THE
ELECTRONICS
HANDBOOK

SECOND EDITION

The Electrical Engineering Handbook Series

Series Editor
Richard C. Dorf
University of California, Davis

Titles Included in the Series

The Handbook of Ad Hoc Wireless Networks, Mohammad Ilyas
The Avionics Handbook, Cary R. Spitzer
The Biomedical Engineering Handbook, Second Edition, Joseph D. Bronzino
The Circuits and Filters Handbook, Second Edition, Wai-Kai Chen
The Communications Handbook, Second Edition, Jerry Gibson
The Computer Engineering Handbook, Vojin G. Oklobdzija
The Control Handbook, William S. Levine
The CRC Handbook of Engineering Tables, Richard C. Dorf
The Digital Signal Processing Handbook, Vijay K. Madisetti and Douglas Williams
The Electrical Engineering Handbook, Second Edition, Richard C. Dorf
The Electric Power Engineering Handbook, Leo L. Grigsby
The Electronics Handbook, Jerry C. Whitaker
The Engineering Handbook, Second Edition, Richard C. Dorf
The Handbook of Formulas and Tables for Signal Processing, Alexander D. Poularikas
The Handbook of Nanoscience, Engineering, and Technology, William A. Goddard, III,
    Donald W. Brenner, Sergey E. Lyshevski, and Gerald J. Iafrate
The Handbook of Optical Communication Networks, Mohammad Ilyas and
    Hussein T. Mouftah
The Industrial Electronics Handbook, J. David Irwin
The Measurement, Instrumentation, and Sensors Handbook, John G. Webster
The Mechanical Systems Design Handbook, Osita D.I. Nwokah and Yidirim Hurmuzlu
The Mechatronics Handbook, Robert H. Bishop
The Mobile Communications Handbook, Second Edition, Jerry D. Gibson
The Ocean Engineering Handbook, Ferial El-Hawary
The RF and Microwave Handbook, Mike Golio
The Technology Management Handbook, Richard C. Dorf
The Transforms and Applications Handbook, Second Edition, Alexander D. Poularikas
The VLSI Handbook, Wai-Kai Chen

Forthcoming Titles

The Electrical Engineering Handbook, Third Edition, Richard C. Dorf
The Electronics Handbook, Second Edition, Jerry C. Whitaker

THE
ELECTRONICS
HANDBOOK

SECOND EDITION

Editor-in-Chief
JERRY C. WHITAKER

Taylor & Francis
Taylor & Francis Group

Boca Raton London New York Singapore

A CRC title, part of the Taylor & Francis imprint, a member of the
Taylor & Francis Group, the academic division of T&F Informa plc.

Published in 2005 by
CRC Press
Taylor & Francis Group
6000 Broken Sound Parkway NW
Boca Raton, FL 33487-2742

International Standard Book Number-10: 0-8493-1889-0 (Hardcover)
International Standard Book Number-13: 978-0-8493-1889-4 (Hardcover)
Library of Congress Card Number 2004057106

Library of Congress Cataloging-in-Publication Data

The electronics handbook / edited by Jerry C. Whitaker. — 2nd ed.
    p. cm. — (Electrical engineering handbook series; v. 34)
  Includes bibliographical references and index.
  ISBN 0-8493-1889-0 (alk. paper)
   1. Electronic circuits–Handbooks, manuals, etc. I. Whitaker, Jerry C. II. Series.

TK7867.E4244 2005
621.381—dc22
                                                        2004057106

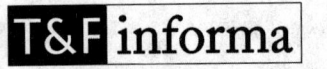

Taylor & Francis Group is the Academic Division of T&F Informa plc.

Visit the Taylor & Francis Web site at
http://www.taylorandfrancis.com

and the CRC Press Web site at
http://www.crcpress.com

Dedication

For Mark Richer
with thanks for the opportunity to contribute to ATSC

Preface

The first edition of *The Electronics Handbook* was published in 1996. Between then and now, tremendous changes have occurred in electronics engineering. During this same period, the value of *The Electronics Handbook* has been recognized by thousands of readers all over the world, for which the editor and authors are very grateful.

The numerous changes in technology over the past few years have led to the publication of a second edition of *The Electronics Handbook*. This new edition builds upon the solid foundation of fundamental theory and practical applications of the original work. All chapters have been reviewed and updated as needed, and many new chapters have been added to explain new developments in electronics engineering.

The Electronics Handbook is intended for engineers and technicians involved in the design, production, installation, operation, and maintenance of electronic devices and systems. This publication covers a broad range of technologies with emphasis on practical applications. In general, the level of detail provided is limited to that necessary to design electronic systems based on the interconnection of operational elements and devices. References are provided throughout the handbook to direct readers to more detailed information on important subjects.

The purpose of *The Electronics Handbook* is to provide in a single volume a comprehensive reference for the practicing engineer in industry, government, and academia. The book is divided into 23 chapters that encompass the field of electronics. The goal is to provide the most up-to-date reference on subjects ranging from classical devices and circuits to emerging technologies and applications.

The fundamentals of electronics have evolved to include a wide range of knowledge, empirical data, and a broad range of practice. The focus of this handbook is on the key concepts, models, and equations that enable the engineer to analyze, design, and predict the behavior of complex electrical devices, circuits, instruments, and systems. The reader will find the key concepts of each subject defined, described, and illustrated; where appropriate, practical applications are given as well.

The level of conceptual development of each topic is challenging, but tutorial and relatively fundamental. Each chapter is written to enlighten the expert, refresh the knowledge of the experienced engineer, and educate the novice.

The information contained in this work is organized into 23 chapters, building a foundation from theory to materials to components to circuits to applications. The Handbook concludes with important chapters on reliability, safety, and engineering management.

At the conclusion of most chapters of the Handbook are three important entries of particular interest to readers:

- Defining Terms, which identifies key terms applicable to the field and their most common definitions
- References, which lists the papers and other resources used in the development of the chapter
- Further Information, which directs the reader to additional sources of in-depth information on the subject matter

These features, a trademark of the CRC Press Electrical Engineering Handbook Series, are a valuable aid to both experienced and novice engineers.

In a publication as large as this, locating the information desired in a rapid manner is important. Numerous aids are provided in this regard. A complete table of contents is given at the beginning of the book. In addition, an individual table of contents precedes each of the 23 chapters. A comprehensive subject index is also provided.

The Electronics Handbook is designed to provide answers to most inquiries and to direct the reader to further sources and references as needed. It is our hope that this publication will continue to serve you—the reader—with important, useful information for years to come.

Jerry C. Whitaker
Editor-in-Chief

Editor-in-Chief

Jerry C. Whitaker is Vice President of Standards Development at the Advanced Television Systems Committee (ATSC). Whitaker supports the work of the various ATSC technology and implementation committees and assists in the development of ATSC standards, recommended practices, and related documents. The ATSC is an international, nonprofit organization developing voluntary standards for digital television.

Whitaker is a Fellow of the Society of Broadcast Engineers and a Fellow of the Society of Motion Picture and Television Engineers. He is also the author and editor of more than 30 books on technical topics. His current CRC titles include:

- *The RF Transmission Systems Handbook*
- *The Electronic Systems Maintaining Handbook*
- *AC Power Systems Handbook*, 2nd edition
- *The Power Vacuum Tubes Handbook*

Whitaker is the former editorial director and associate publisher of *Broadcast Engineering* and *Video Systems* magazines.

Contributors

Samuel O. Agbo
California Polytechnic State
  University
San Luis Obispo, California

Oktay Alkin
School of Engineering
Southern Illinois University
Edwardsville, Illinois

Brent Allen
Bell Northern Research
Ottawa, Canada

William F. Ames
School of Mathematics
Georgia Institute of Technology
Atlanta, Georgia

**Constantine N.
  Anagnostopoulos**
Microelectronics Technical
  Division
Eastman Kodak Company
Rochester, New York

Peter Aronhime
Department of Electrical and
  Computer Engineering
University of Louisville
Louisville, Kentucky

Praveen Asthana
IBM Corporation
San Jose, California

Fred Baumgartner
TCI Technology Ventures
Denver, Colorado

Terrence M. Baun
Criterion Broadcast Services
Milwaukee, Wisconsin

Melvin L. Belcher, Jr.
Radar Systems Analysis
Georgia Tech Research Institute
Symrna, Georgia

Carl Bentz
Intertec Publishing
Overland Park, Kansas

David F. Besch
University of the Pacific
Stockton, California

Ashoka K.S. Bhat
Department of Electrical and
  Computer Engineering
University of Victoria
Victoria, Canada

Glenn R. Blackwell
Department of Electrical and
  Computer Engineering
  Technology
Purdue University
West Lafayette, Indiana

Bruce W. Bomar
Department of Electrical and
  Computer Engineering
University of Tennessee Space
  Institute
Tullahoma, Tennessee

Iuliana Bordelon
CALCE Electronic Packaging
  Research Center
University of Maryland
College Park, Maryland

Jerome R. Breitenbach
Department of Electrical
  Engineering
California Polytechnic State
  University
San Luis Obispo, California

John R. Brews
University of Arizona
Tucson, Arizona

Isidor Buchmann
Cadex Electronics, Inc.
Richmond, Canada

Stuart K. Burgess
University of Southern
  California
Los Angeles, California

George Cain
School of Mathematics
Georgia Institute of Technology
Atlanta, Georgia

Paulo Cardieri
University of Campinas
São Paulo, Brazil

Clifford G. Carter
Naval Undersea Warfare Center
Newport, Rhode Island

Jonathon A. Chambers
Cardiff School of Engineering
Cardiff University
Wales, United Kingdom

Curtis J. Chan
Chan and Associates
Fullerton, California

Ken A. Chauvin
Corning Cable Systems
Hickory, North Carolina

Tom Chen
Department of Electrical
    Engineering
Colorado State University
Fort Collins, Colorado

John Choma, Jr.
University of Southern
    California
San Dimas, California

Badrul H. Chowdhury
Department of Electrical
    Engineering
University of Wyoming
Laramie, Wyoming

Michael D. Ciletti
University of Colorado
Colorado Springs, Colorado

Almon H. Clegg
CCi
Cumming, Georgia

George I. Cohn
California State University
Flintridge, California

James G. Cottle
Hewlett-Packard
San Francisco, California

Leon W. Couch, II
University of Florida
Gainesville, Florida

Charles H. Cox, III
Research Laboratory of
    Electronics
Massachusetts Institute of
    Technology
Cambridge, Massachusetts

Gilles A. Daigle
Institute for Microstructural
    Sciences
Ottawa, Canada

Edward J. Delp, III
ECE Department
Purdue University
West Lafayette, Indiana

Kenneth R. Demarest
University of Kansas
Lawrence, Kansas

Gene DeSantis
DeSantis Associates
New Milford, New Jersey

William E. DeWitt
School of Technology
Purdue University
West Lafayette, Indiana

Daniel F. DiFonzo
Planar Communications
    Corporation
Rockville, Maryland

Dennis F. Doelitzsch
3-D Communications
    Corporation
Marion, Illinois

Barry G. Douglass
Department of Electrical
    Engineering
Texas A&M University
College Station, Texas

Thomas F. Edgar
Department of Chemical
    Engineering
University of Texas
Austin, Texas

Ezz I. El-Masry
Department of Electrical
    Engineering
Technical Institute of Nova
    Scotia
Halifax, Canada

Yariv Ephraim
Department of Electrical and
    Computer Engineering
George Mason University
Fairfax, Virginia

Steve Epstein
Video Systems Magazine
Overland Park, Kansas

Eugene D. Fabricius
EL/EE Department
California Polytechnic State
    University
San Luis Obispo, California

Anthony J. Ferraro
Department of Electrical
    Engineering
Pennsylvania State University
University Park, Pennsylvania

Clifford D. Ferris
University of Wyoming
Laramie, Wyoming

Robert J. Feugate, Jr.
College of Engineering and
    Technology
University of Arizona
Flagstaff, Arizona

Igor M. Filanovsky
Department of Electrical and
    Computer Engineering
University of Alberta
Edmonton, Canada

Paul D. Franzon
Department of Electrical and
    Computer Engineering
North Carolina State University
Raleigh, North Carolina

Susan A. Garrod
Department of Electrical
  Engineering
Purdue University
West Lafayette, Indiana

James E. Goldman
Purdue University
West Lafayette, Indiana

Juergen Hahn
Department of Chemical
  Engineering
Texas A&M University
College Station, Texas

Jerry C. Hamann
Department of Electrical
  Engineering
University of Wyoming
Laramie, Wyoming

Margaret H. Hamilton
Hamilton Technologies, Inc.
Cambridge, Massachusetts

William F. Hammett
Hammett and Edison, Inc.
Sonoma, California

Cecil Harrison
University of Southern
  Mississippi
Brandon, Mississippi

Robert D. Hayes
RDH Incorporated
Marietta, Georgia

Robert J. Hofinger
School of Technology
Purdue University
Columbus, Indiana

James M. Howell
Raytheon Corporation
Woodstock, Georgia

Timothy P. Hulick
Acrodyne Industries, Inc.
Schwenksville, Pennsylvania

Thomas K. Ishii
Department of Electrical and
  Computer Engineering
Marquette University
Milwaukee, Wisconsin

David Jernigan
National Instruments
Austin, Texas

T. S. Kalkur
Department of Electrical and
  Computer Engineering
University of Colorado
Colorado Springs, Colorado

Rangachar Kasturi
Department of Computer
  Science
Pennsylvania State University
State College, Pennsylvania

Hagbae Kim
Langley Research Center
National Aeronautics and
  Space Administration
Hampton, Virginia

Ravindranath Kollipara
LSI Logic Corporation
Palo Alto, California

Kurt L. Kosbar
Department of Electrical
  Engineering
University of Missouri
Rolla, Missouri

David A. Kosiba
Pennsylvania State University
State College, Pennsylvania

Robert Kubichek
Department of Electrical
  Engineering
University of Wyoming
Laramie, Wyoming

Gopal Lakhani
Department of Computer
  Science
Texas Tech University
Lubbock, Texas

Paul P.K. Lee
Microelectronics Technical
  Division
Eastman Kodak Company
Rochester, New York

Élvio João Leonardo
University of Campinas
São Paulo, Brazil

Honoch Lev-Ari
Department of Electrical and
  Computer Engineering
Northeastern University
Boston, Massachusetts

Francis Long
University of Denver
Denver, Colorado

Shih-Lien Lu
Department of Electronics and
  Computer Engineering
Oregon State University
Corvallis, Oregon

Melissa S. Mattmuller
Department of Engineering
  Technology
Purdue University
West Lafayette, Indiana

Edward McConnell
National Instruments
Austin, Texas

John E. McInroy
Department of Electrical
  Engineering
University of Wyoming
Laramie, Wyoming

Bernard E. McTaggart
Naval Undersea Warfare Center
Baltic, Connecticut

Patricia F. Mead
Department of Mechanical
    Engineering
University of Maryland
College Park, Maryland

**Álvaro Augusto Machado
    Medeiros**
University of Campinas
São Paulo, Brazil

Victor Meeldijk
Network Processing Group
Intel Corporation
Parsippany, New Jersey

Sanjay K. Mehta
Naval Undersea Warfare Center
Newport, Rhode Island

John D. Meyer
Printing Technologies
    Department
Hewlett-Packard Co.
Palo Alto, California

Jan H. Mooij
Ceramic Innovation Center
Philips Components
Roermond, Netherlands

Gibson Morris, Jr.
Department of Electrical
    Engineering
University of Southern
    Alabama
Mobile, Alabama

Wayne Needham
Intel Corporation
Chandler, Arizona

John Norgard
University of Colorado
Colorado Springs, Colorado

Martin R. Parker
Department of Electrical
    Engineering
University of Alabama
Tuscaloosa, Alabama

Eugene T. Patronis, Jr.
School of Physics
Georgia Institute of Technology
Atlanta, Georgia

Michael Pecht
CALCE Electronic Products
    and Systems Center
University of Maryland
College Park, Maryland

Benjamin B. Peterson
U.S. Coast Guard Academy
Waterford, Connecticut

John W. Pierre
Department of Electrical
    Engineering
University of Wyoming
Laramie, Wyoming

Fabrizio Pollara
Jet Propulsion Lab
California Institute of
    Technology
Pasadena, California

Roy W. Rising
ABC - TV
Valley Village, California

David E. Rittenhouse
Siecor Corporation
Hickory, North Carolina

William J.J. Roberts
Atlantic Coast Technologies, Inc.
Silver Spring, Maryland

Richard Rudman
KFWB Radio
Los Angeles, California

Stanley Salek
Hammett and Edison, Inc
Sonoma, California

C. Sankaran
Electro-Test Inc.
Shoreline, Washington

James A. Scheer
Department of GTRI - Sensors
    Electronics
Georgia Tech Research Institute
Atlanta, Georgia

Jerry E. Sergent
BBS PowerMod, Inc.
Corbin, Kentucky

Ken Seymour
AT&T Wireless Services
Sprynet
Santa Rosa, California

James F. Shackelford
University of California
Davis, California

E. A. G. Shaw
National Research Council
    of Canada
Ottawa, Canada

Joy S. Shetler
Computer Engineering
    Program
California Polytechnic State
    University
San Luis Obispo, California

Carol Smidts
Reliability Engineering
  Program
University of Maryland
College Park, Maryland

Sidney Soclof
California State University
San Gabriel, California

David Stafford
Quantegy, Inc.
Opelika, Alabama

Michael Starling
National Public Radio
Washington, DC

Zbigniew J. Staszak
Department of Electronic
  Circuits
Technical University of Gdansk
Gdansk, Poland

Michael B. Steer
Department of Electronics and
  Computer Engineering
North Carolina State University
Raleigh, North Carolina

Michel R. Stinson
Institute for Microstructural
  Sciences
Ottawa, Canada

Gerhard J. Straub
Hammett and Edison, Inc.
La Plata, Maryland

Robert A. Surette
Shively Labs
Bridgton, Maine

Sawasd Tantaratana
Department of Electrical and
  Computer Engineering
University of Massachusetts
Amherst, Massachusetts

Jeffrey P. Tate
Department of Electronic
  Engineering and Technology
Florida A&M University
Tallahassee, Florida

Stuart K. Tewksbury
Department of Electrical and
  Computer Engineering
Stevens Institute of Technology
Hoboken, New Jersey

Floyd E. Toole
Harman International
  Industries, Inc.
Northridge, California

William H. Tranter
Department of Electrical
  Engineering
Virginia Polytechnic Institute
  and State University
Blacksburg, Virginia

Vijai K. Tripathi
Oregon State University
Corvallis, Oregon

Sadrul Ula
Department of Electrical
  Engineering
University of Wyoming
Laramie, Wyoming

Ardie D. Walser
Department of Electrical
  Engineering
City College of New York
New York, New York

William E. Webb
Department of Electrical
  Engineering
University of Alabama
Tuscaloosa, Alabama

Robert R. Weirather
Harris Corporation
Quincy, Illinois

Douglas H. Werner
Pennsylvania State University
University Park, Pennsylvania

Pingjuan L. Werner
Pennsylvania State University
State College, Pennsylvania

Jerry C. Whitaker
Advance Television Systems
  Committee
Washington, DC

Allan White
Langley Research Center
National Aeronautics and
  Space Administration
Hampton, Virginia

Donald White
Don White Consultants, Inc.
Warrenton, Virginia

Bogdan M. Wilamowski
Department of Electrical and
  Computer Engineering
Auburn University
Auburn, Alabama

Tsong-Ho Wu
Transtech Networks, Inc.
Iselin, New Jersey

Fred Wylie
Audio Processing Technology,
  Ltd.
Belfast, Northern Ireland

Michel D. Yacoub
University of Campinas
Pathumthaui, Brazil

Harry E. Young
MTA-EMCI
Atlanta, Georgia

Paul Kit-Lai Yu
Department of Electrical and
  Computer Engineering
University of California, San
  Diego
La Jolla, California

Rodger E. Ziemer
University of Colorado
Colorado Springs, Colorado

Contents

Chapter 3   Properties of Passive Components   239

Chapter 4   Passive Electrical Circuit   309

Chapter 5   Electron Vacuum Devices   353

Chapter 10   Power Supplies and Regulation   1013

Chapter 11   Packaging Electronic Systems   1259

Chapter 12 Communication Principles   1353

Chapter 13 Electromagnetic Radiation   1483

1

Fundamental Electrical Theory

John Norgard
University of Colorado

Michael D. Ciletti
University of Colorado

Eugene T. Patronis, Jr.
Georgia Institute of Technology

Barry G. Douglass
Texas A&M University

George I. Cohn
California State University

Floyd E. Toole
Harman International Industries, Inc.

E. A. G. Shaw
National Research Council of Canada

Gilles A. Daigle
Institute for Microstructural Sciences

Michel R. Stinson
Institute for Microstructural Sciences

Jerry C. Whitaker
Editor-in-Chief

1.1 Electromagnetic Spectrum

John Norgard

1.1.1 Introduction

The **electromagnetic** (EM) **spectrum** consists of all forms of EM radiation, for example, EM waves propagating through space, from direct current (DC) to **light** to **gamma rays**. The EM spectrum can be arranged in order of frequency or wavelength into a number of regions, usually wide in extent, within which the EM waves have some specified common characteristics, for example, those characteristics relating to the production or detection of radiation. Note that in this section, specific frequency ranges are called bands; several contiguous frequency bands are called spectrums; and subfrequency ranges within a band are called segments. A common example is the spectrum of the radiant energy in white light, as dispersed by a prism, to produce a rainbow of its constituent colors.

The EM spectrum can be displayed as a function of frequency (or wavelength), as shown schematically in Fig. 1.1. In air, frequency and wavelength are inversely proportional ($f = c/\lambda$). The meter-kilogram-second (MKS) unit of frequency is the hertz (1 Hz = 1 cycle per second); the MKS unit of wavelength is the meter.

Frequency is also measured in the following subunits.

kilohertz   [1 kHz = 10^3 Hz]
megahertz   [1 MHz = 10^6 Hz]
gigahertz   [1 GHz = 10^9 Hz]
terahertz   [1 THz = 10^{12} Hz]

or, for very high frequencies,

electron volt   [1 eV $\sim 2.41 \times 10^{14}$ Hz]

Note that $\mathcal{E} = hf = qV$ where $h = 6.63 \times 10^{-34}$ Js (Plancks constant) and $q = 1.602 \times 10^{-19}$ C.
Wavelength is also measured in the following subunits.

centimeters   [1 cm = 10^{-2} m]
millimeters   [1 mm = 10^{-3} m]
micrometers   [1 μm = 10^{-6} m] (microns)
nanometers   [1 nm = 10^{-9} m]
Ångstroms   [1 Å = 10^{-10} m]
picometers   [1 pm = 10^{-12} m]
femtometers   [1 fm = 10^{-15} m]
attometers   [1 am = 10^{-18} m]

1.1.2 Spectral Subregions

In this section, the EM spectrum is divided for convenience into three main subregions: (1) the optical spectrum, (2) the DC to light spectrum, and (3) the light to gamma ray spectrum.

The main subregions of the EM spectrum are now discussed, starting with the optical spectrum and continuing with the DC to light and the light to gamma ray spectrums. Note that the boundaries between

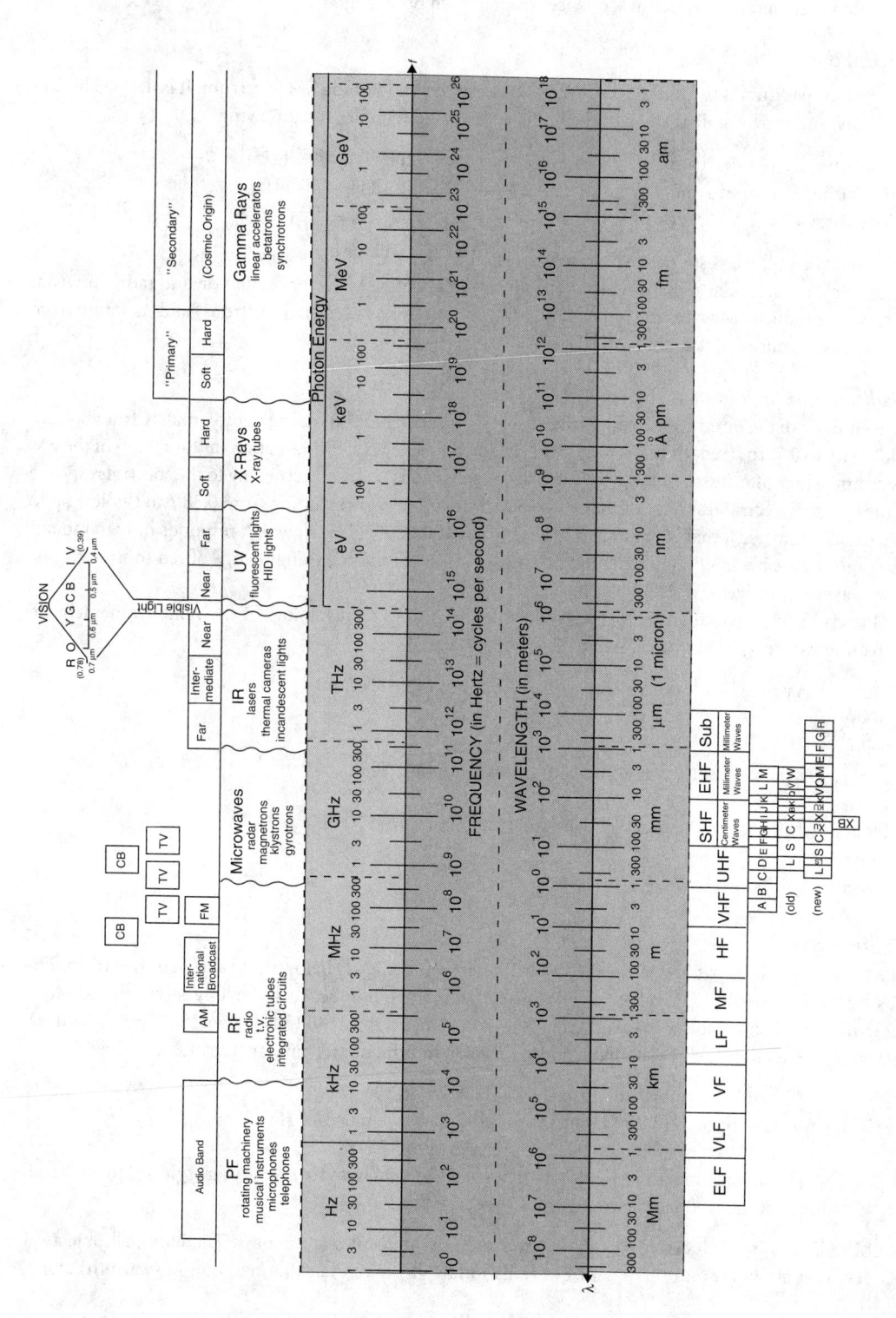

FIGURE 1.1 The electromagnetic spectrum.

some of the spectral regions are somewhat arbitrary. Some spectral bands have no sharp edges and merge into each other, and some spectral sequences overlap each other slightly.

Optical Spectrum

The optical spectrum is the middle frequency/wavelength region of the EM spectrum. It is defined here as the visible and near-visible regions of the EM spectrum and includes the following.

The infrared (IR) spectrum:	1 mm–0.7 μm	(300 GHz–429 THz)
The visible light spectrum:	0.7 μm–0.4 μm	(429 THz–750 THz)
The ultraviolet (UV) spectrum:	0.4 μm–circa 3 nm	(750 THz–circa 300 eV)

These regions of the EM spectrum are usually described in terms of their wavelengths.

Atomic and molecular radiation produce radiant light energy. Molecular radiation and radiation from hot bodies produce EM waves in the IR band. Atomic radiation (outer shell electrons) and radiation from arcs/sparks produce EM waves in the UV band.

Visible Light Spectrum

In the middle of the optical spectrum is the visible light spectrum, extending approximately from 0.4 μm (violet) up to 0.7 μm (red), that is, from 750 THz down to 429 THz. EM radiation in this region of the EM spectrum, when entering the eye, gives rise to visual sensations (colors), according to the spectral response of the eye [the spectral response of the eye is sometimes quoted as extending from 0.38 μm (violet) up to 0.75 or 0.78 μm (red), that is, from 790 THz down to 400 or 385 THz)], which responds only to radiant energy in the visible light band extending from the extreme long wavelength edge of red to the extreme short wavelength edge of violet.

This visible light spectrum is further subdivided into the various colors of the rainbow, namely (in decreasing wavelength/increasing frequency):

red	a primary color; peak intensity at 700.0 nm (429 THz)
orange	
yellow	
green	a primary color; peak intensity at 546.1 nm (549 THz)
cyan	
blue	a primary color; peak intensity at 435.8 nm (688 THz)
indigo	
violet	

IR Spectrum

The IR spectrum is the region of the EM spectrum lying immediately below the visible light spectrum. The IR spectrum consists of EM radiation with wavelengths extending between the longest visible red (circa 0.7 μm) and the shortest microwaves (circa 300–1000 μm, i.e., from 400 THz down to 1 THz–300 GHz).

The IR spectrum is further subdivided into the near, intermediate,* and far IR bands as follows:

1. *Near* IR band: 0.7 μm up to 3 μm (429 THz down to 100 THz)
2. *Intermediate* IR band: 3 μm up to 10 μm (100 THz down to 30 THz)
3. *Far* IR band: 10 μm up to 100 μm (30 THz down to 3 THz)
   The submillimeter region of wavelengths is sometimes included in the very far region of the IR band.
4. Submillimeter: 100 μm up to 1 mm (3 THz down to 300 GHz)

EM radiation is produced by oscillating and rotating molecules and atoms. Therefore, all objects at temperatures above absolute zero emit EM radiation by virtue of their thermal motion (warmth) alone.

---

*Note some reference texts use 2.5 μm (120 THz) as the breakpoint between the near and the intermediate IR bands.

Objects near room temperature emit most of their radiation in the IR band. Even relatively cool objects, however, emit some **IR radiation**; hot objects, such as incandescent filaments, emit strong IR radiation.

IR radiation is sometimes incorrectly called radiant heat, because warm bodies emit IR radiation and bodies that absorb IR radiation are warmed. However, IR radiation is not itself heat. This EM radiation is called black body radiation. Such waves are emitted by all material objects. For example, the background cosmic radiation (2.7 K) emits microwaves; room temperature objects (295 K) emit IR rays; the sun (6000 K) emits yellow light; the solar corona 10^6 K emits **X rays**.

IR astronomy uses the 1 μm–1 mm part of the IR band to study celestial objects by their IR emissions. IR detectors are used in night vision systems, intruder alarm systems, weather forecasting, and missile guidance systems. IR photography uses multilayered color film, with an IR sensitive emulsion in the wavelengths between 700 and 900 nm, for medical and forensic applications and for aerial surveying.

UV Spectrum

The UV spectrum is the region of the EM spectrum lying immediately above the visible light spectrum. The UV spectrum consists of EM radiation with wavelengths extending between the shortest visible violet (circa 0.4 μm) and the longest X rays (circa 3 nm), that is, from 750 THz (circa 3 eV) up to 125 eV (some reference texts use 4, 5, or 10 nm as the upper edge of the UV band.)

The UV spectrum is further subdivided into the near and the far UV bands as follows:

Near UV band: 0.4 μm down to 100 nm   (3 eV up to 10 eV)
Far UV band: 100 nm down to circa 3 nm   (10 eV up to circa 300 eV)

The far UV band is also referred to as the vacuum UV band, since air is opaque to all UV radiation in this region.

UV radiation is produced by electron transitions in atoms and molecules, as in a mercury discharge lamp. UV radiation from the sun causes tanning of the skin. Radiation in the UV range can cause florescence in some substances, can produce photographic and ionizing effects, and is easily detected.

In UV astronomy, the emissions of celestial bodies in the wavelength band between 50 and 320 nm are detected and analyzed to study the heavens. The hottest stars emit most of their radiation in the UV band.

DC to Light Spectrum

Below the IR spectrum are the lower frequency (longer wavelength) regions of the EM spectrum, subdivided generally into the following spectral regions (by frequency/wavelength).

Microwave spectrum: 300 GHz down to 300 MHz   (1 mm up to 1 m)
Radio frequency (RF) spectrum: 300 MHz down to 10 kHz   (1 m up to 30 km)
Power/telephony spectrum: 10 kHz down to DC   (30 km up to ∞)

Note that some reference works define the lower edge of the microwave spectrum at 1 GHz. The three regions of the EM spectrum are usually described in terms of their frequencies.

Radiations having wavelengths of the order of millimeters and centimeters are called **microwaves**; those still longer are called **radio waves** (or Hertzian waves).

Radiation from electronic devices produces EM waves in both the microwave and RF bands. Power frequency energy is generated by rotating machinery. Direct current is produced by batteries or rectified alternating current (AC).

Microwave Spectrum

The microwave spectrum is the region of wavelengths lying between the far IR/submillimeter regions and the conventional RF region. The boundaries of the microwave spectrum have not been definitely fixed, but it is commonly regarded as the region of the EM spectrum extending from about 1 mm to 1 m in wavelengths, that is, 300 GHz down to 300 MHz.

The microwave spectrum is further subdivided into the following segments.

Millimeter waves: 300 GHz down to 30 GHz   (1 mm up to 1 cm) extremely high-frequency
(EHF) band

Centimeter waves: 30 GHz down to 3 GHz   (1 cm up to 10 cm) super high-frequency
(SHF) band

Note that some reference articles consider the top edge of the millimeter region to stop at 100 GHz. The microwave spectrum usually includes the ultra high-frequency (UHF) band from 3 GHz down to 300 MHz (10 cm up to 1 m).

Microwaves are used in radar, in communication links spanning moderate distances, as radio carrier waves in radio broadcasting, for mechanical heating, and cooking in microwave ovens.

Radio Frequency Spectrum

The RF range of the EM spectrum is the wavelength band suitable for utilization in radio communications extending from 10 kHz to 300 MHz (some authors consider the RF band as extending from 10 kHz to 300 GHz, with the microwave band as a subset of the RF band from 300 MHz to 300 GHz.) Some of the radio waves serve as the carriers of the low-frequency audio signals; other radio waves are modulated by video and digital information. The amplitude modulated (AM) broadcasting band uses waves with frequencies between 550 and 1640 kHz; the frequency modulated (FM) broadcasting band uses waves with frequencies between 88 and 108 MHz.

In the U.S., the Federal Communications Commission (FCC) is responsible for assigning a range of frequencies, for example, a frequency band in the RF spectrum, to a broadcasting station or service. The International Telecommunications Union (ITU) coordinates frequency band allocation and cooperation on a worldwide basis.

Radio astronomy uses a radio telescope to receive and study radio waves naturally emitted by objects in space. Radio waves are emitted from hot gases (thermal radiation), from charged particles spiraling in magnetic fields (synchrotron radiation), and from excited atoms and molecules in space (spectral lines), such as the 21-cm line emitted by hydrogen gas.

Power Frequency/Telephone Spectrum

The power frequency (PF) range of the EM spectrum is the wavelength band suitable for generating, transmitting, and consuming low-frequency power, extending from 10 kHz down to DC (zero frequency). In the U.S., most power is generated at 60 Hz (some military applications use 400 Hz); in other countries, for example, in Europe, power is generated at 50 Hz.

Frequency Bands

The combined microwave, RF (Hertzian waves), and power/telephone spectra are subdivided into the following specific bands.

Extremely low-frequency (ELF) band:	30 Hz up to 300 Hz	(10 Mm down to 1 Mm)
Voice-frequency (VF) band:	300 Hz up to 3 kHz	(1 Mm down to 100 km)
Very low-frequency (VLF) band:	3 kHz up to 30 kHz	(100 km down to 10 km)
Low-frequency (LF) band:	30 kHz up to 300 kHz	(10 km down to 1 km)
Medium-frequency (MF) band:	300 kHz up to 3 MHz	(1 km down to 100 m)
High-frequency (HF) band:	3 MHz up to 30 MHz	(100 m down to 10 m)
Very high-frequency (VHF) band:	30 MHz up to 300 MHz	(10 m down to 1 m)
Ultra high-frequency (UHF) band:	300 MHz up to 3 GHz	(1 m down to 10 cm)
Super high-frequency (SHF) band:	3 GHz up to 30 GHz	(1 cm down to 1 cm)
Extremely high-frequency (EHF) band:	30 GHz up to 300 GHz	(1 cm down to 1 mm)

The upper portion of the UHF band, the SHF band, and the lower part of the EHF band are further subdivided into the following bands.

L band:	1 GHz up to 2 GHz	(30 cm down to 15 cm)
S band:	2 GHz up to 4 GHz	(15 cm down to 7.5 cm)
C band:	4 GHz up to 8 GHz	(7.5 cm down to 3.75 cm)
X band:	8 GHz up to 12 GHz	(3.75 cm down to 2.5 cm)
Ku band:	12 GHz up to 18 GHz	(2.5 cm down to 1.67 cm)
K band:	18 GHz up to 26.5 GHz	(1.67 cm down to 1.13 cm)
Ka band:	26.5 GHz up to 40 GHz	(1.13 cm down to 7.5 mm)
Q band:	32 GHz up to 50 GHz	(9.38 mm down to 6 mm)
U band:	40 GHz up to 60 GHz	(7.5 mm down to 5 mm)
V band:	50 GHz up to 75 GHz	(6 mm down to 4 mm)
W band:	75 GHz up to 100 GHz	(4 mm down to 3.33 mm)

An alternate and more detailed subdivision of the UHF, SHF, and EHF bands is as follows:

L band:	1.12 GHz up to 1.7 GHz	(26.8 cm down to 17.6 cm)
LS band:	1.7 GHz up to 2.6 GHz	(17.6 cm down to 11.5 cm)
S band:	2.6 GHz up to 3.95 GHz	(11.5 cm down to 7.59 cm)
C(G) band:	3.95 GHz up to 5.85 GHz	(7.59 cm down to 5.13 cm)
XN(J, XC) band:	5.85 GHz up to 8.2 GHz	(5.13 cm down to 3.66 cm)
XB(H, BL) band:	7.05 GHz up to 10 GHz	(4.26 cm down to 3 cm)
X band:	8.2 GHz up to 12.4 GHz	(3.66 cm down to 2.42 cm)
Ku(P) band:	12.4 GHz up to 18 GHz	(2.42 cm down to 1.67 cm)
K band:	18 GHz up to 26.5 GHz	(1.67 cm down to 1.13 cm)
V(R, Ka) band:	26.5 GHz up to 40 GHz	(1.13 cm down to 7.5 mm)
Q(V) band:	33 GHz up to 50 GHz	(9.09 mm down to 6 mm)
M(W) band:	50 GHz up to 75 GHz	(6 mm down to 4 mm)
E(Y) band:	60 GHz up to 90 GHz	(5 mm down to 3.33 mm)
F(N) band:	90 GHz up to 140 GHz	(3.33 mm down to 2.14 mm)
G(A) band:	140 GHz up to 220 GHz	(2.14 mm down to 1.36 mm)
R band:	220 GHz up to 325 GHz	(1.36 mm down to 0.923 mm)

The upper portion of the VHF band, the UHF and SHF bands, and the lower part of the EHF band have been more recently divided into the following bands.

A band:	100 MHz up to 250 MHz	(3 m down to 1.2 m)
B band:	250 MHz up to 500 MHz	(1.2 m down to 60 cm)
C band:	500 MHz up to 1 GHz	(60 cm down to 30 cm)
D band:	1 GHz up to 2 GHz	(30 cm down to 15 cm)
E band:	2 GHz up to 3 GHz	(15 cm down to 10 cm)
F band:	3 GHz up to 4 GHz	(10 cm down to 7.5 cm)
G band:	4 GHz up to 6 GHz	(7.5 cm down to 5 cm)
H band:	6 GHz up to 8 GHz	(5 cm down to 3.75 cm)
I band:	8 GHz up to 10 GHz	(3.75 cm down to 3 cm)
J band:	10 GHz up to 20 GHz	(3 cm down to 1.5 cm)
K band:	20 GHz up to 40 GHz	(1.5 cm down to 7.5 mm)
L band:	40 GHz up to 60 GHz	(7.5 mm down to 5 mm)
M band:	60 GHz up to 100 GHz	(5 mm down to 3 mm)

Several other frequency bands of interest (not exclusive) are now listed.
In the power spectrum:

Audio band:	10 Hz–10 kHz

In the RF spectrum:

Longwave broadcasting band:	150–290 kHz
AM broadcasting band:	550–1640 kHz (1.640 MHz) (107 Channels, 10-kHz separation)
International broadcasting band:	3–30 MHz
Shortwave broadcasting band:	5.95–26.1 MHz (8 bands)
VHF television (channels 2–4):	54–72 MHz
VHF television (channels 5–6):	76–88 MHz
FM broadcasting band:	88–108 MHz
VHF television (channels 7–13):	174–216 MHz
UHF television (channels 14–83):	470–890 MHz

In the *Microwave* spectrum (up to 40 GHz):

Aeronavigation:	0.96–1.215 GHz
Global positioning system (GPS) down link:	1.2276 GHz
Military communications (COM)/radar:	1.35–1.40 GHz
Miscellaneous COM/radar:	1.40–1.71 GHz
L-band telemetry:	1.435–1.535 GHz
GPS downlink:	1.57 GHz
Military COM (troposcatter/telemetry):	1.71–1.85 GHz
Commercial COM and private line of sight (LOS):	1.85–2.20 GHz
Microwave ovens:	2.45 GHz
Commercial COM/radar:	2.45–2.69 GHz
Instructional television:	2.50–2.69 GHz
Military radar (airport surveillance):	2.70–2.90 GHz
Maritime navigation radar:	2.90–3.10 GHz
Miscellaneous radars:	2.90–3.70 GHz
Commercial C-band satellite (SAT) COM downlink:	3.70–4.20 GHz
Radar altimeter:	4.20–4.40 GHz
Military COM (troposcatter):	4.40–4.99 GHz
Commercial microwave landing system:	5.00–5.25 GHz
Miscellaneous radars:	5.25–5.925 GHz
C-band weather radar:	5.35–5.47 GHz
Commercial C-band SAT COM uplink:	5.925–6.425 GHz
Commercial COM:	6.425–7.125 GHz
Mobile television links:	6.875–7.125 GHz
Military LOS COM:	7.125–7.25 GHz
Military SAT COM downlink:	7.25–7.75 GHz
Military LOS COM:	7.75–7.9 GHz
Military SAT COM uplink:	7.90–8.40 GHz
Miscellaneous radars:	8.50–10.55 GHz
Precision approach radar:	9.00–9.20 GHz
X-band weather radar (and maritime navigation radar):	9.30–9.50 GHz
Police radar:	10.525 GHz
Commercial mobile COM [LOS and electronic news gathering (ENG)]:	10.55–10.68 GHz
Common carrier LOS COM:	10.70–11.70 GHz
Commercial COM:	10.70–13.25 GHz
Commercial Ku-band SAT COM downlink:	11.70–12.20 GHz
Direct broadcast satellite (DBS) downlink and private LOS COM:	12.20–12.70 GHz
ENG and LOS COM:	12.75–13.25 GHz
Miscellaneous radars and SAT COM:	13.25–14.00 GHz
Commercial Ku-band SAT COM uplink:	14.00–14.50 GHz

Military COM (LOS, mobile, and Tactical):	14.50–15.35 GHz
Aeronavigation:	15.40–15.70 GHz
Miscellaneous radars:	15.70–17.70 GHz
DBS uplink:	17.30–17.80 GHz
Common carrier LOS COM:	17.70–19.70 GHz
Commercial COM (SAT COM and LOS):	17.70–20.20 GHz
Private LOS COM:	18.36–19.04 GHz
Military SAT COM:	20.20–21.20 GHz
Miscellaneous COM:	21.20–24.00 GHz
Police radar:	24.15 GHz
Navigation radar:	24.25–25.25 GHz
Military COM:	25.25–27.50 GHz
Commercial COM:	27.50–30.00 GHz
Military SAT COM:	30.00–31.00 GHz
Commercial COM:	31.00–31.20 GHz
Navigation radar:	31.80–33.40 GHz
Miscellaneous radars:	33.40–36.00 GHz
Military COM:	36.00–38.60 GHz
Commercial COM:	38.60–40.00 GHz

Light to Gamma Ray Spectrum

Above the UV spectrum are the higher frequency (shorter wavelength) regions of the EM spectrum, subdivided generally into the following spectral regions (by frequency/wavelength)

X-ray spectrum:	circa 300 eV up to circa 30 keV	(circa 3 nm down to circa 30 pm)	(circa 1×10^{17} Hz up to 1×10^{19} Hz)
Gamma ray spectrum:	circa 30 keV up to ∞	(30 pm down to 0 m)	(1×10^{19} Hz up to ∞)

These regions of the EM spectrum are usually described in terms of their photon energies in electron volts.

Note that **cosmic rays** (from astronomical sources) are not EM waves (rays) and, therefore, are not part of the EM spectrum. Cosmic rays are high-energy-charged particles (electrons, protons, and ions) of extraterrestrial origin moving through space that may have energies as high as 10^{20} eV. Cosmic rays have been traced to cataclysmic astrophysical/cosmological events, such as exploding stars and black holes. Cosmic rays are emitted by supernova remnants, pulsars, quasars, and radio galaxies. Cosmic rays collide with molecules in the Earth's upper atmosphere producing secondary cosmic rays and gamma rays of high energy. These gamma rays are sometimes called cosmic or secondary gamma rays. Cosmic rays are a useful source of high-energy particles for experiments. They also contribute to the natural background radiation.

Radiation from atomic inner shell excitations produces EM waves in the X-ray spectrum. Radiation from naturally radioactive nuclei produces EM waves in the gamma ray spectrum.

X-Ray Spectrum

The X-ray spectrum is further subdivided into the following segments.

Soft X rays:	circa 300 eV up to 10 keV	(circa 3 nm down to 10 nm)	(circa 1×10^{17} Hz up to 3×10^{18} Hz)
Hard X rays:	10 keV up to circa 30 keV	(10 nm down to circa 30 pm)	(3×10^{18} Hz up to circa 3×10^{19} Hz)

Because the physical nature of these rays was first unknown, this radiation was called X rays. The more powerful X rays are called hard X rays and are of high frequencies and, therefore, are more energetic; less powerful X rays are called soft X rays and have lower energies.

X rays are produced by transitions of electrons in the inner levels of excited atoms or by rapid deceleration of charged particles—Brehmsstrahlung breaking radiation. An important source of X rays is synchrotron radiation. X rays can also be produced when high-energy electrons from a heated filament cathode strike the surface of a target anode (usually tungsten) between which a high alternating voltage (approximately 100 kV) is applied.

X rays are a highly penetrating form of EM radiation and applications of X rays are based on their short wavelengths and their ability to easily pass through matter. X rays are very useful in crystallography for determining crystalline structure and in medicine for photographing the body. Since different parts of the body absorb X rays to a different extent, X rays passing through the body provide a visual image of its interior structure when striking a photographic plate. X rays are dangerous and can destroy living tissue and can cause severe skin burns; however, X rays are useful in the diagnosis and nondestructive testing of products for defects.

Gamma Ray Spectrum

The gamma ray spectrum is subdivided into the following segments.

Primary gamma rays:	circa 30 keV up to 3 MeV	(circa 30 pm down to 300 fm)	(circa 1×10^{19} Hz up to 1×10^{21} Hz)
Secondary gamma rays:	3 MeV up to ∞	(300 fm down to 0 m)	(1×10^{21} Hz up to ∞)

The primary gamma rays are further subdivided into the following segments.

Soft gamma rays:	circa 30 keV up to circa 300 keV	(circa 30 pm down to circa 3 pm)	(circa 1×10^{19} Hz up to circa 1×10^{20} Hz)
Hard gamma rays:	circa 300 keV up to 3 MeV	(circa 3 pm down to 300 fm)	(circa 1×10^{20} Hz up to 1×10^{21} Hz)

Secondary gamma rays are created from collisions of high-energy cosmic rays with particles in the Earth's upper atmosphere.

Gamma rays are essentially very energetic X rays. The distinction between the two is based on their origin. X rays are emitted during atomic processes involving energetic electrons; gamma rays are emitted by excited nuclei or other processes involving subatomic particles.

Gamma rays are emitted by the nucleus of radioactive material during the process of natural radioactive decay as a result of transitions from high-energy excited states to low-energy states in atomic nuclei. Cobalt 90 is a common gamma ray source (with a half-life of 5.26 years). Gamma rays are also produced by the interaction of high-energy electrons with matter. Cosmic gamma rays cannot penetrate the Earth's atmosphere.

Applications of gamma rays are used both in medicine and in industry. In medicine, gamma rays are used for cancer treatment, diagnoses, and prevention. Gamma ray emitting radioisotopes are used as tracers. In industry, gamma rays are used in the inspection of castings, seams, and welds.

Defining Terms

Cosmic rays: Highly penetrating particle rays from outer space. Primary cosmic rays that enter the Earth's upper atmosphere consist mainly of protons. Cosmic rays of low energy have their origin in the sun, those of high energy in galactic or extragalactic space, possibly as a result of supernova explosions. Collisions with atmospheric particles result in secondary cosmic rays (particles) and secondary gamma rays (EM waves).

Electromagnetic spectrum: EM radiant energy arranged in order of frequency or wavelength and divided into regions within which the waves have some common specified characteristics, for example, the waves are generated, received, detected, or recorded in a similar way.

Gamma rays: Electromagnetic radiation of very high energy (greater than 30 keV) emitted after nuclear reactions or by a radioactive atom when its nucleus is left in an excited state after emission of alpha or beta particles.

Infrared (IR) radiation: Electromagnetic radiations having wavelengths in the range, 0.7 nm (the long-wavelength limit of visible red light) to 1 mm (the shortest microwaves). A convenient subdivision is as follows: near, 0.7 μm to 2–5 μm; intermediate, 2–5 μm to 10 μm; far, 10 μm to 1 mm.

Light: White light, when split into a spectrum of colors, is composed of a continuous range of merging colors: red, orange, yellow, green, cyan, blue, indigo, and violet.

Microwaves: An electromagnetic wave that has a wavelength between approximately 0.3 cm (or 1 mm) and 30 (or 10) cm, corresponding to frequencies between 1 GHz (or 300 MHz) and 100 (or 300) GHz. Note that there are no well-defined boundaries distinguishing microwaves from infrared and radio and waves.

Radio waves: Electromagnetic radiation suitable for radio transmission in the range of frequencies from about 10 kHz to about 300 MHz.

Ultraviolet (UV) radiation: Electromagnetic radiations having wavelengths in the range from 0.4 nm (the shortest wavelength limit of visible violet light) to 3 nm (the longest X rays). A convenient subdivision is as follows: near, 0.4 μm to 100 nm; far, 100 nm to 3 nm.

X rays: Electromagnetic radiation of short wavelengths (circa 3 nm to 30 pm) produced when cathode rays impinge on matter.

References

Cambridge Encyclopedia. 1990. Cambridge University Press, New York.

Collocott, T.C. and Dobson, A.B., Eds. *Dictionary of Science & Technology.* W & R Chambers.

Columbia Encyclopedia. 1993. Columbia University Press, New York.

Handbook of Physics. 1958. McGraw-Hill, New York.

Judd, D.B. and Wyszecki, G. *Color in Business, Science and Industry*, 3rd ed. Wiley, New York.

Kaufman, Ed. *IES Illumination Handbook.* Illumination Engineering Society.

Lapedes, D.N., Ed. *The McGraw-Hill Encyclopedia of Science & Technology*, 2nd ed. McGraw-Hill, New York.

Stemson, A. *Photometry and Radiometry for Engineers.* Wiley, New York.

Webster's New World Encyclopedia. 1992. Prentice-Hall, Englewood Cliffs, NJ.

Wyszecki, G. and Stiles, W.S. *Color Science, Concepts and Methods, Quantitative Data and Formulae*, 2nd Ed. Wiley, New York.

Further Information

Kobb, B.Z. *Spectrum Guide, Radio Frequency Allocations in the United States, 30 MHz–300 GHz.* New Signal Press.

IEEE Standard 521.

U.S.A.F. Regulation 55–44.

1.2   Resonance

Michael D. Ciletti

1.2.1   Introduction

Resonant circuits play an important role in filters and tuning circuits for communications, radar, and many other electronics systems. In a typical application a resonant circuit is used to tune a radio or television to a particular station while blocking the signals that it receives simultaneously from other stations and sources of signals. Resonant circuits effect a frequency-dependent attenuation of signals, thus passing certain signals and rejecting others, including noise. Widespread use of resonant circuits in consumer, industrial, and defense electronics warrants their study. We will examine resonant circuits from the point of view of their frequency-domain and time-domain properties.

1.2.2 Frequency-Domain Description of Resonance

When a sinusoidal source is applied to a stable linear circuit all of the steady-state node voltages and branch currents in the circuit will be sinusoids having the same frequency as the input. A sinusoidal input signal produces a sinusoidal steady-state output (response). The steady-state response of a given node voltage or branch current, however, may have different amplitude and phase than those of the input signal. This relationship is illustrated by the block diagram in Fig. 1.2, in which the input signal $A_{in} \sin(\omega t + \phi_{in})$ produces a steady-state output signal (i.e., a voltage or current) $y_{ss}(t) = A_{out} \sin(\omega t + \phi_{out})$.

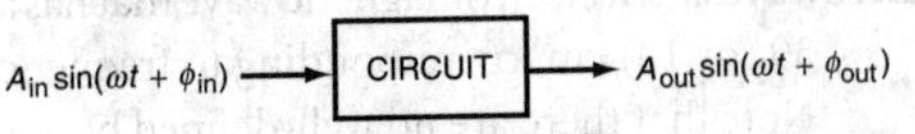

$$A_{in} \sin(\omega t + \phi_{in}) \longrightarrow \boxed{\text{CIRCUIT}} \longrightarrow A_{out} \sin(\omega t + \phi_{out})$$

FIGURE 1.2 Input–output relationship for linear circuits in sinusoidal steady state.

In general, when a linear circuit's input signal is the sinusoid $v_{in}(t) = A_{in} \sin(\omega t + \phi_{in})$ for $t \geq 0$, the steady-state output signal will be a sinusoid described by $y_{ss}(t) = A_{out} \sin(\omega t + \phi_{out})$. The magnitude and phase of the steady-state output signal are related to the magnitude and phase of the circuit's input sinusoidal signal by

$$A_{out} = A_{in} |H(j\omega)|$$

and

$$\phi_{out} = \phi_{in} + \text{ang} H(j\omega)$$

where $|H(j\omega)|$ denotes the magnitude of the complex-valued s-domain input-output transfer function $H(s)$, and ang $H(j\omega)$ denotes the angle of $H(s)$, with both quantities evaluated at $s = j\omega$ in the complex plane. It is important to realize that the steady-state output signal will be a scaled and time-shifted copy of the input signal, as depicted in Fig. 1.3. The time t_{ss} denotes the time at which the circuit is considered to be in the steady state. The magnitude of the steady-state output signal $y_{ss}(t)$ is created by scaling the magnitude of the input signal by $|H(j\omega)|$; the phase angle

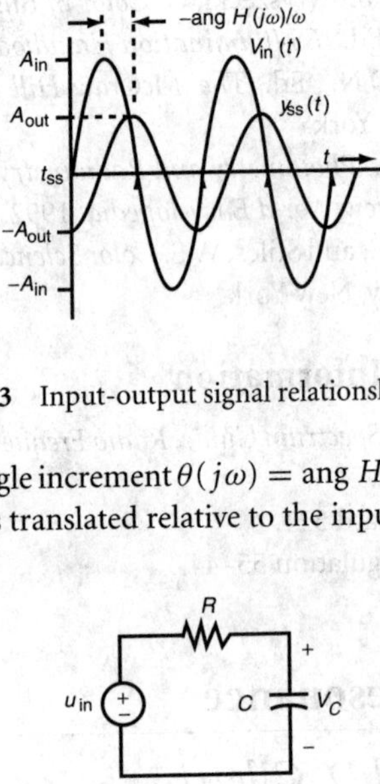

FIGURE 1.3 Input-output signal relationships.

of the steady-state output signal is created by adding the phase angle increment $\theta(j\omega) = \text{ang} H(j\omega)$ to the phase angle of the input signal. The steady-state output signal is translated relative to the input signal on the time axis by an amount $\tau = -\text{ang} H(j\omega)/\omega$.

The relationship between a circuit's sinusoidal input signal and its steady-state output signal can be represented in the frequency domain by the Bode magnitude and phase responses of the circuit [DeCarlo and Lin, 1995; Irwin 1995]. For example, the capacitor voltage in the simple RC circuit shown in Fig. 1.4 has the Bode magnitude $|H(j\omega)|$ and Bode phase response, $\theta(j\omega)$ shown in Fig. 1.5.

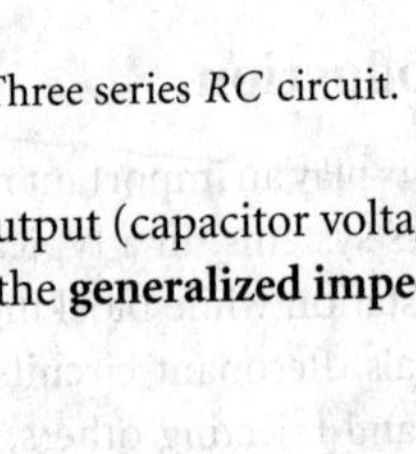

FIGURE 1.4 Three series RC circuit.

The s-domain input-output transfer function $H(s)$ relating the output (capacitor voltage) to the input source voltage in Fig. 1.4 is obtained by using voltage division with the **generalized impedances** (Ciletti, 1988) in the series RC circuit

$$H(s) = Z_C(s)/[Z_R(s) + Z_C(s)]$$

where $Z_C(s) = 1/(sC)$ and $Z_R(s) = R$. Making these substitutions leads to

$$H(s) = 1/(1 + sRC)$$

and

$$H(j\omega) = 1/(1 + j\omega RC)$$

with

$$|H(j\omega)| = 1/\sqrt{(1 + \omega^2 R^2 C^2)}$$

and

$$\theta(j\omega) = \text{ang}[H(j\omega)] = -\tan^{-1}(\omega RC)$$

(Note: Generalized (s-domain) impedances $Z(s)$ and admittances $Y(s)$ obey the same algebraic laws of series and parallel combination as do resistors, thereby simplifying circuit analysis.)

The Bode magnitude response shown in Fig. 1.5 for the response of the capacitor voltage in the RC circuit in Fig. 1.4 said to be a low-pass response, indicating that sinusoidal sources with low frequencies will be less attenuated in steady state than those with relatively high frequency. In fact, the graph of $|H(j\omega)|$ is considered to be relatively flat for $\omega \leq \omega_c$, with ω_c the so-called cutoff frequency of the filter.

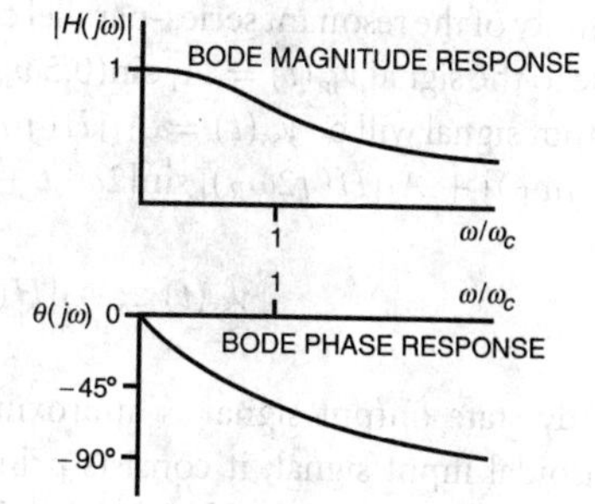

The cutoff frequency of a filter is determined by the value of the circuit's components; here, $\omega_c = 1/(RC)$. The cutoff frequency of a low-pass filter has the significance that a sinusoid signal the frequency of which is outside of the passband of the

FIGURE 1.5 Bode magnitude and phase responses for a simple RC lowpass filter.

filter contributes less than 50% of the power to the output than would a DC signal having the same input amplitude. At low frequencies the output signal's amplitude will be a close approximation to that of the input. We note that $|H(j0)| = 1$ and $|H(j\omega_c)| = 0.707$ for this circuit.

1.2.3   Series-Parallel RLC Resonant Filter

A circuit's ability to selectively attenuate signals at certain frequencies is determined by its topology and by the value of its physical components. For certain values of its components, a circuit's Bode magnitude response might be much sharper in shape than for other choices of components. When a circuit's Bode magnitude response exhibits a sharp characteristic the circuit is said to be in **resonance**.

The Bode magnitude response of the simple RC circuit will always have the shape shown in Fig. 1.5 and can never exhibit resonance. On the other hand, the capacitor voltage in the series or parallel RLC circuit shown in Fig. 1.6 has the sharp Bode magnitude response shown in Fig. 1.7 when $R = 10\ \Omega$, $L = 1$ H, and $C = 1$ F (frequency scaling can be

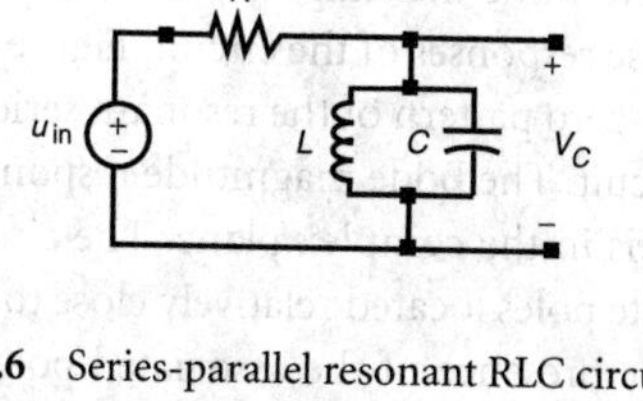

FIGURE 1.6   Series-parallel resonant RLC circuit.

used to obtain more realistic component values for a given cutoff frequency). This circuit is distinguished by its sharp Bode magnitude response and is said to be a resonant circuit. We note, however, that for other choices of its component values the same circuit might not exhibit the sharp Bode magnitude response that is characteristic of resonance (e.g., $R = 1\,\Omega, L = 1\,H, C = 1\,F$). The resonant frequency of this circuit, ω_R, can be shown to be given by $\omega_R = 1/\sqrt{(LC)}$.

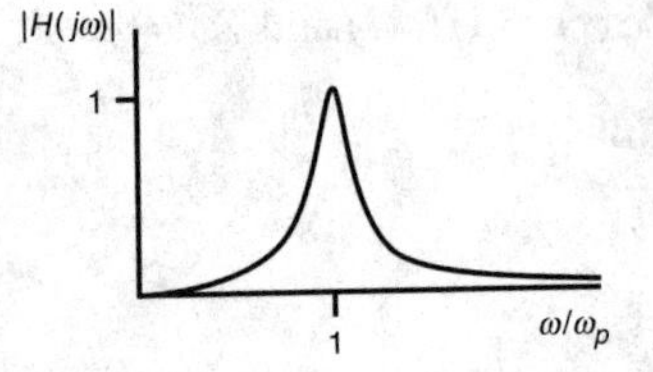

The **transfer function** relating the output voltage (across L and C) to the input voltage of the series/parallel RLC circuit is obtained as follows:

FIGURE 1.7 Bode magnitude response for a resonant series-parallel RLC circuit.

$$H(s) = Z_{LC}(s)/[R + Z_{LC}(s)]$$
$$= \{Z_L(s)Z_C(s)/[Z_L(s) + Z_C(s)]\}/\{R + Z_L(s)Z_C(s)/[Z_L(s) + Z_C(s)]\}$$
$$= \{(sL)[1/(sC)]/[sL + 1/(sC)]\}/\{R + (sL)[1/(sC)]/[sL + 1/(sC)]\}$$
$$= [s/(RC)]/[s^2 + s/(RC) + 1/(LC)]$$

The utility of the resonant series-parallel circuit is demonstrated by considering the circuit's **steady-state response** to the signal $v_{in}(t) = A_1 \sin(0.5\omega_R t + \phi_1) + A_2 \sin(\omega_R t + \phi_2) + A_3 \sin(2\omega_R t + \phi_3)$. The steady-state output signal will be $y_{ss}(t) = A_1|H(j0.5\omega_R)| \sin[0.5\omega_R t + \phi_1 + \theta_1(j\omega_R)] + A_2|H(j\omega_R)| \sin[\omega_R t + \phi_2 + \theta_2(j\omega_R)] + A_3|H(j2\omega_R)| \sin[2\omega_R t + \phi_3 + \theta_3(j\omega_R)]$, which can be approximated by

$$y_{ss}(t) \simeq A_2|H(j\omega_R)| \sin[\omega_R t + \phi_2 + \theta_2(j\omega_R)]$$

The steady-state output signal is approximately a scaled and time-translated (phase-shifted) copy of the sinusoidal input signal; it consists primarily of a sinusoidal signal having frequency $\omega = \omega_R$. The components of the input signal at frequencies $0.5\omega_R$ rad/s and $2\omega_R$ rad/s will be effectively filtered by the circuit and make minimal contribution to the output voltage.

1.2.4 The Pole–Zero Pattern Description of Resonance

The pole-zero pattern of the input/output transfer function, $H(s)$, of a circuit is formed by plotting the location of the roots of the polynomials comprising the numerator and denominator of $H(s)$, when $H(s)$ is expressed as a ratio of two polynomials. The roots of the numerator polynomial are called the *zeros* of $H(s)$, and those of the denominator are called the *poles* of $H(s)$. The key relationship is that the location of the poles and zeros in the s plane determine the shape of the Bode magnitude and phase responses of the circuit. Figure 1.8 shows the pole-zero pattern of the resonant series-parallel

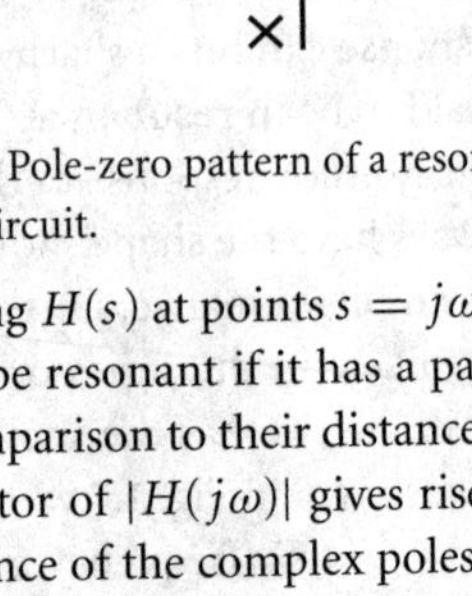

FIGURE 1.8 Pole-zero pattern of a resonant series-parallel RLC circuit.

RLC circuit. The Bode magnitude response is obtained by evaluating $H(s)$ at points $s = j\omega$ on the imaginary axis in the complex plane. The circuit's Bode response will be resonant if it has a pair of complex conjugate poles located relatively close to the imaginary axis in comparison to their distance from the real axis. The presence of the associated pole factors in the denominator of $|H(j\omega)|$ gives rise to the sharp resonant peak feature in the Bode magnitude response. (The distance of the complex poles from the real axis determines the frequency of the circuit's damped frequency of oscillation, and the distance of the pole pair determines the decay factor of the oscillation. The same is true of higher-order circuits having multiple repeated poles located near the j axis.

1.2.5 Time-Domain Description of Resonance

Time-domain methods are often used to characterize linear circuits, and can also be used to describe resonance. When an electrical circuit exhibits an undamped oscillatory or slightly damped behavior it is said to be in *resonance*, and the waveforms of the voltages and currents in the circuit can oscillate indefinitely.

The response $y(t)$ of a linear circuit to an input signal $u_{in}(t)$ comprises the sum of two parts. The *zero-input response* $y_{ZIR}(t)$ is that part of $y(t)$ due solely to the energy that was stored in the circuit's capacitors and inductors before the input signal was applied. The *zero-state response* $y_{ZSR}(t)$ is the response that the circuit exhibits to the input signal when no energy is initially stored in the circuit. When a circuit has no initial stored energy it is said to be *initially relaxed*. For a linear circuit, $y(t) = y_{ZIR}(t) + y_{ZSR}(t)$, and $y(t)$ is called the *initial state response* of the circuit, denoted by $y_{ISR}(t)$. That is, the response of a linear circuit to an input signal is the sum of its response to its initial stored energy with the circuit's input signal set to zero and its response to the input signal when the circuit is initially relaxed.

The time-domain behavior of the zero-input response of a circuit is related to the frequency-domain property of resonance. In the case of a second-order circuit, its zero-input response will be overdamped, critically damped, or underdamped, depending on the value of the circuit's components. If the components are such that the response is highly underdamped, the circuit is said to be in resonance, and its zero-input response will be oscillatory in nature and will not decay rapidly. The relative proximity of the poles of the circuit's transfer function $H(s)$ to the j axis accounts for this oscillatory behavior. To see this, note that each distinct pair of complex poles in $H(s)$ contributes to $y_{ZIR}(t)$ a term having the following form:

$$y(t) = Ke^{-\alpha t}\sin(\omega_d t + \psi).$$

[If $H(s)$ has repeated (multiple) complex poles at the same location, the expression for $y(t)$ also includes a polynomial factor in the variable t.] The damping factor α determines the time constant of the decay of $y(t)$, with $\tau = 1/|\alpha|$, and the damped frequency of oscillation ω_d determines the frequency of the oscillation. When $\omega_d \gg |\alpha|$ the circuit is said to be resonant, and the period of oscillation is small compared to the time constant of decay. The time-domain waveform of the response is said to exhibit ringing. The complex poles associated with ringing are relatively closer to the j axis than to the real axis.

1.2.6 Resonance and Energy Storage in Inductors and Capacitors

The physical phenomena of resonance is due to an exchange of energy between electric and magnetic fields. In passive RLC circuits, the energy can be stored in the electrical field of a capacitor and transferred to the magnetic field of an inductor, and vice versa. In an active circuit, such as an op-amp bandpass filter with no inductors, energy can be exchanged between capacitors.

The energy stored in a capacitor when charged to a voltage v is $W_c = 1/2Cv^2$; the energy stored in an inductor carrying a current i is $W_L = 1/2Li^2$. In a passive RLC circuit, resonance involves an alternating exchange between the inductor current and the capacitor voltage, with inductor current decreasing from a maximum value to a minimum value and capacitor voltage increasing from a minimum value to a maximum value. When this exchange occurs with relatively little dissipation of energy, the circuit is in resonance.

The Ideal Parallel LC Resonant Circuit

The ideal (lossless) LC circuit shown in Fig. 1.9 illustrates the physical nature of resonance in circuits. The circuit is assumed to consist of an ideal inductor and capacitor, that is, the inductor has no associated series resistance and the capacitor has no associated shunt leakage conductance.

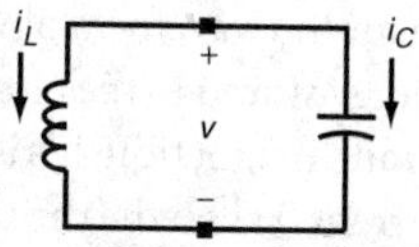

In the configuration shown in Fig. 1.9, the capacitor and inductor share a common current and

FIGURE 1.9   Ideal LC resonant circuit.

have the same voltage across their terminals. The capacitor current is described by $i_C = C dv/dt$, and the inductor voltage is described by $v = L di_L/dt$. The circuit topology imposes the condition that $i_L = -i_C$. These three relationships lead to the following differential equation model of the time-domain behavior of the circuit:

$$\frac{d^2v}{dt^2} + \frac{1}{LC}v = 0$$

The solution to this equation is a sinusoidal waveform having the parametric description

$$v(t) = K \sin(\omega t + \phi)$$

This can be verified by substituting the expression for $v(t)$ into the differential equation model and performing the indicated operations. The fact that $v(t)$ can be shown to have this form indicates that it is possible for this circuit to sustain oscillatory voltage and current waveforms indefinitely. When the parametric expression for $v(t)$ is substituted into the differential equation model the value of ω that is compatible with the solution of the equation is revealed to be $\omega = 1/\sqrt{(LC)}$. This is an example of the important fact that the frequency at which an electrical circuit exhibits resonance is determined by the physical value of its components. The remaining parameters of $v(t)$, K, and ϕ are determined by the initial energy stored in the circuit (i.e., the boundary conditions for the solution to the differential equation model of the behavior of the circuit's voltage).

Let v_0 and i_0 be the initial capacitor voltage and inductor current, respectively, and consider $t = 0^+$. Then the solution for $v(t)$ evaluated at $t = 0^+$ gives

$$v(0^+) = v_0 = K \sin \phi$$

and the capacitor current gives

$$i_C(0^+) = C dv(0^+)/dt = -i_L(0^+) = -i_0 = C[K\omega \cos(\phi)]$$

These two conditions must be satisfied simultaneously. The values of K and ϕ can be determined by considering three cases of the values of v_0 and i_0. Case 1: If $v_0 = 0$ then K must be nonzero for a nontrivial solution, and ϕ can be 0 or π. We choose $\phi = 0$ to satisfy the physical condition that $i_0 > 0 \Rightarrow C dv(0^+)/dt < 0$. Also, $K = -i_0/(C\omega)$. Thus, for this case, $v_C(t) = -i_0/(\omega C) \sin(\omega t)$ and $i_L(t) = i_0 \cos(\omega t)$. The capacitor voltage and the inductor current are 90° out of phase. Case 2: Similarly, if $i_0 = 0$ then $v_C(t) = v_0 \sin(\omega t + \pi/2)$ and $i_L(t) - C dv/dt = -C v_0 \omega \cos(\omega t + \pi/2) = C v_0 \omega \sin \omega t$. Case 3: When $v_0 \neq 0$ and $i_0 \neq 0$, we get

$$\phi = -\tan^{-1}(\omega C v_0 / i_0)$$

and

$$K = v_0 / \sin \phi$$

Two values of ϕ satisfy the first equation; the value for which the sign of $C dv(0^+)/dt$ is compatible with the sign of i_0 must be chosen. This leads to the solution for $v(t)$

$$v(t) = \sqrt{[v_0^2 + i_0^2/(\omega^2 C^2)]} \sin\left(\omega t - \tan^{-1}(\omega C v_0 / i_0)\right)$$

with $\omega = 1/\sqrt{(LC)}$.

The solution for the waveform of $v(t)$ depends on the initial capacitor voltage and inductor current, that is, the initial energy stored in the circuit's electric and magnetic fields. The exchange of energy between the circuit's electric and magnetic fields is evident from the phase relationship between the capacitor voltage and the inductor current. When the capacitor voltage is at a maximum or minimum value the inductor current is at a minimum or maximum value, and vice versa. When the energy stored in the capacitor's electric field is a maximum or minimum, the energy stored in the inductor's magnetic field is a zero, and vice versa.

The Ideal Series LC Resonant Circuit

The ideal (lossless) series LC section shown in Fig. 1.10 can exhibit resonance when imbedded in a circuit operating at the frequency $\omega = 1/\sqrt{(LC)}$. To see this, we question whether it is possible for the inductor and capacitor voltages to exactly cancel each other. Consider

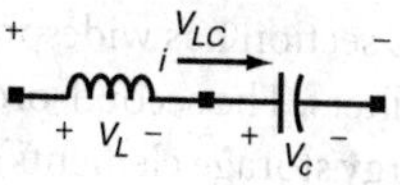

FIGURE 1.10   Ideal series LC section.

$$0 = v_L(t) + v_C(t)$$
$$= L\,di/dt + v_C(t)$$

Taking the derivative and multiplying the terms by the factor C gives

$$0 = LC d^2i/dt + C dv_c/dt$$
$$0 = LC d^2i/dt + C dv_c/dt$$
$$0 = LC d^2i/dt + i(t)$$

The parametric form of the solution of this second-order differential equation is $i(t) = K \sin(\omega t + \phi)$, which is the same form that was found for the parallel resonant LC section. Substituting this expression into the equation and canceling terms gives $\omega^2 = 1/(LC)$, or $\omega = 1/\sqrt{(LC)}$, the frequency at which the cancellation of v_L by v_C occurs. When the series LC section is imbedded in a circuit whose sinusoidal source has the frequency $\omega = 1/\sqrt{(LC)}$, the steady-state voltage across the section $v_{LC}(t)$ will be zero. The resonant series LC section appears to be a short circuit. This feature can be exploited in a filter whose output voltage is measured as $v_{LC}(t)$ and which must reject a signal at or very near the frequency $\omega = 1/\sqrt{(LC)}$.

Parallel LC Section Impedance

The utility of the parallel resonant LC section shown in Fig. 1.11 can be understood by examining the impedance that the circuit presents to the circuit in which it is imbedded. We have $Z_{LC}(s) = Z_L Z_c/(Z_L + Z_C) = (s/C)/[s^2 + 1/(LC)]$. $Z_{LC}(s)$ has a pair of poles on the j axis in the s plane. In resonance, the circuit presents infinite impedance (i.e., an open circuit) to the external circuit in which it is imbedded, that is, when the source frequency

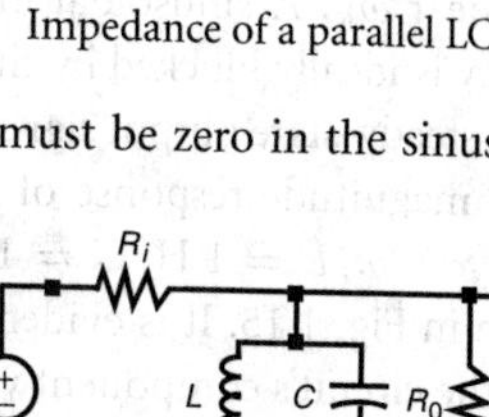

FIGURE 1.11   Impedance of a parallel LC section.

is $\omega = 1/[\sqrt{(LC)}]$. The physical consequence of this is that i_{LC} must be zero in the sinusoidal steady state.

When the resonant LC section is imbedded within the circuit shown in Fig. 1.12, and the source frequency is the same as the resonant frequency of the section, the circuit looks like the voltage source is connected to a series circuit consisting of R_i in series with R_0. All of the current from the source is delivered to the output resistor. At other frequencies, the impedance of the LC section provides a path for

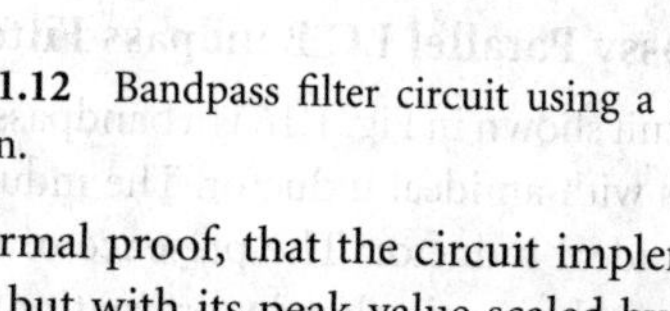

FIGURE 1.12   Bandpass filter circuit using a resonant LC section.

current to bypass R_0. From this we can conclude, without formal proof, that the circuit implements a bandpass Bode magnitude characteristic like that in Fig. 1.7, but with its peak value scaled by voltage division to be $R_0/(R_i + R_0)$.

Parallel LC Bandstop Filter

The parallel LC section has widespread application in a variety of filters. The second-order (i.e., two independent energy storage elements) bandstop filter in Fig. 1.13 exploits the fact that the LC section is capable of blocking all current through it from the external circuit when it is operating at its resonant frequency.

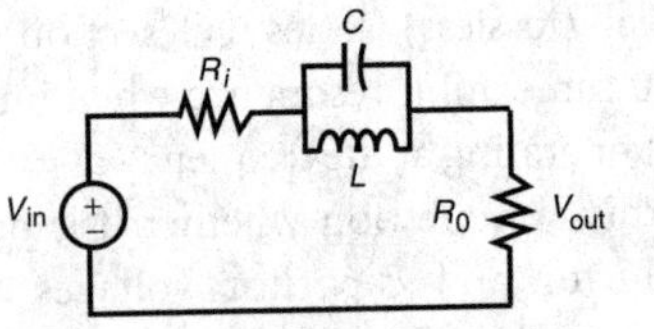

FIGURE 1.13 Bandstop filter using a resonant parallel resonant LC section.

When the frequency of the sinusoidal source, v_{in}, is the same frequency as the resonant frequency of the circuit, the LC section effectively blocks the steady-state flow of current to the load resistor R_0. This can be confirmed by examining the s-domain transfer function between the source and the load. Using s-domain models for inductor impedance $[Z_L(s) = sL]$ and capacitor impedance $[Z_C(s) = 1/(sC)$, we apply Kirchhoff's current and voltage laws to the circuit and, using series/parallel reduction as we would for resistors, find the following expression for $H(s)$:

$$H(s) = V_0(s)/V_{in}(s) = \frac{R_0}{R_i + R_0} \frac{s^2 + \dfrac{1}{LC}}{s^2 + \dfrac{s}{(R_i + R_0)C} + \dfrac{1}{LC}}$$

which fits the generic form of the transfer function for a second-order bandstop filter [Ciletti, 1988]

$$H(s) = \frac{s^2 + \omega_R^2}{s^2 + s\omega_b + \omega_R^2}$$

with $\omega_b = 1/[(R_i + R_0)C]$ and $\omega_R = 1/\sqrt{(LC)}$. The sharpness of the filter's Bode plot can be described in terms of the *resonant quality factor* $Q = \omega_R/\omega_b$. A large value of Q implies a highly resonant circuit. The pole-zero pattern of the generic second-order bandstop filter with $R_i = 10\ \Omega$, $L = 1$ H, $C = 1$ F, and $R_0 = 1\ \Omega$ is shown in Fig. 1.14; it has a pair of zeros located on the j axis in the s domain at $\pm j\omega_R$. The poles of the generic bandstop filter are located at $s = 1/2 - \omega_b \pm j\sqrt{(\omega_b^2 - 4\omega_R^2)}$. When Q is high, the poles are very close to the zeros of $H(s)$ and the filter sharply rejects signals with frequencies near ω_R. A sinusoidal input signal at this frequency is ideally blocked by the filter in the sense that its steady-state output signal will be zero.

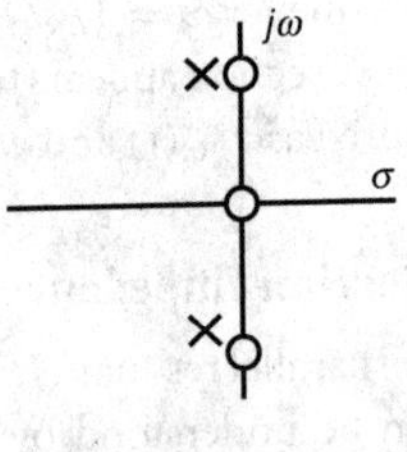

FIGURE 1.14 Pole-zero pattern of a second order bandstop filter.

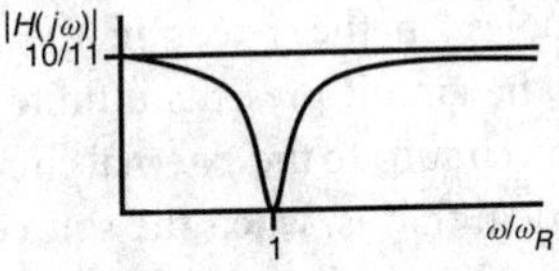

FIGURE 1.15 Bode magnitude response of a second-order bandstop filter.

The Bode magnitude response of the bandstop filter for $R_i = 10\omega$, $L = 1$ H, $C = 1$ F, and $R_0 = 1\ \Omega$ is shown in Fig. 1.15. It is evident that careful selection of the circuit's component values will result in a selective bandstop filter, one that rejects signals in a very narrowband of frequencies.

The Lossy Parallel LC Bandpass Filter

The circuit shown in Fig. 1.16 is a bandpass filter. It has a physical inductor, that is, one that has a resistance in series with an ideal inductor. The inductor in a physical circuit will always have a resistance in series with an ideal inductor. The presence of the series resistor has the effect of moving the poles of $H(s)$ away from the j axis, thereby preventing the transfer function from exhibiting a perfect open circuit at resonance.

This can be verified by deriving the impedance of the lossy parallel LC section: $Z_{LC}(s) = (1/C)(s + r/L)/[s^2 + sr/L + 1/(LC)]$. Note that the denominator of $Z_{LC}(s)$ cannot have poles on the j axis unless $r = 0$. The physical implication of this condition is that $Z_{LC}(s)$ will always have a nonzero value of I_{LC} in steady state, and $Z_{LC}(s)$ cannot behave like an ideal open circuit. In practice, when r is very small the circuit behaves for all practical purposes as an ideal LC section.

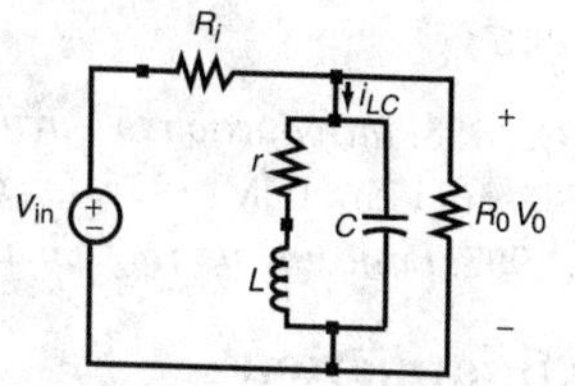

FIGURE 1.16 Lossy parallel LC filter.

Active Resonant Bandpass Filter

The op-amp circuit in Fig. 1.17 implements a bandpass filter without inductors. The input-output transfer function of this circuit can be obtained by using Kirchhoff's current law at the node common to R_1, R_2, C_1, and C_2, and by noting that C_2 and R_3 share a common current. Furthermore, the output voltage appears across R_3. The transfer function $H(s)$, between the input voltage source and the output node of the filter is given by

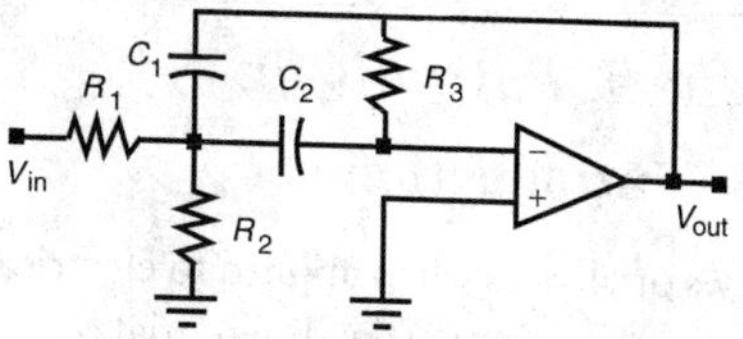

FIGURE 1.17 Op-amp bandpass filter.

$$H(s) = \frac{-s/(R_1 C_1)}{s^2 + s\omega_b + \omega_p^2}$$

where $\omega_b = (C_1 + C_2)/(R_3 C_1 C_2)$ and $\omega_p^2 = (R_1 + R_2)/(R_1 R_2 R_3 C_1 C_2)$. The bandwidth of the filter is ω_b and the resonant frequency is ω_p. The design of a sharply resonant (high-Q) active bandpass filter will have $Q \gg 1$.

1.2.7 Physical Hazards with Resonant Circuits

The voltages and currents in a resonant circuit can be very high, and therefore warrant caution. Under the condition of resonance, the circuit in Fig. 1.11 (parallel LC) is such that the LC section looks like an open circuit to the circuit in which it is imbedded. However, the current that circulates between the inductor and the capacitor in the LC section itself can be quite large. This has implications for the size of the conductor that must conduct the resonant current and dissipate heat. Likewise, when the operating conditions of the series LC circuit in Fig. 1.10 place the circuit in resonance, the voltage v_{LC} must be zero, but v_L and v_C can be very large, especially in high-Q circuits, thereby posing a hazard to the careless handler of the circuit.

Defining Terms

Generalized impedance: An s-domain transfer function in which the input signal is current source and the output signal is a voltage in the circuit.

Resonance: A condition in which a circuit exhibits a highly selective frequency-dependent attenuation of sinusoidal input signals.

Steady-state response: The response of a circuit after a sufficient time has elapsed to allow the value of the transient response to become insignificant.

Transfer function: An s-domain function that determines the relationship between an exponential forcing function and the particular solution of a circuit's differential equation model. It also describes a relationship between the Laplace transform s-domain spectral representation of a circuit's input signal and the Laplace transform s-domain spectral description of its output signal when the circuit is initially at rest (no stored energy in its capacitors and inductors) and excited by the source.

References

Ciletti, M.D. 1988. *Introduction to Circuit Analysis and Design.* Holt, Rinehart, and Winston, New York.
DeCarlo, R.A. and Lin, P.-M. 1995. *Linear Circuit Analysis.* Prentice-Hall, New York.
Irwin, J.D. 1995. *Basic Engineering Circuit Analysis,* 5th ed. Macmillan, New York.

Further Information

For further information on the basic concepts of resonance in linear circuits, *see Circuits, Devices, and Systems* by R.J. Smith and R.C. Dorf. For an introductory discussion of the design of active filters see the above-cited reference *Introduction to Circuit Analysis and Design* by Ciletti.

1.3 Electroacoustics

Eugene T. Patronis, Jr.

1.3.1 Introduction

The laws of physics when applied to electrical, mechanical, acoustical, or other systems amount to expressions of ordinary or partial differential equations. The work of solving these equations can be tedious, and techniques have been developed over a period of time to both shorten and simplify the procedure in certain classes of problems. In particular, the area of electrical circuit analysis, early on the concept of lumped circuit elements each possessing an impedance property was introduced. This impedance property is the voltage-to-current ratio of the circuit element in question and was identifiable with a corresponding differential operator in the governing differential equation. The representation of each lumped circuit element by its associated impedance in effect converts the differential equation into a simpler algebraic equation. In electroacoustics this electrical circuit analysis technique is extended to include mechanical as well as acoustical circuit elements. There remain, however, certain situations in which this approach is not viable such as extended or distributed systems whose dimensions are comparable to or exceed the sound wavelength and those systems involving distant radiation phenomena. These systems must be handled by a more traditional approach. Some of the more important results of the most basic of these systems will be presented, first in a form that leads naturally to a completion of the description employing the techniques of electroacoustics.

1.3.2 Linear Acoustics

The equations of linear acoustics stem from Euler's equations for a perfect compressible fluid.

Newton's second law

$$\rho_0 \frac{\partial u}{\partial t} + \nabla p = 0$$

Conservation of mass

$$\frac{\partial \rho}{\partial t} + \rho_0 \nabla \cdot u = 0$$

from which we find the wave equations

$$\frac{\partial^2 p}{\partial t^2} = c^2 \nabla^2 p$$

$$\frac{\partial^2 \rho}{\partial t^2} = c^2 \nabla^2 \rho$$

$$\frac{\partial^2 \varphi}{\partial t^2} = c^2 \nabla^2 \varphi$$

where

$$u = -\nabla \varphi \qquad c^2 = \left(\frac{dp}{d\rho}\right)_0$$

When the fluid medium is air, the dependent variables in these equations are:

p = **acoustic pressure**, which is the variation above or below static atmospheric pressure, Pa

ρ = air density where ρ_0 is the static air density, kgm/m^3

u = air particle velocity, m/s

φ = scalar potential function from which the vector particle velocity is derivable, m^2/s

c = sound phase velocity, m/s

1.3.3 Radiation Models

The classic starting point for treating acoustic radiation is the pulsating sphere of nominal radius a and surface area S that is alternately expanding and contracting such that the velocity of the spherical surface is directed along radial lines and has a sinusoidal dependence on the time such that $u = u_m \cos(\omega t)$, where u is the instantaneous surface velocity (m/s), u_m the surface velocity amplitude (m/s), and ω the angular frequency (rad/s). The wave equation can be solved quite readily for this simple geometry. Additionally, if the nominal radius of the sphere, a, is small compared with the wavelength, $a < l$, and the observation point is very distant from the center of the sphere, $r \gg l$, the radiation is isotropic with the acoustic pressure varying inversely with the radial distance r, as depicted in Fig. 1.18. In Fig. 1.18, p represents the acoustic pressure phasor at the observation point, except that the phase lag associated with wave propagation along the path from the source has been omitted. In addition to the pulsating sphere, Fig. 1.18 displays the results for an acoustic dipole composed of two identical pulsating spheres arranged so that

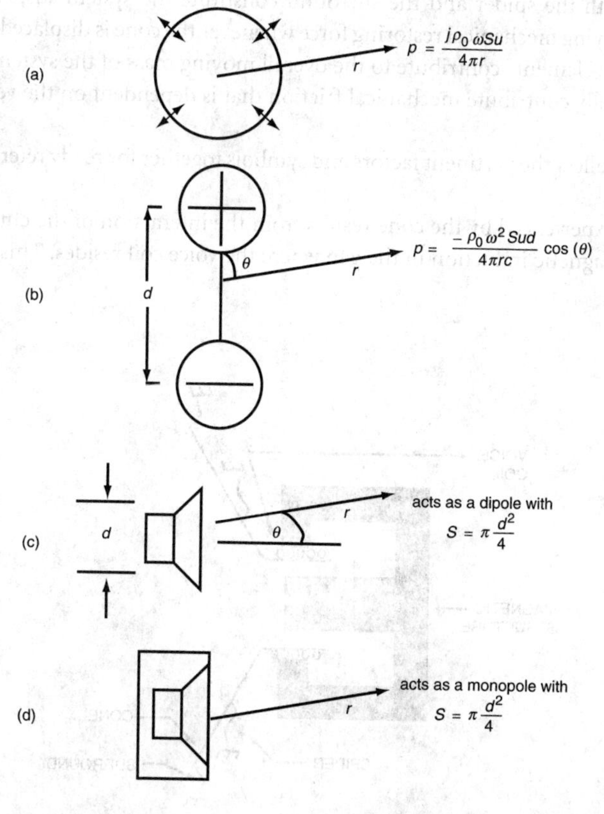

(a) $p = \dfrac{j\rho_0\,\omega S u}{4\pi r}$

(b) $p = \dfrac{-\rho_0\,\omega^2 S u d}{4\pi r c}\cos(\theta)$

(c) acts as a dipole with $S = \pi\dfrac{d^2}{4}$

(d) acts as a monopole with $S = \pi\dfrac{d^2}{4}$

FIGURE 1.18 Radiation models: (a) pulsating sphere [acoustic monopole], (b) two out of polarity pulsation spheres [acoustic dipole], (c) bare loudspeaker, (d) back enclosed loudspeaker.

when one sphere expands the other contracts by an equal amount. Two practical sound sources and their properties at low frequencies are also displayed in Fig. 1.18. In each instance it is found that the acoustic pressure is directly proportional to the surface velocity of the radiator whether it be a theoretical sphere or a loudspeaker diaphragm. Attention will now be turned to a determination of this surface velocity for a typical loudspeaker structure.

1.3.4 Dynamic Low-Frequency Loudspeaker

Analysis will be made of a low-frequency transducer of the conventional type that consists of a piston in the form of a truncated cone of some lightweight but hopefully rigid material flush mounted in a sealed cabinet and arranged to radiate into a half-space. Half-space radiation doubles the acoustic pressure for a given velocity and is assumed because such devices are usually positioned close to at least one large plane surface such as a floor. To describe the piston motion and its subsequent radiation, knowledge is required not only of the acoustical behavior of the device but also of its purely mechanical and electrical properties as well. Figure 1.19 illustrates in a simplified form the features of the device construction that impact on its motional behavior.

The voice coil is a helical spiral of wire wound on a hollow cylinder with the entire assembly attached to the cone, which acts as a piston. The magnetic structure is arranged such that the magnetic flux passes radially through the cylindrical surface about which the voice coil is wound. This is achieved by having a permanent magnet attached to a cylindrical pole piece that is placed within and concentric with the voice coil former. A cylindrically symmetric magnetic return path straddles the outer surface of the voice coil such that the voice coil resides symmetrically in the magnetic air gap. The *spider* centers the voice coil within the magnetic gap whereas the *surround* centers the base of the cone in the overall supporting frame. Both the spider and the surround constitute the system suspension and behave as linear springs by supplying mechanical restoring force whenever the cone is displaced from its equilibrium position. Each of these elements contribute to the overall moving mass of the system and the suspension components additionally contribute mechanical friction that is dependent on the velocity of the moving system.

It is convenient to collect the pertinent factors and symbols together for ready reference, and this is done in Table 1.1.

The driving force experienced by the cone results from the interaction of the current in the voice coil conductor with the magnetic induction in the gap where the voice coil resides. This is an axially directed

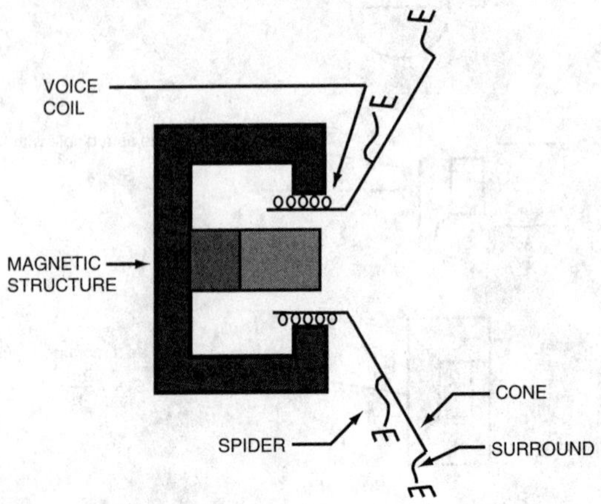

FIGURE 1.19 Simplified loudspeaker construction.

TABLE 1.1 Loudspeaker Descriptive Factors

Factor Description	Symbol	Typical Value
Piston radius, m	a	0.10
Piston area, m^2	$S = \pi a^2$	0.0314
Mechanical moving mass, kgm	M	0.03
Suspension stiffness, N/m	K	3000
Suspension resistance, kgm/s	R_m	2.5
Electrical voice coil resistance, Ω	R_e	6
Electrical voice coil inductance, H	L_e	0.001
Magnetic induction in gap, T	B	1
Length of conductor in gap, m	l	10
Instantaneous voice coil current, A	i	
Enclosure interior air volume, m^3	V_0	0.045

force given by Bli. A restriction to sinusoidal time-dependent currents will greatly simplify the analysis without loss of generality. In this way time will enter as $e^{j\omega t}$, actually as the real part of $e^{j\omega t}$, and because the system is linear, differentiation with respect to time is replaced by multiplication by $j\omega$ and integration with respect to time is replaced by division by $j\omega$. Newton's second law of motion, which equates acceleration to the totality of forces divided by the moving mass, can now be written

$$j\omega u = F_t/M$$

where

u = piston velocity, m/s
F_t = total force acting on piston, N
M = total moving mass, kgm

It is necessary at this point to delineate the individual forces acting on the piston.

Aside from the applied force F_a of the voice coil current, the piston experiences a restoring force F_s from the suspension whenever the piston is displaced from its equilibrium position, a frictional force F_f resulting from flexure of the spider and surround whenever the piston is in motion, a restoring force F_b resulting from compression or rarefaction of the air trapped in the sealed cabinet behind the loudspeaker cone whenever the piston is displaced from its equilibrium position, and a force of radiation F_r exerted by the air in front of the piston whenever the piston is moving so as to generate sound waves in the air. Table 1.2 collects all of these forces and indicates their relationship to the piston velocity which is to be considered the dependent variable in the following analysis.

TABLE 1.2 Forces Acting on Loudspeaker Cone

Force	Origin	Value
F_a	Applied	Bli
F_s	Suspension	$\dfrac{-Ku}{j\omega}$
F_b	Air in box	$\dfrac{-\rho_0 c^2 S^2 u}{V_0 j\omega}$
F_f	Motional friction	$-R_m u$
F_r	Radiation	$-\rho_0 c S \left[R_1\left(\dfrac{2\omega a}{c}\right) + jX_1\left(\dfrac{2\omega a}{c}\right) \right] u$

Source: Kinsler, L.E. and Frey, A.R. 1962. *Fundamentals of Acoustics.* Wiley, New York.

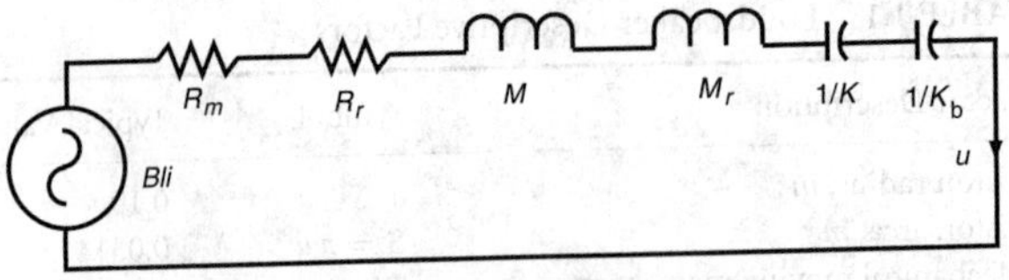

FIGURE 1.20 Equivalent circuit for an acoustical system.

It is now possible to write the equation of motion as a statement of Newton's second law in the form

$$Mj\omega u = Bli - \frac{Ku}{j\omega} - \frac{\rho_0 c^2 S^2 u}{V_0 j\omega} - R_m u - \rho_0 c S\left[R_1\left(\frac{2\omega a}{c}\right) + jX_1\left(\frac{2\omega a}{c}\right)\right]u \qquad (1.1)$$

The negative terms in this equation are forces that are oppositely directed to the piston velocity and, hence, oppose the motion. The term representing the force on the piston resulting from radiation is obscure and will be discussed later. For the moment, it is necessary to solve this equation for the piston velocity to obtain

$$u = \frac{Bli}{R_m + \rho_0 c S R_1\left(\frac{2\omega a}{c}\right) + j\left[\omega M + \rho_0 c S X_1\left(\frac{2\omega a}{c}\right) - \frac{K}{\omega} - \frac{\rho_0 c^2 S^2}{V_0 \omega}\right]} \qquad (1.2)$$

or

$$u = \frac{Bli}{Z_m} \qquad (1.3)$$

At this point it is possible to draw an analogy with an electrical system. In the electrical case the current is the ratio of the applied voltage to the complex circuit impedance. In this instance, if the velocity is identified with electrical current and the applied force with the applied voltage, the denominator of the velocity equation should be identified as a complex **mechanical impedance**. Additionally, in an electrical circuit consisting of several elements connected in series, the current is common to all elements and the applied voltage exists across series combination. In this mechanical circuit the velocity is the same for the voice coil, suspension, cone, and the air, and the force F_a exists across the combination. A possible analogous mechanical circuit thus appears as given in Fig. 1.20.

In Fig. 1.20 the positive mechanical reactances are symbolized by inductors and negative mechanical reactances are symbolized by capacitors as would be true in the analogous electrical situation. The constant applied force generator is symbolized as a voltage source. Also in Fig. 1.20, R_r is identified as $\rho_0 c S R_1(2\omega a/c)$, whereas K_b is identified as $\rho_0 c^2 S^2/V_0$ and M_r as $\rho_0 c S X_1(\frac{2\omega a}{c})/\omega$. Unlike the common electrical situation, however, those elements that stem from radiation phenomena, such as R_r and M_r, are themselves frequency dependent. This complication will be dealt with after the model has been more fully developed. The circuit of Fig. 1.20, though useful, can only be an intermediate result as it does not involve any of the purely electrical properties of the loudspeaker such as the voice coil resistance or self-inductance. In addition, whenever the voice coil is in motion it has induced in it a back electromotive force (emf) of size Blu that opposes the current in the voice coil. This situation is represented schematically in Fig. 1.21. The current in the circuit of Fig. 1.21 is given by

$$i = \frac{(E - Blu)}{(R_e + j\omega L_e)}$$

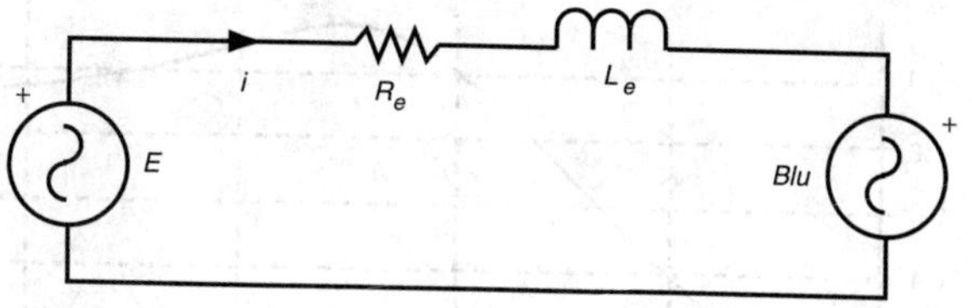

FIGURE 1.21 Equivalent circuit of a voice coil in motion.

whereas from the former equation of motion we have the piston velocity [Equation (1.2)]. When i is eliminated from these two equations the result becomes

$$u = \frac{\dfrac{Ebl}{R_e + j\omega L_e}}{Z_m + \dfrac{B^2 l^2 (R_e - j\omega L_e)}{R_e^2 + \omega^2 L_e^2}}$$

In this result all of the electrical as well as acoustical properties have been transformed into equivalent mechanical terms, and the appropriate circuit appears as Fig. 1.22. In Fig. 1.22 all of the system properties have been expressed in mechanical terms. The generator on the left is a force generator supplying the driving force expressed in Newtons. All of the impedances, whether from electrical, mechanical, or acoustical origin, have been transformed to equivalent mechanical ohms, that is, kilograms/second. To proceed further it is necessary to finally address the complicated frequency-dependent behavior of the radiation impedance. The real part of the radiation impedance accounts for the fact that some mechanical energy is removed from the loudspeaker cone and is carried away by subsequent wave propagation in the surrounding air. The imaginary part of the radiation impedance accounts for the fact that there exists a cyclic lossless exchange of energy between the loudspeaker cone and the air in the immediate vicinity of the loudspeaker. The radiation resistance, in mechanical ohms, can be expressed as

$$R_r = \rho_0 c S R_1 \left(\frac{2\omega a}{c} \right)$$

whereas the radiation reactance can be expressed as

$$X_r = \rho_0 c S X_1 \left(\frac{2\omega a}{c} \right)$$

The details of the behavior of the functions R_1 and X_1 are presented in Fig. 1.23.

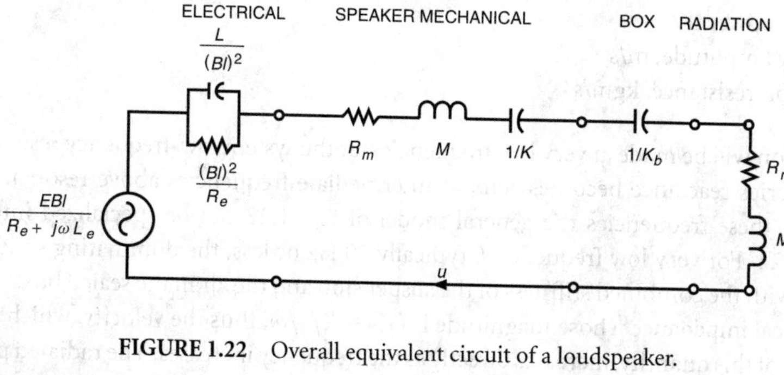

FIGURE 1.22 Overall equivalent circuit of a loudspeaker.

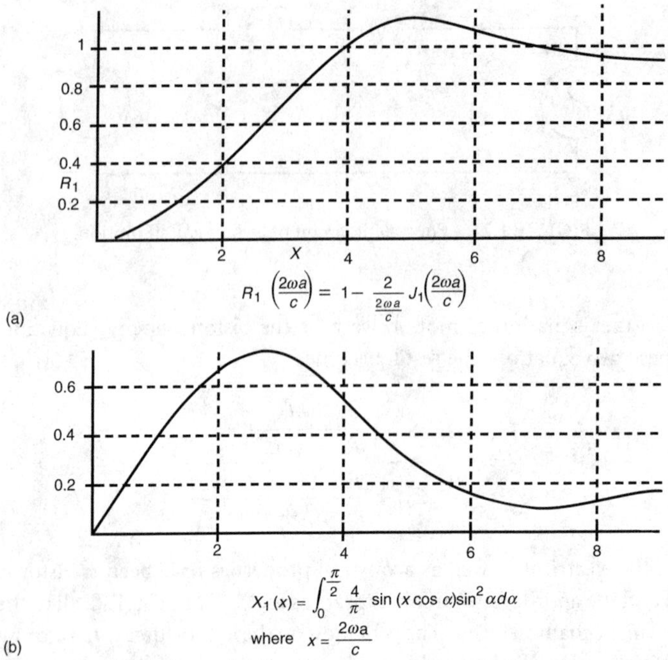

$$R_1\left(\frac{2\omega a}{c}\right) = 1 - \frac{2}{\frac{2\omega a}{c}} J_1\left(\frac{2\omega a}{c}\right)$$

(a)

$$X_1(x) = \int_0^{\frac{\pi}{2}} \frac{4}{\pi}\sin(x\cos\alpha)\sin^2\alpha\,d\alpha$$

where $x = \frac{2\omega a}{c}$

(b)

FIGURE 1.23   Piston impedance functions: (a) resistance function, (b) reactance function.

From Fig. 1.23, R_1 at low frequencies has a parabolic dependence on the frequency and, thus, grows as the square of the frequency, whereas at low frequencies X_1 has a linear dependence on the frequency. At high frequencies R_1 becomes constant at a value of 1, whereas X_1 slowly approaches 0. As a consequence, regardless of the frequency, both $R_r = \rho_0 c S R_1$ and $X_r = \rho_0 c S X_1$ will have a magnitude of $\rho_0 c S$ or very much lower.

1.3.5   Radiated Power

The model that has been developed can now usefully be employed to calculate the acoustical power radiated by the loudspeaker over a wide range of frequencies. The average radiated acoustical power can be expressed as

$$P = \frac{1}{2}u_m^2 R_r$$

where
   P = average power, W
   u_m = velocity amplitude, m/s
   R_r = radiation resistance, kgm/s

An examination will be made at very low frequencies, at the system low-frequency resonance frequency where the net series reactance becomes zero, at intermediate frequencies above resonance, and at high frequencies. For these frequencies the general model of Fig. 1.22 can be specialized into four cases as shown in Fig. 1.24. For very low frequencies, typically 20 Hz or less, the dominating series impedance is that associated with the combined stiffness of the suspension and the air in the sealed box. This presents an overall mechanical impedance whose magnitude is $(K + K_b)/\omega$, thus the velocity, which is proportional to the reciprocal of this quantity, increases directly as the frequency increases. The radiated power, however,

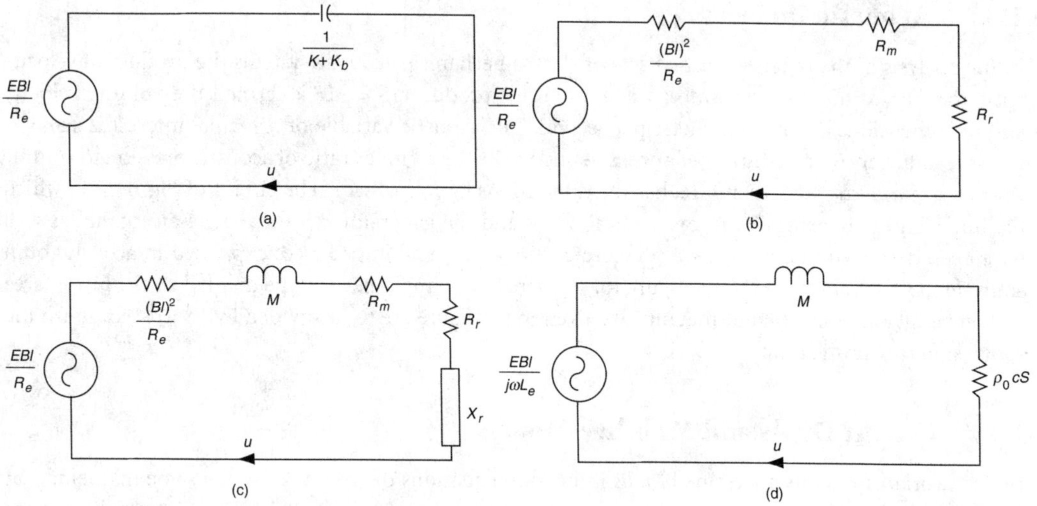

FIGURE 1.24 Specialized classes of the general model: (a) very low frequencies, (b) low frequency system resonance, (c) intermediate frequencies above resonance, (d) high frequency.

depends on the product of the velocity squared with the radiation resistance. The radiation resistance itself grows as the square of the frequency in this frequency range and, hence, the radiated power increases with the fourth power of the frequency or 12 dB per octave. In the Fig. 1.24(b) the operating frequency is such that the positive reactances of the moving mass and that contributed by radiation reactance are just annulled by the negative reactances of the suspension and the air in the box. For the typical values in Table 1.1 this occurs at about 70 Hz. The Fig. 1.24(c) describes a broad frequency range above resonance where the mass and radiation reactances dominate the circuit behavior. Both of these reactances are increasing linearly with frequency and, hence, the cone velocity decreases inversely with the frequency. The radiation resistance still increases as the square of the frequency in this range and, hence, the product $u_m^2 R_r$ will be a constant independent of frequency, as consequently will be the radiated power. This range extends from just above the resonance frequency to a few thousand hertz, that is, more for small pistons and less for large ones. Finally, in the high-frequency range of the Fig. 1.24(d), the radiation resistance has become the constant $\rho_0 c S$ while the output of the force generator falls inversely with the frequency and the mass reactance increases directly with the frequency. As a consequence, the velocity decreases as the square of the frequency and the radiated power falls as the fourth power of the frequency or -12 dB per octave. These results are shown in Fig. 1.25. In Fig. 1.25 the ordinate has been taken as $10\,\mathrm{dB}\log P/P_{\mathrm{mid}}$, where P_{mid} is the power radiated at the midband of the device.

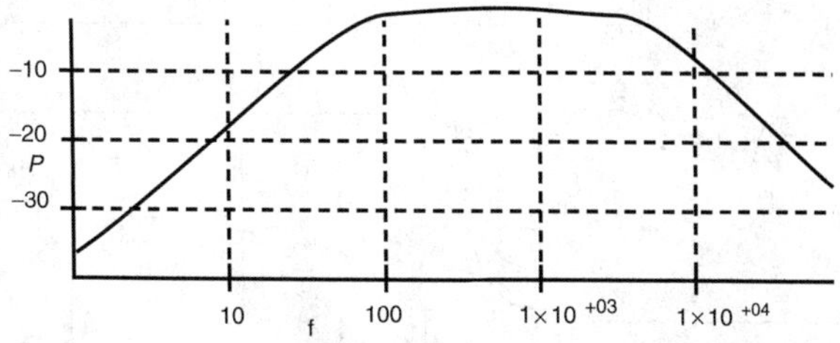

FIGURE 1.25 Loudspeaker power response by circuit analysis.

1.3.6 Acoustic Impedance

In the analysis of the purely acoustic parts of electromechanical acoustic systems the product of velocity with area is a commonly encountered quantity. This product $uS = U$ is termed the **volume velocity** and its dimensions are in cubic meter per second. The acoustic variable of principal interest is acoustic pressure and, hence, **acoustic impedance** is defined to be the complex ratio of acoustic pressure to volume velocity. An *acoustic ohm*, therefore, has the dimensions kgm/(m^4 sec). The circuit of Fig. 1.22 in which the impedances are expressed in mechanical ohms and the generator is a force generator could as well have been described as employing a pressure generator with all impedances expressed in acoustic ohm equivalents. The value given in the figure for the generator simply needs to be divided by the piston area S whereas all of the impedance magnitudes given in the figure are to be divided by S^2 to accomplish the conversion to acoustic ohms.

1.3.7 Circuit Duals and Mobility Models

Two networks or circuits are termed *duals* if the loop equations of the one have the same mathematical form as the node equations of the other. In the electrical case, in going from a given circuit to its dual, series connections are replaced by parallel connections, resistance by conductance, inductance by capacitance, voltage by current, impedance by admittance, and voltage source by current source. Duality also applies to analogous circuit models because the basis for the construction of the circuit model is the mathematical form of the governing equation of the physical system being represented by the model in analogous form.

The mechanical and acoustic parts of the loudspeaker circuit of Fig. 1.22 can be represented by its dual as displayed in Fig. 1.26(a). In this circuit the applied force Bli flows in the circuit and the piston velocity u appears across the circuit. The complex ratio of the piston velocity to the applied force is the **mechanical** admittance or **mobility**, which is just the reciprocal of the former mechanical impedance, as

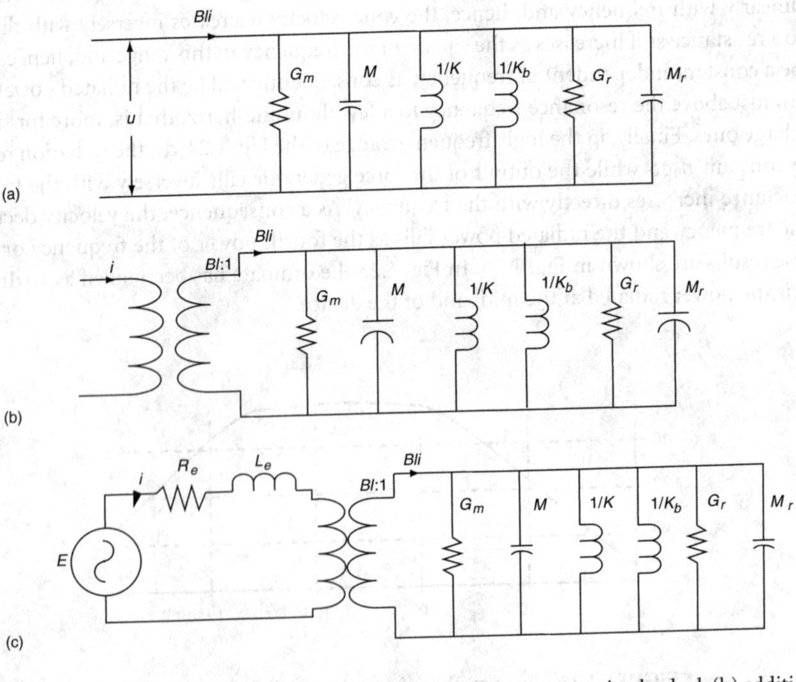

FIGURE 1.26 Mechanical and acoustical parts of the loudspeaker: (a) representation by dual, (b) addition of an ideal transformer, (c) incorporation of electrical aspects.

will be shown shortly. It should be observed that all former inductors have been replaced by capacitors, all former capacitors by inductors, and all former resistances by conductances of equal value, that is, G_m is R_m, etc.

In Fig. 1.26(a) the flow in each element is the quantity applied across the element multiplied by the admittance of the element. Hence, the total flow Bli is given by

$$Bli = u\left[G_m + j\omega M - \left(\frac{jK}{\omega}\right) - \left(\frac{jK_b}{\omega}\right) + G_r + j\omega M_r\right]$$

When the appropriate values are substituted for G_m and G_r and when K_b and M_r are replaced by the identifications given earlier, this equation becomes

$$Bli = u\left\{R_m + \rho_0 cSR_1\left(\frac{2\omega a}{c}\right) + j\left[\omega M + \rho cSX_1\left(\frac{2\omega a}{c}\right) - \frac{K}{\omega} - \frac{\rho_0 c^2 S^2}{V_0\omega}\right]\right\}$$

which when solved for u yields the former equation of motion (1.1) from which

$$\frac{u}{Bli} = \frac{1}{Z_m}$$

The advantage of this approach becomes apparent on examining Fig. 1.26(b) and Fig. 1.26(c). In Fig. 1.26(b) a hypothetical ideal transformer has been added that has a turns ratio of $Bl : 1$. With a secondary current of Bli, the primary current of such a transformer will be the secondary current divided by the turns ratio or

$$\frac{Bli}{Bl} = i$$

This allows the natural incorporation of the purely electrical aspects of the loudspeaker system, as shown in Fig. 1.26(c). In Fig. 1.26(c) the voltage generator of emf E supplies the normal actual voice coil current i. The electrical impedance as viewed from the vantage point of this generator is the series combination of the voice coil impedance, $R_e + j\omega L_e$, with the impedance presented by the primary of the ideal transformer. This latter impedance is the square of the turns ratio multiplied by the impedance connected to the secondary of the transformer. This product is

$$\frac{(Bl)^2}{Z_m}$$

Therefore, the electrical generator sees a total electrical impedance given by

$$Z_e = R_e + j\omega L_e + \frac{(Bl)^2}{Z_m}$$

Recall from Fig. 1.21, Equation (1.3) and from the original equation of motion, Equation (1.2). When u is eliminated from these two equations, i is found to be

$$i = \frac{E}{R_e + j\omega L_e + \dfrac{(Bl)^2}{Z_m}}$$

The conclusion to be drawn is that the models of Fig. 1.22, as well as Fig. 1.26(c), both accurately describe the physical loudspeaker system, though in different terms. The choice of which model to employ in a given situation is really a matter of preference as they both yield the same results. Scientists with electrical backgrounds will probably prefer that of Fig. 1.26(c), however, as the terminology is more familiar.

Defining Terms

Acoustic impedance: The complex ratio of acoustic pressure to volume velocity.
Acoustic pressure: The pressure variation above or below static atmospheric pressure.
Mechanical impedance: The complex ratio of applied force to resulting velocity.
Mechanical mobility: The complex ratio of velocity to applied force.
Volume velocity: The product of surface area with surface velocity.

References

Beranek, L.L. 1954. *Acoustics.* McGraw-Hill, New York.
Kinsler, L.E. and Frey, A.R. 1962. *Fundamentals of Acoustics.* Wiley, New York.
Skilling, H.H. 1965. *Electrical Engineering Circuits.* Wiley, New York.

Further Information

Journal of the Acoustical Society of America. Acoustical Society of America, 500 Sunnyside Blvd. Woodbury, NY 11797. A scholarly journal dealing with acoustics and related topics.

Journal of the Audio Engineering Society. Audio Engineering Society, 60 East 42nd St., New York, NY 10165. Current articles in the audio engineering field.

1.4   Thermal Noise and Other Circuit Noise

Barry G. Douglass

1.4.1   Introduction

All electronic circuits are affected by many factors, which cause their performance to be degraded from the ideal assumed in simple component models, in ways that can be controlled but not eliminated entirely. One limitation is the failure of the model to properly account for the real behavior of the components, either due to the oversimplified model description or variations in manufacture. Usually by careful design the engineer can work around the limitations of the model and produce a circuit the operation of which is very close to predictions. One source of performance degradation, which cannot be easily overcome, however, is called **noise.** When vacuum tubes first came into use in the early part of the 20th century, they were extensively used in radios as signal amplifiers and for other related signal conditioning functions. Thus, the measure of performance that was of greatest importance to electronic circuit designers was the quality of the sound produced at the radio speaker. It was immediately noticed that the sound coming from speakers not only consisted of the transmitted signal but also of popping, crackling, and hissing noises, which seemed to have no pattern and were distinctly different than the sounds that resulted from other interfering signal sources such as other radio stations using neighboring frequencies. This patternless or random signal was immediately labeled noise. This has become the standard term for signals that are random and that are combined with the circuit signal to affect the overall performance of the system.

As electronic engineers worked to discover the sources of this new phenomenon, they realized that part of it seems to come from the airwaves, in other words, it is part of the signal arriving at the receiver antenna, and part of it seems to be created internally by the circuit itself. The first type of internally generated noise to be detected is called **thermal noise.** This was discovered in vacuum tube triodes which were being used as signal amplifiers. Vacuum tube triodes have three filaments or plates in a single glass vacuum tube similar to a light bulb, and one of these filaments is purposely heated to a high temperature (several thousand degrees Fahrenheit) by passing a current through it just like the filament in a light bulb. This is necessary for proper conduction of electrons from the heated surface to either of the other two plates in the tube. In studying the noise problem, engineers discovered that as the element is heated, the noise level increases, even though the level of the incoming signal (the one being amplified) does not. If the noise were part of the incoming signal, it could only change as the amplified signal changed;

therefore, it must be created by the circuit itself. Since it is dependent on temperature, it is called thermal noise.

As the study of noise has progressed engineers have come to realize that there are many sources of noise in circuits, and not all of these are temperature dependent. The term thermal noise has come to mean any source of noise generated within a circuit that depends on temperature. More recently as digital logic circuits have acquired great importance in a wide variety of circuit applications, many of them well beyond the traditional field of data processing computers, some new manifestations of the noise phenomenon have appeared. In analyzing **digital circuits** it is often common to talk about **error** as opposed to noise. The distinction between error and noise is one of perspective. Traditional computer applications involve a large amount of processing of electronically stored data, with a very high standard of accuracy required in the results. Just as in other circuits occasional failures in the circuits occur; these being neither due to damage to the components, nor due to poor circuit design, but rather to the presence of noise in the circuit. Since a very high degree of accuracy is both required and achieved in computers, individual instances of error are notable and various techniques to detect and correct for such errors have been developed. Indeed, one of the reasons computerized circuits are coming into greater use in applications where other circuits have previously been used is the fact that these error correction techniques, which can only be practically implemented in digital logic systems, have become so sophisticated as to allow a much higher level of circuit performance in the presence of noise than is possible with other circuits.

In the following sections, we will study each type of internally generated noise in circuits and see how this noise impacts the performance of both digital and other (analog) circuits. Much of the analysis of noise is equally applicable to noise that originates before the input to a circuit; however, we are only interested here in noise that originates within a circuit and the way in which various circuit components alter the noise as it passes through the system.

1.4.2 Thermal Noise

By definition thermal noise is internally generated noise, which is temperature-dependent. Although first observed in vacuum tube devices because their amplifying capabilities tend to bring out thermal noise, it is also observed in semiconductor devices. It is a phenomenon due to the ohmic resistance of devices, which dissipate the energy lost in them as heat. Heat consists of random motion of molecules, which are more energetic as temperature increases. Since the motion is random, it is to be expected that as electrons pass through the ohmic device incurring resistance, there should be some random deviations in the rate of energy loss. This fluctuation has the effect of causing variation in the resulting current, which is noise. As the temperature of the device increases, the random motion of the molecules increases, and so does the corresponding noise level; therefore, it is known as thermal noise. In semiconductor diodes, bipolar junction transistors (BJTs), and field effect transistors (FETs), this noise is evident whenever the device is turned on and conducting current. Note that these devices also have their noiseless performance dependent on temperature; noise is strictly due to the variance in the current, which occurs at any given temperature.

From quantum mechanics it is found that an ohmic resistance R operating at absolute temperature T (in degrees kelvin) has a noise power spectral density $G(f)$ (in volts squared per hertz), which can be expressed as a function of frequency f (in hertz) as

$$G(f) = \frac{2Rhf}{e^{hfkT} - 1} \tag{1.4}$$

where e is the Euler number, k is Boltzmann's constant $= 1.37 \times 10^{-23}$ J/K, h is Planck's constant $= 6.62 \times 10^{-34}$ Js. If room temperature, 295 K, is inserted for the operating temperature in Equation (1.4), it is found that until the infrared frequency range, which is well beyond the operating capacity of ordinary electronic circuit components, the noise power $G(f)$ remains nearly independent of frequency and can be adequately approximated as

$$G(f) = 2RkT \tag{1.5}$$

Note that $G(f)$ has units of volts squared per hertz. When delivered at a matched load with resistance R, the available noise power spectral density $G_a(f)$ becomes

$$G_a(f) = \frac{2RkT}{4R} = \frac{kT}{2} \tag{1.6}$$

This noise power spectral density expressed in Equation (1.5) and Equation (1.6) increases linearly with absolute temperature, and is independent of frequency. By solving Equation (1.6) for the temperature T, it is possible to determine it from the available noise spectral density. This can be extended to define a noise temperature T_N for any source of **white noise**, even nonthermal noise, in order to allow convenient calculations

$$T_N = \frac{2G_a(f)}{k} \tag{1.7}$$

This noise temperature may not correspond to any physical temperature, since a source of nonthermal white noise may produce far more white noise than would correspond to the thermal noise at its actual operating temperature. The fact that Equation (1.6) and Equation (1.7) are independent of resistance makes them easily extendable to transistors and diodes, which have varying resistances depending on the current they carry. Since source and load devices would be carrying the same current the formulas can be applied.

1.4.3 Shot Noise

Semiconductor devices display a type of noise that is temperature independent and is due to the fact that the current carrying mechanizm involves randomly distributed doping impurities creating either excess electrons or positive charge-carrying holes in the n-doped or p-doped semiconductor. Because of the random distribution of the impurities, and even more because of the random way in which the carriers are swept across a p-n junction and then recombine, there is a variance in the operation of the junctions, which is called **shot noise**. It is similar to thermal white noise except that it is independent of temperature, and tends to fall off at frequencies above $1/t$, which is the reciprocal of the charge carrier lifetime t before recombining. Bipolar junction transistors and diodes tend to produce this kind of noise because of the fact that both devices have active p-n junctions while carrying current. In contrast, field effect transistors, which have either alternating p-n-p regions or n-p-n regions, are activated by converting the middle region between the two junctions from n to p or p to n by adding charge through a capacitor plate next to the region. In this way the junctions disappear and the random recombination does not take place; thus, FETs are relatively immune to shot noise.

In low-frequency applications, typically below a few kilohertz, some transistors exhibit so-called *burst* noise and *flicker* noise, which tend to fall off quickly with increasing frequency. These are not significant in applications above a few kilohertz, whereas at very high frequencies, transmission line effects and **capacitive coupling noise** begins to become very significant, especially in digital logic circuits, as will be discussed subsequently. The relationship between white noise and frequency-dependent burst and flicker noise is illustrated in Fig. 1.27.

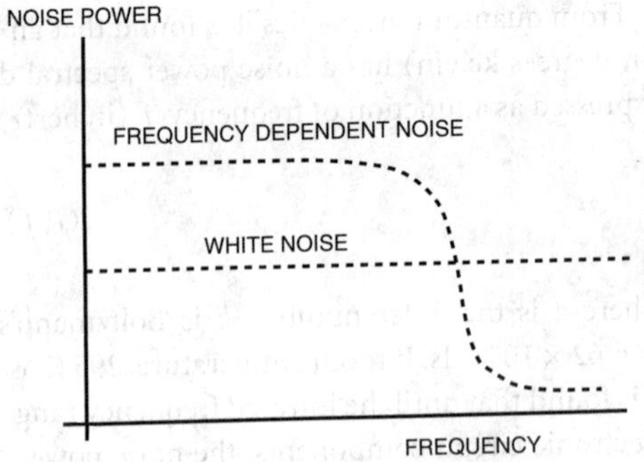

FIGURE 1.27 A plot of signal power against frequency for white noise and frequency limited noise.

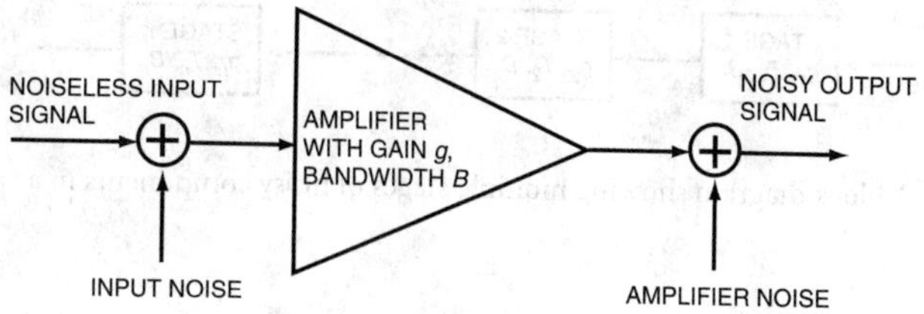

FIGURE 1.28   A block diagram modeling how noise is introduced to a signal during amplification.

1.4.4   Noise in Systems of Cascaded Stages

The noise level produced by thermal noise sources is not necessarily large; however, since source signal power may also be low it is usually necessary to amplify the source signal. Since noise is combined with the source signal, and both are then amplified, with more noise added at each successive stage of amplification, noise can become a very noticeable phenomenon. Consider a system as illustrated in Fig. 1.28, having a source signal power S_s with source noise power spectral density $n_s(f) = kT_s$ as input to an amplifier with gain g (the ratio of the amplified output to the input), and internal amplifier noise spectral density $n_i(f)$, operating over frequency bandwidth B. The effective noise temperature T_e of the amplifier is given as

$$T_e = \frac{1}{gkB} \int_0^\infty n_i(f)df \qquad (1.8)$$

Thus, the output noise power is

$$N_o = gkB(T_s + T_e) \qquad (1.9)$$

Now an amplifier with gain g and input signal power S_s will produce an output signal power $S_o = g S_s$. Hence, the output signal-to-noise ratio will be

$$\frac{S_o}{N_o} = \frac{S_s}{kB(T_s + T_e)} \qquad (1.10)$$

The signal-to-noise ratio of the source within the same bandwidth is

$$\frac{S_s}{N_s} = \frac{S_s}{kBT_s} \qquad (1.11)$$

The noise factor n_F is defined as the ratio of the source signal-to-noise ratio divided by the output signal-to-noise ratio, when the source noise temperature is set to ambient temperature T_0. Combining Equation (1.10) and Equation (1.11) we get

$$n_F = 1 + \frac{T_e}{T_0} \qquad (1.12)$$

or

$$T_e = T_0(n_F - 1) \qquad (1.13)$$

Figure 1.29 illustrates a system of cascaded amplifier stages. In analog systems amplifiers with fixed bandwidth are a good model for all of the components in a cascaded system of C stages, so the preceding analysis for a single filtering amplifier stage can be easily extended so that Equation (1.12) and

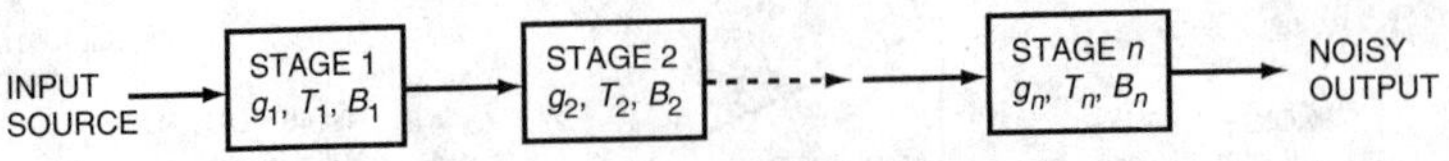

FIGURE 1.29 A block diagram showing multiple stages of noisy components in a cascaded system.

Equation (1.13) become

$$
\begin{aligned}
n_F &= 1 + \frac{T_1}{T_0} + \frac{T_2}{g_1 T_0} + \cdots + \frac{T_C}{g_{C-1} T_0} \\
&= n_{F1} + \frac{n_{F2} - 1}{g_1} + \cdots + \frac{n_{F2} - 1}{g_1 \cdots g_{C-1}}
\end{aligned}
\tag{1.14}
$$

and

$$
T_e = T_1 + \frac{T_2}{g_1} + \cdots + \frac{T_2}{g_1 \cdots g_{C-1}}
\tag{1.15}
$$

Equation (1.15), which is called *Frii's formula*, combined with Equation (1.6) allows calculation of the internally generated noise power at the output to a system of cascaded stages. Note that this analysis applies only to analog systems; digital systems do not pass noise from one stage to the next in the same way as analog circuits. Typical values of effective noise temperature T_e in amplifiers range from 400 K up to 2000 K in integrated circuit amplifiers.

1.4.5 Noise-Induced Error in Digital Circuits

As described in the Introduction, digital logic circuits respond differently from analog circuits to the presence of noise, since any noise signal that falls below the threshold necessary to alter a single discreet bit of information to another (erroneous) discreet level will be completely rejected by the circuit. This provides the possibility of extremely high noise immunity for digital logic circuits. Furthermore, the extraordinary ease with which complex digital circuits can be constructed to perform a wide variety of functions has made them increasingly popular replacements for analog systems. In applications where the inputs and outputs of the system are to be analog, the digital logic circuits are preceded by an analog-to-digital converter (ADC) and followed by a digital-to-analog converter (DAC). The ADC encodes the analog signal values (usually voltage) as binary numbers and the DAC uses the binary numerical outputs of the digital system to create analog signal values.

To design complex circuits, which favor the digital approach, it is necessary to mount many extremely small devices on a single semiconductor chip substrate. These devices are closely packed together and, further, because of their size are extremely sensitive to small stray currents. Because of the dense packing and the associated network of interlaced wires, the circuits are susceptible to crosstalk between the various components. This crosstalk is primarily due to capacitive coupling between adjacent wires, since at the dimensions of chip circuitry inductance is normally too small to be significant. At the board level inductive coupling can be a problem; boards must be carefully designed so as to minimize the effects of inductance, such as by providing a ground plane next to the wires. At the board level, as circuit switching speed increases into the gigahertz range, the length of wires on boards approaches the wavelength of the electromagnetic waves of the switching pulses (as a rule of thumb, an electromagnetic wave travels about 6 in/ns in a wire, corresponding to a 1-GHz pulse duration). Therefore, line reflections from chip input terminations and sharp bends in wire runs can cause false pulses to become a problem. Care in terminating wires and rounding corners in wires is necessary to minimize this problem.

All of these concerns about capacitive coupling, inductive coupling, and transmission line effects can be managed in order to ensure that they do not become a major problem in operating digital logic. Nevertheless, their random nature, which results from the unpredictability of extremely complex systems,

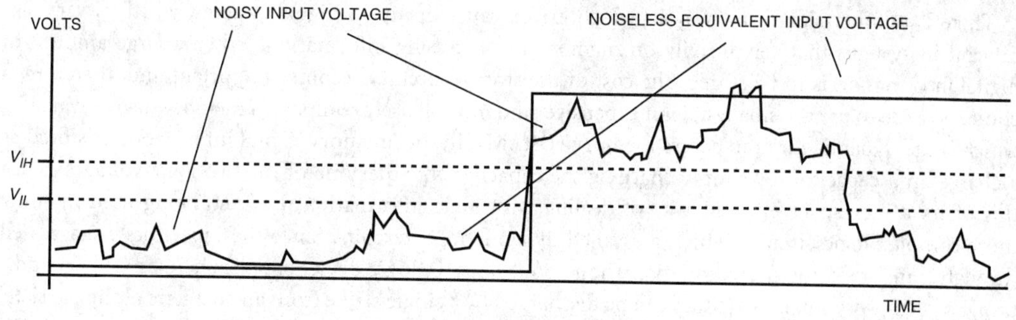

VOLTS NOISY INPUT VOLTAGE NOISELESS EQUIVALENT INPUT VOLTAGE

V_{IH}
V_{IL}

TIME

FIGURE 1.30 A depiction of a noisy input signal to a digital circuit and the noiseless equivalent input as interpreted by the circuit.

means that it is never possible to guarantee that occasional errors due to these effects will not occur. These effects, in addition to the random thermal and other noise associated with semiconductor devices, conspire to create significant levels of noise at the inputs to logic circuit components. For this reason logic circuits must have **noise margins** built into their operating characteristics. For binary systems, where there are two logic levels, high and low, the devices have associated output voltages for each logic level or state. V_{OH} is the output voltage corresponding to a logic high state, and V_{OL} is the output voltage corresponding to a logic low state. At the input to each device is a component that measures the voltage and determines whether this is to be interpreted as a logic low input or a logic high input. The minimum voltage that is recognized by the device as a logic high input is called V_{IH}, whereas V_{IL} is the maximum voltage recognized as a logic low. These are depicted in Fig. 1.30. Obviously $V_{IL} < V_{IH}$. For the region of voltage in between these two voltages, that is to say if the input voltage V_I is $V_{IL} < V_I < V_{IH}$, then the circuit will behave erratically and produce erroneous output values. It is said that in this region the logic value is *indeterminate*. The noise margin high NM_H and noise margin low NM_L are determined as

$$NM_H = V_{OH} - V_{IH} \qquad (1.16)$$

and

$$NM_L = V_{IL} - V_{OL} \qquad (1.17)$$

and the overall noise margin NM is the minimum of NM_H and NM_L, since this will determine the noise level where an error can occur. This noise margin allows a certain level of noise to exist on the inputs to logic circuits without the noise affecting the correctness of the outputs of the devices, and so errors do not occur. However, it is important to note that noise, because it is a random process, has a distribution that because of the central limit theorem can be assumed, in general, to be Gaussian. Thus, regardless of how large a noise margin is designed into a digital logic device, some noise will occasionally create error; the noise margin can only be increased in order to reduce the error rate. Nevertheless, the fact that a mechanism exists to reduce the level of noise passing through the system represents a powerful tool for noise reduction, which is not available to the analog circuit designer. Note in Equation (1.9) that in an analog amplifier the noise is passed to the output amplified along with the input signal. In a digital logic circuit as long as the signal has a magnitude predictably larger than that of the noise, then it can be separated from the noise by the choice of a noise margin.

We have seen that noise on the input to a logic circuit can create errors in the output of the device. Individual errors may or may not be significant in a system, but given a continuous error rate passing to the system output the impact on the overall system performance will be similar to an equivalent output noise level in an analog system. The error rate (number of erroneous bits per unit time) as a fraction of the data rate (total number of data bits per unit time) gives a good indication of the impact of errors in a digital logic system.

There is another significant way in which errors can crop up in digital systems, which is very significant in systems that rely heavily on memory in processing information. When a large amount of digital information is to be stored, the cost of the storage circuits becomes important and, therefore, a choice is made to use the smallest, least expensive, and most densely compacted memory device available, which is the dynamic random access memory (DRAM). In this memory digital information is stored as a charge on a capacitor, by simply charging the capacitor to either one of the two logic voltages. Since this offers a very space-efficient way of dealing with logic information it is also being used in some more complex logic circuits, which are called dynamic logic circuits. Since the capacitors are accessed through transistors, the preceding discussion about noise in logic devices applies here as well. In addition, however, very small capacitors can be discharged by a single strike from an energetic **alpha particle**, which can originate externally to the system from cosmic rays and, also, which can occur naturally as a small amount of radiation in packaging and even the semiconductor substrate itself. Figure 1.31 illustrates how an alpha particle can penetrate through a charged surface region on a silicon chip, which can be the bottom plate of a semiconductor capacitor, and by charging the substrate directly below the surface region can cause the capacitor to lose its charge into the substrate. Since this discharging does not significantly damage the device, only causing loss of information, it is termed a *soft error*. Since this type of alpha particle occurs due to radioactivity in materials, care must be taken to use packaging with a minimum amount of radioactive material. Nevertheless, some radioactivity is unavoidable, and even with special techniques to avoid the effects of alpha particle impacts some soft errors will occur.

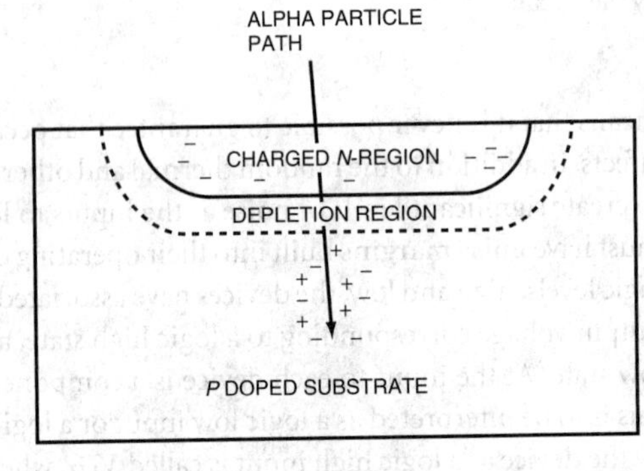

FIGURE 1.31 A depiction of an alpha particle crossing a MOSFET circuit component into the chip substrate.

As stated previously, any single bit of information can be used repeatedly, having a large impact on the overall error level in the system, and so the fact that such errors may be relatively uncommon can nevertheless not be counted on to make such errors insignificant.

The solution to dealing with errors in digital logic systems after they have occurred is to employ error-correcting coding techniques. Although the subject of error-correcting coding is well beyond the scope of this chapter, it is worth pointing out that these techniques are extremely powerful in reducing the impact of errors in overall system performance, and since they can be incorporated as software operations in a digital processing unit, they lend themselves very well to digital logic implementation. The strategy in error-correcting coding is to use extra information within the data stream to pin down errors by comparing copies of the data, which should be consistent for inconsistencies due to error. If, for example, two identical copies of a data stream become different, then one is in error. Note that if an error can be detected in transmission, it may be possible to request a retransmission, or if the error location can be pinned down exactly, then in a binary system an erroneous bit can be corrected since it can only have one other value besides the incorrect one. The references contain further reading on the subject.

1.4.6 Noise in Mixed Signal Systems

In the previous section it was shown how a noisy input signal can affect a digital logic circuit to become errors. Since in some applications the final output of an electronic circuit may be in digital form, these errors may be of great importance in themselves. It is also common, however, as stated earlier, for electronic systems to begin and end as analog circuits, but in between have digital logic subsections. Such a system is called a mixed signal system, and noise is a concern both at the input to the system and at the output. Noise at the input is converted into errors within the digital logic circuit; the data then picks up further errors as it is processed within the digital logic circuits, and when the conversion from the digital to the analog domain is finally made, these accumulated errors become noise once again at the digital logic circuit's output. There are some mixed signal systems that combine digital and analog circuits on a single semiconductor chip substrate, and these can produce crosstalk, which originates within the digital logic circuits and creates noise within the analog circuit section. The most common of these are power supply noise and **ground bounce**.

When analog and digital circuits are closely linked to common ground and power supplies, it is possible to have temporary spikes occur in the power supply or even shifts in the grounding level through the chip substrate due to large power draw during switching of the digital portion of the circuit. Digital circuits are normally designed to switch in response to timing signals called clock pulses, and since the switching of digital circuits is random there is a probability in many cases that a large portion of the components will be switching state between high and low outputs, all in response to the same clock pulse. This can cause sudden temporary drain on the power supply and disturb the grounding of the substrate, to the point where neighboring analog components will pick up these disturbances as noise.

Further noise problems can occur due to errors in converting analog inputs into digital form. The circuits that perform these functions, called quantizing, rely on a process of comparing the analog input to a reference value created by the digital circuit and then converted back to analog. By such successive comparisons the digital circuit determines which digital value most closely matches the analog input. But digital-to-analog converters used in this process are subject to all of the noise problems of analog circuits since they are themselves analog devices, and rely on resistive components, which have thermal noise limitations; thus, the digital information necessarily enters the digital system with noise present. The quantizer merely converts the noisy signal to binary numbers, including the noise values unless they are below the quantizer precision. When the outputs are converted back to analog, they once again are subjected to noise from the digital-to-analog converter, so more noise is introduced. This places a limitation on the accuracy of digital-to-analog as well as analog-to-digital systems, and contributes to the overall noise level of mixed signal systems. This is distinct from quantization error, which is caused by the limited precision of the quantization process, which converts the continuous analog stream into discrete levels of digitally represented signal strength. There is a connection between the two, in that noise affects the accuracy of the digital-analog conversion processes and therefore dictates the highest level of meaningful precision that can be obtained, thereby establishing the number of digital data bits that will be used and the precision of the quantizer. Once the digital data is converted back to analog form, the quantization error becomes **quantization noise**, since the noise level cannot be less than the precision of the quantized data stream from which the analog output is generated. The total quantization noise level is determined as the greater of the quantized data precision and the inaccuracy due to the noise level occurring at the quantizer. Since there is no benefit in using quantization which is more precise than accurate, in general the quantization noise is determined by the quantizing precision. Thus, quantization noise is the noise level which equals the limit of precision of the digital data representation.

Determining the error rate of an individual digital circuit component can be straightforward once the noise distribution and noise margin of the component are known. The error probability for each input value is taken from the distribution of noise that exceeds the noise margin. Computing the effect of individual errors in a complex digital system on the overall output error rate is extremely difficult and can be intractable, so that designers normally rely on general estimates based on experience and testing. This results in a paradox that although digital systems offer unique capability to reduce the effect of noise in degrading the performance of electronic systems, one of their weakest points is the extreme difficulty of properly characterizing the level of noise.

1.4.7 Conclusions

This chapter has dealt exclusively with internally generated noise and its effects on electronic circuits, both digital and analog. This covers a large portion of the sources of circuit noise, though another equally important source is external natural phenomena such as cosmic rays, the solar wind, and lightning discharges. Fortunately the methods for analyzing the effects of these noise sources on the operation of circuits is largely the same as for internal noise, once the extent of these sources has been characterized. To some degree all electronic circuits can be viewed as open systems with the environment forming either an integral part of the system as in the case of radio communication systems, or a peripheral influence in the case of nominally closed systems such as embedded controllers.

The treatment of the very important and extensive field of noise in electronic circuits has necessarily been introductory, and readers are urged to consult the suggestions for further reading to fill in the gaps left by this chapter. The areas of error-correcting coding and optical data transmission, while highly specialized, are fast increasing in importance, and they present their own unique challenges. Readers interested in acquiring a greater understanding of noise in the field of communications systems are urged to follow developments in these important areas.

Defining Terms

Alpha particle noise: This type of noise occurs exclusively in small semiconductor capacitors, when an energetic alpha particle, either from cosmic rays or from the packaging or substrate itself, traverses the capacitor, discharging it, thereby creating an error in the stored charge. Such an accumulation of errors in a digital system has the effect of creating a noise signal.

Analog circuits: These are circuits that process the input signal as a continuous range of values from some minimum to some maximum extremes. Active components (those with their own power source) in such circuits usually are used to amplify input signals without altering their shape (i.e., linear amplification). Thus noise at the input is faithfully reproduced at the output except amplified. Other components in analog circuits, usually involving capacitors and inductors, are used to filter out unwanted frequency ranges.

Capacitive coupling noise: In digital systems most circuit components consist of very small elements, which are very sensitive to correspondingly small voltage and current pulses such as those caused by crosstalk between neighboring wires, which are capacitively coupled. Although this can be regarded as a type of **interference**, typically the larger number of crisscrossing wires in a large system, in conjunction with independence of the various signals on the lines, results in a completely unpredictable crosstalk having all the characteristics of noise. It is generated internally within the circuit and is temperature independent. Analog circuits suffer much less from this type of noise because each component is normally handling much larger signals, and so small capacitive couplings are relatively insignificant.

Digital circuits: These are circuits that only allow a limited number of fixed signal levels at their inputs to be recognized as valid, and produce the same fixed signal levels at their outputs. The most common of these by far are binary circuits, which allow only two levels, such as a high voltage (e.g., 5 V) and a low voltage (e.g., 0 V). All other voltages are truncated to the nearest valid voltage level. Thus, noise at an input will be suppressed unless it is sufficient to move the signal from one valid level to another.

Error: A single error refers to a single event of a degraded signal propagating through a digital circuit as an incorrect voltage level. The collective noun error is sometimes used to refer to a pattern of repeated errors occurring in a digital circuit, in much the same way as noise occurs.

Ground bounce noise: Ground bounce occurs when a large number of semiconductor circuit components are mounted on a common semiconductor chip substrate, so that they are imperfectly insulated from each other. In normal operation the substrate should act as an insulator; however, during certain unusual fluctuations in signal levels the system power and ground connections can experience fluctuations, which affect the performance of each component in a random way that has the characteristics of noise, much like capacitive coupling.

Interference: This is the name given to any predictable, periodic signal that occurs in an electronic circuit in addition to the signal that the circuit is designed to process. This is distinguished from a noise signal by the fact that it occupies a relatively small frequency range, and since it is predictable it can often be filtered out. Usually interference comes from another electronic system such as an interfering radio source.

Noise: This is the name given to any random (i.e., without pattern) signal that occurs in an electronic circuit in addition to the signal that the circuit is designed to process.

Noise margin: In digital circuits this is the difference between the nominal noiseless signal value for a given allowable level and the threshold signal value, which will just barely be distinguishable from the next allowable signal level. For example, a typical MOSFET transistor-based logic device might have a nominal input voltage of 4.5 V for a logic high value, and a 3.5 V minimum voltage, which is still recognized as a logic high. Thus the logic high noise margin is $4.5 - 3.5 = 1$ V.

Quantization noise: This is a type of noise that occurs when an analog signal is converted into digital form by measuring and encoding the range of analog input signal values as a set of numbers, usually in the binary system, for manipulation in digital circuits, possibly followed by reconversion to analog form. Quantization noise is an example of how noise such as thermal noise can become part of a digital signal.

Shot noise: This is a type of circuit noise, which is not temperature dependent, and which is not white noise in the sense that it tends to diminish at higher frequencies. This usually occurs in components whose operation depends on a mean particle residence time for the active electrons within the device. The cutoff frequency above which noise disappears is closely related to the inverse of this characteristic particle residence time. It is called shot noise because in a radio it can make a sound similar to buckshot hitting a drum, as opposed to white noise, which tends to sound more like hissing (due to the higher-frequency content).

Thermal noise: This is any noise that is generated within a circuit and that is temperature dependent. It usually is due to the influence of temperature directly on the operating characteristics of circuit components, which because of the random motion of molecules as a result of temperature, in turn creates a random fluctuation of the signal being processed.

White noise: This is noise that has its energy evenly distributed over the entire frequency spectrum, within the frequency range of interest (typically below frequencies in the infrared range). Because noise is totally random it may seem inappropriate to refer to its frequency range, it is not really periodic in the ordinary sense. Nevertheless, by examining an oscilloscope trace for white noise, one can verify that every trace is different, as the noise never repeats itself, and yet each trace looks the same. Similarly, a television set, which is tuned to a dead frequency, displays never-ending snow, which always looks the same, yet clearly is always changing. There is a strong theoretical foundation to represent the frequency content of such signals as covering the frequency spectrum evenly. In this way the impact on other periodic signals can be analyzed. The term white noise arises from the fact that, similar to white light which has equal amounts of all light frequencies, white noise has equal amounts of noise at all frequencies within circuit operating ranges.

References

Bakoglu, H.B. 1990. *Circuits, Interconnections, and Packaging for VLSI*. Addison-Wesley, New York.

Buckingham, M.J. 1983. *Noise in Electronic Devices and Systems*. Ellis Horwood, Ltd., Chichester, England (distributed in U.S. through Wiley).

Carlson, A.B. 1986. *Communication Systems, An Introduction to Signals and Noise in Electrical Communication*, 3rd ed. McGraw-Hill, New York.

Engberg, J. and Larsen, T. 1995. *Noise Theory of Linear and Non-Linear Circuits*. Wiley, New York.

Fabricius, E.D. 1990. *Introduction to VLSI Design*. McGraw-Hill, New York.

Franca, J.E. and Tsividis, Y., eds. 1994. *Design of Analog-Digital VLSI Circuits for Telecommunications and Signal Processing*, 2nd ed. Prentice-Hall, Englewood Cliffs, NJ.

Glasser, L.A. and Doberpuhl, D.W. 1985. *The Design and Analysis of VLSI Circuits*. Addison-Wesley, New York.

Jacobsen, G. 1994. *Noise in Digital Optical Transmission Systems*. Artech House, Boston, MA.

Mardiguian, M. 1992. *Controlling Radiated Emissions by Design*. Van Nostrand Reinhold, New York.

Motchenbacher, C.D. and Connelly, J.A. 1993. *Low Noise Electronic Systems Design*. Wiley, New York.

Rhee, M.Y. 1989. *Error Correcting Coding Theory.* McGraw-Hill, New York.

Shwartz, M. 1980. *Information Transmission, Modulation, and Noise: A Unified Approach to Communications Systems*, 3rd ed. McGraw-Hill, New York.

Smith, D.C. 1993. *High Frequency Measurement and Noise in Electronic Circuits.* Van Nostrand Reinhold, New York.

Weste, N.H.E. and Eshraghian, K. 1992. *Principles of CMOS VLSI Design, A Systems Perspective*, 2nd ed. Addison-Wesley, New York.

Further Information

Readers who wish to study the subject of circuit noise in more depth will find excellent coverage of this extensive topic in Buckingham: *Noise in Electronic Devices and Systems* and Engberg and Larsen: *Noise Theory of Linear and Non-Linear Circuits,* including thorough discussions of thermal noise, shot noise, and other types of internal noise. For in-depth discussions of methods for designing circuits to reduce the effects of noise readers are referred to Motchenbacher and Connelly: *Low Noise Electronic Systems Design* and Mardiguian: *Controlling Radiated Emissions by Design.* Readers who are interested in exploring methods for measuring noise in circuits should consult Smith: *High Frequency Measurement and Noise in Electronic Circuits.* For a thorough discussion of circuit noise in the newest form of electronic circuit components, namely, digital optical devices and circuits, readers are directed to Jacobsen: *Noise in Digital Optical Transmission Systems.* A very complete (though technically advanced) introduction to the extensive field of error correcting coding theory is found in Rhee: *Error Correcting Coding Theory.* Readers who are just interested in a text that covers the field of communications theory, including the study of noise, should consult Shwartz: *Information Transmission, Modulation, and Noise: A Unified Approach to Communications Systems* or Carlson: *Communication Systems, An Introduction to Signals and Noise in Electrical Communication.* For descriptions of noise sources in digital circuits including alpha particle noise, capacitive coupling crosstalk, and other digital effects, readers are directed to Weste and Eshraghian: *Principles of CMOS VLSI Design, A Systems Perspective,* Fabricius: *Introduction to VLSI Design,* Bakoglu: *Circuits, Interconnections, and Packaging for VLSI,* and Glasser and Doberpuhl: *The Design and Analysis of VLSI Circuits.* Recent developments in the study of noise and errors are reported in the *IEEE Journal of Solid State Circuits* and the *IEEE Transactions on Communications,* both published through the Institute of Electrical and Electronic Engineers (*IEEE*).

1.5   Logic Concepts and Design

George I. Cohn

1.5.1   Introduction

Digital logic deals with the representation, transmission, manipulation, and storage of digital information. A digital quantity has only certain discrete values in contrast with an analog quantity, which can have any value in an allowed continuum. The enormous advantage digital has over analog is its immunity to degradation by noise, if that noise does not exceed a tolerance threshold.

1.5.2   Digital Information Representation

Information can be characterized as qualitative or quantiative. Quantitative information requires a number system for its representation. Qualitative does not. In either case, however, digitalized information is represented by a finite set of different characters. Each character is a discrete quanta of information. The set of characters used constitutes the alphabet.

1.5.3   Number Systems

Quantitative information is represented by a number system. A character that represents quantitative information is called a **digit**. The number of different values which a digit may have is called the **radix**, designated

TABLE 1.3   Notation for Numbers

	Juxtaposition	Polynomial
Integer	$N = N_{n-1} N_{n-2} \cdots N_1 N_0$	$N = \sum\limits_{k=0}^{n-1} N_k R^k$
Fraction	$F = \cdot F_{-1} F_{-2} \cdots F_{-m+1} F_{-m}$	$F = \sum\limits_{k=-m}^{-1} F_k R^k$
Real	$X = X_{n-1} X_{n-2} \cdots X_1 X_0 \cdot X_{-1} X_{-2} \cdots X_{-m+1} X_{-m}$	$X = \sum\limits_{k=-m}^{n-1} X_k R^k$

by R. The symbols that designate the different values a digit can have are called numeric characters. The most conventionally used numeric characters are $0, 1, 2, \ldots$, etc., with 0 representing the smallest value. The largest value that a digit may have in a number system is the **reduced radix**, $r = R - 1$. Different radix values characterize different number systems: with R different numeric character values the number system is Rnary, with 2 it is binary, with 3 it is ternary, with 8 it is octal, with 10 it is decimal, and with 16 it is hexadecimal.

Any value that can be expressed solely in terms of digits is an **integer**. A negative integer is any integer obtained by subtracting a positive integer from a smaller integer. Any number obtained by dividing a number by a larger number is a **fraction**. A number that has both an integer part and a fraction part is a **real number**.

All of the digits in a *number system* have the same radix. The radix is the **base** of the number system. Presumably, the possession of 10 fingers has made the decimal number system the most convenient for humans to use. The characters representing the 10 values a decimal digit can have are: 0, 1, 2, 3, 4, 5, 6, 7, 8, and 9. The binary number system is most natural for digital electronic systems because a desired reliability for a system can be most economically achieved using elements with two stable states. The characters normally used to represent the two values a binary digit may have are 0 and 1. The hexadecimal number system $(0, 1, 2, 3, 4, 5, 6, 7, 8, 9, A, B, C, D, E,$ and $F; R = 16)$ is of importance because it shortens by a factor of four the string of digits representing the binary information stored and manipulated in digital computers.

1.5.4   Number Representation

Numbers that require more than one digit can be represented in different formats, as shown in Table 1.3. Different formats facilitate execution of different procedures. Arithmetic is most conveniently done with the juxtaposition format. Theoretical developments are facilitated by the polynomial format.

1.5.5   Arithmetic

The most common arithmetic processes, addition, subtraction, multiplication, and division are conveniently implemented using juxtaposition notation. Development of formulation procedure is facilitated using the polynomial notation. Since the numbers are digital representations, the logic used to maniputate the numbers is *digital logic*. However, this is different than the logic of boolean algebra, which is what is usually meant by the term digital logic. The logic of the former is implemented in hardware by using the logic of the latter. The four basic arithmetic operations can be represented as functional procedures in equation form or in arithmetic manipulation form, as shown in Table 1.4.

The arithmetic processes in the binary system are based on the binary addition and multiplication tables given in Table 1.5.

Table 1.6 gives binary examples for each of the basic arithmetic operations.

1.5.6   Number Conversion from One Base to Another

The method of using series polynomial expansions for converting numbers from one base to another is illustrated in Table 1.7.

TABLE 1.4 Arithmetic Operations

	Algebraic Form	Arithmetic Form
Addition	Sum = Augend + Addend	Augend + Addend Sum
Subtraction	Difference = Minuend − Subtrahend	Minuend − Subtrahend Difference
Multiplication	Product = Multiplicand × Multiplier	Multiplicand × Multiplier Product
Division	Dividend/Divisor = Quotient + Remainder/Divisor	Quotient Remainder Divisor ⟌Dividend

TABLE 1.5 Single Digit Binary Arithmetic Tables

(a) Addition

	0	1
0	0	1
1	1	10

(b) Multiplication

	0	1
0	0	0
1	0	1

TABLE 1.6 Binary Arithmetic Operation Examples

Addition:

```
  1 1 0 0    carries
  1 1 1 0 0  augend
+ 1 1 0 1    addend
1 0 1 0 0 1  sum
```

Multiplication:

```
      1 1 0 1    multiplicand
    × 1 1 0      multiplier
      0 0 0 0    partial product 1
      1 1 0 1    partial product 2
    1 1 0 1      partial product 3
  1 0 0 1 1 1 0  product
```

Subtraction, borrow method:

```
  1 0      borrows
  1̸ 0̸ 1    minuend
− 1 0      subtrahend
  1 1      difference
```

Division, with fraction remainder:

```
                                   remainder
                                      1 1 1
            quotient                 ───────
              1 0 0 1 0              1 0 1 1
      1 0 1 1⟌1 1 0 0 1 1 0 1  dividend
divisor 1 0 1 1
          0 1 1 1 0
          1 0 1 1
          0 1 1 1
```

Subtraction, payback method:

```
  1 0      borrows
  1 0 1    minuend
− 1        payback
  − 1 0    subtrahend
  1 1      difference
```

Division, with remainder in quotient:

```
              quotient
              .1 0 1 0 0 0 1 ...
      1 0 1 1⟌1 1 1.0 0 0 0 0 0 0  dividend
divisor 1 0 1 1
          1 1 0 0
          1 0 1 1
          0 1 0 0 0 0
              1 0 1 1
```

TABLE 1.7 Series Polynomial Method for Converting Numbers Between Bases

	Sample Conversion From:	
Method	A Lower to a Higher Base	A Higher to a Lower Base
1. Express number in polynomial form in the given base	$101.1_2 = 1 \times 2^2 + 0 \times 2^1 + 1 \times 2^0 + 1 \times 2^{-1}$	$36.5_{10} = 3 \times 10^1 + 6 \times 10^0 + 5 \times 10^{-1}$
2. Convert radix and coefficients to the new base	$= 1 \times 4 + 0 \times 2 + 1 \times 1 + 1 \times 0.5$	$= 11 \times 1010^1 + 110 \times 1010^0 + 101 \times 1010^{-1}$
3. Evaluate terms in the new base	$= 4 + 0 + 1 + 0.5$	$= 11 \times 1010 + 110 + 101/1010$ $= 11110 + 110 + .1$
4. Add the terms	$101.1_2 = 5.5_{10}$	$36.5_{10} = 100100.1_2$

TABLE 1.8 Nested Polynomials

(a) Nested Polynomial via Iterated Factoring	(b) Lower Order Polynomials
$N = N_{n-1}R^{n-1} + N_{n-2}R^{n-2} + \cdots + N_2R^2 + N_1R + N_0$	$N = N^{(1)}R + N_0$
$N = (N_{n-1}R^{n-2} + N_{n-2}R^{n-3} + \cdots + N_2R + N_1)R + N_0$	$N^{(1)} = N^{(2)}R + N_1$
$N = ((N_{n-1}R^{n-3} + N_{n-2}R^{n-4} + \cdots + N_2)R + N_1)R + N_0$	$N^{(2)} = N^{(3)}R + N_2$
$\vdots$	$\vdots$
	$N^{(n-2)} = N^{(n-3)}R + N_{n-2}$
$N = (\cdots(N_{n-1})R + N_{n-2})R + \cdots + N_2)R + N_1)R + N_0$	$N^{(n-1)} = N_{n-1}$

Evaluation of polynomials is more efficiently done with the *nested form*. The nested form is obtained from the series form by successive factoring of the variable from all terms in which it appears as shown in Table 1.8. The number of multiplications to evaluate the nested form increases linearly with the order of the polynomial, whereas the number of multiplications to evaluate the series form increases with the square of the order.

Conversion of integers between bases is more easily done using the lower order polynomials (Table 1.8(b)) obtained by nesting. The least significant digit of the number in the new base is the remainder obtained after dividing the number in the old base by the new radix. The next least significant digit is the remainder obtained by dividing the first reduced polynomial by the new radix. The process is repeated until the most significant digit in the new base is obtained as the remainder when the new radix no longer fits into the last reduced polynomial. This is more compactly represented with the arithmetic notation shown in Table 1.9 along with the same examples used to illustrate the polynomial series method.

TABLE 1.9 Radix Divide Method for Converting Numbers Between Bases

		Sample Conversion From:	
Method		A Higher to a Lower Base	A Lower to a Higher Base
$R\underline{/\,N}$		$2\underline{/\,36_{10}}$	
$R\underline{/\,N^{(1)}}$	N_0	$2\underline{/\,18}\quad 0$	
$R\underline{/\,N^{(2)}}$	N_1	$2\underline{/\,9}\quad 0$	
$R\underline{/\,N^{(3)}}$	N_2	$2\underline{/\,4}\quad 1$	$12\underline{/\,2012_3}$
$\vdots$		$2\underline{/\,2}\quad 0$	$12\underline{/\,102}\quad 11$
		$2\underline{/\,1}\quad 0$	$12\underline{/\,2}\quad 1$
$R\underline{/\,N^{(n-1)}}$	N_{n-2}	$0\quad 1$	$12\underline{/\,0}\quad 2$
$R\underline{/\,0}$	N_{n-1}	$36_{10} = 100100_2$	$2012_3 = 214_5$

TABLE 1.10 Radix Multiply Number Conversion Method (Terminating Case)

Formalism		Sample Conversion Between Bases	
Algebraic	Arithmetic	Higher to Lower	Lower to Higher
$F = \cdot F_{-1}F_{-2}F_{-3}\cdots F_{-m}$	$\begin{array}{r} \cdot F_{-1}F_{-2}F_{-3}\cdots F_{-m} \\ \times R \end{array}$	$\begin{array}{r} 0.125_{10} \\ \times 2 \end{array}$	$\begin{array}{r} 0.100101_2 \\ \times 1010 \end{array}$
$R^* F = F_{-1}\cdot F_{-2}F_{-3}\cdots F_{-m}$ $= F_{-1}\cdot F^{(1)}$	$\begin{array}{r} \underline{F_{-1}}\mid\cdot F_{-2}F_{-3}\cdots F_{-m} \\ \times R \end{array}$	$\begin{array}{r} \underline{0}\mid.25 \\ \times 2 \end{array}$	$\begin{array}{r} \underline{101}\mid.11001 \\ \times 1010 \end{array}$
$R^* F^{(1)} = F_{-2}\cdot F_{-3}F_{-4}$ $\cdots F_{-m} = F_{-2}\cdot F^{(2)}$	$\begin{array}{r} \underline{F_{-2}}\mid\cdot F_{-3}\cdots F_{-m} \\ \times R \end{array}$	$\begin{array}{r} \underline{0}\mid.5 \\ \times 2 \end{array}$	$\begin{array}{r} \underline{111}\mid.1101 \\ \times 1010 \end{array}$
$R^* F^{(2)} = F_{-3}\cdot F_{-4}F_{-5}$ $\cdots F_{-m} = F_{-2}\cdot F^{(3)}$	$\vdots$	$\begin{array}{r} \underline{1}\mid.0 \\ \end{array}$	$\begin{array}{r} \underline{1000}\mid.001 \\ \times 1010 \end{array}$
$\vdots$			$\begin{array}{r} \underline{1}\mid.10 \\ \times 1010 \end{array}$
$R^* F^{(m-2)} = F_{-m+1}\cdot F_{-m}$ $= F_{-m+1}\cdot F^{(m-1)}$	$\begin{array}{r} \underline{F_{-m+1}}\mid\cdot F_{-m} \\ \times R \end{array}$		$\begin{array}{r} \underline{10}\mid.1 \\ \times 1010 \end{array}$
$R^* F^{(m-1)} = F_{-m}\cdot$	$\underline{F_{-m}}\mid\cdot$	$.125_{10} = .001_2$	$\begin{array}{r} \underline{101}\mid. \\ .100101_2 = .578125_{10} \end{array}$

Conversion of a fraction from one base to another can be done by successive multiplications of the fraction by the radix in the base to which the fraction is to be converted. Each multiplication by the radix gives a product that has the digits shifted to the left by one position. This moves the most significant digit of the fraction to the left of the radix point placing that digit in the integer portion of the product, thereby isolating it from the fraction. This process is illustrated in algebraic form in the left column of Table 1.10 and in arithmetic form in the next column. Two sample numeric conversions are shown in the next two columns of Table 1.10.

Table 1.10 deals only with terminating fractions, that is, the remaining fractional part vanishes after a finite number of steps. For a nonterminating case the procedure is continued until a sufficient number of digits have been obtained to give the desired accuracy. A nonterminating case is illustrated in Table 1.11. A set of digits, which repeat *ad infinitum*, are designated by an underscore as shown in Table 1.11.

Conversion to base 2 from a base which is an integer power of 2 can be most simply accomplished by independent conversion of each successive digit, as illustrated in Table 1.12(a). Inversely, conversion from base 2 to a base 2^k can be simply accomplished by grouping the bits into sets of k bits each starting with the least significant bit for the integer portion and starting with the most significant bit for the fraction portion, as shown by the examples in Table 1.12(b).

TABLE 1.11 Nonterminating Fraction Conversion Example

$$0.1_{10} = 0.000110011\ldots_2$$

$$\begin{array}{r} \times 2 \\ \hline \underline{0}\mid.2 \\ \times 2 \\ \hline \underline{0}\mid.4 \\ \times 2 \\ \hline \underline{0}\mid.8 \\ \times 2 \\ \hline \underline{1}\mid.6 \\ \times 2 \\ \hline \underline{1}\mid.2 \\ \times 2 \\ \hline \underline{0}\mid.4 \\ \times 2 \\ \hline \underline{0}\mid.8 \\ \times 2 \\ \hline \underline{1}\mid.6 \end{array}$$

or more compactly

$$0.1_{10} = 0.0\underline{0011}_2$$

1.5.7 Complements

Each number system has two conventionally used **complements**:

$$\text{radix complement of } N = N^{RC} = R^n - N$$
$$\text{reduced radix complement of } N = N^{rC} = N^{RC} - 1$$

TABLE 1.12 Conversions between Systems Where One Base Is an Integer Power of the Other Base

(a) Conversion from high base to lower base:

$B2 \cdot C5_{16} = 1011 \quad 0010 \quad .1100 \quad 0101_2$

$62.75_8 = 110 \quad 010 \quad .111 \quad 101_2$

(b) Conversion from lower base to high base:

$11 \quad 0010 \quad 0100.0001 \quad 1100 \quad 01_2 = 324.1C4_{16}$

$10 \quad 110 \quad 001.011 \; 111 \quad 01_2 = 261.372_8$

where R is the radix and n is the number of digits in the number N. These equations provide complements for numbers having the magnitude N.

A positive number can be represented by a code in the two character machine language alphabet, 0 and 1, which is simply the positive number expressed in the base 2, that is, the code for the number is the number itself. A negative number requires that the sign be coded in the binary alphabet. This can be done by separately coding the sign and the magnitude or by coding the negative number as a single entity. Table 1.13 illustrates four different code types for negative numbers. Negative numbers can be represented in the sign magnitude form by using the leftmost digit as the code for the sign ($+$ for 0 and $-$ for 1) and the rest of the digits as the code for the magnitude. Complements and biasing provide the means for coding the negative number as a single entity instead of having a discrete separate coding for the sign. The use of complements provides for essentially equal ranges of values for both positive and negative numbers. The biased representation can also provide essentially equal ranges for positive and negative values by choosing the biasing value to be essentially half of the largest binary number that could be represented with the available number of digits. The bias code is obtained by subtracting the biasing value from the code considered as a positive number, as shown in the rightmost column of Table 1.13.

Complements enable subtraction to be done by addition of the complement. If the result fits into the available field size the result is automatically correct. A diagnostic must be provided to show that the result is incorrect if **overflow** occurs, that is, the number does not fit in the available field. Table 1.14 illustrates arithmetic operations with and without complements. The two rightmost columns illustrate cases where the result overflows the 3-b field size for the magnitude. A condition can be represented in terms of two

TABLE 1.13 Number Representations

Available Codes	Positive Numbers	Signed Numbers via the Following Representations:			
		Sign Magnitude	One's Complement	Two's Complement	111 bias
1111	1111	−111	−000	−001	000
1110	1110	−110	−001	−010	+111
1101	1101	−101	−010	−011	+110
1100	1100	−100	−011	−100	+101
1011	1011	−011	−100	−101	+100
1010	1010	−010	−101	−110	+011
1001	1001	−001	−110	−111	+010
1000	1000	−000	−111	−000	+001
0111	0111	+111	+111	+111	000
0110	0110	+110	+110	+110	−001
0101	0101	+101	+101	+101	−010
0100	0100	+100	+100	+100	−011
0011	0011	+011	+011	+011	−100
0010	0010	+010	+010	+010	−101
0001	0001	+001	+001	+001	−110
0000	0000	+000	+000	+000	−111

TABLE 1.14 Comparison of Arithmetic With and Without Complements

	$N = 7 - 5$ $= 2$	$N = 5 - 7$ $= -2$	$N = 5 + 7$ $= 12$	$N = -5 - 7$ $= -12$
Sample Illustrations				
Pencil and Paper Arithmetic (without complements)	111 -101 $\overline{10}$	(-111) $-(-101)$ $\overline{-10}$	101 $+111$ $\overline{1100}$	(-101) $+(-111)$ $\overline{-1100}$
Computer Arithmetic with 2's complement	0111 $+1011$ $\cancel{1}0010$	0101 $+1001$ $\overline{1110}$	0101 $+0111$ $\overline{1100}$	1011 $+1001$ $\cancel{1}0100$
4 binary digit working field (accommodates 3-b magnitude)	↓ designates positive number	↓ designates negative number	↓ designates negative number	↓ designates positive number
Result Veracity	$+010$ True	-010 True	-100 False	$+100$ False
Significant Carries	$C_0 = 1$ $C_1 = 1$	$C_0 = 0$ $C_1 = 0$	$C_0 = 0$ $C_1 = 1$	$C_0 = 1$ $C_1 = 0$
Veracity Condition	$C_0 \equiv C_1$ $C_0 \odot C_1 = 1$	$C_0 \equiv C_1$ $C_0 \odot C_1 = 1$	$C_0 \not\equiv C_1$ $C_0 \odot C_1 = 0$	$C_0 \not\equiv C_1$ $C_0 \odot C_1 = 0$

carry parameters:

- C_0 the output carry from the leftmost digit position
- C_1, the output carry from the second leftmost digit position (the output carry from the magnitude field if sign magnitude representation is used)

If both of these carries are coincident (i.e., have the same value) the result fits in the available field and, hence, is correct. If these two carries are not coincident the result is incorrect.

1.5.8 Codes

Various types of **codes** have been developed for serving different purposes. There are codes that enable characters in an alphabet to be individually expressed in terms of codes in a smaller alphabet. For example, the alphabet of decimal numeric symbols can be expressed in terms of the binary alphabet by the **binary coded decimal** (**BCD**) or 8421 code shown in Table 1.15. The 8421 designation represents the weight given to each of the binary digits in the coding process.

TABLE 1.15 Sample Codes

Decimal Digits	BCD 8421	2421	2-out-of-5	Parity Even	Parity Odd	Gray 1-bit	Gray 2-bit	Gray 3-bit
0	0000	0000	00011	0000 0	0000 1	0	00	000
1	0001	0001	00101	0001 1	0001 0	1	01	001
2	0010	0010	00110	0010 1	0010 0		11	011
3	0011	0011	01001	0011 0	0011 1		10	010
4	0100	0100	01010	0100 1	0100 0			110
5	0101	1011	01100	0101 0	0101 1			111
6	0110	1100	10001	0100 1	0110 0			101
7	0111	1101	10010	0111 1	0111 0			100
8	1000	1110	10100	1000 1	1000 0			
9	1001	1111	11000	1001 0	1001 1			

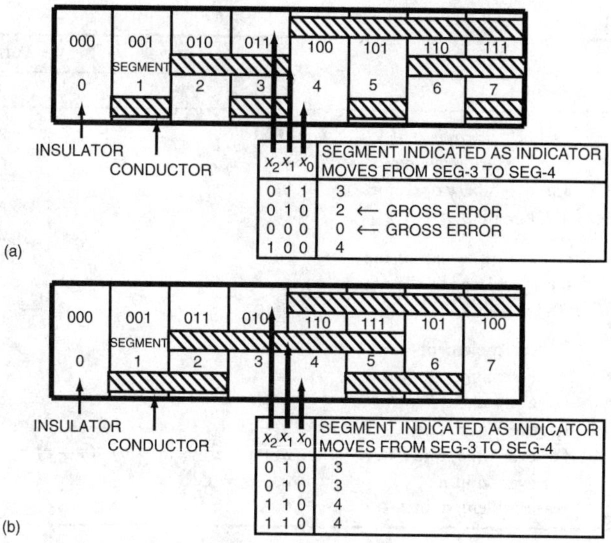

FIGURE 1.32 Eight-segment position sensor with slightly misaligned contacts: (a) binary code physical configuration, (b) Gray code physical configuration.

There are codes that facilitate doing arithmetic. The 2421 code can also be used to represent the decimal numeric symbols. The 2421 code has the advantage that the code for the reduced radix complement is the same as the reduced radix complement of the code, and this is not true of the BCD code. Thus the 2421 code facilitates arithmetic with multiple individually coded digit numbers.

There are codes designed to detect errors that may occur in storage or transmission. Examples are the even and odd parity codes and the 2-out-of-5 code shown in Table 1.15. The 2-out-of-5 error detection code is such that each decimal value has exactly two high digit values. Parity code attaches an extra bit having a value such that the total number of high bits is odd if odd parity is used and the total number of bits is even if even parity is used. An even number of bit errors are not detectable by a single bit parity code. Hence, single bit parity codes are adequate only for sufficiently low bit error rates. Including a sufficient number of **parity bits** enables the detection and correction of all errors.

There are codes designed to prevent measurement misrepresentation due to small errors in sensor alignment. **Gray codes** are used for this purpose. A Gray code is one in which the codes for physically adjacent positions are also **logically adjacent**, that is, they differ in only one digit. Gray codes can be generated for any number of digits by reflecting the Gray code for the case with one less digit as shown in Table 1.15 for the case of 1, 2, and 3-bit codes. The advantage of a Gray scale coded lineal position sensor is illustrated in Fig. 1.32 for the eight-segment case.

1.5.9 Boolean Algebra

Boolean algebra provides a means to analyze and design binary systems and is based on the seven postulates given in Table 1.16. All other Boolean relationships are derived from these seven postulates. Expressed in graphical form, called *Venn diagrams*, the postulates appear more natural and logical. This benefit results from the two-dimensional pictorial representation freeing the expressions from the one-dimensional constraints imposed by lineal language format.

The OR and AND operations are normally designated by the arithmetic operator symbols $+$ and $\cdot$ and referred to as sum and product operators in basic digital logic literature. However, in digital systems that perform arithmetic operations this notation is ambiguous and the symbols $\vee$ for OR and $\wedge$ for AND eliminates the ambiguity between arithmetic and boolean operators. Understanding the conceptual meaning of these boolean operations is probably best provided by set theory, which uses the union operator $\cup$ for OR and the intersection operator $\cap$ for AND. An element in a set that is the union of sets is a member

TABLE 1.16 Boolean Postulates

Postulate	Name	Meaning	Forms (a)	(b)
1	Definition	$\exists$ a set $\{K\} = \{a, b, \ldots\}$	**OR**	**AND**
		of two or more elements	$+$	$\cdot$
		and two binary operators	$\vee$	$\wedge$
		$\ni \{K\} = \{a, b, a+b, a \cdot b, \ldots\}$	$\cup$	$\cap$
2	Substitution Law	expression$_1$ = expression$_2$ If one replaced by the other does not alter the value		
3	Identity Element	$\exists$ identity elements for each operator	$a + 0 = a$	$a \cdot 1 = a$
4	Commutativity	For every a and b in K	$a + b = b + a$	$a \cdot b = b \cdot a$
5	Associativity	For every a, b, and c in K	$a + (b + c) = (a + b) + c$	$a \cdot (b \cdot c) = (a \cdot b) \cdot c$
6	Distributivity	For every a, b, and c in K	$a + (b \cdot c) = (a + b) \cdot (a + c)$	$a \cdot (b + c) = (a \cdot b) + (a \cdot c)$
7	Complement	For every a in K $\exists$ a complement in $K \ni$	$a + \bar{a} = 1$	$a \cdot \bar{a} = 0$

of one set OR another of the sets in the union. An element in a set that is the intersection of sets is a member of one set AND a member of the other sets in the intersection.

A set of theorems derived from the postulates facilitates further developments. The theorems are summarized in Table 1.17. Use of the postulates is illustrated by the proof of a theorem in Fig. 1.33.

1)	$x + x = x + x$	IDENTITY
2)	$= (x + x) \cdot 1$	P3b IDENTITY ELEMENT EXISTENCE
3)	$= (x + x) \cdot (x + \bar{x})$	P7a COMPLEMENT EXISTENCE
4)	$= x + x \cdot \bar{x}$	P6a DISTRIBUTIVITY
5)	$= x + 0$	P7b COMPLEMENT EXISTENCE
6)	$= x$	P3a IDENTITY ELEMENT EXISTENCE

FIGURE 1.33 Proof of Theorem 8: Idempotency (a): $x + x = x$.

1.5.10 Boolean Functions

Boolean functions can be defined and represented in terms of boolean expressions and in terms of **truth tables** as illustrated in Fig. 1.34(a) and Fig. 1.34(c). Each form can be converted into the other form. The function values needed for the construction of the truth table can be obtained by evaluating the function as illustrated in Fig. 1.18(b). The reverse conversion will be illustrated subsequently.

TABLE 1.17 Boolean Theorems

Theorem		Forms (a)	(b)
8	Idempotency	$a + a = a$	$a \cdot a = a$
9	Complement Theorem	$a + 1 = 1$	$a \cdot 0 = 0$
10	Absorption	$a + ab = a$	$a(a + b) = a$
11	Extra Element Elimination	$a + \bar{a}b = a + b$	$a(\bar{a} + b) = ab$
12	De Morgan's Theorem	$\overline{a + b} = \bar{a} \cdot \bar{b}$	$\overline{ab} = \bar{a} + \bar{b}$
13	Concensus	$ab + \bar{a}c + bc = ab + \bar{a}c$	$(a + b)(\bar{a} + c)(b + c) = (a + b)(\bar{a} + c)$
14	Complement Theorem 2	$ab + a\bar{b} = a$	$(a + b)(a + \bar{b}) = a$
15	Concensus 2	$ab + a\bar{b}c = ab + ac$	$(a + b)(a + \bar{b} + c) = (a + b)(a + c)$
16	Concensus 3	$ab + \bar{a}c = (a + c)(\bar{a} + b)$	$(a + b)(\bar{a} + c) = ac + \bar{a}b$

For a given number of variables there are a finite number of boolean functions. Since each boolean variable can have two values, 0 or 1, a set of n variables has 2^n different values. A boolean function has a specific value for each of the possible values that the independent variables can have. Since there are two possible values for each value of the independent variables there are 2^{2^n} different boolean functions of two variables. The number of functions increases very rapidly with the number of independent variables as shown in Table 1.18.

	A	B	C	f(A, B, C)
	0	0	0	0
	0	0	1	1
	0	1	0	0
	0	1	1	1
	1	0	0	1
	1	0	1	0
	1	1	0	1
	1	1	1	1

(a) $\quad f(A, B, C) = AB + A\bar{C} + \bar{A}C$

(b) $\quad f(0, 0, 1) = 0 \cdot 0 + 0 \cdot \bar{1} + \bar{0} \cdot 1$
$\quad\quad\quad\quad\quad = 0 \cdot 0 + 0 \cdot 0 + 1 \cdot 1$
$\quad\quad\quad\quad\quad = 0 + 0 + 1$
$\quad\quad\quad\quad\quad = 1$

(c)

FIGURE 1.34 Example of forms for defining boolean functions: (a) boolean expression definition, (b) boolean expression evaluation, (c) truth table definition.

The 16 different boolean functions of two independent variables are defined in algebraic form in Table 1.19 and in truth table form in Table 1.20.

TABLE 1.18 Number of Different Boolean Functions

Variables n	Arguments 2^n	Functions 2^{2^n}
0	1	2
1	2	4
2	4	16
3	8	256
4	16	65,536
5	32	4,194,304
⋮	⋮	⋮

TABLE 1.19 Functions of Two Variables Defined as Boolean Expressions

	Name	Expression	Circuit Representation
0	ALWAYS	1	
1	NEVER	0	
2	1st Var	a	
3	2nd Var	b	
4	NOT 1st Var	$\bar{a}$	
5	NOT 2nd Var	$\bar{b}$	
6	MIN-0/NOR	$\bar{a} \wedge \bar{b} = a \downarrow b$	
7	MIN-1	$a \wedge \bar{b}$	
8	MIN-2	$\bar{a} \wedge b$	
9	MIN-3/AND	$a \wedge b$	
10	MAX-0/OR	$a \vee b$	
11	MAX-1	$a \vee \bar{b}$	
12	MAX-2	$\bar{a} \vee b$	
13	MAX-3/NAND	$\bar{a} \vee \bar{b} = a \uparrow b$	
14	EXOR	$A \oplus b = a \wedge \bar{b} \vee \bar{a} \wedge b$	
15	COIN	$a \Theta b = \overline{a \oplus b}$	

TABLE 1.20   Truth Tables for Two Variable Functions

a	b	NOR m_0	m_1	m_2	AND m_3	OR M_0	M_1	M_2	NAND M_3	XOR	COIN	a	b	$\bar{a}$	$\bar{b}$	LO	HI
0	0	1	0	0	0	0	1	1	1	0	1	0	0	1	1	0	1
0	1	0	1	0	0	1	0	1	1	1	0	0	1	1	0	0	1
1	0	0	0	1	0	1	1	0	1	1	0	1	0	0	1	0	1
1	1	0	0	0	1	1	1	1	0	0	1	1	1	0	0	0	1

1.5.11   Switching Circuits

Boolean functions can be performed by digital circuits. Circuits that perform complicated boolean functions can be subdivided into simpler circuits that perform simpler boolean functions. The circuits that perform the simplest boolean functions are taken as basic elements, called *gates* and are represented by specialized symbols. The circuit symbols for the gates that perform functions of two independent variables are shown in Table 1.19. The gates are identified by the adjective representing the operation they perform. The most common gates are the AND, OR, NAND, NOR, XOR and COIN gates. The only nontrivial single input gate is the invertor or NOT gate. Gates are the basic elements from which more complicated digital logic circuits are constructed.

A logic circuit whose steady-state outputs depend only on the present steady-state inputs (and not on any prior inputs) is called a *combinational logic circuit*. To depend on previous inputs would require memory, thus a combinational logic circuit has no memory elements.

Boolean algebra allows any combinational logic circuit to be constructed solely with AND, OR, and NOT gates. Any combinational logic circuit may also be constructed solely with NAND gates, as well as solely with NOR gates.

1.5.12   Expansion Forms

The **sum of products (SP)** is a basic form in which all boolean functions can be expressed. The **product of sums (PS)** is another basic form in which all boolean functions can be expressed. An illustrative example is given in Fig. 1.35(b) and Fig. 1.35(c) for the example given in Fig. 1.35(a).

Minterms are a special set of functions, none of which can be expressed in terms of the others. Each minterm has each of the variables in the complemented or the uncomplemented form ANDed together. An SP expansion in which only minterms appear is a canonical SP expansion. Figure 1.35(d) shows the development of the canonical SP expansion for the previous example. The canonical SP expansion may also be simply expressed by enumerating the minterms as shown in Fig. 1.35(f). Comparison of the truth table with the minterm expansion shows that each function value of 1 represents a minterm of the function and vice versa. All other function values are 0.

Maxterms are a special set of functions, none of which can be expressed in terms of the others. Each maxterm has each of the variables in the complemented or the uncomplemented form ORed together. A PS expansion in which only maxterms appear is a canonical PS expansion. Figure 1.35(e) shows the development of the canonical PS expansion for the previous example. The canonical PS expansion may also be simply expressed by enumerating the maxterms as shown in Fig. 1.35(g). Comparison of the truth table with the maxterm expansion shows that each function value of 0 represents a maxterm of the function and vice versa. All other function values are 1.

1.5.13   Realization

The different types of boolean expansions provide different circuits for implementing the generation of the function. A function expressed in the SP form is directly realized as an AND-OR **realization** as illustrated in Fig. 1.36(a). A function expressed in the PS form is directly realized as an OR-AND realization as illustrated

TRUTH TABLE

A	B	C	f(A, B, C)
0	0	0	0
0	0	1	1
0	1	0	0
0	1	1	1
1	0	0	1
1	0	1	0
1	1	0	1
1	1	1	1

$f(A, B, C) = A(B + \overline{C}) + (\overline{A} + B)C$

(a)

(b) P6b $\rightarrow$ $f(A, B, C) = AB + A\overline{C} + \overline{A}C + BC$
    T13a $\rightarrow$ $f(A, B, C) = AB + A\overline{C} + \overline{A}C$

(c) P7b $\rightarrow$ $f(A, B, C) = A(\overline{A} + B + \overline{C}) + (\overline{A} + B + \overline{C})C$
    P6b $\rightarrow$ $f(A, B, C) = (A + C)(\overline{A} + B + \overline{C})$

(d) P7a $\rightarrow$
$f(A, B, C) = AB(C + \overline{C}) + A(B + \overline{B})\overline{C} + \overline{A}(B + \overline{B})C$
P6b $\rightarrow$
$f(A, B, C) = ABC + AB\overline{C} + AB\overline{C} + A\overline{B}\overline{C} + \overline{A}BC + \overline{A}\overline{B}C$
T8a $\rightarrow$
$f(A, B, C) = ABC + AB\overline{C} + A\overline{B}\overline{C} + \overline{A}BC + \overline{A}\overline{B}C$

(e) P7b $\rightarrow$ $f(A, B, C) = (A + C + B\overline{B})(\overline{A} + B + \overline{C})$
    P6a $\rightarrow$ $f(A, B, C) = (A + B + C)(A + \overline{B} + C)(\overline{A} + B + \overline{C})$

(f) $f(A, B, C) = m_{111} + m_{110} + m_{100} + m_{011} + m_{001}$
    $f(A, B, C) = \Sigma m(1, 3, 4, 6, 7)$

(g) $f(A, B, C) = M_{000} + M_{010} + M_{101}$
    $f(A, B, C) = \Pi M(0, 2, 5)$

FIGURE 1.35 Examples of converting boolean functions between forms: (a) given example, (b) conversion to SP form, (c) conversion to PS form, (d) conversion to canonical SP form, (e) conversion to canonical PS form, (f) minterm notation/canonical SP form, (g) maxterm notation/canonical PS form.

in Fig. 1.36(b). By using involution and deMorgan's theorem the SP expansion can be expressed in terms of NAND- NAND and the PS expansion can be expressed in terms of NOR-NOR as shown in Fig. 1.36(c) and Fig. 1.36(d). The variable inversions specified in the inputs can be supplied by either NAND or NOR gates as shown in Fig. 1.36(g) and Fig. 1.36(h), which then provide the NAND-NAND-NAND and the NOR-NOR-NOR circuits shown in Fig. 1.36(i) and Fig. 1.36(j).

1.5.14 Timing Diagrams

Timing diagrams are of two major types. A **microtiming diagram** has a time scale sufficiently expanded in space to display clearly the gate delay, such as shown in Fig. 1.37(a) for an AND gate. A **macrotiming diagram** has a time scale sufficiently contracted in space so that the gate delay is not noticeable, as shown in Fig. 1.37(b) for an AND gate.

The advantage of the macrotiming diagram is that larger time intervals can be represented in a given spatial size and they can be more quickly developed. The disadvantage is that they do not display the information required for speed limitation considerations.

1.5.15 Hazards

The variation in signal delays through different circuit elements in different paths may cause the output signal to fluctuate from that predicted by non-time-dependent truth tables for the elements. This fluctuation can cause an undesired result and, hence, is a *hazard*. This is illustrated in Fig. 1.38.

1.5.16 *K*-Map Formats

In a truth table the values of a boolean function are displayed in a one-dimensional array. A *K*-**map** contains the same information arranged in as many effective dimensions as there are independent variables in the function. The special form of the representation provides a simple procedure for minimizing the expression and, hence, the number of components required for realizing the function in a given form. The function

(c) INVOLUTION GIVES
$f(a, b, c, d) = \overline{a\bar{c} + bcd + ad}$
DEMORGAN'S THEOREM GIVES
$f(a, b, c, d) = \overline{a\bar{c}} \cdot \overline{bcd} \cdot \overline{ad}$
OR IN TERMS OF THE NAND OPERATOR
$f(a, b, c, d) = (a{\uparrow}\bar{c}){\uparrow}(b{\uparrow}c{\uparrow}d){\uparrow}(a{\uparrow}d)$

(d) INVOLUTION GIVES
$f(a, b, c, d) = \overline{(a + b)(\bar{a} + c)(b + c + d)}$
DEMORGAN'S THEOREM GIVES
$f(a, b, c, d) = \overline{(a + b)} + \overline{(\bar{a} + c)} + \overline{(b + c + d)}$
OR IN TERMS OF THE NOR OPERATOR
$f(a, b, c, d) = (a{\downarrow}b){\downarrow}(\bar{a}{\downarrow}c){\downarrow}(b{\downarrow}c{\downarrow}d)$

FIGURE 1.36 Examples of realizations based on various expansion forms: (a) AND-OR realization of $f(a, b, c, d) = a\bar{c} + bcd + ad$, (b) OR-AND realization of $f(a, b, c, d) = (a + b)(\bar{a} + c)(b + c + d)$, (c) AND-OR conversion to NAND-NAND, (d) OR-AND conversion to NOR-NOR, (e) NAND-NAND realization of $f(a, b, c, d) = a\bar{c} + bcd + ad$, (f) NOR-NOR realization of $f(a, b, c, d) = (a + b)(\bar{a} + c)(b + c + d)$, (g) NAND gate realization of NOT gate, (h) NOR gate realization of NOT gate, (i) NAND-NAND-NAND realization of $f(a, b, c, d) = a\bar{c} + bcd + ad$, (j) NOR-NOR-NOR realization of $f(a, b, c, d) = (a + b)(\bar{a} + c)(b + c + d)$.

is represented in a space that in a Venn diagram is called the *universal set*. The K-map is a special form of Venn diagram. The space is divided into two halves for each of the independent variables. The division of the space into halves is different for each independent variable. For one independent variable the space is divided into two different identical size regions, each of which represents a minterm of the function. For n independent variables the space is divided into 2^n different identical size regions, one for each of the 2^n minterms of the function. This and associated considerations are illustrated in a sequence of figures from Fig. 1.39 to Fig. 1.46.

FIGURE 1.37 Timing diagrams for the AND gate circuit: (a) microtiming diagram, (b) macrotiming diagram.

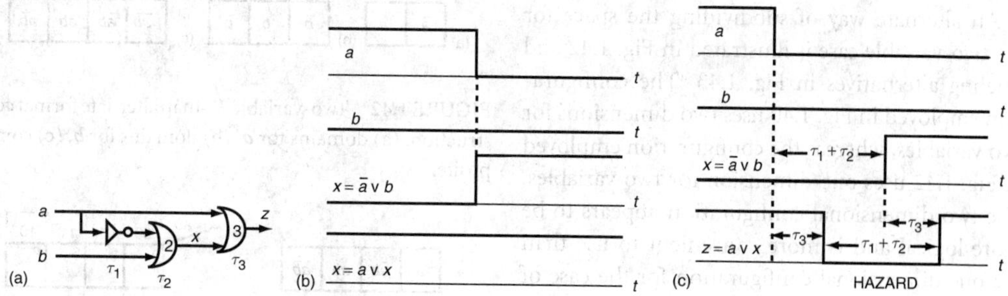

FIGURE 1.38 Example of a hazard (output variation) caused by unequal delay paths: (a) circuit for illustrating a hazard, (b) ideal case, no delays, $\tau_1 = \tau_2 = \tau_3 = 0$, no hazard introduced; (c) signal paths with different delays, $\tau_1 + \tau_2 + \tau_3 > \tau_3$, hazard introduced.

Figure 1.39 shows one-variable K-map formats. Figure 1.39(a) shows the space divided into two equal areas, one for each of the two minterms possible for a single variable. The squares could also be identified by the variable placed external to the space to designate the region that is the domain of the variable, with the unlabeled space being the domain of the complement of the variable, as shown in Fig. 1.39(b). Another way of identifying the regions is by means of the minterm number the area is for, as shown in Fig. 1.39(c). Still another way is to place the values the variable can have as a scale alongside the space, as shown in Fig. 1.39(d). The composite labeling, shown in Fig. 1.39(e), appears redundant but is often useful because of the different modes of thought used with the different label types. Putting the actual minterm expressions inside each square is too cluttering and is rarely used except as an aid in teaching the theory of K-maps. The use of minterm numbers, although widely used, also clutters up the diagram, and the methodology presented here makes their use superfluous once the concepts are understood.

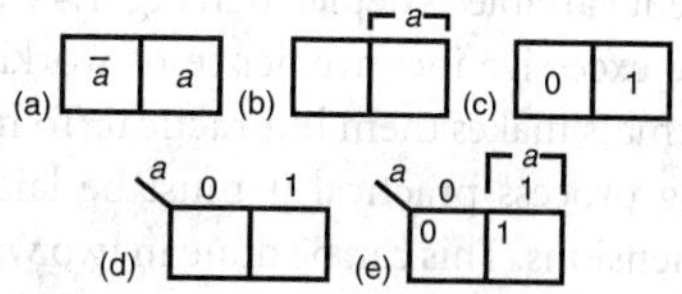

The organization of a two-variable K-map format is illustrated in Fig. 1.40. The space is subdivided vertically into two domains for the variable a and its complement, and is subdivided horizontally for the variable b and its complement as shown in Fig. 1.40(a) and Fig. 1.40(b). The composite subdivisions for both variables together with the expressions for the two-variable minterms are shown in Fig. 1.40(c).

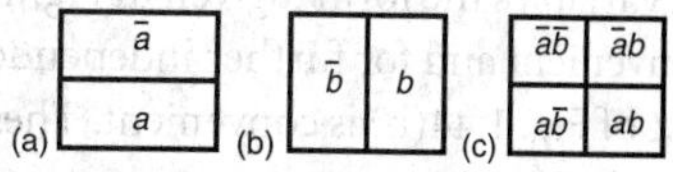

FIGURE 1.39 One variable K-map forms: (a) internal minterm labels, (b) external domain label, (c) internal minterm number labels, (d) external scale label, (e) composite labeling.

FIGURE 1.40 Two-variable K-map format construction: (a) domains for a, (b) domains for b, composite.

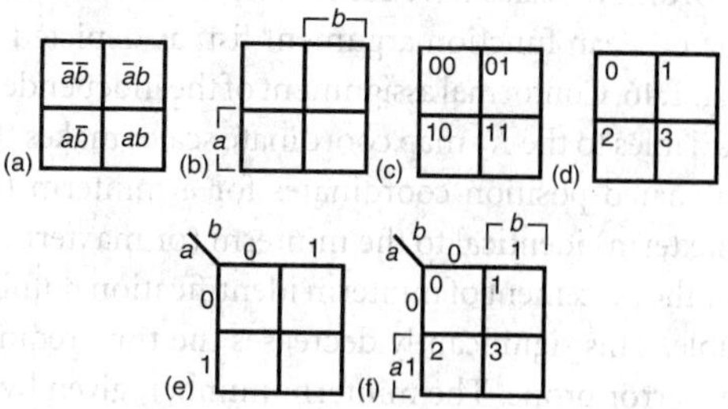

FIGURE 1.41 Two variable K-map formats: (a) minterm labels, (b) domain labels, (c) minterm binary labels, (d) minterm decimal labels, (e) scale labels, (f) composite labeling.

The different formats for identifying the areas for the two-variable case are shown in Fig. 1.41. Of particular interest is the comparison of the binary and decimal minterm number labels. The binary minterm number is simply the **catenation** of the vertical and horizontal scale number for the position. It is the use of this identity that makes internal labels in the square totally superfluous.

An alternate way of subdividing the space for the two-variable case is illustrated in Fig. 1.42 and labeling alternatives in Fig. 1.43. The configuration employed in Fig. 1.40 uses two dimensions for two variables, whereas the configuration employed in Fig. 1.42 uses one-dimension for two variables. The two dimensional configuration appears to be more logical and is more convenient to use than the one dimensional configuration for the case of two variables. For the case of a larger number of variables the configuration in Fig. 1.43 offers special advantages, as will be shown.

The organization of three-variable K-map formats is illustrated in Fig. 1.44. It is logical to introduce an additional space dimension for each additional independent variable as depicted in Fig. 1.44(a); however, the excessive inconvenience of working with such formats makes them impractical. To make the mapping process practical it must be laid out in two dimensions. This can be done in two ways. One way is to take the individual slices of the three-dimensional configuration and place them adjacent to each other as illustrated in Fig. 1.44(b). Another way is to use the one-dimensional form for two variables, illustrated in Fig. 1.43, as shown in Fig. 1.44(c). For the case of three and four independent variables the format given in Fig. 1.44(c) is more convenient and for further independent variables that of Fig. 1.44(b) is convenient. These are all illustrated in Fig. 1.46.

Labeling for three independent variables is given in Fig. 1.45.

The independent boolean variables in **conformal coordinate scales** have exactly the same order as in the boolean function argument list, as depicted in Fig. 1.46. Conformal assignment of the independent variables to the K-map coordinate scales makes the catenated position coordinates for a minterm (or maxterm) identical to the minterm (or maxterm) number. Utilization of this identity eliminates the need for the placement of minterm identification numbers in each square or for a separate position identification table. This significantly decreases the time required to construct K-maps and makes their construction less error prone. The minterm number, given by the catenation of the vertical and horizontal coordinate numbers, is obvious if the binary or octal number system is used.

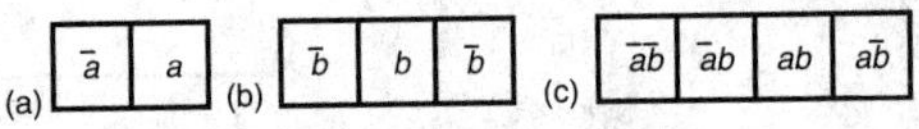

FIGURE 1.42   Two-variable K-map alternate format construction: (a) domains for **a**, (b) domains for **b**, (c) composite.

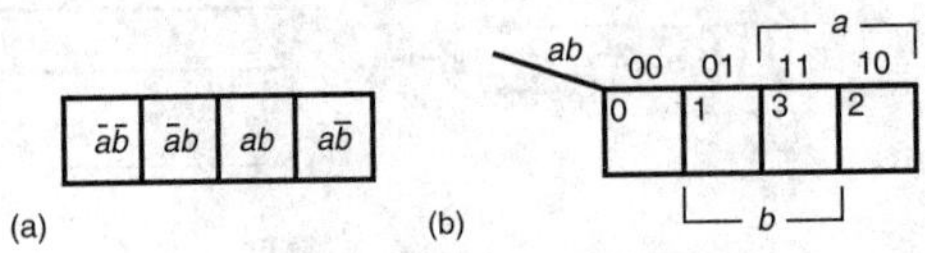

FIGURE 1.43   Two-variable K-map alternate format: (a) minterm labels, (b) composite labeling.

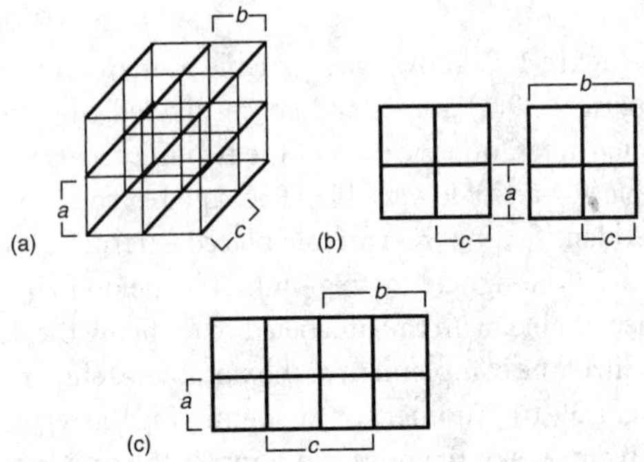

FIGURE 1.44   Three-variable K-map formats: (a) three dimensional, (b) three-dimensional left and right halves, (c) two dimensional.

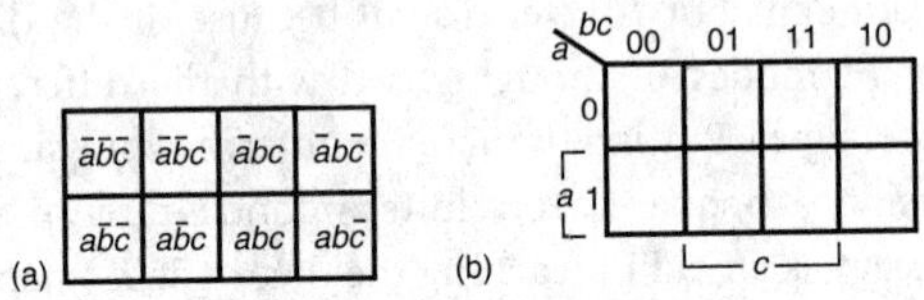

FIGURE 1.45   Three variable two dimensional K-map formats: (a) minterm labels, (b) composite labels.

1.5.17   *K-Maps and Minimization*

A function is mapped into the K-map format by entering the value for each of the minterms in the space for that minterm. The function values can be obtained in various ways such as from the truth table for the function, the boolean expression for the function, or from other means by which the function may be defined. An example is given for the truth table given in Fig. 1.35(a), which is repeated here in Fig. 1.47(a) and whose K-map is shown in various forms in Fig. 1.47(b) to Fig. 1.47(d).

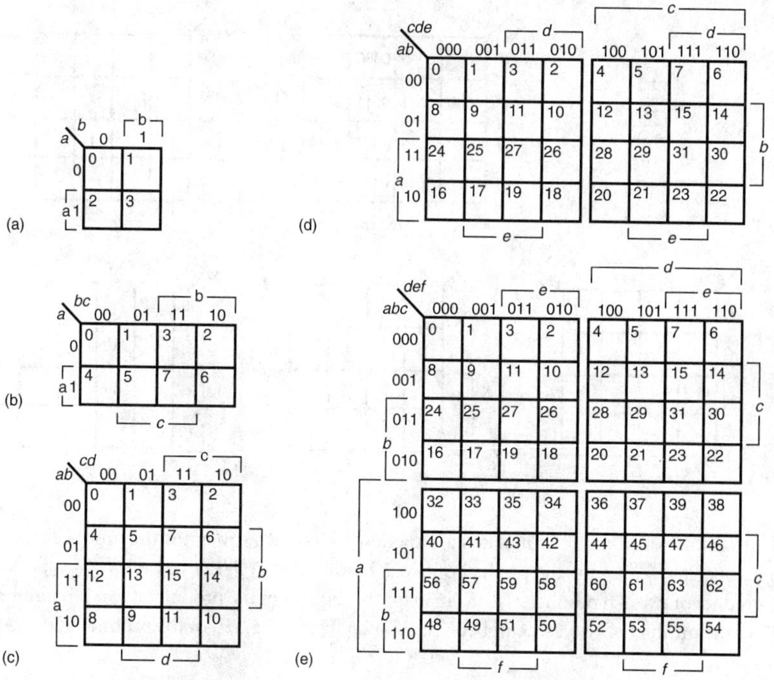

FIGURE 1.46 Formats for K-maps for functions of 2–6 independent variables with conformal coordinate scales: (a) two-variable case f(a, b), three-variable case f(a, b, c), (c) four-variable case f(a, b, c, d), (d) five-variable case f(a, b, c, d, e), (e) six-variable case f(a, b, c, d, e, f).

The function can also be mapped into a conformally scaled K-map directly from the canonical expansion, this being essentially the same process as entering the minterms (or maxterms) from the truth table. The function may also be directly mapped from any PS or SP expansion form. Another means of obtaining the K-map is to formulate it as a function table, as illustrated for the multiplication of 1- and 2-b numbers in Fig. 1.48.

1.5.18 Minimization with K-Maps

The key feature of K-maps that renders them convenient for minimization is that minterms that are spatially adjacent in the horizontal or vertical directions are logically adjacent. Logically adjacent minterms are identical in all variables except one. This allows the two minterms to be combined into a single terms with one less variable. This is illustrated in Fig. 1.49. Two adjacent minterms combine into a first-order **implicant**.

A	B	C	F(A, B, C)
0	0	0	0
0	0	1	1
0	1	0	0
0	1	1	1
1	0	0	1
1	0	1	0
1	1	0	1
1	1	1	1

(a)

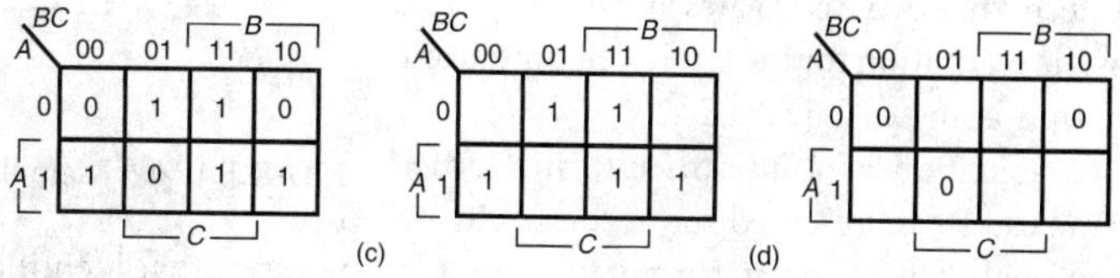

FIGURE 1.47 Three-variable K-map example: (a) truth table, (b) K-map with all values shown, (c) minterm K-map, (d) maxterm K-map.

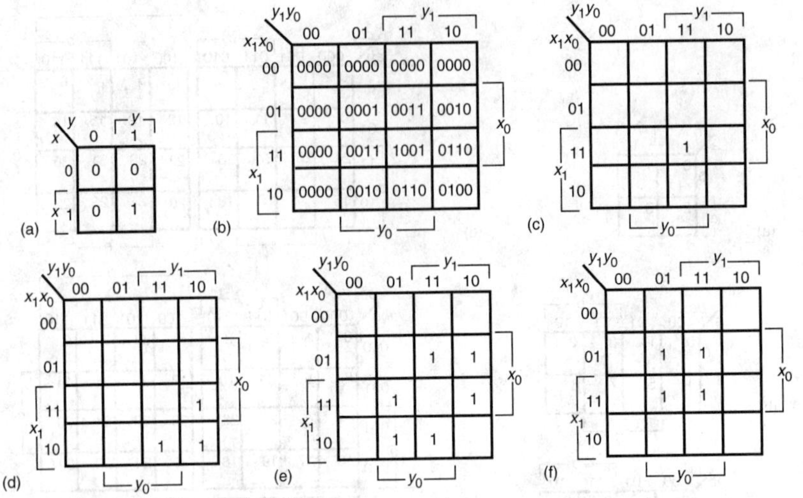

FIGURE 1.48 Examples of K-maps formulated as function tables: (a) K-map for the product of two 1-b numbers $P = x^*y$; (b) composite K-map for the product of two 2-b numbers, $P_3 P_2 P_1 P_0 = x_1 x_0 * y_1 y_0$; (c) K-map for the digit P_3 of the product of two 2-b numbers; (d) K-map for the digit P_2 of the product of two 2-b numbers; (e) K-map for the digit P_1 of the product of two 2-b numbers; (f) K-map for the digit P_0 of the product of two 2-b numbers.

(b) $F(x_1, x_0, y_1, y_0) = \bar{x}_1 x_0 y_1 y_0 + x_1 x_0 y_1 y_0$

FACTORING GIVES $F(x_1, x_0, y_1, y_0) = (\bar{x}_1 + x_1) x_0 y_1 y_0$
(c) HENCE $F(x_1, x_0, y_1, y_0) = x_0 y_1 y_0$

(d) $F = m_{0111} + m_{1111}$
      $F = l_{-111}$

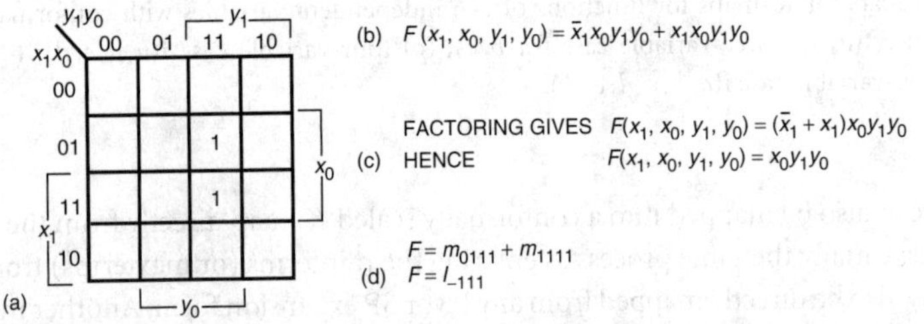

FIGURE 1.49 Example of minimization with a K-map: (a) sample K-map, (b) expression in minterms of function definition in (a), (c) simplification of the expression in (b), (d) simplification of the expression in (b) using single symnbol minterm and implicant notation.

A first-order implicant is the combination of all of the independent variables but one. In this example the first-order implicant expressed in terms of minterms contains eight literals but the minimized expression contains only three literals. The circuit realization for the OR combination of the two minterms has two AND gates and one OR gate, whereas the realization for the equivalent implicant requires only a single AND gate.

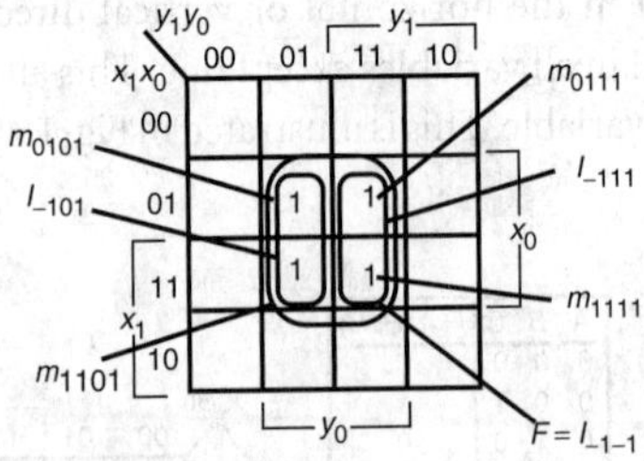

The combination of minterms into first-order implicants can be represented more compactly by

FIGURE 1.50 Example of minimization with K-map.

using the single symbol minterm notation with the subscript that identifies the particular minterm expressed in binary, as illustrated in Fig. 1.49(d).

Two adjacent first-order implicants can be combined into a second-order implicant as illustrated in Fig. 1.50. A second-order implicant contains all of the independent variables except two. In general, an *n*th-order implicant contains all of the variables except *n* and requires an appropriately grouped set of 2^n minterms.

Minterms that are at opposite edges of the same row or column are logically adjacent since they differ in only one variable. If the plane space is rolled into a cylinder with opposite edges touching, then the logically adjacent edge pairs become physically adjacent. For larger numbers of variables using *K*-maps with parallel sheets the corresponding positions on different sheets are logically adjacent. If the sheets are considered as overlaid the corresponding squares are physically adjacent. The minimized expression is obtained by covering the all of the minterms with the fewest number of largest possible implicants. A minterm is a zero-order implicant. Figure 1.51 illustrates a variety of examples. A **don't care** is a value that never occurs or if it does occur it is not used, and hence, it does not matter what its value is. Don't cares are also included to illustrate that they can be used to simplify expressions by taking their values to maximize the order of the implicants.

Maxterm *K*-maps can also be utilized to obtain minimized expressions by combining maxterms into higher order implicants, as illustrated for the example in Fig. 1.52.

$$F = bef + \bar{a}b\bar{c}\bar{f} + \bar{a}bd\bar{e}$$

FIGURE 1.51 Example of minimization with six-variable *K*-map.

$$F = B(A + C)$$

FIGURE 1.52 Three-variable maxterm *K*-map example.

GIVEN $F(A, B, C, D) = \Sigma m(2, 6, 7, 8) + d(0, 4, 5, 12, 13)$ and $G(A, B, C, D) = \Sigma m(2, 4, 5) + d(6, 7, 8, 10)$

ZERO-ORDER IMPLICANT LIST.					FIRST-ORDER IMPLICANT LIST.				SECOND-ORDER IMPLICANT LIST			
NO. OF 1s	MIN-TERM	CODE ABCD	FLAGS	PI	IMPLICANTS	CODE ABCD	FLAGS	PI	IMPLICANTS	CODE ABGD	FLAGS	PI
0	0	0000	F-	√	0, 2	00-0	F-	√	0, 2, 4, 6	0-0	F-	1
	2	0010	FG	√	0, 4	0-00	F-	√	0, 4, 8, 12	-00	F-	2
1	4	0100	FG	√	0, 8	-000	F-	√	4, 5, 6, 7	01-	FG	3
	8	1000	FG	√	2, 6	0-10	FG	5	4, 5, 12, 13	-10-	F-	4
	5	0101	FG	√	2, 10	-010	-G	6				
2	6	0110	FG	√	4, 5	010-	FG	√				
	10	1010	-G	√	4, 6	01-0	FG	√				
	12	1100	F-	√	4, 12	-100	F-	√				
	7	0111	FG	√	8, 10	10-0	-G	7				
3	13	1101	F-	√	8, 12	1-00	F-	√				
					5, 7	01-1	FG	√				
					5, 13	-101	F-	√				
					6, 7	011-	FG	√				
					12, 13	110-	F-	√				

PRIME IMPLICANT CHART

	F				G		
PI \ m	2	6	7	8	2	4	5
1	X	X			X		
2				X			
3		X	X			X	X
4							
5	X	X					X
6							
7							

MINIMUM SP EXPANSIONS

$$F = PI_2 + PI_3 + PI_5$$
$$= \overline{CD} + \overline{A}B + A\overline{C}\overline{D}$$

$$G = PI_3 + PI_5$$
$$= \overline{A}B + \overline{A}C\overline{D}$$

FIGURE 1.53 Illustration of the Quine–McClusky method of simultaneous minimization.

1.5.19   Quine–McCluskey Tabular Minimization

The *K*-map minimization method is too cumbersome for more than six variables and does not readily lend itself to computerization. A tabular method, which can be implemented for any number of variables and which lends itself to computer program implementation, consists of the following steps:

1. List all the minterms in the boolean function (with their binary code) organized into groups having the same number of 1s. The groups must be listed in consecutive order of the number of 1s.
2. Construct the list of first-order implicants. Use flags to indicate which minterms, don't cares, or implicants go with which functions. (Only minterms in adjacent groups have the possibility of being adjacent and, hence, this ordering method significantly reduces the labor of compiling the implicants.)
3. Construct the list of second-order implicants and the lists of all higher order implicants until no higher order implicants can be constructed.
4. Construct the *prime implicant chart*. The prime implicant chart shows what prime implicants cover which minterms.
5. Select the minimum number of largest prime implicants that cover the minterms.

This procedure is illustrated in Fig. 1.53 for the simultaneous minimization of two boolean functions.

Defining Terms

Base: The number a digit must be multiplied by to move it one digit to the left, also called the radix.

Binary coded decimal (BCD): Each decimal digit is expressed individually in binary form.

Catenation: Symbols strung together to form a larger sequence, as the characters in a word and the digits in a number.

Code: The representation in one alphabet of something in another (usually larger) alphabet.

Complement: The quantity obtained by subtracting a number from the largest quantity that can be expressed in the specified number of digits in a given number system.

Conformal: The same arrangement of a set of quantities in two different contexts.

Digit: A character that represents quantitative information.

Don't care: A function that can be taken either as a minterm or a maxterm at the convenience of the user.

Gray code: A set of codes having the property of logical adjacency.

Implicant: A first-order implicant is a pair of logically adjacent minterms. A second-order implicant is a set of logically adjacent first-order implicants and so on.

Integer: Any number that can be expressed solely in terms of digits.

Fraction: Any number divided by a larger number.

K-map: An arrangement of space into equal size units, each of which represents a minterm (or maxterm) such that each physically adjacent square is also logically adjacent.

Logically adjacent: Any two codes having the same number of digits for which they differ in the value of only one of the digits.

Macrotiming diagram: A graphical display showing how the waveforms vary with time but with a time scale that does not have sufficient resolution to display the delays introduced by the individual basic elements of the digital circuit.

Maxterm: A function of a set of boolean variables that has a low value for only one combination of variable values and has a high value for all other combinations of the variable values.

Microtiming diagram: A graphical display showing how the waveforms vary with time but with a time scale that has sufficient resolution to display clearly the delays introduced by the individual basic elements of the digital circuit.

Minterm: A function of a set of boolean variables that has a high value for only one combination of variable values and has a low value for all other combinations of the variable values.

Overflow: That part of a numerical operation that does not fit into the allocated field.

Parity bit: An extra bit catenated to a code and given a value such that the total number of high bits is even for even parity and odd for odd parity.

Product of sums (PS): The AND combination of terms, which are OR combinations of boolean variables.

Prime implicant: An implicant that is not part of a larger implicant.

Radix: The number of different values that a digit can have in a number system.

Reduced radix: The largest value a digit can have in a number system. It is one less than the radix.

Real number: A number that has a fractional part and an integer part.

Realization: A circuit that can produce the value of a function.

Sum of products (SP): The OR combination of terms, which are AND combinations of boolean variables.

Truth table: The table of values that a Boolean function can have for which the independent variables considered as a multidigit number are arranged in consecutive order.

References

Hayes, J.P. 1993. *Introduction of Digital Logic Design*. Addison-Wesley, Reading, MA.

Humphrey, W.S., Jr. 1958. *Switching Circuits with Computer Applications*. McGraw-Hill, New York.

Hill and Peterson. 1974. *Introduction to Switching Theory and Logical Design*, 2nd ed. Wiley, New York.

Johnson and Karim. 1987. *Digital Design a Pragmatic Approach*. Prindle, Weber and Schmidt, Boston.

Karnaugh, M. 1953. The map method for synthesis of combinational logic circuits. *AIEE Trans. Comm. Elec.* 72 (Nov.): 593–599.

Mano, M.M. 1991. *Digital Design*. Prentice-Hall, Englewood Cliffs, NJ.

McClusky, E.J. 1986. *Logic Design Principles*. Prentice-Hall, Englewood Cliffs, NJ.

Mowle, F.J. 1976. *A Systematic Approach to Digital Logic Design*. Addison-Wesley, Reading, MA.

Nagle, Carrol, and Irwin. 1975. *An Introduction to Computer Logic*, 2nd ed. Prentice-Hall, Englewood Cliffs, NJ.

Pappas, N.L. 1994. *Digital Design*. West, St. Paul, MN.

Roth, C.H., Jr. 1985. *Fundamentals of Logic Design*, 3rd ed. West, St. Paul, MN.

Sandige, R.S. 1990. *Modern Digital Design*. McGraw-Hill, New York.

Shaw, A.W. 1991. *Logic Circuit Design*. Saunders, Fort Worth, TX.

Wakerly, 1990. *Digital Design Principles and Practices*. Prentice-Hall, Englewood Cliffs, NJ.

Further Information

Further information on basic logic concepts and combinational logic design can be found in occasional articles in the following journals:

Lecture Notes in Computer Science (annual)
International Journal of Electronics (monthly)
IEEE Transactions on Education (quarterly)
IEEE Transactions on Computers (monthly)
IEEE Transactions on Software Engineering (monthly)
IEEE Transactions on Circuits and Systems 1. Fundamental Theory and Applications (monthly)

1.6 Digital Logic and Sequential Logic Circuits

George I. Cohn

1.6.1 Combinational and Sequential Logic Circuits

A **combinational logic** circuit contains logic gates but does not contain storage elements. A symbolic representation of a multiple-input multiple-output combinational logic circuit is shown in Fig. 1.54(a). The outputs of a combinational logic circuit depend only on the present inputs to the circuit. A sequential

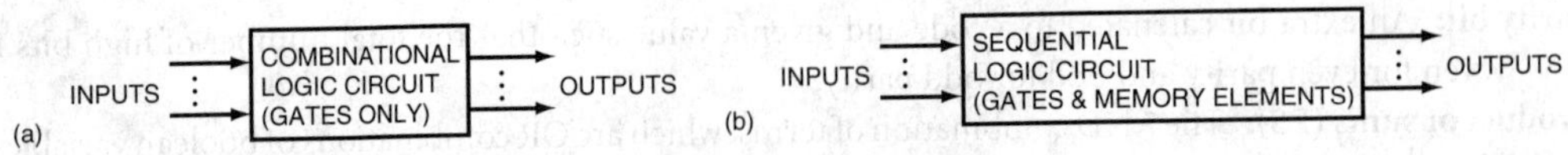

FIGURE 1.54   Logic circuits: (a) combinational, (b) sequential.

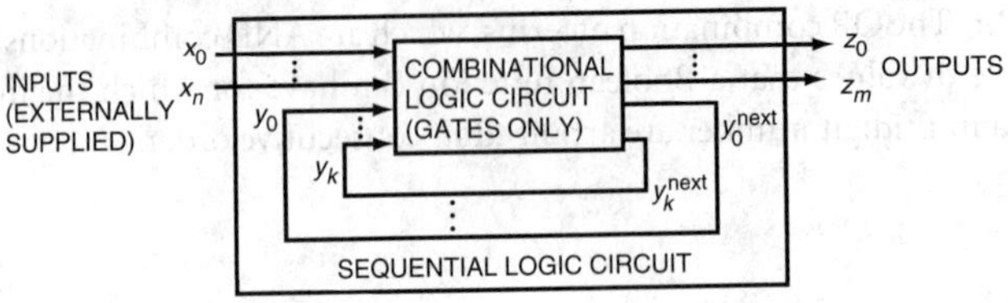

FIGURE 1.55   A combinatioal logic circuit with stable feedback is a sequential logic circuit.

logic circuit contains storage elements in addition to gates. A symbolic representation of a multiple-input multiple-output **sequential logic** circuit is shown in Fig. 1.54(b). The outputs of a sequential logic circuit depend on prior inputs as well as current inputs.

Combinational logic circuits with **feedback**, as shown in Fig. 1.55, with stable behavior can store information. Feedback provides one or more closed loops through the circuit. The outputs fed back as inputs may cause the outputs to change, thereby changing the inputs again, and so on. If the outputs keep changing without reaching a fixed equilibrium **state** the system is **unstable**. If the outputs settle to a fixed unchanging equilibrium state the system is **stable**. A combinational logic circuit with closed-loop feedback that is stable can store information and, hence, be used as a memory element. The current inputs produce the next outputs at a later time due to the propagation delays through the circuit. If the next outputs, which are fed back, are equal to the previous inputs due to feedback the system is in a stable state. If they are unequal they are in an unstable state since it is undergoing a transient variation. These conditions are shown in Table 1.21.

The system may eventually reach a stable state or it may not. Because of the variation in the delay time through the various elements of the system the state may not be predictable from the elementary tables employed for representation of the elements and the circuit. This may happen when more than one input is changed before an equilibrium condition is reached. This situation is avoided when the operation restriction is employed that only one input is changed at a time and that there is no further change in any input until equilibrium is established. In practice, this restraint must be removed or the design must be implemented so as to remove ambiguities. If two or more state variables change in a process the changes may not take place simultaneously. The phenomena can be considered a **race** as to which takes place first and this is referred to as a *race condition*. If the final state depends on which variable changes first the phenomena is a **critical race**. If the final state is independent of the order of change the phenomena is a **noncritical race**.

TABLE 1.21   System State

State	Condition	
Stable	$y_i^{next} = y_i$;	$i = 0{:}k$
Unstable	$y_i^{next} \neq y_i$;	$i = 0{:}k$

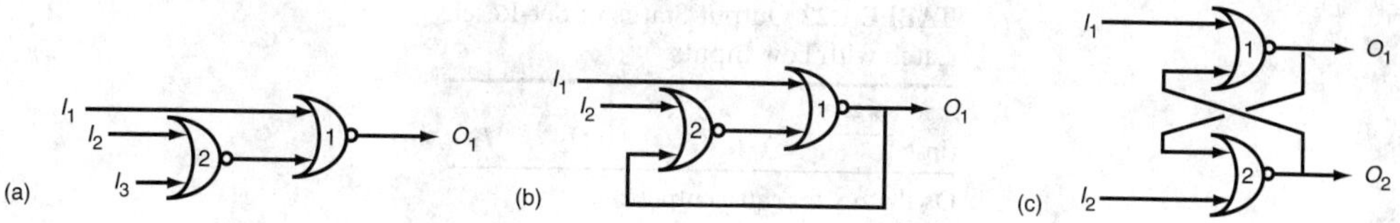

FIGURE 1.56 Combinational logic circuit example: (a) without feedback, (b) with feedback, (c) with feedback but symmetrically configured.

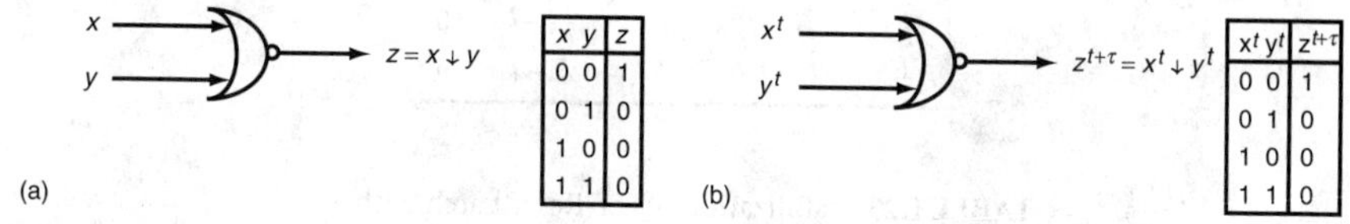

FIGURE 1.57 NOR gate and NOR gate truth table: (a) for equilibrium conditions, (b) with gate delay time represented.

1.6.2 Set-Reset Latch

A simple combinational logic circuit with two NOR gates is shown in Fig. 1.56(a). This circuit with a feedback path providing a closed loop as shown in Fig. 1.56(b) has two stable states and, hence, can be used to store one binary digit of information. Since the gates function symmetrically it is convenient to draw the circuit in symmetric form as shown in Fig. 1.56(c), to convey the perception of symmetric behavior.

The circuit element characteristics are required to analyze this circuit to determine the conditions under which the behavior is stable and the conditions under which it is unstable. The NOR gate characteristics normally specified by the truth table in Fig. 1.57(a) correctly give the steady-state equilibrium characteristics. However, signals presented to the input terminals produce effects at the output terminals at a later time. The time delay for the effect to propagate through the device (called the *gate delay time* and represented by the symbol tau τ) has an important effect on the behavior of the circuit. Figure 1.57(b) shows the truth table labeled so that this delay is explicitly designated.

1.6.3 Latch Analysis with Difference Equations

The **latch** circuit shown in Fig. 1.56(c) is redrawn in Fig. 1.58 with the outputs represented in terms of the inputs, with the gate delay explicitly shown, and with the outputs expressed as first-order difference equations in terms of the inputs. The case for general inputs is shown in Fig. 1.58(a) and the special case for low inputs in Fig. 1.58(b). With low inputs the equations in Fig. 1.57(b) show that if the two outputs are complements, $O_2^t = \overline{O_1^t}$, that the system is stable, that is, $O_1^{t+\tau} = O_1^t$ and $O_2^{t+\tau} = O_2^t$. It also shows that if the two outputs are equal $O_2^t = O_1^t$ the system is unstable, that is, $O_1^{t+\tau} = \overline{O_1^t}$ and $O_2^{t+\tau} = \overline{O_2^t}$. Table 1.22 illustrates the development for the low input cases.

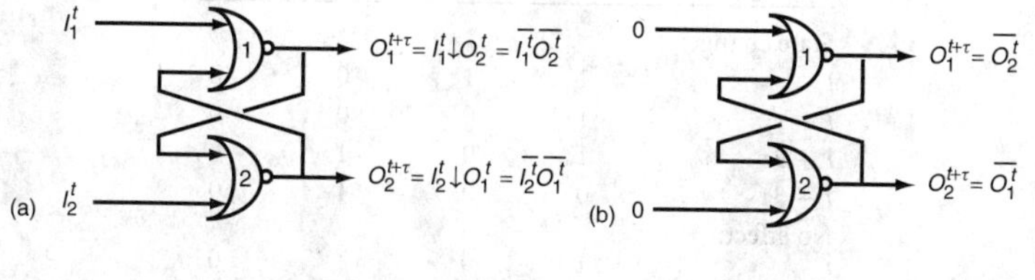

FIGURE 1.58 Two NOR gate latchs: (a) arbitrary inputs, (b) low inputs.

TABLE 1.22 Output Status of Set-Reset Latch with Low Inputs

time $\downarrow$ signal $\rightarrow$	I_1	O_1	O_2	I_2
Oscillatory for equal outputs:				
t	0	n	n	0
$t + \tau$	0	$1-n$	$1-n$	0
$t + 2\tau$	0	n	n	0
$t + 3\tau$	0	$1-n$	$1-n$	0
Stable for complementary outputs:				
t	0	k	$1-k$	0
$t + \tau$	0	k	$1-k$	0

TABLE 1.23 Stabilization of Set-Reset Latch with a High Input

time $\downarrow$ signal $\rightarrow$	I_1	O_1	O_2	I_2
t	1	k	k	0
$t + \tau$	1	0	$1-k$	0
$t + 2\tau$	1	0	1	0
$t + 3\tau$	0	0	1	0

A high signal into either input stabilizes the circuit. Because of the circuit symmetry it is only necessary to demonstrate this for one case. This behavior is shown by the output status chart of Table 1.23.

A **set-reset** (SR) latch in a stable state is changed to a different stable state by a high external input to the gate, which has the initially high output. The new state remains stable with the new value and continues to remain stable when the external input is removed. If the external input to a gate is less than the other gate's output there is no change in state. This output status is charted in Table 1.24.

With both inputs high, both outputs are driven low and will remain low as long as both high inputs are in place. The circuit becomes oscillatory one gate delay after the pair of high inputs are removed.

1.6.4 Microtiming Diagram Construction

Timing diagrams were introduced in Chapter 1.5, Section 1.5.14. The circuit example in Fig. 1.58 has four waveforms in the timing diagram, one for each distinct wire. The signals supplied from external sources, that is, **excitations** or driving functions, are the independent electrical variables, which have to be

TABLE 1.24 Effect of a High Input on a Stable Set-Reset Latch

time $\downarrow$ signal $\rightarrow$	I_1	O_1	O_2	I_2
State change:				
t	1	1	0	0
$t + \tau$	1	0	0	0
$t + 2\tau$	1	0	1	0
$t + 3\tau$	0	0	1	0
No effect:				
t	1	0	1	0
$t + \tau$	1	0	1	0

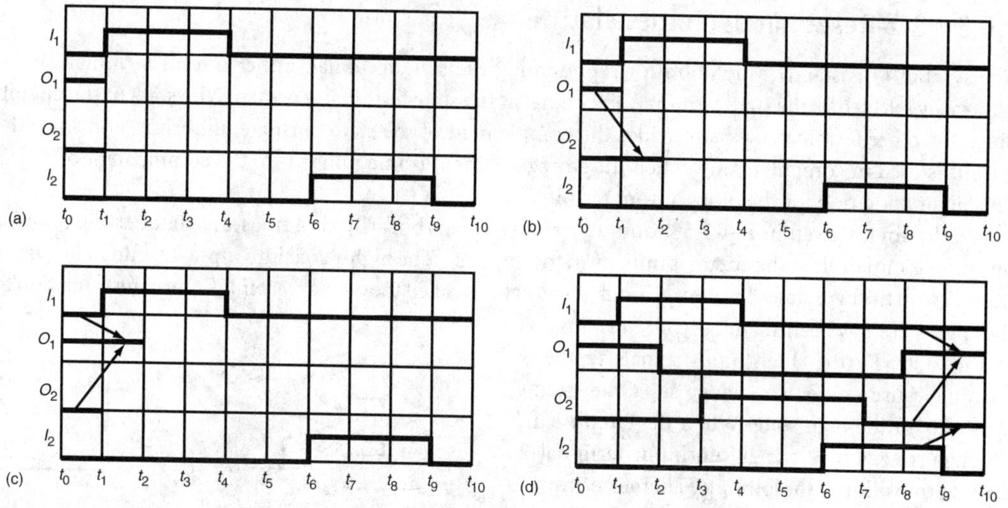

FIGURE 1.59 Microtiming diagram example for the set-reset latch: (a) specified driving and initial conditions, (b) output-2 advanced to t_2, (c) output-1 advanced to t_2, (d) outputs advanced to t_{10}.

prespecified for the time duration of interest in order to construct the timing diagram. The waveforms on all of the wires that are not directly connected to the input terminal are the dependent electrical variables that have to be determined for the timing diagram. The initial conditions, that is, the initial values of the dependent variables, must also be specified prior to the construction of the timing diagram.

Construction of a microtiming diagram for the circuit given in Fig. 1.58 is illustrated. If the initial conditions of the dependent variable waveforms are not specified, then it would be necessary to develop the timing diagrams for each of the possible sets of initial conditions in order to have the complete set of possible responses of the system to the specified driving functions. The construction process is illustrated for the initial conditions given in Fig. 1.59(a). Normally, the set of driving waveforms would be given in one group and the dependent variable waveforms in a subsequent group. However, the arrangement given in Fig. 1.59(a) facilitates the development of the microtiming diagram for this example.

Physically the waveforms develop simultaneously with time. The procedure for reasoning out the dependent variable waveforms is facilitated by using discrete time steps having a quantized value equal to the gate delay $\tau = t_n - t_{n-1}$ since it takes a gate delay for a cause at the input terminals of a gate to produce an effect at the output terminal, shown by the gate equations in Fig. 1.58.

Since the effect waveform is due to the cause waveforms a gate delay earlier, the causal arrows in Fig. 1.59(b) and Fig. 1.59(c) show the waveform values at the arrow tails that cause the waveform values at the arrow tip. The advancement of output-2 by one gate delay is shown in Fig. 1.59(b). Since any high input to a NOR gate drives the output low it is only necessary to consider a single high input to determine a low gate output. If all of the inputs to a NOR gate are low the output is driven high one gate delay later as shown in Fig. 1.59(c) for the advancement of output-1 by one gate delay. This process is continued until the timing diagram has been constructed for the time range of interest as shown in Fig. 1.59(d).

The microtiming diagram shows that whenever the outputs are in an equilibrium state they are complements of each other. The noncomplementary condition occurs only for a portion of the interval during which the transient transition between states takes place. In the timing diagram shown a transient transition between states occurs when a high input signal is supplied to the gate which has the high output.

1.6.5 Set-Reset Latch Nomenclature

The symbol Q is standard notation for representing the state terminal. Either output terminal can be arbitrarily selected as the state terminal. The signal at the state terminal represents the state of the circuit. Since the other terminal always provides the complement of the state during equilibrium and since the output is used in computers only when the storage element is in equilibrium, the complement of Q is a satisfactory notation for the other output terminal.

For the circuit given in Fig. 1.58, output-terminal-2 is arbitrarily chosen as the state terminal. Since output-2 is initially low, the circuit is initially in the low state. The high signal into input-terminal-1 causes a transition to the high state. Therefore, input-terminal-1 is called the S or *set terminal*. Once the transition to the high state has been made, the circuit remains in this state. Further high input signals to the S terminal cause no further change since the circuit is already in the high state. When the circuit is in the high state, a high signal into input-terminal-2 causes a transition to the low state. Therefore, input-terminal-2 is called the R or *reset terminal*. With this standard nomenclature for the terminals the *set-reset latch* circuit is shown in Fig. 1.60(a). The standard *single-block symbol* (SBS) representation

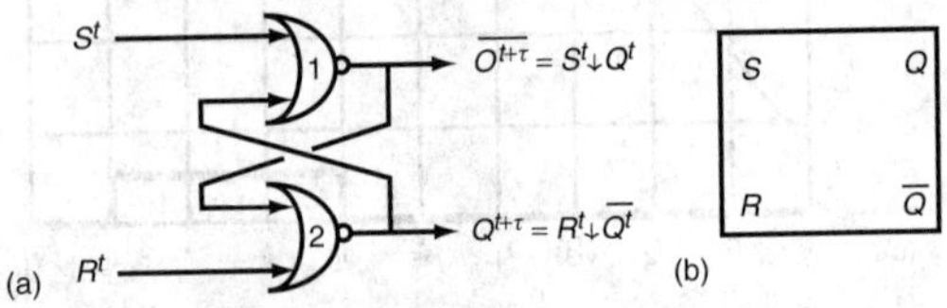

FIGURE 1.60   The set-reset (SR) latch: (a) circuit diagram, (b) SBS representation.

for a set-reset latch is shown in Fig. 1.60(b). In the SBS notation the Q terminal is placed opposite the S terminal rather than opposite to the R terminal as in the circuit diagram.

1.6.6 Set-Reset Latch Truth Table

The current state and the input signals are the independent variables that determine the next state, which is the *dependent variable*. The input signals are the driving functions that command the device to change its state. A high input signal to the S terminal commands the device to be set, that is, to go to the high state. A high input signal to the R terminal commands the device to be reset, that is, to go to the low state. The SR latch operation is given in **state diagram** form in Fig. 1.61. A complete set of transition command combinations is given in Table 1.25.

1.6.7 Set-Reset Latch Macrotiming Diagram

To better observe the differences between micro- and macrotiming diagrams, corresponding examples are directly compared. The microtiming diagram in Fig. 1.59(d) is redrawn in Fig. 1.62(a) with the input and output waveforms separately grouped. Figure 1.62(b) gives the corresponding macrotiming diagram, which neglects the propagation delay through the gates.

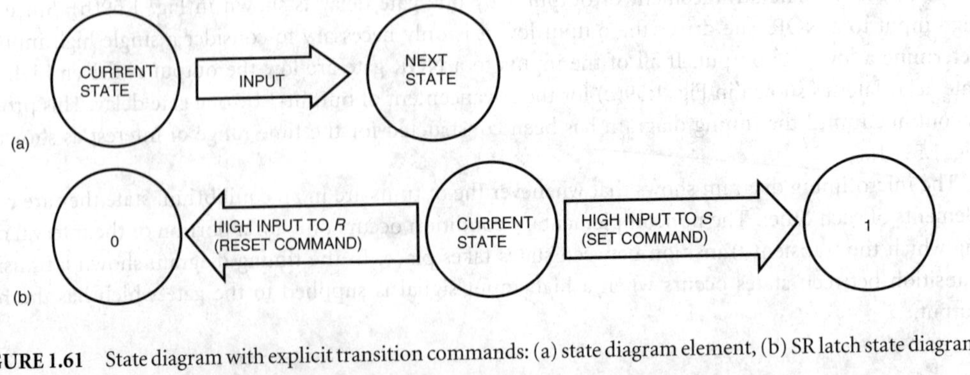

FIGURE 1.61   State diagram with explicit transition commands: (a) state diagram element, (b) SR latch state diagram.

TABLE 1.25 Asynchronous Set-Reset Latch Truth Table

S	R	Q	Q^{next}	
0	0	0	0	No
0	0	1	1	change
0	1	0	0	Clear
0	1	1	0	
1	0	0	1	Set
1	0	1	1	
1	1	0	?	Do not
1	1	1	?	use

1.6.8 *J K* Latch

The SR latch's undesirable behavior resulting when both inputs are high can be eliminated by using AND gates in the input leads to block the simultaneous application of both high input signals, as shown in Fig. 1.63(a). The input AND gate layer is controlled by the state and complementary state signals. This causes the state to complement whenever both inputs are high. The single block symbol for the *J K* latch is shown in Fig. 1.63(b). The truth table for the *J K* latch differs from that of the SR latch only in the last two entries as shown in Fig. 1.63(c).

1.6.9 *T* Latch

Combining the two inputs of a *J K* latch into a single input gives a latch that changes state when a high signal is applied. This action is called *toggling* and, hence, it is called a toggle or *T latch*. The circuit, symbol, and truth table are shown in Fig. 1.64.

FIGURE 1.62 Example timing diagrams for an asynchronous SR latch: (a) microtiming diagram, (b) macrotiming diagram.

J	K	Q	Q^{next}	
0	0	0	0	NO
0	0	1	1	CHANGE
0	1	0	0	CLEAR
0	1	1	0	
1	0	0	1	SET
1	0	1	1	
1	1	0	1	COMP-
1	1	1	0	LEMENT

FIGURE 1.63 The *JK* latch: (a) *JK* latch circuit, (b) *JK* latch SBS representation, (c) *JK* latch truth table.

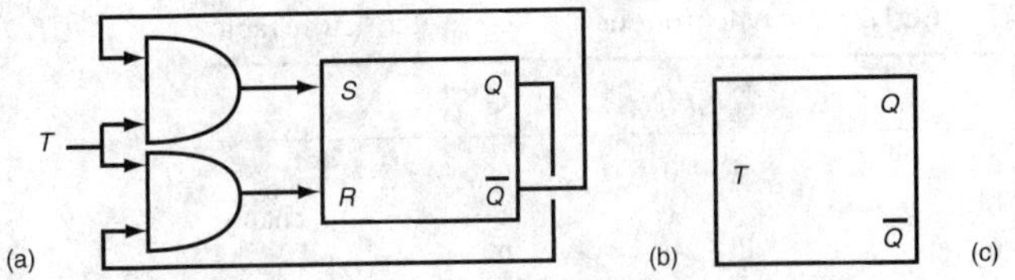

T	K	Q_{next}
0	0	0
0	1	1
1	0	1
1	1	0

FIGURE 1.64 The *T* latch: (a) *T* latch circuit, (b) *T* latch SBS representation, (c) *T* latch truth table.

1.6.10 *D* Latch

A latch that stores the signal on an input data line is a data or *D latch.* A circuit that can accomplish this is shown in Fig. 1.65 along with its symbol representation and truth table. The *D* latch requires a second input, called a control signal and is sometimes a clock pulse, to control when the input signal is placed in storage.

1.6.11 Synchronous Latches

The circuit shown in Fig. 1.60 is an **asynchronous** set-reset latch, it responds whenever the input signal is applied. In a system using a multiplicity of such latches there may be significant variations in the arrival time of the input signals. In some applications it is desirable to have the latches respond simultaneously even though there is a variation in the input signal arrival times. Simultaneity is obtained by synchronizing the actions. This is done by the use of a common synchronizing signal, usually the *clock signal.* Latches that require a synchronizing signal for their activation are referred to as **synchronous** latches such as depicted in Fig. 1.66. It is advantageous to use the SBS for a latch that is an element in a more complex circuit. A macrotiming diagram example for a synchronous SR latch is shown in Fig. 1.67.

The complete truth table for the synchronous SR latch is given in Fig. 1.68. A truth table is more than doubled in size (twice the length plus one column) by the incorporation of each additional input variable. Since there are no changes that occur in the absence of a clock pulse, the abbreviated truth table contains all of the required information and is less than half the size.

1.6.12 Master-Slave Flip-Flops

It is often desirable to process information stored in memory elements and then restore the information in the memory element. If the memory element is a latch, such as shown in Fig. 1.69, the feedback would cause a problem. Signals can propagate around such a feedback loop an unknown number of times causing unknown results at the time the control pulse is turned off, breaking the feedback loop.

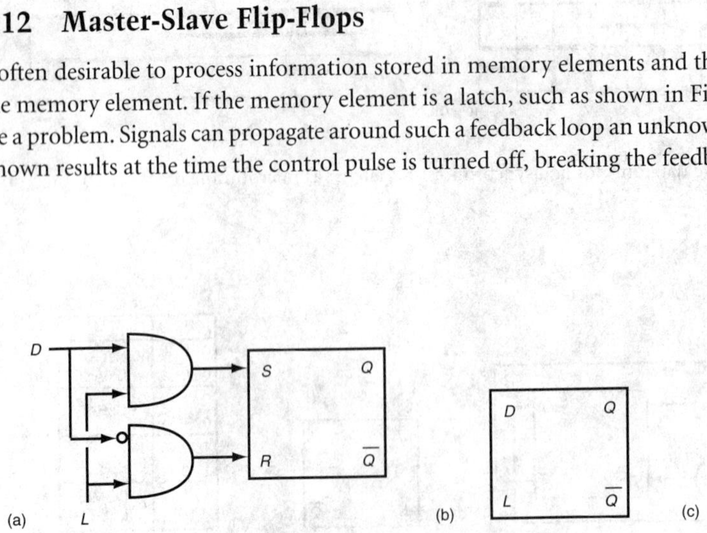

D	L	Q	Q_{next}	
0	0	0	0	NO CHANGE
0	0	1	1	
0	1	0	0	LOAD 0
0	1	1	0	
1	0	0	0	NO CHANGE
1	0	1	1	
1	1	0	1	LOAD 1
1	1	1	1	

FIGURE 1.65 The *D* latch: (a) *D* latch circuit, (b) *D* latch SBS representation, (c) *D* latch truth table.

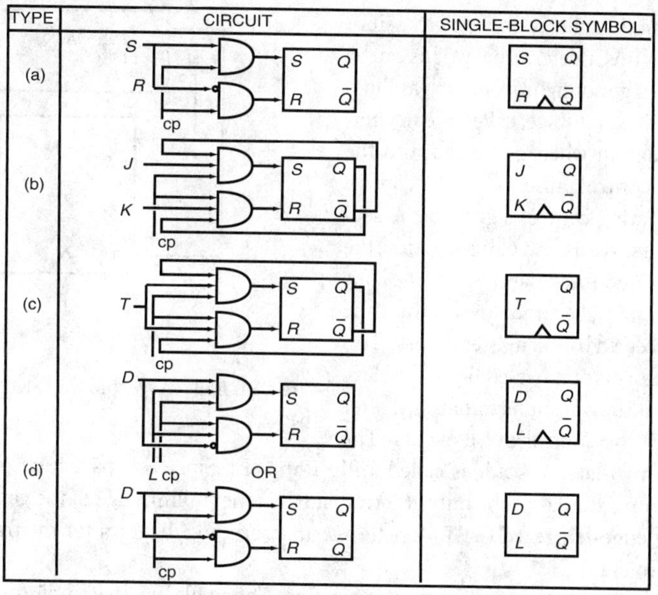

FIGURE 1.66 Four types of synchronous latches: (a) SR, (b) JK, (c) T, (d) D.

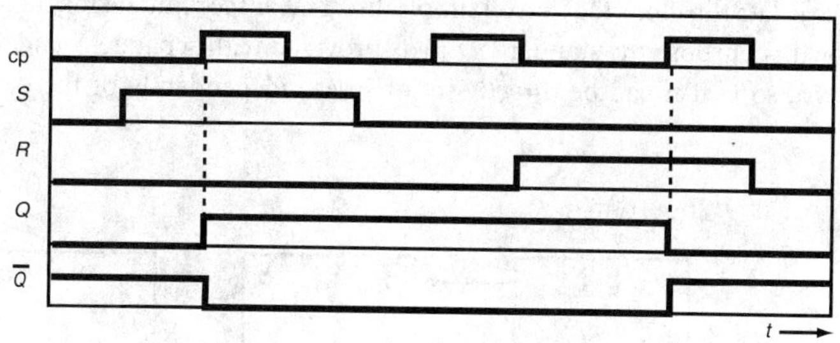

FIGURE 1.67 Macrotiming diagram for a synchronous SR latch.

(a)

cp	S	R	Q	Q^{next}	
0	0	0	0	0	
0	0	0	1	1	
0	0	1	0	0	
0	0	1	1	1	NO
0	1	0	0	0	CHANGE
0	1	0	1	1	
0	1	1	0	0	
0	1	1	1	1	
1	0	0	0	0	NO
1	0	0	1	1	CHANGE
1	0	1	0	0	
1	0	1	1	0	CLEAR
1	1	0	0	1	
1	1	0	1	1	SET
1	1	1	0	?	
1	1	1	1	?	DO NOT USE

(b)

S	R	Q	Q^{next}	
0	0	0	0	NO CHANGE
0	0	1	1	
0	1	0	0	CLEAR
0	1	1	0	
1	0	0	1	SET
1	0	1	1	
1	1	0	?	DO NOT USE
1	1	1	?	

FIGURE 1.68 Synchronous SR latch truth-tables: (a) complete, (b) abbreviated.

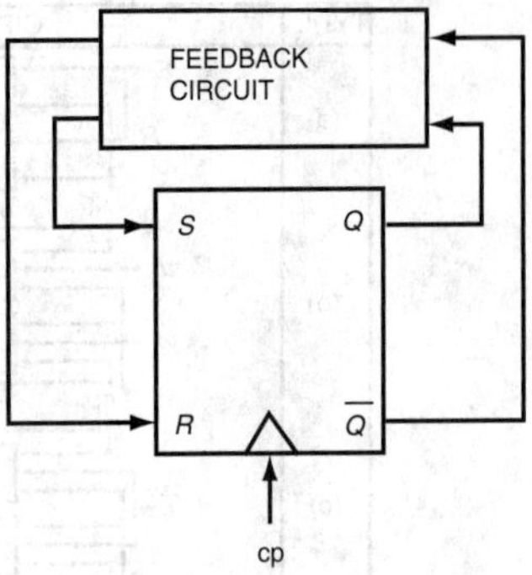

A storage element that allows only one transit of the signal eliminates this indeterminacy. This can be provided by a two-stage configuration such as illustrated in Fig. 1.70. The first stage, called the **master**, receives the information when it is gated into the master stage by the control pulse. When the control pulse is high, the gating control signal to the second stage, called the **slave**, is low. Consequently, the slave cannot receive new information until the control pulse goes off, that is, becomes low. At this time the information received by the master is passed on to the slave. The information passed on to the slave becomes available at the output terminals just after the trailing edge of the control pulse occurs. The

FIGURE 1.69 A synchronous SR latch with external feedback.

master-slave (MS) two-latch cascade is called a **flip-flop** (FF). Since this particular design provides the information gated into storage at the output terminal when the trailing edge of the gating pulse occurs, it is called a **trailing-edge-triggered** (TET) flip-flop. A macrotiming diagram for the trailing-edge-trigged flip-flop is shown in Fig. 1.71.

Each type of latch has a corresponding type of flip-flop. The single block symbols for the corresponding FFs are given in Fig. 1.72.

1.6.13   Standard Master-Slave Data Flip-Flop

It is convenient to have data flip-flops that have a simple direct means for initializing them to either the high or the low state. For this purpose the standard D FF is provided with separate S and R input terminals, as shown in Fig. 1.73, so that it can be directly set or reset independently of the data input terminal.

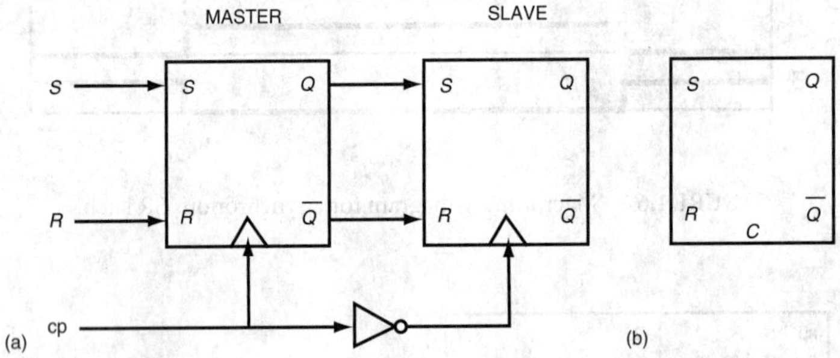

FIGURE 1.70   Trailing-edge triggered master-steve (MS) SR-FF: (a) circuit diagram, (b) SBS representation.

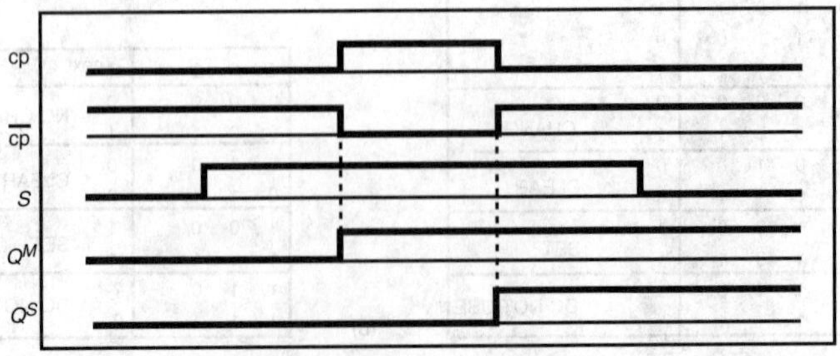

FIGURE 1.71   Macrotiming diagram for a trailing-edge-triggered MS SR-FF.

The standard D FF, thus, can be either used as an asynchronous SR latch by using terminals S and R (without signals to terminals D and C), or can be used as a D FF by using terminals D and C (without signals to terminals R and S).

1.6.14 Sequential Logic System Description

Sequential logic circuits differs from combinational logic circuits by containing memory. A sequential logic circuit output consequently depends on prior inputs as well as the current inputs. For the purpose of digital systems the state of a wire (or terminal) is the value of the signal on it. The state of a circuit is the set of signals on the set of wires in the circuit. Since each wire is connected to an output terminal of an element in the circuit, the set of signals at the output terminals of the elements comprising the circuit represent the state of the circuit. If the element is a gate, then the output is given by the current inputs of the gate. Hence, a sufficient description of the system is given by the set of values of the signals at the output terminals of the memory elements in the circuit. Hence, the state of the circuit is represented by the collected states of the set of memory elements comprising the circuit.

If the output is a function of the input as well as the state, the system is referred to as a **Mealy model**, Fig. 1.74(a). If the output is only a function of the state the system, it is referred to as a **Moore model**, Fig. 1.74(b).

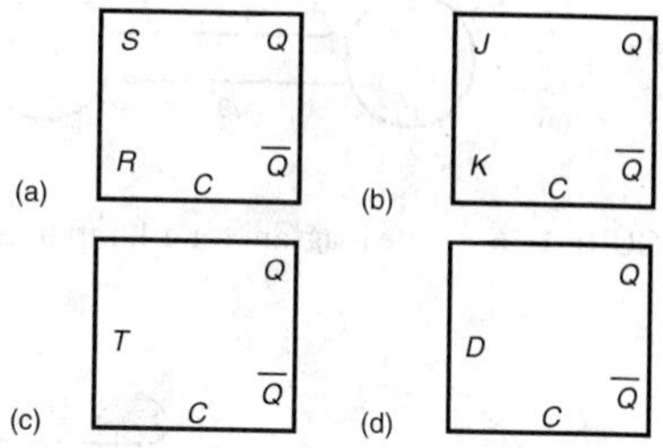

FIGURE 1.72 Flip-flop symbols: (a) SR FF, (b) JK FF, (c) T FF, (d) D FF.

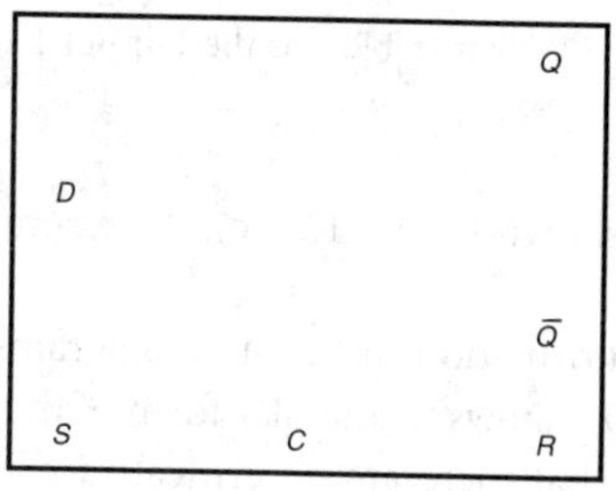

FIGURE 1.73 Standard D FF with set amd reset input terminals.

The behavior of a system can be represented by its sequence of states induced and the outputs produced in response to the inputs. This can be represented by a state diagram, which shows the possible states, and the transitions between the states, and the outputs produced in response to the inputs. The basic elements of the state diagram are shown in Fig. 1.75(a). For the Mealy model the output is designated in the transition arrow as shown in Fig. 1.75(b). For the Moore model the output is designated in the state

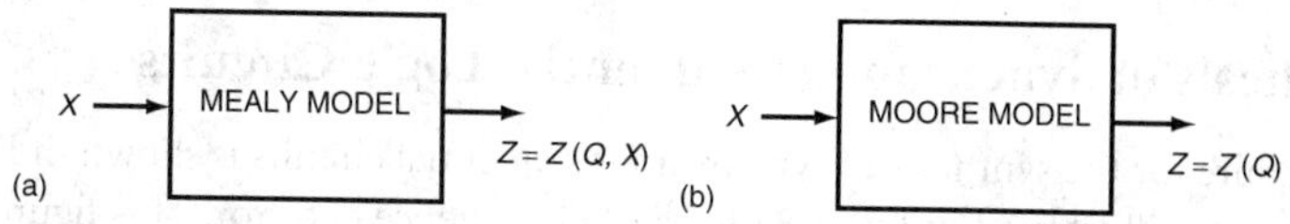

FIGURE 1.74 System models: (a) Mealy, (b) Moore.

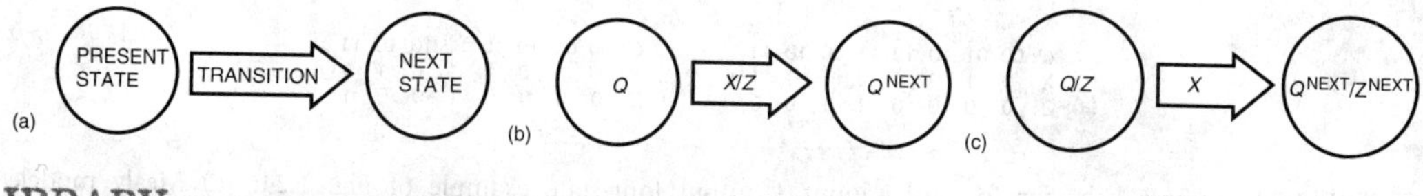

FIGURE 1.75 State diagram elements: (a) prototype, (b) Mealy model, (c) Moore model.

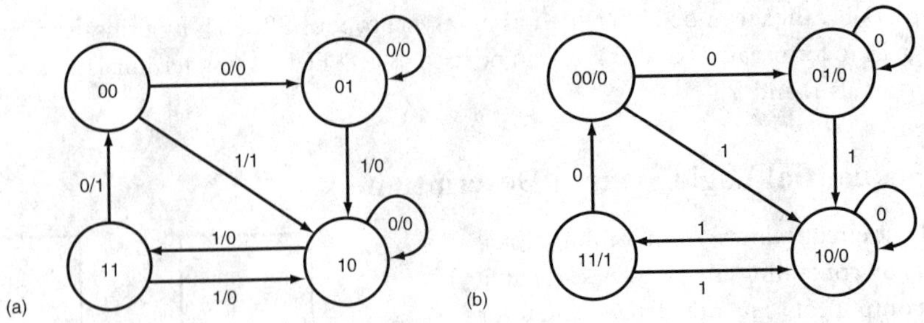

FIGURE 1.76 State diagram for a 1-input 1-output four-state example: (a) Mealy model, (b) More model.

(a)

Q \ X	0	1
00	01/0	10/1
01	01/0	10/0
11	00/1	10/0
10	10/0	11/0

(b)

Q \ X	0	1	Z
00	01	10	0
01	01	10	0
11	00	10	1
10	10	11	0

FIGURE 1.77 State tables for the 1-input 1-output four-state example of Figure 1.76: (a) Mealy model, (b) Moore model.

circle as shown in Fig. 1.75(c). An example of a four-state system is illustrated in the state diagrams in Fig. 1.76.

The information in the state diagrams can also be expressed in a tabular form called a **state table.** Figure 1.77 shows the tabular form of the state diagrams in Fig. 1.76.

Given that there are no critical race conditions, that only one input at a time is changed, and that no changes are initiated unless the system is in equilibrium, the state table or state diagram allows the prediction of the behavior of the system for the specification of the input sequence and the initial state. An example is given in Fig. 1.78.

1.6.15 Analysis of Synchronous Sequential Logic Circuits

The general procedure for analyzing synchronous sequential circuits is given by the flow chart in Fig. 1.79. As an example of analysis consider the circuit shown in Fig. 1.80(a). The excitation K-maps, the output K-map, the **excitation table**, the FF truth tables, the **transition table**, the state table, the state diagram, and an output sequence example are shown in the successive parts of Fig. 1.80.

1.6.16 Synthesis of Synchronous Sequential Logic Circuits

The general procedure for the synthesis of synchronous sequential circuits is shown in Fig. 1.81.

The design process is illustrated in Fig. 1.82 for a 0111 sequence detector. This figure displays each of the steps in the synthesis process starting with the state diagram specification of the sequence detector in Fig. 1.82(a). Letters are initially assigned to represent the states. The considerations required for assigning binary code representations for the states require considerations that are covered later in this chapter. The machine is initialized to be in state D. If the first input is a 1 the machine remains in state D. The system

(a)

Q:	00	01	10	10	11	00	10	11
X:	0	1	0	1	0	1	1	1
Z:	0	0	0	0	1	1	0	0

(b)

Q:	00	01	10	10	11	10	10	11
X:	0	1	0	1	1	0	1	1
Z:	0	0	0	0	1	0	0	1

FIGURE 1.78 Sample behavior for the 1-input 1-output four-state example of Fig. 1.24: (a) Mealy model, (b) Moore model.

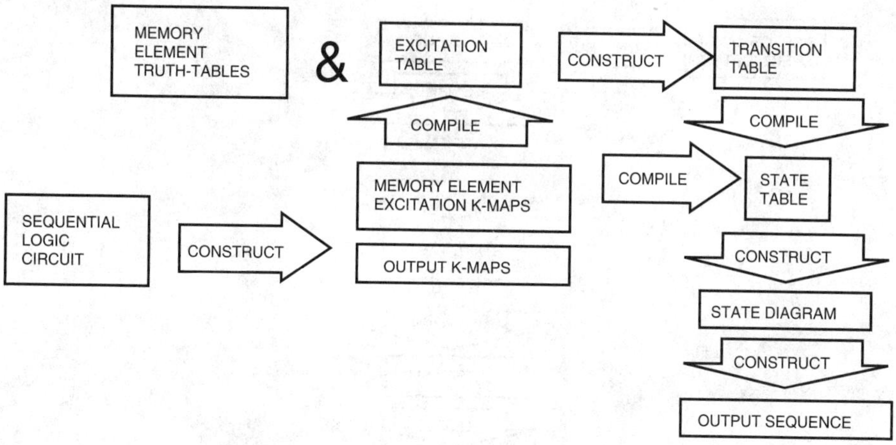

FIGURE 1.79 Analysis procedure for sequential logic circuits.

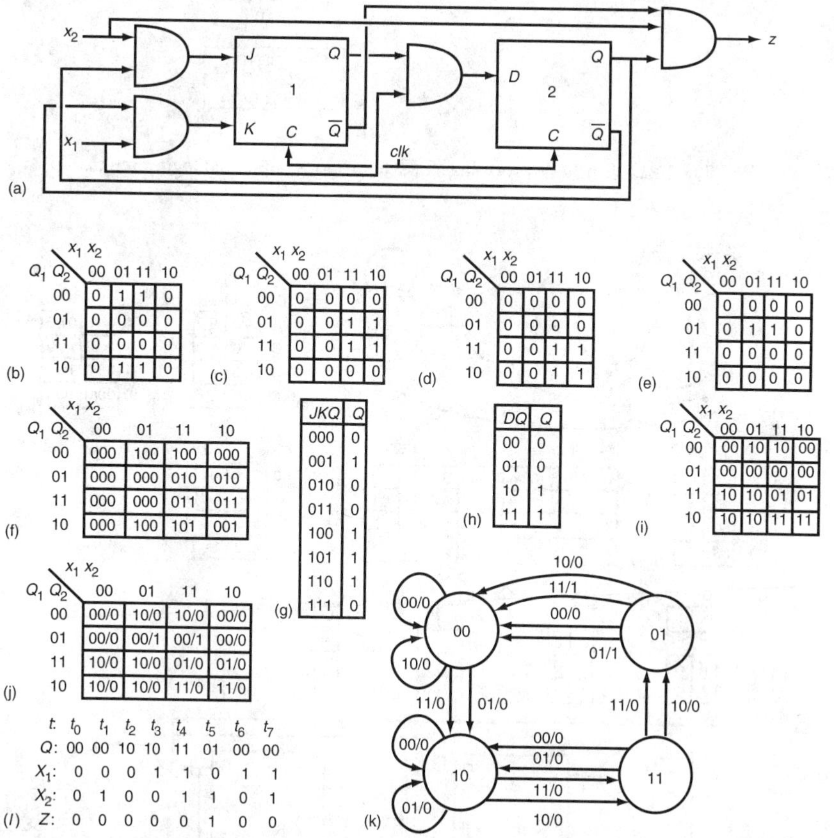

FIGURE 1.80 Synchronous 1-output 2-input four-state sequential logic circuit analysis example: (a) circuit diagram, (b) K-map $J = x_2 \bar{Q}_2$, (c) K-map $K = x_1 Q_2$, (d) K-map $D = x_1 Q_1$, (e) K-map $z = x_2 \bar{Q}_1 Q_2$, (f) JKD excitation table, (g) JK truth table, (h) D truth table, (i) transition table, (j) state table, (k) state diagram, (l) output sequence example.

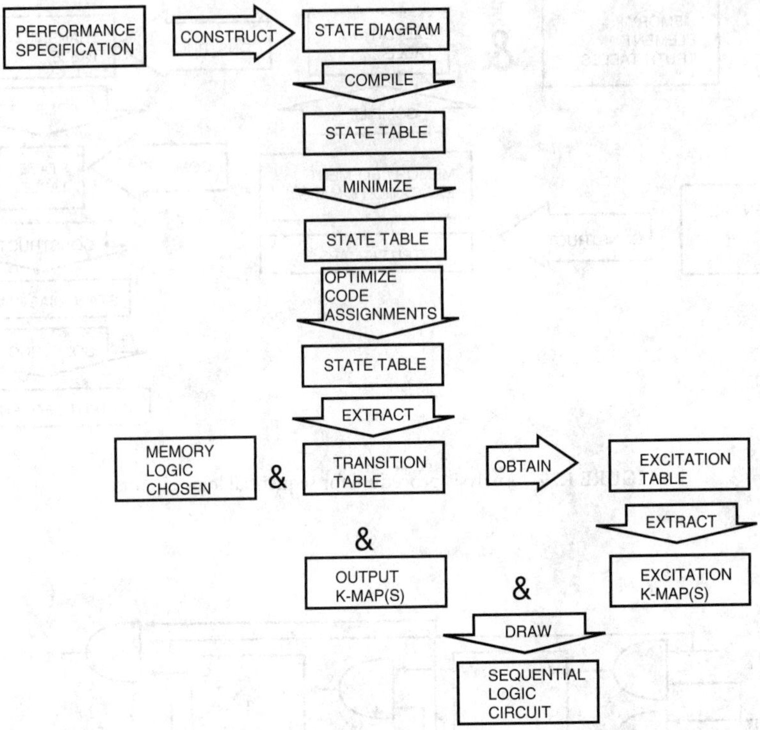

FIGURE 1.81 Synthesis procedure for synchronous sequential logic circuits.

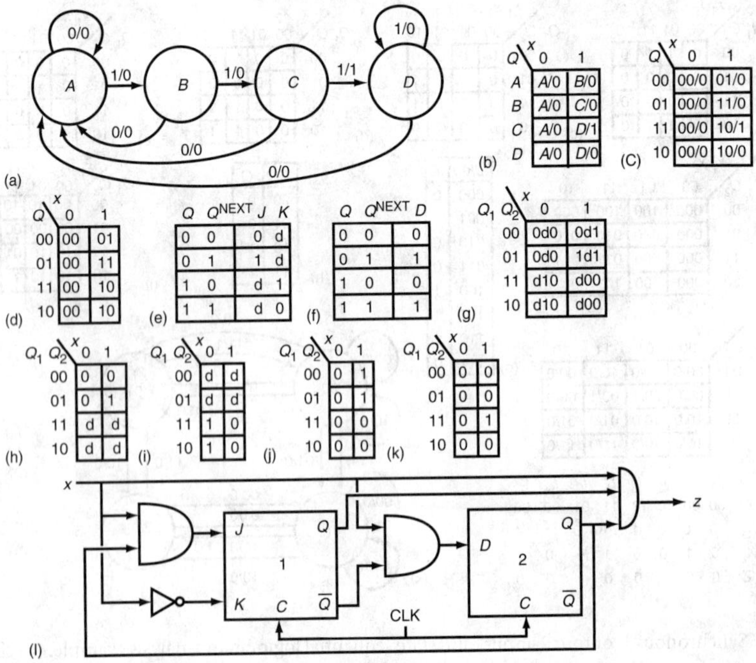

FIGURE 1.82 Development for a 0111 sequence detector: (a) state diagram, (b) character state table, (c) binary state table, (d) transition table, (e) $J K$ FF input table, (f) D FF input table, (g) $J K D$ excitation table, (h) K-map $J = x Q_2$, (i) K-map $K = \bar{x}$, (j) K-map $D = x \bar{Q}_1$ (k) K-map $z = x Q_1 Q_2$, (l) synchronous sequential logic circuit.

is placed in state A to designate that a 0 has been received, in state B to designate that the sequence 01 has been received, state C to designate the sequence 011 has been received, and state D when the sequence 0111 has been received. The system outputs a 1 when the transition to state D is made to signify that the desired sequence has been detected, otherwise the output is 0. The character state and the binary state diagrams are constructed next, as shown in Fig. 1.82(b) and Fig. 1.82(c). Since there are four states the number of storage elements required is two. This first example will illustrate the procedure without considering the problems of minimizing the number of states or of making optimum code assignments. Hence, the Gray scale code is assigned without changing the state table configuration as shown in Fig. 1.82(c). The transition table is extracted from the binary state table and is shown in Fig. 1.82(d). For illustrative purposes these will be chosen as $J\,K$ and D flip-flops. To determine the excitation table it is necessary to have the input tables for these flip-flops given in Fig. 1.82(e) and Fig. 1.82(f). Consequently, the excitation table is that shown in part Fig. 1.82(g). The excitations are then separated into individual K-maps shown in Fig. 1.82(h) to Fig. 1.82(j). The output K-map is taken from Fig. 1.82(c) and shown in Fig. 1.82(k). The resulting circuit diagram specified by the K-maps is shown in Fig. 1.82(l).

1.6.17 Equivalent States

Minimizing the number of states minimizes the number of storage elements required and, hence, simplifies the circuit, which reduces its cost, increases its reliability, and increases its speed. States that are **equivalent** to other **states** can be eliminated. Two states are equivalent if and only if for every possible input:

- they produce the same output
- the next states are equivalent.

If starting from two different states a sequence of k inputs produces the same sequence of outputs the two states are k *equivalent*. If the states are k equivalent for all k, then the states are equivalent. Three techniques for determining equivalency are:

- Inspection
- Partitioning
- Implication table

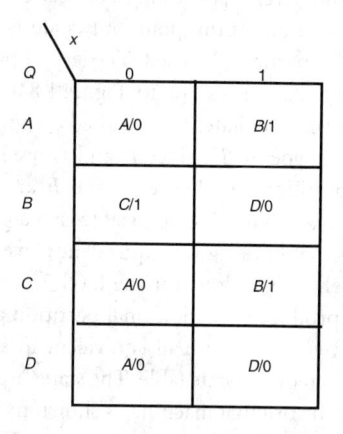

FIGURE 1.83 State table for a machine, which has equivalent states that are apparent by inspection.

Inspection can easily be done for cases in which single step comparisons can be made. For example, if two states have the same next states as well as the same output they are the same state. This is illustrated in Fig. 1.83, which shows that states A C are the same.

1.6.18 Partitioning

Partitioning provides a straightforward procedure for determining equivalency for any amount of complexity. Successive partitioning steps produce smaller partitions. If the next step does not yield any smaller partitions no further steps will yield any smaller partitions and, hence, the partitioning process is then complete. All states that are in the same partition after k steps are k equivalent. All states that are in the same partition when no further partitioning can be accomplished are equivalent. States that are not in the same final partition are not equivalent. The **partition table** resulting from the partioning process, shown

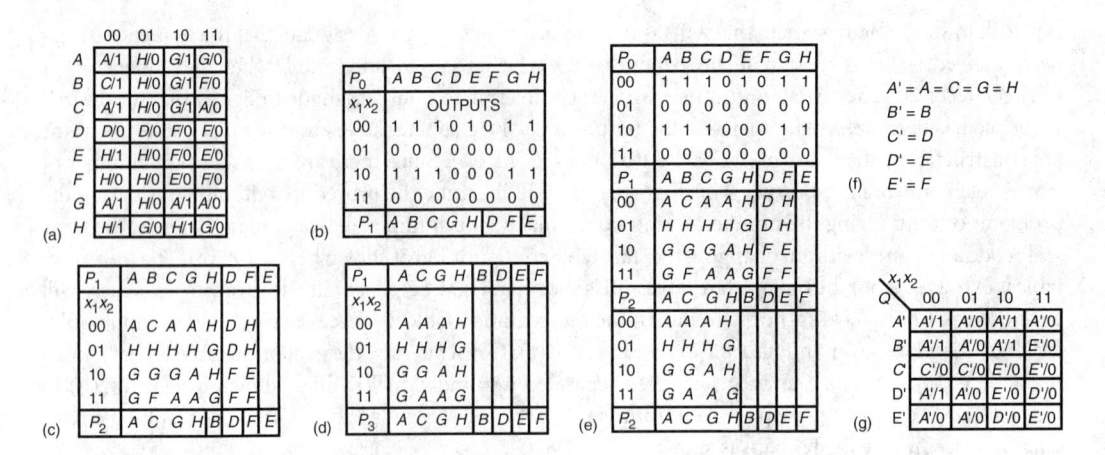

FIGURE 1.84 Partition table construction process: (a) sample machine state table, (b) first partition step, (c) second partition step, (d) third partition step, (e) composite partition table, (f) reduced state assignments, (g) reduced state table.

in Fig. 1.84, is illustrated in Fig. 1.84(e) for the machine whose state table is shown in Fig. 1.84(a). The first partitioning step is based on the output. If two states have the same outputs for all possible inputs they are *1-equivalent*. Figure 1.84(b) displays the outputs for all possible inputs organized into a format that facilitates the required comparisons. States A, B, C, G, and H all have the same outputs and, hence, belong in the same level-1 partition. Likewise, D and F have the same outputs and, hence, belong in the same level-1 partition. At this point only state E is shown not to be equivalent to any other state and, hence, is in a level-1 partition by itself. The level-2 partitions are based on comparisons of the next states that each state can make a transition to. Figure 1.84(c) displays the next states for all possible inputs organized into a format that facilitates the required comparisons. In the level-1 partition /D, F/ with a 00 input D goes to D but F goes to H. Since D and H are not in the same level-1 partition, D and F are not in the same level-2 partition and, hence, D and F are not equivalent. Similarly, B is in a different level-2 partition from A, C, G, and H. Once a state is in a partition by itself, there are no other states equivalent to it.

States that have a given equivalence level with other states have to be examined for partitioning at the next level. This is done for the level-2 partition /A, C, G, H/ in Fig. 1.84(d). The level-3 partitioning process produced no additional partition and, hence, no further partitioning is possible. Consequently states C, G, and H are all equivalent to state A and, hence, can be eliminated from the state table to produce a reduced state table. The states are lettered consecutively but distinguished with primes from the states of the original machine as shown in Fig. 1.84(f). The reduced state table is given in Fig. 1.84(g). In practice the individual partitions are not done in separate tables, but in one composite table as shown in Fig. 1.84(e).

1.6.19 Implication Table

An implication table contains a triangular array of squares such that there is one square for each pair of states as shown in Fig. 1.85. Notation is placed in each square to indicate that the pair of states are not equivalent, may be equivalent, or are equivalent. The first step in the process consists of placing an X in those cells representing states that are not equivalent because they do not have identical outputs for the same inputs, as shown in Fig. 1.86(a) for the machine defined in Fig. 1.84(a). The second step in the process is to list in each cell the pairs of states that must be equivalent if the pair represented by that cell is to be equivalent, as shown in Fig. 1.86(b). If a pair required for equivalence is not equivalent then the lack of equivalence is designated by a / drawn across the square, as shown in Fig. 1.86(c). The next sweep rules out one additional pair, B, G, as shown in Fig. 1.86(d). The array is successively scanned to check

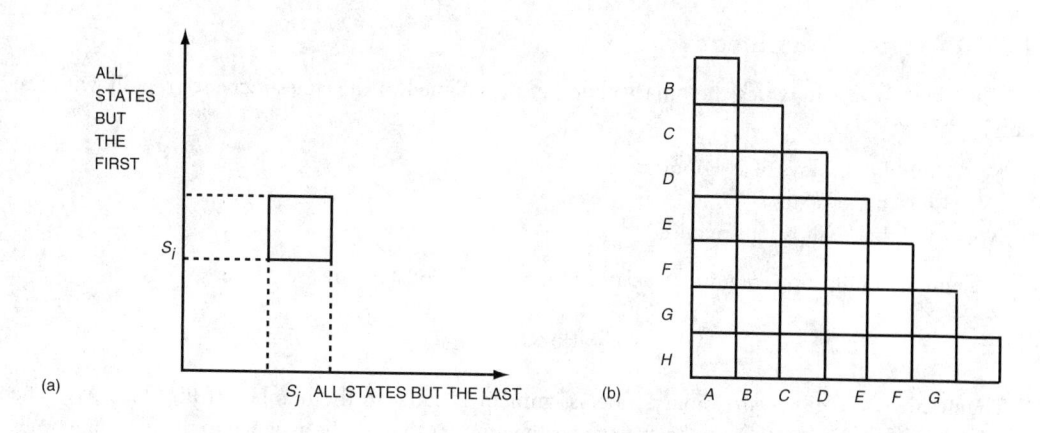

FIGURE 1.85 Implication table format: (a) square for state pair S_i, S_j, (b) implication table format for an eight-state machine.

each remaining possible equivalence to see what pairs can be ruled out. Further sweeps uncover no further nonequivalencies. Consequently, the squares that have not been ruled out represent pairs of equivalent states. The equivalencies are, consequently, (A, C), (A, G), (A, H), (C, G), (C, H), and (G, H). These equivalencies can be more compactly expressed as the set (A, C, G, H). This is the same result as obtained using the partition table.

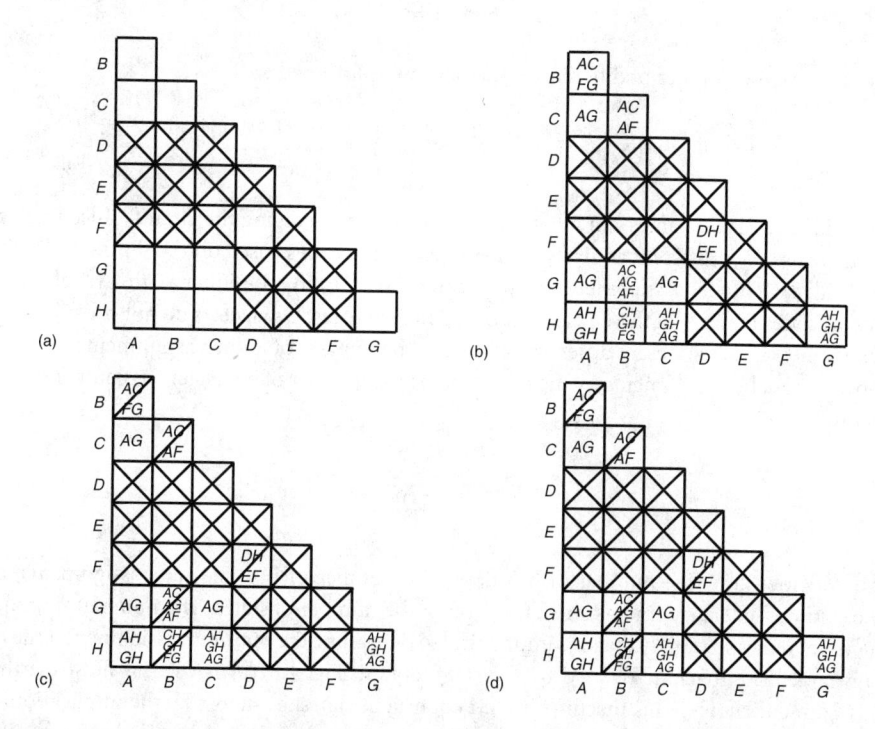

FIGURE 1.86 Implication table development for the machine specified by the state table in Fig. 1.84(a): (a) partial implication table with state pairs Xed out that are not equivalent by virtue of different outputs, (b) partial implication table with the remaining squares having state pair entries, which must be equivalent for that state pair to be equivalent, (c) partial implication table with state pairs /ed out on the first sweep because the next state pairs are not equivalent, (d) implication table after last sweep rules out no additional state pairs.

1.6.20   State Assignment

The number of ways in which the binary code can be assigned to the states increases rapidly with the number of states. Let

N_{SA} = number of ways of assigning states
N_S = number of states
N_{FF} = number of flip-flops required

The number of flip-flops required is related to the number of states by

$$N_{FF} = (\log_2 N_S)_{\text{truncated up}}$$

The number of digits in the binary code assignment is equal to the number of flip-flops N_{FF}. The number of different codes that can be provided is the number that can be assigned to the first state times the number that can be assigned to the second state, and so on down to the number that can be provided to the last state. The number of codes that can be provided to the first state is

$$N_1 = 2^{N_{FF}}$$

Once a code has been assigned to the first state, the choice of codes for the second state is one less,

$$N_2 = N_1 - 1$$

the number of codes available for assigning to the third state is

$$N_3 = N_2 - 1 = N_1 - 2$$

etc. Consequently, the number of different codes that can be assigned to states is

$$N_{SA} = N_1(N_1 - 1)(N_1 - 2)\cdots(N_1 - N_S + 1) = \frac{N_1!}{(N_1 - N_S)!} = \frac{2^{N_{FF}}!}{(2^{N_{FF}} - N_S)!}$$

The individual variable names can be interchanged without changing the circuit. Since there are N_{FF} flip-flops there are $N_{FF}!$ different equivalent ways of assigning codes based on the flip-flop identification numbers. With each of these assignments the particular code digit may be either a 0 or a 1; hence, there are $2^{N_{FF}}$ different codes for each of these permutations. Since these last two factors do not constitute different unique assignments (i.e., represent different hardware), the number of unique assignments is smaller than the number of possible assignments by the factor that is the number of equivalent assignments, $2^{N_{FF}} N_{FF}!$, consequently,

$$N_{UA} = \frac{N_{SA}}{N_{FF}! 2^{N_{FF}}}$$

Table 1.26 shows how the number of flip-flops, the number of possible state assignments, and the number of unique state assignments, are a function of the number of states. The number of unique state assignments increases so rapidly that it is impractical to determine the optimum assignment if the number of states is large. The machine defined by the state table in Fig. 1.87(a) is used to illustrate the effect of code assignment choice. This machine has no equivalent states, as shown by the implication table in Fig. 1.87(b). With seven independent states, three flip-flops are required and, hence, there are 840 unique code assignments. Implementations of the required combinational logic circuits using $J K$ flip-flops for the two code assignments specified in Fig. 1.87(c) are given in Fig. 1.87(d). The gate counts required for these implementations are summarized in Fig. 1.87(e).

TABLE 1.26   State Assignments

N_S	N_{FF}	N_{SA}	N_{UA}
2	1	2	1
3	2	24	3
4	2	24	3
5	3	6,720	140
6	3	20,160	420
7	3	40,320	840
8	3	40,320	840
9	4	4,151,347,200	10,810,800
.	.	.	.
.	.	.	.
.	.	.	.

TABLE 1.27   State Table Based State Assignment Guidelines.

Rule No.	Give logically adjacent assignments to states that:
1	have the same next state for a given input
2	are the same next state of a present state

1.6.21   State Assignment Guidelines

An exhaustive study of all possible implementations is generally impractical; hence, it is desirable to have some guidelines for making a near optimum choice. Complexity is reduced by having realizations with higher order prime implicants. This is provided by codings that have large clusters of minterms. The arrangement of 1s on the K-map are improved by using the logically adjacent assignments rules specified in Table 1.27.

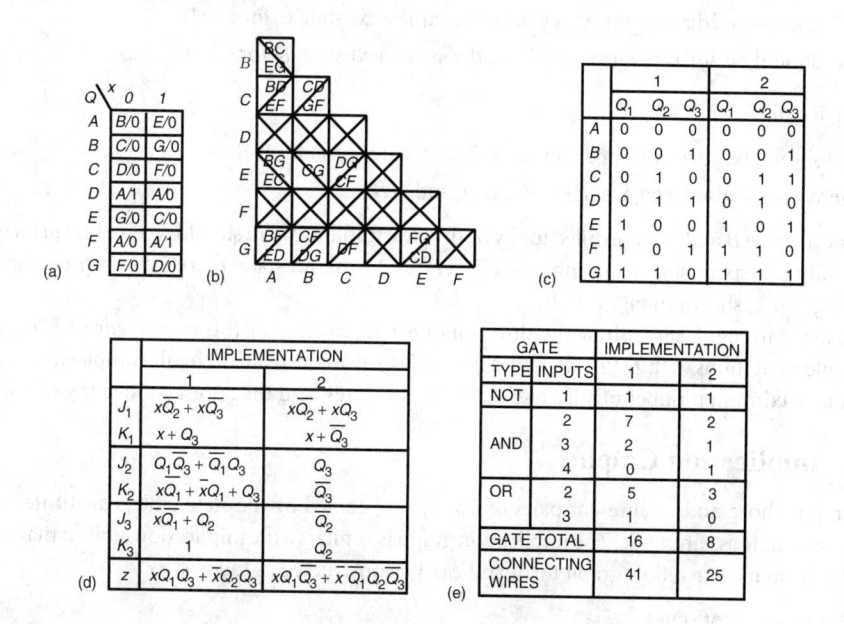

FIGURE 1.87   Comparison of implementations for two code assignments: (a) state table, (b) implication table, (c) selected code assignments, (d) combinational logic circuit implementations, (e) relative complexity of the two implementations.

$Q \backslash x$	0	1
A	C/0	D/0
B	C/0	A/0
C	B/0	D/0
D	A/1	B/1

(a)

ASSIGNMENT			
Q	1	2	3
A	00	00	00
B	01	11	10
C	11	01	01
D	10	10	11

(b)

$Q_1 \backslash Q_2$ 0 1

	0	1
0	A	B
1	C	D

(c) ASSIGNMENT 1

	0	1
0	A	D
1	C	B

ASSIGNMENT 2

	0	1
0	A	D
1	B	C

ASSIGNMENT 3

ADJACENCY	DESIRED ADJACENCIES				NO. OF ADJACENCIES
	AB	AC	BD	CD	
1	Y	Y	Y	Y	4
2	N	Y	Y	N	2
3	Y	N	N	Y	2

(d)

	IMPLEMENTATION		
	1	2	3
D_1	$\bar{x}\bar{Q_1} + x\bar{Q_2}$	$\bar{x}Q_1 + \bar{x}\bar{Q_2} + x\bar{Q_1}Q_2$	$\bar{x}Q_1 + \bar{x}\bar{Q_2} + x\bar{Q_1}Q_2$
D_2	$xQ_1 + x\bar{Q_2} + Q_1\bar{Q_2}$	$xQ_1 + x\bar{Q_2} + Q_1\bar{Q_2}$	$\bar{x}\bar{Q_2} + x\bar{Q_1}Q_2 + \bar{Q_1}Q_2$
z	Q_1Q_2	$\bar{Q_1}Q_2$	$\bar{Q_1}Q_2$

(e)

GATE TYPE	INPUTS	IMPLEMENTATION		
		1	2	3
NOT	1	1	1	1
AND	2	5	6	4
	3	0	1	2
OR	2	1	0	0
	3	1	2	2
GATE TOTAL		8	10	9
CONNECTING WIRES		19	25	24

(f)

FIGURE 1.88 Example used to illustrate rules for state assignments: (a) example machine, (b) unique state assignments, (c) unique state assignment K-maps, (d) adjacencies provided by the various orderings, (e) combinational logic circuit implementations, (f) relative complexity of the three implementations.

These are illustrated in Fig. 1.88 with the machine defined by the state table in Fig. 1.88(a). According to rule 1:

- AB should be adjacent since they have the same next state C for $x = 0$.
- AC should be adjacent since they have the same next state D for $x = 1$.

According to rule 2:

- CD should be adjacent since they are the next states of state A.
- there should also be adjacencies AC, BD, and AB.

Since adjacencies AB and AC are called for by both rules, fulfilling them should receive first priority within the constraint of maximizing the number of adjacencies. With four states there are only three unique ways of assigning codes, shown in Fig. 1.88(b).

The K-maps in Fig. 1.88(c) show the three unique assignments for this case. Figure 1.88(e) gives the circuit implementations with D flip-flops and Fig. 1.88(f) gives the relative circuit complexity. Assignment one has the maximum number of adjacencies, the fewest gates, and the fewest connecting wires.

1.6.22 Implication Graph

A flow graph whose nodes represent pairs of states, as specified by the state table, constitutes a tool for selecting good state assignments. The implication graph is similar to the implication table in that it contains the same information in a flow graph form and can be constructed as follows:

- choose a pair of states
- go to the next pair of states for each input and enter as a new node, but omit if the pair is a single state
- repeat until no new pairs can be generated

An illustrative example is given in Fig. 1.89.

TABLE 1.27 Continuation: Implication-Graph-Based State Assignment Guidelines

Rule No.	Give logically adjacent assignments to:
3	state pairs on closed or contiguous subgraphs

An implication graph is complete if it contains all possible pairs. The example in Fig. 1.90 is a complete implication graph. Normally, cases are not complete. A subgraph is part of a complete graph. A closed subgraph is one in which all outgoing arcs for each node within the subgraph terminate on nodes within the subgraph. This is important for state assignments. The example in Fig. 1.89 has no closed subgraphs.

The implication graph should be used to guide the rearrangement of the state table to make:

- logically adjacent states physically adjacent
- the next-state-pair of two current-adjacent states adjacent and to follow rule 3 in the continuation of Table 1.27.

Rule 3 dictates the same choice for the example represented in Fig. 1.89 as rules 1 and 2 do for the same example applied directly to the state table. For other and more complicated examples the various rules may not be in agreement, in which case preference is recommended for rules 1 and 3. In general, the rules should be followed as closely as possible to obtain on an average a more optimum solution than would be obtained on an average by a purely random code assignment choice.

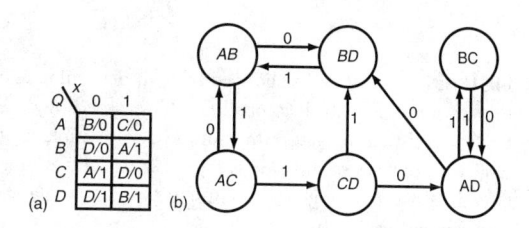

FIGURE 1.89 Example to illustrate an implication graph: (a) state table example, (b) implication graph.

1.6.23 Incompletely Specified Circuits

Incompletely specified circuits may have a combination of unspecified outputs and next states. These unspecified quantities can be treated as *don't cares*. However, the presence of don't cares complicates the process of minimizing the number of states required. With don't cares the concept of **compatible** states replaces that of equivalent states. Two states that are equivalent to a third state are equivalent to each other. However, two states that are compatible with a third state are not necessarily compatible with each other because different choices for the don't cares may be evoked by the different equivalences. As a result the incompletely specified circuits have an added complexity for which merger diagrams are introduced to facilitate treatment. An illustrative example of the incomplete specification is given in Fig. 1.90.

Incompatible states of an incompletely specified machine are caused by different outputs or by incompatible next states. An incompatible class is a set of mutually incompatible states. A maximal incompatible class is one for which there are no other incompatible states that are incompatible with the states already in the class. The maximal incompatibles can be constructed as illustrated for this example in Fig. 1.91(e). Starting with the rightmost column of the implication table the compatible state pairs are cumulatively listed. Whenever compatible states can be combined in a larger compatible class this is done on the next line. The incompatible sets are given in Fig. 1.91(f).

FIGURE 1.90 An incompletely specified circuit.

(a)

(b)

(c)

(d)

(e)

F   (FG)
E   (FG)(EG)
D   (FG)(EG)(DG)
C   (FG)(EG)(DG)(CD)(CE)(CF)(CG)
C   (CDG)(CEG)(CFG)
B   (CDG)(CEG)(CFG)(BC)(BG)
B   (BCG)(CDG)(CEG)(CFG)
A   (AEG)(BCG)(CDG)(CEG)(CFG)

(f)

F   —
E   (EF)
D   (EF)(DE)(DF)
D   (DEF)
C   (DEF)
B   (DEF)(BD)(BE)(BF)
B   (BDEF)
A   (BDEF)(AB)(AC)(AD)(AF)
A   (BDEF)(ABDF)(AC)

FIGURE 1.91 Use of implication table to determine compatible states for the example in Fig. 1.90(a): (a) partial implication table with incompatible states caused by different outputs designated by an *X*, (b) partial implication table with listings of states which must be compatible for the present states to be compatible, (c) partial implication table with first sweep through listings of states with a / to designate incompatibles, (d) implication table for machine after final sweep through listings of states with a \ to designate last found incompatibles, (e) compatibility classes, (f) incompatibility classes.

Determination of the maximal compatible sets is facilitated by a *merger diagram*. The merger diagram is formed by representing the states as dots spaced around a circles as shown in Fig. 1.92(a). A straight line is then drawn between each pair of compatible states. Each maximal set is obtained from the merger diagram by noting those sets of points for which every point in the set is connected to every other point in the set. For the example shown there are five maximal sets: (AEG), (BCG), (CDG), (CEG), and (CFG).

The minimization of an incompletely specified state table is as follows. A set of compatibility classes must be selected that meet the following conditions:

- *Completeness.* All of the states in the original machine must be contained in the set of compatibility classes chosen.
- *Consistency.* The next states must be in some compatibility class within the set, that is, the set of compatibility classes chosen must be closed.
- *Minimality.* Choose the smallest set of compatibility classes that meets the preceding criteria.

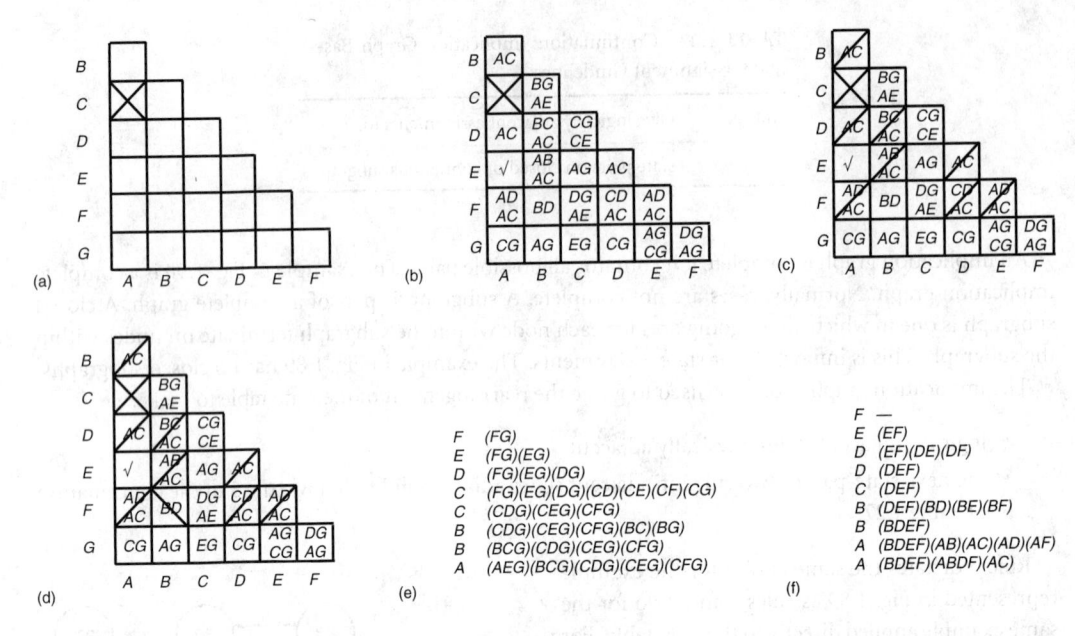
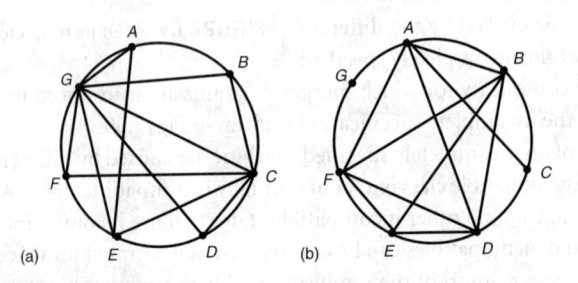

(a)   (b)

FIGURE 1.92 Merger diagram for the machine defined in Fig. 1.90: (a) merger diagram for maximal compatibles, (b) merger diagram for maximal incompatibles.

Each compatibility class corresponds to a state in the reduced machine. Since the selection of the compatibility classes is a trial-and-error process, it is helpful to bound the number of states required S_M in the realization of the minimal-state machine. The upper bound on the number of states is given by

$S_U = \text{minimum } (N_{MC}, S_O)$
$S_O = $ number of states in the original machine
$N_{MC} = $ number of sets of maximum compatibles

Since

$$N_{MC} \leq S_O$$

it follows that

$$S_U = N_{MC}$$

The lower bound on the number of states in the minimal circuit is given by

$S_L = \text{maximum}(N_{MI1}, N_{MI2}, \ldots, N_{MIk}, \ldots)$
$N_{MIk} = $ number of states in kth maximal incompatibles group of the original circuit

For the illustrative sample $S_L = 4$ and $S_U = 5$. The set of maximal compatibles is always complete and consistent but may not be minimal. Search for a minimal set of compatibility classes by examining sets of maximal incompatibles in groups of S_L.

The example defined in Fig. 1.90 has a closure table and reduced state table given in Fig. 1.93(a) and Fig. 1.93(b). When more than one state is listed for the next state it is in essence a restricted don't care in that it does not matter which of the listed states is the next state, whereas a don't care is unrestricted as to which state is the next state.

	0	1
(AEG)	AG	CG
(BCG)	BG	AEG
(CDG)	CG	CEG
(CEG)	AG	CEG
(a) (CFG)	DG	AEG

	0	1
A'	A'/1	B',C',D',F'/1
B'	B'/-	A'/0
C'	B',C',D',F'/1	D'/0
D'	A'/1	D'/0
(b) E'	C'/-	A'/0

FIGURE 1.93 Incompletely specified example: (a) closure table, (b) reduced state table.

1.6.24 Algorithmic State Machines

A state machine is a sequential logic circuit. An **algorithmic state machine (ASM)** is a state machine that is designed directly from a hardware algorithm, which is the procedure to be directly implemented in hardware. The procedure consists of two major portions:

- A circuit to process data, that is, produce the output data from the input data
- Another circuit to produce the control signals that initiate the successive steps in the production process

The control logic receives status signals from the processor and external signals that dictate the sequence of control signals produced as shown in Fig. 1.94(a). The objective of this approach is to provide a more logically straightforward and more easily and quickly executable design methodology rather than providing the most optimum circuit with the fewest components.

An algorithmic state machine chart is a specialized form of flow chart, which represents the timing relationships, as well as the sequence of operations and the conditions required for their performance. The ASM chart has three basic elements:

- State box
- Decision box
- Conditional box

These are represented by rectangles and diamonds, both common to conventional flow charts, and boxes with rounded ends, respectively. The state box is represented by labeling it with the state name and

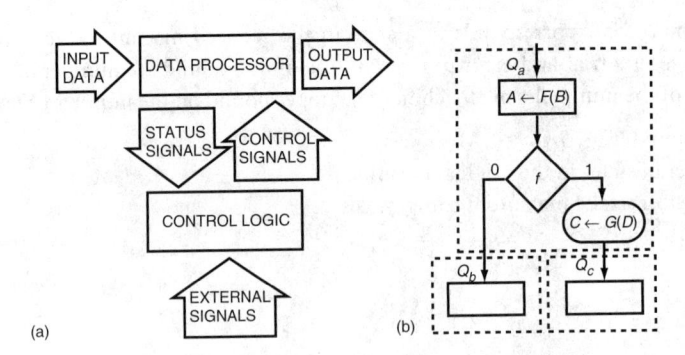

FIGURE 1.94 Algorithmic state machine: (a) hardware algorithm implementation, (b) illustrative ASM chart.

placing the operation that is executed when the machine goes to that state. The decision boxes contain the expression that can have more than one value with a different output for each value. The conditional boxes contain the expression for the operation that is executed if the condition designated by the decision box leading to it is true. The conditional operations are performed when the machine is in the specified state. These elements are illustrated in Fig. 1.94(b). When more than a few storage elements are needed, the previous design procedures get out of hand. The algorithmic state machine is convenient to use for the design of large systems.

1.6.25 Asynchronous Sequential Machines

The basic memory elements, latches, previously introduced are asynchronous sequential machines. Analysis of asynchronous machines with difference equations and timing diagrams rapidly gets out of hand with an increase in complexity. Utilizing the state table, transition table, and excitation table methodology is generally simpler but requires that the system be restricted to the fundamental mode of operation to make the process valid. In the fundamental mode of operation only one input at a time is applied and the system must reach equilibrium before another input is applied. Problems that can occur when not operating in the fundamental mode are described in the discussion of races and cycles.

Transition Tables

Transition tables for asynchronous circuits can convey significant information more quickly by placing circles about the next states that are stable, unstable next states are uncircled. The output and feedback variables have been labeled Q to identify them as state variables. This identifies the transition table for the asynchronous circuit with the transition table of the synchronous circuit. There is usually at least one next state entry that is, the same as the current state in each row. If this were not true then that state would not be stable. The procedure for obtaining a transition table is summarized in Table 1.28.

TABLE 1.28 Procedure for Obtaining an Asynchronous Circuit Transition table

No.	Step
1.	Let Q_k designate the input provided by feedback loop k.
2.	Let Q_k^n designate the output provided by feedback loop k.
3.	Derive the boolean expression for Q_k^n in terms of the inputs.
4.	Construct the K-map for the outputs in terms of the inputs.
5.	Combine the individual K-maps into a composite.
6.	Circle the next state values that are stable.

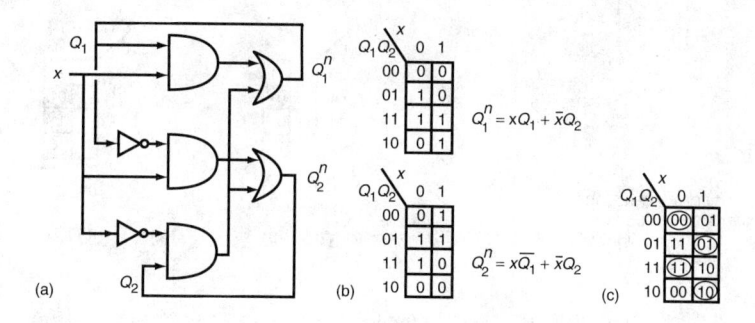

FIGURE 1.95   Asynchronous sequential circuit example: (a) circuit, (b) output K-maps, (c) transition table.

An example of an asynchronous sequential circuit is shown in Fig. 1.95(a). The K-maps for the outputs are shown in Fig. 1.95(b). The composite K-maps show how the signals change upon passage through the circuit and constitute the transition table shown in Fig. 1.95(c).

Flow Table

In the design process the optimum codes are not known in advance; hence, it is necessary to designate the states by nonnumeric symbols. A flow table, in essence, is a transition table in which the states are designated by letter symbols. Figure 1.96(a) shows a flow table in which the output is included for the case of a two-state circuit with two inputs and one output. The table for this example shows that if the input $x_1 = 0$ the circuit is in state A independent of the value of x_2. If the inputs $x_1 x_2 = 11$ the circuit may be in either state A or state B. If the inputs changed from 01 to 11 the circuit remains in state A, but if the inputs changed from 10 to 11 the circuit remains in state B. A change in input, from 00 to 11, is not allowed in the fundamental mode. To illustrate problems that can occur with the nonfundamental mode of operation, consider the circuit that results with the code assignment $A = 0$ and $B = 1$, which gives the transition table, output K-map, and circuit shown in Fig. 1.96(b), Fig. 1.96(c), and Fig. 1.96(d).

Race Conditions

When a change in the input causes two or more variables to change nonsimultaneously, due to unequal delays, one variable reaches its stable state before the others, that is, there is a race between the variables.

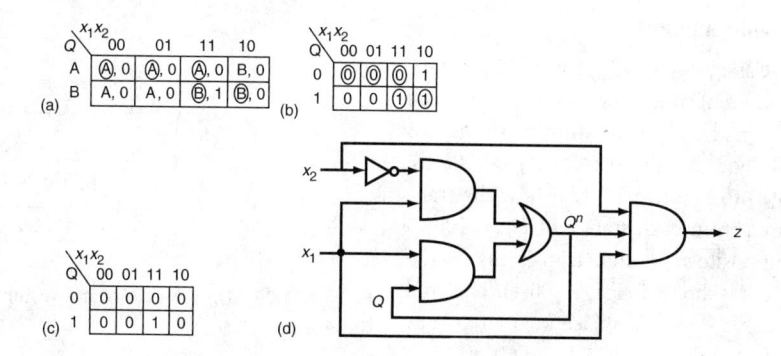

FIGURE 1.96   Asynchronous two-state two-input one-output sequential circuit example specified by a flow table: (a) flow table, (b) transition table, (c) output K-map, (d) circuit diagram.

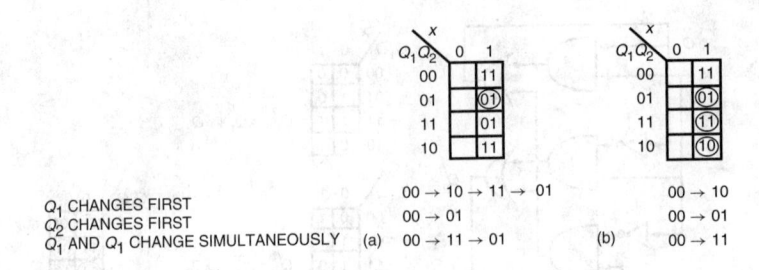

Q_1 CHANGES FIRST
Q_2 CHANGES FIRST
Q_2 AND Q_1 CHANGE SIMULTANEOUSLY

FIGURE 1.97   Race Examples: (a) uncritical, (b) critical.

This may cause state variables to change in an unpredictable manner. Consider the examples shown in Fig. 1.97. If the state changes from 00 to 11 and there are unequal delay times, then the system may go through the intermediate state 01 or 10 depending on whether Q_2 or Q_1 changes first. The race is critical if the final state depends on the order in which the state variables change, otherwise it is noncritical. Critical races must be avoided for reliable operation. Race conditions can be avoided by making binary state assignments in a way that only one state variable can change at a time when a state transition occurs.

Cycles

A cycle is a unique sequence of unstable states. A cycle will keep repeating if it does not end in a stable state. Examples are given in Fig. 1.98.

Stability

The utility of a digital circuit depends on its being able to remain indefinitely in the same state unless driven by an input signal to another state where it remains indefinitely. Figure 1.98(b) illustrates a case where an input signal 1 causes the states to change cyclically through three different states and, hence, is unstable. Figure 1.99 is an example of a two-state two-input that oscillates between 0 and 1 when the both inputs are 1.

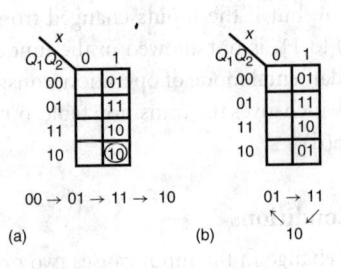

FIGURE 1.98   Cycle examples: (a) single transit, (b) unstable.

Race-Free Code Assignments

Race-free code assignments are provided by assigning logically adjacent codes to states between which transitions are made. The procedure for doing this is illustrated by the example in Fig. 1.100(a). Figure 1.100(b) to Figure 1.100(d) show the transitions required from each state and Fig. 1.100(e) shows the composite set of state transitions. A trial code assignment is shown in Fig. 1.100(f). This trial assignment makes A and B adjacent, B and C adjacent, but not A and C.

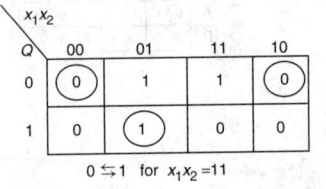

FIGURE 1.99   Unstable two-state two-input example transition table.

The insertion of an extra row in the flow table of this example, as shown in Fig. 1.101, allows the elimination of critical races by the introduction of **cycles**. This is done by changing an unstable

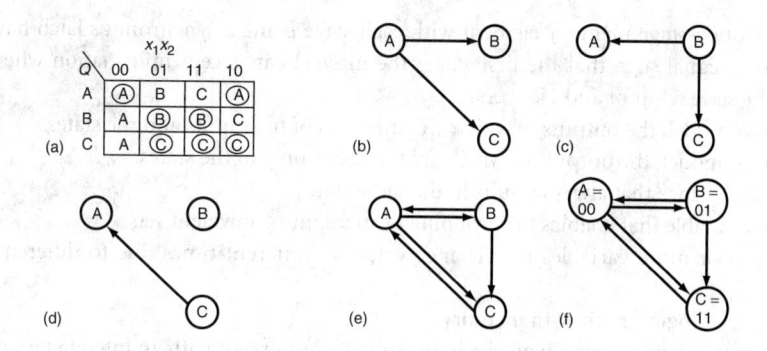

FIGURE 1.100 Example illustrating a method of obtaining race-free assignments: (a) transition table, (b) transition required for row A, (c) transition required for row B, (d) transition required for row C, (e) all transitions required, (f) trial code assignement.

destination state to the added state and then assigning the destination state for the added state to be the original destination state before the added state was introduced. A restriction on the assigned destination states for the added state is that it not be itself in order not to introduce an extra stable state. In Fig. 1.101(a) slashes are used to indicate that the next state may be any state except the current state. Also, in this example the fourth state can be added without increasing the number of binary variables in the state description.

More involved examples may require the introduction of an extra binary variable in the state code in order to provide the extra states needed to introduce the cycles required for elimination of critical races.

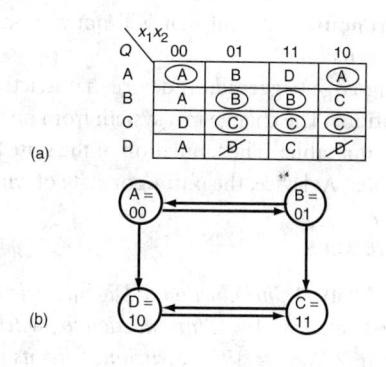

FIGURE 1.101 Modification of the example in Fig. 1.100 giving a race-free assignment: (a) transition table modification, (b) transition diagram.

Defining Terms

Algorithmic state machine (ASM): A sequential logic circuit whose design is directly specified by the algorithm for the task the machine is to accomplish.

Asynchronous: A circuit whose response is directly initiated by the input without depending on a synchronizing signal.

Combinational logic: A logic circuit without memory elements.

Compatible: Two states that can be treated as the same because of the presence of a *don't care*.

Critical race: A race that results in different states depending on the relative delays involved.

Equivalent states: States that produce the same outputs and whose next states are equivalent.

Excitation: Any input to a circuit, but more specifically an input to a storage element.

Excitation table: A table of inputs required to produce a specified set of transitions.

Feedback: An output of a unit connected to the unit's input.

Flip-flop: A two-stage latch in a master-slave configuration.

Implication table: A triangular table having squares identified by state pairs containing information to enable the determination of state equivalency.

Latch: A memory element that does not have immunity to external feedback.

Leading-edge triggered: A device that is activated by the leading edge of a signal.

Master-slave: A two-stage memory element with each stage being a synchronous latch having the synchronizing signal such that the first stage (the master) can accept information when the second stage (the slave) cannot and vice versa.

Mealy model: A model, the outputs of which are functions of the inputs and the states.

Moore model: A model, the outputs of which are functions only of the states.

Noncritical race: A race that always results in the same state.

Partition table: A table that enables the grouping of states into equivalent sets.

Race: When two or more variables may change values at different times due to differences in circuit delays.

Sequential logic: A logic circuit with memory.

Stable: A device, the output or state of which does not change except with an intended excitation.

State: The condition in which a system exists.

State diagram: A figure showing the set of states a system may have and the transitions that can occur between the states and the outputs which are produced.

State table: A tabular display of the same information contained in a state diagram.

Synchronous: A circuit that will not respond to the input without the occurrence of a synchronizing signal.

Trailing-edge triggered: A device that is activated by the trailing edge of a signal.

Transition: The change in a system from one state to another.

Transition table: The tabulation of the state transitions a system has.

Unstable: A device, the output or state of which changes without an intended excitation.

References

Hayes, J.P. 1993. *Introduction to Digital Logic Design*. Addison-Wesley, Reading, MA.

Hill and Peterson. 1974. *Introduction to Switching Theory and Logical Design*, 2nd ed. Wiley, New York.

Humphrey, W.S., Jr. 1958. *Switching Circuits with Computer Applications*. McGraw-Hill, New York.

Johnson and Karim. 1987. *Digital Design A Pragmatic Approach*. Prindle, Weber, and Schmidt, Boston.

Karnaugh, M. 1953. The map method for synthesis of combinational logic circuits. *AIEE Trans. Comm. Elec.* 72 (Nov):593–599.

Mano, M.M. 1991. *Digital Design*. Prentice-Hall, Englewood Cliffs, NJ.

McCluskey, E.J. 1986. *Logic Design Principles*. Prentice-Hall, Englewood Cliffs, NJ.

Mowle, F.J. 1976. *A Systematic Approach to Digital Logic Design*. Addison-Wesley, Reading, MA.

Nagle, Carrol, and Irwin. 1975. *An Introduction to Computer Logic*, 2nd ed. Prentice-Hall, Englewood Cliffs, NJ.

Pappas, N.L. 1994. *Digital Design*. West, St. Paul, MN.

Roth, C.H., Jr. 1985. *Fundamentals of Logic Design*, 3rd ed. West, St. Paul, MN.

Sandige, R.S. 1990. *Modern Digital Design*. McGraw-Hill, New York.

Shaw, A.W. 1991. *Logic Circuit Design*. Saunders, Fort Worth, TX.

Wakerly, 1990. *Digital Design Principles and Practices*. Prentice-Hall, Englewood Cliffs, NJ.

Further Information

Further information on sequential logic design and analysis can be found in occasional articles in the following journals:

Lecture Notes in Computer Science (annual)
International Journal of Electronics (monthly)
IEEE Transactions on Education (quarterly)
IEEE Transactions on Computers (monthly)
IEEE Transactions on Software Engineering (monthly)
IEEE Transactions on Circuits and Systems 1. Fundamental Theory and Applications (monthly)

1.7 The Physical Nature of Sound

Floyd E. Toole, E. A. G. Shaw, Gilles A. Daigle and Michel R. Stinson

1.7.1 Introduction[1]

Sound is a physical disturbance in the medium through which it is propagated. Although the most common medium is air, sound can travel in any solid, liquid, or gas. In air, sound consists of localized variations in pressure above and below normal atmospheric pressure (*compressions* and *rarefactions*).

Air pressure rises and falls routinely, as environmental weather systems come and go, or with changes in altitude. These fluctuation cycles are very slow, and no perceptible sound results, although it is sometimes evident that the ears are responding in a different way to these *infrasonic* events. At fluctuation frequencies in the range from about 20 cycles per second up to about 20,000 cycles per second the physical phenomenon of sound can be perceived as having pitch or tonal character. This generally is regarded as the *audible* or *audio-frequency range*, and it is the frequencies in this range that are the concern of this chapter. Frequencies above 20,000 cycles per second are classified as *ultrasonic*.

1.7.2 Sound Waves

The essence of sound waves is illustrated in Fig. 1.102, which shows a tube with a piston in one end. Initially, the air within and outside the tube is all at the prevailing atmospheric pressure. When the piston moves quickly inward, it compresses the air in contact with its surface. This energetic compression is rapidly passed on to the adjoining layer of air, and so on, repeatedly. As it delivers its energy to its neighbor, each layer of air returns to its original uncompressed state. A longitudinal sound pulse is moving outward through the air in the tube, causing only a passing disturbance on the way. It is a pulse because there is only an isolated action, and it is longitudinal because the air movement occurs along the axis of sound propagation. The rate at which the pulse propagates is the speed of sound. The pressure rise in the compressed air is proportional to the velocity with which the piston moves, and the perceived loudness of the resulting sound pulse is related to the incremental amplitude of the pressure wave above the ambient atmospheric pressure.

Percussive or impulsive sounds such as these are common, but most sounds do not cease after a single impulsive event. Sound waves that are repetitive at a regular rate are called *periodic*. Many musical sounds are periodic, and they embrace a very wide range of repetitive patterns. The simplest of periodic sounds is a pure tone, similar to the sound of a tuning fork or a whistle. An example is presented when the end of the tube is driven by a loudspeaker reproducing a recording of such a sound (Fig. 1.103). The pattern of displacement vs. time for the loudspeaker diaphragm, shown in Fig. 1.103(b), is called a *sine wave* or *sinusoid*.

If the first diaphragm movement is inward, the first event in the tube is a pressure compression, as seen previously. When the diaphragm changes direction, the adjacent layer of air undergoes a *pressure rarefaction*. These cyclic compressions and rarefactions are repeated, so that the sound wave propagating down the tube has a regularly repeated, periodic form. If the air pressure at all points along the tube were measured at a specific instant, the result would be the graph of air pressure vs. distance shown in Fig. 1.103(c). This reveals a smoothly sinusoidal waveform with a repetition distance along the tube symbolized by λ, the *wavelength* of the periodic sound wave.

If a pressure-measuring device were placed at some point in the tube to record the instantaneous changes in pressure at that point as a function of time, the result would be as shown in Fig. 1.103(d). Clearly, the curve has the same shape as the previous one except that the horizontal axis is time instead of distance. The periodic nature of the waveform is here defined by the time period T, known simply as the *period*

---

[1]From: Jerry Whitaker and K. Blair Benson, *Standard Handbook of Audio and Radio Engineering*, 2nd edition, McGraw-Hill, New York, NY, 2001. Used with permission.

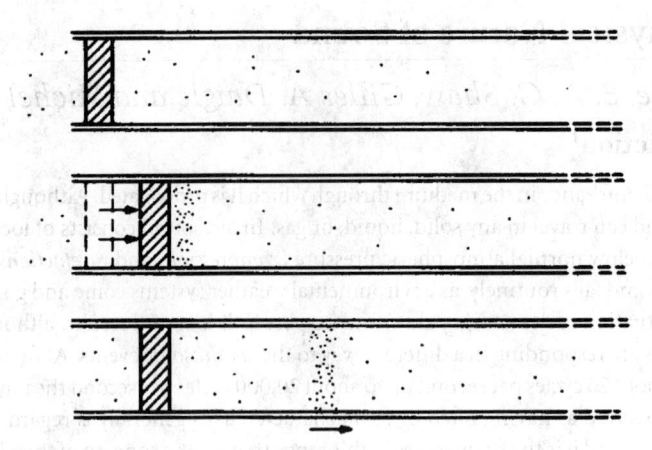

FIGURE 1.102 Generation of a longitudinal sound wave by the rapid movement of a piston in the end of a tube, showing the propagation of the wave pulse at the speed of sound down the length of the tube.

of the sound wave. The inverse of the period, $1/T$, is the *frequency* of the sound wave, describing the number of repetition cycles per second passing a fixed point in space. An ear placed in the path of a sound wave corresponding to the musical tone middle C would be exposed to a frequency of 261.6 cycles per second or, using standard scientific terminology, a frequency of 261.6 Hz. The perceived loudness of the tone would depend on the magnitude of the pressure deviations above and below the ambient air pressure.

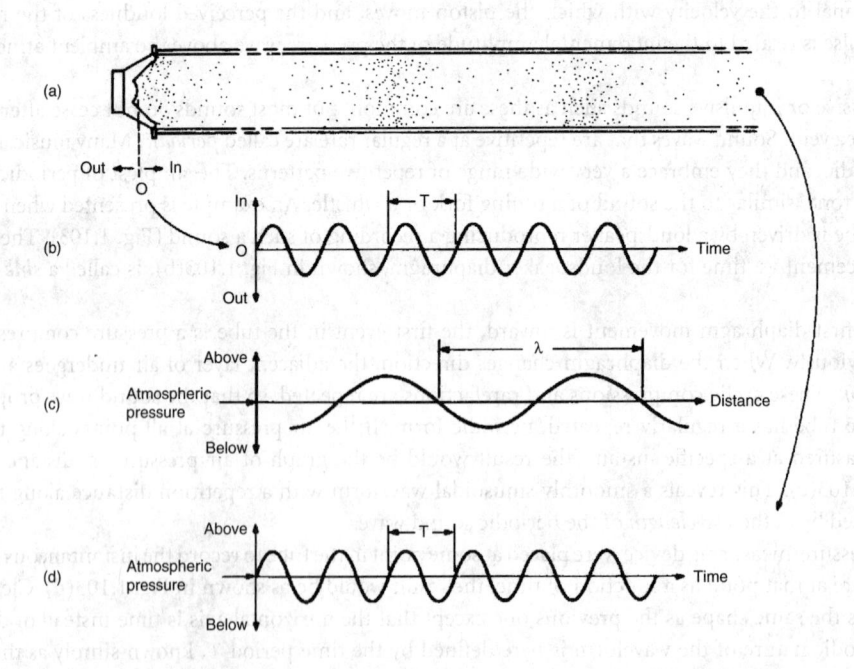

FIGURE 1.103 Characteristics of sound waves: (a) a periodic sound wave, a sinusoid in this example, is generated by a loudspeaker placed at the end of a tube. (b) waveform showing the movement of the loudspeaker diaphragm as a function of time: displacement vs. time. (c) waveform showing the instantaneous distribution of pressure along a section of the tube: pressure vs. distance. (d) waveform showing the pressure variation as a function of time at some point along the tube: pressure vs. time.

The parameters discussed so far are all related by the *speed of sound*. Given the speed of sound and the duration of one period, the wavelength can be calculated as

$$\lambda = cT \tag{1.18}$$

where
  λ = wavelength
  c = speed of sound
  T = period

By knowing that the frequency $f = 1/T$, the following useful equation and its variations can be derived:

$$\lambda = \frac{c}{f} \quad f = \frac{c}{\lambda} \quad c = f\lambda \tag{1.19}$$

The speed of sound in air at a room temperature of 22°C (72°F) is 345 m/s (1131 ft/s). At any other ambient temperature, the speed of sound in air is given by the following approximate relationships [1, 2]:

$$c(m/s) = 331.29 + 0.607t(°C) \tag{1.20}$$

or

$$c(m/s) = 1051.5 + 1.106t(°F) \tag{1.21}$$

where t = ambient temperature.

The relationships between the frequency of a sound wave and its wavelength are essential to understanding many of the fundamental properties of sound and hearing. The graph of Fig. 1.104 is a useful quick reference illustrating the large ranges of distance and time embraced by audible sounds. For example, the tone middle C with a frequency of 261.6 Hz has a wavelength of 1.3 m (4.3 ft) in air at 20°C. In contrast, an organ pedal note at Cl, 32.7 Hz, has a wavelength of 10.5 m (34.5 ft), and the third-harmonic overtone of C8, at 12,558 Hz, has a wavelength of 27.5 mm (1.1 in). The corresponding periods are, respectively, 3.8, 30.6, and 0.08 ms. The contrasts in these dimensions are remarkable, and they result in some interesting and troublesome effects in the realms of perception and audio engineering. For the discussions that follow it is often more helpful to think in terms of wavelengths rather than in frequencies.

Complex Sounds

The simple sine waves used for illustration reveal their periodicity very clearly. Normal sounds, however, are much more complex, being combinations of several such pure tones of different frequencies and perhaps additional transient sound components that punctuate the more sustained elements. For example, speech is a mixture of approximately periodic vowel sounds and staccato consonant sounds. Complex sounds can also be periodic; the repeated wave pattern is just more intricate, as is shown in Fig. 1.105(a). The period identified as T_1 applies to the *fundamental frequency* of the sound wave, the component that normally is related to the characteristic pitch of the sound. Higher-frequency components of the complex wave are also periodic, but because they are typically lower in amplitude, that aspect tends to be disguised in the summation of several such components of different frequency. If, however, the sound wave were analyzed, or broken down into its constituent parts, a different picture emerges: Fig. 1.105(b), (c), and (d). In this example, the analysis shows that the components are all *harmonics*, or whole-number multiples, of the fundamental frequency; the higher-frequency components all have multiples of entire cycles within the period of the fundamental.

To generalize, it can be stated that all *complex periodic waveforms* are combinations of several harmonically related sine waves. The shape of a complex waveform depends upon the relative amplitudes of the various harmonics and the position in time of each individual component with respect to the others. If one of the harmonic components in Fig. 1.105 is shifted slightly in time, the shape of the waveform

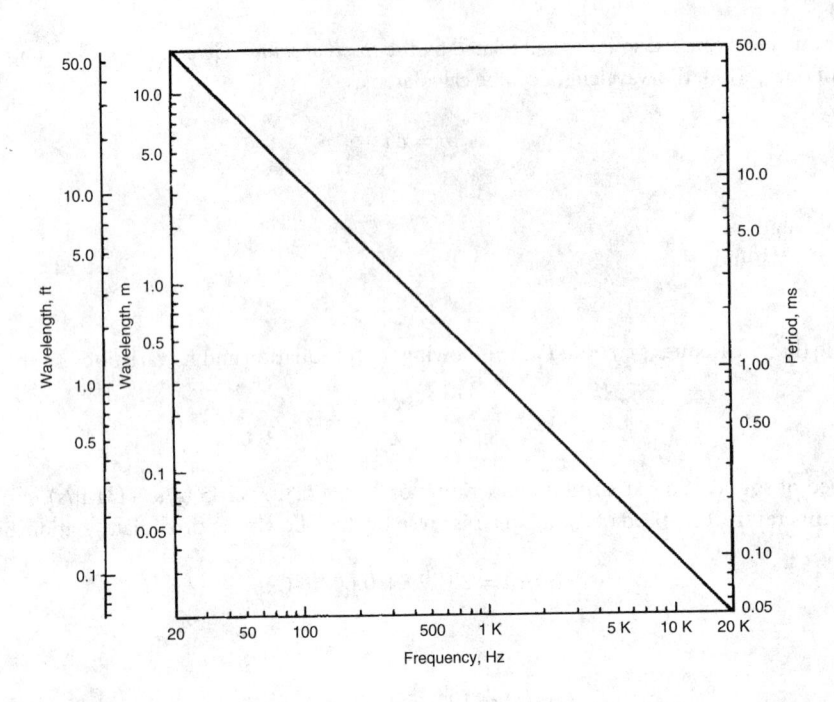

FIGURE 1.104   Relationships between wavelength, period, and frequency for sound waves in air.

is changed, although the frequency composition remains the same (Fig. 1.106). Obviously a record of the time locations of the various harmonic components is required to completely describe the complex waveform. This information is noted as the *phase* of the individual components.

Phase

Phase is a notation in which the time of one period of a sine wave is divided into 360°. It is a relative quantity, and although it can be defined with respect to any reference point in a cycle, it is convenient to start (0°) with the upward, or positive-going, zero crossing and to end (360°) at precisely the same point at the beginning of the next cycle (Fig. 1.107). *Phase shift* expresses in degrees the fraction of a period or wavelength by which a single-frequency component is shifted in the time domain. For example, a phase shift of 90° corresponds to a shift of one-fourth period. For different frequencies this translates into different time shifts. Looking at it from the other point of view, if a complex waveform is time-delayed, the various harmonic components will experience different phase shifts, depending on their frequencies.

A special case of phase shift is a *polarity reversal*, an inversion of the waveform, where all frequency components undergo a 180° phase shift. This occurs when, for example, the connections to a loudspeaker are reversed.

Spectra

Translating time-domain information into the frequency domain yields an *amplitude-frequency spectrum* or, as it is commonly called, simply a *spectrum*. Figure 1.108(a) shows the spectrum of the waveform in Fig. 1.104 and Fig. 1.106, in which the height of each line represents the amplitude of that particular component and the position of the line along the frequency axis identifies its frequency. This kind of display is a *line spectrum* because there are sound components at only certain specific frequencies. The phase information is shown in Fig. 1.108(b), where the difference between the two waveforms is revealed in the different *phase-frequency spectra*.

The equivalence of the information presented in the two domains—the waveform in the time domain and the amplitude- and phase-frequency spectra in the frequency domain—is a matter of considerable

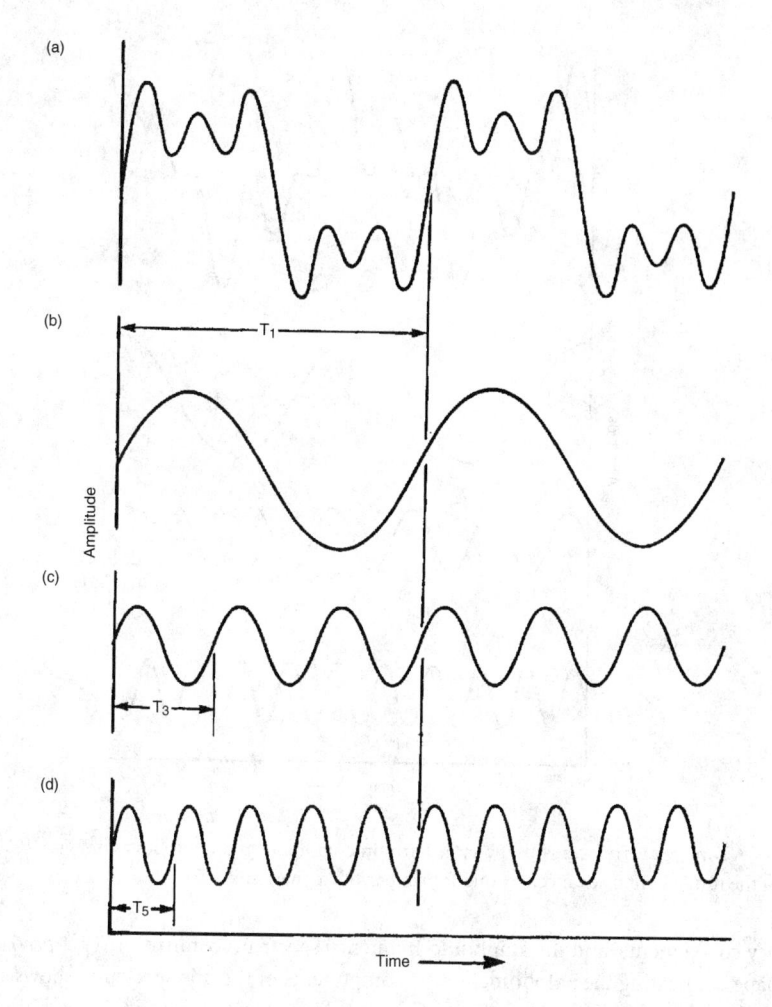

FIGURE 1.105 A complex waveform constructed from the sum of three harmonically related sinusoidal components, all of which start at the origin of the time scale with a positive-going zero crossing. Extending the series of odd-harmonic components to include those above the fifth would result in the complex waveform progressively assuming the form of a square wave. (a) complex waveform, the sum of b, c, and d (b) fundamental frequency (c) third harmonic (d) fifth harmonic.

importance. The proofs have been thoroughly worked out by the French mathematician Fourier, and the well-known relationships bear his name. The breaking down of waveforms into their constituent sinusoidal parts is known as *Fourier analysis*. The construction of complex waveshapes from summations of sine waves is called *Fourier synthesis*. *Fourier transformations* permit the conversion of time-domain information into frequency-domain information, and vice versa. These interchangeable descriptions of waveforms form the basis for powerful methods of measurement and, at the present stage, provide a convenient means of understanding audio phenomena. In the examples that follow, the relationships between time-domain and frequency-domain descriptions of waveforms will be noted.

Figure 1.109 illustrates the sound waveform that emerges from the larynx, the buzzing sound that is the basis for vocalized speech sounds. This sound is modified in various ways in its passage down the vocal tract before it emerges from the mouth as speech. The waveform is a series of periodic pulses, corresponding to the pulses of air that are expelled, under lung pressure, from the vibrating vocal cords. The spectrum of this waveform consists of a harmonic series of components, with a fundamental frequency, for this male talker, of 100 Hz. The gently rounded contours of the waveform suggest the absence of strong

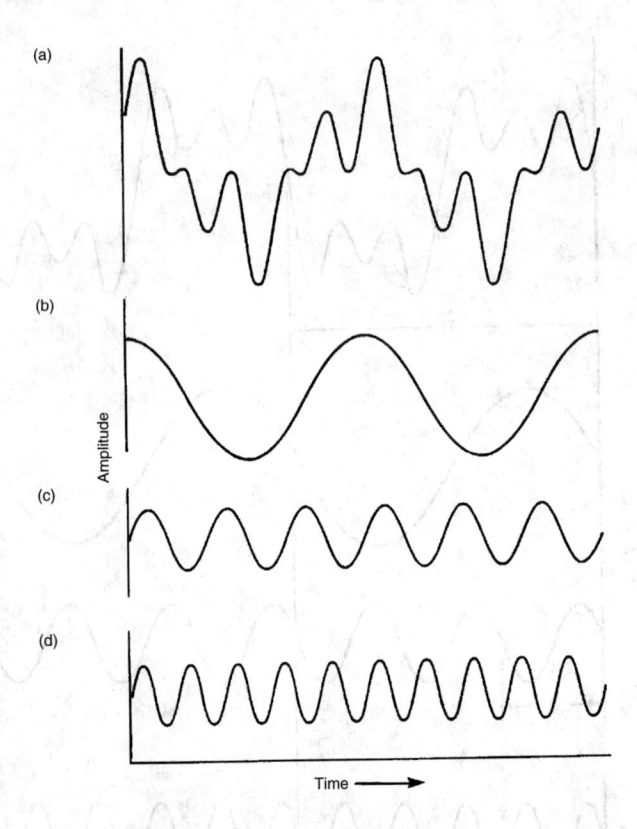

FIGURE 1.106 A complex waveform with the same harmonic-component amplitudes as in Fig. 1.105, but with the starting time of the fundamental advanced by one-fourth period: a phase shift of 90°.

high-frequency components, and the amplitude-frequency spectrum confirms it. The *spectrum envelope*, the overall shape delineating the amplitudes of the components of the line spectrum, shows a progressive decline in amplitude as a function of frequency. The amplitudes are described in *decibels*, abbreviated dB. This is the common unit for describing sound-level differences. The rate of this decline is about −12 dB per octave (an *octave* is a 2:1 ratio of frequencies).

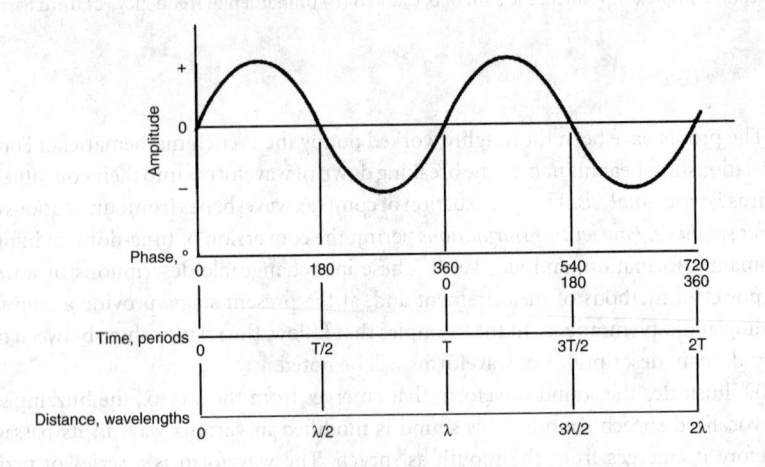

FIGURE 1.107 The relationship between the period T and wavelength λ of a sinusoidal waveform and the phase expressed in degrees. Although it is normal to consider each repetitive cycle as an independent 360°, it is sometimes necessary to sum successive cycles starting from a reference point in one of them.

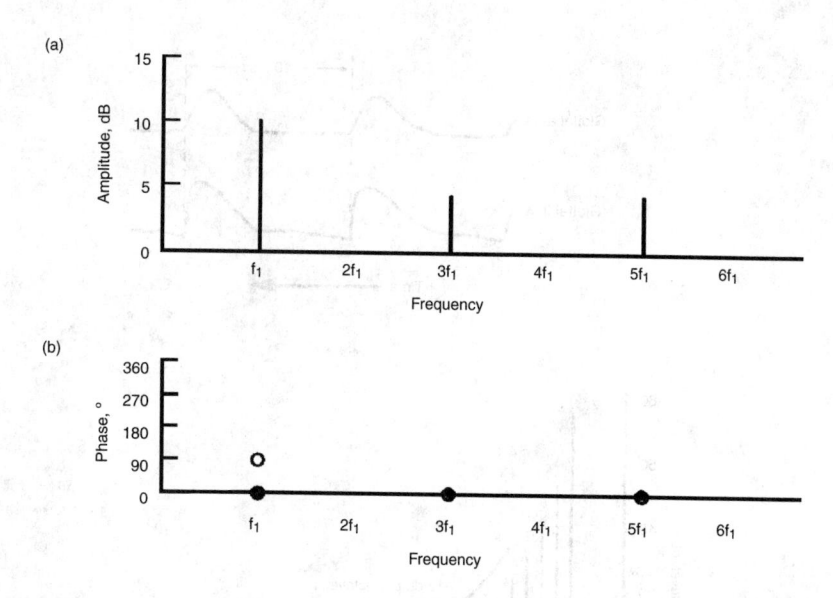

FIGURE 1.108 The amplitude-frequency spectra (a) and the phase-frequency spectra (b) of the complex waveforms shown in Fig. 1.105 and Fig. 1.106. The amplitude spectra are identical for both waveforms, but the phase-frequency spectra show the 90° phase shift of the fundamental component in the waveform of Fig. 1.106. Note that frequency is expressed as a multiple of the fundamental frequency f_1. The numerals are the harmonic numbers. Only the fundamental f_1 and the third and fifth harmonics (f_3 and f_5) are present.

Increasing the pitch of the voice brings the pulses closer together in time and raises the fundamental frequency. The harmonic-spectrum lines displayed in the frequency domain are then spaced farther apart but still within the overall form of the spectrum envelope, which is defined by the shape of the pulse itself. Reducing the pitch of the voice has the opposite effect, increasing the spacing between pulses and reducing the spacing between the spectral lines under the envelope. Continuing this process to the limiting condition, if it were possible to emit just a single pulse, would be equivalent to an infinitely long period, and the spacing between the spectral lines would vanish. The discontinuous, or *aperiodic*, pulse waveform therefore yields a *continuous* spectrum having the form of the spectrum envelope.

Isolated pulses of sound occur in speech as any of the variations of consonant sounds, and in music as percussive sounds and as transient events punctuating more continuous melodic lines. All these aperiodic sounds exhibit continuous spectra with shapes that are dictated by the waveforms. The leisurely undulations of a bass drum waveform contain predominantly low-frequency energy, just as the more rapid pressure changes in a snare drum waveform require the presence of higher frequencies with their more rapid rates of change. A technical waveform of considerable use in measurements consists of a very brief impulse which has the important feature of containing equal amplitudes of all frequencies within the audio-frequency bandwidth. This is moving toward a limiting condition in which an infinitely short event in the time domain is associated with an infinitely wide amplitude-frequency spectrum.

1.7.3 Dimensions of Sound

The descriptions of sound in the preceding section involved only pressure variation, and while this is the dimension that is most commonly referred to, it is not the only one. Accompanying the pressure changes are temporary movements of the air "particles" as the sound wave passes (in this context a particle is a volume of air that is large enough to contain many molecules while its dimensions are small compared with the wavelength). Other measures of the magnitude of the sound event are the displacement amplitude of the air particles away from their rest positions and the velocity amplitude of the particles during the movement cycle. In the physics of sound, the *particle displacement* and the *particle velocity* are useful

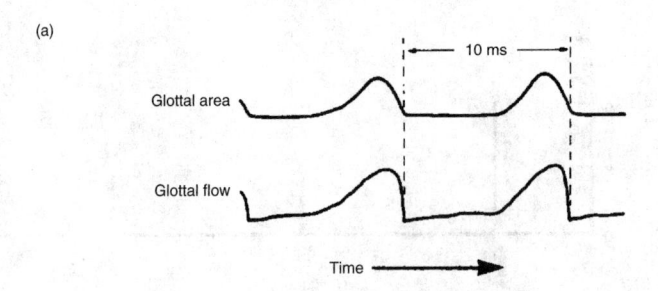

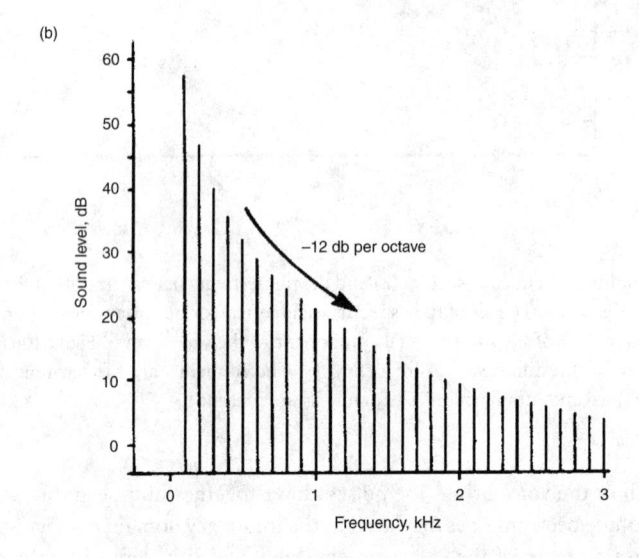

FIGURE 1.109 Characteristics of speech (a) waveforms showing the varying area between vibrating vocal cords and the corresponding airflow during vocalized speech as a function of time (b) the corresponding amplitude-frequency spectrum, showing the 100-Hz fundamental frequency for this male speaker. (*From* [3]. *Used with permission*).

concepts, but the difficulty of their measurement limits their practical application. They can, however, help in understanding other concepts.

In a normally propagating sound wave, energy is required to move the air particles; they must be pushed or pulled against the elasticity of the air, causing the incremental rises and falls in pressure. Doubling the displacement doubles the pressure change, and this requires double the force. Because the work done is the product of force times distance and both are doubled, the energy in a sound wave is therefore proportional to the square of the particle displacement amplitude or, in more practical terms, to the square of the sound pressure amplitude.

Sound energy spreads outward from the source in the three dimensions of space, in addition to those of amplitude and time. The energy of such a sound field is usually described in terms of the energy flow through an imaginary surface. The sound energy transmitted per unit of time is called *sound power*. The sound power passing through a unit area of a surface perpendicular to a specified direction is called the *sound intensity*. Because intensity is a measure of energy flow, it also is proportional to the square of the sound pressure amplitude.

The ear responds to a very wide range of sound pressure amplitudes. From the smallest sound that is audible to sounds large enough to cause discomfort, there is a ratio of approximately 1 million in sound pressure amplitude, or 1 trillion (10^{12}) in sound intensity or power. Dealing routinely with such large numbers is impractical, so a logarithmic scale is used. This is based on the *bel*, which represents a ratio of 10:1 in sound intensity or sound power (the power can be acoustical or electrical). More commonly the

TABLE 1.29 Various Power and Amplitude Ratios and Their Decibel Equivalents*

Sound or Electrical Power Ratio	Decibels	Sound Pressure, Voltage, or Current Ratio	Decibels
1	0	1	0
2	3.0	2	6.0
3	4.8	3	9.5
4	6.0	4	12.0
5	7.0	5	14.0
6	7.8	6	15.6
7	8.5	7	16.9
8	9.0	8	18.1
9	9.5	9	19.1
10	10.0	10	20.0
100	20.0	100	40.0
1,000	30.0	1,000	60.0
10,000	40.0	10,000	80.0
100,000	50.0	100,000	100.0
1,000,000	60.0	1,000,000	120.0

* Other values can be calculated precisely by using Equation (1.18) and Equation (1.19) or estimated by using this table and the following rules:

Power ratios that are multiples of 10 are converted into their decibel equivalents by multiplying the appropriate exponent by 10. For example, a power ratio of 1000 is 10^3, and this translates into $3 \times 10 = 30$ dB. Since power is proportional to the square of amplitude, the exponent of 10 must be doubled to arrive at the decibel equivalent of an amplitude ratio.

Intermediate values can be estimated by combining values in this table by means of the rule that the multiplication of power or amplitude ratios is equivalent to adding level differences in decibels. For example, increasing a sound level 27 dB requires increasing the power by a ratio of 500 (20 dB is a ratio of 100, and 7 dB is a ratio of 5; the product of the ratios is 500). The corresponding increase in sound pressure or electrical signal amplitude is a factor of just over 20 (20 dB is a ratio of 10, and 7 dB falls between 6.0 and 9.5 and is therefore a ratio of something in excess of 2); the calculated value is 22.4. Reversing the process, if the output from a power amplifier is increased from 40 to 800 W, a ratio of 20, the sound pressure level would be expected to increase by 13 dB (a power ratio of 10 is 10 dB, a ratio of 2 is 3 dB, and the sum is 13 dB). The corresponding voltage increase measured at the output of the amplifier would be a factor of between 4 and 5 (by calculation, 10.5).

decibel, one-tenth of a bel, is used. A difference of 10 dB therefore corresponds to a factor-of-10 difference in sound intensity or sound power. Mathematically this can be generalized as

$$\text{Level difference} = \log \frac{P_1}{P_2} \text{ bels} \tag{1.22}$$

or

$$\text{Level difference} = 10 \log \frac{P_1}{P_2} \text{ decibels} \tag{1.23}$$

where P_1 and P_2 are two levels of power.

For ratios of sound pressures (analogous to voltage or current ratios in electrical systems) the squared relationship with power is accommodated by multiplying the logarithm of the ratio of pressures by 2, as follows:

$$\text{Level difference} = 10 \log \frac{P_1^2}{P_2^2} = 20 \log \frac{P_1}{P_2} \text{ dB} \tag{1.24}$$

where P_1 and P_2 are sound pressures.

TABLE 1.30 Typical Sound Pressure Levels and Intensities for Various Sound Sources*

Sound Source	Sound Pressure Level, dB	Intensity W/m^2	Listener Reaction
	160		Immediate damage
Jet engine at 10 m	150	10^3	
	140		Painful feeling
	130		
SST takeoff at 500 m	120	1	Discomfort
Amplified rock music	110		
Chain saw at 1 m	100		fff
Power mower at 1.5 m	90	10^{-3}	
75-piece orchestra at 7 m	80		f
City traffic at 15 m	70		
Normal speech at 1 m	60	10^{-6}	p
Suburban residence	50		
Library	40		ppp
Empty auditorium	30	10^{-9}	
Recording studio	20		
Breathing	10		
	0†	10^{-12}	Inaudible

*The relationships illustrated in this table are necessarily approximate because the conditions of measurement are not defined. Typical levels should, however, be within about 10 dB of the stated values.

†O-dB sound pressure level (SPL) represents a reference sound pressure of 0.0002 μbar, or 0.00002 N/m^2.

The relationship between decibels and a selection of power and pressure ratios is given in Table 1.29. The footnote to the table describes a simple process for interpolating between these values, an exercise that helps to develop a feel for the meaning of the quantities.

The representation of the relative magnitudes of sound pressures and powers in decibels is important, but there is no indication of the absolute magnitude of either quantity being compared. This limitation is easily overcome by the use of a universally accepted reference level with which others are compared. For convenience the standard reference level is close to the smallest sound that is audible to a person with normal hearing. This defines a scale of *sound pressure level* (SPL), in which 0 dB represents a sound level close to the hearing-threshold level for middle and high frequencies (the most sensitive range). The SPL of a sound therefore describes, in decibels, the relationship between the level of that sounds and the reference level. Table 1.30 gives examples of SPLs of some common sounds with the corresponding intensities and an indication of listener reactions. From this table it is clear that the musically useful range of SPLs extend from the level of background noises in quiet surroundings to levels at which listeners begin to experience auditory discomfort and nonauditory sensations of feeling or pain in the ears themselves.

While some sound sources, such as chain saws and power mowers, produce a relatively constant sound output, others, like a 75-piece orchestra, are variable. The sound from such an orchestra might have a *peak factor* of 20 to 30 dB; the momentary, or peak, levels can be this amount higher than the long-term average SPL indicated [4].

The sound power produced by sources gives another perspective on the quantities being described. In spite of some impressively large sounds, a full symphony orchestra produces only about 1 acoustic watt when working through a typical musical passage. On crescendos with percussion, though, the levels can be of the order of 100 W. A bass drum alone can produce about 25 W of acoustic power of peaks. All these levels are dependent on the instruments and how they are played. Maximum sound output from cymbals might be 10 W; from a trombone, 6 W; and from a piano, 0.4 W [5]. By comparison, average speech generates about 25 μW, and a present- day jet liner at takeoff between 50 and 100 kW. Small gasoline engines produce from 0.001 to 1.0 acoustic watt, and electric home appliances less than 0.01 W [6].

References

1. Beranek, Leo L: *Acoustics*, McGraw-Hill, New York, NY, 1954.
2. Wong, G. S. K: Speed of sound in standard air, *J. Acoust. Soc. Am.*, vol. 79, pp. 1359–1366, 1986.
3. Pickett, J. M: *The Sounds of Speech Communications*, University Park Press, Baltimore, MD, 1980.
4. Ehara, Shiro: Instantaneous pressure distributions of orchestra sounds, *J. Acoust. Soc. Japan*, vol. 22, pp. 276–289, 1966.
5. Stephens, R. W. B. and A. E. Bate: *Acoustics and Vibrational Physics*, 2nd ed., E. Arnold (ed.), London, 1966.
6. Shaw, E. A. G.: "Noise pollution—what can be done?" *Phys. Today*, vol. 28, no. 1, pp. 46–58, 1975.

1.8 Principles of Light, Vision, and Photometry

Jerry C. Whitaker

1.8.1 Introduction

Vision results from stimulation of the eye by light and consequent interaction through connecting nerves with the brain.[2] In physical terms, light constitutes a small section in the range of electromagnetic radiation, extending in wavelength from about 400–700 nm or billionths (10^{-9}) of a meter. (See Fig. 1.110.)

Under ideal conditions, the human visual system can detect:

- Wavelength differences of 1 nm (10 Å, 1 angstrom unit $= 10^{-8}$ cm)
- Intensity differences of as little as 1%
- Forms subtending an angle at the eye of 1 arc minute, and often smaller objects

Although the range of human vision is small compared with the total energy spectrum, human discrimination—the ability to detect differences in intensity or quality—is excellent.

1.8.2 Sources of Illumination

Light reaching an observer usually has been reflected from some object. The original source of such energy typically is radiation from molecules or atoms resulting from internal (atomic) changes. The exact type of emission is determined by

- The ways in which the atoms or molecules are supplied with energy to replace that which they radiate
- The physical state of the substance, whether solid, liquid, or gaseous

The most common source of radiant energy is the thermal excitation of atoms in the solid or gaseous state.

The Spectrum

When a beam of light traveling in air falls upon a glass surface at an angle, it is *refracted* or bent. The amount of **refraction** depends on the wavelength, its variation with wavelength being known as **dispersion**. Similarly when the beam, traveling in glass, emerges into air, it is refracted (with dispersion). A glass prism provides a refracting system of this type. Because different wavelengths are refracted by different amounts, an incident white beam is split up into several beams corresponding to the many wavelengths contained in the composite white beam. Thus is obtained the spectrum.

If a spectrum is allowed to fall upon a narrow slit arranged parallel to the edge of the prism, a narrow band of wavelengths passes through the slit. Obviously the narrower the slit, the narrower the band of

---

[2]Portions of this chapter were adapted from Whitaker, J.C. 2000. VIDEO *Displays*: McGraw-Hill, New York. Used with permission.

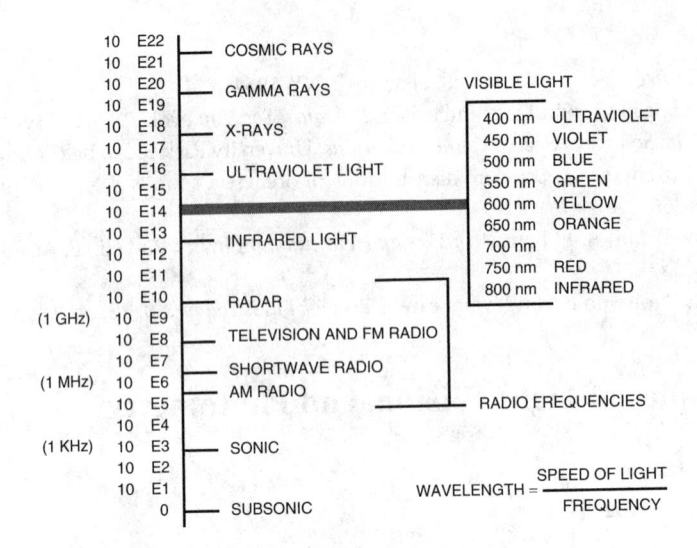

FIGURE 1.110   The electromagnetic spectrum.

wavelengths or the sharper the spectral line. Also, more dispersion in the prism will cause a wider spectrum to be produced, and a narrower spectral line will be obtained for a given slit width.

It should be noted that purples are not included in the list of spectral colors. The purples belong to a special class of colors; they can be produced by mixing the light from two spectral lines, one in the red end of the spectrum, the other in the blue end. Purple (magenta is a more scientific name) is therefore referred to as a *nonspectral color*.

A plot of the power distribution of a source of light is indicative of the watts radiated at each wavelength per nanometer of wavelength. It is usual to refer to such a graph as an **energy distribution curve**.

Individual narrow bands of wavelengths of light are seen as strongly colored elements. Increasingly broader bandwidths retain the appearance of color, but with decreasing purity, as if white light had been added to them. A very broad band extending generally throughout the visible spectrum is perceived as white light. Many white light sources are of this type, such as the familiar tungsten-filament electric light bulb (see Fig. 1.111). Daylight also has a broad band of radiation, as illustrated in Fig. 1.112. The energy distributions shown in Fig. 1.111 and Fig. 1.112 are quite different and, if the corresponding sets of radiation were seen side by side, would be different in appearance. Either one, particularly if seen alone, however, would represent a very acceptable white. A sensation of white light can also be induced by light sources that do not have a uniform energy distribution. Among these is fluorescent lighting, which exhibits sharp peaks of energy through the visible spectrum. Similarly, the light from a monochrome (black-and-white) video cathode ray tube is not uniform within the visible spectrum, generally exhibiting peaks in the yellow and blue regions of the spectrum; yet it appears as an acceptable white (see Fig. 1.113).

1.8.3   Monochrome and Color Vision

The color sensation associated with a light stimulus can be described in terms of three characteristics:

- Hue
- Saturation
- Brightness

The spectrum contains most of the principal hues: red, orange, yellow, green, blue, and violet. Additional hues are obtained from mixtures of red and blue light. These constitute the purple colors. *Saturation* pertains to the strength of the hue. Spectrum colors are highly saturated. White and grays have no hue and,

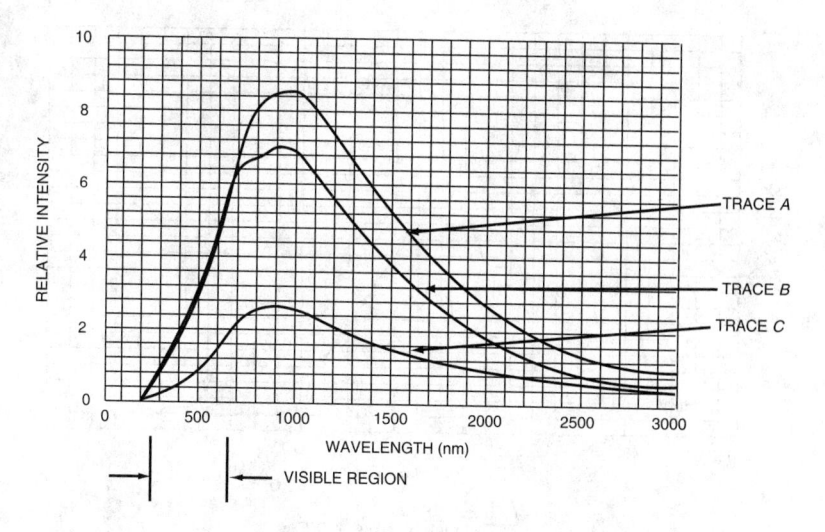

FIGURE 1.111 The radiating characteristic of tungsten: (trace *A*) radiant flux from 1 cm² of a blackbody at 3000 K, (trace *B*) radiant flux from 1 cm² of tungsten at 3000 K, (trace *C*) radiant flux from 2.27 cm² of tungsten at 3000 K (equal to curve *A* in the visible region). (*Source:* Adapted from IES. 1981. *IES Lighting Handbook.* Illuminating Engineering Society of North America, New York.)

therefore, have zero saturation. Pastel colors are of low or intermediate saturation. *Brightness* pertains to the intensity of the stimulation. If a stimulus has high intensity, regardless of its hue, it is said to be bright.

The psychophysical analogs of hue, saturation, and brightness are:

- Dominant wavelength
- Excitation purity
- Luminance

This principle is illustrated in Table 1.31.

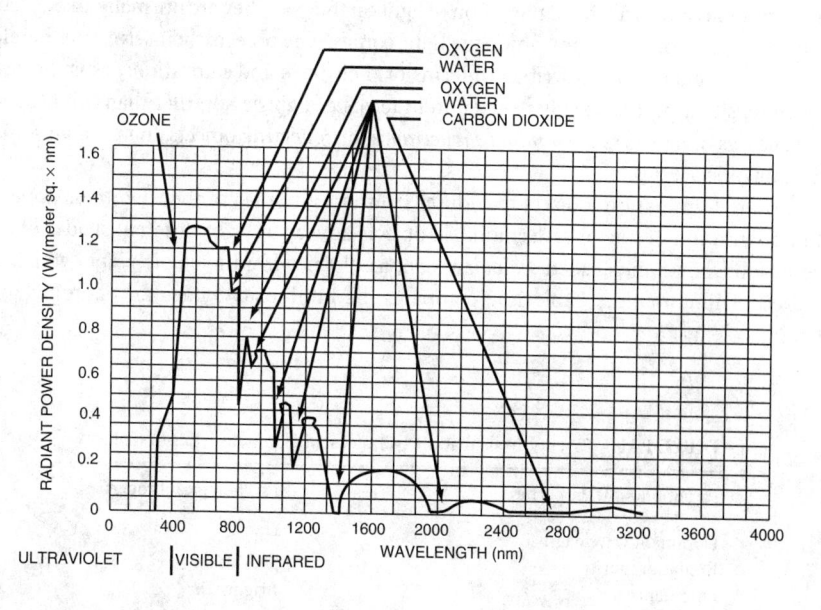

FIGURE 1.112 Spectral distribution of solar radiant power density at sea level, showing the ozone, oxygen, and carbon dioxide absorption bands. (*Source:* Adapted from IES. 1981. *IES Lighting Handbook.* Illuminating Engineering Society of North America, New York.)

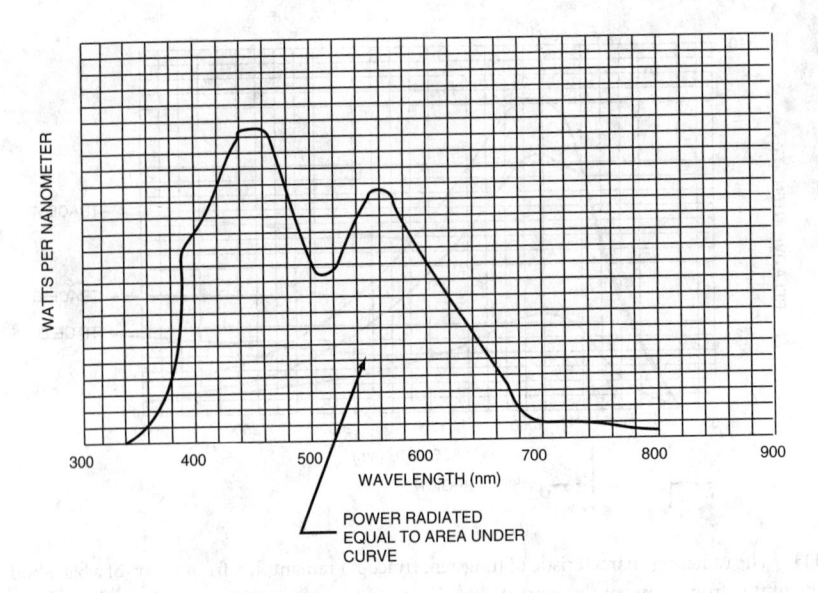

FIGURE 1.113   Power distribution of a monochrome video picture tube light source. (*Source:* Adapted from Fink, D.G. 1952. *Television Engineering,* 2nd ed. McGraw-Hill, New York.)

By means of definitions and standard response functions, which have received international acceptance through the International Commission on Illumination, the dominant wavelength, purity, and luminance of any stimulus of known spectral energy distribution may be determined by simple computations. Although roughly analogous to their psychophysical counterparts, the psychological attributes of hue, saturation, and brightness pertain to observer responses to light stimuli and are not subject to calculation. These sensation characteristics as applied to any given stimulus depend in part on other visual stimuli in the field of view and on the immediately preceding stimulations.

Color sensations arise directly from the action of light on the eye. They are normally associated, however, with objects in the field of view from which the light comes. The objects themselves are therefore said to have color. *Object colors* may be described in terms of their hues and saturations, as is the case for light stimuli. The intensity aspect is usually referred to in terms of lightness, rather than brightness. The psychophysical analogs of lightness are *luminous reflectance* for reflecting objects and *luminous transmittance* for transmitting objects.

At low levels of illumination, objects may differ from one another in their lightness appearances but give rise to no sensation of hue or saturation. All objects appear to be of different shades of gray. Vision at low levels of illumination is called *scotopic vision,* as distinct from *photopic vision,* which takes place at higher levels of illumination. Table 1.32 compares the luminosity values for photopic and scotopic vision.

TABLE 1.31   Psychophysical and Psychological Characteristics of Color

Psychophysical Properties	Psychological Properties
Dominant wavelength	Hue
Excitation purity	Saturation
Luminance	Brightness
Luminous transmittance	Lightness
Luminous reflectance	Lightness

TABLE 1.32 Relative Luminosity Values for Photopic and Scotopic Vision

Wavelength, nm	Photopic Vision	Scotopic Vision
390	0.00012	0.0022
400	0.0004	0.0093
410	0.0012	0.0348
420	0.0040	0.0966
430	0.0116	0.1998
440	0.023	0.3281
450	0.038	0.4550
460	0.060	0.5670
470	0.091	0.6760
480	0.139	0.7930
490	0.208	0.9040
500	0.323	0.9820
510	0.503	0.9970
520	0.710	0.9350
530	0.862	0.8110
540	0.954	0.6500
550	0.995	0.4810
560	0.995	0.3288
570	0.952	0.2076
580	0.870	0.1212
590	0.757	0.0655
600	0.631	0.0332
610	0.503	0.0159
620	0.381	0.0074
630	0.265	0.0033
640	0.175	0.0015
650	0.107	0.0007
660	0.061	0.0003
670	0.032	0.0001
680	0.017	0.0001
690	0.0082	
700	0.0041	
710	0.0021	
720	0.00105	
730	0.00052	
740	0.00025	
750	0.00012	
760	0.00006	

Only the rods of the retina are involved in scotopic vision; the cones play no part. As the *fovea centralis* is free of rods, scotopic vision takes place outside the fovea. Visual acuity of scotopic vision is low compared with photopic vision.

At high levels of illumination, where cone vision predominates, all vision is color vision. Reproducing systems such as black-and-white photography and monochrome video, however, cannot reproduce all three types of characteristics of colored objects. All images belong to the series of grays, differing only in their relative brightness.

The relative brightness of the reproduced image of any object depends primarily on the luminance of the object as seen by the photographic or video camera. Depending on the camera pickup element or the film, however, the dominant wavelength and purity of the light may also be of consequence. Most films and video pickup elements now in use exhibit sensitivity throughout the visible spectrum and, consequently, marked distortions in luminance as a function of dominant wavelength and purity are not encountered.

Their spectral sensitivities seldom conform exactly to that of the human observer, however, so that some brightness distortions do exist.

Visual Requirements for Video

The objective in any type of visual reproduction system is to present to the viewer a combination of visual stimuli that can be readily interpreted as representing or having close association with a real viewing situation. It is by no means necessary that the light stimuli from the original scene be duplicated. There are certain characteristics in the reproduced image, however, that are necessary and others that are highly desirable. Only a general qualitative discussion of such characteristics will be given here.

In monochrome video, images of objects are distinguished from one another and from their backgrounds as a result of luminance differences. In order that detail in the picture be visible and that objects have clear, sharp edges, it is necessary that the video system be capable of rapid transitions from areas of one luminance level to another. This degree of resolution need not match that possible in the eye itself, but too low an effective **resolution** results in pictures with a fuzzy appearance and lacking fineness of detail.

Luminance range and the transfer characteristic associated with luminance reproduction are likewise of importance in monochrome video. Objects seen as white usually have minimum reflectances of approximately 80%. Black objects have reflectances of approximately 4%. This gives a luminance ratio of 20/1 in the range from white to black. To obtain the total luminance range in a scene, the reflectance range must be multiplied by the illumination range. In outdoor scenes, the illumination ratio between full sunlight and shadow may be as high as 100/1. The full luminance ranges involved with objects in such scenes cannot be reproduced in normal video reproduction equipment. Video systems must be capable of handling illumination ratios of at least 2, however, and ratios as high as 4 or 5 are desirable. This implies a luminance range on the output of the system of at least 40, with possible upper limits as high as 80 or 100.

Monochrome video transmits only luminance information, and the relative luminances of the images should correspond at least roughly to the relative luminances of the original objects. Red objects, for example, should not be reproduced markedly darker than objects of other hues but of the same luminance. Exact luminance reproduction, however, is by no means a necessity. Considerable distortion as a function of hue is acceptable in many applications. Luminance reproduction is probably of primary consequence only if detail in some hues becomes lost.

Images in monochrome video are transmitted one point, or small area, at a time. The complete picture image is repeatedly scanned at frequent intervals. If the frequency of scan is not sufficiently high, the picture appears to flicker. At frequencies above a **critical frequency** no flicker is apparent. The critical frequency changes as a function of luminance, being higher for higher luminance. The basic requirement for monochrome video is that the **field frequency** (the rate at which images are presented) be above the critical frequency for the highest image luminances.

Images of objects in color video are distinguished from one another by luminance differences or by differences in hue or saturation. Exact reproduction in the image of the original scene differences is not necessary or even attainable. Nevertheless, some reasonable correspondence must prevail because the luminance gradation requirements for color are essentially the same as those for monochrome video.

Luminous Considerations in Visual Response

Vision is considered in terms of physical, psychophysical, and psychological quantities. The primary stimulus for vision is radiant energy. The study of this radiant energy in its various manifestations, including the effects on it of reflecting, refracting, and absorbing materials, is a study in physics. The response part of the visual process embodies the sensations and perceptions of seeing. Sensing and perceiving are mental operations and therefore belong to the field of psychology. Evaluation of radiant-energy stimuli in terms of the observer responses they evoke is within the realm of psychophysics. Because observer response sensations can be described only in terms of other sensations, psychophysical specifications of stimuli are made according to sensation equalities or differences.

1.8.4 Photometric Measurements

Evaluation of a radiant-energy stimulus in terms of its brightness-producing capacity is a photometric measurement. An instrument for making such measurements is called a *photometer*. In visual photometers, which must be used in obtaining basic photometric measurements, the two stimuli to be compared are normally directed into small adjacent parts of a viewing field. The stimulus to be evaluated is presented in the *test field*; the stimulus against which it is compared is presented in the *comparison field*. For most high-precision measurements the total size of the combined test and comparison fields is kept small, subtending about 2° at the eye. The area outside these fields is called the *surround*. Although the surround does not enter directly into the measurements, it has adaptation effects on the retina and thus affects the appearances of the test and comparison fields. It also influences the precision of measurement.

1.8.5 Luminosity Curve

A *luminosity curve* is a plot indicative of the relative brightnesses of spectrum colors of different wavelength or frequency. To a normal observer, the brightest part of a spectrum consisting of equal amounts of radiant flux per unit wavelength interval is at about 555 nm. Luminosity curves are therefore commonly normalized to have a value of *unity* at 555 nm. If, at some other wavelength, twice as much radiant flux as at 555 nm is required to obtain brightness equality with radiant flux at 555 nm, the luminosity at this wavelength is 0.5. The luminosity at any wavelength λ is, therefore, defined as the ratio P_{555}/P_λ, where P_λ denotes the amount of radiant flux at the wavelength λ, which is equal in brightness to a radiant flux of P_{555}.

The luminosity function that has been accepted as standard for photopic vision is given in Fig. 1.114. Tabulated values at 10-nm intervals are given in Table 1.32. This function was agreed on by the International Commission on Illumination (CIE) in 1924. It is based on considerable experimental work that was conducted over a number of years. Chief reliance in arriving at this function was based on the step-by-step equality-of-brightness method. Flicker photometry provided additional data.

In the scotopic range of intensities, the luminosity function is somewhat different from that of the photopic range. The two curves are compared in Fig. 1.115. Values are listed in Table 1.32. The two curves are similar in shape, but there is a shift for the scotopic curve of about 40 nm to the shorter wavelengths.

Measurements of luminosity in the scotopic range are usually made by the *threshold-of-vision* method. A single stimulus in a dark surround is used. The stimulus is presented to the observer at each of a number

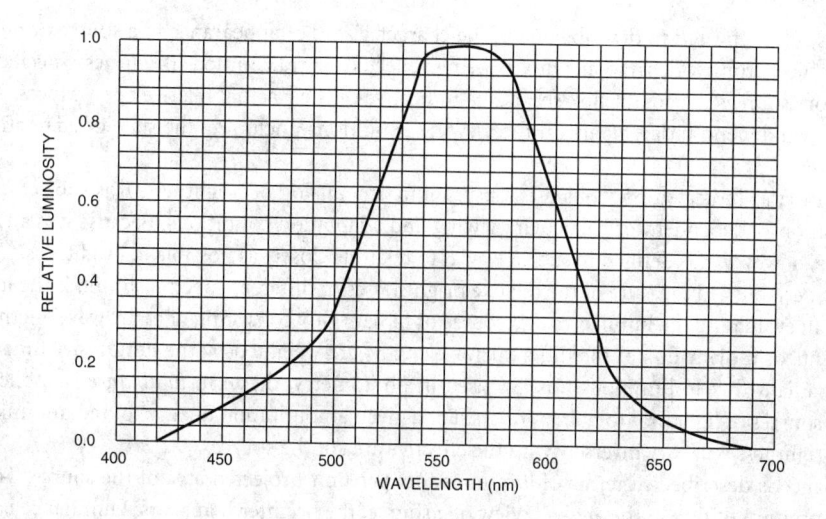

FIGURE 1.114 The photopic luminosity function. (*Source:* Adapted from Fink, D.G. 1952. *Television Engineering,* 2nd ed. McGraw-Hill, New York.)

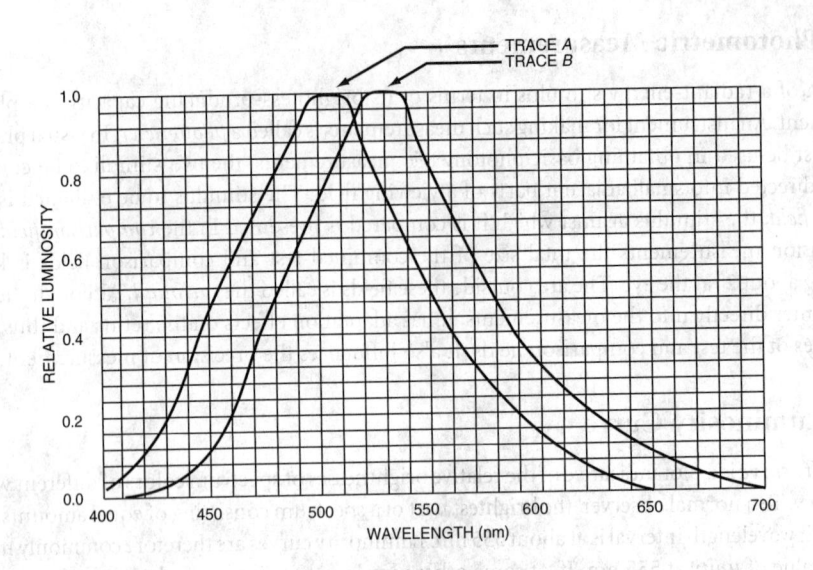

FIGURE 1.115   Scotopic luminosity function (trace *A*) as compared with photopic luminosity function (trace *B*). (*Source:* Adapted from Fink, D.G. 1952. *Television Engineering*, 2nd ed. McGraw-Hill, New York.)

of different intensities, ranging from well below the threshold to intensities sufficiently high to be definitely visible. Determinations are made of the amount of energy, at each chosen wavelength, that is reported visible by the observer a certain percentage of the time, such as 50%. The reciprocal of this amount of energy determines the relative luminosity at the given wavelength. The wavelength plot is normalized to have a maximum value of 1.00 to give the scotopic luminosity function.

In the intensity region between scotopic and photopic vision, called the *Purkinje* or *mesopic region*, the measured luminosity function takes on sets of values intermediate between those obtained for scotopic and photopic vision. Relative luminosities of colors within the mesopic region will therefore vary, depending on the particular intensity level at which the viewing takes place. Reds tend to become darker in approaching scotopic levels; greens and blues tend to become relatively lighter.

Luminance

Brightness is a term used to describe one of the characteristics of appearance of a source of radiant flux or of an object from which radiant flux is being reflected or transmitted. Brightness specifications of two or more sources of radiant flux should be indicative of their actual relative appearances. These appearances will depend in large part on the viewing conditions, including the state of adaptation of the observer's eye.

Luminance, as indicated previously, is a psychophysical analog of brightness. It is subject to physical determination, independent of particular viewing and adaptation conditions. Because it is an analog of brightness, however, it is defined in such a way as to relate as closely as possible to brightness.

The best established measure of the relative brightnesses of different spectral stimuli is the luminosity function. In evaluating the luminance of a source of radiant flux consisting of many wavelengths of light, the amounts of radiant flux at the different wavelengths are weighted by the luminosity function. This converts radiant flux to luminous flux. As used in photometry, the term *luminance* is applied only to extended sources of light, not to point sources. For a given amount (and quality) of radiant flux reaching the eye, brightness will vary inversely with the effective area of the source.

Luminance is described in terms of luminous flux per unit projected area of the source. The greater the concentration of flux in the angle of view of a source, the brighter it appears. Luminance is therefore expressed in terms of amounts of flux per unit solid angle or *steradian*.

In considering the relative luminances of various objects of a scene to be captured and reproduced by a video system, it is convenient to normalize the luminance values so that the white in the region of principal illumination has a relative luminance value of 1.00. The relative luminance of any other object then becomes the ratio of its luminance to that of the white. This white is an object of highly diffusing surface with high and uniform reflectance throughout the visible spectrum. For purposes of computation it may be idealized to have 100% reflectance and perfect diffusion.

Perception of Fine Detail

Detail is seen in an image because of brightness differences between small adjacent areas in a monochrome display or because of brightness, hue, or saturation differences in a color display. Visibility of detail in a picture is important because it determines the extent to which small or distant objects of a scene are visible, and because of its relationship to the **sharpness** appearance of the edges of objects.

Picture definition is probably the most acceptable term for describing the general characteristic of crispness, sharpness, or image-detail visibility in a picture. Picture definition depends on characteristics of the eye, such as visual acuity, and on a variety of characteristics of the picture-image medium, including its resolving power, luminance range, contrast, and image edge gradients.

Visual acuity may be measured in terms of the visual angle subtended by the smallest detail in an object that is visible. The *Landolt ring* is one type of test object frequently employed. The ring, which has a segment cut from it, is shown in any one of four orientations, with the opening at the top or bottom or on the right or left side. The observer identifies the location of this opening. The visual angle subtended by the opening that can be properly located 50% of the time is a measure of visual acuity.

Test-object illuminance, contrast between the test object and its background, time of viewing, and other factors greatly affect visual-acuity measurements. Up to a visual distance of about 20-ft (6-m) acuity is partially a function of distance, because of changes in shape of the eye lens in focusing. Beyond 20 ft it remains relatively constant. Visual acuity is highest for foveal vision, dropping off rapidly for retinal areas outside the fovea.

A black line on a light background is visible if it has a visual angle no greater than 0.5″. This is not, however, a true measure of visual acuity. For visual-acuity tests of the type described, normal vision, corresponding to a Snellen 20/20 rating, represents an angular discrimination of about 1′. Separations between adjacent cones in the fovea and resolving power limitations of the eye lens give theoretical visual-acuity values of about this same magnitude.

The extent to which a picture medium, such as a photographic or a video system, can reproduce fine detail is expressed in terms of *resolving power* or *resolution*. Resolution is a measure of the distance between two fine lines in the reproduced image that are visually distinct. The image is examined under the best possible conditions of viewing, including magnification.

Two types of test charts are commonly employed in determining resolving power, either a wedge of radial lines or groups of parallel lines at different pitches for each group. For either type of chart, the spaces between pairs of lines usually are made equal to the line widths. Figure 1.116 shows a test signal electronically generated by a video measuring test set.

Resolution in photography is usually expressed as the maximum number of lines (counting only the black ones or only the white ones) per millimeter that can be distinguished from one another. Measured values of resolving power depend on a number of factors in addition to the photographic material itself. The most important of these typically are:

- Density differences between the black and the white lines of the test chart photographed
- Sharpness of focus of the test-chart image during exposure
- Contrast to which the photographic image is developed
- Composition of the developer

Sharpness of focus depends on the general quality of the focusing lens, image and object distances from the lens, and the part of the projected field in which the image lies. In determining the resolving power

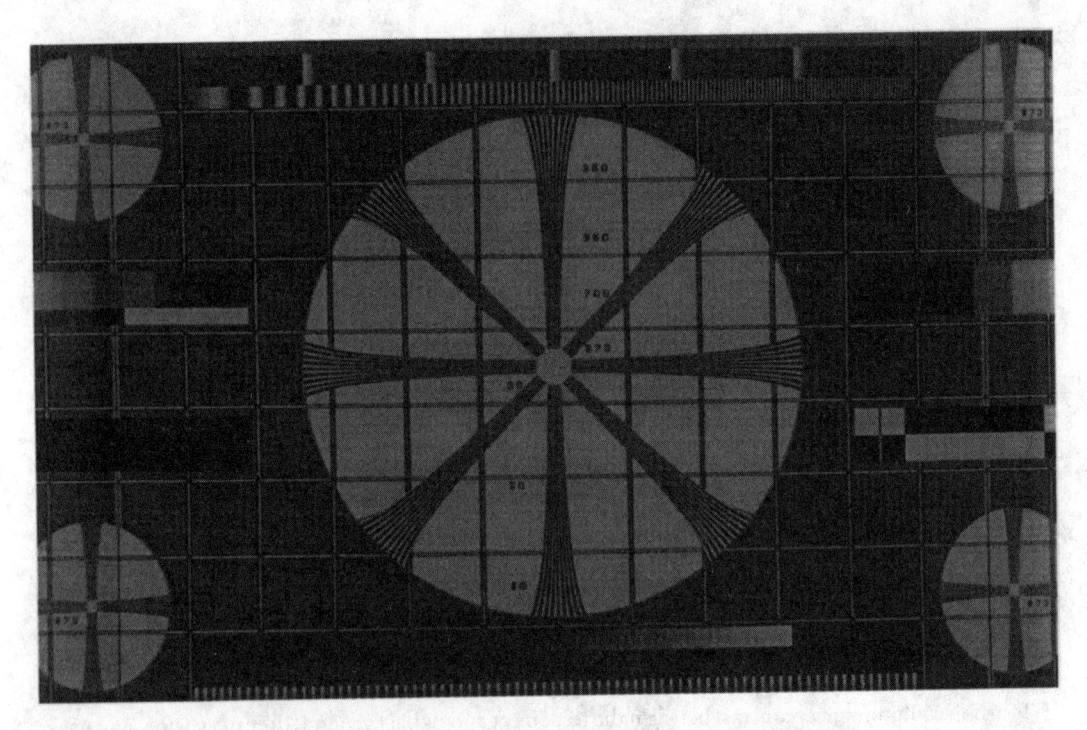

FIGURE 1.116 Test chart for high-definition television applications produced by a signal waveform generator. The electronically produced pattern is used to check resolution, geometry, bandwidth, and color reproduction. (*Source: Tektronix.*)

of a photographic negative or positive material, a test chart is generally employed that has a high-density difference, such as 3.0, between the black-and-white lines. A high-quality lens is used, the projected field is limited, and focusing is critically adjusted. Under these conditions, ordinary black-and-white photographic materials generally have resolving powers in the range of 30–200 line-pairs/mm. Special photographic materials are available with resolving powers greater than 1000 line-pairs/mm.

Resolution in a video system is expressed in terms of the maximum number of lines, counting both black and white, that are discernible in viewing a test chart. The value of horizontal (vertical lines) or vertical (horizontal lines) resolution is the number of lines equal to the vertical dimension of the raster. Vertical resolution in a well-adjusted system equals the number of scanning lines, roughly 500 in conventional television. In normal broadcasting and reception practice, however, typical values of vertical resolution range from 350–400 lines. The theoretical limiting value for horizontal resolution (R_H) in a 525 line, 30-Hz frame rate system is given by

$$R_H = \frac{2 \times 0.75 \times \Delta f}{30 \times 525}$$

$$R_H = 0.954 \times 10^{-4} \times \Delta f$$

Where Δf is the available bandwidth frequency in Hz. The constants 30 and 525 represent the frame and line frequencies, respectively in the conventional U.S. vision system. A factor of 2 is introduced because in one complete cycle both a black and a white line are obtainable. The factor 0.75 is necessary because of the receiver aspect ratio; the picture height is three-fourths of the picture width. There is an additional reduction of about 15% (not included in the equation) in the theoretical value because of horizontal blanking time during which retrace takes place. A bandwidth of 4.25 MHz—typically that of the conventional terrestrial television system—thus makes possible a maximum resolution of about 345 lines.

Sharpness

The appearance evaluation of a picture image in terms of the edge characteristics of objects is called sharpness. The more clearly defined the line that separates dark areas from lighter ones, the greater the sharpness of the picture. Sharpness is, naturally, related to the transient curve in the image across an edge. The average gradient and the total density difference appear to be the most important characteristics. No physical measure has been devised, however, that in all cases will predict the sharpness (appearance) of an image.

Picture resolution and sharpness are to some extent interrelated but they are by no means perfectly correlated. Pictures ranked according to resolution measures may be rated somewhat differently on the basis of sharpness. Both resolution and sharpness are related to the more general characteristic of picture definition. For pictures in which, under particular viewing conditions, effective resolution is limited by the visual acuity of the eye rather than by picture resolution, sharpness is probably a good indication of picture definition. If visual acuity is not the limiting factor, however, picture definition depends to an appreciable extent on both resolution and sharpness.

1.8.6   Human Visual System

The human visual system (HVS) is powerful and exceeds the performance of artificial visual systems in almost all areas of comparison. Researchers have, therefore, studied the HVS extensively to ascertain the most efficient and effective methods of communicating information to the eye. An important component of this work has been the development of models of how humans see in an effort to improve image processing systems.

1.8.7   A Model for Image Quality

The classic approach to image quality assessment involves the presentation to a group of test subjects visual test material for evaluation and rating. The test material may include side-by-side display comparisons, or a variety of perception-threshold presentations. One common visual comparison technique is called the *pair-comparison* method. A number of observers are asked to view a specified number of images at two or more distances. At each distance, the subjects are asked to rank the order of the images in terms of overall quality, clearness, and personal preference.

An image acquisition, storage, transmission, and display system need not present more visual information to the viewer than the viewer can process. For this reason, image quality assessment is an important element in the development of any new video system. For example, in the development of a video compressing algorithm, the designer needs to know at what compression point impairments are visible to the average viewer.

Evaluation by human subjects, although an important part of this process, is also expensive and time consuming. Numerous efforts have been made to reduce the human visual system and its interaction with a display device to one or more mathematical models. (Some of this research is cited in Grogan (1992), Barten (1992), Daly (1992), Reese (1992), and Martin (1992).) The benefits of this research to design engineers is obvious: more timely feedback on new product offerings. In the development of advanced displays, it is important to evaluate the visual performance of the system well in advance of the expensive fabrication process. Display design software, therefore, is valuable. Because image quality is the primary goal in many display designs, tools that permit the engineer to analyze and emulate the displayed image assist in the design process. Such tools provide early feedback on new display techniques and permit a wider range of prospective products to be evaluated.

After the system or algorithm has successfully passed the minimum criteria established by the model, it can be subjected to human evaluation. The model simulation requires the selection of many interrelated parameters. A series of experiments is typically conducted to improve the model to more closely approximate human visual perception.

Defining Terms

Brightness: A term used to describe one of the characteristics of appearance of a source of radiant flux or of an object from which radiant flux is being reflected or transmitted.

Critical frequency: The rate of picture presentation, as in a video system or motion picture display, above which the presented image ceases to give the appearance of flickering. The critical frequency changes as a function of luminance, being higher for higher luminance.

Dispersion: The variation of refraction as a function of wavelength.

Energy distribution curve: A plot of the power distribution of a source of light giving the energy radiated at each wavelength per nanometer of wavelength.

Field frequency: The rate at which images in an electronic imaging system are presented. In conventional terrestrial television (National Televison Systems Committee (NTSC)), two fields are presented to make up one *frame* (a complete picture).

Refraction: The bending of light as it passes from one medium to another, such as from air to glass, at an angle.

Resolution: The extent to which an imaging system, such as a photographic or a video system, can reproduce fine detail.

Sharpness: The appearance evaluation of a picture image in terms of the edge characteristics of objects contained therein. The more clearly defined the lines separating dark areas from light ones, the greater the sharpness of the picture.

References

Barten, P.G.J. 1992. Physical model for the contrast sensitivity of the human eye. In *Human Vision, Visual Processing, and Digital Display III*, ed. B.E. Rogowitz, pp. 57–72. Proc. SPIE 1666.

Benson, K.B. and Whitaker, J.C. 1991. *Television Engineering Handbook*, revised ed. McGraw-Hill, New York.

Boynton, R.M. 1979. *Human Color Vision*. Holt, New York.

Committee on Colorimetry. 1953. *The Science of Color*. Optical Society of America, New York.

Daly, S. 1992. The visible differences predictor: An algorithm for the assessment of image fidelity. In *Human Vision, Visual Processing, and Digital Display III*, ed. B.E. Rogowitz, pp. 2–15. Proc. SPIE 1666.

Davson, H. 1980. *Physiology of the Eye*, 4th ed. Academic, New York.

Evans, R.M., Hanson, W.T., Jr., and Brewer, W.L. 1952. *Principles of Color Photography*. Wiley, New York.

Fink, D.G. 1952. *Television Engineering*, 2nd ed. McGraw-Hill, New York.

Grogan, T.A. 1992. Image evaluation with a contour-based perceptual model. *Human Vision, Visual Processing, and Digital Display III*, ed. B.E. Rogowitz, pp. 188–197. Proc. SPIE 1666.

IES. 1981. *IES Handbook*, Illuminating Society of North America, New York.

Kingslake, R., ed. 1965. *Applied Optics and Optical Engineering*, Vol. 1. Academic, New York.

Martin, R.A., Ahumanda, A.J., Jr., James O., and Larimer, J.O. 1992. Color matrix display simulation based upon luminance and chromatic contrast sensitivity of early vision. *Human Vision, Visual Processing, and Digital Display III*, ed. B.E. Rogowitz, pp. 336–342. Proc. SPIE 1666.

Polysak, S.L. 1941. *The Retina*. University of Chicago Press, Chicago, IL.

Reese, G. 1992. Enhancing images with intensity-dependent spread functions. *Human Vision, Visual Processing, and Digital Display III*, ed. B.E. Rogowitz, pp. 253–261. Proc. SPIE 1666.

Richards, C.J. 1973. *Electronic Displays and Data Systems: Constructional Practice*. McGraw-Hill, New York.

Schade, O.H. 1948. Electro-optical characteristics of television systems. *RCA Review* 9:5–37, 245–286, 490–530, 653–686.

Whitaker, J.C. 1993. *Electronic Displays: Technology, Design and Applications*. McGraw-Hill, New York.

Wright, W.D. 1947. *Researches on Normal and Defective Colour Vision*. Mosby, St. Louis, MO.

Wright, W.D. 1969. *The Measurement of Colour*, 4th ed. Adam Hilger, London.

Further Information

The SPIE (Bellingham, Washington) offers a number of publications examining the characteristics of the human visual system. The organization also conducts technical seminars on various topics relating to optics and the application of optical technologies.

The following books are recommended for further reading:

Jerry Whitaker and K. BlairBenson: *Standard Handbook of Video and Television Engineering*, 4th ed., McGraw-Hill, New York, N.Y., 2003.

Jerry Whitaker: *Video Displays*, McGraw-Hill, New York, N.Y., 2000.

2

Properties of Materials and Components

John Choma, Jr.
University of Southern California

Stuart K. Burgess
University of Southern California

Stuart K. Tewksbury
Stevens Institute of Technology

David F. Besch
University of the Pacific

Jan H. Mooij
Philips Components

Martin R. Parker
University of Alabama

William E. Webb
University of Alabama

Igor M. Filanovsky
University of Alberta

James F. Shackelford
University of California

2.1 Circuit Fundamentals

John Choma, Jr. and Stuart K. Burgess

2.1.1 The Electrical Circuit

An **electrical circuit** or *electrical network* is an array of interconnected elements wired so as to be capable of conducting current. The fundamental *two-terminal elements* of an electrical circuit are the **resistor**, the **capacitor**, the **inductor**, the **voltage** *source*, and the **current** *source*. The circuit schematic symbols of these elements, together with the algebraic symbols used to denote their respective general values, appear in Fig. 2.1(a) through Fig. 2.1(e).

The value of a resistor is its resistance R, the dimensional of which units are *ohms*. The case of a wire used to interconnect the terminals of two electrical elements corresponds to the special case of a resistor, the resistance of which is ideally zero ohms; that is, $R = 0$. For the capacitor in Fig. 2.1(b), the capacitance C, has units of *farads*, and in Fig. 2.1(c), the value of an inductor is its inductance L, the dimensions of which are *henries*. In the case of the voltage sources depicted in Fig. 2.1(d), a constant, time-invariant source of voltage, or *battery*, is distinguished from a voltage source, the voltage value of which varies with time. The latter type of voltage source is often referred to as a *time-varying signal* or simply, a *signal*. In either case, the value of the battery voltage, E, and the time-varying signal, $v(t)$, is in units of *volts*. Finally, the current source of Fig. 2.1(e) has a value I, the value of which is expressed in *amperes*, which is typically abbreviated to *amps*.

Elements having three, four, or more than four terminals can also appear in practical electrical networks. The discrete component *bipolar junction transistor* (BJT), which is schematically portrayed in Fig. 2.2(a), is an example of a three-terminal element, where the three terminals at issue are the collector, the base, and the emitter. On the other hand, the monolithic *metal-oxide-semiconductor field-effect transistor* (MOSFET) depicted in Fig. 2.2(b) has four terminals: the drain, the gate, the source, and the bulk substrate.

Multiterminal elements appearing in circuits are routinely represented, or *modeled*, by equivalent subcircuits formed of only interconnected two-terminal elements. Such a representation is always possible, provided that the list of two-terminal elements itemized in Fig. 2.1 is appended by an additional type of two-terminal element known as the **controlled source**, or *dependent generator*. There are four types of controlled sources; two are voltage sources and two are current sources. In Fig. 2.3(a), the dependent generator is a *voltage-controlled voltage source* (VCVS) in that the voltage $v_o(t)$, developed from terminal 3 to terminal 4, is a function of, and is therefore dependent on, the voltage $v_i(t)$ established elsewhere in

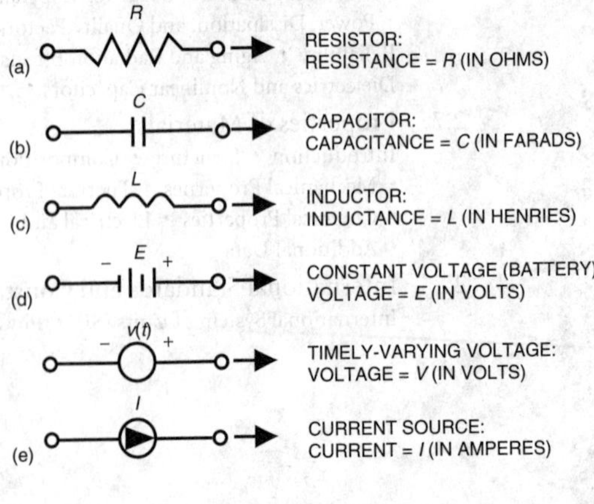

FIGURE 2.1 Circuit schematic symbol and corresponding value notation for: (a) resistor, (b) capacitor, (c) inductor, (d) voltage source, and (e) current source. Note that a constant voltage source, or battery, is distinguished from a voltage source the value of which varies with time.

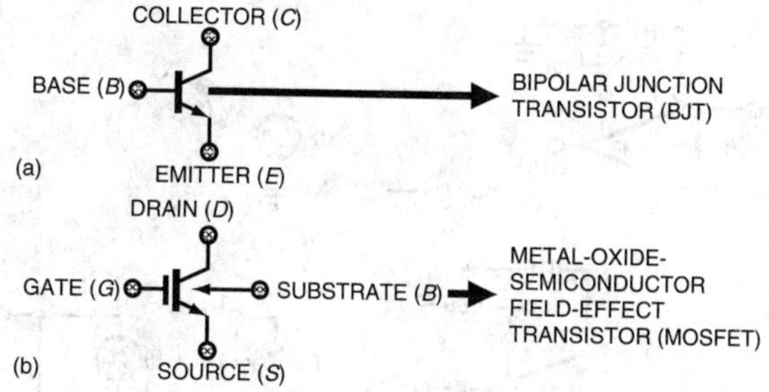

FIGURE 2.2 Circuit schematic symbol for: (a) discrete component bipolar junction transistor (BJT), (b) monolithic metal-oxide-semiconductor field-effect transistor (MOSFET).

the network from terminal 1 to terminal 2. The *controlled voltage* $v_o(t)$, as well as the *controlling voltage* $v_i(t)$, can be constant or time varying. Regardless of the time-domain nature of these two voltages, the value of $v_o(t)$ is not an independent number. Instead, its value is determined by $v_i(t)$ in accordance with a prescribed functional relationship, say,

$$v_o(t) = f[v_i(t)] \qquad (2.1)$$

If the function $f(\cdot)$ is linearly related to its argument, then Eq. (2.1) collapses to the form

$$v_o(t) = f_\mu v_i(t) \qquad (2.2)$$

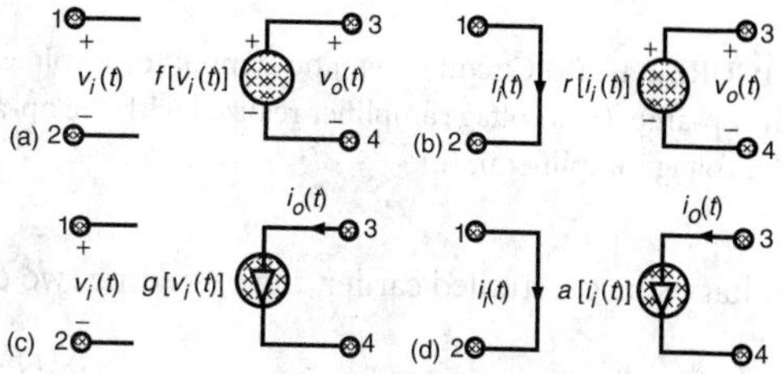

FIGURE 2.3 Circuit schematic symbol for: (a) voltage-controlled voltage source (VCVS), (b) current-controlled voltage source (CCVS), (c) voltage-controlled current source (VCCS), (d) current-controlled current source (CCCS).

where f_μ is a constant, independent of either $v_o(t)$ or $v_i(t)$. When the function on the right-hand side of Eq. (2.1) is linear, the subject VCVS becomes known as a *linear voltage-controlled voltage source*.

The second type of controlled voltage source is *the current-controlled voltage source* (CCVS) depicted in Fig. 2.3(b). In this dependent generator, the controlled voltage $v_o(t)$ developed from terminal 3 to terminal 4 is a function of the *controlling current* $i_i(t)$, flowing in the network between terminals 1 and 2, as indicated. In this case, the generalized functional dependence of $v_o(t)$ on $i_i(t)$ is expressible as

$$v_o(t) = r[i_i(t)] \qquad (2.3)$$

which reduces to

$$v_o(t) = r_m i_i(t) \qquad (2.4)$$

when $r(\cdot)$ is a linear function of its argument.

The two types of dependent current sources are diagrammed symbolically in Fig. 2.3(c) and Fig. 2.3(d). Figure 2.3(c) depicts a *voltage-controlled current source* (VCCS), for which the controlled current $i_o(t)$ flowing in the electrical path from terminal 3 to terminal 4 is determined by the controlling voltage $v_i(t)$ established across terminals 1 and 2. The controlled current can, therefore, be written as

$$i_o(t) = g[v_i(t)] \qquad (2.5)$$

In the *current-controlled current source* (CCCS) of Fig. 2.3(d),

$$i_o(t) = a[i_i(t)] \qquad (2.6)$$

where the controlled current $i_o(t)$ flowing from terminal 3 to terminal 4 is a function of the controlling current $i_i(t)$ flowing in the circuit from terminal 1 to terminal 2. As is the case with the two controlled

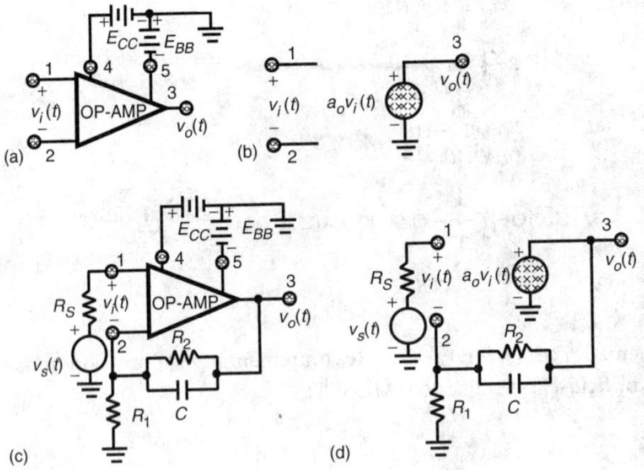

FIGURE 2.4 (a) Circuit schematic symbol for a voltage mode operational amplifier, (b) first order linear model of the op-amp, (c) a voltage amplifier realized with the op-amp functioning as the gain element, (d) equivalent circuit of the voltage amplifier in (c).

voltage sources studied earlier, the preceding two equations collapse to the linear relationships,

$$i_o(t) = g_m v_i(t) \tag{2.7}$$

and

$$i_o(t) = a_\alpha i_i(t) \tag{2.8}$$

when $g(\cdot)$ and $a(\cdot)$, are linear functions of their arguments, respectively.

The immediate implication of the controlled source concept is that the definition for an electrical circuit given at the beginning of this subsection can be revised to read: an electrical circuit or electrical network is an array of *interconnected two-terminal elements* wired in such a way as to be capable of conducting current. Implicit to this revised definition is the understanding that the two-terminal elements allowed in an electrical circuit are the resistor, the capacitor, the inductor, the voltage source, the current source, and any of the four possible types of dependent generators.

In an attempt to reinforce the engineering utility of the foregoing definition, consider the voltage mode *operational amplifier (op-amp)*, whose circuit schematic symbol is shown in Fig. 2.4(a). Observe that the op-amp is a five-terminal element. Two terminals, labeled 1 and 2, are provided to receive input signals that derive either from external signal sources or from the output terminals of subcircuits that feed back a designable fraction of the output signal established between terminal 3 and the system ground. Battery voltages, identified as E_{CC} and E_{BB} in the figure, are applied to the remaining two op-amp terminals (terminals 4 and 5) with respect to ground to *bias* or *activate* the op-amp for its intended application. When E_{CC} and E_{BB} are selected to ensure that the subject operational amplifier behaves as a linear circuit element, the linearized, or small-signal, characteristics of the op-amp in Fig. 2.4(a) can be modeled by the electrical circuit appearing in Fig. 2.4(b). Thus, the voltage amplifier of Fig. 2.4(c), which interconnects two batteries, a signal source voltage, three resistors, a capacitor, and an op-amp, can be represented by the network given in Fig. 2.4(d). Note that the latter configuration uses only two terminal elements, one of which is a voltage-controlled voltage source.

2.1.2 Current and Current Polarity

The concept of an electrical *current* is implicit to the definition of an electrical circuit in that a circuit is said to be an array of two terminal elements that are connected in such a way as to permit the *conduction of current*. Current flow through an element that is capable of current conduction requires that a net nonzero charge, say, $q(t)$, be transferred over finite time across any cross-sectional area of the element. The current $i(t)$ that actually flows is the time rate of change of this transferred charge; that is,

$$i(t) = \frac{dq(t)}{dt} \qquad (2.9)$$

where the MKS unit of charge is the *coulomb*, time t is measured in seconds, and the resultant current is in units of amperes. Note that zero current does not necessarily imply a lack of charge at a given cross section of a conductive element. Rather, zero current implies only that the subject charge is not changing with time; that is, the charge is not moving through the elemental cross section.

Electrical charge can be negative, as in the case of *electrons* transported through a cross section of a conductive element such as aluminum or copper. A single electron has a charge of $-(1.602 \times 10^{-19})$ C. Thus, Eq. (2.9) implies a need to transport an average of (6.242×10^{18}) electrons in 1 s through a cross section of aluminum if the aluminum element is to conduct a constant current of 1 A. Charge can also be positive, as in the case of *holes* transported through a cross section of a semiconductor such as germanium or silicon. To first order, the electrical charge of a hole is the negative of the charge of an electron, which implies that the charge of a hole is $+(1.602 \times 10^{-19})$ C.

A positive charge, $q(t)$, transported from the left of the cross section to the right of the cross section in the element abstracted in Fig. 2.5(a) gives rise to a positive current, $i(t)$, that also flows from left to right across the indicated cross section. Assume that prior to the transport of such charge, the volumes to the left and to the right of the cross section are electrically neutral; that is, these volumes have zero initial net charge. Then, the transport of a positive charge, say, q_o, from the left side to the right side of the element charges the right side to $+q_o$ and the left side to $-q_o$.

Alternatively, suppose a negative charge in the amount of $-q_o$ is transported from the right side of the element to its left side, as suggested in Fig. 2.5(b). Then, the left side charges to $-q_o$ and the right side charges to $+q_o$, which is identical to the electrostatic condition incurred by the transport of a positive charge in the amount of q_o from left- to right-hand sides. Resultantly, the transport of a net negative

FIGURE 2.5 (a) Transport of a positive charge from the left-hand side to the right-hand side of an arbitrary cross section of a conductive element, (b) transport of a negative charge from the right-hand side to the left-hand side of an arbitrary cross section of a conductive element, (c) transport of positive or negative charges from either side to the other side of an arbitrary cross section of a conductive element.

charge from right to left produces a positive current, $i(t)$, flowing from left to right, just as positive charge transported from left- to right-hand sides induces a current flow from left to right.

Assume, as portrayed in Fig. 2.5(c), that a positive or a negative charge, say, $q_1(t)$, is transported from the left side of the indicated cross section to the right side. Simultaneously, a positive or a negative charge in the amount of $q_2(t)$ is directed through the cross section from right to left. If $i_1(t)$ is the current arising from the transport of the charge $q_1(t)$, and if $i_2(t)$ denotes the current corresponding to the transport of the charge $q_2(t)$, the net effective current $i_e(t)$ flowing from the left side of the cross section to the right side of the cross section is

$$i_e(t) = \frac{d}{dt}[q_1(t) - q_2(t)] = i_1(t) - i_2(t) \qquad (2.10)$$

where the charge difference, $(q_1(t) - q_2(t))$, represents the net charge transported from left to right. Observe that if $q_1(t) \equiv q_2(t)$, the net effective current is zero even though conceivably large numbers of charges are transported to and fro across the junction.

2.1.3 Energy and Voltage

The preceding section highlights the fact that the flow of current through a conductive electrical element mandates that a net charge be transported over finite time across any arbitrary cross section of that element. The electrical effect of this charge transport is a net positive charge induced on one side of the element in question and a net negative charge (equal in magnitude to the aforementioned positive charge) mirrored on the other side of the element. But this ramification conflicts with the observable electrical properties of an element in equilibrium. In particular, an element sitting in free space, without any electrical connection to a source of *energy*, is necessarily in equilibrium in the sense that the net positive charge in any volume of the element is precisely counteracted by an equal amount of charge of opposite sign in said volume. Thus, if none of the elements abstracted in Fig. 2.5 is connected to an external source of energy, it is impossible to achieve the indicated electrical charge differential that materializes across an arbitrary cross section of the element when charge is transferred from one side of the cross section to the other.

The energy commensurate with sustaining current flow through an electrical element derives from the application of a voltage, say, $v(t)$, across the element in question. Equivalently, the application of electrical energy to an element manifests itself as a voltage developed across the terminals of an element to which energy is supplied. The amount of applied voltage $v(t)$ required to sustain the flow of current, $i(t)$, as diagrammed in Fig. 2.6(a), is precisely the voltage required to offset the electrostatic implications of the differential charge induced across the element through which $i(t)$ flows. This is to say that without the connection of the voltage $v(t)$ to the element in Fig. 2.6(a), the element cannot be in equilibrium. With $v(t)$ connected, equilibrium for the entire system comprising element and voltage source is re-established by allowing for the conduction of the current $i(t)$.

Instead of viewing the delivery of energy to an electrical element as a voltage source applied to the element, this energy delivery may be interpreted as the upshot of a current source used to excite the element, as depicted in Fig. 2.6(b). This interpretation follows from the fact that energy must be applied in an amount that effects charge transport at a desired time rate of change. It follows that the application of a current source in the amount of the desired current is necessarily in one-to-one correspondence with the voltage required to offset the charge differential manifested by the charge transport that yields the subject current. To be sure, a voltage source is a physical entity, whereas a current source is not. But the mathematical modeling of energy delivery to an electrical element can be accomplished through either a voltage source or a current source.

FIGURE 2.6 (a) The application of energy in the form of a voltage applied to an element that is made to conduct a specified current, the applied voltage $v(t)$ causes the current $i(t)$ to flow, (b) the application of energy in the form of a current applied to an element that is made to establish a specified terminal voltage, the applied current $i(t)$ causes the voltage $v(t)$ to be developed across the terminals of the electrical element.

In Fig. 2.6, the terminal voltage $v(t)$ corresponding to the applied energy, say, $w(t)$, that is required to transfer an amount of charge $q(t)$ across an arbitrary cross section of the element is

$$v(t) = \frac{dw(t)}{dq(t)} \tag{2.11}$$

where $v(t)$ is in volts when $q(t)$ is in coulombs and $w(t)$ is in joules. Thus, if 1 J of applied energy results in the transport of 1 C of charge through an element, the elemental terminal voltage manifested by the 1 J of applied energy is 1 V.

It should be understood that the derivative on the right-hand side of Eq. (2.11), and thus the terminal voltage demanded of an element that is transporting a certain amount of charge through its cross section, is a function of the properties of the type of material from which the element undergoing study is fabricated. For example, an insulator, such as paper, air, or silicon dioxide, is incapable of current conduction and, hence, intrinsic charge transport. Thus, $q(t)$ is essentially zero in an insulator and by Eq. (2.11), an infinitely large terminal voltage is mandated for even the smallest possible current. In a conductor, such as aluminum, iron, or copper, large amounts of charge can be transported for very small applied energies. Accordingly, the requisite terminal voltage for even very large currents approaches zero in ideal conductors. The electrical properties of semiconductors, such as germanium, silicon, and gallium arsenide, lie between the extremes of those for an insulator and a conductor. In particular, semiconductor elements behave as insulators when their terminals are subjected to small voltages, whereas progressively larger terminal voltages render the electrical behavior of semiconductors akin to conductors. This conditionally conductive property of a semiconductor explains why semiconductor devices and circuits must be biased to appropriate voltage levels before these devices and circuits can function in accordance with their design requirements.

2.1.4 Power

The foregoing material underscores the fact that the flow of current through a two terminal element requires that charge be transported over time through any cross section of that element or network. In turn, such charge transport requires that energy be supplied to the network, usually through the application of an external voltage source. The time rate of change of this applied energy is the **power** delivered by the external voltage or current source to the network in question. If $p(t)$ denotes this power in watts,

$$p(t) = \frac{dw(t)}{dt} \tag{2.12}$$

where, of course, $w(t)$ is the energy supplied to the network in joules. By rewriting Eq. (2.12) in the form

$$p(t) = \frac{dw(t)}{dq(t)} \frac{dq(t)}{dt} \tag{2.13}$$

and applying Eqs. (2.9) and (2.11), the power supplied to the two terminals of an element or a network becomes the more expedient relationship,

$$p(t) = v(t)i(t) \tag{2.14}$$

Equation (2.14) expresses the power delivered to an element as a simple product of the voltage applied across the terminals of the element and the resultant current conducted by that element. But care must be exercised when applying Eq. (2.14) to practical circuits.

To this end, it is useful to revisit the simple abstraction of Fig. 2.6(a), which is redrawn in slightly modified form in Fig. 2.7. In this circuit, a signal source voltage $v_s(t)$ is applied across the two terminals, terminals 1 and 2, of an element, which responds by conducting a current $i(t)$ from terminal 1 to terminal 2 and developing a corresponding terminal voltage $v(t)$, as shown. If the wires that connect the signal source to the element are ideal zero resistance conductors, the voltage $v(t)$ is identical to $v_s(t)$. Moreover, since the current is manifested by the application of the signal source, which thereby establishes a closed electrical path for current conduction, the element current, $i(t)$, is necessarily the same as the current, $i_s(t)$, that flows through $v_s(t)$.

If attention is focused on only the element in Fig. 2.7, it is natural to presume that the current conducted by the element actually flows from terminal 1 to terminal 2 when, as shown, the voltage

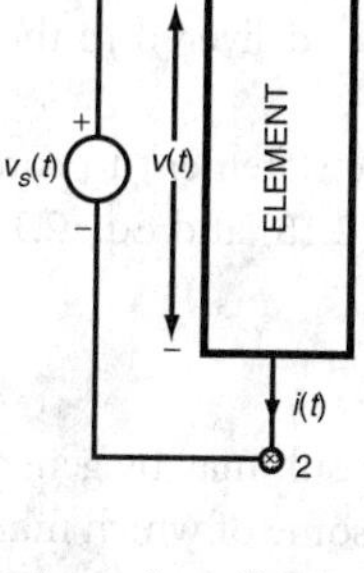

FIGURE 2.7 Circuit used to illustrate power calculations and the associated reference polarity convention.

developed across the element is positive at terminal 1 with respect to terminal 2. This assertion might be rationalized qualitatively by noting that the positive voltage nature at terminal 1 acts to repel positive charges from terminal 1 to terminal 2, where the negative nature of the developed voltage $v(t)$ tends to attract the repulsed positive charges. Similarly, the positive nature of the voltage at terminal 1 serves to attract negative charges from terminal 2, where the negative nature of $v(t)$ tends to repel such negative charges. Since current flows in the direction of transported positive charge and opposite to the direction of transported negative charge, either interpretation gives rise to an elemental current $i(t)$ that flows from terminal 1 to terminal 2. In general, if current is indicated as flowing from the *high* (+) voltage terminal to the *low* (−) voltage terminal of an element, the current conducted by the element and the voltage developed across the element to cause this flow of current are said to be in **associated reference polarity**. When the element current $i(t)$ and the corresponding element voltage $v(t)$, as exploited in the defining power relationship of Eq. (2.14), are in associated reference polarity, the resultantly computed power is a positive number and is said to represent the *power delivered to* the element. In contrast, $v(t)$ and $i(t)$ are said to be in *disassociated reference polarity* when $i(t)$ flows from the low voltage terminal of the element to its high voltage terminal. In this case, the voltage-current product in Eq. (2.14) is a negative number. Rather than stating that the resultant negative power is delivered to the element, it is more meaningful to assert that the computed negative power is a *positive power that is generated by* the element in question.

At first glance, it may appear as though polarity disassociation between element voltage and current variables is an impossible circumstance. Not only is polarity disassociation possible, it is absolutely necessary if electrical circuits are to subscribe to the fundamental principle of *conservation of power*. This principle states that the net power dissipated by a circuit must be identical to the net power supplied to that circuit. A confirmation of this basic principle derives from a further consideration of the topology in Fig. 2.7. The electrical variables, $v(t)$ and $i(t)$, pertinent to the element delineated in this circuit are in associated reference polarity. Accordingly, the power $p_e(t)$ dissipated by this element is positive and given by Eq. (2.14); namely,

$$p_e(t) = v(t)i(t) \tag{2.15}$$

However, the voltage and current variables, $v_s(t)$ and $i_s(t)$, relative to the signal source voltage are in disassociated polarity. It follows that the power $p_s(t)$ *delivered to the signal source* is

$$p_s(t) = -v_s(t)i_s(t) \tag{2.16}$$

Since, as previously stated, $v_s(t) = v(t)$ and $i_s(t) = i(t)$ for the circuit at hand, Eq. (2.16) can be written as

$$p_s(t) = -v(t)i(t) \tag{2.17}$$

The last result implies that the

$$\text{power delivered by the signal source} = +v(t)i(t) \equiv p_e(t) \tag{2.18}$$

that is, the power delivered to the element by the signal source is equal to the power dissipated by the element.

An alternative statement to conservation of power, as applied to the circuit in Fig. 2.7, derives from combining Eq. (2.15) and Eq. (2.17) to arrive at

$$p_s(t) + p_e(t) = 0 \tag{2.19}$$

The foregoing result may be generalized to the case of a circuit comprising an electrical interconnection of N elements, some of which may be voltage and current sources. Let the voltage across the kth element be $v_k(t)$, and let the current flowing through this kth element, in associated reference polarity with $v_k(t)$, be $i_k(t)$. Then, the power, say, $p_k(t)$, delivered to the kth electrical element is $v_k(t)i_k(t)$. By conservation of power,

$$\sum_{k=1}^{N} p_k(t) = \sum_{k=1}^{N} v_k(t)i_k(t) = 0 \tag{2.20}$$

The satisfaction of this expression requires that at least one of the $p_k(t)$ be negative or, equivalently, at least one of the N elements embedded in the circuit at hand must be a source of energy.

2.1.5 Circuit Classifications

It was pointed out earlier that the relationship between the current that is made to flow through an electrical element and the applied energy, and thus voltage, that is required to sustain such current flow is dictated by the material from which the subject element is fabricated. The elemental material and the associated manufacturing methods exploited to realize a particular type of circuit element determine the mathematical nature between the voltage applied across the terminals of the element and the resultant current flowing through the element. To this end, electrical elements and circuits in which they are embedded are generally codified as linear or nonlinear, **active** or **passive**, time varying or time invariant, and **lumped** or **distributed**.

Linear Versus Nonlinear

A *linear two-terminal circuit element* is one for which the voltage developed across, and the current flowing through, are related to one another by a linear algebraic or a linear integrodifferential equation. If the relationship between terminal voltage and corresponding current is nonlinear, the element is said to be *nonlinear*. A linear circuit contains only linear circuit elements, whereas a circuit is said to be nonlinear if at least one of its embedded electrical elements is nonlinear.

All practical circuit elements, and thus all practical electrical networks, are inherently nonlinear. But over suitably restricted ranges of applied voltages and corresponding currents, the volt–ampere characteristics of these elements and networks emulate idealized linear relationships. In the design of an electronic linear signal processor, such as an amplifier, an implicit engineering task is the implementation of biasing subcircuits that constrain the voltages and currents of internal semiconductor elements to ranges that ensure linear elemental behavior over all possible operating conditions.

The voltage–current relationship for the linear resistor offered in Fig. 2.8(a) is

$$v(t) = Ri(t) \tag{2.21}$$

where the voltage $v(t)$ appearing across the terminals of the resistor and the resultant current $i(t)$ conducted by the resistor are in associated reference polarity. The resistance R is independent of either $v(t)$ or $i(t)$. From Eq. (2.14), the dissipated resistor power, which is manifested in the form of heat, is

FIGURE 2.8 Circuit schematic symbol and corresponding voltage and current notation for a: (a) linear resistor, (b) linear capacitor, (c) linear inductor.

$$p_r(t) = i^2(t)R = \frac{v^2(t)}{R} \tag{2.22}$$

The linear capacitor and the linear inductor, whose schematic symbols appear, respectively, in Figs. 2.8(b) and 2.8(c), store energy, as opposed to dissipating power. Their volt–ampere equations are the linear relationships,

$$i(t) = C\frac{dv(t)}{dt} \tag{2.23}$$

for the capacitor, whereas for the inductor in Fig. 2.8(c),

$$v(t) = L\frac{di(t)}{dt} \tag{2.24}$$

Observe from Eqs. (2.23) and (2.14) that the power $p_c(t)$ delivered to the linear capacitor is

$$p_c(t) = v(t)i(t) = Cv(t)\frac{dv(t)}{dt} \tag{2.25}$$

From Eq. (2.12), this power is related to the energy, say, $w_c(t)$, stored in the form of charge deposited on the plates of the capacitor by

$$Cv(t)dv(t) = dw_c(t) \tag{2.26}$$

It follows that the energy delivered to, and hence stored in, the capacitor from time $t = 0$ to time t is

$$w_c(t) = \frac{1}{2}Cv^2(t) \tag{2.27}$$

It should be noted that this stored energy, like the energy associated with a signal source or a battery voltage, is available to supply power to other elements in the network in which the capacitor is embedded. For example, if very little current is conducted by the capacitor in question, Eq. (2.23) implies that the voltage across the capacitor is essentially constant. Recall that an element whose terminal voltage is a constant and in which energy is stored and, therefore, available for use, behaves as a battery.

If the preceding analysis is repeated for the inductor of Fig. 2.8(c), it can be shown that the energy $w_l(t)$ stored in the inductive element from time $t = 0$ to time t is

$$w_l(t) = \frac{1}{2}Li^2(t) \tag{2.28}$$

Whereas an energized capacitor conducting almost zero current functions as a voltage source, an energized inductor supporting almost zero terminal voltage emulates a constant current source.

Active Versus Passive

An electrical element or network is said to be passive if the power delivered to it, defined in accordance with Eq. (2.14), is nonnegative. This definition exploits the requirement that the terminal voltage $v(t)$ and the element current $i(t)$ appearing in Eq. (2.14) be in associated reference polarity. In contrast, an element or network to which the delivered power is negative is said to be active; that is, an active element or network generates power instead of dissipating it.

Conventional two-terminal resistors, capacitors, and inductors are passive elements. It follows that networks formed of interconnected two-terminal resistors, capacitors, and inductors are passive networks. Two-terminal voltage and current sources generally behave as active elements. However, when more than one source of externally applied energy is present in an electrical network, it is possible for one or more of these sources to behave as passive structures. Comments similar to those made in conjunction with two-terminal voltage and current sources apply equally well to each of the four possible dependent generators. Accordingly, multiterminal configurations, whose models exploit dependent sources, can behave as either passive or active networks.

Time Varying Versus Time Invariant

The elements of a circuit are defined electrically by an identifying parameter, such as resistance, capacitance, inductance, and the gain factors associated with dependent voltage or current sources. An element, the identifying parameter of which changes as a function of time is said to be a time-varying element. If said parameter is a constant over time, the element in question is time invariant. A network containing at least one time-varying electrical element is said to be a time-varying network. Otherwise, the network is time-invariant. Excluded from the list of elements whose electrical character establishes the time variance or time invariance of a considered network are externally applied voltage and current sources. Thus, for example, a network the internal elements of which are exclusively time-invariant resistors, capacitors, inductors, and dependent sources but which is excited by a sinusoidal signal source is nonetheless a time-invariant network.

Although some circuits, and particularly electromechanical networks, are purposely designed to exhibit time-varying volt–ampere characteristics, parametric time variance is generally viewed as a parasitic phenomenon in the majority of practical circuits. Unfortunately, a degree of parametric time variance is unavoidable in even those circuits that are specifically designed to achieve input/output response properties that closely approximate time-invariant characteristics. For example, the best of network elements exhibit a slow aging phenomenon that shifts the values of its intrinsic physical parameters. The upshot of these shifts is electrical circuits the overall performance of which deteriorates with time.

Lumped Versus Distributed

Electrons in conventional conductive elements are not transported instantaneously across elemental cross sections. But their transport velocities are very high. In fact, these velocities approach the speed of light, say, c, which is (3×10^8) m/s or about 982 ft/μs. Electrons and holes in semiconductors are transported at somewhat slower speeds, but generally no less than an order of magnitude or so smaller than the speed of light. The time required to transport charge from one terminal of a two-terminal electrical element to its other terminal, compared with the time required to propagate energy uniformly through the element, determines whether an element is lumped or distributed. In particular, if the time required to transport charge through an element is significantly smaller than the time required to propagate the energy through the element that is required to incur such charge transport, the element in question is said to be lumped. On the other hand, if the charge transport time is comparable to the energy propagation time, the element is said to be distributed.

The feature size of an element, in relation to the wavelength of energy propagated through the element, determines whether it should be viewed as a lumped or as a distributed element. If the feature size is significantly smaller than the wavelength of applied energy, the element is lumped; otherwise, it is distributed. Antennas used to capture high-frequency transmissions are excellent examples of distributed circuit elements. Practical resistors, capacitors, and inductors used in most discrete component, hybrid, and integrated circuits are lumped circuit elements.

Defining Terms

Active element: A circuit element that generates power.

Associated reference polarity: A polarity convention that fixes the direction of current flow through a circuit element from the positive voltage side of the element to the negative voltage side of said element.

Capacitor: A two-terminal electrical element that satisfies a prescribed algebraic relationship in the charge–voltage (q–V) plane.

Controlled source: A current or voltage branch element whose value depends on a branch current or voltage determined elsewhere in the network.

Current: The time rate of change of net charge passing through an arbitrary cross section of an electrical element.

Distributed element: A circuit element whose feature size is comparable to, or significantly larger than, the wavelength of applied energy.

Electrical circuit: An array of interconnected elements wired so as to be capable of conducting current.

Inductor: A two-terminal electrical element that satisfies a prescribed algebraic relationship in the flux–current (ϕ–I) plane.

Lumped element: A circuit element whose feature size is significantly smaller than the wavelength of applied energy.

Passive element: A circuit element that dissipates power or *stores* energy.

Power: The time rate of change of energy applied to an electrical element.

Resistor: A two-terminal electrical element that satisfies a prescribed algebraic relationship in the voltage–current (V–I) plane.

Voltage: The energy applied to an element per unit of charge transported through the element.

References

Balabanian, N. and Bickart, T.A. 1985. *Electrical Network Theory*. Robert E. Krieger, Malabar, FL.

Chua, L.O. 1969. *Nonlinear Network Theory*. McGraw-Hill, New York.

Cunningham, D.A. and Stuller, J.A. 1995. *Circuit Analysis*. Houghton-Mifflin, Boston, MA.

Dorf, R.C. and Svoboda, J.A. 1996. *Introduction to Electric Circuits*. Wiley, New York.

Huelsman L.P., ed. 1970. *Active Filters: Lumped, Distributed, Integrated, Digital, and Parametric*. McGraw-Hill, New York.

Further Information

The *Circuits and Devices Magazine*, published by the Institute of Electrical and Electronic Engineers (IEEE), is a good entry level periodical that focuses on applications of the various concepts introduced in this chapter.

The IEEE also publishes two advanced relevant scholarly journals, namely, the *IEEE Transactions on Circuits and Systems, Part I* and the *IEEE Transactions on Circuits and Systems, Part II*.

2.2 Semiconductor Materials

Stuart K. Tewksbury

2.2.1 Introduction

A semiconductor material has a resistivity lying between that of a conductor and that of an insulator. In contrast to the granular materials used for resistors, however, a semiconductor establishes its conduction properties through a complex quantum mechanical behavior within a periodic array of semiconductor atoms, that is, within a crystalline structure. For appropriate atomic elements, the crystalline structure leads to a disallowed energy band between the energy level of electrons bound to the crystal's atoms and the energy level of electrons free to move within the crystalline structure (i.e., not bound to an atom). This energy gap fundamentally impacts the mechanisms through which electrons associated with the crystal's atoms can become free and serve as conduction electrons. The resistivity of a semiconductor is proportional to the free carrier density, and that density can be changed over a wide range by replacing a very small portion (about 1 in 10^6) of the base crystal's atoms with different atomic species (*doping atoms*). The majority carrier density is largely pinned to the net dopant impurity density. By selectively changing the crystalline atoms within small regions of the crystal, a vast number of small regions of the crystal can be given different conductivities. In addition, some dopants establish the electron carrier density (*free electron density*), whereas others establish the *hole* carrier density (holes are the dual of electrons within semiconductors). In this manner, different types of semiconductor (n type with much higher electron carrier density than the hole density and p type with much higher hole carrier density than the electron carrier density) can be located in small but contacting regions within the crystal.

By applying electric fields appropriately, small regions of the semiconductor can be placed in a state in which all of the carriers (electron and hole) have been expelled by the electric field and that electric field sustained by the exposed dopant ions. This allows electric switching between a conducting state (with a settable resistivity) and a nonconducting state (with conductance vanishing as the carriers vanish).

This combination of localized regions with precisely controlled resistivity (dominated by *electron conduction* or by *hole conduction*) combined with the ability to electronically control the flow of the carriers (electrons and holes) leads to the semiconductors being the foundation for contemporary electronics. This foundation is particularly strong because a wide variety of atomic elements (and mixtures of atomic elements) can be used to tailor the semiconductor material to specific needs. The dominance of silicon semiconductor material in the electronics area (e.g., the very large-scale integrated (VLSI) digital electronics area) contrasts with the rich variety of semiconductor materials widely used in optoelectronics. In the latter case, the ability to adjust the **bandgap** to desired wavelengths of light has stimulated a vast number of optoelectronic components, based on a variety of technologies. Electronic components also provide a role for nonsilicon semiconductor technologies, particularly for very high bandwidth circuits that can take advantage of the higher speed capabilities of semiconductors using atomic elements similar to those used in optoelectronics. This rich interest in nonsilicon technologies will undoubtedly continue to grow, due to the rapidly advancing applications of optoelectronics, for the simple reason that silicon is not suitable for producing an efficient optical source.

2.2.2 Crystalline Structures

Basic Semiconductor Materials Groups

Most semiconductor materials are crystals created by atomic bonds through which the **valence band** of the atoms are filled with eight electrons through sharing of an electron from each of four nearest neighbor atoms. These materials include semiconductors composed of a single atomic species, with the basic atom having four electrons in its valence band (supplemented by covalent bonds to four neighboring atoms to complete the valence band). These elemental semiconductors, therefore, use atoms from group IV of the atomic chart. Other semiconductor materials are composed of two atoms, one from group N ($N < 4$) and the other from group M ($M > 4$) with $N + M = 8$, filling the valence bands with eight electrons. The major categories of semiconductor material are summarized in the following sections.

Elemental (IV–IV) Semiconductors

Elemental semiconductors consist of crystals composed of only a single atomic element from group IV of the periodic chart, that is, germanium (Ge), silicon (Si), carbon (C), and tin (Sn). Silicon is the most commonly used electronic semiconductor material and is also the most common element on Earth. Table 2.1 summarizes the naturally occurring abundance of some elements used for semiconductors, including nonelemental (compound) semiconductors.

Figure 2.9(a) illustrates the covalent bonding (sharing of outer shell, valence band electrons by two atoms) through which each group IV atom of the crystal is bonded to four neighboring group IV atoms, creating filled outer electron bands of eight electrons.

In addition to crystals composed of only a single group IV atomic species, one can also create semiconductor crystals consisting of two or more atoms, all from group IV. For example, silicon carbide (SiC) has been investigated for high-temperature applications. $Si_x Ge_{1-x}$ semiconductors are presently under study to achieve bandgap engineering within the silicon system. In this case, a fraction x ($0 < x < 1$) of the atoms in an otherwise silicon crystal is silicon whereas a fraction $1 - x$ has been replaced by germanium. This ability to replace a single atomic element with a combination of two atomic elements from the same column of the periodic chart appears in the other categories of semiconductor described subsequently (and is particularly important for optoelectronic devices).

TABLE 2.1 Abundance (Fraction of Elements Occurring on Earth) of Common Elements Used for Semiconductors

Element	Abundance
Si	0.28
Ga	1.5×10^{-5}
As	1.8×10^{-6}
Ge	5×10^{-6}
Cd	2×10^{-7}
In	1×10^{-7}

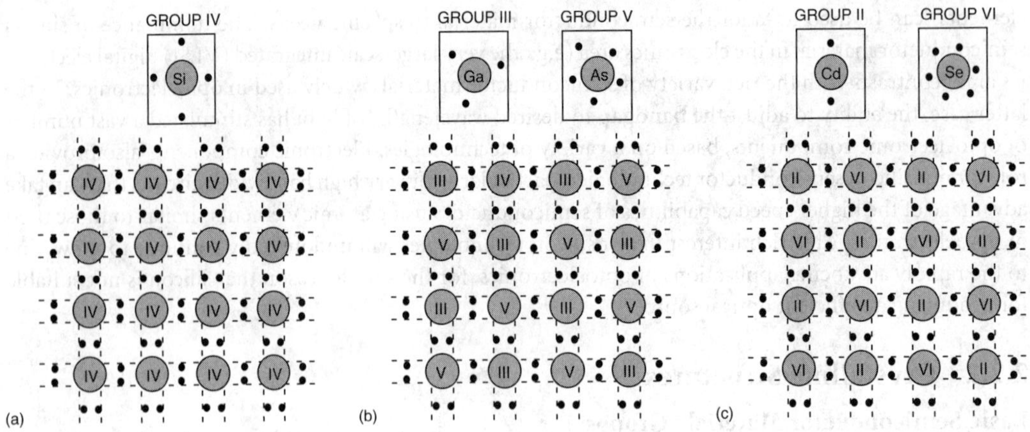

FIGURE 2.9 Bonding arrangements of atoms in semiconductor crystals: (a) elemental semiconductor such as silicon, (b) compound III–V semiconductor such as GaAs, (c) compound II–VI semiconductor such as CdS.

Compound III–V Semiconductors

The III–V semiconductors are prominent (and will gain in importance) for applications of optoelectronics. In addition, III–V semiconductors have a potential for higher speed operation than silicon semiconductors in electronics applications, with particular importance for areas such as wireless communications. The **compound semiconductors** have a crystal lattice constructed from atomic elements in different groups of the periodic chart. The III–V semiconductors are based on an atomic element A from group III and an atomic element B from group V. Each group III atom is bound to four group V atoms, and each group V atom is bound to four group III atoms, giving the general arrangement shown in Fig. 2.9(b). The bonds are produced by sharing of electrons such that each atom has a filled (eight electron) valence band. The bonding is largely covalent, though the shift of valence charge from the group V atoms to the group III atoms induces a component of ionic bonding to the crystal (in contrast to the elemental semiconductors that have purely covalent bonds). Representative III–V compound semiconductors are GaP, GaAs, GaSb, InP, InAs, and InSb.

GaAs is probably the most familiar example of III–V compound semiconductors, used for both high-speed electronics and for optoelectronic devices. Optoelectronics has taken advantage of ternary and quaternary III–V semiconductors to establish optical wavelengths and to achieve a variety of novel device structures. The **ternary semiconductors** have the general form $(A_x, A'_{1-x})B$ (with two group III atoms used to fill the group III atom positions in the lattice) or $A(B_x, B'_{1-x})$ (using two group V atoms in the group V atomic positions in the lattice). The **quaternary semiconductors** use two group III atomic elements and two group V atomic elements, yielding the general form $(A_x, A'_{1-x})(B_y, B'_{1-y})$. In such constructions, $0 \leq x \leq 1$. Such ternary and quaternary versions are important since the mixing factors (x and y) allow the bandgap to be adjusted to lie between the bandgaps of the simple compound crystals with only one type of group III and one type of group V atomic element. The adjustment of wavelength allows the material to be tailored for particular optical wavelengths, since the wavelength λ of light is related to energy (in this case the gap energy E_g) by $\lambda = hc/E_g$, where h is Planck's constant and c is the speed of light. Table 2.2 provides examples of semiconductor laser materials and a representative optical wavelength for each, providing a hint of the vast range of combinations that are available for optoelectronic applications. Table 2.3, on the other hand, illustrates the change in wavelength (here corresponding to color in the visible spectrum) by adjusting the mixture of a ternary semiconductor.

In contrast to single element elemental semiconductors (for which the positioning of each atom on a lattice site is not relevant), III–V semiconductors require very good control of stoichiometry (i.e., the ratio of the two atomic species) during crystal growth. For example, each Ga atom must reside on a Ga (and

TABLE 2.2 Semiconductor Optical Sources and Representative Wavelengths

Material Layers Used	Wavelength, nm
ZnS	454
AlGaInP/GaAs	580
AlGaAs/GaAs	680
GaInAsP/InP	1580
InGaAsSb/GaSb	2200
AlGaSb/InAsSb/GaSb	3900
PbSnTe/PbTe	6000

not an As) site and vice versa. For these and other reasons, large III–V crystals of high quality are generally more difficult to grow than a large crystal of an elemental semiconductor such as Si.

Compound II–VI Semiconductors

These semiconductors are based on one atomic element from group II and one atomic element from group VI, each type being bonded to four nearest neighbors of the other type, as shown in Fig. 2.9(c). The increased amount of charge from group VI to group II atoms tends to cause the bonding to be more ionic than in the case of III–V semiconductors. II–VI semiconductors can be created in ternary and quaternary forms, much like the III–V semiconductors. Although less common than the III–V semiconductors, the II–VI semiconductors have served the needs of several important applications. Representative II–VI semiconductors are ZnS, ZnSe, and ZnTe (which form in the zinc blende lattice structure discussed subsequently); CdS and CdSe (which can form in either the zinc blende or the wurtzite lattice structure); and CdTe (which forms in the wurtzite lattice structure).

Three-Dimensional Crystal Lattice

The two-dimensional views illustrated in the preceding section provide a simple view of the sharing of valence band electrons and the bonds between atoms. The full three-dimensional lattice structure, however, is considerably more complex than this simple two-dimensional illustration. Fortunately, most semiconductor crystals share a common basic structure, developed as follows.

The crystal structure begins with a cubic arrangement of eight atoms as shown in Fig. 2.10(a). This cubic lattice is extended to a *face-centered cubic* (fcc) lattice, shown in Fig. 2.10(b), by adding an atom to the center of each face of the cube (leading to a lattice with 14 atoms). The *lattice constant* is the side dimension of this cube.

The full lattice structure combines two of these fcc lattices, one lattice interpenetrating the other (i.e., the corner of one cube is positioned within the interior of the other cube, with the faces remaining parallel), as illustrated in Fig. 2.10(c). For the III–V and II–VI semiconductors with this fcc lattice foundation, one fcc lattice is constructed from one type of element (e.g., type III) and the second fcc lattice is constructed from the other type of element (e.g., group V). In the case of ternary and quaternary semiconductors, elements from the same atomic group are placed on the same fcc lattice. All bonds between atoms occur between atoms in different fcc lattices. For example, all Ga atoms in the GaAs crystal are located on

TABLE 2.3 Variation of x to Adjust Wavelength in $GaAs_x P_{1-x}$ Semiconductors

Ternary Compound	Color
$GaAs_{0.14}P_{0.86}$	Yellow
$GaAs_{0.35}P_{0.65}$	Orange
$GaAs_{0.6}P_{0.4}$	Red

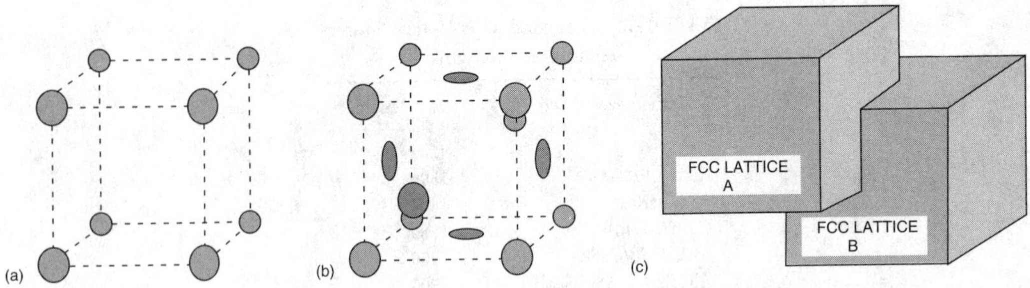

FIGURE 2.10 Three-dimensional crystal lattice structure: (a) basic cubic lattice, (b) face-centered cubic (fcc) lattice, (c) two interpenetrating fcc lattices. In this figure, the dashed lines between atoms are not atomic bonds but, instead, are used merely to show the basic outline of the cube.

one of the fcc lattices and are bonded to As atoms, all of which appear on the second fcc lattice. The interatomic distances between neighboring atoms is, therefore, less than the lattice constant. For example, the interatomic spacing of Si atoms is 2.35 Å but the lattice constant of Si is 5.43 Å.

If the two fcc lattices contain elements from different groups of the periodic chart, the overall crystal structure is called the *zinc blende* lattice. In the case of an elemental semiconductor such as silicon, silicon atoms appear in both fcc lattices, and the overall crystal structure is called the *diamond* lattice (carbon crystallizes into a diamond lattice creating true diamonds, and carbon is a group IV element). As in the discussion regarding III–V semiconductors, the bonds between silicon atoms in the silicon crystal extend between fcc sublattices.

Although the common semiconductor materials share this basic diamond/zinc blende lattice structure, some semiconductor crystals are based on a hexagonal close-packed (hcp) lattice. Examples are CdS and CdSe. In this example, all of the Cd atoms are located on one hcp lattice whereas the other atom (S or Se) is located on a second hcp lattice. In the spirit of the diamond and zinc blende lattices, the complete lattice is constructed by interpenetrating these two hcp lattices. The overall crystal structure is called a *wurtzite* lattice. Type IV–VI semiconductors (PbS, PbSe, PbTe, and SnTe) exhibit a narrow bandgap and have been used for infrared detectors. The lattice structure of these example IV–VI semiconductors is the simple cubic lattice (also called an NaCl lattice).

Crystal Directions and Planes

Crystallographic directions and planes are important in both the characteristics and the applications of semiconductor materials since different crystallographic planes can exhibit significantly different physical properties. For example, the surface density of atoms (atoms per square centimeter) can differ substantially on different crystal planes. A standardized notation (the so-called *Miller indices*) is used to define the crystallographic planes and directions normal to those planes.

The general crystal lattice defines a set of unit vectors (a, b, and c) such that an entire crystal can be developed by copying the unit cell of the crystal and duplicating it at integer offsets along the unit vectors, that is, replicating the basis cell at positions $n_a a + n_b b + n_c c$, where n_a, n_b, and n_c are integers. The unit vectors need not be orthogonal in general. For the cubic foundation of the diamond and zinc blende structures, however, the unit vectors are in the orthogonal x, y, and z directions.

Figure 2.11 shows a cubic crystal, with basis vectors in the x, y, and z directions. Superimposed on this lattice are three planes (Fig. 2.11(a), Fig 2.11(b), and Fig. 2.11(c)). The planes are defined relative to the crystal axes by a set of three integers (h, k, l) where h corresponds to the plane's intercept with the x axis, k corresponds to the plane's intercept with the y axis, and l corresponds to the plane's intercept with the z axis. Since parallel planes are equivalent planes, the intercept integers are reduced to the set of the three smallest integers having the same ratios as the described intercepts. The (100), (010), and (001) planes correspond to the faces of the cube. The (111) plane is tilted with respect to the cube faces, intercepting the x, y, and z axes at 1, 1, and 1, respectively. In the case of a negative axis intercept, the corresponding

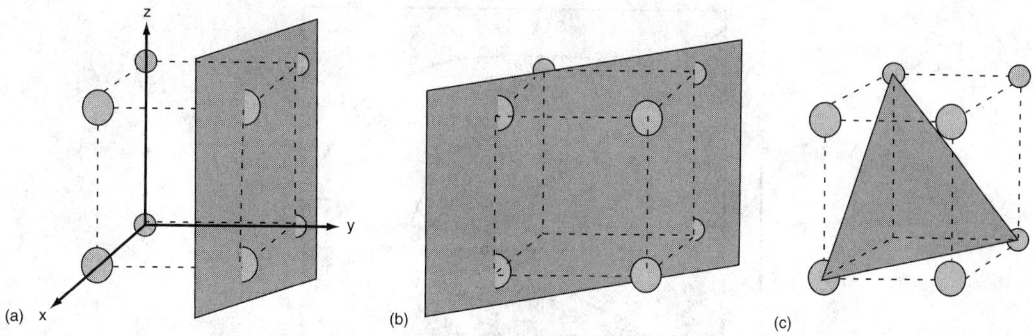

FIGURE 2.11 Examples of crystallographic planes within a cubic lattice organized semiconductor crystal: (a) (010) plane, (b) (110) plane, (c) (111) plane.

Miller index is given as an integer and a bar over the integer, for example, $(\bar{1}00)$, that is, similar to the (100) plane but intersecting the x axis at -1.

Additional notation is used to represent sets of planes with equivalent symmetry and to represent directions. For example, {100} represents the set of equivalent planes (100), $(\bar{1}00)$, (010), $(0\bar{1}0)$, (001), and $(00\bar{1})$. The direction normal to the (hkl) plane is designated $[hkl]$. The different planes exhibit different behavior during device fabrication and impact electrical device performance differently. One difference is due to the different reconstructions of the crystal lattice near a surface to minimize energy. Another is the different surface density of atoms on different crystallographic planes. For example, in Si the (100), (110), and (111) planes have surface atom densities (atoms per square centimeter) of 6.78×10^{14}, 9.59×10^{14}, and 7.83×10^{14}, respectively.

2.2.3 Energy Bands and Related Semiconductor Parameters

A semiconductor crystal establishes a periodic arrangement of atoms, leading to a periodic spatial variation of the potential energy throughout the crystal. Since that potential energy varies significantly over interatomic distances, quantum mechanics must be used as the basis for allowed energy levels and other properties related to the semiconductor. Different semiconductor crystals (with their different atomic elements and different interatomic spacings) lead to different characteristics. The periodicity of the potential variations, however, leads to several powerful general results applicable to all semiconductor crystals. Given these general characteristics, the different semiconductor materials exhibit properties related to the variables associated with these general results. A coherent discussion of these quantum mechanical results is beyond the scope of this chapter and, therefore, we take those general results as given.

In the case of materials that are semiconductors, a central result is the energy-momentum functions defining the state of the electronic charge carriers. In addition to the familiar electrons, semiconductors also provide holes (i.e., positively charged particles) that behave similarly to the electrons. Two energy levels are important: one is the energy level (**conduction band**) corresponding to electrons that are not bound to crystal atoms and that can move through the crystal, and the other energy level (valence band) corresponds to holes that can move through the crystal. Between these two energy levels, there is a region of forbidden energies (i.e., energies for which a free carrier cannot exist). The separation between the conduction and valence band minima is called the *energy gap* or bandgap. The energy bands and the energy gap are fundamentally important features of the semiconductor material.

Conduction and Valence Band

In quantum mechanics, a particle is represented by a collection of plane waves $(e^{j(\omega t - k \cdot x)})$ where the frequency ω is related to the energy E according to $E = \hbar\omega$ and the momentum p is related to the wave vector by $p = \hbar k$. In the case of a classical particle with mass m moving in free space, the energy

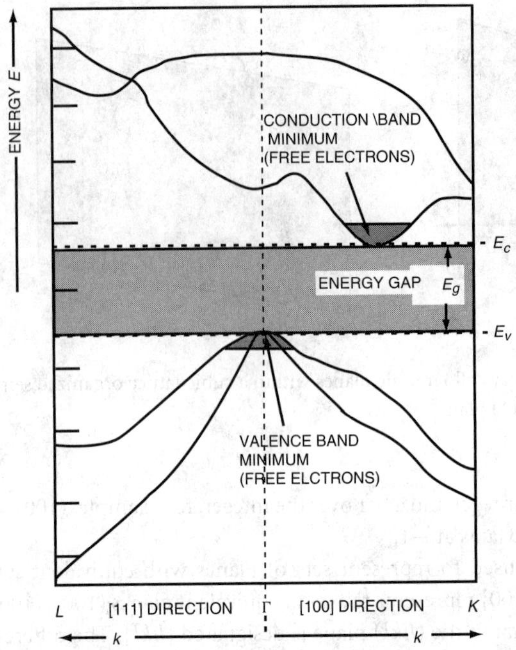

FIGURE 2.12   General structure of conduction and valence bands.

and momentum are related by $E = p^2/(2m)$ that, using the relationship between momentum and wave vector, can be expressed as $E = (\hbar k)^2/(2m)$. In the case of the semiconductor, we are interested in the energy/momentum relationship for a free electron (or hole) moving in the semiconductor, rather than moving in free space. In general, this $E-k$ relationship will be quite complex, and there will be a multiplicity of $E-k$ states resulting from the quantum mechanical effects. One consequence of the periodicity of the crystal's atom sites is a periodicity in the wave vector $\mathbf{k}$, requiring that we consider only values of k over a limited range (with the $E-k$ relationship periodic in k).

Figure 2.12 illustrates a simple example (not a real case) of a conduction band and a valence band in the energy-momentum plane (i.e., the E vs. k plane). The E vs. k relationship of the conduction band will exhibit a minimum energy value and, under equilibrium conditions, the electrons will favor being in that minimum energy state. Electron energy levels above this minimum (E_c) exist, with a corresponding value of momentum. The E vs. k relationship for the valence band corresponds to the energy–momentum relationship for holes. In this case, the energy values increase in the direction toward the bottom of the page and the minimum valence band energy level E_v is the maximum value in Fig. 2.12. When an electron bound to an atom is provided with sufficient energy to become a free electron, a hole is left behind. Therefore, the energy gap $E_g = E_c - E_v$ represents the minimum energy necessary to generate an electron-hole pair (higher energies will initially produce electrons with energy greater than E_c, but such electrons will generally lose energy and fall into the potential minimum).

The details of the energy bands and the bandgap depend on the detailed quantum mechanical solutions for the semiconductor crystal structure. Changes in that structure (even for a given semiconductor crystal such as Si) can therefore lead to changes in the energy band results. Since the thermal coefficient of expansion of semiconductors is nonzero, the bandgap depends on temperature due to changes in atomic spacing with changing temperature. Changes in pressure also lead to changes in atomic spacing. Though these changes are small, the are observable in the value of the energy gap. Table 2.4 gives the room temperature value of the energy gap E_g for several common semiconductors, along with the rate of change of E_g with temperature (T) and pressure (P) at room temperature.

TABLE 2.4 Variation of Energy Gap with Temperature and Pressure

Semiconductor	E_g, 300 K	dE_g/dT, meV/K	dE_g/dP, meV/kbar
Si	1.110	−0.28	−1.41
Ge	0.664	−0.37	5.1
GaP	2.272	−0.37	10.5
GaAs	1.411	−0.39	11.3
GaSb	0.70	−0.37	14.5
InP	1.34	−0.29	9.1
InAs	0.356	−0.34	10.0
InSb	0.180	−0.28	15.7
ZnSe	2.713	−0.45	0.7
ZnTe	2.26	−0.52	8.3
CdS	2.485	−0.41	4.5
CdSe	1.751	0.36	5
CdTe	1.43	−0.54	8

Source: Adapted from Böer, K.W. 1990. *Survey of Semiconductor Physics, Vol. 1: Electrons and Other Particles in Bulk Semiconductors.* Van Nostrand, New York.

The temperature dependence, though small, can have a significant impact on carrier densities. A heuristic model of the temperature dependence of E_g is $E_g(T) = E_g(0 \text{ K}) - \alpha T^2/(T + \beta)$. Values for the parameters in this equation are provided in Table 2.5. Between 0 and 1000 K, the values predicted by this equation for the energy gap of GaAs are accurate to about 2×10^{-3} eV.

Direct Gap and Indirect Gap Semiconductors

Figure 2.13 illustrates the energy bands for Ge, Si, and GaAs crystals. In Fig. 2.13(b), for silicon, the valence band has a minimum at a value of $\boldsymbol{k}$ different than that for the conduction band minimum. This is an **indirect gap**, with generation of an electron-hole pair requiring an energy E_g and a change in momentum (i.e., k). For direct recombination of an electron-hole pair, a change in momentum is also required. This requirement for a momentum change (in combination with energy and momentum conservation laws) leads to a requirement that a phonon participate with the carrier pair during a direct recombination process generating a photon. This is a highly unlikely event, rendering silicon ineffective as an optoelectronic source of light. The direct generation process is more readily allowed (with the simultaneous generation of an electron, a hole, and a phonon), allowing silicon and other **direct gap semiconductors** to serve as optical detectors.

In Fig. 2.13(c), for GaAs, the conduction band minimum and the valence band minimum occur at the same value of momentum, corresponding to a direct gap. Since no momentum change is necessary during direct recombination, such recombination proceeds readily, producing a photon with the energy of the initial electron and hole (i.e., a photon energy equal to the bandgap energy). For this reason, direct gap semiconductors are efficient sources of light (and use of different direct gap semiconductors with different E_g provides a means of tailoring the wavelength of the source). The wavelength λ corresponding to the gap energy is $\lambda = hc/E_g$.

TABLE 2.5 Temperature Dependence Parameters for Common Semiconductors

	E_g (0 K), eV	$\alpha(\times 10^{-4})$	β	E_g(300 K), eV
GaAs	1.519	5.405	204	1.42
Si	1.170	4.73	636	1.12
Ge	0.7437	4.774	235	0.66

Source: Adapted from Sze, S.M. 1981. *Physics of Semiconductor Devices,* 2nd ed. Wiley-Interscience, New York.

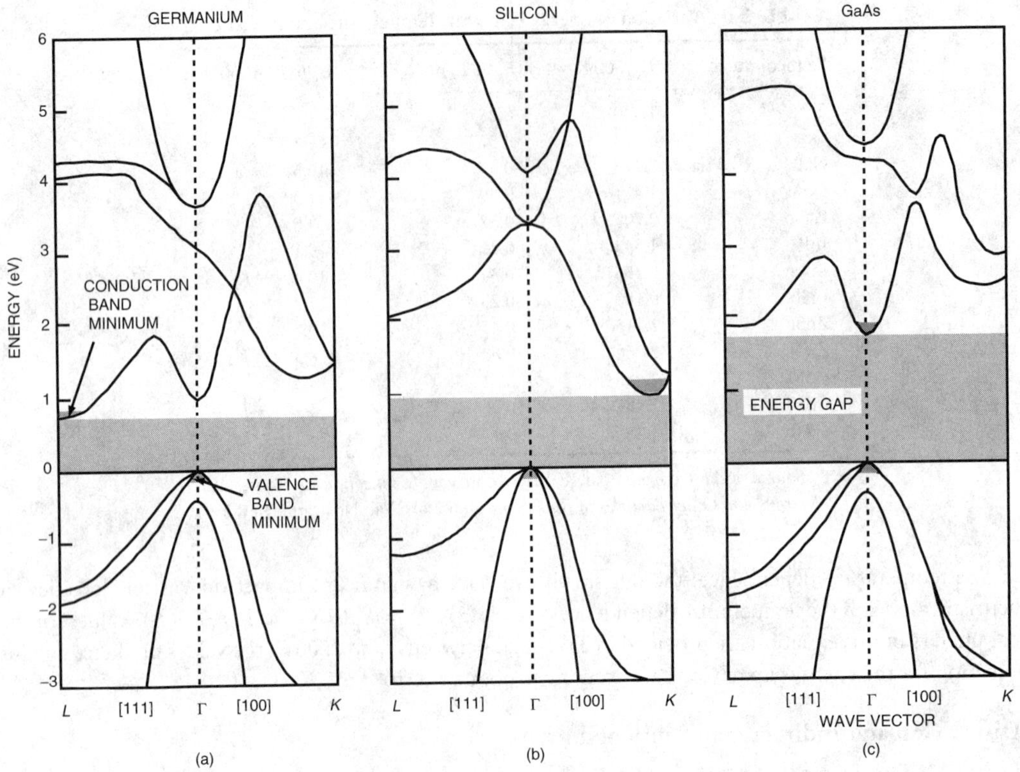

FIGURE 2.13 Conduction and valence bands for (a) Ge, (b) Si, (c) GaAs. (*Source:* Adapted from Sze, S.M. 1981. *Physics of Semiconductor Devices*, 2nd ed. Wiley Interscience, New York.)

Figure 2.13(c) also illustrates a second conduction band minimum with an indirect gap, but at a higher energy than the minimum associated with the direct gap. The higher conduction band minimum can be populated by electrons (which are in an equilibrium state of higher energy) but the population will decrease as the electrons gain energy sufficient to overcome that upper barrier.

Effective Masses of Carriers

For an electron with energy close to the minimum of the conduction band, the energy vs. momentum relationship along each momentum axis is approximately given by $E(k) = E_0 + a_2(k - k^*)^2 + a_4(k - k^*)^4 + \cdots$. Here, $E_0 = E_c$ is the *ground state energy* corresponding to a free electron at rest and k^* is the wave vector at which the conduction band minimum occurs. Only even powers of $k - k^*$ appear in the expansion of $E(k)$ around k^* due to the symmetry of the E–k relationship around $k = k^*$. This approximation holds for sufficiently small increases in E above E_c. For sufficiently small movements away from the minimum (i.e., sufficiently small $k - k^*$), the terms in $k - k^*$ higher than quadratic can be ignored and $E(k) \approx E_0 + a_2 k^2$, where we have taken $k^* = 0$. If, instead of a free electron moving in the semiconductor crystal, we had a free electron moving in free space with potential energy E_0, the energy-momentum relationship would be $E(k) = E_0 + (\hbar k)^2/(2m_0)$, where m_0 is the mass of an electron. By comparison of these results, it is clear that we can relate the curvature coefficient a_2 associated with the parabolic minimum of the conduction band to an **effective mass** m_e^*, that is, $a_2 = (\hbar^2)/(2m_e^*)$ or

$$\frac{1}{m_e^*} = \frac{2}{\hbar^2} \cdot \frac{\partial^2 E_c(k)}{\partial k^2}$$

Similarly for holes, an effective mass m_h^* of the holes can be defined by the curvature of the valence band minimum, that is,

$$\frac{1}{m_h^*} = \frac{2}{\hbar^2} \cdot \frac{\partial^2 E_v(k)}{\partial k^2}$$

Since the energy bands depend on temperature and pressure, the effective masses can also be expected to have such dependencies, though the room temperature and normal pressure value is normally used in device calculations.

This discussion assumes the simplified case of a scalar variable k. In fact, the wave vector k has three components (k_1, k_2, k_3), with directions defined by the unit vectors of the underlying crystal. Therefore, there are separate masses for each of these vector components of k, that is, masses m_1, m_2, m_3. A scalar mass m^* can be defined using these directional masses, the relationship depending on the details of the directional masses. For cubic crystals (as in the diamond and zinc blende structures), the directions are the usual orthonormal directions and $m^* = (m_1 \cdot m_2 \cdot m_3)^{1/3}$. The three directional masses effectively reduce to two components if two values are equal (e.g., $m_1 = m_2$), as in the case of *longitudinal* and *transverse* effective masses (m_l and m_t, respectively) seen in silicon and several other semiconductors. In this case, $m^* = [(m_t)^2 \cdot m_l]^{1/3}$. If all three values of m_1, m_2, m_3 are equal, then a single value m^* can be used.

An additional complication is seen in the valence band structures in Fig. 2.13. Here, two different E–k valence bands have the same minima. Since their curvatures are different, the two bands correspond to different masses, one corresponding to *heavy holes* with mass m_h and the other to *light holes* with mass m_l. The effective scalar mass in this case is $m^* = (m_h^{3/2} + m_l^{3/2})^{2/3}$. Such light and heavy holes occur in several semiconductors, including Si.

Values of effective mass are given in Tables 2.8 and 2.13.

Intrinsic Carrier Densities

The density of free electrons in the conduction band depends on two functions. One is the density of states $D(E)$ in which electrons can exist and the other is the energy distribution function $F(E, T)$ of free electrons.

The energy distribution function (under thermal equilibrium conditions) is given by the Fermi-Dirac distribution function

$$F(E) = \left[1 + \exp\left(\frac{E - E_f}{k_B T}\right)\right]^{-1}$$

which, in most practical cases, can be approximated by the classical Maxwell-Boltzmann distribution. These distribution functions are general functions, not dependent on the specific semiconductor material.

The density of states $D(E)$, on the other hand, depends on the semiconductor material. A common approximation is

$$D_n(E) = M_c \frac{\sqrt{2}}{\pi^2} \frac{(E - E_c)^{1/2}}{\hbar^3} \frac{(m_e^*)^3}{2}$$

for electrons and

$$D_p(E) = M_v \frac{\sqrt{2}}{\pi^2} \frac{(E_v - E)^{1/2}}{\hbar^3} \frac{(m_h^*)^3}{2}$$

for holes. Here, M_c and M_v are the number of equivalent minima in the conduction band and valence band, respectively. Note that, necessarily, $E \geq E_c$ for free electrons and $E \leq E_v$ for free holes due to the forbidden region between E_c and E_v.

The density n of electrons in the conduction band can be calculated as

$$n = \int_{E=E_c}^{\infty} F(E, T) D(E) dE$$

TABLE 2.6 N_c and N_v at 300 K

	N_c ($\times 10^{19}/\text{cm}^3$)	N_v ($\times 10^{19}/\text{cm}^3$)
Ge	1.54	1.9
Si	2.8	1.02
GaAs	0.043	0.81
GaP	1.83	1.14
GaSb	0.021	0.62
InAs	0.0056	0.62
InP	0.052	1.26
InSb	0.0043	0.62
CdS	0.224	2.5
CdSe	0.11	0.74
CdTe	0.13	0.55
ZnSe	0.31	0.87
ZnTe	0.22	0.078

Source: Adapted from Böer, K.W. 1990. *Survey of Semiconductor Physics, Vol. 1: Electrons and Other Particles in Bulk Semiconductors.* Van Nostrand, New York.

For Fermi levels significantly (more than a few $k_B T$) below E_c and above E_v, this integration leads to the results

$$n = N_c e^{-(E_c - E_f)/k_b T}$$

and

$$p = N_v e^{-(E_f - E_v)/k_b T}$$

where n and p are the densities of free electrons in the conduction band and of holes in the valence band, respectively. N_c and N_v are *effective densities of states* that vary with temperature (slower than the exponential in the preceding equations), effective mass, and other conditions. Table 2.6 gives values of N_c and N_v for several semiconductors. Approximate expressions for these densities of state are $N_c = 2(2\pi m_e^* k_B T/\hbar^2)^{3/2} M_c$ and $N_v = 2(2\pi m_e^* k_B T/\hbar^2)^{3/2} M_v$. These effective densities of states are fundamental parameters used in evaluating the electrical characteristics of semiconductors. The preceding equations for n and p apply both to **intrinsic semiconductors** (i.e., semiconductors with no impurity dopants) as well as to semiconductors that have been doped with donor and/or acceptor impurities. Changes in the interrelated values of n and p through introduction of dopant impurities can be represented by changes in a single variable, the Fermi level E_f.

The product of n and p is independent of Fermi level and is given by

$$n \cdot p = N_c \cdot N_v e^{-E_g/k_B T}$$

where the energy gap $E_g = E_c - E_v$. Again, this holds for both intrinsic semiconductors and for doped semiconductors. In the case of an intrinsic semiconductor, charge neutrality requires that $n = p \equiv n_i$, where n_i is the *intrinsic carrier concentration* and

$$n_i^2 = N_c \cdot N_v e^{-E_g/k_B T}$$

Since, under thermal equilibrium conditions $np \equiv n_i^2$ (even under impurity doping conditions), knowledge of the density of one of the carrier types (e.g., of p) allows direct determination of the density of the other (e.g., $n = n_i^2/p$). Values of n_i vary considerably among semiconductor materials: $2 \times 10^{-3}/\text{cm}^3$ for CdS, $3.3 \times 10^6/\text{cm}^3$ for GaAs, $0.9 \times 10^{10}/\text{cm}^3$ for Si, $1.9 \times 10^{13}/\text{cm}^3$ for Ge, and 9.1×10^{14} for PbS.

Since there is appreciable temperature dependence in the effective density of states, the equations do not accurately represent the temperature variations in n_i over wide temperature ranges. Using the approximate expressions for N_c and N_v already given,

$$n_i = 2\left(2\pi k_B T/\hbar^2\right)^{3/2}\left(m_e^* m_h^*\right)^{3/4}\sqrt{M_c M_v}\,e^{-E_g/k_B T}$$

exhibiting a $T^{3/2}$ temperature dependence superimposed on the exponential dependence on $1/T$. For example, at 300 K

$$n_i(T) = 1.76 \times 10^{16}\,T^{3/2}e^{-4550/T}\ \text{cm}^{-3}\quad \text{for Ge}$$

and

$$n_i(T) = 3.88 \times 10^{16}\,T^{3/2}e^{-7000/T}\ \text{cm}^{-3}\quad \text{for Si}$$

Substitutional Dopants

An intrinsic semiconductor material contains only the elemental atoms of the basic material (e.g., silicon atoms for Si, gallium and arsenic atoms for GaAs, etc.). The resistivity is quite high for such intrinsic semiconductors and doping is used to establish a controlled lower resistivity material and to establish *pn* junctions (interface between *p*-type and *n*-type regions of the semiconductor). Doping concentrations are generally in the range 10^{14}–$10^{17}/\text{cm}^3$, small relative to the density of atoms in the crystal (e.g., to the density 5×10^{22} atoms/cm^3 of silicon atoms in Si crystals). Table 2.7 lists a variety of dopants and their energy levels for Si and GaAs.

Figure 2.14(a) illustrates acceptor dopants and donor dopants in silicon. In the case of acceptor dopants, group III elements of the periodic chart are used to substitute for the group IV silicon atoms in the crystal. This acceptor atom has one fewer electron in its outer shell than the silicon atom and readily captures a free electron to provide the missing electron needed to complete the outer shells (eight electrons) provided by the covalent bonds. The acceptor atom, with a captured electron, becomes a negative ion. The electron captured from the outer shell of a neighboring silicon atom leaves behind a hole at that neighboring silicon atom (i.e., generates a free hole when ionized). By using acceptor doping densities N_A substantially greater than n_i, a net hole density $p \gg n_i$ is created. With $np = n_i^2$ a constant, the electron density n decreases as p increases above n_i. The resulting semiconductor material becomes p type. In the case of donor dopants, group V elements are used to substitute for a silicon atom in the crystal. The donor atom has one extra electron in its outer shell, compared to a silicon atom, and that extra electron can leave the donor site and become a free electron. In this case, the donor becomes a positive ion, generating a free electron. By using donor doping densities N_D substantially greater than n_i, a net electron density $n \gg n_i$ is created, and p decreases substantially below n_i to maintain the np product $np = n_i^2$. An n type semiconductor is produced.

TABLE 2.7 Acceptor and Donor Impurities Used in Si and GaAs

	Donor	$E_c - E_d$, eV	Acceptor	$E_a - E_v$, eV
Si crystal	Sb	0.039	B	0.045
	P	0.045	Al	0.067
	As	0.054	Ga	0.073
GaAs crystal	S	0.006	Mg	0.028
	Se	0.006	Zn	0.031
	Te	0.03	Cd	0.035
	Si	0.058	Si	0.026

Source: Adapted from Tyagi, M.S. 1991. *Introduction to Semiconductor Materials.* Wiley, New York.

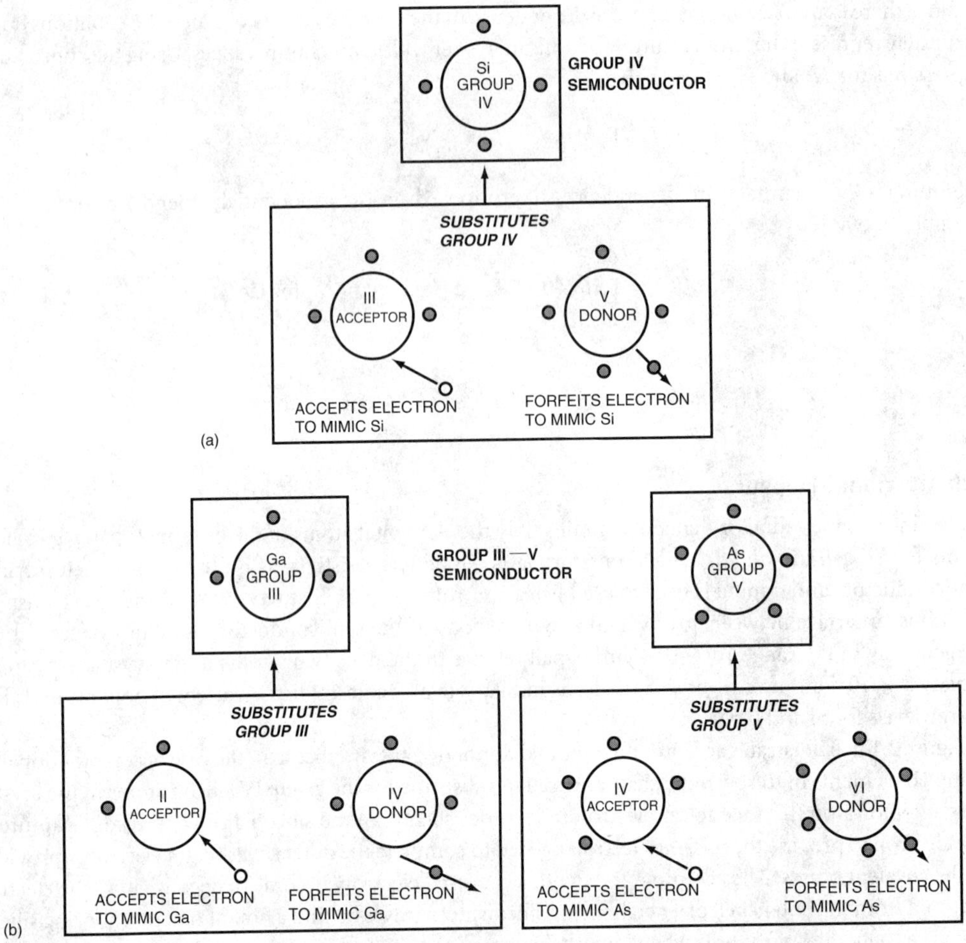

FIGURE 2.14 Substitution of dopant atoms for crystal atoms: (a) IV–IV semiconductors (e.g., silicon), (b) III–V semiconductors (e.g., GaAs).

Figure 2.14(b) illustrates the alternative doping options for a III–V semiconductor (GaAs used as an example). Replacement of a group III element with a group II element renders that group II element an acceptor (one fewer electron). Replacement of a group V element with a group VI element renders that group VI element a donor (one extra electron). Group IV elements such as silicon can also be used for doping. In this case, the group IV element is a donor if it replaces a group III element in the crystal and is an acceptor if it replaces a Group V element in the crystal. Impurities which can serve as either an acceptor or as a donor within a crystal are called **amphoteric impurities**.

Acceptor and donor impurities are most effectively used when the energy required to generate a carrier is small. In the case of small ionization energies (in the crystal lattice), the energy levels associated with the impurities lie within the bandgap close to their respective bands (i.e., donor ionization energies close to the conduction band and acceptor ionization energies close to the valence band). If the difference between the ionization level and the corresponding valence/conduction band is less than about $3k_B T$ (≈ 0.075 eV at 300 K), then the impurities (called **shallow energy level dopants**) are essentially fully ionized at room temperature. The dopants listed in Table 2.7 are such shallow energy dopants. A semiconductor doped with $N_A \gg n_i$ acceptors then has a hole density $p \approx N_A$ and a much smaller electron density $n \approx n_i^2/N_A$. Similarly, a semiconductor doped with $N_D \gg n_i$ donors has an electron density $n \approx N_D$ and a much smaller hole density $p = n_i^2/N_D$. From the results given earlier for carrier concentrations as a function of Fermi level, the Fermi level is readily calculated (given the majority carrier concentration).

Most semiconductors can be selectively made (by doping) either n type or p type, in which case they are called **ambipolar semiconductors.** Some semiconductors can be selectively made only n type or only p type. For example, ZnTe is always p type whereas CdS is always n type. Such semiconductors are called **unipolar semiconductors.**

Dopants with energy levels closer to the center of the energy gap (i.e., the so-called **deep energy level dopants**) serve as electron-hole recombination sites, impacting the minority carrier lifetime and the dominant recombination mechanism in indirect bandgap semiconductors.

2.2.4   Carrier Transport

Currents in semiconductors arise both due to movement of free carriers in the presence of an electric field and due to diffusion of carriers from high, carrier density regions into lower, carrier density regions. Currents due to electric fields are considered first.

In earlier discussions, it was noted that the motion of an electron in a perfect crystal can be represented by a free electron with an effective mass m_e^* somewhat different than the real mass m_e of an electron. In this model, once the effective mass has been determined, the atoms of the perfect crystal can be discarded and the electron viewed as moving within free space. If the crystal is not perfect, however, those deviations from perfection remain after the perfect crystal lattice has been discarded and act as scattering sites within the free space seen by the electron in the crystal.

Substitution of a dopant for an element of the perfect crystal leads to a distortion of the perfect lattice from which electrons can scatter. If that substitutional dopant is ionized, the electric field of that ion adds to the scattering. Impurities located at interstitial sites (i.e., between atoms in the normal lattice sites) also disrupt the perfect crystal and lead to scattering sites. Crystal defects (e.g., a missing atom) disrupt the perfect crystal and appears as a scattering site in the free space seen by the electron. In useful semiconductor crystals, the density of such scattering sites is small relative to the density of silicon atoms. As a result, removal of the silicon atoms through use of the effective mass leaves a somewhat sparsely populated space of scattering sites.

The perfect crystal corresponds to all atomic elements at the lattice positions and not moving, a condition which can occur only at 0 K. At temperatures above absolute zero, the atoms have thermal energy that causes them to move away from their ideal site. As the atom moves away from the nominal, equilibrium site, forces act to return it to that site, establishing the conditions for a classical oscillator problem (with the atom oscillating about its equilibrium location). Such oscillations of an atom can transfer to a neighboring atom (by energy exchange), leading to the oscillation propagating through space. This wavelike disturbance is called a *phonon* and serves as a scattering site (*phonon scattering*, also called *lattice scattering*), which can appear anywhere in the crystal.

Low Field Mobilities

The dominant scattering mechanisms in silicon are ionized impurity scattering and phonon scattering, though other mechanisms such as mentioned previously do contribute. Within this free space contaminated by scattering centers, the free electron moves at a high velocity (the thermal velocity v_{therm}) set by the thermal energy $(k_B T)$ of the electron, with $2k_B T/3 = 0.5\,m_e^* v_{therm}^2$ and, therefore, $v_{therm} = \sqrt{4k_B T/3m_e^*}$. At room temperature in Si, the thermal velocity is about 1.5×10^7 cm/s, substantially higher than most velocities that will be induced by applied electric fields or by diffusion. Thermal velocities depend reciprocally on effective mass, with semiconductors having lower effective masses displaying higher thermal velocities than semiconductors with higher effective masses. At these high thermal velocities, the electron will have an average mean time τ_n between collisions with the scattering centers during which it is moving as a free electron in free space. It is during this period between collisions that an external field acts on the electron, creating a slight displacement of the orbit of the free electron. Upon colliding, the process starts again, producing again a slight displacement in the orbit of the free electron. This displacement divided by the mean free time τ_n between collisions represents the velocity component induced by the external electric field. In the absence of such an electric field, the electron would be scattered in random directions and

TABLE 2.8 Conductivity Effective Masses for Common Semiconductors

	Ge	Si	GaAs
m_e^*	$0.12m_0$	$0.26m_0$	$0.063m_0$
m_h^*	$0.23m_0$	$0.38m_0$	0.53

display no net displacement in time. With the applied electric field, the electron has a net drift in the direction set by the field. For this reason, the induced velocity component is called the *drift velocity* and the thermal velocities can be ignored.

By using the standard force equation $F = eE = m_e^* dv/dt$ with velocity $v = 0$ at time $t = 0$ and with an acceleration time τ_n, the final velocity v_f after time τ_n is then simply $v_f = e\tau_n E/m_e^*$ and, letting $v_{\text{drift}} = v_f/2$, $v_{\text{drift}} = e\tau_n E/(2m_e^*)$. Table 2.8 gives values of the electron and hole effective masses for semiconductors.

The drift velocity v_{drift} in an external field E is seen to vary as $v_{\text{drift}} = \mu_n E$, where the electron's **low field mobility** μ_n is given approximately by $\mu_n \approx e\tau_n/(2m_e^*)$. Similarly, holes are characterized by a low field mobility $\mu_p \approx e\tau_p/(2m_h^*)$, where τ_p is the mean time between collision for holes.

This simplified mobility model yields a mobility that is inversely proportional to the effective mass. The effective electron masses in GaAs and Si are $0.09m_e$ and $0.26m_e$, respectively, suggesting a higher electron mobility in GaAs than in Si (in fact, the electron mobility in GaAs is about three times greater than that in Si). The electron and hole effective masses in Si are 0.26 and $0.38m_e$, respectively, suggesting a higher electron mobility than hole mobility in Si (in Si, $\mu_n \approx 1400 \text{ cm}^2/\text{V} \cdot \text{s}$ and $\mu_p \approx 500 \text{ cm}^2/\text{V} \cdot \text{s}$). The simplified model for μ_n and μ_p is based on the highly simplified model of the scattering conditions encountered by carriers moving with thermal energy. Far more complex analysis is necessary to obtain theoretical values of these mobilities. For this reason, the approximate model should be regarded as a guide to mobility variations among semiconductors and not as a predictive model.

The linear dependence of the mobility μ_n (μ_p) on τ_n (τ_p), suggested by the simplified development given previously, also provides a qualitative understanding of the mobility dependence on impurity doping and on temperature. As noted earlier, phonon scattering and ionized impurity scattering are the dominant mechanisms controlling the scattering of carriers in most semiconductors. At room temperature and for normally used impurity doping concentrations, phonon scattering typically dominates ionized impurity scattering. As the temperature decreases, the thermal energy of the crystal atoms decreases, leading to a decrease in the phonon scattering and an increase in the mean free time between phonon scattering events. The result is a mobility μ_{phonon} that increases with decreasing temperature according to $\mu_{\text{phonon}} \approx B_1 T^{-\alpha}$ where α is typically between 1.6 and 2.8 at 300 K. Table 2.9 gives the room temperature mobility and temperature dependence of mobility at 300 K for Ge, Si, and GaAs. In the case of ionized impurity scattering, the lower temperature leads to a lower thermal velocity, with the electrons passing ionized impurity sites more slowly and suffering a stronger scattering effect. As a result, the mobility μ_{ion} decreases with decreasing temperature. Starting at a sufficiently high temperature that phonon scattering dominates, the overall mobility will initially increase with decreasing temperature until the ionized

TABLE 2.9 Mobility and Temperature Dependence at 300 K

	Ge		Si		GaAs	
	μ_n	μ_p	μ_n	μ_p	μ_n	μ_p
Mobility, $\text{cm}^{-2}/\text{V} \cdot \text{s}$	3900	1900	1400	470	8000	340
Temperature	$T^{-1.66}$	$T^{-2.33}$	$T^{-2.5}$	$T^{-2.7}$	—	$T^{-2.3}$

Source: Adapted from Wang, S. 1989. *Fundamentals of Semiconductor Theory and Device Physics.* Prentice-Hall, Englewood Cliffs, NJ.

impurity scattering becomes dominant, leading to subsequent decreases in the overall mobility with decreasing temperature.

In Si, Ge, and GaAs, for example, the mobility increases by a factor of roughly 7 at 77 K relative to the value at room temperature, illustrating the dominant role of phonon scattering in these semiconductors under normal doping conditions.

Since scattering probabilities for different mechanisms add to yield the net scattering probability that, in turn, defines the overall mean free time between collisions, mobilities (e.g., phonon scattering mobility and lattice scattering mobility) due to different mechanisms are combined as $\mu^{-1} = \mu_{\text{phonon}}^{-1} + \mu_{\text{ion}}^{-1}$, that is, the smallest mobility dominates.

The mobility due to lattice scattering depends on the density of ionized impurity sites, with higher impurity densities leading to shorter distances between scattering sites and smaller mean free times between collisions. For this reason, the mobility shows a strong dependence on impurity doping levels at temperatures for which such scattering is dominant. As the ionized impurity density increases, the temperature at which the overall mobility becomes dominated by impurity scattering increases. In Si at room temperature, for example, $\mu_e \approx 1400$ and $\mu_p \approx 500$ for dopant concentrations below $\approx 10^{15}$ cm^3, decreasing to approximately 300 and 100, respectively, for concentrations $> 10^{18}$ cm^3.

These qualitative statements can be made substantially more quantitative by providing detailed plots of mobility vs. temperature and vs. impurity density for the various semiconductor materials. Examples of such plots are provided in several references (e.g., Tyagi, 1991; Nicollian and Brews, 1982; Shur, 1987; Sze, 1981; Böer, 1990, 1992; Haug, 1975; Wolfe, Holonyak, and Stillman, 1989; Smith, 1978; Howes and Morgan, 1985).

Diffusion results from a gradient in the carrier density. For example, the flux of electrons due to diffusion is given by $F_e = -D_n dn/dx$, with $F_p = -D_p dp/dx$ for holes. The diffusion constants D_n and D_p are related to the mobilities given by $D_n = \mu_n(k_b T/e)\mu_e$ and $D_p = (k_B T/e)\mu_p$, the so-called *Einstein relations*. In particular, the mean time between collisions t_{col} determines both the mobility and the diffusion constant.

Saturated Carrier Velocities

The mobilities discussed previously are called low field mobilities since they apply only for sufficiently small electric fields. The low field mobilities represent the scattering from distortions of the perfect lattice, with the electron gaining energy from the electric field between collisions at such distortion sites. At sufficiently high electric fields, the electron gains sufficient energy to encounter inelastic collisions with the elemental atoms of the perfect crystal. Since the density of such atoms is very high (i.e., compared to the density of scattering sites), this new mechanism dominates the carrier velocity at sufficiently high fields and causes the velocity to become independent of the field (i.e., regardless of the electric field strength, the electron accelerates to a peak velocity at which the inelastic collisions appear and the energy of the electron is lost). The electric field at which velocities become saturated is referred to as the *critical field* E_{cr}. Table 2.10 summarizes the **saturated velocities** and critical fields for electrons and holes in Si. The saturated velocities in GaAs and Ge are about 6×10^6 cm/s, slightly lower than the saturated velocity in Si.

TABLE 2.10 Saturated Velocity and Critical Field for Si at Room Temperature

	Saturated Velocity, cm/s	Critical Electric Field, V/cm
Electrons	1.1×10^7	8×10^3
Holes	9.5×10^6	1.95×10^4

Source: Adapted from Tyagi, M.S. 1991. *Introduction to Semiconductor Materials.* Wiley, New York.

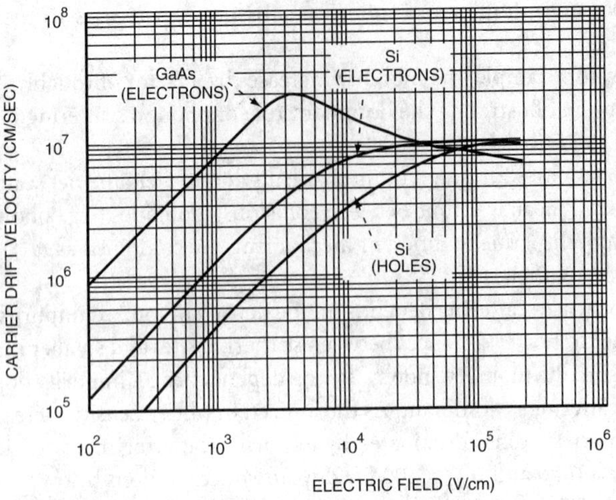

FIGURE 2.15 Velocity vs. electric field for silicon and GaAs semiconductors. (*Source:* Adapted from Sze, S.M. 1981. *Physics of Semiconductor Devices*, 2nd ed. Wiley Interscience, New York with permission.)

Figure 2.15 shows representative velocity vs. electric field characteristics for electrons in silicon and in GaAs. The linear dependence at small fields illustrates the low field mobility, with GaAs having a larger low field mobility than silicon. The saturated velocities v_{sat}, however, do not exhibit such a large difference between Si and GaAs. Also, the saturated velocities do not exhibit as strong a temperature dependence as the low field mobilities (since the saturated velocities are induced by large electric fields, rather than being perturbations on thermal velocities).

Figure 2.15 also exhibits an interesting feature in the case of GaAs. With increasing electric field, the velocity initially increases beyond the saturated velocity value, falling at higher electric fields to the saturated velocity. This *negative differential mobility* region has been discussed extensively as a potential means of achieving device speeds higher than would be obtained with fully saturated velocities. Table 2.11 summarizes the peak velocities for various semiconductors.

As device dimensions have decreased, the saturated velocities have become a more severe limitation. With critical fields in silicon about 10^4 V/cm, one volt across one micron leads to saturated velocities. Rather than achieving a current proportional to applied voltage (as in the case of the low field mobility condition), currents become largely independent of voltage under saturated velocity conditions. In addition, the emergence of saturated velocities also compromises the speed advantages implied by high mobilities for some semiconductors (e.g., GaAs vs. Si). In particular, although higher low field velocities generally lead to higher device speeds under low field conditions, the saturated velocities give similar speed performance at high electric fields.

TABLE 2.11 Saturated Velocity and
Critical Field for Various Semiconductors

Semiconductor	Peak Velocity, cm/s
AsAs	6×10^6
AlSb	7×10^6
GaP	1.2×10^7
GaAs	2×10^7
PbTe	1.7×10^7
InP	4×10^7
InSb	7×10^7

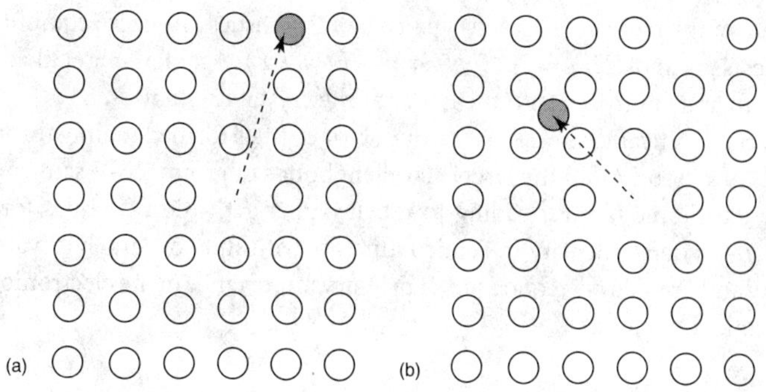

FIGURE 2.16 Point defects in semiconductors: (a) Schottky defects, (b) Frenkle defects.

2.2.5 Crystalline Defects

A variety of defects occur in semiconductor crystals, many of which lead to degradations in performance and require careful growth and fabrication conditions to minimize their occurrence. Other defects are relatively benign. This section summarizes several of the defect types.

Point Defects

The point defects correspond to a lattice atom missing from its position. Two distinct cases appear, as shown in Fig. 2.16. *Schottky defects*, shown in Fig. 2.16(a), result when an atom is missing from a lattice site. Typically the atom is assumed to have migrated to the surface (during crystal growth) where it takes a normal position at a surface lattice site. An energy E_s is required to move a lattice atom to the crystal surface, that energy serves as an activation energy for Schottky defects. At temperature T, the equilibrium concentration N_{sd} of Schottky defects is given by $N_{sd} = N_L \exp(-E_{sd}/kT)$, where N_L is the density of lattice atoms, and there is necessarily some nonzero concentration of such defects appearing while the crystal is at the high temperatures seen during growth. The high-temperature defect densities are frozen into the lattice as the crystal cools.

Frenkle defects result when an atom moves away from a lattice site and assumes a position between lattice sites (i.e., takes an interstitial position), as shown in Fig. 2.17(b). The Frenkle defect, therefore, corresponds to a pair of defects, namely, the empty lattice site and the extra interstitially positioned atom. The activation

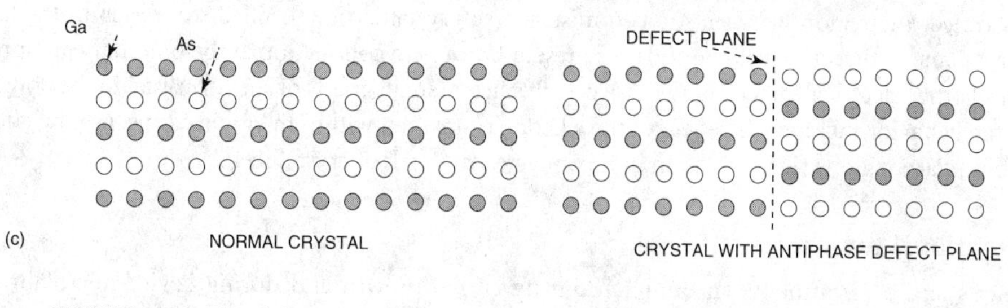

FIGURE 2.17 Line defects: (a) edge dislocation, (b) screw dislocation, (c) antiphase defect.

energy E_{fd} required for formation of this defect pair again establishes a nonzero equilibrium concentration N_{fd} of Frenkle defects given by $N_{fd} = \sqrt{N_L N_I} \exp(-E_{fd}/kT)$. Again, the Frenkle defects tend to form during crystal growth and are frozen into the lattice as the crystal crystallizes.

These point defects significantly impact semiconductor crystals formed mainly through ionic bonding. The covalent bonds of group IV and the largely covalent bonds of group III–V semiconductors, however, are much stronger than ionic bonds, leading to much higher activation energies for point defects and a correspondingly lower defect density in semiconductors with fully or largely covalent bonds. For this reason, the Schottky and Frenkle defects are not of primary importance in the electronic behavior of typical IV–IV and III–V semiconductors.

Line Defects

Three major types of line defects (*edge dislocations, screw dislocations*, and *antiphase defects*) are summarized here. The line defects are generally important factors in the electrical behavior of semiconductor devices since their influence (as trapping centers, scattering sites, etc.) extends over distances substantially larger than atomic spacings.

Edge dislocations correspond to an extra plane inserted orthogonal to the growth direction of a crystal, as illustrated in Fig. 2.17(a). The crystalline lattice is disrupted in the region near where the extra plane starts, leaving behind a line of *dangling bonds*, as shown. The dangling bonds can lead to a variety of effects. The atoms at the start of the dislocation are missing a shared electron in their outer band, suggesting that they may act as traps, producing a linear chain of trap sites with interatomic spacings (vastly smaller than normally encountered between intentionally introduced acceptor impurities. In addition to their impact on the electrical performance, such defects can compromise the mechanical strength of the crystal.

Screw dislocations result from an extra plane of atoms appearing at the surface. In this case (Fig. 2.17(b)), the growth process leads to a spiral structure growing vertically (from the edge of the extra plane). The change in lattice structure over an extended region as the crystal grows can introduce a variety of changes in the etching properties of the crystal, degraded junctions, etc. In addition, the dangling bonds act as traps, impacting electrical characteristics.

Antiphase defects occur in compound semiconductors. Figure 2.17(c) illustrates a section of a III–V semiconductor crystal in which the layers of alternating Ga and As atoms on the right side is one layer out of phase relative to the layers on the left side. This phase error leads to bonding defects in the plane joining the two sides, impacting the electrical performance of devices. Such defects require precision engineering of the growth of large diameter crystals to ensure that entire planes form simultaneously.

Stacking Faults and Grain Boundaries

Stacking faults result when an extra, small area plane (platelet) of atoms occurs within the larger crystal. The result is that normal planes are distorted to extend over that platelet (Fig. 2.18(a)). Here, a stack of defects appears above the location of the platelet. Electrical properties are impacted due to traps and other effects.

Grain boundaries appear when two growing microcrystals with different crystallographic orientations merge, leaving a line of defects along the plane separating the two crystalline lattices as shown in Fig. 2.18(b). As illustrated, the region between the two misoriented crystal lattices is filled from each of the lattices, leaving a plane of defects. Grain boundaries present barriers to conduction at the boundary and produce traps and other electrical effects. In the case of polysilicon (small grains of silicon crystal in the polysilicon area), the behavior includes the effect of the silicon crystal seen within the grains as well as the effects of grain boundary regions (i.e., acts as an interconnected network of crystallites).

Unintentional Impurities

Chemicals and environments encountered during crystal growth and during device microfabrication contain small amounts of impurities that can be incorporated into the semiconductor crystal lattice.

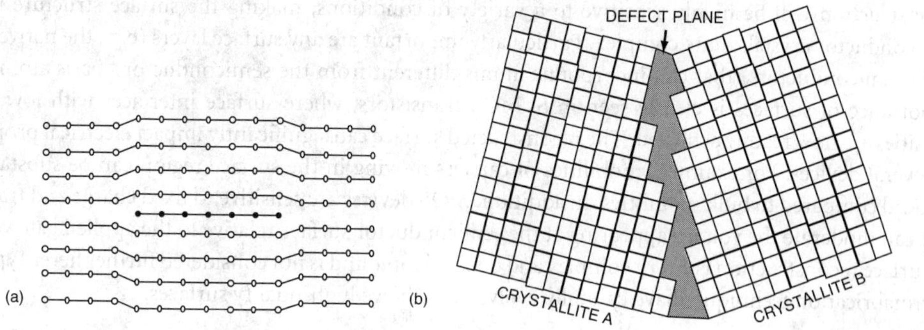

FIGURE 2.18 (a) Stacking faults, (b) grain boundary defects.

Some unintentional impurities replace atoms at lattice sites and are called **substitutional impurities.** Others take positions between lattice sites and are called **interstitial impurities**. Some impurities are benign in the sense of not strongly impacting the behavior of devices. Others are beneficial, improving device performance. Examples include hydrogen, which can compensate dangling bonds and elements that give rise to deep energy trapping levels (i.e., energy levels near the center of the band gap). Such deep level trapping levels are important in indirect gap semiconductors in establishing recombination time constants. In fact, gold was deliberately incorporated in earlier silicon bipolar transistors to increase the recombination rate and achieve faster transistors. Finally, other impurities are detrimental to the device performance.

Optoelectronic devices are often more sensitive to unintentionally occurring impurities. A variety of characteristic trap levels are associated with several of the defects encountered. Table 2.12 summarizes several established trap levels in GaAs, some caused by unintentional impurities and others caused by defects. The strategy for semiconductor material growth and microfabrication is, therefore, a complex strategy, carefully minimizing those unintentional impurities (e.g., sodium in the gate oxides of silicon metal oxide semiconductor (MOS) transistors), selectively adding impurities at appropriate levels to advantage, and maintaining a reasonable strategy regarding unintentional impurities which are benign.

Surface Defects: The Reconstructed Surface

If one imagines slicing a single crystal along a crystallographic plane, a very high density of atomic bonds is broken between the two halves. Such dangling bonds would represent the surface of the crystal, if the crystal structure extended fully to the surface. The atomic lattice and broken bonds implied by the slicing, however, do not represent a minimum energy state for the surface. As a result, there is a reordering of bonds and atomic sites at the surface to achieve a minimum energy structure. The process is called *surface reconstruction* and leads to a substantially different lattice structure at the surface. This surface

TABLE 2.12   Trap States (with Energy E_t) in GaAs

Type	$E_c - E_t$	Name	Type	$E_t - E_v$	Name
Shallow donor	≈5.8 meV		Shallow acceptor	≈10 meV	
Oxygen donor	0.82 eV	EL2	Tin acceptor	0.17 eV	
Chromium acceptor	0.61 eV	EL1	Copper acceptor	0.42 eV	
Deep acceptor	0.58 eV	EL3	Hole trap	0.71 eV	HB2
Electron trap	0.90 eV	EB3	Hole trap	0.29 eV	HB6
Electron trap	0.41 eV	EB6	Hole trap	0.15 eV	HC1

Source: Adapted from Shur, M. 1987. *GaAs Devices and Circuits.* Plenum Press, New York.

reconstruction will be highly sensitive to a variety of conditions, making the surface structure in real semiconductor crystals quite complex. Particularly important are any surface layers (e.g., the native oxide on Si semiconductors) that can incorporate atoms different from the semiconductor's basis atoms. The importance of surfaces is clearly seen in Si MOS transistors, where surface interfaces with low defect densities are now readily achieved. The reconstructed surface can significantly impact electrical properties of several devices. For example, mobilities of carriers moving in the surface region can be substantially reduced compared to bulk mobilities. In addition, MOS devices are sensitive to fixed charge and trap sites that can moderate the voltage appearing at the semiconductor surface relative to the applied gate voltage.

Surface reconstruction is a very complex and detailed topic and is not considered further here. Typically, microfabrication techniques have carefully evolved to achieve high-quality surfaces.

2.2.6 Summary

A rich diversity of semiconductor materials, led by the extraordinarily advanced device technologies for silicon microelectronics and III–V optoelectronics, has been explored for a wide range of applications. Much of the computing, information, and consumer electronics revolutions expected over the next decade will rest of the foundation of these important crystalline materials. As established semiconductor materials such as silicon continue to define the frontier for advanced fabrication of very small devices, new materials are emerging or being reconsidered as possible additional capabilities for higher speed, lower power, and other advantages. This chapter has provided a broad overview of semiconductor materials. Many fine and highly readable books, listed in the references, provide the additional depth which could not be provided in this brief chapter. Tables of various properties are provided throughout the chapter, with much of the relevant information provided in Table 2.13.

Defining Terms

Ambipolar semiconductors: Semiconductors that can be selectively doped to achieve either *n* type or *p* type material.

Amphoteric impurities: Doping impurities that may act as either a donor or an acceptor.

Bandgap: Energy difference between the conduction band and the valence band.

Compound semiconductors: Semiconductor crystals composed of two or more atomic elements from different groups of the periodic chart.

Conduction band: Energy level for free electrons at rest in the semiconductor crystal.

Deep energy level impurities: Doping impurities or other impurities whose energy level lies toward the center of the bandgap. Important for carrier recombination in indirect gap semiconductors.

Direct gap semiconductor: Semiconductor whose conduction band minimum and valance band minimum appear at the same wave vector (same momenta); important for optical sources.

Effective mass: Value of carrier mass used to represent motion of a carrier in the semiconductor as though the motion were in free space.

Elemental semiconductors: Semiconductor crystals composed of a single atomic element.

Indirect gap semiconductor: Semiconductor whose conduction band minimum and valence band minimum appear at different wave vectors (different momenta).

Intrinsic semiconductors: Semiconductors with no intentional or nonintentional impurities (dopants).

Low field mobility: Proportionality constant between carrier velocity and electric field.

Quaternary semiconductors: Compound semiconductors with two atomic elements from one group and two atomic elements from a second group of the periodic chart.

Saturated velocities: Maximum carrier velocity induced by electric field (due to onset of inelastic scattering).

Shallow energy level dopants: Doping impurities whose energy level lies very close to the conduction (valence) band for donors (acceptors).

Substitutional impurities: Impurities that replace the crystal's base atom at that base atom's lattice position.

TABLE 2.13 Properties of GaAs and Si Semiconductors at Room Temperature (300 K), CdS and CdSe Both Can Appear in Either Zinc Blende or Wurtzite Lattice Forms

	Lattice	Lattice Constant, Å	Energy Gap, eV	Electron Effective Mass	Hole Effective Mass	Dielectric Constant	Electron Mobility, cm²/V-s	Hole Mobility, cm²/V-s
C	Diamond	3.5668	5.47 (I)[a]	0.2	0.25	5.7	1,800	1,200
Si	Diamond	5.4310	1.11 (I)	m_l: 0.98	m_l: 0.16	11.7	1,350	480
				m_t: 0.19	m_h: 0.5			
Ge	Diamond	5.6461	0.67 (I)	m_l: 1.58	m_l: 0.04	16.3	3,900	1,900
				m_t: 0.08	m_h: 0.3			
AlP	Zinc blende	5.4625	2.43	0.13		9.8	80	
AlAs	Zinc blende	5.6605	2.16 (I)	0.5	m_l: 0.49	12.0	1,000	180
					m_h: 1.06			
AlSb	Zinc blende	6.1355	1.52 (I)	0.11	m_l: 0.39	11	900	400
GaN	Wurtzite	3.189	3.4	0.2	0.8	12	300	
GaP	Zinc blende	5.4506	2.26 (I)	0.13	0.67	10	300	150
GaAs	Zinc blende	5.6535	1.43 (D)[b]	0.067	m_l: 0.12	12.5	8,500	400
					m_h: 0.5			
GaSb	Zinc blende	6.0954	0.72 (D)	0.045	0.39	15.0	5,000	1,000
InAs	Zinc blende	6.0584	0.36 (D)	0.028	0.33	12.5	22,600	200
InSb	Zinc blende	6.4788	0.18 (D)	0.013	0.18	18	100,000	1,700
InP	Zinc blende	5.8687	1.35 (D)	0.077	m_l: 0.12	12.1	4,000	600
					m_h: 0.60			
CdS	Zinc blende	5.83	2.42 (D)	0.2	0.7	5.4	340	50
CdSe	Zinc blende	6.05	1.73 (D)	0.13	0.4	10.0	800	
CdTe	Zinc blende	6.4816	1.50 (D)	0.11	0.35	10.2	1,050	100
PbS	NaCl	5.936	0.37 (I)	0.1	0.1	17.0	500	600
PbSe	NaCl	6.147	0.26 (I)	m_l: 0.07	m_l: 0.06	23.6	1,800	930
				m_t: 0.039	m_h: 0.03			
PbTe	NaCl	6.45	0.29 (I)	m_l: 0.24	m_l: 0.3	30	6,000	4,100
				m_t: 0.02	m_h: 0.02			

[a] Indirect bands.
[b] Direct bands.

Source: Adapted from Wolfe, C.M., Holonyak, N., and Stillman, G.E. 1989. *Physical Properties of Semiconductors.* Prentice-Hall, New York.

Ternary semiconductors: Compound semiconductors with two atomic elements from one group and one atomic element from a second group of the periodic chart.

Unipolar semiconductors: Semiconductors that can be selectively doped to achieve only *n* type or only *p* type material.

Valence band: Energy for bound electrons and for free holes at rest in the semiconductor crystal.

References

Beadle, W.E., Tsai, J.C.C., and Plummer, R.D., eds., 1985. *Quick Reference Manual for Silicon Integrated Circuit Technology.* Wiley, New York.

Böer, K.W. 1990. *Survey of Semiconductor Physics, Vol. 1: Electrons and Other Particles in Bulk Semiconductors.* Van Nostrand, New York.

Böer, K.W. 1992. *Survey of Semiconductor Physics, Vol. 2: Barriers, Junctions, Surfaces, and Devices.* Van Nostrand, New York.

Capasso, F. and Margaritondo, G., eds. 1987. *Heterojunction and Band Discontinuities.* North Holland, Amsterdam.

Haug, A. 1975. *Theoretical Solid State Physics, Vols. 1 and 2.* Pergamon Press, Oxford, England.

Howes, M.J. and Morgan, D.V. 1985. *Gallium Arsenide: Materials, Devices, and Circuits.* Wiley, New York.

Irvine, S.J.C., Lum, B., Mullin, J.B., and Woods, J., eds. 1982. *II–VI Compounds 1982.* North Holland, Amsterdam.

Lannoo, M. and Bourgoin, J. 1981. *Point Defects in Semiconductors.* Springer-Verlag, Berlin.

Loewrro, M.H., ed. 1985. *Dislocations and Properties of Real Materials.* Inst. of Metals, London.

Moss, T.S. and Balkanski, M., eds. 1980. *Handbook on Semiconductors, Vol. 2: Optical Properties of Solids.* North Holland, Amsterdam.

Moss, T.S. and Keller, S.P., eds. 1980. *Handbook on Semiconductors, Vol. 3: Material Properties and Preparation.* North Holland, Amsterdam.

Moss, T.S. and Paul, W. 1982. *Handbook on Semiconductors, Vol. 1: Band Theory and Transport Properties.* North Holland, Amsterdam.

Nicollian, E.H. and Brews, J.R. 1982. *MOS Physics and Technology.* Wiley, New York.

Pantelides, S.T. 1986. *Deep Centers in Semiconductors.* Gordon & Breach Science, New York.

Shur, M. 1987. *GaAs Devices and Circuits.* Plenum Press, New York.

Smith, R.A. 1978. *Semiconductors.* Cambridge University Press, London.

Sze, S.M. 1981. *Physics of Semiconductor Devices,* 2nd ed. Wiley Interscience, New York.

Tyagi, M.S. 1991. *Introduction to Semiconductor Materials.* Wiley, New York.

Wang, S. 1989. *Fundamentals of Semiconductor Theory and Device Physics.* Prentice-Hall, Englewood Cliffs, New Jersey.

Willardson, A.K. and Beer, A.C., eds. 1985. *Semiconductors and Semimetals, Vol. 22.* Academic Press, New York.

Wilmsen, C.W., ed. 1985a. *Physics and Chemistry of III–V Compounds.* Plenum Press, New York.

Wilmsen, C.W., ed. 1985b. *Physics and Chemistry of III–V Compound Semiconductor Interfaces.* Plenum Press, New York.

Wolfe, C.M., Holonyak, N., and Stillman, G.E. 1989. *Physical Properties of Semiconductors.* Prentice-Hall, New York.

Further Information

The list of references focuses on books that have become popular in the field. The reader is advised to refer to such books for additional detail on semiconductor materials. The Institute for Electrical and Electronics Engineers (IEEE, Piscataway, NJ) publishes several journals that provide information on recent advances in materials and their uses in devices. Examples include the *IEEE Trans. Electron Devices*, the *IEEE Solid State Circuits Journal*, the *IEEE Journal on Quantum Electronics*, and the *IEEE Journal on Lightwave Technology*. The American Physical Society also publishes several journals focusing on semiconductor materials. One providing continuing studies of materials is the *Journal of Applied Physics*. The Optical Society of America (OSA, Washington, DC) publishes *Applied Optics*, a journal covering many practical issues as well as new directions. The Society of Photo-Optical Instrumentation Engineers (SPIE, Bellingham, WA) sponsors a broad range of conferences, with an emphasis on optoelectronic materials but also covering other materials. The SPIE can be contacted directly for a list of its extensive set of conference proceedings.

2.3 Thermal Properties

David F. Besch

2.3.1 Introduction

The rating of an electronic or electrical device depends on the capability of the device to dissipate heat. As miniaturization continues, engineers are more concerned about heat dissipation and the change in properties of the device and its material makeup with respect to temperature.

The following section focuses on heat and its result. Materials may be categorized in a number of different ways. In this chapter, materials will be organized in the general classifications according to their resistivities:

- Insulators
- Semiconductors
- Conductors

It is understood that with this breakdown, some materials will fit naturally into more than one category. Ceramics, for example, are insulators, yet with alloying of various other elements, can be classified as semiconductors, resistors, a form of conductor, and even conductors. Although, in general, the change in resistivity with respect to temperature of a material is of interest to all, the design engineer is more concerned with how much a resistor changes with temperature and if the change drives the circuit parameters out of specification.

2.3.2 Fundamentals of Heat

In the commonly used model for materials, heat is a form of energy associated with the position and motion of the material's molecules, atoms, and ions. The position is analogous with the state of the material and is potential energy, whereas the motion of the molecules, atoms, and ions is kinetic energy. Heat added to a material makes it hotter and vice versa. Heat also can melt a solid into a liquid and convert liquids into gases, both changes of state. Heat energy is measured in calories (cal), British thermal units (Btu), or joules (J). A calorie is the amount of energy required to raise the temperature of one gram (1 g) of water one degree Celsius (1°C) (14.5 to 15.5°C). A Btu is a unit of energy necessary to raise the temperature of one pound (1 lb) of water by one degree Fahrenheit (1°F). A joule is an equivalent amount of energy equal to work done when a force of one newton (1 N) acts through a distance of one meter (1 m). Thus heat energy can be turned into mechanical energy to do work. The relationship among the three measures is: 1 Btu = 251.996 cal = 1054.8 J.

Temperature

Temperature is a measure of the average kinetic energy of a substance. It can also be considered a relative measure of the difference of the heat content between bodies. Temperature is measured on either the Fahrenheit scale or the Celsius scale. The Fahrenheit scale registers the freezing point of water as 32°F and the boiling point as 212°F. The Celsius scale or centigrade scale (old) registers the freezing point of water as 0°C and the boiling point as 100°C.

The Rankine scale is an absolute temperature scale based on the Fahrenheit scale. The Kevin [Kelvin] scale is an absolute temperature scale based on the Celsius scale. The absolute scales are those in which zero degree corresponds with zero pressure on the hydrogen thermometer. For the definition of temperature just given, zero °R and zero K register zero kinetic energy.

The four scales are related by the following:

$$°C = 5/9(°F - 32)$$
$$°F = 9/5(°C) + 32$$
$$K = °C + 273.16$$
$$°R = °F + 459.69$$

Heat Capacity

Heat capacity is defined as the amount of heat energy required to raise the temperature of one mole or atom of a material by 1°C without changing the state of the material. Thus it is the ratio of the change in heat energy of a unit mass of a substance to its change in temperature. The heat capacity, often called thermal capacity, is a characteristic of a material and is measured in cal/g per °C or Btu/lb per °F,

$$c_p = \frac{\partial H}{\partial T}$$

Specific Heat

Specific heat is the ratio of the heat capacity of a material to the heat capacity of a reference material, usually water. Since the heat capacity of water is 1 Btu/lb and 1 cal/g, the specific heat is numerically equal to the heat capacity.

Thermal Conductivity

Heat transfers through a material by conduction resulting when the energy of atomic and molecular vibrations is passed to atoms and molecules with lower energy. In addition, energy flows due to free electrons,

$$Q = kA\frac{\partial T}{\partial l}$$

where
Q = heat flow per unit time
k = thermal conductivity
A = area of thermal path
l = length of thermal path
T = temperature

The coefficient of thermal conductivity k is temperature sensitive and decreases as the temperature is raised above room temperature.

Thermal Expansion

As heat is added to a substance the kinetic energy of the lattice atoms and molecules increases. This, in turn, causes an expansion of the material that is proportional to the temperature change, over normal temperature ranges. If a material is restrained from expanding or contracting during heating and cooling, internal stress is established in the material.

$$\frac{\partial l}{\partial T} = \beta_L l \quad \text{and} \quad \frac{\partial V}{\partial T} = \beta_V V$$

where
l = length
V = volume
T = temperature
β_L = coefficient of linear expansion
β_V = coefficient of volume expansion

Solids

Solids are materials in a state in which the energy of attraction between atoms or molecules is greater than the kinetic energy of the vibrating atoms or molecules. This atomic attraction causes most materials to form into a crystal structure. Noncrystalline solids are called *amorphous*, including glasses, a majority of plastics, and some metals in a semistable state resulting from being cooled rapidly from the liquid state. Amorphous materials lack a long range order.

Crystalline materials will solidify into one of the following geometric patterns:

- Cubic
- Tetragonal

- Orthorhombic
- Monoclinic
- Triclinic
- Hexagonal
- Rhombohedral

Often the properties of a material will be a function of the density and direction of the lattice plane of the crystal.

Some materials will undergo a change of state while still solid. As it is heated, pure iron changes from body centered cubic to face centered cubic at 912°C with a corresponding increase in atomic radius from 0.12 to 0.129 nm due to thermal expansion. Materials that can have two or more distinct types of crystals with the same composition are called *polymorphic*.

Liquids

Liquids are materials in a state in which the energies of the atomic or molecular vibrations are approximately equal to the energy of their attraction. Liquids flow under their own mass. The change from solid to liquid is called *melting*. Materials need a characteristic amount of heat to be melted, called the *heat of fusion*. During melting the atomic crystal experiences a disorder that increases the volume of most materials. A few materials, like water, with **stereospecific** covalent bonds and low packing factors attain a denser structure when they are thermally excited.

Gases

Gases are materials in a state in which the kinetic energies of the atomic and molecular oscillations are much greater than the energy of attraction. For a given pressure, gas expands in proportion to the absolute temperature. For a given volume, the absolute pressure of a gas varies in proportion to the absolute pressure. For a given temperature, the volume of a given weight of gas varies inversely as the absolute pressure. These three facts can be summed up into the Gas law:

$$PV = RT$$

where
- P = absolute pressure
- V = specific volume
- T = absolute temperature
- R = universal gas constant t

Materials need a characteristic amount of heat to transform from liquid to solid, called the *heat of vaporization*.

2.3.3 Other Material Properties

Insulators

Insulators are materials with resistivities greater than about $10^7 \, \Omega \cdot \text{cm}$. Most ceramics, plastics, various oxides, paper, and air are all insulators. Alumina (Al_2O_3) and beryllia (BeO) are ceramics used as substrates and chip carriers. Some ceramics and plastic films serve as the dielectric for capacitors.

Dielectric Constant

A capacitor consists of two conductive plates separated by a dielectric. Capacitance is directly proportional to the dielectric constant of the insulating material. Ceramic compounds doped with barium titanate have high dielectric constants and are used in capacitors. Plastics, such as mica, polystyrene, polycarbonate, and polyester films also serve as dielectrics for capacitors. Capacitor values are available with both positive and negative changes in value with increased temperature. See the first subsection in Sec. 2.3.4 for a method to calculate the change in capacitor values at different temperatures.

Resistivity

The resistivity of insulators typically decreases with increasing temperature.

Semiconductors

Semiconductors are materials that range in resistivity from approximately 10^{-4} to 10^{+7} $\Omega \cdot$ cm. Silicon (Si), Germanium (Ge), and Gallium Arsenide (GaAs) are typical semiconductors. The resistivity and its inverse, the conductivity, vary over a wide range, due primarily to doping of other elements. The conductivity of intrinsic Si and Ge follows an exponential function of temperature,

$$\sigma = \sigma_0 e^{\frac{E_g}{2kT}}$$

where
  σ = conductivity
  σ_0 = constant t
  E_g = 1.1 eV for Si
  k = Bolzmann's constant t
  T = temperature $^\circ$K

Thus, the electrical conductivity of Si increases by a factor of 2400 when the temperatures rises from 27 to 200 K.

Conductors

Conductors have resistivity value less than 10^{-4} $\Omega \cdot$ cm and include metals, metal oxides, and conductive nonmetals. The resistivity of conductors typically increases with increasing temperature as shown in Fig. 2.19

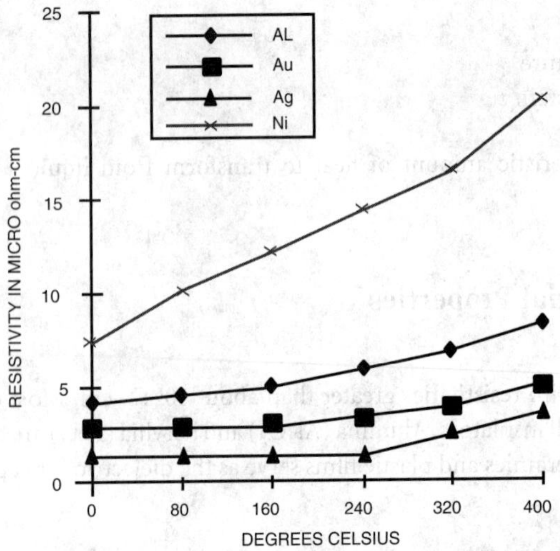

FIGURE 2.19 Resistivity in a function of temperature.

TABLE 2.14 Characteristics of certain alloys

Sn (%)	Pb (%)	Ag (%)	°C
60	40	—	190
60	38	2	192
10	90	—	302
90	10	—	213
95	5	5	230

Melting Point

Solder is an important material used in electronic systems. The tin-lead solder system is the most used solder compositions. The system's equilibrium diagram shows a typical eutectic at 61.9% Sn. Alloys around the eutectic are useful for general soldering. High Pb content solders have up to 10% Sn and are useful as high-temperature solders. High Sn solders are used in special cases such as in high corrosive environments. Some useful alloys are listed in Table 2.14.

2.3.4   Engineering Data

Graphs of resistivity and dielectric constant vs. temperature are difficult to translate to values of electronic components. The electronic design engineer is more concerned with how much a resistor changes with temperature and if the change drives the circuit parameters out of specification. The following defines the commonly used terms for components related to temperature variation.

Temperature Coefficient of Capacitance

Capacitor values vary with temperature due to the change in the dielectric constant with temperature change. The temperature coefficient of capacitance (TCC) is expressed as this change in capacitance with a change in temperature.

$$\text{TCC} = \frac{1}{C}\frac{\partial C}{\partial T}$$

where
  TCC = temperature coefficient of capacitance
  C = capacitor value
  T = temperature

The TCC is usually expressed in parts per million per degree Celsius (ppm/°C). Values of TCC may be positive, negative, or zero. If the TCC is positive, the capacitor will be marked with a P preceding the numerical value of the TCC. If negative, N will precede the value. Capacitors are marked with NPO if there is no change in value with a change in temperature. For example, a capacitor marked N1500 has a $-1500/1,000,000$ change in value per each degree Celsius change in temperature.

Temperature Coefficient of Resistance

Resistors change in value due to the variation in resistivity with temperature change. The temperature coefficient of resistance (TCR) represents this change. The TCR is usually expressed in parts per million per degree Celsius (ppm/°C).

$$\text{TCR} = \frac{1}{R}\frac{\partial R}{\partial T}$$

TABLE 2.15   TCR values of common resistors

Resistor Type	TCR, ppm/°C		
Carbon composition	+500	to	+2000
Wire wound	+200	to	+500
Thick film	+20	to	+200
Thin film	+20	to	+100
Base diffused	+1500	to	+2000
Emitter diffused	+600		
Ion implanted	±100		

where
TCR = temperature coefficient of resisitance
R   = resistance value
T   = temperature

Values of TCR may be positive, negative, or zero. TCR values for often used resistors are shown in Table 2.15. The last three TCR values refer to resistors imbedded in silicon monolithic integrated circuits.

Temperature Compensation

Temperature compensation refers to the active attempt by the design engineer to improve the performance and stability of an electronic circuit or system by minimizing the effects of temperature change. In addition to utilizing optimum TCC and TCR values of capacitors and resistors, the following components and techniques can also be explored.

- Thermistors
- Circuit design stability analysis
- Thermal analysis

Thermistors

Thermistors are semiconductor resistors that have resistor values that vary over a wide range. They are available with both positive and negative temperature coefficients and are used for temperature measurements and control systems, as well as for temperature compensation. In the latter they are utilized to offset unwanted increases or decreases in resistance due to temperature change.

Circuit Analysis

Analog circuits with semiconductor devices have potential problems with bias stability due to changes in temperature. The current through junction devices is an exponential function as follows:

$$i_D = I_S \left(e^{\frac{q v_D}{nkT}} - 1 \right)$$

where
i_D = junction current
I_S = saturation current
v_D = junction voltage
q = electron charge
n = emission coefficient
k = Boltzmann's constant
T = temperature, in 0 K

Junction diodes and bipolar junction transistor currents have this exponential form. Some biasing circuits have better temperature stability than others. The designer can evaluate a circuit by finding its fractional temperature coefficient,

$$TC_F = \frac{1}{v(T)} \frac{\partial v(T)}{\partial T}$$

where

$v(T)$ = circuit variable
TC_F = temperature coefficient
T = temperature

Commercially available circuit simulation programs are useful for evaluating a given circuit for the result of temperature change. SPICE, for example, will run simulations at any temperature with elaborate models included for all circuit components.

Thermal Analysis

Electronic systems that are small or that dissipate high power are subject to increases in internal temperature. Thermal analysis is a technique in which the designer evaluates the heat transfer from active devices that dissipate power to the ambient.

Defining Terms

Eutectic: Alloy composition with minimum melting temperature at the intersection of two solubility curves.
Stereospecific: Directional covalent bonding between two atoms.

References

Guy, A.G. 1967. *Elements of Physical Metallurgy*, 2nd ed., pp. 255–276. Addison-Wesley, Reading, MA.
Incropera, F.P. and Dewitt, D.P. 1990. *Fundamentals of Heat and Mass Transfer*, 3rd ed., pp. 44–66. Wiley, New York.

Further Information

Additional information on the topic of thermal properties of materials is available from the following sources:

Banzhaf, W. 1990. *Computer-Aided Circuit Analysis Using Psice*. Prentice-Hall, Englewood Cliffs, NJ.
Smith, W.F. 1990. *Principles of Material Science and Engineering*. McGraw-Hill, New York.
Van Vlack, L.H. 1980. *Elements of Material Science and Engineering*. Addison-Wesley, Reading, MA.

2.4 Resistive Materials

Jan H. Mooij

2.4.1 Introduction

Resistance was defined by Ohm as the ratio between the potential difference between the ends of a conductor and the current flowing through the conductor. The standard unit of resistance is the *Absolute Ohm*, generally denoted Ω. It is defined as the resistance over a conductor that has a potential drop of 1 V when a current of 1 A is flowing through it.

Resistance is by no means a constant. It may depend on voltage or current, as well as temperature, humidity and other parameters. Every device that is not superconducting has a resistance. This is true even for interconnections, which usually have a very low resistance, or insulators where it is extremely high. It is difficult to give a good definition of a resistor. Almost any device may be considered a resistor. Here a resistor may be defined as an electronic device that is primarily meant to provide a specified resistance under specified conditions. This resistance may be constant, it may depend on some external parameter or it can also be adjustable.

This definition is rather vague, which is in accordance with the practical situation. Devices such as heating elements and incandescent lamps are made for other purposes, but under certain circumstances they can be used as a resistor. It is rather easy to measure a resistance, and so sensors based on resistance variations are popular. On the other hand, standard resistors are sometimes used as heating elements or sensors.

The requirements that must be met by resistors are extremely variable. Therefore, many types of resistors exist. A first division may be in:

- **Fixed resistors**, which in first approximation should have a constant resistance, independent of temperature, voltage, etc.
- **Variable resistors**, for which the resistance value can be changed by the user, but must be independent of other outside influences.
- **Nonlinear resistors**, which have been specially designed to depend strongly on a parameter such as temperature or voltage.

2.4.2   Fixed Resistors

By far the largest number of resistors are fixed resistors. Traditional fixed resistors usually are cylindrical with solderable leads. Here several groups may be distinguished.

Wire Wound Resistors

As the name indicates, for wire wound resistors the resistive element consists of a wire. To get sufficient length this wire is wound around a ceramic rod or tube. The ends of the wire are welded or soldered to caps or rings, which are attached to the rod. In turn, these caps or rings are attached to the leads, which may be wires or tags.

Because of the mechanical requirements of winding, in practice all resistive wires are made of metal, usually alloys based on CuNi or NiCr. The useful **resistivity** range in metals is small and, therefore, the resistance can primarily be varied by variation of the length and the diameter of the wire. Restrictions for maximum dimensions limit the resistance range for the standard dimensions and a good **temperature coefficient of resistance (TCR)** to 1–100,000 Ω. A lower value can be reached by using a wire with a low resistivity, but this will result in a higher TCR. Higher values need larger resistors, which can accommodate more wire length.

When insulation is provided with an organic coating the maximum hot spot temperature may not exceed 300°C. Even then a special coating must be used to prevent degradation of the coating. With an enamel coating, higher temperatures up to 450°C may be used.

As a result of the high allowed surface temperature the maximum load of wire wound resistors is relatively high. Their behavior under pulse loads is also good, and they have a low current noise. Because of the windings they tend to have a high inductance. Special resistors are produced in which two layers of wire are wound in opposite directions to minimize the inductance.

Carbon Composition Resistors

The resistive element of carbon composition resistors is a rod that consists of a mixture of carbon powder and organic binder. The outside of the rod is covered with an insulation layer. Wires are introduced into the ends of the conductive rod. It follows from the dimensions of the rod that the resistivity of the material must be high to get the usual resistance values. Such materials have rather bad stability at elevated temperatures; the TCR is also very negative. They tend to be replaced by film resistors in newer designs. One advantage is their good pulse behavior, especially for single pulses.

Carbon composition resistors are provided in dimensions ranging from 2 to 5 mm diameter. Typically, they have a tolerance of $\pm 20\%$ and a TCR ranging from -500 (low values) to -2000 ppm/$°$C (high values).

Film Resistors

Film resistors are made with many different kinds of films, but their basic design is rather constant. A cylindrical ceramic core is covered with a film. Caps that are pressed over this film are used as contacts. To adjust the value of the resistor, a spiral is cut or lasered in the film. During trimming the resistance is measured continuously, and trimming is stopped when the required value has been reached. In this way the initial tolerance of film resistors can be low. Wires are welded to the caps. Insulation and protection against moisture and pollution is provided by an organic coating. Such a coating is very important, as the resistive films can be extremely thin. Such films can easily be damaged by chemical attack, especially under humid conditions.

The marking of film resistors is usually done with a color code. This consists of a number of colored rings around the resistor. For resistors with a tolerance down to $\pm 5\%$ four rings are used: numbers one and two give the first two digits of the resistance value, the third one gives a multiplication factor, and the fourth one indicates the rated tolerance of the resistance value. For resistors with a lower tolerance, five rings are used; here the first three rings give the first three digits of the resistance value. Sometimes an extra ring or dot is used to indicate the TCR of the resistor. The meanings of the different colors are given in Fig. 2.20.

The properties of film resistors are primarily determined by the film. Many resistive materials have been used. They each have their weak and strong points and these determine the suitable applications.

Cracked Carbon Film Resistors

Traditionally cracked carbon film resistors have been used for general purpose applications. The films are made by cracking a hydrocarbon in the absence of oxygen on the rods. The sheet resistance is only partly determined by the thickness of the film. The resistivity of the material can vary strongly, depending on the proportion of amorphous carbon and the amount of contaminants in the film. The TCR usually becomes more negative when the resistivity increases.

FIGURES	MULTIPLIER (Ω)	TOLERANCE	TEMP. COEFF. ($.10^{-6}$/K)	
	0.01	10%		SILVER
	0.1	5%		GOLD
0	1		200	BLACK
1	10	1%	100	BROWN
2	100	2%	50	RED
3	1 K		15	ORANGE
4	10 K		25	YELLOW
5	100 K	0.5%		GREEN
6	1 M	0.25%	10	BLUE
7	10 M	0.1%	5	VIOLET
8			1	GREY
9				WHITE

FIGURE 2.20 Color code for fixed resistors in accordance with International Electrotechnical Commission (IEC) publication 62.

Carbon film resistors are available over a resistance range from 1 Ω to 10 MΩ. As already stated, their TCR may vary; usually it ranges from −400 ppm/°C for the low values to up to −2000 ppm/°C for very high values. The smallest ones have a length of 3 mm and a maximum diameter of 1.3 mm, with a maximum power rating of 0.25 W. Most resistors are somewhat larger, with a diameter of 1.8 mm. The use of large resistors with a power rating up to 2 W has decreased to a negligible level.

Metal Film Resistors

Metal film resistors are used for many applications. Very low-resistance values can be made with films of nickel phosphorous. Such films can be deposited on ceramic by a chemical process and, therefore, it is relatively cheap to make them very thick. They can have a TCR below ±50 ppm/°C.

Most other metal films are deposited by physical vapor deposition, usually evaporation or sputtering. Both processes require a high vacuum and are relatively expensive. Sputtering is the most expensive method, but it allows a much better control of the film composition. As a result of mechanization of physical vapor deposition, the deposition price of metal films is now so low that application for general purpose resistors is possible.

Most alloys for metal film resistors are based on nickel-chromium. For lower values, alloys based on copper-nickel also are used. For very high values, a higher resistivity is needed, and some silicides can be used. For resistors with a high-power rating, special alloys or metal oxide films may be employed.

The TCR of metal film resistors is usually better than ±200 ppm/°C. The stability is very good as long as the maximum working temperature is not exceeded. For most standard resistors this maximum temperature is around 175°C; special alloys may allow temperatures around 300°C. A tolerance of ±5% is standard.

Precision resistors are also made with metal films. Basically these resistors are identical to standard metal film resistors, but special care is taken to improve the TCR, tolerance, and stability. A TCR down to ±5 ppm/°C and a tolerance better than ±0.02% can be delivered.

Cermet or Metal Glaze Resistors

For most materials a very high sheet resistance can only be obtained by using an extremely thin film. By using a composite material such as a metal glaze, a thick and therefore more stable high ohmic film can be made. Such a film is produced by dipping the ceramic rod in a paste and drying and firing the layer.

As cylindrical resistors cermet or metal glaze resistors are mostly used for high-resistance values and high voltages. The available resistance range is 100 kΩ to 100 MΩ. The standard TCR is ±200 ppm/K but a TCR better than ±25 ppm/°C is possible. The stability is good and the current noise acceptable for such high values.

Surface Mounted Device (SMD) Resistors

A large and increasing part of the fixed resistor market is taken by surface mounted device (SMD) resistors. These resistors are mounted directly on the printed circuit board. Electrical contacts are provided at the sides of the resistor.

For most SMD resistors the substrate is a rectangular piece of ceramic, usually alumina. At the upper side conductors are deposited along two edges. The resistive film is deposited between and partly over these electrodes. The resistance value is adjusted by laser trimming. A protective coating is applied over the resistive film. Contacts are made over the edges; these usually extend over the sides of the ceramic and even some distance over the lower side to improve the soldering quality.

For the resistive layer in general purpose resistors thick film technology is normally used. In most cases the contacts are made by dipping a conductive paste. Sometimes evaporated or sputtered contacts are used. To ensure good soldering the contacts are covered with a solder or tin layer. Standard SMD resistors have a tolerance of ±5% and a TCR of 200 ppm/°C. With special thick-film techniques and appropriate contacts, tolerances of ±1% and a TCR of ±50 ppm/°C can be made.

For precision SMD resistors metal films are used. These films are made of the same alloys as used for cylindrical metal film resistors. Tolerances down to ±0.1% and a TCR of 25 ppm/°C are available.

The dimensions of SMD resistors are usually given in units of 0.01 in. Most are in the ranges 1206 (3 × 1.5 mm) to 0603 (1.5 × 0.75 mm). There is a trend to smaller dimensions, 0402 and even 0201 are available commercially. Larger dimensions (e.g., 1218 size, for 1 W) are provided for applications that require a higher power rating.

Metal electrode facebonding (MELF) resistors are special type of SMD device. Their resistive elements are identical to those of cylindrical film resistors. Instead of wires, they have solderable caps, which are soldered directly on the printed circuit board. Different kinds of resistive films are used, and their properties are comparable to those of similar conventional film resistors.

Resistor Networks

Mounting of resistors is often more expensive than the components themselves. The use of the smallest SMD components is often limited by the difficulty of handling them. The amount of handling and mounting can be reduced by combining a number of resistors with their interconnections to a network on one substrate. This also makes it easier to use smaller resistors.

These networks are mostly made on ceramic or glass substrates. Both thick-film and thin-film technologies may be used. The specifications of these networks are comparable to those of discrete film resistors made with the same technology. Thin films are used when the requirements on tolerance or TCR are high or when very small resistor dimensions are needed. By the use of lithography small devices can be made in thin film. Small dimensions not only save space and weight but they also have advantages when high frequencies are used.

Other components such as capacitors or coils may be integrated in the network. As this requires an extra processing such components are also often mounted on the network, usually as SMDs. The network may also be used to mount integrated circuits and so provide complete circuit blocks.

Resistors on Integrated Circuits

Integrated circuits (IC) themselves may also contain resistors. These resistors are integrated in the IC. It is convenient to make these resistors in the same process steps as the rest of the IC. Therefore, the exact way in which the resistors are made depends on which process is used to produce the IC.

Diffused resistors are used often because they can be formed together with other circuit elements and no extra steps are needed. The available square resistance depends on the doping concentration and the diffusion depth. Further resistance variation can be obtained by using the appropriate number of squares in the design. The square resistances that can be obtained range from 10 to 600 Ω per square, the TCR lies around −1000 ppm/°C. The resistance tolerance is within ±20%. With special structures a higher square resistance, up to 3000 Ω per square, can be obtained, but such resistors have a much more negative TCR and their tolerance is higher.

It is possible to use sputtered resistive films with a better TCR on an IC. However, this may lead to contamination of the production line. Therefore, this is only used when it is absolutely necessary.

2.4.3 Variable Resistors

In variable resistors the resistance value can be modified at will, but after adjustment this value must remain constant. The usual way to obtain this is a sliding contact that can move over the surface of the resistive element. The variable resistance can be used as such. In this case two connections are used, one at the start of the resistive element and one connected with the sliding contact.

A variable resistor can also be used as a voltage divider. In this case three connections are used. The total voltage is put over the resistive element; the required voltage is obtained by adjusting the position of the sliding contact. Variable resistors that are used in this way are also called potentiometers.

The resistance increment can be proportional to the displacement of the sliding contact, but also different resistance tracks are possible. The latter are, for instance, used for volume control in audio equipment.

As the sensitivity of the human ear is not linear, a nonlinear track may be better suited. Usually such potentiometers have a low-resistance increment at the start and an ever higher one further on. Therefore, they are sometimes called logarithmic potentiometers. The effect is obtained by varying the **sheet resistivity** and/or the track width over the resistive element in film potentiometers.

Adjustment of variable resistors may be done by turning a knob or by shifting along a line. In the first case, the sliding contact is rotated along an axis and the resistive element is part of a circle. In the second case, a linear resistive element is used. Apart from the mechanics, the technology of both designs is identical. In most cases the one used is a matter of design preference. Another design uses a linear resistive element, but the sliding contact is driven by a spindle that can be rotated. In this way it is easier to achieve small movement of the sliding contact, and so a more precise adjustment can be obtained.

In wire wound variable resistors the resistive element is a wire that is wound around a support. The contact slides over one side of the wires. The wire metals are the same as used for fixed resistors, but care should be taken that no thick oxide layers are present as this may cause a high contact resistance. Wire wound resistors can be made with good tolerance; they also have excellent stability and can have high rated power. They are, therefore, nearly always used when a high-power rating is needed. Their disadvantages include rather bad resolution and high price.

Variable film resistors are made with carbon composition films, usually on a paper laminate substrate, or with metal glaze films on a ceramic substrate. It is complicated to trim these resistive elements. Therefore, the tolerance on the resistance of film variable resistors is usually high, in the order of $\pm 20\%$. For most applications, however, this is no problem.

Carbon composition films have a rather high negative TCR, moderate stability, and sensitivity to humid conditions. Their major advantage is that they are inexpensive.

Metal glaze films are very stable, and their TCR can be much better than carbon composition. Besides being used in professional equipment they are also used in consumer electronics where conductive carbon films are not good enough, such as high-voltage trimmers in television sets. For specific applications, variable resistors may be combined with fixed resistors in a network.

In the usual design, the surface of the resistive element cannot have a protective coating, as this would impede the use of a sliding contact. For use in an aggressive environment special types are made. They depend mostly on hermetic sealing of the whole resistor.

Some variable resistors are only used for trimming circuits during manufacture or repair. They may be replaced by fixed resistors that are trimmed by the customer. Although the components are cheaper, the expensive trimming equipment makes this only attractive in special cases. Other devices are used many times during the operation of the equipment. For the latter types, minimum wear of the resistive element and the wiper is important.

An important property of variable resistors is the contact resistance between the sliding contact and the resistive element because it contributes to the resistance value. In the case of use as a potentiometer, it may be important if a current is flowing through the sliding contact. This may be especially important for volume or balance potentiometers in audio equipment. Here, variations in the contact resistance will give a crackling noise during the movement of the sliding contact. To reduce the contact resistance special sliding contacts are often used, which have multiple contact points or a custom soft contact material.

The *resolution* of the adjustment is the minimum relative resistance adjustment that can be made. For a wire wound variable resistor this is usually determined by the number of windings: each new winding that is contacted causes a step in resistance value or, in the case of use as a potentiometer, a step in potential. If the resistive element is a film, irregularities in the film may determine the resolution, but usually mechanical problems related with slip/stick performance are more important.

2.4.4 Nonlinear Resistors

Nonlinear resistors have a resistance that depends strongly on some external parameter. For any resistor the resistance depends on many parameters, but nonlinear resistors are specially made to show such a dependence. Usually the parameters are voltage or temperature. The major kinds are

- Resistors with a positive temperature coefficient (PTC)
- Resistors with a negative temperature coefficient (NTC)
- Voltage dependent resistors (VDR)

Nonlinear resistors consist of a piece of ceramic material with two metal connections at opposing sides. Usually the device is insulated with some kind of organic coating. The shape and dimensions may vary widely depending on type, manufacturer, and specification. Within each kind of nonlinear resistor special types are available for many specific applications.

There is a growing demand for SMD types of nonlinear resistors.

PTC

Positive temperature coefficient resistors are characterized by a resistance that is only moderately dependent on temperature below a certain temperature, but increases very strongly above that temperature. This increase may be as high as 15% per degree Celsius, and the total increase over some tens of degrees may be several orders of magnitude. A typical example of a resistance/temperature characteristic is given in Fig. 2.21.

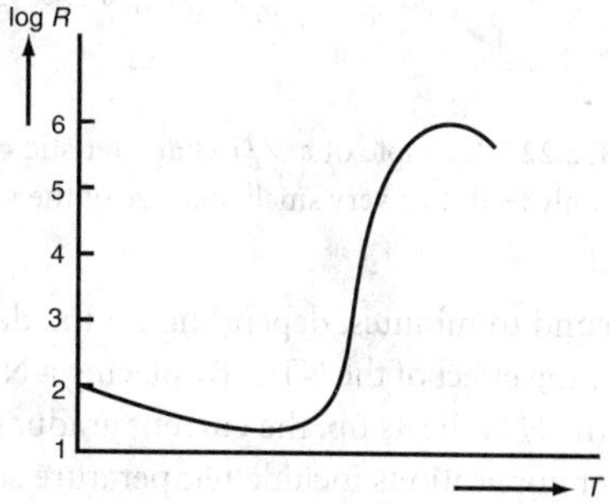

The ceramic of PTCs is based on $BaTiO_3$ with certain additions. Normal grains of ceramic are surrounded by barrier layers of strongly oxidized material. The barrier potential is inversely proportional to the dielectric constant of the crystals. Therefore,

FIGURE 2.21   Typical resistance/temperature curve for a PTC resistor.

the resistance increases sharply when the ferroelectric Curie temperature of the material is exceeded. It is possible to make materials with a specific Curie temperature by variations of the processing conditions. Above a certain temperature the TCR again becomes negative; this means that PTCs can only be used in a limited temperature range. As the barrier layers are thin, their resistance is shunted by a high capacitance. This means that they become strongly frequency dependent above 5 MHz.

Positive temperature coefficients are often used as protective devices in equipment to avoid overheating. By placing the PTC on the hot spot in series with the feeding circuit, overheating leads immediately to a reduction of the current. Protection against overcurrent may be provided by the self-heating of the PTC: when the current becomes too high the PTC limits it to a safe value. Another possibility is to use the PTC as a heating element. As the threshold temperature is approached the current and so the power is reduced, independent of the supply voltage. A PTC can also be used as a temperature sensor.

NTC

Negative temperature coefficient resistors, also called *thermistors,* have a negative temperature dependence in the order of several percent per degree Celsius. They are made from polycrystalline semiconductors, consisting of oxides of chromium, manganese, iron, cobalt, and nickel. The resistance of a NTC resistor can be expressed as

$$R = Ae^{B/T}$$

where A and B are constants and T is the absolute temperature in $K \cdot B$ depends on the material, it lies between 2000 and 5500 K. This means that around 25°C a temperature change of 1°C gives a resistance change between about 2 and 6%. Most NTCs are specified with the resistance value at 25°C and a constant B value between 20 and 85°C.

In measuring an NTC, care should be taken that the dissipated power remains small. When a sufficiently high voltage is applied, so much heat is generated that the resistor temperature increases. This leads to a lower resistance, a higher current, and a further increase of temperature. After some time (from a fraction

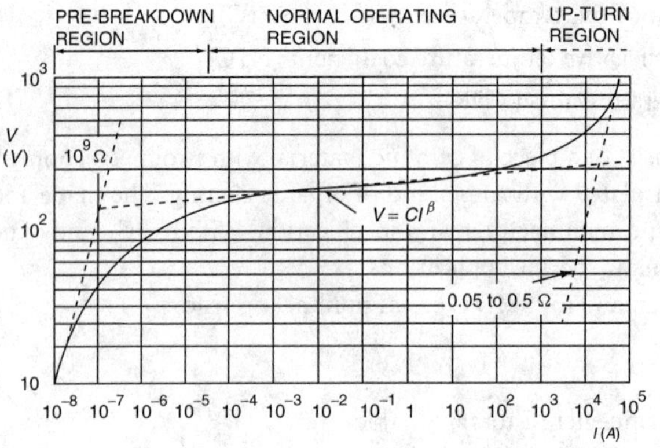

FIGURE 2.22　Example of a V/i characteristic of a varistor. In the normal operating region a very large change in current only leads to a very small increase of the voltage.

of a second to minutes, depending on the design) an equilibrium is reached. Many applications use this self-heating effect of the NTC. By placing a NTC in series with another device, the start current is reduced. When the NTC heats up, the current gradually increases to its final value.

Other applications include temperature sensing and compensation of the temperature dependence of other devices.

VDR

In VDR resistors (also called varistors) the effective resistance decreases strongly with increasing voltage. The major materials for VDRs are zincoxide, strontium titanate, and silicon carbide. They consist of many particles with a rectifying action at their contact points. When a certain voltage over a contact is exceeded, the current strongly increases. The number of contacts in series determines the voltage rating, and the number of parallel connections determines the maximum current.

A typical voltage/current curve for a varistor is given in Fig. 2.22. The voltage in the protection zone may vary between some tens of volts to thousands of volts.

ZnO varistors are typically used to suppress voltage surges. Such surges may be generated by lightning, the switching of inductive loads, or electrostatic discharges. The resistor is connected parallel to the load. In normal conditions a negligible current is flowing. When a voltage surge occurs the current through the PTC strongly increases and the surge is suppressed. Another application is in micromotors, where strontium titanate varistors are normally used.

2.4.5　Fundamental Laws of Resistive Elements

The best-known law on resistance is Ohm's law, which states that

$$V = i \times R$$

where
  V = voltage over the resistor
  i　= current
  R = resistance

This law actually includes the definition of a resistance. It can also be written as $i = V/R$ or $R = V/i$.

Resistance is a property of a specific conducting element. It depends on the dimensions and the material of the conductor. In most cases the resistance R can be given as

$$R = \rho \times l/A$$

where

   l = length of the conductor
   A = area of cross section
   ρ = resistivity of the material

For resistive films, often the sheet resistivity is used. This is the resistance of a square area on the resistor surface.

   Application of the last formula gives

$$R = \rho \times 1/t$$

where t is the thickness of the film. This value does not depend on the size of the square, as a greater length is compensated by the greater width.

   The standard unit of resistivity is the ohm meter, but the ohm centimeter or the micro-ohm centimeter is also used. Between different materials the ratio between the highest and lowest resistivities at ambient temperature is on the order of 10^{25}. A still higher ratio is found at low temperatures, even when we forget about superconductors, for which the resistivity is essentially zero. The TCR is defined as

$$\frac{1}{R} \times \frac{dR}{dT}$$

It is usually given in parts per million per degree celsius (ppm/°C). In most cases the TCR depends also on temperature. The TCR, therefore, must be specified over a temperature region. Sometimes only the average between 20 and 70°C is given. Other specifications include values for different temperature regions.

Conduction Mechanisms

Electrical conduction is usually separated into two mechanisms. One is metallic conduction, in which the charge carriers in the conduction band are free to move through the lattice at any temperature. Here the number of charge carriers does not depend on temperature, so the temperature dependence is only caused by the variation of the carrier mean free path, that is, the mean distance a charge carrier will travel without being scattered.

   In a perfect lattice at absolute zero temperature no scattering of the carriers occurs. At higher temperatures vibrations will disturb the lattice and scattering does occur. This leads to a resistivity that increases with temperature. For most metals this increase is linear at temperatures above approximately 40 K.

   In practice, metal lattices are not perfect. Lattice imperfections may include missing atoms, dislocations, and impurities. These imperfections also scatter the charge carriers. It turns out that usually the contributions of thermal vibrations and lattice imperfections can be added up. This is known as Matthiessen's Rule.

   Alloying of pure metals increases the amount of disorder in the lattice. Therefore, the resistivity always increases when a pure metal is alloyed, even when the added metal has a lower resistivity in its pure form. At first the resistivity increases linearly with the amount of second metal, but for higher percentages the increase goes slower. A maximum is usually reached around a 50% composition, as higher contents may be considered an alloy of the first metal in the second one.

   Quantum mechanics show that the mean free path of the electrons in a metal cannot be smaller than the interatomic distance. For many concentrated alloys of transition metals this limit is approached. Near this limit the resistivity becomes insensitive to a further increase of lattice disorder, and thus the TCR becomes low. It turns out that for these alloys the TCR is approximately zero when the resistivity is around 150 $\mu\Omega \cdot$ cm, positive when it is lower and negative when it is higher. Their low TCR makes these alloys useful for fixed resistors.

   Ideal semiconductors have no electrical conduction at absolute zero temperature as there are no free charge carriers. At higher temperatures free charge carriers are created by thermal excitation. The electrical conduction is determined by both the number of charge carriers and their mean free path. The change in carrier density is dominant, and so the TCR of semiconductors is negative.

Because of their strongly negative TCR, ideal semiconductors are not appropriate for fixed resistors as long as only a small part of the charge carriers are in the conduction band. If a large part of the carriers is activated, which may be obtained by heavy doping or a low gap energy, this condition will improve. A special case is graphitic carbon. Here the atoms are arranged in planes. Within the planes the electrical conduction is metallike, but between the planes it is semiconducting.

It is tempting to mix metals and semiconductors to combine their respective positive and negative TCR to achieve a zero TCR. If a simple mixture is made, the conductivity and the TCR are determined by a percolation process. This implies that in a very small composition region the TCR suddenly changes from positive to negative. Therefore, for such materials the reproducibility and the spread of both the resistivity and the TCR are unacceptable.

Useful mixed materials contain thin filaments of conducting material that are incorporated in an insulating matrix. The TCR is almost completely determined by the conductor, but the effective resistivity is very high. The matrix increases the adhesion to the substrate and it may protect the conductor, so such materials have rather good stability. Many cermet or metal glaze materials are based on this principle. Another possibility is the use of grains of a good conductor surrounded by a thin layer of semiconductor. Again the reproducibility is the major problem here.

Maximum Load

The generation of heat is an essential property of any resistance. The power P that is dissipated in a resistor is

$$P = i \times V$$

with i = current and V = voltage. By combination with ohm's law this may also be written as

$$P = i^2 \times R$$

or

$$P = V^2/R$$

The second formula indicates that for a constant current the dissipated power is proportional to the resistance. According to the third formula, for a constant voltage the power is inversely proportional to the resistance.

The dissipated heat will cause an increase in temperature of the resistor and its surroundings. In this way the temperature may become so high that problems occur. Usually the maximum power is determined by either the hottest spot on the resistor or the temperature of the connection between the resistor and the printed circuit. As the latter temperature is usually the lowest one it becomes more important when the temperature of the surroundings increases.

When a part of a resistor becomes too hot a resistance drift occurs. The magnitude of this drift depends on the resistive material and its geometry. The maximum allowed temperature depends also on the specified maximum drift; therefore, for precision resistors a lower temperature is allowed.

The magnitude of the temperature increase is determined by the amount of dissipated energy and the heat resistance to its surroundings. Heat can be dissipated by three processes, transfer to air moving along the resistor, radiation, and conduction.

The heat transfer to air depends on the temperature difference, the surface of the resistor, and the speed and direction of the air flow. Air flow along the resistor may be due to external causes but it can also be caused by natural convection. The air that is heated by the resistor becomes lighter than the surrounding air and, therefore, it tends to rise. The actual formulas for the transfer of heat to air are complicated as they depend strongly on the geometry of the resistor and its surroundings.

The heat transfer by radiation is given by

$$E \times A \times \left(T_R^4 - T_S^4 \right)$$

Here E is the emission coefficient of the surface, A is the surface area, T_R and T_S are the absolute temperatures of the resistor and its surroundings, respectively. It is clear that the contribution of radiation increases strongly for a higher resistor temperature. The emission coefficient lies between 0 and 1; for organic coatings it has almost the maximum value in the relevant wavelength area.

The equation for heat transfer by conduction is

$$Q = \lambda * A / l$$

Here Q is the heat flow in 1 s, l and A are again the length and the area of the heat conductor, and λ is the heat conduction coefficient. The heat conduction coefficient is much less variable than the electrical conductivity; for solids its value varies around ambient temperature between 1 W(m · K) for some glasses to 500 W(m · K) for diamond. Metals generally have a high heat conduction, between 420 W/(m · K) for pure silver to 10 W/(m · K) for some concentrated alloys. For air the heat conduction under standard conditions is so low that it can be neglected relative to convection. Heat conduction is also important for the internal heat flow in the resistor body.

The relative importance of the contributions of the various ways of heat transmission depends on the construction and the geometry of the resistor and its surroundings. Some rules can be given, as most resistors fall into one of two categories.

For small resistors (length up to 5 mm with a maximum surface temperature under 200°C) almost all heat transport occurs by conduction. The area of such resistors is simply too small to dissipate a significant part of the power. This applies for most low-power resistors and all SMD resistors. The nominal rating of such resistors is typically around 0.5 W. In this case the temperature of the resistor depends strongly on the heat resistance of the printed circuit board. Good heat conduction is provided by the connectors on the print. Often in practice the maximum rating is not determined by the hottest spot on the resistor but by the temperature of the soldering spot. It should be realized that the nominal power rating of a resistor is based on a single resistor on a circuit board with a rather good heat conduction. For applications where more resistors are used on one board, a derating may be necessary.

Large resistors that can reach a surface temperature well over 200°C lose most of the produced heat by convection or radiation. In fact, this applies to all normal resistors with a rating over 1 W. For such resistors special measures may be necessary to make sure that the soldering point does not become too hot. Iron wires, with a low-heat conductivity or extra long wires can be used.

Resistors with a very high-power rating cannot be used on a printed circuit board. For these resistors a special mounting must be provided. Of course, it is advisable to keep such resistors away from components that are sensitive to heat. It may also be necessary to provide ventilation slots or a ventilator to dissipate the heat outside the equipment.

Pulse Load

It is well known that a resistor can withstand a higher power than usual when the load is applied for only a short time. Although several mechanisms may determine the maximum allowed power during a pulse, in by far the most cases the maximum temperature turns out to be the limiting factor. Under continuous load a resistor dissipates all energy to its environment. At the start, however, energy also is needed to increase the temperature of the resistor. As the resistor needs time to reach its maximum temperature a lower temperature will be reached when the load is only applied for a short time. Therefore, a higher load can be applied without damage to the resistor.

The maximum allowed pulse power depends on the pulse duration, the repetition frequency, and the properties of the resistor. This dependence may be quite complex, but some general rules always hold. For resistors of a similar type, larger ones always can handle a higher pulse. Also, for a shorter pulse, the allowed pulse power is at least as high as for a longer one. An important rule for repetitive pulses is that the average power may never exceed the maximum allowed power under continuous load. This rule seems obvious, but is easily forgotten.

Under typical pulse load conditions the dissipation of energy to the environment of the resistor during a pulse can be neglected. Therefore, the rate of temperature increase is determined by the pulse power divided by the effective heat capacity of the resistor. During the pulse this effective heat capacity is not constant. All heat is produced in the resistive element. If the pulse is very short, the heat capacity of the resistive element determines the temperature increase as long as the proportion of the energy that can diffuse to the rest of the resistor is relatively small. Here the temperature is proportional to the pulse time. In this limit good behavior under pulse load can be obtained by using a large heat capacity of the resistive element itself and a resistive material that can withstand high temperatures.

In the course of time, part of the pulse energy diffuses into the rest of the resistor. The thickness of the effective zone d is given by

$$d = \sqrt{\frac{\lambda}{\rho \times C} \times t}$$

where
  λ = heat conductivity of the substrate material
  ρ = density of the material
  C = specific heat capacity of the material
  t = pulse time

All mentioned material properties are those of the material into which the heat is diffusing.

In a first approximation the energy diffusion to the rest of the resistor becomes important when the pulse time is so long that d approaches the thickness of the resistive element. This critical time will vary from around 1 s for large wire wound resistors with 1-mm wires to 10^{-10} s for film resistors with a film thickness of 10 nm. Under this regime the temperature of the resistive element increases with the square root of time.

For even longer times, d becomes on the order of the dimensions of the resistor. In this case also, heat loss to the environment becomes important. Here a gradual transition to continuous load, with a temperature that is independent of time, occurs.

The good pulse behavior of wire wound resistors is primarily based on the relatively large heat capacity of the wires. For pulse behavior, a large wire diameter should be used. As for a given resistance value, a thinner wire also implies a shorter wire; the maximum pulse rating for different wire wound resistors may be quite different.

For thin-film resistors the heat diffusion into the substrate is dominant for all but extremely short pulses. The effective heat capacity is proportional to the surface area; thus, large resistors here also are at an advantage. Breakdown usually occurs at spots where the current density is high, for instance at the end of trimming groove. Irregularities in the film may complicate matters even more. Nevertheless, it is possible to make mathematic models of the pulse behavior of such resistors that accurately predict the maximum allowed pulse power under specified conditions. An example of such a calculated curve is given in Fig. 2.23.

The behavior of resistors with inhomogeneous resistive elements is complicated. The relatively large volume may give a rather good pulse behavior, especially for carbon composition resistors. Very short pulses, however, may reveal the fact that in such materials the actual current carrying part is thin and for such pulses the acceptable pulse energy may be surprisingly small.

Noise

All resistors exhibit *thermal noise*. This depends only on the resistance value; its energy does not depend on current or frequency.

In addition, resistors have *current noise*. This has the character of resistance variations, so that the noise is proportional to the total voltage over the resistor. Its magnitude is different for different resistors. The

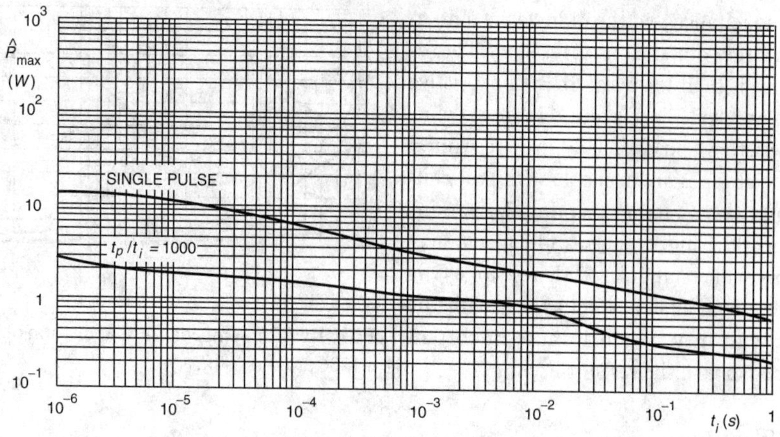

FIGURE 2.23 Maximum permissible peak pulse power (P_{max}) for a 1206 SMD resistor as a function of pulse duration for single pulses and for repetitive pulses with a pulse frequency that is 1000 times the pulse duration.

noise energy drops with frequency. Usually it is inversely proportional to the frequency f but higher powers of f may occur.

The noise energy depends on the number of charge carriers in the resistive element. Resistors with many charge carriers have a low current noise level. Therefore, wire wound resistors are extremely good in this respect. For film resistors, metal films are better than carbon films and metal glaze. Carbon composition resistors have an extremely high noise level. Within one type, low values are better than high ones and big resistors better than small ones.

The specified noise of a resistor always only applies to the current noise. The noise spectrum involves the whole frequency range, but for practical purposes noise is usually specified as measured for a band from 680 to 1680 Hz.

2.4.6 Long-Term Damp Heat

Resistance changes under humid conditions are due to electro-chemical transport of resistive material. In other words, this only happens when liquid water is present in combination with a sufficiently high potential difference. Liquid water may form when the relative humidity is near 100%. When the relative humidity is somewhat below 100%, liquid water may also form when slits or pollution is present. On the other hand, no water will be found on resistors that have a significantly higher temperature than their surroundings.

Humidity problems are almost exclusively found for high-value resistors that are operated far below their power rating. They may combine a sufficiently high voltage with a low heating effect. In addition, film resistors often have very thin films and so even small material transport may cause interruption. This is one reason that thick-film resistors are preferred for very high values.

Frequency Dependence

An ideal resistor has only a resistance, but in practice resistors always also have a capacitance and an inductance component. The simplest equivalent circuit for a practical resistor is given in Fig. 2.24.

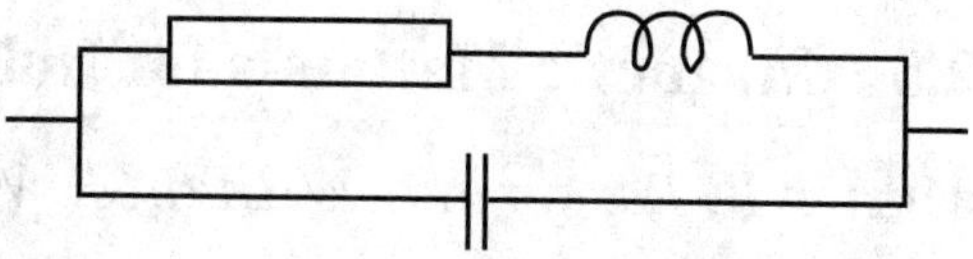

As a result, at high frequencies the impedance of low-value resistors increases, whereas for high value resistors it decreases. The changeover usually takes

FIGURE 2.24 Simple equivalent circuit for a resistor.

place at nominal resistance values around 100 Ω. An example of the dependence of impedance on frequency is given in Fig. 2.25.

Both the capacitance and the inductance tend to increase with larger dimensions of resistors. The series inductance is high for resistors with a resistive element such as a coil. This is found in most cylindrical film resistors and especially in wire wound resistors. Special types are available with two coils wound in opposite direction to minimize the inductance. Very low inductances and capacitances are found for small rectangular SMD resistors. One should remember that here interactions with the rest of the circuit may be much more important than the impedance of the resistor itself.

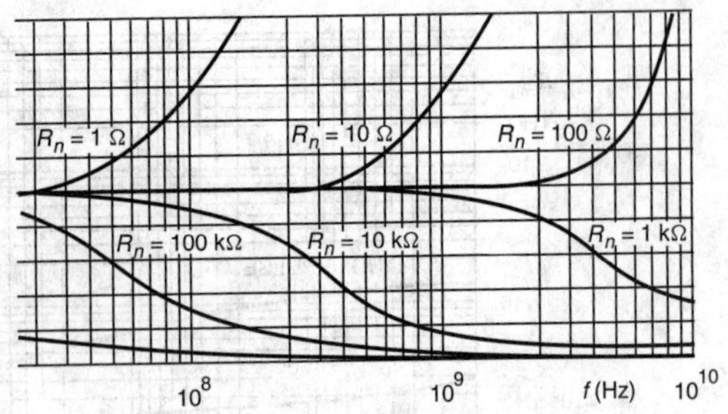

FIGURE 2.25 Impedance as a function of frequency for a 1206 SMD resistor.

Defining Terms

Fixed resistor: A resistor that, in first approximation, should have a constant resistance, independent of temperature, voltage, and other operating parameters.

Nonlinear resistor: A resistor with a value that depends strongly on some external parameter such as temperature (thermistor) or voltage (varistor).

Resistivity: The resistance of a cube with specified dimensions made of a given material. Usual units are the $\Omega \cdot m$ (resistance of a cube with sides of one meter) and the $\Omega \cdot cm$.

Sheet resistivity: The resistance of a square made of a thin film. This resistance depends only on the film material and the thickness, not on the size of the square.

Temperature coefficient of resistance (TCR): Defined as

$$\frac{1}{R} \times \frac{dR}{dT}$$

It is usually given in parts per million per degree celsius (ppm/°C).

Variable resistor: A resistor for which the resistance value can be changed by the user, but must be independent of other outside influences.

References

Dummer, G.W.A. 1967. *Fixed Resistors.* Sir Isaac Pitman & Sons, London.

Huissoon, P.M. 1994. Pulse loading, a hot issue for resistors. *Electronic Design* (Sept.):99–104.

Maissel, L.I. 1970. Electrical properties of metallic thin films. In *Handbook of Thin Film Technology*, eds. L.I. Maissel and R. Glang. McGraw-Hill, New York.

Further Information

The best source for current information on resistive components can be found in the various applications catalogs of resistor manufacturers. Application notes are available on a wide range of subjects.

2.5 Magnetic Materials for Inductive Processes

Martin R. Parker and William E. Webb

2.5.1 Introduction

Inductance is, primarily, a geometrical property of a current-carrying element in an electrical circuit. A circuit element with this property may be termed an inductor. The magnitude and, for that matter, the frequency dependence of inductance, also depend on the material environment of that element. Similar remarks to these could, of course, be used to define capacitance and, as may be seen throughout this book, inductance and capacitance are invariably intimately related in electronic circuits. One reason is that both are (electrical) energy storage devices in time-varying electronic systems. Capacitance is, however, a measure of the capability of a (potential) circuit element to store charge and, thereby, electric field energy; inductance, by contrast, is a measure of a circuit element's ability to store magnetic field energy. Since magnetic field is derived from current flow, inductance is always associated with current-carrying circuit elements. This dichotomy extends also to the matter of frequency dependence. At zero signal frequency, for example, an (ideal) capacitor has infinite impedance whereas the inductor has zero impedance; at infinite frequency the opposite is true.

The concept of inductance, or, more correctly, **self-induc- tance** is, perhaps, best illustrated by the archetypal example of the current-carrying conducting coil of N circular turns (Fig. 2.26). When a current I flows through this circuit, as described, a magnetic field is generated inside and around the coil turns, in accordance with Ampere's law, as shown schematically in Fig. 2.26. (A detailed description of these field lines is a nontrivial matter, but is of little importance to the issue in question, which is how much **flux linkage** there is with the coil). Flux linkage Λ is defined by how much of the total **magnetic flux** ϕ threads all of the turns of the coil, multiplied by the total numbers of turns N. Thus, in simple algebraic terms

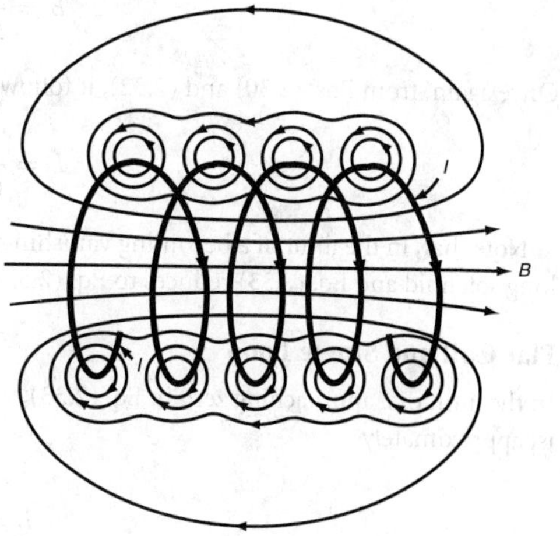

FIGURE 2.26 A magnetic field produced by a current through a coil. (Adapted from Plonus, M.A. 1978. *Applied Electromagnetics*. McGraw-Hill, New York.)

$$\Lambda = N\phi = N \iint_A B \cdot dA (\cong NBA)$$

where, because of incomplete linkage by the magnetic flux (Fig. 2.26), in actuality A will typically be slightly less than the cross-sectional area of the coils. The inductance L of this coil is defined as the flux linkage per unit current and is defined as (Plonus, 1978)

$$L = \frac{\Lambda}{I} = \frac{N\left(\iint_A B \cdot dA\right)}{I} \tag{2.29}$$

The SI unit of inductance is the henry (symbol H) if I is in amperes, Λ in webers, A in square meters, and B in teslas (webers/square meter). The inductance of some current-carrying elements of comparatively simple geometry is described in approximate form next.

2.5.2 Coils

Long Solenoid

The long solenoid is a longer, more tightly wound version of the coil of Fig. 2.26, with a relatively large number of turns (Fig. 2.27). Using Eq. (2.30) and the result that the B-field inside a long solenoid is fairly uniform and given by

$$\frac{\mu_0 NI}{\ell} \tag{2.30}$$

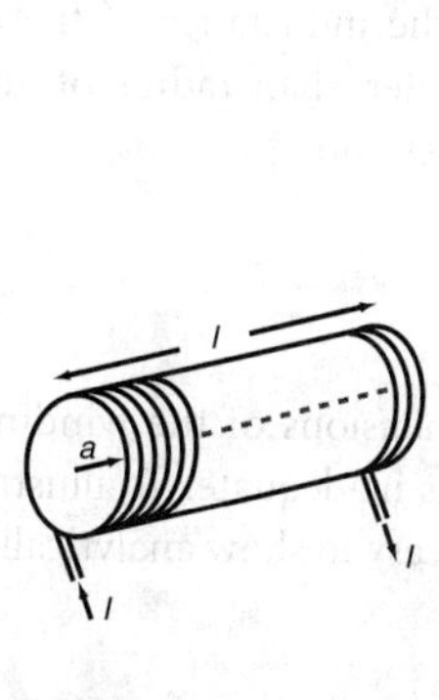

FIGURE 2.27 A short solenoid.

where ℓ is its length, means that the inductance is given by

$$L = \frac{\mu_0 N^2 A}{\ell} \tag{2.31}$$

where $m_0 (= 4p \times 10^{-7} \text{ H/m})$ is the **permeability** of free space.

Short Solenoid

Similarly, for a short solenoid of N turns, also of length ℓ and of radius a (Fig. 2.27), it can be shown that

$$B = \frac{\mu_0 N I}{(\ell^2 + 4a^2)^{\frac{1}{2}}} \tag{2.32}$$

Once again, from Eqs. (2.30) and (2.32), it follows that

$$L = \frac{\mu_0 N^2 A}{(\ell^2 + 4a^2)^{\frac{1}{2}}} \tag{2.33}$$

Note that, in the limit of a becoming vanishingly small, the short solenoid takes on the appearance of a long solenoid and Eq. (2.33) reduces to Eq. (2.31).

Flat Coil and Single Loop

In the limit of ℓ approaching zero in Eq. (2.33), we have a flat coil of N turns, radius a, whose inductance is, approximately

$$L = \frac{\mu_0 N^2 a}{2} \tag{2.34}$$

where A (Eq. (2.33)) has been replaced by a. Note that the special case of a single loop is derived from Eq. (2.34), simply by making $N = \ell$, leading to

$$L = \frac{\mu_0 a}{2} \tag{2.35}$$

The Toroid

Another simple classic geometry (Fig. 2.28) is the toroid. This doughnut-shaped winding is a close relative of the long solenoid. Indeed, it may be viewed as a long solenoid whose ends have been joined. Accordingly, the B-field inside the toroid may be given, approximately, by (Eq. 2.31), with the modification that the length ℓ is replaced by $2r$, where r is the (mean) radius of the toroid. A crude approximation for the inductance of the toroid, when r is much greater than radius of the individual windings is, therefore

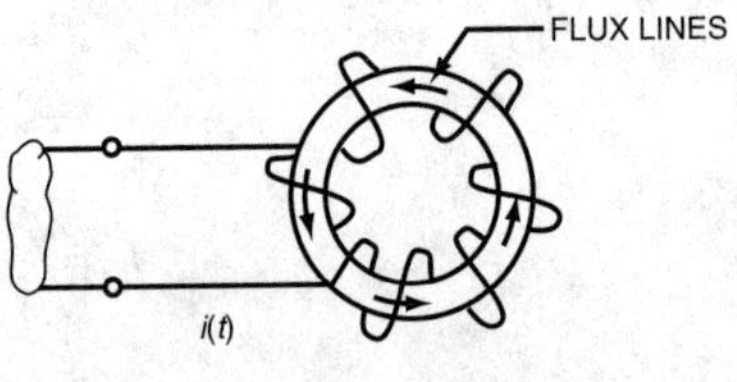

FIGURE 2.28 A toroid inductor. (Adapted from Irwin, J.D. 1995. *Basic Engineering Circuit Analysis*, 5th ed. Macmillan, NY.)

$$L = \frac{\mu_0 N^2 A}{2r} \tag{2.36}$$

When the dimensions of the windings compare in size with the toroid radius, the simple expression of Eq. (2.36) is inadequate. To illustrate the point, consider the rectangularly cross-sectioned toroid of Fig. 2.29. It is easy to show analytically (Cheng, 1993) that, for this geometry

$$L = \frac{\mu_0 N^2 h(\ell n b/a)}{2} \tag{2.37}$$

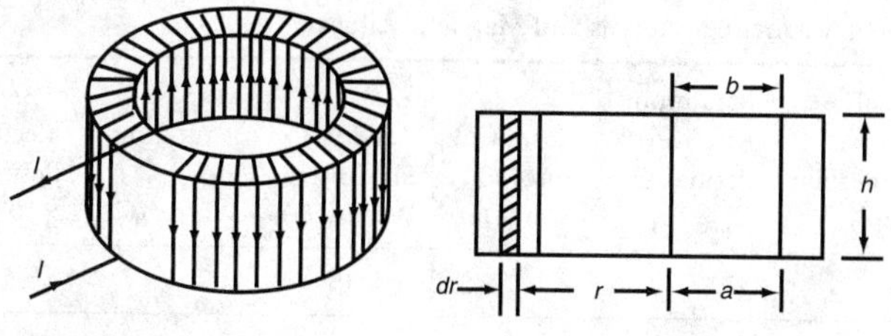

FIGURE 2.29 A closely wound toroidal coil. (Adapted from Cheng, D.A. 1993. *Fundamentals of Engineering Electromagnetics*. Addison-Wesley, NY.)

Circuit Description of Self-Inductance

It is a relatively simple matter to describe the voltage drop across a simple self-inductance such as the coil of Fig. 2.30. For a steady-state current I, in the coils, the voltage drop is simply IR, as dictated by Ohm's Law, where R is the coil resistance. On the other hand, if the current $i(t)$ is time varying, then it follows quite simply from a combination of Faraday's Law of electromagnetic induction and equations (see, for example, Eqs. (2.29) and (2.31)) for magnetic flux and flux density that the (time-varying) voltage drop v across a self-inductance is

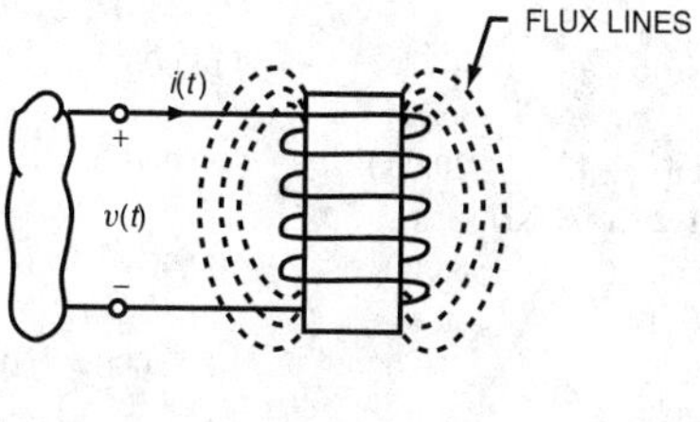

FIGURE 2.30 A circuit illustrating self-inductance. (Adapted from Irwin, J.D. 1995. *Basic Engineering Circuit Analysis*, 5th ed. Macmillan, NY.)

$$v = L\frac{di}{dt} + Ri \qquad (2.38)$$

Given that magnetic field energy density is expressible (Plonus, 1978) as

$$w_m = B \cdot H \qquad (2.39)$$

it follows that for any of the simple inductive windings described, the total magnetic field energy (that is, the stored energy of any inductor) is

$$W_m = \frac{1}{2}Li^2 \qquad (2.40)$$

Note that L may be defined from Eq. (2.40), thus,

$$L = \left(\frac{2W_m}{i^2}\right)^{\frac{1}{2}} \qquad (2.40a)$$

2.5.3 Magnetic Materials

All of the preceding described current elements were characterized by air-filled windings. In fact the formulas of Eq. (2.31) are essentially valid (better than 1% accuracy) if the windings encompass any condensed matter other than a magnetic (i.e., ferromagnetic/ferrimagnetic, etc.) material. If, however, any of the described windings are constructed around a **magnetic** material of **permeability** μ, then in each of the formulas of Eqs. (2.31–2.37), μ_0 is replaced by μ. Now, for magnetic materials μ is, typically, much greater than μ_0. The actual value of μ depends crucially on (1) the current signal frequency and (2) the chemical identity and temperature of the magnetic material comprising the inductive element as well as on the details of its microstructure.

TABLE 2.16　Properties of Magnetic Materials and Magnetic Alloys

Material (Composition)	Initial Relative Permeability, μ_i/μ_0	Maximum Relative Permeability, μ_{max}/μ_0	Coercive Force H_c, A/m (Oe)	Residual Field B_r, Wb/m^2(G)	Saturation Field B_s, Wb/m^2(G)	Electrical Resistivity ρ $\times 10^{-8}$ $\Omega \cdot$ m	Uses
			Soft				
Commercial iron (0.2 imp.)	250	9,000	≈80 (1)	0.77 (7,700)	2.15 (21,500)	10	Relays
Purified iron (0.05 imp.)	10,000	200,000	4 (0.05)	—	2.15 (21,500)	10	
Silicon-iron (4 Si)	1,500	7,000	20 (0.25)	0.5 (5,000)	1.95 (19,500)	60	Transformers
Silicon-iron (3 Si)	7,500	55,000	8 (0.1)	0.95 (9,500)	2 (20,000)	50	Transformers
Silicon-iron (3 Si)	—	116,000	4.8 (0.06)	1.22 (12,200)	2 (20,100)	50	Transformers
Mu metal (5 Cu, 2 Cr, 77 Ni)	20,000	100,000	4 (0.05)	0.23 (2,300)	0.65 (6,500)	62	Transformers
78 Permalloy (78.5 Ni)	8,000	100,000	4(0.05)	0.6 (6,000)	1.08 (10,800)	16	Sensitive relays
Supermalloy (79 Ni, 5 Mo)	100,000	1,000,000	0.16 (0.002)	0.5 (5,000)	0.79 (7,900)	60	Transformers
Permendur (50 Cs)	800	5,000	160 (2)	1.4 (14,000)	2.45 (24,500)	7	Electromagnets
Mn-Zn ferrite	1,500	2,500	16 (0.2)	—	0.34 (3,400)	20 × 10^6	Core material
Ni-Zn ferrite	2,500	5,000	8 (0.1)	—	0.32 (3,200)	10^{11}	for coils

Source: After Plonus, M.A. 1978. *Applied Electromagnetics*. McGraw-Hill, New York.

In inductive devices, one will, typically, find *soft* magnetic materials employed as the core medium. Here, a soft magnetic material is one in which the **relative permeability** μ_r defined as

$$\mu_r = \frac{\mu}{\mu_0} \qquad (2.41)$$

is, numerically, $\gg 1$. (See Table 2.16.) For the magnetic materials of Table 2.16 and the like, μ_r is (1) a nonanalytical function of magnetic field intensity (see, for example Fig. 2.31), (2) a complex quantity of the form

$$\mu = \mu' + j\mu'' \qquad (2.42)$$

FIGURE 2.31　Magnetization curve of commercial iron. Permeability is given by the ratio B/H. (*Source:* Plonus, M.A. 1978. *Applied Electromagnetics*. McGraw-Hill, New York.)

where μ'' is, typically, significant only at high frequencies (see, for example Fig. 2.32) and is a consequence of one or more of the lossy mechanisms outlined in Sec. 2.5.5.

2.5.4　Self-Inductance of Wires and Transmission Lines

The concept of inductance is by no means limited to the multiturned windings of Sec. 2.5.1. Any current-carrying element, even one with the most basic geometry, has at least marginal inductive characteristics. This is, perhaps, best illustrated with the case of a simple, straight copper wire. By integrating the self-(magnetic) field energy density over all internal radial values for a given wire current I (and applying

Eq. (2.40a) as a definition for L), it follows that the self-inductance of a wire/unit length, $L'(= L/\ell)$, is

$$L' = \frac{\mu_0}{8\pi} \text{ (H/m)} \qquad (2.43)$$

Note that the result is independent of wire radius.

By a similar approach, it can be shown (Plonus, 1978; Cheng, 1993) that the inductance/unit length of a coaxial cable (Fig. 2.33(a)) is

$$L' = \left(\frac{\mu_0}{8\pi}\right) + \left(\frac{\mu_0}{2\pi}\right) \ell n \left(\frac{b}{a}\right) \text{ (H/m)} \qquad (2.44)$$

Likewise, a rather more tedious calculation (Plonus, 1978; Cheng, 1993) shows that for the parallel-wire transmission line (Fig. 2.33(b)), the corresponding result is

$$L' = \frac{\mu_0}{\pi} \left(\frac{1}{4} + \ell n \frac{d}{a}\right) \text{ (H/m)} \qquad (2.45)$$

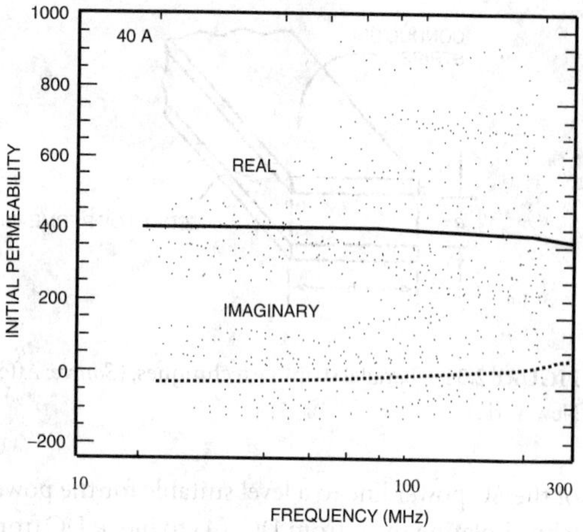

FIGURE 2.32 Frequency dependence of the initial permeability of a nanocrystalline NiFeCo/Cu/NiFeCo/FeMn thin film spin valve structure. (Courtesy of B. Pirkle.)

Finally, the growth of ultra-high-frequency, large-scale integrated circuitry places ever-increasing importance on parallel strip lines of the type shown in Fig. 2.34(a) and Fig. 2.34(b). It is a rather simple matter (Plonus, 1978) to show that the inductance/unit length for both of these structures is

$$L' = \frac{\mu a}{b} \text{ (H/m)} \qquad (2.46)$$

where μ is the permeability of the medium between the strips.

2.5.5 Transformers

A transformer is a device that converts variations in the current in one circuit to variations in the current in one or more other circuits by means of magnetic coupling between two or more coils. In electronic circuits, transformers are used (1) to change the magnitude of an electrical signal, (2) to convert the voltage

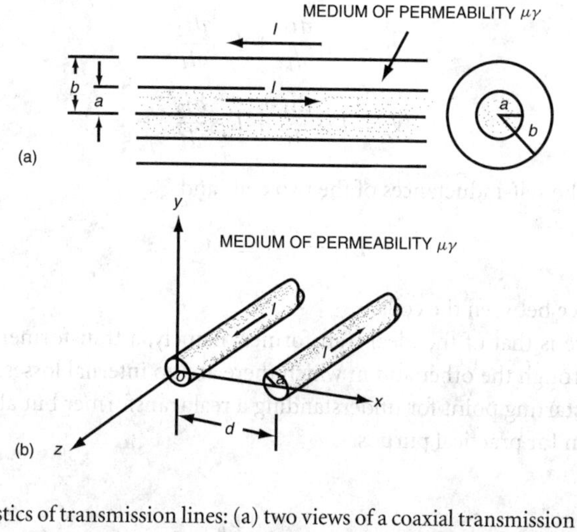

FIGURE 2.33 Characteristics of transmission lines: (a) two views of a coaxial transmission line, (b) a two-wire transmission line. (*Source:* Cheng, D.A. 1993. *Fundamentals of Engineering Electromagnetics*. Addison-Wesley, New York.)

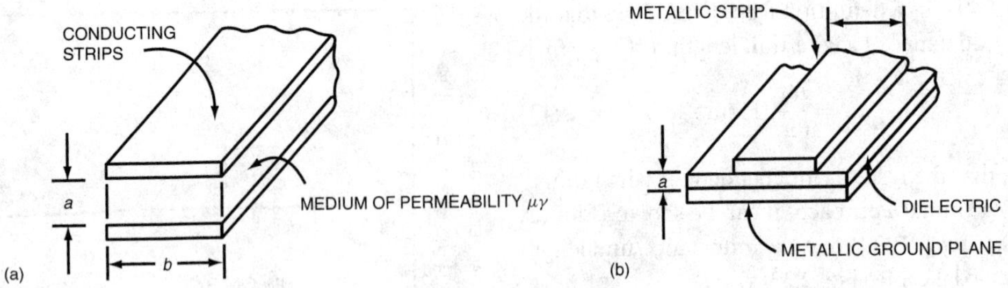

FIGURE 2.34 Parallel stripline techniques. (*Source:* After Plonus, M.A. 1978. *Applied Electromagnetics.* McGraw-Hill, New York.)

of the AC power line to a level suitable for the power supply, (3) to pass electrical signals between circuits while isolating them from DC, (4) to block DC from devices, such as speakers, whose operation would be adversely affected by large DC currents, (5) to match the impedance of a source to that of a load in order to achieve maximum power transfer and minimize distortion, and (6) used in connection with capacitors they can be part of a resonant circuit to provide narrow-band amplifications.

A transformer consist of two coils placed close together so that most of the magnetic flux generated by a current in one coil passes through the other (Fig. 2.35). To insure that this is the case usually both coils are wound on a common core of magnetic material (see Table 2.16) although, occasionally, transformers intended for high-frequency application have air cores. One coil, the primary, is connected to a time-varying voltage source while the other, the secondary, is connected to a load. The current in the primary creates a time-varying flux in the core. Since this flux also passes through the secondary winding a voltage is generated in the secondary.

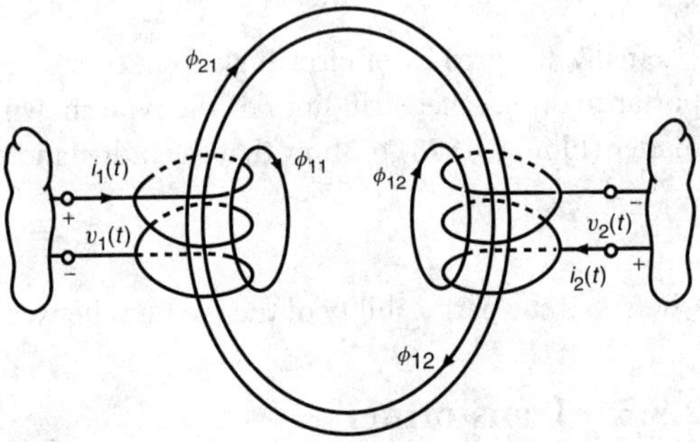

In general, the voltages V_1 and V_2 in each coil is given in terms of the currents i_1 and i_2

FIGURE 2.35 Flux relationships for mutually coupled coils. (Adapted from Irwin, J.D. 1995. *Basic Engineering Circuit Analysis*, 5th ed. Macmillan, New York.)

$$v_1 = L_{11}\frac{di_1}{dt} + L_{12}\frac{di_2}{dt}$$

$$v_2 = L_{21}\frac{di_1}{dt} + L_{22}\frac{di_2}{dt} \qquad (2.47)$$

where L_{11} and L_{22} are the self-inductances of the two coils and

$$L_{12} = L_{21} = M \qquad (2.48)$$

is the **mutual inductance** between the coils.

A useful concept here is that of the ideal transformer; namely, a transformer in which all of the flux from one coil passes through the other and in which there are no internal losses. Not only does the ideal transformer provide a starting point for understanding a real transformer but also in many cases it is an adequate approximation for practical purposes.

From Faraday's Law

$$e_p = N_p\frac{\partial \Phi_p}{dt} \qquad (2.49)$$

and

$$e_s = N_s \frac{\partial \phi_s}{dt} \tag{2.50}$$

where e is the voltage across the coil, N is the number of turns, and Φ is the flux through the coil; the subscripts p and s refer to the primary and secondary, respectively. Since for an ideal transformer $\phi_p = \phi_s$, we have

$$\frac{e_s}{e_p} = \frac{N_s}{N_p} = \frac{1}{a} \tag{2.51}$$

where a is the turns ratio. Thus, if the secondary winding has more turns than the primary, the output voltage is greater than the input voltage and the transformer is said to be a *step-up* transformer. Conversely, if the primary has the most turns, the output voltage is lower than the input voltage and we have a *step-down* transformer.

Since the ideal transformer is lossless the input and output power must be the same. Hence, we can write

$$e_p i_p = e_s i_s \tag{2.52}$$

where i_p and i_s are the currents flowing in the primary and secondary, respectively. Eliminating the ratio e_p/e_s from Eqs. (2.51) and (2.52) yields

$$\frac{i_s}{i_p} = \frac{N_p}{N_s} = a \tag{2.53}$$

Clearly, a transformer which increases the voltage decreases the current and vice versa.

A salient point to remember is that the preceding equations apply only to the time-varying components of the voltages and currents. Since static magnetic fields produce no induced voltages, the primary and secondary windings are isolated as far as DC voltages are concerned.

The maximum power transfer theorem states that maximum power is transferred from a source to a load when the internal impedance of the source is equal to the load impedance. In addition, when signals are transmitted over transmission lines it is, in general, necessary that the source and load impedances be matched to the characteristic impedance of the line to avoid reflections and distortion. As mentioned previously transformers are often used to accomplish the necessary impedance matching. The impedance (Z) looking into the primary and secondary terminals of a transformer are

$$Z_p = \frac{e_p}{i_p} \quad \text{and} \quad Z_s = \frac{e_s}{i_s} \tag{2.54}$$

Hence,

$$\frac{Z_s}{Z_p} = \frac{e_s}{i_s} \frac{i_p}{e_p} = \frac{1}{a^2} \tag{2.55}$$

that is, the ratio of input to output impedance varies as the square of the turns ratio.

The principle units of magnetic properties relating to transformer design are listed in Table 2.17.

TABLE 2.17   Units

Quantity	Symbol	Unit (S.I.)
Magnetic flux density	B	T (or Wb/m^2)
Magnetic field intensity	H	A/m
Magnetic flux	ϕ'	Wb
Magnetic flux linkage	Λ	Wb-turns
Self, mutual inductance	L, M	H
Magnetic permeability	μ	H/m

The polarity of the voltage can be determined by examining the direction in which the coils are wound on the core. For example, consider the transformer shown in Fig. 2.36(a). With the coils wound as shown, a current flowing into the upper terminal of the primary in the direction of the magnetic flux induced in the core can be determined from the right-hand rule to be in the clockwise direction as shown. The current in the secondary will be in the direction that will produce a flux that opposes the flux that induced it. Therefore, the current in the

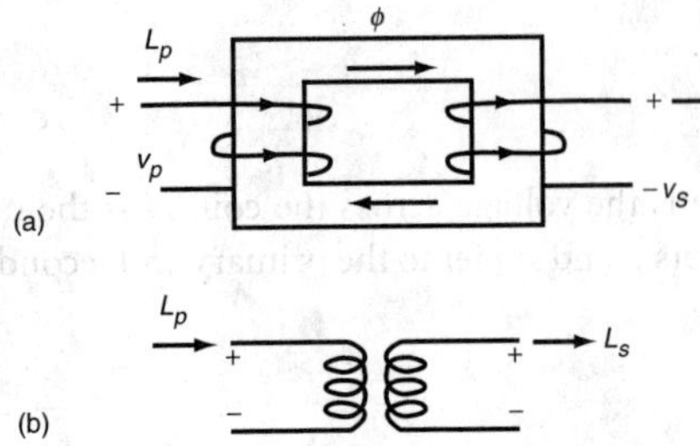

FIGURE 2.36   Voltage polarity of transformers.

secondary winding is out of the upper terminal. If the direction of either winding is reversed, then the polarity of that coil is also reversed. To indicate the polarity of a transformer dots are placed on the circuit symbol, as shown in Fig. 2.36(b). The manufacturer also sometimes places dots on the case of the transformer. The dot convention states: (1) if current flows into the dotted end of one coil it flows out the dotted end of the other and (2) the polarity of the dotted ends of both coils are the same.

Transformer Loss

A real transformer differs from an ideal transformer in two important ways. First, in a real transformer only part of the flux from one coil passes through the other. The part of the flux from the primary that does not pass through the secondary is called the leakage flux. The magnetic coupling between the primary and secondary windings is expressed in terms of the leakage constant (k), which varies between 1 for an ideal transformer and ϕ for two coils that have no magnetic coupling between them. A well-designed ferrite core transformer may have a k very near to 1, whereas the typical value of k for an air core transformer may be as low as 0.5.

In addition to imperfect flux leakage a real transformer has internal power losses that must be considered:

- *Copper loss.* Since the wire from which the coils are core wound has a finite resistance there is a power loss due to ohmic heating of the wire.
- *Eddy current loss.* The time-varying flux not only induces current in the secondary winding of the transformer but also induces circulating currents in the core as well (Fig. 2.37(a) and Fig. 2.37(b)). These eddy currents, as they are called, cause ohmic heating of the core and hence a loss of power. Power transformers normally have laminated cores made of alternate layers of iron and insulating material to coil erupt the flow of the eddy currents and thus reduce the eddy current loss (Fig. 2.37(c)).
- *Hysteresis loss.* Work is also done in rotating the magnetic domains in the core, thus loss is termed hysteresis loss.

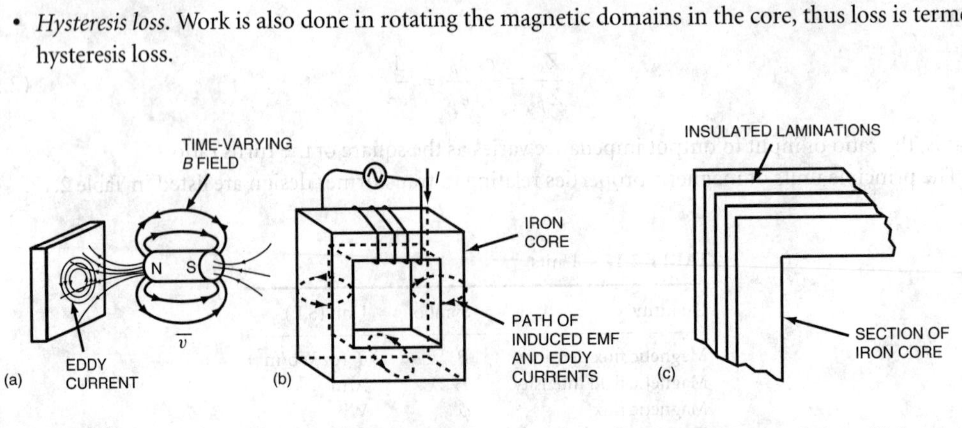

FIGURE 2.37   (a) A magnet moving toward a conducting sheet will induce eddy currents as shown; (b) formation of emfs, which cause eddy currents; (c) the reduction of eddy currents by laminating the core. (*Source:* After Plonus, M.A. 1978. *Applied Electromagnetics.* McGraw-Hill, New York.)

- *Loss due to energy being converted to sound in the core.* This phenomenon, called magnetostriction, is of importance only in very large power transformers. The combination of the eddy current and hystersis losses is called the *iron loss*.

In addition to these effects, real transformers often have strong capacitance both between the turns of the windings and between the primary and secondary windings. These capacitances may become significant at high frequencies.

Networks containing transformers can be analyzed by introducing an equivalent circuit. Let

L_p = inductance of the primary winding
L_s = inductance of the secondary winding
R_p = inductance of the primary winding
R_s = inductance of the secondary winding
Z = the equivalent impedance of the load
R_0 = hypothetical resistance introduced to
            account for core losses
M = mutual inductance between the primary and
            secondary windings

By definition the mutual inductance is given by

$$M = k\sqrt{L_p L_s} \tag{2.56}$$

We also let

$$L_1 = L_p - aM \tag{2.57}$$

and

$$L_2 = L_s - M/a \tag{2.58}$$

The real transformer of Fig. 2.37(a) can be replaced by the circuit containing an ideal transformer shown in Fig. 2.37(b). It is easily shown that the circuits of Fig. 2.37(a) and Fig. 2.37(b) are described by the same set of loop equations and are, therefore, equivalent. This deviation is beyond the scope of this chapter but can be found in standard circuits texts.

The ideal transformer's only effects are (1) to change the voltage and current by a factor a and ℓ/a, respectively, and (2) to transform the impedances by a factor of a^2 (See Fig. 2.38). We can therefore eliminate the transformer by replacing the secondary voltage and current by scaled quantities i'_1 and i'_2 where

$$i'_s = \frac{i_s}{a} \tag{2.59}$$

$$v_s = av_s \tag{2.60}$$

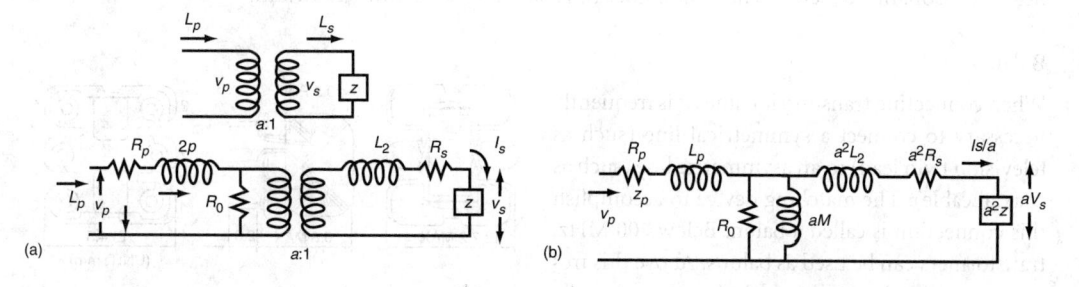

(a)   (b)

FIGURE 2.38 Transformer characteristics: (a) real transformer, (b) circuit model with ideal transformer.

and by multiply in all impedances on the secondary side of the transformer by a^2 to obtain the network of Fig. 2.37(a). This network can be solved for V_s' and i_s' using standard circuit analysis techniques. Equations (2.59) and (2.60) can then be used to obtain V_s and i_s.

By this procedure it is said that the secondary quantities have been referred to the primary side. It is also possible to refer the primary quantities to the secondary side. In this case the multiplier would be ℓ/a.

Autotransformer

An **autotransformer** consists of a single coil, usually wound on a magnetic core, as shown in Fig. 2.39. The part of the coil common to both input and output sides of the transformer is called the common winding and the rest of the coil is called the series winding. We denote the number of turns on each winding by N_c and N_s, respectively. Usually N_s is much smaller than N_c. Symbols for the various voltages and currents are defined in Fig. 2.39. Either side of the autotransformer can be used as the primary side. If the high-voltage (left-hand) side of the autotransformer in Fig. 2.39 is the primary side

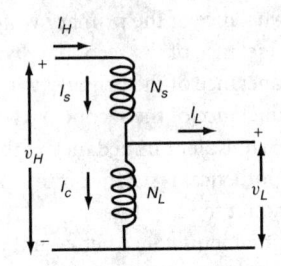

FIGURE 2.39   An auto-transformer. Current directions are shown for a step-down autotransformer.

it is a step-down transformer, whereas if the low-voltage (right-hand) side is the primary it is a step-up transformer. In the latter case, the direction of all four currents are reversed.

For an ideal autotransformer the relations between the high- and low-side voltages and currents are given by Eq. (2.61), which is valid for both step-up and step-down configurations.

$$\frac{v_L}{v_H} = \frac{I_H}{I_L} = \frac{N_c}{N_c + N_s} \qquad (2.61)$$

Autotransformers are frequently wound on a cylindrical core, and the output taken from a movable contact as shown in Fig. 2.40. Such a variable autotransformer could provide a voltage from 0 V to about 10% above the line voltage.

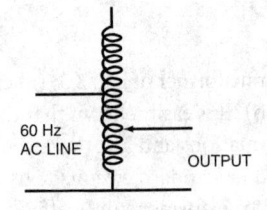

FIGURE 2.40   Variable autotransformer.

Intermediate-Frequency Transformer

Transformers are often used to couple the intermediate-frequency (IF) stages of a superheterodyne radio. These IF transformers, or *IF-cans* as they are often called, are parallel by capacitors, as shown in Fig. 2.40. Each coil has a moveable ferrite core that may be screwed into or out of the coil changing the inductance of the coil and, hence, the resonant frequency of the tank circuit. The IF transformer will thus pass signals near the resonant frequency. The IF amplifier thus have a very narrow bandwidth.

Baluns

When connecting transmission lines it is frequently necessary to connect a symmetrical line (such as television twin lead) to an asymmetric line (such as coaxial cable). The matching device to accomplish this connection is called a balun. Below 200 MHz, transformers can be used as baluns. Above this frequency, transmission line techniques are usually used. A balun is shown in Fig. 2.41.

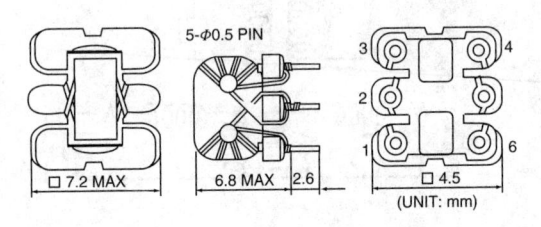

FIGURE 2.41   A balun transformer. (Courtesy of Toko.)

Distortion

Transformers can be a major source of distortion in electronic circuits. This distortion arises from two major sources. First, the transformer consists of inductive elements whose impedance is frequency dependent. Transformers, therefore, display harmonic distortion. Second, the magnetization curve of ferromagnetic material (Fig. 2.31) is very nonlinear. This introduces distortion into signals coupled by iron or ferrite core transformers unless the signal level is kept very small.

Defining Terms

Autotransformer: A transformer in which the input and output sides have a common winding.
Flux linkage: Magnetic flux multiplied by the number of turns.
Magnetic flux: Magnetic flux density integrated over cross-sectional area.
Magnetic permeability: Magnetic analog of electrical conductivity, describing the degree to which a magnetic material is capable of sustaining magnetic flux. Defined by ratio of magnetic induction field over magnetic field intensity.
Mutual inductance: The magnetic flux linkage common to two magnetic circuit elements per unit current flowing through the element in question.
Relative permeability: The magnetic permeability of a material divided by the magnetic permeability of free space.
Self-inductance: The magnetic flux linkage per unit current flowing through the circuit element.

References

Cheng, D.A. 1993. *Fundamentals of Engineering Electromagnetics*. Addison-Wesley, New York.
Plonus, M.A. 1978. *Applied Electromagnetics*. McGraw-Hill, New York.

Further Information

For further information the following books are recommended:
    David K. Cheng, *Field and Wave Electromagnetics*, 2nd ed. Addison-Wesley, NY.
    Irwin, J. D. 1995. *Basic Engineering Circuit Analysis*, 5th ed. Macmillan, NY.
See, also, the following journals:
    IEEE Transactions on Magnetics.
    Spectrum (The Institute of Electrical and Electronics Engineers, Inc.).

2.6   Capacitance and Capacitors

Igor M. Filanovsky

2.6.1   Capacitor and Capacitance in Linear Electromagnetic Field

A system of two conducting bodies (which are frequently determined as plates) located in the electromagnetic field and having equal charges of opposite signs $+Q$ and $-Q$ can be called a **capacitor**. The **capacitance** C of this system is equal to the ratio of the charge Q (here the charge absolute value is taken) to the voltage V (again, the absolute value is important) between the bodies, that is

$$C = \frac{Q}{V} \tag{2.62}$$

Capacitance C depends on the size and shape of the bodies and their mutual location. It is proportional to the dielectric permittivity ε of the media where the bodies are located. The capacitance is measured in **farads** (F) if the charge is measured in coulombs (C) and the voltage in volts (V). One farad is a very big unit; the practical capacitors have the capacitances that are measured in micro- (μF, or 10^{-6} F), nano- (nF, or 10^{-9} F), and picofarads (pF, or 10^{-12} F).

The calculation of capacitance requires knowledge of the electrostatic field between the bodies. The following two theorems (Stuart, 1965) are most important in these calculations. The integral of the flux density D over a closed surface is equal to the charge Q enclosed by the surface (the Gauss theorem), that is

$$\oint D\,ds = Q \tag{2.63}$$

This result is valid for linear and nonlinear dielectrics. For a linear and isotropic media $D = \varepsilon E$, where E is the electric field, and this case will be assumed here (unless otherwise specified). The magnitude E of the field is measured in volt per meter, the magnitude D of the flux in coulomb per square meter, and the dielectric permittivity has the dimension of farad per meter. The dielectric permittivity is usually represented as $\varepsilon = \varepsilon_0 K_d$ where ε_0 is the permittivity of air ($\varepsilon_0 = 8.86 \cdot 10^{-12}$ F/m) and K_d is the dielectric constant.

The electric field is defined by an electric potential ϕ. The directional derivative of the potential taken with the minus sign is equal to the component of the electric field in this direction. The voltage V_{AB} between the points A and B, having the potentials ϕ_A and ϕ_B, respectively (the potential is also measured in volts), is equal to

$$V_{AB} = \int_A^B E\,dl = \phi_A - \phi_B \tag{2.64}$$

This result is the second basic relationship. The left hand side of Eq. (2.64) is a line integral. At each point of the line AB we have two vectors: E defined by the field and dl that defines the direction of the line at this point.

Let us consider some examples of application of these two basic relationships.

Parallel Plate Capacitor

Assume (Fig. 2.42(a)) that we have two plates of area S each. The plates have the potentials ϕ_A and ϕ_B. The distance d between the plates is small, so that one can neglect the distortions of the electric field along the capacitor perimeter.

The last statement needs some comments. In the general case the parallel plate capacitor has the field that is not uniform (Fig. 2.42(b)). Yet, if the plates are cut so that the inner part has the guard rings, which have also the potentials ϕ_A and ϕ_B, then the inner part will have a uniform electric field (Fig. 2.42(c)). We will assume that Fig. 2.42(a) represents this inner part of Fig. 2.42(c).

Taking a path from the point A to the point B along one of the field lines

$$V_{AB} = \int_A^B E\,dl = E \int_A^B dl = Ed \tag{2.65}$$

if this uniform field has the magnitude E. If one takes a surface (Fig. 2.42(c)) enveloping one of the plates and closely located to it, then

$$Q = \oint \varepsilon E\,ds = \varepsilon E S \tag{2.66}$$

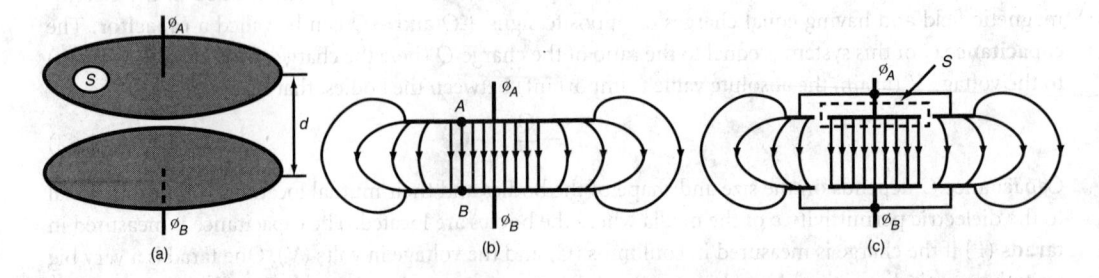

FIGURE 2.42 Parallel plate capacitor.

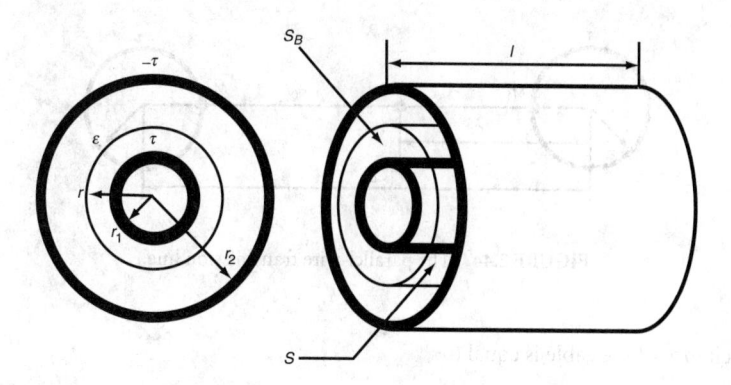

FIGURE 2.43 Characteristics of coaxial cable.

because the field exists inside the capacitor only. From Eqs. (2.65) and (2.66) one immediately obtains that

$$C = \frac{Q}{V} = \frac{Q}{V_{AB}} = \frac{\varepsilon S}{d} \tag{2.67}$$

The construction of nearly all practical capacitors can be reduced to this simple two-plate configuration. And when the area S increases (proportionally to the square of the plate diameter), the role of the perimeter (proportional to the first degree of diameter) field becomes smaller, and the guard rings could be omitted.

Coaxial Cable Capacitance

Let us consider a piece of a coaxial cable having the length l. Thus, we are calculating the capacitance of two concentric cylinders, the radius of the inner cylinder being r_1 and the inner radius of the external cylinder being r_2 (Fig. 2.43). The inner cylinder carries a positive charge Q, the dielectric between these two cylinders is characterized by the permittivity ε, and the voltage between the cylinders is V.

If the condition $l \gg r_2$ is satisfied, one can neglect the distortions of the electric field near the ends of the coaxial cable and consider that the charge is distributed uniformly, with the linear density $\tau = Q/l$ (τ is measured in coulomb per meter) and the vectors of the electric field E and the electric flux density D are directed along the radial lines passing from one cylinder to another. Then one can apply the Gauss theorem to a cylindric surface having the radius r and located between the cylinders. The radial direction of the electric field allows us to neglect the flux through the bottoms S_B and take into consideration the flux through the cylinder side surface S only. Then, for a given r, the electric flux density D is constant on the side surface and one can write

$$\oint D\,ds = \int_S D\,ds = \int_S D\,ds = D\int_S ds = D2\pi rl = Q\tau l \tag{2.68}$$

From this result one obtains that $D = \tau/(2\pi r)$ and

$$E = \frac{\tau}{2\pi \varepsilon r} \tag{2.69}$$

Notice that Eq. (2.69) does not depend on the radius r_1; this means that the electric field outside the inner cylinder is the same as the field of a line coinciding with the cylinder axis and carrying the same charge.

The voltage at the cable is

$$V = \int_{r_1}^{r_2} E\,dr = \int_{r_1}^{r_2} \frac{\tau}{2\pi \varepsilon r}\,dr = \frac{\tau}{2\pi \varepsilon}\ln\frac{r_2}{r_1} \tag{2.70}$$

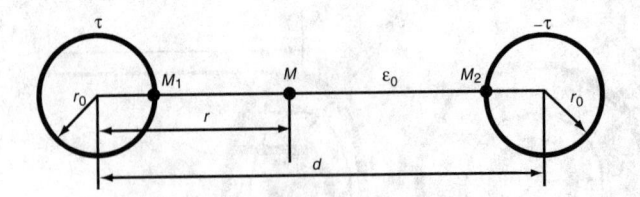

FIGURE 2.44   The parallel-wire transmission line.

Thus, the capacitance of the cable is equal to

$$C = \frac{2\pi \varepsilon l}{\ell n (r_2/r_1)} \tag{2.71}$$

The capacitance of tubular capacitors can also be calculated using Eq. (2.71).

Capacitance of a Two-Wire Transmission Line

The cross section of this line is shown in Fig. 2.44. The wires have the charges of opposite signs when a voltage is applied to the line. Again, it is assumed that the length l of wires is much larger than the distance between the wires, that is, $l \gg d$. This allows us to neglect the distortion of electric field near the ends of the transmission line and one can consider that the field is the same in any plane that is perpendicular to the wires axes. Assume also that the line is in air and far enough from ground so that the influence of this conducting ground can be neglected. Let the voltage between the wires be V, the charge at each wire is $Q = \tau l$, where τ is the charge linear density, and each wire has the radius r_0.

In the coaxial cable the distribution of the electric charge on the inner wire was uniform along the surface of this wire. In the considered case, the charge density on the wires is not uniform; it is higher on the inner parts of the wire surfaces due to attraction force between the charges of opposite sign. Hence, the distance between the electric axes of the wires is less than the distance between their geometrical axes. This difference become smaller when the ratio d/r becomes larger. If $d \gg r$ one can consider that the electric and geometric axes of the wires are coinciding. Then, at the point M the electric field can be found as a result of superposition of the fields from the left and right wires, that is

$$E = \frac{\tau}{2\pi \varepsilon_0 r} + \frac{\tau}{2\pi \varepsilon_0 (d - r)} \tag{2.72}$$

and the voltage V between the wires can be calculated as

$$V = \int_{M_1}^{M_2} E\, dr = \frac{\tau}{2\pi \varepsilon_0} \int_{r_0}^{d-r_0} \left(\frac{1}{r} + \frac{1}{d-r} \right) dr = \frac{\tau}{\pi \varepsilon_0} \ell n \frac{d - r_0}{r_0} \tag{2.73}$$

From Eq. (2.73) one finds that the capacitance of the two-wire line

$$C = \frac{Q}{V} = \frac{\pi \varepsilon_0 l}{\ell n [(d - r_0)/r_0]} \approx \frac{\pi \varepsilon_0 l}{\ell n (d/r_0)} \tag{2.74}$$

The capacitance of a wire located at a distance $d/2$ from the ground plane can be obtained using the doubled value given by Eq. (2.74).

2.6.2   Mutual Capacitance

In many technical problems one has to consider an electric field created by a system of some pairs of the bodies with the charges of opposite sign in each pair. The charges of these pairs and the voltages between them are connected by a system of linear equations.

In a system of two capacitors (bodies 1 and 1' and 2 and 2') shown in Fig. 2.45 all four bodies create their common electric field, and the charge of each capacitor depends on both voltages V_1 and V_2. The field of Fig. 2.45 corresponds to the case when the potential of the body 1 is larger than the potential of the bodies 2 and 2' (with the potential of the body 2 larger than the potential of the body 2'), and the potential of the body 2' is larger than the potential of the body 1'. In this case the charge of one capacitor due to the voltage on another capacitor coincides with the charge created by the voltage on the capacitor itself. Hence, the total charges for the capacitors is

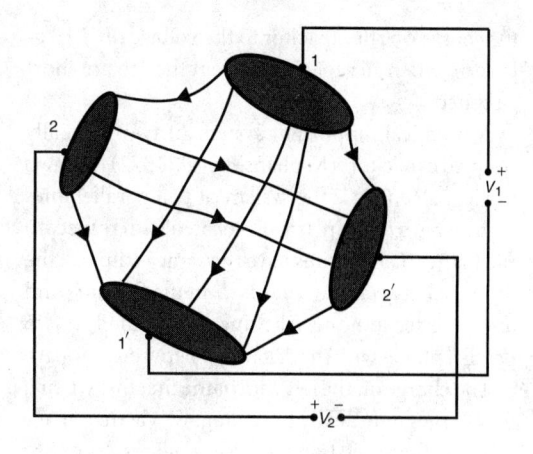

FIGURE 2.45   System of two capacitors.

$$Q_1 = C_{11} V_1 + C_{12} V_2$$
$$Q_2 = C_{21} V_1 + C_{22} V_2 \tag{2.75}$$

where C_{11} and C_{22} can be called self-capacitances of the first and second capacitor, correspondingly, and C_{12} and C_{21} can be called **mutual capacitances**.

The self-capacitances

$$C_{11} = \left.\frac{Q_1}{V_1}\right|_{V_2=0} \quad \text{and} \quad C_{22} = \left.\frac{Q_2}{V_2}\right|_{V_1=0} \tag{2.76}$$

are determined by all four bodies of the system and is different from the capacitances C_1 and C_2 of each of these capacitors when the other is absent. When one calculates the capacitance C_{11}, the condition $V_2 = 0$ simply means that the plates of another capacitor are connected by a wire, yet under the influence of the first capacitor the plates of this another capacitor obtain equal charges of opposite signs. The self-capacitance C_{22} is calculated in a similar way.

The mutual capacitances

$$C_{12} = \left.\frac{Q_1}{V_2}\right|_{V_1=0} \quad \text{and} \quad C_{21} = \left.\frac{Q_2}{V_1}\right|_{V_2=0} \tag{2.77}$$

are determined by the ratios of the charge on the plates of shortened capacitor to the voltage at the plates of another capacitor, when this charge is induced by this voltage.

The mutual capacitance is determined by the geometric configuration of the bodies, their mutual location, and the dielectric permittivity of the media in which the bodies are immersed. Contrary to the self-capacitance, it can be positive or negative. The mutual capacitance is also measured in farads. Also, it is possible to show that

$$C_{12} = C_{21} \tag{2.78}$$

(this property is analogous to the reciprocity condition in passive networks (Scott, 1960)).

In the general case of a system of many capacitors, the charge Q_i of the ith capacitor is determined by all voltages, that is

$$Q_i = C_{ii} V_i + \sum_{i \neq j} C_{ij} V_j \tag{2.79}$$

where C_{ii} is the self-capacitance of the capacitor i (in the presence of other capacitances), and $C_{ij} = C_{ji}$ is the mutual capacitance between the capacitors i and j. This mutual capacitance is equal to the ratio of the charge on ith capacitor to the voltage on jth capacitor, when all capacitors except the jth are short circuited.

A practical, important system of two mutually coupled capacitors (Kapliansky et al., 1972) is shown in Fig. 2.46. This is the system of two parallel lines of the length l symmetrically located with respect to each other. Let all wires have the same radius r_0; the distance between the wires is d_1 in the first line and d_2 in the second one. The lines are located in two

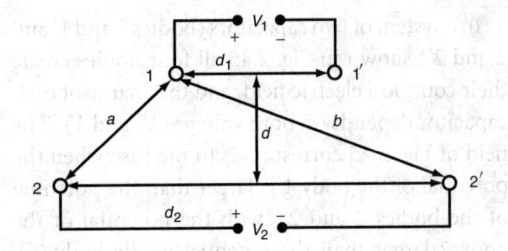

FIGURE 2.46 Capacitively coupled transmission lines.

parallel planes with the distance d between them. Assume as well that $l \gg r_0, d_1 \gg r_0, d_2 \gg r_0$, and $d \gg r_0$.

The charge on the left wire of the first line when it is shortened ($V_1 = 0$) can be found as the difference of the charges induced by the voltage V_2 via the capacitance between the wires 1 and 2 and via the capacitance between the wires 1 and 2'. Using the results of Sec. 2.6.1 one can write that

$$Q_1|_{V_1=0} = \frac{\pi \varepsilon_0 l}{2\ell n(a/r_0)} V_2 - \frac{\pi \varepsilon_0 l}{2\ell n(b/r_0)} V_2 = \frac{\pi \varepsilon_0 l}{2} \frac{\ell n(b/a)}{\ell n(a/r_0)\ell n(b/r_0)} V_2 \tag{2.80}$$

Here $a = (\frac{1}{2})\sqrt{[(d_1 - d_2)^2 + 4d^2]}$ and $b = (\frac{1}{2})\sqrt{[(d_1 + d_2)^2 + 4d^2]}$. Notice also the multiplier of $1/2$ in Eq. (2.80) that arises due to the symmetry of the picture: one can assume that wire 2 has the potential $V_1/2$ and wire 2' has the potential $-V_1/2$ with respect to the ground. Thus, one finds that the mutual capacitance is equal to

$$C_{12} = \frac{\pi \varepsilon_0 l}{2} \frac{\ell n(b/a)}{\ell n(a/r_0)\ell n(b/r_0)} \tag{2.81}$$

and the self-capacitance of the first line, taking into consideration the presence of the second line, is equal to

$$C_{11} = \frac{\pi \varepsilon_0 l}{\ell n(d_1/r_0)} + \frac{\pi \varepsilon_0 l}{2} \frac{\ell n(b/a)}{\ell n(a/r_0)\ell n(b/r_0)} \tag{2.82}$$

The problem of calculating self-capacitances and mutual capacitances is important, for example, when it is required to determine the level of interference of the transmission line 2 with the communication line 1. This interference can be determined, for example, by the ratio of the charge $C_{12} V_2$ of the mutual capacitance to the charge $C_{11} V_1$ of the self-capacitance.

2.6.3 Energy Stored in the System of Charged Capacitors

Let n conducting bodies be located in a dielectric with permittivity ε with the potential of each body ϕ_k and the charge of each body equal to Q_k. It is possible to show (Stuart, 1965) that the energy stored in the electric field of this system is equal to

$$W = \frac{1}{2} \sum_k \phi_k Q_k \tag{2.83}$$

Hence, for one only body this energy will be

$$W = \frac{1}{2}\phi Q = \frac{C\phi^2}{2} = \frac{Q^2}{2C} \tag{2.84}$$

where $C = Q/\phi$ is the body capacitance. For a capacitor, that is the system of two bodies with $Q_1 = Q_2 = Q$ and the voltage $V = \phi_1 - \phi_2$ between the bodies, the energy will be

$$W = \frac{1}{2}\phi_1 Q_1 + \frac{1}{2}\phi_2 Q_2 = \frac{1}{2}Q(\phi_1 - \phi_2) = \frac{QV}{2} = \frac{CV^2}{2} = \frac{Q^2}{2C} \tag{2.85}$$

and it is a quadratic function of the charge Q or the voltage V.

Thus, capacitance can be considered as the concept of energy storage in an electric field. And the particular result (Eq. (2.85)) can be obtained (Stuart, 1965) easily if one accepts Eq. (2.62) as an experimental law of a linear capacitor. It is also known from experiment that work must be done in order to charge the capacitor and, therefore, the capacitor stores energy. This can be calculated as follows. If a current i is flowing into a capacitor with a voltage V across it, then the source is supplying power Vi. Hence, in time dt the energy supplied by the source is $Vi\,dt$. If there are no resistive elements in the system and the energy is not dissipated, then this energy must be stored in the capacitor. But $i = dQ/dt$, and thus the energy dW is given by $V\,dQ$. One also has the law (2.62) relating V and Q and, hence

$$dW = \frac{Q}{C}dQ \tag{2.86}$$

If the initial charge on the capacitor is zero, the final charge is Q, and the final voltage is V, then, indeed

$$W = \int_0^Q \frac{Q}{C}dQ = \frac{1}{2}\frac{Q^2}{C} = \frac{1}{2}CV^2 \tag{2.87}$$

This is the energy of the field created in the space around the capacitor plates.

The energy of two coupled capacitors (four bodies) will be

$$W = \frac{Q_1 V_1}{2} + \frac{Q_2 V_2}{2} = \frac{(C_{11}V_1 + C_{12}V_2)V_1}{2} + \frac{(C_{21}V_1 + C_{22}V_2)V_2}{2}$$

$$= \frac{C_{11}V_1^2}{2} + \frac{C_{22}V_2^2}{2} + C_{12}V_1 V_2 \tag{2.88}$$

Hence, this energy includes the sum of energies stored in each capacitor and the term of mutual energy $C_{12}V_1 V_2$. This last energy can be positive or negative, and the energy of the whole system can be larger or smaller than the sum of the energies stored in each capacitor. Yet, the whole energy of the system will be always positive.

In the same way one can write that for a system of coupled capacitors the stored energy is

$$W = \sum_k \frac{C_k V_k^2}{2} + \sum_{p \neq k} C_{kp} V_k V_p \tag{2.89}$$

and one can show that it is always positive.

2.6.4 Capacitor as a Circuit Component

In an electric circuit the capacitor appears as a two-terminal element described by the following relationship between the current and voltage:

$$i(t) = C\frac{dv(t)}{dt} \tag{2.90}$$

where C is a positive real constant. The result (2.90) follows, of course, from the fact that the capacitor energy is stored as the energy of an electrostatic field and is given by Eq. (2.87). But dealing with an electric circuit one simply defines that if Eq. (2.90) is valid, this circuit element is a capacitor, and C is the capacitance. This element is denoted then as shown in Fig. 2.47(a).

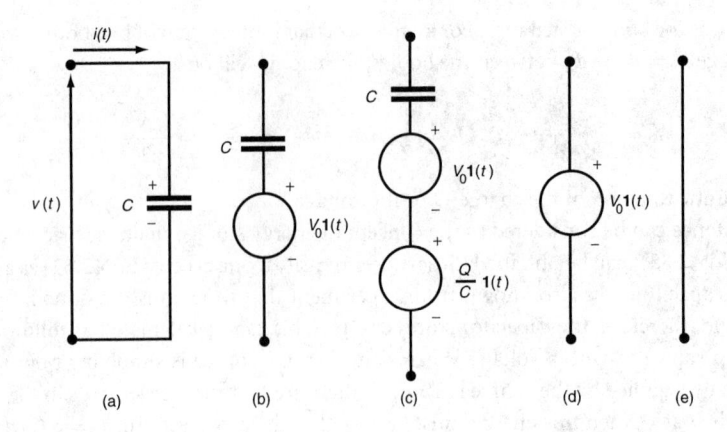

FIGURE 2.47   Charged capacitor and its equivalent circuits in transients.

One can formulate some properties of this element that follow directly from this definition. If t_0 is an instant of time when the voltage on the capacitor is zero, and $p(t) = i(t)v(t)$ is the instant power in the capacitor, then the energy stored in the capacitor is

$$w(t) = \int_{t_0}^{t} p(t)\, dt = \int_{t_0}^{t} C \frac{dv(t)}{dt} v(t)\, dt = \frac{1}{2} C v^2(t) \tag{2.91}$$

This is the same result (2.87) that was obtained from the capacitor electric field consideration, except that the voltage is considered not as a constant but as a variable. One can see that the constant C is the capacitance and check that it is measured in farads. Indeed, the dimension of this unit, as follows from Eq. (2.90), is (1 A/1 V) 1 s = 1 C/1 V = 1 F and coincides with the dimension followed from Eq. (2.62).

For any positive real C

$$\frac{1}{2} C v^2(t) \geq 0 \tag{2.92}$$

and the energy stored in the capacitor in the time interval t_1, t_2 is equal to

$$W = \frac{1}{2} C \left[v^2(t_2) - v^2(t_1) \right] \tag{2.93}$$

Hence, if $v(t_2) = v(t_1)$ the energy stored in the capacitor in this time interval is equal to zero for any arbitrary variation of voltage $v(t)$ between t_1 and t_2. This means that if $v(t_3) > v(t_1)$ (here $t_1 < t_3 < t_2$) then the energy stored in the capacitor in the interval t_1, t_3 will be returned in the circuit during the interval t_3, t_2; the capacitor transforms the energy without dissipation. In other words, the energy stored in the capacitor equals zero at all instants when $v(t) = 0$, and in all other instants it is positive. Of course, this energy is stored as the energy of electrostatic field inside the capacitor; the capacitor does not have any stored energy if this electrostatic field is absent. The capacitor is a *definite* component. This means that if $v(t)$ is known the result (2.90) immediately provides the corresponding current $i(t)$. From the other side, if $i(t)$ is given, the voltage $v(t)$ can be found by an integration procedure. Yet, these two cases need some additional remarks (Biorci, 1975) if the relationship (2.90) is considered as a differential relationship that describes the input–output variables of a linear system.

If the voltage $v(t)$ is considered as an input variable, and the current $i(t)$ as an output variable, then in Eq. (2.90) the behavior of the current in a certain time interval is completely determined by the behavior of the voltage in this interval. In this case the solution of the differential equation (2.90) has the forced component only, it is not possible to determine a state variable, and the current is the function that is more impulsive than the voltage (e.g., if the voltage includes a step, the current should include a pulse). Hence, the particular solution of the differential equation (2.90) coincides in this case with the full solution. The Laplace transform applied to Eq. (2.90) gives the relationship

$$I(s) = Cs\,V(s) \tag{2.94}$$

If the current $i(t)$ is considered as an input variable and $v(t)$ as an output variable, the relationship (2.90) can be rewritten as

$$C\frac{dv(t)}{dt} = i(t) \tag{2.95}$$

In this differential equation the right part does not include any derivatives. In this case the solution, that is, the voltage $v(t)$, should be less impulsive than the current. Yet, the voltage will be known in a given time interval if the value of $v(t)$ is known at the initial point of this interval, and it is not necessary to know any derivatives of $v(t)$ at this instant. The voltage $v(t)$ in this case is called a *state variable*. This assumes that if $i(t)$ is equal to zero for all $t \le 0$ and that if at $t = 0^-$ the state variable is zero, that is, $v(0^-) = 0$ (the capacitor is initially discharged) the solution of the differential equation (2.95) obtained using the Laplace transform gives the relationship

$$Cs\,V(s) = I(s) \tag{2.96}$$

The relationships (2.94) and (2.96) show that considering the input-output relationships one can introduce two transfer functions, namely, the capacitor impedance

$$Z(s) = \frac{1}{Cs} \tag{2.97}$$

and the capacitor admittance

$$Y(s) = \frac{1}{Z(s)} = sC \tag{2.98}$$

Then in the stationary condition $s \to 0$ and $Z \to \infty$, and $Y \to 0$.

In the sinusoidal regime $s = j\omega = 2\pi f$, where f is the frequency, in hertz, and

$$Z(j\omega) = \frac{1}{j\omega C} = -j\frac{1}{\omega C} \tag{2.99}$$

and

$$Y(j\omega) = j\omega C \tag{2.100}$$

In this case $\dot{V} = (1/j\omega C)\dot{I}$ (here $\dot{V}$ and $\dot{I}$ are phasors) and $\dot{I} = j\omega C\dot{V}$. Thus, in the sinusoidal regime the capacitor is characterized by a reactance

$$X_C = \frac{1}{\omega C} \tag{2.101}$$

measured in ohms, and the current will lead the voltage by 90°.

When one is solving Eq. (2.95) (the input variable is voltage) the particular integral (forced response) coincides with the total response; this is not the case when one is solving Eq. (2.99) (the input variable is current). But even in this case, one can write, as it was in the case of zero initial conditions, that

$$v(t) = \frac{1}{C}\int_{0-}^{t} i(t)\,dt \tag{2.102}$$

for $t \geq 0$. If the current $i(t)$ has a pulse component $Q\delta(t)$ at $t = 0$, then one can divide the integral in two parts (one is from $t = 0^-$ to $t = 0^+$, another from $t = 0^+$ to t) and write

$$v(t) = \left[\frac{1}{C} \int_{0^+}^{t} i(t)\,dt \right] 1(t) + \frac{Q}{C} 1(t) \tag{2.103}$$

(here the expression under integral does not include pulse components). Finally, if the capacitor initial condition is different from zero (nonzero state), and $v(0^-) = V_0$ one can write

$$v(t) = \frac{1}{C} \int_{0^-}^{t} i(t)\,dt + v(0^-) \tag{2.104}$$

which can be represented in two forms

$$v(t) = \left[\frac{1}{C} \int_{0^-}^{t} i(t)\,dt \right] 1(t) + V_0 1(t) \tag{2.105}$$

and

$$v(t) = \left[\frac{1}{C} \int_{0^-}^{t} i(t)\,dt \right] 1(t) + \frac{Q}{C} 1(t) + V_0 1(t) \tag{2.106}$$

depending on whether $i(t)$ is free of a pulse component at $t = 0$ or includes it.

It is easy to see from these expressions that $v(0^-) = v(0^+)$ excluding the case when $i(t)$ has a pulse component at $t = 0$. This is a formalization of the statement that the capacitor does resist to a change of voltage. If, in a circuit with this capacitor, at an instant $t = 0$ a perturbation occurs that is not able to produce a current pulse, then the capacitor preserves in $t = 0^+$ the voltage that it has at $t = 0^-$.

The Eq. (2.105) assumes that in calculation of transients an initially charged capacitor (Fig. 2.47(a)) can be substituted for by the equivalent circuit shown in Fig. 2.47(b) where the capacitor C is discharged at $t = 0^-$. Thus, the result of voltage variations preceding the instant $t = 0$ is summarized by the voltage $v(0^-) = V_0$. The Eq. (2.106) (much less useful) shows the capacitor equivalent circuit (Fig. 2.47(c)) if it is known *a priori* that the input includes a pulse current that appears at $t = 0$. If one considers the capacitance only at the instant $t = 0^+$, it can be substituted for by a voltage source with the voltage $v(0^-) = V_0$ (Fig. 2.47(d)) or simply by a short circuit (Fig. 2.47(e)) if the capacitor was initially discharged (assuming that there were no pulses in the current at $t = 0$).

Finally, if $v(t)$ is constant, the current $i(t)$ should be permanently zero. Or, it is better to say that if $v(t) = V_0 + v_1(t)$ the capacitor current is the same as if only the component $v_1(t)$ was applied. This particular property is important when the capacitor is used as a coupling element in electronic circuits.

Let us return to the analysis of the sinusoidal regime again. If

$$v(t) = \sqrt{2} V \cos \omega t \tag{2.107}$$

then, as follows from Eq. (2.90)

$$i(t) = -\sqrt{2}\omega C V \sin \omega t = -\sqrt{2} I \sin \omega t \tag{2.108}$$

where $I = V\omega C$, and the instantaneous power is given by

$$p(t) = v(t)i(t) = -2VI \cos \omega t \sin \omega t = -VI \sin 2\omega t \tag{2.109}$$

Hence, the instantaneous power is a sinusoidal variable, the frequency of which is two times larger than the frequency of the current. This power is plotted in Fig. 2.48 (which summarizes the capacitor properties). On the first half-cycle the power is negative, representing a flow of energy from the capacitance to the source. It does not mean that the capacitance can give up energy to the capacitance source before it gets

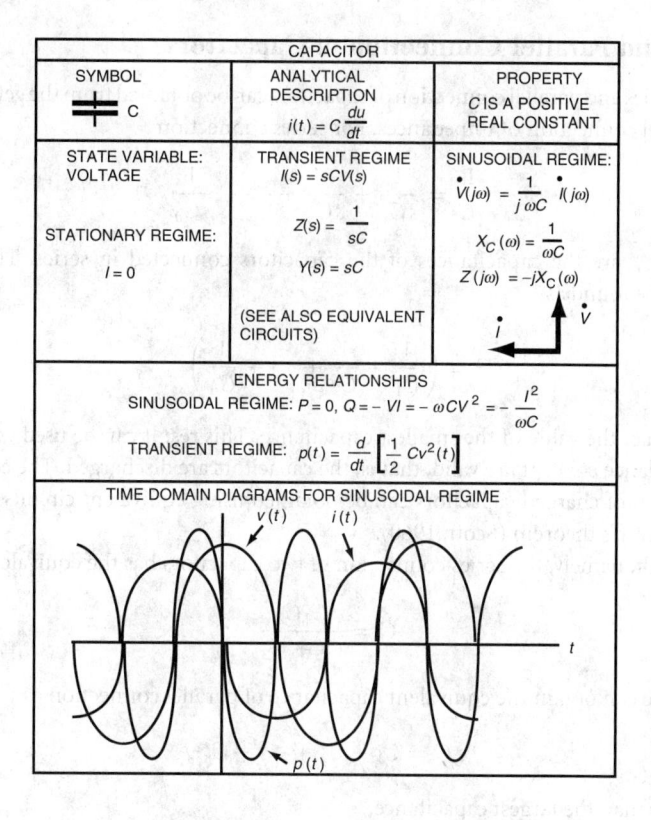

FIGURE 2.48 Summary of capacitor properties.

energy from the source. The system in the steady-state condition, and the original energy got into the capacitance during the original transient of the circuit, while the steady-state was being established. Yet, there is a general agreement to consider the instantaneous power in the capacitor with negative sign (Scott, 1960). The average value of this instantaneous power is zero. The active power is usually determined as the average value of the instantaneous power; hence, the capacitor does not have any active power. The instantaneous power in the capacitor is called *instantaneous reactive power* and its maximal value (with negative sign) is called *reactive power* and is denoted by Q. Thus

$$\text{reactive power} = Q = -VI = -\omega C V^2 = -\frac{I^2}{\omega C} = -\frac{V^2}{X_C} = -X_C I^2 \qquad (2.110)$$

Again, there is a relationship between the reactive power and the average energy stored in the capacitance. The instantaneous energy is

$$w(t) = C\frac{v^2(t)}{2} = C V^2 \cos^2 \omega t = \frac{C V^2}{2}(1 + \cos 2\omega t) \qquad (2.111)$$

and the average value of the energy stored is

$$W = \frac{C V^2}{2} \qquad (2.112)$$

It is apparent that for a capacitor

$$Q = -2\omega W \qquad (2.113)$$

2.6.5 Series and Parallel Connection of Capacitors

The formulas for series and parallel connection of capacitors can be obtained from the general consideration of series and parallel connection of impedances. For series connection

$$\frac{1}{sC} = \frac{1}{sC_1} + \frac{1}{sC_2} + \cdots + \frac{1}{sC_n} \tag{2.114}$$

where $C_1, C_2, \ldots, C_n$ are the capacitances of the capacitors connected in series. Then the equivalent capacitance C can be found as

$$C = \left(\frac{1}{C_1} + \frac{1}{C_2} + \cdots + \frac{1}{C_n} \right)^{-1} \tag{2.115}$$

and is always less than the value of the smallest capacitance. This result can be used when the conditions for using the impedance concept are valid, that is, the capacitors are discharged. The equivalent circuit of the series connection of charged capacitors can be obtained using equivalent circuits shown in Fig. 2.47 and applying Thevenin's theorem (Scott, 1960).

A particular result, namely, the series connection of two capacitors has the equivalent capacitance

$$C = \frac{C_1 C_2}{C_1 + C_2} \tag{2.116}$$

In a similar way one can obtain the equivalent capacitance of parallel connection as

$$C = C_1 + C_2 + \cdots + C_n \tag{2.117}$$

and is always larger than the largest capacitance.

Furthermore, the following two results are useful (for initially discharged capacitors).

- The voltage applied to two capacitors connected in series is divided between these two capacitors inversely proportionally to their capacitance (thus, the larger part of the voltage is on the smaller capacitance).

- The current applied to two capacitors connected in parallel is divided between these two capacitors proportionally to their capacitance (thus, the larger part of the current goes through the larger capacitance).

2.6.6 Construction of Practical Capacitors

The properties of the capacitors that are used as discrete elements of electronic circuits will be described in the following sections (RIFA, 1990/91; Lindquist, 1977). Most of these capacitors are obtained by making contacts with a system of metal plates separated by a dielectric material.

One can divide these capacitors in three broad classes. The first class includes low-loss capacitors with good capacitance stability. These are (in accordance with the dielectric) mica, glass, some ceramic (of low K_d), and low-loss plastic film (polypropylene, polystyrene) capacitors. They are more expensive and should be used for critical precision circuitry (e.g., active and passive telecommunication filters). The second class are capacitors of medium loss and medium stability designed to operate over a wide range of DC and AC voltages. These are paper (oil and wax impregnated), plastic film, and ceramic (medium and high K_d) capacitors. Their DC applications include coupling, decoupling, bypass, smoothing and power separating filters, and energy storage, to mention a few. Their AC applications are no less wide and include motor start, lightning, interference suppression, power line applications (switching and measurement equipment), etc. The third class are aluminum and tantalum electrolytic capacitors providing high capacitance in a tolerable size; however, they do have drawbacks, which are discussed later. For the same rated voltage tantalum capacitors have as much as three times larger capacitance per volume unit as aluminum (for tantalum oxide $K_d = 27.6$, for aluminum oxide $K_d = 8$), they are more reliable and have longer service life.

The electrolytics are designed for DC and polarized voltage applications. Less expensive aluminum capacitors are used in radio and television equipment; more expensive tantalum capacitors are destined for military and harsh environmental applications.

There is a small group of fixed value capacitors (mica, glass, oil, gas and vacuum) designed for high-voltage (up to 35-kV peak), high-current (up to 200 A) transmitter applications. They and variable capacitors are not considered here.

Film and Foil Capacitors

In many cases the capacitor manufacturing process starts by forming one plate using metallization of one side of a flexible dielectric film. A foil (usually aluminum) is used as another plate. The film and foil combination (the nonmetallized film side meets the foil) is rolled on a plastic core with alternate layers slightly offset or extended (Fig. 2.49(a)). The wound roll is then heat treated to harden the film. Then the end disks (tinned copper) are connected. They are soldered to the extended ends of the foil using induction heating. At the metallization foil-ends side they are press fitted using conductive pastes. Subsequently, axial or radial-lead wires (tinned copper or nickel) are welded to the end caps (in some technologies the wires are welded before soldering). Then the capacitor is sealed. It may be dipped in a liquid dielectric, which afterwards dries out (or polymerized) wrapped in a plastic film and end filled, or may be encapsulated in a case or jacket of molded plastic.

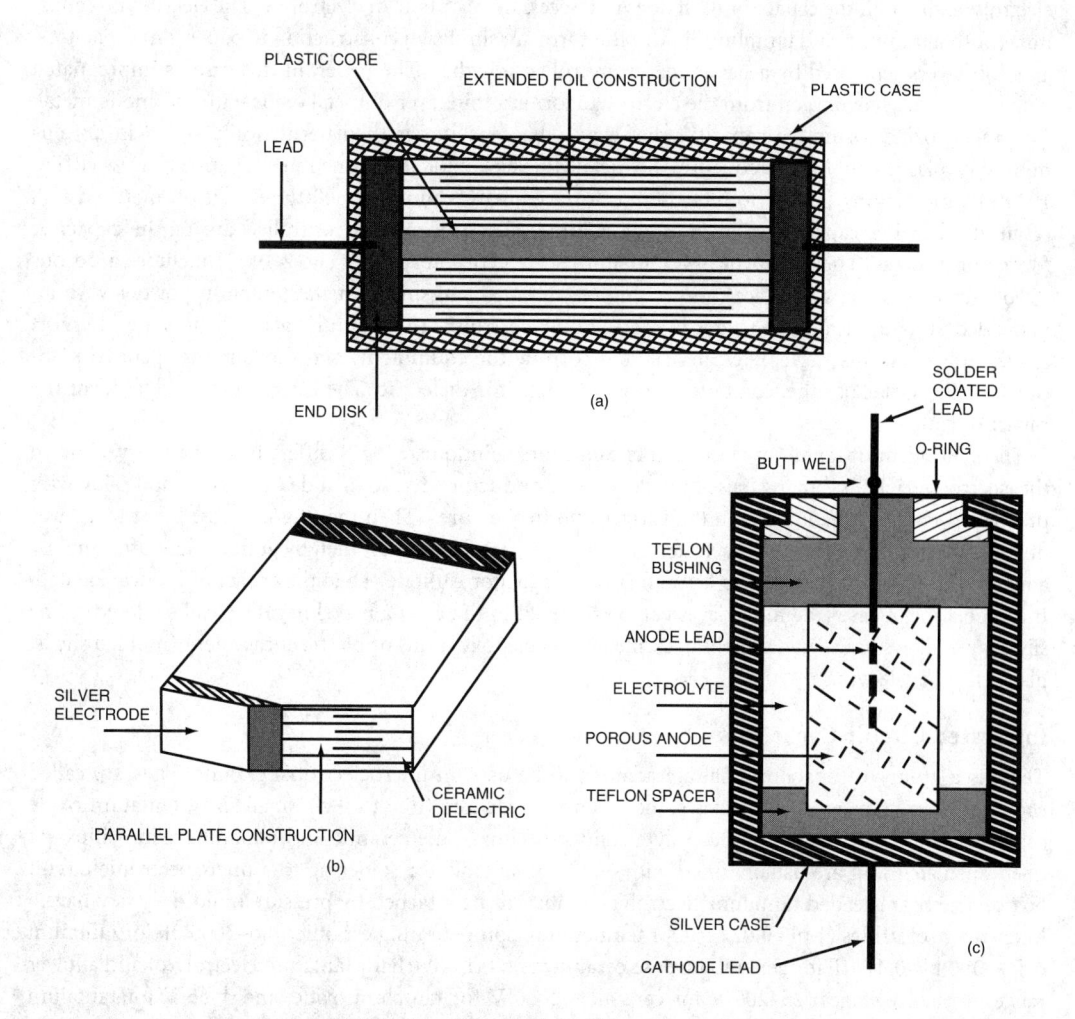

FIGURE 2.49 Construction of discrete capacitors.

The variations of this basic process depend on the dielectric film used. For example, in some capacitors two foil layers can be divided by a plain film (polystyrene capacitors) (RIFA, 1990/91) or paper (impregnated with oil); two metallized offseted films can be used as well (polypropylene capacitors). There are many other sealing procedures as well.

Solid Dielectric Capacitors

The capacitors with brittle dielectric (mica, glass, porcelain) and ceramic capacitors are constructed in stacks (Fig. 2.49(b)).

The dielectric layer is metallized at one side (e.g., a silver electrode is screened on the mica insulators in mica capacitors), then cut into rectangular pieces. The pieces are stacked with an offset of the alternative layers. The end caps are connected using press fitting and conductive pastes. Then the leads are connected and one of the sealing procedures follows.

The dielectric layer for ceramic capacitors is formed by preparing a barium titanate paste that is rolled into a flexible film. The film is covered by conducting ink to form the electrode layer. Then the film is cut, stacked into layers, and the procedure is completed in a manner similar to that already described.

Electrolytic Capacitors

The term *electrolytic capacitor* is applied to any capacitor in which the dielectric layer is formed by an electrolytic method; the capacitor itself does not necessarily contain an electrolyte. The electrolytic capacitors (both aluminum and tantalum) in dry foil form are similar in construction to paper film capacitors. Two foil layers separated by a paper spacer are rolled together. The paper, in this case, is impregnated with electrolyte. The oxygen from the electrolyte forms a thin layer of metal oxide with the anode metal. This metal oxide forms the capacitor dielectric. The capacitor is then hermetically sealed in an aluminum or plastic can. More frequently the metal (thicker) oxide is obtained in a separate process (RIFA, 1990/91) and etching is used to increase the surface area (as much as 300 times that of unetched foil, resulting in higher capacitance in the same volume). The thickness is controlled during the capacitor formation process. The wet form of a tantalum electrolytic capacitor ((Fig. 2.49(c)) includes a porous (to obtain a large active surface) anode slug obtained by pressing of metal tantalum powder with an imbedded tantalum lead. The slug is immersed into electrolyte (e.g., sulfuric acid which is used in most stable capacitors; this electrolyte can leak and corrode the equipment) and the tantalum pentoxide located on the surface of the slug pores forms the capacitor dielectric. The case and electrolyte form the capacitor cathode.

The solid form of tantalum electrolythic capacitors (Lindquist, 1977) differs from the wet version in the electrolyte, which is manganese dioxide (a semiconductor). It also includes a porous anode slug with preformed tantalum pentoxide on the surface and in the pores. Then the slug is covered first by a layer of manganese dioxide (which penetrates into pores in dissolved state), then by graphite and after that by a metal (silver) layer. The last two layers form the capacitor cathode. Then the capacitor is encapsulated. It can be a metal case (the metal layer can be soldered to the case; a nickel negative lead soldered to the silver layer allows to use metal shells backfilled with an epoxy resin) or plastic encasement (molded shells, plastic film, or sleeving).

Integrated Circuit Capacitors

There is a group of capacitors that are adapted to be used in microelectronic circuits. They are called *chip capacitors*. This group includes some ceramic capacitors (due to their small size), tantalum oxide solid capacitors, and tantalum electrolyte solid capacitors. The ceramic and tantalum oxide chips are unencapsulated and are usually fitted with end caps suitable for soldering to a microelectronic circuit board. The beam-leaded tantalum electrolytic chips can be attached by pressure bonding. They have a lower range of values (1 pF–0.027 μF for temperature compensating ceramic, 100–3000 pF for tantalum oxide, 390 pF–0.47 μF for general purpose ceramic, and 0.1–10 μF for tantalum electrolyte) and a lower range of rated voltages (25–200 V for ceramic, 12–50 V for tantalum oxide, and 3–35 V for tantalum electrolyte). Other parameters are the same as for ordinary discrete capacitors.

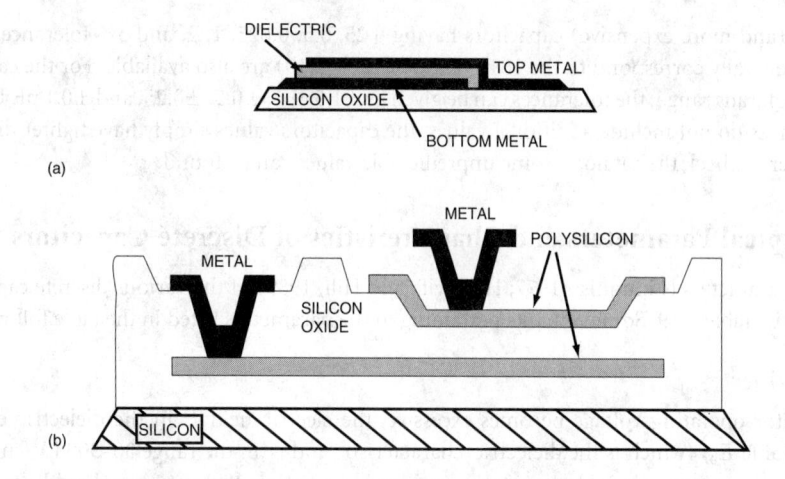

FIGURE 2.50 Integrated circuit capacitors.

Another group can be found in the microelectronic circuits themselves. These monolayer capacitors appear, first of all, as thin-film tantalum oxide capacitors in some postprocessing technologies of integrated circuits. In this case a tantalum conducting layer is deposited on the top insulating layer of an integrated circuit. Then the bottom capacitor plate is formed using a photo resist-mask-stripping procedure. A dielectric layer covering the bottom plate is formed. In the case of the tantalum bottom plate, this is usually tantalum oxide, which is well controlled by oxidization. Then the top plate (usually aluminum) is deposited.

The plates of the capacitors of integrated circuits (Fig. 2.50(b)) in most modern technologies (Allstot and Black, 1983) are formed by two heavily doped polysilicon layers. The dielectric is usually a thin layer of silicon oxide. The whole capacitor is formed on a layer of thick (field) oxide. The capacitors are temperature stable, their temperature coefficient is about 20 ppm/°C. A detailed comparative evaluation of the different metal oxide semiconductor (MOS) capacitor structures can be found in Allstot and Black (1983).

Other dielectrics, in sensor applications, can be used as well (Boltshauser and Baltes, 1991). Such devices usually incorporate a dielectric with given sensitivity (to humidity, force, etc.) properties. The metallization layer formed on the top of dielectric frequently has a shape providing access of the measured physical variable to the dielectric.

2.6.7 Standard Values and Tolerances of Discrete Capacitors

The first two significant figures (for each of the standard tolerances 5, 10, and 20%) of the standard capacitor values are given in Table 2.18. Capacitors values have bimodal distribution, that is, the values tend to be grouped close to, but within, their tolerance values.

TABLE 2.18 Standard 5, 10, and 20% Capacitor Values

5%	10%	20%	5%	10%	20%
10	10	10	33	33	33
11			36		
12	12		39	39	
13			43		
15	15	15	47	47	47
16			51		
18	18		56	56	
20			62		
22	22	22	68	68	68
24			75		
27	27		82	82	
30			91		

Precision (and more expensive) capacitors having 0.25, 0.5, 0.625, 1, 2, and 3% tolerances (with the values that generally correspond to the 10% values of Table 2.19) are also available. For the capacitors in the small picofarads range, the tolerances can be given as ± 1.5, ± 1, ± 0.5, ± 0.25, and ± 0.1 pF. Usually low tolerance ranges do not include additional values, the capacitors values simply have tighter distribution. Yet, it is better to check the catalogs; some unpredictable values can be found.

2.6.8   Typical Parameters and Characteristics of Discrete Capacitors

The typical parameters (Lindquist, 1977; Horowitz and Hill, 1989) of the various discrete capacitors are summarized in Table 2.19. Some remarks pertaining to the parameters listed in the table follow.

Voltage Ratings

If the capacitor operating voltage becomes excessive, the electric field E in the dielectric exceeds the breakdown value E_{br} (which is the dielectric characteristic and is in the range 30–300 kV/cm), and the dielectric becomes permanently damaged. In continuous operation, the capacitor should always operate at a voltage that is less than its rated voltage V_R (which, of course, is less than the breakdown voltage V_{br}). Furthermore, this increases the capacitor's reliability and expected lifetime. Most capacitors have standard voltage ratings of 50, 100, 200, 400, and 600 V DC. The tantalum and electrolytic capacitors have ratings of 6, 10, 12, 15, 20, 25, 35, 50, 75, 100 V and higher.

A breakdown usually starts at the plate imperfections, in the points of more dense electric field. Yet, for a chosen technology it is possible to assume that $V_{br} = E_{br}d$ and, hence, is proportional to the dielectric thickness d. To increase the voltage rating of a capacitor, d should be increased. To maintain the same capacitance value $C = \varepsilon_0 K_d S/d$, the area must also be increased. One can consider that the capacitor volume increases proportionally to the square of the voltage rating.

The *surge voltage* (or rated peak nonrepetitive voltage) specification of a capacitor determines its ability to withstand high transient voltages induced by a switching or any other disturbances of equipment operation. The allowable (for a limited number of times) surge voltage levels are generally 10% above the rated voltage for electrolytic capacitors (for aluminum capacitors with rated voltage lower than 300 V this figure can be as high as 50% (RIFA, 1990/91)). For ceramic and mica capacitors this voltage is called the *dielectric withstanding voltage* and can be as high as 250% of rated DC voltage.

For electrolytic capacitors the rated *reverse voltage* is specified as well. For aluminum capacitors (undefinite duration) this figure is 1.5 V (usually in the full temperature range). Higher reverse voltages can be applied for very short periods only; this causes some change in capacitance but not capacitor failure. The foil and wet tantalum capacitors have a rated reverse voltage of 3 V at $+85°C$ and 2 V at $+125°C$. Solid-electrolyte tantalum electrolytic capacitors can typically withstand a reverse voltage up to 15% of their rated voltage.

Temperature Coefficient

The temperature characteristic of capacitors depends primarily on the temperature dependence of the dielectric constant K_d. A low value of temperature coefficient is obtained for capacitors with dielectrics of low K_d.

Film, mica, glass, porcelain, and paper capacitors have K_d in the range 2–7, aluminum and tantalum electrolytic capacitors have K_d in the range 7–25. The ceramic capacitors are divided in three classes: capacitors with low K_d in the range 25–100, with medium K_d in the range 300–1800, and with high K_d in the range 2500–15,000.

Polycarbonate, polystyrene, polysulfone, teflon, mica, and glass capacitors have the smallest variation with temperature; for the last three types the temperature coefficient (TC) is practically constant in the whole temperature range of their operation.

The temperature behavior of ceramic capacitors depends on the K_d range, and the variation of K_d increases with its value. Yet, the temperature characteristics of ceramic capacitors can be tailored using special additions to the dielectric (barium titanate) and carefully controlled technology. A wide variety

TABLE 2.19 Parameters and Characteristics of Discrete Capacitors

Capacitor Type	Range	Rated Voltage, V_R	TC ppm/°C	Tolerance, ±%	Insulation Resistance, MΩ/μF	Dissipation Factor, %	Dielectric Absorption, %	Temperature Range, °C	Comments, Applications	Cost
Polycarbonate	100 pF–30 μF	50–800	±50	10	5×10^5	0.2	0.1	−55/+125	High quality, small, low TC	High
Polyester/Mylar	1000 pF–50 μF	50–600	+400	10	10^5	0.75	0.3	−55/+125	Good, popular	Medium
Polypropylene	100 pF–50 μF	100–800	−200	10	10^5	0.2	0.1	−55/+105	High quality low absorption	High
Polystyrene	10 pF–2.7 μF	100–600	−100	10	10^6	0.05	0.04	−55/+85	High quality, large, low TC, signal filters	Medium
Polysulfone	1000 pF–1 μF	100–600	+80	5	10^5	0.3	0.2	−55/+150	High temperature	High
Parylene	5000 pF–1 μF		±100	10	10^5	0.1	0.1	−55/+125	High temperature	High
Kapton	1000 pF–1 μF		+100	10	10^5	0.3	0.3	−55/+220	High temperature	High
Teflon	1000 pF–2 μF	50–200	−200	10	5×10^6	0.04	0.04	−70/+250	High temperature lowest absorption	High
Mica	5 pF–0.01 μF	100–600	−50	5	2.5×10^4	0.001	0.75	−55/+125	Good at RF, low TC	High
Glass	5 pF–1000 pF	100–600	+140	5	10^6	0.001		−55/+125	Excellent long-term stability	High
Porcelain	100 pF–0.1 μF	50–400	+120	5	5×10^5	0.10	4.2	−55/+125	Good long-term stability	High
Ceramic (NPO)	100 pF–1 μF	50–400	±30	10	5×10^3	0.02	0.75	−55/+125	Active filters, low TC	Medium
Ceramic	10 pF–1 μF	50–30,000	±800	10	5×10^3		2.5	−55/+125	Small, very popular selectable TC	Low
Paper	0.01 μF–10 μF	200–1600		10		1.0	2.5	−55/+125	Motor capacitors	Low
Aluminum	0.1 μF–1.6 F	3–600	+2500	−10/+100	100	10	8.0	−40/+85	Power supply filters short life	High
Tantalum (Foil)	0.1 μF–1000 μF	6–100	+800	−10/+100	20	4.0	8.5	−55/+85	High capacitance small size, low inductance	High
Thin-film	10 pF–200 pF	6–30	+100	10	10^6	0.01		−55/+125		High
Oil	0.1 μF–20 μF	200–10,000				0.5			High voltage filters, large, long life	
Vacuum	1 pF–1000 pF	2000–3600							Transmitters	

of negative TCs and a few positive TCs are available. The capacitors are marked in accordance with their temperature coefficient; the high-stability types have values ranging from P100 (positive, 100 ppm/°C) to N080 (negative, 80 ppm/°C). The most stable capacitors in this group are denoted as NPO type; they have temperature coefficient (nonspecified) ±30 ppm/°C. The so-called temperature compensating types (they are frequently used in timing circuits, to compensate the positive temperature coefficient of a resistor) have values in the range from N150 to N750. The TC of high-stability and compensating capacitors is practically constant in the temperature range from −55 to +125°C. Monolithic ceramic capacitors are capable of withstanding and operating at temperatures up to +260°C when properly designed and manufactured for these applications (ceramic cased capacitors); of course, in this case the temperature coefficient is not constant and the capacitor value itself can drop by as much as 60% (or even more when an additional DC voltage is superimposed).

Tantalum and aluminum capacitors have extremely large variations with temperature. At low temperatures the electrolyte can freeze; at high temperatures it can dry out.

Aluminum capacitors are usually designed for the range from −55 to +125°C (with the characteristics guaranteed for the range from −40 to +105°C (RIFA, 1990/91)); at −55°C the capacitance can drop down to 30% of its nominal value; the increase at +125°C is about 20%.

Tantalum capacitors can be usually designed for the range from −55 to +200°C (with the characteristics guaranteed for the range from −55 to +125°C); at −55°C the capacitance can drop down to 60% of its nominal value (to 20–30% for large values of capacitance); the increase at +125°C is also about 20%. The capacitor change is less for devices with higher rated voltage (for example, the change for a capacitor of 150 V is approximately three times less than that of the capacitor of 6-V rated voltage).

Insulation Resistance

Capacitors are not ideal circuit elements. The equivalent circuit of a capacitor (valid from DC up to the low radio frequency range) is shown in Fig. 2.51(a). The shunt resistance R_p (which, in general, is the function of frequency) at DC and low frequencies is referred as the **insulation resistance** of the capacitor. It accounts for dielectric leakage current and dielectric power losses. When the capacitor is charged to $V(0)$ volts and allowed to discharge through itself, the capacitor voltage is

$$v(t) = V(0)[1 - \exp(-t/R_p C)] \qquad (2.118)$$

If the time constant $R_p C$ is constant, it can be calculated measuring the voltage $V(0)$ and the voltage $V(T_e)$ that resides on the capacitor when the time T_e has elapsed. Then

$$R_p C = -T_e \ln[1 - V(T_e)/V(0)] \qquad (2.119)$$

The time T_e is referred to as the *time of electrification* and is usually chosen to be one (RIFA, 1990/91) or two minutes (Lindquist, 1977). Since for a flat capacitor $C = \varepsilon S/d$ and leakage resistance $R_p = \rho d/S$, where ρ is the resistivity of the dielectric, then $R_p C = \rho \varepsilon$. Hence, the time constant $R_p C$ is a function of the dielectric properties and depends on the capacitor type only. This allows the use of $R_p C$ (measured in megaohm microfarad or in ohm farad) as a figure of merit (rather than R_p). This figure is referred to as *isolation resistance* in Table 2.19.

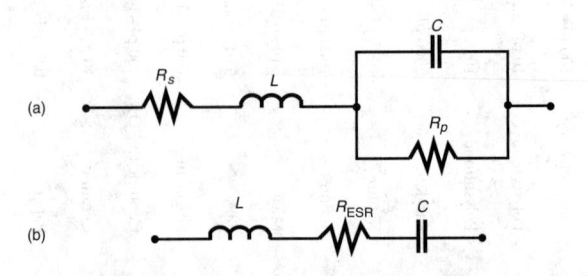

FIGURE 2.51   Capacitor equivalent circuits.

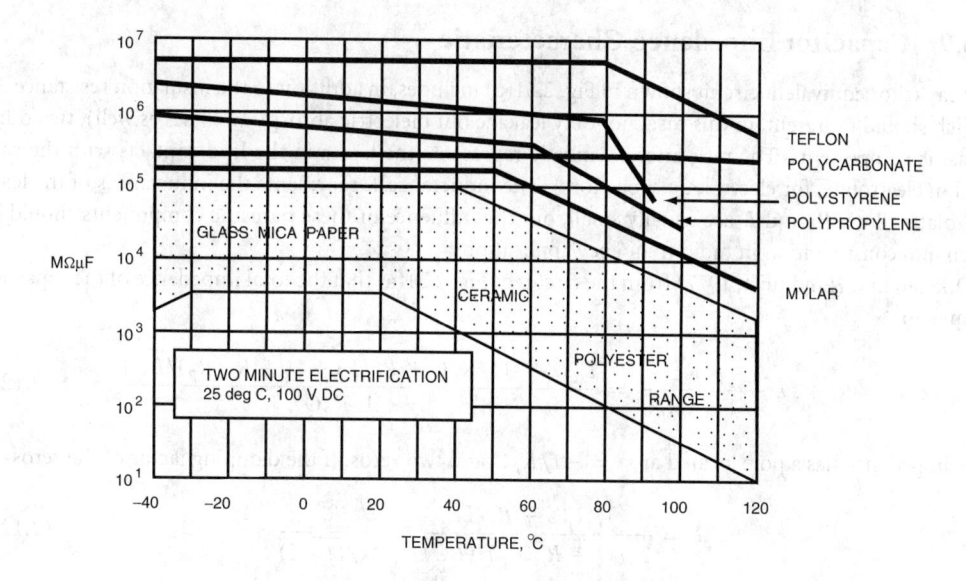

FIGURE 2.52 Temperature dependence of the insulation resistance.

The R_pC product decreases (nearly exponentially for many dielectrics) when the temperature increases (Fig. 2.52). It is approximately halved (RIFA, 1990/91) for each 7°C of temperature rise; the temperature where this decrease starts is different for different dielectrics. The R_pC product also can vary with electrification time (due to dielectric absorption effects) and can increase (up to two orders of magnitude) in subsequent measurements.

In reality the R_pC product depends on the capacitor package as well. For larger capacitor values the decrease with temperature starts earlier. Many manufacturers divide capacitors in groups and also give the R_p value for each group. This is especially valid for electrolytic capacitors; for these capacitors very frequently the rated leakage current I_{RL} is given. This current can be described by an empiric dependence

$$I_{RL} = kCV_R + 4 \tag{2.120}$$

where I_{RL} is in microamperes, V_R is the capacitor rated voltage in volts, C is the capacitance in microfarad, and $0.003 < k < 0.01$. The leakage current is also approximately proportional to temperature.

Dielectric Absorption

Dielectric absorption occurs in all solid dielectric capacitors. If a capacitor is charged for some prolonged period of time, then quickly discharged and allowed to stay open circuited for a certain period of time, a *recovery voltage* can be measured across the capacitor. This recovery voltage is due to residual energy stored in the dielectric that is not released during the original discharge period (the only perfect dielectric from that the whole of the stored energy may be recovered is a perfect vacuum). The dielectric absorption coefficient is determined as

$$DA = \frac{\text{recovery voltage}}{\text{charging voltage}} \tag{2.121}$$

The values listed in Table 2.19 are determined by charging the capacitor for 1 h at the rated voltage, discharging it for 10 s through 5 Ω, and measuring the voltage 15 min after the resistor is disconnected. The dielectric absorption coefficient represent the losses in the capacitor that occur at pulse and high-frequency operation.

2.6.9 Capacitor Impedance Characteristic

The capacitor equivalent circuit shown in Fig. 2.51(a) includes, in addition to the insulation resistance R_p (which should represent, in this case, not only leakage but dielectric absorption losses as well), two other parasitic components. The resistance R_s mostly represents the losses in the lead contacts with the caps (and in electrolyte, for electrolytic capacitors). The inductance L represents the inductances of the leads and plates. Both R_s and L are usually small, but the influence of these parasitic components should be taken into consideration in radio frequency applications.

One can find (Lindquist, 1977) from the circuit of Fig. 2.51(a) that the total impedance of the capacitor is equal to

$$Z(s) = R_s + sL + \frac{R_p}{1 + s R_p C} = \frac{L[s^2 + s(1/R_pC + R_s/L) + (1 + R_s/R_p)/LC]}{s + 1/R_pC} \tag{2.122}$$

This impedance has a pole located at $s_p = -1/R_pC$ and two zeros. If the damping factor of the zeros

$$\zeta = \frac{1/R_pC + R_s/L}{2\sqrt{(1 + R_s/R_p)/(LC)}} \approx \frac{R_s}{2\sqrt{(L/C)}} \tag{2.123}$$

is small (low loss, $\zeta < 1$), the zeros are complex conjugate, having a resonant frequency $\omega_z = \sqrt{[(1 + R_s/R_p)/LC]} \approx 1/\sqrt{(LC)}$ since $R_s \ll R_p$ in all capacitors. A noticeable resonance effect occurs. The magnitude frequency characteristic for this case is shown in Fig. 2.53(a). If the damping factor is large (high loss, $\zeta > 1$) the zeros are real with values $s_{z1} = -1/R_sC$ and $s_{z2} = -R_s/L$. The magnitude frequency characteristic does not show any resonance (Fig. 2.53(b)). It is clear that the capacitor has a maximum usable frequency range between $1/R_pC$ and $1/\sqrt{(LC)}$ (low loss) or between $1/R_pC$ and $1/R_sC$ (high loss). The input resistance of the capacitor in the usable range is between R_p and R_s.

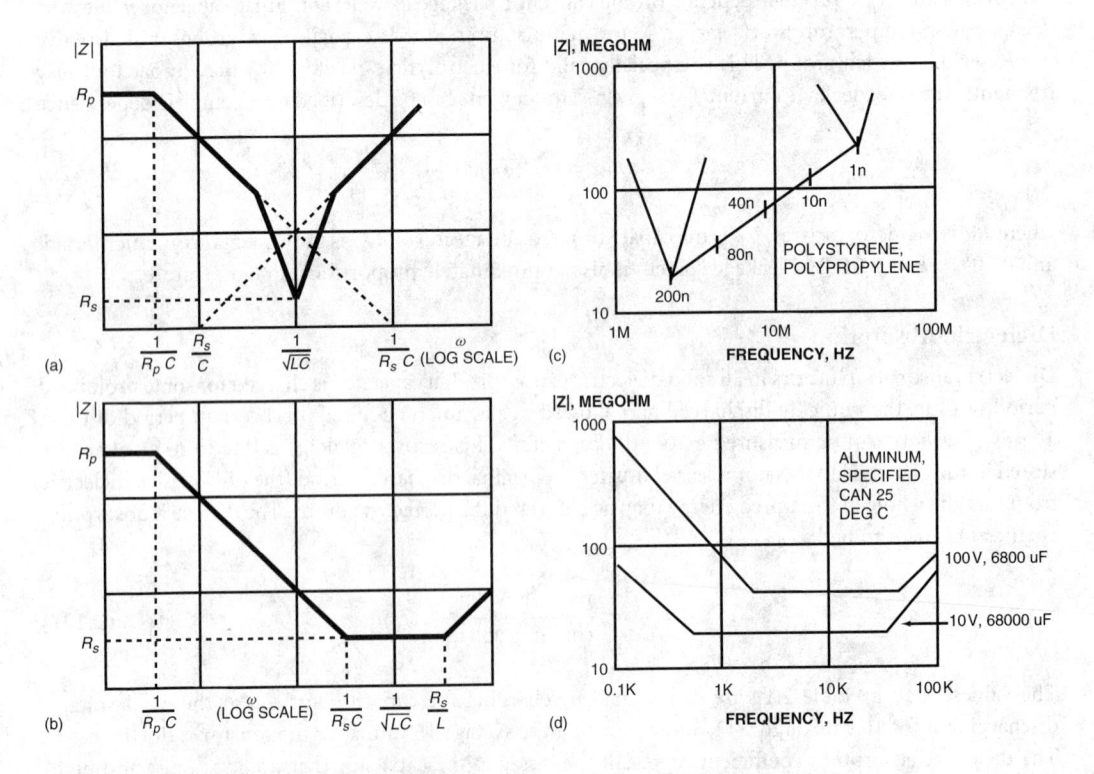

FIGURE 2.53 Capacitor impedance characteristics.

As was mentioned before, both R_p and R_s are functions of frequency, and it would be difficult to describe the capacitor impedance in a wide range of frequencies. Usually, a small part of the magnitude characteristic is given. For precision capacitors one can usually find the characteristics as shown in Fig. 2.53(c) and for the electrolytic capacitances as shown in Fig. 2.53(d). In both cases the characteristics are around the capacitor resonant frequency $\omega_z \approx 1/\sqrt{(LC)}$.

In this vicinity the equivalent circuit can be simplified. If the condition $\omega^2 R_p C \gg 1$ is satisfied, then

$$Z(j\omega) = R(\omega) + jX(\omega) = R_s + j\omega L + \frac{R_p}{1 + j\omega R_p C}$$

$$\approx R_s + \frac{1}{R_p \omega^2 C^2} + j\left(\omega L - \frac{1}{\omega C}\right) = R_{\mathrm{ESR}} + j(X_L - X_C) \quad (2.124)$$

as it follows from Eq. (2.122). This equation allows us to obtain the parameters of the simplified equivalent circuit of the capacitor (Fig. 2.51(b)), where the effective series resistor $R_{\mathrm{ESR}} = R_s + 1/(R_p\omega^2 C^2)$ represents all of the losses in R_p and R_s. For electrolytic capacitances the inductance L is usually 3–4 nH. For these capacitors R_{ESR} sharply increases at freezing temperatures (RIFA, 1990/91).

2.6.10 Power, Dissipation, and Quality Factors

If the capacitor impedance is represented at AC steady-state operation as

$$Z(j\omega) = R(\omega) + jX(\omega) \quad (2.125)$$

then several figures of merit can be used to describe the losses in the capacitors. The first one is the *power factor* (PF), which is defined as

$$\mathrm{PF} = \frac{\text{power loss}}{\text{apparent power in}} = \frac{I^2 R}{I^2 |Z|} = \frac{R}{|Z|} = \cos\theta \quad (2.126)$$

Two other factors are the **dissipation factor** (DF) defined as

$$\mathrm{DF} = \frac{\text{power loss}}{\text{reactive power stored}} = \frac{I^2 R}{I^2 X} = \frac{R}{X} = \cot\theta = \tan\delta \quad (2.127)$$

and its inverse, the capacitor *quality factor* $Q_C = 1/\mathrm{DF} = X/R = \tan\theta$ (sometimes the DF is determined as the ratio of R_{ESR}/X_C, this approximation is satisfactory in the vicinity of the minimum value of DF magnitude). The DF and PF are essentially equal when the PF is 10% or less. The value δ is called the *angle of losses.*

Substituting $s = j\omega$ in Eq. (2.120) and finding the real and imaginary parts for the impedance $Z(j\omega)$ one can find that

$$\mathrm{DF}(\omega) = \frac{\omega^2 R_s/L + (1 + R_s/R_p)/R_p L C^2}{\omega[\omega^2 - (1/LC - 1/(R_p C)^2]} \approx \frac{R_s(\omega^2 - 1/R_p R_s C^2)}{\omega(\omega^2 - 1/LC)} \quad (2.128)$$

The magnitude of this frequency-dependent value is shown in Fig. 2.54(a). One can see that, over the maximum useful frequency range of the capacitor, the dissipation factor has a minimum. This minimum occurs in the vicinity of $\omega_{\min} = 1/C\sqrt{(R_p R_s)}$ and it has the value of $\mathrm{DF}_{\min} \approx 2\sqrt{(R_s R_p)}$.

The standard measurements of DF are made at the frequency of 1 kHz with a 1V rms voltage applied to the capacitor. This frequency is usually close to $f_{\min} = \omega_{\min}/2\pi$ and the manufacturer frequently provides the plot of dissipation factor in the region of $f_{\min}$ (Fig. 2.54(b)) for different groups of capacitors. The dissipation factor at any given frequency is a function of the temperature. Usually, a capacitor is designed in such a way that this function is nearly flat in the industrial range of temperatures with an increase toward the ends of the full temperature range.

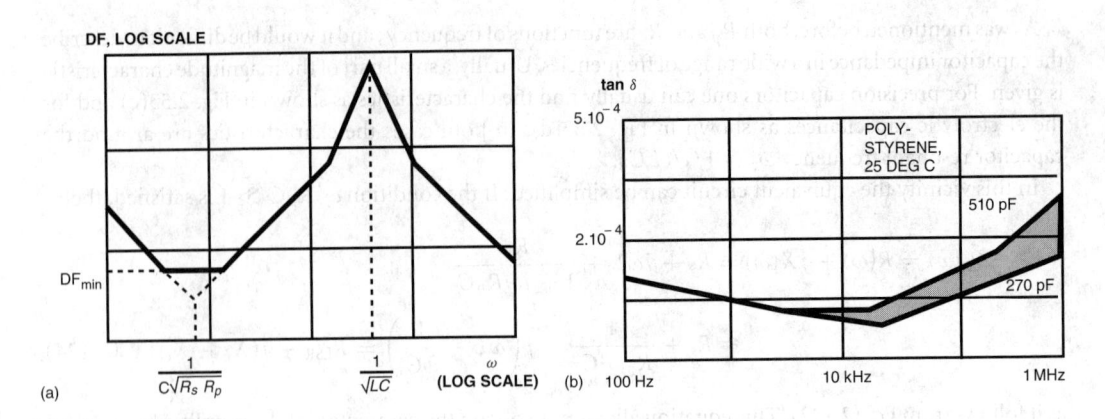

FIGURE 2.54 Capacitor dissipation factor.

The frequency dependence of the quality factor Q_C (this figure of merit usually applies to capacitors used in tuned circuits) and its maximal value can be found by inverting the DF(ω) characteristic. The standard measurements of Q_C for the tuned circuit capacitors are made at the frequency of 1 MHz.

2.6.11 Capacitor Reliability

About one-quarter of the part failures in electronic equipment are capacitor failures. One-half of these failures are caused by improper selection and application. Some typical causes of capacitor failure are: (1) voltage overload, (2) current overload, (3) high temperature, (4) frequency effects, (5) shock and vibration, (6) humidity, and (7) pressure.

Voltage overload produces an excessive electric field in the dielectric that, as was mentioned before, results in the breakdown and permanent damage of the dielectric.

Rapid voltage variations result in current transients. If these currents are of sufficient amplitude and duration, they can permanently damage the dielectric, deform the capacitor, and permanently change the capacitor value. Pulse current limits are commonly expressed in the form of maximal dV/dt in volts per microsecond. The figure is usually given when pulse operation is implied. The values (RIFA, 1990/91) are strongly dependent on the design and dielectric used; thus, for the metallized propylene pulse capacitors the maximal dV/dt can be from 20 to 120 V/μs; for the same metallized propylene power capacitors it will be from 100 to 500 V/μs; and for metallized paper interference supressor capacitors this figure can be from 150 to 2000 V/μs. For electrolytic capacitors frequently used in power supply rectifiers the figure of rated ripple current (in milliamperes) at operating frequency (usually 100 or 120 Hz) and operating temperature (usually 100°C) is provided. Current overload can also occur when the capacitor operates at high frequencies.

High temperatures and overheating are also common causes of capacitor failure. Excessive temperature accelerates the dielectric aging, makes the plastic film brittle, and can cause cracks in hermetic seals.

Moisture and humidity cause corrosion, reduce the dielectric strength, and lower insulation resistance. Pressure variation, shock, and vibration can cause mechanical damages (seal cracks, etc.) that result in electrical failures. It is interesting to note that for electrolytic capacitors three parameters, namely, capacitance, leakage current, and the equivalent series resistor R_{ESR} are considered as the basic ones, and the failure of the capacitor after a test is established comparing the values of these parameters with the specified ones.

In calculation of capacitor reliability (Lindquist, 1977) it is assumed that capacitors have lifetimes that are exponentially distributed. In this case the distribution is characterized by one parameter only, namely, the mean time to failure (MTTF) τ_0. The inverse of the MTTF is the failure rate of the component,

TABLE 2.20 Typical Failure Rates for Capacitors

Capacitor Type	Failure Rate (failures/10^6 h)			T_{max}, °C
	Low	Average	High	
Paper/Film	0.0006	0.002	0.66	125
Ceramic (gen. purpose)	0.017	0.11	0.92	150
Ceramic (temp. compensating)	0.0006	0.008	0.41	100
Mica	0.004	0.013	1.0	125
Glass	0.0001	0.0042	0.49	125
Tantalum (gen. purpose)	0.042	0.15	1.7	85
Tantalum (established reliability)	0.0033	0.011	0.35	85
Aluminum (gen. purpose)	0.0096	0.047	1.8	85
Aluminum (established reliability)	0.0072	0.03	0.59	85

$\lambda = 1/\tau_0$. Some typical failure rates are listed in Table 2.20. In capacitors the thermal stress (temperature) and electrical stress (voltage) essentially determine the failure rate. The results in this table are based on voltage stress-temperature combinations: $(0.1 V_{max}, 0°C)$ for low rate, $(0.5 V_{max}, T_{max}/2°C)$ for average rate, and $(V_{max}, T_{max}°C)$ for high rates. The contribution of overvoltage or overheating in each particular case varies from one dielectric to another, the combined stress is an acceptable compromise.

Table 2.20 also provides the so-called basic failure rate λ_0. To obtain the failure rate that considers the capacitor environmental conditions, the basic rate λ_0 is multiplied by a factor π_E that increases from 1 for the case of nearly zero environmental stress with optimum engineering operation and maintenance to 15 for the case of severe conditions of noise, vibration, and other environmental stresses related to missile launch and space vehicle operation.

The failure rate depends on quality control as well. The failure rate of lower quality capacitors is obtained by multiplying the basic rate λ_0 by a multiplier π_Q. The value of this multiplier increases from 1 for high-quality capacitors to 4 for low-grade ones.

2.6.12 Aging and Radiation Effects

Capacitors exhibit aging effects (Lindquist, 1977) of varying degrees depending on their type. The data must be obtained from the manufacturer.

For frequently used ceramic capacitors the aging consists of a constant decrease of capacitance with time. The change is exponential and is expressed in percentage change per decade of time. The rate of change increases with increase of the dielectric constant K_d (Fig. 2.55). Little aging takes place after 1000 h (about 1.4 month). Usually preaging allowances are made by a manufacturer for the initial aging to maintain shelflife tolerances. Aging also reduces the capacitor insulation resistance.

Radiation gradually deteriorates certain plastic film and ceramic dielectrics (Lindquist, 1977). Organic dielectrics are more susceptible than inorganic dielectrics. Radiation reduces the insulation resistance and decomposes impregnates in paper

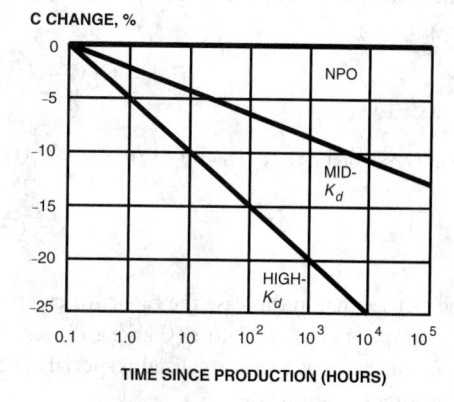

FIGURE 2.55 Aging of ceramic capacitors.

capacitors. The radiation susceptibility for some capacitors is shown in Fig. 2.56. Glass, mica, and ceramic capacitors are the least susceptible. Film and electrolytic capacitors are average (tantalums are better than aluminums), and paper capacitors are the most susceptible.

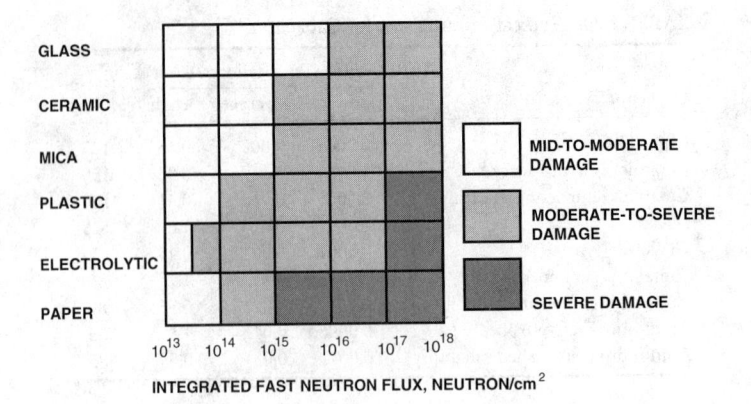

GLASS
CERAMIC
MICA
PLASTIC
ELECTROLYTIC
PAPER

MID-TO-MODERATE DAMAGE

MODERATE-TO-SEVERE DAMAGE

SEVERE DAMAGE

10^{13} 10^{14} 10^{15} 10^{16} 10^{17} 10^{18}

INTEGRATED FAST NEUTRON FLUX, NEUTRON/cm^2

FIGURE 2.56   Radiation damage of capacitors.

2.6.13   Nonlinear Dielectrics and Nonlinear Capacitors

The dielectric constant K_d of linear dielectrics does not depend on the electric field E. The electric flux density $D = \varepsilon E = \varepsilon_0 K_d E$ in such dielectrics is proportional to the field and

$$\varepsilon = \frac{D}{E} = \text{const.} \qquad (2.129)$$

Most of the dielectrics that have small K_d (1–7) are practically linear, and the dielectrics with high K_d are usually linearized (as in ceramic capacitors). As a result one has linear capacitances. Yet, nonlinear capacitors exist. The charge $Q = g(V)$ residing on this capacitor is a nonlinear function of the voltage V. The current in this case can be described by the equation

$$i(t) = \frac{dQ}{dt} = \frac{dg(v)}{dv}\frac{dv}{dt} = C(v)\frac{dv}{dt} \qquad (2.130)$$

If the voltage v includes a bias component V_0 and a slowly changeable control component $V_C(t)$, then

$$i(t) = C[V_0 + V_C(t) + v]\left[\frac{dv}{dt} + \frac{dV_C(t)}{dt}\right] \qquad (2.131)$$

If $V_C(t) \ll V_0, v \ll V_C(t)$, and $dv/dt \gg dV_C(t)/dt$, then

$$i(t) = C[V_0 + V_C(t)]\frac{dv}{dt} \qquad (2.132)$$

One can see that in this case the capacitor can be controlled. The circuit with this capacitor can be used, for example, in FM oscillators (Clarke and Hess, 1971).

Nonlinear capacitors are realized as special hyper-abrupt p–n junctions. The devices are generally known as *varicaps* or *varactors*.

Defining Terms

Capacitance: Equal to the ratio $C = Q/V$ of the charge Q to the voltage V between two aforementioned conducting bodies.

Capacitor: A system of two conducting bodies located in the electromagnetic field and having equal charges $+Q$ and $-Q$ of opposite signs.

Dielectric absorption: Equal to the ratio of recovery voltage (the voltage which appears at the quickly discharged capacitor if it is allowed to stay open-circuited for one hour) to the initial charging voltage.

Dissipation factor: Equal to the ratio of power loss to the reactive power stored when a sinusoidal voltage is applied to the capacitor.

Farad: The measurement unit of capacitance; this is a very big unit, practical capacitors have the capacitances that are measured in nano-, micro-, and picofarads.

Insulation resistance: Equal to the time constant of the discharge of the initially charged capacitor when the capacitor is discharged through itself.

Mutual capacitance: Equal to the ratio of the charge on the plates of ith capacitor to the voltage at the plates of another jth capacitor, when this charge is induced by this voltage and when all capacitors except jth are short circuited.

References

Allstot, D.J. and Black, W.C., Jr. 1983. Technological design considerations for monolitic MOS switched-capacitor filtering systems. *Proc. IEEE* 71(8):967–986.

Biorci, G. 1975. *Fondamenti di elettrotecnica: circuiti*. UTET, Torino, Italy.

Boltshauser, T. and Baltes, H. 1991. Capacitive humidity sensors in SACMOS technology with moisture absorbing photosensitive polyimide. *Sensors and Actuators*, Ser. A, 25-27:509–512.

Clarke, K.K. and Hess, D.T. 1971. *Communication Circuits*. Addison-Wesley, Reading, MA.

Horowitz, P. and Hill, W. 1989. *The Art of Electronics*, 2nd ed. Cambridge University Press, Cambridge, England.

Kapliansky, A.E., Lisenko, A.P., and Polotovsky, L.S. 1972. *Fundamentals of Electrical Engineering* (in Russian). Visshaya Shkola Pub. House, Moscow.

Lindquist, C.S. 1977. *Active Network Design*. Steward & Sons, Long Beach, CA.

RIFA Capacitors. Catalog 1990/91. Kalmar, Sweden.

Scott, R.E. 1960. *Linear Circuits*. Addison-Wesley, Reading, MA.

Stuart, R.D. 1965. *Electromagnetic Field Theory*. Addison-Wesley, Reading, MA.

Further Information

IEEE Transactions on Microwave Theory and Techniques (capacitance calculation for different geometric configurations).

IEEE Transactions on Circuits and Systems (capacitor as a component of electric and electronic circuits).

IEEE Transactions on Components, Hybrids, and Manufacturing Technology (capacitors characteristics and parameters, measurements of parameters, technologies).

IEEE Journal of Solid-State Circuits (capacitor as a component of solid-state electronic circuits).

Sensors and Actuators (application of capacitors in sensors and sensor systems).

The Electrical Engineering Handbook, R.C. Dorf, editor-in-chief, CRC Press, Boca Raton, FL, 1993 (good diagram reviewing variations in characteristics of aluminum electrolytic capacitors, references on electrolytic capacitors).

Reference Data for Engineers: Radio, Electronics, Computer and Communication, E.C. Jordan, editor-in-chief, 7th ed. H.W. Sams & Co, Inc., 1985, Indianapolis, IN (data on high-quality film capacitors, deciphering of color code for ceramic capacitors, the type designation for mica capacitors, EIA and MIL standards for mica capacitors).

Electronics Engineers Handbook, D.G. Fink, D. Christiansen, eds., 3rd ed. McGraw-Hill, New York, 1989 (data on microelectronic chip capacitors, fixed transmitter capacitors, and variable capacitors).

Handbook of Modern Electronics and Electrical Engineering, C. Belove, editor-in-chief. Wiley, New York, 1986 (useful diagrams for selection of capacitors, tabulated stability data, more detailed tabulated temperature coefficient).

2.7 Properties of Materials

James F. Shackelford

2.7.1 Introduction

The term *materials science and engineering* refers to that branch of engineering dealing with the processing, selection, and evaluation of solid-state materials (Shackelford, 1996). As such, this is a highly interdisciplinary field. This chapter reflects the fact that engineers outside of the specialized field of materials science and engineering largely need guidance in the selection of materials for their specific applications. A comprehensive source of property data for engineering materials is available in *The CRC Materials Science and Engineering Handbook* (Shackelford, Alexander, and Park, 1994). This brief chapter will be devoted to defining key terms associated with the properties of engineering materials and providing representative tables of such properties. Because the underlying principle of the fundamental understanding of solid-state materials is the fact that atomic- or microscopic-scale structure is responsible for the nature of materials properties, we shall begin with a discussion of structure. This will be followed by a discussion of the importance of specifying the chemical composition of commercial materials. These discussions will be followed by the definition of the main categories of materials properties.

2.7.2 Structure

A central tenet of materials science is that the behavior of materials (represented by their **properties**) is determined by their structure on the atomic and microscopic scales (Shackelford, 1996). Perhaps the most fundamental aspect of the structure–property relationship is to appreciate the basic skeletal arrangement of atoms in **crystalline** solids. Table 2.21 illustrates the fundamental possibilities, known as the 14 **Bravais lattices**. All crystalline structures of real materials can be produced by decorating the unit cell patterns of Table 2.21 with one or more atoms and repetitively stacking the unit cell structure through three-dimensional space.

2.7.3 Composition

The properties of commercially available materials are determined by chemical composition as well as structure (Shackelford, 1996). As a result, extensive numbering systems have been developed to label materials, especially metal **alloys**. Table 2.22 gives an example for gray cast irons.

TABLE 2.21   The 14 Bravais Lattices

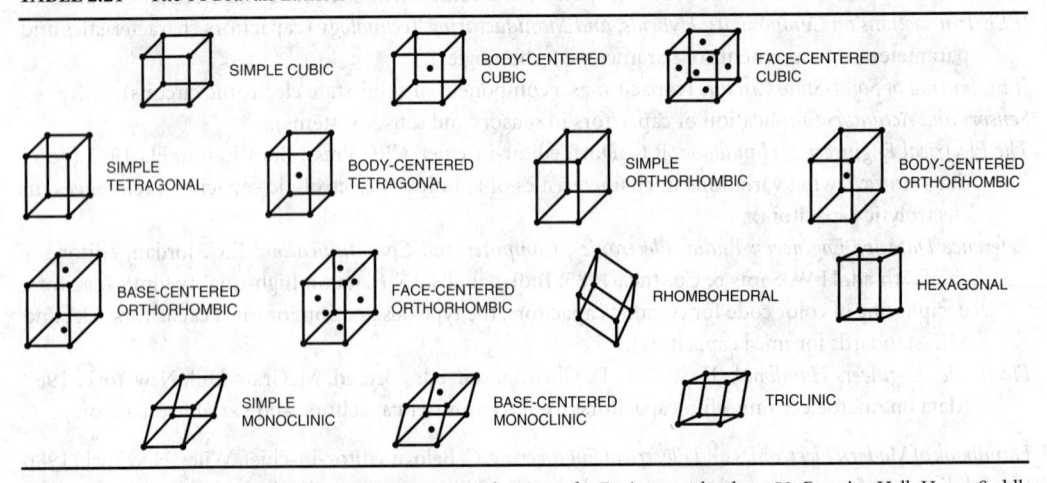

Source: Shackelford, J.F. 1996. *Introduction to Materials Science for Engineers*, 4th ed., p. 58. Prentice-Hall, Upper Saddle River, NJ.

TABLE 2.22 Composition Limits of Selected Gray Cast Irons (%)

UNS	SAE Grade	C	Mn	Si	P	S
F10004	G1800	3.40 to 3.70	0.50 to 0.80	2.80 to 2.30	0.15	0.15
F10005	G2500	3.20 to 3.50	0.60 to 0.90	2.40 to 2.00	0.12	0.15
F10009	G2500	3.40 min	0.60 to 0.90	1.60 to 2.10	0.12	0.12
F10006	G3000	3.10 to 3.40	0.60 to 0.90	2.30 to 1.90	0.10	0.16
F10007	G3500	3.00 to 3.30	0.60 to 0.90	2.20 to 1.80	0.08	0.16
F10010	G3500	3.40 min	0.60 to 0.90	1.30 to 1.80	0.08	0.12
F10011	G3500	3.50 min	0.60 to 0.90	1.30 to 1.80	0.08	0.12
F10008	G4000	3.00 to 3.30	0.70 to 1.00	2.10 to 1.80	0.07	0.16
F10012	G4000	3.10 to 3.60	0.60 to 0.90	1.95 to 2.40	0.07	0.12

Source: Data from ASM. 1984. *ASM Metals Reference Book*, 2nd ed., p. 166. American Society for Metals, Metals Park, OH.

2.7.4 Physical Properties

Among the most basic and practical characteristics of engineering materials are their physical properties. Table 2.23 gives the **density** of a wide range of materials in units of mega gram per cubic meter (= gram per cubic centimeter), whereas Table 2.24 gives the **melting points** for several common metals and ceramics.

2.7.5 Mechanical Properties

Central to the selection of materials for structural applications is their behavior in response to mechanical loads. A wide variety of mechanical properties are available to help guide materials selection (Shackelford, Alexander, and Park, 1994). The most basic of the mechanical properties are defined in terms of the **engineering stress** and the **engineering strain**. The engineering stress σ is defined as

$$\sigma = \frac{P}{A_0}$$

(2.133)

TABLE 2.23 Density of Selected Materials, Mg/m^3

Metal		Ceramic		Glass		Polymer	
Ag	10.50	Al_2O_3	3.97–3.986	SiO_2	2.20	ABS	1.05–1.07
Al	2.7	BN (cub)	3.49	SiO_2 10 wt% Na_2O	2.291	Acrylic	1.17–1.19
Au	19.28	BeO	3.01–3.03	SiO_2 19.55 wt% Na_2O	2.383	Epoxy	1.80–2.00
Co	8.8	MgO	3.581	SiO_2 29.20 wt% Na_2O	2.459	HDPE	0.96
Cr	7.19	SiC(hex)	3.217	SiO_2 39.66 wt% Na_2O	2.521	Nylon, type 6	1.12–1.14
Cu	8.93	Si_3N_4 (α)	3.184	SiO_2 39.0 wt% CaO	2.746	Nylon 6/6	1.13–1.15
Fe	7.87	Si_3N_4 (β)	3.187			Phenolic	1.32–1.46
Ni	8.91	TiO_2 (rutile)	4.25			Polyacetal	1.425
Pb	11.34	UO_2	10.949–10.97			Polycarbonate	1.2
Pt	21.44	ZrO_2 (CaO)	5.5			Polyester	1.31
Ti	4.51	Al_2O_3 MgO	3.580			Polystyrene	1.04
W	19.25	$3Al_2O_3$ $2SiO_2$	2.6–3.26			PTFE	2.1–2.3

Source: Selected data from Shackelford, J.F., Alexander, W., and Park, J.S., ed. 1994. *CRC Materials Science and Engineering Handbook*, 2nd ed. CRC Press, Boca Raton, FL.

TABLE 2.24   Melting Point of Selected Metals and Ceramics

Metal	M.P. (°C)	Ceramic	M.P. (°C)
Ag	962	Al_2O_3	2049
Al	660	BN	2727
Au	1064	B_2O_3	450
Co	1495	BeO	2452
Cr	1857	NiO	1984
Cu	1083	PbO	886
Fe	1535	SiC	2697
Ni	1453	Si_3N_4	2442
Pb	328	SiO_2	1723
Pt	1772	WC	2627
Ti	1660	ZnO	1975
W	3410	ZrO_2	2850

Source: Selected data from Shackelford, J.F., Alexander, W., and Park, J.S., ed. 1994. *CRC Materials Science and Engineering Handbook,* 2nd ed. CRC Press, Boca Raton, FL.

where P is the load on the sample with an original (zero stress) cross-sectional area A_0. The engineering strain ε is defined as

$$\varepsilon = \frac{[l - l_0]}{l_0} = \frac{\Delta l}{l_0} \tag{2.134}$$

where l is the sample length at a given load and l_0 is the original (zero stress) length. The maximum engineering stress that can be withstood by the material during its load history is termed the *ultimate tensile strength*, or simply **tensile strength** (TS). An example of the tensile strength for selected wrought (meaning worked, as opposed to cast) aluminum alloys is given in Table 2.25. The stiffness of a material is indicated by the linear relationship between engineering stress and engineering strain for relatively small levels of load application. The **modulus of elasticity** E, also known as **Young's modulus**, is given by the ratio

$$E = \frac{\sigma}{\varepsilon} \tag{2.135}$$

TABLE 2.25   Tensile Strength of Selected Wrought Aluminum Alloys

Alloy	Temper	TS (MPa)
1050	0	76
1050	H16	130
2024	0	185
2024	T361	495
3003	0	110
3003	H16	180
5050	0	145
5050	H34	195
6061	0	125
6061	T6, T651	310
7075	0	230
7075	T6, T651	570

Source: Selected data from Shackelford, J.F., Alexander, W., and Park, J.S., ed. 1994. *CRC Materials Science and Engineering Handbook,* 2nd ed. CRC Press, Boca Raton, FL.

TABLE 2.26   Young's Modulus of Selected Glasses, GPa

Type	E
SiO_2	72.76–74.15
SiO_2 20 mol % Na_2O	62.0
SiO_2 30 mol % Na_2O	60.5
SiO_2 35 mol % Na_2O	60.2
SiO_2 24.6 mol % PbO	47.1
SiO_2 50.0 mol % PbO	44.1
SiO_2 65.0 mol % PbO	41.2
SiO_2 60 mol % B_2O_3	23.3
SiO_2 90 mol % B_2O_3	20.9
B_2O_3	17.2–17.7
B_2O_3 10 mol % Na_2O	31.4
B_2O_3 20 mol % Na_2O	43.2

Source: Selected data from Shackelford, J.F., Alexander, W., and Park, J.S., ed. 1994. *CRC Materials Science and Engineering Handbook*, 2nd ed. CRC Press, Boca Raton, FL.

Table 2.26 gives values of Young's modulus for selected compositions of glass materials. The ductility of a material is indicated by the percent elongation at failure ($= 100 \times \varepsilon_{failure}$), representing the general ability of the material to be plastically (i.e., permanently) deformed. The percent elongation at failure for selected polymers is given in Table 2.27.

2.7.6   Thermal Properties

Many applications of engineering materials depend on their response to a thermal environment. The **thermal conductivity** k is defined by **Fourier's law**

$$k = \frac{-\left[\dfrac{dQ}{dt}\right]}{\left[A\left(\dfrac{dT}{dx}\right)\right]} \tag{2.136}$$

TABLE 2.27   Total Elongation at Failure of Selected Polymers

Polymer	Elongation
ABS	5–20
Acrylic	2–7
Epoxy	4.4
HDPE	700–1000
Nylon, type 6	30–100
Nylon 6/6	15–300
Phenolic	0.4–0.8
Polyacetal	25
Polycarbonate	110
Polyester	300
Polypropylene	100–600
PTFE	250–350

Source: Selected data from Shackelford, J.F., Alexander, W., and Park, J.S., eds. 1994. *CRC Materials Science and Engineering Handbook*, 2nd ed. CRC Press, Boca Raton, FL.

TABLE 2.28 Thermal Conductivity and Thermal Expansion of Alloy Cast Irons

Alloy	Thermal Conductivity, W/(m K)	Thermal Expansion Coefficient, 10^{-6} mm/(mm °C)
Low-C white iron	22[a]	12[b]
Martensitic nickel-chromium iron	30[a]	8–9[b]
High-nickel gray iron	38–40	8.1–19.3
High-nickel ductile iron	13.4	12.6–18.7
Medium-silicon iron	37	10.8
High-chromium iron	20	9.3–9.9
High-nickel iron	37–40	8.1–19.3
Nickel-chromium-silicon iron	30	12.6–16.2
High-nickel (20%) ductile iron	13	18.7

[a]Estimated.
[b]10–260°C. (*Source:* Data from ASM. 1984. *ASM Metals Reference Book*, 2nd ed., p. 172. American Society for Metals, Metals Park, OH.)

where dQ/dt is the rate of heat transfer across an area A due to a temperature gradient dT/dx. It is also important to note that the dimensions of a material, in general, increases with temperature. Increases in temperature lead to greater thermal vibration of the atoms and an increase in average separation distance of adjacent atoms. The **linear coefficient of thermal expansion** α is given by

$$\alpha = \frac{dl}{l\,dT} \tag{2.137}$$

with α having units of millimeter per millimeter degree Celsius. Examples of thermal conductivity and thermal expansion coefficient for alloy cast irons are given in Table 2.28.

2.7.7 Chemical Properties

A wide variety of data is available to characterize the nature of the reaction between engineering materials and their chemical environments (Shackelford, Alexander, and Park, 1994). Perhaps no such data are more fundamental and practical than the **electromotive force series** of metals shown in Table 2.29. The voltage associated with various half-cell reactions in standard aqueous environments are arranged in order, with

TABLE 2.29 Electromotive Force Series of Metals

Metal	Potential, V	Metal	Potential, V	Metal	Potential, V
Anodic or Corroded End					
Li	−3.04	Al	−1.70	Pb	−0.13
Rb	−2.93	Mn	−1.04	H	0.00
K	−2.92	Zn	−0.76	Cu	0.52
Ba	−2.90	Cr	−0.60	Ag	0.80
Sr	−2.89	Cd	−0.40	Hg	0.85
Ca	−2.80	Ti	−0.33	Pd	1.0
Na	−2.71	Co	−0.28	Pt	1.2
Mg	−2.37	Ni	−0.23	Au	1.5
Be	−1.70	Sn	−0.14		
				Cathodic or Noble Metal End	

Source: Data compiled by J.S. Park from Bolz, R.E. and Tuve, G.L., eds. 1973. *CRC Handbook of Tables for Applied Engineering Science.* CRC Press, Boca Raton, FL.

the materials associated with more anodic reactions tending to be corroded in the presence of a metal associated with a more cathodic reaction.

2.7.8 Electrical and Optical Properties

To this point, we have concentrated on various properties dealing largely with the structural applications of engineering materials. In many cases the electromagnetic nature of the materials may determine their engineering applications. Perhaps the most fundamental relationship in this regard is **Ohm's law,** which states that the magnitude of current flow I through a circuit with a given resistance R and voltage V is related by

$$V = IR \tag{2.138}$$

where V is in units of volts, I is in amperes, and R is in ohms. The resistance value depends on the specific sample geometry. In general, R increases with sample length l and decreases with sample area A. As a result, the property more characteristic of a given material and independent of its geometry is **resistivity** ρ defined as

$$\rho = \frac{[RA]}{l} \tag{2.139}$$

The units for resistivity are ohm meter ($\Omega \cdot$ m). Table 2.30 gives the values of electrical resistivity for various materials, indicating that metals typically have low resistivities (and correspondingly high electrical conductivities) and ceramics and polymers typically have high resistivities (and correspondingly low conductivities).

An important aspect of the electromagnetic nature of materials is their optical properties. Among the most fundamental optical characteristics of a light transmitting material is the **index of refraction** n defined as

$$n = \frac{v_{\text{vac}}}{v} \tag{2.140}$$

where v_{vac} is the speed of light in vacuum (essentially equal to that in air) and v is the speed of light in a transparent material. The index of refraction for a variety of polymers is given in Table 2.31.

TABLE 2.30   Electrical Resistivity of Selected Materials

Metal (Alloy Cast Iron)	$\rho, \Omega \cdot$ m	Ceramic	$\rho, \Omega \cdot$ m	Polymer	$\rho, \Omega \cdot$ m
Low-C white cast iron	$0.53 \cdot 10^{-6}$	Al_2O_3	$>10^{13}$	ABS	$2\text{--}4 \cdot 10^{13}$
Martensitic Ni-Cr iron	$0.80 \cdot 10^{-6}$	B_4C	$0.3\text{--}0.8 \cdot 10^{-2}$	Acrylic	$>10^{13}$
High-Si iron	$0.50 \cdot 10^{-6}$	BN	$1.7 \cdot 10^{11}$	HDPE	$>10^{15}$
High-Ni iron	$1.4\text{--}1.7 \cdot 10^{-6}$	BeO	$>10^{15}$	Nylon 6/6	$10^{12}\text{--}10^{13}$
Ni-Cr-Si iron	$1.5\text{--}1.7 \cdot 10^{-6}$	MgO	$1.3 \cdot 10^{13}$	Phenolic	$10^{7}\text{--}10^{11}$
High-Al iron	$2.4 \cdot 10^{-6}$	SiC	$1\text{--}1 \cdot 10^{10}$	Polyacetal	10^{13}
Medium-Si ductile iron	$0.58\text{-}0.87 \cdot 10^{-6}$	Si_3N_4	10^{11}	Polypropylene	$>10^{15}$
High-Ni (20%) ductile iron	$1.02 \cdot 10^{-6}$	SiO_2	10^{16}	PTFE	$>10^{16}$

Source: Selected data from Shackelford, J.F., Alexander, W., and Park, J.S., eds. 1994. *CRC Materials Science and Engineering Handbook,* 2nd ed. CRC Press, Boca Raton, FL.

TABLE 2.31　Refractive Index of Selected Polymers

Polymer	n
Acrylic	1.485–1.500
Cellulose Acetate	1.46–1.50
Epoxy	1.61
HDPE	1.54
Polycarbonate	1.586
PTFE	1.35
Polyester	1.50–1.58
Polystyrene	1.6
SAN	1.565–1.569
Vinylidene Chloride	1.60–1.63

Source: Data compiled by J.S. Park from Lynch, C.T., ed. 1975. *CRC Handbook of Materials Science, Vol. 3.* CRC Press, Inc., Boca Raton, FL; and ASM. 1988. *Engineered Materials Handbook, Vol. 2, Engineering Plastics.* ASM International, Metals Park, OH.

2.7.9　Additional Data

The following, more detailed tabular data are reprinted from: Dorf, R.C. 1993. *The Electrical Engineering Handbook.* CRC Press, Boca Raton, FL, pp. 2527–2544.

Electrical Resistivity

Electrical Resistivity of Pure Metals

The first part of this table gives the electrical resistivity, in units of $10^{-8}\ \Omega \cdot m$, for 28 common metallic elements as a function of temperature. The data refer to polycrystalline samples. The number of significant figures indicates the accuracy of the values. However, at low temperatures (especially below 50 K) the electrical resistivity is extremely sensitive to sample purity. Thus the low-temperature values refer to samples of specified purity and treatment.

The second part of the table gives resistivity values in the neighborhood of room temperature for other metallic elements that have not been studied over an extended temperature range.

Electrical Resistivity in $10^{-8}\ \Omega \cdot m$

T/K	Aluminum	Barium	Beryllium	Calcium	Cesium	Chromium	Copper
1	0.000100	0.081	0.0332	0.045	0.0026		0.00200
10	0.000193	0.189	0.0332	0.047	0.243		0.00202
20	0.000755	0.94	0.0336	0.060	0.86		0.00280
40	0.0181	2.91	0.0367	0.175	1.99		0.0239
60	0.0959	4.86	0.067	0.40	3.07		0.0971
80	0.245	6.83	0.075	0.65	4.16		0.215
100	0.442	8.85	0.133	0.91	5.28	1.6	0.348
150	1.006	14.3	0.510	1.56	8.43	4.5	0.699
200	1.587	20.2	1.29	2.19	12.2	7.7	1.046
273	2.417	30.2	3.02	3.11	18.7	11.8	1.543
293	2.650	33.2	3.56	3.36	20.5	12.5	1.678
298	2.709	34.0	3.70	3.42	20.8	12.6	1.712
300	2.733	34.3	3.76	3.45	21.0	12.7	1.725

Electrical Resistivity in 10^{-8} $\Omega \cdot$ m (Continued)

T/K	Aluminum	Barium	Beryllium	Calcium	Cesium	Chromium	Copper
400	3.87	51.4	6.76	4.7		15.8	2.402
500	4.99	72.4	9.9	6.0		20.1	3.090
600	6.13	98.2	13.2	7.3		24.7	3.792
700	7.35	130	16.5	8.7		29.5	4.514
800	8.70	168	20.0	10.0		34.6	5.262
900	10.18	216	23.7	11.4		39.9	6.041

T/K	Gold	Hafnium	Iron	Lead	Lithium	Magnesium	Manganese
1	0.0220	1.00	0.0225		0.007	0.0062	7.02
10	0.0226	1.00	0.0238		0.008	0.0069	18.9
20	0.035	1.11	0.0287		0.012	0.0123	54
40	0.141	2.52	0.0758		0.074	0.074	116
60	0.308	4.53	0.271		0.345	0.261	131
80	0.481	6.75	0.693	4.9	1.00	0.557	132
100	0.650	9.12	1.28	6.4	1.73	0.91	132
150	1.061	15.0	3.15	9.9	3.72	1.84	136
200	1.462	21.0	5.20	13.6	5.71	2.75	139
273	2.051	30.4	8.57	19.2	8.53	4.05	143
293	2.214	33.1	9.61	20.8	9.28	4.39	144
298	2.255	33.7	9.87	21.1	9.47	4.48	144
300	2.271	34.0	9.98	21.3	9.55	4.51	144
400	3.107	48.1	16.1	29.6	13.4	6.19	147
500	3.97	63.1	23.7	38.3		7.86	149
600	4.87	78.5	32.9			9.52	151
700	5.82		44.0			11.2	152
800	6.81		57.1			12.8	
900	7.86					14.4	

T/K	Molybdenum	Nickel	Palladium	Platinum	Potassium	Rubidium	Silver
1	0.00070	0.0032	0.0200	0.002	0.0008	0.0131	0.00100
10	0.00089	0.0057	0.0242	0.0154	0.0160	0.109	0.00115
20	0.00261	0.0140	0.0563	0.0484	0.117	0.444	0.0042
40	0.0457	0.068	0.334	0.409	0.480	1.21	0.0539
60	0.206	0.242	0.938	1.107	0.90	1.94	0.162
80	0.482	0.545	1.75	1.922	1.34	2.65	0.289
100	0.858	0.96	2.62	2.755	1.79	3.36	0.418
150	1.99	2.21	4.80	4.76	2.99	5.27	0.726
200	3.13	3.67	6.88	6.77	4.26	7.49	1.029
273	4.85	6.16	9.78	9.6	6.49	11.5	1.467
293	5.34	6.93	10.54	10.5	7.20	12.8	1.587
298	5.47	7.12	10.73	10.7	7.39	13.1	1.617
300	5.52	7.20	10.80	10.8	7.47	13.3	1.629
400	8.02	11.8	14.48	14.6			2.241
500	10.6	17.7	17.94	18.3			2.87
600	13.1	25.5	21.2	21.9			3.53
700	15.8	32.1	24.2	25.4			4.21
800	18.4	35.5	27.1	28.7			4.91
900	21.2	38.6	29.4	32.0			5.64

(Continued)

Electrical Resistivity in $10^{-8}\ \Omega \cdot m$ (Continued)

T/K	Sodium	Strontium	Tantalum	Tungsten	Vanadium	Zinc	Zirconium
1	0.0009	0.80	0.10	0.000016		0.0100	0.250
10	0.0015	0.80	0.102	0.000137	0.0145	0.0112	0.253
20	0.016	0.92	0.146	0.00196	0.039	0.0387	0.357
40	0.172	1.70	0.751	0.0544	0.304	0.306	1.44
60	0.447	2.68	1.65	0.266	1.11	0.715	3.75
80	0.80	3.64	2.62	0.606	2.41	1.15	6.64
100	1.16	4.58	3.64	1.02	4.01	1.60	9.79
150	2.03	6.84	6.19	2.09	8.2	2.71	17.8
200	2.89	9.04	8.66	3.18	12.4	3.83	26.3
273	4.33	12.3	12.2	4.82	18.1	5.46	38.8
293	4.77	13.2	13.1	5.28	19.7	5.90	42.1
298	4.88	13.4	13.4	5.39	20.1	6.01	42.9
300	4.93	13.5	13.5	5.44	20.2	6.06	43.3
400		17.8	18.2	7.83	28.0	8.37	60.3
500		22.2	22.9	10.3	34.8	10.82	76.5
600		26.7	27.4	13.0	41.1	13.49	91.5
700		31.2	31.8	15.7	47.2		104.2
800		35.6	35.9	18.6	53.1		114.9
900			40.1	21.5	58.7		123.1

Electrical Resistivity of Other Metallic Elements in the Neighborhood of Room Temperature

Element	T/K	Electrical Resistivity $10^{-8}\ \Omega \cdot m$	Element	T/K	Electrical Resistivity $10^{-8}\ \Omega \cdot m$
Antimony	273	39	Polonium	273	40
Bismuth	273	107	Praseodymium	290–300	70.0
Cadmium	273	6.8	Promethium	290–300	75
Cerium	290–300	82.8	Protactinium	273	17.7
Cobalt	273	5.6	Rhenium	273	17.2
Dysprosium	290–300	92.6	Rhodium	273	4.3
Erbium	290–300	86.0	Ruthenium	273	7.1
Europium	290–300	90.0	Samarium	290–300	94.0
Gadolinium	290–300	131	Scandium	290–300	56.2
Gallium	273	13.6	Terbium	290–300	115
Holmium	290–300	81.4	Thallium	273	15
Indium	273	8.0	Thorium	273	14.7
Iridium	273	4.7	Thulium	290–300	67.6
Lanthanum	290–300	61.5	Tin	273	11.5
Lutetium	290–300	58.2	Titanium	273	39
Mercury	273	94.1	Uranium	273	28
Neodymium	290–300	64.3	Ytterbium	290–300	25.0
Niobium	273	15.2	Yttrium	290–300	59.6
Osmium	273	8.1			

Electrical Resistivity of Selected Alloys

Values of the resistivity are given in units of $10^{-8}\ \Omega \cdot m$. General comments in the preceding table for pure metals also apply here.

	273 K	293 K	300 K	350 K	400 K
Alloy—Aluminum-Copper					
Wt % Al					
99[a]	2.51	2.74	2.82	3.38	3.95
95[a]	2.88	3.10	3.18	3.75	4.33
90[b]	3.36	3.59	3.67	4.25	4.86
85[b]	3.87	4.10	4.19	4.79	5.42
80[b]	4.33	4.58	4.67	5.31	5.99
70[b]	5.03	5.31	5.41	6.16	6.94
60[b]	5.56	5.88	5.99	6.77	7.63
50[b]	6.22	6.55	6.67	7.55	8.52
40[c]	7.57	7.96	8.10	9.12	10.2
30[c]	11.2	11.8	12.0	13.5	15.2
25[f]	16.3[aa]	17.2	17.6	19.8	22.2
15[h]	—	12.3	—	—	—
19[g]	1.8[aa]	11.0	11.1	11.7	12.3
5[e]	9.43	9.61	9.68	10.2	10.7
1[b]	4.46	4.60	4.65	5.00	5.37
Alloy—Aluminum-Magnesium					
99[c]	2.96	3.18	3.26	3.82	4.39
95[c]	5.05	5.28	5.36	5.93	6.51
90[c]	7.52	7.76	7.85	8.43	9.02
85	—	—	—	—	—
80	—	—	—	—	—
70	—	—	—	—	—
60	—	—	—	—	—
50	—	—	—	—	—
40	—	—	—	—	—
30	—	—	—	—	—
25	—	—	—	—	—
15	—	—	—	—	—
10[b]	17.1	17.4	17.6	18.4	19.2
5[b]	13.1	13.4	13.5	14.3	15.2
1[a]	5.92	6.25	6.37	7.20	8.03
Alloy—Copper-Gold					
Wt % Cu					
99[c]	1.73	1.86[aa]	1.91[aa]	2.24[aa]	2.58[aa]
95[c]	2.41	2.54[aa]	2.59[aa]	2.92[aa]	3.26[aa]
90[c]	3.29	4.42[aa]	3.46[aa]	3.79[aa]	4.12[aa]
85[c]	4.20	4.33	4.38[aa]	4.71[aa]	5.05[aa]
80[c]	5.15	5.28	5.32	5.65	5.99
70[c]	7.12	7.25	7.30	7.64	7.99
60[c]	9.18	9.13	9.36	9.70	10.05
50[c]	11.07	11.20	11.25	11.60	11.94
40[c]	12.70	12.85	12.90[aa]	13.27[aa]	13.65[aa]
30[c]	13.77	13.93	13.99[aa]	14.38[aa]	14.78[aa]
25[c]	13.93	14.09	14.14	14.54	14.94
15[c]	12.75	12.91	12.96[aa]	13.36[aa]	13.77
10[c]	10.70	10.86	10.91	11.31	11.72
5[c]	7.25	7.41[aa]	7.46	7.87	8.28
1[c]	3.40	3.57	3.62	4.03	4.45

(*Continued*)

Electrical Resistivity of Selected Alloys (Continued)

	273 K	293 K	300 K	350 K	400 K
Alloy—Copper-Nickel					
Wt % Cu					
99[c]	2.71	2.85	2.91	3.27	3.62
95[c]	7.60	7.71	7.82	8.22	8.62
90[c]	13.69	13.89	13.96	14.40	14.81
85[c]	19.63	19.83	19.90	2032	20.70
80[c]	25.46	25.66	25.72	26.12[aa]	26.44[aa]
70[i]	36.67	36.72	36.76	36.85	36.89
60[i]	45.43	45.38	45.35	45.20	45.01
50[i]	50.19	50.05	50.01	49.73	49.50
40[c]	47.42	47.73	47.82	48.28	48.49
30[i]	40.19	41.79	42.34	44.51	45.40
25[c]	33.46	35.11	35.69	39.67[aa]	42.81[aa]
15[c]	22.00	23.35	23.85	27.60	31.38
10[c]	16.65	17.82	18.26	21.51	25.19
5[c]	11.49	12.50	12.90	15.69	18.78
1[c]	7.23	8.08	8.37	10.63[aa]	13.18[aa]
Alloy—Copper-Palladium					
Wt % Cu					
99[c]	2.10	2.23	2.27	2.59	2.92
95[c]	4.21	4.35	4.40	4.74	5.08
90[c]	6.89	7.03	7.08	7.41	7.74
85[c]	9.48	9.61	9.66	10.01	10.36
80[c]	11.99	12.12	12.16	12.51[aa]	12.87
70[c]	16.87	17.01	17.06	17.41	17.78
60[c]	21.73	21.87	21.92	22.30	22.69
50[c]	27.62	27.79	27.86	28.25	28.64
40[c]	35.31	35.51	35.57	36.03	36.47
30[c]	46.50	46.66	46.71	47.11	47.47
25[c]	46.25	46.45	46.52	46.99[aa]	47.43[aa]
15[c]	36.52	36.99	37.16	38.28	39.35
10[c]	28.90	29.51	29.73	31.19[aa]	32.56[aa]
5[c]	20.00	20.75	21.02	22.84[aa]	24.54[aa]
1[c]	11.90	12.67	12.93[aa]	14.82[aa]	16.68[aa]
Alloy—Copper-Zinc					
Wt % Cu					
99[b]	1.84	1.97	2.02	2.36	2.71
95[b]	2.78	2.92	2.97	3.33	3.69
90[b]	3.66	3.81	3.86	4.25	4.63
85[b]	4.37	4.54	4.60	5.02	5.44
80[b]	5.01	5.19	5.26	5.71	6.17
70[b]	5.87	6.08	6.15	6.67	7.19
60	—	—	—	—	—
50	—	—	—	—	—
40	—	—	—	—	—
30	—	—	—	—	—
25	—	—	—	—	—
15	—	—	—	—	—
10	—	—	—	—	—
5	—	—	—	—	—
1	—	—	—	—	—

Electrical Resistivity of Selected Alloys (Continued)

	273 K	293 K	300 K	350 K	400 K
Alloy—Gold-Palladium					
Wt % Au					
99[c]	2.69	2.86	2.91	3.32	3.73
95[c]	5.21	5.35	5.41	5.79	6.17
90[i]	8.01	8.17	8.22	8.56	8.93
85[b]	10.50[aa]	10.66	10.72[aa]	11.100[aa]	11.48[aa]
80[b]	12.75	12.93	12.99	13.45	13.93
70[c]	18.23	18.46	18.54	19.10	19.67
60[b]	26.70	26.94	27.02	27.63[aa]	28.23[aa]
50[a]	27.23	27.63	27.76	28.64[aa]	29.42[aa]
40[a]	24.65	25.23	25.42	26.74	27.95
30[b]	20.82	21.49	21.72	23.35	24.92
25[b]	18.86	19.53	19.77	21.51	23.19
15[a]	15.08	15.77	16.01	17.80	19.61
10[a]	13.25	13.95	14.20[aa]	16.00[aa]	17.81[aa]
5[a]	11.49[aa]	12.21	12.46[aa]	14.26[aa]	16.07[aa]
1[a]	10.07	10.85[aa]	11.12[aa]	12.99[aa]	14.80[aa]
Alloy—Gold-Silver					
Wt % Au					
99[b]	2.58	2.75	2.80[aa]	3.22[aa]	3.63[aa]
95[a]	4.58	4.74	4.79	5.19	5.59
90[i]	6.57	6.73	6.78	7.19	7.58
85[j]	8.14	8.30	8.36[aa]	8.75	9.15
80[j]	9.34	9.50	9.55	9.94	10.33
70[j]	10.70	10.86	10.91	11.29	11.68[aa]
60[j]	10.92	11.07	11.12	11.50	11.87
50[j]	10.23	10.37	10.42	10.78	11.14
40[j]	8.92	9.06	9.11	9.46[aa]	9.81
30[a]	7.34	7.47	7.52	7.85	8.19
25[a]	6.46	6.59	6.63	6.96	7.30[aa]
15[a]	4.55	4.67	4.72	5.03	5.34
10[a]	3.54	3.66	3.71	4.00	4.31
5[i]	2.52	2.64[aa]	2.68[aa]	2.96[aa]	3.25[aa]
1[b]	1.69	1.80	1.84[aa]	2.12[aa]	2.42[aa]
Alloy—Iron-Nickel					
Wt % Fe					
99[a]	10.9	12.0	12.4	—	18.7
95[c]	18.7	19.9	20.2	—	26.8
90[c]	24.2	25.5	25.9	—	33.2
85[c]	27.8	29.2	29.7	—	37.3
80[c]	30.1	31.6	32.2	—	40.0
70[b]	32.3	33.9	34.4	—	42.4
60[c]	53.8	57.1	58.2	—	73.9
50[d]	28.4	30.6	31.4	—	43.7
40[d]	19.6	21.6	22.5	—	34.0
30[c]	15.3	17.1	17.7	—	27.4
25[b]	14.3	15.9	16.4	—	25.1
15[c]	12.6	13.8	14.2	—	21.1
10[c]	11.4	12.5	12.9	—	18.9
5[c]	9.66	10.6	10.9	—	16.1[aa]
1[b]	7.17	7.94	8.12	—	12.8

(*Continued*)

Electrical Resistivity of Selected Alloys (Continued)

	273 K	293 K	300 K	350 K	400 K
Alloy—Silver-Palladium					
Wt % Ag					
99[b]	1.891	2.007	2.049	2.35	2.66
95[b]	3.58	3.70	3.74	4.04	4.34
90[b]	5.82	5.94	5.98	6.28	6.59
85[k]	7.92[aa]	8.04[aa]	8.08	8.38[aa]	8.68[aa]
80[k]	10.01	10.13	10.17	10.47	10.78
70[k]	14.53	14.65	14.69	14.99	15.30
60[i]	20.9	21.1	21.2	21.6	22.0
50[k]	31.2	31.4	31.5	32.0	32.4
40[m]	42.2	42.2	42.2	42.3	42.3
30[b]	40.4	40.6	40.7	41.3	41.7
25[k]	36.67[aa]	37.06	37.19	38.1[aa]	38.8[aa]
15[i]	27.08[aa]	26.68[aa]	27.89[aa]	29.3[aa]	30.6[aa]
10[i]	21.69	22.39	22.63	24.3	25.9
5[b]	15.98	16.72	16.98	18.8[aa]	20.5[aa]
1[a]	11.06	11.82	12.08[aa]	13.92[aa]	15.70[aa]

[a] Uncertainty in resistivity is ±2%.
[b] Uncertainty in resistivity is ±3%.
[c] Uncertainty in resistivity is ±5%.
[d] Uncertainty in resistivity is ±7% below 300 K and ±5% at 300 and 400 K.
[e] Uncertainty in resistivity is ±7%.
[f] Uncertainty in resistivity is ±8%.
[g] Uncertainty in resistivity is ±10%.
[h] Uncertainty in resistivity is ±12%.
[i] Uncertainty in resistivity is ±4%.
[j] Uncertainty in resistivity is ±1%.
[k] Uncertainty in resistivity is ±3% up to 300 K and ± 4% above 300 K.
[m] Uncertainty in resistivity is ±2% up to 300 K and ± 4% above 300 K.
[a] Crystal usually a mixture of α-hcp and fcc lattice.
[aa] In temperature range where no experimental data are available.

Electrical Resistivity of Selected Alloy Cast Irons

Description	Electrical Resistivity, $\mu\Omega \cdot m$
Abrasion-resistant white irons	
Low-C white iron	0.53
Martensitic nickel–chromium iron	0.80
Corrosion-resistant irons	
High-silicon iron	0.50
High-chromium iron	
High-nickel gray iron	1.0[a]
High-nickel ductile iron	1.0[a]
Heat-resistant gray irons	
Medium-silicon iron	
High-chromium iron	
High-nickel iron	1.4–1.7
Nickel-chromium–silicon iron	1.5–1.7
High-aluminum iron	2.4
Heat-resistant ductile irons	
Medium-silicon ductile iron	0.58–0.87
High-nickel ductile (20 Ni)	1.02
High-nickel ductile (23 Ni)	1.0[a]

[a] Estimated. (*Source:* Data from 1984. *ASM Metals Reference Book*, 2nd ed. American Society for Metals, Metals Park, OH.)

Resistivity of Selected Ceramics (Listed by Ceramic)

Ceramic	Resistivity, $\Omega \cdot$ cm
Borides	
Chromium diboride (CrB_2)	21×10^{-6}
Hafnium diboride (HfB_2)	$10–12 \times 10^{-6}$ at room temp.
Tantalum diboride (TaB_2)	68×10^{-6}
Titanium diboride (TiB_2) (polycrystalline)	
85% dense	$26.5–28.4 \times 10^{-6}$ at room temp.
85% dense	9.0×10^{-6} at room temp.
100% dense, extrapolated values	$8.7–14.1 \times 10^{-6}$ at room temp.
	3.7×10^{-6} at liquid air temp.
Titanium diboride (TiB_2) (monocrystalline)	
Crystal length 5 cm, 39 and 59 deg. orientation with respect to growth axis	$6.6 \pm 0.2 \times 10^{-6}$ at room temp.
Crystal length 1.5 cm, 16.5 and 90 deg. orientation with respect to growth axis	$6.7 \pm 0.2 \times 10^{-6}$ at room temp.
Zirconium diboride (ZrB_2)	9.2×10^{-6} at 20°C
	1.8×10^{-6} at liquid air temp.
Carbides: boron carbide (B_4C)	$0.3–0.8$

Dielectric Constants

Dielectric Constants of Solids: Temperatures in the Range 17–22°C

Material	Frequency (Hz)	Dielectric Constant	Material	Frequency (Hz)	Dielectric Constant
Acetamide	4×10^8	4.0	Phenanthrene	4×10^8	2.80
Acetanilide	—	2.9	Phenol (10°C)	4×10^8	4.3
Acetic acid (2°C)	4×10^8	4.1	Phosphorus, red	10^8	4.1
Aluminum oleate	4×10^8	2.40	Phosphorus, yellow	10^8	3.6
Ammonium bromide	10^8	7.1	Potassium aluminum		
Ammonium chloride	10^8	7.0	sulfate	10^6	3.8
Antimony trichloride	10^8	5.34	Potassium carbonate		
Apatite $\perp$ optic axis	3×10^8	9.50	(15°C)	10^8	5.6
Apatite $\parallel$ optic axis	3×10^8	7.41	Potassium chlorate	6×10^7	5.1
Asphalt	$<3 \times 10^6$	2.68	Potassium chloride	10^4	5.03
Barium chloride (anhyd.)	6×10^7	11.4	Potassium chromate	6×10^7	7.3
Barium chloride ($2H_2O$)	6×10^7	9.4	Potassium iodide	6×10^7	5.6
Barium nitrate	6×10^7	5.9	Potassium nitrate	6×10^7	5.0
Barium sulfate(15°C)	10^8	11.4	Potassium sulfate	6×10^7	5.9
Beryl $\perp$ optic axis	10^4	7.02	Quartz $\perp$ optic axis	3×10^7	4.34
Beryl $\parallel$ optic axis	10^4	6.08	Quartz $\parallel$ optic axis	3×10^7	4.27
Calcite $\perp$ optic axis	10^4	8.5	Resorcinol	4×10^8	3.2
Calcite $\parallel$ optic axis	10^4	8.0	Ruby $\perp$ optic axis	10^4	13.27
Calcium carbonate	10^6	6.14	Ruby $\parallel$ optic axis	10^4	11.28
Calcium fluoride	10^4	7.36	Rutile $\perp$ optic axis	10^8	86
Calcium sulfate ($2H_2O$)	10^4	5.66	Rutile $\parallel$ optic axis	10^8	170
Cassiterite $\perp$ optic axis	10^{12}	23.4	Selenium	10^8	6.6
Cassiterite $\parallel$ optic axis	10^{12}	24	Silver bromide	10^6	12.2
d-Cocaine	5×10^8	3.10	Silver chloride	10^6	11.2
Cupric oleate	4×10^8	2.80	Silver cyanide	10^6	5.6
Cupric oxide (15°C)	10^8	18.1	Smithsonite $\perp$ optic axis	10^{12}	9.3
Cupric sulfate (anhyd.)	6×10^7	10.3			
Cupric sulfate ($5H_2O$)	6×10^7	7.8	Smithsonite $\parallel$ optic axis	10^{10}	9.4
Diamond	10^8	5.5			
Diphenylmethane	4×10^8	2.7	Sodium carbonate (anhyd.)	6×10^7	8.4

(Continued)

Dielectric Constants of Solids: Temperatures in the Range 17–22°C (Continued)

Material	Frequency (Hz)	Dielectric constant	Material	Frequency (Hz)	Dielectric Constant
Dolomite ⊥ optic axis	10^8	8.0			
Dolomite ‖ optic axis	10^8	6.8	Sodium carbonate (10H$_2$O)	6×10^7	5.3
Ferrous oxide (15°C)	10^8	14.2			
Iodine	10^8	4	Sodium chloride	10^4	6.12
Lead acetate	10^6	2.6	Sodium nitrate	—	5.2
Lead carbonate (15°C)	10^8	18.6	Sodium oleate	4×10^8	2.75
Lead chloride	10^6	4.2	Sodium perchlorate	6×10^7	5.4
Lead monoxide (15°C)	10^8	25.9	Sucrose (mean)	3×10^8	3.32
Lead nitrate	6×10^7	37.7	Sulfur (mean)	—	4.0
Lead oleate	4×10^8	3.27	Thallium chloride	10^6	46.9
Lead sulfate	10^6	14.3	*p*-Toluidine	4×10^8	3.0
Lead sulfide (15°)	10^6	17.9	Tourmaline ⊥ optic axis	10^4	7.10
Malachite (mean)	10^{12}	7.2			
Mercuric chloride	10^6	3.2	Tourmaline ‖ optic axis	10^4	6.3
Mercurous chloride	10^6	9.4			
Naphthalene	4×10^8	2.52	Urea	4×10^8	3.5
			Zircon ⊥, ‖	10^8	12

Dielectric Constants of Ceramics

Material	Dielectric Constant, 10^6 Hz	Dielectric Strength V/mil	Volume Resistivity $\Omega \cdot$ cm (23°C)	Loss Factor[a]
Alumina	4.5–8.4	40–160	10^{11}–10^{14}	0.0002–0.01
Corderite	4.5–5.4	40–250	10^{12}–10^{14}	0.004–0.012
Forsterite	6.2	240	10^{14}	0.0004
Porcelain (dry process)	6.0–8.0	40–240	10^{12}–10^{14}	0.0003–0.02
Porcelain (wet process)	6.0–7.0	90–400	10^{12}–10^{14}	0.006–0.01
Porcelain, zircon	7.1–10.5	250–400	10^{13}–10^{15}	0.0002–0.008
Steatite	5.5–7.5	200–400	10^{13}–10^{15}	0.0002–0.004
Titanates (Ba, Sr, Ca, Mg, and Pb)	15–12.000	50–300	10^8–10^{15}	0.0001–0.02
Titanium dioxide	14–110	100–210	10^{13}–10^{18}	0.0002–0.005

[a]Power factor × dielectric constant equals loss factor.

Dielectric Constants of Glasses

Type	Dielectric Constant at 100 MHz (20°C)	Volume Resistivity (350°C M $\Omega \cdot$ cm)	Loss Factor[a]
Corning 0010	6.32	10	0.015
Corning 0080	6.75	0.13	0.058
Corning 0120	6.65	100	0.012
Pyrex 1710	6.00	2,500	0.025
Pyrex 3320	4.71	—	0.019
Pyrex 7040	4.65	80	0.013
Pyrex 7050	4.77	16	0.017
Pyrex 7052	5.07	25	0.019
Pyrex 7060	4.70	13	0.018
Pyrex 7070	4.00	1,300	0.0048
Vycor 7230	3.83	—	0.0061
Pyrex 7720	4.50	16	0.014
Pyrex 7740	5.00	4	0.040
Pyrex 7750	4.28	50	0.011
Pyrex 7760	4.50	50	0.0081

Dielectric Constants of Glasses (Continued)

Type	Dielectric Constant at 100 MHz (20°C)	Volume Resistivity (350°C M Ω · cm)	Loss Factor[a]
Vycor 7900	3.9	130	0.0023
Vycor 7910	3.8	1600	0.00091
Vycor 7911	3.8	4000	0.00072
Corning 8870	9.5	5000	0.0085
G.E. Clear (silica glass)	3.81	4000–30,000	0.00038
Quartz (fused)	3.75–4.1 (1 MHz)	—	0.0002 (1 MHz)

[a]Power factor × dielectric constant equals loss factor.

Properties of Semiconductors

Semiconducting Properties of Selected Materials

Substance	Minimum Energy Gap, eV R.T.	0 K	$\frac{dE_g}{dT}$ ×10⁴, eV/°C	$\frac{dE_g}{dP}$ ×10⁶, eV · cm²/kg	Density of States Electron Effective Mass m_{d_n} (m_o)	Electron Mobility and Temperature Dependence μ_n, cm²/V·s	−x	Density of States Hole Effective Mass m_{d_p}, (m_o)	Hole Mobility and Temperature Dependence μ_p, cm²/V · s	−x
Si	1.107	1.153	−2.3	−2.0	1.1	1900	2.6	0.56	500	2.3
Ge	0.67	0.744	−3.7	±7.3	0.55	3800	1.66	0.3	1820	2.33
αSn	0.08	0.094	−0.5		0.02	2500	1.65	0.3	2400	2.0
Te	0.33				0.68	1100		0.19	560	
III–V Compounds										
AlAs	2.2	2.3				1200			420	
AlSb	1.6	1.7	−3.5	−1.6	0.09	200	1.5	0.4	500	1.8
GaP	2.24	2.40	−5.4	−1.7	0.35	300	1.5	0.5	150	1.5
GaAs	1.35	1.53	−5.0	+9.4	0.068	9000	1.0	0.5	500	2.1
GaSb	0.67	0.78	−3.5	+12	0.050	5000	2.0	0.23	1400	0.9
InP	1.27	1.41	−4.6	+4.6	0.067	5000	2.0		200	2.4
InAs	0.36	0.43	−2.8	+8	0.022	33,000	1.2	0.41	460	2.3
InSb	0.165	0.23	−2.8	+15	0.014	78,000	1.6	0.4	750	2.1
II–VI Compounds										
ZnO	3.2		−9.5	+0.6	0.38	180	1.5			
ZnS	3.54		−5.3	+5.7		180			5(400°C)	
ZnSe	2.58	2.80	−7.2	+6		540			28	
ZnTe	2.26			+6		340			100	
CdO	2.5 ± 0.1		−6		0.1	120				
CdS	2.42		−5	+3.3	0.165	400		0.8		
CdSe	1.74	1.85	−4.6		0.13	650	1.0	0.6		
CdTe	1.44	1.56	−4.1	+8	0.14	1200		0.35	50	
HgSe	0.30				0.030	20,000	2.0			
HgTe	0.15		−1		0.017	25,000		0.5	350	
Halite Structure Compounds										
PbS	0.37	0.28	+4		0.16	800		0.1	1,000	2.2
PbSe	0.26	0.16	+4		0.3	1500		0.34	1500	2.2
PbTe	0.25	0.19	+4	−7	0.21	1600		0.14	750	2.2
Others										
ZnSb	0.50	0.56			0.15	10				1.5
CdSb	0.45	0.57	−5.4		0.15	300			2,000	1.5
Bi₂S₃	1.3					200			1100	
Bi₂Se₃	0.27					600			675	
Bi₂Te₃	0.13		−0.95		0.58	1200	1.68	1.07	510	1.95

(*Continued*)

Semiconducting Properties of Selected Materials (Continued)

Substance	Minimum Energy Gap, eV R.T.	0 K	$\frac{dE_g}{dT}$ $\times 10^4$ eV/°C	$\frac{dE_g}{dP}$ $\times 10^6$ eV · cm²/kg	Density of States Electron Effective Mass m_{d_n} (m_o)	Electron Mobility and Temperature Dependence μ_n, cm²/V·s $-x$		Density of States Hole Effective Mass m_{d_p}, (m_o)	Hole Mobility and Temperature Dependence μ_p, cm²/V·s $-x$	
Mg₂Si		0.77	−6.4		0.46	400	2.5		70	
Mg₂Ge		0.74	−9			280	2		110	
Mg₂Sn	0.21	0.33	−3.5		0.37	320			260	
Mg₃Sb₂		0.32				20			82	
Zn₃As₂	0.93					10	1.1		10	
Cd₃As₂	0.55				0.046	100,000	0.88			
GaSe	2.05		3.8						20	
GaTe	1.66	1.80	−3.6			14	−5			
InSe	1.8					9000				
TlSe	0.57		−3.9		0.3	30		0.6	20	1.5
CdSnAs₂	0.23				0.05	25,000	1.7			
Ga₂Te₃	1.1	1.55	−4.8							
α-In₂Te₃	1.1	1.2			0.7				50	1.1
β-In₂Te₃	1.0								5	
Hg₅In₂Te₈	0.5								11,000	
SnO₂									78	

Band Properties of Semiconductors

Part A. Data on Valence Bands of Semiconductors (Room Temperature)

Substance	Band Curvature Effective Mass (expressed as fraction of free electron mass)			Energy Separation of Split-Off Band, eV	Measured (light) Hole Mobility cm²/V · s
	Heavy Holes	Light Holes	Split-off Band Holes		
Semiconductors with Valence Band Maximum at the Center of the Brillouin Zone ("F")					
Si	0.52	0.16	0.25	0.044	500
Ge	0.34	0.043	0.08	0.3	1820
Sn	0.3				2400
AlAs					
AlSb	0.4			0.7	550
GaP				0.13	100
GaAs	0.8	0.12	0.20	0.34	400
GaSb	0.23	0.06		0.7	1400
InP				0.21	150
InAs	0.41	0.025	0.083	0.43	460
InSb	0.4	0.015		0.85	750
CbTe	0.35				50
HgTe	0.5				350

Substance	**Semiconductors with Multiple Valence Band Maxima** Band Curvature Effective Masses				Measured (light) Hole Mobility, cm²/V · s
	Number of Equivalent Valleys and Direction	Longitudinal m_L	Transverse m_T	Anisotropy, $K = m_L/m_T$	
PbSe	4 "L" [111]	0.095	0.047	2.0	1,500
PbTe	4 "L" [111]	0.27	0.02	10	750
Bi₂Te₃	6	0.207	~0.045	4.5	515

Part B. Data on conduction Bands of Semiconductors (Room Temperature Data)

Single Valley Semiconductors

Substance	Energy Gap, eV	Effective Mass (m_o)	Mobility cm^2/V · s	Comments
GaAs	1.35	0.067	8500	3 (or 6?) equivalent [100] valleys 0.36 eV above this maximum with a mobility of ~50
InP	1.27	0.067	5000	3 (or 6?) equivalent [100] valleys 0.4 eV above this minimum
InAs	0.36	0.022	33,000	Equivalent valleys ~1.0 eV above this minimum
InSb	0.165	0.014	78,000	
CdTe	1.44	0.11	1000	4 (or 8?) equivalent [111] valleys 0.51 eV above this minimum

Multivalley Semiconductors

		Band Curvature Effective Mass		
Substance	Energy Gap	Number of Equivalent Valleys and Direction	Longitudinal m_L	Transverse m_T	Anisotropy $K = m_L/m_T$
Si	1.107	6 in [100] Δ	0.90	0.192	4.7
Ge	0.67	4 in [111] at L	1.588	0.0815	19.5
GaSb	0.67	as Ge (?)	~1.0	~0.2	~5
PbSe	0.26	4 in [111] at L	0.085	0.05	1.7
PbTe	0.25	4 in [111] at L	0.21	0.029	5.5
Bi$_2$Te$_3$	0.13	6			~0.05

Resistivity of Semiconducting Minerals

Mineral	$\rho, \Omega \cdot$ m	Mineral	$\rho, \Omega \cdot$ m
Diamond (C)	2.7	Gersdorffite, NiAsS	1 to 160 $\times 10^{-6}$
Sulfides		Glaucodote, (Co, Fe)AsS	5 to 100 $\times 10^{-6}$
Argentite, Ag$_2$S	1.5 to 2.0 $\times 10^{-3}$	Antimonide	
Bismuthinite, Bi$_2$S$_3$	3 to 570	Dyscrasite, Ag$_3$Sb	0.12 to 1.2 $\times 10^{-6}$
Bornite, Fe$_2$S$_3 \cdot n$Cu$_2$S	1.6 to 6000 $\times 10^{-6}$	Arsenides	
Chalcocite, Cu$_2$S	80 to 100 $\times 10^{-6}$	Allemonite, SbAs$_2$	70 to 60,000
Chalcopyrite, Fe$_2$S$_3 \cdot$ Cu$_2$S	150 to 9000 $\times 10^{-6}$	Lollingite, FeAs$_2$	2 to 270 $\times 10^{-6}$
Covellite, CuS	0.30 to 83 $\times 10^{-6}$	Nicollite, NiAs	0.1 to 2 $\times 10^{-6}$
Galena, PbS	6.8 $\times 10^{-6}$ to 9.0 $\times 10^{-2}$	Skutterudite, CoAs$_3$	1 to 400 $\times 10^{-6}$
Haverite, MnS$_2$	10 to 20	Smaltite, CoAs$_2$	1 to 12 $\times 10^{-6}$
Marcasite, FeS$_2$	1 to 150 $\times 10^{-3}$	Tellurides	
Metacinnabarite, 4HgS	2 $\times 10^{-6}$ to 1 $\times 10^{-3}$	Altaite, PbTe	20 to 200 $\times 10^{-6}$
Millerite, NiS	2 to 4 $\times 10^{-7}$	Calavarite, AuTe$_2$	6 to 12 $\times 10^{-6}$
Molybdenite, MoS$_2$	0.12 to 7.5	Coloradoite, HgTe	4 to 100 $\times 10^{-6}$
Pentlandite, (Fe, Ni)$_9$S$_8$	1 to 11 $\times 10^{-6}$	Hessite, Ag$_2$Te	4 to 100 $\times 10^{-6}$
Pyrrhotite, Fe$_7$S$_8$	2 to 160 $\times 10^{-6}$	Nagyagite, Pb$_6$Au(S, Te)$_{14}$	20 to 80 $\times 10^{-6}$
Pyrite, FeS$_2$	1.2 to 600 $\times 10^{-3}$	Sylvanite, AgAuTe$_4$	4 to 20 $\times 10^{-6}$
Sphalerite, ZnS	2.7 $\times 10^{-3}$ to 1.2 $\times 10^4$	Oxides	
Antimony-sulfur compounds		Braunite, Mn$_2$O$_3$	0.16 to 1.0
Berthierite, FeSb$_2$S$_4$	0.0083 to 2.0	Cassiterite, SnO$_2$	4.5 $\times 10^{-4}$ to 10,000
Boulangerite, Pb$_5$Sb$_4$S$_{11}$	2 $\times 10^3$ to 4 $\times 10^4$	Cuprite, Cu$_2$O	10 to 50
Cylindrite, Pb$_3$Sn$_4$Sb$_2$S$_{14}$	2.5 to 60	Hollandite, (Ba, Na, K)Mn$_8$O$_{16}$	2 to 100 $\times 10^{-3}$
Franckeite, Pb$_5$Sn$_3$Sb$_2$S$_{14}$	1.2 to 4	Ilmenite, FeTiO$_3$	0.001 to 4
Hauchecornite, Ni$_9$(Bi, Sb)$_2$S$_8$	1 to 83 $\times 10^{-6}$	Magnetite, Fe$_3$O$_4$	52 $\times 10^{-6}$
Jamesonite, Pb$_4$FeSb$_6$S$_{14}$	0.020 to 0.15	Manganite, MnO $\cdot$ OH	0.018 to 0.5
Tetrahedrite, Cu$_3$SbS$_3$	0.30 to 30,000	Melaconite, CuO	6000

(Continued)

Resistivity of Semiconducting Minerals (Continued)

Mineral	$\rho, \Omega \cdot m$	Mineral	$\rho, \Omega \cdot m$
Arsenic-sulfur compounds		Psilomelane, $KMnO \cdot MnO_2 \cdot nH_2O$	0.04 to 6000
Arsenopyrite, FeAsS	20 to 300 $\times 10^{-6}$	Pyrolusite, MnO_2	0.007 to 30
Cobaltite, CoAsS	6.5 to 130 $\times 10^{-3}$	Rutile, TiO_2	29 to 910
Enargite, Cu_3AsS_4	0.2 to 40 $\times 10^{-3}$	Uraninite, UO	1.5 to 200

Source: Carmichael, R.S., ed. 1982. *Handbook of Physical Properties of Rocks, Vol. I.* CRC Press, Boca Raton, FL.

Properties of Magnetic Alloys

Name	Composition,[a] Weight Percent					Remanence, B_r, G	Coercive Force, H_e, O	Maximum Energy Product, $(BH)_{max}$, G–O $\times 10^{-6}$
	Al	Ni	Co	Cu	Other			
U.S.								
Alnico I	12	20–22	5			7100	440	1.4
Alnico II	10	17	12.5	6		7200	540	1.6
Alnico III	12	24–26		3		6900	470	1.35
Alnico IV	12	27–28	5			5500	700	1.3
Alnico V[b]	8	14	24	3		12,500	600	5.0
Alnico V DG[b]	8	14	24	3		13,100	640	6.0
Alnico VI[b]	8	15	24	3	1.25 Ti	10,500	750	3.75
Alnico VII[b]	8.5	18	24	3	5 Ti	7200	1050	2.75
Alnico XII	6	18	35		8 Ti	5800	950	1.6
Chromium steel					1 Mn 0.9 C 3.5 Cr 0.9 C	10,000 9700	50 65	0.2 0.3
Cobalt steel			17		0.3 Mn 2.5 Cr 8 W 0.75 C	9500	150	0.65
Cunico		21	29	50		3400	660	0.80
Cunife		20		60		5400	550	1.5
Ferroxdur 1		$BaFe_{12}O_{19}$				2200	1800	1.0
Ferroxdur 2		$BaF_{12}O_{19}$ (oriented)				3840	2000	3.5
Platinum-cobalt			23		77 Pt	6000	4300	7.5
Remalloy			12		17 Mo	10,500	250	1.1
Silmanol	4.4				86.6 Ag 8.8 Mn	550	6000	0.075
Tungsten steel					5 W 0.3 Mn 0.7 C	10,000	70	0.32
Vicalloy I			52		10 V	8800	300	1.0
Vicalloy II (wire)			52		14 V	10,000	510	3.5
Germany								
Alni 90	12	21				8000	350	1.2
Alni 120	13	27				6000	570	1.2
Alnico 130	12	23	5			6300	620	1.4
Alnico 160	11	24	12	4		6200	700	1.6
Alnico 190	12	21	15	4		7000	700	1.8
Alnico 250	8	19	23	4	6 Ti	6500	1000	2.2
Alnico 400[b]	9	15	23	4		12,000	650	4.8
Alnico 580[b] (semicolumnar)	9	15	23	4		13,000	700	6.0

Properties of Magnetic Alloys (Continued)

Name	Composition,[a] Weight Percent					Remanence, B_r, G	Coercive Force, H_e, O	Maximum Energy Product, $(BH)_{max}$, G–O $\times 10^{-6}$
	Al	Ni	Co	Cu	Other			
Oerstit 800	9	18	19	4	4 Ti	6600	750	1.95
Great Britain								
Alcomax I	7.5	11	25	3	1.5 Ti	12,000	475	3.5
Alcomax II	8	11.5	24	4.5		12,400	575	4.7
Alcomax IISC (semicolumnar)	8	11	22	4.5		12,800	600	5.15
Alcomax III	8	13.5	24	3	0.8 Nb	12,500	670	5.10
Alcomax IIISC (semicolumnar)	8	13.5	24	3	0.8 Nb	13,000	700	5.80
Alcomax IV	8	13.5	24	3	2.5 Nb	11,200	750	4.30
Alcomax IVSC (semicolumnar)	8	13.5	24	3	2.5 Nb	11,700	780	5.10
Alni, high B_r	13	24		3.5		6200	480	1.25
Alni, normal						5600	580	1.25
Alni, high H_e	12	32			0–0.5Ti	5000	680	1.25
Alnico, high B_r	10	17	12	6		8000	500	1.70
Alnico, normal						7250	560	1.70
Alnico, high H_e	10	20	13.5	6	0.25 Ti	6600	620	1.70
Columax (columnar)	similar to Alcomax III or IV					13,000–14,000	700–800	7.0–8.5
Hycomax	9	21	20	1.6		9500	830	3.3

[a]Remainder of unlisted composition is either iron or iron plus trace impurities.
[b]Cast anisotropic. Unmarked are cast isotropic.

Properties of Magnetic Alloys: High-Permeability Magnetic Alloys

Name	Composition[a], Weight Percent	Sp. gr., g/cm^3	Tensile Strength		Remark	Use
			kg/mm^{2b}	Form		
Silicon iron AISI M 15	Si 4	7.68–7.64	44.3	—	Annealed 4 h 802–1093°C	Low core losses
Silicon iron AISI M 8	Si 3	7.68–7.64	44.2	Grain oriented	Annealed 4 h 802–1204°C	
45 Permalloy	Ni 45; Mn 0.3	8.17	—	—	—	Audio transformer, coils, relays
Monimax	Ni 47; Mo 3	8.27	—	—	—	High-frequency coils
4–79 Permalloy	Ni 79; Mo 4; Mn 0.3	8.74	55.4	—	H$_2$ annealed 1121°C	Audio coils, transformers, magnetic shields
Sinimax	Ni 43; Si 3	7.70	—	—	—	High-frequency coils
Nu-metal	Ni 75; Cr 2; Cu 5	8.58	44.8	—	H$_2$ annealed 1221°C	Audio coils, magnetic shields, transformers
Supermalloy	Ni 79; Mo 5; Mn 0.3	8.77	—	—	—	Pulse transformers, magnetic amplifiers, coils
2-V Permendur	Co 40; V 2	8.15	46.3	—	—	DC electromagnets, pole tips

[a]Iron is additional alloying metal.
[b]kg/mm^2 $\times$ 1422.33 = lbs/in^2.

Properties of Magnetic Alloys: Cast Permanent Magnetic Alloys

Alloy Name (country of manufacture[a])	Composition[b] Weight Percent	Sp.gr., g/cm^3	Thermal Expansion		Tensile Strength			
			$Cm \times 10^{-6}$ cm$\times°$ C	Between °C	kg/ mm^{2c}	Form	Remark[d]	Use
Alnico I (US)	Al 12; Ni 20–22; Co 5	6.9	12.6	20–300	2.9	Cast	i	Permanent magnets
Alnico II (US)	Al 10; Ni 17; Cu 6; Co 12.5	7.1	12.4	20–300	2.1 45.7	Cast Sintered	i	Temperature controls magnetic toys and novelties
Alnico III (US)	Al 12; Ni 24–26; Cu 3	6.9	12	20–300	8.5	Cast	i	Tractor magnetos
Alnico IV (US)	Al 12; Ni 27–28; Co 5	7.0	13.1	20–300	6.3 42.1	Cast Sintered	i	Application requiring high coercive force
Alnico V (US)	Al 8; Ni 14; Co 24; Cu 3	7.3	11.3	—	3.8 35	Cast Sintered	a	Application requiring high energy
Alnico V DG (US)	Al 8; Ni 14; Co 24; Cu 3	7.3	11.3	—	—	—	a, c	
Alnico VI (US)	Al 8; Ni 15; Co 24; Cu 3; Ti 1.25	7.3	11.4	—	16.1	Cast	a	Application requiring high energy
Alnico VII (US)	Al 8.5; Ni 18; Cu 3; Co 24; Ti 5	7.17	11.4	—	—	—	a	
Alnico XII (US)	Al 6; Ni 18; Co 35; Ti 8	7.2	11	20–300	—	—	—	Permanent magnets
Comol (US)	Co 12; Mo 17	8.16	9.3	20–300	88.6	—	—	Permanent magnets
Cunife (US)	Cu 60; Ni 20	8.52	—	—	70.3	—	—	Permanent magnets
Cunico (US)	Cu 50; Ni 21	8.31	—	—	70.3	—	—	Permanent magnets
Barium ferrite Feroxdur (US)	$BaFe_{12}O_{19}$	4.7	10	—	70.3	—	—	Ceramics
Alcomax I (GB)	Al 7.5; Ni 11; Co 25; Cu 3; Ti 1.5	—	—	—	—	—	a	Permanent magnets
Alcomax II (GB)	Al 8; Ni 11.5; Co 24; Cu 4.5	—	—	—	—	—	a	Permanent magnets
Alcomax II SC (GB)	Al 8; Ni 11; Co 22; Cu 4.5	7.3	—	—	—	—	a, sc	
Alcomax III (GB)	Al 8; Ni 13.5; Co 24; Nb 0.8	7.3	—	—	—	—	a	Magnets for motors, loudspeakers
Alcomax IV (GB)	Al 8; Ni 13.5; Cu 3; Co 24; Nb 2.5	—	—	—	—	—	—	Magnets for cycle-dynamos
Columax (GB)	Similar to Alcomax III or IV	—	—	—	—	—	a, sc	Permanent magnets, heat treatable
Hycomax (GB)	Al 9; Ni 21; Co 20; Cu 1.6	—	—	—	—	—	a	Permanent magnets
Alnico (high H_c) (GB)	Al 10; Ni 20; Co 13.5; Cu 6; Ti 0.25	7.3	—	—	—	—	i	
Alnico (high B_r) (GB)	Al 10; Ni 17; Co 12; Cu 6	7.3	—	—	—	—	i	
Alni (high H_c) (GB)	Al 12; Ni 32; Ti 0–0.5	6.9	—	—	—	—	i	

Properties of Magnetic Alloys: Cast Permanent Magnetic Alloys (Continued)

Alloy Name (country of manufacture[a])	Composition[b] Weight Percent	Sp.gr., g/cm³	Thermal Expansion		Tensile Strength	Form	Remark[d]	Use
			$Cm \times 10^{-6}$ $cm \times °C$	Between °C	kg/ mm²[c]			
Alni (high B_r) (GB)	Al 13; Ni 24; Cu 3.5	—	—	—	—	—	i	
Alnico 580 (Ger)	Al 9; Ni 15; Co 23; Cu 4	—	—	—	—	—	i	
Alnico 400 (Ger)	Al 9; Ni 15; Co 23; Cu 4	—	—	—	—	—	a	
Oerstit 800 (Ger)	Al 9; Ni 18; Co 19; Cu 4; Ti 4	—	—	—	—	—	i	Permanent magnets
Alnico 250 (Ger)	Al 8; Ni 19; Co 23; Cu 4; Ti 6	—	—	—	—	—	i	
Alnico 190 (Ger)	Al 12; Ni 21; Cu 4; Co 15	—	—	—	—	—	i	
Alnico 160 (Austria)	Al 11; Ni 24; Co 12; Cu 4	—	—	—	—	—	i	Permanent magnets, sintered
Alnico 130 (Ger)	Al 12; Ni 23; Co 5	—	—	—	—	—	i	
Alni 120 (Ger)	Al 13; Ni 27	—	—	—	—	—	i	
Alni 90 (Ger)	Al 12; Ni 21	—	—	—	—	—	i	

[a]US, United States; GB, Great Britain; Ger, Germany.
[b]Iron is the additional alloying metal for each of the magnets listed.
[c] kg/mm² × 1422.33 = lb/in².
[d]i. = isotropic, a. = anisotropic, c. = columnar, sc. = semicolumnar.

Properties of Antiferromagnetic Compounds

Compound	Crystal Symmetry	θ_N[a] K	θ_P[b] K	$(P_A)_{eff}$[c] μ_B	P_A[d] μ_B
$CoCl_2$	Rhombohedral	25	−38.1	5.18	3.1 ± 0.6
CoF_2	Tetragonal	38	50	5.15	3.0
CoO	Tetragonal	291	330	5.1	3.8
Cr	Cubic	475			
Cr_2O_3	Rhombohedral	307	485	3.73	3.0
$CrSb$	Hexagonal	723	550	4.92	2.7
$CuBr_2$	Monoclinic	189	246	1.9	
$CuCl_2 \cdot 2H_2O$	Orthorhombic	4.3	4–5	1.9	
$CuCl_2$	Monoclinic	~70	109	2.08	

(*Continued*)

Properties of Antiferromagnetic Compounds (Continued)

Compound	Crystal Symmetry	θ_N[a] K	θ_P[b] K	$(P_A)^c_{eff}$ μ_B	P_A[d] μ_B
$FeCl_2$	Hexagonal	24	−48	5.38	4.4 ± 0.7
FeF_2	Tetragonal	79–90	117	5.56	4.64
FeO	Rhombohedral	198	507	7.06	3.32
α-Fe_2O_3	Rhombohedral	953	2940	6.4	5.0
α-Mn	Cubic	95			
$MnBr_2 \cdot 4H_2O$	Monoclinic	2.1	$\begin{Bmatrix} 2.5 \\ 1.3 \end{Bmatrix}$	5.93	
$MnCl_2 \cdot 4H_2O$	Monoclinic	1.66	1.8	5.94	
MnF_2	Tetragonal	72–75	113.2	5.71	5
MnO	Rhombohedral	122	610	5.95	5.0
β-MnS	Cubic	160	982	5.82	5.0
MnSe	Cubic	~173	361	5.67	
MnTe	Hexagonal	310–323	690	6.07	5.0
$NiCl_2$	Hexagonal	50	−68	3.32	
NiF_2	Tetragonal	78.5–83	115.6	3.5	2.0
NiO	Rhombohedral	533–650	~2000	4.6	2.0
$TiCl_3$		100			
V_2O_3		170			

[a] θ_N = Néel temperature, determined from susceptibility maxima or from the disappearance of magnetic scattering.

[b] θ_P = a constant in the Curie-Weiss law written in the form $\chi_A = C_A/(T + \theta_P)$, which is valid for antiferromagnetic material for $T > \theta_N$.

[c] $(P_A)_{eff}$ = effective moment per atom, derived from the atomic Curie constant $C_A = (P_A)^2_{eff}(N^2/3R)$ and expressed in units of the Bohr magneton, $\mu_B = 0.9273 \times 10^{-20}$ erg G^{-1}.

[d] P_A = magnetic moment per atom, obtained from neutron diffraction measurements in the ordered state.

Properties of Magnetic Alloys: Saturation Constants and Curie Points of Ferromagnetic Elements

Element	σ_s[a] (20°C)	M_s[b] (20°C)	σ_s (0 K)	n_B[c]	Curie point, °C
Fe	218.0	1,714	221.9	2.219	770
Co	161	1,422	162.5	1.715	1,131
Ni	54.39	484.1	57.50	0.604	358
Gd	0	0	253.5	7.12	16

[a] σ_s = saturation magnetic moment/gram.

[b] M_s = saturation magnetic moment/cm^3, in cgs units.

[c] n_B = magnetic moment per atom in Bohr magnetons.

(*Source:* 1963. *American Institute of Physics Handbook.* McGraw-Hill, New York.)

Magnetic Properties of Transformer Steels

Ordinary Transformer Steel

B(Gauss)	H(Oersted)	Permeability $= B/H$
2,000	0.60	3,340
4,000	0.87	4,600
6,000	1.10	5,450
8,000	1.48	5,400
10,000	2.28	4,380
12,000	3.85	3,120
14,000	10.9	1,280
16,000	43.0	372
18,000	149	121

High Silicon Transformer Steels

B	H	Permeability
2,000	0.50	4,000
4,000	0.70	5,720
6,000	0.90	6,670
8,000	1.28	6,250
10,000	1.99	5,020
12,000	3.60	3,340
14,000	9.80	1,430
16,000	47.4	338
18,000	165	109

Initial Permeability of High Purity Iron for Various Temperatures

L. Alberts and B.J. Shepstone

Temperature °C	Permeability (Gauss/oersted)
0	920
200	1,040
400	1,440
600	2,550
700	3,900
770	12,580

Saturation Constants for Magnetic Substances

Substance	Field Intensity	Induced Magnetization	Substance	Field Intensity	Induced Magnetization
	(For Saturation)			(For Saturation)	
Cobalt	9,000	1,300	Nickel, hard	8,000	400
Iron, wrought	2,000	1,700	annealed	7,000	515
cast	4,000	1,200	Vicker's steel	15,000	1,600
Manganese steel	7,000	200			

Magnetic Materials: High-Permeability Materials

Material	Form	Approximate % Composition					Typical Heat Treatment, °C	Permeability at $B = 20$, G	Maximum Permeability	Saturation flux density B, G	Hysteresis[a] loss, W_h, ergs/cm²	Coercive[a] Force H_a, O	Resistivity $\mu \cdot \Omega$cm	Density, g/cm³
		Fe	Ni	Co	Mo	Other								
Cold rolled steel	Sheet	98.5	—	—	—	—	950 Anneal	180	2,000	21,000	—	1.8	10	7.88
Iron	Sheet	99.91	—	—	—	—	950 Anneal	200	5,000	21,500	5,000	1.0	10	7.88
Purified iron	Sheet	99.95	—	—	—	—	1480 H$_2$ + 880	5,000	180,000	21,500	300	0.05	10	7.88
4% Silicon-iron	Sheet	96	—	—	—	4 Si	800 Anneal	500	7,000	19,700	3,500	0.5	60	7.65
Grain oriented[b]	Sheet	97	—	—	—	3 Si	800 Anneal	1,500	30,000	20,000	—	0.15	47	7.67
45 Permalloy	Sheet	54.7	45	—	—	0.3 Mn	1050 Anneal	2,500	25,000	16,000	1,200	0.3	45	8.17
45 permalloy[c]	Sheet	54.7	45	—	—	0.3 Mn	1200 H$_2$ Anneal	4,000	50,000	16,000	—	0.07	45	8.17
Hipernik	Sheet	50	50	—	—	—	1200 H$_2$ Anneal	4,500	70,000	16,000	220	0.05	50	8.25
Monimax	Sheet	—	50	—	—	—	1125 H$_2$ Anneal	2,000	35,000	15,000	—	0.1	80	8.27
Sinimax	Sheet	—	—	—	—	—	1125 H$_2$ Anneal	3,000	35,000	11,000	—	—	90	—
78 Permalloy	Sheet	21.2	78.5	—	—	0.3 Mn	1050 + 600 Q[d]	8,000	100,000	10,700	200	0.05	16	8.60
4–79 Permalloy	Sheet	16.7	79	—	4	0.3 Mn	1100 + Q	20,000	100,000	8,700	200	0.05	55	8.72
Mu metal	Sheet	18	75	—	—	2 Cr, 5 Cu	1175 H$_2$	20,000	100,000	6,500	—	0.05	62	8.58
Supermalloy	Sheet	15.7	79	—	5	0.3 Mn	1300 H$_2$+ Q	100,000	800,000	8,000	—	0.002	60	8.77
Permendur	Sheet	49.7	—	50	—	—	800 Anneal	800	5,000	24,500	12,000	2.0	7	8.3
2 V Permendur	Sheet	49	—	49	—	2 V	800 Anneal	800	4,500	24,000	6,000	2.0	26	8.2
Hiperco	Sheet	64	—	34	—	Cr	850 Anneal	650	10,000	24,200	—	1.0	25	8.0
2-81 Permalloy	Insulated powder	17	81	—	2	—	650 Anneal	125	130	8,000	—	<1.0	10^6	7.8
Carbonyl iron	Insulated powder	99.9	—	—	—	—	—	55	132	—	—	—	—	7.86
Ferroxcube III	Sintered powder	MnFe$_2$O$_4$ + ZnFe$_2$O$_4$					—	1,000	1,500	2,500	—	0.1	10^8	5.0

[a] At saturation.
[b] Properties in direction of rolling.
[c] Similar properties for Nicaloi, 4750 alloy, Carpenter 49, Armco 48.
[d] Q, quench or controlled cooling.

Magnetic Materials: Permanent Magnet Alloys

Material	% Composition (remainder Fe)	Heat Treatment[a] (temperature, °C)	Magnetizing Force H_{max}, O	Coercive Force H_c, O	Residual Induction B_r, G	Energy Product $BH_{max} \times 10^{-6}$	Method of Fabrication[b]	Mechanical Properties[c]	Weight lb/In.³
Carbon steel	1 Mn, 0.9 C	Q 800	300	50	10,000	0.20	HR, M, P	H, S	0.280
Tungsten steel	5 W, 0.3 Mn, 0.7 C	Q 850	300	70	10,300	0.32	HR, M, P	H, S	0.292
Chromium steel	3.5 Cr, 0.9 C, 0.3 Mn	Q 830	300	65	9,700	0.30	HR, M, P	H, S	0.280
17% Cobalt steel	17 Co, 0.75 C, 2.5 Cr, 8 W	–	1,000	150	9,500	0.65	HR, M, P	H, S	–
36% Cobalt steel	36 Co, 0.7 C, 4 Cr, 5 W	Q 950	1,000	240	9,500	0.97	HR, M, P	H, S	0.296
Remalloy or Comol	17 Mo, 12 Co	Q 1200, B 700	1,000	250	10,500	1.1	HR, M, P	H	0.295
Alnico I	12 Al, 20 Ni, 5 Co	A 1200, B 700	2,000	440	7,200	1.4	C, G	H, B	0.249
Alnico II	10 Al, 17 Ni, 2.5 Co, 6 Cu	A 1200, B 600	2,000	550	7,200	1.6	C, G	H, B	0.256
Alnico II (sintered)	10 Al, 17 Ni, 2.5 Co, 6 Cu	A 1300	2,000	520	6,900	1.4	Sn, G	H	0.249
Alnico IV	12 Al, 28 Ni, 5 Co	Q 1200, B 650	3,000	700	5,500	1.3	Sn, C, G	H	0.253
Alnico V	8 Al, 14 Ni, 24 Co, 3 Cu	AF 1300, B 600	2,000	550	12,500	4.5	C, G	H, B	0.264
Alnico VI	8 Al, 15 Ni, 24 Co, 3 Cu, 1 Ti	–	3,000	750	10,000	3.5	C, G	H, B	0.268
Alnico XII	6 Al, 18 Ni, 35 Co, 8 Ti	–	3,000	950	5,800	1.5	C, G	H, B	0.26
Vicalloy I	52 Co, 10 V	B 600	1,000	300	8,800	1.0	C, CR, M, P	D	0.295
Vicalloy II (wire)	52 Co, 14 V	CW + B 600	2,000	510	10,000	3.5	C, CR, M, P	D	0.292
Cunife (wire)	60 Cu, 20 Ni	CW + B 600	2,400	550	5,400	1.5	C, CR, M, P	D, M	0.311
Cunico	50 Cu, 21 Ni, 29 Co	–	3,200	660	3,400	0.80	C, CR, M, P	D, M	0.300
Vectolite	30Fe$_2$O$_3$, 44Fe$_3$O$_4$, 26Co$_2$O$_3$	–	3,000	1,000	1,600	0.60	Sn, G	W	0.113
Silmanal	86.8Ag, 8.8Mn, 4.4Al	–	20,000	6,000[d]	550	0.075	C, CR, M, P	D, M	0.325
Platinum-cobalt	77 Pt, 23 Co	Q 1200, B 650	15,000	3,600	5,900	6.5	C, CR, M	D	0.325
Hyflux	Fine powder	–	2,000	390	6,600	0.97	–	–	0.176

[a] Q, quenched in oil or water; A, air cooled; B, baked; F, cooled in magnetic field; CW, cold worked.
[b] HR, hot rolled or froged; CR, cold rolled or drawn; M, machined; G, must be ground; P, punched; C, cast; Sn, sintered.
[c] H, hard; B, brittle; S, strong; D, ductile; M, malleable; W, weak.
[d] Value given is intrinsic H_c.

Resistance of Wires

The following table gives the approximate resistance of various metallic conductors. The values have been computed from the resistivities at 20°C, except as otherwise stated, and for the dimensions of wire indicated. Owing to differences in purity in the case of elements and of composition in alloys, the values can be considered only as approximations.

The following dimensions have been adopted in the computation.

B. & S. gauge	mm	Diameter mils (1 mil = 0.001 in)	B. & S. gauge	mm	Diameter mils (1 mil = 0.001 in)
10	2.588	101.9	26	0.4049	15.94
12	2.053	80.81	27	0.3606	14.20
14	1.628	64.08	28	0.3211	12.64
16	1.291	50.82	30	0.2546	10.03
18	1.024	40.30	32	0.2019	7.950
20	0.8118	31.96	34	0.1601	6.305
22	0.6438	25.35	36	0.1270	5.000
24	0.5106	20.10	40	0.07987	3.145

Advance[a] $(0\,°C)\ \varrho = 48. \times 10^{-6}$ $\Omega\cdot cm$

B. & S. No.	Ω/cm	Ω/ft
10	0.000912	0.0278
12	0.00145	0.0442
14	0.00231	0.0703
16	0.00367	0.112
18	0.00583	0.178
20	0.00927	0.283
22	0.0147	0.449
24	0.0234	0.715
26	0.0373	1.14
27	0.0470	1.43
28	0.0593	1.81
30	0.0942	2.87
32	0.150	4.57
34	0.238	7.26
36	0.379	11.5
40	0.958	29.2

Aluminum $\varrho = 2.828 \times 10^{-6}$ $\Omega\cdot cm$

B. & S. No.	Ω/cm	Ω/ft
10	0.0000538	0.00164
12	0.0000855	0.00260
14	0.000136	0.00414
16	0.000216	0.00658
18	0.000344	0.0105
20	0.000546	0.0167
22	0.000869	0.0265
24	0.00138	0.0421
26	0.00220	0.0669
27	0.00277	0.0844
28	0.00349	0.106
30	0.00555	0.169
32	0.00883	0.269
34	0.0140	0.428
36	0.0223	0.680
40	0.0564	1.72

Brass $\varrho = 7.00 \times 10^{-6}$ $\Omega\cdot cm$

B. & S. No.	Ω/cm	Ω/ft
10	0.000133	0.00406
12	0.000212	0.00645
14	0.000336	0.0103
16	0.000535	0.0163
18	0.000850	0.0259
20	0.00135	0.0412
22	0.00215	0.655
24	0.00342	0.104
26	0.00543	0.166
27	0.00686	0.209
28	0.00864	0.263
30	0.0137	0.419
32	0.0219	0.666
34	0.0348	1.06
36	0.0552	1.68
40	0.140	4.26

Climax $\varrho = 87 \times 10^{-6}$ $\Omega\cdot cm$

No.	Ω/cm	Ω/ft
10	0.00165	0.0504
12	0.00263	0.0801
14	0.00418	0.127
16	0.00665	0.203
18	0.0106	0.322
20	0.0168	0.512
22	0.0267	0.815
24	0.0425	1.30
26	0.0675	2.06
27	0.0852	2.60
28	0.107	3.27
30	0.171	5.21
32	0.272	8.28
34	0.432	13.2
36	0.687	20.9
40	1.74	52.9

Constantan $(0\,°C)\ \varrho = 44.1 \times 10^{-6}$ $\Omega\cdot cm$

No.	Ω/cm	Ω/ft
10	0.000838	0.0255
12	0.00133	0.0406
14	0.00212	0.0646
16	0.00337	0.103
18	0.00536	0.163
20	0.00852	0.260
22	0.0135	0.413
24	0.0215	0.657
26	0.0342	1.04
27	0.0432	1.32
28	0.0545	1.66
30	0.0866	2.64
32	0.138	4.20
34	0.219	6.67
36	0.348	10.6
40	0.880	26.8

Copper, annealed $\varrho = 1.724 \times 10^{-6}$ $\Omega\cdot cm$

No.	Ω/cm	Ω/ft
10	0.0000328	0.000999
12	0.0000521	0.00159
14	0.0000828	0.00253
16	0.000132	0.00401
18	0.000209	0.00638
20	0.000333	0.0102
22	0.000530	0.0161
24	0.000842	0.0257
26	0.00134	0.0408
27	0.00169	0.0515
28	0.00213	0.0649
30	0.00339	0.103
32	0.00538	0.164
34	0.00856	0.261
36	0.0136	0.415
40	0.0344	1.05

[a]Trademark.

Resistance of Wires (Continued)

Eureka[a] (0°C) ρ = 47. × 10⁻⁶ Ω·cm

B. & S.

No.	Ω/cm	Ω/ft
10	0.000893	0.0272
12	0.00142	0.0433
14	0.00226	0.0688
16	0.00359	0.109
18	0.00571	0.174
20	0.00908	0.277
22	0.0144	0.440
24	0.0230	0.700
26	0.0365	1.11
27	0.0460	1.40
28	0.0580	1.77
30	0.0923	2.81
32	0.147	4.47
34	0.233	7.11
36	0.371	11.3
40	0.938	28.6

Excello ρ = 92. × 10⁻⁶ Ω·cm

B. & S.

No.	Ω/cm	Ω/ft
10	0.00175	0.0533
12	0.00278	0.0847
14	0.00442	0.135
16	0.00703	0.214
18	0.0112	0.341
20	0.0178	0.542
22	0.0283	0.861
24	0.0449	1.37
26	0.0714	2.18
27	0.0901	2.75
28	0.114	3.46
30	0.181	5.51
32	0.287	8.75
34	0.457	13.9
36	0.726	22.1
40	1.84	56.0

German silver ρ = 33 × 10⁻⁶ Ω·cm

B. & S.

No.	Ω/cm	Ω/ft
10	0.000627	0.0191
12	0.000997	0.304
14	0.00159	0.0483
16	0.00252	0.0768
18	0.00401	0.122
20	0.00638	0.194
22	0.0101	0.309
24	0.0161	0.491
26	0.0256	0.781
27	0.0323	0.985
28	0.0408	1.24
30	0.0648	1.97
32	0.103	3.14
34	0.164	4.99
36	0.260	0.794
40	0.659	20.1

Gold ρ = 2.44 × 10⁻⁶ Ω·cm

No.	Ω/cm	Ω/ft
10	0.0000464	0.00141
12	0.0000737	0.00225
14	0.000117	0.00357
16	0.000186	0.00568
18	0.000296	0.00904
20	0.000471	0.0144
22	0.000750	0.0228
24	0.00119	0.0363
26	0.00189	0.0577
27	0.00239	0.728
28	0.00301	0.0918
30	0.00479	0.146
32	0.00762	0.232
34	0.0121	0.369
36	0.0193	0.587
40	0.0487	1.48

Iron ρ = 10 × 10⁻⁶ Ω·cm

No.	Ω/cm	Ω/ft
10	0.000190	0.00579
12	0.000302	0.00921
14	0.000481	0.0146
16	0.000764	0.0233
18	0.00121	0.0370
20	0.00193	0.0589
22	0.00307	0.0936
24	0.00489	0.149
26	0.00776	0.237
27	0.00979	0.299
28	0.0123	0.376
30	0.0196	0.598
32	0.0312	0.952
34	0.0497	1.51
36	0.789	2.41
40	0.200	6.08

Lead ρ = 22 × 10⁻⁶ Ω·cm

No.	Ω/cm	Ω/ft
10	0.000418	0.0127
12	0.000665	0.0203
14	0.00106	0.0322
16	0.00168	0.0512
18	0.00267	0.0815
20	0.00425	0.130
22	0.00676	0.206
24	0.0107	0.328
26	0.0171	0.521
27	0.0215	0.657
28	0.0272	0.828
30	0.0432	1.32
32	0.0687	2.09
34	0.109	3.33
36	0.174	5.29
40	0.439	13.4

Magnesium ρ = 4.6 × 10⁻⁶ Ω·cm

No.	Ω/cm	Ω/ft
10	0.0000874	0.00267
12	0.000139	0.00424
14	0.000221	0.00674
16	0.000351	0.0107
18	0.000559	0.0170
20	0.000889	0.0271
22	0.00141	0.0431
24	0.00225	0.0685
26	0.00357	0.109
27	0.00451	0.137
28	0.00568	0.173
30	0.00903	0.275
32	0.0144	0.438
34	0.0228	0.696
36	0.0363	1.11
40	0.0918	2.80

Manganin ρ = 44 × 10⁻⁶ Ω·cm

No.	Ω/cm	Ω/ft
10	0.000836	0.0255
12	0.00133	0.0405
14	0.00211	0.0644
16	0.00336	0.102
18	0.00535	0.163
20	0.00850	0.259
22	0.0135	0.412
24	0.0215	0.655
26	0.0342	1.04
27	0.0431	1.31
28	0.0543	1.66
30	0.0864	2.63
32	0.137	4.19
34	0.218	6.66
36	0.347	10.6
40	0.878	26.8

Molybdenum ρ = 5.7 × 10⁻⁶ Ω·cm

No.	Ω/cm	Ω/ft
10	0.000108	0.00330
12	0.000172	0.00525
14	0.000274	0.00835
16	0.000435	0.0133
18	0.000693	0.0211
20	0.00110	0.0336
22	0.00175	0.0534
24	0.00278	0.0849
26	0.00443	0.135
27	0.00558	0.170
28	0.00704	0.215
30	0.0112	0.341
32	0.0178	0.542
34	0.0283	0.863
36	0.0450	1.37
40	0.114	3.47

[a]Trademark.

Resistance of Wires (Continued)

Monel Metal $\varrho = 42 \times 10^{-6}$ $\Omega \cdot$ cm

B. & S. No.	Ω/cm	Ω/ft
10	0.000798	0.0243
12	0.00127	0.0387
14	0.00202	0.0615
16	0.00321	0.0978
18	0.00510	0.156
20	0.00811	0.247
22	0.0129	0.393
24	0.0205	0.625
26	0.0326	0.994
27	0.0411	1.25
28	0.0519	1.58
30	0.0825	2.51
32	0.131	4.00
34	0.209	6.36
36	0.331	10.1
40	0.838	25.6

ᵃNichrome $\varrho = 150 \times 10^{-6}$ $\Omega \cdot$ cm

B. & S. No.	Ω/cm	Ω/ft
10	0.0021281	0.06488
12	0.0033751	0.1029
14	0.0054054	0.1648
16	0.0085116	0.2595
18	0.0138383	0.4219
20	0.0216218	0.6592
22	0.0346040	1.055
24	0.0548088	1.671
26	0.0875760	2.670
28	0.1394328	4.251
30	0.2214000	6.750
32	0.346040	10.55
34	0.557600	17.00
36	0.885600	27.00
38	1.383832	42.19
40	2.303872	70.24

Nickel $\varrho = 7.8 \times 10^{-6}$ $\Omega \cdot$ cm

B. & S. No.	Ω/cm	Ω/ft
10	0.000148	0.00452
12	0.000236	0.00718
14	0.000375	0.0114
16	0.000596	0.0182
18	0.000948	0.0289
20	0.00151	0.0459
22	0.00240	0.0730
24	0.00381	0.116
26	0.00606	0.185
27	0.00764	0.233
28	0.00963	0.294
30	0.0153	0.467
32	0.0244	0.742
34	0.0387	1.18
36	0.0616	1.88
40	0.156	4.75

Platinum $\varrho = 10 \times 10^{-6}$ $\Omega \cdot$ cm

10	0.000190	0.00579
12	0.000302	0.00921
14	0.000481	0.0146
16	0.000764	0.0233
18	0.00121	0.0370
20	0.00193	0.0589
22	0.00307	0.0936
24	0.00489	0.149
26	0.00776	0.237
27	0.00979	0.299
28	0.0123	0.376
30	0.0196	0.598
32	0.0312	0.952
34	0.0497	1.51
36	0.0789	2.41
40	0.200	6.08

Silver (18°C) $\varrho = 1.629 \times 10^{-6}$ $\Omega \cdot$ cm

10	0.0000310	0.000944
12	0.0000492	0.00150
14	0.0000783	0.00239
16	0.000124	0.00379
18	0.000198	0.00603
20	0.000315	0.00959
22	0.000500	0.0153
24	0.000796	0.0243
26	0.00126	0.0386
27	0.00160	0.0486
28	0.00201	0.0613
30	0.00320	0.0975
32	0.00509	0.155
34	0.00809	0.247
36	0.0129	0.392
40	0.325	0.991

Steel, piano wire (0°C) $\varrho = 11.8 \times 10^{-6}$ $\Omega \cdot$ cm

10	0.000224	0.00684
12	0.000357	0.0109
14	0.000567	0.0173
16	0.000901	0.0275
18	0.00143	0.0437
20	0.00228	0.0695
22	0.00363	0.110
24	0.00576	0.176
26	0.00916	0.279
27	0.0116	0.352
28	0.0146	0.444
30	0.0232	0.706
32	0.0368	1.12
34	0.0586	1.79
36	0.0931	2.84
40	0.236	7.18

Steel, Invar (35% Ni) $\varrho = 81 \times 10^{-6}$ $\Omega \cdot$ cm

10	0.00154	0.0469
12	0.00245	0.0746
14	0.00389	0.119
16	0.00619	0.189
18	0.00984	0.300
20	0.0156	0.477
22	0.0249	0.758
24	0.0396	1.21
26	0.0629	1.92
27	0.0793	2.42
28	0.100	3.05
30	0.159	4.85
32	0.253	7.71
34	0.402	12.3
36	0.639	19.5
40	1.62	49.3

Tantalum $\varrho = 15.5 \times 10^{-6}$ $\Omega \cdot$ cm

10	0.000295	0.00898
12	0.000468	0.0143
14	0.000745	0.0227
16	0.00118	0.0361
18	0.00188	0.0574
20	0.00299	0.0913
22	0.00476	0.145
24	0.00757	0.231
26	0.0120	0.367
27	0.0152	0.463
28	0.0191	0.583
30	0.0304	0.928
32	0.0484	1.47
34	0.0770	2.35
36	0.122	3.73
40	0.309	9.43

Tin $\varrho = 11.5 \times 10^{-6}$ $\Omega \cdot$ cm

10	0.000219	0.00666
12	0.000348	0.0106
14	0.000553	0.0168
16	0.000879	0.0268
18	0.00140	0.0426
20	0.00222	0.0677
22	0.00353	0.108
24	0.00562	0.171
26	0.00893	0.272
27	0.0113	0.343
28	0.0142	0.433
30	0.0226	0.688
32	0.0359	1.09
34	0.0571	1.74
36	0.0908	2.77
40	0.230	7.00

ᵃTrademark.

Resistance of Wires (Continued)

B & S. No.	Ω/cm	Ω/ft	B & S. No.	Ω/cm	Ω/ft
Tungsten $\varrho = 5.51 \times 10^{-6}\,\Omega \cdot cm$			Zinc (0°C) $\varrho = 5.75 \times 10^{-6}\,\Omega \cdot cm$		
10	0.000105	0.00319	10	0.000109	0.00333
12	0.000167	0.00508	12	0.000174	0.00530
14	0.000265	0.00807	14	0.000276	0.00842
16	0.000421	0.0128	16	0.000439	0.0134
18	0.000669	0.0204	18	0.000699	0.0213
20	0.00106	0.0324	20	0.00111	0.0339
22	0.00169	0.0516	22	0.00177	0.0538
24	0.00269	0.0820	24	0.00281	0.0856
26	0.00428	0.130	26	0.00446	0.136
27	0.00540	0.164	27	0.00563	0.172
28	0.00680	0.207	28	0.00710	0.216
30	0.0108	0.330	30	0.0113	0.344
32	0.0172	0.524	32	0.0180	0.547
34	0.0274	0.834	34	0.0286	0.870
36	0.0435	1.33	36	0.0454	1.38
40	0.110	3.35	40	0.115	3.50

Defining Terms

Alloy: Metal composed of more than one element.

Bravais lattice: One of the 14 possible arrangements of points in three-dimensional space.

Crystalline: Having constituent atoms stacked together in a regular, repeating pattern.

Density: Mass per unit volume.

Electromotive force series: Systematic listing of half-cell reaction voltages.

Engineering strain: Increase in sample length at a given load divided by the original (stress-free) length.

Engineering stress: Load on a sample divided by the original (stress-free) area.

Fourier's law: Relationship between rate of heat transfer and temperature gradient.

Index of refraction: Ratio of speed of light in vacuum to that in a transparent material.

Linear coefficient of thermal expansion: Material parameter indicating dimensional change as a function of increasing temperature.

Melting point: Temperature of transformation from solid to liquid upon heating.

Ohm's law: Relationship between voltage, current, and resistance in an electrical circuit.

Property: Observable characteristic of a material.

Resistivity: Electrical resistance normalized for sample geometry.

Tensile strength: Maximum engineering stress during a tensile test.

Thermal conductivity: Proportionality constant in Fourier's law.

Young's modulus (modulus of elasticity): Ratio of engineering stress to engineering strain for relatively small levels of load application.

References

ASM. 1984. *ASM Metals Reference Book,* 2nd ed. American Society for Metals, Metals Park, OH.

ASM. 1988. *Engineered Materials Handbook,* Vol. 2, *Engineering Plastics.* ASM International, Metals Park, OH.

Bolz, R.E. and Tuve, G.L., eds. 1973. *CRC Handbook of Tables for Applied Engineering Science.* CRC Press, Boca Raton, FL.

Lynch, C.T., ed. 1975. *CRC Handbook of Materials Science,* Vol. 3. CRC Press, Boca Raton, FL.

Shackelford, J.F. 1996. *Introduction to Materials Science for Engineers,* 4th ed. Prentice-Hall, Upper Saddle River, NJ.

Shackelford, J.F., Alexander, W., and Park, J.S., eds. 1994. *The CRC Materials Science and Engineering Handbook*, 2nd ed. CRC Press, Boca Raton, FL.

Further Information

A general introduction to the field of materials science and engineering is available from a variety of introductory textbooks. In addition to Shackelford (1996), readily available references include:

Askeland, D.R. 1994. *The Science and Engineering of Materials*, 3rd ed. PWS–Kent, Boston.
Callister, W.D. 1994. *Materials Science and Engineering-An Introduction*, 3rd ed. Wiley, New York.
Flinn, R.A. and Trojan, P.K. 1990. *Engineering Materials and Their Applications*, 4th ed. Houghton Mifflin, Boston.
Smith, W.F. 1990. *Principles of Materials Science and Engineering*, 2nd ed. McGraw-Hill New York.
Van Vlack, L.H. 1989. *Elements of Materials Science and Engineering*, 6th ed. Addison-Wesley Reading, MA.

2.8   International Standards and Constants[1]

2.8.1   International System of Units (SI)

The International System of units (SI) was adopted by the 11th General Conference on Weights and Measures (CGPM) in 1960. It is a coherent system of units built from seven *SI base units*, one for each of the seven dimensionally independent base quantities: they are the meter, kilogram, second, ampere, kelvin, mole, and candela, for the dimensions length, mass, time, electric current, thermodynamic temperature, amount of substance, and luminous intensity, respectively. The definitions of the SI base units are given subsequently. The *SI derived units* are expressed as products of powers of the base units, analogous to the corresponding relations between physical quantities but with numerical factors equal to unity.

In the International System there is only one SI unit for each physical quantity. This is either the appropriate SI base unit itself or the appropriate SI derived unit. However, any of the approved decimal prefixes, called *SI prefixes*, may be used to construct decimal multiples or submultiples of SI units.

It is recommended that only SI units be used in science and technology (with SI prefixes where appropriate). Where there are special reasons for making an exception to this rule, it is recommended always to define the units used in terms of SI units. This section is based on information supplied by IUPAC.

Definitions of SI Base Units

Meter: The meter is the length of path traveled by light in vacuum during a time interval of 1/299 792 458 of a second (17th CGPM, 1983).

Kilogram: The kilogram is the unit of mass; it is equal to the mass of the international prototype of the kilogram (3rd CGPM, 1901).

Second: The second is the duration of 9 192 631 770 periods of the radiation corresponding the transition between the two hyperfine levels of the ground state of the cesium-133 atom (13th CGPM, 1967).

Ampere: The ampere is that constant current which, if maintained in two straight parallel conductors of infinite length, of negligible circular cross section, and placed 1 m apart in vacuum, would produce between these conductors a force equal to 2×10^{-7} newton per meter of length (9th CGPM, 1948).

Kelvin: The kelvin, unit of thermodynamic temperature, is the fraction 1/273.16 of the thermodynamic temperature of the triple point of water (13th CGPM, 1967).

Mole: The mole is the amount of substance of a system that contains as many elementary entities as there are atoms in 0.012 kilogram of carbon-12. When the mole is used, the elementary entities must

---

[1]Material herein was reprinted from the following sources:
    Lide, D.R., ed. 1992. *CRC Handbook of Chemistry and Physics*, 76th ed. CRC Press, Boca Raton, FL: International System of Units (SI), symbols and terminology for physical and chemical quantities, classification of electromagnetic radiation.
    Zwillinger, D., ed. 1996. *CRC Standard Mathematical Tables and Formulae*, 30th ed. CRC Press, Boca Raton, FL: Greek alphabet, physical constants.

be specified and may be atoms, molecules, ions, electrons, or other particles, or specified groups of such particles (14th CGPM, 1971).

Examples of the use of the mole:

1 mol of H_2 contains about 6.022×10^{23} H_2 molecules, or 12.044×10^{23} H atoms

1 mol of HgCl has a mass of 236.04 g

1 mol of Hg_2Cl_2 has a mass of 472.08 g

1 mol of Hg_2^{2+} has a mass of 401.18 g and a charge of 192.97 kC

1 mol of $Fe_{0.91}S$ has a mass of 82.88 g

1 mol of e^- has a mass of 548.60 μg and a charge of -96.49 kC

1 mol of photons whose frequency is 10^{14} Hz has energy of about 39.90 kJ

Candela: The candela is the luminous intensity, in a given direction, of a source that emits monochromatic radiation of frequency 540×10^{12} hertz and that has a radiant intensity in that direction of $(1/683)$ watt per steradian (16th CGPM, 1979).

Names and Symbols for the SI Base Units

Physical Quantity	Name of SI Unit	Symbol for SI Unit
Length	meter	m
Mass	kilogram	kg
Time	second	s
Electric current	ampere	A
Thermodynamic temperature	kelvin	K
Amount of substance	mole	mol
Luminous intensity	candela	cd

SI Derived Units with Special Names and Symbols

Physical Quantity	Name of SI Unit	Symbol for SI Unit	Expression in Terms of SI Base Units
Frequency[a]	hertz	Hz	s^{-1}
Force	newton	N	$m\,kg\,s^{-2}$
Pressure, stress	pascal	Pa	$N\,m^{-2} = m^{-1}\,kg\,s^{-2}$
Energy, work, heat	joule	J	$N\,m = m^2\,kg\,s^{-2}$
Power, radiant flux	watt	W	$J\,s^{-1} = m^2\,kg\,s^{-3}$
Electric charge	coulomb	C	$A\,s$
Electric potential, electromotive force	volt	V	$J\,C^{-1} = m^2\,kg\,s^{-3}\,A^{-1}$
Electric resistance	ohm	Ω	$V\,A^{-1} = m^2\,kg\,s^{-3}\,A^{-2}$
Electric conductance	siemens	S	$\Omega^{-1} = m^{-2}\,kg^{-1}\,s^3\,A^2$
Electric capacitance	farad	F	$C\,V^{-1} = m^{-2}\,kg^{-1}\,s^4\,A^2$
Magnetic flux density	tesla	T	$V\,s\,m^{-2} = kg\,s^{-2}\,A^{-1}$
Magnetic flux	weber	Wb	$V\,s = m^2\,kg\,s^{-2}\,A^{-1}$
Inductance	henry	H	$V\,A^{-1}\,s = m^2\,kg\,s^{-2}\,A^{-2}$
Celsius temperature[b]	degree Celsius	°C	K
Luminous flux	lumen	lm	cd sr
Illuminance	lux	lx	$cd\,sr\,m^{-2}$
Activity (radioactive)	becquerel	Bq	s^{-1}
Absorbed dose (of radiation)	gray	Gy	$J\,kg^{-1} = m^2 s^{-2}$
Dose equivalent (dose equivalent index)	sievert	Sv	$J\,kg^{-1} = m^2\,s^{-2}$
Plane angle	radian	rad	$1 = m\,m^{-1}$
Solid angle	steradian	sr	$1 = m^2\,m^{-2}$

[a] For radial (circular) frequency and for angular velocity the unit $rad\,s^{-1}$, or simply s^{-1}, should be used, and this may not be simplified to Hz. The unit Hz should be used only for frequency in the sense of cycles per second.

[b] The Celsius temperature θ is defined by the equation:

$$\theta/°C = T/K - 273.15$$

The SI unit of Celsius temperature interval is the degree Celsius, °C, which is equal to the kelvin, K. °C should be treated as a single symbol, with no space between the ° sign and the letter C. (The symbol °K and the symbol °, should no longer be used.)

Units in Use Together with the SI

These units are not part of the SI, but it is recognized that they will continue to be used in appropriate contexts. SI prefixes may be attached to some of these units, such as milliliter, ml; millibar, mbar; megaelectronvolt, MeV; kilotonne, ktonne.

Physical Quantity	Name of Unit	Symbol for Unit	Value in SI Units
Time	minute	min	60 s
Time	hour	h	3600 s
Time	day	d	86 400 s
Plane angle	degree	°	$(\pi/180)$ rad
Plane angle	minute	′	$(\pi/10\ 800)$ rad
Plane angle	second	″	$(\pi/648\ 000)$ rad
Length	ångstrom[a]	Å	10^{-10} m
Area	barn	b	$10^{-28}\ \text{m}^2$
Volume	litre	l, L	$\text{dm}^3 = 10^{-3}\text{m}^3$
Mass	tonne	t	$\text{Mg} = 10^3$ kg
Pressure	bar[a]	bar	$10^5\ \text{Pa} = 10^5\ \text{N m}^{-2}$
Energy	electronvolt[b]	$\text{eV} (= e \times V)$	$\approx 1.60218 \times 10^{-19}$ J
Mass	unified atomic mass unit[b,c]	$u (= m_a(^{12}\text{C})/12)$	$\approx 1.66054 \times 10^{-27}$ kg

[a]The ångstrom and the bar are approved by CIPM for temporary use with SI units, until CIPM makes a further recommendation. However, they should not be introduced where they are not used at present.

[b]The values of these units in terms of the corresponding SI units are not exact, since they depend on the values of the physical constants e (for the electronvolt) and N_A (for the unified atomic mass unit), which are determined by experiment.

[c]The unified atomic mass unit is also sometimes called the dalton, with symbol Da, although the name and symbol have not been approved by CGPM.

Greek Alphabet

	Greek Letter	Greek Name	English Equivalent		Greek Letter	Greek Name	English Equivalent
A	α	Alpha	a	N	ν	Nu	n
B	β	Beta	b	Ξ	ξ	Xi	x
Γ	γ	Gamma	g	O	o	Omicron	ŏ
Δ	δ	Delta	d	Π	π	Pi	p
E	ϵ	Epsilon	ĕ	P	ρ	Rho	r
Z	ζ	Zeta	z	Σ	σ ς	Sigma	s
H	η	Eta	ē	T	τ	Tau	t
Θ	θ ϑ	Theta	th	Υ	υ	Upsilon	u
I	ι	Iota	i	Φ	ϕ φ	Phi	ph
K	κ	Kappa	k	X	χ	Chi	ch
Λ	λ	Lambda	l	Ψ	ψ	Psi	ps
M	μ	Mu	m	Ω	ω	Omega	ō

2.8.2 Physical Constants

General

Equatorial radius of the Earth = 6378.388 km = 3963.34 miles (statute).

Polar radius of the Earth, 6356.912 km = 3949.99 miles (statute).

1 degree of latitude at 40° = 69 miles.

1 international nautical mile = 1.15078 miles (statute) = 1852 m = 6076.115 ft.

Mean density of the Earth = 5.522 g/cm³ = 344.7 lb/ft³

Constant of gravitation $(6.673 \pm 0.003) \times 10^{-8}/\text{cm}^3\text{gm}^{-1}\text{s}^{-2}$.

Acceleration due to gravity at sea level, latitude 45° = 980.6194 cm/s² = 32.1726 ft/s².

Length of seconds pendulum at sea level, latitude 45° = 99.3575 cm = 39.1171 in.

1 knot (international) = 101.269 ft/min = 1.6878 ft/s = 1.1508 miles (statute)/h.

1 micron = 10^{-4} cm.

1 ångstrom = 10^{-8} cm.

Mass of hydrogen atom = $(1.67339 \pm 0.0031) \times 10^{-24}$ g.

Density of mercury at 0°C = 13.5955 g/ml.

Density of water at 3.98°C = 1.000000 g/ml.

Density, maximum, of water, at 3.98°C = 0.999973 g/cm^3.

Density of dry air at 0°C, 760 mm = 1.2929 g/1.

Velocity of sound in dry air at 0°C = 331.36 m/s -1087.1 ft/s.

Velocity of light in vacuum = $(2.997925 \pm 0.000002) \times 10^{10}$ cm/s.

Heat of fusion of water 0°C = 79.71 cal/g.

Heat of vaporization of water 100°C = 539.55 cal/g.

Electrochemical equivalent of silver 0.001118 g/s international amp.

Absolute wavelength of red cadmium light in air at 15°C, 760 mm pressure = 6438.4696 Å.

Wavelength of orange-red line of krypton 86 = 6057.802 Å.

π Constants

$\pi = 3.14159$	26535	89793	23846	26433	83279	50288	41971	69399	37511
$1/\pi = 0.31830$	98861	83790	67153	77675	26745	02872	40689	19291	48091
$\pi^2 = 9.8690$	44010	89358	61883	44909	99876	15113	53136	99407	24079
$\log_e \pi = 1.14472$	98858	49400	17414	34273	51353	05871	16472	94812	91531
$\log_{10} \pi = 0.49714$	98726	94133	85435	12682	88290	89887	36516	78324	38044
$\log_{10} \sqrt{2\pi} = 0.39908$	99341	79057	52478	25035	91507	69595	02099	34102	92128

Constants Involving e

$e = 2.71828$	18284	59045	23536	02874	71352	66249	77572	47093	69996
$1/e = 0.36787$	94411	71442	32159	55237	70161	46086	74458	11131	03177
$e^2 = 7.38905$	60989	30650	22723	04274	60575	00781	31803	15570	55185
$M = \log_{10} e = 0.43429$	44819	03251	82765	11289	18916	60508	22943	97005	80367
$1/M = \log_e 10 = 2.30258$	50929	94045	68401	79914	54684	36420	76011	01488	62877
$\log_{10} M = 9.63778$	43113	00536	78912	29674	98645	-10			

Numerical Constants

$\sqrt{2} = 1.41421$	35623	73095	04880	16887	24209	69807	85696	71875	37695
$\sqrt[3]{2} = 1.25992$	10498	94873	16476	72106	07278	22835	05702	51464	70151
$\log_e 2 = 0.69314$	71805	59945	30941	72321	21458	17656	80755	00134	36026
$\log_{10} 2 = 0.30102$	99956	63981	19521	37388	94724	49302	67881	89881	46211
$\sqrt{3} = 1.73205$	08075	68877	29352	74463	41505	87236	69428	05253	81039
$\sqrt[3]{3} = 1.44224$	95703	07408	38232	16383	10780	10958	83918	69253	49935
$\log_e 3 = 1.09861$	22886	68109	69139	52452	36922	52570	46474	90557	82275
$\log_{10} 3 = 0.47712$	12547	19662	43729	50279	03255	11530	92001	28864	19070

Symbols and Terminology for Physical and Chemical Quantities: Classical Mechanics

Name	Symbol	Definition	SI Unit
Mass	m		kg
Reduced mass	μ	$\mu = m_1 m_2/(m_1 + m_2)$	kg
Density, mass density	ρ	$\rho = m/V$	kg m^{-3}
Relative density	d	$d = \rho/\rho^\theta$	1
Surface density	ρ_A, ρ_S	$\rho_A = m/A$	kg m^{-2}
Specific volume	v	$v = V/M = 1/\rho$	m^3 kg^{-1}
Momentum	$\boldsymbol{p}$	$\boldsymbol{p} = mv$	kg ms^{-1}

(Continued)

Symbols and Terminology for Physical and Chemical Quantities: Classical Mechanics (Continued)

Name	Symbol	Definition	SI Unit
Angular momentum, action	L	$L = r \times p$	J s
Moment of inertia	I, J	$I = \sum m_i r_i^2$	kg m^2
Force	F	$F = dp/dt = ma$	N
Torque, moment of a force	$T, (M)$	$T = r \times F$	N m
Energy	E		J
Potential energy	E_p, V, Φ	$E_p = -\int F \cdot ds$	J
Kinetic energy	E_k, T, K	$E_k = (1/2)mv^2$	J
Work	W, w	$W = \int F \cdot ds$	J
Hamilton function	H	$H(q, p)$ $= T(q, p) + V(q)$	J
Lagrange function	L	$L(q, \dot{q})$ $= T(q, \dot{q}) - V(q)$	J
Pressure	p, P	$p = F/A$	Pa, N m^{-2}
Surface tension	γ, σ	$\gamma = dW/dA$	N m^{-1}, J m^{-2}
Weight	$G, (W, P)$	$G = mg$	N
Gravitational constant	G	$F = Gm_1m_2/r^2$	N m^2 kg^{-2}
Normal stress	σ	$\sigma = F/A$	Pa
Shear stress	τ	$\tau = F/A$	Pa
Linear strain, relative elongation	ε, e	$\varepsilon - \Delta l/l$	1
Modulus of elasticity, Young's modulus	E	$E = \sigma/\varepsilon$	Pa
Shear strain	γ	$\gamma = \Delta x/d$	1
Shear modulus	G	$G = \tau/\gamma$	Pa
Volume strain, bulk strain	θ	$\theta = \Delta V/V_0$	1
Bulk modulus, compression modulus	K	$K = -V_0(dp/dV)$	Pa
Viscosity, dynamic viscosity	η, μ	$\tau_{x,z} = \eta(dv_x/dz)$	Pa s
Fluidity	ϕ	$\phi = 1/\eta$	m kg^{-1}s
Kinematic viscosity	v	$v = \eta/\rho$	m^2 s^{-1}
Friction coefficient	$\mu, (f)$	$F_{frict} = \mu F_{norm}$	1
Power	P	$P = dW/dt$	W
Sound energy flux	P, P_a	$P = dE/dt$	W
Acoustic factors			
Reflection factor	ρ	$\rho = P_r/P_0$	1
Acoustic absorption factor	$\alpha_a, (\alpha)$	$\alpha_a = 1 - \rho$	1
Transmission factor	τ	$\tau = P_{tr}/P_0$	1
Dissipation factor	δ	$\delta = \alpha_a - \tau$	1

Symbols and Terminology for Physical and Chemical Quantities: Electricity and Magnetism

Name	Symbol	Definition	SI Unit
Quantity of electricity, electric charge	Q		C
Charge density	ρ	$\rho = Q/V$	C m^{-3}
Surface charge density	σ	$\sigma = Q/A$	C m^{-2}
Electric potential	V, ϕ	$V = dW/dQ$	V, J C^{-1}
Electric potential difference	$U, \Delta V, \Delta\phi$	$U = V_2 - V_1$	V
Electromotive force	E	$E = \int (F/Q) \cdot ds$	V
Electric field strength	E	$E = F/Q = -\text{grad } V$	V m^{-1}
Electric flux	Ψ	$\Psi = \int D \cdot dA$	C
Electric displacement	D	$D = \varepsilon E$	C m^{-2}
Capacitance	C	$C = Q/U$	F, C V^{-1}
Permittivity	ε	$D = \varepsilon E$	F m^{-1}
Permittivity of vacuum	ε_0	$\varepsilon_0 = \mu_0^{-1} c_0^{-2}$	F m^{-1}
Relative permittivity	ε_r	$\varepsilon_r = \varepsilon/\varepsilon_0$	1

Symbols and Terminology for Physical and Chemical Quantities: Electricity and Magnetism (Continued)

Name	Symbol	Definition	SI Unit
Dielectric polarization (dipole moment per volume)	$\boldsymbol{P}$	$\boldsymbol{P} = \boldsymbol{D} - \varepsilon_0 \boldsymbol{E}$	$\mathrm{C\,m^{-2}}$
Electric susceptibility	χ_e	$\chi_e = \varepsilon_r - 1$	1
Electric dipole moment	$\boldsymbol{p}, \boldsymbol{\mu}$	$\boldsymbol{p} = Q\boldsymbol{r}$	C m
Electric current	I	$I = dQ/dt$	A
Electric current density	$\boldsymbol{j}, \boldsymbol{J}$	$I = \int \boldsymbol{j} \cdot d\boldsymbol{A}$	$\mathrm{A\,m^{-2}}$
Magnetic flux density, magnetic induction	$\boldsymbol{B}$	$\boldsymbol{F} = Q\boldsymbol{v} \times \boldsymbol{B}$	T
Magnetic flux	$\boldsymbol{\Phi}$	$\boldsymbol{\Phi} = \int \boldsymbol{B} \cdot d\boldsymbol{A}$	Wb
Magnetic field strength	$\boldsymbol{H}$	$\boldsymbol{B} = \mu \boldsymbol{H}$	$\mathrm{A\,M^{-1}}$
Permeability	μ	$\boldsymbol{B} = \mu \boldsymbol{H}$	$\mathrm{N\,A^{-2}}$, $\mathrm{H\,m^{-1}}$
Permeability of vacuum	μ_0		$\mathrm{H\,m^{-1}}$
Relative permeability	μ_r	$\mu_r = \mu/\mu_0$	1
Magnetization (magnetic dipole moment per volume)	$\boldsymbol{M}$	$M = \boldsymbol{B}/\mu_0 - \boldsymbol{H}$	$\mathrm{A\,m^{-1}}$
Magnetic susceptibility	$\chi, \kappa, (\chi_m)$	$\chi = \mu_r - 1$	1
Molar magnetic susceptibility	χ_m	$\chi_m = V_m \chi$	$\mathrm{m^3\,mol^{-1}}$
Magnetic dipole moment	$\boldsymbol{m}, \boldsymbol{\mu}$	$E_p = -\boldsymbol{m} \cdot \boldsymbol{B}$	$\mathrm{A\,m^2}$, $\mathrm{J\,T^{-1}}$
Electrical resistance	R	$R = U/I$	Ω
Conductance	G	$G = 1/R$	S
Loss angle	δ	$\delta = (\pi/2) + \phi_I - \phi_U$	1, rad
Reactance	X	$X = (U/I)\sin\delta$	Ω
Impedance (complex impedance)	Z	$Z = R + iX$	Ω
Admittance (complex admittance)	Y	$Y = 1/Z$	S
Susceptance	B	$Y = G + iB$	S
Resistivity	ρ	$\rho = E/j$	Ω m
Conductivity	κ, γ, σ	$\kappa = 1/\rho$	$\mathrm{S\,m^{-1}}$
Self-inductance	L	$E = -L(dI/dt)$	H
Mutual inductance	M, L_{12}	$E_1 = L_{12}(dI_2/dt)$	H
Magnetic vector potential	$\boldsymbol{A}$	$\boldsymbol{B} = \boldsymbol{\nabla} \times \boldsymbol{A}$	$\mathrm{Wb\,m^{-1}}$
Poynting vector	$\boldsymbol{S}$	$\boldsymbol{S} = \boldsymbol{E} \times \boldsymbol{H}$	$\mathrm{W\,m^{-2}}$

Symbols and Terminology for Physical and Chemical Quantities: Electromagnetic Radiation

Name	Symbol	Definition	SI Unit
Wavelength	λ		m
Speed of light in vacuum	c_0		$\mathrm{m\,s^{-1}}$
in a medium	c	$c = c_0/n$	$\mathrm{m\,s^{-1}}$
Wavenumber in vacuum	$\tilde{\nu}$	$\tilde{\nu} = \nu/c_0 = 1/n\lambda$	$\mathrm{m^{-1}}$
Wavenumber (in a medium)	σ	$\sigma = 1/\lambda$	$\mathrm{m^{-1}}$
Frequency	ν	$\nu = c/\lambda$	Hz
Circular frequency, pulsatance	ω	$\omega = 2\pi\nu$	$\mathrm{s^{-1}}$, $\mathrm{rad\,s^{-1}}$
Refractive index	n	$n = c_0/c$	1
Planck constant	h		J s
Planck constant/2π	$\hbar$	$\hbar = h/2\pi$	J s
Radiant energy	Q, W		J
Radiant energy density	ρ, w	$\rho = Q/V$	$\mathrm{J\,m^{-3}}$
Spectral radiant energy density			
in terms of frequency	ρ_ν, w_ν	$\rho_\nu = d\rho/d\nu$	$\mathrm{J\,m^{-3}Hz^{-1}}$
in terms of wavenumber	$\rho_{\tilde{\nu}}, w_{\tilde{\nu}}$	$\rho_{\tilde{\nu}} = d\rho/d\tilde{\nu}$	$\mathrm{J\,m^{-2}}$
in terms of wavelength	ρ_λ, w_λ	$\rho_\lambda = d\rho/d\lambda$	$\mathrm{J\,m^{-4}}$

(Continued)

Symbols and Terminology for Physical and Chemical Quantities: Electromagnetic Radiation (Continued)

Name	Symbol	Definition	SI Unit
Einstein transition probabilities			
Spontaneous emission	A_{nm}	$dN_n/dt = -A_{nm}N_n$	s^{-1}
Stimulated emission	B_{nm}	$dN_n/dt = -\rho_{\bar{v}}(\bar{v}_{nm}) \times B_{nm}N_n$	$s\,kg^{-1}$
Stimulated absorption	B_{mn}	$dN_n/dt = \rho_{\bar{v}}(\bar{v}_{nm})B_{mn}N_m$	$s\,kg^{-1}$
Radiant power, Radiant energy per time	Φ, P	$\Phi = dQ/dt$	W
Radiant intensity	I	$I = d\Phi/d\Omega$	$W\,sr^{-1}$
Radiant exitance (emitted radiant flux)	M	$M = d\Phi/dA_{source}$	$W\,m^{-2}$
Irradiance (radiant flux received)	$E, (I)$	$E = d\Phi/dA$	$W\,m^{-2}$
Emittance	ε	$\varepsilon = M/M_{bb}$	1
Stefan-Boltzman constant	σ	$M_{bb} = \sigma T^4$	$W\,m^{-2}\,K^{-4}$
First radiation constant	c_1	$c_1 = 2\pi hc_0^2$	$W\,m^2$
Second radiation constant	c_2	$c_2 = hc_0/k$	$K\,m$
Transmittance, transmission factor	τ, T	$\tau = \Phi_{tr}/\Phi_0$	1
Absorptance, absorption factor	α	$\alpha = \Phi_{abs}/\Phi_0$	1
Reflectance, reflection factor	ρ	$\rho = \Phi_{refl}/\Phi_0$	1
(decadic) absorbance	A	$A = \lg(1 - \alpha_i)$	1
Napierian absorbance	B	$B = \ln(1 - \alpha_i)$	1
Absorption coefficient			
(linear) decadic	a, K	$a = A/l$	m^{-1}
(linear) napierian	α	$\alpha = B/l$	m^{-1}
molar (decadic)	ε	$\varepsilon = a/c = A/cl$	$m^2\,mol^{-1}$
molar napierian	κ	$\kappa = \alpha/c = B/cl$	$m^2\,mol^{-1}$
Absorption index	k	$k = \alpha/4\pi\bar{v}$	1
Complex refractive index	$\hat{n}$	$\hat{n} = n + ik$	1
Molar refraction	R, R_m	$R = \dfrac{n^2 - 1}{n^2 + 2}V_m$	$m^3\,mol^{-1}$
Angle of optical rotation	α		1, rad

Symbols and Terminology for Physical and Chemical Quantities: Solid State

Name	Symbol	Definition	SI Unit
Lattice vector	R, R_0		m
Fundamental translation vectors for the crystal lattice	$a_1; a_2; a_3,$ $a; b; c$	$R = n_1a_1 + n_2a_2 + n_3a_3$	m
(circular) reciprocal lattice vector	G	$G \cdot R = 2\pi m$	m^{-1}
(circular) fundamental translation vectors for the reciprocal lattice	$b_1; b_2; b_3,$ $a^*; b^*; c^*$	$a_i \cdot b_k = 2\pi\delta_{ik}$	m^{-1}
Lattice plane spacing	d		m
Bragg angle	θ	$n\lambda = 2d\sin\theta$	1, rad
Order of reflection	n		1
Order parameters			
short range	σ		1
long range	s		1
Burgers vector	b		m
Particle position vector	r, R_j		m
Equilibrium position vector of an ion	R_0		m

Symbols and Terminology for Physical and Chemical Quantities: Solid State (Continued)

Name	Symbol	Definition	SI Unit
Equilibrium position vector of an ion	$\boldsymbol{R}_0$		m
Displacement vector of an ion	$\boldsymbol{u}$	$\boldsymbol{u} = \boldsymbol{R} - \boldsymbol{R}_0$	m
Debye–Waller factor	B, D		1
Debye circular wavenumber	q_D		m^{-1}
Debye circular frequency	ω_D		s^{-1}
Grüneisen parameter	γ, Γ	$\gamma = \alpha V / \kappa C_V$	1
Madelung constant	$\alpha, \mathcal{M}$	$E_{\mathrm{coul}} = \dfrac{\alpha N_A z_+ z_- e^2}{4\pi \varepsilon_0 R_0}$	1
Density of states	N_E	$N_E = dN(E)/dE$	$J^{-1}\,m^{-3}$
(spectral) density of vibrational modes	N_ω, g	$N_\omega = dN(\omega)/d\omega$	$s\,m^{-3}$
Resistivity tensor	ρ_{ik}	$E = \rho \cdot \boldsymbol{j}$	$\Omega\,m$
Conductivity tensor	σ_{ik}	$\sigma = \rho^{-1}$	$S\,m^{-1}$
Thermal conductivity tensor	λ_{ik}	$J_q = -\lambda \cdot \operatorname{grad} T$	$W\,m^{-1}\,K^{-1}$
Residual resistivity	ρ_R		$\Omega\,m$
Relaxation time	τ	$\tau = l/v_F$	s
Lorenz coefficient	L	$L = \lambda/\sigma T$	$V^2\,K^{-2}$
Hall coefficient	A_H, R_H	$E = \rho \cdot \boldsymbol{j} + R_H(\boldsymbol{B} \times \boldsymbol{j})$	$m^3\,C^{-1}$
Thermoelectric force	E		V
Peltier coefficient	Π		V
Thomson coefficient	$\mu, (\tau)$		$V\,K^{-1}$
Work function	Φ	$\Phi = E_\infty - E_F$	J
Number density, number concentration	$n, (p)$		m^{-3}
Gap energy	E_g		J
Donor ionization energy	E_d		J
Acceptor ionization energy	E_a		J
Fermi energy	E_F, ε_F		J
Circular wave vector, propagation vector	$\boldsymbol{k}, \boldsymbol{q}$	$k = 2\pi/\lambda$	m^{-1}
Bloch function	$u_k(\boldsymbol{r})$	$\psi(\boldsymbol{r}) = u_k(\boldsymbol{r})\exp(i\boldsymbol{k} \cdot \boldsymbol{r})$	$m^{-3/2}$
Charge density of electrons	ρ	$\rho(\boldsymbol{r}) = -e\psi^*(\boldsymbol{r})\psi(\boldsymbol{r})$	$C\,m^{-3}$
Effective mass	m^*		kg
Mobility	μ	$\mu = v_{\mathrm{drift}}/E$	$m^2\,V^{-1}\,s^{-1}$
Mobility ratio	b	$b = \mu_n/\mu_p$	1
Diffusion coefficient	D	$dN/dt = -DA(dn/dx)$	$m^2\,s^{-1}$
Diffusion length	L	$L = \sqrt{D\tau}$	m
Characteristic (Weiss) temperature	ϕ, ϕ_W		K
Curie temperature	T_C		K
Néel temperature	T_N		K

Letter Designations of Microwave Bands

Frequency, GHz	Wavelength, cm	Wavenumber, cm^{-1}	Band
1–2	30–15	0.033–0.067	L-band
1–4	15–7.5	0.067–0.133	S-band
4–8	7.5–3.7	0.133–0.267	C-band
8–12	3.7–2.5	0.267–0.4	X-band
12–18	2.5–1.7	0.4–0.6	Ku-band
18–27	1.7–1.1	0.6–0.9	K-band
27–40	1.1–0.75	0.9–1.33	Ka-band

3

Properties of Passive Components

Jeffrey P. Tate
Florida A&M University

Patricia F. Mead
University of Maryland

Ardie D. Walser
City College of New York

C. Sankaran
Electro-Test Inc.

Roy W. Rising
ABC-TV

Michael Starling
National Public Radio

3.1 Crystal Oscillators

Jeffrey P. Tate and Patricia F. Mead

3.1.1 Introduction

Crystal oscillators were first developed in the early 1920s and have found wide spread application in radio and television communications. Today, these oscillators remain popular in precision **frequency** applications including timing device circuits and radio frequency filters. The oscillators are typically described as *voltage controlled oscillators* (VCXOs) or *oven controlled oscillators* (OCXOs). These devices offer frequency precision and stability over large temperature ranges, exhibiting stability tolerances better than $+/-0.005\%$ for sources with over 100-MHz oscillation.

3.1.2   Piezoelectric Effect

Basic Physical Principles

Materials that are used for crystal oscillator applications exhibit the **piezoelectric** effect. This phenomenon produces an electric field within a material when a mechanical force is applied. Conversely, if an electrical drive signal is applied to the crystal, a mechanical vibration results. If the driving signal is periodic in nature, then resulting vibration will also be periodic. Very accurate electric signal frequencies can be produced by certain materials with the appropriate shape, electrode geometry, and ambient conditions. For this reason, crystal **resonators** are used in applications such as television, cellular radio communications, and electronic test equipment where accurate synthesis of signal frequencies is required throughout.

The relationship between the electric field, electric displacement, electric polarization, and mechanical stress and strain is given by what are called *constituitive relations*. The electrical quantities are related in the following manner:

$$D = \epsilon_0 E + P$$

where D is the electrical displacement, E is the electric field and P is the electric polarization. The term ϵ_0 is the free-space permittivity. The polarization depends on the applied electric field. In a piezoelectric material an electric polarization can also result from an applied stress or strain. This is called the *direct piezoelectric effect*. Therefore the polarization is proportional to the stress (T) or strain (S) that is

$$P = dT \quad \text{or} \quad P = eS$$

In the converse piezoelectric effect, the stress or strain forces are generated by the electric field applied to the crystal. Thus, we have

$$S = dE \quad \text{or} \quad T = eE$$

The constants of proportionality d and e are called piezoelectric stress and strain coefficients. The stress and strain forces are represented by matrix quantities, and the coefficients are tensor quantities. A tensor mathematically represents the fact that the polarization can depend on the stress or strain in more than one direction. This is also true for the relationship between the stress or strain and the electric field. Many other physical properties in crystals also exhibit this nature, which is called *anisotropy*. Thus when a property is anisotropic, its value depends on the direction of orientation in the crystal. For the direct piezoelectric effect, the total polarization effect is the sum of these two contributions, an applied electric field and applied mechanical force. Based on the relationship between the electric displacement and the electric polarization it is then possible to write equations that relate the displacement D to the applied stress or strain. Electric displacement is the quantity that is preferred in experiment and engineering.

Excitation of Wave Modes in Crystals

A variety of electric polarization effects can be produced in piezoelectric crystals by squeezing, bending, shearing, or torsion forces. Conversely, a variety of simple and coupled mechanical vibrations can be produced by applied electric fields. When an electric field is applied, the orientation of the field results in different mechanical vibrations or elastic wave motions in the crystal. These motions are called **modes**, and they depend on the placement of the electrodes. The electrode shape and the shape and cut of the crystal also affects the mode spectrum. Some shapes considered include thick plates, thin plates, and narrow bars. Plates can be square, rectangular, or circular in cross section. In Fig. 3.1 an example of a bar with different electrode placements for exciting longitudinal vibrations is shown. In case of Fig. 3.1(a) the electric field is acting along the bar thickness and in case of Fig. 3.1(b) the electric field is perpendicular to the bar thickness. Another example is a flexing mode in a narrow bar. In this case the electric field is transverse to the bar thickness and parallel to its bending axis. It is possible to generate both even and odd flexing modes. Mechanical bending of the bar conversely produces opposing electric fields and currents. Other mechanical vibrations that are possible include torsional, face-shear, and thickness-shear modes.

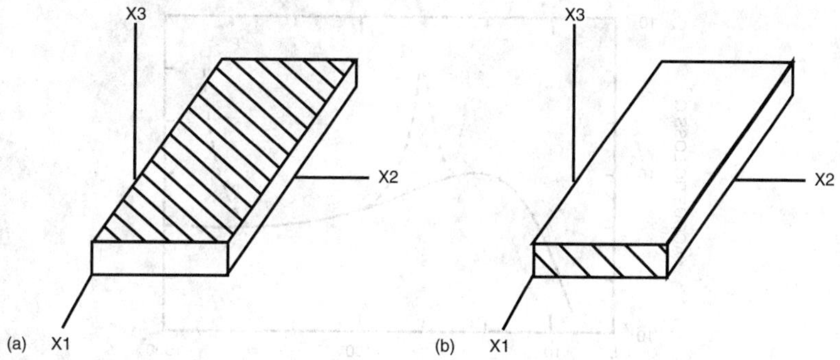

FIGURE 3.1 Arrangement of electrodes on a narrow bar for excitation of longitudinal vibrations: (a) electrodes on planes perpendicular to axis X3, (b) electrodes on planes perpendicular to axis X1. (From Zelenka, J. 1986. *Piezoelectric Resonators and Their Application*, Chap. 3. Elsevier, Amsterdam.)

A description of the propagation of elastic wave modes in piezoelectric resonators requires the use of continuum mechanics and the Maxwell equations for the electromagnetic field. When a driving electric field is applied to a crystal, the frequency of the field determines the amplitude of the elastic wave vibration. Conversely, a harmonically applied stress/strain determines what electrical signal frequencies are produced. The maximum vibration amplitude occurs when the frequency of the driving electrical signal coincides with a mechanical resonance of the crystal.

3.1.3 Crystal Resonators

Resonant Frequency and Loss Mechanisms

The resonant frequency of the crystal depends on the physical properties and dimensions of the crystal. For applications requiring precision frequency control these properties and dimensions must be very stable. When stability is an issue, compensation techniques may be required to achieve the desired frequency control. Crystal resonators are typically plates with finite dimensions. The mechanical resonant frequencies for these bounded crystal plates are not easy to obtain through exact calculation. For circular and rectangular plates three-dimensional vibrations are possible, but they are difficult to model exactly. Approximations, however, are used to describe the complex modes of vibration in these systems. Methods for employing these approximations are discussed in detail in Zelenka (1986).

Another important concern for piezoelectric resonator applications is the amount of energy lost due to internal damping forces. A system capable of resonant oscillations will have, associated with it, a figure of merit called the material quality **factor Q**. The quality factor relates the energy stored to the energy dissipated per cycle. Mechanical losses can arise from intrinsic mechanisms such as thermal phonons (lattice vibrations) or defects in crystals. As the temperature increases, the thermal phonon population increases and the intrinsic phonon loss will also increase. The presence of this loss causes the stress and strain to differ in phase angle. A measure of this deviation is the relationship

$$\tan \delta = \frac{1}{Q}$$

where δ is the phase deviation between stress and strain and Q is the material quality factor. For crystals used in precision frequency control applications the value of the quality factor will be very high, indicating that the losses are minimal. Values of Q from 10^5 to 10^7 are typical.

Some of the potential sources of loss in piezo-electronic crystals include defect centers (e.g., AL-OH−, AL-Na+, Al-Li+, and Al-hole centers) and the liberation of interstitial alkali ions. An example of the measured loss of an **AT-cut** crystal resonator is shown in Fig. 3.2 where the two curves indicate the acoustic loss before and after the application of hydrogen sweeping. This sweeping process uses an electric field to

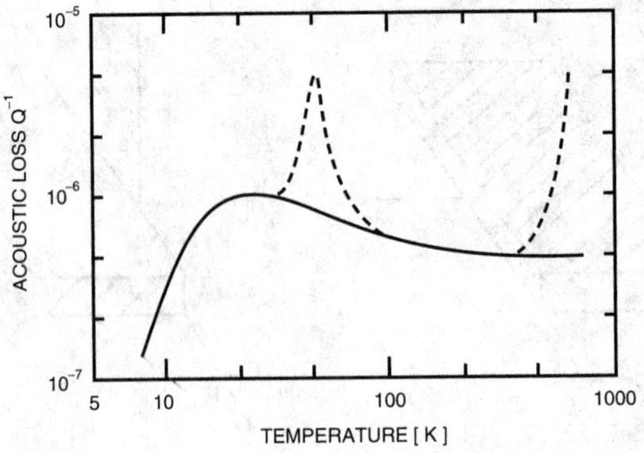

FIGURE 3.2 Acoustic loss vs temperature spectrum of an AT-cut resonator in the as-grown state (dashed curve) and hydrogen sweeping (solid curve). (After Halliburton, L.E. et al. 1985. In *Precision Frequency Control Vol. 1*, eds. E. Gerber and A. Ballato, Chap. 1. Academic Press, Orlando, FL.)

sweep out interstitial cation impurities such as hydrogen from the crystal. Significant loss peaks are present when the crystal is used prior to the sweeping procedure.

Another important parameter for piezoelectronic systems is the electrochemical coupling coefficient κ, which is given by the formula

$$k = \sqrt{\frac{W_p}{W_e}}$$

where W_p is the stored elastic energy and W_e is the stored electric energy. For crystals used in oscillator applications the value for κ is quite low. For an AT-cut quartz plate the value is about 0.0001. A low value for the coupling coefficient means that the electrical and mechanical systems are loosely coupled. If the crystal has a high Q factor, however, the presence of a small driving signal is sufficient to generate a response that maintains the excitation.

Calculation of Resonant Frequency

The resonant frequency of the quartz crystal is largely influenced by the shape of the cavity resonator which is composed of the piezoelectric quartz material. A quartz crystal resonator does not behave as a one-dimensional system; however, the basic principles behind three-dimensional resonance calculations are described in a one-dimensional picture. The equation describing free vibration in a one-dimensional system is

$$\frac{\partial^2 \Psi(x,t)}{\partial t^2} = \frac{v^2 \partial^2 \Psi(x,t)}{\partial x^2}$$

where v is the speed at which a vibrational disturbance can propagate through the medium and $\Psi(x,t)$ describes the displacement of a point in the material.

If we assume that the wave function Ψ may be factored into a product of independent functions of time t and position x, and if we apply the boundary conditions

$$\Psi(x,0) = \Psi(0,t) = \Psi(L,t) = 0$$

it may be shown that

$$\Psi(x,t) = A_n \sin\left[\frac{n\pi vt}{L}\right] \sin\left[\frac{n\pi x}{L}\right]$$

From this expression we may readily recognize the resonant frequencies

$$f_n = \frac{nv}{2L}$$

The fundamental frequency occurs for $f_1 = v/2L$, and the second harmonic occurs for $f_2 = v/L$.

TABLE 3.1 Elastic Coefficients of Alpha Quartz

Parameter	Bechmann	Mason
c_{11}	86.74	$86.05 \times 10^9\,\mathrm{N/m^2}$
c_{12}	6.99	5.05
c_{13}	11.91	10.45
c_{14}	−17.91	18.25
c_{33}	107.2	107.1
c_{44}	57.94	58.65
c_{66}	39.88	40.5

Sources: See Bechmann, R. 1958. *Phys. Rev.* 110(5): 1060; Mason, W.P. 1950. *Piezoelectric Crystals and Their Applications to Ultrasonics*, p. 84. Van Nostrand Reinhold, New York, NY.

Extending to the three-dimensional case, we find for a body of width w, length l, and thickness e, the fundamental frequencies are expressed as

$$f_{nmp} = \frac{v}{2}\left[\left(\frac{n}{e}\right)^2 + \left(\frac{m}{l}\right)^2 + \left(\frac{p}{w}\right)^2\right]^{\frac{1}{2}}$$

but if the thickness is much less than the length or width of the system, the expression reduces to the one-dimensional expression with $f_n = nv/2L$. Further, the velocity of propagation in quartz crystals is determined from the direction of vibration through the elastic coefficients c_{ij}. Substituting an explicit expression for velocity into the resonant frequency expression, we find for an X-cut and Y-cut crystal,

$$f_{nx} = n/2e\sqrt{\frac{c_{11}}{\rho}}$$

$$f_{ny} = n/2e\sqrt{\frac{c_{66}}{\rho}}$$

where ρ is mass density. Values for the elastic coefficients of **alpha quartz** is given in Table 3.1.

The mode structure discussed previously has been developed assuming rectangular-shaped electrodes. Most high-frequency resonators, however, use flat or spherical circular electrode shapes. The electrode shapes affect the boundary conditions for the mode vibrations, and the resulting resonant frequencies must now be described by mathematical functions that are able to satisfy those boundary conditions. For example, with flat circular electrodes, Bessel functions are used to describe the resonant mode behavior. Spherical contoured plates are also routinely applied to resonator designs.

Oscillation of Overtones and Spurious Modes

Under normal conditions, the fundamental frequency oscillates in the quartz resonator. However, the characteristics of the external network can be utilized to promote oscillation in a higher order *overtone* mode. The exact overtone frequency is typically not an exact harmonic of the fundamental frequency. However, the overtone is normally close to a harmonic value. The external circuit is tuned to a frequency near the desired overtone frequency.

Spurious modes are inharmonic modes that may also appear in the vibrational spectrum of the quartz resonator. The spurious modes usually occur in high-resistance modes, and they generally occur at frequencies that are higher than the desired oscillation frequency. The occurrence of spurious modes is increased for crystals operating in an overtone mode as shown in Fig. 3.3.

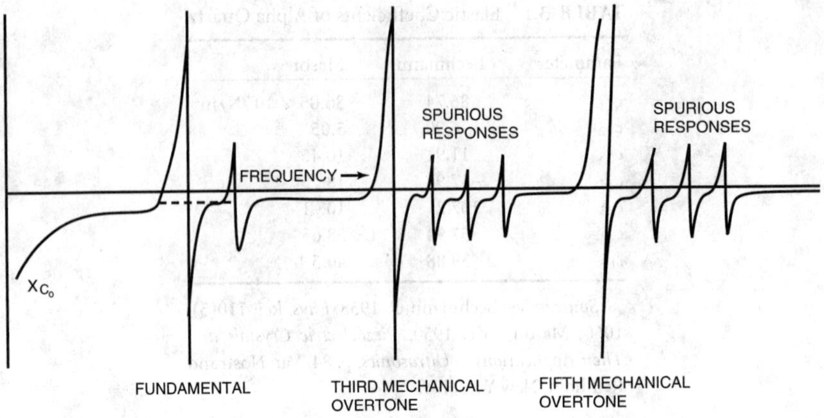

FIGURE 3.3 Overtone and spurious frequency response of a quartz crystal. (*Source:* Frerking, M.E. 1978. *Crystal Oscillator Design and Temperature Compensation*, Chap. 4. Van Nostrand Reinhold, New York, NY.)

Resonator Cut Types

Control of the resonant frequency response of crystal oscillators is also influenced by a variety of crystal cavity configurations. These configurations also influence the coefficients that determine the sensitivity of the resonant frequency to temperature changes or applied stress fields. Additionally, the application of the crystal electrodes also affects these coefficients.

Quartz crystals have a trigonal symmetry so that physical features repeat every 120°. A diagram showing the natural faces of a quartz crystal is shown in Fig. 3.4. The structure shows 6 m-faces, which are parallel to the z axis. The cap faces include two r-faces, called *minor rhomb* faces; an R-face, called the *major rhomb* face; and the rarely occurring s- and x-faces. Note that the crystals shown in Fig. 3.4 are also mirror images of each other. Quartz crystals occur in one or the other of the varieties shown. These are called left quartz (Fig. 3.4(a)) and right quartz (Fig. 3.4(b)). Also, the two varieties shown are found both in naturally occurring and cultured (man-made) quartz.

X-, Y-, and Z-Cuts

The convention for making an X-cut is simply to saw the quartz crystal along a plane that is parallel to the X axis. For a Y-cut, the crystal is sawed in a plane that is parallel to the Y axis. Finally, for a Z-cut, the crystal is sawed in a plane that is parallel to the Z axis.

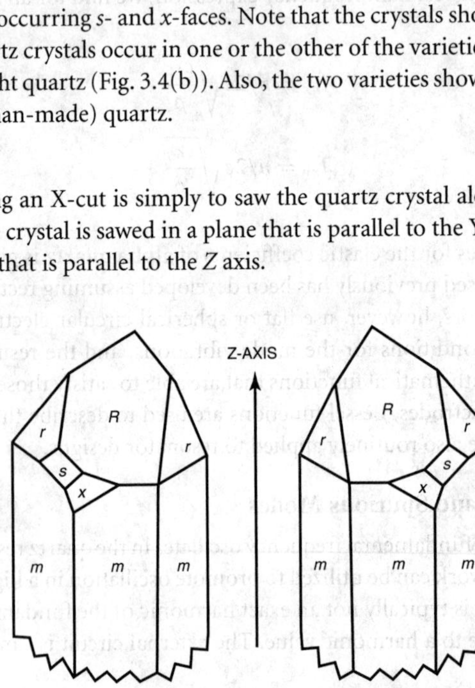

FIGURE 3.4 Face geometry of quartz crystal: (a) left quartz, (b) right quartz. (*Source:* Bottom, V. 1982. *Introduction to Quartz Crystal Unit Design.* Van Nostrand Reinhold, New York, NY.)

AT- and BT-Cuts

Circuit parameter values of the crystal are largely influenced by the angle at which the quartz is cut from the raw quartz slab (precut form). One of the most popular cuts is the AT-cut, which typically is operated in a thickness shear mode between 1 and 200 MHz. Also, above $\sim$25 MHz, the AT-cut quartz crystal typically operates in an overtone mode. The AT-cut is usually made from a Y-bar, which simply indicates that the maximum dimension of the crystal is in the Y direction. Also, the actual cut is defined as a cut with angle, $\theta = 35°15'$ with respect to the z axis. Another popular cut is the **BT-cut**; this is a cut made at an angle of $-49°$ with respect to the z axis. The BT-cut crystals are usually Y-cut also, and the BT-cut crystals are more likely to be applied in an overtone mode, although fundamental mode oscillation is also common.

CT- and DT-Cuts

CT-cuts and DT-cuts represent a family of cuts where the centerpoint of the electrode face is placed at a mechanical node. The modes created by the CT or DT family of cuts are two-dimensional face shear modes. The two dimesional vibration is achieved by supporting the structure at the position of the node, so that movement is then restricted. The most significant feature of the CT- and DT-cuts lies in the fact that they are achieved by rotating the Y-cut about the X axis into two positions at which the temperature dependence of the face shear modes are virtually independent of temperature. Crystals with these cuts are largely used in the frequency range of 100–1000 kHz.

Compensational Cut Formations

The resonant frequency of the quartz cavity has been observed to be highly dependent on several environmental factors, including temperature and external or internal stresses applied to the quartz material. These factors can result in frequency sensitivities on the order of 100 ppm in some cases. This type of sensitivity would make precision frequency control applications extremely difficult without the use of fairly sophisticated control networks that would stabilize temperature and isolate the quartz oscillator from external vibration noise. Fortunately, however, a variety of crystal cuts have been discovered that result in a much lower sensitivity to temperature and stress effects. The stress compensated SC-cut (EerNisse, 1975) is described in Sec. 3.1.7.

Much of the important theoretical development involving compensation cuts for quartz crystals has been done by EerNisse (1975, 1976). EerNisse has been able to show that the frequency-stress coefficient is determined by the Φ cut angle, and the first-order frequency-temperature coefficient is determined almost entirely by the θ-cut angle. Several standard cut types, such as the thermal-transient-compensated (TTC-) cut (Klusters et al., 1977), TS-cut (Holland, 1964), and SC-cut all refer to double rotated cuts with $\theta \sim 34°$ and $\Phi \sim 22°$. A similar cut is the IT-cut by Bottom and Ives (1951). The IT-cut has $\theta = 34°17'$ and $\Phi = 19°06'$. The frequency-temperature coefficient is zero for the IT-cut.

3.1.4 Equivalent Circuit of a Quartz Resonator

The discussion of this section follows the development presented by Frerking (1978). Because of the piezoelectric effect, the quartz resonator behaves physically as a vibrating object when driven by a periodic electric signal near a resonant frequency of the cavity. This resonant behavior may be characterized by equivalent electrical characteristics that may then be used to accurately determine the response and performance of the electromechanical system.

Dye and Butterworth (1926) developed the equivalent lump electrical network that describes an electrostatic signal driving a mechanically vibrating system (electromechanical transducers). The equivalent electric network that models the piezoresonator was later developed by Van Dyke (1928), Butterworth, and others, and is shown in Fig. 3.5. In the equivalent network, the circuit parameters

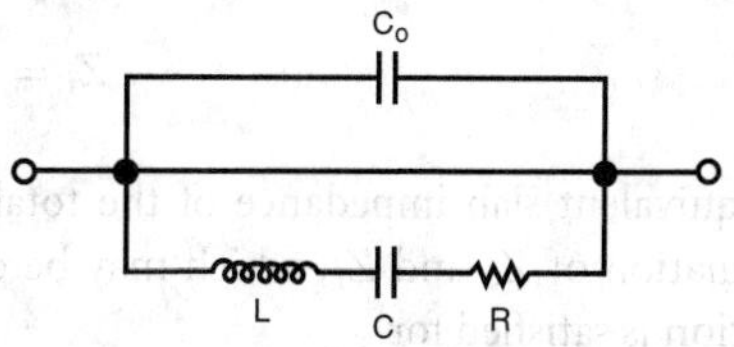

FIGURE 3.5 Equivalent network of piezoelectric resonator.

TABLE 3.2 Typical Equivalent Circuit Parameter Values for Quartz Crystals

Parameter	200 kHz Fundamental	2 MHz	30 MHz 3rd o/tone	90 MHz 5th o/tone
R_1	2 kΩ	100 Ω	20 Ω	40 Ω
L_1	27 H	520 mH	11 mH	6 mH
C_1	0.024 pF	0.012 pF	0.0026 pF	0.0005 pF
C_0	9 pF	4 pF	6 pF	4 pF
Q	18×10^3	18×10^3	18×10^3	18×10^3

R_i, L_i, and C_i represent parameters associated with the quartz crystal, whereas C_0 represents the shunt capacitance of the resonator electrodes in parallel with the container that packages the crystal.

The inductance L_i is associated with the mass of the quartz slab, and the capacitance C_i is associated with the stiffness of the slab. Finally, the resistance R_i is determined by the loss processes associated with the operation of the crystal. Each of the equivalent parameters can be measured using standard network measurement techniques (see Sec. 3.1.6). Table 3.2 lists typical parameter values.

Application of an electric field across the electrodes of the crystal results in both face and thickness shear waves developing within the crystal resonator. For a Y-cut crystal where the thickness of the cut is along the Y-direction, face shear modes occur in the XZ plane, whereas thickness shear modes develop in the XY plane.

The vibration is enhanced for electric signals near a resonant frequency of the resonator. Low-frequency resonances tend to occur in the face shear modes, whereas high-frequency resonances occur in thickness shear modes. In either case, when the applied electric signal is near a resonant frequency, the energy required to maintain the vibrational motion is very small.

Nonresonant Behavior
If the applied electric signal is not near a resonant frequency of the resonator, the quartz crystal behaves as a parallel plate capacitor with capacitance

$$C_0 = \frac{k e_0 A}{d}$$

where A is the electrode area, d is the thickness of the quartz slab, and k is the relative dielectric constant of the resonator material. Because of the anisotropic nature of quartz crystals, the value of k varies with the direction of the applied electric signal. Finally, the impedance of the nonresonant quartz crystal device is described by

$$Z_0 = \frac{j}{\omega C_0}$$

where ω is the angular frequency of the applied electric signal.

Resonant Behavior
The characteristic series combination impedance of the quartz slab may be expressed as

$$Z_i = R_i + j\left[\omega L_i - \frac{1}{\omega C_i}\right]$$

The equivalent slab impedance of the total parallel network is determined by calculating the parallel combination of Z_0 and Z_i, which may be defined as Z_p. At resonance, Z_p is purely resistive, and this condition is satisfied for

$$\omega = \sqrt{\left(\frac{1}{L_i C_i} + \frac{1}{2 L_i C_0} - R_i^2 \frac{1}{2 L_i^2}\right) \pm \left(\frac{1}{2 L_i C_0} - \frac{R_i^2}{2 L_i^2}\right)}$$

with

$$\left(\frac{1}{2 L_i C_0} - \frac{R_i^2}{2 L_i^2} \right)^2 \gg \frac{R_i^2}{L_i^3 C_i}$$

These two frequencies correspond to the series resonance f_s (obtained using the minus sign) and the parallel resonance f_a.

The series resonance is given by

$$f_s = \frac{1}{2\pi} \sqrt{\left(\frac{1}{L_i C_i} \right)}$$

and when $C_i / C_0 \ll 1$, the parallel resonance (also known as the *antiresonance*) is given by

$$f_a = f_s \left[1 + \frac{C_i}{2 C_0} \right]$$

The crystal is normally operated at frequencies located between the series resonance and the parallel or so-called antiresonance frequencies. This defines a useful bandwidth for the crystal relative to the series resonance and may be expressed as

$$\Delta f = f_s \frac{C_i}{2 C_0}$$

For the preceding expressions, the quartz resonator is not connected to an external driving network. The effect of the external network can be accounted for by a load capacitance C_L in parallel with C_0. In the presence of an external load capacitance C_L, the antiresonance frequency of the loaded quartz resonator is expressed as

$$f_L = f_S \left[1 + \frac{C_i}{2(C_0 + C_L)} \right]$$

and the frequency shift away from the series resonant frequency is

$$\Delta f_L = \frac{C_i}{2(C_0 + C_L)}$$

3.1.5 Experimental Determination of the Crystal Oscillator Parameters

Determination of Resonant Frequency

The discussions of the first two subsections of Sec. 3.1.5 follow the derivation of Zelenka (1986). The circuit for measurement of the resonant frequency is shown in Fig. 3.6. This is a passive π network that is also useful for obtaining equivalent circuit parameters, such as dynamic capacitance, crystal resistance, and inductance. The measurement of resonant frequency requires a signal generator at the input and a voltmeter or phase meter connected to the output. The resonant frequency can be measured under two distinct circuit conditions: at the minimum resonator impedance (f_{im}) or at zero resonator reactance (f_{ir}).

The first condition is observed when the output voltage of the π network is a maximum. The second condition occurs when the phase shift between the input and output voltages is zero. This case requires the use of a phase meter for the measurement. For the ith resonance, the two frequencies are equal to

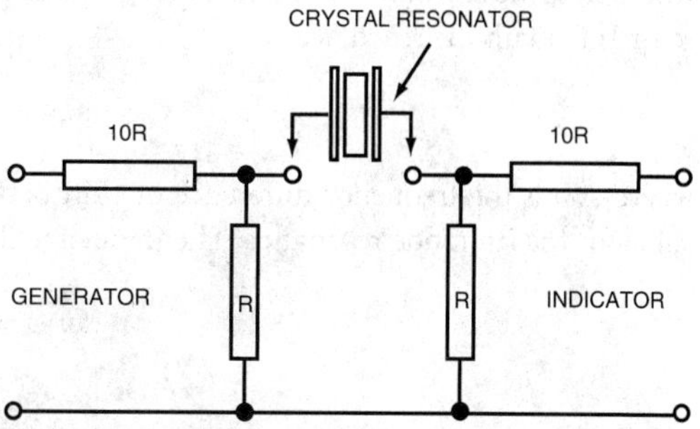

FIGURE 3.6 Passive π network for measurement of equivalent circuit parameters. (From Zelenka, J. 1986. *Piezoelectric Resonators and Their Applications*, Chap. 5. Elsevier, Amsterdam.)

the series resonant frequency defined previously if the resonator losses are neglilgible (i.e., resonator resistance R_i is close to zero). If the losses cannot be neglected, then more complex formulas that relate the measured frequencies to the series resonant frequency are required. They are given as

$$f_{im} = f_{is}\left[1 - \frac{r_i}{Q_i^2} + \frac{1}{2}\left(r_i^3 + \frac{3r_i^2}{4}\right)\frac{1}{Q_i^4} + \cdots\right]$$

and

$$f_{ir} = f_{is}\left[1 + \frac{r_i}{Q_i^2} + \frac{1}{2}\left(r_i^3 + \frac{3r_i^2}{4}\right)\frac{1}{Q_i^4} + \cdots\right]$$

where r_i is the ratio C_0/C_i and Q_i is the quality factor (for the ith resonator mode).

Measurement of the Equivalent Circuit Parameters

The dynamic capacitance of the resonator C_i is determined by measuring two frequency detunings. The different detunings result from connection of two different external capacitances C_{s1} and C_{s2} in series with the resonator. The measurement method is described in detail by Zelenka (1986). If the resonator Q factor satisfies the inequality

$$Q_i > 20\sqrt{r_i}$$

then the magnitude of the dynamic capacitance is same for both f_{im} and f_{ir} resonant frequencies. If this condition is not satisfied, then the capacitance differs for the two cases. In situations where applications are not as demanding, the dynamic capacitance can be obtained from the expression

$$C_i = \frac{2\Delta f_i}{f_{is}}(C_0 + C_{si})$$

where Δf_1 is the detuning that results when the capacitor C_{s1} is connected. This formulation is less accurate because the parasitic capacitances included in C_0 are not canceled out. The magnitude of these spurious capacitance effects cannot always be determined with sufficient accuracy.

Once the resonant frequency and the dynamic capacitance are determined, the inductance L_i is found using a measurement of the phase shift between the input and output voltages of the π network. The resistance R_i is given for the crystal of the resonator. The phase shift is given by

$$\tan \phi = \frac{X_i}{R_i + R}$$

where X_i is the ith-mode resonator reactance and R is the output resistance of the π network. For a series resonant circuit the reactance is

$$X_i = \omega L_i + \frac{1}{\omega C_i} \simeq 2\Delta\omega L_i$$

where $\Delta\omega$ is the frequency difference or shift between the operating frequency and the series resonant value for the ith mode resonance. The frequency shift can be related to the phase shift by

$$\tan \phi = \frac{\Delta f}{f_{is}}2Q_{\text{eff}}$$

$$Q_{\text{eff}} = \frac{w_{is}L_i}{R_i + R}$$

A measurement of the phase shift and the determination of the frequency ratio $\Delta f/f_{is}$ allows us to calculate Q_{eff}. Using a value of R_i provided for the crystal of interest, the L_i can be obtained. The value of the resonator resistance R_{ir} at series resonance can also be determined by measuring the input and output voltages at resonance (V_{1r}, V_{2r}) and when the crystal resonator is short-cirtcuited (V_1, V_2) using the formula

$$R_{ir} = R\left(\frac{V_2}{V_1}\frac{V_{1r}}{V_{2r}} - 1\right)$$

The expression that relates R_{ir} to R_i is

$$R_{ir} = R_i\left[1 + \frac{C_0^2}{C_i^2 Q_i^2}\right]$$

which can be used to obtain R_i if the resistance data for a specific resonator are not available.

Crystal Quality Factor

It is possible to determine the electrical quality factor of the crystal resonator indirectly using the equivalent circuit parameters and the formula

$$Q_i = \frac{1}{2\pi f_{is} R_i C_i}$$

The infrared quality factor or *material Q* can be obtained with a measurement of the extinction coefficient α. A detailed discussion of the theory behind this measurement is provided in an article by Brice (1985). An approximation to the extinction coefficient called α_v is obtained for several wave numbers of interest (wave number is reciprocal of wavelength). The value of Q is then calculated from the formula

$$Q = \frac{C 10^5}{\alpha_v^*}$$

where C is a constant that depends on wave number. Typical wave numbers are 3410, 3500, and 3585 cm^{-1}. Values of Q have been extensively determined using this formulation with statistical deviation on the order of 10%, when Q is in the range $0.3 < Q \times 10^{-6} < 3.0$.

3.1.6 Piezoelectric Materials

Piezoelectric materials that have found significant application in frequency control applications include quartz, lithium niobate, lithium tantalate, bismuth germanium oxide, aluminum phosphate (berlinite), and langasite. These materials have been covered extensively in the literature, which includes the reference texts by Zelenka (1986), Gerber and Ballato (1985, Vol. 1), and Bottom (1982) and many journal publications (Ballato and Gualtieri, 1995; Wall and Vig, 1994; Brice, 1985). Quartz is most commonly used for crystal resonator applications and the subject is covered in great detail in many of these works. Materials have undergone considerable changes over the last two decades in terms of the quality and types of materials available. An extensive summary of the material properties such as elastic and piezoelectric coefficients for some representative samples are given in the text by Zelenka (1987) and also the work edited by Gerber and Ballato (1985).

Defects and Impurities

As discussed in Sec. 3.1.3 the presence of defects and impurities can promote loss in piezoelectric crystals. Defects are divided into two main groups, *point* and *extended defects*. The point defects include the aluminum-related centers as well as oxygen-vacancy centers. Aluminum ions can easily substitute for silicon in quartz; however, charge compensation is required. Aluminum has a +3 charge whereas the silicon valence is +4. An additional positive charge is required that can be supplied by H+, Li+, Na+, or holes at interstitial sites in the crystal lattice. Iron-related defects are also possible since iron is also a trivalent (+3) ion.

Oxygen-vacancy centers, which are also called E′ centers, are also possible. These centers have been extensively studied in quartz crystals and can take several forms. These point defects have not yet been

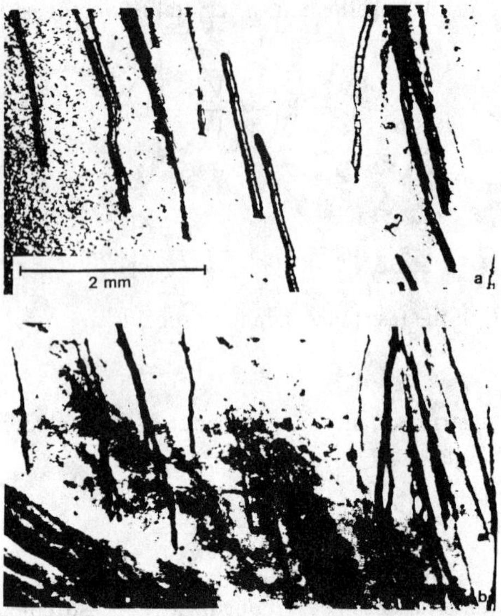

FIGURE 3.7 Correspondence between etch tunnels and X-ray topographic region in +X region: (a) Optical photograph, (b) X-ray topograph. (From Halliburton, L.E. et al. 1985 in *Precision Frequency Control Vol. 1*, eds. E. Gerber and A. Ballato, Chap. 1. Academic Press, Orlando, FL.)

observed to cause acoustic loss peaks in quartz. An example of an oxygen-vacancy center is the E_1' center that is formed when an oxygen vacancy traps an unpaired elecion.

In quartz materials grown from seed crystals, the dominant impurity near the seed is molecular H_2O. This gives way to OH− as you move away from the seed. It is the presence of such water molecules that is primarily responsible for the hydrolytic weakening in quartz.

Extended defects such as dislocations and fault surfaces can also arise in piezoelectric crystals. These defects have been studied extensively using a number of experimental techniques. One type of extended defect, called an *etch tunnel*, is shown in Fig. 3.7. It is believed that the growth of such tunnels depends on the control of the initial seed–crystal interface and maintenance of a uniform growth rate. It is also possible that interstitial action impurities such as those mentioned previously are involved in etch tunnel formation.

Piezoelectric Ceramic and Polymers

Some ceramic and polymer materials also display piezoelectric properties. For ceramics these include polycrystal line materials based on solid solutions of $Pb(Zr, Ti)O_3$. The ratio of zirconium and titanium constituents help to determine the crystal symmetry structure that results after heating and then cooling of the samples. Modern piezoelectric ceramics are referred to as PZT ceramics. Other materials such as barium, strontium, calcium, and/or trivalent rare earth elements can be added to the ceramics in small quantities to modify the piezoelectric properties.

Polymers that exhibit the piezoelectric effect include polyvinyl chloride, polyvinyl fluoride, and difluor polyethylene. These polymers acquire their properties through technological processing. The thin plastic foil samples are exposed to strong electric fields and then cooled to room temperature. This process results in a polarization of the material.

Motional Time Constant

Ballato and Gualtieri (1994) employ an intrinsic parameter $\tau_1^{(m)}$ (the **motional time constant**) to make comparisons between different piezoelectric materials. This constant is inversely proportional to the product of frequency and quality factor for the mth resonator mode. It is possible to define the motional time

TABLE 3.3 Motional Time Constants and Characteristic Lengths for Some Piezoelectric Bulk Resonators

Type/Cut	N_0, m/s	Q ($\times 10^3$)	τ_1 ($\times 10^{-15}$ s)	k^2, %	L_0, nm
Quartz (AT)	1,661		11.8	0.77	39.24
Quartz (BT)	2,536		4.9	0.32	60.99
Quarts (SC)	1,797		11.7	0.25	130.9
Quartz (STW)		3.3	24.0		
Quartz (miniBVA)		2500.0	13.0		
Quartz (TC)		1600.0	10.0		
LiNbO$_3$ (Z)	1,787		2.94	0.13	60.70
La$_3$Ga$_5$SiO$_{14}$ (Y)	1,384		6.30	2.89	4.68
La$_3$Ga$_5$Si$_{14}$ (X)	2,892		4.20	0.64	29.42
Li$_2$Ba$_4$O$_7$ (TA)	2,550	450	7.80	1.32	23.36

Source: See Ballato, J.G. and Gualtieri, J.G. 1994. Advances in high-Q piezoelectric resonator materials and devices. *IEEE Trans. on Ultrasonics, Ferroelectrics, and Frequency Control* 41(6): 834–844.

constant in several ways. One of these definitions is based on the equivalent circuit model. In this case the quality factor is the electrical quality factor. It is also believed that the motional time constant can be related to the material properties directly, for example, by relating it to the ratio of viscosity and the elastic stiffness. In Table 3.3, the motional time constant for a variety of materials used in bulk resonator applications is shown. Other parameters included in the table are quality factor Q, characteristic length L_0, square of the piezoelectric coupling coefficient k^2, and a frequency constant N_0. The motional time constant is inversely proportional to the maximum operating frequency for a given material and overtone. The relationship between L_0, N_0, and operating frequency is given by the formula from Ballato and Gualtieri (1994) as

$$f = N_0 (L_0 T^2)^{\frac{-1}{3}}$$

where T is the resonator thickness and f is the operating frequency at the maximum overtone.

3.1.7 Temperature and Stress Compensation Techniques and Aging Effects

Thermal Properties of Crystal Resonators

The property that most directly affects the performance of a crystal resonator is the thermal expansion of the crystal. This can result in shifts in the resonant frequency due to stresses occurring in the crystal and at the crystal–electrode interfaces. The temperature characteristics of the elastic constants are important to the frequency-temperature characteristic. These constants also depends in part on the cut of the crystal. An example is the common AT-cut for quartz that has a limiting frequency stability of $\pm 0.002\%$ over the range from -55 to $+105°C$. This stability limit can be further improved by reducing the operating temperature range. For applications requiring more enhanced stability performance, methods of temperature control or compensation can be employed.

The general relationship between frequency and temperature for crystal resonators is given by

$$f = A_1(t - t_0) + A_2(t - t_0)^2 + A_3(t - t_0)^3$$

where t_0 is the room temperature and t is the operating temperature, and A_1, A_2, and A_3 are constants. This relationship accounts for the resonator dimensions, size, and thickness of the electrodes and the electrode material. Another important factor in frequency-temperature stability is the overtone order of the operating frequency. It has been shown experimentally with AT-cut resonators that the frequency-temperature characteristic can be minimized for higher overtone operation (e.g., the third overtone).

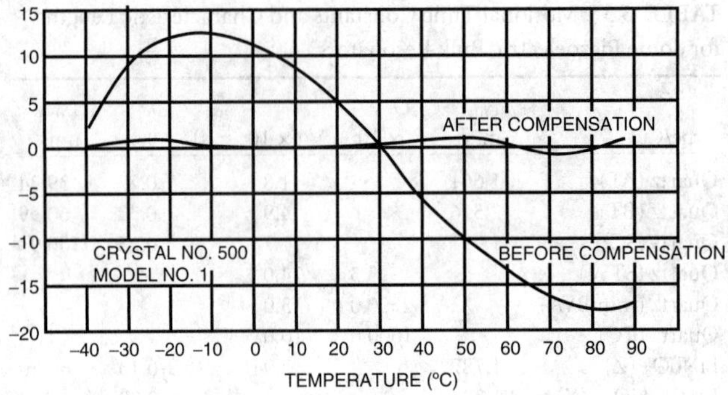

FIGURE 3.8 Frequency vs. temperature characteristics for a typical temperature compensated crystal oscillator. (From Frerking, M.E. 1978. *Crystal Oscillator Design and Temperature Compensation*, Chap. 4. Van Nostrand Reinhold, New York)

Temperature Control and Compensation and Aging

An oven is typically used for the purpose of controlling temperature. Crystal resonators reside inside the oven with the oscillator control circuitry either internal or external to the oven. When the circuitry is externally connected, stabilities in frequency of ± 5 parts in 10^7 are possible. For internal circuitry, stabilities of several parts per billion have been obtained in proportionally controlled ovens. Improved frequency-temperature stability can also be obtained with two-stage ovens. Usage of ovens is limited, however, by the disadvantages that arise such as required warmup time, high-power consumption, large volume, and reduced component reliability when frequent on/off cycling of the oven occurs.

Compensation for temperature-related frequency shifts can be accomplished by connecting a reactance (e.g., a variable voltage capacitor called a varactor) in series with the resonator. The effect of this reactance is to pull the frequency to cancel the effect of frequency drift caused by temperature changes. An example of a frequency vs. temperature characteristic both before and after compensation is shown in Fig. 3.8. Notice that the general form for the characteristic before compensation follows the cubic variation given earlier in this section.

Several techniques for temperature compensation are possible. They include analog temperature compensation, hybrid analog–digital compensation, and digital temperature compensation. For analog compensation a varactor is controlled by a voltage divider network that contains both thermistors and resistors to control the compensation. Stability values are in the range from 0.5 to 10 ppm. The lowest value requires that individual components in the voltage divider network be adjusted under controlled temperature conditions. This is difficult over large temperature ranges for analog networks because of the tight component value tolerances required and because of interactions between different analog segments. Analog networks can be segmented to increase the independence of the component adjustments required to improve stability.

These problems can be alleviated with the use of some form of hybrid or digital compensation. In this scheme, the crystal oscillator is connected to a varactor in series as in the analog case. This capacitor provides the coarse compensation (about ± 4 parts in 10^7). The fine compensation is provided by the digital network to further improve the stability. The hybrid analog–digital method is then capable of producing stability values of ± 5 parts in 10^8 over the range -40 to $+80°C$. For a detailed discussion of these temperature compensation techniques see Frerking (1978).

Stress Compensation

Stress compensation is employed to stabilize the crystal resonator frequency against shifts caused by mechanical stresses. These result from temperature changes or from electrode changes, due to differences in the expansion coefficients that can occur over time. The frequency variations are caused by alteration of the crystal elastic coefficients with stress bias. The method developed for stress compensation is a special crystal cut called the *stress-compensated* or SC-cut. This cut was developed to obtain a resonator, the frequency of which would be independent of the mentioned stresses. It was determined that a cut at angles near $\phi = 22.5°$ and $\theta = 34.5°$ on a doubly rotated quartz plate would be free of frequency shifts caused by stress bias in the plane of the plate. This cut was theoretically predicted by EerNisse, who named it.

The SC-cut has some advantages over the more common AT- and BT-cuts. These include a flatter frequency-temperature response, especially at the turning points, and less susceptibility to aging due to changes in the mounting stresses. The lower sensitivity to frequency changes means that the SC-cut resonators can be frequency-temperature tested more rapidly than their AT- and BT-cut counterparts. A disadvantage of the SC-cut is that it is sensitive to air damping and must be operated in a vacuum environment to obtain optimum Q factors. The frequency-temperature curve is also roughly 70° higher for the SC-cut over the more common AT-cut, which could also be a disadvantage in some applications.

Aging Effects

Aging in crystal oscillators generally refers to any change over time that affects the frequency characteristics of the oscillator or the physical parameters that describe the device, for example, motional time constant and equivalent circuit parameters. Some factors that influence aging include surface deterioration, surface contamination, electrode composition, and environmental conditions.

Deterioration on the surface contributes to strain in the crystal, and this type of strain can impact the operating frequency of the oscillator. More important, the surface strain is extremely sensitive to humidity and temperature early in the life of the crystal. One method of combating this surface degradation mechanism is to etch away the strained surface layers. HF or NH_4F_2 can be used to remove the strained surface layers. The surface is etched until the change in operating frequency increases only linearly as the etch time is increased. The initial slope of the frequency change vs. etch time curve is exponential. Once the strained surface layers have been removed, however, the slope becomes linear.

Another surface issue is contamination from foreign materials. The presence of foreign substances loads the oscillator with an additional mass, and this can be reflected in the equivalent circuit parameters. The contaminants can be introduced during unclean fabrication processes, or from out gassing of other materials after the device has been packaged or from contaminants in the environment during storage. The contamination process is often increased in higher temperature environments and temperature cycling leads to stability problems for the device.

Aging problems associated with the device electrode include delamination, leading to increased resistance or corrosion effects. Typical electrode materials include gold, silver, and aluminum. Of these, silver is the most popular material. Because of silver oxidation processes, however, the electrode mass can significantly affect the operational frequency. Reactions typically involve either oxygen or other elements that may be classified as halides. Occasionally, silver electrodes are exposed to iodine to produce a layer of silver iodide. Silver iodide is a largely inactive compound that protects against further increases in the corrosion process. Alternatively, gold is largely inactive and easy to deposit, so that a gold electrode may not require additional coating materials. On the other hand, gold only loosely binds with the quartz substrate so that a layer of chromium is often added to provide a surface for adhesion to gold. Without the chrome layer, delaminations may develop at the gold–quartz interface. Aluminum is a popular electrode material for high-speed operations. Aluminum, however, is highly reactive with oxygen. The growth of aluminum oxide layers is usually very slow, and after a layer thickness of $\sim$50 Å has been reached, the growth is reduced to near zero. Unfortunately, the reaction rate is not strongly influenced by temperature so that accelerated procedures involving high-temperature operations are largely ineffective. One technique to combat this aging effect is anodic oxidation. This process leaves a thick oxide layer on the electrode, but the growth rate is proportional to the electrode voltage. The process can, therefore, be accelerated to an acceptable time frame. Additional discussion of aging effects is presented Bottom (1982).

Defining Terms

Alpha quartz: The crystalline form of SiO_2 at temperatures below 573°.

AT-cut: A quartz plate cut from Y-bar quartz at an angle 35° with respect to the z axis.

BT-cut: A quartz plate cut from Y-bar quartz at an angle $-49°$ with respect to the z axis.

Crystal: A solid material composed of atoms, ions, or molecules that are arranged in a structured three-dimensional pattern.

Frequency: A measure of the number of cycles repeated per second.

Mode: An allowed oscillating frequency.

Motional time constant: A material parameter that is inversely proportional to product of the frequency and Q-factor for a particular mode.

Piezoelectric: A material effect whereby an applied electric field produces a mechanical strain, and an applied mechanical strain produces an electric field.

Q-factor: A figure of merit that represents that ratio of stored to dissipated energy per cycle.

Resonator: A cavity or chamber which has been cut to permit oscillation at a particular frequency.

References

Ballato, A. and Gualtieri, J.G. 1994. Advances in high-Q piezoelectric resonator materials and devices. *IEEE Trans. Ultrasonics, Ferroelectrics and Frequency Control* 41(6):834–844.

Bottom, V. 1982. *Introduction to Quartz Crystal Unit Design*. Van Nostrand Reinhold, New York.

Brendel, R., Marianneau, G., Blin, T., and Brunet, M. 1995. Computer-aided design of quartz crystal oscillators. *IEEE Trans. Ultrasonics, Ferroelectronic and Frequency Control* 42(4):700–707.

Brice, J.C. 1985. Crystals for quartz resonators. *Rev. of Modern Phys.* 57(1):105–146.

Dye, D.W. 1926. *Proc. Phys. Soc.* 38:399–457.

Eccles, M. 1994. Applying crystals. *Elec. World and Wireless World*, pp. 659–663.

EerNisse, E.P. 1975. *Proceedings of 29th Annual Symposium on Frequency Control*, pp. 1–4.

EerNisse, E.P. 1976. *Proceedings of 30th Annual Symposium on Frequency Control*, pp. 8–10.

Frerking, M.E. 1978. *Crystal Oscillator Design and Temperature Compensation*. Van Nostrand Reinhold, New York.

Gerber, E.A. and Ballato, A., eds. 1985. *Precision Frequency Control Vol. 1 and 2*. Academic Press, Orlando, FL.

Gottlieb, I.M. 1987. *Understanding Oscillators*. TAB Books, Blue Ridge Summit.

Walls, F. and Vig, J.R. 1995. Fundamental limits on the frequency stabilities of crystal oscillators. *IEEE Trans. Ultrasonics, Ferroelectronic and Frequency Control* 42(4):576–587.

Zelenka, J. 1986. *Piezoelectric Resonators and Their Applications*. Elsevier, Amsterdam.

Further Information

IEEE Transactions on Ultrasonics, Ferroelectronics, and Frequency Control. Proceedings of the Annual Symposium on Frequency Control.

3.2 Surface Acoustic Wave (SAW) Devices

Ardie D. Walser

3.2.1 Introduction

Acousto-optic modulators control the transmission of light by local changes in the index of refraction of the transmission medium. This modulation or change occurs by means of a traveling sound wave that induces a stress-related modification of the local index. The sound wave can be transverse (*coplanar device*), longitudinal (*colinear*), or a combination of both as in the case of a *surface acoustic wave* (SAW).

Optical wave diffraction can be produced by either bulk acoustic waves traveling in the volume of the material, or by surface acoustic waves propagating on the surface of the medium (that is, in the form of a planar waveguide). The penetration depth of the surface acoustic wave is within one acoustic wavelength (Λ) of the surface. The energy of the surface acoustic field is concentrated in the same spatial region as **optical waveguides**, making SAW devices well suited for integrated optic applications.

Whether bulk or surface acoustic waves are used, two fundamentally different types of optical modulations are possible. The type of modulation is dictated by the interaction length L of the optical beam with the acoustic wave, where L is approximately the width of the acoustic wave front (or acoustic beam spot size). In the Raman–Nath type modulator the optical beam transverses the acoustic beam, and the interaction length of the optical path (i.e., the width of the acoustic wave) is so short that the optical waves only undergo a phase grating diffraction, producing many interference peaks in the far field.

Increasing the interaction length by widening the acoustic beam causes the optical waves to undergo multiple diffractions before leaving the acoustic field. This is similar to the diffraction of X rays by atomic planes observed by Bragg. In such a situation the optical beam is incident at a specific angle (i.e., Bragg angle) to the acoustically induced index grating displaying only one intense diffraction lobe in the far-field pattern.

3.2.2 Photoelastic Effect

A material's optical index of refraction is coupled to its mechanical strain via the **photoelastic effect**, traditionally expressed by a fourth-rank tensor, $\mathbf{p}_{ijkl}$

$$\Delta \eta_{ij} = \Delta \left(\frac{1}{n^2} \right)_{ij} = \mathbf{p}_{ijkl} \mathbf{S}_{kl} \tag{3.1}$$

where $\Delta \eta_{ij}$ (or $\Delta(1/n^2)_{ij}$) is the change in the optical impermeability tensor and $\mathbf{S}_{kl}$ is the strain tensor, defined as

$$\mathbf{S}_{kl}(r) = \frac{1}{2} \left[\frac{\partial u_k(r)}{\partial x_l} + \frac{\partial u_l(r)}{\partial x_k} \right] \tag{3.2}$$

where u_k characterizes the mechanical displacement along the direction x_l, of an arbitrary point $P(x_1, x_2, x_3)$ in the body of a uniform elastic solid under stress. The variables $\mathbf{p}_{ijkl}$ form the strain-optic tensor. For a crystal in the presence of a strain field the index ellipsoid is given by

$$(\eta_{ij} + \mathbf{p}_{ijkl} \mathbf{S}_{kl}) x_i x_j = 1 \tag{3.3}$$

Exploiting the symmetry of the optical impermeability tensor η_{ij}, the strain tensor $\mathbf{S}_{kl}$ and the strain-optic tensor $\mathbf{p}_{ijkl}$, contracted indices can be used to abbreviate the notation. The equation for the change in the optical impermeability tensor then becomes

$$\Delta \left(\frac{1}{n^2} \right) = \mathbf{p}_{ij} \mathbf{S}_j i j, \quad i, j = 1, 2, \dots, 6 \tag{3.4}$$

where S_j are the strain components. The strain field alters the orientation and dimensions of the index ellipsoid. This change is dependent on the applied strain field and the strain optic coefficients $\mathbf{p}_{ijkl}$. Only the orientation or form of the strain-optic coefficients $\mathbf{p}_{ijkl}$ can be derived from the symmetry of the material, which is usually of crystalline form. Point group symmetry determines which of the 36 coefficients are zero, as well as the relationships that may exist between nonzero coefficients. For a more comprehensive analysis of photoelastic tensors, refer to Nye (1957).

An acoustic wave can produce a strain causing a change in the index of refraction via the photoelastic effect (Dixon, 1967). It has been shown (Pinnow, 1970) that the change in the index of refraction is related to the acoustic power P_a by the expression

$$\Delta n = \sqrt{\frac{n^6 p^2 P_a}{2\rho v^3 A}} \tag{3.5}$$

TABLE 3.4 Common Materials Used in Acousto-optic Interactions

Material	λ, μm	n	ρ, g/cm^2	v (10^3)/s	M, (10^{-15})
Fused quartz	0.63	1.46	2.2	5.95	1.51
GaP	0.63	3.31	4.13	6.32	44.6
GaAs	1.15	3.37	5.34	5.15	104
TiO$_2$	0.63	2.58	4.6	7.86	3.93
LiNbO$_3$	0.63	2.2	4.7	6.57	6.99
YAG	0.63	1.83	4.2	8.53	0.012
LiTaO$_3$	0.63	2.18	7.45	6.19	1.37
ADP	0.63	1.58	1.803	6.15	2.78
KDP	0.63	1.51	2.34	5.50	1.91

Sources: Adapted from Yariv, A. and Yeh, P. 1994. *Optical Waves in Crystals*, p. 346. Wiley, New York.

where n is the index of refraction in the unstrain medium, p the appropriate element of the photoelastic tensor, P_a the acoustic power in watts, ρ the mass density, v the acoustic velocity, and A the cross-sectional area that the acoustic wave travels through. This expression is commonly written as

$$\Delta n = \sqrt{\frac{M P_a}{2A}} \tag{3.6}$$

where M is defined by

$$M = \frac{n^6 p^2}{\rho v^3} \tag{3.7}$$

the acousto-optic figure of merit. M represents the relative magnitude of the index change produced by an elastic wave of constant power, it is therefore a convenient figure for comparing various acousto-optic materials. The value of M depends on the material and its orientation. Values of M for some common acousto-optic materials are shown in Table 3.4.

3.2.3 Acousto-Optic Effect

We will establish the basic principles that govern the behavior of all acousto-optic devices whether of bulk or waveguide (SAW) construction. An acousto-optic modulator is composed of an acoustic medium (such as water, glass, lithium niobate, rutile, etc.) and a transducer. The transducer converts electrical signals into sound waves propagating in the acoustic medium with an acoustic frequency spectrum that is limited by the bandwidth of the transducer that matches the electrical excitation. The sound wave causes a perturbation in the index of refraction of the material, setting up a refractive index grating of the form

$$n(\mathbf{r}, t) = n + \text{Re}[\Delta n \exp\{j(\Omega t - \mathbf{K} \cdot \mathbf{r})\}] \tag{3.8}$$

which has a grating period Λ equal to the sound wavelength in the material, and moves along the direction of the sound wave with a velocity

$$v = \frac{\Omega}{|\mathbf{K}|} = \frac{\Omega}{2\pi}\Lambda \tag{3.9}$$

equal to the sound velocity in the material. Since the velocity of light ($c = 3 \times 10^8$ m/s) is approximately five orders of magnitude greater than the velocity of sound (several times 10^3 m/s) the index grating can be viewed as stationary.

For clarity and ease of calculation we shall examine the scattering of an electromagnetic wave by an acoustical wave in an isotropic medium. Consider an acoustic wave that is launched from the bottom of a cell containing an isotropic fluid (water, etc.). A periodic strain distribution is established inducing a traveling index grating in the medium.

Raman–Nath Diffraction

When the cell is thin (Fig. 3.9) and the acoustic wave is weak, the optical beam direction change is very small on transmission through the cell. For Raman–Nath type diffraction, the interaction length must be so short that multiple diffractions do not take place. This condition occurs when

$$L \ll \frac{\Lambda^2}{\lambda} \qquad (3.10)$$

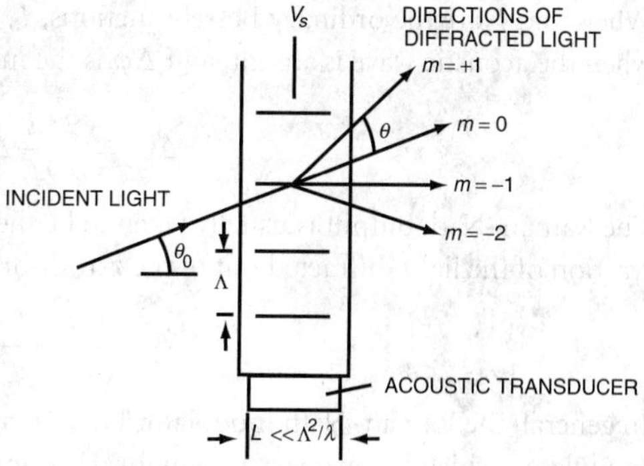

is satisfied, where λ is the optical wavelength within the material. Equation (3.10) establishes a relationship between the interaction length L and the acoustic wavelength Λ. The phase change of the emerging optical beam after passing through the acoustic wave is

FIGURE 3.9 An acousto-optic deflector in the Raman–Nath regime ($L \ll \Lambda^2/\lambda$), where the cell can be considered a phase grating spatially modulating the emerging beam.

$$\Delta\varphi = \frac{\Delta n 2\pi L}{\lambda} \sin\left(\frac{2\pi y}{\Lambda}\right) \qquad (3.11a)$$

and using Eq. (3.6)

$$\Delta\varphi = \frac{2\pi}{\lambda} \sqrt{\frac{MPL}{2a}} \sin\left(\frac{2\pi y}{\Lambda}\right) \qquad (3.11b)$$

where Δn is the index change caused by the sound wave, and A equals L multiplied by the thickness of the acoustic beam a. The spatial component $(\sin(2\pi y/\Lambda))$ adds a corrugated structure to the phase front of the transmitted optical beam. The index variation produced by the acoustic wave can be considered a phase diffraction grating, in which the transmitted light reflects off the grating into many orders (Fig. 3.9). The angle θ of the direction of a maximum of the emergent wave is given by

$$\sin\theta - \sin\theta_0 = m\frac{\lambda}{\Lambda}, \quad m = 0, \pm 1, \pm 2, \ldots \qquad (3.12)$$

where θ_0 is the angle of incidence, Λ is the wavelength of the acoustic wave, and m is an integer representing the order of a diffraction maxima. For normal incidence ($\theta_0 = \pi/2$) Eq. (3.12) becomes

$$\sin\theta = m\frac{\lambda}{\Lambda}, \quad m = 0, \pm 1, \pm 2, \ldots \qquad (3.13)$$

Hence, for Raman–Nath diffraction several discrete peaks are observed in the emerging optical beam. The intensity of these orders are given by (Hammer, 1982)

$$\frac{I}{I_o} = \frac{1}{2}[J_m(\Delta\varphi')]^2, \quad |m| > 0$$

$$\frac{I}{I_o} = [J_0(\Delta\varphi)]^2, \quad m = 0 \qquad (3.14)$$

Where the J are the ordinary Bessel functions, I_o is the intensity of the optical beam that is transmitted when the acoustic wave is absent, and $\Delta\varphi$ is the maximum value of $\Delta\varphi$ written as

$$\Delta\varphi' = \frac{2\pi L}{\lambda}\Delta n = \frac{2\pi}{\lambda}\sqrt{\frac{MPL}{2a}} \tag{3.15}$$

The Raman–Nath output is usually taken to be the zeroth-order mode. The modulation index equals the fraction of the light diffracted out of the zeroth-order and is expressed as

$$\frac{[I_o - I(m=0)]}{I_o} = 1 - [J_0(\Delta\varphi')]^2 \tag{3.16}$$

In general, the Raman–Nath modulator has a smaller modulation index than that of a comparable Bragg modulator, which is a primary reason for why the Raman–Nath modulator is not as popular as the Bragg type.

Bragg Diffraction

When the interaction length between the optical and acoustic beams are relatively long, so that multiple diffraction can occur, as is the case for thick cells, Bragg type diffraction takes place. The relation expressing this condition is given by

$$L \geq \frac{\Lambda^2}{\lambda} \tag{3.17}$$

where L is the interaction length and Λ is the acoustic wavelength defined earlier. The cell can be interpreted as a stack of reflectors spaced by the wavelength of the acoustic wave (Fig. 3.10). This is analogous to the treatment of X-ray diffraction by layers of atoms in a lattice, first studied by Bragg. When the path difference between rays reflected from adjacent layers equals an integral multiple of the optical wavelength (Fig. 3.11), the rays reflected from each layer interfere constructively and the total reflected light intensity displays a peak. The Bragg diffraction condition is

$$2\Lambda\sin\theta = \frac{\lambda}{n} \tag{3.18}$$

with n equal to the index of refraction of the medium and θ the angle of incidence. For a modulator, generally the zeroth-order beam is taken as the output. The modulation depth is then given as

$$\frac{I_O - I}{I_O} = \sin^2\left(\frac{\Delta\varphi'}{2}\right) = \sin^2\left[\frac{\pi}{\lambda}\sqrt{\frac{MPL}{2a}}\right] \tag{3.19}$$

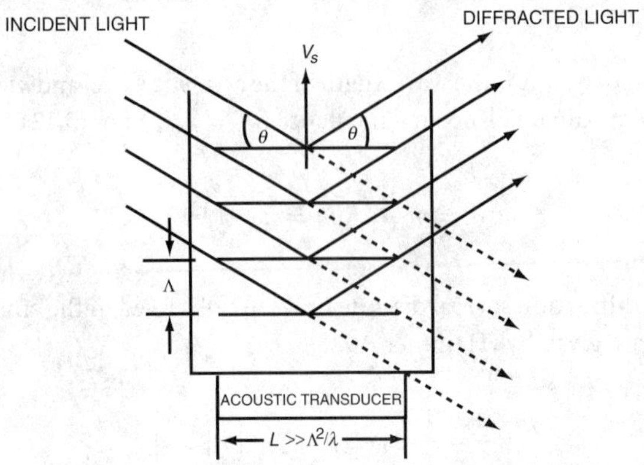

FIGURE 3.10   An acousto-optic deflector that is in the Bragg regime ($L \gg \Lambda^2/\lambda$); the cell can be viewed as a stack of reflecting layers with a period equal to the acoustic wavelength (Λ).

where I_O is the transmitted intensity when the acoustic wave is not present, I is the zeroth-order intensity in the presence of the acoustic beam, and $\Delta\varphi'$ is the maximum phase shift of the optical beam produced by the acoustic wave.

The Bragg diffraction condition shown in Eq. (3.18) is the result of assuming that the traveling acoustic wave is stationary relative to the optical beam. The motion of the acoustic wave can be accounted for by considering the Doppler shift undergone by an optical beam reflecting from a mirror moving at the velocity of sound v at an angle that satisfies the Bragg condition. The Doppler frequency shift of a wave reflected from a moving surface (such as the wave front of a traveling sound wave) is

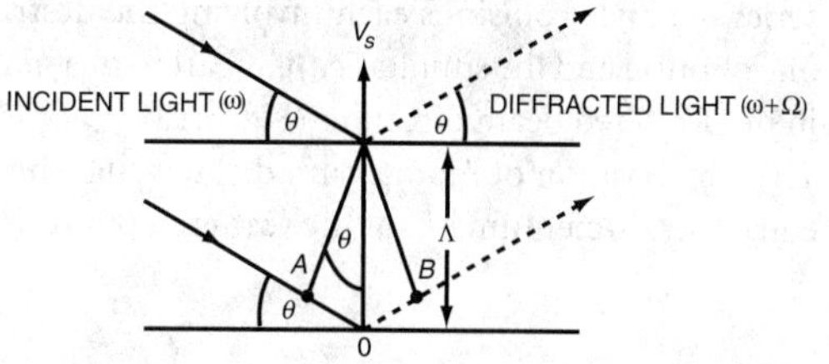

FIGURE 3.11 The reflections of an incident optical beam from two adjacent crests (separated by the acoustic wavelength Λ) of an acoustical wave moving with the velocity v_s. The reflections from the two crests will interfere constructively only if the optical path difference AO + OB equals an optical wavelength, thus, satisfying the Bragg condition ($2L \sin\theta = \lambda/n$).

$$\Delta\omega = 2\omega \frac{v_\parallel}{c/n} \tag{3.20}$$

where $\Delta\omega$ is the optical frequency and $v_\parallel$ is the projection of the surface velocity vector along the optical wave propagation direction. We can see from Fig. 3.12 that $v_\parallel = v \sin\theta$ substituting it into Eq. (3.20)

$$\Delta\omega = 2\omega \frac{v \sin\theta}{c/n} \tag{3.21}$$

Using the expression for the Bragg diffraction, Eq. (18) condition we can write

$$\Delta\Omega = \frac{2\pi v}{\Lambda} = \Omega \tag{3.22}$$

where the frequency of the reflected optical wave is up shifted by Ω (Fig. 3.11). If the direction of propagation of the sound wave were reversed, then the optical wave would be down shifted by Ω.

Particle Description of Acousto-Optic Interactions

Taking advantage of the dual particle-wave nature of light and of sound we can deduce many of the features of the diffraction of light by sound. In accordance with the quantum theory of light, an optical wave of angular frequency ω can be described as a stream of particles (photons), each of energy $\hbar\omega$ and momentum $\hbar k$. Similarly, an acoustic wave of angular frequency Ω can be thought of as made up of particles (phonons) with momentum $\hbar K$ and energy $\hbar\Omega$. The diffraction of light by an acoustic wave can be described as a

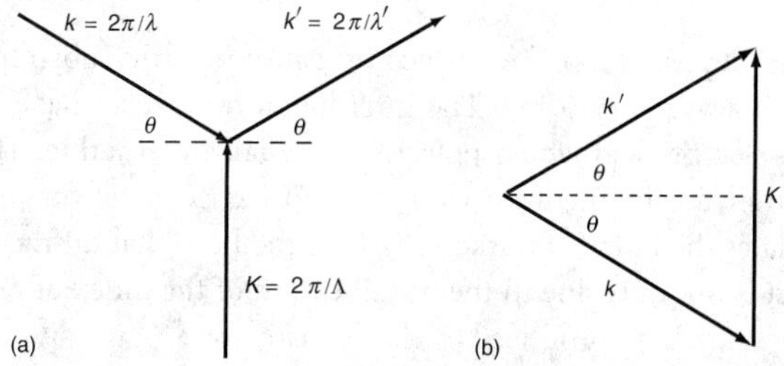

FIGURE 3.12 (a) Wavevector diagram of an incident optical beam (k) being scattered by an acoustic beam (K) into a diffracted beam (k'); (b) the vectorial description of the conservation of momentum that is used to derive the Bragg condition for the diffraction of an optical beam by an acoustic wave front.

series of single collisions each involving the destruction or annihilation of an incident photon at ω and one phonon and the simultaneous creation of a new diffracted photon at a frequency ω', which propagates in the scattered beam direction (Fig. 3.12).

The conservation of momentum dictates that the total momentum $\hbar(k + K)$ of the colliding particules equal the momentum $\hbar k'$ of the scattered photon, hence

$$k' = k + K \tag{3.23}$$

The conservation of energy requires that

$$\omega' = \omega + \Omega \tag{3.24}$$

the Doppler shift formula, where the diffracted optical waves frequency shift is equal to the sound frequency. The conservation of momentum condition is an alternate expression of Braggs condition (Fig. 3.12). To demonstrate this, consider Fig. 3.11 where the value of the sound frequency Ω is much less than that of the incident optical frequency ω. We can express this condition as follows

$$\omega \gg \Omega \tag{3.25}$$

From Eq. (3.24) and Eq. (3.25) we have

$$\omega' \approx \omega, \quad \text{and} \quad k' \approx k \tag{3.26}$$

Therefore, the magnitudes of the incident and diffracted optical wavevectors are equal, and the magnitude of the acoustic wavevector is

$$K = 2k \sin\theta \tag{3.27}$$

the Bragg condition.

3.2.4 Devices and Applications

The use of acousto-optic interaction offers the possibility of manipulating laser light at high speeds since the use of mechanical moving parts is not involved. Advances in the acousto-optic device applications stem from the availability of lasers, which produce intense, coherent light; from the development of efficient broadband transducers, which generate elastic waves up to the microwave frequencies; and from the discovery of materials with excellent elastic and optical properties. The **acousto-optic effect** can be used to build beam deflectors, light modulators, tunable filters, signal processors, and spectral analyzers.

Our discussion on the acousto-optic effect applies to both bulk and thin-film materials. In this section we will study devices using SAW. The SAW device is lighter and more compact, and for a given acoustic energy its diffraction efficiency is greater than that of corresponding bulk-effect devices.

SAW Construction

An interdigital transducer deposited on the surface of a piezoelectric substrate (i.e., lithium niobate) launches a surface acoustic wave (Fig. 3.13). The interdigital transducer has a comblike structure that establishes an alternating electric field whose polarity is spatially alternating. This alternation produces an electrically induced strain in the medium via the piezoelectric effect that excites an acoustic wave. The acoustic wave modulates the index of refraction of the medium that diffracts the optical beam. Note that this diffraction effect is not only due to the modification of the index of refraction, but also to the physical deformation of the medium produced by the acoustic wave.

Transducer

For a SAW waveguide modulator, an interdigitated, single periodic pattern of metal fingers are deposited directly onto the waveguide surface, as shown in (Fig. 3.14), where the transducer has a design

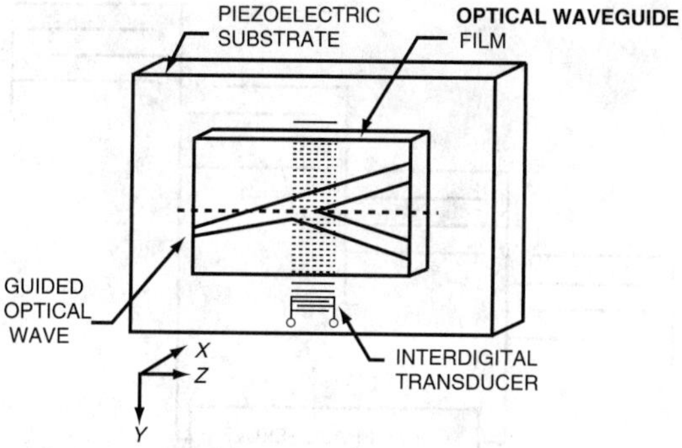

FIGURE 3.13 Waveguide acousto-optic Bragg modulator. The incident guided optical wave is deflected through an angle of 2θ by a surface acousitc wave (SAW) launched by an interdigital transducer.

constant referred to as the *aspect ratio L/Λ*. The interdigital transducer (IDT) launches a SAW in the direction normal to the finger pattern. Frequencies up to 1 GHz are possible using photolithographic techniques for this simple patterning. When designing a SAW waveguide modulator, wide bandwidth and high diffraction efficiency (which implies low acoustic driver power) are desired attributes. By decreasing the interactive length (acoustic aperture) L of a modulator, a larger bandwidth can be

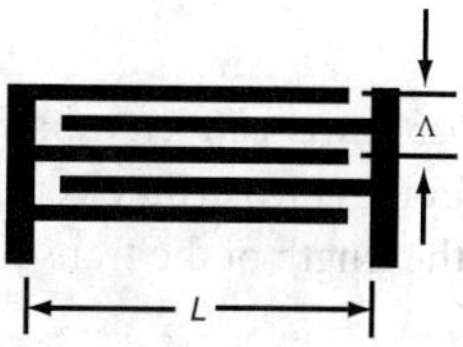

FIGURE 3.14 The standard interdigital transducer for SAW excitation, where the aspect ratio or relative aperture (L/Λ) is a design parameter.

achieved. However, this would be at the expense of the diffraction efficiency as can be seen in Eq. (3.19). To satisfy both conflicting conditions more elaborate transducer designs have been employed. Wide bandwidths have been accomplished by such schemes as the following

- *The multiple tilted-array transducer* (Lee, Kim, and Tsai, 1979), where an array of standard IDTs with staggered center frequencies are tilted toward each other in order to satisfy the Bragg condition at each frequency (Fig. 3.15).
- The *Phase-array transducer* (Nguyen and Tsai, 1977) that consists of several IDT segments with one period displacement (phase difference) between adjacent segments, using frequency control SAW beam steering (Fig. 3.16).

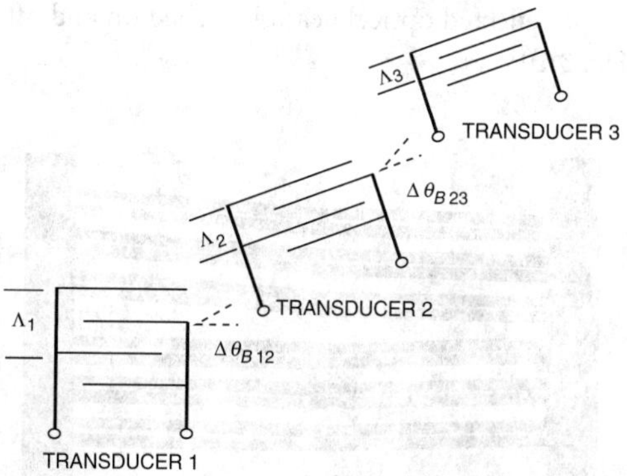

FIGURE 3.15 Multiperiodic transducer in a tilted array for increased bandwidth. The array of simple transducers have staggered center frequencies and are tilted toward each other in order to satisfy the Bragg condition at each center frequency. This increases the acoustic aperture and the frequency range of the transducer.

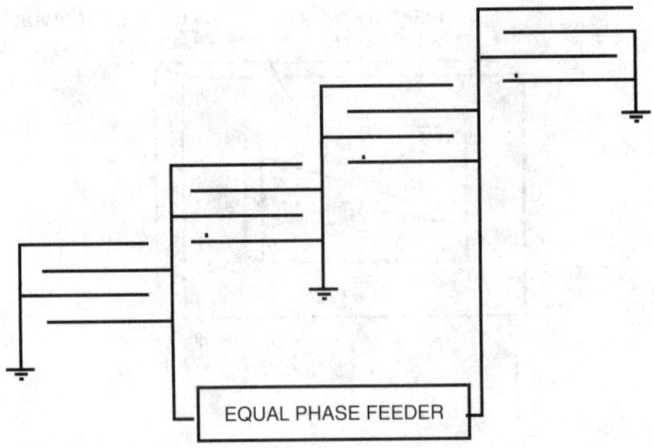

FIGURE 3.16 A phase-array transducer. By placing a variable phase shift of the RF signal between the individual transducers, a scanning acoustic wave front is generated, producing a wide aperture acoustic beam that tracks the Bragg condition over a large frequency bandwidth.

- The *tilted-finger chirped transducer* (Joseph and Chen, 1979), which broadens the excitation bandwidth by the chirp in the electrode period (Fig. 3.17). The interdigitated spacing is gradually changed along the length of the transducer in the direction of the SAW propagation.

Beam Deflectors and Switches (Bragg Type)

A number of light modulators using the acousto-optic effect can be fabricated. Both amplitude and frequency modulation can be achieved. The modulator can operate in either the Raman–Nath regime or the Bragg regime. However, the Raman–Nath modulators are limited to low frequency and have limited bandwidth. This is due to the short interaction length permitted by the criterion $L \ll \Lambda^2/\lambda$. Even at the maximum allowable interaction length L, for high RF (small Λ), excessive acoustic power would have to be used, causing Raman–Nath modulators to be impractical. We will concentrate on devices that operate in the Bragg regime, since these types of devices have wide bandwidth and can operate at high modulation frequencies.

The intensity of the diffracted beam of a Bragg cell is propotional to the amplitude of the acoustic wave. Modulating the amplitude of the acoustic wave causes a corresponding modulation on the diffracted optical beam. This device can be used as a linear light modulator (Fig. 3.18).

Increasing the amplitude of the acoustic wave eventually causes saturation to occur and almost complete diffraction of the optical beam is achieved. The modulator now behaves like an optical switch. By turning the acoustic wave on and off, the reflected optical beam is turned on and off and the transmitted optical beam is turned off and on (Fig. 3.19).

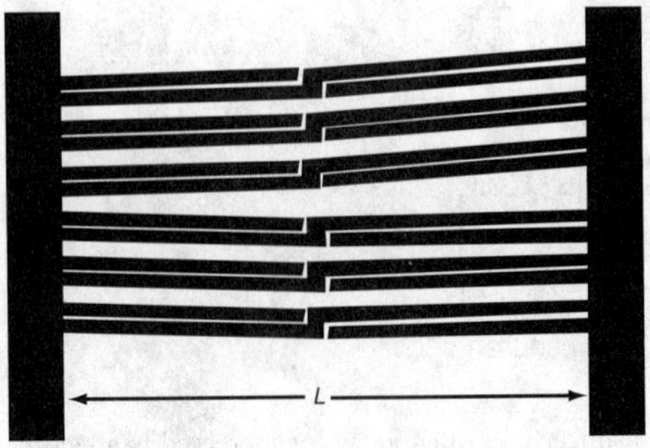

FIGURE 3.17 Tilted-finger chirped transducer.

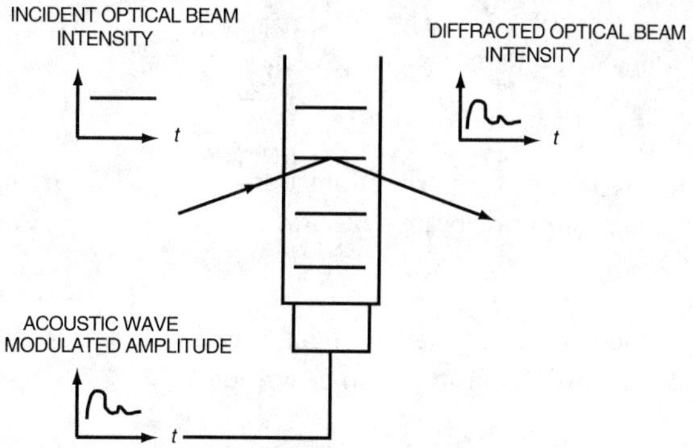

FIGURE 3.18 The intensity of the diffracted optical beam is proportional to the amplitude of the acoustic wave, hence the device is an acousto-optic modulator.

Important Parameters of SAW Devices

Next we shall define some of the important parameters that describe the characteristics of Bragg type acousto-optic devices.

Deflection Angle

An acousto-optic Bragg diffraction occurs only when the incident beam is at the Bragg angle θ_B with respect to the acoustic wave. Using the Bragg condition

$$K = 2k \sin \theta_B \tag{3.28}$$

we can write

$$\theta_B = \sin^{-1} \frac{\lambda}{2n\Lambda} = \sin^{-1} \frac{\lambda f}{2nv} \tag{3.29}$$

where λ and n are the wavelength and effective index of the optical guided wave, respectively and Λ, f, and v are the SAW wavelength, frequency, and phase velocity, respectively. In practice the Bragg angle is

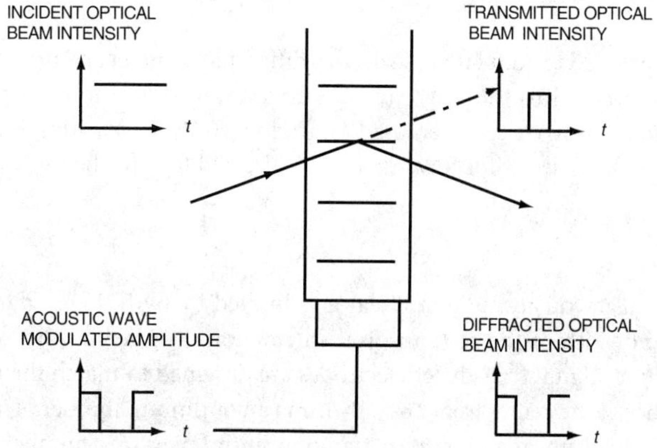

FIGURE 3.19 Increasing the power of the acoustic wave will eventually cause saturation, where total diffraction occurs and the modulator performs as an acousto-optic switch. Turning the sound on causes the optical beam to deflect, and turning the sound off allows the optical beam to pass unperturbed.

small so we can write

$$2\theta_B = \frac{\lambda}{n\Lambda} = \frac{\lambda f}{nv} \tag{3.30}$$

Thus, the deflection angle $2\theta_B$ is approximately proportional to the SAW frequency, which is the basic principle of operation of an acousto-optic beam deflector.

Diffraction Efficiency

The diffraction efficiency (the fraction of incident light energy diffracted) at the Bragg condition for SAW-propagation width L and SAW intensity I_a can be written as

$$\eta = \frac{I_{\text{diffracted}}}{I_{\text{incident}}} = \sin^2\left(\frac{\pi L}{\lambda\sqrt{2}\cos\theta_B}\sqrt{MI_a}\right) \tag{3.31}$$

The acoustic intensity I_a required for total conversion of the incident light is given by

$$I_a = \frac{\lambda^2 \cos^2\theta_B}{2ML^2} \tag{3.32}$$

Noting that a SAW is localized in a layer of thickness nearly equal to the acoustic wavelength Λ, the power per unit area averaged in this layer (intensity I_a) can be approximated by $P_a/\Lambda L$. Using this relation with Eq. (3.32) the acoustic power required, P_a, is

$$P_a = \Lambda L I_a = \frac{\lambda^2 \cos^2\theta_B}{2M}\left(\frac{\Lambda}{L}\right) \tag{3.33}$$

It is desirable according to Eq. (3.33) to have a small aspect ratio (Λ/L) for efficient operation of a modulator. In the low diffraction efficiency region ($n < 0.5$, low acoustic power levels) Eq. (3.31) can be written as

$$\eta = \frac{I_{\text{diffracted}}}{I_{\text{incident}}} = \frac{\pi^2}{2\lambda^2 \cos^2\theta_B}\left(\frac{L}{\Lambda}\right) MP_a \tag{3.34}$$

or using $I_a = p_a/\Lambda L$

$$\eta = \frac{I_{\text{diffracted}}}{I_{\text{incident}}} = \frac{\pi^2}{2\lambda^2 \cos^2\theta_B}\left(\frac{L}{\Lambda}\right) MP_a \tag{3.35}$$

Thus, if the acoustic power is modulated, so is the diffracted light beam intensity. The SAW power P_a is the RF power that is fed to the SAW transducer multiplied by the conversion efficiency of the transducer. The interdigital transducer often used to produce SAW propagation is a bidirectional device. The maximum theoretical efficiency for SAW excitation directed toward the interaction region is -3 dB, which is approached in practice.

Bandwidth

The bandwidth of a Bragg acousto-optic modulator is limited by both the excitation bandwidth of the transducer and the bandwidth of the acousto-optic interaction. In practice, the aspect ratio (or relative aperture) L/Λ of the interdigital transducer electrodes are designed to match the radiation resistance to the RF source impedance at the center frequency. By using an optimum number of interdigital finger pairs N_{opt}, the excitation bandwidth is maximized. For an optimum finger pair number N_{opt} the relative SAW excitation bandwidth $2\Delta f_T/f$ is given by

$$\frac{2\Delta f_T}{f} \approx \frac{1}{N_{\text{opt}}} \tag{3.36}$$

The interaction bandwidth $2\Delta f_{AO}/f$ is determined by the angular selectivity $(2\Delta\theta_i)$ of a Bragg diffraction. The SAW transducer is oriented such that the direction of propagation of the acoustical and optical waves satisfy the Bragg condition at the center frequency. A change in the frequency dictates a corresponding change in the Bragg angle θ_B and a deviation from the Bragg condition. This, in turn, causes a decrease in the diffraction efficiency through the angular selectivity of the acousto-optic grating. From Eq. (3.30) we have

$$\frac{\Delta f_{AO}}{f} = \frac{2\Delta\theta_b nv}{\lambda f} \tag{3.37}$$

Using

$$2\Delta\theta_i \approx \frac{5}{QL} \approx \frac{\Lambda}{L} \tag{3.38}$$

the expression for the angular selectivity (a parameter associated with the decline in the diffraction efficiency due to angular deviation from the Bragg condition) where Q is the magnitude acoustic wave vector $2\pi/\Lambda$, we can write Eq. (3.37) as

$$\frac{2\Delta f_{AO}}{f} = \frac{10nv}{\lambda 2\pi f}\left(\frac{\Lambda}{L}\right) \tag{3.39}$$

Note that the interaction bandwidth is inversely proportional to the acoustic frequency f and proportional to the acoustic wavelength Λ for a given aspect ratio L/Λ. The smaller value of $2\Delta f_{AO}/f$ and $2\Delta f_T/f$ determines the overall bandwidth of the Bragg cell. At low frequencies the SAW excitation bandwidth limits the overall bandwidth. At high frequencies the SAW is limited by the interaction bandwidth.

The modulation bandwidth Δf of an acousto-optic modulator is determined primarily by the angular spread of the optical beam. The finite angular distribution of the wave vector allows the Bragg diffraction to occur over a finite range of acoustic frequencies. By differentiating Eq. (3.30), the expression for the bandwidth is

$$\Delta f = \frac{2nv}{\lambda}\Delta\theta_B \tag{3.40}$$

Response Speed

The SAW transit time over the interaction region determines the Bragg cells response speed τ. Where D represents the guide wave width (or the interaction region) the response time is written as

$$\tau = \frac{D}{v} \tag{3.41}$$

To insure that the input and output guided waves spatially overlap the SAW propagation width L, width D must satisfy the following condition:

$$D > 2L\tan\theta_B \approx \frac{\lambda L}{n\Lambda} \tag{3.42}$$

This relationship places a lower limit on D, where one might seek to reduce it in order to increase the response time of the Bragg cell.

Time Bandwidth Product

The time bandwidth product is a nondimensional parameter defined as the product of the bandwidth $2\Delta f$ and the SAW transit time τ. For a Bragg cell not only is size of the absolute deflection important but so is the number of resolvable spots. For a guided wave with a width D the beam divergence angle is

$$2\delta\theta = 0.88\lambda/nD \tag{3.43}$$

due to diffraction. Using Eq. (3.40) and Eq. (3.41) the number N of resolvable spots can be written

$$N = \frac{\Delta\theta_b}{\delta\theta} = \frac{D}{0.88v}(2\Delta f) \approx \tau 2\Delta f \tag{3.44}$$

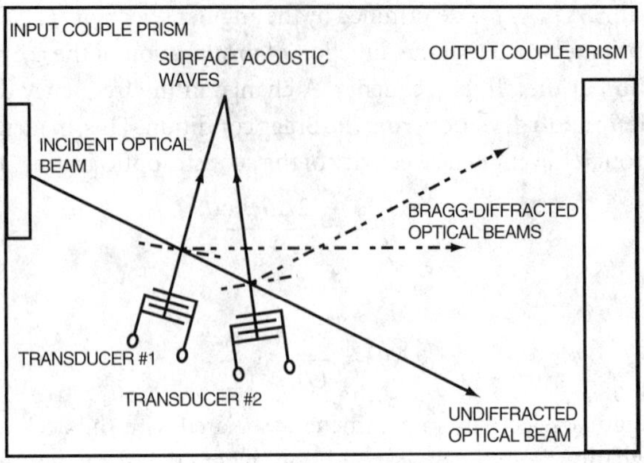

FIGURE 3.20   Schematic drawing of a guided-wave acousto-optic Bragg deflector, using two tilted surface acoustic waves.

Surface Acoustic Wave Beam Deflectors

The use of $LiNbO_3$ thin-film waveguides for Bragg interactions between surface acoustic waves and optical guided waves have been demonstrated (Kuhn et al., 1970). Since the diffraction efficiency η depends on the acoustic intensity I_a, with the confinement of the acoustic power near the surface (around Λ), only a low modulation or switching power is needed to alter the guided optical wave. Shown in Fig. 3.20 is an experimental setup in which both the surface wave and optical beam are guided in the waveguide. Transducer 1 and transducer 2 diffract the incident beam into two different diffraction optical beams that are spatially displaced. Multiple tilted transducers are very effective because they permit the Bragg condition to be satisfied over a large acoustic aperture (interaction length) and over a wide frequency range.

Acousto-Optic Tunable Filters (AOTF)

Spectral tunable filters can be fabricated exploiting the acousto-optic interaction. In general tunable filters employ codirectional (large-angle) acousto-optic interaction; however, a small angle scheme can be used. In either case a strong diffraction (or reflection) of the incident optical wave occurs only when the Bragg condition is satisfied. If the incident optical beam contains several frequencies (or consists of multiple spectral components), only the optical frequency that satisfies the Bragg condition at a given acoustic frequency will be diffracted. Therefore, by varying the acoustic frequency, the frequency of the diffracted optical beam can also be varied.

Illustrated in Fig. 3.21 is an acousto-optic tunable filter in the noncollinear geometry. These filters offer advantages such as the freedom of choosing propagation directions for various applications and the

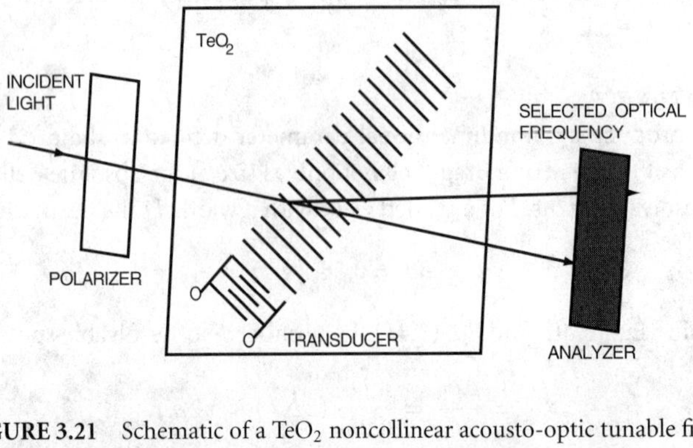

FIGURE 3.21   Schematic of a TeO_2 noncollinear acousto-optic tunable filter.

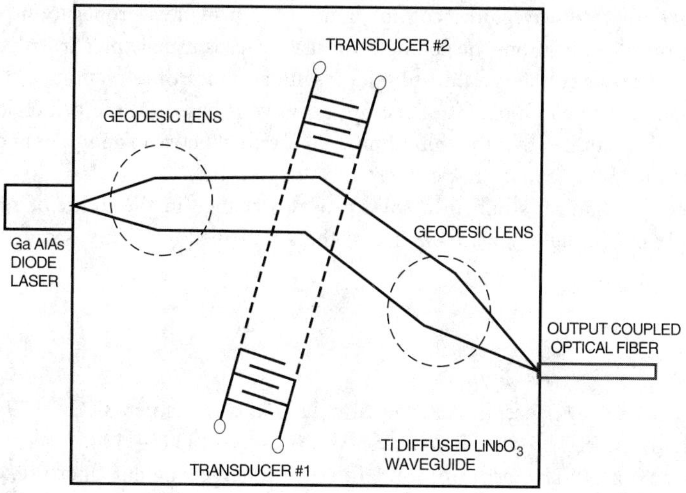

FIGURE 3.22 Schematic for a standing wave surface acoustic wave optical modulator (SWSAWOM) for optical frequency shifting.

freedom of choosing the operation acoustic frequency for a given crystal. A TeO_2 noncollinear acousto-optic tunable filter was constructed that could tune from 7000 to 4500 A by changing the acoustic frequency from 100 to 180 MHz (Chang, 1970).

Acousto-Optic Frequency Shifter

As stated earlier the frequency shifting effect of an acousto-optic device is usually considered neglible when the device is used as a modulator or beam shifter. In certain applications, however, this small shift in frequency can be useful. Illustrated in Fig. 3.22, is a standing wave surface acoustic wave optical modulator (SWSAWOM) that can produce frequency shifts of an optical carrier wave that are large enough to permit frequency multiplexing of a number of information signal channels onto one optical beam (Ih et al., 1988). In this device a Ti diffused waveguide is formed in a $LiNbO_3$ substrate, where the output of a semiconductor laser is butt coupled into the waveguide. The guided optical beam is then collimated by a **geodesic lens** before passing through a SWSAWOM. The SWSAWOM consist of two transducers each generating a SAW in opposite directions, producing an upward and downward frequency shift of the optical waves. Thus the diffracted optical beam has frequency components at $f_o + f_m$ and $f_o - f_m$, where f_m is the RF signal at which the transducers are driven. The diffracted optical beam is focused into a square-law detector by a second geodesic lens. The two spectral components of the optical beam beat together to produce a modulated subcarrier frequency of $f_c = 2 f_m$. Several of these units could be constructed, each with its own particular information impressed upon its unique subcarrier frequency f_{ci}. The signals can be sorted out at the receiver via microwave filters.

The applications of SAW devices are numerous, their implementation as deflectors, tunable filters, and frequency shifters, as described, are just a few of their possible uses. Other important device applications are time multiplexer, pulse compression of chirped signal, correlators, spectrum analyzer, and isolators.

Defining Terms

Acousto-optic effect: Mechanical strain produced in a material by the passage of an acoustic wave, causing a change in the index of refraction in the material via the photoelastic effect.

Geodesic lens: A lens composed of circularly symmetric aspherical depressions (or dimples) in the surface of a thin-film waveguide. In this type of lens, light waves are confined in the waveguide and follow the longer curved path through the lens region. Waves propagating near the center of the lens travel a longer path than waves near the edge, modifing the wave front so that focusing can occur.

Optical waveguides: Optical waveguides confine, guide, and provide a propagation path for light. For integrated optical applications the geometry of the device is usually planar with a rectangular slab structure. The device contains a thin-film layer with an index of refraction (n_g) within which the light propagates. The confinement of the optical wave is based on the principle of total internal reflection, which requires that the thin film be sandwiched between an upper and lower layer with indices of refraction n_u and n_c, respectively, of values less than n_g.

Photoelastic effect: Mechanical strain in a solid causes a change in the index of refraction that can affect the phase of a light wave traveling in the strain medium. This is called the photoelastic effect.

References

Barnoski, M.K., Chen, B.-U., Joseph, T.R., Ya-Min Lee, J., and Ramer, O.G. 1979. Integrated-optic spectrum analyzer. *IEEE Trans. Cir. and Syst.* Vol. Cas-26, 12:1113–1124.

Chang, I.C. 1974. Noncollinear acousto-optic filter with large angular aperture. *Appl. Phys. Lett.* 25:370–372.

Dixon, R.W. 1967. Photoelastic properties of selected materials and their relevance for applications to acoustic light modulators and scanners, *J. of App. Phys.* (Dec.) 38:5149.

Hammer, J.M. 1982. Modulation and switching of light in dielectric waveguides. In *Integrated Optics*, 2nd ed., ed. T. Tamir, Topic Appl. Phys., Vol. 7, p. 155. Springer–Verlag, Berlin.

Kuhn, L., Dakss, M.L., Heidrich, P.F., and Scott, B.A. 1970. Deflection of an optical guided wave by a surface acoustic wave. *Appl. Phys. Lett.* 17:265.

Lee, C.C., Liao, K.-Y., Lang, C.L., and Tasi, C.S. 1979. Wide-band guided wave acoustooptic Bragg deflector using a titled-finger chirp transducer. *IEEE J.* QE15(10):1166–1170.

Nguyen, L.T. and Tsai, C.S. 1977. Efficient wideband guided-wave acoustooptic Bragg diffraction using phased surface acoustic wave array in LiNbO$_3$ waveguides. *Appl. Opt.* 16:1297–1304.

Nye, J.F. 1957. *Physical Properties of Crystals.* p. 241. Oxford Clarendon Press, London.

Pinnow, D. 1970. Guidelines for the selection of acoustooptic materials. *IEEE J.* QE6:223–238.

Tsai, C.S. 1979. Guided-wave acoustooptic Bragg modulators for wide-band integrated optic communications and signal processing. *IEEE Trans. Cir. and Syst.* Vol. Cas-12, 12:1072–1123.

Further Information

IEEE Journal of Quantum Electronics: Published monthly by the IEEE Lasers and Electro-Optics Society. 445 Hoes Lane, P.O. Box 1331, Piscataway, NJ 08855-1331.

Laser Focus World: Published monthly by PennWell Publications. 10 Tara Blvd., 5th Floor, Nashua, NH 03062.

Optics Letters: Published monthly by the Optical Society of America. 2010 Massachusetts Ave., NW, Washington, DC 20036.

Phototonics Spectra: Published monthly by Laurin Publications. Berkshire Common, P.O. Box 4949, Pittsfield, MA 01202.

Banerjee, P.P. and Poon, T. 1991. *Principles of Applied Optics.* Richard D. Irwin, Inc. and Aksen Associates, Inc. Homewood, IL.

Buckman, A.B. 1992. *Guided-Wave Photonics.* Jovanich College, Harcourt Brace, Fort Worth, TX.

Hunsperger, R.G. 1982, 1984, 1991. *Integrated Optics*, 3rd ed. Theory and Technology, Springer Series in Optical Sciences, Vol. 33. Springer–Verlag. Berlin.

Lizuka, K. 1985, 1987. *Engineering Optics*, 2nd ed. Springer Series in Optical Sciences, Vol. 35. Springer–Verlag, Berlin.

Nishihara, H., Haruna, M., and Suhara, T. 1989. *Optical Integrated Circuits.* McGraw–Hill, New York.

Pollock, C.R. 1995. *Fundamentals of Optoelectronics.* Richard D. Irwin, Inc., Chicago, IL.

Yariv, A. and Pochi Yeh. 1984. *Optical Waves in Crystals.* Wiley, New York.

3.3 Electromechanical Devices

C. Sankaran

3.3.1 Introduction

Electricity has long been recognized as a force with unlimited potential. Uncontrolled, electricity can perform no useful function, but can cause serious damage to property and personnel. Electromechanical devices allow us to control and harness the vast potential of electricity. While discussing electromechanical equipment, electricity and magnetism must be given equal status. Most electromechanical devices require interaction between electricity and magnetism to perform the required task. Because of the interdependency of electricity and magnetism, knowledge of the basic relationship between the two is vital to understand how electromechanical devices operate. The following laws emphasize the relationship between the two physical properties.

3.3.2 Ampere's Law

The law states that the **magnetic field intensity** around a closed magnetic circuit is equal to the total current linkage of the electromagnetic circuit,

$$\oint H dL = Ni,$$

where H is the magnetic field intensity, N the number of turns in the electrical winding, and dL is an elementary length of the magnetic circuit. In Fig. 3.23, $L = 2\pi R$ is the mean length of the toroid magnetic circuit. From Ampere's law, $HL = Ni$. The field intensity $H(\text{A/m}) = Ni/L$.

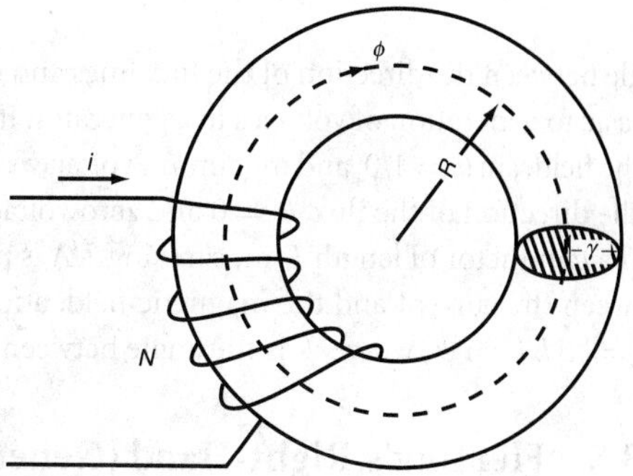

FIGURE 3.23 Ampere's law.

The magnetic field intensity is responsible for producing a magnetic **flux** Ø in the magnetic medium. The **flux density** or the amount of flux lines per unit area of cross section of the magnetic medium is given by $B = \mu H$, where μ is known as the absolute permeability of the magnetic medium. Also, $\mu = \mu_0 \mu_r$, where μ_0 is the permeability of free space equal to $4\pi\,10^{-7}$ Wb/A/m and μ_r is the **relative permeability** of the magnetic medium. Permeability is the measure of ease with which a magnetic medium is made to carry magnetic flux. The higher the permeability, the smaller is the magnetizing force necessary to produce a given quantity of flux.

3.3.3 Faraday's Law

The law states that a time-varying magnetic flux field induces a voltage in a single conductor given by $e = -d\varphi/dt$ where $d\varphi/dt$ is the rate of change of flux with time and e is the induced voltage in the conductor. The negative sign indicates that the direction of the induced voltage is in a direction that would oppose the direction of the changing field. If N number of conductors are subjected to the changing magnetic field, then $e = -N d\varphi/dt$. This equation states that if the expression for the flux with respect to time is known, then the expression for the induced voltage is obtained by differentiating φ with respect to time.

3.3.4 Generator and Motor Laws

A voltage is induced in a conductor situated in a magnetic field if the conductor is moved across the flux field so as to create relative motion between the conductor and the flux lines. Generators utilize this principle for generating electricity. The induced voltage in a conductor of length L m, moving across a magnetic field of density B T with a velocity of V m/s is given by $e = BLV \sin \theta$, where θ is the relative

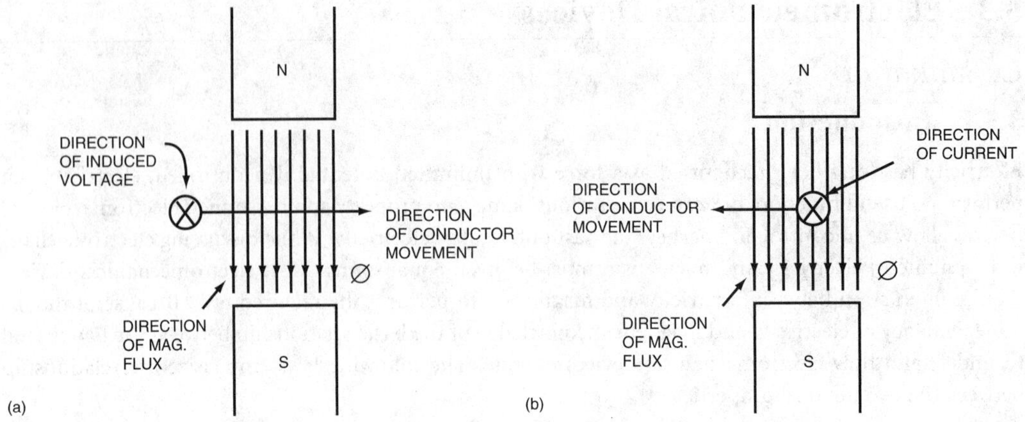

FIGURE 3.24   (a) Generator action, (b) motor action.

angle between the direction of the flux lines and direction of movement of the conductor. This expression is basic to generation of voltages in a generator. It is to be noted that if the conductors move perpendicular to the field, $\sin \theta = 1.0$, and maximum voltage is induced in the conductor. If the conductor moves parallel to the direction of the flux, $\theta = 0$ and zero voltage is induced in the conductor.

If a conductor of length L m, carrying I A is placed in a magnetic field of density B, due to interaction between the current and the magnetic field, a force is exerted on the conductor. The force is given by F (N) $= BLI \sin \psi$, where ψ is the angle between the direction of the flux and the axis of the conductor.

3.3.5   Fleming's Right-Hand (Generator) and Left-Hand (Motor) Rules

The directions of the magnetic field, the direction of movement of the conductor, and the direction of the induced voltage during generator action are governed by Fleming's right-hand rule. The rule states, with the index finger, middle finger, and the thumb of the right hand oriented at right angle to each other, if the index finger and the thumb represent the directions of the flux lines and the direction of movement of the conductor, the middle finger points in the direction of the induced voltage. Figure 3.24(a) illustrates Fleming's right-hand generator rule.

The direction of the magnetic field, the direction of the current, and the direction of motion experienced by a conductor during motor action are governed by Fleming's left-hand rule. The rule states, with the index finger, the middle finger and the thumb of the left hand placed at right angles to each other, if the index finger and the middle finger, represent the direction of the flux lines and the direction of the current in a conductor, the thumb points in the direction of movement of the conductor. Figure 3.24(b) illustrates the left-hand motor rule.

3.3.6   Switch Mechanisms and Relays

Switches are devices that are used to make or break an electrical circuit. Switches are classified into manually operated and electrically operated types. Electrically operated switches are usually larger and rated to interrupt large magnitudes of currents. During interruption of current in an electrical circuit, arcs are produced. The temperature of arcs can exceed $5000°$C. If the arc is not rapidly extinguished, substantial contact erosion would occur. This would shorten the life of the switch considerably. To withstand high-temperatures and the mechanical forces associated with contact operation, the contacts in large switches are made of high-temperature, high-strength alloys of copper and silver. Periodic inspection and cleaning of the contacts must be part of maintenance of the switches. Irregularities associated with arcing and pitting and carbon deposits must be cleaned using a fine steel wool pad. Push buttons and toggle switches are usually low-current devices, rated 10 A or less used in low-energy electrical control or power circuits rated less than 600 V. These are classified into maintained contact or momentary contact devices depending on if the change in contact status is to be momentary or maintained for the circuit to operate. These devices are provided with a *make* and a *break* rating. Since arcing accompanies current disruption, the break ratings of these devices are lower than the make ratings. Normally, push buttons and relays are tested for number of operations exceeding 100,000 and perhaps as high as one million. These devices are sometimes provided with DC ratings. DC ratings are lower than AC ratings since current interruptions are more difficult to accomplish for direct currents due to the absence of zero crossover points of the current, which facilitate current interruption.

Power disconnect switches are used in high-energy electrical power systems for opening and closing electrical circuits rated up to 4000 A at 480 V. Smaller disconnect switches are manually operated, whereas larger switches use spring charged mechanisms to achieve fast contact operation. Arc chutes and arc chambers are provided in the power switches to divide, divert, and extinguish the arc. Maintenance procedures of power disconnect switches must include examination and cleaning of the arc chutes.

Relays are electromechanical devices that open or close an electrical circuit in response to a signal, which may be voltage, current, power, frequency, speed, temperature, pressure, flow, level, or any other physical phenomenon. Electromagnetic relays operate by the forces due to magnetic fields on an armature. Electromagnetic relays consist of an operating coil located on an electromagnet in which an armature is free to slide. When a current is circulated in the coil, a magnetic field is produced in the core of the electromagnet. The action of the magnetic field on the armature causes the armature to slide, which is the actuating action in a relay. Although it takes only a few milliamps of current to keep a relay sealed in, several amperes may be needed to initiate the movement of the armature. An important consideration for application of relays in control circuits is the selection of control transformers, which supply the relays. Excessive voltage drop in the transformer windings during initial relay operation could cause the relay to not operate. Solid-state relays are more versatile and operate within a broader voltage range. Some solid-state relays may be used for AC or DC operation. Solid-state relays are also more easily integrated into process controller schemes due to their versatility.

3.3.7 Fuses

Fuses are electrothermal devices that are used to interrupt current flow through an electrical circuit if the heating effect of the current exceeds the thermal rating of the fusible element within the fuse. As current flows through the fuse element, heat, which is proportional to the square of the current, is produced. If the current exceeds the rating of the fuse, then the heat generated by the current melts the fusible element thereby interrupting the current flow. Fuses have unique current/time (I/T) characteristics based on their design. Fast acting fuses clear the overload condition in less time than time-lag fuses. Figure 3.25 shows typical melt times of a family of time-lag fuses. Selection of fuses must be based on the thermal withstand capabilities of the load being protected. Fuse coordination is a term that refers to a fuse application process, which assigns priorities to the order in which fuses must open while protecting an electrical system. It is always preferred that the fuse immediately upstream of the protected circuit open first in case of overcurrents and the fuse nearest to the source should be the last device to open. This selectivity is required to maintain power to rest of the electrical system in case of abnormal conditions in one circuit.

Current limiting fuses are designed to open an electrical circuit before the current waveform can reach its first peak. Figure 3.26 compares energy levels that would be absorbed by the faulted circuit with and without the current limiting fuse. Without fault current limiting, considerable damage to equipment would occur. Dual-element fuses are designed to withstand transient currents, which may be present during motor starting or transformer turn on. Dual-element fuses have time-delay characteristics, which allow them to withstand short durations of current surges. These fuses are provided with a short-circuit element and an overload, element. The overload element operates under prolonged overloads, whereas the short circuit element acts to open the circuit in case of short circuits.

Fuses are classified according to their voltage and current ratings and according to their ability to interrupt fault currents. The interrupting rating of fuses varies according to their design. Failure to provide

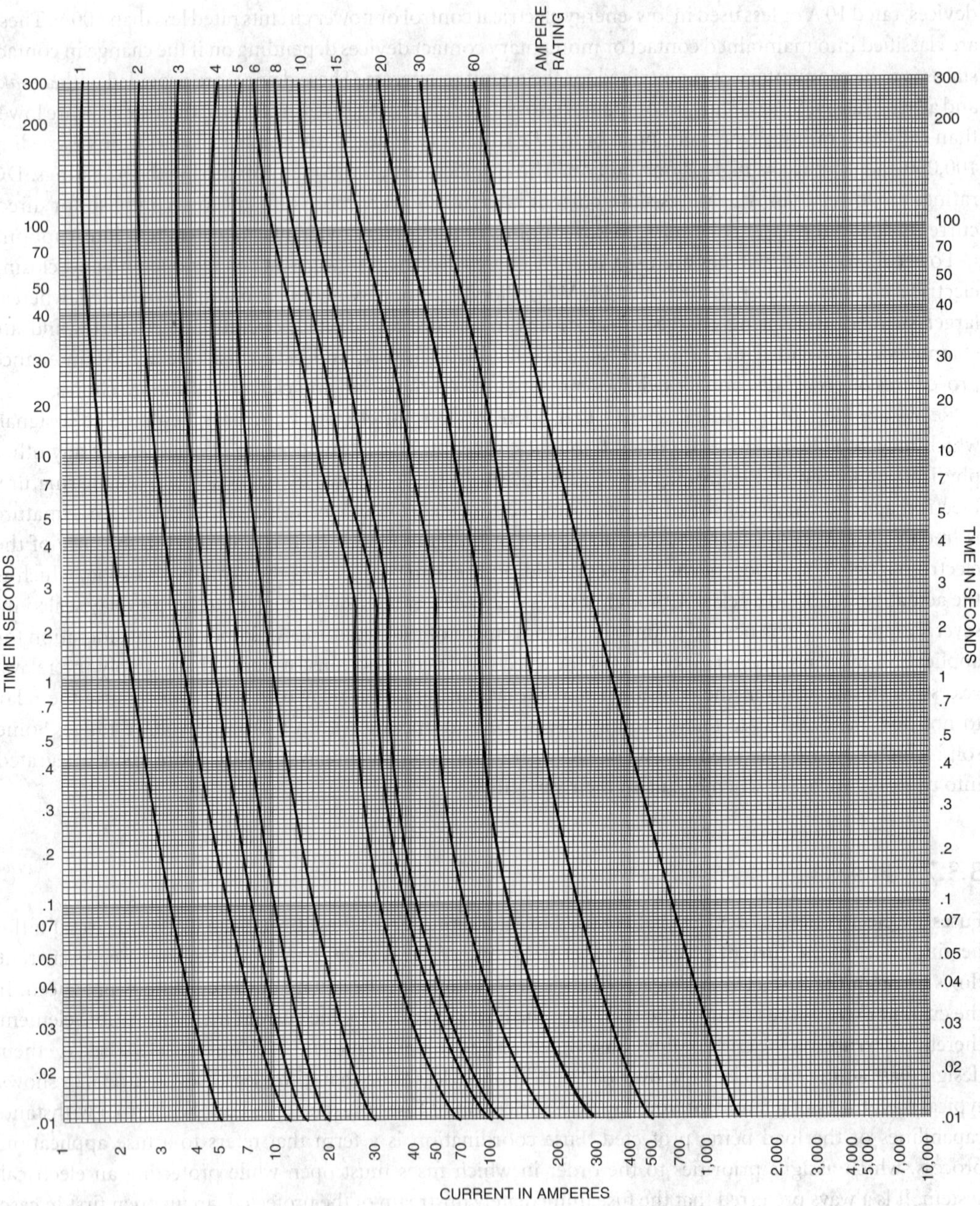

FIGURE 3.25 Average total clearing time-current characteristic curves.

fuses with adequate interrupting ratings can result in considerable damage to equipment and possible injury to personnel.

3.3.8 Circuit Breakers

Circuit breakers are electromechanical devices used for the interruption of electrical currents due to overloads or short circuits. Most circuit breakers are thermal magnetic devices. The thermal element in a circuit breaker is a bimetal strip, which is heated by the current. The deflection of the bimetal strip due to overloads causes the breaker to trip. The bimetal strip provides inverse time trip protection as the amount

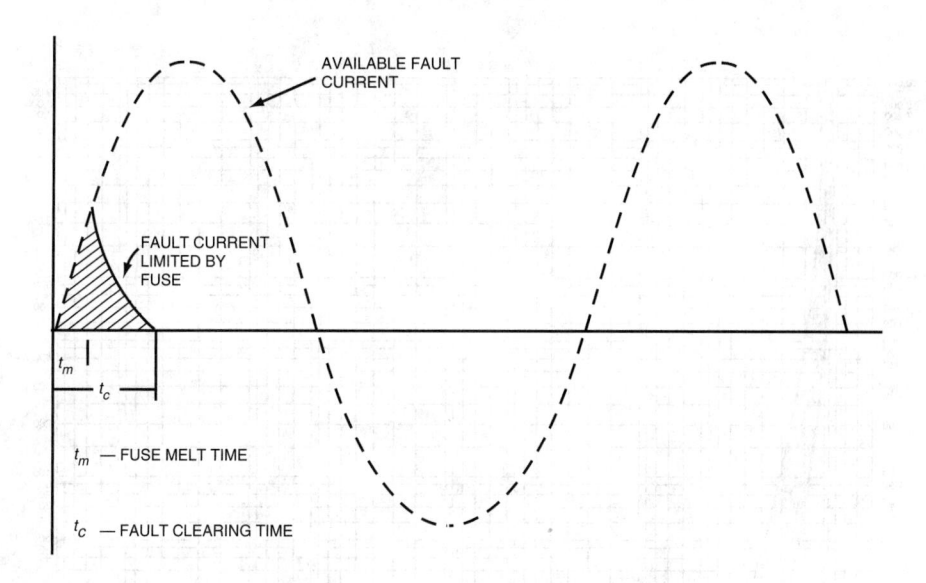

FIGURE 3.26 Current limiting fuse action.

of heat produced in the strip is proportional to the product of the square of the current and time (I^2t). Protection during short circuit conditions is provided by a magnetic element, which is actuated by the large currents that are generated during a short circuit. Typical operating characteristics of a thermal magnetic circuit breaker is shown in Fig. 3.27. Conventional thermal magnetic circuit breakers are being replaced by electronic circuit breakers. These breakers incorporate circuitry to sense the current and operate the circuit breaker in accordance with its design algorithm. These breakers incorporate potentiometer adjustments to change the overcurrent protection levels and operating time. In addition, changing the rating of the circuit breaker is readily accomplished by changing the rating plug provided on the breaker. Electronic circuit breakers also provide effective protection against ground faults. Ground fault occurs when an energized electrical wire contacts a grounded metal structure. Depending on the impedance of the ground fault, the fault currents may not be high enough to cause a conventional thermal magnetic circuit breaker to open. The electronic circuit breakers have special circuitry to detect low current ground faults. Figure 3.28 shows typical adjustments available on an electronic circuit breaker.

Circuit breakers are rated according to their voltage, current, and symmetrical current interrupting ratings. It is vital that the ratings of the breaker match the requirements of the circuit being protected.

3.3.9 Indicating Devices

Indicating devices are used to detect and measure electrical parameters such as voltage, current, frequency, real and reactive power, power factor, and energy. Most indicating devices use the principle of interaction between magnetic fields for their operation. In a moving coil meter, the coil carrying the signal to be measured is placed across the open poles of a permanent magnet iron core. The interaction between the permanent magnet field and the field due to the signal causes a rotating torque proportional to the signal. Since reversal of current in the coil produces opposite torque, moving coil indicators are only used to measure DC signals. Iron vane-type instruments have two iron vanes, which are magnetized by the coil carrying the signal. The two vanes are magnetized to opposite polarities producing a repulsive torque, which is proportional to square of the signal. Iron vane instruments can be used to measure DC or AC signals. Dynamometer-type instruments are used to measure power. These contain two field coils and a moving coil, which is suspended in the resultant field of the field coils. The field coils carry the voltage signal and the moving coil carries the current signal. The torque produced is proportional to the product of the voltage and the current and the phase angle between the two.

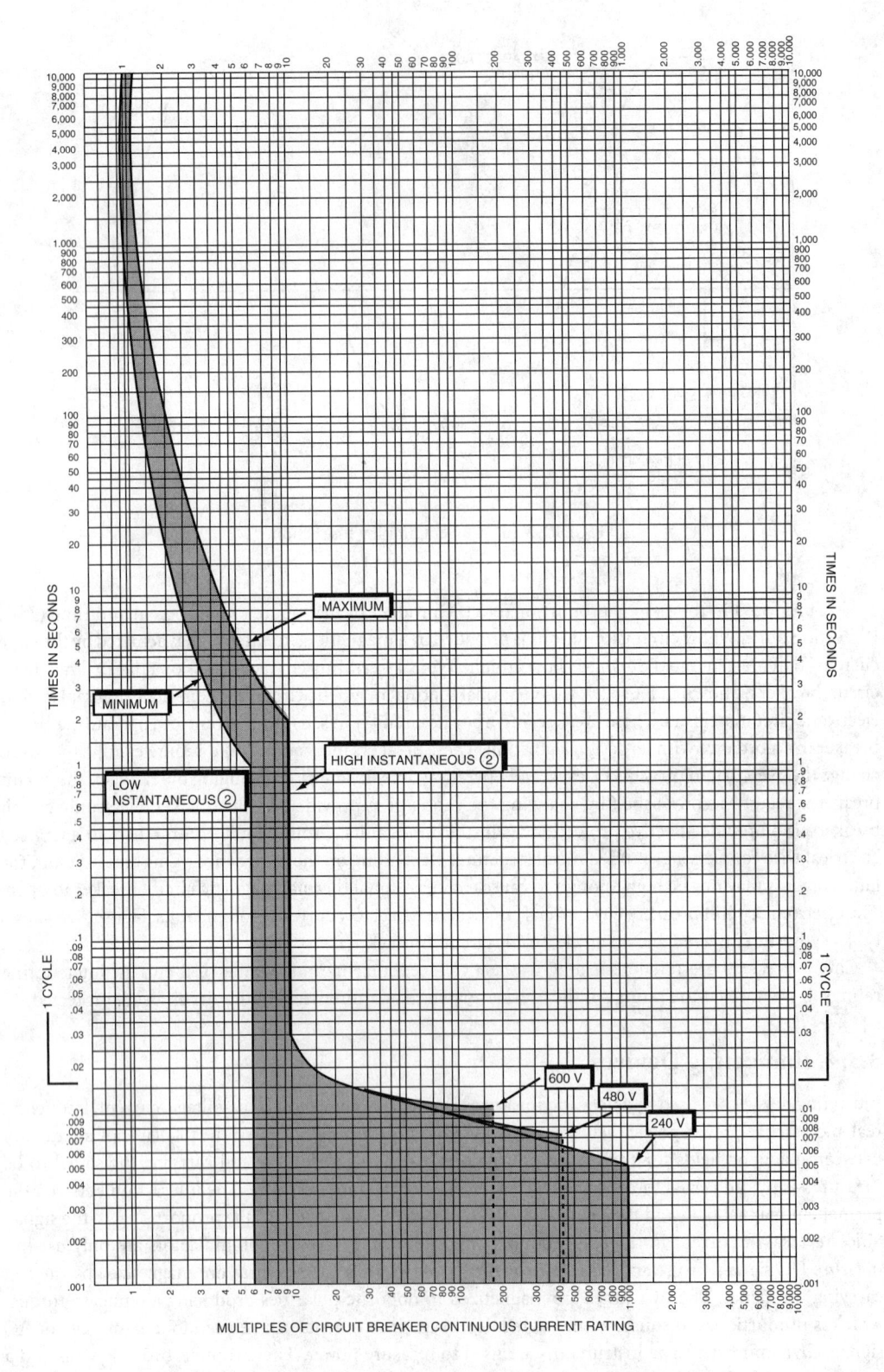

FIGURE 3.27   Operating characteristics of thermal magnetic circuit breaker.

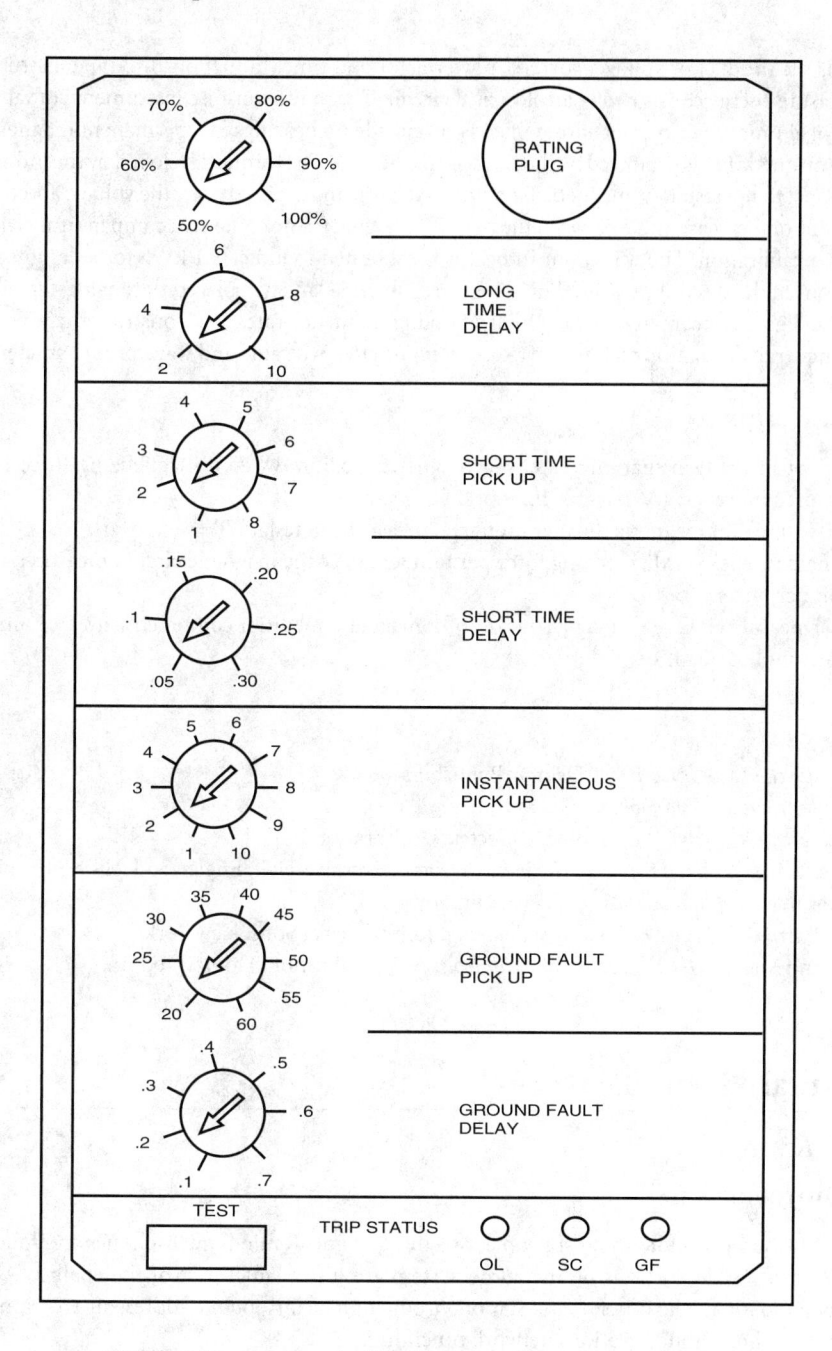

FIGURE 3.28 Front panel adjustment on an electronic circuit breaker.

Rectifier-type instruments rectify the AC signal to DC for detection in a moving coil instrument. Since the moving coil instrument produces deflection proportional to the average value of the signal, the meter readings are scaled by 1.11 to indicate RMS values of the measured AC signal. However, if the AC signal contains substantial waveform distortion due to harmonics, large errors are introduced in the measurement.

Thermocouple-type instruments measure the temperature produced when a precision resistor is heated by the signal. The electrical current produced by the thermocouple is used in a moving coil instrument to measure the signal. Since the heat produced in a resistance is proportional to the RMS value of the signal, accuracy is maintained while measuring waveforms containing high-frequency components.

Energy is the product of voltage, current, powerfactor and time. Induction-disc-type instruments are used to measure energy consumed in an electrical circuit. The induction-disc instrument contains a metal disc made of aluminum or copper alloy, which is supported on bearings in a permanent magnet core and exposed to magnetic fields produced by the voltage and current. The torque developed by the induction disc is proportional to the resultant magnetic field induced in the induction disc by the voltage and the current and the cosine of the phase angle between the two. The magnet frame contains a compensating winding for power factor calibration. The watt-hour indicator is calibrated to indicate 1 kWh for a designed number of revolutions of the disc. At present, digital instruments have largely replaced the analog-type indicating devices. The digital instruments employ high sampling frequency rates to reconstruct the waveform and use frequency transformation techniques to ascertain its effective value and waveform characteristics.

Defining Terms

Flux: Lines of magnetism that circulate in a magnetic medium when a magnetizing force is applied, measured in weber; 1 Wb = 10^8 lines of flux.

Flux density: Lines of magnetic flux per unit area, measured in tesla; 1 T = 1 Wb/m^2.

Magnetic field intensity: Magnetizing force per unit length of the magnetic circuit, measured in ampere turns per meter.

Relative permeability: Ease of magnetization of a magnetic medium compared with free air. This is a dimensionless quantity.

References

Alerich, W. *Electric Motor Controls*. Delmar Publishers.
American Electricians Handbook. McGraw-Hill.
Applied Protective Relaying. Westinghouse Electric Corporation.
Burgseth, F.R. and Venkata, S.S. *Electric Energy Devices*. Prentice-Hall, Englewood Cliffs, NJ.
Electrical Protection Handbook. BUSS Fuse Company.
Mason, C.R. *Art and Science of Protective Relaying*. John Wiley & Sons, New York.
Rockis, G. and Mazur, G. *Electrical Motor Controls*. ATP Publication, Homewood, IL.

3.4   Aural Pickup Devices

Roy W. Rising

3.4.1   Introduction

When you say the word audio to most people, two things come to mind: microphones and loudspeakers. These two most visible portions of any audio system are just a small part of the chain of equipment that makes audio for a radio or television station, production facility, public address/intercommunication system, or any other communications network function.

The selection of equipment for a professional audio environment is a critical part of facility planning. Good planning can make the difference between a system that is capable of handling whatever jobs may arise and one that serves as a barrier to getting projects done. Every part of the audio chain is important. There are no insignificant links. Key elements of an audio facility include:

- Source origination equipment, including microphones, tape recorders and disk-based systems
- Audio control and mixing equipment
- Distribution and buffer amplifiers
- Patch bay systems
- Audio equalization and special effects equipment

3.4.2 Aural Pickup Devices

Every electronic system has a starting point. For audio, that point usually is a microphone. Advancements in recording technology have, however, given operators a multitude of options with regard to programming. The source origination equipment available for a well-equipped audio system ranges from traditional microphones to recorders of various types including digital tape and disk-based recording and playback systems. Technology has given audio programmers a new menu of sources.

Microphone Types

Microphones are transducers. No matter how large or small, elaborate or simple, expensive or economical a microphone might be, it has only one basic function: to convert acoustical energy to electrical energy. With this fundamental in mind, you might wonder why microphones exist in such a wide array of sizes, shapes, frequency-response characteristics, and element types. The answer is that although the basic function of all microphones is the same, they have to work in many different applications and under various conditions. Choosing the right microphone for a particular job might seem easy, but it is a decision that deserves considerable thought. Just as no two production situations are alike, the microphone requirements also are varied.

Microphone manufacturers try to offer a selection of products to match any application. With a good working knowledge of the various designs, choosing the right microphone for the job becomes an easier task. The learning process begins with a look at some of the microphones commonly in use today.

Microphone types can be roughly divided into two basic classes: those intended for professional applications, such as broadcast stations and recording studios, and those designed for industrial applications, such as communications equipment and special purpose audio pickup where faithful audio reproduction is not the primary concern.

Professional Microphones

A wide variety of microphones is available for professional applications. The requirements of the professional audio/video industry have led to the development of products intended for specialized applications. In general, three basic classes of microphone elements are used by professionals: dynamic, condenser, and ribbon types.

Apart from the basic microphone elements, professional users also have a number of product classes from which to choose. The selection includes:

- Stand mounted or handheld
- Tiny clip-on
- Shotgun
- Wireless
- Parabolic
- Line level

Each has its place in the professional sound recorder's bag of technical tricks.

Dynamic Microphones

Dynamic microphones are the real workhorses of the professional world. They are rugged and reliable, not particularly temperamental, and they do not require a powered impedance-conversion circuit (as condenser types do). Dynamic microphones also provide excellent sound quality, some matching the detail and precision of the finest, most expensive condensers.

The basic mechanical design of a dynamic microphone is shown in Fig. 3.29. The dynamic transducer consists of a lightweight, bobbinless coil (the **voice coil**) attached to a rigid aluminum or mylar dome diaphragm and suspended in the gap between the poles of a permanent magnet. When sound waves move the diaphragm, the coil produces an output voltage. The cartridge is shock mounted in the case, which normally includes a foam filter screen for dirt and breath pop protection. Some products include a non-moving *hum bucking* winding whose inverted output phase cancels magnetically induced hum.

The source impedance of a dynamic microphone is mostly resistive, so that the output is not substantially affected by electrical loading. These microphones are capable of wide dynamic range with low noise and minimal distortion. Dynamic microphones can be readily designed for unidirectional or omnidirectional response patterns.

Although dynamic microphones can be made fairly small and lightweight, performance tends to suffer as the size is decreased. For this reason, condenser-type transducers have taken over the clip-on body microphone market. These devices, however, are easily damaged by moisture.

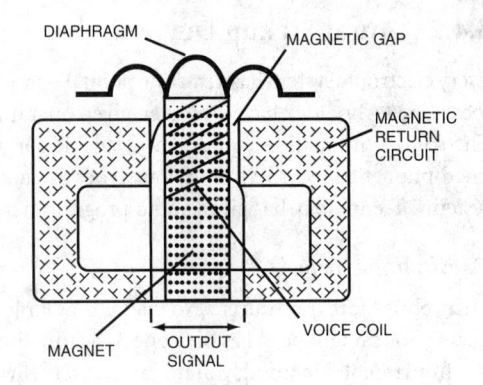

FIGURE 3.29   Basic construction of a dynamic microphone.

Condenser Microphones

Condenser types have become popular for good reasons. They can be small and lightweight, and they can deliver accurate sound reproduction. In the world of microphones there are tradeoffs. To get the advantages of a condenser microphone you have to put up with some disadvantages. These include the following:

- The need to power an impedance-converting circuit
- A high degree of sensitivity to wind noise
- A dynamic range that is limited by the impedance converter
- The need for shielding against RF and electromagnetic interference

Condenser microphones also are more sensitive to humidity and physical shock than dynamic types.

The transducer element of a condenser microphone consists of a lightweight metal or metalized plastic diaphragm located very near a metal backplate. This forms a capacitor, as shown in Fig. 3.30. When sound waves strike the diaphragm, they cause a change in spacing between it and the backplate, varying the capacitance. With the application of a DC bias, the capacitance variations are translated into variations in electrical voltage. Extremely small capacitance gives condenser microphones a high impedance, making it necessary to add an active circuit that converts the element impedance to a lower value compatible with established standards.

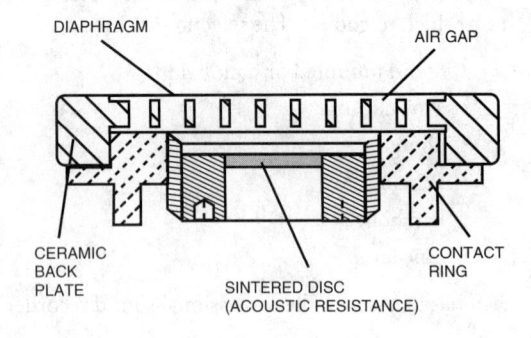

FIGURE 3.30   Mechanical construction of a condenser microphone.

A polarizing voltage of between 9 and 48 V DC is applied to the diaphragm by an external power supply. This often is built into the audio mixing console and fed as a phantom current through the microphone cable. Figure 3.31 shows the circuit for delivering standard 48 V DC **phantom power**. Some microphones do not work correctly at voltages below 36 V DC.

Because the diaphragm of a condenser microphone is not loaded down with the mass of a voice coil, it can respond quickly and accurately to incident sound. Condensers generally have excellent audio char acteristics, making them a good choice for professional audio applications. Photomicrographs of condenser diaphragms at work have revealed why these microphones have varying personalities. Instead of the expected resonating drum head appearance, the surface may look like crinkled aluminum foil (L. Burroughs, 1969; personal communication). This results in uneven high-frequency response sometimes

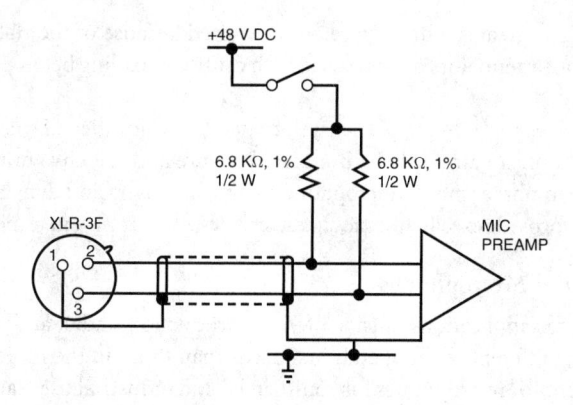

FIGURE 3.31   Typical circuit for 48-VDC phantom power.

called peakiness. Manufacturers' graphs usually average this unevenness as a rise in the presence range, about 6–10 kHz.

Another type of condenser microphone is the **electret** (more accurately called an **electret-biased** condenser microphone). In this case, the condenser's DC bias is supplied by an electret material, rather than a battery or power supply. The electret material, generally a fluorocarbon polymer, can be a part of the diaphragm or the backplate. Its electrostatic charge lasts almost indefinitely. A well-designed electret capsule retains its charge and sensitivity for 100 years or so. After more than 30 years of use, the sensitivity may drop by no more than 3 dB.

Electrets still require an impedance converter. It normally is housed within the microphone case, powered by either an internal battery or external phantom power. The amplifier, usually a field-effect transistor (FET) stage with high-input impedance, buffers the capsule from standard preamp input impedances that otherwise would resemble a short-circuit.

Electrets are most common as tiny clip-on microphones. The extremely small diaphragm in these units eliminates most of the physical shortcomings found in large diaphragm microphones. They provide very high accuracy, approaching the theoretically perfect point pickup. Electrets are available for both professional sound recording and laboratory applications.

Ribbon Microphones

Ribbon devices are not nearly as common as condenser and dynamic microphones, largely because of their reputation for fragility. However, they do provide good frequency response, excellent sound quality, and low handling noise.

The construction of a typical ribbon microphone is shown in Fig. 3.32. Basically, the ribbon microphone produces an output signal the same way as the dynamic microphone: a conductor moves in a magnetic field inducing a voltage. Instead of a voice coil, the ribbon microphone uses a thin corrugated strip of aluminum foil suspended between the poles of a permanent magnet. The ribbon, which may be as thin as 0.001 in, acts like a 1/2-turn coil and also serves as the diaphragm.

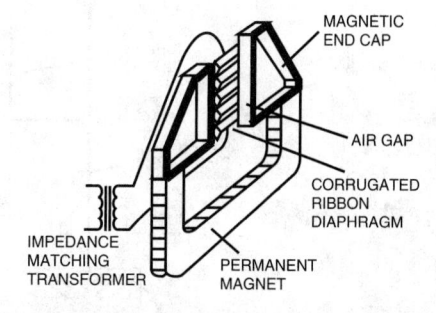

FIGURE 3.32   Basic construction of a ribbon microphone.

The output voltage of the ribbon diaphragm is very small, and the impedance is very low. Therefore, ribbon microphones include a transformer to boost the signal voltage. The highly inductive source impedance of ribbon microphones is sensitive to loading. The nominal (named) input impedance of a microphone preamp usually is 150 Ω. The actual value should be 2000 Ω or greater. Although seldom required because of the ribbon microphone's low-output level, passive input attenuators must be used with caution. Loading below 2000 Ω will compromise frequency response.

A ribbon microphone responds to the particle velocity of the atmosphere. Dynamic and condenser microphones react to variations in atmospheric pressure. Pressure and velocity always are 90° out of phase, and so it is not possible to phase ribbon microphones with condensers and dynamics.

Ribbon microphones provide excellent sonic characteristics, good transient response, and low self-noise.

Commercial/Industrial Microphones

Microphones intended for applications outside of video production, radio, and music recording or performance are built for a different set of operational needs than those in the professional audio industry. The primary requirements of microphones for commercial and industrial users are ruggedness and intelligibility under adverse conditions. Sonic purity is less important.

Microphones in this group include hand-held noise-canceling units for mobile communications and industrial plant operations, handset and headset-boom microphones for intercom and order wire circuits, and public address microphones. Each user group has its own particular set of needs, met through variations on a few basic designs.

Carbon Microphones

Carbon microphones are low cost, rugged, reliable, and produce a high-output signal. They are used by the millions in telephone handsets. However, their output signal is inconsistent, somewhat distorted, and useful only over a limited frequency range.

The carbon microphone consists of a metal diaphragm, a fixed backplate, and carbon granules sandwiched in between (see Fig. 3.33). When sound waves move the diaphragm they alter the pressure on the granules, changing the resistance between diaphragm and backplate. Like a condenser microphone, this modulates an externally supplied voltage. The output of the carbon button is balanced by a transformer also serving to block DC from the preamplifier input.

Piezoelectric Microphones

Piezoelectric devices (also known as **crystal** or **ceramic** microphones) provide fairly broad frequency response and high output, especially at low frequencies. They generally are inexpensive and reliable, finding

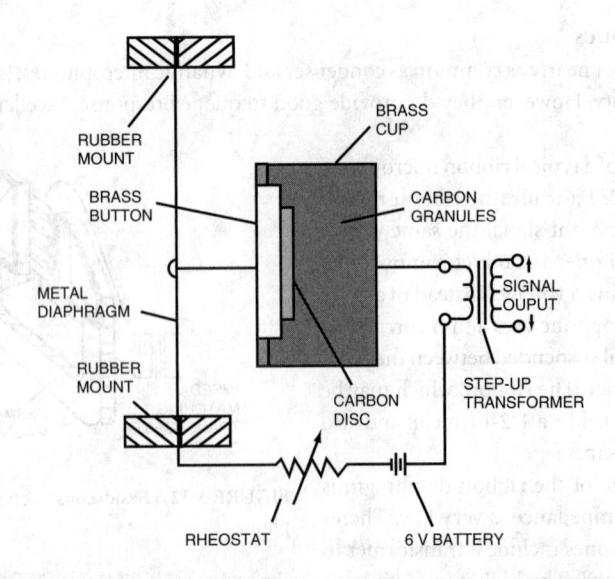

FIGURE 3.33 Construction and external electrical elements of a carbon microphone.

applications in communication microphones and some sound measurement devices. The main disadvantages of piezoelectric microphones are high-output impedance, which makes them susceptible to electrical noise, and sensitivity to mechanical vibration.

Figure 3.34 shows the basic construction of a piezoelectric microphone. The equivalent circuit is that of a capacitor in series with a voltage generator. When mechanically stressed, the element generates a voltage across its opposing surfaces. One end of the piezoelectric element is attached to the center of a diaphragm (usually made of aluminum foil) and the other end of the element is clamped to the microphone frame. When sound energy strikes the diaphragm, it vibrates the crystal, deforming it slightly. In response, the crystal generates a voltage proportional to this bending, delivering an electrical representation of the sound.

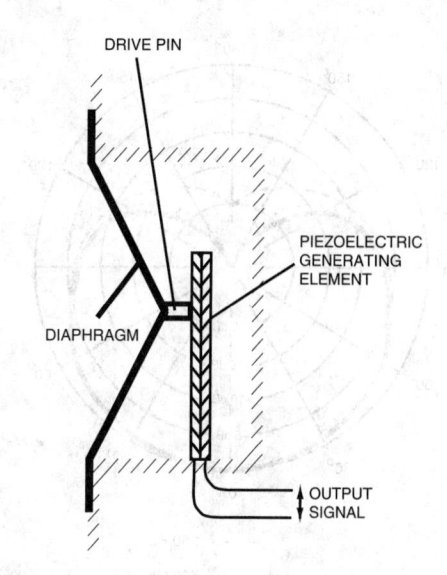

FIGURE 3.34 Mechanical layout of a piezoelectric microphone.

Controlled Magnetic Microphones

A controlled magnetic microphone consists of a strip of magnetically permeable material suspended in a coil of wire, with one end placed between the poles of a permanent magnet. The center of the diaphragm is attached to the suspended end of the armature. Movement of the armature in the magnet's gap induces a voltage in the surrounding coil. The voltage is proportionate to the armature swing and constitutes the output signal. Electrically the transducer is equivalent to a voltage generator in series with a resistor and an inductor. Controlled magnetic microphones are also known as magnetic, variable-reluctance, moving-armature, or balanced-armature devices, depending on the manufacturer.

Controlled magnetic microphones commonly are used for communications and paging because of their sensitivity, ruggedness, and dependability.

3.4.3   Directional Patterns

Every microphone type has a specific directional or polar pattern. That is, each microphone responds in a specific way to sounds arriving from different directions. The polar pattern describes graphically the directionality of a given microphone. Although many different polar patterns are possible, the most common are **omnidirectional**, **bidirectional**, and varieties of **unidirectional**. Some products provide selection among these patterns and a few offer intermediate choices as well. An omnidirectional microphone responds uniformly to sound arriving from any direction. A unidirectional microphone is most sensitive to sounds arriving at the front of the microphone; it is less sensitive to sounds coming from other directions. Bidirectional microphones are most sensitive to sounds on either side of the pickup element. The pickup pattern for a bidirectional microphone is a figure eight.

The main disadvantages of an omnidirectional microphone, as compared with a unidirectional, are lack of rejection of background sounds or noise and a susceptibility to feedback in sound-reinforcement applications.

Omnidirectional microphones are widely used in broadcast and other professional applications for several reasons. First, it is difficult to make a unidirectional microphone very small. As a result, most on-camera clip-on microphones are omnidirectional. This does not create much of a problem, however, because most broadcast environments where body microphones are employed can be controlled to minimize background noise and other extraneous sounds. Also, a body microphone can be located closer to the mouth than a boom microphone at the edge of the frame.

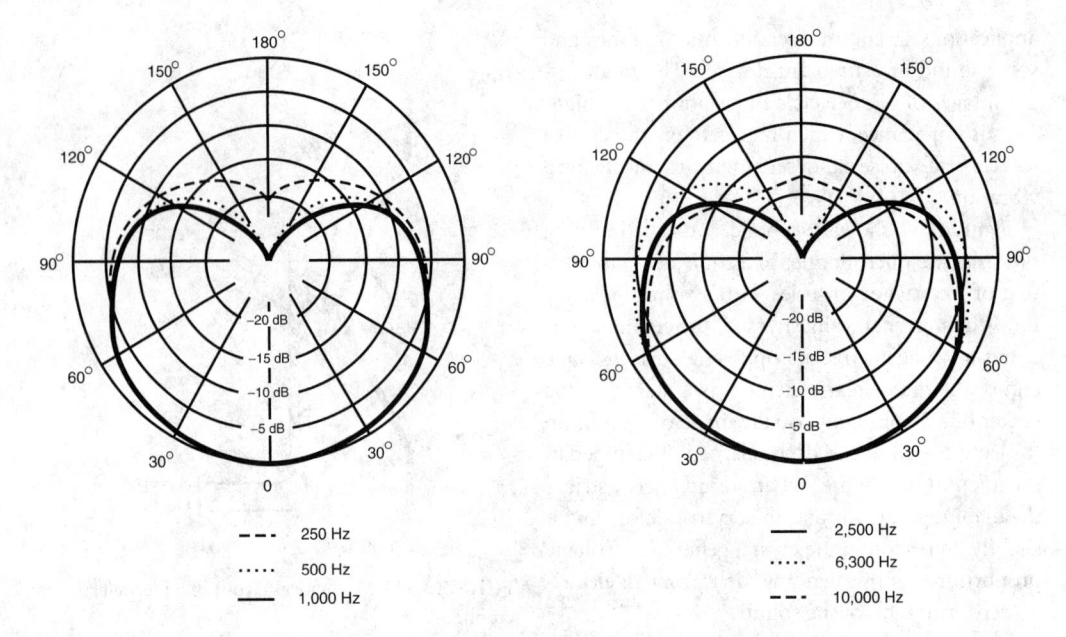

FIGURE 3.35 Typical cardioid polar pattern, plotted at six discrete frequencies. The graphics plot the output in decibels as a function of the microphone's angle relative to the sound source.

Many of the microphones used by radio and television reporters are omnidirectional because it is often desirable to pick up some ambient sounds in the field to add to the broadcast's sense of liveness. Another plus is the omnidirectional microphone's ability to pick up room ambiance and natural reverberation to alleviate the aural flatness a unidirectional microphone might produce.

There are many applications, however, where a unidirectional pickup pattern is desirable. The most obvious is sound reinforcement, where the need to avoid feedback is a primary concern. By reducing the pickup of extraneous sounds, a unidirectional microphone makes available more overall system gain.

The most common form of unidirectional microphone is the **cardioid** type, whose polar pattern is graphically depicted as heart shaped (see Fig. 3.35). The area at the bottom of the heart corresponds to the on-axis front of the microphone, the area of greatest sensitivity. The notch at the top of the heart corresponds to the back of the microphone, the area of least receptivity (referred to as the null). By contrast, the omnidirectional microphone's polar pattern is (ideally) depicted as a circle.

Special Pattern Microphones

Other varieties of unidirectional microphones include the **supercardioid**, **hypercardioid**, and ultradirectional types such as the line/gradient **shotgun** microphone. Supercardioid and hypercardioid microphones are variations of the bidirectional pattern. The front lobe is enlarged as the back lobe is reduced. The null points are moved from the sides toward the rear of the pattern. The supercardioid polar pattern is shown in Fig. 3.36. Also note the variation in polar response for the four frequencies shown.

Hypercardioid microphones offer somewhat greater rejection of off-axis sound, but usually at the expense of frequency response and/or polar pattern smoothness. A few units do provide uniform frequency response from all directions, but sometimes the truth they tell about ambient conditions is less desirable. Lack of polar pattern smoothness can be a hazard in live performance foldback systems. An unexpected off-axis lobe may be pointed at a nearby high-level loudspeaker. Some foldback applications have demonstrated higher gain before feedback using omnidirectional microphones with extremely smooth polar frequency response.

The line/gradient shotgun microphone is a special type of ultradirectional design. The shotgun offers an extremely tight polar response pattern sometimes referred to as club shaped. However, most shotguns come

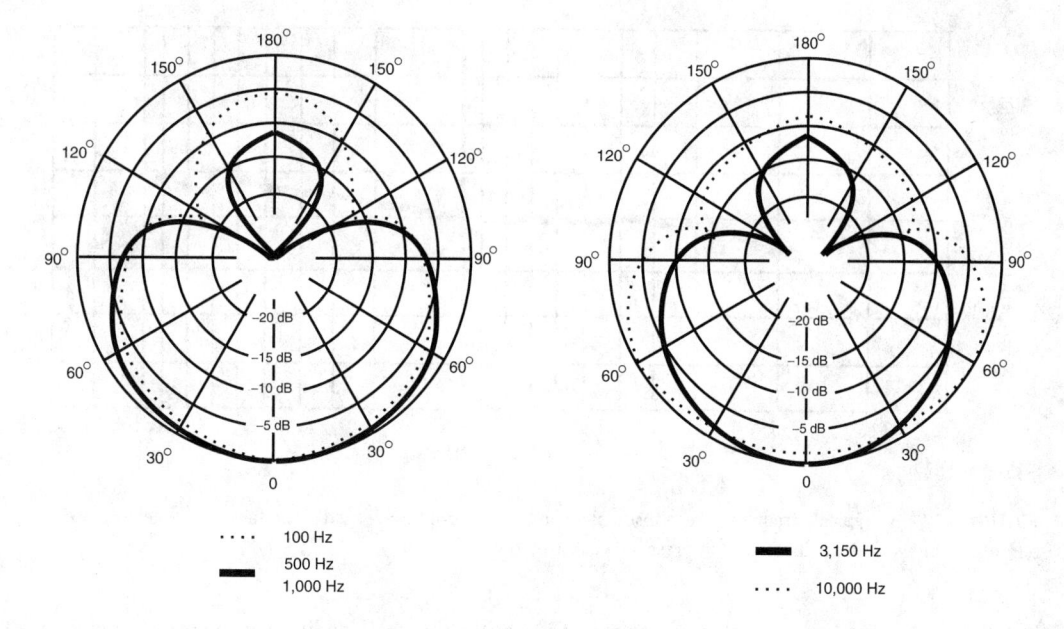

FIGURE 3.36 Typical supercardioid polar pattern, plotted at four discrete frequencies. Note the nulls that appear at the sides of the microphone, rather than at the back.

with a dizzying array of unpredictable side lobes. For interior applications, these often result in deteriorated quality because of reflection off the walls and other hard surfaces. Outdoors, shotgun microphones seem to reach farther than other directional microphones. In truth, they simply reject background noise better. Keep in mind that microphones do not reach; sound travels. The shotgun can pick out sounds at a distance, while ignoring ambient noise. Because the shotgun has a narrow working angle, the microphones must be kept on axis while in use. This is in contrast to the wider acceptance of other directional microphones, the null of which sometimes can be directed toward a problem source without losing the intended source. Generally speaking, the longer the body, the more directional a shotgun microphone will be.

New shotgun designs feature a combination of cartridge tuning and interference tube tuning to achieve smoother frequency response and polar patterns. Compromises in the tuning factors dictate what the microphone sounds like, and how it behaves.

Stereo Pickup Techniques

Stereo microphones contain dual elements. A front facing cardioid or super cardioid picks up the direct monophonic signal. A side-facing figure eight element captures the stereo space surrounding the microphone. Internal addition and subtraction of the two signals yields left and right information. Known as *mid-side* (MS) microphones, these systems also offer direct output from the elements for recording and subsequent processing into left and right signals with adjustable stereo spatiality. Some products provide built-in left/right selection of mono, partial stereo, and full stereo.

Frequency Response

The frequency response of a microphone describes the sensitivity of the device as a function of frequency. The **range of response** defines the highest and lowest frequencies the microphone can successfully reproduce. The shape of a microphone's response curve indicates how it responds within this range. Figure 3.37 shows a typical microphone frequency-response plot.

Frequency response often is affected by the polar pattern of the microphone. As mentioned previously, the frequency response may vary depending on the direction of the sound source and the distance from the sound source to the microphone. Most directional microphones exhibit proximity effect, an unnatural

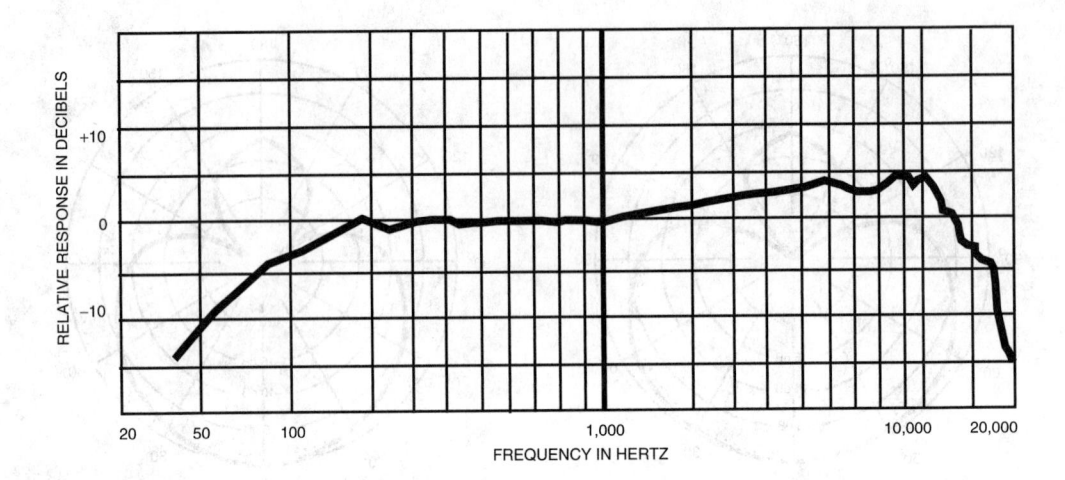

FIGURE 3.37 A typical frequency-response plot for a microphone used in stage and general sound-reinforcement work. Note the high-end presence peak and low-end rolloff.

boost of low frequencies when placed less than 1 m from the source. In addition, some microphones have built-in equalization controls that can alter their frequency response.

The ideal frequency range for a microphone may be 16 Hz–20 kHz with ruler-flat response and no peakiness or other colorations. Sound from such a microphone can be adjusted by careful use of filters and equalizers for specific results. In practice, the best frequency response for a given microphone depends on the application and the personal taste of the user. For precise, accurate, lifelike studio recording, a microphone with flat response usually is desirable. For most sound reinforcement applications and other voice-oriented uses, a presence boost in the upper midrange and rolloff on the low end adds clarity and intelligibility. Many handheld microphones used for field reporting and interviews have response tailored for voice intelligibility. The range from 150 Hz to 12 kHz is effective for this purpose.

3.4.4 Selecting Microphone Types

The handheld microphone, probably the most popular type of microphone, is available in many shapes and sizes. Manufactured in both directional and nondirectional versions, the handheld microphone provides wide frequency response, low handling noise, and a wide choice of characteristic sounds. Because adequate space is available, a shock-mount system is incorporated into most professional handheld microphones. Just holding a microphone or dragging the cable across a floor can cause low-frequency noise to be transmitted to the cartridge. A shock-mount system minimizes such noise.

The clip-on body microphone also is in high demand today. Its small size and wide frequency response offer professional users what appears to be the best of all worlds. Small size translates into minimum visual distraction on-camera or before an audience. Wide frequency response assures good audio quality. There are other points to consider, however, before a clip-on microphone is chosen for an application.

The smallest clip-ons available are omnidirectional. This makes the talent's job more comfortable because staying on-microphone is easier. Directional clip-ons are available, but they have performance tradeoffs. The most obvious is size. To make a microphone directional, a back entry must be added to the housing so that sound can reach the back of the capsule. The distance from the back port to the element determines the wave length above which (hence, the frequency below which) the pattern becomes omnidirectional. This translates into a larger housing for the microphone capsule. Although not as large as a handheld microphone, an effective unidirectional clip-on must be noticeably larger than its omnidirectional counterpart. To keep the size under control, shock mounting of the directional capsule may be compromised. This results in a microphone that exhibits more handling noise than a comparable omnidirectional microphone.

Windscreens for clip-on microphones usually are needed for outdoor use. Even a soft breeze can cause damage to the audio track. The culprit is turbulence, caused by wind hitting the grille or case of the microphone. The sharper the edges, the greater the turbulence. A good windscreen helps to break up the flow of air around the microphone and reduce turbulence. Windscreens work best when fitted loosely around the grille of any microphone. A windscreen that has been jammed down on a microphone serves only to close off part of the normal acoustic path from the sound source to the diaphragm. The end result is attenuated high-frequency response and reduced wind protection.

The task of applying a microphone to a sound source engages an interesting process of elimination. If the source may be expected to be constant within defined limits, the selection range may be narrowed. If the unexpected must be anticipated, however, more generic specifications should be considered.

Defining Terms

Bidirectional pattern: A microphone pickup pattern resembling a figure eight, in which the device is most sensitive to sounds on either side of the pickup element.

Cardioid pattern: A microphone pickup pattern whose polar plot is graphically depicted as heart shaped.

Electret (or electret-biased) microphone: A variation on the basic condenser microphone in which the DC bias is supplied by an electret material, rather than a battery or power supply.

Phantom power: The practice of powering the element of a condenser microphone via a low-value current on the microphone cable.

Range of responses: The highest and lowest frequencies that a microphone can successfully reproduce.

Voice coil: The bobbinless coil transducer element of a dynamic microphone.

References

Tremaine, H.M. 1969. Microphones. In *Audio Cyclopedia*, 2nd ed., p. 147. Howard W. Sams & Co., Indianapolis, IN.

Further Information

Numerous publications are available from the Audio Engineering Society (New York, NY), which serves the professional audio industry. AES also sponsors yearly trade shows and technical seminars that address various areas of audio technology.

Recommended trade magazines that address professional audio applications include *Video Systems*. Intertec Publishing, Overland Park, KS, and *Broadcast Engineering*, PRIMEDIA Publishing, Overland Park, KS.

Recommended books include:

Alkin, G. 1989. *Sound Techniques for Video & TV*. Focal Press, Stoneham, MA.
Davis, G. and Jones, R. 1987. *Sound Reinforcement Handbook*. Yamaha Music Corp., Buena Park, CA.
Whitaker, J. C. and Benson, K. B. 2001. eds. *Standard Handbook of Audio and Radio Engineering*, 2nd ed. McGraw-Hill, New York, New York.

3.5   Aural Reproduction Devices

Michael Starling

3.5.1   Introduction

The most frequently employed aural reproduction device is the loudspeaker. Loudspeakers are *electromechano-acoustic* transducers that transform electrical energy into acoustic energy, through the mechanical coupling of a diaphragm to the air mass. Aural reproduction design consists of a series of resonance, damping, matching, and energy transfer tradeoffs.

Many acousticians believe that the loudspeaker is the single most critical component in the recording and playback chain. Even the most subtle, and sometimes nonquantifiable nonlinearity of the loudspeaker system can seriously color sound quality. The physics of the electromechanoacoustic process presents a seemingly infinite number of opportunities for reproduction error. Moreover, these anomalies are further complicated by radically differing effects when reproduced stereophonically.

Various techniques and approaches have been devised to deal with the distortions and errors that accumulate through each stage of the reproducing system. These anomalies are best understood by reducing the elements of electromechanoacoustic transduction to core principles.

3.5.2 Sound Theory

Wavelength and frequency are related inversely as illustrated by $\lambda = v/f$, where the wavelength (in meters or feet) is equal to the velocity (meters or feet per second)/frequency in hertz. Velocity at sea level and 25°C is 1130 f/s or 339 m/s. Temperature and other effects on velocity will be shown later, but can be assumed to be minimal under normal indoor conditions. See Fig. 3.38 (Eargle, 1994, 5) to determine the wavelength for any given frequency.

Point Source Radiation

The inverse square law states that sound field intensity drops roughly 6 dB for each doubling of distance in *spherical space*. This is shown by

$$Lp \text{ (measuring point)} = Lp \text{ (reference point)} + 20 \log Dr/Dm$$

Spherical space is an ideal, often referred to as *free space* and limited in practice to elevated sound sources in outdoor settings and in the characteristics of anechoic chambers.

As noted by D. Davis and C. Davis (1987), the sound intensity drops by only 3 dB per doubling of distance in cylindrical space. This occurs when encountering an infinite line source, such as can be approached by

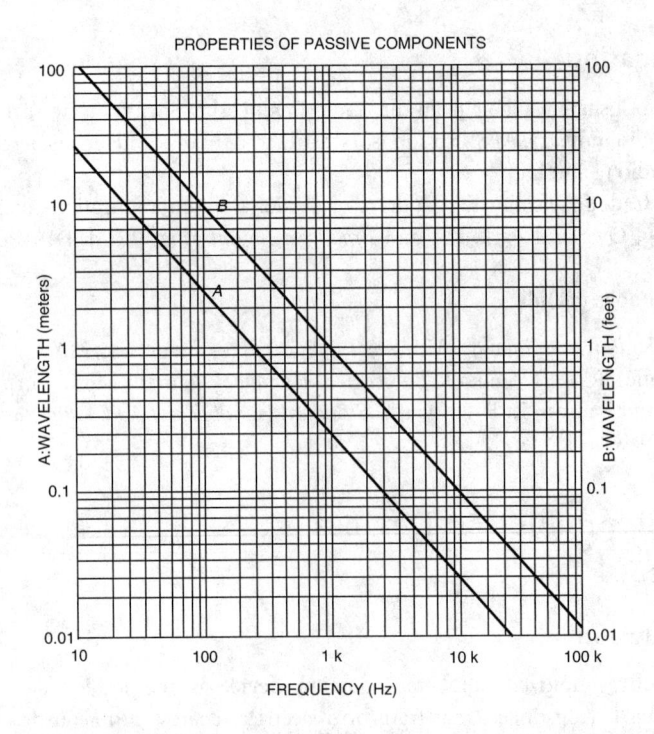

FIGURE 3.38 Frequency and wavelength in air. (*Source:* Eargle, J.M. 1994. *Electroacoustical Reference Data*, p. 5. Van Nostrand Reinhold, New York.)

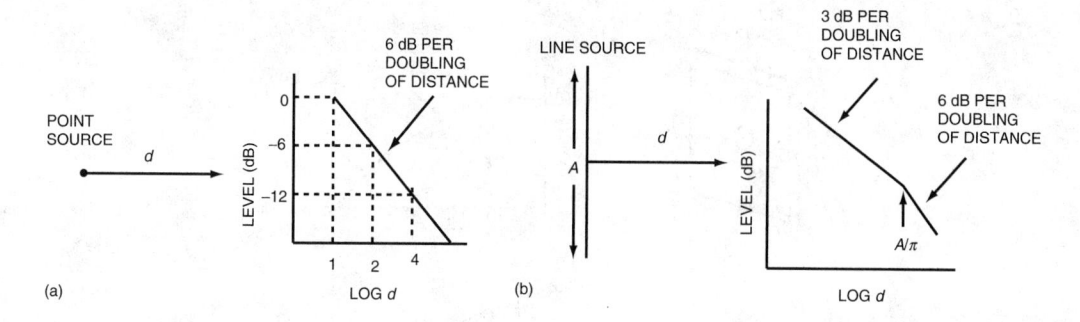

FIGURE 3.39 Attenuation with distance from plane and line sources in a free field. (*Source:* Eargle, J.M. 1994. *Electroacoustical Reference Data*. Van Nostrand Reinhold, New York.)

the ambient noise of race cars evenly spread. Beyond a distance of roughly one-third of the length of the line source, however, the loss will approach 6 dB per doubling of distance. This is illustrated by Fig. 3.39 (Eargle, 1994, 9).

One example of an effort to simulate a point source radiator is shown in Fig. 3.40. (Tremaine 1973, 1111).

By connecting the voice coils 180° out of phase and perforating the diaphragms, the internal air resistance is minimized, and the surrounding air mass is modulated coherently. Such speaker systems have been installed by hanging them from ceilings. The lack of ideal free space and infinite line sources in most settings is readily apparent. Free field inverse square losses can be calculated using the nomograph in Fig. 3.41 (Eargle, 1994, 7).

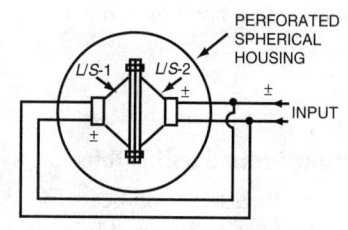

FIGURE 3.40 Simulated spherical loud-speaker. The housing is perforated with many small holes. The voice coils are connected 180° out of phase. (*Source:* Termaine, H.M. 1973. *Audio Cyclopedia*, 2nd ed., p. 111. Howard Sams.)

Plane Source Radiation in an Infinite Baffle

Plane source radiation loss patterns are predicted based on the length of the sides of the plane radiator. The plane radiator is assumed to present a uniform emanated wave with the surface of the radiating **piston** vibrating in phase at all points.

No discernible loss is noted until a distance of roughly one-third of the length of the shorter side of the plane radiator is reached. Once a distance from the plane radiator is reached that is roughly one-third of the distance of the shorter side, the sound field falls off at roughly 3 dB per doubling of distance. Once a distance that is roughly one-third of the length of the longer side of the plane radiator is reached, the attenuation approaches 6 dB per doubling of distance.

The actual distances are given as A/π and B/π. When the ratio of the sides is within two or three of each other, the three rate of loss regions are difficult to discern. However, once A and B are in a ratio of six or higher, the three regions become more apparent. See Fig. 3.42 (Eargle, 1994, 9).

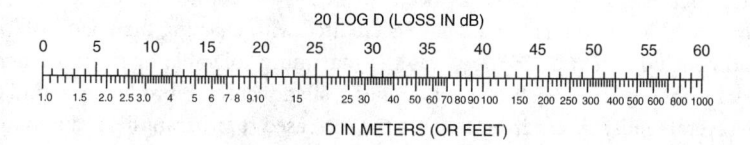

FIGURE 3.41 Inverse square losses in a free field. (From Eargle, J.M. 1994. *Electroacoustical Reference Data*, p. 7. Van Nostrand Reinhold, New York.)

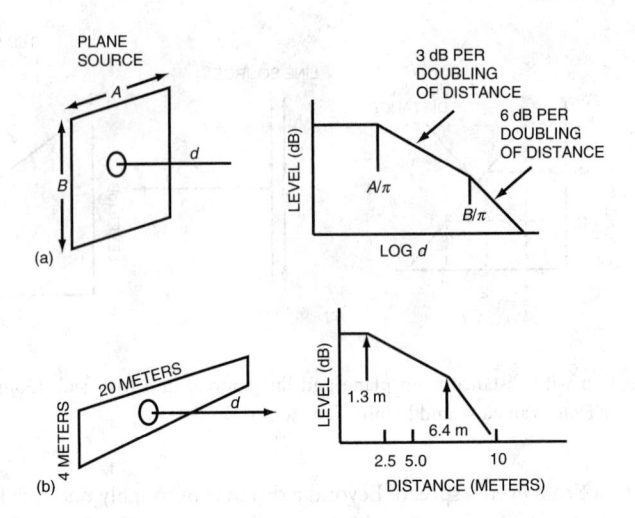

FIGURE 3.42 Attenuation with distance from plane and line sources in a free field. (*Source:* Eargle, J.M. 1994. *Electroacoustical Reference Data*, p. 9. Van Nostrand Reinhold, New York.)

Diffraction Around Solid Objects

The environmental setting of the reproducing system greatly affects objective (measurable by test instruments) and subjective (discernible to trained listeners) frequency response and imaging location. Comb filtering is especially apparent when studying the diffraction patterns where sound waves from planal sources impinge the object. Not unexpectedly, objects with minimal sharp transitions produce the least serious off-axis colorations.

Note that spherical objects produce the most uniform frequency response. Also note that where $\theta = 150°$ the most serious diffraction occurs. For the cylinder and cube this angle places a very sharp transition between the point of radiation and the measuring point at $0°$. Since a variety of objects are found in most listening areas, even in recording studios, very careful design of the ambient monitoring environment is essential to transparent sound reproduction. See Fig. 3.43 (Eargle, 1994, 29).

Reflection

Reflection occurs whenever a sound wave hits an object that is larger than 1/4 wavelength. Reflections from large objects often impairs speech intelligibility. However, reflections with short time delays can be very beneficial because they can be designed to constructively combine with the incident sound to produce an apparently much louder sound (see the **Haas effect**, discussed subsequently). For reflections occurring beyond 50 ms, however, loss of intelligibility can result. Diffuse reverberation combines to impair intelligibility once the direct-to-noise ratio falls below 10–15 dBA. An ideal direct signal-to-noise ratio for maximum intelligibility is considered 25 dBA.

With any significant reflection, where reverberation time exceeds 1.5 s, the intelligibility of speech is severely affected. Because syllables typically occur some 3–4 times per second, the effect of reverberant overhang is considerably less under 1.5 s. Peutz is credited with charting the relationship of the loss of consonants and concluded that a 15% loss was the maximum tolerable by most listeners. It has been noted that for large indoor areas, such as sports arenas, a single central reproducing sound cluster is more favorable than several loudspeakers groups due to the increased reverberant field caused by the dispersed arrays. Moreover, the remaining problem of sound coverage in areas not within the field of the main cluster (walkways, galleries, and under balconies) is solved by low-level distributed systems, with time delays synchronized to that of the main cluster.

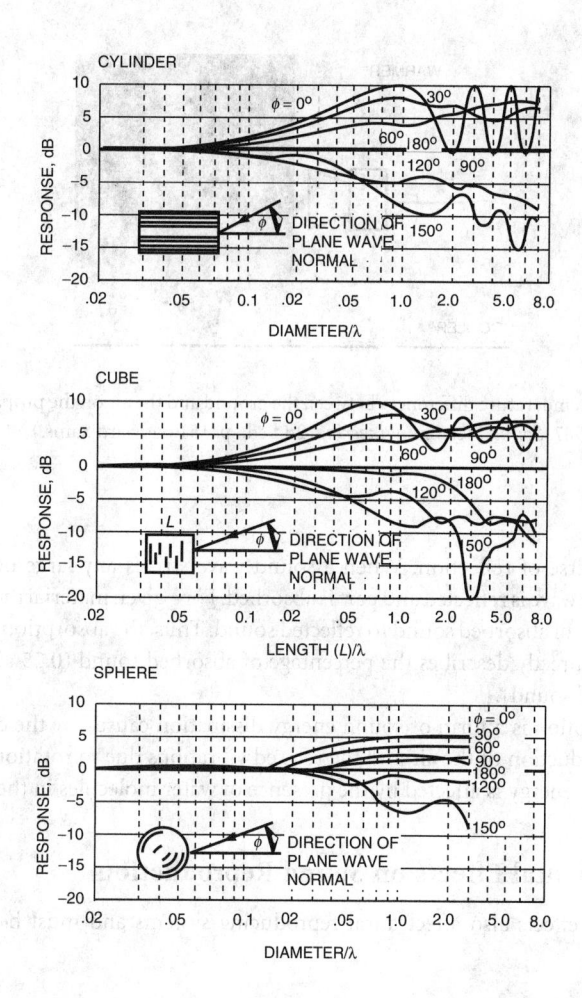

FIGURE 3.43 Diffraction of sound by a cylinder, a cube, and a sphere. (*Source:* Eargle, J.M. 1994. *Electroacoustical Reference Data*, p. 29. Van Nostrand Reinhold, New York.)

Refraction

Refraction occurs whenever sound passes between different transmission media. This commonly occurs when sound passes between layers of air of differing temperature. The velocity of sound increases with increasing temperature, and differing wind velocities encountered at different elevations can also cause sound bending. D. Davis and C. Davis (1987) have noted that the effect of temperature differences is particularly apparent over frozen surfaces on a sunny day. This is because the warmer air above the surface of the cold surface travels faster than the cooler air below, thus bending the sound wave back toward the ground and resulting in tremendous transmission range. See Fig. 3.44 and Fig. 3.45 (D. Davis and C. Davis, 1987, 151).

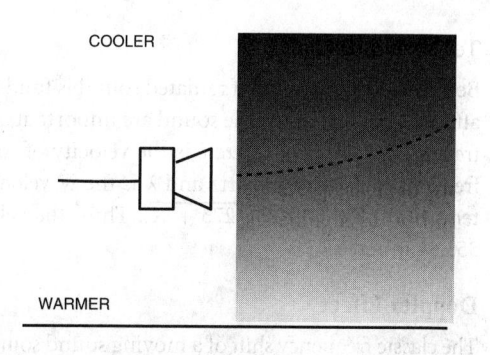

FIGURE 3.44 Effect of temperature differences between the ground and the air on the propagation of sound. (*Source:* Davis, D. and Davis, C. 1987. *Sound System Engineering*, 2nd ed., p. 151. Howard Sams.)

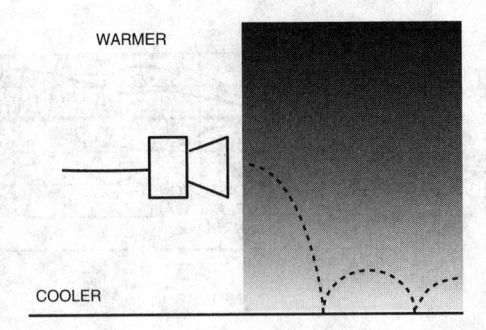

WARMER

COOLER

FIGURE 3.45   Effect of temperature differences between the ground and the air on the propagation of sound. (*Source:* Davis, D. and Davis, C. 1987. *Sound System Engineering*, 2nd ed., p. 151. Howard Sams.)

Absorption

Absorption is the inverse of reflection. When a sound wave strikes any large object (greater than 1/4 wavelength) part of the wave is reflected and part is absorbed. For a given material the absorption coefficient is expressed as the ratio of absorbed sound to reflected sound. Thus, the absorption coefficient is a number between 0 and 1 and directly describes the percentage of absorbed sound (0.25 absorption coefficient is equal to 25% absorbed sound).

Atmospheric absorption is a form of sound energy dissipation caused by the combined action of the viscosity and heat conduction of the air and the relaxed vibrations due to rotational energy effects of air molecules. Vibrational energy is affected by the presence of water molecules in the air.

3.5.3   Environmental Effects on Sound Reproduction

Other environmental effects also affect aural reproducing systems and must be considered by careful designers.

Humidity

The greater air density of higher humidities actually improves (lessens) the normally increased attenuation of higher audio frequencies. At 20% relative humidity the atmospheric absorption begins to deviate from the absorption rate of 80% humidity above 3 kHz. At a distance of 40 m (131 ft) the 10-kHz frequency absorption is 10 dB greater with only 20% relative humidity. This factor is important for large outdoor sound reinforcement settings. See Fig. 3.46 (Eargle, 1994, 15).

Temperature

Because the wavelength of radiated sound is fundamental to interactions in the ambient field, factors also affecting the velocity of the sound are important. As we have seen, velocity is a product of wavelength and frequency. $\lambda = c/f$, where c is the velocity of sound in feet per second or meters per second, f is the frequency of sound in hertz and λ is the wavelength in feet or meters. To calculate velocity at a specific temperature, $c = 20.6\sqrt{273 + °C}$. Thus, the velocity of sound varies between 340.36 m/s at 0°C and 355.61 m/s at 25°C.

Doppler Effect

The classic frequency shift of a moving sound source that approaches or recedes from the listener is also an important component of reproducing systems. The compression and expansion of wavelength has been shown by D. Davis and C. Davis (1987) for a car or train passing the listener at 60 mi/h. The shift of a 1000-Hz tone shifts from 1184 to 928 Hz. This effect is also at work due to the low-frequency excursion of a loudspeaker diaphragm while simultaneously reproducing a high-frequency signal. Calculation of

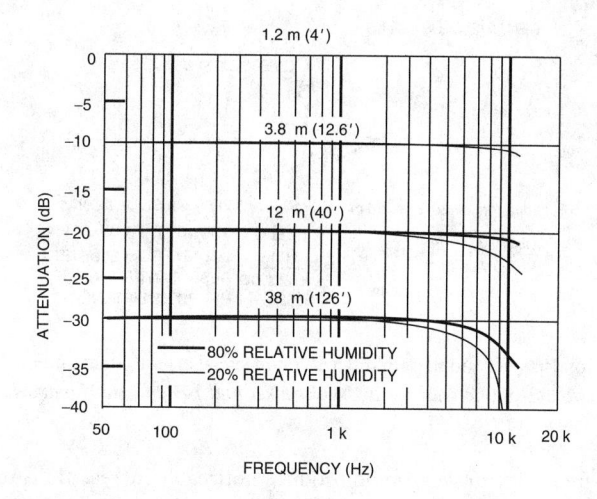

FIGURE 3.46 Atmospheric absorption due to inverse square losses and relative humidity. (*Source:* Eargle, J.M. 1994. *Electroacoustical Reference Data*, p. 15. Van Nostrand Reinhold, New York.)

Doppler shift is illustrated by

$$F = \frac{V +/- V_L}{V +/- V_s} F_s$$

where

F = frequency heard
V = velocity of sound
V_L = velocity of the listener
V_s = velocity of the source
F_s = frequency of the sound source

3.5.4 Characteristics of Human Hearing

Equal Loudness Contours

As early as 1933, Fletcher and Munson demonstrated that human hearing was subject to diminished response, particularly to lower frequencies, at lower sound pressure levels. This effect is commonly compensated for by the familiar loudness buttons on most consumer amplifiers. See Fig. 3.47 (Borwick, 1994, 2).

3.5.4.1 High-Frequency Acuity and Aging

High-frequency response generally diminishes with age, and this loss usually affects men more than women. D. Davis and C. Davis (1987) recommend that design engineers should not compensate for the hearing loss. They state three compelling reasons:

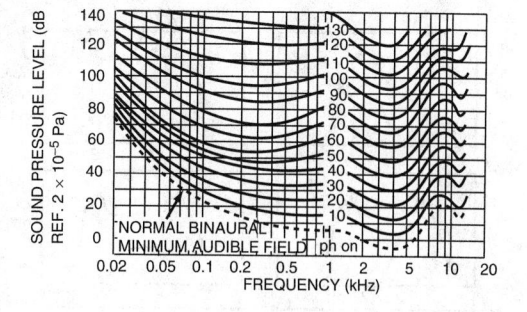

FIGURE 3.47 Equal loudness contours for human hearing in free-field listening conditions. (*Source:* Ford. 1994. Principles of sound radiation. In *Loudspeaker and Headphone Handbook*, ed. J. Borwick, 2nd ed. p. 2. Focal.)

- Young people with normal hearing will be annoyed.
- Older people who have made mental compensations for the gradual onset of the loss would also be annoyed.

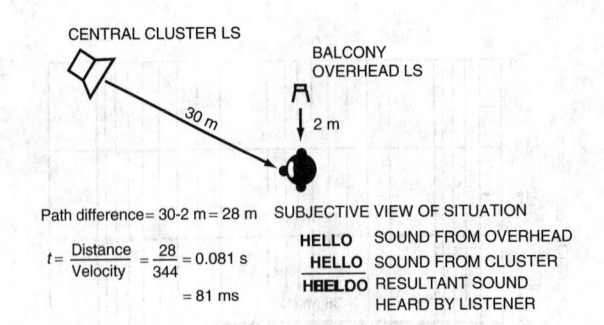

CENTRAL CLUSTER LS

BALCONY
OVERHEAD LS

30 m

2 m

Path difference= 30-2 m = 28 m

$$t = \frac{\text{Distance}}{\text{Velocity}} = \frac{28}{344} = 0.081 \text{ s}$$
$$= 81 \text{ ms}$$

SUBJECTIVE VIEW OF SITUATION

HELLO   SOUND FROM OVERHEAD
HELLO   SOUND FROM CLUSTER
HBELDO   RESULTANT SOUND
            HEARD BY LISTENER

FIGURE 3.48   Effect of loudspeaker sound path differences on signal arrival times and intelligibility. (*Source:* Mapp. 1994. Sound reinforcement and public address. In *Loudspeaker and Headphone Handbook*, ed. J. Borwick, 2nd ed., p. 355. Focal.)

- Available high-frequency drivers would produce noticeably increased distortion due to the high-frequency boost.

Since spoken word intelligibility is dependent on accurate consonant perception, designers of specialized systems, such as hearing aids or radio reading services for the elderly, should make appropriate system compensations.

Haas Effect

The effect of time delays between direct and reflected sounds can cause apparent spatial shifts in the sound source and also affect loudness and intelligibility. These effects were documented by Haas in 1951. As shown by Mapp, reflections can result in the listening phenomena illustrated in Fig. 3.48 (Borwick, 1994, 355).

Delays of as little as a few milliseconds are what constitutes location information in the integration abilities of humans. Two synchronized sound sources placed before a listener, as is standard with stereo loudspeakers, appear to shift from being at the midpoint location to emanating toward the first in time source until delayed by 20–30 ms. Once the delay exceeds 20–30 ms the sources will diverge and be clearly discernible as two separately emanating sounds.

Haas' work was primarily concerned with the effect of delayed sound from a common source. He noted that with up to 20–30 ms of delay, the delayed sound would be masked and listeners would detect only a single louder direct sound source. See Fig. 3.49 (Borwick, 1994, 356).

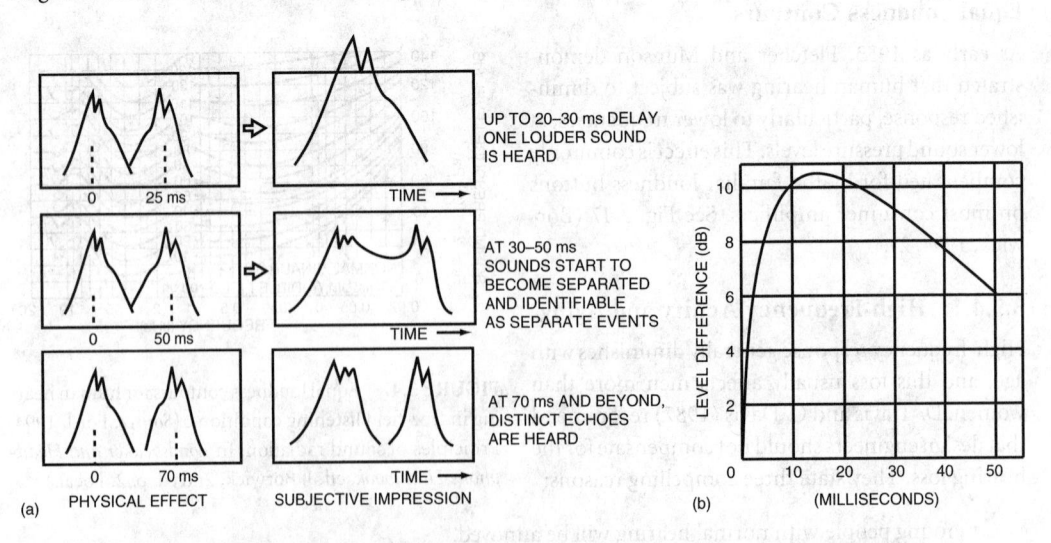

UP TO 20–30 ms DELAY
ONE LOUDER SOUND
IS HEARD

AT 30–50 ms
SOUNDS START TO
BECOME SEPARATED
AND IDENTIFIABLE
AS SEPARATE EVENTS

AT 70 ms AND BEYOND,
DISTINCT ECHOES
ARE HEARD

(a)   PHYSICAL EFFECT   SUBJECTIVE IMPRESSION

(b)   (MILLISECONDS)

LEVEL DIFFERENCE (dB)

FIGURE 3.49   Haas effect curve. (*Source:* Mapp. 1994. Sound reinforcement and public address. In *Loudspeaker and Headphone Handbook*, ed. J. Borwick, 2nd ed., p. 356. Focal.)

3.5.5 Elements of Aural Transduction Systems

Horn Systems

Horn loudspeakers may be viewed as an impedance matching device that mechanically couples the diaphragm to the air mass at the mouth of the horn. Horn-type loudspeakers were the first loudspeakers, and to this day their efficiency is unmatched by other loudspeaker types. Typical horn efficiency is between 25 and 50%, compared to less than 5% efficiency for most direct radiating loudspeakers. Any system employing only a horn is known as a *projector type* system.

Over the course of many years of experimentation with various rates of flare, it was discovered that the exponential and hypex flares produced the highest throat resistances, and also the flattest response. A hypex flare is a hybrid between a hyperbolic and exponential flare. See Fig. 3.50 (Tremaine, 1973, 1099).

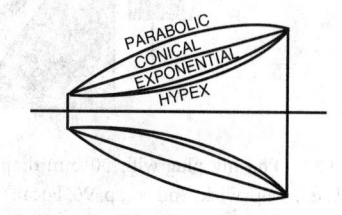

FIGURE 3.50 Comparison of the different rates of flare used in the design of loudspeaker horns. (*Source:* Tremaine, H.M. 1973. *Audio Cyclopedia*, 2nd ed., p. 1099. Howard Sams.)

The impedance and frequency response of various horn flares is shown in Fig. 3.51 (Borwick, 1994, 18).

Lower frequency response of horn radiators generally falls off at 6 dB/octave, provided that throat impedance is resistive (above the horn cutoff frequency). High-frequency response is limited by the mass reactance of the moving system, but can be extended several octaves by designing the cavity volume to resonate with the mass of the diaphragm at a frequency above cutoff impedance of the moving system and reflected throat resistance. The response above this resonance point falls off at 12 dB/octave.

A special problem at high frequencies, when the diaphragm radius approaches one-quarter wavelength, results in a lack of phase coherence between the radiating edge and the center of the diaphragm. Response irregularities result, but are often compensated for by use of a phasing plug. The plug is fitted between the diaphragm and throat and presents

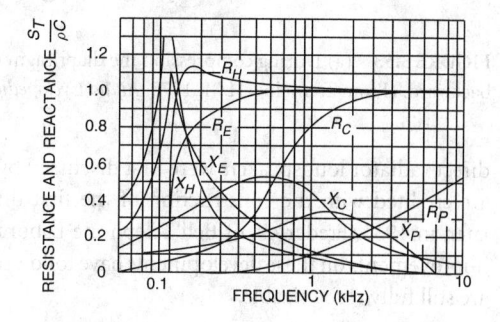

FIGURE 3.51 The acoustic impedance at the throats of infinitely long exponential, hyperbolic, conical and parabolic horns (after Dinsdale and Olson). (*Source:* Ford. 1994. In *Loudspeaker and Headphone Handbook*, ed. J. Borwick, 2nd ed., p. 18. Focal.)

several identical air paths between these two elements. This results in a constant phase delay across the diaphragm and smoothes the response irregularities at the higher frequencies. See Fig. 3.52 (Borwick, 1994, 96).

An unusual side effect of horn radiator efficiency and directional air mass coupling is that air overload can be encountered in horn systems above 120 dB. At these levels the air itself becomes nonlinear. The most troubling obstacle in large high-output systems are the harmonic distortions due to the compression volume being less than the volume displaced during rarefaction. See Fig. 3.53 (Tremaine, 1973, 1093).

For optimum efficiency, Kelly has recommended that the diaphragm area must be larger than the throat area. He notes that for a given distortion level, the power output is directly proportional to the area of the throat entrance port. See Fig. 3.54 (Borwick, 1994, 99).

Direct Radiators

Direct radiators are the most widely employed form of loudspeaker design. These are speakers in which a diaphragm is directly coupled to the air mass and radiates directly into the listening area. Although

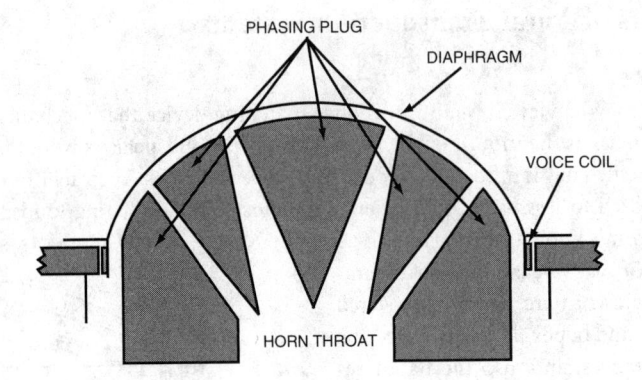

FIGURE 3.52 Phasing plug with 50-mm diaphragm and 20-mm throat. (*Source:* In *Loudspeaker and Headphone Handbook*, ed. J. Borwick, 2nd ed., p. 96. Focal.)

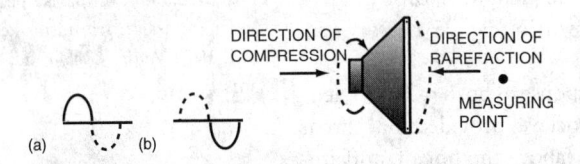

FIGURE 3.53   (a) During compression the diaphragm moves forward, (b) during rarefaction the diaphragm moves backward. (From Tremaine, H.M. 1973. *Audio Cyclopedia*, 2nd ed., p. 1093. Howard Sams.)

direct radiator loudspeakers were first discussed by Lord Rayleigh in 1877, Chester Rice and E.W. Kellogg are credited with the construction of the first direct radiator dynamic loudspeaker in 1925. Six years later, in 1931, Frederick, of Bell Telephone Laboratories, demonstrated a two-way system. Innumerable improvements on these developments have followed over the ensuing decades, yet the fundamental designs are still fully recognizable.

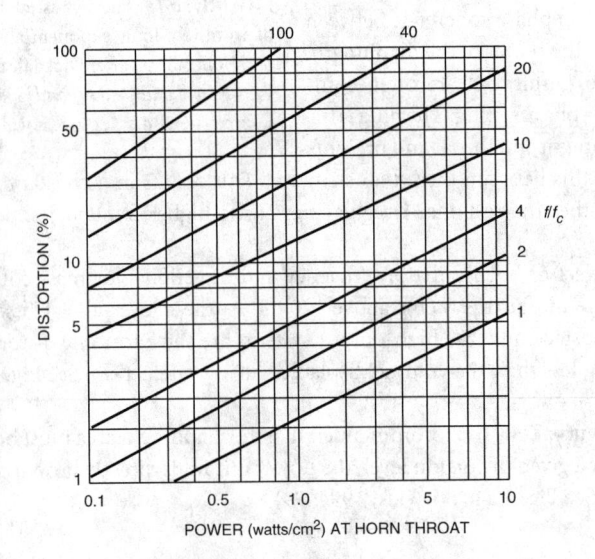

FIGURE 3.54 Percentage second harmonic distortion as a function of power at horn throat, with the ratio of the frequency f to the cutoff frequency f_c. (From Kelly. 1994. In *Loudspeaker and Headphone Handbook*, ed. J. Borwick, 2nd ed., p. 99. Focal.)

Despite their common usage, direct radiators achieve their modern success through the accumulated improvement of many decades of experience with such diverse problems as voice coil heat buildup, diaphragm nodes, enclosure types, and the complications of multiple radiator/crossover networks.

The objective of these interrelated tasks is illusory: to recreate for the listener the original incident sound field with a flat direct frequency response and a flat reverberant field. As noted by Colloms (1991): "[N]early all box-type direct radiator loudspeakers have the opposite characteristic to that which is required. For most loudspeakers their power response reduces with frequency, worsened by the natural increase in room absorption with increasing frequency."

The obstacles to successful direct radiator systems cross diverse branches of physics and continue to provide fertile ground for radical design departures as new approaches to old problems are documented before professional societies.

Figure 3.55 provides a cross sectional view of the basic dynamic loudspeaker direct radiator (Tremaine, 1973, 1079, and 1080).

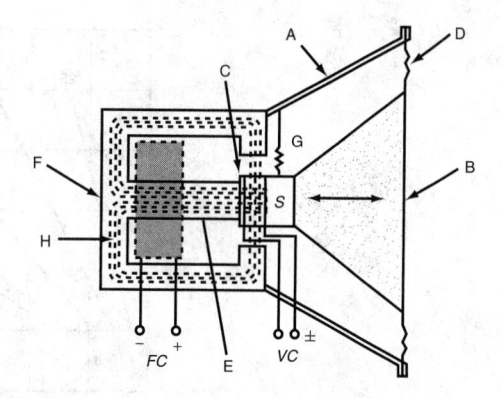

FIGURE 3.55 Cross-sectional construction view of a dynamic loudspeaker. Item A is the frame or basket, supporting a diaphragm B, with a voice coil C attached to the apex of the diaphragm. The outer edges of the diaphragm are supported by a flexible edge D. The voice coil consists of several turns of small copper or aluminum wire, centered over a central pole piece E in a magnetic structure F built in the form of the letter E. The voice-coil assembly is centered over the pole piece by means of spider G. The magnetic lines of flux are indicated by the dotted lines H. The central pole piece E becomes the south pole of the magnetic field and the outer edges of the magnetic structure are the north pole of the magnetic field. (From Tremaine, H.M. 1973. *Audio Cyclopedia*, 2nd ed., pp. 1079, 1080. Howard Sams.)

Voice Coil

The voice coil is the heart of modern loudspeakers. In its simplest form the voice coil is a coil of wire, through which electrical waves are applied, suspended within a fixed magnetic field, and directly coupled to a movable diaphragm. It is the physical movement of the voice coil due to the interaction of its magnetic current flow and the magnetic field surrounding the coil that causes the diaphragm to vibrate in response to the applied audio waves and thereby produce sound.

Temperature Rise

The voice coil itself is inherently nonlinear due to heat buildup that occurs over prolonged activity, especially at higher volume levels. Power handling of early loudspeakers was largely limited by the paper formers then in use. Those voice coils could only tolerate a temperature rise to about 100°C, which limited a typical 10-in diaphragm unit to about 35 W of continuous input power. Modern formers using a polyamide (Nomex) or polyimide (Kapton) can operate with up to 350° temperature rises.

Temperature rise immunity is critical since an increased input power raises both temperature and resistance. It has been generally noted that for a rise of 250°C resistance of the coil doubled, similar to the *apparent power* handling capacity of the loudspeaker. In reality, of course, the operative coupling force of the loudspeaker is proportional to current only and under these conditions the overall efficiency of the system is halved. See Fig. 3.56 (Borwick, 1994, 74).

Gap Placement and Dimensional Configuration

For maximum efficiency, maximizing the magnetic field flux across the voice coil requires the smallest coil to pole piece gap dimension possible based on peak voice coil excursion and power dissipation. As a practical matter this gap must be increased somewhat to compensate for dimensional expansion as operating temperature increases.

For a nominal 50-mm voice coil operating at 250°C gap clearance must be increased by 0.4 mm.

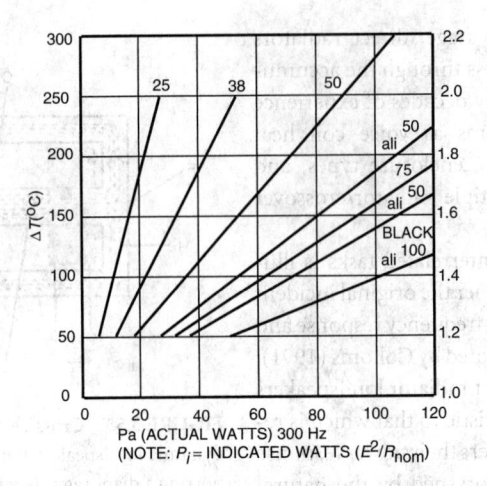

FIGURE 3.56 Voice-coil temperature vs. input power. (*Source:* In *Loudspeaker and Headphone Handbook*, ed. J. Borwick, 2nd ed., p. 74. Focal.)

Three basic voice coil/magnetic gap ratios are in common use: overhanging, short coil, and equal coil and gap. See Fig. 3.57. (Borwick, 1994, 76). The equal gap ratio is the most cost-effective configuration. Distortion due to asymmetry of the leakage flux is minimal, provided the top plate thickness of the magnetic assembly is four times greater than the peak voice coil displacement. This requirement is readily met and is the most commonly employed method in small excursion midrange and high-frequency systems.

The oversized voice coil configuration is best for wide excursion units, such as subwoofers and other high-power, low-frequency systems. The main disadvantage of this configuration is that considerable distortion results from any asymmetry in the flux leakage between the compression and rarefaction swing. This flux leakage distortion is best minimized by the oversized top plate assembly, but is also the most costly.

FIGURE 3.57 The three basic voice-coil/magnet gap configurations: (a) overhanging coil, (b) short coil, (c) equal-length coil and gap. (*Source:* In *Loudspeaker and Headphone Handbook*, 1994. ed. J. Borwick, 2nd ed., p. 76. Focal.)

Flux distribution and distortion are additionally affected by whether the effective center pole of the voice coil is flush or extended above the top plate of the magnetic structure. The most cost-effective manufacturing places the units flush but results in asymmetrical flux leakage. By lengthening the center pole of the voice coil above the top plate of the magnetic structure, more symmetrical flux leakage is achieved and second harmonic distortion is dramatically decreased.

The Magnetic Structure

The earliest dynamic loudspeakers employed an electrodynamic design whereby the field coil was energized by a DC current. When the field coil was energized the audio frequency signal applied to the voice coil caused the voice coil to swing inward or outward depending on the instantaneous polarity. Modern designs have dispensed with the mechanical offset this required and now rely exclusively on permanent-magnet systems for the loudspeaker's fixed field.

From the earliest designs of carbon steel with magnetic flux (B) and magnetic force (H) products of about 1.6 kJ/m^3, Alnico was developed by 1934 (cobalt-iron-nickel-aluminum-copper) with a BH product of

12.8 kJ/m^3. Fully columnar magnets, heated above the Curie point (1200°C), cooled in a strong unidirectional flux field resulted in anisotropic magnets with a BH maximum of 60 kJ/m^3. However, during the 1970s the world supply of cobalt was destabilized due to mining problems. Earlier work based on using nonmetallic ferrite cores to reduce eddy currents was accelerated. Ferrite cores are still far more economical, but require much greater volume to achieve flux densities approaching that of Alnico cores.

Operating temperature of the magnet has a substantial effect on voice coil assembly performance. Modernly, **ferrofluids** (generally MOFe$_2$O$_3$) are commonly employed in the gap between the voice coil and the magnet surface. The fluid is held in intimate contact with the magnetic assembly surface and the voice coil itself. Over short throw applications the ferrofluid is directionally secure and can even be stabilized for low-frequency high-excursion systems by proper venting of the dust cap.

The ferrofluid results in greatly improved heat transfer, which allows the voice coil assembly to handle much greater transient peaks. Temperature recovery is substantially improved. A peak of 8 W above nominal program for 1 min would cause the voice coil to rise from 80° to 100°C with the return to nominal operating temperature in 10 s. Without ferrofluid treatment the voice coil would rise to 213°C.

Also note that the normally resulting increase in mechanical damping is generally transparent, even at low signal levels, due to the well-stabilized magnetic moment of the ferrofluid itself. See Fig. 3.58 (Borwick, 1994, 92).

In many direct radiators a centering spider connects between the voice coil and basket frame or magnet as a means of stabilizing the coil's position relative to the pole piece. A second spider is sometimes employed for greater ruggedness, but with a commensurate effect on mechanical complexity of the voice coil/magnet motor and, thus, on frequency response and efficiency. Modern spiders are generally constructed of impregnated cloth for rigidity.

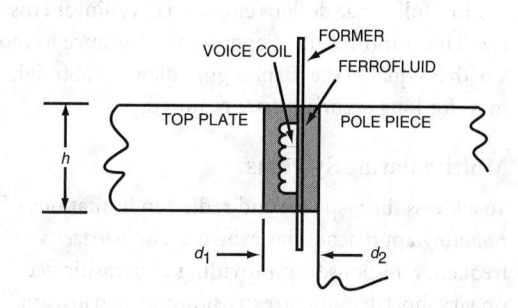

FIGURE 3.58 Schematic showing ferrofluid in the magnet gap. (*Source: In Loudspeaker and Headphone Handbook,* ed. J. Borwick, 2nd ed., p. 92. Focal.)

The term spider evolved from the first 15-in loudspeaker developed by RCA which used heavy threads to center the voice coil. Unlike modern spiders these early true spider systems required frequent recentering.

Diaphragm

The acoustic impedance of the diaphragm varies with factors such as material construction, baffle design, surface compliance, suspension damping, applied frequency, and even operating temperature and humidity. Traditional cone diaphragms were made of paper construction. Modern musical instrument loudspeakers still employ paper felted cones due to high stiffness-to-mass and ruggedness. Polypropylene is popular for consumer applications, but generally is limited to applications below 7 kHz.

Although heat buildup and physical distortion of the voice coil are a central reproducing system design problem, the diaphragm itself is perhaps the most challenging hurdle simply because the physical dimensions required to move any appreciable air volume deviate significantly from the ideal uniform piston in which all radiating mechanical parts vibrate with the same phase, magnitude, and velocity.

Direct radiator systems frequently exhibit nonlinearities due to the flexure nodes caused by standing waves across the diaphragmatic surface. To minimize flexure nodes a given diaphragm should be designed to cover no more than one decade (e.g., 20–200 Hz, 200 Hz–2 kHz, and 2–20 kHz). Many proprietary techniques are employed by various manufacturers that seek to achieve maximum stiffness and minimum mass of the diaphragm. Specific formulations of pulp, slurry, binders, and coefficient additives are often closely guarded trade secrets.

Hybrid techniques such as corrugating the cone with secondary suction and coating can achieve successful decoupling at higher frequencies, thereby minimizing **node flexure**. Nonetheless, some degree of node flexure is evident in most direct radiators. Extreme examples are shown in Fig. 3.59 (Borwick, 1994, 49) that illustrate the deleterious effects of such uniform piston breakup.

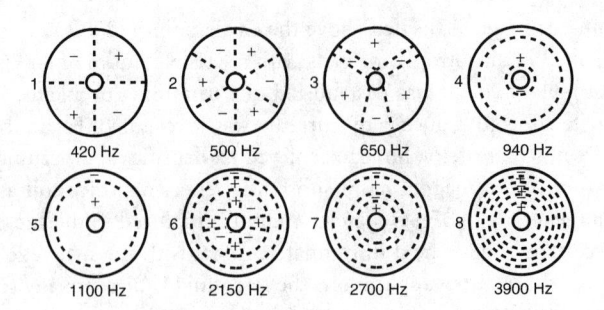

FIGURE 3.59   Nodal patterns of the cone. (*Source:* In *Loudspeaker and Headphone Handbook,* ed. J. Borwick, 2nd ed., p. 49. Focal.)

Most full-range designs employ a curvilinear cross sectional design following exponential or hyperbolic law. This improves the diaphragm's resistance to resonance above the piston band limit where the wavelength is equal to the diaphragm's diameter. Straight sided designs are usually employed only where extra mass for long excursions are required.

Multiradiating Systems

To address the response and radiation limitations of single diaphragm systems a variety of unique *multi-radiating* approaches has evolved. The *whizzer* was among the earliest modifications designed to extend frequency response by providing a parasitic secondary short diaphragm extrusion attached just outside the cone-former junction. See Fig. 3.60 (Tremain, 1973, 1090).

Where a large radiation angle is required to cover a wide listening area, the task cannot be accomplished by a single direct radiator. And where low distortion for a wide variety of program types is required, the task cannot be accomplished by a single direct radiator.

In accomplishing an extended smooth frequency response, two-way and three-way loudspeaker systems are the norm, with each diaphragmatic radiator being suited for a specific portion of the audio frequency spectrum. To prevent distortion due to unintended power leakage reaching a radiator designed for a different range of the spectrum electrical crossover networks are required ahead of the electromotive assemblies.

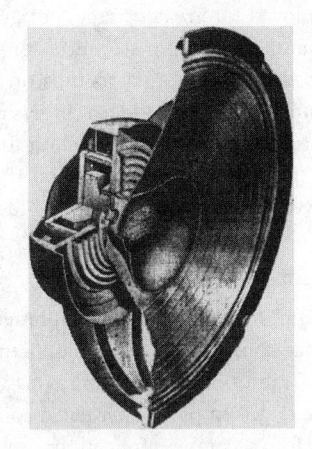

FIGURE 3.60   Electrovoice dynamic speaker unit with a whizzer attached to diaphragm to increase the midrange and high-frequency response. (*Source:* Tremaine, H.M. 1973. *Audio Cyclopedia,* 2nd ed., p. 1090. Howard Sams.)

Because usable filter designs do not have an infinite rolloff at crossover, careful overlap design is essential. This requires smooth response for each drive unit across the transition region as well as precision design to match the complex impedance characteristics of the target driver and present a stable impedance to the source amplifier.

Two categories of crossover networks are employed, high level and low level. High-level crossover networks feed the target drivers through passive components. The primary practical advantage of the low-level crossover is that matching to the complex driver unit load impedance is handled by separate power amplifiers rather than the network itself. The cost implications for low-level crossover systems are obvious and are commonly found only in high-performance applications. See Fig. 3.61 (Borwick, 1994, 198).

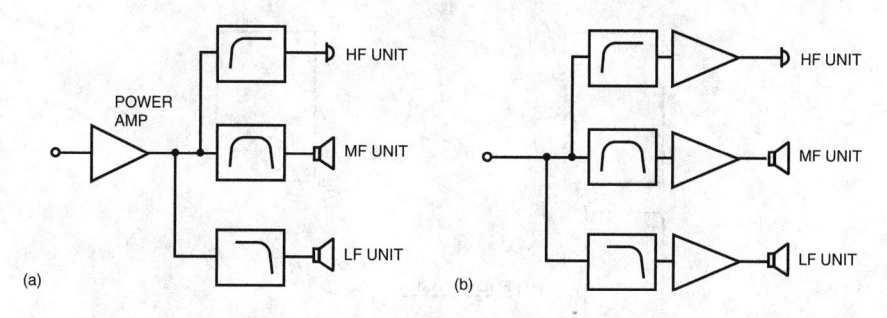

FIGURE 3.61 Crossover network with high level filters (a), and low-level filters (b). (*Source:* In *Loudspeaker and Headphone Handbook*, ed. J. Borwick, 2nd ed., p. 198. Focal.)

Modern crossover design for high-level systems requires matching the complex electroacoustic response of the driver to one of the commonly executable filter section types. This desired filter shape is referred to as the *target function.* These target filter functions are commonly based on first order Butterworth derivations, including Linkwitz–Riley cascades. The phase-delay characteristics of crossover networks are especially problematic and frequently result in sharply irregular polar plots for the multidriver systems. See Fig. 3.62 (Borwick, 1994, 210). The Linkwitz–Riley target shapes are especially desirable due to their in-phase relationship. The reference axis of the drive units is not altered with this crossover target shape.

Reference axis plots are especially important to multidriver units. By knowing the polar response of the multidriver system, proper vertical mounting of the frequency-dependent radiators can be determined. Depending on anticipated listener location, the reference axis can be vertically aimed by use of sloping or stepped baffles, as well as by inverting typical driver layout. See Fig. 3.63 (Borwick, 1994, 213).

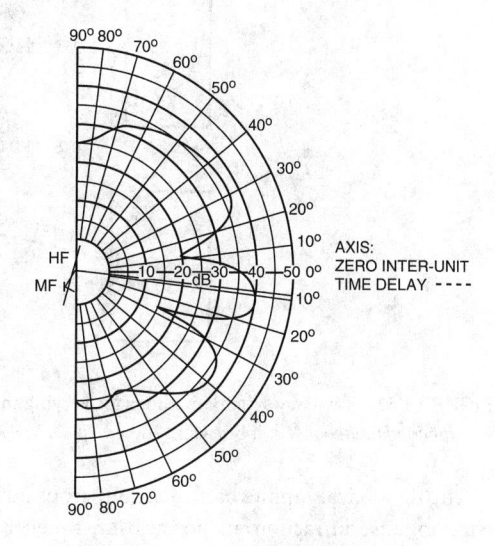

FIGURE 3.62 Vertical polar characteristic at crossover frequency; two-way system with unit target functions; third-order Butterworth. (*Source:* In *Loudspeaker and Headphone Handbook*, ed. J. Borwick, 2nd ed., p. 210. Focal.)

Enclosures

At an early date open baffle systems were realized to result in unwanted reinforcement and cancellation effects due to radiation interference between the front and rear diaphragm surface. To the ensuing box type enclosure, a vent was quickly added, although it was originally envisioned as a mere pressure equalizer. Later work showed that careful vent design could appreciably improve low-frequency response of the radiating system. Yet, in addition to improved low-frequency response, the vented box affects higher frequencies due to a variety of mechanical and acoustic resonances and diffraction.

In 1957 Olson of RCA laboratories showed that the measured frequency response of direct radiating loudspeakers was demonstrably affected by the type of enclosure in which it was mounted. The unmistakable conclusion of these experiments was that shapes which avoided sharp transitions provided significantly flatter response. In fact, the response was close to ideal only for spherical enclosures. The truncated pyramidal shapes also presented little opportunity for response coloration as shown by the graphs in Fig. 3.64 (Eargle, 1994, 31).

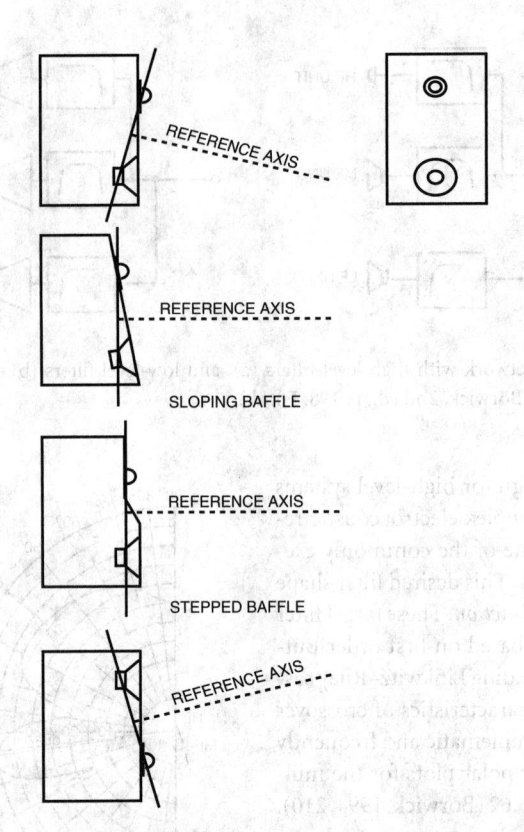

FIGURE 3.63 Two-way system: methods for varying the direction of the reference axis. (*Source:* In *Loudspeaker and Headphone Handbook*, ed. J. Borwick, 2nd ed., p. 213. Focal.)

The theoretical infinite baffle has numerous advantages over real-world designs. There are no baffled edges to cause diffraction and no enclosure to generate undesirable resonances. Moreover, with the forward plane completely isolated from the rear radiation, a solid angle of dipole radiation is produced. Although the finite baffle (open baffle) shares the figure-8 radiation of the infinite baffle, directivity will be induced as the plane surface approaches one wavelength.

The enclosed box, although the nearest practical embodiment of the infinite baffle, produces deleterious resonance, and diffraction and distorted radiation angle effects.

It is the vented enclosure, sometimes referred to as a *bass reflex* or *ported* box that couples rear radiation over a narrow range of low frequencies to direct radiation. The aperture forms a Helmholtz resonator coupling the air in the vent to the compliance of the air in the box. Various varieties of vents involving multiple chambers, transmission line, and bandpass designs are employed depending on specific frequencies to be reinforced. The dimension, shape, position, and air flow of the vent requires careful consideration. In general the minimum area and diameter is illustrated by (as shown in Borwick (1994, 274)):

$$S_v >/ = 0.8 f_B V_D$$

and

$$d_v >/ = \sqrt{f_B V_D}$$

where

S_v = area of the vent
d_v = diameter of a circular vent
V_D = peak displacement volume of the driver
f_B = resonant frequency of the vented enclosure

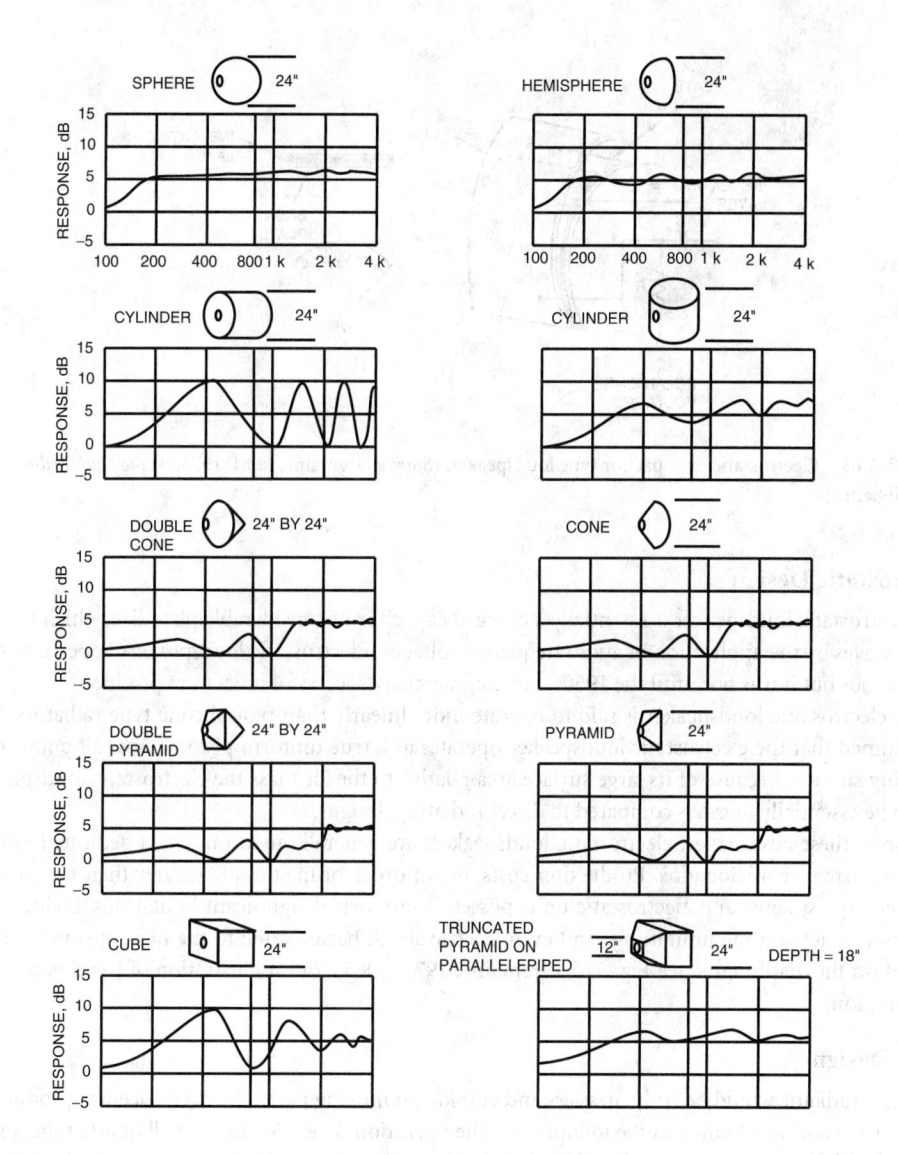

FIGURE 3.64 Response curves showing diffraction by 10 objects of different shape. (*Source:* Eargle, J.M. 1994. *Electroacoustical Reference Data,* p. 31. Van Nostrand Reinhold, New York.)

The cabinet itself will vibrate and reradiate into the sound field to some extent for most enclosure types. In general, the more dense the enclosure material, the less susceptible the unit will be to reradiation. Nonetheless, some modern loudspeakers employ aluminum, synthetic resins, and even foamed plastic for enclosure construction. Various techniques such as bracing, damping, and layered enclosure construction can be employed with good success in suppressing cabinet radiation.

It is imperative that corner and edge joints be fully rigid and secure. Since the 1970s virtually all wooden cabinet designs employ an all-glued joint construction with cabinet access via the bass unit mounting hole. This construction has substantially enhanced low-frequency reinforcement.

3.5.6 Unique Approaches to Loudspeaker Design and Assessment

Despite the near ubiquity of direct radiators, the foregoing discussion illustrates a host of mechanical and electrical obstacles unique to this design class. Unique approaches to moving the air mass for sound reproduction have attempted to overcome such difficulties.

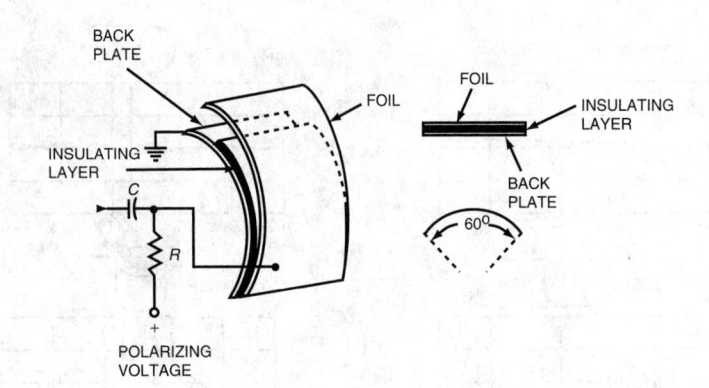

FIGURE 3.65   Electrostatic or capacitor type loudspeaker. (*Source:* Tremaine, H.M. 1973. *Audio Cyclopedia,* 2nd ed. Howard Sams.)

Electrostatic Design

The electrostatic loudspeaker is essentially a charged capacitor of considerable plate size, which produces sound waves by the application of audio frequency voltages. Electrostatic loudspeakers appeared as early as the 1930s but it was not until the 1950s that commercially successful units were produced.

The electrostatic loudspeaker is said to operate more linearly than typical cone type radiators. Thus, it is claimed that the electrostatic loudspeaker operates as a true uniform piston across all points of the radiating surface. Because of its large surface area relative to the air mass the electrostatic loudspeaker is said to be essentially *massless* compared to direct radiating designs.

Despite these advantages, electrostatic loudspeakers are generally used only by a dedicated group of high-performance aficionados. Production costs are an order of magnitude greater than for comparable cone type systems and electrostatic units possess a number of significant limitations, including low efficiency, strict volume limitations, and even electric shock hazards due to the high operating charges present on the diaphragm. See Fig. 3.65 (Tremaine, 1973, 1085) for an illustration of basic electrostatic construction.

Ionic Design

The ideal radiator would be truly massless and employ no moving parts. Such a device was produced for a short time and was known as the ionophone. The operation depended on a small quartz tube, open at one end with the other connected to a Kanthal electrode. The electrode is connected to a radio frequency oscillator of high-voltage output causing ionic discharge to occur in the air within the quartz tube. See Fig. 3.66 (Borwick, 1994, 37).

By amplitude modulating the RF oscillator at an audio frequency, the glow discharge in the tube varies accordingly. Pressure waves are set up within the tube at the audio modulating frequency, and by connecting the open end of the tube to the throat of an exponential horn fairly high sound pressure levels near 100 SPL can be generated. Nonetheless, overall efficiency is low due to the substantial voltages required to produce ionic discharge, and the horn coupling places strict limits on low-frequency response.

Despite the theoretical promise of electrostatic and ionic design approaches, the direct radiators, due primarily to the good success and low cost of modern construction continue to remain the most commonly used among all reproducing systems, even for the most demanding professional requirements.

Loudspeaker Assessments

Most loudspeaker manufacturers periodically engage in various forms of subjective loudspeaker assessments. Such subjective assessments can produce statistically significant results that are extremely difficult, if not impossible, to quantify using objective testing techniques. Since subjective tests involve relatively expensive testing devices (humans, usually on the clock) such tests are infrequently employed. Because of

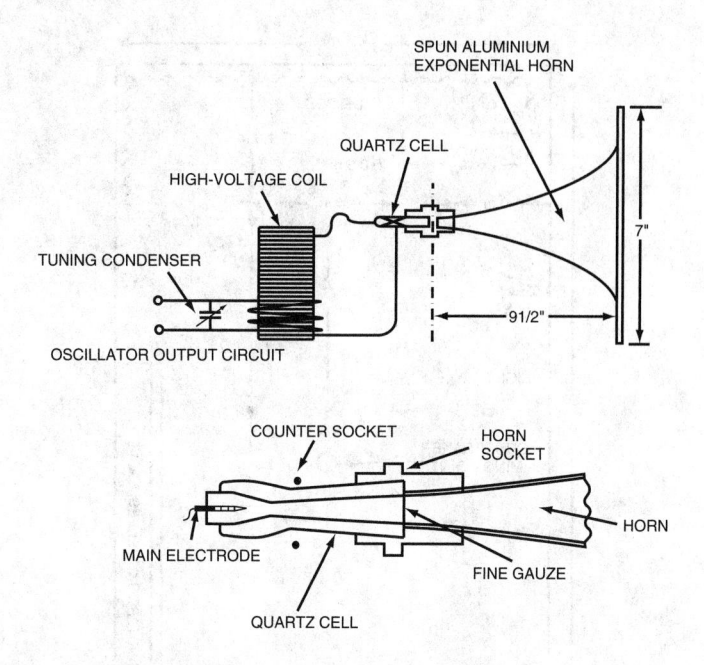

FIGURE 3.66 Detail of the ionophone assembly with the output circuit of the oscillator. (*Source:* In *Loudspeaker and Headphone Handbook,* ed. J. Borwick, 2nd ed., p. 37. Focal.)

the many variables inherent in subjective testing, as well as the costs required to produce a scientifically valid study, increasing attention has been paid to such methodology in recent years.

So called *double blind* testing, employing A/B/X methodology has been particularly successful in producing verifiable results of listener preferences and discernment. The reference, source, and random sampling of either is used under A/B/X testing to achieve good correlation in statistical validity. Commonly utilized analysis of variance (ANOVA) is frequently employed, and the test administrators are blind to the testing material, the specific parameter under test, and often, even the subjects.

Of course many variables will affect subjective tests and the International Electrotechnical Commission (IEC) has produced a recommended listening room configuration for conducting tests on loudspeakers. See Fig. 3.67 (Toole).

Not surprisingly listeners exhibit biases related to hearing ability, and to a lesser extent, age. Broadband sensitivity is apparently the most important factor in the reliability of most test subjects evaluating loudspeaker systems. It is estimated that approximately 60% of the population possess acceptable broadband sensitivity to be reliable test subjects. It has been noted that stereo assessments tend to obscure slightly the discernment of typical listeners. Monophonic tests of the same loudspeaker systems revealed a more heightened sensitivity to colorations than when listeners heard the systems in stereo.

A sample listener rating grid is shown in Fig. 3.68, as utilized by the Communications Research Centre in Ottawa during evaluations of proposed digital radio broadcast systems.

3.5.7 Headphones and Earphones

Several unique factors render headphone and earphone design even more complex than loudspeaker systems. Most specialized communications headphones rely on restricted frequency response tailored to maximize speech intelligibility. This discussion will cover the more common issues associated with high-fidelity headphone reproduction.

A fundamental problem of binaural listening is known as *in-head localization* (IHL) and is apparently due to the lack of a dispersed wavefront striking the aural cavity. Other problems such as bone conduction (BC), the frequency response and resonance challenges of small aperture assemblies, variable leakage

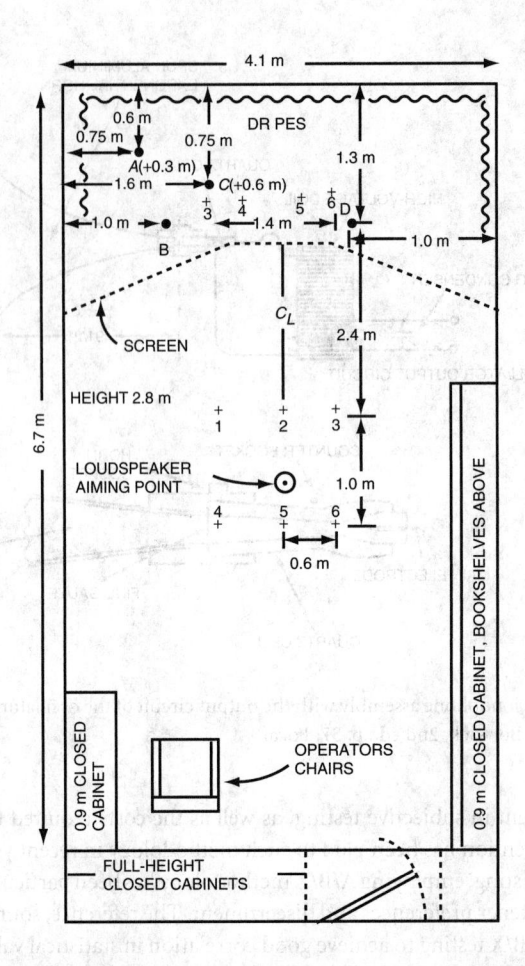

FIGURE 3.67 A plain view of the listening room of the IEC recommended configuration. In monophonic tests, loudspeakers can be located at positions 3, 4, 5, and 6, with listeners at positions 1, 3, and 5. In stereophonic tests, loudspeakers are located at positions B and D, and listeners at positions 2 and 5. (*Source:* Toole, F.E. Loudspeaker measurements and their relationship to listener preferences. *JAES* 34 (Pts 1 and 2).

effects, the *cone of confusion*, the *occlusion effect*, the *missing 6-dB effect*, and even factors such as cushion compliance add additional complexity to headphone design.

In-Head Localization

Headphones, unlike ambient listening conditions, offer a consistent and stable aural environment, do not normally affect others in the vicinity, and offer excellent sound quality at a far lower price than comparable loudspeaker systems. In-head localization is perhaps the greatest single obstacle to ubiquitous use of headphone systems.

It is the lack of frontal spatial localization that most vexes headphone designers. Substantial progress has been made at improving not only median plane front and rear cues, but also elevation information can now be added to headphone listening. Through much study it has been determined that specific spectral bands hold the key to significant location information. Two directional bands are associated with front localization, 260–550 Hz and 2.5–6 kHz. Rear localization is achieved by boosting 700 Hz–1.8 kHz and 10–13 kHz. Upward elevation localization is achieved by boosting 7–10 kHz. Additional cues have been successfully added to replicate the dynamic mechanism of head movements by synchronously varying interaural time delays (IDT) and interaural amplitude delays (IAD). Directional shifts of up to 60° have been reported using these simple techniques, yet full frontal localization remains elusive.

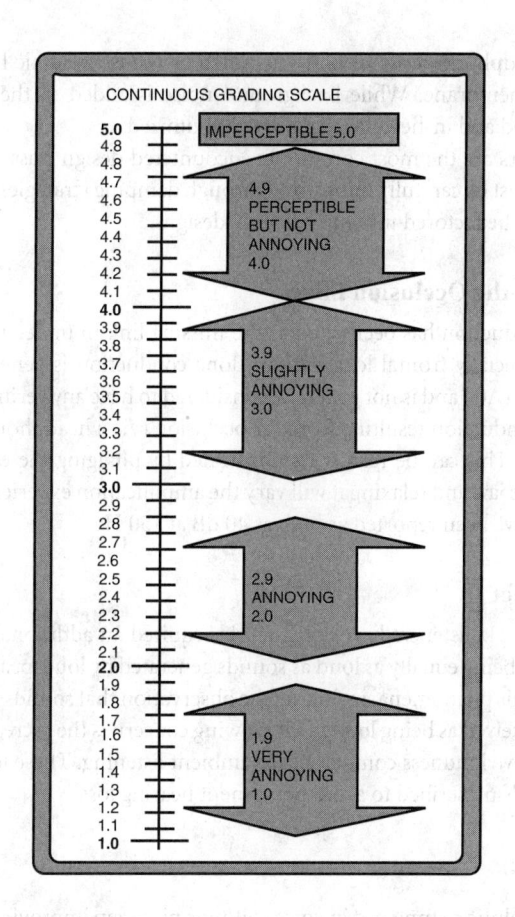

CONTINUOUS GRADING SCALE

5.0	IMPERCEPTIBLE 5.0
4.8	
4.8	
4.7	4.9
4.6	PERCEPTIBLE
4.5	BUT NOT
4.4	ANNOYING
4.3	4.0
4.2	
4.1	
4.0	
3.9	
3.8	3.9
3.7	SLIGHTLY
3.6	ANNOYING
3.5	3.0
3.4	
3.3	
3.2	
3.1	
3.0	
2.9	
2.8	
2.7	
2.6	
2.5	2.9
2.4	ANNOYING
2.3	2.0
2.2	
2.1	
2.0	
1.9	
1.8	
1.7	
1.6	1.9
1.5	VERY
1.4	ANNOYING
1.3	1.0
1.2	
1.1	
1.0	

FIGURE 3.68 Image of the 5-grade impairment scale.

One interesting experiment reported by Poldy involved making a binaural recording using the listener's own head for placement of the electret microphones in the ear canals. This process was repeated for other listeners. When playing the recordings back using symmetrical headphones each listener easily recognized his own recording and heard the sounds in front. The recordings made in other person's ears were elevated or in-head localized. Thus, although progress is being made in this area, each listener's unique interaural cues to frontal localization may provide an insurmountable challenge to solving this long standing problem. Real-time signal processing power may eventually lead to more specific, and certainly more costly, results.

Related to the problem of missing frontal information is the so-called cone of confusion which describes the ambiguity of lateral localization when extreme left or right information is reproduced in the headphone assembly. See Fig. 3.69.

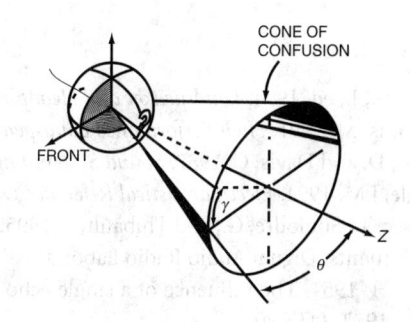

FIGURE 3.69 The cone of confusion for a spherical head model.

Small Aperture Assemblies

Basic headphone designs are based on circumaural, supra-aural, closed, and open. In circumaural construction the headphone surrounds the ear, for supra-aural design the headphones rests on the ear. In general, the closed and circumaural designs have better low-frequency response, which is largely controlled by leakage. The open headphone design is best illustrated by the isodynamic headphone, which uses an acoustically permeable membrane. While background noise is added to the reproduced sound, low-frequency response is good and in-head localization is minimized.

Moving coil transducers are the most commonly encountered design class but, as with loudspeakers, membrane resonances must be carefully minimized through damping strategies. Cushion compliance and cup resonances must also be factored into high-fidelity designs.

Bone Conduction and the Occlusion Effect

Low-frequency bone conduction has been cited as the missing link in understanding headphone reproduction phenomena, especially frontal localization. Bone conduction is generally masked some 50 dB below *acoustic conduction* (AC) and is not generally considered to have any verifiable effect on localization. Amplification of bone conduction resulting from ear occlusion (via a headphone, for example) occurs up to approximately 2 kHz. This can be readily demonstrated by plugging the ears while humming a low sweep tone. Clenching the jaw and relaxing it will vary the amplification experienced. Sound pressure level shifts due to occlusion have been reported as high as 40 dB at 150 Hz.

The Missing 6 dB Effect

At frequencies below 300 Hz listeners have consistently required an additional 6–10 dB in SPL to judge sounds in headphones as being equally as loud as sounds generated by loudspeakers. Various explanations have been proffered for this phenomena, including the observation that sounds that emanate from a larger sound source may be perceived as being louder. Of growing concern is the increased SPL levels required by headphone users to achieve loudness comparable to ambient listening. These levels may more frequently pass the danger thresholds prescribed to avoid permanent hearing loss.

Defining Terms

Ferrofluid: Iron based solution employed in voice coil/pole piece gap improving magnetic flux and power
handling capacity.
Flexure nodes: Standing wave patterns in loudspeaker membrane caused by applied energy.
Haas effect (law of the first wavefront): The first sound heard will mask subsequent short delay arriving
sounds, the combination appearing as a louder source.
Piston: Ideal loudspeaker radiating diaphragm with all segments of the diaphragm moving uniformly.

References

Borwick, J., ed. 1994. *Loudspeaker and Headphone Handbook*, 2nd ed. Focal Press.
Colloms, M. 1991. *High Performance Loudspeakers*, 4th ed. Pentech Press, London.
Davis, D. and Davis, C. 1987. *Sound System Engineering*, 2nd ed. Howard Sams.
Eargle, J.M. 1994. *Electroacoustical Reference Data*. Van Nostrand Reinhold, New York.
Grusec, T., Soulodre, G., and Thibault, L. 1995. Subjective tests of audio quality and transmission impairments. Digital Audio Radio Laboratory Tests, EIA, Washington, D.C.
Haas, H. 1951. The influence of a single echo on the audibility of speech. *Acoustica* I(49). Reprinted in 1972, *JAES* 20.
Peutz, U.M.A. 1971. Articulation loss of consonants as a criterion for speech transmission in a room. *JAES* 19.
Rice, C.W. and Kellogg, E.W. 1925. Notes on the development of a new type of hornless loudspeaker. *JAIEE* 12: 461–480.
Toole, F.E. 1985. Subjective measurements of loudspeaker sound quality and listener performance. *JAES* 33.
Tremaine, H.M. 1973. *Audio Cyclopedia*, 2nd ed. Howard Sams.

Further Information

The Howard M. Tremaine *Audio Cyclopedia* in the references is out of print, but worth searching speciality bookstores.

AES. 1980. Loudspeakers: An Anthology Vol. 1, 2nd ed. Audio Engineering Society.

Beranek, L.L. 1986. *Acoustics*. American Institute of Physics, New York.

Bullock, R.M. 1982. Loudspeaker crossover systems: an optimal crossover choice. *JAES* 30, No. 7/8, 486.

ISO R226-1961 (1961). Normal equal loudness contours for pure tones and normal threshold of hearing under free field conditions.

Olson, H.F. 1957. *Acoustical Engineering*. Van Nostrand Co. Reprinted. Professional Audio Journals, Philadelphia (1991).

4

Passive Electrical Circuit

Jerry C. Whitaker
Editor-in-Chief

Kenneth R. Demarest
University of Kansas

Michael B. Steer
North Carolina State University

Cecil Harrison
University of Southern Mississippi

4.1 Coaxial Transmission Lines

Jerry C. Whitaker

4.1.1 Introduction

The components that connect, interface, transfer, and filter RF energy within a given system or between systems are critical elements in the operation of vacuum tube devices. Such hardware, usually passive, determines to a large extent the overall performance of the RF generator. To optimize the performance of power vacuum devices, it is first necessary to understand and optimize the components upon which the tube depends.

The mechanical and electrical characteristics of the transmission line, waveguide, and associated hardware that carry power from a power source (usually a transmitter) to the load (usually an antenna) are critical to proper operation of any RF system. Mechanical considerations determine the ability of the components to withstand temperature extremes, lightning, rain, and wind, that is, they determine the overall reliability of the system.

Skin Effect

The effective resistance offered by a given conductor to radio frequencies is considerably higher than the ohmic resistance measured with direct current. This is because of an action known as the *skin effect*, which causes the currents to be concentrated in certain parts of the conductor and leaves the remainder of the cross-section to contribute little or nothing toward carrying the applied current.

When a conductor carries an alternating current, a magnetic field is produced that surrounds the wire. This field continually expands and contracts as the ac wave increases from zero to its maximum positive value and back to zero, then through its negative half-cycle. The changing magnetic lines of force cutting the conductor induce a voltage in the conductor in a direction that tends to retard the normal flow of current in the wire. This effect is more pronounced at the center of the conductor. Thus, current within the conductor tends to flow more easily toward the surface of the wire. The higher the frequency, the greater the tendency for current to flow at the surface. The depth of current flow d is a function of frequency and is determined from the following equation:

$$d = \frac{2.6}{\sqrt{\mu f}} \tag{4.1}$$

where d is the depth of current in mils, μ is the permeability (copper $= 1$, steel $= 300$), and f is the frequency of signal in MHz. It can be calculated that at a frequency of 100 kHz, current flow penetrates a conductor by 8 mils. At 1 MHz, the skin effect causes current to travel in only the top 2.6 mils in copper, and even less in almost all other conductors. Therefore, the series impedance of conductors at high frequencies is significantly higher than at low frequencies. Figure 4.1 shows the distribution of current in a radial conductor.

When a circuit is operating at high frequencies, the skin effect causes the current to be redistributed over the conductor cross-section in such a way as to make most of the current flow where it is encircled by the smallest number of flux lines. This general principle controls the distribution of current, regardless of the shape of the conductor involved. With a flat-strip conductor, the current flows primarily along the edges, where it is surrounded by the smallest amount of flux.

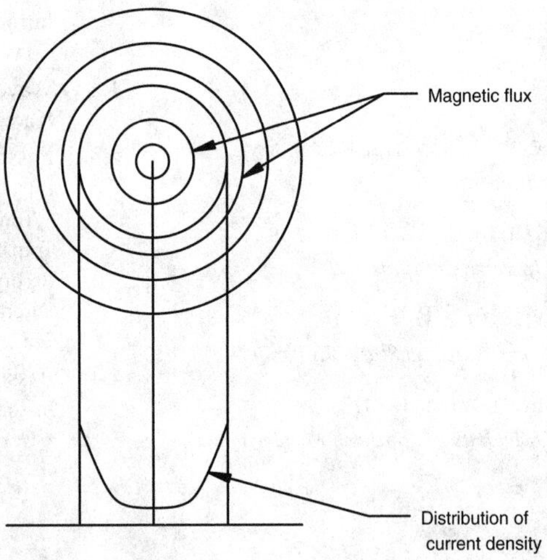

FIGURE 4.1 Skin effect on an isolated round conductor carrying a moderately high frequency signal.

4.1.2 Coaxial Transmission Line

Two types of coaxial transmission line are in common use today: rigid line and corrugated (semiflexible) line. Rigid coaxial cable is constructed of heavy-wall copper tubes with Teflon or ceramic spacers. (Teflon^TM is a registered trademark of DuPont.) Rigid line provides electrical performance approaching an ideal transmission line, including

- High power-handling capability
- Low loss
- Low VSWR (voltage standing wave ratio)

Rigid transmission line is, however, expensive to purchase and install.

The primary alternative to rigid coax is semiflexible transmission line made of corrugated outer and inner conductor tubes with a spiral polyethylene (or Teflon) insulator. The internal construction of a semiflexible line is shown in Fig. 4.2. Semiflexible line has four primary benefits:

- It is manufactured in a continuous length, rather than the 20-foot sections typically used for rigid line.

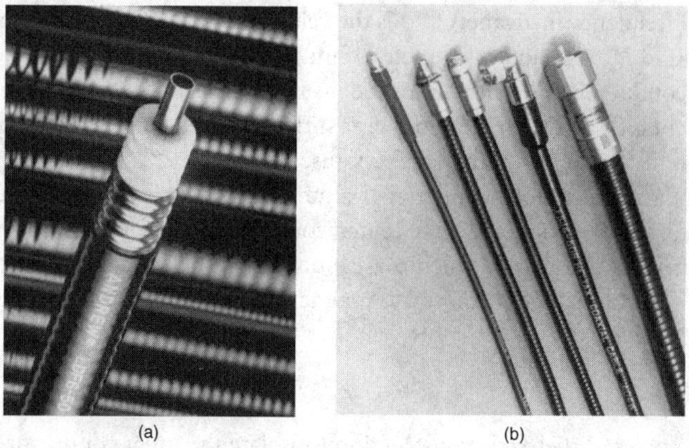

FIGURE 4.2   Semiflexible coaxial cable: (a) section of cable showing the basic construction, and (b) cable with various terminations. (Courtesy of Andrew Corporation, Orland Park, IL.)

- Because of the corrugated construction, the line can be shaped as required for routing from the transmitter to the antenna.
- The corrugated construction permits differential expansion of the outer and inner conductors.
- Each size of line has a minimum bending radius. For most installations, the flexible nature of corrugated line permits the use of a single piece of cable from the transmitter to the antenna, with no elbows or other transition elements. This speeds installation and provides for a more reliable system.

Electrical Parameters

A signal traveling in free space is unimpeded; it has a free-space velocity equal to the speed of light. In a transmission line, capacitance and inductance slow the signal as it propagates along the line. The degree to which the signal is slowed is represented as a percentage of the free-space velocity. This quantity is called the relative velocity of propagation and is described by

$$V_p = \frac{1}{\sqrt{LC}} \tag{4.2}$$

where L is the inductance in henrys per foot, C is the capacitance in farads per foot, and

$$V_r = \frac{V_p}{c} \times 100\% \tag{4.3}$$

where V_p is the velocity of propagation, c is 9.842×10^8 feet per second (free-space velocity), and V_r is the velocity of propagation as a percentage of free-space velocity.

Transverse Electromagnetic Mode

The principal mode of propagation in a coaxial line is the *transverse electromagnetic mode* (TEM). This mode will not propagate in a waveguide, and that is why coaxial lines can propagate a broad band of frequencies efficiently. The cutoff frequency for a coaxial transmission line is determined by the line dimensions. Above cutoff, modes other than TEM can exist and the transmission properties are no longer defined. The cutoff frequency is equivalent to

$$F_c = \frac{7.50 \times V_r}{D_i D_o} \tag{4.4}$$

where F_c is the cutoff frequency in gigahertz, V_r is the velocity (percent), D_i is the inner diameter of outer conductor in inches, and D_o is the outer diameter of inner conductor in inches.

At dc, current in a conductor flows with uniform density over the cross-section of the conductor. At high frequencies, the current is displaced to the conductor surface. The effective cross-section of the conductor decreases and the conductor resistance increases because of the skin effect.

Center conductors are made from copper-clad aluminum or high-purity copper and can be solid, hollow tubular, or corrugated tubular. Solid center conductors are found on semiflexible cable with $1/2$-inch or smaller diameter. Tubular conductors are found in $7/8$-inch or larger-diameter cables. Although the tubular center conductor is used primarily to maintain flexibility, it can also be used to pressurize an antenna through the feeder.

Dielectric

Coaxial lines use two types of dielectric construction to isolate the inner conductor from the outer conductor. The first is an air dielectric, with the inner conductor supported by a dielectric spacer and the remaining volume filled with air or nitrogen gas. The spacer, which may be constructed of spiral or discrete rings, typically is made of Teflon or polyethylene. Air-dielectric cable offers lower attenuation and higher average power ratings than foam-filled cable but requires pressurization to prevent moisture entry.

Foam-dielectric cables are ideal for use as feeders with antennas that do not require pressurization. The center conductor is completely surrounded by foam-dielectric material, resulting in a high dielectric breakdown level. The dielectric materials are polyethylene-based formulations that contain antioxidants to reduce dielectric deterioration at high temperatures.

Impedance

The expression *transmission line impedance* applied to a point on a transmission line signifies the vector ratio of line voltage to line current at that particular point. This is the impedance that would be obtained if the transmission line were cut at the point in question, and the impedance looking toward the receiver were measured.

Because the voltage and current distribution on a line are such that the current tends to be small when the voltage is large (and vice versa), as shown in Fig. 4.3, the impedance will, in general, be oscillatory in the same manner as the voltage (large when the voltage is high and small when the voltage is low). Thus, in the case of a short-circuited receiver, the impedance will be high at distances from the receiving end that are odd multiples of $1/4$-wavelength, and will be low at distances that are even multiples of $1/4$-wavelength.

The extent to which the impedance fluctuates with distance depends on the *standing wave ratio* (ratio of reflected to incident waves), being less as the reflected wave is proportionally smaller than the incident wave. In the particular case where the load impedance equals the characteristic impedance, the impedance of the transmission line is equal to the characteristic impedance at all points along the line.

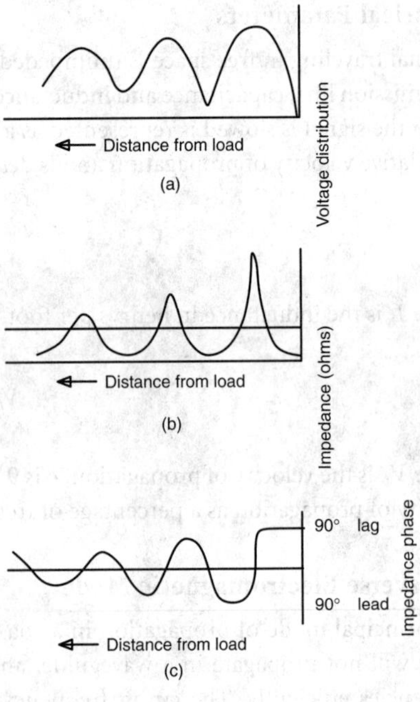

FIGURE 4.3 Magnitude and power factor of line impedance with increasing distance from the load for the case of a short-circuited receiver and a line with moderate attenuation: (a) voltage distribution, (b) impedance magnitude, and (c) impedance phase.

The *power factor* of the impedance of a transmission line varies according to the standing waves present. When the load impedance equals the characteristic impedance, there is no reflected wave and the power factor of the impedance is equal to the power factor of the characteristic impedance. At radio frequencies, the power factor under these conditions is accordingly resistive. However, when a reflected wave is present, the power factor is unity (resistive) only at the points on the line where the voltage passes through a maximum or a minimum. At other points the power factor will be reactive, alternating from leading to lagging at intervals of $^1/_4$-wavelength. When the line is short-circuited at the receiver, or when it has a resistive load less than the characteristic impedance so that the voltage distribution is of the short-circuit type, the power factor is inductive for lengths corresponding to less than the distance to the first voltage maximum. Thereafter, it alternates between capacitive and inductive at intervals of $^1/_4$-wavelength. Similarly, with an open-circuited receiver or with a resistive load greater than the characteristic impedance so that the voltage distribution is of the open-circuit type, the power factor is capacitive for lengths corresponding to less than the distance to the first voltage minimum. Thereafter, the power factor alternates between capacitive and inductive at intervals of $^1/_4$-wavelength, as in the short-circuited case.

Resonant Characteristics

A transmission line can be used to perform the functions of a resonant circuit. For example, if the line is short-circuited at the receiver, at frequencies in the vicinity of a frequency at which the line is an odd number of $^1/_4$-wavelengths long, the impedance will be high and will vary with frequency in the vicinity of resonance. This characteristic is similar in nature to a conventional parallel resonant circuit. The difference is that with the transmission line, there are a number of resonant frequencies, one for each of the infinite number of frequencies that make the line an odd number of $^1/_4$-wavelengths long. At VHF, the parallel impedance at resonance and the circuit Q obtainable are far higher than can be realized with lumped circuits. Behavior corresponding to that of a series resonant circuit can be obtained from a transmission line that is an odd number of $^1/_4$-wavelengths long and open-circuited at the receiver.

Transmission lines can also be used to provide low-loss inductances or capacitances if the proper combination of length, frequency, and termination is employed. Thus, a line short-circuited at the receiver will offer an inductive reactance when less than $^1/_4$-wavelength, and a capacitive reactance when between $^1/_4$- and $^1/_2$-wavelength. With an open-circuited receiver, the conditions for inductive and capacitive reactances are reversed.

4.1.3 Electrical Considerations

VSWR (voltage standing wave ratio), attenuation, and power-handling capability are key electrical factors in the application of coaxial cable. High VSWR can cause power loss, voltage breakdown, and thermal degradation of the line. High attenuation means less power delivered to the antenna, higher power consumption at the transmitter, and increased heating of the transmission line itself.

VSWR is a common measure of the quality of a coaxial cable. High VSWR indicates nonuniformities in the cable that can be caused by one or more of the following conditions:

- Variations in the dielectric core diameter
- Variations in the outer conductor
- Poor concentricity of the inner conductor
- Nonhomogeneous or periodic dielectric core

Although each of these conditions may contribute only a small reflection, they can add up to a measurable VSWR at a particular frequency.

Rigid transmission line is typically available in a standard length of 20 ft, and in alternative lengths of 19.5 and 19.75 ft. The shorter lines are used to avoid VSWR buildup caused by discontinuities resulting from the physical spacing between line section joints. If the section length selected and the operating frequency have a $^1/_2$-wave correlation, the connector junction discontinuities will add. This effect is known as flange buildup. The result can be excessive VSWR. The *critical frequency* at which a $^1/_2$-wave relationship exists

TABLE 4.1   Representative Specifications for Various Types of Flexible Air-Dielectric Coaxial Cable

Cable Size (in.)	Maximum Frequency (MHz)	Velocity (%)	Peak Power 1 MHz (kW)	Average Power		Attenuation[a]	
				100 MHz (kW)	1 MHz (kW)	100 MHz (dB)	1 MHz (dB)
$1^5/_8$	2.7	92.1	145	145	14.4	0.020	0.207
3	1.64	93.3	320	320	37	0.013	0.14
4	1.22	92	490	490	56	0.010	0.113
5	0.96	93.1	765	765	73	0.007	0.079

[a] Attenuation specified in dB/100 ft.

is given by

$$F_{cr} = \frac{490.4 \times n}{L} \tag{4.5}$$

where F_{cr} is the critical frequency, n is any integer, and L is the transmission line length in feet. For most applications, the critical frequency for a chosen line length should not fall closer than ± 2 MHz of the passband at the operating frequency.

Attenuation is related to the construction of the cable itself and varies with frequency, product dimensions, and dielectric constant. Larger-diameter cable exhibits lower attenuation than smaller-diameter cable of similar construction when operated at the same frequency. It follows, therefore, that larger-diameter cables should be used for long runs.

Air-dielectric coax exhibits less attenuation than comparable-sized foam-dielectric cable. The attenuation characteristic of a given cable is also affected by standing waves present on the line resulting from an impedance mismatch. Table 4.1 shows a representative sampling of semiflexible coaxial cable specifications for a variety of line sizes.

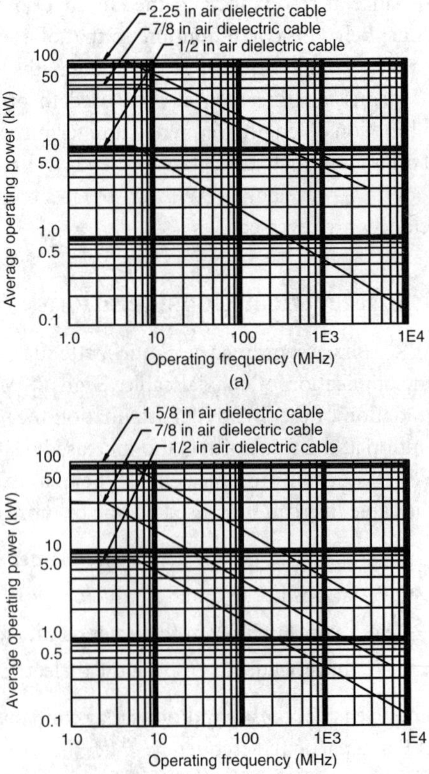

4.1.4   Coaxial Cable Ratings

Selection of a type and size of transmission line is determined by a number of parameters, including power-handling capability, attenuation, and phase stability.

Power Rating

Both peak and average power ratings are required to fully describe the capabilities of a transmission line. In most applications, the peak power rating limits the low frequency or pulse energy, and the average power rating limits high-frequency applications, as shown in Fig. 4.4. Peak power ratings are usually

FIGURE 4.4   Power rating data for a variety of coaxial transmission lines: (a) 50 Ω line, and (b) 75 Ω line.

stated for the following conditions:

- VSWR = 1.0
- Zero modulation
- One atmosphere of absolute dry air pressure at sea level

The peak power rating of a selected cable must be greater than the following expression, in addition to satisfying the average-power-handling criteria:

$$E_{pk} > P_t \times (1 + M)^2 \times \text{VSWR} \tag{4.6}$$

where E_{pk} is the cable peak power rating in kilowatts, P_t is the transmitter power in kilowatts, M is the amplitude modulation percentage expressed decimally (100% = 1.0), and VSWR is the voltage standing wave ratio. From this equation, it can be seen that 100% amplitude modulation will increase the peak power in the transmission line by a factor of 4. Furthermore, the peak power in the transmission line increases directly with VSWR.

The peak power rating is limited by the voltage breakdown potential between the inner and outer conductors of the line. The breakdown point is independent of frequency. It varies, however, with the line pressure (for an air-dielectric cable) and the type of pressurizing gas.

The average power rating of a transmission line is limited by the safe, long-term operating temperature of the inner conductor and the dielectric. Excessive temperatures on the inner conductor will cause the dielectric material to soften, leading to mechanical instability inside the line.

The primary purpose of pressurization of an air-dielectric cable is to prevent the ingress of moisture. Moisture, if allowed to accumulate in the line, can increase attenuation and reduce the breakdown voltage between the inner and outer conductors. Pressurization with high-density gases can be used to increase both the average power and the peak power ratings of a transmission line. For a given line pressure, the increased power rating is more significant for peak power than for average power. High-density gases used for such applications include Freon 116 and sulfur hexafluoride. Figure 4.5 illustrates the effects of pressurization on cable power rating.

An adequate safety factor is necessary for peak and average power ratings. Most transmission lines are tested at two or more times their rated peak power before shipment to the customer. This safety factor is intended as a provision for transmitter transients, lightning-induced effects, and high-voltage excursions resulting from unforeseen operating conditions.

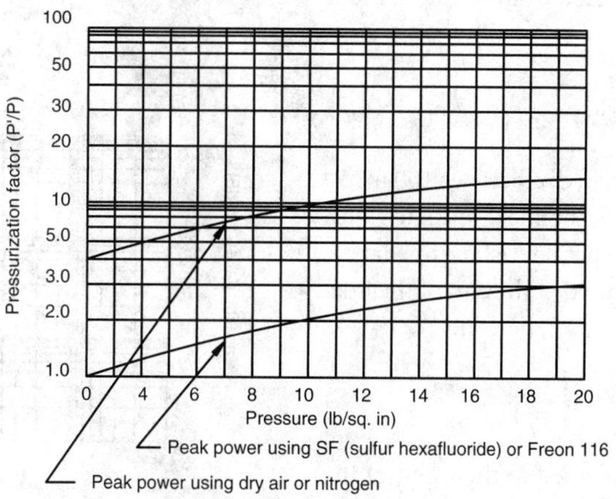

FIGURE 4.5 Effects of transmission line pressurization on peak power rating. Note that $P' =$ rating of the line at the increased pressure and $P =$ rating of the line at atmospheric pressure.

TABLE 4.2   Electrical Characteristics of Common RF Connectors

Connector Type	DC Test Voltage (kW)	Peak Power (kW)
SMA	1.0	1.2
BNC, TNC	1.5	2.8
N, UHF	2.0	4.9
GR	3.0	11
HN, $^7/_{16}$	4.0	20
LC	5.0	31
$^7/_8$ EIA, F Flange	6.0	44
$1^5/_8$ EIA	11.0	150
$3^1/_8$ EIA	19.0	4

Connector Effects

Foam-dielectric cables typically have a greater dielectric strength than air-dielectric cables of similar size. For this reason, foam cables would be expected to exhibit higher peak power ratings than air lines. Higher values, however, usually cannot be realized in practice because the connectors commonly used for foam cables have air spaces at the cable/connector interface that limit the allowable RF voltage to "air cable" values.

The peak-power-handling capability of a transmission line is the smaller of the values for the cable and the connectors attached to it. Table 4.2 lists the peak power ratings of several common RF connectors at standard conditions.

Attenuation

The attenuation characteristics of a transmission line vary as a function of the operating frequency and the size of the line itself. The relationships are shown in Fig. 4.6.

The efficiency of a transmission line dictates how much power output from the transmitter actually reaches the antenna. Efficiency is determined by the length of the line and the attenuation per unit length.

The attenuation of a coaxial transmission line is defined by

$$\alpha = 10 \times \log \left\{ \frac{P_1}{P_2} \right\} \qquad (4.7)$$

where α is the attenuation in decibels per 100 m, P_1 is the input power into a 100-m line terminated with the nominal value of its characteristic impedance, and P_2 is the power measured at the end of the line. Stated in terms of efficiency (E, percent),

$$E = 100 \times \left\{ \frac{P_0}{P_i} \right\} \qquad (4.8)$$

where P_i is the power delivered to the input of the transmission line and P_o is the power delivered to the antenna. The relationship between efficiency and loss in decibels (insertion loss) is illustrated in Fig. 4.7.

FIGURE 4.6 Attenuation characteristics for a selection of coaxial cables: (a) 50 Ω line, and (b) 75 Ω line.

Manufacturer-supplied attenuation curves are typically guaranteed to within approximately ±5%. The values given usually are rated at 24°C (75°F) ambient temperature. Attenuation increases slightly with higher temperature or applied power. The effects of ambient temperature on attenuation are illustrated in Fig. 4.8.

Loss in connectors is negligible, except for small (SMA and BNC) connectors at frequencies of several gigahertz and higher. Small connectors used at high frequencies typically add 0.1 dB of loss per connector.

When a transmission line is attached to a load such as an antenna, the VSWR of the load increases the total transmission loss of the system. This effect is small under conditions of low VSWR. Figure 4.9 illustrates the interdependence of these two elements.

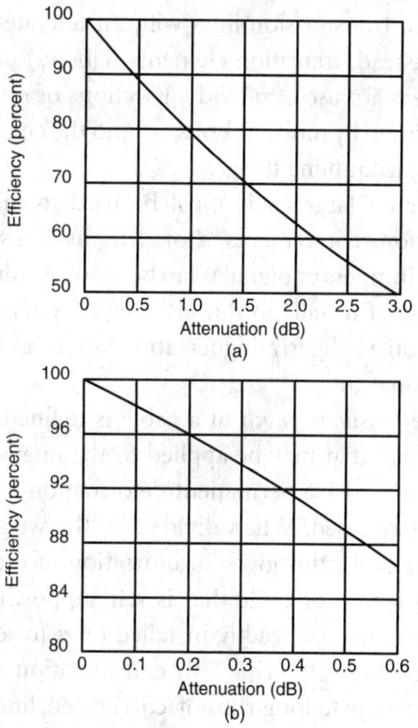

Phase Stability

A coaxial cable expands as the temperature of the cable increases, causing the electrical length of the line to increase as well. This factor results in phase changes that are a function of operating temperature. The phase relationship can be described by

$$\theta = 3.66 \times 10^{-7} \times P \times L \times T \times F \qquad (4.9)$$

where θ is the phase change in degrees, P is the phase-temperature coefficient of the cable, L is the length of coax in feet, T is the temperature range (minimum-to-maximum operating temperature), and F is the frequency in MHz. Phase changes that are a function of temperature are important in systems utilizing multiple transmission lines, such as a directional array fed from a single phasing source. To maintain proper operating parameters, the phase changes of the cables must be minimized. Specially designed coaxial cables that offer low-phase-temperature characteristics are available. Two types of coax are commonly used for this purpose: (1) phase-stabilized cables, which have undergone extensive temperature cycling until such time as they exhibit their minimum phase-temperature coefficient, and (2) phase-compensated cables, in which changes in the electrical length have been minimized through adjustment of the mechanical properties of the dielectric and inner/outer conductors.

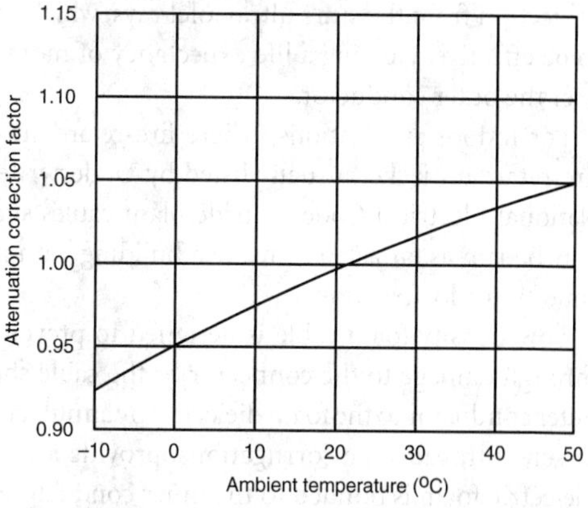

FIGURE 4.7 Conversion chart showing the relationship between decibel loss and efficiency of a transmission line: (a) high-loss line, and (b) low-loss line.

FIGURE 4.8 The variation of coaxial cable attenuation as a function of ambient temperature.

Mechanical Parameters

Corrugated copper cables are designed to withstand bending with no change in properties. Low-density foam- and air-dielectric cables generally have a minimum bending radius of ten times the cable diameter. Super flexible versions provide a smaller allowable bending radius.

Rigid transmission lines will not tolerate bending. Instead, transition elements (elbows) of various sizes are used. Individual sections of rigid line are secured by multiple bolts around the circumference of a coupling flange.

When a large cable must be used to meet attenuation requirements, short lengths of a smaller cable (jumpers or pigtails) can be used on either end for ease of installation in low-power systems. The trade-off is slightly higher attenuation and some additional cost.

The *tensile strength* of a cable is defined as the axial load that may be applied to the line with no more than 0.2% permanent deformation after the load is released. When divided by the weight per foot of cable, this gives an indication of the maximum length of cable that is self-supporting and therefore can be readily installed on a tower with a single hoisting grip. This consideration usually applies only to long runs of corrugated line; rigid line is installed one section at a time.

(a) 6 dB normal line attenuation
(b) 3 dB normal line attenuation
(c) 2 dB normal line attenuation
(d) 1 dB normal line attenuation
(e) 0.5 dB normal line attenuation

FIGURE 4.9 The effect of load VSWR on transmission line loss.

The *crush strength* of a cable is defined as the maximum force per linear inch that can be applied by a flat plate without causing more than a 5% deformation of the cable diameter. Crush strength is a good indicator of the ruggedness of a cable and its ability to withstand rough handling during installation.

Cable jacketing affords mechanical protection during installation and service. Semiflexible cables typically are supplied with a jacket consisting of low-density polyethylene blended with 3% carbon black for protection from the sun's ultraviolet rays, which can degrade plastics over time. This approach has proved to be effective, yielding a life expectancy of more than 20 years. Rigid transmission line has no covering over the outer conductor.

For indoor applications, where fire-retardant properties are required, cables can be supplied with a fire-retardant jacket, usually listed by Underwriters Laboratories. Note that under the provisions of the National Electrical Code, outside plant cables such as standard black polyethylene-jacketed coaxial line may be run as far as 50 ft inside a building with no additional protection. The line can also be placed in conduit for longer runs.

Low-density foam cable is designed to prevent water from traveling along its length, should it enter through damage to the connector or the cable sheath. This is accomplished by mechanically locking the outer conductor to the foam dielectric by annular corrugations. Annular or ring corrugations, unlike helical or screw-thread-type corrugations, provide a water block at each corrugation. Closed-cell polyethylene dielectric foam is bonded to the inner conductor, completing the moisture seal.

A coaxial cable line is only as good as the connectors used to tie it together. The connector interface must provide a weatherproof bond with the cable to prevent water from penetrating the connection. This is ensured by the use of O-ring seals. The cable connector interface must also provide a good electrical bond that does not introduce a mismatch and increase VSWR. Good electrical contact between the connector and the cable ensures that proper RF shielding is maintained.

Bibliography

Andrew Corporation, Broadcast Transmission Line Systems, Technical Bulletin 1063H, Orland Park, IL, 1982.

Benson, K. B. and J. C. Whitaker, *Television and Audio Handbook for Technicians and Engineers*, McGraw-Hill, New York, 1989.

Cablewave Systems, The Broadcaster's Guide to Transmission Line Systems, Technical Bulletin 21A, North Haven, CT, 1976.

Cablewave Systems, Rigid Coaxial Transmission Lines, Cablewave Systems Catalog 700, North Haven, CT, 1989.

Crutchfield, E. B., Ed., *NAB Engineering Handbook*, 8th ed., National Association of Broadcasters, Washington, D.C., 1992.

Fink, D. and D. Christiansen, Eds., *Electronics Engineers' Handbook*, 3rd ed., McGraw-Hill, New York, 1989.

Jordan, Edward C., Ed., *Reference Data for Engineers: Radio, Electronics, Computer and Communications*, 7th ed., Howard W. Sams, Indianapolis, IN, 1985.

Perelman, R. and T. Sullivan, Selecting flexible coaxial cable, in *Broadcast Engineering*, Intertec Publishing, Overland Park, KS, May 1988.

Terman, F. E., *Radio Engineering*, 3rd ed., McGraw-Hill, New York, 1947.

Whitaker, Jerry C., G. DeSantis, and C. Paulson, *Interconnecting Electronic Systems*, CRC Press, Boca Raton, FL, 1993.

Whitaker, Jerry C., *Radio Frequency Transmission Systems: Design and Operation*, McGraw-Hill, New York, 1990.

4.2 Waveguide

Kenneth R. Demarest

4.2.1 Waveguide Modes

Any structure that guides electromagnetic waves can be considered a **waveguide**. Most often, however, this term refers to closed metal cylinders that maintain the same cross-sectional dimensions over long distances. Such a structure is shown in Fig. 4.10, which consists of a metal cylinder filled with a dielectric. When filled with low-loss dielectrics (such as air), waveguides typically exhibit lower losses than transmission lines, which makes them useful for transporting RF energy over relatively long distances. They are most often used for frequencies ranging from 1 to 150 GHz.

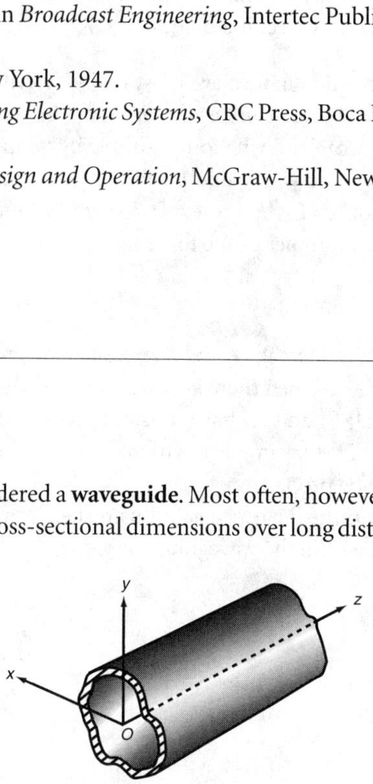

Every type of waveguide has an infinite number of distinct electromagnetic field configurations that can exist inside it. Each of these configurations is called a **waveguide mode**. The characteristics of these modes depend upon the cross-sectional dimensions of the conducting cylinder, the type of dielectric material inside the waveguide, and the frequency of operation.

FIGURE 4.10   A uniform waveguide with arbitrary cross section.

Waveguide modes are typically classed according to the nature of the electric and magnetic field components E_z and H_z. These components are called the longitudinal components of the fields. Several types of modes are possible in waveguides.

- *TE modes.* Transverse-electric modes, sometimes called H modes. These modes have $E_z = 0$ at all points within the waveguide, which means that the electric field vector is always perpendicular (i.e., transverse) to the waveguide axis. These modes are always possible in waveguides with uniform dielectrics.

- *TM modes.* Transverse-magnetic modes, sometimes called E modes. These modes have $H_z = 0$ at all points within the waveguide, which means that the magnetic field vector is perpendicular to the waveguide axis. Like TE modes, they are always possible in waveguides with uniform dielectrics.

- *EH modes.* EH modes are hybrid modes in which neither E_z nor H_z are zero, but the characteristics of the transverse fields are controlled more by E_z than H_z. These modes are often possible in waveguides with inhomogeneous dielectrics.

- *HE modes.* HE modes are hybrid modes in which neither E_z nor H_z are zero, but the characteristics of the transverse fields are controlled more by H_z than E_z. Like EH modes, these modes are often possible in waveguide with inhomogeneous dielectrics.

- *TEM modes.* Transverse-electromagnetic modes, often called transmission line modes. These modes can exist only when a second conductor exists within the waveguide, such as a center conductor on a coaxial cable. Because these modes cannot exist in single, closed conductor structures, they are not waveguide modes.

Waveguide modes are most easily determined by first computing the longitudinal field components, E_z and H_z, that can be supported by the waveguide. From these, the transverse components (such as E_x and E_y) can easily be found simply by taking spatial derivatives of the longitudinal fields (Collin, 1992).

When the waveguide properties are constant along the z axis, E_z and H_z vary in the longitudinal direction as $E_z, H_z \propto \exp(\omega t - \gamma z)$, where $\omega = 2\pi f$ is the radian frequency of operation and γ is a complex number of the form

$$\gamma = \alpha + j\beta \tag{4.10}$$

The parameters γ, α, and β are called the propagation, attenuation, and phase constants, respectively, and $j = \sqrt{-1}$. When there are no metal or dielectric losses, γ is always either purely real or imaginary. When γ is real, E_z and H_z have constant phase and decay exponentially with increasing z. When γ is imaginary, E_z and H_z vary in phase with increasing z but do not decay in amplitude. When this occurs, the fields are said to be propagating.

When the dielectric is uniform (i.e., homogeneous), E_z and H_z satisfy the scalar wave equation at all points within the waveguide

$$\nabla_t^2 E_z + h^2 E_z = 0 \tag{4.11}$$

and

$$\nabla_t^2 H_z + h^2 H_z = 0 \tag{4.12}$$

where

$$h^2 = (2\pi f)^2 \mu\epsilon + \gamma^2 = k^2 + \gamma^2 \tag{4.13}$$

Here, μ and ϵ are the permeability and permittivity of the dielectric media, respectively, and $k = 2\pi f \sqrt{\mu\epsilon}$ is the wave number of the dielectric. The operator ∇_t^2 is called transverse Laplacian operator. In Cartesian coordinates

$$\nabla_t^2 = \frac{\partial^2}{\partial x^2} + \frac{\partial^2}{\partial y^2}$$

Most of the properties of the allowed modes in real waveguides can usually be found by assuming that the walls are perfectly conducting. Under this condition, $E_z = 0$ and $\partial H_z / \partial p = 0$ at the waveguide walls, where p is the direction perpendicular to the waveguide wall. When these conditions are imposed upon the general solutions of Eqs. (4.11) and (4.12), it is found that only certain values of h are allowed. These values are called the *modal eigenvalues* and are determined by the cross-sectional shape of the waveguide. Using Eq. (4.13), the propagation constant γ for each mode varies with frequency according to

$$\gamma = \alpha + j\beta = h\sqrt{1 - \left(\frac{f}{f_c}\right)^2} \tag{4.14}$$

where

$$f_c = \frac{h}{2\pi\sqrt{\mu\epsilon}} \tag{4.15}$$

The modal parameter f_c is in hertz and is called the **cutoff frequency** of the mode it is associated with. According to Eq. (4.14), when $f > f_c$, the propagation constant γ is imaginary, and thus the mode is propagating. On the other hand, when $f < f_c$, then γ is real, which means that the fields decay exponentially with increasing values of z. Modes operated at frequencies below their cutoff frequency are not able to propagate energy over long distances and are called evanescent modes.

The dominant mode of a waveguide is the one with the lowest cutoff frequency. Although higher order modes are often useful for a variety of specialized uses of waveguides, signal distortion is usually minimized when a waveguide is operated in the frequency range where only the dominant mode exists. This range of frequencies is called the *dominant range* of the waveguide.

The distance over which the fields of propagating modes repeat themselves is called the **guide wavelength** λ_g. From Eq. (4.14), it can be shown that γ_g always varies with frequency according to

$$\lambda_g = \frac{\lambda_o}{1 - \left(\dfrac{f_c}{f}\right)^2} \tag{4.16}$$

where $\lambda_o = 1/(f\sqrt{\mu\epsilon})$ is the wavelength of a plane wave of the same frequency in an infinite sample of the waveguide dielectric. For $f \gg f_c$, $\lambda_g \approx \lambda_o$. Also, $\lambda_g \to \infty$ as $f \to f_c$, which is one reason why it is usually undesirable to operate a waveguide mode near modal cutoff frequencies.

Although waveguide modes are not plane waves, the ratio of their transverse electric and magnetic field magnitudes is constant throughout the cross section of the waveguide, just as for plane waves. This ratio is called the modal **wave impedance** and has the following values for TE and TM modes:

$$Z_{\text{TE}} = \frac{E_T}{H_T} = \frac{j\omega\mu}{\gamma} \tag{4.17}$$

and

$$Z_{\text{TM}} = \frac{E_T}{H_T} = \frac{\gamma}{j\omega\epsilon} \tag{4.18}$$

where E_T and H_T are the magnitudes of the transverse electric and magnetic fields, respectively. In the limit as $f \to \infty$, both Z_{TE} and Z_{TE} approach $\sqrt{\mu/\epsilon}$, which is the intrinsic impedance of the dielectric medium. On the other hand, as $f \to f_c$, $Z_{\text{TE}} \to \infty$ and $Z_{\text{TM}} \to 0$, which means that the transverse electric fields are dominant in TE modes near cutoff and vice versa for TM modes.

Rectangular Waveguides

A rectangular waveguide is shown in Fig. 4.11. The conducting walls are formed such that the inner surfaces form a rectangular cross section, with dimensions a and b along the x and y coordinate axes, respectively.

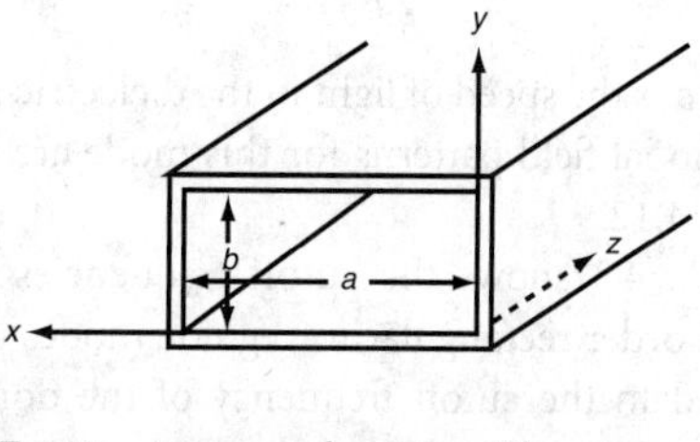

If the walls are perfectly conducting and the dielectric material is lossless, the field components for

FIGURE 4.11 A rectangular waveguide.

the TE$_{mn}$ modes are given by

$$E_x = H_0 \frac{j\omega\mu}{h_{mn}^2} \left(\frac{n\pi}{b}\right) \cos\left(\frac{m\pi}{a}x\right) \sin\left(\frac{n\pi}{b}y\right) \exp(j\omega t - \gamma_{mn}z) \tag{4.19}$$

$$E_y = -H_0 \frac{j\omega\mu}{h_{mn}^2} \left(\frac{m\pi}{a}\right) \sin\left(\frac{m\pi}{a}x\right) \cos\left(\frac{n\pi}{b}y\right) \exp(j\omega t - \gamma_{mn}z) \tag{4.20}$$

$$E_z = 0$$

$$H_x = H_0 \frac{\gamma_{mn}}{h_{mn}^2} \left(\frac{m\pi}{a}\right) \sin\left(\frac{m\pi}{a}x\right) \cos\left(\frac{n\pi}{b}y\right) \exp(j\omega t - \gamma_{mn}z) \tag{4.21}$$

$$H_y = H_0 \frac{\gamma_{mn}}{h_{mn}^2} \left(\frac{n\pi}{b}\right) \cos\left(\frac{m\pi}{a}x\right) \sin\left(\frac{n\pi}{b}y\right) \exp(j\omega t - \gamma_{mn}z) \tag{4.22}$$

$$H_z = H_0 \cos\left(\frac{m\pi}{a}x\right) \sin\left(\frac{n\pi}{b}y\right) \exp(j\omega t - \gamma_{mn}z) \tag{4.23}$$

where

$$h_{mn} = \left(\frac{m\pi}{a}\right)^2 + \left(\frac{n\pi}{b}\right)^2 = 2\pi f_{c_{mn}}\sqrt{\mu\epsilon} \tag{4.24}$$

For the TM$_{mn}$ modes, m and n can be any positive integer value, including zero, as long as both are not zero. The field components for the TM$_{mn}$ modes are

$$E_x = -E_0 \frac{\gamma_{mn}}{h_{mn}^2} \left(\frac{m\pi}{a}\right) \cos\left(\frac{m\pi}{a}x\right) \sin\left(\frac{n\pi}{b}y\right) \exp(j\omega t - \gamma_{mn}z) \tag{4.25}$$

$$E_y = -E_0 \frac{\gamma_{mn}}{h_{mn}^2} \left(\frac{n\pi}{b}\right) \sin\left(\frac{m\pi}{a}x\right) \cos\left(\frac{n\pi}{b}y\right) \exp(j\omega t - \gamma_{mn}z) \tag{4.26}$$

$$E_z = E_0 \sin\left(\frac{m\pi}{a}x\right) \sin\left(\frac{n\pi}{b}y\right) \exp(j\omega t - \gamma_{mn}z) \tag{4.27}$$

$$H_x = E_0 \frac{j\omega\epsilon}{h_{mn}^2} \left(\frac{n\pi}{b}\right) \sin\left(\frac{m\pi}{a}x\right) \cos\left(\frac{n\pi}{b}y\right) \exp(j\omega t - \gamma_{mn}z) \tag{4.28}$$

$$H_y = -E_0 \frac{j\omega\epsilon}{h_{mn}^2} \left(\frac{m\pi}{a}\right) \cos\left(\frac{m\pi}{a}x\right) \sin\left(\frac{n\pi}{b}y\right) \exp(j\omega t - \gamma_{mn}z) \tag{4.29}$$

$$H_z = 0 \tag{4.30}$$

where the values of h_{mn} and $f_{c_{mn}}$ are given by Eq. (4.24). For the TM$_{mn}$ modes, m and n can be any positive integer value except zero.

The dominant mode in a rectangular waveguide is the TE$_{10}$ mode, which has a cutoff frequency

$$f_{c_{10}} = \frac{1}{2a\sqrt{\mu\epsilon}} = \frac{c}{2a} \tag{4.31}$$

where c is the speed of light in the dielectric media. The modal field patterns for this mode are shown in Fig. 4.12.

Table 4.3 shows the cutoff frequencies of the lowest order rectangular waveguide modes (as referenced to the cutoff frequency of the dominant mode) when $a/b = 2.1$.

FIGURE 4.12 Field configurations for the TE$_{10}$ (dominant) mode of a rectangular waveguide. Solid lines, E; dashed lines, H. (*Source:* Adapted from Marcuvitz, N. 1986. *Waveguide Handbook*, 2nd ed., p. 63. Peter Peregrinus, Ltd., London. With permission.)

TABLE 4.3 Cutoff Frequencies of the Lowest Order Rectangular Waveguide Modes (Referenced to the Cutoff Frequency of the Dominant Mode) for a Rectangular Waveguide with $a/b = 2.1$

$f_c/f_{c_{10}}$	Modes
1.0	TE_{10}
2.0	TE_{20}
2.1	TE_{01}
2.326	TE_{11}, TM_{11}
2.9	TE_{21}, TM_{21}
3.0	TE_{30}
3.662	TE_{31}, TM_{31}
4.0	TE_{40}

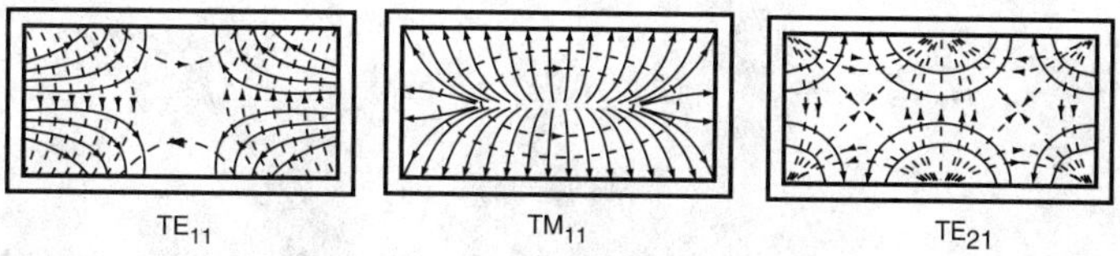

FIGURE 4.13 Field configurations for the TE_{11}, TM_{11}, and the TE_{21} modes: solid lines, E; dashed lines, H. (*Source:* Adapted from Marcuvitz, N. 1986. *Waveguide Handbook*, 2nd ed., p. 59. Peter Peregrinus, Ltd., London. With permission.)

The modal field patterns for several lower order modes are shown in Fig. 4. 13.

Circular Waveguides

A circular waveguide with inner radius a is shown in Fig. 4.14. Here the axis of the waveguide is aligned with the z axis of a circular-cylindrical coordinate system, where ρ and ϕ are the radial and azimuthal coordinates, respectively. If the walls are perfectly conducting and the dielectric material is lossless, the equations for the TE_{nm} modes are

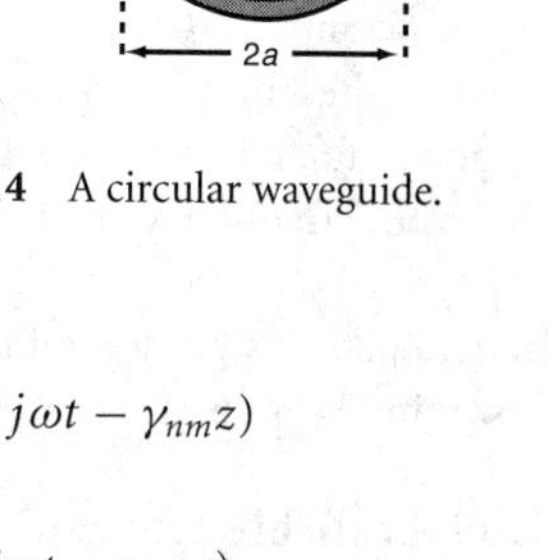

FIGURE 4.14 A circular waveguide.

$$E_\rho = H_0 \frac{j\omega\mu n}{h_{nm}^2 \rho} J_n(h_{nm}\rho) \sin(n\phi)\exp(j\omega t - \gamma_{nm}z) \tag{4.32}$$

$$E_\phi = H_0 \frac{j\omega\mu}{h_{nm}} J_n'(h_{nm}\rho) \cos(n\phi)\exp(j\omega t - \gamma_{nm}z) \tag{4.33}$$

$$E_z = 0 \tag{4.34}$$

$$H_\rho = -H_0 \frac{\gamma_{nm}}{h_{nm}} J_n'(h_{nm}\rho) \cos(n\phi)\exp(j\omega t - \gamma_{nm}z) \tag{4.35}$$

$$H_\phi = H_0 \frac{\gamma_{nm}}{h_{nm}^2 \rho} J_n(h_{nm}\rho) \sin(n\phi)\exp(j\omega t - \gamma_{nm}z) \tag{4.36}$$

$$H_z = H_0 J_n(h_{nm}) \cos(n\phi)\exp(j\omega t - \gamma_{nm}z) \tag{4.37}$$

where n is any positive valued integer, including zero, and $J_n(x)$ and $J'_n(x)$ are the regular Bessel functions of order n and its first derivative, respectively. The allowed values of the modal eigenvalues h_{nm} satisfy

$$J'_n(h_{nm}a) = 0 \tag{4.38}$$

where m signifies the root number of Eq. (4.38). By convention, $1 < m < \infty$, where $m = 1$ indicates the smallest root.

The equations that define the TM_{nm} modes in circular waveguides are

$$E_\rho = -E_0 \frac{\gamma_{nm}}{h_{nm}} J'_n(h_{nm}\rho) \cos(n\phi) \exp(j\omega t - \gamma_{nm}z) \tag{4.39}$$

$$E_\phi = E_0 \frac{\gamma_{nm}}{h_{nm}^2 \rho} J_n(h_{nm}\rho) \sin(n\phi) \exp(j\omega t - \gamma_{nm}z) \tag{4.40}$$

$$E_z = E_0 J_n(h_{nm}\rho) \cos(n\phi) \exp(j\omega t - \gamma_{nm}z) \tag{4.41}$$

$$H_\rho = -E_0 \frac{j\omega \epsilon n}{h_{nm}^2 \rho} J_n(h_{nm}\rho) \sin(n\phi) \exp(j\omega t - \gamma_{nm}z) \tag{4.42}$$

$$H_\phi = -E_0 \frac{j\omega \epsilon}{h_{nm}} J'_n(h_{nm}\rho) \cos(n\phi) \exp(j\omega t - \gamma_{nm}z) \tag{4.43}$$

$$H_z = 0 \tag{4.44}$$

where n is any positive valued integer, including zero. For the TM_{nm} modes, the values of the modal eigenvalues are solutions of

$$J_n(h_{nm}a) = 0 \tag{4.45}$$

where m signifies the root number of Eq. (4.35). As in the case of the TE modes, $1 < m < \infty$.

The dominant mode in a circular waveguide is the TE_{11} mode, which has a cutoff frequency given by

$$f_{c_{11}} = \frac{0.293}{a\sqrt{\mu\epsilon}} \tag{4.46}$$

The configuration of the electric and magnetic fields of this mode is shown in Fig. 4.15.

Table 4.4 shows the cutoff frequencies of the lowest order modes for circular waveguides, referenced to the cutoff frequency of the dominant mode.

The modal field patterns for several lower order modes are shown in Fig. 4.16.

FIGURE 4.15 Field configuration for the TE_{11} (dominant) mode of a circular waveguide: Solid lines, E; dashed lines, H. (Source: Adapted fom Marcuvitz, N. 1986. Waveguide Handbook, 2nd ed., p. 68. Peter Peregrinus, Ltd., London. With permission.)

Commercially Available Waveguides

The dimensions of standard rectangular waveguides are given in Table 4.5.

In addition to rectangular and circular waveguides, several other waveguide types are commonly used in microwave applications. Among these are ridge waveguides and elliptical waveguides. The modes of elliptical waveguides can be expressed in terms of Mathieu functions (Kretzschmar, 1970) and are similar to those of circular waveguides but are less perturbed by minor twists and bends of the waveguide. This property makes them attractive for coupling to antennas.

Single-ridge and double-ridge waveguides are shown in Fig. 4.17. The modes of these waveguides bear similarities to those of rectangular guides, but can only be derived numerically (Montgomery, 1971). Ridge

TABLE 4.4 Cutoff Frequencies of the Lowest Order Circular Waveguide Modes, Referenced to the Cutoff Frequency of the Dominant Mode

$f_c/f_{c_{11}}$	Modes
1.0	TE_{11}
1.307	TM_{01}
1.66	TE_{21}
2.083	TE_{01}, TM_{11}
2.283	TE_{31}
2.791	TM_{21}
2.89	TE_{41}
3.0	TE_{12}

TM_{01} TM_{21} TE_{01}

FIGURE 4.16 Field configurations for the TM_{01}, TE_{21}, and the TE_{01} circular waveguide modes: Solid lines, E; dashed lines, H. (*Source:* Adapted from Marcuvitz, N. 1986. *Waveguide Handbook*, 2nd ed., p. 71. Peter Peregrinus, Ltd., London. With permission.)

TABLE 4.5 Standard Rectangular Waveguides

EIA[a] Designation WR[b]()	Physical Dimensions				Cutoff Frequency for Air-Filled Waveguide, GHz	Recommended Frequency Range for TE_{10} Mode, GHz
	Inside, cm(in.)		Outside, cm(in.)			
	Width	Height	Width	Height		
2300	58.420 (23.000)	29.210 (11.500)	59.055 (23.250)	29.845 (11.750)	0.257	0.32–0.49
2100	53.340 (21.000)	26.670 (10.500)	53.973 (21.250)	27.305 (10.750)	0.281	0.35–0.53
1800	45.720 (18.000)	22.860 (9.000)	46.350 (18.250)	23.495 (9.250)	0.328	0.41–0.62
1500	38.100 (15.000)	19.050 (7.500)	38.735 (15.250)	19.685 (7.750)	0.394	0.49–0.75
1150	29.210 (11.500)	14.605 (5.750)	29.845 (11.750)	15.240 (6.000)	0.514	0.64–0.98
975	24.765 (9.750)	12.383 (4.875)	25.400 (10.000)	13.018 (5.125)	0.606	0.76–1.15
770	19.550 (7.700)	9.779 (3.850)	20.244 (7.970)	10.414 (4.100)	0.767	0.96–1.46
650	16.510 (6.500)	8.255 (3.250)	16.916 (6.660)	8.661 (3.410)	0.909	1.14–1.73
510	12.954 (5.100)	6.477 (2.500)	13.360 (5.260)	6.883 (2.710)	1.158	1.45–2.20
430	10.922 (4.300)	5.461 (2.150)	11.328 (4.460)	5.867 (2.310)	1.373	1.72–2.61

(*Continued*)

TABLE 4.5   (*Continued*)

EIA[a] Designation WR[b]()	Physical Dimensions				Cutoff Frequency for Air-Filled Waveguide, GHz	Recommended Frequency Range for TE$_{10}$ Mode, GHz
	Inside, cm(in.)		Outside, cm(in.)			
	Width	Height	Width	Height		
340	8.636 (3.400)	4.318 (1.700)	9.042 (3.560)	4.724 (1.860)	1.737	2.17–3.30
284	7.214 (2.840)	3.404 (1.340)	7.620 (3.000)	3.810 (1.500)	2.079	2.60–3.95
229	5.817 (2.290)	2.908 (1.145)	6.142 (2.418)	3.233 (1.273)	2.579	3.22–4.90
187	4.755 (1.872)	2.215 (0.872)	5.080 (2.000)	2.540 (1.000)	3.155	3.94–5.99
159	4.039 (1.590)	2.019 (0.795)	4.364 (1.718)	2.344 (0.923)	3.714	4.64–7.05
137	3.485 (1.372)	1.580 (0.622)	3.810 (1.500)	1.905 (0.750)	4.304	5.38–8.17
112	2.850 (1.122)	1.262 (0.497)	3.175 (1.250)	1.588 (0.625)	5.263	6.57–9.99
90	2.286 (0.900)	1.016 (0.400)	2.540 (1.000)	1.270 (0.500)	6.562	8.20–12.50
75	1.905 (0.750)	0.953 (0.375)	2.159 (0.850)	1.207 (0.475)	7.874	9.84–15.00
62	1.580 (0.622)	0.790 (0.311)	1.783 (0.702)	0.993 (0.391)	9.494	11.90–18.00
51	1.295 (0.510)	0.648 (0.255)	1.499 (0.590)	0.851 (0.335)	11.583	14.50–22.00
42	1.067 (0.420)	0.432 (0.170)	1.270 (0.500)	0.635 (0.250)	14.058	17.60–26.70
34	0.864 (0.340)	0.432 (0.170)	1.067 (0.420)	0.635 (0.250)	17.361	21.70–33.00
28	0.711 (0.280)	0.356 (0.140)	0.914 (0.360)	0.559 (0.220)	21.097	26.40–40.00
22	0.569 (0.224)	0.284 (0.112)	0.772 (0.304)	0.488 (0.192)	26.362	32.90–50.10
19	0.478 (0.188)	0.239 (0.094)	0.681 (0.268)	0.442 (0.174)	31.381	39.20–59.60
15	0.376 (0.148)	0.188 (0.074)	0.579 (0.228)	0.391 (0.154)	39.894	49.80–75.80
12	0.310 (0.122)	0.155 (0.061)	0.513 (0.202)	0.358 (0.141)	48.387	60.50–91.90
10	0.254 (0.100)	0.127 (0.050)	0.457 (0.180)	0.330 (0.130)	59.055	73.80–112.00
8	0.203 (0.080)	0.102 (0.040)	0.406 (0.160)	0.305 (0.120)	73.892	92.20–140.00
7	0.165 (0.065)	0.084 (0.033)	0.343 (0.135)	0.262 (0.103)	90.909	114.00–173.00
5	0.130 (0.051)	0.066 (0.026)	0.257 (0.101)	0.193 (0.076)	115.385	145.00–220.00
4	0.109 (0.043)	0.056 (0.022)	0.211 (0.083)	0.157 (0.062)	137.615	172.00–261.00
3	0.086 (0.034)	0.043 (0.017)	0.163 (0.064)	0.119 (0.047)	174.419	217.00–333.00

[a]Electronics Industry Association.
[b]Rectangular waveguide. (*Source:* Liao, S.Y. 1990. *Microwave Devices and Circuits*, 3rd ed., p. 118. Prentice-Hall, Englewood Cliffs, NJ. With permission.)

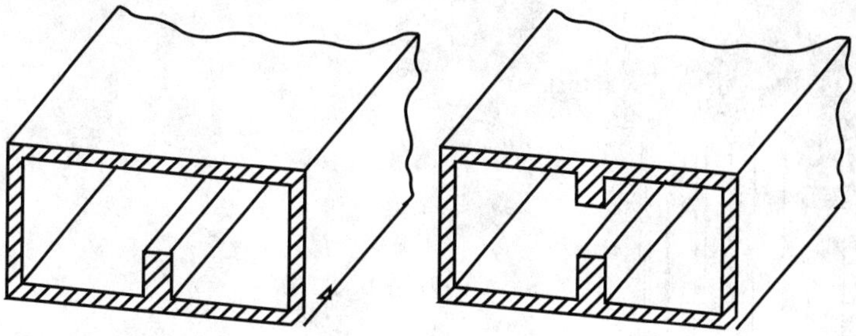

FIGURE 4.17   Single- and double-ridged waveguides.

waveguides are useful because their dominant ranges exceed those of reactangular waveguides. However, this range increase is obtained at the expense of higher losses.

Waveguides are also available in a number of constructions, including rigid, semirigid, and flexible. In applications where it is not necessary for the waveguide to bend, rigid construction is always the best since it exhibits the lowest loss. In general, the more flexible the waveguide construction, the higher the loss.

Waveguide Losses

There are two mechanisms that causes losses in waveguides: dielectric losses and metal losses. In both cases, these losses cause the amplitudes of the propagating modes to decay as $\exp(-\alpha z)$, where α is the attenuation constant, measured in neper/meter. Typically, the attenuation constant is considered as the sum of two components: $\alpha = \alpha_{\text{die}} + \alpha_{\text{met}}$, where α_{die} and α_{met} are the dielectric and metal attenuation constants, respectively.

The attenuation constant α_{die} can be found directly from Eq. (4.14) simply by generalizing the dielectric wave number k to include the effect of the dielectric conductivity σ. For a lossy dielectric, the wave number is given by $k^2 = \omega^2 \mu \epsilon (1 + (\sigma/j\omega\epsilon))$. Thus, from Eqs. (4.13) and (4.14) the attenuation constant α_{die} due to dielectric losses is given by

$$\alpha_{\text{die}} = \text{real} \left[\sqrt{h^2 - \omega^2 \mu \epsilon \left(1 + \frac{\sigma}{j\omega\epsilon} \right)} \right] \tag{4.47}$$

where the allowed values of h are given by Eq. (4.24) for rectangular modes and Eqs. (4.38) and (4.45) for circular modes.

The metal loss constant α_{met} is usually obtained by assuming that the wall conductivity is high enough to have only a negligible effect on the transverse properties of the modal field patterns. Using this assumption, the power loss in the walls per unit distance along the waveguide can then be calculated to obtain α_{met} (Marcuvitz, 1986). Figure 4.18 shows the metal attenuation constants for several circular waveguide modes, each normalized to the resistivity R_s of the walls, where $R_s = \sqrt{(\pi f \mu/\sigma)}$ and where μ and σ are the permeability and conductivity of the metal walls, respectively. As can be seen from this figure, the TE_{0m} modes exhibit particularly low loss at frequencies significantly above their cutoff frequencies, making them useful for transporting microwave energy over large distances.

Mode Launching

When coupling electromagnetic energy into a waveguide, it is important to ensure that the desired modes are excited and that reflections back to the source are minimized. Similar concerns must be considered when coupling energy from a waveguide to a transmission line or circuit element. This is achieved by using (or coupling) structures that allow strong coupling between the desired modes on both structures.

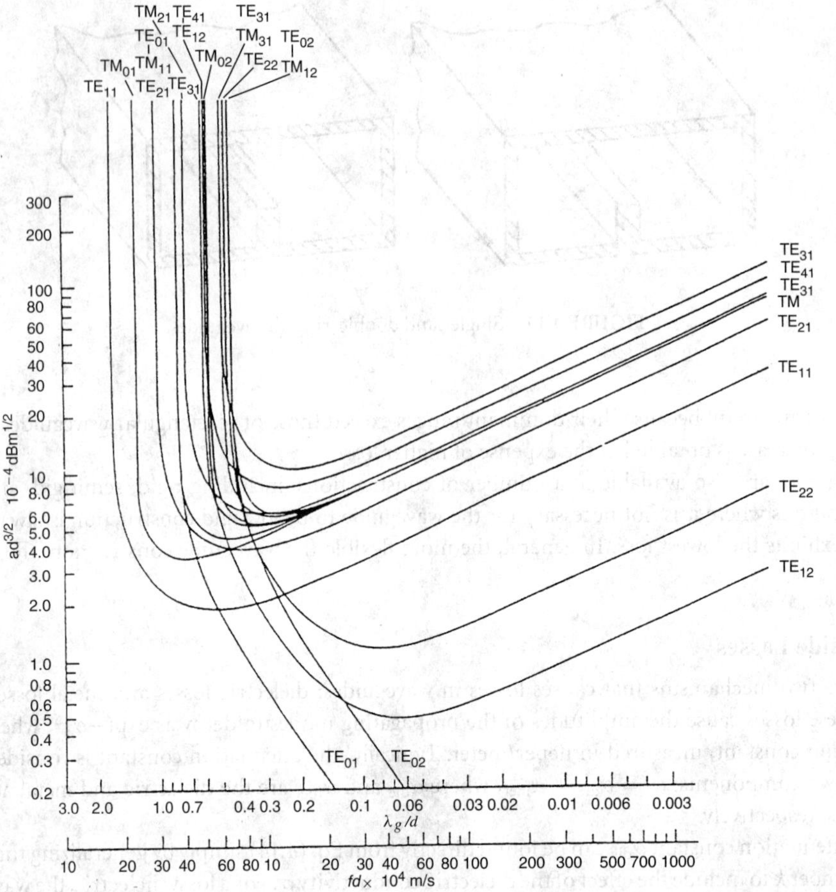

FIGURE 4.18 Values of metallic attenuation constant α for the first few waveguide modes in a circular waveguide of diameter d, plotted aganist normalized wavelength. (*Source:* Baden Fuller, A.J. 1979. *Microwaves*, 2nd ed., p. 138. Pergamon, New York. With permission.)

Figure 4.19 shows a mode launching structure for coaxial cable to rectangular waveguide transitions. This structure provides good coupling between the TEM (transmission line) mode on a coaxial cable and the TE_{10} mode in the waveguide because the antenna probe excites a strong transverse electric field in the center of the waveguide, directed between the broad walls. The distance between the probe and the short circuit back wall is chosen to be approximately $\lambda/4$, which allows the TE_{10} mode launched in this direction to reflect off the short circuit and arrive in phase with the mode launched toward the right.

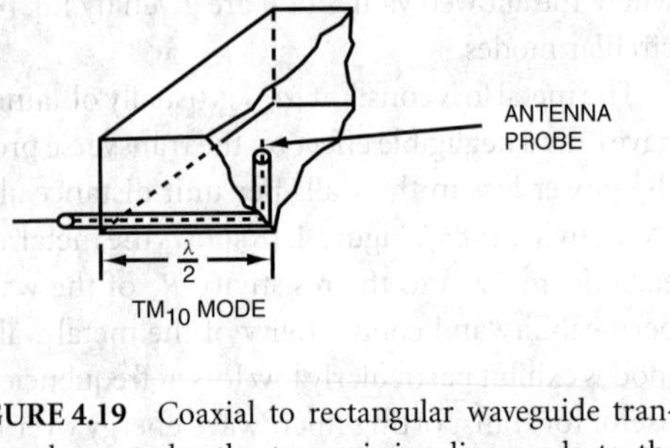

Launching structures can also be devised to launch higher order modes. Mode launchers that couple the transmission line mode on a coaxial cable to the TM_{11} and TM_{21} waveguide mode are shown in Fig. 4.20.

FIGURE 4.19 Coaxial to rectangular waveguide transition that couples the transmission line mode to the dominant waveguide modes. (*Source:* Liao, S.Y. 1990. *Microwave Devices and Circuits*, 3rd ed., p. 117. Prentice-Hall, Englewood Cliffs, NJ. With permission.)

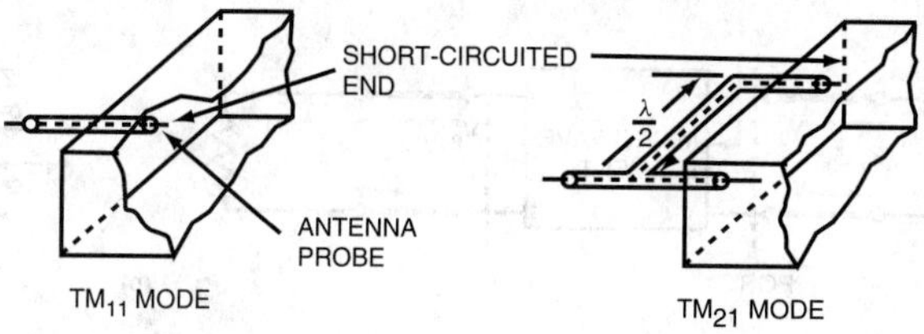

FIGURE 4.20   Coaxial to rectangular waveguide transitions that couples the transmission line mode to the TM_{11} and TM_{21} waveguide modes. (*Source:* Liao, S.Y. 1990. *Microwave Devices and Circuits*, 3rd ed., p. 117. Prentice-Hall, Englewood Cliffs, NJ. With permission.)

Defining Terms

Cutoff frequency: The minimum frequency at which a waveguide mode will propagate energy with little or no attenuation.

Guide wavelength: The distance over which the fields of propagating modes repeat themselves in a waveguide.

Waveguide: A closed metal cylinder, filled with a dielectric, used to transport electromagnetic energy over short or long distances.

Waveguide modes: Unique electromagnetic field configurations supported by a waveguide that have distinct electrical characteristics.

Wave impedance: The ratio of the transverse electric and magnetic fields inside a waveguide.

References

Baden Fuller, A.J. 1979. *Microwaves*, 2nd ed. Pergamon, New York.

Collin, R.E. 1992. *Foundations for Microwave Engineering*, 2nd ed. McGraw-Hill, New York.

Kretzschmar, J. 1970. Wave propagation in hollow conducting elliptical waveguides. *IEEE Trans. Micro. Theory Tech.* MTT-18(9):547–554.

Liao, S.Y. 1990. *Microwave Devices and Circuits*, 3rd ed. Prentice-Hall, Englewood Cliff, NJ.

Marcuvitz, N. 1986.*Waveguide Handbook*, 2nd ed. Peter Peregrinus, Ltd., London.

Montgomery, J. 1971. On the complete eigenvalue solution of ridged waveguide. *IEEE Trans. Micro. Theory Tech.* MTT-19(6):457–555.

Further Information

There are many textbooks and handbooks that cover the subject of waveguides in great detail. In addition to the references cited, others include:

Collin, R. E. 1991. *Field Theory of Guided Waves*, 2nd ed. IEEE Press, Piscataway, NJ.

Gardiol, F. 1984. *Introduction to Microwaves*. Artech House, Dedham, MA.

Lewin, L. 1975. *Theory of Waveguides*. Wiley, New York.

Ramo, S., Whinnery, J., and Van Duzer, T. 1965. *Fields and Waves in Communication Electronics*. Wiley, New York.

Sams, H.W. 1975. *Reference Data for Radio Engineers*.

4.3   Passive Microwave Devices

Michael B. Steer

4.3.1   Introduction

Wavelengths in air at microwave and millimeter-wave frequencies range from 1 m at 300 MHz to 1 mm at 300 GHz and are comparable to the physical dimensions of fabricated electrical components. For this reason circuit components commonly used at lower frequencies, such as resistors, capacitors, and

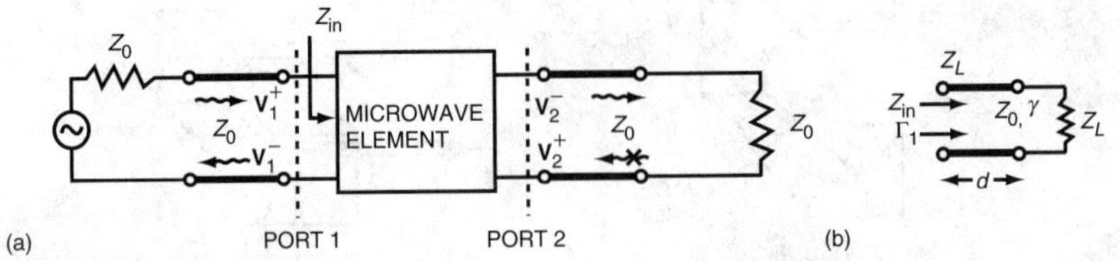

FIGURE 4.21 Incident, reflected, and transmitted traveling voltage waves at: (a) A passive microwave element, (b) a transmission line.

inductors, are not readily available. The relationship between the wavelength and physical dimensions enables new classes of distributed components to be constructed that have no analogy at lower frequencies. Components are realized by disturbing the field structure on a transmission line, resulting in energy storage and thus reactive effects. Electric (E) field disturbances have a capacitive effect and the magnetic (H) field disturbances appear inductive. Microwave components are fabricated in waveguide, coaxial lines, and strip lines. The majority of circuits are constructed using strip lines as the cost is relatively low and they are highly reproducible due to the photolithographic techniques used. Fabrication of waveguide components requires precision machining but they can tolerate higher power levels and are more easily realized at millimeter-wave frequencies (30–300 GHz) than either coaxial or microstrip components.

4.3.2 Characterization of Passive Elements

Passive microwave elements are defined in terms of their reflection and transmission properties for an incident wave of electric field or voltage. In Fig. 4.21(a) a traveling voltage wave with phasor $\mathbf{V}_1^+$ is incident at port 1 of a two-port passive element. A voltage $\mathbf{V}_1^-$ is reflected and $\mathbf{V}_2^-$ is transmitted. In the absence of an incident voltage wave at port 2 (the voltage wave $\mathbf{V}_2^-$ is totally absorbed by Z_0), at port 1 the element has a voltage reflection coefficient

$$\Gamma_1 = \frac{\mathbf{V}_1^-}{\mathbf{V}_1^+} \tag{4.48}$$

and transmission coefficient

$$T = \frac{\mathbf{V}_2^-}{\mathbf{V}_1^+} \tag{4.49}$$

More convenient measures of reflection and transmission performance are the **return loss** and **insertion loss** as they are relative measures of power in transmitted and reflected signals. In decibels

$$\text{return loss} = -20 \log \Gamma_1 \quad \text{insertion loss} = -20 \log T \tag{4.50}$$

The input impedance at port 1, Z_{in}, is related to Γ_1 by

$$Z_{in} = Z_0 \frac{1 + \Gamma_1}{1 - \Gamma_1} \tag{4.51}$$

The reflection characteristics are also described by the **voltage standing wave ratio (VSWR)**, a quantity which is more easily measured. The VSWR is the ratio of the maximum voltage amplitude on a transmission line ($|\mathbf{V}_1^+| + |\mathbf{V}_1^-|$) to the minimum voltage amplitude ($|\mathbf{V}_1^+| - |\mathbf{V}_1^-|$). Thus,

$$\text{VSWR} = \frac{1 + |\Gamma_1|}{1 - |\Gamma_1|} \tag{4.52}$$

These quantities will change if the loading conditions are changed. For this reason scattering parameters are used which are defined as the reflection and transmission coefficients with a load referred to as a

reference impedance. Simple formulas relate the S parameters to other network parameters (Vendelin et al., 1990, pp. 16–17). Thus,

$$S_{11} = \Gamma_1 \qquad S_{21} = T \tag{4.53}$$

S_{11} and S_{12} are similarly defined when a voltage wave is incident at port 2. For a multiport

$$S_{pq} = \frac{\mathbf{V}_p^-}{\mathbf{V}_q^+} \tag{4.54}$$

with all of the ports terminated in the reference impedance. S parameters are the most convenient network parameters to use with distributed circuits as a change in line length results in a phase change. Also, they are the only network parameters that can be measured directly at microwave and millimeter-wave frequencies. Most passive devices, with the notable exception of ferrite devices, are reciprocal and so $S_{pq} = S_{qp}$. A loss-less passive device also satisfies the unitary condition: $\Sigma_q |S_{pq}|^2 = 1$, which is a statement of power conservation indicating that all power is either reflected or transmitted. A passive element is fully defined by its S parameters together with its reference impedance, here Z_0. In general, the reference impedance at each port can be different.

Circuits are designed to minimize the reflected energy and maximize transmission at least over the frequency range of operation. Thus the return loss is high, and the VSWR $\simeq 1$ for well-designed circuits. Individual elements may have high reflection, and the interaction of elements is used in design.

A terminated transmission line such as that in Fig. 4.21(b) has an input impedance

$$Z_{\text{in}} = Z_0 \frac{Z_L + j Z_0 \tanh \gamma d}{Z_0 + j Z_L \tanh \gamma d} \tag{4.55}$$

Thus a short section ($\gamma d \ll 1$) of short-circuited ($Z_L = 0$) transmission line looks like an inductor, and a capacitor if it is open circuited ($Z_L = \infty$). When the line is a half-wavelength long, an open circuit is presented at the input to the line if the other end is short circuited.

4.3.3 Transmission Line Sections

The simplest microwave circuit element is a uniform section of transmission line which can be used to introduce a time delay or a frequency-dependent phase shift. Other line segments for interconnections include bends, corners, twists, and transitions between lines of different dimensions. (See Fig. 4.22.) The dimensions and shapes are designed to minimize reflections and so maximize return loss and minimize insertion loss.

4.3.4 Discontinuities

The waveguide discontinuities shown in Fig. 4.23(a) to Fig. 4.23(f) illustrate most clearly the use of E and H field disturbances to realize capacitive and inductive components. An E-plane discontinuity (Fig. 4.23(a)) can be modeled approximately by a frequency-dependent capacitor. H-plane discontinuities (Fig. 4.23(b) and Fig. 4.23(c)) resemble inductors as does the circular iris of Fig. 4.23(d). The resonant waveguide iris of Fig. 4.23(e) disturbs both the E and H fields and can be modeled by a parallel LC resonant circuit near the frequency of resonance. Posts in waveguide are used both as reactive elements (Fig. 4.23(f)) and to mount active devices (Fig. 4.23(g)). The equivalent circuits of microstrip discontinuities (Fig. 4.23(k) to Fig. 4.23(o)) are again modeled by capacitive elements if the E field is interrupted and by inductive elements if the H field (or current) is interrupted. The stub shown in Fig. 4.23(j) presents a short circuit to the through transmission line when the length of the stub is $\lambda_g/4$. When the stubs are electrically short ($\ll \lambda_g/4$) they introduce shunt capacitances in the through transmission line.

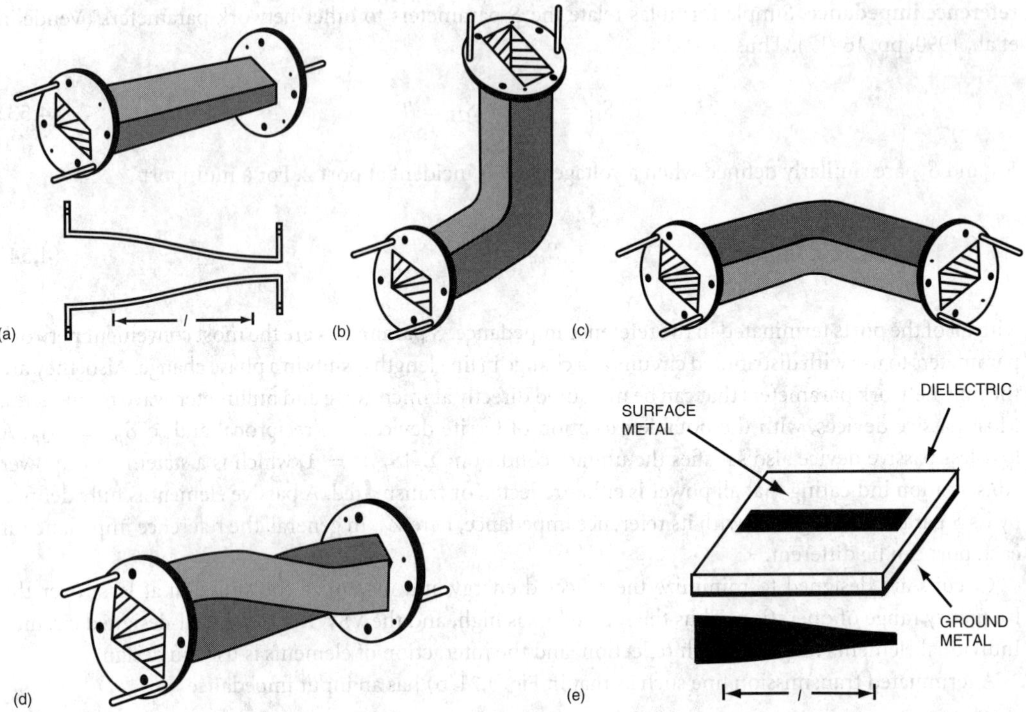

FIGURE 4.22   Sections of transmission lines used for interconnecting components: (a) waveguide tapered section, (b) waveguide E-plane bend, (c) waveguide H-plane bend, (d) waveguide twist, (e) microstrip taper.

4.3.5   Impedance Transformers

Impedance transformers are used to interface two sections of line with different **characteristic impedances**. The smoothest transition and the one with the broadest bandwidth is a tapered line as shown in Fig. 4.22(a) and Fig. 4.22(e). This element tends to be very long and so step terminations called quarter-wave impedance transformers (see Fig. 4.23(h) and Fig. 4.23(i)) are sometimes used although their bandwidth is relatively small centered on the frequency at which $l = \lambda_g/4$. Ideally, $Z_{0,2} = \sqrt{Z_{0,1} Z_{0,3}}$.

4.3.6   Terminations

In a termination, power is absorbed by a length of lossy material at the end of a shorted piece of transmission line (Fig. 4.24(a) and Fig. 4.23(c)). This type of termination is called a matched load as power is absorbed and reflections are very small irrespective of the characteristic impedance of the transmission line. This is generally preferred as the characteristic impedance of transmission lines varies with frequency, particularly so for waveguides. When the characteristic impedance of a line does not vary much with frequency, as is the case with a coaxial line, a simpler smaller termination can be realized by placing a resistor to ground (Fig. 4.24(b)).

4.3.7   Attenuators

Attenuators reduce the level of a signal traveling along a transmission line. The basic construction is to make the line lossy but with a characteristic impedance approximating that of the connecting lines so as to reduce reflections. The line is made lossy by introducing a resistive vane in the case of a waveguide (Fig. 4.24(d)), replacing part of the outer conductor of a coaxial line by resistive material (Fig. 4.24(e)), or covering the line by resistive material in the case of a microstrip line (Fig. 4.24(f)). If the amount of lossy material introduced into the transmission line is controlled, a variable attenuator is achieved, for example, Fig. 4.24(d).

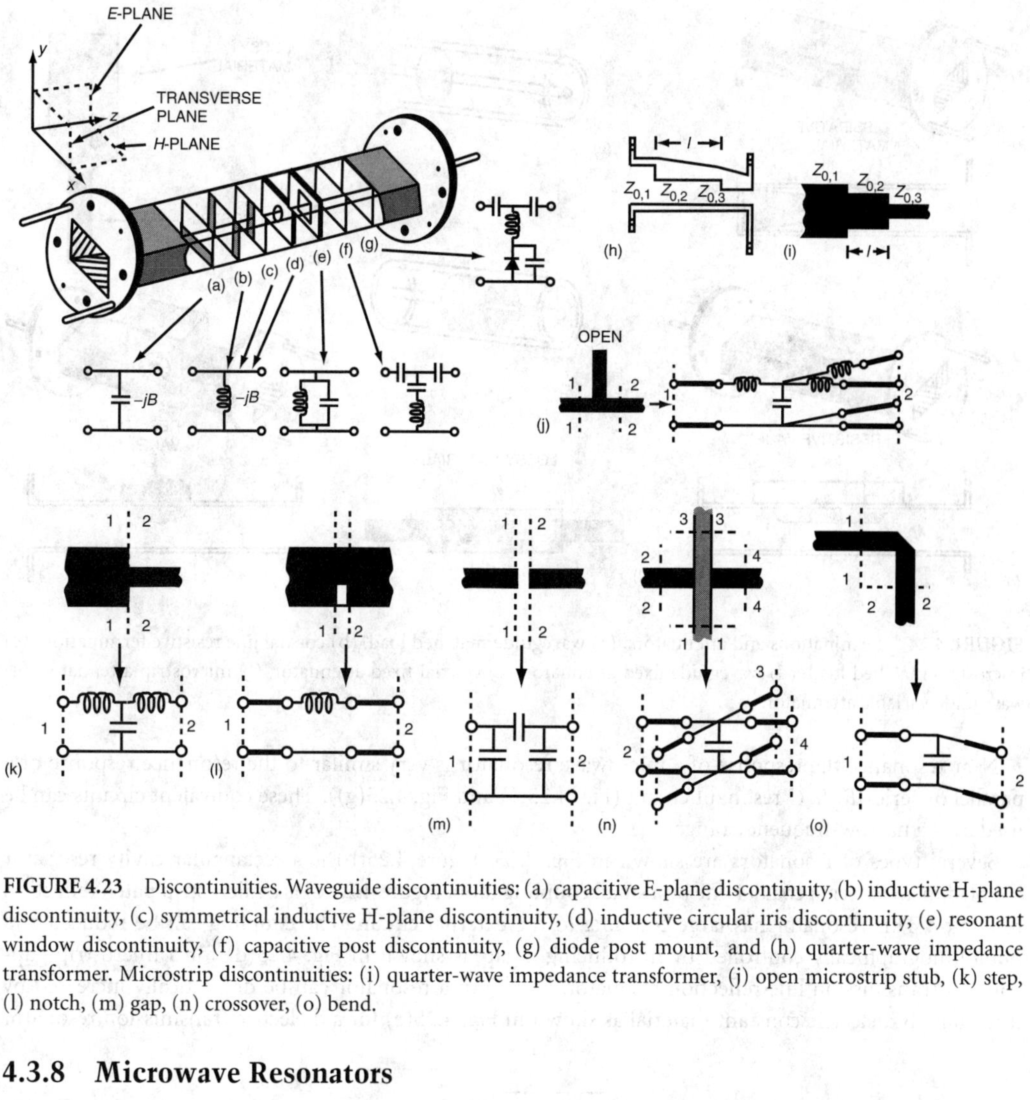

FIGURE 4.23 Discontinuities. Waveguide discontinuities: (a) capacitive E-plane discontinuity, (b) inductive H-plane discontinuity, (c) symmetrical inductive H-plane discontinuity, (d) inductive circular iris discontinuity, (e) resonant window discontinuity, (f) capacitive post discontinuity, (g) diode post mount, and (h) quarter-wave impedance transformer. Microstrip discontinuities: (i) quarter-wave impedance transformer, (j) open microstrip stub, (k) step, (l) notch, (m) gap, (n) crossover, (o) bend.

4.3.8 Microwave Resonators

In a lumped element resonant circuit, stored energy is transferred between an inductor which stores magnetic energy and a capacitor which stores electric energy, and back again every period. Microwave resonators function the same way, exchanging energy stored in electric and magnetic forms but with the energy stored spatially. Resonators are described in terms of their quality factor

$$Q = 2\pi f_0 \left(\frac{\text{maximum energy stored in the resonator at } f_0}{\text{power lost in the cavity}} \right) \tag{4.56}$$

where f_0 is the resonant frequency. The Q is reduced and thus the resonator bandwidth is increased by the power lost due to coupling to the external circuit so that the loaded Q

$$Q_L = 2\pi f_0 \left(\frac{\text{maximum energy stored in the resonator at } f_0}{\text{power lost in the cavity and to the external circuit}} \right) \tag{4.57}$$

$$= \frac{1}{1/Q + 1/Q_{\text{ext}}}$$

where Q_{ext} is called the external Q. Q_L accounts for the power extracted from the resonant circuit and is typically large. For the simple response shown in Fig. 4.25(a) the half-power (3-dB) bandwidth is f_0/Q_L.

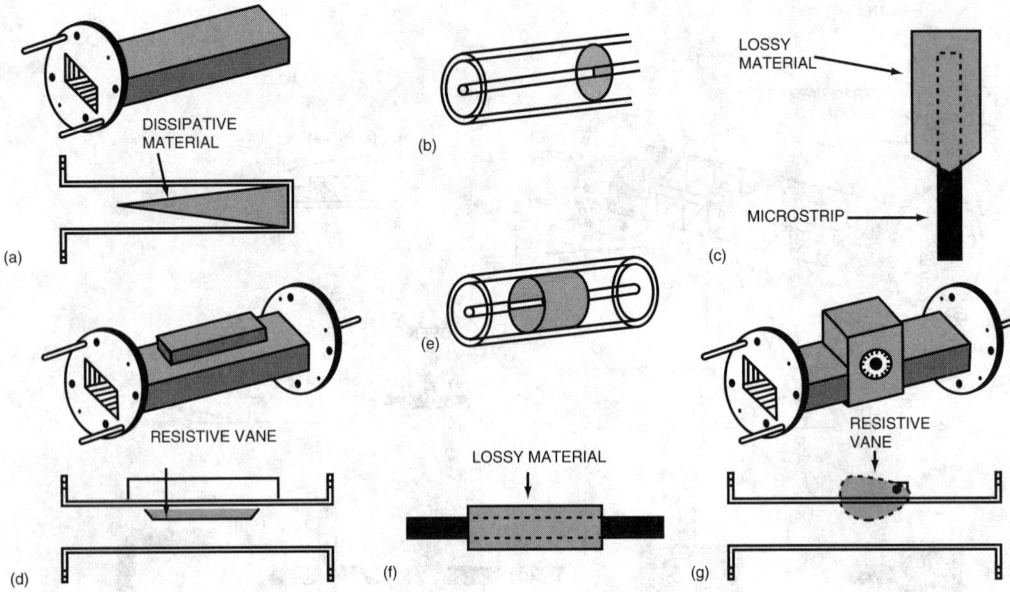

FIGURE 4.24 Terminations and attenuators: (a) waveguide matched load, (b) coaxial line resistive termination, (c) microstrip matched load, (d) waveguide fixed attenuator, (e) coaxial fixed attenuator, (f) microstrip attenuator, (g) waveguide variable attenuator.

Near resonance the response of a microwave resonator is very similar to the resonance response of a parallel or series R, L, C resonant circuit (Fig. 4.25(f) and Fig. 4.25(g)). These equivalent circuits can be used over a narrow-frequency range.

Several types of resonators are shown in Fig. 4.25. Figure 4.25(b) is a rectangular cavity resonator coupled to an external coaxial line by a small coupling loop. Figure 4.25(c) is a microstrip patch reflection resonator. This resonator has large coupling to the external circuit. The coupling can be reduced and photolithographically controlled by introducing a gap as shown in Fig. 4.25(d) for a microstrip gap-coupled transmission line reflection resonator. The Q of a resonator can be dramatically increased by using a high dielectric constant material as shown in Fig. 4.25(e) for a dielectric transmission resonator

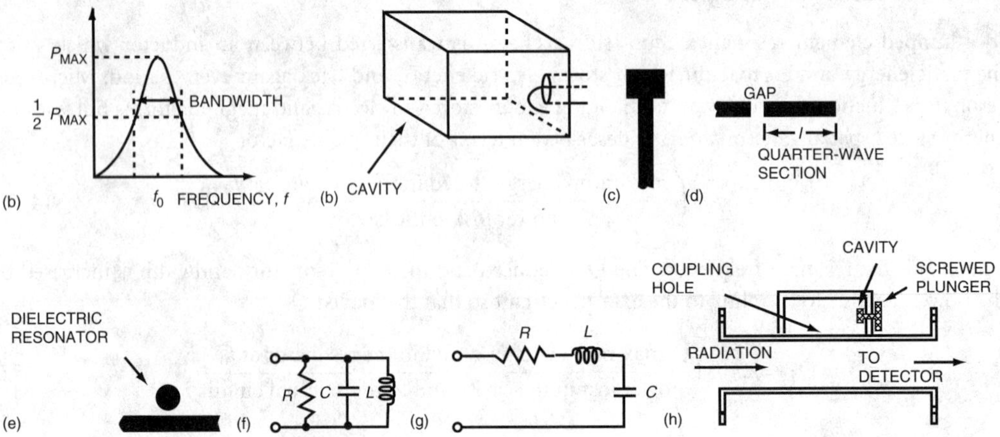

FIGURE 4.25 Microwave resonators: (a) resonator response, (b) rectangular cavity resonator, (c) microstrip patch resonator, (d) microstrip gap-coupled reflection resonator, (e) transmission dielectric transmission resonator in microstrip, (f) parallel equivalent circuits, (g) series equivalent circuits, (h) waveguide wavemeter.

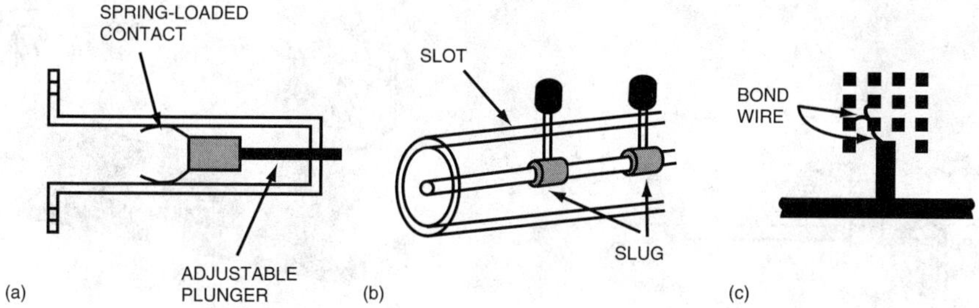

FIGURE 4.26 Tuning elements: (a) waveguide sliding short circuit, (b) coaxial line slug tuner, (c) microstrip stub with tuning pads.

in microstrip. One simple application of a cavity resonator is the waveguide wavemeter (Fig. 4.25(h)). Here the resonant frequency of a rectangular cavity is varied by changing the physical dimensions of the cavity with a null of the detector indicating that the frequency corresponds to the resonant cavity frequency.

4.3.9 Tuning Elements

In rectangular waveguide the basic adjustable tuning element is the sliding short shown in Fig. 4.26(a). Varying the position of the short will change resonance frequencies of cavities. It can be combined with hybrid tees to achieve a variety of tuning functions. The post in Fig. 4.23(f) can be replaced by a screw to obtain a screw tuner, which is commonly used in waveguide filters. Sliding short circuits can be used in coaxial lines and in conjunction with branching elements to obtain stub tuners. Coaxial slug tuners are also used to provide adjustable matching at the input and output of active circuits. The slug is movable and changes the characteristic impedance of the transmission line. It is more difficult to achieve variable tuning in passive microstrip circuits. One solution is to provide a number of pads as shown in Fig. 4.26(c), which, in this case, can be bonded to the stub to obtain an adjustable stub length. Variable amounts of phase shift can be inserted by using a variable length of line called a line stretcher, or by a line with a variable propagation constant. One type of waveguide variable phase shifter is similar to the variable attenuator of Fig. 4.24(d) with the resistive material replaced by a low-loss dielectric.

4.3.10 Hybrid Circuits and Directional Couplers

Hybrid circuits are multiport components that preferentially route a signal incident at one port to the other ports. This property is called directivity. One type of hybrid is called a directional coupler, the schematic of which is shown in Fig. 4.27(a). Here the signal incident at port 1 is coupled to ports 2 and 3, whereas very little is coupled to port 4. Similarly, a signal incident at port 2 is coupled to ports 1 and 4 but very little power appears at port 3. The feature that distinguishes a directional coupler from other types of hybrids is that the power at the output ports (here ports 2 and 3) is different. The performance of a directional coupler is specified by three parameters:

$$\text{Coupling factor} = P_1/P_3$$
$$\text{Directivity} = P_3/P_4 \qquad (4.58)$$
$$\text{Isolation} = P_1/P_4$$

Microstrip and waveguide realizations of directional couplers are shown in Fig. 4.27(b) and Fig. 4.27(c) where the microstrip coupler couples in the backward direction and the waveguide coupler couples in the forward direction. The powers at the output ports of the hybrids shown in Fig. 4.28 and Fig. 4.29 are equal and so the hybrids serve to split a signal into two equal half power signals, as well as having directional sensitivity.

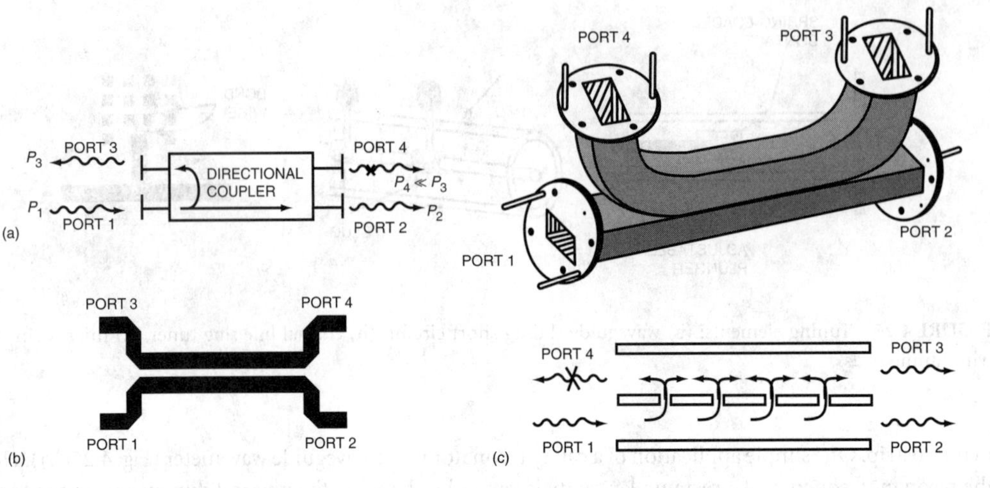

FIGURE 4.27 Directional couplers: (a) schematic, (b) backward-coupling microstrip directional coupler, (c) forward-coupling waveguide directional coupler.

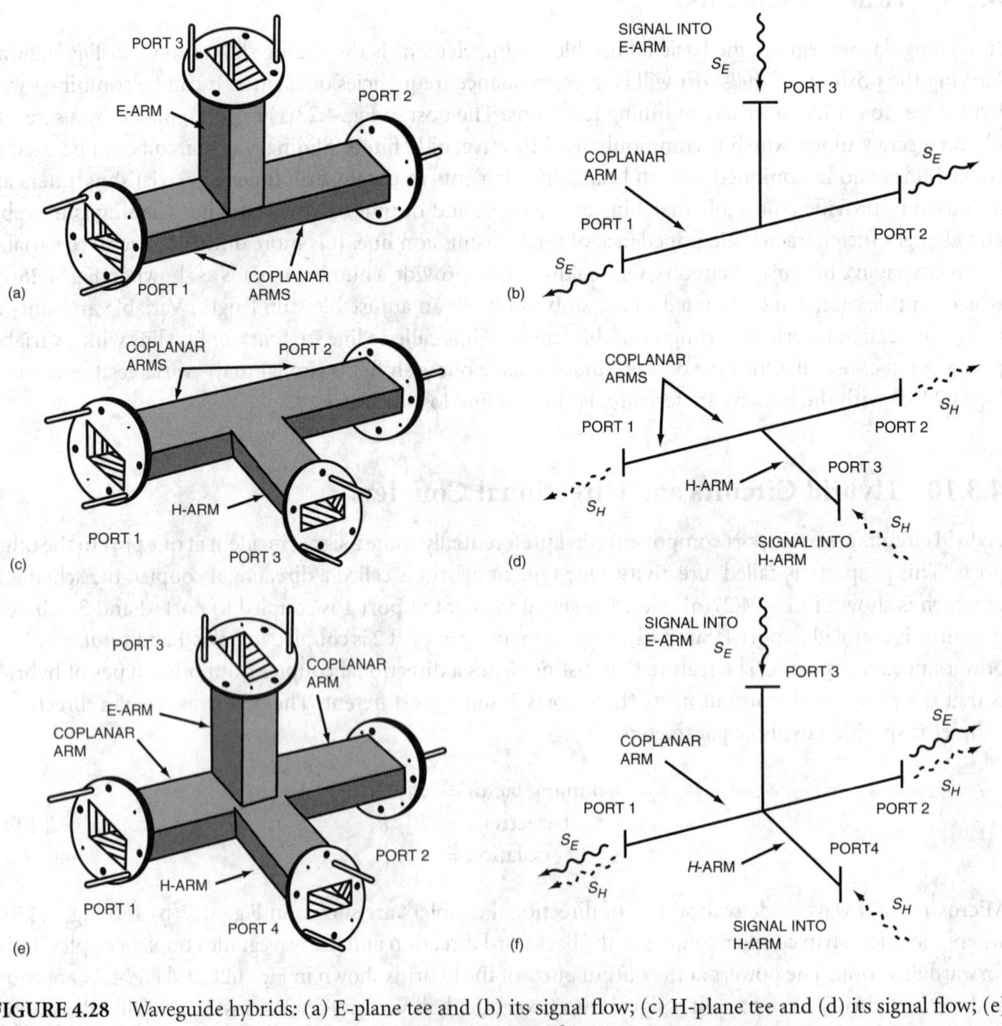

FIGURE 4.28 Waveguide hybrids: (a) E-plane tee and (b) its signal flow; (c) H-plane tee and (d) its signal flow; (e) magic tee and (f) its signal flow. The negative sign indicates $180^{\to 0}$ phase reversal.

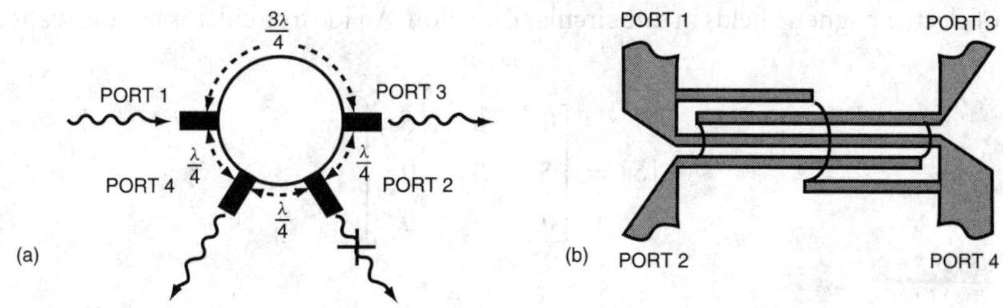

FIGURE 4.29 Microstrip hybrids: (a) Rat race hybrid, (b) Lange coupler.

4.3.11 Filters

Filters are combinations of microwave passive elements designed to have a specified frequency response. Typically, a topology of a filter is chosen based on established lumped element filter design theory. Then computer-aided design techniques are used to optimize the response of the circuit to the desired response.

4.3.12 Ferrite Components

Ferrite components are nonreciprocal in that the insertion loss for a wave traveling from port A to port B is not the same as that from port B to port A.

Circulators and Isolators

The most important type of ferrite component is a circulator (Fig. 4.30(a) and Fig. 4.30(b)). The essential element of a circulator is a piece of ferrite, which when magnetized becomes nonreciprocal, preferring

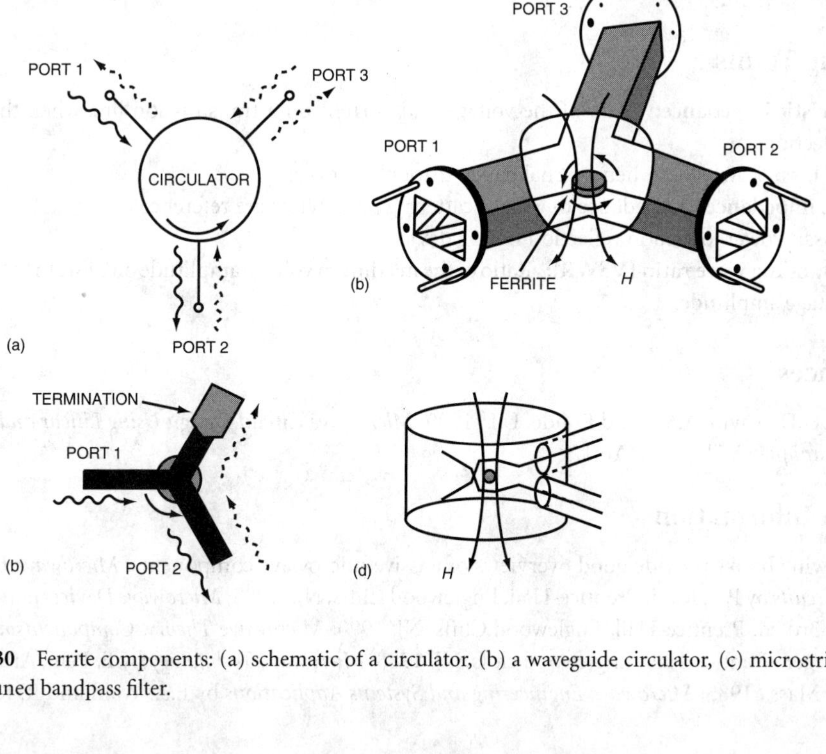

FIGURE 4.30 Ferrite components: (a) schematic of a circulator, (b) a waveguide circulator, (c) microstrip isolator, (d) a YIG tuned bandpass filter.

progression of electromagnetic fields in one circular direction. An ideal circulator has the scattering matrix

$$[S] = \begin{bmatrix} 0 & 0 & S_{13} \\ S_{21} & 0 & 0 \\ 0 & S_{32} & 0 \end{bmatrix} \tag{4.59}$$

In addition to the insertion and return losses, the performance of a circulator is described by its isolation, which is its insertion loss in the undesired direction. An isolator is just a three-port circulator with one of the ports terminated in a matched load as shown in the microstrip realization of Fig. 4.30(c). It is used in a transmission line to pass power in one direction but not in the reverse direction. It is commonly used to protect the output of equipment from high reflected signals. The heart of isolators and circulators is the nonreciprocal element. Electronic versions have been developed for microwave monolithic integrated circuits (MMICs). A four-port version is called a duplexer and is used in radar systems and to separate the received and transmitted signals in a transceiver.

YIG Tuned Resonator

A magnetized YIG (yttrium iron garnet) sphere, shown in Fig. 4.30(d), provides coupling between two lines over a very narrow bandwidth. The center frequency of this bandpass filter can be adjusted by varying the magnetizing field.

4.3.13 Passive Semiconductor Devices

A semiconductor diode can be modeled by a voltage-dependent resistor and capacitor in shunt. Thus an applied DC voltage can be used to change the value of a passive circuit element. Diode optimized to produce a voltage variable capacitor are called varactors. In detector circuits a diode voltage variable resistance is used to achieve rectification and, through design, produce a DC voltage proportional to the power of an incident microwave signal. A controllable variable resistance is used in a *p-i-n* diode to realize an electronic switch.

Defining Terms

Characteristic impedance: Ratio of the voltage and current on a transmission line when there are no reflections.

Insertion loss: Power lost when a signal passes through a device.

Reference impedance: Impedance to which scattering parameters are referenced.

Return loss: Power lost upon reflection from a device.

Voltage standing wave ratio (VSWR): Ratio of the maximum voltage amplitude on a line to the minimum voltage amplitude.

References

Vendelin, G.D., Pavio, A.M., and Rohde, U.L. 1990. *Microwave Circuit Design Using Linear and Nonlinear Techniques.* Wiley, New York.

Further Information

The following books provide good overviews of passive microwave components: *Microwave Engineering Passive Circuits* by P.A. Rizzi, Prentice-Hall, Englewood Cliffs, NJ., 1988; *Microwave Devices and Circuits* by S.Y. Liao, 3rd ed. Prentice-Hall, Englewood Cliffs, NJ. 1990; *Microwave Theory, Components and Devices* by J.A. Seeger, Prentice-Hall, Englewood cliffs, NJ. 1986; *Microwave Technology* by E. Pehl, Artech House, Dedham, Mass., 1985; *Microwave Engineering and Systems Applications* by E.A. Wolff and R. Kaul, Wiley,

New York, 1988; and *Microwave Engineering* by T.K. Ishii, 2nd ed. Harcourt Brace Jovanovich, Orlando, FL. 1989. *Microwave Circuit Design Using Linear and Nonlinear Techniques* by G.D. Vendelin, A.M. Pavio, and U.L. Rohde, Wiley, New York, 1990, provides a comprehensive treatment of computer-aided design techniques for both passive and active microwave circuits.

The monthly journals *IEEE Transactions on Microwave Theory Techniques, IEEE Microwave and Guided Wave Letters*, and *IEEE Transactions on Antennas and Propagation* publish articles on modeling and design of microwave passive circuit components. Articles in the first two journals are more circuit and component oriented, whereas the third focuses on field theoretic analysis. These are published by the Institute of Electrical and Electronics Engineers, Inc. For subscription or ordering contact: IEEE Service Center, 445 Hoes Lane, P.O. Box 1331, Piscataway, NJ 08855-1331.

Articles can also be found in the biweekly magazine *Electronics Letters* and the bimonthly magazine *IEE Proceedings Part H—Microwave, Optics and Antennas*. Both are published by the Institute of Electrical Engineers and subscription inquiries should be sent to IEE Publication Sales, P.O. Box 96, Stenage, Herts. SG1 2SD, United Kingdom. Telephone number (0438) 313311.

The *International Journal of Microwave and Millimeter-Wave Computer-Aided Engineering* is a quarterly journal devoted to the computer-aided design aspects of microwave circuits and has articles on component modeling and computer-aided design techniques. It has a large number of review-type articles. For subscription information contact John Wiley & Sons, Inc., Periodicals Division, P.O. Box 7247-8491, Philadelphia, Pennsylvania 19170-8491.

4.4 Passive Filters

Cecil Harrison

4.4.1 Introduction

A **filter** is a multiport network designed specifically to respond differently to signals of different frequency. This definition excludes networks which behave as filters incidentally, sometimes to the detriment of their main purpose. **Passive filters** are constructed exclusively with passive elements (i.e., resistors, inductors, and capacitors).

Filters are generally categorized by

- Type
- Alignment (or class)
- Order

4.4.2 Filter Type

Filters are categorized by **type**, according to the magnitude of the frequency-response, as

- Lowpass (LP)
- Highpass (HP)
- Bandpass (BP)
- Bandstop (BS)

The terms *band reject* or *notch* are also used as descriptive of the BS filter. The term *all pass* is sometimes applied to a filter whose purpose is to alter the *phase angle* without affecting the magnitude of the frequency response. Ideal and practical interpretations of filter types and the associated terminology are illustrated in Fig. 4.31.

In general, the voltage gain of a filter in the **stopband** (or *attenuation band*) is less than $\sqrt{2}/2 (\approx 0.707)$ times the maximum voltage gain in the **pass band**. In logarithmic terms, the gain in the stopband is at least 3.01 dB less than the maximum gain in the pass band. The **cutoff** (or *break* or *corner*) **frequency** separates the pass band from the stop band. In BP and BS filters, there are two cutoff frequencies, sometimes referred

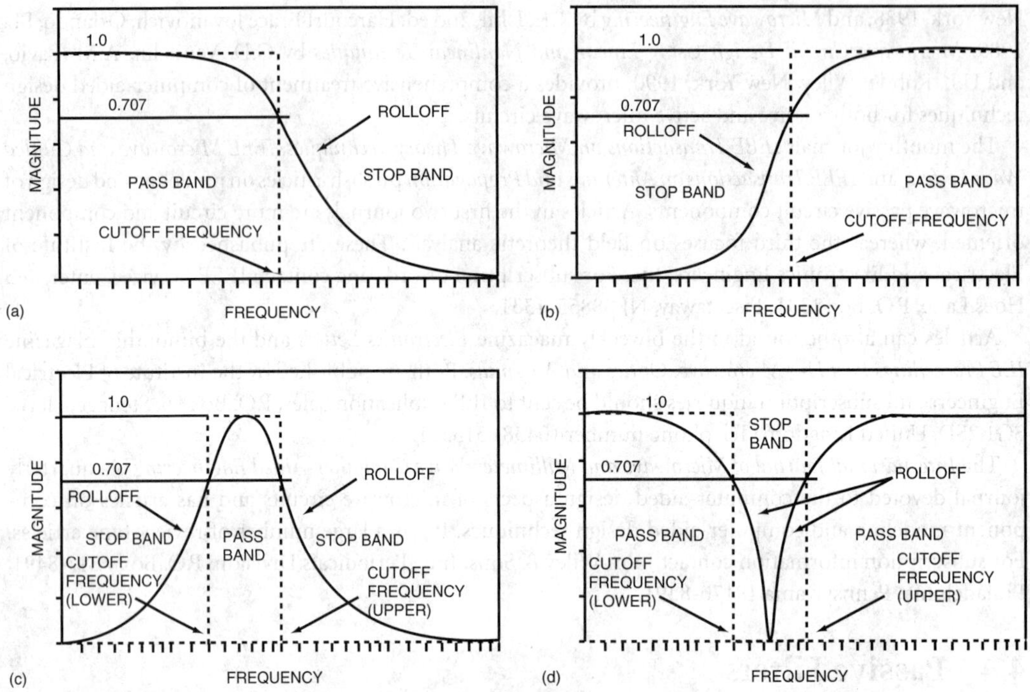

FIGURE 4.31 Filter characteristics by type: (a) low-pass, (b) high-pass, (c) bandpass, (d) bandstop.

to as lower and upper cutoff frequencies. Another term for cutoff frequency is *half-power frequency,* since the power delivered to a resistive load at cutoff frequency is half the maximum power delivered to the same load in the pass band. For BP and BS filters, the **center frequency** is the frequency of maximum or minimum response magnitude, respectively, and **bandwidth** is the difference between upper and lower cutoff frequencies. **Rolloff** is the transition from pass band to stop band and is specified in gain unit per frequency unit (e.g., dB/decade, dB/octave, etc.)

4.4.3 Filter Alignment

The *alignment* (or *class*) of a filter refers to the shape of the frequency response. Fundamentally, **filter alignment** is determined by the coefficients of the filter network transfer function, and so there are an infinite number of filter alignments, some of which may not be realizable. The more common alignments are:

- **Butterworth**
- **Chebyshev**
- Bessel
- Inverse Chebyshev
- Elliptic (or Cauer)

Each filter alignment has a frequency response with a characteristic shape (Fig. 4.32), which provides some particular advantage. For example, filters with Butterworth, Chebyshev, or Bessel alignment are called all-pole filters because their lowpass transfer functions have no zeros. Table 4.6 summarizes the characteristics of the standard filter alignments.

4.4.4 Filter Order

The **order** of a filter is equal to the number of poles in the filter network transfer function. For a lossless L–C (inductance- capacitance) filter with resistive (nonreactive) termination, the number of reactive elements

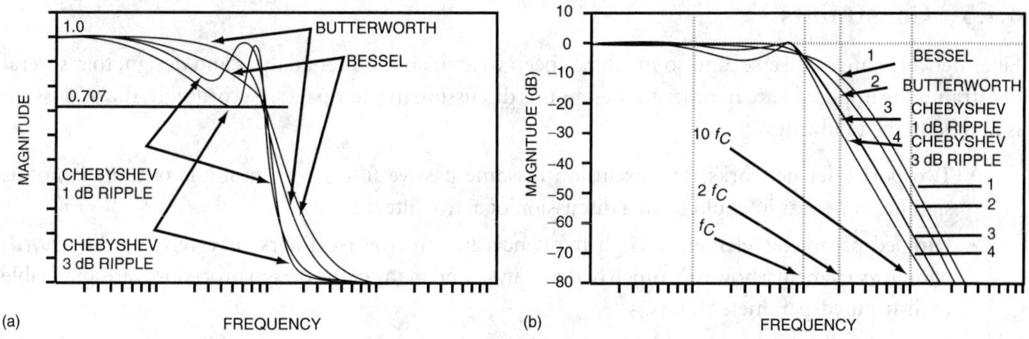

FIGURE 4.32 Filter characteristics by alignment, third-order, all-pole filters: (a) magnitude, (b) magnitude in decibel.

(inductors or capacitors) required to realize a LP or HP filter is equal to the order of the filter. Twice the number of reactive elements is required to realize a BP or a BS filter of the same order. In general, the order of a filter determines the slope of the rolloff: the higher the order, the steeper the rolloff. At frequencies greater than approximately one octave above cutoff (i.e., $f > 2f_C$), the rolloff for all-pole filters is $20n$ dB/decade (or approximately $6n$ dB/octave), where n is the order of the filter (Fig. 4.33). In the vicinity of f_C, both filter alignment (Fig. 4.32) and filter order determine rolloff.

TABLE 4.6   Summary of Standard Filter Alignments

Alignment	Pass-Band Description	Stop Band Description	Comments
Butterworth	Monotonic	Monotonic	All-pole; maximally flat
Chebyshev	Rippled	Monotonic	All-pole
Bessel	Monotonic	Monotonic	All-pole; constant phase shift
Inverse Chebyshev	Monotonic	Rippled	
Elliptic (or Cauer)	Rippled	Rippled	

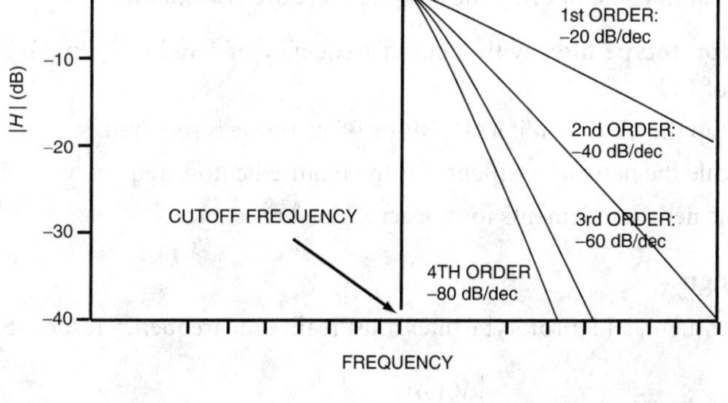

FIGURE 4.33   Effect of filter order on rolloff (Butterworth alignment).

4.4.5   Constraints

Filter networks are so diverse, and so much has been written about filter analysis and design, that several constraints are imposed here in order to present this discussion in a few pages. Accordingly, the discussion is limited to the following:

- Two-port filter networks. The exclusion of some passive filters with multiple outputs is not as significant here as it would be in a discussion of active filters.

- Lumped-parameter elements. High-frequency and microwave filters may rely on *distributed-parameter* elements; however, much of the analysis and many of the design procedures are applicable to distributed-parameter filters.

- Low-order filters. The unavoidable *insertion loss* associated with a passive filter increases with the number of elements in the filter. Consequently, it is usually more practical to use *active filters,* which can be designed to offset insertion loss (or even to provide insertion gain) for applications requiring higher than a third- or fourth-order transfer function.

- Magnitude response. Although a filter alters both the magnitude and the phase angle of the input signal, the predominant applications are those in which the magnitude of the signal is of greater interest.

- Single, resistive termination. The transfer function for a filter depends on both the load and the source impedances; hence, a filter is said to be *doubly terminated.* To avoid complications, the discussion here assumes that source impedance is negligible (i.e., the filter is *singly terminated*) and that the load impedance is nonreactive.

- All-pole filters with Butterworth or Chebyshev alignment. The Bessel alignment is primarily of interest in for *phase filters,* which will not be discussed here, and the techniques for obtaining transfer functions for the other alignments are too lengthy to be treated here. Once the transfer functions are obtained, however, the design procedures presented are generally applicable to all filters of all alignments.

The principal advantage of the Butterworth alignment is the flatness of the amplitude response within the pass band. The principal disadvantage is that the rolloff is not as steep in the vicinity of cutoff as that of some other filter designs of the same complexity (number of poles). The Chebyshev filter has a frequency response with a steeper rolloff in the vicinity of cutoff than that of the Butterworth filter, but *ripples* are present in the pass band. The Chebyshev filter is sometimes called an *equiripple* filter because the ripples have the same peak-to-peak amplitude throughout the pass band.

4.4.6   Summary of Design Procedures

The design procedures suggested here are based on what is sometimes called *modern filter theory.* Modern filter theory (as contrasted with *classical filter theory* relies more on *ad hoc* design calculations and less on the use of tabulated parameters. In brief, the design procedure is as follows:

- Design a LP **prototype filter** with a cutoff frequency of 1 rad/s ($1/2\pi$ Hz) and a resistive load termination of 1 Ω.

- **Frequency map** the LP design if a HP, BP, or BS response is required.

- **Frequency scale** the network elements for the required cutoff frequency.

- **Load scale** the network elements for the actual load resistance.

LP Prototype Design

For Butterworth alignment, a LP prototype filter must realize the frequency response magnitude function

$$\left| \frac{V_2(\omega)}{V_1(\omega)} \right| = \frac{1}{\sqrt{1 + \omega^{2n}}} \tag{4.60}$$

TABLE 4.7 Frequency Response
Magnitude Functions for Butterworth
LP Prototype Filters

Order n	$\left\| \dfrac{V_2(\omega)}{V_1(\omega)} \right\|$ Eq. (22.1)
2	$\dfrac{1}{\sqrt{1+\omega^4}}$
3	$\dfrac{1}{\sqrt{1+\omega^6}}$
4	$\dfrac{1}{\sqrt{1+\omega^8}}$

where

ω = frequency, rad/s
n = filter order
$|V_2|$ = magnitude of the output voltage
$|V_1|$ = magnitude of the input voltage (usually 1 V in experimental and computational procedures)

The magnitude functions are given in Table 4.7 for Butterworth LP prototype filters of second, third, and fourth order.

For Chebyshev alignment, a LP prototype filter must realize the frequency response magnitude function

$$\left| \frac{V_2(\omega)}{V_1(\omega)} \right| = \frac{1}{\sqrt{1 + \epsilon^2 T_n^2(\omega)}} \tag{4.61}$$

where

$T_n(\omega)$ = nth-order Chebyshev polynomial of the first kind (Table 4.8)
ϵ = ripple parameter for the allowable pass-band ripple [Eq. (4.62)]

$$\text{ripple} = 1 - \frac{1}{\sqrt{1+\epsilon^2}}$$

$$\text{ripple (dB)} = 10 \log_{10}(1 + \epsilon^2)$$

$$\epsilon^2 = 10^{0.1 \times \text{ripple(dB)}} - 1 \tag{4.62}$$

The magnitude functions for selected Chebyshev LP prototype filters are given in Table 4.8.

TABLE 4.8 Frequency Response Magnitude Functions for Chebyshev LP Prototype Filters

Order n	Chebyshev Polynomial $T_n(\omega)$	$\left\| \dfrac{V_2(\omega)}{V_1(\omega)} \right\|$ Eq. (4.61)
2	$2\omega^2 - 1$	$\dfrac{1}{\sqrt{4\epsilon^2\omega^4 - 4\epsilon^2\omega^2 + (\epsilon^2 + 1)}}$
3	$4\omega^3 - 3\omega$	$\dfrac{1}{\sqrt{16\epsilon^2\omega^6 - 24\epsilon^2\omega^4 + 9\epsilon^2\omega^2 + 1}}$
4	$8\omega^4 - 8\omega^2 + 1$	$\dfrac{1}{\sqrt{64\epsilon^2\omega^8 - 128\epsilon^2\omega^6 + 80\epsilon^2\omega^4 - 16\epsilon^2\omega^2 + (\epsilon^2 + 1)}}$

TABLE 4.9 s-Domain Transfer Functions for LP Protoype Filters

$$\frac{V_2(s)}{V_1(s)} = \frac{c_0}{c_n s^n + c_{n-1} s^{n-1} + \cdots + c_1 s + c_0}$$

Alignment	Order n	c_4	c_3	c_2	c_1	c_0
Butterworth	2			1.000	1.414	1.000
	3		1.000	2.000	2.000	1.000
	4	1.000	2.613	3.414	2.613	1.000
Chebyshev	2			1.000	2.372	3.314
ripple:	3		1.000	1.939	2.630	1.638
0.1 dB	4	1.000	1.804	2.627	2.026	0.829
$\epsilon = 0.1526$						
Chebyshev	2			1.000	1.098	1.103
ripple:	3		1.000	0.988	1.238	0.491
1.0 dB	4	1.000	0.953	1.454	0.743	0.276
$\epsilon = 0.5088$						
Chebyshev	2			1.000	0.645	0.708
ripple:	3		1.000	0.597	0.928	0.251
3.0 dB	4	1.000	0.582	1.169	0.405	0.177
$\epsilon = 0.9976$						

To design L–C networks to realize the magnitude functions in Tables 4.7 or 4.8, it is necessary to know the s-domain voltage transfer functions from which the magnitude functions are derived. The general form of these transfer functions is

$$\frac{V_2(s)}{V_1(s)} = \frac{c_0}{c_n s^n + c_{n-1} s^{n-1} + \cdots + c_1 s + c_0} \tag{4.63}$$

where

s = Laplace transform domain variable, s^{-1}
c_i = polynomial coefficients

Filter alignment is determined by the coefficients c_i in the denominator polynomial of the transfer function. Table 4.9 provides the coefficients for Butterworth and Chebyshev alignment for second-, third-, and fourth-order filters.

Phase response is not tabulated here, but may be determined by

$$\phi(\omega) = -\tan^{-1} \frac{c_1 \omega - c_3 \omega^3 + c_5 \omega^5 - \cdots}{c_0 - c_2 \omega^2 + c_4 \omega^4 - \cdots} \tag{4.64}$$

where $\phi(\omega)$ is the phase angle in radian.

Passive filters are very often (but not always) realized with an L–C ladder network, which takes on one of the forms shown in Fig. 4.34. The configuration of Fig. 4.34(a) is used for odd-order filters, and the configuration of Fig. 4.34(b) is used for even-order filters. In either case, the inductors provide low impedance between input and output at low frequencies and high impedance at high frequencies, and the capacitors provide low admittance to ground at low frequencies and high admittance at high frequencies. This discussion of filter realization is limited to *lossless* L–C ladder networks; i.e., the capacitors are assumed to have no dielectric conductance and the inductors are assumed to have no winding resistance. (The network configuration in Fig. 4.34 is known as aa *Cauer network* after Wilhelm Cauer, one of the preeminent network theorists of the early 20th century.)

When terminated by a load $Z_L(s)$, the entire circuit consisting of L–C ladder network filter and load (Fig. 4.35) must have the transfer function $V_2(s)/V_1(s)$, as specified in Eq. (4.63). The L–C ladder network

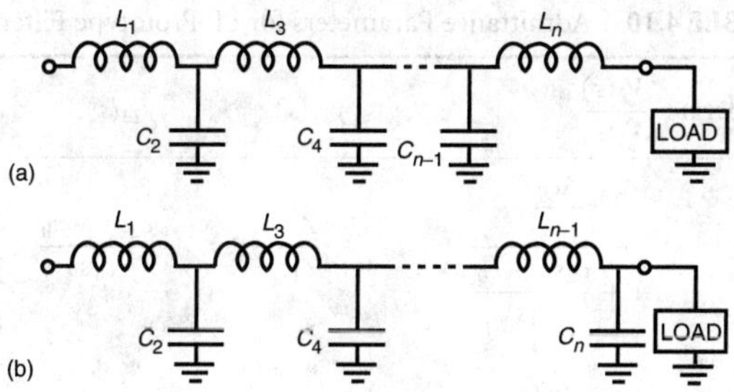

FIGURE 4.34 Passive LP filters: (a) lossless L–C ladder network, odd order, (b) lossless L–C ladder network, even order.

can be characterized by its network admittance parameters

$$y_{11} = \left.\frac{I_1}{V_1}\right|_{V_2=0} \qquad y_{12} = \left.\frac{I_1}{V_2}\right|_{V_1=0} \qquad y_{21} = \left.\frac{I_2}{V_1}\right|_{V_2=0} \qquad y_{22} = \left.\frac{I_2}{V_2}\right|_{V_1=0} \tag{4.65}$$

such that

$$I_1 = y_{11}V_1 + y_{12}V_2 \tag{4.66}$$

and

$$I_2 = y_{21}V_1 + y_{22}V_2 \tag{4.67}$$

When the filter network is terminated by $Z_L(s)$

$$I_2 = -V_2 Z_L \tag{4.68}$$

and so Eq. (4.67) becomes

$$-V_2 Z_L = y_{21}V_1 + y_{22}V_2 \tag{4.69}$$

or

$$\frac{V_2}{V_1} = \frac{-y_{21}}{Z_L + y_{22}} \tag{4.70}$$

In a prototype filter, $Z_L = 1\,\Omega$ (resistive); thus

$$\frac{V_2}{V_1} = \frac{-y_{21}}{1 + y_{22}} \tag{4.71}$$

The voltage transfer function in Eq. (4.63) can be manipulated into the form of Eq. (4.71) by separating the denominator terms into one group of odd-degree terms and another group of even-degree terms. When the numerator and denominator are both divided by the group of odd-degree terms, the result is as follows.

Even order:

$$\frac{V_2(s)}{V_1(s)} = \frac{\dfrac{c_0}{c_{n-1}s^{n-1} + c_{n-3}s^{n-3} + \cdots + c_1 s}}{1 + \dfrac{c_n s^n + c_{n-2}s^{n-2} + \cdots + c_0}{c_{n-1}s^{n-1} + c_{n-3}s^{n-3} + \cdots + c_1 s}} \tag{4.72a}$$

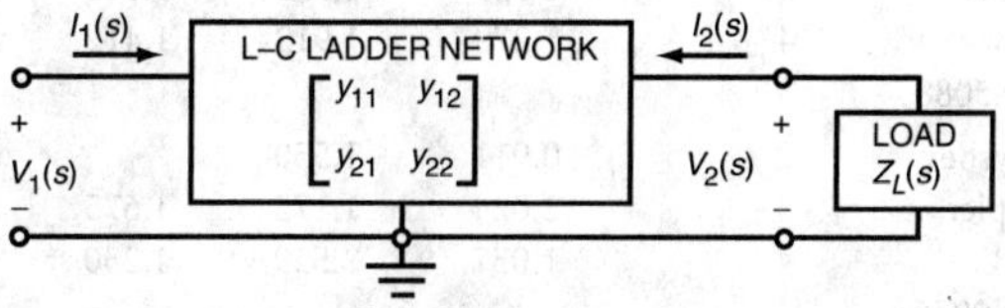

FIGURE 4.35 L–C ladder network with load.

TABLE 4.10 Admittance Parameters for LP Prototype Filters

Order n	$\dfrac{V_2(s)}{V_1(s)}$	$y_{21}(s)$	$y_{22}(s)$
2	$\dfrac{\dfrac{Ac_0}{c_1 s}}{1 + \dfrac{c_2 s^2 + c_0}{c_1 s}}$	$-\dfrac{Ac_0}{c_1 s}$	$\dfrac{c_2 s^2 + c_0}{c_1 s}$
3	$\dfrac{\dfrac{Ac_0}{c_3 s^3 + c_1 s}}{1 + \dfrac{c_2 s^2 + c_0}{c_3 s^3 + c_1 s}}$	$-\dfrac{Ac_0}{c_3 s^3 + c_1 s}$	$\dfrac{c_2 s^2 + c_0}{c_3 s^3 + c_1 s}$
4	$\dfrac{\dfrac{Ac_0}{c_3 s^3 + c_1 s}}{1 + \dfrac{c_4 s^4 + c_2 s^2 + c_0}{c_3 s^3 + c_1 s}}$	$-\dfrac{Ac_0}{c_3 s^3 + c_1 s}$	$\dfrac{c_4 s^4 + c_2 s^2 + c_0}{c_3 s^3 + c_1 s}$

Odd order:

$$\frac{V_2(s)}{V_1(s)} = \frac{\dfrac{c_0}{c_n s^n + c_{n-2} s^{n-2} + \cdots + c_1 s}}{1 + \dfrac{c_{n-1} s^{n-1} + c_{n-3} s^{n-3} + \cdots + c_0}{c_n s^n + c_{n-2} s^{n-2} + \cdots + c_1 s}} \tag{4.72b}$$

The expressions for y_{21} and y_{22} are obtained by matching terms from Eq. (4.71) with terms from Eq. (4.72a) or (4.72b). These results are provided in Table 4.10 for LP filters.

LP Prototype Design Example

The next step in designing an LP prototype filter network is to select inductors and capacitors so that the admittance parameters y_{21} and y_{22} are as specified in Table 4.10. The procedure is demonstrated for a representative fourth-order filter (Chebyshev with 3-dB ripple), and the results are given in Table 4.11 for

TABLE 4.11 L–C Elements for Prototype LP Filter

Alignment	Order n	L_1, H	C_2, F	L_3, H	C_4, F
Butterworth	2	1.414	0.707		
	3	1.500	1.333	0.500	
	4	1.531	1.577	1.082	0.383
Chebyshev	2	0.716	0.442		
ripple:	3	1.090	1.086	0.516	
0.1 dB	4	1.245	1.458	1.199	0.554
$\epsilon = 0.1526$					
Chebyshev	2	0.995	0.911		
ripple:	3	1.509	1.333	1.012	
0.1 dB	4	1.279	1.912	1.412	1.049
$\epsilon = 0.5088$					
Chebyshev	2	0.911	1.550		
ripple:	3	2.024	1.175	1.675	
3.0 dB	4	1.057	2.529	1.230	1.718
$\epsilon = 0.9976$					

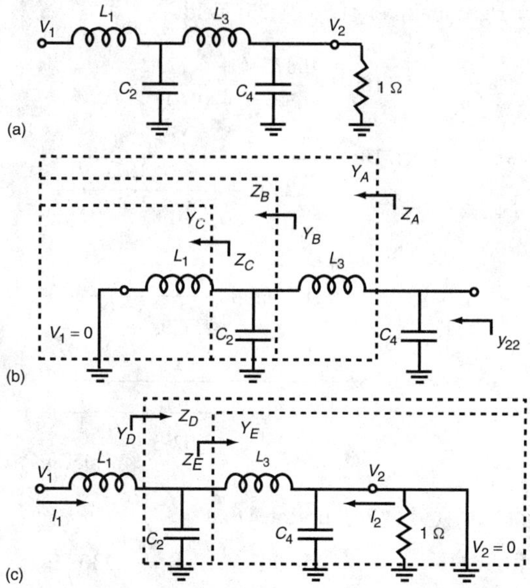

FIGURE 4.36 Design example, fourth-order LP prototype filter: (a) singly terminated LP filter, (b) network model for obtaining y_{22}, (c) network model for obtaining y_{21}.

second-, third-, and fourth-order Butterworth and selected Chebyshev filters. This procedure is based on satisfying the requirements for y_{22}, which necessarily satisfies the requirements for y_{21}.

1. The singly terminated filter network is shown in Fig. 4.36(a).
2. To obtain the network parameter y_{22}, port 1 is shorted (i.e., $V_1 = 0$), which makes the admittance at port 2 equivalent to y_{22} (Fig. 4.36(b)).
3. To simplify analysis, the network is divided into subnetworks A, B, and C, as shown, which leads to

$$y_{22} = Y_4 + Y_A = C_4 s + \frac{1}{Z_A}$$

$$Z_A = Z_3 + Z_B = L_3 s + \frac{1}{Y_B}$$

$$Y_B = Y_2 + Y_C = C_2 s + \frac{1}{Z_C}$$

$$Z_C = Z_1 = L_1 s$$

(4.73)

4. Back substitution in Eq. (4.73) yields

$$y_{22} = C_4 s + \cfrac{1}{L_3 s + \cfrac{1}{C_2 s + \cfrac{1}{L_1 s}}}$$

(4.74)

5. The form of Eq. (4.74) suggests that the values of L_1, C_2, L_3, and C_4 can be determined by comparing Eq. (4.74) to a long division of the required y_{22}. From Tables 4.9 and 4.10

$$y_{22} = \frac{s^4 + 1.169 s^2 + 0.177}{0.582 s^3 + 0.405 s}$$

(4.75)

6. Long division of the right member of Eq. (4.75) yields

$$y_{22} = 1.718s + \frac{0.473s^2 + 0.177}{0.582s^3 + 0.405}$$

$$= 1.718s + \cfrac{1}{1.230s + \cfrac{0.187s}{0.473s^2 + 0.177}}$$

$$= 1.718s + \cfrac{1}{1.230s + \cfrac{1}{2.529s + \cfrac{0.177s}{0.187}}} \tag{4.76}$$

$$= 1.718s + \cfrac{1}{1.230s + \cfrac{1}{2.529s + \cfrac{1}{1.057s}}}$$

7. Comparison of Eq. (4.74) and (4.76) indicates that

$$L_1 = 1.057 \text{ H}; \qquad C_2 = 2.529 \text{ F}; \qquad L_3 = 1.230 \text{ H}; \qquad C_4 = 1.718 \text{ F}$$

These values are prototype values and must be appropriately scaled for the actual cutoff frequency (other than 1 rad/s) and the actual load resistance (other than 1 Ω).

8. Confirmation that the prototype values for L_1, C_2, L_3, and C_4 yield the required value of y_{21} (as obtained from Tables 4.9 and 4.10) is not necessary, but serves as a test of the design procedure. In Fig. 4.36(c)

$$I_1 = \frac{V_1}{Z_1 + Z_D} = \frac{V_1}{L_1s + \cfrac{1}{Y_D}}$$

$$Y_D = Y_2 + Y_E = C_2s + \frac{1}{Z_E} \tag{4.77}$$

$$Z_E = L_3s$$

Back substitution in Eq. (4.77) yields

$$I_1 = \cfrac{V_1}{L_1s + \cfrac{1}{C_2s + \cfrac{1}{L_3s}}} \tag{4.78}$$

Because of current division

$$I_2 = -I_1 \frac{Y_E}{Y_2 + Y_E} = -I_1 \cfrac{\cfrac{1}{L_3s}}{C_2s + \cfrac{1}{L_3s}} \tag{4.79}$$

9. The admittance parameter y_{21} is obtained by combining Eqs. (4.78) and (4.79)

$$y_{21} = -\frac{I_2}{V_1} = -\frac{1}{L_1C_2L_3s^3 + (L_1 + L_3)s}$$

$$= -\frac{1}{3.288s^3 + 2.287s} = -\frac{0.177}{0.582s^3 + 0.405s} \tag{4.80}$$

which confirms y_{21} as given in Tables 4.9 and 4.10.

TABLE 4.12 Frequency Mapping of Network Elements

Type	Series Branch(es)	Shunt Branch(es)
Low pass	L_{LP}	C_{LP}
High pass	$C_{HP} = \dfrac{1}{L_{LP}}$	$L_{HP} = \dfrac{1}{C_{LP}}$
Bandpass	$L_{BP} = \dfrac{L_{LP}}{B}$ $C_{BP} = \dfrac{B}{L_{LP}}$	$C_{BP} = \dfrac{L_{LP}}{B}$ $L_{BP} = \dfrac{B}{L_{LP}}$
Bandstop	$L_{BS} = BL_{LP}$ $C_{BS} = \dfrac{1}{BL_{LP}}$	$C_{BS} = BC_{LP}$ $L_{BS} = \dfrac{1}{BC_{LP}}$

Frequency Mapping

Once a LP prototype filter has been designed, a prototype HP, BP, or BS L–C filter is realized by replacing the LP network elements according to the scheme shown in Table 4.12. This technique is called frequency mapping. Recall that prototype means that the cutoff frequency of a frequency-mapped HP filter or the center frequency of a frequency-mapped BP or BS filter is 1 rad/s (Fig. 4.37) when operating with a 1-Ω load. The bandwidth parameter B in Table 4.12 expresses bandwidth of a BS or BP filter as the ratio of bandwidth to center frequency. If, for example, the bandwidth and center frequency requirements for a BP or BS filter are 100 Hz and 1 kHz, respectively, then the prototype bandwidth parameter B is 0.1. The general network configurations for HP, BP, and BS L–C filters are shown in Fig. 4.38.

The mathematical basis for frequency mapping is omitted here; however, the technique involves replacing the LP transfer function variable (s or ω) with the mapped variable, as presented in Table 4.13. The mapped amplitude response for a Butterworth filter or a Chebyshev filter is obtained by performing the appropriate operation from column two of Table 4.13 on Eq. (4.60) or (4.61), respectively. The mapped phase response

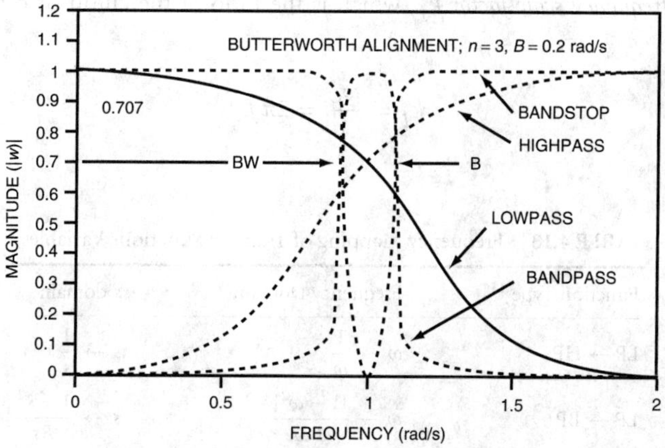

FIGURE 4.37 Mapped frequency responses.

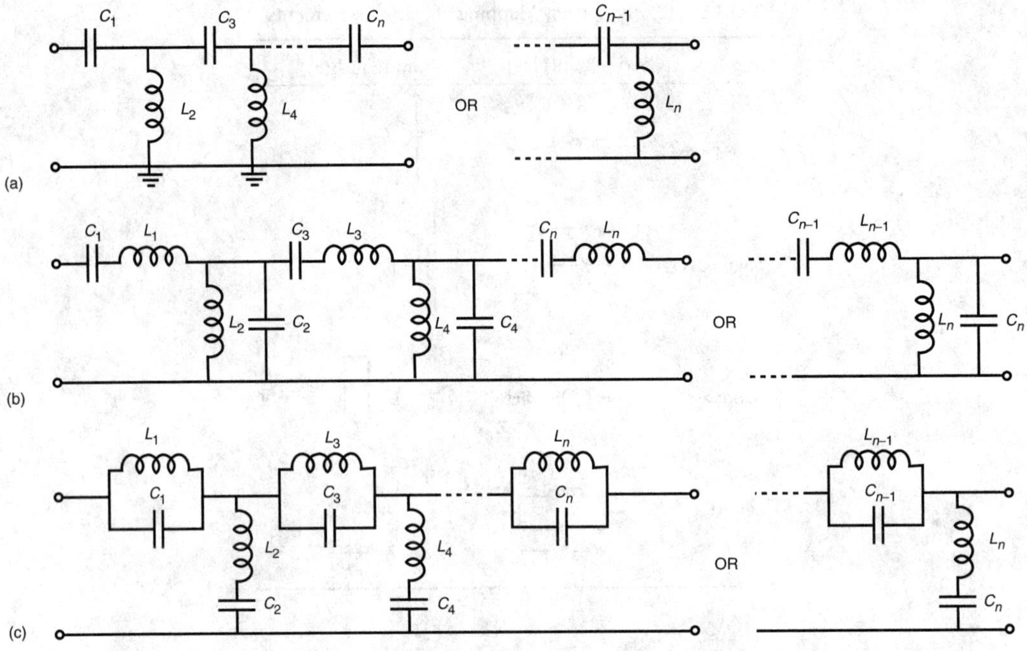

FIGURE 4.38 Mapped filter networks: (a) high pass, (b) bandpass, (c) bandstop.

is obtained by performing the same indicated operation on Eq. (4.64). The mapped s-domain transfer function is realized, when required, by performing the appropriate operation from column 3 of Table 4.13 on Eq. (4.63).

The highest degree term in the denominator of a mapped BP or BS transfer function is twice that of the original LP transfer function. This means that the s-domain transfer function must have twice as many poles, and the L–C ladder network must correspondingly have twice as many elements (Fig. 4.38).

Frequency and Load Scaling

A prototype filter is designed to operate at a cutoff or center frequency of 1 rad/s ($\approx$0.159 Hz) and with a 1-Ω load, and so the inductor and capacitor values for the prototype must be *scaled* for operation of the filter at a specified cutoff or center frequency f_C and with a specified load resistance R_L. Frequency scaling produces a *frequency scale factor* k_f, which is the ratio of the cutoff or center frequency f_C to 1 rad/s

$$k_f = \frac{\omega_C}{1} = 2\pi f_C \tag{4.81}$$

TABLE 4.13 Frequency Mapping of Transfer Functions Variables

Function Type	Frequency Domain	s domain
LP $\rightarrow$ HP	$\omega \rightarrow \dfrac{1}{\omega}$	$s \rightarrow \dfrac{1}{s}$
LP $\rightarrow$ BP	$\omega \rightarrow \dfrac{\lvert 1 - \omega^2 \rvert}{B\omega}$	$s \rightarrow \dfrac{1 + s^2}{Bs}$
LP $\rightarrow$ BS	$\omega \rightarrow \dfrac{B\omega}{\lvert 1 - \omega^2 \rvert}$	$s \rightarrow \dfrac{Bs}{1 + s^2}$

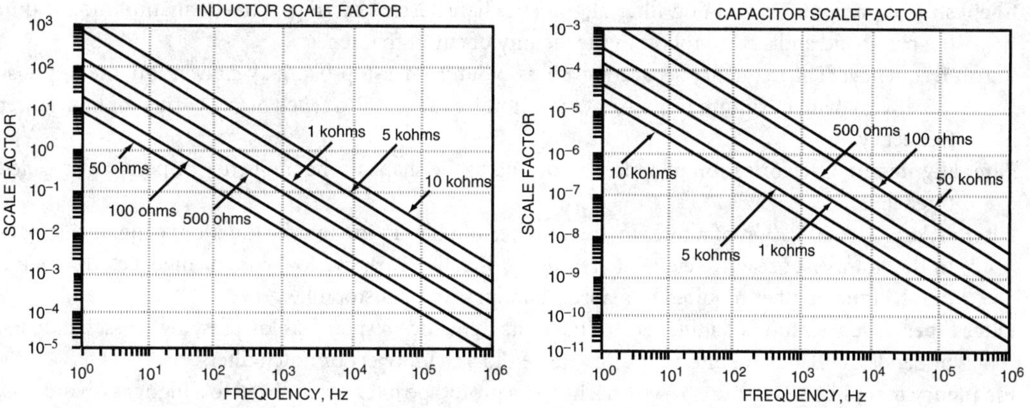

FIGURE 4.39 Scale factor.

and load scaling produces a *load scale factor* k_R, which is the ratio of the load resistance R_L to $1\ \Omega$

$$k_R = \frac{R_L}{1} = R_L \tag{4.82}$$

In practice, the frequency and impedance scale factors are combined to produce a single scale factor K_L for inductors

$$K_L = \frac{k_R}{k_f} = \frac{R_L}{2\pi f_C} \tag{4.83}$$

and a single scale factor K_C for capacitors

$$K_C = \frac{1}{k_f k_R} = \frac{1}{2\pi f_C R_L} \tag{4.84}$$

The inductor and capacitor scale factors are used as multipliers for the prototype values

$$L = K_L L_P \tag{4.85}$$
$$C = K_C C_P \tag{4.86}$$

where

L = scaled inductance
L_P = corresponding inductance in prototype filter
C = scaled capacitance
C_P = corresponding capacitance in prototype filter

The graphs in Fig. 4.39 can be used for quick estimates of the scaled values of inductance and capacitance.

Defining Terms

Bandwidth: The absolute difference between the lower and upper cutoff frequencies of a bandpass or bandstop filter.

Butterworth alignment: A common filter alignment characterized by a maximally flat, monotonic frequency response.

Center frequency: The frequency of maximum or minimum response for a bandpass or a bandstop filter, respectively; often taken as the geometric mean of the lower and upper cutoff frequencies.

Chebyshev alignment: A common filter alignment characterized by ripples of equal amplitude within the pass-band and a steep rolloff in the vicinity of cutoff frequency.

Cutoff frequency: The frequency separating a pass-band from a stop-band; by convention, the response is 3 dB less than maximum response at the cutoff frequency; also called break, corner, or half-power frequency.

Filter alignment: Categorization of a filter according to the shape of the frequency response; also called filter class.

Filter order: Categorization of a filter by the number of poles in its transfer function; generally equates to the number of reactive elements required to realize a passive low-pass or high-pass filter, and one-half the number required to realize a bandpass or bandstop filter.

Filter type: Categorization of a filter according to its frequency response as lowpass, highpass, bandpass, or bandstop. Bandstop filters are also referred to as bandreject or notch filters.

Frequency mapping: Procedure by which a lowpass prototype filter is converted to a highpass, bandpass, or bandstop prototype.

Frequency scaling: Procedure by which the values of prototype filter elements are adjusted to allow the filter to operate at a specified cutoff frequency (other than 1 rad/s).

Load scaling: Procedure by which the values of prototype filter elements are adjusted to allow the filter to operate with a specified load (other than 1 Ω).

Pass-band: A range of frequency over which the filter response is, by convention, within 3 dB of its maximum response.

Passive filter: An electrical network comprised entirely of passive elements and designed specifically to respond differently to signals of different frequency.

Prototype filter: A filter designed to operate with a cutoff frequency of 1 rad/s and with a load termination of 1 Ω.

Rolloff: Transition from pass-band to stop-band.

Stop-band: A range of frequency over which the filter response is, by convention, at least 3 dB less than its maximum response; also called the attenuation band.

References

Chirlian, P.M. 1969. *Basic Network Theory.* McGraw-Hill, New York.

Ciletti, M.D. 1988. *Introduction to Circuit Analysis and Design.* Holt, Rinehart, and Winston, New York.

Dorf, R.C., ed. 1993. *The Electrical Engineering Handbook.* CRC Press, Boca Raton, FL.

Harrison, C.A. 1990. *Transform Methods in Circuit Analysis.* Saunders College Pub., Philadelphia, PA.

Johnson, D.E. 1976. *Introduction to Filter Theory.* Prentice-Hall, Englewood Cliffs, NJ.

Johnson, D.E. and Hilburn, J.L. 1975. *Rapid Practical Designs of Active Filters.* Wiley, New York.

Nilsson, J.W. 1986. *Electric Circuits.* Addison-Wesley, Reading, MA.

Tuttle, D.F., Jr. 1965. *Electric Networks: Analysis and Synthesis.* McGraw-Hill, New York.

Further Information

The textbook by D.E. Johnson, *Introduction to Filter Theory*, mentioned in the references, is recommended as a compact but thorough seminal reference for the design of lumped-parameter passive filters. In addition, this subject has been the primary focus of a number of other textbooks and is included in virtually every basic circuit theory textbook.

5
Electron Vacuum Devices

5.1 Electron Tube Fundamentals

Clifford D. Ferris

5.1.1 Introduction

Although widely used in many electronic applications until the 1960s, the traditional electron vacuum tube has now largely been replaced by semiconductor devices. Notable exceptions include cathode-ray tubes used in oscilloscopes and video displays, power grid tubes used in radio and television broadcast transmitters, and various classes of vacuum microwave devices. This chapter presents an introduction to the operation of electron tubes. Subsequent chapters are concerned with specific vacuum devices and their applications.

The elementary vacuum-tube structure is illustrated in Fig. 5.1. An evacuated volume is enclosed by an envelope, which may be glass, metal, ceramic, or a combination of these materials. Metallic electrodes contained within the envelope form the working device, and external connections are made via metal wires that pass through vacuum seals in the envelope. The three basic internal structures are the cathode (K), the anode (A) (often called the **plate**), which are solid surfaces, and a single open wire-mesh structure that is the **grid**. Under typical operating conditions, the anode is held at a more positive potential than the cathode, and the grid is held at a voltage that is negative relative to the cathode. In this regard, operation is analogous to that of an *n*-channel junction field-effect transistor (JFET). The cathode, grid, and plate are analogous to the source, gate, and drain, respectively.

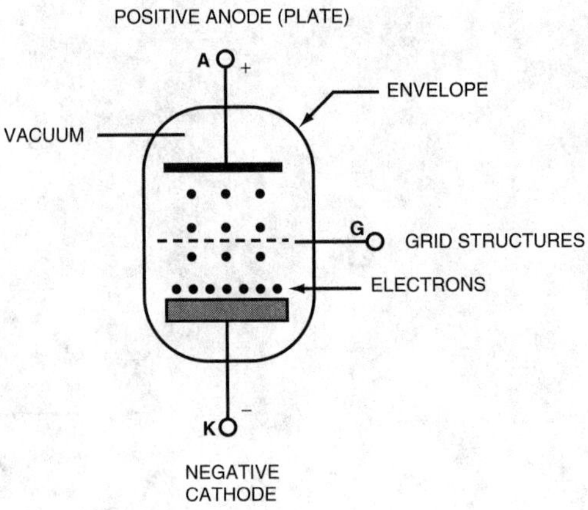

The function of the cathode is to provide a source of free electrons, and this is accomplished by heating the cathode to a high temperature, a process known as **thermionic emission**. Because the electrons are negatively charged and the cathode is lower in potential than the plate, the emitted electrons are accelerated toward the positively charged plate. The electron flow from the cathode to the plate can be controlled by adjusting the electric potential (grid-to-cathode voltage) applied between the grid and cathode. If the grid is made sufficiently negative relative to the cathode, the electron flow to the plate can be cut off completely. As the grid becomes less negative, electrons pass through the grid structure to the plate. The grid potential, however, must not increase to the point that it collects electrons in place of the plate. The grid is composed of fine wires that are not designed to support high current and is easily damaged by any significant electron collection.

FIGURE 5.1   The basic vacuum tube.

5.1.2   Cathodes and Thermionic Emission

There are two fundamental cathode structures, directly heated (filament) and indirectly heated. The filament type of cathode is similar to the tungsten filament in an incandescent light bulb. To increase the efficiency of electron emission, thorium is added to the tungsten (1 or 2% by weight of thorium oxide), thus forming a **thoriated** tungsten filament (after an activation process that converts the thorium oxide to metallic thorium). The actual mechanical configuration or geometry of the filament varies widely among vacuum tubes depending on the output-power rating of the vacuum tube. Typically, the filament (or heater, as it is often called) is supported by the lead wires that pass through the envelope.

Figure 5.2 represents a vacuum diode that contains a directly heated cathode (filament). The filament supply voltage may be either AC or DC (DC is shown), and historically is called the *A supply*. Depending on the service for which a vacuum tube has been designed, filament (heater) supply voltages typically range from 2.5 to 12.6 V. Filament currents vary from as low as 150 mA to hundreds of amperes. A DC potential (the plate supply voltage) is required between the plate and cathode. Under many conditions, the cathode is at 0 V DC so that the plate-to-cathode voltage is the same as the plate voltage.

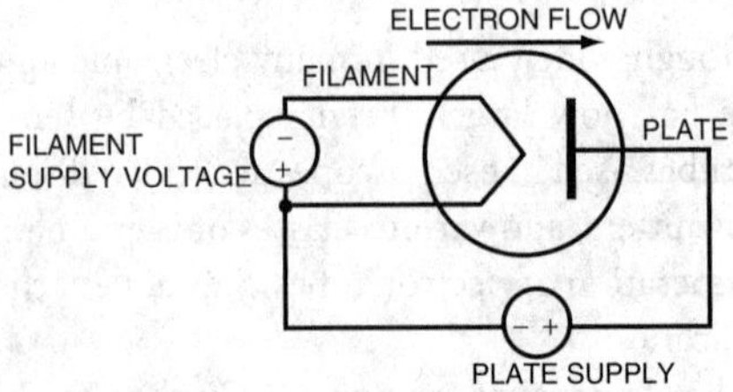

FIGURE 5.2   Vacuum diode for directly heated cathode discussion.

The thermionic-emission current I_{th}, in amperes, generated by a heated filament is described by the Richardson–Dushman equation

$$I_{th} = AD_0 T^2 \exp(-b_0/T) \qquad (5.1)$$

where

A = surface (emitting) area of the filament, m^2
D_0 = material constant for a given filament type, A/m^2-K^2
T = absolute temperature, K
b_0 = $11,600 E_w$, K
E_w = the work function of the emitting surface, eV

Some typical values for thoriated tungsten are $D_0 = 3.0 \times 10^4$ A/m^2-K^2, $T = 1900$ K, $b_0 = 30,500$ K, and $E_w = 2.63$ eV. Cathode efficiency is defined in terms of the number of amperes (or milliamperes) produced per square meter of emitting surface per watt of applied filament power. For a thoriated tungsten filament with plate voltages in the range from 0.75 to 5 kV, emission efficiency ranges from 50 to 1000 A/m^2/W. Thoriated tungsten filaments are normally incorporated into power grid tubes that generate output power levels from a few hundred watts to thousands of kilowatts. These devices are used in high-power radio and television broadcasting service and in industrial applications such as induction heating.

Oxide-coated cathodes that are indirectly heated are more efficient ($E_w \approx 1$ eV) than thoriated tungsten filaments, but they cannot sustain the high-voltage high-current operating conditions associated with power grid tubes. When the maximum output power required is below a few hundred watts, they are the device of choice and find application in low-power high-frequency transmitters. They are also widely used by audiophiles who feel that vacuum-tube power amplifiers produce a quality of sound output that is superior to that produced by semiconductor devices.

A typical geometry for an oxide-coated cathode is a nickel or Konal-metal sleeve or cylinder that surrounds a heating filament (the heater). The heater wire is usually tungsten or a tungsten-molybdenum alloy that is coated with a ceramic insulating material (alundum, aluminum oxide or beryllium oxide) to isolate the heater electrically from the cathode sleeve. The outside coating of the cathode is applied as a mixture of carbonates that are subsequently converted to oxides during final processing of the electron tube. Barium and strontium oxides are widely used. Typical heater temperatures are on the order of $1000°C$, which produce an orange to yellow-orange glow from the heater. Cathode temperatures are on the order of $850°C$. Maximum heater-power requirements are on the order of 12 W for oxide-coated cathode devices, whereas thoriated tungsten filaments may require many kilowatts of heater input power.

Figure 5.3 illustrates two of many possible heater configurations. The twisted configuration (Fig. 5.3(a)) and the folded linear arrangement (Fig. 5.3(b)) are commonly used. The ceramic oxide

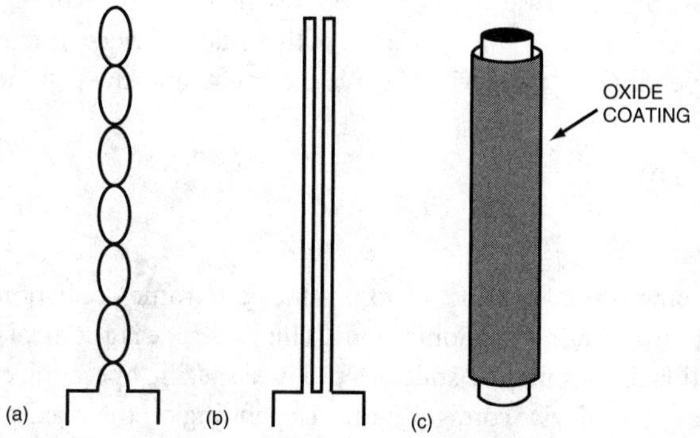

OXIDE
COATING

(a) (b) (c)

FIGURE 5.3 Heater and cathode structures: (a) Twisted filament, (b) folded filament, (c) cathode sleeve.

coating that is applied to the wires prevents electrical short circuits between the wire segments. The heater is inserted into the cathode sleeve, and the ceramic coating applied to the heater surface insulates the heater electrically from the cathode sleeve. This ceramic layer cannot sustain a high potential difference between the heater and the cathode, and 200 V is a typical figure for the maximum heater-cathode voltage. For this reason, indirectly heated cathodes are not appropriate for power grid tube service.

Before we move on to specific vacuum-tube configurations, a final comment is required concerning electron emission from a cathode. To simplify mathematical derivations, it is often assumed that all of the electrons are emitted with zero initial velocity. If some of the electrons have initial velocities, a negative potential gradient is established at the cathode, which then produces a potential-energy barrier and corresponding space-charge region. Electrons emitted with very small initial velocities are then repelled toward the cathode. Equilibrium is established between emission of electrons and this space-charge region. In effect, the electrons that are accelerated toward the plate (the emission current) are drawn from the space-charge region.

5.1.3 Thermionic Diode

The fundamental thermionic diode has been illustrated in Fig. 5.2, and Fig. 5.4 illustrates the representation of a vacuum diode that has an oxide cathode. The plate current (I_P) vs plate-to-cathode voltage (V_P) description of diode operation is provided by the Langmuir–Child law that states that the plate current is proportional to the plate voltage raised to the three-halves power.

FIGURE 5.4 Vacuum diode with indirectly heated cathode.

$$I_P = k_0 V_P^{1.5} \tag{5.2}$$

The proportionality constant k_0 depends on the physical dimensions of the plate and cathode, their respective geometries, and the ratio of electronic charge to mass. It is sometimes called the *permeance*. It must be noted that the plate current cannot exceed the thermionic emission current, which provides an upper limit to the value of I_P in Eq. (5.2). A typical characteristic curve is presented in Fig. 5.5, in which $k_0 = 8.94 \times 10^{-6}$ (A/V$^{1.5}$).

Vacuum diodes are used as circuit elements in the same manner as modern semiconductor diodes. They may serve in rectifier circuits, and as voltage clippers, clamps, and DC restorers for high-voltage applications. From a circuit viewpoint, the only difference is the requirement of a heater voltage when a thermionic diode is used. The forward-voltage drop associated with vacuum diodes, however, is considerably larger than the fraction of a volt associated with a silicon device. It may range from a few volts to more than 100 V depending on the device, the plate current, and the plate-to-cathode voltage (plate voltage).

5.1.4 Grid Tubes

To achieve voltage and power amplification, or to produce electronic oscillation, at least one additional electrode must be added to the basic thermionic diode. This electrode is a control grid. As will be discussed subsequently, two additional grids may be added to produce specific operating characteristics of the grid tube. Figure 5.6 illustrates four device configurations. Depending on the maximum power dissipation of the device, either of the two cathode types shown may be used. For graphical convenience, the heater connections have been omitted in Fig. 5.6(a) to Fig. 5.6(d).

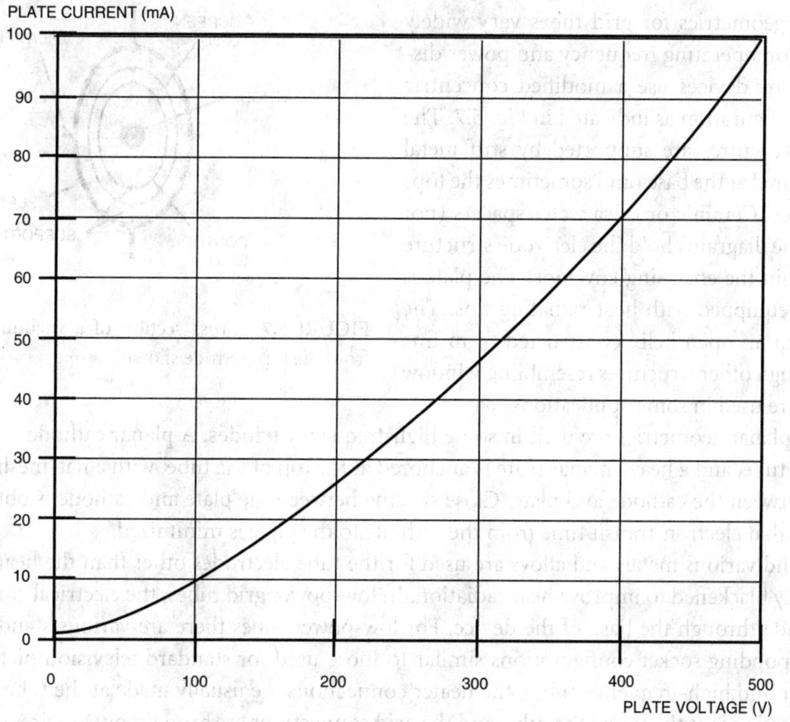

FIGURE 5.5 Vacuum diode characteristic for $k_o = 8.94 \times 10^{-6}$ (A/V$^{1.5}$).

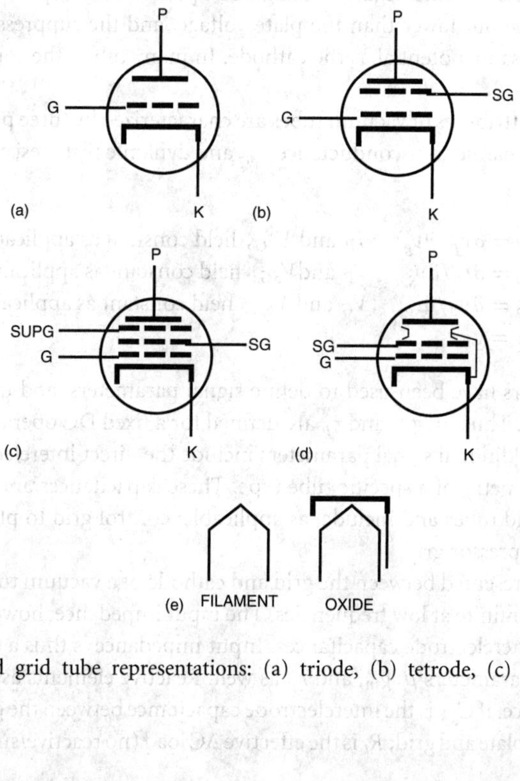

FIGURE 5.6 Four standard grid tube representations: (a) triode, (b) tetrode, (c) pentode, (d) beam power, (e) cathode types.

Physical geometries for grid tubes vary widely depending on operating frequency and power dissipated. Many devices use a modified concentric cylinder configuration as indicated in Fig. 5.7. The electrode structures are supported by stiff metal posts anchored at the base (and sometimes the top) of the device. Ceramic or mica wafer spacers (not shown in the diagram) hold the electrode structure rigidly within the enclosing envelope. The plate is frequently equipped with heat radiating fins. The grid is often an open helix constructed from fine wire, although other structures resembling window screening are used in some applications.

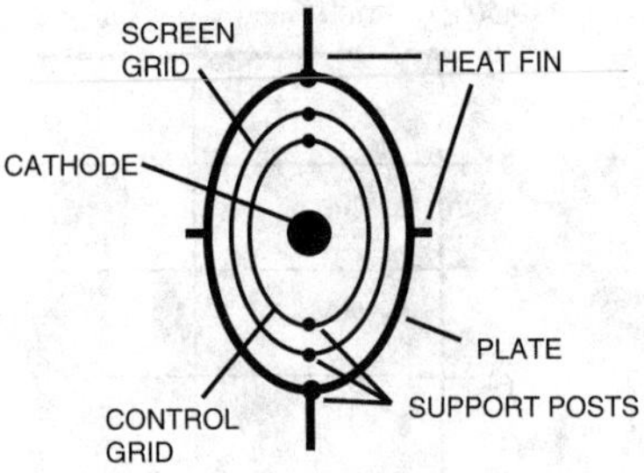

FIGURE 5.7 Cross section of a vacuum tube viewed from the top (tetrode shown).

Parallel planar geometries are used in some high-frequency **triodes**. A planar cathode is placed at the base of the tube, and a heavy planar plate is anchored at the top of the tube, with a flat-mesh control grid inserted between the cathode and plate. Close spacing between the plate and cathode is obtained in this manner so that electron-transit time from the cathode to the plate is minimized.

Nickel and various metals and alloys are used for the tube electrodes other than the heater. The plate is frequently blackened to improve heat radiation. In low-power grid tubes, the electrical connections are usually made through the base of the device. For low-power tubes there are various standard base-pin and corresponding socket configurations similar to those used for standard television picture tubes. In high-power and high-frequency tubes, the heater connections are usually made at the tube base, but the plate connection is at the top of the tube, and the grid connections are brought out as wires or rings along the side of the device. Power grid tubes are described in a separate chapter in this handbook.

By general convention, the cathode potential is assumed to be close to or at zero volts. Thus, references to grid volts, screen volts, and plate volts (rather than grid-to-cathode, screen-to-cathode, or plate-to-cathode) are standard in vacuum-tube literature. In low-power tubes, the control grid (G in Fig. 5.6) is operated at either zero volts or some negative voltage. The plate voltage is positive, the screen grid voltage (SG in Fig. 5.6) is positive but lower than the plate voltage, and the suppressor grid voltage (SUPG in Fig. 5.6) is usually at the same potential as the cathode. In many tubes, the suppressor grid is connected internally to the cathode.

The signal processing attributes of vacuum tubes are characterized by three primary parameters, amplification factor μ (**mu**), dynamic transconductance g_m, and dynamic plate resistance r_p, which are defined as follows:

$$\mu = \partial v_p / \partial v_g; \quad I_P \text{ and } V_{SG}, \text{ held constant as applicable}$$
$$g_m = \partial i_p / \partial v_g; \quad V_P \text{ and } V_{SG}, \text{ held constant as applicable}$$
$$r_p = \partial v_p / \partial i_p; \quad V_G \text{ and } V_{SG}, \text{ held constant as applicable}$$
$$\mu = g_m r_p$$

Note that lowercase letters have been used to define signal parameters, and uppercase letters have been used to define DC values. Thus, μ, g_m, and r_p are defined for a fixed DC operating point (Q point) of the corresponding device. Additional signal parameters include the direct interelectrode capacitances, which are a function of the geometry of a specific tube type. These capacitances are generally a few picofarads at most in low-power grid tubes and include, as applicable: control grid to plate; control grid to heater, cathode, screen grid, suppressor grid.

The input resistance presented between the grid and cathode of a vacuum tube is extremely high and is generally assumed to be infinite at low frequencies. The input impedance, however, is not infinite because of the influence of the interelectrode capacitances. Input impedance is thus a function of frequency, and it depends on the tube parameters μ, g_m, and r_p as well. Reactive elements associated with the load also influence input impedance. If C_{gk} is the interelectrode capacitance between the grid and cathode, C_{pg} is the capacitance between the plate and grid, R_l is the effective AC load (no reactive/susceptive components), and

ω is the radian frequency of operation for an amplifier tube, then input *admittance* (Y_i) is, approximately, for triode

$$Y_i \approx j\omega\{C_{gk} + [\mu R_l/(r_p + R_l)C_{pg}]\} \tag{5.3a}$$

and **tetrode** and **pentode**

$$Y_i \approx j\omega C_{gk} \tag{5.3b}$$

At very high frequencies, lead inductances and stray capacitances associated with the tube itself and its external connections modify the expressions shown.

With the exception of cathode **emissivity**, the operating parameters of vacuum tubes depend solely on geometry and the placement of the electrodes within an evacuated envelope. If mechanical tolerances are rigidly held during manufacture, then there is very little variation in operating parameters across vacuum tubes of the same type number. Cathode manufacture can also be controlled to high tolerances. For this reason, when an operating point is specified, single numbers define μ, g_m, and r_p in vacuum tubes. The designer does not have to contend with the equivalent of beta spread in bipolar-junction transistors or transconductance spread in field-effect devices. To ensure reliable operation over the life of the tube, a barium *getter* is installed before the tube structure is sealed into an envelope and evacuated. After the tube has been evacuated and the final seal has been made, the getter is flashed (usually using a radio frequency induction-heating coil) and elemental barium is deposited on a portion of the inside of the envelope. The function of the getter is to trap gas molecules that are generated from the tube components with time. In this manner the integrity of the vacuum is preserved. Heater burnout or oxide-layer deterioration (in indirectly heated cathode tubes) are the usual causes of failure in vacuum tubes.

Triode

Historically, the triode shown in Fig. 5.6(a) was the first vacuum tube used for signal amplification. As noted in the introductory section (Fig. 5.1), the voltage applied between the grid (G) and the cathode (K) can be used to control the electron flow and, hence, the plate current. The dynamic operation of a triode is defined approximately by a modified form of the Langmuir–Child law as follows

$$i_p = k_0(v_g + v_p/\mu)^{1.5} \tag{5.4}$$

The quantity k_0 is the permeance (as defined in the diode discussion) of the triode, and its value is specific to a particular triode type. The plate characteristic curves generated by Eq. (5.4), for a triode in which $k_0 = 0.0012$, are illustrated in Fig. 5.8. As the plate characteristic curves indicate, the triode is approximately a voltage-controlled voltage source.

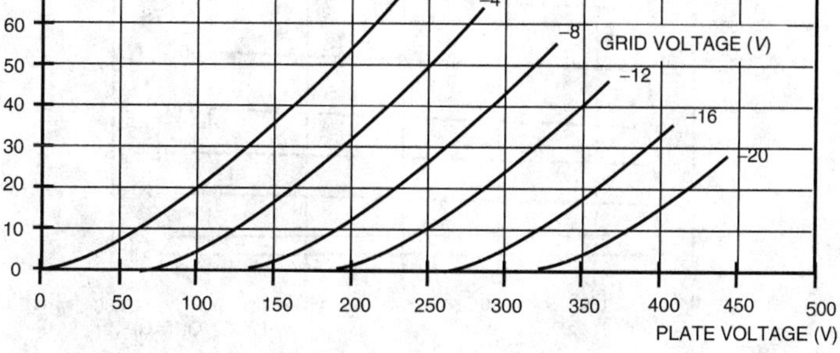

FIGURE 5.8 Triode plate characteristics.

The value of μ in triodes typically ranges from about 4 (low-mu triode) to 120 (high-mu triode). Typical ranges for r_p and g_m are, respectively, from 800 Ω to 100 kΩ, and from 1 to 5 mS.

Tetrode

In general, the triode exhibits a high direct capacitance (relatively speaking) between the control grid and the plate. To provide electrostatic shielding, a second grid (the screen grid) is added between the control grid and the plate. The screen grid must be configured geometrically to provide the amount of shielding desired, and arranged so that it does not appreciably affect electron flow. Insertion of the screen grid, however, does significantly change the plate characteristic curves, as is shown in Fig. 5.9. The negative-resistance region below 100 V may lead to unstable operation and care must be taken to assure that the plate voltage remains above this region. In the early days of radio, the negative-resistance of the tetrode was used in conjunction with tuned circuits to produce sinusoidal oscillators.

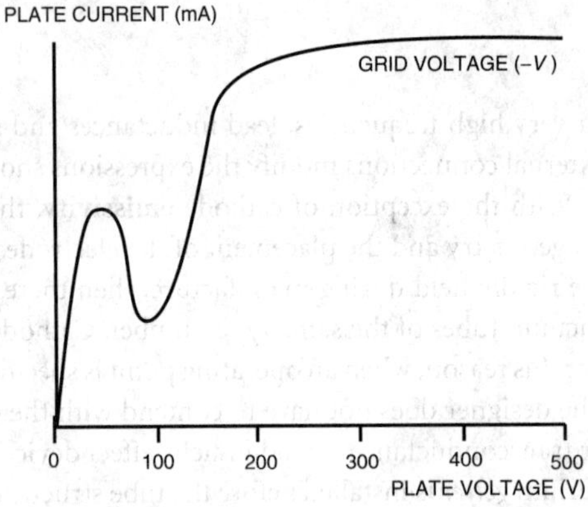

FIGURE 5.9 Tetrode plate characteristic for a single grid voltage (at constant screen voltage).

Pentode

To suppress the negative-resistance characteristic generated by the tetrode, a third coarse grid structure (the suppressor grid) is inserted between the screen grid and the plate. In this manner, the plate characteristic curves are smoothed and take the form shown in Fig. 5.10.

The dynamic operation of a pentode is described approximately by the relation

$$i_p = k_0(v_g + v_p/\mu_1 + v_{sg}/\mu_2)^{1.5} \tag{5.5}$$

where μ_1 is the amplification factor for the plate and control grid and μ_2 is the amplification factor for the plate and screen grid. As the plate characteristic curves indicate, the pentode is approximately a voltage-controlled current source, and the curves are similar to those generated by an n-channel depletion-mode metal oxide semiconductor field effect transistor (MOSFET). The screen grid draws an appreciable current that may amount to as much as 20–30% of the plate current.

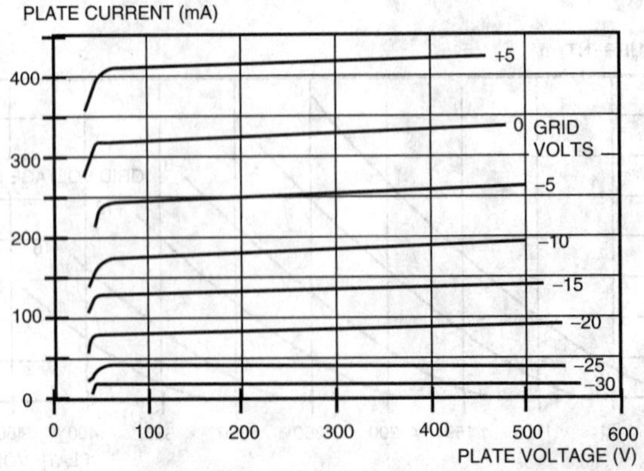

FIGURE 5.10 Typical pentode plate characteristics ($V_S = 150$ V).

The value of μ is not normally provided on the specifications sheets for pentodes, and only the values for g_m and r_p are listed. Transconductance values are typically on the order of 5–6 mS, and plate resistance values vary from 300 kΩ to over 1 MΩ.

Pentodes are sometimes operated in a triode connection, in which the screen grid is connected to the plate. In the 1950s, this mode of operation was popular in high-fidelity vacuum-tube audio power amplifiers.

Beam Power Tube

The beam power tube is a modified form of the tetrode or pentode that utilizes directed electron beams to produce a substantial increase in the output power. Beam-forming electrodes are used to shape the electron beam into the form of a fan (as indicated by the dotted lines in Fig. 5.11). An optional suppressor grid may be included. The control and screen grids are exactly aligned such that the electron beam is broken into dense fan-shaped layers or sheets. Beam power tubes are characterized by high-power output, high-power sensitivity, and high efficiency. The operating screen current ranges from approximately 5–10% of the plate current, as opposed to pentodes in which the screen current is on the order of 20–30% of the plate current.

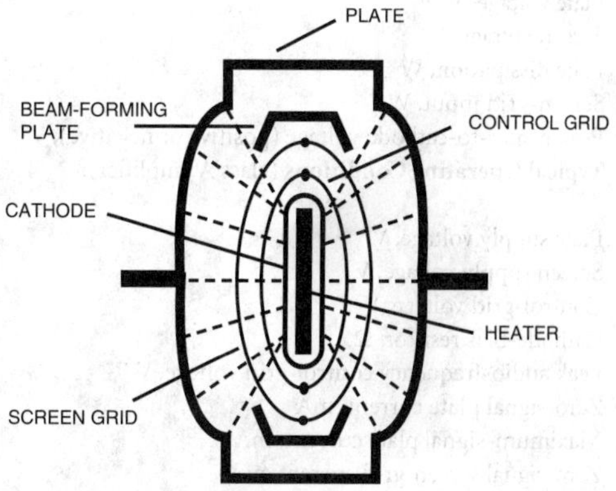

FIGURE 5.11 Cross-section of a beam power tetrode as viewed from the top.

5.1.5 Vacuum Tube Specifications

Data sheets for vacuum tubes typically provide the information shown in Table 5.1 for a hypothetical low-power beam power tube. Some of the parameters and operating conditions specified in the table have yet to be addressed, and these matters are discussed subsequently.

5.1.6 Biasing Methods to Establish a DC Operating Point

As is the case with any electronic active device that is used in amplifier service, vacuum tubes must be biased for a DC operating point (Q point). Figure 5.12 illustrates this process graphically for a triode when the plate supply voltage is 400 V and the grid potential is −8 V. The DC load line is $R_L = (400 \text{ V})/(80 \text{ mA}) = 5 \text{ k}\Omega$.

In fixed-bias operation, the control grid is supplied from a negative fixed-voltage bias supply (historically called the C supply). The plate is operated from a high-voltage positive supply. The circuit diagram for a resistor–capacitor- (RC-) coupled triode audio amplifier operating under fixed-bias conditions is shown in Fig. 5.13(a). The DC load line is achieved by using a 5-kΩ resistor in the plate circuit. The cathode is connected directly to ground, and the control grid is connected through a series resistor to the negative 8-V bias supply. The value of the grid resistor is generally not critical. It is inserted simply to isolate the applied input signal from the bias supply. Based on the Q-point location in Fig. 5.12, the plate current is approximately 28 mA and the voltage at the plate is approximately 260 V. Thus the Q-point parameters are: $V_G = -8$ V, $V_P = 260$ V, $I_P = 28$ mA. Based on the data (not included here) used to generate the characteristic curves, $\mu = 16$, $g_m = 4.5$ mS, and $r_p = 3550 \ \Omega$.

Figure 5.13(b) illustrates the same audio amplifier configured for self-bias operation. The IR drop across a cathode resistor is used to place the cathode at the positive value of the grid potential above ground, whereas the grid is returned to ground through a resistor. Traditionally the grid resistor is called the *grid-leak* resistor, because its function is to bleed away from the grid any static charge that might build up

TABLE 5.1   Operating Specifications for Hypothetical Beam Power Pentode Single-Tube Operation

General:		
Heater voltage (AC/DC), V	6.3	
Heater current, A	0.9	
Direct interelectrode capacitances		
Control grid to plate, pF	0.4	
Control grid to cathode, pF	10	
Plate to cathode, heater, screen, suppressor, pF	12	
Maximum Ratings:		
Plate voltage, V	400	
Screen voltage, V	300	
Plate dissipation, W	20	
Screen-grid input, W	2.5	
Peak heater-to-cathode voltage (positive or negative), V	180	
Typical Operating Conditions (class A amplifier):		
	Fixed bias	Cathode bias
Plate supply voltage, V	350	300
Screen supply voltage, V	250	200
Control-grid voltage, V	−18	—
Cathode-bias resistor, Ω	—	220
Peak audio-frequency control-grid voltage, V	18	13
Zero-signal plate current, mA	54	51
Maximum-signal plate current, mA	66	55
Zero-signal screen-grid current, mA	2.5	3
Maximum-signal screen-grid current, mA	7	4.5
Plate resistance, Ω	33000	—
Transconductance, mS	5.2	
Load resistance, Ω	4200	4500
Total harmonic distortion, %	15	10
Maximum-signal power output, W	11	7
Maximum Ratings (Triode Connection: Screen Grid Connected to Plate):		
Plate voltage, V		300
Plate dissipation, W		20
Peak heater-to-cathode voltage (positive or negative), V		180
Typical Operating Conditions (Class A Amplifier):		
	Fixed bias	Cathode bias
Plate supply voltage, V	250	250
Control-grid voltage, V	−20	—
Cathode-bias resistor, Ω	—	470
Peak audio-frequency control-grid voltage, V	20	20
Zero-signal plate current, mA	40	40
Maximum-signal plate current, mA	45	43
Plate resistance, Ω	1700	—
Amplification factor	8	—
Transconductance, mS	4.7	
Load resistance, Ω	5000	6000
Total harmonic distortion, %	5	6
Maximum-signal power output, W	1.4	1.3

on the grid structure. Since in normal operation the grid draws no current, the grid voltage rests at ground potential or zero volts. The cathode is biased above ground to provide the required value of grid-to-cathode potential. In the example case, if the cathode is at +8 V and the grid at 0 V, the effective grid-to-cathode voltage is −8 V. The value of the cathode resistor is calculated as $(8\ \text{V})/(28\ \text{mA}) = 286\ \Omega$. The closest standard value is 270 Ω, as is shown in Fig. 5.13(b). To achieve maximum signal gain from the amplifier, the cathode resistor is bypassed by an appropriate value of capacitance.

A graphical approach may also be applied to evaluate the amplification properties of the circuits shown in Fig. 5.13. The AC load line (R_l) for the circuits shown is the parallel combination of the 5-kΩ plate

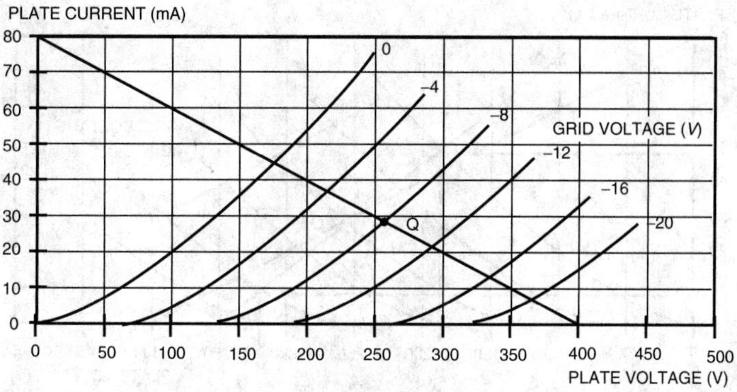

FIGURE 5.12 Triode plate characteristics with superimposed DC load line for Q-point calculations.

resistor and the external signal load. Let us assume that $R_l = 3.3\ \text{k}\Omega$ and we now add it to Fig. 5.12 with the new plot shown in Fig. 5.14. The AC and DC load line intersect at the Q point. The amplifier gain is evaluated from the plot as follows. Apply to the grid a sinusoidal signal (v_i) with a peak value of 8 V. Thus, the grid voltage will swing from the Q-point value of -8 V to 0 V and -16 V, as indicated by the double-headed arrow. The corresponding plate voltages are indicated by the vertical arrows. When $v_i = +8$ V, $v_g = 0$ V and the plate voltage is ≈ 190 V; when $v_i = -8$ V, $v_g = -16$ V and the plate voltage

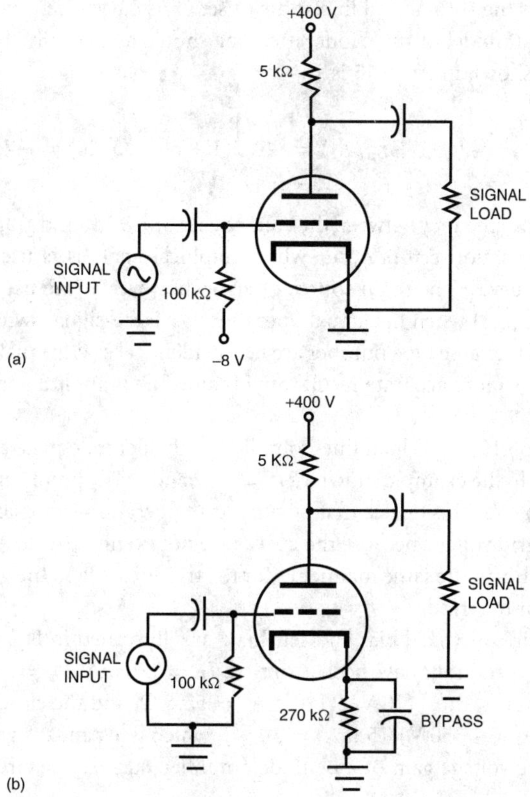

FIGURE 5.13 Triode audio amplifier circuits: (a) audio amplifier with fixed bias, (b) audio amplifier with cathode bias (self-bias).

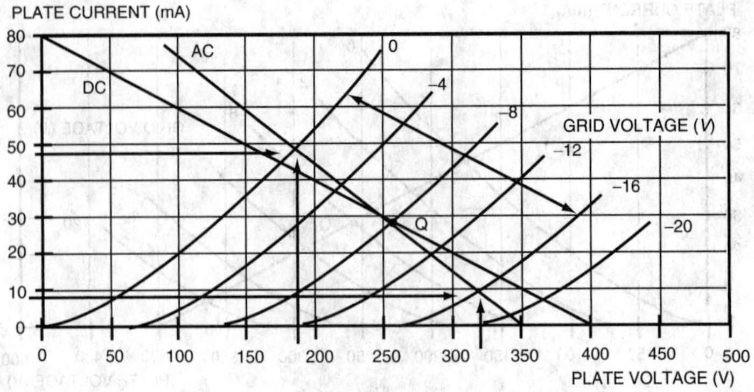

FIGURE 5.14 Load-line constructions to evaluate amplifier voltage gain.

is $\approx$320 V. Note that plate voltage decreases when grid voltage increases, thus the output signal is 180° out of phase with the input signal. The voltage gain is, thus, $A_v = -(320 - 190)/16 = -8.1$.

The output current has the same phase as the grid voltage, and the peak values are indicated by the horizontal arrows. When the grid voltage is 0 V, the plate current is $\approx$49 mA, and when the grid voltage is -16 V, the plate current is $\approx$9 mA. Note that the output-voltage swing about the Q point is not quite symmetric. The Q-point value of the plate voltage is $\approx$260 V. The positive swing in plate voltage is $320 - 260 = 60$ V, and the negative swing is $260 - 190 = 70$ V. This asymmetry is caused by the fact that the AC load line intersects the -16-V grid line at the onset of its curvature.

Based on a small-signal model of the triode, it can be shown analytically that the voltage gain of the amplifier configurations shown in Fig. 5.13 is

$$A_v = -\mu/(1 + r_p/R_l) = -16/(1 + 3550/3300) = -7.7 \qquad (5.6)$$

Note that there is a slight discrepancy between the voltage gain of -8.1 found graphically and the calculated value of -7.7. Some estimation errors occur when graphical analysis is used, and the solution is to use expanded graphical curves. The main source of error, however, is the use of a small-signal analysis result as predicted by Eq. (5.6) when the actual operation is a large-signal swing along the AC load line. Because vacuum tubes of the same type number are nearly identical and the plate characteristic curves are meaningful, in many cases more accurate results are obtained by graphical analysis than by small-signal calculations.

As is illustrated in Fig. 5.15, a DC load line is applied to the pentode plate characteristics in the same manner as for the triode. In the example shown, $R_L = 1$ kΩ and the Q-point values are $V_P = 300$ V, $I_P = 200$ mA, $V_G \approx -9$ V. Note for this device that positive-grid operation is possible and that nonlinearities occur for large negative grid voltage because the curves are not evenly spaced. An AC load is added to the pentode plate characteristics in the same manner as for the triode case, and the graphical analysis proceeds in the same manner as for the triode.

Two audio amplifier circuits (fixed bias and self-bias) are illustrated in Fig. 5.16. In the self-bias case, Fig. 5.16(b), the Q-point conditions have been assumed: $I_P = 180$ mA, $V_G = -10$ V, $I_{SG} = 5$ mA, $V_{SG} = 350$ V. The cathode resistor is, thus, $(10 \text{ V})/(180 \text{ mA}) = 55.6$ Ω, and the closest standard value is 56 Ω. The screen resistor is $(500 \text{ V} - 350 \text{ V})/(5 \text{ mA}) = 30$ kΩ, which is a standard value.

At low frequencies, the voltage gain of a pentode amplifier, such as the circuits shown in Fig. 5.16 is given by

$$A_v = -g_m R_l \qquad (5.7)$$

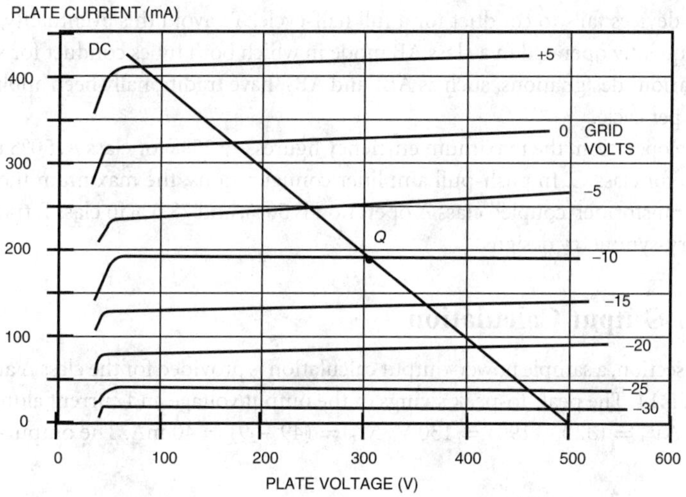

FIGURE 5.15 Pentode plate characteristics with superimposed DC load line for Q-point calculations.

5.1.7 Power Amplifier Classes

Power amplifiers may be divided into various classes, designated by letters of the alphabet, according to how long the power grid tube conducts during one complete cycle of a sinusoidal input signal. Classes A, B, and C apply to analog operation, whereas additional designations (D–S) apply to pulsed operation. Pulsed-mode classes are addressed in other chapters.

In class A operation (sometimes designated as A_1), the tube is biased as in Fig. 5.12 and Fig. 5.15, and conduction occurs for a full cycle. Because the tube is always on and drawing quiescent power ($P_Q = V_{PQ}I_{PQ}$), operating efficiency is low. In class B operation, the tube is biased such that it conducts for exactly one-half cycle. In class C operation, the tube is biased such that it conducts for less than one-half cycle but more than one-quarter of a cycle. Class C is the most efficient mode of these classes, and it is used in radio-telephony power amplifiers in which the associated high-Q circuits restore the missing portions of each cycle.

Class B operation is adopted for push-pull amplifier design in which the two tubes conduct during alternate half-cycles, thus reproducing the complete input signal. Harmonic distortion is introduced if

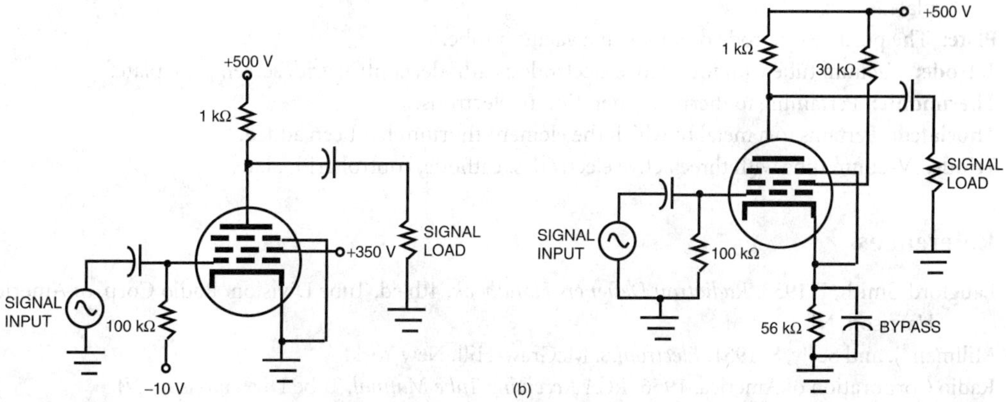

FIGURE 5.16 Pentode audio amplifier circuits: (a) audio amplifier with fixed bias, (b) audio amplifier cathode bias (self-bias).

either of the two devices fails to conduct for a full half-cycle. To avoid this eventuality, push–pull power amplifiers are frequently operated in a class AB mode in which both tubes conduct for slightly more than one-half cycle. Various designations, such as AB_1 and AB_2, have traditionally been applied to indicate the conduction time per cycle.

For single-tube operation, the maximum efficiency figures are 25% for class A, 50% for class B, and on the order of 70% for class C. In push-pull amplifier configurations, the maximum theoretical efficiency obtainable with transformer-coupled class A operation is 50%, and 78.5% in class B transformer-coupled or complementary-symmetry designs.

5.1.8 Power-Output Calculation

To complete this section, a sample power-output calculation is provided for the class A amplifier operation described by Fig. 5.14. The peak-to-peak swings of the output voltage and current along the AC load line are, respectively, $\Delta v_p = (320 - 190) = 130$ V, $\Delta i_p = (49 - 9) = 40$ mA. The output-signal power (P_0) is, thus

$$P_0 = \Delta v_p \Delta i_p / 8 = 0.125 \times 130 \text{ V} \times 0.04 \text{ A} = 650 \text{ mW}$$

In power triodes, output distortion is primarily generated by a second harmonic component, and the percentage distortion may be calculated from the empirical relation

$$\% \text{ distortion} = 100 \times [0.5(i_{p\,min} + i_{p\,max}) - I_{PQ}]/\Delta i_p \qquad (5.8)$$

In our example, $I_{PQ} = 28$ mA $= 0.028$ A, $i_{p\,min} = 9$ mA $= 0.009$ A, $i_{p\,max} = 49$ mA $= 0.049$ A, $\Delta i_p = 40$ mA $= 0.04$ A, therefore

$$\% \text{ distortion} = 100 \times (0.5(0.058) - 0.028)/0.04 = 2.5\%$$

The quiescent (zero-signal) power (P_Q) is 260 V $\times$ 28 mA $= 7.28$ W. Thus the percent efficiency is

$$\% \text{ efficiency} = 100 \times P_0/P_Q = 100 \times 0.65/7.28 = 8.9\%$$

Defining Terms

Cathode: The heated electrode in a vacuum tube that emits electrons.
Emissivity: A measure of the degree of emission of thermionic electrons from a hot cathode or filament.
Mu: Dimensionless voltage amplification factor of a vacuum tube.
Pentode: Vacuum tube with five active electrodes: cathode, control grid, screen grid, suppressor grid, plate.
Plate: The positive electrode or anode in a vacuum tube.
Tetrode: Vacuum tube with four active electrodes: cathode, control grid, screen grid, plate.
Thermionic: Pertaining to thermal generation of electrons.
Thoriated: Pertains to a metal to which the element thorium has been added.
Triode: Vacuum tube with three active electrodes: cathode, control grid, plate.

References

Langford-Smith, F. 1953. *Radiotron Designers Handbook*, 4th ed. Tube Division, Radio Corp. of America, Harrison, NJ.
Millman, J. and Seely, S. 1951. *Electronics*. McGraw-Hill, New York.
Radio Corporation of America. 1956. *RCA Receiving Tube Manual*. Tube Div. Harrison, NJ.
Seely, S. 1956. *Electronic Engineering*. McGraw-Hill, New York.
Spangenberg, K.R. 1948. *Vacuum Tubes*. McGraw-Hill, New York.
Spangenberg, K.R. 1957. *Fundamentals of Electron Devices*. McGraw-Hill, New York.

Further Information

For additional information on the design, construction, and application of vacuum tubes to electronic circuits, the following books are recommended.

Chance, B., Hughes, V., MacNichol, E.F., Sayre, D., and Williams, F.C. 1949. *Waveforms*. McGraw-Hill, New York.

Elmore, W.C. and Sands, M. 1949. *Electronics, Experimental Techniques*. McGraw-Hill, New York.

Gray, T.S. 1954. *Applied Electronics*. Wiley, New York.

Millman, J. 1958. *Vacuum-Tube and Semiconductor Electronics*. McGraw-Hill, New York.

Seely, S. 1956. *Radio Electronics*. McGraw-Hill, New York.

Terman, F.E. 1955. *Electronic and Radio Engineering*. McGraw-Hill, New York.

Valley, G.E., Jr. and Wallman, H. 1948. *Vacuum Tube Amplifiers*. McGraw-Hill, New York.

5.2 Power Grid Tubes

Jerry C. Whitaker

5.2.1 Introduction

A power-grid tube is a device using the flow of free electrons in a vacuum to produce useful work.[1] It has an emitting surface (the cathode), one or more grids that control the flow of electrons, and an element that collects the electrons (the anode). As discussed in the previous chapter, power tubes can be separated into groups according to the number of electrodes (grids) they contain. The physical shape and location of the grids relative to the plate and cathode are the main factors that determine the *amplification factor* μ and other parameters of the device. The physical size and types of material used to construct the individual elements determines the power capability of the tube. A wide variety of tube designs are available to commercial and industrial users. By far the most common are triodes and tetrodes.

5.2.2 Vacuum Tube Design

Any particular power vacuum tube may be designed to meet a number of operating parameters, the most important of which are usually high operating efficiency and high-gain/bandwidth properties. Above all, the tube must be reliable and provide long operating life. The design of a new power tube is a lengthy process that involves computer-aided calculations and modeling. The design engineers must examine a laundry list of items, including

- *Cooling:* how the tube will dissipate heat generated during normal operation. A high-performance tube is of little value if it will not provide long life in typical applications. Design questions include whether the tube will be air cooled or water cooled, the number of fins the device will have, and the thickness and spacing of the fins.

- *Electro-optics:* how the internal elements line up to achieve the desired performance. A careful analysis must be made of what happens to the electrons in their paths from the cathode to the anode, including the expected power gain of the tube.

- *Operational parameters:* what the typical interelectrode capacitances will be, and the manufacturing tolerances that can be expected. This analysis includes: spacing variations between elements within the tube, the types of materials used in construction, the long-term stability of the internal elements, and the effects of thermal cycling.

---

[1] Portions of this chapter were adapted from Varian. 1982. *Care and Feeding of Power Grid Tubes*. Varian/Eimac, San Carlos, CA. Used with permission.

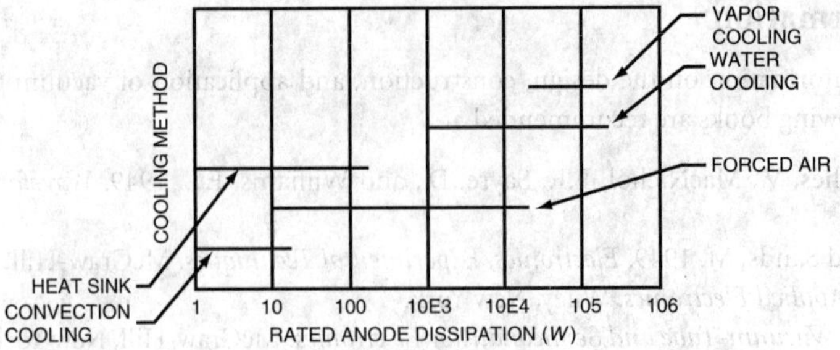

FIGURE 5.17 The relationship between anode dissipation and cooling method.

Device Cooling

The first factor that separates tube types is the method of cooling used: air, water, or vapor. *Air-cooled* tubes are common at power levels below 50 kW. A *water-cooling* system, although more complicated, is more effective than air cooling—by a factor of 5–10 or more—in transferring heat from the device. Air cooling at the 100-kW level is not normally used because it is virtually impossible to physically move sufficient air through the device (if the tube is to be of reasonable size) to keep the anode sufficiently cool. **Vapor cooling** provides an even more efficient method of cooling a power amplifier (PA) tube than water cooling. Naturally, the complexity of the external blowers, fans, ducts, plumbing, heat exchangers, and other hardware must be taken into consideration in the selection of a cooling method. Figure 5.17 shows how the choice of cooling method is related to anode dissipation.

Air Cooling

A typical air-cooling system for a transmitter is shown in Fig. 5.18. Cooling system performance for an air-cooled device is not necessarily related to airflow volume. The cooling capability of air is a function of its mass, not its volume. An appropriate airflow rate within the equipment is established by the manufacturer, resulting in a given resistance to air movement.

The altitude of operation is also a consideration in cooling system design. As altitude increases, the density (and cooling capability) of air decreases. To maintain the same cooling effectiveness, increased airflow must be provided.

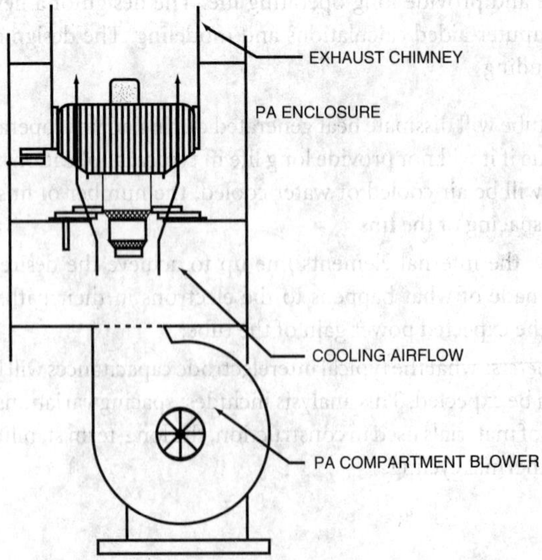

FIGURE 5.18 A typical transmitter PA stage cooling system.

Water Cooling

Water cooling is usually preferred over air cooling for power outputs above about 50 kW. Multiple grooves on the outside of the anode, in conjunction with a cylindrical jacket, force the cooling water to flow over the surface of the anode, as illustrated in Fig. 5.19.

Because the water is in contact with the outer surface of the anode, a high degree of purity must be maintained. A resistivity of 1 mΩ/cm (at 25°C) is typically specified by tube manufacturers. Circulating water can remove about 1 kW/cm² of effective internal anode area. In practice, the temperature of water leaving the tube must be limited to 70°C to prevent the possibility of spot boiling.

After leaving the anode, the heated water is passed through a heat exchanger where it is cooled to 30–40°C before being pumped back to the tube.

FIGURE 5.19   Water cooled anode with grooves for controlled water flow.

Vapor-Phase Cooling

Vapor cooling allows the permissible output temperature of the water to rise to the boiling point, giving higher cooling efficiency compared with water cooling. The benefits of vapor-phase cooling are the result of the physics of boiling water. Increasing the temperature of one gram of water from 40 to 70°C requires 30 cal of energy. However, transforming one gram of water at 100°C into steam vapor requires 540 cal. Thus, a vapor-phase cooling system permits essentially the same cooling capacity as water cooling, but with greatly reduced water flow. Viewed from another perspective, for the same water flow, the dissipation of the tube may be increased significantly (all other considerations being the same).

A typical vapor-phase cooling system is shown in Fig. 5.20. A tube incorporating a specially designed anode is immersed in a boiler filled with distilled water. When power is applied to the tube, anode dissipation heats the water to the boiling point, converting the water to steam vapor. The vapor passes to a condenser, where it gives up its energy and reverts to a liquid state. The condensate is then returned to the boiler, completing the cycle. Electric valves and interlocks are included in the system to provide for operating safety and maintenance. A vapor-phase cooling system for a transmitter with multiple PA tubes is shown in Fig. 5.21.

Special Applications

Power devices used for research applications must be designed for transient overloading, requiring special considerations with regard to cooling. Oil, heat pipes, refrigerants (such as Freon), and, where high-voltage holdoff is a problem, gases (such as sulfahexafluoride) are sometimes used to cool the anode of a power tube.

Cathode Assembly

The ultimate performance of any vacuum tube is determined by the accuracy of design and construction of the internal elements. The requirements for a successful tube include the ability to operate at high temperatures and withstand physical shock. Each element is critical to achieving this objective.

The cathode used in a power tube obtains the energy required for electron emission from heat. The cathode may be directly heated (filament type) or indirectly heated. The three types of emitting surfaces most commonly used are

- **Thoriated tungsten**
- Alkaline-earth oxides
- Tungsten barium aluminate-impregnated emitters

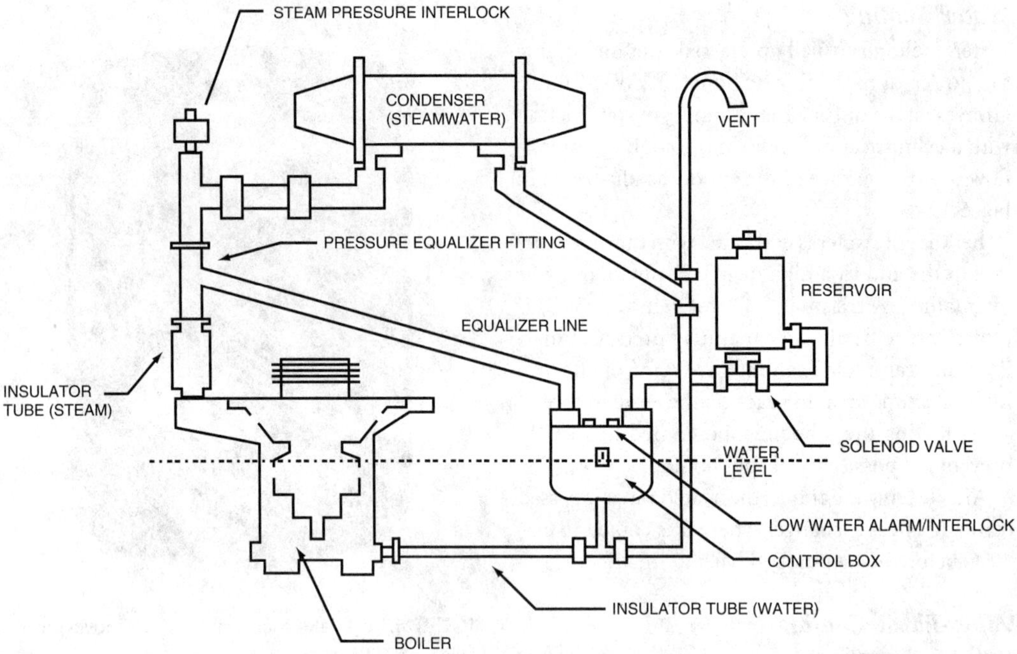

FIGURE 5.20   Typical vapor-phase cooling system.

The thoriated tungsten and tungsten-impregnated cathodes are preferred in power tube applications because they are more tolerant to *ion bombardment*. The characteristics of the three emitting surfaces are summarized in Table 5.2.

A variety of materials may be used as a source of electrons in a vacuum tube. Certain combinations of materials are preferred, however, for reasons of performance and economics.

Oxide Cathode

The conventional production-type oxide cathode consists of a coating of barium and strontium oxides on a base metal such as nickel. Nickel alloys, in general, are stronger, tougher, and harder than most nonferrous alloys and many steels. The most important property of nickel alloys is their ability to retain strength and toughness at elevated temperatures. The oxide layer is formed by first coating a nickel structure (a can or disc) with a mixture of barium and strontium carbonates, suspended in a binder material. The mixture is approximately 60% barium carbonate and 40% strontium carbonate.

During vacuum processing of the tube, these elements are *baked* at high temperatures. As the binder is burned away, the carbonates are subsequently reduced to oxides. The cathode is then *activated* and will emit electrons.

An oxide cathode operates CW at 700–820°C and is capable of an average emission density of 100–500 mA/cm^2. High-emission current capability is one of the main advantages of the oxide cathode. Other advantages include high peak emission for short pulses and low operating temperature. As shown in Table 5.2, peak emission of up to 20 A/cm^2 is possible from an oxide cathode. A typical device is shown in Fig. 5.22.

Although oxide-coated emitters provide more peak emission per watt of heating power than any other type, they are not without their drawbacks. Oxide emitters are more easily damaged or poisoned than other emitters, and also deteriorate more rapidly when subjected to bombardment of high-energy particles.

The oxide cathode material will evaporate during the life of the tube, causing free barium to migrate to other areas within the device. This evaporation can be minimized in the design stage by means of a high-efficiency cathode that runs as cool as possible but still is not emission limited at the desired heater

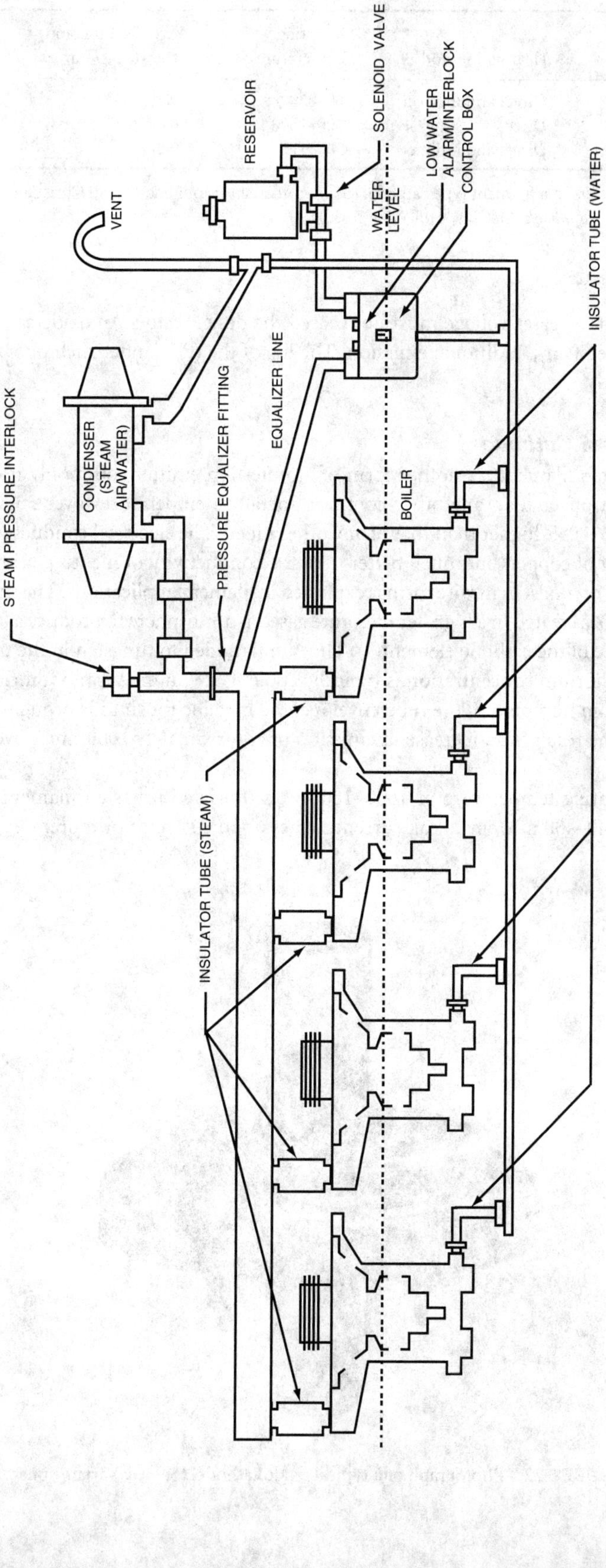

FIGURE 5.21 Vapour-phase cooling system for a 4-tube transmitter using a common waer supply.

TABLE 5.2   Characteristics of Common Thermonic Emitters

Emitter	Heating Method[a]	Operating Temperature, °C	Average Emission Density, A/cm^2	Peak Emission Density, A/cm^2
Oxide	Direct and indirect	700–820	0.10–0.50	0.10–20
Thoriated tungsten	Direct	1600–1800	0.04–0.43	0.04–10
Impregnated tungsten	Direct and indirect	900–1175	0.5–8.0	0.5–12

[a]Directly heated refers to a filament type cathode. (*Source:* Adapted from Fink, D. and Christiansen, D., eds. 1989. *Electronics Engineers' Handbook.* McGraw-Hill, New York.)

voltage. In the field, the heater voltage must not exceed the design value. An oxide cathode that is over-heated produces little, if any, additional emission. The life of the tube under such operation, however, is shortened significantly.

Thoriated Tungsten Cathode

The thoriated tungsten filament is another form of atomic-film emitter commonly used in power grid tubes. Tungsten is stronger than any other common metal at temperatures over 3500°F. The melting point of tungsten is 6170°F, higher than that of any other metal. The electrical conductivity of tungsten is approximately $\frac{1}{3}$ that of copper, but much better than the conductivity of nickel, platinum, or iron-base alloys. The resistivity of tungsten in wire form is exploited in filament applications. The thoriated-tungsten filament (or cathode) is created in a high-temperature gaseous atmosphere to produce a layer of *ditungsten carbide* on the surface of the cathode element(s). Thorium is added to tungsten in the process of making tungsten wire. The thorium concentration is typically about 1.5%, in the form of thoria. By proper processing during vacuum pumping of the tube envelope, the metallic thorium is brought to the surface of the filament wire. The result is an increase in emission of approximately 1000 times over a conventional cathode.

At a typical operating temperature of 1600–1800°C, a thoriated tungsten filament will produce an average emission of 40–430 mA/cm^2. Peak current ranges up to 10 A/cm^2 or more.

FIGURE 5.22   Photograph of a typical oxide cathode. (*Source:* Varian/Eimac.)

One of the advantages of a thoriated tungsten cathode over an oxide cathode is the ability to operate the plate at higher voltages. Oxide cathodes are susceptible to deterioration caused by ion bombardment. To achieve reasonable life, plate voltage must be limited. A thoriated tungsten cathode is more tolerant of ion bombardment, and so higher plate voltages can be safely applied.

The end of useful life for a thoriated tungsten tube occurs when most of the carbon has evaporated or has combined with residual gas, depleting the carbide surface layer. Theoretically, a 3% increase in filament voltage will result in a 20 K increase in cathode temperature, a 20% increase in peak emission, and a 50% decrease in tube life because of carbon loss. This cycle works in reverse, too. For a small decrease in temperature and peak emission, the life of the carbide layer—and hence, the tube—may be increased.

Tungsten-Impregnated Cathode

The tungsten-impregnated cathode typically operates at 900–1175°C and provides the highest average emission density of the three types of cathodes discussed in this chapter (500 mA/cm^2–8 A/cm^2). Peak power performance ranges up to 12 A/cm^2.

Tungsten as an element is better able than other emitters to withstand bombardment by high-energy positive ions without having its emission impaired. These positive ions are always present in small numbers in vacuum tubes as a result of ionization by collision with the residual gas.

Cathode Construction

Power tube filaments can be assembled in several different configurations. Figure 5.23 shows a spiral-type filament, and Fig. 5.24 shows a bar-type design. The spiral filament is used extensively in low-power tubes. As the size of the tube increases, mechanical considerations dictate a bar-type filament with spring loading to compensate for thermal expansion. A mesh filament can be used for both small and large tubes. It is

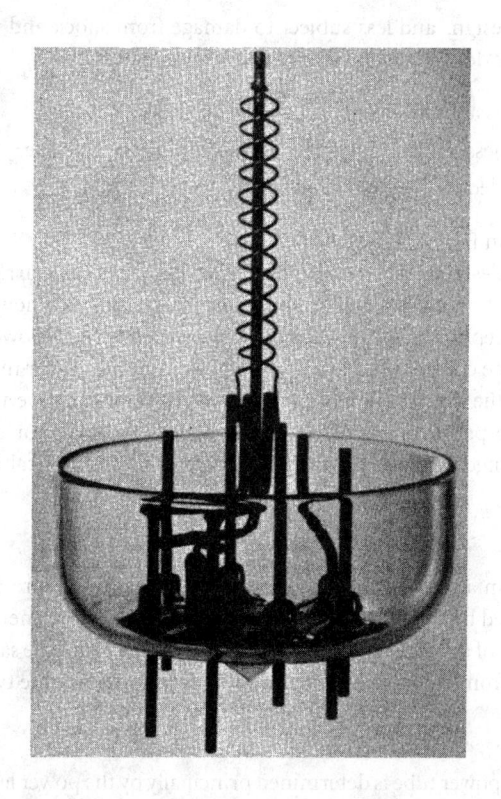

FIGURE 5.23 Spiral-type tungsten filament. (*Source:* Varian/Eimac.)

FIGURE 5.24   Bar-type tungsten filament. (*Source:* Varian/Eimac.)

more rugged than other designs and less subject to damage from shock and vibration. The rigidity of a cylindrical mesh cathode is determined by the following parameters:

- The diameter of the cathode
- The number, thickness, and length of the wires forming the cathode
- The fraction of welded to total wire crossings

A mesh cathode is shown in Fig. 5.25.

Some power tubes are designed as a series of electron gun structures arranged in a cylinder around a centerline. This construction allows large amounts of plate current to flow and to be controlled with a minimum of grid interception. With reduced grid interception, less power is dissipated in the grid structures. In the case of the control grid, less driving power is required for the tube.

In certain applications, the construction of the filament assembly can have an effect on the performance of the tube itself, and the performance of the RF system as a whole. For example, filaments built in a basket-weave mesh arrangement usually offer lower distortion in critical high-level AM modulation circuits.

Velocity of Emission

The electrons emitted from a hot cathode depart with a velocity that represents the difference between the kinetic energy possessed by the electron just before emission and the energy that must be given up to escape. Because the energy of different electrons within the emitter is not the same, the velocity of emission will vary as well, ranging from zero to a maximum value determined by the type and style of emitter.

Grid Structures

The type of grid used for a power tube is determined principally by the power level and operating frequency required. For most medium-power tubes (5–25 kW dissipation) welded wire construction is common.

FIGURE 5.25   Mesh tungsten filament. (*Source:* Varian/Eimac.)

At higher power levels, laser-cut **pyrolytic** graphite **grids** may be found. The grid structures of a power tube must maintain their shape and spacing at elevated temperatures. They must also withstand shock and vibration.

Wire Grids

Conventional wire grids are prepared by operators that wind the assemblies using special *mandrels* (forms) that include the required outline of the finished grid. The operators spot weld the wires at intersecting points, shown in Fig. 5.26. Most grids of this type are made with tungsten or *molybdenum*, which exhibit stable physical properties at elevated temperatures. On a strength basis, pure molybdenum is generally considered the most suitable of all refractory metals at temperatures between 1600 and 3000°F. The thermal conductivity of molybdenum is more than three times that of iron and almost half that of copper.

Grids for higher power tubes are typically built using a bar-cage type of construction. A number of vertical supports are fastened to a metal ring at the top and to a base cone at the bottom. The lower end of the assembly is bonded to a contact ring. The construction of the ring, metal base cone, and cylindrical metal base give the assembly low lead inductance and low RF resistance.

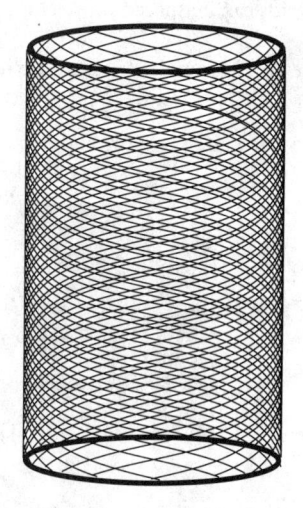

FIGURE 5.26   Mesh grid structure. (*Source:* Varian/Eimac.)

The external loading of a grid during operation and the proximity of the grid to the hot cathode impose severe demands on both the mechanical stability of the structure and the physical characteristics of its surface. The grid absorbs a high proportion of the heat radiated by the cathode. It also intercepts the electron beam, converting part of its kinetic energy into heat. Further, high-frequency capacitive currents flowing in the grid results in additional heat.

The end result is that grids are forced to work at temperatures as high as 1500°C. Their primary and secondary emission must, however, be low. To prevent grid emission, high electron affinity must be ensured throughout the life of the tube, even though it is impossible to prevent material evaporated from the cathode from contaminating the grid surface.

In tubes with oxide cathodes, grids made of tungsten or molybdenum wire are coated with gold to reduce primary emission caused by deposition. The maximum safe operating temperature for gold plating, however, is limited (about 550°C). Special coatings have, therefore, been developed for high-temperature applications that are effective in reducing grid emission. In tubes with thoriated tungsten cathodes, grids made of tungsten or molybdenum are coated with proprietary compounds to reduce primary emission.

Primary grid emission is usually low in a thoriated tungsten cathode device. In the case of an oxide cathode, however, free barium can evaporate from the cathode coating material and find its way to the control and screen grids. The rate of evaporation is a function of cathode temperature and time. A grid contaminated with barium will become another emitting surface. The hotter the grid, the greater the emissions.

Pyrolytic Grid

Pyrolytic grids are a high-performance alternative to wire or bar grid assemblies. Used primarily for high-power devices, pyrolytic grids are formed by laser-cutting a graphite cup of the proper dimensions. The computer-controlled laser cuts numerous holes in the cup to simulate a conventional-style grid. Figure 5.27 shows a typical pyrolytic-type grid before and after laser processing.

Pyrolytic (or oriented) graphite is a form of crystallized carbon produced by the decomposition of a hydrocarbon gas at high temperatures in a controlled environment. A layer of pyrolytic graphite is deposited on a special form. The thickness of the layer is proportional to the time that deposition is allowed to continue. The structure and mechanical properties of the deposited graphite depend on the imposed conditions.

Pyrolytic grids are ideal vacuum-tube elements because they do not expand like metal. Their small coefficient of expansion prevents movement of the grids inside the tube at elevated temperatures. This preserves the desired electrical characteristics of the device. Because tighter tolerances can be maintained, pyrolytic grids can be spaced more closely than conventional wire grids. Additional benefits include

- The grid is a single structure having no weld points.
- The grid has a thermal conductivity in two of the three planes nearly that of copper.
- The grid can operate at high temperatures with low vapor pressure.

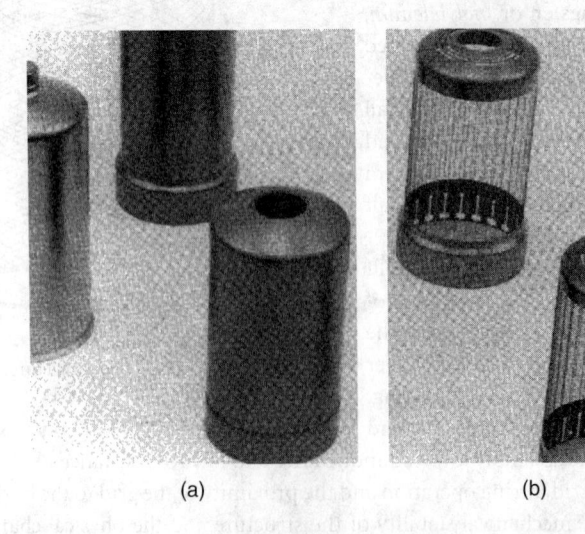

(a)   (b)

FIGURE 5.27   Pyrolytic graphite grid: (a) before laser processing, (b) completed assembly. (*Source:* Varian/Eimac.)

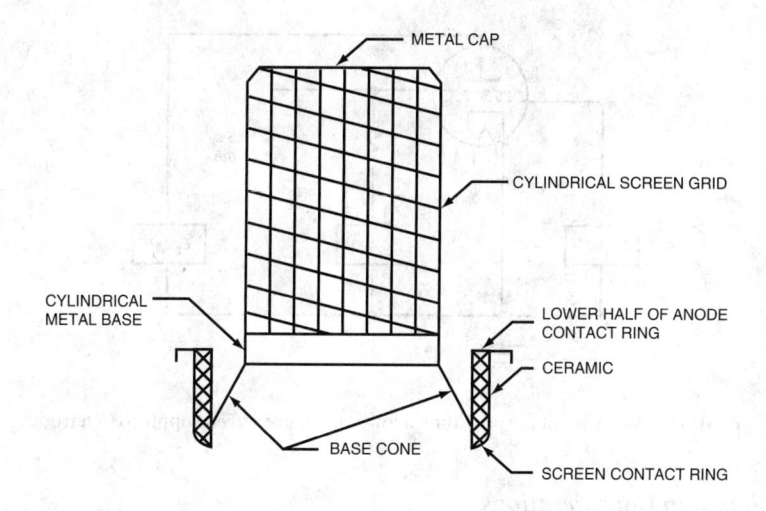

FIGURE 5.28 The screen grid assembly of a typical tetrode PA tube.

Grid Physical Structure

The control, screen, and suppressor grids are cylindrical and concentric. Each is slightly larger than the previous grid, as viewed from the cathode. Each is fastened to a metal base cone, the lower end of which is bonded to a contact ring. Figure 5.28 shows the construction of a typical screen grid assembly. Figure 5.29 illustrates a cutaway view of a tetrode power tube.

The shape of the control grid and its spacing from the cathode defines, in large part, the operating characteristics of the tube. For best performance, the grid must be essentially transparent to the electron path from the cathode to the plate. In a tetrode, the control and screen grids must be precisely aligned to minimize grid current. For pentode tubes, the previous two conventions apply, in addition to the requirement for precise alignment and minimum beam interception for the suppressor grid.

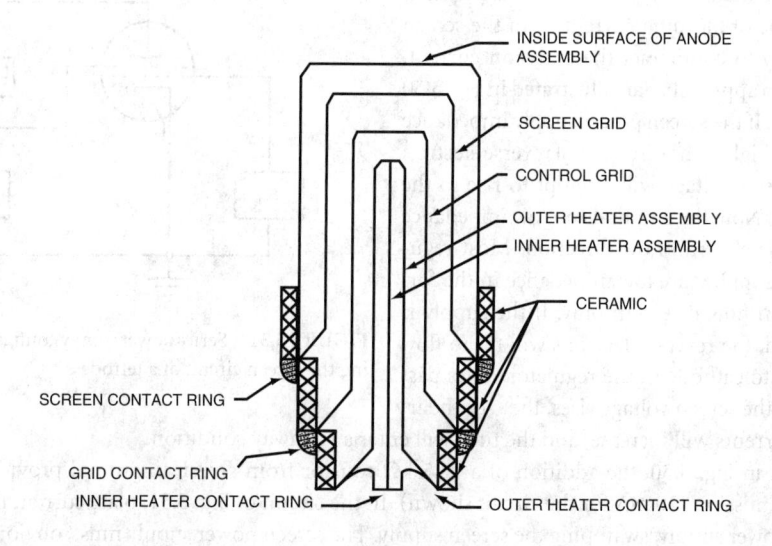

FIGURE 5.29 The internal arrangement of the anode, screen, control grid, and cathode assemblies of a tetrode power tube.

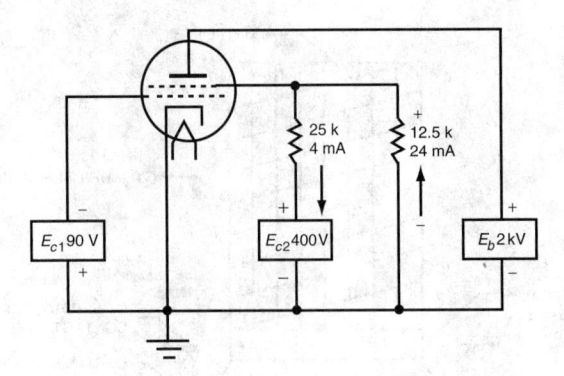

FIGURE 5.30   Circuit for providing a low impedance screen supply for a tetrode.

Secondary Emission Considerations

The relationship of the properties of secondary electrons to the grid structures and other elements of a vacuum tube must be carefully considered in any design. As the power capability of a tube increases, the physical size of the elements also increase. This raises the potential for secondary emission from the control, screen, and suppressor grids. Secondary emission can occur regardless of the type of cathode used. The yield of secondary electrons may be reduced through the application of surface treatments.

In a tetrode, the screen is operated at a relatively low potential, necessary to accelerate the electrons emitted from the cathode. Not all electrons pass through the screen on their way to the plate. Some are intercepted by the screen grid. As the electrons strike the screen, other low-energy electrons are emitted. If these electrons have a stronger attraction to the screen, they will fall back to that element. If, however, they pass into the region between the screen and the plate, the much higher anode potential will attract them. The result is electron flow from screen to plate.

Because of the physical construction of a tetrode, the control grid will have virtually no control over screen-to-plate current flow as a result of secondary electrons. During a portion of the operating cycle of the device, it is possible that more electrons will leave the screen grid than will arrive. The result will be a reverse electron flow on the screen element, a condition common to high-power tetrodes. A low-impedance path for reverse electron flow must be provided.

Tube manufacturers typically specify the recommended value of bleeding current from the screen power supply to counteract the emission current. Two common approaches are illustrated in Fig. 5.30 and Fig. 5.31. If the screen power supply impedance is excessively high in the direction of reverse electron flow, the screen voltage will attempt to rise to the plate voltage. Note the emphasis on low impedance in the reverse electron flow direction. Most regulated power supplies are low impedance in the forward electron flow direction only. If the supply is not well bled, the reverse electrons will try to flow from anode to cathode in the regulator series pass element. As the screen voltage rises, the secondary

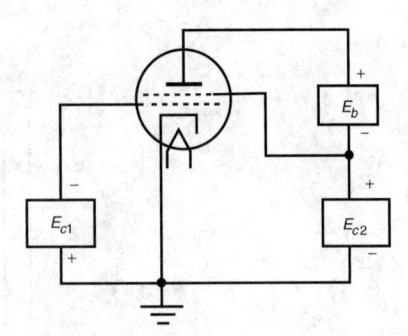

FIGURE 5.31   Series power supply configuration for swamping the screen circuit of a tetrode.

and plate currents will increase, and the tube will enter a runaway condition.

As shown in Fig. 5.30, the addition of a 12.5-kΩ resistor from screen to ground provides a path for screen grid emission (20 mA for the circuit shown). In the circuit of Fig. 5.31, plate current flows through the screen power supply, swamping the screen supply. The screen power supply must, obviously, carry the normal screen and plate currents. This scheme is used extensively in circuits where the screen is operated at DC ground potential. The plate-to-cathode voltage is then the sum of the E_b and E_{c2} power supplies.

The suppressor grid of a pentode reduces the effects of secondary emission in the device. This attribute, thus, reduces the requirement to provide a reverse electron flow path for the screen grid power supply. The screen current requirement for a pentode may, however, be somewhat higher than for a tetrode of the same general characteristics.

The designer must also consider the impedance of the control grid circuit. Primary grid emission can result in operational problems if the grid circuit impedance is excessively high. Primary grid emission, in the case of an oxide cathode tube, will increase with tube life.

The size and power of gridded tubes dictate certain characteristics of electrical potential. As this geometry increases in electrical terms, significant secondary emission from the control grid can occur. Control grid secondary emission can exist whether the cathode is a thoriated tungsten or oxide emitter, and can occur in a triode, tetrode, or pentode. A typical curve of grid current as a function of grid voltage for a high-power thoriated tungsten filament tetrode is shown in Fig. 5.32. As shown in the figure, grid current decreases and eventually takes a reverse direction as the grid voltage increases. This reduction and reversal of grid current can be explained by the normal secondary emission characteristics of the metals used in the grid structure. The secondary emission characteristics of common metals are presented in curve form in Fig. 5.33. The ratio of secondary-to-primary electron current is given as a function of primary electron potential. An examination of the chart will show the region between 200 and 600 V to be rather critical as far as secondary emission is concerned. Any power grid tube that normally operates with 200–600 V on the grid can exhibit the negative resistance characteristic of decreasing grid current with increasing grid voltage when another electrode, such as the anode in

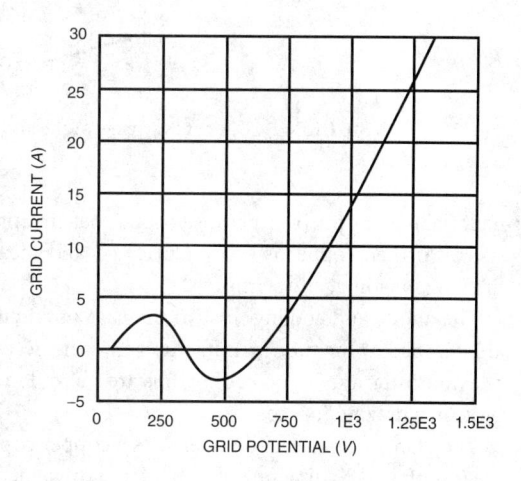

FIGURE 5.32 Typical curve of grid current as a function of control grid voltage for a high power thoriated tungsten filament tetrode.

a triode or the screen grid in a tetrode, is at a sufficiently high potential to attract the emitted electrons from the control grid. A driver stage that works into such a nonlinear load normally must be designed in such a manner as to tolerate this condition. One technique involves swamping the driver so that changes in load resulting from secondary grid emission is a small percentage of the total load the driver works into.

Plate Assembly

The plate assembly of a power tube is typically a collection of many smaller parts that are machined and assembled to tight specifications. Copper is generally used to construct the anode. It is an excellent

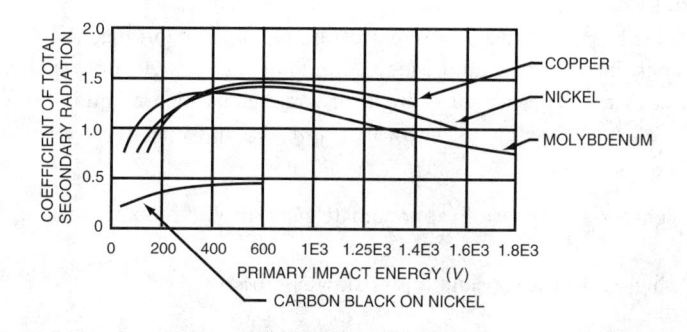

FIGURE 5.33 Secondary emission characteristics of various metals under ordinary conditions.

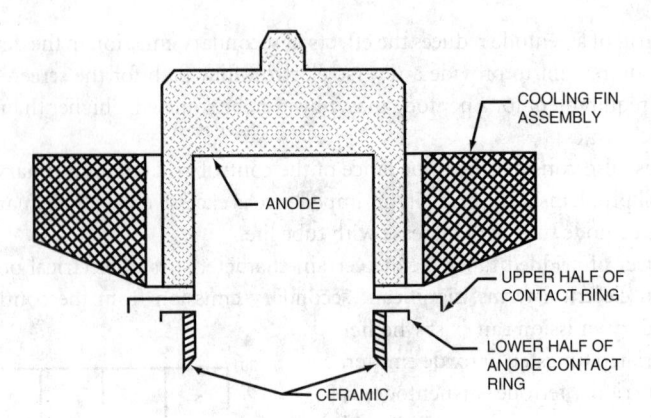

FIGURE 5.34   A cutaway view of the anode structure of an RF power amplifier tube.

material for this purpose because copper has the highest electrical conductivity of any metal except pure silver. Further, copper is easily fabricated and ideally suited to cold-forming operations such as deep drawing, bending, and stamping.

The anode and cooling fins (in the case of an air-cooled device) begin as flat sheets of copper. They are stamped by the tube manufacturer into the necessary sizes and shapes. After all of the parts have been machined, the anode and cooling fins are stacked in their proper positions, clamped, and brazed into one piece in a brazing furnace.

The plate of a power tube resembles a copper cup with the upper-half of a plate contact ring welded to the mouth and cooling fins silver soldered or welded to the outside of the assembly. The lower-half of the anode contact ring is bonded to a base ceramic spacer. At the time of assembly, the two halves of the ring are welded together to form a complete unit, as shown in Fig. 5.34.

In most power tubes, the anode is a part of the envelope, and, because the outer surface is external to the vacuum, it can be cooled directly. Glass envelopes were used in older power tubes. Most have been replaced, however, with devices that use ceramic as the envelope material.

Ceramic Elements

Ceramics are an integral part of modern power vacuum tubes. Three types of ceramics are in common usage in the production of vacuum devices:

- *Aluminum oxide*-based ceramics
- *Beryllium oxide*- (BeO-) based ceramics
- *Aluminum nitride*- (AIN-) based ceramics

Aluminum Oxide Ceramics

Aluminum oxide-based ceramic insulators are a common construction material for a wide variety of electrical components, including vacuum tubes. Aluminum oxide is 20 times higher in thermal conductivity than most oxides. The flexure strength of commercial high-alumina ceramics is 2–4 times greater than that of most oxide ceramics. There are drawbacks, however, to the use of alumina ceramics, including

- Relatively high thermal expansion (approximately 7 ppm/°C), compared to other ceramic materials, such as BeO
- Moderately high dielectric constant (approximately 10)

Aluminas are fabricated from aluminum oxide powders with various percentages of *sintering promoters,* which produce the so-called *glassy phase*. The later additives reduce the densification temperatures to between 1500 and 1600°C. Based on the final application, the powders may be pressed, extruded, or prepared in slurries for slip casting or tape casting. The surface finish of aluminas is typically 3–25 μm/in as a result of normal processing. For very smooth finishes (2 μm/in) the surfaces can be lapped or polished.

Alumina ceramics for vacuum tube applications are rarely used apart from being bonded to metals. The means by which this is accomplished frequently dictate the processing technique. Metallization of aluminas is usually accomplished by either high-temperature firing or low-temperature thick-film processing. Fired, shaped aluminas are usually postmetallized by refiring the formed article after coating with a slurry of molybdenum and manganese powder or tungsten metal powder. Based on the purity of the alumina, glass powder may be added to the metal powder. The mechanism of metallization requires an internal glassy phase or the added glassy phase for proper bonding. This is accomplished by firing in slightly reducing and moist atmospheres at temperatures above 1500°C. The resulting metallization is usually plated electrochemically with nickel or copper, or both.

Thick-film processing of alumina ceramics has traditionally been performed in oxidizing atmospheres at moderate temperatures (800–1000°C). Precious metals are used, such as gold, silver, platinum, palladium, and their combinations. A glassy phase is usually incorporated in the thick-film paste.

Each approach to metallization has advantages and disadvantages. The advantage of the high-temperature metallization schemes with molybdenum or tungsten is their moderate cost. The disadvantage is the high resistivity of the resulting films. These relative merits are reversed for thick-film materials. Another advantage of thick-film metallization is that the process is often applied by the device fabricator, not the ceramic vendor. This can allow for greater efficiency in design modification.

Beryllium Oxide Ceramics

Beryllium oxide-based ceramics are in many ways superior to alumina-based ceramics. The major drawback is the toxicity of BeO. Beryllium and its compounds are a group of materials that are potentially hazardous and must be handled properly. With the necessary safeguards, BeO has been used successfully in many tube designs.

Beryllium oxide materials are particularly attractive for use in power vacuum tubes because of their electrical, physical, and chemical properties. The thermal conductivity of BeO is approximately 10 times higher than that of alumina-based materials. Figure 5.35 compares the thermal conductivity of BeO to that of alumina and some alternative materials. As the chart illustrates, BeO has a lower dielectric constant and coefficient of thermal expansion than alumina. It is, however, also slightly lower in strength.

Beryllia materials are fabricated in much the same way as alumina compounds, although the toxic properties of the powders mandate that they be processed in laboratories equipped to handle them safely.

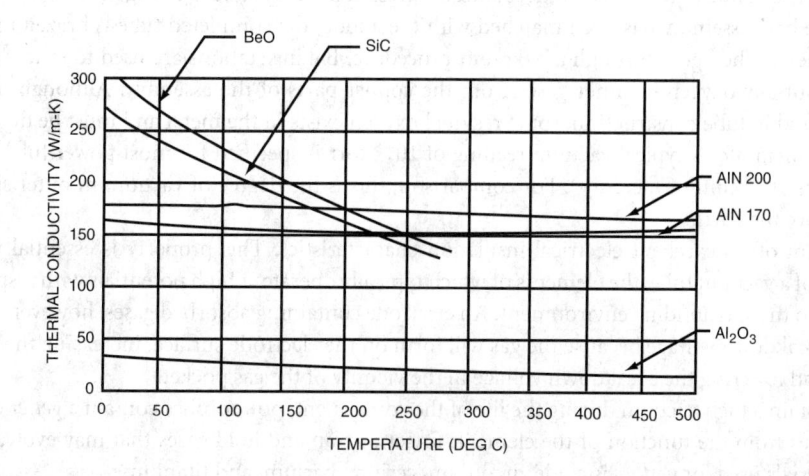

FIGURE 5.35 Thermal conductivity of common ceramics. (*Source:* Adapted from Harpter, C.A. 1991. *Electronic Packaging and Interconnection Handbook*, pp. 1–64. McGraw-Hill, New York.)

Simple cutting, drilling, and postmetallization are also handled by vendors with special equipment. Reliable thick-film systems may be applied to BeO substrates. Such coatings usually necessitate less elaborate safety precautions and are often applied by the device fabricator.

Other Ceramics

Aluminum nitride- (AlN-) based ceramics have been developed as an alternative to the toxicity concerns of BeO-based materials. As shown in Fig. 5.35, the thermal conductivity of AlN is comparable to that of BeO but deteriorates less with temperature. The dielectric constant of AlN is comparable to that of alumina (a drawback) but its thermal expansion is low (4 ppm/°C).

Other ceramic materials that find some use in vacuum devices include silicon carbide- (SiC-) based substances and boron nitride ceramics.

Tube Construction

Each type of power grid tube is unique insofar as its operating characteristics is concerned. The basic physical construction, however, is common to most devices. A vacuum tube is built in essentially two parts:

- The base, which includes the filament and supporting stem, control grid, screen grid, and suppressor grid (if used)
- The anode, which includes the heat-dissipating fins made in various machining steps

The base subassembly is welded using a *tungsten-inert gas* (TIG) process in an oxygen-free atmosphere (a process sometimes referred to as *Heliarc* welding) to produce a finished base unit.

The ceramic elements used in a vacuum tube are critical elements of the device. Assembled in sections, each element builds on the previous one to form the base of the tube. The ceramic-to-metal seals are created using a material that is *painted* onto the ceramic and then heated in a brazing oven. After preparation in a high-temperature oven, the painted area provides a metallic structure that is molecularly bonded to the ceramic, and provides a surface suitable for brazing.

This process requires temperature sequences that dictate completion of the highest temperature stages first. As the assembly takes form, lower oven temperatures are used so that completed bonds will not be damaged.

Despite all of the advantages that make ceramics one of the best insulators for a tube envelope, their brittleness is a potential cause of failure. The smallest cracks in the ceramic, not in themselves damaging, can cause the ceramic to break when mechanically stressed by temperature changes.

After the base assembly has been matched with the anode, the completed tube is brazed into a single unit. The device then goes through a **bake-out** procedure. Baking stations are used to evacuate the tube and bake out any oxygen or other gases from the copper parts of the assembly. Although oxygen-free copper is used in tube construction, some residual oxygen exists in the metal and must be driven out for long component life. A typical vacuum reading of 10^{-8} torr is specified for most power tubes (formerly expressed as 10^{-8} mm of mercury). For comparison, this is the degree of vacuum in outer space about 200 mi above the Earth.

A vacuum offers excellent electrical insulation characteristics. This property is essential for reliable operation of a vacuum tube, the elements of which typically operate at high potentials with respect to each other and to the surrounding environment. An electrode containing absorbed gases, however, will exhibit reduced breakdown voltage because the gas will form on the electrode surface, increasing the surface gas pressure and lowering the breakdown voltage in the vicinity of the gas pocket.

To maintain a high vacuum during the life of the component, power tubes contain a *getter* device. The name comes from the function of the element: to get or trap and hold gases that may evolve inside the tube. Materials used for getters include zirconium, cerium, barium, and titanium.

The operation of a vacuum tube is an evolving chemical process. End of life in a vacuum tube is generally caused by loss of emission.

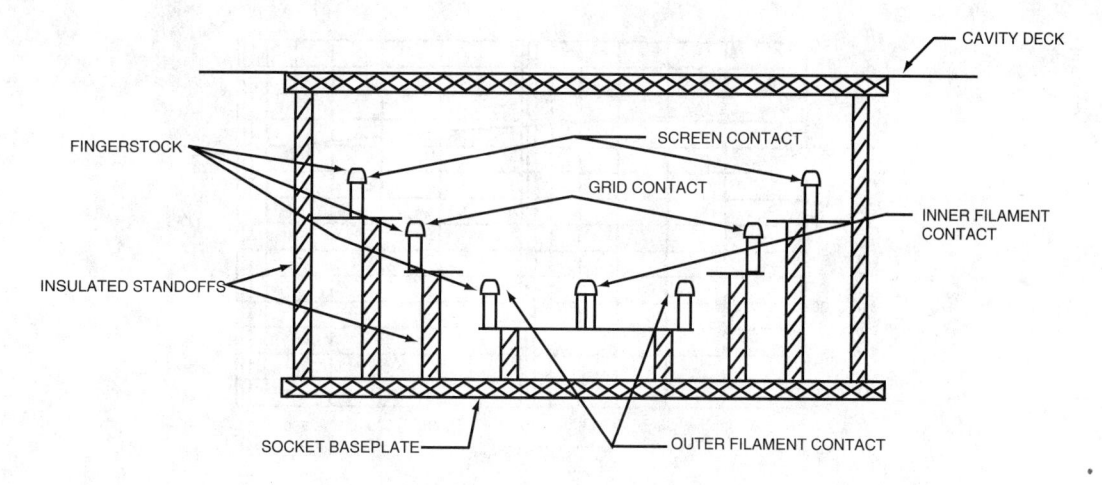

FIGURE 5.36 Cross section of the base of a tetrode showing the connection points. (*Source:* Varian/Eimac.)

Connection Points

The high-power levels and frequencies at which vacuum tubes operate place stringent demands on the connectors used to tie the outside world to the inside elements. Figure 5.36 shows a cutaway view of the base of a tetrode. Tubes are designed to be mounted vertically on their electrical connectors. The connectors provide a broad contact surface and mechanical support for the device.

The cathode and grids are mounted on ring-shaped *Kovar*™ bases, which also serve as contact rings for the external connections. Kovar is an iron-nickel-cobalt alloy whose coefficient of thermal expansion is comparable with that of aluminum oxide ceramic. The different diameters of the various contact rings allow them to be grouped coaxially. The concentric tube/connector design provides for operation at high frequencies. Conductivity is improved by silver plating.

Tube Sockets

Any one tube design may have several possible socket configurations, depending on the frequency of operation. If the tube terminals are large cylindrical surfaces, the contacting portions of the socket consist of either spring **collets** or an assembly of spring finger stock. Usually, these multiple-contacting surfaces are made of beryllium copper to preserve spring tension at the high temperatures present at the tube terminals. The fingers are silver plated to reduce RF resistance.

If the connecting fingers of a power-tube socket fail to provide adequate contact with the tube element rings, a concentration of RF currents will result. Depending on the extent of this concentration, damage may result to the socket. After a connecting finger loses its spring action, the heating effect is aggravated and tube damage is possible.

A series of specialized power tubes is available with no sockets at all. Intended for cathode-driven service, the grid assembly is formed into a flange that is bolted to the chassis. The filament leads are connected via studs on the bottom of the tube. Such a configuration completely eliminates the tube socket. This type of device is useful for low-frequency applications, such as induction heating.

5.2.3 High Frequency Operating Limits

Like most active devices, performance of a given vacuum tube deteriorates as the operating frequency is increased beyond its designed limit. **Electron transit time** is a significant factor in the upper frequency limitation of electron tubes. A finite time is taken by electrons to traverse the space from the cathode, through the grid, and on to the plate. As the operating frequency increases, a point is reached at which the electron transit time effects become significant. This point depends on the accelerating voltages at the grid and anode, and their respective spacings. Tubes with reduced spacing in the grid-to-cathode region exhibit reduced transit time effects.

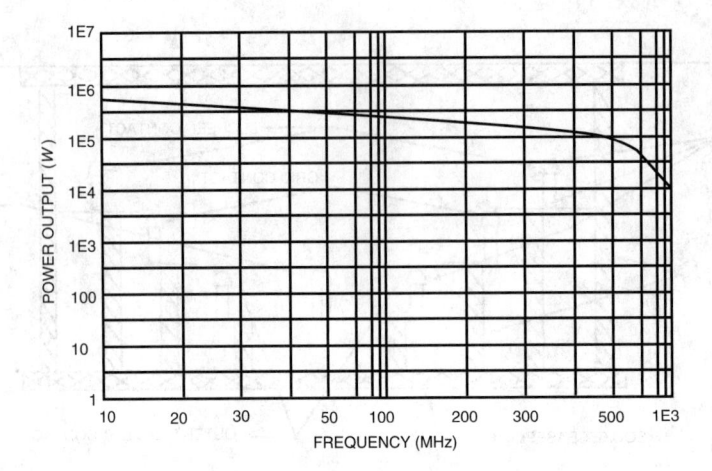

FIGURE 5.37   Continuous-wave output power capability of a gridded vacuum tube.

There is also a power limitation that is interrelated with the high-frequency limit of a device. As the operating frequency is increased, closer spacing and smaller-sized electrodes must be used. This reduces the power handling capability of the tube. Figure 5.37 illustrates the relationship.

Gridded tubes at all power levels for frequencies up to about 1 GHz are invariably cylindrical in form. At higher frequencies, planar construction is almost universal. As the operating frequency is increased beyond design limits, output power and efficiency both decrease. Figure 5.38 illustrates the relationship.

Transit Time Effects

When class C, class B, or similar operations are carried out at frequencies sufficiently high that the transit time of the electrons is not a negligible fraction of the waveform cycle, the following complications are observed in grid-based vacuum tubes:

- **Back heating** of the cathode
- Loading of the control grid circuit as a result of energy transferred to electrons that do not necessarily reach the grid to produce a DC grid current

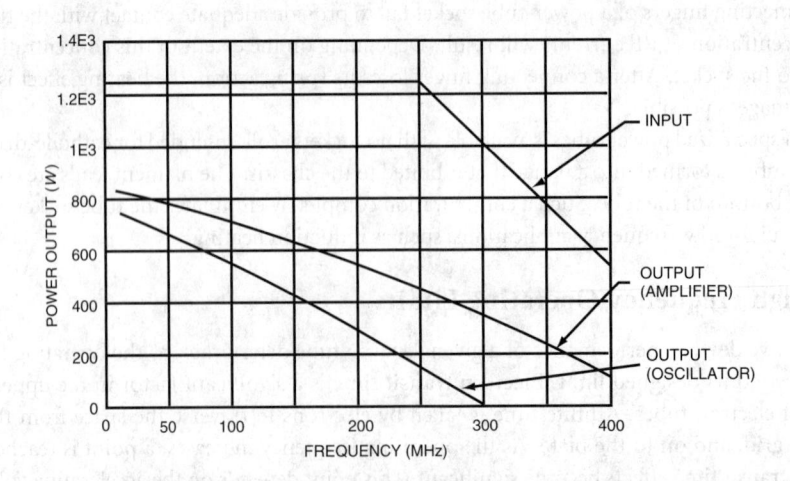

FIGURE 5.38   Performance of a class C amplifier as the operating frequency is increased beyond the design limits of the vacuum tube.

- Debunching of plate current pulses
- Phase differences between the plate current and the exciting voltage applied to the control grid

Back heating of the cathode occurs when the transit time in the grid-cathode space is sufficiently great to cause an appreciable number of electrons to be in transit at the instant the plate current pulse would be cut off in the case of a low-frequency operation. A considerable fraction of the electrons thus trapped in the interelectrode space are returned to the cathode by the negative field existing in the grid-cathode space during the cutoff portion of the cycle. These returning electrons act to heat the cathode. At very high frequencies this back heating is sufficient to supply a considerable fraction of the total cathode heating required for normal operation. Back heating may reduce the life of the cathode as a result of electron bombardment of the emitting surface. It also causes the required filament current to depend on the conditions of operation within the tube.

Energy absorbed by the control grid as a result of input loading is transferred directly to the electron stream in the tube. Part of this stream is then used by the electrons in producing back heating of the cathode. The remainder affects the velocity of the electrons as they arrive at the anode of the tube. This portion of the energy is not necessarily all wasted. In fact, a considerable percentage of it may, under favorable conditions, appear as useful output in the tube. To the extent that this is the case, the energy supplied by the exciting voltage to the electron stream is simply transferred directly from the control grid to the output circuits of the tube without amplification.

An examination of the total time required by electrons to travel from the cathode to the anode in a triode, tetrode, or pentode operated as a class C amplifier reveals that the resulting transit times for electrons at the beginning, middle, and end of the current pulse will differ as the operating frequency is increased. In general, electrons traversing the distance during the first segment of the pulse will have the shortest transit time, whereas those near the middle and end of the pulse will have the longest transit times, as illustrated in Fig. 5.39. The first electrons in the pulse have a short transit time because they approach the plate before the plate potential is at its minimum value. Electrons near the middle of the pulse approach the plate with the instantaneous plate potential at or near minimum and, consequently, travel less rapidly in the grid-plate space. Finally, those electrons that left the cathode late in the current pulse (those just able to escape being trapped in the control grid-cathode space and returned toward the cathode) will be slowed as they approach the grid, and so have a large transit time. The net effect is to cause the pulse of plate current to be longer than it would in operation at a low frequency. This causes the efficiency of the amplifier to drop at high frequencies, because a longer plate current pulse increases plate losses.

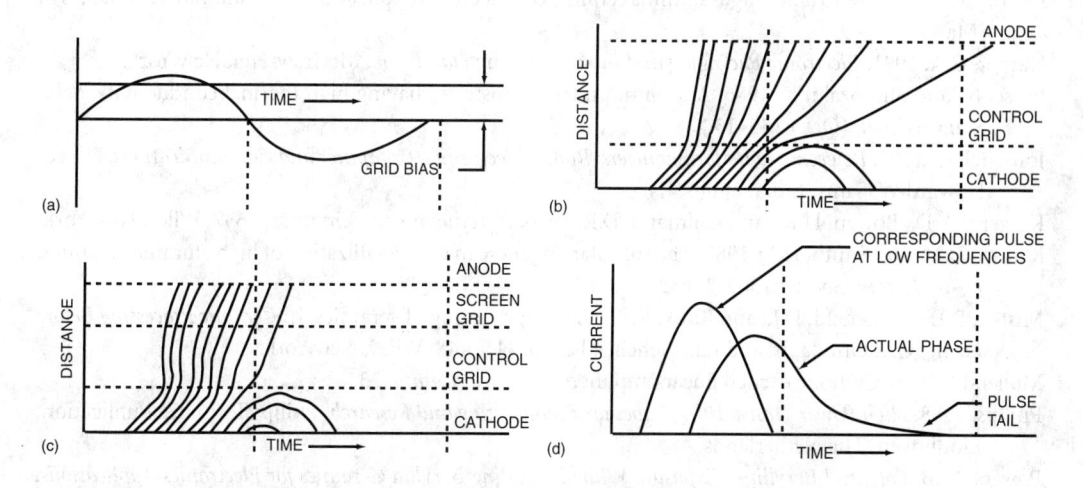

FIGURE 5.39 Transit time effects in a class C amplifier: (a) control grid voltage, (b) electron position as a function of time (triode case), (c) electron position as a function of time (tetrode case), (d) plate current (triode case). (*Source:* Adapted from Terman, F.E. 1947. *Radio Engineering*, p. 405. McGraw-Hill, New York.)

Defining Terms

Back heating: The effect on the cathode of a power vacuum tube when the transit time in the grid-cathode space is sufficiently great to cause, under certain conditions, an appreciable number of electrons to be trapped in the interelectrode space and returned to the cathode, thereby heating the cathode.

Bake-out: A process during tube manufacture in which the device is heated to a high temperature while the envelope is evacuated.

Collet: A circular spring fingerstock connection element for a power vacuum tube.

Electron transit time: The finite time taken by electrons to traverse the space from the cathode, through the grid(s), and on to the plate of a vacuum tube.

Pyrolytic grid: A grid structure made of pyrolytic (oriented) graphite that is laser-cut to the proper geometry for use in a power vacuum tube.

Thoriated tungsten: A form of atomic-film emitter commonly used in power tubes. Through proper processing, the hybrid material is made to provide high-emissions capabilities and long life.

Vapor cooling: A cooling technique for power vacuum tubes utilizing the conversion of hot water to steam as a means of safely conducting heat from the device and to a heat sink.

References

Block, R. 1988. CPS microengineers new breed of materials. *Ceramic Ind.* (April):51–53.

Brush. Ceramic products. Brush Wellman, Cleveland, OH.

Buchanan, R.C. 1986. *Ceramic Materials for Electronics.* Marcel Dekker, New York.

Coors. Coors ceramics: Materials for tough jobs. Coors Data Sheet K.P.G.-2500-2/87 6429.

Cote, R.E. and Bouchard, R.J. 1988. Thick film technology. In *Electronic Ceramics,* L.M. Levinson, ed., pp. 307–370. Marcel Dekker, New York.

CPL. 1984. Combat boron nitride, solids, powders, coatings. Carborundum Product Literature, Form A-14, 011, Sept.

Dettmer, E.S., Charles, H.K., Jr., Mobley, S.J., and Romenesko, B.M. 1988. Hybrid design and processing using aluminum nitride substrates, pp. 545–553. *ISHM 88 Proc.*

Dettmer, E.S. and Charles, H.K., Jr. 1989. AlN and SiC substrate properties and processing characteristics. *Advances in Ceramics,* Vol. 31. Am. Ceramic Soc., Columbus, OH.

Fink, D. and Christiansen, D., eds. 1989. *Electronics Engineer's Handbook,* 3rd ed. McGraw-Hill, New York.

Floyd, J.R. 1969. How to tailor high-alumina ceramics for electrical applications. *Ceramic Ind.* (Feb.):44–77, (March):46–49.

Harper, C.A. 1991. *Electronic Packaging and Interconnecting Handbook.* McGraw-Hill, New York.

Iwase, N. and Shinozaki, K. 1986. Aluminum nitride substrates having high thermal conductivity. *Solid State Technol.* (Oct.):135–137.

Jordan, E.C. 1985. *Reference Data for Engineers: Radio, Electronics, Computer and Communications,* 7th ed. Howard W. Sams, Indianapolis, IN.

Kingery, W.D., Bowen, H.K., and Uhlmann, D.R. 1976. *Introduction to Ceramics,* p. 637. Wiley, New York.

Mattox, D.M. and Smith, H.D. 1985. The role of manganese in the metallization of high alumina ceramics. *J. Am. Ceram. Soc.* 64:1363–1369.

Mistler, R.E., Shanefield, D.J., and Runk, R.B. 1978. Tape casting of ceramics. In *Ceramic Processing Before Firing,* G.Y. Onoda, Jr. and L.L. Hench, eds., pp. 411–448. Wiley, New York.

Muller, J.J. 1946. Cathode excited linear amplifiers. *Electrical Comm.* 23.

Philips. 1988. *High Power Transmitting Tubes for Broadcasting and Research.* Philips Technical Publication, Eindhoven, The Netherlands.

Powers, M.B. *Potential Beryllium Exposure While Processing Beryllia Ceramics for Electronics Applications.* Brush Wellman, Cleveland, OH.

Sawhill, H.T., Eustice, A.L., Horowitz, S.J., Gar-El, J., and Travis, A.R. 1986. Low temperature co-fireable ceramics with co-fired resistors. pp. 173–180. In *Proc. Int. Symp. on Microelectronics.*

Schwartz, B. 1988. Ceramic packaging of integrated circuits. In *Electronic Ceramics,* L.M. Levinson, ed., pp. 1–44. Marcel Dekker, New York.

Terman, F.E. 1947. *Radio Engineering,* 3rd ed. McGraw-Hill, New York.

Varian. 1982. *Care and Feeding of Power Grid Tubes.* Laboratory Staff, Varian/Eimac, San Carlos, CA.

Whitaker, J.C. 1991. *Radio Frequency Transmission Systems: Design and Operation.* McGraw-Hill, New York.

Whitaker, J.C. 1999. *Power Vacuum Tubes Handbook,* 2nd ed. CRC Press, Boca Raton, FL.

Further Information

Specific information on the application of power vacuum tubes can be obtained from the manufacturers of those devices. More general application information can be found in the following publications:

Whitaker, J.C. 1999. *Power Vacuum Tubes Handbook,* 2nd ed. CRC Press, Boca Raton, FL.

Golio, M. 2000. *RF and Microwave Handbook,* CRC Press, Boca Raton, FL.

5.3   Neutralization Techniques

Jerry C. Whitaker

5.3.1   Introduction

An RF power amplifier must be properly neutralized to provide acceptable performance in most applications.[2] The means to accomplish this end can vary considerably from one design to another. An RF amplifier is neutralized when two conditions are met:

- The **interelectrode capacitance** between the input and output circuits is canceled.
- The inductance of the screen grid and cathode assemblies (in a tetrode) is canceled.

Cancellation of these common forms of coupling between the input and output circuits of vacuum tube amplifiers prevents self-oscillation and the generation of spurious products.

5.3.2   Circuit Analysis

Figure 5.40 illustrates the primary elements that effect **neutralization** of a vacuum tube RF amplifier operating in the VHF band. (Many of the following principles also apply to lower frequencies.) The feedback elements include the residual grid-to-plate capacitance C_{gp}, plate-to-screen capacitance C_{ps}, and screen grid lead inductance L_s. The RF energy developed in the plate circuit E_p causes a current I to flow through the plate-to-screen capacitance and the screen lead inductance. The current through the screen inductance develops a voltage $-E$ with a polarity opposite that of the plate voltage E_p. The $-E$ potential is often used as a method of neutralizing tetrode and pentode tubes operating in the VHF band.

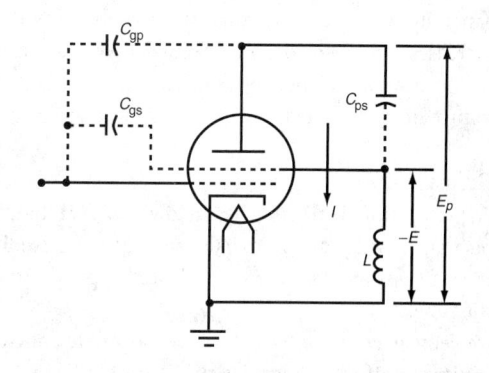

FIGURE 5.40   The elements involved in the neutralization of a tetrode PA stage.

Figure 5.41 graphically illustrates the electrical properties at work. The circuit elements of the previous figure have been arranged so that the height above or below the zero potential line represents magnitude

---

[2]Portions of this chapter were adapted from: Varian. 1982. *Care and Feeding of Power Grid Tubes.* Varian/Eimac, San Carlos, CA. Used with permission.

and polarity of the RF voltage for each part of the circuit with respect to ground (zero). For the purposes of this illustration, assume that all of the circuit elements involved are pure reactances. The voltages represented by each, therefore, are either in phase or out of phase and can be treated as positive or negative with respect to each other.

The voltages plotted in the figure represent those generated as a result of the RF output circuit voltage E_p. No attempt is made to illustrate the typical driving current on the grid of the tube. The plate P has a high positive potential above the zero line, established at the ground point. Keep in mind that the distance above the baseline represents increasing positive potential. The effect of the out-of-phase screen potential developed as a result of inductance L_s is shown, resulting in the generation of $-E$.

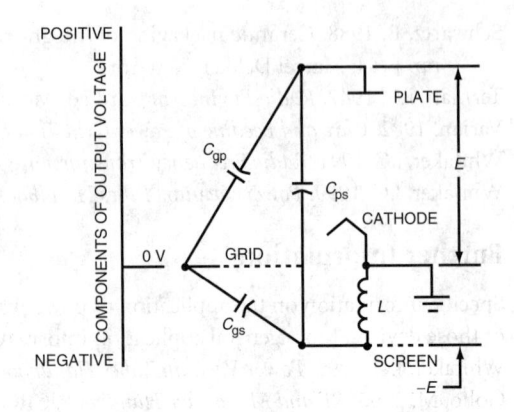

FIGURE 5.41 A graphical representation of the elements involved in the neutralization of a tetrode RF stage when self-neutralized.

As depicted, the figure constitutes a perfectly neutralized circuit. The grid potential rests at the zero baseline. The grid operates at filament potential insofar as any action of the output circuit on the input circuit is concerned.

The total RF voltage between plate and screen is made up of the plate potential and screen lead inductance voltage, $-E$. This total voltage is applied across a divider circuit that consists of the grid-to-plate capacitance and grid-to-screen capacitance (C_{gp} and C_{gs}). When this potential divider is properly matched for the values of plate RF voltage (E_p) and screen lead inductance voltage ($-E$), the control grid will exhibit zero voltage difference with respect to the filament as a result of E_p.

Circuit Design

There are a variety of methods that may be used to neutralize a vacuum tube amplifier. Generally speaking, a grounded-grid, cathode-driven triode can be operated into the VHF band without external neutralization components. The grounded-grid element is sufficient to prevent spurious oscillations. Tetrode amplifiers generally will operate through the MF band without neutralization. However, as the gain of the stage increases, the need to cancel feedback voltages caused by tube interelectrode capacitances and external connection inductances becomes more important. At VHF frequencies and above, it is generally necessary to provide some form of stage neutralization.

Below VHF

For operation at frequencies below the VHF band, neutralization for a tetrode typically employs a capacitance bridge circuit to balance out the RF feedback caused by residual plate-to-grid capacitance. This method assumes that the screen is well bypassed to ground, providing the expected screening action inside the tube.

Neutralization of low-power push-pull tetrode or pentode tubes can be accomplished with cross neutralization of the devices, as shown in Fig. 5.42. Small-value neutralization capacitors are used. In some cases, neutralization can be accomplished with a simple wire connected to each side of the grid circuit and brought through the chassis deck. Each wire is positioned to look at the plate of the tube on the opposite-half of the circuit. Typically, the wire (or a short rod) is spaced a short distance from the plate of each tube. Fine adjustment is accomplished by moving the conductor in or out from its respective tube.

A similar method of neutralization can be used for a cathode-driven symmetrical stage, as shown in Fig. 5.43. Note that the neutralization capacitors C_n are connected from the cathode of one tube to the plate of the opposite tube. The neutralizing capacitors have a value equal to the internal cathode-to-plate capacitance of the PA tubes.

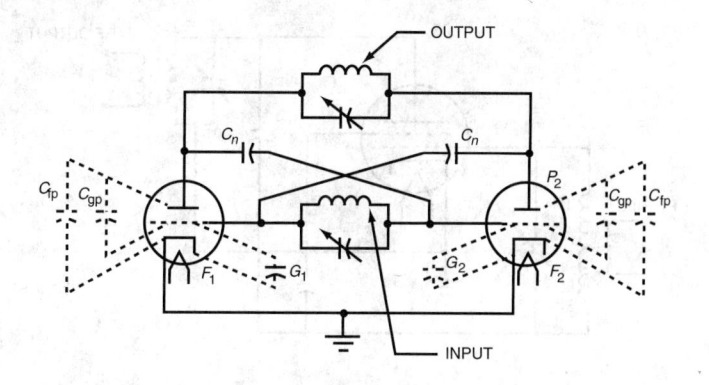

FIGURE 5.42 Push-pull grid neutralization.

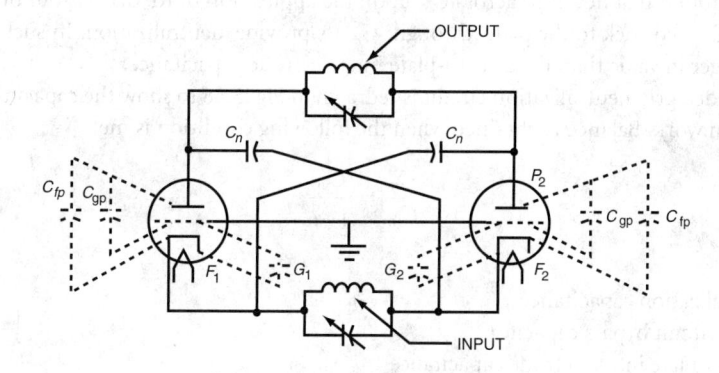

FIGURE 5.43 Symmetrical stage neutralization for a grounded-grid circuit.

In the case of a single-ended amplifier, neutralization can be accomplished using either a push-pull output or push-pull input circuit. Figure 5.44 shows a basic push-pull grid neutralization scheme that provides the out-of-phase voltage necessary for proper neutralization. It is usually simpler to create a push-pull network in the grid circuit than the plate because of the lower voltages present. The neutralizing capacitor C_n is small and may consist of a simple feed-through wire (described previously). Padding capacitor C_p is often added to maintain the balance of the input circuit while tuning. C_p is generally equal in size to the input capacitance of the tube.

Single-ended tetrode and pentode stages can be neutralized using the method shown in Fig. 5.45. The input resonant circuit is placed above ground by a small amount because of the addition of capacitor C_{in}.

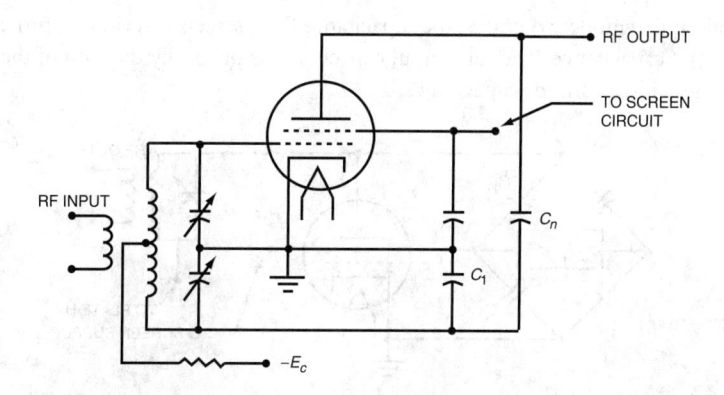

FIGURE 5.44 Push-pull grid neutralization in a single-ended tetrode stage.

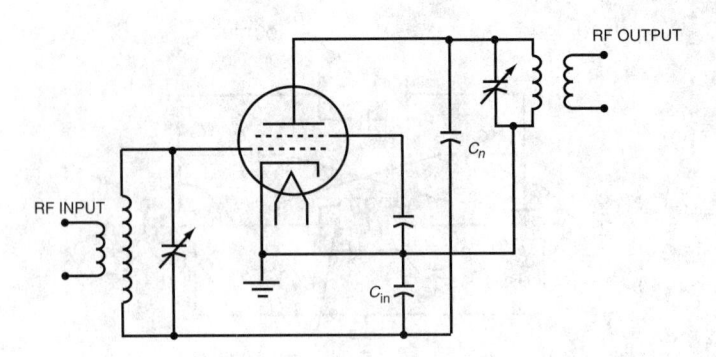

FIGURE 5.45   Single-ended grid neutralization for a tetrode.

The voltage to ground that develops across C_{in} upon the application of RF drive is out of phase with the grid voltage, and is fed back to the plate through C_n to provide neutralization. In such a design, C_n is considerably larger in value than the grid-to-plate interelectrode capacitance.

The single-ended grid neutralization circuit is redrawn in Fig. 5.46 to show the capacitance bridge that makes the design work. Balance is obtained when the following condition is met

$$\frac{C_n}{C_{\text{in}}} = \frac{C_{\text{gp}}}{C_{gf}}$$

where

C_n = neutralization capacitance
C_{in} = input circuit bypass capacitor
C_{gp} = grid-to-plate interelectrode capacitance
C_{gf} = total input capacitance, including tube and stray capacitance

A single-ended amplifier also can be neutralized by taking the plate circuit slightly above ground and using the tube capacitances as part of the neutralizing bridge. This circuit differs from the usual RF amplifier design in that the plate bypass capacitor is returned to the screen side of the screen bypass capacitor, as shown in Fig. 5.47. The size of screen bypass capacitor C_s and the amount of stray capacitances in C_p are chosen to balance the voltages induced in the input by the internal tube capacitances, grid-to-plate C_{gp} and screen-to-grid C_{sg}. This circuit is redrawn in Fig. 5.48 in the usual bridge form. Balance is obtained when the following condition is met

$$\frac{C_p}{C_s} = \frac{C_{\text{gp}}}{C_{\text{sg}}}$$

In usual tetrode and pentode structures, the capacitance from screen-to-grid is approximately half the published tube input capacitance. The tube input capacitance is primarily the sum of the grid-to-screen capacitance and the grid-to-cathode capacitance.

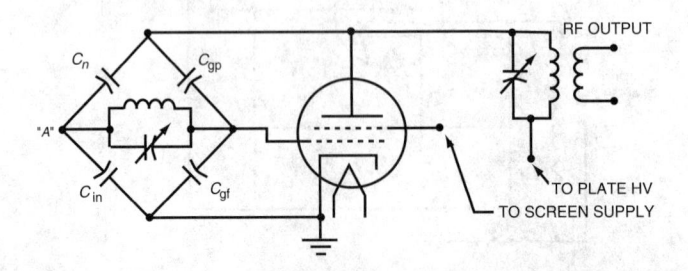

FIGURE 5.46   The previous figure redrawn to show the elements involved in neutralization.

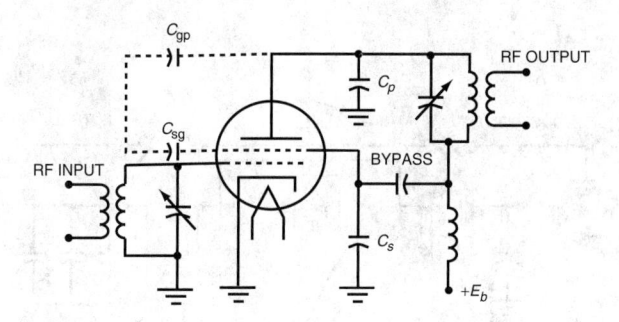

FIGURE 5.47   Single-ended plate neutralization.

It should be noted that in the examples given, it is assumed that the frequency of operation is low enough so that inductances in the socket and connecting leads can be ignored. This is basically true in MF applications and below. At higher bands, however, the effects of stray inductances must be considered, especially in single-ended tetrode and pentode stages.

VHF and Above

Neutralization of power-grid tubes operating at VHF frequencies provides special challenges and opportunities to the design engineer. At VHF frequencies and above, significant RF voltages can develop in the residual inductance of the screen, grid, and cathode elements. If managed properly, these inductances can be used to accomplish neutralization in a simple, straightforward manner.

At VHF and above, neutralization is required to make the tube input and output circuits independent of each other with respect to reactive currents. Isolation is necessary to ensure independent tuning of the input and output. If variations in the output voltage of the stage produce variations of phase angle of the input impedance, phase modulation will result.

As noted previously, a circuit exhibiting independence between the input and output circuits is only half of the equation required for proper operation at RF frequencies. The effects of incidental inductance of the control grid must also be canceled for complete stability. This condition is required because the suppression of coupling by capacitive currents between the input and output circuits is not, by itself, sufficient to negate the effects of the output signal on the cathode-to-grid circuit. Both conditions, input and output circuit independence and compensation for control grid lead inductance, must be met for complete stage stability at VHF and above.

Figure 5.49 shows a PA stage employing stray inductance of the screen grid to achieve neutralization. In this grounded-screen application, the screen is tied to the cavity deck using six short connecting straps. Two additional adjustable ground straps are set to achieve neutralization.

Triode amplifiers operating in a grounded-grid configuration offer an interesting alternative to the more common grounded-cathode system. When the control grid is operated at ground potential, it serves to

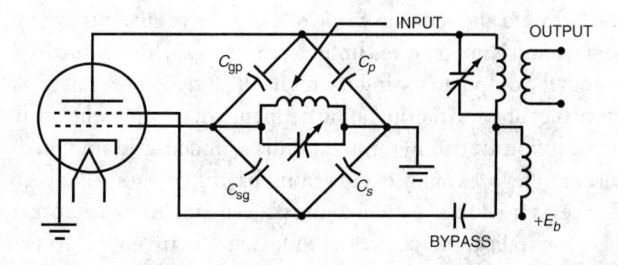

FIGURE 5.48   Single-ended plate neutralization showing the capacitance bridge present.

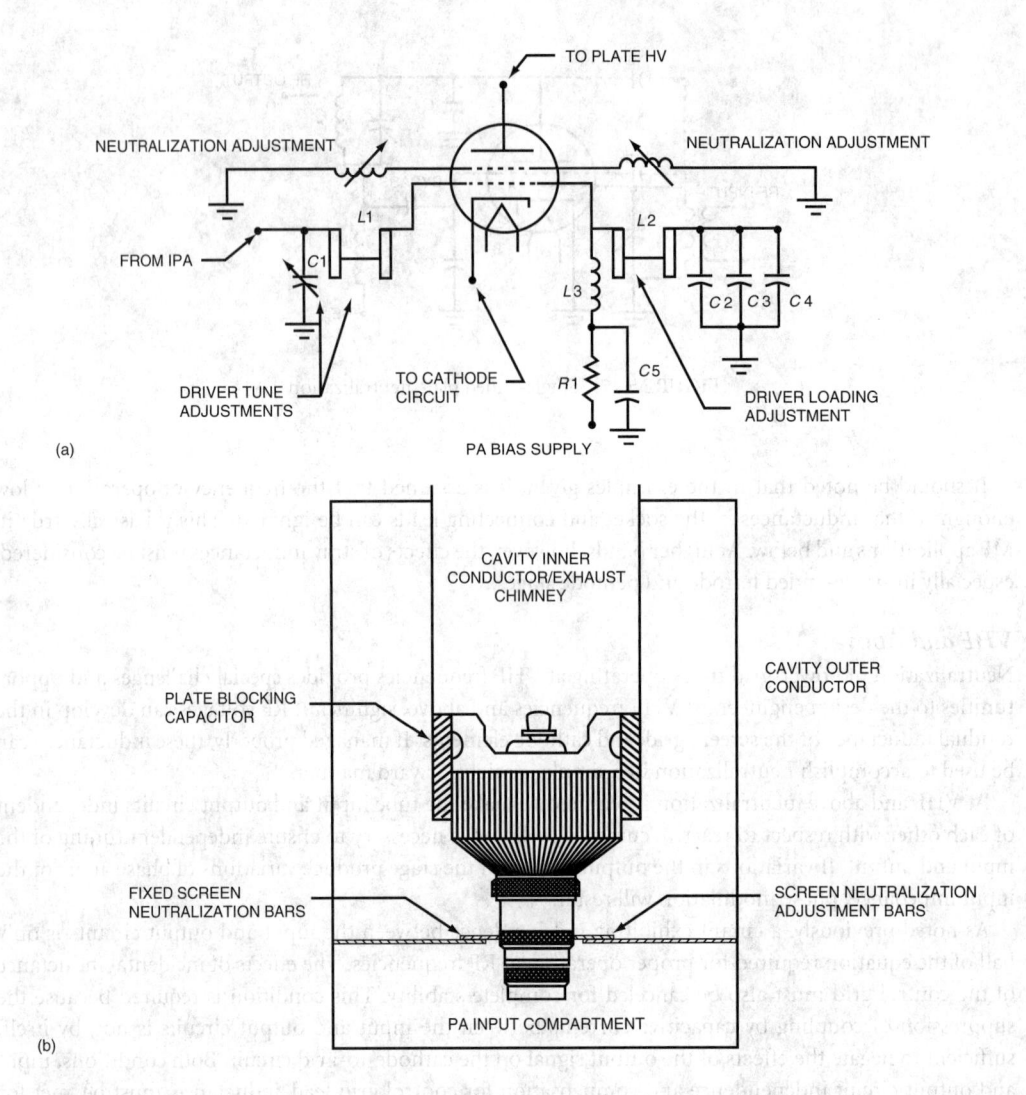

FIGURE 5.49 A grounded-screen PA stage neutralized through the use of stray inductance between the screen connector and the cavity deck: (a) circuit diagram, (b) mechanical implementation.

shield capacitive currents from the output to the input circuit. Typically, provisions for neutralization are not required until the point at which grid lead inductance becomes significant.

Grounded-Grid Amplifier Neutralization

Grounded-grid amplifiers offer an attractive alternative to the more common grid-driven circuit. The control grid is operated at RF ground and serves as a shield to capacitive currents from the output to the input circuit. Generally, neutralization is not required until the control grid lead inductive reactance becomes significant. The feedback from the output to the input circuit is no longer the result of plate-to-filament capacitance. The physical size of the tube and the operating frequency determine when neutralization is required.

Two methods of neutralization are commonly used with grounded-grid amplifiers. In the first technique, the grids of a push-pull amplifier are connected to a point having zero impedance to ground, and a bridge of neutralizing capacitances is used that is equal to the plate-filament capacitances of the tubes.

The second method of neutralization requires an inductance between the grid and ground or between the grids of a push-pull amplifier of a value that will compensate for the coupling between input and output circuits resulting from the internal capacitances of the tubes.

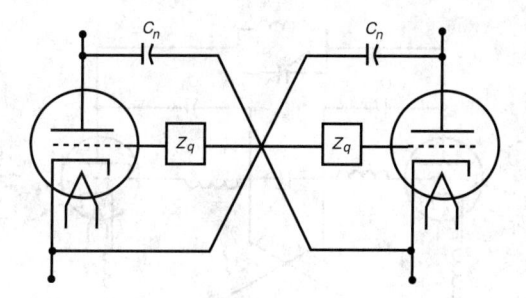

FIGURE 5.50   Circuit of grounded-grid amplifier having grid impedance and neutralized for reactive currents.

Behavior of these two circuits is quite different. They may be considered as special forms of the more general case in which the neutralizing capacitors have values differing from the internal capacitances of the tubes, and in which an appropriate reactance is connected between the grids. Under these conditions, the value of neutralizing capacitance permits continuous variation of power amplification, stability, and negative feedback.

The purpose of neutralization is to make the input and output circuits independent of each other with respect to reactive currents. The input current must be independent of the output voltage, and vice versa. This condition is necessary to permit independent tuning of the input and output circuits, so that variations in output voltage do not produce variations of phase angle of the input impedance, resulting in phase modulation.

The condition of independence between input and output circuits, which may be called the neutralized condition, does not necessarily imply stability. This is because the suppression of coupling by capacitive currents between input and output circuits is not sufficient to remove the effect of the output voltage on the cathode-to-grid voltage. A second condition, distinct from neutralization, must be met for complete stability: the effect of the control grid lead inductance must be canceled.

Grid Impedance

In the special case of a grounded-grid amplifier having a grid impedance and the reactive currents neutralized, the following equations apply (see Fig. 5.50):

$$C_n = C_{fp} - \frac{C_{fg}}{\mu}$$

$$Z_g = \frac{1}{j\omega C_{fg} + C_{gp}(1+\mu)}$$

If in solving the equation for C_n the sign is negative, this indicates that in-phase neutralization is required. Conversely, if the sign of C_n is positive, then out-of-phase neutralization is needed. A negative value of Z_g indicates capacitive reactance required, and a positive value indicates that inductive reactance is to be used.

Application Example

If the grids of a push-pull cathode-driven amplifier are not at ground potential because the inductance of the leads is not negligible, coupling may exist between the input and output circuits through the plate-grid capacitances, cathode-grid capacitances, and grid-to-grid inductance. One method of reducing this coupling is to insert between the grids a series tuned circuit that has zero reactance at the operating frequency. This technique is illustrated in Fig. 5.51. This neutralization scheme is useful only for the case where no grid current flows. If grid current flows, a grid resistance will appear in parallel with the grid-to-filament capacitance. If the resistance is small in comparison to the reactance of this grid-to-filament capacitance, phase modulation will result.

Another important property of the preceding neutralization scheme is that power amplification is a function of the neutralizing capacitance, whereas the independence of cathode and plate circuits from

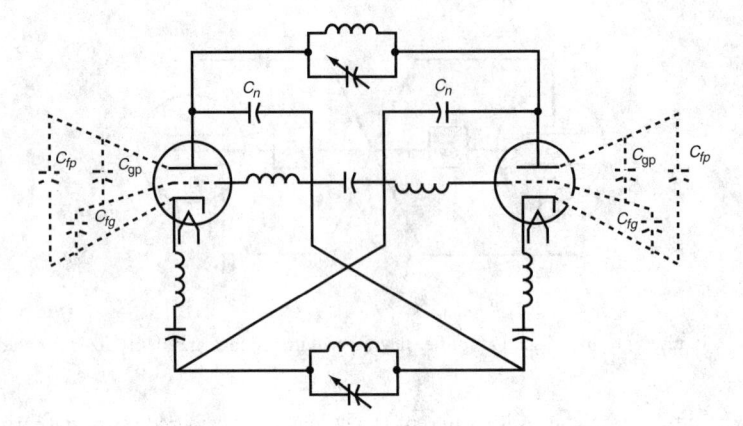

FIGURE 5.51   Neutralization by cross-connected capacitors of a symmetrical cathode-excited amplifier with compensation of lead inductance.

the viewpoint of reactive currents may be obtained with any value of neutralizing capacitance. If the neutralizing capacitance is less than the plate-to-filament capacitance of the tube, the stage will operate with low-excitation power and high-power amplification.

If the neutralizing capacitance is greater than the plate-to-filament capacitance, the power amplification would be quite low, but the total output power possibly would be increased.

5.3.3   Self-Neutralizing Frequency

The voltage dividing action between the plate-to-grid capacitance C_{pg} and the grid-to-screen capacitance C_{gs} will not change with changes in operating frequency. The voltage division between the plate and screen, and screen and ground caused by the charging current I, will however, vary significantly with frequency. There will be a particular frequency, therefore, where this potential dividing circuit will effectively place the grid at filament potential insofar as the plate is concerned. This point is known as the **self-neutralizing frequency** illustrated in Fig. 5.52.

At the self-neutralizing frequency, the tetrode or pentode is inherently neutralized by the circuit elements within the device itself, and external screen inductance to ground. When a device is operated

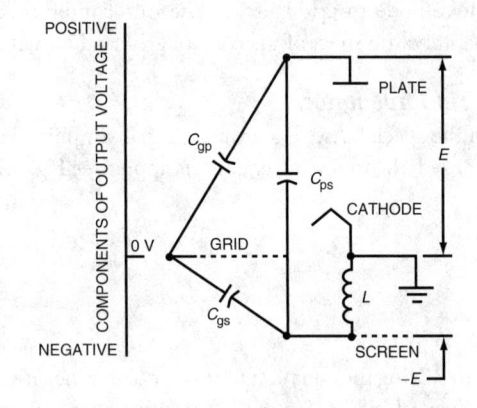

FIGURE 5.52   Graphical representation of the elements of a tetrode when self-neutralized.

below its self-neutralizing frequency, the normal cross-neutralization circuits apply. When the operating frequency is above the self-neutralizing frequency, the voltage developed in the screen lead inductance is too large to give the proper voltage division between the internal capacitances of the device. One approach to neutralization in this case involves adjusting the inductive reactance of the screen lead to ground so as to lower the total reactance. In the example shown in Fig. 5.53, this is accomplished with a series variable capacitor.

Another approach is shown in Fig. 5.54, in which the potential divider network made up of the tube capacitance is changed. In the example, additional plate-to-grid capacitance is added external to the tube. The external capacitance C_{ext} can take the form of a small wire or rod positioned adjacent to the plate of the tube. This approach is similar to the one described previously for conventional neutralization, except that in this case the neutralizing probe is connected to the grid of the tube, rather than to an opposite polarity in the circuit.

If the RF power amplifier is operating above the self-neutralizing frequency of the tube and must be tuned over a range of frequencies, it is probably easier to use the screen series tuning capacitor method and make this control available to the operator. If operation is desired over a range of frequencies including the self-neutralizing frequency of the tube, this circuit is also desirable because the incidental lead inductance in the variable capacitor lowers the self-neutralizing frequency of the circuit so that the neutralizing series capacitor can be made to operate over the total desired frequency range. If this range is too great, switching of neutralizing circuits will be required. A small 50–100 pF variable capacitor in the screen lead has often been found to be satisfactory. Another method of changing the self-neutralizing frequency of a tetrode or pentode can be fashioned from the general bypassing arrangement of the screen and filament shown in Fig. 5.55. The screen lead is bypassed with minimum inductance to the filament terminal of the tube. Some inductance is introduced in the common filament and screen ground leads. The grid is shown below the zero voltage or chassis potential, indicating that the voltage developed in the screen lead inductance to chassis is excessive. If the filament is tapped on this inductance, a point can be found where the voltage difference between the grid and filament is zero, as far as the components of plate voltage are concerned. This arrangement will be found to self-neutralize at a higher frequency than if the filament and screen were separately bypassed to

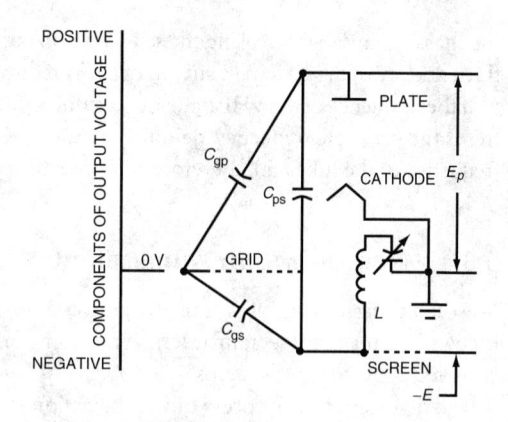

FIGURE 5.53 Components of the output voltage of a tetrode when neutralized by added series screen-lead capacitance.

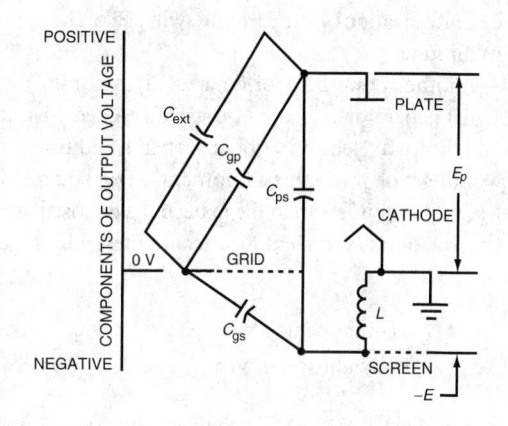

FIGURE 5.54 Components of the output voltage of a tetrode when neutralized by added external grid-to-plate capacitance.

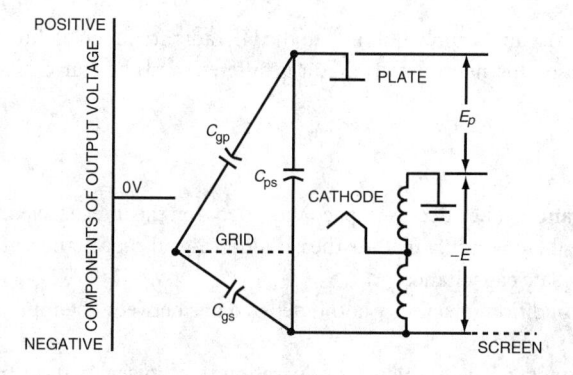

FIGURE 5.55 Components of the output voltage of a tetrode neutralized by adding inductance common to the screen and cathode return.

the chassis. Thus, by increasing the self-neutralizing frequency of the tube and screen bypass arrangement, the tendency of the VHF parasitic to occur is reduced.

If the frequency of the VHF parasitic is reduced by increasing the inductance of the plate lead (presuming this is the principle frequency-defining circuit), the circuit can be made to approach the self-neutralizing frequency of the tube and, therefore, suppress the parasitic.

5.3.4 Neutralization Adjustment

Most neutralization circuits must be adjusted for operation at a given frequency. The exact procedure followed to make these adjustments vary from one circuit to the next. The following generalizations, however, apply to most systems.

The first step in the process of neutralization is to break the DC connections to the plate voltage and screen voltage supplies, leaving the RF circuits intact. If the DC current path is not broken, some current can flow in either one of these circuits even though the voltages are zero. The presence of this current causes the amplifier to work in the normal manner, generating RF power in the plate circuit. It will then be incorrect to adjust for zero power in the plate circuit. Sufficient RF grid drive must be applied to provide some grid current or to cause a sensitive RF meter coupled to the plate to give an indication of feedthrough power. When the plate circuit is tuned through resonance, the grid current will dip when the circuit is out of neutralization or the RF meter will peak. The neutralization adjustments are made until the indication is minimum.

Another powerful tool for roughly neutralizing an RF amplifier is to feed the power output from a signal generator into the grid circuit. A sensitive RF detector is inserted between the output connector and the load. Neutralization can then be adjusted for minimum feedthrough. This technique is useful in working with prototype equipment. Actual qualitative measurements can be made. If the insertion loss of the amplifier is less than the expected gain, oscillation will occur. Circuit modifications can be made until the isolation is sufficient to warrant a test with high voltages applied. The advantages of this cold system test include

- No components are subjected to unusual stress if the amplifier is unstable.
- Circuit adjustments can be made safely because no high voltages are present.

For final trimming of the neutralization adjustment, the stage should be returned to operating condition at reduced power (similar to that used when testing for parasitic oscillations), or under the final loaded operating conditions. At the higher frequencies and in the VHF region, it will be found that a small additional trimming adjustment of the neutralization circuit is usually required. When the plate circuit is tuned through resonance, minimum plate current and maximum control grid current should occur simultaneously. In the case of the tetrode and pentode, the DC screen current should be maximum at the same time.

These neutralizing procedures apply not only to the HF radio frequencies, but also in the VHF or UHF regions. In the latter cases, the neutralizing circuit is different and the conventional cross-neutralization schemes may not apply.

Defining Terms

Interelectrode capacitance: The effective capacitances between the internal elements of a power vacuum tube. In a triode these typically include the cathode-to-grid capacitance, grid-to-plate capacitance, and cathode-to-plate capacitance.

Neutralization: The condition of stability and independence between the input and output circuits of a power grid tube stage.

Self-neutralizing frequency: The frequency of operation of a power vacuum tube stage where the potential dividing circuit of the inherent device capacitances will effectively place the grid at filament potential insofar as the plate is concerned.

References

Fink, D. and Christiansen, D., eds. 1989. *Electronics Engineer's Handbook,* 3rd ed. McGraw-Hill, New York.

Jordan, E.C. 1985. *Reference Data for Engineers: Radio, Electronics, Computer and Communications,* 7th ed. Howard W. Sams, Indianapolis, IN.

Muller, J.J. 1946. Cathode excited linear amplifiers. *Electrical Comm.* 23.

Philips. 1988. *High Power Transmitting Tubes for Broadcasting and Research.* Philips Technical Publication, Eindhoven, The Netherlands.

Terman, F.E. 1947. *Radio Engineering,* 3rd ed. McGraw-Hill, New York.

Varian. 1982. *Care and Feeding of Power Grid Tubes.* Varian/Eimac, Laboratory Staff, San Carlos, CA.

Whitaker, J.C. 1991. *Radio Frequency Transmission Systems: Design and Operation.* McGraw-Hill, New York.

Whitaker, J.C. 1999. *Power Vacuum Tubes Handbook.* 2nd ed. CRC Press, Boca Raton, FL.

Further Information

Specific information on neutralization considerations for power vacuum tubes can be obtained from the manufacturers of those devices. More general application information can be found in the following publications:

Whitaker, J.C. 1999. *Power Vacuum Tubes Handbook.* 2nd ed. CRC Press, Boca Raton, FL.

Golio, M. 2000. *RF and Microwave Handbook,* CRC Press, Boca Raton, FL.

5.4 Amplifier Systems

Jerry C. Whitaker

5.4.1 Introduction

Any number of configurations may be used to generate RF signals using vacuum tubes.[3] Circuit design is dictated primarily by the operating frequency, output power, type of modulation, duty cycle, and available power supply. Tube circuits can be divided generally by their **operating class** and type of modulation employed. The angle of plate current flow determines the class of operation:

- Class A = 360° conduction angle
- Class B = 180° conduction angle
- Class C = conduction angle less than 180°
- Class AB = conduction angle between 180° and 360°

The class of operation has nothing to do with whether the amplifier is grid driven or cathode driven. A cathode-driven amplifier, for example, can be operated in any desired class. The class of operation is only a function of the plate current conduction angle. The efficiency of an amplifier is also a function of the plate current conduction angle.

The efficiency of conversion of DC to RF power is one of the most important characteristics of a vacuum tube amplifier circuit. The DC power that is not converted into useful output energy is, for the most part, converted to heat. This heat represents wasted power; the result of low efficiency is increased operating cost for energy. Low efficiency also compounds itself. This wasted power must be dissipated, requiring increased cooling capacity. The efficiency of the amplifier must, therefore, be carefully considered, consistent with the other requirements of the system. Figure 5.56 shows the theoretical efficiency attainable with a tuned or resistive load assuming that peak AC plate voltage is equal to the plate supply voltage.

---

[3] Adapted from: Whitaker, J.C. 1999. *Power Vacuum Tubes Handbook,* 2nd ed. CRC Press, Boca Raton, FL. Used with permission.

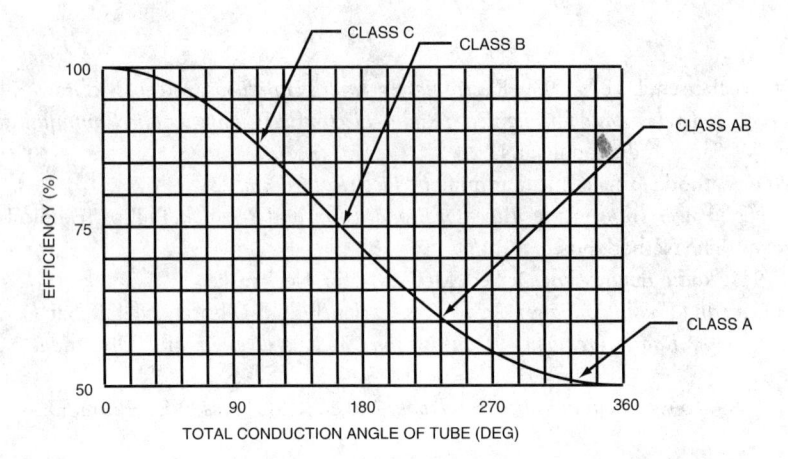

FIGURE 5.56 Plate efficiency as a function of conduction angle for an amplifier with a tuned load. (*Source:* Martin, T.L. Jr. 1955. *Electronic Circuits*, p. 452. Prentice-Hall, Englewood Cliffs, NJ.)

Class A Amplifier

A class A amplifier is used in applications requiring low harmonic distortion output power. A class A amplifier can be operated with low intermodulation distortion in linear RF amplifier service. Typical plate efficiency for a class A amplifier is about 30%. Power gain is high because of the low drive power required. Gains as high as 30 dB are typical.

Class B and AB Amplifiers

A class AB power amplifier is capable of generating more power—using the same tube—than the class A amplifier, but more intermodulation distortion will also be generated. A class B RF linear amplifier will generate still more intermodulation distortion, but is acceptable in certain applications. The plate efficiency is typically 66%, and stage gain is about 20–25 dB.

Class C Amplifiers

A class C power amplifier is used where large amounts of RF energy need to be generated with high efficiency. Class C RF amplifiers must be used in conjunction with tuned circuits or cavities, which restore the amplified waveform through the flywheel effect.

The grounded cathode class C amplifier is the building block of RF technology. It is the simplest method of amplifying CW, pulsed, and FM signals. The basic configuration is shown in Fig. 5.57. Tuned input and output circuits are used for impedance matching and to resonate the stage at the desired operating frequency. The cathode is bypassed to ground using low-value capacitors. Bias is applied to the grid as shown. The bias power supply may be eliminated if a self-bias configuration is used. The typical operating efficiency of a class C stage ranges from 65 to 85%.

Figure 5.58 illustrates the application of a zero-bias triode in a grounded-grid arrangement. Because the grid operates at RF ground potential, this circuit offers stable performance without the need for neutralization (at medium frequency (MF) and below). The input signal is coupled to the cathode through a matching network. The output of the triode feeds a pi network through a blocking capacitor.

5.4.2 Principles of RF Power Amplification

In an RF power amplifier, a varying voltage is applied to the control grid (or cathode in the case of a grounded-grid circuit) from a driver stage whose output is usually one of the following:

- Carrier frequency signal only
- Modulation (intelligence) signal only
- Modulated carrier signal

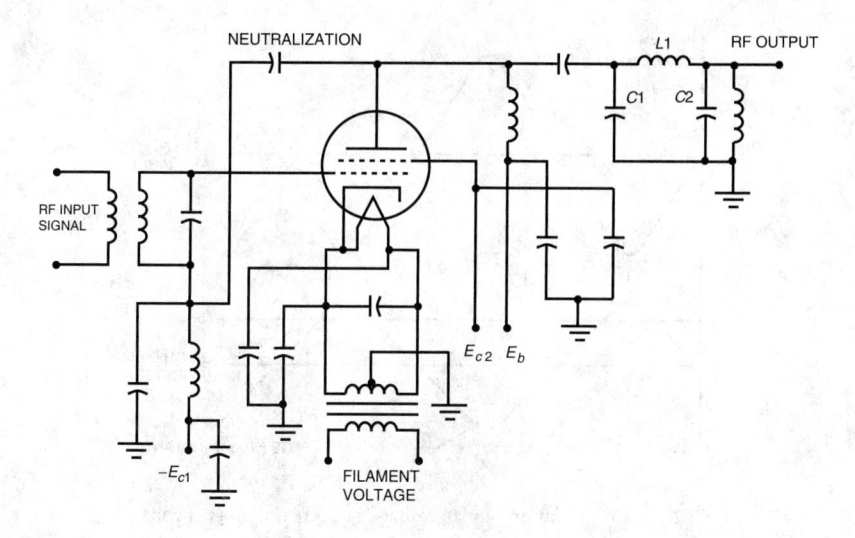

FIGURE 5.57 Basic grounded-cathode RF amplifier circuit.

Simultaneous with the varying control grid signal, the plate voltage will vary in a similar manner, resulting from the action of the amplified current flowing in the plate circuit. In RF applications with resonant circuits, these voltage changes are smooth sinewave variations, 180° out of phase with the input. The relationship is illustrated in Fig. 5.59. Note how these variations center about the DC plate voltage and the DC control grid bias. In Fig. 5.60 the variations have been indicated next to the plate voltage and grid voltage scales of a typical constant current curve. At some instant in time, shown as t on the time scales, the grid voltage has a value denoted e_g on the grid voltage sinewave.

Any point on the operating line (when drawn on constant current curves as illustrated in Fig. 5.60) indicates the instantaneous values of plate current, screen current, and grid current that must flow when

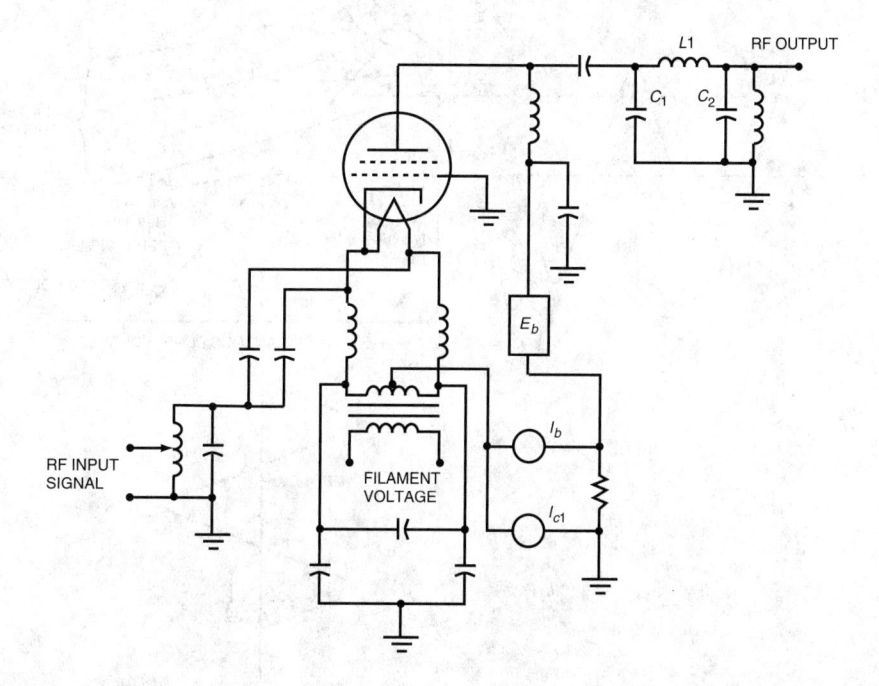

FIGURE 5.58 Typical amplifier circuit using a zero-bias triode. Grid current is measured in the return lead from ground to the filament.

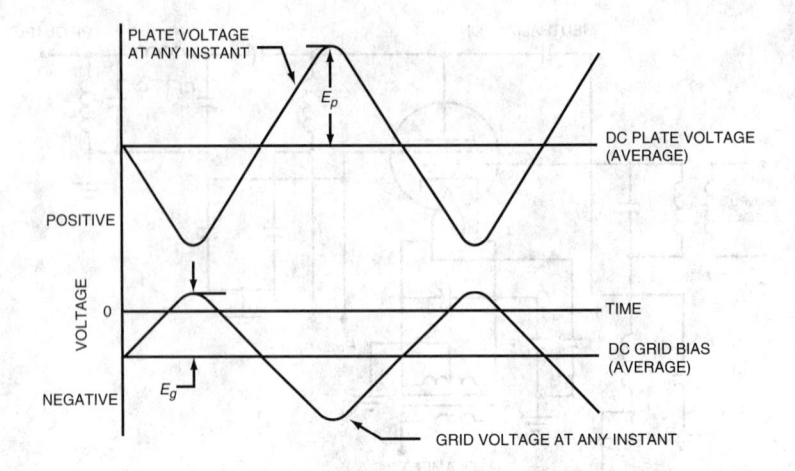

FIGURE 5.59   Variation of plate voltage as a function of grid voltage.

these particular values of grid and plate voltage are applied to the tube. Thus, by plotting the values of plate and grid current as a function of time t, a curve of instantaneous values of plate and grid current can be produced. Such plots are shown in Fig. 5.61.

By analyzing the plate and grid current values it is possible to predict with accuracy the effect on the plate circuit of a change at the grid. It follows that if a properly loaded resonant circuit is connected to

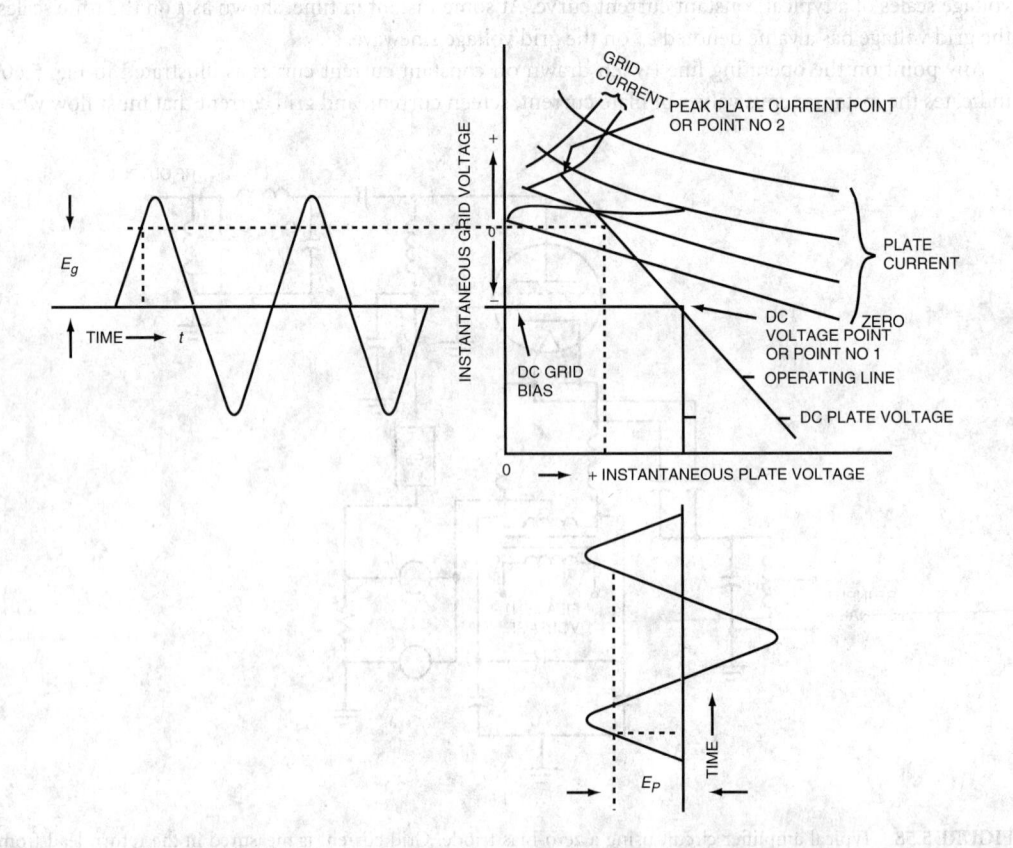

FIGURE 5.60   Relationship between grid voltage and plate voltage plotted on a constant-current curve.

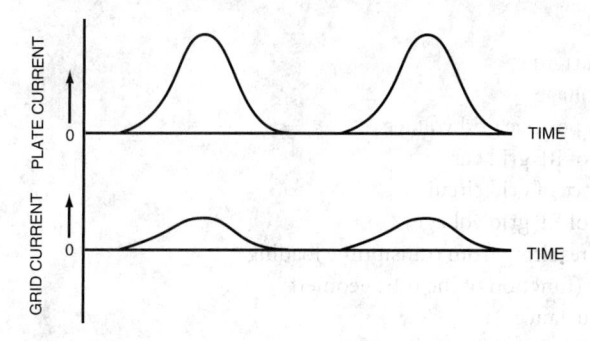

FIGURE 5.61 Instantaneous values of plate and grid current as a function of time.

the plate, a certain amount of RF power will be delivered to that circuit. If the resonant circuit is tuned to the fundamental frequency (the same frequency as the RF grid voltage), the power delivered will be that of the fundamental or principal RF component of plate current. If the circuit is tuned to a harmonic of the grid voltage frequency, the power delivered will be the result of a harmonic component of the plate current.

Drive Power Requirements

The technical data sheet for a given tube type provides the approximate drive power required for various operating modes. As the frequency of operation increases and the physical size of the tube structure becomes large with respect to this frequency, the drive power requirements will also increase.

The drive power requirements of a typical grounded-cathode amplifier consist of six major elements

- The power consumed by the bias source, given by

$$P_1 = I_{c1} \times E_{c1}$$

- The power dissipated in the grid as a result of rectified grid current

$$P_2 = I_{c1} \times e_{cmp}$$

- The power consumed in the tuned grid circuit

$$P_3 = I_{c,\mathrm{rms}}^2 \times R_{\mathrm{rf}}$$

- The power loss as a result of transit time effects

$$P_4 = \left[\frac{e_{c,\mathrm{rms}}}{R_t} \right]^2$$

R_t is that part of the resistive component of the tube input impedance resulting from transit time effects, and is given by

$$R_t = \frac{1}{K g_m f^2 T^2}$$

- The power consumed in that part of the resistive component of the input impedance resulting from cathode lead inductance

$$P_5 = \frac{e_g^2}{R_s}$$

- Input resistance resulting from the inductance of the cathode leads is found from the following

$$R_s = \frac{1}{\omega^2 g_m L_k C_{gk}}$$

- The power dissipated in the tube envelope because of dielectric loss

$$P_6 = 1.41 f E_1^2 v$$

where

I_{c1} = DC grid current
E_{c1} = DC grid voltage
E_{cmp} = maximum positive grid voltage
$I_{c,\text{rms}}$ = rms value of RF grid current
R_{rf} = RF resistance of grid circuit
$e_{c,\text{rms}}$ = rms value of RF grid voltage
R_t = resistance resulting from transit time loading
K = a constant (function of the tube geometry)
g_m = transconductance
f = frequency, Hz
T = transit time, cathode-to-grid
R_s = cathode lead inductance input resistance loading
ω = $2\pi f$
L_k = cathode lead inductance, H
C_{gk} = grid-to-cathode capacitance, F
E_1 = voltage gradient, kV/in, rms
v = loss factor of dielectric materials

The total drive power P_t is, then, equal to

$$P_t = P_1 + P_2 + P_3 + P_4 + P_5 + P_6$$

Particular attention must be given to grid dissipation when a tube is operated in the VHF and UHF regions. The total driving power required for a given output may be greater than the grid dissipation capability of the device.

Operational Considerations for VHF and UHF

When operating a tube in the VHF and UHF regions, several techniques may be applied to minimize the driving power without appreciably affecting plate conversion efficiency. These techniques include

- Use the minimum DC control bias. Frequently, it is advisable to bring the bias down to approximately cutoff.
- Maintain a high value of DC screen voltage, even though it appears to increase the fraction of the cycle during which plate current flows.
- Use the minimum RF excitation voltage necessary to obtain the desired plate circuit performance, even though the DC grid current is considerably lower than would be expected at lower frequencies.
- Keep the cathode lead inductance to the output and input circuits as low as possible. This can be accomplished by (1) using short and wide straps, (2) using two separate return paths for the input and output circuits, or (3) proper choice of cathode bypass capacitor(s).

These techniques do not necessarily decrease the plate efficiency significantly when the circuit is operated at VHF and UHF. The steps should be tried experimentally to determine whether the plate circuit efficiency is appreciably affected. It is usually acceptable—and even preferable—to sacrifice some plate efficiency for improved tube life when operating at VHF and UHF.

Optimum power output at these frequencies is obtained when the loading is greater than would be used at lower frequencies. Fortunately, the same condition reduces driving power and screen current (for the tetrode and pentode cases), and improves tube life expectancy in the process.

Mechanical and Electrical Considerations

To maintain proper isolation of the output and input circuits, careful consideration must be given to the location of the component parts of the amplifier. All elements of the grid or input circuit and any earlier stages must be kept out of the plate circuit compartment. Similarly, plate circuit elements must be kept out of the input compartment. It should be noted, however, that in the case of the tetrode and pentode, the screen lead of the tube and connections via the socket are common to both the output and input resonant circuits. Because of the plate-to-screen capacitance of a tetrode or pentode, the RF plate voltage (developed in the output circuit) causes an RF current to flow out the screen lead to the chassis. In the case of a push-pull stage, this current may flow from the screen terminal of one tube to the screen terminal of the other tube. Similarly, because of the grid-to-screen capacitance of the tube, the RF voltage in the input circuit will cause an RF current to flow in this same screen lead to the chassis, or to the opposite tube of the push-pull circuit.

The inductance of the lead common to both the output and input circuits has the desirable feature of providing voltage of opposite polarity to neutralize the feedback voltage of the residual plate to control grid capacitance of the tube. It should be noted, however, that the mutual coupling from the screen lead to the input resonant circuit may be a possible source of trouble, depending on the design.

Power Supply Considerations

The power supply requirements for a triode are straightforward. The degree of regulation and ripple depends on the requirements of the system. In the case of a linear RF amplifier, it is important to have good plate power supply regulation. Without tight regulation, the plate voltage will drop during the time the plate is conducting current heavily. This drop will cause **flat topping** and will appear as distortion in the output. In push-pull applications where grid current flows, it is important to keep the grid circuit resistance to a minimum. If this is not done, positive peak clipping will occur.

In the case of the tetrode and pentode, the need for screen voltage introduces some new considerations and provides some new possibilities. Voltage for the screen grid of a low-power tetrode or pentode can readily be taken from the power supply used for the plate of the tube. In this case, a series resistor, or potential dividing resistor, is chosen so that with the intended screen current flowing the voltage drop through the resistor is adequate to give the desired screen voltage. The potential dividing resistor is the preferred technique for those tubes with significant secondary screen emission.

It is possible to take the screen voltage from a separate low-voltage supply. A combination scheme may also be employed, where a dropping resistor is used in conjunction with a low-voltage or intermediate-voltage supply. Frequently a combination of series resistor and voltage source can be chosen so that the rated screen dissipation will not be exceeded regardless of variations in screen current. With a fixed screen supply, there are advantages in using an appreciable amount of fixed grid bias so as to provide protection against the loss of **excitation**, or for cases where the driver stage is being keyed.

If the screen voltage is taken through a dropping resistor from the plate supply, there is usually little point in using a fixed grid bias because an unreasonable amount of bias would be required to protect the tube if the excitation failed. Under operating conditions with normal screen voltage, the cutoff bias is low (screen voltage divided by the screen μ). When a stage loses excitation and runs statically, the screen current falls to nearly zero. (See the static curves of the tube in question.) If the screen voltage is obtained through a simple dropping resistor from the plate supply, the screen voltage will then rise close to full plate voltage. Because the cutoff bias required is proportional to screen voltage, the grid bias required will be much greater than the amount of bias desired under normal operating conditions. When a screen dropping resistor is used, most of the bias is normally supplied through a grid resistor and other means are used for tube protection.

The power output from a tetrode or pentode is sensitive to screen voltage. For this reason, any application requiring a high degree of linearity through the amplifier requires a well-regulated screen power supply. A screen dropping resistor from the plate supply is not recommended in such applications.

The suppressor grid power supply requirements are similar to the control grid power supply. The suppressor grid intercepts little current and, therefore, a low power supply may be used. Any variation in suppressor voltage as a result of ripple or lack of regulation will appear at the output of the amplifier because of suppressor grid modulation of the plate current.

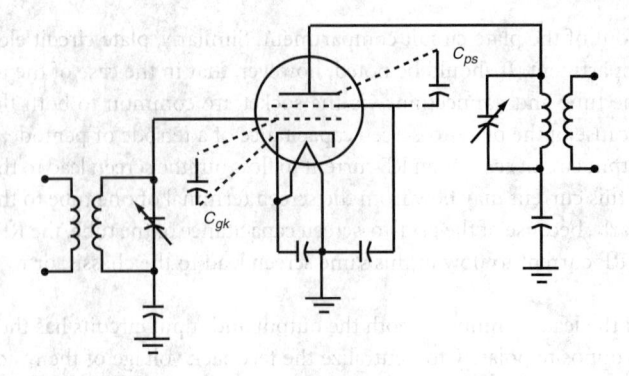

FIGURE 5.62   Interelectrode capacitances supporting VHF parasitic oscillation in an HF RF amplfier.

Parasitic Oscillations

Most self-oscillations in RF power amplifiers using gridded tubes have been found to fall in the following three classes:

- Oscillation at VHF from about 40 to 200 MHz, regardless of the operating frequency of the amplifier.
- Self-oscillation on the fundamental frequency of the amplifier.
- Oscillation at a low radio frequency below the normal frequency of the amplifier.

Low-frequency oscillation in an amplifier usually involves the RF chokes, especially when chokes are used in both the output and input circuits. Oscillation near the fundamental frequency involves the normal resonant circuits, and brings up the question of neutralization. When a parasitic self-oscillation is found on a very high frequency, the interconnecting leads of the tube, tuning capacitor, and bypass capacitors are typically involved.

VHF oscillation occurs commonly in amplifiers where the radio frequency circuits consist of coils and capacitors, as opposed to cavity designs. As illustrated in Fig. 5.62, the tube capacitances effectively form a tuned-plate tuned-grid oscillator.

The frequency of a VHF parasitic is typically well above the self-neutralizing frequency of the tube. However, if the self-neutralizing frequency of the device can be increased and the frequency of the parasitic lowered, complete suppression of the parasitic may result, or its suppression by resistor-inductor parasitic suppressors may be made easier.

It is possible to predict with the use of a grid dip wave meter the parasitic frequency to be expected in a given circuit. The circuit should be complete and with no voltages on the tube. Couple the meter to the plate or screen lead and determine the resonant frequency.

Elimination of the VHF parasitic oscillation may be accomplished using one of the following techniques:

- Place a small coil and resistor combination in the plate lead between the plate of the tube and the tank circuit, as shown in Fig. 5.63(a). The resistor-coil combination is usually made up of a noninductive resistor of about 25–100 Ω, shunted by three or four turns approximately 1/2-in diameter and frequently wound around the resistor. In some cases it may be necessary to use such a suppressor in both the plate and grid leads. The resistor-coil combination operates on the principle that resistor loads the VHF circuit but is shunted by the coil for the lower fundamental frequency. In the process of adjusting the resistor-coil combination, it is often found that the resistor runs hot. This heat is usually caused by the dissipation of fundamental power in the resistor, and is an indication of too many turns in the suppressor coil. Just enough turns should be used to suppress the parasitic and no more. Once the parasitic has been suppressed there will be no parasitic voltage or current present. Therefore, there is no parasitic power to be dissipated.
- Use small parasitic chokes in the plate lead as shown in Fig. 5.63(b). The size of the coil will vary considerably depending on the tube and circuit layout. A coil of 4–10 turns measuring approximately 1/2-in diameter is typical. The presence of the choke in the frequency determining part of the circuit

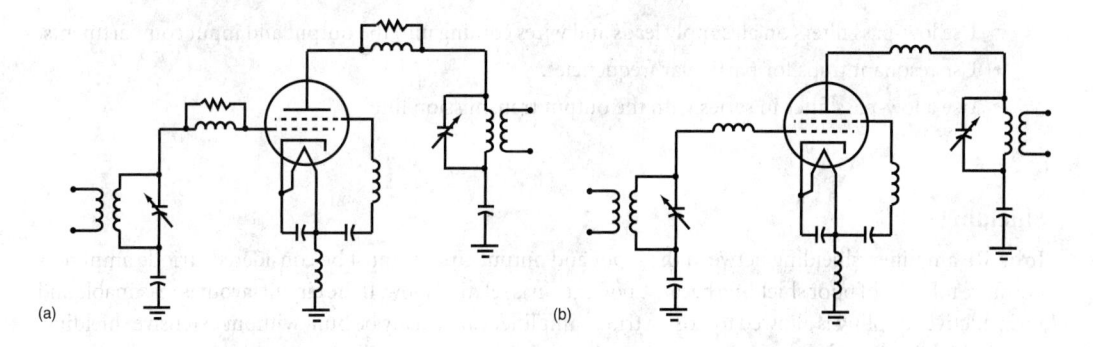

FIGURE 5.63 Placement of parasitic suppressors to eliminate VHF parasitic oscillations in an HF RF amplifier: (a) resistor-coil combination, (b) parasitic choke.

> lowers the frequency of a possible VHF parasitic so that it falls near the self-neutralizing frequency of the tube and bypass leads. In addition to varying the size of the suppressor choke, the amount of inductance common to the screen and filament in the filament grounding strap may be a factor. This parameter can be varied simultaneously with the suppressor choke.

Of the two methods outlined for suppressing VHF parasitic oscillations, the first is probably the simpler to use and has been widely employed.

Dynatron Oscillation

Another form of commonly encountered self-oscillation is known as *dynatron* oscillation. Dynatron oscillation is caused when any electrode in a vacuum tube has negative resistance. At times there may be more electrons leaving the screen grid than are arriving. If the screen voltage is allowed to increase under these conditions, even more electrons will leave the grid. This phenomenon implies a negative resistance characteristic. If there is high alternating current impedance in the circuit from the screen grid through the screen grid power supply, and from the plate power supply to the plate, dynatron oscillation may be sustained.

Dynatron oscillation typically occurs in the region of 1–20 Hz. This low-frequency oscillation is usually accompanied by another oscillation in the 1–2 kHz region. Suppression of these oscillations can be accomplished by placing a large bypass capacitor (1000 μF) across the output of the screen grid power supply. The circuit supporting the oscillation can also be detuned by a large inductor. Increasing the circuit losses at the frequency of oscillation is also effective.

Harmonic Energy

It is generally not appreciated that the pulse of grid current contains energy on harmonic frequencies and that control of these harmonic energies may be important. The ability of the tetrode and pentode to isolate the output circuit from the input circuit over a wide range of frequencies is important in avoiding feed through of harmonic voltages from the grid circuit. Properly designed tetrode and pentode amplifiers provide for complete shielding in the amplifier layout so that coupling external to the tube is prevented.

In RF amplifiers operating either on the fundamental or a desired harmonic frequency, the control of unwanted harmonics is important. The following steps permit reduction of the unwanted harmonic energies present in the output circuit:

- Keep the circuit impedance between the plate and cathode low for the high harmonic frequencies. This requirement may be achieved by having some or all of the tuning capacitance of the resonant circuit close to the tube.
- Completely shield the input and output compartments.
- Use inductive output coupling from the resonant plate circuit and possibly a capacitive or Faraday shield between the coupling coil and the tank coil, or a high-frequency attenuating circuit such as a π or π-L network.

- Use low-pass filters on all supply leads and wires coming into the output and input compartments.
- Use resonant traps for particular frequencies.
- Use a low-pass filter in series with the output transmission line.

Shielding

In an RF amplifier, shielding between the input and output circuits must be considered. Triode amplifiers are more tolerant of poor shielding because power gain is relatively low. If the circuit layout is reasonable and no inductive coupling is allowed to exist, a triode amplifier can usually be built without extensive shielding. Even if shielding is not necessary to prevent fundamental frequency oscillation, it will aid in eliminating any tendency toward parasitic oscillation. The higher the gain of an amplifier the more important the shielding.

Pierced Shields

Tetrode and pentode amplifiers require comprehensive shielding to prevent input-to-output circuit coupling. It is advisable to use nonmagnetic materials such as copper, aluminum, or brass in the RF fields to provide the shielding. Quite often a shield must have holes through it to allow the passage of cooling air. In the LF and part of the HF range, the presence of small holes will not impair shield effectiveness. As the frequency is increased, however, the RF currents flowing around the hole in one compartment cause fields to pass through the hole into another compartment. Currents are, therefore, induced on the shield in the other compartment. This problem can be eliminated by using holes that have significant length. A piece of pipe with a favorable length-to-diameter ratio as compared to the frequency of operation will act as a *waveguide beyond cutoff* attenuator. If more than one hole is required to pass air, a material resembling a honeycomb may be used. This material is commercially available and provides excellent isolation with a minimum air pressure drop. A section of honeycomb shielding is shown in Fig. 5.64. Some tube sockets incorporate a waveguide beyond cutoff air path. These sockets allow the tube in the amplifier to operate at high gain and up through VHF.

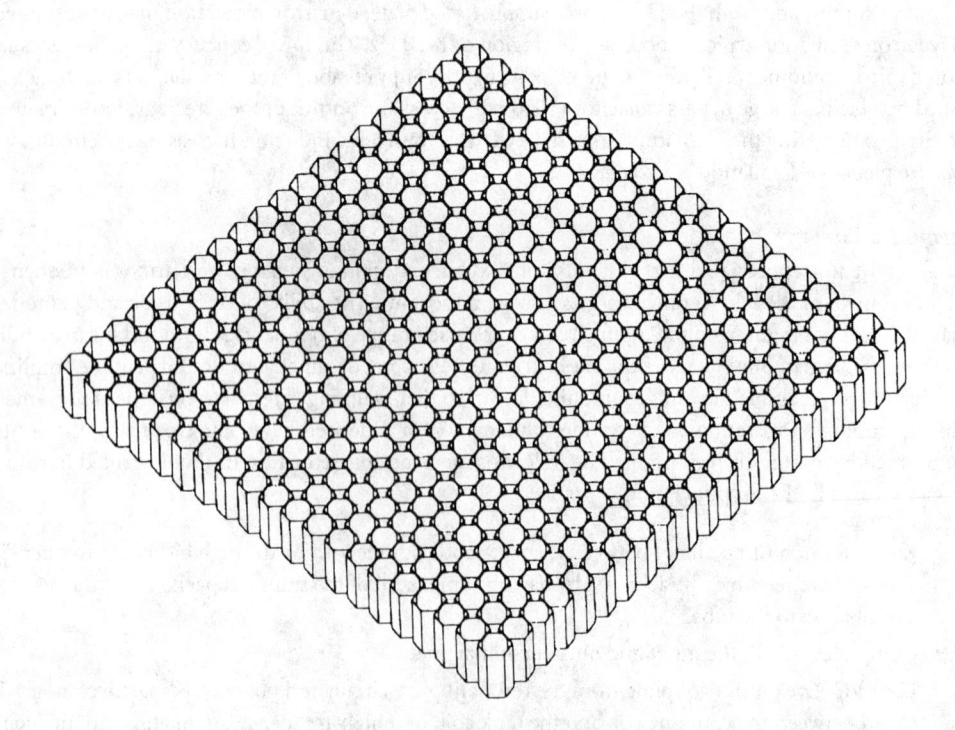

FIGURE 5.64   A section of honeycomb shielding material used in an RF amplifier.

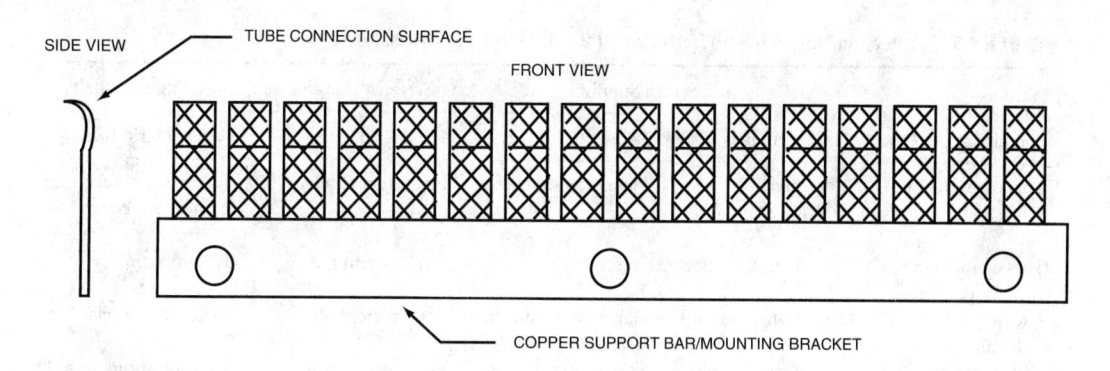

FIGURE 5.65   A selection of common finger stock.

Compartments

By placing the tube and related circuits in completely enclosed compartments and properly filtering incoming supply wires, it is possible to prevent coupling radio frequency energy out of the circuit by means other than the desired output coupling. Such filtering prevents the coupling of energy that may be radiated or fed back to the input section or earlier stages in the amplifier chain. Energy fed back to the input circuit causes undesirable interaction in tuning and/or self-oscillation. If energy is fed back to earlier stages, significant operational problems may result because of the large-power gain over several stages.

In the design of an RF amplifier, doors or removable panels must typically be used. The requirement for making a good, low-resistance joint at the discontinuity must be met. There are several materials available commercially for this purpose. Finger stock, shown in Fig. 5.65, has been used for many years.

Protection Measures

Power grid tubes are designed to withstand considerable abuse. The maximum ratings for most devices are conservative. For example, the excess anode dissipation resulting from detuning the plate circuit will have no ill effects on most tubes if not applied for periods of time sufficient to overheat the envelope and the seal structure.

Similarly, the control, screen, and suppressor grids will stand some excess dissipation. Typically, the maximum dissipation for each grid indicated on the data sheet should not be exceeded except for time intervals of less than 1 s. The maximum dissipation rating for each grid structure is usually considerably above typical values used for maximum output so that ample operating reserve is provided. The time of duration of overload on a grid structure is necessarily short because of the small heat storage capacity of the grid wires. Furthermore, grid temperatures cannot be measured or seen, and so no warning of accidental overload is apparent.

The type and degree of protection required in an RF amplifier against circuit failure will vary with the type of screen and grid voltage supply. Table 5.3 lists protection criteria for tetrode and pentode devices. The table provides guidelines on the location of a suitable relay that should act to remove the principal supply voltage from the stage or transmitter to prevent damage to the tube.

For designs where screen voltage taken through a dropping resistor from the plate supply, a plate relay provides almost universal protection. For the case of a fixed screen supply, a relay provides protection in most cases. For protection against excessive antenna loading and consequent high plate dissipation, a screen undercurrent relay may also be used in some services.

The plate, screen, and bias voltages may be applied simultaneously to a tetrode. The same holds true for a pentode, plus the application of the suppressor voltage. In a grid-driven amplifier, grid bias and excitation can usually be applied alone to the tube, especially if a grid leak resistor is used. Plate voltage can be applied to the tetrode and pentode before the screen voltage with or without excitation to the control grid. Never apply screen voltage before the plate voltage. The only exception would be when the tube is cut off so that no space current (screen or plate current) will flow, or when the excitation and screen voltage are low. If screen voltage is applied before the plate voltage and screen current can

TABLE 5.3   Protection Guidelines for Tetrode and Pentode Devices

	Fixed Screen Supply		Screen Voltage through Dropping Resistor	
Circuit Failure Type	Fixed Grid Bias	Resistor Grid Bias	Fixed Grid Bias	Resistor Grid Bias
Loss of excitation	No protection required	Plate current relay	Plate current relay	Plate current relay or screen control circuit
Loss of antenna loading	Screen current relay	Screen current relay	Grid current relay	No protection required
Excess antenna loading	Screen undercurrent relay	Screen undercurrent relay	Plate current relay	Plate current relay
Failure of plate supply	Screen current relay	Screen current relay	Grid current relay	No protection required
Failure of screen supply	Grid current relay	No protection required	Does not apply	Does not apply
Failure of grid bias supply	Plate current relay or screen current relay	Does not apply	Plate current relay or grid current relay	Does not apply

flow, the maximum allowable screen dissipation will almost always be exceeded and tube damage will result.

Table 5.4 lists protection guidelines for a triode. The table covers the grid-driven triode amplifier and the high-μ (zero-bias) cathode-driven triode amplifier. Drive voltage must never be applied to a zero-bias triode amplifier without plate voltage. The table indicates the recommended location of a suitable relay that should act to remove the principal supply voltage from the stage or transmitter to prevent damage to the tube or transmitter.

5.4.3 Cavity Amplifier Systems

Power grid tubes are ideally suited for use as the power generating element in a **cavity amplifier**. Because of the physical dimensions involved, cavity designs are typically limited to VHF frequencies and above. Lower frequency RF generators utilize discrete L and C devices to create the needed tank circuit. Two types of cavity amplifiers are commonly used: 1/4-wavelength and 1/2-wavelength systems.

In a cavity amplifier, the tube becomes part of a resonant transmission line. The stray interelectrode and distributed capacity and inductance of the tube are used to advantage as part of the resonant line. This resonant line is physically larger than the equivalent lumped constant LRC resonant circuit operating at the same frequency, and this larger physical size aids in solving the challenges of high-power operation, skin effect losses, and high-voltage standoff concerns.

A shorted 1/4-wavelength transmission line has a high, purely resistive input impedance. Electrically, it appears as a parallel resonant circuit, as shown in Fig. 5.66. When the physical length of the line is less than

TABLE 5.4   Protection Guidelines for a Triode

	Triode		
Circuit Failure Type	Fixed Grid Bias	Resistor Grid Bias	Zero-Bias Triode
Loss of excitation	No protection required	Plate overcurrent relay	No protection required
Loss of antenna loading	RF output detector and relay	RF output detector and relay	Grid overcurrent relay
Excess antenna loading	RF output detector and relay	RF output detector and relay	RF output detector and relay
Failure of plate supply	No protection required	No protection required	Grid overcurrent relay
Failure of grid bias supply	Plate overcurrent relay	Does not apply	Does not apply

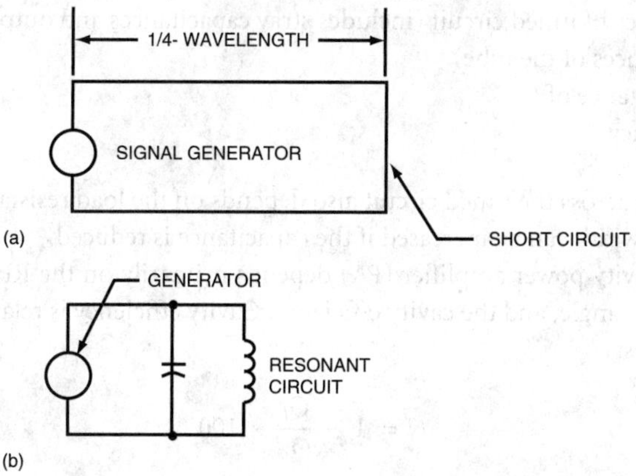

FIGURE 5.66   Shorted 1/4-wavelength line: (a) physical circuit, (b) electrical equivalent.

1/4-wavelength, the impedance will be lower and the line will appear inductive, as illustrated in Fig. 5.67. This inductance is used to resonate with the capacitive reactance in the tube and the surrounding circuit.

Bandwidth and Efficiency

Power amplifier bandwidth has a significant effect on modulation performance. Available bandwidth determines the amplitude response, phase response, and group delay response. Performance tradeoffs must often be considered in the design of a cavity amplifier, including bandwidth, gain, and efficiency.

Power amplifier bandwidth is restricted by the equivalent load resistance across the parallel tuned circuits in the stage. Tuned circuits are necessary to cancel the low reactive impedance presented by the relatively high input and output capacitances of the amplifying device. The bandwidth for a single tuned circuit is proportional to the ratio of capacitive reactance X_c to load resistance R_l appearing across the tuned circuit

$$\text{BW} \approx \frac{K}{2\pi f_c C R_l} \approx \frac{K X_c}{R_l}$$

where
$\quad \text{BW} = $ bandwidth between half-power (−3 dB) points
$\quad K \quad = $ proportionality constant
$\quad R_l \quad = $ load resistance appearing across tuned circuit

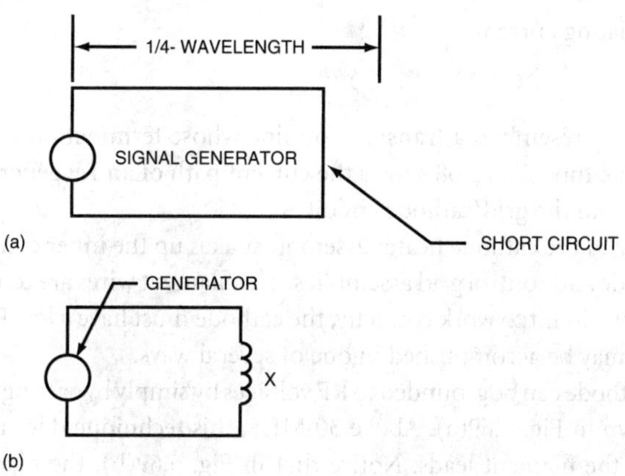

FIGURE 5.67   Shorted line less than 1/4-wavelength: (a) physical circuit, (b) electrical equivalent.

C = total capacitance of tuned circuit (includes stray capacitances and output or
        input capacitances of the tube)
X_c = capacitive reactance of C
f_c = carrier frequency

The RF voltage swing across the tuned circuit also depends on the load resistance. For the same power and efficiency, the bandwidth can be increased if the capacitance is reduced.

The efficiency of a cavity power amplifier (PA) depends primarily on the RF plate voltage swing, the plate current conduction angle, and the cavity efficiency. Cavity efficiency is related to the ratio of loaded to unloaded Q as follows:

$$N = 1 - \frac{Q_l}{Q_u} \times 100$$

where
N = cavity efficiency,
Q_l = cavity loaded Q
Q_u = cavity unloaded Q

The loaded Q is determined by the plate load impedance and output circuit capacitance. Unloaded Q is determined by the cavity volume and the RF resistivity of the conductors resulting from the skin effect. For best cavity efficiency, the following conditions are desirable:

- High unloaded Q
- Low loaded Q

As the loaded Q increases, the bandwidth decreases. For a given tube output capacitance and power level, loaded Q decreases with decreasing plate voltage or with increasing plate current. The increase in bandwidth at reduced plate voltage occurs because the load resistance is directly related to the RF voltage swing on the tube anode. For the same power and efficiency, the bandwidth can also be increased if the output capacitance is reduced. Power tube selection and minimization of stray capacitance are important considerations when designing for maximum bandwidth.

Current Paths

The operation of a cavity amplifier is an extension of the current paths inside the tube. Two elements must be examined in this discussion:

- The input circulating currents
- The output circulating currents

Input Circuit

The grid/cathode assembly resembles a transmission line whose termination is the RF resistance of the electron stream within the tube. Fig. 5.68 shows the current path of an RF generator (the RF driver stage output) feeding a signal into the grid/cathode circuit.

The outer contact ring of the cathode heater assembly makes up the inner conductor of a transmission line formed by the cathode and control grid assemblies. The filament wires are returned down the center of the cathode. For the input circuit to work correctly, the cathode must have a low RF impedance to ground. This cathode bypassing may be accomplished in one of several ways.

Below 30 MHz, the cathode can be grounded to RF voltages by simply bypassing the filament connections with capacitors, as shown in Fig. 5.69(a). Above 30 MHz, this technique does not work well because of the stray inductance of the filament leads. Notice that in Fig. 5.69(b), the filament leads appear as RF chokes, preventing the cathode from being placed at RF ground potential. This causes negative feedback and reduces the efficiency of the input and output circuits.

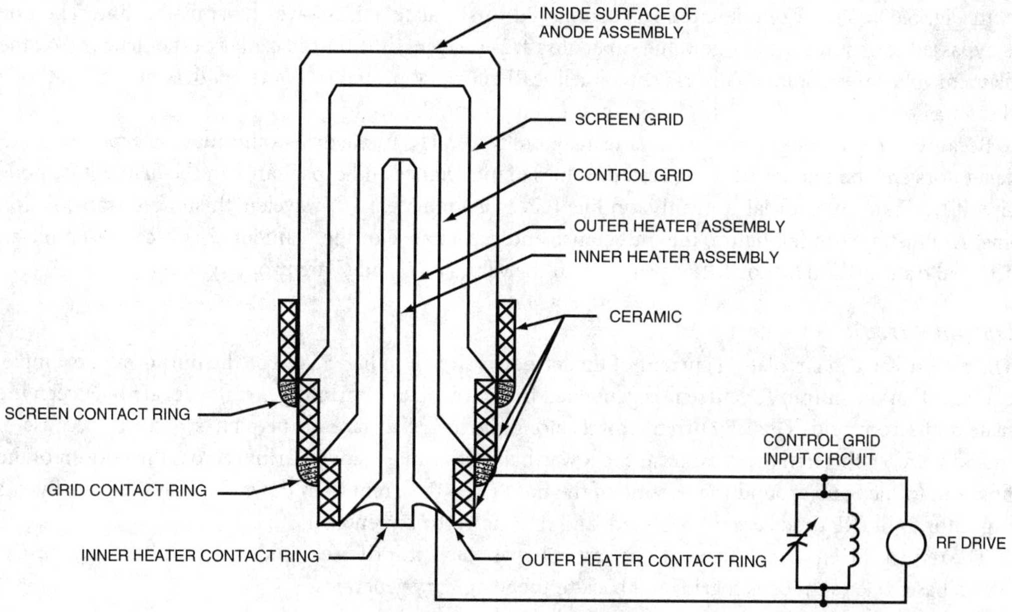

FIGURE 5.68 A simplified representation of the grid input circuit of a PA tube.

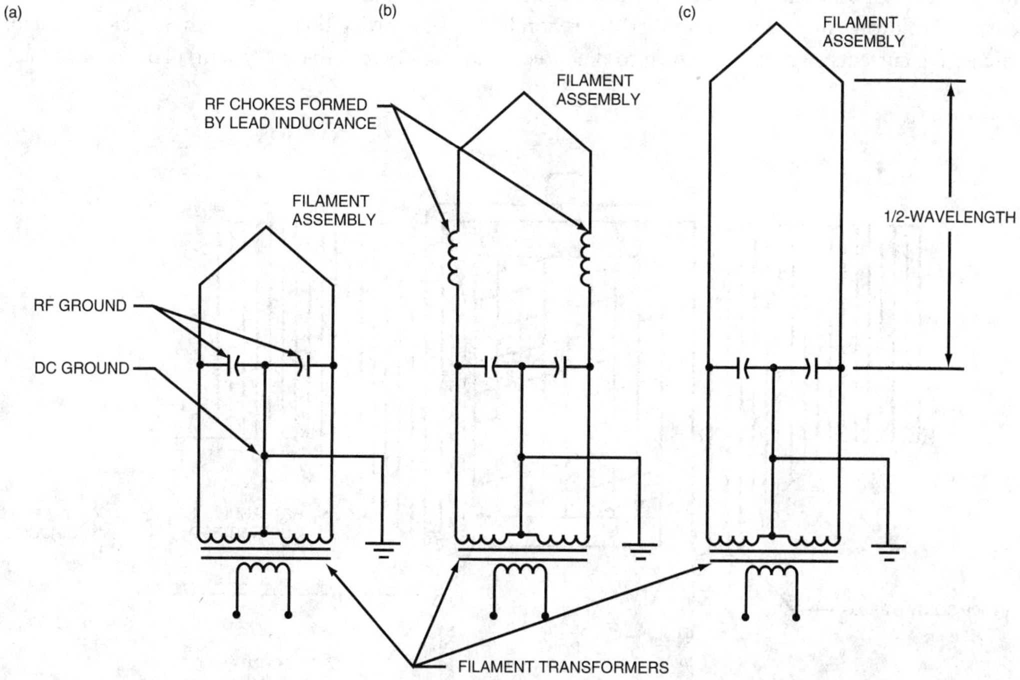

FIGURE 5.69 Three common methods for RF bypassing of the cathode of a tetrode PA tube: (a) grounding the cathode below 30 MHz, (b) grounding the cathode above 30 MHz, (c) grounding the cathode via a 1/2-wave transmission line.

In Fig. 5.69(c) the cathode circuit is configured to simulate a 1/2-wave transmission line. The line is bypassed to ground with large value capacitors 1/2-wavelength from the center of the filament (at the filament voltage feed point). This transmission line RF short is repeated 1/2-wavelength away at the cathode (heater assembly) and effectively places it at ground potential.

Because 1/2-wavelength bypassing is usually bulky at VHF frequencies (and may be expensive), RF generators are often designed using certain values of inductance and capacitance in the filament/cathode circuit to create an artificial transmission line that will simulate a 1/2-wavelength shorted transmission line. As illustrated in the figure, the inductance and capacitance of the filament circuit can resemble an artificial transmission line of 1/2- wavelength, if the values of L and C are properly selected.

Output Circuit

The plate-to-screen circulating current of the tetrode is shown in Fig. 5.70. For the purposes of example, consider that the output RF current is generated by an imaginary current generator located between the plate and screen grid. The RF current travels along the inside surface of the plate structure (because of the skin effect), through the ceramic at the lower-half of the anode contact ring, across the bottom of the fins, and to the band around the outside of the fins. The RF current then flows through the plate bypass capacitor to the RF tuned circuit and load, and returns to the screen grid.

The return current travels through the screen bypass capacitor (if used) and screen contact ring, up the screen base cone to the screen grid, and back to the imaginary generator.

The screen grid has RF current returning to it, but because of the assembly's low impedance, the screen grid is effectively at RF ground potential. The RF current generator, therefore, appears to be feeding an open-ended transmission line consisting of the anode (plate) assembly and the screen assembly. The RF voltage developed by the anode is determined by the plate impedance (Z_p) presented to the anode by the resonant circuit and its load.

When current flows on one conductor of a transmission line cavity circuit, an equal magnitude current flows in the opposite direction on the other conductor. This means that a large value of RF circulating current is flowing in a cavity amplifier outer conductor (the cavity box). All of the outer conductor circulating currents start at and return to the screen grid (in a tetrode-based system). The front or back

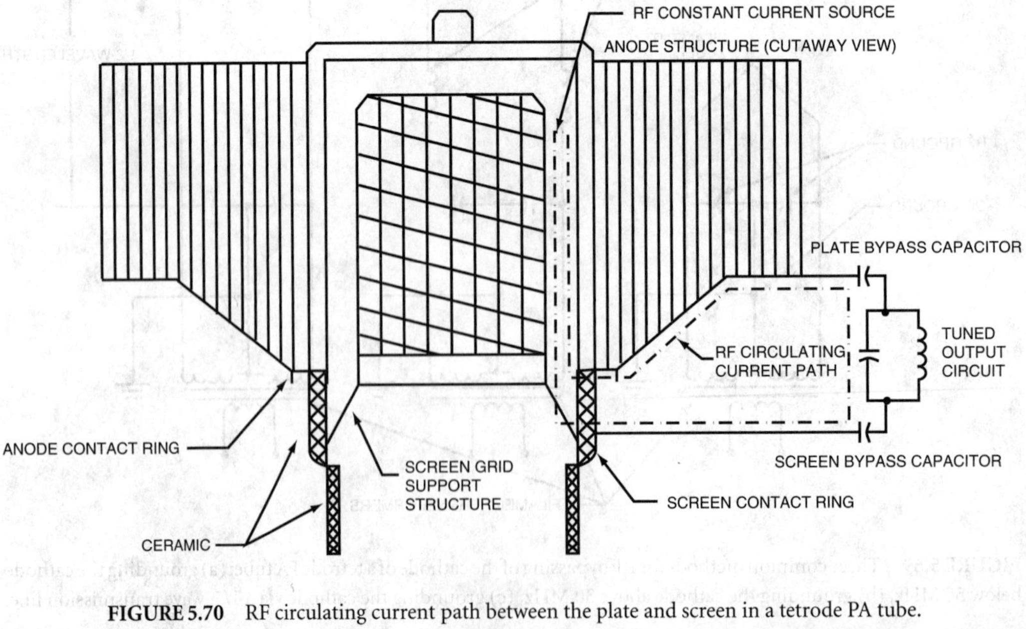

FIGURE 5.70   RF circulating current path between the plate and screen in a tetrode PA tube.

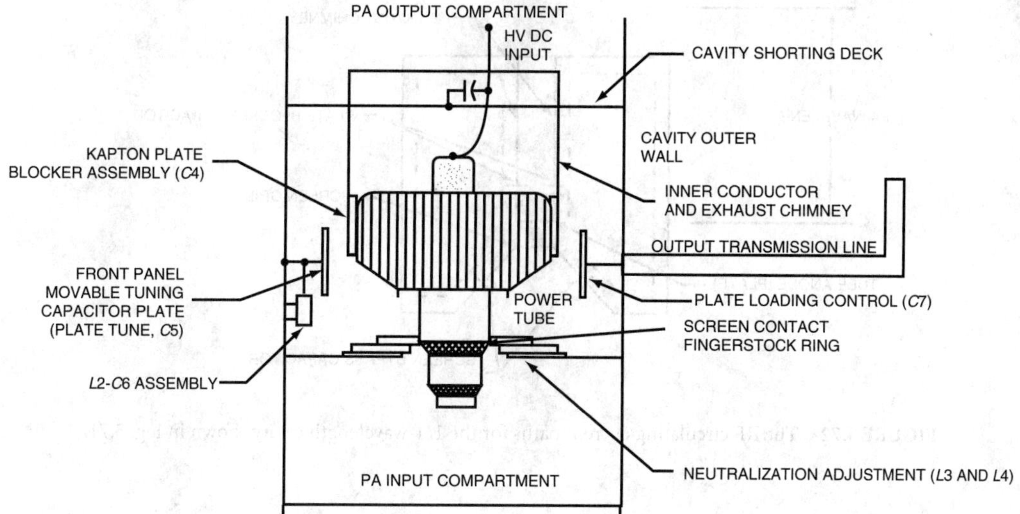

FIGURE 5.71   The physical layout of a common type of 1/4-wavelength PA cavity designed for VHF operation.

access panel (door) of the cavity is part of the outer conductor and large values of circulating current flow into it, through it, and out of it. A mesh contact strap is generally used to electrically connect the access panel to the rest of the cavity.

1/4-Wavelength Cavity

The 1/4-wavelength PA cavity is common in transmitting equipment. The design is simple and straightforward. A number of variations can be found in different RF generators, but the underlying theory of operation is the same.

A typical 1/4-wave cavity is shown in Fig. 5.71. The plate of the tube connects directly to the inner section (tube) of the plate-blocking capacitor. The exhaust chimney/inner conductor forms the other plate of the blocking capacitor. The cavity walls form the outer conductor of the 1/4-wave transmission line circuit. The DC-plate voltage is applied to the PA tube by a cable routed inside the exhaust chimney and inner tube conductor. In the design shown in the figure, the screen-contact fingerstock ring mounts on a metal plate that is insulated from the grounded cavity deck by an insulating material. This hardware makes up the screen-blocking capacitor assembly. The DC-screen voltage feeds to the fingerstock ring from underneath the cavity deck through an insulated feedthrough assembly.

A grounded-screen configuration may also be used in this design in which the screen-contact fingerstock ring is connected directly to the grounded cavity deck. The PA cathode then operates at below ground potential (in other words, at a negative voltage), establishing the required screen voltage to the tube.

The cavity design shown in the figure is set up to be slightly shorter than a full 1/4-wavelength at the operating frequency. This makes the load inductive and resonates the tube's output capacity. Thus, the physically foreshortened shorted transmission line is resonated and electrically lengthened to 1/4-wavelength.

Figure 5.72 illustrates the paths taken by the RF-circulating currents in the circuit. RF energy flows from the plate, through the plate-blocking capacitor, along the inside surface of the chimney/inner conductor (because of the skin effect), across the top of the cavity, down the inside surface of the cavity box, across the cavity deck, through the screen-blocking capacitor, over the screen-contact fingerstock and into the screen grid.

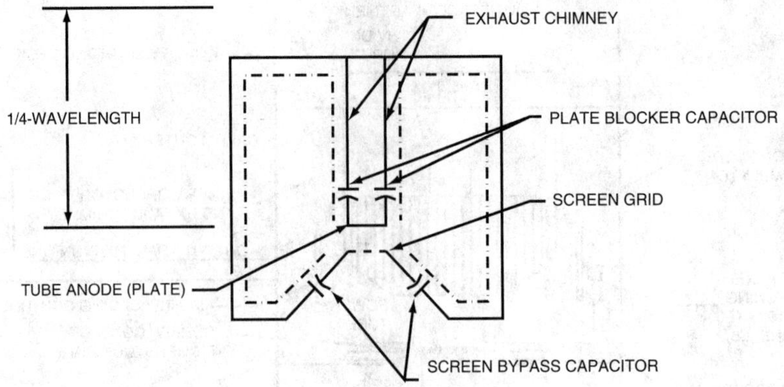

FIGURE 5.72   The RF circulating current paths for the 1/4-wavelength cavity shown in Fig. 5.71.

Figure 5.73 shows a graph of RF current, voltage, and impedance for a 1/4-wavelength coaxial transmission line. It shows that infinite impedance, zero RF current, and maximum RF voltage occur at the feed point. This would not be suitable for a practical PA circuit because arcing would result from the high RF voltage, and poor efficiency would be caused by the mismatch between the tube and the load.

Notice, however, the point on the graph marked at slightly less than 1/4-wavelength. This length yields an impedance of 600–800 Ω and is ideal for the PA-plate circuit. It is necessary, therefore, to physically foreshorten the shorted coaxial transmission-line cavity to provide the correct plate impedance. Shortening the line also is a requirement for resonating the tube's output capacity, because the capacity shunts the transmission line and electrically lengthens it.

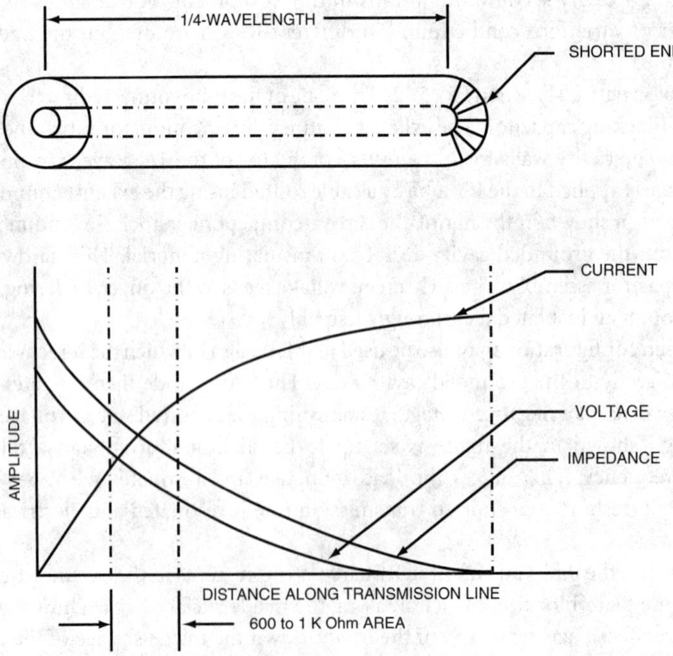

FIGURE 5.73   Graph of the RF current, RF voltage, and RF impedance for a 1/4-wavelength shorted transmission line. At the feed point RF current is zero, RF voltage is maximum, and RF impedance is infinite.

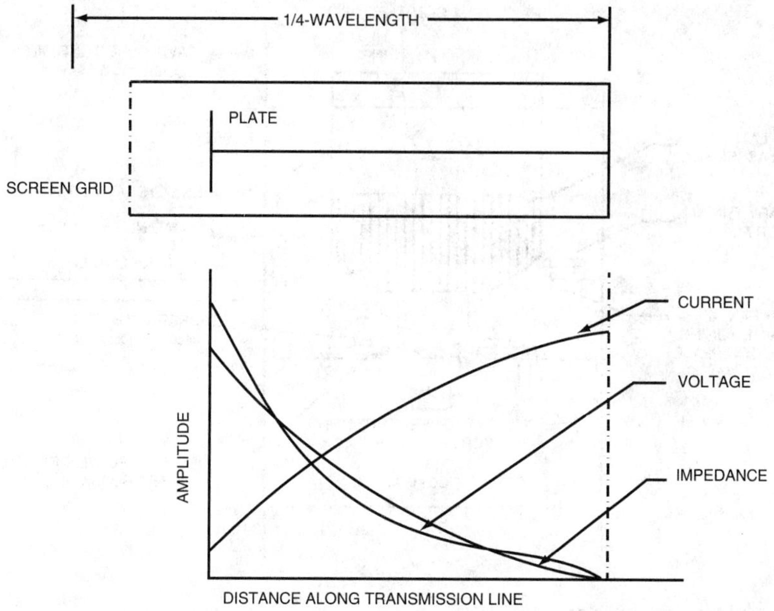

FIGURE 5.74 Graph of the RF current, RF voltage, and RF impedance produced by the physically foreshortened coaxial transmission line cavity.

Figure 5.74 shows a graph of the RF current, voltage, and impedance presented to the plate of the tube as a result of the physically foreshortened line. This plate impedance is now closer to the ideal 600–800 Ω value required by the tube anode.

Tuning the Cavity

Coarse tuning of the cavity is accomplished by adjusting the cavity length. The top of the cavity (the cavity shorting deck) is fastened by screws or clamps and can be raised or lowered to set the length of the cavity for the particular operating frequency. Fine tuning is accomplished by a variable-capacity plate-tuning control built into the cavity. In a typical design, one plate of the tuning capacitor, the stationary plate, is fastened to the inner conductor just above the plate-blocking capacitor. The movable tuning plate is fastened to the cavity box, the outer conductor, and is mechanically linked to the front-panel tuning control. This capacity shunts the inner conductor to the outer conductor and is used to vary the electrical length and resonant frequency of the cavity.

The 1/4-wavelength cavity is inductively coupled to the output port. This coupling is usually on the side opposite the cavity access door. The inductive pickup loop can take several forms. In one design it consists of a half-loop of flat copper bar stock that terminates in the loading capacitor at one end and feeds the output transmission line inner conductor at the other end. This configuration is shown in Fig. 5.75. The inductive pickup ideally would be placed at the maximum current point in the 1/4-wavelength cavity. However, this point is located at the cavity shorting deck and when the deck is moved for coarse tuning, the magnetic coupling will be changed. A compromise in positioning, therefore, must be made. The use of a broad, flat copper bar for the coupling loop adds some capacitive coupling to augment the reduced magnetic coupling.

Adjustment of the loading capacitor couples the 50-Ω transmission-line impedance to the impedance of the cavity. Heavy loading lowers the plate impedance presented to the tube by the cavity. Light loading reflects a higher load impedance to the amplifier plate.

Methods commonly used to increase the operating bandwidth of the cavity include minimizing added tuning capacitance. The ideal case would be to resonate the plate capacitance alone with a perfect inductor. Practical cavities, however, require either the addition of a variable capacitor or a variable inductor using

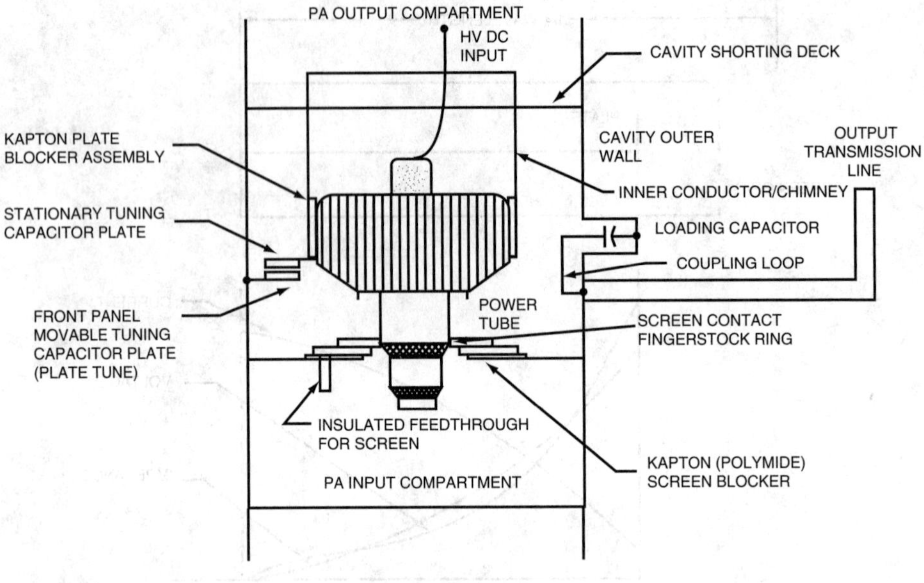

FIGURE 5.75   Basic loading mechanism for a VHF 1/4-wave cavity.

sliding contacts for tuning. Other inherent mechanical and electrical compromises include:

- The requirement for a plate DC blocking capacitor.
- The presence of maximum RF current at the grounded end of the simulated transmission line where the conductor may be nonhomogeneous. This can result in accompanying losses in the contact resistance.

Wideband Cavity

The cavity systems discussed so far in this section provide adequate bandwidth for many applications. Some uses, however, dictate an operating bandwidth beyond what a conventional cavity amplifier can provide. One method of achieving wider operating bandwidth involves the use of a double-tuned overcoupled amplifier, shown in Fig. 5.76. The system includes four controls to accomplish tuning:

- *Primary tune*, which resonates the plate circuit and tends to tilt the response and slide it up and down the bandpass.
- *Coupling*, which sets the bandwidth of the PA. Increased coupling increases the operating bandwidth and lowers the PA efficiency. When the coupling is adjusted, it can tilt the response and change the center of the bandpass, necessitating readjustment of the plate tune control.
- *Secondary tune*, which resonates the secondary cavity and will tilt the response if so adjusted. The secondary tune control typically can be used to slide the response up and down the bandpass, as the primary tune control.
- *Loading*, which determines the value of ripple in the bandpass response. Adjustment of the loading control usually tilts the response and changes the bandwidth, necessitating readjustment of the secondary tune and coupling controls.

Output Coupling

Coupling is the process by which RF energy is transferred from the amplifier cavity to the output transmission line. Wideband cavity systems use coupling to transfer energy from the primary cavity to the secondary cavity. Coupling in tube type power amplifiers usually transforms a high (plate or cavity) impedance to a lower output (transmission line) impedance. Both capacitive (electrostatic) and inductive (magnetic) coupling methods are used in cavity RF amplifiers. In some designs, combinations of the two methods are used.

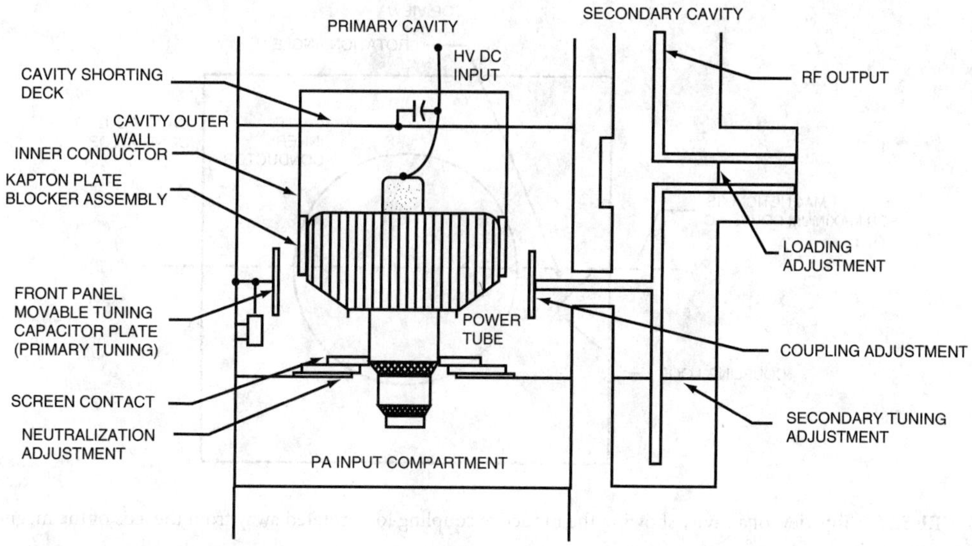

FIGURE 5.76 Double-tuned overcoupled **broadband cavity amplifier**. (*Source:* Adapted from Weirather, R., Best, G., and Plonka, R. 1992. TV transmitters. In *NAB Engineering Handbook*, ed. E.B. Crutchfield, 8th ed. p. 524. National Association of Broadcasters, Washington, DC.)

Inductive Coupling

Inductive (magnetic) coupling employs the principles of transformer action. The efficiency of the coupling depends on three conditions:

- The cross-sectional area under the coupling loop, compared to the cross-sectional area of the cavity (see Fig. 5.77). This effect can be compared to the turns ratio principle of a transformer.
- The orientation of the coupling loop to the axis of the magnetic field. Coupling from the cavity is proportional to the cosine of the angle at which the coupling loop is rotated away from the axis of the magnetic field, as illustrated in Fig. 5.78.
- The amount of magnetic field that the coupling loop intercepts (see Fig. 5.79). The strongest magnetic field will be found at the point of maximum RF current in the cavity. This is the area where maximum inductive coupling is obtained. Greater magnetic field strength also is found closer to the center conductor of the cavity. Coupling, therefore, is inversely proportional to the distance of the coupling loop from the center conductor.

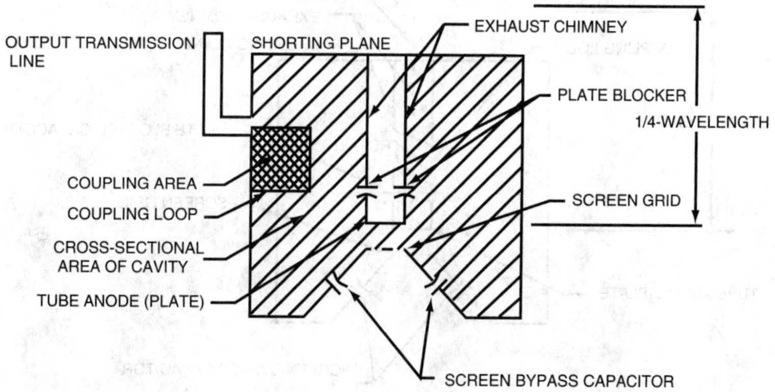

FIGURE 5.77 The use of inductive coupling in a 1/4-wavelength PA stage.

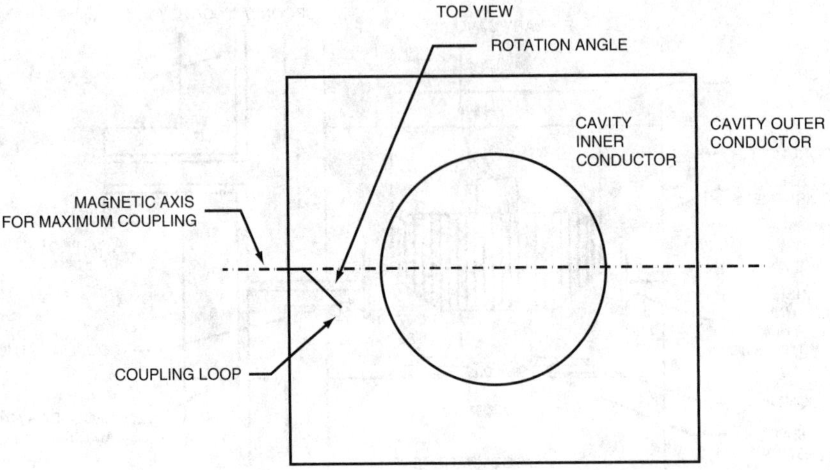

FIGURE 5.78 Top view of a cavity showing the inductive coupling loop rotated away from the axis of the magnetic field of the cavity.

In 1/4-wavelength cavities, the coupling loop generally feeds a 50-Ω transmission line (the load). The loop is in series with the load and has considerable inductance at VHF frequencies. This inductance will reduce the RF current that flows into the load, thus reducing power output. This effect can be overcome by placing a variable capacitor in series with the output coupling loop. The load is connected to one end of the coupling loop and the variable capacitor ties the other end of the loop to ground. The variable capacitor cancels some or all of the loop inductance. It functions as the PA-stage loading control.

Maximum loop current and output power occurs when the loading capacitor cancels all of the inductance of the loading loop. This lowers the plate impedance and results in heavier loading.

Light loading results if the loading capacitance does not cancel all of the loop inductance. The loop inductance that is not canceled causes a decrease in load current and power output, and an increase in plate impedance.

Capacitive Coupling

Capacitive (electrostatic) coupling, which physically appears to be straightforward, often baffles the applications engineer because of its unique characteristics. Figure 5.80 shows a cavity amplifier with a capacitive-coupling plate positioned near its center conductor. This coupling plate is connected to the

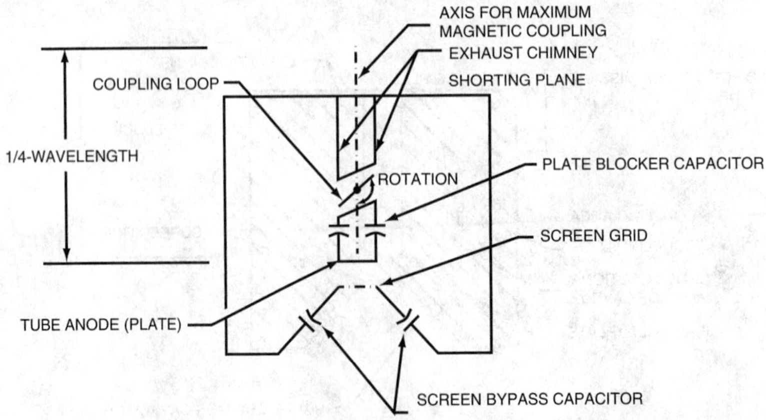

FIGURE 5.79 Cutaway view of the cavity showing the coupling loop rotated away from the axis of the magnetic field.

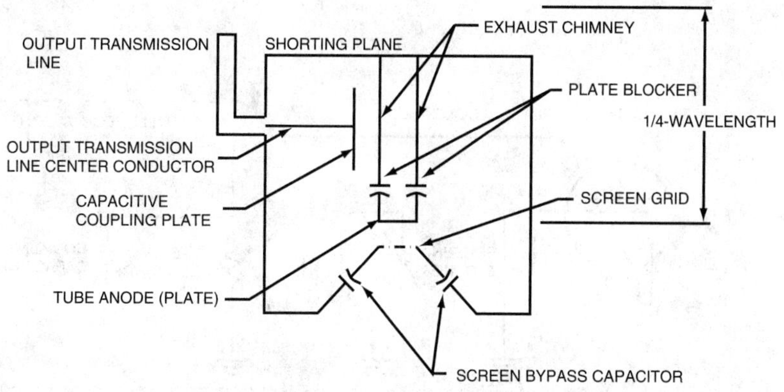

FIGURE 5.80 The 1/4-wavelength cavity with capacitive coupling to the output load.

output load, which can be a transmission line or a secondary cavity (for wideband operation). The parameters that control the amount of capacitive coupling include

- The area of the coupling-capacitor plate (the larger the area, the greater the coupling)
- The distance from the coupling plate to the center conductor (the greater the distance, the lighter the coupling)

Maximum capacitive coupling occurs when the coupling plate is at the maximum voltage point on the cavity center conductor.

To understand the effects of the capacitive coupling, the equivalent circuit of the cavity must be observed. Fig. 5.81 shows the PA tube, cavity (functioning as a parallel resonant circuit), and output section. The plate-blocking capacitor isolates the tube's DC voltage from the cavity. The coupling capacitor and output load are physically in series, but electrically they appear to be in parallel, as shown in Fig. 5.82. The resistive component of the equivalent parallel circuit is increased by the coupling reactance. The equivalent parallel coupling reactance is absorbed into the parallel resonant circuit. This explains the necessity to retune the plate after changing the PA stage coupling (loading). The coupling reactance can be a series capacitor or inductor.

The series-to-parallel transformations are described by the following formulas

$$R_p = \frac{R_s^2 + X_s^2}{R_s}$$

$$X_p = \frac{R_s^2 + X_s^2}{X_s}$$

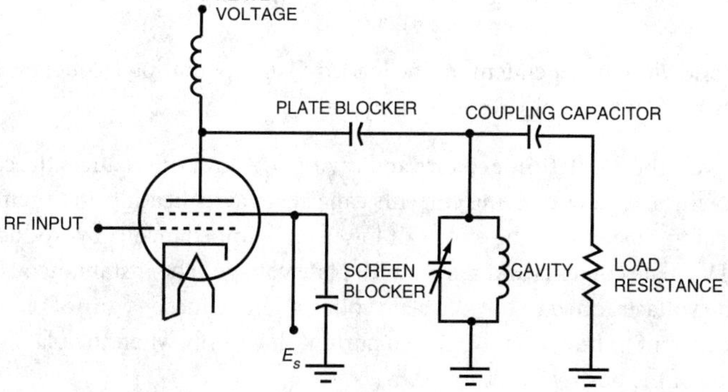

FIGURE 5.81 The equivalent circuit of a 1/4-wavelength cavity amplifier with capacitive coupling.

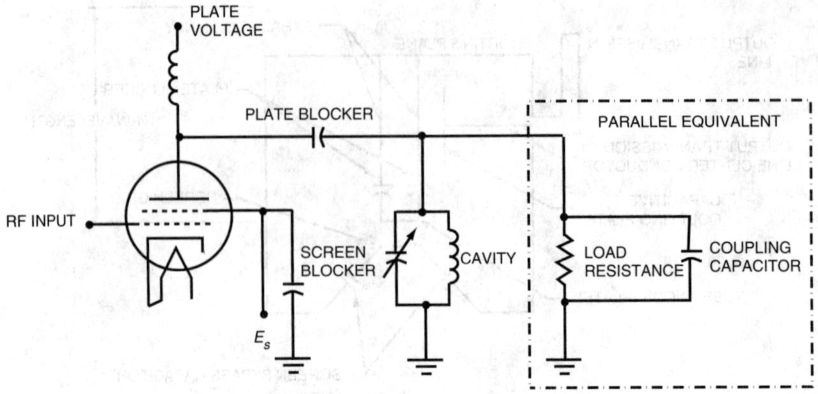

FIGURE 5.82 The equivalent circuit of a 1/4-wavelength cavity amplifier showing how series capacitive coupling appears electrically as a parallel circuit to the PA tube.

where

R_p = effective parallel resistance
R_s = actual series resistance
X_s = actual series reactance
X_p = effective parallel reactance

PA Loading

Proper **loading** of a cavity PA stage to the output transmission line is critical to dependable operation and optimum efficiency of the overall system. Light coupling produces light loading and results in a high plate impedance; conversely, heavy coupling results in heavier loading and a lower plate impedance. Maximum output power, coinciding with maximum efficiency and acceptable dissipation, dictates a specific plate impedance for a cavity of given design. This plate impedance is also dependent upon the values of DC plate voltage (E_p) and plate current (I_p).

Plate impedance dictates the cavity parameters of loaded Q, RF circulating current, and bandwidth. The relationships can be expressed as follows:

- Loaded Q is directly proportional to the plate impedance and controls the other two cavity parameters. Loaded $Q = Z_p/X_l$, where Z_p is the cavity plate impedance and X_l is the cavity inductive reactance.
- Circulating current in the cavity is much greater (by a factor of the loaded Q) than the RF current supplied by the tube. Circulating current = $Q \times I_p$, where I_p is the RF current supplied to the cavity by the tube.
- The cavity bandwidth is dependent on the loaded Q and operating frequency. Bandwidth is equal to F_r/Q, where F_r is the cavity resonant frequency.

Heavy loading lowers the PA plate impedance and cavity Q. A lower Q reduces the cavity RF circulating currents. In some cavities, high circulating currents can cause cavity heating and premature failure of the plate or screen blocking capacitors. The effects of lower plate impedance, a byproduct of heavy loading, are higher RF and DC plate currents and reduced RF plate voltage. The instantaneous plate voltage is the result of the RF plate voltage added to the DC plate voltage. The reduced swing of plate voltage causes less positive DC screen current to flow. Positive screen current flows only when the plate voltage swings close to or below the value of the positive screen grid.

Light loading raises the plate impedance and cavity Q. A higher Q will increase the cavity circulating currents, raising the possibility of component overheating and failure. The effects of higher plate impedance are reduced RF and DC plate current and increased RF and DC plate voltage excursions. The higher cavity RF or peak DC voltages may cause arcing in the cavity.

There is one value of plate impedance that will yield optimum output power, efficiency, dissipation, and dependable operation. It is dictated by cavity design and the values of the various DC and RF voltages and currents supplied to the stage.

Depending on the cavity design, light loading may seem deceptively attractive. The DC plate voltage is constant (set by the power supply), and the lower DC plate current resulting from light loading reduces the tube's overall DC input power. The RF output power may change with light loading, depending on the plate impedance and cavity design, while efficiency will probably increase or, at worst, remain constant. Caution must be exercised with light loading, however, because of the increased RF voltages and circulating currents that such operation creates. Possible adverse effects include cavity arcing and overheating of cavity components, such as capacitors. The manufacturer's recommendations on PA tube loading should, therefore, be carefully observed.

Although there are many similarities among various cavity designs, each one imposes its own set of operational requirements and limitations. No two cavity systems will tune up in exactly the same fashion. Given proper maintenance, a cavity amplifier will provide years of reliable service.

Mechanical Design

Understanding the operation of a cavity amplifier is usually difficult because of the nature of the major component elements. It is often difficult to relate the electrical schematic diagram to the mechanical assembly that exists within the transmitter. Consider the PA cavity schematic diagram shown in Fig. 5.83. The grounded-screen stage is of conventional design. Decoupling of the high-voltage power supply is accomplished by $C1, C2, C3$, and $L1$. Capacitor $C3$ is located inside the PA chimney (cavity inner conductor). The RF sample lines provide two low-power RF outputs for a modulation monitor or other test instrument. Neutralization inductors $L3$ and $L4$ consist of adjustable grounding bars on the screen grid ring assembly.

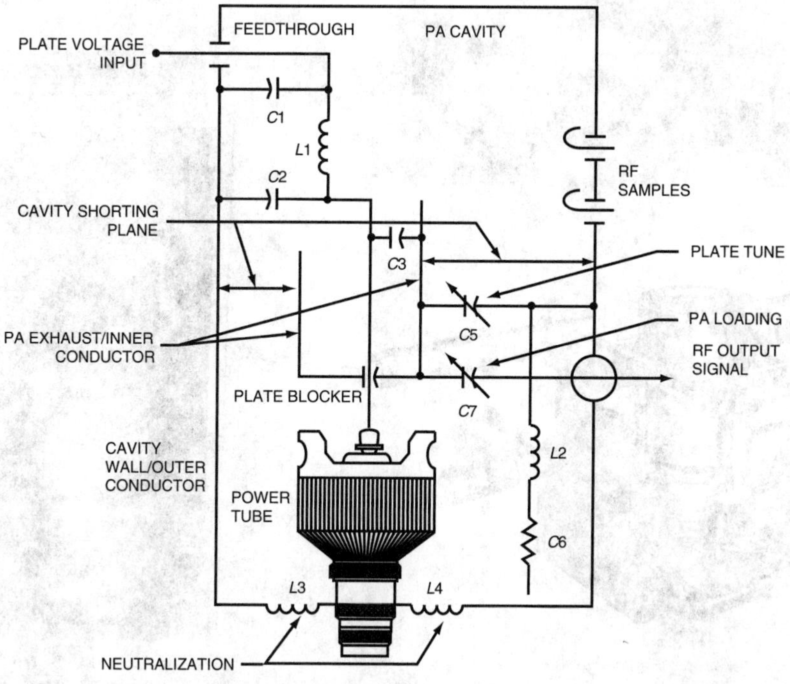

FIGURE 5.83 Typical VHF cavity amplifier.

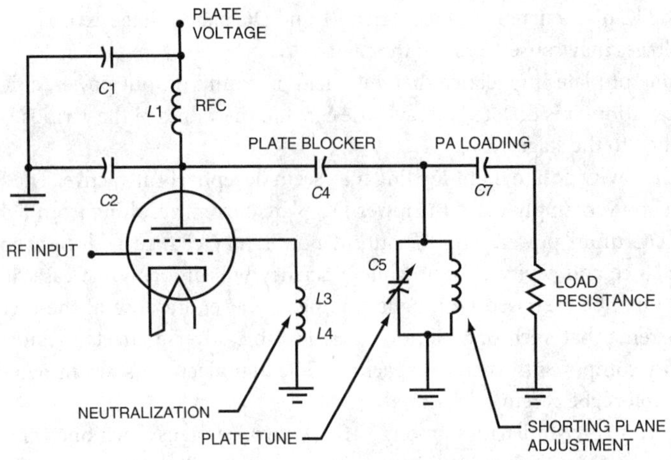

FIGURE 5.84   Electrical equivalent of the cavity amplifier shown in Fig. 5.83.

Figure 5.84 shows the electrical equivalent of the PA cavity schematic diagram. The 1/4-wavelength cavity acts as the resonant tank for the PA. Coarse tuning of the cavity is accomplished by adjustment of the shorting plane. Fine tuning is performed by the PA tuning control, which acts as a variable capacitor to bring the cavity into resonance. The PA loading control consists of a variable capacitor that matches the cavity to the load. The assembly made up of $L2$ and $C6$ prevents spurious oscillations within the cavity.

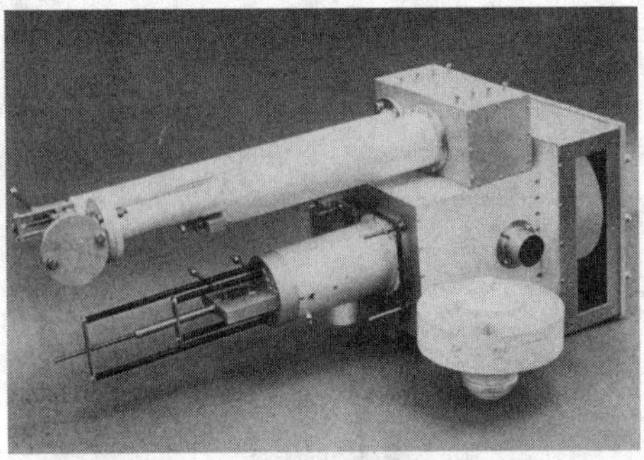

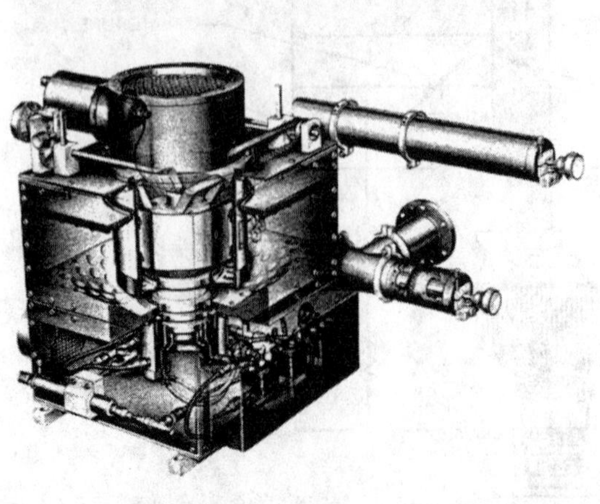

FIGURE 5.85   VHF cavity amplifiers: (a) cross-sectional view of a broadband design, (b) cavity designed for television service, (c) FM broadcast cavity amplifier.

Blocking capacitor $C4$ is constructed of a roll of *Kapton*^{TM} insulating material sandwiched between two circular sections of aluminum. (Kapton is a registered trademark of Du Pont.) PA plate tuning control $C5$ consists of an aluminum plate of large surface area that can be moved in or out of the cavity to reach resonance. PA loading control $C7$ is constructed much the same as the PA tuning assembly, with a large-area paddle feeding the harmonic filter, located external to the cavity. The loading paddle may be moved toward the PA tube or away from it to achieve the required loading. The $L2$-$C6$ damper assembly consists of a 50-Ω noninductive resistor mounted on the side of the cavity wall. Component $L2$ is formed by the inductance of the connecting strap between the plate tuning paddle and the resistor. Component $C6$ is the equivalent stray capacitance between the resistor and the surrounding cavity box.

It can be seen that cavity amplifiers involve as much mechanical engineering as they do electrical engineering. The photographs of Fig. 5.85 show graphically the level of complexity that a cavity amplifier may involve. Figure 5.85(a) depicts a VHF power amplifier (Philips) with broadband input circuitry. Figure 5.85(b) shows a wideband VHF amplifier intended for television service (Varian). Figure 5.85(c) illustrates a VHF cavity amplifier designed for FM broadcast service (Varian).

Defining Terms

Broadband cavity amplifier: A cavity amplifier made to operating over a broad range of frequencies, typically through the addition of a secondary cavity that is coupled to the primary.

Cavity amplifier: A vacuum tube-based amplifying stage in which the physical elements of the output resonating circuit consist of the tube chimney and enclosing box. These and related components in the output circuit typically simulate a 1/4- or 1/2-wavelength transmission line.

Excitation: The input signal (typically to the grid of a power vacuum tube) that controls the operation of the stage, usually an amplifier or oscillator.

Flat topping: A distortion mechanism in an amplifying stage in which the stage is unable to faithfully reproduce the positive peaks of the output signal. Reasons for such problems include poor regulation of the plate voltage supply.

Loading: The process of extracting energy from a cavity amplifier through either inductive or capacitive coupling.

Operating class: A division of operating modes for an amplifying device. In the case of power vacuum tubes, the operating class is determined by the angle of plate current flow as follows: class A = 360° conduction angle, class B = 180° conduction angle, class C = conduction angle less than 180°, and class AB = conduction angle between 180 and 360°.

References

Crutchfield, E.B., ed. 1992. *NAB Engineering Handbook*, 8th ed. National Association of Broadcasters, Washington, DC.

Gray, T.S. 1954. *Applied Electronics*. MIT Press, Cambridge, MA.

Heising, R.A. 1921. Modulation in radio telephony. *Proceedings of the IRE* 9(3):305.

Heising, R.A. 1928. Transmission system. U.S. Patent 1,655,543, Jan.

Jordan, E.C. 1985. *Reference Data for Engineers: Radio, Electronics, Computer and Communications*, 7th ed. Howard W. Sams, Indianapolis, IN.

Martin, T.L., Jr. 1955. *Electronic Circuits*. Prentice-Hall, Englewood Cliffs, NJ.

Mendenhall, G.N., Shrestha, M.B., and Anthony, E.J. 1992. FM broadcast transmitters. In *NAB Engineering Handbook*, ed. E.B. Crutchfield, 8th ed., pp. 429–474. National Association of Broadcasters, Washington, DC.

Shrestha, M.B. 1990. Design of tube power amplifiers for optimum FM transmitter performance. *IEEE Trans. Broadcasting* 36(1):46–64.

Terman, F.E. and Woodyard, J.R. 1938. A high efficiency grid-modulated amplifier. *Proceedings of the IRE* 26(8): 929.

Terman, F.E. 1947. *Radio Engineering*. McGraw-Hill, New York.

Varian. 1984. *The Care and Feeding of Power Grid Tubes*. Varian/EIMAC, San Carlos, CA.

Weirather, R.R., Best, G.L., and Plonka, R. 1992. TV transmitters. In *NAB Engineering Handbook*, ed. E.B. Crutchfield, 8th ed., pp. 499–551. National Association of Broadcasters, Washington, DC.

Whitaker, J.C. 1991. *Radio Frequency Transmission Systems: Design and Operation*. McGraw-Hill, New York.

Whitaker, J.C. 1999. *Power Vacuum Tubes Handbook*. 2nd ed. CRC Press, Boca Raton, FL.

Woodard, G.W. 1992. AM transmitters. In *NAB Engineering Handbook*, ed. E.B. Crutchfield, 8th ed., pp. 353–381. National Association of Broadcasters, Washington, DC.

Further Information

Specific information on the design of power amplifier systems using vacuum tubes can usually be obtained from the manufacturers of those devices. More general application information can be found in the following publications:

Whitaker, J.C. 1999. *Power Vacuum Tubes Handbook*. 2nd ed. CRC Press, Boca Raton, FL.

Golio, M. 2000. *RF and Microwave Handbook*. CRC Press, Boca Raton, FL.

5.5 Image Capture Devices

Steve Epstein

5.5.1 Introduction

The invention of image capture tubes, and cathode ray tube (CRT) technology for display, is at the heart of what eventually became television. Near the end of the 19th century, when early work was being done on these devices, generating artificial images (i.e., test patterns) simply was not practical. Image capture devices allowed real world images to be captured and turned into an electronic signal. Early attempts were mechanical, followed later by electronic vacuum tube devices. As tube technology improved, picture **resolution** increased to the point where, during the 1960s and 1970s, a variety of tubes were found in professional video cameras throughout the world.

Ultimately tube technology improved to the point where reasonably high-resolution images (>750 TV lines) could be obtained with compact, lightweight devices. These units found their way into the electronic news gathering (ENG) cameras of the mid-1970s. These cameras, along with portable videocassette recorders, changed the face of television news forever. No longer burdened by the requirement of film, television news became more immediate. Stories shot within minutes of newscasts could be edited and aired.

By the 1980s, solid-state charge-coupled devices (CCDs) were finding their way into cameras. Early CCDs, like early tubes, had their problems. Today, however, CCDs offer high resolution, long-life, high **sensitivity**, and negligible maintenance costs. Tube devices did see a short resurrection during the early design phase of high-definition television, however, increased resolution CCDs became available and high definition development work quickly moved toward the CCD. The majority of the image pickup tubes available in quantity today are of the Plumbicon™ variety.

5.5.2 The Plumbicon

The development of the Plumbicon in 1965 (Philips Components) was an enormous technological achievement. At a time when color television was just getting off the ground, the Plumbicon paved the way for its explosive growth. The Plumbicon and similar tubes, including the Leddicon™ (English Electric Value Company, Ltd. (EEV)), used a lead-oxide photoconductive layer. Lead oxide is a solid-state semiconductor, a porous vaporgrown crystalline layer of lead monoxide.

The basic physical construction of the Plumbicon is shown in Fig. 5.86. The photosensitive PbO layer is deposited on a transparent electrode of SnO_2 on the window of the tube. This conducting layer is the target electrode and is typically biased at about 40 V positive with respect to the cathode. The remainder of the tube structure is designed to provide a focused, low-energy electron beam for scanning the target.

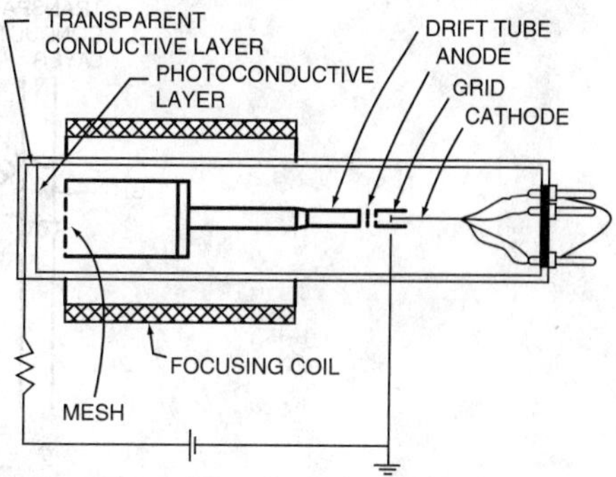

Figure 5.87 provides a more detailed schematic of the operation of the Plumbicon. The side of the PbO layer facing the gun is reduced to cathode potential by the charge deposited by the scanning beam. Thus, the full target voltage appears across the layer. In the absence of an optical signal (lower-half of figure), there is no mechanism for discharging, and the surface charges to such a potential that further charge deposition, often called *beam landing*, is prevented. At this point, there is no further current induced in the external circuit. When light is incident as in the

FIGURE 5.86 Basic geometry of the Plumbicon tube. (*Source:* Stupp, E.H. 1968. *Physical Properties of the Plumbicon.*)

upper-half of the figure, photocarriers are generated. Under the influence of the electric field across the layer, these carriers move in the appropriate direction to discharge the stored charge on the surface. As the scanning beam passes across the varied potential of the surface, an additional charge can be deposited on the higher positive potential regions. The amount of charge deposited is equal to that discharged by the light. The attendant current in the external circuit is the photosignal.

In the Plumbicon, the lead oxide layer acts as a *p-i-n* diode. The *n*-region is formed at the $PbO : SnO_2$ interface (window side). The *p*-region is a thin layer on the electron-gun side of the layer. The bulk of the material behaves as if it were intrinsic, the *i*-layer. Thus, the electron beam landing on the *p*-region charges it negatively with respect to the *n*-region. The *p-i-n* layer is thus biased in the reverse or low conduction direction, and in the absence of illumination, little or no current flows.

Figure 5.88 shows a simplified diagram of a reverse biased *p-i-n* diode. Virtually all of the applied voltage *V* appears across the *i*-region. The **dark current** of this device is determined by the inverse currents through the *p-i* and *i-n* junctions and the thermal generation of carriers in the *i*-region. To ensure that the dark current in the tube due to these sources is sufficiently low at ambient temperatures, a semiconductor of bandgap greater than 1.0 eV is required. For the somewhat elevated temperatures that are likely to be encountered in a camera, a still larger bandgap is needed. The bandgap of the tetragonal lead oxide used in a Plumbicon is about 1.9 eV, and extremely low dark currents result.

When the target structure is illuminated, the amount of current is dependent on photon energy and wavelength. Both blue and green light generate reasonable amounts of current. Plumbicons, however, are not particularly responsive to wavelengths in the red region. To improve the red response photoconductors

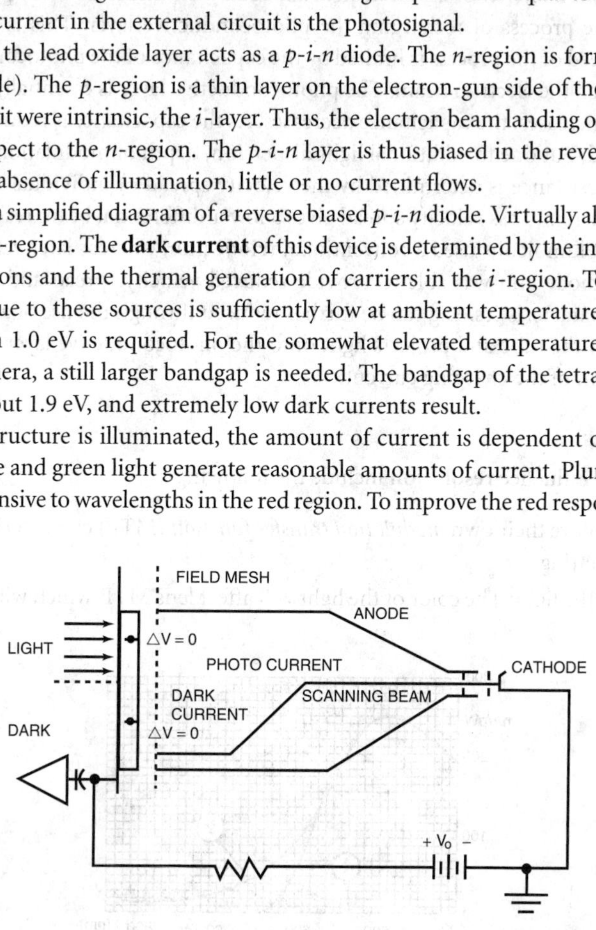

FIGURE 5.87 Schematic equivalent of the Plumbicon. (*Source:* Stupp, E.H. 1968. *Physical Properties of the Plumbicon.*)

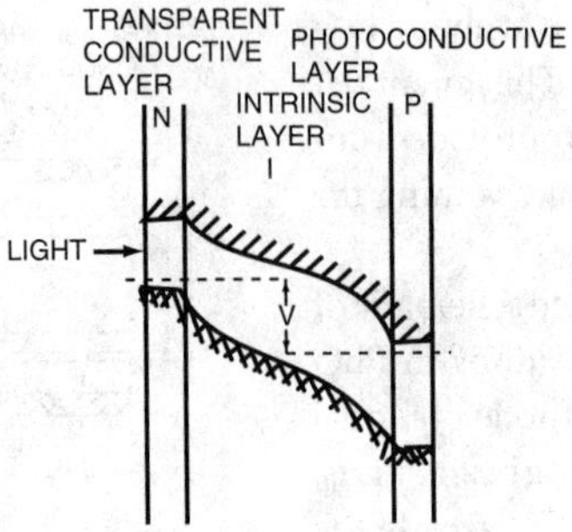

FIGURE 5.88 Simplified energy band characteristics of a reverse biased *p-i-n* diode.

are doped, usually with sulfur. These specially prepared tubes, often called *extended reds*, are then used in the red channel to improve camera performance (Fig. 5.89).

5.5.3   Operating Characteristics

For camera tubes, two important parameters to consider are **modulation depth** and **lag**. Modulation depth is determined by the spot diameter of the electron beam and by the thickness of the photoconductive layer. Lag is governed by the process of recharging the photoconductive layer. This is accomplished by the electron beam. The speed of the process is limited by the capacitance of the photoconductive layer and the differential resistance of the electron beam. The photoconductive layer is essentially capacitive in nature; as the layer thickness increases, the capacitance decreases. As the capacitance is reduced, response time for recharging is improved, but due to additional light dispersion, depth of modulation decreases.

Differential beam resistance is determined by the velocity spread of individual electrons in the beam and by the charging current. Less velocity spread gives a lower differential beam resistance and, therefore, faster response. Early Plumbicons used a conventional **triode gun**, whereas later versions took advantage of a **diode gun**. The diode gun was able to provide reduced velocity spread and, therefore, offered an improvement in lag characteristics. Lag can be further reduced through the use of a *bias light*. A bias light is placed at the base of the tube and projects light onto the photoconductive layer. This has the effect of offsetting the zero point of the charging current.

Designing for Tubes

For tubes, factors that influence **resolution** include the following.

- Lenses: Lenses have their own *modulation transfer function* (MTF) characteristics that change with aperture of *f* setting.
- Wavelength of the light: The color of the light will affect lens MTF, which will influence resolution.

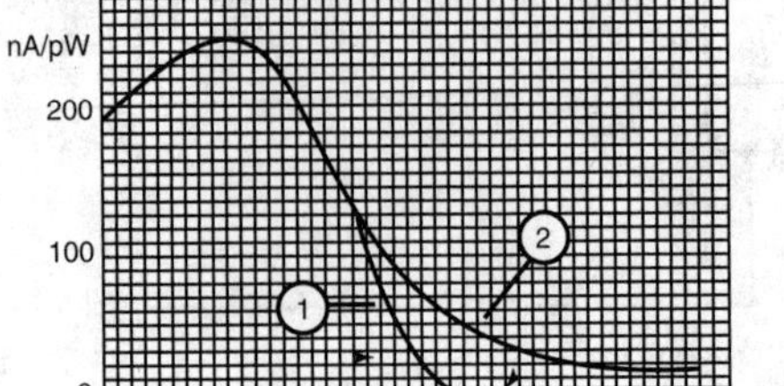

FIGURE 5.89   Differences in the light sensitivity of (1) standard and (2) extended red Plumbicons.

- Imaging tube properties: The photosensitive layer and the beam spot size will affect maximum resolution.
- Yoke properties: The predominant function of the yoke is to provide the scanning action of the beam and to focus the beam. The focus field, along with the G_2, G_3, and G_4 electrodes, forms an electromagnetic lens. This lens influences the beam shape at the target. Focus field strength in the vicinity of the target strongly influences spot size, as does the voltage ratio of G_4 to G_3.
- Flare or veiling glare: Flare can reduce resolution by softening edges. Various methods are employed to reduce flare including special coatings on lens surfaces and antihalation pills on the front surface of Plumbicons.
- Signal preamps: Preamp frequency response directly effects output resolution. Preamps should be swept with a device that accurately simulates the tube to verify that high-frequency **noise** is minimal and that overall response is flat.
- Noise: Noise is the electrical equivalent of attempting to view objects through a snowstorm. Resolution is reduced by the noise. Poor signal-to-noise response results in pictures with apparently poor resolution due to the noise inherent in the image.
- Beam alignment: If beam alignment is not optimized, the beam will not land perpendicular to the center of the target, thereby reducing center resolution.

When designing circuitry for tubes, it is best to consult the manufacturer for data specific to the device to be used. For general purposes, the specifications of a generic 1.5-in tube will be discussed here so that designers will be familiar with the concepts involved. Because a 1.5-in tube is fairly large, voltage and current specifications will generally be higher than what would be required in smaller tubes. Like most tubes, imaging tubes require relatively high voltage and significant amounts of current and, generally, produce heat that must be dissipated. Because circuit requirements and output characteristics can vary with temperature, dissipation must be considered at the design stage.

Imaging tubes generally have seven or eight pins on their base, arranged around the center. Many times a slot is left for a ninth pin, which is used as an alignment reference. Other tubes have a registration mark on the side. This mark is used to align the tube within the deflection yokes. This orients the internal tube structures within the external magnetic fields such that tube performance is maximized. Bias lights are internal on some tubes, others have external units that fit over the base pins and remain in place when the socket is installed. Most bias lights use the heater supply as their power source, adding approximately 100 mA each to the power supply requirements. In existing cameras adapted for operation with a bias light, the added current is not usually a problem and modifications are not usually required. Additional heating is negligible.

Imaging tubes require heaters that normally draw less than 1 A at 6 V. Cathode voltage is generally 0 V; however, most are driven positive during blanking periods. Current requirements are usually less than 250 nA. The target is typically biased at +40 V for Plumbicons and up to +125 V for vidicons. The signal electrode is attached to the target and is generally brought out of the glass envelope near the front of the unit rather than through the base at the rear. Because of the small currents involved, the target ring and its mating connector within the camera head must be kept clean.

Up to four internal grid structures are connected to pins located on the base along with the heater and cathode pins. Grid 1 (G_1) is used to blank the beam when necessary, and is typically biased from 0 (beam on) to −100 V (beam off). G_2 is used to accelerate the beam and is biased at +300 V or thereabouts. G_3 is used to focus the beam and is normally set from +800 to +1000 V. G_4 is the mesh assembly located adjacent to the target and is typically biased at +1400 V.

External structures are also required for proper tube operation. These structures fall into two categories, mechanical and electrical. Mechanical structures include the mounting assembly and its associated adjustments. In a typical application, a locking unit holds the tube and its electromagnetic coils rigidly in place. When loosened, the tube and coils can be adjusted within a small range. One adjustment moves the tube back and forth to obtain optimal optical focus, the other rotates the assembly to align the tube with the horizontal and vertical axes of the image. An additional unit, usually a locking ring, is used to hold the tube in place within the deflection coil assembly.

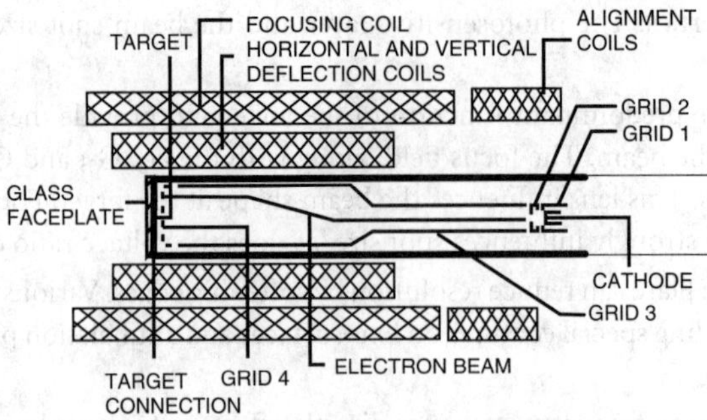

FIGURE 5.90   Internal and external structures used in the design, operation, and alignment of a Plumbicon tube.

External electrical structures include the deflection coils, external focusing coils, and alignment coils (Fig. 5.90). For the focusing coil, field strength should be of the order of 4–5 mT. Alignment coils are used to correct slight misalignments that result from tube and yoke manufacturing. Typical field strengths required are less than 0.5 mT. Two coils are generally used: one to correct horizontally, the other to correct vertically. Deflection coils are used to deflect the beam in such a manner as to produce a raster. For a typical national television system committee (NTSC) raster, coil current for horizontal deflection will be on the order of 250 mA, with another 50 mA required for vertical deflection.

Other considerations include the ability to *beam down* and cap the tubes when they are not in use. Tubes have a finite lifetime; eventually cathode emissions fall to a point where the tube becomes unusable. Circuitry that turns off the beam when not in use can extend tube life considerably. A capping mechanism, either in the camera's filter wheel or a physical lens cap, should be used to prevent highlights from reaching the tube when the camera is not in operation.

Tubes should not be subjected to vibration or shock when oriented vertically with the base in the uppermost position. Under these circumstances, particles can be dislodged within the tube. These loose particles can land on the target, causing spots or blemishes. Large particles landing on the target within the image area can effectively ruin an otherwise good tube. Optical quality glass is used on the surface of the tube and care should be taken to ensure it is kept clean and scratch free.

Defining Terms

Anticomet tail (ACT): A special type of electron gun designed to handle highlights by increasing beam current with a defocused beam during line retrace.

Blooming: An area of the target that is unstable due to insufficient beam current. The area normally appears as a white puddle without definition. Insufficient beam current may be the result of a low beam control setting.

Dark current: Current flow from the photosensitive layer in the absence of incident light. This could be compared to leakage current through a reverse-biased diode.

Diode gun Plumbicon: A Plumbicon tube with an electron gun that operates with a positive voltage applied to G_1 with respect to the cathode. The diode gun principle provides a finer beam spot size and lower beam temperature. This results in higher resolution and improved lag performance compared to triode gun tubes. The diode gun also provides a much higher current reserve for highlight handling when used in conjunction with a **dynamic beam control** (DBC) circuit.

Dynamic beam control: A circuit in a camera designed to instantaneously increase beam current to handle highlights in a scene. This is an alternative to the ACT solution of highlight control.

Dynamic resolution: The ability to distinguish fine detail in a moving object in a scene.

High stability (HS) diode gun Plumbicon: A diode gun Plumbicon tube, with electrostatic focus and magnetic deflection, which uses a high stability electrode structure evaporated and bonded to the tube envelope.

Image retention: A short-term after-image created by a high-contrast stationary scene within the beaming capabilities of the tube.

Lag: The inability of an imaging tube to respond to instantaneous changes in light. For measurement purposes, lag has two components: *rise lag* is the response time from dark to light, whereas *decay lag* is the response time from light to dark. Lag is a very short-term effect, and should not be confused with **image retention**, image burn, or **sticking**.

Low output capacitance (LOC) Plumbicon tube: The LOC Plumbicon tube is designed to reduce the capacitance of the target to ground, resulting in an improved signal-to-noise ratio.

MS diode gun Plumbicon: A diode gun Plumbicon tube with magnetic focus and electrostatic deflection.

Modulation depth: The ratio of the peak-to-peak amplitudes of, for example, 0.5–5 MHz (40–400 television lines) signal as measured on a waveform monitor. When measuring modulation depth, consideration of the frequency response of the preamplifiers well as the MTF of the lens must be taken into account.

Noise: The lower limit of useable dynamic range of the camera. Signal-to-noise is defined as the ratio of the peak value of the maximum signal [100 Institute of Radio Engineers (IRE) units] to the RMS noise voltage when the camera is capped.

Picture quality: Cleanliness of the scanned raster area with respect to spots and blemishes.

Resolution: The ability to distinguish fine detail in a video scene. The resolution of a camera tube is commonly expressed in terms of modulation depth.

Sensitivity: The signal current developed per unit of incident light, measured in microamps per lumen.

Shading: The variation of sensitivity across the target with reference to image center.

Stabilized beam current: The amount of beam current required to stabilize the target when a given amount of light is incident on the target. The beam current is normally set at two times picture white (commonly referred to as one *F* stop past 100 IRE or one *F* stop of head room).

Sticking: Very short-term image retention created in a high-contrast moving or stationary scene. Sticking takes on the appearance of lag except that it occurs under normal lighting conditions, whereas lag is generally a low light level phenomenon.

Triode gun: An electron gun that operates with a negative voltage applied to G_1 with respect to the cathode.

References

Benson, K.B. and Whitaker, J.C., ed. 1990. *Television and Audio Handbook*. McGraw-Hill, New York.

Crutchfield, E.B., ed. 1988. *NAB Engineering Handbook*, 7th ed. National Association of Broadcasters, Washington, DC.

Crutchfield, E.B., ed. 1993. *NAB Engineering Handbook*, 8th ed. National Association of Broadcasters, Washington, DC.

Inglis, A.F. 1990. *Behind the Tube*. Focal Press, London.

Levitt, R.S. 1968. *Operating Characteristics of the Plumbicon.*

National Association of Broadcasters. 1985. *NAB Engineering Handbook*. Washington, DC.

Philips Components. Plumbicon Application Bulletin 43. Jan. 1985, Slatersville, RI.

Rao, N.V. 1990. Development of high resolution camera tube for 2000 line TV system. SPIE, *Electron Image Tubes and Image Intensifiers*, vol. 1243, pp. 80–86.

Steen, R. 1991. CCDs vs. camera tubes: a comparison. *Broadcast Engineering*. Intertec Publishing, Overland Park, KS, May.

Stupp, E.H. 1968. *Physical Properties of the Plumbicon.*

In addition, the following books are recommended.

Benson, K.B. and Whitaker, J. 1990. *Television and Audio Handbook for Technicians and Engineers*. McGraw-Hill, New York.

Benson, K.B. and Whitaker, J., ed. 1991. *Television Engineering Handbook*, rev. ed. McGraw-Hill, New York.

Further Information

Additional information on the topic of image capture devices is available from the following sources.

Broadcast Engineering magazine is a monthly periodical dealing with television and radio technology. The magazine, published in Overland Park, KS, is free to qualified subscribers.

The Society of Motion Picture and Television Engineers (SMPTE) publishes a monthly *Journal*, and holds an annual technical conference in February and a convention in the Fall. The SMPTE is headquartered in White Plains, NY.

The Society of Broadcast Engineers (SBE) holds technical conferences throughout the year. The SBE is located in Indianapolis, IN.

The National Association of Broadcasters (NAB) holds an annual engineering conference and trade show in the Spring. The NAB is headquartered in Washington, D.C.

5.6   CRT Devices and Displays

Jerry C. Whitaker

5.6.1   Introduction

The cathode ray tube (CRT) is the dominant display technology for a wide range of applications—both consumer and professional.[4] As the requirements for greater resolution and color purity have increased, improvements have also been made in the design and manufacture of CRT devices and signal-driving circuits. Improvements to the basic monochrome and/or color CRT have been pushed within the last 10 years by the explosion of the personal computer industry and the increased resolution demanded by end users. Display size has also been a key element in CRT development. Consumer interest in large-screen home television has been strong within the last decade, and the roll-out of high-definition television has accelerated this trend.

5.6.2   Basic Operating System

The CRT produces visible or ultraviolet radiation by bombardment of a thin layer of phosphor material by an energetic beam of electrons. Nearly all commercial applications involve the use of a sharply focused electron beam directed time sequentially toward relevant locations on the phosphor layer by means of externally controlled electrostatic or electromagnetic fields. In addition, the current in the electron beam can be controlled or modulated in response to an externally applied varying electric signal. A generalized CRT consists of the following elements:

- An electron beam-forming system
- Electron-beam deflecting system (electrostatic or electromagnetic)
- Phosphor screen
- Evacuated envelope

Figure 5.91 shows the basic design of a monochrome CRT. The electron beam is formed in the electron gun, where it is modulated and focused. The beam then travels through the deflection region, where it is directed toward a specific spot or sequence of spots on the phosphor screen. At the phosphor screen the electron beam gives up some of the energy of the electrons in producing light or other radiation, some in generating secondary electrons, and the remainder in producing heat.

---

[4]This chapter was adapted from: Whitaker, J.C. 1994. *Electronic Displays: Technology, Design and Applications.* McGraw-Hill, New York. Used with permission.

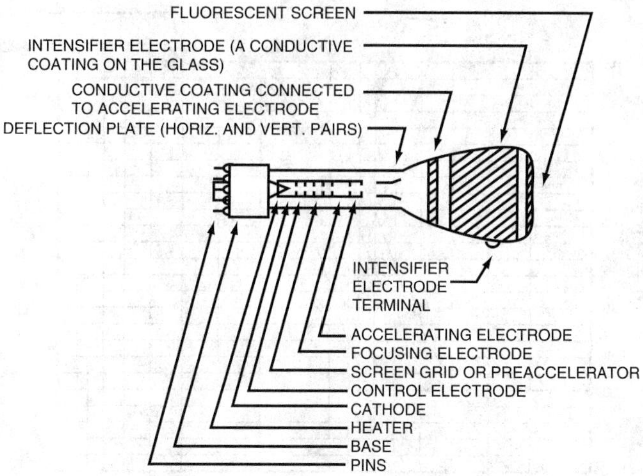

FLUORESCENT SCREEN
INTENSIFIER ELECTRODE (A CONDUCTIVE COATING ON THE GLASS)
CONDUCTIVE COATING CONNECTED TO ACCELERATING ELECTRODE
DEFLECTION PLATE (HORIZ. AND VERT. PAIRS)

INTENSIFIER ELECTRODE TERMINAL
ACCELERATING ELECTRODE
FOCUSING ELECTRODE
SCREEN GRID OR PREACCELERATOR
CONTROL ELECTRODE
CATHODE
HEATER
BASE
PINS

FIGURE 5.91 A generalized schematic of a cathode ray tube using electrostatic deflection.

Classification of CRT Devices

Tubes may be classified in terms of bulb parameters and screen/gun geometry. The principle categories that separate one class of device from another include

- Tube size. Conventionally, tube size is measured as the screen diagonal dimension in rounded inch units. This number is included in tube-type numbers.
- Neck diameter (OD). The gun, yoke, neck hardware, and socketing are affected by this dimension (typically given in millimeter). Common neck sizes are 36.5, 29, and 22.5 mm.
- Deflection angle. This parameter is calculated from the rated full-screen diagonal and glassware drawings, using the yoke reference plane as an assumed center of deflection. Angles in common use include 90, 100, and 110°. Higher deflection angles enable shorter tubes but entail other tradeoffs.
- Other characteristics, including gun type (**delta** or **in-line**), screen structure (stripes or dots) for color CRTs, and flat or curved face plates.

The CRT Envelope

The cathode ray tube envelope consists of the faceplate, bulb, funnel, neck, base press, base, faceplate safety panels, shielding, and potting. (Not all CRTs incorporate each of these components.) The faceplate is the most critical component of the envelope because the display on the phosphor must be viewed through it. Most faceplates are pressed in molds from molten glass and trimmed and annealed before further processing. Some specialized CRTs for photographic recording or flying-spot scanning use optical-quality glass faceplates sealed to the bulb section in such a way as to produce minimum distortion.

To minimize the return scattering of ambient light from the white phosphor, many CRT types use a neutral-gray-tinted faceplate. While the display information is attenuated as it makes a single pass through this glass, ambient light will be attenuated both going in and coming out, thus squaring the attenuation ratio and increasing contrast.

Certain specialized CRTs have faceplates made wholly or partially of fiber optics, which may have extraordinary characteristics, such as high ultraviolet transmission. A fiber optic region in the faceplate permits direct-contact exposure of photographic or other sensitive film without the necessity for external lenses or physical space for optical projection.

The bulb section of the CRT is the transition element necessary to enclose the full deflection volume of the electron beam between the deflection region and the phosphor screen on the faceplate. In most

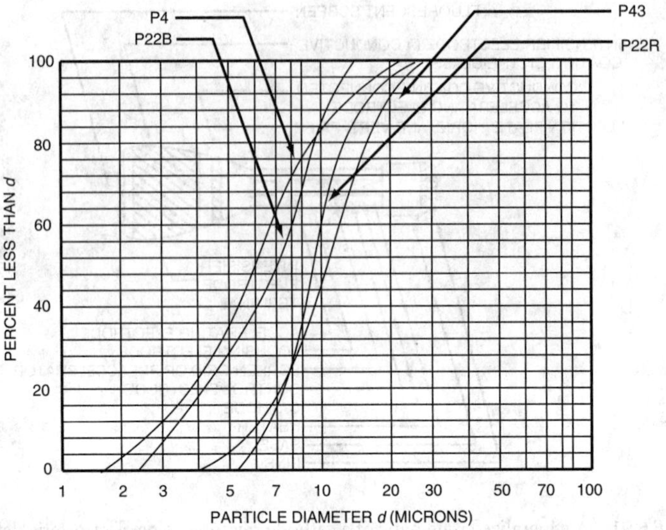

FIGURE 5.92   Particle size distribution of four representative phosphors.

CRTs, this is a roughly cone-shaped molded-glass component. Instead of an ordinary glass bulb, many large CRTs include metal cone sections made of a glass-sealing iron alloy. The metal cones are generally lighter in weight than the corresponding glass sections.

The junction region of the bulb of a CRT with the neck section is critical to the geometry of the device. Tubes made with these separate sections are intended for electromagnetic deflection, and this region is where the deflection yoke is located.

The neck diameter of a CRT depends to a great extent on the type of deflection used and the intended application of the CRT. In general, electrostatic deflection CRTs have large neck diameters, whereas the electromagnetic-deflection devices have small diameters.

Phosphor and Screen Characteristics

Many materials, naturally occurring and synthetic, organic, and inorganic, have the ability to give off light. For video applications, the materials of interest are crystalline inorganic solids that are stable under cathode-ray tube fabricating and operating conditions. These materials are generally powders having average particle sizes in the range of 5–15 μm. Figure 5.92 shows some typical particle size distributions. Because of defects and irregularities in the crystal lattice structure, these materials have the ability to absorb incident energy and convert this energy into visible light. This process involves the transfer of energy from the electron beam to electrons in the phosphor crystal. The phosphor electrons are thereby excited or raised to levels higher than the ground state. Light is emitted when the electrons return to the more stable states. Figure 5.93 illustrates these changes in energy levels.

Phosphors are composed of a host crystal that comprises the bulk of the material and one or more *activators*, which may be present in amounts from parts per million to a few mole percent. Either the host or activator can determine the luminescent properties of a phosphor system. For example, in the zinc sulfide/cadmium sulfide: silver (ZnS:Ag/CdS:Ag) system, the emitted color ranges from blue at zero cadmium through green, to yellow, and into red as the cadmium content is increased.

In the commercial preparation of phosphors, the highly purified host and the required amount of activator are intimately mixed, normally with a flux, such as an alkali or alkaline earth halide or phosphate, which supplies a low-temperature melting phase. The flux controls the particle development and aids in the diffusion of the activator into the lattice. This mixture is then fired at high temperature, 1472–2192°F (800–1200°C) on a prescribed schedule in order to develop the desired physical and luminescent properties.

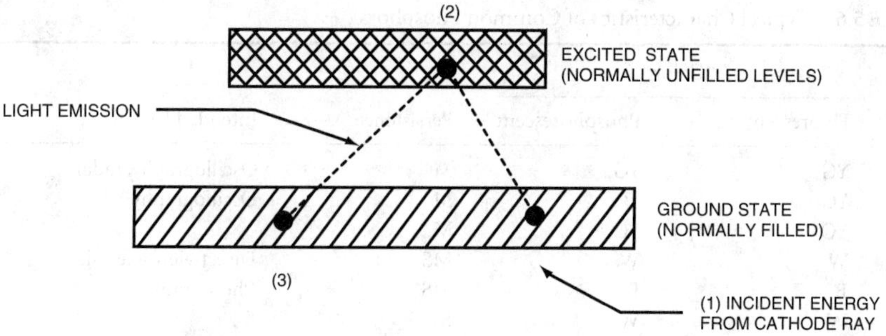

FIGURE 5.93   Energy transitions of electrons leading to luminescence.

In some cases the firing is carried on under specific controlled atmospheres. After firing, the resultant cake is broken up, residual soluble materials are removed by washing, and any required coatings are applied.

Phosphor Screen Types

A phosphor screen is used to convert electron energy into radiant energy in a CRT display device. The screen is composed of a thin layer of luminescent crystals, phosphors, that emit light when bombarded by electrons. This property is referred to as **cathodoluminescence**. It occurs when the energy of the electron beam is transferred to electrons in the phosphor crystal. The property of light emission during excitation is termed **fluorescence**, and that immediately after excitation is removed is termed **phosphorescence**.

Standardization of phosphor types has been accomplished by registration of the various phosphors with the *Joint Electron Device Engineering Council* (JEDEC) of the Electronic Industries Association (EIA). Registered phosphors are designated by a number series known as *P numbers*, as shown in Table 5.5.

Luminescent Properties

Phosphors are commercially available with cathodoluminescent emission over the entire visible band, including ultraviolet and near-infrared. Table 5.6 lists the important characteristics of the most common phosphors used in video display applications. Absolute phosphor efficiency is measured as the ratio of total absolute energy emitted to the total excitation energy applied. When evaluating or comparing picture

TABLE 5.5   Characteristics of Standard Phosphor Types

EIA no.	Worldwide Designation	Typical Application	Composition	Relative Efficiency, %	x^a	y^a	u^a	v^a	Decay
P-4	WW	Black and white television	$ZnS:Ag +$ $Zn_{(1-x)}Cd_xS:Cu, Al$	100	0.270	0.300	0.178	0.446	Medium short
P-1	GJ	Projection green	$ZnSiO_4:Mn$	130	0.218	0.712	0.079	0.577	Medium
P-43	GY	Projection green	$Gd_2O_2S:Tb$	155	0.333	0.556	0.148	0.556	Medium
P22R	X	Red direct-view, projection	$Y_2O_3:Eu$ $Y_2O_2S:Eu$	65	0.640 0.625	0.340 0.340	0.441 0.429	0.528 0.525	Medium short
P-22G	X	Green direct-view, projection	$Zn_{(1-x)}Cd_xS:Cu,Al$ $ZnS:Cu,Al$	180	0.340 0.285	0.595 0.600	0.144 0.119	0.566 0.561	Medium
P-22B	X	Blue direct-view, projection	$ZnS:Ag$	25	0.150	0.065	0.172	0.168	Medium short

After Skolnik, M.I., ed. 1970. *Radar Handbook*, pp. 6–8. McGraw-Hill, New York.

TABLE 5.6 Typical Characteristics of Common Phosphors

Type	Color[a] Fluorescent	Phosphorescent	Persistence[b]	Intended Use
P1	YG	YG	M	Oscillography, radar
P2	YG	YG	M	Oscillography
P3	YO	YO	M	
P4	W	W	MS	Direct view television
P5	B	B	MS	Photographic
P6	W	W	S	
P7	B	Y	MS(B), L(Y)	Radar
P8	Replaced by P7			
P9	Not registered			
P10	Dark trace screen	VL	Radar	
P11	B	B	MS	Photographic
P12	O	O	L	Radar
P13	RO	RO	M	Radar
P14	B	YO	MS(B), M(YO)	Radar
P15	UV	G	UV(VS), G(S)	Flying-spot scanner
P16	UV	UV	VS	Flying-spot scanner, photographic
P17	B	Y	S(B), L(Y)	Oscillography, radar
P18	W	W	M-MS	Projection television
P19	O	O	L	Radar
P20	YG	YG	M-MS	Storage tube
P21	RO	RO	M	Radar
P22	W(R,G,B)	W(R,G,B)	MS	Tricolor video (television)
P23	W	W	MS	Direct-view television
P24	G	G	S	Flying-spot scanner
P25	O	O	M	Radar
P26	O	O	VL	Radar
P27	RO	RO	M	Color TV monitor
P28	YG	YG	L	Radar
P29	P2 and P25 stripes		Radar, indicators	
P30	Canceled			
P31	G	G	MS	Oscillography, bright video
P32	PB	YG	L	Radar
P33	O	O	VL	Radar
P34	BG	YG	VL	Radar, oscillography
P35	G	B	MS	Oscillography
P36	YG	YG	VS	Flying-spot scanner
P37	B	B	VS	Flying-spot scanner, photographic
P38	O	O	VL	Radar
P39	YG	YG	L	Radar
P40	B	YG	MS(B), L(YG)	Low repetition rate (P12 and P16)
P41	UV	O	VS(UV), L(O)	Radar with light trigger

[a]Color: B = blue, P = purple, G = green, O = orange, Y = yellow, R = red, W = white, UV = ultraviolet.
[b]Persistence to 10% level: VS = less than 1 ms, S = 1–10 ms, MS = 10–1 ms, M = 1–100 ms, L = 100 ms–1 s.

tubes, it is more meaningful to measure luminescence in footlambert using a system the response of which matches the eye. In addition to the use of a suitable detector, a number of other parameters must be controlled if meaningful measurements are to be obtained. Most important of these is the total energy to the screen, which includes

- Anode voltage
- Cathode current
- Raster size

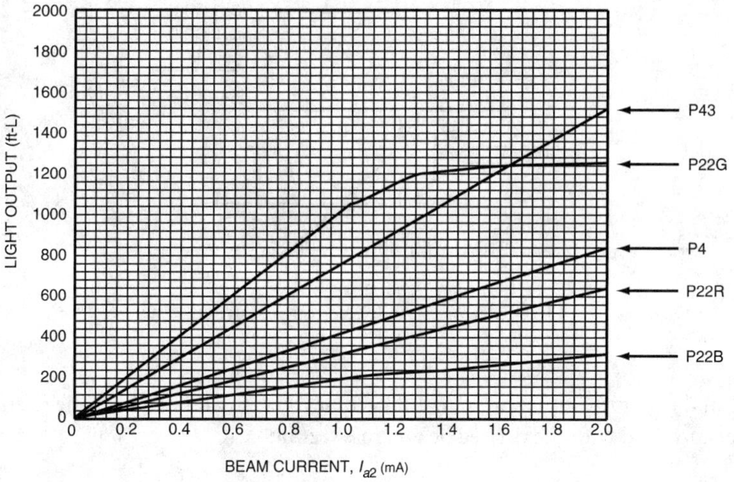

FIGURE 5.94 Linearity characteristics of five common phosphors. Conditions: $E_{a2} = 25$ kV (1 fL = 3426 cd/m^2).

Under normal CRT operating conditions, the luminescence of phosphors is essentially proportional to the beam current applied. However, when high beam currents are employed, some phosphors saturate and depart from linear behavior. The same effect can be observed in a direct-view color video display in areas of highlight brightness and with electron guns having small spot size. If linearity of the three primary phosphors is not closely matched, noticeable shift in white field color can result in highlight areas. Figure 5.94 shows some typical light output/ beam current curves.

Thermal quenching is another mechanism that may result in a loss of phosphor efficiency. In most phosphors the energy transitions become less efficient as screen temperature increases. This phenomenon can be quite pronounced in projection systems where the high-power loading can be responsible for a large increase in screen temperature. Figure 5.95 shows some examples of thermal quenching. Quenching is a transient condition, and the screen returns to normal efficiency after being cooled.

Electron Gun

The electron gun is basic to the structure and operation of any cathode-ray device, specifically display devices. In its simplest schematic form, an electron gun may be represented by the diagram in Fig. 5.96, which shows a triode gun in cross section. Electrons are emitted by the cathode, which is heated by the filament to a temperature sufficiently high to release the electrons. Because this stream of electrons emits from the cathode as a cloud rather than a beam, and it is necessary to accelerate, focus, deflect, and

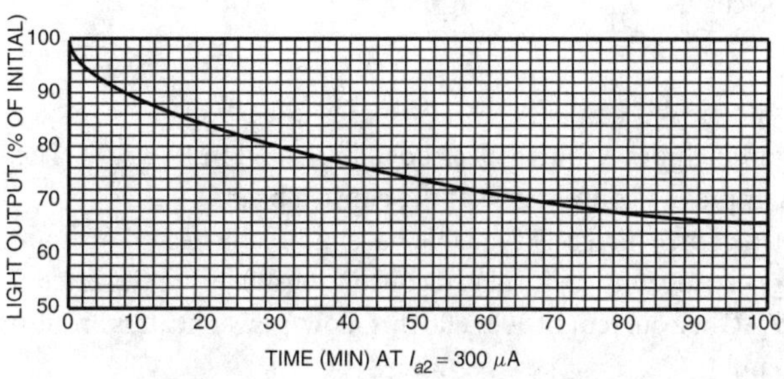

FIGURE 5.95 The loss in phosphor efficiency as a display screen heats at high-current operation (rare-earth green).

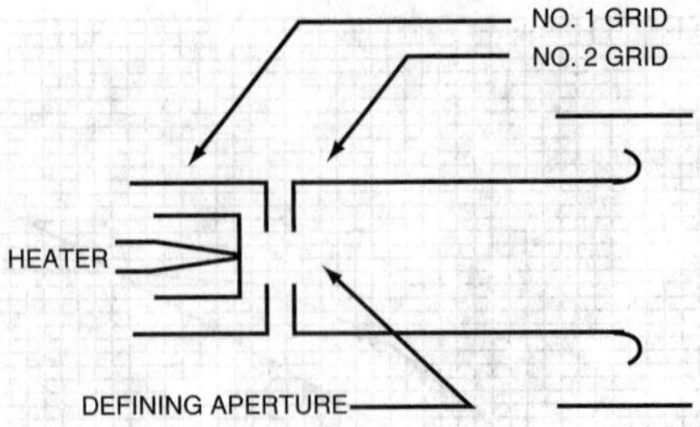

FIGURE 5.96 Triode electron gun structure.

otherwise control the electron emission so that it becomes a beam and can be made to strike a phosphor at the proper location and with the desired beam cross section.

Electron Motion

The properties of motion for an electron in a uniform electrostatic field are obtained from Newton's second law. The velocity of an emitted electron is given by the following:

$$V = \left(\frac{2\,eV}{m}\right)^{\frac{1}{2}} \tag{5.9}$$

where

$e = 1.6 \times 10^{-19}$ C

$m = 9.1 \times 10^{-28}$ g

$V = -Ex$, the potential through which the electron has fallen

When practical units are substituted for the values in the previous equation, the following results:

$$V = 5.93 \times 10^5 V^{1/2} \text{ m/s} \tag{5.10}$$

This expression represents the velocity of the electron. If the electron velocity is at the angle θ to the potential gradient in a uniform field, the motion of the electron is specified by

$$y = \frac{Ex^2}{4V_0 \sin^2 \theta} + \frac{x}{\tan \theta} \tag{5.11}$$

where θ = the electron potential at initial velocity

This equation defines a parabola. The electron trajectory is illustrated in Fig. 5.97, in which the following conditions apply:

y_m = maximum height

x_m = x displacement at the maximum height

a = the slope of the curve

Tetrode Gun

The tetrode electron gun includes a fourth electrode, illustrated in Fig. 5.98. The main advantage of the additional electrode is improved convergence of the emitted beam.

Nearly all CRT electron guns currently available have indirectly heated cathodes in the form of a small capped nickel sleeve or cylinder with an insulated coiled tungsten heater inserted from the back end. Most heaters operate at 6.3 V AC at a current of 300–600 mA. Low-power heaters are also available that operated at 1.5 V (typically, 140 mA).

The cathode assembly is mounted on the axis of the modulating or *control grid* cylinder (or simply *grid*), which is a metal cup of low permeability or stainless steel about 0.5-in in diameter and 0.375–0.5 in long.

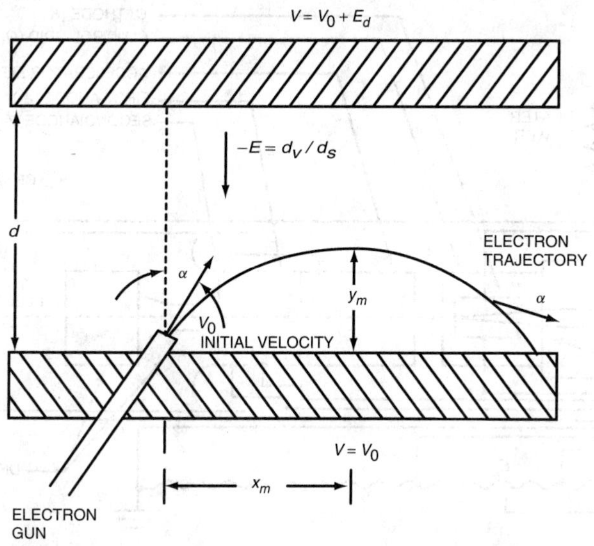

FIGURE 5.97   The electron trajectory from an electron gun using the parameters specified in the text.

A small aperture on the order of 10 mils diameter is punched or drilled in the cap. The grid (G_1) voltage is usually negative with respect to the cathode (K).

To obtain electron current from the cathode through the grid aperture, there must be another electrode beyond the aperture at a positive potential great enough for its electrostatic field to penetrate the aperture to the cathode surface. Figure 5.99 illustrates a typical *accelerating electrode* (G_2) in relation to the cathode structure. The accelerating electrode may be implemented in any given device in a number of ways. In a simple accelerating lens in which successive electrodes have progressively higher voltages, the electrode may also be used for focusing the electron beam upon the phosphor, in which case it may be designated the *focusing* (or *first*) anode (A_1). This element is usually a cylinder, longer than its diameter and probably containing one or more disk apertures.

Electron Beam Focusing

The general principles involved in focusing the electron beam can be best understood by initially examining optical lenses and then establishing the parallelism between them and electrical focusing techniques.

Figure 5.100 shows a simplified diagram of an electrostatic lens. An electron emitted at zero velocity enters the V_1 region and moves in that region at a constant velocity (because the region has a constant potential). The velocity of the electron in that region is defined by Eq. (5.10) for the straightline component,

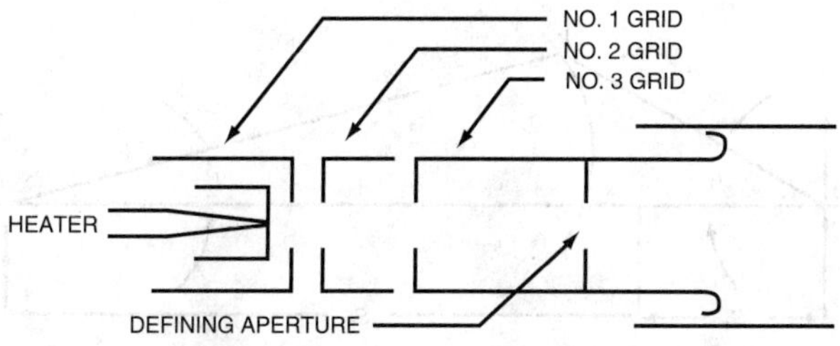

FIGURE 5.98   Basic structure of the tetrode electron gun.

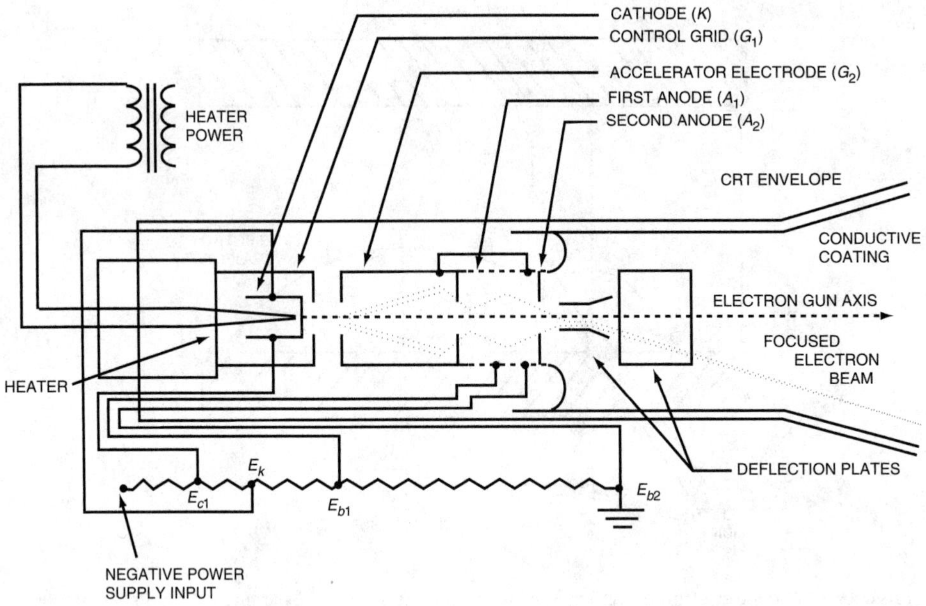

FIGURE 5.99 Generalized schematic of a CRT grid structure and accelerating electrode in a device using electrostatic deflection.

with V_1 replacing V in the equation. After passing through the surface into the V_2 region, the velocity changes to a new value determined by V_2. The only component of the velocity that is changed is the one normal to the surface, so that the following conditions are true:

$$V_t = V_1 \sin I_1$$
$$V_1 \sin I_1 = V_2 \sin I_2$$

Snell's law, also known as the law of refraction, has the form:

$$N_1 \sin I_1 = N_2 \sin I_2$$

where
N_1 = index of refraction for the first medium
N_2 = index of refraction for the second medium
I_1 = angle of the incident ray with the surface normal
I_2 = angle of the refracted ray with the surface normal

The parallelism between the optical and the electrostatic lens is apparent if the appropriate substitutions

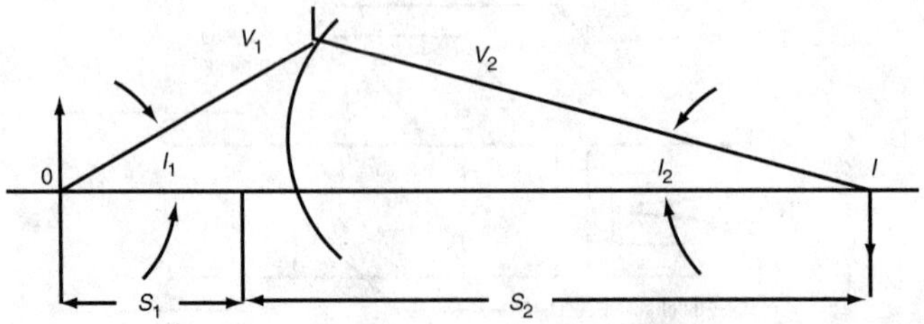

FIGURE 5.100 The basic principles of electron optics.

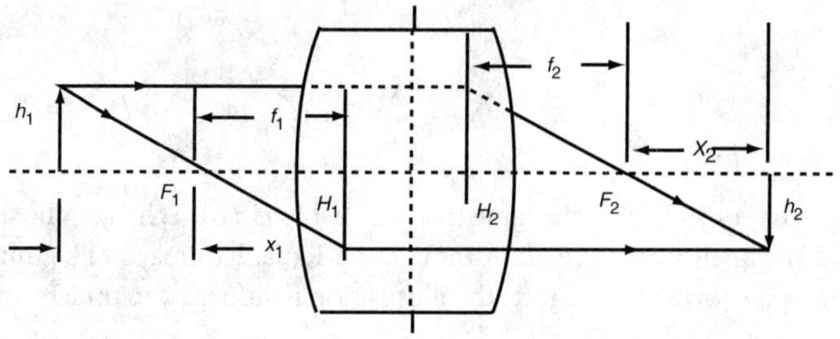

FIGURE 5.101 A unipotential lens. (After Sherr, S. 1979. *Electronic Displays*, pp. 67–70. Wiley, New York.)

are made,

$$V_1 \sin I_1 = V_2 \sin I_2$$
$$\frac{\sin I_1}{\sin I_2} = \frac{V_2}{V_1}$$

For Snell's law, the following applies:

$$\frac{\sin I_1}{\sin I_2} = \frac{N_2}{N_1}$$

The analogy between the optical lens and the electrostatic lens is, thus, apparent. The magnification of the electrostatic lens is given by the following:

$$m = \frac{(V_1/V_2)^{\frac{1}{2}} S_2}{S_1}$$

The condition of a thin, unipotential lens, where $V_1 = V_2$, is illustrated in Fig. 5.101. The following applies:

$$m = \frac{h_2}{h_1} = \frac{f_2}{X_1} = -\frac{X_2}{f_2}$$

The shape of the electron beam under the foregoing conditions is shown in Fig. 5.102. If the potential at the screen is the same as the potential at the anode, the crossover is imaged at the screen with the magnification

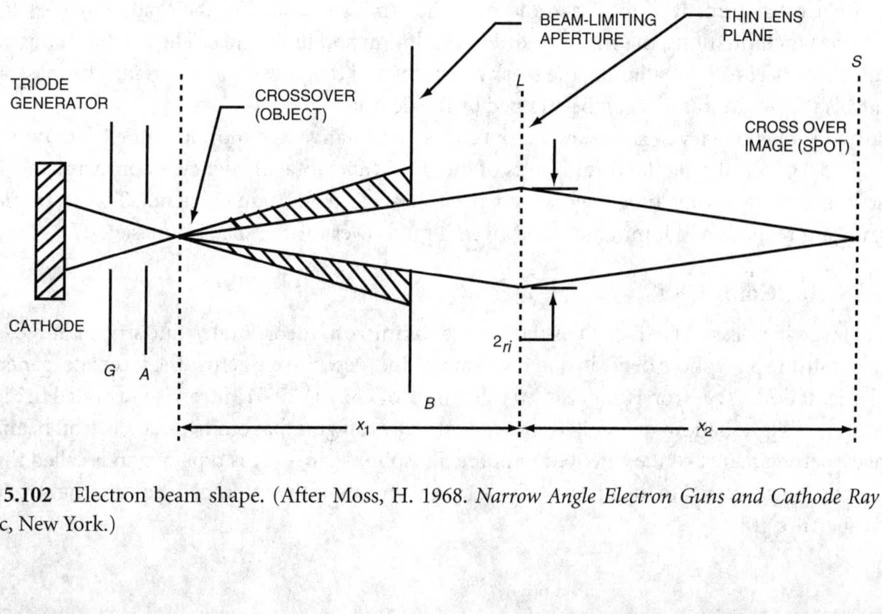

FIGURE 5.102 Electron beam shape. (After Moss, H. 1968. *Narrow Angle Electron Guns and Cathode Ray Tubes*. Academic, New York.)

given by

$$m = \frac{X_2}{X_1}$$

The magnification can be controlled by changing this ratio, which in turn changes the size of the spot. This is one way to control the quality of the focus. Although the actual lens may not be thin, and in general is more complicated than is shown in the figure, the illustration is sufficient to understand the operation of electrostatic focus.

The size of the spot can be controlled by changing the ratio of V_1 to V_2 or of X_2 to X_1 in the previous equations. Because X_1 and X_2 are established by the design of the CRT, the voltage ratio is the parameter available to the circuit designer to control the size or focus of the spot. It is by this means that focusing is achieved for CRTs using electrostatic control.

5.6.3   Color CRT Devices

The shadow-mask CRT is the most common type of color display device. As illustrated in Fig. 5.103, it utilizes a cluster of three electron guns in a wide neck, one gun for each of the colors red, green, and blue. All of the guns are aimed at the same point at the center of the shadow-mask, which is an iron-alloy grid with an array of perforations in triangular arrangement, generally spaced 0.025-in between centers for entertainment television. For high-resolution studio monitor or computer graphic monitor applications, color CRTs with shadow mask aperture spacing of 0.012-in center-to-center or less are readily available. This triangular arrangement of electron guns and shadow-mask apertures is known as the delta-gun configuration. Phosphor dots on the faceplate just beyond the shadow mask are arranged so that after passing through the perforations, the electron beam from each gun can strike only the dots emitting one color.

All three beams are deflected simultaneously by a single large-diameter deflection yoke, which is usually permanently bonded to the CRT envelope by the tube manufacturer. The three phosphors together are designated P-22, individual phosphors of each color being denoted by the numbers P-22R, P-22G, and P-22B. Most color CRTs are constructed with rare-earth-element-activated phosphors, which offer superior color and brightness compared with previously used phosphors.

Because of the close proximity of the phosphor dots to each other and the strict dependence on angle of penetration of the electrons through the apertures, tight control over electron optics must be maintained. Close attention is also paid to shielding the CRT from extraneous ambient magnetic fields and to degaussing of the shield and shadow mask (usually carried out automatically when the equipment is switched on or off).

Even if perfect alignment of the mask and phosphor triads is assumed, the shadow-mask CRT is still subject to certain limitations, mainly in regard to resolution and luminance. The resolution restriction is the result of the necessity for aligning the mask apertures and the phosphor dot triads; the mask aperture size controls the resolution that can be attained by the device.

Electron beam efficiency in a shadow-mask tube is low, relative to a monochrome CRT. Typical beam efficiency is 9%; considering the three beams of the color tube, total efficiency is approximately 27%. By comparison, a monochrome tube may easily achieve 80% electron beam efficiency. This restriction leads to a significant reduction in luminance for a given input power for the shadow-mask CRT.

Parallel-Stripe Color CRT

The parallel-stripe class of CRT, such as the popular **Trinitron**, incorporates fine stripes of red-, green-, and blue-emitting phosphors deposited in continuous lines repetitively across the faceplate, generally in a vertical orientation. (Trinitron is a registered trademark of Sony.) The Trinitron is illustrated in Fig. 5.104. This device, unlike a shadow-mask CRT, uses a single electron gun that emits three electron beams across a diameter perpendicular to the orientation of the phosphor stripes. This type of gun is called the in-line gun. Each beam is directed to the proper color stripe by means of the internal beam-aiming structure and a slitted aperture grille.

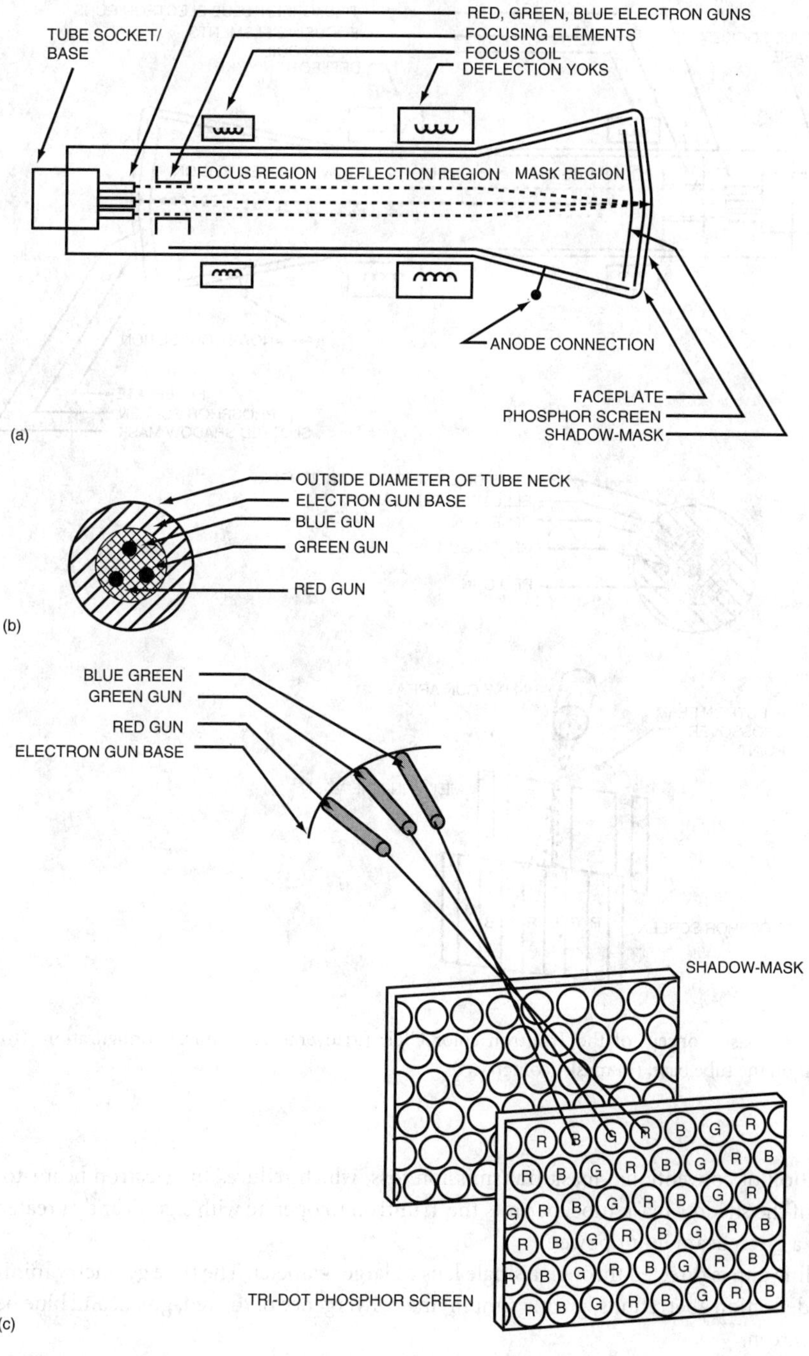

FIGURE 5.103 Basic concept of a shadow-mask color CRT: (a) overall mechanical configuration, (b) delta gun arrangement on the tube base, (c) shadow-mask geometry.

The Trinitron phosphor screen is built in parallel stripes of alternating red, green, and blue elements. A grid structure placed in front of the phosphors, relative to the CRT gun, is used to focus and deflect the beams to the appropriate color stripes. Because the grid spacing and stripe width can be made smaller than the shadow-mask apertures and phosphor dot triplets, higher resolutions may be attained with the Trinitron system.

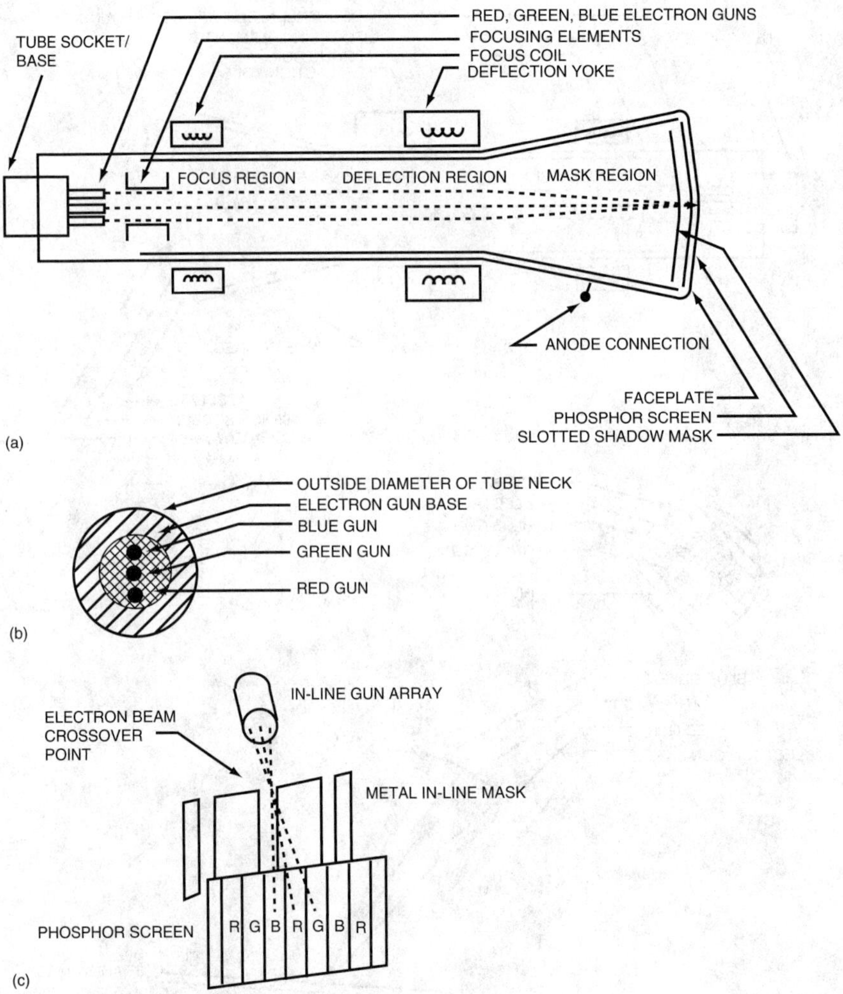

FIGURE 5.104   Basic concept of the Trinitron color CRT: (a) overall mechanical configuration, (b) in-line gun arrangement on the tube base, (c) mask geometry.

Elimination of conventional mask transmission loss, which reduces the electron beam-to-luminance efficiency of the shadow mask tube, permits the Trinitron to operate with significantly greater luminance output for a given beam input power.

The in-line gun is directed through a single lens of large diameter. The tube geometry minimizes beam focus and deflection aberrations, greatly simplifying convergence of the red, green, and blue beams on the phosphor screen.

Basics of Color CRT Design

The shadow-mask CRT is the workhorse of the video display industry. Used in the majority of color video displays since the introduction of color television in the early 1950s, the shadow-mask technique has been refined to yield greater performance and lower manufacturing cost.

Figure 5.105 illustrates the shadow-mask geometry for a tube at face center using in-line guns and a shadow mask of round holes. As an alternative, the shadow-mask may consist of vertical slots, as shown in Fig. 5.106. The three guns and their undeflected beams lie in the horizontal plane. The beams are

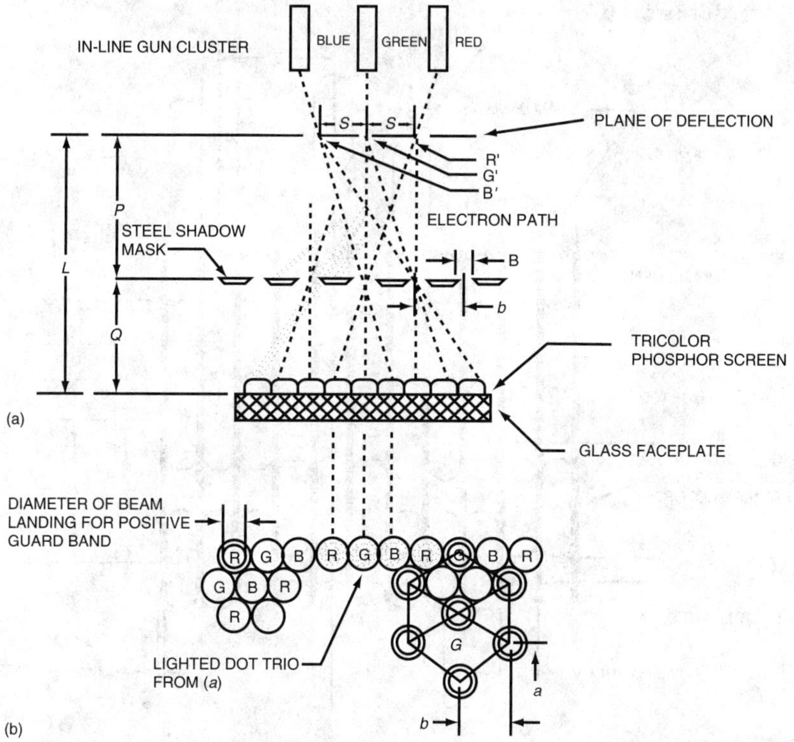

FIGURE 5.105 Shadow-mask CRT using in-line guns and round mask holes: (a) overall tube geometry, (b) detail of phosphor dot layout.

shown converged at the mask surface. The beams may overlap more than one hole and the holes are encountered only as they happen to fall in the scan line. By convention, a beam in the figure is represented by a single straight line projected backward at the incident angle from an aperture to an apparent *center of deflection* located in the *deflection plane*. In Fig. 5.105, the points B′, G′, and R′, lying in the deflection plane, represent such apparent centers of deflections for blue, green, and red beams striking an aperture under study. (These deflection centers move with varying deflection angles.) Extending the rays forward to the facepanel denotes the printing location for the respective colored dots (or stripes) of a tricolor group. Thus, centers of deflection become color centers with a spacing S in the deflection plane. The distance S projects in the ratio Q/P as the dot spacing within the trio. Figure 5.105 also shows how the mask hole horizontal pitch b projects as screen horizontal pitch in the ratio L/P. The same ratio applies for projection of mask vertical pitch a. The Q-space (mask to panel spacing) is optimized to obtain the largest possible theoretical dots without overlap. At panel center, the ideal screen geometry is then a mosaic of equally spaced dots (or stripes).

The stripe screen shown in Fig. 5.106 is used extensively in color CRTs. One variation of this stripe (or line) screen uses a cylindrical faceplate with a vertically tensioned grill shadow-mask without tie bars. Prior to the stripe screen, the standard construction was a tridot screen with a delta gun cluster as shown in Fig. 5.107.

Guard Band

The use of **guard bands** is a common feature for aiding purity in a CRT. The guard band, where the lighted area is smaller than the theoretical tangency condition, may be either *positive* or *negative*. In Fig. 5.105, the leftmost red phosphor exemplifies a positive guard band; the lighted area is smaller than the

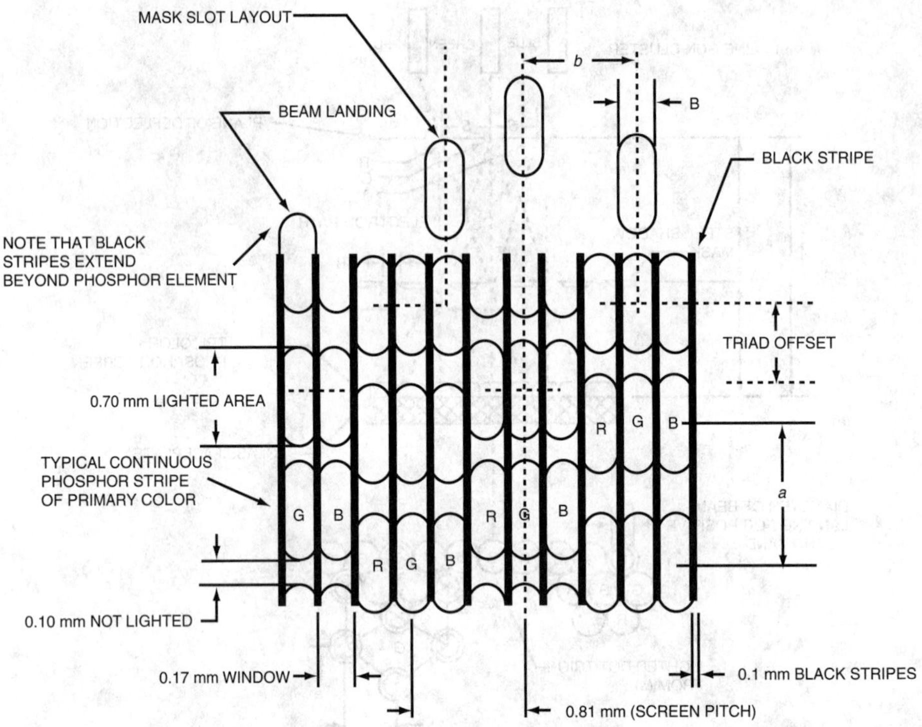

FIGURE 5.106   Shadow-mask CRT using in-line guns and vertical stripe mask holes.

actual phosphor segment, accomplished by mask hole diameter design. Figure 5.106, on the other hand, shows negative guard band (NGB) (or *window-limited*) construction for stripe screens. Vertical black stripes about 0.1 mm (0.004-in) wide separate the phosphor stripes forming windows to be lighted by a beam wider than the window opening by about 0.1 mm. Figure 5.107 shows NGB construction of a tridot screen.

Electron Gun Classifications

Electron guns for color tubes can be classified according to the main lens configuration, which include

- Unipotential
- Bipotential
- Tripotential
- Hybrid lenses

The unipotential gun, discussed previously in this chapter as it applies to monochrome operation, is the simplest of all designs. This type of gun is rarely used for color applications, except for small-screen sizes. The system suffers from a tendency toward arcing at high anode voltage, and relatively large low current spots.

The bipotential lens is the most commonly used gun in shadow-mask color tubes. The arrangement of gun electrodes is shown in Fig. 5.108 along with a computer-generated plot of equipotential lines and electron trajectories. The main lens of the gun is formed in the gap between grid 3 and grid 4. When grid 3 operates at 18–22% of the grid-voltage, the lens is referred to as a *low bipotential* configuration, often called simply *LoBi*. When grid 3 operates at 26–30% of the grid-4 voltage, the lens is referred to as a *high bipotential* or *HiBi* configuration.

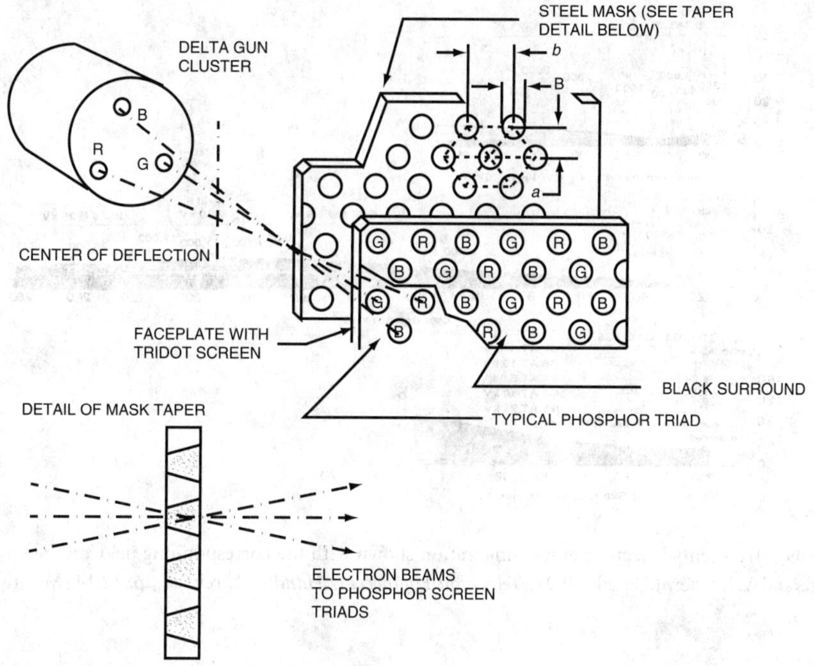

FIGURE 5.107 Delta-gun, round-hole mask negative guard band tridot screen. The taper on the mask holes is shown in the detail drawing only.

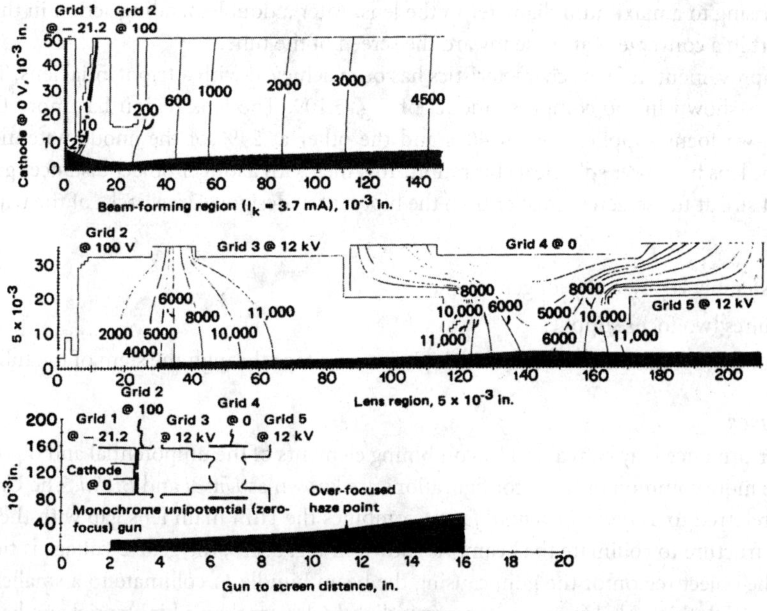

FIGURE 5.108 Bipotential electron gun configuration, shown with the corresponding field and beam plots. (After Benson, K.B. and Whitaker, J.C., eds. 1991. *Television Engineering Handbook*, rev. ed., p. 12.13. McGraw-Hill, New York.)

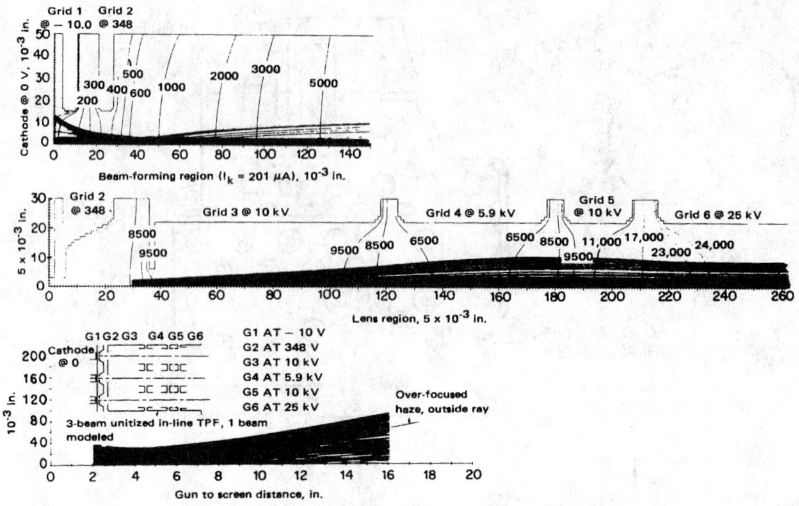

FIGURE 5.109 Tripotential electron gun configuration shown with the corresponding field and beam plots. (After Benson, K.B. and Whitaker, J.C., eds. 1991. *Television Engineering Handbook*, rev. ed., p. 12.14. McGraw-Hill, New York.)

The LoBi configuration has the advantages of a short grid 3 and shorter overall length, with parts assembly generally less critical than the HiBi configuration. However, with its shorter object distance (grid-3 length), the lens suffers from somewhat larger spot size than the HiBi. The HiBi, on the other hand, with a longer grid-3 object distance, has improved spot size and resolution. The focus voltage supply for the LoBi also can be less expensive.

It can be seen in the plot Fig. 5.108 that the bipotential beam starts with a crossover of the beam near the cathode, rising to a maximum diameter in the lens. After a double bending action in the lens region the rays depart in a convergent attitude toward the screen of the tube.

Further improvement in focus characteristics has been achieved with a tripotential lens. The electrode arrangement is shown in the computer model of Fig. 5.109. The lens region has more than one gap and requires two focus supplies, one at 40% and the other at 24% of the anode potential. With this refinement the lens has lower spherical aberration. Together with a longer object distance (grids 3–5), the resulting spot size at the screen is smaller than the bipotential designs. Drawbacks of the tripotential gun include

- The assembly is physically longer.
- It requires two focus supplies.
- It requires a special base to deliver the high focus voltage through the stem of the tube.

Hybrid Lenses

Improved performance may be realized by combining elements of the unipotential and bipotential lenses in series. The more common of these configurations are known as *UniBi* and *BiUni*. The UniBi structure (sometimes referred to as *quadripotential focus*) combines the HiBi main lens gap with the unipotential type of lens structure to collimate the beam. As shown in Fig. 5.110, the grid-2 voltage is tied to a grid 4 inserted in the object region of the gun, causing the beam bundle to collimate to a smaller diameter in the main lens. With this added focusing, the gun is slightly shorter than a bipotential gun having an equal focus voltage.

The BiUni structure, illustrated in Fig. 5.111, achieves a similar beam collimation by tying the added element to the anode, rather than to grid 2. The gun structure is shorter because of the added focusing early in the device. With three high gradient gaps, arcing can be a problem in the BiUni configuration.

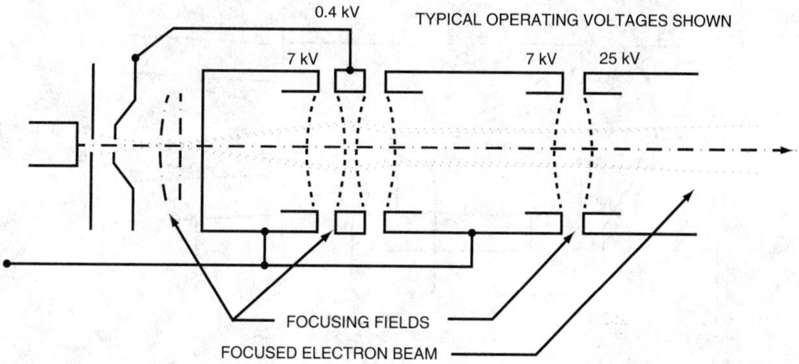

FIGURE 5.110 Electrode arrangement of UniBi gun.

Trinitron

The Trinitron gun consists of three beams focused through the use of a single large main focus lens of unipotential design. Figure 5.112 shows the three in-line beams mechanically tilted to pass through the center of the lens, then reconverged toward a common point on the screen. The gun is somewhat longer than other color guns, and the mechanical structuring of the device requires unusual care and accuracy in assembly.

Gun Arrangements

Prior to 1970 most color gun clusters used a delta arrangement. Since that time the design trend has been toward the three guns in-line on the horizontal axis of the tube.

The electron optical performance of a delta gun is superior to in-line, in its basic form, by virtue of a larger lens diameter, resulting from more efficient use of the available area inside the neck. Larger neck diameters improve lens diameter, but at the expense of higher deflection power and more difficult convergence problems over the face of the tube. Typical delta gun lens diameters include

- 12 mm (0.5 in) for a 51-mm (2-in) neck OD
- 9 mm (0.4 in) for a 36-mm (1.4-in) neck
- 7 mm (0.3 in) for a 29-mm (1.1-in) neck

Delta guns use individual cylinders to form the electron lenses, and the three guns are tilted toward a common point on the screen. The individual cylinders are subject to random errors in position that can

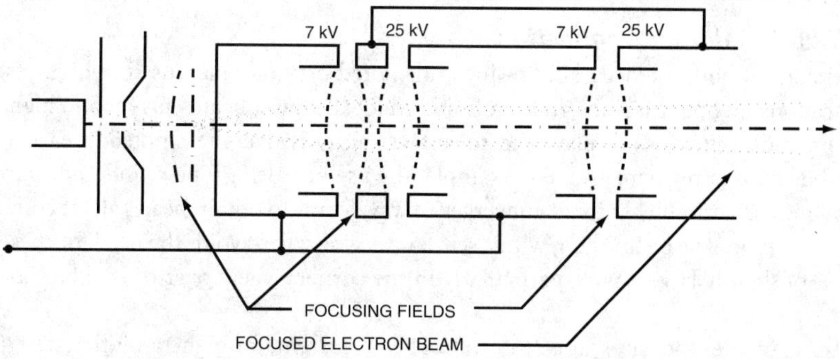

FIGURE 5.111 Electrode arrangement of a BiUni gun.

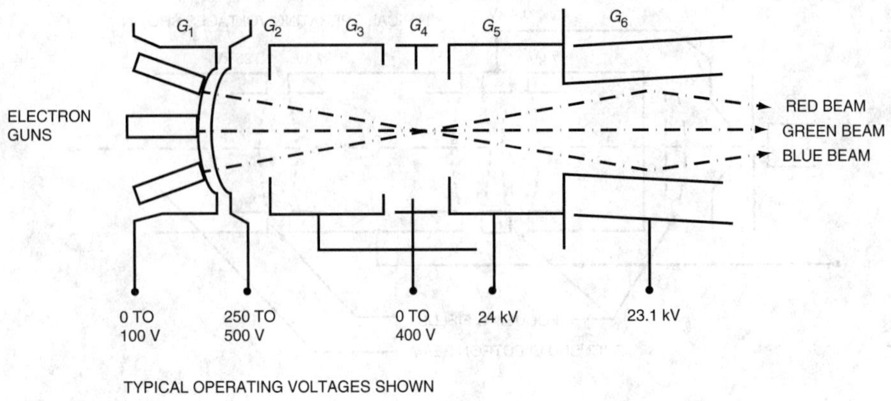

FIGURE 5.112   Electrode arrangement of the Trinitron gun. (After Morrell, A. et al. 1974. *Color Picture Tubes*, pp. 91–98. Academic, New York.)

cause misconvergence of the three undeflected spots. Further, carefully tailored current waveforms are applied to magnetic pole pieces on the separate guns to dynamically converge the three beams over the full face of the tube.

In-line guns enjoy one major advantage that more than offsets their less efficient use of available neck area. That is, with three beams on the horizontal axis, a deflection yoke can be built that will maintain dynamic convergence over the full face of the tube, without the need for correcting waveforms. With this major simplification in needed circuitry almost all guns now being built for commercial video applications are of the in-line design. In-line guns are available for 36-, 29-, and 22.5-mm neck diameters (among others). Typical lens diameters for these three guns are 7.5, 5.5, and 3.5 mm, respectively.

Unitized Construction

Most in-line guns are of *unitized construction* in which the three apertures or lens diameters are formed from a single piece of metal. With this arrangement the apertures and lens diameters are accurately fixed with respect to each other so that beam landings at the screen are more predictable than with the cylinder guns. Also, self-converging deflection yokes need the more accurate positioning of the beams in the yoke area. The only trio of gun elements not electrically tied together are the cathodes. Therefore, all varying voltages controlling luminance and color must be applied to the cathodes only. It should be noted that the three beams in unitized guns travel parallel until they reach the final lens gap. Here, an offset, or tilted, lens on the two outboard beams bends the two outer beams toward the center beam at the screen. Any change in the strength of this final lens gap, such as a focus voltage adjustment, will cause a slight change in the undeflected convergence pattern at the screen.

Guns for High-Resolution Applications

Improved versions of both the delta and in-line guns noted previously are used in high-resolution data display applications. In both cases the guns are adjusted for the lower beam current and higher resolution needed in data and/or graphics display. The advantages and disadvantages noted in earlier sections for delta and in-line guns apply here also. For example, the use of a delta-type cylinder, or *barrel-type* gun, requires as many as 20 carefully tailored convergence waveforms to obtain near-perfect convergence over the full face of the tube. The in-line gun, with a self-converging yoke, avoids the need for these waveforms, at the expense of slightly larger spots, particularly in the corners where overfocused haze tails can cause problems.

Figure 5.113 compares spot sizes, at up to 1 mA of beam current, for high-resolution designs compared with commercial receiver-type devices. Both delta and in-line 13- and 19-in vertical (13 V and 19 V)

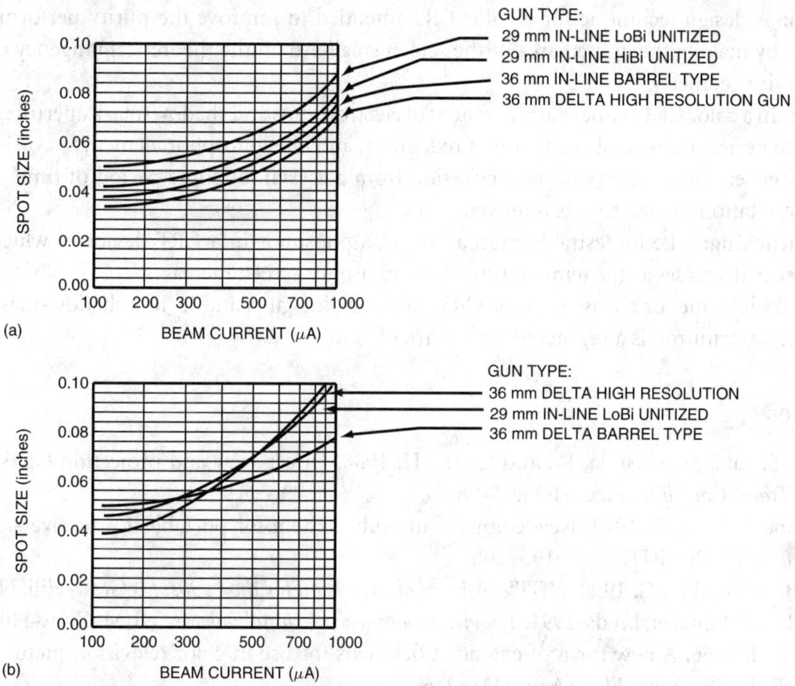

FIGURE 5.113 Spot-size comparison of high-resolution guns vs commercial television guns: (a) 13-in vertical display, (b) 19-in vertical display. (After Benson, K.B. and Whitaker, J.C., eds. *Television Engineering Handbook*, rev. ed., p. 12.20. McGraw-Hill, New York.)

devices are shown. Note the marked improvement in spot size at current levels below 500 μA for the high-resolution devices.

Deflecting Multiple Electron Beams

Deflection of the electron beams in a color CRT represents a difficult technical exercise. The main problems that occur when the three beams are deflected by a common deflection system are associated with the spot distortions that occur in single-beam tubes. However, the effect is intensified by the need for the three beams to cross over and combine as a spot on the shadow mask. The two most significant effects are curvature of field and astigmatism.

Misalignment and misregistration of the three beams of the color CRT will lead to loss of purity for colors produced by combinations of the primary colors. Such reproduction distortions can also result in a reduction in luminance output because a smaller part of the beams are passing through the apertures. Additional errors that must be considered include

- Deflection-angle changes in the yoke-deflection center
- Stray electromagnetic fields

Defining Terms

Cathodoluminescence: The property of luminescent crystals (phosphors) to emit visible light with bombarded by electrons.

Delta gun: In a color CRT, a triangular arrangement of electron guns and shadow-mask apertures positioned so as to permit individual excitation of red, green, and blue phosphor elements.

Fluorescence: The property of light emission from a crystal during excitation by electrons.

Guard band: A design technique for a color CRT intended to improve the purity performance of the device by making the lighted area of the screen smaller than the theoretical tangency condition of the device geometry.

In-line gun: In a color CRT, a linear arrangement of electron guns and shadow-mask apertures positioned so as to permit individual excitation of red, green, and blue phosphor elements.

Phosphorescence: The property of light emission from a crystal for some period of time immediately after excitation by electrons is removed.

Thermal quenching: The undesirable mechanism of a phosphor in a CRT device in which phosphor efficiency decreases as the temperature of the phosphor screen increases.

Trinitron: Trade name for a class of color CRTs using in-line guns and in-line shadow-mask/phosphor geometry. Trinitron is a registered trademark of Sony.

References

Ashizaki, S., Suzuki, Y., Mitsuda, K., and Omae, H. 1988. Direct-view and projection CRTs for HDTV. *IEEE Trans. Consumer Elec.* 34(1):91–98.

Barbin, R. and Hughes, R. 1972. New color picture tube system for portable TV receivers. *IEEE Trans. Broadcast TV Rec.* BTR-18(3):193–200.

Benson, K.B. and Fink, D.G. 1990. *HDTV: Advanced Television for the 1990s.* McGraw-Hill, New York.

Benson, K.B. and Whitaker, J., eds. 1991. *Television Engineering Handbook,* rev. ed. McGraw-Hill, New York.

Blacker, A. et al. 1966. A new form of extended field lens for use in color television picture tube guns. *IEEE Trans. Consumer Elec.* (Aug.):238–246.

Blaha, R. 1972. Degaussing circuits for color TV receivers. *IEEE Trans. Broadcast TV Rec.* BTR-18(1):7–10.

Carpenter, C. et al. 1955. An analysis of focusing and deflection in the post-deflection focus color kinescope. *IRE Trans. Elec. Dev.* 2:1–7.

Casteloano, J.A. 1992. *Handbook of Display Technology.* Academic, New York.

Chang, I. 1980. Recent advances in display technologies. *Proc. SID* 21(2):45.

Chen, H. and Hughes, R. 1980. A high performance color CRT gun with an asymmetrical beam forming region. *IEEE Trans. Consumer Elec.* CE-26(Aug.):459–465.

Chen, K.C., Ho, W.Y., and Tseng, C.H. 1992. Invar mask for color cathode ray tubes. In *Display Technologies,* Chen, S.H. and Wu, S.T., eds. *Proc. SPIE 1815,* pp. 42–48, Bellingham, WA.

Davis, C. and Say, D. 1979. High performance guns for color TV—A comparison of recent designs. *IEEE Trans. Consumer Elec.* CE-25(Aug.).

Donofrio, R. 1972. Image sharpness of a color picture tube by MTF techniques. *IEEE Trans. Broadcast TV Rec.* BTR-18(1):1–6.

EIA. 1975. CRTs: *Glossary of Terms and Definitions.* Publication TEP92, Electronic Industries Association, Washington, DC.

Fink, D., ed. 1957. *Television Engineering Handbook.* McGraw-Hill, New York.

Fink, D. and Christiansen, D., eds. 1989. *Electronics Engineers Handbook,* 3rd ed. McGraw-Hill, New York.

Fiore, J. and Kaplin, S. 1982. A second generation color tube providing more than twice the brightness and improved contrast. *IEEE Trans. Consumer Elec.* CE-28(1):65–73.

Godfrey, R. et al. 1968. Development of the permachrome color picture tube. *IEEE Trans. Broadcast TV Rec.* BTR-14(1).

Gow, J. and Door, R. 1954. Compatible color picture presentation with the single-gun tri color chromatron. *Proc. IRE* 42(1):308–314.

Hockenbrock, R. 1989. New technology advances for brighter color CRT displays. In *Display System Optics II,* ed. H.M. Assenheim, *Proc. SPIE 1117,* pp. 219–226, Bellingham, WA.

Hoskoshi, K. et al. 1980. A new approach to a high performance electron gun design for color picture tubes.1980 *IEEE Chicago Spring Conf. Consumer Electronics.*

Hu, C., Yu, Y., and Wang, K. 1992. Antiglare/antistatic coatings for cathode ray tube based on polymer system, In *Display Technologies.* Chen, S.H. and Wu, S.T., eds. *Proc. SPIE 1815,* pp. 42–48.

Law, H. 1951. A three-gun shadow mask color kinescope. *Proc. IRE* 39(Oct.):1186–1194.

Masterson, W. and Barbin, R. 1971. Designing out the problems of wide-angle color TV tube. *Electronics* (April 26):60–63.

Mokhoff, N. 1981. A step toward perfect resolution. *IEEE Spectrum* 18(7):56–58.

Morrell, A. 1981. Color picture tube design trends. *Proc. SID* 22(1):3–9.

Morrell, A. et al. 1971. *Color Television Picture Tubes.* Academic, New York.

Moss, H. 1968. *Narrow Angle Electron Guns and Cathode Ray Tubes.* Academic, New York.

Palac, K. 1976. Method for manufacturing a color CRT using mask and screen masters. U.S. Patent 3,989,524. 1976.

Say, D. 1977. Picture tube spot analysis using direct photography. *IEEE Trans. Consumer Elec.* CE-23(Feb.)32–37.

Sherr, S. 1979. *Electronic Displays.* Wiley, New York.

Tong, H.S. 1992. HDTV Display–A CRT approach. In *Display Technologies*, Chen, S.H. and Wu, S.T., eds. *Proc. SPIE 1815*, pp. 2–8, Bellingham, WA.

U.S. Patent. 1980. Electron gun with astigmatic flare-reducing beam forming regcon. U.S. Patent 4,234,814, Nov. 18.

Whitaker, J.C. 1993. *Electronic Displays: Technology, Design and Applications.* McGraw-Hill, New York.

Yoshida, S. et. al. 1973. A wide deflection angle (114°) Trinitron color picture tube. *IEEE Trans. Elec. Dev.* 19(4):231–238.

Yoshida, S. et al. 1974. 25-V inch 114-degree Trinitron color picture tube and associated new development. *Trans. BTR* (Aug.):193–200.

Yoshida. S. et al. 1982. The Trinitron—A new color tube. *IEEE Trans. Consumer Elec.* CE-28(1):56-64.

Further Information

The SPIE (the international optical technology society) offers a wide variety of publications on the topic of display systems engineering. The organization also holds technical seminars on this and other topics several times each year. SPIE is headquartered in Bellingham, WA. The Society for Information Display (SID) also holds regular technical seminars and offers a number of publications on the topic of display systems. SID is headquartered in Santa Ana, CA. Furthermore, two books on the topic of display systems are recommended:

Sherr, S. 1994. *Fundamentals of Display System Design*, 2nd ed. Wiley-Interscience, New York.
Whitaker, J.C. 2001. *Video Displays.* McGraw-Hill, New York.

The Electronic Industries Association (EIA), Washington, DC, also offers several publications on the topic of display systems as they relate to consumer products.

5.7 Projection Systems

Jerry C. Whitaker

5.7.1 Introduction

The use of large-screen projection displays continues to grow at a rapid rate as the need to present high-resolution video and graphics information steadily increases.[5] High definition television (HDTV) will requires large screens (greater than 40-in diagonal) to provide effective presentation. Some form of projection is, therefore, the only practical solution. The role of HDTV and film in future theaters is also being explored, along with performance criteria for effective large-screen video presentations.

---

[5]Portions of this chapter were adapted from: Whitaker, J.C. 1994. *Electronic Displays: Technology, Design and Application.* McGraw-Hill, New York.

New projection system hardware is taking advantage of liquid crystal (LC) and thin film transistor (TFT) technology originally developed for direct-view flat panel displays. Also, new deformable-membrane light valves used in Schlieren systems are being developed to update oil film light-valve projectors, which have been the mainstay of large-screen projection technology for many years.

The number of specific applications for large-screen displays continues to grow. Military command centers provide a stringent test where high resolution, brightness, and reliability are critical. Flight simulation presents perhaps some of the greatest challenges, as large images must be presented on domes with high brightness, contrast, and resolution. Success in this application may lead to new entertainment vehicles, where domed entertainment theaters will combine traditional presentations with dynamic, high-definition video enhancements.

Extensive developmental efforts are being conducted into large-screen display systems. Much of this research is aimed at advancing new technologies such as plasma, electroluminescent and LCD, lasers, and new varieties of CRTs. Because this section is concerned with vacuum tube technologies, this chapter will focus on CRT-based projection systems. Light valve and laser technologies will also be examined. Solid-state projection systems are covered in Chapter 41.

Display Types

Large-screen projectors fall into four broad classes, or grades:

- *Graphics.* Graphics projectors are the highest quality, and generally most expensive, projectors. These systems are capable of the highest operating frequency and resolution. They can genlock to almost any computer or image source and typically offer resolutions of 2000 × 2000 pixels and greater with horizontal sweep rates up to 89 kHz or more.
- *Data.* Data projectors are less expensive than graphics projectors and are suitable for use with common computer image generators, such as PCs equipped with EGA, VGA, and Macintosh standard graphics cards.
- *HDTV.* HDTV projectors provide the quality level necessary to take full advantage of high-definition imaging systems. Resolution on the order of 1125 TV lines is provided at an aspect ratio of 16:9.
- *Video.* Video projectors provide resolution suitable for National Television Systems Committee (NTSC) level images. Display performance of 380–480 TV lines is typical.

Computer signal sources can follow many different, and sometimes incompatible, standards. Multisync projectors are common. Generally speaking, projectors are *downward compatible*. In other words, most graphic projectors can function as data projectors, HDTV projectors, and video projectors; most data projectors can also display HDTV and video, and so on.

Displays for HDTV Applications

The most significant difference between a conventional display and one designed for high-definition video is the increased resolution and wider aspect ratio of HDTV. Three technologies are practical for viewing high-definition images of 40-in diagonal and larger:

- **Light valve** *projection display.* Capable of modulating high-power external light sources, light valves are mainly used for large screen displays.
- *CRT projection displays.* Widely use for middle-sized screen displays, CRT projection systems are popular because of their relative ease of manufacturing (and hence competitive cost) and good performance. These displays are likely to be the mainstay technology for HDTV in the near-term future.
- *Flat panel plasma display panel (PDP) and liquid crystal display (LCD).* Flat panel PDP displays are available demonstrated in 40-in and larger sizes.

TABLE 5.7 Performance Requirements for Large-Screen Military Display Applications

Parameter	Performance Requirement
Output luminance[a]:	
Small conference room	50–200 lm
Large conference room	200–600 lm
Control center	600–2000 lm
Addressable resolution	1280×1024 pixels
Visual resolution	15% absolute minimum modulation depth to an alternate pixel input
Video interface	15.75–80 kHz horizontal scan, 60–120 Hz noninterlaced refresh
Video sources	10–20 different sources, autosynchronous lock
Response to new update	Less than 1 s, no smear to dynamic response
Reliability	1000 h MTBF, excluding consumables
Operating cost	Less than $10/h per projector, including AC power
Maintenance	*Level 5* maintenance technician

[a]Ambinent lighting of 20–40 fc is assumed. The values shown are modulated luminance, spatially averaged over the full screen area. (*Source:* Blaha, R.J. 1990. Large screen display technology assessment for military applications. *Large Screen Projection Displays II*, ed. W.P. Bleha, Jr., pp. 80–92. Proc. SPIE 1255.)

Displays for Military Applications

Full-color, large-screen display systems can enhance military applications that require group presentation, coordinated decisions, or interaction between decision makers. Projection display technology already plays an important role in operations centers, simulation facilities, conference rooms, and training centers. Each application requires unique values of luminance, resolution, response time, reliability, and video interface. The majority of large-screen applications involve fixed environments where commanders and their staff interact as a group to track an operation and make mission-critical decisions. As sensor technology improves and military culture drives toward *scene presence*, future applications will require large-screen display systems that offer greater realism and real-time response. Table 5.7 lists the basic requirements for military display applications. The available technologies meeting these demands are illustrated in Fig. 5.114.

5.7.2 Projection System Fundamentals

Video projection systems provide a method of displaying a much larger image than can be generated on a direct-view cathode-ray picture tube. Optical magnification and other techniques are employed to throw an expanded image on a passive viewing surface that may have a diagonal dimension of 75-in or more. Direct-view CRTs are generally restricted to less than 39-in diagonal. A primary factor limiting CRT size is glass strength. To withstand atmospheric pressure on the evacuated envelope, CRT weight increases exponentially with linear dimension.

The basic elements of a projection system, illustrated in Fig. 5.115, include

- Viewing screen
- Optical elements
- Image source
- Drive electronics

The major differences of projection systems from direct-view displays are embodied in the first three areas, whereas the electronics assembly is (typically) essentially the same as for direct-view systems.

Projection Requirements

To provide an acceptable image, a projection system must approach or equal the performance of a direct-view device in terms of brightness, contrast, and resolution. Whereas the first two parameters may be compromised to some extent, large displays must excel in resolution because of the tendency of viewers

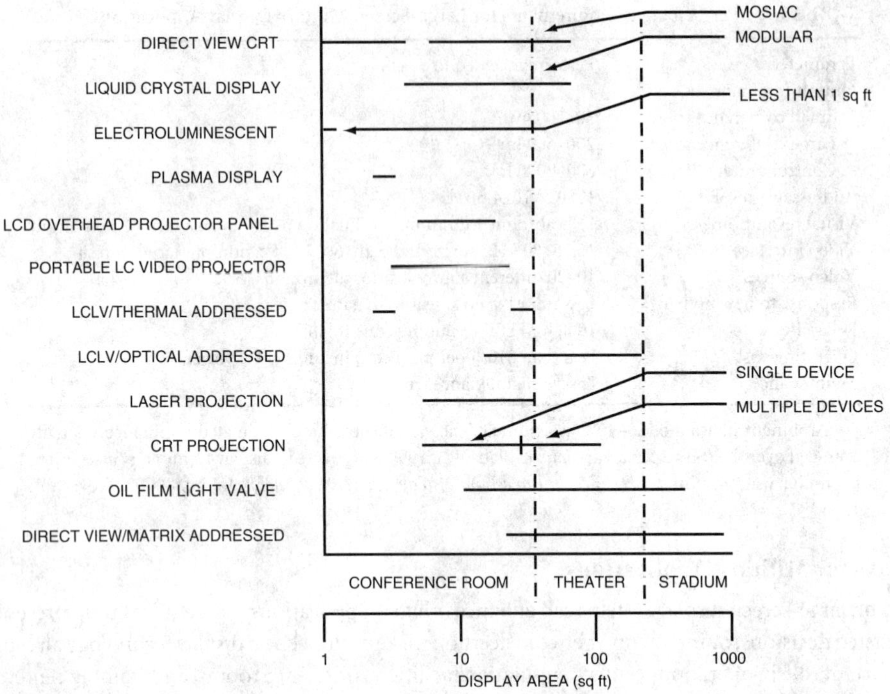

FIGURE 5.114   Basic grading of full-color display technologies vs. display area for military applications.

to be positioned less than the normal relative distance of four to eight times the picture height from the viewing surface. Table 5.8 provides performance levels achieved by direct-view video displays and conventional film theater equipment.

Evaluation of overall projection system brightness B, as a function of its optical components, can be calculated using the following equation:

$$B = \frac{L_g G T R^M D}{4 W_g (f/N)^2 (1+m)^2}$$

where
L_G = luminance of the green source (CRT or other device)
G = **screen gain**
T = lens transmission

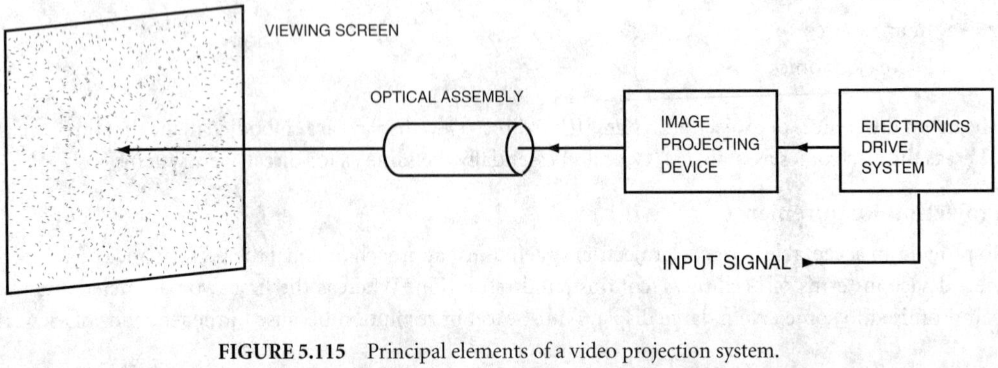

FIGURE 5.115   Principal elements of a video projection system.

TABLE 5.8   Performance Levels of Video and Theater Displays

Display System	Luminous Output (brightness), nits, f L	Contrast Ratio at Ambient Illumination, fc	Resolution (TVL)
Television receiver	200–400, 60–120	30:1 at 5	275
Theater (film projector)	34–69, 10–20[a]	100:1 at 0.1[b]	4800

[a]U.S. standard (PH-22.124-1969); see Luxenberg and Kuehn, 1968, p. 29.
[b]Limited by lens flare.

R = mirror reflectance
M = number of mirrors
D = dichroic efficiency
W_G = green contribution to desired white output, %
f/N = lens f-number
m = magnification

For systems in which dichroics or mirrors are not employed, those terms drop out.

Two basic categories of viewing screens are employed for projection video displays. As illustrated in Fig. 5.116, the systems are

- **Front projection**, where the image is viewed from the same side of the screen as that on which it is projected.
- **Rear projection**, where the image is viewed from the opposite side of the screen as that on which it is projected.

Front projection depends on reflectivity to provide a bright image, whereas rear projection requires high transmission to achieve that characteristic. In either case, screen size influences display brightness inversely as follows:

$$B = \frac{L}{A}$$

where
    B = apparent brightness, cd/m^2
    L = projector light output, lm
    A = screen viewing area, m^2

Thus, for a given projector luminance output, viewed brightness varies in proportion to the reciprocal of the square of any screen dimension (width, height, or diagonal). An increase in screen width from the

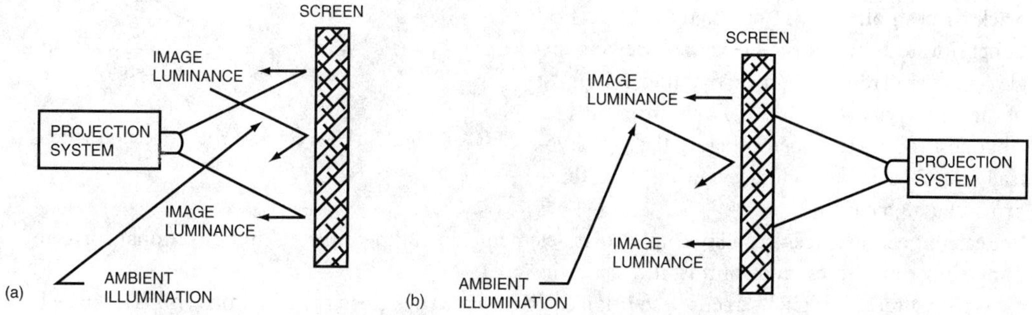

FIGURE 5.116   Projection system screen characteristics: (a) front projection, (b) rear projection.

TABLE 5.9 Screen Gain Characteristics for Various Materials

Screen Type	Gain, cd/m^2
Lambertian (flat-white paint, magnesium oxide)	0.85–0.90
White semigloss	1.5
White pearlescent	1.5–2.5
Aluminized	1–12
Lenticular	1.5–2
Beaded	1.5–3
Ektalite (Kodak)	10–15
Scotch-light (3 M)	Up to 200

conventional aspect ratio of 4:3 (1.33) to an HDTV ratio of 16:9 (1.777) requires an increase in projector light output of approximately 33% for the same screen brightness.

To improve apparent brightness, directional characteristics can be designed into viewing screens. This property is termed *screen gain G*, and the preceding equation becomes

$$B = G \times \frac{L}{A}$$

Gain is expressed as screen brightness relative to a *lambertian surface*. Table 5.9 lists some typical front-projection screens and their associated gains.

Screen contrast is a function of the manner in which ambient illumination is treated. Figure 5.116 illustrates that a highly reflective screen (used in front projection) reflects ambient illumination as well as the projected illumination (image). The reflected light thus tends to dilute contrast, although highly directional screens diminish this effect. A rear-projection screen depends on high transmission for brightness but can capitalize on low reflectance to improve contrast. A scheme for achieving this is equivalent to the black matrix utilized in tricolor CRTs. Illustrated in Fig. 5.117, the technique focuses projected light through lenticular lens segments onto strips of the viewing surface, allowing intervening areas to be coated with a black (non-reflective) material. The lenticular segments and black stripes normally are oriented in the vertical dimension to broaden the horizontal viewing angle. The overall result is a screen that transmits most of the light (typically, 60%) incident from the rear while absorbing a large percentage of the light (typically, 90%) incident from the viewing side, thus providing high contrast.

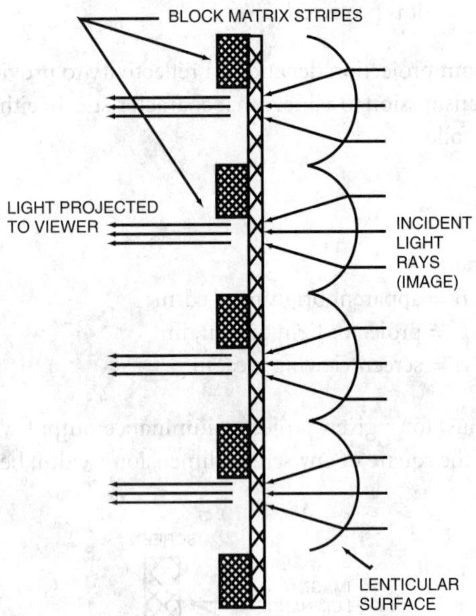

FIGURE 5.117 High contrast rear projection system.

Rear-projection screens usually employ extra elements, including diffusers and directional correctors, to maximize brightness and contrast in the viewing area.

As with direct-view CRT screens, resolution can be affected by screen construction. This is not usually a problem with front-projection screens, although granularity or lenticular patterns can limit image detail. In general, any screen element, such as the matrix arrangement described previously, that quantizes the image (breaks it into discrete segments) limits resolution. For 525- and 625-line video, this factor does not provide the limiting aperture. High-resolution applications, however, may require attention to this parameter.

Projectors for Cinema Applications

Screen brightness is a critical element in providing an acceptable HDTV large-screen display. Without adequate brightness, the impact on the audience is reduced. Theaters have historically utilized front projection systems for 35-mm film. This arrangement provides for efficient and flexible theater seating. Large-screen HDTV is likely to maintain the same arrangement. Typical motion picture theater projection system specifications are as follows:

- Screen width 30 ft
- Aspect ratio 2.35:1
- Contrast ratio 300:1
- Screen luminance 15 fL
- Center-to-edge brightness uniformity 85%

These specifications meet the expectations of motion picture viewers. It follows that for HDTV projectors to be competitive in theatrical display applications, they must meet similar specifications.

Optical Projection Systems

Both refractive and reflective lens configurations have been used for the display of a CRT raster on screens 40-in or more in diagonal width. The first attempts merely placed a lenticular Fresnel lens, or an inefficient $f/1.6$ projection lens in front of a shadow mask direct view tube, as shown in Fig. 5.118(a). The resulting brightness of no greater than 2 or 3 fL was suitable for viewing only in a darkened room. Figure 5.118(b) shows a variation on this basic theme. Three individual CRTs are combined with cross reflecting mirrors and then focused onto the screen. The in-line projection layout is shown in Fig. 5.118(c) using three tubes, each with its own lens. This is the most common system used for multitube displays. Typical packaging to reduce cabinet size for front or rear projection is shown in Fig. 5.118(d) and 5.118(e), respectively.

Because of the off-center positioning of the outboard color channels, the optical paths differ from the center channel, and keystone scanning height modulation is necessary to correct for differences in optical throw from left to right. The problem, illustrated in Fig. 5.119, is more severe for wide-screen formats.

Variables to be evaluated in choosing among the many schemes include the following:

- Source luminance
- Source area
- Image magnification
- Optical-path transmission
- Light collection efficiency (of the lens)
- Cost, weight, and complexity of components and corrective circuitry

The lens package is a critical factor in rendering a projection system cost effective. The package must possess good luminance collection efficiency (small f-number), high transmission, good modulation transfer function (MTF), light weight, and low cost. Table 5.10 compares the characteristics of some available lens complements.

The total light incident upon a projection screen is equal to the total light emerging from the projection optical system, neglecting losses in the intervening medium. The distribution of this light generally is not uniform. Its intensity is less at screen edges than at the center in most projection systems as a result of light ray obliquity through the lens ($\cos^4 \theta$ law) and *vignetting effects* [Luxenberg and Kuehn 1968].

Light output from a lens is determined by collection efficiency and transmittance, as well as source luminance. Typical figures for these characteristics are

- Collection efficiency: 15–25%
- Transmittance: 75–90%

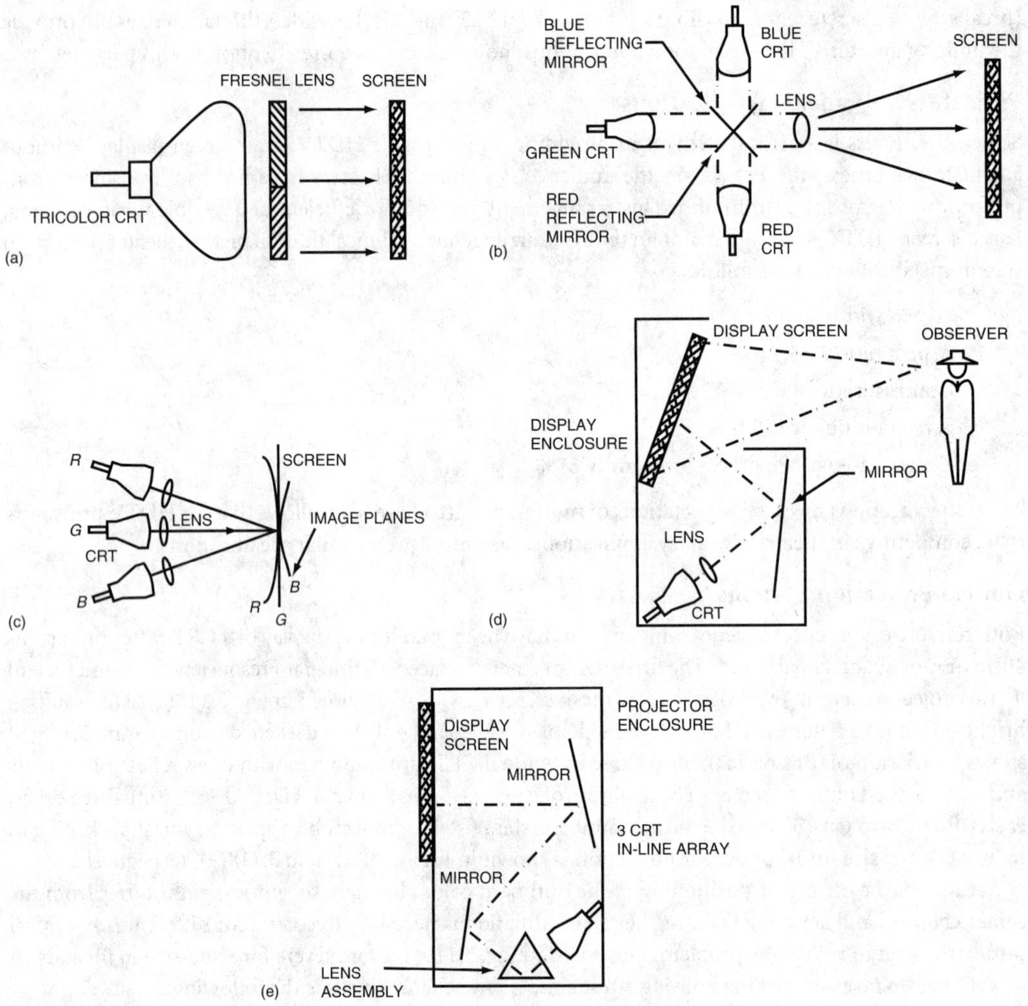

FIGURE 5.118 Optical projection configurations: (a) single tube, single lens rear projection system; (b) crossed-mirror trinescope rear projection; (c) three tube, three lens rear projection systems; (d) folded optics, front projection with three tubes in-line and a dichroic mirror; (e) folded optics, rear projection.

Collection efficiency is partially a function of the light source, and the figure given is typical for a lambertian source (CRT) and lens having a half-field angle of approximately 25°.

Optical Distortions

Optical distortions are important to image geometry and resolution. Geometry is generally corrected electronically, both for pincushion/barrel effects and keystoning, which result from the fact that the three image sources are not coaxially disposed in the common in-line array.

Resolution, however, is affected by lens astigmation, coma, spherical aberration, and chromatic aberration. The first three factors are dependent on the excellence of the lens, but chromatic aberration can be minimized by using line emitters (monochromatic) or narrow-band emitters for each of the three image sources. Because a specific lens design possesses different magnification for each of the three primary colors (the index of refraction varies with wavelength), throw distance for each must also be adjusted independently to attain perfect registration.

In determining final display luminance, transmission, reflectance, and scattering by additional optical elements such as dichroic filters, optical-path folding mirrors, or corrective lenses must also be accounted

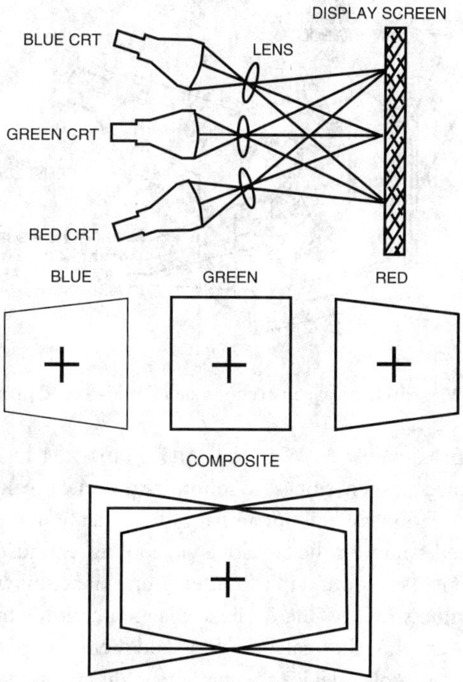

FIGURE 5.119 Three tube in-line array. The optical axis of the outboard (red and blue) tubes intersect the axis of the green tube at screen center. Red and blue rasters show trapezoidal (*keystone*) distortion and result in misconvergence when superimposed on the green raster, thus requiring electrical correction of scanning geometry and focus.

for. Dichroics exhibit light attenuations of 5–30% and mirrors can reduce light transmission by as much as 5% each. Front-surface mirrors exhibit minimum absorption and scattering but are susceptible to damage during cleaning. Contrast is also affected by the number and nature of optical elements employed in the projection system. Each optical interface generates internal reflections and scattering, which dilute contrast and reduce MTF amplitude response. Optical coatings may be utilized to minimize these effects, but their contribution must be balanced against their cost.

Image Devices

CRTs and light valves are the two most common devices for creating images to be optically projected. Each is available in a multitude of variations. Projection CRTs have historically ranged in size from 1 in (2.5 cm) to 13 in (33 cm) diagonal (diameter for round envelope types). Because screen power must increase in proportion to the square of the magnification ratio, it is clear that faceplate dissipation for CRTs used in projection systems must be extremely high. Electrical-to-luminance conversion efficiency for common video phosphors is on the order of 15% (McKechnie, 1981). A 50-in (1.3-m) diagonal screen at 60 fL requires a 5-in (12.7-cm) CRT to emit 6000 fL, exclusive of system optical losses, resulting in

TABLE 5.10 Performance of Various Types of Projection Lenses

Lens type	Aperture, $f/$	Image Diagonal, mm	Focal Length, mm	Magnification	Response at 300 TVL, %
Refractive, glass	1.6	196	170	8	33
Refractive, acrylic	1.0	127	127	10	13
Schmidt .	0.7	76	87	30	15
Fresnel	1.7	305	300	5	6

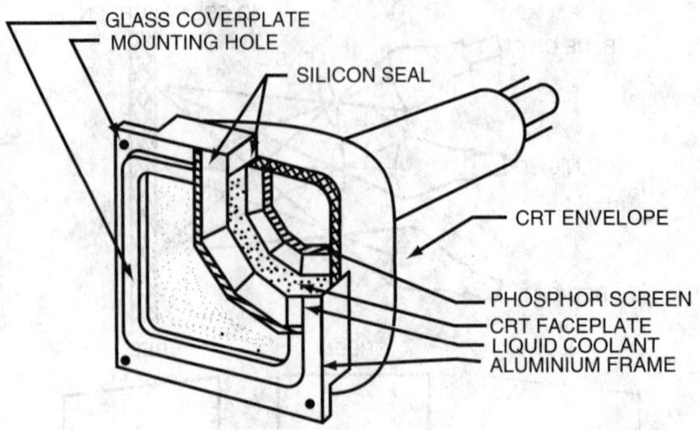

FIGURE 5.120   Mechanical construction of a liquid-cooled projection CRT.

a faceplate dissipation of approximately 20 W in a 3-in (7.6-cm) × 4-in (10.2-cm) raster. A practical limitation for ambient air-cooled glass envelopes (to minimize thermal breakage) is 1 mW/mm^2 or 7.74 W for this size display. Accommodation of this incompatibility must be achieved in the form of improved phosphor efficiency or reduced strain on the envelope via cooling. Because phosphor development is a mature science, maximum benefits are found in the latter course of action with liquid cooling assemblies employed to equalize differential strain on the CRT faceplate. Such implementations produce an added benefit through reduction of phosphor thermal quenching and thereby supply up to 25% more luminance output than is attainable in an uncooled device at equal screen dissipation (Kikuchi et al., 1981).

A liquid-cooled CRT assembly, shown in Fig. 5.120, depends on a large heat sink to carry away and dissipate a substantial portion of the heat generated in the CRT. Large-screen projectors using such assemblies commonly operate CRTs at four to five times their rated thermal capacities. Economic constraints mitigate against the added cost of cooling assemblies, however, and methods to improve phosphor conversion efficiency and optical coupling/transmission efficiencies continue to be investigated.

Concomitant to high power are high voltage and/or high beam current. Each has benefits and penalties. Resolution, dependent on spot diameter, is improved by increased anode voltage and reduced beam current. For a 525- or 625-line display, spot diameter should be 0.006 in (0.16 mm) on the 3-in × 4-in (7.6-cm × 10.2-cm) raster discussed previously. Higher resolving power displays require yet smaller spot diameters, but a practical maximum anode-voltage limit is 30–32 kV when X radiation, arcing, and stray emission are considered. Exceeding 32 kV typically requires special shielding and CRT processing during assembly.

One form of the in-line array benefits from the relatively large aperture and light transmission efficiency of Schmidt reflective optics by combining the electron optics and phosphor screen with the projection optics in a single tube. The principle components of an integral Schmidt system are shown in Fig. 5.121.

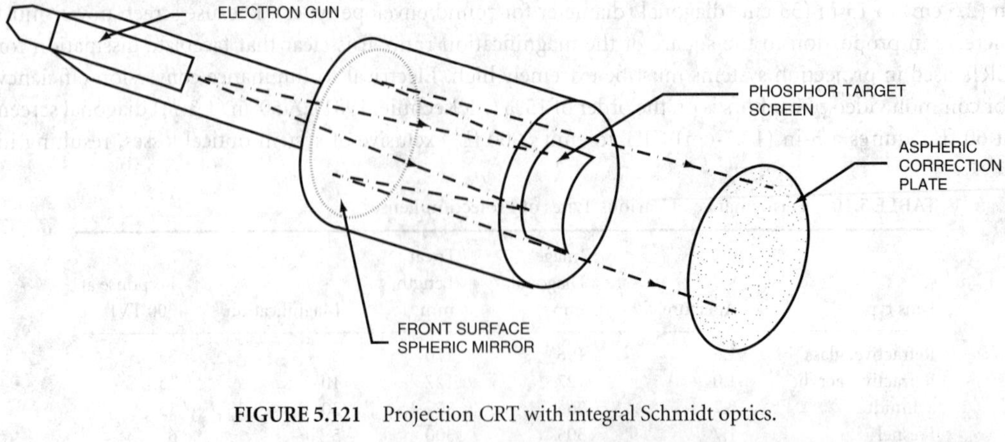

FIGURE 5.121   Projection CRT with integral Schmidt optics.

Electrons emitted from the electron gun pass through the center opening in the spherical mirror of the reflective optical system to scan a metal-backed phosphor screen. Light from the phosphor (red, green, or blue, depending on the color channel) is reflected from the spherical mirror through an *aspheric corrector lens*, which serves as the face of the projection tube. Schmidt reflective optical systems are significantly more efficient than refractive systems because of the lower f characteristic and the reduced attenuation by glass in the optical path.

Discrete-Element Display Systems

The essential characteristic defining discrete-element display devices is division of the viewing surface (volume for three-dimensional displays) into separate segments that are individually controlled to generate an image. Each element, therefore, embodies a dedicated controlling switch or valve as opposed to raster display devices, which employ one (or a small number of) control device(s) for activation of all display elements.

Flood Beam CRT

Large displays (20 m diagonal and larger) have been employed for mass audiences in stadiums to provide special coverage of sporting events. These consist of flood beam CRT arrays in which each device fulfills the function of a single phosphor dot in a delta-gun shadow-mask CRT. Thus, each display element consists of a trio of flood beam tubes, one red, one green, one blue, each with a 1–6 in (2.5–15 cm) diameter. A fully NTSC-capable display would require in excess of 400,000 such tubes (147,000 of each color) to be individually addressed. Practical implementations have employed less than 40,000, thus requiring substantially reduced drive complexity.

Matrix Addressing

Provision for individual control of each element of a discrete-element display implies a proliferation of electrical connections and, therefore, rapidly expanding cost and complexity. These factors are minimized in discrete-element displays through *matrixing* techniques wherein a display element must receive two or more gating signals before it is activated. Geometric arrangements of the matrixing connectors guarantee that only the appropriate elements are activated at a given time.

Matrix addressing requires that the picture elements of a panel be arranged in rows and columns, as illustrated Fig. 5.122 for a simplified 4×4 matrix. Each element in a row or column is connected to all other elements in that row or column by active devices (typically transistors). Thus, it is possible to select the drive signals so that when only a row or column is driven, no elements in that row or column is activated. However, when both the row and column containing a selected element are driven, the sum of the two signals will activate a given element. The resulting reduction in the number of connections required, compared to a discrete-element drive, is significant because the number of lines is reduced from one-per-element to the sum of the number of rows and columns in the panel.

For active matrix addressing to work properly, it is necessary to have a sharp discontinuity in the transfer characteristic of the device at each pixel at the intersection of the row and column address lines. In some cases, the characteristics of the display device itself can be used to provide this discontinuity. The breakdown potential of a gas discharge can be used (for plasma displays) or alternatively, liquid crystals can be made to have a reasonably sharp transfer curve.

Stadium applications require color video displays that can be viewed in sunlight. In such cases, the high brightness required is achieved with a matrix system. A bright source at each pixel is controlled by its own drive circuit, as illustrated in Fig. 5.123. Light sources are typically either color incandescent light bulbs or electron-excited fluorescent phosphors. The switching matrix must control the intensity of each color in each pixel of the display. For video, the system must have a continuous gray scale. Because of the cost and complexity of providing individual control of each pixel light source, such displays usually have somewhat lower resolution than those using light valves.

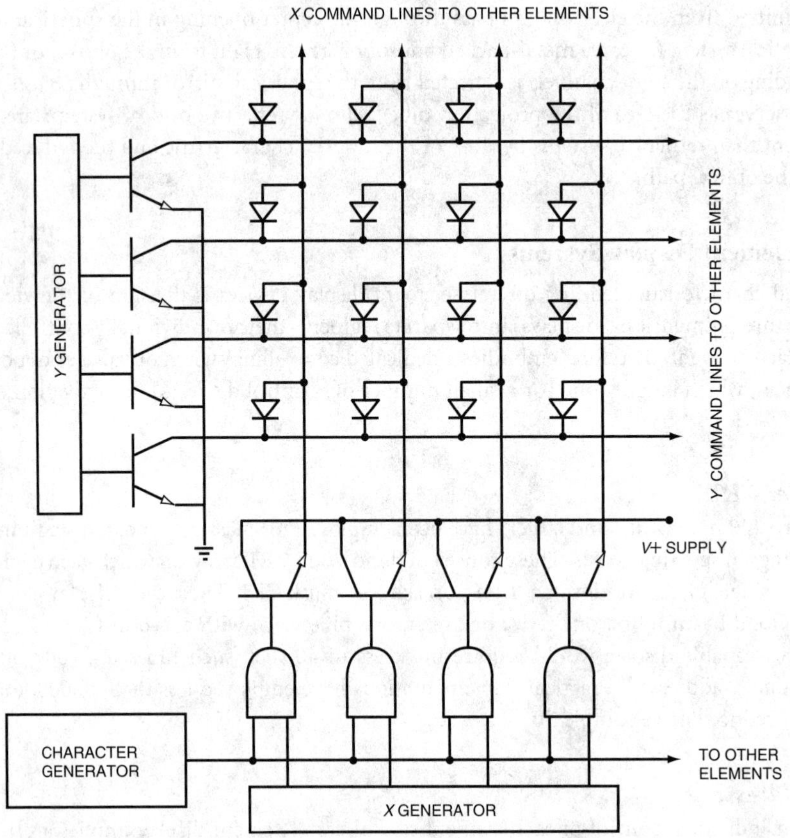

FIGURE 5.122   Matrix addressing for a discrete-element display.

5.7.3 Projection Display Systems

The use of large-screen video displays has been increasing for a wide variety of applications. Because of the greater resolution provided by HDTV and advanced computer graphics systems, higher image quality is being demanded from large-screen display systems. Available technologies include the following:

- CRT-based display systems
- *Eidophor* reflective optical system
- *Talaria* transmissive color system
- Dual light valve system for HDTV
- Laser beam projection scanning system
- LCD projection system

HDTV applications present additional, more stringent requirements than conventional 525/625-line color displays, including

- Increased light output for wide screen presentation
- Increased horizontal and vertical resolution
- Broader gamut of color response to meet system specifications
- Projection image aspect ratio of 16:9 rather than 4:3

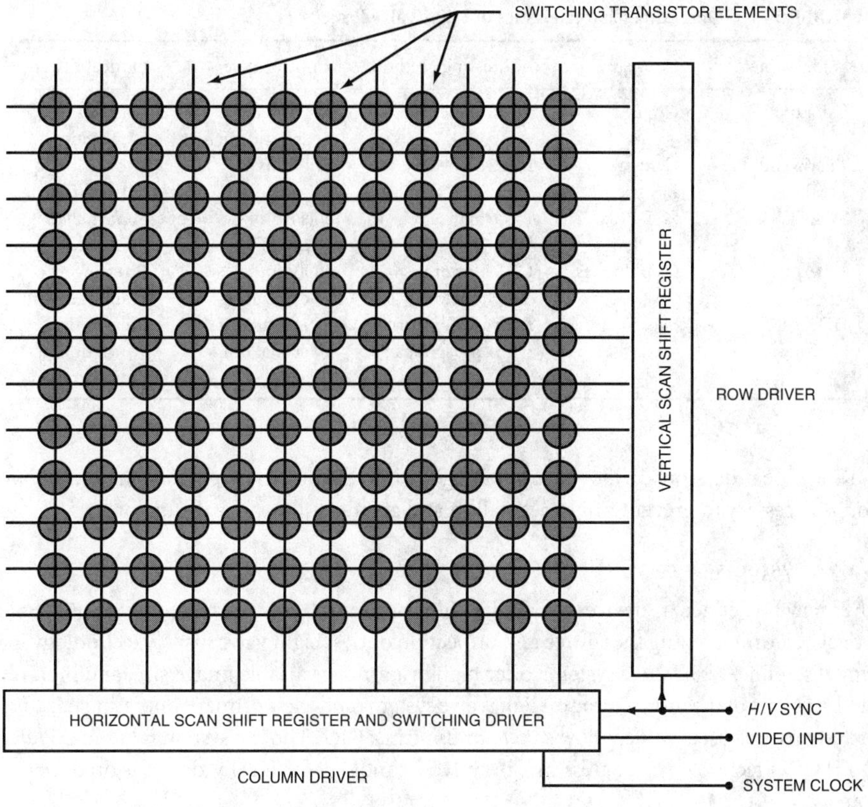

FIGURE 5.123 Principle of active matrix addressing for a large-screen display.

CRT Projection Systems

Large-screen color projection systems typically employ three monochrome CRT assemblies (red, green, blue) and the necessary optics to project full-color images. CRT systems based on 5- and 7-in CRT technology are popular because of their low cost; however, they suffer from low luminance levels (usually less than 50–60 average lumens). Higher luminance levels are provided by 9-in CRTs (160 lumens is typical). The larger CRT surface area permits increased beam energy without sacrificing resolution. Still higher luminance and resolution are offered by 13-in CRT projection systems.

Static and dynamic convergence circuitry is used to align the three beams over the display area. Digital convergence circuits permit convergence files tailored for each video source to be stored within a dedicated microprocessor. Advanced operating capabilities include autosync to a variety of input sources, built-in diagnostics, and automatic setup.

With CRT technology, high luminance and high resolution are conflicting requirements. For a given acceleration voltage, increased beam current will provide increased luminance, but reduced resolution because the spot size tends to increase. High-luminance levels also raise the operating temperature of the device, which may shorten its expected lifetime.

Multiple sets of CRT-based projection systems can be linked to increase luminance and resolution for a given application. The multiple beams overlay each other to yield the improved performance. Convergence of the 6–12 CRT assemblies, however, can become quite complex. Multiple CRT systems are satisfactory for NTSC video sources, but are usually inadequate for the display of computer-generated graphics data with single pixel characters.

The *mosaic* approach to improved luminance and resolution subdivides the screen into segments, each allocated to an individual projection system. As a result, larger images are delivered with higher overall performance. This scheme has been used effectively in simulation training to provide seamless, panoramic

TABLE 5.11   Operating Characteristics of Light Valve Projectors

Trade Name	Company	Addressing Method	Control Layer	Optical Effect
Eidophor	Gretag	Electron beam	Oil film on spheric mirror	Diffraction in reflection
Talaria and MLV	General Electric	Electron beam	Oil film on glass disk	Diffraction in transmission
SVS(TITUS)	SODERN (Philips)	Electron beam	Electro-optic crystal	*Pockels effect* in reflection
Super-projector	GM/Hughes	CRT + photo-conductor	ECB/liquid-crystal cell	Birefringence in reflection
HDTV LCD projector	Sharp	Active TFT matrix in amorphous Si	TN-liquid crystal cell	Polarization rotation in transmission

displays. Real-time processors divide the input image(s) into the proper display subchannels. Careful edge correction is necessary to prevent brightness falloff and geometric linearity distortions.

Light Valve Systems

Light valves may be defined as devices which, like film projectors, employ a fixed light source modulated by an optical-valve intervening the source and projection optics. Light valve display technology is a rapidly developing discipline. Light valve systems offer high brightness, variable image size, and high resolution. Table 5.11 lists some of the more common light valve systems and their primary operating parameters. The first four systems are based on *electron-beam addressing* in a CRT. The last system is based on *liquid crystal light valve* (LCLV) technology. Progress in light valve technology for HDTV depends on developments in two key areas:

- Materials and technologies for light control
- Integrated electronic driving circuits for addressing picture elements

Eidophor Reflective Optical System

Light valve systems are capable of producing images of substantially higher resolution than are required for 525/625-line systems. They are ideally suited to large screen theater displays.

In a manner similar to film projectors, a fixed light source is modulated by an optical value system (Schlieren optics) located between the light source and the projection optics (see Fig. 5.124). In the basic Eidophor system, collimated light typically from a 2-kW xenon source (component 1 in the figure) is directed by a mirror to a viscous oil surface in a vacuum by a grill of mirrored slits (component 3).

The slits are positioned relative to the oil-coated reflective surface so that when the surface is flat, no light is reflected back through the slits. An electron beam scanning the surface of the oil with a video picture raster (components 4–6) deforms the surface in varying amounts, depending on the video modulation of the scanning beam. Where the oil is deformed by the modulated electron scanning beam, light rays from the mirrored slits are reflected at an angle that permits them to pass through the slits to the projection lens. The viscosity of the liquid is high enough to retain the deformation over a period slightly greater than a television field.

Projection of color signals is accomplished through the use of three units, one for each of the red, green, and blue primary colors converged on a screen.

Talaria Transmissive Color System

The Talaria system also uses the principle of deformation of an oil film to modulate light rays with video information. However, the oil film is transmissive rather than reflective. In addition, for full-color displays, only one gun is used to produce red, green, and blue colors. This is accomplished in a single light valve by the more complex Schlieren optical system shown in Fig. 5.125.

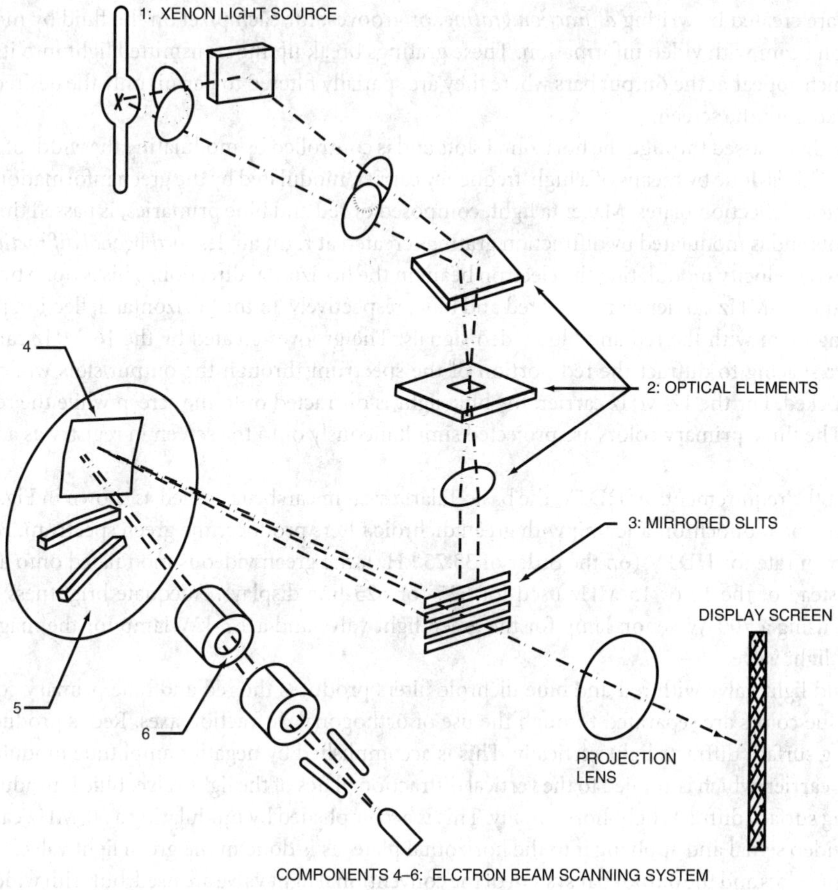

FIGURE 5.124 Mechanical configuration of the Eidophor projector optical system.

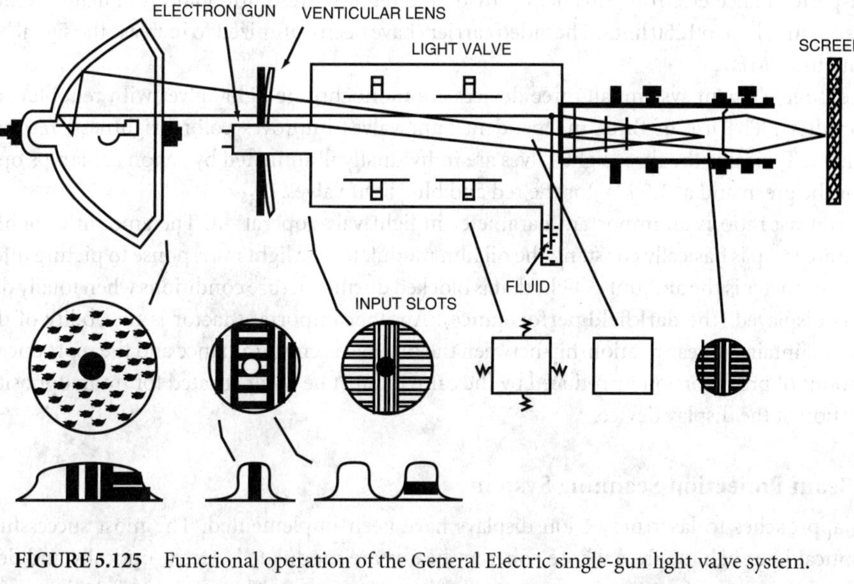

FIGURE 5.125 Functional operation of the General Electric single-gun light valve system.

Colors are created by writing *diffraction grating*, or grooves, for each pixel on the fluid by modulating the electron beam with video information. These gratings break up the transmitted light into its spectral colors, which appear at the output bars where they are spatially filtered to permit only the desired color to be projected onto the screen.

Green light is passed through the horizontal slots and is controlled by modulating the width of the raster scan lines. This is done by means of a high-frequency carrier, modulated by the green information, applied to the vertical deflection plates. Magenta light, composed of red and blue primaries, is passed through the vertical slots and is modulated by diffraction gratings created at right angles (*orthogonal diffraction*) to the raster lines by velocity modulating the electron beam in the horizontal direction. This is done by applying 16-MHz and 12-MHz carrier signals for red and blue, respectively, to the horizontal deflection plates and modulating them with the red and blue video signals. The grooves created by the 16-MHz carrier have the proper spacing to diffract the red portion of the spectrum through the output slots while the blue light is blocked. For the 12-MHz carrier, the blue light is diffracted onto the screen while the red light is blocked. The three primary colors are projected simultaneously onto the screen in registry as a full-color picture.

To meet the requirements of HDTV, the basic Talaria system can be modified as shown in Fig. 5.126. In the system, one monochromatic unit with green dichroic filters produces the green spectrum. Because of the high scan rate for HDTV (on the order of 33.75 kHz), the green video is modulated onto a 30-MHz carrier instead of the 12 or 15 MHz used for 525- or 625-line displays. Adequate brightness levels are produced using a 700-W xenon lamp for the green light valve and a 1.3-kW lamp for the magenta (red and blue) light valve.

A second light valve with red and blue dichroic filters produces the red and blue primary colors. The red and blue colors are separated through the use of orthogonal diffraction axes. Red is produced when the writing surface diffracts light vertically. This is accomplished by negative amplitude modulation of a 120-MHz carrier, which is applied to the vertical diffraction plates of the light valve. Blue is produced when the writing surface diffracts light horizontally. This is accomplished by modulating a 30-MHz carrier with the blue video signal and applying it to the horizontal plate, as is done in the green light valve.

The input slots and the output bar system of the conventional light valve are used, but with wider spacing of the bars. Therefore, the resolution limit is increased. The wider bar spacing is achievable because the red and blue colors do not have to be separated on the same diffraction axis as in the single light valve system. This arrangement eliminates the cross-color artifact present with the single light valve system, and therefore improves the overall colorimetric characteristics, as shown in Fig. 5.127.

High-performance electron guns help provide the required resolution and modulation efficiency for HDTV systems of up to 1250 lines. The video carriers have been optimized to increase the signal bandwidth capability to 30 MHz.

In the three-element system, all three devices are monochrome light valves with red, blue, and green dichroic filters. The use of three independent light valves improves color brightness, resolution, and colorimetry. Typically, the three light valves are individually illuminated by xenon arc lamps operating at 1 kW for the green and at 1.3 kW for the red and blue light valves.

The contrast ratio is an important parameter in light valve operation. The amount of light available from the arc lamp is basically constant; the oil film modulates the light in response to picture information. The key parameter is the amount of light that is blocked during picture conditions when totally dark scenes are being displayed (the **darkfield performance**). Another important factor is the ability of the display device to maintain a linear relationship between the original scene luminance and the reproduced picture. The amount of predistortion introduced by the camera must be compensated for by an opposite amount of distortion at the display device.

Laser Beam Projection Scanning System

Several approaches to laser projection displays have been implemented. The most successful employs three optical laser light sources whose coherent beams are modulated electro-optically and deflected by electromechanical means to project a raster display on a screen. The scanning functions are typically

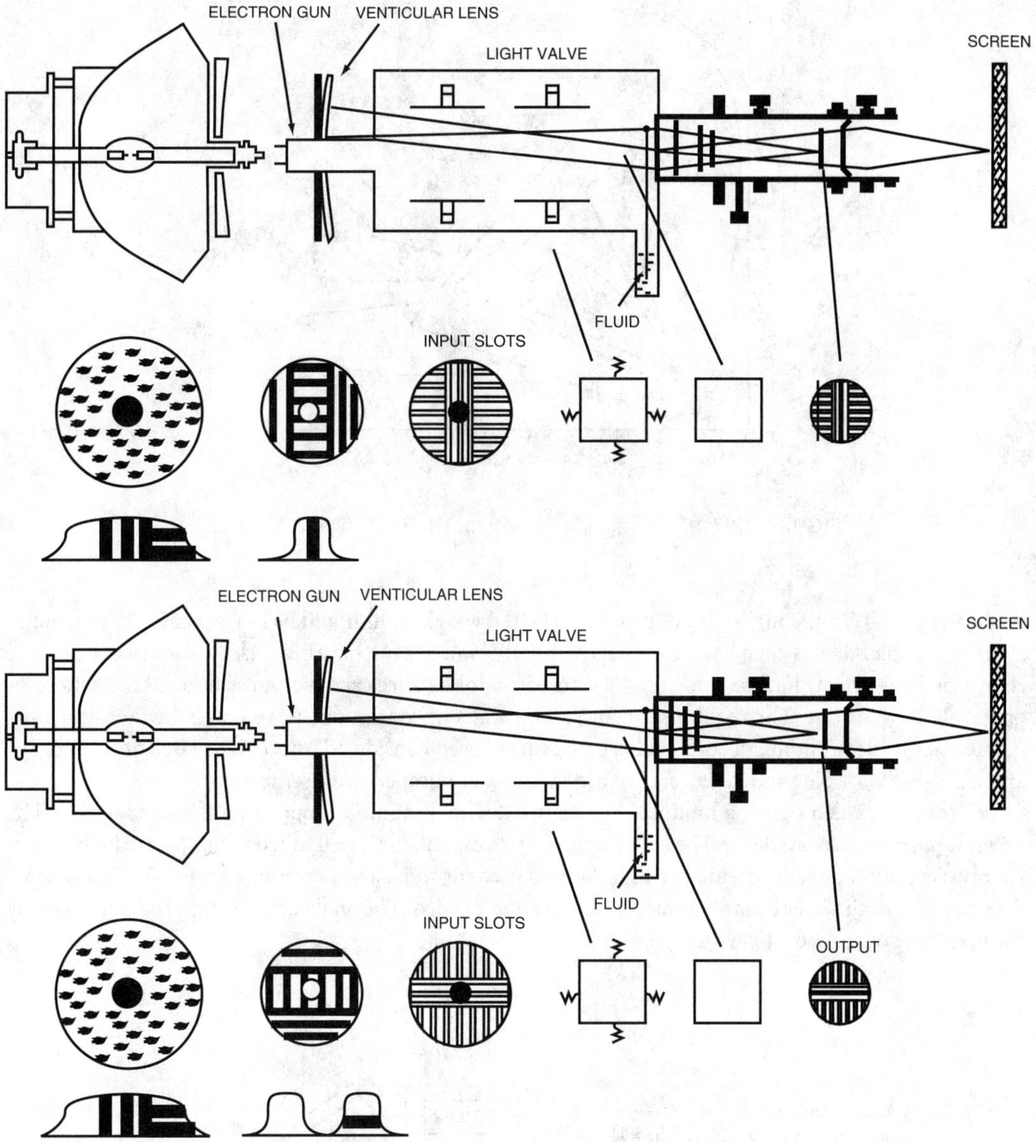

FIGURE 5.126   Functional operation of the two-channel HDTV light valve system.

provided by a rotating polygon mirror and a separate vibrating mirror. A block diagram of the basic system is shown in Fig. 5.128.

The flying spots of light used in this approach are one scan line (or less) in height and a small number of pixels wide. This means that any part of the screen may only be illuminated for a few nanoseconds. The scanned laser light projector is capable of high contrast ratios (as high as 1000:1) in a darkened environment. A laser projector may, however, be subject to a brightness variable referred to as **speckle**. Speckle is a sparkling effect resulting from interference patterns in coherent light. This effect causes a flat, dull projection surface to look as though it had a beaded texture. This tends to increase the perception of brightness, at the expense of image quality.

Figure 5.129 shows the configuration of a laser projector using continuous-wave lasers and mechanical scanners. The requirements for the light wavelengths of the lasers are critical. The blue wavelength must

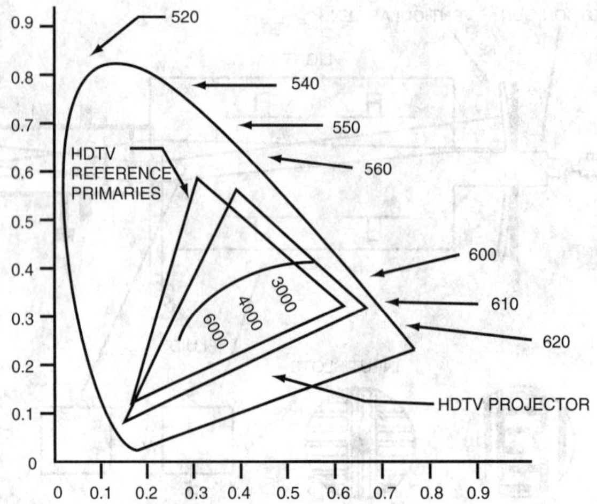

FIGURE 5.127   Color characteristics of the GE HDTV light valve system.

be shorter than 477 nm, but as long as possible. The red wavelength should be longer than 595 nm, but as short as possible. Greens having wavelengths of 510, 514, and 532 nm have been used with success. Because laser projectors display intense colors, small errors in color balance can result in significant distortions in gray scale or skin tone. The requirement for several watts of continuous-wave power further limits the usable laser devices. Although several alternatives have been considered, most color laser projectors use argon ion lasers for blue and green, and an argon ion laser pumped dye laser for red.

To produce a video signal, a modulator is required with a minimum operating frequency of 6 MHz. Several approaches are available. The bulk acousto-opitc modulator is well suited to this task in high-power laser projection. A directly modulated laser diode is used for some low-power operations. The modulator does not absorb the laser beam, but rather deflects it as needed. The angle of deflection is determined by the frequency of the acoustic wave.

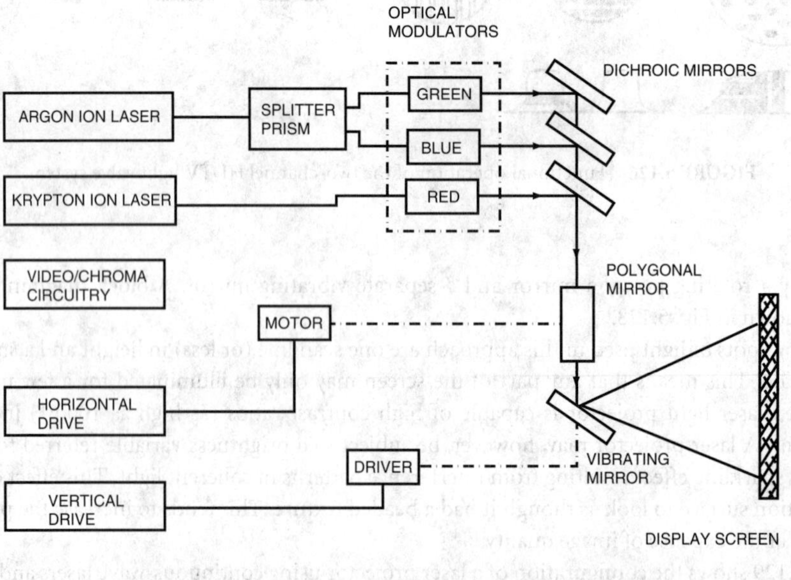

FIGURE 5.128   Block diagram of a laser-scanning projection display.

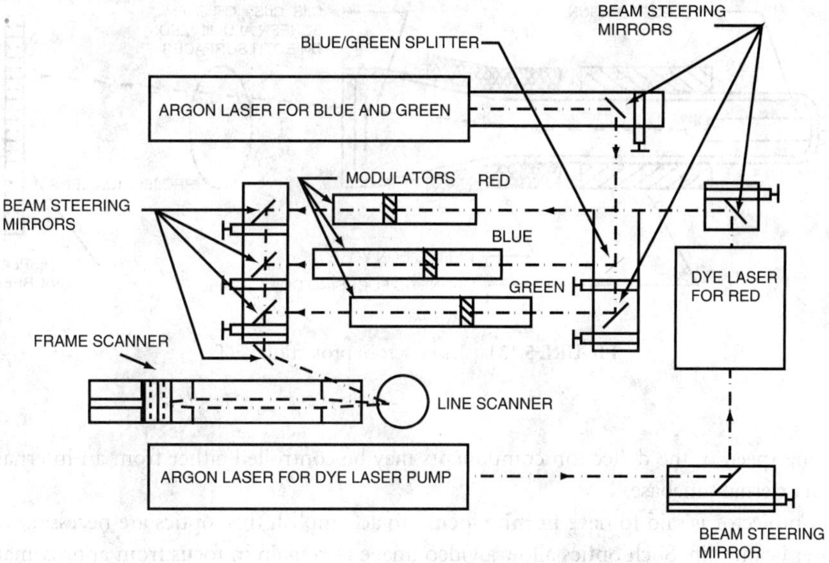

FIGURE 5.129 Configuration of a color laser projector.

The scanner is the component of the laser projector that moves the point of modulated light across and down the image plane. Several types of scanners may be used including mechanical, acousto-optic, and electro-optic. Two categories of scanning devices are used in the system:

- *Line scanner*, which scans the beam in horizontal lines across the screen. Lines are traced at 15,000–70,000 times per second. The rotating polygon scanner is commonly used, featuring 24–60 mirrored facets. Figure 5.130 illustrates a mechanical rotating polygon scanner.

- *Frame scanner*, which scans the beam vertically and forms frames of horizontal scan lines. The frame scanner cycles 50–120 times per second. A galvanometer-based scanner is typically used in conjunction with a mirrored surface. Because the mirror must fly back to the top of the screen in less than a millisecond, the device must be small and light. Optical element are used to force the horizontally scanned beam into a small spot on the frame scanner mirror.

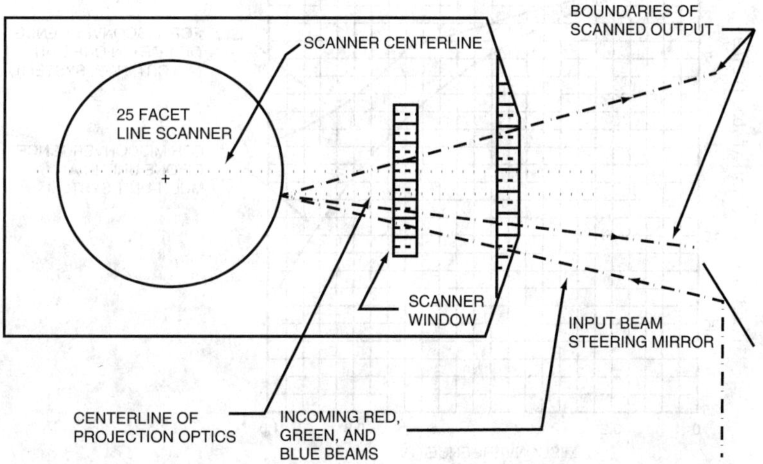

FIGURE 5.130 Rotating polygon line scanner for a laser projector.

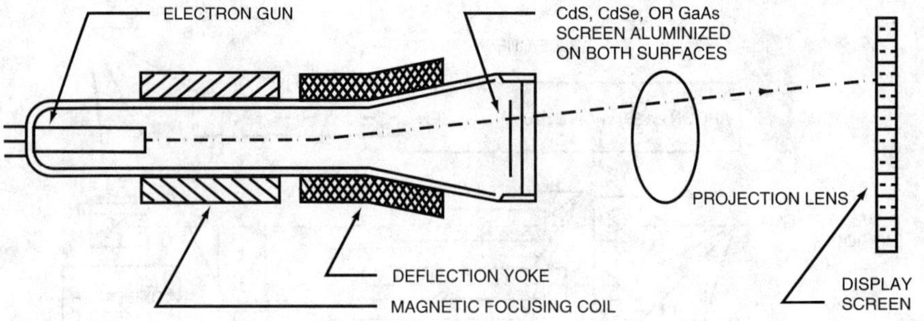

FIGURE 5.131 Laser-screen projection CRT.

The operating speed of the deflection components may be controlled either from an internal clock, or locked to an external timebase.

The laser projector is said to have **infinite focus**. To accomplish this, optics are necessary to keep the scanned laser beam thin. Such optics allow a video image to remain in focus from approximately 4 ft to infinity. For high-resolution applications, the focus is more critical.

There is an undesirable byproduct of high-power laser operation. Most devices are water cooled and use a heat exchanger to dump the waste heat.

Another approach large-screen display employs an electron beam pumped monocrystalline screen in a CRT to produce a 1-in raster. The raster screen image is projected by conventional optics, as shown in Fig. 5.131. This technology promises three optical benefits:

- High luminance in the image plane
- Highly directional luminance output for efficient optical coupling
- Compact, lightweight, and inexpensive projection optics

Full-color operations is accomplished using three such devices, one for each primary color.

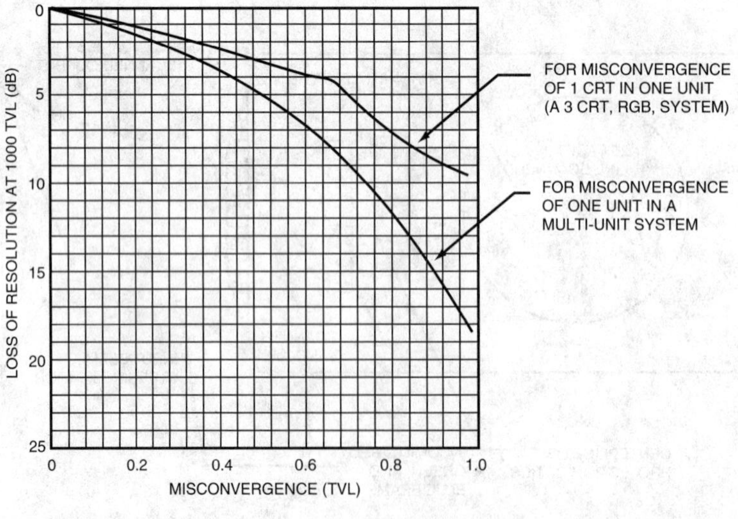

FIGURE 5.132 The relationship between misconvergence and display resolution.

TABLE 5.12 Appearance of Scan Line Judder in a Projection Display

Condition of Adjacent Lines	S/N (p-p)
Clearly overlapped	Less than 69 dB
Just before overlapped	75 dB
No judder	Greater than 86 dB

5.7.4 Operational Considerations

Accurate convergence is critical to display resolution. Figure 5.132 shows the degradation in resolution resulting from convergence error. It is necessary to keep convergence errors to less than half the distance between the scanning lines to hold resolution loss below 3 dB. Errors in convergence also result in color contours in a displayed image. Estimates have put the detectable threshold of color contours at 0.75–0.5 min of arc. This figure also indicates that convergence error must held under 0.5 scanning lines.

Raster stability influences the short-term stability of the display. The signal-to-noise ratio (S/N) and raster stability relationships for deflection circuits are shown in Table 5.12. A S/N equivalent of 1/5 scanning line is necessary to obtain sufficient raster stability. In HDTV applications, this translates to approximately 80 dB. Other important factors are high speed and improved efficiency of the deflection circuit.

A number of projection manufacturers have begun to incorporate automatic convergence systems into their products. These usually take the form of a charge-coupled device (CCD) camera sensor that scans various portions of the screen as test patterns are displayed.

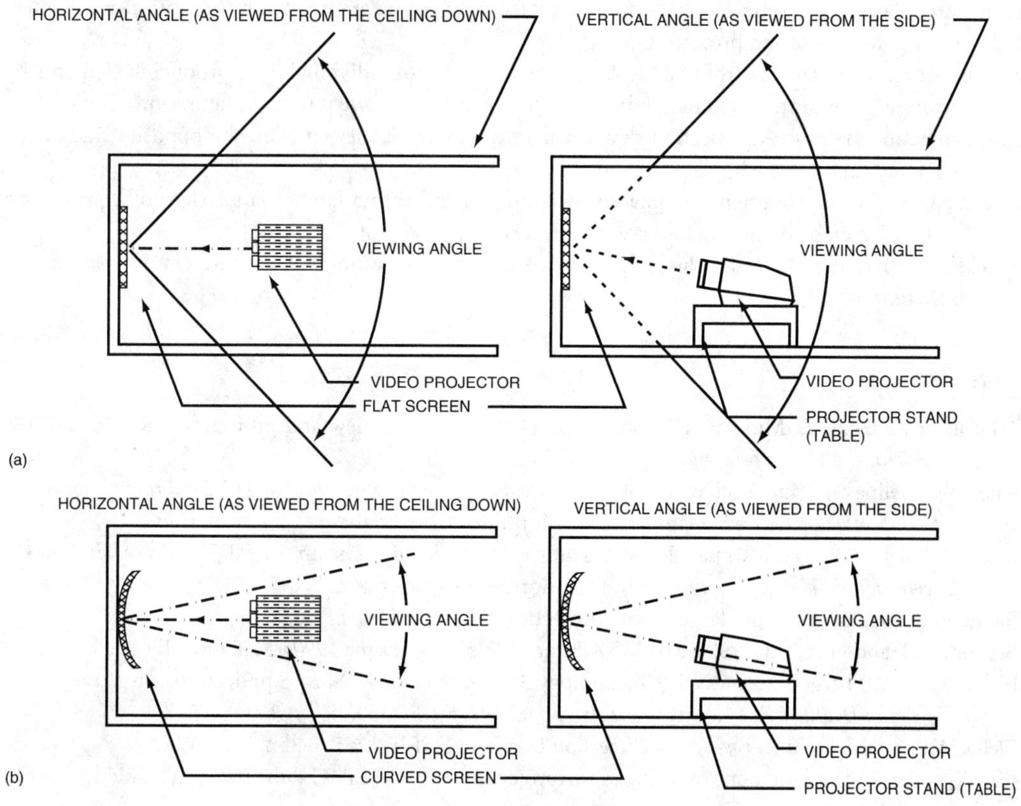

FIGURE 5.133 Relationship between screen type and viewing angle: (a) conventional (flat) screen, (b) high gain (curved) screen.

Screen Considerations

Because most screens reflect both ambient light in the room and the incident light from the projector (in a front projection system), lighting levels are an important consideration. A rear screen system can reduce this problem. Ambient light on the screen's projector side reflects harmlessly back toward the projector, away from the audience. Light on the viewing side of the screen may fall on the screen but the human eye has experience in filtering out this kind of reflection. The result is that a rear screen projector provides an increased contrast ratio (all other considerations being equal); it can, therefore, be used in lighter areas. Special screen coatings can also help scatter ambient light away from the viewers, enhancing the rear screen image. Coatings can also increase the apparent brightness of screens operated from the front. The biggest gains, however, are made by using a screen that is curved slightly. Such screens can increase the apparent brightness of an image. Curved screens, however, also narrow the acceptable field of view. High-gain screens provide a typical viewing angle of ±50° from the centerline of the screen. Conventional (low-gain) screens can provide viewing of up to ±90° from the centerline. Figure 5.133 illustrates the tradeoffs involved.

Defining Terms

Darkfield performance: The amount of light that is blocked in a light valve projector when a totally dark scene is being displayed. This parameter is critical to good contrast ratio.

Front projection system: A projection device where the image is viewed from the same side of the screen as that on which it is projected.

Infinite focus: A characteristic of a laser projector in which the video image remains in focus from approximately four feet to infinity.

Light valve: A projection device that employs a fixed light source modulated by an optical-valve intervening the source and projection optics.

Matrix addressing: The control of a display panel consisting of individual light-producing elements by arranging the control system to address each element in a row/column configuration.

Rear projection system: A projection device where the image is viewed from the opposite side of the screen as that on which it is projected.

Screen gain: The improvement in apparent brightness relative to a lambertian surface of a projection system screen by designing the screen with certain directional characteristics.

Speckle: A sparkling effect in a laser beam projection system resulting from interference patterns in coherent light.

References

Ashizaki, S., Suzuki, Y., Mitsuda, K., and Omae, H. 1988. Direct-view and projection CRTs for HDTV. *IEEE Trans. on Consumer Elec.*

Bates, W., Gelinas, P., and Recuay, P. 1991. Light valve projection system for HDTV. *Proceedings of the International Television Symposium*. ITS, Montreux, Switzerland.

Bates, W., Gelinas, P., and Recuay, P. 1991. Light valve projection system for HDTV. *Proceedings of the International Television Symposium*. ITS, Montreux, Switzerland.

Bauman, E. 1953. The Fischer large-screen projection system. *J. SMPTE* 60:351.

Benson, K.B. and Fink, D.G. 1990. *HDTV: Advanced Television for the 1990s*. McGraw-Hill, New York.

Blaha, R.J. 1990. Large screen display Technology Assessment for Military Applications. In *Large-Screen Projection Displays II*, ed. W.P. Bleha, Jr., pp. 80–92. SPIE 1255, Bellingham, WA.

Bleha, W.P., Jr., ed. 1990. *Large-Screen Projection Displays II*. SPIE 1255, Bellingham, WA.

EIA. X-radiation measurement procedures for projection tubes. TEPAC Pub. 102, Electronic Industries Association, Washington, DC.

Fink, D.G., ed. 1957. *Television Engineering Handbook*. McGraw-Hill, New York.

Fink, D.G. 1986. *Color Television Standards*. McGraw-Hill, New York.

Fink, D.G. et al. 1980. The future of high definition television. *J. SMPTE* 89(Feb./March).

Fujio, T., Ishida, J., Komoto, T., and Nishizawa, T. 1980. High definition television systems—signal standards and transmission. *J. SMPTE* 89(Aug.).

Gerhard-Multhaupt, R. 1991. Light valve technologies for HDTV projection displays: a summary. *Proceedings of the International Television Symposium*. ITS, Montreux, Switzerland.

Glenn, W.E. 1970. Principles of simultaneous color projection using fluid deformation. *J. SMPTE* 79:788.

Glenn, W. 1990. Large screen displays for consumer and theater use. In *Large-Screen Projection Displays II*, ed. W.P. Bleha, Jr., pp. 36–43. SPIE 1255, Bellingham, WA.

Good, W. 1975. Recent advances in the single-gun color television light-valve projector. *Soc. Photo-Optical Instrumentation Engrs.* 59.

Good, W. 1975. Projection television. *IEEE Trans. Consumer Elec.* CE-21(3):206–212.

Gretag A.G. *What You May Want to Know About the Technique of Eidophor*. Regensdorf, Switzerland.

Hardy, A.C. and Perrin, F.H. 1932. *The principles of Optics*. McGraw-Hill, New York.

Howe, R. and Welham, B. 1980. Developments in plastic optics for projection television systems. *IEEE Trans. Consumer Elec.* CE-26(1):44–53.

Hube, D.H. 1988. *Eye, Brain and Vision*. Scientific American Library, New York.

Itah, N. et al. 1979. New color video projection system with glass optics and three primary color tubes for consumer use. *IEEE Trans. Consumer Elec.* CE-25(4):497–503.

Judd, D.B. 1933. The 1931 C.I.E. standard observer and coordinate system for colorimetry. *J. Opt. Soc. Am.* 23.

Kikuchi, M. et al. 1981. A new coolant-sealed CRT for projection color TV. *IEEE Trans. Consumer Elec.* CE-27(3):478–485.

Kingslake, R., ed. 1965. *Applied Optics and Optical Engineering*, vol. 1, chap. 6. Academic, New York.

Kodak. Kodak filters for scientific and technical uses. Eastman Kodak Co., Rochester, NY.

Kurahashi, K. et al. 1981. An outdoor screen color display system. *SID Int. Symp. Digest 7*. Technical Papers, XII (April):132–133.

Luxenberg, H. and Kuehn, R. 1968. *Display Systems Engineering*. McGraw-Hill, New York.

McKechnie, S. 1981. Philips Laboratories (NA) Report, unpublished.

Mitsuhashi, T. 1990. HDTV and large screen display. In *Large-Screen Projection Displays II*, ed. W.P. Bleha, Jr., pp. 2–12. SPIE 1255, Bellingham, WA.

Morizono, M. 1990. Technological trends in high resolution displays including HDTV. *SID International Symposium Digest*. Paper 3.1, May.

Nasibov, A. et al. 1974. Electron-beam tube with a laser screen. *Sov. J. Quant. Electron.* 4(3):300.

Pease, R.W. 1990. An overview of technology for large wall screen projection using lasers as a light source. In *Large-Screen Projection Displays II*, ed. W.P. Bleha, Jr., pp. 93–103. SPIE 1255, Bellingham. WA.

Pfahnl, A. Aging of electronic phosphors in cathode ray tubes. In *Advances in Electron Tube Techniques*, pp. 204–208. Pergamon, New York.

Pitts, K. and Hurst, N. 1989. How much do people prefer widescreen (16 × 9) to standard NTSC (4 × 3)? *IEEE Trans. Consumer Elec.* (Aug.).

Pointer, R.M. 1945. The gamut of real surface colors. *Color Res. App.* 5.

Poorter, T. and deVrijer, F.W. 1959. The projection of color television pictures. *J. SMPTE* 68:141.

Robertson, A. 1979. Projection television—1 review of practice. *Wireless World* 82(Sept.):47–52.

Schiecke, K. 1981. Projection television: Correcting distortions. *IEEE Spectrum* 18(11):40–45.

Sears, F.W. 1946. *Principles of Physics, III. Optics*. Addison-Wesley, Cambridge, MA.

Sherr, S. 1970. *Fundamentals of Display System Design*. Wiley-Interscience, New York.

Taneda, T. et al. 1973. A 1125-scanning line laser color TV display. *SID 1973 Symp. Digest. Technical Papers*, IV (May):86–87.

Wang, S. et al. 1968. Spectral and spatial distribution of X-rays from color television receivers. *Proc. Conf. Detection and Measurement of X-radiation from Color Television Receivers* (Washington, DC), pp. 53–72, March 28–29.

Williams, C.S., and Becklund, O.A. 1972. *Optics: A Short Course for Engineers and Scientists*. Wiley-Interscience, New York.

Whitaker, J.C. 1994. *Electronic Displays: Technology, Design and Applications*. McGraw-Hill, New York.

Yamamoto, Y., Nagaoka, Y., Nakajima, Y., and Murao, T. 1986. Super-compact projection lenses for projection television. *IEEE Trans. Consumer Elec.* (Aug.).

Further Information

The SPIE (the international optical technology society) offers a wide variety of publications on the topic of display systems engineering. The organization also holds technical seminars on this and other topics several times each year. SPIE is headquartered in Bellingham, WA. The Society for Information Display (SID) also holds regular technical seminars and offers a number of publications on the topic of projection display systems. SID is headquartered in Santa Ana, CA.

Two books on the topic of display systems are recommended:

Sherr, S. 1994. *Fundamentals of Display System Design*, 2nd ed. Wiley-Interscience, New York.

Whitaker, J.C. 2001. *Video Displays*. McGraw-Hill, New York.

6

Microwave Vacuum Devices

Jerry C. Whitaker
Editor-in-Chief

Robert R. Weirather
Harris Corporation

Thomas K. Ishii
Marquette University

6.1   Microwave Power Tubes

Jerry C. Whitaker

6.1.1   Introduction

Microwave power tubes span a wide range of applications, operating at frequencies from 300 MHz to 300 GHz with output powers from a few hundred watts to more than 10 MW. Applications range from the familiar to the exotic. The following devices are included under the general description of microwave power tubes:

- Klystron, including the *reflex* and *multicavity* klystron
- *Multistage depressed collector* (MSDC) klystron
- *Klystrode* (IOT) tube
- Traveling-wave tube (TWT)
- **Crossed-field tube**

- *Coaxial magnetron*
- *Gyrotron*
- **Planar triode**
- High-frequency tetrode

This wide variety of microwave devices has been developed to meet a wide range of applications. Some common uses include

- UHF-TV transmission
- Shipboard and ground-based radar
- Weapons guidance systems
- Electronic countermeasure (ECM) systems
- Satellite communications
- Tropospheric scatter communications
- Fusion research

As new applications are identified, improved devices are designed to meet the need. Microwave power tubes manufacturers continue to push the limits of frequency, operating power, and efficiency. Microwave technology is an evolving science. Figure 6.1 charts device type as a function of operating frequency and power output.

Two principal classes of microwave vacuum devices are in common use today:

- **Linear beam** tubes
- Crossed-field tubes

Each class serves a specific range of applications. In addition to these primary classes, some power grid tubes are also used at microwave frequencies.

Linear Beam Tubes

As the name implies, in a linear beam tube the electron beam and the circuit elements with which it interacts are arranged linearly. The major classifications of linear beam tubes are shown in Fig. 6.2. In

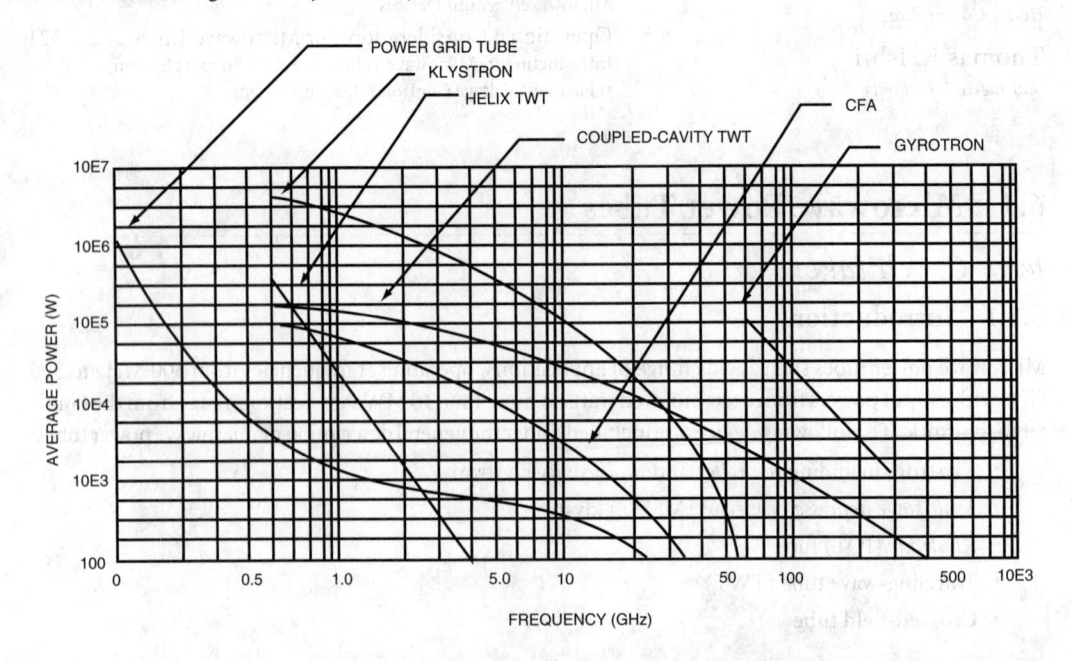

FIGURE 6.1 Microwave power tube type as a function of frequency and output power.

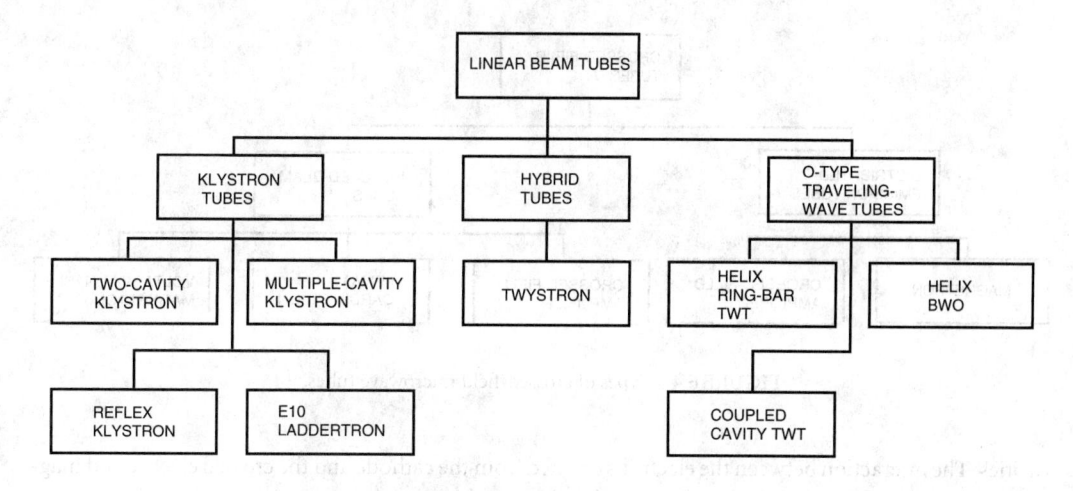

FIGURE 6.2 Types of linear beam microwave tubes.

such a device, a voltage applied to an anode accelerates electrons drawn from a cathode, creating a beam of kinetic energy. Power supply potential energy is converted to kinetic energy in the electron beam as it travels toward the microwave circuit. A portion of this kinetic energy is transferred to microwave energy as RF waves slow down the electrons. The remaining beam energy is either dissipated as heat or returned to the power supply at the collector. Because electrons will respell one another, there usually is an applied magnetic focusing field to maintain the beam during the interaction process. The magnetic field is supplied either by a solenoid or permanent magnets. Figure 6.3 shows a simplified schematic of a linear beam tube.

Crossed-Field Tubes

The magnetron is the pioneering device of the family of crossed-field tubes. The family tree of this class of devices is shown in Fig. 6.4. Although the physical appearance differs from that of linear beam tubes, which are usually circular in format, the major difference is in the interaction physics that requires a magnetic field at right angles to the applied electric field. Whereas the linear beam tube sometimes requires a magnetic field to maintain the beam, the crossed-field tube always requires a magnetic focusing field.

Figure 6.5 shows a cross section of the magnetron, including the magnetic field applied perpendicular to the cathode–anode plane. The device is basically a diode with the anode composed of a plurality of resonant

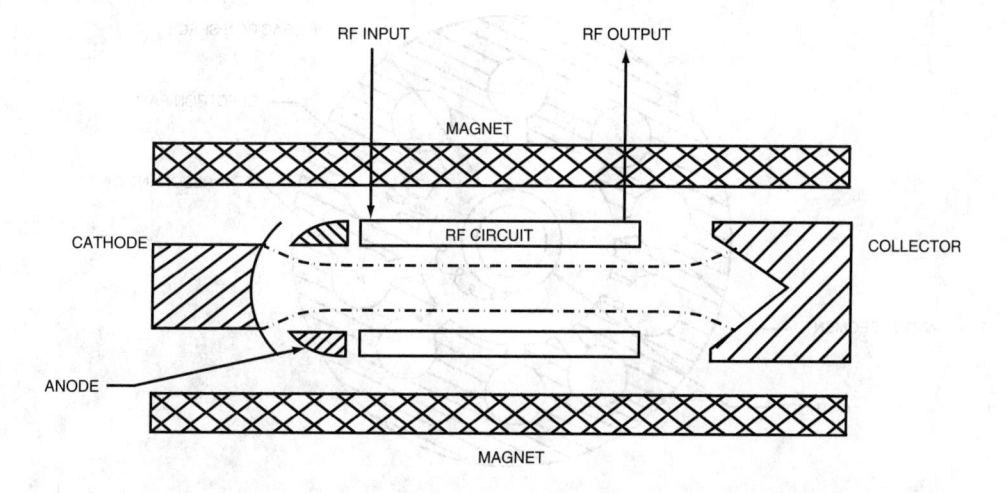

FIGURE 6.3 Schematic diagram of a linear beam tube.

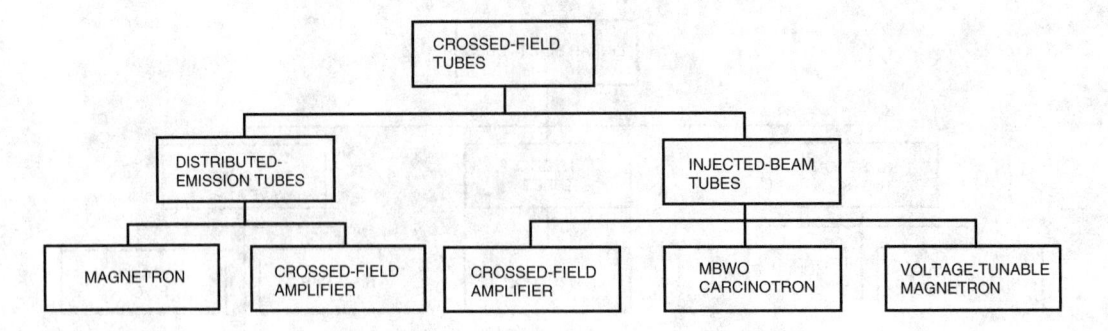

FIGURE 6.4   Types of crossed-field microwave tubes.

cavities. The interaction between the electrons emitted from the cathode and the crossed electric and magnetic fields produces a series of space charge spokes that travel around the anode–cathode space in a manner that transfers energy to the RF signal supported by the multicavity circuit. The mechanism is highly efficient.

Crossed-Field Amplifiers

Figure 6.6 shows the family tree of the crossed-field amplifier (CFA). The configuration of a typical present day distributed emission amplifier is similar to that of the magnetron except that the device has an input for the introduction of RF into the circuit. Current is obtained primarily by secondary emission from the negative electrode that serves as a cathode throughout all or most of the interaction space. The earliest versions of this tube type were called **amplitrons**.

The CFA is deployed in radar systems operating from the UHF to the Ku band and at power levels up to several megawatts. In general, bandwidth ranges from a few percent to as much as 25% of the center frequency.

6.1.2 Grid Vacuum Tubes

The physical construction of a vacuum tube causes the output power and available gain to decrease with increasing frequency. The principal limitations faced by grid-based devices include the following:

- Physical size. Ideally, the RF voltages between electrodes should be uniform; however, this condition cannot be realized unless the major electrode dimensions are significantly less than $\frac{1}{4}$-wavelength

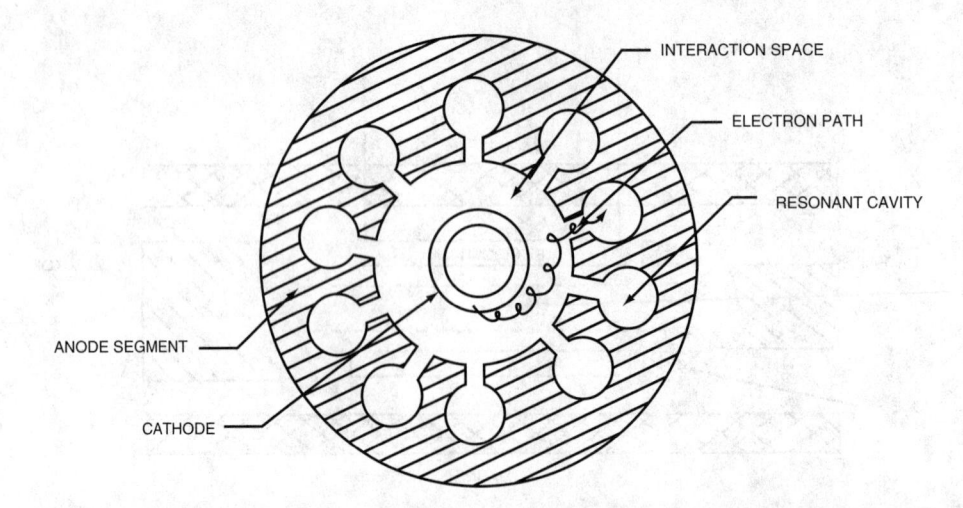

FIGURE 6.5   Magnetron electron path looking down into the cavity with the magnetic field applied.

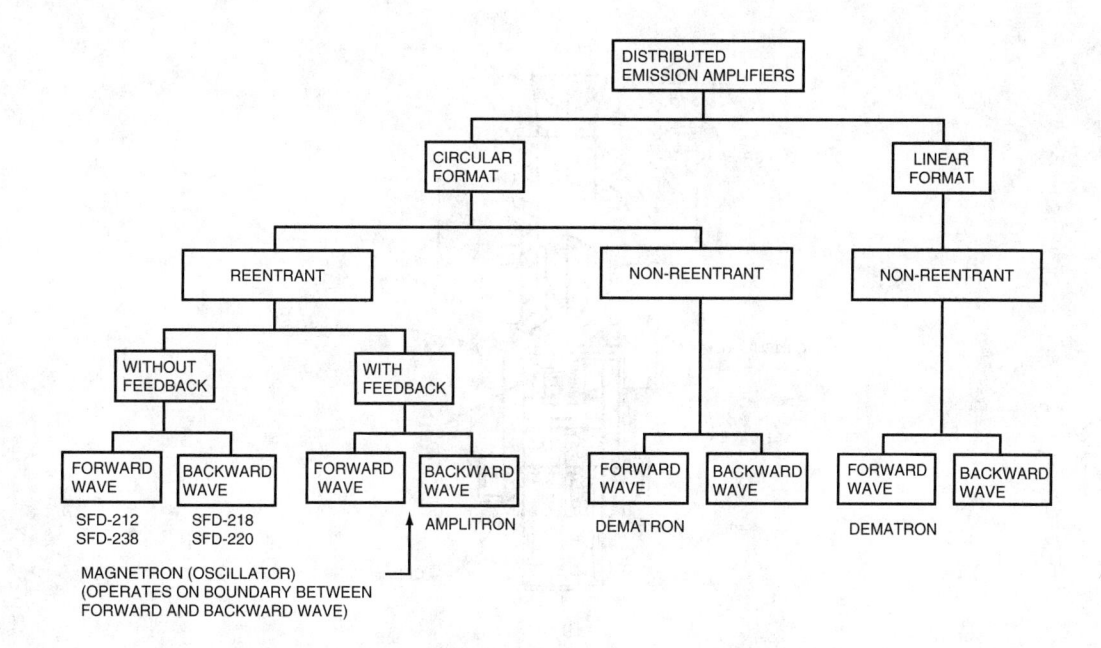

FIGURE 6.6   Family tree of the distributed emission crossed-field amplifier (CFA).

at the operating frequency. This restriction presents no problems at VHF frequencies; however, as the operating frequency increases into the microwave range, severe restrictions are placed on the physical size of individual tube elements.

- Electron transit-time. Interelectrode spacing, principally between the grid and the cathode, must be scaled inversely with frequency to avoid problems associated with electron transit time. Possible adverse conditions include: (1) excessive loading of the drive source, (2) reduction in power gain, (3) back heating of the cathode as a result of electron bombardment, and (4) reduced conversion efficiency.

- Voltage standoff. High-power tubes operate at high voltages. This presents significant problems for microwave vacuum tubes. For example, at 1 GHz the grid-cathode spacing must not exceed a few mils. This places restrictions on the operating voltages that may be applied to individual elements.

- Circulating currents. Substantial RF currents may develop as a result of the inherent interelectrode capacitances and stray inductances/capacitances of the device. Significant heating of the grid, connecting leads, and vacuum seals may result.

- Heat dissipation. Because the elements of a microwave grid tube must be kept small, power dissipation is limited.

Still, some grid vacuum tubes find applications at high frequencies. Planar triodes are available that operate at several gigahertz, with output powers of 1–2 kW in pulsed service. Efficiency (again for pulsed applications) ranges from 30 to 60%, depending on the frequency.

Planar Triode

A cross-sectional diagram of a planar triode is shown in Fig. 6.7. The envelope is made of ceramic, with metal members penetrating the ceramic to provide for connection points. The metal members are shaped either as disks or as disks with cylindrical projections.

The cathode is typically oxide coated and indirectly heated. The key design objective for a cathode is high-emission density and long tube life. Low-temperature emitters are preferred because high cathode temperatures typically result in more evaporation and shorter life.

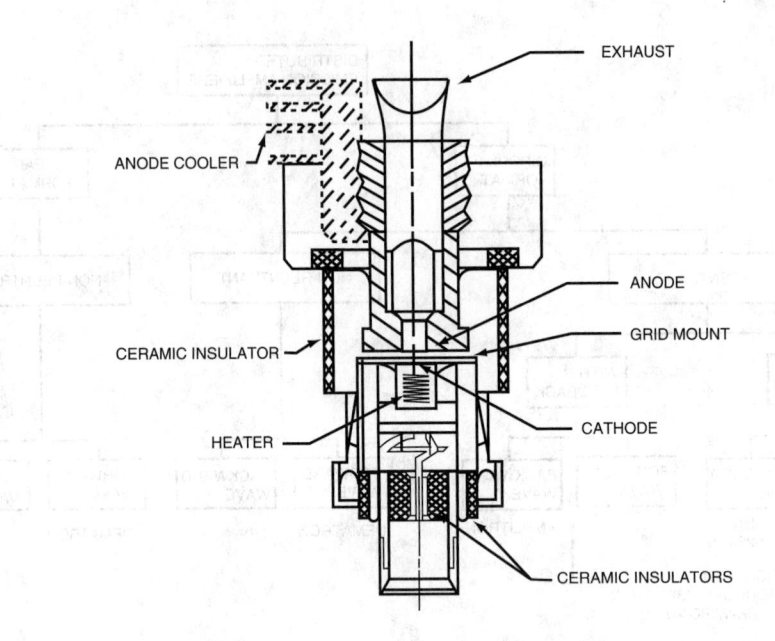

FIGURE 6.7   Cross section of a 7289 planar triode. (*Source:* Varian/Eimac.)

The grid of the planar triode represents perhaps the greatest design challenge for tube manufacturers. Close spacing of small-sized elements is needed, at tight tolerances. Good thermal stability is also required, because the grid is subjected to heating from currents in the element itself, plus heating from the cathode and bombardment of electrons from the cathode.

The anode, usually made of copper, conducts the heat of electron bombardment to an external heat sink. Most planar triodes are air cooled.

Planar triodes designed for operation at 1 GHz and above are used in a variety of circuits. The grounded grid configuration is most common. The plate resonant circuit is cavity based, using waveguide, coaxial line, or stripline. Electrically, the operation of planar triode is much more complicated at microwave frequencies than at low frequencies. Figure 6.8 compares the elements at work for a grounded-grid amplifier operating at (a) low frequencies and (b) at microwave frequencies. The equivalent circuit is made more complicated by

- Stray inductance and capacitance of the tube elements
- Effects of the tube contact rings and socket elements
- Distributed reactance of cavity resonators and the device itself
- Electron transit-time effects, which result in resistive loading and phase shifts

Reasonable gains of 5–10 dB may be achieved with a planar triode. Increased gain is available by cascading stages. Interstage coupling may consist of waveguide or coaxial-line elements. Tuning is accomplished by varying the cavity inductance or capacitance. Additional bandwidth is possible by stagger tuning cascaded stages.

High-Power UHF Tetrode

New advancements in vacuum tube technology have permitted the construction of high-power UHF transmitters based on tetrodes. Such devices are attractive because they inherently operate in an efficient class A–B mode. UHF tetrodes operating at high-power levels provide essentially the same specifications, gain, and efficiency as tubes operating at lower powers. The anode power supply is much lower in voltage than the collector potential of a klystron-based system (8 kV is common). The tetrode also does not require focusing magnets.

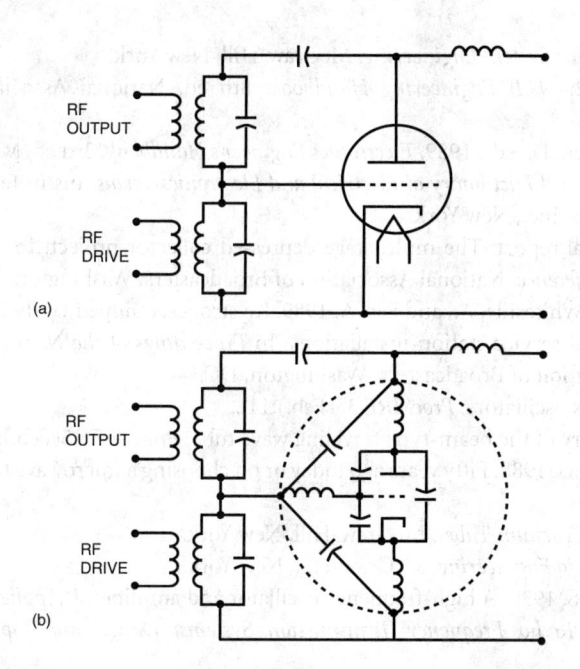

FIGURE 6.8 Grounded-grid equivalent circuits: (a) low-frequency operation, (b) microwave frequency operation. The cathode-heating and grid-bias circuits are not shown.

Efficient removal of heat is the key to making a tetrode practical at high-power levels. Such devices typically use water or vapor-phase cooling. Air cooling at such levels is impractical because of the fin size that would be required. Also, the blower for the tube would have to be quite large, reducing the overall transmitter AC-to-RF efficiency.

The expected lifetime of a tetrode in UHF service is usually shorter than a klystron of the same power level. Typical lifetimes of 8000–15,000 h have been reported.

Defining Terms

Amplitron: A classic crossed-field amplifier in which output current is obtained primarily by secondary emission from the negative electrode that serves as a cathode throughout all or most of the interaction space.

Crossed-field tube: A class of microwave devices where the electron beam and the circuit elements with which it interact are arranged in a nonlinear, usually circular, format. The interaction physics of a crossed-field tube requires a magnetic field at right angles to the applied electric field.

Linear beam tube: A class of microwave devices where the electron beam and the circuit elements with which it interacts are arranged linearly.

Planar triode: A grid power tube whose physical design make it practical for use at microwave frequencies.

References

Badger, G. 1986. The Klystrode: A new high-efficiency UHF-TV power amplifier. In *Proceedings of the NAB Engineering Conference.* National Association of Broadcasters, Washington, DC.

Clayworth, G.T., Bohlen, H.P., and Heppinstall, R. 1992. Some exciting adventures in the IOT business. In *NAB 1992 Broadcast Engineering Conference Proceedings,* pp. 200–208. National Association of Broadcasters, Washington, DC.

Collins, G.B. 1947. *Radar System Engineering.* McGraw-Hill, New York.

Crutchfield, E.B. 1992. *NAB Engineering Handbook,* 8th ed. National Association of Broadcasters, Washington DC.

Fink, D. and Christensen, D., eds. 1989. *Electronics Engineer's Handbook,* 3rd ed. McGraw-Hill, New York.

IEEE. 1984. *IEEE Standard Dictionary of Electrical and Electronics Terms.* Institute of Electrical and Electronics Engineers, Inc., New York.

McCune, E. 1988. Final report: The multi-stage depressed collector project. In *Proceedings of the NAB Engineering Conference.* National Association of Broadcasters, Washington, DC.

Ostroff, N., Kiesel, A., Whiteside, A., and See, A. 1989. Klystrode-equipped UHF-TV transmitters: Report on the initial full service station installations. In *Proceedings of the NAB Engineering Conference.* National Association of Broadcasters, Washington, DC.

Pierce, J.A. 1945. Reflex oscillators. *Proc. IRE* 33(Feb.):112.

Pierce, J.R. 1947. Theory of the beam-type traveling wave tube. *Proc. IRE* 35(Feb.):111.

Pond, N.H. and Lob, C.G. 1988. Fifty years ago today or on choosing a microwave tube. *Microwave Journal* (Sept.):226–238.

Spangenberg, K. 1947. *Vacuum Tubes.* McGraw-Hill, New York.

Terman, F.E. 1947. *Radio Engineering.* McGraw-Hill, New York.

Varian, R. and Varian, S. 1939. A high-frequency oscillator and amplifier. *J. Applied Phys.* 10(May):321.

Whitaker, J.C. 1991. *Radio Frequency Transmission Systems: Design and Operation.* McGraw-Hill, New York.

Whitaker, J.C. *Power Vacuum Tubes Handbook.* 2nd ed. CRC Press, New York.

Further Information

Specific information on the application of microwave tubes can be obtained from the manufacturers of those devices. More general application information can be found in the following publications:

Radio Frequency Transmission Systems: Design and Operation, written by Jerry C. Whitaker, McGraw-Hill, New York, 1991.

Power Vacuum Tubes Handbook 2nd ed., written by Jerry C. Whitaker, CRC Press, New York, 1994.

The following classic texts are also recommended:

Radio Engineering, 2nd ed., written by Frederick E. Terman, McGraw-Hill, New York, 1947.

Vacuum Tubes, written by Karl R. Spangenberg, McGraw-Hill, New York, 1948.

6.2 Klystron

Robert R. Weirather

6.2.1 Introduction

Energy in a microwave system may be viewed as with the passage of traveling electromagnetic waves along a waveguide axis. The generation of these waves can be resolved to one of energy conversion: the source potential energy (DC) is converted to the energy of the electromagnetic fields. At lower radio frequencies one use of the familiar multielectrode vacuum tube is as a valve controlling the flow of electrons in the anode conductor. (RF sources can use semiconductor devices for microwave power but these are generally limited to less than a kilowatt per device.) To generate considerable power at higher frequencies, the vacuum tube is less desirable than other means such as klystrons.

Conventional vacuum tubes are not efficient sources of currents at microwave frequencies. The assumption upon which their efficient design is based, namely, that the electrons pass from **cathode** to **anode** in a time short compared with the period of the anode potential, no longer holds true at 10^9–10^{10} Hz. The transit time of the order of magnitude of 10^{-8} s may be required for an electron to travel from cathode to anode of a conventional vacuum tube at higher frequencies. The anode and grid potentials may alternate from 10 to 100 times during the electron transit. The anode potential during the negative part of its cycle thus removes energy that was given to the electrons during the phase of positive anode potential. The overall result of transit-time effects is that of a reduction in the operating efficiency of the vacuum tube becoming more marked with increasing frequency. Reducing the transit time by decreasing cathode-to-anode spacing and increasing anode potential creates excessive interelectrode capacitance and overheating of the anode. Thus, the necessity for the design of tubes that are specifically suited for operation in the higher frequency range. Consider a tube the electron beam of which interacts with the time-varying fields of that part of the microwave system that is included within the tube. This interaction proceeds in such a way that the velocity of the electrons is lower after the interaction than before, the kinetic energy removed from the electrons can reappear as energy of the electromagnetic fields. Practical tubes in use differ among themselves in the details of the mode of interaction between the electron beam and the rapidly varying fields. this chapter reviews one such tube: the klystron and some of its derivatives.

6.2.2 The Klystron

The three major functions of the high-power linear-beam tube (beam generation, interaction, and dissipation) are accomplished in three separate and well-defined sections of the tube, each of which can be optimized almost independently of the others. A relatively large high-convergence **electron gun** may be used to provide a small beam diameter (compared to wavelength) for interaction with a high-frequency circuit. Beam focusing can be made sufficiently good so that interception of current can often be neglected, and only RF circuit losses need be considered. The beam is usually allowed to expand into a large collector when it leaves the interaction region. Electron guns, beam focusing, **collectors**, and output windows are independent of the type of interaction structure used and are common to all tubes.

6.2.3 Typical Klystron Amplifier

Figure 6.9 shows a cutaway representation of a typical klystron amplifier. Schematically it is very similar to a triode tube in that it includes an electron source, resonant circuits, and a collector (which is roughly equivalent to the plate of a triode). In fact, the klystron amplifier consists of three separate sections—the electron gun for beam generation, the RF section for beam interaction, and the collector section for electron energy dissipation.

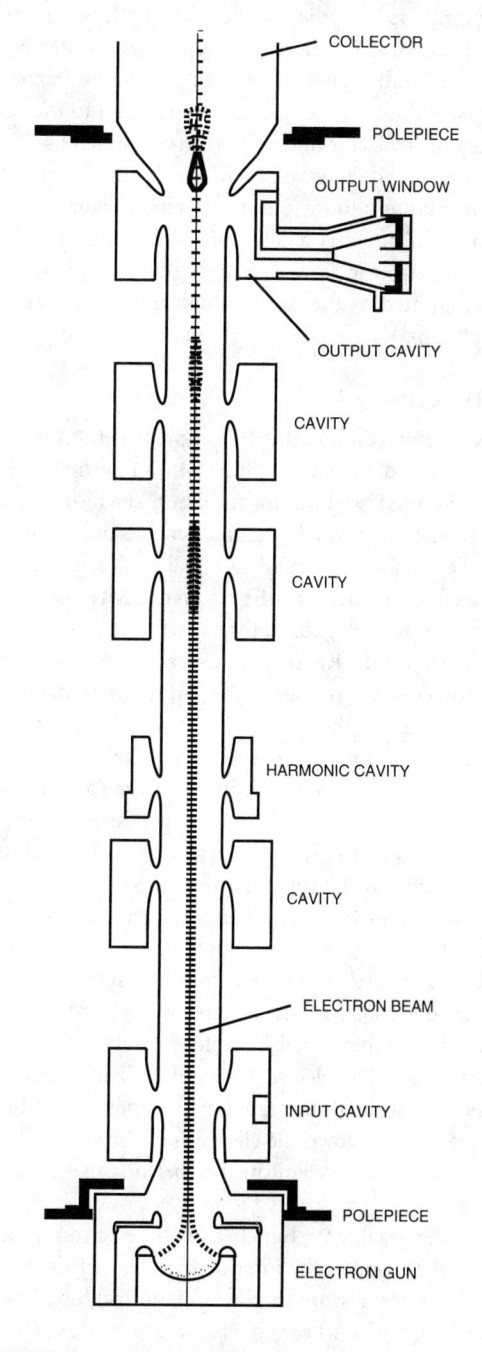

FIGURE 6.9 Sectional view of a klystron showing the principal elements.

The Electron Gun

Let us consider, first, the electron gun structure: As in the triode, the electron gun consists of heater and cathode, a control grid (sometimes), and an anode. Electrons are emitted by the hot cathode surface and are drawn toward the anode, which is operated at a positive potential with respect to the cathode. The electrons are formed into a small, dense beam by either electrostatic or magnetic focusing techniques, similar to the techniques used for beam formation in a cathode ray tube. In some klystron amplifiers, a control grid is used to permit adjustment of the number of electrons that reach the anode region; this control grid may be used to turn the tube completely on or completely off in certain pulsed-amplifier applications. The electron beam is well formed by the time it reaches the anode. It passes through a hole in the anode, passes on to the RF section of the tube, and eventually the electrons are intercepted by the collector. The electrons are returned to the cathode through an external power supply. It is evident that the collector in the klystron acts much like the plate of a triode insofar as collecting of the electrons is concerned, however, there is one important difference; the plate of a triode is normally connected in some fashion to the output RF circuit, whereas, in a klystron amplifier, the collector has no connection to the RF circuitry at all. From the preceding discussion, it is apparent that the klystron amplifier, insofar as the electron flow is concerned, is quite analogous to a stretched-out triode tube in that electrons are emitted by the cathode, controlled in number by the control grid, and collected eventually by the collector.

RF Section

Now let us consider the RF section of a klystron amplifier. This part of the tube is physically quite different from a gridded tube amplifier. One of the major differences is in the physical configuration of the resonant circuit used in a klystron amplifier. The resonant circuit used with a triode circuit, at lower frequencies, is generally composed of an inductance and a capacitor, whereas the resonant circuit used in a microwave tube is almost invariably a metal-enclosed chamber known as a cavity resonator. A very crude analogy can be made between the resonant cavity and a conventional L-C resonant circuit. The gap in the cavity is roughly analogous to the capacitor in a conventional low-frequency resonant circuit in that alternating voltages, at the RF frequencies, can be made to appear across the cavity gap. Circulating currents will flow between the two sides of the gap through the metal walls of the cavity, roughly analogous to the flow of RF current in the inductance of an L-C resonant circuit. Since RF voltages appear across the sides of the cavity gap, it is apparent that an electric field will be present, oscillating at the radio frequency, between the two surfaces of the cavity gap. When a cavity is the correct size, it will resonate. To bring to resonance, a cavity can be tuned by adjusting its size by some mechanical means.

As shown in Fig. 6.9, electrons pass through the cavity in each of the resonators, and pass through cylindrical metal tubes between the various gaps. These metal tubes are called *drift tubes*. In a klystron amplifier the low-level RF input signal is coupled to the first or input resonator, which is called the *buncher* cavity. The signal may be coupled in through either a waveguide or a coaxial connection. The RF input signal will excite oscillating currents in the cavity walls. These oscillating currents will cause the alternate sides of the buncher gap to become first positive and then negative in potential at a frequency equal to the frequency of the RF input signal. Therefore, an electric field will appear across the buncher gap, alternating at the RF frequency. This electric field will, for half a cycle, be in a direction that will tend to speed up the electrons flowing through the gap; on the other half of the cycle the electric field will be in a direction that will tend to slow down the electrons as they cross the buncher gap. This effect is called velocity modulation, and it is the mechanism that permits the klystron amplifier to operate at frequencies higher than the triode.

After leaving the buncher gap the electrons proceed toward the collector in the drift tube region. Ignore for the moment the intermediate resonators shown in Fig. 6.9, and let us consider the simple case of a two-cavity klystron amplifier. In the drift tube region the electrons that have been speeded up by the electric field in the buncher gap will tend to overtake those electrons that have previously been slowed down (by the preceding half of the RF wave across the buncher gap). It is apparent that, since some electrons are tending to overtake other electrons, clumps or bunches of electrons will be formed in the drift tube region.

If the average velocity of the electron stream is correct, as determined by the original voltage between anode and cathode, and if the length of the drift tube is proper, these bunches of electrons will be quite completely formed by the time they reach the catcher gap of the last or output cavity (which is called the catcher). This results in bunches of electrons flowing through the catcher gap periodically, and during the time between these bunches relatively fewer electrons flow through the catcher gap. The time between arrival of bunches of electrons is equal to the time of one cycle of the RF input signal.

These bunches of electrons will induce alternating current flow in the metal walls of the catcher cavity as they pass through the catcher gap. The output cavity will have large oscillating currents generated in its walls. These currents cause electric fields to exist, at the radio frequency, within the output cavity. These electric fields can be coupled from the cavity (to output waveguide, or coaxial transmission lines) resulting in the RF output power from the tube.

Collector

The electron beam energy is coupled in the last or output cavity to provide the desired RF power. The residual electron beam energy is absorbed in the collector. Collector design is based on the shape and energy of the electron beam. Electrons may have considerable energy upon impact on the collector and thus create undesired secondary emissions. Collector design includes coating and geometry considerations to minimize secondary emission. The collector may have to dissipate the full energy of the beam and is most frequently liquid cooled at higher powers.

6.2.4   High-Efficiency Klystron Design

The early analysis of ballistic bunching in a two-cavity klystron led to a maximum predicted efficiency of 58%. Early researchers recognized very early that tuning the penultimate cavity to the high-frequency side of the carrier could increase the efficiency over that which had been predicted. This occurs because it presents an inductive impedance to the beam and creates fields that squeeze the bunch formed at previous cavities. Locating a cavity tuned to the second harmonic of the operating frequency at the point where the second-harmonic content of the beam had increased enough to interact with that cavity, one could form a bunch that contained a higher fundamental component of current than by using conventional fundamental bunching only. The combination of fundamental bunching plus properly phased second harmonic bunching at a subsequent cavity is an approximation to a sawtooth voltage at a single gap. In theory, 100% efficiency would be possible. As we will see, further improvements have led to very efficient DC power to RF power conversion.

External and Internal Cavities

The high-power, multicavity klystron can be manufactured with the cavities either integral with the body or as external to the body. These two types are generally referred to klystron constructions as either with external or internal cavities. The internal (integral) cavities are manufactured in place by the OEM and thus field replacement of a failed tube requires shipment back to the vendor. The advantage with internal cavities is that details of installation are completely under the control of the OEM and less subject to installation error. Additionally, the cooling and heat flow is controlled better. With external cavity models, the cavities can be removed and the tubes replaced. This is a feature that lowers cost and hastens repair.

Multistage Depressed Collector Klystron

Development efforts at NASA to achieve highly efficient microwave transmitters for space applications resulted in achieving multistage depressed collector (MSDC) technology with great potential for efficiency enhancement. The UHF television community recognized the potential of this technology for reducing operating costs, and preliminary analysis indicated that applying MSDC technology to the power klystron used in UHF-TV transmitters could reduce amplifier power consumption by half.

A development program was organized and achieved an MSDC klystron design. A four-stage design was selected that would operate with equal voltage steps per stage (see Fig. 6.10). These MSDC designs are identical to the ordinary klystron except a different collector is used. The collector is segmented by different voltage stages. These different stages collect the various electron energies at lower voltages and thus reduce the overall dissipation resulting in increased efficiency (see Fig. 6.11). Power efficiency varies from 74% with no RF output to 57% at saturated output. The overall efficiency is 71% with no RF output to 57% at saturation output. The overall efficiency is 71% at this point, compared with 55% without the MSDC. For television operation, the input power is reduced from 118 to 51.8 kW by use of the MSDC, consequently, reducing the power costs to less than one-half.

6.2.5 Broadband Klystron Operation

For television service in the U.S., at least a 5.25-MHz bandwidth is required. Some European systems require as much as 8 MHz. At the power levels used in television service, it is necessary to stagger tune the cavities of a klystron to obtain this bandwidth. This is particularly true at the low end of the UHF band, but less so at the high end where the percentage bandwidth is lower.

The stagger tuning of a klystron in order to improve bandwidth is much more complicated, from a theoretical point of view, than is, for example, the stagger tuning of an IF amplifier. In an IF amplifier, the gain functions of the various stages may simply be multiplied together. In the klystron, however, the velocity modulation applied to the beam at a cavity gap will contribute a component to the exciting current at each and every succeeding gap. Thus, the interaction of tuning and the beam shaping for optimum power, bandwidth, efficiency, and gain is a tradeoff. To achieve the desired operation, the OEM must do studies and tests to determine the cavity tuning.

6.2.6 Frequency of Operation

Practical considerations limit klystron designs at the lower frequencies. The length of a klystron would be several meters at 100 MHz, but only 1 m long at

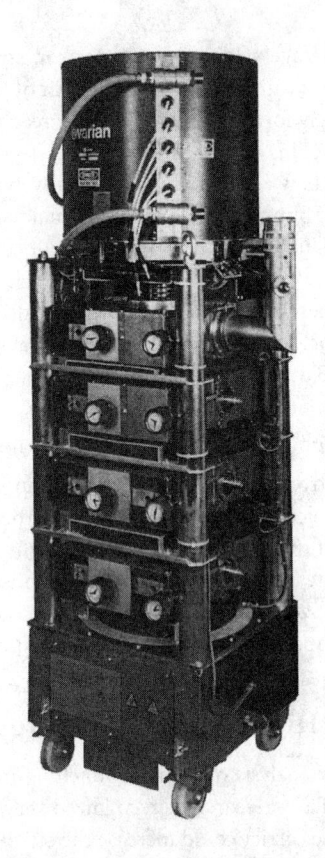

FIGURE 6.10 Multistage depressed collector (MSDC) for a klystron. (*Source:* Courtesy of Varian Associates, San Carlos, CA.)

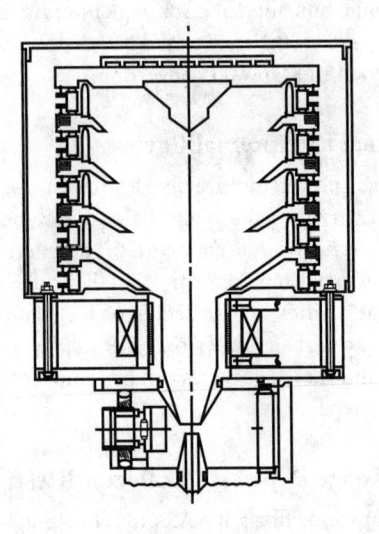

FIGURE 6.11 Cross-sectional view of a klystrode multistage depressed collector with four stages. (*Source:* Courtesy of Varian Associates, San Carlos, CA.)

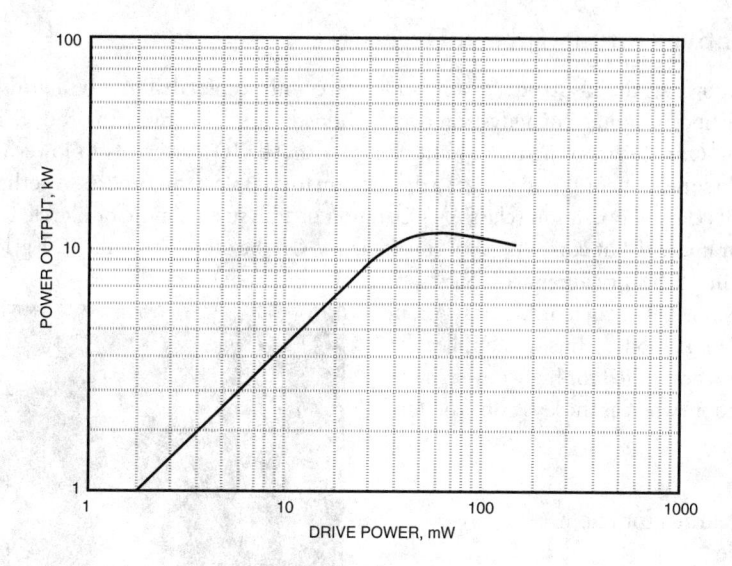

FIGURE 6.12 Typical gain and saturation characteristics of a klystron.

1 GHz. Triodes and tetrodes are more practical tubes at lower frequencies, whereas the klystron becomes practical at 300 MHz and above. At frequencies above 20 GHz, the design of klystrons is limited by the transit time.

6.2.7 Power and Linearity

The power of a klystron is limited to a theoretical 100% of that in the electron beam ($P_v = I_b V_b$). In some areas, efficiencies of 70–80% have been achieved. Normally, efficiency of 40–60% can be expected. RF power bandwidth, of course, is dependent on the number of cavities and their tuning. Typically, klystrons with multiple cavities can be staggered tuned to provide a flat RF passband at some reduction in efficiency and gain. For example, in four cavity tubes, staggered tuning results in a-1-dB bandwidth of 1% with a gain of 40 dB.

Klystron linearity (both amplitude and phase) is considered fair to good. For RF power, several decibels below saturation, the amplitude linearity exhibits a curvature as shown in Fig. 6.12. As drive is increased, the RF output reaches saturation and further increases in drive reduces power output. The linearity curvature and saturation characteristic will generate harmonic and intermodulation products, which may be filtered or precorrected.

The phase shift vs. drive is also nonlinear. As the output power reaches saturation, the phase shift increases and may reach 10° or more. This nonlinearity creates AM to PM conversion. Compensation circuits can be used in the RF drive to correct this condition.

6.2.8 Klystron Summary

When the need for high frequency and high power is indicated, the klystron is one of the first candidates for service. The klystron continues to be workhorse with long life and dependable performance. Klystrons are a powerful, high-frequency answer to many RF problems. Klystrons come in a variety of powers from a 1 kW to several megawatts. Modern designs have years of life, high gains, and improved efficiency. Found in radars, accelerators, television transmitters, and communication systems, these tubes operate in outer space to fixed base locations.

6.2.9   Inductive Output Tube A Derivative

The IOT, also known as the **klystrode**, is a hybrid device with some attributes of a gridded tube and a klystron. The distinct advantage of a klystron over a gridded tube is its ability to produce high powers at high frequencies. One shortcoming of a klystron is its cost. The IOT attempts to lower the cost gap by using a grid to do some bunching and utilizes a drift region to complete this process. Another shortcoming of a klystron is its class of operation (class A). Operation of a klystron conduction cycle is class A, which delivers superior linearity but comes up short in efficiency. On the other hand, the IOT grid can be biased to completely cut off **beam current** for class B or be biased with some beam current for class AB, or even used in class A mode. Thus, the application of the klystrode is best suited for high-power, high-efficiency operation, whereas the klystron is best for high linearity.

History and Names for the IOT Klystrode

The klystrode was invented by Varian in the 1980s. Its name was chosen for its hybrid nature and describes the device quite well. Other manufacturers began work on similar devices and use other descriptions. English Electric Valve, Ltd., of the UK, chose the name inductive output tube (IOT) (see Fig. 6.13). This name was chosen due to a reference in the literature to this device possibility in the 1930s. Thus, it was 50 years before the klystrode (or IOT) was made commercially practical.

IOT Description

The klystron relies on the velocity bunching of the electrons from a heated cathode over a body length of 1 m or more. This is necessary to permit orderly physical spacing of the buncher cavities and the individual electron velocities to form properly. The superiority of the klystron over the gridded tube (typically a tetrode) is its output coupling, a cavity. Contrasted, it is quite difficult to design circuits to couple from tetrodes at higher power due to the increasing capacitance, which lowers the impedance. The superiority of the gridded tube is in its compact size and the general possibility of lower cost.

   The IOT takes advantage of two superior characteristics of the gridded tube and the klystron for high frequencies. A grid is used to provide simple control of the electron beam while a cavity is used to couple the RF energy from the beam. This

FIGURE 6.13   60-kW IOT (Klystrode). (*Source:* Courtesy of EEV.)

combination makes possible a smaller, lower cost, high-frequency, high-power tube. Still required, however, are focus magnets to maintain the beam, liquid cooling at higher powers, high beam supply voltages to accelerate the electrons, and the necessity for an external circuit called a **crowbar**. A crowbar is required

TABLE 6.1 Comparison of Klystron and IOT Operating Parameters

Parameter	Klystron	IOT	Units
Frequency	UHF	UHF	—
Peak power	60	60	kW
Beam voltage	28	32	kV max
Beam current	7	4	A max
Drive power	8	300	W
Efficiency	42	60	%
Collector dis.	150	50	kW max

for a IOT due to the possibility of the tube developing an internal arc. This internal arc creates a surge of current from the stored energy of the power supply, which would damage the grid and thus must be diverted. A crowbar is a device such as a thyratron used in shunt that diverts the powers supply stored energy based on the onset of an internal arc.

Gun, Grid, and Gain

The IOT electron gun is identical to that of the klystron except a grid is inserted between the cathode and the anode to control the electrons created by the cathode. The grid bunches the gun generated electrons, and the beam drifts through the field free region and interacts with the RF field in an output cavity to produce the RF output. The gun resembles a klystron gun but also has a grid much like a tetrode. Thus, control of the beam is like a gridded tube and operation is similar. The grid can be used to turn on or off the beam, modulate with RF, or both. Various classes of operation are possible. The most desired is class B, which is the most efficient and linear. Frequently used is class AB, which permits some beam current with no RF drive thus minimizing the class B crossover distortions.

The IOT gain is not as high as that of the klystron. Gain is typically 20–23 dB whereas the klystron is 35–40 dB. These gains are compared to those of a tetrode, which are typically 15 dB at higher frequencies. The lack of gain is not a severe limit for the IOT since solid-state drivers at 100th the output power (20 dB less) usually can be economical. The even higher gain of the klystron simplifies the driver stages even more.

Typical Operation

The IOT and klystron can be compared in operational parameters by reviewing Table 6.1. As this data clearly shows the klystron and IOT are close cousins in the high-frequency tube family. Not shown with this data is the other important performance parameters of amplitude linearity, phase distortion, bandwidth, size, or expected life. Depending on the application, these parameters may be more important than efficiency or gain.

Acknowledgments

We kindly acknowledge the assistance of Earl McCune of Varian, Robert Symons of Litton, and Michael Kirk of EEV.

Defining Terms

Anode: A positive portion of the tube that creates the field to accelerate the electrons produced by the cathode for passage down the body of the tube.

Beam current: The current that flows from anode to cathode.

Cathode: The most negative terminal of a vacuum tube; the source of electrons.

Collector: The portion of the tube that collects the electrons. The most electrically positive portion of the tube and dissipates residual the beam energy.

Crowbar: A triggered, shunt device that diverts the stored energy of the beam power supply.

Electron gun (gun): An electron emission material that is heated to produce the electron beam. Usually physically shaped and electrically controlled to form a pencil lead sized beam.

Focus coil: Coils of wire encircling the tube forming an electromagnet to focus the beam electrons. The focus coils have DC applied during operation.

Grid or modulating anode: A terminal in the tube used to control the electron beam usually placed near the cathode.

Klystrode: A hybrid tube that uses elements from the klystron and the tetrode.

References

Badger, G. 1985. The klystrode, a new high-efficiency UHF TV power amplifier. Varian, *NAB Proceedings of the Engineering Conference.*

Heppinstall, R. 1986. Klystron operating efficiencies: Is 100 percent realistic? *NAB Engineering Conference*, EEV.

McCune, E.W. 1988. UHF-TV klystron multistage depressed collector development program. NASA Report.

Nelson, B. 1963. *Introduction to Klystron Amplifiers.* Varian Associates, San Carlos, CA, April.

Staprans, A., McCune, E.W., and Ruetz, J.A. High-Power Linear-Beam Tubes. *Proceedings IEEE*, vol. 61.

Symons, R. 1982. Klystrons for UHF television. *Proceedings IEEE* (Nov.).

Symons, R.S. 1986. Scaling laws and power limits for klystrons. *IEDM Digest.*

Symons, R.S. and Vaughan, J.R.M. 1994. The linear theory of the clustered-cavity^{TM} klystron. *IEEE Transactions* (Oct.).

Varian. 1972. Notes on Klystron Operation. Varian Associates, San Carlos, CA, April.

Varian. Klystrons for Ku-band satellite ground transmitters. Varian Associates, San Carlos, CA, Pub. No. 4084.

Further Information

For further information refer to:

Klystrons and Microwave Triodes. Hamilton, Knipp, Kuper, Radiation Laboratory Series.

Practical Theory and Operation of UHF-TV Klystrons. Reginal Perkins, R&L Technical Publishers, Boston, MA.

Very High Frequency Techniques. Radio Research Laboratory, Harvard Univ. Press, Boston, MA.

6.3   Traveling Wave Tubes

Thomas K. Ishii

6.3.1   Definitions (Ishii, 1989, 1990)

A **traveling wave tube** (TWT) is a vacuum device in which traveling microwave electromagnetic fields and an electron beam interact with each other. In most **traveling wave** tubes, the direction of traveling microwave propagation inside the tube is made parallel to and in the same direction with the **electron beam**. Microwaves propagating in the same direction as the electron beam are termed **forward waves**. Traveling wave tubes based on the interaction with forward waves and the electron beam are termed the *forward wave tubes.* Most forward wave tubes are microwave amplifiers.

To make interaction between the traveling microwave electromagnetic fields and the electron beam efficient, in a traveling wave tube, the phase velocity or propagation speed of microwaves in the tube is slowed down to approximately equal speed with the speed of electrons in the electron beam. In most practical traveling wave tubes, the speed of electrons in the beam is approximately 1/10 of the speed of light. In most traveling wave tubes, there is a transmission line in the device. Microwave energy to be amplified is fed to one end of the transmission line to which the electron beam is coupled. The transmission line is designed so that the speed of microwave propagation is also slowed down to approximately 1/10 of speed of light. For the best result, the speed of microwave propagation on the transmission line and the speed of electrons in the coupled electron beam must be approximately equal. Because the speed of microwave propagation is slowed down in the transmission line, the line is termed a **slow-wave structure**.

Thus in a traveling wave tube, microwaves and electrons travel together with approximately the same speed and interact with each other. During the interaction, the kinetic energy of the electrons is transferred or converted into microwave energy. As both microwaves and electrons are traveling together, the electrons lose their kinetic energy and the microwaves gain the energy of electromagnetic fields. Thus microwave electromagnetic fields are amplified while traveling inside the traveling wave tube. Amplified microwave energy is extracted at the other end of the transmission line.

Since a traveling wave tube does not have a resonator, if the input circuit and the output circuit are well impedance matched, a wide operating frequency bandwidth can be expected. In practice, the operating frequency bandwidth usually exceeds an octave (EEV, 1991, 1995).

The gain of a traveling wave tube is, among other things, dependent on the length of the tube; the longer the tube, the higher is the gain. Therefore, the physical shape of a traveling wave tube is long to accommodate the transmission line. The transmission line must have an input coupling circuit to couple the input microwave signal that is to be amplified into the transmission line. At the end of the transmission line, there must be another coupling circuit to take the amplified microwave out to the output circuit.

In a traveling wave tube, the electron beam is generated by an **electron gun** that consists of a hot cathode and electrostatic focusing electrodes. Electrons are generated from the heated cathode by **thermionic emission**. The launched electron beam to couple with the transmission line is focused by longitudinally applied DC magnetic fields. These **focusing DC magnetic fields** are provided by either a stack of permanent magnet rings or a solenoid. Beyond the end of the transmission line, there is a positive electrode to collect the used electron beam. This electrode is termed the **collector**.

A generic configuration of a traveling wave tube is shown in Fig. 6.14. The electron beam is coupled to a slow-wave structure transmission line to which the microwave signal to be amplified is fed through an input waveguide and a coupling antenna. The amplified microwaves are coupled out to the output waveguide through the output coupling antenna and the coupling waveguide circuit. As seen from Fig. 6.14, the electron gun is biased negatively by the use of the DC power supply and radio frequency choke (RFC) against the waveguide, the transmission line, and the collector. During the interaction between the electron beam and microwaves, the electron beam loses kinetic energy, and the lost energy is transferred into the electromagnetic field energy of the microwaves traveling in the transmission line.

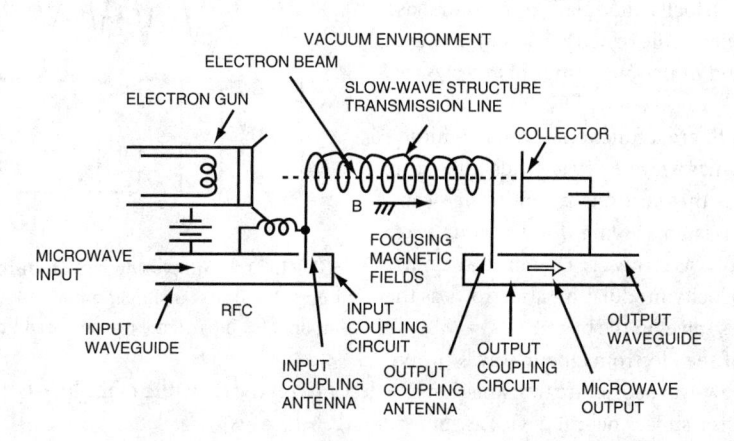

FIGURE 6.14 Generic configuration of a traveling wave tube.

Usually a traveling wave tube means a forward wave tube. In the vacuum tube, the electron beam and forward waves interact with each other. There is a vacuum tube in which the electron beam interacts with **backward waves**. This type of tube is termed the **backward wave tube**. The backward wave tube is inherently highly regenerative (built-in positive feedback); therefore, it is usually an oscillator. Such an oscillator is termed a **backward wave oscillator (BWO)**. The BWO has a **traveling wave** structure inside, but usually it is not called a traveling wave tube. Details of a BWO are presented in Chap. 6.4.3. The oscillation frequency of a BWO is dominated by the electron speed, which is determined by the anode voltage. Therefore, a BWO is a **microwave frequency voltage controlled oscillator** (VCO).

6.3.2 Design Principles (Ishii, 1990)

By design, in a traveling wave tube, the speed of electrons in the electron bean u (m/s) and the **longitudinal phase velocity of** microwave electromagnetic field in the slow-wave structure transmission line v (m/s) are made approximately equal. When an electron comes out of the electron gun, the speed of an electron is

$$u_0 = \sqrt{\frac{2q\,V_a}{m}} \tag{6.1}$$

where q is the electric charge of an electron, 1.602×10^{-19} coulombs; m is the mass of an electron, 9.1085×10^{-31} kg; and V_a is the acceleration anode voltage.

But the speed of electrons changes as it travels within the beam due to interaction with traveling electromagnetic waves in the transmission line slow-wave structure. In the beginning, the electrons are velocity modulated. Some are accelerated and some are decelerated by the microwave electromagnetic field in the slow-wave transmission line. The result of **velocity modulation** is **electron bunching**. In a traveling wave tube, bunched electrons are traveling with the traveling electromagnetic waves. The speed of bunched electrons is given by

$$u = u_0 + u_m \epsilon^{-\alpha_e z} \cdot \epsilon^{-j(\beta_e z - \omega t)} \tag{6.2}$$

In this equation, a one-dimensional coordinate system z is laid along the direction of traveling waves and the electron beam. It is assumed that the interaction of electrons and traveling electromagnetic waves started at $z = 0$ and the time $t = 0$. In Eq. (6.2), u_m is the magnitude of velocity modulation at the initial location, α_e the attenuation constant of the velocity modulation, β_e the phase constant of the electronic velocity modulation waves in the beam, and ω the operating microwave frequency.

Both α_e and β_e are a function of the magnitude of interacting microwave electric fields and time, which change as the electrons travel in the beam. When the TWT is amplifying, the traveling wave electric field grows as it travels. Therefore, the magnitude of the velocity modulation also grows as the electronic waves travel in the beam.

The speed of the electromagnetic waves propa-

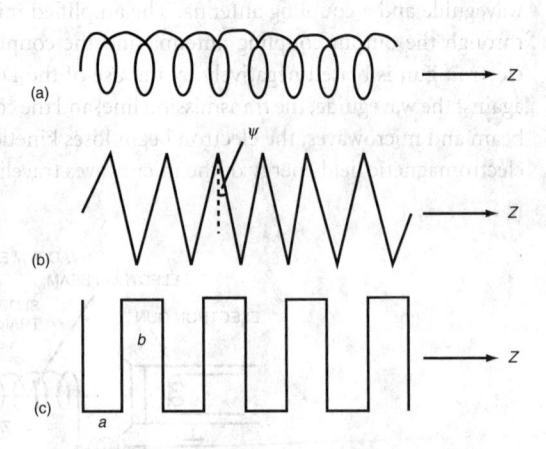

FIGURE 6.15 Simple slow-wave transmission line structures: (a) helical transmission line, (b) zig-zag transmission line, (c) interdigital transmission line.

gating on the slow-wave structure transmission line depends on the structure of the line itself. For example, if the structure is a simple helical device as shown in Fig. 6.15(a) is

$$v_z = c \sin \psi \tag{6.3}$$

where c is the speed of light in the vacuum and ψ the pitch angle of the helix. If the slow-wave structure is a simple zig-zag transmission with the angle of the zig-zag at 2ψ (as shown in Fig. 6.2(b)), the longitudinal velocity of electromagnetic wave propagation in the z direction is

$$v_z = c \sin \psi \tag{6.4}$$

If the slow-wave structure is a simple interdigital circuit with the pitch dimensions a and b, as shown in Fig. 6.2(c),

$$v_z = \frac{a}{a+b} c \tag{6.5}$$

These equations assume that the coupling between the electron beam and the transmission line is light. It is further assumed that the waves propagate along the conducting wire slow-wave transmission line with the speed c. In reality, the electron beam will load these transmission lines. Then the speed of the electrons in the beam will be slightly different from these equations. In an electron beam coupled generic transmission line

$$v_z = \frac{\omega}{\beta_c - \frac{1}{2}b} \tag{6.6}$$

where ω is the operating angular frequency, β_c is the phase constant of the slow-wave transmission line with the series impedance $\dot{Z}$ (Ω/m) and the shunt admittance of $\dot{Y}$ (S/m), which is

$$\beta_c = I_m \sqrt{\dot{Z}\dot{Y}} \tag{6.7}$$

and

$$b = (\dot{\gamma} - \dot{\gamma}_c)\epsilon^{j\frac{5\pi}{6}} \tag{6.8}$$

where

$$\dot{\gamma}_c = \sqrt{\dot{Z}\dot{Y}} \tag{6.9}$$

and

$$\dot{\gamma} = \left(\alpha_c - \frac{\sqrt{3}}{2}b\right) + j\left(\beta_c - \frac{1}{2}b\right) \tag{6.10}$$

$$\alpha_c = R_e \sqrt{\dot{Z}\dot{Y}} \tag{6.11}$$

and β_c is given in Eq. (6.7).

The **voltage gain** per meter of propagation in **neper** per meter and **decibel** per meter is

$$\alpha = \alpha_c - \frac{\sqrt{3}}{2}b \tag{6.12}$$

$$\alpha = 8.686\left[\alpha_c - \frac{\sqrt{3}}{2}b\right] \tag{6.13}$$

If the interaction transmission line length of the TWT is ℓ m long, then the voltage gain is

$$G = \epsilon^{\alpha\ell} \tag{6.14}$$

The **frequency bandwidth** is calculated by the following procedure. At the midband frequency ω_0

$$\epsilon^{\alpha(\omega_0)} = \epsilon \exp\left\{\alpha_c(\omega_0) - \frac{\sqrt{3}}{2}b(\omega_0)\right\} \tag{6.15}$$

At the lower band edge ω^-

$$\epsilon \exp\left\{\alpha_c(\omega^-) - \frac{\sqrt{3}}{2}b(\omega)\right\} = \frac{1}{\sqrt{2}}\epsilon^{\alpha(\omega_0)} \tag{6.16}$$

At the upper band edge ω^+

$$\epsilon \exp\left\{\alpha_c(\omega^+) - \frac{\sqrt{3}}{2}b(\omega^+)\right\} = \frac{1}{\sqrt{2}}\epsilon^{\alpha(\omega_0)} \tag{6.17}$$

Solving Eq. (6.16) ω^- is found. Solving Eq. (6.17) ω^+ is found. Then the frequency bandwidth is

$$\Delta\omega = \omega^+ - \omega^- \tag{6.18}$$

$$\Delta f = \frac{\omega^+ - \omega^-}{2\pi} \tag{6.19}$$

The **noise figure** (NF) of a TWT ℓ-m-long interaction transmission line can be calculated by the following procedure:

$$\mathrm{NF} = \frac{(kT\Delta f G_p + \int_0^\ell (P_s + P_p + P_f + kT_N\Delta f)dz + kT\Delta f)}{kT\Delta f G_p} \tag{6.20}$$

where T is the TWT circuit temperature of the input port and the output port, Δf the frequency bandwidth, G_p the **power gain** and

$$G_p = \epsilon^{2\alpha\ell} \tag{6.21}$$

and k is Boltzmann constant, 1.38054×10^{-23} J/K; P_s is the **shot noise** power per meter; P_p is the **partition noise** power per meter; P_f is the **flicker noise** per meter of propagation in the interaction transmission line section; and T_N is the **noise temperature** per meter of the interaction transmission line.

6.3.3 Electron Gun

The electron gun of a traveling wave tube is a device that supplies the electron beam to the tube. A schematic diagram of a generic electron gun is shown in Fig. 6.16. A generic electron gun consists of a hot cathode heated by an electric heater, a negatively biased **focusing electrode** or **focuser** and positively biased accelerating anode. Figure 6.16 is a cross-sectional view, and the structure can be a two-dimensional or three-dimensional coaxial structure (Liao, 1988; Sims and Stephenson, 1953). If it is two dimensional, it is for a ribbon-shaped electron beam and if it is coaxial, it is for a cylindrical electron beam. The focusing of the electron beam and the amount of the beam current can be adjusted by adjusting the heater voltage V_H and the bias voltage for the cathode V_c, for the focusing electrode or focuser V_f and the accelerator anode voltage V_a.

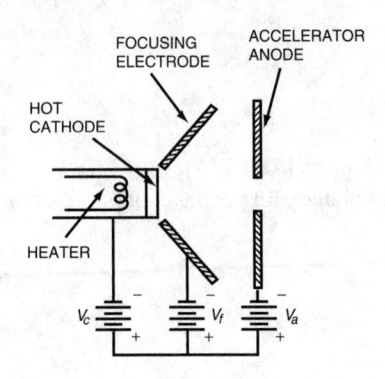

FIGURE 6.16   A schematic diagram of a generic electron gun.

Electron emission current density (A/m^2) can be calculated by the following **Richardson–Dushman equation** (Chodorow and Susskind, 1964):

$$J = 6.02 \times 10\,T^2\epsilon^{-b/T} \tag{6.22}$$

where T is the absolute temperature of the cathode in kelvin, and b also in kelvin is (Chodorow and Susskind, 1964)

$$b = 1.1605 \times 10^4\phi \tag{6.23}$$

where ϕ is the **work function** of the cathode material. If the cathode material is tungsten, $\phi = 4.5$ V. If it is tantalum, $\phi = 4.1$ V. If it is an oxide coated cathode by B_aO, S_rO or C_aO, $\phi = 1$–2 V. If it is a combination of B_aO and S_rO, then $\phi = 1.0$ V. If it is thoriated tungsten, $\phi = 2.6$ V. The operating temperature of the cathode is about 1300~1600 K. Once electrons are thermally emitted from the cathode, electrons are pushed toward the center axis of the structure by the **coulomb force**

$$F_r = q\,E_r aa \tag{6.24}$$

where q is the electric charge of an electron 1.602×10^{-19} C and E_r is the radial electric field intensity (V/m). E_r is determined by the geometry of various electrodes and the bias voltages V_c, V_f, and V_a. However, V_f has the most direct influence on it. This negative electrode pushes the electron toward the center axis. The electrons are also pulled along the center axis by the positive **accelerator** anode. The longitudinal coulomb force acting on the electron is

$$F_z = q\,E_z \tag{6.25}$$

where E_z is the longitudinal electric field intensity in volt per meter.

E_z is determined by the geometry of the electrodes and the bias voltages V_c, V_f, and V_a. However, V_a has the most direct effect on pulling electrons out of the hot cathode.

The procedure to obtain the electron trajectory in the electron gun is to solve **Newton's equations of motion** for the electron. In a generic electron motion, a tangential force may be involved, if it is not balanced. It is

$$F_\phi = q\,E_\phi \tag{6.26}$$

where E_ϕ is a tangential electric field. If the electron gun is structured axially circularly symmetrical, then E_ϕ does not exist. Otherwise, E_ϕ must be considered. Newton's equations of motion of an electron in the electron gun are

$$m\frac{d^2r}{dt^2} = q\,E_r \tag{6.27}$$

$$m\frac{d^2z}{dt^2} = q\,E_z \tag{6.28}$$

$$m\frac{d^2(r\phi)}{dt^2} = q\,E_\phi \tag{6.29}$$

E_r, E_z, and E_ϕ are obtainable from the electric potential V within the electron gun

$$E_r = -\frac{\partial V}{\partial r} \tag{6.30}$$

$$E_z = -\frac{\partial V}{\partial z} \tag{6.31}$$

$$E_\phi = -\frac{1}{r}\frac{\partial V}{\partial \phi} \tag{6.32}$$

The electric potential V within the electron gun is obtainable by solving a Laplace equation with specifically given boundary conditions,

$$\frac{1}{r}\frac{\partial}{\partial r}\left(r\frac{\partial V}{\partial r}\right) + \frac{1}{r^2}\frac{\partial^2 V}{\partial \phi^2} + \frac{\partial^2 V}{\partial z^2} = 0 \qquad (6.33)$$

An axially symmetrical solid cylindrical electron beam is produced by the gun structure illustrated in Fig. 6.16, if the structure is axially cylindrically symmetrical. If the middle of the hot cathode is made nonemitting and only the edge of the cathode is emitting, then the cathode becomes an **annular cathode**. The annular cathode produces a **hollow beam**. The annular electron beam can save the electron beam current for equal microwave output.

If the gun structure shown in Fig. 6.16 is two dimensional, then it produces a ribbon-shaped electron beam. A ribbon-shaped beam is used for a TWT of a two-dimensional structure.

If the angle of the focusing electrode against the axis of the electron beam is 67.5° and the anode is also tilted forward to produce a rectilinear flow, which is an electron flow parallel to the z axis in Fig. 6.16, then, such electron gun is termed the **Pierce gun**.

In practice the hot cathode surface is curved as shown in Fig. 6.17 to increase the electron emitting surface and to obtain a high-density electron beam.

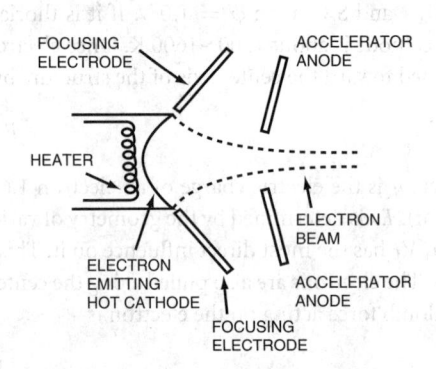

FIGURE 6.17   Cross-sectional view of an electron gun with curved hot cathode.

6.3.4   Electron Beam Focusing

Electrons in an electron beam mutually repel each other by the electron's own coulomb force due to their own negative charge. In addition, the electron beam usually exists in proximity to the positively biased slow-wave structure as seen in Fig. 6.14. Therefore, the electron beam tends to diverge. The process of confining the electron beam within the designed trajectory against the mutual repulsion and diverging force from the slow-wave structure is termed *electron beam focusing*.

The electron beam in a TWT is usually focused by a DC magnetic flux applied parallel to the direction of the electron beam, which is coaxial to the slow-wave transmission line as illustrated in Fig. 6.14. If there is no focusing magnetic flux only at the output of the electron gun, then such electron flow in the beam is termed **Brillouin flow** (Hutter, 1960). If the electron gun itself is exposed to and unshielded from the focusing para-axial longitudinal magnetic flux, then the electron flow produced is termed the **immersed flow** (Hutter, 1960). If the electron gun is not shielded from focusing magnetic flux, which is not para-axial, then the electron flow produced in such a focusing system is neither Brillouin flow nor immersed flow, but it is a *generic flow*.

In a generic flow device, the focusing force or equation of motion of an electron in the electron beam is

$$m\frac{\partial^2 r}{\partial t^2} = mr\left(\frac{\partial \phi}{\partial t}\right)^2 + qr\frac{\partial \phi}{\partial t}B_z + \frac{qr_0^2\rho_0}{2\epsilon_0 r} + qE_r \qquad (6.34)$$

where r_0 is the radius of a part of the electron beam that is repelling the electron in question, (r, ϕ, z) is the cylindrical coordinate of the electron in question, q is the charge of the electron, which is 1.602×10^{-19} C. B_z is the z component of applied magnetic flux density, m is the mass of the electron, which is 9.109×10^{-31} kg. Also, ρ_0 is the electronic charge density of the electron beam, and ϵ_0 is the **permittivity** of the vacuum, which is 8.854×10^{-12} F/m.

The angular velocity of the electron trajectory $\partial\phi/\partial t$ may be calculated from the following equation of **Busch's theorem** (Hutter, 1960; Chodorow and Susskind, 1964; Harman, 1953):

$$\frac{\partial\phi}{\partial t} = \frac{\eta}{2\pi r^2}(\psi - \psi_c) \tag{6.35}$$

where η is a constant q/m, which is 1.759×10^{11} C/kg,

$$\psi = \pi r^2 B_z \tag{6.36}$$
$$\psi_c = \pi r_c^2 B_z \tag{6.37}$$

where r_c is the radial position of the electron in question where it first encountered the focusing magnetic flux.

If the electron beam is the immersed flow or Brillouin flow, then

$$m\frac{\partial^2 r}{\partial t^2} = mr\left(\frac{\partial\phi}{\partial t}\right)^2 + qr\frac{\partial\phi}{\partial t}B_z + \frac{qr_0^2\rho_0}{2\epsilon_0 r} + qE_r \tag{6.38}$$

Focusing conditions are

$$\left.\frac{\partial r}{\partial t}\right|_{r=a} = 0 \tag{6.39}$$

$$\left.\frac{\partial^2 r}{\partial t^2}\right|_{r=a} = 0 \tag{6.40}$$

where a is the radius of the electron beam.

In practice, the focusing magnetic flux is created by either a solenoid or a structure of permanent magnetic rings.

6.3.5 Electron Velocity Modulation (Ishii, 1990)

If there are no microwaves on the slow-wave transmission line of a traveling wave tube, the velocity of an electron in the electron beam emitted from the electron gun is constant and uniform. As seen from Fig. 6.1, there is practically no longitudinal DC electric field in the transmission line section. Therefore, the electrons are just drifting with the speed given by the electron gun, which is

$$u_0 = \sqrt{\frac{2qV_a}{m}} \tag{6.41}$$

where q is the charge of the electron, m is the mass of the electron, and V_a is the accelerating electric potential of the electron gun. Now, if microwaves are launched on the slow-wave transmission line from the input as illustrated in Fig. 6.14 then there will be interaction between electrons in the electron beam and microwave electromagnetic fields on the slow-wave transmission line. In a traveling wave tube, it is designed that the electron velocity u_0 and the longitudinal propagation velocity of microwaves on the slow wave structure v_z are made approximately equal to each other.

$$u_0 \approx v_z \tag{6.42}$$

Therefore, electrons under the acceleration field due to traveling microwaves are kept accelerating and other electrons under the decelerating electric field of the traveling microwaves are kept decelerating. The longitudinal electric field component of traveling microwaves on the slow-wave transmission line is expressed by

$$E_z = E_0 \epsilon^{-\alpha z}\epsilon^{-j(\beta z - \omega t)} \tag{6.43}$$

where z is a one-dimensional coordinate set along the slow-wave transmission line and $z = 0$ is the point where microwaves are launched. E_0 is the magnitude of the longitudinal component of the microwave electric field at $z = 0$, α is the **attenuation constant** of the line, and β is the phase constant of the line. Depending on the phase angle of E_z at specific time t and location z, which is $(\beta z - \omega t)$, the electron may be kept accelerating or may be kept decelerating. This acceleration and deceleration of electrons due to applied microwave electric fields is termed the velocity modulation of an electron beam. The modulated electron velocity is

$$u = u_0 + u_m \exp\{-\alpha_e z - j(\beta_e z - \omega t)\} \tag{6.44}$$

where u_m is the magnitude of velocity modulation of the electron velocity. Here, α_e is the attenuation constant and β_e is the phase constant of the propagation of the velocity modulation on the electron beam.

6.3.6 Electron Bunching (Ishii, 1989)

If the speed of electrons in a focused electron beam is velocity modulated, as seen in Eq. (6.44), then the electron flow is no longer uniform and smooth. Some electrons in the beam are faster or slower than others, and as a result electrons tend to *bunch* together as they travel from the electron gun in the beginning to the electron **collector** at the end. The traveling electron charge density in the velocity modulated electron beam is expressed by

$$\rho = \rho_0 + \rho_m \epsilon^{-\alpha_e z} \epsilon^{-j(\beta_e z - \omega t)} \tag{6.45}$$

where ρ_0 is the electron charge density of an unmodulated electron beam, ρ_m is the magnitude of charge density modulation as the result of the velocity modulation of the electrons, α_e is the attenuation constant of the electron charge wave, and β_e is the phase constant of the wave.

The effect of electron bunching or propagation of electron charge waves in the electron beam is observed in a form of an electron beam current. If the cross-sectional area of the electron beam is S, (in square meter), then the electron beam current that contains bunched electrons is

$$I = I_0 + I_1 \epsilon^{-\alpha_e z} \cdot \epsilon^{-j(\beta_e z - \omega t)} + I_2 \epsilon^{-2\alpha_e z} \cdot \epsilon^{-2j(\beta_e z - \omega t)} \tag{6.46}$$

where I_0 is the DC beam current or unmodulated electron beam current, I_1 is the magnitude of beam current modulation at the frequency of microwaves ω, and I_2 is the magnitude of beam current modulation at the second harmonic frequency of operating frequency,

$$I_0 = S\rho_0 u_0 \tag{6.47}$$

$$I_1 = S(\rho_0 u_m + \rho_m u_0) \tag{6.48}$$

and

$$I_2 = S\rho_m u_m \tag{6.49}$$

6.3.7 Electron Beam Coupling

As seen from Fig. 6.14, to give electrons velocity modulation, microwave energy at or in proximity to the input must couple to the electron beam. To extract microwave energy from the bunched electrons at or in proximity to the output, the energy of the bounced electrons must be coupled to the slow-wave transmission line. This coupling of energy between the electron beam and the slow-wave transmission line is termed the **electron beam coupling**.

The strength of the beam coupling is represented by the beam coupling coefficient, which is defined as

$$k = \frac{I_c}{I_1} \tag{6.50}$$

where I_c is the microwave current in the slow-wave transmission line at operating frequency and I_1 is the magnitude of microwave beam current in the electron beam at the operating frequency ω. The beam coupling coefficient for a helical slow-wave line is obtained by

$$k = \frac{\pi u_0^2}{\omega \left(u_m \dfrac{\rho_0}{\rho_m} + u_0 \right) \sin \psi} \tag{6.51}$$

If the slow-wave transmission line is not a helix line, then the slow-wave ratio (less than 1) must be used instead of $\sin \psi$ where ψ is the **pitch angle of the helix**.

6.3.8 Power Gain

As microwaves propagate along the slow wave transmission line, they receive energy from the bunched electron beam. The loss of kinetic energy due to the microwave deceleration is the gain of microwave energy. Microwave power increase per meter of propagation is given by

$$\frac{dP}{dz} = \frac{d}{dz} \left(\frac{k^2}{2} mu^3 NS \right) = 2\alpha P \tag{6.52}$$

where P is microwave power flow on the transmission line, z is a one-dimensional coordinate set up along the slow-wave transmission line, k the beam coupling coefficient, m the mass of an electron, u the velocity of the electron, N the density of electrons in the beam, S the cross-sectional area of the beam, and α the gain parameter in neper/m that represents the neper power increase on the transmission line per meter of propagation.

The **gain parameter** α in neper per meter is given by (Hutter, 1960)

$$\alpha = \frac{\sqrt{3}}{2} \beta_e \left\{ Z_0 \frac{I_0}{4V_a} \right\}^{\frac{1}{3}} \tag{6.53}$$

where Z_0 is the **characteristic impedance** of the slow-wave transmission line, I_0 the DC beam current, V_a the DC acceleration anode voltage, and β_e the phase constant of the electron charge wave.

The output power P_0 of a traveling wave tube that is the length of the interaction of the slow-wave transmission line ℓ m long is

$$P_0 = P_i \epsilon^{2\alpha\ell} \tag{6.54}$$

where P_i is the input power to the slow wave transmission line. Therefore, the power gain in number is

$$G = \frac{P_0}{P_i} = \epsilon^{2\alpha\ell} \tag{6.55}$$

The power gain in neper is

$$G = \ell n \frac{P_0}{P_i} = 2\alpha\ell \tag{6.56}$$

The power gain in decibel is

$$G = 10 \log_{10} \frac{P_0}{P_i} = 10 \log_{10} \epsilon^{2\alpha\ell} \\ = 8.686\,\alpha\ell \tag{6.57}$$

6.3.9 Frequency Bandwidth

The gain parameter of a traveling wave tube is frequency independent. As seen in Eq. (6.53)

$$\beta_e = \frac{\omega}{u_0} \tag{6.58}$$

where u_0 is the DC electron velocity, and

$$Z_0 = \sqrt{\frac{R + j\omega L}{G + j\omega C}} = \sqrt{\frac{Z}{Y}} \epsilon^{j\frac{\phi - \psi}{2}} \tag{6.59}$$

where R, L, G, and C are resistance, inductance, conductance, and capacitance per meter of the slow-wave transmission line and

$$Z = \sqrt{R^2 + (\omega L)^2} \tag{6.60}$$

$$Y = \sqrt{G^2 + (\omega C)^2}$$

$$\phi = \tan^{-1}\frac{\omega L}{R} \tag{6.61}$$

$$\psi = \tan^{-1}\frac{\omega C}{G}$$

and

$$\alpha = \frac{\sqrt{3}}{2}\frac{\omega}{u_0}\left\{\frac{Z(\omega)}{Y(\omega)}\frac{I_0}{4V_a}\right\}^{\frac{1}{3}}\cos\frac{\phi(\omega) - \psi(\omega)}{6} \tag{6.62}$$

If the maximum gain occurs at $\omega = \omega_0$, then the maximum power gain is

$$G(\omega_0) = \epsilon^{2\alpha(\omega_0)\ell} \tag{6.63}$$

If the power falls to $\frac{1}{2}\epsilon^{2\alpha(\omega_0)\ell}$ at $\omega = \omega_+$ and $\omega = \omega_-$, then

$$G(\omega_\pm) = \frac{1}{2}\epsilon^{2\alpha(\omega_0)\ell} \tag{6.64}$$

Solving Eq. (6.64) for both ω_+ and ω_-, the frequency bandwidth is obtained as

$$\Delta f = \frac{\omega_+ - \omega_-}{2\pi} \tag{6.65}$$

Equation (6.64) is usually written as transcendental equation:

$$\exp\left[\frac{\sqrt{3}\omega_\pm}{u_0}\left\{\frac{Z(\omega_\pm)}{Y(\omega_\pm)}\frac{I_0}{4V_a}\right\}^{\frac{1}{3}}\cos\frac{\phi(\omega_\pm) - \psi(\omega_\pm)}{6}\right]$$

$$= \frac{1}{2}\exp\left[\frac{\sqrt{3}\omega_0}{u_0}\left\{\frac{Z(\omega_0)}{Y(\omega_0)}\frac{I_0}{4V_a}\right\}^{\frac{1}{3}}\cos\frac{\phi(\omega_0) - \psi(\omega_0)}{6}\right] \tag{6.66}$$

6.3.10 Noise

As seen from Eq. (6.20), the **noise** output of a traveling wave tube with the interaction length of the electron beam and the slow-wave transmission line ℓ, power gain G_p, frequency bandwidth Δf, and temperature of both the input and the output circuits $T(K)$ is

$$P_N = kT\Delta f G_p + \int_0^\ell (P_s + P_p + P_F + kT_N\Delta f)dz + kT\Delta f \tag{6.67}$$

Other symbols are as defined after Eq. (6.20). Here, $kT\Delta f$ is the thermal noise input power, also, $kT\Delta f$ is the **thermal noise** generated at the output circuit. Boltzmann's constant k is 1.38×10^{-23} J/K.

The shot noise power in the beam coupled out to the transmission line per meter is

$$P_s = 2\xi q I_0 \Delta f \tag{6.68}$$

where q is the electron charge, I_0 the DC beam current, and ξ the power coupling coefficient between the beam and the transmission line.

The flicker noise P_F is due to random variation of **emissivity** of the cathode and

$$P_F = \eta \frac{1}{f} \tag{6.69}$$

where f is the operating frequency and η a proportionality constant.

The partition noise P_p is generated by electron hitting the slow-wave transmission line and is proportional to the speed of the electron u and the beam current I_0,

$$P_p = \zeta I_0 u \tag{6.70}$$

The noise figure of a traveling wave tube is presented with Eq. (6.20).

6.3.11 Sensitivity

The amount of the input power that makes the signal-to-noise ratio at the output of a traveling wave tube unity is termed the **tangential sensitivity** of the traveling wave tube. The output power of the traveling wave tube is

$$P_0 = P_i G_p \tag{6.71}$$

where P_i is the input power and G_p is the power gain of the traveling wave tube.

For the tangential sensitivity

$$P_0 = P_N \tag{6.72}$$

where P_N is the noise output power presented by Eq. (6.67). For the tangential sensitivity P_s, $P_i = P_s$ and

$$P_i = P_s \quad \text{and} \quad P_s G_p = P_N \tag{6.73}$$

Combining Eq. (6.73) with Eq. (6.67),

$$P_S = \frac{P_N}{G_P} = kT\Delta f + \frac{1}{G_p}\left[\int_0^\ell (P_s + P_p + P_F + kT_N\Delta f)dz + kT\Delta f\right] \tag{6.74}$$

For a generic **sensitivity**

$$P_0 = nP_N \tag{6.75}$$

where n is an arbitrary specified number.

Since

$$P_0 = nP_N \tag{6.76}$$

$$P_S = \frac{P_N}{G_P} = kT\Delta f + \frac{1}{G_P}\left[\int_0^\ell (P_s + P_p + P_F + kT_N\Delta f)dz + kT\Delta f\right] \tag{6.77}$$

The value of n can be

$$n \gtrless 1 \tag{6.78}$$

6.3.12   Collector

As seen from Fig. 6.14, the electron beam emitted from the electron gun goes through the interaction region with the microwave slow-wave transmission line and hits the electron collector at the end. Various configurations of the electron collector are shown schematically in Fig. 6.18. Collectors are hit by high-energy electrons of several thousand electron volts or more (EEV 1991, 1995). Collectors, therefore, often need forced air or water cooling. In practical traveling wave tubes, collectors and transmission lines are biased to the ground potential or near ground potential and the electron gun or cathode are biased to a high negative potential for operational security reasons.

For low-power traveling wave tubes, a flat collector as shown in Fig. 6.18(a) is adequate. To prevent electrons from returning to the transmission line, the collector is sometimes biased slightly positively. If electrons have sufficient kinetic energy and no chance of going back to the interaction region, then the collector may be biased negatively to make the electrons soft landing on the collector to prevent overheating the collector.

Regardless, the flat corrector is not adequate for a high-power TWT. The cone-shaped collector as shown in Fig. 6.18(b) has more electron collection area than the flat collector shown in Fig. 6.18(a). Therefore, it reduces collector current density on the collector. This reduces the heating problem of the collector. A larger electron collection area can be obtained by making the collector configuration a curved cone shape, shown in Fig. 6.18(c), or a cylindrical shape, shown in Fig. 6.18(d) to reduce the heating of the collector by electron bombardment.

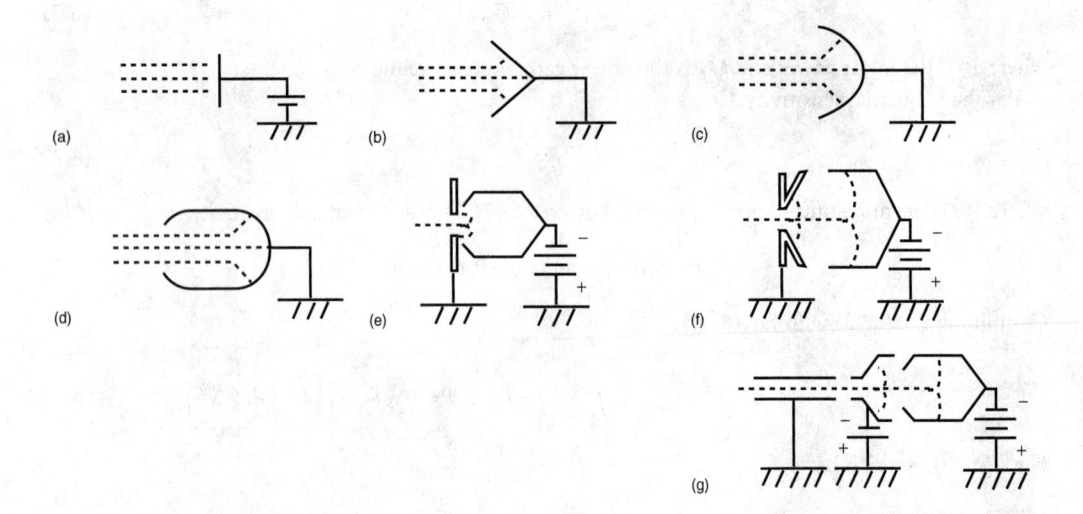

FIGURE 6.18   Schematic cross-sectional view of various collector configurations for a TWT: (a) plate, (b) cone, (c) curved cone, (d) cylinder, (e) depressed potential cylinder, (f) two stage, and (g) three stage.

For a high-power TWT, the electron energy may be on the order of tens of thousand electron volts (EEV 1991, 1995). Typically, electrons in the collector region are gradually slowed down for soft landing to prevent excess heating of the collector structure. In Fig. 6.18(e) (Liao, 1988) the cylindrical collector is negatively biased and electrons are pushed back and collected by a grounded plate at the end of the interaction region. The electron collection region may be expanded further as illustrated in Fig. 6.18(f) (Liao, 1988). The electron collection is accomplished in two stages, as seen from the figure. For an extremely high-power operation a three-stage collector, as shown in Fig. 6.18(g) (Liao, 1988), is employed. Each stage is progressively negatively biased for soft landing of electrons to prevent excess heating of the collector structure.

The energy of an electron at the collector is approximately,

$$w = qV_a \tag{6.79}$$

where q is the charge of an electron and V_a is the acceleration voltage. Therefore, for complete soft landing, the collector must provide

$$w_c = -qV_a = qV_c \tag{6.80}$$

where V_c is the collector voltage

$$V_c = -V_a \tag{6.81}$$

For any specified soft landing

$$V_c = -bV_a \tag{6.82}$$

where b is a specified constant that is less than unity.

6.3.13 Practical Examples

Traveling wave tubes are amplifiers. They are used for microwave communications, radar, industrial and biomedical sensing, and scientific research, as exemplified by the radio telescope. Traveling wave tube are used wherever wide frequency band microwave amplification is needed. They are widely used for multichannel telephone and television channel transmission. They are also useful for short pulsed radar amplification and high data rate data transmission. A traveling wave tube can be used as transmitter amplifier of high power, or it can be used as a receiver amplifier of high sensitivity. A traveling wave tube may be for continuous wave operation, or for pulsed operation. Depending on the application, there are a great variety of traveling wave tubes in terms of power, voltage, current, mode of operation, size and weight.

Defining Terms

Accelerator: A positive electrode in a vacuum tube to accelerate emitted electrons from its cathode by coulomb force in a desired direction.

Annular cathode: A cathode of a vacuum tube with the shape of the emitting surface of the cathode is annular. The annular cathode can produce a hollow electron beam.

Attenuation constant: A constant representing attenuation of magnitude of propagating quantity such as the electromagnetic waves and acoustic waves.

Backward wave: Electromagnetic wave that travels in the direction opposite to the direction of electrons in the coupled electron beam.

Backward wave oscillator (BWO): In which the electron beam and microwaves travel in opposite direction to each other.

Backward wave tube: A backward wave traveling wave tube; the microwave input is located in proximity to the collector and the microwave output is located in proximity to the electron gun. The direction of traveling microwave and direction of electron motion in the beam are opposite to one another.

Brillouin flow: A stream of electron beam emitted from an electron gun that is not exposed to a focusing magnetic field.

Busch's theorem: Angular velocity of an electron in a magnetically focused electron beam is proportional to the difference of magnetic flux at the electron beam cross section, where the electron first encounters the focusing magnetic field, and at the cross section that includes the observation point. It is inversely proportional to the beam cross-sectional area at the observation point. The proportionality constant is 0.879×10^{11} C/kg.

Characteristic impedance: Ratio of a transmission line voltage of a forward voltage traveling wave to the accompanying transmission line current of a forward current traveling wave.

Collector: An electron collector; a positive electrode against the cathode. It is located at the final termination of the electron beam in a vacuum tube to collect electrons finished by the electronic process of the vacuum tube.

Coulomb force: Electric force exerted on an electrically charged body, which is proportional to the amount of the charge and the electric field strength in which the charged body is placed.

Decibel (dB): Ten times the common logarithm of the power ratio or 20 times the common logarithm of voltage ratio.

Duty cycle: In a periodical on-off system, ratio of the on time duration to the entire period of time of repetition.

Electron beam: Collimated and focused stream of electrons.

Electron beam coupling: Electromagnetic coupling between an electron beam and an electric circuit or a transmission line.

Electron bunching: Gathering of electrons due to both the velocity modulation and the drifting of electrons.

Electron gun: An electronic structure that contains a thermionic cathode, an accelerator, and a focusing electrode to emit an electron beam.

Electronic countermeasure: Electronic equipment, system, or procedure to sense, detect, disturb, or destroy a transmitter by detecting the transmitter signals.

Emissitivity: Ratio of emission current to the anode voltage under the given temperature of the cathode.

Focuser: A focusing electrode for an electron steam in a vacuum tube.

Focusing electrode: A special purpose electrode or an assembly of electrodes negatively charged to focus the electron beam by pushing electrons in desirable directions.

Focusing magnetic fields: Moving electrons are pushed by applied magnetic force or Lortentz force, when the direction and strength of magnetic fields are properly adjusted. It is possible to focus an electron beam in a desired direction. Such magnetic fields are termed the focusing magnetic fields.

Forward waves: In a traveling wave tube, microwaves propagate in the same direction with electrons in the electron beam. Electromagnetic waves travel in the positive direction of the coordinate system.

Frequency bandwidth: A range of frequencies within which the variation of the gain of the device is less than 3 dB.

Flicker noise: Electrical noise current in an electron stream due to stochastic variation of a heated cathode surface; it is inversely proportional to operating frequency of interest, the $1/f$ noise.

Gain parameter: Neper gain constant per meter of propagation.

Hollow beam: An electron beam with hollow core.

Immersed flow: A flow of electrons emitted from an electron gun exposed to the focusing magnetic fields.

Longitudinal phase velocity: Phase velocity of microwave propagation in the axial direction of a slow-wave structure of a traveling wave tube.

Microwave voltage controlled oscillator (MVCO): A backward wave oscillator (BWO) is an example.

Neper: A natural logarithm of a ratio.

Newton's equation of motion: The force is equal to a product of the mass and the acceleration.

Noise: Electrical noise that is unwanted and/or undesirable electrical variations.

Noise temperature: For a passive device, it is the ratio of noise power to the product of Boltzmann's constant and operating frequency bandwidth. For an amplifier, it is the ratio of output noise power to a product of Boltzmann's constant, power gain, and operating frequency bandwidth.

Noise figure: Ratio of actual noise output of an actual two-port network to fictitious ideal noise output of an ideal network that does not produce noise within the circuit but has transferable noise from the input. The ratio of the input S/N ratio to the output S/N ratio.

Permittivity: A ratio of the electric flux density to the electric field strength.

Partition noise: Electrical noise generated within a vacuum tube when the electron stream hits obstacles and is divided.

Pierce gun: An electron gun with which the focusing electrode is tilted 67.5° from the axis of the electron beam and the accelerator anode is also tilted forward to produce a rectilinear flow of an electron beam.

Pitch angle of a helix: An angle between a tangent to the helix and another tangent to a cylinder that contains the helix and is perpendicular to the cylinder axis at a common tangential point on the helix.

Power gain: A ratio of the output power to the input power. The ratio can be expressed by a number, in decibel, or in neper.

Richardson–Dushman equation: An equation of electron saturation emission current density from a heated cathode. It states that the emission electron current density is proportional to the square of the cathode temperature and an exponential of negative constant times a reciprocal of the absolute temperature of the cathode. The proportionality constant is 6.02×10^{-3} and the negative constant depends on the cathode material.

Sensitivity: Amount of input signal power to a two-port network that produces a designed signal-to-noise ratio at the output.

Shot noise: Electrical noise presents an electron current due to stochastic modulation of electronic emission from the source.

Slow-wave structure: A short microwave transmission line in a traveling wave tube in which the longitudinal phase velocity of traveling microwave is slowed down to almost equal speed of electrons in the interacting electron beam of the tube.

Tangential sensitivity: Amount of input signal power to a two-port network to produce the output signal-to-noise power ratio unity.

Thermal noise: Electric noise power that is due to heat. It is equal to the product of Boltzmann's constant, the absolute temperature of the subject, and the operating frequency bandwidth.

Thermionic emission: Electron emission from a heated cathode.

Traveling wave: Electromagnetic waves propagating into a specified direction coherently.

Traveling wave tube (TWT): A microwave amplifier vacuum tube that is based on interaction between the traveling microwave and the electron beam.

Velocity modulation: A process that induces variation of velocities to electrons in an electron beam with otherwise uniform velocity.

Voltage gain: Ratio of the output voltage of an amplifier to the input voltage to the amplifier. The voltage gain is expressed by a number, in decibel, or in Neper.

Work function: Amount of energy necessary to take out an electron from a material.

References

Chodorow, M. and Susskind, C. 1964. *Fundamentals of Microwave Electronics*. McGraw-Hill, New York.

EEV. 1991. *Microwave Products*. Chelmsford, England.

EEV. 1995. *Traveling Wave Tubes and Amplifiers*. Chelmsford, England.

Harman, W.W. 1953. *Fundamentals of Electronic Motion*. McGraw-Hill, New York.

Hutter, R.G.E. 1960. *Beam and Wave Electronics in Microwave Tubes*. D. Van Nostrand, Princeton, NJ.

Ishii, T.K. 1989. *Microwave Engineering.* Harcourt Brace Jovanovich, San Diego, CA.
Ishii, T.K. 1990. *Practical Microwave Electron Devices.* Academic Press, San Diego, CA.
Liao, S.Y. 1988. *Microwave Electron Tube Devices.* Prentice-Hall, Englewood Cliffs, NJ.
Sims, G.D. and Stephenson, I.M. 1963. *Microwave Tubes and Semiconductor Devices.* Interscience, London.

Further Information

Additional information on the topic of traveling wave tubes is available from the following sources.

IEEE Electron Device Meeting Digest.
IEEE Transactions on Electron Devices, published by the Institute of Electrical and Electronic Engineers Inc., 345 East 47th St., New York, NY 10017.
Manufacturers' catalogs such as from EEV, Hughes, Thomson Tubes Electroniques, Varian, and NEC.

6.4   Other Microwave Vacuum Devices

T.K. Ishii

6.4.1   Introduction

Thus far in Sections 6.1–6.3 planar microwave tubes, **klystrons**, and traveling wave tubes in microwave vacuum devices have been covered. There are a great number of other microwave devices available, and it is almost impossible to cover every available microwave vacuum device in detail. Therefore, in this chapter, some selected microwave vacuum devices are presented in some detail. These microwave devices include **magnetrons, backward wave oscillators (BWO), carcinotrons, twystrons, laddertrons, ubitrons, gyrotrons, helical beam tubes, amplitrons,** and **platinotrons** (Ishii, 1989).

6.4.2   Magnetrons

Magnetrons are used for high-power applications. The device is widely used for radar transmitters and as a power source for microwave heating processes. Though one of the most common types of magnetron is a radial magnetron, the linear magnetrons and inverted magnetrons may also be used depending on the applications. Both magnetrons and inverted magnetrons are oscillators, and the linear magnetron is an amplifier.

Schematic diagrams of cross-sectional views of linear magnetrons are shown in Fig. 6.19(a) and Fig. 6.19(b). In the **O-type linear magnetron** in Fig. 6.19(a), the **electron beam** emitted from the **electron gun** is focused by a longitudinally applied DC magnetic flux density B as is the case in the traveling wave tube. In Fig. 6.19(a), both the anode and the **collector** are electropositive against the electron gun and the **sole**. The sole is a nonemitting cathode. In practice, for safety reasons, the anode block is grounded and the cathode and the sole are biased to a high negative voltage. In this biasing arrangement, the elementary trajectory of an electron in an O-type linear magnetron is a helix. But as a whole, it is considered to be a uniform flow electron beam if there is no microwave excitation. As seen from Fig. 6.19(a), the anode block has a number of slots. These slots are cut one-quarter wavelength deep. The slots are quarter-wavelength cavity resonators. The structure is a series of microwave **cavity resonators** coupling to an electron beam. This is similar to the multicavity klystron presented in Chap. 31. Therefore, the principles of a **linear magnetron** are similar to the principles of a multicavity klystron. The input microwaves impart velocity modulation to the electron beam and distributes microwave energy to the first cavity and following cavities. The **velocity modulated** electrons are bunched, and the tightly bunched electrons induce a large amount of amplified microwave energy at the output cavity. The energy in the output is coupled to the output circuit. Usually a linear magnetron is an amplifier of high gain, but it is also an amplifier of a narrow frequency bandwidth.

When the orientation of the electron gun and the direction of focusing magnetic flux density B are changed, as shown in Fig. 6.19(b), the magnetron is termed as **M-type linear magnetron.** In this arrangement, the elementary trajectory of an electron in the beam is a cycloid.

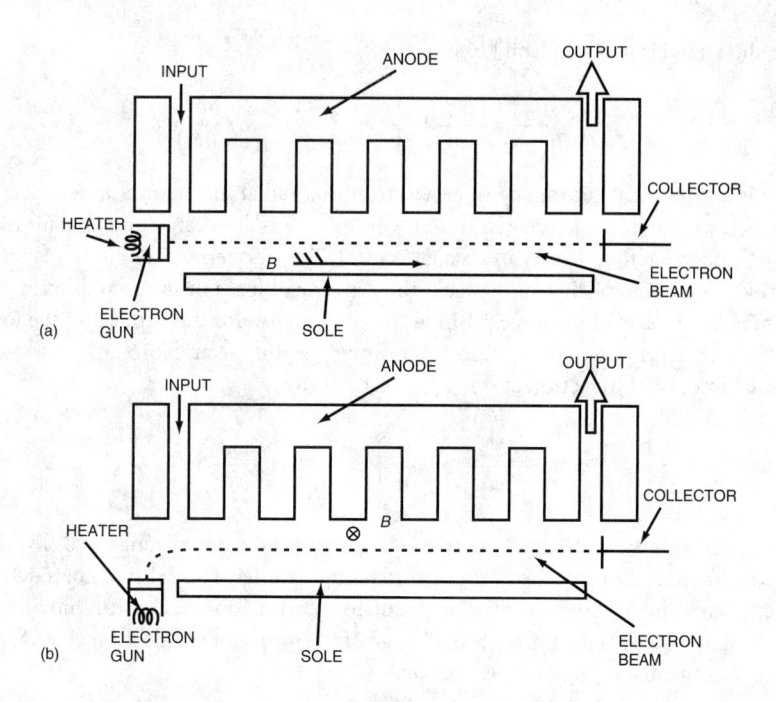

FIGURE 6.19 Cross-sectional view of linear magnetrons: (a) O-type linear magnetron, (b) M-type linear magnetron.

If the slots in the anode are not resonating, the tube is no longer a linear magnetron. If many nonresonating slots are cut in the anode block, then the device is a traveling wave tube. This type of traveling wave tube is termed a **crossed field amplifier (CFA)** or **crossed field device (CFD)**.

If the anode block of M-type linear magnetron is bent and made re-entrant, in which the end of the structure is connected to the beginning of the structure as shown in Fig. 6.20, then the magnetron is termed a **radial magnetron**. In the radial magnetron, there is no nonemitting sole. The cathode is an emitting hot cathode. Electrons emitted from the hot cathode are pulled toward the anode by the radial electric field. Therefore, the radial force is expressed in a cylindrical coordinate coaxially set up with the radial

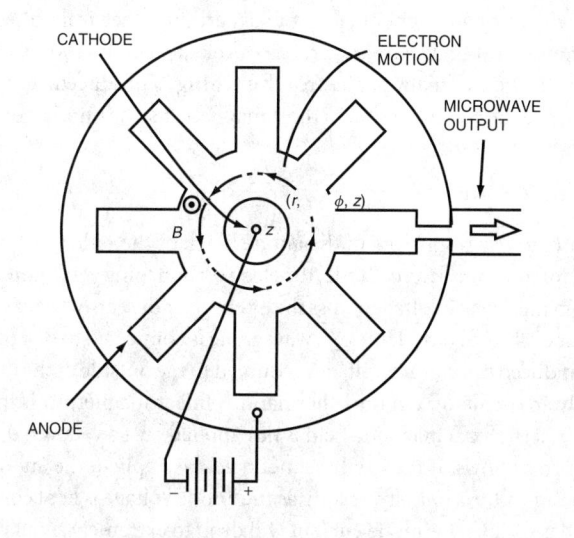

FIGURE 6.20 Cross-sectional view of a radial linear magnetron.

magnetron as shown in Fig. 6.20. In Ishii (1989)

$$m\left\{\frac{\partial^2 r}{\partial t^2} - r\left(\frac{\partial \phi}{\partial t}\right)^2\right\} = q\left\{\frac{\partial V}{\partial y} - Br\frac{\partial \phi}{\partial t}\right\} \tag{6.83}$$

where (r, ϕ, z) is a cylindrical coordinate of an electron in a radial magnetron, m is an electronic mass 9.1091×10^{-31} kg, q is the electronic charge 1.6021×10^{-19} C, V is the electric potential in volts at (r, ϕ, z), and B is the DC magnetic flux density in tesla at (r, ϕ, z). The first term of Eq. (6.83) is the radial force toward the anode, which inwardly counteracted the centripetal force in the second term of Eq. (6.83). The third term is the electric radial force pushing the electron toward the anode and the fourth term of Eq. (6.83) is the magnetic force counteracting inwardly against the electric force.

The tangential force on the electron at (r, ϕ, z) is (Ishii, 1989)

$$r\left\{q\frac{1}{r}\frac{\partial V}{\partial \phi} + qB\frac{\partial r}{\partial t}\right\} = \frac{\partial}{\partial t}\left(rmr\frac{\partial \phi}{\partial t}\right) \tag{6.84}$$

The first term on the left-hand side of Eq. (6.84) is the torque due to the tangential electric force of an electron at (r, ϕ, z) against the center axis. The second term on the left-hand side of Eq. (6.84) is the torque against the center axis due to the tangential magnetic force acting on the electron moving radially. The right-hand side of Eq. (6.84) is the time rate of change of the torque of the electron at (r, ϕ, z) against the center axis due to tangential motion of the electron.

The energy of the electron at (r, ϕ, z) is (Ishii, 1989)

$$\frac{1}{2}m\{\dot{r}^2 + (r\dot{\phi})^2\} = qV \tag{6.85}$$

The first term on the left-hand side is the radial **kinetic energy** and the second term is the tangential kinetic energy. The right-hand side of Eq. (6.85) is the electric potential energy of the electron at the specific location (r, ϕ, z). Therefore, the electric potential V in Eqs. (6.83–6.85) should contain a solution of **Poisson's equation** and a solution of **Helmhortz's wave equation**. Since the work is very involved, however, most designers use a solution of **Laplace's** equation only for the first approximation. Solving Eqs. (6.83–6.85), the trajectory of an electron in a radial magnetron will be known. As a result, electrons emitted from the cathode surface in all directions circle around the cathode forming an **electron cloud**. The relationship between the electron cloud and the cavities in the anode of a radial magnetron is similar to the electron beam in the M-type linear magnetron to its cavity resonators. A small microwave voltage induced at the opening of the cavity resonator generated from background electrical noise, such as **thermal noise**, **shot noise**, **partition noise**, and **flicker noise**, produces velocity modulation to the circulating electron cloud. Because of this velocity modulation, **electron bunching** is produced. The bunched electrons in a radial magnetron are termed the **electron pole**. In most magnetrons, the number of electron poles formed is

$$n = \frac{N}{2} \tag{6.86}$$

where N is the number of cavity resonators in the anode block of the radial magnetron.

As the electron poles rotate around the cathode, the electron bunching gets tighter. The high-density electron pole produces a high microwave voltage across the opening or the gap of the cavity resonator. Thus, if an output transmission line such as a coaxial line or a waveguide is connected to one of the cavity resonators as shown in Fig. 6.20, the induced microwave voltage is coupled to the outside of the magnetron. Thus, a radial magnetron is inherently an oscillator. On the other hand, a linear magnetron is inherently an amplifier.

A radial magnetron is, if the axial magnetic field is not applied, just a vacuum diode. If an anode voltage is applied, the anode current flows. If the axial magnetic field is applied, the anode current keeps flowing approximately the same amount as initially, as long as the anode voltage is kept constant. Then, at a certain value of magnetic flux density B_c, the anode current will drop to extremely small value. This indicates that the magnetic force has become so large that the **electron orbit** has been bent excessively. Electrons cannot reach the anode with this amount of magnetic force. The magnetic flux density that causes the anode current to drop sharply is termed the **cutoff magnetic flux density**. Magnetron oscillation is frequently found near the cutoff magnetic flux density. Oscillation under extremely small anode current is one of the features of a radial magnetron that provides high efficiency. The relationship among the anode voltage V_a, the cutoff magnetic flux density B_c, the radius of the cathode r_c, and the radius of the anode V_a is given by (Ishii, 1989)

$$B_c = \sqrt{45.5 V_a} \frac{r_a}{r_a^2 - r_c^2} \tag{6.87}$$

The high efficiency of radial magnetrons stems from this low operating anode current as well as the **re-entrant use** of electron poles. In practice, whenever the high efficiency and high-power generation of microwave power is considered, magnetrons are the first to be considered. Radar and industrial and domestic microwave ovens are examples of applications of radial magnetrons.

In the radial magnetron shown in Fig. 6.20, the output waveguide is connected to only one of the cavity resonators in the anode block. The output transmission line can be connected to all of the cavity resonators if the cavity resonators form a coaxial configuration with the outer jacket to which the output transmission line is connected, as shown in Fig. 6.21(a). This type of magnetron is termed the **coaxial magnetron**. Microwaves generated by the rotating electron poles at the anode vane cavity resonators are coupled to the output coaxial cavity through intervane tip gaps and slits cut on the anode ring between the vane cavities, as shown in Fig. 6.21(a). All vane cavity resonators are thus connected to a coaxial cavity consisting of the conducting outer jacket and the anode ring. Thus, if a transmission line is coupled to this coaxial cavity, generated microwaves are coupled to the output circuit through the output transmission line in a side way as shown in Fig. 6.21(a). An alternate coupling configuration to the output is in the direction perpendicular to the plane of the drawing. The outer jacket is tapered to an outer conductor of the output coaxial line and the slitted anode ring is tapered into the center conductor of the output coaxial line. The side opposite to the output is terminated by a conducting endplate at an impedance matching location, which is approximately one-quarter wavelength from the active part of the coaxial magnetron.

To obtain a large emission current for higher power, the electron emitting hot cathode can be the inside wall of the outermost cylinder as shown in Fig. 6.21(b). This type of coaxial magnetron is termed the **inverted coaxial magnetron**. Microwave cavity resonators formed from a number of anode vanes are tied together with a cylindrical slitted anode ring. Through the slits in the slitted anode ring, microwave power induced in the vane cavity resonators is coupled to the central coaxial cavity, which is formed from the slitted anode ring and the center conductor (which is not emitting electrons). Therefore, if the center conductor is tapered to a center conductor of an output coaxial line and the slitted anode ring is tapered into an outer conductor of the output coaxial line, then the output will be taken from the coaxial line in the direction perpendicular to the plane of the drawing. The side opposite to the output coaxial line of the inverted coaxial magnetron is terminated by a conducting endplate at an impedance matching location, which is approximately one-quarter wavelength away from the active part of the inverted coaxial magnetron.

6.4.3 Backward Wave Oscillators

In a traveling wave tube, if the microwave signal to be amplified is propagating in the slow-wave structure backwardly to the direction of the electron beam, as shown in Fig. 6.22(a), the device is termed a **backward wave oscillator** (BWO). When the electron beam is focused through the use of longitudinal magnetic flux density as shown in Fig. 6.22(a), the device is *O*-type *BWO*. Regardless the direction of microwave propagation in the slow-wave structure, electrons are velocity modulated in the neighborhood of the electron gun and bunched as the electrons travel toward the collector, and the bunched electrons induce a large microwave voltage in the slow-wave structure at a location near the collector. Microwaves traveling backward carry positive feedback energy toward the electron gun for stronger velocity modulation and bunching. Thus, the system is inherently an oscillator rather than a stable amplifier. It can be used as a high-gain amplifier, but usually it is used as a microwave voltage controlled oscillator (MVCO). The input

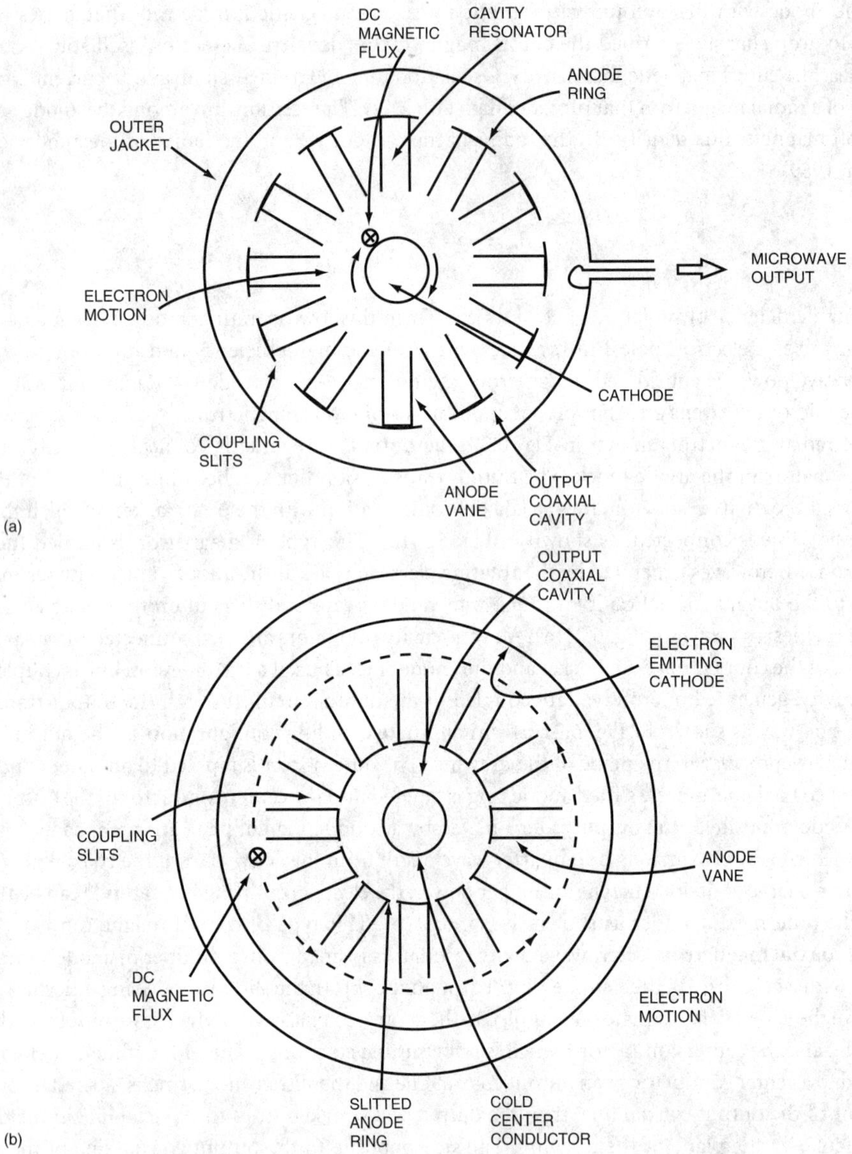

FIGURE 6.21 Schematic diagrams of a co-axial magnetron and an inverted coaxial magnetron: (a) coaxial magnetron, (b) inverted coaxial magnetron.

is typically terminated by an impedance-matched reflectionless termination, as shown in Fig. 6.22. The oscillation frequency is determined by the speed of electrons and the time constant of the feedback. The speed of electrons is controlled by the anode voltage.

The BWO can take a form of an *M-type radial BWO* as shown in Fig. 6.22(b). The direction of electron pole motion and the direction of microwave propagation along the annular re-entrant type slow-wave structure are opposite to each other. It should be noted that the depth of the slits cut in the inner surface of the anode is very shallow and it is much less than one-quarter wavelength deep. In other words, the slits are not in resonance or the slits are not cavity resonators as in the case of magnetrons; rather the slits are not resonating, as is in the case of traveling wave tubes. In the M-type radial BWO, the electron beam is focused by a magnetic flux density applied perpendicular to the electron beam, as seen from Fig. 6.22(b). Again

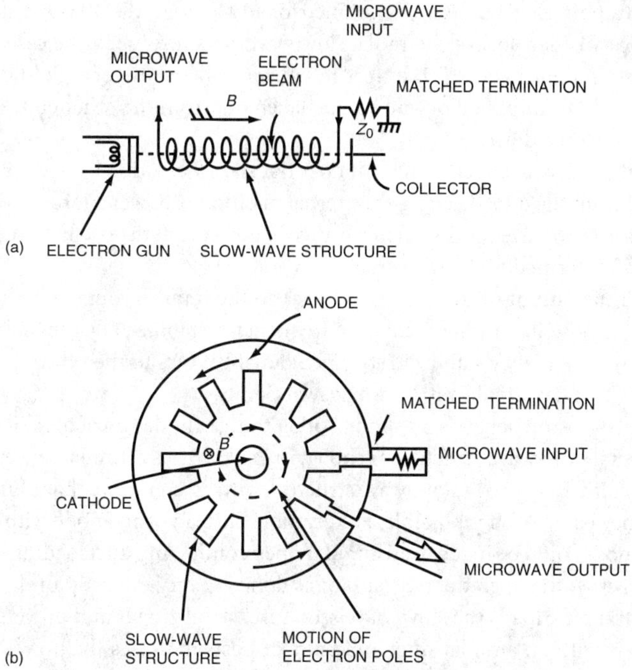

FIGURE 6.22 Schematic diagrams of backward wave oscillators: (a) O-type linear BWO, (b) M-type radial BWO.

the input terminal of a M-type radial BWO is usually terminated by an impedance-matched reflectionless termination. An M-type radial BWO is sometime termed the carcinotron by a trade name. A key feature of the carcinotron is its wide voltage tunability over a broad frequency range.

6.4.4 Strap-Fed Devices

In a radial magnetron, sometimes every other pole of the anode resonators is conductively tied for microwave potential equalization, as shown in Fig. 6.23(a). These conducting tie rings are termed **straps**; this technique of using strap rings is termed *strapping*. Strapping ensures good synchronization of microwaves in the magnetron's resonators with the rotation of electron poles.

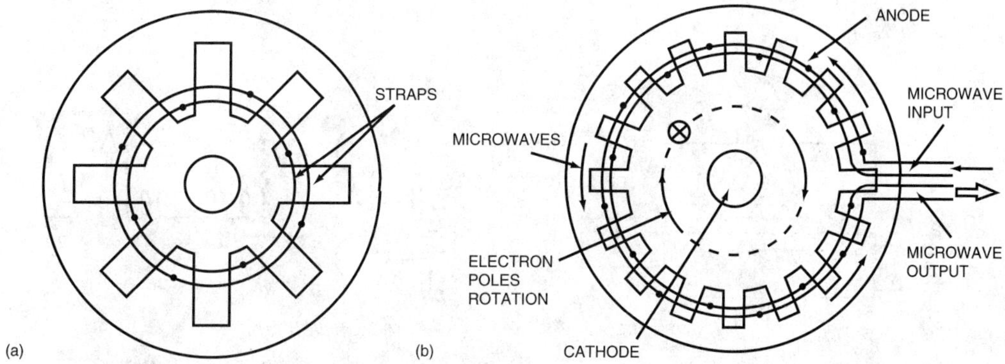

FIGURE 6.23 Strapped devices: (a) strapped radial magnetron, (b) nonre-entrant strapping of a BWO.

The technique of strapping is extended and modified for an M-type radial BWO, as shown in Fig. 6.23(b). Strapping rings tie every other pole of the radial slow wave structure, as in the case of a strapped radial magnetron, but the strapping rings are no longer re-entrant. Microwave energy to be amplified is fed to the strap at one end and the amplified output can be taken out from the other end. This type of electron device is termed the strap-fed device.

If an M-type radial BWO is strapped, usually it does not start oscillation by itself. But if microwaves are fed through the strap from the outside using an external microwave power source to the microwave input, then the oscillation starts and even if the exciter source is turned off, the oscillation continues. This type of M-type radial BWO is termed the **platinotron** (Ishii, 1989).

In a platinotron, if the output of the tube is fed back to the input through a high-Q cavity resonator, it becomes a self-starting oscillator. The oscillation frequency is stabilized by the high-Q cavity resonator. This type of high-power frequency stabilized strapped radial BWO is termed the *stabilotron* (Ishii, 1989). The operating power levels are at kilowatt and megawatt levels.

Performance of a platinotron depends on, among other things, the design of the slow-wave structure. For example, the interdigital slow-wave structure as shown in Fig. 6.23 has a limited power handling capability and frequency bandwidth. Design of a slow-wave structure with greater power handling capacity and stable with broader frequency bandwidth is possible. For example, instead of an anode with an interdigital slow-wave structure, the anode could be made of an annular open conducting duct loading with number of pairs of conducting posts across the open duct. Strapping is done at every other tip of the pairs of conducting posts. This type of strapping loads the slow-wave structure, stabilizing it and preventing oscillation. The structure of the anode with an annular duct and pairs of posts increases the power handling capability. This type of loaded radial BWO is termed the *amplitron* (Ishii, 1989; Liao, 1988). The amplitron is capable of amplifying high-power microwave signals with pulses and **continuous waves**. It is used for long-range pulsed radar transmitter amplifiers and industrial microwave heating generators. The operating power levels are at kilowatt and megawatt levels.

6.4.5 Modified Klystrons

Instead of using a multiple number of cavity resonators as in a multicavity klystron, the cavities may be combined to a single cavity, as shown in Fig. 6.24(a). Multiple cavities are hard to fabricate at millimeter wavelengths. A multimode single cavity as shown in Fig. 6.24(a) is more convenient. Beam coupling is accomplished through the use of ladder-shaped grids, as shown by the dashed lines with the single cavity in Fig. 6.24(a). This type of klystron is termed the laddertron (Ishii, 1989). There will be a standing wave on the ladder line. Velocity modulation is performed in the ladder structure near the electron gun and bunching begins near the middle of the electron stream. Catching is done in the ladder structure near the collector. Since this is only a one-cavity resonator, there is a built-in positive feedback, and the oscillation starts like a reflex klystron, which has a single cavity. At millimeter wavelengths, a dominant mode cavity resonator becomes very small and loses its power handling capability. In a laddertron, the cavity resonator

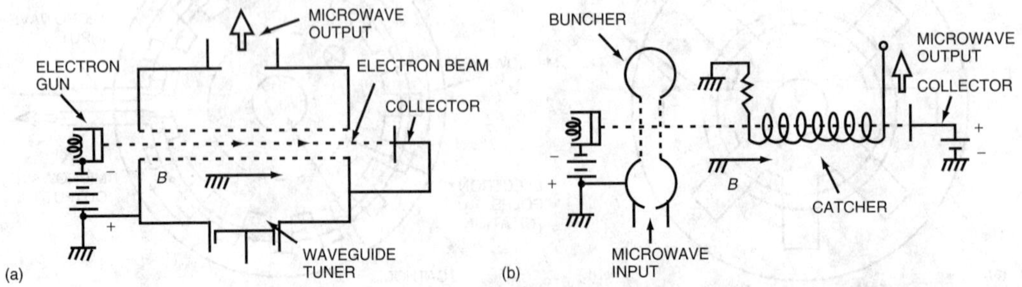

FIGURE 6.24 Schematic diagrams of modified klystrons: (a) laddertron, (b) twystron.

is a multiple higher mode resonator. Therefore, the quality factor Q of the cavity is high and the size of the cavity resonator is reasonably large. It can handle tens of watts of power at frequency as high as 50 GHz.

The catcher of a two-cavity klystron amplifier can be replaced by the slow-wave structure of a traveling wave tube, as shown in Fig. 6.24(b). This type of tube is termed the *twystron* (Ishii, 1989). The slow-wave structure provides a broader frequency bandwidth than a regular two-cavity klystron. A microwave input fed into the buncher cavity produces velocity modulation to the electron beam. The electron beam is bunched while drifting, and the bunched electrons induce a microwave voltage in the microwave slow-wave structure. The electron beam is focused by the use of longitudinally applied magnetic flux density B.

Commercial twystrons operate in the **S-band** (2.6–3.95 GHz) and **C-band** (3.95–5.85 GHz), and are used for pulsed radar transmitter power amplifiers. The pulsed peak output ranges from 1 to 7 MW with a pulse width of 10–50 μs. This results in an average output power in the range of 1–28 k W with efficiency of 30–35%. The tube requires a beam voltage of 80–117 kV and a beam current of 45–150 A. The microwave input power to drive the amplifier twystron is in a range of 0.3–3.0 kW (Liao, 1988).

6.4.6   Electron Gyration Devices

As seen in an O-type traveling wave tube, electron trajectory of an electron in an electron beam focused by a longitudinally applied magnetic field is a helix. If the electron velocity, **electron injection angle** and applied longitudinal magnetic flux density are varied, then an electron beam of helical form with different size and pitch will be formed. A coil-shaped electron beam will be produced by adjusting the acceleration voltage, applied magnetic flux density, and the electron injection angle to the focusing magnetic field. The coil of the electron beam can be a simple single coils; or, depending on the adjustment of the aforementioned three parameters, it can be an electron beam of a double coil, or a large coil made of thin small coils. In the case of the double-coil trajectory, the large coil-shaped trajectory is termed the **major orbit** and smaller coil trajectory is termed the **minor orbit**.

If a single coil-shaped electron beam is launched in a waveguide as shown in Fig. 6.25, then microwaves in the waveguide interact with the helical beam. This type of vacuum tube is termed the helical beam tube (Ishii, 1989). A single-coil helical beam is launched into a TE_{10}-mode (Transverse Electric one-zero mode) rectangular waveguide. Inside of the waveguide, microwaves travel from right to left and the helical beam travels in an opposite direction. Therefore, the microwaves–electron beam interaction is the **backward wave interaction**. If the

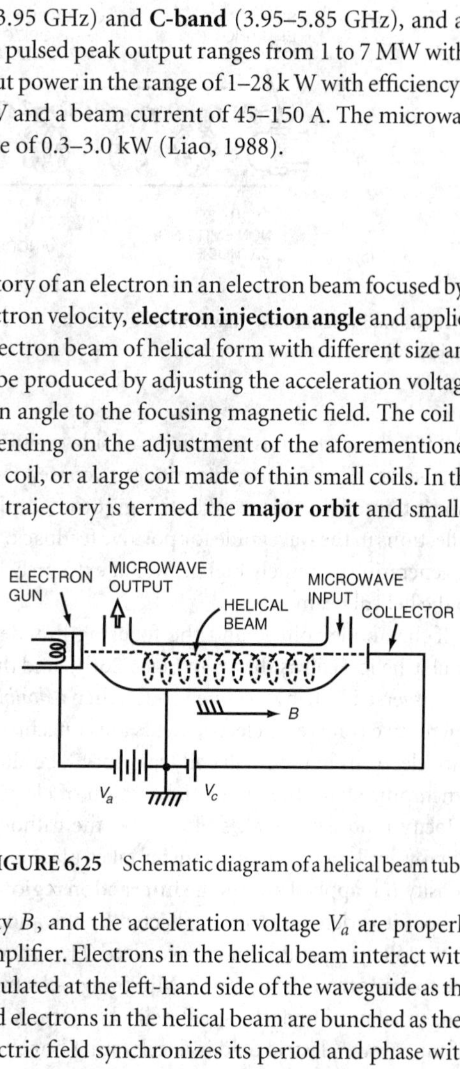

FIGURE 6.25   Schematic diagram of a helical beam tube.

microwave frequency, the focusing magnetic flux density B, and the acceleration voltage V_a are properly adjusted, this device will function as a backward wave amplifier. Electrons in the helical beam interact with the transverse microwave electric fields and velocity modulated at the left-hand side of the waveguide as the beam enters into the wave-guide. The velocity modulated electrons in the helical beam are bunched as they travel toward the right. If the alternating microwave electric field synchronizes its period and phase with the helical motion of bunched electrons so that the electrons always receive retardation from microwave transverse electric fields, then the electrons loose their kinetic energy and microwave gains in the electric field energy by the **kinetic energy conservation principles**. Thus, the amplified microwave power emerges at the waveguide output at the left, since the microwaves travel backward. The amplified microwaves at the left give strong velocity modulation to the helical electron beam at the left.

In Fig. 6.25, if the microwave input port and the output port are interchanged with each other, then the system becomes a forward wave amplifier. Such a forward wave amplifier is termed a peniotron (Ishii, 1990).

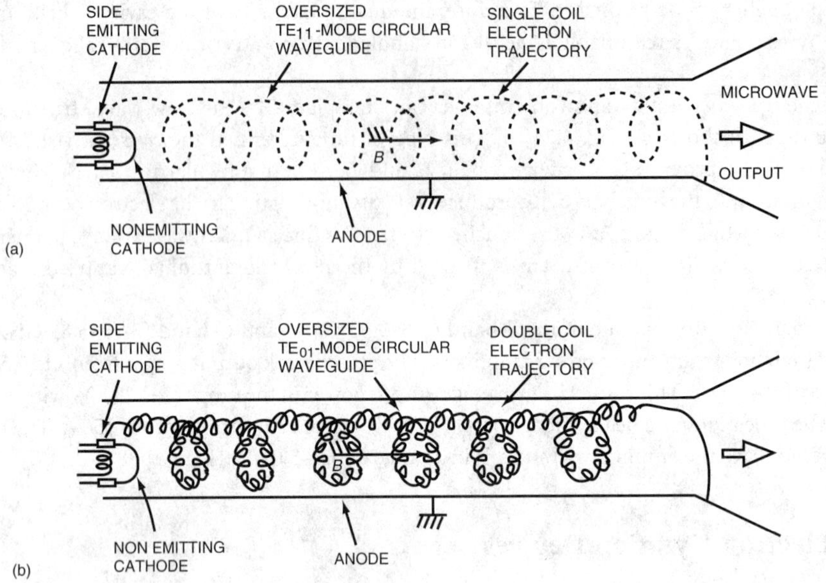

FIGURE 6.26 Schematic diagram of gyrotrons: (a) single coiled helical beam gyrotron, (b) electron trajectory of double-coil helical beam gyrotron.

If the electron gun is modified to a side emitting cathode and the waveguide is changed to **TE$_{11}$-mode** (Transverse Electric one-one mode) oversized circular waveguide as shown in Fig. 6.26(a), this type of tube is termed the gyrotron. In this gyrotron, both ends of the waveguide are open and there are sufficient reflections in the waveguide for positive feedback. The gyrotron is the **forward wave oscillator**. It is capable of generating extremely high power at extremely high frequency. The principles of operation are identical to the helical beam tubes.

If the anode voltage and the focusing flux density are readjusted, the electron beam is made into a double helical coil as shown in Fig. 6.26(b), and the oversized circular waveguide is operated in **TE$_{01}$** mode (Transverse Electric zero-one mode), then a *double-coil helical beam gyrotron* is formed. In the TE$_{01}$ mode, microwave transverse electric fields are concentric circles. Therefore, the tangential electric fields interact with electrons in the small coil trajectory. The alternating tangential microwave electric fields are made to synchronize with the tangential motion of electrons in the minor coil-shaped trajectory. Thus, electron velocity modulation takes place near the cathode and bunching takes place in the middle of the tube. Microwave kinetic energy transfer takes place as beam approaches the right. The focusing magnetic flux density B is applied only in the **interaction region**. Therefore, if electron beam comes out of the interaction region, it is defocused and collected by the anode waveguide as depicted in Fig. 6.26(b). If the circular waveguide is operated in an oversized TE$_{11}$ mode, with the double-coil helical beam, then it is termed the **tornadotron** (Ishii, 1989). Microwave–electron interaction occurs between the parallel component of tangential motion of the small helical trajectory and the TE$_{11}$-mode microwave electric field. If the phase of microwave electric field is decelerating bunched electrons, then the lost kinetic energy of the bunched electrons is transferred to the microwave signal and the oscillation starts.

In a single-coiled helical beam gyrotron as shown in Fig. 6.26(a), frequency of the helical motion of electrons is the **cyclotron frequency** and it is given by (Ishii, 1990)

$$f_c = \frac{qB}{2\pi m} \tag{6.88}$$

where q is the charge of the electron, 1.6021×10^{-19} C; B is the longitudinal magnetic flux density in tesla; and m is the mass of an electron, 9.1091×10^{-31} kg. The amount of necessary magnetic flux density

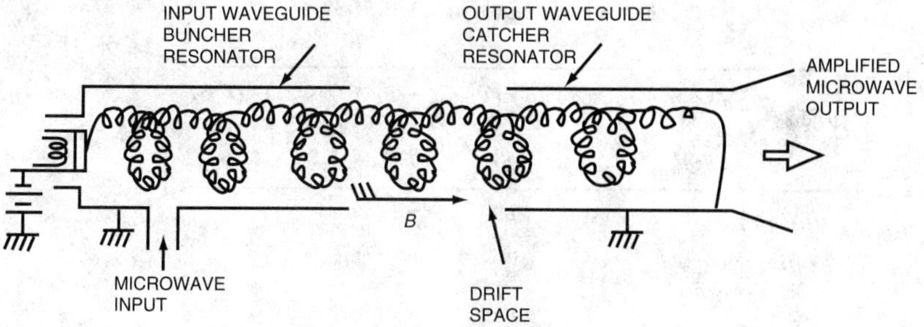

FIGURE 6.27 The gyroklystron amplifier.

to produce desired cyclotron frequency f_c is, then (Ishii, 1990)

$$B = \frac{2\pi m f_c}{q} \qquad (6.89)$$

The radius of curvature r of the helical motion is given by (Ishii, 1990)

$$r = \frac{u_\phi}{2\pi f_c} \qquad (6.90)$$

where u_ϕ is the tangential velocity of the electron and f_c is the cyclotron-frequency of the electron.

In the double coiled helical beam gyrotron, the cycloidal frequency f of the double coiled trajectory is calculated to be

$$f = \frac{qB}{2\pi mc} \qquad (6.91)$$

This cycloidal frequency must be made equal to the operating microwave frequency.

Gyrotrons are capable of producing high microwave power on the order of 100 kW at microwave frequency on the order of 100 GHz.

There are a family of gyrotrons. Gyrotrons with a single waveguide, as shown in Fig. 6.26 are termed **gyromonotron**. There are many other variations in gyrotrons (Coleman, 1982; McCune, 1985).

When the circular waveguide is split as shown in Fig. 6.27, the tube is termed the **gyroklystron** amplifier. Both waveguides resonate to the input frequency. There are strong standing waves in both waveguide resonators. The input microwave signal to be amplified is fed through a side opening to the input waveguide resonator. This is the buncher resonator as in the case of a klystron. This buncher resonator gives velocity modulation to gyrating electrons in the double helical coil-shaped electron beam. There is a **drift space** between the buncher resonator and the **catcher** resonator at the output. While drifting electrons bunch and bunched electrons enter into the output waveguide catcher resonator, electron speed is adjusted in such a way that electrons are decelerated by the resonating microwave electric field. The lost kinetic energy in bunched electrons is transformed into microwave energy and microwaves in catcher resonator are amplified. The amplified power appears at the output of the tube.

If the waveguide is an unsplit one piece waveguide that is impedance matched and not resonating, as shown in Fig. 6.28, the tube is termed the **gyrotron traveling wave tube amplifier**. In this tube the input microwaves are fed through an opening in the waveguide near the electron gun, as shown in Fig. 6.28. Microwaves in the waveguide are amplified gradually as they travel toward the output port by interacting with the double coiled helical electron beam, which is velocity modulated and bunched. There are no significant standing waves in the waveguide. Microwaves grow gradually in the waveguide as they travel toward the output port by interaction with electrons.

If the electron gun is moved to the side of the waveguide and the microwave power is extracted from the wave-guide opening in proximity to the electron gun, as shown in Fig. 6.29, then it is termed a **gyrotron**

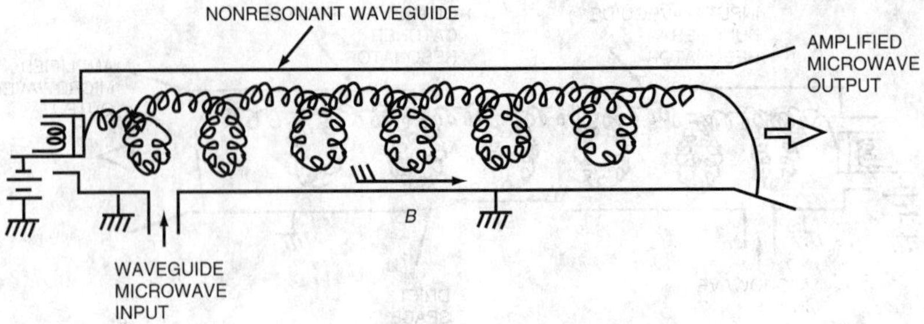

FIGURE 6.28   Gyrotron traveling wave tube amplifier.

backward oscillator. The principle is similar to the backward wave oscillator; the process of velocity modulation, drifting, bunching, and catching are similar to the klystron. Microwave energy induced in the waveguide travels in both directions. The circuit is adjusted to emphasize the waves traveling in a backward direction. The backward waves become the output of the tube and, at the same time, they carry the positive feedback energy to the electrons just emitted and to be velocity modulated. Thus, the system goes into oscillation. This is a backward wave oscillator in a gyrotron form.

If the waveguide is split into two again, but this time the input side waveguide is short and the output side waveguide is long, as shown in Fig. 6.30, then the tube is termed a **gyrotwystron** amplifier. The input side waveguide resonator is the same as the input resonator wave-guide of a gyroklystron, as shown in Fig. 6.27. There are strong standing waves in the input bunched **waveguide** resonator. There is no drift space between the two waveguides. The output side waveguide is a long impedance-matched waveguide and there is no microwave standing wave in the waveguide. This is a traveling wave waveguide. As microwaves travel in the waveguide, they interact with bunched electrons and the microwaves grow as they travel toward the output port. This is a combination of the gyroklystron and gyrotron traveling wave tube amplifier. Therefore it is termed the gyrotwystron amplifier.

6.4.7   Quasiquantum Devices

In a gyrotron, if electrons are accelerated by extremely high voltage, electrons become *relativistic*, or the mass of an electron becomes a function of the velocity as

$$m = \frac{m_0}{\sqrt{1 - \left(\frac{c}{v}\right)^2}} \tag{6.92}$$

where m is the relativistic mass of an electron, m_0 the static mass of an electron, c the speed of light in vacuum, and v the speed of the electron in question.

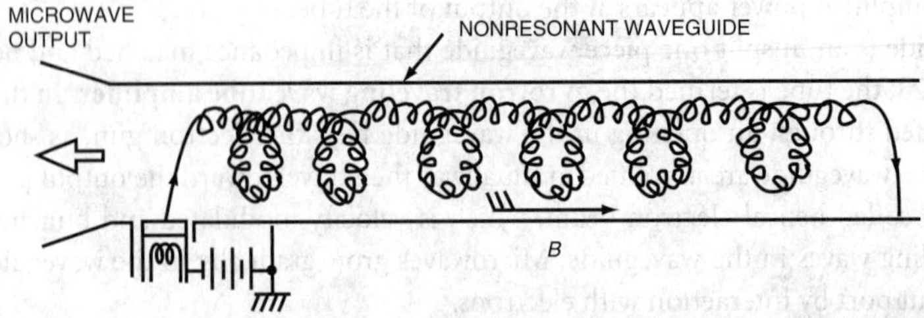

FIGURE 6.29   Gyrotron backward oscillator.

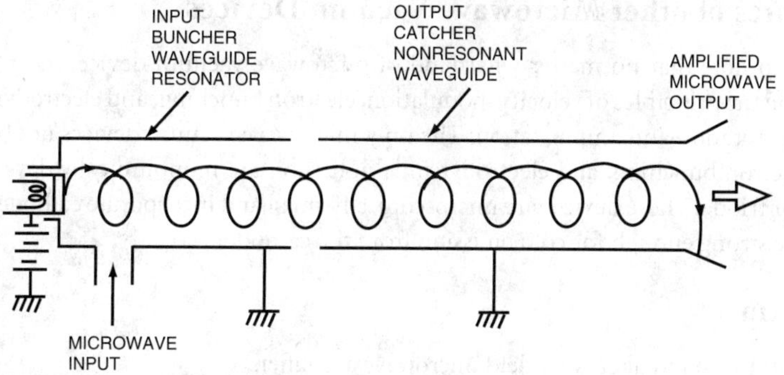

FIGURE 6.30 Gyrotwystron amplifier.

The energy states of electrons are specified by a set of **quantum numbers,** and the transition of energy state occurs only between the energy state described by a set of quantum numbers. The set of quantum numbers specify a high-energy orbit of an electron and a low-energy orbit of an electron. With the interaction with the waveguide or waveguide resonator at the cyclotron frequency, if the electrons are stimulated to induce a downward transition of high-energy orbit electrons to the low-energy orbit, then microwave radiation results. This is a *cyclotron frequency maser* (Ishii, 1989).

Another **quasiquantum** vacuum **device** is the *free electron laser,* or the *ubitron.* A schematic diagram of a ubitron is shown in Fig. 6.31. A high-speed relativistic electron beam is emitted from an electron gun focused by a longitudinally applied DC magnetic flux density B and periodically deflected by deflection magnetics. These repetitive deflections create among relativistic electrons high-energy states and low-energy states. If the high-energy electrons in the deflection waveguides are stimulated by the resonance frequency of the waveguide resonator, then the downward transition (which is the transition of an energy state of electrons in the high-energy state to the low-energy state) occurs. Then, there will be microwave emissions due to the energy transition at the stimulation frequency. The emission-frequency is

$$f = \frac{\Delta E}{h} \tag{6.93}$$

where ΔE is the energy difference between the high-energy state and the low-energy state in Joule and h is Planck's constant, which is 6.6256×10^{-34} J-s. The operating frequency of ubitrons are in millimeter wave or submillimeter wave frequencies and operate at a high-power level.

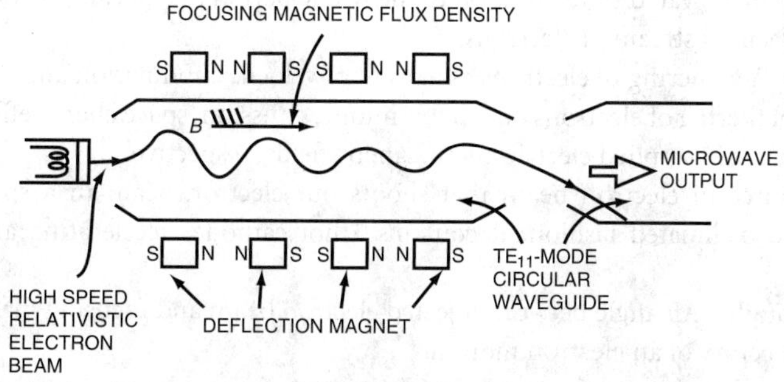

FIGURE 6.31 Schematic diagram of a ubitron.

6.4.8   Features of other Microwave Vacuum Devices

It is interesting to note that no matter what kind of microwave vacuum device, most electron devices function based on the principles of velocity modulation, electron bunching, and electrodynamic induction for microwave generation and amplification. The only microwave vacuum devices not based on velocity modulation, electron bunching, and electrodynamic induction are monolithically developed solid-state vacuum diode or triode. These devices are microscopically small and incorporate cathodes. The operating electric fields are strong enough for cold emission from the cathode.

Defining Terms

Amplitron: A strap fed radial crossed field microwave amplifier.

Backward wave interaction: Interaction between backward propagating microwave electric fields against an electron stream and the electron in the electron beam. The direction of propagating microwaves and the direction of motion of electrons in the beam are opposite each other.

Backward wave oscillators (BWO): A microwave oscillator tube that is based on a backward wave interaction.

Carcinotron: A forward radial traveling wave amplifier in which microwave signals are fed to the radial slow-wave structure.

Catcher: A cavity resonator of a multicavity klystron proximite to the collector to catch microwave energy from the bunched electrons.

Cavity resonator: A resonating space surrounded by conducting walls. An electromagnetic wave energy storage device.

C-band: Microwave frequency range, 3.95–5.85 GHz.

Coaxial magnetron: A radial magnetron where the anode and cathode are gradually transformed into a coaxial line.

Continuous wave: A wave of a constant and uninterrupted amplitude, unmodulated waves, carrier waves.

Collector: An electrode at the end of the electron beam in a microwave vacuum tube to collect electrons finished with the electronic processes.

Crossed field amplifier (CFA): Radial or linear traveling wave amplifier where radial or transverse DC accelerating electric fields are perpendicular to axial or longitudinal DC magnetic fields, respectively.

Crossed field devices (CFD): Radial or linear forward traveling wave amplifier or backward wave oscillator where radial or transverse DC accelerating electric fields are perpendicular to axial or longitudinal DC magnetic fields, respectively.

Cutoff magnetic flux density: In magnetrons operated under a constant anode voltage, a value of magnetic flux density at which the anode current decreases abruptly.

Cyclotron frequency: A frequency of electron oscillation in a cyclotron. A frequency of circular motion of an electron under magnetic fields applied perpendicularly to the plane of the circular motion.

Drift space: Space where electrons move only due to their inertia.

Duty cycles: A ratio of activated time duration to the whole period in a periodically activated system.

Electron beam: A focused stream of electrons.

Electron bunching: A gathering of electrons as a result of velocity modulation and drifting.

Electron cloud: A gathering of electrons due to thermionic emission, space charge effect, Coulomb force, and Lorentz force by applied electric and magnetic fields, respectively.

Electron gun: A source of electron beam that shoots out electron beam in a specified direction in a focused and collimated fashion. It contains a hot cathode, accelerating anode, and focusing electrodes.

Electron injection angle: An angle between injected electron beam and focusing magnetic fields.

Electron orbit: Trajectory of an electron motion.

Electron pole: In a radial magnetron, gathering of electrons formed by tangential velocity modulation and the electromagnetic fields. It rotates around the cathode.

Flicker noise: An electrical noise current contained in a thermal emission current due to stochastic change on conditions of cathode surface. For example in case of the $1/f$ noise, the noise power is inversely proportional to the operating frequency of interest.

Forward wave oscillator: A traveling wave amplifier with a positive feedback.

Gyroklystron: A gyrotron with a catcher waveguide and a buncher waveguide.

Gyromonotron: A gyrotron with a single resonating waveguide.

Gyrotron: A high-power millimeter wave generator based on transverse interaction on helically gyrating electrons against the millimeter wave electric field that is traveling longitudinally in a waveguide.

Gyrotron backward wave oscillator: A gyrotron in which the gyrating helical electron beam and traveling millimeter waves in a waveguide are in opposite directions.

Gyrotron traveling wave amplifier: A gyrotron with a matched waveguide. The waveguide carried only forward waves. The gyrating electrons interact with forward millimeter waves only.

Gyrotwystron: A gyrotron with a resonating buncher waveguide and a nonresonating traveling wave catcher waveguide.

Helical beam tube: A backward wave amplifier based on interaction between a forward helical electron beam launched into a waveguide and backward traveling microwave in the waveguide.

Helmhortz' wave equation: A second-order linear differential equation of electric potential, magnetic potential, electric field, magnetic field, or any other field quantities associated with propagating waves and/or standing waves.

Interaction region: A region where an electron beam and microwaves transfer energy to each other.

Inverted magnetron: A radial magnetron in which the anode structure is at inside and the cathode structure is at outside.

Kinetic energy: Energy of motion.

Kinetic energy conservation principle: Any change of kinetic energy of an electron is transformed into a change in electric potential energy with which the electron is interacting, and vice versa.

Klystron: A microwave vacuum tube amplifier or oscillator that contains buncher cavity or cavities, drift space, and a catcher cavity, among other components.

Laddertron: A microwave vacuum tube oscillator with a slow-wave structure coupled to a single-cavity resonator.

Laplace's equation: A second-order linear homogeneous differential equation of electric potential or any other function.

Linear magnetron: A magnetron in which the cavity resonator loaded anode, a sole (nonemitting cathode), and the electron beam are arranged in rectilinear fashion.

Magnetron: A microwave vacuum tube oscillator in which microwaves in a number of cavity resonators interact with bunched electrons in motion rotating or rectilinearly. The basic electron motion is formed by the applied accelerating electric fields and the DC magnetic fields.

Major orbit: A larger helical orbit of an electron beam in a gyrotron.

Minor orbit: A smaller helical orbit of electron beam in a double-coil electron beam gyrotron.

M-type linear magnetron: A magnetron in which the cavity loaded anode, the sole (nonemitting cathode), and the electron beam are arranged in a rectilinear fashion. Focusing DC magnetic fields are applied perpendicularly to both the electron beam and the DC electric fields so that the individual electron orbit is a cycloid.

O-type linear magnetron: A magnetron in which the cavity loaded anode, the sole (nonemitting cathode), and the electron beam are arranged in a rectilinear fashion. Focusing DC magnetic fields are applied parallel to the electron beam and perpendicularly to the DC electric fields so that the individual electron orbit is a helix.

Partition noise: Electrical noise current in an electron stream in an electron device due to separation of electron stream owing to the electron stream striking an obstacle or obstacles in the device.

Platinotron: A strap feed starting radial M-type backward wave oscillator.

Poisson's equation: Second-order linear inhomogeneous differential equation of electric potential or other electrical or magnetic functions.

Quantum numbers: A group of integrating constants in a solution of Schroedinger's equation.

Quasiquantum devices: Electron devices in which the principle of the device function is more easily explainable by quantum mechanics than by Newtonian classical dynamics.

Radial magnetron: A magnetron in which the anode and the cathode are arranged in a coaxial radial fashion.

Re-entrant use: Operation of a device in which a majority of the output of the device re-enters the input of the device.

S-band: A range of microwave frequency band, 2.6–3.95 GHz.

Shot noise: Electrical noise current in an electron stream in an electron device due to stochastic emission of electrons from the cathode.

Sole: A nonemitting cathode.

Strap: A conducting ring that ties tips of poles of magnetron or magnetronlike devices in a specified fashion for microwave potential and phase equalization.

Strap fed devices: Strapped magnetronlike devices that operate by microwaves fed through the strapping such as amplitron amplifies and platinotron oscillators.

TE$_{01}$ mode: A mode of circular or rectangular waveguide excitation in which all microwave electric fields are transverse to the direction of microwave propagation and coaxially concentric.

TE$_{11}$ mode: A mode of a circular or rectangular waveguide excitation in which all microwave electric fields are transverse to the direction of microwave propagation and approximately parallel to each other ending at the waveguide wall.

Thermal noise: Electric noise power of a device or component that is due to thermal vibration of atoms, molecules, and electrons. It is proportional to a product of the frequency bandwidth of interest and the absolute temperature of the device or component. The proportionality constant is Boltzmann's constant.

Tornadotron: A gyrotron with a simple helical beam.

Ubitron: A millimeter wave high-power quasiquantum generator with relativistically high-speed electron beam. Millimeter waves are generated due to quantum transition between two energy states of electrons and amplified due to the velocity modulation and kinetic energy transfer principles.

Velocity modulation: Production of variation of electron speed in a stream of electrons.

Waveguide: A transmission line that guides electromagnetic waves.

References

Coleman, J.T. 1982. *Microwave Devices*. Reston Pub., Reston, VA.

EEV Ltd. 1991. *EEV Microwave Products*. EEV Ltd., Chelmsford, England.

Ishii, T.K. 1989. *Microwave Engineering*. Harcourt Brace Jovanovich, San Diego, CA.

Ishii, T.K. 1990. *Practical Microwave Electron Devices*. Academic Press, San Diego, CA.

Liao, S.Y. 1988. *Microwave Electron-Tube Devices*. Prentice-Hall, Englewood Cliffs, NJ.

McCune, E.W. 1985. Fusion plasma heating with high power microwave and millimeter wave tubes. *Journal of Microwave Power*, vol. 20 (April) pp. 131–136.

Richardson Electronics Ltd. 1994. *Industrial Microwave Catalog*, June, LaFox, IL.

Veley, V.F. *Modern Microwave Technology*. Prentice-Hall, Englewood Cliffs, NJ.

Further Information

Additional information on the topic of other microwave vacuum devices may be obtained from the following sources:

IEEE Electron Device Meeting Digest.

IEEE Transactions on Electron Devices, published by the Institute of Electrical and Electronics Engineers Inc., 345 East 47th St., New York, NY 10017.

Manufacturers' catalog, application notes, and other technical publications such as from EEV, Hughes, Thomson Tubes Electroniques, Varian, and NEC.

6.5   Operational Considerations for Microwave Tubes

Jerry C. Whitaker

6.5.1   Introduction

Long-term reliability of a microwave power tube requires detailed attention to the operating environment of the device, including supply voltages, load point, and cooling. Optimum performance of the system can only be achieved when all elements are functioning within specified parameters.

6.5.2   Microwave Tube Life

Any analysis of microwave tube life must first identify the parameters that define *life*. The primary *wear out* mechanism in a microwave power tube is the electron gun at the cathode. In principle, the cathode will eventually evaporate the activating material and cease to produce the required output power. Tubes, however, rarely fail because of low emission, but for a variety of other reasons that are usually external to the device.

Power tubes designed for microwave applications provide long life when operated within their designed parameters. The point at which the device fails to produce the required output power can be predicted with some accuracy, based on design data and in-service experience. Most power tubes, however, fail because of mechanisms other than predictable chemical reactions inside the device itself. External forces, such as transient overvoltages caused by lightning, cooling system faults, and improper tuning more often than not lead to the failure of a microwave tube.

6.5.3   Life-Support System

Transmitter control logic is usually configured for two states of operation:

- An *operational level*, which requires all of the life-support systems to be present before the high-voltage (HV) command is enabled
- An *overload level*, which removes HV when one or more fault conditions occur

The cooling system is the primary life-support element in most RF generators. The cooling system should be fully operational before the application of voltages to the tube. Likewise, a cool-down period is usually recommended between the removal of beam and filament voltages and shut-down of the cooling system.

Most microwave power tubes require a high-voltage removal time of less than 100 ms from the occurrence of an overload. If the trip time is longer, damage may result to the device. **Arc detectors** are often installed in the cavities of high-power tubes to sense fault conditions, and shut down the high-voltage power supply before damage can be done to the tube. Depending on the circuit parameters, arcs can be sustaining, requiring removal of high voltage to squelch the arc. A number of factors can cause RF arcing, including

- Overdrive condition
- Mistuning of one or more cavities
- Poor cavity fit (applies to external types only)
- Under coupling of the output to the load
- Lightning strike at the antenna
- High voltage standing wave ratio (VSWR)

Regardless of the cause, arcing can destroy internal elements or the vacuum seal if drive and/or high voltage are not removed quickly. A lamp is usually included with each arc detector photocell for test purposes.

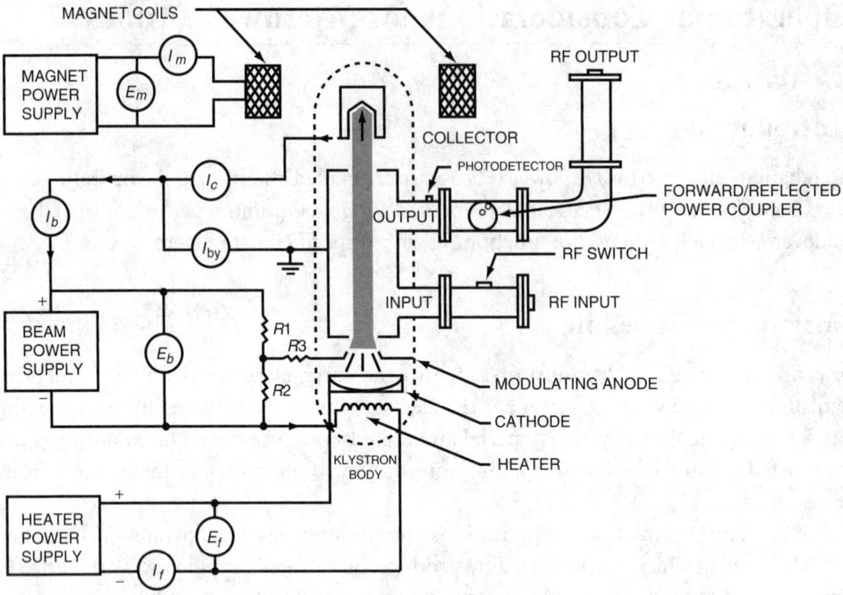

FIGURE 6.32　Protection and metering system for a klystron amplifier.

Protection Measures

A microwave power tube must be protected by control devices in the amplifier system. Such devices offer either visual indications, aural alarm warnings, or actuate interlocks within the system. Figure 6.32 shows a klystron amplifier and the basic components associated with its operation, including metering for each of the power supplies. Other types of microwave power devices use similar protection schemes. Sections of coaxial transmission line, representing essential components, are shown in the figure attached to the RF input and RF output ports of the tube. A single magnet coil is shown to represent any coil configuration that may exist; its position in the drawing is for convenience only and does not represent the true position in the system.

Heater Supply

The heater power supply can be either AC or DC. If it is DC, the positive terminal must be connected to the common heater-cathode terminal and the negative terminal to the heater terminal. The amount of power supplied to the heater is important because it establishes the cathode operating temperature. The temperature must be high enough to provide ample electron emission but not so high that emission life is jeopardized. The test performance sheet accompanying each tube lists the proper operating values of heater voltage and current for that tube. Heater voltage should be measured at the terminals of the device using a true-rms-reading voltmeter.

　　Because the cathode and heater are connected to the negative side of the beam supply, they must be insulated to withstand the full beam potential.

Beam Supply

The high-voltage **beam supply** furnishes the DC input power to the klystron. The positive side of the beam supply is connected to the body and collector of the klystron. The negative terminal is connected to the common heater-cathode terminal. Never connect the negative terminal of the beam supply to the heater-only terminal because the beam current will then flow through the heater to the cathode and cause premature heater failure. The voltmeter, E_b in Fig. 6.32, measures the beam voltage applied between the cathode and the body of the klystron.

Current meter I_c measures collector current, typically 95% or more of the total device current. Current meter I_{by} measures **body current**. An interlock should interrupt the beam supply if the body current exceeds a specified maximum value.

The sum of the body current I_{by} and collector current I_c is equal to the beam current I_b, which should stay constant as long as the beam voltage and modulating anode voltage are held constant.

Magnet Supply

Electrical connections to the DC magnet include two meters, one for measuring current through the circuit (I_m in Fig. 6.32) and one for measuring voltage (E_m). When the device is first installed in its magnet assembly, both parameters should be measured and recorded for future reference. If excessive body current or other unusual symptoms occur, these data will be valuable for system analysis.

Undercurrent protection should be provided to remove beam voltage if the magnetic circuit current falls below a preset value. The interlock should also prevent the beam voltage from being applied if the magnetic circuit is not energized. This scheme will not, however, provide protection if the coils are shorted. Shorted conditions can be determined by measuring the normal values of voltage and current and recording them for future reference.

The body-current overload protection should actuate if the magnetic field is reduced for any reason.

RF Circuits

In Fig. 6.32, monitoring devices are shown on the RF input and output of the device. These monitors protect the tube should a failure occur in the RF output circuit. Two directional couplers and a photodetector are attached to the output. These components and an RF switching device on the input form a protective network against output transmission line mismatch. The RF switch is activated by the photodetector or the reflected power monitor and must be capable of removing RF drive power from the klystron in less than 10 ms (typically).

In the RF output circuit, the forward power coupler is used to measure the relative power output of the klystron. The reflected power coupler is used to measure the RF power reflected by the output circuit components, or antenna. Damaged components or foreign material in the RF line will increase the RF reflected power. The amount of reflected power should be no more than 5% of the actual forward RF output power of the device in most applications. An interlock monitors the reflected power and removes RF drive to the klystron if the reflected energy reaches an unsafe level. To protect against arcs occurring between the monitor and the output window, a photodetector is placed between the monitor and the window. Light from an arc will trigger the photodetector, which actuates the interlock system and removes RF drive before the window is damaged.

6.5.4 Filament Voltage Control

Extending the life of a microwave tube begins with accurate adjustment of filament voltage. The filament should not be operated at a reduced voltage in an effort to extend tube life. Unlike a thoriated tungsten grid tube, reduced filament voltage may cause uneven emission from the surface of the cathode with little or no improvement in cathode life.

Voltage should be applied to the filament for a specified warm-up period before the application of beam current to minimize thermal stress on the cathode/gun structure. However, voltage normally should not be applied to the filaments for extended periods (2 h or more) if no beam voltage is present. The net rate of evaporation of emissive material from the cathode surface is greater without beam voltage. Subsequent condensation of material on gun components may lead to voltage standoff problems.

6.5.5 Cooling System

The cooling system is vital to any RF generator. In a high-power microwave transmitter, the cooling system may need to dissipate as much as 70% of the input AC power in the form of waste heat. For vapor phase-cooled devices, pure (distilled or demineralized) water must be used. Because the collector is usually only several volts above ground potential, it is generally not necessary to use deionized water.

TABLE 6.2 Variation of Electrical and Thermal Properties of Common Insulators as a Function of Temperature

	20°C	120°C	260°C	400°C	538°C
Thermal Conductivity[a]					
99.5% BeO	140	120	65	50	40
99.5% Al_2O_3	20	17	12	7.5	6
95.0% Al_2O_3	13.5				
Glass	0.3				
Power Dissipation[b]					
BeO	2.4	2.1	1.1	0.9	0.7
Electrical Resistivity[c]					
BeO	10^{16}	10^{14}	5×10^{12}	10^{12}	10^{11}
Al_2O_3	10^{14}	10^{14}	10^{12}	10^{12}	10^{11}
Glass	10^{12}	10^{10}	10^{8}	10^{6}	
Dielectric Constant[d]					
BeO	6.57	6.64	6.75	6.90	7.05
Al_2O_3	9.4	9.5	9.6	9.7	9.8
Loss Tangent[d]					
BeO	0.00044	0.00040	0.00040	0.00049	0.00080

[a] Heat transfer in $Btu/ft^2/h/°F$.
[b] Dissipation in $W/cm/°C$.
[c] Resistivity in $\Omega \cdot cm$.
[d] At 8.5 GHz.

Excessive dissipation is perhaps the single greatest cause of catastrophic failure in a power tube. The critical points of almost every tube type are the metal-to-ceramic junctions or seals. At temperatures below 250°C these seals remain secure, but above that temperature, the bonding in the seal may begin to disintegrate. Warping of internal structures also may occur at temperatures above the maximum operating level of the device. The result of prolonged overheating is shortened tube life or catastrophic failure. Several precautions are usually taken to prevent damage to tube seals under normal operating conditions. Air directors or sections of tubing may be used to provide spot cooling at critical surface areas of the device. Airflow and waterflow sensors typically prevent operation of the RF system in the event of a cooling system failure.

Temperature control is important for microwave tube operation because the properties of many of the materials used to build a device change with increasing temperature. In some applications, these changes are insignificant. In others, however, such changes can result in detrimental effects, leading to, in the worst case, catastrophic failure. Table 6.2 details the variation of electrical and thermal properties with temperature for various substances.

Water Cooling Systems

A water-cooled tube depends on an adequate flow of fluid to remove heat from the device and transport it to an external heat sink. The recommended flow as specified in the technical data sheet should be maintained at all times when the tube is in operation. Inadequate water flow at high temperature may cause the formation of steam bubbles at the collector surface where the water is in direct contact with the collector. This condition can contribute to premature tube failure.

Circulating water can remove about $1.0 \, kW/cm^2$ of effective internal collector area. In practice, the temperature of water leaving the tube is limited to 70°C to preclude the possibility of spot boiling. The water is then passed through a heat exchanger where it is cooled to 30–40°C before being pumped back to the device.

Cooling System Design

A liquid cooling system consists of the following principal components:

- A source of coolant
- Circulation pump

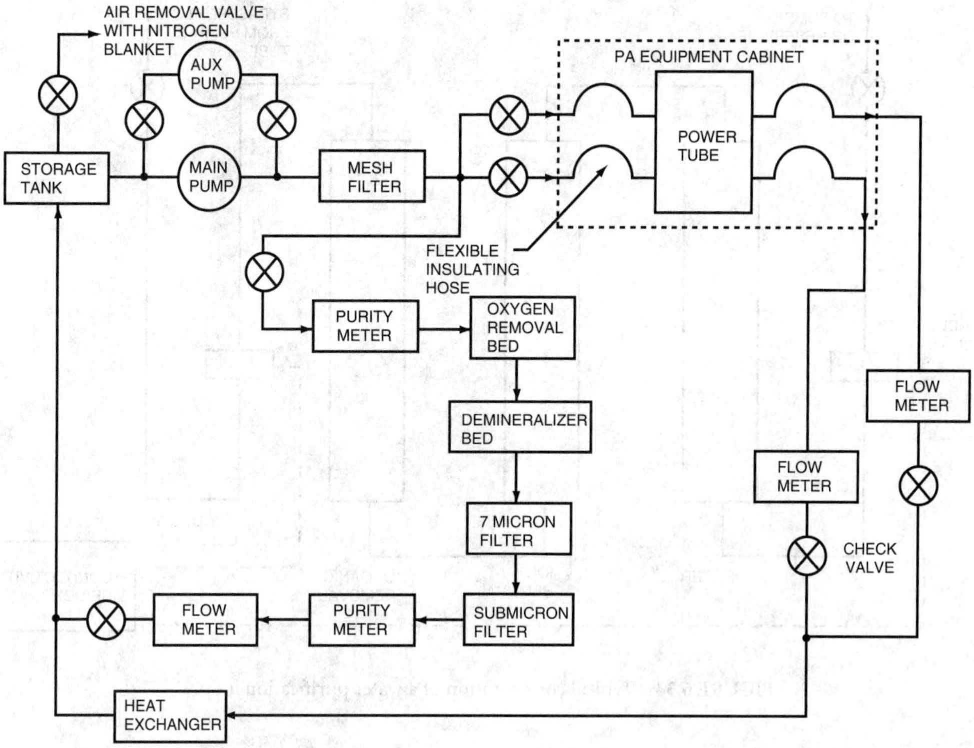

FIGURE 6.33 Functional schematic of a water-cooling system for a high-power amplifier or oscillator.

- Heat exchanger
- Coolant purification loop
- Various connection pipes, valves, and gauges
- Flow interlocking devices (required to ensure coolant flow anytime the equipment is energized)

Such a system is shown schematically in Fig. 6.33. In most cases the liquid coolant will be water, however, if there is a danger of freezing, it will be necessary to use an antifreeze solution such as ethylene glycol. In these cases, coolant flow must be increased or plate dissipation reduced to compensate for the poorer heat capacity of the ethylene glycol solution. A mixture of 60% ethylene glycol to 40% water by weight will be about 75% as efficient as pure water as a coolant at 25°C. Regardless of the choice of liquid, the system volume must be maintained above the minimum required to insure proper cooling of the vacuum tube(s).

The main circulation pump must be of sufficient size to ensure necessary flow and pressure as specified on the tube data sheet. A filter screen of at least 60 mesh is usually installed in the pump outlet line to trap any circulating debris that might clog coolant passages within the tube.

The heat exchanger system is sized to maintain the outlet temperature such that the outlet water from the tube at full dissipation does not exceed 70°C. Supplementary coolant courses may be connected in parallel or series with the main supply as long as the maximum outlet temperature is not exceeded.

Valves and pressure meters are installed on the inlet lines to the tube to permit adjustment of flow and measurement of pressure drop, respectively. A pressure meter and check valve are employed in the outlet line. In addition, a flow meter, sized in accordance with the tube data sheet, and a thermometer are included in the outlet line of each coolant course. These flow meters are equipped with automatic interlock switches, wired into the system electrical controls, so that the tube will be completely deenergized in the event of a loss of coolant flow in any one of the coolant passages. In some tubes, filament power alone is sufficient to damage the tube structure in the absence of proper water flow.

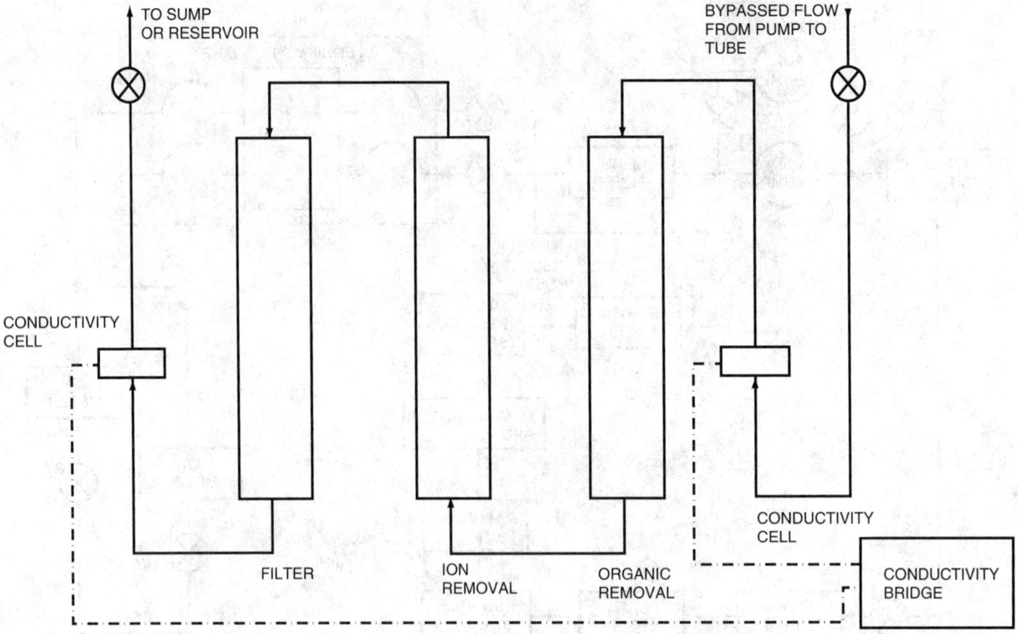

FIGURE 6.34   Typical configuration of a water purification loop.

The lines connecting the plumbing system to the inlet and outlet ports of tube are made of a flexible insulating material configured so as to avoid excessive strain on the tube flanges. The coolant lines must be of sufficient length to keep electrical leakage below approximately 4 mA at full operating power.

The hoses are coiled or otherwise supported so they do not contact each other or any conducting surface between the high voltage end and ground. Conducting hose *barbs*, connected to ground, are provided at the low-potential end so that the insulating column of water is broken and grounded at the point it exits the equipment cabinet.

Even if the cooling system is constructed using the recommended materials, and is filled with distilled or deionized water, the solubility of the metals, carbon dioxide, and free oxygen in the water make the use of a coolant purification **regeneration loop** essential. The integration of the purification loop into the overall cooling system is shown in Fig. 6.34. The regeneration loop typically taps 5–10% of the total cooling system capacity, circulating it through oxygen scavenging and deionization beds and submicron filters before returning it into the main system. The purification loop can theoretically process water to 18-MΩ resistivity at 25°C. In practice, resistivity will be somewhat below this value.

Figure 6.34 shows a typical purification loop configuration. Packaged systems such as the one illustrated are available from a number of manufacturers. In general, such systems consist of replaceable cartridges that perform the filtering, ion exchange, and organic solid removal functions. The system will usually include flow and pressure gauges and valves and conductivity cells for continuous evaluation of the condition of the water and filters.

Water Purity and Resistivity

The purity of the cooling water is an important operating parameter for any water-cooled amplifier or oscillator. The specific **resistivity** typically must be maintained at 1 M$\Omega \cdot$ cm minimum at 25°C. Distilled or deionized water should be used and the purity and flow protection periodically checked to ensure against degradation.

Oxygen and carbon dioxide in the coolant will form copper oxide, reducing cooling efficiency, and electrolysis may destroy the coolant passages. In addition, a filter screen should be installed in the tube inlet line to trap any circulating debris that might clog coolant passages within the tube.

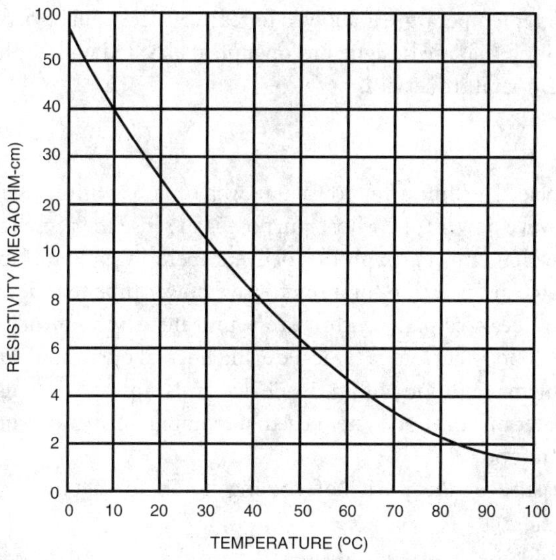

FIGURE 6.35 The effect of temperature on the resistivity of ultrapure water.

After normal operation for an extended period, the cooling system should be capable of holding 3–4 $M\Omega \cdot cm$ until the filter beds become contaminated and must be replaced. The need to replace the filter bed resins is indicated when the purification loop output water falls below 5 $M\Omega \cdot cm$. Although the resistivity measurement is not a test for free oxygen in the coolant, the oxygen filter bed should always be replaced when replacing the deionizing bed.

The resistivity of the coolant is also affected by the temperature of the water. The temperature dependence of the resistivity of pure water is charted in Fig. 6.35.

Generally speaking, it is recommended that the coolant water be circulated at all times. This procedure provides the following benefits:

- Maintains high resistivity
- Reduces bacteria growth
- Minimizes oxidation resulting from coolant stagnation

If it is undesirable to circulate the coolant at the regular rate when the tube is deenergized, a secondary circulating pump can be used to move the coolant at a lower rate to purge any air that might enter the system and to prevent stagnation. Typical recommended minimum circulation rates within the coolant lines are as follows:

- 2 m/s during normal operation
- 30 cm/s during standby

The regeneration loop is typically capable of maintaining the cooling system such that the following maximum contaminant levels are not exceeded:

- Copper: 0.05 PPM by weight
- Oxygen: 0.5 PPM by weight
- CO_2: 0.5 PPM by weight
- Total solids: 3 PPM by weight

These parameters represent maximum levels; if the precautions outlined in this section are taken, actual levels will be considerably lower.

If the cooling system water temperature is allowed to reach 50°C, it will be necessary to use cartridges in the coolant regeneration loop that are designed to operate at elevated temperatures. Ordinary cartridges will decompose in high temperature service.

Defining Terms

Arc detector: A device placed within a microwave power tube or within one or more of the external cavities of a microwave power tube whose purpose is to sense the presence of an overvoltage arc.

Beam supply: The high-voltage power supply that delivers operating potential to a microwave power tube.

Body current: A parameter of operation for a microwave power tube that describes the amount of current passing through leakage paths within or around the device, principally through the body of the tube. Some small body current is typical during normal operation. Excessive current, however, can indicate a problem condition within the device or its support systems.

Regeneration loop: A water purification system used to maintain proper conditions of the cooling liquid for a power vacuum tube.

Water resistivity: A measure of the purity of cooling liquid for a power tube, typically measured in $M\Omega \cdot cm$.

References

Crutchfield, E.B. 1992. *NAB Engineering Handbook*, 8th ed. National Association of Broadcasters, Washington DC.

Harper, C.A. 1991. *Electronic Packaging and Interconnection Handbook*. McGraw-Hill, New York.

Kimmel, E. 1983. Temperature sensitive indicators: how they work, when to use them. *Heat Treating* (March).

Spangenberg, K. 1947. *Vacuum Tubes*. McGraw-Hill, New York.

Terman, F.E. 1947. *Radio Engineering*. McGraw-Hill, New York.

Varian. n.d. *Integral Cavity Klystrons for UHF-TV Transmitters*. Varian Associates, Palo Alto, CA.

Varian. 1971. Foaming test for water purity. *Application Engineering Bulletin* No. AEB-26. Varian Associates, Palo Alto, CA, Sept.

Varian. 1977. Water purity requirements in liquid cooling systems. *Application Bulletin* No. 16. Varian Associates, San Carlos, CA, July.

Varian. 1978. Protecting integral-cavity klystrons against water leakage. *Technical Bulletin* No. 3834. Varian Associates, Palo Alto, CA, July.

Varian. 1982. Cleaning and flushing klystron water and vapor cooling systems. *Application Engineering Bulletin* No. AEB-32. Varian Associates, Palo Alto, CA, Feb.

Varian. 1984. *Care and Feeding of Power Grid Tubes*. Varian Associates, San Carlos, CA.

Varian. 1984. Temperature measurements with eimac power tubes. *Application Bulletin* No. 20. Varian Associates, San Carlos, CA, Jan.

Varian. 1991. Water purity requirements in water and vapor cooling systems. *Application Engineering Bulletin* No. AEB-31. Varian Associates, Palo Alto, CA, Sept.

Whitaker, J.C. 1991. *Radio Frequency Transmission Systems: Design and Operation*. McGraw-Hill, New York.

Whitaker, J.C. 1992, *Maintaining Electronic Systems*. CRC Press, Boca Raton, FL.

Whitaker, J.C. *Power Vacuum Tubes Handbook* 2nd ed., CRC Press, Boca Raton, FL.

Further Information

Specific information on the application of microwave tubes can be obtained from the manufacturers of those devices. More general application information can be found in the following publications:

Maintaining Electronic Systems, written by Jerry C. Whitaker, CRC Press, Boca Raton, FL, 1990.

Radio Frequency Transmission Systems: Design and Operation, written by Jerry C. Whitaker, McGraw-Hill, New York, 1991.

Power Vacuum Tubes Handbook 2nd ed., written by Jerry C. Whitaker, CRC Press, Boca Raton, FL.

7

Semiconductor Devices and Circuits

Sidney Soclof
California State University

John R. Brews
University of Arizona

Edward J. Delp, III
Purdue University

Jerry C. Whitaker
Editor-in-Chief

Timothy P. Hulick
Acrodyne Industries, Inc.

Peter Aronhime
University of Louisville

Ezz I. El-Masry
Technical Institute of Nova Scotia

Victor Meeldijk
Intel Corporation

7.1 Semiconductors

Sidney Soclof

7.1.1 Introduction

Transistors form the basis of all modern electronic devices and systems, including the integrated circuits used in systems ranging from radio and television to computers. Transistors are solid-state electron devices made out of a category of materials called semiconductors. The mostly widely used semiconductor for transistors, by far, is silicon, although gallium arsenide, which is a compound semiconductor, is used for some very high-speed applications.

Semiconductors

Semiconductors are a category of materials with an electrical conductivity that is between that of conductors and insulators. Good conductors, which are all metals, have electrical resistivities down in the range of 10^{-6} Ω-cm. Insulators have electrical resistivities that are up in the range from 10^6 to as much as about 10^{12} Ω-cm. Semiconductors have resistivities that are generally in the range of 10^{-4}–10^4 Ω-cm. The resistivity of a semiconductor is strongly influenced by impurities, called **dopants**, that are purposely added to the material to change the electronic characteristics.

We will first consider the case of the pure, or **intrinsic semiconductor**. As a result of the thermal energy present in the material, electrons can break loose from covalent bonds and become free electrons able to move through the solid and contribute to the electrical conductivity. The covalent bonds left behind have an electron vacancy called a **hole**. Electrons from neighboring covalent bonds can easily move into an adjacent bond with an electron vacancy, or hole, and thus the hold can move from one covalent bond to an adjacent bond. As this process continues, we can say that the hole is moving through the material. These holes act as if they have a positive charge equal in magnitude to the electron charge, and they can also contribute to the electrical conductivity. Thus, in a semiconductor there are two types of mobile electrical charge carriers that can contribute to the electrical conductivity, the free electrons and the holes. Since the electrons and holes are generated in equal numbers, and recombine in equal numbers, the free electron and hole populations are equal.

In the **extrinsic** or **doped semiconductor**, impurities are purposely added to modify the electronic characteristics. In the case of silicon, every silicon atom shares its four valence electrons with each of its four nearest neighbors in covalent bonds. If an impurity or dopant atom with a valency of five, such as phosphorus, is substituted for silicon, four of the five valence electrons of the dopant atom will be held in covalent bonds. The extra, or fifth electron will not be in a covalent bond, and is loosely held. At room temperature, almost all of these extra electrons will have broken loose from their parent atoms, and become free electrons. These pentavalent dopants thus donate free electrons to the semiconductor and are called **donors**. These donated electrons upset the balance between the electron and hole populations, so there are now more electrons than holes. This is now called an **N-type semiconductor**, in which the electrons are the **majority carriers**, and holes are the **minority carriers**. In an N-type semiconductor the free electron concentration is generally many orders of magnitude larger than the hole concentration.

If an impurity or dopant atom with a valency of three, such as boron, is substituted for silicon, three of the four valence electrons of the dopant atom will be held in covalent bonds. One of the covalent bonds will be missing an electron. An electron from a neighboring silicon-to-silicon covalent bond, however, can easily jump into this electron vacancy, thereby creating a vacancy, or hole, in the silicon-to-silicon covalent bond. Thus, these trivalent dopants accept free electrons, thereby generating holes, and are called **acceptors**. These additional holes upset the balance between the electron and hole populations, and so there are now more holes than electrons. This is called a **P-type semiconductor**, in which the holes are the majority carriers, and the electrons are the minority carriers. In a P-type semiconductor the hole concentration is generally many orders of magnitude larger than the electron concentration.

Figure 7.1 shows a single crystal chip of silicon that is doped with acceptors to make it P-type on one side, and doped with donors to make it N-type on the other side. The transition between the two sides is

called the *PN junction*. As a result of the concentration difference of the free electrons and holes there
will be an initial flow of these charge carriers across
the junction, which will result in the N-type side
attaining a net positive charge with respect to the

FIGURE 7.1 PN junction.

P-type side. This results in the formation of an electric potential *hill* or barrier at the junction. Under
equilibrium conditions the height of this potential hill, called the **contact potential** is such that the flow
of the majority carrier holes from the P-type side up the hill to the N-type side is reduced to the extent
that it becomes equal to the flow of the minority carrier holes from the N-type side down the hill to
the P-type side. Similarly, the flow of the majority carrier free electrons from the N-type side is reduced to
the extent that it becomes equal to the flow of the minority carrier electrons from the P-type side. Thus,
the net current flow across the junction under equilibrium conditions is zero.

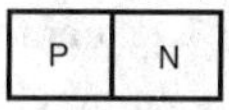

7.1.2 Diodes

In Fig. 7.2 the silicon chip is connected as a *diode*
or two-terminal electron device. The situation in
which a bias voltage is applied is shown. In Fig. 7.2(a)
the bias voltage is a **forward bias**, which produces
a reduction in the height of the potential hill at the
junction. This allows for a large increase in the flow
of electrons and holes across the junction. As the
forward bias voltage increases, the **forward current**

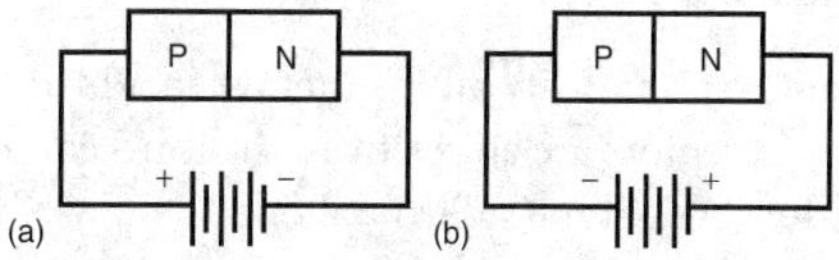

FIGURE 7.2 Biasing of a diode: (a) forward bias, (b)
reverse bias.

will increase at approximately an exponential rate, and can become very large. The variation of forward
current flow with forward bias voltage is given by the diode equation as

$$I = I_0(\exp(qV/nkT) - 1)$$

where
- I_0 = reverse saturation current, constant
- q = electron charge
- n = dimensionless factor between 1 and 2
- k = Boltzmann's constant
- T = absolute temperature, K

If we define the **thermal voltage** as $V_T = kT/q$, the diode equation can be written as

$$I = I_0(\exp(V/nV_T) - 1)$$

At room temperature $V_T \cong 26\ mV$, and n is typically around 1.5 for silicon diodes.

In Fig. 7.2(b) the bias voltage is a **reverse bias**, which produces an increase in the height of the potential
hill at the junction. This essentially chokes off the flow of electrons from the N-type side to the P-type side,
and holes from the P-type side to the N-type side. The only thing left is the very small trickle of electrons
from the P-type side and holes from the N-type side. Thus the **reverse current** of the diode will be very
small.

In Fig. 7.3 the circuit schematic symbol for the
diode is shown, and in Fig. 7.4 a graph of the current
vs. voltage curve for the diode is presented. The
P-type side of the diode is called the **anode**, and the
N-type side is the **cathode** of the diode. The forward
current of diodes can be very large, in the case of
large power diodes, up into the range of 10–100 A.

FIGURE 7.3 Diode symbol.

The reverse current is generally very small, often down in the low nanoampere, or even picoampere range. The diode is basically a one-way voltage-controlled current switch. It allows current to flow in the forward direction when a forward bias voltage is applied, but when a reverse bias is applied the current flow becomes extremely small. Diodes are used extensively in electronic circuits. Applications include rectifiers to convert AC to DC, wave shaping circuits, peak detectors, DC level shifting circuits, and signal transmission gates. Diodes are also used for the demodulation of amplitude-modulated (AM) signals.

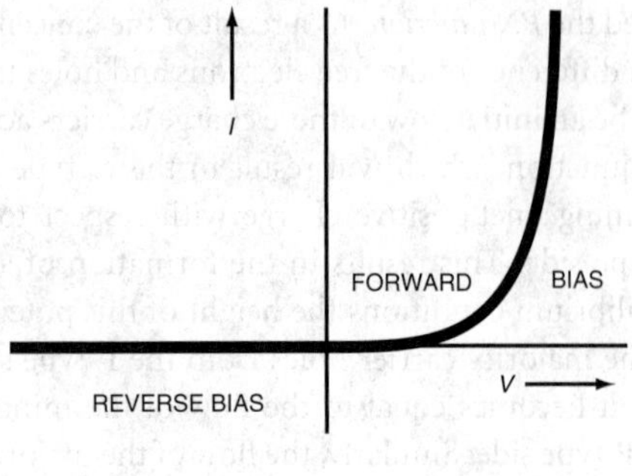

FIGURE 7.4   Current vs. voltage curve for a diode.

Defining Terms

Acceptors: Impurity atoms that when added to a semiconductor contribute holes. In the case of silicon, acceptors are atoms from the third column of the periodic table, such as boron.

Anode: The P-type side of a diode.

Cathode: The N-type side of a diode.

Contact potential: The internal voltage that exists across a PN junction under thermal equilibrium conditions, when no external bias voltage is applied.

Donors: Impurity atoms that when added to a semiconductor contribute free electrons. In the case of silicon, donors are atoms from the fifth column of the periodic table, such as phosphorus, arsenic, and antimony.

Dopants: Impurity atoms that are added to a semiconductor to modify the electrical conduction characteristics.

Doped semiconductor: A semiconductor that has had impurity atoms added to modify the electrical conduction characteristics.

Extrinsic semiconductor: A semiconductor that has been doped with impurities to modify the electrical conduction characteristics.

Forward bias: A bias voltage applied to the PN junction of a diode or transistor that makes the P-type side positive with respect to the N-type side.

Forward current: The large current flow in a diode that results from the application of a forward bias voltage.

Hole: An electron vacancy in a covalent bond between two atoms in a semiconductor. Holes are mobile charge carriers with an effective charge that is opposite to the charge on an electron.

Intrinsic semiconductor: A semiconductor with a degree of purity such that the electrical characteristics are not significantly affected.

Majority carriers: In a semiconductor, the type of charge carrier with the larger population. For example, in an N-type semiconductor, electrons are the majority carriers.

Minority carriers: In a semiconductor, the type of charge carrier with the smaller population. For example, in an N-type semiconductor, holes are the minority carriers.

N-type semiconductor: A semiconductor that has been doped with donor impurities to produce the condition that the population of free electrons is greater than the population of holes.

P-type semiconductor: A semiconductor that has been doped with acceptor impurities to produce the condition that the population of holes is greater than the population of free electrons.

Reverse bias: A bias voltage applied to the PN junction of a diode or transistor that makes the P-type side negative with respect to the N-type side.

Reverse current: The small current flow in a diode that results from the application of a reverse bias voltage.

Thermal voltage: The quantity kT/q where k is Boltzmann's constant, T is absolute temperature, and q is electron charge. The thermal voltage has units of volts, and is a function only of temperature, being approximately 25 mV at room temperature.

References

Comer, D.J. and Comer, D.T., 2002. *Advanced Electronic Circuit Design*. John Wiley & Sons, New York, NY.

Hambley, A.R. 2000. *Electronics*, 2nd ed. Prentice-Hall, Englewood Cliffs, NJ.

Jaeger, R.C. and Travis, B. 2004. *Microelectronic Circuit Design with CD-ROM*. McGraw-Hill, New York.

Mauro, R. 1989. *Engineering Electronics*. Prentice-Hall, Englewood Cliffs, NJ.

Millman, J. and Grabel, A. 1987. *Microelectronics*, 2nd ed. McGraw-Hill, New York.

Mitchell, F.H., Jr. and Mitchell, F.H., Sr. 1992. *Introduction to Electronics Design*, 2nd ed. Prentice-Hall, Englewood Cliffs, NJ.

Martin, S., Roden, M.S., Carpenter, G.L., and Wieserman, W.R. 2002. *Electronic Design*, Discovery Press, Los Angeles, CA.

Neamen, D. 2001. Electronic Circuit Analysis with CD-ROM. McGraw-Hill, New York, NY.

Sedra, A.S. and Smith, K.C. 2003. *Microelectronics Circuits*, 5th ed. Oxford University Press, Oxford.

Spencer, R. and Mohammed G. 2003. *Introduction to Electronic Circuit Design*. Prentice-Hall, Englewood Cliffs, NJ.

Further Information

An excellent introduction to the physical electronics of devices is found in Ben G. Streetman, *Solid State Electronic Devices*, 4th ed. Prentice-Hall, Englewood Cliffs, NJ, 1995. Another excellent reference on a wide variety of electronic devices is Kwok K. Ng, *Complete Guide to Semiconductor Devices*, 2002. Wiley-IEEE Computer Society Press, New York, NY. A useful reference on circuits and applications is Donald A. Neamen, *Electronic Circuit Analysis and Design*, 2nd ed., 2001, Irwin, Chicago, IL.

7.2 Bipolar Junction and Junction Field-Effect Transistors

Sidney Soclof

7.2.1 Bipolar Junction Transistors

A basic diagram of the bipolar junction transistor (BJT) is shown in Fig. 7.5. Whereas the diode has one PN junction, the BJT has two PN junctions. The three regions of the BJT are the emitter, base, and collector. The middle, or base region, is very thin, generally less than 1 μm wide. This middle electrode, or base, can be considered to be the control electrode that controls the current flow through the device between emitter and collector. A small voltage applied to the base (i.e., between base and emitter) can produce a large change in the current flow through the BJT.

FIGURE 7.5 Bipolar junction transistor.

BJTs are often used for the amplification of electrical signals. In these applications the emitter-base PN junction is turned on (forward biased) and the collector-base PN junction is off (reverse biased). For the NPN BJT as shown in Fig. 7.5, the emitter will emit electrons into the base region. Since the P-type base region is so thin, most of these electrons will survive the trip across the base and reach the collector-base junction. When the electrons reach the collector-base junction they will roll downhill into the collector, and thus be collected by the collector to become the collector current I_C. The emitter and collector currents will be approximately equal, so $I_C \cong I_E$. There will be a small base current, I_B, resulting from the emission of holes from the base across the emitter-base junction into the emitter. There will also be a small component of the base current due to the recombination of electrons and holes in the base. The ratio of collector current to base current is given by the parameter β or h_{FE}, is $\beta = I_C/I_B$, and will be very large, generally up in the range of 50–300 for most BJTs.

In Fig. 7.6(a) the circuit schematic symbol for the NPN transistor is shown, and in Fig. 7.6(b) the corresponding symbol for the PNP transistor is given. The basic operation of the PNP transistor is similar to that of the NPN, except for a reversal of the polarity of the algebraic signs of all DC currents and voltages.

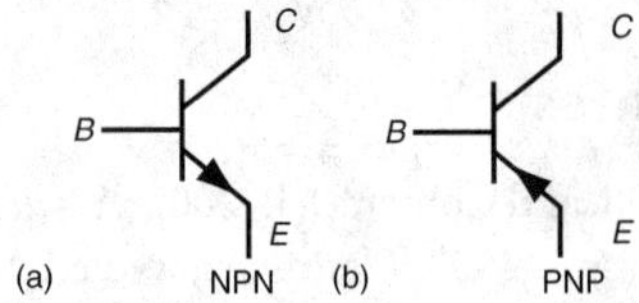

In Fig. 7.7 the operation of a BJT as an amplifier is shown. When the BJT is operated as an amplifier the emitter-base PN junction is turned on (forward biased) and the collector-base PN junction is off (reverse biased). An AC input voltage applied between base and emitter, $v_{in} = v_{be}$, will produce an AC component, i_c, of the collector current. Since i_c flows through a load resistor, R_L, an AC voltage, $v_o = v_{ce} = -i_c \cdot R_L$ will be produced at the collector. The AC small-signal voltage gain is $A_V = v_o/v_{in} = v_{ce}/v_{be}$.

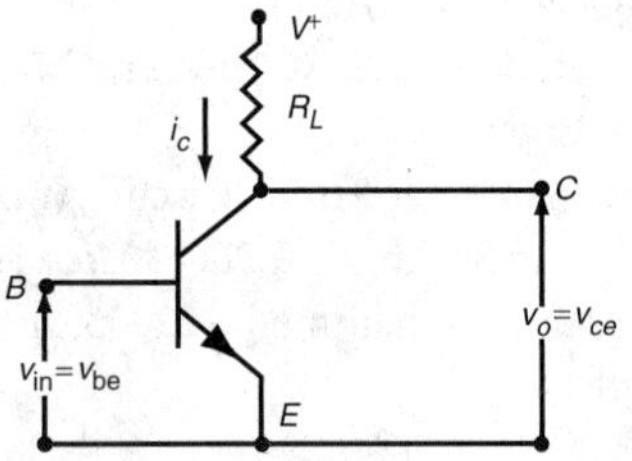

The collector current I_C of a BJT when operated as an amplifier is related to the base-to-emitter voltage V_{BE} by the exponential relationship $I_C = I_{CO} \cdot \exp(V_{BE}/V_T)$, where I_{CO} is a constant, and $V_T = thermal\ voltage = 25$ mV. The rate of change of I_C with respect to V_{BE} is given by the *transfer conductance*, $g_m = dI_C/dV_{BE} = I_C/V_T$. If the net load driven by the collector of the transistor is R_L, the AC small-signal voltage gain is $A_V = v_{ce}/v_{be} = -g_m \cdot R_L$. The negative sign indicates that the output voltage will be an amplified, but inverted, replica of the input signal. If, for example, the transistor is biased at a DC collector current level of $I_C = 1$ mA and drives a net load of $R_L = 10$ kΩ, then $g_m = I_C/V_T = 1$ mA/25 mV = 40 mS, and $A_V = v_c/v_{be} = -g_m \cdot R_L = -40$ mS $\cdot 10$ k$\Omega = -400$. Thus we see that the voltage gain of a single BJT amplifier stage can be very large, often up in the range of 100, or more.

The BJT is a three electrode or **triode** electron device. When connected in a circuit it is usually operated as a two-port, or two-terminal, pair device as shown in Fig. 7.8. Therefore, one of the three electrodes of the BJT must be common to both the input and output ports. Thus, there are three basic BJT configurations, common emitter (CE), common base (CB), and common collector (CC), as shown in Fig. 7.8. The most often used configuration, especially for amplifiers, is the common-emitter (CE), although the other two configurations are used in some applications.

FIGURE 7.6   BJT schematic symbols: (a) NPN BJT, (b) PNP BJT.

FIGURE 7.7   A BJT amplifier.

7.2.2   Amplifier Configurations

We will first compare the common-emitter circuit of Fig. 7.8(b) to the common-base circuit of Fig. 7.8(c). The AC small-signal voltage gain of the common-emitter circuit is given by $A_V = -g_m R_{NET}$ where

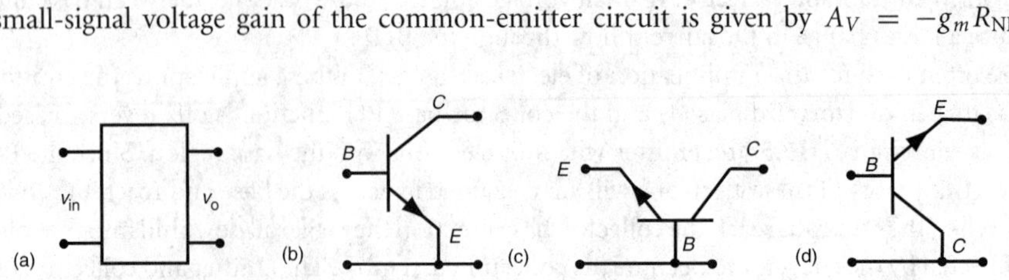

FIGURE 7.8   The BJT as a two-port device: (a) block representation, (b) common emitter, (c) common base, (d) common collector.

g_m is the dynamic forward transfer conductance as given by $g_m = I_C/V_T$, and R_{NET} is the net load resistance driven by the collector of the transistor. Note that the common-emitter circuit is an inverting amplifier, in that the output voltage is an amplified, but inverted, replica of the input voltage. The AC small-signal voltage gain of the common-base circuit is given by $A_V = g_m R_{NET}$, so we see that the common-base circuit is a noninverting amplifier, in that the output voltage is an amplified replica of the input voltage. Note that the magnitude of the gain is given by the same expression for both amplifier circuits.

The big difference between the two amplifier configurations is in the input resistance. For the common-emitter circuit the AC small-signal input resistance is given by $r_{IN} = n\beta V_T/I_C$, where n is the ideality factor, which is a dimensionless factor between 1 and 2, and is typically around 1.5 for silicon transistors operating at moderate current levels, in the 1–10 mA range. For the common-base circuit the AC small-signal input resistance is given by $r_{IN} = V_T/I_E \cong V_T/I_C$. The input resistance of the common-base circuit is smaller than that of the common-emitter circuit by a factor of approximately $n\beta$. For example, taking $n = 1.5$ and $\beta = 100$ as representative values, at $I_C = 1.0$ mA we get for the common-emitter case $r_{IN} = n\beta V_T/I_C = 1.5 \cdot 100 \cdot 25$ mV/1 mA $= 3750\ \Omega$; whereas for the common-base case, we get $r_{IN} \cong V_T/I_C = 25$ mV/1 mA $= 25\ \Omega$. We see that r_{IN} for the common emitter is 150 times larger than for the common base. The small input resistance of the common-base case will severely load most signal sources. Indeed, if we consider cascaded common-base stages with one common-base stage driving another operating at the same quiescent collector current level, then we get $A_V = g_m R_{NET} \cong g_m \cdot r_{IN} \cong (I_C/V_T)\cdot(V_T/I_C) = 1$, so that no net voltage gain is obtained from a common-base stage driving another common-base stage under these conditions. For the cascaded common-emitter case, we have $A_V = -g_m R_{NET} \cong -g_m r_{IN} = (I_C/V_T) \cdot (n\beta V_T/I_C) =$, $n\beta$, so that if $n = 1.5$ and $\beta = 100$, a gain of about 150 can be achieved. It is for this reason that the common-emitter stage is usually chosen.

The common-base stage is used primarily in high-frequency applications due to the fact that there is no direct capacitive feedback from output (collector) to input (emitter) as a result of the common or grounded base terminal. A circuit configuration that is often used to take advantage of this, and at the same time to have the higher input impedance of the common-emitter circuit is the *cascode* configuration, as shown in Fig. 7.9. The cascode circuit is a combination of the common-emitter stage directly coupled to a common-base stage. The input impedance is that of the common-emitter stage, and the grounded base of the common-base stage blocks the capacitive feedback from output to input.

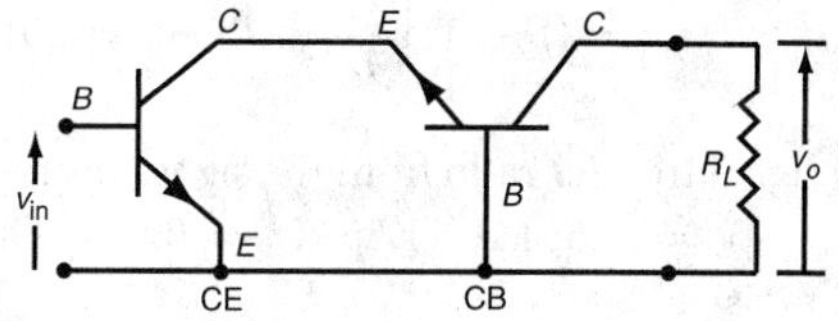

FIGURE 7.9 A cascode circuit.

The common-collector circuit of Fig. 7.10 will now be considered. The common-collector circuit has an AC small-signal voltage gain given by

$$A_V = \frac{g_m R_{NET}}{[1 + g_m R_{NET}]} = \frac{R_{NET}}{[R_{NET} + V_T/I_C]}$$

The voltage gain is positive, but will always be less than unity, although it will usually be close to unity. For example if $I_C = 10$ mA and $R_{NET} = 50\ \Omega$ we obtain

$$A_V = \frac{R_{NET}}{\left[R_{NET} + \dfrac{V_T}{I_C}\right]} = \frac{50}{\left[50 + \dfrac{25\ \text{mV}}{10\ \text{mA}}\right]} = \frac{50}{50 + 2.5} = \frac{50}{52.5} = 0.952$$

Since the voltage gain for the common-collector stage is positive and usually close to unity, the AC voltage at the emitter will rather closely follow the voltage at the base, hence, the name *emitter-follower* that is usually used to describe this circuit.

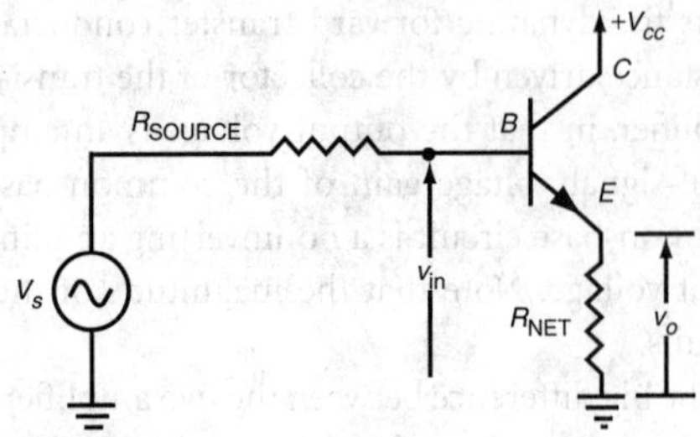

FIGURE 7.10   Emitter-follower circuit.

We have seen that the common-collector or emitter-follower stage will always have a voltage gain that is less than unity. The emitter follower is nevertheless a very important circuit because of its impedance transforming properties. The AC small-signal input resistance is given by $r_{IN} = (\beta + 1)((V_T/I_C) + R_{NET})$, where R_{NET} is the net AC load driven by the emitter of the emitter follower. We see that looking into the base, the load resistance R_{NET} is transformed up in value by a factor of $\beta + 1$. Looking from the load back into the emitter, the AC small-signal output resistance is given by

$$r_O = \frac{1}{g_m} + \frac{R_{SOURCE}}{(\beta + 1)} = \frac{V_T}{I_C} + \frac{R_{SOURCE}}{(\beta + 1)}$$

where R_{SOURCE} is the net AC resistance that is seen looking out from the base toward the signal source. Thus, as seen from the load looking back into the emitter follower, the source resistance is transformed down by a factor of $\beta + 1$. This impedance transforming property of the emitter follower is useful for coupling high-impedance sources to low-impedance loads. For example, if a 1-kΩ source is coupled directly to a 50-Ω load, the transfer ratio will be $T = R_{LOAD}/(R_{LOAD} + R_{SOURCE}) = 50/1050 \cong 0.05$. If an emitter follower with $I_C = 10$ mA and $\beta = 200$ is interposed between the signal source and the load, the input resistance of the emitter follower will be

$$r_{IN} = (\beta + 1)\left[\frac{V_T}{I_C} + R_{NET}\right] = 201 \cdot (2.5\ \Omega + 50\ \Omega) = 201 \cdot 52.5\ \Omega = 10.55\ k\Omega$$

The signal transfer ratio from the signal source to the base of the emitter follower is now $T = r_{IN}/(r_{IN} + R_{SOURCE}) = 10.55\,k\Omega/11.55\,k\Omega = 0.913$. The voltage gain through the emitter follower from base to emitter is

$$A_V = \frac{R_{NET}}{\left[R_{NET} + \dfrac{V_T}{I_C}\right]} = \frac{50}{\left[50 + \dfrac{25\ mV}{10\ mA}\right]} = \frac{50}{[50 + 2.5]} = \frac{50}{52.5} = 0.952$$

and so the overall transfer ratio is $T_{NET} = 0.913 \cdot 0.952 = 0.87$. Thus, there is a very large improvement in the transfer ratio.

As a second example of the usefulness of the emitter follower, consider a common-emitter stage operating at $I_C = 1.0$ mA and driving a 50 Ω load. We have

$$A_V = -g_m R_{NET} = -\frac{1\ mA}{25\ mV} \cdot 50\ \Omega = -40\ mS \cdot 50\ \Omega = -2$$

If an emitter follower operating at $I_C = 10$ mA is interposed between the common-emitter stage and the load, we now have for the common-emitter stage a gain of

$$A_V = -g_m R_{NET} = -\frac{1\ mA}{25\ mV} \cdot r_{IN} = -40\ mS \cdot 10.55\ k\Omega = -422$$

The voltage gain of the emitter follower from base to emitter is 0.952, so the overall gain is now $-422 \cdot 0.952 = -402$, as compared to the gain of only -2 that was available without the emitter follower.

The BJT is often used as a switching device, especially in digital circuits, and in high-power applications. When used as a switching device, the transistor is switched between the *cutoff region* in which both junctions are off, and the *saturation region* in which both junctions are on. In the *cutoff region* the collector current is reduced to a small value, down in the low nanoampere range, and so the transistor looks essentially like an open circuit. In the saturation region the voltage drop between collector and emitter becomes small, usually less than 0.1 V, and the transistor looks like a small resistance.

7.2.3 Junction Field-Effect Transistors

A junction field-effect transistor or JFET is a type of transistor in which the current flow through the device between the drain and source electrodes is controlled by the voltage applied to the gate electrode. A simple physical model of the JFET is shown in Fig. 7.11. In this JFET an N-type conducting channel exists between *drain* and *source*. The gate is a heavily doped P-type region (designated as P^+), that surrounds the N-type channel. The gate-to-channel PN junction is normally kept reverse biased. As the reverse bias voltage between gate and channel increases, the depletion region width increases, as shown in Fig. 7.12. The depletion region extends mostly into the N-type channel because of the heavy doping on the P^+ side. The depletion region is depleted of mobile charge carriers and, thus, cannot contribute to the conduction of current between drain and source. Thus, as the gate voltage increases, the cross-sectional area of the N-type channel available for current flow decreases. This reduces the current flow between drain and source. As the gate voltage increases, the channel becomes further constricted, and the current flow gets smaller. Finally, when the depletion regions meet in the middle of the channel, as shown in Fig. 7.13, the channel is pinched off in its entirety, all of the way between

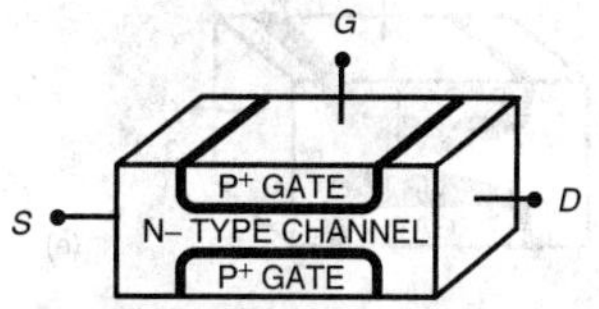

FIGURE 7.11 Model of a JFET device.

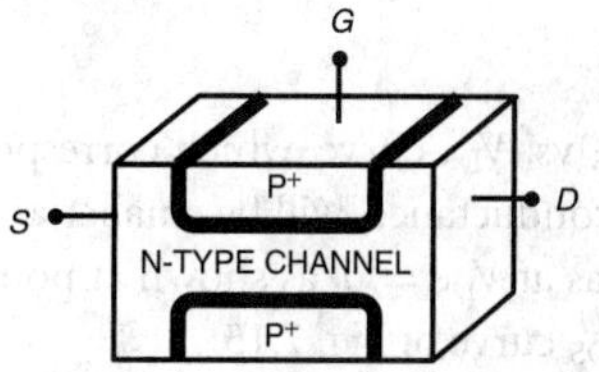

FIGURE 7.12 JFET with increased gate voltage.

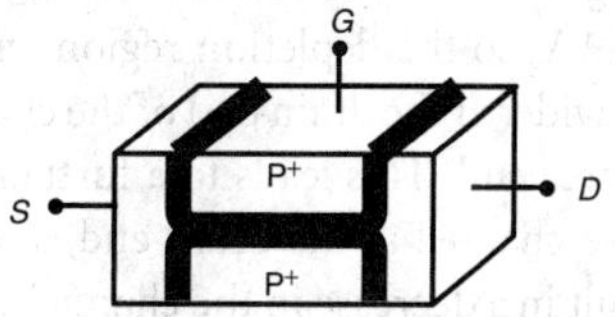

FIGURE 7.13 JFET with pinched-off channel.

the source and the drain. At this point the current flow between drain and source is reduced to essentially zero. This voltage is called the **pinch-off voltage** V_P. The pinch-off voltage is also represented as V_{GS} (OFF), as being the gate-to-source voltage that turns the drain-to-source current I_{DS} off. We have been considering here an N-channel JFET. The complementary device is the P-channel JFET, which has a heavily doped N-type (N^+) gate region surrounding a P-type channel. The operation of a P-channel JFET is the same as for an N-channel device, except the algebraic signs of all DC voltages and currents are reversed.

We have been considering the case for V_{DS} small compared to the pinch-off voltage such that the channel is essentially uniform from drain to source, as shown in Fig. 7.14(a). Now let us see what happens as V_{DS} increases. As an example, assume an N-channel JFET with a pinch-off voltage of $V_P = -4$ V. We will see what happens for the case of $V_{GS} = 0$ as V_{DS} increases. In Fig. 7.14(a) the situation is shown for the case of $V_{DS} = 0$ in which the JFET is fully on and there is a uniform channel from source to drain. This is at point A on the I_{DS} vs. V_{DS} curve of Fig. 7.15. The drain-to-source conductance is at its maximum value of g_{ds} (ON), and the drain-to-source resistance is correspondingly at its minimum value of r_{ds}(ON). Now, consider the case of $V_{DS} = +1$ V as shown in Fig. 7.14(b). The gate-to-channel bias voltage at the source end is still $V_{GS} = 0$. The gate-to-channel bias voltage at the drain end is $V_{GD} = V_{GS} - V_{DS} = -1$ V, so the depletion region will be wider at the drain end of the channel than at the source end. The channel will, thus, be narrower at the drain end than at the source end, and this will result in a decrease in the channel conductance g_{ds}, and correspondingly, an increase in the channel resistance r_{ds}. Thus, the slope

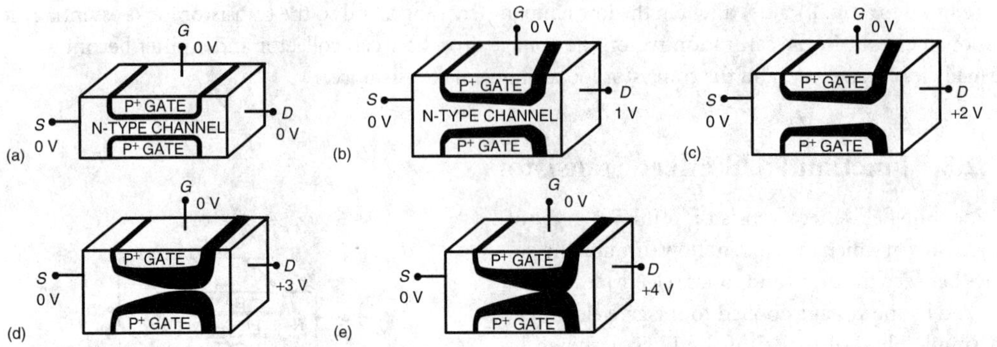

FIGURE 7.14　JFET operational characteristics: (a) Uniform channel from drain to source, (b) depletion region wider at the drain end, (c) depletion region significantly wider at the drain, (d) channel near pinchoff, (e) channel at pinchoff.

of the I_{DS} vs. V_{DS} curve, which corresponds to the channel conductance, will be smaller at $V_{DS} = 1$ V than it was at $V_{DS} = 0$, as shown at point B on the I_{DS} vs. V_{DS} curve of Fig. 7.15.

In Fig. 7.14(c) the situation for $V_{DS} = +2$ V is shown. The gate-to-channel bias voltage at the source end is still $V_{GS} = 0$, but the gate-to-channel bias voltage at the drain end is now $V_{GD} = V_{GS} - V_{DS} = -2$ V, so the depletion region will be substantially wider at the drain end of the channel than at the source end. This leads to a further constriction of the channel at the drain end, and this will

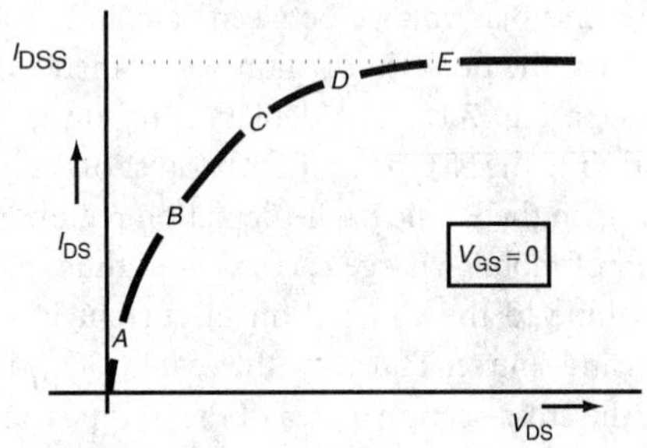

FIGURE 7.15　I_{DS} vs. V_{DS} curve.

again result in a decrease in the channel conductance g_{ds}, and correspondingly, an increase in the channel resistance r_{ds}. Thus the slope of the I_{DS} vs. V_{DS} curve will be smaller at $V_{DS} = 2$ V than it was at $V_{DS} = 1$ V, as shown at point C on the I_{DS} vs. V_{DS} curve of Fig. 7.15.

In Fig. 7.14(d) the situation for $V_{DS} = +3$ V is shown, and this corresponds to point D on the I_{DS} vs. V_{DS} curve of Fig. 7.15.

When $V_{DS} = +4$ V the gate-to-channel bias voltage will be $V_{GD} = V_{GS} - V_{DS} = 0 - 4\,V = -4\,V = V_P$. As a result the channel is now pinched off at the drain end, but is still wide open at the source end since $V_{GS} = 0$, as shown in Fig. 7.14(e). It is important to note that channel is pinched off just for a very short distance at the drain end, so that the drain-to-source current I_{DS} can still continue to flow. This is not at all the same situation as for the case of $V_{GS} = V_P$ wherein the channel is pinched off in its entirety, all of the way from source to drain. When this happens, it is like having a big block of insulator the entire distance between source and drain, and I_{DS} is reduced to essentially zero. The situation for $V_{DS} = +4\,V = -V_P$ is shown at point E on the I_{DS} vs. V_{DS} curve of Fig. 7.15.

For $V_{DS} > +4$ V, the current essentially saturates, and does not increase much with further increases in V_{DS}. As V_{DS} increases above $+4$ V, the pinched-off region at the drain end of the channel gets wider, which increases r_{ds}. This increase in r_{ds} essentially counterbalances the increase in V_{DS} such that I_{DS} does not increase much. This region of the I_{DS} vs. V_{DS} curve in which the channel is pinched off at the drain end is called the **active region**, also known as the *saturated region*. It is called the active region because when the JFET is to be used as an amplifier it should be biased and operated in this region. The saturated value of drain current up in the active region for the case of $V_{GS} = 0$ is called I_{DSS}. Since there is not really a true saturation of current in the active region, I_{DSS} is usually specified at some value of V_{DS}. For most JFETs, the values of I_{DSS} fall in the range of 1–30 mA. In the current specification, I_{DSS}, the third subscript S refers to I_{DS} under the condition of the gate *shorted* to the source.

The region below the active region where $V_{DS} < +4 \text{ V} = -V_P$ has several names. It is called the **non-saturated region**, the **triode region**, and the **ohmic region**. The term triode region apparently originates from the similarity of the shape of the curves to that of the vacuum tube triode. The term ohmic region is due to the variation of I_{DS} with V_{DS} as in Ohm's law, although this variation is nonlinear except for the region of V_{DS}, which is small compared to the pinch-off voltage, where I_{DS} will have an approximately linear variation with V_{DS}.

The upper limit of the active region is marked by the onset of the breakdown of the gate-to-channel PN junction. This will occur at the drain end at a voltage designated as $BV_{DG} = BV_{DS}$, since $V_{GS} = 0$. This breakdown voltages is generally in the 30–150 V range for most JFETs.

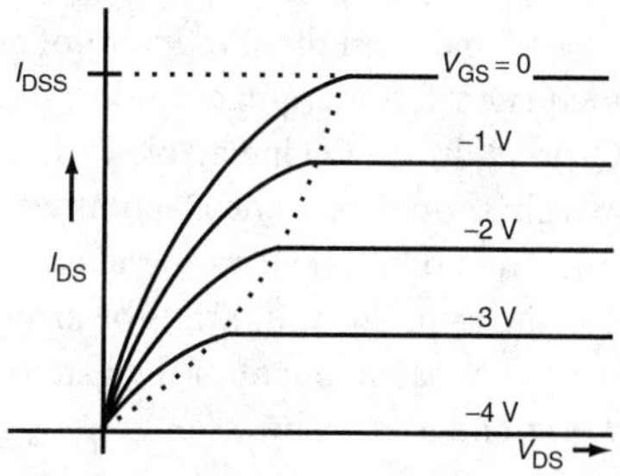

Thus far we have looked at the I_{DS} vs. V_{DS} curve only for the case of $V_{GS} = 0$. In Fig. 7.16 a family of curves of I_{DS} vs. V_{DS} for various constant values of V_{GS} is presented. This is called the *drain characteristics*, and is also known as the **output characteristics**, since the output side of the JFET is usually the drain side. In the active region where I_{DS} is relatively independent of V_{DS}, there is a simple approximate equation relating I_{DS} to V_{GS}. This is the square law **transfer equation** as given by $I_{DS} = I_{DSS}[1-(V_{GS}/V_P)]^2$. In Fig. 7.17 a graph of the I_{DS} vs. V_{GS} *transfer characteristics* for the JFET is presented. When $V_{GS} = 0$, $I_{DS} = I_{DSS}$ as expected, and as $V_{GS} \to V_P$, $I_{DS} \to 0$. The lower boundary of the active region is controlled by the condition that the channel be pinched off at the drain end. To meet this condition the basic requirement is that the gate-to-channel bias voltage at the drain end of the channel, V_{GD}, be greater than the pinch-off voltage V_P. For the example under consideration with $V_P = -4$ V, this means that $V_{GD} = V_{GS} - V_{DS}$ must be more negative than -4 V. Therefore, $V_{DS} - V_{GS} \geq +4$ V. Thus, for $V_{GS} = 0$, the active region will begin at $V_{DS} = +4$ V. When $V_{GS} = -1$ V, the active region will begin at $V_{DS} = +3$ V, for now $V_{GD} = -4$ V.

FIGURE 7.16 JFET drain characteristics.

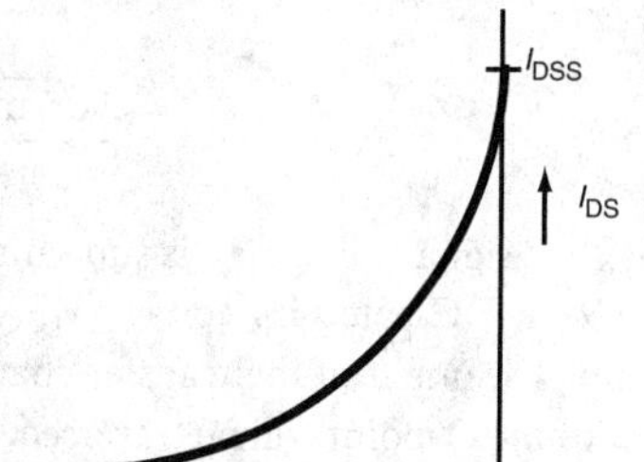

FIGURE 7.17 JFET transfer characteristic.

When $V_{GS} = -2$ V, the active region begins at $V_{DS} = +2$ V, and when $V_{GS} = -3$ V, the active region begins at $V_{DS} = +1$ V. The dotted line in Fig. 7.16 marks the boundary between the nonsaturated and active regions. The upper boundary of the active region is marked by the onset of the avalanche breakdown of the gate-to-channel PN junction. When $V_{GS} = 0$, this occurs at $V_{DS} = BV_{DS} = BV_{DG}$. Since $V_{DG} = V_{DS} - V_{GS}$, and breakdown occurs when $V_{DG} = BV_{DG}$, as V_{GS} increases the breakdown voltages decreases as given by $BV_{DG} = BV_{DS} - V_{GS}$. Thus, $BV_{DS} = BV_{DG} + V_{GS}$. For example, if the gate-to-channel breakdown voltage is 50 V, the V_{DS} breakdown voltage will start off at 50 V when $V_{GS} = 0$, but decreases to 46 V when $V_{GS} = -4$ V. In the nonsaturated region I_{DS} is a function of both V_{GS} and I_{DS}, and in the lower portion of the nonsaturated region where V_{DS} is small compared to V_P, I_{DS} becomes an approximately linear function of V_{DS}. This linear portion of the nonsaturated region is called the *voltage-variable resistance* (VVR) region, for in this region the JFET acts like a linear resistance element between source and drain. The resistance is variable in that it is controlled by the gate voltage.

JFET as an Amplifier: Small-Signal AC Voltage Gain

Consider the common-source amplifier circuit of Fig. 7.18. The input AC signal is applied between gate and source, and the output AC voltage between be taken between drain and source. Thus the source electrode of this triode device is common to input and output, hence the designation of this JFET as a common-source (CS) amplifier.

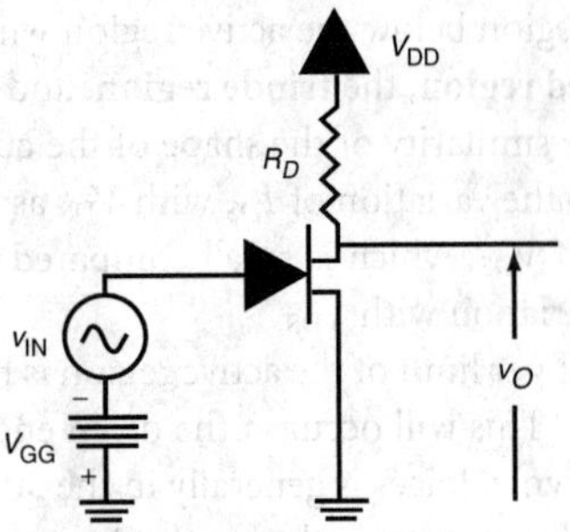

A good choice of the DC operating point or quiescent point (Q-point) for an amplifier is in the middle of the active region at $I_{DS} = I_{DSS}/2$. This allows for the maximum symmetrical drain current swing, from the quiescent level of $I_{DSQ} = I_{DSS}/2$, down to a minimum of $I_{DS} \cong 0$ and up to a maximum of $I_{DS} = I_{DSS}$. This choice for the Q-point is also a good one from the standpoint of allowing for an adequate safety margin for the location of the

FIGURE 7.18   A common source amplifier.

actual Q-point due to the inevitable variations in device and component characteristics and values. This safety margin should keep the Q-point well away from the extreme limits of the active region, and thus insure operation of the JFET in the active region under most conditions. If $I_{DSS} = +10$ mA, then a good choice for the Q-point would thus be around $+5.0$ mA. The AC component of the drain current, i_{ds} is related to the AC component of the gate voltage, v_{gs} by $i_{ds} = g_m \cdot v_{gs}$, where g_m is the *dynamic transfer conductance*, and is given by

$$g_m = \frac{2\sqrt{I_{DS} \cdot I_{DSS}}}{-V_P}$$

If $V_p = -4$ V, then

$$g_m = \frac{2\sqrt{5\,\text{mA} \cdot 10\,\text{mA}}}{4\,\text{V}} = \frac{3.54\,\text{mA}}{V} = 3.54\,\text{mS}$$

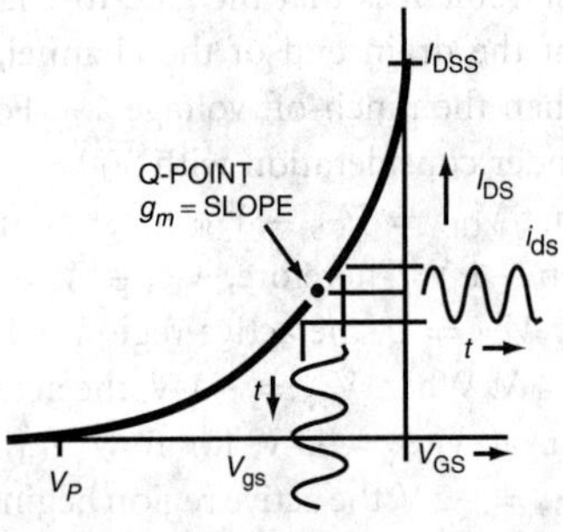

If a small AC signal voltage v_{gs} is superimposed on the quiescent DC gate bias voltage $V_{GSQ} = V_{GG}$, only a small segment of the transfer characteristic adjacent to the Q-point will be traversed, as shown in Fig. 7.19. This small segment will be close to a straight line, and as a result the AC drain current i_{ds}, will have a waveform close to that of the AC voltage applied to the gate. The ratio of i_{ds} to v_{gs} will be the slope of the transfer curve as given by

$$\frac{i_{ds}}{v_{gs}} \cong \frac{d\,I_{DS}}{d\,V_{GS}} = g_m$$

FIGURE 7.19   JFET transfer characteristic.

Thus $i_{ds} \cong g_m \cdot v_{gs}$. If the net load driven by the drain of the JFET is the drain load resistor R_D, as shown in Fig. 7.18, then the AC drain current i_{ds} will produce an AC drain voltage of $v_{ds} = -i_{ds} \cdot R_D$. Since $i_{ds} = g_m \cdot v_{gs}$, this becomes $v_{ds} = -g_m v_{gs} \cdot R_D$. The AC small-signal voltage gain from gate to drain thus becomes

$$A_V = \frac{v_O}{v_{IN}} = \frac{v_{ds}}{v_{gs}} = -g_m \cdot R_D$$

The negative sign indicates signal inversion as is the case for a common-source amplifier.

If the DC drain supply voltage is $V_{DD} = +20$ V, a quiescent drain-to-source voltage of $V_{DSQ} = V_{DD}/2 = +10$ V will result in the JFET being biased in the middle of the active region. Since $I_{DSQ} = 5$ mA, in the example under consideration, the voltage drop across the drain load resistor R_D is 10 V. Thus $R_D = 10$ V/5 mA $= 2$ kΩ. The AC small-signal voltage gain, A_V, thus becomes

$$A_V = -g_m \cdot R_D = -3.54\,\text{mS} \cdot 2\,\text{k}\Omega = -7.07$$

Note that the voltage gain is relatively modest, as compared to the much larger voltage gains that can be obtained with the bipolar-junction transistor common-emitter amplifier. This is due to the lower transfer conductance of both JFETs and metal-oxide-semiconductor field-effect transistors (MOSFETs) as compared to BJTs. For a BJT the transfer conductance is given by $g_m = I_C/V_T$ where I_C is the quiescent collector current and $V_T = kT/q \cong 25$ mV is the **thermal voltage**. At $I_C = 5$ mA, $g_m = 5$ mA/25 mV $= 200$ mS for the BJT, as compared to only 3.5 mS for the JFET in this example. With a net load of 2 kΩ, the BJT voltage gain will be -400 as compared to the JFET voltage gain of only 7.1. Thus FETs do have the disadvantage of a much lower transfer conductance and, therefore, lower voltage gain than BJTs operating under similar quiescent current levels; but they do have the major advantage of a much higher input impedance and a much lower input current. In the case of a JFET the input signal is applied to the *reverse-biased* gate-to-channel PN junction, and thus sees a very high impedance. In the case of a common-emitter BJT amplifier, the input signal is applied to the *forward-biased* base-emitter junction and the input impedance is given approximately by $r_{IN} = r_{BE} \cong 1.5 \cdot \beta \cdot V_T/I_C$. If $I_C = 5$ mA and $\beta = 200$, for example, then $r_{IN} \cong 1500$ Ω. This moderate input resistance value of 1.5 kΩ is certainly no problem if the signal source resistance is less than around 100 Ω. However, if the source resistance is above 1 kΩ, there will be a substantial signal loss in the coupling of the signal from the signal source to the base of the transistor. If the source resistance is in the range of above 100 kΩ, and certainly if it is above 1 MΩ, then there will be severe signal attenuation due to the BJT input impedance, and a FET amplifier will probably offer a greater overall voltage gain. Indeed, when high impedance signal sources are encountered, a multistage amplifier with a FET input stage, followed by cascaded BJT stages is often used.

JFET as a Constant Current Source

An important application of a JFET is as a constant current source or as a current regulator diode. When a JFET is operating in the active region, the drain current I_{DS} is relatively independent of the drain voltage V_{DS}. The JFET does not, however, act as an ideal constant current source since I_{DS} does increase slowly with increases in V_{DS}. The rate of change of I_{DS} with V_{DS} is given by the drain-to-source conductance $g_{ds} = dI_{DS}/dV_{DS}$. Since I_{DS} is related to the channel length L by $I_{DS} \propto 1/L$, the drain-to-source conductance g_{ds} can be expressed as

$$g_{ds} = \frac{dI_{DS}}{dV_{DS}} = \frac{dI_{DS}}{dL} \cdot \frac{dL}{dV_{DS}}$$

$$= \frac{-I_{DS}}{L} \cdot \frac{dL}{dV_{DS}} = I_{DS}\left(\frac{-1}{L}\right)\left(\frac{dL}{dV_{DS}}\right)$$

The channel length modulation coefficient is defined as

$$\text{channel length modulation coefficient} = \frac{1}{V_A} = \frac{-1}{L}\left(\frac{dL}{dV_{DS}}\right)$$

where V_A is the JFET *early voltage*. Thus we have that $g_{ds} = I_{DS}/V_A$. The early voltage V_A for JFETs is generally in the range of 20–200 V.

The *current regulation* of the JFET acting as a constant current source can be expressed in terms of the fractional change in current with voltage as given by

$$\text{current regulation} = \left(\frac{1}{I_{DS}}\right)\frac{dI_{DS}}{dV_{DS}} = \frac{g_{ds}}{I_{ds}} = \frac{1}{V_A}$$

For example, if $V_A = 100$ V, the current regulation will be $1/(100$ V$) = 0.01/$V $= 1\%/$V, so I_{DS} changes by only 1% for every 1 V change in V_{DS}.

In Fig. 7.20 a diode-connected JFET or *current regulator diode* is shown. Since $V_{GS} = 0$, $I_{DS} = I_{DSS}$. The current regulator diode can be modeled as an ideal constant current source in parallel with a resistance r_O

as shown in Fig. 7.21. The *voltage compliance range* is the voltage range over which a device or system acts as a good approximation to the ideal constant current source. For the JFET this will be the extent of the active region. The lower limit is the point where the channel just becomes pinched off at the drain end. The requirement is, thus, that $V_{DG} = V_{DS} - V_{GS} > -V_P$. For the case of $V_{GS} = 0$, this occurs at $V_{DS} = -V_P$. The upper limit of the voltage compliance range is set by the breakdown voltage of the gate-to-channel PN junction, BV_{DG}. Since V_P is typically in the 2–5 V range, and BV_{DG} is typically >30 V, this means that the voltage compliance range will be relatively large. For example, if $V_P = -3$ V and $BV_{DG} = +50$ V, the voltage compliance range

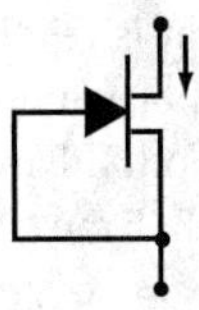

FIGURE 7.20 A current regulator diode

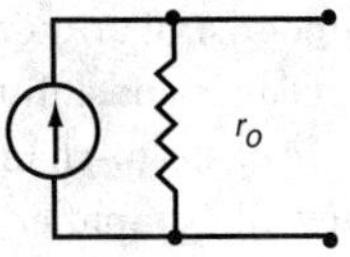

FIGURE 7.21 Model of the current regulator diode.

will extend from $V_{DS} = +3$ V up to $V_{DS} = +50$ V. If $V_A = 100$ V and $I_{DSS} = 10$ mA, the current regulator dynamic output conductance will be

$$g_O = \frac{dI_O}{dV_O} = \frac{dI_{ds}}{dV_{ds}} = g_{ds} = \frac{10\ \text{mA}}{100\text{V}} = 0.1\ \text{mA/V} = 0.1\ \text{mS}$$

The current regulator dynamic output resistance will be $r_O = r_{ds} = 1/g_{ds} = 10$ kΩ. Thus, the current regulator diode can be represented as a 10-mA constant current source in parallel with a 10-kΩ dynamic resistance.

In Fig. 7.22 a current regulator diode is shown in which a resistor R_S is placed in series with the JFET source in order to reduce I_{DS} below I_{DSS}. The current I_{DS} flowing through R_S produces a voltage drop $V_{SG} = I_{DS} \cdot R_S$. This results in a gate-to-source bias voltage of $V_{GS} = -V_{SG} = -I_{DS} \cdot R_S$. From the JFET transfer equation, $I_{DS} = I_{DSS}(1 - (V_{GS}/V_P))^2$ we have that $V_{GS} = V_P(1 - \sqrt{I_{DS}I_{DSS}})$. From the required value of I_{DS} the corresponding value of V_{GS} can be determined, and from that the value of R_S can be found.

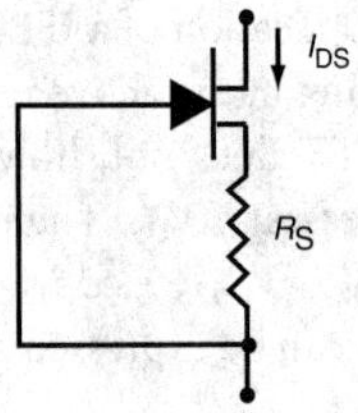

FIGURE 7.22 A current regulator diode for $I_{DS} < I_{DSS}$.

With R_S present, the dynamic output conductance, $g_O = dI_O/dV_O$, becomes $g_O = g_{ds}/(1 + g_m R_S)$. The current regulation now given as

$$\text{current regulation} = \left(\frac{1}{I_O}\right)\frac{dI_O}{dV_O} = \frac{g_O}{I_O}$$

Thus R_S can have a beneficial effect in reducing g_O and improving the current regulation. For example, let $V_P = -3$ V, $V_A = 100$ V, $I_{DSS} = 10$ mA, and $I_O = I_{DS} = 1$ mA. We now have that

$$V_{GS} = V_{GS} = V_P\left[1 - \sqrt{\frac{I_{DS}}{I_{DSS}}}\right] = -3\,\text{V}\left[1 - \sqrt{\frac{1\,\text{mA}}{10\,\text{mA}}}\right] = -2.05\,\text{V}$$

and so $R_S = 2.05$ V/1 mA $= 2.05$ kΩ. The transfer conductance g_m is given by

$$g_m = \frac{2\sqrt{I_{DS} \cdot I_{DSS}}}{-V_P} = \frac{2\sqrt{1\,\text{mA} \cdot 10\,\text{mA}}}{3\,\text{V}} = 2.1\,\text{mS}$$

and so $g_m R_S = 2.1$ mS $\cdot 2.05$ kΩ $= 4.32$. Since $g_{ds} = 1$ mA/100 V $= 10\ \mu$S, we have

$$g_O = \frac{g_{ds}}{1 + g_m R_S} = \frac{10\ \mu\text{S}}{5.32} = 1.9\ \mu\text{S}$$

The current regulation is thus $g_O/I_O = 1.9\ \mu\text{S}/1\ \text{mA} = 0.0019/\text{V} = 0.19\%/\text{V}$. This is to be compared to the current regulation of $1\%/\text{V}$ obtained for the case of $I_{DS} = I_{DSS}$.

Any JFET can be used as a current regulating diode. There are, however, JFETs that are especially made for this application. These JFETs have an extra long channel length, which reduces the channel length modulation effect and, hence, results in a large value for V_A. This in turn leads to a small g_{ds} and, hence, a small g_O and, thus, good current regulation.

Operation of a JFET as a Voltage-Variable Resistor

A JFET can be used as voltage-variable resistor in which the drain-to-source resistance r_{ds} of the JFET can be varied by variation of V_{GS}. For values of $V_{DS} \ll V_P$ the I_{DS} vs. V_{DS} characteristics are approximately linear, and so the JFET looks like a resistor, the resistance value of which can be varied by the gate voltage.

The channel conductance in the region where $V_{DS} \ll V_P$ is given by $g_{ds} = A\sigma/L = WH\sigma/L$, where the channel height H is given by $H = H_0 - 2W_D$. In this equation W_D is the depletion region width and H_0 is the value of H as $W_D \to 0$. The depletion region width is given by $W_D = K\sqrt{V_J} = K\sqrt{V_{GS} + \phi}$ where K is a constant, V_J is the junction voltage, and ϕ is the PN junction **contact potential** (typical around 0.8–1.0 V). As V_{GS} increases, W_D increases and the channel height H decreases as given by $H = H_0 - 2K\sqrt{V_{GS} + \phi}$. When $V_{GS} = V_P$, the channel is completely pinched off, so $H = 0$.

The drain-to-source resistance r_{ds} is given approximately by $r_{ds} \cong r_{ds}(\text{ON})/[1 - \sqrt{V_{GS}/V_P}]$. As $V_{GS} \to 0, r_{ds} \to r_{ds}(\text{ON})$ and as $V_{GS} \to V_P, r_{ds} \to \infty$. This latter condition corresponds to the channel being pinched off in its entirety, all of the way from source to drain. This is like having big block of insulator (i.e., the depletion region) between source and drain. When $V_{GS} = 0, r_{ds}$ is reduced to its minimum value of $r_{ds}(\text{ON})$, which for most JFETs is in the 20–4000 Ω range. At the other extreme, when $V_{GS} > V_P$, the drain-to-source current I_{DS} is reduced to a small value, generally down into the low nanoampere, or even picoampere range. The corresponding value of r_{ds} is not really infinite, but is very

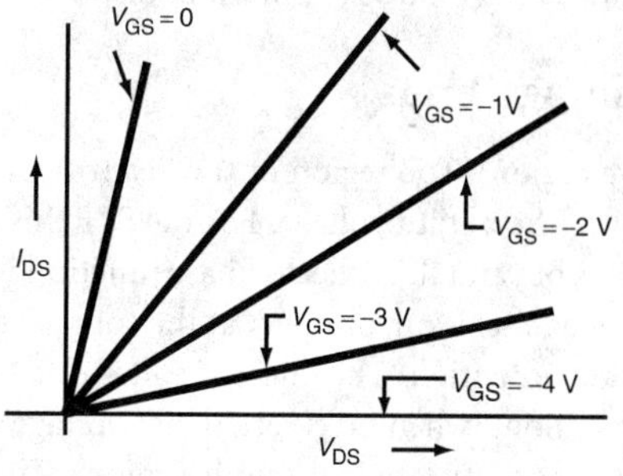

FIGURE 7.23 Voltage-variable resistor characteristics of the JFET.

large, generally well up into the gigaohm (1000 MΩ) range. Thus, by variation of V_{GS}, the drain-to-source resistance can be varied over a very wide range. As long as the gate-to-channel junction is reverse biased, the gate current will be very small, generally down in the low nanoampere, or even picoampere range, so that the gate as a control electrode draws little current. Since V_P is generally in the 2–5 V range for most JFETs, the V_{DS} values required to operate the JFET in the VVR range is generally < 0.1 V. In Fig. 7.23 the VVR region of the JFET I_{DS} vs. V_{DS} characteristics is shown.

Voltage-Variable Resistor Applications

Applications of VVRs include automatic gain control (AGC) circuits, electronic attenuators, electronically variable filters, and oscillator amplitude control circuits.

When using a JFET as a VVR it is necessary to limit V_{DS} to values that are small compared to V_P to maintain good linearity. In addition, V_{GS} should preferably not exceed $0.8V_P$ for good linearity, control, and stability. This limitation corresponds to an r_{ds} resistance ratio of about 10:1. As V_{GS} approaches V_P, a small change in V_P can produce a large change in r_{ds}. Thus, unit-to-unit variations in V_P as well as changes in V_P with temperature can result in large changes in r_{ds} as V_{GS} approaches V_P.

The drain-to-source resistance r_{ds} will have a temperature coefficient TC due to two causes: (1) the variation of the channel resistivity with temperature and (2) the temperature variation of V_P. The TC of the channel resistivity is positive, whereas the TC of V_P is positive due to the negative TC of the contact potential ϕ. The positive TC of the channel resistivity will contribute to a positive TC of r_{ds}. The negative TC of V_P will contribute to a negative TC of r_{ds}. At small values of V_{GS}, the dominant contribution to the TC is the positive TC of the channel resistivity, and so r_{ds} will have a positive TC. As V_{GS} gets larger, the negative TC contribution of V_P becomes increasingly important, and there will be a value of V_{GS} at which the net TC of r_{ds} is zero and above this value of V_{GS} the TC will be negative. The TC of r_{ds}(ON) is typically +0.3%/°C for N-channel JFETs, and +0.7%/°C for P-channel JFETs. For example, for a typical JFET with an r_{ds}(ON) = 500 Ω at 25°C and V_P = 2.6 V, the zero temperature coefficient point will occur at V_{GS} = 2.0 V. Any JFET can be used as a VVR, although there are JFETs that are specifically made for this application.

Example of VVR Application

A simple example of a VVR application is the electronic gain control circuit of Fig. 7.24. The voltage gain is given by $A_V = 1 + (R_F/r_{ds})$. If, for example, R_F = 19 kΩ and $r_{ds}(ON)$ = 1 kΩ, then the maximum gain will be $A_{VMAX} = 1 + [R_F/r_{ds}(ON)] = 20$. As V_{GS} approaches V_P, the r_{ds} will increase and become very large such that $r_{ds} \gg R_F$, so that A_V will decrease to a minimum value of close to unity. Thus, the gain can be varied over a 20:1 ratio. Note that $V_{DS} \cong V_{IN}$, and so to minimize distortion, the input signal amplitude should be small compared to V_P.

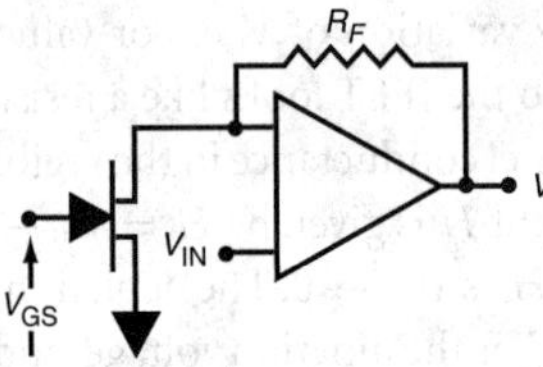

FIGURE 7.24   An electronic gain control circuit.

Defining Terms

Active region: The region of transistor operation in which the output current is relatively independent of the output voltage. For the BJT this corresponds to the condition that the emitter-base junction is on, and the collector-base junction is off. For the FETs this corresponds to the condition that the channel is on, or open, at the source end, and pinched off at the drain end.

Contact potential: The internal voltage that exists across a PN junction under thermal equilibrium conditions, when no external bias voltage is applied.

Ohmic, nonsaturated, or triode region: These three terms all refer to the region of FET operation in which a conducting channel exists all of the way between source and drain. In this region the drain current varies with both the gate voltage and the drain voltage.

Output characteristics: The family of curves of output current vs. output voltage. For the BJT this will be curves of collector current vs. collector voltage for various constant values of base current or voltage, and is also called the collector characteristics. For FETs this will be curves of drain current vs. drain voltage for various constant values of gate voltage, and is also called the drain characteristics.

Pinch-off voltage, V_P: The voltage that when applied across the gate-to-channel PN junction will cause the conducting channel between drain and source to become pinched off. This is also represented as $V_{GS(OFF)}$.

Thermal voltage: The quantity kT/q where k is Boltzmann's constant, T is absolute temperature, and q is electron charge. The thermal voltage has units of volts, and is a function only of temperature, being approximately 25 mV at room temperature.

Transfer conductance: The AC or dynamic parameter of a device that is the ratio of the AC output current to the AC input voltage. The transfer conductance is also called the *mutual transconductance*, and is usually designated by the symbol g_m.

Transfer equation: The equation that relates the output current (collector or drain current) to the input voltage (base-to-emitter or gate-to-source voltage).

Triode: A three-terminal electron device, such as a bipolar junction transistor or a field-effect transistor.

References

Bogart, T.F., Beasley, J.S., and Rico, G. 2001. *Electronic Devices and Circuits*, 5th ed. Prentice-Hall, Upper Saddle River, NJ.

Boylestad, R. and Nashelsky, L. 2000. *Electronic Devices and Circuit Theory*, 8th ed. Prentice Hall, Upper Saddle River, NJ.

Cathey, J. 2002. *Electronic Devices and Circuits*, McGraw-Hill, New York.

Comer, D.J. and Comer, D.T. 2003. *Advanced Electronic Circuit Design*, John Wiley & Sons, New York.

Comer, D.J. and Comer, D.T. 2003. *Fundamentals of Electronic Circuit Design*, John Wiley & Sons, New York.

Dailey, D. 2001. *Electronic Devices and Circuits: Discrete and Integrated*, Prentice-Hall, Upper Saddle River, NJ.

Fleeman, S. 2003. *Electronic Devices*, 7th ed. Prentice-Hall, Upper Saddle River, NJ.

Floyd, T. 2002. *Electronic Devices*, 6th ed. Prentice-Hall, Upper Saddle River, NJ.

Hassul, M. and Zimmerman, D.E. 1996. *Electronic Devices and Circuits*, Prentice-Hall, Upper Saddle River, NJ.

Horenstein, M.N. 1996. *Microelectronic Circuit and Devices*, 2nd ed. Prentice-Hall, Upper Saddle River, NJ.

Kasap, S. 2002. *Principles Of Electronic Materials And Devices With CD-ROM*, 2nd ed. McGraw-Hill, New York.

Malvino, Albert P. 1999. *Electronic Principles*, 6th ed. McGraw-Hill, New York.

Mauro, R. 1989. *Engineering Electronics*, Prentice-Hall, Upper Saddle River, NJ.

Millman, J. and Grabel, A. 1987. *Microelectronics*, 2nd ed. McGraw-Hill, New York.

Mitchell, F.H., Jr. and Mitchell, F.H., Sr. 1992. *Introduction to Electronics Design*, 2nd ed. Prentice-Hall, Upper Saddler River, NJ.

Neamen, D. 2001. *Electronic Circuit Analysis With CD-ROM With E-Text*, 2nd ed. McGraw-Hill, New York.

Schuler, C.A. 2003. *Electronics: Principles and Applications, Student Text with MultiSIM CD-ROM*, 6th ed. McGraw-Hill, New York.

Sedra, A.S. and Smith, K.C. 2003. *Microelectronic Circuits*, 5th ed. Oxford University Press, New York.

Shur, M. 1996. *Introduction to Electronic Devices*, John Wiley & Sons, New York.

Singh, J. 2001. *Semiconductor Devices: Basic Principles*, John Wiley & Sons, New York.

Spencer, R. and Ghausi, M. 2003. *Introduction to Electronic Circuit Design*, Prentice-Hall, Upper Saddle River, NJ.

Further Information

An excellent introduction to the physical electronics of devices is found in Ben G. Streetman, *Solid State Electronic Devices*, 4th ed. Prentice-Hall, Englewood Cliffs, NJ. 1995. Another excellent reference on a wide variety of electronic devices is Kwok K. Ng, *Complete Guide to Semiconductor Devices*, McGraw-Hill, New York, 1995. A useful reference on circuits and applications is Donald A. Neamen, *Electronic Circuit Analysis and Design*, Irwin, Chicago, IL, 1996.

7.3   Metal-Oxide-Semiconductor Field-Effect Transistor

John R. Brews

7.3.1   Introduction

The metal-oxide-semiconductor field-effect transistor (MOSFET) is a transistor that uses a control electrode, the **gate**, to capacitively modulate the conductance of a surface **channel** joining two end contacts, the **source** and the **drain**. The gate is separated from the semiconductor **body** underlying the gate by a thin *gate insulator*, usually silicon dioxide. The surface channel is formed at the interface between the semiconductor body and the gate insulator, see Fig. 7.25.

The MOSFET can be understood by contrast with other field-effect devices, like the junction field-effect transistor (JFET) and the metal-semiconductor field-effect transistor (MESFET) (Hollis and Murphy, 1990). These other transistors modulate the conductance of a *majority-carrier* path between two *ohmic* contacts by capacitive control of its cross section. (Majority carriers are those in greatest abundance in

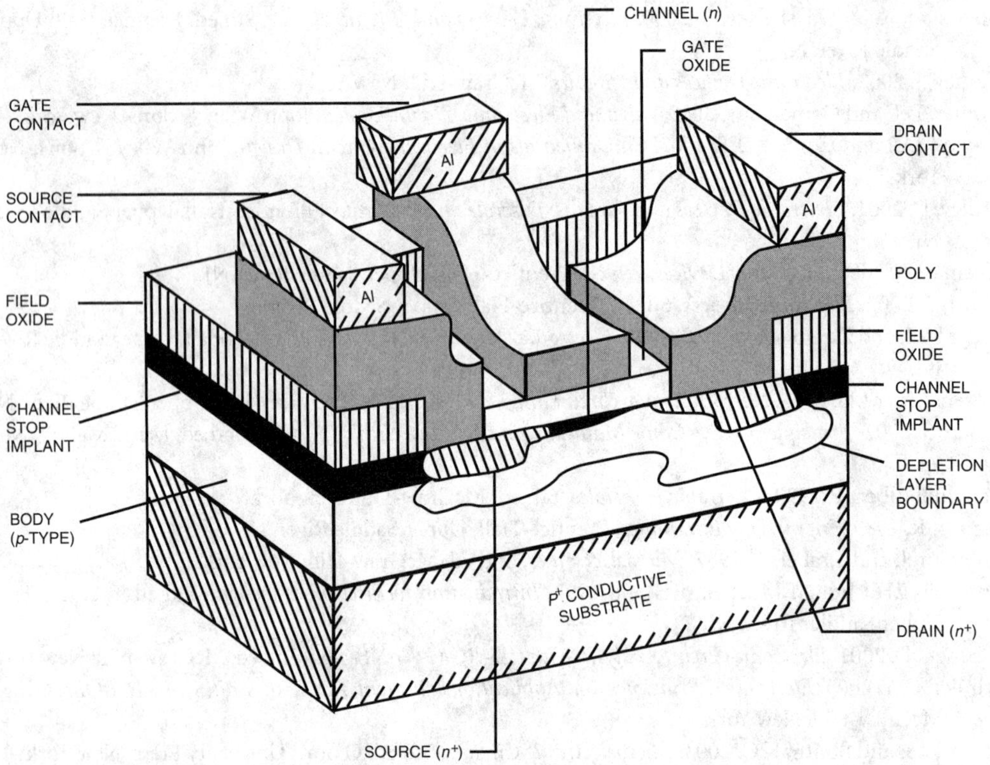

FIGURE 7.25 A high-performance *n*-channel MOSFET. The device is isolated from its neighbors by a surrounding thick *field oxide* under which is a heavily doped *channel stop* implant intended to suppress accidental channel formation that could couple the device to its neighbors. The drain contacts are placed over the field oxide to reduce the capacitance to the body, a parasitic that slows response times. These structural details are described later. (*Source:* After Brews, J.R. 1990. The submicron MOSFET. In *High-Speed Semiconductor Devices*, ed. S.M. Sze, pp. 139–210. Wiley, New York.)

field-free semiconductor, electrons in *n*-type material and holes in *p*-type material.) This modulation of the cross section can take place at any point along the length of the channel, and so the gate electrode can be positioned anywhere and need not extend the entire length of the channel.

Analogous to these field-effect devices is the *buried-channel, depletion-mode,* or *normally on* MOSFET, which contains a surface layer of the same doping type as the source and drain (opposite type to the semiconductor body of the device). As a result, it has a built-in or normally on channel from source to drain with a conductance that is reduced when the gate depletes the majority carriers.

In contrast, the true MOSFET is an *enhancement-mode* or *normally off* device. The device is normally off because the body forms *p–n* junctions with both the source and the drain, so no majority-carrier current can flow between them. Instead, *minority-carrier* current can flow, provided minority carriers are available. As discussed later, for gate biases that are sufficiently attractive, above **threshold**, minority carriers are drawn into a surface channel, forming a conducting path from source to drain. The gate and channel then form two sides of a capacitor separated by the gate insulator. As additional attractive charges are placed on the gate side, the channel side of the capacitor draws a balancing charge of minority carriers from the source and the drain. The more charges on the gate, the more populated the channel, and the larger the conductance. Because the gate *creates* the channel, to insure electrical continuity the gate must extend over the entire length of the separation between source and drain.

The MOSFET channel is created by attraction to the gate and relies on the insulating layer between the channel and the gate to prevent leakage of minority carriers to the gate. As a result, MOSFETs can be made only in material systems that provide very good gate insulators, and the best system known is the silicon-silicon dioxide combination. This requirement for a good gate insulator is not as important for JFETs and MESFETs where the role of the gate is to *push away* majority carriers, rather than to *attract* minority carriers. Thus, in GaAs systems where good insulators are incompatible with other device or fabricational requirements, MESFETs are used.

A more recent development in GaAs systems is the heterostructure field-effect transistor (HFET) (Pearton and Shah, 1990) made up of layers of varying compositions of Al, Ga, and As or In, Ga, P, and As. These devices are made using molecular beam epitaxy or by organometallic vapor phase epitaxy. HFETs include a variety of structures, the best known of which is the modulation doped FET (MODFET). HFETs are field-effect devices, not MOSFETs, because the gate simply modulates the carrier density in a pre-existent channel between ohmic contacts. The channel is formed spontaneously, regardless of the quality of the gate insulator, as a condition of equilibrium between the layers, just as a depletion layer is formed in a *p–n* junction. The resulting channel is created very near to the gate electrode, resulting in gate control as effective as in a MOSFET.

The silicon-based MOSFET has been successful primarily because the silicon-silicon dioxide system provides a stable interface with low trap densities and because the oxide is impermeable to many environmental contaminants, has a high breakdown strength, and is easy to grow uniformly and reproducibly (Nicollian and Brews, 1982). These attributes allow easy fabrication using lithographic processes, resulting in integrated circuits (ICs) with very small devices, very large device counts, and very high reliability at low cost. Because the importance of the MOSFET lies in this relationship to high-density manufacture, an emphasis of this chapter is to describe the issues involved in continuing miniaturization.

An additional advantage of the MOSFET is that it can be made using either electrons or holes as channel carrier. Using both types of devices in so-called complementary MOS (CMOS) technology allows circuits that draw no DC power if current paths include at least one series connection of both types of device because, in steady state, only one or the other type conducts, not both at once. Of course, in exercising the circuit, power is drawn during switching of the devices. This flexibility in choosing *n*- or *p*-channel devices has enabled large circuits to be made that use low-power levels. Hence, complex systems can be manufactured without expensive packaging or cooling requirements.

7.3.2 Current-Voltage Characteristics

The derivation of the current-voltage characteristics of the MOSFET can be found in (Annaratone, 1986; Brews, 1981; and Pierret, 1990). Here a qualitative discussion is provided.

Strong-Inversion Characteristics

In Fig. 7.26 the source-drain current I_D is plotted vs. drain-to-source voltage V_D (the I–V curves for the MOSFET). At low V_D the current increases approximately linearly with increased V_D, behaving like a simple resistor with a resistance that is controlled by the gate voltage V_G: as the gate voltage is made more attractive for channel carriers, the channel becomes stronger, more carriers are contained in the channel, and its resistance R_{ch} drops. Hence, at larger V_G the current is larger.

At large V_D the curves flatten out, and the current is less sensitive to drain bias. The MOSFET is said to be in *saturation*. There are different reasons for this behavior, depending on the field along the channel caused by the drain voltage. If the source-drain

FIGURE 7.26 Drain current I_D vs. drain voltage V_D for various choices of gate bias V_G. The dashed-line curves are for a *long-channel* device for which the current in saturation increases quadratically with gate bias. The solid-time curves are for a *short-channel* device that is approaching *velocity saturation* and thus exhibits a more linear increase in saturation current with gate bias, as discussed in the text.

separation is short, near, or below a micrometer, the usual drain voltage is sufficient to create fields along the channel of more then a few $\times 10^4$ V/cm. In this case the carrier energy is sufficient for carriers to lose energy by causing vibrations of the silicon atoms composing the crystal (optical phonon emission). Consequently, the carrier velocity does not increase much with increased field, saturating at a value $v_{sat} \approx 10^7$ cm/s in silicon MOSFETs. Because the carriers do not move faster with increased V_D, the current also saturates.

For longer devices the current-voltage curves saturate for a different reason. Consider the potential along the insulator-channel interface, the surface potential. Whatever the surface potential is at the source end of the channel, it varies from the source end to a value larger at the drain end by V_D because the drain potential is V_D higher than the source. The gate, on the other hand, is at the same potential everywhere. Thus, the difference in potential between the gate and the source is larger than that between the gate and drain. Correspondingly, the oxide field at the source is larger than that at the drain and, as a result, less charge can be supported at the drain. This reduction in attractive power of the gate reduces the number of carriers in the channel at the drain end, increasing channel resistance. In short, we have $I_D \approx V_D/R_{ch}$ but the channel resistance $R_{ch} = R_{ch}(V_D)$ is increasing with V_D. As a result, the current-voltage curves do not continue along the initial straight line, but bend over and saturate.

Another difference between the current-voltage curves for short devices and those for long devices is the dependence on gate voltage. For long devices, the current level in saturation $I_{D,sat}$ increases quadratically with gate bias. The reason is that the number of carriers in the channel is proportional to $V_G - V_{TH}$ (where V_{TH} is the *threshold voltage*) as is discussed later, the channel resistance $R_{ch} \propto 1/(V_G - V_{TH})$, and the drain bias in saturation is approximately V_G. Thus, $I_{D,sat} = V_D/R_{ch} \propto (V_G - V_{TH})^2$, and we have quadratic dependence. When the carrier velocity is saturated, however, the dependence of the current on drain bias is suppressed because the speed of the carriers is fixed at v_{sat}, and $I_{D,sat} \propto v_{sat}/R_{ch} \propto (V_G - V_{TH})v_{sat}$, a linear gate-voltage dependence. As a result, the current available from a short device is not as large as would be expected if we assumed it behaved like a long device.

Subthreshold Characteristics

Quite different current-voltage behavior is seen in **subthreshold**, that is, for gate biases so low that the channel is in *weak inversion*. In this case the number of carriers in the channel is so small that their charge does not affect the potential, and channel carriers simply must adapt to the potential set up by the electrodes and the dopant ions. Likewise, in subthreshold any flow of current is so small that it causes no potential drop along the interface, which becomes an equipotential.

As there is no lateral field to move the channel carriers, they move by diffusion only, driven by a gradient in carrier density setup because the drain is effective in reducing the carrier density at the drain end of the channel. In subthreshold the current is then independent of drain bias once this bias exceeds a few tens of millivolts, enough to reduce the carrier density at the drain end of the channel to near zero.

In short devices, however, the source and drain are close enough together to begin to share control of the potential with the gate. If this effect is too strong, a drain-voltage dependence of the subthreshold characteristic then occurs, which is undesirable because it increases the MOSFET off current and can cause a drain-bias dependent threshold voltage.

Although for a well-designed device there is no drain-voltage dependence in subthreshold, gate-bias dependence is exponential. The surface is lowered in energy relative to the semiconductor body by the action of the gate. If this *surface potential* is ϕ_S below that of the body, the carrier density is enhanced by a Boltzmann factor $\exp(q\,\phi_S/kT)$ relative to the body concentration, where kT/q is the thermal voltage, ≈ 25 mV at 290 K. As ϕ_S is roughly proportional to V_G, this exponential dependence on ϕ_S leads to an exponential dependence on V_G for the carrier density and, hence, for the current in subthreshold.

7.3.3 Important Device Parameters

A number of MOSFET parameters are important to the performance of a MOSFET. In this section some of these parameters are discussed, particularly from the viewpoint of digital ICs.

Threshold Voltage

The threshold voltage is vaguely defined as the gate voltage V_{TH} at which the channel begins to form. At this voltage devices begin to switch from off to on, and circuits depend on a voltage swing that straddles this value. Thus, threshold voltage helps in deciding the necessary supply voltage for circuit operation and it also helps in determining the leakage or off current that flows when the device is in the off state.

We now will make the definition of threshold voltage precise and relate its magnitude to the doping profile inside the device, as well as other device parameters such as oxide thickness and flatband voltage.

Threshold voltage is controlled by oxide thickness d and by body doping. To control the body doping, ion implantation is used, so that the dopant-ion density is not simply a uniform extension of the bulk, background level N_B ions/unit volume, but has superposed on it an implanted ion density. To estimate the threshold voltage, we need a picture of what happens in the semiconductor under the gate as the gate voltage is changed from its off level toward threshold.

If we imagine changing the gate bias from its off condition toward threshold, at first the result is to repel majority carriers, forming a surface *depletion layer*, refer to Fig. 7.25. In the depletion layer there are almost no carriers present, but there are dopant ions. In n-type material these dopant ions are positive donor impurities that cannot move under fields because they are locked in the silicon lattice, where they have been deliberately introduced to replace silicon atoms. In p-type material these dopant ions are negative acceptors. Thus, each charge added to the gate electrode to bring the gate voltage closer to threshold causes an increase in the depletion-layer width sufficient to balance the gate charge by an equal but opposite charge of dopant ions in the silicon depletion layer.

This expansion of the depletion layer continues to balance the addition of gate charge until threshold is reached. Then this charge response changes: above threshold any additional gate charge is balanced by an increasingly strong inversion layer or channel. The border between a depletion-layer and an inversion-layer response, threshold, should occur when

$$\frac{dq\,N_{inv}}{d\phi_S} = \frac{dQ_D}{d\phi_S} \qquad (7.1)$$

where $d\phi_S$ is the small change in surface potential that corresponds to our incremental change in gate charge, $q\,N_{inv}$ is the inversion-layer charge/unit area, and Q_D the depletion-layer charge/unit area. According to Eq. (7.1), the two types of response are equal at threshold, so that one is larger than the other on either side of this condition. To be more quantitative, the rate of increase in $q\,N_{inv}$ is exponential, that is, its rate of change is proportional to $q\,N_{inv}$, and so as $q\,N_{inv}$ increases, so does the left side of Eq. (7.1). On the other hand, Q_D has a square-root dependence on ϕ_S, which means its rate of change becomes smaller as Q_D increases. Thus, as surface potential is increased, the left side of Eq. (7.1) increases $\propto q\,N_{inv}$ until, at threshold, Eq. (7.1) is satisfied. Then, beyond threshold, the exponential increase in $q\,N_{inv}$ with ϕ_S swamps Q_D, making change in $q\,N_{inv}$ the dominant response. Likewise, below threshold, the exponential decrease in $q\,N_{inv}$ with decreasing ϕ_S makes $q\,N_{inv}$ negligible and change in Q_D becomes the dominant response. The abruptness of this change in behavior is the reason for the term threshold to describe MOSFET switching.

To use Eq. (7.1) to find a formula for threshold voltage, we need expressions for N_{inv} and Q_D. Assuming the interface is held at a lower energy than the bulk due to the charge on the gate, the minority-carrier density at the interface is larger than in the bulk semiconductor, even below threshold. Below threshold and even up to the threshold of Eq. (7.1), the number of charges in the channel/unit area N_{inv} is given for n-channel devices approximately by (Brews, 1981)

$$N_{inv} \approx d_{INV}\frac{n_i^2}{N_B}e^{q(\phi_S-V_S)/kT} \qquad (7.2)$$

where the various symbols are defined as follows: n_i is the intrinsic carrier density/unit volume $\approx 10^{10}/\text{cm}^3$ in silicon at 290 K and V_S is the body reverse bias, if any. The first factor, d_{INV}, is an effective depth of minority carriers from the interface given by

$$d_{\text{INV}} = \frac{\varepsilon_s kT/q}{Q_D} \tag{7.3}$$

where Q_D is the depletion-layer charge/unit area due to charged dopant ions in the region where there are no carriers and ε_s is the dielectric permittivity of the semiconductor.

Equation (7.2) expresses the net minority-carrier density/unit area as the product of the bulk minority-carrier density/unit volume n_i^2/N_B, with the depth of the minority-carrier distribution d_{INV} multiplied in turn by the customary Boltzmann factor $\exp(q(\phi_S - V_S)/kT)$ expressing the enhancement of the interface density over the bulk due to lower energy at the interface. The depth d_{INV} is related to the carrier distribution near the interface using the approximation (valid in weak inversion) that the minority-carrier density decays exponentially with distance from the oxide-silicon surface. In this approximation, d_{INV} is the *centroid* of the minority-carrier density. For example, for a uniform bulk doping of 10^{16} dopant ions/cm^3 at 290 K, using Eq. (7.2) and the surface potential at threshold from Eq. (7.7) ($\phi_{\text{TH}} = 0.69$ V), there are $Q_D/q = 3 \times 10^{11}$ charges/cm^2 in the depletion layer at threshold. This Q_D corresponds to a $d_{\text{INV}} = 5.4$ nm and a carrier density at threshold of $N_{\text{inv}} = 5.4 \times 10^9$ charges/cm^2.

The next step in using the definition of threshold, Eq. (7.1), is to introduce the depletion-layer charge/unit area Q_D. For the ion-implanted case, Q_D is made up of two terms (Brews, 1981)

$$Q_D = q N_B L_B (2(q\phi_{\text{TH}}/kT - m_1 - 1))^{\frac{1}{2}} + q D_I \tag{7.4}$$

where the first term is Q_B, the depletion-layer charge from bulk dopant atoms in the depletion layer with a width that has been reduced by the first moment of the implant, namely, m_1 given in terms of the centroid of the implant x_C by

$$m_1 = \frac{D_I x_C}{N_B L_B^2} \tag{7.5}$$

The second term is the additional charge due to the implanted-ion density within the depletion layer, D_I/unit area. The Debye length L_B is defined as

$$L_B^2 \equiv \left[\frac{kT}{q}\right]\left[\frac{\varepsilon_s}{q N_B}\right] \tag{7.6}$$

where ε_s is the dielectric permittivity of the semiconductor. The Debye length is a measure of how deeply a variation of surface potential penetrates into the body when $D_I = 0$ and the depletion layer is of zero width.

Approximating $q N_{\text{inv}}$ by Eq. (7.2) and Q_D by Eq. (7.4), Eq. (7.1) determines the surface potential at threshold ϕ_{TH} to be

$$\phi_{\text{TH}} = 2(kT/q)\ell_n(N_B/n_i) + (kT/q)\ell_n\left[1 + \frac{q D_I}{Q_B}\right] \tag{7.7}$$

where the new symbols are defined as follows: Q_B is the depletion-layer charge/unit area due to bulk body dopant N_B in the depletion layer, and $q D_I$ is the depletion-layer charge/unit area due to implanted ions in the depletion layer between the inversion-layer edge and the depletion-layer edge. Because even a small increase in ϕ_S above ϕ_{TH} causes a large increase in $q N_{\text{inv}}$, which can balance a rather large change in gate charge or gate voltage, ϕ_S does not increase much as $V_G - V_{\text{TH}}$ increases. Nonetheless, in strong inversion $N_{\text{inv}} \approx 10^{12}$ charges/cm^2, and so in **strong inversion** ϕ_S will be about $7kT/q$ larger than ϕ_{TH}.

Equation (7.7) indicates for uniform doping (no implant, $D_I = 0$) that threshold occurs approximately for $\phi_S = \phi_{\text{TH}} = 2(kT/q)\ell_n(N_B/n_i) \equiv 2\phi_B$, but for the nonuniformly doped case, a larger surface potential is needed, assuming the case of a normal implant where D_I is positive, increasing the dopant density. The implant increases the required surface potential because the field at the surface is larger, narrowing the inversion layer and reducing the channel strength for $\phi_S = 2\phi_B$. Hence, a somewhat larger surface potential is needed to increase $q\,N_{inv}$ to the point that Eq. (7.1) is satisfied. Equation (7.7) would not apply if a significant fraction of the implant were confined to lie within the inversion layer itself. No realistic implant can be confined within a distance comparable to an inversion-layer thickness (a few tens of nanometers), however, and so Eq. (7.7) covers practical cases.

With the surface potential ϕ_{TH} known, the potential on the gate at threshold Φ_{TH} can be found if we know the oxide field F_{ox} by simply adding the potential drop across the semiconductor to that across the oxide. That is, $\Phi_{TH} = \phi_{TH} + F_{ox}d$, where d is the oxide thickness and F_{ox} is given by Gauss' law as

$$\varepsilon_{ox}F_{ox} = Q_D \tag{7.8}$$

There are two more complications in finding the threshold voltage. First, the gate *voltage* V_{TH} usually differs from the gate *potential* Φ_{TH} at threshold because of a work-function difference between the body and the gate material. This difference causes a spontaneous charge exchange between the two materials as soon as the MOSFET is placed in a circuit allowing charge transfer to occur. Thus, even before any voltage is applied to the device, a potential difference exists between the gate and the body due to spontaneous charge transfer. The second complication affecting threshold voltage is the existence of charges in the insulator and at the insulator–semiconductor interface. These nonideal contributions to the overall charge balance are due to traps and fixed charges incorporated during the device processing.

Ordinarily interface-trap charge is negligible ($< 10^{10}/\text{cm}^2$ in silicon MOSFETs) and the other nonideal effects on threshold voltage are accounted for by introducing the *flatband voltage* V_{FB}, which corrects the gate bias for these contributions. Then, using Eq. (7.8) with $F_{ox} = (V_{TH} - V_{FB} - \phi_{TH})/d$ we find

$$V_{TH} = V_{FB} + \phi_{TH} + Q_D\frac{d}{\varepsilon_{ox}} \tag{7.9}$$

which determines V_{TH} even for the nonuniformly doped case, using Eq. (7.7) for ϕ_{TH} and Q_D at threshold from Eq. (7.4). If interface-trap charge/unit area is not negligible, then terms in the interface-trap charge/unit area Q_{IT} must be added to Q_D in Eq. (7.9).

From Eqs. (7.4) and (7.7), the threshold voltage depends on the implanted dopant-ion profile only through two parameters, the net charge introduced by the implant in the region between the inversion layer and the depletion-layer edge $q\,D_I$, and the centroid of this portion of the implanted charge x_C. As a result, a variety of implants can result in the same threshold, ranging from the extreme of a δ-function spike implant of dose D_I/unit area located at the centroid x_C, to a box type rectangular distribution with the same dose and centroid, namely, a rectangular distribution of width $x_W = 2x_C$ and volume density D_I/x_W. (Of course, x_W must be no larger than the depletion-layer width at threshold for this equivalence to hold true, and x_C must not lie within the inversion layer.) This weak dependence on the details of the profile leaves flexibility to satisfy other requirements, such as control of off current.

As already stated, for gate biases $V_G > V_{TH}$, any gate charge above the threshold value is balanced mainly by inversion-layer charge. Thus, the additional oxide field, given by $(V_G - V_{TH})/d$, is related by Gauss' law to the inversion-layer carrier density approximately by

$$\varepsilon_{ox}(V_G - V_{TH})/d \approx q\,N_{inv} \tag{7.10}$$

which shows that channel strength above threshold is proportional to $V_G - V_{TH}$, an approximation often used in this chapter. Thus, the switch in balancing gate charge from the depletion layer to the inversion layer causes N_{inv} to switch from an exponential gate-voltage dependence in subthreshold to a linear dependence above threshold.

For circuit analysis Eq. (7.10) is a convenient *definition* of V_{TH} because it fits current-voltage curves. If this definition is chosen instead of the charge-balance definition Eq. (7.1), then Eqs. (7.1) and (7.7) result in an *approximation* to ϕ_{TH}.

Driving Ability and $I_{D,\text{sat}}$

The driving ability of the MOSFET is proportional to the current it can provide at a given gate bias. One might anticipate that the larger this current, the faster the circuit. Here this current is used to find some response times governing MOSFET circuits.

MOSFET current is dependent on the carrier density in the channel, or on $V_G - V_{TH}$, see Eq. (7.10). For a long-channel device, driving ability depends also on channel length. The shorter the channel length L, the greater the driving ability, because the channel resistance is directly proportional to the channel length. Although it is an oversimplification, let us suppose that the MOSFET is primarily in saturation during the driving of its load. This simplification will allow a clear discussion of the issues involved in making faster MOSFETs without complicated mathematics. Assuming the MOSFET to be saturated over most of the switching period, driving ability is proportional to current in saturation, or to

$$I_{D,\text{sat}} = \frac{\varepsilon_{\text{ox}} Z \mu}{2dL}(V_G - V_{TH})^2 \tag{7.11}$$

where the factor of two results from the saturating behavior of the I–V curves at large drain biases, and Z is the width of the channel normal to the direction of current flow. Evidently, for long devices driving ability is quadratic in $V_G - V_{TH}$ and inversely proportional to d.

The result of Eq. (7.11) holds for long devices. For short devices, as explained for Fig. 7.26, the larger fields exerted by the drain electrode cause *velocity saturation* and, as a result, $I_{D,\text{sat}}$ is given roughly by (Einspruch and Gildenblat, 1989)

$$I_{D,\text{sat}} \approx \frac{\varepsilon_{\text{ox}} Z \upsilon_{\text{sat}}}{d} \frac{(V_G - V_{TH})^2}{V_G - V_{TH} + F_{\text{sat}}L} \tag{7.12}$$

where υ_{sat} is the carrier saturation velocity, about 10^7 cm/s for silicon at 290 K, and F_{sat} is the field at which velocity saturation sets in, about 5×10^4 V/cm for electrons and not well established as $\gtrsim 10^5$ V/cm for holes in silicon MOSFETs. For Eq. (7.12) to agree with Eq. (7.11) at long L, we need $\mu \approx 2\upsilon_{\text{sat}}/F_{\text{sat}} \approx 400$ cm^2/V · s for electrons in silicon MOSFETs, which is only roughly correct. Nonetheless, we can see that for devices in the submicron channel length regime, $I_{D,\text{sat}}$ tends to become independent of channel length L and becomes more linear with $V_G - V_{TH}$ and less quadratic, see Fig. 7.26. Equation (7.12) shows that velocity saturation is significant when $(V_G - V_{TH})/L \gtrsim F_{\text{sat}}$, for example, when $L \lesssim 0.5 \,\mu$m if $V_G - V_{TH} = 2.3$ V.

To relate $I_{D,\text{sat}}$ to a gate response time τ_G, consider one MOSFET driving an identical MOSFET as load capacitance. Then the current from (Eq. 7.12) charges this capacitance to a voltage V_G in a gate response time τ_G given by (Shoji, 1988)

$$\tau_G = C_G V_G / I_{D,\text{sat}}$$
$$= \left[\frac{L}{\upsilon_{\text{sat}}}\right]\left[1 + \frac{C_{\text{par}}}{C_{\text{ox}}}\right]\frac{V_G(V_G - V_{TH} + F_{\text{sat}}L)}{(V_G - V_{TH})^2} \tag{7.13}$$

where C_G is the MOSFET gate capacitance $C_G = C_{\text{ox}} + C_{\text{par}}$, with $C_{\text{ox}} = \varepsilon_{\text{ox}} ZL/d$ the MOSFET oxide capacitance, and C_{par} the parasitic component of the gate capacitance (Chen, 1990). The parasitic capacitance C_{par} is due mainly to overlap of the gate electrode over the source and drain and partly to fringing-field and channel-edge capacitances. For short-channel lengths, C_{par} is a significant part of C_G, and keeping C_{par} under control as L is reduced is an objective of gate-drain alignment technology. Typically, $V_{TH} \approx V_G/4$, so that

$$\tau_G = \left[\frac{L}{\upsilon_{\text{sat}}}\right]\left[1 + \frac{C_{\text{par}}}{C_{\text{ox}}}\right]\left[1.3 + 1.8\frac{F_{\text{sat}}L}{V_G}\right] \tag{7.14}$$

Thus, on an intrinsic level, the gate response time is a multiple of the transit time of an electron from source to drain, which is L/υ_{sat} in velocity saturation. At shorter L, a linear reduction in delay with L is predicted, whereas for longer devices the improvement can be quadratic in L, depending on how V_G is scaled as L is reduced.

The gate response time is not the only delay in device switching, because the drain-body p–n junction also must charge or discharge for the MOSFET to change state (Shoji, 1988). Hence, we must also consider a *drain* response time τ_D. Following Eq. (7.13), we suppose that the drain capacitance C_D is charged by the supply voltage through a MOSFET in saturation so that

$$\tau_D = C_D V_G / I_{D,\text{sat}} = \left[\frac{C_D}{C_G}\right]\tau_G \tag{7.15}$$

Equation (7.15) suggests that τ_D will show a similar improvement to τ_G as L is reduced, provided that C_D/C_G does not increase as L is reduced. However, $C_{\text{ox}} \propto L/d$, and the major component of C_{par}, namely, the overlap capacitance contribution, leads to $C_{\text{par}} \propto L_{\text{ovlp}}/d$ where L_{ovlp} is roughly three times the length of overlap of the gate over the source or drain (Chen, 1990). Then $C_G \propto (L + L_{\text{ovlp}})/d$ and, to keep the C_D/C_G ratio from increasing as L is reduced, either C_D or oxide thickness d must be reduced along with L.

Clever design can reduce C_D. For example, various *raised-drain* designs reduce the drain-to-body capacitance by separating much of the drain area from the body using a thick oxide layer. The contribution to drain capacitance stemming from the sidewall depletion-layer width next to the channel region is more difficult to handle, because the sidewall depletion layer is deliberately reduced during miniaturization to avoid *short-channel* effects, that is, drain influence on the channel in competition with gate control. As a result, this sidewall contribution to the drain capacitance tends to increase with miniaturization unless junction depth can be shrunk.

Equations (7.14) and (7.15) predict reduction of response times by reduction in channel length L. Decreasing oxide thickness leads to no improvement in τ_G, but Eq. (7.15) shows a possibility of improvement in τ_D. The *ring oscillator,* a closed loop of an odd number of inverters, is a test circuit the performance of which depends primarily on τ_G and τ_D. Gate delay/stage for ring oscillators is found to be near 12 ps/stage at 0.1-μm channel length, and 60 ps/stage at 0.5 μm.

For circuits, interconnection capacitances and fan out (multiple MOSFET loads) will increase response times beyond the device response time, even when parasitics are taken into account. Thus, we are led to consider interconnection delay τ_{INT}. Although a lumped model suggests, as with Eq. (7.15), that $\tau_{\text{INT}} \approx (C_{\text{INT}}/C_G)\tau_G$, the length of interconnections requires a *distributed* model. Interconnection delay is then

$$\tau_{\text{INT}} = R_{\text{INT}}\, C_{\text{INT}}/2 + R_{\text{INT}}\, C_G + (1 + C_{\text{INT}}/C_G)\tau_G \tag{7.16}$$

where R_{INT} is the interconnection resistance, C_{INT} is the interconnection capacitance, and we have assumed that the interconnection joins a MOSFET driver in saturation to a MOSFET load, C_G. For small R_{INT}, the τ_{INT} is dominated by the last term, which resembles Eqs. (7.13) and (7.15). Unlike the ratio C_D/C_G in Eq. (7.15), however, it is difficult to reduce or even maintain the ratio C_{INT}/C_G in Eq. (7.16) as L is reduced. Remember, $C_G \propto Z(L + L_{\text{ovlp}})/d$. Reduction of L, therefore, tends to increase C_{INT}/C_G, especially because interconnect cross sections cannot be reduced without impractical increases in R_{INT}. What is worse, along with reduction in L, chip sizes usually increase, making line lengths longer, increasing R_{INT} even at constant cross section. As a result, interconnection delay becomes a major problem as L is reduced. The obvious way to keep C_{INT}/C_G under control is to increase the device width Z so that $C_G \propto Z(L + L_{\text{ovlp}})/d$ remains constant as L is reduced. A better way is to cascade drivers of increasing Z (Chen, 1990; Shoji, 1988). Either solution requires extra area, however, reducing the packing density that is a major objective in decreasing L in the first place. An alternative is to reduce the oxide thickness d, a major technology objective.

Transconductance

Another important device parameter is the small-signal transconductance g_m (Malik, 1995; Sedra and Smith, 1991; Haznedar, 1991) that determines the amount of output current swing at the drain that results from a given input voltage variation at the gate, that is, the small-signal gain

$$g_m = \left.\frac{\partial I_D}{\partial V_G}\right|_{V_D=\text{const}} \tag{7.17}$$

Using the chain rule of differentiation, the transconductance in saturation can be related to the small-signal *transition* or *unity-gain frequency* that determines at how high a frequency ω the small-signal current gain $|\iota_{\text{out}}/\iota_{\text{in}}| = g_m/(\omega C_G)$ drops to unity. Using the chain rule

$$g_m = \frac{\partial I_{D,\text{sat}}}{\partial Q_G} \frac{\partial Q_G}{\partial V_G} = \omega_T C_G \tag{7.18}$$

where C_G is the oxide capacitance of the device, $C_G = \partial Q_G/\partial V_G|_{V_D}$ where Q_G is the charge on the gate electrode. The frequency ω_T is a measure of the small-signal, high-frequency speed of the device, neglecting parasitic resistances. Using Eq. (7.12) in Eq. (7.18) we find that the transition frequency also is related to the transit time L/υ_{sat} of Eq. (7.14), so that both the digital and small-signal circuit speeds are related to this parameter.

Output Resistance and Drain Conductance

For small-signal circuits the output resistance r_o of the MOSFET (Malik, 1995; Sedra and Smith, 1991) is important in limiting the gain of amplifiers. This resistance is related to the small-signal drain conductance in saturation by

$$r_o = \frac{1}{g_D} = \frac{\partial V_D}{\partial I_{D,\text{sat}}}\bigg|_{V_G=\text{const}} \tag{7.19}$$

If the MOSFET is used alone as a simple amplifier with a load line set by a resistor R_L, the gain becomes

$$\left|\frac{\upsilon_o}{\upsilon_{\text{in}}}\right| = g_m \frac{R_L r_o}{R_L + r_o} \leq g_m R_L \tag{7.20}$$

showing how gain is reduced if r_o is reduced to a value approaching R_L.

As devices are miniaturized, r_o is decreased and g_D increased, due to several factors. At moderate drain biases, the main factor is channel-length modulation, the reduction of the channel length with increasing drain voltage that results when the depletion region around the drain expands toward the source, causing L to become drain-bias dependent. At larger drain biases, a second factor is drain control of the inversion-layer charge density that can compete with gate control in short devices. This is the same mechanism discussed later in the context of subthreshold behavior. At rather high drain bias, carrier multiplication further lowers r_o.

In a digital inverter, a lower r_o widens the voltage swing needed to cause a transition in output voltage. This widening increases power loss due to current spiking during the transition and reduces noise margins (Annaratone, 1986). It is not, however, a first-order concern in device miniaturization for digital applications. Because small-signal circuits are more sensitive to r_o than digital circuits, MOSFETs designed for small-signal applications cannot be made as small as those for digital applications.

7.3.4   Limitations on Miniaturization

A major factor in the success of the MOSFET has been its compatibility with processing useful down to very small dimensions. Today channel lengths (source-to-drain spacings) of 0.5 μm are manufacturable, and further reduction to 0.1 μm has been achieved for limited numbers of devices in test circuits, such as ring oscillators. In this section some of the limits that must be considered in miniaturization are outlined (Brews, 1990).

Subthreshold Control

When a MOSFET is in the off condition, that is, when the MOSFET is in subthreshold, the off current drawn with the drain at supply voltage must not be too large in order to avoid power consumption and discharge of ostensibly isolated nodes (Shoji, 1988). In small devices, however, the source and drain are closely spaced, and so there exists a danger of direct interaction of the drain with the source, rather than an interaction mediated by the gate and channel. In an extreme case, the drain may draw current directly from the source, even though the gate is off (*punchthrough*). A less extreme but also undesirable case occurs when the drain and gate jointly control the carrier density in the channel (*drain-induced barrier lowering*, or drain control of threshold voltage). In such a case, the on-off behavior of the MOSFET is not controlled by the gate alone, and switching can occur over a range of gate voltages dependent on the drain voltage. Reliable circuit design under these circumstances is very complicated, and testing for design errors is prohibitive. Hence, in designing MOSFETs, a drain-bias independent subthreshold behavior is necessary.

A measure of the range of influence of the source and drain is the depletion-layer width of the associated *p–n* junctions. The depletion layer of such a junction is the region in which all carriers have been depleted, or pushed away, due to the potential drop across the junction. This potential drop includes the applied bias across the junction and a spontaneous built-in potential drop induced by spontaneous charge exchange when *p*- and *n*-regions are brought into contact. The depletion-layer width W of an abrupt junction is related to potential drop V and dopant-ion concentration/unit volume N by

$$W = \left[\frac{2\varepsilon_s V}{qN} \right]^{\frac{1}{2}} \tag{7.21}$$

To avoid subthreshold problems, a commonly used rule of thumb is to make sure that the channel length is longer than a minimum length $L_{\min}$ related to the junction depth r_j, the oxide thickness d, and the depletion-layer widths of the source and drain W_S and W_D by (Brews, 1990)

$$L_{\min} = A[r_j d (W_S W_D)^2]^{\frac{1}{3}} \tag{7.22}$$

where the empirical constant $A = 0.88 \text{ nm}^{-1/3}$ if r_j, W_S, and W_D are in micrometers and d is in nanometers.

Equation (7.22) shows that smaller devices require shallower junctions (smaller r_j), or thinner oxides (smaller d), or smaller depletion-layer widths (smaller voltage levels or heavier doping). These requirements introduce side effects that are difficult to control. For example, if the oxide is made thinner while voltages are not reduced proportionately, then oxide fields increase, requiring better oxides. If junction depths are reduced, better control of processing is required, and the junction resistance is increased due to smaller cross sections. To control this resistance, various *self-aligned contact* schemes have been developed to bring the source and drain contacts closer to the gate (Brews, 1990; Einspruch and Gildenblat, 1989), reducing the resistance of these connections. If depletion-layer widths are reduced by increasing the dopant-ion density, the *driving ability* of the MOSFET suffers because the threshold voltage increases. That is, Q_D increases in Eq. (7.9), reducing $V_G - V_{TH}$. Thus, increasing V_{TH} results in slower circuits.

As secondary consequences of increasing dopant-ion density, channel conductance is further reduced due to the combined effects of increased scattering of electrons from the dopant atoms and increased oxide fields that pin carriers in the inversion layer closer to the insulator-semiconductor interface, increasing scattering at the interface. These effects also reduce driving ability, although for shorter devices they are important only in the linear region (that is, below saturation), assuming that mobility μ is more strongly affected than saturation velocity v_{sat}.

Hot-Electron Effects

Another limit on how small a MOSFET can be made is a direct result of the larger fields in small devices. Let us digress to consider why proportionately larger voltages, and thus larger fields, are used in smaller devices. First, according to Eq. (7.14), τ_G is shortened if voltages are increased, at least so long as $V_G/L \lesssim F_{\text{sat}} \approx 5 \times 10^4$ V/cm. If τ_G is shortened this way, then so are τ_D and τ_{INT}, Eqs. (7.15) and (7.16). Thus, faster response is gained by increasing voltages into the velocity saturation region. Second, the fabricational control of smaller devices has not improved proportionately as L has shrunk, and so there is a larger percentage variation in device parameters with smaller devices. Thus, disproportionately larger voltages are needed to insure all devices operate in the circuit, to overcome this increased fabricational *noise*. Thus, to increase speed and to cope with fabricational variations, fields get larger in smaller devices.

As a result of these larger fields along the channel direction, a small fraction of the channel carriers have enough energy to enter the insulating layer near the drain. In silicon-based *p*-channel MOSFETs, energetic holes can be trapped in the oxide, leading to a positive oxide charge near the drain that reduces the strength of the channel, degrading device behavior. In *n*-channel MOSFETs, energetic electrons entering the oxide create interface traps and oxide wear out, eventually leading to gate-to-drain shorts (Pimbley et al., 1989).

To cope with these problems *drain engineering* has been tried, the most common solution being the *lightly doped drain* (Chen, 1990; Einspruch and Gildenblat, 1989; Pimbley et al., 1989; Wolf, 1995). In this design, a lightly doped extension of the drain is inserted between the channel and the drain proper. To keep the field moderate and reduce any peaks in the field, the lightly doped drain extension is designed to spread the drain-to-channel voltage drop as evenly as possible. The aim is to smooth out the field at a value close to F_{sat} so that energetic carriers are kept to a minimum. The expense of this solution is an increase in drain resistance and a decreased gain. To increase packing density, this lightly doped drain extension can be stacked vertically alongside the gate, rather than laterally under the gate, to control the overall device area.

Thin Oxides

According to Eq. (7.22), thinner oxides allow shorter devices and, therefore, higher packing densities for devices. In addition, driving ability is increased, shortening response times for capacitive loads, and output resistance and transconductance are increased. There are some basic limitations on how thin the oxide can be made. For instance, there is a maximum oxide field that the insulator can withstand. It is thought that the intrinsic break-down voltage of SiO_2 is of the order of 10^7 V/cm, a field that can support $\approx 2 \times 10^{13}$ charges/cm^2, a large enough value to make this field limitation secondary. Unfortunately, as they are presently manufactured, the intrinsic breakdown of MOSFET oxides is much less likely to limit fields than defect-related leakage or breakdown, and control of these defects has limited reduction of oxide thicknesses in manufacture to about 5 nm to date.

If defect-related problems could be avoided, the thinnest useful oxide would probably be about 3 nm, limited by direct tunneling of channel carriers to the gate. This tunneling limit is not well established and also is subject to oxide-defect enhancement due to tunneling through intermediate defect levels. Thus, defect-free manufacture of thin oxides is a very active area of exploration.

Dopant-Ion Control

As devices are made smaller, the precise positioning of dopant inside the device is critical. At high temperatures during processing, dopant ions can move. For example, source and drain dopants can enter the channel region, causing position-dependence of threshold voltage. Similar problems occur in isolation structures that separate one device from another (Primbley et al., 1989; Einspruch and Gildenblat, 1989; Wolf, 1995).

To control these thermal effects, process sequences are carefully designed to limit high-temperature steps. This design effort is shortened and improved by the use of computer modeling of the processes. Dopant-ion movement is complex, however, and its theory is made more difficult by the growing trend to use *rapid thermal processing* that involves short-time heat treatments. As a result, dopant response is not steady state, but transient. Computer models of transient response are primitive, forcing further advance in small device design to be more empirical.

Other Limitations

Besides limitations directly related to the MOSFET, there are some broader difficulties in using MOSFETs of smaller dimension in chips involving even greater numbers of devices. Already mentioned is the increased delay due to interconnections that are lengthening due to increasing chip area and increasing complexity of connection. The capacitive loading of MOSFETs that must drive signals down these lines can slow circuit response, requiring extra circuitry to compensate.

Another limitation is the need to isolate devices from each other (Brews, 1990; Chen, 1990, Einspruch and Gildenblat, 1989; Pimbley et al., 1989; Wolf, 1995), so that their actions remain uncoupled by parasitics. As isolation structures are reduced in size to increase device densities, new parasitics are discovered. A solution to this problem is the manufacture of circuits on insulating substrates, *silicon-on-insulator* technology (Colinge, 1991). To succeed, this approach must deal with new problems, such as the electrical quality of the underlying silicon-insulator interface, and the defect densities in the silicon layer on top of this insulator.

Acknowledgments

The author is pleased to thank R.D. Schrimpf and especially S.L. Gilbert for suggestions that clarified the manuscript.

Defining Terms

Channel: The conducting region in a MOSFET between source and drain. In an *enhancement*-mode, or normally off MOSFET the channel is an inversion layer formed by attraction of minority carriers toward the gate. These carriers form a thin conducting layer that is prevented from reaching the gate by a thin *gate-oxide* insulating layer when the gate bias exceeds *threshold*. In a *buried-channel* or *depletion*-mode, or normally on MOSFET, the channel is present even at zero gate bias, and the gate serves to increase the channel resistance when its bias is nonzero. Thus, this device is based on majority-carrier modulation, like a MESFET.

Gate: The control electrode of a MOSFET. The voltage on the gate capacitively modulates the resistance of the connecting *channel* between the *source* and *drain*.

Source, drain: The two output contacts of a MOSFET, usually formed as *p–n* junctions with the *substrate* or *body* of the device.

Strong inversion: The range of gate biases corresponding to the on condition of the MOSFET. At a fixed gate bias in this region, for low drain-to-source biases the MOSFET behaves as a simple gate-controlled resistor. At larger drain biases, the channel resistance can increase with drain bias, even to the point that the current *saturates,* or becomes independent of drain bias.

Substrate or body: The portion of the MOSFET that lies between the *source* and *drain* and under the gate. The gate is separated from the body by a thin *gate insulator,* usually silicon dioxide. The gate modulates the conductivity of the body, providing a gate-controlled resistance between the source and drain. The body is sometimes DC biased to adjust overall circuit operation. In some circuits the body voltage can swing up and down as a result of input signals, leading to body-effect or back-gate bias effects that must be controlled for reliable circuit response.

Subthreshold: The range of gate biases corresponding to the off condition of the MOSFET. In this regime the MOSFET is not perfectly off, but conducts a leakage current that must be controlled to avoid circuit errors and power consumption.

Threshold: The gate bias of a MOSFET that marks the boundary between on and off conditions.

References

The following references are not to the original sources of the ideas discussed in this article, but have been chosen to be generally useful to the reader.

Annaratone, M. 1986. *Digital CMOS Circuit Design.* Kluwer Academic, Boston, MA.

Brews, J.R. 1981. Physics of the MOS Transistor. In *Applied Solid State Science, Supplement 2A,* ed. D. Kahng, pp. 1–20. Academic Press, New York.

Brews, J.R. 1990. The Submicron MOSFET. In *High-Speed Semiconductor Devices,* ed. S.M. Sze, pp. 139–210. Wiley, New York.

Chen, J.Y. 1990. *CMOS Devices and Technology for VLSI.* Prentice-Hall, Englewood Cliffs, NJ.

Colinge, J.-P. 1991. *Silicon-on-Insulator Technology: Materials to VLSI.* Kluwer Academic, Boston, MA.

Haznedar, H. 1991. *Digital Microelectronics.* Benjamin/Cummings, Redwood City, CA.

Hollis, M.A. and Murphy, R.A. 1990. Homogeneous Field-Effect Transistors. In *High-Speed Semiconductor Devices,* ed. S.M. Sze, pp. 211–282. Wiley, New York.

Einspruch, N.G. and Gildenblat, G.S., eds. 1989. *Advanced MOS Device Physics,* Vol. 18, *VLSI Microstructure Science.* Academic, New York.

Malik, N.R. 1995. *Electronic Circuits: Analysis, Simulation, and Design.* Prentice-Hall, Englewood Cliffs, NJ.

Nicollian, E.H. and Brews, J.R. 1982. *MOS Physics and Technology,* Chap. 1. Wiley, New York.

Pearton, S.J. and Shah, N.J. 1990. Heterostructure Field-Effect Transistors. In *High-Speed Semiconductor Devices,* ed. S.M. Sze, pp. 283–334. Wiley, New York.

Pierret, R.F. 1990. *Field Effect Devices,* 2nd ed., Vol. 4, *Modular Series on Solid State Devices.* Addison-Wesley, Reading, MA.

Pimbley, J.M., Ghezzo, M., Parks, H.G., and Brown, D.M. 1989. *Advanced CMOS Process Technology,* ed. N.G. Einspruch, Vol. 19, *VLSI Electronics Microstructure Science.* Academic Press, New York.

Sedra, S.S. and Smith, K.C. 1991. *Microelectronic Circuits,* 3rd ed. Saunders College Publishing, Philadelphia, PA.

Shoji, M. 1988. *CMOS Digital Circuit Technology.* Prentice-Hall, Englewood Cliffs, NJ.

Wolf, S. 1995. *Silicon Processing for the VLSI Era: Volume 3—The Submicron MOSFET.* Lattice Press, Sunset Beach, CA.

Further Information

The references given in this section have been chosen to provide more detail than is possible to provide in the limited space of this article. In particular, Annaratone (1986) and Shoji (1988) provide much more detail about device and circuit behavior. Chen (1990), Pimbley (1989), and Wolf (1995) provide many technological details of processing and its device impact. Haznedar (1991), Sedra and Smith (1991), and Malik (1995) provide much information about circuits. Brews (1981) and Pierret (1990) provide good discussion of the derivation of the device current-voltage curves and device behavior in all bias regions.

7.4 Image Capture Devices

Edward J. Delp, III

7.4.1 Image Capture

A digital image is nothing more than a matrix of numbers. The question is how does this matrix represent a real image that one sees on a computer screen?

Like all imaging processes, whether they are analog or digital, one first starts with a sensor (or transducer) that converts the original imaging energy into an electrical signal. These sensors, for instance, could be the photomultiplier tubes used in an X-ray system that convert the X-ray energy into a *known* electrical voltage. The transducer system used in ultrasound imaging is an example where sound pressure is converted to electrical energy; a simple TV camera is perhaps the most ubiquitous example. An important fact to note is that the process of conversion from one energy form to an electrical signal is not necessarily a *linear* process. In other words, a proportional change in the input energy to the sensor will not always cause the same proportional change in the output electrical signal. In many cases calibration data are obtained in the laboratory so that the relationship between the input energy and output electrical signal is known. These data are necessary because some transducer performance characteristics change with age and other usage factors.

The sensor is not the only thing needed to form an image in an imaging system. The sensor must have some spatial extent before an image is formed. By spatial extent we mean that the sensor must not be a simple point source examining only one location of energy output. To explain this further, let us examine two types of imaging sensors used in imaging: a charge-coupled device (CCD) video camera and the ultrasound transducer used in many medical imaging applications.

The CCD camera consists of an *array* of light sensors known as charge-coupled devices. The image is formed by examining the output of each sensor in a preset order for a finite time. The electronics of the system then forms an electrical signal, which produces an image that is shown on a cathode-ray tube (CRT) display. The image is formed because there is an array of sensors, each one examining only one spatial location of the region to be sensed.

The process of **sampling** the output of the sensor array in a particular order is known as *scanning*. Scanning is the typical method used to convert a two-dimensional energy signal or image to a one-dimensional electrical signal that can be handled by the computer. (An image can be thought of as an energy field with spatial extent.) Another form of scanning is used in ultrasonic imaging. In this application there is only one sensor instead of an array of sensors. The ultrasound transducer is moved or steered (either mechanically or electrically) to various spatial locations on the patient's chest or stomach. As the sensor is moved to each location, the output electrical signal of the sensor is sampled and the electronics of the system then form a television-like signal, which is displayed. Nearly all of the transducers used in imaging form an image by either using an array of sensors or a single sensor that is moved to each spatial location.

One immediately observes that both of the approaches discussed are equivalent in that the energy is sensed at various spatial locations of the object to be imaged. This energy is then converted to an electrical signal by the transducer. The image formation processes just described are classical analog image formation, with the distance between the sensor locations limiting the spatial resolution in the system. In the array sensors, resolution is also determined by how close the sensors are located in the array. In the single-sensor approach, the spatial resolution is limited by how far the sensor is moved. In an actual system spatial resolution is also determined by the performance characteristics of the sensor. Here we are assuming for our purposes *perfect* sensors.

In digital image formation one is concerned about two processes: *spatial sampling* and **quantization**. Sampling is quite similar to scanning in analog image formation. The second process is known as *quantization* or *analog-to-digital conversion*, whereby at each spatial location a *number* is assigned to the amount of energy the transducer observes at that location. This number is usually proportional to the electrical signal at the output of the transducer. The overall process of sampling and quantization is known as *digitization*. Sometimes the digitization process is just referred to as analog-to-digital conversion, or A/D conversion; however, the reader should remember that digitization also includes spatial sampling.

The digital image formulation process is summarized in Fig. 7.27. The spatial sampling process can be considered as overlaying a grid on the object, with the sensor examining the energy output from each grid box and converting it to an electrical signal. The quantization process then assigns a number to the electrical signal; the result, which is a *matrix* of numbers, is the digital representation of the image. Each spatial location in the image (or grid) to which a number is assigned is known as a *picture element* or **pixel** (or pel). The size of the sampling grid is usually given by the number of pixels on each side of the grid, for examples, $256 \times 256, 512 \times 512, 640 \times 480$.

The quantization process is necessary because all information to be processed using computers must be represented by numbers. The quantization process can be thought of as one where the input energy to the transducer is represented by a finite number of energy values. If the energy at a particular pixel location does not take on one of the finite energy values, it is assigned to the closest value. For instance, suppose that we assume *a priori* that only energy values of 10, 20, 50, and 110 will be represented (the units are of no concern in this example). Suppose at one pixel an energy of 23.5 was observed by the transducer. The A/D converter would then assign this pixel the energy value of 20 (the closest one). Notice that the quantization process makes mistakes; this error in assignment is known as *quantization error* or *quantization noise*.

In our example, each pixel is represented by one of four possible values. For ease of representation of the data, it would be simpler to assign to each pixel the index value 0, 1, 2, 3, instead of 10, 20, 50, 110. In fact, this is typically done by the quantization process. One needs a simple table to know that a pixel assigned the value 2 corresponds to an energy of 50. Also, the number of possible energy levels is typically some integer power of two to also aid in representation. This power is known as the number of *bits* needed to represent the energy of each pixel. In our example each pixel is represented by 2 b.

One question that immediately arises is how accurate the digital representation of the image is when one compares the digital image with a corresponding analog image. It should first be pointed out that after the digital image is obtained one requires special hardware to convert the matrix of pixels back to an image that can be viewed on a CRT display. The process of converting the digital image back to an image that can be viewed is known as *digital-to-analog conversion*, or *D/A conversion*.

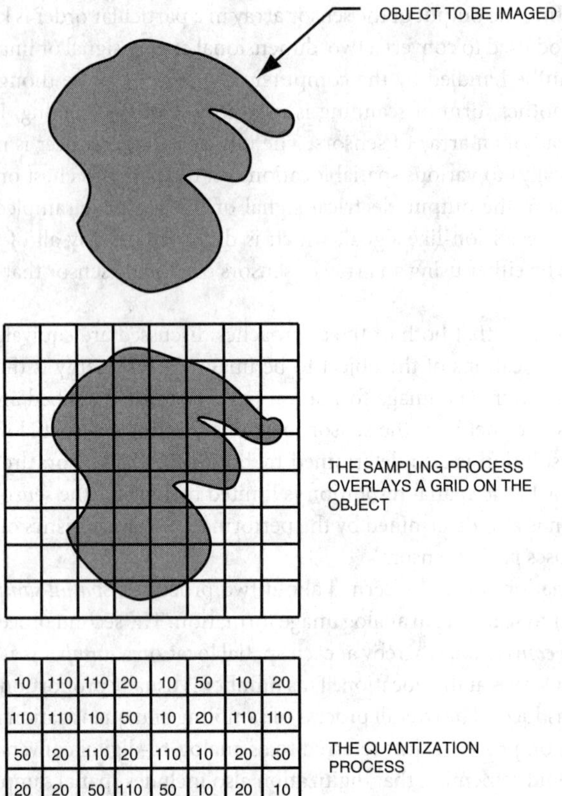

FIGURE 7.27   Digital image formation: sampling and quantization.

The quality of representation of the image is determined by how close spatially the pixels are located and how many levels or numbers are used in the quantization, that is, how coarse or fine is the quantization. The sampling accuracy is usually measured in how many pixels there are in a given area and is cited in pixels/unit length, e.g., pixels/cm. This is known as *the spatial sampling rate*. One would desire to use the lowest rate possible to minimize the number of pixels needed to represent the object. If the sampling rate is too low, then obviously some details of the object to be imaged will not be represented very well. In fact, there is a mathematical theorem that determines the lowest sampling rate possible to preserve details in the object. This rate is known as the *Nyquist* sampling rate. The theorem states that the sampling rate must be *twice* the highest possible detail one expects to image in the object. If the object has details closer than, say, 1 mm, one must take at least 2 pixels/mm. (The Nyquist theorem actually says more than this, but a discussion of the entire theorem is beyond the scope of this chapter). If we sample at a lower rate than the theoretical lowest limit, the resulting digital representation of the object will be distorted. This type of distortion or sampling error is known as *aliasing* errors. Aliasing errors usually manifest themselves in the image as moire patterns (Fig. 7.28). The important point to remember is that there is a *lower limit* to the spatial sampling rate such that object detail can be maintained. The sampling rate can also be stated as the total number of pixels needed to represent the digital image, that is, the matrix size (or grid size). One often sees these sampling rates cited as $256 \times 256, 512 \times 512$, and so on. If the same object is imaged with a large matrix size, the sampling rate has obviously increased. Typically, images

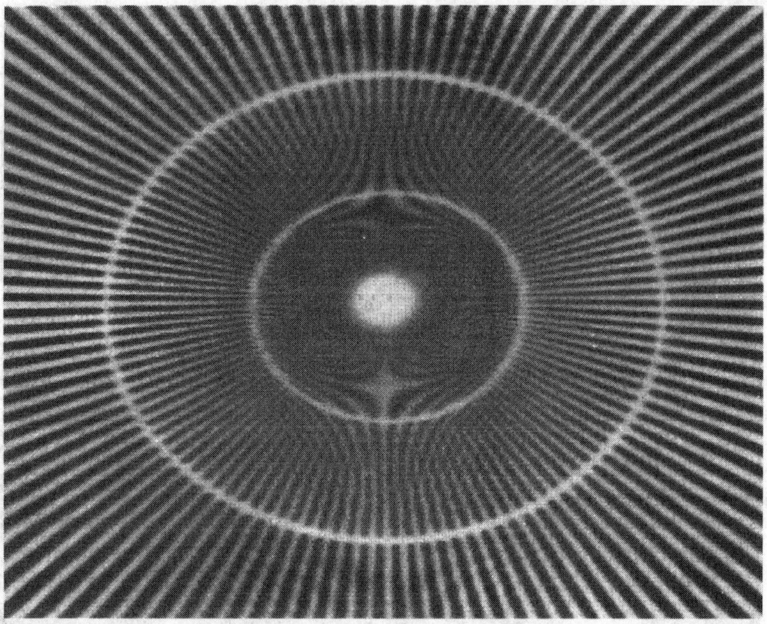

FIGURE 7.28   This image shows the effects of aliasing due to sampling the image at too low a rate. The image should be straight lines converging at a point. Because of undersampling, it appears as if there are patterns in the lines at various angles. These are known as moiré patterns.

are sampled on $256 \times 256, 512 \times 512$, or 1024×1024 grids, depending on the application and type of modality. One immediately observes an important issue in digital representation of images: that of the large number of pixels needed to represent the image. A 256×256 image has 65,536 pixels and a 512×512 image has 262,144 pixels. We shall return to this point later when we discuss processing or storage of these images.

The quality of the representation of the digital image is also determined by the number of levels or shades of gray that are used in the quantization. If one has more levels, then fewer mistakes will be made in assigning values at the output of the transducer. Figure 7.29 demonstrates how the number of gray levels affects the digital representation of an artery. When a small number of levels are used, the quantization is coarse and the quantization error is large. The quantization error usually manifests itself in the digital image by the appearance of *false contouring* in the picture. One usually needs at least 6 b or 64 gray levels to represent an image adequately. Higher quality imaging systems use 8 b (256 levels) or even as many as 10 b (1024 levels) per pixel. In most applications, the human observer cannot distinguish quantization error when there are more than 256 levels. (Many times the number of gray levels is cited in bytes. One byte is 8 b, that is, high-quality monochrome digital imaging systems use one byte per pixel.)

One of the problems briefly mentioned previously is the large number of pixels needed to represent an image, which translates into a large amount of digital data needed for the representation. A 512×512 image with 8 b/pixel (1 byte/pixel) of gray levels representation requires 2,097, 152 b of computer data to describe it. A typical computer file that contains 1000 words usually requires only about 56,000 b to describe it. The 512×512 image is 37 times larger! (A picture is truly worth more than 1000 words.) This data requirement is one of the major problems with digital imaging, given that the storage of digital images in a computer file system is expensive. Perhaps another example will demonstrate this problem. Many computers and word processing systems have the capability of transmitting information over telephone lines to other systems at data rates of 14,400 b/s. At this speed it would require nearly 2.5 min to transmit a 512×512 image! Moving objects are imaged digitally by taking *digital snapshots* of them, that is, digital video. True digital imaging would acquire about 30 images/s to capture all of the important motion in a scene.

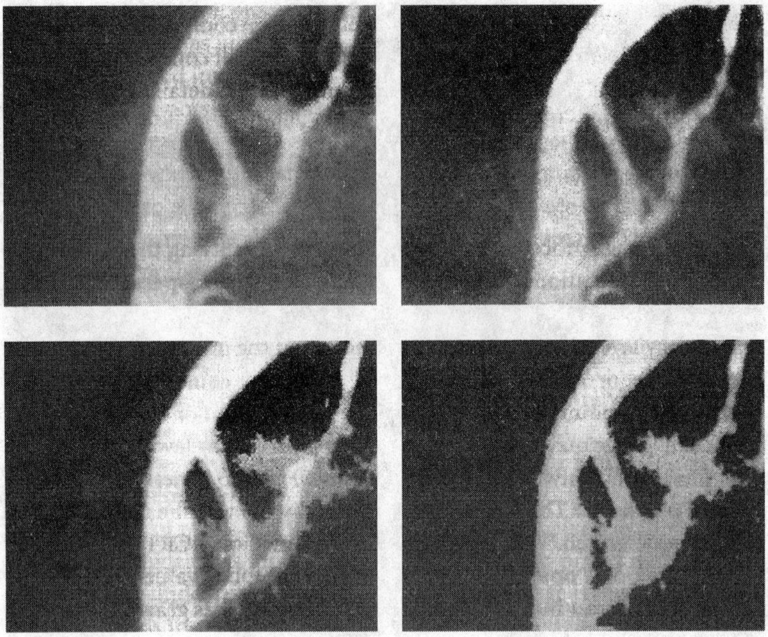

FIGURE 7.29   This image demonstrates the effects of quantization error. The upper left image is a coronary artery image with 8 b (256 levels or shades of gray) per pixel. The upper right image has 4 b/pixel (16 levels). The lower left image has 3 b/pixel (8 levels). The lower right image has 2 b/pixel (4 levels). Note the false contouring in the images as the number of possible levels in the pixel representation is reduced. This false contouring is the quantization error, and as the number of levels increases the quantization error decreases because fewer mistakes are being made in the representation.

At 30 images/s, with each image sampled at 512×512 and with 8 b/pixel, the system must handle 62,914,560 b/s. Only very expensive acquisition systems are capable of handling these large data rates.

The greatest advantage of digital images is that they can be processed on a computer. Any type of operation that one can do on a computer can be done to a digital image. Recall that a digital image is just a (huge) matrix of numbers. Digital image processing is the process of using a computer to extract useful information from this matrix. Processing that cannot be done optically or with analog systems (such as early video systems) can be easily done on computers. The disadvantage is that a large amount of data needs to be processed, and on some small computer systems this can take a long time (hours).

7.4.2   Point Operations

Perhaps the simplest image processing operation is that of modifying the values of individual pixels in an image. These operations are commonly known as **point operations**. A point operation might be used to highlight certain regions in an image. Suppose one wished to know where all of the pixels in a certain gray level region were *spatially* located in the image. One would modify all those pixel values to 0 (black) or 255 (white) such that the observer could see where they were located.

Another example of a point operation is *contrast enhancement* or *contrast stretching*. The pixel values in a particular image may occupy only a small region of gray level distribution. For instance, the pixels in an image may only take on values between 0 and 63, when they could nominally take on values between 0 and 255. This is sometimes caused by the way the image was digitized and/or by the type of transducer used. When this image is examined on a CRT display the contrast looks washed out. A simple point operation that multiplies each pixel value in the image by four will increase the apparent contrast in the image; the new image now has gray values between 0 and 252. This operation is shown in Fig. 7.30. Possibly the most

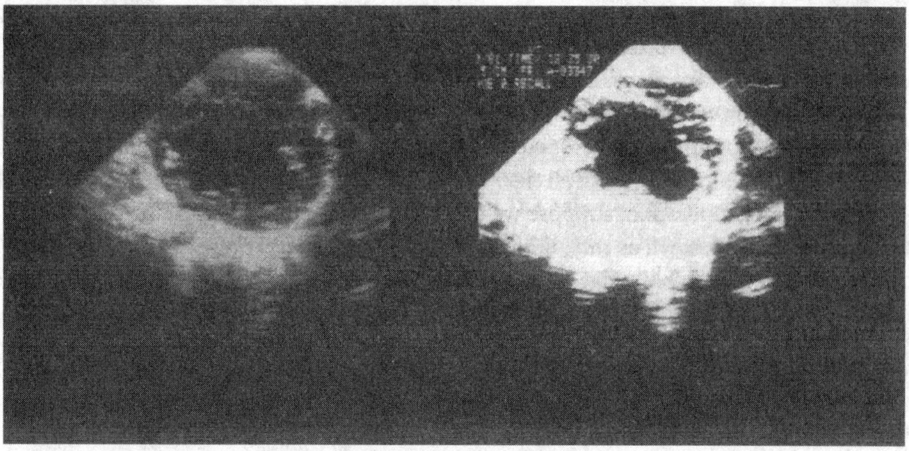

FIGURE 7.30 Contrast stretching. The image on the right has gray values between 0 and 63, causing the contrast to look washed out. The image on the right has been contrast enhanced by multiplying the gray levels by four.

widely used point operation in medical imaging is *pseudecoloring*. In this point operation all of the pixels in the image with a particular gray value are assigned a *color*. Various schemes have been proposed for appropriate pseudocolor tables that assign the gray values to colors. It should be mentioned that point operations are often cascaded, that is, an image undergoes contrast enhancement and then pseudocoloring.

The operations just described can be thought of as operations (or *algorithms*) that modify the range of the gray levels of the pixels. An important feature that describes a great deal about an image is the *histogram* of the pixel values. A histogram is a table that lists how many pixels in an image take on a particular gray value. These data are often plotted as a function of the gray value. Point operations are also known as *histogram modification* or *histogram stretching*. The contrast enhancement operation shown in Fig. 7.30 modifies the histogram of the resultant image by stretching the gray values from a range of 0–63 to a range of 0–252. Some point operations are such that the resulting histogram of the processed image has a particular shape. A popular form of histogram modification is known as *histogram equalization*, whereby the pixel are modified such that the histogram of the processed image is almost flat, that is, all of the pixel values occur equally.

It is impossible to list all possible types of point operations; however, the important thing to remember is that these operations process one pixel at a time by modifying the pixel based only on its gray level value and not where it is distributed spatially (i.e., location in the pixel matrix). These operations are performed to enhance the image, to make it easier to see certain structures or regions in the image, or to force a particular shape to the histogram of the image. They are also used as initial operations in a more complicated image processing algorithm.

7.4.3 Image Enhancement

Image enhancement is the use of image processing algorithms to remove certain types of distortion in an image. The image is enhanced by removing noise, by making the edge structures in the image stand out, or by any other operation that makes the image *look* better.[1] Point operations just discussed are generally

---

[1] Image enhancement is often confused with *image restoration*. Image enhancement is the *ad hoc* application of various processing algorithms to enhance the appearance of the image. Image restoration is the application of algorithms that use knowledge of the degradation process to enhance or restore the image, that is, deconvolution algorithms used to remove the effect of the aperture point spread function in blurred images. A discussion of image restoration is beyond the scope of this section.

considered to be enhancement operations. Enhancement also includes operations that use groups of pixels and the spatial location of the pixels in the image.

The most widely used algorithms for enhancement are based on pixel functions that are known as *window operations*. A window operation performed on an image is nothing more than the process of examining the pixels in a certain region of the image, called the window region, and computing some type of mathematical function derived from the pixels in the window. In most cases the windows are square or rectangle, although other shapes have been used. After the operation is performed, the result of the computation is placed in the center pixel of the window. As an example of a window operation, suppose we compute the average value of the pixels in the window. This operation is known as *smoothing* and will tend to reduce noise in the image, but unfortunately it will also tend to blur edge structures in the image.

It is possible to compute a nonlinear funtion of the pixels in the window. One of the more powerful nonlinear window operations is that of *median filtering*. In this operations all of the pixels in the window are listed in descending magnitude and the middle, or *median*, pixel is obtained. The median pixel then is used to replace the center pixel. The median filter is used to remove noise from an image and at the same time preserve the edge structure in the image. More recently, there has been a great deal of interest in *morphological operators*. These are also nonlinear window operations that can be used to extract or enhance shape information in an image.

Defining Terms

Image enhancement: An image processing operation that is intended to improve the visual quality of the image or to emphasize certain features.

Pixel: A single sample or picture element in the digital image, which is located at specific spatial coordinates.

Point operation: An image processing operation in which individual pixels are mapped to new values irrespective of the values of any neighboring pixels.

Quantization: The process of converting from a continuous-amplitude image to an image that takes on only a finite number of different amplitude values.

Sampling: The process of converting from a continuous-parameter image to a discrete-parameter image by discretizing the spatial coordinate.

References

Gonzalez, R.C. and Wintz, P. 1991. *Digital Image Processing.* Addison-Wesley, Reading, MA.

Jain, A.K. 1989. *Fundamentals of Digital Image Processing.* Prentice-Hall, Englewood Cliffs, NJ.

Macovski, A. 1983. *Medical Imaging Systems.* Prentice-Hall, Englewood Cliffs, NJ.

McFarlane, M.D. 1972. Digital pictures fifty years ago. *Proc. IEEE* (July):768–770.

Pratt, W.K. 1991. *Digital Image Processing.* Wiley, New York.

Rosenfeld, A. and Kak, A. 1982. *Digital Picture Processing* Vols. 1 and 2. Academic, San Diego, CA.

Further Information

A number of textbooks are available that cover the broad area of image processing and several that focus on more specialized topics within this field. The texts by Gonzalez and Wintz (1991), Jain (1989), Pratt (1991), and Rosenfeld and Kak (1982, Vol. 1) are quite broad in their scope. Gonzalez and Wintz's treatment is written at a somewhat lower level than that of the other texts. Current research and applications of image processing are reported in a number of journals. Of particular note are the *IEEE Transactions on Image Processing;* the *IEEE Transactions on Pattern Analysis and Machine Intelligence;* the *IEEE Transactions on Geoscience and Remote Sensing;* the *IEEE Transactions on Medical Imaging; the Journal of the Optical Society of America, A: Optical Engineering;* the *Journal of Electronic Imaging;* and *Computer Vision, Graphics, and Image Processing.*

7.5 Image Display Devices

Jerry C. Whitaker

7.5.1 Introduction

Information display is a key element in the advancement of electronics technology. The continued rapid growth of computer applications and embedded computer controllers in a wide variety of products have pushed forward the requirements for display components of all types. The basic display technologies in common use include the following:

- Cathode ray tube (CRT)
- CRT-based large screen projection
- Light valve-based large screen projection
- Light emitting diode (LED)
- Plasma display panel (PDP)
- Electroluminescence (EL)
- Liquid crystal display (LCD)

These technologies find use in a broad spectrum of applications, including

- Consumer television
- Computer display
- Medical imaging
- Industrial process control
- Test and measurement
- Automotive instrumentation
- Aerospace instrumentation
- Military devices and systems
- Camera imaging
- Printing
- Textiles
- Office equipment
- Telecommunications systems
- Advertising and point-of-purchase

CRT-based display technologies are covered in Chapter 28. Solid-state display systems will be addressed in this chapter.

Flat Panel Display Systems

Flat panel display systems are of interest to design engineers because of their favorable operational characteristics relative to CRT devices. These advantages include

- Portability
- Low occupied volume
- Low weight
- Modest power requirements

Within the last few years, a great deal of progress has been made on improving the performance of FPDs.

Advanced Display Applications

The applications for information display technologies are wide and varied. Aerospace is one area where different display technologies must be merged and where significant development is being conducted on an on-going basis. More information must be presented in the cockpit than ever before, and in greater varieties as well. The glass cockpit has already become a reality, with electronic displays replacing electromechanical devices. With the need for more detailed information, the size and resolution of the display devices must increase. Full-color plates and multilevel gray scales are a definite requirement. Reliability, naturally, is of the utmost importance.

The cockpit will grow into an interactive space that may include a heads-up display and/or helmet visor display. Alternatively, a big picture approach has been suggested in which the entire cockpit display area is one large reconfigurable element that integrates system readouts and controls. Reconfiguration is important where flight mode determines the cockpit display needs.

Multisensor Environment

Many different types of imaging sensors exist, each sensitive to a different portion of the electromagnetic spectrum (Foyle, 1992). Passive sensors, which collect energy transmitted by or reflected from a source, include television (visible light), night-vision devices (intensified visible and near-infrared light), passive millimeter wave sensors, and thermal imaging (infrared sensors). Active sensors, when objects are irradiated and the energy reflected from those objects is collected, include the various types of radar.

Each of these systems was developed because of their ability to increase the probability of identification, or detection of objects under difficult environment conditions. To best present this information to an operator, image processing algorithms have been developed that *fuse* the information into a single coherent image containing information from more than one sensor. Such systems are referred to as *multisensor displays*. The process of combining the display data is known an **sensor fusion**. New sensor integration hardware and software can be evaluated both relatively, by determining which combinations are better than others, and absolutely, by comparing operator performance to theoretical expectations.

Typically, as the environmental conditions change in which an individual sensor operators, so does the information content of that image. The information content of the image can further be *scaled* by the operator's ability to perform a target identification or discrimination task. It is reasonable to expect task performance with a sensor fusion display formed from two low-information content (and, hence, poor performance) images to still be relatively poor. Similarly, two high-information content (high-performance) sensor images should yield good performance when combined into a sensor fusion display. Assuming that there was some independent information in the two individual sensor images, it is logical to expect performance with the sensor fusion display to be better than either of the two individual sensors alone. This property results in a three-dimensional performance space.

The greater the number of sensor inputs, the greater the overall performance of the fusion display. However, as the number of inputs grows, the task of designing a display that provides coherent information to the operator is significantly increased. This area of display technology represents perhaps the greatest single technical challenge for the display industry. Although a wealth of sensor data is available in a tactical situation, the ability of the operator to integrate the data on a split-second basis remains limited. As the computer industry has already discovered, gathering data is relative easy—determining how to display the data so that the operator can understand it is the real challenge.

7.5.2 LCD Projection Systems

Liquid crystal displays have been widely employed in high-information content systems and high-density projection devices. The steady progress in active-matrix liquid crystal displays, in which each pixel is addressed by means of a thin film transistor (TFT), has led to the development of full-color video TFT-addressed liquid crystal light valve (LCLV) projectors. Compared with conventional CRT-based projectors.

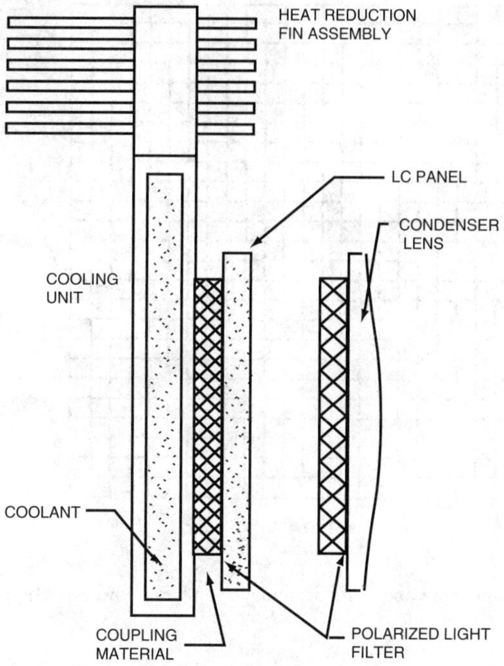

FIGURE 7.31   Structure of a liquid-cooled LC panel.

LCLV systems have a number of advantages, including

- Compact size
- Light weight
- Low cost
- Accurate color registration

Total light output (brightness) and contrast ratio, however, have remained a challenge. Improvements in the transmittance of polarizer elements have helped increase display brightness. By arrangement of the direction of the polarizers, the LCD can be placed in either the *normally black* (NB) or *normally white* (NW) mode. In the NW mode, light will be transmitted through the cell at V_{off} and will be blocked at V_{on}. The opposite situation applies for the NB mode. A high contrast display cannot be obtained without a satisfactory dark state.

Cooling the LC panels represents a significant technical challenge as the light output of projection systems increases. The contrast ratio of the displayed image will decrease as the temperature of the LC panels rises. Furthermore, long-term operation under conditions of elevated temperature will result in shortened life for the panel elements.

The conventional approach to cooling has been to circulate forced air over the LC panels. This approach is simple and works well for low- to medium-light-output systems. Higher power operation, however, may require liquid cooling. One approach incorporates a liquid-cooled heat exchanger to maintain acceptable contrast ratio on a high-output HDTV projector. The structure of the cooling system is shown in Fig. 7.31. The cooling unit, mounted behind the LC panel, is made of frame glass and two glass panels. The coolant is a mixture of water and ethylene glycol, which prevents the substance from freezing at low temperatures.

Figure 7.32 compares the temperature in the center of the panel for conventional air cooling and for liquid cooling. Because LC panel temperature is a function of the brightness distribution of the light source, the highest temperature is at the center of the panel, where most of the light is usually concentrated. As the amount of light from the lamp increases, the cooling activity the liquid-based system accelerates. Waste heat is carried away from the projector by directing cooling air across the heat reduction fins.

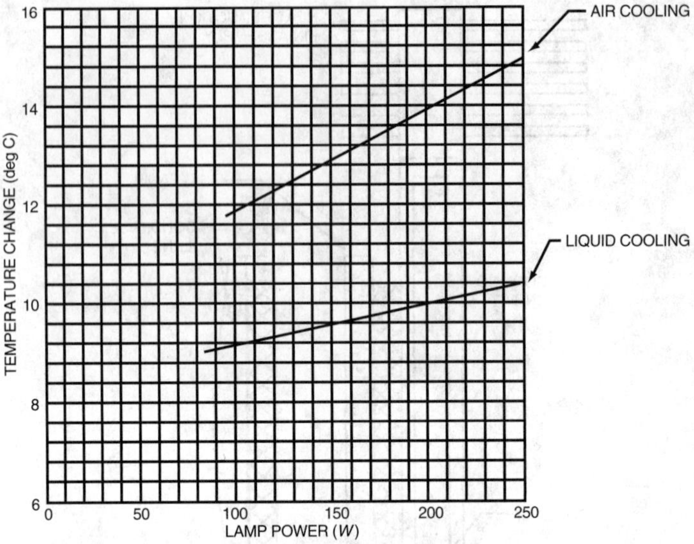

FIGURE 7.32   LC panel temperature dependence on lamp power in a liquid crystal projector.

Application Examples

The numerous recent improvements in LC display technology have permitted significant advances in LC projection techniques. One common approach incorporates three 5.5-in. 1.2 million pixel TFT active matrix LCD panels as light valves to project onto a 110-in. (or larger) screen. A simplified block diagram of the system is shown in Fig. 7.33. A short-arc, single jacket type 250-W high-intensity metal halide lamp is used as the light source. A parabolic cold mirror is utilized that reflects only the visible spectrum of the light beam to the front of the projector, allowing the infrared and ultraviolet spectrum rays to pass behind the mirror. Such rays have a harmful effect on the LC panels. The white light from the lamp is separated into the primary colors by dichroic mirrors. Each of the three primary color beams are regulated as they pass through their respective LC panels.

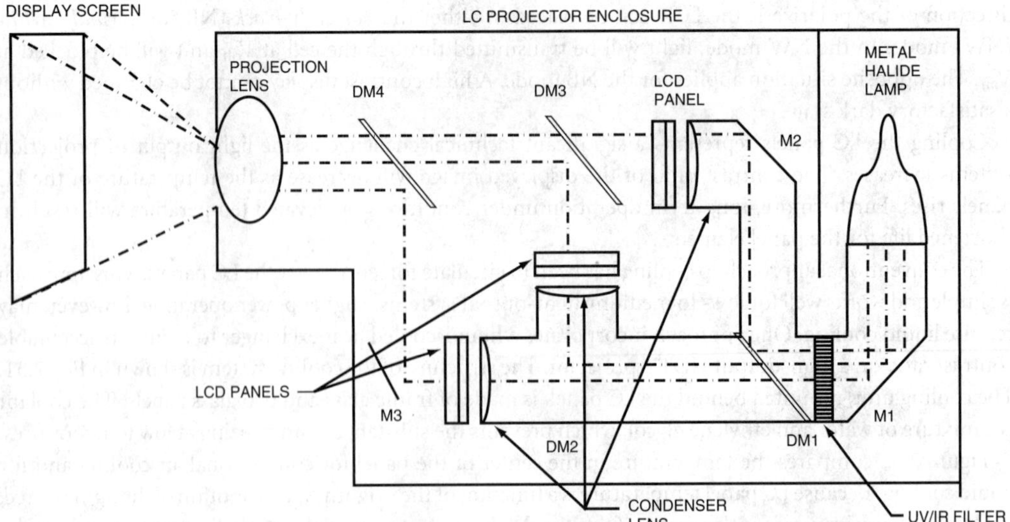

FIGURE 7.33   LC projection system for HDTV applications.

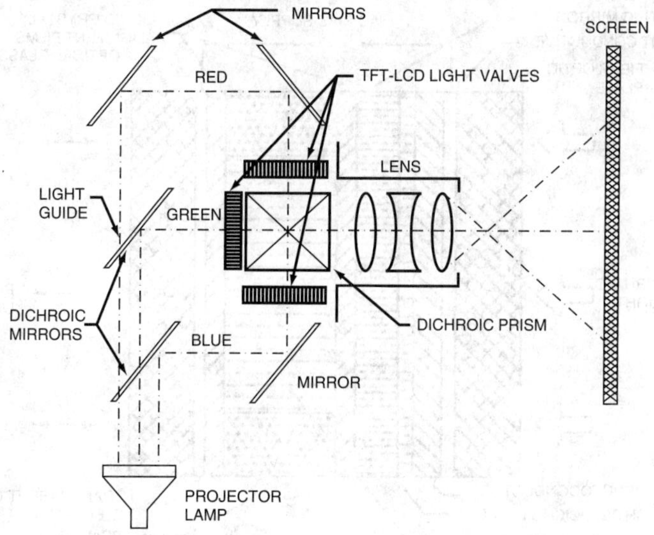

FIGURE 7.34   Optical system for an NTSC LC projector.

The LC system offers good signal clarity, compared to other projection systems. A single lens optical system is used; the primary colors are combined before entering the projection lens. The convergence of the picture projected, therefore, is unaffected by changes in the distance between the projector and the screen. Furthermore, changes in the viewing position relative to the center axis of the display screen will not adversely affect color purity.

Figure 7.34 illustrates the optical system of an LC projector designed for National Television Systems Committee (NTSC) applications. The TFT active matrix LC panels contain 211,000 pixels. Vertical resolution of 440 television lines is achieved through the use of noninterlaced progressive scanning. Interpolative picture processing techniques render the scanning lines essentially invisible. As shown in the figure, white light is separated into red, green, and blue (RGB) light by dichroic mirrors. Light of each color is directed to its corresponding RGB LC imaging panel. The LC panels are mounted to a dichroic prism for fixed convergence. The RGB images are combined into a single full-color image in the dichroic prism and projected by the lens onto the screen.

Homeotropic LCLV

The *homeotropic* (perpendicular alignment) LCLV[2] (Hughes) is based on the optical switching element shown in Fig. 7.35. Sandwiched between two transparent idium-tin oxide electrodes are a layer of cadmium sulfide photoconductor, a cadmium telluride light blocking layer, a dielectric mirror, and a 6-μm-thick layer of liquid crystal. An AC bias voltage connects the transparent electrodes. The light blocking layer and the dielectric mirror are thin and have high dielectric constants, and so the AC field is primarily across the photoconductor and the liquid crystal layer. When there is no writing light impinging on the photoconductor, the field or voltage drop is primarily across the photoconductor and not across the liquid crystal. When a point on the phtoconductor is activated by a point of light from an external writing CRT, the impedance at that point drops and the AC field is applied to the corresponding point in the liquid crystal layer. The lateral impedances of the layers are high, and so the photocarriers generated at the point of exposure do not spread to adjacent points. Thus, the image from the CRT that is exposing

---

[2]This section is based on: Fritz, V.J. 1990. Full-color liquid crystal light valve projector for shipboard use. In *Large Screen Projection Displays II*, ed. W.P. Bleha, Jr., pp. 59–68. SPIE 1255.

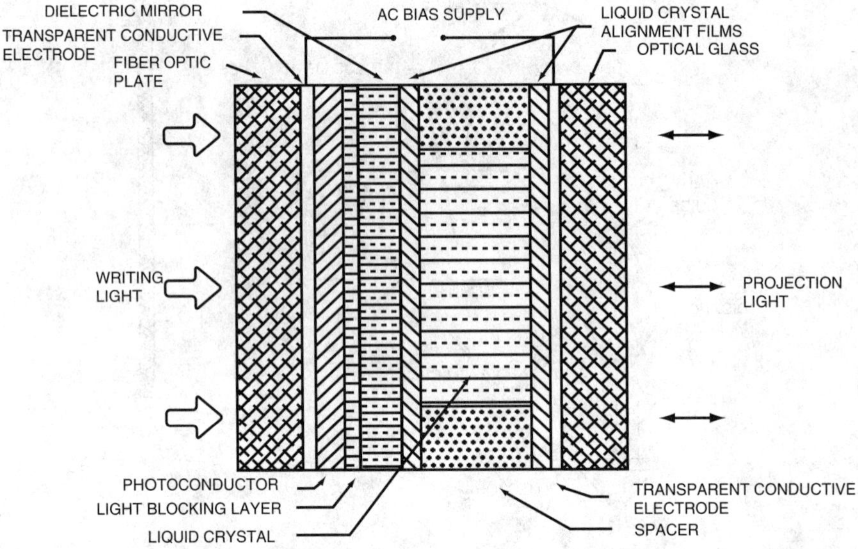

FIGURE 7.35   Layout of a homeotropic liquid crystal light valve.

the photoconductor is reproduced as a voltage image across the liquid crystal. The electric field causes the liquid crystal molecules to rotate the plane of polarization of the projected light that passes through the liquid crystal layer.

The homeotropic LCLV provides a perpendicular alignment in the *off* state (no voltage across the liquid crystal) and a pure optical birefringence effect of the liquid crystal in the *on* state (voltage applied across the liquid crystal). To implement this electro-optical effect, the liquid crystal layer is fabricated in a perpendicular alignment configuration; the liquid crystal molecules at the electrodes are aligned with their long axes perpendicular to the electrode surfaces. In addition, they are aligned with their long axes parallel to each other along a preferred direction that is fabricated into the surface of the electrodes.

A conceptual drawing of an LCLV projector using this technology is shown in Fig. 7.36. Light from a xenon arc lamp is polarized by a polarized beamsplitting prism. The prism is designed to reflect *S* polarized light (the polarization axis, the *E*-field vector, is perpendicular to the plane of incidence) and transmit

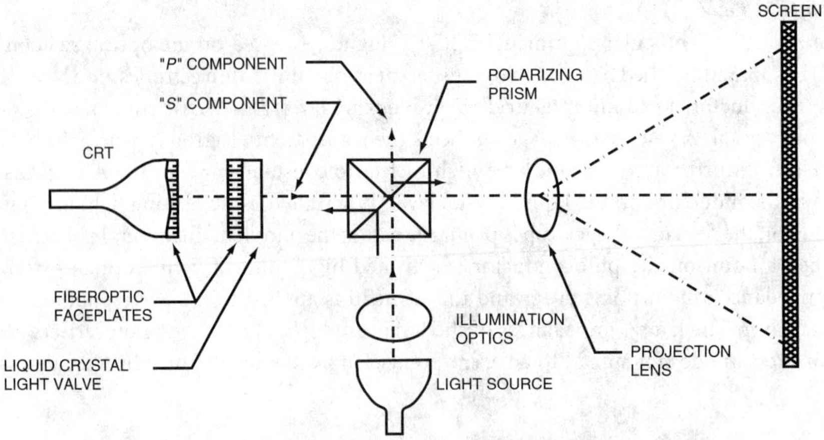

FIGURE 7.36   Block diagram of an LCLV system.

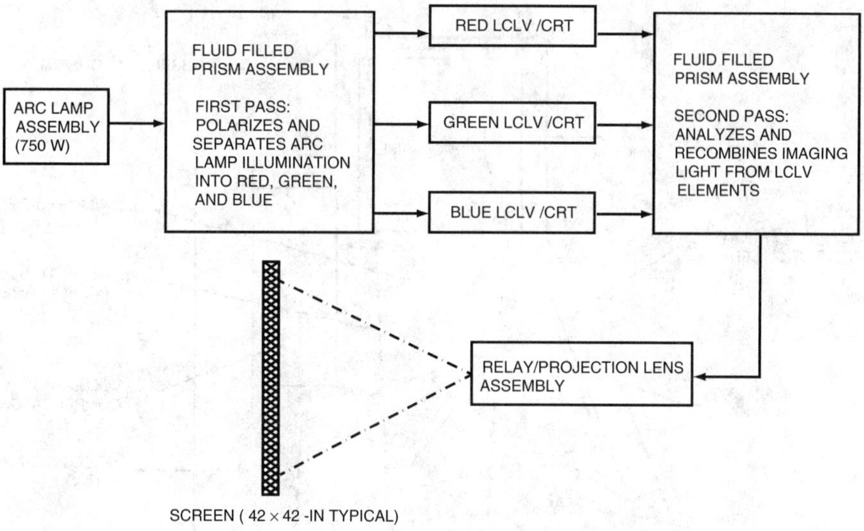

FIGURE 7.37 Functional optical system diagram of an LCLV projector.

P polarized light (the polarization axis, the *E*-field vector, is parallel to the plane of incidence). The *S* polarization component of the light is reflected to the light valve. When the light valve is activated by an image from the CRT, the reflecting polarized light is rotated 90° and becomes *P* polarized. The *P* polarization component transmits through the prism to the projection lens. In this type of system the prism functions as both a polarizer and analyzer. Both monochrome and full-color versions of this design have been implemented. Figure 7.37 shows a block diagram of the color system.

The light valve is coupled to the CRT assembly by a fiber optic backplate on valve and a fiber optic faceplate on the CRT.

The heart of this system is the fluid filled prism assembly. This device contains all of the fillers and beamsplitter elements that separate the light into different polarizations and colors (for the full-color model). Figure 7.38 shows the position of the beamsplitter elements inside the prism assembly.

The light that exits the arc lamp housing enters the prism through the ultraviolet (UV) filter. The UV filter is a combination of a UV reflectance dichroic on a UV absorption filter. The UV energy below 400 nm can be destructive to the life of the LCLV. Two separate beamsplitters *prepolarize* the light: the green prepolarizer and the red/blue prepolarizer. Prepolarizers are used to increase the extinction ratio of the two polarization states. The green prepolarizer is designed to reflect only the *S* polarization component of green light. The *P* polarization component of green and both the *S* and *P* polarizations of red and blue are transmitted through the green prepolarizer. The green prepolarizer is oriented 90° with respect to the other beamsplitters in the prism. This change of orientation means that the transmitted *P* component of the green light appears to be *S* polarized relative to the other beamsplitters in the prism assembly. The red/blue prepolarizer is designed to reflect the *S* polarization component of red and blue. The remaining green that is now *S* polarized and the red and blue *P* polarized components are transmitted to the main beamsplitter.

The main beamsplitter is a broadband polarizer designed to reflect all visible *S* polarized light and transmit all visible *P* polarized light. The *S* polarized green light is reflected by the main beamsplitter to the green mirror. The green mirror reflects the light to the green LCLV. The *P* polarized component of red and blue transmits through the main polarizer. The red/blue separator reflects the blue light and transmits the red to the respective LCLVs.

Image light is rotated by the light valve and is reflected to the prism. The rotation of the light or phase change (*P* polarization to *S* polarization) allows the image light from the red and blue LCLVs to reflect off the main beamsplitter toward the projection lens. When the green LCLV is activated, the green image light is transmitted through the main beamsplitter.

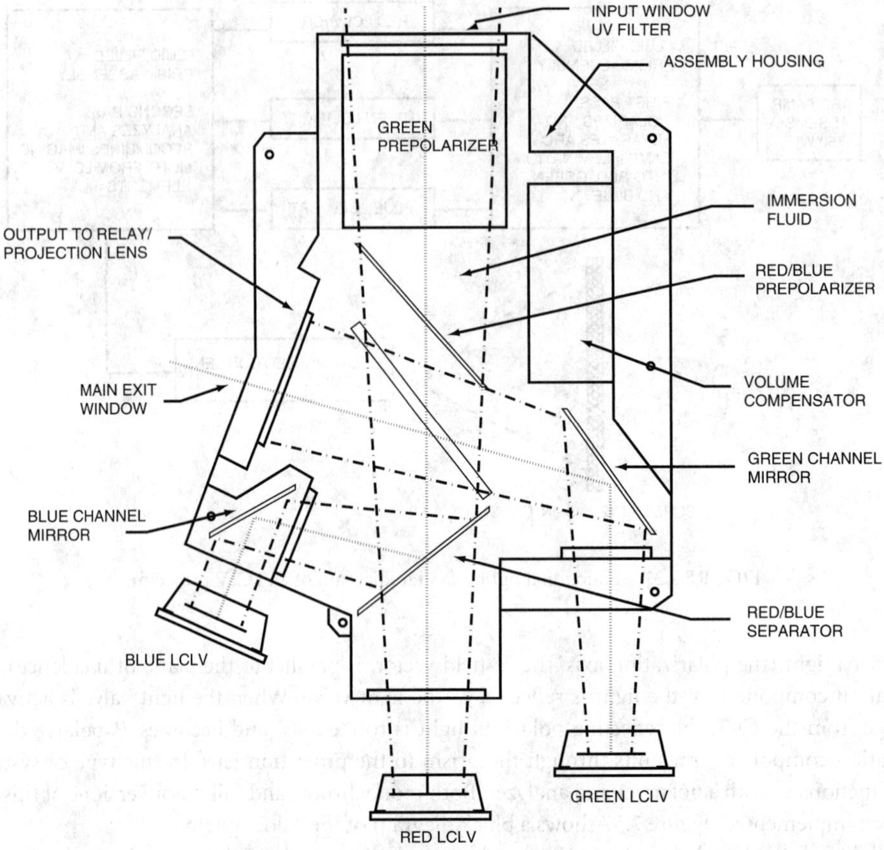

FIGURE 7.38   Layout of the fluid filled prism assembly for an LCLV projector.

The fluid filled prism assembly is constructed with thin-film coatings deposited on quartz plates and immersed in an index matching fluid. Because polarized light is used, it is necessary to eliminate stress birefringence in the prism, which, if present, would cause contrast degradation and background nonuniformity in the projected image. Using a fluid filled prism effectively eliminates mechanically and thermally induced stress birefringence as well as any residual stress birefringence. Expansion of the fluid with temperature is compensated for by using a bellows arrangement.

The color projector described in this section is capable of light output of over 1300 lumens with a contrast ratio of 200:1 or greater. The projector uses a 750-W xenon arc lamp as the light source. The video resolution is 1600 (H) × 1200 (V).

Laser-Addressed LCLV

The laser-addressed liquid crystal light valve is a variation on the CRT-addressed LCLV discussed in the previous section. Instead of using a CRT to write the video information on the LC element, a laser is used.[3] The principle advantages of this approach is the small spot size possible with laser technology, thereby increasing the resolution of the displayed image. In addition, a single beam from a laser scanner can be directed to three separate RGB light valves, reducing potential convergence problems by having only one source for the image.

---

[3] Portions of this section are based on: Phillips, T.E. et al. 1992. 1280 × 1024 video rate laser-addressed liquid crystal light valve color projection display. *Optical Engineering* 31(11):2300–2312.

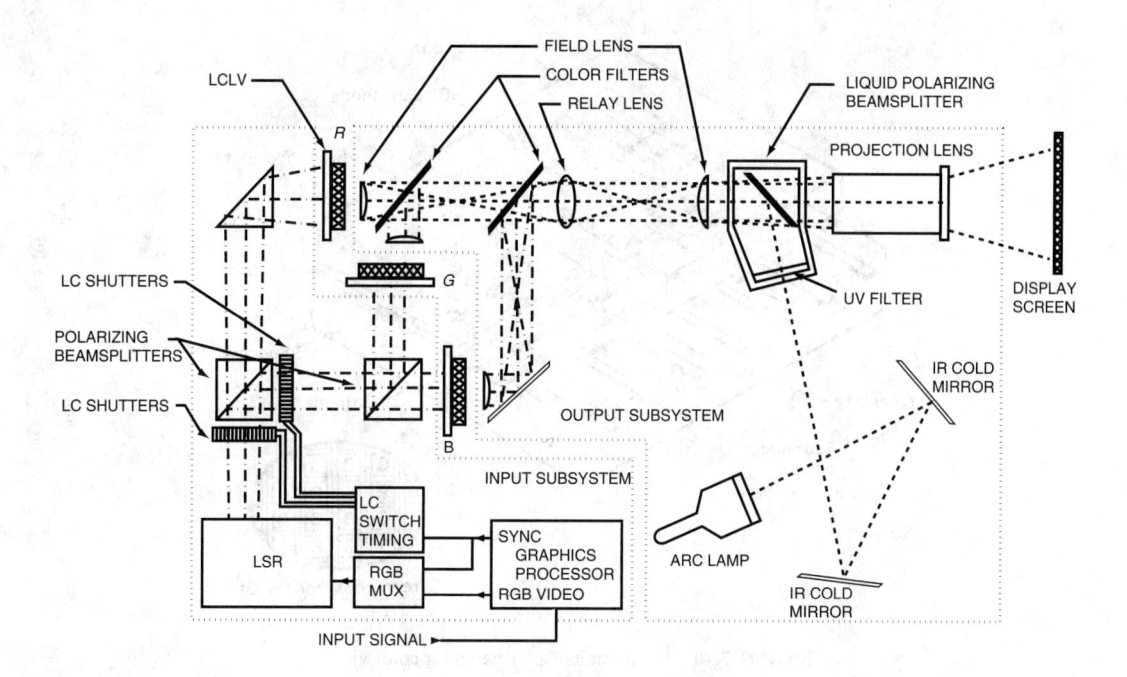

FIGURE 7.39 Block diagram of a color laser-addressed light valve display system.

A block diagram of the laser-addressed LCLV is shown in Fig. 7.39. Major components of the system include

- Laser
- Video raster scanner
- Polarizing multiplexing switches
- LCLV assemblies
- Projection optics

Projector operation can be divided into two basic subsystems:

- Input subsystem, which includes the laser, scanning mechanism, and polarizing beamsplitters
- Output subsystem, which includes the LCLV assemblies, projection optics, and light source

The output subsystem is similar in nature to the homeotropic LCLV discussed in the previous section.

The heart of the input subsystem is the *laser raster scanner* (LRS). The X-Y deflection system incorporated into the LRS uses four acousto-optic devices to achieve all solid-state modulation and scanning. A diode-pumped Nd:YAG, doubled to 535 nm (18 mW at 40:1 polarization ratio), is used as the laser source for the raster scan system.

Because only a single-channel scanner is used to drive the LCLV assemblies, a method is required to sequence the RGB video fields to their respective light valves. This sequencing is accomplished through the use of LC polarization switches. These devices act in a manner similar to the LCLV in that they change the polarization state of incident light. In this case the object is to rotate the polarization by 0° or 90° and exploit the beam steering capabilities of polarizing beamsplitters to direct the image field to the proper light valve at the proper time.

As shown in Fig 7.39, each switch location actually consists of a pair of individual switches. This was incorporated to achieve the necessary fall time required by the LRS to avoid overlapping of the RGB fields. Although these devices have a short rise time (approximately 100 μs), their decay time is relatively long (approximately 1 ms). By using two switches in tandem, compensation for the relatively slow relaxation

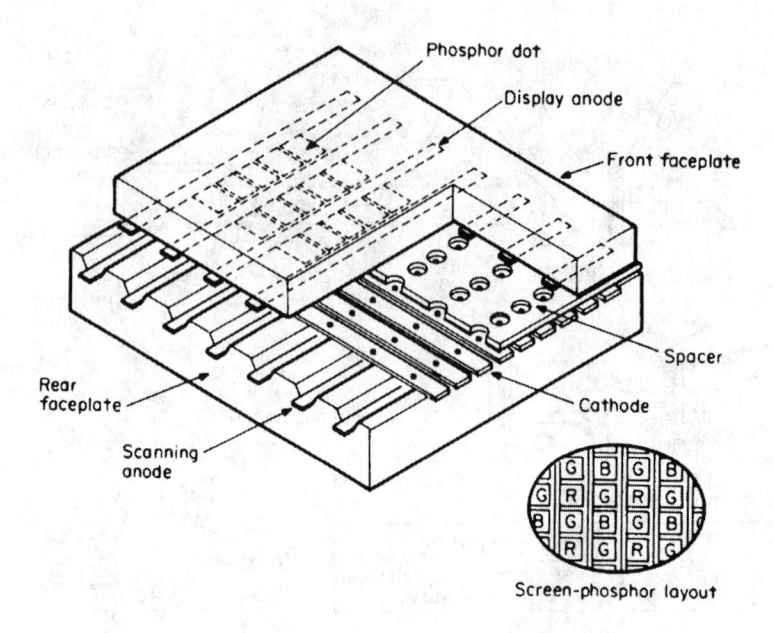

FIGURE 7.40 DC plasma display panel for color video.

time is accomplished by biasing the second switch to relax in an equal, but opposite, polarization sense. This causes the intermediate polarization state that is present during the switch-off cycle to be nulled out, allowing light to pass through the LC switch during this phase of the switching operation.

Luminescent Panels

The prevalent technology of large screen luminescent panels employs plasma excitation of phosphors overlying each display cell. The plasma is formed in low-pressure gas that may be excited by either AC or DC fields. Figure 7.40 depicts a color DC plasma panel. Such devices have been constructed in a variety of sizes and cell structures. Color rendition requires red/green/ blue trios that are excited preferentially or proportionally, as in a color CRT, which triples the number of actual cells (three cells comprise one picture element).

Because of their favorable size and weight considerations, PDP systems are well suited to large screen (40-in. diagonal) full-color emissive display applications.

Both AC and DC PDPs have been used to produce full-color, flat-panel large area dot-matrix devices. PDP systems operate in a memory mode. Memory operation is achieved through the use of a thick-film current-limiting series capacitor at each cell site. The internal cell memory capability eliminates the need for a refresh scan. The cell memory holds an image until it is erased. This eliminates flicker because each cell operates at a duty cycle of one. The series cell capacitor in an AC-PDP is fabricated using a thick-film screen-printed dielectric glass, as illustrated in Fig. 7.41. The gap separation between the two substrates is typically 4 mil. The surface of the thick-film dielectric is coated with a thin-film dielectric material such as magnesium oxide.

Matrix Addressing

Provision for individual control of each element of a discrete element display implies a proliferation of electrical connections and, therefore, rapidly expanding cost and complexity. These factors are minimized in discrete-element displays through *matrixing* techniques wherein a display element must receive two or more gating signals before it is activated. Geometric arrangements of the matrixing connectors guarantee that only the appropriate elements are activated at a given time.

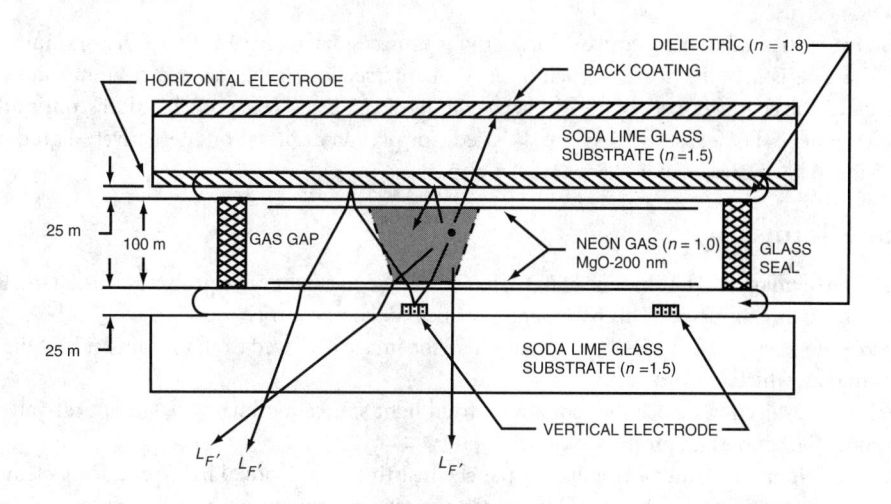

FIGURE 7.41 Cross section of an AC-PDP cell capacitor.

Matrix addressing requires that the picture elements of a panel be arranged in rows and columns, as illustrated in Fig. 7.42 for a simplified 4 × 4 matrix. Each element in a row or column is connected to all other elements in that row or column by active devices (typically transistors). Thus, it is possible to select the drive signals so that when only a row or column is driven, no elements in that row or column will be activated. However, when both the row and column containing a selected element are driven, the sum of the two signals will activate a given element. The resulting reduction in the number of connections required, compared to a discrete-element drive, is significant because the number of lines is reduced from one-per-element to the sum of the number of rows and columns in the panel.

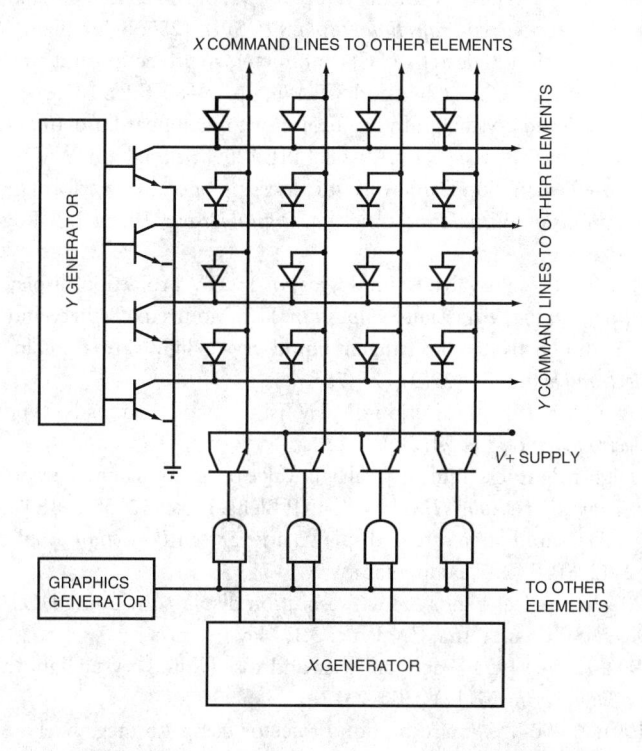

FIGURE 7.42 Matrix addressing for flat-panel display.

For active matrix addressing to work properly, it is necessary to have a sharp discontinuity in the transfer characteristic of the device at each pixel at the intersection of the row and column address lines. In some cases, the characteristics of the display device itself can be used to provide this discontinuity. The breakdown potential of a gas discharge can be used (for plasma displays) or alternatively, liquid crystals can be made to have a reasonably sharp transfer curve.

Defining Terms

Darkfield performance: The amount of light that is blocked in a light valve projector when a totally dark scene is being displayed. This parameter is critical to good contrast ratio.

Front projection system: A projection device where the image is viewed from the same side of the screen as that on which it is projected.

Light valve: A projection device that employs a fixed light source modulated by an optical-valve intervening the source and projection optics.

Matrix addressing: The control of a display panel consisting of individual light-producing elements by arranging the control system to address each element in a row/column configuration.

Rear projection system: A projection device where the image is viewed from the opposite side of the screen from that on which it is projected.

Sensor fusion: The combining of multiple sensor inputs into a single output display that is optimized for operator interpretation through manipulation of data and, if applicable, pre-programmed decision making processes.

References

Bates, W., Gelinas, P., and Recuay, P. 1991. Light valve projection system for HDTV. *Proceedings of the International Television Symposium*. ITS, Montreux, Switzerland.

Blaha, R.J. 1990. Large screen display technology assessment for military applications. In *Large-Screen Projection Displays II*, ed. William P. Bleha, Jr. pp. 80–92. SPIE 1255, Bellingham, WA.

Bleha, W.P., Jr., ed. 1990. *Large-Screen Projection Displays II*. SPIE 1255, Bellingham, WA.

Cheng, J.S. et al. 1992. The optimum design of LCD parameters in projection and direct view applications. In *Display Technologies*, eds. S.H. Chen and S.T. Wu, pp. 69–80. SPIE 1815.

Fritz, V.J. 1990. Full-color liquid crystal light valve projector for shipboard use. In *Large-Screen Projection Displays II*, ed. William P. Bleha, Jr., pp. 59–68. SPIE 1255, Bellingham, WA.

Foyle, D.C. 1992. Proposed evaluation framework for assessing operation performance with multisensor displays. In *Human Vision, Visual Processing, and Digital Display III*, ed. B.E. Rogowitz, pp. 514–525. SPIE 1666.

Gerhard-Multhaupt, R. 1991. Light valve technologies for HDTV projection displays: A summary. *Proceedings of the International Television Symposium*. ITS, Montreux, Switzerland.

Grinberg, J. et al. 1975. Photoactivated birefringent liquid-crystal light valve for color symbology display. *IEEE Trans. Electron Devices* ED-22(9):775–783.

Lakatos, A.I. and Bergen, R.F. 1977. Projection display using an amorphous-Se-type ruticon light valve. *IEEE Trans. Electron Devices* ED-24(7):930–934.

Maseo, I. et al. 1990. High-brightness liquid crystal light valve projector using a new polarization converter. In *Large-Screen Projection Displays II*, ed. William P. Bleha, Jr., pp. 52–58. SPIE 1255, Bellingham, WA.

Mitsuhashi, T. 1990. HDTV and large screen display. *Large-Screen Projection Displays II*, ed. William P. Bleha, Jr., pp. 2–12. SPIE 1255, Bellingham, WA.

Morizono, M. 1990. Technological trends in high resolution displays including HDTV. *SID International Symposium Digest*. SID, Santa Ana, CA, Paper 3.1, May.

Phillips, T.E. et al. 1992. 1280 × 1024 Video rate laser-addressed liquid crystal light valve color projection display. *Optical Engineering* 31(11):2300–2312.

Takeuchi, K. et al. 1991. A 750-TV-line resolution projector using 1.5 megapixel a-Si TFT LC modules. *SID 91 Digest*. SID, Santa Ana, CA, pp. 415–418.

Tomioka, M. and Hayshi, Y. 1991. Liquid crystal projection display for HDTV. *Proceedings of the International Television Symposium*. ITS, Montreux, Switzerland.

Tsuruta, M. and Neubert, N. 1992. An advanced high resolution, high brightness LCD color video projector. *J. SMPTE* (June):399–403.

Wedding, D.K., Sr. 1990. Large area full color AC plasma display monitor. In *Large-Screen Projection Displays II*, ed. William P. Bleha, Jr., pp. 29–35. SPIE 1255, Bellingham, WA.

Whitaker, J.C. 2001. *Video Displays*. McGraw-Hill, New York.

Further Information

The SPIE (the international optical technology society) offers a wide variety of publications on the topic of display systems engineering. The organization also holds technical seminars on this and other topics several times each year. SPIE is headquartered in Bellingham, WA. The Society for Information Display (SID) also holds regular technical seminars and offers a number of publications on the topic of projection display systems. SID is headquartered in Santa Ana, CA.

Two books on the topic of display systems are recommended:

Whitaker, Jerry C.: "Video Display Engineering," McGraw-Hill, New York, N.Y., 2001.

Whitaker, Jerry C., and K. Blair Benson: "Standard Handbook of Video and Television Engineering," McGraw-Hill, New York, N.Y., 2003.

7.6   Solid-State Amplifiers

Timothy P. Hulick

7.6.1   Introduction

Amplifiers are used in all types of electronic equipment and form the basis from which other active circuits are derived. Designing an oscillator would be impossible without an amplifier built in. Without amplifiers, any generated signal (if a meaningful signal could be generated without one) would only suffer loss after loss as it passed through a circuit to the point where it would become immeasurably small and useless. In fact, amplifiers are designed and built to counter loss so that deterioration of a signal as it passes through a medium is restored to its former or even greater level while allowing it to do something to the circuit that it just passed through. An amplified signal does not have to be a greater reproduction of itself, although this may be desirable, but only has to be bigger in some way than the one it came from. The form of an amplified signal may not resemble the signal from which it came, but must be controlled by it. The controlling of a big signal by a smaller one defines the notion of **gain**. All useful amplifiers have a gain magnitude equal to or greater than one which counters the effect of gain less than one (or loss) by passive elements of a circuit. Gain of an amplifier is defined to be the ratio of output power level to a load, divided by input power level driving the amplifier. In general, those devices that have a gain magnitude greater than one are considered to be *amplifiers*, whereas those of the other case are considered to be *attenuators*. The universally accepted electronic symbol for an amplifier is the triangle, however oriented, and it may or may not have more than one input or output or other connection(s) made to it. A simple amplifier is shown in Fig. 7.43.

Technically speaking, active logic gates found in the various types of logic families are also amplifiers since they offer power gain in their ability to drive relatively low-impedance loads, while not acting as a significant load themselves.

Amplifiers discussed in this chapter are limited to those that use solid-state devices as the active element since other types are discussed in Sections 7.7 and 7.8. In addition, amplifiers discussed in this section are presented in a way leading the reader to think that it is a section on *RF* amplifiers. This could not be further from the truth. Examples and presentations made here are meant to be *general*, using the general case where carrier frequencies are not zero. The same rules apply to amplifiers, whether they are used for baseband purposes, in which case the carrier frequency is zero, or are used as RF amplifiers, in which case the carrier frequency is not zero.

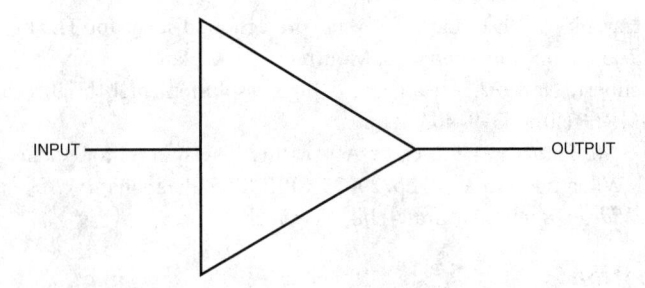

FIGURE 7.43   Any amplifier is universally symbolically represented as a triangle.

7.6.2   Linear Amplifiers and Characterizing Distortion

Amplifiers overcome loss and may or may not preserve the amplitude-time-frequency attributes of the signal to be amplified. Those that do come under the broad category known as **linear** amplifiers, whereas those that do not come under the category called *nonlinear* amplifiers. A linear amplifier has a frequency domain transfer function so that each point along the time-domain curve that defines the input signal perfectly maps into the output signal in direct proportion with constant time delay. This can only happen if a plot of output vs. input is a straight line, hence, the term linear. Since a straight line function represents only the ideal case, there are degrees of linearity that define the usefulness of such a thing in a particular application. What is considered linear in one application may not be linear enough in another. Linearity is measured in the amount of distortion of many different kinds, again, according to the application.

Compression

For an amplifier to be considered linear, the amplified signal output must be a faithful reproduction of the input signal. Without any *fixes* applied, such as negative feedback, the basis linear amplifier has limitations that are functions of its load impedance, supply voltage, input bias current (or voltage depending on device type), and its ability to just make the power asked of it. With sufficiently large drive, any device can be driven beyond its limits to the point of burnout, and so it is important to know the limitations of any amplifier. The onset of output signal compression is the first sign that a linear amplifier is being driven beyond its linear limits, that is, the transfer function curve is beginning to slope away from a straight line such that the output *compresses* compared to what it should be. Figure 7.44 illustrates the normal straight line path compared to the **compression** taking place when the amplifier is being driven too hard.

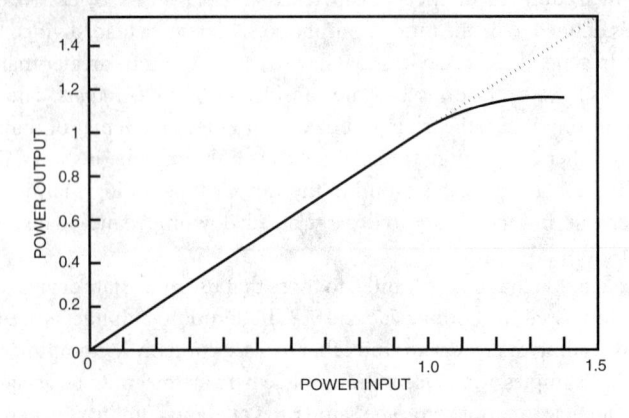

FIGURE 7.44   In the normalized plot of power output vs. power input, the amplifier is seen to be linear in the region from 0 to 1.0. Beyond this point, the output becomes compressed and finally flattens.

Linear amplifiers are characterized according to their 1-dBm power gain compression point. Simply put, the 1-dBm compression point is that power output from which a 9-dBm decrease is realized when the drive level is decreased by 10 dBm. A single-frequency continuous wave (CW) signal is usually used as the driving signal. When performing this test it is best to begin with a low-level signal far away from the compression region. The drive level is increased in 1-dBm steps while monitoring the linear tracking of input and output powers. Steps taken in 1-dBm increments are adequate and beneficial when looking for less than a 1-dBm increase in output power as the compression region is reached. At the onset of compression, the drive should be increased by 1 dBm more. At this point drive may be decreased by 10 dBm. If the output power decreases by more than 9 dBm, the 1-dBm compression point is still higher. Increase drive by a few tenths of a decibel and try again. Finding the 1-dBm compression point should be done by decreasing drive, not increasing it. Not knowing where it is, an increase this large could put the amplifier well into saturation (compression) and perhaps permanent damage may result.

The 1-dBm compression point should be noted. Generally, excellent linear operation will be 10 dBm or more below this point, but more power may be gotten from an amplifier if the compression point is approached more closely and compensated for by a low-level precorrector (expander) circuit before the input to the amplifier.

Odd Order Intermodulation Distortion

Another very important measure of linearity is the relative level of generated odd order intermodulation products in the amplifier. For intermodulation to take place, two or more signals must exist at the same time as a combined input signal. They must be at different frequencies, but near enough to each other so that they both fit within the passband of the amplifier. A suitable example input signal is of the form of Eq. (7.23)

$$S_i(t) = A\cos(2\pi f_1 t) + B\cos(2\pi f_2 t) \tag{7.23}$$

where A and B are the amplitudes of the independent input signals of frequencies f_1 and f_2, respectively.

In the absence of any nonlinearity in the amplifier transfer function, the output is of the same form as the input, and there is no change except that A and B are multiplied by the gain of the amplifier G

$$S_0(t) = G A\cos(2\pi f_1 t) + G B\cos(2\pi f_2 t) \tag{7.24}$$

If the transfer function is not perfectly linear, the input components are also multiplied together within the amplifier and appear at the output. The amplifier actually behaves like an ideal amplifier shunted (input to output) by a mixer, as shown in Fig. 7.45. The expression showing all frequency components out of the amplifier is

$$S_0(t) = G A\cos(2\pi f_1 t) + G B\cos(2\pi f_2 t) + \sum_{m=\text{int}} \sum_{n=\text{int}} c_{mn} A B\cos(2\pi m f_1 t)\cos(2\pi n f_2 t)$$

$$S_0(t) = G A\cos(2\pi f_1 t) + G B\cos(2\pi f_2 t) \tag{7.25}$$

$$+ \sum_{m=\text{int}} \sum_{n=\text{int}} \left[\frac{1}{2} c_{mn} A B\cos 2\pi(m f_1 + n f_2)t + \frac{1}{2} c_{mn} A B\cos 2\pi(m f_1 - n f_2)t \right]$$

where c_{mn} is the amplitude of the product component at frequency $f_{mn} = (m f_1 + n f_2)$ and is determined by the degree of nonlinearity.

Note that m and n can be any positive or negative integer and that the frequency components f_{mn} closest to the original signal frequencies are for m or n negative, but not both, $|m| + |n| = \text{odd}$ and m and n differ only by one. Generally, the amplitudes of the multiplication (intermodulation product) frequency components at f_{mn} are greater for $|m| + |n| = 3$ than for $|m| + |n| = 5$. Also the magnitudes of the $|m| + |n| = 5$ components are greater than the $|m| + |n| = 7$ components. The difference frequencies for the sum of m and n being odd are special because these frequency components may be difficult to filter out. Even order components, that is, when $|m| + |n| = \text{even}$ and the sum frequencies are far away

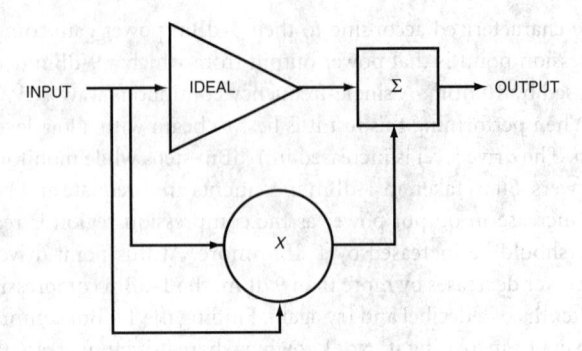

FIGURE 7.45   For intermodulation distortion analysis, the amplifier under test may be thought of as an ideal amplifier shunted by a mixer that creates the intermodulation product components. Multiple frequency signal components make up the inputs to the mixer.

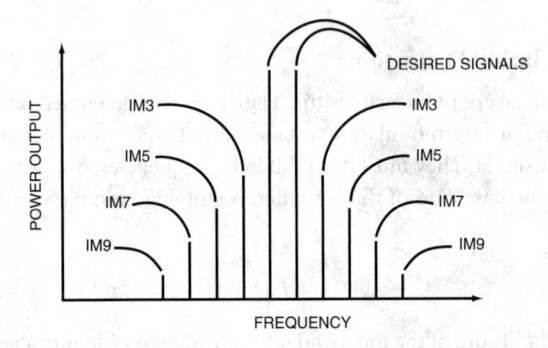

FIGURE 7.46   Example output spectral components are shown for a real linear amplifier with odd order intermodulation distortion. A two-frequency (two-tone) input signal is shown amplified at the output along with odd order difference frequencies generated within the amplifier.

in frequency, therefore, are easy to filter and generally are of no consequence. Because of this, odd order **intermodulation distortion** components are of importance to the amplifier linearity whereas even order components are not. Specifically, the third-order components are of concern because they are the closest in frequency to the original signals and are of the greatest amplitude. Figure 7.46 is a view of the relative positions and amplitudes of odd order intermodulation products around the original signals; their frequencies are listed in Table 7.1.

The CW signal used to find the 1-dBm compression point sets the maximum **peak envelope power** (**PEP**) with which the amplifier should be driven. The PEP of Eq. (7.23) is $(A + B)^2$, whereas the PEP of each frequency component is A^2 or B^2. The CW signal used in the compression test would be, for

TABLE 7.1   Some Odd Order Intermodulation Distortion Components to the Ninth Order

Third Order	Fifth Order	Seventh Order	Ninth Order	Comment
$2f_1 - f_2$	$3f_1 - 2f_2$	$4f_1 - 3f_2$	$5f_1 - 4f_2$	Near
$2f_2 - f_1$	$3f_2 - 2f_1$	$4f_2 - 3f_1$	$5f_2 - 4f_1$	Near
$f_1 - 2f_2$	$2f_1 - 3f_2$	$3f_1 - 4f_2$	$4f_1 - 5f_2$	Negative
$f_2 - 2f_1$	$2f_2 - 3f_1$	$3f_2 - 4f_1$	$4f_2 - 5f_1$	Negative
$f_1 + 2f_2$	$2f_1 + 3f_2$	$3f_1 + 4f_2$	$4f_1 + 5f_2$	Far away
$f_2 + 2f_1$	$2f_2 + 3f_1$	$3f_2 + 4f_1$	$4f_2 + 5f_1$	Far away

example, C^2. Using a *two-tone* signal with equal amplitudes such that $A = B$, the PEP becomes $4A^2$ or $4B^2$. The PEP of the two-tone signal should not exceed the PEP of the CW signal, or the output will enter the compressed region more than anticipated. If $4A^2 = C^2$, then the PEP needed in each of the two tones is one-fourth of the equivalent CW PEP of the single-tone compression test signal, or A^2. For this reason the two-tone test signals are individually 6 dBm below the compression test limit chosen.

The frequencies of the two tones are spaced such that the third-order intermodulation products commonly known as the IM3 products fall within the passband of the amplifier. They are not to be suppressed by means of any filters or bandwidth shaping circuits. As the amplifier is driven by the two-tone test signal, the output is observed on a spectrum analyzer. The transfer function of one of the desired signals is plotted, whereas the level of one of the IM3 signals is also plotted directly below that. The IM3 point plotted in the *power output vs. power input* space indicates that it could have come from an input signal that already had the IM3 component in it and it was just amplified and that the amplifier was not the source of the IM3 component at all. Of course, this is not true, but the IM3 curve, as it is derived along with the real transfer function as points are plotted at other power levels, begins to show itself as an imaginary transfer function. At increasing power levels, the loci of the two curves may be linearly extended to a crossover point where the desired signal output power equals the IM3 level. An example plot is shown in Fig. 7.47. Linear amplifiers are never driven this hard, but this point is specified (usually in (dBm)) for every linear amplifier and used along with the gain (slope of the real transfer function) to recreate a plot such as that of Fig. 7.47.

The level of IM3 may be related to the PEP of the two tones, in which case it is referred to as $(X)\mathrm{dB_{PEP}}$, or it may be related to the PEP of either of the two tone levels that are, of course, 6 dB lower than the PEP value. When expressed this way, it is referred to as $(X)\mathrm{dB_c}$, referenced to one of the carrier or tone levels. It is important to know which is being used.

Although the two-tone test has become a standard to determine the third-order intercept point, it does not tell the whole story. The two-tone test applies where only two tones are expected in the signal to be amplified. In the case of an analog television signal, for example, there are three tones of constant amplitude: the picture carrier, commonly called the *visual carrier*; the **color subcarrier**; and the sound carrier, commonly called the **aural subcarrier**. Their respective equivalent CW amplitudes in most television formats around the world are -8, -17, and -10 dB, respectively, referenced to the peak of the synchronization (sync) level. The PEP level is the PEP of the peak of sync and is used as the reference, but for common visual/aural amplification, the combined peak of sync value is an additional 73% greater. If the peak of sync level is considered to be normalized to one, and the aural subcarrier is -10 dBm down from that,

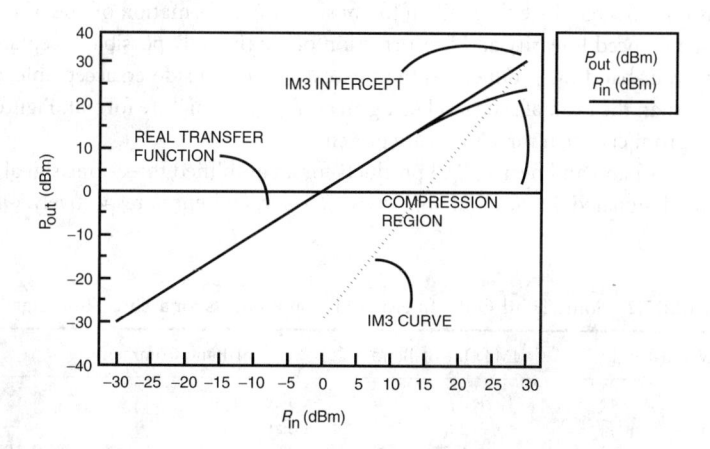

FIGURE 7.47 The output power where the IM3 line and the real transfer function line intersect is called the IM3 intercept point. The slope of the IM3 line is twice that of the transfer function so that for every decibel increase in output power, the IM3 level increases by 3 dB.

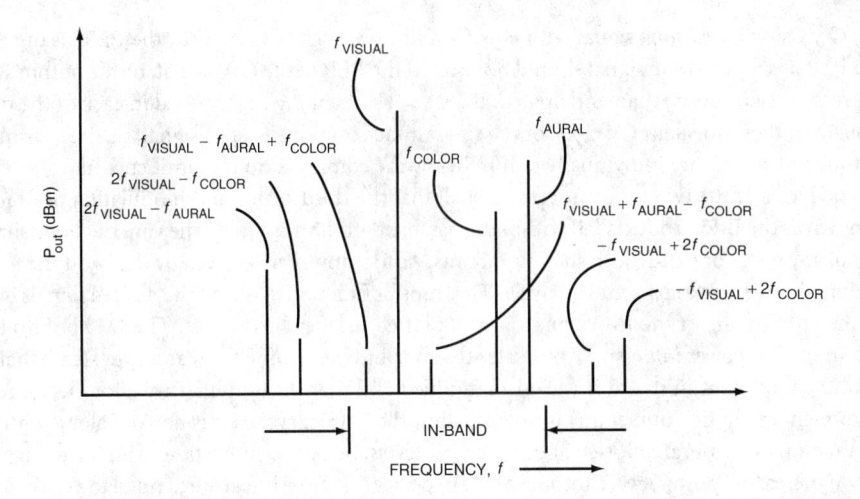

FIGURE 7.48   A three-tone test is used when evaluating a linear amplifier for television applications. The tones are set to the level of the carriers making up the signal, for example, −8-dB visual, −17-dB color, and −10-dB aural, relative to peak of sync. Some IM3 products fall in between the carriers and, therefore, are in-band and impossible to filter out.

it follows that

$$P_{PEP} = \left(\sqrt{P_{PEP\ VISUAL}} + \sqrt{P_{AURAL}}\right)^2$$

$$P_{PEP} = \left(\sqrt{1} + \sqrt{0.1}\right)^2 = 1.73$$

(7.26)

The additional PEP of the color subcarrier is insignificant at 17 dBm down from the peak of sync value. In television, however, the level of intermodulation is related to the peak of sync value as if the aural subcarrier were not present. A three-tone test is commonly used in this industry with the tone levels set to their respective values and placed at the proper frequencies. The visual carrier frequency is taken as the reference, whereas the color subcarrier is about 3.579 MHz higher and the aural subcarrier is 4.5 MHz above the visual carrier. When three tones are used for an intermodulation test, it is important to realize that some of the IM products will fall in between the original three signals. This is not the case when only two tones are used. For IM distortion products in between desired signals, filtering these components to an acceptable level is not possible since they fall on top of sideband information of one or more subcarriers. Concentrating on enhanced linearity and precorrection offers the only possible acceptable performance solutions. Generally, in-band IM products, as they are called, are considered acceptable when they are at least −52 dB lower than the peak of sync level using an average gray picture format. Figure 7.48 shows the placement of IM3 products when three tones are present.

Table 7.2 lists the in-band and nearby IM3 products for an amplified three-tone signal where the three carriers (tones) are designated f_V, f_C, and f_A for *visual*, *color*, and *aural*, respectively. One peculiarity is

TABLE 7.2   Some Third-Order Intermodulation Products for a Three-Tone Signal

IM3 In-band	IM3 Out-of-Band	Out-of-Band Comment
$f_V + f_A - f_C$	$f_V + 2f_A$	Far away
$f_V - f_A + f_C$	$f_V + 2f_C$	Far away
	$f_V - 2f_A$	Negative, only exists mathematically
	$f_V - 2f_C$	Negative, only exists mathematically
	$2f_V - f_A$	Sideband Regeneration
	$2f_V - f_C$	Sideband Regeneration

that $2f_v - f_a$ and $2f_v - f_c$ are also IM3 products, but appear to look as if they were lower sideband images of the legitimate signals at f_a and f_c, respectively. Although these components arise due to intermodulation, they were originally present when the color and aural subcarriers modulated the visual carrier before they were filtered out by a vestigial sideband filter to conform to television standards. This reappearance of the lower sidebands is sometimes called *sideband regeneration* because of where they are placed even though they are third-order intermodulation products.

Communications systems, such as those used by the telephone industry, make use of large numbers of carriers through one amplifier placing very demanding requirements on intermodulation performance. For a two-tone application, it was shown that the PEP of the individual carriers is 6 dB down from the PEP of the combined carriers. IM3 levels are 6 dB higher when referenced to either of these two carriers compared to the combined PEP of the two tones. For n equal amplitude carriers of power P, the PEP of the combined carriers is shown to be

$$PEP = n^2 P.$$

or that each carrier's peak envelope power contribution to the total PEP becomes

$$P = \frac{PEP}{n^2} \tag{7.27}$$

If 16 carriers are used to drive an amplifier capable of 100-W PEP (50 dBm), the individual carriers cannot exceed about 0.4 W (26 dBm). This is a 24-dB degradation in IM3 level relative to the individual carriers. If Δ is the relative increase in IM3 in dB due to a multicarrier signal and n is the number of carriers, it follows that

$$\Delta = 20 \log n$$

For this amplifier passing 16 carriers, requiring the IM3 levels to be 60 dB_c to prevent cross talk, the IM3 level relative to total PEP must be $60 + 20 \log 16 = 84$ dB_{PEP}.

Other Distortions

Odd order intermodulation distortion and 1-dBm gain compression go a long way to characterize the linearity of a linear amplifier, but they are by no means the only ones. Other distortion types tend to be application specific relating to their importance. For video signals other distortion type include the following:

- Differential gain
- Differential phase
- High-frequency to low-frequency delay
- Group delay
- Incidental carrier phase modulation
- Cross modulation (one carrier modulation to another)
- RF harmonic distortion
- Total harmonic distortion

Each of these is briefly discussed, but for a more in-depth understanding, the reader is referred to more specialized sources of information (Craig, 1994).

Differential gain distortion is a distortion of the amplitude information on a subcarrier in the presence of amplitude modulation of another subcarrier or main carrier. *Differential phase distortion* is a distortion of the phase information on a subcarrier in the presence of amplitude modulation of another subcarrier or main carrier. The former is measured in percent whereas the latter in degrees.

High-frequency to low-frequency delay is a distortion caused by the difference in time delay presented to high-frequency sidebands compared to low-frequency sidebands when it is important that their time relationship not be changed. A distortion of this type causes misalignment of subcarrier information with main carrier or other subcarriers and is measured in time.

Group delay distortion has the same effect as high-frequency to low-frequency delay, but is caused by associated filters (notch, bandpass, etc.) that may be built into an amplifier. Sudden changes of transfer function of a filter such as that near a notch filter frequency will cause sudden changes in group delay for amplifier frequency components nearby. It is also measured in time.

Incidental carrier phase modulation (ICPM) could be characterized as AM to PM conversion taking place within an amplifier. As an amplifier passes an amplitude modulated signal, dynamic capacitances within the amplifying device itself act in conjunction with fixed impedances in the amplifier circuit to cause dynamic time constants leading to carrier phase shift with amplitude modulation. ICPM is measured in degrees relative to some clamped amplitude level.

Cross modulation is a distortion caused by the modulation of one carrier through the amplifier onto another subcarrier or carrier. A cause of cross modulation is the mixer or multiplier behavior of real amplifiers modeled in Fig. 7.45. If two or more carriers are multiplied together, at least one of which is modulated with AM, its modulation sidebands may appear on the other. This is different from intermodulation distortion where sum and difference carrier frequencies arise whether they are modulated or not, but the mechanism is the same. Cross modulation occurs in linear amplifiers when the signal levels at the input are simply too high in amplitude for the amplifier to pass without *mixing*. Cross modulation is also caused when the amplifier modulates the power supply voltage as supply current changes with amplitude modulated signal level. If an AM signal and a frequency modulated (FM) or phase modulated (PM) signal are being amplified simultaneously, the AM signal causes a modulated current draw from the supply in certain classes of linear amplifiers (see Class B and Class AB amplifiers later in this chapter). This in turn causes a corresponding modulated voltage drop across the internal resistance of the supply to the amplifier causing synchronous amplitude modulation of the FM or PM signal. Cross modulation is measured in terms of percent AM in the case where AM is the result, or, frequency deviation in the case where FM is the result.

When the peaks of the instantaneous RF carrier enter the compression region, the RF carrier sinusoidal waveform is compressed to something other than a perfect sinusoid. When this happens, a Fourier series of harmonic components is generated leading to RF *harmonic distortion*. These are far away in frequency and easy to filter out, but may cause significant degradation of power efficiency since it is power that cannot be used and frequently is as high as 10% of the total RF power coming out of the amplifier. With no harmonic filter, a frequency insensitive power meter may indicate 100 W, but only 90 of these 100 W may be at the fundamental frequency.

In RF amplifiers, harmonics of the carrier frequencies are located at integer multiples of those frequencies and are easy to filter out. Harmonics of the modulating (baseband) signals may fall within the passband of the amplifier. Distortion by the appearance of baseband harmonics will arise out of a real amplifier when they did not exist going in. For example, a typical AM radio amplifier needs a passband of 30 kHz to faithfully pass all of the sidebands out to 15 kHz for quality amplification of a music modulated carrier. If just a single note at 1 kHz is the modulating signal, there is room for 15 harmonics to be passed along should the harmonic distortion create them. What an amplifier does to baseband signals is characterized by *total harmonic distortion*, which is expressed as a percent of total harmonic power to fundamental power. It is an envelope or baseband distortion, not an RF distortion.

7.6.3 Nonlinear Amplifiers and Characterizing Distortion

Any amplifier that significantly degrades amplitude modulation is a nonlinear amplifier. Linear amplifiers are supposed to be perfect, whereas nonlinear amplifiers are deliberately not this way. Usually a nonlinear amplifier is optimized in one or more of its characteristics, such as efficiency or ability to make power no matter how distorted. Nonlinear amplifiers have their place where a single tone input signal or one that is not amplitude modulated (except for the case where the modulation is a full on or full off signal) is the signal to be amplified. An example of this is a signal that is only frequency modulated. Specifically, an RF carrier that is deviated in frequency does not require *amplitude* linearity. Even though a FM signal may be carrying beautiful music, it is always only a single carrier that is moving around in frequency, hopefully symmetrically about the carrier frequency or *rest point*. Instantaneously it is only at one frequency. This is important to understand and often thought of only casually. The instantaneous frequency of a deviated carrier is singular even if there are subcarriers as part of the baseband signal providing the modulation. The fact remains that instantaneously, there is simply no other carrier to intermodulate with. For this reason, intermodulation cannot exist and so there is no intermodulation distortion. There is no distortion of the modulating signal because of amplitude compression. There is even no distortion of the frequency modulation if there is simultaneous amplitude modulation as long as the AM is not near 100% where the carrier is made to *pinch off* or disappear at times. Phase modulation amplification is affected in the same way as frequency modulation amplification through nonlinear amplifiers. Nonlinear amplifiers are also very good at making high power much more efficiently than linear amplifiers. Industrial RF heating applications also do not care about linearity since the purpose is to make high-power RF at a particular frequency in the most efficient way possible. If all these good things are attributable to nonlinear amplifier's amplifying frequency or phase modulated signals why are they not used universally instead of AM? References on the subject of modulation methods will indicate that theoretically FM and PM require infinite bandwidth and practically five or six times the bandwidth of an equivalent AM signal carrying the same information. Amplitude modulation methods have their place, with their linearity requirements, whereas nonamplitude modulation methods have theirs.

Distortions

Just as there are those conditions that cause distortion in linear amplifiers, there are other conditions that cause distortion in nonlinear amplifiers. For example, it was stated that the bandwidth requirement for FM or PM is much greater than for AM. In fact, infinite bandwidth is the ideal case, when in practice, of course, this is not possible. The bandwidth, however, must be wide enough to pass enough of the sideband components so that harmonic distortion of the recovered baseband signal at the receiving end is below a specified limit. Therefore, distortion of the modulated signal can only be avoided by ensuring that the bandwidth of the nonlinear amplifier is sufficient. The bandwidth required is estimated by the well-known *Carson's Rule* (Taub and Schilling, 1986) which states that

$$B = 2(\Delta f + f_m) \tag{7.28}$$

where B is the bandwidth that will carry 98% of the sideband power, Δf is the frequency deviation of the carrier, and f_m is the highest frequency component of the modulating signal.

Other distortions that affect an FM or PM signal take place in the modulator or source of the modulated signal to be amplified, not in the amplifier, thus they will not be discussed in this chapter.

Linear amplifiers are used wherever the integrity of the signal to be amplified must be maintained to the greatest extent possible. Minimizing the distortion types presented in Sec. 7.6.2 will optimize a linear amplifier so that it can be used for its intended purpose, but an amplifier must be classified as a linear amplifier in the first place or it will be impossible to achieve *acceptable linear performance*. Linear is a relative term implying that distortion is imperceivable for the application. Relative linearity is classified alphabetically from A to Z with A being the most linear. Not all letters are used, thus there are enough of them available for future classification as new amplifier types are discovered. The most widely used linear classes of amplifiers are Class A, Class AB, and Class B. Others are classes BD, G, H, and S.

Often while perusing through the literature, two opposing phrases may appear even within the same paragraph. They are *small signal* and *large signal*. Nebulous words such as these are used among amplifier design engineers and appear to be unscientific in a most unlikely place, but they do have definition. These terms are used because active solid-state amplifying devices have a small signal behavior and a large signal behavior. *Small signal* behavior means simply that the signal going through a transistor is small enough in magnitude so that the transistor may be modeled in terms of equivalent linear elements. The signal is not large enough to be noticed by the modeled elements of the transistor to cause them to change value. Using the model provides predictable behavior. In addition, the signal is not large enough to cause departure from the user's definition of what is linear enough. Best possible linearity requires that the compression region shown in Fig. 7.44 or Fig. 7.47 be avoided, but if it is acceptable, the small signal domain may be extended into this region of the transfer function. The term small signal is confined to Class A amplifiers and has no meaning in the other classes. The dividing line between small and large signals is at the power level below which the transistor equivalent elements are linear components and the amplifier is still considered linear according to the needs of the application.

7.6.4 Linear Amplifier Classes of Operation

Class A Amplifier

The term Class A does not refer to whether an amplifier is used in a small or large signal application even though all small signal amplifiers are Class A amplifiers. A Class A amplifier could also be used for large signals if the linear model no longer applies. Class A refers to the conduction angle of the amplifying device with respect to the driving signal and how it is biased to achieve this. For an amplifying device to be operating in Class A, it must be biased so that it is conducting current throughout the entire swing of the signal to be amplified. In other words, it is biased so that extremes of the amplitude of the driving signal never cause the device to go to zero **current** draw (cut off) and never cause the voltage across the output of the device to go to zero (**saturation**). In this manner, the device is always active to follow the waveform of the driving signal. Intuitively, it has the best chance at being linear compared to other bias conditions to be presented later. Notice that the word *device* is used when addressing the part that does the amplifying. **Class of operation** has nothing to do with the device type. The requirements for Class A operation are the same whether the device is a tetrode vacuum tube, bipolar junction transistor, field effect transistor, klystron, or anything else, and the mathematical behavior and theoretical limits of a device operated in Class A are the same regardless of the device type. It is true, however, that some devices are better than others in practical linearity, which allows variations in power efficiency and other performance parameters, but not in theoretical limits. The more general statement is that class of operation has nothing to do with device type, but only how it is biased, which, in turn, determines the **conduction angle**.

The operation of the Class A amplifier is easily understood if it is analyzed in graphical form. A bipolar junction NPN transistor is assumed. Figure 7.49 is a plot of collector current I_c vs. collector to emitter voltage V_{ce} for various values of constant base current I_b for a common emitter configuration. There is no resistor in the collector bias circuit so that the transistor quiesces at the supply voltage V_{CC}. To allow a symmetrical current and voltage swing about the quiescent point, the quiescent point is chosen at $\frac{1}{2}(2V_{CC})$ or V_{CC} and $\frac{1}{2}\hat{I}_c$, the peak value of collector current. I_{CQ} is the quiescent collector current. Collector-emitter saturation voltage is assumed to be zero. The negative reciprocal of the correct *load*

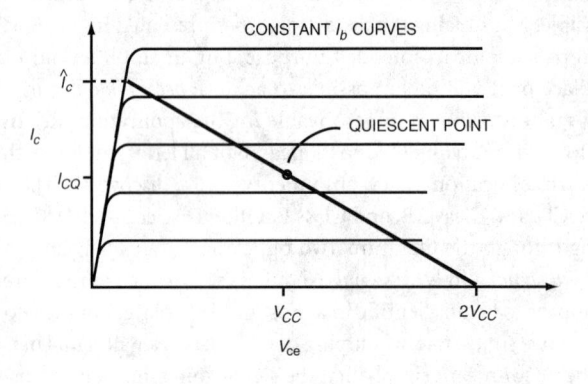

FIGURE 7.49 The I_c vs. V_{ce} plot for a bipolar junction transistor biased for Class A operation is shown with the staticquiescent point halfway between voltage and current limits for symmetrical excursions along the loadline. Strict Class A operation requires that the ends of the loadline not be exceeded.

resistance is the slope of the loadline plotted to allow maximum power transfer to the load. The loadline must, therefore, connect $2V_{CC}$ at $I_c = 0$ with $\hat{I}_c$ at $V_{ce} = 0$. Class A operation dictates that the ends of the loadline cannot be exceeded, but approaching the ends while maintaining linearity is possible if the harmonic power generated is filtered out so that it does not reach the load. If the peak of the AC component of the collector-emitter voltage is $\hat{V}_{ac}$ and the load resistance is R, then the output power P_{out}, may be calculated to be

$$P_{out} = \frac{\hat{V}_{ac}^2}{2R} \tag{7.29}$$

The DC input power from the power supply P_{in}, is calculated from

$$P_{in} = V_{CC} I_{CQ} = \frac{V_{CC}^2}{R} \tag{7.30}$$

Comparing Eq. (7.29) with Eq. (7.30), it is obvious that

$$P_{out} = \frac{1}{2} P_{in} \tag{7.31}$$

when $\hat{V}_{ac} = V_{CC}$. This occurs when the collector current and collector-emitter voltage are driven to the limits. The maximum theoretical power efficiency η_{max} is

$$\eta_{max} = \frac{P_{out}}{P_{in}} = \frac{1}{2} \tag{7.32}$$

And, of course, this number decreases from the maximum value as $\hat{V}_{ac}$ is decreased from the end limits of the loadline.

In practice, the limits imposed on $\hat{V}_{ac}$ are usually much less than that to provide an efficiency of 50%. A Class A amplifier used as a power amplifier typically operates at 10–20% efficiency in order to maintain third-order intermodulation distortion down to required levels for most communications applications. Since the Class A stage is usually used as the driver in a cascaded series of power amplifiers, the Class A stage must be very clean. The IM3 distortion products that the Class A stage produces are amplified by follow-on stages and the distortion only gets worse as more IM3 is produced. At UHF, or more it is practical to design and construct Class A power amplifiers using one bipolar junction NPN transistor that will provide 13 dBm of gain at 40-W PEP with −30 dBm IM3 performance using the two-tone test. Of course, any number of these may be power combined to produce hundreds of watts. After combining four or eight Class A stages, it is no longer practical to continue, since power combining losses become significant and adding stages only serves to cancel out losses.

Looking at the amplifier as a whole and not just the transistor, the input and output impedances are designed to be some value: 50 Ω for power RF amplifiers, 75 Ω for video, 600 Ω for audio or some other standard value for interfacing with the outside world. Whatever value the amplifier is designed for, it must be a good resistive load to whatever is driving it, and it must work into a load designed to be its load resistance. The actual impedances of the transistor input and the load presented to its output usually are much different than the external impedances, called the **standard**. In fact, for a power amplifier these actual impedances could be a fraction of an ohm. Input and output matching circuits connect these very low-impedance values to the standard. Looking into either port of a Class A amplifier with an impedance measuring device such as an impedance bridge or network analyzer will allow the impedance standard to be measured when the amplifier supply voltages are applied and the transistor is operating at its quiescent point. A return loss of 20 dBm or more across the frequency band of interest signifies a *good* amplifier. In the case of Class A, the entire amplifier may be modeled to be a resistor of the chosen standard value at the input $R_{StandardIn}$ and a current controlled current source shunted by the standard resistance value $R_{StandardLoad(SL)}$ at the output. These are the measured values. A model of the Class A transistor (neglecting the effects of capacitances and stray inductances that are *tuned out* by the matching circuits) is presented

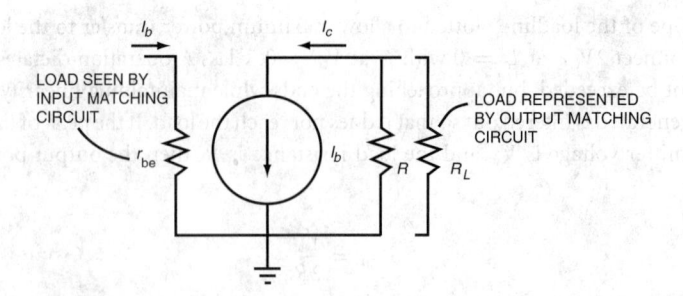

FIGURE 7.50   The Class A amplifier transistor can be modeled as a current controlled current source with internal output resistance R. When $R = R_L$, half of the collector power is dissipated in R and the other half in R_L, providing a maximum efficiency η_{max} of 50%. This agrees with Eq. (7.31). Not shown are the matching circuits transforming r_{be} and R_L to a standard impedance, such as 50 Ω.

in Fig. 7.50, whereas the model for the transistor amplifier with internal impedance matching circuits modeled is shown in Fig. 7.51.

Class B Amplifier

At high-power levels not practical for Class A, the Class B amplifier is used. Class B is more power efficient than Class A and unlike Class A, a transistor (or other amplifying device) conducts for precisely half of the drive cycle or π radians. It is biased precisely at collector current cutoff. As stated previously, the class of operation has nothing to do with the device type, but rather its bias condition and its conduction angle. Since Class B operation means that the device is only conducting half of the time, a single ended amplifier may only be used at narrow bandwidths where the tuned output network provides the missing half-cycle by storing the energy presented to it by the active half-cycle, then returning it to the circuit when the device is off. For untuned applications, such as audio amplifiers and wideband RF amplifiers, a push-pull pair of devices operate during their respective half-cycles causing the amplifier to be active all of the time. In this situation, the point on the loadline halfway between the ends is at zero supply current for both devices and driven away from zero with drive. Figure 7.52 shows a graphical presentation of the Class B push-pull configuration.

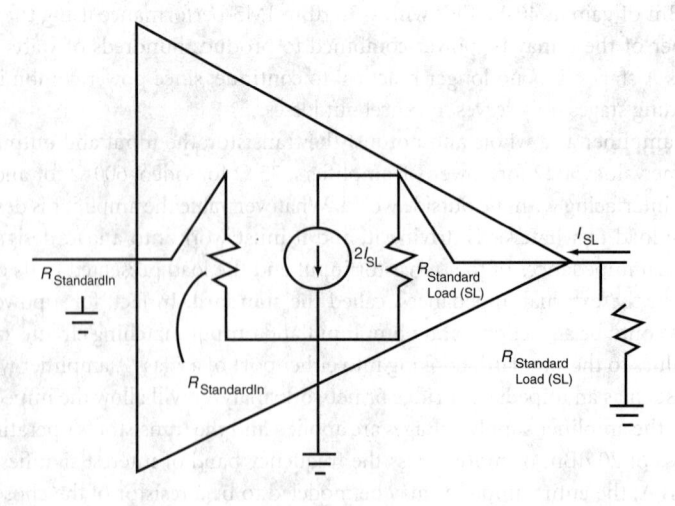

FIGURE 7.51   The Class A transistor amplifier is modeled as a resistive load to its driver whereas the output impedance seen at the output port is equal to the standard load resistance. The input and output impedances may be measured from the outside looking in with a network analyzer or impedance bridge.

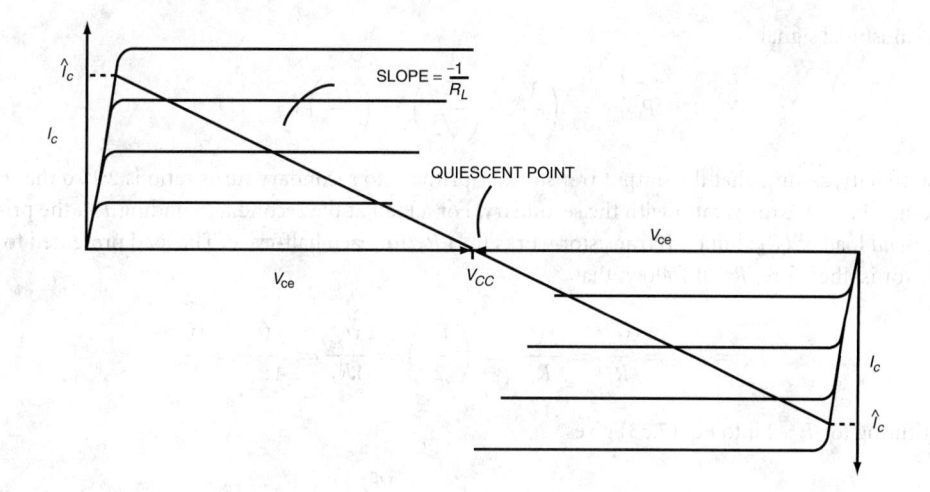

FIGURE 7.52 The graphical solution to the push–pull Class B amplifier configuration is shown. The quiescent point is midway between the ends of the loadline as in Class A, but unlike Class A, the quiescent current is zero. Each transistor output sees R_L at the collector since the center-tapped primary transformer has a 1:1 turns ratio from R_L to each transistor even though it has a 2:1 primary (end-to-end) to secondary turns ratio.

Because the phase of one transistor is of the opposite phase of the other, a center tapped transformer is needed to combine the two halves properly. This may be done without the use of a transformer if complementary pairs of transistors are available (one NPN and one PNP). Both cases are presented in Fig. 7.53.

To find the maximum linear output power and maximum theoretical power efficiency of a Class B push-pull amplifier, the loadline in Fig. 7.52 provides the information for the following analysis. Let

$\overset{\diamond}{V}_{ac}$ = maximum peak-to-peak voltage swing along the loadline, $\leq 2V_{CC}$

$\overset{\diamond}{I}_{ac}$ = maximum peak-to-peak current swing along the loadline, $\leq 2\hat{I}_c$

$\hat{V}_{ac}$ = maximum peak voltage swing along the loadline from the quiescent point, $\leq V_{CC}$

$\hat{I}_{ac}$ = maximum peak current swing along the loadline from the quiescent point, $\leq \hat{I}_c$

It follows that

$$2\hat{V}_{ac} = \overset{\diamond}{V}_{ac} \leq 2V_{CC}$$

$$2\hat{I}_{ac} = \overset{\diamond}{I}_{ac} \leq 2\hat{I}_{ac}$$

FIGURE 7.53 Push–pull amplifier: (a) like transistors are configured in push–pull Class B. An input and an output transformer are needed to invert the phase of one transistor with respect to the other. The turns ratio of T1 is chosen so that the standard load impedance R_{SL} is transformed to the load impedance R_L presented to each collector. Their graphical solution is shown in Fig. 7.52. (b) The phase of one transistor with respect to the other is inverted without transformers since one transistor is NPN whereas the other is PNP.

For a sinusoidal signal

$$P_{\text{out}_{\text{max}}} = \left(\frac{\hat{V}_{\text{ac}}}{\sqrt{2}}\right)\left(\frac{\hat{I}_{\text{ac}}}{\sqrt{2}}\right) = \left(\frac{\hat{I}_{\text{ac}}}{2}\right)\hat{V}_{\text{ac}} \tag{7.33}$$

For simplicity, assume that the output transformer primary to secondary turns ratio is 2:1 so that either active-half has a 1:1 turns ratio with the secondary. For a load at the secondary winding R_L, the primary *end-to-end* load is $4R_L$, but each transistor works into R_L for each half-cycle. The load presented to each transistor is, therefore, R_L. It follows that

$$\overset{\circ}{I}_{\text{ac}} = 2\hat{I}_{\text{ac}} = \frac{\overset{\circ}{V}_{\text{ac}}}{R_L} = \frac{2\hat{V}_{\text{ac}}}{R_L} \Rightarrow \left(\frac{\hat{I}_{\text{ac}}}{2}\right) = \frac{\overset{\circ}{V}_{\text{ac}}}{4R_L} = \frac{2\hat{V}_{\text{ac}}}{4R_L} = \frac{\hat{V}_{\text{ac}}}{2R_L}$$

Substituting for $\hat{I}_{\text{ac}}/2$ into Eq. (7.33) gives

$$P_{\text{out}_{\text{max}}} = \frac{\hat{V}_{\text{ac}}^2}{2R_L} \leq \frac{V_{CC}^2}{2R_L} \tag{7.34}$$

The center tap current of the transformer is half-wave DC at twice the frequency of the drive signal so that

$$I_{\text{ct}}(\theta) = \hat{I}_{\text{ac}} |\sin \theta|$$

The DC component of I_{ct} is found from

$$I_{\text{DC}} = \frac{1}{2\pi} \int_0^{2\pi} \hat{I}_{\text{ac}} |\sin \theta| d\theta = \frac{1}{2\pi} \int_0^{\pi} \hat{I}_{\text{ac}} \sin \theta d\theta - \frac{1}{2\pi} \int_{\pi}^{2\pi} \hat{I}_{\text{ac}} \sin \theta d\theta = \frac{2\hat{I}_{\text{ac}}}{\pi}$$

and

$$I_{\text{DC}} = \frac{2\hat{V}_{\text{ac}}}{\pi R_L} \leq \frac{2V_{CC}}{\pi R_L}$$

The DC input power is found from

$$P_{\text{in}} = I_{\text{DC}} V_{CC} = \frac{2\hat{V}_{\text{ac}} V_{CC}}{\pi R_L} \leq \frac{2V_{CC}^2}{\pi R_L} \tag{7.35}$$

The power efficiency becomes (from Eqs. (7.34) and (7.35))

$$\eta = \frac{P_{\text{out}}}{P_{\text{in}}} = \frac{\hat{V}_{\text{ac}}^2}{2R_L} \frac{\pi R_L}{2\hat{V}_{\text{ac}} V_{CC}} = \frac{\pi \hat{V}_{\text{ac}}}{4V_{CC}} \tag{7.36}$$

whereas the maximum theoretical power efficiency (when $\hat{V}_{\text{ac}} = V_{CC}$) becomes

$$\eta_{\text{max}} = \frac{\pi}{4} = 78.5\% \tag{7.37}$$

The upper limit of Class B power efficiency is considerably higher than that of Class A at 50%. The power dissipated in each of the two transistors P_{Dis}, is half of the total. The total power dissipated in the case of maximum efficiency is

$$2P_{\text{Dis}} = P_{\text{in}} - P_{\text{out}}$$

$$= \frac{2V_{CC}^2}{\pi R_L} - \frac{V_{CC}^2}{2R_L} \tag{7.38}$$

$$2P_{\text{Dis}} = \frac{V_{CC}^2}{R_L}\left(\frac{2}{\pi} - \frac{1}{2}\right)$$

and

$$P_{\text{Dis}} = \frac{V_{CC}^2}{R_L}\left(\frac{1}{\pi} - \frac{1}{4}\right) \tag{7.39}$$

In Fig. 7.52 the loadline passes through the zero collector current point of each transistor where one device is turning off and the other is turning on. At this precise point both amplifier devices could be *off* or *on*. To avoid this *glitch* or *discontinuity*, strict Class B is never really used, but instead, each transistor is biased on to a small value of current to avoid this *crossover distortion*. The choice of bias current has an affect on IM3 so that it may be adjusted for best IM performance under a multitone test condition. The quiescent bias current is usually set for about 1–10% of the highest DC collector current under drive. Biased under this condition, the active devices are not operating in strict Class B, but rather, in a direction toward Class A, and is usually referred to as Class AB.

Since the quiescent point of a Class AB or B amplifier is near or at zero, looking into the input port or output port expecting to measure the standard impedance to which the amplifier was designed is not possible. Large signal behavior comes into play in power amplifiers other than Class A and only exists under drive. Without drive, the amplifier is cut off in the case of Class B or near cut off in the case of Class AB. With drive, the input return loss may be measured as in the case of Class A, but a directional coupler is used to sense the incident and reflected power. Forward output power is also measured with a directional coupler and compared with theoretical efficiency to determine the worth of the output matching circuit. Efficiency should be in the 40–60% range for good linearity, staying away from the ends of the loadline. Frequently, impedance information about the input and output of a transistor used in Class AB or B is for large signal (power) applications and arrived at experimentally by beginning with a best guess set of parameters about the transistor, then designing a matching circuit and optimizing it for best performance under power. When the goal is reached, the transistor is removed, and the input and output impedances of the circuit are measured at the points where the transistor was connected. These points may then be plotted on a Smith chart. If the input impedance to the transistor is called Z_{in}, the complex conjugate of what the circuit measured is Z_{in} of the transistor. If the output impedance of the transistor is called Z_{out}, then this is the complex conjugate of that measured for the output circuit. If, however, the output impedance is called Z_{load}, then it is not the complex conjugate of what was measured, but the load impedance that the transistor is working into. There is some ambiguity about the output impedance among the stated parameters of many manufacturers of devices. Some call it Z_{out} when they really mean Z_{load}. A general guide is that the load that a power transistor wants to work into for best overall performance is capacitive, so if Z_{load} is specified to be $R + jX$, this is probably Z_{out} and not Z_{load}. Z_{load} would be the complex conjugate of this or $R - jX$.

Third-order intermodulation performance for Class AB or B is generally worse than Class A, and harmonic distortion is better in the sense that even harmonics are canceled to a large degree in the primary of the output transformer. These currents flow in opposite directions at the same time and net to zero, whereas odd order harmonic currents do not. An amplifier model for a Class AB or B amplifier is shown in Fig. 7.54. Its topology is the same as that for the Class A in Fig. 7.51 except that $R_{\text{StandardLoad(SL)}}$ *inside* the amplifier is replaced by R_{Internal} or R_{Int}. For a maximum power efficiency, $\eta_{\text{max}} = 78.5\%$,

$$\eta_{\text{max}} = \frac{P_{\text{out}_{\text{max}}}}{P_{\text{in}}} = \frac{R_{\text{Int max}}}{R_{\text{Int max}} + R_{\text{SL}}} \tag{7.40}$$
$$= 0.785$$

and

$$R_{\text{Int max}} = 3.6512 R_{\text{SL}}$$

so that for efficiency figures less than the theoretical maximum

$$R_{\text{Int}} \leq 3.6512 R_{\text{SL}} \tag{7.41}$$

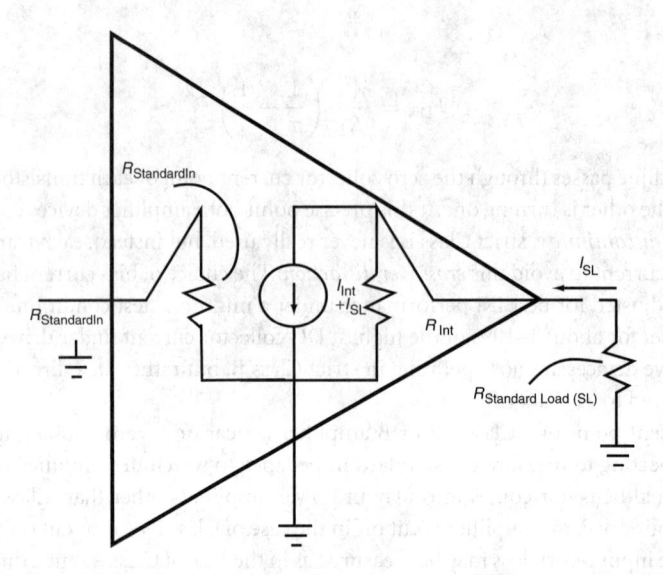

FIGURE 7.54   The Class AB or B amplifier may be modeled as having a measurable input resistance when driven and an internal output resistance that maintains the power efficiency at or below the theoretical maximum of 78.5%. For this, $R_{\text{Int}} \leq 3.6512 R_{\text{SL}}$.

Deleting max from Eq. (7.40)

$$\eta = \frac{R_{\text{Int}}}{R_{\text{Int}} + R_{\text{SL}}} \tag{7.42}$$

R_{Int} may be measured indirectly by knowing the efficiency of the amplifier and the standard load resistance

$$R_{\text{Int}} = R_{\text{SL}} \left(\frac{\eta}{1 - \eta} \right) \tag{7.43}$$

An enhancement to the efficiency of the Class B amplifier makes use of the addition of a third harmonic component of the right amount to the collector-emitter voltage waveform to cause near square-wave flattening when it is near zero (where the collector current is greatest). This modification alters the amplifier enough so that it enjoys a different classification called Class F. Flattening enhances efficiency by as much as one-eighth so that

$$\eta_{\text{ClassF}} = \frac{9}{8} \eta_{\text{ClassB}}$$

From Eq. (7.37)

$$\eta_{\text{ClassFmax}} = \frac{9}{8} \left(\frac{\pi}{4} \right) = 0.884$$

7.6.5   Nonlinear Amplifier Classes of Operation

If the Class A amplifier is biased to cause device current for the entire drive cycle, and Class B is biased to cause device current for precisely half of the cycle, then it is logical to assign the Class C designation to that bias condition that causes less than half-cycle device current. There are two types of Class C amplifiers: nonsaturated and saturated.

Nonsaturated Class C Amplifier

It is not convenient to analyze the Class C amplifier by means of the loadline argument as in the cases of Classes A or B, but rather, a time- or phase-domain analysis proves to be more workable. The reason is that in the other classes conduction time is fixed at full time or half. In Class C, the conduction time is variable so that a general analysis must be performed where time or phase is the independent variable, not collector-emitter voltage. A time or phase analysis for Classes A or B could have been done, but then only one method would be presented here and the intention is to present another way. The results are the same and comparisons presented at the end of this section.

A phase domain diagram of the nonsaturated Class C amplifier is shown in Fig. 7.55 along with a representative schematic diagram. The term nonsaturated means that the active device is not driven to the point where the collector-emitter voltage is at the lowest possible value. The curve of $V_{ce}(\theta)$ in Fig. 7.55(a) does not quite touch the zero line. Another way of saying this is that the transistor is always active or never conducts like a switch. It is active in the same way as Classes A and B, and because of this similarity, it may be modeled in the same way, that is, as a dependent current source.

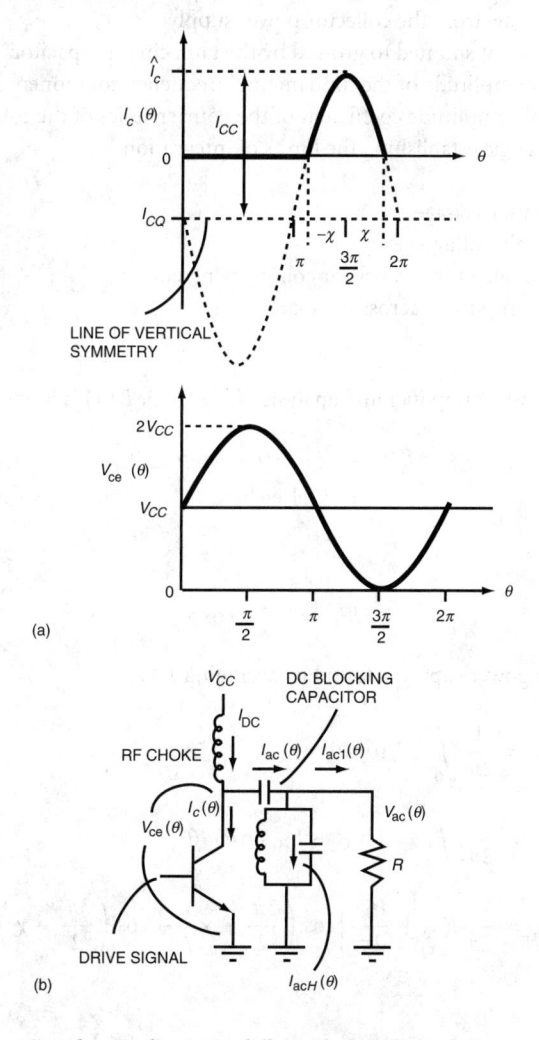

FIGURE 7.55 The phase- or time-domain diagrams of all significant voltages and currents of the nonsaturated Class C amplifier are shown: (a) a representative NPN transistor, (b) schematic diagram.

In an amplifier where the drive signal causes the onset of conduction so that the conduction angle is less than π, it is important to realize the following.

1. The transistor will not conduct until its base-emitter junction self-reverse bias voltage of about 0.7 V is overcome. With no external base bias applied, the conduction angle will be less than π and the quiescent collector current fixed at a negative default value I_{CQ}. If this default value is acceptable in the outcome, then no external bias needs to be applied and the design becomes simpler. Applying a DC base current will allow control over I_{CQ}.
2. The collector current $I_c(\theta)$, is sinusoidal in shape when it exists.
3. The collector-emitter voltage $V_{ce}(\theta)$, is sinusoidal, but π radians out of phase with the collector current.

Some definitions are in order.

$I_c(\theta)$ = instantaneous collector current
$\hat{I}_c$ = peak value of collector current
I_{CQ} = negative self bias collector current
$I_{CC}(\theta)$ = positive net value of $I_c(\theta)$ and I_{CQ}
$I_c(\theta)$ = $I_{CQ} - I_{CC} \sin\theta$ when the net collector current $I_c(\theta) \geq 0$; it is zero elsewhere
I_{DC} = DC current draw from the collector power supply
$I_{acH}(\theta)$ = harmonic current shunted to ground by the tank circuit capacitor
$\hat{I}_{ac1}$ = peak current amplitude of the fundamental frequency component in the load; it is the first harmonic amplitude coefficient of the Fourier series of the collector current with the conduction angle establishing the limits of integration
$I_{ac1}(\theta)$ = $\hat{I}_{ac1} \sin\theta$
$V_{ce}(\theta)$ = collector-emitter voltage
V_{CC} = collector supply voltage
2χ = conduction angle of the transistor collector current
$\hat{V}_{ac1}$ = peak voltage amplitude across the load
$V_{ac1}(\theta)$ = $\hat{V}_{ac1} \sin\theta$

By the definition of the Class C amplifier and analysis of Fig. 7.55, $I_c(\theta)$ is found to be

$$I_c(\theta) = I_{CQ} - I_{CC} \sin\theta \geq 0$$
$$= 0, \text{ elsewhere} \tag{7.44}$$

and

$$I_{CQ} = -I_{CC} \cos\chi \tag{7.45}$$

The DC current from the power supply is found by averaging $I_c(\theta)$

$$I_{DC} = \frac{1}{2\pi} \int_0^{2\pi} I_c(\theta)d\theta$$

$$I_{DC} = \frac{1}{2\pi} \int_{\frac{3\pi}{2}-\chi}^{\frac{3\pi}{2}+\chi} (I_{CQ} - I_{CC}\sin\theta)d\theta$$

$$I_{DC} = \frac{2\chi}{2\pi} I_{CQ} + \frac{I_{CC}}{2\pi}\left[\cos\left(\frac{3\pi}{2}+\chi\right) - \cos\left(\frac{3\pi}{2}-\chi\right)\right]$$

However

$$\cos\left(\frac{3\pi}{2}+\chi\right) = \sin\chi \qquad \text{and} \qquad -\cos\left(\frac{3\pi}{2}-\chi\right) = \sin\chi$$

so that

$$I_{DC} = \frac{\chi}{\pi}(-I_{CC}\cos\chi) + \frac{I_{CC}}{\pi}\sin\chi$$

and

$$I_{DC} = \frac{I_{CC}}{\pi}(\sin\chi - \chi\cos\chi) \tag{7.46}$$

Also,

$$I_{CC} = \frac{\pi I_{DC}}{(\sin\chi - \chi\cos\chi)} \tag{7.47}$$

DC power supplied to the collector is found to be

$$P_{in} = I_{DC}V_{CC} = \frac{I_{CC}V_{CC}}{\pi}(\sin\chi - \chi\cos\chi) \tag{7.48}$$

The peak current value of the fundamental frequency component in the load, $\hat{I}_{ac1}$ is the coefficient of the fundamental frequency component of the Fourier series of the collector current $I_c(\theta)$. All harmonic current is shunted to ground by the tank circuit capacitor and assumed to be zero. The tank circuit is resonant at the fundamental frequency.

$\hat{I}_{ac1}$ is found from

$$\hat{I}_{ac1} = -\frac{1}{\pi}\int_0^{2\pi} I_c(\theta)\sin\theta\,d\theta$$

$$\hat{I}_{ac1} = -\frac{1}{\pi}\int_{\frac{3\pi}{2}-\chi}^{\frac{3\pi}{2}+\chi} (I_{CQ} - I_{CC}\sin\theta)\sin\theta\,d\theta$$

$$\hat{I}_{ac1} = -\frac{I_{CQ}}{\pi}\int_{\frac{3\pi}{2}-\chi}^{\frac{3\pi}{2}+\chi} \sin\theta\,d\theta + \frac{I_{CC}}{\pi}\int_{\frac{3\pi}{2}-\chi}^{\frac{3\pi}{2}+\chi} \sin^2\theta\,d\theta$$

$$\hat{I}_{ac1} = \frac{I_{CQ}}{\pi}\cos\theta\Big|_{\frac{3\pi}{2}-\chi}^{\frac{3\pi}{2}+\chi} + \frac{I_{CC}}{\pi}\left(\frac{\theta}{2} + \frac{\sin 2\theta}{4}\right)\Big|_{\frac{3\pi}{2}-\chi}^{\frac{3\pi}{2}+\chi}$$

$$\hat{I}_{ac1} = \frac{I_{CQ}}{\pi}\left[\cos\left(\frac{3\pi}{2}+\chi\right) - \cos\left(\frac{3\pi}{2}-\chi\right)\right] + \frac{I_{CC}}{\pi}\left[\chi - \frac{-\sin 2\chi - \sin 2\chi}{4}\right]$$

$$\hat{I}_{ac1} = \frac{2I_{CQ}}{\pi}\sin\chi + \frac{I_{CC}}{\pi}\chi + \frac{I_{CC}}{2\pi}\sin 2\chi$$

Since $I_{CQ} = -I_{CC}\cos\chi$

$$\hat{I}_{ac1} = -\frac{2I_{CC}}{\pi}\sin\chi\cos\chi + \frac{I_{CC}}{\pi}\chi + \frac{I_{CC}}{2\pi}\sin 2\chi$$

$$\hat{I}_{ac1} = \frac{I_{CC}}{2\pi}(-4\sin\chi\cos\chi + 2\chi + \sin 2\chi)$$

rearranging and making use of the identity, $\sin 2x = 2\sin x \cos x$

$$\hat{I}_{ac1} = \frac{I_{CC}}{\pi}(\chi - \sin\chi\cos\chi) \tag{7.49}$$

The peak voltage across the load becomes

$$\hat{V}_{ac1} = \frac{RI_{CC}}{\pi}(\chi - \sin\chi\cos\chi) \tag{7.50}$$

The output power in the load is

$$P_{out} = P_{ac1} = \frac{\hat{V}_{ac1}^2}{2R} = \frac{RI_{CC}^2}{2\pi^2}(\chi - \sin \chi \cos \chi)^2 \tag{7.51}$$

Substituting Eq. (7.47) for I_{CC}

$$P_{out} = \frac{RI_{DC}^2}{2}\left(\frac{\chi - \sin \chi \cos \chi}{\sin \chi - \chi \cos \chi}\right)^2 \tag{7.52}$$

Dividing Eq. (7.51) by Eq. (7.52) the power efficiency η is

$$\eta = \frac{P_{out}}{P_{in}} = \frac{RI_{CC}^2}{2\pi^2}\frac{\pi}{I_{CC}V_{CC}}\frac{(\chi - \sin \chi \cos \chi)^2}{(\sin \chi - \chi \cos \chi)}$$

$$\eta = \frac{R}{2\pi}\frac{I_{CC}}{V_{CC}}\frac{(\chi - \sin \chi \cos \chi)^2}{(\sin \chi - \chi \cos \chi)} \tag{7.53}$$

Here η_{max} occurs when $\hat{V}_{ac1} = V_{CC}$. Setting Eq. (7.50) equal to V_{CC}, solving for I_{CC}, and substituting into Eq. (7.53), $\eta \to \eta_{max}$

$$\eta_{max} = \frac{\chi - \sin \chi \cos \chi}{2(\sin \chi - \chi \cos \chi)}$$

$$V_{ce}(\theta) = V_{CC} + \hat{V}_{ac1}\sin \theta \tag{7.54}$$

and in the case of maximum efficiency where $\hat{V}_{ac1} = V_{CC}$

$$V_{ce}(\theta) = V_{CC}(1 + \sin \theta)|_{\eta=\eta_{max}} \tag{7.55}$$

The collector-emitter voltage $V_{ce}(\theta)$ swings from $2V_{CC}$ to zero and differs from the voltage across the load R by the DC component V_{CC}. Stated before, $I_{acH}(\theta)$, the harmonic currents, do not enter the equations because these currents do not appear in the load.

The first thing to note about Eq. (7.54) is that the theoretical maximum efficiency is not limited to a value less than one. The second thing is that it is only a function of conduction angle, 2χ. Here, $\eta_{max} \to 1$ as $\chi \to 0$. For $\chi \to 0$, Eq. (7.54) becomes

$$\lim_{\chi \to 0} \eta_{max} = \frac{\chi - \sin \chi \cos \chi}{2(\sin \chi - \chi \cos \chi)} \tag{7.56}$$

For small χ

$$\sin \chi \to \chi, \qquad \sin 2\chi \to 2\chi, \qquad -4 \sin \chi \to -4\chi \qquad \text{and} \qquad \cos \chi \to (<1)$$

It follows that

$$\lim_{\chi \to 0} \eta_{max} = \frac{2\chi - 2\chi(<1)}{4\chi - 4\chi(<1)} = 1 \tag{7.57}$$

To achieve 100% power efficiency, the collector current pulse $I_c(\theta)$ would become infinitesimal in width, carry no energy, and make no power. Even though the amplifier would produce no power, it would consume no power, and so it would be 100% efficient. In real applications, power efficiency can be in the 80s of percent.

Power dissipated as heat in the transistor is

$$P_{Dis} = (1 - \eta)P_{in} \tag{7.58}$$

Furthermore, the resistance represented by the transistor collector-emitter path, that is, that which is represented by the internal shunt resistance of the current-dependent current source, is found by dividing Eq. (7.58) by I_{DC}^2. This resistance may be called R_{Dis} and shares the input power with the load according

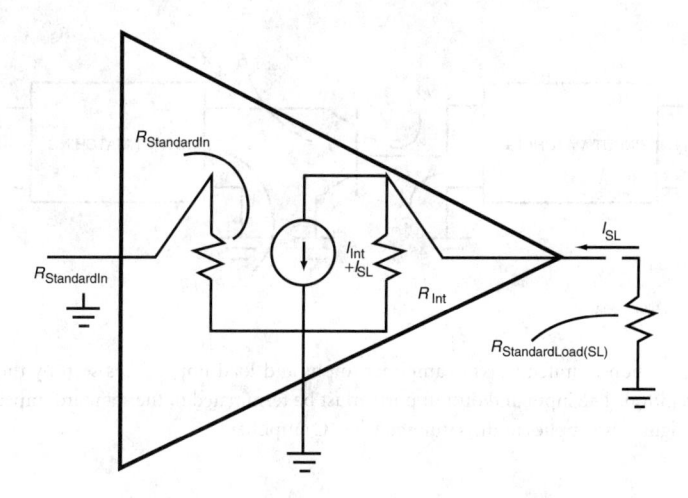

FIGURE 7.56 The nonsaturated Class C amplifier may be modeled as a current controlled current source with a measurable input resistance when driven, and an internal output resistance determined by the power efficiency, $\eta \cdot R_{Int}$ shares power with R_{SL}, but $R_{Int} \to \infty$ when $\eta \to 1$.

to Eq. (7.58)

$$R_{Dis} = \frac{P_{Dis}}{I_{DC}^2} = \frac{V_{CC}}{I_{DC}} - \frac{R}{2}\left(\frac{\chi - \sin\chi\cos\chi}{\sin\chi - \chi\cos\chi}\right)^2 \qquad (7.59)$$

For the amplifier (not the transistor), the equivalent internal resistance may be found referenced to the standard amplifier load resistance by simply scaling $R \to R_{StandardLoad(SL)}$ and

$$R_{int} = R_{Dis}\frac{R_{SL}}{R} \qquad (7.60)$$

The nonsaturated Class C amplifier, like the linear classes, may be modeled as a current controlled current source. Figure 7.56 is a representation of the amplifier and is very similar to Fig. 7.54.

Measuring port impedances without drive and making output power is senseless because the transistor is biased off without drive. Input resistance, $R_{StandardIn}$ may only be measured when the transistor is driven to full power. For nonsaturated Class C, $R_{int} \to \infty$ as $\eta \to 1$.

The transistor impedances within the amplifier may be measured by removing the transistor after matching circuit adjustments are made for best specified performance and connecting the impedance measuring instrument, such as a network analyzer, to the input and output connections left open by the transistor. Figure 7.57 illustrates the same concept as that presented in Fig. 7.53(a) for the Class B amplifier except that the latter is for a push-pull configuration and the former is for the single ended.

A measuring instrument will give the complex conjugate of the transistor input impedance Z_{in}. Measuring the impedance seen at the input of the output matching network will give Z_{load}, which will be of the form $R - jX$. R is the same R used in the equations derived for the transistor. There may be a small jX component for best operation of a particular transistor type.

An $I_c(\theta)$ vs. $V_{ce}(\theta)$ loadline representation of the nonsaturated Class C amplifier is shown in Fig. 7.58 and is based on the derived results of the given equations.

The greatest P_{out} is obtained when the conduction angle 2χ is near π in width, that is, where the breakpoint in the loadline is just below V_{CC}. The greatest power efficiency occurs where the breakpoint is far to the left near $V_{ce}(\theta) = 0$ and the conduction angle is near zero. The load R presented to a Class C amplifier is usually chosen to be

$$R = \frac{V_{CC}^2}{2P_{out}} \qquad (7.61)$$

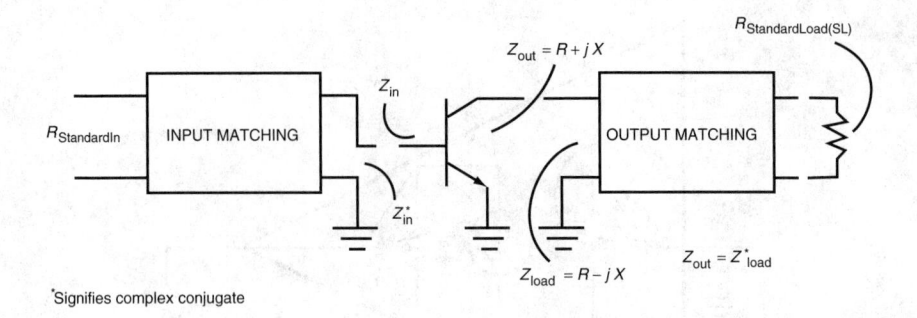

FIGURE 7.57 For the nonsaturated Class C amplifier, input and load impedances seen by the transistor may be measured after its removal. The input and output ports must be terminated in the standard impedance for this static measurement. This figure also applies to the saturated Class C amplifier.

for a practical design giving a high conduction angle and power making capability using the simple default base bias current. This is where the drive signal must overcome the base-emitter reverse bias voltage of about 0.7 V. A base bias source is not needed and the base is simply grounded for DC.

Saturated Class C Amplifier

The saturated Class C amplifier allows the collector **voltage** (in the case of the bipolar junction transistor) to be driven into **saturation** for a portion of the active time. Figures 7.44 and 7.47 show this as the compression region, but in Class C with no regard to amplitude linearity, power efficiency is increased when the transistor is driven into hard saturation so that further increases in drive yield little, if any, increase in output power. The collector-emitter voltage is as low as it can be. The variable in this approach is the length of time or phase that the tran

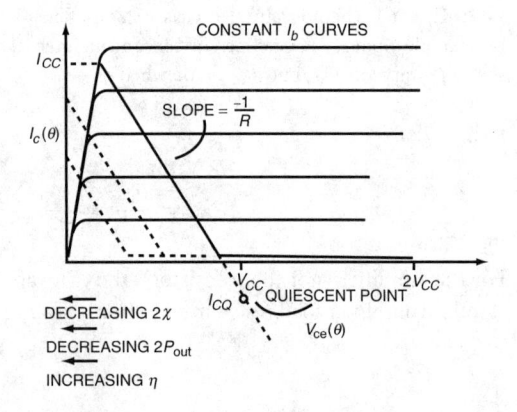

FIGURE 7.58 The nonsaturated Class C amplifier is shown graphically with trends as a function of 2χ, P_{out}, and η. The intersection of the loadline with the $I_c(\theta) = 0$ line at a point lower than V_{CC} accounts for higher efficiency than classes A or B. Quiescent point I_{CQ} is negative.

sistor is saturated compared to the conduction angle. With the nonsaturated Class C transistor, there is a sequence of three changes of collector current behavior, that is, cutoff, linear, and cutoff. With the saturated transistor, there are five: cutoff, linear, saturated, linear, and cutoff. The mathematical expressions for each change are necessary to completely describe the transistor behavior, and this involves the introduction of another variable, **saturation angle** $2\chi_s$ to those described for the nonsaturated case.

When using a bipolar junction transistor in the saturated mode, the saturation voltage must be taken into account. It cannot be ignored as in the previously discussed classifications because, in those, V_{Sat} was never reached. Also when saturated, collector current is not infinite. It is limited by an equivalent **on resistance** so that $I_{cSat} = V_{Sat}/R_{on}$.

In addition to the definitions described for the nonsaturated case, the following are also appropriate for the saturated case:

$2\chi_s$ = saturation angle of the transistor collector current
V_{Sat} = saturation voltage of the transistor collector-emitter path
I_{cSat} = saturated collector current when $V_{ce} = V_{Sat}$
R_{on} = equivalent collector-emitter resistance during saturation

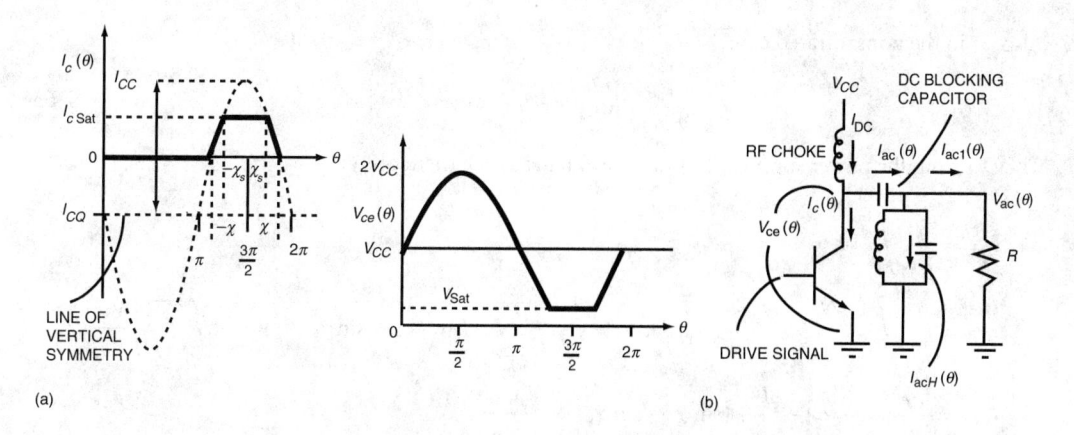

FIGURE 7.59 Phase- or time-domain diagrams of all significant voltages and currents of the saturated Class C amplifier: (a) representative NPN transistor (b) circuit diagram.

Figure 7.59(a) illustrates the collector current vs. collector-emitter voltage behavior of the saturated transistor. Of course, the saturation angle is always contained within the conduction angle. The circuit diagram in Fig. 7.59(b) is the same as Fig. (7.55 b).

Certain assumptions are made regarding the ideal behavior of a transistor in the saturated Class C mode:

1. The transistor has no dynamic collector-emitter or collector-base capacitances associated with it, which in reality it has.
2. The transistor collector current and collector-emitter voltage waveforms are sinusoidal, which they are not.
3. The saturated collector current is constant, which it is not.
4. Saturation on-resistance is constant, which it is not.
5. There is no parasitic inductance or capacitance associated with the circuit, which there is.
6. There are no harmonic currents flowing in the load, which there are.

The precise saturated Class C amplifier is nearly impossible to analyze for a number of reasons:

1. The transistor has dynamically changing capacitances from collector to base and collector to emitter as functions of dynamic voltages.
2. The current and voltage waveforms are anything but sinusoidal.
3. The saturated collector current has peaks and dips.
4. The saturated on-resistance has peaks and dips.
5. There may be significant parasitic inductances and capacitances at VHF and UHF.
6. There are some harmonic currents in the load.

Because of these things, the assumptions are used in the analysis of the saturated Class C amplifier and in the associated figures to make it manageable.

By definition of saturated Class C transistor operation and analysis of Fig. 7.59, $I_c(\theta)$ is found to be

$$
I_c(\theta) = \begin{cases} 0, & \text{when } 0 < \theta < \dfrac{3\pi}{2} - \chi, \text{ and } \dfrac{3\pi}{2} + \chi < \theta < 2\pi \\[2mm] I_{CQ} - I_{CC}\sin\theta, & \text{when } \dfrac{3\pi}{2} - \chi \le \theta \le \dfrac{3\pi}{2} - \chi_s, \text{ and } \dfrac{3\pi}{2} + \chi_s \le \theta \le \dfrac{3\pi}{2} + \chi, \quad (7.62) \\[2mm] \dfrac{V_{\text{Sat}}}{R_{\text{on}}}, & \text{when } \dfrac{3\pi}{2} - \chi_s < \theta < \dfrac{3\pi}{2} + \chi_s \end{cases}
$$

The onset of saturation occurs when the collector-emitter voltage equals the saturation voltage. R_{on} is finite allowing V_{Sat} to be greater than zero so that the saturated collector current is finite during saturation.

Also, as in the nonsaturated case

$$I_{CQ} = -I_{CC}\cos\chi \tag{7.63}$$

The DC from the power supply is found by averaging $I_c(\theta)$, therefore

$$I_{DC} = \frac{1}{2\pi}\int_0^{2\pi} I_c(\theta)d\theta$$

$$I_{DC} = \frac{1}{2\pi}\int_{\frac{3\pi}{2}-\chi}^{\frac{3\pi}{2}-\chi_s}(I_{CQ}-I_{CC}\sin\theta)d\theta + \frac{1}{2\pi}\int_{\frac{3\pi}{2}+\chi_s}^{\frac{3\pi}{2}+\chi}(I_{CQ}-I_{CC}\sin\theta)d\theta + \frac{1}{2\pi}\int_{\frac{3\pi}{2}-\chi_s}^{\frac{3\pi}{2}+\chi_s}\frac{V_{Sat}}{R_{on}}d\theta$$

$$I_{DC} = \frac{I_{CQ}}{\pi}(\chi-\chi_s) + \frac{I_{CC}}{\pi}(\sin\chi-\sin\chi_s) + \frac{V_{Sat}\chi_s}{\pi R_{on}}$$

Substituting Eq. (7.63) for I_{CQ}

$$I_{DC} = \frac{I_{CC}}{\pi}[(\sin\chi-\sin\chi_s) - (\chi-\chi_s)\cos\chi] + \frac{V_{Sat}\chi_s}{\pi R_{on}} \tag{7.64}$$

Compare Eq. (7.64) with Eq. (7.46). Solving (7.64) for I_{CC}

$$I_{CC} = \frac{\pi I_{DC} - \dfrac{V_{Sat}\chi_s}{\pi R_{on}}}{(\sin\chi-\sin\chi_s) - (\chi-\chi_s)\cos\chi} \tag{7.65}$$

Compare Eq. (7.65) with Eq. (7.47). DC power supplied to the collector is found to be

$$P_{in} = I_{DC}V_{CC} = \left\{\frac{I_{CC}}{\pi}[(\sin\chi-\sin\chi_s) - (\chi-\chi_s)\cos\chi] + \frac{V_{Sat}\chi_s}{\pi R_{on}}\right\}V_{CC} \tag{7.66}$$

Compare Eq. (7.65) with Eq. (7.66).

As in the nonsaturated case, the peak current value of the fundamental frequency component in the load, $\hat{V}_{ac1}$, is the coefficient of the fundamental frequency component of the Fourier series of the collector current $I_c(\theta)$. All harmonic current is shunted to ground by the tank circuit capacitor and *assumed* to be zero. The tank circuit is resonant at the fundamental frequency. $\hat{I}_{ac1}$ is found from

$$\hat{I}_{ac1} = -\frac{1}{\pi}\int_0^{2\pi} I_c(\theta)\sin\theta d\theta$$

$$\hat{I}_{ac1} = -\frac{1}{\pi}\int_{\frac{3\pi}{2}-\chi}^{\frac{3\pi}{2}-\chi_s}(I_{CQ}-I_{CC}\sin\theta)\sin\theta d\theta$$

$$-\frac{1}{\pi}\int_{\frac{3\pi}{2}-\chi_s}^{\frac{3\pi}{2}+\chi_s}\frac{V_{Sat}}{R_{on}}\sin\theta d\theta - \frac{1}{\pi}\int_{\frac{3\pi}{2}+\chi_s}^{\frac{3\pi}{2}+\chi}(I_{CQ}-I_{CC}\sin\theta)\sin\theta d\theta$$

After much painful integration

$$\hat{I}_{ac1} = \frac{2I_{CQ}}{\pi}(\sin\chi-\sin\chi_s) + \frac{I_{CC}}{\pi}(\chi-\chi_s) - \frac{I_{CC}}{2\pi}(\sin 2\chi_s - \sin 2\chi) + \frac{2V_{Sat}}{\pi R_{on}}\sin\chi_s$$

Rearranging

$$\hat{I}_{ac1} = \frac{2I_{CQ}}{\pi}(\sin\chi-\sin\chi_s) + \frac{I_{CC}}{4\pi}(4\chi - 4\chi_s + 2\sin 2\chi - 2\sin 2\chi_s) + \frac{2V_{Sat}}{\pi R_{on}}\sin\chi_s$$

And again applying Eq. (7.63)

$$\hat{I}_{ac1} = \frac{I_{CC}}{2\pi}[2(\chi - \chi_s) - 4\cos\chi(\sin\chi - \sin\chi_s) + (\sin 2\chi - \sin 2\chi_s)] + \frac{2V_{Sat}}{\pi R_{on}}\sin\chi_s \qquad (7.67)$$

Compare Eq. (7.67) with Eq. (7.49).

Hypothetically, if $\chi = \chi_s$, then the transistor is immediately saturated and there is no linear region. This can only happen if V_{ce} never is higher than V_{Sat}. Having no linear region, the peak current in the load becomes

$$\hat{I}_{ac1} = \frac{2V_{Sat}}{\pi R_{on}}\sin\chi_s$$

The entire term within the brackets in Eq. (7.67) is zero, meaning that I_{CC} could be very large, approaching ∞. To find the voltage across the load R, under this condition, $\hat{I}_{ac1}$ is multiplied by R, so that

$$\hat{V}_{ac1} = \frac{2RV_{Sat}}{\pi R_{on}}\sin\chi_s$$

Now, hypothetically suppose that very hard limiting is maintained to ensure that $\chi = \chi_s$, but that the supply voltage is allowed to increase from V_{Sat} to some higher V_{CC}, so that V_{Sat} is replaced by $(V_{CC} - V_{Sat})$. It follows that

$$\hat{V}_{ac1} = \frac{2R(V_{CC} - V_{Sat})}{\pi R_{on}}\sin\chi_s$$

for very hard limiting. The voltage across the load is proportional to supply voltage and can be much greater than it as a function of R/R_{on} and χ_s, and this is the case as will be shown.

Finding $\hat{V}_{ac1}$ by multiplying Eq. (7.67) by R,

$$\hat{V}_{ac1} = \frac{RI_{CC}}{2\pi}[2(\chi - \chi_s) - 4\cos\chi(\sin\chi - \sin\chi_s) + (\sin 2\chi - \sin 2\chi_s)] + \frac{2RV_{Sat}}{\pi R_{on}}\sin\chi_s \qquad (7.68)$$

Compare Eq. (7.68) with Eq. (7.50).

To make things a bit easier, let

$$\Psi = [2(\chi - \chi_s) - 4\cos\chi(\sin\chi - \sin\chi_s) + (\sin 2\chi - \sin 2\chi_s)] \qquad (7.69)$$

and it has no unit.

Let

$$\Upsilon = \frac{2V_{Sat}}{\pi R_{on}}\sin\chi_s \qquad (7.70)$$

and it has the unit of current.

Let

$$\Gamma = \frac{V_{Sat}}{\pi R_{on}}\chi_s \qquad (7.71)$$

and it has the unit of current.

Let

$$\Lambda = [(\sin\chi - \sin\chi_s) - (\chi - \chi_s)\cos\chi] \qquad (7.72)$$

and it has no unit.

Rewriting Eq. (7.68)

$$\hat{V}_{ac1} = \frac{RI_{CC}}{2\pi}\Psi + \Upsilon R \qquad (7.73)$$

Rewriting Eq. (7.65)

$$I_{CC} = \frac{\pi I_{DC} - \Gamma}{\Lambda}$$ (7.74)

Output power is

$$P_{out} = \frac{\hat{V}_{ac1}^2}{2R} = \frac{1}{2R}\left[\frac{RI_{CC}}{2\pi}\Psi + \Upsilon R\right]^2$$ (7.75)

Substituting Eq. (7.74) into Eq. (7.75), we get

$$P_{out} = \frac{R}{2}\left[\frac{1}{2\pi}\left(\frac{\pi I_{DC} - \Gamma}{\Lambda}\right)\Psi + \Upsilon\right]^2$$ (7.76)

Rewriting Eq. (7.66)

$$P_{in} = \left(\frac{I_{CC}}{\pi}\Lambda + \Gamma\right)V_{CC}$$ (7.77)

Efficiency η is found by dividing Eq. (7.75) by Eq. (7.77), so that

$$\eta = \frac{P_{out}}{P_{in}} = \frac{\left[\frac{RI_{CC}}{2\pi}\Psi + \Upsilon R\right]^2}{2RV_{CC}\left[\frac{I_{CC}}{\pi}\Lambda + \Gamma\right]}$$ (7.78)

Compare Eq. (7.78) with Eq. (7.53).

At this point, η appears to be a function of both I_{CC} and V_{CC}. Equation (7.74) may be used to rid Eq. (7.78) of I_{CC}, but this will only introduce I_{DC} into the equation. Instead, let $\hat{V}_{ac1} = \alpha V_{CC}$, where α is a constant of proportionality greater than one. (This has been shown to be the case for hard saturation.) This will allow expressing Eq. (7.78) only in terms of one voltage αV_{CC}, and no current variable. Being simplified in this way provides for a more meaningful analysis. It also forces the case for hard saturation so that Eq. (7.78) becomes an expression for η_{max}. Setting Eq. (7.73) equal to αV_{CC}, solving for I_{CC}; and substituting into Eq. (7.78) gives

$$\eta_{max} = \frac{1}{4}\frac{\alpha^2 \Psi V_{CC}}{(\alpha V_{CC} - R\Upsilon)\Lambda + \frac{1}{2}R\Gamma\Psi}$$ (7.79)

We dare to expand Eq. (7.79) so that

$$\eta_{max} = \frac{1}{4}\frac{\alpha^2 V_{CC}[2(\chi - \chi_s) - 4\cos\chi(\sin\chi - \sin\chi_s) + (\sin 2\chi - \sin 2\chi_s)]}{\alpha V_{CC}[(\sin\chi - \sin\chi_s) - (\chi - \chi_s)\cos\chi] - \left[\frac{R}{R_{ON}}\frac{2V_{Sat}}{\pi}\sin\chi_s\right][X]}$$

$$\underset{\text{Numerator}}{}$$ (7.80)

$$\frac{}{[(\sin\chi - \sin\chi_s) - (\chi - \chi_s)\cos\chi] + \frac{R}{R_{ON}}\left[\frac{V_{Sat}}{2\pi}\chi_s\right][X]}$$

$$\underset{\text{Numerator}}{}$$

$$\frac{}{[2(\chi - \chi_s) - 4\cos\chi(\sin\chi - \sin\chi_s) + (\sin 2\chi - \sin 2\chi_s)]}$$

Although this equation is lengthy, it offers some intuitive feel for those things that affect efficiency. At first, the casual reader would expect that efficiency would not depend on V_{CC} because the efficiency of the nonsaturating Class C case does not depend on anything but 2χ. There is conduction angle dependency because, in saturation, the transistor collector equivalent circuit is a voltage source, not a current source. Efficiency is high as long as V_{CC} is much larger than V_{Sat}. Intuitively, efficiency is high because the collector current is highest when the collector-emitter is in saturation. As V_{CC} is made to approach V_{Sat}, there is less voltage to appear across the load and more current flows through R_{on}. When V_{CC} is made equal to V_{Sat}, all of the current flows through R_{on}, none to the load, and $\eta \to 0$. Other important conclusions may be drawn by verifying that Eq. (7.80) collapses to η_{max} predicted for the following:

1. Nonsaturated Class C case, where $2\chi_s = 0$ and $\alpha \to 1$
2. Linear Class B case, where $2\chi = \pi$
3. Linear Class A case, where $2\chi = 2\pi$

For the nonsaturated Class C case, Eq. (7.80) reduces to

$$\eta_{max} = \frac{1}{4} \frac{\alpha^2 V_{CC}[2\chi - 4\cos\chi\sin\chi + \sin 2\chi]}{\alpha V_{CC}[\sin\chi - \chi\cos\chi]}$$

and realizing the identity, $\sin 2\chi = 2\sin\chi\cos\chi$

$$\eta_{max} = \frac{1}{2} \frac{(\chi - \cos\chi\sin\chi)}{(\sin\chi - \chi\cos\chi)} \tag{7.81}$$

which agrees with Eq. (7.54)

For Class B, Eq. (7.81) reduces to

$$\eta_{max} = \frac{\dfrac{\pi}{2} - \cos\dfrac{\pi}{2}\sin\dfrac{\pi}{2}}{2\left(\sin\dfrac{\pi}{2} - \dfrac{\pi}{2}\cos\dfrac{\pi}{2}\right)} = \frac{\dfrac{\pi}{2} - 0}{2(1 - 0)} = \frac{\pi}{4} \tag{7.82}$$

which agrees with Eq. (7.37).

For Class A, Eq. (7.81) reduces to

$$\eta_{max} = \frac{\pi - 2\sin\pi\cos\pi}{2(\sin\pi - \pi\cos\pi)} = \frac{\pi - 0}{2(1 - 0)} = \frac{1}{2} \tag{7.83}$$

which agrees with Eq. (7.32).

In saturated Class C, the voltage across the load is proportional to the supply voltage V_{CC}. Because of this it may be *high-level modulated* by connecting a baseband modulation signal voltage, $V_m(\theta)$, in series with the supply line. V_{CC} then becomes $V_{CC}(\varphi)$ so that

$$V_{CC}(\varphi) = V_{CC} + V_m(\varphi)$$

Assuming that $V_m(\varphi)$ is sinusoidal and that $\hat{V}_m$ is made equal to V_{CC}, the amplifier will be modulated 100%, that is, $V_{CC}(\varphi)$ will swing from V_{Sat} to $2V_{CC}$. The amplifier is modeled in Fig. 7.60 and transistor,

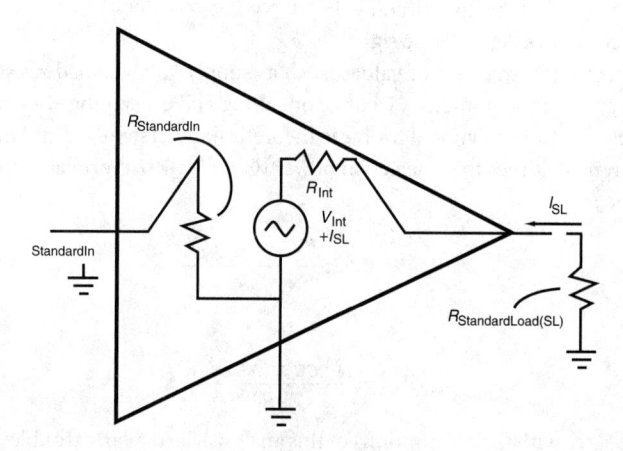

FIGURE 7.60 The saturated Class C amplifier may be modeled as a current controlled voltage source with a measurable input resistance when driven, and an internal output resistance determined by the power efficiency $\eta \cdot R_{Int}$ shares power with R_{SL}, but $R_{Int} \to 0$ when $\eta \to 1$.

circuit, and standard impedances are measured in the same way as the nonsaturated amplifier. The loadline representation is given in Fig. 7.61 and should be compared with Fig. 7.58.

Referring to Fig. 7.60, R_{Int} is contained in

$$P_{\text{out}} = P_{\text{in}} \frac{R_{\text{SL}}}{R_{\text{SL}} + R_{\text{Int}}} \tag{7.84}$$

and along with R_{SL}, determines efficiency. In fact

$$\eta = \frac{R_{\text{SL}}}{R_{\text{SL}} + R_{\text{Int}}} \tag{7.85}$$

and should be compared with the complexity of Eq. (7.80). It is much easier to measure efficiency than it is to calculate it analytically.

Power dissipated in the transistor P_{Dis}, is found from

$$P_{\text{Dis}} = P_{\text{in}} - P_{\text{out}} = P_{\text{in}}\left(1 - \frac{R_{\text{SL}}}{R_{\text{SL}} + R_{\text{Int}}}\right) = P_{\text{in}}(1 - \eta) \tag{7.86}$$

Although Class C amplifiers, saturated or not, cannot perform as linear amplifiers, they are almost always used to amplify phase or frequency modulated signals to higher levels where high-power efficiency is important. Since the saturated Class C amplifier may be V_{CC} modulated, it is often used as the output stage of high-power high-level modulated full carrier AM transmitters. And since the Class C amplifier output tank circuit is high in harmonic current, the amplifier is useful as a frequency multiplier by resonating the tank to a harmonic so that only harmonic energy is passed to the load. An undesirable trait of the Class C is that the transistor collector current *snaps on* as drive appears and cannot be controlled with drive signal if it is saturated. Only V_{CC} will control collector current.

Because of the assumptions made about the saturated Class C amplifier before its mathematical analysis, it must be realized that the final circuit will need *adjusting* to get it to work right. Large signal

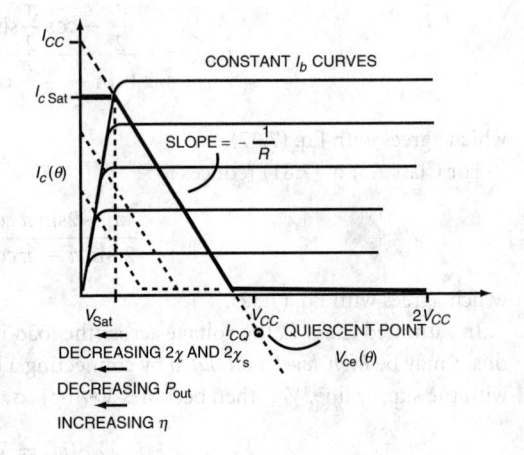

FIGURE 7.61 The saturated Class C amplifier is shown graphically with trends as a function of $2\chi, 2\chi_s, P_{\text{out}}$, and η. The intersection of the loadline with I_{CQ} is negative current, whereas saturation occurs where the loadline meets the transistor curve.

behavior is only defined for a single set of values, such as supply voltage, load resistance, saturated currents, tank circuit Q, and other parameters. Change one thing and everything else changes! The idealized analysis presented here can only be viewed for basic theoretical understanding and initial design and test. Experience has shown many times that the actual power (O'Reilly, 1975) available from a saturated Class C amplifier is closer to

$$P_{\text{out}} \approx 0.625 \frac{(V_{CC} - V_{\text{Sat}})^2}{2R} \tag{7.87}$$

rather than

$$P_{\text{out}} = \frac{(V_{CC} - V_{\text{Sat}})^2}{2R}$$

since reality does not agree with the assumptions of this analysis. Particularly troublesome are the parasitic impedances associated with the emitter ground connection where frequently fractions of an ohm are involved and in bypassing supply lines with real capacitors and decoupling chokes that allow more than zero RF current to flow where in the ideal case, should not.

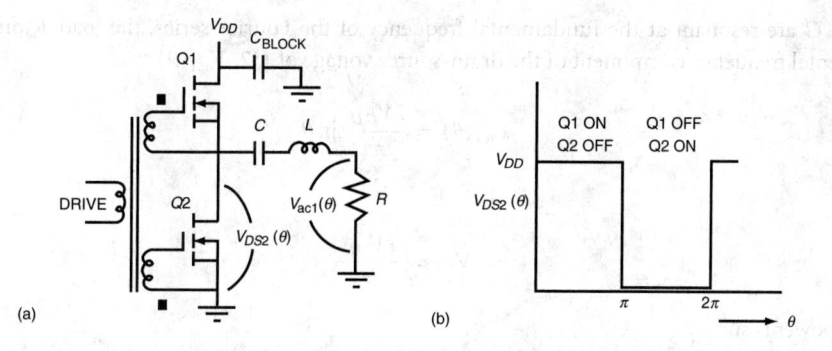

FIGURE 7.62 The half-wave Class D switch-mode pair in (a) produce power at the fundamental component of the switching frequency. Near-100% efficiency may be achieved when the transistors are modeled as ideal switches. The on-resistance of the transistors and drain-source capacitance prevents 100% efficient operation. Drive waveform may be of any shape, but must transition through the on-off threshold region ($+4$ V) very fast to avoid the active region. (b) The drain-source voltage of Q2.

Class D Voltage Switch-Mode Amplifier

Switch-mode amplifiers comprise a special class unlike Classes A, B, or C or their derivatives. Switch mode implies that something is on or off, but not in between. In switch mode there is no linear region as there is in Class A, B, or C. In fact, the active device is modeled as a switch. Insulated gate bipolar junction transistors (IGBJT) are used in switch-mode amplifiers up to a few hundred kilohertz whereas field effect transistors (FETs) are used to about 10 MHz. IGBJTs are advanced in their voltage and current capability, but are slower to switch on and off than FETs. Device development has advanced in these areas over the past few years at such a rapid rate that it is difficult to predict current, voltage, and switching speed limits with any degree of accuracy. For this discussion, the depletion type power FET is used, but it is only representative. Typically, a power FET has several thousand picofarads of gate-source capacitance, a saturation resistance R_{on}, of tenths or hundredths of an ohm, and snaps closed (to R_{on}) whenever gate-source voltage exceeds about $+4$ V for an N-channel type. Likewise, it snaps open whenever gate-source voltage falls below this same threshold. For the following discussion it is assumed that the time to pass through $+4$ V in either direction is zero and independent of drive signal shape. Except for R_{on}, it is close to being an ideal switch. In many ways it is better than a real switch because it is faster, does not bounce, and does not wear out like a real mechanical switch. Moreover, for this discussion, drain-source capacitance when the FET switch is open is also considered to be zero.

Figure 7.62(a) shows the schematic diagram of a half-wave depletion type FET Class D amplifier and the drain-source voltage of Q2 is shown in Fig. 7.62(b). The purpose of the amplifier is to produce a sinusoid output voltage across the load R at the switching frequency. The driving signal may be any shape, but a square voltage drive is preferred so that passage of the gate drive voltage through the threshold voltage is fast. As stated, R_{on} is assumed to be zero, as is any drain-source capacitance.

Because the drain-source voltage of Q2 is square, it contains only odd harmonics according to the nonzero Fourier series coefficients.

$$V_{DS2}(\theta) = V_{DD}\left[\frac{1}{2\pi}\int_0^\pi d\theta + \left(\frac{1}{\pi}\int_0^\pi \sin\theta\, d\theta\right)\sin\theta + \left(\frac{1}{\pi}\int_0^\pi \sin 2\theta\, d\theta\right)\sin 2\theta \right.$$
$$\left. +\cdots\left(\frac{1}{\pi}\int_0^\pi \sin 3\theta\, d\theta\right)\sin 3\theta + \cdots\right] \tag{7.88}$$

$$V_{DS2}(\theta) = V_{DD}\left[\frac{1}{2} - \frac{1}{\pi}\cos\theta\Big|_0^\pi |\sin\theta - \frac{1}{2\pi}\cos 2\theta\Big|_0^\pi |\sin 2\theta - \frac{1}{3\pi}\cos 3\theta\Big|_0^\pi |\sin 3\theta + \cdots\right]$$

$$V_{DS2}(\theta) = V_{DD}\left[\frac{1}{2} + \frac{2}{\pi}\sin\theta + \frac{2}{3\pi}\sin 3\theta + \frac{2}{5\pi}\sin 5\theta + \frac{2}{7\pi}\sin 7\theta + \cdots\right] \tag{7.89}$$

If L and C are resonant at the fundamental frequency of the Fourier series, the load R only sees the fundamental frequency component of the drain-source voltage of $Q2$, $V_{\text{ac}1}(\theta)$

$$V_{\text{ac}1}(\theta) = \frac{2V_{DD}}{\pi}\sin\theta \qquad (7.90)$$

and

$$\hat{V}_{\text{ac}1} = \frac{2V_{DD}}{\pi}$$

Output power is simply

$$P_{\text{out}} = \frac{\hat{V}_{\text{ac}1}^2}{2R} = \frac{4V_{DD}^2}{2\pi^2 R} = \frac{2V_{DD}^2}{\pi^2 R} \qquad (7.91)$$

The current through $Q1$ or $Q2$ is a half-sinusoid since each transistor conducts for only half of the time. The average current through $Q1$ is the DC supply current I_{DC}

$$I_{\text{ac}1}(\theta) = \frac{V_{\text{ac}1}(\theta)}{R} = \frac{2V_{DD}}{\pi R}\sin\theta$$

$$I_{Q1}(\theta) = \begin{cases} I_{\text{ac}1}(\theta), & \text{when } \sin\theta > 0 \\ 0, & \text{when } \sin\theta \le 0 \end{cases}$$

$$I_{\text{DC}} = \frac{1}{2\pi}\int_0^\pi I_{Q1}(\theta)d\theta \qquad (7.92)$$

$$I_{\text{DC}} = \frac{-V_{DD}}{\pi^2 R}\cos\theta|_0^\pi| = \frac{2V_{DD}}{\pi^2 R}$$

The DC input power is found from

$$P_{\text{in}} = I_{\text{DC}}V_{DD} = \frac{2V_{DD}^2}{\pi^2 R} \qquad (7.93)$$

Since Eq. (7.93) equals Eq. (7.91), the efficiency η is equal to one.

A nonzero R_{on} is the only resistance preventing perfect efficiency if L and C in Fig. 7.62 are considered ideal. In reality, power efficiency near 98% is possible with very low on-resistance depletion type FETs. C_{block} is necessary because the current through $Q1$ has a DC component.

The full-wave depletion type FET Class D amplifier may be configured as a pair of transistors such as the push-pull Class B schematic diagram shown in Fig. 7.53(a). Using a center-tapped primary transformer in the output circuit allows both transistors to conduct to the supply instead of one to the supply and one to ground, hence, full wave. A series tuned $L–C$ resonant circuit in series with the load is necessary to allow only fundamental frequency load current. Full wave may also be achieved by using the four transistor *H-bridge*. It is the configuration of choice here in order to illustrate another way. The H-bridge name is obvious and is shown in Fig. 7.63. As before, R_{on} and drain-source capacitance are assumed to be zero.

The voltage across the series RLC output circuit alternates polarity between $+V_{DD}$ and $-V_{DD}$ so that $V_{\text{ac}}(\theta)$ is of the same form as Eq. (7.89). When L and C are tuned to fundamental resonance

$$V_{\text{ac}1}(\theta) = 2V_{DD}\left(\frac{2}{\pi}\sin\theta\right) \qquad (7.94)$$

and

$$\hat{V}_{\text{ac}1} = \frac{4V_{DD}}{\pi}$$

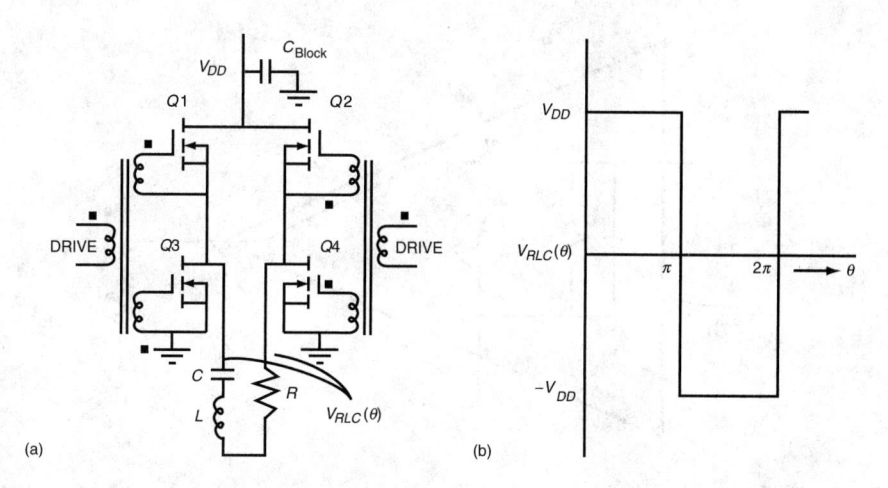

FIGURE 7.63 (a) The H-bridge full-wave Class D amplifier configuration. A diagonal pair of FETs is on, while the opposite pair is off at any instant. The H-bridge provides four times the power as the half-wave configuration. LC is a series resonant circuit at the fundamental frequency so that harmonic currents do not flow in the load R. (b) The waveform is the bridge voltage across the RLC network.

The output power is found from

$$P_{\text{out}} = \frac{\hat{V}_{\text{ac1}}^2}{2R} = \frac{8V_{DD}^2}{\pi^2 R} \tag{7.95}$$

The current through the load is

$$I_{\text{ac1}}(\theta) = \frac{4V_{DD}}{\pi R}\sin\theta$$

and

$$\hat{I}_{\text{ac1}} = \frac{4V_{DD}}{\pi R}$$

The currents through Q1–Q4 are half-sinusoids since complementary pairs (Q1 Q4) and/or (Q2, Q3) only conduct for half of each drive cycle. Over 2π, the pulsating DC from the supply consists of two half-sinusoids so that C_{block} is needed and

$$
\begin{aligned}
I_{\text{DC}} &= 2\left(\frac{1}{2\pi}\int_0^\pi I_{\text{ac1}}(\theta)d\theta\right) \\
&= -\frac{4V_{DD}}{\pi^2 R}\cos\theta|_0^\pi| = \frac{8V_{DD}}{\pi^2 R}
\end{aligned} \tag{7.96}
$$

Input power is

$$P_{\text{in}} = I_{\text{DC}}V_{DD} = \frac{8V_{DD}^2}{\pi^2 R} \tag{7.97}$$

Since Eq. (7.95) equals Eq. (7.97), power efficiency is equal to one as long as the transistors act as ideal switches with $R_{\text{on}} = 0$ and drain-source capacitance is also equal to zero. Four times the power is available from the full-wave H-bridge compared to the two transistor half-wave circuit of Fig. 7.62 because twice V_{DD} is the voltage swing across RLC.

Either the full-wave or half-wave Class D amplifier may be transformer coupled to R. It is only necessary to consider the turns ratio, but P_{out} and P_{in} remain the same.

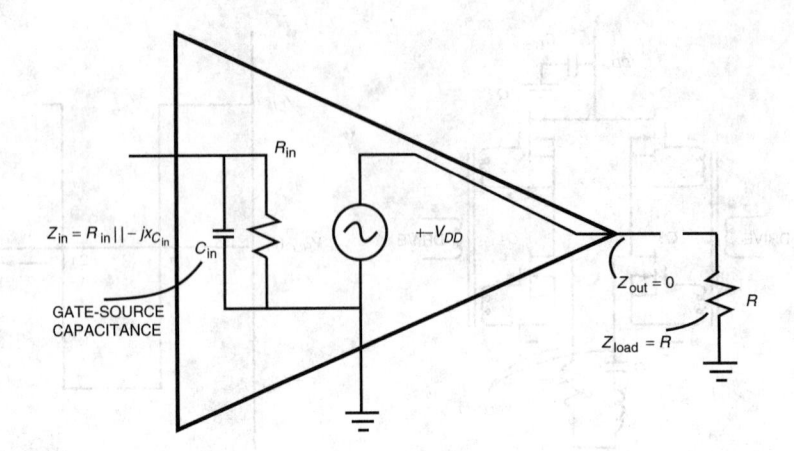

FIGURE 7.64   The full-wave H-bridge is represented by a voltage source with no internal output resistance, assuming that the on-resistance and drain-source capacitance are zero. The input circuit must be a good load to the driving source, but usually is driven by a low impedance source capable of driving the capacitive input. R_{in} is high.

Modeling the entire switch mode amplifier in Fig. 7.64 shows that R_{Int} of the output circuit of the amplifier is zero. It must be since power efficiency is 100%, and the output circuit is a voltage source. It may be a current controlled voltage source if the switching devices are bipolar function transistors (BJTs), or voltage controlled if FETs or IGBJTs are used. Although not much attention is given to the drive input circuit, it must be designed to present itself as some standard impedance at the operating frequency and, perhaps, very broadband if the drive signal is square. With no tuned circuits in the way and a very low source impedance, the drive signal should remain square.

In practice, each transistor in the full-wave or half-wave case is shunted by a *fast recovery diode* placed in the reverse direction from drain to source. The diode provides a path for current to flow when the load R is better represented by a complex impedance, $Z = R \pm jX$. A reactive component in the load causes out-of-phase currents to want to flow backward through any off transistor. Without the diode, the output capacitance of an off transistor will develop a spike of voltage due to the out-of-phase current charging the capacitance, and permanent damage will result. The diodes must be fast enough to conduct or not conduct dictated by the switching phase. Ordinary 60-Hz diodes imitate a resistor at 100 kHz.

In the case of the FET, drive power is derived by charging and discharging the gate-source capacitance. The amount of energy required to charge it is

$$E = \frac{1}{2}CV^2 \tag{7.98}$$

For example, at 10 MHz, the period, τ is 10^{-7} s. A half-period (charge time) is $\tau = 0.5(10)^{-7}$ s. If gate capacitance is 2000 pF, $2(10)^{-9}$ F, for each transistor charging to 10 V, then the energy needed to charge each gate becomes

$$E = \frac{1}{2}(2)(10)^{-9}(10)^2 = (10)^{-7} \text{ J}$$

The same energy must be taken out and dissipated (during the off half-period) at a rate of 10 MHz. The power needed to do this is

$$P = \frac{dE}{dt} = \frac{E}{\frac{\tau}{2}} = 2Ef = 2(10)^{-7}(10)^7 = 2 \text{ watts per gate} \tag{7.99}$$

This may or may not be a large amount of power, but it is not insignificant and must be taken into consideration when designing the overall amplifier. Furthermore, drive power is proportional to carrier frequency.

A potential for disaster exists if a set of on transistors does not turn off as quickly as it should and, instead, remains on while the opposite phase set is turning on. Should this happen, a momentary short circuit appears across the supply, and the transistors are history. A solution is to relatively slowly charge the gates and relatively rapidly discharge them to prevent on-state overlap. The charge path is slow, whereas the discharge path is fast. An effective circuit is shown in Fig. 7.65.

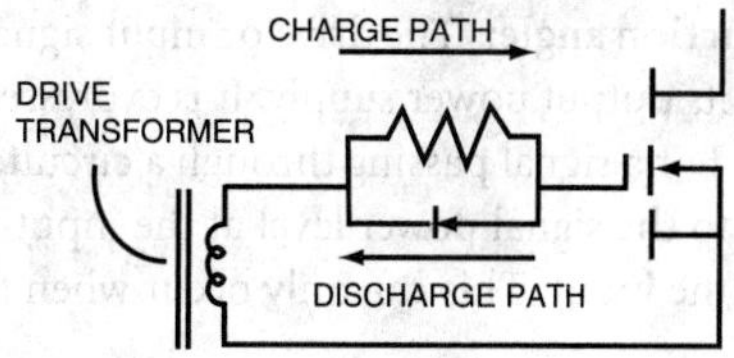

FIGURE 7.65 Slow charge and rapid discharge of the gate-source capacitance prevents possible transistor on-state overlap, which places a short circuit across the supply and destroys transistors.

An enhancement to Class D makes use of the real drain-source capacitance previously ignored. It is a source of inefficiency and takes energy to charge, which is dissipated as heat during discharge much like the gate-source capacitance. In the Class E enhancement mode (modified Class D), the capacitance is used as part of the output network, and an additional external capacitance is selected so that the total is the correct amount to cause the drain voltage to reach zero at the instant the transistor turns off. In this way no discharge of the capacitance takes place and efficiency approaches unity.

The switch-mode Class D amplifier can be used as a baseband linear amplifier by applying a square gate signal, the pulse width (duty cycle) of which is proportional to the amplitude of the signal to be amplified.

Assume that a 25% duty cycle represents the halfway point between off and full on (50%) for each transistor set. Modulating up or down from 25% drives the amplifier to zero and 50% duty cycle for each set (100% total). The average of the output voltage across R will follow the pulse width represented drive signal. The output must be low-pass filtered to obtain the average, which is a reproduction of the signal to be amplified. (Use of the Class D amplifier in this way is sometimes referred to as Class S.)

Applying a pulse width modulated baseband amplifier like this as the power supply for a switch-mode RF amplifier (same configuration) provides a high-level amplitude modulated RF signal. After bandpass filtering at its output, it is suitable for transmission as a full carrier double sideband signal.

There are many variations of the amplifier classes presented in this section and most are *enhancements* of the basic ones.

Defining Terms

Amplifier class of operation: An alphabetical tag associated with an amplifier configuration describing its bias condition (which, in turn, determines its conduction angle) and sometimes its external circuit configuration.

Attenuation: For a signal passing through a circuit, the ratio of the signal power level at the output of the circuit to the signal power level at the input to the circuit when the output signal is of less power than the input. Usually expressed in decibels

$$\text{attenuation} = -10\log\frac{P_{\text{out}}}{P_{\text{in}}} \geq 0$$

Aural subcarrier: In a composite television signal, the frequency division multiplexed carrier placed outside the visual passband that carries the audio modulation. In the NTSC (United States) system, it is placed 4.5 MHz higher than the visual carrier.

Color subcarrier: In a composite television signal, the frequency division multiplexed carrier placed within the visual passband which carries the color modulation. In the National Television Systems Committee (NTSC) (United States) system, it is placed approximately 3.579 MHz higher that the visual carrier.

Compression: For the transfer function of an amplifier, the relatively higher power region where the output power is less than proportional to the input power. The 1-dBm compression point is that output power level that decreases by 9 dBm when the input power decreases by 10 dBm.

Conduction angle: The drive or input signal phase angle over which an amplifier conducts current from its output power supply. It is expressed mathematically by 2χ.

Gain: For a signal passing through a circuit, the ratio of the signal power level at the output of the circuit to the signal power level at the input to the circuit when the output signal is of greater power that the input. This can only occur when the circuit is an amplifier. Usually expressed in decibels

$$\text{gain} = 10\log\frac{P_{\text{out}}}{P_{\text{in}}} \geq 0$$

Intermodulation distortion: The result of two or more time-domain signals being multiplied together when they were intended to be only added together. Undesirable multiplication takes place in every amplifier amplifying multiple signals yielding sum and difference frequency components that are not present at the input to the amplifier.

Linear: Refers to the ability of an amplifier to maintain the integrity of the signal being amplified. A perfectly linear amplifier causes no distortion to the signal while making it greater in amplitude. The output vs. input transfer function plots as a straight line.

On-resistance: The collector-emitter resistance of a bipolar junction transistor or the drain-source resistance of a field effect transistor when driven to saturation. In the case of the BJT, it is equal to the collector-emitter saturation voltage divided by the collector current and is a function of both. In the case of the FET, it is equal to the saturated drain-source voltage divided by the drain current, but is not a function of either one.

Peak envelope power: The average power of the peak value of a signal.

Saturation angle: The drive or input signal phase angle over which an amplifying device is driven into saturation thereby resulting in a flat maximum of amplifying device supply current. Even though drive level may continue to increase, output circuit current does not. It is expressed mathematically by $2\chi_s$.

Saturation current: The collector current of a bipolar junction transistor or the drain current of a field effect transistor during the time of the saturation angle.

Standard load: Refers to the resistance of a load that is a standard established by an organization or committee commissioned to decide standards. Input, output, and interstage resistance in RF circuits is established to be 50 Ω unbalanced. Input, output, and interestage resistance in video circuits is established to be 75 Ω unbalanced, and that for audio circuits is 300 Ω balanced and sometimes 600 Ω balanced or 8 Ω unbalanced.

References

Craig, M. 1994. *Television Measurements NTSC Systems.* Tektronix, Inc., Beaverton, OR.

Daley, J.L., ed. 1962. *Principles of Electronics and Electronic Systems.* U.S. Naval Inst., Annapolis, MD.

Hulick, T.P. Switching power supplies for high voltage. *QEX,* Feb.

Jordan, E.C., ed. 1986. *Reference Data for Engineers: Radio, Electronics, Computer, and Communications,* 7th ed. Howard W. Sams & Co., Indianapolis, IN.

Krauss, H.L., Bostian, C.W., and Raab, F.H. 1980. *Solid State Radio Engineering.* Wiley, New York.

O'Reilly, W.P. 1975. Transmitter power amplifier design II. *Wireless World* 81 (Oct.).

Orr, W.I. 1986. *Radio Handbook,* 23rd ed. Howard W. Sams & Co., Indianapolis, IN.

Schetgen, R., ed. 1995. *The ARRL Handbook,* 72nd ed. American Radio Relay League. Newington, CT.

Taub, H. and Schilling, D.L. 1986. *Principles of Communication Systems,* 2nd ed. McGraw-Hill, New York.

Further Information

Additional information on the topic of amplifiers is available from the following sources:

RF Design magazine, a periodical dealing with RF technology over a broad range of frequencies and uses.

Microwaves & RF magazine, a periodical specializing in rigorous presentations of topics related to the name.

Broadcast Engineering magazine, a periodical dealing with television and radio technology.

The National Association of Broadcasters (NAB), hosts an annual engineering conference and trade show in the Spring. The NAB is headquartered in Washington, D.C.

In addition, the following books are recommended:

Golio, M. ed. 2000. *The RF and Microwave Systems Handbook*, CRC Press, Boca Raton, FL.
Whitaker, J. C. ed. 2002. *The RF Transmission Systems Handbook*, CRC Press, Boca Raton, FL.

7.7 Operational Amplifiers

Peter Aronhime

7.7.1 Introduction

The name *operational amplifier* was originally applied to the high-gain voltage amplifiers that were employed in analog computers. When operational amplifiers were used with feedback circuits they could perform mathematical operations such as addition and integration on input signals. In the context of analog computers, operational amplifiers were most often used to solve differential equations.

The first commercially successful *integrated* operational amplifier (op-amp) was the Fairchild Company's μA702, which was available in 1963. The first practical integrated op-amp was the μA709, which was available in 1965. National Semiconductor Company's LM101 followed soon after. These three op-amps were designed by Bob Widlar (*Electronic Design*, 1992).

Probably the most widely used op-amp, the μA741, was designed by Dave Fullagar and became available in 1967. This op-amp has relatively good characteristics, is easy to use, and is very inexpensive. It is an indication of the commercial success of the 741 that many of the more recently designed op-amps still have the same pin-outs as the 741 so that newer op-amps can replace 741s in a circuit board without redesigning the circuit board.

7.7.2 The Ideal Op-Amp

The ideal op-amp is an abstraction that is useful in the design and analysis of active circuits. A design can be formulated in terms of ideal op-amps, and later, if the design shows promise, the circuit can be analyzed in more detail to make sure it works in all circumstances. Figure 7.66 shows a schematic representation of an ideal op-amp. The presence of the *infinity* on the op-amp symbol indicates that this is an ideal op-amp representation.

Terminal 2 in this figure is denoted as the inverting input terminal, and a minus sign is marked next to this terminal on the op-amp symbol. Terminal 3 is called the noninverting input terminal, and a plus sign is placed next to this terminal. Terminal 6 is the output terminal, and terminals 4 and 7 are used to supply power to the device. Power is provided by a split supply consisting of two DC voltage sources with a ground between them. This ground is the

FIGURE 7.66 A schematic representation of an ideal op-amp shown with required power supplies and bypass capacitors. Not shown is the feedback network. The terminal numbers correspond to those used for the 741 op-amp in the minidip configuration.

electronic ground of the circuit, and all signals are referenced to this ground. Thus, the inverting input port exists between terminals 2 and 2' in Fig. 7.66, the noninverting input port exists between terminals 3 and 3' in the figure, and the output port exists between terminals 6 and 6'. In a physical circuit, terminals 2', 3', and 6' should be electrically the same as the ground point between the power supplies. Although an op-amp will not operate properly and may even be damaged if signals are applied to the op-amp while no

FIGURE 7.67 Ideal op-amp models: (a) model with note stating that $v_a = 0$, (b) model with note that $i = 0$, (c) model consisting of a nullator and a grounded norator.

power connections are made to it, the power connections will not be shown in the schematics that follow. It is assumed that the op-amps shown are properly powered.

Figure 7.66 shows bypass capacitors connected between the power supply leads and ground at the op-amp. Values for these capacitors typically range between 0.1 and 1 μF, and they may be necessary to prevent the op-amp from oscillating or latching up even when negative feedback is connected around it. The phrase *latching up* refers to op-amp behavior characterized by the output voltage heading quickly to the maximum or minimum value that the op-amp can sustain and remaining at that level. Some op-amps may require more elaborate power supply bypass circuits, but generally speaking, individual capacitors connected as shown in the figure will suffice. However, a circuit designer should consult the manufacturer's specification sheet for a specific op-amp.

Ideal op-amps have the following properties:

- The voltage gain from the inverting input port to the output port is $-\infty$ over all frequencies, and the voltage gain from the noninverting port to the output port is $+\infty$.
- No current is drawn by the input leads. Thus, i_2 and i_3 in Fig. 7.66 are zero.
- The voltage between the input leads of the ideal op-amp is zero. Thus, v_a in Fig. 7.66 is zero.
- The output voltage can change in zero time in response to an input voltage that changes in zero time.
- The output resistance is zero. Thus, the output voltage of an ideal op-amp is independent of the current drawn from the output.
- Ideal op-amps generate no noise.

No real op-amp has these properties, but the properties of real op-amps approach the ones listed for many applications.

Three circuit models, which summarize the properties of ideal op-amps, are shown in Fig. 7.67. These models are equivalent, and the choice of which one to use in the analysis of a linear op-amp circuit is merely a matter of convenience. The model shown in Fig. 7.67(a) includes the note $v_a = 0$. The connection between terminals 2 and 3 in this figure is sometimes called a **virtual connection** or, if terminal 3 is connected to ground, a *virtual ground*. No current flows through this connection, and there is zero voltage between terminals 2 and 3.

In Fig. 7.67(b), the virtual connection between terminals 2 and 3 is realized by a short circuit with the additional note $i = 0$. In the third model, the virtual connection is represented by a **nullator**, which is an element defined to have zero current through it and zero volts across it (Bruton, 1980).

The output circuits of the models in Fig. 7.67(a) and Fig. 7.67(b) are ideal voltage sources, and the output circuit of the model in Fig. 7.67(c) is a **norator** with one end grounded. A norator is an element in which both the current through it and the voltage across it are arbitrary (Bruton, 1980), where arbitrary in the context of circuit analysis means that the current and voltage are determined by external circuitry.

The use of the model in Fig. 7.67(b) is illustrated in Fig. 7.68. Figure 7.68(a) shows an amplifier circuit consisting of an ideal op-amp and two resistors. The ideal op-amp symbol is replaced by a model in Fig. 7.68(b). A node equation written at node X yields

$$\frac{v_i}{R_1} + \frac{v_O}{R_2} = 0 \tag{7.100}$$

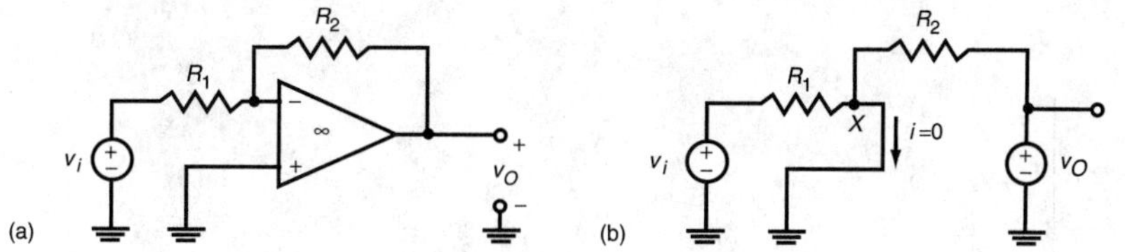

FIGURE 7.68 Inverting amplifier: (a) schematic, (b) a model for the ideal op-amp has been substituted into the circuit.

from which we obtain

$$\frac{v_O}{v_i} = -\frac{R_2}{R_1} \tag{7.101}$$

This amplifier configuration is called an inverting amplifier. The input resistance of this amplifier is easily determined from Fig. 7.68(b) to be R_1.

In most applications, external circuit elements such as resistors and capacitors must be connected to the terminals of an op-amp to obtain a useful circuit. Very few practical applications exist in which the op-amp is used alone.

7.7.3 A Collection of Functional Circuits

A medley of op-amp circuits is presented in this section. Table 7.3 shows some basic op-amp circuits and lists the chief characteristics of each one. The power supply connections are not shown for these circuits (except for circuits 10 and 11), but it is understood that they are present. Power supply bypass capacitors are also not shown but are assumed to be connected. Circuit 2 in the table is a voltage buffer. It is an example of one of the few applications that require no external components except for a wire connecting the output to the inverting input terminal of the op-amp.

The adder, shown as circuit 4 in Table 7.3, can be expanded into an adder/subtractor as depicted in Fig. 7.69. In this circuit, we have $(n + Z)$ voltage sources labeled $v_1, v_2, \ldots, v_n$ and $v_A, v_B, \ldots, v_Z$. The output voltage of the circuit is

$$v_O = -\sum_{i=1}^{n} M_i v_i + \sum_{k=A}^{Z} M_k v_k \tag{7.102}$$

where the M are gain magnitudes. The variable S in the figure is

$$S = S_n - S_p \tag{7.103}$$

where

$$S_n = 1 + \sum_{i=1}^{n} M_i = 1 + |(\text{sum of negative gains})|$$

$$S_p = \sum_{k=A}^{Z} M_k = \text{sum of positive gains}$$

We require that $S \geq 0$ so that the resistor R_F/S is not negative. As an example, suppose we have four voltage sources labeled v_1, v_2, v_A, and v_B, and we want

$$v_O = -6v_1 - 3v_2 + 5v_A + \frac{1}{2}v_B$$

TABLE 7.3 Op-amp Circuits

No.	Type of Circuit	Schematic	Circuit Gain or Variable of Interest	Input Resistance or Input Currents or Voltages	Special Requirements or Remarks
1	Noninverting amplifier		$\dfrac{v_O}{v_i} = 1 + \dfrac{R_2}{R_1}$	$R_{\text{in}} = \infty$ (ideally)	
2	Buffer		$\dfrac{v_O}{v_i} = 1$	$R_{\text{in}} = \infty$ (ideally)	Special case of circuit 1
3	Difference amplifier		$v_O = \dfrac{R_2}{R_1}(v_2 - v_1)$	$i_1 = \dfrac{v_1 - v_2 \dfrac{R_b}{(R_a + R_b)}}{R_1}$ $i_2 = \dfrac{v_2}{R_a + R_b}$	$\dfrac{R_1}{R_2} = \dfrac{R_a}{R_b}$
4	Adder		$v_O = -\left\{ v_1\dfrac{R_f}{R_1} + v_2\dfrac{R_f}{R_2} + \cdots + v_n\dfrac{R_f}{R_n} \right\}$	$i_1 = \dfrac{v_1}{R_1}$ $i_2 = \dfrac{v_2}{R_2}$ $\cdots$ $i_n = \dfrac{v_n}{R_n}$	

#					
5	Variable gain circuit		$$\frac{v_O}{v_i} = (2Kx - K)$$ $$0 \le x \le 1, \quad K > 1$$	$$i = \frac{v_i}{R_3} + \frac{Kv_i(1-x)}{R}$$	Potentiometer R_3 adjusts the gain over the range $-K$ to $+K$.
6	Voltage-to-current converter		$$i = \frac{v_i}{R_1}$$		The current through R_L is independent of R_L
7	Voltage-to-current converter with grounded load		$$i = \frac{v_i}{R}$$	$$i_s = \frac{v_i}{R}\left(1 - \frac{R_L}{R}\right)$$	$v_O = v_i(2R_L/R)$ The current i is independent of R_L Circuit has wide band-width for $R_L \ll R$.
8	Current-to-voltage converter		$$v_O = -Ri_i$$	$$v = 0$$	The voltage v_O is independent of R_L and R_i

(Continued)

TABLE 7.3 Op-amp Circuits (Continued)

No.	Type of Circuit	Schematic	Circuit Gain or Variable of Interest	Input Resistance or Input Currents or Voltages	Special Requirements or Remarks
9	Current-to-voltage converter		$v_O = -2i R_1 \dfrac{R_4}{R_3}$	$v = 0$	
10	Inverting amplifier with single supply		$v_O = 7.5 - v_i \dfrac{R_2}{R_1}$		$R = 3.9\ \mathrm{k\Omega}$
11.	Noninverting amplifier with single supply		$v_O = 7.5 + v_i \left(1 + \dfrac{R_2}{R_1}\right)$		$R = 3.9\ \mathrm{k\Omega}$

12	Integrator		$v_O = -V(0) - \dfrac{1}{RC}\displaystyle\int_0^t v_i(t)\,dt$ $V(0)$ is the initial voltage across the capacitor. RC is very large.	$i = \dfrac{v_i}{R}$ Negative feedback is required at DC. A large value of R_C can be used or a feedback path can be established through an external circuit.
13.	De Boo integrator		$v_O = 2V(0) + \dfrac{2}{RC}\displaystyle\int_0^t v_i(t)\,dt$	$i = \dfrac{v_i}{R} - \dfrac{v_O}{2R}$ One end of capacitor is physically grounded.
14.	Differentiator		$v_O = -RC\dfrac{dv_i}{dt}$	$i = C\dfrac{dv_i}{dt}$ Differentiators are usually avoided in the design of circuits because they accentuate noise.
15.	Generalized impedance converter (GIC)		$Z_{in} = \dfrac{Z_1 Z_3 Z_5}{Z_2 Z_4}$	

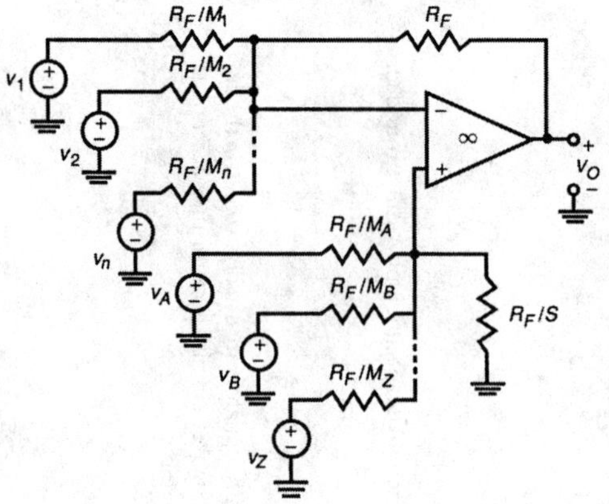

FIGURE 7.69   Single op-amp adder/subtractor circuit.

We note that $S_n = 1 + (6 + 3) = 10$ and $S_p = 5.5$ so that $S = 4.5$, which is not negative. The resulting circuit is shown in Fig. 7.70.

However, suppose we require a different circuit in which

$$v_O = -v_1 - 2v_2 + 3v_A + 4v_B$$

In this case, S is negative. So we append a fictitious source v_3 and require that the gain from v_3 to v_O be -3. Now $S_n = 1 + (1 + 2 + 3) = 7$. The sum of positive gains, S_p, remains equal to 7. Thus $S = 0$, and the resistor R_F/S in Fig. 7.69 is replaced by an open circuit. Of course, we set $v_3 = 0$. The resulting circuit is shown in Fig. 7.71.

Another useful circuit, based on a single op-amp, is shown in Fig. 7.72. The schematic depicts a rendering of the Burr-Brown RCV420 4–20 mA current loop receiver, a circuit which converts a 4–20 mA input signal to a 0–5 V output signal.

The circuit realizes a transresistance R_T given by

$$R_T = \frac{\Delta V_O}{\Delta I_N} = \frac{5\,\text{V}}{16\,\text{mA}} = 0.3125\,\text{V/mA}$$

Also, the output voltage of the circuit must be offset by -4 mA $(0.3125\,\text{V/mA}) = -1.25$ V so that 0 V will be obtained when 4 mA is applied to the input. Thus, the output voltage is given by

$$V_O = -1.25\,\text{V} + I_N(\text{mA})\,[0.3125\,(\text{V/mA})] \tag{7.104}$$

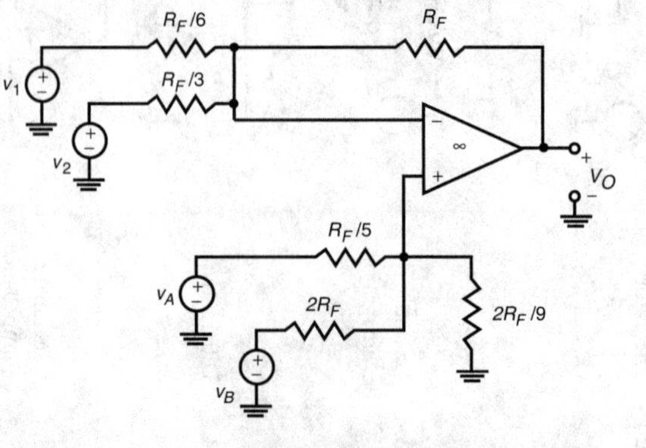

FIGURE 7.70   Adder/subtractor that provides gains: -6, -3, $+5$, and $+\frac{1}{2}$.

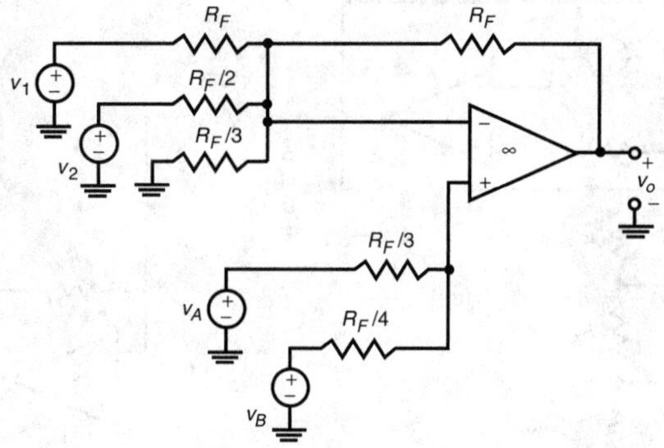

FIGURE 7.71 Adder/subtractor with gains equal to $-1, -2, +3$, and $+4$.

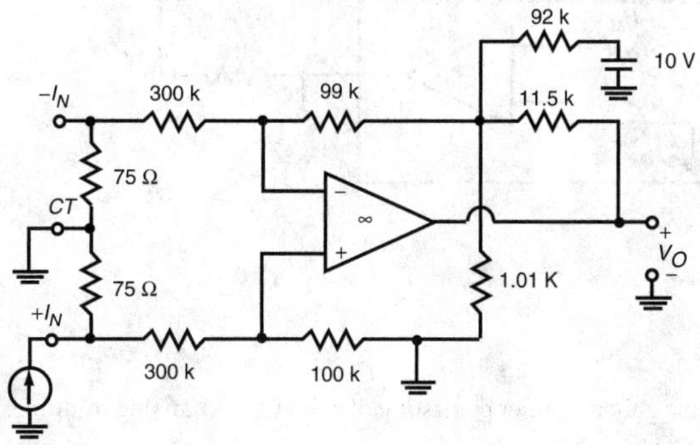

FIGURE 7.72 A 4–20 mA current loop receiver shown with a positive input current signal. This signal returns to ground through terminal CT. The terminal labeled $-I_N$ is left open. If the signal is a negative current signal, then it is applied to the terminal labeled $-I_N$ and the terminal $+I_N$ is left open.

To verify Eq. (7.104), we reduce the circuit in Fig. 7.72 to that shown in Fig. 7.73(a) using the techniques described in Aronhime (1995). Further reduction yields the circuit in Fig. 7.73(b) from which we obtain Eq. (7.104). Note that if $+I_N$ were zero (open circuit) in Fig. 7.72, then from Eq. (7.105), V_O would be -1.25 V. This voltage can be easily sensed and used as an indicator that the signal line is open.

Op-amps can be advantageously combined with analog switches to create innovative and useful circuits. For example, Fig. 7.74 shows a representation of a programmable-gain amplifier composed of a dual 4-to-1 analog multiplexer and a difference amplifier. The positions of the switches in the multiplexer are selected digitally by means of leads A_0 and A_1 and yield amplifier gains according to Table 7.4.

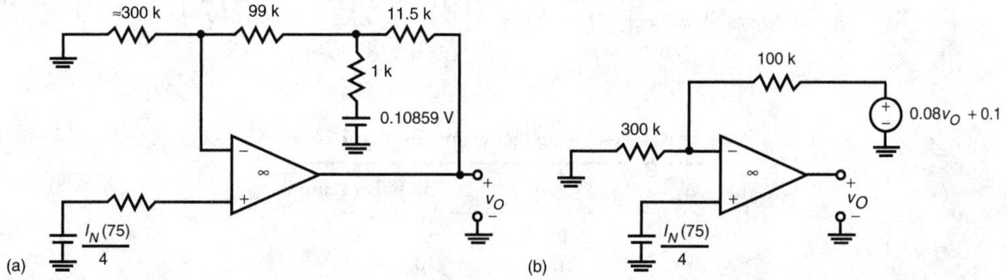

FIGURE 7.73 Steps in the analysis of the current loop receiver.

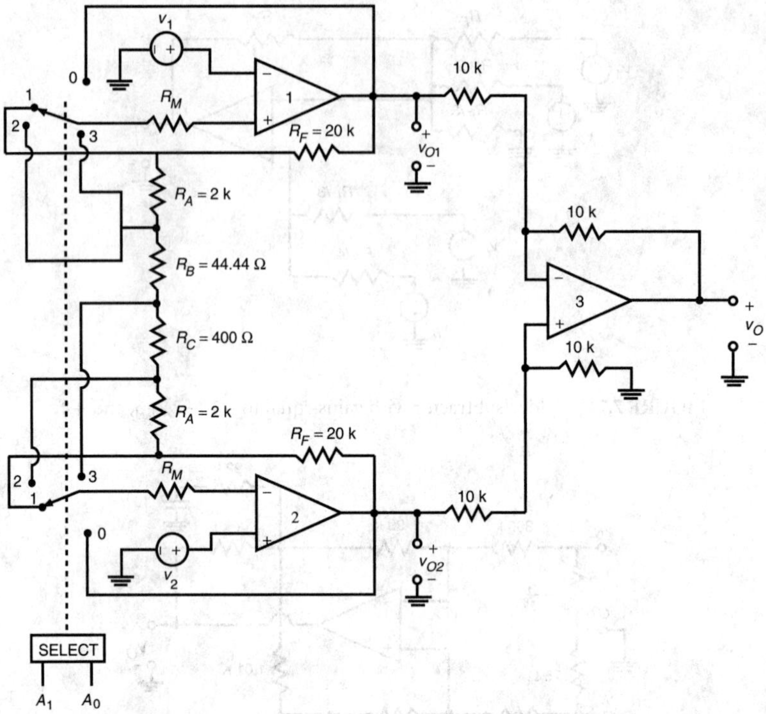

FIGURE 7.74   Programmable gain amplifier consisting of a dual 4-to-1 analog multiplexer and a difference amplifier. The op-amps are ideal.

The resistors labeled R_M represent parasitic resistances of the analog switches used in the multiplexer. These resistances can range in value from 35 Ω to several kilohms.

To verify the second entry in Table 7.4, suppose A_1 is low and A_0 is high. The resulting connections to op-amps 1 and 2 are shown in Fig. 7.75. Since the op-amps are considered to be ideal, no currents flow through the resistors labeled R_M, and so these resistors do not enter into the calculations. Using superposition and the results previously given for the ideal inverting amplifier, we can write

$$v_{O1} = v_1 \left(1 + \frac{R_F}{2R_A + R_B + R_C} \right) - v_2 \left(\frac{R_F}{2R_A + R_B + R_C} \right) \tag{7.105}$$

and

$$v_{O2} = -v_1 \left(\frac{R_F}{2R_A + R_B + R_C} \right) + v_2 \left(1 + \frac{R_F}{2R_A + R_B + R_C} \right) \tag{7.106}$$

TABLE 7.4   Amplifier Gains (for Fig. 7.74)

A_1	A_0	Amplifier Gain
0	0	1
0	1	10
1	0	100
1	1	1000

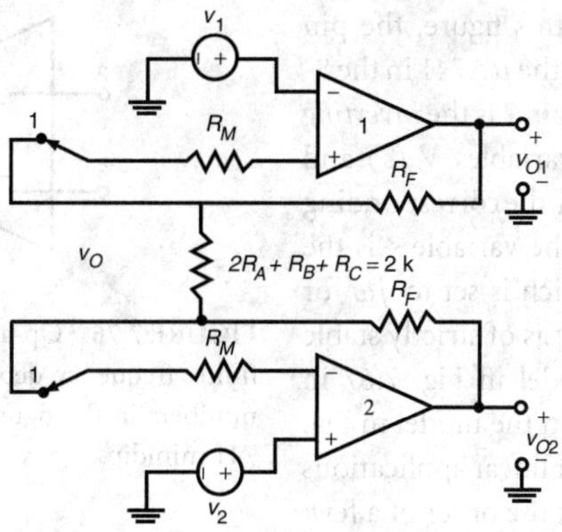

FIGURE 7.75 Reduced circuit corresponding to the programmable gain amplifier in Fig. 7.74 with both switches set to position 1 ($A_1 = 0$ and $A_0 = 1$). Op-amp 3 and its associated resistors are not shown.

The difference of the two voltages v_{O1} and v_{O2} is formed by the difference amplifier composed of op-amp 3 and the four 10-kΩ resistors in Fig. 7.74 (see circuit 3 in Table 7.3). Thus, v_O is given by

$$v_O = v_{O2} - v_{O1} = (v_2 - v_1)\left(1 + \frac{2R_F}{2R_A + R_B + R_C}\right) = 10.0(v_2 - v_1) \qquad (7.107)$$

At low frequencies, the resistances of the switches used to create the analog multiplexers have no effect on the output voltage, but they may affect the results at higher frequencies if parasitic capacitances are not kept to low values. However, the properties of real or tangible op-amps (as opposed to ideal op-amps) will most likely impose tighter limits on the range of frequencies over which this circuit and any of the circuits in this section can be used. Furthermore, real op-amp characteristics will affect gain accuracies. In the next section, we discuss the characteristics of real op-amps and show how these characteristics can be modeled.

7.7.4   Op-Amp Limitations

Real, tangible op-amps differ from ideal op-amps in the following ways:

- The gain of a real op-amp is neither infinite nor constant. It is a function of frequency.
- A real op-amp has an offset voltage and nonzero bias current.
- Input resistances are not infinite and output resistance is not zero. Input capacitance is not zero as well.
- The output voltage is limited in amplitude, and it is slew-rate limited. The output current is also limited in amplitude.
- A tangible op-amp has nonzero **common-mode gain**.
- Real op-amps generate noise.

In this section we discuss these properties of real op-amps and compare them with the specifications of the commercial version of the μA741 op-amp. Although the list above is somewhat lengthy, in many applications only one or two properties of real op-amps matter. For this reason, we develop a series of models for the real op-amp. Each model accounts for only one or two departures from ideal behavior. Thus, the effects of a particular property of an op-amp on a circuit can be ascertained.

Op-Amp Gain

For many applications, the most important departure from ideal behavior is the finite, frequency-dependent gain of real op-amps. Figure 7.76 shows an op-amp model which is useful for determining the effects of nonideal gain on a circuit. In this figure, the pin numbers correspond to those of the $\mu A741$ in the 8-pin minidip (see Fig. 7.66), and pin 2 is the *inverting* terminal of the op-amp. The variables $V_a(s)$ and $V_O(s)$ are Laplace transforms of the corresponding functions of time v_a and v_O. The variable s is the Laplace transform variable, which is set to $j\omega$ for the sinusoidal steady-state analysis of strictly stable networks. Unlike the ideal model in Fig. 7.67 in which v_a is zero, v_a is not zero in the model in Fig. 7.76, although it is quite small for linear applications of an op-amp. Typically, v_a is of the order of a few millivolts.

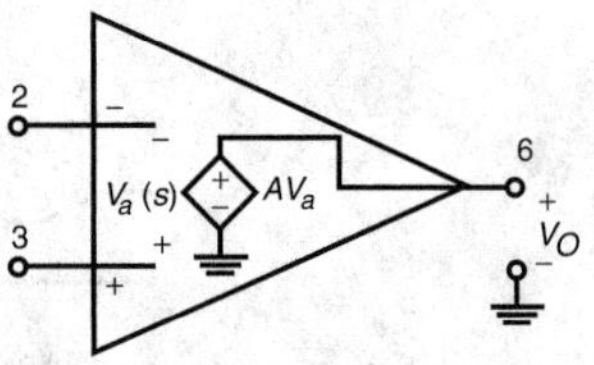

FIGURE 7.76 Op-amp model which accounts for the finite, frequency-dependent gain of a real op-amp. Pin numbers in the figure correspond to those used on the 741 minidip.

$A(s)$ represents the gain of the op-amp. For most modern general-purpose op-amps, this frequency-dependent gain is closely approximated by

$$A(s) = \frac{A_o \omega_a}{(s + \omega_a)} \tag{7.108}$$

where A_o is the gain of the op-amp at DC and ω_a is the (radian) frequency at which the magnitude of the gain has decreased 3.01 dB from its DC value. The frequency-dependent gain given in Eq. (7.108) is termed the **open-loop gain** of the op-amp, and the schematic in Fig. 7.76 with $A(s)$ given by Eq. (7.108) is termed the **one-pole model** or the one-pole rolloff model.

A_o is very large. Even an inexpensive op-amp such as the $\mu A741$ has a gain at DC of 100 dB or more. However, ω_a is only 10–20 rad/s, which corresponds to only a few hertz. Figure 7.77 shows sketches of the magnitude and phase of Eq. (7.108). Note that the magnitude of the **open-loop gain** equals one at $\omega = A_o \omega_a$.

If A_o is 105 dB and ω_a is 10 rad/s, the magnitude of the gain is reduced to about 85 dB at $\omega = 100$ rad/s. The question arises: Why are op-amps designed to exhibit the characteristics of this one-pole model? A brief answer to this question begins with the fact that an op-amp consists of many transistors and, therefore, there are many parasitic capacitances, which in turn create many poles in $A(s)$. Suppose the op-amp is used to make an amplifier such as the one shown as circuit 1 in Table 7.3. As the gain of this amplifier is reduced by, say, decreasing R_2, the poles of the transfer function of the *amplifier* move in the s plane. Some of these poles move toward the imaginary axis and may even cross over into the right-half s plane if the gain is reduced to a low value. This would cause the amplifier to be unstable. The poles of the amplifier exist because $A(s)$ has poles (and zeros), but the poles of the amplifier are different from the poles of $A(s)$. To ensure that the amplifier remains stable while providing a gain that is accurately modeled by $1 + (R_2/R_1)$, it can be shown that $A(s)$

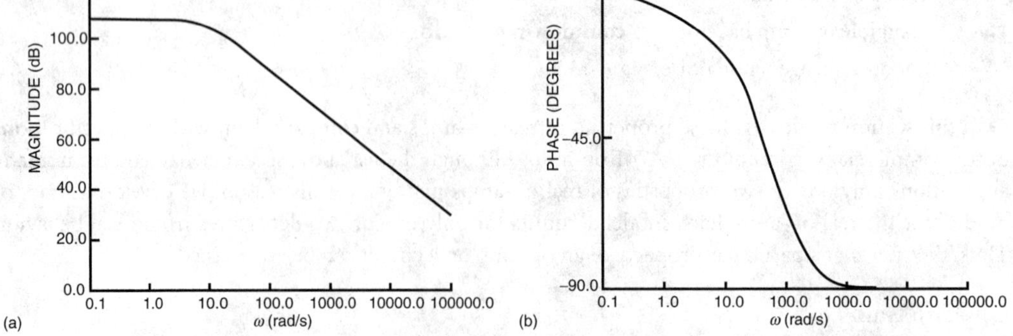

FIGURE 7.77 (a) Magnitude and (b) phase of the open-loop gain $A(s)$ of an op-amp.

should have one dominant pole close to the origin of the s plane (Budak, 1974; Geiger, Allen, and Strader, 1990; Franco, 1988). This discussion has assumed that the feedback around the op-amp is purely resistive and has ignored stray capacitance that might be associated with a load on the op-amp. If the feedback network around the op-amp is not purely resistive and/or there are capacitors (or inductors) in the load, then an analysis must be performed to determine whether the amplifier is stable or not. The one-pole model developed here can be used for this analysis.

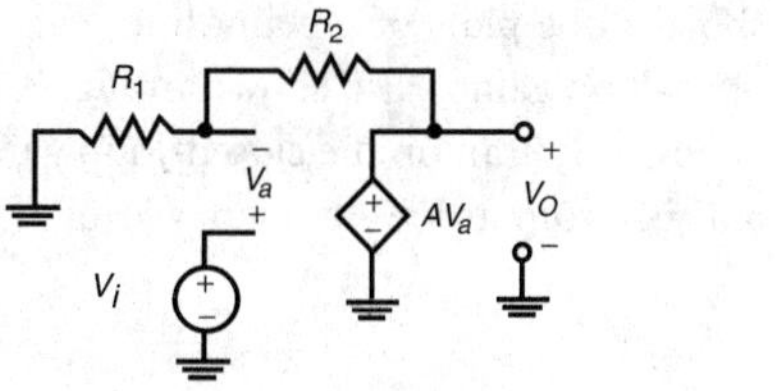

FIGURE 7.78 Noninverting amplifier schematic including one-pole model of Fig. 7.76.

The **gain-bandwidth product** GB of the op-amp is $GB = A_o \omega_a$, and the **op-amp time constant** is denoted as $\tau = 1/GB$. For analyses in which the frequencies of interest are much greater than ω_a, $A(s)$ is usually approximated as

$$A(s) \approx \frac{GB}{s} = \frac{1}{s\tau} \tag{7.109}$$

Ideally, GB is infinite and τ is zero.

Applying the one-pole model to the noninverting amplifier shown as circuit 1 in Table 7.3, we obtain the circuit shown in Fig. 7.78.

Since $V_O = A(s)V_a$, we obtain

$$V_i - \frac{V_O}{A(s)} = V_O \frac{R_1}{R_1 + R_2}$$

Using $A(s) = GB/s$ and letting $K = 1 + R_2/R_1$, we get

$$\frac{V_O}{V_i} = K \left[\frac{\dfrac{GB}{K}}{s + \dfrac{GB}{K}} \right] \tag{7.110}$$

The gain of the amplifier in Eq. (7.110) is referred to as the closed-loop gain. Examination of Eq. (7.110) shows that the gain consists of the ideal gain K (gain obtained with an ideal op-amp) multiplied by a first-order function of the Laplace variable s. This first-order function causes the magnitude of the amplifier gain to fall off for higher frequencies, a behavior that is readily observed in the laboratory. Figure 7.79

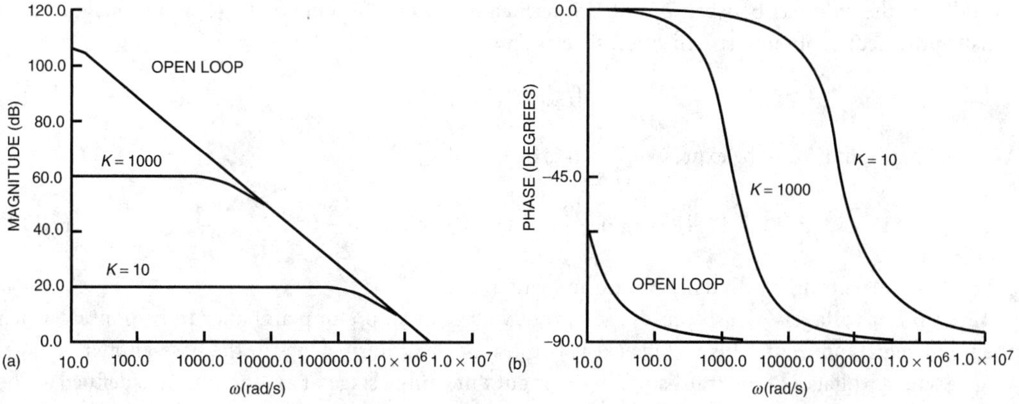

FIGURE 7.79 (a) Bode plots of the closed-loop gain of the noninverting amplifier for $K = 1000$ and for $K = 10$. The magnitude of the op-amp gain is also shown. (b) Plots of phase are shown for $K = 1000$ and $K = 10$, as well as for the op-amp gain.

shows Bode plots of the amplifier gain and phase for two different gain settings. The magnitude of the open-loop gain is also shown in Fig. 7.79(a), and we see that the magnitude of the open-loop gain sets a bandwidth limit on the closed-loop gain. Let ω_{CL} be the frequency at which the closed loop gain is down 3.01 dB from its low-frequency level of 20 log (K). From Eq. (7.110), we have $\omega_{CL} = GB/K$. Thus

$$\omega_{CL} K = A_o \omega_a \qquad (7.111)$$

The quantity $\omega_{CL} K$ is the **closed-loop gain-bandwidth product**, and so Eq. (7.111) shows that the closed-loop gain-bandwidth product for this amplifier equals the open-loop gain-bandwidth product.

The finite gain-bandwidth product of the op-amp also manifests itself in other ways in addition to causing the gain of op-amp amplifiers to fall off at higher frequencies. For example, it can cause the pole and zero placements in active filter circuits to be inaccurate, it can cause active filters to be unstable, and it can limit the **rise time** of an amplifier's response to a step. For example, suppose the buffer shown as circuit 2 in Table 7.3 is excited by a step of voltage of small amplitude V. Small amplitude is needed so that we avoid slew rate limiting which is discussed in a later section. The rise time t_r of the response is defined as the time required for the output voltage to go from 10 to 90% of the final value V. If we use the one-pole model given in Fig. 7.76 in the buffer circuit and solve for t_r, we obtain

$$t_r = \tau \ell_n 9 = 2.2\tau$$

where τ is the time constant of the op-amp. Thus, in response to the step excitation, the output voltage has a nonzero rise time. Clearly, the rise time is improved for smaller τ.

Offset Current and Voltage

An op-amp must be permitted to conduct DC currents through its input leads in order to operate properly. These currents are small and range from picoamps to nanoamps for an IGFET-input op-amp such as a CA3140. For a μA741, they range from nanoamps to a microamp. In an ideal op-amp, they are taken as zero. But these small currents cause the output voltage of an op-amp amplifier to be nonzero when the signals applied to the amplifier are zero. In applications such as instrumentation amplifiers or a track-and-hold circuits, these currents can cause appreciable error.

As in Fig. 7.66, we denote these currents as I_2 and I_3. The term **bias current** or *input bias current* on the specification sheets of op-amps from most semiconductor manufacturers is the *average* of I_2 and I_3. That is, the bias current I_B is

$$I_B = \frac{I_2 + I_3}{2} \qquad (7.112)$$

In addition, the *difference* between I_2 and I_3 is termed the **input offset current** I_{OS}, and its magnitude is usually provided on op-amp specification sheets. Thus

$$|I_{OS}| = |I_2 - I_3| \qquad (7.113)$$

We note that I_2 and I_3 can be expressed in terms of I_B and I_{OS} as

$$I_2 = I_B + \frac{I_{OS}}{2} \quad \text{and} \quad I_3 = I_B - \frac{I_{OS}}{2} \qquad (7.114)$$

All of these currents drift with temperature and with time.

Input offset voltage V_{OS} also causes the output voltage of an op-amp amplifier to be nonzero when the input signals are zero. This voltage exists because of small mismatches in the base-emitter voltages or gate-source voltages of the transistors in the input differential stage of an op-amp. It is defined as the voltage that must be applied to the input of a noninverting amplifier to drive the output voltage to zero with no load on the output. It is assumed that the feedback resistors are chosen so that negligible current is drawn from the output, and currents flowing in the input leads of the op-amp cause insignificant voltages. Offset voltage also drifts with temperature and time.

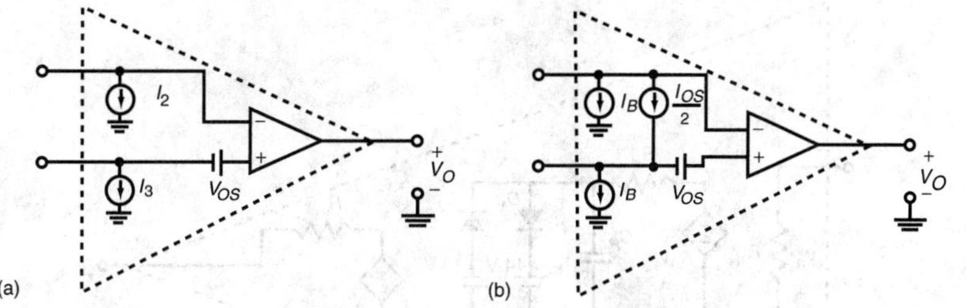

(a)   (b)

FIGURE 7.80 (a) Model showing DC input currents and DC offset voltage. (b) Model which incorporates specification sheet parameters for the input currents.

Figure 7.80(a) shows a model for an op-amp that incorporates I_2, I_3, and V_{OS}. Using Eq. (7.114) we obtain the model shown in Fig. 7.80(b) in which the currents are given in terms of I_B and I_{OS}. For both schematics in Fig. 7.80, the nonideal op-amp, depicted with dotted lines, is modeled with independent sources and an op-amp, shown with solid lines, which has zero input currents and zero offset voltage.

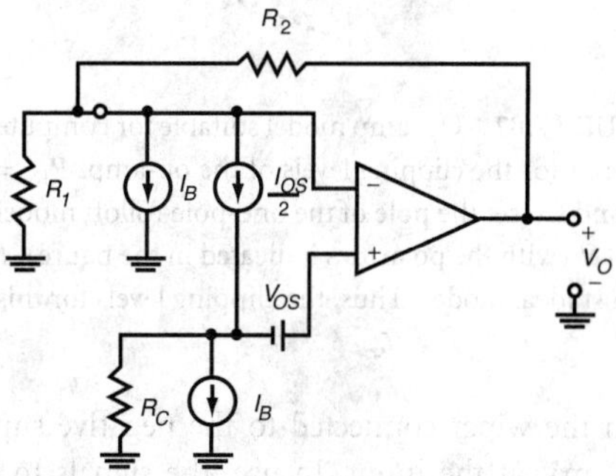

To assess the effects of nonzero I_B, I_{OS}, and V_{OS} on the output voltage of the noninverting amplifier, we employ the circuit shown in Fig. 7.81. Note that no external signals are applied and that an extra resistor R_c has been included in the noninverting input lead of the amplifier. The DC output voltage can be written as

FIGURE 7.81 Noninverting amplifier schematic incorporating the op-amp model shown in Fig. 7.80(b).

$$V_O = I_B \left[R_2 - R_C \left(1 + \frac{R_2}{R_1} \right) \right] + \frac{I_{OS}}{2} \left[R_2 + R_C \left(1 + \frac{R_2}{R_1} \right) \right] + V_{OS} \left(1 + \frac{R_2}{R_1} \right) \qquad (7.115)$$

Ideally, V_O should be zero. If

$$R_C = \frac{R_2}{1 + \dfrac{R_2}{R_1}} = R_1 \| R_2$$

then V_O becomes independent of I_B and can be expressed as

$$V_O = I_{OS} R_2 + V_{OS} \left(1 + \frac{R_2}{R_1} \right)$$

Since I_B is typically 5–10 times larger than I_{OS}, the resistor R_C should be included if the voltage $I_B R_2$ is significant in comparison with the expected range of input signal amplitudes.

Although R_C can be selected to make the output of the noninverting amplifier (and other amplifiers as well) independent of I_B, it cannot be used to eliminate the effects of the other error terms in Eq. (7.115). The reason for this is that the polarities of I_{OS} and V_{OS}, shown with specific polarities in Fig. 7.80(b), are not really known in practice. In fact, these offsets may change polarities as temperature is changed. On the other hand, most integrated op-amps have provision for the connection of an external circuit that can be used to adjust the effects of the offset sources to zero. For a μA741, the external circuit consists of a 10-kΩ potentiometer connected between pins 1 and 5 (on the minidip)

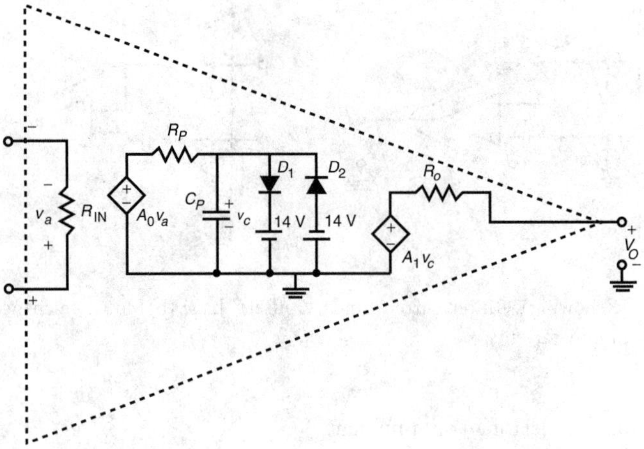

FIGURE 7.82 Op-amp model suitable for computer simulations. The two batteries in the output circuit of the model account for the clipping levels of the op-amp. $R_{IN} = 2$ MΩ, $A_o = 200{,}000$, $R_p = 10$ kΩ, $C_p = 5.0$ μF, and $A_1 = 1$. R_p and C_p set the pole of the one-pole rolloff model. For the values indicated, ω_a is 20. The two batteries are each set to 14 V (with the polarities indicated in the figure). N for the diodes (emission coefficient) is set to 0.001, which yields almost ideal diodes. Thus, the clipping levels for this model are set to ± 14 V. R_o is 75 Ω.

with the wiper connected to the negative supply. The range of adjustment for this circuit is typically ± 15 mV at the input. In use, the signals to the op-amp amplifier are set to zero, and the output of the amplifier is then adjusted to zero with the potentiometer. As ambient temperature changes or if the power to the amplifier is removed and then reapplied, this adjustment procedure should be performed again.

Output Voltage and Current Limits

If an amplifier containing a typical μA741 op-amp is driven by a sine wave of sufficient amplitude to cause clipping and the op-amp is powered by ± 15-V power supplies, the voltage amplitudes at which clipping occurs are approximately $+14.3$ V and -13.8 V for a light load ($R_L \geq 10$ kΩ). Thus, the clipping levels are slightly unsymmetrical. If the load is increased, the magnitudes of the clipping levels are decreased. During the time the op-amp is clipping the output signal, the input voltage $v_a(t)$ is not held to approximately 0 V, and the models shown in Fig. 7.67 are not valid. Figure 7.82 shows a model that accounts for clipping levels, incorporates the one-pole open-loop gain model, and includes input and output resistances. This model is suitable for use in computer simulations of op-amp circuits. Input and output resistances are discussed in a later section.

The output current is also limited. The output current for a μA741 op-amp is limited to about 25 mA by internal protection circuitry. Thus, the current demanded from the output of this op-amp should be kept below 25 mA to avoid distortion of the output waveform. In fact, the output current demand should be kept small for best results because most parameters of an op-amp, such as gain-bandwidth product, degrade as output current demand is increased. Also, the op-amp output must supply current to the feedback network, and this current is included as part of the output current demand. For example, the op-amp in the circuit shown in Fig. 7.83 supplies current to both 1-kΩ resistors. This op-amp will current limit if $|v_{O1}| \geq 12.5$ V.

If more current is needed, a power op-amp or a power buffer can be employed.

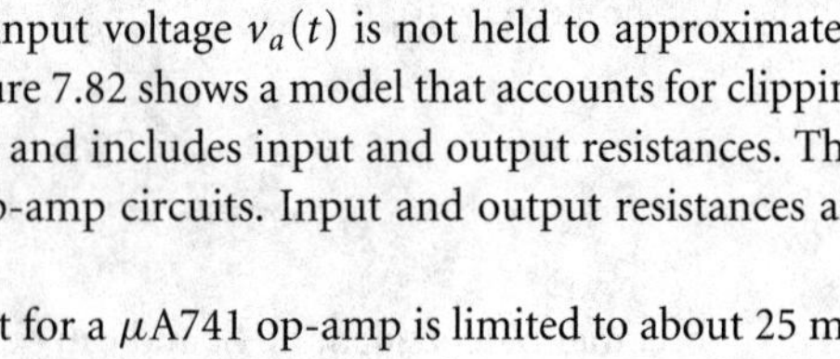

FIGURE 7.83 Op-amp 1 supplies current from its output terminal to both 1-kΩ resistors shown.

Slew Rate

For large signal excitations, a nonlinear phenomenon called **slew rate** (*SR*) limiting imposes an upper limit on how fast the output voltage can change with time. If the input voltage signal is a sine wave given by $V \sin(\omega t)$, and if the gain of the amplifier is G, then the output voltage becomes slew rate limited if

$$\frac{d[GV\sin(\omega t)]}{dt}\bigg/_{\max} > SR \qquad (7.116)$$

Since the maximum slope of the output voltage occurs when the sine wave goes through zero, we have slew rate limiting for

$$GV\omega > SR$$

Thus, if the product of the output voltage amplitude, GV, and frequency ω is too large, then the output waveform becomes distorted as shown in Fig. 7.84. To avoid slew rate distortion for a given op-amp, one must reduce either the amplitude or the frequency of the sinusoidal input waveform so that the product is less than the *SR* of the op-amp.

Slew rate limiting begins abruptly as the amplitude and frequency of the output voltage are increased. The cause of slew rate limiting is illustrated by the circuit diagram shown in Fig. 7.85. This diagram represents an op-amp as a three-stage amplifier. The compensation capacitor, which sets the position of the dominant pole, is represented by C_c. The first stage is a transconductance amplifier and its maximum output current is limited to $I_{O1\max}$. Stage 2 is a voltage amplifier with a high-input resistance and low-output resistance, and stage 3 is a voltage amplifier with a gain approximately equal to one. For a μA741, $I_{O1\max}$ is approximately 18 μA and C_c is about 30 pF. If the input signal is large enough in frequency or amplitude to cause *SR* limiting, then I_{O1} is constant at $I_{O1\max}$. From $I = C \, dV/dt$, we obtain $SR = 0.6 \times 10^6$ V/s or 0.6 V/μs. The typical value for *SR* for a μA741 is given as 0.5 V/μs, a result that compares favorably with the outcome of this simple calculation. Op-amps are available having slew rates greater than hundreds of volts per microsecond.

The magnitude of the *SR* of the μA741 is approximately the same whether the output voltage is increasing or decreasing. Some op-amps slew faster in one direction than in the other. As a result, when the output

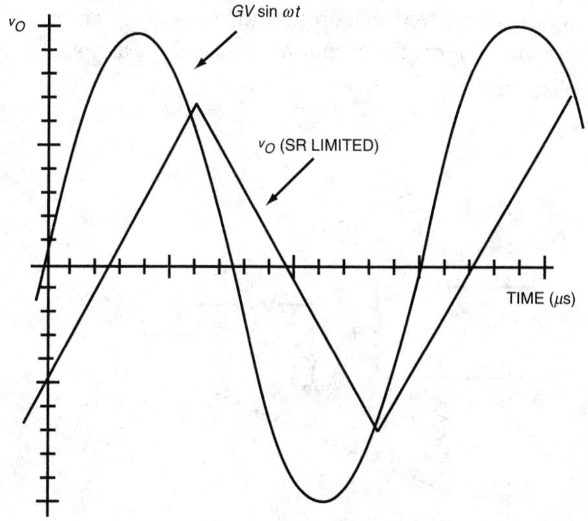

FIGURE 7.84   Desired output and slew rate limited output for an op-amp amplifier.

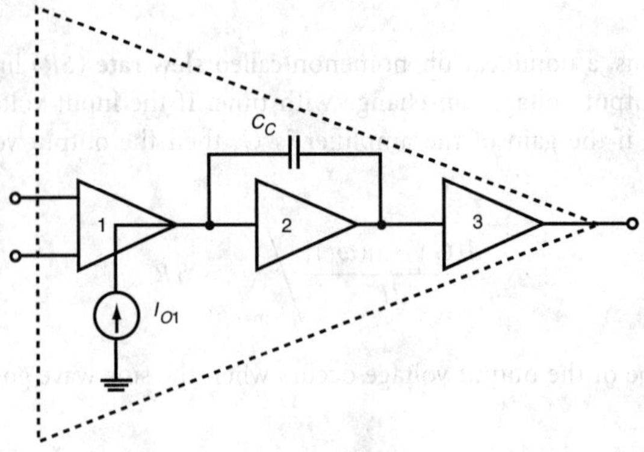

FIGURE 7.85 A three-stage representation for an op-amp that emphasizes the causes of slew rate limiting.

voltage is slew rate limited, a DC offset voltage is created, and the magnitude of this DC voltage increases as the input voltage amplitude and frequency are increased.

Another specification for op-amps that pertains to SR is **full power bandwidth**. Suppose the sinusoidal output voltage of an op-amp amplifier is to be, say, 10-V peak. How high can the frequency ω be increased while avoiding distortion due to SR? This question is answered by the calculation: $10\omega \leq SR$. If SR is 0.5 V/μs, then $\omega = 2\pi f \leq 0.05$ rad/μs or $f \leq 8$ kHz.

Input and Output Impedances

An ideal op-amp has infinite-input and zero-output impedances. Figure 7.86 shows a model for an op-amp, which accounts for the finite- (but large) input resistances and the low- (but not zero) output resistance of a real op-amp. For a μA741, the differential input resistance R_{id} is about 2 MΩ, the common-mode input resistance R_{iCM} is 100 MΩ, the input capacitance C_i is 4–7 pF (including the capacitance of the op-amp package), and the output resistance R_o is about 75 Ω. There are other parasitic capacitances such as capacitance between input and output that are not shown, but for many applications, C_i is the most significant capacitance.

Although the input resistances are in the megohm range and the output resistance is 75 Ω, it appears that the input and output resistances depart considerably from the ideal values of infinity and zero, respectively. However, negative feedback around the op-amp in an op-amp voltage amplifier increases the input resistance and reduces the output resistance. For example, let us determine the output

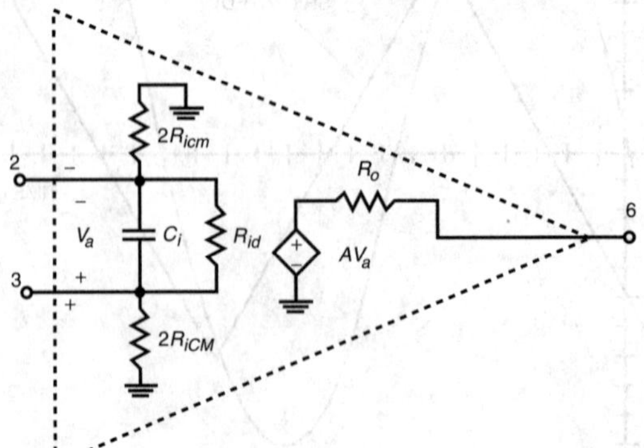

FIGURE 7.86 A model for the op-amp which accounts for input impedances and output resistance.

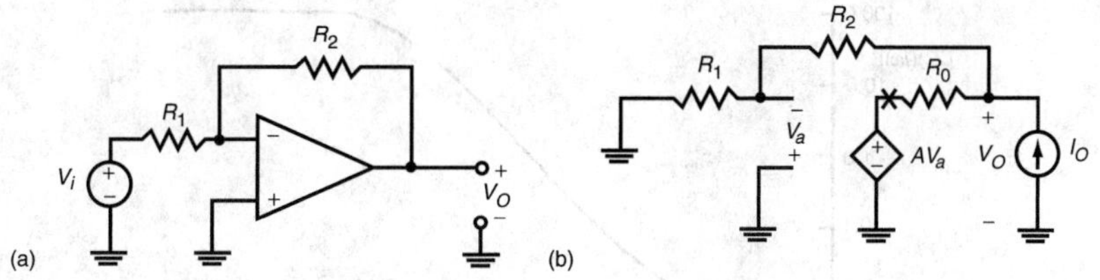

FIGURE 7.87 (a) Inverting amplifier whose output impedance Z_o is to be found. (b) Amplifier schematic prepared for analysis.

impedance of the inverting amplifier shown in Fig. 7.87(a). Figure 7.87(b) shows the resulting schematic after:

- The model in Fig. 7.86 has been substituted into Fig. 7.87(a). The resistances and C_i at the input of the model have been neglected since they are in parallel with R_1.
- The independent source v_i has been set to 0 V.
- An independent source $I_O(s)$ where s is the Laplace variable has been applied to the output port.

The output impedance is given by $Z_O = V_O/I_O$, and the result is

$$Z_O = \frac{R_o(R_1 + R_2)}{R_o + R_1 + R_2} \left[\frac{1}{1 + \dfrac{AR_1}{R_o + R_1 + R_2}} \right] \tag{7.117}$$

Thus, the output impedance is given by $R_o \,\|\, (R_1 + R_2) = R_{pas}$ modified by the terms in the braces. Note that R_{pas} is merely the output resistance of the passive network that is obtained when A is set to zero. We also note that Z_O approaches zero as GB of the op-amp approaches infinity and that for noninfinite GB, Z_O is a function of frequency since A is given by Eq. (7.108).

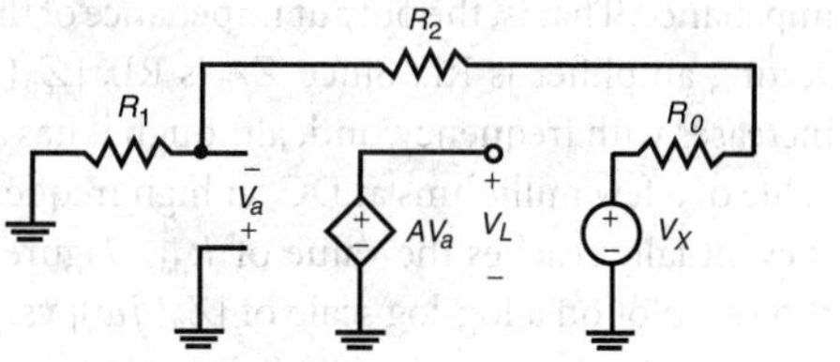

The output impedance can be expressed in terms of loop gain. If the circuit in Fig. 7.87(b) is broken at point X, the current source I_O removed, and the

FIGURE 7.88 Finding the loop gain of the inverting amplifier shown in Fig. 7.87.

circuit excited with a voltage source V_X, we obtain the circuit shown in Fig. 7.88. The output of the circuit is V_L and the loop gain is V_L/V_X. The loop gain is found as

$$\frac{V_L}{V_X} = \text{loop gain} = \frac{-AR_1}{R_o + R_1 + R_2} \tag{7.118}$$

Thus

$$Z_O = \frac{R_{pas}}{1 - \text{loop gain}} \tag{7.119}$$

If $R_o = 75\ \Omega$, $R_1 = 1\ \text{k}\Omega$, $R_2 = 10\ \text{k}\Omega$, $A_O = 200{,}000$, and $\omega_a = 20$ rad/s, then substituting the expression for A from Eq. (7.108) into Eq. (7.117) and setting $s = 0$ yields the output resistance at DC as

$$Z_O\big|_{s=0} \approx 4.13\ \text{m}\Omega$$

This value is very near the ideal value of zero.

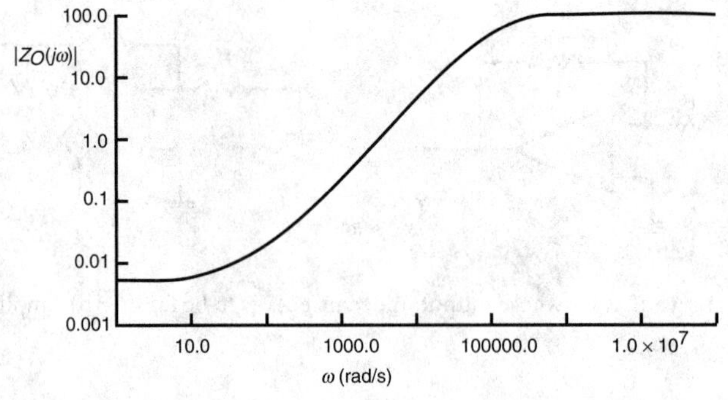

FIGURE 7.89 Plot of $|Z_o(j\omega)|$ of an inverting amplifier vs. ω on a log-log scale.

Z_O is a function of frequency. Again, substituting the expression for A from Eq. (7.108) into Eq. (7.117) and neglecting ω_a in comparison with $A_o\omega_a R_1/(R_o + R_1 + R_2)$, we obtain

$$Z_O \approx R_{\text{pas}} \left[\frac{s + \omega_a}{s + \dfrac{A_o\omega_a R_1}{R_o + R_1 + R_2}} \right]$$

We note that $Z_O(s)$ has one zero and one pole, that both the zero and the pole are located on the negative real axis of the s plane, and that the zero is closer to the origin than the pole for reasonable values of R_1 and R_2. Thus $Z_O(s)$ is an RL driving point impedance. That is, the output impedance of the inverting amplifier is RL. Since Z_O is RL, $|Z_o(j\omega)|$ increases with frequency, and, although it has a low value of a few milliohms at DC, at high frequencies it eventually reaches the value of R_{pas}. Figure 7.89 shows a plot on a log-log scale of $|Z_o(j\omega)|$ vs. ω.

Next, we examine the input impedance of the noninverting amplifier shown as circuit 1 in Table 7.3. Ideally, this impedance should be infinite. Figure 7.90 shows the noninverting amplifier with the model shown in Fig. 7.86 substituted for the op-amp. The upper resistor $2R_{i\text{CM}}$ in Fig. 7.90 is ignored since it is parallel with R_1 and $R_1 \ll 2R_{i\text{CM}}$. The lower resistor $2R_{i\text{CM}}$ is temporarily removed because it is in parallel with the impedance of the remainder of the circuit. Let the impedance of the remainder of the circuit be denoted as Z' so that the input impedance of the noninverting amplifier is $Z_{\text{IN}} = (2R_{i\text{CM}}) \| Z'$. Figure 7.91 shows the circuit used to determine $Z' = V_i/I_i$. The results are

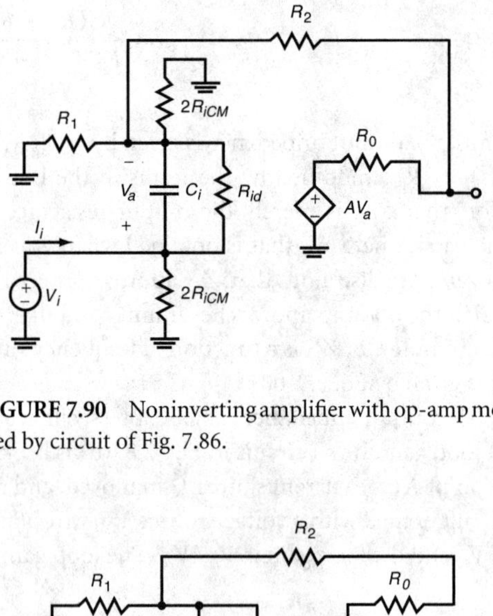

FIGURE 7.90 Noninverting amplifier with op-amp modeled by circuit of Fig. 7.86.

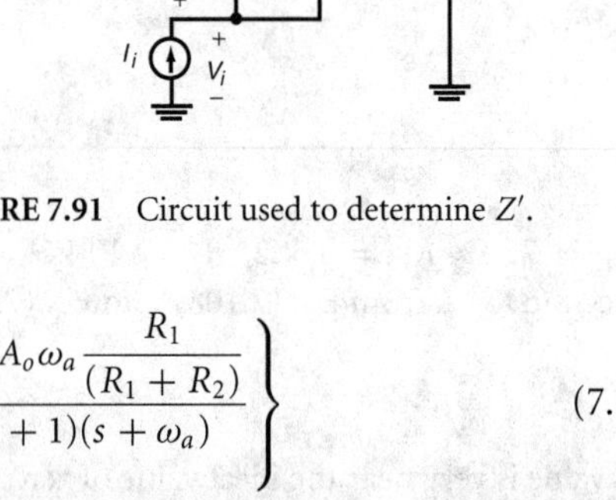

FIGURE 7.91 Circuit used to determine Z'.

$$Z' = R_1 \| R_2 + R_{id} \left\{ \frac{s + \omega_a + A_o\omega_a \dfrac{R_1}{(R_1 + R_2)}}{(sC_i R_{id} + 1)(s + \omega_a)} \right\} \tag{7.120}$$

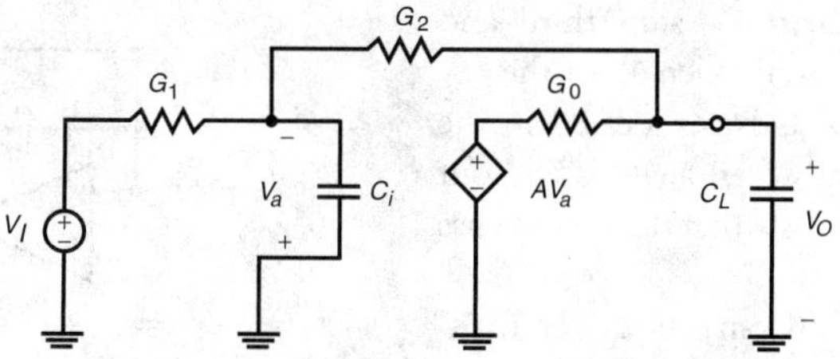

FIGURE 7.92 Inverting amplifier with the op-amp modeled by the essential components in the model in Fig. 7.86.

If Eq. (7.120) is rewritten with a common denominator, it is noticed that the poles and zeros of the results are located on the negative real axis of the s plane, and the lowest critical frequency is a pole (at $s = -\omega_a$). But the poles and zeros may or may not be interlaced depending on the values chosen. Thus, Z' may be an RC or an RLC driving point impedance. Nevertheless, $|Z'|$ falls off more or less uniformly until it levels out at high frequency. At DC, Z' is given by

$$Z' \mid_{s=0} = R_1 \| R_2 + R_{id}\left(1 + \frac{A_o R_1}{R_1 + R_2}\right) \approx R_{id}\left(\frac{A_o R_1}{R_1 + R_2}\right) \tag{7.121}$$

If $A_o = 200,000$, $R_1 = 1$ kΩ, $R_2 = 10$ kΩ, and $R_{id} = 2$ MΩ, then $Z' = 3.6 \times 10^{10}$ Ω for $s = 0$. Thus, $Z_{IN} = 2R_{iCM}$ at DC. Although Z_{IN} is about 200 MΩ at low frequencies, its magnitude decreases with increasing frequency toward a value of $R_1 \| R_2$.

The model in Fig. 7.86 can explain an instability that sometimes occurs when an op-amp amplifier drives a capacitive load. Figure 7.92 shows the inverting amplifier with the op-amp replaced by the essential components of the model in Fig. 7.86. For convenience, the resistors of the circuit have been replaced by their corresponding conductances. Thus, $G_1 = 1/R_1$, $G_o = 1/R_o$, etc. We have neglected the resistors ($2R_{iCM}$) as well as R_{id} because they are large. However, we have included C_i. C_L at the output of the circuit represents a capacitive load. The transfer function of the circuit is given by

$$\frac{V_O}{V_i} = \frac{-G_1(G_o - s\tau G_2)}{s^3\tau C_i C_L + s^2\tau(C_L(G_1 + G_2) + C_i(G_o + G_2)) + s\tau(G_o(G_1 + G_2) + G_1 G_2) + G_o G_2} \tag{7.122}$$

where $\tau = 1/GB$. We have employed the simplified expression for A given in Eq. (7.109) for this analysis. Note that if $\tau = 0$, then Eq. (7.122) reduces to

$$\frac{V_O}{V_i} = -\frac{G_1}{G_2} = -\frac{R_2}{R_1}$$

as we would expect for an ideal op-amp.

The characteristic polynomial of the amplifier, given by the denominator of Eq. (7.122), is third order, and it is not difficult to find a set of values of the components and parameters of the amplifier that causes two roots of the denominator to be in the right-half s plane. That is, sets of reasonable component and parameter values can be found that make the circuit unstable. The instability in this circuit arises from a combination of nonzero τ and nonzero C_i, R_o, and C_L. We have already shown that when $\tau = 0$, the transfer function reduces to a constant, which has no poles. Setting the other components to zero one at a time reduces the characteristic polynomial to one of second order in which all of the coefficients are positive. For example, if $R_o = 0$ ($G_o =$ infinity), then the transfer function reduces to

$$\frac{V_O}{V_i} = \frac{-G_1}{s^2\tau C_i + s\tau(G_1 + G_2) + G_2}$$

A second-order polynomial (unlike some third- and higher-order polynomials) in which all of the coefficients are positive has its roots located in the open left-half s plane, and so an analysis without R_o would indicate erroneously that the amplifier is always stable.

If a load is capacitive, how can it be driven without risking amplifier instability? Figure 7.93 shows a commonly used solution to this problem (Franco, 1988). The inverting amplifier has been modified in two ways. First, a resistor R, which is approxim-

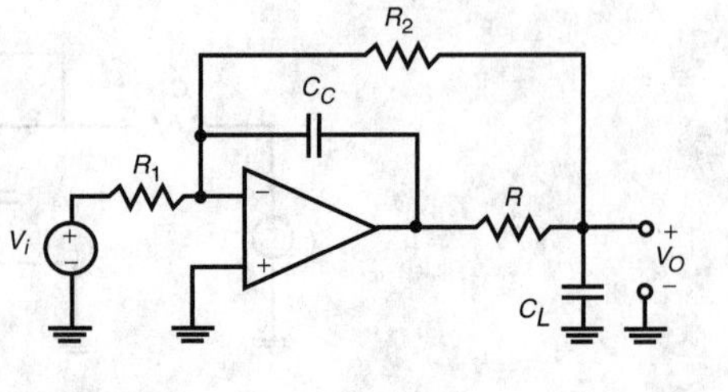

FIGURE 7.93   Driving a capacitive load with a modified inverting amplifier.

ately the same value as R_o, is connected to the output of the op-amp, and, second, a small capacitor C_c is connected around the op-amp as shown. The value of this capacitor is determined by $C_c = (R_1/R_2)C_i$.

Common-Mode Rejection Ratio

Common-mode rejection ratio (CMRR) and its companion, common-mode gain, are concepts developed for difference amplifiers. They are important in the design of amplifiers for sensors and transducers since, in many such applications, it is necessary to determine the difference of two voltages. Consider the generic difference amplifier represented in Fig. 7.94. This amplifier consists of one or more op-amps and associated external resistors. Circuit number three in Table 7.3 is an example of

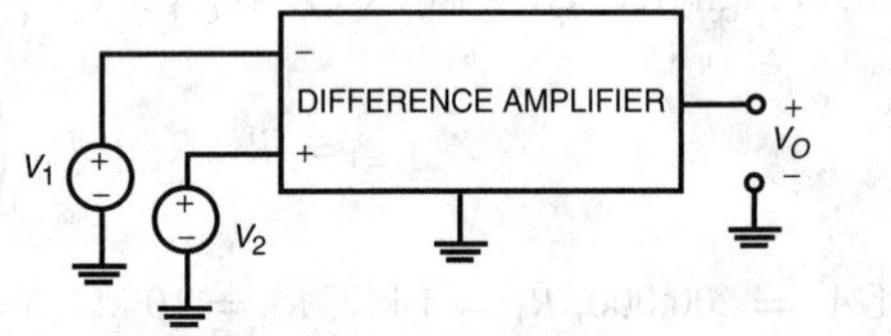

FIGURE 7.94   Generic difference amplifier with arbitrary excitations V_1 and V_2.

a difference amplifier that uses one op-amp and four external resistors.

The input voltages V_1 and V_2 are assumed to be of reasonable amplitude and frequency, but otherwise they are arbitrary. These arbitrary signals can be expressed in terms of common-mode and difference-mode excitations. The arithmetic average of the two excitations is the common-mode voltage. That is

$$V_{CM} = \frac{V_1 + V_2}{2} \tag{7.123}$$

The difference of the two arbitrary excitations is the difference-mode voltage V_D. That is

$$V_D = V_1 - V_2 \tag{7.124}$$

Thus, the excitations in Fig. 7.94 can be written as

$$V_1 = \frac{V_D}{2} + V_{CM} \qquad V_2 = -\frac{V_D}{2} + V_{CM} \tag{7.125}$$

and Fig. 7.94 can be redrawn as shown in Fig. 7.95. If $V_D = 0$, the output voltage of the difference amplifier V_O should be zero. But due to the amplifier's common-mode gain G_{CM}, there is an output voltage that depends on V_{CM}. Ideally, G_{CM} should be zero, but mismatches among the external components as well as the common-mode gain of the op-amps contribute to nonzero G_{CM}.

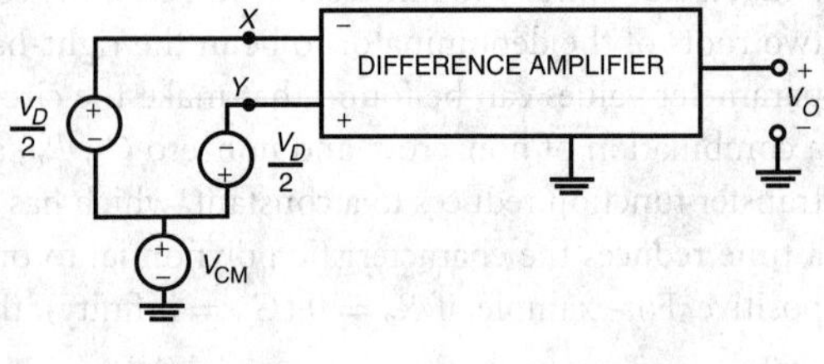

Let the voltage gain from port X (terminal X to ground in Fig. 7.95) to the output port be $-G_X$, and let the gain from port Y to the output port be

FIGURE 7.95   Difference amplifier with excitations expressed as difference-mode and common-mode voltages.

G_Y. Then we can express V_O as

$$V_O = -V_D \left(\frac{G_X + G_Y}{2} \right) + V_{CM}(G_Y - G_X) \tag{7.126}$$

Nonzero common-mode gain is caused by the two gains being unequal. If $G_X = G_Y = G$, then $V_O = -GV_D$, and there is no common-mode contribution to the output voltage. The term $(G_Y - G_X) = G_{CM}$, and $(G_X + G_Y)/2$ is the **difference-mode gain** of the amplifier. The ratio of the difference-mode gain to the common-mode gain is the common-mode rejection ratio. Thus, CMRR can be expressed as

$$\text{CMRR} = \frac{G_X + G_Y}{2(G_Y - G_X)}$$

and ideally, this ratio should be infinity. CMRR is a useful parameter because it permits us to rewrite Eq. (7.126) in the form

$$V_O = \frac{G_X + G_Y}{2} \left(-V_D + \frac{V_{CM}}{\text{CMRR}} \right)$$

which emphasizes the error caused by V_{CM} and noninfinite CMRR.

An op-amp is also a difference amplifier although the box representing a difference amplifier in Fig. 7.94 cannot be replaced directly by an op-amp because there would be no negative feedback around the op-amp. Thus, the op-amp would not operate properly. Figure 7.96 shows a specific difference amplifier based on a single op-amp. We assume the resistors R_1, R_2, R_a, and R_b are identical. Thus, error in the output voltage is caused by the op-amp alone. For the purpose of developing a *general* model that accounts for the op-amp's common-mode gain, we assume the op-amp has no imperfections except for common-mode gain and noninfinite open-loop

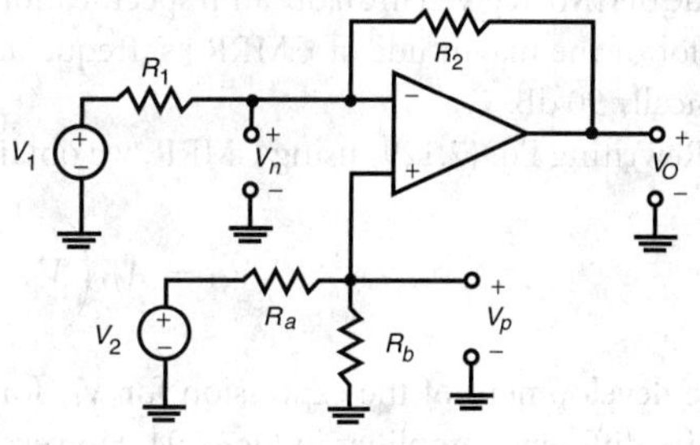

FIGURE 7.96 A specific difference amplifier.

gain. The voltages at the inverting and noninverting input leads of the op-amp are labeled V_n and V_p, respectively. Let the gain of the op-amp from the inverting input **port** to the output port be A_1, and let A_2 be the gain from the noninverting input port to the output port. The op-amp exhibits nonzero common-mode gain because the magnitudes of A_1 and A_2 are not exactly equal, the phase of A_1 is not exactly π, and the phase of A_2 is not zero.

The output voltage of the op-amp can be expressed as

$$V_O = V_n A_1 + V_p A_2 \tag{7.127}$$

If $A_1 = -A_2 = -A$, then $V_O = (V_p - V_n)A_2 = V_a A$. That is, the results correspond to the op-amp model shown in Fig. 7.76.

The common-mode voltage applied to the op-amp in Fig. 7.96 is

$$V_{CM} = \frac{V_n + V_p}{2} \tag{7.128}$$

and the difference-mode voltage is $V_D = V_p - V_n$. Then Eq. (7.127) can be rewritten as

$$V_O = V_D \left(\frac{A_2 - A_1}{2} \right) + V_{CM}(A_1 + A_2) \tag{7.129}$$

The difference-mode gain is

$$A_D = \frac{V_O}{V_D}\bigg/_{V_{CM}=0} = \frac{A_2 - A_1}{2} \tag{7.130}$$

and the common-mode gain is

$$A_c = \frac{V_O}{V_{CM}}\bigg/_{V_D=0} = A_1 + A_2 \tag{7.131}$$

Common-mode rejection ratio for the op-amp is given by

$$\text{CMRR} = \frac{A_D(s)}{A_c(s)} \tag{7.132}$$

Often, CMRR is expressed in decibels as

$$\text{CMRR}_{dB} = 20 \log \left| \frac{A_D(j\omega)}{A_c(j\omega)} \right| \tag{7.133}$$

CMRR for an op-amp is a function of frequency, and its magnitude decreases as frequency increases. The value of CMRR given in an op-amp specification sheet is usually the low-frequency (large) value, although a plot of the magnitude of CMRR vs. frequency is sometimes provided. CMRR for a μA741 is listed as typically 90 dB.

Rewriting Eq. (7.129) using CMRR, we obtain

$$V_O = A_D \left(V_p - V_n + \frac{V_n + V_p}{2} \frac{1}{\text{CMRR}} \right) \tag{7.134}$$

The development of the expression for V_O for the op-amp has paralleled the derivation of V_O for the entire difference amplifier in Fig. 7.94. However, by means of the negative feedback around the op-amp, the voltages V_n and V_p are nearly equal. Thus, we approximate $(V_n + V_p)/2$ as V_p. We also approximate A_D as the open-loop gain $A(s)$ given in Eq. (7.107). With these approximations, Eq. (7.134) becomes

$$V_O \approx A \left[V_p \left(1 + \frac{1}{\text{CMRR}} \right) - V_n \right] \tag{7.135}$$

Equation (7.135) indicates that to account for noninfinite CMRR, we increase the input signal at the noninverting input lead of the op-amp by V_p/CMRR as shown in Fig. 7.97.

As an example of the application of this model, we analyze the difference amplifier in Fig. 7.96 with $V_1 = 5.000$ and $V_2 = 5.002$ V DC. Let us assume the resistors are equal and find V_O for infinite CMRR and for CMRR = 1000. Let us also assume $A(s)$ is infinite and that op-amp offsets are zero. For infinite

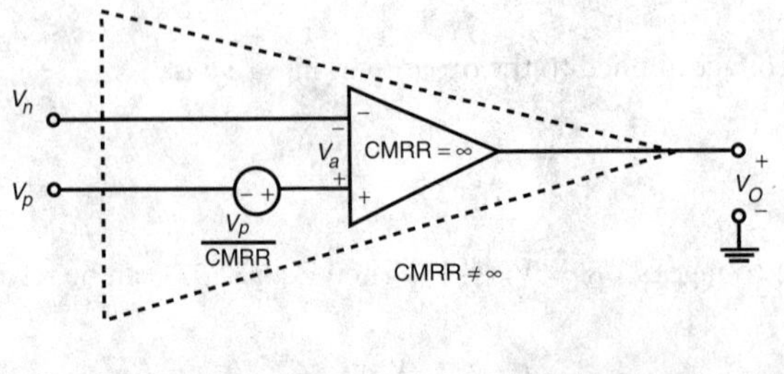

FIGURE 7.97   Model for op-amp which accounts for noninfinite CMRR.

CMRR, we immediately have

$$V_O = V_1\left(-\frac{R_2}{R_1}\right) + V_2\frac{R_b}{R_a + R_b}\left(1 + \frac{R_2}{R_1}\right)$$

If the resistors are equal, $V_O = 2$ mV, and the difference amplifier is able to discern a 2-mV difference in the two input signals. However, if CMRR $= 1000$, using Fig. 7.97 and Eq. (7.135), we have

$$V_O = -5.000 + \frac{5.002}{2}\left(1 \pm \frac{1}{\text{CMRR}}\right)2 = (2 \pm 5) \text{ mV DC} \tag{7.136}$$

Thus, the desired output voltage is obscured by the error introduced by the low CMRR of the op-amp. Note that the sign of the CMRR term in Eq. (7.136) is not really known, and so a $\pm$ is used.

Suppose the input voltages to the difference amplifier in Fig. 7.96 are 100 and 102 mV DC. Again, we take CMRR to be 1000. In this case, we have $V_O = (2 \pm 0.1)$ mV DC. Thus, the output voltage is very nearly the desired value with only $\pm 5\%$ error. The reason for this improvement is that the common-mode excitation has been reduced from about 5 V to 100 mV.

There is one other consideration that must be kept in mind when contending with common-mode gain. That is, the specification sheets for most op-amps specify a *range* of V_{CM} over which the published CMRR specifications apply. For a μA741, this range is ± 12 V when the op-amp is powered by ± 15-V power supplies. If the input voltage is outside this range, the CMRR specification does not apply.

Noise

An op-amp generates noise currents and voltages, and the resulting noise at the output of an op-amp amplifier degrades and may entirely obscure a useful output signal. In this section, we briefly review some noise concepts and provide a noise model for the op-amp.

Noise is a random phenomenon; its instantaneous value cannot be known. Thus, statistical concepts are employed to determine the effects of noise on a circuit. If $e_n(t)$ is a noise voltage, then its root-mean-square (rms) value E_n is given by

$$E_n = \left[\frac{1}{T}\int_0^T e_n^2(t)dt\right]^{\frac{1}{2}} \tag{7.137}$$

where T is a specified time interval over which the average of $e_n^2(t)$ is taken. E_n^2, denoted as the mean square value of the noise, represents the average power a 1-Ω resistor dissipates when $e_n(t)$ is applied across it. If E_n is known and if its amplitude distribution is Gaussian, then it is common engineering practice to calculate the peak-to-peak value of noise as $6E_n$. That is, the **crest factor** of Gaussian noise is considered to be 3 where crest factor is defined as the ratio of the peak value of the noise to its rms value.

If we have two noise sources $e_{n1}(t)$ and $e_{n2}(t)$, to determine the rms value of the sum, we have

$$E_n = \left[\frac{1}{T}\int_0^T [e_{n1}(t) + e_{n2}(t)]^2 dt\right]^{\frac{1}{2}}$$
$$= \left[E_{n1}^2 + E_{n2}^2 + \frac{2}{T}\int_0^T [e_{n1}(t)e_{n2}(t)]dt\right]^{\frac{1}{2}} \tag{7.138}$$

The average of the product of two uncorrelated noise sources is zero. Thus, if $e_{n1}(t)$ and $e_{n2}(t)$ are uncorrelated, then

$$E_n = \left(E_{n1}^2 + E_{n2}^2\right)^{\frac{1}{2}} \tag{7.139}$$

Since noise power E_n^2 is spread over all parts of the spectrum, we must specify the frequency band over which we are performing measurements or making calculations. In general, the noise power depends on the width of the frequency band specified and on the location of the band within the frequency spectrum. A useful concept here is **noise power density**, which is the rate of change of noise power with frequency. It is denoted as $e_n^2(f)$, its units are volts squared per hertz, and it is defined as

$$e_n^2(f) = \frac{dE_n^2}{df} \qquad (7.140)$$

Noise power density represents the average noise power over a 1-Hz bandwidth as a function of frequency. A companion quantity is *noise spectral density* denoted by $e_n(f)$ with units given by volts/$\sqrt{\text{hertz}}$.

If the noise power density is known over a specified frequency band, then the rms value of the noise can be obtained by integrating Eq. (7.140). Thus

$$E_n = \left[\int_{f_L}^{f_U} e_n^2(f)df \right]^{\frac{1}{2}} \qquad (7.141)$$

where f_L and f_U are the lower and upper limits of the frequency band, respectively.

Two common varieties of noise power densities are white noise and $1/f$ noise. The spectral density of white noise is a constant e_{nW}, and its rms value, from Eq. (7.141), is

$$E_n = e_{nW}\sqrt{f_U - f_L} \qquad (7.142)$$

From Eq. (7.142) we see that the power of white noise, E_n^2, is proportional to the bandwidth no matter where this bandwidth is located in the frequency spectrum.

The noise power density of $1/f$ noise has the form $e_n^2(f) = K_v^2/f$ where K_v is a constant representing the extrapolated value of the noise spectral density at $f = 1$ Hz. The spectral density of $1/f$ noise is $K_v/\sqrt{f}$ indicating that on a log-log plot the spectral density has a slope of -0.5 decades/decade. The rms value of $1/f$ noise, obtained from Eq. (7.141), is

$$E_n = K_v \left[\ell n \frac{f_U}{f_L} \right]^{\frac{1}{2}} \qquad (7.143)$$

and the power of $1/f$ noise is given by

$$E_n^2 = K_v^2 \ell n \left(\frac{f_U}{f_L} \right) \qquad (7.144)$$

which indicates that the power is proportional to the log ratio of the band-edge frequencies independent of the band's placement in the frequency spectrum.

The noise sources in the noise model for an opamp are composed of a mixture of white and $1/f$ noise as shown for a voltage noise source in Fig. 7.98. At low frequencies, $1/f$ noise dominates, and at high frequencies, white noise dominates. The boundary is called the *corner frequency* and is denoted as f_{cv} in Fig. 7.98. A similar plot applies to the noise spectral density of current noise sources in the opamp model. The corner frequency for a noise current source is denoted as f_{ci}.

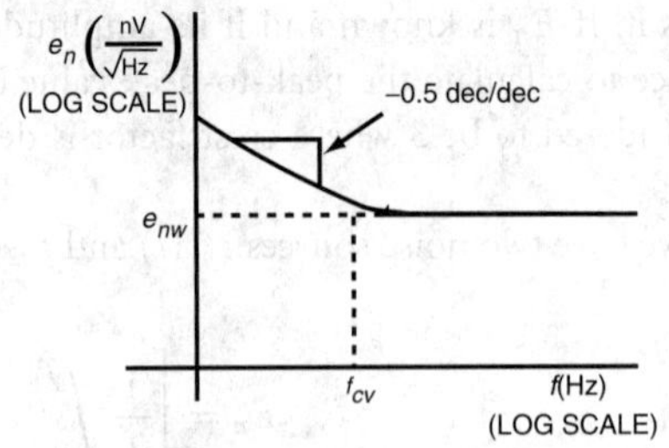

FIGURE 7.98 Spectral noise density plot of noise consisting of a mixture of white and $1/f$ noise.

The power density of the noise represented in Fig. 7.98 can be expressed analytically as

$$e_n^2 = e_{nW}^2 \left(\frac{f_{cv}}{f} + 1 \right) \qquad (7.145)$$

and the rms value corresponding to Eq. (7.145), from Eq. (7.141), is

$$E_n = e_{nW} \left[f_{cv} \ell_n \frac{f_U}{f_L} + f_U - f_L \right]^{\frac{1}{2}} \tag{7.146}$$

The corresponding rms value for a noise current source is

$$I_n = i_{nW} \left[f_{ci} \ell_n \frac{f_U}{f_L} + f_U - f_L \right]^{\frac{1}{2}}$$

For a μA741 op-amp, $e_{nW} \approx 20\,\mathrm{nV}/\sqrt{\mathrm{Hz}}$, $f_{cv} \approx 200\,\mathrm{Hz}$, $i_{nW} \approx 0.5\mathrm{pA}/\sqrt{\mathrm{Hz}}$, and $f_{ci} \approx 2\,\mathrm{kHz}$. As another point of comparison, a low-noise op-amp such as an OP-27F has typical noise specifications given by $e_{nW} = 3.0\,\mathrm{nV}/\sqrt{\mathrm{Hz}}$ and $i_{nW} = 0.5\,\mathrm{pA}/\sqrt{\mathrm{Hz}}$ with corner frequencies $f_{cv} = 2.7\,\mathrm{Hz}$ and $f_{ci} = 140\,\mathrm{Hz}$.

Figure 7.99 shows the op-amp noise model. A noisy op-amp is modeled by three noise generators and a noiseless op-amp. Polarities and reference directions are not shown since noise generators do not have true polarity or reference direction assignments. The spectral densities of the current sources are listed as i_{nn} and i_{np} in the model in order to keep track of the effects of these generators. Op-amp data sheets list information for or provide graphs for a single current spectral density i_n. At the end of our analysis, we set $i_{nn} = i_{np} = i_n$ and consider the sources to be uncorrelated.

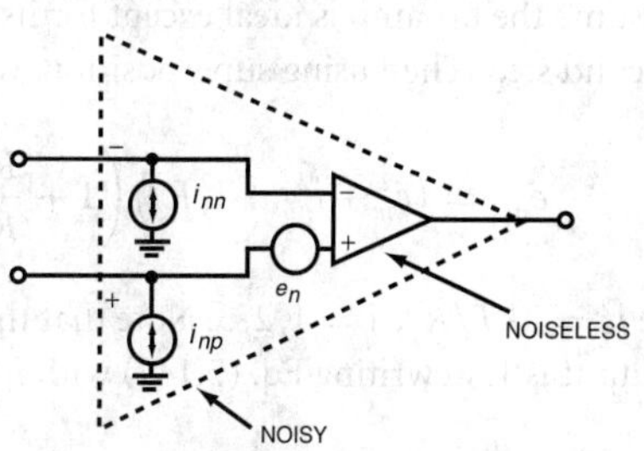

To analyze an op-amp circuit with resistive feedback, we also need noise models for resistors. Two

FIGURE 7.99 Noise model for an op-amp consisting of two current noise generators, a voltage noise generator, and a noiseless op-amp.

models for a noisy resistor are shown in Fig. 7.100. The thermal noise (Johnson noise) of the resistor is modeled by a voltage noise source with spectral density e_R as shown in Fig. 7.100(b). The power density of this source is $e_R^2 = 4kTR$ V^2/Hz where R is in ohms, k is Boltzmann's constant $= 1.38 \times 10^{-23}$ J/K, and T is temperature in degrees Kelvin. Thus, thermal noise is white. At $25°$C, the noise spectral density is approximately $e_R \approx 4\sqrt{R}$ nV/$\sqrt{\mathrm{Hz}}$ where R is expressed in kilohms. Figure 7.100(c) shows an equivalent resistor noise model that includes a current noise source having power density $i_R^2 = 4kT/R$ A^2/Hz.

As an example of the application of the op-amp noise model we find the noise power output of the circuit shown in Fig. 7.101. The steps followed to perform this analysis are

- Substitute noise models for the op-amp and resistors. This step yields six noise sources for this circuit.
- Reduce the number of sources where possible by combining power densities.
- If the circuit is linear and the noise sources are uncorrelated, then superposition can be used. Thus, the contribution to the output power density is found by multiplying each source power density

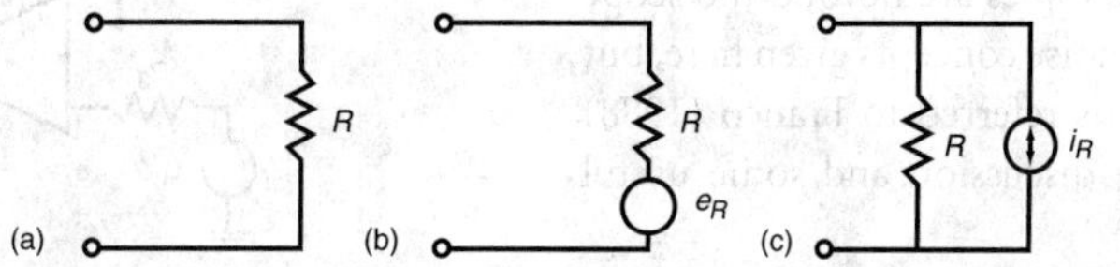

FIGURE 7.100 (a) The noisy resistor R is modeled by (b) a voltage noise source in series with a noiseless resistor or by (c) a current noise source in parallel with a noise-free resistor.

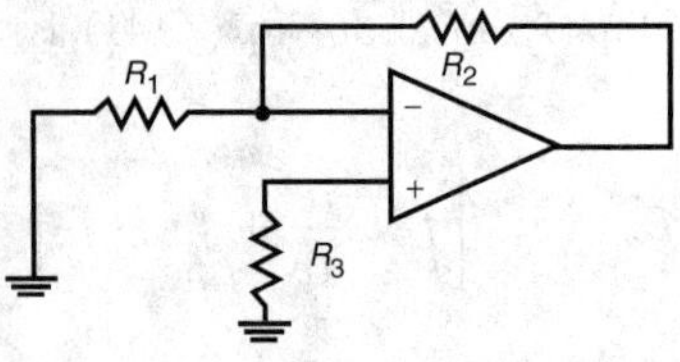

FIGURE 7.101   Example circuit for noise calculations.

by the magnitude squared of the transfer function from that source to the output. The total output voltage noise power density is obtained by adding all the contributions together.

- The rms value of the output noise can be found by integrating over a specified bandwidth and taking the square root of the result.

We assume the op-amp is ideal except for its noise sources. Figure 7.102 shows the results obtained after the second step. Then using superposition, we have

$$e_{nO}^2 = \left(e_n^2 + i_{np}^2 + i_3^2 R_3^2\right)\left(1 + \frac{R_2}{R_1}\right)^2 + \left(i_1^2 + i_{nn}^2\right)R_1^2\left(\frac{R_2}{R_1}\right)^2 + i_2^2 R_2^2(1)^2 \tag{7.147}$$

where $i_j^2 = 4kT/R_j$, $j = 1, 2, 3$. Note that the transfer function from the generator associated with R_2 to the output is 1. Rewriting Eq. (7.147) with $i_{nn} = i_{np} = i_n$, we have

$$e_{nO}^2 = \left(1 + \frac{R_2}{R_1}\right)^2 \left\{ \begin{array}{l} e_n^2 + i_n^2\left[R_3^2 + (R_1 \,\|\, R_2)^2\right] \\ +4kT[R_3 + (R_1 \,\|\, R_2)] \end{array} \right\} \tag{7.148}$$

Equation (7.148) has the form

$$e_{nO}^2 = |A_n|^2 \; e_{ni}^2 \tag{7.149}$$

where A_n is the noise gain of the circuit and e_{ni} is the equivalent input noise spectral density. This result can be represented by the schematic shown in Fig. 7.103 in which the op-amp and the resistors are noiseless.

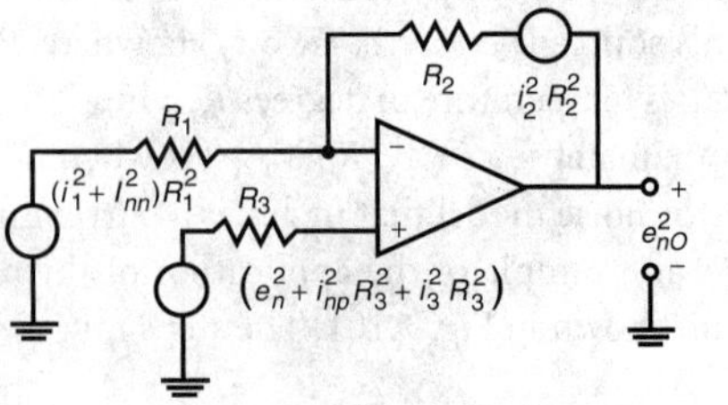

Using Eq. (7.145) for e_n^2 (and a corresponding expression for i_n^2), we can integrate e_{nO}^2 as in Eq. (7.141) to obtain the rms value of the output noise. An estimate of the peak-to-peak output noise is found by multiplying the results of Eq. (7.141) by 6.

Purely reactive elements do not generate noise, but they affect frequency characteristics of the noise gain of the circuit as does the open-loop gain of the op-amp. These topics are beyond the scope of the brief review of noise concepts given here, but the interested reader is referred to Franco (1988) for a more complete discussion and some useful examples.

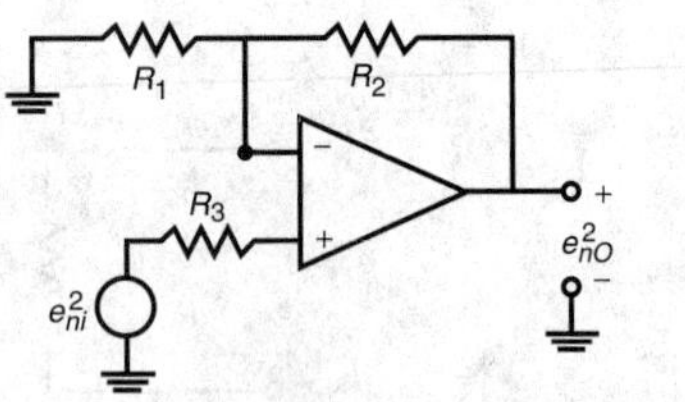

This completes the discussion of op-amp limitations. Not every op-amp departure from the ideal has been discussed, but the ones presented here are

FIGURE 7.102   Results after the second step: combining noise sources.

FIGURE 7.103   Equivalent representation for circuit shown in Fig. 7.102.

important in many applications. There is ample literature on op-amps, and the reader is urged to consult the listings in Further Information following Section 7.8.

Types and Values of Elements

In this section we discuss briefly the ranges of values and types of passive components that are normally used in discrete design. As a rule of thumb, resistors most often employed in op-amp circuits range between a few hundred ohms to several hundred kilohms. Resistors of value below a few hundred ohms may be too much of a load, and most op-amps are low-power devices whose parameters degrade as they are required to supply more current to load and feedback circuits. At the high end of the range, resistors larger in value than several hundred kilohms generate more noise and allow stray pickup signals to couple into the circuit more easily. This rule of thumb is probably broken as often as it is observed, but it serves notice that a circuit requiring resistors outside the range indicated should be examined carefully.

Carbon composition, carbon film, metal oxide film, and metal film resistors are the usual resistors employed at the design and prototyping stage. Wire-wound resistors should be avoided because of their inductive traits which can affect the frequency characteristics of circuits even at low frequencies. Resistors used in finished products must be flame retardant.

Mica and plastic film (polycarbonate, polyester, or polystyrene) are most often the types of capacitors used to determine the frequency characteristics or timing behavior of discrete op-amp circuits. Ceramic disc or electrolytic capacitors should generally not be used in these applications. Ceramic disc capacitors are sensitive to temperature and humidity, and electrolytic capacitors are sensitive to temperature and conduct small amounts of DC current. However, ceramic disc and electrolytic capacitors are used for power supply bypassing duties.

Values of capacitors used in discrete design range from tens of picofarads to a few microfarads. Capacitor values should not be so small that parasitic capacitances, which may range up to 10 pF or more, will have an appreciable effect.

Plastic film capacitors for electronic applications can be obtained with values up to about 10 μF, but these capacitors are bulky and expensive, and the use of capacitors this large in value should be avoided if possible. Film capacitors that range in value from nanofarads to microfarads perform nicely in most op-amp circuits.

Acknowledgments

The author wishes to express his gratitude to Dr. Jacek Zurada, Mr. Zbigniew J. Lata, Mr. Damon Miller, Mr. A. Wisen, Mr. Peichu (Peter) Sheng, and Mr. KeChang Wang for their help and encouragement.

Defining Terms

Bias Current (I_B): The arithmetic average of the currents that flow in the input leads of an op-amp.
Closed loop gain-bandwidth product: The product of the magnitude of low-frequency gain and the 3.01 dB bandwidth of an op-amp amplifier.
Common-mode gain: The gain given to a common-mode input signal by a difference amplifier. Ideally, this gain should be zero.
Common-mode rejection ratio: The ratio of the difference-mode gain to the common-mode gain. Ideally, this ratio should be infinite. This quantity is often expressed in decibels.
Common-mode signal: If two arbitrary signals v_1 and v_2 are applied to the inputs of a difference amplifier, then the common-mode signal is the arithmetic average of the two signals. That is, the common-mode signal is $(v_1 + v_2)/2$.
Crest factor: The ratio of the peak value of a signal to its rms value.
Difference-mode gain: The gain given to a difference-mode input signal by a difference amplifier.
Difference-mode signal: If two arbitrary signals v_1 and v_2 are applied to the inputs of a difference amplifier, then the difference-mode signal is the difference of the two signals.
Driving point function: The ratio of the transform of the time-domain response of a linear network to the transform of the excitation causing the response. Both the excitation and the response are at the same

port of the network. In determining a driving point function, all independent sources associated with the network are set to zero except the excitation. Dependent sources remain operating.

Dynamic range: The dynamic range of a filter is the ratio of the largest undistorted output signal amplitude and the noise floor.

Full power bandwidth: The frequency at which a sinusoidal signal having large amplitude begins to become distorted at the output of an op-amp due to the slew rate limitation. Full power bandwidth, ω_{FB} in rad/s, is related to SR by $\omega_{FB} = SR/V_{Omax}$, where V_{Omax} is the peak value of output voltage.

Gain-bandwidth product (GB): The product of the magnitude of op-amp gain at DC and its open loop bandwidth.

Input offset current (I_{OS}): The difference between the currents that flow in the input leads of an op-amp. Usually only the magnitude of this difference is known.

Input offset voltage (V_{OS}): The small DC voltage that must be applied at the input of a noninverting amplifier to drive the output voltage to zero with no load on the output. It is assumed that the feedback resistors are chosen so that no appreciable current is drawn from the output, and currents flowing in the input leads of the op-amp cause insignificant voltages.

Loop gain: If a feedback network is cut and a signal is applied at one side of the cut so as to maintain impedance levels, then the loop gain is the gain experienced by that signal as it traverses the circuit to the other side of the cut.

Noise power density: The rate of change of noise power with frequency. The units are volt squared per hertz or ampere squared per hertz.

Norator: An idealized two-terminal network element for which the voltage across it and the current through it are determined by the network to which it is connected.

Nullator: An idealized two-terminal network element that conducts no current and yet maintains zero volts across itself. This element is often used to represent a virtual connection.

One-pole model: An approximation for the open-loop gain of an op-amp. This approximation represents the frequency behavior of most general purpose op-amps fairly well and has only one pole.

Op-amp time constant (τ): The reciprocal of the gain-bandwidth product of the op-amp.

Open-loop gain [$A(s)$]: In the context of this chapter, this term refers to the gain of the op-amp itself without a feedback network.

Port: Two terminals of a network that are considered to be associated.

Rise time (t_r): The time required for the response of a network to go from 10% to 90% of its final value when the network is excited by a step.

Slew rate (SR): A nonlinear phenomenon in which the rate of change of output voltage in response to a large-amplitude step of input voltage is internally limited.

Small signal: A level of signal applied to an amplifier or filter so that the circuit retains linear behavior.

Virtual connection: A representation of the circuit between the input leads of an ideal op-amp. The voltage across and the current through a virtual connection are both zero. If one input lead of an ideal op-amp is connected to ground, the virtual connection is often termed a virtual ground.

References

Aronhime, P.B., 1995. Frequency domain methods. In *The Circuits and Filters Handbook*, ed. W.-K. Chen, Chap. 20, pp. 660–701. CRC Press, Boca Raton, FL.

Bruton, L.T. 1980. *RC-Active Circuits*. Prentice-Hall, Englewood Cliffs, NJ.

Budak, A. 1974. *Passive and Active Network Analysis and Synthesis*. Houghton Mifflin, Boston, MA; reissued Waveland Press, 1991, Prospect Heights, IL.

Electronic Design, 1992. Vol. 40, p. 96. Penton, Nov. 25.

Franco, S. 1988. *Design with Operational Amplifiers and Analog Integrated Circuits*. McGraw-Hill, New York.

Geiger, R., Allen, P., and Strader, N. 1990. *VLSI Design Techniques for Analog and Digital Circuits*, Chap. 6, pp. 476–485. McGraw-Hill, New York.

Further Information

For sources of more detailed information on the subject of operational amplifiers, see the Further Information section at the end of Section 7.8.

7.8 Applications of Operational Amplifiers

Peter Aronhime

7.8.1 Instrumentation Amplifiers

Instrumentation amplifiers are, generally, high-performance differential input voltage amplifiers that have high-gain accuracy, high-input resistance, low-output resistance, high common-mode rejection ratio (CMRR), low noise, and low offsets with low offset drift with temperature. Differential input current instrumentation amplifiers have low-input resistance and high-output resistance. Most applications of instrumentation amplifiers are based on amplifying the signal of a transducer to an amplitude that can be accurately converted to a digital signal without introducing significant errors. Thus, instrumentation amplifiers are used with strain gauge bridges, with thermocouples and other types of thermosensors, and with precision semiconductor gas sensors. In some applications such as medical instrumentation, an instrumentation amplifier may be required to operate with high common-mode signals. Gains of instrumentation amplifiers are often preset by internal, precision trimmed resistors.

Circuit three in Table 7.3 (see Section 7.7) is a simple example of an instrumentation amplifier that utilizes four resistors and a single op-amp. It has many drawbacks, not the least of which is that the input currents are generally different and depend on the input voltages. This defect could be remedied by using gain-of-one buffers (circuit two in Table 7.3) between the signals and the input leads of the difference amplifier, but this modification would require two more op-amps to fix one problem. Figure 7.104 shows a difference amplifier that employs two op-amps and four resistors (Wojslaw and Moustakas, 1986). The input resistances are, ideally, infinite. The output voltage of the circuit is given by

$$v_O = \left(1 + \frac{R_4}{R_3}\right)\left[v_2 - v_1\left(1 + \frac{R_2}{R_1}\right)\left(\frac{R_4}{R_3 + R_4}\right)\right]$$

(See Section 7.7 for definitions of parameters.) If we impose the requirement that $R_2/R_1 = R_3/R_4$, then the output voltage of the amplifier can be expressed as

$$v_O = \left(1 + \frac{R_1}{R_2}\right)(v_2 - v_1)$$

For a gain of 101, typical values for resistors are $R_1 = R_4 = 100 \text{ k}\Omega$ and $R_2 = R_3 = 1 \text{ k}\Omega$.

The most common three-op-amp configuration is shown in Fig. 7.105. The CMRR of this instrumentation amplifier can be quite large, over 110 dB. Minus and plus signs in parentheses at the inputs denote inverting and noninverting inputs of the amplifier. Op-amps (1) and (2) are best matched, and a Linear Technology LT1002 can be used for this application. To analyze the circuit, we let $R_{2A} + R_{2B} = R_2$, and

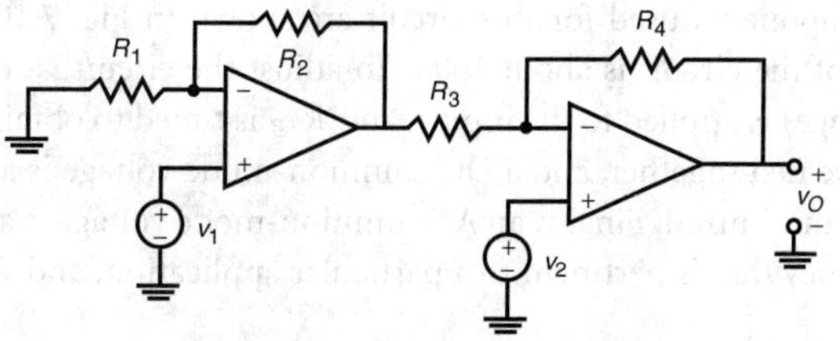

FIGURE 7.104 Difference amplifier which ideally draws no current from the input signal sources.

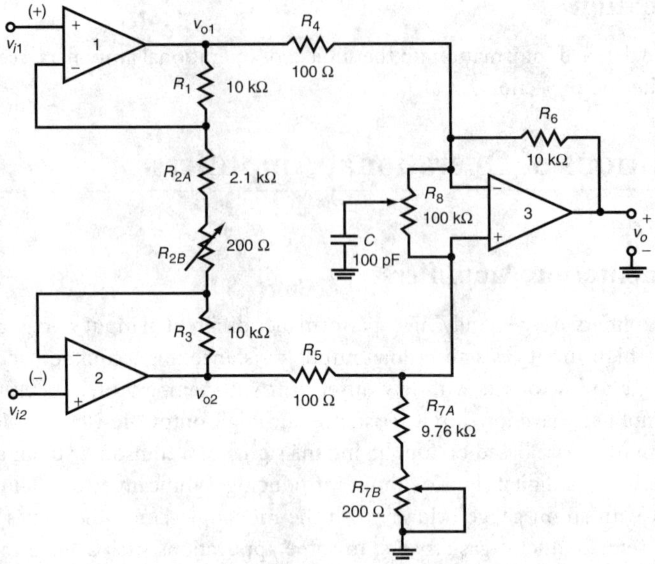

FIGURE 7.105 Instrumentation amplifier with high CMRR.

$R_{7A} + R_{7B} = R_7$, and ignore C and R_8. Also, we assume the op-amps are ideal. Using superposition and making liberal use of the virtual connections between the input leads of op-amps (1) and (2), we obtain

$$v_{o1} = v_{i1}\left(1 + \frac{R_1}{R_2}\right) + v_{i2}\left(-\frac{R_1}{R_2}\right) \tag{7.150}$$

and

$$v_{o2} = v_{i1}\left(-\frac{R_3}{R_2}\right) + v_{i2}\left(1 + \frac{R_3}{R_2}\right) \tag{7.151}$$

Op-amp (3) and its associated resistors R_5–R_7 are recognized as the simple difference amplifier from Table 7.3 (see Section 7.7). Here, it forms the difference between v_{o1} and v_{o2}. The result is

$$v_O = v_{o1}\left(-\frac{R_6}{R_4}\right) + v_{o2}\left(\frac{R_7}{R_5 + R_7}\right)\left(1 + \frac{R_6}{R_4}\right) \tag{7.152}$$

Normally, $R_4 = R_5$ and $R_6 = R_7$ so that we have

$$v_O = (v_{o2} - v_{o1})\frac{R_6}{R_4}$$

$$= (v_{i2} - v_{i1})\left(1 + \frac{R_1 + R_3}{R_2}\right)\frac{R_6}{R_4} \tag{7.153}$$

Typical values of the components used for this circuit are shown in Fig. 7.105. With these values, the differential voltage gain of the circuit is about 1000. To adjust the circuit, a low-frequency signal (with low-common-mode voltage) is applied to the inputs, and R_{2B} is tuned to obtain a gain of 1000. Then the inputs of the amplifier are tied together, and a DC common-mode voltage is applied. R_{7B} is adjusted so that the output voltage is minimized. Finally, an AC common-mode voltage is applied having a frequency of 10–20 kHz or a frequency that is pertinent to a particular application, and R_8 is adjusted to minimize the AC output voltage.

As an example of the application of an instrumentation amplifier we consider the bridge circuit shown in Fig. 7.106(a), which is formed by two 350-Ω strain gauges and two other 350-Ω precision resistors, each

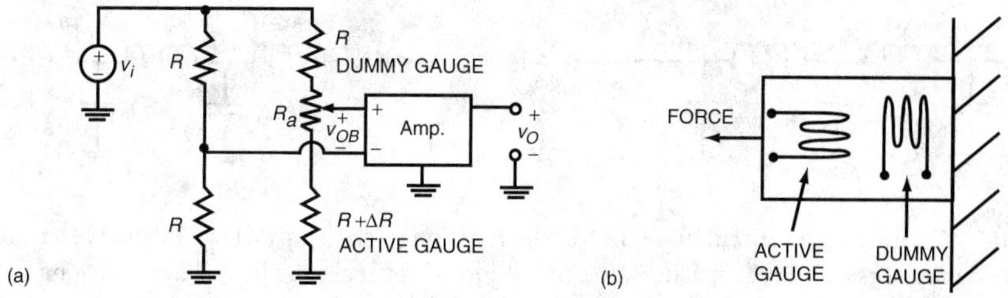

FIGURE 7.106 (a) Bridge circuit with instrumentation amplifier, (b) location of strain gauges on a bar which is subjected to uniaxial force F in the direction shown.

labeled R. A 2-Ω potentiometer R_a is included in the bridge between the gauges with the wiper connected to the noninverting input of the amplifier. One gauge is a dummy gauge, which is included to compensate for environmental factors such as temperature. It is mounted as shown in Fig. 7.106(b) so that it is not subjected to strain. The other strain gauge is termed the active gauge. An excitation v_i, which may be DC or AC, is applied, and the bridge is balanced using R_a with no force applied to the bar.

A uniaxial force is then applied to the bar. The output voltage of the bridge v_{OB} is amplified by an instrumentation amplifier with gain $G = 1000$. What is the strain measured in the bar if v_O of the amplifier equals v_i and the **gauge factor (GF)** of the strain gauge is 2.05? The resistance of the active gauge increases to $R + \Delta R$ when force is applied to the bar. We assume that R_a is zero and obtain

$$v_{OB} = v_i \left(\frac{R + \Delta R}{2R + \Delta R} - \frac{1}{2} \right) = v_i \left(\frac{\dfrac{\Delta R}{R}}{4 + 2\left(\dfrac{\Delta R}{R}\right)} \right)$$

For $\Delta R \ll 2R$,

$$v_{OB} \approx \frac{v_i}{4} \frac{\Delta R}{R}$$

Thus, $\Delta R / R = 0.004$. The gauge factor is defined as

$$\text{GF} = \frac{\dfrac{\Delta R}{R}}{\dfrac{\Delta L}{L}} \tag{7.154}$$

where L is the active length of the gauge, and $\Delta L / L$ is the strain ϵ. Since GF for these strain gauges is 2.05, we have $\epsilon = 0.002$ in/in.

In this application, the instrument amplifier is required to have high-common-mode rejection ratio because the common-mode voltage at its input is $v_i/2$. If the bridge excitation is DC, low-offset voltage is required because v_{OB} is about 1 mV. The large gain of the amplifier is an aid in initially balancing the bridge by adjusting R_a.

7.8.2 Active Filter Circuits

A primary incentive for the development of active filters is the elimination of inductors that are required if a filter is to be realized only with passive elements. Especially for low frequencies, the inductors required for filter realizations are large, heavy, expensive, and not standardized. Other advantages of active filters are that they are versatile, easily constructed, easily tuned, and inexpensive. Over the years several approaches to the design of active filters have been developed. Each approach has its advantages and disadvantages, and these will become evident for the design approaches presented in this section. The common theme for the

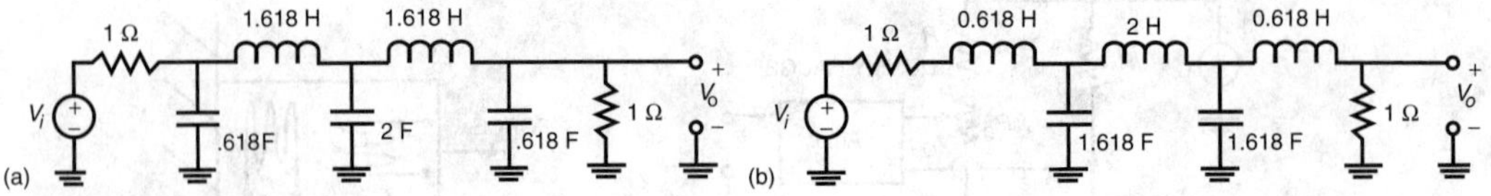

FIGURE 7.107 Passive circuit realizations of the transfer function given in Eq. (7.155): (a) double terminated circuit employing two inductors and three capacitors, (b) circuit formed by replacing the voltage source by a short circuit, forming the dual, and then reinserting the voltage source.

filters discussed in this section is that they are continuous time as opposed to digital or switched-capacitor filters. Continuous-time filters are often referred to as analog filters.

The **transfer function** $T(s)$ of a fifth-order low-pass Butterworth filter with (3.01-dB) cutoff frequency $\omega_c = 1$ rad/s is

$$T(s) = \frac{1}{s^5 + 3.2361s^4 + 5.2361s^3 + 5.2361s^2 + 3.2361s + 1} \qquad (7.155)$$

If the denominator of Eq. (7.155) is factored, we have

$$T(s) = \frac{1}{(s+1)(s^2 + 0.618s + 1)(s^2 + 1.618s + 1)} \qquad (7.156)$$

Figure 7.107 shows double-terminated LC ladder realizations of $T(s)$. One circuit can be obtained from the other by replacing the voltage excitation by a short circuit, forming the dual, and then inserting the voltage source using the pliers technique.

Cascade Design

The cascade design procedure begins with a transfer function in factored form such as is shown in Eq. (7.156) and realizes the transfer function as a product of second-order transfer functions and, if the order is odd, a first-order transfer function. These constituent transfer functions are termed *sections*. For the transfer function in Eq. (7.156), we have three sections so that $T = T_1 T_2 T_3$ where $T_1 = 1/(s+1)$, and T_2 and T_3 have the form $1/(s^2 + s(\omega_0/Q) + \omega_0^2) = 1/((s+\alpha)^2 + \beta^2)$. In the expressions for T_2 and T_3, ω_0 and Q are used to describe the locations of the poles in a polar manner, and α and β describe pole locations in a rectangular manner. The quantity ω_0 is termed the *undamped natural frequency* and is the radius of the circle centered at the origin of the s-plane on which the pair of complex poles lie. To have poles that are complex and in the open left-half s-plane, $1/2 <$ $Q < \infty$. A circuit representation of the product of transfer functions is shown in Fig. 7.108; the individual blocks representing T_j, $j = 1, 2, 3$, are said to be connected in cascade. It is assumed that there is no loading between the blocks in the figure. From a purely mathematical point of view, the order of the blocks in the figure does not make a difference. But from a practical view, the order of placement may be important. If the input signal to the filter is of

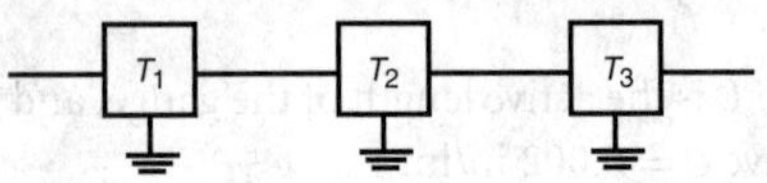

FIGURE 7.108 Cascade connection of circuits realizing transfer function T_j, $j = 1, 2, 3$. The overall transfer function of the circuit is $T = \Pi T_j$. It is assumed there is no loading between the blocks shown.

sufficient amplitude, then, usually, the sections are cascaded in the order of increasing Q in order to prevent clipping in the high Q stages. However, if the input signal is small, a higher Q stage may be placed at the beginning of the cascade in order to amplify the signal over some portion of its frequency range.

The issues in the cascade design of an active filter, once a desired overall transfer function has been determined, are the selection of appropriate circuits to realize each section, especially the second-order sections, and the determination of the values for the resistors and capacitors for each section. Factors that influence the selection of active **filter sections** are the sensitivity of the section to the gain-bandwidth product (GB) of the op-amp(s), the sensitivities to changes in values of the passive elements of the section, and the number of components required. In the following, we provide a representative listing and design information for active filter second-order sections.

Figure 7.109 shows a low-pass Rauch section (Nichols and Rauch, 1956). Its transfer function, assuming the op-amp is ideal, is given by

$$\frac{V_O}{V_i}(s) = \frac{-\left(\dfrac{1}{C_1 C_2 R_2 R_3}\right)}{s^2 + s\dfrac{1}{C_1}\left[\dfrac{1}{R_1} + \dfrac{1}{R_2} + \dfrac{1}{R_3}\right] + \dfrac{1}{C_1 C_2 R_1 R_2}} \tag{7.157}$$

For the Rauch section

$$\omega_0 = \frac{1}{\sqrt{C_1 C_2 R_1 R_2}} \qquad \frac{1}{Q} = \sqrt{\frac{C_2}{C_1}}\left[\sqrt{\frac{R_2}{R_1}} + \sqrt{\frac{R_1}{R_2}} + \frac{\sqrt{R_1 R_2}}{R_3}\right] \tag{7.158}$$

The sensitivities of ω_0 and Q to the passive elements, where the sensitivity of Q and of ω_0 to element x are defined as

$$\mathop{S}_{x}^{Q} = \frac{x}{Q}\frac{\delta Q}{\delta x} \qquad \mathop{S}_{x}^{\omega_0} = \frac{x}{\omega_0}\frac{\delta \omega_0}{\delta x} \tag{7.159}$$

are given by

$$\mathop{S}_{R_1}^{\omega_0} = \mathop{S}_{R_2}^{\omega_0} = \mathop{S}_{C_1}^{\omega_0} = \mathop{S}_{C_2}^{\omega_0} = -\frac{1}{2}; \qquad \mathop{S}_{R_3}^{\omega_0} = 0; \qquad \mathop{S}_{C_1}^{Q} = -\mathop{S}_{C_2}^{Q} = \frac{1}{2}$$

and

$$\mathop{S}_{R_1}^{Q} = -\frac{1}{2} + Q\sqrt{\frac{C_2 R_2}{C_1 R_1}}; \qquad \mathop{S}_{R_2}^{Q} = -\frac{1}{2} + Q\sqrt{\frac{C_2 R_1}{C_1 R_2}}; \qquad \mathop{S}_{R_3}^{Q} = Q\sqrt{\frac{C_2}{C_1}}\frac{\sqrt{R_1 R_2}}{R_3}$$

To realize a second-order section, values must be chosen for the five unknowns, R_1, R_2, R_3, C_1, and C_2. There are only two known quantities, ω_0 and Q, and so, usually, the resistors are chosen to be equal. In this case, the sensitivities to the passive elements are quite low and rival those of a passive RLC circuit realization of a filter section. However, the Rauch circuit is not appropriate for the realization of a high Q section. For example, even to realize $Q = 10$ requires $C_1 = 900C_2$, which is a relatively large spread of capacitor values. Also, for $Q > 5$, the

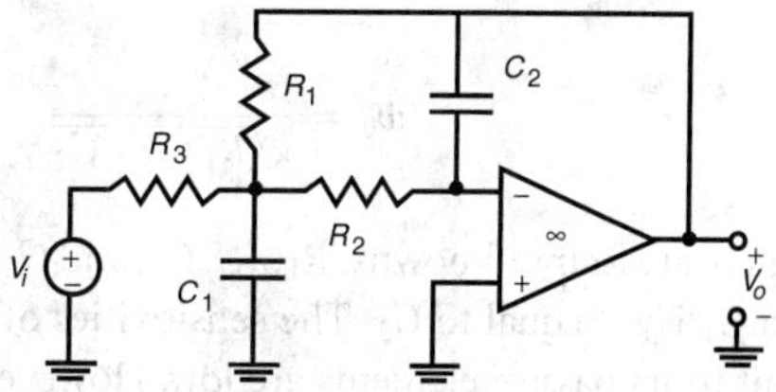

FIGURE 7.109 Low-pass Rauch filter section.

Rauch section is sensitive to GB of the op-amp (Budak, 1974). However, for low Q sections, such as would be needed to realize the transfer function in Eq. (7.156), the Rauch section is a good choice. Finally, we note that if ω_0 is set, then Q can be adjusted using R_3. However, changing R_3 also changes the gain constant for the section.

Figure 7.110 shows a cascade design for the fifth-order low-pass filter transfer function in Eq. (7.156). Two Rauch sections and a first-order section are utilized. All the resistors are 1 Ω, and the values of the capacitors are shown in the figure. These values for the resistors and capacitors are not practical, but the elements would be **magnitude scaled** (resistors multiplied by a scaling factor K_m and all capacitors divided by K_m where K_m can be of the order of 10^5) and **frequency scaled** (all capacitors divided by another scaling factor K_f but the resistors left unchanged) to meet given specifications for the filter. Buffers are employed at the input and output of the circuit so that source and load impedances do not change the frequency characteristics of the filter. The first-order section is placed last in the cascade of sections because if it were placed earlier in the circuit, an additional buffer would be needed.

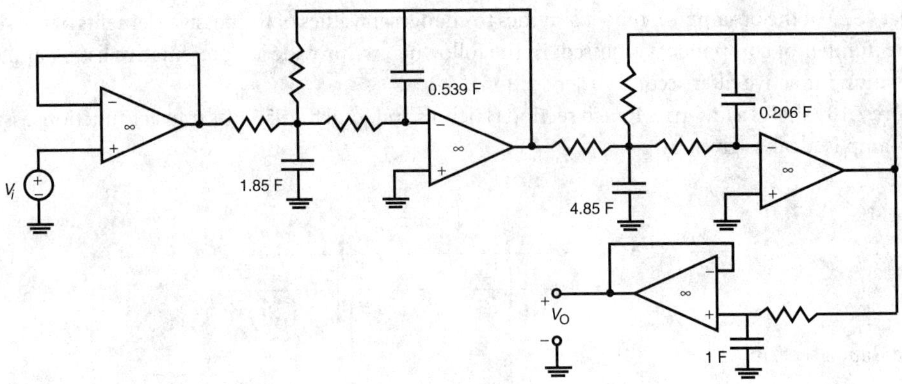

FIGURE 7.110 Realization for the transfer function in Eq. (7.156). All resistors are 1 Ω. These element values are impractical, but they can be magnitude and frequency scaled to obtain practical values.

Figure 7.111 shows a second-order bandpass section that corresponds to the low-pass Rauch section. The circuit in Fig. 7.111 realizes a transfer function of the form

$$T(s) = \frac{Ks}{s^2 + s\dfrac{\omega_0}{Q} + \omega_0^2} \tag{7.160}$$

Specifically, the transfer function is

$$\frac{V_O}{V_i} = \frac{-\dfrac{s}{R_1 C_1}}{s^2 + \dfrac{s}{R_2}\left(\dfrac{1}{C_1} + \dfrac{1}{C_2}\right) + \dfrac{1}{R_1 R_2 C_1 C_2}} \tag{7.161}$$

and the expressions for ω_0 and $1/Q$ for this circuit are

$$\omega_0 = \frac{1}{\sqrt{R_1 R_2 C_1 C_2}} \qquad \frac{1}{Q} = \sqrt{\frac{R_1}{R_2}}\left[\sqrt{\frac{C_1}{C_2}} + \sqrt{\frac{C_2}{C_1}}\right] \tag{7.162}$$

There are four unknowns, R_1, R_2, C_1, and C_2 and often C_1 is set equal to C_2. The sensitivities of this circuit to its passive elements are low. However, as with the low-pass version, the circuit is not appropriate for the realization of high Q sections because of high element value spread and sensitivity to op-amp GB.

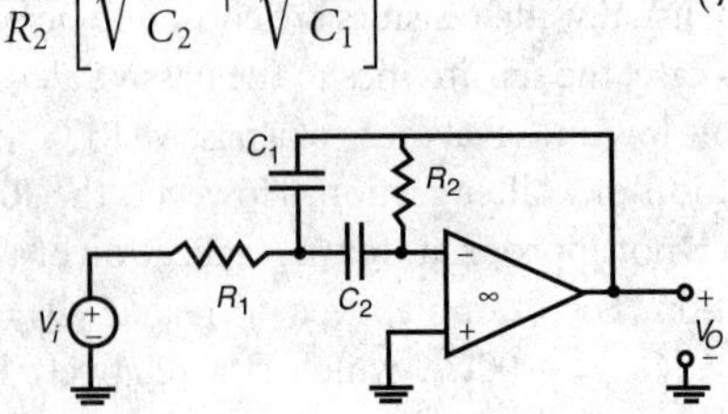

Figure 7.112 shows a high-pass version of the Rauch section. Its transfer function is

FIGURE 7.111 Bandpass filter section corresponding to the Rauch low-pass section.

$$\frac{V_O}{V_i} = \frac{-s^2\dfrac{C_3}{C_1}}{s^2 + \dfrac{s}{R_2}\left(\dfrac{1}{C_1} + \dfrac{1}{C_2} + \dfrac{C_3}{C_1 C_2}\right) + \dfrac{1}{R_1 R_2 C_1 C_2}} \tag{7.163}$$

with ω_0 and $1/Q$ given by

$$\omega_0 = \frac{1}{\sqrt{R_1 R_2 C_1 C_2}} \qquad \frac{1}{Q} = \sqrt{\frac{R_1}{R_2}}\left(\sqrt{\frac{C_1}{C_2}} + \sqrt{\frac{C_2}{C_1}} + \frac{C_3}{\sqrt{C_1 C_2}}\right) \tag{7.164}$$

As with the low-pass and bandpass versions, this high-pass circuit can only realize second-order sections with moderate Q.

A notch filter or a second-order allpass filter can be obtained using the circuit shown in Fig. 7.113. It is based on the fact that a notch or second-order allpass can be realized by subtracting a second-order bandpass transfer function from an appropriate constant. Setting $C_1 = C_2 = C$ in Fig. 7.111 and adding V_i by means of R_A and R_B as shown in Fig. 7.113, we obtain, using superposition,

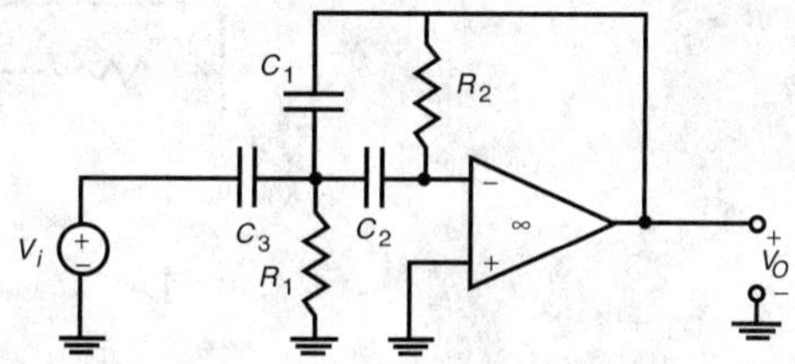

FIGURE 7.112 High-pass Rauch section.

$$V_O = V_i T_{BP} + k V_i (1 - T_{BP}) \qquad (7.165)$$

where $k = R_B/(R_A + R_B)$ and T_{BP} is given by Eq. (7.161). Thus, the transfer function for the circuit in Fig 7.113 can be rewritten as

$$\frac{V_O}{V_i} = k \left[\frac{s^2 + s\omega_0 \left(\frac{1}{Q} + 2Q \left(1 - \frac{1}{k} \right) \right) + \omega_0^2}{s^2 + s\frac{\omega_0}{Q} + \omega_0^2} \right] \qquad (7.166)$$

and a notch filter is realized with

$$k = \frac{2Q^2}{2Q^2 + 1} \qquad (7.167)$$

A second-order allpass with complex poles is obtained for

$$k = \frac{Q^2}{Q^2 + 1} \qquad (7.168)$$

This circuit is easily adjusted, but it is best used for moderate and low Q sections.

Many other filter sections that employ only one op-amp have been used over the years. Particular mention should be made of the Sallen-Key circuits (Budak, 1974) and J.J. Friend's single-amplifier-biquad (SAB) (Ghausi and Laker, 1981). The Friend SAB can realize second-order notch sections in which the undamped natural frequency of the zeros can differ from that of the poles. This is a feature that is needed if a higher-order bandstop filter is to be realized by cascading second-order notch sections.

The Rauch low-pass section and the other sec-

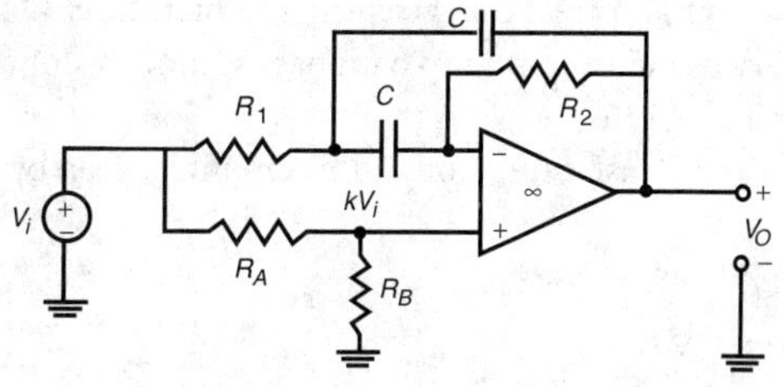

FIGURE 7.113 Circuit used to obtain a second-order notch or a second-order allpass filter.

tions derived from it are limited in the Q they can realize. In some instances, a scheme called *Q-boosting* can be employed to increase Q without additionally burdening the existing op-amp. This scheme is illustrated with the bandpass section of Fig. 7.111. An additional op-amp amplifier is connected into the original circuit as shown in Fig. 7.114. The overall feedback loop through R_5 introduces positive feedback into the circuit. The output voltage of the two-op-amp circuit can be written as

$$V_O = -T_{BP} \left[V_i \frac{R_4}{R_3} + V_O \frac{R_4}{R_5} \right] \qquad (7.169)$$

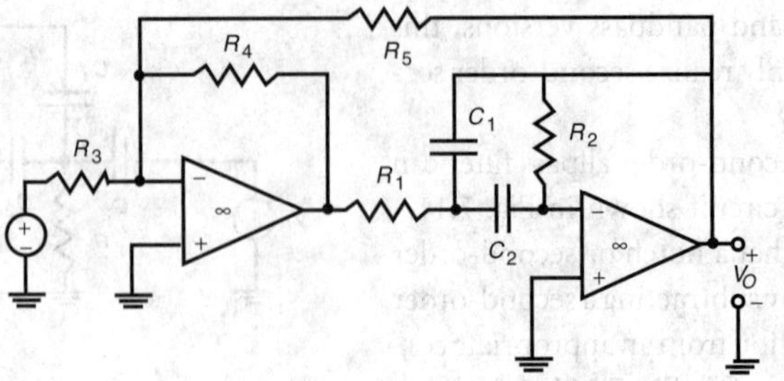

FIGURE 7.114   Q-boosting applied to the bandpass section in Fig. 7.111.

where T_{BP} is given in Eq. (7.161). After substituting Eq. (7.161) into Eq. (7.167) and rearranging the results we obtain

$$\frac{V_O}{V_i} = \frac{s\,\dfrac{R_4}{R_1 C_1 R_3}}{s^2 + s\left[\dfrac{1}{R_2}\left(\dfrac{1}{C_1}+\dfrac{1}{C_2}\right) - \dfrac{1}{R_1 C_1}\dfrac{R_4}{R_5}\right] + \dfrac{1}{R_1 R_2 C_1 C_2}} \tag{7.170}$$

Let Q' be the Q of the complex poles in Eq. (7.170). Then it is easy to show that

$$\frac{1}{Q'} = \frac{1}{Q} - \frac{R_4}{R_5}\sqrt{\frac{R_2 C_2}{R_1 C_1}}$$

and, therefore, $Q' > Q$, and a higher Q can be obtained without as much spread in element values by means of the ratio R_5/R_4. Also note that ω_0 stays the same as before. However, a penalty in the form of increased sensitivities is exacted when Q is increased by inserting additional terms into the s term in the denominator. This scheme is limited to Q-boosts up to a factor of about five or ten.

As with single op-amp sections, there are many two op-amp sections given in the literature. One very useful two op-amp bandpass section is shown in Fig. 7.115.

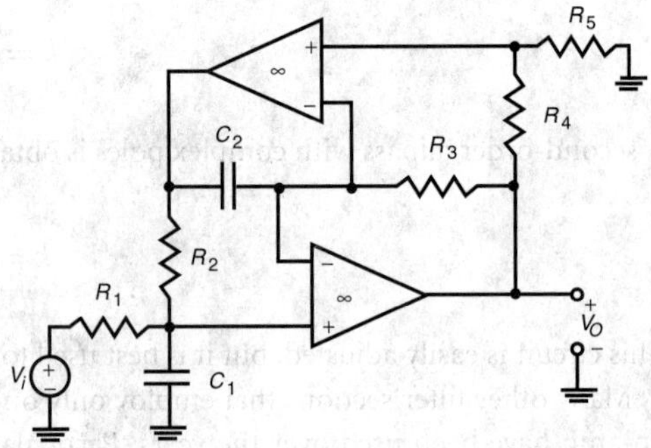

The transfer function of this circuit is given by

FIGURE 7.115   Two op-amp bandpass section.

$$\frac{V_O}{V_i} = \frac{s\,\dfrac{1+K_R}{C_1 R_1}}{s^2 + \dfrac{s}{C_1 R_1} + \dfrac{K_R}{C_1 C_2 R_2 R_3}} \tag{7.171}$$

where $K_R = R_4/R_5$ and

$$\omega_0 = \sqrt{\frac{K_R}{C_1 C_2 R_2 R_3}} \qquad Q = \sqrt{K_R}\sqrt{\frac{C_1}{C_2}}\left(\frac{R_1}{\sqrt{R_2 R_3}}\right)$$

Note that the coefficient of the s term and the constant term in the denominator of Eq. (7.171) can be changed independently. Thus, the coefficient of the s term can be adjusted by R_1 without affecting the constant term, and the constant term can be adjusted by, say, R_3 without affecting the s term. For best results, the two op-amps should be matched (Franco, 1988).

As a final example of a second-order section that can be used in cascade design, we examine the Akerberg-Mossberg low-pass section shown in Fig. 7.116. This section has three op-amps, and, if the op-amps are

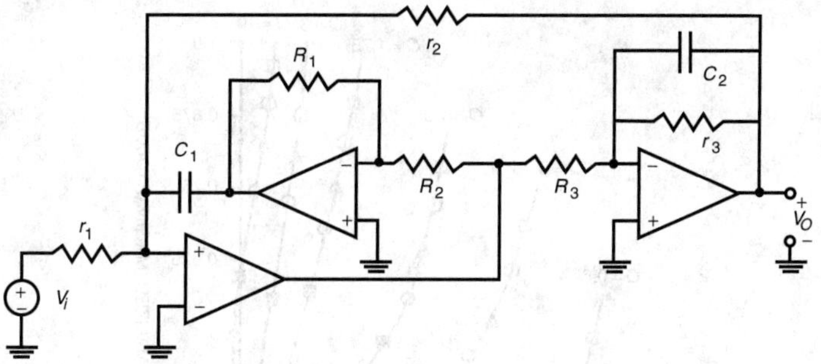

FIGURE 7.116   Akerberg-Mossberg low-pass filter section.

ideal, has a transfer function given by

$$\frac{V_O}{V_i} = \frac{-\dfrac{R_2}{C_1 C_2 r_1 R_1 R_3}}{s^2 + \dfrac{s}{C_2 r_3} + \dfrac{R_2}{C_1 C_2 R_1 r_2 R_3}} \tag{7.172}$$

The quantities ω_0 and Q for this circuit are

$$\omega_0 = \sqrt{\frac{R_2}{C_1 C_2 R_1 r_2 R_3}} \qquad Q = r_3 \sqrt{\frac{C_2}{C_1}} \sqrt{\frac{R_2}{R_1 r_2 R_3}} \tag{7.173}$$

From Eq. (7.172) and (7.173), it is seen that ω_0 can be adjusted by r_2, and then Q can be adjusted using r_3 without affecting ω_0. Finally, the gain constant can be adjusted using r_1 without affecting the previously set variables. This circuit is said to feature orthogonal adjustments. Several three op-amp circuits have this attribute. All five basic second-order filter sections can be realized with modifications to the low-pass circuit (Schaumann, Ghausi, and Laker, 1990).

The Akerberg-Mossberg section is relatively insensitive to op-amp GB compared with, say, the lossy integrator three op-amp circuit (Aronhime, 1977). To provide a comparison, the effects of op-amp GB are now determined for the Akerberg-Mossberg low-pass section. The one-pole rolloff model given in Eq. (41.109) (see Section 7.7) is used for each op-amp. For the sake of simplicity, we assume GB of the op-amps are equal, that $R_1 = R_2 = R_3 = r_1 = r_2 = R$, and that $C_1 = C_2 = C$. Thus, the expressions for ω_0 and Q simplify to $\omega_0 = 1/(CR)$ and $Q = r_3/R$. Then the characteristic polynomial $D(s)$ of the filter can be written as

$$D(s) = s^2 + s\frac{\omega_0}{Q} + \omega_0^2 + \frac{s}{GB}\left[2s^2 + s\left(3\omega_0 + 2\frac{\omega_0}{Q}\right) + 2\frac{\omega_0^2}{Q}\right]$$

$$+ \frac{s^2}{GB^2}\left[3s^2 + s\left(7\omega_0 + 3\frac{\omega_0}{Q}\right) + \omega_0^2\left(2 + \frac{6}{Q}\right)\right] \tag{7.174}$$

$$+ \frac{s^3}{GB^3}\left[2s^2 + s\left(6\frac{\omega_0}{Q} + 2\omega_0\right) + 4\omega_0^2\left(1 + \frac{1}{Q}\right)\right]$$

Next, we let $s_n = s/\omega_0$ and $GB_n = GB/\omega_0$ where the subscript n indicates normalization with respect to ω_0. Upon collecting terms, we have

$$GB_n^3 \frac{D_n}{2} = s_n^5 + s_n^4\left[\frac{3}{2}GB_n + \frac{3}{Q} + 1\right] + s_n^3\left[GB_n^2 + GB_n\left(\frac{7}{2} + \frac{3}{2Q}\right) + 2\left(1 + \frac{1}{Q}\right)\right]$$

$$+ s_n^2\left[\frac{GB_n^3}{2} + GB_n^2\left(\frac{3}{2} + \frac{1}{Q}\right) + GB_n\left(1 + \frac{3}{Q}\right)\right] + s_n\left[\frac{GB_n^3}{2Q} + \frac{GB_n^2}{Q}\right] + \frac{GB_n^3}{2} \tag{7.175}$$

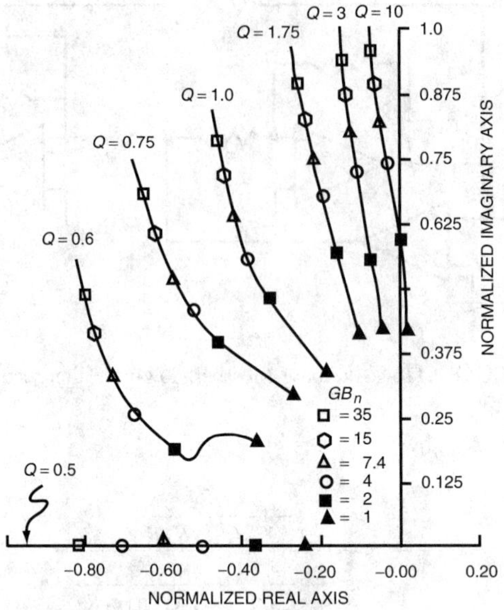

FIGURE 7.117 Loci of second quadrant dominant pole of Eq. (7.175) for values of Q (ideal) from 10 to 0.5. These loci show how the position of the dominant poles of the Akerberg–Mossberg section are influenced by finite op-amp GB. Each locus corresponds to a single value of the ideal Q. The loci are generated by factoring Eq. (7.175) for values of GB_n in the range $\infty > GB_n \geq 1$, where $GB_n = GB/\omega_0$ and ω_0 is the ideal magnitude of the (dominant) poles. The points on the loci for $GB_n = \infty$ are obtained from the denominator of Eq. (7.172).

From Eq. (7.174), we see that the ideal terms obtained with infinite GB have been corrupted by additional terms caused by finite GB. To see the effects graphically, we select several values of Q and then factor the polynomial in Eq. (7.175) for many values of GB_n in order to draw loci. The polynomial in Eq. (7.175) is fifth order, but three of the roots are in the far left-half normalized s plane. The dominant poles are a pair of complex poles that correspond to the ideal poles in Eq. (7.172) but are shifted because of finite GB. Figure 7.117 shows the family of loci generated, one locus for each value of Q selected. Only the loci of the dominant second quadrant pole are plotted. The other dominant pole is the conjugate.

To interpret the results, let us choose the locus for $Q = 10$. The ideal pole location is at the top end of the locus and corresponds to infinite GB. However, if GB_n equals 35, then the pole, instead of being at the upper end of the locus, is at the point on the locus marked with an unshaded square. This means that if the op-amp GB is 35 times the ω_0 of the pair of complex poles desired, then the actual ω_0 will be less than the desired value by the amount indicated on the locus. As another example, on the same locus, if the GB of the op-amp is only two times the desired ω_0, then the pole location will be on the imaginary axis at the shaded square, and the filter will be on the verge of oscillation. This set of loci can be compared to the loci of other three op-amp sections given in Aronhime (1977) and to loci for single op-amp sections provided in Budak (1974).

Advantages of cascade design are: each section can be tuned to some extent so that its poles and zeros are at the correct locations, circuit realizations for each section can be selected depending on the Q of the poles to be realized and other factors, and the filters, if properly designed, can be relatively easily tested. However, the main advantage of cascade design, namely, easy tuning, is lost if thousands of filters have to be produced that must meet a particular set of design requirements accurately. Manual tuning of filter sections in this case would be too expensive and time consuming. Many approaches have been employed to eliminate the need to tune filter sections. Since the most important factor causing actual pole and zero locations to differ from the ideal location is finite op-amp GB, there have been several schemes that attempt to compensate for this problem. Predistortion, a technique used with passive filters to account for parasitic resistance in inductors and capacitors, can be applied. If it is known that finite *GB* will lower ω_0 and increase *Q* for a particular section, the section can be designed to have poles with a larger magnitude and lower *Q*. Then when real op-amps are used, the poles, hopefully, will be near the correct locations. The problem with this and other schemes is that, if they are to be accurate, the user must have some idea of the values of *GB* of the op-amps being used in the section. But *GB* is not a controlled parameter of an op-amp.

Automatic tuning is another scheme (Schaumann, Ghausi, and Laker, 1990), but it adds circuit complexity. The best solution is to eliminate the need for tuning without the requirement for knowing op-amp *GB*. An approach that employs composite op-amps accomplishes this goal.

Composite Op-Amps

Given accurate passive components, the primary reason that the poles and zeros realized by active filter sections are not located where they are supposed to be is the finite *GB* of the op-amps. How can op-amps be modified, without knowing their *GB*, so that filter sections are insensitive to op-amp *GB*? To answer this question, we consider the circuit shown in Fig. 7.118 which represents an active filter section consisting of an RC network and an op-

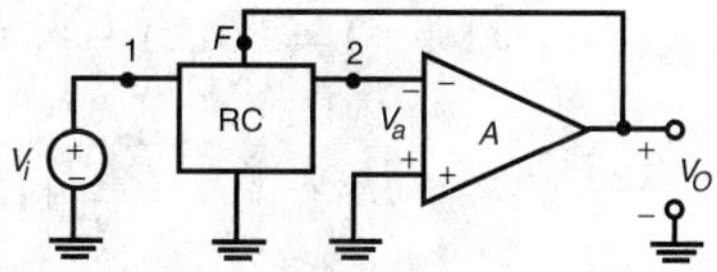

FIGURE 7.118 Active filter section consisting of an RC network and an op-amp.

amp (Budak, Wullink, and Geiger, 1980; Geiger and Budak, 1981). The open loop gain of the op-amp is given in Section 7.7 by Eq. (4.109) and is repeated here for convenience:

$$\frac{V_O}{V_a} = A(s) \approx \frac{GB}{s} = \frac{1}{\tau s}$$

The transfer function of the entire network in Fig. 7.118 is given by

$$\frac{V_O}{V_i} = T(s) = \frac{-T_{21}}{T_{2F} + \dfrac{1}{A}} \tag{7.176}$$

where T_{21} and T_{2F} are transfer functions of the RC network given by

$$T_{21} = \frac{-V_a}{V_i}\bigg/_{V_O=0} \qquad \text{and} \qquad T_{2F} = \frac{-V_a}{V_O}\bigg/_{V_i=0}$$

To see the effect of the op-amp on the filter section, we examine

$$\frac{\delta T(s)}{\delta \tau} = \frac{T_{21}\dfrac{\delta\left(\dfrac{1}{A}\right)}{\delta \tau}}{\left[T_{2F} + \dfrac{1}{A}\right]^2} \tag{7.177}$$

If the section is to be insensitive to the time constant of the op-amp, we must have

$$\frac{\delta T(s)}{\delta \tau}\bigg/_{\tau=0} = 0$$

and from Eq. (7.177) this requires

$$\frac{\delta\left(\dfrac{1}{A}\right)}{\delta \tau}\bigg/_{\tau=0} = 0 \tag{7.178}$$

But $1/A = \tau s$, and so Eq. (7.178) yields s instead of zero. The conclusion is that no active filter section employing only a single op-amp (modeled by the one-pole rolloff model) can be made insensitive to the time constant of the op-amp. However, it is possible to devise a composite op-amp consisting of *two* individual op-amps that allows us to synthesize a filter transfer function $T(s)$ that is first-order insensitive to the op-amp time constants. A circuit

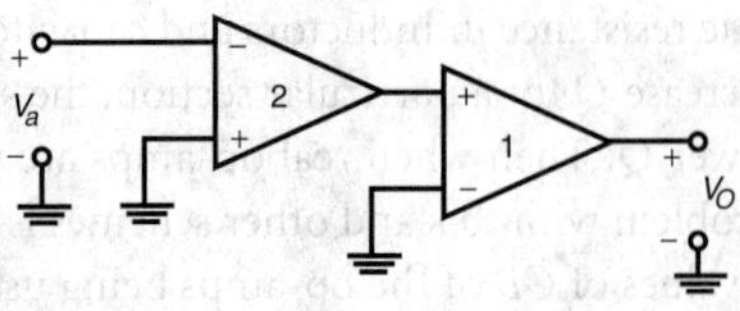

FIGURE 7.119 A composite op-amp composed of two individual op-amps.

realized with such a composite op-amp has a transfer function T with Maclaurin series expansion,

$$T(\tau_1, \tau_2) = T(0,0) + \left[\left(\tau_1 \frac{\delta}{\delta \tau_1} + \tau_2 \frac{\delta}{\delta \tau_2} \right) T(\tau_1, \tau_2) \Big/ \right]_{\tau_1 = \tau_2 = 0}$$
$$+ \cdots + \frac{1}{n!} \left[\left(\tau_1 \frac{\delta}{\delta \tau_1} + \tau_2 \frac{\delta}{\delta \tau_2} \right)^n T(\tau_1, \tau_2) \Big/ \right]_{\tau_1 = \tau_2 = 0} + \cdots \tag{7.179}$$

where $T(0,0)$ corresponds to the ideal transfer function of the circuit. To make the overall transfer function $T(s)$ insensitive to first-order op-amp time constant effects, we devise composite op-amps for which the first partials in Eq. (7.179) are zero. Figure 7.119 shows an example of such an op-amp. The gain of this op-amp is given by

$$\frac{V_O}{V_a} = A(s) = \frac{-1}{s^2 \tau_1 \tau_2} \tag{7.180}$$

Clearly, the composite op-amp in Fig. 7.119 is unstable, but it is not meant to be used by itself. The issue is whether the filter it is used in is stable. Also note that

$$\frac{\delta \frac{1}{A}}{\delta \tau_1} = s^2 \tau_2 \quad \text{and} \quad \frac{\delta \frac{1}{A}}{\delta \tau_2} = s^2 \tau_1 \tag{7.181}$$

Thus, when τ_1 and τ_2 are zero, both partials in Eq. (7.181) are zero. Although the ideal values of τ_1 and τ_2 are zero, the values of these quantities are not required to be equal in the actual composite op-amp. In other words, matching of the time constants of the individual op-amps is not required. If the op-amp in Fig. 7.118 is replaced by the composite amplifier in Fig. 7.119 then $T(s)$ will be, to a first-order approximation, insensitive to the op-amp time constants as they take on values that differ from zero. The values for the passive components of the filter section are obtained assuming the op-amp is ideal. Then, instead of constructing the section with a single op-amp, a composite op-amp is used.

Figure 7.120 shows a representation for a composite op-amp (or a composite amplifier) employing a resistor network and two one-pole rolloff op-amps. The transfer function of this circuit is

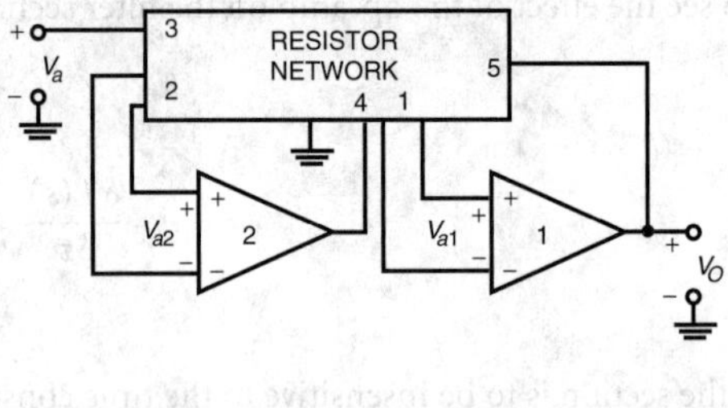

FIGURE 7.120 A representation for a composite op-amp or a composite amplifier composed of a resistor network and two individual one-pole rolloff op-amps.

$$\frac{V_O}{V_a} = A(s) = \frac{s \tau_2 t_{13} + t_{14} t_{23} - t_{13} t_{24}}{s^2 \tau_1 \tau_2 - s \tau_1 t_{24} - s \tau_2 t_{15} + t_{15} t_{24} - t_{14} t_{25}} \tag{7.182}$$

where τ_1 and τ_2 are the time constants of the two op-amps, and t_{ij}, $i = 1, 2$ and $j = 3, 4, 5$, are transfer functions of the resistor network. Since the transfer functions of the resistor network are real numbers, we rewrite Eq. (7.182) as

$$A(s) = \frac{s\tau_2 a_1 + a_0}{s^2\tau_1\tau_2 + s\tau_1 b_1 + s\tau_2 b_2 + b_0} \tag{7.183}$$

The partial derivatives of $1/A$ with respect to the time constants are

$$\frac{\delta\frac{1}{A}}{\delta\tau_1} = \frac{s^2\tau_2 + sb_1}{s\tau_2 a_1 + a_0} \qquad \frac{\delta\frac{1}{A}}{\delta\tau_2} = \frac{s^2\tau_1(a_0 - a_1 b_1) + s(a_0 b_2 - a_1 b_0)}{(s\tau_2 a_1 + a_0)^2} \tag{7.184}$$

To make the partial derivative with respect to τ_1 equal to zero when the time constants are zero requires that $b_1 = 0$ and $a_0 \neq 0$. For the other derivative to be zero requires $(a_0 b_2 - a_1 b_0) = 0$ and, again $a_0 \neq 0$. Imposing these requirements on $A(s)$ we obtain

$$A = \frac{s\tau_2 a_1 + a_0}{s^2\tau_1\tau_2 + \dfrac{b_0}{a_0}(s\tau_2 a_1 + a_0)}, \qquad a_0 \neq 0 \tag{7.185}$$

which has the expected form

$$A = \frac{\text{polynomial}}{K_1 s^2 + K_2(\text{polynomial})} \tag{7.186}$$

Table 7.5 shows several composite devices which meet the requirements given in Eq. (7.184) and thus have the form of Eqs. (7.185) and (7.186). Circuit 1 in this table is a composite op-amp in which b_0 in Eq. (7.185) is zero.

The other two circuits in the table are positive gain, infinite input resistance amplifiers. Note that circuit 3 is obtained from circuit 2 by switching the input leads of op-amp 1. Although circuits two and three in

TABLE 7.5 Composite Op-Amps and Amplifiers

No.	Type of Circuit	Schematic	Gain $A(s)$	Remarks
1	Op-amp		$\dfrac{V_O}{V_a} = \dfrac{s\tau_2 + a}{s^2\tau_1\tau_2}$	$a = \dfrac{R_B}{R_A + R_B}$
2	Amplifier (positive gain)		$\dfrac{V_O}{V_i} = -\dfrac{s\tau_2 + \dfrac{m}{m+1}}{s^2\tau_1\tau_2 - s\tau_2\dfrac{1}{m+1} - \dfrac{m}{(m+1)^2}}$	$m = \dfrac{R_B}{R_A}$
3	Amplifier (positive gain)		$\dfrac{V_O}{V_i} = \dfrac{s\tau_2 + \dfrac{m}{m+1}}{s^2\tau_1\tau_2 + s\tau_2\dfrac{1}{m+1} + \dfrac{m}{(m+1)^2}}$	$m = \dfrac{R_B}{R_A}$

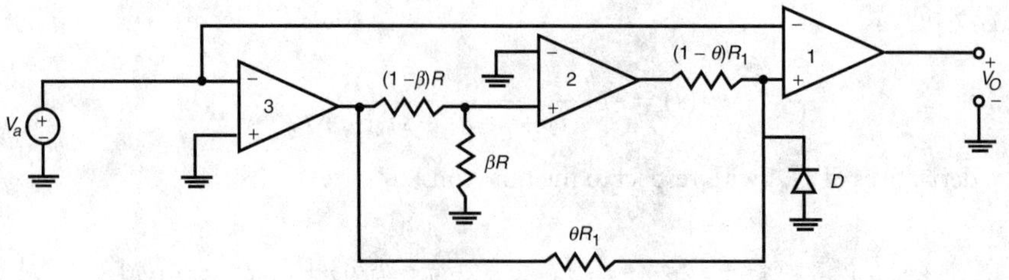

FIGURE 7.121   Composite op-amp consisting of three individual op-amps.

Table 7.5 require the matching of resistor ratios, none of the circuits require that op-amp time constants be matched.

These results can be extended so that first and second partials of $1/A$ with respect to the time constants of the op-amps can be made zero when evaluated for values of the time constants equal to zero. Figure 7.121 shows a three op-amp composite operational amplifier. Its transfer function is given by

$$\frac{V_O}{V_a} = -\frac{s^2\tau_2\tau_3 + s\tau_2(1-\theta) + \theta\beta}{s^3\tau_1\tau_2\tau_3} \tag{7.187}$$

Note that this transfer function has the form of Eq. (7.186). Not only are the first partial derivatives of Eq. (7.187) with respect to the op-amp time constants zero when evaluated at zero, but also, the second partials are zero,

$$\left.\frac{\delta^2\frac{1}{A}}{\delta\tau_j^2}\right|_{\tau_1=\tau_2=\tau_3=0} = 0, \quad j = 1,2$$

and

$$\left.\frac{\delta^2\frac{1}{A}}{\delta\tau_j\delta\tau_k}\right|_{\tau_1=\tau_2=\tau_3=0} = 0, \quad j,k = 1,2,3, \quad j \neq k$$

However, as more individual op-amps are combined, the resulting composite op-amp becomes less forgiving. Circuit layout, lead dress, and second-order effects become much more important. The diode D in Fig. 7.121 is included to prevent a large unstable mode of operation due to op-amp nonlinearities that may be excited when power is first applied or when large noise spikes exist (Budak, Wullink, and Geiger, 1980). Normally, the diode is off, and it is not included in the analysis.

The composite op-amp in Fig. 7.121 consists of three one-pole rolloff op-amps, and if it is used in a (nominally) second-order filter section, there are five poles that must be accounted for. Thus, a careful stability analysis must be performed to ensure that the dominant pair of complex poles realized are the ones desired. The resistor ratios in Fig. 7.121 are used as an aid in achieving stability in this application. Usually, the product $\theta\beta$ is set to be less than one, such as 0.2. In this case, θ can be set to 0.2 and β can be set to 1 (no β resistor divider needed).

We end this brief description of composite op-amps and amplifiers with a rule of thumb. To realize second-order filter sections with $Q \leq 10$ using μA741 op-amps so that f_o is within about 2% of the ideal value and Q is within about 5% of its ideal value *with no tuning*, then for $f_o \leq 2$ kHz, only one op-amp or op-amp amplifier is needed; for $f_o \leq 10$ kHz, a first-order zero sensitivity composite op-amp or amplifier is needed; and for $f_o \leq 50$ kHz, a second-order zero sensitivity composite op-amp or amplifier is required. With higher performance op-amps such as an LF356s, the design of active filter sections that require no tuning can readily be pushed to a value of f_o beyond 150 kHz.

LC Ladder Simulations

Passive circuit realizations of filters have very low sensitivities to element values. Lossless filters designed to maximize power transfer between source and load have the lowest possible sensitivities in the passband (Schaumann, Ghausi, and Laker, 1990). These networks are realized as double-terminated LC ladders as shown in Fig. 7.107, and the corresponding active filter realizations based on ladder simulations begin with the passive filter schematic rather than with the transfer function.

Ladder simulations can be classed into two groups: operational simulations and element simulations. In other words, one class of simulations is focussed on the *equations* of the passive network, and the other class concentrates on the *elements*, especially the inductors.

We first examine an operational simulation. Consider the doubly terminated ladder network shown in Fig. 7.122. This network represents a low-pass filter that has been frequency denormalized (frequency scaled up to the frequencies of interest). Interior node voltage V_1 and branch currents I_1 and I_2 have been identified in the figure. The first step in obtaining the operational simulation of this network is to write the network equations using Kirchhoff's and Ohm's laws. The equations are

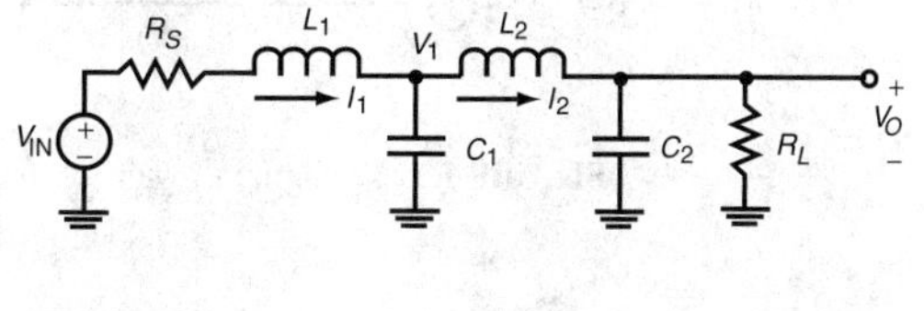

FIGURE 7.122 A doubly terminated ladder low-pass filter.

$$I_1 = \frac{K V_{IN} - V_1}{s L_1 + R_S} \tag{7.188a}$$

$$V_1 = \frac{I_1 - I_2}{s C_1} \tag{7.188b}$$

$$I_2 = \frac{V_1 - V_O}{s L_2} \tag{7.188c}$$

$$V_O = I_2 \frac{R_L}{s C_2 R_L + 1} \tag{7.188d}$$

In Eq. (7.188a), V_{IN} has been multiplied by a constant K so that the active circuit can realize the required transfer function with an arbitrary gain factor.

The next step in realizing the operational simulation is to construct a *voltage* signal flow graph. This is done by first converting all currents in Eqs. (7.188) to voltages by the use of an arbitrary scaling resistor R. Thus, the equations in Eqs. (7.188) become

$$I_1 R = \frac{K V_{IN} R - V_1 R}{s L_1 + R_S} \tag{7.189a}$$

$$V_1 = \frac{I_1 R - I_2 R}{s C_1 R} \tag{7.189b}$$

$$I_2 R = \frac{V_1 R - V_O R}{s L_2} \tag{7.189c}$$

$$V_O = \frac{I_2 R}{s C_2 R + \dfrac{R}{R_L}} \tag{7.189d}$$

The original network has two inductors and two capacitors, and the equations in Eqs. (7.189) can be considered as describing four integrators. Two integrators incorporate a source or a load resistor and are lossy. The other two integrators are lossless. In other words, for an nth-order doubly terminated low-pass LC ladder, the network equations will have the form of n integrators of which two will be lossy, and the remaining integrators will be lossless. To construct the signal flow graph corresponding to Eqs. (7.189), we follow the convention that the voltages associated with the circuit, such as input and output voltages

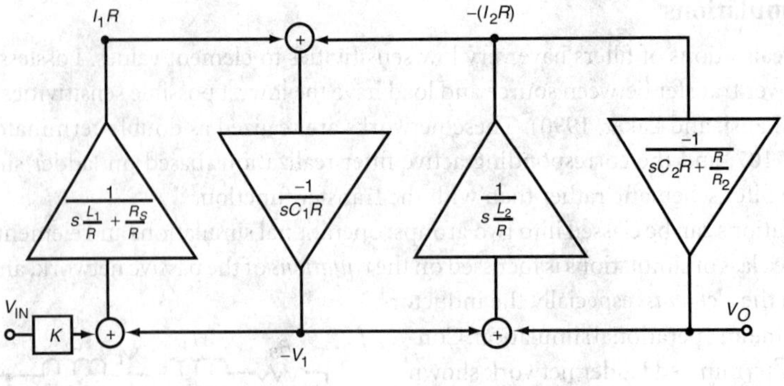

FIGURE 7.123   Signal flow graph for the low-pass ladder circuit in Fig. 7.122.

and voltage V_1 at the internal node, are represented by nodes at the bottom of the graph. Voltages that represent the branch currents I_1 and I_2 are placed at nodes at the top of the graph. We also adopt the rule: Employ inverting integrators in branches corresponding to the ladder capacitors, and use noninverting intergrators to simulate the ladder inductors. The resulting signal flow graph is depicted in Fig. 7.123. We note that the gain around each loop in the figure is negative and independent of R.

The next step in obtaining an operational realization is to substitute appropriate integrators (or summer-integrators) into the signal flow graph. The usual integrators used are the Miller integrator and the noninverting integrator shown in Fig. 7.124(a) and Fig. 7.124(b), respectively. For lossless integration, resistors R_C in the circuits are set open. The output voltages of these circuits can be expressed as

$$V_O = \mp \left[\frac{V_A}{sCR_A + \dfrac{R_A}{R_C}} + \frac{V_B}{sCR_B + \dfrac{R_B}{R_C}} \right] \tag{7.190}$$

where the minus sign is used for the Miller integrator, and the plus sign is employed for the noninverting circuit. When the integrators are substituted into the signal flow graph of Fig. 7.123, there will be many more unknown elements than equations. Thus, usually the capacitors are set equal. The resulting operational realization is shown in Fig. 7.125. However, even with equal capacitors, there still remain 10 unknown elements consisting of 9 resistors (excluding r and R) and 1 capacitor value. There are 6 known values in Fig. 7.122. The available degrees of freedom can be used to maximize the dynamic range of the filter or to achieve other design objectives. Here, we will be content merely to equate the time constants in Eq. (7.189) with those of the circuit in Fig. 7.125. Equating CR_3 to L_1/R_s (from Eq. (7.189a)), we obtain

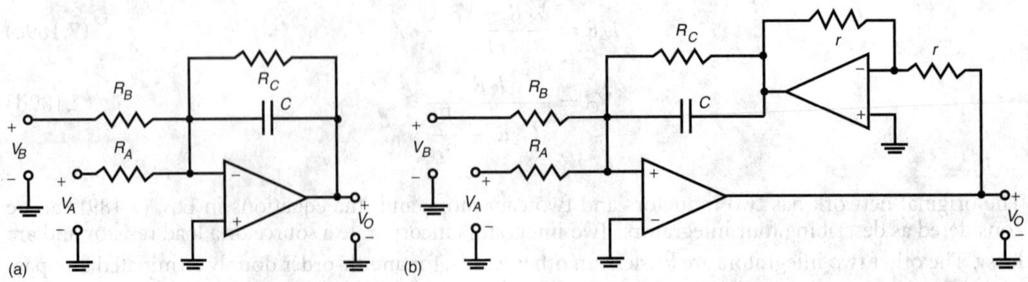

FIGURE 7.124   (a) Lossy inverting integrator (Miller integrator), (b) noninverting lossy integrator, sometimes referred to as the phase lead integrator.

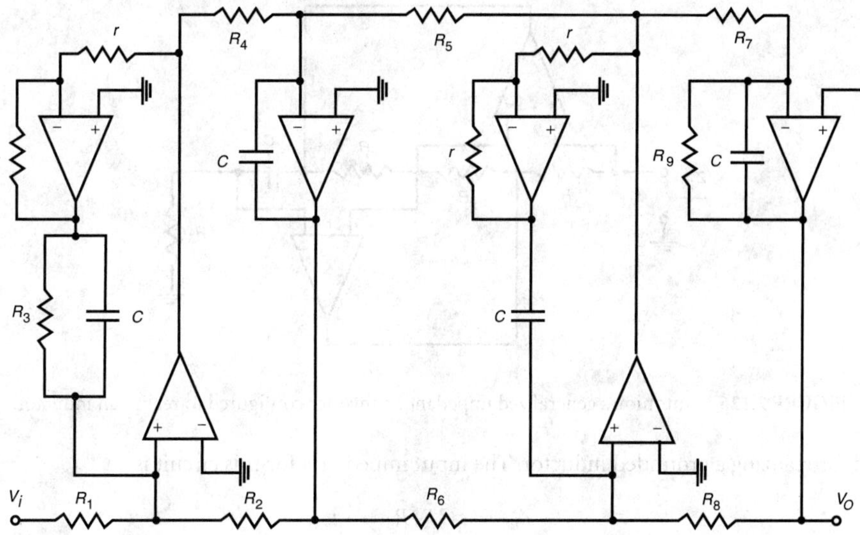

FIGURE 7.125 Operational simulation of ladder network in Fig. 7.122.

$R_3 = L_1/(C R_s)$. Also from Eq. (7.189a) at DC, we obtain $R_3/R_1 = K R/R_s$ and $R_3/R_2 = R/R_s$. Thus

$$R_1 = R_3 \frac{R_s}{KR} = \frac{L_1}{CKR} \quad \text{and} \quad R_2 = R_3 \frac{R_s}{R} = \frac{L_1}{CR}$$

From Eq. (7.189b), we require that $C_1 R = C R_4 = C R_5$. Thus

$$R_4 = R_5 = R \frac{C_1}{C}$$

In the same manner from Eq. (7.189c), we have $L_2/R = C R_6 = C R_8$. Thus

$$R_6 = R_8 = \frac{L_2}{CR}$$

Equation (7.189d) yields $C_2 R = C R_7$, which produces

$$R_7 = R \frac{C_2}{C}$$

Also, at DC, Eq. (7.189d) yields $R_7/R_9 = R/R_L$ so that

$$R_9 = R_7 \frac{R_L}{R} = R_L \frac{C_2}{C}$$

The scaling resistor R and the capacitor C are arbitrary and can be selected to obtain practical values for the circuit components. In this section, we have obtained an active operational simulation of a double-terminated LC ladder filter. This simulation mimics the equations of the passive network and inherits its low element sensitivities. Although we have only examined a low-pass realization, the basic idea can be extended to rather general LC ladder circuits (Schaumann, Ghausi, and Laker, 1990), and so it can be used with some modifications for the other types of basic filter functions (highpass, bandpass, bandstop) as well as for general filter and signal processing circuits.

We now examine a method of ladder simulation that utilizes element substitution. This is a direct approach in which Antoniou's *generalized impedance converter* (GIC), listed as circuit number 15 in Table 7.3 (see Section 7.7), is used to realize grounded inductors in the passive network. Figure 7.126 shows the GIC

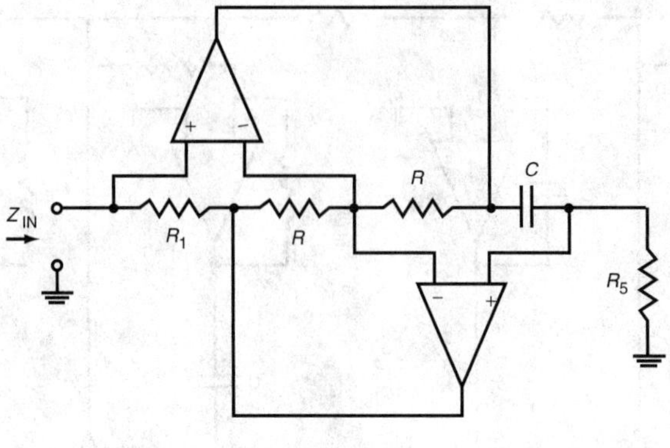

FIGURE 7.126 Antoniou's generalized impedance converter configured to realize an inductor.

arranged for realizing a grounded inductor. The input impedance for this circuit is

$$Z_{IN} = sCR_1R_5 = sL$$

It can be shown (Schaumann, Ghausi, and Laker, 1990) that inductor errors are minimized if $R_5 = 1/(\omega_e C)$ where ω_e is a special frequency of interest such as a filter passband edge. Then the value of R_1 is found from the desired value of the inductor as $R_1 = \omega_e L$.

Figure 7.127(a) shows a passive high-pass ladder network to be simulated. The resulting active network is shown in Fig. 7.127(b) in which the grounded inductors have been replaced by GICs so that $L_1 = R_1 R_5 C_A$ and $L_2 = R_6 R_{10} C_B$. The resistors and capacitors remain unchanged.

This direct method of ladder network simulation is best employed if the inductors to be replaced are grounded. If floating inductors are to be replaced, then $1\frac{1}{2}$–2 times as many op-amps are needed depending

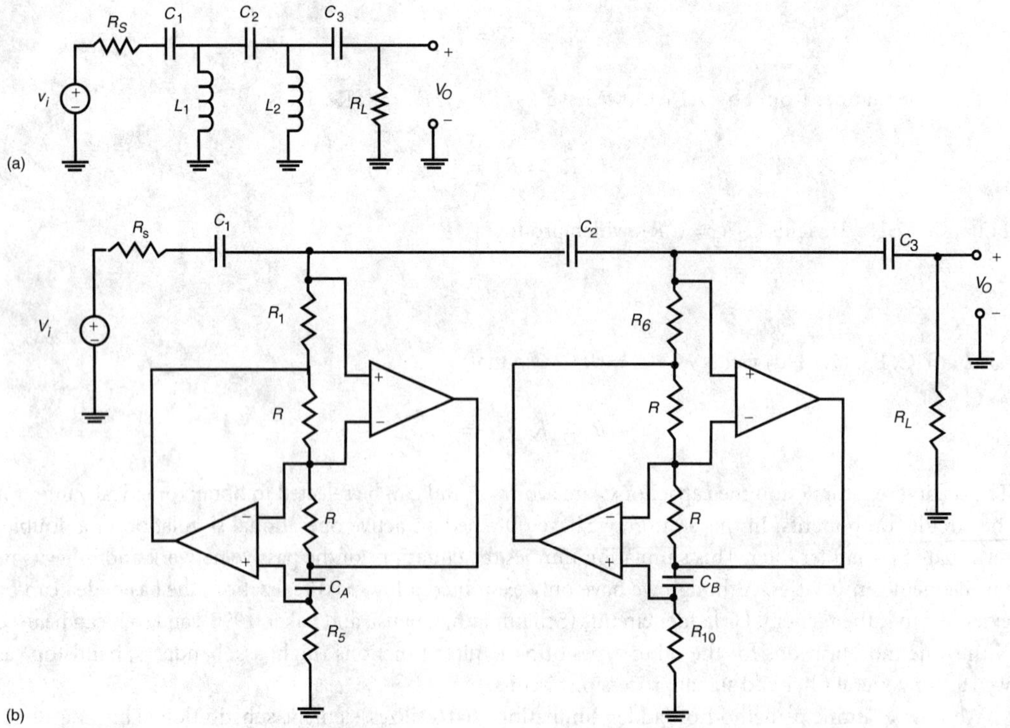

FIGURE 7.127 (a) Passive ladder network with grounded inductors, (b) active circuit simulation of (a).

TABLE 7.6 Bruton's Transformation

Impedance in Passive Network	Impedance in Transformed Network
$Z = R$	$Z' = R/(ks) = 1/(sC')$
$Z = sL$	$Z' = L/k = R'$
$Z = 1/(sC)$	$Z' = 1/(s^2Ck) = 1/(s^2D')$

on the circuits used to realize the inductors. Also, parasitic capacitors and resistors have significant effects on the resulting networks. Instead, if a network with floating inductors has only grounded capacitors, then Bruton's transformation can be used to advantage. To apply this transformation, each impedance of the passive network is scaled by $1/(ks)$. This amounts to a magnitude scaling of the network, and so the overall transfer function remains unaffected, but the impedances of the resistors, inductors, and capacitors are transformed according to Table 7.6. Primed quantities refer to impedances and element values in the transformed network. We see from Table 7.6 that resistors are transformed to capacitors, inductors become resistors, and capacitors convert to *frequency-dependent negative resistors* (FDNRs) having impedances $Z'(j\omega) = -1/(\omega^2 D')$. The constant k is arbitrary and can be used to obtain practical values of components. Antoniou's GIC is used to realize grounded FDNRs.

Figure 7.127(a) shows the network obtained by applying Bruton's transformation to the low-pass passive network in Fig. 7.122. Note that the FDNRs are denoted in this schematic by a symbol consisting of four parallel bars. The active realization of the network in Fig. 7.128(a) is given in Fig. 7.128(b). Each GIC in this figure requires two capacitors to realize an FDNR, but a smaller number of op-amps is needed than if the floating inductors in Fig. 7.122 were realized directly. The values of the two FDNRs in Fig. 7.128(b) are $D'_1 = C_A^2 R_A$ and $D'_2 = C_B^2 R_B$. It can be shown (Schaumann, Ghausi, and Laker, 1990) that a good

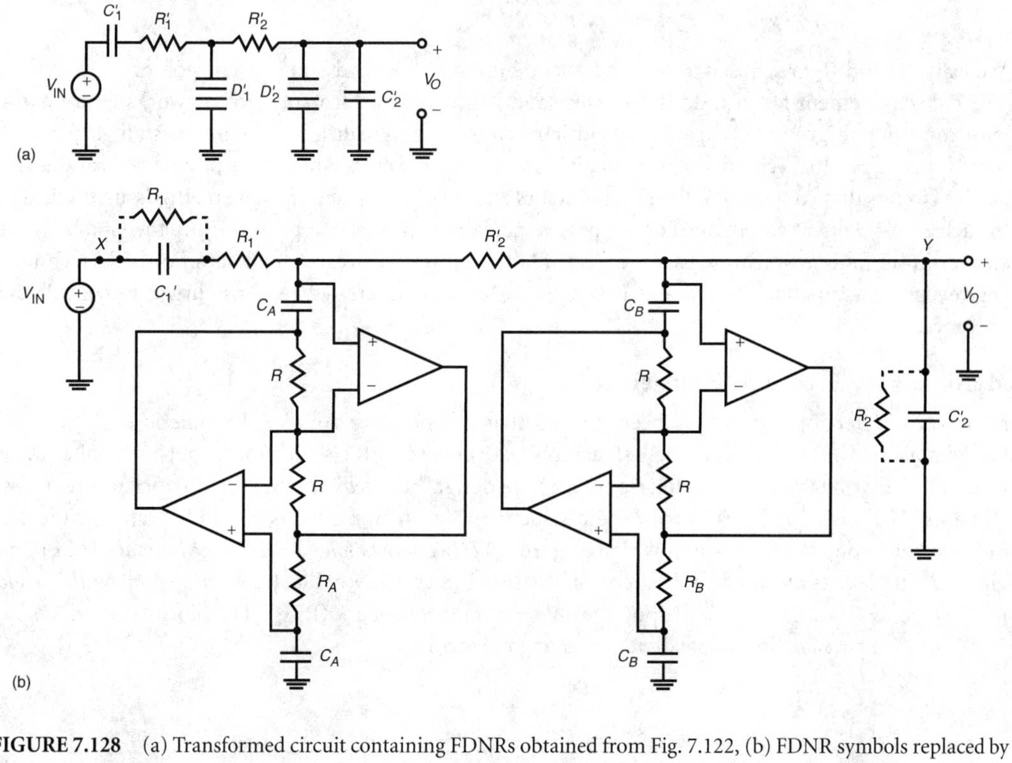

FIGURE 7.128 (a) Transformed circuit containing FDNRs obtained from Fig. 7.122, (b) FDNR symbols replaced by GIC realizations.

choice for R_A is $1/(\omega_e C_A)$ where ω_e is a frequency of interest such as a filter passband edge. This choice minimizes the errors in D' caused by finite GB of the op-amps. The same remarks apply to R_B and C_B.

Examination of the active circuit structure in Fig. 7.128(b) reveals two problems. The first is that, since the entire ladder is transformed, the active structure does not have a true source and load resistor. If the source and load resistors are specified, then gain-of-one buffers should be inserted into the circuit at points X and Y. Then the specified resistors can be connected without changing the overall transfer function.

The second problem is that DC current cannot flow through the noninverting inputs of the upper op-amps unless the extra resistors R_1 and R_2, shown with dotted lines, are added to the circuit. Recall that a DC current must be allowed to flow through the input leads for an op-amp to operate properly. Thus, there must be a path for DC current to ground, to the output terminal of an op-amp, or to a (DC coupled) input voltage source. The added resistors R_1 and R_2 in Fig. 7.128(b) provide this path at the expense of inaccuracy at low frequency in the transfer function realized. From the figure at DC, we have

$$T(j0) = \frac{R_2}{R_1 + R_1' + R_2' + R_2}$$

where $T(j\omega)$ is the transfer function of the overall circuit. From the original circuit in Fig. 7.122, we have at DC

$$T(j0) = \frac{R_L}{R_L + R_S}$$

Thus

$$R_2 = \frac{R_L}{R_S}(R_1 + R_1' + R_2')$$

The added resistors R_1 and R_2 represent parasitic inductors in the transformed circuit. To minimize their effects, it can be shown that each resistor should be chosen to be much larger than $R_1' + R_2'$. As a result

$$\frac{R_2}{R_1} = \frac{R_L}{R_S}$$

Typically, R_1 and R_2 are chosen to be from several hundred thousand ohms to a megohm.

Direct replacement for grounded inductors and Bruton's transformation for networks with floating inductors but only grounded capacitors yield active, inductorless realizations for low-pass, high-pass, and some types of bandpass circuits. These circuits inherit the low sensitivities of the passive realizations. For passive circuits that contain *both* floating inductors and floating capacitors, these methods are inefficient. In such cases, element simulations of the passive network can be accomplished using the Gorski-Popiel embedding technique which also uses a version of Antoniou's GIC. An excellent description of the technique is given in (Schaumann, Ghausi, and Laker, 1990), and the interested reader is urged to consult that reference.

Multiple-Loop Feedback Topologies

Many active filter topologies have been devised that can be classed under the generic term *multiple-loop feedback* (MLF) topologies. These structures were created in the search for higher-order filter active realizations that retained the modularity of cascade design but yielded lower sensitivities to the passive elements. Figure 7.129 shows four of these structures (feedforward paths are added later). Note that each feedback loop includes a minus sign. Figure 7.129(a) shows *follow-the-leader feedback* (FLF), and Fig. 7.129(b) depicts *modified leapfrog feedback* (MLF). Figure 7.129(c) illustrates *inverse follow-the-leader feedback* (IFLF), and Fig. 7.129(d) shows *primary resonator feedback* (PRB). The blocks labeled T_i, $i = 1, 2, 3$, in the figure are, in general, biquads having the form

$$T_i = K_i \frac{s^2 + s\dfrac{\omega_{Zi}}{Q_{Zi}} + \omega_{Zi}^2}{s^2 + s\dfrac{\omega_0}{Q} + \omega_0^2}$$

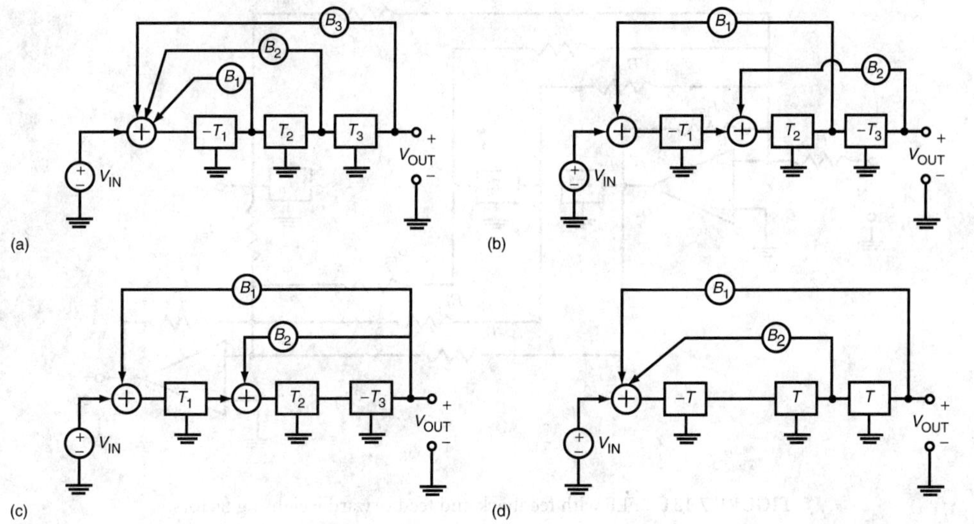

FIGURE 7.129 Multiple-loop feedback topologies: (a) follow-the-leader feedback (FLF), (b) modified leapfrog feedback (MLF), (c) inverse follow-the-leader feedback, (d) primary resonator block feedback (PRB).

except for the PRB in which all of the transfer functions are the same. The circles labeled B_i, $i = 1, 2, 3$, are feedback weighting factors. The configurations shown realize overall sixth-order transfer functions, but the diagrams can be easily extended for the realizations of higher-order filters.

Figure 7.130 illustrates two types of feedforward topologies that can be utilized with any of the four diagrams in Fig. 7.129 with the addition of summers where needed. The individual transfer function blocks in Fig. 7.129 are labeled T_A, T_B, and T_C for generality, and the F_i are feedforward weighting factors. A common pairing of feedback and feedforward topologies is shown in the FLF structure depicted in Fig. 7.131. In this diagram, the feedback and feedforward weighting factors have been combined with the summers. The summer at the input of the diagram also supplies the minus sign needed at the first biquad block. The overall transfer function T is given by

$$T = \frac{V_{OUT}}{V_{IN}} = B_0 \frac{F_0 + \sum_{i=1}^{n}\left[F_i \prod_{j=1}^{i} T_j\right]}{1 + \sum_{i=1}^{n}\left[B_i \prod_{j=1}^{i} T_j\right]} \tag{7.191}$$

where $B_i = R_{B_i}/R_1$ and $F_i = R_2/R_{F_i}$, $i = 0, 1, \ldots, n$. From Eq. (7.191), it can be seen that the poles and zeros of the individual transfer function blocks affect the poles and zeros of the overall transfer function T.

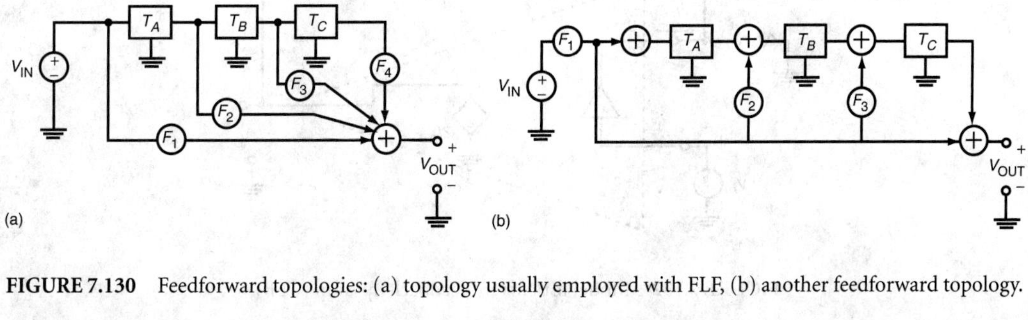

FIGURE 7.130 Feedforward topologies: (a) topology usually employed with FLF, (b) another feedforward topology.

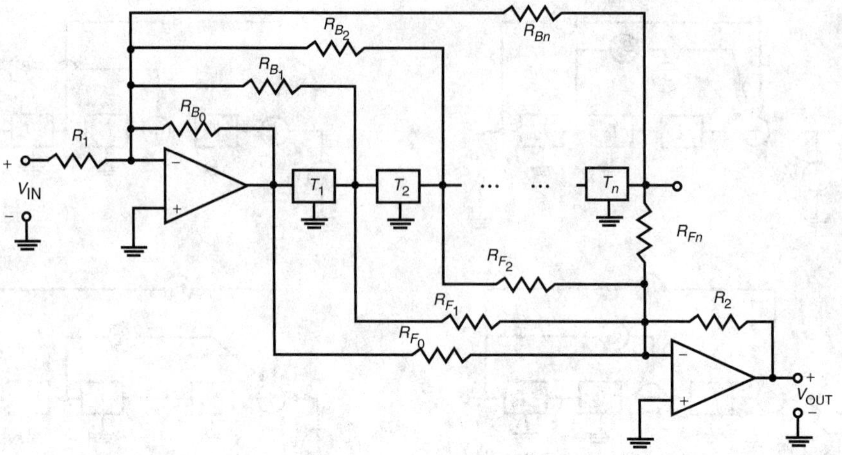

FIGURE 7.131	FLF with feedback and feedforward weighting factors.

To realize a general high-order filter with arbitrary zeros using MLF circuit to achieve low sensitivities is difficult. Some simplifications can be made in the design of geometrically symmetrical bandpass filters, but we will leave these procedures to the references (Ghausi and Laker, 1981; Schaumann, Ghausi, and Laker, 1990).

## 7.8.3	Other Operational Elements

In previous sections, we described the conventional op-amp and examined some of its applications. In this section, we survey other important operational elements—the *current feedback op-amp* (CFB), the *operational transconductance amplifier* (OTA), and the *current conveyor* (CC). All of these other operational elements are available commercially, and all offer advantages and have some disadvantages in comparison with conventional op-amps such as the μA741. Unlike the conventional op-amp, which is a *voltage-controlled voltage source* (VCVS) operational element, current is the main variable of interest at either an input or an output port of all the other operational elements.

Current Feedback Operational Amplifier

CFBs have large closed-loop bandwidths that range from 3×10^6 rad/s to over 10^9 rad/s, and the bandwidth is relatively independent of closed-loop gain (Soclof, 1991; Franco, 1991). Commercial examples of CFBs include the Analog Devices AD844 and AD9611, the Comlinear CLC220A, and the Burr-Brown OPA603.

Figure 7.132 depicts a small signal model of a CFB together with external feedback resistors R_F and R_1 and an input voltage signal V_i. As is the case for the conventional op-amp, inverting and noninverting

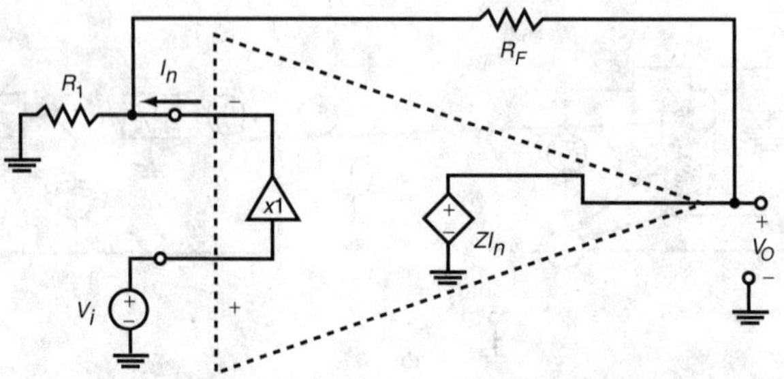

FIGURE 7.132	Model for current feedback op-amp (CFB).

terminals are indicated at the inputs to the CFB. However, the structure of the CFB differs from that of the conventional op-amp. A unity-gain voltage buffer is located between the two input leads which forces the voltage at the inverting input port to equal the voltage at the noninverting input port. Ideally, the input resistance at the noninverting input port is infinite so that no current flows in the lead. However, the input resistance at the inverting port is ideally zero. At the output port of the CFB, the model has an ideal voltage source which depends on the current I_n. The coefficient of the dependent source is Z so that

$$V_O = ZI_n \tag{7.192}$$

where Z is the transimpedance of the CFB. Generally, Z is designed to have a large magnitude at low frequencies, ranging from 300 kΩ to over 3 MΩ. Its magnitude rolls off with a one-pole rolloff characteristic so that we can express Z as

$$Z(j\omega) = \frac{r_T}{1 + j\dfrac{\omega}{\omega_a}} \tag{7.193}$$

where $r_T = Z(0)$ and ω_a is the frequency at which the magnitude of this impedance has decreased to 0.707 of its low-frequency value. This frequency can also be written as

$$\omega_a = \frac{1}{r_T C_T} \tag{7.194}$$

where C_T is termed the transcapacitance of the CFB. C_T is held to as low a value as possible so that larger values of the feedback resistor R_F can be used while retaining a fast response. Values of C_T typically range from 0.5 to 5 pF, and a typical value of ω_a is about $2\pi(12 \times 10^3)$ rad/s. However, as we shall see, the closed-loop bandwidth is much larger than ω_a.

Consider the noninverting amplifier in Fig. 7.132. The current I_n can be expressed as

$$I_n = \frac{V_i}{R_1 \| R_F} - \frac{V_O}{R_F} \tag{7.195}$$

and it is seen from Eq. (7.195) that the feedback signal V_O/R_F is in the form of a current. Then employing Eq. (7.192) and Eq. (7.193) in Eq. (7.195) and solving for V_O/V_i, we obtain

$$\frac{V_O}{V_i} = G = \frac{1 + \dfrac{R_F}{R_1}}{1 + \dfrac{R_F}{r_T}\left(1 + j\dfrac{\omega}{\omega_a}\right)} \tag{7.196}$$

where G is the closed-loop gain of the amplifier. Since R_F/r_T is much less than 1, we can write

$$G \approx \frac{1 + \dfrac{R_F}{R_1}}{1 + j\dfrac{R_F \omega}{r_T \omega_a}} \tag{7.197}$$

from which we obtain the closed-loop 3.01-dB bandwidth of the amplifier, BW_{CL} in radians per second, as

$$BW_{\text{CL}} = \frac{\omega_a r_T}{R_F} \tag{7.198}$$

For example, if $R_F = 1000 \, \Omega$, $R_1 = 111 \, \Omega$, $r_T = 3 \, \text{M}\Omega$, and $\omega_a = 2\pi(12 \times 10^3)$ rad/s, then $G(0) = 10$, and $BW_{\text{CL}} = 22.6 \times 10^7$ rad/s. The resulting closed-loop gain-bandwidth product is 22.6×10^8 rad/s corresponding to 360 MHz.

A remarkable aspect of these results is that the closed-loop bandwidth does not depend, in this somewhat idealized case, on the closed-loop gain $G(0)$. Figure 7.133 shows plots of the magnitude of the gain vs.

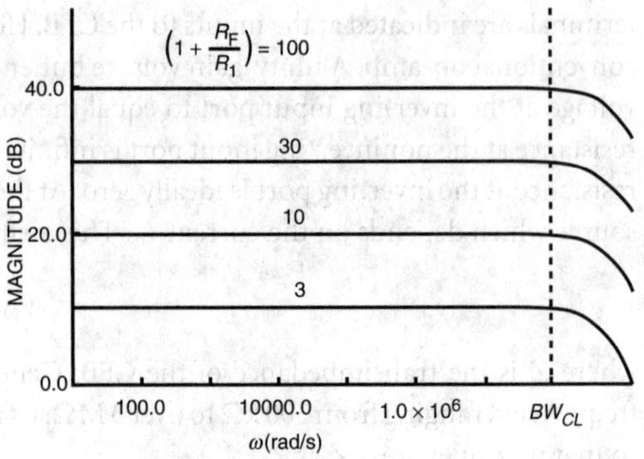

ω obtained from Eq. (7.197). These plots should be compared with the corresponding plots shown in Fig. 7.79 (see Section 7.197), for an amplifier based on a conventional op-amp.

Another interesting aspect of these results concerns the current I_n. Since $I_n = V_O/Z$, the current in the inverting input lead is very small. In the limit as $|Z|$ approaches infinity, I_n approaches zero. Thus, the CFB's output voltage ideally takes on whatever value is required to make I_n equal zero even though the output resistance of the voltage buffer between the input leads ranges from 3 to 50 Ω. As a result, the conditions that apply for an ideal conventional op-amp, namely, zero volts between the two input leads and zero currents drawn by these leads, also hold for the CFB, although for different reasons (Franco, 1991).

FIGURE 7.133 Magnitudes of gain vs. ω obtained from Eq. (7.147) for noninverting CFB amplifier.

CFBs have very large slew rates because of the small size of C_T and because the current available to charge C_T is not limited by the maximum current of a current source. Slew rates of CFBs are one or two orders of magnitude greater than the slew rates of conventional op-amps. Slew rates greater than 6000 V/μs are not uncommon (Soclof, 1991).

However, no CFB is ideal, and small departures from ideal characteristics have adverse effects on the performance of these operational elements. For example, the output resistance of the buffer between the input leads causes some dependence of the closed-loop bandwidth on the closed-loop gain. Figure 7.134 shows an amplifier that includes a CFB model that accounts for this resistance which is denoted as r. With r included, the closed-loop gain of the noninverting amplifier becomes

$$\frac{V_O}{V_i} = G = \frac{1 + \dfrac{R_F}{R_1}}{1 + \dfrac{R_F}{Z}\left[1 + \dfrac{r}{R_1 \| R_F}\right]} \tag{7.199}$$

Substituting for Z from Eq. (7.193) we have

$$G(j\omega) = \frac{G(0)}{1 + \dfrac{R_F}{r_T}\left(1 + j\dfrac{\omega}{\omega_a}\right)\left(1 + \dfrac{r}{R_1 \| R_F}\right)} \tag{7.200}$$

$$\approx \frac{G(0)}{1 + j\dfrac{\omega R_F}{\omega_a r_T}\left(1 + \dfrac{r}{R_1 \| R_F}\right)}$$

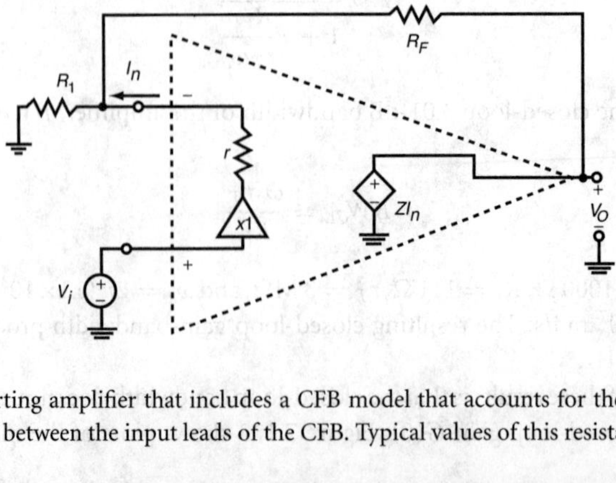

FIGURE 7.134 Noninverting amplifier that includes a CFB model that accounts for the output resistance r of the gain-of-one voltage buffer between the input leads of the CFB. Typical values of this resistance range from 3 to 50 Ω.

TABLE 7.7 Closed-Loop Gains and Bandwidths

Gains [$G(0)$]	Bandwidths (BW_{CL} in M rad/s)
3	377.4
10	333.3
30	250.0
100	133.3

where $G(0) = 1 + R_F/R_1$. The closed-loop bandwidth is

$$BW_{CL} = \frac{\omega_a \left(\dfrac{r_T}{R_F} \right)}{1 + \dfrac{r}{R_F} G(0)} \tag{7.201}$$

Here we see that $G(0)$ is present in the denominator of BW_{CL}, and so there is a small decrease in bandwidth as the gain is increased. For example, suppose a CFB has $r = 30$ Ω, $R_F = 1.5$ kΩ, $r_T = 3$ MΩ, and $\omega_a = 200 \times 10^3$ rad/s. The closed-loop bandwidths for various gains are given in Table 7.7, and plots of the magnitudes of the gains vs. frequency are shown in Fig. 7.135. We see that in changing the gain from 10 to 100, the bandwidth is reduced by a factor of 2.5 for the CFB of this example. But the bandwidth of the noninverting amplifier shown as circuit 1 in Table 7.3 (see Section 7.7) would be reduced by a factor of 10.

Equation (7.201) can be used to modify the external resistor of the CFB amplifier of Fig. 7.134 to compensate for bandwidth reduction caused by r. For required closed-loop bandwidth BW_{CL} and low-frequency gain $G(0) = 1 + R_F/R_1$, we obtain

$$R_F = \frac{\omega_a r_T}{BW_{CL}} - r G(0)$$

Suppose a low-frequency gain $G(0)$ equal to 30 is required with a bandwidth of 300×10^6 rad/s. We obtain $R_F = 1100$ Ω, and $R_1 = 40$ Ω.

CFBs can be used to realize inverting and noninverting amplifiers, current-to-voltage and voltage-to-current converters, and summing and differencing amplifiers. Thus, they can be utilized in a variety of active filters such as Sallen-Key active filters. However, careful stability analysis must be performed for circuits in which CFBs have reactive components in their feedback networks. For example, the usual Miller

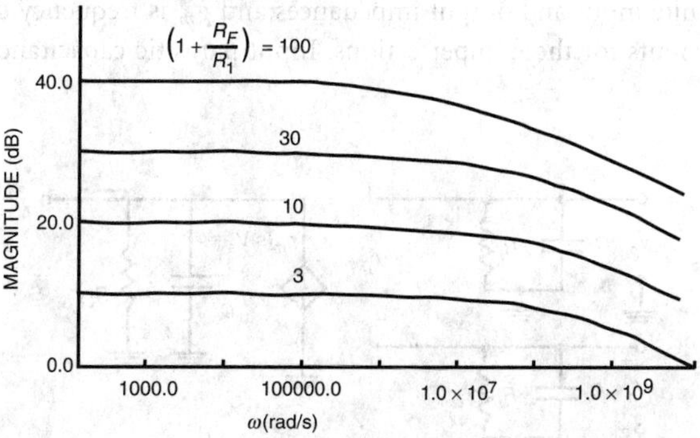

FIGURE 7.135 Plots of the magnitudes of closed loop gains vs. frequency which show dependence of the bandwidth on r.

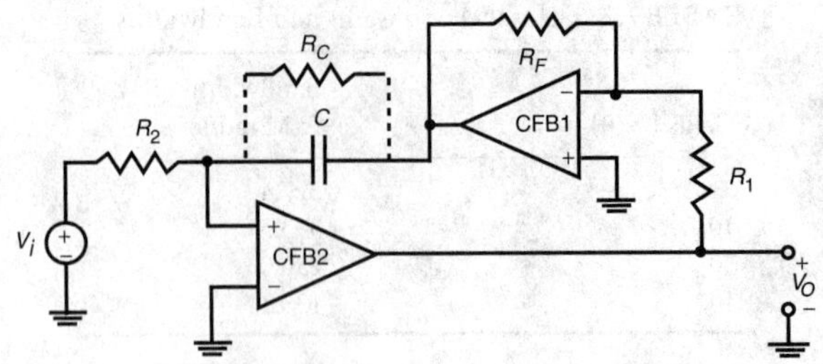

FIGURE 7.136   A CFB integrator.

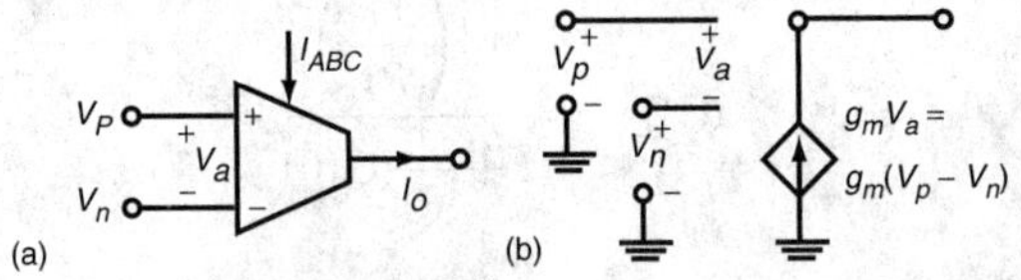

FIGURE 7.137   Operational transconductance amplifier: (a) schematic representation, (b) ideal model.

integrator is not often realized with a CFB, although the interested reader should refer to (Mahattanakul and Toumazou, 1996) for a and more thorough discussion of this circuit. Instead, the integrator employed in the Akerberg-Mossberg active filter section, Fig. 7.116, can be used. This integrator is shown in Fig. 7.136. Ideally its transfer function is $V_O/V_i = R_1/(sCR_FR_2)$, and it can be made a lossy integrator by shunting C with a resistor R_C as shown in dotted lines in the figure. Thus, a variety of active filter sections and topologies can be constructed with CFBs.

Operational Transconductance Amplifier

An OTA is an approximation to an ideal voltage-controlled current source. Figure 7.137(a) shows a representation for the device, and Fig. 7.137(b) depicts an ideal model. In Fig. 7.137(a), the DC current I_{ABC} is termed the amplifier bias current. It controls the transconductance, g_m, of the OTA and can vary g_m over several decades. Usually, g_m is proportional to I_{ABC} so that

$$g_m = kI_{ABC} \tag{7.202}$$

where $k = q/(2KT)$ and $KT/q = 26$ mV at 25°C. The typical range of I_{ABC} is 0.1 μA to 1 mA.

Ideally, the OTA has infinite input and output impedances, and g_m is assumed to be constant. However, real OTAs have noninfinite input and output impedances and g_m is frequency dependent. Figure 7.138 shows a model that accounts for these imperfections. Input parasitic capacitance C_i typically ranges in

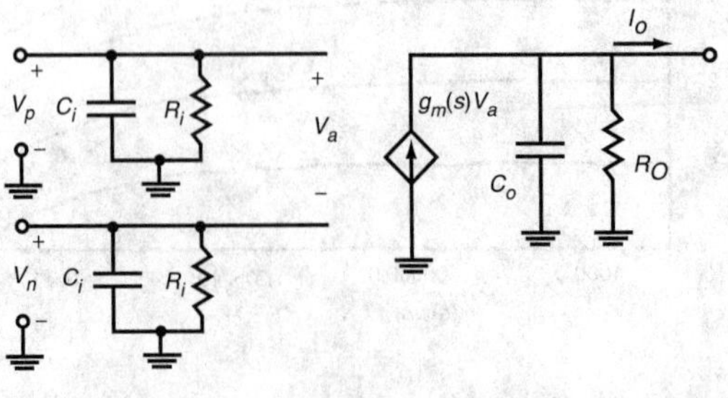

FIGURE 7.138   Model for nonideal OTA.

value from 0.5 to 4 pF, and the output capacitance is typically from 1 to 6 pF. Discrete OTAs such as the CA3080 and the National LM13600 have parasitic capacitances at the higher end of these ranges whereas OTAs that are incorporated on an integrated circuit chip have capacitances at the lower end. The capacitances across the input ports are assumed to be equal. To show the frequency dependence of the transconductance, g_m can be expressed as

$$g_m = \frac{g_{mo}\omega_a}{s + \omega_a} \tag{7.203}$$

where g_{mo} is the low-frequency value of g_m and ω_a is the frequency in radians per second where the transconductance has decreased to 0.707 of its low-frequency value. As I_{ABC} is changed from 0.1 μA to 1 mA, g_{mo} changes from about 2 μS to 20×10^4 μS. The frequency ω_a is about $2\pi(2 \times 10^6)$ rad/s for a CA3080 operating with $I_{ABC} = 500$ μA. The parasitic resistors R_i in Fig. 7.138 are typically several megohms for lower values of I_{ABC} and R_O typically ranges from several megohms to over 100 MΩ. But all of these parasitic resistances are heavily dependent on I_{ABC}. As I_{ABC} is increased so that g_m is increased, R_i and R_O decrease. For example, if I_{ABC} is 500 μA, R_i is about 30 kΩ for a CA3080. But if I_{ABC} is 5 μA, R_i is about 1.5 MΩ. Also, the peak output current that can be drawn is approximately equal to I_{ABC}. Another aspect of OTAs that must be kept in mind is that $|V_a|$ is limited to about 25-mV peak in order to preserve linearity in the CA3080. However, this limit is being broadened in newer OTAs. Finally, it should be observed that OTAs must be allowed to conduct DC current in their input leads. It is clear that careful attention must be given to the manufacturer's specification sheets when discrete versions of OTAs are used.

Despite the many departures from ideal behavior exhibited by OTAs, their usefulness for design is not as meager as it first appears. The advantages afforded by OTAs include control of g_m by means of I_{ABC}, good high-frequency performance, greater adaptability to monolithic integrated circuit design than conventional op-amps and elimination of the requirement for negative feedback in order to operate properly, although feedback is often employed to realize particular circuit functions. Also, some commercial OTAs have buffers included on the chip so that a load may be driven without changing the overall transfer function of the circuit.

We next examine some of the basic OTA building blocks for linear signal processing applications. Figure 7.139(a) shows an OTA circuit that realizes a grounded resistor with a value that can be controlled by I_{ABC}. The input impedance is

$$\frac{V_i}{I_i} = Z_{IN} = \frac{1}{g_m + Y_i + Y_O} \tag{7.204}$$

where $Y_i = sC_i + 1/R_i$ and $Y_O = sC_O + 1/R_O$. For frequencies low enough and parasitic resistances large enough, $Z_{IN} \approx 1/g_m$. If the input leads of the OTA are interchanged, the circuit yields a short circuit stable negative resistor with $Z_{IN} \approx -1/g_m$. Figure 7.139(b) shows a circuit that can realize a floating resistor. Using the ideal model and solving for I_1 and I_2, we obtain $I_1 = (V_1 - V_2)g_{m1}$ and $I_2 = (V_2 - V_1)g_{m2}$. If $g_{m1} = g_{m2} = g_m$, then $I_2 = -I_1$, and the circuit realizes a floating resistor of value $1/g_m$. The two circuits in Fig. 7.139 permit the construction of active filters without the use of resistors. Thus, transconductance-C filters can be realized, and such filters are attractive for monolithic integrated circuit realizations (Schaumann, Ghausi, and Laker, 1990).

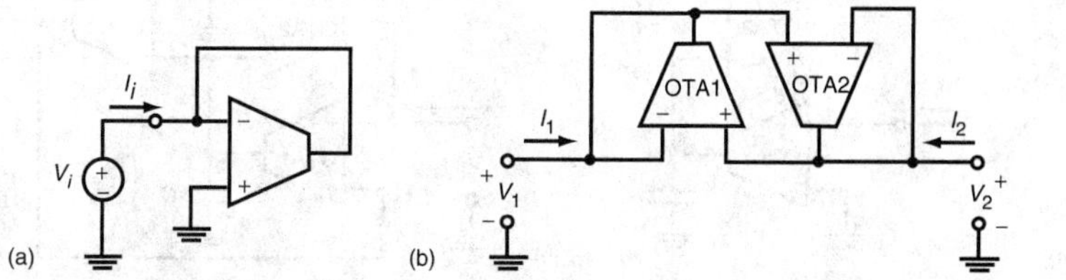

FIGURE 7.139 OTA circuits: (a) OTA realization of a grounded resistor, (b) floating resistor realization.

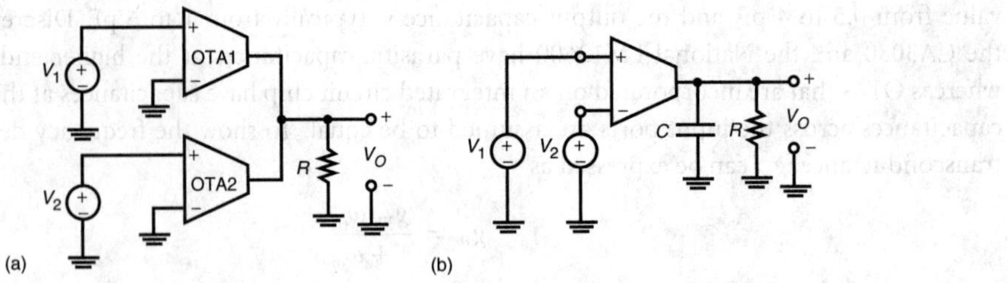

FIGURE 7.140   OTA circuits: (a) OTA summer, (b) OTA lossy integrator.

For active filters, summers and integrators are also needed. Figure 7.140(a) shows a summer circuit for which the output voltage is given by

$$V_O = V_1 \frac{g_{m1}}{G} + V_2 \frac{g_{m2}}{G}$$

where $G = 1/R$. This circuit is versatile because the resistor R can be replaced by the OTA grounded resistor circuit in Fig. 7.139(a) to obtain an all-OTA circuit, more input voltages can be accommodated simply by adding more OTAs at the input, and interchanging the input leads of one of the OTAs at the input side of the summer changes the sign of the corresponding term in the output.

Figure 7.140(b) shows a lossy differential input OTA integrator. The output voltage is

$$V_O = (V_1 - V_2) \frac{g_m R}{sCR + 1}$$

and, again, the resistor can be replaced by the circuit in Fig. 7.139(a). Removal of the resistor results in a lossless integrator.

Grounded and floating realizations of an inductor are shown in Fig. 7.141(a) and Fig. 7.141(b), respectively. These are OTA gyrator realizations of inductors. The first circuit realizes

$$\frac{V_i}{I_i} = Z_{IN} = \frac{sC}{g_{m1}g_{m2}} \tag{7.205}$$

For the second circuit, we can write

$$I_1 - g_{m1}g_{m2}(V_2 - V_1)\frac{1}{sC} = 0$$

$$I_2 + g_{m1}g_{m3}(V_1 - V_2)\frac{1}{sC} = 0 \tag{7.206}$$

If $g_{m2} = g_{m3} = g_m$, then $I_2 = -I_1$, and the floating inductor realized ideally is $L = C/(g_m g_{m1})$. However, the parasitic resistors and capacitors can seriously influence the floating inductor realized. Unwanted resistance and capacitance will exist across both ports, the inductor itself will be lossy, the

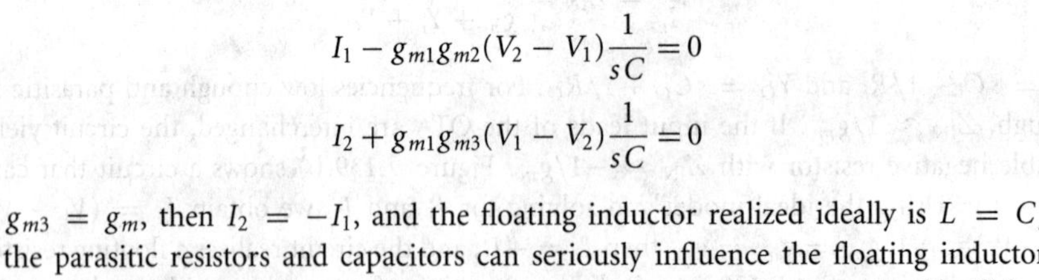

FIGURE 7.141   OTA circuits: (a) OTA realization of a grounded inductor, (b) floating inductor realization.

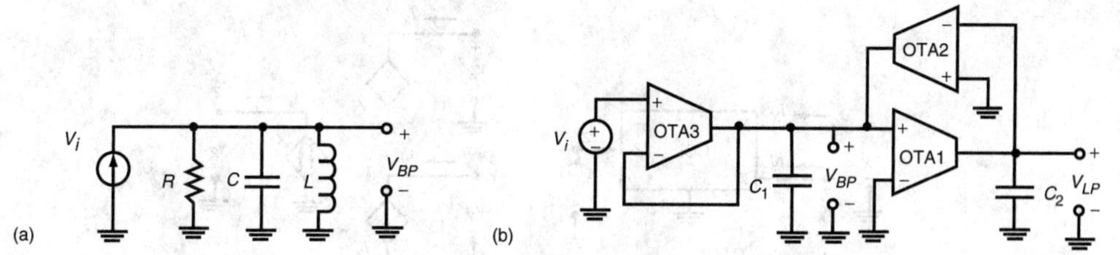

FIGURE 7.142 OTA circuits: (a) RLC bandpass circuit, (b) OTA realization.

value of the inductor will be changed from its ideal value, and a current source will be included in the overall equivalent circuit of the inductor (Schaumann, Ghausi, and Laker, 1990). Whether these departures from ideal behavior will prevent the OTA realization from being utilized depends on the particular application.

As a final example of a useful OTA circuit, we examine a bandpass/low-pass OTA circuit based on the passive RLC tuned circuit shown in Fig. 7.142(a). The active circuit, containing three OTAs, is shown in Fig. 7.142(b). Using ideal models for the OTAs, we obtain

$$\frac{V_{BP}}{V_i} = \frac{s\dfrac{g_{m3}}{C_1}}{s^2 + s\dfrac{g_{m3}}{C_1} + \dfrac{g_{m1}g_{m2}}{C_1 C_2}} \tag{7.207}$$

As a bonus, if we take an output across C_2, we obtain a low-pass transfer function given by

$$\frac{V_{LP}}{V_i} = \frac{-\dfrac{g_{m1}g_{m3}}{C_1 C_2}}{s^2 + s\dfrac{g_{m3}}{C_1} + \dfrac{g_{m1}g_{m2}}{C_1 C_2}} \tag{7.208}$$

In the active circuit, OTA3 converts the input voltage signal to a current source in parallel with a resistor $1/g_{m3}$. Capacitor C_1 realizes the capacitor of the RLC network, and the remaining two OTAs together with C_2 realize the inductor L. This circuit has the advantage that all parasitic capacitors are in parallel with C_1 or C_2. They do not increase the order of the circuit and can be accounted for in the values chosen for C_1 and C_2. Also, the output conductances of OTAs 2 and 3 are in parallel with the simulated conductance g_{m3} and can be absorbed into this value. Only the output conductance of OTA1 cannot be lumped directly with a design element of the circuit. Its main effect is to lower the Q and increase ω_0 of the circuit.

This brief catalog of circuits provides an indication of the versatility of OTAs and their usefulness especially in the design of linear integrated circuit signal processors.

Current Conveyor (CC)

Current conveyors (CCs) were first introduced by Smith and Sedra (1968) (see also Sedra (1989)) and have been used in a wide variety of electronic applications.

The current conveyor is a three-port device described by the hybrid equations

$$\begin{bmatrix} i_Y \\ v_X \\ i_Z \end{bmatrix} = \begin{bmatrix} 0 & a & 0 \\ 1 & 0 & 0 \\ 0 & \pm 1 & 0 \end{bmatrix} \begin{bmatrix} v_Y \\ i_X \\ v_Z \end{bmatrix} \tag{7.209}$$

in which the symbols for the port variables represent total instantaneous quantities. If $a = 1$, the first generation current conveyor, denoted as CCI, is described, and if $a = 0$, the second generation current conveyor, CCII, is characterized. Also, if the plus sign is employed in the equation for i_Z in Eq. (7.209), the corresponding current conveyor is termed a *positive* current conveyor, denoted as CCI+ or CCII+. If the minus sign is chosen, a negative current conveyor is described, indicated by CCI− or CCII−.

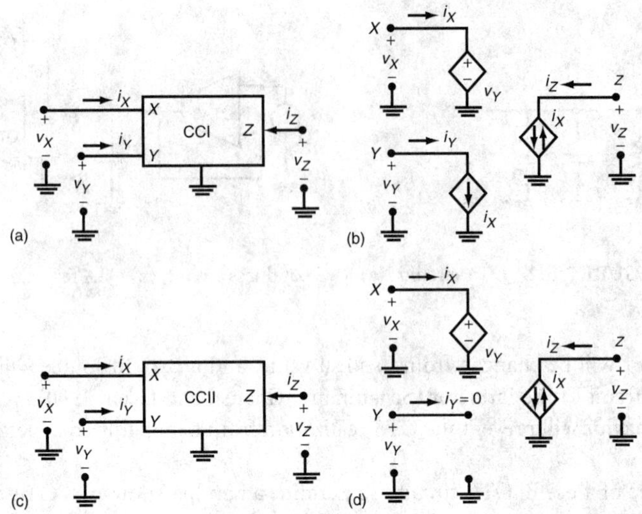

FIGURE 7.143   (a) Representation for a CCI. (b) Ideal model for a CCI. The downward-pointing arrow in the output current source is used for a CCI+. (c) Representation for a CCII. (d) Ideal model for a CCII. The downward-pointing arrow is used to model a CCII+.

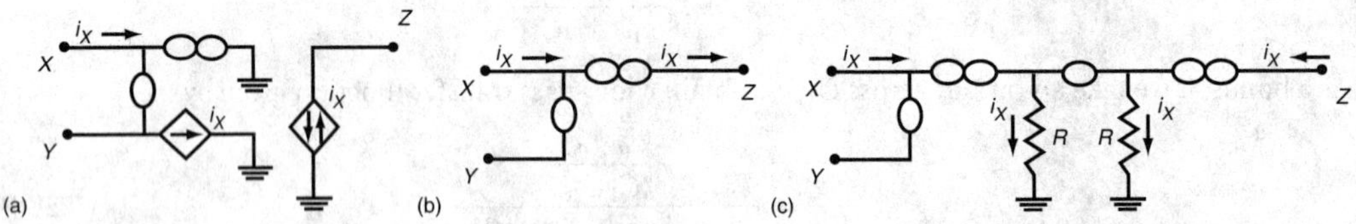

FIGURE 7.144   Ideal models for current conveyors that incorporate nullators and norators: (a) model for a CCI, (b) model for a CCII−, (c) model for a CCII+.

Figure 7.143 shows schematics and ideal models for CCIs and CCIIs. The current generators driving port Z in the models shown in Fig. 7.43(b) and 7.43(d) have two arrows. The downward-pointing arrows are used for positive CCs, and the upward-pointing arrows are employed for negative CCs. In these models, current is conveyed from the low-impedance input at port X to the high-impedance output at port Z. Note that the CCI has three dependent sources in its model, and the CCII has only two. Because it is easier to make, the CCII is more commonly encountered. If the input to the CCII at port Y is connected to ground, the CCII becomes a gain-of-one current buffer.

Figure 7.144 shows ideal models that incorporate nullors. A CCI is modeled in Fig. 7.144(a), and a CCII− model is shown in Fig. 7.144(b). The simplicity of the model for the CCII− makes it attractive for design, and it should be noted that, unlike the nullor model for the conventional op-amp, the norator is not required to be grounded. Figure 7.144(c) shows the slightly more complicated model for the CCII+.

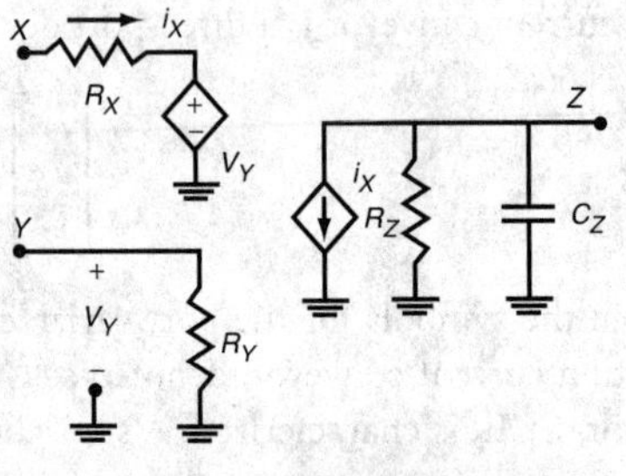

Figure 7.145 shows a model for a CCII+ that accounts for some of the imperfections of this operational element (Svoboda, 1994). R_X typically ranges from 3 to 50 Ω depending on which commercial current conveyor is being used. R_Y is typically 2 to 10M Ω, R_Z is typically 3 MΩ, and C_Z can range from 3 to 11 pF. Also, i_X and i_Z are not exactly

FIGURE 7.145   Nonideal model for a CCII+.

equal. That is $i_Z = (1 \pm \Delta)i_X$ where Δ is a positive number $\ll 1$. Typically, $\Delta = 0.005$. Finally, the open-loop bandwidth in hertz for a current conveyor with $R_L = 1$ kΩ at port Z can range from 5 to over 20 MHz depending on the current conveyor used. Similar models apply for the CCII+ and for CCIs.

The input side of the CCII− in Fig. 7.145 is the same as for a CFB (see Fig. 7.134). Thus, a CFB can be obtained by connecting the output port of a CCII− to a voltage buffer to obtain V_O. Then Z for the resulting CFB is $R_Z/(sC_ZR_Z + 1)$. The close relationship between CFBs and CCIIs and the fact that the *input* to the voltage buffer at the output is brought out to a pin are the reasons an AD844 CFB can be used as a current conveyor.

Table 7.8 lists some circuit building blocks that are based on current conveyors. In the first four circuits listed, the CCII− can be replaced with a CCII+ yielding an extra minus sign in the result. Note that the results for circuit 4 in the table are the same whether a negative or positive CCII is used for the second current conveyor in the voltage amplifier.

Current conveyors are particularly suited for **current-mode circuits**. Current-mode circuits are circuits in which the internal processing of signals is performed on currents rather than voltages. When the voltages at the nodes of a circuit change, the ever-present parasitic capacitances must be charged or discharged, and this is one cause of reduced circuit bandwidth. However, if signals are processed in the form of currents such that voltage amplitudes in the circuit change relatively little, then an improvement in the bandwidth of the circuit can be obtained if stray inductance is minimized.

Second-order cascadable current-mode filter sections can be constructed from CCIs or CCIIs (Aronhime and Dinwiddie, 1991; Sedra and Roberts, 1990). However, it is often convenient to derive current-mode circuits directly from **voltage-mode circuits** using the **adjoint network** concept. Given a network N, an adjoint network N' can be obtained having a current ratio transfer function that is the same as the voltage ratio transfer function of the original network N. Figure 7.146 depicts the relations between the transfer functions of N and N'.

To construct N' given N, voltage-controlled-voltage-sources in N are replaced by current-controlled-current-sources in N' with input terminal pairs of the sources exchanged with output terminal pairs as shown in Fig. 7.147. Passive elements R, L, and C are left unchanged. The resulting networks are termed *interreciprocal networks*.

As an example, consider the voltage-mode circuit shown in Fig. 7.148(a) in which two gain-of-one voltage amplifiers are used. The corresponding adjoint network is depicted in Fig. 7.148(b), and its

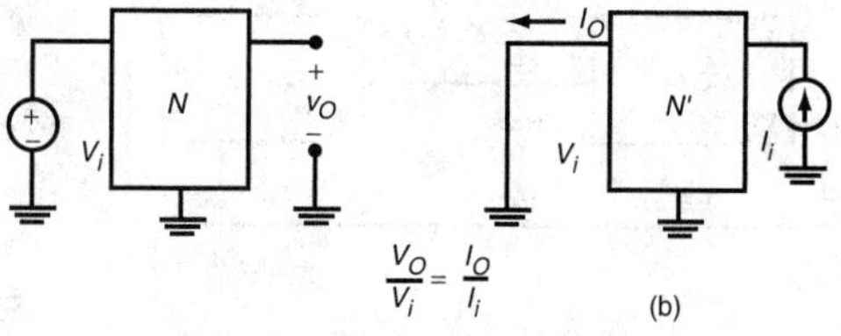

$$\frac{V_O}{V_i} = \frac{I_O}{I_i}$$

FIGURE 7.146 (a) Given network N with transfer function V_o/V_i, (b) adjoint network N' with transfer function I_o/I_i. The two transfer functions are equal.

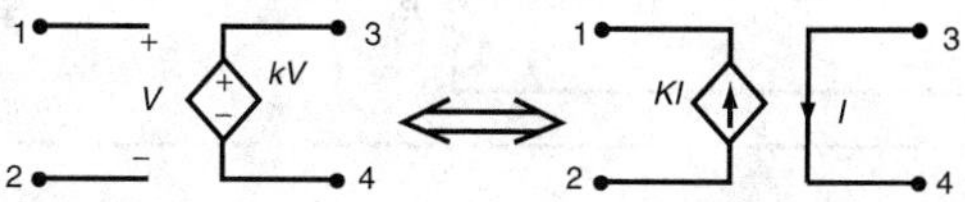

FIGURE 7.147 Voltage-controlled-voltage-sources are replaced by current-controlled-current-sources with input and output terminal pairs exchanged to obtain N' from given network N.

TABLE 7.8 Applications of Current Conveyors

No.	Type of Circuit	Schematic	Gain, Output Current, or Impedance
1	Current amplifier		$\dfrac{i_z}{i_i} = \dfrac{R_1}{R_2}$
2	Current adder subtractor		$i_z = (i_A + \cdots + i_C)R\left(\dfrac{1}{R_1} + \cdots + \dfrac{1}{R_n}\right)$ $- (i_1 + \cdots + i_n)$
3	Current integrator		$I_z = \dfrac{I_i}{sCR}$
4	Volatge amplifier		$\dfrac{v_O}{v_i} = -\dfrac{R_1}{R_2}$
5	Grounded inductor		$Z_{IN} = sCR^2$
6	Floating FDNR		$Z_{IN} = \dfrac{R_3}{s^2 C_1 C_2 R_1 R_2}$

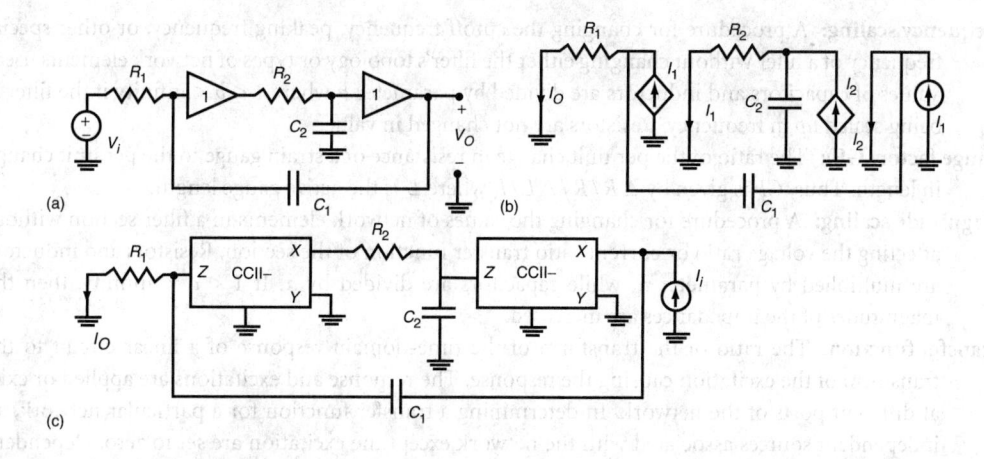

FIGURE 7.148 (a) A voltage-mode filter circuit, (b) adjoint network corresponding to (a), (c) realization of adjoint network using current conveyors.

current conveyor realization is shown in Fig. 7.148(c). The transfer functions for the two networks are given by

$$\frac{V_O}{V_i} = \frac{I_O}{I_i} = \frac{\dfrac{1}{R_1 C_1 R_2 C_2}}{s^2 + \dfrac{s}{R_1 C_1} + \dfrac{1}{R_1 C_1 R_2 C_2}}$$

Thus, the transfer function for the two networks are the same except that one is a voltage ratio, and the other is a current ratio.

In this example, a voltage-mode low-pass filter section that contained gain-of-one voltage amplifiers was converted to a current-mode filter that could be realized by current conveyors. However, conversion of voltage-mode circuits with voltage amplifiers having gains greater than one requires current amplifiers with corresponding equal gains in the adjoint network. Such current amplifiers can be realized if the required gain is not great by ratioing the areas of the transistors in the output current mirror in the usual realization of current conveyors. In some cases, the current amplifier in Table 7.8 can be used.

The list of operational elements described in this section is by no means a complete list of elements available. Devices such as Norton operational amplifiers, four-terminal floating nullors, and operational floating conveyors are objects of study in the literature (Sedra and Roberts, 1990).

Acknowledgments

The author wishes to express his gratitude to Jacek Zurada, Zbigniew J. Lata, Damon Miller, Peichu (Peter) Sheng, KeChang Wang and A. Wisen for their help and encouragement.

Defining Terms

Adjoint network: A network having a nodal admittance matrix which is the transpose of that of a given network. This concept is useful in deriving current-mode networks from voltage-mode networks.

Critical frequencies: A collective term for the poles and zeros of a transfer function or a driving point function of a network.

Current-mode circuits: Signal processing circuits in which the primary variables of interest are currents.

Filter section: An active circuit with a transfer function that is usually second order that can be used in cascade design to realize higher-order filters.

Frequency scaling: A procedure for changing the cutoff frequency, peaking frequency, or other special frequency of a filter without changing either the filter's topology or types of network elements used. Values of capacitors and inductors are divided by parameter b where $1 < b <$ infinity if the filter is being scaled *up* in frequency. Resistors are not changed in value.

Gauge factor (GF): The ratio of the per unit change in resistance of a strain gauge to the per unit change in length. Thus, *GF* is given by $\Delta R/R \big/ \Delta L/L$ where L is the active gauge length.

Magnitude scaling: A procedure for changing the values of network elements in a filter section without affecting the voltage ratio or current ratio transfer function of the section. Resistors and inductors are multiplied by parameter a, while capacitors are divided by a. If $1 < a <$ infinity, then the magnitudes of the impedances are increased.

Transfer function: The ratio of the transform of the time-domain response of a linear circuit to the transform of the excitation causing the response. The response and excitations are applied or exist at different ports of the network. In determining a transfer function for a particular network, all independent sources associated with the network except the excitation are set to zero. Dependent sources remain operating.

Voltage-mode circuits: Signal processing circuits in which the primary variables of interest are voltages.

References

Aronhime, P.B. 1977. Effects of finite gain-bandwidth product on three recently proposed quadratic networks. *IEEE Trans. on Circuits and Systems* CAS-24(Nov.):657–660.

Aronhime, P.B. and Dinwiddie, A. 1991. Biquadratic current-mode filters using a single CCI. *Intl. J. Elec.*, 70:1063–1071.

Budak, A. 1974. *Passive and Active Network Analysis and Synthesis*. Houghton Mifflin, Boston. Reissued 1991. Waveland Press, Prospect Heights, IL.

Budak, A., Wullink, G., and Geiger, R.L. 1980. Active filters with zero transfer function sensitivity with respect to the time constants of operational amplifiers. *IEEE Trans. on Circuits and Systems* CAS-27 (Oct.):849–854.

Franco, S. 1988. *Design with Operational Amplifiers and Analog Integrated Circuits*. McGraw-Hill, New York.

Franco, S. 1991. Current-feedback amplifiers. In *Analog Circuit Design*, ed. J. Williams. Butterworth-Heinemann, Boston.

Geiger, R.L. and Budak, 1981. Design of active filters independent of first- and second-order operational amplifier time constant effects. *IEEE Trans. on Circuits and Systems* CAS-28(Aug.):749–757.

Ghausi, M.S. and Laker, K.R. 1981. *Modern Filter Design*. Prentice-Hall, Englewood Cliffs, NJ.

Mahattanakul, J., and Toumazou, C. 1996. A theoretical study of the stability of high frequency current feedback op-amp integrators. *IEEE Trans. on Circuits and Systems I*, Vol. 43, No. 1, pp. 2–12.

Nichols, M.H. and Rauch, L.L. 1956. *Radio Telemetry*, pp. 137–138. Wiley, New York.

Schaumann, R., Ghausi, M.S., and Laker, K.R. 1990. *Design of Analog Filters*. Prentice-Hall, Englewood Cliffs, NJ.

Sedra, A.S. 1989. The current conveyor: History and progress. In *Proceedings of the IEEE International Symposium on Circuits and Systems* (Portland, OR), pp. 1567–1571. IEEE Press, New York.

Sedra, A. and Roberts, G.W. 1990. Current conveyor theory and practice. In *Analogue IC Design: The Current-Mode Approach*, eds. C. Toumazou, F.J. Lidgey, and D.G. Haigh. Peter Peregrinus Ltd., London.

Smith, K.C. and Sedra, A.S. 1968. The current conveyor—A new circuit building block. *Proc. IEEE* 56(Aug.) 1368–1369.

Soclof, S. 1991. *Design and Applications of Analog Integrated Circuits*. Prentice-Hall, Englewood Cliffs, NJ.

Svoboda, J.A. 1994. Comparison of RC op-amp and RC current conveyor filters. *Intl. J. Elec.* 76:615–626.

Wojslaw, C.F. and Moustakas, E.A. 1986. *Operational Amplifiers*. J. Wiley, New York.

Further Information

Supplementary data and material about operational amplifiers and their uses can be obtained from the following sources:

EDN magazine, a twice monthly periodical that has a general electronics orientation. The section titled "Design Ideas" is noteworthy. The magazine is published by Cahners Publishing Co., Highlands Ranch, CO, and is free to qualified subscribers.

Electronic Design magazine, a twice monthly periodical that discusses many aspects of electronics. Particularly noteworthy are the column "Pease Porridge" by Bob Pease and the section "Ideas for Design." This magazine is published by Penton Publishing, Inc., and is free to qualified subscribers.

Applications notes from manufacturers provide another valuable source of information. An alphabetical listing of companies that manufacture integrated operational amplifers and that provide applications notes follows:

Analog Devices	Maxim Integrated Products
Burr Brown	Motorola
Harris Semiconductor	National Semiconductor
Linear Technology	Texas Instruments

Articles in technical journals are another source of information. New developments and innovative op-amp applications are often first made known in these publications.

Published by the Institute of Electrical and Electronics Engineers, NY:

Circuits and Devices Magazine
Transactions on Circuits and Systems I: Fundamental Theory and Applications
Transactions on Circuits and Systems II: Analog and Digital Signal Processing
Transactions on Instrumentation and Measurements

Published by the Institution of Electrical Engineers, London, England:

Electronics Letters
Proceedings of the IEE—Part G

Published by Kluwer Academic Publishers, Boston:

Analog Integrated Circuits and Signal Processing

Other technical journals that carry articles of importance on op-amps and their applications are:

Electronics and Wireless World
International Journal of Circuit Theory and Applications
International Journal of Electronics
Journal of Circuits, Systems, and Computers
Journal of the Franklin Institute

The proceedings of two conferences are also sources of useful information on op-amps and applications. These proceedings are available from the Institute of Electrical and Electronics Engineers:

Proceedings of the International Symposium on Circuits and Systems
Proceedings of the Midwest Symposium on Circuits and Systems

The following bibliography also lists excellent sources for additional information on operational amplifiers and applications.

Allen, P.E. and Holberg, D.R. 1987. *CMOS Analog Circuit Design*. Holt, Reinhart and Winston, New York.

Barna, A. and Porat, D.I. 1989. *Operational Amplifiers*, 2nd ed. Wiley, New York.

Blinchikoff, H.J. and Zverev, A.I. 1976. *Filtering in the Time and Frequency Domains*. Wiley, New York. Reprinted 1987, Krieger, Malabar, FL.

Bruton, L.T. 1980. *RC-Active Circuits Theory and Design*. Prentice-Hall, Englewood Cliffs, NJ.

Chirlian, P.M. 1967. *Integrated and Active Network Analysis and Synthesis*. Prentice-Hall, Englewood Cliffs, NJ.

Chua, L.O. and Lin, P.-M. 1975. *Computer-Aided Analysis of Electronic Circuits*. Prentice-Hall, Englewood Cliffs, NJ.

Clayton, G.B. 1979. *Operational Amplifiers*, 2nd ed. Butterworths, London.

Coughlin, R.F. and Driscoll, F.F. 1982. *Operational Amplifiers and Linear Integrated Circuits*, 2nd ed. Prentice-Hall, Englewood Cliffs, NJ.

Daniels, R.W. 1974. *Approximation Methods for Electronic Filter Design*. McGraw-Hill, New York.

Daryanani, G. 1976. *Principles of Active Network Synthesis and Design*. Wiley, New York.

De Pian, L. 1962. *Linear Active Network Theory*. Prentice-Hall, Englewood Cliffs, NJ.

Graeme, J.G. 1977. *Designing with Operational Amplifiers*. McGraw-Hill, New York.

Gray, P.R. and Meyer, R.G. 1993. *Analysis and Design of Analog Integrated Circuits*, 3rd ed. Wiley, New York.

Grebene, A.B. 1984. *Bipolar and MOS Analog Integrated Circuit Design*. Wiley, New York.

Horowitz, P. and Hill, W. 1989. *The Art of Electronics*, 2nd ed. Cambridge Univ. Press, Cambridge, England.

Huelsman, L. 1968. *Digital Computations in Basic Circuit Theory*. McGraw-Hill, New York.

Huelsman, L. 1968. *Theory and Design of Active RC Circuits*. McGraw-Hill, New York.

Huelsman, L. 1972. *Basic Circuit Theory with Digital Computations*. Prentice-Hall, Englewood Cliffs, NJ.

Huelsman, L. 1993. *Active and Passive Analog Filter Design*. McGraw-Hill, New York.

Huelsman, L. and Allen, P.E. 1980. *Introduction to the Theory and Design of Active Filters*. McGraw-Hill, New York.

Humpherys, D.S. 1970. *The Analysis, Design, and Synthesis of Electrical Filters*. Prentice-Hall, Englewood Cliffs, NJ.

Irvine, R.G. 1994. *Operational Amplifiers Characteristics and Applications*, 3rd ed. Prentice-Hall, Englewood Cliffs, NJ.

Ismail, M. and Fiez, T. 1994. *Analog VLSI Signal and Information Processing*. McGraw-Hill, New York.

Johnson, D.E., Johnson, J.R., and Moore, H.P. 1980. *Handbook of Active Filters*. Prentice-Hall, Englewood Cliffs, NJ.

Kennedy, E.J. 1988. *Operational Amplifier Circuits*. Holt, Reinhart, and Winston, New York.

Laker, K.R. and Sansen, W. 1994. *Design of Analog Integrated Circuits and Systems*. McGraw-Hill, New York.

Lam, H.Y.-F. 1979. *Analog and Digital Filters: Design and Realization*. Prentice-Hall, Englewood Cliffs, NJ.

Michael, S. and Mikhael, W.B. 1987. Inverting integrator and active filter applications of composite operational amplifiers. *IEEE Trans. on Circuits and Systems* CAS-34(May):461–470.

Mikhael, W.B. and Michael, S. 1987. Composite operational amplifiers: Generation and finite-gain applications. *IEEE Trans. on Circuits and Systems* CAS-34(May):449–460.

Mitra, S. 1969. *Analysis and Synthesis of Linear Active Networks*. Wiley, New York.

Mitra, S. 1980. *An Introduction to Digital and Analog Integrated Circuits and Applications*. Harper & Row, New York.

Moschytz, G.S. 1975. *Linear Integrated Networks: Fundamentals* Van Nostrand Reinhold, New York.

Moschytz, G.S. 1975. *Linear Integrated Networks: Design* Van Nostrand Reinhold, New York.

Roberge, J.K. 1975. *Operational Amplifiers: Theory and Practice*. Wiley, New York.

Sedra, A.C. and Brackett, P.O. 1978. *Filter Theory and Design: Active and Passive*. Matrix, Portland, OR.

Seidman, A. 1983. *Integrated Circuits Applications Handbook*. Wiley, New York.

Sheingold, D.H. 1980. *Transducer Interfacing Handbook, Analog Devices*. Norwood, MA.

Smith, J.I. 1971. *Modern Operational Circuit Design*. Wiley, New York.

Stanley, W.D. 1994. *Operational Amplifiers with Linear Integrated Circuits*, 3rd ed. Merrill, New York.

Stephenson, F.W. 1985. *RC Active Filter Design Handbook*. Wiley, New York.

Temes, G. and LaPatra, J. 1977. *Introduction to Circuit Synthesis and Design*. McGraw-Hill, New York.

Van Valkenburg, M.E. 1982. *Analog Filter Design*. Holt, Reinhart, and Winston, Fort Worth, TX.

Vlach, J. and Singhal, K. 1994. *Computer Methods for Circuit Analysis and Design*, 2nd ed. Van Nostrand Reinhold, New York.

Wait, J.V., Huelsman, L., and Korn, G.A. 1992. *Introduction to Operational Amplifier Theory and Applications*, 2nd ed. McGraw-Hill, New York.

Weinberg, L. 1962. *Network Analysis and Synthesis*. McGraw-Hill, New York.

Wierzba, G. 1986. Op-amp relocation: A topological active network synthesis. *IEEE Trans. on Circuits and Systems* CAS-33(5):469–475.

Williams, A.B. and Taylor, F. 1995. *Electronic Filter Design Handbook*, 3rd ed. McGraw-Hill, New York.

Wobschall, D. 1979. *Circuit Design for Electronic Instrumentation*. McGraw-Hill, New York.

Wong, Y.J. and Ott, W.E. 1976. *Function Circuits*. McGraw-Hill, New York.

7.9 Switched-Capacitor Circuits

Ezz I. El-Masry

7.9.1 Introduction

Implementation of electrical systems has evolved dramatically over the years as a result of numerous developments in technology. Originally, electrical systems had been implemented using passive components such as resistors (R), capacitors (C), and inductors. Because inductors are bulky, noisy, lossy, and unsuitable for miniaturization, tremendous efforts to simulate and/or replace them by circuits containing R, C, and **operational amplifiers** (**op amps**) had begun following the development of the monolithic op amp by Widlar in 1967 (Graeme, 1973). Most of the resulting analog RC–op amp circuits were constructed in hybrid form with monolithic op amps and chip capacitors soldered on a board containing thick-film resistors.

Practically, to achieve full integration in silicon chips, designer would have to choose between the bipolar or **metal oxide semiconductor** (**MOS**) **integrated circuit** (**IC**) technologies (Grey and Meyer, 1984; Grebene, 1984). Each technology has advantages and disadvantages. Bipolar technology is faster, offers higher transconductance values, and has lower turn-on values (0.5 V for silicon). On the other hand, MOS technology has the ability to store signal-carrying charges at a node for a relatively long period of time, to move the charge between different nodes (under the control of a digital clock), and to continuously, nondestructively, sense these charges. Prior to about 1977, there existed a clear separation of bipolar and MOS technologies, according to the function required. MOS technology, with its superior device density, was used mostly for digital logic and memory applications; on the other hand, bipolar ICs, such as op amps, were used to implement analog filters.

Strong motivation for analog circuits designers to develop techniques that are compatible with the MOS technology came with development of the MOS large scale integration (LSI) microprocessor and other low-cost digital ICs. Currently, MOS technology is the most widely used IC technology for digital **very large scale integration** (**VLSI**) circuits. In addition, new developments have occurred in communication technology (such as digital telephony, data transmission via telephone lines, etc.) that require mixed analog and digital signal processing circuitry. Therefore, problems of compatibility with digital systems, as well as cost, reliability, weight, and size added greater incentives to the development of analog circuits that can be fabricated using MOS technologies.

There has been long-standing difficulty in producing fully integrated MOS RC–op amp circuits with precise characteristics. This is because the MOS fabrication process does not result in sufficiently precise control over the absolute values of resistance and capacitance (and, hence, the RC product). In addition, MOS integrated (diffused) resistors have poor temperature and linearity characteristics, as well as requiring a large silicon area. Ordinarily the RC values cannot be controlled to better than 20%. This limitation was soon overcome by the development of circuit techniques wherein the resistor is simulated by a capacitor and MOS switches. Hence, the circuit's frequency response depends on the ratios of MOS capacitors. It has been shown that MOS technology has the capability of realizing capacitor ratios within 0.1% accuracy. This class of circuits is known as **switched-capacitor** (SC) circuits, and they have received considerable attention. In addition to being very accurate, the SC circuit characteristic can be controlled by a precision digital clock. In addition, absolute capacitance values can be sufficiently reduced while maintaining the same capacitance ratios, resulting in circuits that consume a small chip area.

The principle of SC is not new. In fact, it was mentioned in 1873 by Maxwell in his fundamental book on electricity and magnetism. Many years later, Fudem (1984) pointed out that the SC approach to circuit synthesis was introduced in 1962 as was documented in his Ph.D. thesis in 1965 (Fudem is the first to use the name *capacitor switch*). The development of the resonant transfer principle (Boite and Thiran, 1968; Fettweis, 1968) had a great impact on the SC network theory. The use of capacitor and MOS switches to realize digital filters was demonstrated by Kuntz in 1968. More recently, Fried (1972) demonstrated the way to implement active filters using only capacitors, MOS transistors, and a pulse generator. Later, the idea of replacing resistors in RC–op amp filters by capacitors and MOS switches was fully investigated and implemented in the design of **SC filters** by Caves et al. (1977) and Hosticka, Broderson, and Gray (1977). Since then, SC circuit design techniques have made rapid progress and are widely used in many communications and signal processing applications. For almost the last two decades, the most successful realizations of analog filters in MOS technologies have been based on the well-established SC circuit techniques (Allan and Sanchez-Sinencio, 1984; Unbenhauen and Cichocki, 1989; Moschytz, 1984; Ghausi and Laker, 1981; Gregorian and Temes, 1986; Nakayama and Kuraishi, 1987).

7.9.2 Switched-Capacitor Simulation of a Resistor

The recognition that a resistor could be simulated with the combination of a capacitor and MOS switches was the basic theory underlying the operation of SC filters. To illustrate this principle let us consider the SC configuration shown in Fig. 7.149 where the switch is moving back and forth between positions 1 and 2 with frequency f_c Hz. In the analysis to follow, switch position 1 or 2 will be referred to as switch-1 or switch-2, respectively. Assume that v_1 and v_2 are voltage sources (or virtual ground or op amp's output); that is, the values of v_1 and v_2 are not affected by the switch actions. When the switch is in position 1

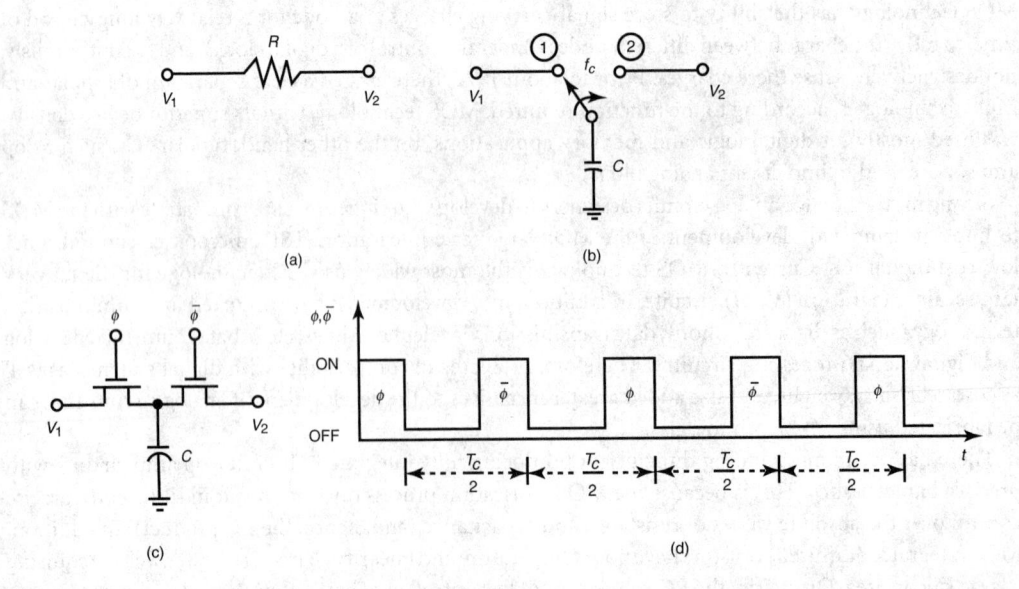

FIGURE 7.149 A switched capacitor circuit: (a) resistor to be simulated by SC, (b) SC simulation of R, (c) MOS implementation of (b), (d) waveforms of the clock signals.

(switch-1 is on) as shown, the capacitor is charging to q_1 given by

$$q_1 = Cv_1 \tag{7.210}$$

It is assumed that the switch stays at that position long enough to fully charge the capacitor. Now if the switch moves to position 2 (switch-2 is on), the capacitor will be charging (discharging); if $v_2 > v_1 (v_2 < v_1)$ at voltage v_2, then the new charge q_2 is given by

$$q_2 = Cv_2 \tag{7.211}$$

Therefore, the transfer of charge Δq is

$$\Delta q = q_1 - q_2 = C(v_1 - v_2) \tag{7.212}$$

and this will be accomplished in $T_c = 1/f_c$ s. The electric current will be, on an average

$$i(t) = \frac{\partial q}{\partial t} = \lim_{\Delta t \to 0} \frac{\Delta q}{\Delta t} \cong \frac{C(V_1 - V_2)}{T_c}, \qquad T_c \ll 1 \tag{7.213}$$

The size of an equivalent resistor that yields the same value of current is then

$$R = \frac{(V_1 - V_2)}{i} \cong \frac{T_c}{C} = \frac{1}{Cf_c} \tag{7.214}$$

For this approximation to be valid, it is necessary that the switching frequency f_c be much larger than the frequencies of v_1 and v_2. This can be easily achieved in the case of audio-frequency filtering, and the SC may be regarded as a direct replacement for the resistor in RC–op amp circuits. This, however, is only valid under the assumption that v_1 and v_2 are not affected by the switch closures. Although this is in effect true in RC–op amp filters that are based on integrators, it is not true in more general circuits. In these cases simple replacement of a resistor by a SC can yield the wrong results. The practicality and technological design considerations of this approach have been discussed in details by many authors (Brodersen, Gray, and Hodges, 1979; Allstot and Black, 1983). MOS realization of the SC resistor, together with the pertinent clock signals, is shown in Fig. 7.149(c) and Fig. 7.149(d).

7.9.3   Switched-Capacitor Integrators

SC integrators are the key elements in the realization of SC filters. A straightforward realization of a SC integrator, Fig. 7.150, is obtained by applying the SC simulation of a resistor, Fig. 7.149, to the RC–op amp inverting integrator. Analysis of this circuit using the preceding approach yields approximate results. Exact results, however, can be achieved by applying a sampled-data analysis technique. Because of the sampled-data nature of SC filters, they are more accurately described, as well as designed, by the use of the z transformation (Jury, 1964). In this formulation the charge distribution in the SC circuit, for one clock

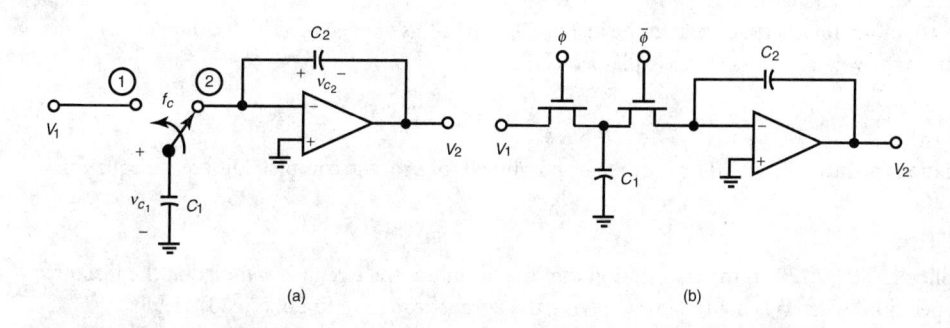

(a)   (b)

FIGURE 7.150   Realization of a SC integrator: (a) SC integrator, (b) MOS implementation of (a).

cycle, is described by a difference equation. This difference equation is then converted into an algebraic equation employing the z transformation. Moreover, the application of this transformation yields a rational function in z that describes the behavior of the SC circuit. Begin by assuming that switch-1, in Fig. 7.150, is closed (ϕ is high) and C_1 is charged to a value

$$q_1 = C_1 v_{c_1} = C_1 v_1 \tag{7.215}$$

The concept of *discrete time* (DT) is introduced at this point. The simplest way of looking at this concept is to visualize the input voltage as being sampled at some discrete point in time. Mathematically, one can state that the signal is uniformly sampled at times $t + nT_c$ where T_c is the clock period and n is an integer. The choice of values for n is arbitrary and, thus, the value $n = -1$ is chosen for this first phase. Equation (7.215) can now be rewritten as

$$q_1(t - T_c) = C_1 v_1(t - T_c) \tag{7.216}$$

Assume that the input voltage remains constant during this clock phase and C_1 charges instantly to the value in Eq. (7.216) and holds that value for the duration of that phase. This assumption is valid as long as the frequency of the input voltage is smaller than the clock frequency. During the same time interval, C_2 is holding a charge

$$q_2(t - T_c) = C_2 v_2(t - T_c) \tag{7.217}$$

Since both switch-2 and the inverting input of the op amp present infinite impedance, no current is allowed to flow through C_2. This requires that the voltage across C_2 remains constant or, equivalently, that any charge stored in C_2 during the previous clock phase is held there during this phase. Therefore, v_2 remains constant, due to the virtual ground at the inverting input terminal of the op amp. After half-clock cycle (at the instant $t - (T_c/2)$) switch-1 is opened and switch-2 is closed (ϕ is low) connecting the top plate of C_1 directly to the inverting input of the op amp and disconnecting the input from the rest of the circuit. Because of the virtual ground, C_1 is completely discharged and all of the charge is transferred to C_2. The net effect of this charge transfer is most easily understood with the aid of the following charge equations for this time instant:

$$q_1[t - (T_c/2)] = 0$$
$$q_2[t - (T_c/2)] = q_2(t - T_c) + C_1\{v_{c_1}[t - (T_c/2)] - v_{c_1}(t - T_c)\} \tag{7.218a}$$

Thus

$$C_2 v_2[t - (T_c/2)] = C_2 v_2(t - T_c) + C_1\{v_{c_1}[t - (T_c/2)] - v_{c_1}(t - T_c)\} \tag{7.218b}$$

with $v_{c_1}[t - (T_c/2)] = 0$ and $v_{c_1}(t - T_c) = v_1(t - T_c)$.

Equations (7.218) make use of the *charge conservation principle* that states that charge is neither created nor destroyed in a closed system. Once the charge transfer has taken place, all currents go to zero, and no further voltage change occur.

After another half-clock cycle (at the instant t), switch-2 is opened (ϕ is high), therefore, no change will occur in the voltage across C_2, as explained earlier. Thus

$$v_2(t) = v_2[(t - (T_c/2)] \tag{7.219}$$

Equations (7.218) and (7.219) may then be combined to form the complete difference equation

$$C_2 v_2(t) = C_2 v_2(t - T_c) - C_1 v_1(t - T_c) \tag{7.220}$$

Implicit in Eq. (7.220) is the assumption that the output of the circuit is sampled at the instants $t + nT_c$ of operation (when ϕ is high). Now applying the z transformation to Eq. (7.220) yields

$$C_2 V_2(z) = C_2 z^{-1} V_2(z) - C_1 z^{-1} V_1(z) \tag{7.221}$$

The transfer function of the integrator is then obtained as

$$\frac{V_2}{V_1}(z) = -\left(\frac{C_1}{C_2}\right)\frac{z^{-1}}{1 - z^{-1}} \tag{7.222}$$

The reader familiar with numerical integration methods can easily recognize that Eq. (7.222) is the forward-Euler numerical integration scheme. Therefore, the integrator in Fig. 7.150 with the output signal sampled when switch-1 is closed (ϕ is high) is known as the **forward-Euler discrete integrator** (FEDI).

7.9.4 Parasitic Capacitors

A major problem with the integrator of Fig. 7.150 is that it suffers from the effect of the inevitable **parasitic capacitors** that are associated with the switches and capacitors due to the MOS fabrication process. These parasitic capacitors have considerable effect on the frequency response of SC filters. It is difficult, if not impossible, to compensate for these parasitics since they do not have specific values but rather they assume a range of values that are poorly controlled. The sources of these parasitics are briefly described next (Ghausi and Laker, 1981; Gregouian and Temes, 1986).

Parasitic Capacitors Associated with the MOS Switch

On a chip associated with every MOS transistor are some undesired capacitances. This is inherent in the MOS technology. Figure 7.151 shows an NMOS switch together with the major parasitic capacitors associated with it. C_{ol} are the gate-diffusion overlap capacitances due to errors in the mask alignment. They are about 0.005 pF each and couple the clock signals into the filter. This clock feedthrough appears at the output as an additional DC offset. These parasitic capacitors can be greatly reduced by using self-alignment MOS technologies that result in very small overlaps between the gate and both the source and drain of a transistor. C_{pn} are the source- (drain-) substrate capacitance due to the reversed-biased pn junctions between the source (drain) and the substrate. They are about 0.02 pF each.

Parasitic Capacitors Associated with the MOS Capacitor

Figure 7.152 shows a cross section of a typical MOS capacitor using double polysilicon MOS technology together with the associated parasitic capacitors and an equivalent circuit model. C_{p1} is a sizable parasitic capacitance that exists from the bottom plate of the MOS capacitor C to the substrate. This is the capacitance of the silicon dioxide layer under the first layer of polysilicon, as shown in Fig. 7.152. Typically, C_{p1} can have a value in the range of 5–20% of C. C_{p2} is a small capacitance from the top plate of C to the substrate,

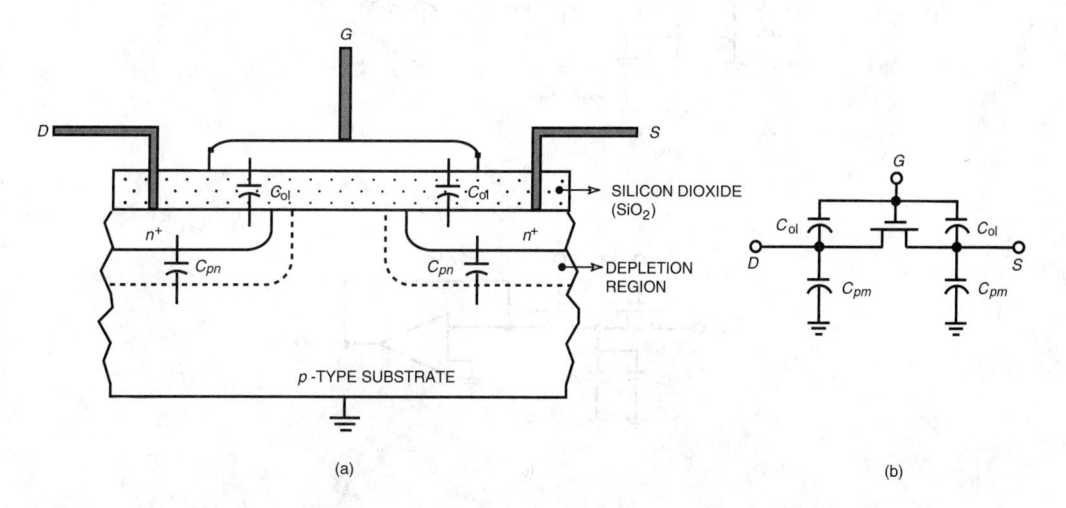

(a)　　　　　　　　　　　　　　　　　　　(b)

FIGURE 7.151 NMOS switch and the associated parasitic capacitances: (a) CMOS implementation, (b) symbolic representation.

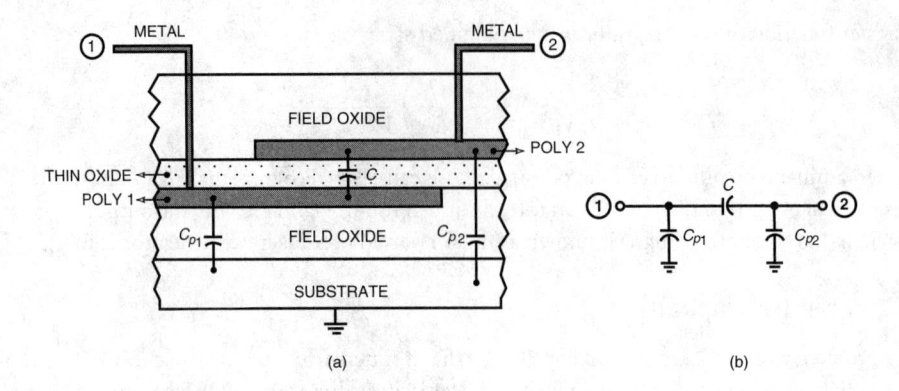

FIGURE 7.152 Double polysilicon CMOS capacitor and the associated parasitic capacitances: (a) CMOS implementation, (b) symbolic representation

and it is due to the interconnection of C to other circuitry. Typically, values of C_{p2} can be from 0.1–1% of C, depending on the layout technique.

Recognizing that the range of capacitance values of the capacitors used as passive components in SC filters are typically from 0.01 to 100 pF, therefore, the effects of this parasitic on the frequency response should not be ignored. The design of SC filters must be done in such a way that the parasitic capacitors do not degrade the performance of the filter, as will be seen later.

The effect of the parasitic capacitors is shown in Fig. 7.153 for the FEDI. Most of the top and bottom plate parasitic capacitors, shown in Fig. 7.153(a), have no effect on V_2 since they do not deliver any charge to C_2 (C_{p12}, C_{pn22}, and C_{p21} are shorted assuming that the open-loop gain of the op amp is sufficiently high, C_{pn11} and C_{p22} are voltage driven and, therefore, they are insignificant). The overlap capacitance

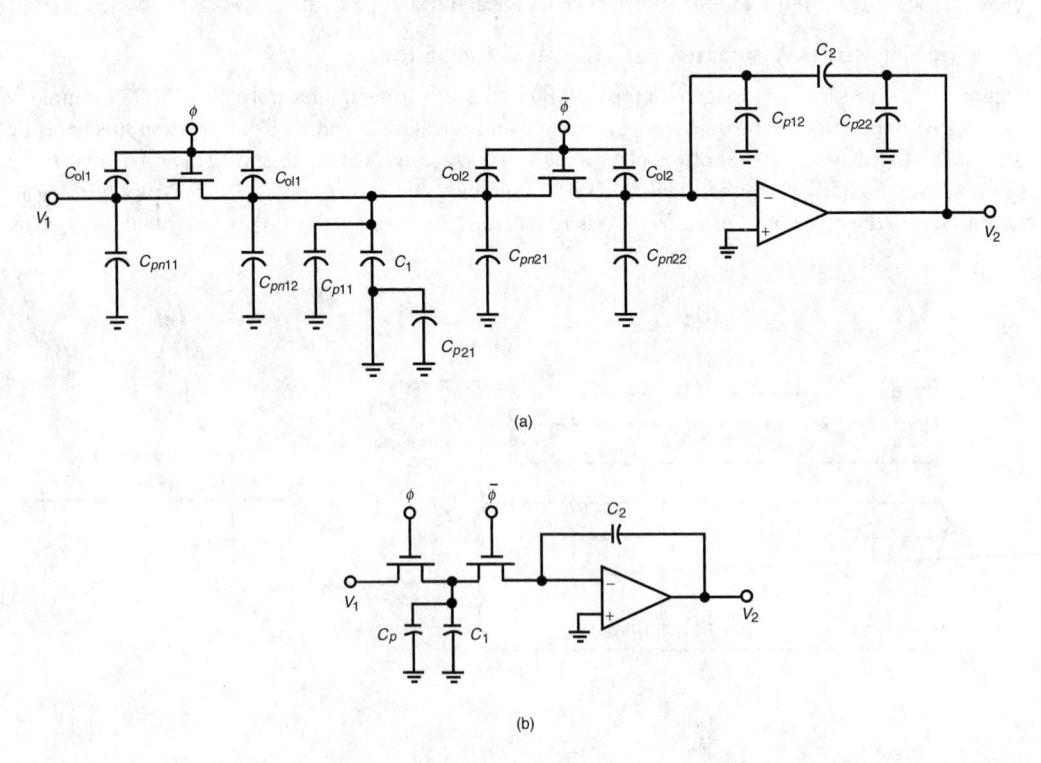

FIGURE 7.153 Management of parasitic capacitances: (a) SC integrator and the associated parasitic capacitances, (b) effect of the parasitics on the performance of SC integrator.

(C_{ol1} and C_{ol2}) can be drastically reduced in the fabrication process (using the self-alignment techniques), therefore, they are irrelevant. The parasitic capacitors that have direct effect on the output of the integrator are C_{pn12}, C_{p11}, and C_{pn21}. Their parallel combination is represented by C_p as shown in Fig. 7.153(b). C_p is poorly controlled and may be as large as 0.05 pF, making the net value of C_1 uncertain. If, for example, 1% accuracy is required for C_1, it should have value greater than 5 pF. For practical reasons, the integrating capacitance $C_2 \gg C_1$; therefore, the integrator will consume a large chip area.

7.9.5 Parasitic-Insensitive Switched-Capacitor Integrators

The effect of the parasitic capacitors can be eliminated by the use of the parasitic-insensitive integrator configurations (Hosticka, Brodersen, and Gray, 1977; Mertin and Sedra, 1979), shown in Fig. 7.154. The circuit in Fig. 7.154(a) is the parasitic-insensitive version of the FEDI, where C_1 is charging through the top plate and discharging through the bottom plate. Therefore, the transfer function of this integrator, with v_2 is sampled when ϕ is high, is a noninverting one. It should be noted that if v_1 is sampled and held for a full-clock cycle, then v_2 will also be held constant for a full cycle. The integrator shown in Fig. 7.154(c) has an inverting transfer function. Applying the charge equation method, previously described in Sec. 7.9.3, to the integrator of Fig. 7.154(c) and assuming v_2 is sampled when ϕ is high, one can easily derive the following transfer function:

$$\frac{V_2}{V_1}(z) = -\left(\frac{C_1}{C_2}\right)\frac{1}{1-z^{-1}} \tag{7.223}$$

It should be noted that the numerical integration in Eq. (7.223) is similar to the backward-Euler numerical integration scheme. Therefore, the integrator in Fig. 7.154(c), with the output signal sampled when ϕ is high, is known as the *backward-Euler discrete integrator (BEDI)*.

The MOS implementations along with the relevant parasitic capacitors, C_{p1} and C_{p2}, are also given in Fig. 7.154(b) and Fig. 7.154(d). Note that C_{p1} and C_{p2} are periodically switched between v_1 ground and the op amp's virtual ground-ground, respectively. Therefore, they do not deliver any charge to C_2.

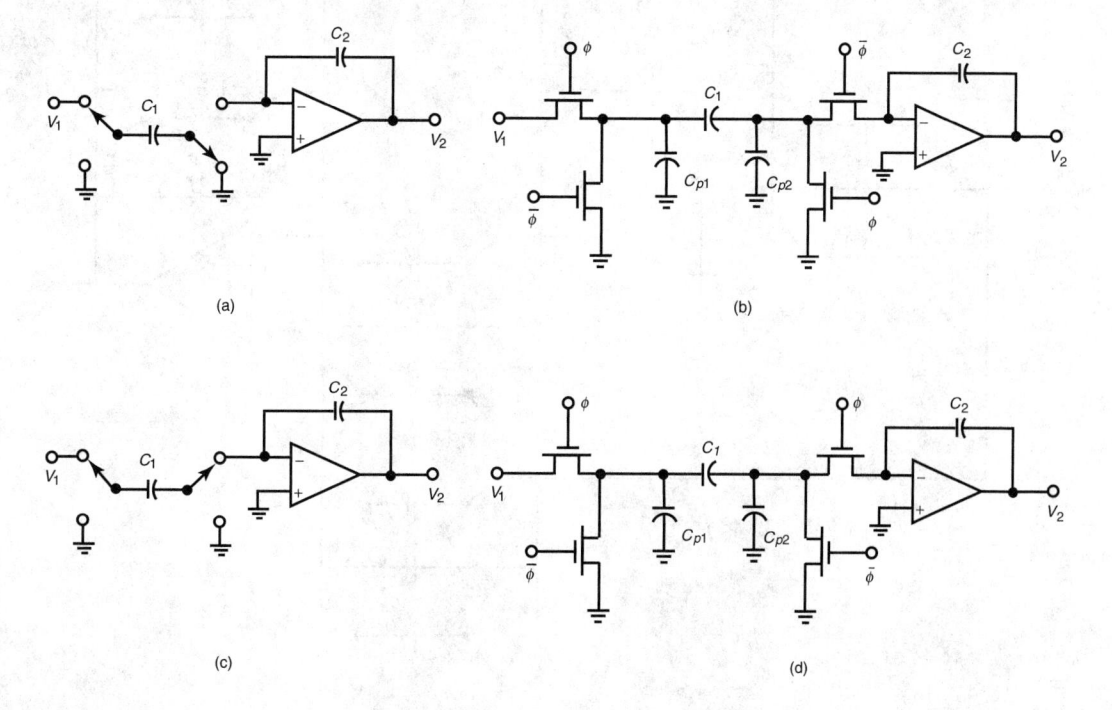

FIGURE 7.154 (a) Parasitic-insensitive FEDI, (b) MOS realization of (a), (c) parasitic-insensitive BEDI, (d) MOS realization of (c).

Thus, the integrator's output voltage v_2 is insensitive to these parasitic capacitors. Also, as a result of this insensitivity, the capacitance values C_1 and C_2 need not be much larger than the parasitic. Therefore, the size of the integrator can be made smaller than that of the parasitic-sensitive realization.

7.9.6 Switched-Capacitor Biquad

Switched-capacitor circuits that realize the DT **biquadratic transfer function** of Eq. (7.224) are known in the literature as SC **biquad**. They are useful circuits to filter designers because they form a fundamental building block for the construction of higher order filters in the form of cascade or multiple-loop feedback topologies. In this section we will introduce a SC biquad that will be used later to realize higher order filters. The SC circuit shown in Fig. 7.155 realizes the general discrete-time biquadratic transfer function (El-Masry, 1980)

$$T(z) = \frac{\alpha_0 + \alpha_1 z^{-1} + \alpha_2 z^{-2}}{1 + \beta_1 z^{-1} + \beta_2 z^{-2}} \tag{7.224}$$

with no exceptions. The circuit is based on the two-integrator loop topology and implements the parasitic-insensitive inverting and noninverting BEDI and FEDI, respectively.

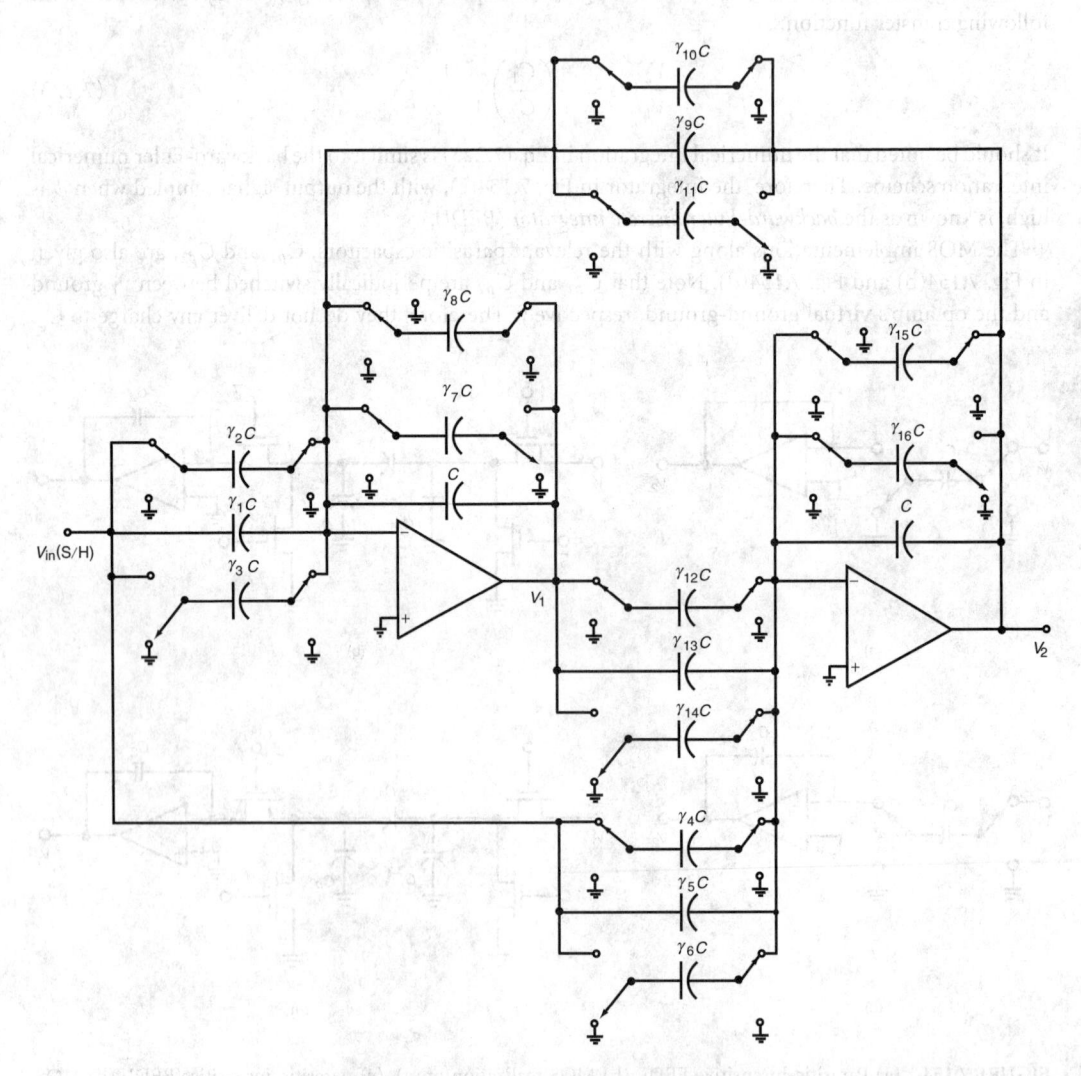

FIGURE 7.155 General SC biquad circuit.

Damping is provided, in general, by the SC $\gamma_8 C$ and $\gamma_{15} C$ (resistive damping) and by the SC $\gamma_{10} C$ and $\gamma_{12} C$ (capacitive damping). Usually only one of these four SC schemes is used. The SC $\gamma_7 C$, $\gamma_{11} C$, and $\gamma_{16} C$ realize positive feedback and the SC $\gamma_7 C$ and $\gamma_{16} C$ provide local positive feedback. Typically, only one of these capacitors is used. Transmission zeros are realized via the *feedforward approach*; that is, the input signal is fed to the input terminal of both integrators through capacitors $\gamma_1 C$ and $\gamma_5 C$ and SCs $\gamma_2 C$, $\gamma_3 C$, $\gamma_4 C$, and $\gamma_6 C$. Assuming that the switches are operated by nonoverlapping clocks and the input signal is sampled and held for the full period T_c, then the voltage-ratio transfer functions of the circuit are obtained as (El-Masry, 1980, 1983)

$$T_1(z) = \frac{V_1}{V_{in}}(z) = \frac{a_0 + a_1 z^{-1} + a_2 z^{-2}}{c_0 + c_1 z^{-1} + c_2 z^{-2}} \tag{7.225a}$$

and

$$T_2(z) = \frac{V_2}{V_{in}}(z) = \frac{b_0 + b_1 z^{-1} + b_2 z^{-2}}{c_0 + c_1 z^{-1} + c_2 z^{-2}} \tag{7.225b}$$

where

$$
\begin{aligned}
a_0 &= (\gamma_4 + \gamma_5)(\gamma_9 + \gamma_{10}) - (\gamma_1 + \gamma_2)(1 + \gamma_{15}) \\
a_1 &= (\gamma_1 + \gamma_2)(1 + \gamma_{16}) + (\gamma_1 + \gamma_3)(1 + \gamma_{15}) \\
&\quad -(\gamma_4 + \gamma_5)(\gamma_9 + \gamma_{11}) - (\gamma_5 + \gamma_6)(\gamma_9 + \gamma_{10}) \\
a_2 &= (\gamma_5 + \gamma_6)(\gamma_9 + \gamma_{11}) - (\gamma_1 + \gamma_3)(1 + \gamma_{16})
\end{aligned}
$$

$$
\begin{aligned}
b_0 &= (\gamma_1 + \gamma_2)(\gamma_{12} + \gamma_{13}) - (\gamma_4 + \gamma_5)(1 + \gamma_8) \\
b_1 &= (\gamma_4 + \gamma_5)(1 + \gamma_7) + (\gamma_5 + \gamma_6)(1 + \gamma_8) \\
&\quad -(\gamma_1 + \gamma_2)(\gamma_{13} + \gamma_{14}) - (\gamma_1 + \gamma_3)(\gamma_{12} + \gamma_{13}) \\
b_2 &= (\gamma_1 + \gamma_3)(\gamma_{13} + \gamma_{14}) - (\gamma_5 + \gamma_6)(1 + \gamma_7)
\end{aligned}
\tag{7.225c}
$$

$$
\begin{aligned}
c_0 &= (1 + \gamma_8)(1 + \gamma_{15}) - (\gamma_9 + \gamma_{10})(\gamma_{12} + \gamma_{13}) \\
c_1 &= (\gamma_9 + \gamma_{11})(\gamma_{12} + \gamma_{13}) + (\gamma_9 + \gamma_{10})(\gamma_{13} + \gamma_{14}) \\
&\quad -(1 + \gamma_7)(1 + \gamma_{15}) - (1 + \gamma_8)(1 + \gamma_{16}) \\
c_2 &= (1 + \gamma_7)(1 + \gamma_{16}) - (\gamma_9 + \gamma_{11})(\gamma_{13} + \gamma_{14})
\end{aligned}
$$

By using the bilinear z transformation

$$S = 2 f_c [(1 - z^{-1})/(1 + z^{-1})] \tag{7.226}$$

the equivalent analog pole frequency ω_0 and its Q-factor are obtained as

$$\omega_0 = 2 f_c \sqrt{\frac{c_0 + c_1 + c_2}{c_0 - c_1 - c_2}} \quad \text{and} \quad Q = \frac{\sqrt{(c_0 + c_1 + c_2)(c_0 - c_1 + c_2)}}{2(c_0 - c_1)} \tag{7.227}$$

It should be noted that as a result of the sampled and held input, the output voltages V_1 and V_2 will also be held constant for the full period T_c. This feature is very useful for the realization of higher order SC filters in cascade or multiple-loop feedback topologies.

Other popular SC biquads based on the two-integrator loop topology can easily be generated from the circuit of Fig. 7.155. For example, the biquad introduced by Fleisher and Laker (1979) can be obtained from the circuit of Fig. 7.155 by letting $\gamma_1 = \gamma_5 = \gamma_7 = \gamma_8 = \gamma_{11} = \gamma_{12} = \gamma_{13} = \gamma_{16} = 0$. Also, the biquad by Martin and Sedra (1980) can be obtained from Fig. 7.155 by letting $\gamma_3 = \gamma_4 = \gamma_6 = \gamma_7 = \gamma_8 = \gamma_{11} = \gamma_{12} = \gamma_{13} = \gamma_{15} = \gamma_{16} = 0$.

Defining Terms

Biquad: An electronic circuit that realizes a biquadratic transfer function.

Biquadratic transfer function: A rational function that comprises a ratio of second-order numerator and denominator polynomials in the frequency variable.

Forward-Euler discrete integrator (FEDI): A discrete-time integrator based on the forward-Euler numerical integration scheme.

Integrated circuit (IC): A combination of interconnected circuit elements inseparably associated on or within a continuous substrate.

Metal oxide semiconductor (MOS): A popular type of integrated circuit technology used to fabricate chips using silicon. It refers specifically to the layer sequence: metal, silicon-dioxide, and silicon.

Operational amplifier (op amps): An amplifier with large gain that has two input terminals (inverting and noninverting) and normally one output terminal.

Parasitic capacitor: An unwanted capacitor that is unavoidable adjunct of a wanted circuit element (a capacitor or a transistor). It results from the integrated circuit fabrication process.

Switched-capacitor filters: An electronic circuit (fabricated from operational amplifiers, transistors, and capacitors) designed to transmit some range of signal frequencies while discriminating others.

Very large scale integration (VLSI): A very dense integrated circuit that contains more than 1000 elements.

References

Allan, P.E. and Sanchez-Sinencio, E. 1984. *Switched Capacitor Circuits*. Van Nostrand Reinhold, New York.

Allstot, D.J. and Black, W.C. Jr. 1983. Technological design considerations for monolithic MOS switched-capacitor filtering systems. *IEEE Proc.* 71(967).

Boite, R. and Thiran, J.P.V. 1968. Synthesis of filters with capacitances switches, and regenerating devices. *IEEE Trans. on Circuits Theory* CT-15 (447).

Brodersen, R.W., Gray, P.R., and Hodges, D.A. 1979. MOS switched-capacitor filters. *IEEE Proc.* 67(61).

Caves, J.T., Copeland, M.A. , Rahim, C.F., and Rosenbaum, S.D. 1977. Sampling analog filtering using switched capacitors as resistor equivalents. *IEEE J. Solid-State Circuits* SC-12:592.

El-Masry, E.I. 1980. Strays-insensitive active switched-capacitor biquad. *Elec. Lett.* 16:480.

El-Masry, E.I. 1983. Strays-insensitive state-space switched-capacitor filters. *IEEE Trans. on Circuits and Systems* CAS-30:474.

Fettweis, A. 1968. Theory of resonant-transfer circuits and their application to the realization of frequency selective networks. Proc. Summer School on Circuit Theory. Czechoslovak Academy of Science, Prague.

Fleischer, P.E. and Laker, K.R. 1979. A family of active switched-capacitor building blocks. *Bell Sys. Tech. Journal* 58:2235.

Fried, D.L. 1972. Analog sampled-data filters. *IEEE J. Solid-State Circuits* SC-7 (Aug.):302–304.

Fudim, E.F. 1965. Capacitor-switch approach and its use in pneumatics. (Ph.D.) Technical Sciences Dissertation, Inst. of Control Sciences, Moscow, USSR, April 8.

Fudim, E.F. 1984. Fundamentals of the switched-capacitor approach to circuit synthesis. *IEEE Circuits and Systems Mag.* 6(12).

Ghausi, M.S. and Laker, K.R. 1981. *Modern Filter Design*, Chap. 6. Prentice-Hall, Englewood Cliffs, NJ.

Graeme, J.G. 1973. *Operational Amplifier: Third-generation Technologies*. McGraw-Hill, New York.

Gray, P.R. and Meyer, R.G. 1984. *Analysis and Design of Analog Integrated Circuits*. Wiley, New York.

Grebene, A.B. 1984. *Bipolar and MOS Analog Integrated Circuits Design*. Wiley, New York.

Gregorian, R. and Temes, G.C. 1986. *Analog MOS Integrated Circuits for Signal Processing*, Chap. 5. Wiley, New York.

Hosticka, B.J., Brodersen, R.W., and Gray, P.R. 1977. MOS sampled-data recursive filters using switched capacitor integrators. *IEEE J. Solid-State Circuits* SC-12:600.

Jury, E.I. 1964. *Theory and Application of the Z-Transform Method*. Wiley, New York.

Kuntz, W. 1968. A new sample and hold device and its application to the realization of digital filters. *IEEE Proc.* 56:2092.

Martin, K. and Sedra, A.S. 1979. Strays-insensitive switched-capacitor filters based on the bilinear *z*-transformation. *Elec. Lett.* 15:365.

Martin, K. and Sedra, A.S. 1980. Exact design of switched-capacitor band-pass filters using coupled-biquad structures. *IEEE Trans. on Circuits and Systems* CAS-27:469.

Maxwell, J.C. 1873. *A Treatise of Electricity and Magnetism*, p. 347. Clarendon Press, Oxford, England.

Moschytz, G.S., ed. 1984. *MOS Switched-Capacitor Filters: Analysis and Design*. IEEE Press, New York.

Nakayama, K. and Kuraishi, Y. 1987. Present and future applications of switched-capacitor circuits. *IEEE Circuits and Devices Mag.* 10(Sept.).

Unbenhauen, R. and Cichocki, A. 1989. *MOS Switched-Capacitor and Continuous-Time Integrated Circuits and Systems*. Springer-Verlag, New York.

Further Information

Additional information on the topic of switched-capacitor circuits is available from the following sources.

IEEE Transactions on Circuits and Systems, a monthly periodical published by the Institute of Electrical and Electronics Engineers. It deals with the analysis and design aspects of electrical circuits and systems.

IEEE Journal Solid-State Circuits, a monthly periodical published by the Institute of Electrical and Electronics Engineers. It deals with design aspects of electronics circuits.

IEEE Circuits and Devices Magazine, a bi-monthly magazine published by the Institute of Electrical and Electronics Engineers.

IEE Proceedings Circuits, Devices and Systems, a monthly periodical published in the U.K. by the Institute of Electrical Engineers. It deals with topics related to electrical circuits, electronics devices, and microelectronics systems.

IEE Electronics Letters, a bi-weekly periodical published in the U.K. by the Institute of Electrical Engineers and deals with various topics related to electrical/electronics circuits, electronics devices, and microelectronics systems.

IEEE International Symposium on Circuits and Systems is an annual symposium dealing with aspects of electrical and electronics circuits and systems.

The Midwest Symposium on circuits and systems is an annual symposium dealing with aspects of electrical and electronics. In addition the following books and tutorial papers are recommended.

Gray, P.R., Hodges, D.A., and Broderson, R.W. 1980. *Analog MOS Integrated Circuits*. IEEE Press, New York.

Gregorian, R., Martin, K.W., and Temes, G.C. 1983. Switched-capacitor circuit design. *IEEE Proc.* 71(8).

Tsividis, Y. and Antognetti, P., eds. 1985. *Design of MOS VLSI Circuits for Telecommunications*. Prentice-Hall, Englewood Cliffs, NJ.

Tsividis, Y. 1983. Principles of operation and analysis of switched-capacitor circuits. *IEEE Proc.* 71(8).

7.10 Semiconductor Failure Modes[4]

Victor Meeldijk

7.10.1 Discrete Semiconductor Failure Modes

Failures of semiconductor devices in storage, or dormant applications, are the result of latent manufacturing defects that were not detected during device screening tests. For discrete semiconductors, such as transistors, a large percentage of failures are the result of **die** and **wire bonding** defects and contamination. A common failure mode of spring loaded diodes is the contact material losing its compression strength, or slipping off the die, resulting in an open circuit.

---

[4]Portions of this chapter were adapted from: Meeldijk, V. 1995. *Electronic Components: Selection and Application Guidelines*, Chaps. 10 and 11. Wiley Interscience, New York. Used with permission.

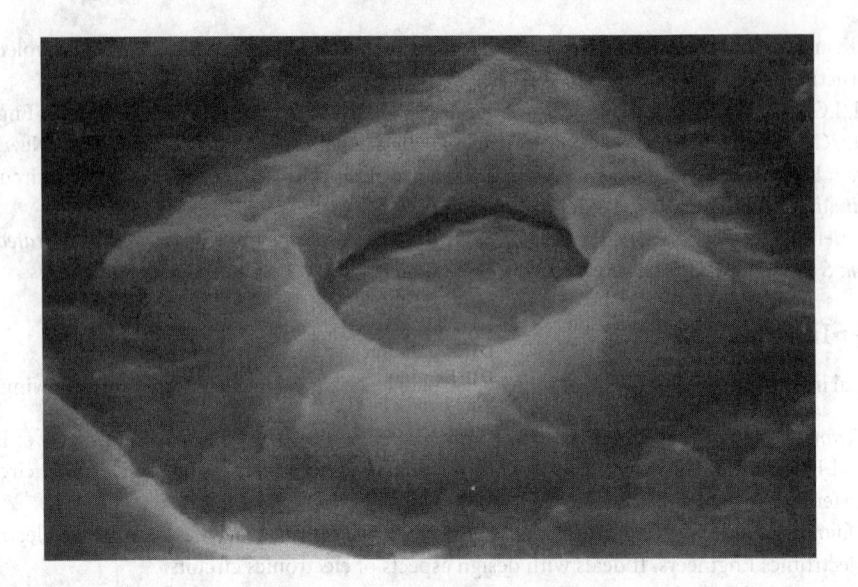

FIGURE 7.156   Excessive thinning or metallization.

Failure mechanisms can be grouped into three categories:

- Those independent of environment (oxide defects, diffusion defects).
- Failures dependent on environment (wire bond or **metallization** defects, see Fig. 7.156 and Fig. 7.157).
- Failure mechanisms that are time and environment dependent (metal migration, corrosion, inter-metallic compound formulations, such as caused by dissimilar metal use).

Table 7.9 and Sec. 7.10.6 discusses the various screening tests to induce failures prior to components being incorporated in assemblies.

Common failure modes of discrete semiconductors are zener diode: 50% short, 50% open; junction diode: 60–70% high reverse, 20–25% open, 10–15% short; silicon controlled rectifier (SCR): 2% open, 98% short; and transistor: 20% high leakage, 20% low gain, 30% open circuit, 30% short circuit (Fig. 7.158).

Interpretation of failure data by the Reliability Analysis Center (RAC, 1991), Rome, New York (used with permission), has the following failure mode breakdowns:

- General purpose diode: 49% short, 36% open, 15% parameter change
- Diode rectifier: 51% short, 29% open, 20% parameter change
- Small signal diode: 18% short, 24% open, 58% parameter change
- Microwave diode: 50% open, 23% parameter change (drift), 10% short, 17% intermittent
- Zener diode reference: 68% parameter change (drift), 17% open, 13% shorted, 2% intermittent
- Zener diode regulator: 45% open, 35% parameter change (drift), 20% shorted
- Optoelectronic LED: 70% open, 30% short
- Optoelectronic sensor: 50% short, 50% open
- Thyristor: 45% failed off, 40% short, 10% open, 5% failed on
- Triac: 90% failed off, 10% failed on

Further breakdown on transistor failure modes by the Reliability Analysis Center (RAC) are

- Bipolar transistor: 73% shorted, 27% open

(Note these new data contrast with that published in January 1976 by the U.S. Army Material Command (AMCP-706-196), which had 59% high collector to base leakage current, 37% low collector to emitter

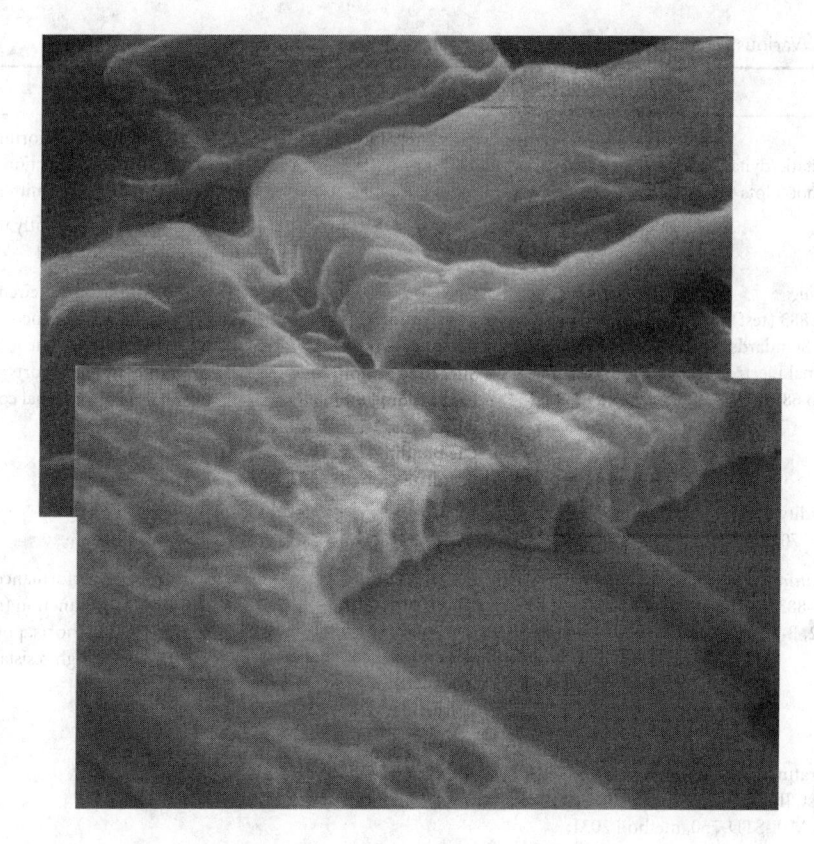

FIGURE 7.157 Separation/depression of metallization and thinness of metallization. The second photograph shows an acceptable metallization step.

breakdown (Bvceo), and 4% open circuited).

- Field effect transistor (FET): 51% shorted, 22% low output, 17% parameter change, 5% open, 5% output high
- GaAs FET transistor: 61% open, 26% shorted, 13% parameter change
- Radio frequency (RF) transistor: 50% parameter change, 40% shorted, 10% open

7.10.2 Integrated Circuit Failure Modes

Most IC failures, as for semiconductors, are related to manufacturing defects (Fig. 7.159 is the inside view of a typical acceptable IC). A breakdown of typical IC failure mechanisms is 40.7% wire bond interconnects, 23.4% misapplication/misuse, 4.2% **masking**/etching defect, 3.3% die mechanical damage, 1.4% cracked die, 0.9% die metallization corrosion, 0.9% die contamination, and 24.8% other causes.

These mechamisms result in the following typical failure modes:

- Digital devices: 40% stuck high, 40% stuck low, 20% loss of logic
- Linear devices: 10–20% drift, 10 output high or low, 70–80% no output

Interpretation of failure data in 1991 by the Reliability Analysis Center (RAC) found the following failure mode percentages:

- Digital bipolar: 28% output stuck high, 28% output stuck low, 22% input open, 22% output open
- Digital MOS: 8% output stuck high, 9% output stuck low, 36% input open, 36% output open, 12% supply open

TABLE 7.9 Various Screening Tests Used to Detect Defects

Test	Process Defect	Failure Mode
Electrical: DC tests, static/dynamic tests, functional tests, Schmoo plots	Slice Preparation Etching Wire Bonding	Improper electrical performance, opens, shorts, degraded junction characteristics, intermittents
	Improper marking	Device operates differently than expected, or not at all
Thermal cycling: MIL-STD-883 (test) method 1010; Japanese Industrial Standards Office (JIS) C 7022 A-4; International Electrotechnical Commission (IEC) PUB 68; Test Na, Nb	Slice preparation **Passivation** (Fig. 7.162) Etching; Diffusions Metallization **DIE separation** **DIE bonding** Wire bonds; Final seal Fatigue cracks	Potential short or open circuits, low breakdown voltage, increased leakage, poor performance due to faulty components, near shorts, near opens, high resistance internal connections
(Temp/humidity cycling MIL-STD-883 method 1004; JIS C 7022 A-5; IEC PUB 68; Test Z/AD)		
High temperature storage: MIL-STD-883, method 1008; JIS C 7022, B-3	Etching Slice preparation Diffusions; Metallizations Lead bonds; Final seal Creep; Corrosion Electromigration; Surface charge spreading	Lowered device performance due to degradation of junction (soldering characteristics, shorts, potential shorts, opens, high resistance)
High temperature/high humidity bias test and storage test, JIS C 7022, B-5; IEC PUB 68; Test C (heat test, MIL-STD-750, method 2031; JIS C 7022; A-1; IEC PUB 68; Test Tb)		
(High temperature operation, MIL-STD-883 method 1005; JIS C 7022 B-1)		
(Low temperature storage, JIS C 7022, B-4; IEC PUB 68; Test A)		
Reverse bias voltage:	Slice preparation Final seal	Degradation of junction characteristics, performance degradation
Operational life test:	Slice preparation Passivation; Etching Diffusions; Metallization Die bonding; Final seal Dendrite growth	Degradation of junction characteristics, short circuits, leakage, low insulation resistance, performance degradation from heating or faulty or unstable internal components, opens, near opens, near shorts, high resistance internal connections
Mechanical: Variable freq. vibration MIL-STD-883 method 2007; JIS C 7022; A-10; IEC PUB 68; Test Fc	Die separation Die bonding; Wire bonds; Die cracks; Lead shape; irregularities package; irregularities creep	Performance degradation caused by overheating, opens or potential opens, shorts or intermittent shorts
Solderability: MIL-STD-883; METHOD 2003; JIS C 7022; A-2;	Lead finish/plating	Intermittent, open or high resistance solder connections between the part and the circuit board
Salt fog (atmosphere) MIL-STD-883 method 1009; JIS C 7022; A-12; IEC PUB 68; Test Ka		Tests resistance to corrosion in a salt atmosphere

TABLE 7.9 Various Screening Tests Used to Detect Defects (*Continued*)

Test	Process Defect	Failure Mode
Thermal shock: MIL-STD-883; method 1011; JIS C 7022; A-3; IEC PUB 68; Test Nc	Die separation Die bonding	Performance degradation caused by overheating, opens or potential opens, shorts or intermittent shorts
X Ray:	Die bonding; Wire bonding Final seal Lead frame shape irregularities	Performance degradation caused by overheating, intermittents, shorts or opens, intermittent short circuits caused by loose conductive particles in the package.
Acceleration: Constant acceleration; MIL-STD-883 method 2001; JIS C 7022; A-9; IEC PUB 68; Test Ga	Die bonding Final seal	Performance degradation caused by overheating, weak wires will be open or intermittent, intermittent shorts caused by loose conductive particles in the package
Hi-voltage test:	Passivation Final seal	Open or short circuits, increased device leakage, low insulation breakdown
ESD testing: MIL-STD-883; method 3015 Electronic Industries Association of Japan (EIAJ) IC 121:20	Design susceptibility to ESD	Degraded device performance or device failure.
Vibration/PIND test:	Die bonding Wire bonding	Shorts or intermittent shorts caused by loose conductive particles in the package, performance degradation caused by overheating, opens and intermittents
Lead fatigue: Lead integrity; JIS C 7002; A-11; IEC PUB 68; Test U	Final seal	Open circuits, external lead loose
Leak tests: Hermetic seal tests; MIL-STD-883 method 1014; JIS C 7022; A-6; IEC PUB 68; Test Q [Pressure cooker test (PCT) to evaluate moisture resistance in a short period of time; EIAJ IC-121: 18]	Final seal	Performance degradation, shorts or opens caused by chemical corrosion or moisture, intermittents
Precap visual:	Slice preparation Passivation; Masking Etching; Metallizations Die separation Die bonding; Wire bonding	Increased leakage, low insulation breakdown, opens, shorts, intermittents, potential shorts or opens, high internal resistances, intermittent shorts
Visual:	Final seal	Open circuits; package cracks; package seal problems

- Digital programmable array logic (PAL): 80% failed truth table, 20% shorted
- Interface IC: 58% output stuck low, 16% input open, 16% output open, 10% supply open
- Linear IC: 50% degraded/improper output, 41% no output, 3% shorted, 2% open, 2% drift
- Linear-op amps: 69% degraded (unstable, clipped output, drifted etc.), 13% intermittent, 10% shorted, 6% overstressed by transients, 3% no output
- Bipolar memory: 79% slow transfer of data, 21% data bit loss
- MOS memory: 34% data bit loss, 26% short, 23% open, 17% slow transfer of data
- Digital memory: 30% single bit error, 25% column error, 25% row error, 10% row and column error, 10% complete failure

FIGURE 7.158   A failed transistor.

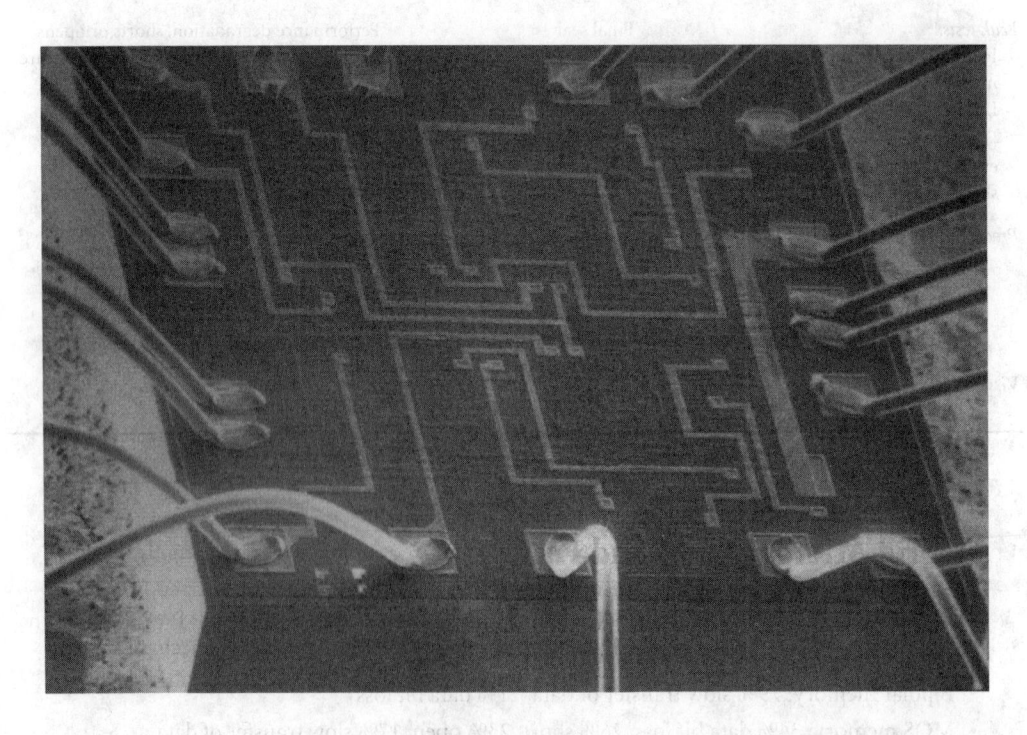

FIGURE 7.159   Inside view of an acceptable IC showing **bond pads** and metallization.

TABLE 7.10 Hybrid Microcircuits—Failure Accelerating Environments

Accelerating Environment	Environments Failure Mechanism	Failure Mode
Mechanical stress (thermal shock, vibration)	Substrate bonding; Cracked/broken substrate; Faulty bonds	Open circuits
High temperature	Damaged resistors	Open or out of tolerance
Thermal cycling	Cracked resistors; Various film defects Excess bonding time Thermal coefficient of expansion mismatch between film and substrate	Open or out of tolerance
High voltage and temperature test	Various film defects; Excess bonding time Thermal coefficient of expansion mismatch between film and substrate	Out of tolerance
Thermal and mechanical stress	Cracked dice; Shorted wires Faulty bonds; Bond separation	Open circuits; Short circuits

- Digital memory RAM: 25% no operation at cold temperatures, 17% parameter change, 16% shorted, 13% open, 13% incorrect data, 7% contaminated (particles inside cavity of hermetic sealed unit)
- Ultraviolet erasable programmable read only memory (UVEPROM): 94% open (unprogrammable) bit locations, 6% would not erase
- Hybrid devices: 51% open circuit (caused by resistor/capacitor opens, internal bond pad corrosion and electromigration), 26% degraded/improper output (distorted or slow response time output), 17% short circuit (**electrostatic discharge (ESD)** overstress, fractures, shorting wire bonds), 6% no output

The analysis data, which can be interpreted by the reader for various applications, is in their report (RAC, 1991). Most vendors have reliability reports on their products that discuss the results of accelerated life tests and provide a device failure rate.

7.10.3 Hybrid Microcircuits and Failures

Hybrid devices are a combination of active components (microcircuits or ICs) and various passive discrete parts mounted on a substrate and interconnected by conductive film traces. Hybrid failure mechanisms, like those of discrete and integrated circuits, are primarily due to manufacturing defects. Failures will occur whether the device is operating or dormant. Various accelerating environments, such as temperature or vibration can reveal the failure mechanisms before parts are shipped (or used in circuits). Table 7.10 lists the accelerating environments and the problems they reveal.

7.10.4 Memory IC Failure Modes

Memory ICs failures result in the inability to read or write data, erroneous data storage, unacceptable output levels, and slow access time. The specific failures that cause these problems are as follows:

- Open and short circuits: can cause various problems from a single bit error to a catastrophic failure of the whole device.
- Open decoder circuits: cause addressing problems to the memory.
- *Multiple writes:* writing one cell actually writes to that cell and other cells (multiple write and address uniqueness problems).
- Pattern sensitivity: the contents of a cell become complemented due to read and write operations in electrically adjacent cells (cell disturbances, adjacent cell disturbances, column disturbances, adjacent column disturbances, row disturbance, adjacent row disturbance).

- *Write recovery:* the access time of the device may be slower than specified when each read cycle is preceded by a write cycle, or a number of write cycles.

- Address decoder switching time: this time can vary depending on the state of the address decoder prior to switching and on the state to which the decoder is switching.

- Sense amplifier sensitivity: memory information may be incorrect after reading a long series of similar data bits followed by a single transition of the opposite data value.

- *Sleeping sickness:* the memory loses information in less than the specified hold time (for DRAMs).

- Refresh problems (may be static or dynamic): the static refresh test checks the data contents after the device has been inactive during a refresh. In dynamic refresh, the device remains active and some cells are refreshed. All cells are then tested to check whether the data in the nonrefreshed cells is still correct.

Special tests are needed for memory devices to insure the integrity of every memory bit location. These functional tests are composed of four basic tests, pattern tests, background tests, timing test, and voltage level tests. A summary of these tests is contained in Table 7.11.

Memory tests cannot be specified to test each part 100% as a RAM can contain any one of 2^N different data patterns and can be addressed in N factorial ($N!$) address sequences without using the same address twice. Test times shown may be extremely long for some high-density memory devices; a GALPAT test for a 4 M DRAM would take 106 days to execute. It is easy to see how impractical it is to overspecify device testing. Any test plan should be developed along with the supplier of the memory devices you are to purchase.

System level tests may use bit-map graphics where the status of the memory cells is displayed on the cathode ray tube (CRT) display of the system (i.e., failing bits show up as a wrong color). Various patterns are read into the memory and the CRT may be configured to show 1 cell equivalent to one pixel (a picture element) or compressed to show 4, 16, or 64 cells per pixel depending on the number of pixels available and the memory size. Zoom features can be used to scale the display for close examination of failing areas of the memory.

Note, in large memory intensive systems, consider using error correction codes to correct hard (permanent failure) and soft errors (i.e., alpha particle upsets, a single upset that does not occur again). One vendor of video RAMs estimates a soft error rate of a 1 Meg part of 3.9 FITs (at a 500-ns cycle time with Vcc at 4.5 V, and a 4 Meg DRAM error rate of 41 FITs (at a 90% confidence level, 5 Vcc operation with 15.625-μs cycle (refresh) rate). Soft error rates are dependent on cycle time and operating voltage; the lower the voltage and faster the cycle time, the higher the failure in time rate. One thousand FITs equals 1 PPM (1 failure in 10^6 h), which equals 0.1%/1000 (0.1% failures every 1000 h). When calculating soft error rates, add the refresh mode rate and the active mode rate, which can be determined from acceleration curves provided by the manufacturer.

7.10.5 IC Packages and Failures

Devices can be classified by package style, which can either be **hermetically** sealed (**ceramic** or metal cans) or nonhermetic (epoxy, silicones, phenolics or plastic encapsulated). Materials employed in most microcircuits will change very slowly if stored in a dry environment at a constant low temperature.

Hermetic packages, glass, ceramic, or metal, protect the device from environmental atmospheric effects. Depending on the metal package it can take anywhere from 50 to well over 100 years for the inside of the IC package to reach 50% of the humidity of the outside air. An epoxy package can take anywhere from minutes to a few days. Therefore, equipment designed for harsh environments, such as might be encountered in military applications, use hermetically sealed devices. Plastic packaged devices are low cost, offer mechanical shock and vibration resistance, are free from loose particles inside the device as they do not have a cavity, and are available in smaller package sizes. All plastics, however, contain some moisture and are permeable to moisture to some degree. (Salt environments have also been shown to

TABLE 7.11   Tests for Memory Circuits

Test	Comments
March	An N pattern test that checks for functional cell failures (opens and shorts and address unique faults) and data sensitivity. In this test a background of 0s is written into memory. Then a 1 is written into each sequential cell. Each cell is then tested and changed back to 0 in reverse order until the first address is reached. The same sequence is then repeated by loading in a background of 1s.
Masest	An N pattern test that checks for address uniqueness and address decoder problems (open decode lines). This is an alternating Multiple-Address Selection test that writes a background of 0s to the memory device at the min address, then the max address, then the min $+1$ address, then the max -1 address, etc. All cells are then read and verified and a complementary pattern of 1s is read into the memory device and the test is repeated.
Refresh	Any one of several patterns designed to test for an acceptable refresh time. Usually an N pattern test
Sliding diagonal	An $N^{3/2}$ pattern that checks for address decoder problems, sense amplifier faults, and sense amplifier recovery. This test is a pattern that generates a diagonal of data in a field of complement data, and slides the diagonal of data in the x direction, creating a barber pole effect.
Surround disturbance	An N pattern test that checks for adjacent cell disturbance. Variations include a row disturbance and column disturbance.
Galloping tests	This is a sequence test where a base cell, containing complement data from the rest of the cells in its set (i.e., the base cell is 1 but the other cells are 0). A set may be a column, row, device or adjacent cells to the base cell.
Walking tests	Walking tests are similar to galloping tests except that the cells in the set with the base cell are read sequentially instead of alternately with the base cell.
Galloping/walking tests: GALPAT rows (galloping rows) Rows	An $N^{3/2}$ pattern test that checks adjacent cell walking disturbance and row/column disturbance.
GALPAT columns (galloping columns) Walking columns	An $N^{3/2}$ test that checks for row/column disturbances.
GALPAT (galloping pattern test)	A $4N^2$ pattern test that checks for address decoder problems, adjacent cell disturbances, row/column disturbances and sense amplifier recovery. This is a long execution time test. Into a background of 0s the first cell is complemented (to 1) and then read alternately with every other cell in memory. This sequence is repeated until every memory cell is tested. The test is then repeated with a background of 1s with the cell changed to 0.
GALDIG (galloping diagonal) pattern test	This test finds unsatisfactory address transitions between each cell and the positions in the cell's diagonal. It also finds slow sense amplifier recovery and destruction of stored data due to noise coupling among cells in a column. In this test a diagonal of 1s is written into a background of 0s. All cells are read and verified and the diagonal is then shifted horizontally. This keeps repeating until the diagonal has been shifted throughout the memory. The test is then repeated with complemented data.
GALWREC	A galloping write recovery test, an N^2 pattern test

(continued)

TABLE 7.12   Tests for Memory Circuits (*Continued*)

Test	Comments
WALKPAT (walking pattern)	An N^2 test that checks for address uniqueness and sense amplifier recovery. This is a long execution time test. In a background of 0s, the first cell is complemented and then all the other cells are read sequentially. The cell is then changed back to zero and the next cell is complemented and all the other cells are then read sequentially again. This repeats for all of the memory cells and then repeats again with a background of 1s with each cell changed to a 0. (A single 1 moves through a field of 0s and a single 0 is moved through a field of 1s.)
Dual walking columns	An $N^{3/2}$ pattern test
MOVI (moving inversion)	This test checks for address propagation failures. It is similar to a march but in addition uses each address line as the least significant bit in a counting sequence.
Scan test	Each cell is written, or read, sequentially. The addressing sequence can count up from the minimum address (minmax) or down from the maximum address (maxmin). Either the x or y axis can be the fast axis (rowfast or columnfast, respectively).
Address complement	An addressing sequence in which all address lines, except one, are changed in each cycle.
Background tests (may be used with the preceding tests)	
Checkerboard inverted checkerboard	An N pattern test that checks for functional cell failures and data sensitivity (to noise). Each cell is checked to verify that it can be forced to a 1 or a 0. Each cell is surrounded by cells with complementary data (a logical 1 stored in a cell would be surrounded by cells that have logical 0 stored in them). Variations include double checkerboard and complemented double checkerboard which have the same execution time.
Pseudocheckerboard	This checkerboard pattern is generated by exclusive OR-ing data with the least significant row address bit and the least significant column address bit (data XOR X0 XOR Y0). This may not generate a true checkerboard background in parts that have folded digit lines, and so a different equation may be necessary for these parts.
Zeros/ones	An N pattern test. A background of either solid 1s or 0s is stored in each cell.
Row stripes	Background data where every row is surrounded by its complement. Therefore a row with logical 1 stored in it would be surrounded by rows with logical 0 in them.
Column stripes	Like row stripes but background data relates to columns instead of rows.
Double variations	The size of the background is doubled, two rows, two columns, or a double checkerboard with two rows by two columns.
Parity	Background data that are generated when the parity of the x and y addresses is even or odd.
Bit map graphics (test that can be used to check memory at system level)	Used to examine patterns of failing cells, characterize the operating regions of memory devices, verify device specifications. This test identifies mask misalignment, diffusion anomalies, contamination defects, design sensitivities and semiconductor defects.

effect silicone encapsulated devices but leave epoxy novalac parts relatively unaffected). The mismatch in thermal coefficients of the plastic vs. the die also reduces the operating temperature of the device from 0 to 70°C vs. hermetic packages, which can operate from −55 to +125°C.

Studies have been done by the Microelectronics and Computer Technology Corporation (Austin, Texas) and Lehigh University on replacing heremetic packages with more effective lightweight protective coatings. Their program, *reliability without hermeticity* (RwoH), was initiated by a research contract awarded to them by the U.S. Air Force Wright Laboratory. The RwoH is an industry working group, initiated in 1988, to address nonhermetic electronic packaging issues, specifically, coating to provide environmental protection for semiconductor devices and multichip modules.

The Joint Electron Device Engineering Council (JEDEC) specification A112 defines six levels of moisture sensitivity with level 1 being the highest rating, or not moisture sensitive. The other levels are reduced levels, such as level 3, which ensures product integrity (i.e., from **popcorning** from entraped moisture during soldering) when stored at 30°C, 85% Rh for 168 h. Level 4 indicates package integrity is ensured for 84 h, and level 6 only permits 6 h of exposure. Plastic parts shipped in sealed bags with desiccant usually are designed for a 12-mo storage and should only be opened when the parts are to be used. Parts stored for longer than this time, especially plastic quad flatpack (PQFP) packaged devices, should be baked to remove the moisture that has entered the package. Entrapped moisture can vaporize during rapid heating, such as in a solder reflow process, and these stresses can cause package cracking. Subsequent high-temperature and moisture exposures can allow contaminants to enter the IC and cause failure at a later time due to corrosion.

Note Hughes Missile Systems reported in 1995 of conditions where organic-based (such as organic acid (OA)) water-soluble fluxes lead to dendrite growth and short circuits. It was found that component seals of high lead glass **frit** (glass frit is used in cerdip and ceramic flat pack components) and tin and lead on the device **leads** results in the growth of lead dendrites on the surface of the glass seals. The lead comes from the high lead oxide content of the soft glass itself. High temperatures accelerate the dendrite growth, with shorts occurring in as little as 5 min with 140°F warm flux. Prevention of this problem includes using different fluxes and minimizing the application of fluxes to the component side of the PC board. Dendrite growth has also been reported for kovar flatpacks stored in foil-lined cardboard packaging. The chlorine dendrite growth, resulting in a high electrical leakage from pins to case, resulted from the chemical impurities in the cardboard and the device. Fiberboard and paperboard packaging is no longer used for device unit packaging. **Tin whisker** growth also occurs with parts having pure tin plating on the leads.

Specific IC packages have definite failures associated with their use. These include

- *Ball grid array* (BGA) or *Overmolded pad-array carrier* (OMPAC): In manufacture, the primary failure mode for this package is short circuits (98% vs. 2% open circuits).
- *Flip chip:* In manufacture, the primary failure mode for a flip chip on an MCM module is a short circuit.
- PQFP: Packages are susceptible to moisture-induced cracking in applications requiring reflow soldering.
- Quad flatpack (QFP): The most common quad flatpack manufacturing failure mode is a short circuit (82% short circuits, 2% open circuits, and 16% misalignment problems).
- Tape carrier package (TCP): Tape automated bonding (TAB) package has an open circuit (98% vs. 2% misalignment problems) failure mode.

Table 7.9 uses the environmental conditions as screening tests to detect parts that may prematurely fail because of a problem or manufacturing defect. The table shows the various screening tests that can be used to detect the various defects that effect microcircuit reliability. The screens that are rated as good to excellent in identifying component defects, and that are low in cost, are high-temperature storage, temperature cycling, thermal shock, nitrogen bomb test (this tests hermetic package seals), and gross leak test (which also tests hermetic package seals). Screening tests are done prior to putting devices in stock for future use. Proper ESD procedures and controls should be followed, and because of ESD degradation effects the parts should not be periodically handled or tested; only electrically test devices prior to use.

Component failures, such as pin to pin leakage, can also be caused by contaminating films that are on the seal and embedment glass of the package. One such failure mode reported for SRAMs was caused by lead sulfide from antistatic ABS/PVC (acrylonitrile-butadiene-styrene/polyvinyl chloride) plastic trays used to store parts during assembly and postseal operations. The lead sulfide exhibited itself as a shiny film over the seal and embedment glass and resulted in more than 20 μA (at 5 V) leakage. PVC plastic that uses Thioglycolate (or Dibutyltin bis or Iso Octylthioglicolate, heat stabilizers), a common additive, can cause sulfur-based contamination (and corrosion) on parts with lead-based embedment glass. Because of this, PVC manufacturers are pursuing alternate stabilizer materials, and products that can have performance degradations are often stored and processed using metal trays.

7.10.6   Lead Finish

Reducing lead in the environment is an environmental health and safety issue that is of prime importance today. While a typical microprocessor has approximately 0.2 g of lead and a computer motherboard has 2 to 3 g of lead, and the whole USA electronic interconnection market uses less than 2% of the world's lead production, to address environmental concerns, there are various environmental directives limiting the use of lead and other hazardous substances. These directives, formulated by the European Union in a Directive of the European parliament and of the Council of the European Union are:

- End of Life Vehicles (ELV) Directive 2000/53/EC in force since October 21, 2000, requires that products be free of heavy metals such as mercury, cadmium, hexavalent chromium and lead. It requires a recycling system be established by vehicle manufacturers for vehicles made after July 1, 2002 and for all vehicles regardless of date of manufacture by July 1, 2007. Lead can still be used as an alloying additive in copper and in solderable applications.
- The WEEE Directive, Waste Electrical and Electronic Equipment (WEEE) Directive 2002/96/EC, expands the recycling requirements of the ELV Directive to include a broad range of electronic and electrical appliances and equipment. WEEE went into effect on February 13, 2003. It is to be scheduled to become European national law by August 13, 2004, and be applicable to consumer use products by August 13, 2005. Article 2(3) however states "Equipment which is connected with the protection of the essential interests of the security of Member States, arms, munitions and war material shall be excluded from this Directive. This does not, however, apply to products which are not intended for specifically military purposes."
- RoHS Directive, The Restriction of Hazardous Substances in Electrical and Electronic Equipment, this Directive 2002/95/EC establishes standards and limits for the hazardous material content in electronic and electrical equipment. The Directive went into effect on February 13, 2003. It is scheduled to become European national law by August 13, 2004 for be in force for products by July 1, 2006. Banned or restricted substances include lead, mercury, cadmium, hexavalent chromium, certain brominated flame retardants (PBBs), and polybrominated diphenyl ethers (PBDEs).

The recommended lead-free solder formulation is Sn-Ag-Cu for board assembly but there are other formulations such as Nickel-Palladium (NiPd), or Nickel-Palladium with Gold flash (NiPdAu). Passive components, to be compatible with a lower temperature Lead process (which is 215°C for 50/50 Tin/Lead formulations and 230°C for 40/60 formulations) and the higher lead-free process of up to 260°C, use pure matte Tin for their contacts. The use of lead in solder is partially based on several potential reliability issues. Pure Tin component leads have been shown to result in inter-metallic migration in the termination of the electronic component and the growth of tin whiskers which could cause short circuits (which is why there is a exemption for military use (only) components).

The National Electronics Manufacturing Initiative (NEMI) has addressed the problem of "tin whiskers" in lead-free assemblies. A tin whisker is defined by them as

A spontaneous columnar or cylindrical filament, which rarely branches, of monocrystalline tin emanating from the surface of a plating finish. Furthermore, tin whiskers may have the following characteristics:

- an aspect ratio (length/width) greater than 2
- can be kinked, bent, twisted
- generally have consistent cross-sectional shape
- may have striations or rings around it

Their recommended test method is

- temperature cycling ($-55°C$ to $+85°C$, approximately 3 cycles/hour)
- temperature humidity tests at $60°C/93\%$ RH
- ambient storage (air-conditioned facility)

(Note: Tin will undergo a phase transformation, becoming a powdery form (called Tin Pest) if Tin plated parts are stored for more than 1 week at temperatures below $13°C$).

7.10.7 Screening and Rescreening Tests

In the 1970s, the Navy instituted a policy requiring devices be rescreened on receipt (by the contractor or an independent test laboratory) because there was evidence that Joint Army Navy (JAN) qualified devices (including JANTX, JANTXV discrete component) were not of a quality level to be used in military hardware (DoD directive 4245.7-M Transition from Development to Production, which was signed in 1984 and saw wide implementation by the end of 1986).

Rescreening tests often imposed included the following:

Destructive physical analysis (DPA) examination tests, where two pieces of each lot of parts (as a minimum) were cut open and examined to determine the workmanship of the parts. If the workmanship was judged to be poor, the whole lot of parts were rejected (a lot of parts is defined as parts from one manufacturer from one assembly line and one date code).

Particle impact noise detection (PIND) tests, where each hybrid part or IC with a cavity where the die was unglassivated (insulated with a glass coating, Fig. 7.160 shows an IC with cracks in this coating), was vibrated and transducers, mounted on the part would detect if there were any particles rattling around. Parts with loose pieces in them were rejected from the shipment.

Go-no go electrical tests, and static and dynamic tests were required to be performed at low, ambient ($25°C$) and high temperatures.

Hermeticity testing was required to test the integrity of the package seal.

To review the reasons for component rescreening we need to examine test data for the time period involved. In 1981 it was reported that the Naval Weapons Support Center (NWSC) in Crane, Indiana, found part defect rates of 15% for ICs and 17% for discrete semiconductors. In a January 1983 report, one independent test laboratory found reject rates of 16.7% for linear ICs, 8.4% for digital ICs, and 9.2% for CMOS ICs. By October 1984, this same laboratory found defect rates of 5.5% for linear ICs, 3.7% for digital ICs, 8.2% for transistors, and 3.8% for optoelectronics. In 1983, the Defense Electronic Center found semiconductor defect rates of 8% for nonmilitary specification parts, with a less than 1% defect rate for military specification devices. In 1986, a military contractor estimated rescreening failures of 0.9% for ICs and 1.5% for transistors. In 1989, a manufacturer of a special purpose computer workstation for the U.S. Navy reported the following rescreening results: IC/semiconductors tested: 8312 (127 types of parts); total rejects PIND/DPA: 25 (16 PIND failures on hybrid components, 6 electrical failures, 3 mechanical failures—broken leads); reject rate: 0.30%.

Breaking down the hybrid failures, of 99 military qualified oscillators tested, 6 were rejected for a rejection rate of 6.1%; 53 nonmilitary qualified oscillators were tested of which 10 were rejected, for a rejection rate of 18.8%.

The Semiconductor Industry of America conducted a quality statistics program to monitor and report on industry data on various quality control indices and parameters of microcircuits from 1985 to the early

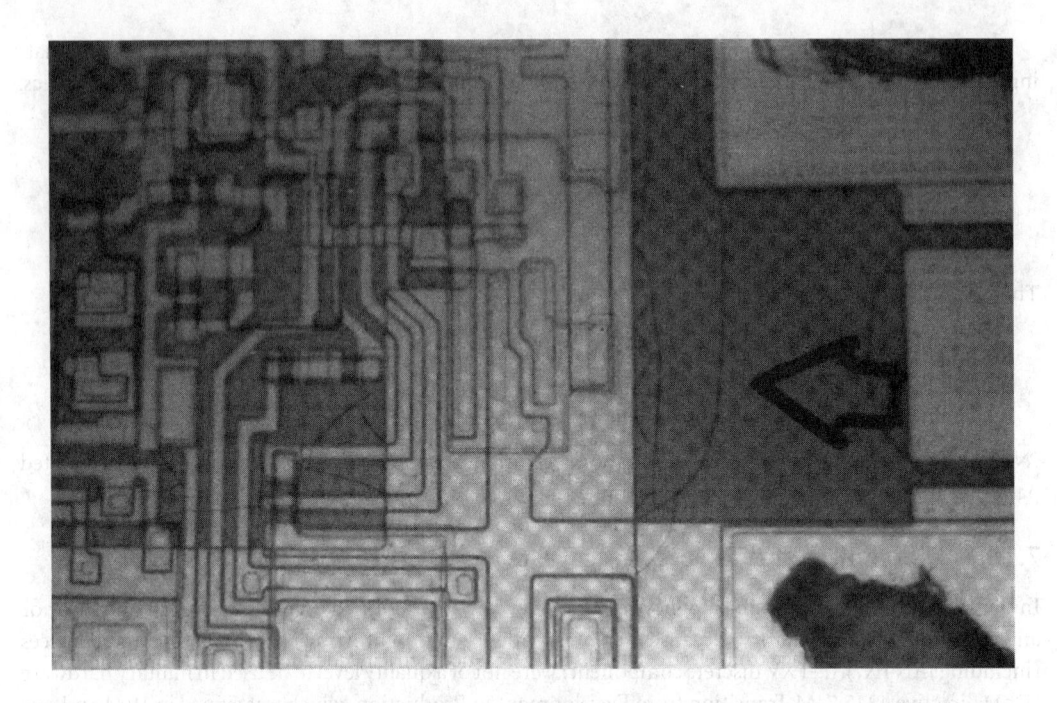

FIGURE 7.160   An IC with cracks in the die glass coating.

1990s. In 1991, the analysis of the data showed that for every 10,000 parts shipped, there is on average only one part with electrical defects, or a 100 **parts per million (PPM)** defect rate. The data were reported using JEDEC standard 16, Assessment of Microcircuit Outgoing Quality Levels in Parts Per Million and represents data on JAN qualified parts, DESC drawing parts, MIL-STD-883C screened devices, and devices procured to source control drawings (SCDs). Over 150 million devices were sampled with the reporting companies supplying about 80% of all military microcircuits.

The breakdown of defects based upon part type are linear parts: 100 PPM, bipolar digital 50 PPM (10 PPM in the first quarter of 1991), MOS digital: <100 PPM, and MOS memory: 160 PPM (with defects falling an average 30 PPM/quarter). If all the data are combined, the data indicate that for every 25,000 parts shipped, there averages only 1 part with electrical defects. The study also shows that there is a steady and continual decline in the level of electrically defective parts. This analysis shows a significant improvement in part quality since the 1970s, when component rescreening was instituted, and rescreening is probably not necessary except for very high-reliability applications (such as space applications).

Semiconductor, Integrated Circuit, and Hybrid Device Screening Tests

Table 7.9 is a summary of the various screening tests that can be used to detect semiconductor failure mechanisms. Failure mechanisms independent of application environment (oxide or diffusion defects) can be accelerated by operating the device, i.e., performance of a **burn-in** test.

Failure mechanisms dependent on the application environment (bond or metallization defects) can be accelerated by temperature or mechanical stresses (vibration).

Time and environment dependent failure mechanisms (metal migration, corrosion, intermetallic compound formulations caused by dissimilar use) can be accelerated by device operation (i.e., elevated temperature burn-in testing).

7.10.8   Electrostatic Discharge Effects

ICs are susceptible to damage from electrostatic discharge. Static electricity does not have to be large enough to cause a visible spark for an IC to be damaged. Figure 7.161 is an ESD failure in a D/A converter circuit. Figure 1.162 shows a passivation fault extending under the metallization.

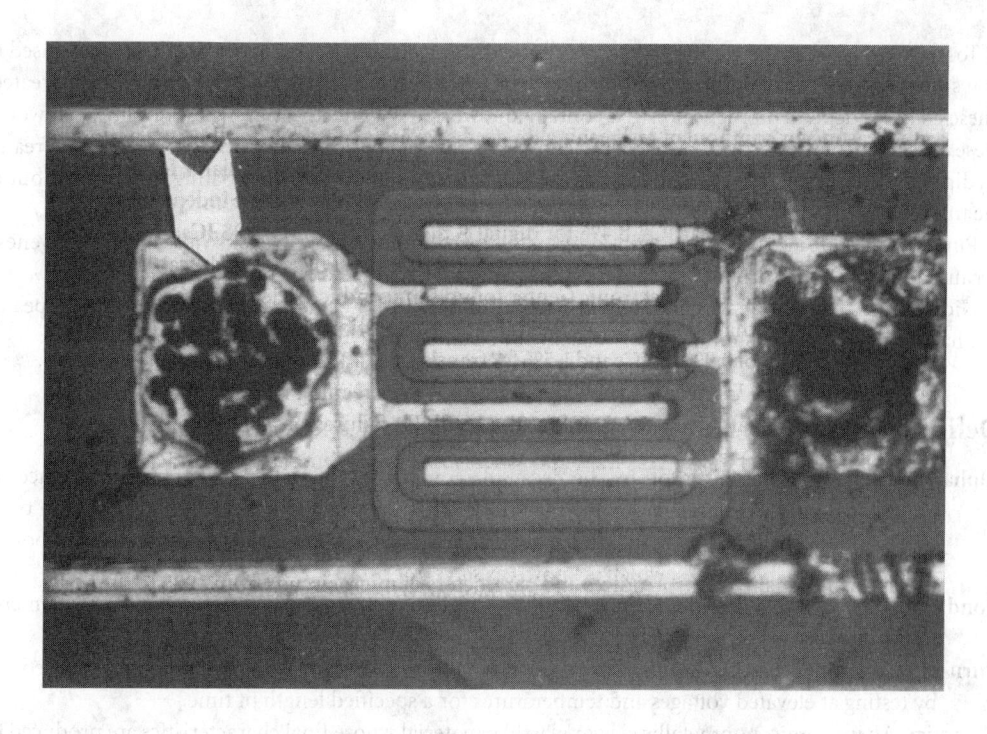

FIGURE 7.161   An ESD failure in a D/A converter circuit.

FIGURE 7.162   Passivation fault extending under the metallization.

To prevent ESD damage, various materials are used. Topical antistats, such as those that may be used to coat some plastic IC dip shipping tubes, include detergent substances, which wet the surface being treated. These coatings depend on their moisture content and work most effectively at high humidities. However, these coatings do wear out and treated surfaces must be checked periodically with a static detector. Treated IC dip tubes should be retreated after use as the IC leads can scrape away the coating as they slide out of the tube.

Pink poly bags are also used to contain components or assemblies. These bags also lose their effectiveness, because they rely on absorbed moisture within the material, and must be checked periodically.

Nickel coated bags may have conductivity discontinuities due to the thin coating cracking after repeated bag handling. These bags should be periodically tested and examined for physical damage.

Defining Terms

Alpha particle: A product given off by the decay of radioactive material (usually emitted by traces of radioactivity in IC ceramic packaging materials). This particle has a positive charge equal to twice that of an electron and is emitted at a very high velocity. Alpha particles can cause temporary memory upsets in DRAMs, which is known as *soft error rate*.

Bond pad: Areas of metallization on the IC die that permit the connection of fine wires or circuit elements to the die. (See also wire bond.)

Burn-in: Component testing where infant mortality failures (defective or weak parts) are screened out by testing at elevated voltages and temperatures for a specified length of time.

Ceramic: An inorganic, nonmetallic clay or glasslike material whose final characteristics are produced by subjection to high temperatures.

DC test: Tests that measure a static parameter, for example, leakage current.

Defense Electric Supply Center (DESC) part: Used to denote a Standard Military drawing part that has been approved by the DESC, in Columbus (formerly in Dayton), Ohio.

Die: A single IC, also known as a chip.

Die information exchange (DIE): A specification created by chip makers and software vendors to provide basic die information in a standard format.

DIE bonding: The attachment of an IC chip (or DIE) to a substrate or a header.

DIE separation: Refers to the separation of the actual microcircuit chip from the inside of the package.

Destructive physical analysis (DPA): Devices are opened and analyzed for process integrity and workmanship.

Electrostatic discharge (ESD): The instantaneous transfer of charges accumulated on a nonconductor to a conductor, into ground.

Failure in time (FIT): A rating equal to the number of failures in one billion (10^9) h.

Frit: A relatively low softening point material of glass composition.

Function test: A check for correct device operation generally by truth table verification.

Hermetic: Sealed so that the object is gas tight (usually to a rate of less than 1×10^{-6} cc/s of helium.)

Joint Army Navy (JAN): When used when referring to microcircuits indicated a part fully qualified to the requirements of MIL-M-38510 for ICs (now replaced by MIL-M-38535) and MIL-S-19500 for semiconductors. JAN class B microcircuit level of the standard military drawing (SMD) program is the preferred level for design in new weapons systems.

JANTX: A prefix denoting that the military specification device has received extra screening and testing, such as an 100% 168-h burn-in.

JANTXV: A JANTX part with an added precapsulation visual requirement.

Joint Electron Device Engineering Council (JEDEC): A part of the EIA.

Lead: A conductive path, usually self-supporting, the portion of an electrical component that connects it to outside circuitry.

Lead frame: The metallic portion of the device package that makes electrical connections from the die to other circuitry.

Mask: The stencil of circuit elements through which light is shown to expose that circuit pattern onto a photoresist coating on the chip (die). The exposed areas are stripped away leaving a pattern.

Metallization: The deposited thin metallic coating layer on a microcircuit or semiconductor.

Passivation: The process in which an insulating dielectric layer is formed over the surface of the die. Passivation is normally achieved by thermal oxidation of the silicon and a thin layer of silicon dioxide is obtained in this manner (a combination of PECVD oxide and PECVD nitride deposited at lower temperature (below 450°C)). Other passivation dielectric coatings may also be applied (used), such as silicon glass (silicon oxynitride).

Particle impact noise detection testing (PIND): A test where cavity devices are vibrated and monitored for the presence of loose particles inside the device package via the noise the material makes. These loose particles may be conductive (such as gold flake particles from gold eutectic die attachment operations) and could result in short circuits. This is not required on a 100% basis for military class B devices. It is required for all class S, or devices generally required for spacecraft application use.

Popcorning: A plastic package crack or delamination that is caused by the phase change and expansion of internally condensed moisture in the package during reflow soldering, which results in stress the plastic package can not withstand.

Parts per million (PPM): The number of failures in one million parts. A statistical estimation of the number of defective devices, usually calculated at a 90% confidence level.

Schmoo plot: An $X–Y$ plot giving the pass/fail region for a specific test while varying the parameters in the X and Y coordinates.

Soft error: An error, or upset in the output of a part (usually applies to memory devices for a single bit output error), which does not reoccur (i.e., the device performs to specifications when tested after the failure occurred).

Substrate: The supporting material upon which the microcircuit (IC) is fabricated, or in hybrids the part to which the IC (and other components, etc.) is attached.

Tin whisker: A hairlike single crystal growth formed on the metallization surface.

Wire bond: A wire connection between the semiconductor die bond pad and the leadframe or terminal.

Acknowledgement

I would like to thank Ron Kalakuntla of ISSI (Integrated Silicon Solution Inc.) and Mr. Lawrence Taccour, of Technical Marketing Group, for their assistance in reviewing this chapter.

References

Arnold, H.D. 1981. Institute of Environmental Sciences Proceedings. ESSEH (Environmental Stress Screening of Electronic Hardware), Institute of Environmental Sciences (Mt. Prospect, IL) Conf., Sept. 21.

Bley, W. 1981. Testing high-speed bipolar memories. Fairchild Camera and Instrument Corp., Mountain View, CA, Nov.

CALCE (Computer Aided Life Cycle Engineering) News, Jan. 1993. Workshop On Temperature Effects, University of Maryland, College Park, MD.

Chester, M. 1986. DOD's rescreening of chips stirs controversy. *Electronic Products,* Oct. 15, pp. 83–84.

Costlow, T. 1995. MCM substrates mixed. *Electronic Engineering Times,* Jan. 16.

Ellis, M. 1984. Non-Mil Defects Surpass Mil. *Electronic Buyers' News,* Nov. 26, p. 6.

Frye, M.A. 1994. Extension of the implementation date of MIL-I-38535 and MIL-S-19500 regarding the prohibition of pure tin as a plating material. Defense Logistics Agency, Defense Electronics Supply Center, Dayton, OH (Letter dated Jan. 21).

GIDEP. Government and Industry Data Exchange Program (Corona, CA), Problem Advisory, 7G-P-95-01 (Jan. 1995), Microcircuits, Flux, Soldering, Liquid; Problem Advisory,G4-P-93-01, Packaging, Cardboard, Flatpack, Microcircuit; Problem Advisory, ZW-P- 9301A (Jan. 1993), Storage Container, Material, Contamination; Problem Advisory, S4-P-93-01 (Jan. 1993), Transistor, Tin Plating, Whisker Growth.

Gulley, D.W. 1992. Texas Instruments, mean time between events, a discussion of device failures in DRAMs. 9(5), Sept.

Hamilton, H.E. 1984. Electronics test. Micro Control Co. Minneapolis, MN, April.

Hitachi. 1991. Reliability report, Multiport Video RAM. HM534251, MC-883, Hitachi, Ltd., Tokyo, Japan, Oct. 14.

Hnatek, E.R. 1983. The case for component rescreening. *Test & Measurement World*, Jan., pp. 18–24.

Hnatek, E.R. 1984. ICs for military and aerospace show dramatic jump in quality, reliability, military/space electronics design. Viking Labs., Inc. *Military/Space Electronics Design* (McGraw-Hill), Oct., pp. 27–30.

Hu, J.M., Barker, D., Dasgupta, A., and Arora, A. 1993. Role of failure-mechanism identification in accelerated testing. *Journal of the IES* (July/Aug.), Institute of Environmental Sciences (Mt. Prospect, IL).

ESSEH (Environmental Stress Screening of Electronic Hardware) 1985. Environmental stress screening for parts. Institute of Environmental Sciences Proceedings, *ESSEH*, Sept.

Klinger, D.J., Nakada, Y., and Memendez, M.A. *AT&T Reliability Manual*, Van Nostrand Reinhold,1990.

Lawrence, J.D. Jr. 1983. *Parallel testing of memory devices*. Reliability Inc., Houston, TX, Oct.

RADC. 1988. *Reliability/maintainability/testability design for dormancy*. Lockheed Electronics Co., Rome Air Development Center, Rept. RADC-TR-88-110, May. (Available from the Defense Technical Information Center (document #AD-A202-704)), Defense Logistics Agency, Department of Defense.

Meeldijk, V. 1995. *Electronic Components: Selection and Application Guidelines*. Wiley Interscience, New York, Chap. 10 and 11.

Meeldijk, V. 1990. Effects of storage and dormancy on components. *Electronic Servicing and Technology Magazine*, Dec., pp. 6–11.

Micron Technology, Inc., Boise, ID. 1991. *Quality/Reliability Handbook*, 4/91, Reliability Monitor, 4 M DRAM, rev. 10/91 and 1 MEG DRAM book.

MIL-HDBK. 1988. *Electronic Reliability Design Handbook*. MIL-HDBK-338 Military Handbook, Oct. 12.

MIL-STD. 1995. NASA standard electrical, electronic, and electromechanical (EEE) parts list. MIL-STD-975, March 17.

MIL-STD. 1992. Electronic parts, materials, and processes for space and launch vehicles. MIL-STD-1547, Dec. 1.

MIL-STD-19500. 1994. General specification for semiconductor devices, April 15.

Motorola, Phoenix, AZ. 1982. Dynamic RAM quality and reliability report. MCM6664A/6665A 64K.

Murray, J. 1994. MCMs pose many production problems. *Electronic Engineering Times*, June 11.

Navsea Systems Command. 1991. Washington, D.C. *Parts application and reliability information manual for Navy electronic equipment*. TE000-AB-GTP-010, Navsea Systems Command (stock number 0910-LP-494-5300), March.

O'Connor, P.T.D. 1985. *Practical Reliability Engineering*, 2nd ed. Wiley, New York.

RAC. 1991. *Failure mode/mechanism distributions*. FMD-91, Reliability Analysis Center, Rome, NY.

School, R. 1985. *Effective screening techniques for dynamic RAMs*. Pacific Reliability Corp. presented at Electro., Institute of Electrical and Electronic Engineers.

Schaefer, S. 1994. DRAM soft error rate calculations. *Design Line*, 3(1).

Semiconductor Industry Association (San Jose, CA). 1991. *Quality statistics report on military integrated circuits*. Government Procurement Committee.

Smith, D.J. 1985. *Reliability and Maintainability in Perspective*, 2nd ed. Halsted Press, Wiley, New York.

Somos, I.L., Eriksson, L.O., and Tobin, W.H. 1986. Understanding di/dt ratings and life expectancy for thyristors. *PCIM Magazine*, Feb.

U.S. Army Material Command AMCP-706-196.

Willoughby, W.J. Jr. 1980. Military electronics/countermeasures: View from the top (interview), Aug., pp. 14–20, 60–61.

Yalamanchili, P., Gannamani, R., Munamarty, R., McClusky, P., and Christou, A. 1995. Optimum processing prevents PQFP popcorning. CALCE Electronic Packaging Research Center, MD, *SMT*, May.

Further Information

The following sources can be referenced for additional data on failure modes:

The Reliability Analysis Center (RAC), Rome, NY, Failure Mode/Mechanism Distributions, 1991, FMD-91.

AT&T Reliability Manual, by David J. Klinger, Yoshinao Nakada, and Maria A. Memendez, published by Van Nostrand Reinhold, 1990.

8

Microelectronics

Tom Chen
Colorado State University

Samuel O. Agbo
California Polytechnic State University

Eugene D. Fabricius
California Polytechnic State University

Robert J. Feugate, Jr.
University of Arizona

Shih-Lien Lu
Oregon State University

James G. Cottle
Hewlett-Packard

Susan A. Garrod
Purdue University

Constantine N. Anagnostopoulos
Eastman Kodak Company

Paul P.K. Lee
Eastman Kodak Company

Jonathon A. Chambers
Cardiff University

Sawasd Tantaratana
University of Massachusetts

Bruce W. Bomar
University of Tennessee Space Institute

Paul D. Franzon
North Carolina State University

Wayne Needham
Intel Corporation

Victor Meeldijk
Intel Corporation

8.1 Integrated Circuits

Tom Chen

8.1.1 Introduction

Transistors and their fabrication into *very large scale integrated (VLSI)* circuits are the invention that has made modern computing possible. Since its inception, integrated circuits have been advancing rapidly from a few transistors on a small silicon die in the early 1960s to 4 millions of transistors integrated on to a single large silicon substrate. The dominant type of transistor used in today's integrated circuits is the metal-oxide-semiconductor (MOS) type transistor. The rapid technological advances in integrated circuit (IC) technology accelerated during and after the 1980s, and one of the most influential factors for such a rapid advance is the technology scaling, that is, the reduction in MOS transistor feature sizes. The MOS feature size is typically measured by the MOS transistor channel length. The smaller the transistors, the more dense the integrated circuits in terms of the number of transistors packed on to a unit area of silicon substrate, and the faster the transistor can switch. Not only can we pack more transistors onto a unit silicon area, the chip size has also increase. As the transistor gets smaller and silicon chip size gets bigger, the transistor's driving capability decreases and the interconnect parasitics (interconnect capacitance and resistance) increases. Consequently, the entire VLSI system has to be designed very carefully to meet the speed demands of the future. Common design issues include optimal gate design and transistor sizing, minimization of clock skew and proper timing budgeting, and realistic modeling of interconnect parasitics.

8.1.2 High-Speed Design Techniques

A modern VLSI device typically consists of several *megacells*, such as memory blocks and data-path arithmetic blocks, and a lot of basic MOS logic gates, such as inverters and NAND/NOR gates. **Complementary MOS (CMOS)** is one of the most widely used logic families, mainly because of its low-power consumption and high-noise margin. Other logic families include NMOS and PMOS logic. Because of its popularity, only the CMOS logic will be discussed. Many approaches to high-speed design discussed here are equally applicable to other logic families.

Optimizing a VLSI device for high-speed operation can be carried out at the system level, as well as at the circuit and logic level. To achieve the maximum operating speed at the circuit and logic levels for a given technology, it is essential to properly set the size of each transistor in a logic gate to optimally drive the output load. If the output load is very large, a string of drivers with geometrically increasing sizes is needed. The size of transistors in a logic gate is also determined by the impact of the transistors as a load to be driven by their preceding gates.

Optimization of Gate Level Design

To optimize the gate level design, let us look at the performance of a single CMOS inverter as shown in Fig. 8.1. Delay of a gate is typically defined as the time difference between input transition and output transition at 50% of supply voltage. The inverter gate delay can be analytically expressed as

$$T_d = C_l(A_n/\beta_n + A_p/\beta_p)/2$$

FIGURE 8.1 Gate delay in a single inverter.

where C_l is the load capacitance of the inverter; β_n and β_p are the forward current gains of n-type and p-type transistors, respectively, and are proportional to the transistor's channel width and inversely proportional to the transistor's channel length; A_n and A_p are process related parameters for a given supply voltage and they are determined by

$$A_n = [2n/(1-n) + \ell_n((2(1-n) - V_0)/V_0)][V_{dd}(1-n)]$$
$$A_n = [-2p/(1+p) + \ell_n((2(1+p) - V_0)/V_0)][V_{dd}(1+p)]$$

where $n = V_{thn}/V_{dd}$ and $p = V_{thp}/V_{dd}$. V_{thn} and V_{thp} are gate threshold voltages for n-channel and p-channel transistors, respectively. This expression does not take the input signal slope into account. Otherwise, the expression would become more complicated. For more complex CMOS gates, an equivalent inverter structure is constructed to reflect the effective strength of their p-tree and n-tree in order to apply the inverter delay model.

In practice, CMOS gate delay is treated in a simple fashion. The delay of a logic gate can be divided into two parts: the intrinsic delay D_{ins}, and the load-related delay D_{load}. The gate intrinsic delay is determined by the internal characteristics of the gate including the implementing technology, the gate structure, and the transistor sizes. The load-related delay is a function of the total load capacitance at the gate's output. The total gate delay can be expressed as

$$T_d = D_{ins} + C_l^* S$$

where C_l is the total load capacitance and S is the factor for gate's driving strength. $C_l^* S$ represents the gate's load-related delay. In most CMOS circuits using leading-edge submicron technologies, the total delay of a gate can be dominated by the load-related delay. For an inverter in a modern submicron CMOS technology of around 0.5-μm feature size, D_{ins} can range from 0.08 to 0.12 ns and S can range from 0.00065 to 0.00085 ns/fF depending on specifics in the technology and the minimum transistor feature size. For other more complex gates such as NAND and NOR gates, D_{ins} and S generally increase.

To optimize a VLSI circuit for its maximum operating speed, **critical paths** must be identified. A critical path in a circuit is a signal path with the longest time delay from a primary input to a primary output. The time delay on the critical path in a circuit determines the maximum operating speed of the circuit. The time delay of a critical path can be minimized by altering the size of the transistors on the critical path. Using the lumped resistor-capacitor (RC) delay model, the problem of transistor sizing can be formulated to an optimization problem with a convex relationship between the path delay and the sizes of the transistors on the path. This optimization problem is simple to solve. The solutions often have 20–30% deviation, however, compared to the **SPICE** simulation results. Realistic modeling of gate delay taking some second-order variables, such as input signal slope, into consideration has shown that the relationship between the path delay and the sizes of the transistors on the path is not convex. Such detailed analysis led to more sophisticated transistor sizing algorithms. One of these algorithms suggested using genetic methods to search for an optimal solution and has shown some promising results.

Clocks and Clock Schemes in High-Speed Circuit Design

Most of the modern electronic systems are **synchronous systems.** The clock is a central pace setter in a synchronous system to step the desired system operations through various stages of the computation. Latches are often used to facilitate catching the output data at the end of each clock cycle. Figure 8.2 shows the typical synchronous circuit with random logic clusters as computational blocks and latches as pace setting devices. When there exist feedbacks, as shown in Fig. 8.2, the circuit is referred to as *sequential circuit.*

A latch is also called a *register* or a *flip-flop.* The way a latch catches data depends on how it is triggered by the clock signal. Generally, there are level-triggered and edge-triggered latches, the former can be further subdivided according to the triggering polarity as positive or negative level or edge-triggered latches. The performance of a digital circuit is often determined by the maximum clock frequency the circuit can run. For a synchronous digital circuit to function properly, the longest delay through any combinational clusters must be less than the clock cycle period. Therefore, the following needs to be done for high-speed design:

- Partition the entire system so that the delays of all of the combinational clusters are as balanced as possible.
- Design the circuit of the combinational clusters so that the delay of critical paths in the circuit is minimized and less than the desired clock cycle period.
- Use a robust clock scheme to ensure that the entire system is free of race conditions and has minimal tolerable clock skew.

The first item listed is beyond the scope of this section. The second item was discussed in the preceding subsection.

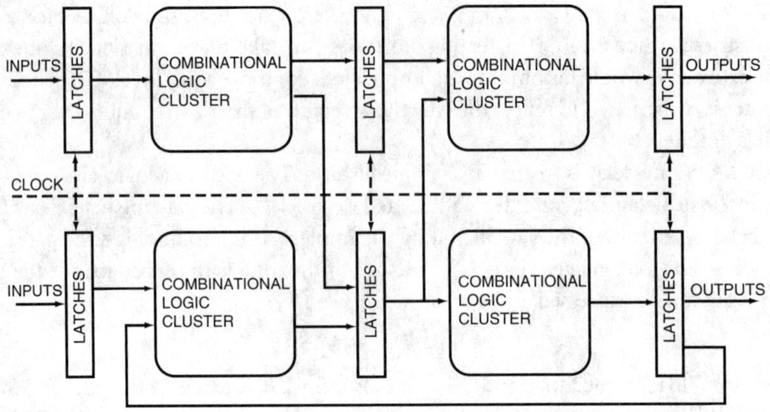

FIGURE 8.2   A typical synchronous circuit with combinational logic clusters and latches.

Race conditions typically occur with the use of level triggered latches. Figure 8.3 shows a typical synchronous system based on level triggered latches. Because of delays on the clock distribution network, such as buffers and capacitive parasitics on the interconnect, the timing difference caused by such a distribution delay is often referred to as **clock skew** which is modeled by a delay element in Fig. 8.3. For the system to operate properly, at the positive edge each latch is supposed to capture its input data from the previous clock cycle. If the clock skew is severe shown as skewed clock clk'', however, it could be possible that the delay from $Q1$ to $D2$ becomes short enough that $D1$ is not only caught by the clk, but also caught by clk'. The solution to such a race condition caused by severe clock skew is as follows:

- Change the latch elements from level triggered to edge triggered or pseudoedge triggered latches such as latches using two-phase, nonoverlapping clocks.
- Resynthesize the system to balance to the critical path delay of different combinational clusters.
- Reduce the clock skew.

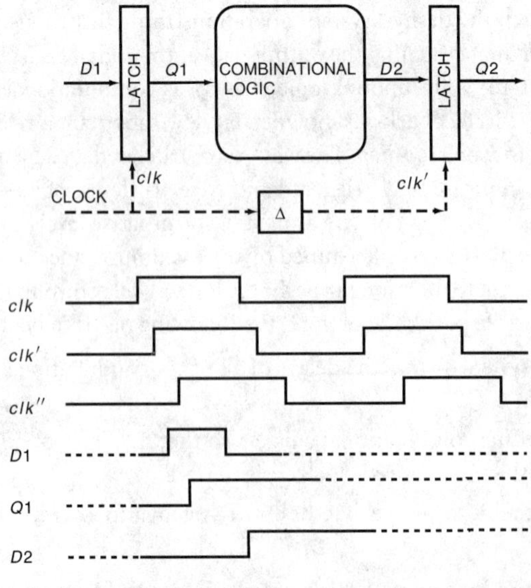

FIGURE 8.3   An example of race condition caused by a severe clock skew in the system.

Clock skew can also cause other types of circuit malfunction if it is not attended properly. In dynamic logic, severe clock skew may result in different functional blocks being in different stages of precharging or evaluating. Severe clock skew eats away a significant amount of precious cycle time in VLSI systems. Therefore, reducing clock skew is an important problem in high-speed circuit design.

As discussed earlier, clock skew is mainly caused by the imbalance of the clock distribution network. Such an imbalance can be the result of distance differences from the nearest clock driver, different functional blocks driven by different clock drivers with different driving strengths, temperature difference on the same die, device characteristic differences on the same die due to process variation, etc. Two general approaches are often taken to minimize the clock skew. The first approach deals with the way the clock signal is distributed. The geometric shape of the clock distribution network is a very important attribute. Depending on the type of system operation, several popular distribution network topologies are illustrated in Fig. 8.4. Among these topologies, **H-tree** presents least amount of clock skew and, therefore, is widely used in high-performance systems.

The second approach deals with employing additional on-chip circuits to force the clock signals of two different functional blocks to be aligned, or to force the on-chip clock signal to be aligned with the global clock signal at the system level. The widely used circuits for this purpose include **phase-locked loop** (PLL) and **delay-locked loop** (DLL). Figure 8.5 shows a simple phase-locked loop. A simple PLL consists of four components: a digital phase detector, a charge pump, a low-pass filter, and a voltage controlled oscillator (VCO). The phase detector accepts reference clock, CLK_ref, and a skewed clock, CLK_out, and compares the phase difference of the two clocks to charge or discharge the charge pump. The low-pass filter is used to convert the phase difference between the reference frequency and the skewed frequency to a voltage level. This voltage is then fed into the VCO to reduce the difference of the reference and skewed clocks until they are locked to each other.

One of the most important design parameters for PLL is the output jitter. The output jitter is demonstrated by the random deviation of the output clock's phase from the reference clock signal. Significant peak-to-peak jitter will effectively reduce the clock period. The main contributor of the output jitter is the noise on the input of the VCO. Additional jitter can be induced by the noise on power supply rails that are common to high-speed VLSI circuits. Furthermore, acquisition time of PLL, in the several microsecond range, is often longer than desirable. This is mainly attributed to the response time of the VCO. In a typical scenario where clock skew is caused by the imbalance of the distribution network, the skewed clock often has the correct frequency. What needs to be corrected is the relative phase of the clock signals. Therefore, there is no need to have a VCO. Rather, a simple delay logic can be used to modify the clock signal's phase. This type of simplified phase correct circuit is referred to as a delay-locked loop. By replacing the VCO with a simple programmable delay line, DLL is simpler, yet exhibits less jitter than its PLL counterpart.

Asynchronous Circuits and Systems

Clock distribution within large VLSI chips is becoming more and more of a problem for high-speed digital systems. Such a problem may be surmountable using state-of-the-art computer aided design (CAD) tools and on-chip PLL/DLL circuits. Asynchronous circuits have, nevertheless, gained a great deal of attention lately. An asynchronous circuit does not require an external clock to get it through the computation. Instead, it works on the principle of handshaking between functional blocks. Therefore, execution of a computational operation is totally dependent on the readiness of all of the input variables of the functional block. The biggest advantage of asynchronous circuits over synchronous circuits is that the correct behavior of asynchronous circuits is independent of the speed of their components or signal interconnect delays.

In a typical asynchronous circuit, functional blocks will have two more signals, *request* and *complete*, apart from their input and output signals as shown in Fig. 8.6. These two binary signals are necessary and sufficient for handshaking purposes. Even though asynchronous circuits are speed independent, the order of computation is still maintained by connecting the complete signal from one block to the request signal to another block. When the request signal is active for a functional block, indicating the computation of the preceding functional block is completed, the current functional block starts computation evaluation using its valid inputs from the preceding functional block. Once the evaluation is completed, the current

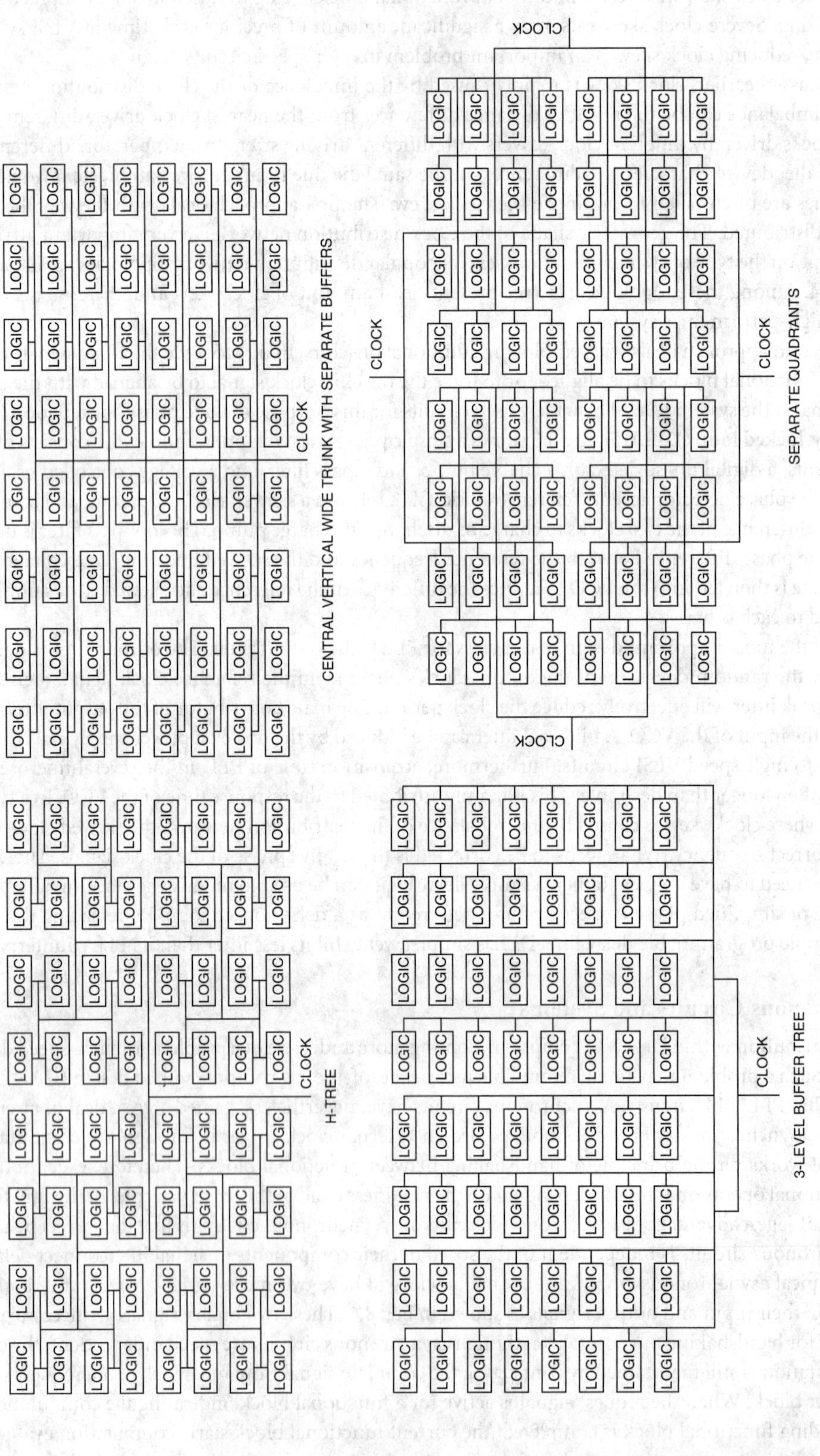

FIGURE 8.4 Various clock distribution structures.

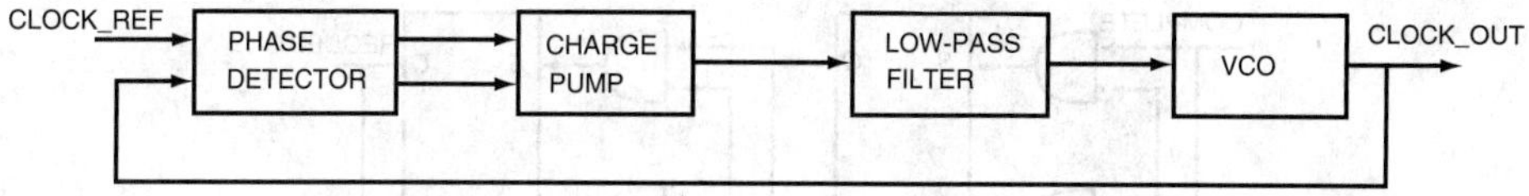

FIGURE 8.5   A simple phase-locked loop structure.

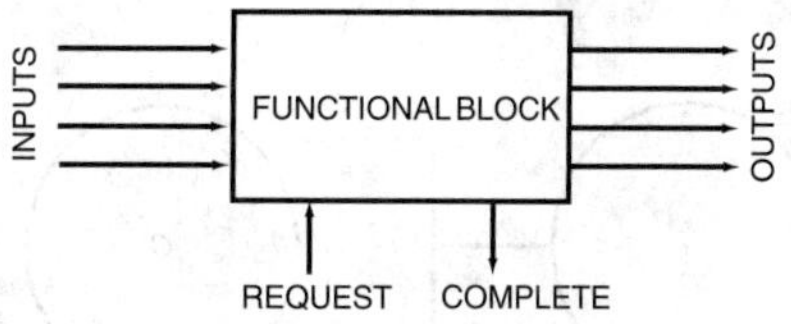

FIGURE 8.6   The single request and complete are two extra signals for a typical asynchronous circuit.

functional block sets the complete signal to active to activate other functional blocks for computation. Figure 8.7 shows a schematic of such a communication protocol where block A and block B are connected in a pipeline.

To ensure that asynchronous circuits function correctly regardless of individual block speed, the *request* signal of a functional block should only be activated if the functional block has already completed the current computation. Otherwise, the current computation would be overwritten by incoming computation requests. To prevent this situation from happening, an interconnection block is required with an *acknowledge* signal from the current functional block to the preceding functional block. An active acknowledge signal indicates to the preceding function block that the current block is ready to accept new data from it. This two-way communication protocol with request and acknowledge is illustrated in Fig. 8.7. The interconnect circuit is unique to asynchronous circuits. It is often referred to as a **C-element**. Figure 8.8 shows a design of the C-element.

In recent years, much effort has been spent on applying the asynchronous circuits to real-world applications. Several totally asynchronous designs of microprocessors have demonstrated their commercial feasibility. Several issues that still need to be addressed with regard to asynchronous circuits include acceptable amounts of silicon overhead, power efficiency, and performance as compared to their synchronous counterparts.

Interconnect Parasitics and Their Impact on High-Speed Design

On-chip interconnects present parasitic capacitance and resistance as loads to active circuits. Such parasitic loads had little impact on earlier ICs because the intrinsic gate delay dominated the total gate delay.

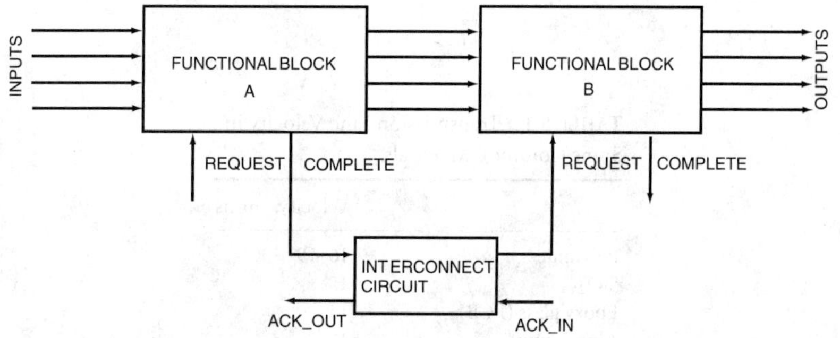

FIGURE 8.7   The schematic of a communication protocol in asynchromous systems.

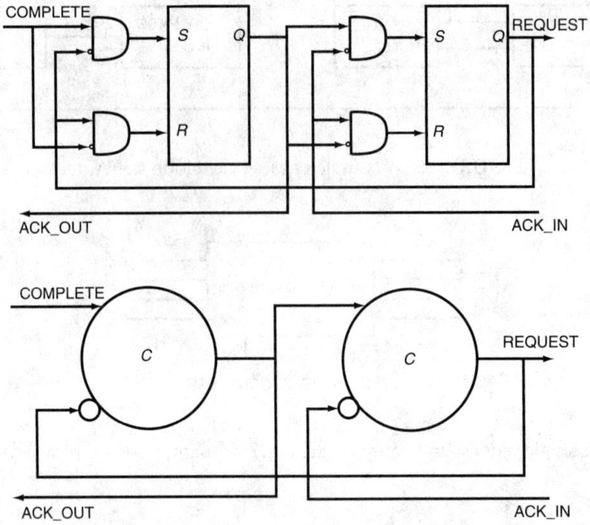

FIGURE 8.8 A schematic design of the C-element.

With aggressive scaling in the VLSI process, the gate intrinsic delay decreases dramatically. The interconnect parasitics does not scale proportionally, however, and the wire resistance tends to increase, resulting in the delay caused by the interconnect load parasitics gradually becoming a dominant factor in the total gate delay. The problem is further exacerbated by the fact that when the operating speed reaches several hundred megahertz, the traditional lumped RC model is no longer accurate. It has been suggested that such a lumped RC model should be modified to include a grounded resistor and an inductor. The RLC interconnect model includes nonequilibrium initial conditions and its response waveform may be nonmonotonic. Such a model may be more accurate because the existence of the inductance reduces the rate of increase in current and, therefore, increases the signal transition time. When the operating speed increases further such that the rise time of a signal is much less than the signal transmission time from point A to point B, a transmission line model should be used. On-chip interconnect, thus, is typically modeled as a microstrip. The characteristics of a transmission line are determined by its relative dielectric constant and magnetic permeability. Table 8.1 shows the signal transmission velocity in some common materials used in VLSI. As a rule of thumb, the transmission line phenomena become significant when

$$t_r < 2.5^* t_f$$

where t_r is the rise time of a signal and t_f is the signal transmission time, which is the interconnect length divided by the signal traveling velocity in the given material. The interconnect can be treated as a lumped RC network when

$$t_r > 5^* t_f$$

TABLE 8.1 Transmission Line Velocity in Some Common Materials

	Velocity, cm/ns
Polymide	16–19
SiO_2	15
Epoxy glass (PCB)	13
Aluminium	10

The signal rise time depends on the driver design and the transmission line's characteristic impedance Z_0. In MOS ICs, the load device at the receiving end of the transmission line can always be treated as an open circuit. Therefore, driver design is a very important aspect of high-speed circuit design. The ideal case is to have the driver's output impedance match the transmission line's characteristic impedance. Driving an unterminated transmission line (the MOS IC case) with its output impedance lower than the line's characteristic impedance, however, can increase driver's settling time due to excess ringing and, therefore, is definitely to be avoided. Excess ringing at the receiving end could also cause the load to switch undesirably. Assuming MOS transistor's threshold is 0.6–0.8 V, to ensure that no undesirable switching takes place, the output impedance of the drive should be at least a third of the charactertstic impedance of the transmission line. When the output impedance is higher than the line's characteristic impedance, multiple wave trips of the signal may be required to switch the load. To ensure that only one wave trip is needed to switch the load, the output impedance of the driver should be within 60% of the characteristic impedance of the transmission line.

For a lossy transmission line due to parasitic resistance of on-chip interconnects, an exponential attenuating transfer function can be applied to the signal transfer at any point on the transmission line. The rate of the attenuation is proportional to the unit resistance of the interconnect. When operating frequency increases beyond a certain level, the on-chip transmission media exhibits the *skin effect* in which the time-varying currents concentrate near the skin of the conductor. Therefore, the unit resistance of the transmission media increases dramatically.

Defining Terms

Application-specific integrated circuit (ASIC): Device designed specifically for a particular application.

Application-specific standard product (ASSP): Device designed specifically for one area of applications, such as graphics and video processing.

Asynchronous system: A system in which the progress of a computation is driven by the readiness of all the necessary input variables for the computation through a handshaking protocol. Therefore, no central clock is needed.

C-element: A circuit used in an asynchronous as an interconnect circuit. The function of this circuit is to facilitate the handshaking communication protocol between two functional blocks.

Clock skew: A phase difference between two clock signals at different part of a chip/system due to imbalance of the distribution media and the distribution network.

Complementary metal-oxide silicon (CMOS): It is a very popular integrated circuit type in use today.

Critical path: A signal path from a primary input pin to a primary output pin with the longest delay time in a logic block.

Delay-locked loop (DLL): It is similar to PLL except that it has better jitter suppression capability.

Digital signal processor (DSP): A processing device specialized in popular math routines used by signal processing algorithms.

Field programmable gate array (FPGA): A popular device which can be tailored to a particular application by loading a customizing program on to the chip.

H-tree: A popular clock distribution tree topologically that resembles the H shape. It introduces the least amount of clock skew compared to other distribution topologies.

Phase-locked loop (PLL): A circuit that can detect the phase difference of two signals and reduce the difference in the presence of the phase difference.

Programmable logic devices (PLD): A class of IC products which are easy to customize for a particular application.

SPICE: A popular circuit level simulation program to perform detailed analysis of circuit behavior.

Synchronous system: A system in which a computation is divided into unit periods defined by a central clock signal. Signal transfer within the system typically occurred at the transition edge of the clock signal.

References

Bakoglu, H.B. 1991. *Circuits, Interconnections, and Packaging for VLSI*. Addison-Wesley, Reading, MA.

Dill, D.L. 1989. *Trace Theory for Automatic Hierarchical Verification of Speed-Independent Circuits*. MIT Press, Cambridge, MA.

Gardner, F.M. 1979. *Phaselock Techniques*, 2nd ed. Wiley, New York.

Jeong, D. et al. 1987. Design of PLL-based clock generation circuits. *IEEE J. Solid-State Circuits* SC-22(2): 255–261.

Johnson, M. and Hudson, E. 1988. A variable delay line PLL for CPU-coprocessor synchronization. *IEEE J. Solid-State Circuits* (Oct.):1218–1223.

Meng, T.H. 1991. *Synchronization Design for Digital Systems*. Kluwer Academic, Norwell, MA.

Rosenstark, S. 1994. *Transmission Lines in Computer Engineering*. McGraw-Hill, New York.

Sapatnekar, S., Rao, V., and Vaidya, P. 1992. A convex optimization approach to transistor sizing for CMOS circuits. *Proc. ICCAD*, pp. 482–485.

Wang, X. and Chen, T. 1995. Performance and area optimization of VLSI systems using genetic algorithms. *Int. J. of VLSI Design* 3(1):43–51.

Weste, N. and Eshraghian, K. 1993. *Principle of CMOS VLSI Design: A Systems Perspective*, 2nd ed. Addison-Wesley, Reading, MA.

Further Information

For general information on the VLSI design process and various design issues, consult several excellent reference books, two of which are listed in the reference section, including Mead and Conway's *Introduction to VLSI Systems*, Glasser and Dobberpuhl's *The Design and Analysis of VLSI Circuits*, and Geiger's *VLSI Design Techniques for Analog and Digital Circuits*. *IEEE Journal of Solid-State Circuits* provides an excellent source for the latest development of novel and high-performance VLSI devices.

Some of the latest applications of PLLs and DLLs can be found in the *Proceedings of International Solid-State Circuit Conference*, the *Symposium on VLSI Circuits*, and the *Custom Integrated Circuit Conference*.

For information on modeling of VLSI interconnects and their transmission line treatment, consult the *Proceedings of Design Automation Conference* and *International Conference on Computer-Aided Design*. *IEEE Transactions on CAD* is also an excellent source of information on the subject.

8.2   Integrated Circuit Design

Samuel O. Agbo and Eugene D. Fabricius

8.2.1   Introduction

Integrated circuits (ICs) are classified according to their levels of complexity: small-scale integration (SSI), medium-scale integration (MSI), large-scale integration (LSI) and very large-scale integration (VLSI). They are also classified according to the technology employed for their fabrication (bipolar, N metal oxide semiconductor (NMOS), complementary metal oxide semiconductor (CMOS), etc.). The design of integrated circuits needs to be addressed at the SSI, MSI, LSI, and VLSI levels. Digital SSI and MSI typically consist of gates and combinations of gates. Design of digital SSI and MSI is presented in Sec. 8.2.3, and consists largely of the design of standard gates. These standard gates are designed to have large noise margins, large fan out, and large load current capability, in order to maximize their versatility.

In principle, the basic gates are sufficient for the design of any digital integrated circuit, no matter how complex. In practice, modifications are necessary in the basic gates and MSI circuits like flip-flops, registers, adders, etc., when such circuits are to be employed in LSI or VLSI design. For example, circuits to be interconnected on the same chip can be designed with lower noise margins, reduced load driving capability, and smaller logic swing. The resulting benefits are lower power consumption, greater circuit density, and improved reliability. On the other hand, several methodologies have emerged in LSI and VLSI design that are not based on interconnections or modification of SSI and MSI circuits. Both approaches to LSI and VLSI design are presented in the following sections.

8.2.2 An Overview of the IC Design Process

The effort required for the design of an integrated circuit depends on the complexity of the circuit. The requirement may range from several days effort for a single designer to several months work for a team of designers. **Custom design** of complex integrated circuits is the most demanding. By contrast, **semicustom design** of LSI and VLSI that utilize preexisting designs, such as standard cells and gate arrays, requires less design effort.

IC design is performed at many different levels and Fig. 8.9 is a nonunique depiction of these levels. Level 1 presents the design in terms of subsystems (standard cells, gate arrays, custom subcircuits, etc.) and their interconnections. Design of the system **layout** begins with the floor plan of level 3. It does not involve the layout of individual transistors and devices, but is concerned with the geometric arrangement and interconnection of the subsystems. Level 4 involves the circuit design of the subsystems. Levels 2 and 5 involve system and subcircuit simulations, respectively, which may lead to modifications in levels 1 and/or 4.

Discussion here will focus primarily on the system design of level 1 and the subsystem circuit design of level 4. Lumped under the fabrication process of level 7 are many tasks, such as **mask** generation, process simulation, wafer fabrication, testing, etc. Broadly speaking, floor plan generation is a part of layout. For large ICs, layout design is often relevant to system and circuit design.

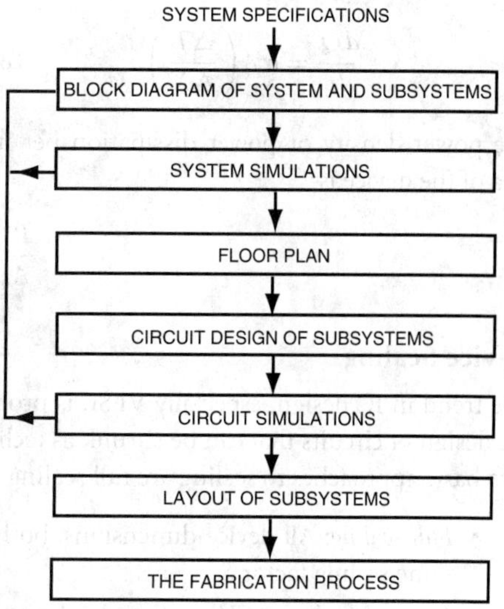

FIGURE 8.9 Different levels in IC design.

8.2.3 General Considerations in IC Design

Layout or actual arrangement of components on an IC is a design process involving design tradeoffs. Circuit design often influences layout and vice versa, especially in LSI and VLSI. Thus, it is helpful to outline some general considerations in IC design:

- *Circuit components*: Chip area is crucial in IC design. The usual IC components are transistors, resistors, and capacitors. Inductors are uneconomical in area requirements and are hardly ever used in ICs except in certain microwave applications. Transistors require small chip areas and are heavily utilized in ICs. Resistor area requirements increase with resistor values. IC resistors generally range between 50 Ω and 100 kΩ. Capacitors are area intensive and tend to be limited to 100 pF.

- *Isolation regions*: Usually, different components would be placed in different isolation regions. The number of isolation regions, however, should be minimized in the interest of chip area economy by placing more than one component in an isolation region whenever possible. For example, several resistors could share one isolation region.

- *Design rules*: Geometric design rules that specify minimum dimensions for features and separation between features must be followed for a given IC process.

- *Power dissipation*: Chip layout should allow for adequate power dissipation and avoid overheating or development of hot spots on the chip. In low-power circuits such as CMOS ICs, device size is determined by lithographic constraints. In circuits with appreciable power dissipation, device size is determined by thermal constraints and may be much larger than lithographic constraints would allow.

Device size determination from power-density considerations is illustrated with the aid of Fig. 8.10. The device resides close to the surface of the substrate. Thus, heat flow is approximately one-dimensional, from the device to the substrate, although it is actually a three-dimensional flow. Assuming an infinite heat sink of ambient temperature T_A, substrate thickness ΔX, thermal conductivity α, device of surface area A, and temperature $T_A + \Delta T$, the rate of heat flow toward the heat sink, dQ/dt, is given by

$$\frac{dQ}{dt} = \alpha A \left(\frac{\Delta T}{\Delta X} \right) \qquad (8.1)$$

FIGURE 8.10   Heat flow in an IC device.

The power density or power dissipation per unit area of the device is

$$\frac{P}{A} = \alpha \left(\frac{\Delta T}{\Delta X} \right) \qquad (8.2)$$

Device Scaling

The trend in IC design, especially VLSI, is progressively toward smaller sizes. Scaling techniques permit the design of circuits that can be shrunk as technological developments allow progressively smaller sizes. Two basic approaches to scaling are full scaling and constant voltage scaling.

- *Full scaling*: All device dimensions, both surface and vertical, and all voltages are reduced by the same scaling factor S.
- *Constant voltage (CV) scaling*: All device dimensions, both surface and vertical, are reduced by the same scaling factor S, but voltages are not scaled. Voltages are maintained at levels compatible with transistor-transistor logic (TTL) supply voltages and logic levels.

Scaling of device dimensions has implications for other device parameters. Full scaling tends to maintain constant electric field strength and, hence, parameters that are no worse off, as device dimensions are reduced, but does not ensure TTL voltage compatibility. Table 8.2 compares effect on device parameters

TABLE 8.2   Effect of Full Scaling and Constant Voltage Scaling on IC Device Parameters

Parameter	Full Scaling	CV Scaling
Channel length, L	$1/S$	$1/S$
Channel width, W	$1/S$	$1/S$
Oxide thickness, t_{OX}	$1/S$	$1/S$
Supply voltage, V_{DD}	$1/S$	1
Threshold voltage, V_{TD}	$1/S$	1
Oxide capacitances, C_{OX}, C_{SW}, C_{FOX}	S	S
Gate capacitance, $C_g = C_{OX} WL$	$1/S$	$1/S$
Transconductances, K_N, K_p	S	S
Current, I_D	$1/S$	$1/S$
Power dissipation per dense, P	$1/S^2$	S
Power dissipation per unit area, P/A	1	S^3
Packing density	S^2	S^2
Propagation delay, t_p	$1/S$	$1/S^2$
Power-delay product, $P t_p$	$1/S^3$	$1/S$

of the two scaling approaches. A compromise that is often adopted is to use full scaling for all internal circuits and maintain TTL voltage compatibility at the chip input/output (I/O) pins.

Although many scaling relationships are common to MOS field effect transistors (MOSFETs) and bipolar ICs (Keyes, 1975), the scaling relationships of Table 8.2 more strictly applies to MOSFETs. Bipolar doping levels, unlike MOSFETs, are not limited by oxide breakdown. Thus, in principle, miniaturization can be carried further in bipolar processing technology. However, bipolar scaling is more complex. One reason is that the junction voltages required to turn on a bipolar junction transistor (BJT) does not scale down with dimensions.

Geometric Design Rules

Design rules specify minimum device dimensions, minimum separation between features and maximum misalignment of features on an IC. Such rules tend to be process and equipment dependent. For example, a design rule for a 2-μm process may not be appropriate for a 0.5-μm process. Design rules should protect against fatal errors such as short circuits due to excessive misalignment of features, or open circuits due to too narrow a metal or polysilicon conductive path.

Generalized design rules that are portable between processes and scalable to facilitate adaptation to shrinking minimum geometries as processes evolve are desirable. Other advantages of generalized design rules include increased design efficiency due to fewer levels and fewer rules, automatic translation to final layout, layout-rule and electrical-rule checking, simulation, verification, etc. The Mead-Conway approach (1980) to generalized design rules is to define a scalable and process-dependent parameter, lambda (λ), as the maximum misalignment of a feature from its intended position on a wafer or half the maximum misalignment of two features on different mask layers. Table 8.3 shows a version of the Mead-Conway scalable design rules for NMOS (Fabricius, 1990).

CMOS ICs utilize both NMOS and PMOS devices. Starting with a p-substrate, the NMOS would be fabricated on the p-substrate and the PMOS in an n-well, and vice versa. With the addition of an n-well, p-well or twin tub process, CMOS fabrication is similar to that for NMOS, although more complex. Table 8.4 (Fabricius, 1990) shows the Mead-Conway scalable CMOS design rules. The dimensions are given in multiples of λ and the rules are specified by the MOS Implementation System (MOSIS) of the

TABLE 8.3   MOS Implementation System (MOSIS) NMOS Design Rules

Mask Level	Feature	Size (times λ)
N^+ Diffusion	diffusion width	2
	diffusion spacing	3
Implant mask	implant-gate overlap	2
	implant to gate spacing	1.5
Buried contact mask	contact to active device	2
	overlap with diffusion	1
	contact to poly spacing	2
	contact to diffusion spacing	2
Poly mask	poly width	2
	poly spacing	2
	poly to diffusion spacing	1
	gate extension beyond diffusion	2
	diffusion to poly edge	2
Contact mask	contact width	2
	contact-diffusion overlap	1
	contact-poly overlap	1
	contact to contact spacing	2
	contact to channel	2
	contact-metal overlap	1
Metal mask	metal width	3
	metal spacing	3

TABLE 8.4 MOSIS Portable CMOS Design Rules

Mask Level	Feature	Size (times λ)
n-well and p-well	well width	6
	well to well spacing	6
n^+, p^+ active diffusion or implant	active width	3
	active to active spacing	3
	source/drain to well edge	6
	substrate/well contact	3
	active to well edge	3
Poly mask	poly width or spacing	2
	gate overlap of active	2
	active overlap of gate	2
	field poly to active	1
p-select, n-select	select-space (overlap) to (of) channel	3
	select-space (overlap) to (of) active	2
	select-space (overlap) to (of) contact	1
Simpler contact to poly	contact size	2 × 2
	active overlap of contact	2
	contact to contact spacing	2
	contact to gate spacing	2
Denser contact to poly	contact size	2 × 2
	poly overlap of contact	1
	contact spacing on same poly	2
	contact spacing on different poly	5
	contact to non-contact poly	4
	space to active short run	2
	space to active long run	3
Simpler contact to active	contact size	2 × 2
	active overlap of contact	2
	contact to contact spacing	2
	contact to gate spacing	2
Denser contact to active	contact size	2 × 2
	active overlap of contact	1
	contact spacing on same active	2
	contact spacing on different active	6
	contact to different active	5
	contact to gate spacing	2
	contact to field poly short run	2
	contact to field poly long run	3
Metal 1	width	3
	metal 1 to metal 1	3
	overlap of contact to poly	1
	overlap of contact to active	1
via	size	2 × 2
	via-to-via separation	2
	metal 1/via overlap	1
	space to poly or active edge	2
	via to contact spacing	2
Metal 2	width	3
	metal 2 to metal 2 spacing	4
	metal overlap of via	1
Overglass	bonding pad (with metal 2 undercut)	$100 \times 100 \ \mu m$
	probe pad	$75 \times 75 \ \mu m$
	pad to glass edge	$6 \ \mu m$

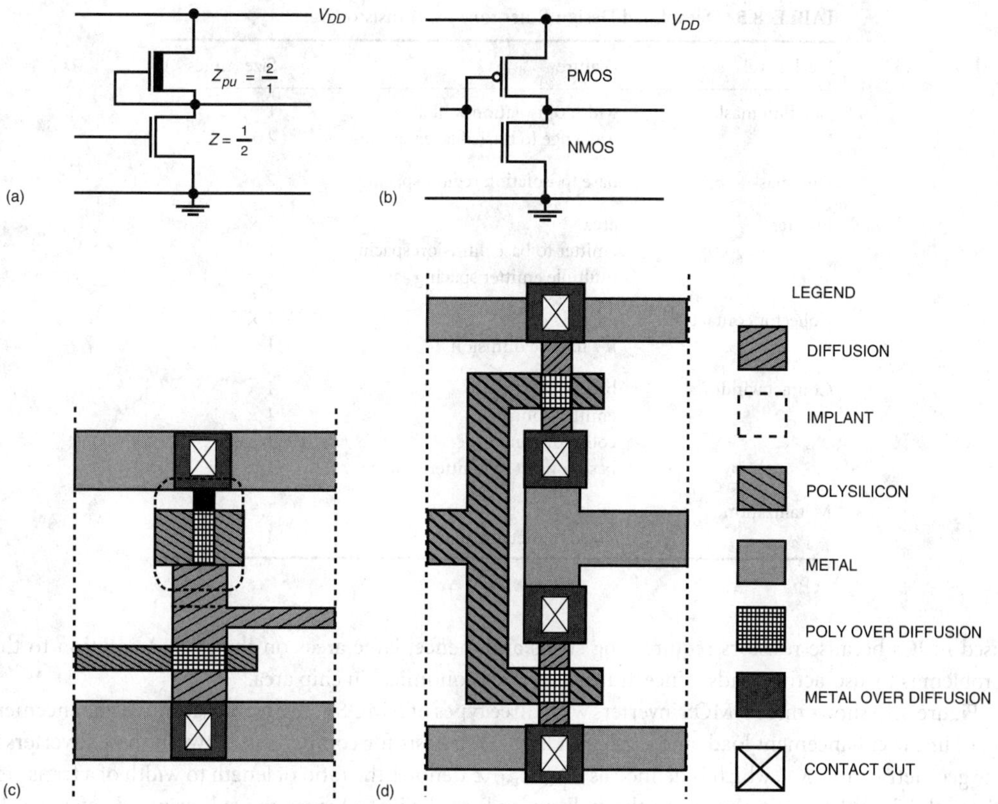

FIGURE 8.11 Layout design for NMOS inverter with depletion load and a CMOS inverter: (a) NMOS inverter, (b) CMOS inverter, (c) NMOS inverter layout, (d) CMOS inverter layout.

University of Southern California Information Sciences Institute. Further clarification of the rules may be obtained from MOSIS manuals (USCISI, 1984, 1988).

Figure 8.11 illustrates the layout of an NMOS inverter with depletion load and a CMOS inverter, employing the NMOS and CMOS scalable design rules just discussed. Figure 8.11(a) and Figure 8.11(b) show the circuit diagrams for the NMOS and CMOS, respectively, with the aspect ratios Z defined as the ratio of the length to the width of a transistor gate. Figure 8.11(c) and Figure 8.11(d) give the layouts for the NMOS and CMOS, respectively.

A simplified design rule for a bipolar *npn* comparable to those previously discussed for NMOS and CMOS is presented in Table 8.5. As before, the minimum size of a feature in the layout is denoted by lambda (λ). The following six masks are required for the fabrication: n^+ buried-layer diffusion, p^+ isolation diffusion, p base region diffusion, n^+ emitter and collector diffusions, contact windows, and metalization.

8.2.4 Design of Small-Scale and Medium-Scale Integrated Circuits

Gates are the basic building blocks in digital integrated circuits. Small-scale integrated circuits are essentially gate circuits, and medium-scale integrated circuits are circuits employing several gates. Gates, in turn, are based on inverters and can be realized from inverter circuits with some modifications, especially those modifications that allow for multiple inputs. This section will start with a discussion of inverters and gates.

NMOS Inverters

A resistive load NMOS inverter, its output characteristics, and its voltage transfer characteristic are shown in Fig. 8.12. The loadline is also shown on the output characteristics. Resistive load inverters are not widely

TABLE 8.5   Simplified Design Rules for *npn* Transistor

Mask Level	Feature	Size (times λ)
Isolation mask	width of isolation wall	1
	wall edge to buried layer spacing	2.5
Base mask	base to isolation region spacing	2.5
Emitter	area	2×2
	emitter to base diffusion spacing	1
	multiple emitter spacing	
Collector contact	area	1×5
	n^+ to base diffusion	1
Contact windows	base contact	1×2
	emitter contact	1×1
	collector contact	1×2
	base contact to emitter spacing	1
Metallization	width	1.5
	metal to metal spacing	1

used in ICs because resistors require long strips and, hence, large areas on the chip. A solution to this problem is to use active loads, since transistors are economical in chip area.

Figure 8.13 shows three NMOS inverters with three types of NMOS active loads: saturated enhancement load, linear enhancement load, and depletion load. One basis for comparison between these inverters is the geometric ratio K_R, which is defined as Z_{pu}/Z_{pd}. Z denotes the ratio of length to width of a transistor channel. The subscript pu refers to the pull-up or load device, whereas the subscript pd refers to the pull-down or driving transistor.

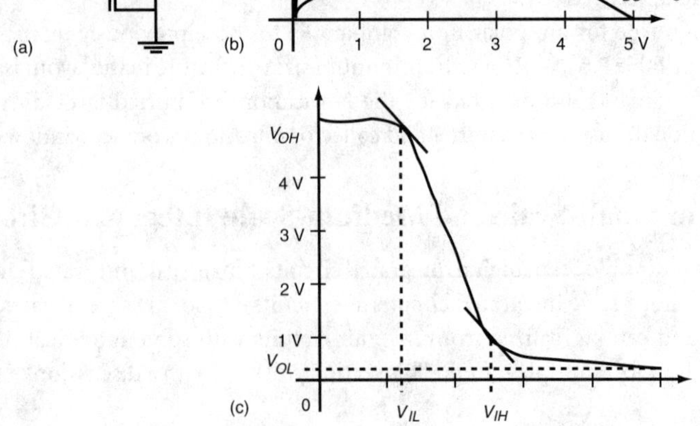

FIGURE 8.12   NMOS resistive load inverter: (a) resistive load NMOS inverter, (b) output characteristics, (c) transfer characteristic.

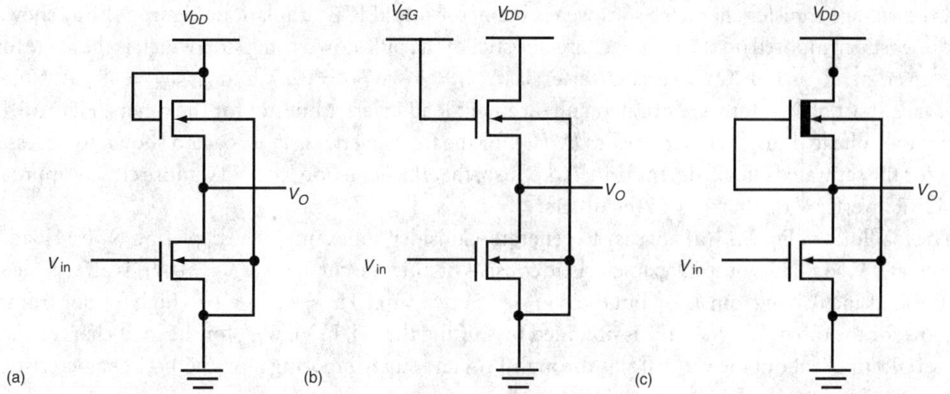

FIGURE 8.13 NMOS inverters with different types of active loads: (a) saturated enhancement load, (b) linear enhancement load, (c) depletion load.

The saturated enhancement load inverter overcomes much of the area disadvantage of the resistive load inverter. When carrying the same current and having the same pull-down transistor as the resistive inverter, however, K_R is large for the saturated enhancement load inverter, indicating load transistor area minimization is still possible. This configuration yields a smaller logic swing relative to the resistive load inverter, however, because the load transistor stops conducting when its $V_{GS} = V_{DS}$ decreases to V_T. Thus, for this inverter, $V_{OH} = V_{DD} - V_T$.

In Fig. 8.13(b), because V_{GG} is greater than $V_{DD} + V_T$, V_{DS} is always smaller than $V_{GS} - V_T$; thus, the load always operates in the linear region. This results in a linear enhancement load NMOS inverter. The high value of V_{GG} also ensures that V_{GS} is always greater than V_T, so that the load remains on and V_{OH} pulls up to V_{DD}. The linear enhancement load configuration, however, requires a load transistor of larger area relative to the saturated enhancement load inverter, and requires additional chip area for the V_{GG} contact.

In the depletion NMOS load inverter of Fig. 8.13(c), $V_{GS} = 0$, thus the load device is always on and V_{OH} pulls all the way to V_{DD}. This configuration overcomes the area disadvantage without incurring a voltage swing penalty. It is, therefore, the preferred alternative. The performance of the NMOS inverters with the four different types of loads are graphically compared in Fig. 8.14(a) and Fig. 8.14(b). Both the loadlines

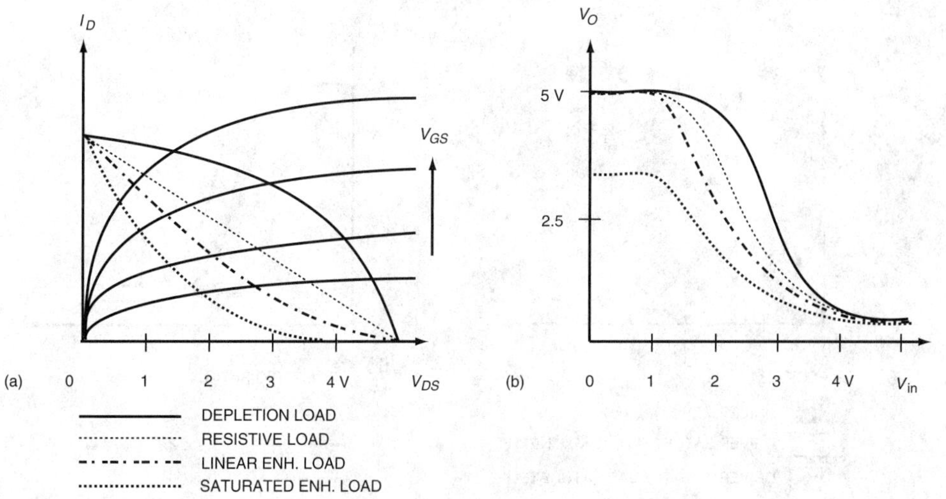

—————— DEPLETION LOAD
··············· RESISTIVE LOAD
— ·· — ·· — LINEAR ENH. LOAD
·················· SATURATED ENH. LOAD

FIGURE 8.14 Performance of NMOS inverters with different types of loads: (a) output characteristics and load lines, (b) voltage transfer characteristics.

and the voltage transfer characteristics were obtained from SPICE simulation. Figure 8.14(a) shows the loadlines superimposed on the output characteristics of the pull-down transistor, which is the same for the four inverters. R_L is 100 kΩ and each inverter has $V_{DD} = 5$ V, $V_{OL} = 0.2$ V and $I_{D\,max} = 48\,\mu$A. Note that V_{OH} falls short of V_{DD} for the saturated enhancement load inverter but not for the others. Figure 8.14(b) shows the voltage transfer characteristics (VTC) for the four inverters. V_{OH} is again shown to be less than V_{DD} for the saturated enhancement load. Note, also, that the depletion load VTC more closely approaches the ideal inverter VTC than any of the others.

The loadlines of Fig. 8.13(a) are easy to generate. Consider, for example the depletion NMOS load. V_{GS} is fixed at 0 V, so that its output characteristic consists of only the curve for $V_{GS} = 0$. I_D is always the same for the load and driving transistor, but their V_{DS} add up to V_{DD}. Thus, when V_{DS} is high for one transistor, it is low for the other. The loadline is obtained by shifting the origin for V_{DS} for the load characteristic to V_{DD}, reflecting it about the vertical axis through V_{DD} and superimposing it on the $V–I$ characteristics for the driving inverter.

The voltage transfer characteristics are best generated by computer simulation. Useful insights, however, can be gained from an analysis yielding the critical voltages V_{OH}, V_{OL}, V_{IH}, V_{IL}, and V_O for any specified V_{in}. The NMOS currents hold the key to such an analysis. Threshold voltages are different for enhancement and depletion NMOS transistors, but the drain current equations are the same. The drain current is given in the linear region and the saturated region, respectively, by

$$I_D = K_n \left[2(V_{GS} - V_T)V_{DS} - V_{DS}^2\right]; \quad V_{DS} \leq V_{GS} - V_T \tag{8.3}$$

$$I_D = K_n(V_{GS} - V_T)^2; \quad V_{DS} \geq V_{GS} - V_T \tag{8.4}$$

where
V_T = threshold voltage
$K_n = \frac{\mu_n C_{ox}}{2}\left(\frac{w}{L}\right)$ = transconductance
μ_n = electron channel mobility
C_{ox} = gate capacitance per unit area

Similar definitions apply to PMOS transistors.

Consider the VTC of Fig. 8.15(a) for a depletion load NMOS inverter. For the region $0 < V_{in} < V_T$ the driving transistor is off, so $V_{OH} = V_{DD}$. At A, V_{in} is small; thus, for the driving transistor, $V_{DS} = V_O > V_{in} - V_T = V_{GS} - V_T$. For the load $V_{DS} = V_{DD} - V_O$ is small. Hence, the driver is saturated and load

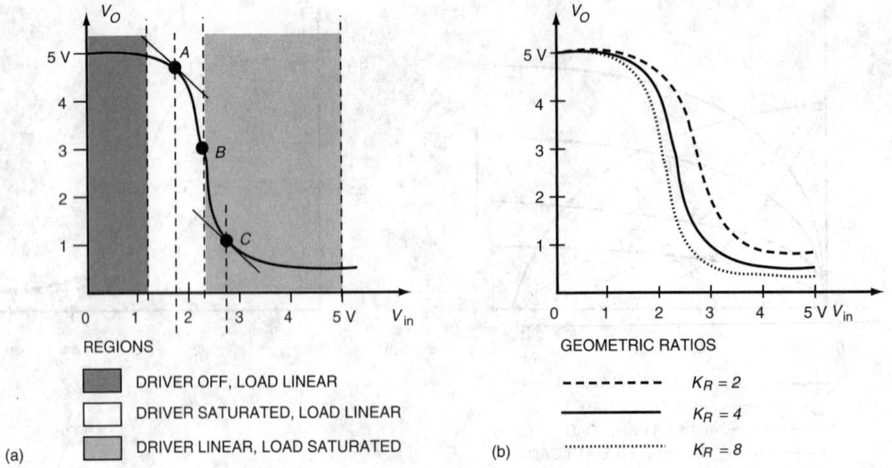

(a)

REGIONS

▪ DRIVER OFF, LOAD LINEAR
☐ DRIVER SATURATED, LOAD LINEAR
▪ DRIVER LINEAR, LOAD SATURATED

(b)

GEOMETRIC RATIOS

-------- $K_R = 2$
——— $K_R = 4$
············ $K_R = 8$

FIGURE 8.15 Regions of the VTC and VTCs for different geometric ratios for the depletion load NMOS inverters: (a) depletion load NMOS inverter VTC and its regions, (b) VTCs for different geometric ratios.

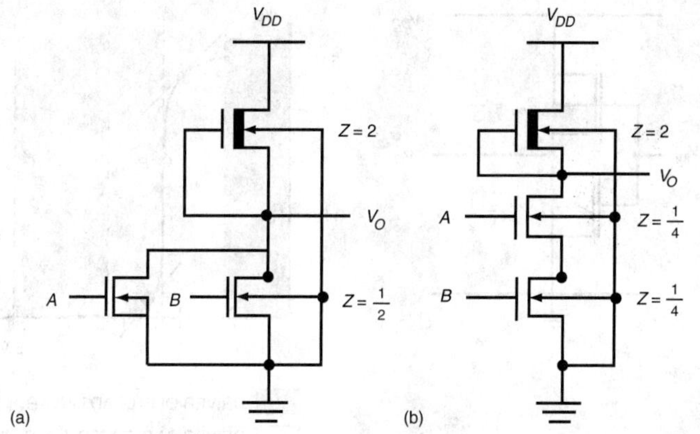

FIGURE 8.16 NMOS gates: (a) NMOS NOR gate, (b) NMOS NAND gate.

is linear. Similar considerations lead to the conclusions as to the region in which each device operates, as noted in the figure. To find V_{IL} and V_{IH}, the drain currents for the appropriate region of operation for points A and C, respectively, are equated, for both transistors. Differentiating the resulting equations with respect to V_{in} and applying the condition that $dV_O/dV_{in} = -1$ yields the required critical voltages. Equating drain currents for saturated load and linear driver at $V_{in} = V_{DD}$ and solving yields V_{OL}. The output voltage V_O may be found at any value of V_{in} by equating the drain currents for the two transistors, appropriate for that region of V_{in}, and solving for V_O at the given V_{in}.

NMOS Gates

Only NOR and NAND need be considered because these are more economical in chip area than OR and AND, and any logic system can be implemented entirely with either NOR or NAND. By connecting driving transistors in parallel to provide the multiple inputs, the NMOS inverter is easily converted to a NOR gate as shown in Fig. 8.16(a). By connecting driving transistors in series as in Fig. 8.16(b), an NAND gate is obtained.

The NMOS, NOR, and NAND gates are essentially modifications of the depletion load NMOS inverter of Fig. 8.13(c). They all have the same depletion load, and their performance should be similar. For the same value of V_{DD}, each has $V_{OH} = V_{DD}$, and they should have the same V_{OL} and the same drain current when $V_O = V_{OL}$. In Fig. 8.13(c), the depletion load inverter has $Z_{pu} = 2$, $Z_{pd} = \frac{1}{2}$ and $K_R = 4$. Thus, Z_{pu} is 2 for the NOR and the NAND gate. With only one driving transistor on in the NOR gate, the drain current should be sufficient to ensure that $V_O = V_{OL}$. Thus, for each of the driving transistors, $Z_I = \frac{1}{2}$, as for the depletion load inverter. For the NAND gate, the equivalent series combination of Z_{pd} (proportional to drain-source resistance) should also be $\frac{1}{2}$ to allow the same value of V_O, leading to $Z_I = \frac{1}{4}$ for each driving transistor in the NAND gate. Thus, K_R is 4 for the inverter, 4 for the NOR gate, and 8 for the NAND. As the number of inputs increases, K_R increases for the NAND but not for the NOR. It is clear that NMOS NAND gates are wasteful of chip area relative to NMOS NOR. Hence, NOR gates are preferred (and the NOR is the standard gate) in NMOS.

CMOS Inverters

As shown in Fig. 8.17(a), the CMOS inverter consists of an enhancement NMOS as the driving transistor, and a complementary enhancement PMOS load transistor. The driving transistor is off when V_{in} is low, and the load transistor is off when V_{in} is high. Thus, one of the two series transistors is always off (equivalently, drain current and power dissipation are zero) except during switching, when both transistors are momentarily on. The resulting low-power dissipation is an important CMOS advantage and makes it an attractive alternative in VLSI design.

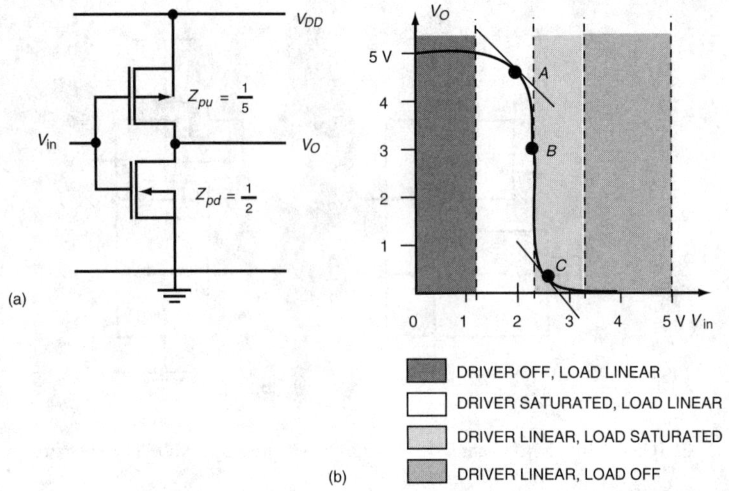

FIGURE 8.17 The CMOS inverter and its voltage transfer characteristics: (a) CMOS inverter, (b) CMOS transfer characteristic and its regions.

NMOS circuits are ratioed in the sense that the pull up never turns off, and V_{OL} is determined by the inverter ratio. CMOS is ratioless in this sense, since V_{OL} is always the negative rail. If one desires equal sourcing and sinking currents, however, the pull-up device must be wider than the pull-down device by the ratio of the electron-to-hole mobilities, typically about 2.5 to 1. This also gives a symmetrical voltage transfer curve, with the voltage at which $V_{in} = V_O$ having a value of $V_{DD}/2$. This voltage is referred to as the inverter voltage V_{inv}.

The voltage transfer for the CMOS inverter is shown in Fig. 8.17(b). Note that the voltage transfer characteristic approaches that of the ideal logic inverter. These characteristics are best obtained with computer circuit simulation programs. As with the depletion load NMOS inverter, useful insights may be gained by performing an analytical solution. The analysis proceeds as previously described for the depletion load NMOS inverter. Note that the VTC of Fig. 8.17(b) has been divided into regions as in Fig. 8.15(a). In each region, the appropriate expressions for the load and driving transistor drain currents are equated so that V_O can be computed for any given V_{in}. To find V_{IL} and V_{IH}, the condition that $dV_O/dV_{in} = -1$ at such critical voltages is applied to the drain current equation. Note that the drain current equations for the PMOS are the same as for NMOS (Eqs. 8.3 and 8.4), except for reverse voltage polarities for the PMOS.

CMOS Gates

CMOS gates are based on simple modifications to the CMOS inverter. Figure 8.18(a) and Figure 8.18(b) show that the CMOS NOR and NAND gates are essentially CMOS inverters in which the load and driving transistor are replaced by series or parallel combinations (as appropriate) of PMOS and NMOS transistors, respectively.

Suppose the NOR gate of Fig. 8.18(a) is to have the same V_{DD} and V_{inv} as the CMOS inverter of Fig. 8.17(a), then the equivalent Z_{pu} and Z_{pd} for the NOR gate should equal those for the inverter. Since only one of the parallel pull-down transistors needs be on in the NOR to ensure $V_O = 0$ V, $Z_I = Z_{pd} = \frac{1}{2}$, as for the inverter. For the series load, however, $Z_L = \frac{1}{10}$ to give equivalent $Z_{pu} = \frac{1}{5}$. If the NAND gate of Fig. 8.18(b) is to have the same V_{inv} as the said inverter, similar arguments lead to $Z_I = \frac{1}{4}$ and $Z_L = \frac{1}{5}$ for the NAND. Thus, $K_R = 0.4$ for the inverter, 0.2 for the NOR, and 0.8 (closer to unity) for NAND. Hence, NAND is the standard gate in CMOS. Another way of putting this is that for the given Z values, if the channel length L is constant, then the widths of the loads for the inverter, NOR, and NAND are in the ratio 1:2:1. Thus, the NOR requires more chip area, and this larger area requirement increases with the number of inputs.

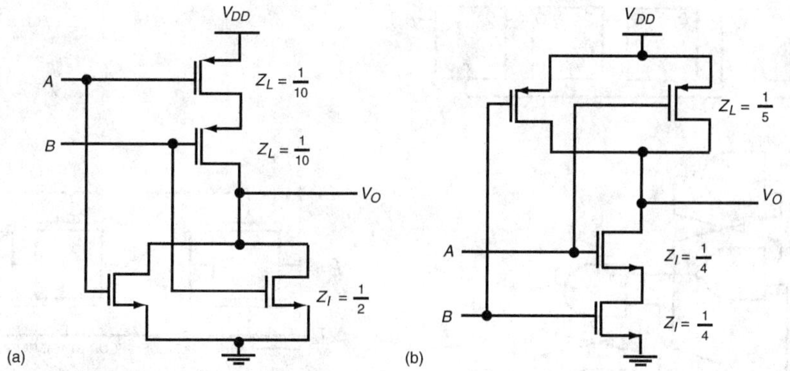

FIGURE 8.18 CMOS gates: (a) CMOS NOR gate, (b) CMOS NAND gate.

Bipolar Gates

The major bipolar digital logic families are TTL, emitter-coupled logic (ECL), and integrated injection logic (I^2L). Within each logic family, there are subclassifications, for example, the Schottky transistor logic (STL), and the integrated Schottky logic (ISL), which developed from the basic I^2L. Bipolar gates have faster switching speeds but greater power dissipation than CMOS gates. The most popular bipolar gate in SSI is the low-power Schottky TTL, which has moderate power dissipation and propagation delay. The fastest switching bipolar family is the ECL, but it has relatively high-power dissipation. The highest packing density is achieved with I^2L and its relatives with low-power dissipation and moderate switching speeds. A better comparison of logic families should be based on the power-delay product, which takes into account both power dissipation and propagation delay.

Medium-Scale Integrated Circuits

MSI circuits have between 10 and 100 transistors per chip. They are built from inverters and basic logic gates with hardly any modifications. They require minimal design effort beyond putting together and interconnecting logic gates. Examples of MSI circuits are flip-flops, counters, registers, adders, multiplexers, demultiplexers, etc.

8.2.5 LSI and VLSI Circuit Design

Semicustom design is a heavily utilized technique in LSI and VLSI design. In this technique, largely predesigned subcircuits or cells are interconnected to form the desired, larger circuit. Such subcircuits are usually highly regular in nature, so that the technique leads to highly regular circuits and layouts.

Multiphase Clocking

Multiphase clocking is an important technique that can be used to reduce device count in LSI and VLSI circuits. To illustrate the savings that can be realized with the technique, device count is compared for a conventional design of a 4-b shift register employing D flip-flops based on CMOS NAND gates and a 4-b shift register employing two-phase clocks and CMOS technology.

Both designs are shown in Fig. 8.19. Figure 8.19(a) shows the conventional design for the shift register, which employs a single phase clock signal, whereas Fig. 8.19(b) shows the circuit realization of each D flip-flop with CMOS NAND gates (Taub and Schilling, 1977). The device count for each in this design is obtained as follows:

- 5 two-input CMOS NAND gates, 4 transistors each: 20 transistors
- 1 three-input CMOS NAND gate: 6 transistors
- Number of transistors per D flip-flop: 26
- Total number of transistors for the 4-b register: 104

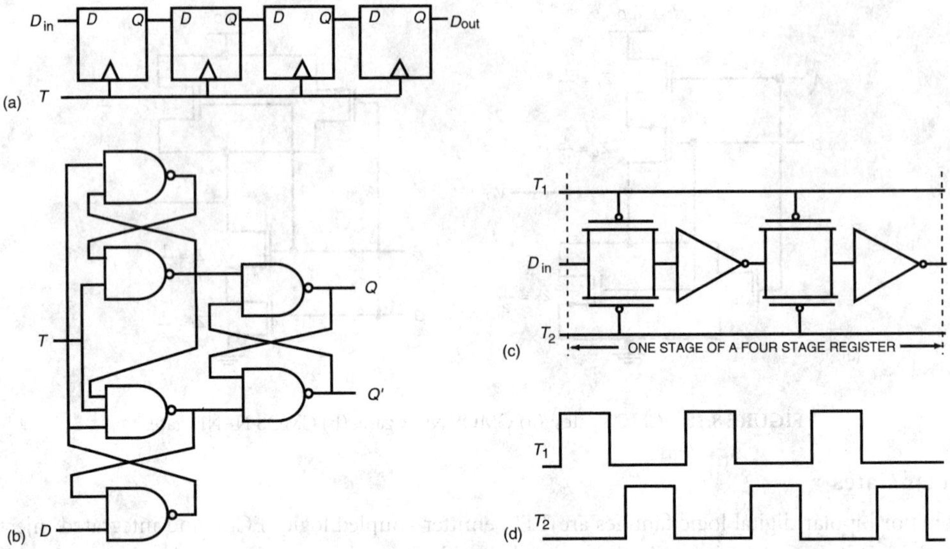

FIGURE 8.19 Conventional (static) and dynamic 4-b shift register: (a) conventional static shift register, (b) D flip-flop, (c) dynamic shift register, (d) two-phase clock pulses.

The second design, which employs two-phase clocking, is shown in 8.19(c), whereas the nonoverlapping clock signals are shown in Fig. 8.19(d). Note that each flip-flop now consists of two CMOS transmission gates and two CMOS inverters. Thus, there are 8 transmission gates and 8 inverters in the 4-b shift register. Device count for this design is as follows:

- Number of transistors for 8 transmission gates: 16
- Number of transistors for 8 CMOS inverters: 16
- Total number of transistors for the 4-b register: 32

In the preceding example, employing two-phase clocking helped to reduce device count to less than one-third of the requirement in the conventional static design. This gain, however, is partly offset by the need for more complex clocking and the fact that the shift register is now dynamic. To avoid loss of data due to leakage through off transistors, the clock must run above a minimum frequency. The times required to charge and discharge capacitive loads determine the upper clock frequency.

8.2.6 Increasing Packing Density and Reducing Power Dissipation in MOS Circuits

CMOS gates have much lower power dissipation than NMOS gates. This is a great advantage in LSI and VLSI design. Standard CMOS gates, however, require two transistors per input and, therefore, have higher device count than NMOS gates that require one driving transistor per input, plus one depletion load transistor, irrespective of the number of inputs (Mavor, Jack, and Denyer, 1983). This NMOS feature is put to advantage in applications such as semiconductor memories and programmable logic arrays, which will be discussed later. In addition to requiring a higher device count, it is necessary to isolate the PMOS and NMOS transistors in the CMOS and to employ metal interconnection between their drains, which are of opposite conductivity. Consequently, gate count per chip for NMOS is about half that of CMOS, using the same design rules.

Figure 8.20 shows a CMOS domino logic circuit in which clocking is employed in an unconventional CMOS circuit to provide both high density and low-power dissipation. When T is low, Q_1 is off, so there is no path to ground irrespective of the logic levels at the inputs A, B, C, D and E. Q_2 is on, so that the

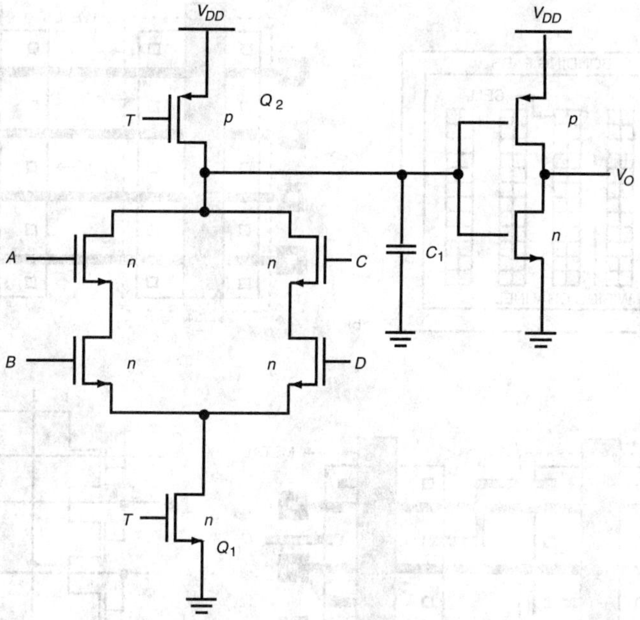

FIGURE 8.20   CMOS domino AND-OR logic.

parasitic capacitance C_1 charges to V_{DD}. When T is high, Q_2 is off and Q_1 is on. Thus if both A and B, or both C and D, or all of A, B, C, and D are high, a path exists from C_1 to ground, and it discharges. Otherwise, C_1 remains high (but for slow leakage), and the valid logic $(A\ B) + (C + D)$ appears at the output F. Note that this circuit has only two load PMOS transistors, and only one driving transistor is required for each additional input. Thus, device count is minimized by using complex instead of simple logic functions. Each transistor, except those in the output inverter, may be minimum size, since they are required only to charge or discharge C_1. Power dissipation is low as for standard CMOS, because no steady-state current flows.

8.2.7   Gate Arrays

Gate arrays are a category of semicustom integrated circuits typically containing 100 to several thousand gate cells arranged in rows and columns on a chip. The gate cell may be a NAND, NOR, or other gate. Often, each gate cell is a set of components that could be interconnected to form the desired gate or gates. Identical gate cell pattern is employed, irrespective of chip function. Consequently, gate arrays can be largely processed in advance (Reinhard, 1987). Less effort is required for design with gate arrays since only the masks needed for interconnection are required to customize a chip for a particular application.

Figure 8.21 illustrates a gate array in various levels of detail and possible interconnections within a cell. The floor plan of Fig. 8.21(a) shows that there are 10 columns of cells with 10 cells per column, for a total of 100 cells in the chip. The cell layout of Fig. 8.21(b) shows that there are 4 NMOS and 4 PMOS transistors per cell. Thus there are a total of 800 transistors in the chip. The transistor channels are under the polysilicon and inside the diffusion areas. Figure 8.21(c) shows the cell layout with interconnection to form an NAND gate, whereas Fig. 8.21(d) shows the circuit equivalent of a cell.

Because of their simplicity, a significant amount of wiring is required for interconnections in gate arrays. Good computer software is essential for designing interconnections. In practice, wiring channels tend to fill up, so that it is difficult to utilize more than 70% of the cells on a chip (Alexander, 1985). The standard cell approach discussed next reduces this problem, to some extent, by permitting use of more complex logic functions or cells.

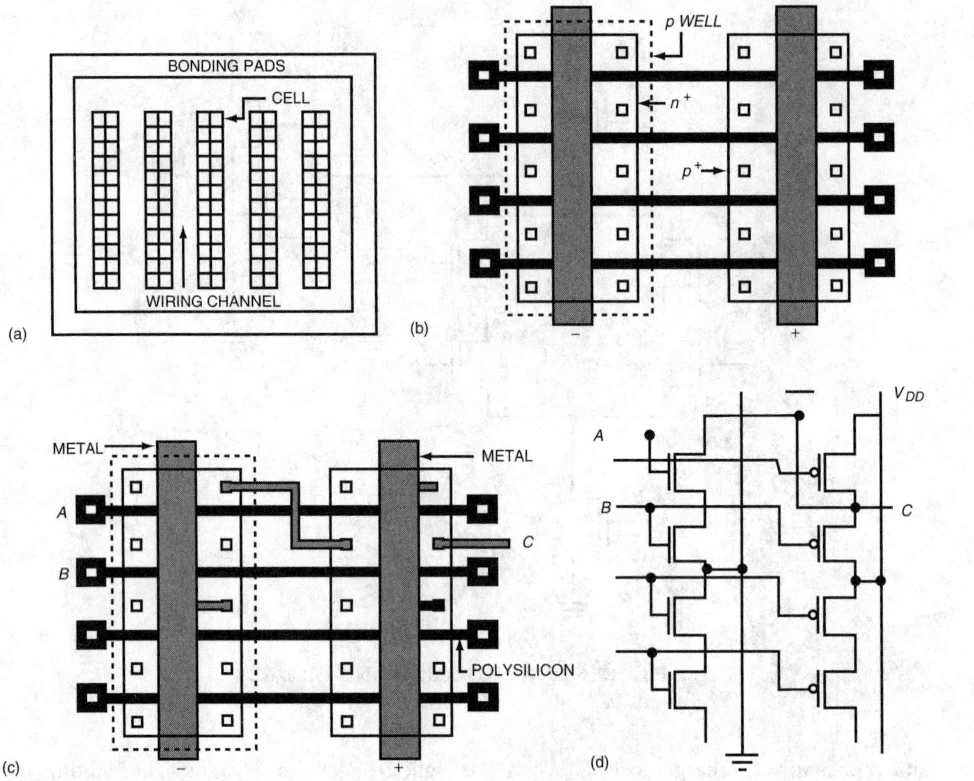

FIGURE 8.21 Gate array at various levels of detail: (a) cell structure, (b) transistor structure, (c) NAND gate interconnection, (d) equivalent circuit.

8.2.8 Standard Cells

In the **standard cell** approach, the IC designer selects from a library of predefined logic circuits or cells to build the desired circuit. In addition to the basic gates, the **cell library** usually includes more complex logic circuits such as exclusive-OR, AND-OR-INVERT, flip-flops, adders, read only memory (ROM), etc.

The standard cell approach to design is well suited to automated layout. The process consists of selecting cells from the library in accordance with the desired circuit functions, the relative placement of the cells, and their interconnections. The floor plan for a chip designed by this method is similar to the floor plan for a gate array chip as shown in Fig. 8.21(a). Note, however, that the designer has control over the number and width of wiring channels in this case. Layout for a cell is always the same each time the cell is used, but the cells used and their relative placement is unique to a chip. Thus, every mask level is unique in this approach, and fabrication is more involved and more costly than in the gate array approach (Hodges and Jackson, 1988).

8.2.9 Programmable Logic Devices

Programmable logic devices (PLDs) are a class of circuits widely used in LSI and VLSI design to implement two-level, sum-of-products, boolean functions. Multilevel logic can be realized with Weinberger arrays or gate matrices (Fabricius, 1990; Weinberger, 1967). Included among PLDs are programmable logic arrays (PLAs), programmable array logic (PAL), and ROM. The AND-OR structure of the PLA, which can be used to implement any two-level function, is the core of all PLDs. The AND-OR function is often implemented with NOR-NOR or NAND-NAND logic.

PLDs have the advantage of leading to highly regular layout structure. The PLD consists of an AND plane followed by an OR plane. The logic function is determined by the presence or absence of contacts or

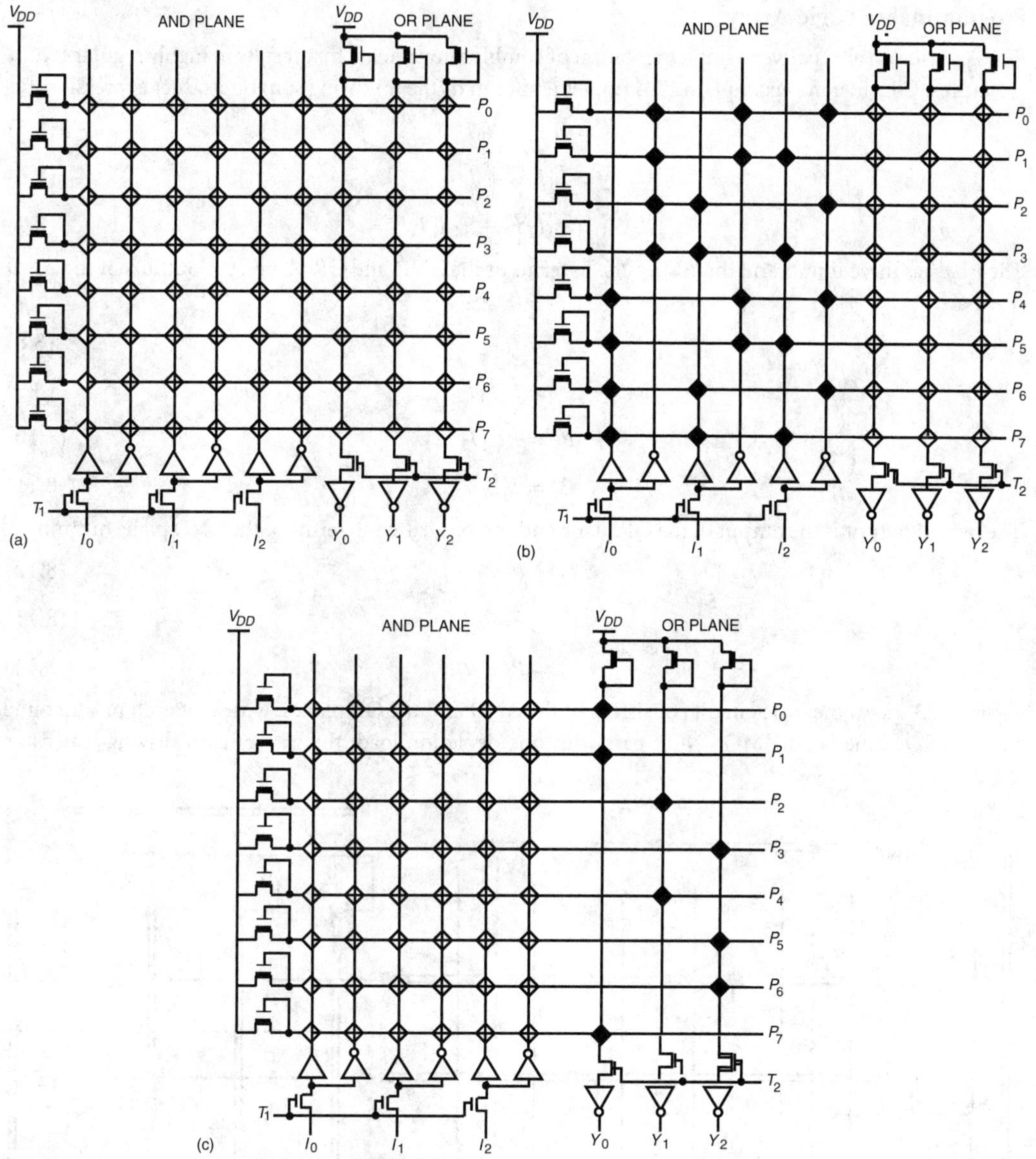

FIGURE 8.22 Types of programmable logic devices: (a) programmable logic array, (b) programmable read only memory, (c) programmable array logic.

connections at row and column intersections in a single conducting layer. Programming or establishment of appropriate contacts may be accomplished during fabrication. Alternatively, the PLDs may be user programmable by means of fuse links.

Figure 8.22 shows the three types of PLDs. Hollow diamonds at row/column intersections in an AND or OR plane indicates that the plane is programmable. Presence of solid diamonds in some row/column intersections indicate that the logic for that plane is already defined and fixed. The PLD is a PLA if both the AND and OR planes are programmable, a PAL if only the AND plane is programmable, and a ROM if only the OR plane (the decoder in this case) is programmable. Because PLAs are programmable in both planes, they permit more versatile logic realizations than PALs. Also, the PAL can be considered a special case of the PLA. Thus, only the PLA is discussed further.

Programmable Logic Array

PLAs provide an alternative to implementation of combinational logic that results in highly regular layout structures. Consider, for example, a PLA implementation of the following sum of product expressions:

$$Y_1 = \bar{I}_0 \bar{I}_1 + I_0 \bar{I}_2 \tag{8.5}$$

$$Y_2 = \bar{I}_0 I_1 I_2 + I_0 \bar{I}_2 \tag{8.6}$$

$$Y_3 = \bar{I}_0 \bar{I}_1 \bar{I}_2 + \bar{I}_0 I_1 I_2 \tag{8.7}$$

The PLA has three inputs and three outputs. In terms of the AND and OR planes, the outputs of the AND plane are

$$P_1 = \bar{I}_0 \bar{I}_2 \tag{8.8}$$

$$P_2 = \bar{I}_0 I_1 \bar{I}_2 \tag{8.9}$$

$$P_3 = \bar{I}_0 I_1 I_2 \tag{8.10}$$

$$P_4 = I_0 \bar{I}_2 \tag{8.11}$$

The overall output is the output of the OR plane and can be written in terms of the AND plane outputs as

$$Y_1 = P_1 + P_4 \tag{8.12}$$

$$Y_2 = P_3 + P_4 \tag{8.13}$$

$$Y_3 = P_2 + P_3 \tag{8.14}$$

Figure 8.23 shows the logic circuit consisting of the AND and the OR planes. Note that each product line in the AND plane is an NMOS NOR gate with one depletion load; the gate of each driving transistor

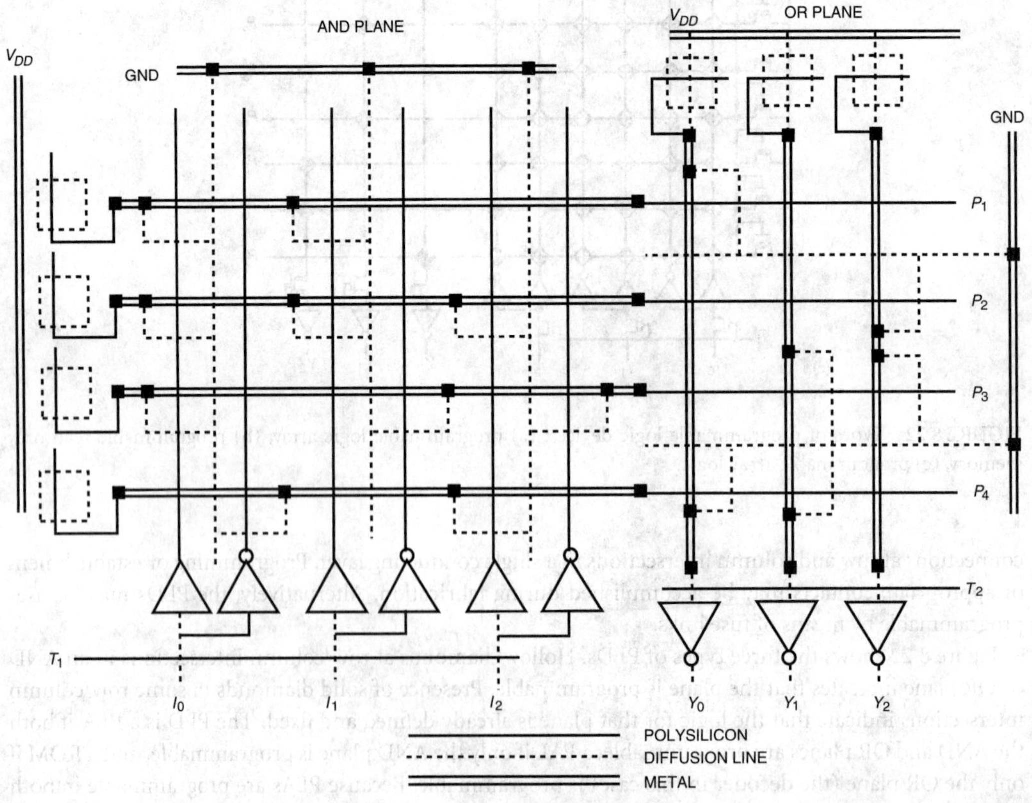

FIGURE 8.23   An NMOS PLA.

is controlled by an input line. Likewise, each output line in the OR plane is an NMOS NOR gate with driving transistors whose gates are controlled by the product lines. Thus, the PLA employs a NOR–NOR implementation.

The personality matrix for a PLA (Lighthart, Aarts, and Beenker, 1986) gives a good description of the PLA and how it is to be programmed. The personality matrix Q of the PLA of Fig. 8.23 is given by Eq. (8.15).

$$Q = \begin{bmatrix} 0 & 0 & x & 1 & 0 & 0 \\ 0 & 0 & 0 & 0 & 0 & 1 \\ 0 & 1 & 1 & 0 & 1 & 1 \\ 1 & x & 0 & 1 & 1 & 0 \end{bmatrix} \tag{8.15}$$

The first three columns comprise the AND plane of the matrix, whereas the last three columns comprise the OR plane of the three-input, three-output PLA. In the AND plane, element $q_{ij} = 0$ if a transistor is to link the product line P_i to the input line I_i; $q_{ij} = 1$ if a transistor is to link P_i to $\bar{I}_i$, and q_{ij} is a *don't care* if neither input is to be connected to P_i. In the OR plane, $q_{ij} = 1$ if product line P_i is connected to output Y_j and 0 otherwise.

Figure 8.24 shows the stick diagram layout of the PLA circuit of Fig. 8.23, and illustrates how the regular structure of the PLA facilitates its layout. The input lines to each plane are polysilicon, the output lines from each plane are metal, and the sources of driving transistors are connected to ground by diffused lines. The connecting transistors are formed by grounded, crossing diffusion lines.

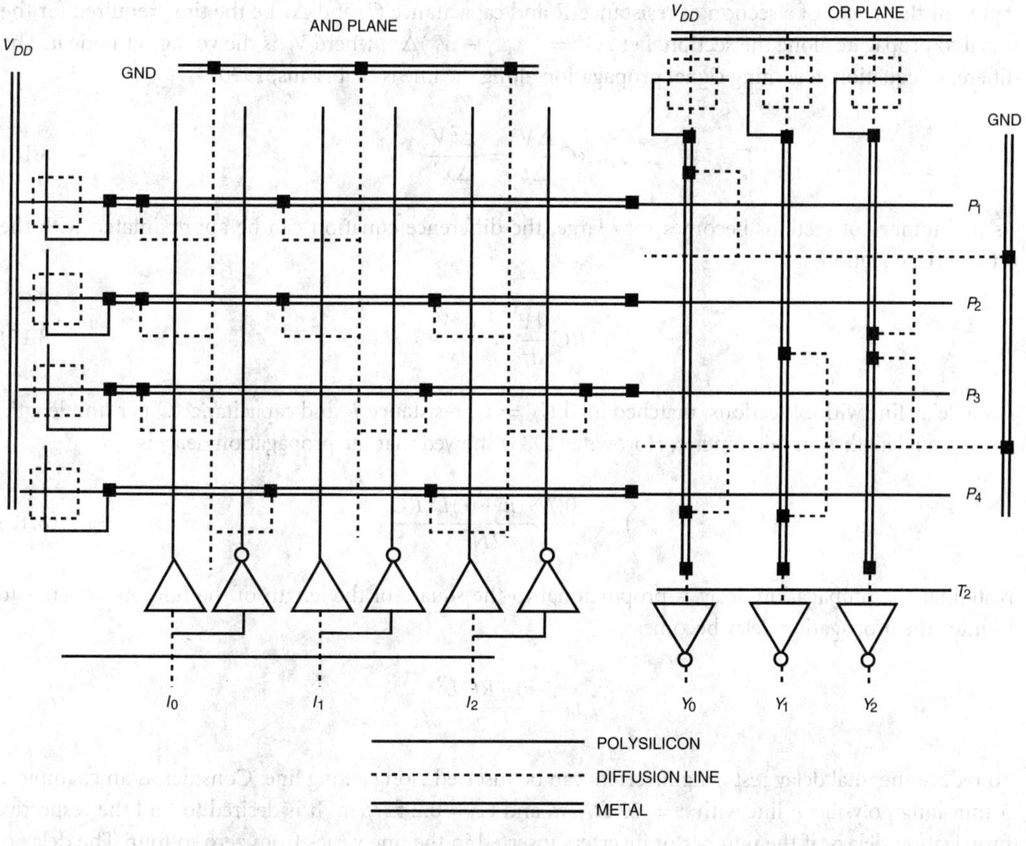

FIGURE 8.24 Stick diagram layout of the PLA shown in Fig. 8.23.

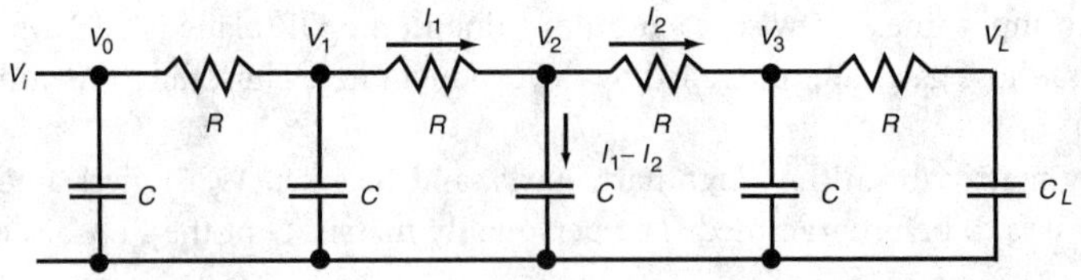

FIGURE 8.25 Lumped circuit model of a long polysilicon line.

8.2.10   Reducing Propagation Delays

Large capacitive loads are encountered in many ways in large integrated circuits. Bonding pads are required for interfacing the chip with other circuits, whereas probe pads are often required for testing. Both present large capacitive loads to their drivers. Interconnections within the chip are by means of metal or polysilicon lines. When long, such lines present long capacitive loads to their drivers. Although regular array structures, such as those of gate arrays, standard cells, and PLAs, are very convenient for semicustom design of LSI and VLSI, they have an inherent drawback with regard to propagation delay. Their row and column lines contact many devices and, hence, are very capacitive. The total delay of a long line may be reduced by inserting buffers along the line to restore the signal. Superbuffers are used for interfacing between small gates internal to the chip and large pad drivers and for driving highly capacitive lines.

Resistance-Capacitance (RC) Delay Lines

A long polysilicon line can be modeled as a lumped RC transmission line as shown in Fig. 8.25. Let Δx represent the length of a section of resistance R and capacitance C, and Δt be the time required for the signal to propagate along the section. Let $\Delta V = (V_{n-1} - V_n)\Delta x$, where V_n is the voltage at node n. The difference equation governing signal propagation along the line is (Fabricius, 1990)

$$RC\frac{\Delta V}{\Delta t} = \frac{\Delta^2 V}{\Delta x^2} \tag{8.16}$$

As the number of sections becomes very large, the difference equation can be approximated with the differential equation

$$RC\frac{dV}{dt} = \frac{d^2 V}{dx^2} \tag{8.17}$$

For a delay line with N sections, matched load $C_L = C$, resistance R and capacitance C per unit length, and overall length L in micrometer, Horowitz (1983) showed that the propagation delay is

$$t_d = \frac{0.7N(n+1)L^2 RC}{2N^2} \tag{8.18}$$

Note that the propagation delay is proportional to the square of the length of the line. As N tends to infinity, the propagation delay becomes

$$t_d = \frac{0.7RCL^2}{2} \tag{8.19}$$

To reduce the total delay restoring inverters can be inserted along a long line. Consider as an example, a 5-mm-long polysilicon line with $r = 20\ \Omega/\mu m$ and $C = 0.2\ fF/\mu m$. It is desired to find the respective propagation delays, if the number of inverters inserted in the line varies from zero to four. The delay of each inverter is proportional to the length of the segment it drives and is given by $t_I = 0.4$ ns when it is

TABLE 8.6 Improvement in Propagation Delay with Increase in
Number of Line Buffered

No. of Inverters K	Total Delay t_d, ns	% of Unbuffered Line Delay
0 (unbuffered)	35.0	100
1	18.5	52.86
2	13.0	37.14
3	10.25	29.29
4	8.6	24.57

driving a 1-mm-long segment. In each case, the inverters used are spaced uniformly along the line. Let

K = number of inverters used
$K + 1$ = number of sections
ℓ = length per section, $\frac{5\text{mm}}{k+1}$
t_d = total delay

then

$$t_d = \left[(k+1)\left(\frac{0.7RC\ell^2}{2} \right) + 0.4\ell K \right] ns \qquad (8.20)$$

From Eq. (8.20), the propagation delays can be calculated. The delay for each number of inverters as a percentage of the unbuffered line delay is also computed. The results are tabulated in Table 8.6.

The results in the table show that the propagation delay decreases as the number of inverters is increased. The improvement in propagation delay, however, is less dramatic for each additional inverter than the one preceding it. The designer would stop increasing the number of inverters when the incremental gain no longer justifies an additional inverter. If the number of inverters is even, there is no inversion of the overall signal.

Superbuffers

Propagation delays can be reduced without excessive power consumption by using superbuffers. These are inverting or noninverting circuits that can source and sink larger currents and drive large capacitive loads faster than standard inverters. Unlike ratioed NMOS inverters in which the pull-up current drive capability is much less than the pull-down capability, superbuffers have symmetric drive capabilities. A superbuffer consists of a push-pull or totem pole output inverter driven by a conventional inverter. In an inverting superbuffer, the gates of both pull-down transistors in the driving and the totem pole inverters are driven by the input signal whereas the gate of the pull-up transistor in the output totem pole inverter is driven by the complement of the input signal.

An inverting and a noninverting NMOS superbuffer is shown in Fig. 8.26. By designing for an inverter ratio (K_R) of 4, and driving the totem pole pull-up with twice the gate voltage of a standard depletion mode pull-up, the NMOS superbuffer can be rendered essentially ratioless. In standard NMOS inverters, the pull-up transistor has the slower switching speed. Consider the inverting superbuffer of Fig. 8.26(a). When the input voltage goes low, the output voltage of the standard inverter and the gate voltage of Q_4 goes high rapidly since the only load it sees is the small gate capacitance of Q_4. Thus, the totem pole output switches rapidly. Similar considerations shows that the noninverting super buffer also results in fast switching speeds.

The improvement in drive current capability of the NMOS superbuffer, relative to the standard (depletion load) NMOS inverter, can be estimated by comparing the average, output pull-up currents (Fabricius, 1990). The depletion load in the standard NMOS inverter is in saturation for $V_O < 2$ V and in linear region for $V_O > 2$ V. For the pull-up device, $V_{DS} = 5\,\text{V} - V_O$. Thus, the pull-up transistor is in saturation when it has $3\,\text{V} < V_{DS} < 5$ V and is in the linear region when $0\,V < V_{DS} < 3$ V. The average current

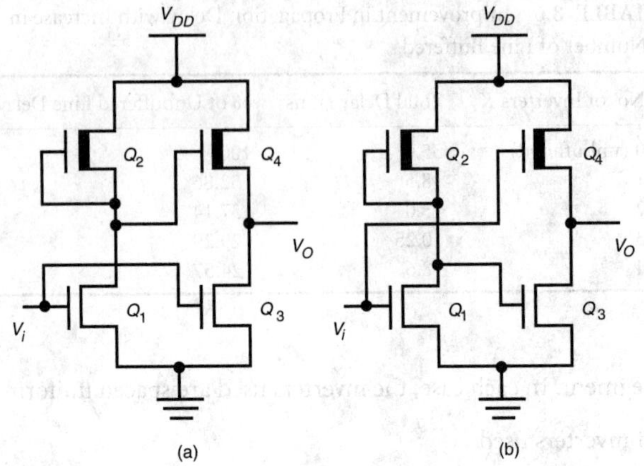

FIGURE 8.26 NMOS superbuffers: (a) inverting superbuffer, (b) noninverting superbuffer.

will be estimated by evaluating $I_D(\text{sat})$ at $V_{DS} = 5$ V and $I_{D(\text{lin})}$ at $V_{DS} = 2.5$ V. Let $V_{TD} = -3$ V for the depletion mode transistor. Then for the standard NMOS inverter

$$I_{D(\text{sat})} = K_{\text{pu}}(V_{GS} - V_{\text{TD}})^2 = K_{\text{pu}}(0+3)^2 = 9K_{\text{pu}} \tag{8.21}$$

$$I_{D(\text{lin})} = K_{\text{pu}}\big(2(V_{GS} - V_{\text{TD}})V_{DS} - V_{DS}^2\big) = K_{\text{pu}}(2(0+3)2.5 - 2.5^2) = 8.76K_{\text{pu}} \tag{8.22}$$

Thus the average pull-up current for the standard NMOS inverter is approximately $8.88K_{\text{pu}}$. For the totem pole output of the NMOS superbuffer, the average pull-up current is also estimated from drain currents at $V_{DS} = 5$ V and 2.5 V. Note that in this case, the pull-up transistor has $V_G = V_{DD} = 5$ V when it is on. Thus, $V_{GS} = V_{DS} = 5$ V so that it always operates in the linear region. The currents are

$$I_D(5\text{ V}) = K_{\text{pu}}[2(5+3)5 - 5^2] = 55K_{\text{pu}} \tag{8.23}$$

$$I_D(2.5\text{ V}) = K_{\text{pu}}(2(2.5+3)2.5 - 2.5^2) = 10.62K_{\text{pu}} \tag{8.24}$$

The average pull-up current for the totem pole output is $38.12K_{\text{pu}}$. The average totem pole pull-up current is approximately 4.3 times the average NMOS pull-up current. Consequently, the superbuffer is roughly ratioless if designed for an inverter ratio of 4.

8.2.11 Output Buffers

Internal gates on a VLSI chip have load capacitances of about 50 fF or less and typical propagation delays of less than 1 ns. However, the chip output pins have to drive large capacitive loads of about 50 pF or more (Hodges and Jackson, 1988). For MOSFETs, the propagation delay is directly proportional to load capacitance. Thus, using a typical gate on the chip to drive an output pin would result in too long a propagation delay. Output buffers utilize a cascade of inverters of progressively larger drive capability to reduce the propagation delay.

An N-stage output buffer is illustrated in Fig. 8.27. Higher drive capability results from employing transistors of increasing channel width. As the transistor width increases from stage to stage by a factor of f, so does the current drive capability and the input capacitance. If C_G is the input or gate capacitance of the first inverter in the buffer, then the second inverter has an input capacitance of fC_G and the Nth inverter has an input capacitance of $f^{N-1}C_G$ and a load capacitance of $f^N C_G$, which is equal to C_L, the load capacitance at the output pin. The inverter on the left in the figure is a typical inverter on the chip with an input or gate capacitance of C_G and a propagation delay of τ. The first inverter in the buffer has an input capacitance of fC_G, but it has a current driving capability f times larger than the on chip inverter.

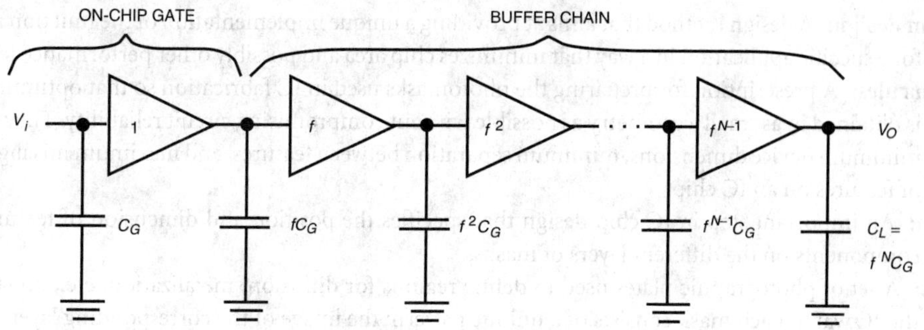

FIGURE 8.27 An N-stage output buffer chain.

Thus, it has a propagation delay of $f\tau$. The second inverter in the buffer has an input capacitance of $f^2 C_G$ and an accumulated delay of $2f\tau$ at its output. The Nth inverter has an input capacitance of $f^N C_G$, which is equal to the load capacitance at the output pin, and an accumulated propagation delay of $Nf\tau$, which is the overall delay of the buffer.

Let

$$\text{load capacitance} = Y C_G = f^N C_G$$
$$\text{total delay} = t_B = Nf\tau$$

Then

$$N = \frac{\ln Y}{\ln f} \tag{8.25}$$

and

$$t_B = \frac{\ln Y}{\ln f} f\tau \tag{8.26}$$

By equating to zero the first derivative of t_B with respect to f, it is found that t_B is minimum at $f = e = 2.72$, the base of the natural logarithms. This is not a sharp minimum (Moshen and Mead, 1979), and values of f between 2 and 5 do not greatly increase the time delay.

Consider an example in which $C_G = 50\text{ fF}$ and $\tau = 0.5$ ns for a typical gate driving an identical gate on the chip. Suppose this typical gate is to drive an output pin with load capacitance $C_L = 55$ pF, instead of an identical gate. If an output buffer is used

$$Y = \frac{C_L}{C_G} = \frac{55\text{ pF}}{50\text{ fF}} = 1100$$

$$N = \ln Y = 7$$

$$t_B = 7e\tau = 9.5\text{ ns}$$

If the typical chip gate is directly connected to the output pin, the propagation delay is $Y\tau = 550$ ns, which is extremely large compared with the 9.5 ns delay obtained when the buffer is used. This example illustrates the effectiveness of the buffer.

Defining Terms

Cell library: A collection of simple logic elements that have been designed in accordance with a specific set of design rules and fabrication processes. Interconnections of such logic elements are often used in semicustom design of more complex IC chips.

Custom design: A design method that aims at providing a unique implementation of the function needed for a specific application in a way that minimizes chip area and possibly other performance features.

Design rules: A prescription for preparing the photomasks used in IC fabrication so that optimum yield is obtained in as small a geometry as possible without compromising circuit reliability. They specify minimum device dimensions, minimum separation between features, and maximum misalignment of features on an IC chip.

Layout: An important step in IC chip design that specifies the position and dimension of features and components on the different layers of masks.

Masks: A set of photographic plates used to define regions for diffusion, metalization, etc., on layers of the IC wafer. Each mask consists of a unique pattern: the image of the corresponding layer.

Standard cell: A predefined logic circuit in a cell library designed in accordance with a specific set of design rules and fabrication processes. Standard cells are typically employed in semicustom design of more complex circuits.

Semicustom design: A design method in which largely predesigned subcircuits or cells are interconnected to form the desired more complex circuit or part of it.

References

Alexander, B. 1985. MOS and CMOS arrays. In *Gate Arrays: Design Techniques and Application.* ed. J.W. Read. McGraw-Hill, New York.

Fabricius, E.D. 1990. *Introduction to VLSI Design.* McGraw-Hill, New York.

Hodges, A.D. and Jackson, H.G. 1988. *Analysis and Design of Digital Integrated Circuits.* McGraw-Hill, New York.

Horowitz, M. 1983. Timing models for MOS pass networks. *Proceedings of the IEEE Symposium on Circuits and Systems*, pp. 198–201.

Keyes, R.W. 1975. Physical limits in digital electronics. *Proc. of IEEE* 63:740–767.

Lighthart, M.M., Aarts, E.H.L., and Beenker, F.P.M. 1986. Design for testability of PLAs using statistical cooling. *Proceedings of the 23rd ACM/IEEE Design Automation Conference*, pp. 339–345, June 29–July 2.

Mavor, J., Jack, M.A., and Denyer, P.B. 1983. *Introduction to MOS LSI Design.* Addison-Wesley, Reading, MA.

Mead, C.A. and Conway, L.A. 1980. *Introduction to VLSI Systems.* Addison-Wesley, Reading, MA.

Moshen, A.M. and Mead, C.A. 1979. Delay time optimization for driving and sensing signals on high capacitance paths of VLSI systems. *IEEE J. of Solid State Circ.* SC-14(2):462–470.

Pucknell, D.A. and Eshroghian, K. 1985. *Basic VLSI Design, Principles and Applications.* Prentice-Hall, Englewood Cliffs, NJ.

Reinhard, D.K. 1987. *Introduction to Integrated Circuit Engineering.* Houghton-Mifflin, Boston, MA.

Taub, H. and Schilling, D. 1977. *Digital Integrated Circuits.* McGraw-Hill, New York.

USC Information Sciences Inst. 1984. MOSIS scalable NMOS process, version 1.0. Univ. of Southern California, Nov., Los Angeles, CA.

USC Information Sciences Inst. 1988. MOSIS scalable and generic CMOS design rules, revision 6. Univ. of Southern California, Feb., Los Angeles, CA.

Weinberger, A. 1967. Large scale integration of MOS complex logic: a layout method. *IEEE J. of Solid-State Circ.* SC-2(4):182–190.

Further Information

IEEE Journal of Solid State Circuits.

IEEE Proceedings of the Custom Integrated Circuits Conference.
IEEE Transactions on Electron Devices.
Proceedings of the European Solid State Circuits Conference (ESSCIRC).
Proceedings of the IEEE Design Automation Conference.

8.3 Digital Logic Families

Robert J. Feugate, Jr.

8.3.1 Introduction

Digital devices are constrained to two stable operating regions (usually voltage ranges) separated by a transition region through which the operating point may pass but not remain (see Fig. 8.28). Prolonged operation in the transition region does not cause harm to the devices, it simply means that the resulting outputs are unspecified. For the inverter shown in the figure, the output voltage is guaranteed to be greater than V_{OH} as long as the input voltage is below the specified V_{IH}. Note that the circuit is designed so that the input high voltage is lower than the output high voltage (vice versa for logic low voltages). This difference, called **noise margin**, permits interfering signals to corrupt the logic voltage within limits without producing erroneous operation. Logical conditions and binary numbers can be represented physically by associating one of the stable voltage ranges with one logic state or binary value and identifying the other stable voltage with the opposite state or value. By extension, then, it is possible to design electronic circuits that physically perform logical or arithmetic operations. As detailed examination of logic design is beyond the scope of this chapter, the reader is referred to any of the large number of digital logic textbooks.[1]

FIGURE 8.28 Logic voltage levels.

Digital logic components were the earliest commercially produced integrated circuits. *Resistor-transistor logic* (RTL) and a speedier variant *resistor-capacitor-transistor logic* (RCTL) were introduced in the early 1960s by Fairchild Semiconductor *Diode-transistor logic* (DTL) was introduced a few years later by Signetics Corporation. Although these families are often discussed in electronics texts as ancestors of later logic families, they have been obsolete for many years and are of historical interest only.

A primary performance characteristic of different logic families is their *speed-power product*, that is, the average propagation delay of a basic gate multiplied by the average power dissipated by the gate. Table 8.7

TABLE 8.7 Speed-Product Comparison for Popular Discrete Logic Families[a]

| | Power Dissipation/Gate | | | |
Logic Family	Static at 100 kHz, mW	Typical Delay, mW	Product at 100 kHz, ns	pJ
Metal-gate CMOS	0.001	0.1	75	7.5
Silicon-gate CMOS	0.0000025	.17	10	1.7
Standard TTL	10	10	10	100
Low Power Schottky TTL	2	2	8	16
Schottky TTL	19	19	4.5	85.5
Adv. Schottky TTL	8.5	8.5	1	8.5
ALS TTL	1	1	2.5	2.5
10 K ECL	24	24	0.2	4.8

[a]Load capacitance is 50 pF, load resistance 500 Ω, except ECL is driving 50-Ω transmission.

---

[1]Two excellent recent logic design texts are *Modern Digital Design*, by Richard Sandige, McGraw-Hill, New York, 1990 and *Contemporary Logic Design*, by Randy Katz, Benjamin Cummings, Redwood City, CA, 1994.

lists the speed-power product for several popular logic families. Note that propagation delays specified in manufacturer handbooks may be measured under different loading conditions, which must be taken into account in computing speed-power product.

8.3.2   Transistor-Transistor Logic

First introduced in the 1960s, transistor-transistor logic (TTL) was the technology of choice for discrete logic designs into the 1990s, when complementary metal oxide semi-conductor (CMOS) equivalents gained ascendancy. Because TTL has an enormous base of previous designs, is available in a rich variety of small-scale and medium-scale building blocks, is electrically rugged, offers relatively high-operating speed, and has well-known characteristics, transistor-transistor logic continues to be an important technology for some time. Texas Instruments Corporation's 54/7400 TTL series became a *de facto* standard, with its device numbers and pinouts used by other TTL (and CMOS) manufacturers.

There are actually several families of TTL devices having circuit and semiconductor process variations that produce different speed-power characteristics. Individual parts are designated by a scheme combining numbers and letters to identify the exact device and the TTL family. Parts having 54 as the first digits have their performance specified over the military temperature range from −55 to +125°C, whereas 74-series parts are specified from 0 to +70°C. Letters identifying the family come next, followed by a number identifying the part's function. For example, 7486 identifies a package containing four, 2-input exclusive-OR gates from the standard TTL family, whereas 74ALS86 indicates the same function in the low-power Schottky family. Additional codes for package type and, possibly, circuit revisions may also be appended. Generally speaking, devices from different TTL families can be intermixed, although attention must be paid to **fanout** (discussed subsequently) and noise performance.

The basic building block of standard TTL is the NAND gate shown in Fig. 8.29. If either input *A* or *B* is connected to a logic-low voltage, multiple-emitter transistor Q1 is saturated, holding the base of transistor Q2 low and keeping it cut off. Consequently, Q4 is starved of base current and is also off, while Q3 receives base drive current through resistor R2 and is turned on, pulling the output voltage up toward V_{cc}. Typical high output voltage for standard TTL circuits is 3.4 V. On the other hand, if both of the inputs are high, Q1 moves into the reverse-active region and current flows out of its collector into the base of Q2.

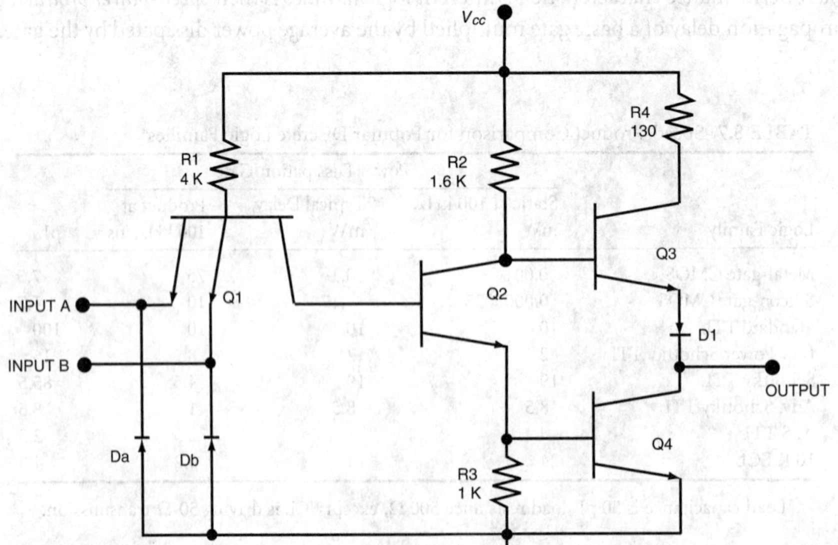

FIGURE 8.29   Two input standard TTL NAND gate.

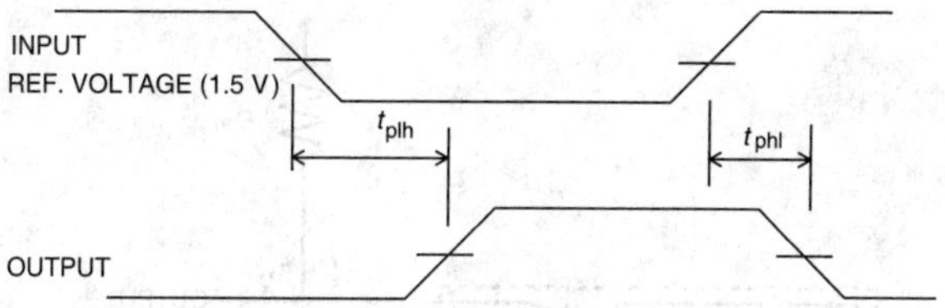

FIGURE 8.30 Propagation delay times.

Q2 turns on, raising Q4's base voltage until it is driven into saturation. At the same time, the Ohm's law voltage drop across R2 from Q2's collector current lowers the base voltage of Q3 below the value needed to keep D1 and Q3's base-emitter junction forward biased. Q3 cuts off, with the net result that the output is drawn low, toward ground. In normal operation, then, transistors Q2, Q3, and Q4 must move between saturation and cutoff. Saturated transistors experience an accumulation in minority carriers in their base regions, and the time required for the transistor to leave saturation depends on how quickly excess carriers can be removed from the base. Standard TTL processes introduce gold as an impurity to create *trapping sites* in the base that speed the annihilation of minority electrons. Later TTL families incorporate circuit refinements and processing enhancements designed to reduce excess carrier concentrations and speed removal of accumulated base electrons.

Because the internal switching dynamics differ and because the pull-up and pull-down transistors have different current drive capabilities, TTL parts show a longer propagation time when driving the output from low to high than from high to low (see Fig. 8.30). Although the average propagation delay $((t_{\text{phl}} + t_{\text{plh}})/2)$ is often used in speed comparisons, conservative designers use the slower propagation delay in computing circuit performance. NAND gate delays for standard TTL are typically 10 ns, with guaranteed maximum t_{phl} of 15 ns and t_{phl} of 22 ns.

When a TTL gate is producing a low output, the output voltage is the collector-to-emitter voltage of pull-down transistor Q4. If the collector current flowing in from the circuits being driven increases, the output voltage rises. The output current, therefore, must be limited to ensure that output voltage remains in the specified logic low range (less than V_{OL}). This places a maximum on the number of inputs that can be driven low by a given output stage; in other words, any TTL output has a maximum low-level fanout. Similarly, there is a maximum fanout for high logic levels. Since circuits must operate properly at either logic level, the smaller of the two establishes the maximum overall fanout. That is

$$\text{fanout} = \text{minimum}(-I_{OH,\text{max}}/I_{IH,\text{max}}, -I_{OL,\text{max}}/I_{IL,\text{max}})$$

where

$I_{OH,\text{max}}$ = specified maximum output current for $V_{OH,\text{min}}$

$I_{OL,\text{max}}$ = specified maximum output current for $V_{OL,\text{max}}$

$I_{IH,\text{max}}$ = specified maximum input current, high level

$I_{IL,\text{max}}$ = specified maximum input current, low level

The minus sign in the fanout computations arises because both input and output current reference directions are defined into the device. Most parts in the standard TTL family have a fanout of 10 when driving other standard TTL circuits. Buffer circuits with increased output current capability are available for applications such as clock signals that must drive an unusually large number of inputs. TTL families other than standard TTL have different input and output current specifications; fanout computations should be performed whenever parts from different families are mixed.

One of the transistors in the conventional *totem pole* TTL output stage is always turned on, pulling the output pin toward either a logic high or low. Consequently, two or more outputs cannot be connected to the same node; if one output were attempting to drive a low logic voltage and the other a high, a low-resistance path is created from between V_{cc} and ground. Excessive currents may damage the output transistors and, even if damage does not occur, the output voltage settles in the forbidden transition region between V_{OH}

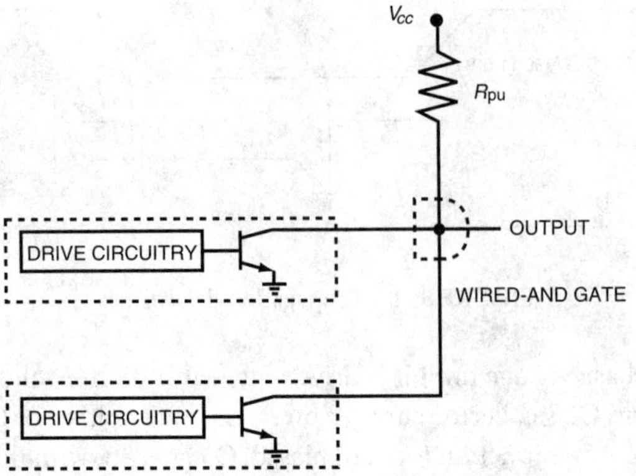

FIGURE 8.31   Wired-AND gate using open collector outputs and a pullup resistor.

and V_{OL}. Some TTL parts are available in versions with open-collector outputs, truncated output stages having only a pull down transistor (see Fig. 8.31). Two or more open-collector outputs can be connected to V_{cc} through a pull-up resistor, creating a wired-AND phantom logic gate. If any output transistor is turned on, the common point is pulled low. It only assumes a high voltage if all output transistors are turned off. The pull-up resistor must be large enough to ensure that the I_{OL} specification is not exceeded when a single gate is driving the output low, but small enough to guarantee that input leakage currents will not pull the output below V_{OH} when all driving gates are off. Since increasing the pull-up resistor value increases the resistance-capacitance (RC) time constant of the wired-AND gate and slows the effective propagation delay, the pull-up resistor is usually selected near its minimum value. The applicable design equations are

$$(V_{cc} - V_{OL,\max})/(I_{OL,\max} + nI_{IL,\max}) < R_{pu} < (V_{cc} - V_{OH,\min})/(mI_{OH,\max} + nI_{IH,\max})$$

where m is the number of open collector gates connected to R_{pu} and n is the number of driven inputs. Reference direction for all currents is into the device pin.

Some TTL parts have three-state outputs: totem pole outputs that have the added feature that both output transistors can be turned off under the control of an output enable signal. When the output enable is in its active state, the part's output functions conventionally, but when the enable is inactive, both transistors are off, effectively disconnecting the output pin from the internal circuitry. Several three-state outputs can be connected to a common bus, and only one output is enabled at a time to selectively place its logic level on the bus.

Over the long history of TTL, a number of variants have been produced to achieve faster speeds, lower power consumption, or both. For example, the 54/74L family of parts incorporated redesigned internal circuitry using higher resistance values to lower the supply current, reducing power consumption to a little as 1/10 that of standard TTL. Speed was reduced as well, with delay times as much as three times longer. Conversely, the 54/74H family offered higher speeds than standard TTL, at the expense of increased power consumption. Both of these TTL families are obsolete, their places taken by various forms of Schottky TTL.

As noted earlier, TTL output transistors are driven well into saturation. Rather than using gold doping to hasten decay of minority carriers, Schottky transistors limit the forward bias of the collector-base junction and, hence, excess base minority carrier accumulation by paralleling the junction with a Schottky (gold-semiconductor) diode, as shown in Fig. 8.32. As the NPN transistor starts to saturate, the Schottky diode becomes forward biased. This clamps the collector-base junction to about 0.3 V, keeping the transistor out of hard saturation. Schottky TTL (S) is about three times faster than standard TTL, while consuming about 1.5 times as much power.

Lower power Schottky TTL (LS) combined an improved input circuit design with Schottky transistors to reduce power consumption to a fifth of standard TTL, while maintaining equal or faster speed.

The redesigned input circuitry of LS logic changes the input logic thresholds, resulting in slightly reduced noise margin.

Still another TTL family is *Fairchild Advanced Schottky* TTL (FAST). The F parts combine circuit improvements with processing techniques that reduce junction capacitances and increase transistor speeds. The result is a fivefold increase in speed compared to standard TTL, with lowered power consumption.

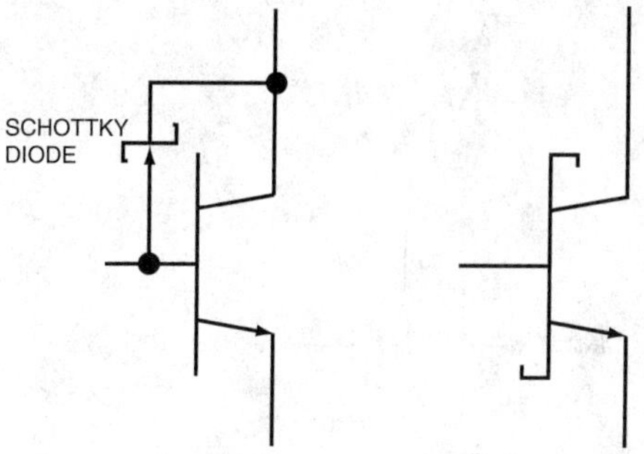

As their names imply, advanced Schottky (AS) and advanced low-power Schottky (ALS) use improved fabrication processes to improve transistor switching speeds. Parts from these families are two

FIGURE 8.32 Schottky transistor construction and symbol.

to three times faster than their S and LS equivalents, while consuming half the power. Although the standard TTL, Schottky (S), and low-power Schottky (LS) families remain in volume production, the newer advanced Schottky families (AS, ALS, F) offer better speed-power products and are favored for new designs.

In making speed comparisons between different logic families, one should keep in mind that manufacturer's published specifications use different load circuits for different TTL families. For example, propagation delays of standard TTL are measured using a load circuit of a 15-pF capacitor paralleled by a 400-Ω resistor, whereas FAST uses 50 pF and 500 Ω. Since effective propagation delay increases approximately linearly with the capacitance of the load circuit, direct comparisons of raw manufacturer data is meaningless. Manufacturer's databooks include correction curves showing how propagation delay changes with changes in load capacitance.

As outputs change from one logic state to another, sharp transients occur on the power supply current. Decoupling capacitors are connected between the V_{cc} and ground leads of TTL logic packages to provide filtering, minimizing transient currents along the power supply lines and reducing noise generation. Guidelines for distribution of decoupling capacitors are found in manufacturer handbooks and texts on digital system design, for example, Barnes (1987).

Circuit boards using advanced TTL with very short **risetimes** (FAST and AS) require physical designs that reduce waveform distortion due to transmission line effects. A brief discussion of such considerations is given in the following section on emitter-coupled logic.

8.3.3   CMOS Logic

In 1962, Frank Wanlass of Fairchild Semiconductor noted that enhancement nMOS and pMOS transistors could be stacked totem pole fashion to form an extraordinarily simple inverter stage (see Fig. 8.33). If the input voltage is near V_{dd}, the nMOS pull-down transistor channel will be enhanced, whereas the pMOS transistor remains turned off. The voltage divider effect of a low nMOS channel resistance and very high pMOS channel resistance produces a low output voltage. On the other hand, if the input voltage is near ground, the nMOS transistor will be turned off, while the pMOS channel is enhanced to pull the output voltage high. Logic gates can be easily created by adding parallel pull-down transistors and series pull-up transistors. When a CMOS gate is in either of its static states, the only current flowing is the very small leakage current through the off transistor plus any current needed by the driven inputs. Since the inputs to CMOS gates are essentially capacitances, input currents can be very small (on the order of 100 pA). Thus, the static power supply current and power dissipation of CMOS gates is far smaller than that of TTL gates. Since the early versions of discrete CMOS logic were much slower than TTL, CMOS was usually relegated to applications where power conservation was more important than performance.

In TTL logic, the high and low output logic levels and the input switching threshold (point at which a particular gate input interprets the voltage as a one rather than a zero) are established largely by forward bias voltages of PN junctions. Since these voltages do not scale with the power supply voltage, TTL operates properly over a rather restricted range (5.0 V $\pm$ 10%). In an all CMOS system, however, output voltage is

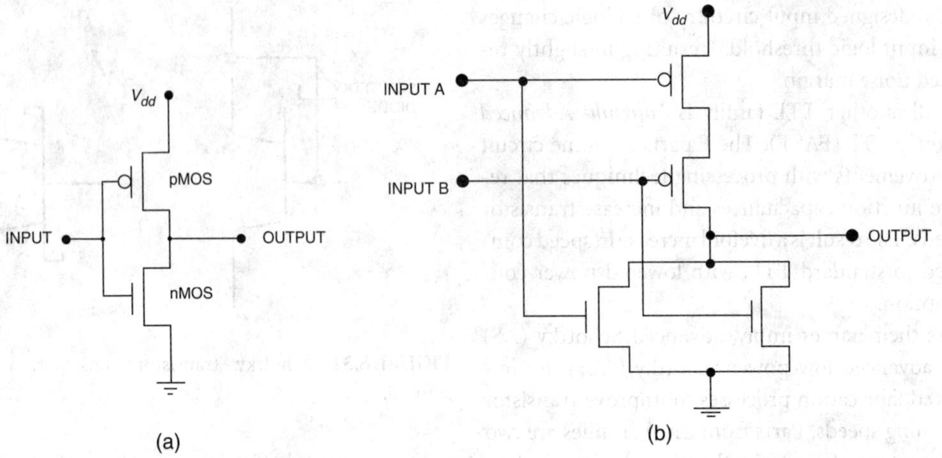

FIGURE 8.33 Elementary CMOS logic circuit: (a) complementary MOS inverter stage, (b) two-input CMOS NOR gate.

determined by the voltage divider effect. Since the off-resistance of the inactive transistor's channel is far larger than the resistance of the active channel, CMOS logic high and low levels both are close to the power supply values (V_{dd} and ground) and scale with power supply voltage. The output transistors are designed to have matching channel resistances when turned on. Thus, as long as the supply voltage remains well above the transistor threshold voltage (about 0.7 V for modern CMOS), the input switching voltage also scales with supply voltage and is about one-half of V_{dd}. Unlike TTL, CMOS parts can operate over a very wide range of supply voltages (3–15 V) although lower voltages reduce noise immunity and speed. In the mid-1990s, integrated circuit manufacturers introduced new generations of CMOS parts optimized for operation in the 3-V range, intended for use in portable applications powered by two dry cell batteries.

Requiring only four transistors, the basic CMOS gate requires much less area than an LS TTL gate with its 11 diodes and bipolar transistors and 6 diffused resistors. This fact, coupled with its much higher static power consumption, disqualifies TTL as a viable technology for very large-scale integrated circuits. In the mid-1980s, CMOS quickly obsoleted the earlier nMOS integrated circuits and has become almost the exclusive technology of choice for complex digital integrated circuits.

Older discrete CMOS integrated circuits (RCA's 4000-series is the standard, with several manufacturers producing compatible devices) use aluminum as the gate material. The metal gate is deposited after creation of the drain and source regions. To guarantee that the gate overlies the entire channel even at maximum interlayer alignment tolerances, there is a substantial overlap of the gate and drain and source. Comparatively large parasitic capacitances result, slowing transistor (hence, gate) speed. In the mid-1980s several manufacturers introduced advanced discrete CMOS logic based on the self-aligning polysilicon gate processes developed for large-scale integrated circuits (ICs). Discrete silicon gate parts are often functional emulations of TTL devices, and have similar numbering schemes (e.g., 74HCxxx). Since silicon gate CMOS is much faster than metal gate CMOS (8 ns typical gate propagation delay vs 125 ns.) while offering similar static power consumption (1 mW per gate vs. 0.6 mW), silicon gate CMOS is the technology of choice for new discrete logic designs. There are two distinct groups of modern CMOS parts: HC high-speed CMOS and AC or C advanced CMOS.

Even though parts from the 74HC and similar families may emulate TTL functions and have matching pin assignments, they cannot be intermingled with TTL parts. As mentioned, CMOS output voltages are much closer to the power supply voltages than TTL outputs. Gates are designed with this characteristic in mind and do not produce solid output voltages when driven by typical TTL high logic levels. Specialized advanced CMOS parts are produced with modified input stages specifically for interfacing with TTL circuits (for example, the HCT family). These parts exhibit slightly higher power consumption than normal silicon-gate CMOS.

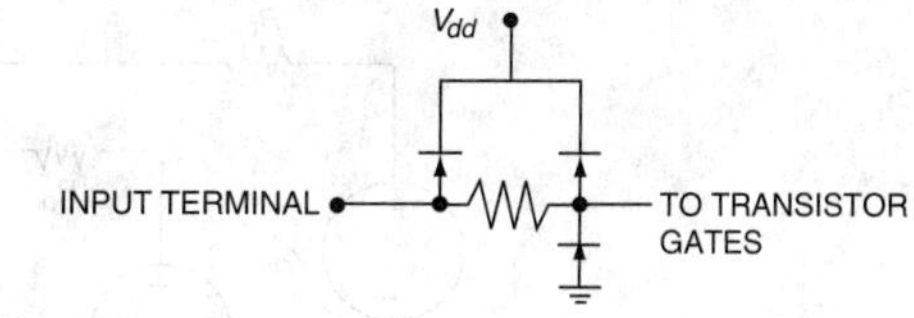

The gate electrodes of CMOS transistors are separated from the channel by a layer of silicon dioxide that may be only a few hundred angstroms thick. Such a thin dielectric layer can be damaged by a potential difference of only 40–100 V. Without precautions, the normal electrostatic potentials that build up on the human body can destroy CMOS devices. Although damage from electrostatic discharge (ESD) can be minimized through proper handling procedures (Matisof, 1986), CMOS parts include ESD protection circuitry on all inputs (see Fig. 8.34). The input diodes serve to limit gate voltages to one diode drop above V_{dd} or below ground, whereas the resistor serves both to limit input current and to slow the rise time of very fast pulses.

FIGURE 8.34 A typical CMOS input protection circuit.

As shown in Fig. 8.35, CMOS inverters and gates inherently have cross-connected parasitic bipolar transistors that form a silicon controlled rectifier (SCR). Suppose that external circuitry connected to the output pin pulls the pin low, below ground level. As this voltage approaches -0.5 V, the base-emitter junction of the parasitic NPN transistor will start to forward bias and the transistor will turn on. The resulting collector current is primarily determined by whatever external circuitry is pulling the output low. That current flows through R_{well} and lowers the base voltage of the parasitic PNP transistor. A large enough current will forward bias the PNP base emitter junction, causing PNP collector current to flow through R_{sub} and helping to maintain the forward bias of the NPN transistor. The SCR then enters a regenerative condition and will quickly assume a stable state in which both transistors remain on even after the initial driving stimulus is removed. In this **latchup** condition, substantial current flows from V_{dd} to ground. Normal operation of the CMOS gate is disrupted and permanent damage is likely. For latchup to occur, the product of the NPN and PNP transistor current gains must be greater than one, the output pin must be driven either below -0.5 V or above $V_{dd} + 0.5$ V by a source that supplies enough current to trigger regenerative operation, and V_{dd} must supply a sustaining current large enough to keep the SCR turned on. Although the potential for latchup cannot be avoided, CMOS manufacturers design input and output circuits that are latchup resistant, using configurations that reduce the gain of the parasitic transistors and that lower the substrate and p-well resistances. Nevertheless, CMOS users must ensure that input and output pins are not driven outside normal operating ranges. Limiting power supply current below the holding current level is another means of preventing latchup damage.

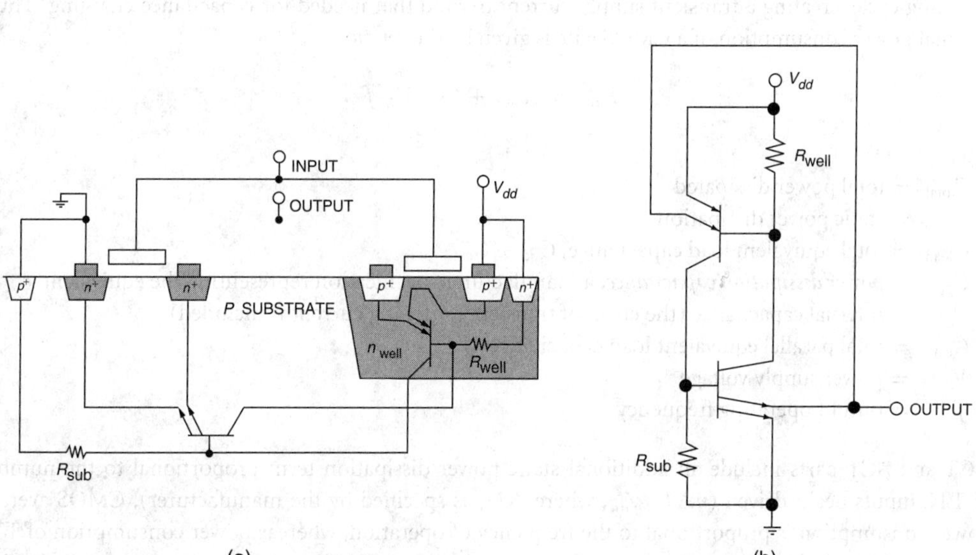

(a)

(b)

FIGURE 8.35 CMOS parasitic SCR: (a) cross-section showing parasitic bipolar transistors, (b) equivalent circuit.

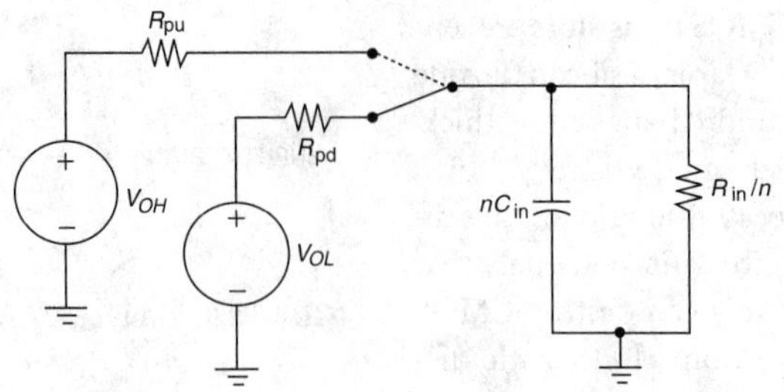

FIGURE 8.36   Equivalent circuit for a CMOS output driving nCMOS inputs.

Fanout of CMOS outputs driving TTL inputs is limited by static current drive requirements and is calculated as for transistor-transistor logic. In all-CMOS systems, the static input current is just the leakage current flowing through the input ESD protection diodes, a current on the order of 100 pA. Consequently, static fanout of CMOS driving CMOS is so large as to be meaningless. Instead, fanout is limited by speed deterioration as the number of driven gates increases (Texas Instruments, 1984). If all driven inputs are assumed identical, an all-CMOS system can be represented by the approximate equivalent circuit of Fig. 8.36. The parallel input resistance R_{in} is on the order of 60 MΩ and can be neglected in later computations. The pull-up resistor R_{pu} can be approximately calculated from manufacturer data as $(V_{dd} - V_{OH})/I_{OH}(\sim 50 \ \Omega)$. The output will charge exponentially from low to high with a time constant of $n R_{pu} C_{in}$, where n is the actual fanout. Since the low output voltage is nearly zero, time required to reach a logic high is given by

$$t = n R_{pu} C_{in} \ln(1 - V_{IH,min} / V_{OH}).$$

Thus, once the maximum allowable risetime (that is, the maximum effective propagation delay) has been specified for an all-CMOS system, the dynamic fanout n can be computed. Although static power consumption of CMOS logic is quite small, outputs that are changing logic states will dissipate power each time they charge and discharge the capacitance loading the output. In addition, both the pull-up and pull-down transistors of the output will be simultaneously partially conducting for a short period on each switching cycle, creating a transient supply current beyond that needed for capacitance charging. Thus, the total power consumption of a CMOS gate is given by the relation

$$P_{total} = P_{static} + C_{load} V_{dd}^2 f$$

where
  P_{total} = total power dissipated
  P_{static} = static power dissipation
  C_{load} = total equivalent load capacitance, $C_{pd} + C_{ext}$
  C_{pd} = *power dissipation capacitance*, a manufacturer specification representing the equivalent internal capacitance (the effect of transient switching current is included)
  C_{ext} = total parallel equivalent load capacitance
  V_{dd} = power supply voltage
  f = output operating frequency

HCT and ACT parts include an additional static power dissipation term proportional to the number of TTL inputs begin driven ($n \Delta I_{dd} V_{dd}$, where ΔI_{dd} is specified by the manufacturer). CMOS average power consumption is proportional to the frequency of operation, whereas power consumption of TTL circuits is almost independent of operating frequency until about 1 MHz, when dynamic power dissipation becomes significant (see Fig. 8.37). Direct comparison of CMOS and TTL power consumption for a digital system is difficult, since not all outputs are changing at the same frequency. Moreover, the relative advantages of CMOS become greater with increasing part complexity; CMOS and TTL power curves for a modestly complicated decoder circuit crossover at a frequency 10 times that of simple gates. Finally, TTL manufacturers do not publish C_{pd} specifications for their parts.

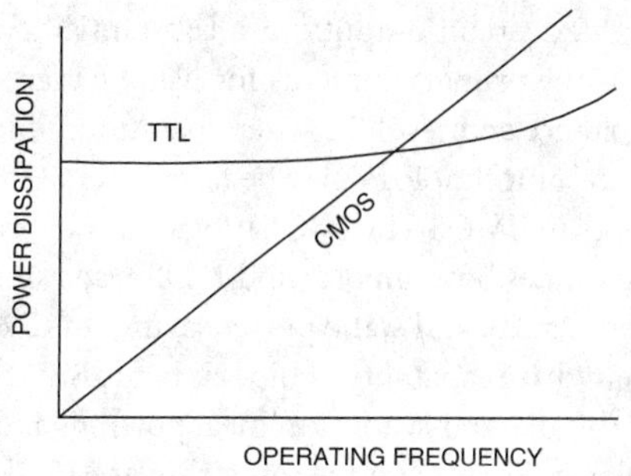

Like TTL parts, CMOS logic requires use of decoupling capacitors to reduce noise due to transient current spikes on power supply lines. Although such decoupling capacitors reduce electromagnetic coupling of noise into parallel signal leads, they do not eliminate an additional noise consideration that arises in both advanced CMOS and advanced

FIGURE 8.37 Power dissipation vs. operating frequency for TTL and CMOS logic devices.

Schottky due to voltages induced on the power supply leads from transient current spikes. Consider the multiple inverter IC package of Fig. 8.38, in which five outputs are switching from low to high simultaneously, while the remaining output stays low. In accord with Faraday's law, the switching current spike will induce transient voltage across L, which represents parasitic inductances from internal package leads and external ground supply traces. That same voltage also lifts the unchanging output above the ground reference. A sufficiently large voltage spike may make signal A exceed the input switching threshold and cause erroneous operation of following logic elements. Careful physical design of the power supply and ground leads can reduce external inductance and limit power supply sag and ground bounce. Other useful techniques include use of synchronous clocking and segregation of high drive signals, such as clock lines, into separate buffer packages, separated from reset and other asynchronous buffers.

As has been suggested several times, effective propagation delay of a logic circuit is directly proportional to the capacitive loading of its output. Systems built from small- and medium-scale logic have many

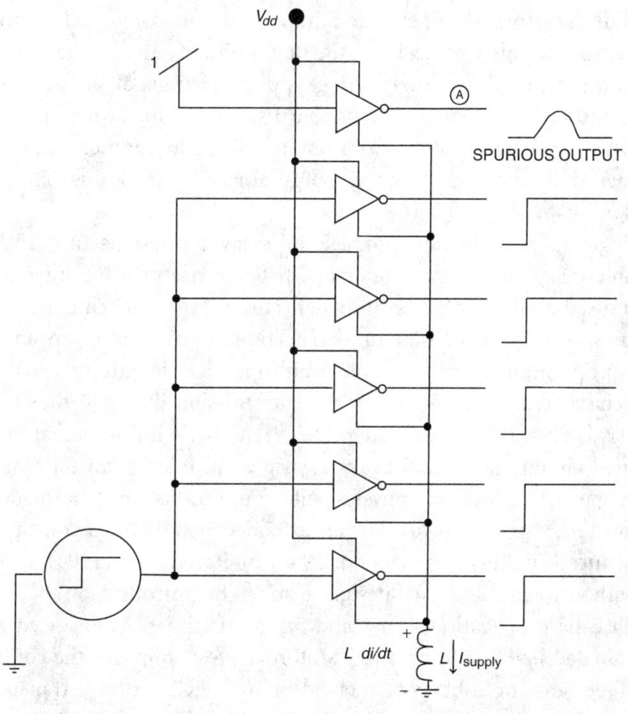

FIGURE 8.38 Output bounce due to ground lead inductance.

integrated circuit outputs with substantial load capacitance from interconnecting printed circuit traces. This is the primary impetus for using larger scale circuits: on-chip interconnections present far smaller capacitance and result in faster operation. Thus, a system built using a highly integrated circuit fabricated in a slower technology may be faster overall that the same system realized using faster small-scale logic.

Bipolar-CMOS (BiCMOS) logic is a hybrid technology for very large-scale integrated circuits that incorporates both bipolar and MOS transistors on the same chip. It is intended to offer the most of the circuit density and static power savings of CMOS while providing the electrical ruggedness and superior current drive capability of bipolar logic. BiCMOS also holds the opportunity for fabricated mixed systems using bipolar transistors for linear portions and CMOS for logic. It is a difficult task to produce high-quality bipolar transistors on the same chip as MOS devices, meaning that biCMOS circuits are more expensive and may have suboptimal performance.

8.3.4 Emitter-Coupled Logic

Emitter coupled logic (ECL) is the fastest variety of discrete logic. It is also one of the oldest, dating back to Motorola's MECL I circuits in 1962. Improved MECL III circuits were introduced in the late 1960s. Today, the 10 and 100 K ECL families remain in use where very high speed is required. Unlike either TTL or CMOS, emitter-coupled logic uses a differential amplifier configuration as its basic topology (see Fig. 8.39). If all inputs to the gate are at a voltage lower than the reference voltage, the corresponding input transistors are cutoff and the entire bias current flows through Q1. The output is pulled up to V_{cc} through resistor R_a. If any input voltage rises above the reference, the corresponding transistor turns on, switching the current from Q1 to the input transistor and lowering the output voltage to $V_{cc} - I_{bias}R_a$. If the collector resistors and bias current are properly chosen, the transistors never enter saturation, and switching time is quite fast. Typical propagation delay for MECL 10 K gates is only 2.0 ns. To reduce the output impedance and improve current drive capability, actual ECL parts include emitter follower outputs (Fig. 8.39(b)). They also include integral voltage references with temperature coefficients matching those of the differential amplifier.

Note that ECL circuits are normally operated with the collectors connected to ground and the emitter current source to a negative supply voltage. From the standpoint of supply voltage fluctuations, the circuit operates as a common base amplifier, attenuating supply variations at the output. The power supply current in both TTL and CMOS circuits exhibits sharp spikes during switching; in ECL gates, the bias current remains constant and simply shifts from one transistor to another. Lower logic voltage swings also reduce noise generation. Although the noise margin for ECL logic is less than TTL and CMOS, ECL circuits generate less noise.

Emitter-coupled logic offers both short propagation delay and fast risetimes. When interconnecting paths introduce a time delay longer than about one-half the risetime, the interconnection's transmission line behavior must be taken into account. Reflections from unterminated lines can distort the waveform with ringing. The need to wait until reflections have damped to an acceptable value increases the effective propagation delay. The maximum open line length depends on the fanout of the driving output, the characteristic impedance of the transmission line, and the velocity of propagation of the line. Motorola databooks indicate that MECL III parts, with an actual fanout of 8 connected with fine-pitch printed circuit traces (100-Ω microstrip), can tolerate an unterminated line length of only 0.1 in. In high-speed ECL designs, physical interconnections must be made using properly terminated transmission lines (see Fig. 8.40). For an extended treatment of transmission lines in digital systems, see manufacturer's publications (Blood, 1988) or Rosenstark (1994). The necessity for using transmission lines with virtually all ECL III designs resulted the introduction in 1971 of the MECL 10 K family of parts, with 2-ns propagation delays and 3.5-ns risetimes. The slowed risetimes meant that much longer unterminated lines could be used (Motorola cites 2.6 in. for the configuration mentioned earlier). Thus, many designs could be implemented in 10 K ECL without transmission line interconnections. This relative ease of use and a wider variety of parts has made 10 K ECL and its variants the dominant ECL family. It should be noted that, although transmission line interconnections were

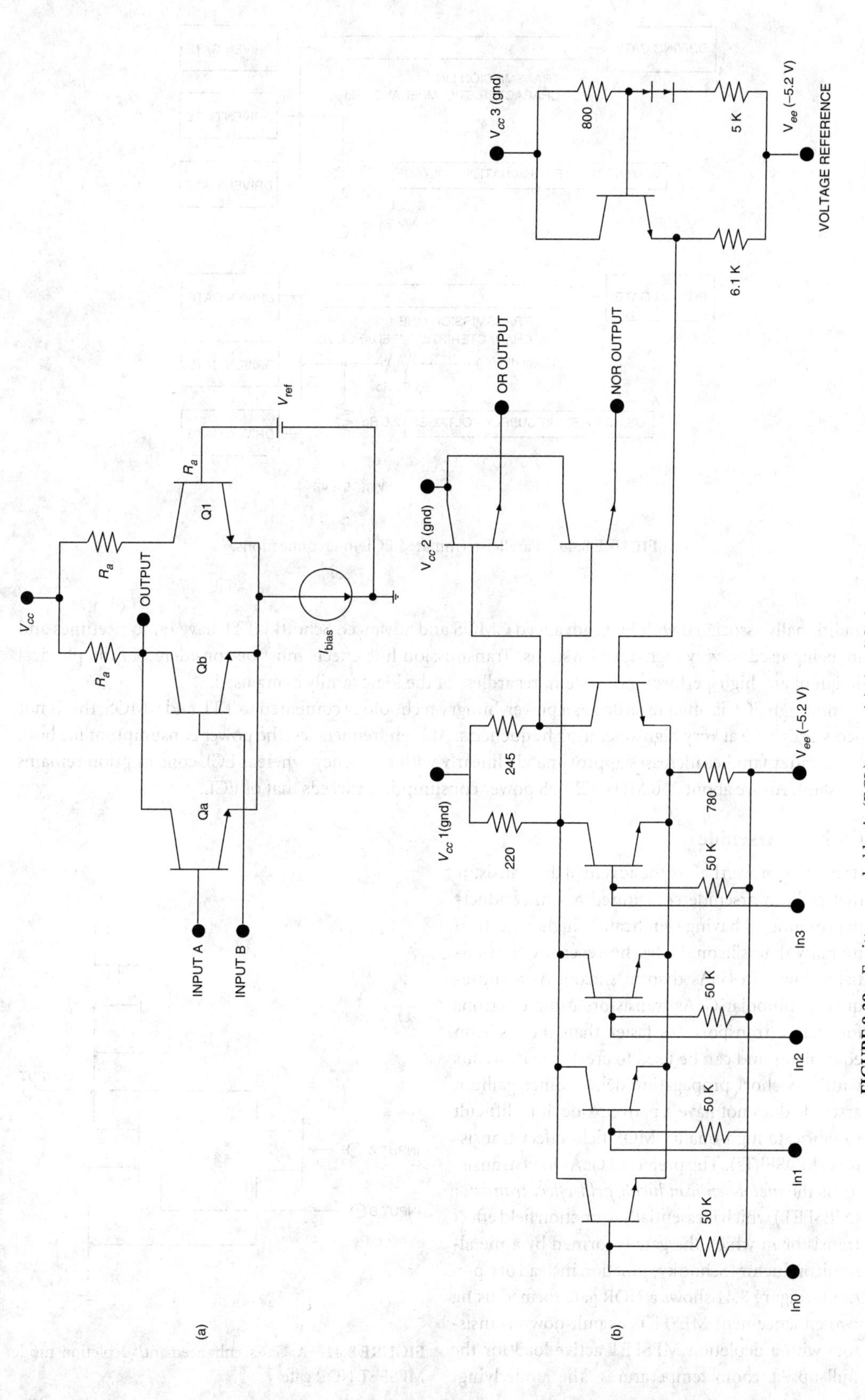

FIGURE 8.39 Emitter coupled logic (ECL) gates: (a) simplified schematic of ECL NOR gate, (b) commercial ECL gate.

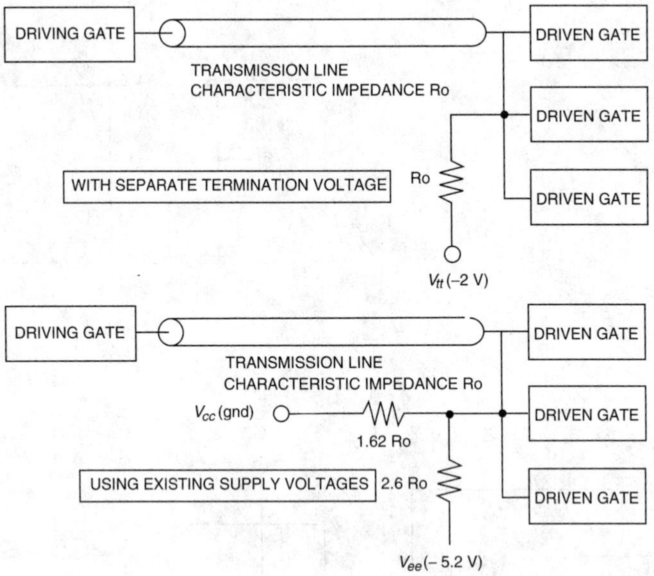

FIGURE 8.40 Parallel-terminated ECL interconnections.

traditionally associated with ECL, advanced CMOS and advanced Schottky TTL have rapid risetimes and are being used in very high-speed systems. Transmission line effects must be considered in the physical design of any high-performance system, regardless of the logic family being used.

Although ECL is often regarded as a power-hungry technology compared to TTL and CMOS, this is not necessarily true at very high-operating frequencies. At high frequencies, the power consumption for both of the latter families increases approximately linearly with frequency, whereas ECL consumption remains constant. Above about 250 MHz, CMOS power consumption exceeds that of ECL.

Gallium Arsenide

It was known even before the advent of the transistor that gallium arsenide constituted a semiconducting compound having significantly higher electron mobility than silicon. Holes, however, are substantially slower in GaAs than in silicon. As a consequence, unipolar GaAs transistors using electrons for charge transport are faster than their silicon equivalents and can be used to create logic circuits with very short propagation delays. Since gallium arsenide does not have a native oxide, it is difficult to fabricate high-quality MOS field effect transistors (MOSFETs). The preferred GaAs logic transistor is the *metal-semiconductor field-effect transistor* (MESFET) which is essentially a junction field effect transistor in which the gate is formed by a metal-semiconductor (Schottky) junction instead of a p–n diode. Figure 8.41 shows a NOR gate formed using two enhancement MESFETs as pull-down transistors with a depletion MESFET active load for the pull-up. At room temperatures, the underlying

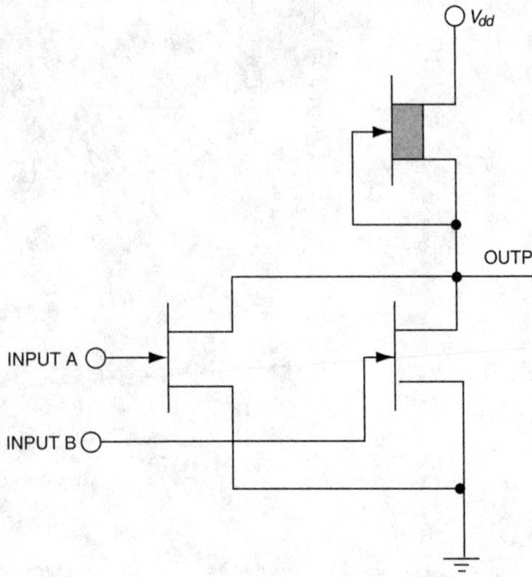

FIGURE 8.41 A GaAs enhancement/ depletion-mode MESFET NOR gate.

GaAs substrate has a very high resistivity (is semi-insulating), providing sufficient transistor-to-transistor isolation without the need for reverse-biased isolation junctions.

Gallium arsenide logic exhibits both short delay and fast risetimes. As noted in the ECL discussion, fast logic signals exhibit ringing if transmission line interconnections are not used. The waveform distortion due to ringing extends the effective propagation delay and can negate the speed advantage of the basic logic circuit. Furthermore, complex fabrication processes and limited volumes make gallium arsenide logic from 10 to 20 times more expensive than silicon. The dual requirements to control costs and to preserve GaAs speed advantage by limiting the number of off-chip interfaces make the technology better suited to large-scale rather small- and medium-scale logic. GaAs digital circuits are available from several vendors in the form of custom integrated circuits or gate arrays. Gate arrays are semifinished large-scale integrated circuits that have large numbers of unconnected gates. The customer specifies the final metallization patterns that interconnect the gates to create the desired logic circuits. As of this writing, vendors were offering application-specific GaAs integrated circuits with more than 20,000 gates and worst-case internal gate delays on the order of 0.07 ns.

8.3.5   Programmable Logic

Although traditional discrete logic remains an important means of implementing digital systems, various forms of programmable logic have become increasingly popular. Broadly speaking, programmable logic denotes large- and very large-scale integrated circuits containing arrays of generalized logic blocks with user-configurable interconnections. In programming the device, the user modifies it to define the functions of each logic block and the interconnections between logic blocks. Depending on the complexity of the programmable logic being used, a single programmable package may replace dozens of conventional small- and medium-scale integrated circuits. Because most interconnections are on the programmable logic chip itself, programmable logic designs offer both speed and reliability gains over conventional logic. Effectively all programmable logic design is done using electronic design automation (EDA) tools that automatically generate device programming information from a description of the function to be performed. The functional description can be supplied in terms of a logic diagram, logic equations, or a hardware description language. EDA tools also perform such tasks as logic synthesis, simplification, device selection, functional simulation, and timing/speed analysis. The combination of electronic design automation and device programmability makes it possible to design and physically implement digital subsystems very quickly compared to conventional logic.

Although the definitions are somewhat imprecise, programmable logic is divided into two broad categories: programmable logic devices (PLDs) and field programmable gate arrays (FPGAs). The PLD category itself is often subdivided into complex PLDs (CPLDs) with numerous internal logic blocks and simpler PLDs. Although the general architecture of small PLDs is fairly standardized, that of FPGAs and CPLDs continues to evolve and differs considerably from one manufacturer to another. Programmable logic devices cover a very wide range, indeed, from simple PALs containing a few gates to field programmable gate arrays providing thousands of usable gate equivalents.

Programmable Array Logic

Programmable array logic (PAL) was first introduced by Monolithic Memories, Inc. In their simplest form, PALs consist of a large array of logic AND gates fed by both the true (uninverted) and complemented (inverted) senses of the input signals. Any AND gate may be fed by either sense of any input. The outputs of the AND gates are hardwired to OR gates that drive the PAL outputs (see Fig. 8.42). By programming the AND inputs, the PAL user can implement any logic function provided that the number of product terms does not exceed the **fanin** to the OR gate. Modern PALs incorporate several logic macrocells that include a programmable inverter following the AND/OR combinatorial logic, along with a programmable flip-flop that can be used when registered outputs are needed. Additional AND gates are used to clock and reset the flip-flop. Macrocells usually also have a programmable three-state buffer driving the block output and programmable switches that can be used to feed macrocell outputs back into the AND–OR array to develop combinatorial logic with many product terms or to implement state machines.

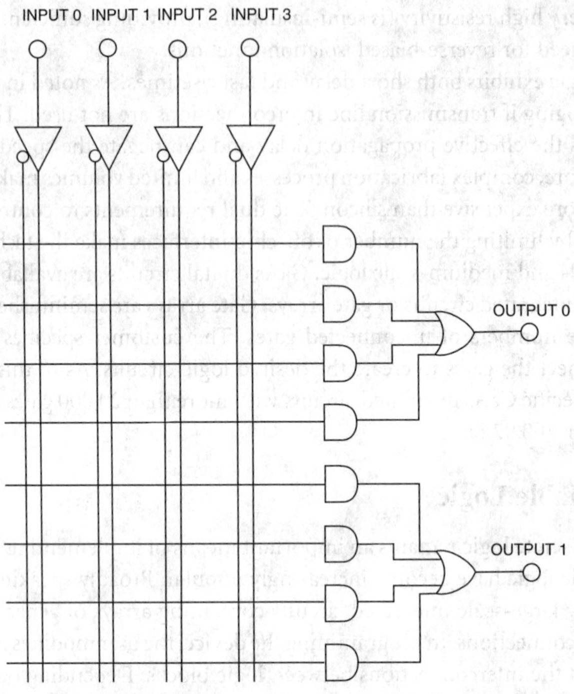

INPUT 0   INPUT 1   INPUT 2   INPUT 3

FIGURE 8.42   A simplified small PAL device.

PLDs are fabricated using either bipolar or CMOS technologies. Although CMOS parts have become quite fast, bipolar PALs offer higher speed (<5 ns input to output combinatorial delay) at the expense of higher power consumption. Different manufacturers accomplish interconnect and logic options programming in different ways. In some instances, the circuit is physically modified by blowing fuses or antifuses (an antifuse is an open circuit until fused, then it becomes a low-resistance circuit). CMOS programmable devices often use interconnections based on erasable programmable read only memory (EPROM) cells. EPROM-based parts can be programmed, then erased and reprogrammed. Although reprogrammability can be useful during the development phase, it is less important in production, and EPROM-based PLDs are also sold in cheaper, one-time programmable versions. Although not a major market factor, electrically reprogrammable PLDs are also available. When used in digital systems, PLDs require fanout computation, just as discrete TTL or CMOS logic. In addition, fast PLDs require the same consideration of transmission line effects.

Although it is possible to hand-develop programming tables for very simple PALs, virtually all design is now done using EDA tools supplied by PLD vendors and third parties. Although computer methods automate the design process to a considerable extent, they usually permit some user control over resource assignments. The user needs to have a thorough knowledge of a specific device's architecture and programming options to optimize the resulting design.

Programmable Logic Arrays

Field programmable logic arrays (FPLAs) resemble PALs with the additional feature that the connections from the AND intermediate outputs to the OR inputs are also programmable, rather than hardwired (Fig. 8.42). Like simpler PALs, FPLAs are produced in registered versions that have flip-flops between the OR gate and the output pin. Even though they are more complex devices, FPLAs were actually introduced before programmable array logic. Programmable AND/OR interconnections make them more flexible than PALs, since product terms (AND gate outputs) can drive more than one OR input and any number of

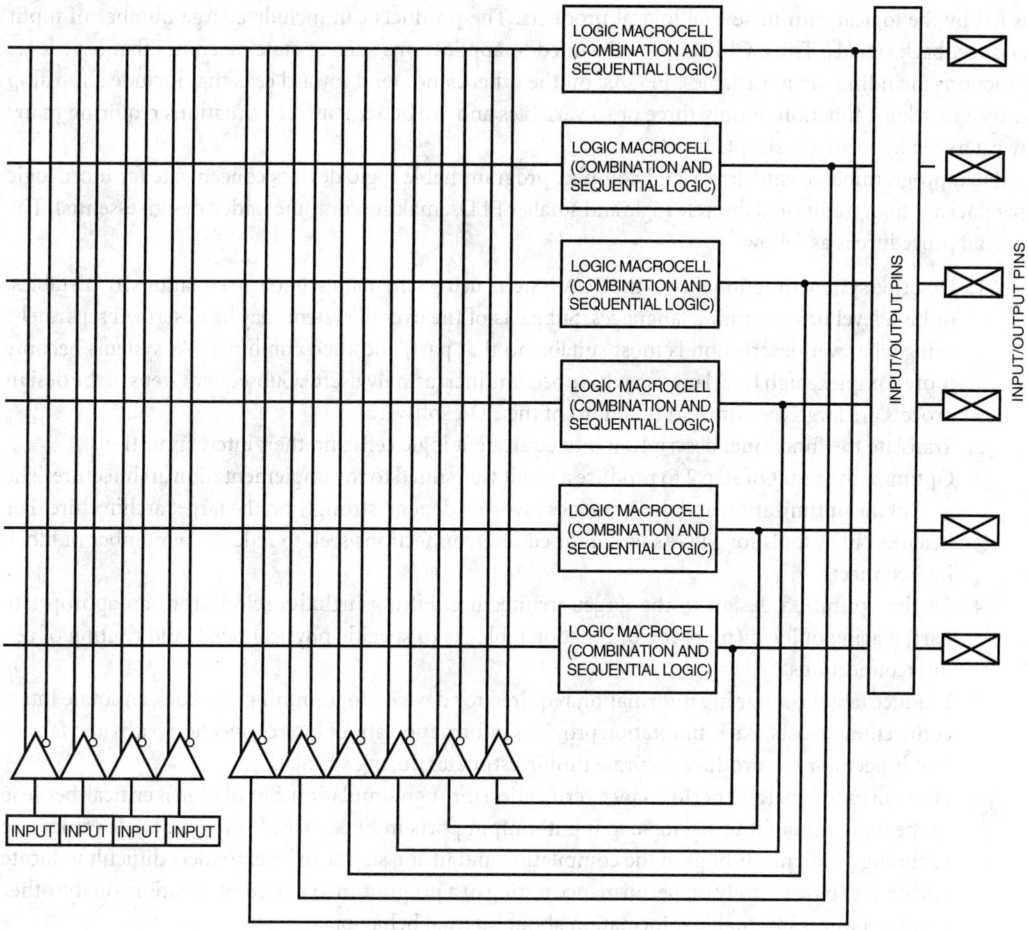

FIGURE 8.43 A simplified complex programmable logic device structure.

product terms can be used to develop a particular output function. The additional layer of programmable interconnects consumes chip area, however, increasing cost. The extra level of interconnections also makes programming more complicated. The most important drawback to FPLAs, however, is lower speed. The programmable interconnections add significant series resistance, slowing signal propagation between the AND and OR gates and making FPLAs sluggish compared to PALs. Since programmable array logic is faster, simpler to program, and flexible enough for most uses, PALs are much more popular than field programmable logic arrays. Most FPLAs marketed today have registered outputs (that is, a flip-flop is placed between the OR gate and the macrocell output) that are fed back into to the AND/OR array. Termed *field programmable logic sequencers*, these parts are intended for implementing small state machines.

Field Programmable Gate Arrays and Complex PLDs

There is no universally accepted dividing line between complex PLDs and FPGAs. Not only do architectures differ from manufacturer to manufacturer, the method of interconnect programming also differs. The term *complex programmable logic device* (CPLD) generally denotes devices organized as large arrays of PAL macrocells, with fixed delay feedback of macrocells outputs into the AND–OR array (Fig. 8.43). CLPD time delays are more predictable than those of FPGAs, which can make them simpler to design with. A more important distinction between CPLDs and FPGAs is the ratio of combinatorial logic to sequential logic in their logic cells. The PAL-like structure of CPLDs means that the flip-flop in each logic macrocell is fed by the logical sum of several logical products. The products can include a large number of input and feedback signals. Thus, CPLDs are well suited to applications such as state machines that have logic functions including many variables. FPGAs, on the other hand, tend toward cells that include a flip-flop driven by a logic function of only three or so variables and are better suited to situations requiring many registers fed by relatively simple functions.

Field programmable gate arrays and complex programmable logic devices concentrate far more logic per package than traditional discrete logic and smaller PLDs, making computer aided design essential. The overall procedure is as follows.

1. Describe system functionality to the EDA system, using schematics, boolean equations, truth tables, or high level programming languages. Subparts of the overall system may be described separately, using whatever description is most suitable to that part, and then combined. As systems become more complex, high level languages have become increasingly useful. Subsequent steps in the design process are largely automated functions of the EDA software.
2. Translate the functional description into equivalent logic cells and their interconnections.
3. Optimize the result of step 2 to produce a result well suited to the implementation architecture. The goal of the optimization and the strategies involved depend strongly on the target architecture. For instance, EDA tools for FPGAs with limited interconnections seek to reduce the number of block interconnects.
4. Fit the optimized design to the target architecture. Fitting includes selection of an appropriate part, placing of logic (mapping of functional blocks to specific physical cells), and routing of cell interconnections.
5. Produce tables containing information required for device programming and back annotate interconnection netlists. Back annotation provides information about interconnection paths and fanout that is necessary to produce accurate timing estimates during simulation.
6. Perform functional and performance verification through simulation. Simulation is critical, because of the high ratio of internal logic to input–output ports in FPGA/CPLD designs. Logic errors due to timing problems or bugs in the compilation and fitting steps can be extremely difficult to locate and correct based purely on performance testing of a programmed device. Simulation, on the other hand, produces invaluable information about internal behavior.

Defining Terms

Fanin: The number of independent inputs to a given logic gate.

Fanout: (1) The maximum number of similar inputs that a logic device output can drive while still meeting output logic voltage specifications. (2) The actual number of inputs connected to a particular output.

Latchup: A faulty operating condition of CMOS circuits in which its parasitic SCRs produce a low resistance path between power supply rails.

Noise margin: The difference between an output logic high (low) voltage produced by a properly functioning output and the input logic high (low) voltage required by a driven input; noise margin provides immunity to a level of signal distortion less than the margin.

Risetime: The time required for a logic signal to transition from one static level to another; usually, risetime is measured from 10 to 90% of the final value.

References

Altera. 1993. *Data Book*. Altera Corp., San Jose, CA.

Alvarez, A. 1989. *BiCMOS Technology and Applications*. Kluwer Academic, Boston, MA.

Barnes, J.R. 1987. *Electronic System Design: Interference and Noise Control Techniques*. Prentice-Hall, Englewood Cliffs, NJ.

Blood, W.R. 1988. *MECL System Design Handbook*. Motorola Semiconductor Products, Phoenix, AZ.

Brown, S.D., Francis, R., Rose, J., and Vranesic, Z. 1992. *Field-Programmable Gate Arrays*. Kluwer Academic, Boston, MA.

Buchanan, J. 1990. *CMOS/TTL Digital System Design*. McGraw-Hill, New York.

Deyhimy, I. 1985. GaAs digital ICs promise speed and lower cost. *Computer Design* 35(11):88–92.

Jenkins, J.H. 1994. *Designing with FPGAs and CPLDs*. Prentice-Hall, Englewood Cliffs, NJ.

Kanopoulos, N. 1989. *Gallium Arsenide Digital Integrated Circuits: A Systems Perspective*. Prentice-Hall, Englewood Cliffs, NJ.

Leigh, B. 1993. Complex PLD & FPGA architectures, *ASIC & EDA*. (Feb.):44–50.

Matisof B. 1986. *Handbook of Electrostatic Controls (ESD)*. Van Nostrand Reinhold, New York.

Matthew, P. 1984. *Choosing and Using ECL*. McGraw-Hill, New York.

Rosenstark, S. 1994. *Transmission Lines in Computer Engineering*. McGraw-Hill, New York.

Scarlett, J.A. 1972. *Transistor–Transistor Logic and Its Interconnections*. Van Nostrand Reinhold, London.

Schilling, D. and Belove, C. 1989. *Electronic Circuits: Discrete and Integrated*, 3rd ed. McGraw-Hill, New York.

Texas Instruments, 1984. *High-Speed CMOS Logic Data Book*. Texas Instruments, Dallas, TX.

Xilinx, 1993. *The Programmable Gate Array Data Book*. Xilinx, Inc., San Jose, CA.

Further Information

For the practitioner, the following technical journals are excellent sources dealing with current trends in both discrete logic and programmable devices.

Electronic Design, published 30 times a year by Penton Publishing, Inc. of Cleveland, OH, covers a very broad range of design-related topics. It is free to qualified subscribers.

EDN, published by Cahners Publishing Company, Newton, MA, appears 38 times annually. It also deals with the entire range of electronics design topics and is free to qualified subscribers.

Computer Design, Pennwell Publishing, Nashua, NH, appears monthly. It focuses more specifically on digital design topics than the foregoing. Like them, it is free to qualified subscribers.

Integrated System Design, a monthly periodical of the Verecom Group, Los Altos, CA, is a valuable source of current information about programmable logic and the associated design tools.

For the researcher, the *IEEE Journal of Solid State Circuits* is perhaps the best single source of information about advances in digital logic. The IEEE sponsors a number of conferences annually that deal in one way or another with digital logic families and programmable logic.

Manufacturer's databooks and application notes provide much current information about logic, especially with regard to applications information. They usually assume some background knowledge, however, and can be difficult to use as a first introduction. Recommended textbooks to provide that background include the following.

Brown, S.D., Francis, R., Rose, J., and Vranesic, Z. 1992. *Field-Programmable Gate Arrays*. Kluwer Academic, Boston, MA.

Buchanan, J. 1990. *CMOS/TTL Digital System Design*. McGraw-Hill, New York.

Jenkins, J.H. 1994. *Designing with FPGAs and CPLDs*. Prentice-Hall, Englewood Cliffs, NJ.

Matthew, P. 1984. *Choosing and Using ECL*. McGraw-Hill, New York.

Rosenstark, S. 1994. *Transmission Lines in Computer Engineering*. McGraw-Hill, New York.

8.4 Memory Devices

Shih-Lien Lu

8.4.1 Introduction

Memory is an essential part of any computation system. It is used to store both the computation instructions and the data. Logically, memory can be viewed as a collection of sequential locations, each with a unique address as its label and capable of storing information. Accessing memory is accomplished by supplying the address of the desired data to the device.

Memory devices can be categorized according to their functionality and fall into two major categories, **read-only-memory (ROM)**, and write-and-read memory or random-access memory (RAM). There is also another subcategory of ROM: mostly read but sometimes write memory or flash ROM memory. Within the RAM category there are two types of memory devices differentiated by storage characteristics, **static RAM (SRAM)** and **dynamic RAM (DRAM)** respectively. DRAM devices need to be refreshed periodically to prevent the corruption of their contents due to charge leakage. SRAM devices, on the other hand, do not need to be refreshed.

Both SRAM and DRAM are volatile memory devices, which means that their contents are lost if the power supply is removed from these devices. Nonvolatile memory, the opposite of volatile memory, retains its contents even when the supply power is turned off. All current ROM devices, including mostly read sometimes write devices are nonvolatile memories. Except for a very few special memories, these devices are all interfaced in a similar way. When an address is presented to a memory device, and sometimes after a control signal is strobed, the information stored at the specified address is retrieved after a certain delay. This process is called a **memory read**. This delay, defined as the time taken from address valid to data ready, is called **memory read access time**. Similarly, data can be stored into the memory device by performing a **memory write**. When writing, data and an address are presented to the memory device with the activation of a write control signal. There are also other control signals used to interface. For example, most of the memory devices in packaged chip format has a chip select (or chip enable) pin. Only when this pin is asserted, the particular memory device gets active. Once an address is supplied to the chip, internal address decoding logic is used to pinpoint the particular content for output. Because of the nature of the circuit structure used in implementing the decoding logic, a memory device usually needs to recover before a subsequent read or write can be performed. Therefore, the time between subsequent address issues is called **cycle time**. Cycle time is usually twice as long as the access time. There are other timing requirements for memory devices. These timing parameters play a very important role in interfacing the memory devices with computation processors. In many situations, a memory device's timing parameters affect the performance of the computation system greatly.

Some special memory structures do not follow the general accessing scheme of using an address. Two of the most frequently used are content addressable memory (CAM), and first-in-first-out (FIFO) memory. Another type of memory device, which accepts multiple addresses and produces several results at different ports, is called multiport memory. There is also a type of memory that can be written in parallel, but is read serially. It is referred to as video RAM or VDRAM since they are used primarily in graphic display applications. We will discuss these in more detail later.

8.4.2 Memory Organization

There are several aspects to memory organization. We will take a top down approach in discussing them.

Memory Hierarchy

The speed of memory devices has been lagging behind the speed of processors. As processors become faster and more capable, larger memory spaces are required to keep up with the every increasing software complexity written for these machines. Figure 8.44(a) illustrates the well-known Moore's law, depicting the exponential growth in central processing unit (CPU) and memory capacity. Although CPUs' speed continues to grow with the advancement of technology and design technique (in particular pipelining), due to the nature of increasing memory size, more time is needed to decode wider and wider addresses and to sense the information stored in the ever-shrinking storage element. The speed gap between CPU and memory devices continues to grow wider. Figure 8.44(b) illustrates this phenomenon.

The strategy used to remedy this problem is called **memory hierarchy**. Memory hierarchy works because of the locality property of memory references due to the sequentially fetched program instructions and the conjugation of related data. In a hierarchical memory system there are many levels of memory hierarchies. A small amount of very fast memory is usually allocated and brought right next to the central processing unit to help match up the speed of the CPU and memory. As the distance becomes greater between the

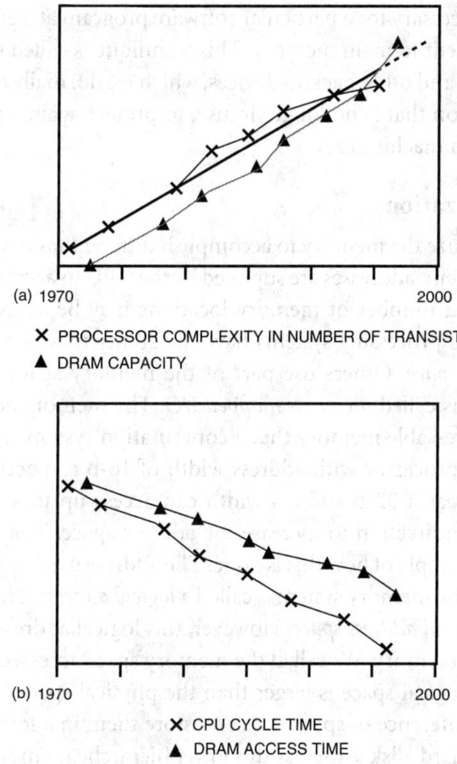

FIGURE 8.44 (a) Processor and memory development trend, (b) speed difference between RAM and CPU.

CPU and memory, the performance requirement for the memory is relaxed. At the same time, the size of the memory grows larger to accommodate the overall memory size requirement. Some of the memory hierarchies are *registers, cache, main memory,* and *disk.* Figure 8.45 illustrates the general memory hierarchy employed in a traditional system. When a memory reference is made, the processor accesses the memory at the top of the hierarchy. If the desired data is in the higher hierarchy, a "bit" is encountered and information is obtained quickly. Otherwise a *miss* is encountered. The requested in-

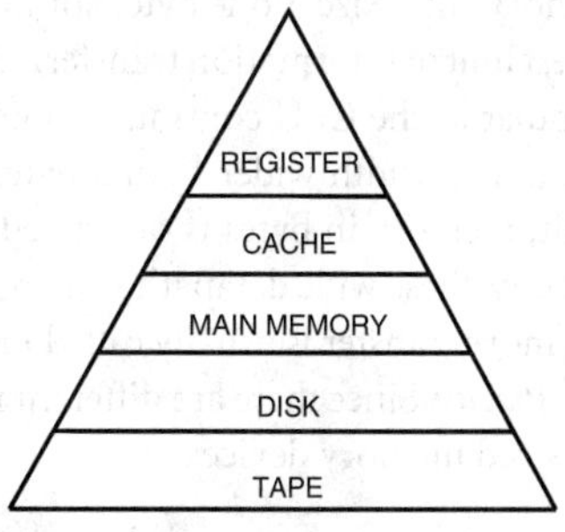

FIGURE 8.45 Traditional memory hierarchy.

formation must be brought up from a lower level in the hierarchy. Usually memory space is divided into blocks so that it can be transferred between levels in groups. At the cache level a chunk is called a *cache block* or a *cache line.* At the main memory level a chunk is referred to as a memory *page.* A miss in the cache is called a *cache miss* and a miss in the main memory is called a *page fault.* When a miss occurs, the whole block of memory containing the requested missing information is brought in from the lower hierarchy as mentioned before. If the current memory hierarchy level is full when a miss occurs, some existing blocks or pages must be removed and sometimes written back to a lower level to allow the new one(s) to be brought in. There are several different *replacement algorithms.* One of the most commonly used is the *least recently used* (LRU) replacement algorithm.

In modern computing systems, there may be several sublevels of cache within the hierarchy of cache. The general principle of memory hierarchy is that the farther away from the CPU it is, the larger its size, slower its speed, and the cheaper its price per memory unit becomes. Because the memory space addressable by the CPU is normally larger than necessary for a particular software program at a given time, disks are used to provide an economical supplement to main memory. This technique is called *virtual memory*. Besides disks there are tapes, optical drives, and other backup devices, which we normally call *backup storage*. They are used mainly to store information that is no longer in use, to protect against main memory and disk failures, or to transfer data between machines.

System Level Memory Organization

On the system level, we must organize the memory to accomplish different tasks and to satisfy the need of the program. In a computation system, addresses are supplied by the CPU to access the data or instruction. With a given address width a fixed number of memory locations may be accessed. This is referred to as the *memory address space*. Some processing systems have the ability to access another separate space called *input/output* (I/O) *address space*. Others use part of the memory space for I/O purposes. This style of performing I/O functions is called *memory–mapped I/O*. The memory address space defines the maximum size of the directly addressable memory that a computation system can access using memory type instructions. For example, a processor with address width of 16-b can access up to 64 K different locations (memory entries), whereas a 32-b address width can access up to 4 Gig different locations. However, sometimes we can use indirection to increase the address space. The method used by 80X86 processors provides an excellent example of how this is done. The address used by the user or programmer in specifying a data item stored in the memory system is called a *logical address*. The address space accessed by the logical address is named *logical address space*. However, this logical address may not necessarily be used directly to index the physical memory. We called the memory space accessed by physical address the *physical address space*. When the logical space is larger than the physical space, then memory hierarchy is required to accommodate the difference of space sizes and store them in a lower hierarchy. In most of the current computing systems, a hard-disk is used as this lower hierarchy memory. This is termed *virtual memory system*. This mapping of logical address to physical address could be either linear or nonlinear. The actual address calculation to accomplish the mapping process is done by the CPU and the memory management unit (MMU). Thus far, we have not specified the exact size of a memory entry. A commonly used memory entry size is one byte. For historical reasons, memory is organized in bytes. A byte is usually the smallest unit of information transferred with each memory access. Wider memory entries are becoming more popular as the CPU continues to grow in speed and complexity. There are many modern systems that have a data width wider than a byte. A common size is a double word (32 b). in current desktop. As a result, memory in bytes is organized in sections of multibytes. However, due to need for backward compatibility, these wide datapath systems are also organized to be byte addressable. The maximum width of the memory transfer is usually called *memory word length*, and the size of the memory in bytes is called *memory capacity*. Since there are different memory device sizes, the memory system can be populated with different sized memory devices.

Memory Device Organization

Physically, within a memory device, cells are arranged in a two-dimensional array with each of the cell capable of storing 1 b of information. This matrix of cells is accessed by specifying the desired row and column addresses. The individual row enable line is generated using an address decoder while the column is selected through a multiplexer. There is usually a sense amplifier between the column bit line and the multiplexer input to detect the content of the memory cell it is accessing. Figure 8.46 illustrates this general memory cell array described with r bit of row address and c bit of column address. With the total number of $r + c$ address bits, this memory structure contains 2^{r+c} number of bits. As the size of memory array increases, the row enable lines, as well as the column bit lines, become longer. To reduce the capacitive load of a long row enable line, the row decoders, sense amplifiers, and column multiplexers are often placed in the middle of divided matrices of cells, as illustrated in Fig. 8.47. By designing the multiplexer differently we are able to construct memory with different output width, for example, ×1, ×8, ×16, etc. In fact, memory designers go to great efforts to design the column multiplexers so that most of the fabrication masks may be shared for memory devices that have the same capacity but with different configurations.

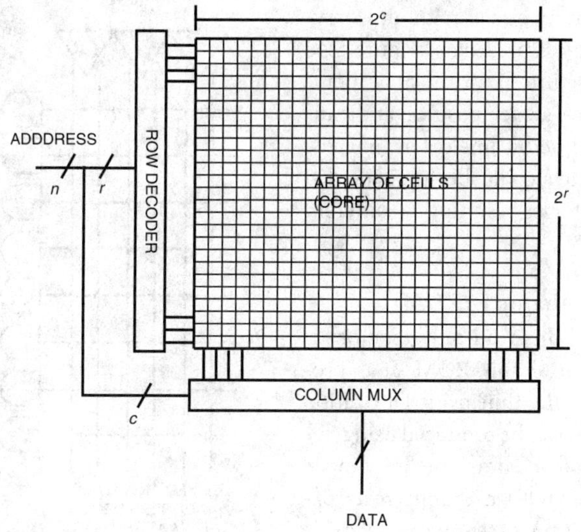

FIGURE 8.46   Generic memory structure.

8.4.3   Memory Device Types

As mentioned before, according to the functionality and characteristics of memory, we may divide memory devices into two major categories: ROM and RAM. We will describe these different type of devices in the following sections.

Read-Only Memory

In many systems, it is desirable to have the system level software (for example, basic input/output system [BIOS]) stored in a read-only format, because these types of programs are seldom changed. Many embedded systems also use read-only memory to store their software routines because these programs also are never changed during their lifetime, in general. Information stored in the read-only memory is permanent. It is retained even if the power supply is turned off. The memory can be read out reliably

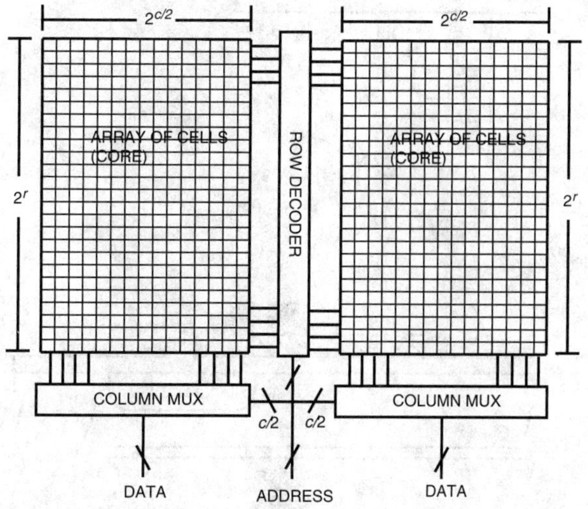

FIGURE 8.47   Divided memory structure.

by a simple current sensing circuit without worrying about destroying the stored data. Figure 8.48 shows the general structure of a read-only memory (ROM). The effective switch position at the intersection of the word-line/bit-line determines the stored value. This switch could be implemented using different technologies resulting in different types of ROMs.

Masked Read-Only Memory (ROM)

The most basic type of this read-only-memory is called masked ROM, or simply ROM. It is programmed at manufacturing time using fabrication processing masks. ROM can be produced using different technologies, bipolar, complementary metal oxide semiconductor (CMOS), n-channel metal oxide semiconductor (nMOS), p-channel metal oxide semiconductor (pMOS), etc. Once they are programmed there are no means to change their contents. Moreover, the programming process is performed at the factory.

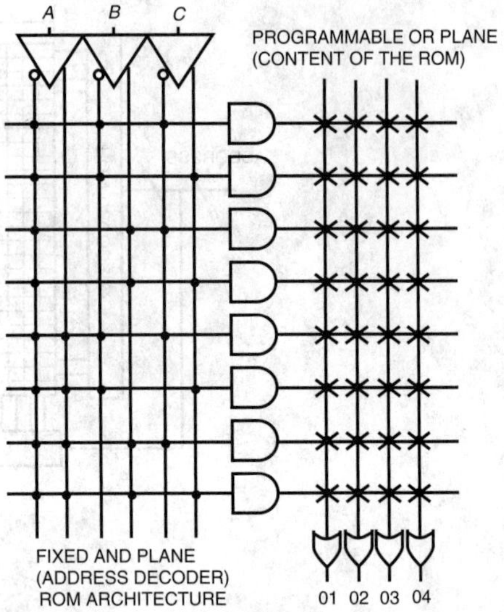

FIGURE 8.48 General structure of a ROM (an 8 × 4 ROM).

Programmable Read-Only Memory (PROM)

Some read-only memory is one-time programmable, but it is programmable by the user at the user's own site. This is called programmable read-only memory (PROM). It is also often referred to as write once memory (WOM). PROMs are based mostly on bipolar technology, since this technology supports it very well. Each of the single transistors in a cell has a fuse connected to its emitter. This transistor and fuse make up the memory cell. When a fuse is blown, no connection can be established when the cell is selected using the row line. Thereby a zero is stored. Otherwise, with the fuse intact, a logic one is represented. The programming is done through a programmer called a PROM programmer or PROM burner. Figure 8.49 illustrates the structure of a bipolar PROM cell and its cross section when fabricated.

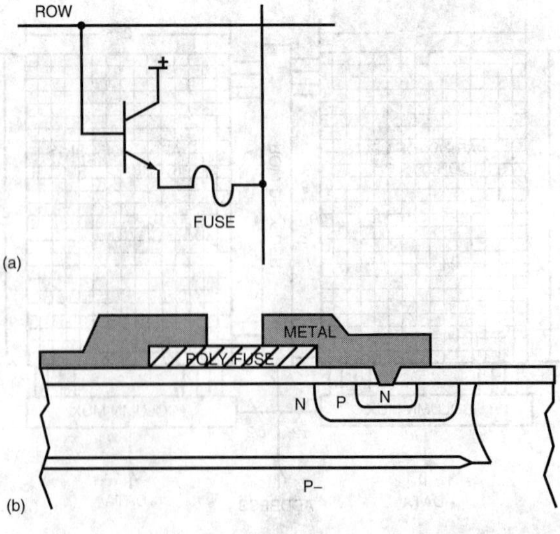

FIGURE 8.49 Bipolar PROM: (a) bipolar PROM cell, (b) cross section of a bipolar PROM cell.

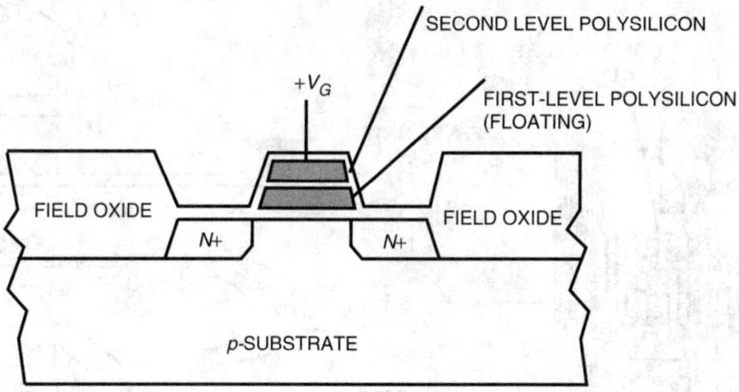

FIGURE 8.50 Cross section of a floating gate EPROM cell.

Erasable Programmable Read-Only Memory (EPROM)

It is sometimes inconvenient to program the ROM only once. Thus, the erasable PROM is designed. This type of erasable PROM is called EPROM. The programming of a cell is achieved by avalanche injection of high-energy electrons from the substrate through the oxide. This is accomplished by applying a high drain voltage, causing the electrons to gain enough energy to jump over the 3.2-eV barrier between the substrate and silicon dioxide thus collecting charge at the floating gate. Once the applied voltage is removed, this charge is trapped on the floating gate. Erasing is done using an ultraviolet (UV) light eraser. Incoming UV light increases the energy of electrons trapped on the floating gate. Once the energy is increased above the 3.2-eV barrier, the electrons leave the floating gate and move toward the substrate and the selected gate. Therefore, these EPROM chips all have windows on their packages where erasing UV light can reach inside the packages to erase the content of cells. The erase time is usually in minutes. The presence of a charge on the floating gate will cause the MOS transistor to have a high threshold voltage. Thus, even with a positive select gate voltage applied at the second level of polysilicon the MOS remains to be turned off. The absence of a charge on the floating gate causes the MOS to have a lower threshold voltage. When the gate is selected the transistor will turn on and give the opposite data bit. Figure 8.50 illustrates the cross section of a EPROM cell with floating gate. EPROM technologies that migrate toward smaller geometries make floating-gate discharge (erase) via UV light exposure increasingly difficult. One problem is that the width of metal bit lines cannot reduce proportionally with advancing process technologies. EPROM metal width requirements limit bit-lines spacing, thus reducing the amount of high-energy photons that reach charged cells. Therefore, EPROM products built on submicron technologies will face longer and longer UV exposure time.

Electrical Erasable Read-Only Memory (EEPROM)

Reprogrammability is a very desirable property. However, it is very inconvenient to use a separate light-source eraser for altering the contents of the memory. Furthermore, even a few minutes of erase time is intolerable. For this reason, an erasable PROM was designed called electrical erasable PROM (EEPROM). EEPROM permits new applications where erasing is done without removing the device from the system it resides in. There are a few basic technologies used in the processing of EEPROMs or electrical reprogrammable ROMs. All of them uses the Fowler-Nordheim tunneling effect to some extent. In this tunneling effect, cold electrons jump through the energy barrier at a silicon-silicon dioxide interface and into the oxide conduction band. This can only happen when the oxide thickness is of 100 Å or less, depending on the technology. This tunneling effect is reversible, allowing the reprogrammable ROMs to be used over and over again. One of the first electrical erasable PROMs is the electrical alterable ROM (EAROM) which is based on metal-nitrite-oxide silicon (MNOS) technology. The other is EEPROM, which is based on silicon floating gate technology used in fabricating EPROMs. The floating gate type of EEPROM is favored because of its reliability and density. The major difference between EPROM and EEPROM is in the way they discharge the charge stored in the floating gate. EEPROM must discharge floating gates electrically as opposed to using an UV light source in an EPROM device where electrons absorb photons from the UV

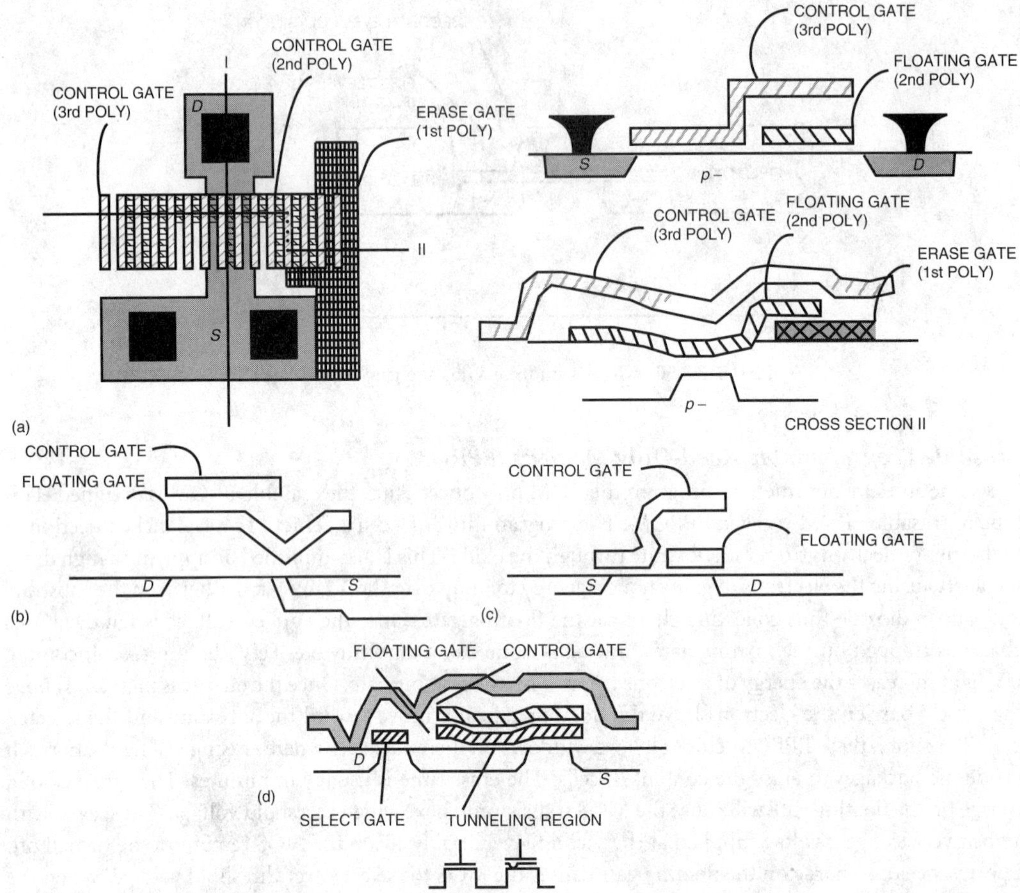

FIGURE 8.51 (a) Triple poly EEPROM cell layout and structure, (b) flotox EEPROM cell structure (source programming), (c) EEPROM with drain programming, (d) another source programming EEPROM.

radiation and gain enough energy to jump the silicon/silicon-dioxide energy barrier in the reverse direction as they return to the substrate. The solution for the EEPROM is to pass low-energy electrons through the thin oxide through high field (10^7 V/cm^2). This is known as the Fowler-Nordheim tunneling, where electrons can pass a short distance through the forbidden gap of the insulator and enter the conduction bank when the field applied is high enough. There are three common types of flash EEPROM cells. One uses the erase gate (three levels of polysilicon), the second and third use source and drain, respectively, to erase. Figures 8.51(a)–8.51(d) illustrate the cross sections of different EEPROMs. To realize a small EEPROM memory cell, the NAND structure was proposed in 1987. In this structure, cells are arranged in series. By using different patterns, an individual cell can be detected whether it is programmed or not. From the user's point of view, EEPROMs differs from RAM only in their write time and number of writes allowed before failure occurs. Early EEPROMs were hard to use because they have no latches for data and address to hold values during the long write operations. They also require a higher programming voltage, other than the operating voltage. Newer EEPROMs use charge pumps to generate the high programming voltage on the chip so the user does not need to provide a separate programming voltage.

Flash-EEPROM

This type of erasable PROM lacks the circuitry to erase individual locations. When you erase them, they are erased completely. By doing so, many transistors may be saved, and larger memory capacities are possible. Note that sometimes you do not need to erase before writing. You can also write to an erased, but unwritten location, which results in an average write time comparable to an EEPROM. Another important thing to know is that writing zeros into a location charges each of the flash EEPROM's memory cells to the same electrical potential so that subsequent erasure drains an equal amount of free charge (electrons) from each cell. Failure to equalize the charge in each cell prior to erasure can result in the overerasure of some cells by dislodging bound electrons in the floating gate and driving them out. When a floating gate is depleted in this way, the corresponding transistor can never be turned off again, thus destroying the flash EEPROM.

Random Access Memory (RAM)

RAM stands for random-access memory. It is really read-write memory because ROM is also random access in the sense that given a random address the corresponding entry is read. RAM can be categorized by content duration. A static RAM's contents is always retained, as long as power is applied. A DRAM, on the other hand, needs to be refreshed every few milliseconds. Most RAMs by themselves are volatile, which means that without the power supply their content will be lost. All of the ROMs mentioned in the previous section are nonvolatile. RAM can be made nonvolatile by using a backup battery.

Static Random Access Memory (SRAM)

Figure 8.52 shows various SRAM memory cells (6T, 5T, 4T). The six transistor (6T) SRAM cell is commonly used SRAM. The crossed coupled inverters in a SRAM cell retain the information indefinitely, as long as the power supply is on since one of the pull-up transistors supplies current to compensate for the leakage current. During a read, bit and bitbar line are precharged while the word enable line is held low. Depending on the content of the cell, one of the lines is discharged slightly causing the precharged voltage to drop when the word enable line is strobed. This difference in voltage between the bit and bitbar lines is sensed by the sense amplifier, which produces the read result. During a write process, one of the bit/bitbar lines is discharged, and by strobing the word enable line the desired data is forced into the cell before the word line goes away.

Figure 8.53 gives the circuit of a complete SRAM circuit design with only one column and one row shown. One of the key design parameters of a SRAM cell is to determine the size of transistors used in the memory cell. We first need to determine the criteria used in sizing transistors in a CMOS 6-transistor/cell SRAM.

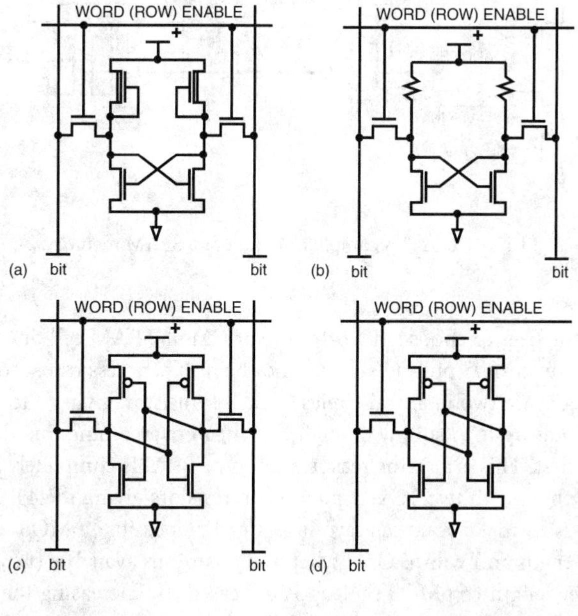

FIGURE 8.52 Different SRAM cells: (a) six-transistor SRAM cell with depletion transistor load, (b) four-transistor SRAM cell with polyresistor load, (c) CMOS six-transistor SRAM cell, (d) five-transistor SRAM cell.

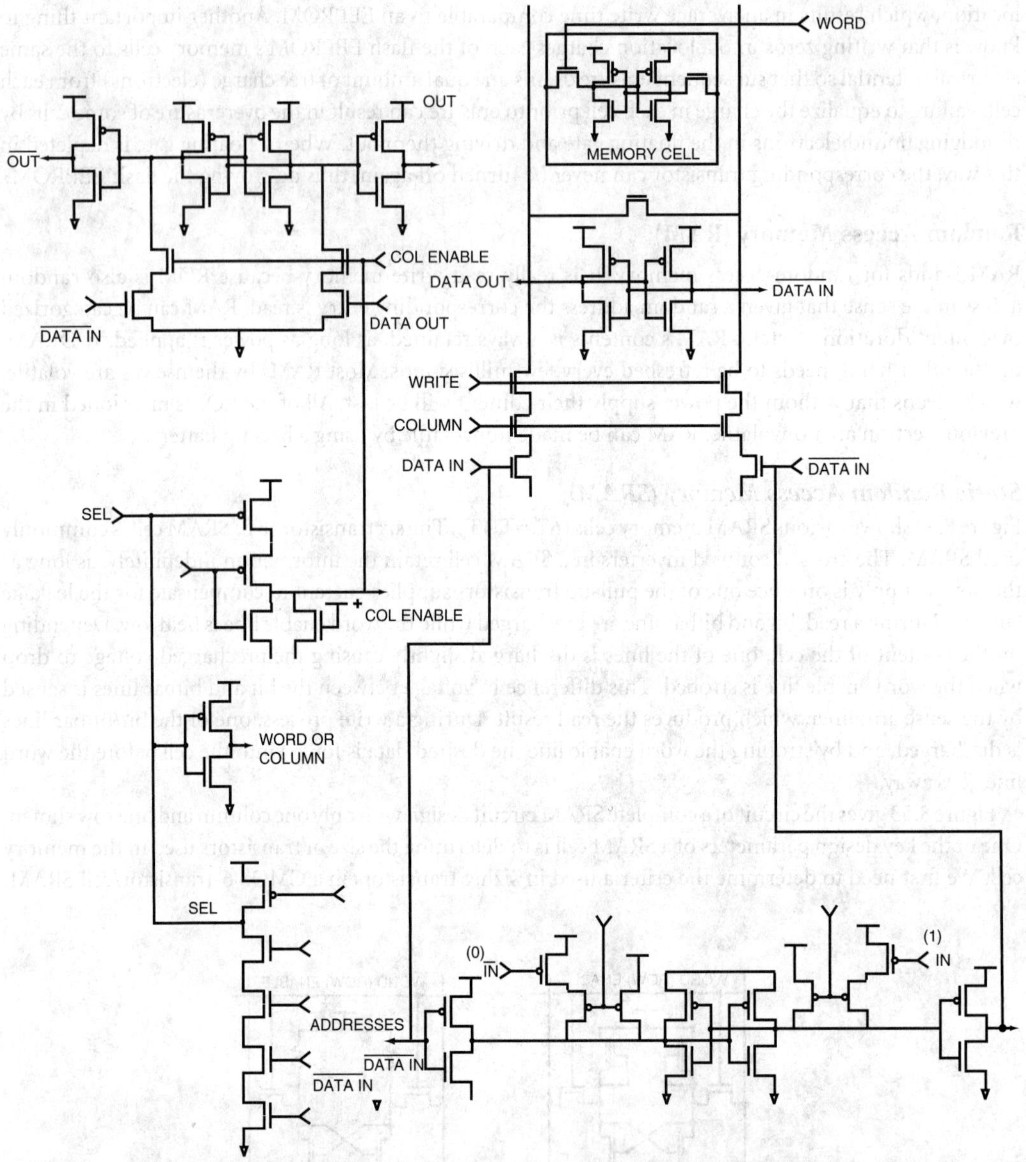

FIGURE 8.53 A complete circuit of an SRAM memory.

There are three transistor sizes to choose in a 6-transistor CMOS SRAM cell due to symmetry. They are the pMOS pull-up size, the nMOS pull-down size, and the nMOS access transistor (also called the pass-transistor gate) size. There are two identical copies of each transistor, giving a total of six. Since the sole purpose of the pMOS pull-up is to supply enough current in overcoming junction current leakage, we should decide this size first. This is also the reason why some SRAMs completely remove the two pMOS transistors and replace them with two 10-GΩ polysilicon resistors giving the 4T cell shown in Fig. 8.52. Since one of the goals is to make your cell layout as small as possible, pMOS pull-up is chosen to be minimum both in its length and width. Only when there is room available (i.e., if it does not increase the cell layout area), the length of pMOS pull-up is increased. By increasing the length of pMOS pull-up transistors, the capacitance on the crossed-coupled inverters output nodes is increased. This helps in protecting against certain soft errors. It also makes the cell slightly easier to write.

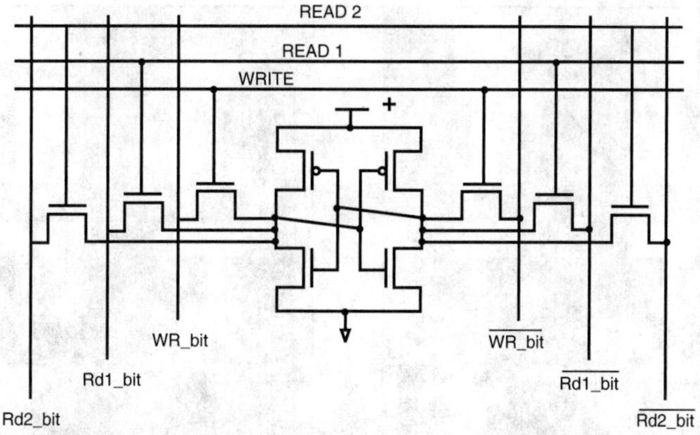

FIGURE 8.54 Multiported CMOS SRAM cell (shown with 2-read and 1-write).

The next step is to choose the nMOS access transistor size. This is a rather complicated process. To begin we need to determine the length of this transistor. It is difficult to choose because, on one hand, we want it also to be minimum in order to reduce the cell layout size. However, on the other hand, a column of n SRAM bits (rows) has to have n of access transistors connected to the bit or bitbar line. If each of the cells leaks just a small amount of current, the leakage is multiplied by n. Thus, the bit or bitbar line, which one might think should be sitting at the bit-line pull-up (or precharge) voltage, is actually pulled down by this leakage. Thus, the bit or bitbar line high level is lower than the intended voltage. When this happens, the voltage difference between the bit and bitbar lines, which is seen by the sense amplifier during a read, is smaller than expected, perhaps catastrophically so. Thus, if the transistors used are not particularly leaky or n is small, a minimum sized nMOS is sufficient. Otherwise, a larger sized transistor should be used. Beside considering leakage, there are three other factors that may affect the transistor sizes of the two nMOSs. They are: (1) cell stability, (2) speed, and (3) layout area. The first factor, cell stability, is a DC phenomenon. It is a measure of the cell's ability to retain its data when reading and to change its data when writing. A read is done by creating a voltage difference between the bit and bitbar lines (which are normally precharged) for the sense amplifier to differentiate. A write is done by pulling one of the bit or bitbar lines down completely. Thus, one must design the size to satisfy the cell stability while achieving the maximum read and write speed and maintaining the minimum layout area.

Much work has been done in writing computer-aided design (CAD) tools that automatically size transistors for arrays of SRAM cells, and then do the polygon layout. Generally, these are known as SRAM macrogenerators or RAM compilers. These generated SRAM blocks are used as drop ins in many application specific intergrated circuits (ASICs). Standard SRAM chips are also available in many different organizations. Common ones are arranged in 4 b, bytes, and double words (32 b) in width.

There is also a special type of SRAM cell used in computers to implement registers. These are called multiple port memories. In general, the contents can be read by many different requests at the same time. Figure 8.54 shows a dual-read port single-write port SRAM cell. When laying out SRAM cells, adjacent cells usually are mirrored to allow sharing of supply or ground lines. Figure 8.55 illustrates the layout of four adjacent SRAM cells using a generic digital process design rules. This block of four cells can be repeated in a two-dimensional array format to form the memory core.

Direct Random Access Memory (DRAM)

The main disadvantage of SRAM is in its size since it takes six transistors (or at least four transistors and two resistors) to construct a single memory cell. Thus, the DRAM is used to improve the capacity. There are different DRAM cell designs. There is the four-transistor DRAM cell, three-transistor DRAM cell, and the one-transistor DRAM cell. Figures 8.56 shows the corresponding circuits for these cells. Data writing is accomplished in a three-transistor cell by keeping the RD line low (see Fig. 8.56(b)) while strobing the WR line with the desired data to be written is kept on the bus. If a 1 is desired to be stored, the gate of T2 is charged turning

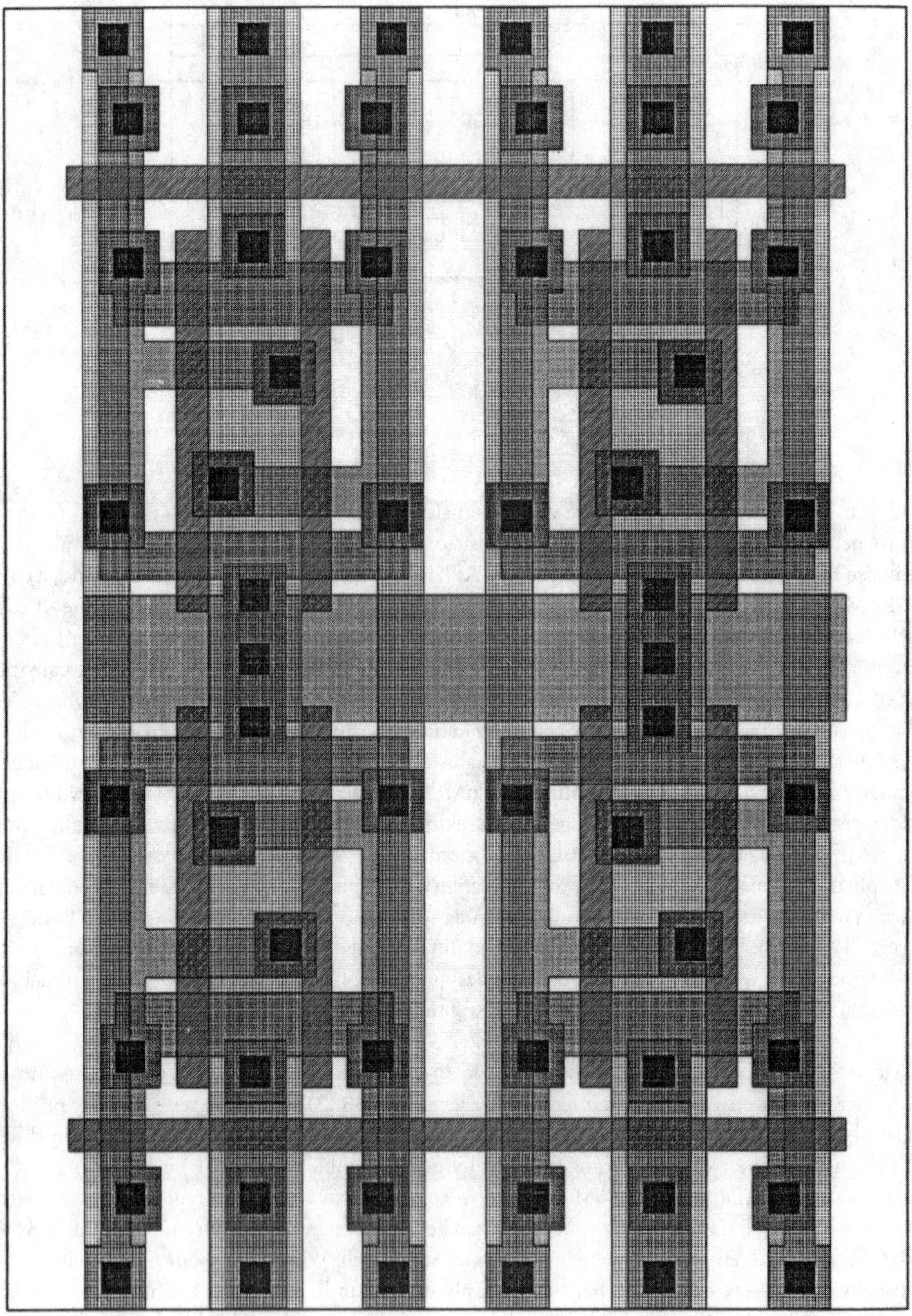

FIGURE 8.55 Layout example of four abutted 6–t SRAM cells.

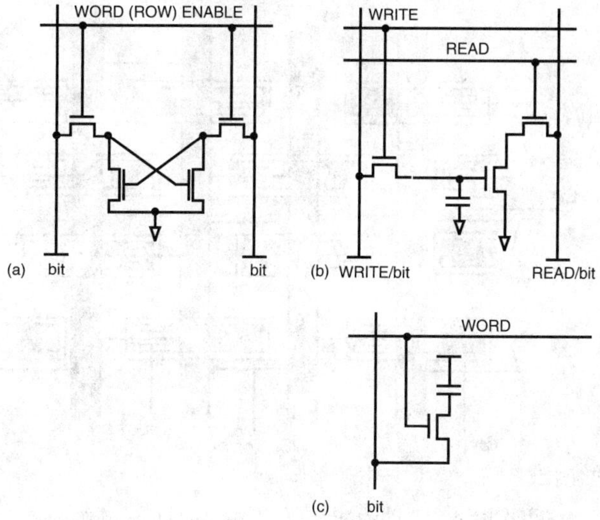

FIGURE 8.56 Different DRAM cells: (a) four-transistor DRAM cell, (b) three-transistor DRAM cell, (c) one-transistor DRAM cell.

on T2. This charge remains on the gate of T2 for a while before the leakage current discharges it to a point where it cannot be used to turn on T2. When the charge is still there, a read can be performed by precharging the bus and strobing the RD line. If a 1 is stored, then both T2 and T3 are on during a read, causing the charge on the bus to be discharged. The lowering of voltage can be picked up by the sense amplifier. If a zero is stored, then there is no direct path from the bus to ground, thus the charge on the bus remains. To further reduce the area of a memory cell, a single transistor cell is often used. Figure 8.56(c) shows the single transistor cell with a capacitor. Usually, two columns of cells are mirror images of each other to reduce the layout area. The sense amplifier is shared, as shown in Fig. 8.57. In this one-transistor DRAM cell, there is a capacitor used to store the charge, which determines the content of the memory. The amount of the charge in the capacitor also determines the overall performance of the memory. A continuing goal is to downscale the physical area of this capacitor to achieve higher and higher density. Usually, as one reduces the area of the capacitor, the capacitance also decreases. One approach is to increase the surface area of the storage electrode without increasing the layout area by employing stacked capacitor structures, such as finned or cylindrical structures. Certain techniques can be used to utilize a cylindrical capacitor structure with hemispherical grains. Figure 8.58 illustrates the cross section of a

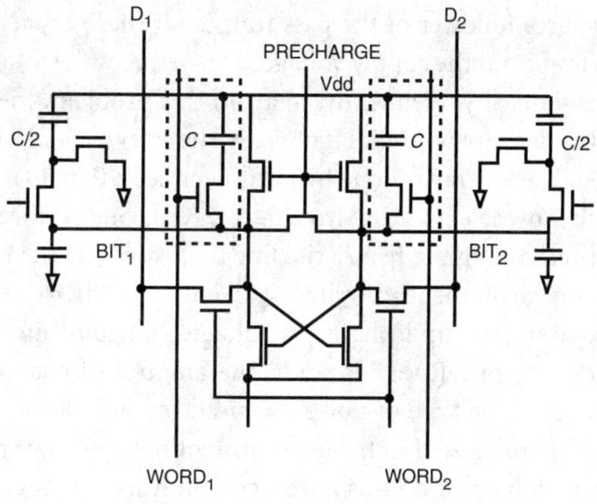

FIGURE 8.57 DRAM sense amplifier with 2 bit lines and 2 cells.

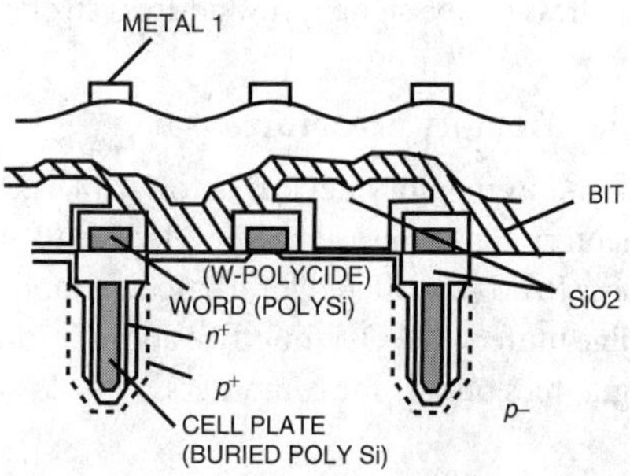

FIGURE 8.58 Cross section view of trenched DRAM cells.

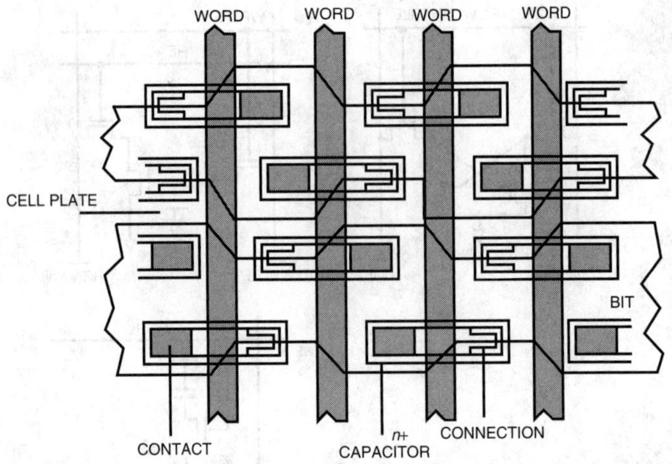

FIGURE 8.59 Physical layout of trenched DRAM cells.

one-transistor DRAM cell with the cylindrical capacitor structure. Since the capacitor is charged by a source follower of the pass transistor, these capacitors can be charged maximally to a threshold voltage drop from the supply voltage. This reduces the total charge stored and affects performance, noise margin, and density. Frequently, to avoid this problem, the word lines are driven above the supply voltage when the data are written. Figure 8.59 shows typical layout of one-transistor DRAM cells.

The writing is done by putting either a 0 or 1 (the desired data to store) on the read/writing line. Then the row select line is strobed. A zero or one is stored in the capacitor as charge. A read is performed with precharging the read/write line then strobing the row select. If a zero is stored due to charge sharing, the voltage on the read/write line decreases. Otherwise, the voltage remains. A sense amplifier is placed at the end to pick up if there is a voltage change or not. DRAM differs from SRAM in another aspect. As the density of DRAM increases, the amount of charge stored in a cell reduces. It becomes more subject to noise. One type of noise is caused by radiation called alpha particles. These particles are helium nuclei that are present in the environment naturally or emitted from the package that houses the DRAM die. If an alpha particle hits a storage cell, it may change the state of the memory. Since alpha particles can be reduced, but not eliminated, some DRAMs institute error detection and correction techniques to increase their reliability. Another difference between DRAMs and SRAMs is in the number of address pins needed for a given size RAM. SRAM chips require all address bits to be given at the same time. DRAMs, however, utilize time-multiplex address lines. Only half of the address bits are given at a given time. They are divided by rows and columns. An extra control signal is thus required. This is the reason why DRAM chips have two address strobe signals: row address strobe (RAS) and column address strobe (CAS).

Special Memory Structures

The trend in memory devices is toward larger, faster and better-performance products. There is a complementary trend toward the development of special purpose memory devices. Several types of special purpose RAMs are offered for particular applications such as content addressable memory for cache memory, line buffers (FIFOs) for office automation machines, frame buffers for TV and broadcast equipment, and graphics buffers for computers.

Content Addressable Memory (CAM)

A special type of memory called content addressable memory (CAM) or associative memory is used in many applications such as cache memory and associative processor. A CAM stores a data item consisting of a tag and a value. Instead of giving an address, a data pattern is given to the tag section of the CAM. This data pattern is matched with the content of the tag section. If an item in the tag section of the CAM matches the supplied data pattern, the CAM outputs the value associated with the matched tag. Figure 8.60 illustrates the basic structure of a CAM. CAM cells must be both readable and writable just like the RAM cell. Figure 8.61 shows a circuit diagram for a basic CAM cell with a match output signal. This output signal may be used as an input for some logic to determine the matching process.

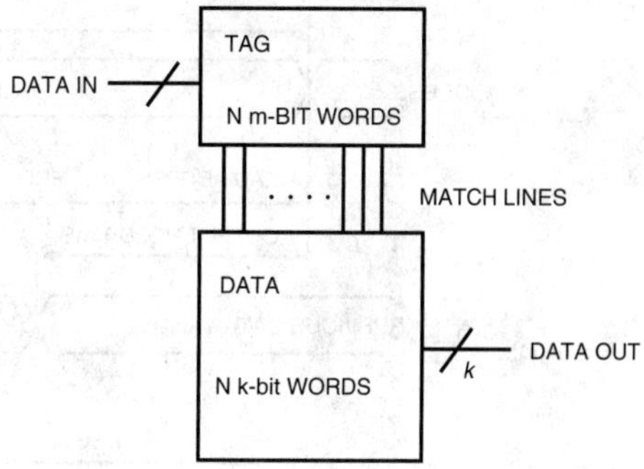

FIGURE 8.60 Functional view of a CAM.

First-In–First-Out/Queue (FIFO/Queue)

A FIFO/queue is used to hold data while waiting. It serves as the buffering region for two systems, which may have different rates of consuming and producing data. FIFO can be implemented using shift registers or RAMs with pointers.

Video RAM: Frame Buffer

There is a rapid growth in computer graphic applications. The technology that is most successful is termed raster scanning. In a raster scanning display system, an image is constructed with series of horizontal lines. Each of these lines are connected pixels of the picture image. Each pixel is represented with bits controlling the intensity. Usually there are three planes corresponding to each of the primary colors: red, green, and blue. These three planes of bit maps are called frame buffer or image memory. Frame buffer architecture affects the performance of a raster scanning graphic system greatly. Since these frame buffers needs to be read out serially to display the image line by line, a special type of DRAM memory called video memory is used. Usually these memory are dual ported with a parallel random access port for writing and a serial port for reading.

8.4.4 Interfacing Memory Devices

Besides capacity and type of devices, other characteristics of memory devices include its speed and the method of access. We mentioned in the Introduction memory access time. It is defined as the time between the address available to the divide and the data available at the pin for access. Sometimes the access time

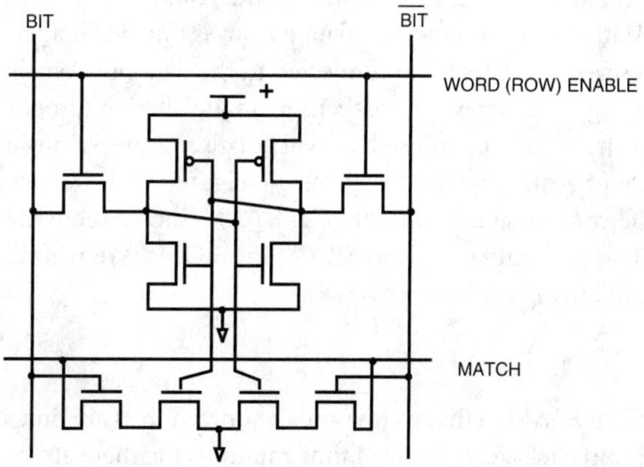

FIGURE 8.61 Static CMOS CAM cell.

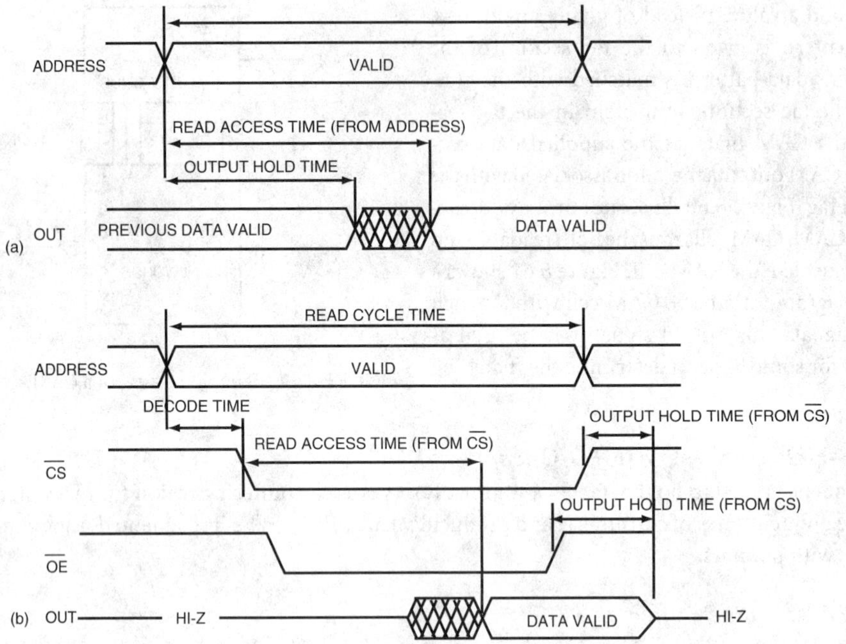

FIGURE 8.62 SRAM read cycle: (a) simple read cycle of SRAM (OE~, CE~ are all asserted and WE~ is low), (b) SRAM read cycle.

is measured from a particular control signal. For example, the time between the read control line ready to data ready is called the read command access time. The memory cycle time is the minimum time between two consecutive accesses. The memory write command time is measured from the write control ready to data stored in the memory. The memory latency time is the interval between the CPU issuing an address to data available for processing. The memory bandwidth is the maximum amount of memory capacity being transferred in a given time. Access is done with address, read/write control lines, and data lines. SRAM and ROMs are accessed similarly during read. Figure 8.62 shows the timing diagram of two SRAM read cycles. In both methods read cycle time is defined as the time period between consecutive read addresses. In the first method, SRAM acts as an asynchronous circuit. Given an address, the output of the SRAM changes and become valid after a certain delay, which is the read access time. The second method uses two control signals, chip select and output enable, to initiate the read process. The main difference is in the data output valid time. With the second method data output is only valid after the output enable signal is asserted, which allows several devices to be connected to the data bus. Writing SRAM and electrically reprogrammable ROM is slightly different. Since there are many different programmable ROMs and their writing processes depend on the technology used, we will not discuss the writing of ROMs here. Figure 8.63 shows the timing diagram of writing typical SRAM chips. Figure 8.63(a) shows the write cycle using the write enable signal as the control signal, whereas Fig. 8.63(b) shows the write cycle using chip enable signals. Accessing DRAM is very different from SRAM and ROMs. We will discuss the different access modes of DRAMs in the following section.

Accessing DRAMs

DRAM is very different from SRAM in that its row and column address are time multiplexed. This is done to reduce the pins of the chip package. Because of time multiplexing there are two address strobe lines for DRAM address, RAS and CAS. There are many ways to access the DRAM. We list the five most common ways.

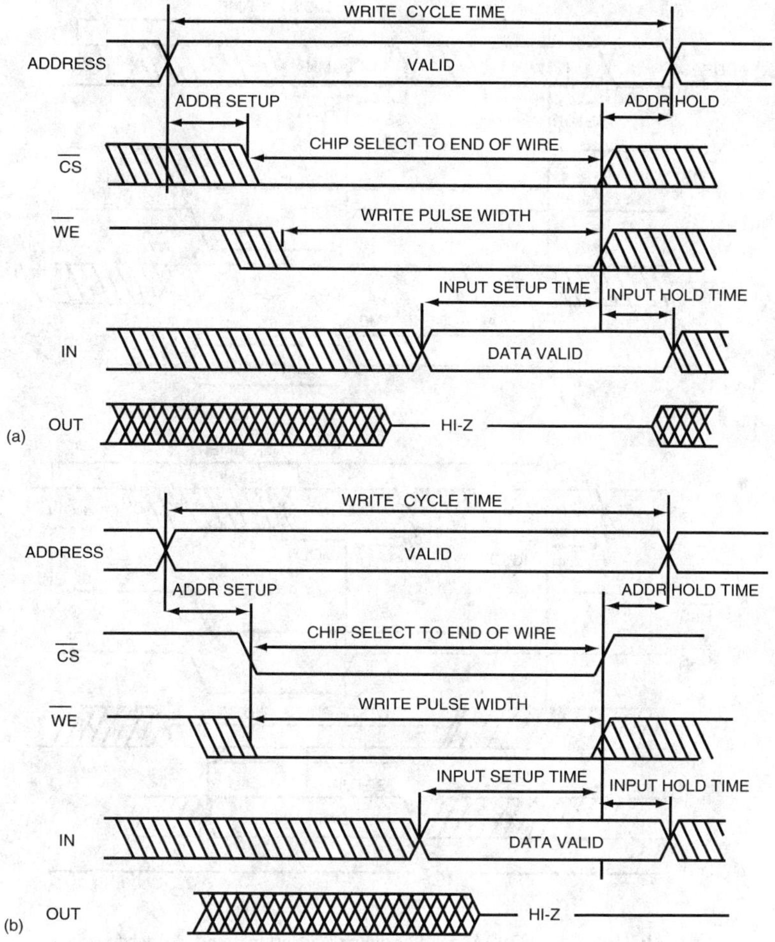

FIGURE 8.63 SRAM write cycles: (a) write enable controlled (b) chip enable controlled.

Normal Read/Write

When reading, a row address is given first, followed by the row address strobe signal RAS. RAS is used to latch the row address on chip. After RAS, a column address is given followed by the column address strobe CAS. After certain delay (read access time) valid data appear on the data lines. Memory write is done similarly to memory read with only the read/write control signal reversed. There are three cycles available to write a DRAM. They are early write, read-modify-write, and late write cycles. Figure 8.64 shows only the early write cycle of a DRAM chip. Other write cycles can be found in most of the DRAM databooks.

Fast Page Mode (FPM) or Page Mode

In page mode (or fast page mode), a read is done by lowering the RAS when the row address is ready. Then, repeatedly give the column address and CAS whenever a new one is ready without cycling the RAS line. In this way a whole row of the two-dimensional array (matrix) can be accessed with only one RAS and the same row address. Figure 8.65 illustrates the read timing cycle of a page mode DRAM chip.

Static Column

Static column is almost the same as page mode except the CAS signal is not cycled when a new column address is given—thus the static column name.

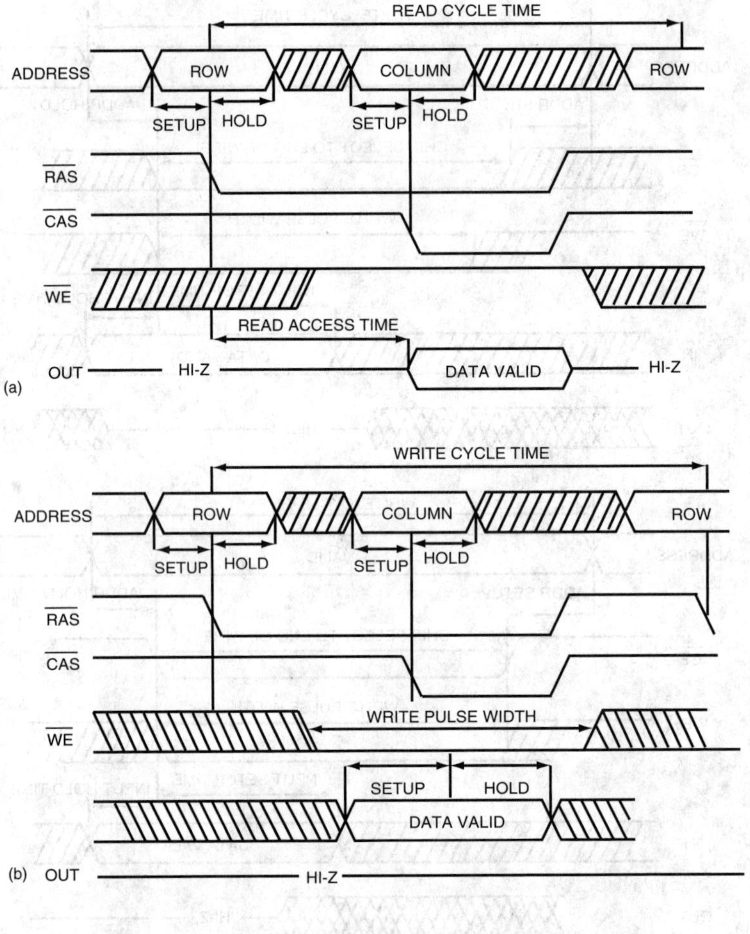

FIGURE 8.64 DRAM read and write cycles: (a) read, (b) write.

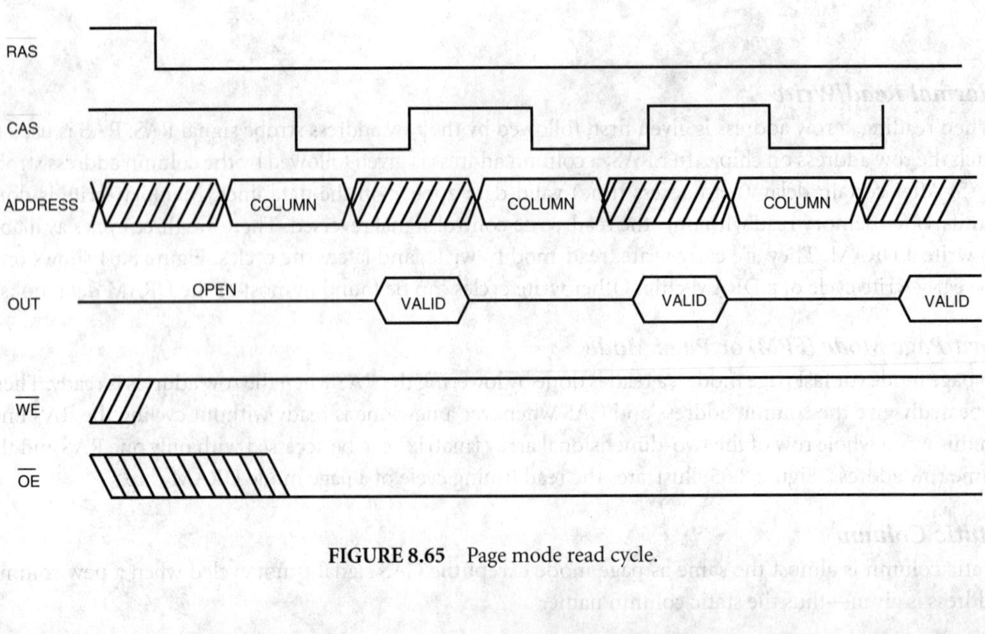

FIGURE 8.65 Page mode read cycle.

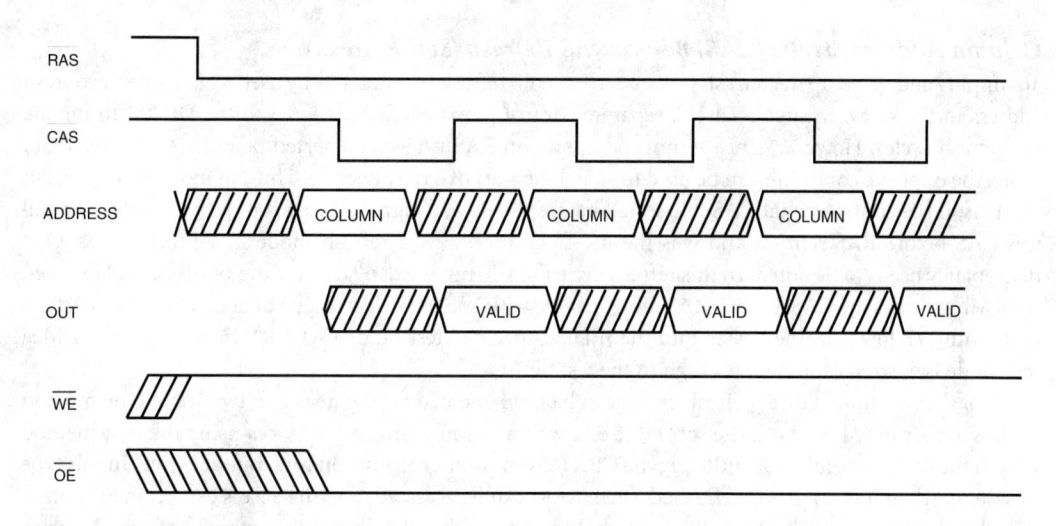

FIGURE 8.66 EDO-page mode read cycle.

Extended Date Output (EDO) Mode

In page mode, CAS must stay low until valid data reach the output. Once the CAS assertion is removed, data are disabled and the output pins goes to open circuit. With EDO DRAM, an extra latch following the sense amplifier allows the CAS line to return to high much sooner, permitting the memory to start precharging earlier to prepare for the next access. Moreover, data are not disabled after CAS goes high. With burst EDO DRAM, not only does the CAS line return to high, it can also be toggled to step though the sequence in burst counter mode, providing even faster data transfer between memory and the host. Figure 8.66 shows a read cycle of an EDO page mode DRAM chip. EDO mode is also called hyper page mode (HPM) by some manufactures.

Nibble Mode

In nibble mode after one CAS with a given column three more accesses are performed automatically without giving another column address (the address is assumed to be incremented from the given address).

Refreshing the DRAM

Row Address Strobe- (RAS-) Only Refresh

This type of refresh is done row by row. As a row is selected by providing the row address and strobing RAS, all memory cells in the row are refreshed in parallel. It will take as many cycles as the number of rows in the memory to refresh the entire device. For example, an 1M × 1 DRAM, which is built with 1024 rows and columns will take 1024 cycles to refresh the device. To reduce the number of refresh cycles, memory arrays are sometimes arranged to have less rows and more columns. The address, however, is nevertheless multiplexed as two evenly divided words (in the case of 1M × 1 DRAM the address word width is 10 b each for rows and columns). The higher order bits of address lines are used internally as column address lines and they are ignored during the refresh cycle. No CAS signal is necessary to perform the RAS-only refresh. Since the DRAM output buffer is enabled only when CAS is asserted, the data bus is not affected during the RAS-only refresh cycles.

Hidden Refresh

During a normal read cycle, RAS and CAS are strobed after the respective row and column addresses are supplied. Instead of restoring the CAS signal to high after the read, several RASs may be asserted with the corresponding refresh row address. This refresh style is called the hidden refresh cycles. Again, since the CAS is strobed and not restored, the output data are not affected by the refresh cycles. The number of refresh cycles performed is limited by the maximum time that CAS signal may be held asserted.

Column Address Strobe (CAS) Before RAS Refresh (Self-Refresh)

To simplify and speed up the refresh process, an on-chip refresh counter may be used to generate the refresh address to the array. In such a case, a separate control pin is needed to signal to the DRAM to initiate the refresh cycles. However, since in normal operation RAS is always asserted before CAS for read and write, the opposite condition can be used to signal the start of a refresh cycle. Thus, in modern self-refresh DRAMs, if the control signal CAS is asserted before the RAS, it signals the start of refresh cycles. We call this CAS-before-RAS refresh, and it is the most commonly used refresh mode in 1-Mb DRAMs. One discrepancy needs to be noted. In this refresh cycle the $\overline{WE}$ pin is a *don't care* for the 1-Mb chips. However, the 4 Mb specifies the CAS_Before_RAS refresh mode with $\overline{WE}$ pin held at high voltage. A CAS-before-RAS cycle with $\overline{WE}$ low puts the 4 Meg into the JEDEC-specified test mode (WCBR). In contrast, the 1 Meg test mode is entered by applying a high to the test pin.

All of the mentioned three refresh cycles can be implemented on the device in two ways. One method utilizes a distributed method, the second uses a wait and burst method. Devices using the first method refresh the row at a regular rate utilizing the CBR refresh counter to turn on rows one at a time. In this type of system, when it is not being refreshed, the DRAM can be accessed, and the access can begin as soon as the self-refresh is done. The first CBR pulse should occur within the time of the external refresh rate prior to active use of the DRAM to ensure maximum data integrity and must be executed within three external refresh rate periods. Since CBR refresh is commonly implemented as the standard refresh, this ability to access the DRAM right after exciting the self-refresh is a desirable advantage over the second method. The second method is to use an internal burst refresh scheme. Instead of turning on rows at regular interval, a sensing circuit is used to detect the voltage of the storage cells to see if they need to be refreshed. The refresh is done with a series of refresh cycles one after another until all rows are completed. During the refresh other access to the DRAM is not allowed.

8.4.5 Error Detection and Correction

Most DRAMs require a read parity bit for two reasons. First, alpha particle strikes disturb cells by ionizing radiation, resulting in lost data. Second, when reading DRAM, the cell's storage mechanism capacitively shares its charge with the bit-line through an enable (select) transistor. This creates a small voltage differential to be sensed during read access. This small voltage difference can be influenced by other close by bit-line voltages and other noises. To have even more reliable memory, error correction code may be used. One of the error correction methods is called the Hamming code, which is capable of correcting any 1-b error.

Defining Terms

Dynamic random access memory (DRAM): This memory is dynamic because it needs to be refreshed periodically. It is random access because it can be read and written.

Memory access time: The time between a valid address supplied to a memory device and data becoming ready at output of the device.

Memory cycle time: The time between subsequent address issues to a memory device.

Memory hierarchy: Organize memory in levels to make the speed of memory comparable to the processor.

Memory read: The process of retrieving information from memory.

Memory write: The process of storing information into memory.

ROM: Acronym for read-only memory.

Static random-access memory (SRAM): This memory is static because it needs not to be refreshed. It is random access because it can be read and written.

References

Alexandridis, N. 1993. *Design of Microprocessor-Based Systems*. Prentice-Hall, Englewood Cliffs, NJ.

Chang, S.S.L. 1980. Multiple-read single-write memory and its applications. *IEEE Trans. Comp.* C-29(8).

Chou, N.J. et. al. 1972. Effects of insulator thickness fluctuations on MNOS charge storage characteristics. *IEEE Trans. Elec. Dev.* ED-19:198.

Denning, P.J. 1968. The working set model for program behavior. *CACM* 11(5).

Flannigan, S. and Chappell, B. 1986. *J. Solid St. Cir.*

Fukuma, M. et al. 1993. Memory LSI reliability. *Proc. IEEE* 81(5). May.

Hamming, R.W. 1950. Error detecting and error correcting codes. *Bell Syst. J.* 29 (April).

Katz, R.H. et al. 1989. Disk system architectures for high performance computing. *Proc. IEEE* 77(12).

Lundstrom, K.I. and Svensson, C.M. 1972. Properties of MNOS structures. *IEEE Trans. Elec. Dev.* ED-19:826.

Masuoka, F. et al. 1984. A new flash EEPROM cell using triple poly-silicon technology. *IEEE Tech. Dig.* IEDM: 464–467.

Micro. *Micro DRAM Databook.*

Mukherjee, S. et al. 1985. A single transistor EEPROM cell and its implementation in a 512 K CMOS EEPROM. *IEEE Tech. Dig.* IEDM:616–619.

NEC. n.d. *NEC Memory Product Databook.*

Pohm, A.V. and Agrawal, O.P. 1983. *High-Speed Memory Systems.* Reston Pub., Reston, VA.

Prince, B. and Gunnar Due-Gundersen, G. 1983. *Semiconductor Memories.* Wiley, New York.

Ross, E.C. and Wallmark, J.T. 1969. Theory of the switching behavior of MIS memory transistors. *RCA Rev.* 30:366.

Samachisa, G. et al. 1987. A 128 K flash EEPROM using double poly-silicon technology. *IEEE International Solid State Circuits Conference*, 76–77.

Sayers. et al. 1991. *Principles of Microprocessors.* CRC Press, Boca Raton, FL.

Scheibe, A. and Schulte, H. 1977. Technology of a new *n*-channel one-transistor EAROM cell called SIMOS. *IEEE Trans. Elec. Dev.* ED-24(5).

Seiichi Aritome, et al. 1993. Reliability issues of flash memory cells. *Proc. IEEE* 81(5).

Shoji, M. 1989. *CMOS Digital Circuit Technology.* Prentice-Hall, Englewood Cliffs, NJ.

Slater, M. 1989. *Design of Microprocessor-based Systems.*

Further Information

More information on basic issues concerning memory organization and memory hierarchy can be found in Pohm and Agrawal (1983). Prince and Due-Gunderson (1983) provides a good background on the different memory devices. Newer memory technology can be found in memory device databooks such as Mukherjee et al. (1985) and the NEC databook. *IEEE Journal on Solid-State Circuits* publishes an annual special issue on the Internation Solid-State Circuits Conference. This conference reports the current state-of-the-art development on most memory devices such as DRAM, SRAM, EEPROM, and flash ROM. Issues related to memory technology can be found in the *IEEE Transactions on Electron Devices*. Both journals have an annual index, which is published at the end of each year (December issue).

8.5   Microprocessors

James G. Cottle

8.5.1   Introduction

In the simplest sense, a *microprocessor* may be thought of as a central processing unit (CPU) on a chip. Technical advances of microprocessors evolve quickly and are driven by progress in ultra large-scale integrated/very large-scale integrated (ULSI/VLSI) physics and technology; fabrication advances, including reduction in feature size; and improvements in architecture.

Developers of microprocessor-based systems are interested in cost, performance, power consumption, and ease of programmability. The latter of these is, perhaps, the most important element in bringing a product to market quickly. A key item in the ease of programmability is availability of program development tools because it is these that save a great deal of time in small system implementation and make life a lot easier for the developer. It is not unusual to find small electronic systems driven by microcontrollers and microprocessors that are far more complex than need be for the project at hand simply because the ease of development on these platforms offsets considerations of cost and power consumption. Development tools are, therefore, a tangible asset to the efficient implementation of systems, utilizing microprocessors.

Often, a more general purpose microprocessor has a companion *microcontroller* within the same family. The subject of microcontrollers is a subset of the broader area of *embedded systems*. A microcontroller contains the basic architecture of the parent microprocessor with additional on-chip special purpose hardware to facilitate easy small system implementation. Examples of special purpose hardware include analog to digital (A/D) and digital to analog (D/A) converters, timers, small amounts of on-chip memory, serial and parallel interfaces, and other input output specific hardware. These devices are most cost efficient for development of the small electronic control system, since all externally needed hardware is already designed onto the chip. Component count external to the microcontroller is therefore kept to a minimum. In addition, the advanced development tools of the parent microprocessor are often available for programming. Therefore, a potential product may be brought to market much faster than if the developer used a more general purpose microprocessor, which often requires a substantial amount of external support hardware.

8.5.2 Architecture Basics

In the early days of mainframe computers, only a few instructions (e.g., addition, load accumulator, store to memory) were available to the programmer and the CPU had to be patched, or its configuration changed, for various applications. Invention of microprogramming, by Maurice Wilkes in 1949, unbound the instruction set. Microprogramming made possible more complex tasks by manipulating the resources of the CPU (its registers, arithmetic logic unit, and internal buses) in a well-defined, but programmable, way. In this type of CPU the microprogram, or *control store*, contained the realization of the semantics of native machine instructions on a particular set of CPU internal resources. Microprogramming is still in strong use today. Within the microprogram resides the routines used to implement more complex instructions and addressing modes. It is this scheme that is still widely used and a key element of the complex instruction set computer(CISC) microprocessor.

Evolution from the original large mainframe computers with hard-wired logical paths and schemes for instruction manipulation toward more flexible and complex instruction sets and addressing modes was a natural one. It was driven by advances in semiconductor fabrication technology, memory design, and device physics. Advances in semiconductor fabrication technology have made it possible to place the CPU on a monolithic integrated circuit along with additional hardware. With the advent of programmable read-only memory, the microprogram was not even bound to the whims of the original designer. It could be reprogrammed at a later date if the manipulation of resources for a particular task needed to be modified or streamlined. In the simplest view, microprocessors became more and more complex with larger instruction sets and many addressing modes. All of these advances were welcomed by compiler writers, and the instruction set complexity was a natural evolution of advances in component reliability and density.

Complex Instruction Set Computer (CISC) and Reduced Instruction Set Computer (RISC) Processors

The strength of the microprocessor is in its ability to perform simple logical tasks at a high rate of speed. In *Complex instruction set computer* (CISC) microprocessors, small register sets, memory to memory operations, large instruction sets (with variable instruction lengths), and the use of microcode are typical. The basic simplified philosophy of the CISC microprocessor is that added hardware can result in an overall increase in speed. The penultimate CISC processor would have each high-level language statement mapped to a single native CPU instruction. Microcode simplifies the complexity somewhat but necessitates the use of multiple machine cycles to execute a single CISC instruction. After the instruction is decoded on a CISC machine, the actual implementation may require 10 or 12 machine cycles depending on the instruction and addressing mode used. The original trend in microprocessor development was toward increased complexity of the instruction set. Although there may be hundreds of native machine instructions, only a handful are actually used. Ironically, the CISC instruction set complexity evolves at a sacrifice in speed because its harder to increase the clock speed of a complex chip. Recently, recognition of this along with demands for increased clock speeds have yielded favor toward the reduced instruction set (RISC) microprocessor.

This trend reversal followed studies in the early 1970s that showed that although the CISC machines had plenty of instructions, only relatively few of these were actually being used by programmers. In fact, 85% of all programming consists of simple assignment instructions) (i.e., A=B).

RISC machines have very few instructions and few machine cycles to implement them. What they do have is a lot of registers and a lot of parallelism. The ideal RISC machine attempts to accomplish a complete instruction in a single machine cycle. If this were the case, a 100-MHz microprocessor would execute 100 million instructions per second. There are typically many registers for moving data to accomplish the goal of reduced machine cycles. There is, therefore, a high degree of parallelism in a RISC processor. On the other hand, CISC machines have relatively few registers, depending on multiple machine cycle manipulation of data by the microprogram. In a CISC machine, the microprogram handles a lot of complexity in interpreting the native machine instruction. It is not uncommon for an advanced microprocessor of this type to have over 100 native machine instructions and a slew of addressing modes.

The two basic philosophies, CISC and RISC, are ideal concepts. Both philosophies have their merits. In practice, microprocessors typically incorporate both philosophical schemes to enhance performance and speed. A general summary of RISC vs CISC is as follows. CISC machines depend on complexity of programming in the microprogram, thereby simplifying the compiler. Complexity in the RISC machine is realized by the compiler itself.

Logical Similarity

Microprocessors may often be highly specialized for a particular application. The external logical appearance of all microprocessors, however, is the same. Although the package itself will differ, it contains connection pins for the various control signals coming from or going to the microprocessor. These are connected to the system power supply, external logic, and hardware to made up a complete computing system. External memory, math-processors, clock generators, and interrupt controllers are examples of such hardware and all must have an interface to the generic microprocessor. A diagram of a typical microprocessor pinout and the common grouping of these control signals is shown in Fig. 8.67 and microprocessor

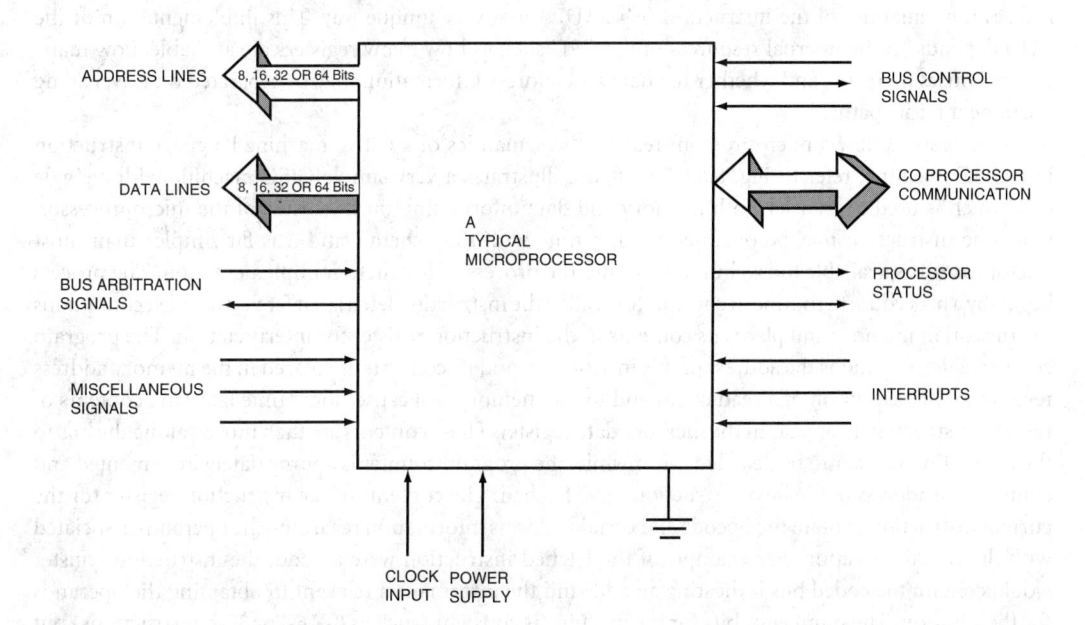

FIGURE 8.67 The pins connecting all microprocessors to external hardware are logically the same. They contain pins for addressing, data bus arbitration and control, coprocessor signals and interrupts, and those pins needed to connect to the power supply and clock generating chips.

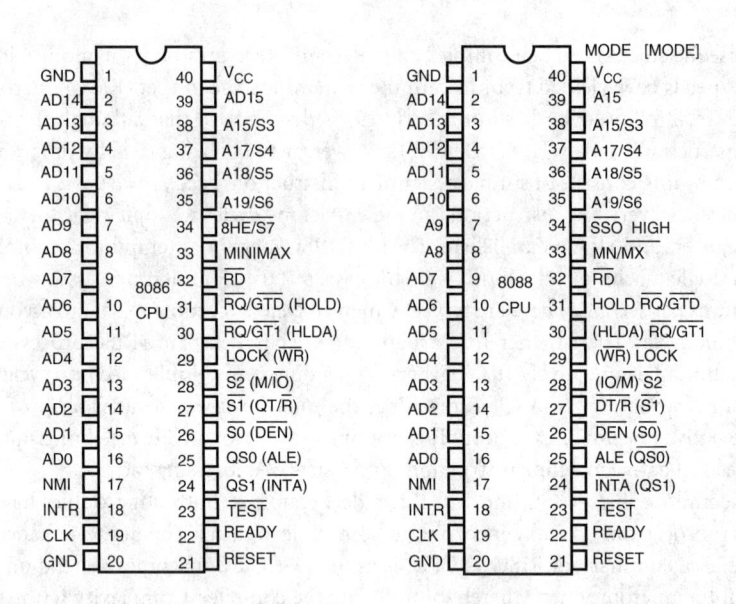

FIGURE 8.68 Microprocessor chip pinouts for the Intel 8086 (left) and the 8088 (right).

printout for Intel® 8086 and 8088 processors are shown Fig. 8.68. Externally, there are signals, which may be grouped as addressing, data, bus arbitration and control, coprocessor signals, interrupts, status, and miscellaneous connections. In some cases, multiple pins are used for connections such as the case for the address and data buses.

Just as the logical connections of the microprocessor are the same from family to family, so are the basic instructions. For instance, all microprocessors have an add instruction, a subtract instruction, and a memory read instruction. Although logically, these instructions all are the same, the particular microprocessor realizes the semantics of the instruction (e.g., ADD) in a very unique way. This implementation of the ADD depends on the internal resources of the CPU such as how many registers are available, how many internal buses there are, and whether the data and address information may be separated on travel along the same internal paths.

To understand how a microprogram realizes the semantics of a native machine language instruction in a CISC machine, refer to Fig. 8.69. The figure illustrates a very simple CISC machine with a single bus, which is used to transfer both memory and data information on the inside of the microprocessor. Only one instruction may be operated on at a time with this scheme, and it is far simpler than most microprocessors available today, but it illustrates the process of instruction implementation. The process begins by a procedure (a routine in control store called the instruction fetch) that fetches the next contiguous instruction in memory and places its contents in the instruction register for interpretation. The program counterregister contains the address of this instruction, and its contents are placed in the memory address register so that execution of a read command to the memory will cause, some time later, the contents of the next instruction to appear in the memory data register. These contents are then moved along the bus to the instruction register to be decoded. Meanwhile, the program counter is appropriately incremented and contains the address of the next instruction to be fetched. The contents of the instruction register for the current instruction contain the opcode and certain address information regarding the operands associated with the coded operation. For example, if the fetched instruction were an add, the instruction register would contain the coded bits indicating an add, and the information relevant to obtaining the operands for the addition. The number of bits for the machine is constant (such as 8-,16-, or 32-b instructions) but the number of bits dedicated to the opcode and operand fields may vary to accommodate instructions with single or multiple operands, different addressing modes, and so forth. The opcode is almost always a coded address referring to a position in the microprogram of the microprocessor. For the case of the add

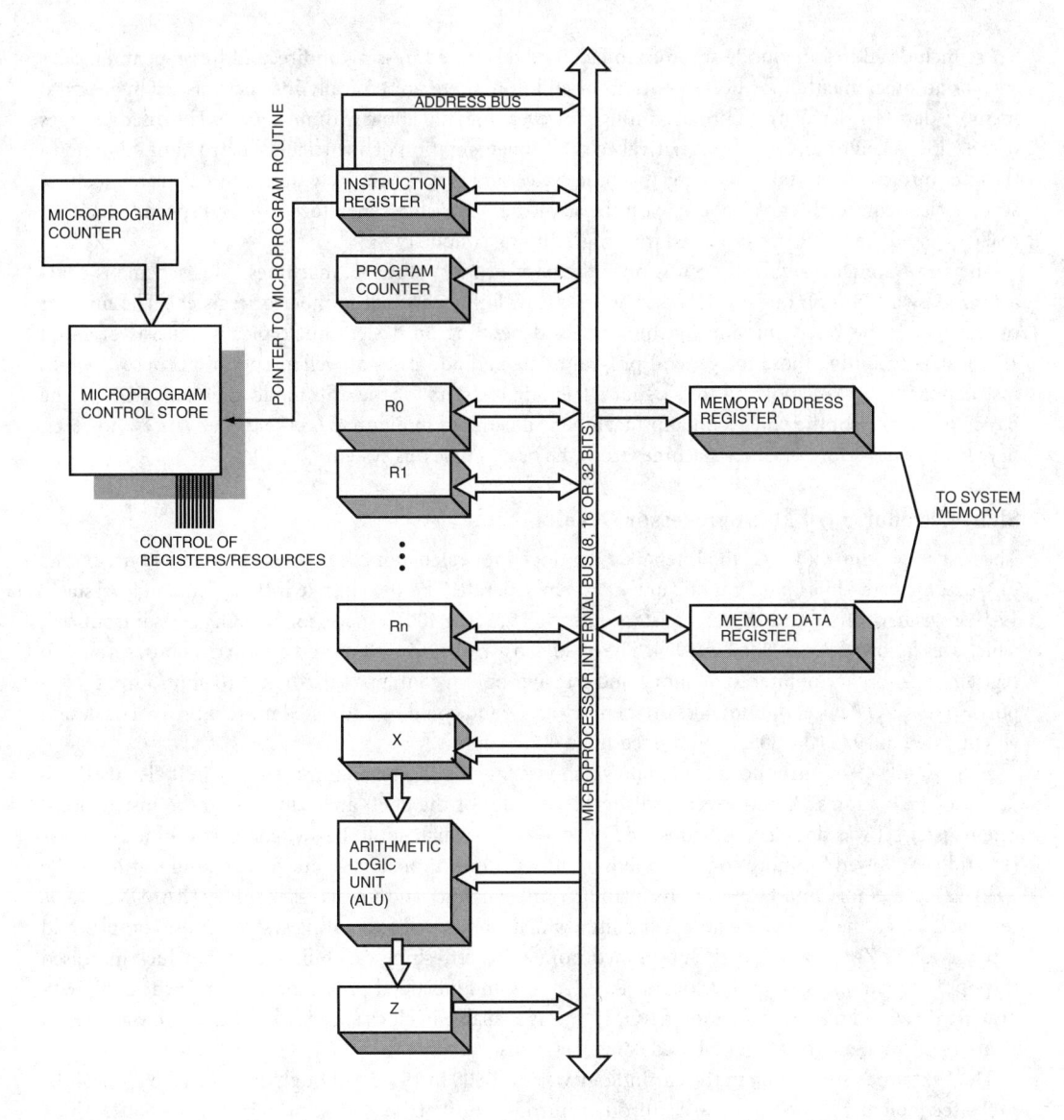

FIGURE 8.69 The basic single bus computer. This CISC machine is characterized by multiple registers linked by a bus, which carries both address and data information to and from the various registers (see text). Control lines gating the input and output of each register are governed by the information contained in the control store.

instruction, the opcode indicates a place in microprogram memory that contains the logic code needed to implement the add. Several machine cycles are necessary for the add instruction, including those steps needed to fetch the instruction itself. The number of these steps depends on the particular microprocessor. That is, all machines have an add instruction. The microprogram contains the realization of the add on the specific microprocessor's architecture with its individual set of hardware resources.

Specific steps (micro-operations), which direct the microprocessor to fetch the next instruction from memory, are included with every native instruction. The instruction fetch procedure is an integral part of every native instruction and represents the minimum number of machine cycles that a microprocessor must go through to implement even the simplest instruction. Even a halt instruction, therefore, requires several machine cycles to implement.

The advantages of the CISC scheme are clear. With a large number of instructions available, the programmer of a compiler or high-level language interpreter has a relatively easy task and a lot of tools at hand.

These include addressing mode schemes to facilitate relative addressing, indirect addressing, and modes such as autoincrementing or decrementing. In addition, there are typically instructions for memory to memory data transfers. Ease of programming, however, does not come without a price. The disadvantages of the CISC scheme are, therefore, relatively clear. Implementation of a particular instruction is bound to the microprogram, may take too many machine cycles, and may be relatively slow. The CISC architectural scheme also requires a lot of real estate on the semiconductor chip. Therefore, size and speed (which are really one and the same) are sacrificed for programming simplicity.

The von Neumann architecture was once the basis of most all CISC machines. This scheme is characterized by a common bus for data and address flow linking a small number of registers. The number of registers in the von Neumann machine varies depending on design, but typically consists of about 10 registers including those for general program data and addresses as well as special purposes such as instruction storage, memory addressing, and latching registers for the arithmetic logic unit (ALU). The basic architecture of the von-Neumann machine is illustrated in Figure 8.68. Recent advances, however, have led to a departure of CISC machines from the basic single bus system.

Short Chronology of Microprocessor Development

The first single chip CPU was the Intel 4004 developed for calculators. It processed data in 4 bits, and its instructions were 8 bits long. Program and data were separate. The 4004 had 46 instructions, a 4 level stack, 12-b program counter, and 16, 4-b registers. Later in 1972, the 4004's successor, the 8008, was introduced, which was followed by the 8080 in 1974. The 8080 had a 16-b address bus and an 8-b data bus, seven, 8-b registers, a 16-b stack pointer to memory, and a 16-b program counter. It also had 256 input/output (I/O) ports so that I/O devices did not take up memory space and could be addressed more directly. The design was updated in 1976 (the 8085) to only require a +5 V supply.

Zilog in July 1976 introduced the Z-80, which was intended to be an improved 8080. It also used 8-b data and 16-b address, could execute all of the opcodes of the 8080 and added 80 more instructions. The register set was doubled and consisted of two banks, which could be switched. Two index registers (IX and IY) allowed for more complex memory instructions. Probably the most successful feature of the Z-80 was its memory interface. Until its introduction, dynamic random-access memory (RAM) had to be refreshed with rather complex external circuitry, which made small computing systems more complex and expensive. The Z-80 was the first chip to incorporate this refreshing capability on-chip, which increased its popularity among system developers. The Z-8 was an embedded processor similar to the Z-80 with on-chip RAM and read-only memory (ROM). It was available in clock speeds to 20 MHz and was used in a variety of small microprocessor-based control systems.

The next processor of note in the chronology was the 6800 in 1975. Although introduced by Motorola, MOS Technologies gained popularity through introducing its 650x series, principally, the 6502, which was used in early desktop computers (Commodores, Apples, and Ataris). The 6502 had very few registers and was principally an 8-b processor with a 16-b address bus. The Apple II, one of the first computers introduced to the mainstream consumer market, incorporated the 6502. Subsequent improvements in the Apple line of micros were downward compatible with the 6502 processor. The extension to the 6502 came in 1977 when Motorola introduced the 6809 with two 8-b accumulators, which could combine mathematics operations in a single 16-b combination. It had 59 instructions. Members of the 6800 family live on in embedded microcontrollers such as the 68HC05 and 68HC11. These microcontrollers are still popular for small control systems. The 68HC11 was extended to 16-b and named the 68HC16. Radiation hardened versions of the 68HC11 have been used in communications satellites.

Advanced micro devices (AMD) introduced a 4-b bit-sliced microprocessor, the Am2901. Bit-sliced processors were modular in that they could be assembled to form larger word sizes. The Am2901 had a 4-b ALU; 16, 4-b registers; and the hardware to connect carry/borrow signals between adjacent modules. In 1979, AMD developed the first floating point coprocessor for microprocessors. The AMD 9511 arithmetic circuit was used in some CP/M, Z-80-based systems and some systems based on the S-100 bus.

Around 1976, competition was heating up for the 16-b microprocessor market. The Texas Instruments (TI) TM9900 was one of the first truly 16-b microprocessors and was designed as a single chip version of the TI 990 minicomputer. The TM9900 had two 16-b registers, good interrupt handling capability, and a decent instruction set for compiler developers. An embedded version (the TMS 9940) was also produced by TI. In 1976, the stage was being set for IBM's choice of a microprocessor for its IBM-PC line of personal computers. Several 16-b microprocessors around at the time had much more powerful features and more straightforward open memory architectures (such as Motorola's 68000). It is rumoured that IBM's own engineers wanted to use the 68000 but at the time IBM had already negotiated the rights to the Intel 8086. Apparently, the choice of the 8-b 8088 was a cost decision because the 8088 could have used the lower cost support chips associated with the 8085, whereas 68000 components were more expensive and not readily available.

Around 1976, Zilog introduced the Z-8000 (shortly after the 8086 by Intel). It was a 16-b microprocessor that had the capability of addressing up to 23-b of address data. The Z-8000 had 16, 16-b registers. The first 8 could be used as 16, 8-b registers, or all 16 could be used as 8, 32-b registers. This offered great flexibility in programming and for arithmetic calculations. Its instruction set included a 32-b multiply and divide instruction. It, also, like the Z-80 had memory refresh circuitry built into the chip. Probably most important, however, in the CPU chronology, is that the Z-8000 was the first microprocessor to incorporate two different modes of operation. One mode was strictly reserved for use by the operating system. The other mode was a general purpose user mode. The use of this scheme improved stability, in that the user could not crash the system as easily, and opened up the possibility of porting the chip towards multitasking, multiuser operating systems such as UNIX.

The Intel Family of Microprocessors

Intel was the first to develop a CPU on a chip in 1971 with the 4004 microprocessor. This chip, along with the 8008, was commissioned for calculators and terminal control. Intel did not expect the demand for these units to be high and several years later, developed a more general purpose microprocessor, the 8080, and a similar chip with more onboard hardware, the 8085. These were the industry's first truly general CPU's available to be integrated into microcomputing systems. The first 16-b chip, the 8086, was developed by Intel in 1978 and was the first industry entry into the realm of the 16-b processors. A companion mathematic coprocessor, the 8087, was developed for calculations requiring higher precision than the 16-b registers the 8086 would offer. Shortly after developing the 8086 and the 16-b address/8-b data version the 8088, IBM chose the 8088 for its IBM PC microcomputers. This decision was a tremendous boon to Intel and its microprocessor efforts. In some ways, it has also made the 80×86 family a victim of its early success, as all subsequent improvements and moves to larger data and address bus CPUs have had to contend with downward compatibility.

The 80186 and 80188 microprocessors were, in general, improvements to the 8086 and 8088 and incorporated more on-chip hardware for input and output support. They were never widely used, however, most likely due to the masking effect of the 8088's success in the IBM PC. The 80186 is architecturally identical to the 8086 but also contains a clock generator, a programmable controlled interrupt, three 16-b programmable timers, two programmable DMA controllers, a chip select unit, programmable control registers, a bus interface unit, and a 6-byte prefetch queue. The 80188 scheme is the same, with the exception that it only has an 8-b external data bus and a 4-byte prefetch queue.

None of the processors of the Intel family up to the beginning of the 1980s had the ability to address more than 1 megabyte of memory. The 80286, a 68-pin microprocessor, was developed to cater to the needs of systems and programs that were evolving, in a large part, due to the success of the 8088. The 80286 increased the available address space of the Intel microprocessor family to 16 megabytes of memory. Also, beginning with the 80286, the data and address lines external to the chip were not shared. In earlier chips, the address pins were multiplexed with the data lines. The internal architecture to accomplish this was a bit cumbersome, however, was kept so as to include downward compatibility with the earlier CPUs. Despite the scheme's unwieldiness, the 80286 was a huge success.

In a decade, the evolution of the microprocessor had advanced from its earliest beginnings (with a 4-b CPU) to a true 16-b microprocessor. Many everday computing tasks were off loaded from mainframes to desktop machines. In 1985 Intel developed a true 32-b processor on a chip, the 80386. It was downward compatible with all object codes back to the 8008 and continued to lock Intel into the rather awkward memory model developed in the 80286. At Motorola, the 68000 in some ways had a far more simple and straightforward open address space and was a serious contender for heavy computing applications being ported to desktop machines from their larger, mainframe cousins. It is for this reason that even today the 68000 is found systems requiring compatibility with UNIX operating systems. The 80386 was, nevertheless, highly successful. The 80386SX was a version of the 80386 developed with an identical package to the 80286 and meant as an upgrade for existing 80286 systems. It did so by upgrading the address but to 32 b but maintained the 16-b data bus of the 80286. A companion mathematic coprocessor (a floating point math unit (FPU)), the 80387, was developed for use with the 80386.

The 80386's success prompted other semiconductor companies (notably AMD and Cyrix) to piggyback on its success by offering clones of the processors and thus alternative sources for its end users and system developers. With Intel's addition of its 80486 in 1989, including full **pipelines**, on-chip **caching** and an integrated rather than separate floating point processor, competition for the chips, popularity was fierce. In late 1993, Intel could no longer protect the next subsequent name in the series (the 80586). It trademarked the Pentium name to its 80586 processor. Because of its popularity, the 80×86 line is the most widely cloned.

The Motorola Family of Microprocessors

Alongside of Intel's development of the 8080, Motorola developed the 6800, 8-b microprocessor. In the early 1970s, the 6800 was used in many embedded industrial control systems. It was not until 1979 however, that Motorola introduced its 16-b entry into the industry, the 68000. The 68000 was designed to be far more advanced than the 8086 microprocessor by Intel in several ways. All internal registers were 32-b wide, and it had the benefit of being able to address all 16 megabytes of external memory without the segmentation schemes utilized in the Intel series. This nonsegmented approach meant that the 68000 had no segment registers, and each of its instructions could address the complete complement of external memory.

The 68000 was the first 16-b microprocessor to incorporate 32-b internal registers. This asset allowed its selection by designers who set out to port sophisticated operating systems to desktop computers. In some ways, the 68000 was ahead of its time. If IBM had chosen the 68000 series as the core chip for its personal computers, the present state-of-the-art of the desktop machine would be radically different. The 68000 was chosen by Apple for its MacIntosh computers. Other computer manufacturers, including Amiga and Atari chose it for its flexibility and its large internal registers.

In 1982, Motorola marketed another chip, the 68008, which was a stripped down version of the 68000 for low-end, low-cost products. The 68008 only had the capability to address 4 megabytes of memory, and its data bus was only 8-b wide. It was never very popular and certainly did not compete well with the 8088 by Intel (which was chosen for the IBM PC computers).

Advanced operating systems were ideal for the 68000 except that the chip had no capability for supporting virtual memory. For this reason, Motorola developed the 68010, which had the capability to continue an instruction after it had been suspended by a bus error. The 68012 was identical to the 68000 except that it had the capability to address 2 gigabytes of memory with its 30 address bus pins.

One of the most successful microprocessors introduced by Motorola was the 68020. It was introduced in 1984 and was the industry's first true 32-b microprocessor. Along with the 32-b registers standard to the 68000 series, it has the capability of addressing 4 gigabytes of memory and a true 32-b wide data bus. It is still widely used.

The 68020 contains an internal 256 byte cache memory. This is an instruction cache, holding up to 64 instructions of the long-word type. Direct access to this cache is not allowed. It serves only as an advance *prefetch queue* to enable the 68020 to execute tight loops of instructions with any further instruction fetches. Since an instruction fetch takes time to process, the presence of the 256-byte instruction cache in the 68020 is a significant speed enhancement.

The cache treatment was expanded in the 68030 to include a 256-byte data cache. In addition, the 68030 includes an onboard paged memory management unit (PMMU) to control access to virtual memory. This is the primary difference in the 68030 and the 68020. The PMMU is available as an extra chip (the 68851) for the 68020 but included on the same chip with the 68030. The 68030 also includes an improved bus interface scheme. Externally, the connections of the 68020 and the 68030 are very nearly the same. The 68030 is available in two speeds, the MC68030RC16 at 16 MHz and the MC68030RC20 with a 20-MHz clock.

The 68000 featured a supervisor and user mode. It was designed for expansion and could fetch the next instruction during an instruction's execution. (This represents 2-stage pipelining.) The 68040 had 6-stages of pipelining. The advances in the 680×0 series continued toward the 68060 in late 1994, which was a *superscalar* microprocessor similar to the Intel Pentium. It truly represents a merging of the two CISC and RISC philosophies of architecture. The 68060 10-stage pipeline translates 680×0 instructions into a decorded RISC-like form and uses a resource renaming scheme to reorder the execution of the instructions. The 68060 includes power saving features that can be shutdown and operates off of a 3.3-V power supply (again similar to the Intel Pentium processors).

RISC Processor Development

The major development efforts for RISC processors were led by the University of California at Berkeley and the Stanford University designs. Sun Microsystems developed the Berkeley version of the RISC processor [scalable processor architecture (SPARC)] for their high-speed workstations. This, however, was not the first RISC processor. It was preceeded by the MIPS R2000 (based on the Stanford University design), the Hewlett Packard PA-RISC CPU, and the AMD 29000.

The AMD 29000 is a RISC design, which follows the lead of the Berkeley scheme. It has a large set of registers spilt into local and global sets. The 64 global registers reduced instruction set processors were developed following the recognition that many of the CISC complex instructions were not being used.

Defining Terms

Cache: Small amount of *fast* memory, physically close to CPU, used as storage of a block of data needed immediately by the processor. Caches exist in a memory hierarchy. There is a small but very fast L1 (level one) cache; if that misses, then the access is passed on to the bigger but slower L2 (level two) cache, and if that misses, the access goes to the main memory (or L3 cache, if it exists).

Pipelining: A microarchitecture technique that divides the execution of an instruction into sequential steps. Pipelined CPUs have multiple instructions executing at the same time but at different stages in the machine. Or, the act of sending out an address before the data is actually needed.

Superscalar: Capable of executing multiple instructions in a given clock cycle. For example, the Pentium processor has two execution pipes (U and V) so it is superscalar level 2. The Pentium Pro processor can dispatch and retire three instructions per clock so it is superscalar level 3.

References

The Alpha 21164A: Continued performance leadership. 1995. *Microprocessor Forum*.

Internal architecture of the Alpha 21164 microprocessor. 1995. *CompCon 95*.

A 300 MHz quad-issue CMOS RISC microprocessor (21164). 1995. In *ISSC 95*, pp. 182–183.

A 200 MHz 64 b dual-issue CMOS microprocessor (21064). 1992. In *ISSC 92*, pp. 106–107.

Hobbit: A high performance, low-power microprocessor. 1993. *CompCon 93*, pp. 88–95.

MIPS R10000 superscalar microprocessor. 1995. *Hot Chips VII*.

The impact of dynamic execution techniques on the data path design of the P6 processor. 1995. *Hot Chips VII*.

A 0.6 μm BiCMOS processor with dynamic execution (P6). 1995. In *ISSC 95*, pp. 176–177.

A 3.3 v 0.6 μm BiCMOS superscalar processor (Pentium). 1994. In *ISSC 94*, pp. 202–203.

An overview of the Intel Pentium processor. 1993. In *CompCon 93*, pp. 60–62.

Superscalar architecture of the P5-×86 next generation processor. 1992. *Hot Chips IV*.

A 93 MHz x86 microprocessor with on-chip L2 cache controller (N586). 1995. In *ISSC 95*, pp. 172–173.

The AMD K5 processor. 1995. *Hot Chips VII*.

The PowerPC620 microprocessor: A high performance superscalar RISC microprocessor. *CompCon 95.*

A new powerPC microprocessor for the low power computing marker (602). 1995. *CompCon 95.*

133 MHz 64 b four-issue CMOS microprocessor (620). In *ISSC 95.* 1995. pp. 174–175.

The powerPC 604 RISC microprocessor. 1994. *IEEE Micro.* (Oct.).

The powerPC user instruction set architecture. 1994. *IEEE Micro.* (Oct.).

PowerPC 604. 1994. *Hot Chips VI.*

The powerPC 603 microprocessor: A low power design for portable applications. 1994. In *CompCon 94,*
      pp. 307–315.

A 3.0 W 75SPECint92 85SPECfp92 superscalar RISC microprocessor (603). 1994. In *ISSC 94,* pp. 212–214.

601 powerPC microprocessor. 1993. *Hot Chips V.*

The powerPC 601 microprocessor. 1993. In *Compcon 93,* pp. 109–116.

      The great dark cloud falls: IBM's choice. In *Great Microprocessors of the Past and Present* (on-line)
      Sec. 3. http:// www.cpu info.berkeley. edu.

Further Information

In the fast changing world of the microprocessor, obsolecence is a fact of life. In fact, the marketing data shows that, with each introduction of a new generation, the time spent ramping up a new technology and ramping down the old gets shorter. There is more need for accuracy in design and development to avoid errors such as Intel experienced with their Pentium Processor with their floating point mathematic computations in 1994. For the small system developer or user of microprocessors, there is an important need to keep abreast of newer, more enhanced chips with better capabilities. Luckily, there is an excellent way to keep up to date.

The CPU Information Center at the University of California, Berkeley maintains an excellent, up to date compilation of microprocessors and microcontrollers, their architectures, and specifications on the World Wide Web (WWW). The site includes chip size and pinout information, tabular comparisons of microprocessor performance and architecture (such as that shown in the tables in this chapter), and references. It is updated regularly and serves as an excellent source for the small systems developer. Some of the information contained in this chapter was obtained and used with permission from that Web site. See http://www. infopad. berkeley.edu for an excellent, up to date summary.

8.6   D/A and A/D Converters

Susan A. Garrod

8.6.1   Introduction

Digital-to-analog (D/A) conversion is the process of converting digital codes into a continuous range of analog signals. *Analog-to-digital (A/D) conversion* is the complementary process of converting a continuous range of analog signals into digital codes. Such conversion processes are necessary to interface real-world systems, which typically monitor continuously varying analog signals, with digital systems that process, store, interpret, and manipulate the analog values.

D/A and A/D applications have evolved from predominately military-driven applications to consumer-oriented applications. Up to the mid-1980s, the military applications determined the design of many D/A and A/D devices. The military applications required very high performance coupled with hermetic packaging, radiation hardening, shock and vibration testing, and military specification and record keeping. Cost was of little concern, and "low power" applications required approximately 2.8 W. The major applications up the mid-1980s included military radar warning and guidance systems, digital oscilloscopes, medical imaging, infrared systems, and professional video.

The applications requiring D/A and A/D circuits today have different performance criteria from those of earlier years. In particular, low power and high speed applications are driving the development of D/A and A/D circuits, as the devices are used extensively in battery-operated consumer products.

8.6.2   D/A and A/D Circuits

D/A and A/D conversion circuits are available as integrated circuits (ICs) from many manufacturers. A huge array of ICs exists, consisting of not only the D/A or A/D conversion circuits, but also closely related circuits such as sample-and-hold amplifiers, analog multiplexers, voltage-to-frequency and frequency-to-voltage converters, voltage references, calibrators, operation amplifiers, isolation amplifiers, instrumentation amplifiers, active filters, DC-to-DC converters, analog interfaces to digital signal processing systems, and data acquisition subsystems. Data books from the IC manufacturers contain an enormous amount of information about these devices and their applications to assist the design engineer.

The ICs discussed in this chapter will be strictly the D/A and A/D conversion circuits. The ICs usually perform either D/A or A/D conversion. There are serial interface ICs, however, typically for digital signal processing applications, that perform both A/D and D/A processes.

D/A and A/D Converter Performance Criteria

The major factors that determine the quality of performance of D/A and A/D converters are *resolution, sampling rate, speed,* and *linearity*.

The *resolution* of a D/A circuit is the smallest change in the output analog signal. In an A/D system, the resolution is the smallest change in voltage that can be detected by the system and can produce a change in the digital code. The resolution determines the total number of digital codes, or *quantization levels*, that is recognized or produced by the circuit.

The *resolution* of a D/A or A/D IC is usually specified in terms of the bits in the digital code or in terms of the least significant bit (LSB) of the system. An n-bit code allows for 2^n quantization levels, or $2^n - 1$ steps between quantization levels. As the number of bits increases, the step size between quantization levels decreases, therefore increasing the accuracy of the system when a conversion is made between an analog and digital signal. The system resolution can be specified also as the voltage step size between quantization levels. For A/D circuits, the resolution is the smallest input voltage that is detected by the system.

The *speed* of a D/A or A/D converter is determined by the time it takes to perform the conversion process. For D/A converters, the speed is specified as the *settling time*. For A/D converters, the speed is specified as the *conversion time*. The settling time for D/A converters varies with supply voltage and transition in the digital code; thus, it is specified in the data sheet with the appropriate conditions stated.

A/D converters have a maximum *sampling rate* that limits the speed at which they can perform continuous conversions. The sampling rate is the number of times per second that the analog signal can be sampled and converted into a digital code. For proper A/D conversion, the minimum sampling rate must be at least two times the highest frequency of the analog signal being sampled to satisfy the Nyquist sampling criterion. The conversion speed and other timing factors must be taken into consideration to determine the maximum sampling rate of an A/D converter. **Nyquist A/D converters** use a sampling rate that is slightly more than twice the highest frequency in the analog signal. **Oversampling A/D converters** use sampling rates of N times rate, where N typically ranges from 2 to 64.

Both D/A and A/D converters require a voltage reference in order to achieve absolute conversion accuracy. Some conversion ICs have internal voltage references, whereas others accept external voltage references. For high-performance systems, an external precision reference is needed to ensure long-term stability, load regulation, and control over temperature fluctuations. External precision voltage reference ICs can be found in manufacturer's data books.

Measurement accuracy is specified by the converter's *linearity*. *Integral linearity* is a measure of linearity over the entire conversion range. It is often defined as the deviation from a straight line drawn between the endpoints and through zero (or the offset value) of the conversion range. Integral linearity is also referred to as *relative accuracy*. The *offset* value is the reference level required to establish the zero or midpoint of the conversion range. *Differential linearity* is the linearity between code transitions. Differential linearity is a measure of the *monotonicity* of the converter. A converter is said to be monotonic if increasing input values result in increasing output values.

The accuracy and linearity values of a converter are specified in the data sheet in units of the LSB of the code. The linearity can vary with temperature, and so the values are often specified at $+25°C$ as well as over the entire temperature range of the device.

D/A Conversion Processes

Digital codes are typically converted to analog voltages by assigning a voltage weight to each bit in the digital code and then summing the voltage weights of the entire code. A general D/A converter consists of a network of precision resistors, input switches, and level shifters to activate the switches to convert a digital code to an analog current or voltage. D/A ICs that produce an analog current output usually have a faster settling time and better linearity than those that produce a voltage output. When the output current is available, the designer can convert this to a voltage through the selection of an appropriate output amplifier to achieve the necessary response speed for the given application.

D/A converters commonly have a fixed or variable reference level. The reference level determines the switching threshold of the precision switches that form a controlled impedance network, which in turn controls the value of the output signal. **Fixed reference D/A converters** produce an output signal that is proportional to the digital input. **Multiplying D/A** converters produce an output signal that is proportional to the product of a varying reference level times a digital code.

D/A converters can produce bipolar, positive, or negative polarity signals. A four-quadrant multiplying D/A converter allows both the reference signal and the value of the binary code to have a positive or negative polarity. The four-quadrant multiplying D/A converter produces bipolar output signals.

D/A Converter ICs

Most D/A converters are designed for general-purpose control applications. Some D/A converters, however, are designed for special applications, such as video or graphic outputs, high-definition video displays, ultra high-speed signal processing, digital video tape recording, digital attenuators, or high-speed function generators.

D/A converter ICs often include special features that enable them to be interfaced easily to microprocessors or other systems. Microprocessor control inputs, input latches, buffers, input registers, and compatibility to standard logic families are features that are readily available in D/A ICs. In addition, the ICs usually have laser-trimmed precision resistors to eliminate the need for user trimming to achieve full-scale performance.

A/D Conversion Processes

Analog signals can be converted to digital codes by many methods, including integration, **successive approximation**, parallel (flash) conversion, **delta modulation, pulse code modulation**, and **sigma–delta conversion**. Two common A/D conversion processes are successive approximation A/D conversion and parallel or **flash A/D** conversion. Very high-resolution digital audio or video systems require specialized A/D techniques that often incorporate one of these general techniques as well as specialized A/D conversion processes. Examples of specialized A/D conversion techniques are pulse code modulation (PCM), and sigma–delta conversion. PCM is a common voice encoding scheme used not only by the audio industry in digital audio recordings but also by the telecommunications industry for voice encoding and multiplexing. Sigma–delta conversion is an oversampling A/D conversion where signals are sampled at very high frequencies. It has very high resolution and low distortion.

Successive approximation A/D conversion is a technique that is commonly used in medium- to high-speed data acquisition applications. It is one of the fastest A/D conversion techniques that requires a minimum amount of circuitry. The conversion times for successive approximation A/D conversion typically range from 10 to 300 μs for 8-b systems.

The successive approximation A/D converter can approximate the analog signal to form an n-bit digital code in n steps. The successive approximation register (SAR) individually compares an analog input voltage to the midpoint of one of n ranges to determine the value of 1 b. This process is repeated a total of n times,

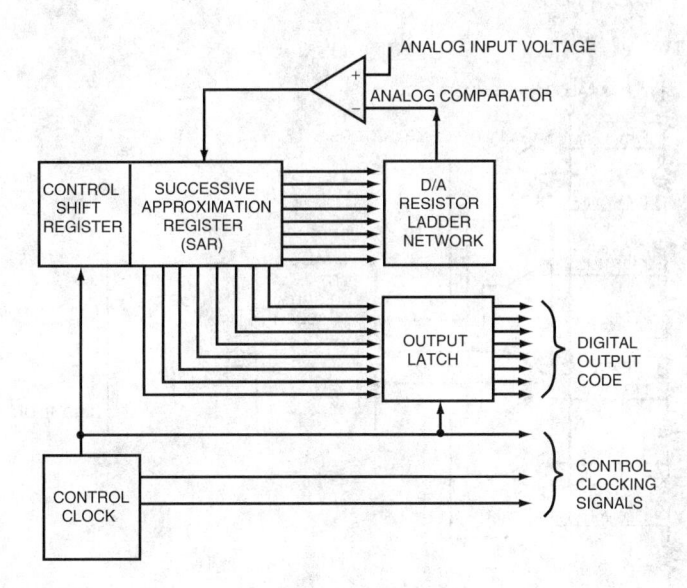

FIGURE 8.70 Successive approximation A/D converter block diagram. (*Source:* Garrod, S. and Borns, R. 1991. *Digital Logic: Analysis, Application, and Design*, p. 919. Copyright ©1991 by Saunders College Publishing, Philadelphia, PA. Reprinted by permission of the publisher.)

using n ranges, to determine the n bits in the code. The comparison is accomplished as follows. The SAR determines if the analog input is above or below the midpoint and sets the bit of the digital code accordingly. The SAR assigns the bits beginning with the most significant bit. The bit is set to a 1 if the analog input is greater than the midpoint voltage, or it is set to a 0 if it is less than the midpoint voltage. The SAR then moves to the next bit and sets it to a 1 or a 0 based on the results of comparing the analog input with the midpoint of the next allowed range. Because the SAR must perform one approximation for each bit in the digital code, an n-bit code requires n approximations.

A successive approximation A/D converter consists of four functional blocks, as shown in Fig. 8.70: the SAR, the analog comparator, a D/A converter, and a clock.

Parallel or flash A/D conversion is used in high-speed applications such as video signal processing, medical imaging, and radar detection systems. A flash A/D converter simultaneously compares the input analog voltage to $2^n - 1$ threshold voltages to produce an n-bit digital code representing the analog voltage. Typical flash A/D converters with 8-b resolution operate at 20–100 MHz.

The functional blocks of a flash A/D converter are shown in Fig. 8.71. The circuitry consists of a precision resistor ladder network, $2^n - 1$ analog comparators, and a digital priority encoder. The resistor network establishes threshold voltages for each allowed quantization level. The analog comparators indicate whether or not the input analog voltage is above or below the threshold at each level. The output of the analog comparators is input to the digital priority encoder. The priority encoder produces the final digital output code that is stored in an output latch.

An 8-b flash A/D converter requires 255 comparators. The cost of high-resolution A/D comparators escalates as the circuit complexity increases and as the number of analog converters rises by $2^n - 1$. As a low-cost alternative, some manufacturers produce modified flash A/D converters that perform the A/D conversion in two steps to reduce the amount of circuitry required. These modified flash A/D converters are also referred to as *half-flash* A/D converters, since they perform only half of the conversion simultaneously.

A/D Converter ICs

A/D converter ICs can be classified as general-purpose, high-speed, flash, and sampling A/D converters. The *general-purpose A/D converters* are typically low speed and low cost, with conversion times ranging

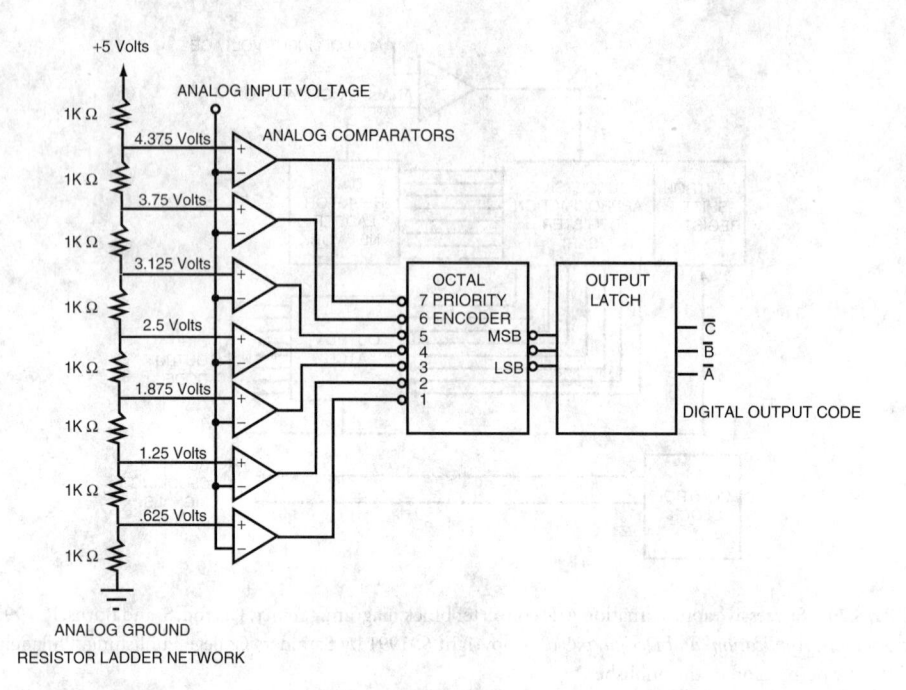

FIGURE 8.71   Flash A/D converter block diagram. (*Source:* Garrod, S. and Borns, R. *Digital Logic: Analysis, Application, and Design*, p. 928. Copyright ©1991 by Saunders College Publishing, Philadelphia, PA. Reprinted by permission of the publisher.)

from 2 μs to 33 ms. A/D conversion techniques used by these devices typically include successive approximation, tracking, and integrating. The general-purpose A/D converters often have control signals for simplified microprocessor interfacing. These ICs are appropriate for many process control, industrial, and instrumentation applications, as well as for environmental monitoring such as seismology, oceanography, meteorology, and pollution monitoring.

High-speed A/D converters have conversion times typically ranging from 400 ns to 3 μs. The higher speed performance of these devices is achieved by using the successive approximation technique, modified flash techniques, and statistically derived A/D conversion techniques. Applications appropriate for these A/D ICs include fast Fourier transform (FFT) analysis, radar digitization, medical instrumentation, and multiplexed data acquisition. Some ICs have been manufactured with an extremely high degree of linearity, to be appropriate for specialized applications in digital spectrum analysis, vibration analysis, geological research, sonar digitizing, and medical imaging.

Flash A/D converters have conversion times ranging typically from 10 to 50 ns. Flash A/D conversion techniques enable these ICs to be used in many specialized high-speed data acquisition applications such as TV video digitizing (encoding), radar analysis, transient analysis, high-speed digital oscilloscopes, medical ultrasound imaging, high-energy physics, and robotic vision applications.

Sampling A/D converters have a sample-and-hold amplifier circuit built into the IC. This eliminates the need for an external sample-and-hold circuit. The throughput of these A/D converter ICs ranges typically from 35 kHz to 100 MHz. The speed of the system is dependent on the A/D technique used by the sampling A/D converter.

A/D converter ICs produce digital codes in a serial or parallel format, and some ICs offer the designer both formats. The digital outputs are compatible with standard logic families to facilitate interfacing to other digital systems. In addition, some A/D converter ICs have a built-in analog multiplexer and therefore can accept more than one analog input signal.

Pulse code modulation (PCM) ICs are high-precision A/D converters. The PCM IC is often refered to as a PCM *codec* with both encoder and decoder functions. The encoder portion of the codec performs the

A/D conversion, and the decoder portion of the codec performs the D/A conversion. The digital code is usually formatted as a serial data stream for ease of interfacing to digital transmission and multiplexing systems.

PCM is a technique where an analog signal is sampled, quantized, and then encoded as a digital word. The PCM IC can include successive approximation techniques or other techniques to accomplish the PCM encoding. In addition, the PCM codec may employ nonlinear data compression techniques, such as companding, if it is necessary to minimize the number of bits in the output digital code. Companding is a logarithmic technique used to compress a code to fewer bits before transmission. The inverse logarithmic function is then used to expand the code to its original number of bits before converting it to the analog signal. Companding is typically used in telecommunications transmission systems to minimize data transmission rates without degrading the resolution of low-amplitude signals. Two standardized companding techniques are used extensively: A-law and μ-law. The A-law companding is used in Europe, whereas the μ-law is used predominantly in the United States and Japan. Linear PCM conversion is used in high-fidelity audio systems to preserve the integrity of the audio signal throughout the entire analog range.

Digital signal processing (DSP) techniques provide another type of A/D conversion ICs. Specialized A/D conversion such as *adaptive differential pulse code modulation* (ADPCM), sigma–delta modulation, *speech subband encoding, adaptive predictive speech encoding*, and *speech recognition* can be accomplished through the use of DSP systems. Some DSP systems require analog front ends that employ traditional PCM codec ICs or DSP interface ICs. These ICs can interface to a digital signal processor for advanced A/D applications. Some manufacturers have incorporated DSP techniques on board the single-chip A/D IC, as in the case of the DSP56ACD16 sigma–delta modulation IC by Motorola.

Integrating A/D converters are used for conversions that must take place over a long period of time, such as digital voltmeter applications or sensor applications such as thermocouples. The integrating A/D converter produces a digital code that represents the average of the signal over time. Noise is reduced by means of the signal averaging, or integration. Dual-slope integration is accomplished by a counter that advances while an input voltage charges a capacitor in a specified time interval, T. This is compared to another count sequence that advances while a reference voltage is discharging across the same capacitor in a time interval, *delta t*. The ratio of the charging count value to the discharging count value is proportional to the ratio of the input voltage to the reference voltage. Hence, the integrating converter provides a digital code that is a measure of the input voltage averaged over time. The conversion accuracy is independent of the capacitor and the clock frequency since they affect both the charging and discharging operations. The charging period, T, is selected to be the period of the fundamental frequency to be rejected. The maximum conversion rate is slightly less than $1/(2\,T)$ conversions per second. While this limits the conversion rate to be too slow for high-speed data acquisition applications, it is appropriate for long-duration applications of slowly varying input signals.

Grounding and Bypassing on D/A and A/D ICs

D/A and A/D converter ICs require correct grounding and capacitive bypassing in order to operate according to performance specifications. The digital signals can severely impair analog signals. To combat the electromagnetic interference induced by the digital signals, the analog and digital grounds should be kept separate and should have only one common point on the circuit board. If possible, this common point should be the connection to the power supply.

Bypass capacitors are required at the power connections to the IC, the reference signal inputs, and the analog inputs to minimize noise that is induced by the digital signals. Each manufacturer specifies the recommended bypass capacitor locations and values in the data sheet. The manufacturers' recommendations should be followed to ensure proper performance.

Selection Criteria for D/A and A/D Converter ICs

Hundreds of D/A and A/D converter ICs are available, with prices ranging from a few dollars to several hundred dollars each. The selection of the appropriate type of converter is based on the application requirements of the system, the performance requirements, and cost. The following issues should be considered in order to select the appropriate converter.

1. What are the input and output requirements of the system? Specify all signal current and voltage ranges, logic levels, input and output impedances, digital codes, data rates, and data formats.
2. What level of accuracy is required? Determine the resolution needed throughout the analog voltage range, the dynamic response, the degree of linearity, and the number of bits encoding.
3. What speed is required? Determine the maximum analog input frequency for sampling in an A/D system, the number of bits for encoding each analog signal, and the rate of change of input digital codes in a D/A system.
4. What is the operating environment of the system? Obtain information on the temperature range and power supply to select a converter that is accurate over the operating range.

Final selection of D/A and A/D converter ICs should be made by consulting manufacturers to obtain their technical specifications of the devices.

Defining Terms

Companding: A process designed to minimize the transmission bit rate of a signal by compressing it prior to transmission and expanding it upon reception. It is a rudimentary "data compression" technique that requires minimal processing.

Delta modulation: An A/D conversion process where the digital output code represents the change, or slope, of the analog input signal, rather than the absolute value of the analog input signal. A 1 indicates a rising slope of the input signal. A 0 indicates a falling slope of the input signal. The sampling rate is dependent on the derivative of the signal, since rapidly changing signal would require a rapid sampling rate for acceptable performance.

Fixed reference D/A converter: The analog output is proportional to a fixed (nonvarying) reference signal.

Flash A/D: The fastest A/D conversion process available to date, also referred to as parallel A/D conversion. The analog signal is simultaneously evaluated by $2^n - 1$ comparators to produce an n-bit digital code in one step. Because of the large number of comparators required, the circuitry for flash A/D converters can be very expensive. This technique is commonly used in digital video systems.

Integrating A/D: The analog input signal is integrated over time to produce a digital signal that represents the area under the curve, or the integral.

Multiplying D/A: A D/A conversion process where the output signal is the product of a digital code multiplied times an analog input reference signal. This allows the analog reference signal to be scaled by a digital code.

Nyquist A/D converters: A/D converters that sample analog signals that have a maximum frequency that is less than the Nyquist frequency. The Nyquist frequency is defined as one-half of the sampling frequency. If a signal has frequencies above the Nyquist frequency, a distortion called *aliasing* occurs. To prevent aliasing, an *antialiasing filter* with a flat passband and very sharp rolloff is required.

Oversampling converters: A/D converters that sample frequencies at a rate much higher than the Nyquist frequency. Typical oversampling rates are 32 and 64 times the sampling rate that would be required with the Nyquist converters.

Pulse code modulation (PCM): An A/D conversion process requiring three steps: the analog signal is sampled, quantized, and encoded into a fixed length digital code. This technique is used in many digital voice and audio systems. The reverse process reconstructs an analog signal from the PCM code. The operation is very similar to other A/D techniques, but specific PCM circuits are optimized for the particular voice or audio application.

Sigma–delta A/D conversion: An *oversampling* A/D conversion process where the analog signal is sampled at rates much higher (typically 64 times) than the sampling rates that would be required with a Nyquist converter. Sigma–delta modulators integrate the analog signal before performing the delta modulation. The integral of the analog signal is encoded rather than the change in the analog signal, as is the case for traditional delta modulation. A digital sample rate reduction filter (also called a digital decimation filter) is used to provide an output sampling rate at twice the Nyquist frequency of the signal. The overall result of oversampling and digital sample rate reduction is greater resolution and less distortion compared to a Nyquist converter process.

Successive approximation: An A/D conversion process that systematically evaluates the analog signal in n steps to produce an n-bit digital code. The analog signal is successively compared to determine the digital code, beginning with the determination of the most significant bit of the code.

References

Analog Devices. 1989. *Analog Devices Data Conversion Products Data Book*. Analog Devices, Inc., Norwood, MA.

Burr–Brown. 1989. *Burr-Brown Integrated Circuits Data Book*. Burr-Brown, Tucson, AZ.

DATEL. 1988. *DATEL Data Conversion Catalog*. DATEL, Inc., Mansfield, MA.

Drachler, W. and Bill, M. 1995. New High-Speed, Low-Power Data-Acquisition ICs. *Analog Dialogue* 29(2):3–6. Analog Devices, Inc., Norwood, MA.

Garrod, S. and Borns, R. 1991. *Digital Logic: Analysis, Application and Design,* Chap. 16. Saunders College Publishing, Philadelphia, PA.

Jacob, J.M. 1989. *Industrial Control Electronics*, Chap. 6. Prentice-Hall, Englewood Cliffs, NJ.

Keiser, B. and Strange, E. 1995. *Digital Telephony and Network Integration*, 2nd ed. Van Nostrand Reinhold, New York.

Motorola. 1989. *Motorola Telecommunications Data Book*. Motorola, Inc., Phoenix, AZ.

National Semiconductor. 1989. *National Semiconductor Data Acquisition Linear Devices Data Book*. National Semiconductor Corp., Santa Clara, CA.

Park, S. 1990. *Principles of Sigma-Delta Modulation for Analog-to-Digital Converters*. Motorola, Inc., Phoenix, AZ.

Texas Instruments. 1986. *Texas Instruments Digital Signal Processing Applications with the TMS320 Family*. Texas Instruments, Dallas, TX.

Texas Instruments. 1989. *Texas Instruments Linear Circuits Data Acquisition and Conversion Data Book*. Texas Instruments, Dallas, TX.

Further Information

Analog Devices, Inc. has edited or published several technical handbooks to assist design engineers with their data acquisition system requirements. These references should be consulted for extensive technical information and depth. The publications include *Analog-Digital Conversion Handbook*, by the engineering staff of Analog Devices, published by Prentice-Hall, Englewood Cliffs, NJ, 1986; *Nonlinear Circuits Handbook, Transducer Interfacing Handbook,* and *Synchro and Resolver Conversion*, all published by Analog Devices Inc., Norwood, MA.

Engineering trade journals and design publications often have articles describing recent A/D and D/A circuits and their applications. These publications include *EDN Magazine, EE Times,* and *IEEE Spectrum*. Research-related topics are covered in *IEEE Transactions on Circuits and Systems* and also *IEEE Transactions on Instrumentation and Measurement.*

8.7 Application-Specific Integrated Circuits

Constantine N. Anagnostopoulos and Paul P.K. Lee

8.7.1 Introduction

Application specific integrated circuits (**ASICs**), also called custom ICs, are chips specially designed to (1) perform a function that cannot be done using standard components, (2) improve the performance of a circuit, or (3) reduce the volume, weight, and power requirement and increase the reliability of a given system by integrating a large number of functions on a single chip or a small number of chips.

ASICs can be classified into the following three categories: (1) full custom, (2) semicustom, and (3) programmable logic devices (PLDs).

The first step of the process toward realizing a custom IC chip is to define the function the chip must perform. This is accomplished during system partitioning, at which time the system engineers and IC designers make some initial decisions as to which circuit functions will be implemented using standard, off the shelf, components and which will require custom ICs. After several iterations the functions that each custom chip has to perform are determined.

There usually are many different ways by which a given function may be implemented. For example, the function may be performed in either the analog or the digital domain. If a digital approach is selected, different execution strategies may be chosen. For example, a delay function may be implemented either by a shift register or by a random access memory. Thus, in the second step of the process, the system engineers and IC designers decide how the functions each of the chips has to perform will be executed. How a function is executed by a given chip constitutes the behavioral description of that chip.

Next, a design approach needs to be developed for each of the custom chips. Depending on the implementation method selected, either a full custom, a semicustom, or a user programmable device approach is chosen.

One important consideration in the design choice is cost and turn around time (Fey and Paraskevopoulos, 1985). Typically, a full custom approach takes the longest to design, and the cost per chip is very high unless the chip volume is also high, normally more than a few hundred thousands chips per year. The shortest design approach is using programmable devices but the cost per chip is the highest. This approach is best for prototype and limited production systems.

8.7.2   Full Custom ASICs

In a typical full custom IC design, every device and circuit element on the chip is designed for that particular chip. Of course, common sense dictates that device and circuit elements proven to work well in previous designs are reused in subsequent ones whenever possible.

A full custom approach is selected to minimize the chip size or to implement a function that is not available or would not be optimum with semicustom or standard ICs. Minimizing chip size increases the fabrication yield and the number of chips per wafer. Both of these factors tend to reduce the cost per chip.

Fabrication yield is given by

$$Y = [(1 - e^{-AD})/AD]^2$$

where A is the chip or die area and D is the average number of defects per square centimeter per wafer. The number of die per wafer is given by

$$N = [\pi(R - A^{1/2})^2]/A$$

where R is the wafer radius and A is, again, the area of the die.

Additional costs incurred involve testing of the die while still on the wafer and packaging a number of the good ones and testing them again. Then good die must be subjected to reliability testing to find out the expected lifetime of the chip at normal operating conditions (Hu, 1992).

Fabrication of full custom ASICs is done at silicon foundries. Some of the foundries are captive, that is, they fabricate devices only for the system divisions of their own company. Others make available some of their lines to outside customers. There also are foundries that exclusively serve external customers. Foundries often provide design services as well. If users are interested in doing their own design, however, then the foundry provides a set of design rules for each of the processes they have available. The design rules describe in broad terms the fabrication technology, whether CMOS, bipolar, BiCMOS, or GaAs and then specify in detail the minimum dimensions that can be defined on the wafer for the various layers, the SPICE [Vladimirescu et al. 1981] parameters of each of the active devices, the range of values for the passive devices and other rules and limitations.

Designing a full custom chip is a complex task and can only be done by expert IC designers. Often a team of people is required both to reduce the design time (Fey and Paraskevopoulos, 1986) and because one person may simply not have all the design expertise required. Also, sophisticated and powerful computer hardware and software are needed. Typically, the more sophisticated the computer aided design (CAD) tools are, the higher the probability that the chip will work the first time. Given the long design and fabrication cycle for a full custom chip and its high cost, it is important that as much as possible of the design be automated, that design rule checking should be utilized, and that the circuit simulation be as complete and accurate as possible.

During the design phase, a substantial effort should be made to incorporate on the chip additional circuitry to help verify that the chip is working properly after it is fabricated, or to help identify the section or small circuit responsible for the chip not working, or verify that an error occurred during the fabrication process. For digital full custom circuits, a large number of testability techniques have been developed (Williams and Mercer, 1993), most of which are automated and easily incorporated in the design, given the proper software. For analog circuits there are no accepted universal testability methods. A unique test methodology for each circuit needs to be developed.

Technology selection for a particular full custom ASIC depends on the functions the chip has to perform, the performance specifications, and its desired cost.

8.7.3 Semicustom ASICs

The main distinguishing feature of semicustom ASICs, compared to full custom ones, is that the basic circuit building blocks, whether analog or digital, are already designed and proven to work. These basic circuits typically reside in libraries within a CAD system. The users simply select from the library the components needed, place them on their circuits, and interconnect them. Circuit simulation is also done at a much higher level than SPICE, and the designer is, therefore, not required to be familiar with either semiconductor or device physics.

Semicustom ICs are designed using either **gate arrays, standard cells**, analog arrays, **functional blocks** and PLDs such as field programmable gate arrays (FPGAs). Note that there does not exist a standard naming convention for these products within the industry. Often different manufacturers use different names to describe what are essentially very similar products. Gate arrays, standard cells, and FPGAs are used for digital designs. Analog arrays are used for analog designs and functional blocks are used for both.

Gate Arrays

A gate array consists of a regular array of transistors, usually arranged in two pairs of n- and p-channel, which is the minimum number required to form a NAND gate, and a fixed number of bonding pads, each incorporating an input/output (I/O) buffer. The major distinguishing feature of gate arrays is that they are partially prefabricated by the manufacturer. The designer customizes only the final contact and metal layers. Partially prefabricating the devices reduces delivery time and cost, particularly of prototype parts.

The layout of a simple 2048-gates CMOS gate array is shown in Fig. 8.72. The device consists of 16 columns of transistors and each column contains 128 pairs of n-channel and p-channel transistors. Between the columns are 18 vertical wiring channels, each containing 21 tracks. There are no active devices in the channels. There are 4 horizontal wiring tracks for each gate for a total of 512 horizontal tracks or *routes* for the whole array. The optimum number R of wiring tracks or routes per gate, of an array of a given number of gates, is given by the empirical formula (Fier and Heikkila, 1982)

$$R = 3CG^{0.124}$$

Where C is the average number of connections per gate and G is the number of gates in the array. For a two-input NAND gate, shown in Fig. 8.73, the number of connections C is 3, the two inputs A and B and

2048 GATES - CMOS GATE ARRAY

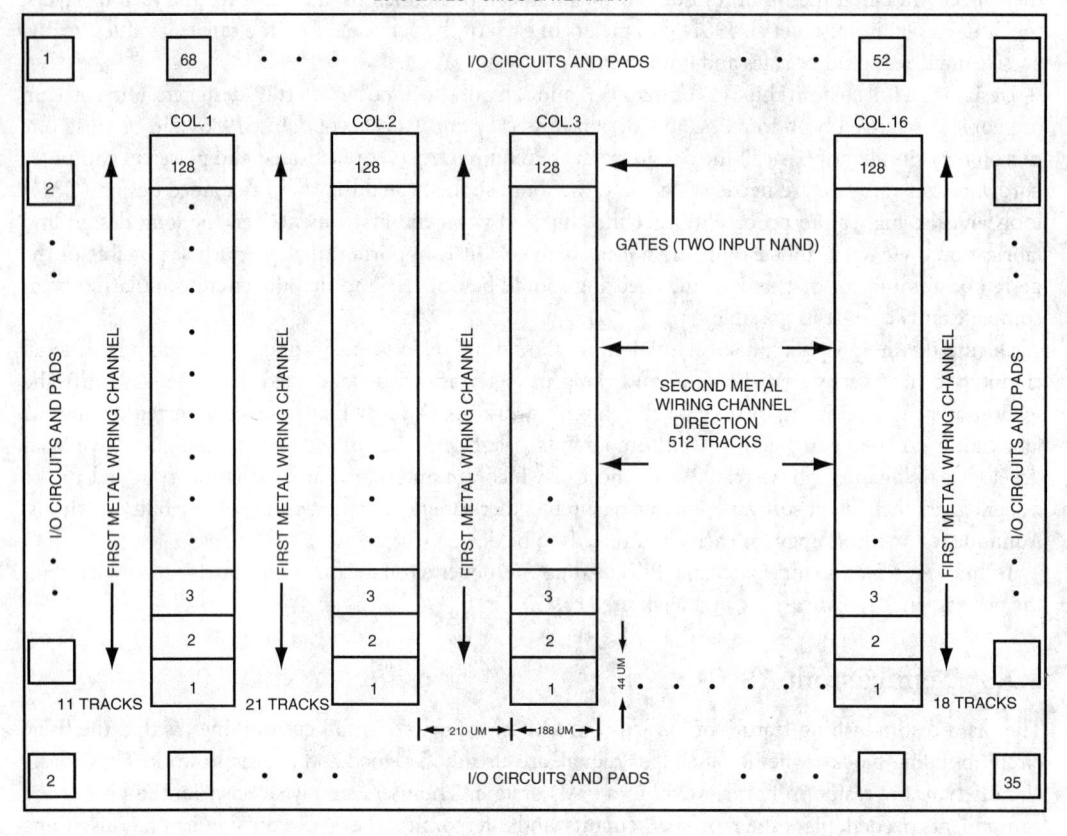

FIGURE 8.72   Schematic layout of a 2048-gates CMOS gate array.

the output AB, and for a 2048-gates array, the preceding formula gives $R = 23$. In this device 25 routes per gate are provided.

In the perimeter of the device 68 bond pads are arranged. Of these, 8 pads are needed for power and ground connections. Another empirical formula (Fier and Heikkila, 1982), based on Rent's rule, specifies that the number of I/O pads required to communicate effectively with the internal gates is given by

$$P = CG^a$$

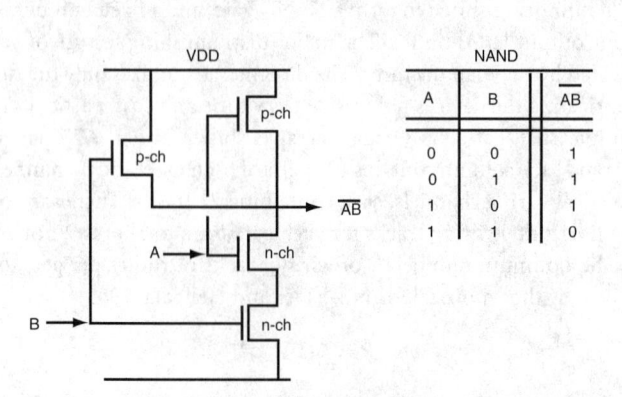

FIGURE 8.73   Electrical schematic of a CMOS NAND logic gate and its truth table.

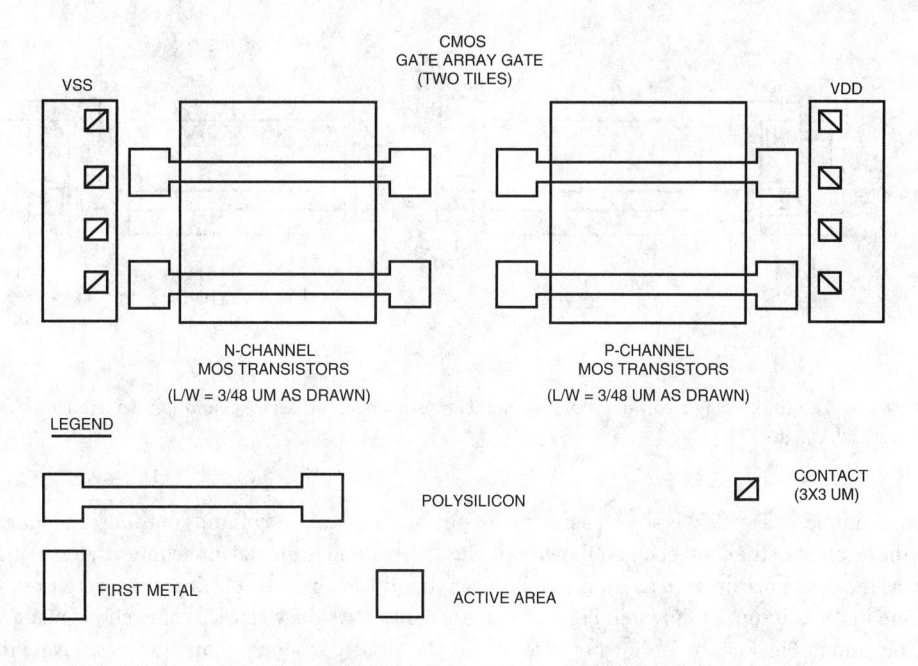

CMOS
GATE ARRAY GATE
(TWO TILES)

VSS

VDD

N-CHANNEL
MOS TRANSISTORS
(L/W = 3/48 UM AS DRAWN)

P-CHANNEL
MOS TRANSISTORS
(L/W = 3/48 UM AS DRAWN)

LEGEND

POLYSILICON

CONTACT
(3X3 UM)

FIRST METAL

ACTIVE AREA

FIGURE 8.74   Schematic layout of one of the 2048 uncommitted gates in a device column.

where P is the sum of input and output pads, C is again the number of connections per gate, G is the number of gates in the array and a is Rent's exponent having a value between 0.5 and 0.7. For large-scale integrated (LSI) circuits, typically $a = 0.46$. Assuming this value for a, then P is equal to 134 for $G = 2048$ and $C = 3$. This gate array, therefore, with only 60 I/O pads should be pad limited for many designs.

As mentioned previously, the process of designing with gate arrays begins with the designer drawing, with the aid of a CAD system, the circuit schematic representing the function the chip must perform. This activity is called **schematic capture**. The schematic, containing circuit elements or *cells*, such as inverters, NAND and NOR gates, flip-flops, adders, etc., is formed by getting these components from a specified library in the computer. Each of the cells has a number of representations associated with it. It has a schematic used, of course, in drawing up the complete circuit schematic. It has a functional description that specifies what the element does, for example, an inverter takes an 1 input and produces a 0 in its output and vice versa. Another file describes its electrical characteristics, such as the time delay between when the inputs reach a certain state and the time the output responds to that state, often referred to as the propagation delay, and it has a physical representation.

In the array shown in Fig. 8.72, before the customized layers are placed on the device, each of the 2048 gates appears schematically as shown in Fig. 8.74. There are two n-channel transistors on the left and two p-channel transistors on the right. There is also a ground or VSS buss line running vertically on the left and a VDD line on the right. Figure 8.75 shows the equivalent electrical schematic of the structure in Fig. 8.74. In Fig. 8.76 is shown only one pair of n- and p-channel transistors and locations within the device area where one of the customizing layers, the contacts, may be placed. The physical representation of a particular library cell contains information of how the uncommitted transistors in the device columns should be connected.

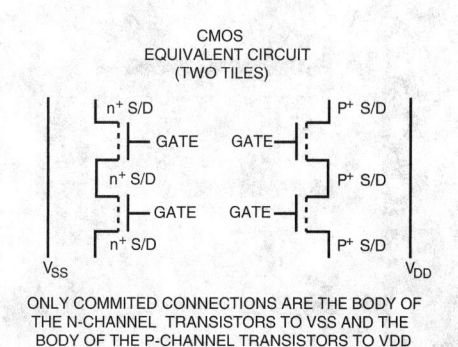

CMOS
EQUIVALENT CIRCUIT
(TWO TILES)

n⁺ S/D   P⁺ S/D
GATE   GATE
n⁺ S/D   P⁺ S/D
GATE   GATE
n⁺ S/D   P⁺ S/D

V_{SS} V_{DD}

ONLY COMMITED CONNECTIONS ARE THE BODY OF
THE N-CHANNEL TRANSISTORS TO VSS AND THE
BODY OF THE P-CHANNEL TRANSISTORS TO VDD

FIGURE 8.75   Equivalent electrical schematic of an uncommitted gate in the device column area of the gate array.

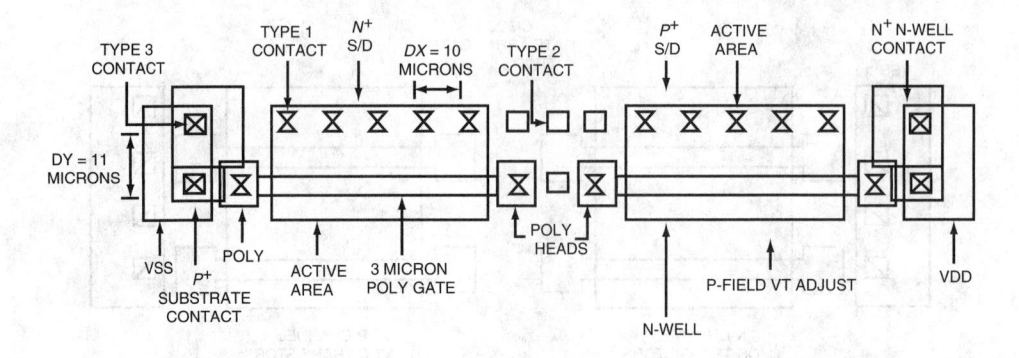

FIGURE 8.76 Layout of a single pair of n- and p-channel transistors in the device column showing how the contact locations are allocated.

As an example, in Fig. 8.77 is shown a portion of one of the 16 device columns in the array. Figure 8.78 shows the location of the contact holes, shown as white squares, and the metal interconnects, for an inverter shown as the dark horizontal bars. An actual inverter circuit is realized by placing this inverter cell along anywhere in the column, as shown in Fig. 8.79. Figure 8.80 shows the electrical connections made in the gate array and in Fig. 8.80(b) is shown, for reference, the electric schematic for an inverter. Note that in Fig. 8.79 and Fig. 8.80, the polysilicon gates of the top and bottom n-channel transistors are connected to VSS

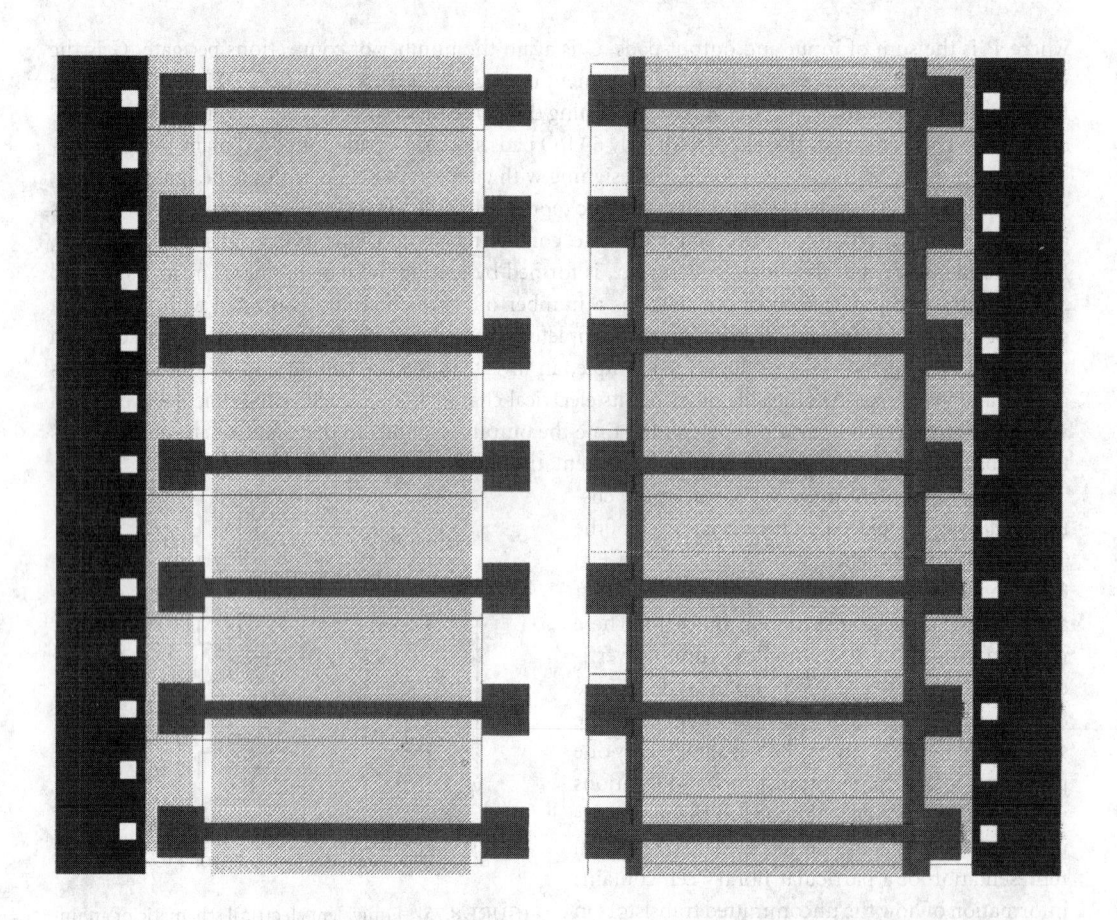

FIGURE 8.77 Layout of a portion of a device column.

FIGURE 8.78 Layout of a single inverter library cell.

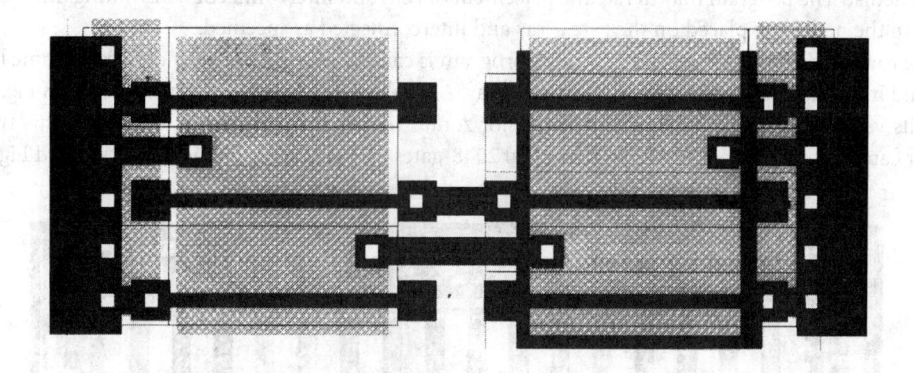

FIGURE 8.79 Completed layout of an inverter in device column.

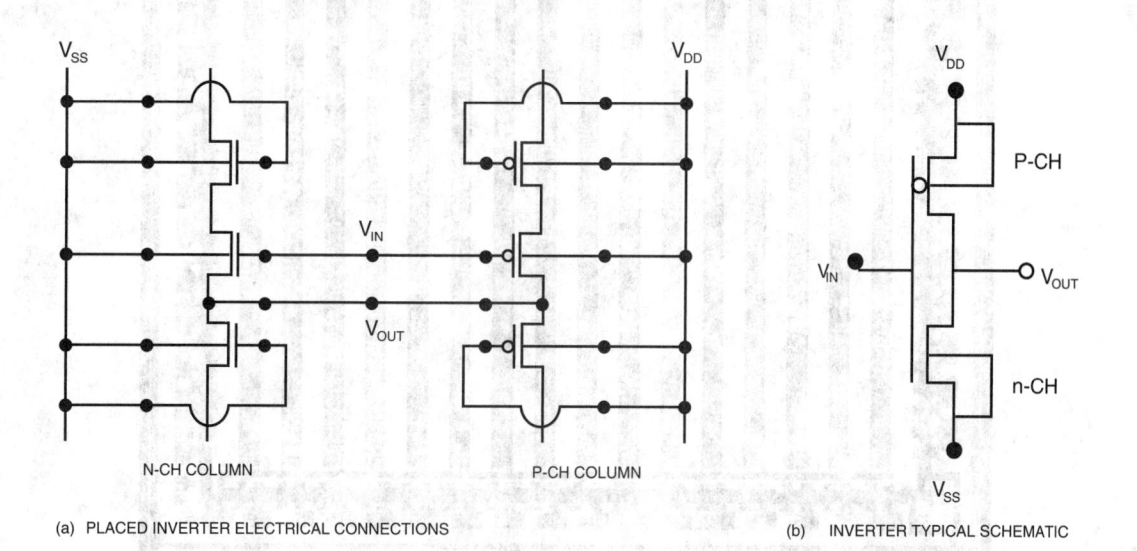

(a) PLACED INVERTER ELECTRICAL CONNECTIONS

(b) INVERTER TYPICAL SCHEMATIC

FIGURE 8.80 (a) Electrical connections of a placed inverter in a device column, (b) electrical schematic of an inverter.

and the corresponding gates of the the p-channel transistors are connected to VDD. This is done to isolate the inverter from interfering electrically with another logic gate that may be placed above or below it.

Returning to the design process, after schematic capture is completed, the designer simulates the entire circuit first to verify that it performs the logic functions the circuit is designed to perform and then using the electrical specifications files, a timing simulation is performed to make sure the circuit will operate at the clock frequency desired. However, the timing simulation is not yet complete. In the electrical characteristics file of each cell, a typical input and load capacitance is assumed. In an actual circuit the load capacitance may be quite different. Therefore, another timing simulation must be done after the actual capacitance values are found for each node.

To do this, first from the circuit schematic, a *netlist* is extracted. This list contains exact information about which node of a given element is connected to which nodes of other elements. This netlist is then submitted to the *place and route* program. This program places all of the elements or cells in the circuit on the gate array, in a fashion similar to what was done in Fig. 8.79, and attempts to connect them as specified in the netlist. The program may iterate the placement of cells and interconnection or routing until all cells listed in the netlist are placed on the gate array and interconnected as specified.

The routing portion of the place and route program is called a *channel router*. It derives its name by the method it uses to interconnect the cells in the array. Also, the gate array architecture, shown in Fig. 8.72, with its vertical wiring channels and a single horizontal wiring channel, is selected so that this type of router can be used. Figure 8.81 shows the actual 2048-gates gate array, before customization and Fig. 8.82

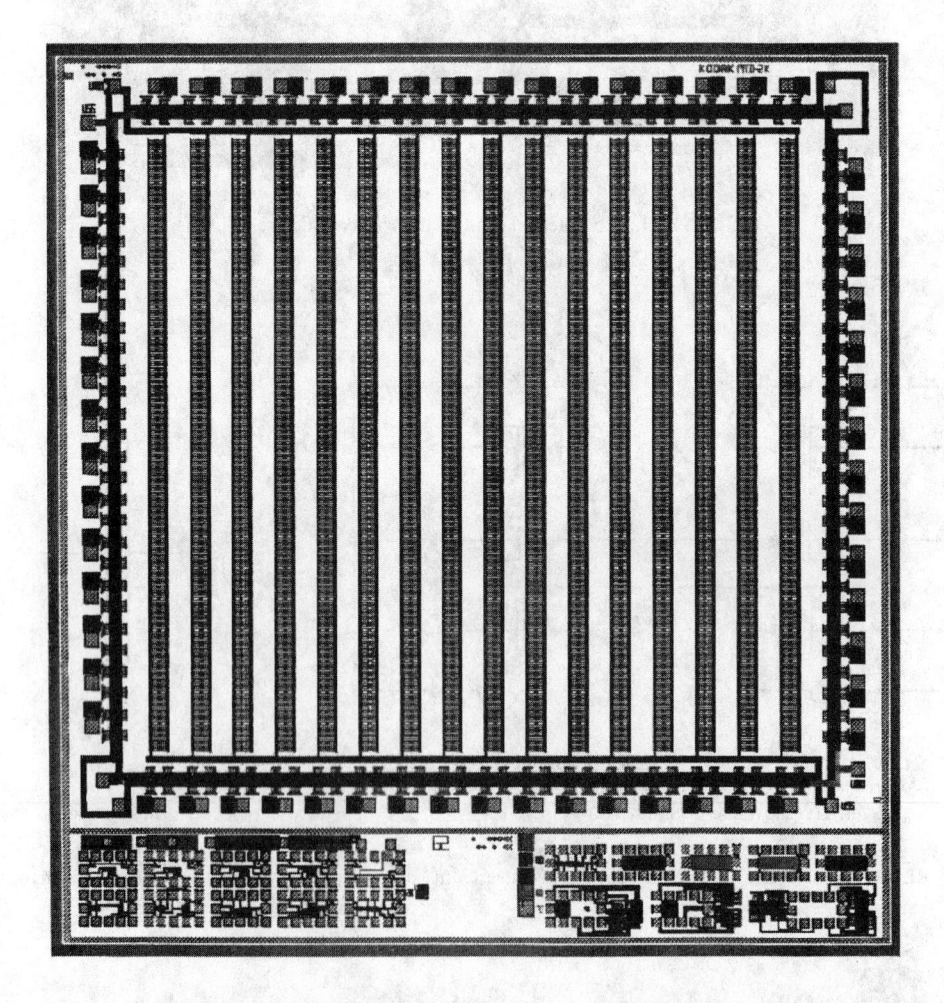

FIGURE 8.81   Layout of the actual 2048-gates gate array before customization.

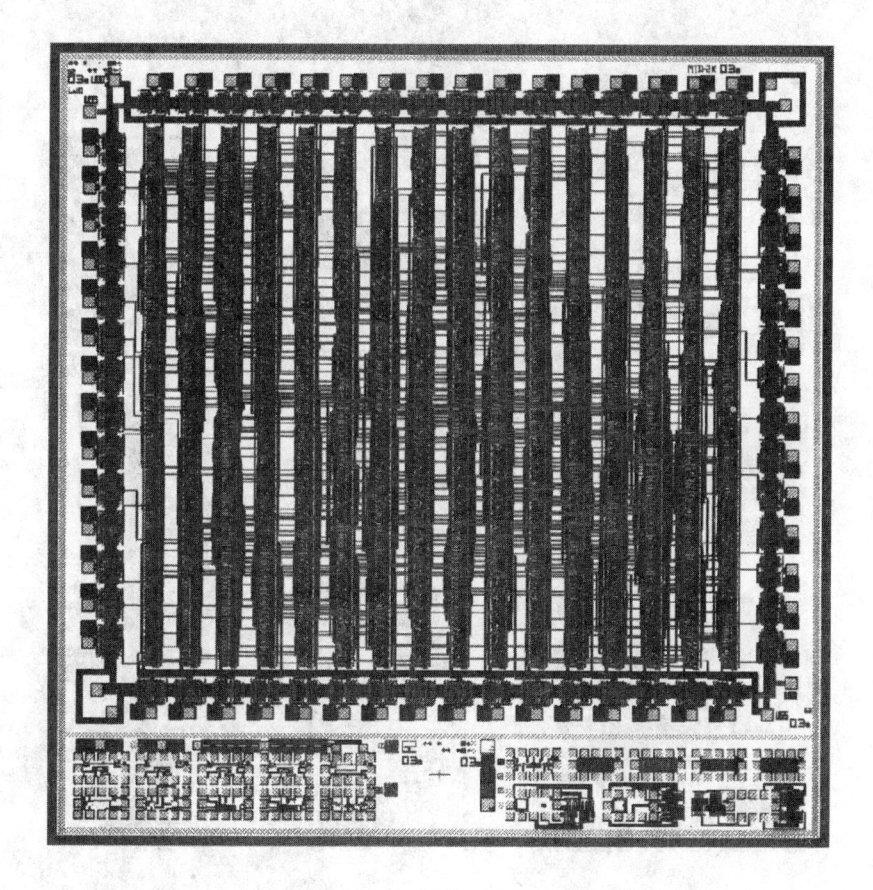

FIGURE 8.82   A custom circuit implementation on the 2048-gates gate array.

shows after it is completed with a custom circuit. In Fig. 8.83 a corner of the array is shown in higher magnification. The vertical darker lines are the first level metal interconnects and the horizontal lighter lines are the second level metal interconnects. This array requires four layers to be customized. These are the first and second level metals, contact, and the via. Vias are contacts exclusively between the two metal layers. Note that all internal wiring in the cells, like the inverter in Fig. 8.78, are made with first level metal and, therefore, the router is not allowed to route first level metal over the device areas.

Once the circuit is placed and routed, an extraction program is applied that does two things. First, it calculates the actual resistance and capacitance seen at each node of the circuit that is then fed back into the netlist for more accurate timing simulation; second, it extracts a new netlist of the circuit on the gate array that can be compared with the initial netlist. The two netlists should, of course, be identical. However, often manual intervention is required to finish either the placement or routing of a particularly difficult circuit and that could cause an inadvertent error to occur in the layout. The error is then detected by comparing this extracted netlist to the original one.

After the design is completed, **test vectors** must be generated. Two sets of test vectors are generated. The input set and the corresponding output set. Each test vector consists of a number of 1s and 0s. The input test vector has as many elements as the number of chip input pins and the output test vector, likewise, has as many elements as the number of output pins.

The input test vector is not, typically, similar to what the chip will see in actual operation. Instead, the goal is to select a set of input vectors that when applied to the chip will cause every internal node to change state at least once. The timing verification program is used for this task. The program keeps track of the nodes toggled, as each input vector is applied, and also saves the output vector. When the finished devices are received, the same input test vectors are applied in sequence, and the resulting output vectors are captured and

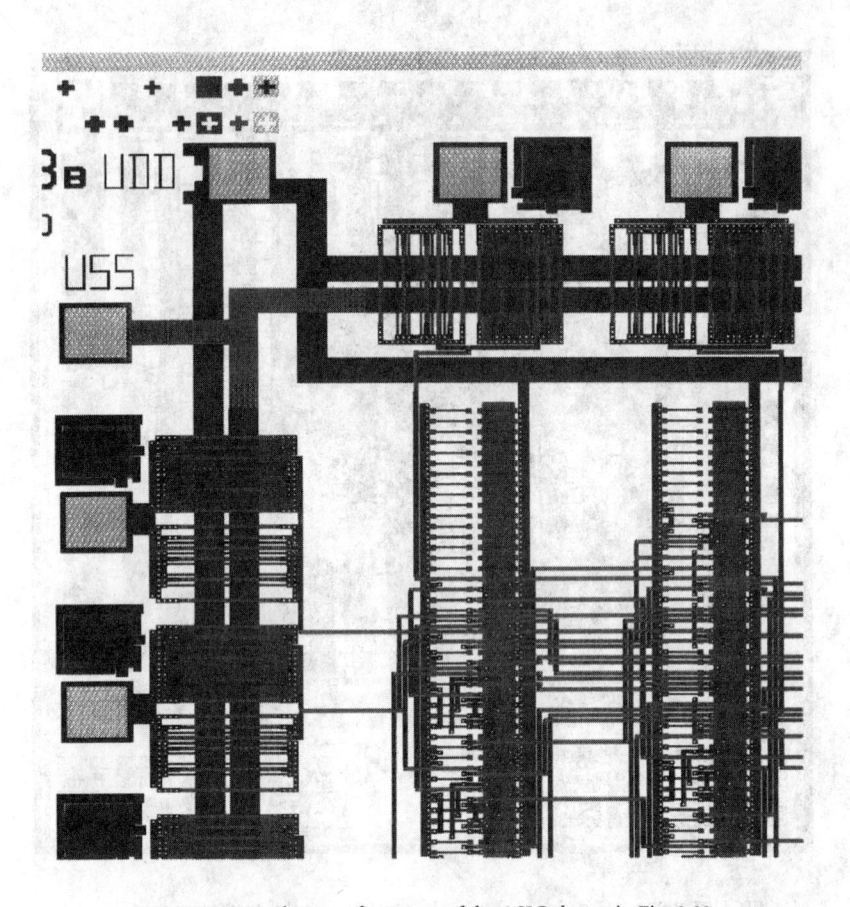

FIGURE 8.83   Closeup of a corner of the ASIC shown in Fig. 8.82.

compared to those obtained from the timing simulation runs. If the two match perfectly, the device is classified as good.

The purpose of this test is to detect whether, because of a defect during fabrication, an internal gate is not operating properly. Usually if a test can cause about 90% of the internal nodes to switch state, it is considered to provide adequate coverage. It would be desirable to have 100% coverage, but often this is not practical because of the length of time it would take for the test to be completed. Finally, because the test is done at the frequency at which the chip will operate, only minimal functional testing is performed.

Just as in the device columns any logic gate can be placed anywhere in any of the columns, each of the pads can serve as either an input, an output, or a power pad, again by customizing the four layers. Figure 8.84 shows an I/O buffer at the end of the prefabrication process. Figure 8.85 shows the contact and metal pattern to make the uncommitted pad an inverting output buffer and Fig. 8.86 shows the two together. Figure 8.87 shows the circuit schematic for the output buffer as well as the schematic when the pad is customized as an input buffer, the layout of which is shown in Fig. 8.88.

The gate array architecture presented here is not the only one. A number of different designs have been developed and are currently used. For each of the architectures, a different place and route algorithm is then developed to take advantage of that particular design. Gate arrays have also taken advantage of developments in silicon technology, which primarily consists of shrinking of the minimum dimensions. As a result, arrays with several million transistors per chip are presently in use.

Standard Cells

Designing with gate arrays is fairly straightforward and is utilized extensively by system engineers. However, gate arrays are quite inefficient in silicon area use since the number of transistors, the number of routing

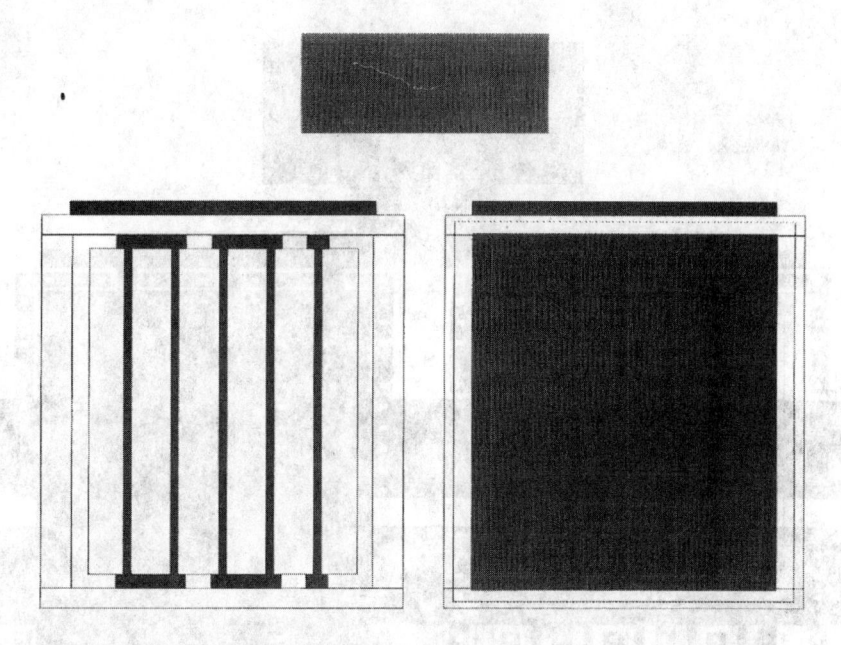

FIGURE 8.84 Layout of an uncommitted I/O buffer. In this figure, the bonding pad is the rectangular pattern at the top, the n-channel transistor is on the left and the p-channel is located on the right.

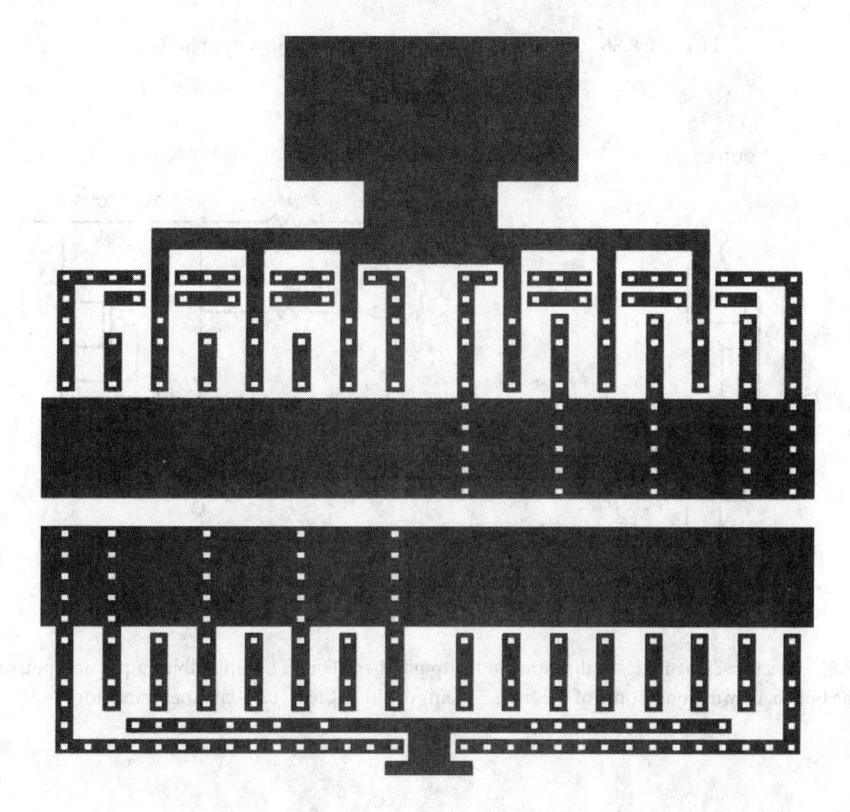

FIGURE 8.85 Layout of the contact and metal pattern of the output buffer library cell.

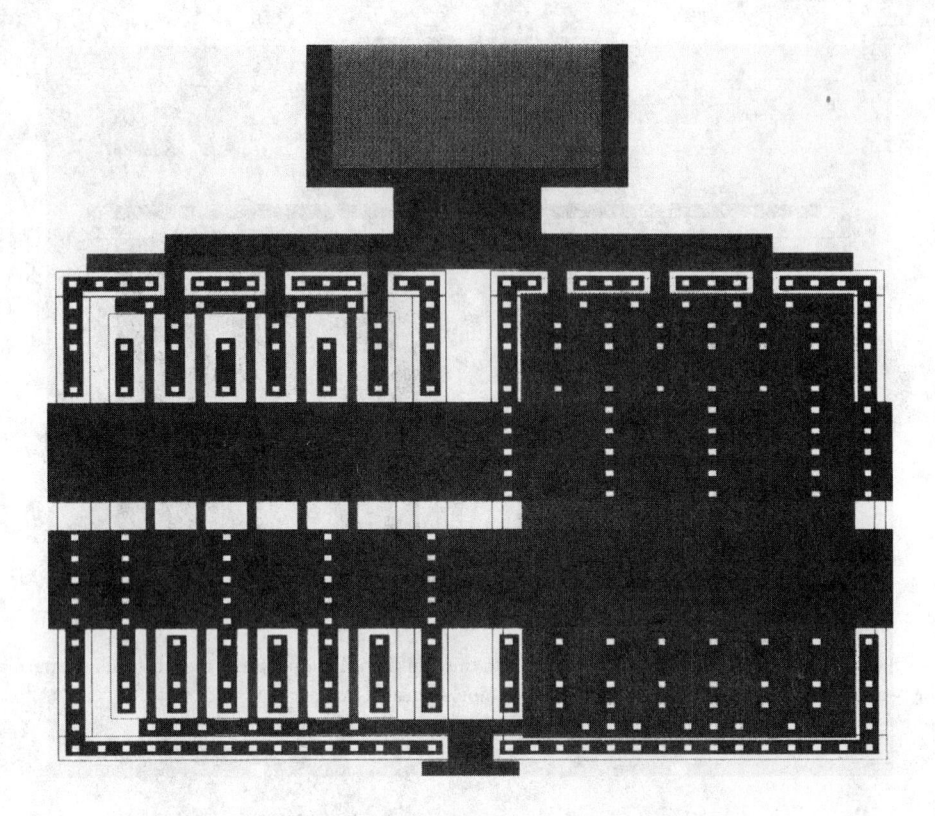

FIGURE 8.86   Completed layout of a full inverting output buffer.

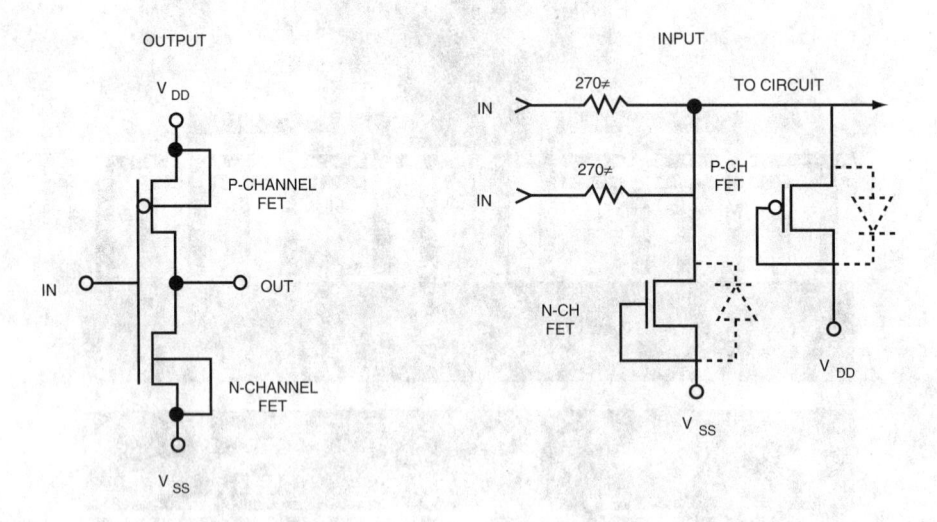

FIGURE 8.87   Electrical schematics of the input and output buffers. For TTL compatible output or input, or for other reasons, the buffers may use only some of the five segments of either the n- or p-channel transistors.

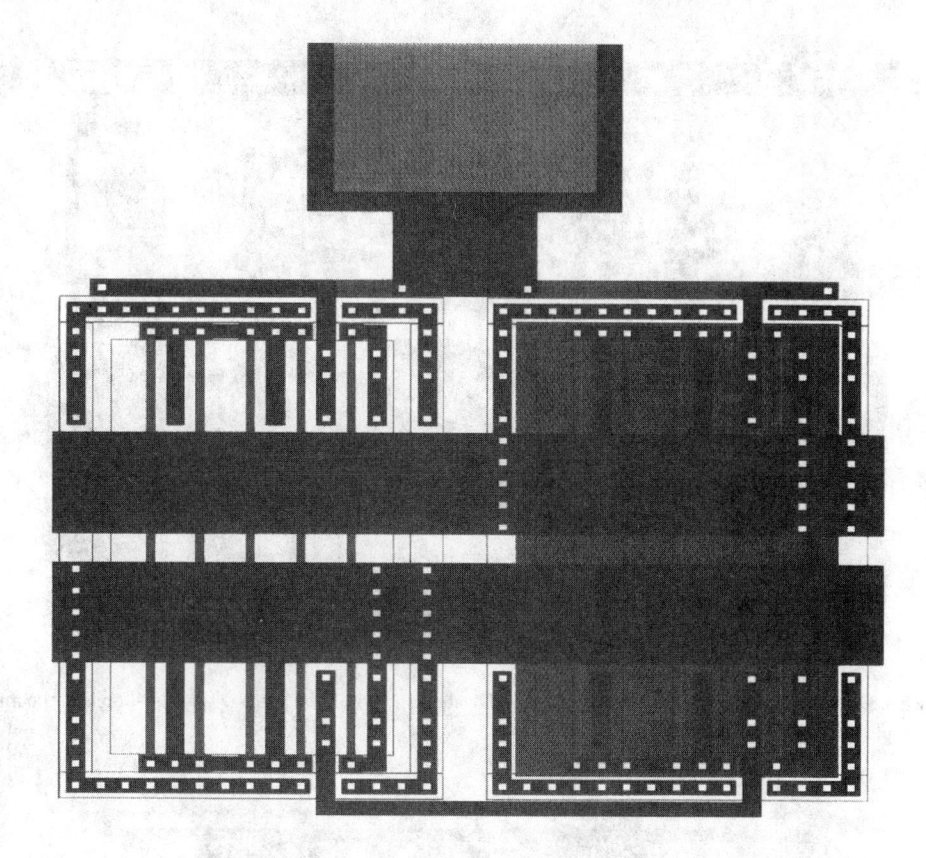

FIGURE 8.88   Completed layout of a simple noninverting input buffer.

tracks, and the number of I/O pads needed by a particular circuit does not always match the number of devices or pads on the array, which are, of course, fixed in number.

The standard cells design methodology was developed to address this problem. As was the case for gate arrays, a predesigned and characterized library of cells exists for the circuit designer to use. But, unlike gate arrays, the chip is layed out containing only the number of transistors and I/O pads specified by the particular circuit, resulting in a smaller, less expensive chip. Fabrication of a standard cells circuit, however, must start from the beginning of the process. This has two major drawbacks. First, the fabrication time for standard cells is several weeks longer than for gate arrays; second, the customer has to pay for the fabrication of the whole lot of wafers, whereas for gate arrays, only a small portion of the lot is processed for each customer. This substantially increases the cost of prototype parts.

Because standard cells are placed and routed automatically, the cells have to be designed with some restrictions, such as uniform height, preassigned locations for clocks and power lines, predefined input and output port locations, and others. Figure 8.89 shows the layout of three common cells from a standard cells library and Fig. 8.90 shows a completed standard cells chip. Its appearance is very similar to that of a gate array, but it contains no unused transistors or I/O pads and the minimum number of routing channels.

Functional Blocks

Advances in place and route software have made possible the functional blocks design methodology. Here the cells are no longer simple logic gates, but complete functions. The cells can be analog or digital and can have arbitrary sizes. Figure 8.91 shows an example of a digital functional block design. In the figure, the three white blocks are placed and routed automatically along with several rows of standard cells. In Fig. 8.92, a small linear charge-coupled device (CCD) image sensor, located in the center of the chip, and

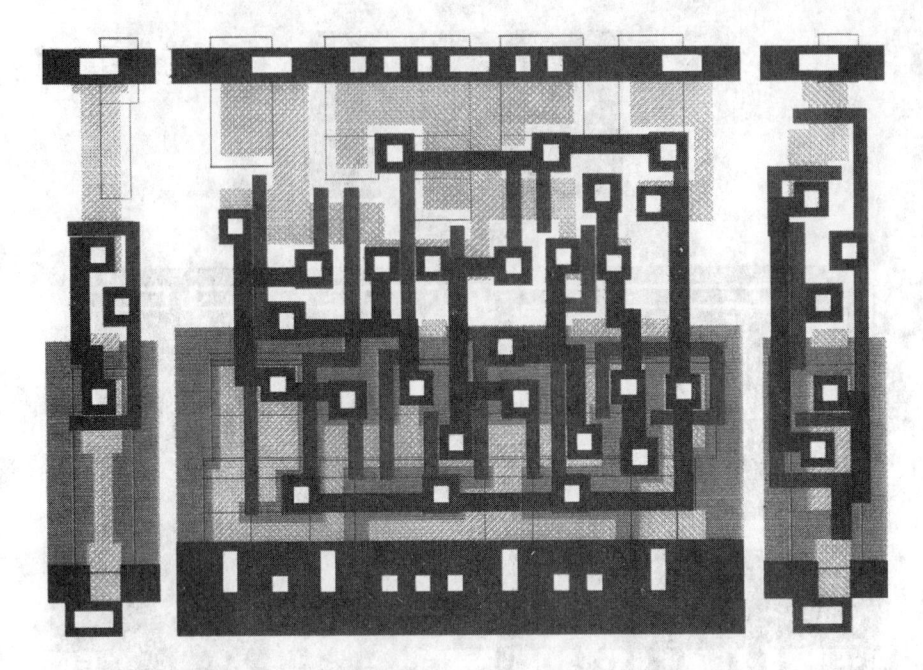

FIGURE 8.89 Three standard cells. A two-input NAND gate is shown on the left, a D type flip-flop is in the middle, and a simple inverter is shown on the right.

FIGURE 8.90 Layout of a completed standard cell ASIC.

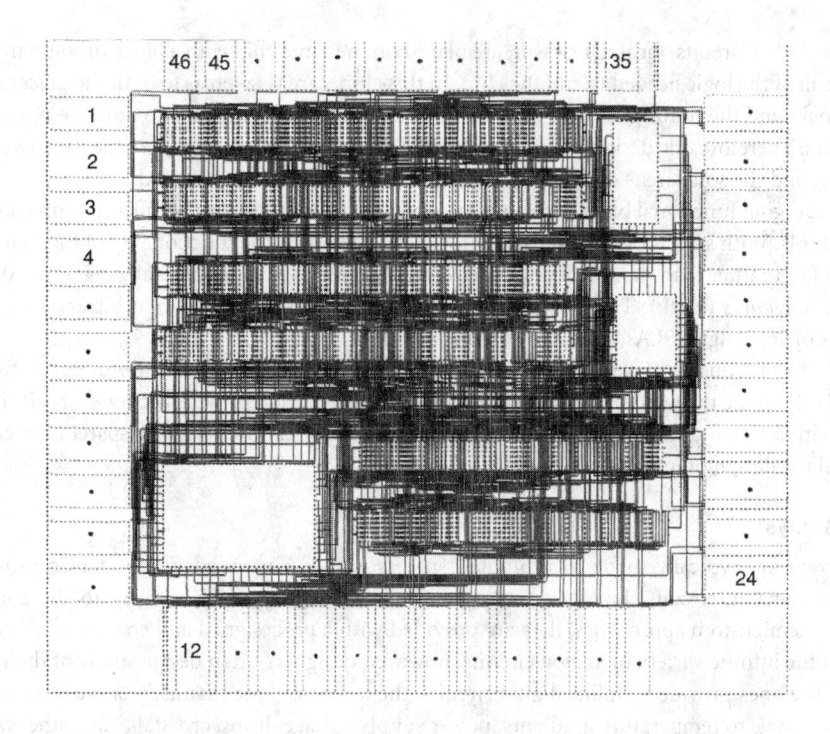

FIGURE 8.91 Layout of a digital functional block designed ASIC.

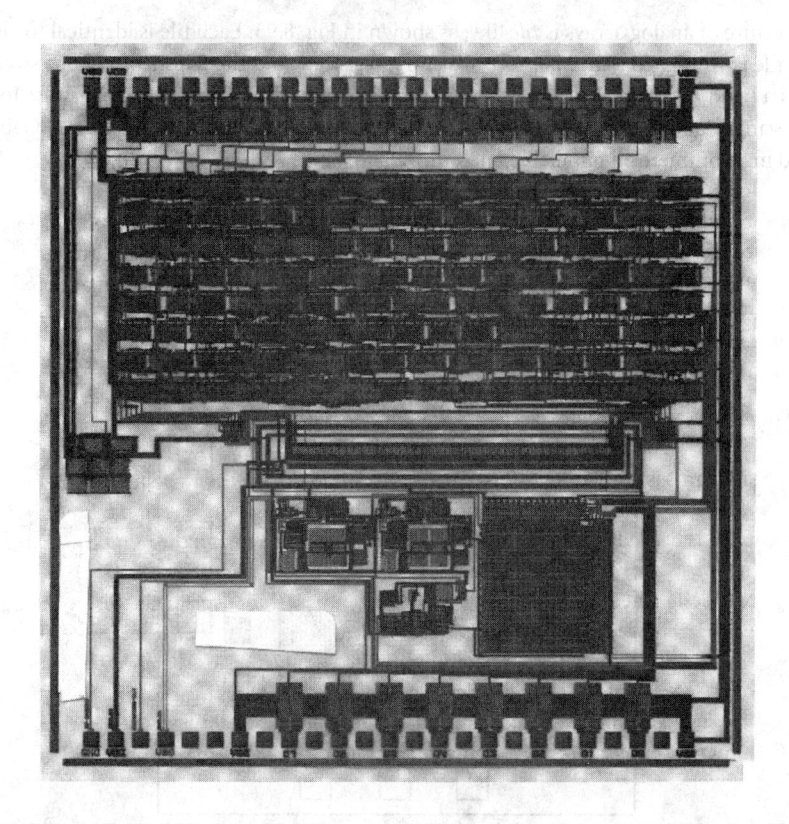

FIGURE 8.92 Layout of a mixed (combined analog and digital circuits on the same chip) functional block designed ASIC.

a few other analog circuits, located below it, are placed on the same chip with a block of standard cells that *C* produce all of the logic needed to run the chip. In this chip, care is taken as to both the placement of the various blocks and the routing because of the sensitivity of the analog circuits to noise, temperature, and other factors. Therefore, the designer had considerable input as to the placement of the various blocks and the routing, and the software is designed to allow this intervention, when needed.

Often the digital functional blocks are built by a class of software programs called **silicon compilers** or **synthesis tools**. With such tools circuit designers typically describe the functionality of their circuit at the behavioral level rather than the gate level, using specially developed computer languages called *hardware description languages* (HDL). These tools can quickly build memory blocks of arbitrary size, adders or multipliers of any length, PLAs, and many other circuits.

It should be obvious by now that the semicustom design approach makes heavy use of software tools. These tools eliminate the tedious and error prone task of hand layout, provide accurate circuit simulation, and help with issues such as testability. Their power and sophistication make transparent the complexity inherent in the design of integrated circuits.

Analog Arrays

Analog arrays are typically made in a bipolar process and are intended for the fabrication of high-performance analog circuits. Like gate arrays, these devices are prefabricated up to the contact level and, like all semicustom approaches, they are provided with a predesigned and characterized cell library. Because of the infinite variety of analog circuits, however, designers often design many of their own cells or modify the ones provided. Unlike digital circuits, the layout is done manually because analog circuits are more sensitive to temperature gradients, power supply voltage drops, crosstalk, and other factors and because the presently available software is not sufficiently sophisticated to take all of these factors into account.

The architecture of analog arrays is *tile* like, as shown in Fig. 8.93. Each tile is identical to all other tiles. Within each tile is contained a number of transistors, resistors, and capacitors of various sizes. The array also contains a fixed number of I/O pads, each of which can be customized with the last four layers to serve either as an input or output buffer. Figure 8.94 shows an actual circuit (Boisvert and Gaboury, 1992) implemented in a commercially available analog array.

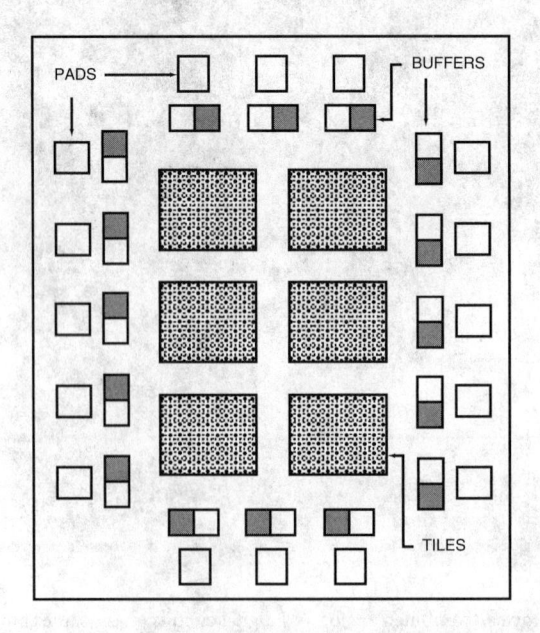

FIGURE 8.93   Architecture of an analog array.

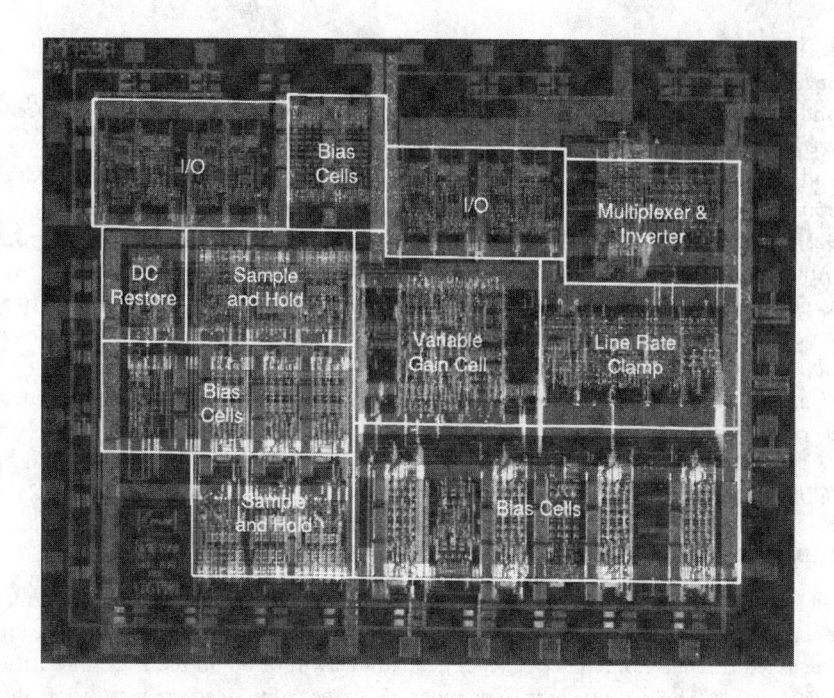

FIGURE 8.94 Example of an ASIC (Boisvert and Gaboury, 1992) implemented in a commercially available analog array.

Defining Terms

ASICs: The acronym of *application specific integrated circuits*; another name for such chips is *custom integrated circuits.*

Functional blocks: ASICs designed using this methodology are more compact than either gate arrays or standard cells because the blocks can perform much more complex functions than do simple logic gates.

Gate arrays: Chips that contain uncommitted arrays of transistors and are prefabricated up to a certain step after which they are customized for the particular application.

Nonreccurring engineering (NRE): Costs the foundry charges the ASIC customer. These costs include engineering time, the cost of making the masks, the cost of fabricating one lot of wafers, and for packaging and testing the prototype parts.

Standard cells: A design methodology for realizing ASICs. Compared to gate arrays, they make more efficient use of silicon.

Schematic capture: The process by which the functionality of the chip is captured in an electrical schematic, usually with the aid of a computer, and using components from a cell library resident in that computer.

Silicon compilers, synthesis tools: Software programs that can construct an ASIC whose functionality is no longer described by a circuit schematic but in a special high level computer languages, generally called hardware description languages or HDL.

Test vectors: A test scheme that consists of pairs of input and output. Each input vector is a unique set of 1s and 0s applied to the chip inputs and the corresponding output vector is the set of 1s and 0s produced at each of the chip's output.

References

Boisvert, D.M. and Gaboury, M.J. 1992. An 8–10-bit, 1–40 MHz analog signal processor with configurable performance for electronic imaging applications. In *Proceedings IEEE International ASIC Conference and Exhibit*, pp. 396–400. Rochester, NY.

Fey, C.F. and Paraskevopoulos, D. 1985. Selection of cost effective LSI design methodologies. In *Proceedings of the IEEE Custom Integrated Circuits Conference*, pp. 148–153. Portland, OR.

Fey, C.F. and Paraskevopoulos, D. 1986. A model of design schedules for application specific ICs. In *Proceedings IEEE Custom Integrated Circuits Conference*, pp. 490–496. Rochester, NY.

Fier, D.F. and Heikkila, W.W. 1982. High performance CMOS design methodologies. In *Proceedings IEEE Custom Integrated Circuits Conference*, pp. 325–328. Rochester, NY.

Hu, C. 1992. IC reliability simulation. *IEEE J. of Solid State Circuits* 27(3):241–246; see also Proceedings of the Annual International Reliability Physics Symposium.

Ting, G., Guidash, R.M., Lee, P.P.K., and Anagnostopoulos, C. 1994. A low-cost, smart-power BiCMOS driver chip for medium power applications. In *Proceedings IEEE International ASIC Conference and Exhibit*, pp. 466–469. Rochester, NY.

Vladimirescu, A. et al. 1981. *SPICE Manual*. Dept. of Electrical Engineering and Computer Sciences, University of California, Berkeley, CA, Oct.

Williams, T.W. and Mercer, M.R. 1993. Testing digital circuits and design for testability. In *Proceedings IEEE International ASIC Conference and Exhibit* (Tutorial Session), p. 10. Rochester, NY.

Further Information

The ASIC field has been expanding rapidly since about 1980. The best source of current information is the two major conferences where the more recent technology developments are reported. These are the IEEE Custom Integrated Circuits Conference (CICC) held annually in May and the IEEE International ASIC Conference and Exhibit, held annually in September. *IEEE Spectrum Magazine* lists these conferences in the calender of events. Apart from the regular technical sessions, educational sessions and exhibits by ASIC vendors are part of these two conferences.

An additional resource is the USC/Information Sciences Institute, 4676 Admiralty Way, Marina del Rey, CA 90292-6695, Telephone (213) 822-1511.

Finally, a selected number of papers from the CICC are published every year, since 1984, in special issues of the *IEEE Journal of Solid State Circuits*.

Additional information can also be found in a number of other conferences and their proceedings, including the Design Automation Conference (DAC) and the International Solid States Circuits Conference (ISSCC).

8.8   Digital Filters

Jonathon A. Chambers, Sawasd Tantaratana and Bruce W. Bomar

8.8.1   Introduction

Digital filtering is concerned with the manipulation of discrete data sequences to remove noise, extract information, change the sample rate, and perform other functions. Although an infinite number of numerical manipulations can be applied to discrete data (e.g., finding the mean value, forming a histogram), the objective of digital filtering is to form a discrete output sequence $y(n)$ from a discrete input sequence $x(n)$. In some manner or another, each output sample is computed from the input sequence—not just from any one sample, but from many, in fact, possibly from all of the input samples. Those filters that compute their output from the present input and a finite number of past inputs are termed *finite impulse response* (FIR), whereas those that use all past inputs are *infinite impulse response* (IIR). This chapter will consider the design and realization of both FIR and IIR digital filters and will examine the effect of finite wordlength arithmetic on implementing these filters.

8.8.2   FIR Filters

A finite impulse response filter is a linear discrete time system that forms its output as the weighted sum of the most recent, and a finite number of past, inputs. Time-invariant FIR filters have finite memory, and their impulse response, namely, their response to a discrete input that is unity at the first sample and otherwise zero, matches the fixed weighting coefficients of the filter. Time-variant FIR filters, on the other hand, may operate at various sampling rates and/or have weighting coefficients that adapt in sympathy with some statistical property of the environment in which they are applied.

Fundamentals

Perhaps the simplest example of an FIR filter is the moving average operation described by the following linear constant-coefficient difference equation:

$$y(n) = \sum_{k=0}^{M} b_k x(n-k) \quad b_k = \frac{1}{M+1}$$

where

  $y(n)$ = output of the filter at integer sample index n
  $x(n)$ = input to the filter at integer sample index n
  b_k = filter weighting coefficients, $k = 0, 1, \ldots, M$
  M = filter order

In a practical application, the input and output discrete-time signals is sampled at some regular sampling time interval, T seconds, denoted $x(nT)$ and $y(nT)$, which is related to the sampling frequency by $f_s = 1/T$ samples per second. For generality, however, it is more convenient to assume that T is unity, so that the effective sampling frequency is also unity and the Nyquist frequency (Oppenheim and Schafer, 1989), namely, the maximum analog frequency that when sampled at f_s will not yield an aliasing distortion, is one-half. It is then straightforward to scale, by multiplication, this normalized frequency range, that is, $(0, 1/2)$, to any other sampling frequency.

    The output of the simple moving average filter is the average of the $M + 1$ most recent values of $x(n)$. Intuitively, this corresponds to a smoothed version of the input, but its operation is more appropriately described by calculating the frequency response of the filter. First, however, the z-domain representation of the filter is introduced in analogy to the s- (or Laplace-) domain representation of analog filters. The z transform of a causal discrete-time signal $x(n)$ is defined by

$$X(z) = \sum_{n=0}^{\infty} x(n) z^{-n}$$

where $X(z)$ is z transform of $x(n)$ and z is complex variable

    The z transform of a delayed version of $x(n)$, namely, $x(n-k)$ with k a positive integer, is found to be given by $z^{-k} X(z)$. This result can be used to relate the z transform of the output $y(n)$ of the simple moving average filter to its input

$$Y(z) = \sum_{k=0}^{M} b_k z^{-k} X(z) \quad b_k = \frac{1}{M+1}$$

The z-domain transfer function, namely, the ratio of the output to input transform, becomes

$$H(z) = \frac{Y(z)}{X(z)} = \sum_{k=0}^{M} b_k z^{-k} \quad b_k = \frac{1}{M+1}$$

Notice the transfer function $H(z)$ is entirely defined by the values of the weighting coefficients $b_k, k = 0, 1, \ldots, M$, which are identical to the discrete impulse response of the filter, and the complex variable z. The finite length of the discrete impulse response means that the transient response of the filter only lasts for $M + 1$ samples, after which steady state is reached. The frequency domain transfer function for the

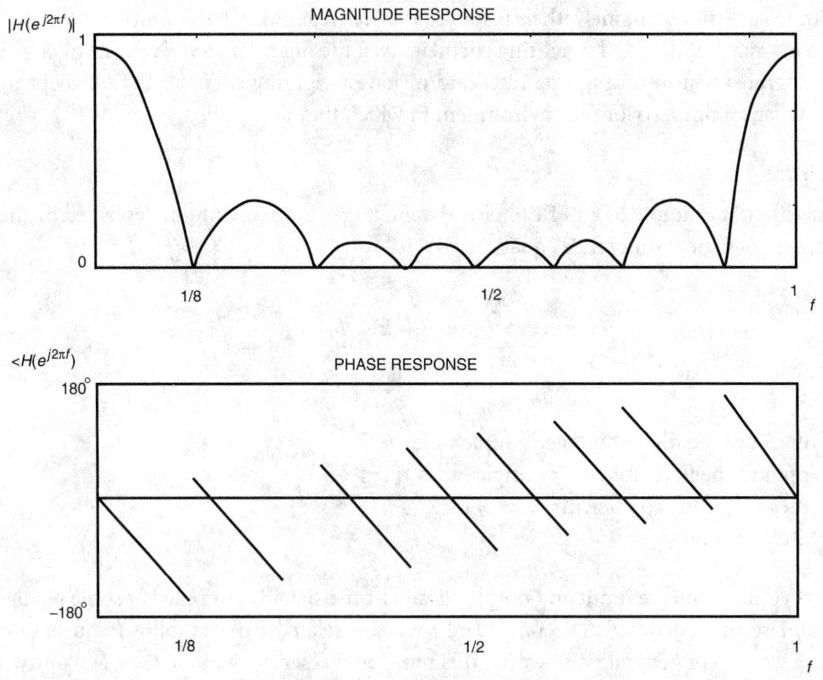

FIGURE 8.95   The magnitude and phase response of the simple moving average filter with $M = 7$.

filter is found by setting $z = e^{j2\pi f}$, where $j = \sqrt{-1}$, and can be written as

$$H(e^{j2\pi f}) = \frac{1}{M+1} \sum_{k=0}^{M} e^{-j2\pi fk} = \frac{1}{M+1} e^{-j\pi fM} \frac{\sin[\pi f(M+1)]}{\sin(\pi f)}$$

The magnitude and phase response of the simple moving average filter, with $M = 7$, are calculated from $H(e^{j2\pi f})$ and shown in Fig. 8.95. The filter is seen clearly to act as a crude low-pass, smoothing filter with a **linear-phase** response. The sampling frequency periodicity in the magnitude and phase response is a property of discrete-time systems. The linear-phase response is due to the $e^{-j\pi fM}$ term in $H(e^{j2\pi f})$ and corresponds to a constant $M/2$ group delay through the filter. A phase discontinuity of $+/- 180°$ is introduced each time the magnitude term changes sign. FIR filters that have center symmetry in their weighting coefficients have this constant, frequency independent group delay property that is very desirable in applications in which time dispersion is to be avoided, for example, in pulse transmission where it is important to preserve pulse shapes (Lee and Messerschmit, 1994).

Another useful interpretation of the z-domain transfer function is obtained by re-writing $H(z)$ as follows:

$$H(z) = \frac{\sum_{k=0}^{M} b_k z^{M-k}}{z^M} = \frac{b_0 z^M + b_1 z^{M-1} + \cdots + b_{M-1} z + b_M}{z^M} = \frac{N(z)}{D(z)}$$

The z-domain transfer function is shown to be the ratio of two Mth-order polynomials in z, namely, $N(z)$ and $D(z)$. The values of z for which $N(z) = 0$ are termed the zeros of the filter, whereas those for which $D(z) = 0$ are the poles. The poles of such an FIR filter are at the origin, that is, $z = 0$, in the z plane. The positions of the zeros are determined by the weighting coefficients, that is, $b_k, k = 0, 1, \ldots, M$. The poles and zeros in the z plane for the simple moving average filter are shown in Fig. 8.96. The zeros, marked with a circle, are coincident with the unit circle, that is, the contour in the z plane for which

$|z| = 1$, and match exactly the zeros in the magnitude response, hence their name; the discontinuities in the phase response are shown in Fig. 8.95. The zeros of an FIR filter may lie anywhere in the z plane because they do not impact on the stability of the filter; however, if the weighting coefficients are real and symmetric, or anti-symmetric, about their center value $M/2$, any complex zeros of the filter are constrained to lie as conjugate pairs coincident with the unit circle or as quartets of roots off the unit circle with the form $(\rho e^{j\theta}, \rho e^{-j\theta},$ $(1/\rho)e^{j\theta}, (1/\rho)e^{-j\theta})$ where ρ and θ are, respectively, the radius and angle of the first zero. Zeros that lie within the unit circle are termed *minimum phase*, whereas those which lie outside the unit circle are called *maximum phase*. This distinction describes the contribution made by a particular zero

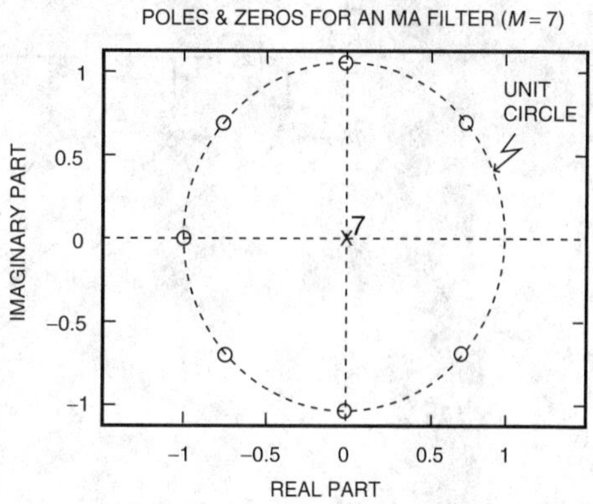

FIGURE 8.96 The pole-zero plot for the simple moving average filter with $M = 7$.

to the overall phase response of a filter. A minimum-phase FIR filter that has all its zeros within the unit circle can have identical magnitude response to a maximum-phase FIR that has all its zeros outside the unit circle for the special case when they have an equal number of zeros with identical angles but reciprocal radii. An example of this would be second-order FIR filters with z-domain transfer functions $H_{\min}(z) = 1 + 0.5z^{-1} + 0.25z^{-2}$ and $H_{\max}(z) = 0.25 + 0.5z^{-1} + z^{-2}$. Notice that the center symmetry in the coefficients is lost, but the minimum- and maximum-phase weighting coefficients are simply reversed. Physically, a minimum-phase FIR filter corresponds to a system for which the energy is rapidly transferred from its input to its output, hence, the large initial weighting coefficients; whereas a maximum-phase FIR filter is slower to transfer the energy from its input to output; as such, its larger coefficients are delayed. Such FIR filters are commonly used for modeling multipath mobile communication transmission paths.

The characteristics of the frequency response of an FIR filter are entirely defined by the values of the weighting coefficients b_k, $k = 0, 1, \ldots, M$, which match the impulse response of the filter, and the order M. Various techniques are available for designing these coefficients to meet the specifications of some application. However, next consider the structures available for realizing an FIR filter.

Structures

The structure of an FIR filter must realize the z-domain transfer function given by

$$H(z) = \sum_{k=0}^{M} b_k z^{-k}$$

where z^{-1} is a unit delay operator. The building blocks for such filters are, therefore, adders, multipliers, and unit delay elements. Such elements do not have the disadvantages of analog components such as capacitors, inductors, operational amplifiers, and resistors, which vary with temperature and age. The direct or tapped-delay line form, as shown in Fig. 8.97, is the most straightforward realization of an FIR filter. The input $x(n)$ is delayed, scaled by the weighting coefficients b_k, $k = 0, 1, \ldots, M$, and accumulated to yield the output.

An equivalent, but transposed, structure of an FIR filter is shown in Fig. 8.98, which is much more modular and well suited to integrated circuit realization. Each module within the structure calculates a partial sum so that only a single addition is calculated at the output stage.

Other structures are possible that exploit symmetries or redundancies in the filter weighting coefficients. For example, FIR filters with linear phase have symmetry about their center coefficient $M/2$; approximately one-half of the coefficients of an odd length, that is, M even, half-band FIR filter, namely, a filter that

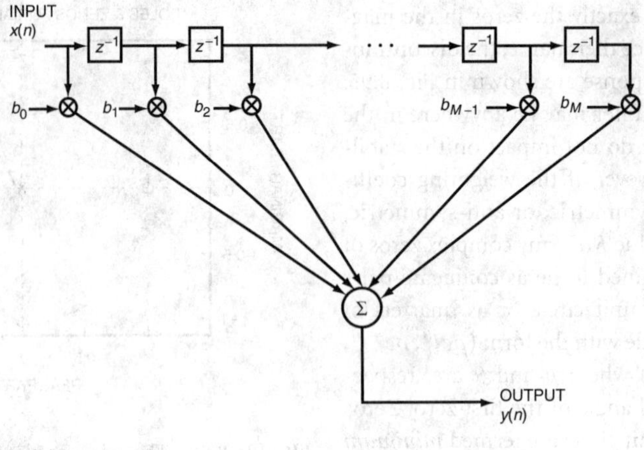

FIGURE 8.97   Direct form structure of an FIR filter.

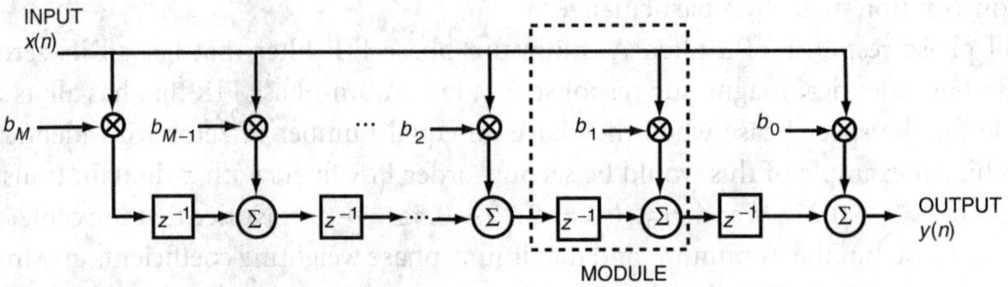

FIGURE 8.98   Modular structure of an FIR filter.

ideally passes frequencies for $0 \leq f \leq 1/4$ and stops frequencies $1/4 < f \leq 1/2$, are zero. Finally, a lattice structure as shown in Fig. 8.99 may be used to realize an FIR filter. The multiplier coefficients within the lattice k_j, $j = 1, \ldots, M$, are not identical to the weighting coefficients of the other FIR filter structures but can be found by an iterative procedure. The attraction of the lattice structure is that it is staightforward to test whether all its zeros lie within the unit circle, namely, the minimum-phase property, and it has low sensitivity to quantization errors. These properties have motivated the use of lattice structures in speech coding applications (Rabiner and Schafer, 1978).

Design Techniques

Linear-phase FIR filters can be designed to meet various filter specifications, such as low-pass, high-pass, band-pass and bandstop filtering. For a low-pass filter, two frequencies are required, namely, the maximum frequency of the pass band below which the magnitude response of the filter is approximately unity, denoted the pass-band corner frequency f_p, and the minimum frequency of the stop band above

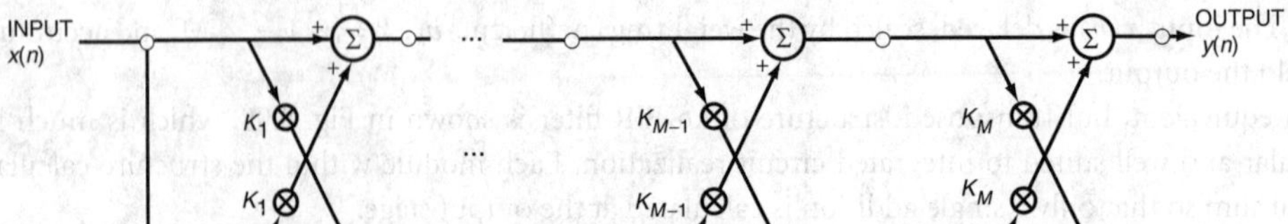

FIGURE 8.99   Lattice structure of an FIR filter.

which the magnitude response of the filter must be less than some prescribed level, named the stop-band corner frequency f_s. The difference between the pass- and stop-band corner frequencies is the *transition-bandwidth*. Generally, the order of FIR filter M required to meet some design specification increases with a reduction in the width of the transition band. There are three established techniques for coefficient design:

- Windowing
- Frequency sampling
- Optimal approximations

The windowing design method calculates the weighting coefficients by sampling the ideal impulse response of an analog filter and multiplying these values by a smoothing window to improve the overall frequency domain response of the filter. The frequency sampling technique samples the ideal frequency domain specification of the filter and calculates the weighting coefficients by inverse transforming these values. However, better results can generally be obtained with the optimal approximations method. The impulse response and magnitude response for a 40th-order optimal half-band FIR low-pass filter designed with the Parks-McClellan algorithm are shown in Fig. 8.100 together with the ideal frequency domain design specification. Notice the zeros in the impulse response. This algorithm minimizes the maximum deviation of the **magnitude response** of the design filter from the ideal magnitude response. The magnitude response of the design filter alternates about the desired specification within the pass band and above the specification in the stop band. The maximum deviation from the desired specification is equalized across the pass and stop bands; this is characteristic of an optimal solution.

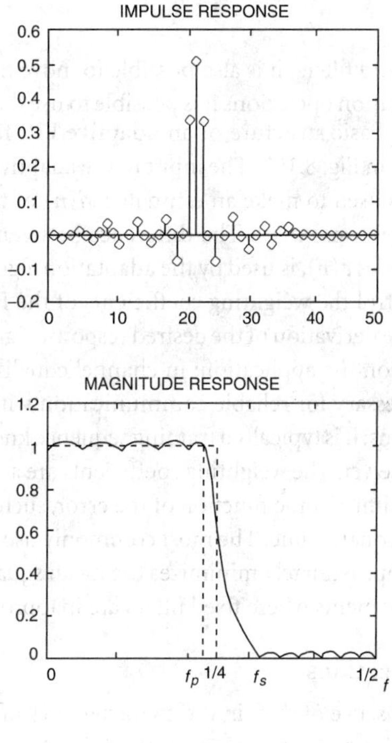

The optimal approximation approach can also be

FIGURE 8.100 Impulse and magnitude response of an optimal 40th-order half-band FIR filter.

used to design *discrete-time differentiators* and *hilbert transformers*, namely, phase shifters. Such filters find extensive application in digital modulation schemes.

Multirate and Adaptive FIR Filters

FIR filter structures can be combined very usefully with multirate processing elements. Such a filtering scheme is, however, time variant and as such may introduce some distortion during processing, but with careful design this can be minimized. Consider the design of a low-pass filter with $f_c = 1/8$; frequencies above this value are essentially to be eliminated. The frequency range between f_c and the Nyquist frequency, that is, 1/2, therefore contains no useful information. The sampling frequency could, provided no aliasing distortion is introduced, therefore be usefully reduced by a factor of 4. Further processing, possibly adaptive, of the signal could then proceed at this reduced rate, hence, reducing the demands on the system design. This operation can be achieved in two ways: A time-invariant low-pass FIR filter, with $f_c = 1/8$, can be designed to operate at the original sampling rate, and its output can be down sampled, termed decimated, by a factor of four; or, more efficiently, the filtering can be performed in two stages, using the same simple half-band filter design twice, each operating at a lower sampling rate. These methods are shown diagrammatically in Fig. 8.101. The second scheme has considerable computational advantages and, because of the nature of

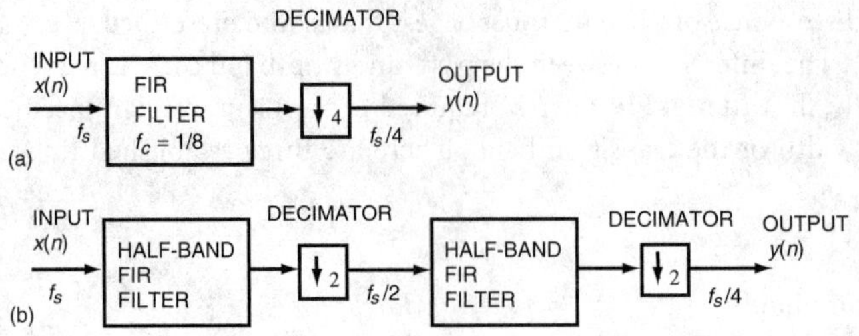

FIGURE 8.101 Multirate **FIR** low-pass **filter** structures: (a) method 1, (b) method 2.

half-band filters, it is also possible to move the decimators in front of the filters. With the introduction of modulation operations it is possible to use the same approach to achieve high-pass and band-pass filtering.

The basic structure of an **adaptive FIR filter** is shown in Fig. 8.102. The input to the adaptive filter $x(n)$ is used to make an estimate $\hat{d}(n)$ of the desired response $d(n)$. The difference between these quantities $e(n)$, is used by the adaptation algorithm to control the weighting coefficients of the FIR filter. The derivation of the desired response signal depends on the application; in channel equalization, as necessary for reliable communication with data modems, it is typically a training sequence known to the receiver. The weighting coefficients are adjusted to minimize some function of the error, such as the

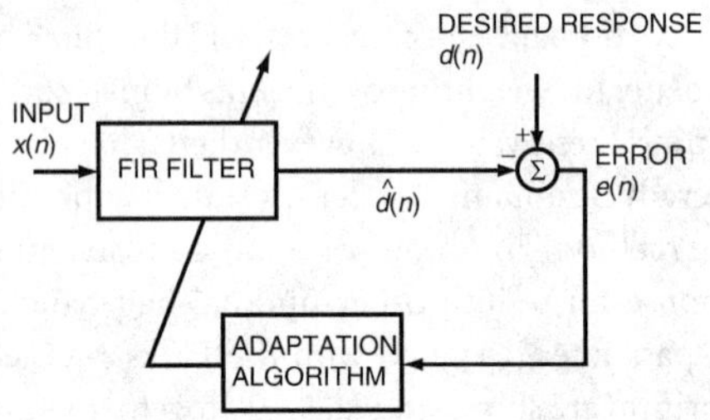

FIGURE 8.102 Adaptive FIR filter structure.

mean square value. The most commonly used adaptive algorithm is the least mean square (lms) algorithm that approximately minimizes the mean square error. Such filters have the ability to adapt to time-varying environments where fixed filters are inappropriate.

Applications

The absence of drift in the characteristics of digitally implemented FIR filters, their ability to reproduce, multirate realizations, and their ability to adapt to time-varying environments has meant that they have found many applications, particularly in telecommunications, for example, in receiver and transmitter design, speech compression and coding, and channel multiplexing. The primary advantages of fixed coefficient FIR filters are their unconditional stability due to the lack of feedback within their structure and their exact linear-phase characteristics. Nonetheless, for applications that require sharp, selective, filtering in standard form, they do require relatively large orders. For some applications, this may be prohibitive and, therefore, recursive; IIR filters are a valuable alternative.

8.8.3 Infinite Impulse Response (IIR) Filters

A digital filter with impulse response having infinite length is called an *infinite impulse response* (IIR) filter. An important class of IIR filters can be described by the difference equation

$$y(n) = b_0 x(n) + b_1 x(n-1) + \cdots + b_M x(n-M) - a_1 y(n-1) - a_2 y(n-2) - \cdots - a_N y(n-N) \quad (8.27)$$

where $x(n)$ is the input, $y(n)$ is the output of the filter, and $(a_1, a_2, \ldots, a_N)$ and $(b_0, b_1, \ldots, b_M)$ are real-value coefficients.

We denote the impulse response by $h(n)$, which is the output of the system when it is driven by a unit impulse at $n = 0$, with the system being initially at rest. The system function $H(z)$ is the z transform of

$h(n)$. For the system in Eq. (8.27), it is given by

$$H(z) = \frac{Y(z)}{X(z)} = \frac{b_0 + b_1 z^{-1} + \cdots + b_M z^{-M}}{1 + a_1 z^{-1} + a_2 z^{-2} + \cdots + a_N z^{-N}} \tag{8.28}$$

where N is called the filter order. Equation (8.28) can be put in the form of poles and zeros as

$$H(z) = b_0 z^{N-M} \frac{(z - q_1)(z - q_2) \cdots (z - q_M)}{(z - p_1)(z - p_2) \cdots (z - p_N)} \tag{8.29}$$

The poles are at $p_1, p_2, \ldots, p_N$. The zeros are at $q_1, q_2, \ldots, q_M$, as well as $N - M$ zeros at the origin.

The frequency response of the IIR filter is the value of the system function evaluated on the unit circle on the complex plane, that is, with $z = e^{j2\pi f}$, where f varies from 0 to 1, or from $-1/2$ to $1/2$. The variable f represents the digital frequency. For simplicity, we write $H(f)$ for $H(z)|_{z=\exp(j2\pi f)}$. Therefore

$$H(f) = b_0 e^{j2\pi(N-M)f} \frac{(e^{j2\pi f} - q_1)(e^{j2\pi f} - q_2) \cdots (e^{j2\pi f} - q_M)}{(e^{j2\pi f} - p_1)(e^{j2\pi f} - p_2) \cdots (e^{j2\pi f} - p_N)} \tag{8.30}$$

$$= |H(f)| e^{j\theta(f)} \tag{8.31}$$

where $|H(f)|$ is the magnitude response and $\theta(f)$ is the phase response.

Compared to an FIR filter, an IIR filter requires a much lower order than an FIR filter to achieve the same requirement of the magnitude response. However, whereas an FIR filter is always stable, an IIR filter can be unstable if the coefficients are not properly chosen. Assuming that the system (8.27) is causal, then it is stable if all of the poles lie inside the unit circle on the z plane. Since the phase of a stable causal IIR filter cannot be made linear, FIR filters are chosen over IIR filters in applications where linear phase is essential.

Realizations

Equation (8.27) suggests a realization of an IIR filter as shown in Fig. 8.103(a), which is called direct form I. By rearranging the structure, we can obtain direct form II, as shown in Fig. 8.103(b). Through transposition, we can obtain transposed direct form I and transposed direct form II, as shown in Fig 8.103(c) and Fig. 8.103(d).

The system function can be put in the form

$$H(z) = \prod_{i=0}^{K} \frac{b_{i0} + b_{i1} z^{-1} + b_{i2} z^{-2}}{1 + a_{i1} z^{-1} + a_{i2} z^{-2}} \tag{8.32}$$

by factoring the numerators and denominators into second-order factors, or in the form

$$H(z) = c_0 + \sum_{i=1}^{K} \frac{b_{i0} + b_{i1} z^{-1}}{1 + a_{i1} z^{-1} + a_{i2} z^{-2}} \tag{8.33}$$

by partial fraction expansion. The value of K is $N/2$ when N is even, and it is $(N+1)/2$ when N is odd. When N is odd, one of a_{i2} must be zero, as well as one of b_{i2} in Eq. (8.32) and one of b_{i1} in Eq. (8.33). According to Eq. (8.32), the IIR filter can be realized by K second-order IIR filters in cascade, as shown in Fig. 8.104(a). According to Eq. (8.33), the IIR filter can be realized by K second-order IIR filters and one scaler (i.e., c_0) in parallel, as depicted in Fig. 8.104(b). Each second-order subsystem can use any of the structures given in Fig. 8.103.

There are other realizations for IIR filters, such as state-space structure, wave structure, and lattice structure. See the references for details. In some situations, it is more convenient or suitable to use software realizations that are implemented by a digital signal processor.

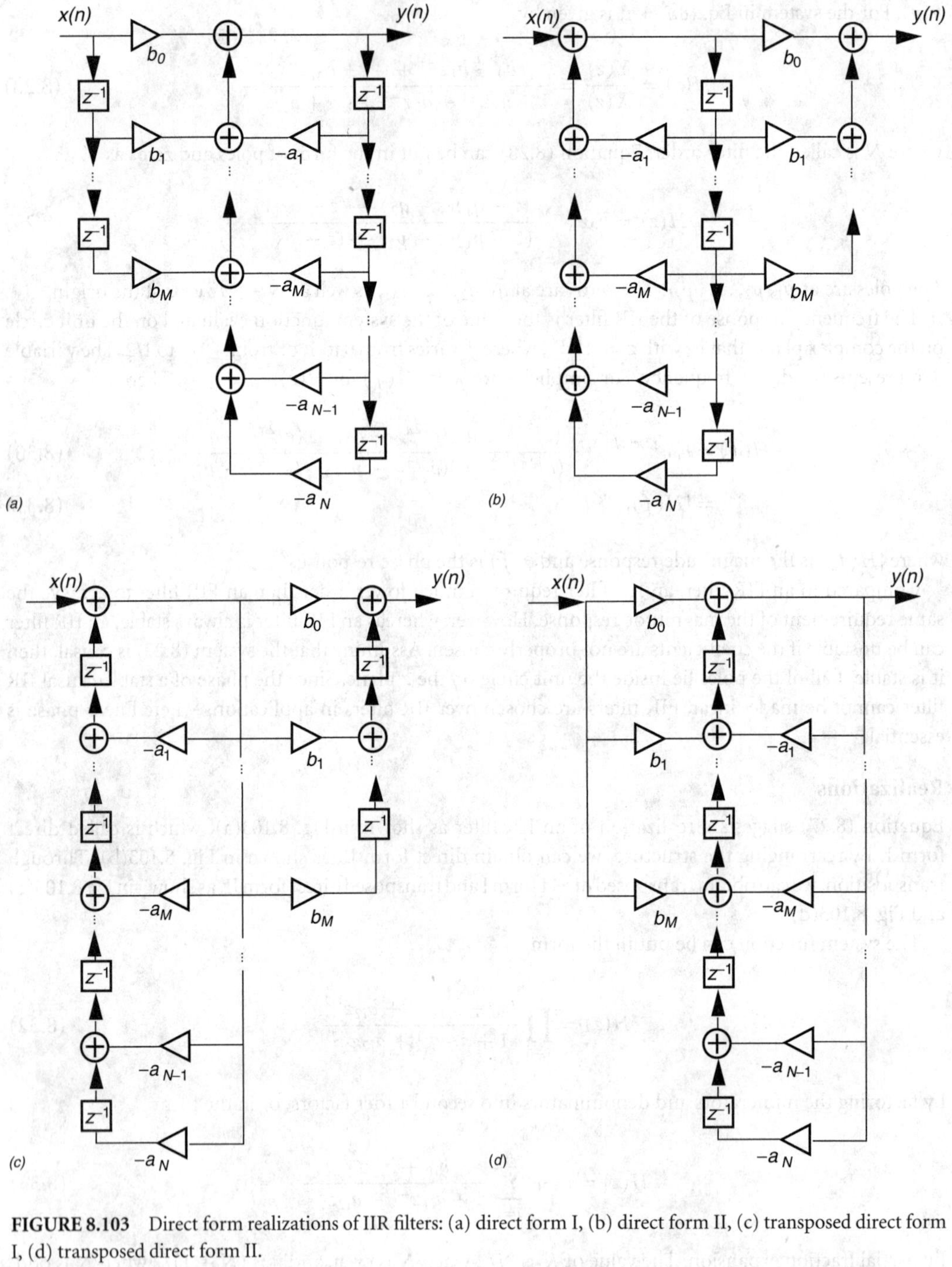

FIGURE 8.103   Direct form realizations of IIR filters: (a) direct form I, (b) direct form II, (c) transposed direct form I, (d) transposed direct form II.

IIR Filter Design

Designing an IIR filter involves choosing the coefficients to satisfy a given specification, usually a magnitude response specification. We assume that the specification is in the form depicted by Fig. 8.105, where the magnitude square must be in the range $(1/(1 + \varepsilon^2), 1)$ in the pass band and it must be no larger than δ^2 in the stop band. The pass-band and the stop-band edges are denoted by f_p and f_s, respectively. No constraint is imposed on the response in the transition band, which lies between a pass band and a stop band.

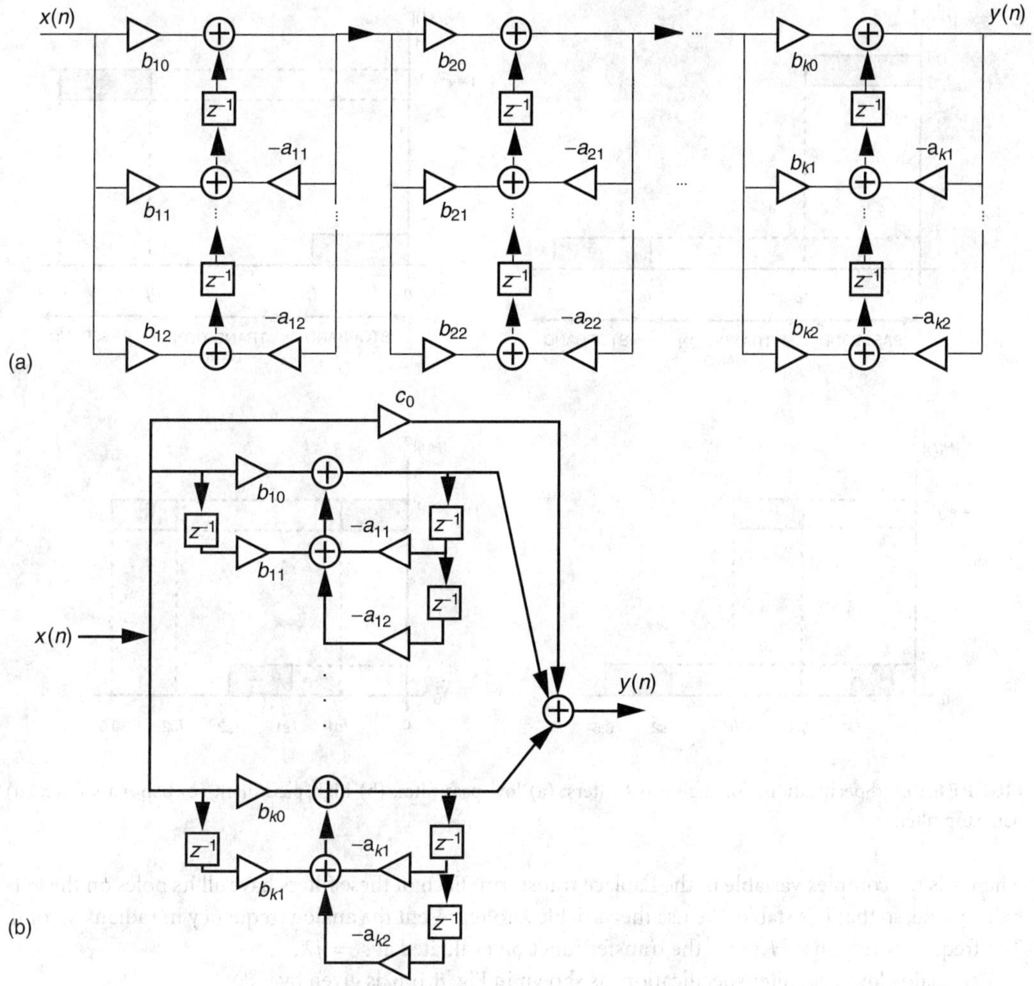

FIGURE 8.104 Realizations of IIR filters: (a) cascade form, (b) parallel form.

There are various IIR filter design methods: design using an analog prototype filter, design using digital frequency transformation, and computer-aided design. In the first method, an analog filter is designed to meet the (analog) specification and the analog filter transfer function is transformed to digital system function. The second method assumes that some digital low-pass filter is available and the desired digital filter is obtained from the digital low-pass filter by a digital frequency transformation. The last method involves algorithms that choose the coefficients so that the response is as close (in some sense) as possible to the desired filter. The first two methods are simple to do, and they are suitable for designing standard filters (low-pass, high-pass, bandpass, and bandstop filters). A computer-aided design requires computer programming, but it can be used to design standard and nonstandard filters. We will focus only on the first method and present some design examples, as well as a summary of some analog filters, in the following sections.

Analog Filters

Here, we summarize three basic types of analog low-pass filters that can be used as prototype for designing IIR filters. For each type, we give the transfer function, its magnitude response, and the order N needed to satisfy the (analog) specification. We use $H_a(s)$ to denote the transfer function of an analog filter,

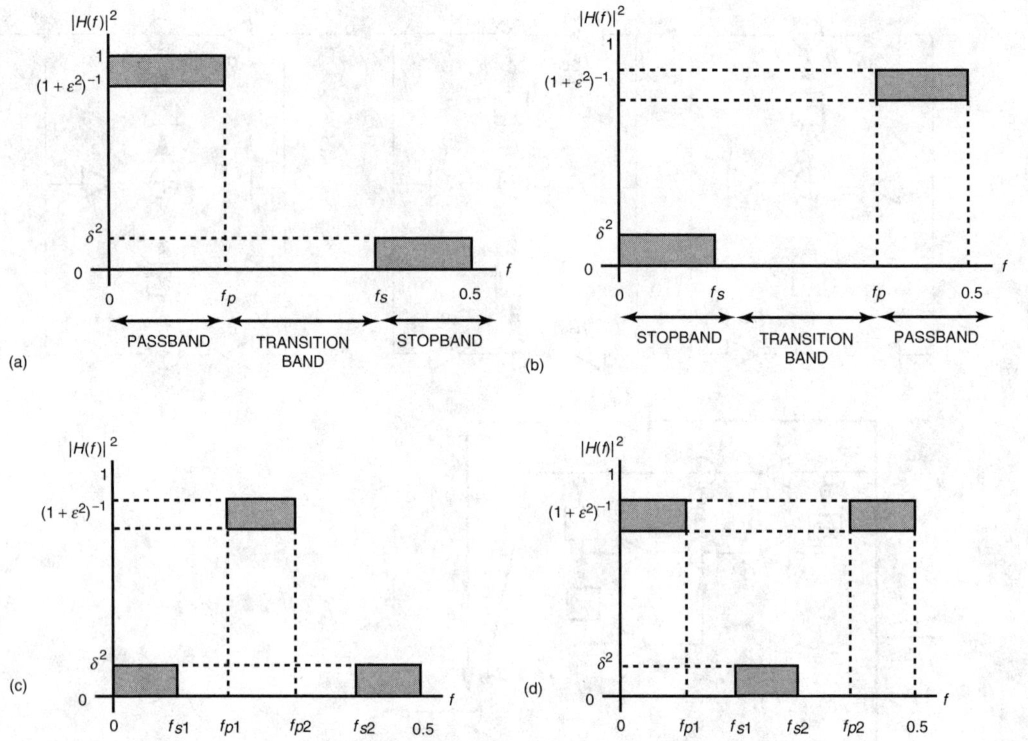

FIGURE 8.105 Specifications for digital IIR filters: (a) low-pass filter, (b) high-pass filter, (c) bandpass filter, (d) bandstop filter.

where s is the complex variable in the Laplace transform. Each of these filters have all its poles on the left-half s plane, so that it is stable. We use the variable λ to represent the analog frequency in radians/second. The frequency response $H_a(\lambda)$ is the transfer function evaluated at $s = j\lambda$.

The analog low-pass filter specification, as shown in Fig. 8.106, is given by

$$(1 + \varepsilon^2)^{-1} \leq |H_a(\lambda)|^2 \leq 1 \quad \text{for } 0 \leq (\lambda/2\pi) \leq (\lambda_p/2\pi)\,\text{Hz}$$
$$0 \leq |H_a(\lambda)|^2 \leq \delta^2 \quad \text{for } (\lambda_s/2\pi) \leq (\lambda/2\pi) < \infty\,\text{Hz} \tag{8.34}$$

where λ_p and λ_s are the pass-band edge and stop-band edge, respectively.

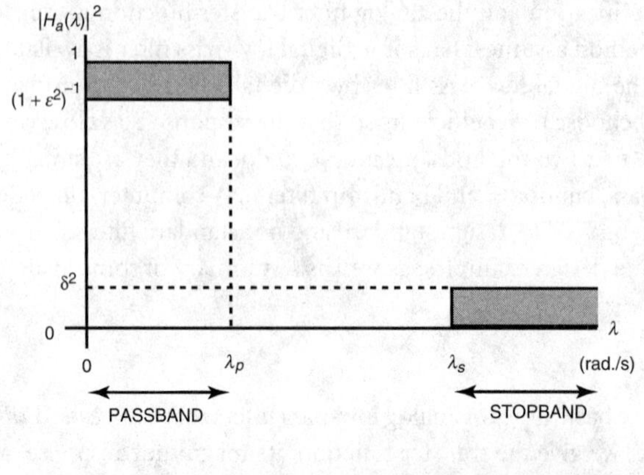

FIGURE 8.106 Specification for an analog low-pass filter.

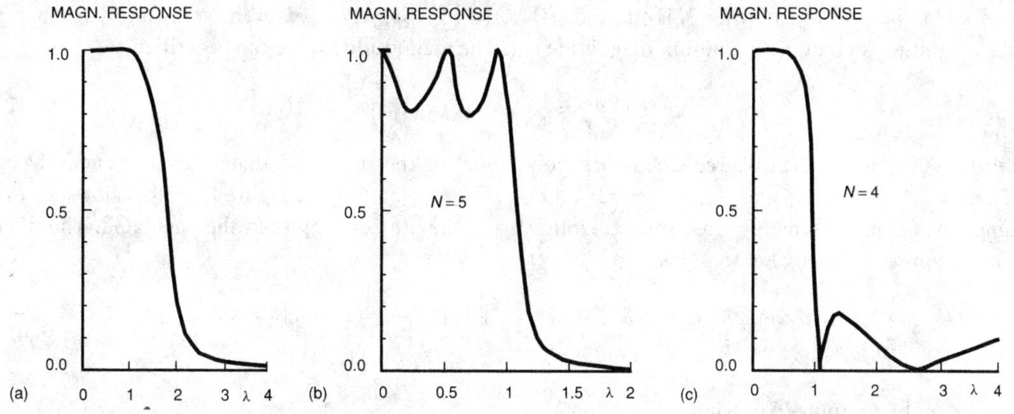

FIGURE 8.107 Magnitude responses of low-pass analog filters: (a) Butterworth filter, (b) Chebyshev filter, (c) inverse Chebyshev filter.

Butterworth Filters

The transfer function of an Nth-order Butterworth filter is given by

$$H_a(s) = \begin{cases} \displaystyle\prod_{i=1}^{N/2} \frac{1}{(s/\lambda_c)^2 - 2Re(s_i)(s/\lambda_c) + 1}, & N = \text{even} \\[2mm] \displaystyle\frac{1}{(s/\lambda_c) + 1} \prod_{i=1}^{(N-1)/2} \frac{1}{(s/\lambda_c)^2 - 2Re(s_i)(s/\lambda_c) + 1}, & N = \text{odd} \end{cases} \tag{8.35}$$

where $s_i = \exp\{j[1 + (2i - 1)/N]\pi/2\}$ and λ_c is the frequency where the magnitude drops by 3 dB. The magnitude response square is

$$|H_a(\lambda)|^2 = [1 + (\lambda/\lambda_c)^{2N}]^{-1} \tag{8.36}$$

Figure 8.107(a) shows the magnitude response $|H_a(\lambda)|$. To satisfy the specification in Eq. (8.34), the filter order is

$$N = \text{integer} \geq \frac{\log\left[\varepsilon/(\delta^{-2} - 1)^{\frac{1}{2}}\right]}{\log[\lambda_p/\lambda_s]} \tag{8.37}$$

and the value of λ_c is chosen from the following range:

$$\lambda_p \varepsilon^{-1/N} \leq \lambda_c \leq \lambda_s(\delta^{-2} - 1)^{-1/(2N)} \tag{8.38}$$

Chebyshev Filters (Type I Chebyshev Filters)

The Nth-order Chebyshev filter has a transfer function given by

$$H_a(s) = C \prod_{i=1}^{N} \frac{1}{(s - p_i)} \tag{8.39}$$

where

$$p_i = -\lambda_p \sinh(\phi)\sin\left(\frac{2i - 1}{2N}\pi\right) + j\lambda_p \cosh(\phi)\cos\left(\frac{2i - 1}{2N}\pi\right) \tag{8.40}$$

$$\phi = \frac{1}{N} \ln\left[\frac{1 + (1 + \varepsilon^2)^{\frac{1}{2}}}{\varepsilon}\right] \tag{8.41}$$

and where $C = -\prod_{i=1}^{N} p_i$ when N is odd and $(1+\varepsilon^2)^{-\frac{1}{2}} \prod_{i=1}^{N} p_i$ when N is even. Note that C normalizes the magnitude so that the maximum magnitude is 1. The magnitude square can be written as

$$|H_a(\lambda)|^2 = \left[1 + \varepsilon^2 T_N^2(\lambda/\lambda_p)\right]^{-1} \tag{8.42}$$

where $T_N(x)$ is the Nth-degree Chebyshev polynomial of the first kind that is given recursively by $T_0(x) = 1, T_1(x) = x$, and $T_{n+1}(x) = 2x T_n(x) - T_{n-1}(x)$ for $n \geq 1$. Figure 8.107(b) shows an example of the magnitude response square. Notice that there are equiripples in the pass band. The filter order required to satisfy Eq. (8.34) is

$$N \geq \frac{\log\left\{\left[(\delta^{-2}-1)^{\frac{1}{2}}/\varepsilon\right] + \left[(\delta^{-2}-1)/\varepsilon^2 - 1\right]^{\frac{1}{2}}\right\}}{\log\{(\lambda_s/\lambda_p) + \left[(\lambda_s/\lambda_p)^2 - 1\right]^{\frac{1}{2}}} \tag{8.43}$$

which can be computed knowing $\varepsilon, \delta, \lambda_p$, and λ_s.

Inverse Chebyshev Filters (Type II Chebyshev Filters)

For inverse Chebyshev filters, the equiripples are inside the stop band, as opposed to the pass band in the case of Chebyshev filters. The magnitude response square of the inverse Chebyshev filter is

$$|H_a(\lambda)|^2 = \left[1 + (\delta^{-2} - 1)/T_N^2(\lambda_s/\lambda)\right]^{-1} \tag{8.44}$$

Figure 8.107(c) depicts an example of Eq. (8.44). The value of $|H_a(\infty)|$ equals 0 if N is odd, and it equals δ if N is even. The transfer function giving rise to Eq. (8.44) is

$$H_a(s) = \begin{cases} C \prod_{i=1}^{N} \dfrac{(s-q_i)}{(s-p_i)}, & N \text{ is even} \\[3mm] \dfrac{C}{(s-p_{(N+1)/2})} \displaystyle\prod_{\substack{i=1 \\ i \neq (N+1)/2}}^{N} \dfrac{(s-q_i)}{(s-p_i)}, & N \text{ is odd} \end{cases} \tag{8.45}$$

where

$$p_i = \frac{\lambda_s}{\alpha_i^2 + \beta_i^2}(\alpha_i - j\beta_i); \qquad q_i = j\frac{\lambda_s}{\cos\left(\dfrac{2i-1}{2N}\pi\right)} \tag{8.46}$$

$$\alpha_i = -\sinh(\phi)\sin\left(\frac{2i-1}{2N}\pi\right); \qquad \beta_i = \cosh(\phi)\cos\left(\frac{2i-1}{2N}\pi\right) \tag{8.47}$$

$$\phi = \frac{1}{N}\cosh^{-1}(\delta^{-1}) = \frac{1}{N}\ln\left[\delta^{-1} + (\delta^{-2}-1)^{\frac{1}{2}}\right] \tag{8.48}$$

and where

$$C = \prod_{i=1}^{N}(p_i/q_i)$$

when N is even and

$$C = -p_{(N+1)/2}\prod_{i=1, i \neq (N+1)/2}^{N}(p_i/q_i)$$

when N is odd. The filter order N required to satisfy Eq. (8.34) is the same as that for the Chebyshev filter, which is given by Eq. (8.43).

Comparison

In comparison, the Butterworth filter requires a higher order than both types of Chebyshev filters to satisfy the same specification. There is another type of filters called elliptic filters (Cauer filters) that have equiripples in the pass band as well as in the stop band. Because of the lengthy expressions, this type of filters is not given here (see the references). The Butterworth filter and the inverse Chebyshev filter have better (closer to linear) phase characteristics in the pass band than Chebyshev and elliptic filters. Elliptic filters require a smaller order than Chebyshev filters to satisfy the same specification.

Design Using Bilinear Transformations

One of the most simple ways of designing a digital filter is by way of transforming an analog low-pass filter to the desired digital filter. Starting from the desired digital filter specification, the low-pass analog specification is obtained. Then, an analog low-pass filter $H_a(s)$ is designed to meet the specification. Finally, the desired digital filter is obtained by transforming $H_a(s)$ to $H(z)$. There are several types of transformation. The all-around best one is the bilinear transformation, which is the subject of this subsection.

In a bilinear transformation, the variable s in $H_a(s)$ is replaced with a bilinear function of z to obtain $H(z)$. Bilinear transformations for the four standard types of filters, namely, low-pass filter (LPF), high-pass filter (HPF), bandpass filter (BPF), and bandstop filter (BSF), are shown in Table 8.8. The second column in the table gives the relations between the variables s and z. The value of T can be chosen arbitrarily without affecting the resulting design. The third column shows the relations between the analog

TABLE 8.8 Bilinear Transformations for Standard Digital Filters

	Transformation (between s and z)	Frequency Relations (between f and λ)	Analog LPF Passband Edge	Analog LPF Stopband Edge
Digital LPF to/from analog LPF	$s = \dfrac{2}{T}\dfrac{1-z^{-1}}{1+z^{-1}}$ $z = \dfrac{(2/T)+s}{(2/T)-s}$	$f = \dfrac{1}{\pi}\tan^{-1}\left(\dfrac{\lambda T}{2}\right)$ $\lambda = \dfrac{2}{T}\tan(\pi f)$	$\lambda_p = \dfrac{2}{T}\tan(\pi f_p)$	$\lambda_s = \dfrac{2}{T}\tan(\pi f_s)$
Digital HPF to/from analog LPF	$s = \dfrac{T}{2}\dfrac{1+z^{-1}}{1-z^{-1}}$ $z = \dfrac{s+(T/2)}{s-(T/2)}$	$f = \begin{cases} -\dfrac{1}{2}+\dfrac{1}{\pi}\tan^{-1}\left(\dfrac{2\lambda}{T}\right), & \lambda \geq 0 \\[2mm] \dfrac{1}{2}+\dfrac{1}{\pi}\tan^{-1}\left(\dfrac{2\lambda}{T}\right), & \lambda \leq 0 \end{cases}$ $\lambda = \dfrac{T}{2}\tan[\pi(f+0.5)], \lambda \geq 0$	$\lambda_p = \dfrac{T}{2}\tan\left[\pi\left(f_p+\dfrac{1}{2}\right)\right]$	$\lambda_s = \dfrac{T}{2}\tan\left[\pi\left(f_s+\dfrac{1}{2}\right)\right]$
Digital BPF to/from analog LPF	$s = \dfrac{\tilde{s}^2 + \tilde{\lambda}_0^2}{W\tilde{s}}$ where $\tilde{s} = \dfrac{2}{T}\dfrac{1-z^{-1}}{1+z^{-1}}$	$\lambda = \dfrac{\tilde{\lambda}^2 - \tilde{\lambda}_0^2}{W\tilde{\lambda}}$ where $\tilde{\lambda} = \dfrac{2}{T}\tan(\pi f)$	$\lambda_p = \max\left\{ \left\|\dfrac{\tilde{\lambda}_0^2 - \tilde{\lambda}_{p1}^2}{W\tilde{\lambda}_{p1}}\right\|, \left\|\dfrac{\tilde{\lambda}_0^2 - \tilde{\lambda}_{p2}^2}{W\tilde{\lambda}_{p2}}\right\| \right\}$ where $\tilde{\lambda}_{p1} = \dfrac{2}{T}\tan(\pi f_{p1})$ $\tilde{\lambda}_{p2} = \dfrac{2}{T}\tan(\pi f_{p2})$	$\lambda_s = \min\left\{ \left\|\dfrac{\tilde{\lambda}_0^2 - \tilde{\lambda}_{s1}^2}{W\tilde{\lambda}_{s1}}\right\|, \left\|\dfrac{\tilde{\lambda}_0^2 - \tilde{\lambda}_{s2}^2}{W\tilde{\lambda}_{s2}}\right\| \right\}$ where $\tilde{\lambda}_{s1} = \dfrac{2}{T}\tan(\pi f_{s1})$ $\tilde{\lambda}_{s2} = \dfrac{2}{T}\tan(\pi f_{s2})$
Digital BSF to/from analog LPF	$s = \dfrac{W\tilde{s}}{\tilde{s}^2 + \tilde{\lambda}_0^2}$ where $\tilde{s} = \dfrac{2}{T}\dfrac{1-z^{-1}}{1+z^{-1}}$	$\lambda = \dfrac{W\tilde{\lambda}}{\tilde{\lambda}_0^2 - \tilde{\lambda}^2}$ where $\tilde{\lambda} = \dfrac{2}{T}\tan(\pi f)$	$\lambda_p = \max\left\{ \left\|\dfrac{W\tilde{\lambda}_{p1}}{\tilde{\lambda}_0^2 - \tilde{\lambda}_{p1}^2}\right\|, \left\|\dfrac{W\tilde{\lambda}_{p2}}{\tilde{\lambda}_0^2 - \tilde{\lambda}_{p2}^2}\right\| \right\}$ where $\tilde{\lambda}_{p1} = \dfrac{2}{T}\tan(\pi f_{p1})$ $\tilde{\lambda}_{p2} = \dfrac{2}{T}\tan(\pi f_{p2})$	$\lambda_s = \min\left\{ \left\|\dfrac{W\tilde{\lambda}_{s1}}{\tilde{\lambda}_0^2 - \tilde{\lambda}_{s1}^2}\right\|, \left\|\dfrac{W\tilde{\lambda}_{s2}}{\tilde{\lambda}_0^2 - \tilde{\lambda}_{s2}^2}\right\| \right\}$ where $\tilde{\lambda}_{s1} = \dfrac{2}{T}\tan(\pi f_{s1})$ $\tilde{\lambda}_{s2} = \dfrac{2}{T}\tan(\pi f_{s2})$

frequency λ and the digital frequency f, obtained from the relations between s and z by replacing s with $j\lambda$ and z with $\exp(j2\pi f)$. The fourth and fifth columns show the required pass-band and stop-band edges for the analog LPF. Note that the allowable variations in the pass band and stop band, or equivalently the values of ε and δ, for the analog low-pass filter remain the same as those for the desired digital filter. Notice that for the BPF and BSF, the transformation is performed in two steps: one for transforming an analog LPF to/from an analog BPF (or BSF), and the other for transforming an analog BPF (or BSF) to/from a digital BPF (or BSF). The values of W and $\tilde{\lambda}_0$ are chosen by the designer. Some convenient choices are: (1) $W = \tilde{\lambda}_{p2} - \tilde{\lambda}_{p1}$ and $\tilde{\lambda}_0^2 = \tilde{\lambda}_{p1}\tilde{\lambda}_{p2}$, which yield $\lambda_p = 1$; (2) $W = \tilde{\lambda}_{s2} - \tilde{\lambda}_{s1}$ and $\tilde{\lambda}_0^2 = \tilde{\lambda}_{s1}\tilde{\lambda}_{s2}$, which yield $\lambda_s = 1$. We demonstrate the design process by the following two examples.

Design Example 1

Consider designing a digital LPF with a pass-band edge at $f_p = 0.15$ and stop-band edge at $f_s = 0.25$. The magnitude response in the pass band is to stay within 0 and -2 dB, although it must be no larger than -40 dB in the stop band. Assuming that the analog Butterworth filter is to be used as the prototype filter, we proceed the design as follows.

1. Compute $\varepsilon, \delta, \lambda_p, \lambda_s$, and the analog filter order. An attenuation of -2 dB means $10 \log_{10}(1+\varepsilon^2)^{-1} = -2$, which yields $\varepsilon = 0.7648$, and attenuation of -40 dB means $10 \log_{10}(\delta^2) = -40$, which gives $\delta = 0.01$. From Table 8.8, the analog pass-band and stop-band edges are $\lambda_p = (2/T)\tan(\pi f_p)$ and $\lambda_s = (2/T)\tan(\pi f_s)$. For convenience, we let $T = 2$, which makes $\lambda_p = 0.5095$ and $\lambda_s = 1.0$. Now, we can calculate the required order of the analog Butterworth filter from Eq. (8.37), which yields $N \geq 7.23$; thus, we choose $N = 8$.

2. Obtain the analog LPF transfer function. From Eq. (8.38), we can pick a value of λ_c between 0.5269 and 0.5623. Let us choose $\lambda_c = 0.54$. With $N = 8$, the transfer function is calculated from Eq. (8.35) as

$$H_a(s) = \frac{7.2302 \times 10^{-3}}{(s^2 + 0.2107s + 0.2916)(s^2 + 0.6s + 0.2916)}$$

$$\bullet \frac{1}{(s^2 + 0.8980s + 0.2916)(s^2 + 1.0592s + 0.2916)} \tag{8.49}$$

3. Obtain the digital filter. Using the transformation in Table 8.8, we transform Eq. (8.49) to the desired digital filter by replacing s with $(2/T)(z - 1)/(z + 1) = (z - 1)/(z + 1)$ (since we let $T = 2$). After substitution and simplification, the resulting digital LPF has a system function given by

$$H(z) = H_a(s)|_{s=(z-1)/(z+1)} = \frac{4.9428 \times 10^{-4}(z^2 + 2z + 1)^4}{(z^2 - 0.9431z + 0.7195)(z^2 - 0.7490z + 0.3656)}$$

$$\bullet \frac{1}{(z^2 - 0.6471z + 0.1798)(z^2 - 0.6027z + 0.0988)} \tag{8.50}$$

The magnitude response $|H(f)|$ is plotted in Fig. 8.108(a). The magnitude is 0.8467 at the pass-band edge $f_p = 0.15$, and it is 0.0072 at the stop-band edge $f_s = 0.25$, both of which are within the specification.

Design Example 2

Now, suppose that we wish to design a digital BSF with pass-band edges at $f_{p1} = 0.15$ and $f_{p2} = 0.30$ and stop-band edges at $f_{s1} = 0.20$ and $f_{s2} = 0.25$. The magnitude response in the pass bands is to stay within 0 and -2 dB, although it must be no larger than -40 dB in the stop band. Let us use the analog type I

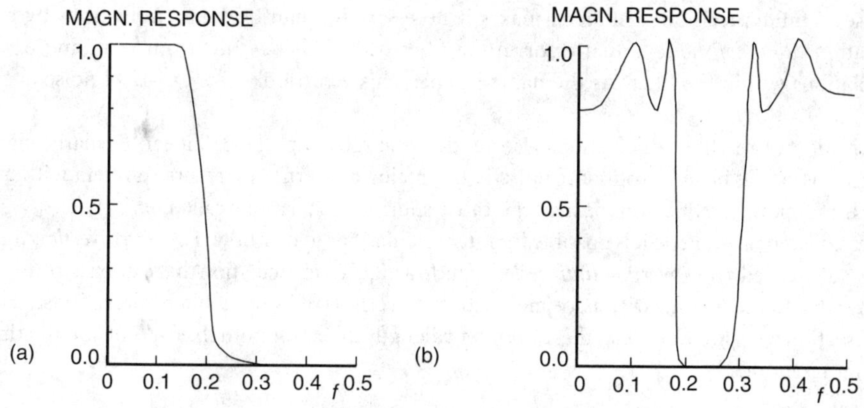

FIGURE 8.108 Magnitude responses of the designed digital filters in the examples: (a) LPF (example 1), (b) BSF (example 2).

Chebyshev filter as the prototype filter. Following the same design process as the first example, we have the following.

1. Compute $\varepsilon, \delta, \lambda_p, \lambda_s$, and the analog filter order. From the first example, we have $\varepsilon = 0.7648$ and $\delta = 0.01$. From Table 8.8 and letting $T = 2$, we compute the analog pass-band and stop-band edges: $\tilde{\lambda}_{p1} = \tan(\pi f_{p1}) = 0.5095, \tilde{\lambda}_{p2} = \tan(\pi f_{p2}) = 1.3764, \tilde{\lambda}_{s1} = \tan(\pi f_{s1}) = 0.7265, \tilde{\lambda}_{s2} = \tan(\pi f_{s2}) = 1.0$. We choose $W = \tilde{\lambda}_{p2} - \tilde{\lambda}_{p1} = 0.8669$ and $\tilde{\lambda}_0^2 = \tilde{\lambda}_{p1}\tilde{\lambda}_{p2} = 0.7013$, so that $\lambda_p = 1.0$ and $\lambda_s = \min\{3.6311, 2.9021\} = 2.9021$. From Eq. (8.43), the required analog filter order is $N \geq 3.22$; thus, we choose $N = 4$.

2. Obtain the analog LPF transfer function. From Eq. (8.39) and with $N = 4$, we find the transfer function of the analog Chebyshev filter as

$$H_a(s) = \frac{1.6344 \times 10^{-1}}{(s^2 + 0.2098s + 0.9287)(s^2 + 0.5064s + 0.2216)} \tag{8.51}$$

3. Obtain the digital filter. Using Table 8.8, we transform Eq. (8.51) to $H(z)$ by replacing s with $W\tilde{s}/(\tilde{s}^2 + \tilde{\lambda}_0^2) = W(z^2 - 1)/[(z-1)^2 + \tilde{\lambda}_0^2(z+1)^2]$ since $T = 2$ and $\tilde{s} = (z-1)/(z+1)$. After substitution and simplification, the resulting digital BSF is

$$H(z) = \frac{1.7071 \times 10^{-1}(z^4 - 0.7023z^3 + 2.1233z^2 - 0.7023z + 1)^2}{(z^4 - 0.5325z^3 + 1.1216z^2 - 0.4746z + 0.8349)}$$
$$\bullet \frac{1}{(z^4 - 0.3331z^3 + 0.0660z^2 - 0.0879z + 0.3019)} \tag{8.52}$$

The magnitude response is plotted in Fig. 8.108(b), which satisfies the specification.

8.8.4 Finite Wordlength Effects

Practical digital filters must be implemented with finite precision numbers and arithmetic. As a result, both the filter coefficients and the filter input and output signals are in discrete form. This leads to four types of finite wordlength effects.

Discretization (quantization) of the filter coefficients has the effect of perturbing the location of the filter poles and zeroes. As a result, the actual filter response differs slightly from the ideal response. This *deterministic* frequency response error is referred to as *coefficient quantization error*.

The use of finite precision arithmetic makes it necessary to quantize filter calculations by rounding or truncation. *Roundoff noise* is that error in the filter output that results from rounding or truncating calculations within the filter. As the name implies, this error looks like low-level noise at the filter output.

Quantization of the filter calculations also renders the filter slightly nonlinear. For large signals this nonlinearity is negligible and roundoff noise is the major concern. For recursive filters with a zero or constant input, however, this nonlinearity can cause spurious oscillations called *limit cycles*.

With fixed-point arithmetic it is possible for filter calculations to overflow. The term *overflow oscillation*, sometimes also called *adder overflow limit cycle*, refers to a high-level oscillation that can exist in an otherwise stable filter due to the nonlinearity associated with the overflow of internal filter calculations.

In this section we examine each of these finite wordlength effects for both fixed-point and floating-point number representions.

Number Representation

In digital signal processing, $(B + 1)$-bit fixed-point numbers are usually represented as two's-complement signed fractions in the format

$$b_0.b_{-1}b_{-2}\cdots b_{-B}$$

The number represented is then

$$X = -b_0 + b_{-1}2^{-1} + b_{-2}2^{-2} + \cdots + b_{-B}2^{-B} \tag{8.53}$$

where b_0 is the sign bit and the number range is $-1 \le X < 1$. The advantage of this representation is that the product of two numbers in the range from -1 to 1 is another number in the same range.

Floating-point numbers are represented as

$$X = (-1)^s m \, 2^c \tag{8.54}$$

where s is the sign bit, m is the *mantissa*, and c is the *characteristic* or *exponent*. To make the representation of a number unique, the mantissa is *normalized* so that $0.5 \le m < 1$.

Fixed-Point Quantization Errors

In fixed-point arithmetic, a multiply doubles the number of significant bits. For example, the product of the two 5-bit numbers 0.0011 and 0.1001 is the 10-bit number 00.00011011. The extra bit to the left of the decimal point can be discarded without introducing any error. However, the least significant four of the remaining bits must ultimately be discarded by some form of quantization so that the result can be stored to five bits for use in other calculations. In the preceding example this results in 0.0010 (quantization by rounding) or 0.0001 (quantization by truncating). When a sum of products calculation is performed, the quantization can be performed either after each multiply or after all products have been summed with double-length precision.

We will examine the case of fixed-point quantization by rounding. If X is an exact value then the rounded value will be denoted $Q_r(X)$. If the quantized value has B bits to the right of the decimal point, the quantization step size is

$$\Delta = 2^{-B} \tag{8.55}$$

Since rounding selects the quantized value nearest the unquantized value, it gives a value which is never more than $\pm\Delta/2$ away from the exact value. If we denote the rounding error by

$$\varepsilon_r = Q_r(X) - X \tag{8.56}$$

then

$$-\frac{\Delta}{2} \le \varepsilon_r \le \frac{\Delta}{2} \qquad (8.57)$$

The error resulting from quantization can be modeled as a random variable uniformly distributed over the appropriate error range. Therefore, calculations with roundoff error can be considered error-free calculations that have been corrupted by additive white noise. The mean of this noise for rounding is

$$m_{\varepsilon_r} = E\{\varepsilon_r\} = \frac{1}{\Delta} \int_{-\frac{\Delta}{2}}^{\frac{\Delta}{2}} \varepsilon_r \, d\varepsilon_r = 0 \qquad (8.58)$$

where $E\{\ \}$ represents the operation of taking the expected value of a random variable. Similarly, the variance of the noise for rounding is

$$\sigma_{\varepsilon_r}^2 = E\{(\varepsilon_r - m_{\varepsilon_r})^2\} = \frac{1}{\Delta} \int_{-\frac{\Delta}{2}}^{\frac{\Delta}{2}} (\varepsilon_r - m_{\varepsilon_r})^2 \, d\varepsilon_r = \frac{\Delta^2}{12} \qquad (8.59)$$

Floating-Point Quantization Errors

With floating-point arithmetic it is necessary to quantize after both multiplications and additions. The addition quantization arises because, prior to addition, the mantissa of the smaller number in the sum is shifted right until the exponent of both numbers is the same. In general, this gives a sum mantissa that is too long and so must be quantized.

We will assume that quantization in floating-point arithmetic is performed by rounding. Because of the exponent in floating-point arithmetic, it is the relative error that is important. The relative error is defined as

$$\varepsilon_r = \frac{Q_r(X) - X}{X} = \frac{\varepsilon_r}{X} \qquad (8.60)$$

Since $X = (-1)^s \, m \, 2^c$, $Q_r(X) = (-1)^s \, Q_r(m) \, 2^c$ and

$$\varepsilon_r = \frac{Q_r(m) - m}{m} = \frac{\varepsilon}{m} \qquad (8.61)$$

If the quantized mantissa has B bits to the right of the decimal point, $|\varepsilon| < \Delta/2$ where, as before, $\Delta = 2^{-B}$. Therefore, since $0.5 \le m < 1$,

$$|\varepsilon_r| < \Delta \qquad (8.62)$$

If we assume that ε is uniformly distributed over the range from $-\Delta/2$ to $\Delta/2$ and m is uniformly distributed over 0.5–1,

$$m_{\varepsilon_r} = E\left\{\frac{\varepsilon}{m}\right\} = 0$$

$$\sigma_{\varepsilon_r}^2 = E\left\{\left(\frac{\varepsilon}{m}\right)^2\right\} = \frac{2}{\Delta} \int_{\frac{1}{2}}^{1} \int_{-\frac{\Delta}{2}}^{\frac{\Delta}{2}} \frac{\varepsilon^2}{m^2} \, d\varepsilon \, dm = \frac{\Delta^2}{6} = (0.167)2^{-2B} \qquad (8.63)$$

From Eq. (8.60) we can represent a quantized floating-point value in terms of the unquantized value and the random variable ε_r using

$$Q_r(X) = X(1 + \varepsilon_r) \qquad (8.64)$$

Therefore, the finite-precision product $X_1 X_2$ and the sum $X_1 + X_2$ can be written

$$fl(X_1 X_2) = X_1 X_2(1 + \varepsilon_r) \qquad (8.65)$$

and

$$fl(X_1 + X_2) = (X_1 + X_2)(1 + \varepsilon_r) \tag{8.66}$$

where ε_r is zero mean with the variance of Eq. (8.63).

Roundoff Noise

To determine the roundoff noise at the output of a digital filter we will assume that the noise due to a quantization is stationary, white, and uncorrelated with the filter input, output, and internal variables. This assumption is good if the filter input changes from sample to sample in a sufficiently complex manner. It is not valid for zero or constant inputs for which the effects of rounding are analyzed from a limit cycle perspective.

To satisfy the assumption of a sufficiently complex input, roundoff noise in digital filters is often calculated for the case of a zero-mean white noise filter input signal $x(n)$ of variance σ_x^2. This simplifies calculation of the output roundoff noise because expected values of the form $E\{x(n)x(n-k)\}$ are zero for $k \neq 0$ and give σ_x^2 when $k = 0$. If there is more than one source of roundoff error in a filter, it is assumed that the errors are uncorrelated so that the output noise variance is simply the sum of the contributions from each source. This approach to analysis has been found to give estimates of the output roundoff noise that are close to the noise actually observed in practice.

Another assumption that will be made in calculating roundoff noise is that the product of two quantization errors is zero. To justify this assumption, consider the case of a 16-bit fixed-point processor. In this case a quantization error is of the order 2^{-15}, whereas the product of two quantization errors is of the order 2^{-30}, which is negligible by comparison.

The simplest case to analyze is a finite impulse response filter realized via the convolution summation

$$y(n) = \sum_{k=0}^{N-1} h(k)x(n - k) \tag{8.67}$$

When fixed-point arithmetic is used and quantization is performed after each multiply, the result of the N multiplies is N times the quantization noise of a single multiply. For example, rounding after each multiply gives, from Eqs. (8.55) and (8.59), an output noise variance of

$$\sigma_o^2 = N \frac{2^{-2B}}{12} \tag{8.68}$$

Virtually all digital signal processor integrated circuits contain one or more double-length accumulator registers that permit the sum-of-products in Eq. (8.67) to be accumulated without quantization. In this case only a single quantization is necessary following the summation and

$$\sigma_o^2 = \frac{2^{-2B}}{12} \tag{8.69}$$

For the floating-point roundoff noise case we will consider Eq. (8.67) for $N = 4$ and then generalize the result to other values of N. The finite-precision output can be written as the exact output plus an error term $e(n)$. Thus

$$y(n) + e(n) = (\{[h(0)x(n)(1 + \varepsilon_1(n)) + h(1)x(n - 1)(1 + \varepsilon_2(n))][1 + \varepsilon_3(n)]$$
$$+ h(2)x(n - 2)(1 + \varepsilon_4(n))\}\{1 + \varepsilon_5(n)\} + h(3)x(n - 3)(1 + \varepsilon_6(n)))(1 + \varepsilon_7(n)) \tag{8.70}$$

In Eq. (8.70) $\varepsilon_1(n)$ represents the error in the first product, $\varepsilon_2(n)$ the error in the second product, $\varepsilon_3(n)$ the error in the first addition, etc. Notice that it has been assumed that the products are summed in the order implied by the summation of Eq. (8.67).

Expanding Eq. (8.70), ignoring products of error terms, and recognizing $y(n)$ gives

$$e(n) = h(0)x(n)[\varepsilon_1(n) + \varepsilon_3(n) + \varepsilon_5(n) + \varepsilon_7(n)] + h(1)x(n-1)[\varepsilon_2(n) + \varepsilon_3(n) + \varepsilon_5(n) + \varepsilon_7(n)]$$
$$+ h(2)x(n-2)[\varepsilon_4(n) + \varepsilon_5(n) + \varepsilon_7(n)] + h(3)x(n-3)[\varepsilon_6(n) + \varepsilon_7(n)] \tag{8.71}$$

Assuming that the input is white noise of variance σ_x^2 so that $E\{x(n)x(n-k)\}$ is zero for $k \neq 0$, and assuming that the errors are uncorrelated,

$$E\{e^2(n)\} = [4h^2(0) + 4h^2(1) + 3h^2(2) + 2h^2(3)]\sigma_x^2\sigma_{\varepsilon_r}^2 \tag{8.72}$$

In general, for any N,

$$\sigma_o^2 = E\{e^2(n)\} = \left[Nh^2(0) + \sum_{k=1}^{N-1}(N+1-k)h^2(k) \right]\sigma_x^2\sigma_{\varepsilon_r}^2 \tag{8.73}$$

Notice that if the order of summation of the product terms in the convolution summation is changed, then the order in which the $h(k)$ appear in Eq. (8.73) changes. If the order is changed so that the $h(k)$ with smallest magnitude is first, followed by the next smallest, etc., then the roundoff noise variance is minimized. Performing the convolution summation in nonsequential order, however, greatly complicates data indexing and so may not be worth the reduction obtained in roundoff noise.

Analysis of roundoff noise for IIR filters proceeds in the same way as for FIR filters. The analysis is more complicated, however, because roundoff noise arising in the computation of internal filter variables (state variables) must be propogated through a transfer function from the point of the quantization to the filter output. This is not necessary for FIR filters realized via the convolution summation since all quantizations are in the output calculation. Another complication in the case of IIR filters realized with fixed-point arithmetic is the need to scale the computation of internal filter variables to avoid their overflow. Examples of roundoff noise analysis for IIR filters can be found in Weinstein and Oppenheim (1969) and Oppenheim and Weinstein (1972) where it is shown that differences in the filter realization structure can make a large difference in the output roundoff noise. In particular, it is shown that IIR filters realized via the parallel or cascade connection of first- and second-order subfilters are almost always superior in terms of roundoff noise to a high-order direct form (single difference equation) realization. It is also possible to choose realizations that are optimal or near optimal in a roundoff noise sense (Mullins and Roberts, 1976; Jackson, Lindgren, and Kim, 1979). These realizations generally require more computation to obtain an output sample from an input sample, however, and so suboptimal realizations with slightly higher roundoff noise are often preferable (Bomar, 1985).

Limit Cycles

A limit cycle, sometimes referred to as a *multiplier roundoff limit cycle*, is a low-level oscillation that can exist in an otherwise stable filter as a result of the nonlinearity associated with rounding (or truncating) internal filter calculations (Parker and Hess, 1971). Limit cycles require recursion to exist and do not occur in nonrecursive FIR filters.

As an example of a limit cycle, consider the second-order filter realized by

$$y(n) = Q_r \left\{ \frac{7}{8}y(n-1) - \frac{5}{8}y(n-2) + x(n) \right\} \tag{8.74}$$

where $Q_r\{\ \}$ represents quantization by rounding. This is a stable filter with poles at $0.4375 \pm j0.6585$. Consider the implementation of this filter with four-bit (three bits and a sign bit) two's complement fixed-point arithmetic, zero initial conditions ($y(-1) = y(-2) = 0$), and an input sequence $x(n) = \frac{3}{8}\delta(n)$

where $\delta(n)$ is the unit impulse or unit sample. The following sequence is obtained:

$$y(0) = Q_r\left\{\frac{3}{8}\right\} = \frac{3}{8} \qquad y(1) = Q_r\left\{\frac{21}{64}\right\} = \frac{3}{8} \qquad y(2) = Q_r\left\{\frac{3}{32}\right\} = \frac{1}{8}$$

$$y(3) = Q_r\left\{-\frac{1}{8}\right\} = -\frac{1}{8} \qquad y(4) = Q_r\left\{-\frac{3}{16}\right\} = -\frac{1}{8} \qquad y(5) = Q_r\left\{-\frac{1}{32}\right\} = 0$$

$$y(6) = Q_r\left\{\frac{5}{64}\right\} = \frac{1}{8} \qquad y(7) = Q_r\left\{\frac{7}{64}\right\} = \frac{1}{8} \qquad y(8) = Q_r\left\{\frac{1}{32}\right\} = 0$$

$$y(9) = Q_r\left\{-\frac{5}{64}\right\} = -\frac{1}{8} \qquad y(10) = Q_r\left\{-\frac{7}{64}\right\} = -\frac{1}{8} \qquad y(11) = Q_r\left\{-\frac{1}{32}\right\} = 0$$

$$y(12) = Q_r\left\{\frac{5}{64}\right\} = \frac{1}{8} \qquad \cdots$$

Notice that although the input is zero except for the first sample, the output oscillates with amplitude $\frac{1}{8}$ and period 6.

Limit cycles are primarily of concern in fixed-point recursive filters. As long as floating-point filters are realized as the parallel or cascade connection of first- and second-order subfilters, limit cycles will generally not be a problem since limit cycles are practically not observable in first- and second-order systems implemented with 32-bit floating-point arithmetic (Bauer, 1993). It has been shown that such systems must have an extremely small margin of stability for limit cycles to exist at anything other than underflow levels, which are at an amplitude of less than 10^{-38} (Bauer, 1993).

There are at least three ways of dealing with limit cycles when fixed-point arithmetic is used. One is to determine a bound on the maximum limit cycle amplitude, expressed as an integral number of quantization steps. It is then possible to choose a wordlength that makes the limit cycle amplitude acceptably low. Alternately, limit cycles can be prevented by randomly rounding calculations up or down (Buttner, 1976). This approach, however, is complicated to implement. The third approach is to properly choose the filter realization structure and then quantize the filter calculations using magnitude truncation (Bomar, 1994). This approach has the disadvantage of slightly increasing roundoff noise.

Overflow Oscillations

With fixed-point arithmetic it is possible for filter calculations to overflow. This happens when two numbers of the same sign add to give a value having magnitude greater than one. Since numbers with magnitude greater than one are not representable, the result overflows. For example, the two's complement numbers 0.101 ($\frac{5}{8}$) and 0.100 ($\frac{4}{8}$) add to give 1.001, which is the two's complement representation of $-\frac{7}{8}$.

The overflow characteristic of two's complement arithmetic can be represented as $R\{\ \}$ where

$$R\{X\} = \begin{cases} X - 2, & X \geq 1 \\ X, & -1 \leq X < 1 \\ X + 2, & X < -1 \end{cases} \tag{8.75}$$

For the example just considered, $R\{\frac{9}{8}\} = -\frac{7}{8}$.

An overflow oscillation, sometimes also referred to as an adder overflow limit cycle, is a high-level oscillation that can exist in an otherwise stable fixed-point filter due to the gross nonlinearity associated with the overflow of internal filter calculations (Ebert, Mazo, and Taylor, 1969). Like limit cycles, overflow oscillations require recursion to exist and do not occur in nonrecursive FIR filters. Overflow oscillations also do not occur with floating-point arithmetic due to the virtual impossibility of overflow.

As an example of an overflow oscillation, once again consider the filter of Eq. (8.74) with four-bit fixed-point two's complement arithmetic and with the two's complement overflow characteristic of Eq. (8.75):

$$y(n) = Q_r\left\{ R\left\{ \frac{7}{8}y(n-1) - \frac{5}{8}y(n-2) + x(n) \right\} \right\}$$ (8.76)

In this case we apply the input

$$x(n) = \left\{ -\frac{3}{4}, -\frac{5}{8}, 0, 0, 0, \ldots \right\}$$ (8.77)

giving the output sequence

$$y(0) = Q_r\left\{ R\left\{ -\frac{3}{4} \right\} \right\} = Q_r\left\{ -\frac{3}{4} \right\} = -\frac{3}{4} \qquad y(1) = Q_r\left\{ R\left\{ -\frac{41}{32} \right\} \right\} = Q_r\left\{ \frac{23}{32} \right\} = \frac{3}{4}$$

$$y(2) = Q_r\left\{ R\left\{ \frac{9}{8} \right\} \right\} = Q_r\left\{ -\frac{7}{8} \right\} = -\frac{7}{8} \qquad y(3) = Q_r\left\{ R\left\{ -\frac{79}{64} \right\} \right\} = Q_r\left\{ \frac{49}{64} \right\} = \frac{3}{4}$$

$$y(4) = Q_r\left\{ R\left\{ \frac{77}{64} \right\} \right\} = Q_r\left\{ -\frac{51}{64} \right\} = -\frac{3}{4} \qquad y(5) = Q_r\left\{ R\left\{ -\frac{9}{8} \right\} \right\} = Q_r\left\{ \frac{7}{8} \right\} = \frac{7}{8}$$

$$y(6) = Q_r\left\{ R\left\{ \frac{79}{64} \right\} \right\} = Q_r\left\{ -\frac{49}{64} \right\} = -\frac{3}{4} \qquad y(7) = Q_r\left\{ R\left\{ -\frac{77}{64} \right\} \right\} = Q_r\left\{ \frac{51}{64} \right\} = \frac{3}{4}$$

$$y(8) = Q_r\left\{ R\left\{ \frac{9}{8} \right\} \right\} = Q_r\left\{ -\frac{7}{8} \right\} = -\frac{7}{8} \qquad \ldots$$

This is a large-scale oscillation with nearly full-scale amplitude.

There are several ways to prevent overflow oscillations in fixed-point filter realizations. The most obvious is to scale the filter calculations so as to render overflow impossible. However, this may unacceptably restrict the filter dynamic range. Another method is to force completed sums-of-products to saturate at ± 1, rather than overflowing (Ritzerfeld, 1989). It is important to saturate only the completed sum, since intermediate overflows in two's complement arithmetic do not affect the accuracy of the final result. Most fixed-point digital signal processors provide for automatic saturation of completed sums if their *saturation arithmetic* feature is enabled. Yet another way to avoid overflow oscillations is to use a filter structure for which any internal filter transient is guaranteed to decay to zero (Mills, Mullis, and Roberts, 1978). Such structures are desirable anyway, since they tend to have low roundoff noise and be insensitive to coefficient quantization (Barnes, 1979).

Coefficient Quantization Error

Each filter structure has its own finite, generally nonuniform grids of realizable pole and zero locations when the filter coefficients are quantized to a finite wordlength. In general the pole and zero locations desired in a filter do not correspond exactly to the realizable locations. The error in filter performance (usually measured in terms of a frequency response error) resulting from the placement of the poles and zeroes at the nonideal but realizable locations is referred to as coefficient quantization error.

Consider the second-order filter with complex-conjugate poles

$$\lambda = r\, e^{\pm j\theta} = \lambda_r \pm j\lambda_i = r\cos(\theta) \pm jr\sin(\theta)$$ (8.78)

and transfer function

$$H(z) = \frac{1}{1 - 2r\cos(\theta)z^{-1} + r^2 z^{-2}}$$ (8.79)

realized by the difference equation

$$y(n) = 2\lambda_r\, y(n-1) - r^2\, y(n-2) + x(n)$$ (8.80)

Quantizing the difference equation coefficients results in a nonuniform grid of realizable pole locations in the z plane. The nonuniform grid is defined by the intersection of vertical lines corresponding to quantization of $2\lambda_r$ and concentric circles corresponding to quantization of $-r^2$. The sparseness of realizable pole locations near $z = \pm 1$ results in a large coefficient quantization error for poles in this region. By contrast, quantizing coefficients of the normal realization (Barnes, 1979) corrresponds to quantizing λ_r and λ_i resulting in a uniform grid of realizable pole locations. In this case large coefficient quantization errors are avoided for all pole locations.

It is well established that filter structures with low roundoff noise tend to be robust to coefficient quantization, and vice versa (Jackson, 1976; Rao, 1986). For this reason, the normal (uniform grid) structure is also popular because of its low roundoff noise.

It is well known that in a high-order polynomial with clustered roots the root location is a very sensitive function of the polynomial coefficients. Therefore, filter poles and zeroes can be much more accurately controlled if higher order filters are realized by breaking them up into the parallel or cascade connection of first- and second-order subfilters. One exception to this rule is the case of linear-phase FIR filters in which the symmetry of the polynomial coefficients and the spacing of the filter zeros around the unit circle usually permits an acceptable direct realization using the convolution summation.

Realization Considerations

Linear-phase FIR digital filters can generally be implemented with acceptable coefficient quantization sensitivity using the direct convolution sum method. When implemented in this way on a digital signal processor, fixed-point arithmetic is not only acceptable but may actually be preferable to floating-point arithmetic. Virtually all fixed-point digital signal processors accumulate a sum of products in a double-length accumulator. This means that only a single quantization is necessary to compute an output. Floating-point arithmetic, on the other hand, requires a quantization after every multiply and after every add in the convolution summation. With 32-bit floating-point arithmetic these quantizations introduce a small enough error to be insignificant for most applications.

When realizing IIR filters, either a parallel or cascade connection of first- and second-order subfilters is almost always preferable to a high-order direct form realization. With the availability of low-cost floating-point digital signal processors, like the Texas Instruments TMS320C32, it is highly recommended that floating-point arithmetic be used for IIR filters. Floating-point arithmetic simultaneously eliminates most concerns regarding scaling, limit cycles, and overflow oscillations. Regardless of the arithmetic employed, a low roundoff noise structure should be used for the second-order sections. The use of a low roundoff noise structure for the second-order sections also tends to give a realization with low coefficient quantization sensitivity. First-order sections are not as critical in determining the roundoff noise and coefficient sensitivity of a realization, and so can generally be implemented with a simple direct form structure.

Defining Terms

Adaptive FIR filter: A finite impulse response structure filter with adjustable coefficients. The adjustment is controlled by an adaptation algorithm such as the least mean square (lms) algorithm. They are used extensively in adaptive echo cancellers and equalizers in communication sytems.

Causality: The property of a system that implies that its output can not appear before its input is applied. This corresponds to an FIR filter with a zero discrete-time impulse response for negative time indices.

Discrete time impulse response: The output of an FIR filter when its input is unity at the first sample and otherwise zero.

Group delay: The group delay of an FIR filter is the negative derivative of the phase response of the filter and is, therefore, a function of the input frequency. At a particular frequency it equals the physical delay that a narrow-band signal will experience passing through the filter.

Linear phase: The phase response of an FIR filter is linearly related to frequency and, therefore, corresponds to constant group delay.

Linear, time invariant (LTI): A system is said to be LTI if superposition holds, that is, its output for an input that consists of the sum of two inputs is identical to the sum of the two outputs that result from the individual application of the inputs; the output is not dependent on the time at that the input is applied. This is the case for an FIR filter with fixed coefficients.

Magnitude response: The change of amplitude, in steady state, of a sinusoid passing through the FIR filter as a function of frequency.

Multirate FIR filter: An FIR filter in which the sampling rate is not constant.

Phase response: The phase change, in steady state, of a sinusoid passing through the FIR filter as a function of frequency.

References

Antoniou, A. 1993. *Digital Filters Analysis, Design, and Applications*, 2nd ed. McGraw-Hill, New York.

Barnes, C.W. 1979. Roundoff noise and overflow in normal digital filters. *IEEE Trans. Circuits Syst.* CAS-26(3):154–159.

Bauer, P.H. 1993. Limit cycle bounds for floating-point implementations of second-order recursive digital filters. *IEEE Trans. Circuits Syst.–II* 40(8):493–501.

Bomar, B.W. 1985. New second-order state-space structures for realizing low roundoff noise digital filters. *IEEE Trans. Acoust., Speech, Signal Processing* ASSP-33(1):106–110.

Bomar, B.W. 1994. Low-roundoff-noise limit-cycle-free implementation of recursive transfer functions on a fixed-point digital signal processor. *IEEE Trans. Industrial Electronics* 41(1):70–78.

Buttner, M. 1976. A novel approach to eliminate limit cycles in digital filters with a minimum increase in the quantization noise. In *Proceedings of the 1976 IEEE International Symposium on Circuits and Systems*, pp. 291–294. IEEE, NY.

Cappellini, V., Constantinides, A.G., and Emiliani, P. 1978. *Digital Filters and their Applications*. Academic Press, New York.

Ebert, P.M., Mazo, J.E., and Taylor, M.G. 1969. Overflow oscillations in digital filters. *Bell Syst. Tech. J.* 48(9): 2999–3020.

Gray, A.H. and Markel, J.D., 1973. Digital lattice and ladder filter synthesis. *IEEE Trans. Acoustics, Speech and Signal Processing* ASSP-21:491–500.

Herrmann, O. and Schuessler, W. 1970. Design of nonrecursive digital filters with minimum phase. *Electronics Letters* 6(11):329–330.

IEEE DSP Committee. 1979. *Programs for Digital Signal Processing*. IEEE Press, New York.

Jackson, L.B. 1976. Roundoff noise bounds derived from coefficient sensitivities for digital filters. *IEEE Trans. Circuits Syst.* CAS-23(8):481–485.

Jackson, L.B., Lindgren, A.G., and Kim, Y. 1979. Optimal synthesis of second-order state-space structures for digital filters. *IEEE Trans. Circuits Syst.* CAS-26(3):149–153.

Lee, E.A. and Messerschmitt, D.G. 1994. *Digital Communications*, 2nd ed. Kluwer, Norwell, MA.

Macchi, O. 1995. *Adaptive Processing: The Least Mean Squares Approach with Applications in Telecommunications*, Wiley, New York.

Mills, W.T., Mullis, C.T., and Roberts, R.A. 1978. Digital filter realizations without overflow oscillations. *IEEE Trans. Acoust., Speech, Signal Processing.* ASSP-26(4):334–338.

Mullis, C.T. and Roberts, R.A. 1976. Synthesis of minimum roundoff noise fixed-point digital filters. *IEEE Trans. Circuits Syst.* CAS-23(9):551–562.

Oppenheim, A.V. and Schafer, R.W. 1989. *Discrete-Time Signal Processing*. Prentice-Hall, Englewood Cliffs, NJ.

Oppenheim, A.V. and Weinstein, C.J. 1972. Effects of finite register length in digital filtering and the fast fourier transform. *Proc. IEEE* 60(8):957–976.

Parker, S.R. and Hess, S.F. 1971. Limit-cycle oscillations in digital filters. *IEEE Trans. Circuit Theory* CT-18(11):687–697.

Parks, T.W. and Burrus, C.S. 1987. *Digital Filter Design*. Wiley, New York.

Parks, T.W. and McClellan, J.H. 1972a. Chebyshev approximations for non recursive digital filters with linear phase. *IEEE Trans. Circuit Theory* CT-19:189–194.

Parks, T.W. and McClellan, J.H. 1972b. A program for the design of linear phase finite impulse response filters. *IEEE Trans. Audio Electroacoustics* AU-20(3):195–199.

Proakis, J.G. and Manolakis, D.G. 1992. *Digital Signal Processing Principles, Algorithms, and Applications*, 2nd ed. MacMillan, New York.

Rabiner, L.R. and Gold, B. 1975. *Theory and Application of Digital Signal Processing*. Prentice-Hall, Englewood Cliffs, NJ.

Rabiner, L.R. and Schafer, R.W. 1978. *Digital Processing of Speech Signals*. Prentice-Hall, Englewood Cliffs, NJ.

Rabiner, L.R. and Schafer, R.W. 1974. On the behavior of minimax FIR digital Hilbert transformers. *Bell Sys. Tech. J.* 53(2):361–388.

Rao, D.B.V. 1986. Analysis of coefficient quantization errors in state-space digital filters. *IEEE Trans. Acoust., Speech, Signal Processing* ASSP-34(1):131–139.

Ritzerfeld, J.H.F. 1989. A condition for the overflow stability of second-order digital filters that is satisfied by all scaled state-space structures using saturation. *IEEE Trans. Circuits Syst.* CAS-36(8): 1049–1057.

Roberts, R.A. and Mullis, C.T. 1987. *Digital Signal Processing*. Addison-Wesley, Reading, MA.

Vaidyanathan, P.P. 1993. *Multirate Systems and Filter Banks*. Prentice-Hall, Englewood Cliffs, NJ.

Weinstein, C. and Oppenheim, A.V. 1969. A comparison of roundoff noise in floating-point and fixed-point digital filter realizations. *Proc. IEEE* 57(6):1181–1183.

Further Information

Additional information on the topic of digital filters is available from the following sources.

IEEE Transactions on Signal Processing, a monthly publication of the Institute of Electrical and Electronics Engineers, Inc, Corporate Office, 345 East 47 Street, NY.

IEEE Transactions on Circuits and Systems—Part II: Analog and Digital Signal Processing, a monthly publication of the Institute of Electrical and Electronics Engineers, Inc, Corporate Office, 345 East 47 Street, NY.

The Institute of Electrical and Electronics Engineers holds an annual conference at worldwide locations called the International Conference on Acoustics, Speech and Signal Processing, ICASSP, Corporate Office, 345 East 47 Street, NY.

IEE Transactions on Vision, Image and Signal Processing, a monthly publication of the Institute of Electrical Engineers, Head Office, Michael Faraday House, Six Hills Way, Stevenage, UK.

Signal Processing, a publication of the European Signal Processing Society, Switzerland, Elsevier Science B.V., Journals Dept., P.O. Box 211, 1000 AE Amsterdam, The Netherlands.

In addition, the following books are recommended.

Bellanger, M. 1984. *Digital Processing of Signals: Theory and Practice*. Wiley, New York.

Burrus, C.S. et al. 1994. *Computer-Based Exercises for Signal Processing Using MATLAB*. Prentice-Hall, Englewood Cliffs, NJ.

Jackson, L.B. 1986. *Digital Filters and Signal Processing*. Kluwer Academic, Norwell, MA, 1986.

Oppenheim, A.V. and Schafer, R.W. 1989. *Discrete-Time Signal Processing*. Prentice-Hall, Englewood Cliffs, NJ.

Widrow, B. and Stearns, S. 1985. *Adaptive Signal Processing*. Prentice-Hall, Englewood Cliffs, NJ.

8.9   Multichip Module Technology

Paul D. Franzon

8.9.1   Introduction

Multichip module (MCM) technology allows bare integrated circuits and passive devices to be mounted together on a common interconnecting substrate. An example is shown in Fig. 8.109. In this photograph, eight chips are **wire-bonded** onto an MCM. MCM technology, however, is not just about packaging chips together; it can lead to new capabilities and unique performance and cost advantages. The purposes of this

FIGURE 8.109   Eight chips wire-bonded into an MCM. (*Source:* MicroModule Systems.)

chapter are as follows: to explain the different multichip module technologies, to show how MCMs can be employed to improve the price/performance of systems, and to indicate some of the more important issues in MCM-system design.

8.9.2   Multichip Module Technology Definitions

In its broadest sense, a multichip module is an assembly in which more than one integrated circuit (IC) are bare mounted on a common substrate. This definition is often narrowed in such a way as to imply that the common substrate provides a higher wiring density than a conventional **printed circuit board (PCB)**. The main physical components are shown in Fig. 8.110 and can be described as follows:

1. The substrate technology provides the interconnect between the chips (ICs or die) and any discrete circuit elements, such as resistors, capacitors, and inductors.
2. The chip connection technology provides the means whereby signals and power are transferred between the substrate and the chips.
3. The MCM package technology is the housing that surrounds the MCM and allows signals, and power and heat to be conducted to the outside world.

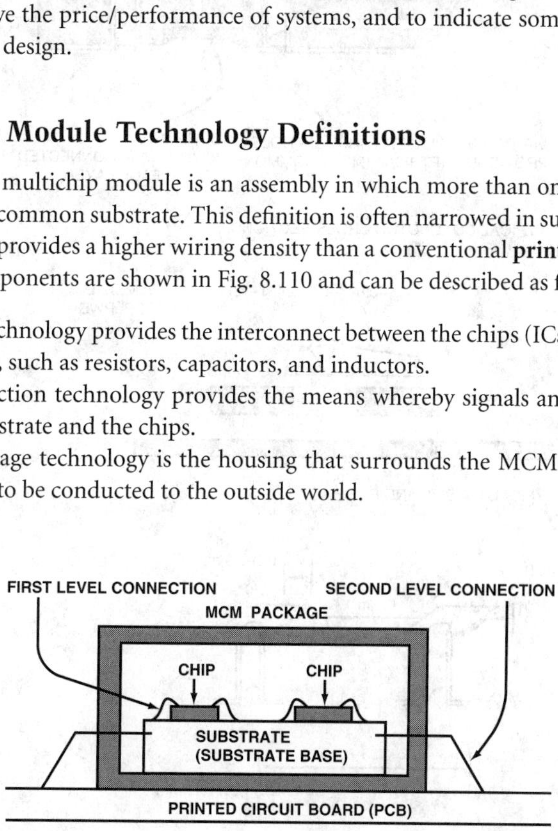

FIGURE 8.110   Physical technology components to an MCM.

What cannot be shown in Fig. 8.110 are several other components that are important to the success of an MCM:

1. The test technology used to ensure correct function of the bare die, the MCM substrate, and the assembled MCM.
2. The repair technology used to replace failed die, if any are detected after assembly.
3. The design technology.

There are many different versions of these technology components. The substrate technology is broadly divided into three categories:

1. **Laminate MCMs (MCM-L)** is shown in (Fig. 8.111). Essentially fine-line PCBs, MCM-Ls are usually constructed by first patterning copper conductors on fiberglass/resin-impregnated sheets as shown in Fig. 8.112. These sheets are laminated together under heat and pressure. Connections between conductors

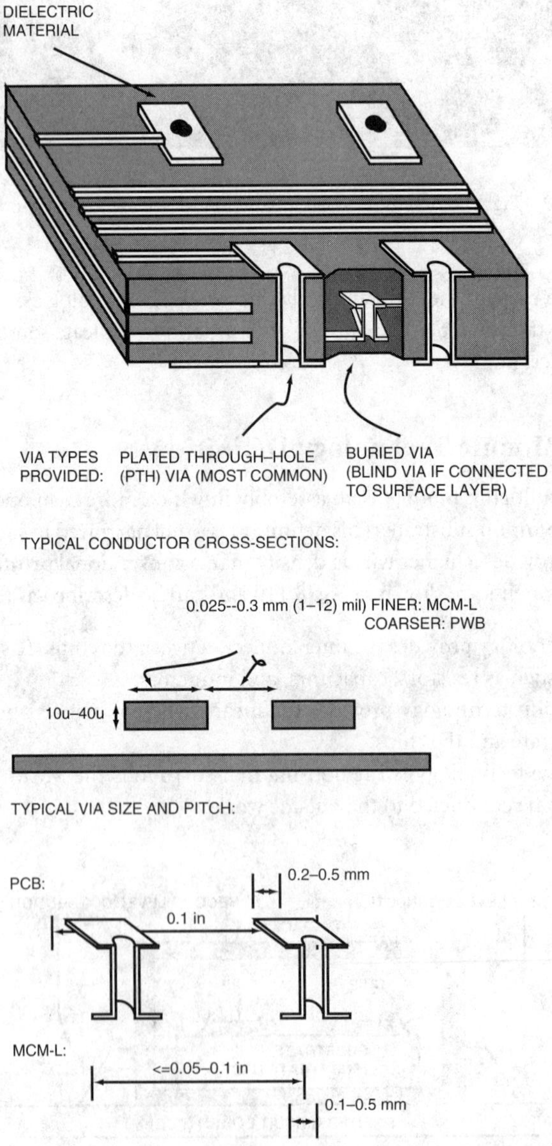

DIELECTRIC
MATERIAL

VIA TYPES   PLATED THROUGH–HOLE   BURIED VIA
PROVIDED: (PTH) VIA (MOST COMMON)   (BLIND VIA IF CONNECTED
                                          TO SURFACE LAYER)

TYPICAL CONDUCTOR CROSS-SECTIONS:

0.025--0.3 mm (1–12) mil) FINER: MCM-L
                          COARSER: PWB

10u–40u

TYPICAL VIA SIZE AND PITCH:

PCB: 0.2–0.5 mm
              0.1 in

MCM-L:
              <=0.05–0.1 in
                          0.1–0.5 mm

FIGURE 8.111 Typical cross sections and feature sizes in printed circuit board and MCM-L technologies.

1. MAKE CORE AND PREPREG LAYERS

FIBER-GLASS CLOTH: FR4 (FIRE RETARDANT
EPOXY -GLASS CLOTH)
IMPREGNATE AND COAT
WITH A POLYMER (PLASTIC)
E.G. RESIN

CORE: FULLY CURE   PRE-PREG: PARTIALLY CURE

('CURING'= HEATING TO MAKE 'SOLID')

2. COPPER CONDUCTOR PROCESSING
COAT ONE OR BOTH SIDES OF CORE LAYERS WITH COPPER

3. PHOTOLITHOGRAPHIC PROCESSING
('SUBTRACTIVE PROCESS')

APPLY PHOTORESIST

MASK
COPPER
CORE

ETCH

CLEAN

4. DRILL BLIND AND BURIED
VIAS

5. PLATE VIA HOLES

6. LAMINATE UNDER HEAT AND
PRESSURE

CORE
PRE-PREG
CORE

7. DRILL AND PLATE THROUGH-HOLE
(PTH) VIAS

8. SURFACE PROCESSING OF COPPER ATTACH PADS
PWB: Pb-Sn SOLDER
(VIA 'SOLDER MASK')

MCM-L: WIRE BOND: SOFT GOLD PLATING
EDGE CONNECTORS: HARD GOLD PLATING
TAB: 60/40 TIN-LEAD SOLDER

FIGURE 8.112 Basic manufacturing steps for MCM-Ls, using fiberglass/resin prepreg sheets as a basis.

on different sheets are made through via holes drilled in the sheets and plated. Recent developments in MCM-L technology have emphasized the use of flexible laminates. Flexible laminates have the potential for permitting finer lines and vias than fiberglass-based laminates.

2. **Ceramic MCMs (MCM-C)** are mostly based on pin grid array (PGA) technology. Typical cross sections and geometries are illustrated in Fig. 8.113 and the basic manufacturing steps are outline in Fig. 8.114. MCM-Cs are made by first casting a uniform sheet of prefired ceramic material, called green tape, then printing a metal ink onto the green tape, then punching and metal filling holes for the vias, and finally cofiring the stacked sheets together under pressure in an oven. In addition to metal, other inks can be used, including ones to print discrete resistors and capacitors in the MCM-C.

3. **Thin-film (deposited) MCMs (MCM-D)** are based on chip metalization processes. MCM-Ds have very fine feature sizes, giving high wiring densities (Fig. 8.115). MCM-Ds are made one layer at a time,

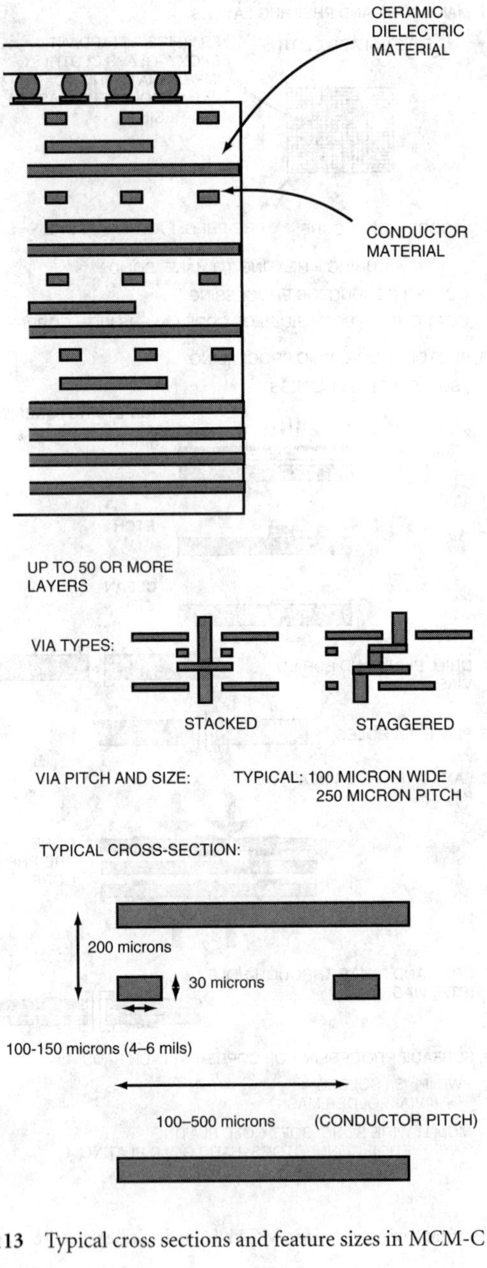

FIGURE 8.113   Typical cross sections and feature sizes in MCM-C technology.

using successive photolithographic definitions of metal conductor patterns and vias and the deposition of insulating layers, usually polyimide (Fig. 8.116). Often MCM-Ds are built on a silicon substrate, allowing capacitors, resistors, and transistors to be built, cheaply, as part of the substrate.

Table 8.9 gives one comparison of the alternative substrate technologies. MCM-Ls provide the lowest wiring and via density but are still very useful for making a small assembly when the total wire count in the design is not too high. MCM-Ds provide the highest wiring density and are useful in designs containing high pin-count chips; however, it is generally the most expensive technology on a per-unit-area basis. MCM-Cs fall somewhere in between, providing intermediate wiring densities at intermediate costs.

Current research promises to reduce the cost of MCM-Ls and MCM-D substrates. MCM-Ls based on flexible board technologies should be both cheaper and provide denser wiring than current fibreglass

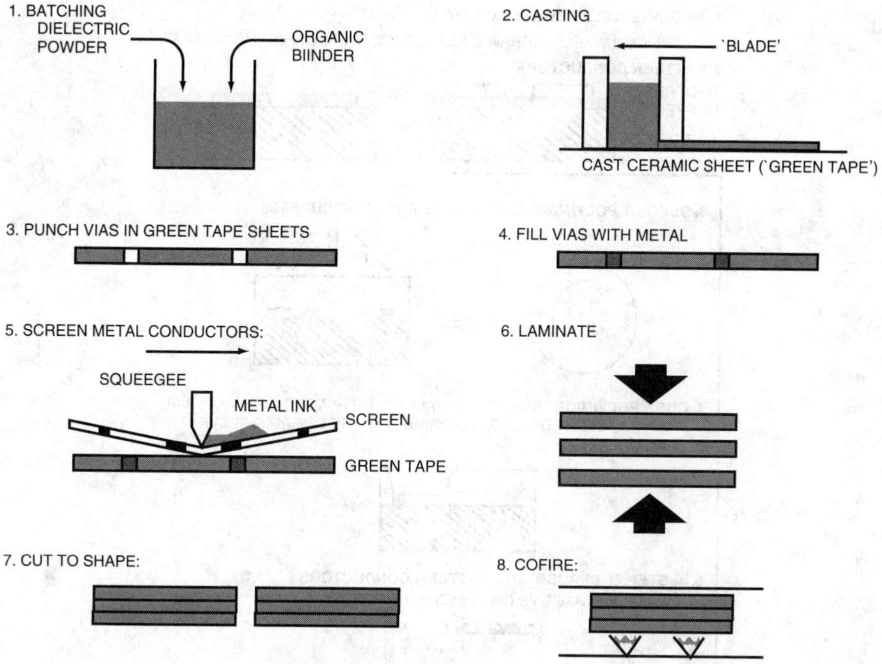

FIGURE 8.114 Basic manufacturing steps for MCM-Cs.

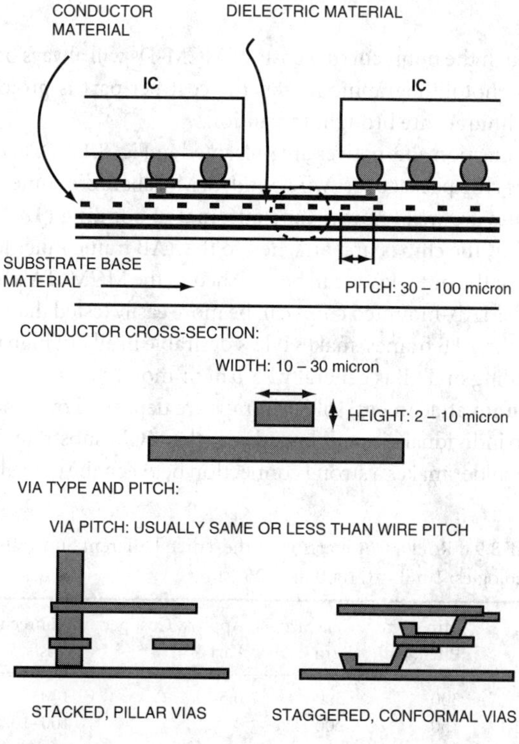

FIGURE 8.115 Typical cross sections and feature sizes in MCM-D technology.

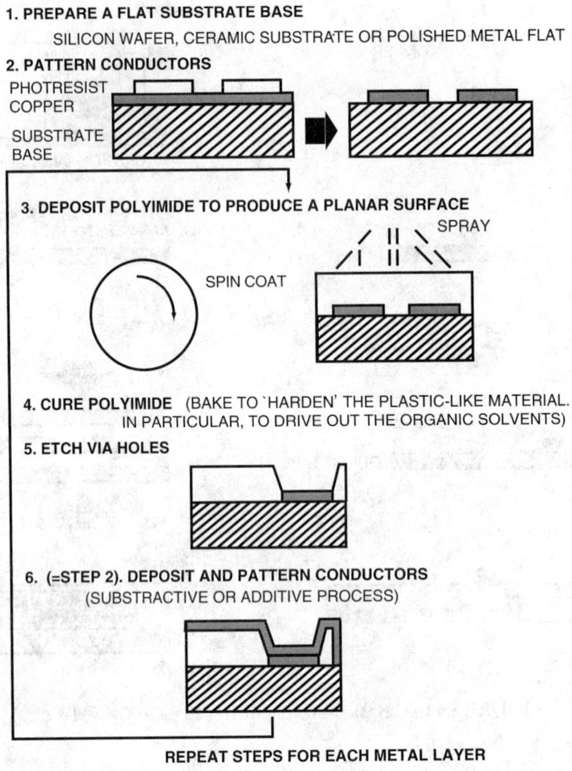

FIGURE 8.116   Basic manufacturing steps for MCM-Cs.

MCM-L technologies. Though the nonrecurring costs of MCM-Ds will always be high, due to the requirement to make fine feature photolithographic masks, the cost per part is projected to decrease as more efficient manufacturing techniques are brought to fruition.

The chip to substrate connection alternatives are presented in Fig. 8.117. Currently, over 95% of the die mounted in MCMs or single chip packages are wire bonded. Most bare die come suitable for wire-bonding, and wire bonding techniques are well known. **Tape automated bonding (TAB)** is an alternative to wire bonding. With TAB assembly, the chip is first attached to the TAB frame inner leads. These leads are then shaped (called *forming*); then the outer leads can be attached to the MCM. TAB has a significant advantage over wire bonding in that the TAB-mounted chips can be more easily tested than bare die. The high tooling costs, however, for making the TAB frames, makes it less desirable in all but high volume production chips. The wire pitch in wire bonding or TAB is generally 75 μm or more.

With **flip-chip solder-bump** attachment, solder bumps are deposited over the area of the silicon wafer. The wafer is then diced into individual die and flipped over the MCM substrate. The module is then placed in a reflow oven, where the solder makes a strong connection between the chip and the substrate. Flip-chip

TABLE 8.9   Rough Comparison of the Three Different Substrate Technologies: 1 mil = 1/1000 in $\approx$ 25 μm

	Min. Wire Pitch, μm	Via Size, μm	Approx. Cost per Part $/in^2	Nonrecurring Costs, $
PCB	300	500	0.3	100
MCM-L	150	200	4	100–10,000
MCM-C	200	100	12	25,000
MCM-D	30	25	20	15,000

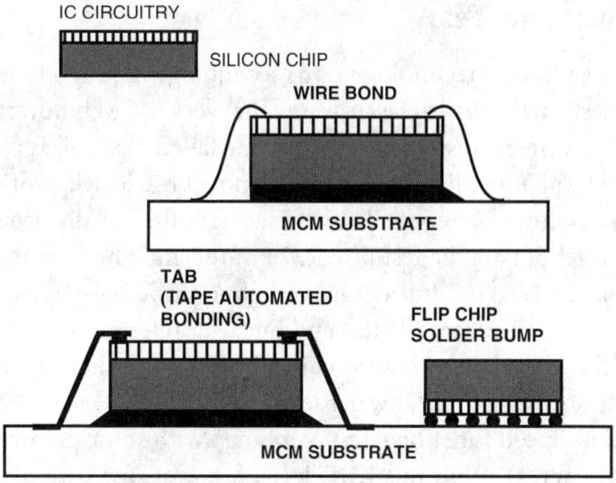

FIGURE 8.117   Chip attach options.

attachment has the following advantages:

- Solder bumps can be placed over the entire area of the chip, allowing chips to have thousands of connections. For example, a 1-cm^2 chip could support 1600 solder bumps (at a conservative 250-μm pitch) but only 533 wire bonds.

- Transistors can be placed under solder bumps but cannot be placed under wire-bond or TAB pads. The reason arises from the relative attachment process. Making a good wire-bond or TAB lead attachment requires compression of the bond into the chip pad, damaging any transistors beneath a pad. On the other hand, a good soldered connection is made through the application of heat only. As a result, the chip can be smaller by an area equivalent to the total pad ring area. For example, consider a 100-mm^2 chip with a 250-μm pad ring. The area consumed by the pad ring would total 10-mm^2, or 10% of the chip area. This area could be used for other functions if the chip were flipped and solder bumped. Alternatively, the chip could be made smaller and cheaper.

- The electrical parasitics of a solder bump are far better than for a wire-bond or TAB lead. The latter generally introduce about 1 nH of inductance and 1 pF of capacitance into a circuit. In contrast, a solder bump introduces about 10 pH and 10 nF. The lower parasitic inductance and capacitance make solder bumps attractive for high-frequency radio applications, for example, in the 5.6-GHz communications band and in high clock rate digital applications.

- The costs of flip-chip are comparable to wire bonding, beyound a certain point.

Flip-chip solder-bump, however, does have some disadvantages. The most significant follows from the requirement that the solder bump pads are larger than wire-bond pads (60–80 μm vs. approximately 50 μm). The solder-bump pads need to be larger in order to make the bumps taller. Taller bumps can more easily absorb the stresses that arise when the module is heated and cooled during assembly. As the thermal coefficient of expansion (TCE) of silicon is usually different from that of the MCM substrate, these two elements expand and contract at different rates, creating stresses in the connecting bumps. For the same reason, it is better to place the bumps in the center of the chip, rather than around the edge. By reducing the distance over which differential expansion and contraction occur, the stresses are reduced.

The larger bump pad sizes requires that the chips be designed specifically for solder-bumping, or that the wafers be postprocessed to distribute solder bumps pads over their surface, and wire (*redistribute*) these pads to the conventional wire-bond pads.

Another potential disadvantage arises from the lead content of the solder bump. Containing radioactive isotopes, most lead is a source of alpha particles that can potentially change the state of nearby transistors not shielded by aluminum. The effects of alpha particles are mainly of concern to dynamic random access memories (DRAMs) and dynamic logic.

8.9.3   Design, Repair, and Test

An important adjunct to the physical technology is the technology required to test and repair the die and modules. An important question that has to be answered for every MCM is how much to test the die before assembly vs. how much to rely on postassembly test to locate failed die. This is purely a cost question. The more the bare die are tested, the more likely that the assembled MCM will work. If the assembled MCM does not work, then it must either be repaired (by replacing a die) or discarded. The question reduces to the one that asks what level of bare die test provides a sufficiently high confidence that the assembled MCM will work, or is the assembled module cheap enough to throw away is a die is faulty.

In general, there are four levels of bare die test and burn-in, referred to as four levels of **known good die (KGD)**. In Table 8.10, these test levels are summarized, along with their impact. The lowest KGD level is to just use the same tests normally done at wafer level, referred to as the *wafer sort tests*. Here the chips are normally subject to a low-speed functional test combined with some parametric measurements (e.g., measurement of transistor curves). With this KGD level, test costs for bare die are limited to wafer test costs only. There is some risk, however, that the chip will not work when tested as part of the MCM. This risk is measured as the *test escape rate*. With conventional packaging, the chips are tested again, perhaps at full speed, once they are packaged, making the test escape rate zero.

If the MCM contains chips that must meet tight timing specifications (e.g., a workstation) or the MCM must meet high reliability standards (e.g., for aerospace), however, then higher KGD test levels are required. For example, a workstation company will want the chips to be speed-sorted. Speed-sorting requires producing a test fixture that can carry high-speed signals to and from the tester. The test fixture type usually used at wafer sort, generally referred to as a *probe card*, is usually not capable of carrying high-speed signals. Instead, it is necessary to build a more expensive, high-speed test fixture, or temporarily mount the bare die in some temporary package for testing. Naturally, these additional expenses increase the cost of providing at-pin speed sorted die.

Some applications require even higher levels of known good die. Aerospace applications usually demand burn-in of the die, particularly for memories, so as to reduce the chance of infant mortality. There are two levels of burned-in KGD. In the lowest level, the die are stressed at high temperatures for a period of time and then tested. In the higher level, the die are continuously tested while in the oven.

How do you decide what test level is appropriate for your MCM? The answer is driven purely by cost; the cost of test vs. the cost of repair. For example, consider a 4-chip MCM using mature ICs. Mature ICs tend to have very high yields, and the process engineers have learned how to prevent most failure modes. As a result, the chances are very small that a mature IC would pass the waferlevel functional test and fail in the MCM. With a test escape rate of 2%, there is only a 8% $(1 - 0.98^4)$ chance that each MCM would need to have a chip replaced after assembly. If the repair costs $30 per replaced chip, then the average excess repair cost per MCM is $2.40. It is unlikely that a higher level of KGD would add only $0.60 to the cost of a chip. Thus, the lowest test level is justified.

On the other hand, consider an MCM containing four immature ICs. The functional and parametric wafer-level tests are poor at capturing speed faults. In addition, the process engineers have not had a chance to learn how to maximize the chip yield and how to best detect potential problems. If the test escape rate was 5%, there would be a 40% chance that a 4-chip MCM would require a chip to be replaced. The average repair cost would be $12 per MCM in this scenario, and the added test cost of obtaining speed-sorted die would be justified.

TABLE 8.10   Levels of Known Good Die and Their Impact

Known Good Die Level	Test Cost Impact	Test Escape Impact
Wafer level functional and parametric	Low	<1–2% fallout for mature ICs possibly >5% fallout for new ICs
At-pin speed sorted	Medium	Min. fallout for new digital ICs
Burned-in	High	Burn-in important for memories
Dynamically burned-in with full testing	Highest	Min. memory fallout

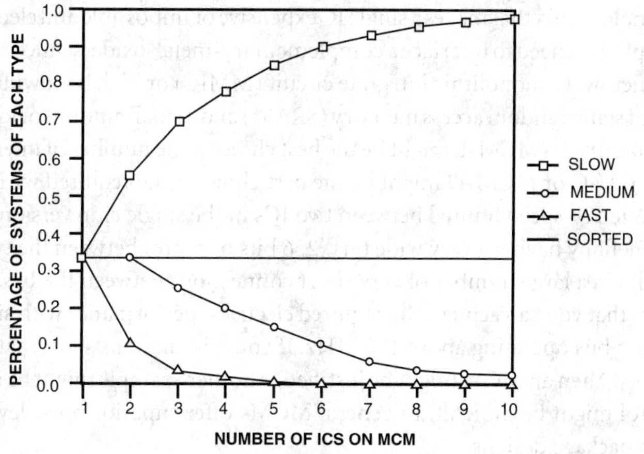

FIGURE 8.118 The effect of speed sorting on MCM performance.

For high-speed systems, speed sorting also greatly influences the speed rating of the module. Today, microprocessors, for example, are graded according to their clock rate. Faster parts command a higher price. In an MCM the entire module will be limited by the slowest chip on the module, and when a system is partitioned into several smaller chips, it is highly likely that a module will contain a slow chip if they are not tested.

For example, consider a set of chips that are manufactured in equal numbers of slow, medium, and fast parts. If the chips are speed sorted before assembly into the MCM, there should be 33% slow modules, 33% medium modules and 33% fast modules. If not speed sorted, these ratios change drastically as shown in Fig. 8.118. For a 4-chip module assembled from unsorted die, there will be 80% slow systems, 19% medium systems, and only 1% fast systems. This dramatic reduction in fast modules also justifies the need for speed-sorted die.

Design technology is also an important component to the MCM equation. Most MCM computer-aided design (CAD) tools have their basis in the PCB design tools. The main differences have been the inclusion of features to permit the use of the wide range of physical geometries possible in an MCM (particularly via geometries), as well as the ability to use bare die. A new format, the *die interchange format*, has been specifically developed to handle physical information concerning bare die (e.g., pad locations).

There is more to MCM design technology, however, than just making small changes to the physical design tools (those tools that actually produce the MCM wiring patterns). Design correctness is more important in an MCM than in a PCB. For example, a jumper wire is difficult to place on an MCM in order to correct for an error. Thus, recent tool developments have concentrated on improving the designers ability to ensure that the multichip system is designed correctly before it is built. These developments include new simulation libraries that allow the designer to simulate the entire chip set before building the MCM, as well as tools that automate the electrical and thermal design of the MCM.

8.9.4 When to Use Multichip Modules

There are a number of scenarios given as follows that typically lead to consideration of an MCM alternative.

1. You must achieve a smaller form factor than is possible with single chip packaging. Often integrating the digital components onto an MCM-L provides the most cost effective way in which to achieve the required size. MCM-D provides the greatest form factor reduction when the utmost smallest size is needed.

2. The alternative is a single die that would either be too large to manufacture or would be so large as to have insufficient yield. The design might be custom or semicustom. In this case, partitioning the die into a number of smaller die and using an MCM to achieve the interconnect performance of the large die is an attractive alternative.

3. You have a technology mix that makes a single IC expensive or impossible and electrical performance is important. For example, you need to interface a complementary-metal-oxide-semiconductor (CMOS) digital IC with a GaAs microwave monolithic integrate circuit (MMIC) or high bandwidth analog IC. Or, you need a large amount of static random access memory (SRAM) and a small amount of logic. In these cases, an MCM might be very useful. An MCM-L might be the best choice if the number of interchip wires are small. A high layer count MCM-C or MCM-D might be the best choice if the required wiring density is large.

4. You are pad limited or speed limited between two ICs in the single chip version. For example, many computer designs benefit by having a very wide bus (256 bits or more) between the two levels of cache. In this case, an MCM allows a large number of very short connections between the ICs.

5. You are not sure that you can achieve the required electrical performance with single chip packaging. An example might be a bus operating above 100 MHz. If your simulations show that it will be difficult to guarantee the bus speed, then an MCM might be justified. Another example might be a mixed signal design in which noise control might be difficult. In general MCMs offer superior noise levels and interconnect speeds to single-chip package designs.

6. The conventional design has a lot of solder joints and reliability is important to you. An MCM design has far fewer solder joints, resulting in less field failures and product returns. (Flip-chip solder bumps have shown themselves to be far more reliable than the solder joints used at board level.)

Though there are many other cases where the use of MCM technology makes sense, these are the main ones that have been encountered so far. If an MCM is justified, the next question might be to decide what ICs need to be placed on the MCM. A number of factors follow that need to be considered in this decision:

1. It is highly desirable that the final MCM package be compatible with single-chip packaging assembly techniques to facilitate manufacturability.
2. Although wires between chips on the MCM are inexpensive, off-MCM pins are expensive. The MCM should contain as much of the wiring as is feasible. On the other hand, an overly complex MCM with a high component count will have a poor final yield.
3. In a mixed signal design, separating the noisy digital components from the sensitive analog components is often desirable. This can be done by placing the digital components on an MCM. If an MCM-D with an integrated decoupling capacitor is used, then the on-MCM noise might be low enough to allow both analog and digital components to be easily mixed. Be aware in this case, however, that the on-MCM ground will have different noise characteristics than the PCB ground.

In short, most MCM system design issues are decided on by careful modeling of the system-wide cost and performance. Despite the higher cost of the MCM package itself, often cost savings achieved elsewhere in the system can be used to justify the use of an MCM.

8.9.5   Issues in the Design of Multichip Modules

The most important issue in the design of multichip modules is obtaining bare die. The first questions to be addressed in any MCM project are the questions as to whether the required bare die are available in the quantities required, with the appropriate test level, with second sourcing, at the right price, etc. Obtaining answers to these questions is time consuming as many chip manufacturers still see their bare die sales as satisfying a niche market. If the manufactured die are not properly available, then it is important to address alternative chip sources early.

The next most important issue is the test and verification plan. There are a number of contrasts with using a printed circuit board. First, as the nonrecurring engineering cost is higher for a multichip module, the desirability for first pass success is higher. Complete prefabrication design verification is more critical when MCMs are being used, so more effort must be spent on logic and electrical simulation prior to fabrication.

It is also important to determine, during design, how faults are going to be diagnosed in the assembled MCM. In a prototype, you wish to be able locate design errors before redoing the design. In a production module, you need to locate faulty die or wire bonds/solder bumps if a faulty module is to be repaired (typically by replacing a die). It is more difficult to physically probe lines on an MCM, however, than on a

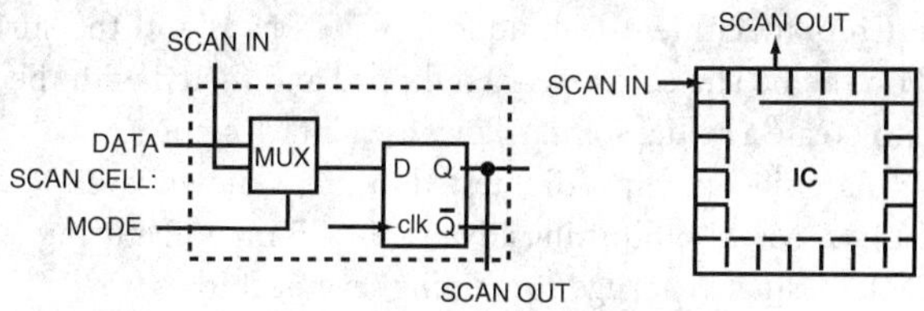

FIGURE 8.119 The use of boundary scan eases fault isolation in an MCM.

PCB. A fault isolation test plan must be developed and implemented. The test plan must be able to isolate a fault to a single chip or chip-to-chip interconnection. It is best to base such a plan on the use of chips with **boundary scan** (Fig. 8.119). With boundary scan chips, test vectors can be scanned in serially into registers around each chip. The MCM can then be run for one clock cycle, and the results scanned out. The results are used to determine which chip or interconnection has failed. If boundary scan is not available, and repair is viewed as necessary, then an alternative means for sensitizing between-chip faults is needed.

The decision as to whether a test is considered necessary is based purely on cost and test-escape considerations. Sandborn and Moreno (1994) and Ng in Donane and Franzon (1992, Chap. 4) provide more information. In general, if an MCM consists only of inexpensive, mature die, repair is unlikely to be worthwhile. The cost of repair is likely to be more than the cost of just throwing away the failed module. For MCMs with only one expensive, low-yielding die, the same is true, particularly if it is confirmed that the cause of most failures is that die. On the other hand, fault diagnosis and repair is usually desirable for modules containing multiple high-value die. You do not wish to throw all of these die away because only one failed.

Thermal design is often important in an MCM. An MCM will have a higher heat density than the equivalent PCB, sometimes necessitating a more complex thermal solution. If the MCM is dissipating more than 1 W, it is necessary to check if any heat sinks and/or thermal spreaders are necessary.

Sometimes, this higher concentration of heat in an MCM can work to the designer's advantage. If the MCM uses one larger heat sink, as compared with the multiple heat sinks required on the single-chip packaged version, then there is the potential for cost savings.

Generally, electrical design is easier for a MCM than for a PCB; the nets connecting the chips are shorter, and the parasitic inductances and capacitances smaller. With MCM-D technology, it is possible to closely space the power and ground planes, so as to produce an excellent decoupling capacitor. Electrical design, however, cannot be ignored in an MCM; 300-MHz MCMs will have the save design complexity as 75-MHz PCBs.

MCMs are often used for mixed signal (mixed analog/RF and digital) designs. The electrical design issues are similar for mixed signal MCMs as for mixed signal PCBs, and there is a definite lack of tools to help the mixed signal designer. Current design practices tend to be qualitative. There is an important fact to remember that is unique to mixed signal MCM design. The on-MCM power and ground supplies are separated, by the package parasitics, from the on-PCB power and ground supplies. Many designers have assumed that the on-MCM and on-PCB references voltages are the same only to find noise problems appearing in their prototypes.

For more information on MCM design techniques, the reader is referred to Doane and Franzon (1992), Tummula and Rymaszewski (1989), and Messner et al. (1992).

Defining Terms

Boundary scan: Technique used to test chips after assembly onto a PCB or MCM. The chips are designed so that test vectors can be scanned into their input/output registers. These vectors are then used to determine if the chips, and the connections between the chips are working.

Ceramic MCM (MCM-C): A MCM built using ceramic packaging techniques.

Deposited MCM (MCM-D): A MCM built using the deposition and thin-film lithography techniques that are similar to those used in integrated circuit manufacturing.

Flip-chip solder-bump: Chip attachment technique in which pads and the surface of the chip have a solder ball placed on them. The chips are then flipped and mated with the MCM or PCB, and the soldered reflowed to create a solder joint. Allows area attachment.

Known good die (KGD): Bare silicon chips (die) tested to some known level.

Laminate MCM (MCM-L): A MCM built using advanced PCB manufacturing techniques.

Multichip module (MCM): A single package containing several chips.

Printed circuit board (PCB): Conventional board found in most electronic assemblies.

Tape automated bonding (TAB): A manufacturing technique in which leads are punched into a metal tape, chips are attached to the inside ends of the leads, and then the chip and lead frame are mounted on the MCM or PCB.

Wire bond: Where a chip is attached to an MCM or PCB by drawing a wire from each pad on the chip to the corresponding pad on the MCM or PCB.

Acknowledgments

The author wishes to thank Andrew Stanaski for his helpful proofreading and for Fig. 8.119 and the comments related to it. He also wishes to thank Jan Vardaman and Daryl A. Doane for providing the tremendous learning experiences that lead to a lot of the knowledge provided here.

References

Dehkordi, P., Ramamurthi, K., Bouldin, D., Davidson, H., and Sandborn, P. 1995. Impact of packaging technology on system partitioning: A case study. In *1995 IEEE MCM Conference*, pp. 144–151.

Doane, D.A. and Franzon, P.D., eds. 1992. *Multichip Module Technologies and Alternatives: The Basics.* Van Nostrand Reinhold, New York.

Franzon, P.D., Stanaski, A., Tekmen, Y., and Banerjia. S. 1996. System design optimization for MCM. *Trans. CPMT.*

Messner, G., Turlik, I., Balde, J.W., and Garrou, P.E., eds. 1992. *Thin Film Multichip Modules.* ISHM.

Sandborn, P.A. and Moreno., H. 1994. *Conceptual Design of Multichip Modules and Systems.* Kluwer, Norwell, MA.

Tummula, R.R. and Rymaszewski, E.J., eds. 1989. *Microelectronics Packaging Handbook.* Van Nostrand Reinhold, Princeton, NJ.

Further Information

Additional information on the topic of multichip module technology is available from the following sources:

The major books in this area are listed in the References.

The primary journals are the *IEEE Transactions on Components, Packaging and Manufacturing Technology, Parts A and B,* and *Advancing Microelectronics,* published by ISHM.

The foremost trade magazine is *Advanced Packaging,* available free of charge to qualified subscribers.

The two main technical conferences are: (1) The IEEE Multichip Module Conference (MCMC), and (2) the ISHM/IEEE Multichip Module Conference.

8.10　Testing of Integrated Circuits

Wayne Needham

8.10.1　Introduction

The goal of the test process for integrated circuits is to separate good devices from bad devices, and to test good devices as good. Bad devices tested as bad become yield loss, but they also represent opportunity for cost reduction as we reduce number of **defects** in the process. When good devices are tested as bad, overkill occurs and this directly impacts cost and profits. Finally, bad devices tested as good signal a quality problem, usually represented as *defects per million* (DPM) devices delivered.

Unfortunately, the test environment is not the same as the operating environment. The test process may reject good devices and may accept defective devices as good. Here lies one of the basic problems of today's test methods, and it must be understood and addressed as we look at how to generate and test integrated circuits. Figure 8.120 shows this miscorrelation of test to use.

FIGURE 8.120   Test miscorrelations.

Bad devices are removed from good devices through a series of test techniques that enable the separation, by either voltage- or current-based testing. The match between the test process and the actual system use will not be addressed in this chapter.

8.10.2   Defect Types

Defects in an integrated circuit process fall into three broad categories. They are:

- Opens or abnormally high-resistance portions of signal lines and power lines that increase the series resistance between connected points. This extra resistance slows the charging of the parasitic capacitance of the line, thus resulting in a time delay. These defects also degrade voltage to a transistor circuit, which reduces levels or causes delays.

- Shorts, bridges, or resistive connections between layers of an integrated circuits, which form a resistance path in a normal insulating area. Given sufficiently low R value, the path may not be able to drive a 1 or a 0.

- Parameter variations such as threshold voltage and saturation current deviations changed by contamination, particles, implant, or process parameter changes. These usually result in speed or drive level changes.

Table 8.11 shows several types of defects and how they are introduced into an integrated circuit. It also describes electrical characteristics.

Defects, although common in integrated circuit processes, are difficult to quantify via test methods. Therefore, the industry has developed abstractions of defects called **faults**. Faults attempt to model the way defects manifest themselves in the test environment.

Traditional Faults and Fault Models

The most widely used fault model is the *single stuck-at fault* model. It is available in a variety of simulation environments with good support tools. The model represents a defect as a short of a signal line to a power supply line. This causes the signal to remain fixed at either a logic one or a zero. Thus, the logical result of an operation is not propagated to an output. Figure 8.121 shows the equivalent of a single stuck-at one or zero fault in a logic circuit. Although this fault model is the most common one, it was developed

TABLE 8.11   Typical IC Defects and Failure Modes

Defect	Source	Electrical Characteristic
Extra metal or interconnect	particle	ohmic connection between lines
Constrained geometry	insufficient or under etching	higher than normal resistance
Foreign material induced in the process in the process	contamination	varying resistance or short

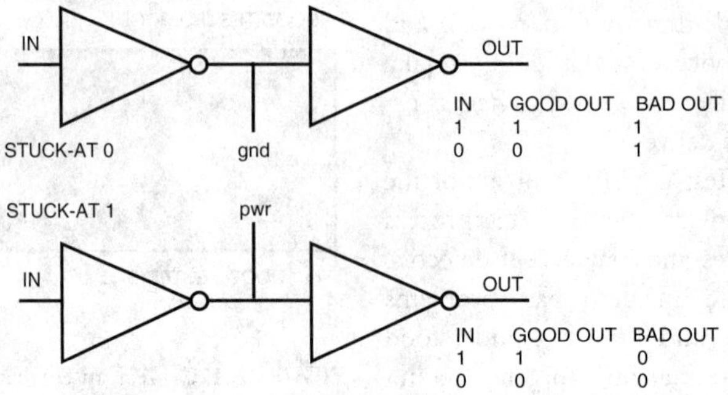

FIGURE 8.121   The single stuck-at model.

over 30 years ago, and does not always reflect today's more advanced processes. Today's very large-scale integrated (VLSI) circuits and submicron processes more commonly have failures that closely approximate bridges between signal lines. These bridges do not map well to the stuck-at fault model.

Other proposed or available fault models include:

- The *open fault model:* This model is for open circuits. Unfortunately in CMOS, missing contacts and transistors can exhibit correct behavior when switching at speed, even if some of the transistors or contacts are missing.

- The *leakage fault model:* This model assumes leakage in transistors and lends itself to an I_{DDQ}[2] test technique (discussed later in this chapter).

- *Timing fault models* (delay, gate and transition faults): All of these models assume a change in the AC characteristics of the circuit. Either a single gate, line, or block of circuitry is slower than needed or simulated, and this slow signal causes a problem with at-speed operation of the integrated circuit.

- *Pseudostuck-at model fault model* for I_{DDQ}: This model assumes nodes look like they are stuck at a one or zero logic level.

- *Voting fault models,* and bridge models: Adjacent lines are assumed connected and the logic result is an X (unknown), or a vote with the highest drive strength circuit dominating the logic level.

- Also for memory circuits.

- *Neighborhood pattern faults*: This fault assumes operations in one memory cell will affect adjacent memory cells (usually limited to the four closest cells).

- *Coupling faults:* This fault model assumes that full rows or columns couple signals into the cells or adjacent rows and columns.

- *Retention faults:* These faults are similar to the open fault for logic (discussed previously). Failures to this fault model do not retain data over an extended period of time.

There are many other fault models.

It is important to realize that the stuck-at model may not necessarily catch failures and/or defects in the fabrication processes. This problem is being addressed by the computer-aided design (CAD) industry. Today, numerous solutions are available, or on the horizon, that address the issues of fault models. Tools currently available can support delay faults. Bridging fault routines have been demonstrated in university research, and should be available from selected CAD vendors within a few years. Soon, we will not be confined to testing solely through the single stuck-at model.

---

[2] I_{DDQ} is the I_{IEEE} symbol for quotient current in a CMOS circuit.

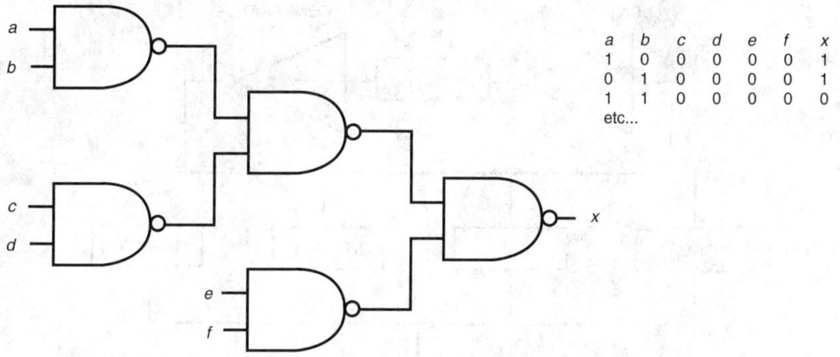

FIGURE 8.122 Logic diagram and test set.

8.10.3 Concepts of Test

The main test method of an integrated circuit is to control and observe nodes. Doing this verifies the logical operation that ensures that the circuit is fault free. This process is usually done by generating stimulus patterns for the input and comparing the outputs to a known good state. Figure 8.122 demonstrates a classic case where a small portion of an integrated circuit is tested with a stimulus pattern set for inputs and an expected set for outputs. Note this test set is logic, not layout, dependent and checks only stuck at faults.

Types of Test

There are three basic types of test for integrated circuits. These are parametric tests, voltage tests, and current tests. Each has a place in the testing process.

Parametrics tests ensure the correct operation of the transistors (voltage and current) of the integrated circuit. Parametric tests include timing, input and output voltage, and current tests. Generally, they measure circuit performance. In all of these tests, numbers are associated with the measurements. Typical readings include: $T_x < 334$ ns, $I_{OH} < 4$ mA, $V_{OI} < 0.4$ V, etc.

The next major category of testing is power supply current tests. Current tests include active switching current or power test, power down test, and standby or stable quiescent current tests. This last test, often called I_{DDQ}, will be discussed later.

The most common method of testing integrated circuits is the voltage-based test. Voltage-based test for logic operations includes functions such as placing a one on the input of an integrated circuit and ensuring that the proper one or zero appears on the output of the IC at the right time. Logical values are repeated, in most cases, at high and low operating voltage. The example in Fig. 8.122 is a voltage-based test.

Test Methods

To implement voltage based testing or any of the other types of tests, initialization of the circuit must take place, and control and observation of the internal nodes must occur. The following are four basic categories of tests:

- External stored response testing. This is the most common form for today's integrated circuits and it relies heavily on large IC **automated test equipment** (**ATE**).
- **Builtin-self-test (BIST).** This is method of defining and building test circuity onto the integrated circuit. This technique provides stimulus patterns and observes the logical output of the integrated circuit.
- **Scan** testing. Some or all of the sequential elements are converted into a shift register for control and observation purposes.
- Parametric tests. The values of circuit parameters are directly measured. This includes the I_{DDQ} test, which is good at detecting certain classes of defects. To implement an I_{DDQ} test, a CMOS integrated

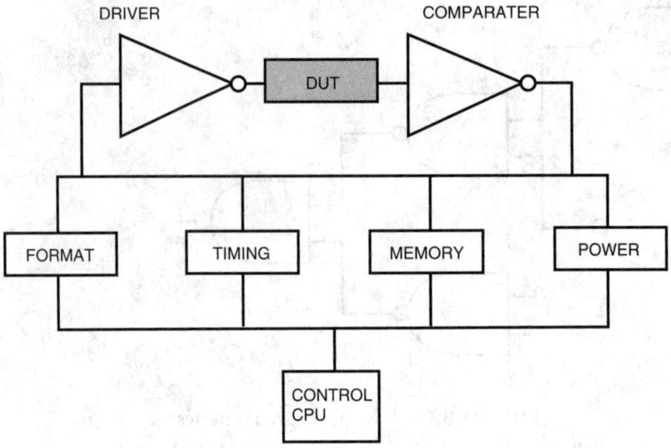

FIGURE 8.123   Stored response tester and DUT.

circuit clock is stopped and the quiescent current is measured. All active circuitry must be placed in a low-current state, including all analog circuitry.

External Stored Response Testing

Figure 8.123 shows a typical integrated circuit being tested by a stored response tester. Note that the patterns for the input and output of the integrated circuit are applied by primary inputs, and the output is compared to known good response. The process of generating stored response patterns is usually done by simulation. Often these patterns are trace files from the original logic simulation of the circuit. These files may be used during simulation for correct operation of the logic and later transferred to a tester for debug, circuit verification, and test. The patterns may take up considerable storage space, but the test pattern can easily detect structural and/or functional defects in an integrated circuit.

To exercise the **device under test (DUT)**, a stored response functional tester relies on patterns saved from the logic simulation. These patterns include input and output states and must account for unknown or X states. The set of patterns for functional test can usually be captured as trace files, or change files, and then ported to the test equipment. Trace or change files capture the logic state of the DUT as its logical outputs change over time. Typically, these patterns represent the logical operation of the device under test. As such, they are good for debugging the operation and design of integrated circuits. They are not the best patterns for fault isolation or yield analysis of the device.

BIST

The most common implementation of BIST is to include a **linear feedback shift register (LFSR)** as an input source. An LFSR is constructed from a shift register with the least significant bit fed back to selected stages with exclusive-OR gates. This feedback mechanism creates a polynomial, which successively divides the quotient and adds back the remainder. Only certain polynomials create pseudorandom patterns. The selection of the polynomial must be carefully considered. The LFSR source, when initialized and clocked correctly, generates a pseudorandom pattern that is used for the stimulus patterns for a block of logic in an integrated circuit.

The common method of output compaction is to use a **multiple input shift register (MISR)**. A MISR is an LFSR with the outputs of the logic block connected to the stages of the LFSR with additional exclusive-OR gates. The block of logic, if operating defect free, would give a single correct signature (see Fig. 8.124). If there is a defect in the block of logic, the resultant logic error would be captured in the output MISR. The MISR constantly divides the output with feedback to the appropriate stages such that an error will remain in the MISR until the results are read at the end of the test sequence. Because the logic states are compacted in a MISR, the results may tend to mask errors if the error repeats itself in the output, or if there are multiple errors in the block. Given a number of errors, the output could contain the correct answer

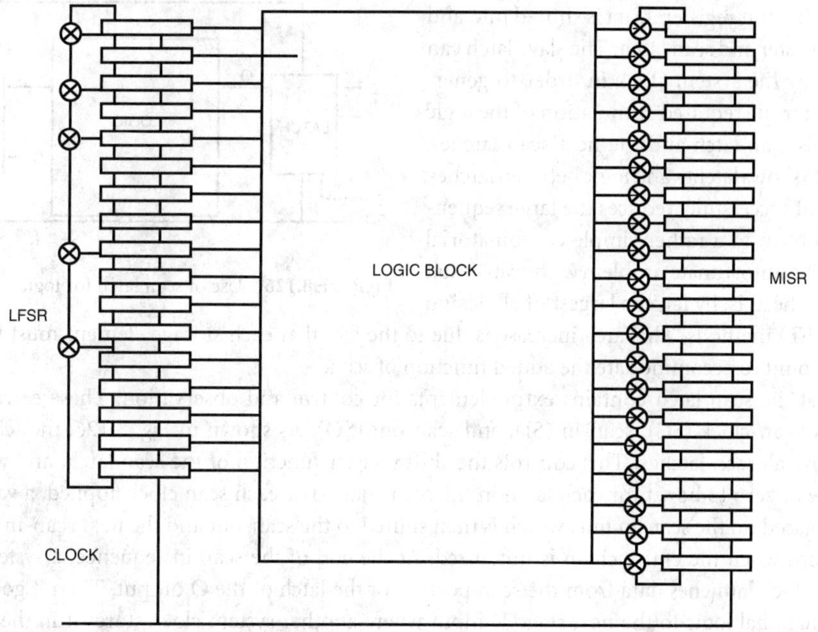

FIGURE 8.124 BIST implementation example with inputs and outputs.

even though the logic block contains errors or defects. This problem is called *aliasing*. For example, the output register shown in Fig. 8.124 is 20-bits long; 2^{20} is approximately one million, which means that there is a very high chance that without an aliasing defect, the output will be the correct state of one out of a million with a fault free integrated circuit.

It should be noted that the example is shown for combination logic only. Sequential logic becomes more difficult as initialization, illegal states, state machines, global feedback, and many other cases must be accounted for in the generation of BIST design and patterns.

Scan

Scan is a technique where storage elements (latches and flip-flops) are changed to dual mode elements. For instance, Fig. 8.125 shows a normal flip-flop of an integrated circuit. Also shown in Fig. 8.125 is a flip-flop converted to a scan register element. During normal operation, the D input and system clock control the output of the flip-flop. During scan operation, the scan clocks are used to control shifting of data into

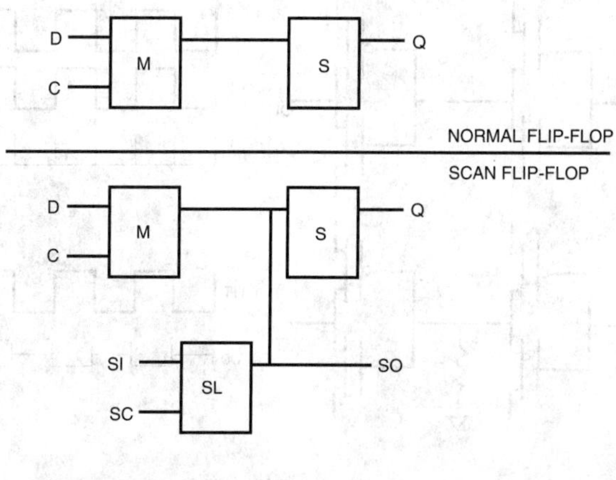

FIGURE 8.125 Example of a scan latch and flip flop.

and out of the shift register. Data is shifted into and out of the master and scan latch. The slave latch can be clocked by the system clock in order to generate a signal for the required stimulation of the logic between this scan latch and the next scan latches. Figure 8.126 shows latches and logic between latches. This method successfully reduces the large sequential test problem to a rather simple combinatorial problem. The unfortunate problem with scan is that the area overhead is, by far, the largest of all **design for test (DFT)** methods. This area increase is due to the fact that each storage element must increase in transistor count to accommodate the added function of scan.

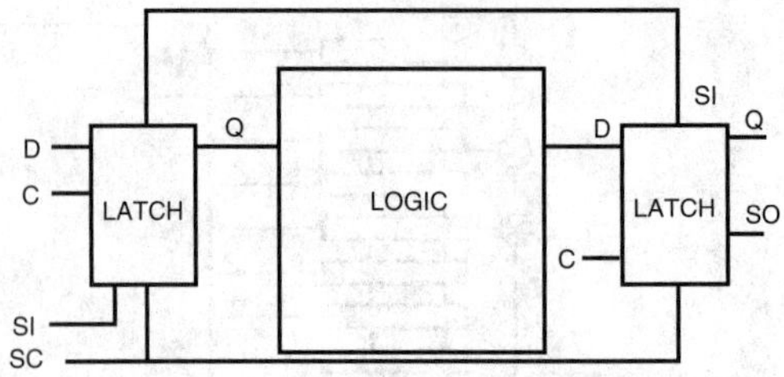

FIGURE 8.126   Use of scan latch for logic.

Note that the scan latch contains extra elements for control and observation. These extra elements include the scan-clock (SC), scan-in (SI), and scan-out (SO). As shown in Fig. 8.126, the scan-clock is used to drive all scan latches. This controls the shift register function of the scan latch, and will allow a known one or zero to be set for each latch in the scan chain. For each scan clock applied, a value of one or zero is placed on the scan-in line, which is then shifted to the scan-out and the next scan-in latch. This shifting stops when the entire chain is initialized. At the end of the scan in sequence, a system clock is triggered, which launches data from the scan portion of the latch of the Q output. Then it goes through the combinational logic to the next stage D input where another system clock stores it in the latch. The next set of scan loads, and shifts out the stored value of the logic operation for comparison. If the logic operation was correct, the scan value will be correct in the shift out sequence.

I_{DDQ}

I_{DDQ} is perhaps the simplest test concept. Figure 8.127 shows a typical dual inverter structure with a potential defect. It also shows the characteristic of I_{DDQ}. The upper circuit is defect free, and the lower circuit has a leakage defect shown as R. I_{DDQ} is shown for each clock cycle. Note the elevated current in the defective circuit during clock cycle A. This elevated current is called I_{DDQ} . It is an easy way to check for most bridging and many other types of defects. The I_{DDQ} test controls and observes approximately half of the transistors in a IC for each test. Therefore, the test is considered massively parallel and is very efficient at detecting bridge defects and partial shorts. Another consideration is the need to design for an I_{DDQ} test. All DC paths, pullups, bus drivers, and contention paths must be designed for zero current when the clock is stopped. One single transistor remaining on, can disable the use of an I_{DDQ} test.

Note that V_{out} and I_{DD} in the defective case may be tolerable if the output voltage is sufficient to drive the next input stage. Not shown here is the timing impact of the degraded voltage. This degraded voltage will cause a time delay in the next stage.

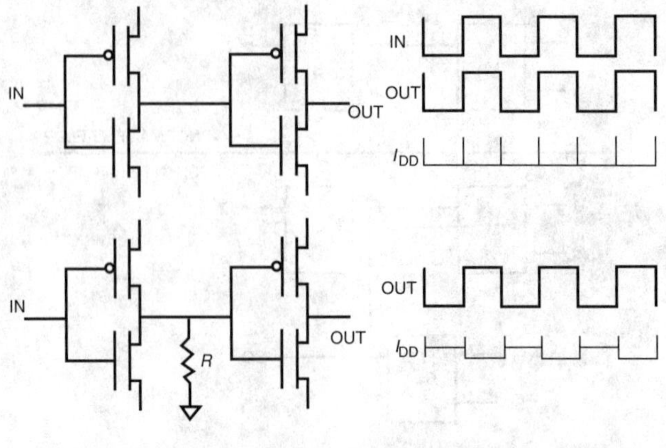

FIGURE 8.127 I_{DDQ} test values.

TABLE 8.12   Attributes of Test

Test Types	Complexity	Illegal States	Aliasing	Silicon Overhead	Test Pattern Generation Time
SCAN	Simple	Controlled	NA	Highest	Shortest
BIST	Complicated with latches	Big problem	Yes	Moderate	Moderate
Stored response	Very complicated	NA	NA	None-little	Longest
I_{DDQ}	Low	None	None	None	Simple

8.10.4   Test Tradeoffs

To determine the best type of test, one must understand the defect types, the logical function of the integrated circuit, and test requirements. Examples of test tradeoffs are included in Table 8.12. The rows in the table correspond to the test types as listed earlier: scan, BIST, stored response, and I_{DDQ}. The columns in the table depict certain attributes of test to be described here.

The first column is the test type. Column number two shows pattern complexity. This is the problem of generating the necessary patterns in order to fully stimulate and control the nodes of the integrated circuit. Column number three is illegal states. Often times the logic in integrated circuits, such as bidirectional bus drivers and signals, must be exclusively set one way or another. Examples include decoders, bus drivers and mutually exclusive line drivers. Tests such as scan and BIST, if not designed and implemented correctly, can damage the integrated circuit during the test process by putting the circuit into an internal contention state. Examples include two bus drivers on simultaneously when one drives high and the other drives low.

Aliasing is a common problem of BIST techniques. This is shown in column four. For example, consider a memory laid out in rows and columns and assume that one column is a complete failure. If there are multiple failures along the column, let us say 256, and if the signature length is 256, the output could possibly show a correct response even though the entire column or row of the memory is defective.

The next column shows overhead. Overhead is the amount of silicon area necessary to implement the test. Each of the areas of overhead relate to the amount of silicon area needed to implement the specific types of test methods.

The final column is the time for test generation. The unfortunate relationship here is that the techniques with the highest area overhead have the most complete control and are easiest to generate tests. Techniques with the lowest area needs have the most complex test generation problem, and take the longest time. This is a complex trade off.

Tradeoffs of Volume and Testability

It is important to forecast the volume of the integrated circuit to be manufactured prior to selecting DFT and test methods. Engineering effort is nonrecurring and can be averaged over every integrated circuit manufactured. The testing of an integrated circuit can be a simple or a complex process. The criteria used for selecting the proper test technique must include volume, expected quality, and expected cost of the integrated circuit. Time to market and/or time to shipments are key factors, as some of the test techniques for very large integrated circuits take an extremely long time.

Defining Terms

Automated test equipment (ATE): Any computer controlled equipment used to test integrated circuits.
Builtin self-test (BIST): An acronym that generally describes a design technique with input stimulus generation circuitry and output response checking circuitry.
Defect: Any error, particle, or contamination of the circuit structure. Defects may or may not result in faulty circuit operation.
Design for test (DFT): A general term which encompasses all techniques used to improve control and observation of internal nodes.
Device under test (DUT): A term used to describe the device being tested by the ATE.

Fault: The operation of a circuit in error. Usually faulty circuits are the result of defects or damage to the circuit.

Linear feedback shift register (LFSR): A method of construction of a register with feedback to generate pseudo random numbers.

Multiple input shift register (MISR): Usually an LFSR with inputs consisting of the previous stage exclusive OR'ed with input data. This method compacts the response of a circuit into a polynomial.

Scan: A design technique where sequential elements (latches, or flipflops) are chained together in a mode that allows data shifting into and out of the latches. In general, scan allows easy access to the logic between latches and greatly reduces the test generation time and effort.

References

Abramovici, M. et al. 1990. *Digital Systems Testing and Testable Design.* IEEE Press, New York.

Needham, W.M. 1991. *Designer's Guide to Testable ASIC Devices.* Van Nostrand Reinhold, New York.

van de Goor, A.J. 1991. *Testing Semiconductor Memories, Theory and Practice.* John Wiley & Sons, New York.

Further Information

IEEE International Test Conference (ITC). This is the largest gathering of test experts, test vendors and researchers in the world.

IEEE Design Automation Conference (DAC) held in various sites throughout the world.

IEEE VLSI Test Symposium.

Test vendors of various types of equipment provide training for operation, programming, and maintenance of ATE.

CAD vendors supply a variety of simulation tools for test analysis and debugging.

8.11 Integrated Circuit Packages[3]

Victor Meeldijk

8.11.1 Introduction

In 1958 the first IC invented had one transistor on it, when the first 8086 microprocessor debuted on June 8, 1978 it contained 29,000 transistors and ran at 5 MHz, the first Intel® Pentium® (1993) had roughly 3.1 million transistors and in 2004 the Intel® Pentium® 4 processor contains 55 million transistors and runs more than 600 times faster than the original 8086 processor at 3.06 GHz. In the near future microprocessors will have more than 1 billion transistors. As semiconductor devices become more complex, the interconnections from the die to the circuit hardware keep evolving. Devices with high clock rates and high-power dissipation, or with multiple die, are leading to various new packages.

The use of bare die in chip on board (COB) and multichip module (MCM) applications is increasing with designers looking for end item size and weight reduction (automotive applications have bond wires going directly from the die to a connector). The use of bare die eliminates the timing delays (caused by stray inductance and capacitance) associated with the leadframes and the device input/outputs (I/Os). For burst mode static RAMs, 1/2–2 ns or a 20% improvement in access time is achieved with bare die product. Small portable devices can also use flip-chip bonding.

IC packaging can be divided into the following categories:

- Surface mount packages (plastic or ceramic)
- Chip-scale packaging
- Bare die

---

[3] Portions of this chapter were adapted from: Meeldijk, V., 1995. *Electronic Components: Selection and Application Guidelines,* Chap. 12. Wiley-Interscience, New York. With permission.

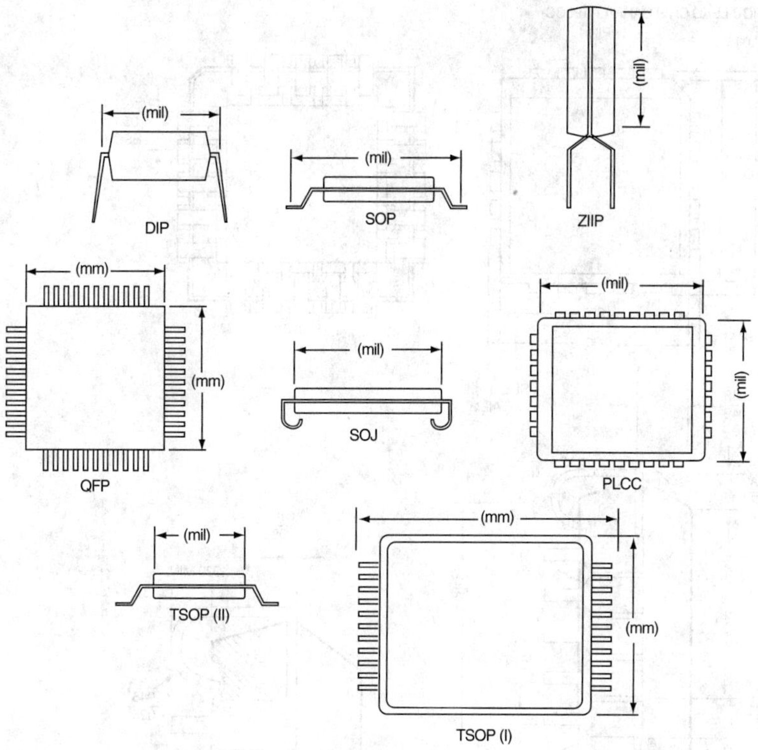

FIGURE 8.128 Some typical IC packages for through-hole applications.

- Through-hole packages
- Module assemblies

In addition to bare die and surface mount techniques, there are still the older device packages for through-hole applications, where the lead of the package goes through the printed wiring board (see Fig. 8.128 and Fig. 8.129). Modules are also available that are assemblies of either packaged parts or die.

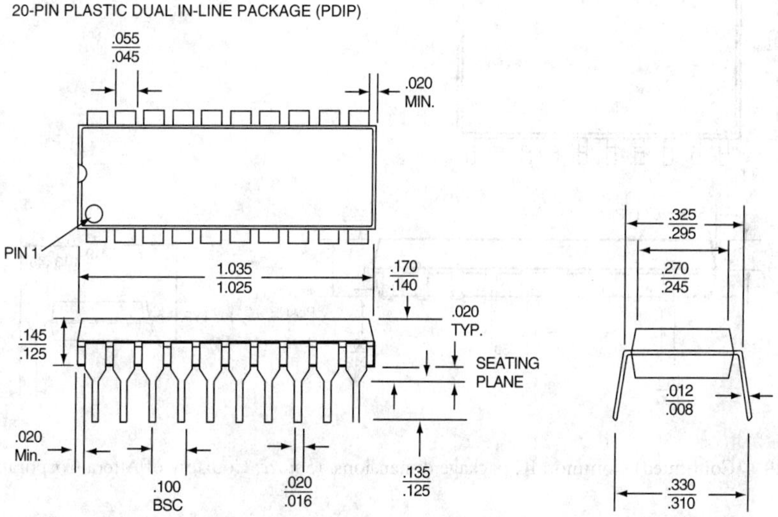

FIGURE 8.129 Common IC package dimensions. (*Source:* Courtesy of Altera Corporation, San Jose, CA.)

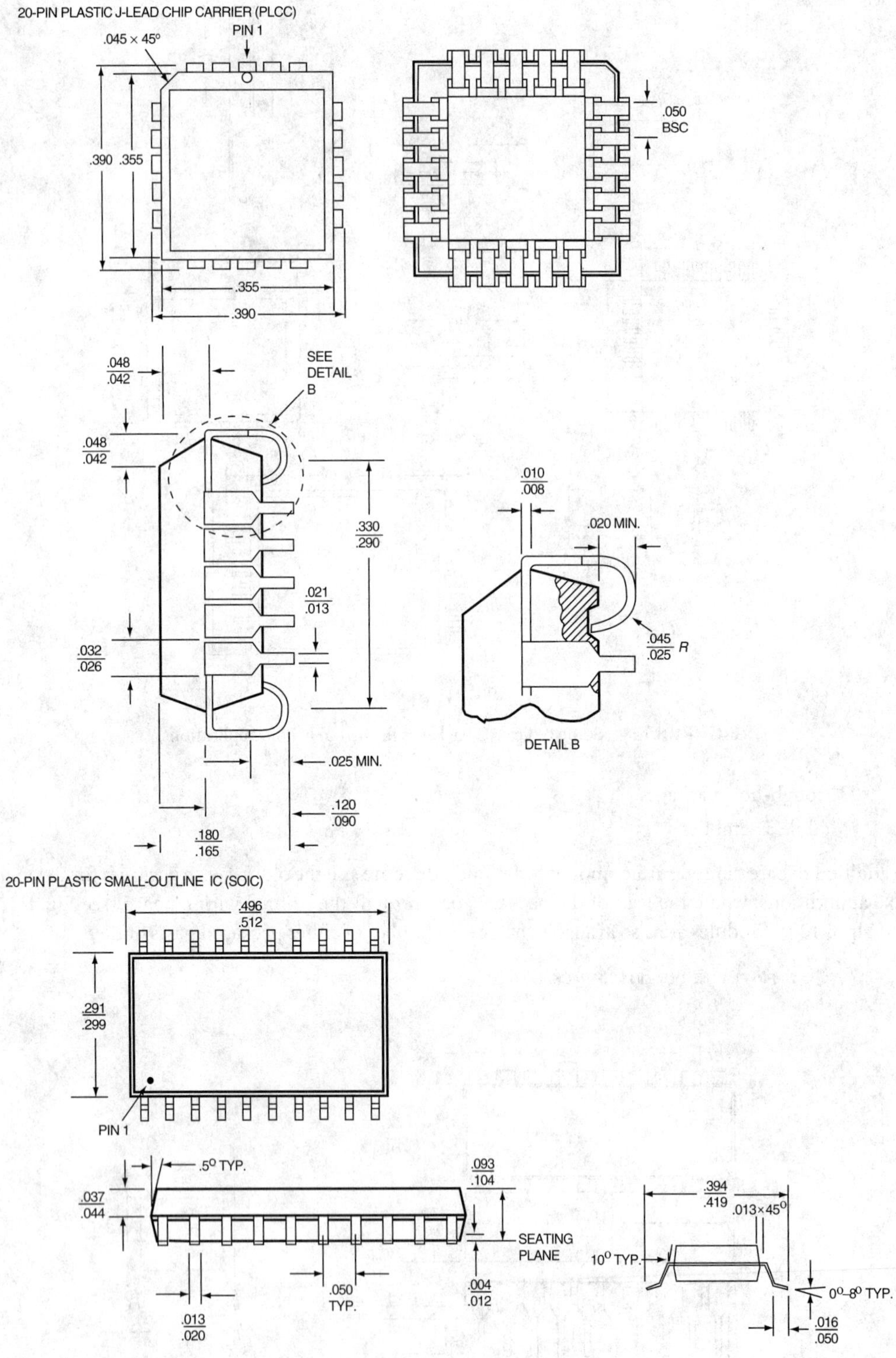

20-PIN PLASTIC J-LEAD CHIP CARRIER (PLCC)

20-PIN PLASTIC SMALL-OUTLINE IC (SOIC)

FIGURE 8.129   (Continued) Common IC package dimensions. (*Source:* Courtesy of Altera Corporation, San Jose, CA.)

28-PIN CERAMIC J-LEAD CHIP CARRIER (JLCC)
FOR MILITARY-QUALIFIED PRODUCT, SEE CASE OUTLINE IN ALTERA MILITARY PRODUCT DRAWING 02D-00194.

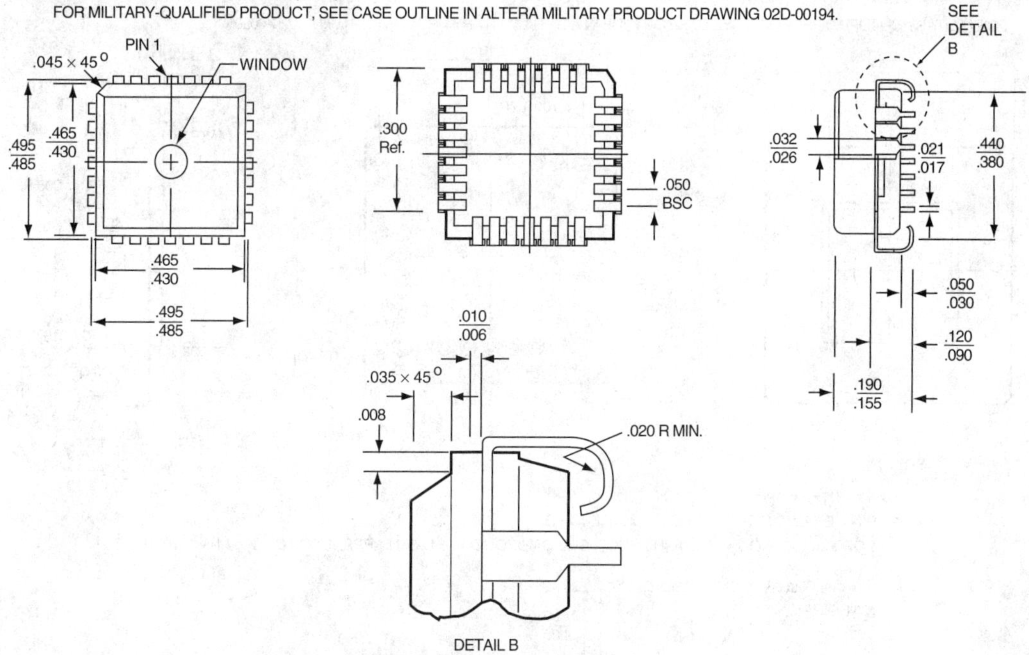

DETAIL B

68-PIN CERAMIC J-LEAD CHIP CARRIER (JLCC)

FOR MILITARY-QUALIFIED PRODUCT, SEE CASE OUTLINE C-J2 IN APPENDIX C OF MIL-M-38510

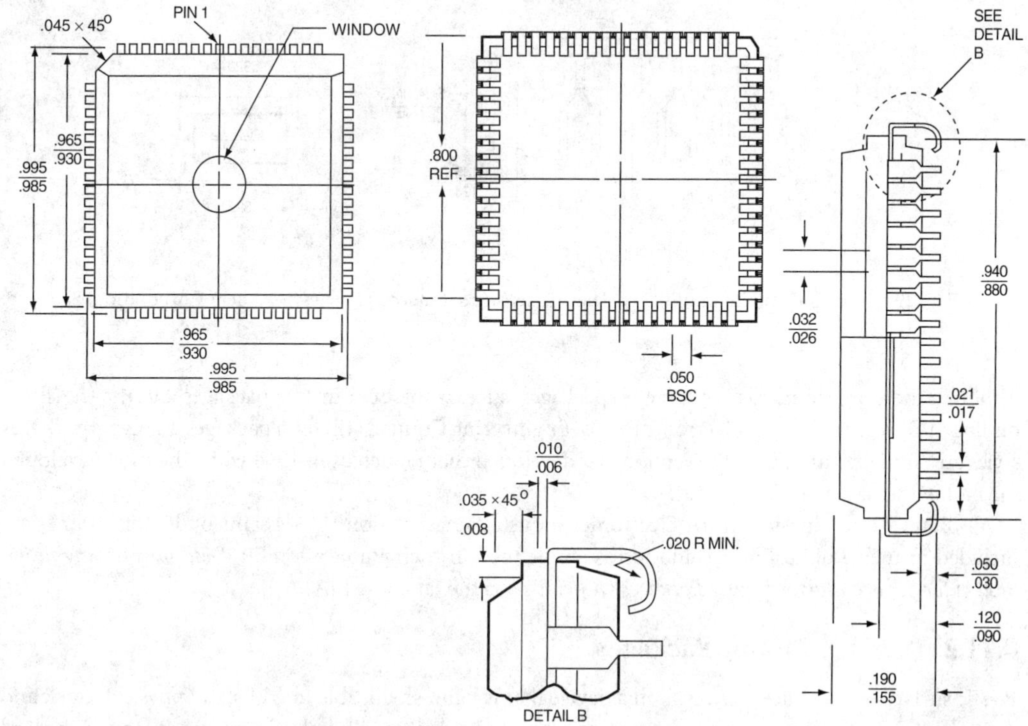

DETAIL B

FIGURE 8.129 (Continued) Common IC package dimensions. (*Source:* Courtesy of Altera Corporation, San Jose, CA.)

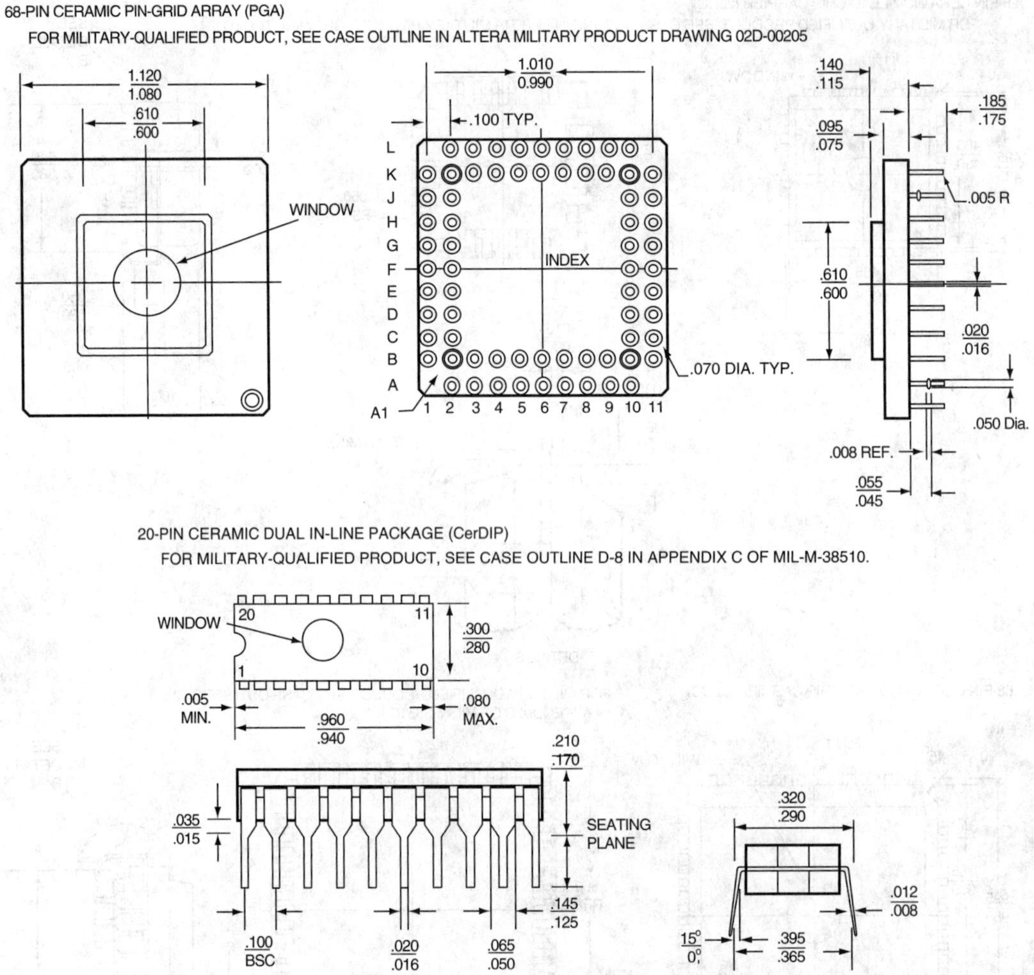

68-PIN CERAMIC PIN-GRID ARRAY (PGA)

FOR MILITARY-QUALIFIED PRODUCT, SEE CASE OUTLINE IN ALTERA MILITARY PRODUCT DRAWING 02D-00205

20-PIN CERAMIC DUAL IN-LINE PACKAGE (CerDIP)

FOR MILITARY-QUALIFIED PRODUCT, SEE CASE OUTLINE D-8 IN APPENDIX C OF MIL-M-38510.

FIGURE 8.129   (Continued) Common IC package dimensions. (*Source:* Courtesy of Altera Corporation, San Jose, CA.)

Although there are standards for some IC packages, such as the dual-in-line packages and the transistor outline (TO) registered Joint Electronic Device Engineering Council (JEDEC) packages, many IC packages styles are (initially) unique to the vendor that developed them (such as in 1990 when the Intel developed the molded plastic quad flat package).

All JEDEC (a subdivision of the Electronics Industries Association [EIA]) semiconductor outlines are included in their Publication 95 and terms are defined in their Publication 30, *Descriptive Designation System for Semiconductor Device Packages* (note: EIAJ is the Japanese EIA).

8.11.2   Surface Mount Packages

Plastic surface mount packages result in a device that is light, small, able to withstand physical shock and g forces, and inexpensive due to a one-step manufacturing process. The plastic is molded around the lead frame of the device. Work is still continuing on developing coatings for the die, such as polyimide, which may resolve hermeticity problems and coefficient of thermal expansion mismatches between the die and the plastic package.

Plastic parts shipped in sealed bags with desiccant are designed for 12 months storage and should only be opened when the parts are to be used. Parts stored for longer than this time, especially plastic quad flatpack (PQFP) packaged devices, should be baked to remove moisture that has entered the package. Plastic packages are hydroscopic and absorb moisture to a level dependent on the storage environment. This moisture can vaporize during rapid heating, such as in a solder reflow process, and these stresses can cause package cracking (known as the popcorn effect). Subsequent high-temperature and moisture exposures can allow contaminants to enter the IC and cause failure at a later time due to corrosion. Consult the device manufacturer for details on proper device handling.

Hermetic packages, such as the older flat pack design or ceramic leadless chip carriers (CLCCs), are used in harsh applications (such as military and space applications) where water vapor and contaminants can shorten the life of the device. Thus they are used in mission critical communication, and navigation and avionic systems. The reliability of CLCC package connections can be further improved by the attachment of leads to the connection pads, to eliminate differences in the thermal coefficient of expansion between the parts and the circuit board material.

Metal packages, with glass seals, provide the highest level of hermetic sealing followed by glasses and ceramics (Fig. 8.130). These parts have a higher temperature range than plastic encapsulated parts (typically, from -55 to $+125°C$ vs. 0 to 70°C, or 85°C, for plastic encapsulated devices). Although aluminum oxide, the most commonly used ceramic material, has a thermal conductivity that is an order of magnitude less than a plastic packaged device, in a ceramic sealed device the circuit die does not come into contact

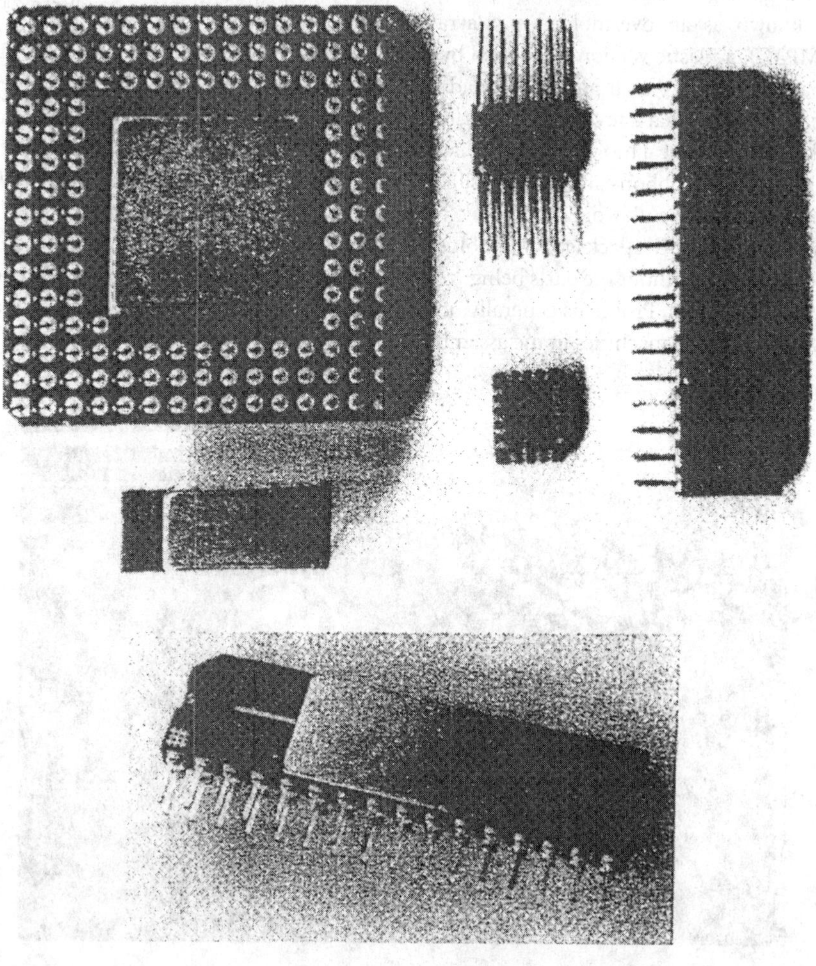

FIGURE 8.130 Typical hermetic IC packages (from the left going clockwise): PGA, flat pack, VIL, LCC and DIP. The bottom photo shows the VIL from a different angle so both rows of pins can be seen.

with the ceramic packaging material. Thus, temperature cycling will not affect the die, and parts can withstand thousands of temperature cycles without damage (1000 temperature cycles would simulate 20 years use on a commercial airliner). Large die plastic encapsulated parts can fail after only 250 temperature cycles. As mentioned earlier, die coatings being developed may resolve this problem. One of the main drawbacks in specifying ceramic packaged devices is the more complicated (than plastic encapsulated parts) manufacturing process. These devices are made up of three ceramic layers, which result in a higher cost device. Availability is limited, and lead times are often longer than standard plastic devices.

Surface mount packages (and related packaging terms) include the following.

First is ball grid array (BGA) (Fig. 8.131 to Fig. 8.133), a packaging technique developed by IBM (also known as an overmolded pad-array

FIGURE 8.131   A 256-pin perimeter BGA (no connections are under the die). (*Source:* Courtesy of AMKOR/ANAM-AMKOR Electronics, Chandler, AZ.)

carrier (OMPAC), a plastic version developed by Motorola, see Fig. 8.134). Leads, or pads, are replaced by solder balls that replace high pin count quad flatpacks (QFPs). Several hundred arrayed leads can be accommodated in areas that a few hundred peripherally arranged leads would occupy.

The package transfers heat more efficiently than a QFP. To further assist heat dissipation there is a metal-lid BGA that uses metal ribbons and thermal silicon to transfer heat to the package's metal lid (pioneered by Samsung Electronics South Korea).

A disadvantage of the BGA package is that solder connections cannot be visually inspected and removed parts cannot be reused without the parts being "re-balled." A major manufacturer however, reports a joint failure rate of only 6 ppm. Problems generally occur only from defective components, bad boards, setup errors, or worn/damaged machines in the assembly line.

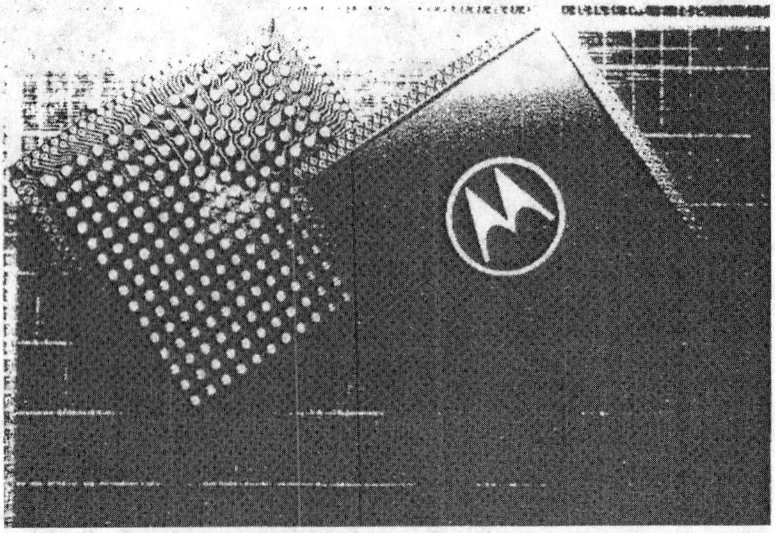

FIGURE 8.132   A 255 bump BGA, 1.5-mm pitch between leads, 27 × 27 mm body size (*Source:* Courtesy of Motorola, Inc., Phoenix, AZ. © Motorola. With permission.)

FIGURE 8.133 A 1.27-mm staggered pitch, 313 bump BGA with distributed vias, 35-mm body size. (*Source:* Courtesy of Motorola, Inc., Phoenix, AZ. © Motorola. With permission.)

Other BGA versions include the perimeter lead BGA, which does not have any solder balls under the chip, uBGA (micro BGA). Reference J-STD-013, first issued by the EIA in 1996 (847-509-7000) for further information on BGA packages.

Other surface mount packages (and related packaging terms) follow:

- Biometric package—A package where the die is exposed but the interconnections are protected by a window frame and plastic encapsulant. Developed by Harris Corporation (Melbourne, FL) in 1996 for a sensor chip used to recognize a person's fingerprint.

- caBGA—Chip array BGA (from Lattice Semiconductor). This package is 87% smaller than a comparable PLCC package.

- Ceramic ball grid array (CBGA).

- Ceramic column grid array (CCGA), invented by IBM, where solder columns replace the solder balls.

- Cerpack, a flatpack composed of a ceramic base and lid. The leadframe is sealed by a glass frit.

- Cerquad, a ceramic equivalent of plastic leaded chip carriers consisting of a glass sealed ceramic package with J leads and ultraviolet window capability.

FIGURE 8.134 A plastic (OMPAC) BGA. (*Source:* Courtesy of Motorola Inc. Phoenix, AZ. © Motorola. With permission.)

FIGURE 8.135   EDQUAD Package. (*Source:* Courtesy of ASAT, Inc., Palo Alto, CA.)

- Chip carrier (CC), a rectangular, or square, package with I/O connections on all four sides.
- Ceramic leaded chip carrier (CLDCC).
- Ceramic leadless chip carrier [CLCC (or CLLCC)], see LCCC.
- Ceramic pin grid array (CPGA).
- Ceramic quad flat package with J leads (CQFJ).
- Ceramic quad flatpack (CQFP), an aluminum ceramic integrated circuit package with four sets of leads extending from the sides and parallel to the base of the IC.
- Ceramic small outline with J leads (CSOJ).
- Chip scale grid array (CSGA), like a BGA, but with a smaller solder ball pitch (0.5–0.65 vs. 1.3–2.5 mm). This is used for devices with 200 or more leads.
- Dual small outline package (DSOP).
- EDQUAD, a trademark of ASAT Inc. for a plastic package that incorporates a heat sink; available as thin QFP (TQFP) and QFP (see Fig. 8.135).
- Enhanced Plastic BGA (EPBGA)—developed by LSI Logic and made available in 1995, this package supports 313 to 750 leads and is designed for high frequency ASIC applications.
- FCBGA, Flip Chip BGA, a design by Fujitsu Microelectronics for use with high frequency, high I/O devices that is large enough to handle the thermal requirements of RISC processors, IP and ATM switching devices.
- FCOB, Flip Chip on Board
- Flatpack (or quad flatpack), one of the oldest surface mount packages, used mainly on military programs. Typically, flatpacks have 14–50 leads on both sides of the body on 0.050-in centers.
- Flat SIP, a single inline package except the leads have a 90° bend.
- Flatpack (FP).
- fpBGA—Fine pitch BGA.
- fpsBGA—Fine pitch super BGA.
- Gull winged leadless chip carrier (GCC). Although easier to inspect, it is weaker than the J lead.
- Gull wing, leads which exit the body and bend downward, resembling a seagull in flight. Gull wings are typically used on small outline (SO) package, but are very fragile, easily bent, and difficult to socket for testing or burn-in. J leads solve these problems.

FIGURE 8.136 An LCC package with C leads attached to compensate for the thermal coefficient of expansion mismatch between the part and the circuit board, thus lowering solder joint stresses. (*Source:* Courtesy of General Dynamics, Information Systems, Formerly Computing Devices International, Bloomington, MN.)

- Hermetic chip carrier (HCC).
- HD-PQFP, originated by Intel for a high density PQFP package with more than 196 leads with a 0.4-mm pitch.
- Heat spreader QFP package (HQFP), package has a copper heat spreader and heat sink.
- HQSOP, a hermetic QSOP package created by Quality Semiconductor in 1995 (Quality Semiconductor has since been acquired by IDT Semiconductor).
- Hermetic small outline transistor (H-SOT) packaged device.
- I lead, IC leads that are formed perpendicular to the printed circuit board making a butt solder joint.
- J leads, leads that are rolled under the body of the package in the shape of the letter J. They are typically used on plastic chip carrier packages.
- J bend leads, leaded chip carrier (JCC).
- J leaded ceramic chip carrier (JLCC).
- Leadless chip carrier (LCC), a chip package with I/O pads on the perimeter of the package. Cavity (lid) up packages, with the backside of the die facing the substrate, have heat dissipated through the substrate. These JEDEC B and C ceramic packages are not suitable for air-cooled systems or for the attachment of heat sinks. Cavity (lid) down parts, (JEDEC A and D) with the die facing away from the substrate, are suitable for air-cooled systems. Rectangular, ceramic leadless E and F are intended for memory devices, with direct attachment in the lid-up position. Computing Devices International devised patented C and S leads that can be attached, as shown in Fig. 8.136.
- Leadless ceramic chip carrier (LCCC), or ceramic leadless chip carrier. JEDEC registered type A must be socketed when on a printed circuit board or a ceramic substrate board, and type B must be soldered. The LCC minipack must be soldered onto printed circuit boards.
- Leaded ceramic chip carrier (LDCC), leaded ceramic chip carrier packages include JEDEC types A, B, C, and D. Leaded type B parts are direct soldered to a substrate. Leaded type A parts can be socketed or direct soldered, and include sub-categories: leaded ceramic, premolded plastic, and postmolded plastic (which are not designated as LDCC devices).
- Land grid array (LGA), an Intel package used for parts such as the 80386L microprocessor. The package is similar to a PGA except that it has gold-plated pads (called *landing pads*) vs. pins.
- Leadless inverted device (LID), a shaped metallized ceramic form used as an intermediate carrier for the semiconductor chip (die). It is especially adapted for attachment to conductor lands of a thick-film or thin-film network by reflow solder bonding.

- Little foot (a trademark of Siliconix), a tiny small outline integrated circuit (SOIC) package.
- Leadless ceramic chip carrier (LLCC).
- Mini-BGA (mBGA), developed by Sandia National Laboratories. Layers of polyimide and metal are applied to the die to redistribute original die pads from a peripheral arrangement into an area array configuration.
- Miniflat, a flat package (MFP) approximately 0.102–0.113 in high (2.6–2.85 mm) that may have leads on two or four sides of the package.
- Multilayer molded (MM) package, an Intel PQFP package with improved high-speed operation due to separate power and ground planes, which reduces device capacitance.
- MM-PQFP, isolated ground and power planes are incorporated within the molded body of the PQFP.
- MQFP—Metal Quad Flatpack. Sometimes used to refer to a plastic metric quad flat pack (such as Harris Corporation, Harris sold its semiconductor operations, now known as Intersil Corp.).
- MQuad, designed by the Olin Corporation, a high thermal dissipation aluminum package available in plastic leaded chip carrier (PLCC) and QFP packages (with 28–300 leads, and clock speeds of 150 MHz) (see Fig. 8.137).
- MSOP—miniature SOP.
- OMPAC, see BGA.
- Pad array carrier (PAD), a surface mount equivalent of the PGA package. This package extends surface mount silicon efficiency from 15 to 40%. A disadvantage of this package is that the solder joints cannot be visually inspected, although X-ray inspection can be used to verify the integrity of the blind solder connections (and to check for solder balls and bridging). Individual solder joints, however, cannot be repaired.
- Plastic (encapsulated) ball grid array (PBGA), see OMPAC.
- Plastic leaded chip carrier [(PLCC) also known as PLDCC or a quad pack by some manufacturers, up to 100 pin packages, commonly 0.050-in pin spacing. Depending on the number of pins, pin 1 is in different locations but is confirmed by a dot in the molding, and the progression of the pin count continues counterclockwise. The leads are J shaped and protrude from the package. These parts

FIGURE 8.137 An MQUAD (trademark OLIN Corporation) package. (*Source:* Courtesy of Hawa Technologies (HTL) formerly Swire Technologies, San Jose, CA.)

can be either surface mounted or socketed. (This package was originally known as a *postmolded type A* leaded device.)

- Plastic quad flatpack (PQFP), this JEDEC approved package is used for devices that have 44–256 I/Os, 0.025-in lead spacing (EIA-JEDEC packages), 0.0256- and 0.0316-in spacing (EIAJ), and 0.0135-in pin spacing. This package has gull wing leads on all four sides and is characterized by bumpers on the corners. The body is slightly thicker than the QFP. PQFP packages are susceptible to moisture-induced cracking in applications requiring reflow soldering.
- Plastic surface mount component (PSMC).
- PSOP—plastic SOP (a term used by the Intel Corporation).
- Quad flat J (QFJ) leaded package.
- QFN—Quad Flatpack, no leads.
- Quad flatpack (QFP), a flatpack with leads on all four sides. The typical JEDEC approved pinouts are 100, 132, and 196, with lead pitch from 0.040 to 0.016 in. The EIAJ types are slightly thinner than the JEDEC equivalents (Fig. 8.138).
- Quarter quad flatpack (QQFT), see SQFT.
- Quad surface mount (QSM).
- Quarter-size small outline package (QSOP).
- Quad pack, see PLCC.
- QVSOP—Quad Very Small Outline Package, introduced by Quality Semiconductor (Note: Acquired by IDT Semiconductor) in late 1993. This package combines the body size of a half width, 14 pin SOIC, with the 25-mill lead pitch of a PQFP (or a 150 × 390 mil, 48 pin package). The package has designation JEDEC MO-154.
- Quad very small outline package (QVSOP), introduced by Quality Semiconductor. This package combines the body size of a half-width, 14-pin SOIC, with the 25-mil lead pitch of a PQFP (or a 150 × 390 mil, 48-pin package). The package has designation JEDEC MO-154.
- Repatterned die, see BGA.
- Square (body) chip carrier (SCC).
- Slam pack, a square ceramic package that looks like an LCC and is always used with a socket.

FIGURE 8.138   A QFP package. (*Source:* Courtesy of ASAT, Inc., Palo Alto, CA.)

- Small outline (SO) (with versions such as SO wide, SO narrow, or SO large), this design originated with the Swiss watch industry in the 1960s (and was reportedly nicknamed SO for Swiss Outline). It was used in the modern electronics industry by N.V. Philips (formerly known as Signetics in the U.S.) in 1971. This package is also known as a *mini-flat* (which has slightly different dimensions than the JEDEC parts by the Japanese). Common packages are SOD-323, SOD-523, SOD-123, see also SOT.

- Small outline IC (SOIC), 8–28 pin packages, commonly 0.050-in pin spacing, with gull wing leads. This package style is about 50–70% of the size of a standard dual inline package (DIP) part (30% as thick).

- Small outline package with J leads on two sides (SOJ). The leads are bent back around the chip carrier.

- Small outline large (SOL), generally refers to a package that is 0.300 mils wide vs 150 mils wide (for an SO package). It is a larger version of the small shrink outline package (SSOP).

- Small outline package (SOP), a package with two rows of narrowly spaced gull-wing leads (same as SOIC).

- Small outline transistor (SOT), a plastic leaded package originally for diodes and transistors but also used for some ICs (such as Hall effect sensors). Some common sizes are SOT-23, SOT-26, SOT-223, and the very small SOT-523.

- Small outline wide (SOW), see SOL.

- Shrink quad flatpack (or flat package) (SQFP), typically a 64-pin small quad flat package (1/4 height of QFT) with a lead pitch of 0.016 in or less. Also known as quarter quad flat pack (QQFT).

- Square surface mounting (SSM).

- Shrink small outline package (SSOP).

- Shrink small outline IC (SSOIC), a plastic package with gull-wing leads on two sides, with a lead pitch equal to or less than 0.025 in.

- Tape ball grid array (TBGA).

- Thin QFP (TQFP). Typical sizes are 1.0- and 1.4-mm body thickness, in sizes ranging from 10×10 mm to 20×20 mm and lead count from 64 to 144 leads.

- Thin small outline package (TSOP). The type-1 plastic surface mount parts have 0.5-mm gull-wing leads on the edges of the parts (the shorter dimension) rather than along the sides (longer dimension). The type-2 parts have the leads on the longer dimension. These packages are generally used for memory ICs.

- Thin shrink (or sometimes called scaled) small outline package (TSSOP), half the height of a standard SOIC. Also known as thin super small outline package by ICT, Inc. (San Jose, CA, see www.anachip.com) and thin small shrink outline, by Mictosemi Integrated products, formerly Linfinity Microelectronics (Garden Grove, CA).

- TVSOP, thin very-small outline package, by Texas Instruments in 1996 for the PCMCIA industry, the package meets the 1.2 mm height for PCMCIA cards. The package has low inductance (for a 14 pin package 2.31 to 2.85 nH and for a 100 lead package 2.58 to 5.51 nH) and capacitance (a 14 pin package had 0.30 to 0.34 pF, with the 100 lead package 0.28 to 0.89 pF). Lead resistance is 37 to 40 mOhm for a 14 pin package and 46 to 66 mOhm for a 100 lead package. Lead spacing is 0.4 mm, with the area of the 14 pin package 23 mm^2, and the 100 lead package 168.5 mm^2.

- UTLP, Ultra Thin Leadless package, developed by Silicon Systems, inc. (Tustin, CA). This package has a mounted height of 0.8 m.

- Vertical mount package (VPAK), a package conceptually like a zig-zag package (ZIP) except instead of through-hole leads it has surface-mount leads (L shaped leads). This package was introduced by Texas Instruments in late 1991 for 4- and 16-Mb direct random access memories (DRAMs).

- Very small (and thin) quad flat package (VQFP). Lead pitch is 0.5 mm, with 32, 48, 64, 80, 100, 128, and 208 pins per package.
- Very small outline (VSO), usually used to denote 25-mil pitch packages with gull-wing leads.
- Very small outline package (VSOP), with 25-mil spaced leads at the ends of the package (also TSOP and SSOP and SSOIC).

8.11.3   Chip-Scale Packaging

This form of packaging technique is designed to have the size and performance of bare die parts but with the handling and testability of packaged devices. The package size is no more than 1.2 times the original chip with various techniques to connect the part to the circuit. Flip chips have enlarged solder pads (bumps or balls). Flip-chip packages include the controlled collapse chip connection (C4), developed by IBM, and the direct chip attachment (DCA), developed by Motorola. In 1964, IBM invented flip-chip interconnection technology to enable production-level joining of discrete transistors. There are various extremely small packages offered by semiconductor vendors, Fairchild has the Chip-Scale MicroPak®, a 6 terminal device is 1.45 square mm, an 8 lead device is 2.5 square mm. An SC-70 device with 6 leads is 4.2 square mm.

General designations are as follows:

- BLP—Bottom Leaded (Plastic) Package developed by LG Semicon (now Hyundai) (Cheonju, South Korea) where the chip leads are molded into the bottom of the plastic package, beneath the chip. This results in a package that is 0.48 to 0.82 mm thick. Packaged devices appeared in 1997.
- Bumped chip, see flip chip.
- FCA—flip chip attach.
- Flip chip—a semiconductor package where the I/O terminations are in the form of bumps on one side of the package (also called *bumped chip*). After the surface of the chip has been passivated, or treated, it is flipped over and attached to a matching substrate. A solder-bumped flip-chip is attached to a printed circuit carrrier substrate by reflow soldering (vs. a ceramic substrate). A polymeric underfill reduces the strain of the thermal coefficient of expansion mismatch between the chip (CTE approximately 2–3) and the PCB substrate (approximately (14–17). CTE Flip Chip (C4) refers to Controlled Collapse Chip Connection. In 1994 flip chip manufacturing techniques only accounted for 2% of the 45 billion ICs produced. In manufacture, the primary failure mode for a flip chip on an MCM module is a short circuit.
- Micro SMT, a peripheral contact package formed using semiconductor fabrication techniques while the IC is in wafer form. This technology was developed by ChipScale, Inc. (San Jose, CA). The resulting device is not much larger than the bare die.
- MiniBGA, uses a predetermined grid array for the solder bumps and is similar to the flip chip.
- Repatterned die, see MiniBGA.
- Slightly larger than IC carrier (SLICC) package.
- MicroBGA (μBGA), also known as *chip-scale packaging* (Mitsubishi); slightly larger than IC or SLICC packaging (Motorola, Inc). A package designed by Tessera, μBGA consists of a flex circuit with gold traces bonded to the die pads. The die pads fan into an array of metal bumps used for the second-level assembly. An elastomeric adhesive is used to attach the flexible circuit to the chip. The package compensates for thermal mismatches between the die and the substrate, and the μBGA can be clamped into a socket for tests before attachment in the final assembly (see Fig. 8.139).

8.11.4   Bare Die

Bare, or unpackaged, parts offer the smallest size, with no signal delays associated with the device package. Current issues related to unpackaged devices are packaging, handling, and testing. The most common

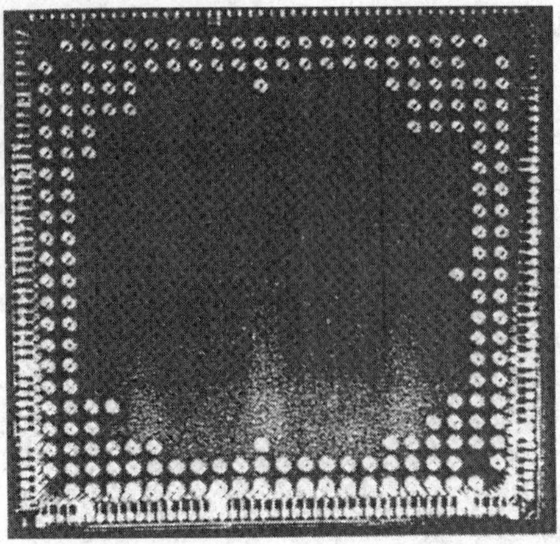

FIGURE 8.139   A μBGA package. (*Source:* Courtesy of Tessera, San Jose, CA.)

bare die and tape packages (bare die mounted on tape) include the following:

- C4PBGA, an IC package developed by Motorola that attaches the IC die to a plastic substrate using the C4 process. It uses a multilayer substrate, unlike the chip-scale or SLICC package.

- Chip scale package (CSP), a Mitsubishi package where the IC is surrounded by a protective covering through which external electrode bumps on the bottom provide electrical contacts. The package, only slightly larger than the chip it houses, has a height about 0.4 mm.

- Chip on board (COB)—the die is mounted directly on the printed circuit substrate (or board). See also tape automated bonding (TAB) and tape carrier packages (TCB).

- Chip on flex (COF), a variation of COB but instead of bonding the bare die to a substrate, it is bonded to a piece of 0.15 in-thick polyimide film (such as FR4) that has a top layer of gold-plated copper. The traces provide bonding ares for wires from the die and a lead pitch similar to QFP packages. It has approximately the same weight and height as COB but failed parts can be removed. It, however, uses about the same board real estate as QFP packages and is more expensive than COB because of the polyimide substrate.

- Demountable tape automated bonding (DTAB), developed by Hewlett–Packard. Mechanical screw and plates align and hold the IC to the board (pressure between the tape and the board provides the contact). The IC can be replaced by removing the screws (vs. a standard TAB package, which is soldered in place).

- GTE—glob top enacpsulation, pertains to the glob of encapsulating material used in MCMs.

- Sealed chips on tape (SCOT), chips mounted on tape supported leads and sealed (usually with a blob of plastic).

- Tape automated bonding (TAB), see TCP.

- Tape carrier packages (TCP), formerly called TAB packages. The chip is mounted to a dielectric film, which has copper foil connection patterns on it. The chip is sealed with a resin compound. This device assembly is mounted directly to a circuit without a plastic or ceramic package. Some tape carrier packages resemble TSOP packages (with gull-wing leads) but are thinner because the bottom of the IC die is exposed. The TCP design is about one-half the volume and one-third the weight of an equivalent pin count TSOP.

- Tape carrier ring (TCR), or a guard ring package, similar to the TCP package but includes a plastic ring to support the outer rings during test, burn-in, and shipment.

- Ultra-high volume density (UHVD), a process developed by General Electric to interconnect bare ICs on prefabricated laminated polyimide film.

8.11.5   Through-Hole Packages

These packages, as the name implies, mount in holes (usually plated through with deposited metal) on the printed circuit board. These device packages include the following:

- Batwing, a package (sometimes a DIP type, but can be surface mounted) with two side tabs used for heat dissipation.
- Ceramic dual inline package (CERDIP). The DIP package was developed by Fairchild Semiconductor and Texas Instruments followed with a metal topped ceramic package that resolved problems with the early ceramic packaged parts.
- Dual inline package (DIL).
- Dual-inline package (DIP), a component with two straight parallel rows of pins or lead wires. The number of leads may be from 8–68 pins (although more than 75% of DIP devices have 14–16 pins), 0.100-in. pin spacing with width anywhere from 0.300-mil centers to 0.900-mil centers. The skinny (or shrink) DIP (SDIP) has 0.300-mil centers (spacing between the rows) vs. 0.600-mil centers. The SDIP usually has 24–28 pins. DIP parts may be ceramic (pins go through a glass frit seal), sidebrazed (pins are brazed onto metal pads on the side of the package), or plastic (where the die is moulded into a plastic package). There is also a shrink DIP, another small package through-hole part.
- Plastic DIP (PDIP).
- Pin grid array (PGA), a plastic or ceramic square package with pins covering the entire bottom surface of the package. Lead pitch is either 0.1 or 0.05 in. perpendicular to the plane of the package. Packages have various pin counts (68 or more). The chip die can be placed opposite the pins (cavity up) or nested in the grid array (cavity array).
- Pinned uncommitted memory array (PUMA II), a PGA package application specific integrated circuit (ASIC) memory array with four 32-pad LCC sites on top of a 66 PGA. Each of the four sites can be individually accessed via a chip select signal thus allowing a user definable configuration (i.e., ×8, ×16, ×24, ×32). It provides for an ASIC memory array without tooling. The substrate is a multilayered cofired alumina substrate with three rows of 11 pins. There is a channel between the pins so that the part can be used with a heat sink rail (or ladder). Onboard decoupling capacitors are mounted in a recess in this channel. This type of device is available from various companies including Mosaic Semiconductor (1.12 × 1.12 in square), Cypress Semiconductor (their 66 pin PGA module, the HG01, is 1.09 × 1.09 max), and Dense-Pac Microsystems (Veraspac or VPAC family, 1.09 × 1.09 max.).
- Plastic pin grid array (PPGA).
- Quad inline package (QIP or QUIL).
- Quad inline package (QUIP)—similar to a DIP except the QUIP has a dual row of pins along the package edge. Row to row spacing is 0.100 in, with adjacent rows aligned directly across from each other.
- Skinny (or shrink) DIP (SDIP), see DIP.
- Shrink DIP, a dual inline package with 24–64 pins with 0.070-in lead spacing.
- Single inline (SIL).
- Single inline module (SIM)—electrical connections are made to a row of conductors along one side.
- Single inline package (SIP), a vertically mounted module with a single row of pins along one edge for through-hole mounting. The pins are 0.100 in. apart. SIPs may also include heatsinks.
- Skinny DIP (SK-DIP), see SDIP.
- Transistor outline-XX (TO-XX), refers to a package style registered with JEDEC.
- Vertical inline package (VIL).

- WOW—Wafer On Wafer, a wafer integration process by FormFactor, Inc, in 1998. A microspring, gold filaments coated with a flexible nickel alloy, is attached to the bonding pads of each die. This enables high speed testing and burn-in at the die level. Microspring contacts can be attached to the PCB so chips can be mounted directly to the PWB.
- Zig-zag inline package (ZIP). This may be either a DIP package that has all the leads on one edge in a staggered zig-zag pattern or a SIP package. Lead spacing is 0.050 in from pin to pin. In modules the leads are on both sides in a staggered zig-zag pattern. Lead spacing is 0.100 in between pins on the same side (or 0.050 in from pin to pin).

8.11.6   Module Assemblies

We use this terminology to refer to packaging schemes that take either packaged parts or bare die and use them to make an assembly. This may be by either mounting the parts to a substrate or printed circuit board or by stacking parts or die to create dense memory modules. This packaging scheme may be surface mount or through-hole. The circuit board assemblies may be through-hole (having pins) or designed for socket mounting (with conductive traces). Variations include the following:

- Dual-inline memory module (DIMM), pioneered by Hitachi, has memory chips mounted on both sides of a PC board (with 168 pins, 84 pins on each side). The DIMM has keying features for 5-, 3.3-, and 2.5-V inputs and for indicating whether it is made with asynchronous DRAMs or flash, static random access memory (SRAM) or DRAM ICs. There is also the SO DIMM or small outline dual in line memory module, which has 72 pins vs. 168 for standard DIMM modules.
- Full stack technology—packing 20–100 dice horizontally, in a loaf-of-bread configuration. This manufacturing method was developed by Irvine Sensors Corporation (see also short stack).
- HDIP module, a hermetic DIP module that has hermetically sealed components mounted on the top and bottom of a ceramic substrate. This package style is generally used in anticipation of a monolithic part (such as a memory device) that will be available at a later time that will fit the same footprint as the module.
- Hermetic vertical DIP (HVDIP) module, a vertically mounted ceramic module with pins along both edges (through-hole mounting). Components used in this module are hermetically sealed. Pins on opposite sides of the module are aligned and are on 0.100-in spacing.
- Leaded multichip module (LMM). An LMMC is an LMM connector.
- Multichip module (MCM), a circuit package with SMT (surface mount) IC chips (minimum of 2) mounted and interconnected via a substrate similar to a multilayer PC board. MCM-L has a substrate that can be FR-4 dielectric but is often polyimide based laminations (first standardized by JEDEC JC-11 committee). Other dielectric substrates are ceramic (MCM-C), which can have passive devices (resistors, capacitors, and inductors) built into the substrate; thin film (i.e., silicon and ceramic construction), with deposited conductors (which can have capacitors built into the substrate) (MCM-D); laminated film (MCM-LF) made up of layers of modified polyimide film; or multichip on a flexible circuit (MCM-F). In 1994, some manufacturers started manufacturing substrates with MCM-D thin-film deposits onto MCM-L laminate substrates in a technique pioneered by IBM. MCM devices inherently offer higher speed and performance at lower cost than conventional devices. Manufacturing issues associated with MCM design include availability of unpackaged die, how to test/inspect the die, and test/rework of the MCM module. MCM modules for military uses fall under MIL-PRF-38534 (see *memory cube*). There is also an MCM-V, with V standing for a vertical, which is a three-dimensional module. This design is also known as Trimrod, by the developer Thomson-CSF (now Thales) and the European Commission (see *memory cube*).
- Memory cube, a three-dimensional module consisting of stacked memory devices such as DRAMs or SRAMs.

- M-RFM, Modified Ring Frame 3-d packaging method developed by Professor Kyung-Wook Paik of the Korea Advanced Institute of Science and Technology (KAIST, Kaejon, Korea) and LG Semicon (now Hyundai) (South Korea) in 1998. In this method the IC wafer is separated into strips vs. individual die. The strip is placed on a polymer adhesive and a ring frame with a matching coefficient of thermal expansion is used to support the strip. A wafer thick layer of polymer adhesive is applied around and on top of the wafer strip. Metal traces are added to route the I/O to the edge of each die, then the die are separated, stacked and solder balls are added.
- NexMod—a module designed for RDRAM by High Connection Dencity, Inc. (HCD, in Sunnyvale, CA), that is a complete RAMBUS channel in one system incorporating termination resistors, clock generator, voltage regulator module, and memory chips.
- Ribcage, a trademark of Staktek Corporation for a three-dimensional memory module design. Also see Uniframe Stakpack Module.
- Short stack, a method of stacking semiconductor dice vertically, where memory ICs are assembled into a three-dimensional thin-film monolithic package with the same footprint as a single SRAM. Recent advances in stacking ICs include forming interconnects at the top of the wafer, adding vias for inner layer interconnections, thinning the wafers to less than 10 micrometers, and then stacking the chips. The via interconnections are made using IC insulating and deposition processes.
- Single inline memory module (SIMM), an assembly containing memory chips. The bottom edge of the SIMM, which is part of the substrate material, acts as an edge card connector. SIMM modules are designed to be used with sockets that may hold the SIMM upright or at an angle, that reduces the height of the module on the circuit board. Typical SIMM parts are 4×9 (4 meg memory by 9), $1 \times 9, 1 \times 8, 256 \times 9$ and 256×8. Also 9-b data width SIMM modules are produced under license to Wang Laboratories, (now Getronics, NV) who developed the SIMM module and socket in the early 1980s as an inexpensive memory expansion for a small workstation.
- SLCC—Stackable Leadless Chip Carrier, developed by DPAC Technologies Corp. (formerly Dense-Pac Microsystems CA) is a multidimensional module) consists of stacked chip carriers. Stacking is accomplished by aligning the packages together and tin dipping each of the four sides. SLCC can achieve a density of 40:1 over conventional packages.
- Uniframe Stakpak Module, a tradename three-dimensional stack of TSOP memory ICs by Staktek Corporation.
- VDIP module, vertically mounted modules with plastic encapsulated components and epoxy encapsulated chips on them. VDIP modules have 0.100 in. spaced pins along both sides of the substrate, with the pins on the alternate sides aligned.
- WLCSP—Wafer level Chip Scale Package.
- Other Definitions:

Plastic encapsulated microcircuit (PEM).
Premolded plastic package (PMP).

8.11.7 Lead Finish

Reducing lead in the environment is an environmental health and safety issue that is of prime importance today. While a typical microprocessor has approximately 0.2 g of lead and a computer motherboard has 2 to 3 g of lead, and the whole U.S. electronic interconnection market uses less than 2% of the world's lead production, to address environmental concerns, there are various environmental directives limiting the use of lead and other hazardous substances. These directives, formulated by the European Union in a Directive of the European parliament and of the Council of the European Union are:

- End of Life Vehicles (ELV) Directive 2000/53/EC: It is in force since October 21, 2000. It states that products be free of heavy metals such as mercury, cadmium, hexavalent chromium, and lead. It requires a recycling system be established by vehicle manufacturers for vehicles made after July 1, 2002 and for all vehicles regardless of date of manufacture by July 1, 2007. Lead can still be used as an alloying additive in copper and in solderable applications.

- The WEEE Directive: Waste Electrical and Electronic Equipment (WEEE) Directive 2002/96/EC, expands the recycling requirements of the ELV Directive to include a broad range of electronic and electrical appliances and equipment. WEEE went into effect on February 13, 2003. It is to be scheduled to become European national law by August 13, 2004, and be applicable to consumer-use products by August 13, 2005. Article 2(3), however, states "Equipment which is connected with the protection of the essential interests of the security of Member States, arms, munitions and war material shall be excluded from this Directive. This does not, however, apply to products which are not intended for specifically military purposes."

- RoHS Directive: The Restriction of Hazardous Substances in Electrical and Electronic Equipment. This Directive 2002/95/EC establishes standards and limits for the hazardous material content in electronic and electrical equipment. The Directive went into effect on February 13, 2003. It is scheduled to become European national law by August 13, 2004 for be in force for products by July 1, 2006. Banned or restricted substances include lead, mercury, cadmium, hexavalent chromium, certain brominated flame retardants (PBBs), and polybrominated diphenyl ethers (PBDEs).

The recommended lead-free solder formulation is Sn-Ag-Cu for board assembly but there are other formulations such as Nickel-Palladium (NiPd), or Nickel-Palladium with Gold flash (NiPdAu). Passive components, to be compatible with a lower temperature Lead process (which is $215°C$ for 50/50 Tin/Lead formulations and $230°C$ for 40/60 formulations) and the higher lead-free process of up to $260°C$, use pure matte Tin for their contacts. The use of lead in solder is partially based on several potential reliability issues. Pure Tin component leads have been shown to result in inter-metallic migration in the termination of the electronic component and the growth of tin whiskers which could cause short circuits (which is why there is a exemption for military use (only) components).

The National Electronics Manufacturing Initiative (NEMI) has addressed the problem of "tin whiskers" in lead-free assemblies. A tin whisker is defined by them as

A spontaneous columnar or cylindrical filament, which rarely branches, of monocrystalline tin emanating from the surface of aplating finish. Furthermore, tin whiskers may have the following characteristics:

- an aspect ratio (length/width) greater than 2
- can be kinked, bent, twisted
- generally have consistent cross-sectional shape
- may have striations or rings around it

Their recommended test method is:

- temperature cycling (-55 to $+85°C$, approximately 3 cycles/h)
- temperature humidity tests at $60°C/93\%$ RH
- ambient storage (air-conditioned facility)

(Note: Tin will undergo a phase transformation, becoming a powdery form called Tin Pest if Tin plated parts are stored for more than 1 week at temperatures below $13°C$).

8.11.8 The Future

While it is difficult to predict the future based upon current trends, making products smaller, lighter, faster, more energy efficient, and less expensive will continue. With these trends will be the requirement to channel the heat away from components. One recent development in 2004 was the Perdue University announcement of a patented technique of a self-ventilating method where microfluidic-like layers pump heat-laden air off-chip using a classic "corona wind" effect. The chip created ions between closely spaced carbon-nanotube electrodes and funneled the resulting air currents down microfluidic channels, allowing the heat to eject out of the sides of the chip.

References

Andrew MacLellan, Electronic Buyers News, December 14, 1998, FormFactor WOWs With Cost Savings, pages 4 and 92.

Baliga, J. Semiconductor International, May 1998, page 46, Flip Chip BGA, IEEE Spectrum, March 2004, pp. 43–47.

Bernard Levine, The Package, Electronic News, August 7, 1995, October 16, 1995, October 23, 1995, May 27, 1996 (with Fred Gunther)

Biancomano, V.J., October 28, 1996, Ball-grid arrays: all in a day's rework, *Electronic Engineering Times*, p. 49.

Bindra, A., Very Small Package Hopes for Big IC Impact, *Electronic Engineering Times*, November 28, 1994; Sandia Shrinks Size of BGA Packages, May 16, 1994; Center for Packaging Research is Proposed, February 13, 1995; *Electronic Engineering Times*, June 3, 1996, TI Eyes New Package Type, p. 56.

Costlow, T. March 6, 1995, LSI rolls enhanced BGA, MCM substrates mixed, *Electronic Engineering Times*; October 10, 1994, BGAs honed to meet diverse needs; September 19, 1994, Chip-scale hits the Fast Track; September 12, 1994, Olin Mines Metal for Ball-Grid Arrays; August 22, 1994, Mitsubishi scales down IC package; July 11, 1994, Amkor trims BGA height; May 9, 1994, MCM design details remain perplexing; April 18, 1994, MCM standard created; April 1, 1996, BGA standard set to ship, *Electronic Engineering Times* p. 90; July 22, 1996, ID sensing presents a touchy situation, *Electronic Engineering Times* p. 64; September 9, 1996, Prolinx photo-imaged vias cut BGA costs, p. 84.

Crum, S. Sept. 1994. Minimal IC packages to be available in high volume. *Electronic Packaging and Production*, News column.

Derman, G. Feb. 28, 1994. Interconnects and packaging. *Electronic Engineering Times*.

Electronic Buyers' News, Briefs Column, September 4, 1995, Briefs Column, Motorola Signs Packaging Deal.

Electronic Express website (Div. of R.S.R. Electronics, Inc. http://www.elexp.com/t_solder.htm

Fairchild Semiconductor website on Lead-Free Implementation http://www.fairchildsemi.com/products/lead_free

Hutchins, C.L. Nov. 1994. Understanding grid array packages. *SMT* (*Surface Mount Technology*.) Hutchins and Associates, Raleigh, NC.

Hwang, J.S., 1995. Chip-Scale Packaging and Assembly, SMT.

Intel Website, http://www.intel.com/research/silicon/leadfree.htm

IPC Website, http://www.ipc.org, IPC Board of Directors Position Statement

Karnezos, M. Feb. 28, 1994. DTAB mounts speed and power threat. *Electronic Engineering Times*.

Lineback, J.R. Nov. 1994. Chip makers try to add quality while removing the package. *Electronic Business Buyer*.

Locke, D. Nov. 28, 1994. Plastic packs pare mil IC pricing. *Electronic Engineering Times*. American Microsystems.

Meeldijk, V. 1995. *Electronic Components: Selection and Application Guidelines.* Wiley-Interscience, New York.

Murray, J. June 11, 1994. MCMs pose many production problems. *Electronic Engineering Times*.

O'Brien, K., Managing ed. May/June 1994. *Advanced Packaging*. μBGA offers flip chip alternative; copper heat sink QFP, dual use sought in Denver. News column.

O'Shea, Paul, Associate Editor, EE—Evaluation Engineering, New Device Packaging Challenges Automated Inspection December 1995.

R. Colin Johnson, Nano work chases self-cooled chips, EE Times, March 29, 2004.

Reuning, K. Aug. 1994. Squeezing the most from BGAs. *U.S. Tech.* deHaart Inc.

Semiconductor International, May 1998, Industry News, pages 20 and 22, 3-D IC Packaging.

Stuart A. *Software for Science*, Vol. 30, Sneak Preview: Intel's Next Generation.

Tadatomo Suga, Lead-free Roadmap 2002—Roadmap 2002 for Commercialization of Lead-free Solder—A report of Lead-Free Soldering Roadmap Committee, JEITA, The University of Tokyo, RCAST, suga@pe.u-tokyo.ac.jp

The National Electronics Manufacturing Initiative's (NEMI's) Tin Whisker Accelerated Test Project Press Release, September 26, 2003—NEMI Team Recommends Three Test Conditions for Evaluating Tin Whisker Growth.

Woolnough, R. April 25, 1994. Thomson 3-D modules ready for prime time. *Electronic Engineering Times.*

Young-Gon Kim (LG Semicon), Novel Lead-Frame Scheme Leads to Small Footprint, Electronic Engineering Times, February 26, 1996 (Note: LG Semicon is now Hyundai).

Further Information

The following sources can be referenced for additional data on IC packaging and related packaging issues (i.e., converting IC packages, lead finish considerations, packaging and failure modes):

CALCE (Computer Aided Life Cycle Engineering) Electronic Packaging Research Center, University of Maryland, College Park, MD, 301-405-5323. CALCE is dedicated to providing a knowledge and resource base to support the development of competitive electronic components, products, and systems. Building 89, Room 1103, University of Maryland College Park, MD 20742 Fax: 301-405-5323, 301-314-9269 webmaster@calce.umd.edu http://www.calce.umd.edu/

Electronic Components: Selection and Application Guidelines, by Victor Meeldijk, John Wiley and Sons Interscience Division, (C)1995. The integrated circuits chapter discusses IC packages, lead material, and failure mechanisms related to packaging.

High Density Packaging Users Group (HDP User Group), a nonprofit organization, formed in 1990, that offers memberships to companies involved in the supply chain of producing products using high density electronic packages. HDP User Group International Inc., Langbrodalsvagen 73, Alvsjo, Sweden.

For General or Management questions:

Executive Director, European Office, Langbrodalsvagen 73, S-12557 Alvsjo, Sweden Phone or Fax: +46 8 86 9868 in Sweden, ruben.bergman@hdpug.se

For Technical or Project questions:

Director of Operations, International Headquarters, 10229 North Scottsdale Road, Suite B, Scottsdale, AZ 85253-1437, USA, Phone: 480-951-7151, Fax: 480-951-1107, Cell: 480-390-4844, Bob@hdpug.org; http://www.hdpug.org/public/1-about-hdp/contact-us.htm

IEEE Components, Packaging, and Manufacturing Technology Society: CPMT Society Executive Office, 445 Hoes Lane, PO Box 1331, Piscataway, NJ, 08855-1331, USA, Phone: 732.562.5529, Fax: 732.981.1769, http://www.ieee.org/portal/index.jsp Email: mtickman@calypso.ieee.org

IMAPS, The International Microelectronics And Packaging Society, 611 2nd Street, N.E., Washington, D.C. 20002 Phone: 202-548-4001; Fax: 202-548-6115 http://www.imaps.org/ E-mail: imaps@imaps.org

International Society for Hybrid Microelectronics (also known as the Microelectronics Society) (ISHM), Reston, VA 800-535-4746, 703-758-1060, http://www.ishm.ee.vt.edu. This technical society was founded in 1967 and covers various microelectronic issues such as MCMs and advanced packaging.

The Institute for Interconnecting and Packaging Electronic Circuits (IPC). This organization issues standards on packaging and printed circuitry.

IPC 2215 SANDERS ROAD NORTHBROOK, IL 60062-6135, Phone: 847-509-9700, Fax: 847-509-9798

IPC Washington Office, 1333 H Street N.W., 11th Floor – West Washington, DC 20005, Phone: 202-962-0460, Fax: 202-962-0464 http://www.ipc.org/ INTERNATIONAL OFFICES: IPC China

Shanghai Centre, Suite 664, 1376 Nanjing West Road Shanghai 200040, PR of China Phone: +86 21 6279-8532 Fax: +86 21 6279-8005

IPC Europe Cabernetlaan 7 6213, GR Maastricht The Netherlands Phone: +31 43 321 2765 Fax: +31 43 344 0975

MEPTEC (Microelectronics Packaging and Test Engineering Council) is a trade association of semiconductor suppliers and manufacturers, committed to enhancing the competitiveness of the back-end portion of the semiconductor business. This organization provides a forum for semiconductor packaging and test professionals to learn and exchange ideas that relate to assembly, test, and handling. It was founded over 25 years ago.

MEPTEC HOME OFFICE 801 W. El Camino Real, No. 258, Mountain View, CA 94040 Bette Cooper, 650-988-7125, Fax: 650-962-8684 info@meptec.org http://www.mepteconline. com/

Microelectronics and Computer Technology Corp. (MMC) http://www.mmc.com. Founded in 1982, this is an R&D consortium whose technology projects have included MEMS (Micro-Electro-mechanical Systems). This organization worked with Lehigh University on a project to replace hermetic packages with more effective lightweight protective coatings. That Reliability without Hermeticity (RwoH) working group (founded in 1988) research contract was awarded by the U.S. Air Force Wright Laboratory.

NEMI, The National Electronics Manufacturing Initiative (NEMI) is an industry-led consortium whose mission is to assure leadership of the global electronics manufacturing supply chain. Its membership includes more than 60 electronics manufacturers, suppliers, associations, government agencies, and universities.

Cynthia Williams cwilliams@nemi.org phone: 207-871-1260 http://www.nemi.org

IMAPS —In 1986, the International Society for Hybrid Microelectronics, now known as ISHM — The Microelectronics Society, formed the ISHM Educational Foundation. The Foundation is dedicated to the long-term goal of promoting development in microelectronics technology, encouraging leadership in the microelectronics industry, and promoting emerging advanced technologies.

Semiconductor Technology Center, Neffs, PA, 610-799-0419. This organization does market research on manufacturing design, holds training seminars, and is involved in the Flip Chip/Tab and Flex Circuit symposiums held each year.

The Surface Mount Technology Association, Triangle Park, NC, 612-920-7682. http://www.smta.org/

The Die Products Consortium (DPC), a program of Dynalog Systems, Inc. has a group of leading microelectronics companies collaborating on enlarging the markets for die products worldwide. The DPC program, is dedicated to providing timely information about insertion of die products into electronic systems. Each year, member companies participate in a variety of projects focusing on die products test, assembly, standardization, infrastructure, and market intelligence.

The DPC is an outgrowth of a close working relationship between two U.S. consortia—MCC and Sematech, to develop technical solutions and business strategies for promoting the die products industry. Dynalog Systems, Inc. started the DPC as a program in 2000. http://www.dieproduct. com/

The Packaging Research Center (PRC) at Georgia Institute of Technology is the largest university-based research and education center focusing on next generation system-level packaging of electronics. The center is sponsored by the National Science Foundation. http://www.ece.gatech.edu/research/ PRC/

Department of Trade and Industry (DTI) http://www.dti.gov.uk/

DTI Web page dedicated to information on the WEEE and ROHS directive including: General information, estimated timetable, provisions, and supporting documents http://164.36.164.20/ sustainability/weee/index.htm

Industry Council for Electronic Equipment Recycling (ICER) http://www.icer.org.uk/ ICER is a source of knowledge and expertise on WEEE issues. It addresses a broad range of measures affecting electrical and electronic equipment. As a cross industry association, it tackles the implications for the whole supply chain.

Intellect provides more information on WEEE and ROHS, it also has related news, articles, and information complimented by sources of advice http://www.intellectuk.org/campaigns/environment/environment.asp

WEEE Recycling Network—The Network is an information service for small businesses covering all aspects of the WEEE Directive. http://www.weeenetwork.com/

Soldertec on Lead-Free Soldering, this site contains the latest information on lead free solutions http://www.tintechnology.biz/soldertec/

The Nordic Electronics Packaging Guideline, A number of microelectronics projects have been carried out in Norway to investigate the technical performance of different packaging and interconnection systems. These projects have resulted in technical knowledge of different packaging techniques, which are contained in the Nordic Packaging Guideline. A guide to help designers in the selection of advanced packaging and interconnection techniques http://extra.ivf.se/ngl/

Semiconductor Technology Center (SemiTech) P.O. Box 38, Neffs, PA 18065, 4525 Spring Hill Drive, Suite #4, Schnecksville, PA 18078, phone: 610-799-0919 Email: info@semitech.com http://www.semitech.com/

Semconductor Technology Center, Inc. has a newsletter "Semiconductor Packaging Update." They also provide client-specific Technology and Market Reports on a variety of semiconductor packaging topics—Equipment, Materials, Processes, Packaging Foundries, Emerging Packaging Trends, Flip Chip Technology, TAB, Substrates, Leadframes, Smart Cards, etc.

The following publications, which offer free subscriptions to qualified readers, cover packaging issues:

EP&P has merged with Semiconductor International magazine. All of the information previously available on epp.com is now available through the Semiconductor International website, at www.semiconductor.net/packaging. If you are not redirected there automatically after a few seconds, please click here: www.semiconductor.net/packaging
http://www.reed-electronics.com/semiconductor/community/3026/Semiconductor+Packaging

Advanced Packaging Advanced Packaging, PO Box 3425, Northbrook, IL 60065-3425 (847) 559-7500 (M-F, 7:30 AM to 6:00 PM CST) Fax: (847) 291-4816 http://ap.pennnet.com/home.cfm subscriptions: http://www.omeda.com/apk/ Email: apk@omeda.com

This publisher also issues Surface Mount Technology.

SMT (Surface Mount Technology) Magazine PO Box 3423, Northbrook, IL 60065-3423, Phone and Fax are the same as for Advanced Packaging. Email: SMT@omeda.com

PennWell publications contact: http://www.omeda.com/custsrv/smt/ Surface Mount Technology (SMT) PennWell 98, Spit Brook Rd. Nashua, NH 03062, Phone: (603) 891-0123

Electronic Engineering Times A CMP Publication 600 Community Drive Manhasset, N.Y. 11030 516-562-5000 516-562-5882 (subscription services) Subscription Inquiries: 516-733-6800 http://www.eetimes.com/ http://www.eetimessubscribe.com/cgi-win/eet.cgi?p=home For customer service help please click here, or we can be reached at 847-291-5215 (7:30 AM–6:00 PM

Electronic Packaging and Production (EP&P) EP&P merged with Semiconductor International magazine. All of the information previously available on epp.com is available through the Semiconductor International website, at www.semiconductor.net/packaging. www.semiconductor.net/packaging

$$9$$

Optoelectronics

Ken A. Chauvin
Corning Cable Systems

Charles H. Cox, III
Massachusetts Institute of Technology

Paul Kit-Lai Yu
University of California, San Diego

David E. Rittenhouse
Siecor Corporation

9.1 Optical Fiber

Ken A. Chauvin

9.1.1 Introduction

Since its beginning, the proliferation of optical fiber has been tremendous by any standard. The uses of optical fibers today are numerous, with new applications continuously emerging. Telecommunications applications of optical fiber and its associated components are widespread, ranging from extensive global networks, to local telephone exchanges, to data centers, and most recently, the long awaited extension of optical fiber all the way to the subscriber premises (i.e., homes and businesses). Nontelecommunications applications for optical fiber are equally as diverse. In medicine, bundles of plastic optical fibers are used to inject light into, and to transmit images from, deep inside the body, allowing doctors to see internal areas without the need for overly invasive exploratory surgical procedures. Specialty optical fibers connected to weapons can support real-time guidance and targeting information using telemetry data provided by the weapon itself. Other applications include uses in control and instrumentation functions aboard automobiles, aircraft, ships, satellites, and in other commercial transportation applications, to name a few. Sections 9.1 and 9.2 will primarily focus on the telecommunications-based uses of optical fiber, fiber optic cable, and the various associated fiber optic components needed to form a complete passive optical system.

Modern day telecommunications involve the transmission of voice, video, and data information over distances ranging from micrometers (1×10^{-6} m) to hundreds, or even thousands, of kilometers (km) using various transmission means. Telecommunications systems continue to incorporate a significant amount of copper transmission media, such as twisted pair and coaxial cables, and the associated electronic equipment, as well as various wireless technologies, to transfer information from point to point. Most current telecommunications systems are complex structures typically comprising several different transmission media types as discussed above (also known as hybrids). This is done in order to take advantage of the strengths of each technology based on the needs of a specific application. However, only the passive fiber optic portions of such systems will be addressed herein.

Section 9.1 covers basic principles of optical fiber operation, fiber manufacturing, and optical fiber design for the major different types of telecommunications-grade optical fibers currently used. Section 9.2 includes information on fiber optic cable designs and applications, basic optical connectivity products, and relevant industry standards and guidelines for fiber optic cable.

Benefits of Optical Fiber Telecommunications Systems

Optical fiber provides many fundamental advantages over alternative transmission technologies for telecommunications applications. The comparatively limited performance of copper conductor based systems forces the use of expensive signal conditioning and regeneration equipment (e.g., amplifiers and repeaters) at much closer intervals than for fiber optic systems. A single line of a voice grade copper system (i.e., 56 kbs) longer than a couple of kilometers requires the use of in-line signal processing for satisfactory performance, and even then is subject to the electromagnetic effects of interfering radio frequency sources such as radio, television, cell phone, and air traffic control broadcasts.

As information throughput requirements increase with the demands of more data-intensive applications at the end-user premises, the spacing between the copper-based repeater points must decrease in order to maintain the same aggregate data rate capability over a given length. Contrast that to all-optical systems in which it is not unusual to transmit 10 gigabits per second data rates over hundreds of kilometers without the need for active signal processing between the transmitter and receiver. Additionally, as it becomes necessary to increase the data transmission capacity or coverage area of a telecommunications system, the diameter and weight of cables for copper conductor systems increase much more rapidly than for optical fiber systems, resulting in a proportionally higher increase in materials, installation, and maintenance related costs.

The small size of optical cables, coupled with readily available components that make efficient use of the optical fiber's transmission capabilities, enable them to be manufactured and installed in much longer lengths than copper cables. The virtually unlimited capacity of optical fiber also alleviates fears of incurring significant long-term costs associated with frequent system upgrades, extensions, or over-builds. The availability of long lengths of individual lightweight fiber optic cables, up to 10 km or more, also make the installation of fiber optic systems much safer, easier, and less expensive, than comparable copper-based systems. Because of their design, fiber optic cables can generally be installed with the same equipment historically used to install twisted pair and coaxial cables, allowing some consideration for the smaller size and lower standard tensile strength properties of fiber optic cable.

More importantly, fiber optic cable design has progressed to the point where it serves as an enabling technology for newer installation methods that are faster, less expensive, and less intrusive to the environment than traditional installation means. Optical cables can be installed in duct system spans of 4000 meters (m) or more depending on the condition, construction, and layout of the duct system, and the details of the installation technique(s) used. Even longer lengths of fiber optic cable can be installed aerially, trenched, or buried in the ground and ocean floor. These extra-long lengths of cable reduce the number of splice points, thereby making the overall installation of optical fiber based telecommunications systems more efficient. The small size of fiber optic cable also saves on valuable conduit space in buried duct applications. This feature becomes even more prevalent when considering some emerging cable types that are specifically designed for use with air-blown or air-assist installation techniques into miniature ducts that are only about one centimeter in diameter.

Another advantage of optical fiber and fiber optic cable is the inherent flexibility in design options, allowing for the development of innovative products for specific applications. Since optical fiber is a man-made composite glass structure, it can be custom designed to meet optimal cost/performance targets in any number of specific applications. As it does not conduct electrical current and is not affected by electromagnetic interference, fiber optic cable can be made all dielectric, making it the ultimate in electromagnetically compatible transmission media. This eliminates such issues as dangerous ground loops, the effects of voltage spikes from the cycling of heavy electrical equipment, and requirements for separate conduits for metallic conductors. It also improves the security of controlled transmission rooms as it is much more difficult to tap a fiber optic line, and much easier to provide security for fiber optic cable.

9.1.2 General Construction

The basic construction of optical fibers used for telecommunications consists of a **core**, an integral **cladding**, and a protective **coating**. Figure 9.1 shows basic cross sections representing the two main types of optical fibers used for telecommunications.

The types of fibers used vary depending on the specific intended application. The most common fibers types used in long-haul, metro, or regional telecommunications systems are single-mode fibers, which can have an effective core diameter in the range of around 8–12 μm depending on the specific design and operating wavelength. Such fibers are capable of carrying data rates measured in terabits per second (1.0×10^{12} bits/sec) for significant distances. For comparatively short-reach applications, multimode fibers are usually preferred because of the relative ease of termination and the low cost of the associated electronics. These fibers find widespread use primarily in applications such as interbuilding local area networks (LANs), storage area networks (SANs) and data centers, intrabuilding backbone and horizontal distribution applications, and the like. The various single-mode and multimode fiber types will be discussed in more detail later in this chapter, in terms of the design and capabilities of each.

Fiber Core

For the common commercially available telecommunications fibers used today, the core is the central region of the fiber and is the area where the majority of the optical energy of a transmitted light pulse is concentrated. It consists of solid silica glass impregnated with dopant materials (mostly Germania), added to the glass structure to optimize its transmission properties by changing the material's **index of refraction**, or to facilitate fiber manufacturing. The distribution of dopants within the core can thus be manufactured-to-suit based on the desired light transmission properties for each particular type of optical fiber application. The resulting refractive characteristic of the core is referred to as its "refractive index profile". Basic types of core size and index profile combinations are discussed in the section on fiber types.

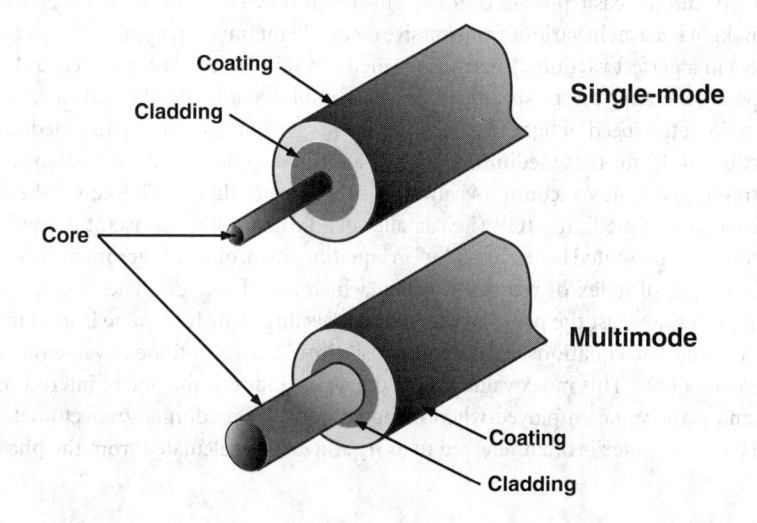

FIGURE 9.1 Multimode and single-mode fiber cross-sections. (Courtesy of Corning Cable Systems LLC and Corning® Inc.)

Fiber Cladding

The cladding is the layer of dielectric material that immediately surrounds the core of an optical fiber and completes the composite structure that is fundamental to the fiber's ability to guide light. The cladding of telecommunications grade optical fiber is also made from silica glass, and is as critical in achieving the desired optical performance properties as the core itself. For optical fiber to work, the core must have a higher index of refraction than the cladding or the light will refract out of the fiber and be lost. Initially multiple cladding diameters were available, but the industry swiftly arrived at a consensus standard cladding diameter of 125 μm, because it was recognized that a common size was needed for intermateability. A cladding diameter of 125 μm is still the most common, although other fiber core and cladding size combinations exist for other applications. Because of their similar physical properties it is possible, and in fact highly desirable, to manufacture the core and cladding as a single piece of glass which cannot be physically separated into the two separate components. It is the refractive index characteristics of the composite core-clad structure that guide the light as it travels down the fiber. The specific materials, design, and construction of these types of optical fibers make them ideally suited for use in transmitting large amounts of data over the considerable distances seen in today's modern telecommunications systems.

Fiber Coating

The third section of an optical fiber is the outer protective coating. The typical diameter of an uncolored coated fiber is 245 μm, but, as with the core and cladding, other sizes are available for certain applications. Coloring fibers for identification increases the final diameter to around 255 μm. The protective coating typically consists of two layers of an ultraviolet (UV) light cured acrylate that is applied during the fiber draw process, by the fiber manufacturer. The inner coating layer is softer to cushion the fiber from stresses that could degrade its performance, while the outer layer is made much harder to improve the fiber's mechanical robustness. This composite coating provides the primary line of physical and environmental protection for the fiber. It protects the fiber surface to preserve the inherent strength of the glass, protects the fiber from bending effects, and simplifies fiber handling. The colored ink layer has properties similar to the outer coating, and is thin enough that its presence does not significantly affect the fiber's mechanical or optical properties.

9.1.3 Basic Principles of Operation

Index of Refraction

In order to understand the basic physics of how light travels down an optical fiber it is first necessary to understand how light behaves in various transmissive media. Light travels at approximately 3.0×10^8 m/sec (186,000 mi/sec) in a perfect vacuum. In any other medium (air, water, glass, plastic, etc.) light travels at some slower speed depending on the specific material(s) employed and the physical structure of the given medium. The ratio of the speed of light in a vacuum to the speed of light in another medium is called the index of refraction (IOR) for that medium. Since it is a ratio of speeds there are no associated units, and because light travels fastest in a vacuum it is always a value greater than 1. The slower the light travels in a medium, the larger that medium's IOR (i.e., smaller denominator for a constant numerator). Index of refraction is typically represented by the letter "n" in equation form, or by the acronym "IOR" in text form.

There are two types of index of refraction values which are of interest. The first is the phase index value, which indicates how fast the phase of the light is traveling. This IOR value is used in essentially all of the related transmission equations, except for transit time. The second index value relates to how fast power travels down a fiber. This index value, called the group index, is the one of interest in determining transit times, and is the value employed when using an optical time domain reflectometer for distance calculations. The group index is often referred to as n_g and can be calculated from the phase index using the equation

$$n_g = n - \lambda \frac{dn}{d\lambda} \tag{9.1}$$

Snell's Law

When light encounters a boundary between two transmissive mediums with differing index of refraction value, it may be reflected back into the first medium at the interface boundary, bent at a different trajectory (i.e., refracted) as it passes into the second medium, or some combination of the two (see Figure 9.2). The actual result depends on the angle the light strikes the interface (angle of incidence) and the wavelength-dependent index of refraction values for the two materials.

As the light passes from one medium to another, the refracted light obeys Snell's law (see Equation 9.2). By convention, the angles used in calculating the light paths are measured from a line drawn normal to the axis of the core-clad boundary or fiber centerline.

$$n_1 \sin \theta_1 = n_2 \sin \theta_2 \qquad (9.2)$$

Based on this relationship, as the angle of incidence (θ_1) increases, the angle of refraction (θ_2) approaches 90°. The angle θ_1 which results in $\theta_2 = 90°$ is called the critical angle. For angles of incidence greater than the critical angle, the light is essentially reflected entirely back into the first medium at an angle equal to the angle of incidence. This condition is called "**total internal reflection**," and it is the basic principle by which optical fibers work. The angle of the reflected light is called the angle of **reflection**.

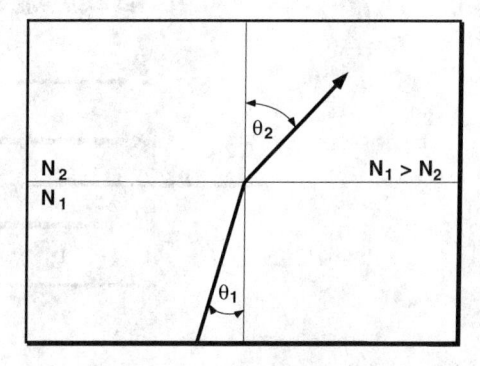

Numerical Aperture

Since light must first be launched into the fiber before it can begin to propagate within it, it is also important to account for the refractive effects at the end of the fiber where it interfaces with a third medium at the transmitter source. This third medium is typically assumed to be air, which has an IOR very close to 1. To ensure that the transmitted light stays within the core, the light must first enter the fiber via the core-air interface (e.g., connector enface) at an angle that will not cause it to subsequently be refracted out of the fiber at the core-clad interface. If the allowable entrance angle for the light at the fiber endface is then rotated 360° around the ring formed by the cylindrical core-clad boundary, it creates a cone that defines the acceptable paths

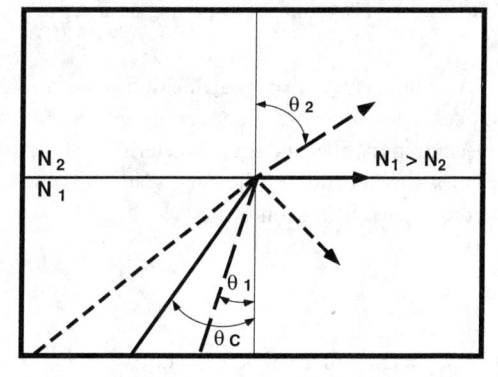

FIGURE 9.2 Refraction of light at boundary between different mediums. (Courtesy of Corning Cable Systems LLC and corning® Inc.)

which light entering the fiber must follow in order to propagate within it (see Figure 9.3).

The size of this "acceptance cone" is a function of the difference between the refractive index values for the core and the cladding. As it is not practical to specify an acceptance cone for a typical optical fiber, the result is usually referred to in terms of a mathematical value called the numerical aperture (NA). The numerical aperture (NA) is related to the acceptance cone by the following:

$$NA = \sin \theta_{NA} = \sqrt{n_1^2 - n_2^2} \qquad (9.3)$$

Where
n_1 = IOR material 1 (transmitting)
n_2 = IOR material 2 (receiving)

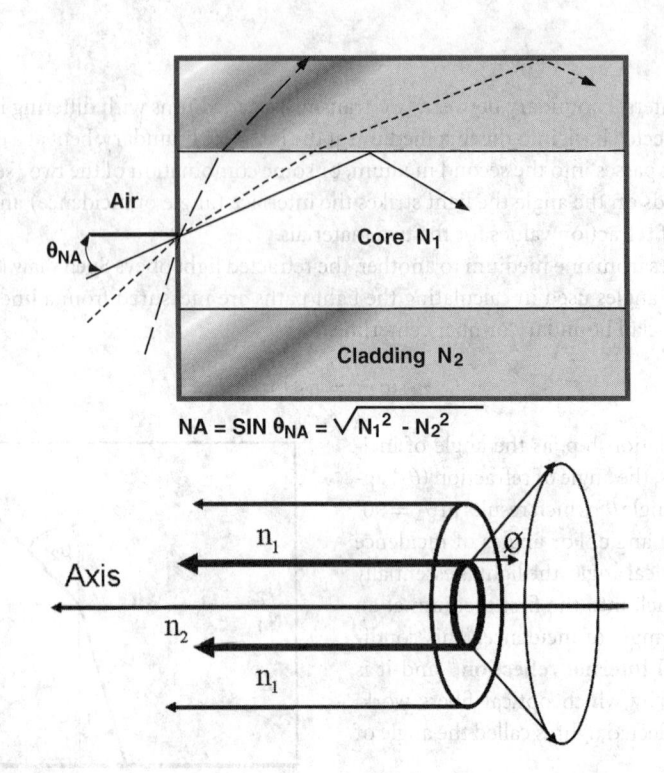

FIGURE 9.3 Fiber Numerical Aperture and acceptance cone. (Courtesy of Corning Cable Systems LLC and Corning® Inc.)

After light enters an optical fiber it travels down the fiber in one or more stable propagation states called modes. The number of modes in an optical fiber can be from one to several hundred depending on the specific type of fiber and the characteristics of the transmitted light. Each **mode** in multimode propagation carries some portion of the light as it traverses the length of fiber. The number of modes (N_m) can be approximated by the following:

$$N_m \cong \frac{\left(D * NA * \frac{\pi}{\lambda} \right)^2}{2} \tag{9.4}$$

Where
λ = operating wavelength (nm)
D = core diameter (nm)
NA = Numerical Aperture

As can be seen from this equation, the number of modes within an optical fiber increases if the core diameter or numerical aperture increase, or if the transmitted wavelength decreases. Since it is somewhat intuitive that larger core fibers can carry more light energy and are easier to connect to, increasing the size of the core, and hence the number of modes, may at first appear to be desirable. However, increasing a fiber's core size significantly can have a number of undesirable consequences depending on the specific application. The effects include a reduction in the aggregate data capacity due to the effects of modal **dispersion**, an increase in intrinsic **attenuation**, and a higher manufacturing cost due to the increased amount of dopant material needed to make the fiber. For multimode fiber, the number of modes also decreases as the wavelength is increased, however; that is not to say that the data rate capacity of the fiber increases as well. This issue is discussed later in more detail. For single-mode fibers there is only one mode present above a certain wavelength, which simplifies the understanding of the transmissive properties of the fiber. However, determining the actual data capacity of a single-mode fiber is more complicated than just accounting for the existence of a single mode. The contributing factors will also be discussed in more detail in the section on key fiber properties.

9.1.4 Key Fiber Properties

Operating Wavelength

The most significant and well-known parameter for any type of fiber is the specified operating wavelength(s). Telecommunications grade optical fibers operate in the near-infrared region of the electromagnetic spectrum (see Figure 9.4). While some custom fiber types designed for use in specialized applications may be optimized for operation at only one wavelength, most standard telecommunications optical fibers used today in fiber optic cables typically have two or more specified operating wavelengths, or wavelength regions. An operating wavelength region for standard fiber types may be specified as a nominal wavelength value with an associated tolerance, or simply as a range of wavelengths with a minimum and a maximum value (also known as an operating band). Examples of single-wavelength specialty fibers are those used in sensing applications, or ones used to carry power from a pump source to an optical amplifier. Specific types of these photonic fibers may be specified for operation at 780, 980, or 1480 nm. Note that it

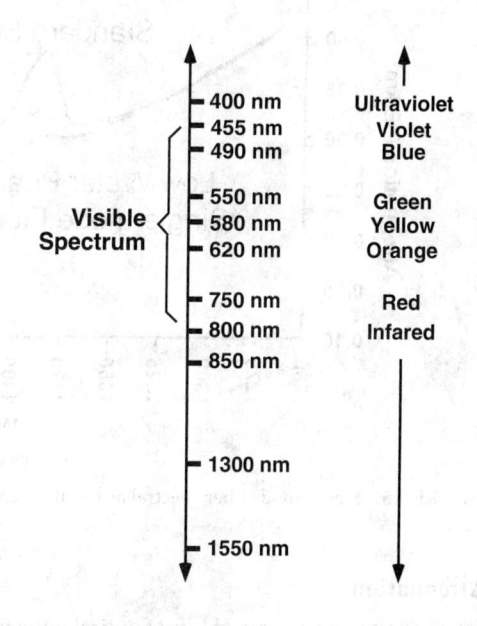

FIGURE 9.4 Electromagnetic spectrum (visible light and telecommunications fiber operation). (Courtesy of Corning Cable Systems LLC and Corning® Inc.)

is physically possible to operate such fibers outside of the specified operating ranges; however, the actual performance in such regions is not as well defined, and may differ significantly from that in specified ones.

The most commonly specified operating wavelengths for multimode silica glass telecommunications fibers are 850 and 1300 nm. Those for single-mode fibers are 1310, 1550, and 1625 nm, although there may be others for specific fiber types or applications. Even though the fibers may physically be able to support transmission over an extended spectral range (e.g., 1260 to 1675 nm), the optical performance is typically referenced to the most commonly used operating wavelengths or regions in order to simplify the processes for specifying and ordering fiber and fiber optic products. The wavelengths used today are primarily driven by the historical operating characteristics of the available sources, such as light emitting diodes (LED) and lasers, and by intrinsic fiber performance attributes, such as attenuation and dispersion characteristics. The basic operating bands for single-mode fiber recommended by the International Telecommunication Union (ITU) are provided in Figure 9.5.

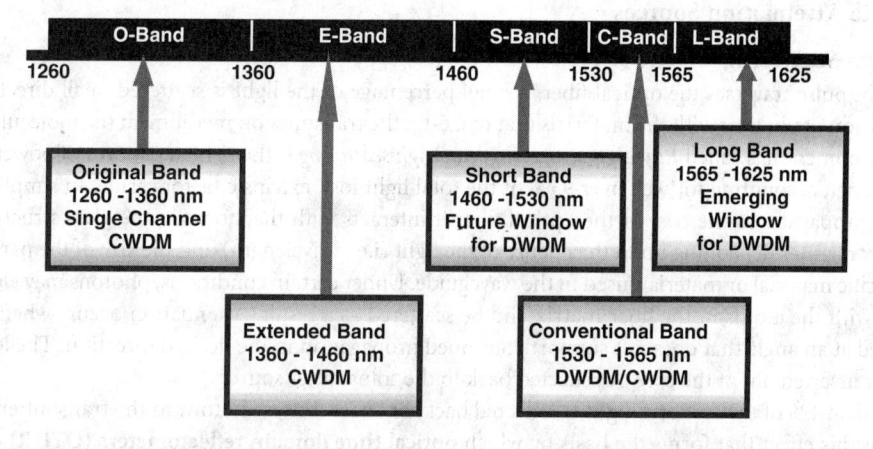

FIGURE 9.5 ITU operating bands. (Courtesy of Corning Cable Systems LLC and Corning® Inc.)

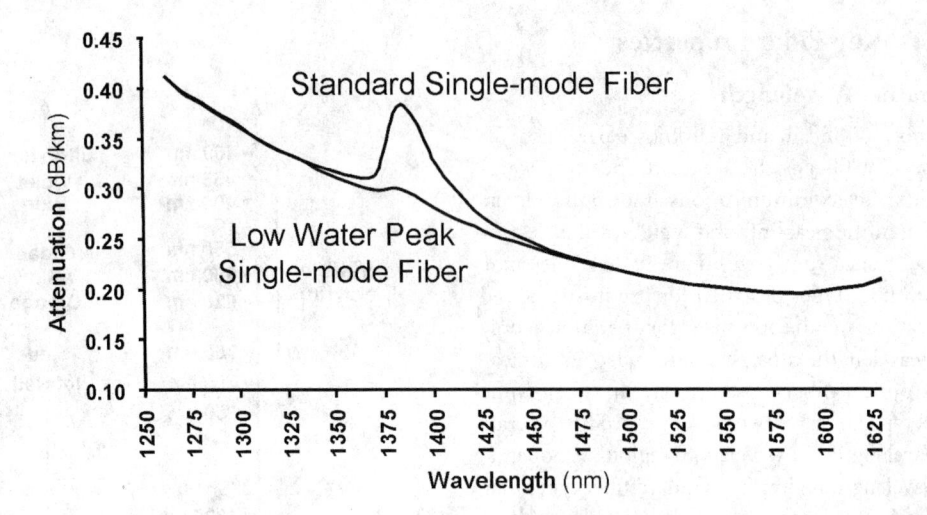

FIGURE 9.6   Single-mode fiber spectral attenuation curve. (Courtesy of Corning Cable Systems LLC and Corning® Inc.)

Attenuation

Attenuation is the combined loss of optical pulse power resulting from internal and external factors as the light propagates within the optical fiber. Mathematically, attenuation is expressed as

$$\text{Attenuation (dB)} = 10 \log \left(\frac{P_{\text{in}}}{P_{\text{out}}} \right) \tag{9.5}$$

By convention optical fiber attenuation is reported as a positive value, and to be useful to the system designer it is usually normalized to a unit length. The resultant value is called the **attenuation coefficient** and it is normally expressed in units of decibels per kilometer (dB/km). The typical attenuation performance of a single-mode fiber as a function of wavelength is shown in Figure 9.6. A characteristic spectral attenuation curve for multimode fiber has a similar shape, but slightly higher attenuation at all wavelengths primarily due to the higher dopant concentration levels present in multimode fibers.

Optical fiber attenuation is caused by several factors both internal and external to the fiber. Fiber attenuation resulting from internal sources is referred to as **intrinsic** attenuation, while that resulting from external sources is referred to as **extrinsic** attenuation. The specific components of intrinsic and extrinsic attenuation are addressed in greater detail in the next section.

Intrinsic Attenuation Sources

Rayleigh Scattering

As a light pulse traverses the optical fiber, a small percentage of the light is scattered in all directions by the collision of photons with the materials that make-up the transmission medium, at the molecular level. This phenomenon is called *Rayleigh scattering*. *Rayleigh scattering* is the primary source of power loss in optical fiber, accounting for well over 95% of the total light lost, extrinsic factors aside. In simple terms, as light propagates in the core of the optical fiber it interacts with the atoms in the glass structure in a complex manner depending upon the energy of the light (i.e., wavelength) and the size of the particles in the specific material or materials used in the waveguide. Under certain conditions, photons may elastically collide with the atoms in the fiber matrix and be scattered as a result. Attenuation occurs when light is scattered at an angle that does not support continued propagation in the desired direction. The lost light is either diverted out of the core or reflected back to the transmitter source.

Less than 1% of the scattered light is reflected back, or "backscattered," toward the transmitter source, and it is this effect that forms the basis by which **optical time domain reflectometers (OTDR)** operate. The OTDR is an important tool in fiber optics and is the piece of equipment most commonly used for

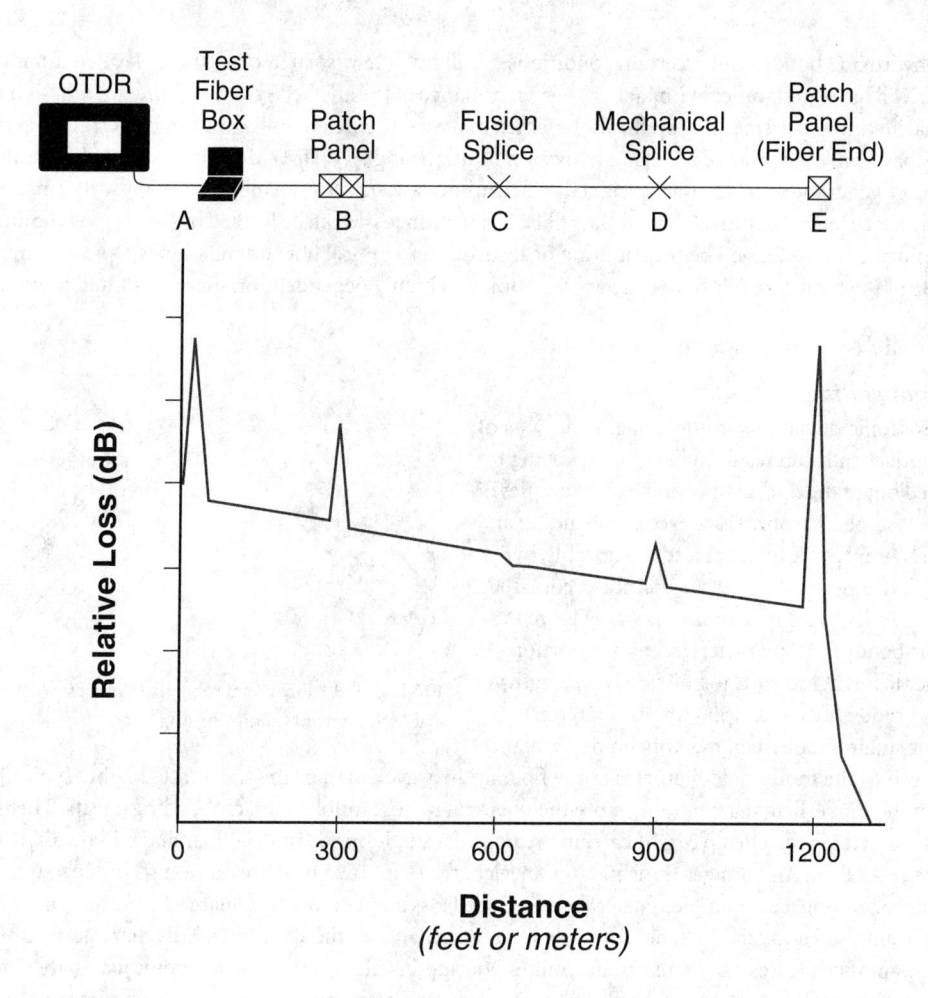

FIGURE 9.7 Sample OTDR trace. (Courtesy of Corning Cable Systems LLC and Corning® Inc.)

testing and troubleshooting fiber optic systems. A characteristic OTDR trace is provided in Figure 9.7. Because the reflected light comes from the light pulse the original pulse power decreases, adding to the total attenuation before it reaches the receiver. The slope of the OTDR trace represents the attenuation coefficient of the fiber for the particular wavelength of light being transmitted. The trace appears essentially straight, barring isolated events, because the overall attenuation coefficient of the fiber, which includes the Rayleigh backscatter coefficient, is assumed to be constant.

Absorption

Some small quantity of impurities can be introduced into the glass structure during the fiber manufacturing process. Large impurities significantly weaken the fiber structure and are screened out during proof testing, but very small impurities can exist which affect primarily the intrinsic optical properties of the fiber. Hydroxide bonds can also exist within the fiber in minute quantities, which, along with other impurities, can absorb light at various wavelengths near the primary operating bands. The light absorbed by the impurities is usually converted to heat, but in any event the original optical signal is attenuated in the process.

Absorption due to hydroxide bonds, most noticeable near 1383 nm (also known as the spectral water-peak region), has historically been the major source of absorption related losses, though recent advances in fiber manufacturing processes have all but eliminated such as an issue. Still, millions of kilometers of early generation single-mode fibers in use today may be measurably impacted by absorption losses associated with hydroxide bonds, under certain conditions. In all but a few isolated cases these elevated attenuation values will have no significant impact on the performance of the affected systems during their design operational lifetimes. For emerging applications which make more efficient use of the total spectral range of optical fiber, such as for course wavelength division multiplexing (CWDM), these issues could be significant, so newer generations of fiber are disigned to accommodate such uses. Unlike Rayleigh scattering, which would exist even in the purest optical glass fiber, absorption losses can be limited by controlling the amount of impurities in the glass. The contribution of absorption to optical fiber attenuation is typically limited to 2–3 % of the total power lost, but the actual amount is highly dependent on the transmitted wavelength.

Extrinsic Attenuation Sources

Macrobending

Macroscopic deviations in the straight-line axis of the optical path can result in light energy being refracted out of the optical fiber and lost. Examples of these macrobend sources are excessive bending and lateral compressive forces placed on the cable, which can then transfer the load to the fibers contained inside under extreme conditions (see Figure 9.8). **Macrobending** is characterized by proportionally higher losses at longer transmission wavelengths in single-mode fiber (i.e., 1550 nm vs. 1310 nm).

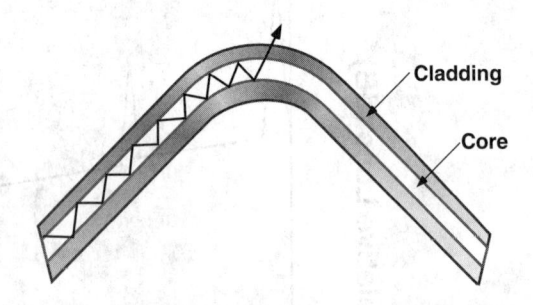

FIGURE 9.8 Fiber macrobend. (Courtesy of Corning Cable Systems LLC and Corning® Inc.)

For single-mode fiber, macrobend performance is linked to the mode-field diameter of the fiber at the wavelength of the transmitted light. Typically, the larger the mode-field diameter, the more the fiber will be susceptible to macrobending effects. The mode-field diameter is the effective light carrying region of a single-mode optical fiber and will be discussed in more detail later. Also, longer transmission wavelengths (e.g., 1625 nm) are more susceptible to the effects of macrobending. For example, a macrobend induced loss may severely attenuate a 1625 nm signal without significantly affecting the 1310 nm transmission performance of the same fiber. This characteristic of macrobend induced losses can be useful in troubleshooting installed systems when point losses are detected. Newer fiber coatings are being developed to improve the macrobend sensitivity performance of optical fibers, especially at the longer wavelengths operated over some single-mode systems. For multimode fiber, **macrobending** losses are essentially wavelength independent, although such fibers are not immune to the effects.

Microbending

Microbending is the result of discrete variations in the refractive index properties of an optical fiber waveguide caused by localized point stresses placed upon it. These stress-induced refractive anomalies result in light pulse energy being refracted out of the optical fiber and lost. Examples of microbend sources are fiber coating defects, fiber optic cable structural defects, or installation related cable damage (e.g., buffer tube kinking). Microbending, however, is an extremely localized phenomenon and the bend source may not be clearly visible upon inspection. Microbending is characterized by proportionally equivalent losses at all transmission wavelengths, making it somewhat akin to a point defect in the fiber's refractive index profile at the core-clad boundary (see Figure 9.9).

Cut-off wavelength-λ_{cc} (Single-mode fibers only)

As explained before, single-mode fiber allows only one mode of transmission if the transmission wavelength is longer than a certain value called the **cutoff wavelength**. Because of the degradation in transmission performance that occurs when small core single-mode fibers are allowed to transmit two or more modes for a significant distance, single-mode fibers are normally operated at wavelengths above this value. However, this value may be affected by cabling the fiber, which can help to strip out undesirable higher order modes, so the practical specified value for this parameter is "cable cutoff wavelength," and represents the value

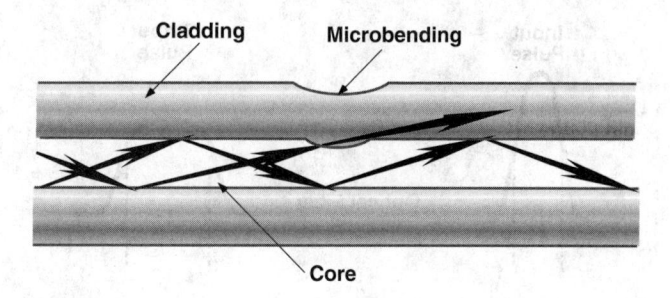

FIGURE 9.9 Fiber microbend. (Courtesy of Corning Cable Systems LLC and Corning® Inc.)

for the fiber in cabled form. The cable cutoff wavelength values for the various fiber types are typically specified to be no greater than 1260 nm for fibers that operate in the 1310 nm operating region (also known as original-band), and 1480 nm for fibers that operate only in the 1550 nm operating region (also known as conventional-band) or higher (e.g., the long-band). See Figure 9.5 for information on the various single-mode fiber operating bands.

Mode-Field Diameter (Single-mode fibers only)

Unlike multimode fibers, which have a core that is well-defined both physically and optically, the light carrying region of a single-mode optical fiber is a function of the characteristic power density distribution of the fiber at a particular wavelength, which follows a Gaussian distribution. In single-mode fibers the optical power distribution extends beyond the physical boundaries of the core such that a portion of the light traversing the fiber travels in the cladding section, nearest to the core. The diameter of the region represented by the effective power distribution in these fibers is called the **mode-field diameter (MFD)**. The MFD is dependent on the transmission wavelength (see Figure 9.10).

Dispersion

In the early years of optical fiber telecommunications, attenuation was generally considered the limiting system performance attribute. As advances in fiber and amplification component designs have been brought to bear on the problem; achievable transmission distances and data rates subsequently became limited by another hurdle, optical signal dispersion. In its simplest form, dispersion causes the optical energy in a light pulse to spread in time (i.e., increase in pulse width) as it traverses the fiber (see Figure 9.11). This spreading reduces the peak pulse signal strength, which decreases the signal (carrier) to noise ratio, and increases the potential that sequential pulses may overlap. In extreme cases the result is a bit error, which occurs when the receiver is not able to discern one or more individual bits. The closer the bits are launched together (i.e., higher data rates), or the longer the system length, the greater the effects of signal dispersion.

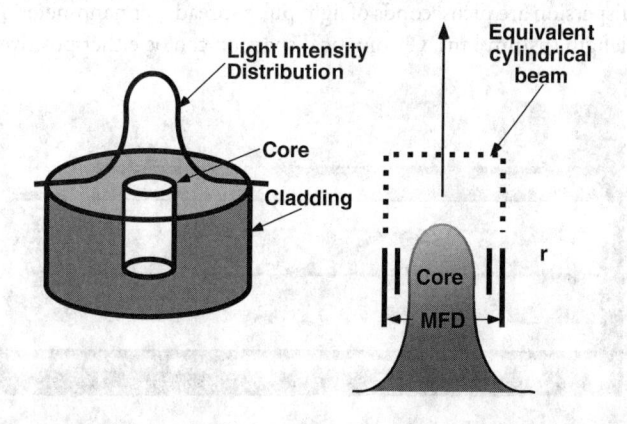

FIGURE 9.10 Mode-field diameter. (Courtesy of Corning Cable Systems LLC and Corning® Inc.)

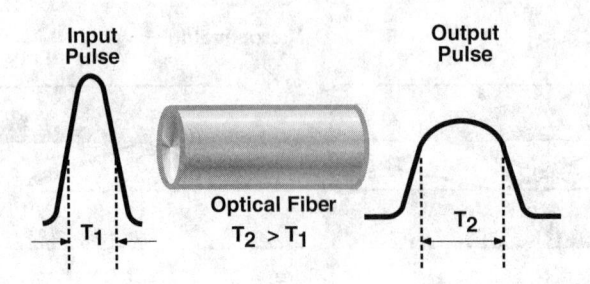

FIGURE 9.11   Pulse dispersion. (Courtesy of Corning Cable Systems LLC and Corning® Inc.)

There are three types optical signal of dispersion to consider and each has its own characteristics and challenges based on the specifics of the operating system. They are modal dispersion, chromatic dispersion, and polarization mode dispersion.

Modal Dispersion

As discussed earlier, light propagation in multimode fibers can take place in a number of stable transmission paths (i.e., modes) depending on the core size, numerical aperture, and system operating wavelength (see Fig 9.12). Because the various stable light paths in a multimode fiber have different total optical lengths, the light following the shortest overall path, near the center of the core, arrives at the receiver end of the fiber before light following the longer outer paths, assuming the speeds of all are constant. This type of dispersion is called modal dispersion. Since an individual pulse of light may comprise hundreds of modes in a multimode fiber, the potential for signal degradation due to modal dispersion is significant. In fact, the issue is so problematic that commercially viable silica-glass step-index multimode optical fibers have had extremely limited use in telecommunications applications. To combat this effect, telecommunications grade multimode optical fibers have a graded-index core that is discussed in more detail in the section on multimode fiber. The nature of modal dispersion is complex and it is not linear. As single-mode fibers have only one stable propagation path for any reasonable distance above the cut-off wavelength, modal dispersion is not a primary consideration in single-mode systems.

Chromatic Dispersion

Chromatic dispersion occurs in optical fiber systems because the pulse of light in the fiber is not truly monochromatic (i.e., it consists of many separate wavelengths). The amount of chromatic dispersion in a fiber is determined by the fiber's material composition and design, and by the light source's operating wavelength and **spectral width**. The spectral width represents the range of different wavelengths present in the transmitter output (see Figure 9.13). A wider spectral width implies a wider range of output wavelengths, which is less desirable than a narrow one with respect to chromatic dispersion effects. The units for chromatic dispersion are picoseconds of light pulse spread, per nanometer of laser spectral width factored to the fiber length (ps/nm·km). Chromatic dispersion can be either positive or negative.

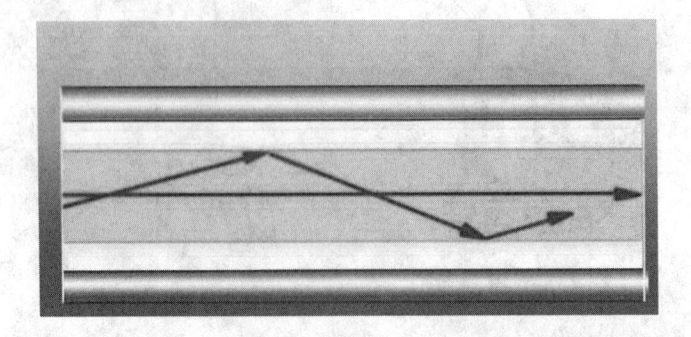

FIGURE 9.12   Multimode Propagation. (Courtesy of Corning Cable Systems LLC and Corning® Inc.)

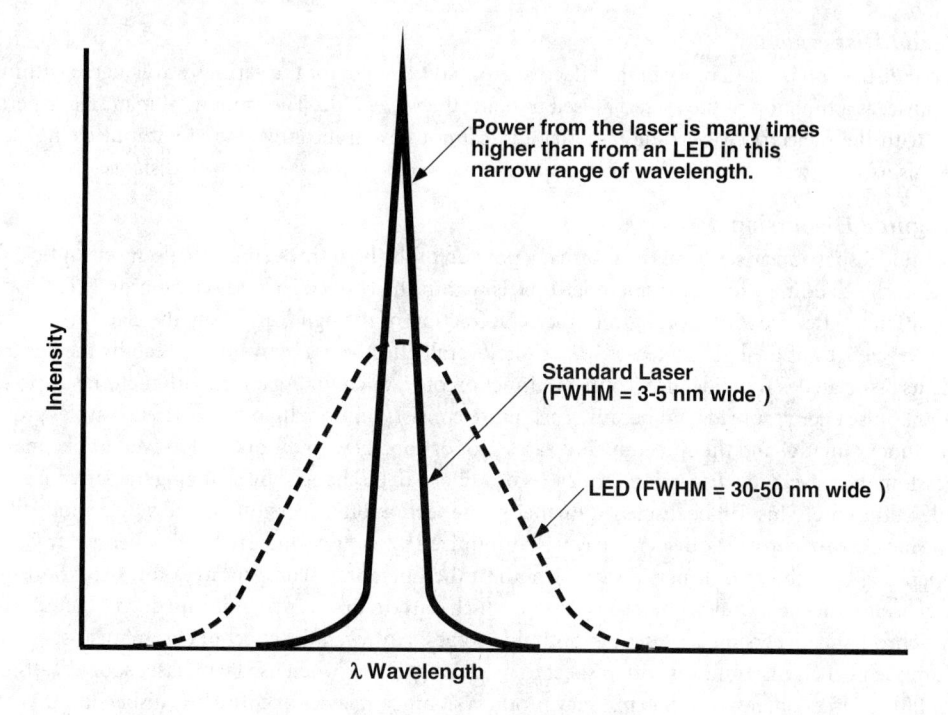

Power from the laser is many times higher than from an LED in this narrow range of wavelength.

Standard Laser (FWHM = 3-5 nm wide)

LED (FWHM = 30-50 nm wide)

Intensity

λ Wavelength

FIGURE 9.13 Spectral width. (Courtesy of Corning Cable Systems LLC and Corning® Inc.)

In a standard nondispersion-shifted single-mode optical fiber the chromatic dispersion is considered to be zero at or around 1310 nm (also known as the zero-dispersion wavelength). This means that signals transmitted near 1310 nm will incur relatively little chromatic dispersion, although the actual zero dispersion wavelength may range from 1302–1322 nm for a specific nondispersion-shifted fiber. At wavelengths longer than the zero dispersion wavelength, the dispersion of these fibers is positive, and at shorter wavelengths it is negative. A positive dispersion value indicates that longer wavelengths in the optical pulse travel faster than the shorter ones. Chromatic dispersion consists of two distinct components: material dispersion and waveguide dispersion (see Figure 9.14).

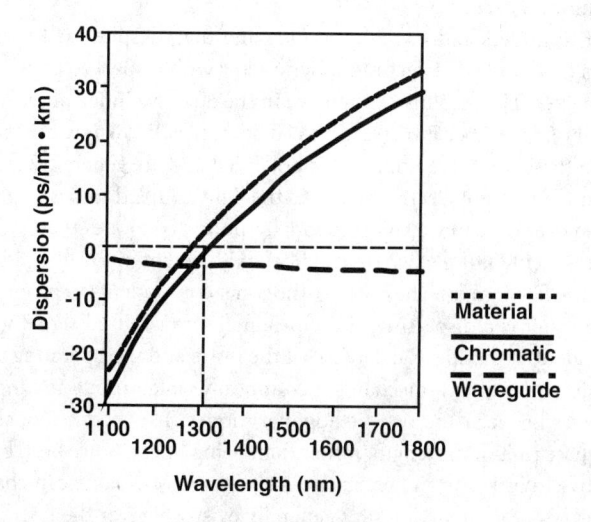

FIGURE 9.14 Single-mode fiber dispersion curve. (Courtesy of Corning Cable Systems LLC and Corning® Inc.)

Material Dispersion

Material dispersion is a difference in the effective group IOR values for the various wavelengths within the light pulse, as a function of the material used to make the waveguide. The amount of material dispersion varies with the dielectric waveguide material and cannot be significantly altered without changing the materials used to make the fiber. Material dispersion increases proportionally with distance.

Waveguide Dispersion

Waveguide dispersion arises from the fact that a portion of the light traversing a single-mode optical fiber travels in the cladding while the remainder travels within the core (refer to discussion on MFD), which has a different effective IOR. For a given fiber, the fraction of the light carried by the cladding increases as the wavelength of the light increases so that the overall effective refractive index seen by the light, and hence its associated waveguide dispersion, is a function of wavelength. Again, as with material dispersion, the total pulse energy spreads out because each pulse consists of more than one discrete wavelength (i.e., not monochromatic) and the spreading increases proportionally with distance. However, unlike material dispersion, the waveguide dispersion properties of a fiber can be changed by changing the refractive index profile of the core. This will be discussed further in the section on dispersion-shifted single-mode fibers.

The net effect of chromatic dispersion can be minimized by using a source with a smaller spectral width, by operating near the zero dispersion wavelength of the fiber, or by compensating for dispersion in the system itself, as needed. Unlike modal dispersion, which only occurs to any significant degree in multimode fibers, chromatic dispersion is present in both fiber types. However, the effect of chromatic dispersion in multimode fibers is insignificant with respect to modal dispersion when used with laser sources, although some LED sources operating at 850 nm may produce significant levels of chromatic dispersion that affect system performance. Industry fiber standards primarily specify modal bandwidth as a measure of the data rate capability of multimode fiber. Equipment manufacturers must then consider the impact of source selection when they subsequently specify fiber modal bandwidth performance requirements for cabling used with the equipment. Conversely, industry standards for single-mode fiber specify chromatic dispersion as the primary parameter to be used in determining the fiber's dispersion performance.

Polarization Mode Dispersion

Polarization mode dispersion (PMD) is comparatively more complex a concept than the other types of optical fiber dispersion discussed earlier, and until relatively recently it wasn't even considered to be a limitation that needed to be specified or addressed. As the effects of system chromatic dispersion have been mitigated by improved, complex fiber designs, and by the availability of dispersion compensating modules, the increased distances of long-haul transmission systems have moved PMD to the forefront as a major performance limiting factor.

It is probably easier to understand PMD by recognizing that, despite the best efforts of fiber manufacturers, optical fibers do not have perfect radial symmetry with respect to physical dimensions, dopant concentrations, and stresses. The result is a difference in the effective index of refraction values along two axes (major and minor) of the fiber cross section. This effect is known as birefringence. Further, light pulses transmitted into these fibers are not linearly polarized and are generally assumed to be random in the orientation of their component electric fields. As the light coupled into a single-mode fiber traverses the length, it is resolved over time into two orthogonally-polarized modes that make up the fundamental mode. These modes are oriented to the two optical fiber axis present in the fiber. PMD occurs because the birefringent nature of the fibers causes these two orthogonal components to propagate at different speeds (see Figure 9.15). The result is a dispersing of the originally transmitted signal which, for simplicity, is assumed to be independent of the spectral content of the pulse and the operating wavelength.

At various points along the fiber a mechanism commonly called mode-mixing occurs, which results in an exchange of power between the two orthogonal modes. It is because of this effect that PMD is proportional to the square root of the length, rather than being linear. Since both birefringence and mode mixing are very sensitive to intrinsic and extrinsic factors, the PMD value can change every time a fiber or system span is measured. Mode mixing sites generally exist wherever the fiber is stressed; though that is not to say that stressing the fiber always produces lower measured PMD values.

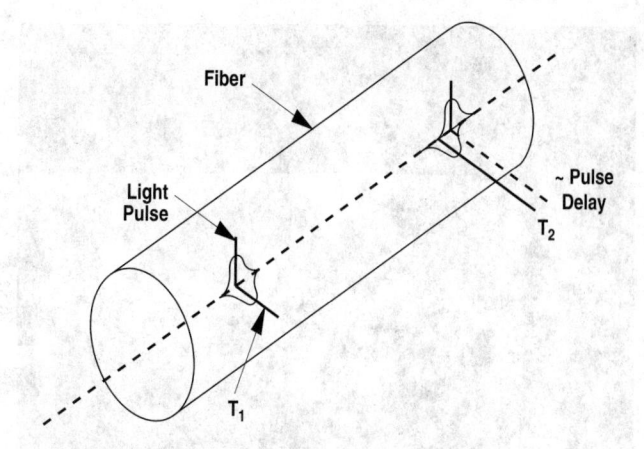

FIGURE 9.15   PMD. (Courtesy of Corning Cable Systems LLC and Corning® Inc.)

PMD is only a concern in single-mode fiber systems because of the relatively higher data rates (10–40 Gb/s) and much longer distances involved. Since both of the polarized components carry a portion of the transmitted power any offset in the arrival time of the two acts to disperse the original signal, which becomes a more significant issue at higher data rates. In addition to limiting the operational transmission capacity of digital systems, it can result in composite second order distortions in analog systems such as those used for cable TV applications.

Although a comprehensive discussion on PMD is beyond the scope of this chapter, it's worthwhile to note that PMD is statistical in nature and is affected by virtually anything that stresses the fiber. Because of this the only way to effectively compensate for the effect is via real-time PMD compensation. Though currently not in widespread use at the time of publication, such devices are reportedly becoming available soon. These devices measure the instantaneous PMD and then utilize a complex array of polarization filters and electronics to provide compensation. Newer generation fibers are also specified to be less PMD sensitive, which is accomplished by reducing fiber manufacturing imperfections. Increasing the coupling between the two orthogonal components can also reduce PMD.

Multimode Fiber Bandwidth

The bandwidth of a multimode fiber is effectively a measure of its modal dispersion, which is the predominant factor that determines the fiber's data carrying capability. It is normalized to a length of fiber and is reported in units of MHz·km. Since cabling does not significantly affect the fiber's modal bandwidth, and because measurements are somewhat complex, it is typically measured by the fiber manufacturer on the bare fiber only. In addition, modal bandwidth does not behave in a linear fashion for systems employing over-filled launch (OFL) conditions (i.e., LED sources) because of the finicky nature of the higher order modes (i.e., those near the core-clad interface) excited by such sources. Therefore special care must be taken when estimating the data rate capacity of such systems. For example, a multimode fiber cut in half does not produce two fiber pieces having double the effective OFL modal bandwidth capability of the original length. Also, the linked OFL bandwidth capability of multiple fibers concatenated together cannot be determined simply by combining the performance of each. Even though there are equations for doing this they are beyond the scope of this chapter.

Newer generation multimode based systems employing vertical cavity surface emitting lasers (VCSELs) or other lasers not only display a higher bandwidth at the operating wavelength, they also display a more linear bandwidth response with respect to transmission distance. In these cases, primarily the lower and intermediate order modes are excited and carry the vast majority of the signal power, thereby mitigating the impact of the higher order modes. This condition is referred to as restricted-mode launch (RML). Effective modal bandwidth (EMB) represents the theoretical worst case restricted-mode launch bandwidth

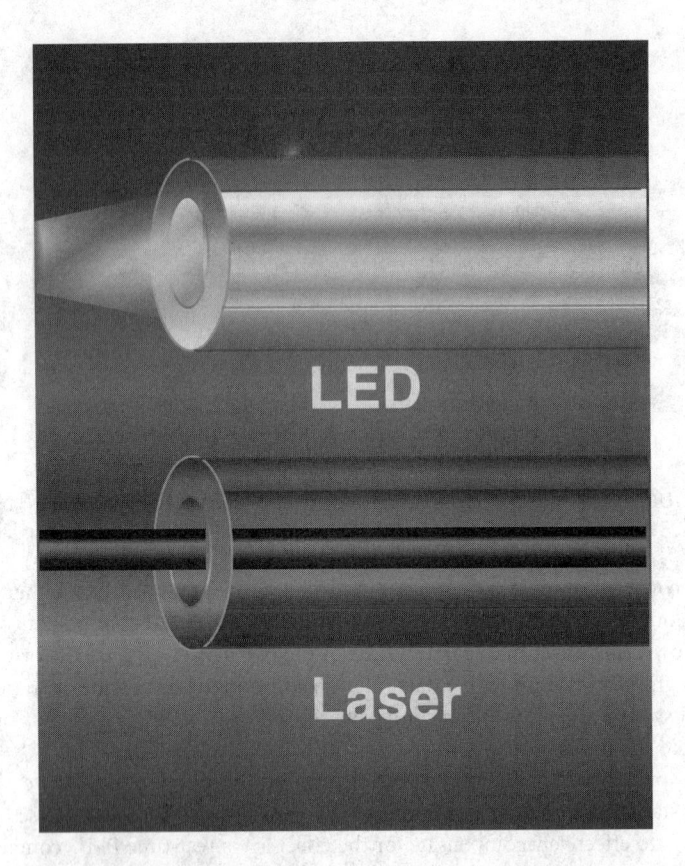

FIGURE 9.16   Transmitter source spot sizes. (Courtesy of Corning Cable Systems LLC and Corning® Inc.)

(RML BW) of a fiber using any VCSEL transmitter meeting certain flux distribution requirements, as established by industry standards. EMB fiber performance is currently only specified for multimode fiber systems operating at 10 Gb/s. Rather than using VCSELs, EMB testing is accomplished by using a narrow beam laser and stepping it across the core of the multimode fiber in small increments (e.g., 2–5 μm). The results are then combined to show pulse delay as a function of radial fiber position at the wavelength on interest (i.e., 850 nm). The delay characteristics of the fiber are then used to ensure that the specified minimum EMB performance of the fiber, as required by industry standards, is met.

The effective launch area of a VCSEL is slightly smaller than the core of the multimode fiber and therefore does not create an over-filled launch condition (see Figure 9.16.) Even then, all current multimode fibers still incorporate a bandwidth value determined using OFL launch conditions (i.e., OFL BW.) So system designers may still need to account for the limitations of individual fibers if an equivalent RML BW or EMB is not included as well. This will change over time and laser-based RML BW performance is expected to eventually become the primary specified value for determining a multimode fiber's transmission capabilities.

Fiber Nonlinearities

As discussed earlier, the information carrying capacity of traditional optical systems is normally limited by attenuation and dispersion (or bandwidth). As the impact of these factors have been mitigated by significant improvements in the design of telecommunication systems and the performance of compensating modules, amplifiers, and other key electronic components, newer systems are experiencing a marked increase in the amount of data they carry, and the distances over which the data are transmitted. The result is an increase in the aggregate optical signal power transmitted over a single fiber, which results from the simultaneous multiplexing of many individual sources at different wavelengths, at several milliwatts each.

These high power operations give rise to an entirely different set of optical issues which are borne from the propagation of light at high intensities in optical fiber. The resulting phenomena are regarded as potentially major system performance-limiting factors, if not properly managed. These effects are collectively termed "fiber nonlinearities," which is derived from the fact that such are dependent on the intensity of light to some higher order power. Fiber nonlinear effects are also length dependent.

To complicate matters there are several factors that affect fiber nonlinear behavior and the impact such effects have on system design and performance. They include the effective size of the light carrying-region of the core, the number of optical channels being transmitted and their relative wavelength spacing, the length of the optical path, and the dispersion characteristics of the fiber. The optical power density within a fiber is a function of the core size, the individual wavelength power and the total input power. The effective length, which depends on both the fiber length and attenuation coefficient, determines the time over which interactions can occur. A high power density and longer available interaction time make for more severe fiber nonlinear behavior.

Raman Scattering

Raman Scattering has two forms, *spontaneous* and *stimulated*. Any light traversing a fiber experiences spontaneous Raman scattering, which is light scattered both forward and backward due to interactions between the optical signal and the vibrations of the silica molecules. Under certain conditions, the atoms in the fiber will absorb some of the light and then quickly re-emit photons with energies equal to the original photons, plus or minus the energy of a characteristic vibration of the atom. This effects both scatters the light and shifts its wavelength. Because of the random nature of the vibrations of the molecules, Spontaneous Raman scattering results in scattered light with a low power and a broad spectral width.

When the spontaneously scattered light gets to a high enough power, or when there is a deliberate signal at an appropriate wavelength, the power transfer to specific target wavelengths is stimulated because of the interaction of the light at those wavelengths and the molecules in the glass. This process results in stimulated Raman scattering (SRS), which is an extremely broadband effect that limits the useable high power bandwidth of optical fibers. SRS production is a function of the number of wavelengths being transmitted and the total number of amplified spans in the system. As no practical methods to fight SRS have been demonstrated to date, the only way to minimize the effect is to maintain the total source power output to levels below the power level that would result in a significant impact on system performance.

Signal degradation due to SRS for single channel systems occurs only at very high powers, typically in the 1 watt range. In optical fibers that simultaneously carry multiple optical channels (i.e., WDM systems), channels operating at shorter wavelengths may provide unwanted gain to channels operating at longer wavelengths. For example, a 1310 nm signal might be scattered and provide gain to a signal over the wavelength range of 1310 to 1430 nm, or longer. The gain carries the signature of the 1310 signal, effectively adding noise to the longer wavelength signals while also reducing the power of the 1310 nm signal. The threshold power required for degradation of longer wavelength channels is reduced as the number of channels carried by the fiber increases. In general, however, SRS is not a significant concern in current optical fiber systems because the threshold power levels are high.

Incidentally, the SRS effect has been successfully exploited with the advent of Raman amplifiers that amplify light signals using this principle. For example, light pumped in at 1480 nm is used to excite the atoms in a section of standard optical fiber forming part of a system link. When a signal pulse (e.g., 1550 nm) subsequently traverses that section of fiber, the excited atoms are stimulated to emit a photon at an energy equivalent to the signal transmission wavelength. Unlike erbium doped fiber amplifiers (EDFAs), which are discrete components, Raman amplifiers can work over kilometers of standard fiber actually forming part of the passive optical fiber span.

Stimulated Brillouin Scattering

Stimulated Brillouin scattering (SBS) occurs at the lowest threshold power levels of the various non-linear effects and can be observed at powers as low as several milliwatts, depending on the transmitter source and fiber characteristics. SBS occurs when signal power reaches a level that results in the generation of acoustic waves in the optical waveguide medium, through a process called electrostriction. The actual process is somewhat complex and involves—the characteristics of the transmitted signal, *Spontaneous Brillouin scattering*, the electrostrictive nature of dielectric waveguides, and the speed of sound in the transmission medium. SBS is primarily an issue with analog system transmission and is not usually a concern for digital systems.

As the generated sound waves travel through the glass material, back toward the source, they induce spatially periodic compressions and rarefactions, which in turn cause localized changes in the IOR value of the transmission medium. These localized fluctuations in the fiber's refractive index scatter the light from subsequent pulses similar to the effect of a Braggs grating, except that the frequency of the back-scattered light is shifted slightly. The reflected light combines with the input signal to reinforce the sound waves, via electrostriction, creating a circular feedback loop. Since the magnitude of this effect increases with increasing input optical power, once the threshold power for SBS is achieved, any subsequent increases in signal strength are essentially back-reflected to the transmitter source. This process effectively creates an upper limit to the power level that can be transmitted down the fiber. The output power of an erbium-doped fiber amplifier (EDFA), given sufficient input signal power, has the capability to produce SBS.

The formation of SBS is more complex than described herein and is aggravated by narrow spectral width transmitters, such as externally modulated lasers that are employed in systems operating at very high data rates, and in systems that employ dense wave division multiplexing (DWDM). Dithering the source output can significantly reduce the SBS threshold power level by delaying the generation of the sound waves in the fiber, but doing so increases the spectral width of the signal, which as we now know can cause other issues. Fiber design can also be used to inhibit the onset of SBS by increasing the threshold power required for SBS to occur. As SBS is related to the speed of the sound in the fiber, changing features that affect this parameter can affect the threshold power level as well. These can include changing the fiber strain and core dopant concentration characteristics.

Self Phase Modulation (SPM)

When a pulse of light passes through an optical fiber it loses energy at the leading edge as it interacts with and aligns the electric fields in the glass. If the pulse power is increased, more energy is transferred to the glass due to these interactions. This power loss causes the light at the front of the pulse to shift to a longer wavelength (red shift). The energy is subsequently transferred back from the glass to the light signal at the trailing edge of the pulse, causing light in that region to be shifted to a shorter wavelength (blue shift). In a fiber with a positive **dispersion coefficient**, the increased in wavelength at the front of the pulse extends the pulse front, while shortened wavelengths at the trailing edge slow down, extending the back of the pulse. Both of these effects therefore result in pulse broadening. The increased spectral width caused by the energy transferred at the pulse edges can also cause significant chromatic dispersion in high power systems. One mitigating factor in self-phase modulation is that, as the pulse spreads, the interaction length for the polarizing reactions with the glass also increases. This reduces the rate at which energy transfers to and from the light signal at the photon level, so pulse spreading does not increase rapidly once it initiates.

In fiber with a negative dispersive coefficient, the wavelength shifting of the light pulse at the leading and trailing edges results in pulse compression. As the longer wavelengths at the front of the pulse slow down, more light at the original signal wavelength interacts with the glass and the process continues. Likewise, at the rear of the pulse, the shorter wavelength light speeds up and moves forward in the pulse. For certain power levels, pulse widths, and fiber dispersion values, this effect can result in nondispersing signals called solitons. Some high data rate, long-length systems have been specifically designed to take advantage of this phenomenon for improved performance.

While SPM can be a problem, the specific combinations of optical power and dispersion needed to cause a significant performance issue are readily accounted for.

Cross Phase Modulation

Since each optical channel in a DWDM system operates at a different wavelength, pulses from the various channels travel through the fiber at different speeds due to chromatic dispersion effects. As a result, pulses at a fast wavelength will catch up to and pass through pulses at slower wavelengths. When two pulses occupy the same section of fiber, an exchange of energy, as described in the section above on SPM, also occurs at the front and back edges of the overlapping area. Under these conditions, the light in both pulses increases in wavelength at the front of the overlap and decreases in wavelength at the back of the overlap. When the difference in the transit times for the two pulses is minimal, such as when the effective dispersion coefficients are close to zero the effect of cross phase modulation (CPM) can be significant. However, the specific combinations of optical power, channel spacing, and dispersion needed to make CPM a problem can be easily managed.

Four Wave Mixing

Another problem which can occur when two signals occupy the same physical location in a fiber is four wave mixing (FWM). FWM occurs when two or more "parent" channels in a WDM system combine to form "daughter" artifacts at the expense of their own power. In a WDM system with many channels at equally spaced frequencies, new wavelengths created by this effect can fall on top of existing channels, resulting in optical crosstalk. If the frequency spacing is not equal, the effect is an increase in system noise that lowers the signal to noise ratio of the operating channels. FWM is a more serious problem for WDM systems running over fibers with low chromatic dispersion values in the operating wavelength region because such a condition increases the amount of time when two or more channels are allowed to interact.

This effect, which is more difficult to account for, coupled with the advent of EDFAs, led to the development of nonzero dispersion-shifted fibers. Requiring a minimum amount of chromatic dispersion in the operating wavelength regions limits the time available for pulses at the various channel wavelengths to interact, which reduces the effects of FWM.

9.1.5 Fiber Manufacturing

There are many potential variations in the processes used to manufacture optical fiber, but the basic premise remains essentially the same for each. This section will introduce the basics of fiber manufacturing and briefly discuss the three primary methods currently employed to support the economical production of optical fiber on a large scale. They are outside vapor deposition (OVD), vapor axial deposition (VAD), and inside vapor deposition (IVD).

All fiber manufacturing processes involve the delivery of oxygen and volatile feed materials in gaseous form to a target reaction area. This reaction produces fine soot that adheres to a bait material or to other deposited soot in a process is generally referred to as "lay-down." Depending on the specific portion of the fiber's core-clad glass structure being created at a given moment, the mixture of reaction materials can be changed to vary the resulting dopant concentration so that the completed fiber has the desired refractive index profile. The more complicated the index profile, the more complex the laydown process. In any event, the laydown process must be tightly controlled to ensure that the completed fiber has the requisite optical and geometric properties.

Variations in the feed materials during the laydown process determine what type of fiber will be manufactured. Changes in the dopant concentration in the core section and the size of the core determine whether the fiber will be intended for multimode or single-mode applications. Also, other materials can be added to the outside of the cladding to enhance certain fiber mechanical properties. For example, carbon can be deposited on the outside of the cladding to hermetically seal the surface of the glass. The carbon layer impedes the migration of hydrogen, present in significant quantities in undersea applications, into the fiber core, which could significantly affect the fiber's attenuation characteristics.

The process for making fiber, including the three primary lay-down process variants, are described in greater detail below

Outside Vapor Deposition

The outside vapor deposition (OVD) process uses a glass mandrel or seed rod placed in a lathe. The feed materials are delivered to the region near the rod surface by a gas-burner type assembly. The burner assembly traverses longitudinally alongside the bait rod as the lathe turns it to build-up the soot thickness

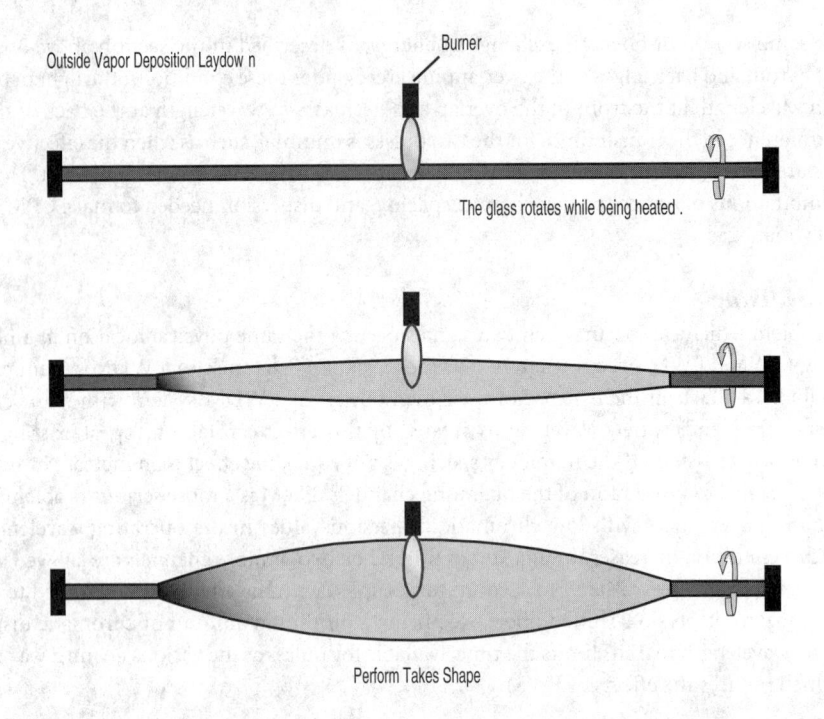

FIGURE 9.17   OVD laydown. (Courtesy of Corning Cable Systems LLC and Corning® Inc.)

in layers (see Figure 9.17). As more layers of porous glass soot are added, the chemical composition of each layer (i.e., dopants) is controlled as necessary to create a refractive index profile specific to the type of fiber being manufactured. Once enough layers have been deposited the seed rod is removed from the center before further processing.

Vapor Axial Deposition

The vapor axial deposition (VAD) process is basically similar to the OVD process, except that the soot is deposited by fixed burners located near the end of a vertically mounted seed rod. This type of process allows for several fixed burners to be used in laydown and the cylindrical soot rod created in this way is drawn upward as the burner assembly deposits additional layers (see Figure 9.18). Again, as in the OVD process, the seed rod is removed before the next step.

Inside Vapor Deposition

The IVD process was the first widely-used, commercially successful laydown process and utilizes a hollow tube vice a solid rod to capture the soot. The seed tube is placed horizontally in a lathe and a gas burner assembly or resonator moves along the outside of the tube surface. The feed materials flow into the seed tube and the soot is laid down on the inside at the desired target area, building inward toward the center of the tube (see Figure 9.19). In most IVD applications the seed tube remains to become the outer part of the glass cladding, so its optical characteristics are not as critical as the rest of the glass structure. Concentrations of the feed materials flowing into the tube are changed as in the other methods to alter the refractive properties of the final product as needed. IVD is in limited use today as the volume of fiber that can be produced is typically much lower, which has an effect on fiber and cable manufacturing efficiencies.

Consolidation

When the laydown process is complete, the resulting composite structure is called a "preform," which is very delicate and has the appearance of a big piece of white chalk. The preform is then placed in a consolidation oven where it is densified into a thick rod of high-purity glass called a "blank." This heating

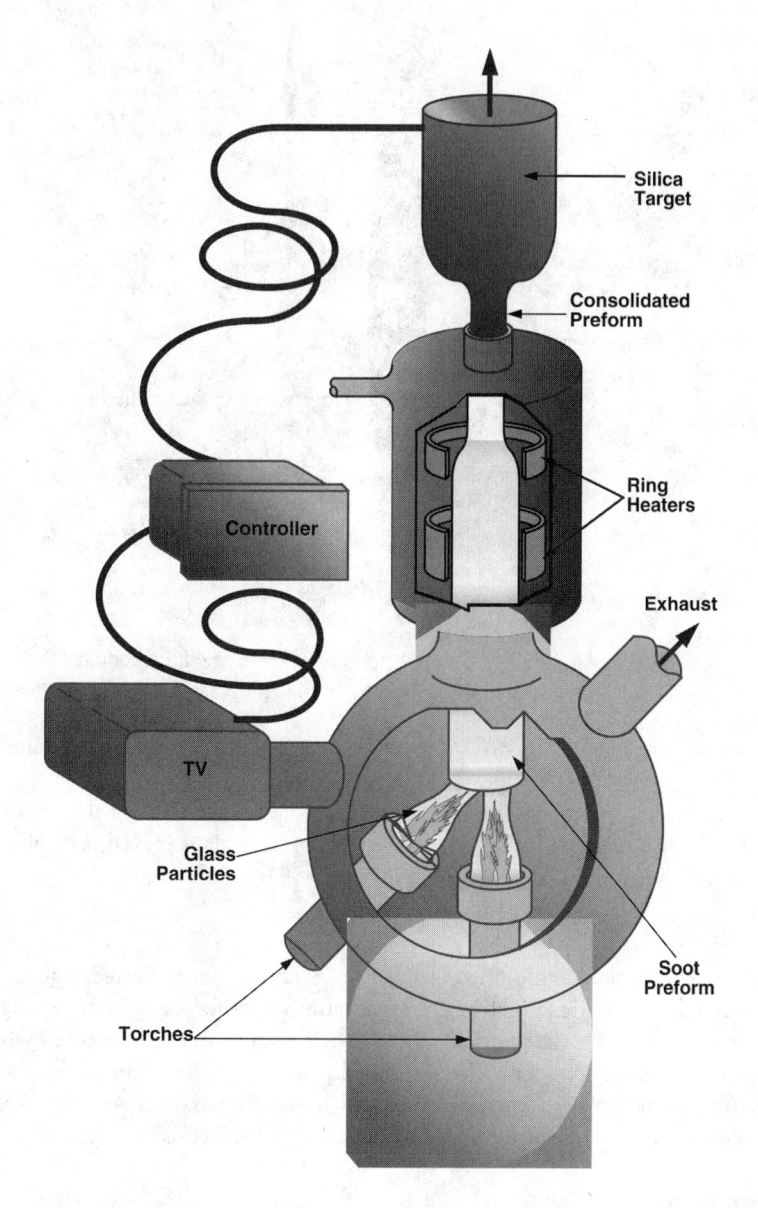

FIGURE 9.18   VAD laydown. (Courtesy of Corning Cable Systems LLC and Corning® Inc.)

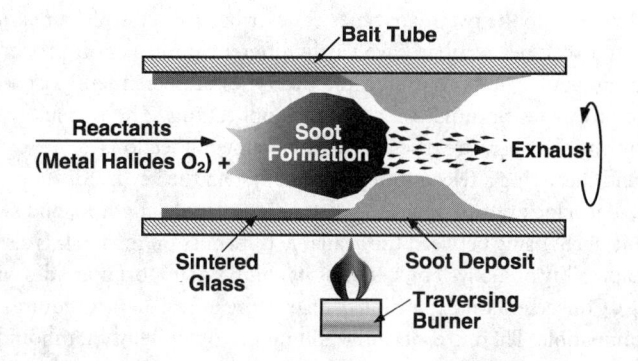

FIGURE 9.19   IVD laydown. (Courtesy of Corning Cable Systems LLC and Corning® Inc.)

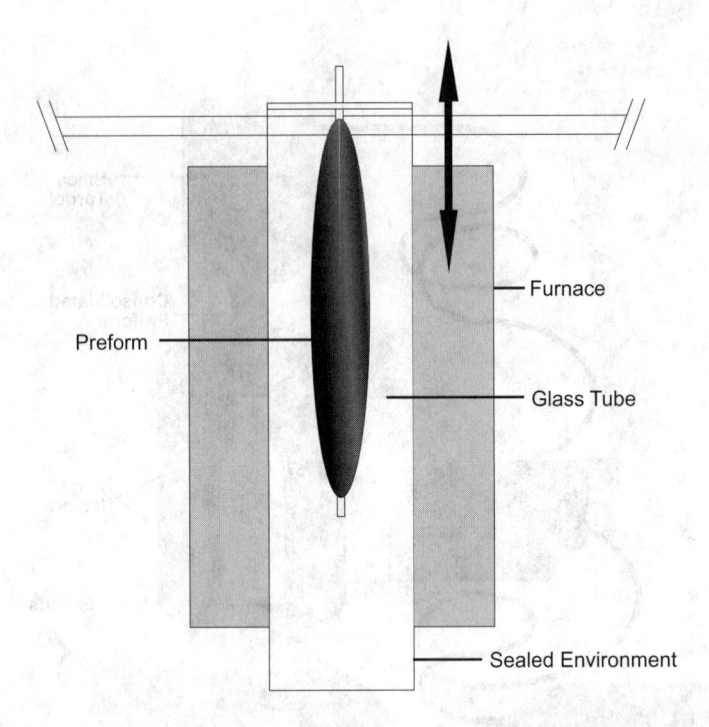

FIGURE 9.20   Fiber consolidation. (Courtesy of Corning Cable Systems LLC and Corning® Inc.)

process, sometimes referred to as sintering, collapses the hole in the center of the preform, if present, forming a solid, ultra-transparent blank rod (see Figure 9.20). This blank is eventually placed in the draw tower which is where the optical fiber is actually produced.

Drawing the Fiber

In the drawing process the blank is mounted vertically in a draw tower and heated to about 2000°C at the bottom end. As the glass rod softens, a gob begins to pull away via the force of gravity, leaving behind a thin strand of fiber (see Figure 9.21). The speed, tension, and temperature of the draw are closely controlled to obtain consistent fiber dimensions. Vibrations, contamination, and other sources of mechanical stresses must be avoided during the draw process to guarantee a high strength glass fiber. An integral part of the draw process is the addition of the protective coating, which is applied and cured in layers as discussed earlier.

9.1.6   Fiber Types

There currently exists a plethora of different fiber designs for a wide range of uses. All telecommunications grade fibers can be divided into the two main types as described earlier, which are known as single-mode and multimode. These two types of fiber have vastly different transmission properties that make each of them ideal for certain applications. Within these two types of fibers there exist a number of different designs, each having properties optimized for specific applications. The various types of commercially available telecommunications fibers are discussed in the following section.

When designing and specifying a telecommunications system the key factors to consider are the transmissions speeds (i.e., data clock rates), operating distances and wavelength(s), and system installed costs, all of which play a role in choosing between the available transmission media. It is also critical to consider the flexibility and adaptability of the system, and thus its ability to support upgrades with minimal expense (i.e., future-proofing). Conversely, the expansion of an existing infrastructure requires that newer products also be backwards-compatible. There are, of course, numerous other items that should factor into the final design and purchase decisions, such as compliance to industry standards, but they will not be discussed in detail herein. With respect to what type of fiber to select, it is usually easy to decide between multimode

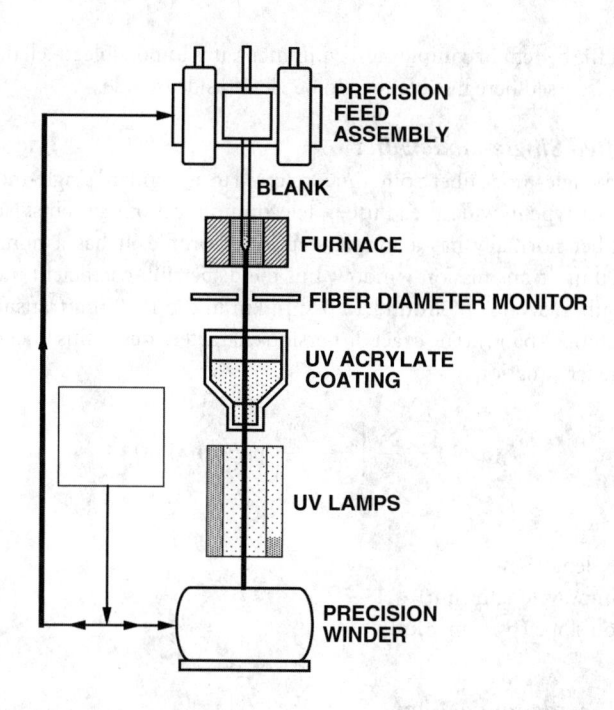

FIGURE 9.21   Fiber draw. (Courtesy of Corning Cable Systems LLC and Corning® Inc.)

and single-mode. Beyond that the selection process becomes somewhat more complicated. The following sections provide general information which can be used to understand the various specific types of single-mode and multimode fibers available. It is not intended to be construed as definitive guidance for selecting a specific fiber type for a particular application.

Single-mode Fibers

Single-mode fibers have historically operated in one of the primary wavelengths regions around either 1310 or 1550 nm, although applications that take advantage of the extended operating spectrum are already commercially available.

The strength of single-mode fibers lay in their inherently low attenuation and dispersion characteristics at the primary operating wavelengths. As a result, single-mode fibers are used in applications to transmit high data rates over distances from tens to thousands of kilometers, where the spacing between costly signal conditioning and amplification equipment needs to be maximized. Single-mode fibers can be generally categorized into two different types that are best differentiated optically by their dispersion characteristics, and physically by the shape of their refractive index profile (see Figure 9.22). Note that under visual

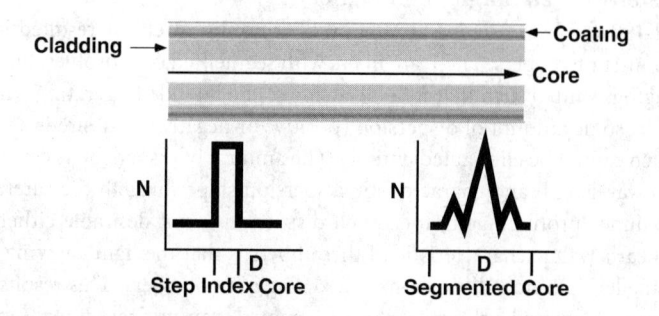

FIGURE 9.22   Single-mode fiber index profiles (Step and segmented-core). (Courtesy of Corning Cable Systems LLC and Corning® Inc.)

inspection, or without the benefit of complex test equipment, it is impossible to tell the difference between any of the fiber types discussed here in were they to be viewed side by side.

Nondispersion-shifted Single-mode Fiber

Nondispersion-shifted single-mode fiber, sometimes referred to as standard single-mode fiber, was the first single-mode optical fiber type in widespread use in telecommunications systems. Nondispersion-shifted single-mode optical fiber normally has a simple step-index profile. It has a nominal zero-dispersion wavelength in the 1300 nm transmission window, but the dispersion coefficient increases as a function of increasing wavelength to a value of around 18 ps/nm·km in the 1550 nm transmission window (also known as conventional- or C-band). The exact dispersion characteristics of this fiber can be approximated by the use of the Sellmeier equation.

$$D(\lambda) \cong \frac{S_0}{4} \left[\lambda - \frac{\lambda_0^4}{\lambda^3} \right] \text{ ps/(nm} \cdot \text{km)} \tag{9.6}$$

Where
λ = operating wavelength (nm)
λ_0 = zero dispersion wavelength (nm)
S_0 = zero dispersion slope (ps/(nm$^2 \cdot$ km))

Dispersion-shifted Single-mode Fiber

Dispersion-shifted fiber is a type of single-mode fiber that has its nominal zero-dispersion wavelength shifted higher in the operating wavelength spectrum in order to optimize the fiber's dispersion properties at longer wavelengths. This was done to support newer generation long-haul telecommunications systems operating at longer wavelengths, where the intrinsic attenuation of the optical fiber is inherently lower. To achieve the desired dispersion properties the index profile of the core must be altered to shift the waveguide component of the fiber's chromatic dispersion coefficient. Recall that the other component, material dispersion, cannot be significantly altered without changing what the fiber is made of. The resulting characteristic index shape for these fibers is called the segmented-core index profile.

The first variants of dispersion-shifted fibers were designed such that the zero dispersion wavelengths were at the lowest point of the spectral attenuation curve for that generation of fibers, which was around 1550 nm. This gave the optimal combination of low attenuation and low dispersion for emerging long-haul applications, or so it was thought. These fibers were an obvious improvement over standard fibers for single-channel haul, high data rate systems, but as we now know their ability to support the multiple narrowly spaced channels that are used in DWDM systems is severely limited. The use of these early dispersion-shifted fibers eventually led to a better understanding of fiber nonlinear effects, and subsequently to the development of improved fiber designs.

Nonzero Dispersion-shifted Single-mode Fiber

In the early to mid-1990s, a better understanding of fiber nonlinear effects resulted in the development of nonzero dispersion-shifted fibers. These are fibers with segmented core profiles, but for which the zero dispersion wavelength is shifted such that it falls outside of the intended operating window, or windows. As mentioned before, some amount of dispersion (positive or negative) is desired to minimize nonlinear effects, although such cannot be eliminated entirely. The amount of dispersion is controlled by managing the zero dispersion wavelength and characteristic dispersion slope through the careful manipulation of the fiber's complex index profile. Again, too much dispersion is not desirable either as it causes other issues as addressed earlier. One characteristic of these fibers is that the light carrying region of the fiber is proportionally smaller than non-dispersion-shifted single-mode fibers. This results in a higher power density at a given source power level, which if severe enough can mitigate improvements in the fiber's nonlinear performance. The latest generation fibers are designed with a larger mode-field diameter to lower the transmitted power density, thereby reducing nonlinear effects.

TABLE 9.1 Key Single-mode Fiber Attributes – Cabled Fiber

Parameter		Non dispersion-shifted Single-Mode	Non-zero Dispersion-shifted Single-Mode
Mode Field	1310 nm	$8.6 - 9.5 \ \mu m$	—
Diameter	1550 nm	$9.6 - 11.2 \ \mu m^a$	$8 - 11 \ \mu m^b$
Numerical Aperture		0.13 – 0.14 (typical)	0.14 (typical)
Cutoff Wavelength		≤ 1260 nm	≤ 1480 nm
Cladding Diameter		$125.0 \pm 1.0 \ \mu m$	$125.0 \pm 1.0 \ \mu m$
Coating Diameter		$245.0 \pm 10.0 \ \mu m$	$245.0 \pm 10.0 \ \mu m$
Attenuation	1310 nm	≤ 0.4 dB/km	NA
	1550 nm	≤ 0.3 dB/km	≤ 0.3 dB/km
Dispersion	O-band	≤ 3.5 ps/nm·km	NA
	C-band	≤ 18 ps/nm·km	$0.1 - 6$ ps/nm·km[2]

[a]Not typically specified by industry standards.
[b]Certain variants of nonzero dispersion shifted single-mode fibers, such as those specifically designed to operate in metropolitan type applications with inexpensive Fabry-Perot laser sources, typically have different specified requirements.

It's worthwhile noting at this point that the possibility of a one-fiber-does-all solution is not on the immediate horizon. The combination revolutionary-evolutionary nature of fiber development has been realized only because of the massive resources dedicated to its understanding. As the requirements placed on them increase, new fiber performance issues continue to be uncovered. While fiber designs advance to meet the challenge, each new operating limitation must be fully understood before design modifications can be made to counter its effects. The biggest challenge facing glass scientists is that altering a fiber design to improve upon one attribute typically comes at the expense of one or more other attributes. This means that current designs are a compromise of competing performance needs in order to optimize for what is most important, given a specific set of expanded operating conditions. As the exact nature of the propagation of electromagnetic fields though dielectric waveguides is very complex, the most likely scenario is that newer generations of fibers will continue to be semicustomized for a specific application or set of applications.

Table 9.1 summarizes some of the parameters for the various single-mode fiber types commonly used today (typical values for cabled fiber are shown). See Section titled "Optical Fiber Standards and Specifications," for additional information on industry standards related to optical fiber.

Multimode Fibers

Multimode fibers are more suited for short distance applications where some distance capability can be sacrificed for a lower installed system cost based on the specific needs. In applications such as LANs, SANs, and data centers, all transmission equipment is located in reasonably close proximity to one another. In many such cases multimode fiber based systems are overall typically more economical because of the availability of comparatively inexpensive connectivity components and low cost electronics, such as LEDs and VCSELs. This allows the total system cost to be lower than an equivalent system based on single-mode fiber and the associated electronics.

Multimode fibers generally operate at one of the two primary wavelengths of 850 and 1300 nm. All telecommunication grade silica-glass multimode fibers are characterized by a graded (i.e., curved) refractive index profile. The core has an index of refraction that is highest at the center, which then falls off approximately in a parabolic shape to where the value is that of the pure glass cladding at the core-clad interface (see Figure 9.23). The primary core sizes for graded index multimode fibers in use today are 50 and 62.5 μm. Other larger core multimode fibers saw limited use early on in telecommunications applications, but these are rare today and current large core multimode fiber designs find service primarily in very-short-length, low data rate applications such as automotive, medical, and similar uses involving imaging or low-intensity lighting.

Recall from the information presented earlier that the dispersion effects from having numerous modes present in fiber could severely limit the data capacity, thereby making it useless for practical

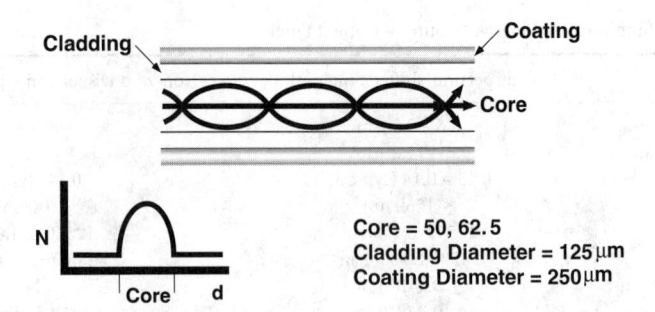

FIGURE 9.23   MM graded-index profile. (Courtesy of Corning Cable Systems LLC and Corning® Inc.)

telecommunications. In order to address this, the refractive index of the core is modified so that the light traveling along the longest stable path (e.g., highest order mode), also travels with the fastest average speed. Near the outside of the core where the index of refraction is lower, the mode travels faster than at the middle of the core. The result is that all the signal modes reach the receiver end of the fiber at essentially the same time. This mode management allows the receiver to differentiate between successive pulses, even though each may consist of hundreds of individual modes. Since the index of refraction is also wavelength dependent, the modal dispersion properties of multimode optical fibers can be optimized for a particular wavelength, such as 1300 nm for 62.5/125 μm fiber, or 850 nm for 850 nm laser-optimized 50/125 μm fiber (see Fig. 19.24). They can also create a balance between the two, such as with standard 50/125 μm fiber.

Table 9.2 summarizes some of the parameters for the various multimode fiber types commonly used today (typical values for cabled fiber are shown). See Section titled, " Optical Fiber Standerds and Specifications," for additional information on industry standards related to optical fiber.

Special Use Fibers

Standard single-mode and multimode telecommunications fibers are not ideally suited for all related applications. Some of the more widely used types of specialty telecommunications fibers are described below.

Chromatic Dispersion Compensating Fiber

As discussed earlier, it is desirable to have some amount of chromatic dispersion at the operating wavelengths for fibers operating in the 1550 nm region or higher in multi-channel systems, in order to minimize fiber nonlinear effects. This causes the dispersion of the system to build up over long distances, to the point

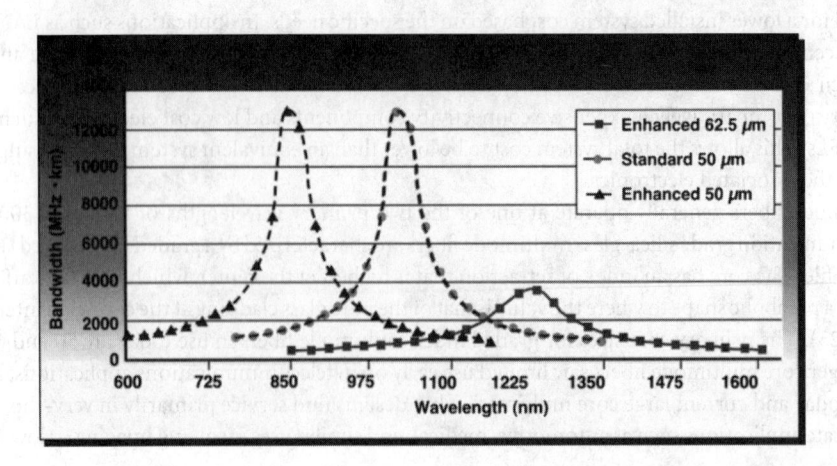

FIGURE 9.24   Multimode Fibre Bandwidth Comparison. (Courtesy of Corning Cable Systems LLC and Corning® Inc.)

TABLE 9.2 Key Multimode Fiber Attributes – Cabled Fiber

Parameter		50/125	62.5/125μm
Core Size		$50 \pm 3.0 \ \mu$m	$62.5 \pm 3.0 \ \mu$m
Numerical Aperture		0.200 ± 0.015	0.275 ± 0.015
Cladding Diameter		$125.0 \pm 2.0 \ \mu$m	$125.0 \pm 2.0 \ \mu$m
Coating Diameter		$245.0 \pm 10.0 \ \mu$m	$245.0 \pm 10.0 \ \mu$m
Attenuation	850 nm	≤ 3.5 dB/km	≤ 3.5 dB/km
	1300 nm	≤ 1.5 dB/km	≤ 1.5 dB/km
OFL BW	850 nm	≥ 500 MHz $\cdot$ km or ≥ 1500 MHz $\cdot$ km[a]	$\geq 160/200$ MHz $\cdot$ km[b]
	1300 nm	≥ 500 MHz $\cdot$ km	≥ 500 MHz $\cdot$ km
EMB 850 nm		≥ 2000 MHz $\cdot$ km[a]	NA

[a] New 50/125μm fibers designed and specified primarily for use in 10 Gb/s
serial systems laser-based systems, but can also be used for legacy applications.
[b] Minimum bandwidth values vary from application to application.

where it can become a performance issue. To counteract this, dispersion compensating fiber can be added at points along the optical link to bring the aggregate system dispersion back down to near zero. This fiber has similar characteristics to other nonzero dispersion-shifted fibers, except that its dispersion coefficient and slope are opposite in sign to those of the system fiber being compensated, and the dispersion values are of a much higher order of magnitude. Dispersion compensating fiber is not typically cabled, but is instead employed in discrete components in the optical path. For long systems it is often necessary to insert dispersion compensating modules (DCMs) at various locations along the system before the accrued dispersion at any one point becomes excessive, to the point where it can no longer be effectively compensated. As these fibers also have an associated attenuation loss, optical amplifiers are often used in tandem with DCMs to boost the output signal strength.

Erbium-Doped Fiber

One of the key enabling technologies for DWDM systems transmitting in the 1550 nm operating window and higher was the advent of erbium-doped fiber amplifiers (EDFAs). EDFAs employ a rare-earth element, erbium, to amplify optical signals traveling through a fiber, including noise, independent of bit rate or protocol. EDFAs are generally referred to as "optical amplifiers" and employ a pump laser operating at 980 or 1480 nm to excite the erbium atoms. When an optical signal passes though the amplifier, the excited erbium atoms are stimulated to emit a photon at the appropriate energy level, similar to the operation of a Raman amplifier.

EDFAs amplify the transmission wavelengths between 1520 and 1630 nm and have output powers ranging from about 10 to 25 dBm. The erbium fibers are not cabled so EDFA modules are discrete components used throughout the system as needed. Optical amplifiers are used most routinely in long-haul Telco systems with one typically installed every 30–80 km along the system route, enabling system distances of 500 km or more before signal conditioning becomes necessary. EDFAs are also used in CATV systems in the links between head ends, or between head ends and hubs.

Polarization Maintaining Fiber

Polarization maintaining fiber (PMF) is designed for use in fiber optic sensor applications (e.g., interferometric sensors) and other polarization sensitive components. In PMF, the light maintains a constant state of polarization as it travels down the fiber due to the intrinsic high birefringence designed into such fiber types. One method of inducing the high birefringence needed is to manufacture a fiber core that is significantly elliptical in shape, while another method is to purposely include stress-inducing rods into the cladding area of the fiber. In these types of fibers, the mode traveling along the minor axis is intentionally attenuated to create the requisite single polarization state of the light.

Submarine Optical Fiber

Due to the great expense involved with the construction, placement, and maintenance of submarine optical fiber systems, component design is critical. As these are the longest telecommunications systems in use anywhere, managing factors such as attenuation and dispersion becomes even more crucial. To account for this submarine fibers have special design parameters to ensure they remain viable over the design life of the installation. These fibers typically have very small cores and correspondingly smaller attenuation coefficients. They are also hermetically sealed by incorporating carbon into the glass structure, as discussed earlier, to prevent the attenuation effects associated with hydrogen migrating into the fiber core region.

In regards to the dispersion issue, the use of DCMs is not viable due to the added system complexity and additional amplification requirements that would result. To address the dispersion issue in submarine systems, a second fiber type must be employed. This fiber has dispersion characteristics that are opposite in sign, but roughly equal in magnitude, to those of the other fiber in the system, at the operating wavelength(s). Unlike terrestrial systems, the dispersion compensating fiber used in submarines systems is cabled and becomes part of the overall system span. Therefore long submarine systems employ these paired fibers, which are alternated along each link to improve the overall system optical characteristics. The two fibers are typically used in equal lengths based on the specific dispersion properties of each. In practice this process is fairly complicated, which tends to increase the associated complexity and cost of such systems.

Radiation Hardened Fiber

Fibers exposed to high radiation levels show an increase in the attenuation coefficient due to a complex interaction between the radiation and some of the materials in the fiber's core. The reaction creates absorption bands at the most common operating wavelengths in the 1310 and 1550 nm regions. A radiation hardened fiber is designed to have reduced radiation sensitivity and also to recover quickly from radiation induced effects. Fibers with a high concentration of Germania as a dopant have been found to operate and recover fairly well in a high radiation environment. Some dopants, such as phosphorous, are more susceptible to long-term radiation induced effects.

Hollow Core Fibers

A newer generation of fiber designs currently under development could potentially eliminate, or significantly reduce, some of the limitations inherent with the current generation of glass optical fibers. Rather than relying on dopants to achieve the requisite core refractive index profile, these fibers employ a matrix of air-filled holes. Some of the primary benefits of these fibers for telecommunications applications are the low dispersion characteristics, the elimination of fiber nonlinearities, and larger cores that support very high power applications. Add in the inherently low attenuation characteristics and materials cost, and these fibers become very attractive. Such fibers also show promise in a number of other useful applications outside of telecommunications, so developmental research continues.

9.1.7   Optical Fiber Standards and Specifications

Optical Fiber Standards Publications

With the plethora of fiber types available, it quickly became necessary early on to develop industry standards to support their use. Industry standards serve a number of functions, the primary ones of which are:

- Establish minimum acceptance criteria to ensure system performance
- Standardize basic design criteria to ensure the interoperability of fibers from the various suppliers
- Educate customers on key performance attributes
- Establish a level playing field for suppliers and manufacturers
- Enable innovation in optical fiber design and performance

Listed in Table 9.3 and Table 9.4 are some of the key telecommunications industry fiber standards employed today for the various optical fiber types. The standards are identified only by document number, publishing organization, and the specific fiber type. Users of industry standards should purchase these from a reliable source to ensure they have the latest applicable version.

TABLE 9.3 Single-mode Fiber Detailed Industry Standards

Single-mode Fiber Type	Industry Forum	Document Number	Fiber Designation
Dispersion Unshifted	TIA[a]	492-CAAA	Class IVa
	IEC[b]	60793-2-50	B1.1
	ITU[C]	G.652	G.652A or B
Non dispersion	TIA	492-CAAB	Class IVa with Low water peak
shifted with Low	IEC	60793-2-50	B1.3
Water-peak	ITU	G.652	G.652C or D
Non-zero	TIA	492-DAAA	Class IVd
Dispersion-	IEC	60793-2-50	B4
shifted	ITU	G.655	G.655A, B or C

[a]TIA[TM] is the Telecommunications Industry Association[TM].
[b]IEC is the International Electrotechnical Commission.
[c]ITU is the International Telecommunication Union.

TABLE 9.4 Multimode Fiber Industry Standards

Multimode Fiber Type	Industry Forum	Document Number	Fiber Designation
50/125 μm	TIA	492-AAAB	Class Ia
	IEC	60793-2-10	A1a.1
	ITU	G.651	—
50/125 μm 850 nm laser-optimized	TIA	492-AAAC	Class Ia
	IEC	60793-2-10	A1a.2
	ITU	NA[a]	NA[a]
62.5/125 μm	TIA	492-AAAA-A	Class Ia
	IEC	60793-2-10	A1b
	ITU	—[b]	NA

[a]ITU standards have not yet been updated to reflect the industry adoption of this fiber type.
[b]ITU is primarily a European organization. 62.5/125μm fiber has not been widely adopted in Europe, so no supporting ITU Standard exists.

Fiber Attribute Specifications

One commonality between the various industry fiber standards are the specified attributes. Some minor differences may exist between the various standards for a particular attribute, but often such differences result primarily from when the documents were last updated, and only to a lesser extent by market preferences or technical factors. Note that each attribute has an associated industry standard test procedure. Refer to the relevant optical fiber Standards listed in Tables 9.3 and 9.4 for information on test procedures.

The key specified performance attributes for the various fiber types an listed in Table 9.5. Other secondary parameters not listed may also be specified, while others only characterized (i.e., informative vs. normative).

TABLE 9.5 Key Industry Standard Specified Fiber Attributes

Attribute	MM	Std SM	NZDS SM
Attenuation coefficient (fiber or cable)	X	X	X
Chromatic dispersion	X	X	X
Cladding diameter	X	X	X
Cladding noncircularity	X	X	X
Core diameter	X	—	—
Core-clad concentricity error	X	X	X
Cut-off wavelength (cable)	—	X	X
Macrobending loss	X	X	X
Modal bandwidth	X	—	—
Mode field diameter	—	1310 nm	1550 nm
Polarization mode dispersion	—	X	X

Defining Terms

Absorption: The attenuation component of an optical fiber which results from the complex interaction of photons with impurities in the glass structure as the light traverses a length of optical fiber. The absorbed optical energy is converted into vibrational energy (i.e., heat), which is typically reemitted at a different energy level.

Attenuation: The combination of optical energy losses from a light pulse traversing a length of optical fiber resulting from absorption and scattering, typically referred to as intrinsic attenuation, and bending losses, typically referred to as extrinsic attenuation, expressed in dB. However, attenuation is often used as a synonym for attenuation coefficient, expressed in dB/km.

Attenuation coefficient: The mathematical expression for optical power attenuation, typically expressed as a change in power measured in decibels (dB), normalized to a unit length (dB/km).

Chromatic Dispersion coefficient: The normalized chromatic dispersion value, normally with units of ps/(nm • km), from which the fiber chromatic dispersion can be calculated for a given fiber length, and spectral width.

Cladding: The layer of dielectric material that immediately surrounds the core of an optical fiber, the combination of which is fundamental to the ability of optical fiber to guide light at certain wavelengths. The cladding material for telecommunications grade optical fibers usually consists of a silica glass that cannot be physically separated from the core.

Core: The central region of an optical fiber consists of a silica-glass structure in which dopants are added to achieve the desired light transmission properties for a particular type of optical fiber. The distribution of dopants alters the index of refraction of the glass, which varies based on dopant concentration. The resulting refractive core properties are referred to as the index profile.

Dispersion: The spreading of the pulse width of a light pulse traversing a length of optical fiber. Pulse spreading from chromatic dispersion results from the interaction of the light pulse and the transmitting fiber medium, at the molecular level, and is highly dependent upon the spectral make-up of the light pulse and the physical structure of the glass both in material make-up and physical design characteristics. Pulse dispersion is generally undesirable as it reduces the signal (carrier) to noise ratio, and if severe enough, can result in bit errors.

Group index of refraction: The mathematical ratio of the power velocity of light in a perfect vacuum (numerator) to its velocity in a given medium (denominator). The group index of refraction is typically denoted by "n_g." As light always travels fastest in a vacuum, the value of n_g for any material is always greater than 1. Caution must be taken that "n" and "n_g" are used appropriately, as "n_g" is used in calculations involving transit time.

Index of refraction: The mathematical ratio of the phase velocity of light in a perfect vacuum (numerator) to its velocity in a given medium (denominator). Index of refraction is typically denoted by "n," or less frequently by the acronym "IOR." As light always travels fastest in a vacuum, the IOR value for any material is always greater than 1 and "n" is used in calculations other than for transit time.

Insertion loss: The measured loss in optical power resulting from the scattering of light at the physical boundary between two optical waveguides (e.g., fibers). The light scattering is caused by the localized disruption in the refractive properties of the optical path due to the insertion of discrete components such as connectors, mechanical splices, couplers, etc. For fusion splices, this loss is typically referred to as "splice loss."

Macrobending: Macroscopic deviations in the straight-line axis of the optical path which result in light pulse energy being refracted out of the optical fiber and lost. Examples of macrobend sources are excessive bending of the protective cable structure, or the placement of excessive compressive forces on the fiber optical cable. Macrobending is characterized by proportionally higher losses at longer transmission wavelengths (i.e., 1550 nm vs. 1310 nm in single-mode fiber).

Microbending: Discrete variations in the refractive index properties of an optical fiber waveguide caused by localized point stresses placed on the fiber that result in light pulse energy being refracted out of the optical fiber and lost. Examples of microbend sources are fiber coating defects, fiber optic cable structural defects, or installation related cable damage. Microbending is characterized by proportionally equivalent losses at all transmission wavelengths.

Mode: A stable path of light in an optical waveguide characterized in complex terms by Maxwell's equations for an electromagnetic field distribution.

Mode field diameter (MFD): A measure of the distribution of the effective light carrying region of a single-mode optical fiber, which extends beyond the fiber core and includes a potion of the integral cladding layer.

Optical time domain reflectometer (OTDR): A device used to characterize the attenuation performance of an optical link by measuring the intensity of light reflected due to Rayleigh scattering as a light pulse traverses the optical path. OTDRs work on the principle that the amount of reflected light is directly proportional to the energy of the light pulse. OTDRs are useful in estimating the attenuation coefficient and in identifying defects and other localized losses, but are not considered a direct loss measurement.

Rayleigh scattering: Light scattering resulting from microscopic variations in the refractive index of an optical path that are proportionally small with respect to the wavelength of the transmitted light. The refractive index variations generally result from changes in material density or composition due to localized stresses or impurities.

Reflection: The redirection of a portion of a light pulse at the boundary of two dissimilar media back into the transmitting medium. The amount of light reflected, vice transmitted into the receiving medium, is a function of the wavelength dependent refractive indices of the two media and the angle at which the light is incident upon the boundary of the two.

Spectral width: The mathematical representation of the extreme wavelengths of the light spectrum that comprises an optical pulse. One method of specifying the spectral line-width of a transmitter is to measure the full-width half-maximum (FWHM) of the source output, or specifically the difference between the wavelengths at which the amplitudes are one-half of the maximum value of the entire pulse.

Total internal reflection: The complete reflection of a light pulse at the boundary of two differing mediums that occurs when the angle at which the light is incident upon the boundary, as measured from the normal, is greater than the critical angle.

References

See all the references of this section at the end of Section 9.2.

9.2 Fiber Optic Cable

Ken A. Chauvin

Fiber optic cables serve a number of purposes, the primary one of which is to house the optical fiber contained inside and protect it from external stresses. The main sources of external stress are those associated with the shipping, handling, installation and termination of the optical fiber, and are relatively short in duration. Other external stresses, such as those experienced over the life of the installation due to the surrounding environment, are more long-term in nature. In addition to protecting the fiber, the cable structure must also organize the fibers and support the identification of individual fibers via printed legends, color coding, groupings, or some combination of the three.

Because the applications of optical fiber and fiber optic cable are diverse, so are the environments in which they are employed. Most environments fall into one of three main categories that are generally referred to as indoor, indoor/outdoor, and outdoor. These environments will be discussed in more detail

TABLE 9.6 Cable Application Considerations

Cable Attribute	Outdoor	Indoor/Outdoor	Indoor
Typical Operating Temperature Range	−40 to 70C	−40 to 70C	−20/0 to 70C
Typical Length	2–12 km	100–2000 m	2–300 m
Typical Fiber Counts	2–432+	2–144	1–24
Mechanical Durability	Excellent	Excellent	Good
Chemical resistance	Excellent	Good	Good
Waterblocking	X	X	—
UV resistance	X	X	—
Fire resistance	—	X	X

in the section on cable designs, but the key differentiating factors are

- Indoor: Resistant to spreading fire and generating smoke
- Outdoor: Resistant to solar radiation, water and moisture, and extreme temperatures and mechanical stresses

Applications in which single cables are used to transition from indoor to outdoor must therefore account for all of the above. Table 9.6 summarizes the primary considerations for the various environments.

For each environment there may be numerous other specific considerations that are germane to a particular application, such as sewer or shipboard environments, or to an installation method, such as aerial self-supporting or jetting into miniature ducts. As with the optical fibers there are a number of unique cable designs optimized for various combinations of application and installation method. The performance criteria for the most common designs are also specified by the various telecommunications industry forums, which will be discussed at the end of this section.

9.2.1 Applications Considerations

Even though the details of a particular cable design depend on the application environment and installation method, there are a few generalities. Various applications expose the cable to different mechanical and environmental conditions that have the potential to stress the fibers contained inside. As optical fiber technology continues to mature end-users are becoming concerned about expected fiber optic cable lifetimes, especially where early generation cables have been already in place for greater than 20 years. To date, no absolute performance model exists to accurately predict all potential failure mechanisms for every application or environment, so the best gauge of the health of a system is typically the optical properties of the cabled fiber.

The ultimate performance of cabled fiber depends on a number of factors including the cable environment, the type of installation, cable design, and the cable materials themselves. Factors which may affect certain installations are structural cable damage caused during or after initial installation, improper cable design, the quality of the splicing and closure work, exposure to periods of severe environmental loading or certain chemicals, or exposure to temperature and humidity conditions outside of specified operating performance ranges. Evidence of performance degradation can be elusive and is largely dependent on the specific characteristics of each installation and the type of cable design deployed.

Environment

Outdoor

Cables designed for placement outdoors are normally installed aerially, directly buried, or placed in underground ducts, using various installation methods. These products normally convey communications signals for extended distances and only penetrate the indoor environment far enough to support a transition to indoor cabling, such as in a telecommunications room. Specific applications range from very long-haul spans running hundreds of kilometers (aerial, buried, or both), all the way to relatively short spans of 100 meters or less, such as from a network distribution point to an enduser business or residence in a fiber-to-the premises (FTTP) system. In each case, the cable is exposed to the various rigors of the outdoor environment, which can be quite extreme for some areas of the world.

The specific type of cable employed in a given application depends primarily upon factors such as fiber type and fiber count, limitations on size, weight or cost, the proposed methods of installation and termination, and perhaps even the need for other specific performance capabilities. The latter may include items such as resistance to extraordinarily high compressive forces, temperatures, or electric field potentials (e.g., aerial installation along high-tension lines), or resistance to various chemicals or other organic hazards such as gnawing rodents.

Fiber-to-the-Home

The drop cable environment is basically a subset of the traditional outside plant environment. As fiber connections get closer to the home, cable designs are changing to adapt to this space. Cable fiber counts for such applications are typically very low (1 to 4 fibers) and the distances involved are much shorter than for traditional outside plant cable installations (hundreds of feet vs. of hundreds of miles). The relative criticality of the cables also is reduced somewhat since drop cables typically connect only one, or a few, residences or businesses vs. linking thousands of individual endusers across hundreds of communications channels, such as is found in long haul cable systems.

Although the installation methods are generally the same as other outdoor cables, the placement of drop cables often involves the use of manual labor to install, obviating the need for heavy equipment and high tensile loads. These cables can either be buried from the curb to the house, or can be hung aerially from the telephone pole to hardware mounted on the house. This unique environment lends itself to a special set of design criteria for such cable types. Since these cables supply only a limited number of endusers, which in many applications may be limited to a single premises, a cable failure is much less critical than a failure in a high density, high traffic, long haul distribution cable. For example, consider how often premises coaxial cables are severed by aggressive gardeners or homebuilders. This defference lets manufactures desing cables that can allow for a limited amount of fiber strain during mechanical and environmental loading, such as could be experienced in self-supporting aerial cable designs. This in turn allows for innovative cable designs and materials selection that can support lower cost product solutions, which is critical in enabling the deployment of affordable fiber-to-the-home systems.

Indoor

Premises cables are generally deployed to support backbone, horizontal, and interconnect applications. Higher fiber count tight buffered cables discribed later, can be used as intrabuilding backbones that connect a main cross-connect to an entrance facility or intermediate cross-connect. Likewise, lower fiber count cables can link an intermediate cross-connect to a telecommunications room (i.e., horizontal cross-connect) feeding multiple workstations. Simplex and duplex interconnect cables are commonly used to patch the optical signal from end equipment to the hardware containing the main distribution lines, in order to complete the passive optical path. These cables can also provide optical service to workstations, or can connect one patch panel to a second patch panel as in a cross-connect application.

In regards to cables designed for indoor-only use, the overriding consideration is to ensure that such cables do not significantly contribute to the spread of flame or the generation of smoke in the event of a fire, although there are other considerations. Indoor cables contain comparatively fewer fibers than outside plant cables, and infrequently go beyond 300 m. Also, because the opto-electronic equipment served by such cables is typically installed in an indoor, climate-controlled environment, these cables are not required to be as mechanically or environmentally robust as cables designed for placement outdoors. Therefore, the design considerations for indoor cables are significantly different than those for outdoor designs.

Although not as severe as those for outside plant cable, there are potential sources of stress for indoor cables that must be accounted for. In addition to stresses related to installation of the cable itself, inside plant cables must be flexible enough to support the tight bends experienced in cable trays and in the optical hardware which supports the overall intra-building cabling infrastructure. Indoor cables might also be placed in close proximity to many other types of cables, including copper cables supporting communications or other low voltage applications. In tall buildings, high fiber count riser cables may go up several stories to reach the various building distribution points. Because of the relatively high number of terminations associated with such short distances, indoor cable must support quick and easy connectorization and be available in variety of fiber counts, ranging from one to two fibers for desktop applications all the way up to 144 fibers or more for building backbone distribution cables.

Indoor/Outdoor

Indoor/outdoor applications involve cables which typically run short distances to make interconnections within and between adjacent buildings. These cables are fully resistant to the typical outside plant environment, but the overall performance requirements are typically not as stringent as those for long-length outside plant communications cables. These cables are used primarily to avoid the need for a costly extra splice point where the telecommunication system cabling transitions from outside to inside, and thus are employed in only relatively short lengths.

Installation Methods

There are a number of diverse installation methods for the various spaces in which fiber optic cable is used. The installation methods discussed below are primarily associated with outside plant applications but may also have some uses in other environments. The relatively benign environment of the inside plant means that installation considerations are somewhat less critical, although this is not to say that they are not important.

Direct Buried

One of the most common and widely used installation methods is the placement of the cable directly into the ground, which is known as direct burial. The primary methods of direct burial are trenching, plowing, or placing cable into narrow grooves cut into hard surfaces, such as roadways or parking lots. The obvious advantage to these types of installations is the inherent measure of safety and stability provided by the cable surroundings, if left undisturbed, but even then these types of applications are not impervious to post-installation damage. Buried cables are subject to damage from dig-ups due to subsequent construction, lightning strikes, chemicals in the soil, gnawing rodents, and ground heaves. Protection against these dangers is somewhat accounted for in the design and construction of the cable itself, which may include the incorporation of a layer of protective metal armoring.

Trenching

In trenching installations, a trench is dug in the earth to an appropriate depth, usually 75 cm or more, and the cable or cables placed into it. The minimum depth of the trench used depends on a number of factors, but deeper is generally better from a protection standpoint. However, deeper trenches are also proportionally more expensive to make. The trench is then backfilled with a variety of materials to protect the cable, such as clean sand or soil. Direct buried installations are typically marked with tapes laid parallel to the cable, but placed closer to the surface, and markers located visibly above the surface, in an effort to prevent dig-ups. Buried cables are typically connected to one another via hermetically sealed closures located in buried hand holes or manholes along the route. These closures house and protect the splices or connectors used to connect sections of cable.

Plowing

For installation by plowing, a groove is plowed into the ground normally using a vibratory plow with the cable laid through a covered trough to the bottom of the plow. As the plow moves forward, the cable is laid in the opening and the soil falls back on top to cover it up. The major benefit to using plows is the relative speed at which the cable can be placed, though such installations are relatively shallow and therefore more likely to be damaged by subsequent construction. For this reason, plowing is generally limited to short lengths of low fiber count cables, such as those used in campus environments or in emerging access market type applications (i.e., drop cable).

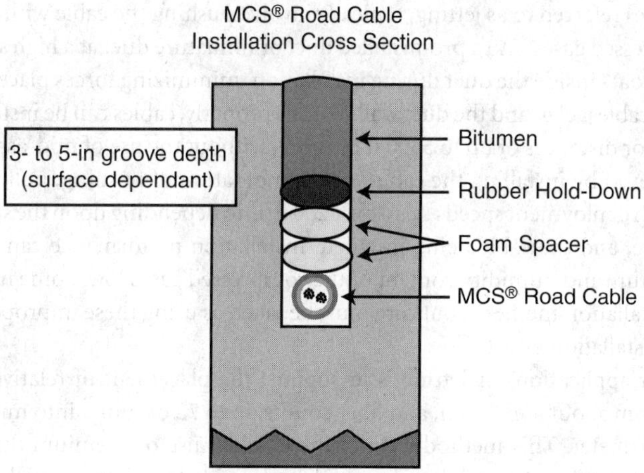

MCS® Road Cable
Installation Cross Section

3- to 5-in groove depth
(surface dependant)

Bitumen

Rubber Hold-Down

Foam Spacer

MCS® Road Cable

FIGURE 9.25   Road cable installation. (Courtesy of Corning Cable Systems LLC and Corning® Inc.)

Roadway Grooves

For cables designed specifically for placement in road surfaces, such as asphalt or concrete, the process for installation involves placing the cable in a narrow groove (about one inch wide) cut into the hard surface of a road or parking lot. The cable can be secured in place with special "hold-down" spacers pressed into the groove on top of the cable. The remainder of the groove is then filled with bitumen—a common road repair material (see Figure 9.25). Since the process can be done as a continuous operation, it minimizes the disruption to traffic. This method also boasts a minimal impact to the useful life of the roadway because the groove is very narrow. Conversely, traditional roadway trenching methods are comparatively more disruptive, costly, and can also significantly reduce the life expectancy of the roadway.

Buried Duct

In duct installations, the duct is first installed below ground usually by trenching. The fiber optic cable is subsequently installed within the duct using a number of methods that are described below. One benefit of using duct is that several ducts comprising a duct bank can be installed simultaneously. As ducting is generally less expensive than the cable itself, this has the potential benefit of reducing the system first-installed cost. Additional fiber optic cables can then subsequently be installed on an as-needed basis to support future growth, vs. all at once. This method minimizes the likelihood of damaging installed cable when new cables are added as each can be installed into a separate protective duct. Since the ducting itself must also be durable to be placed in the ground, it affords the cable some additional measure of mechanical protection. This means that cables designed for installation in ducts need not be as rugged as cable designed for direct-burial applications, which typically equates to a lower cost for these types of cables. Such cables are generally not armored; hence the term duct cable is often used to describe all-dielectric cables designed for placement in such ducts.

The most common method for a duct installation is to pull the cable through the pre-installed duct with a pulling rope or tape. Pulling distances, and thus installed cable lengths, can be increased by the use of pulling lubricants and mid-span assists, thereby allowing one longer cable to be pulled into sequential segments of duct. It is imperative during such operations to ensure that the cable is not damaged by applying excessive tension, by pulling the cable around overly restrictive bends in the conduit or through damaged conduit, or by using guide sheaves or rollers that are too small. In some cases cables can be pulled into duct in groups of two or more, or pulled in on top of other existing cables, but doing so can significantly reduce achievable installation distances and increase the likelihood of damage to the cables themselves.

Another method of installing cable into duct is by moving the cable along the duct length using the action of a high-velocity air or other compressed fluid acting on the cable jacket. Air-assisted installation techniques, sometimes referred to as jetting, typically involve pushing the cable with a tractor mechanism while blowing compressed gases into a pre-installed duct or miniature duct at a high speed. This technique allows the cable to "float" inside the duct during installation, minimizing forces placed on the cable due to friction between the cable jacket and the duct wall. If done properly, cables can be installed into ducts using this type of method for distances of up to 5500 ft or more, without the use of mid-assist machines. Longer distance can be achieved by installing the cable via a combination of pushing, pulling, and jetting. Such installations can reach deployment speeds up to over 200 ft/min depending upon the specific characteristics of the duct, the route, and the cable being installed. Installation performance can also be improved by keeping the temperature and humidity content of the compressed gases low. Some lubricants can be used with jetting-type installation methods, but care must be taken as using these improperly could actually be detrimental to the installation effort.

The most popular application for jetting is to support the placement of relatively small cables with diameters ranging from about 4 to 8 mm, and fiber counts up to 72 or more, into miniature ducts around one centimeter in diameter. This method enables the efficient use of premium duct space by allowing larger traditional ducts, such as 1 to 1-$^1/_2$ in. sizes, to be subdivided into many smaller ones. The result is a honey-comb of miniature ducts that can be populated with cost-reduced miniature cable types specifically designed for installation with high pressure fluid methods.

Since the distances that cables can be placed into ducts may vary significantly depending upon the installation method used and the specific characteristics of the cable and ducting, most cable manufacturers have computer software that estimates achievable distances and cable loading during installation. For accurate calculations these programs require detailed information on the physical and material characteristics of the duct and cable(s) in order to determine an accurate coefficient of friction and cable loading characteristics. Critical duct physical characteristics are the inner diameter, section lengths, changes in elevation, and the number and location of bends, including bend radii and directions. This information is best determined by an accurate route survey conducted on the system prior to the placement of the cable. Key cable characteristics are diameter weight, and jacket material, which can be obtained by contacting the cable manufacturer, and the number of cables being installed.

Aerial Installations

The common methods for placing cables aerially include lashing the cable to a pre-installed steel messenger wire, or employing the use of cable designed to be self-supporting. Self-supporting fiber optic cable designs are usually based on one of two primary designs. Figure-8 self-supporting aerial cables (Figure 9.30(c)) consist of a fiber optic cable core and integrated messenger, typically made from stranded steel. Both the cable and the messenger share a common outer jacket resulting in the characteristic shape that gives these cables their name. All-dielectric, self-supporting (ADSS) aerial cables (Figures 9.30 (a and b)) contain no metallic elements and are usually round or oval in shape, although some are offered in a "Figure-8" design as well. In round cable designs, dielectric aramid yarn is stranded around the cable core or inner jacket to provide the necessary tensile strength for aerial applications.

Aerial applications for fiber optic cable offer unique challenges. Planning for aerial cable installation includes taking into account proper clearances over hazards, the cable types and physical properties, and the mechanical stress loading placed on the cable over the life of the installation, due to the environment. Proper planning requires knowing the "sag and tension" characteristics of the cable once installed. Sag is the amount of vertical distance an aerial cable will hang, or droop from an imaginary straight line drawn between successive attachment points. The tension is the force that the ends of the cable experience under various installation and environmental conditions at the attachment points. Understanding the stresses placed on a cable installed aerially is important to ensure that safe operating conditions are not exceeded, cable lifetime shortened, or that dangers from material failures do not occur.

When planning for the aerial installation of fiber optic cables there are several important factors to consider with respect to the design, specification, installation, and performance of the optical plant. Fiber optic cables used in aerial applications must not only be able to withstand high tensile loads over the life of the installation, but they are also subject to the full range of harsh environmental conditions of the unprotected outside plant. With any aerial installation, factors such as cable design (size, fiber count, tensile properties, etc.), span distance, installation method, installation tension, and environmental loading conditions (temperature, wind and ice) need to be addressed. As with duct installations, major manufactures of fiber optic cables designed to be placed aerially can offer guidance on achievable span distances for a given cable by accounting for the prevailing environmental conditions. The primary limiting conditions for a typical aerial installation are the allowable cable sag, the tensile rating of the cable messenger (if used) or attachment hardware, and the allowable cable strain or fiber attenuation at maximum environmental loading.

Another potential hazard to aerial fiber optic cable installations is that, given certain environmental conditions, the cable could experience additional stresses from oscillating wind-induced motions. One type of cable motion is Aeolian vibration, which can occur with sustained low velocity winds. Aeolian vibrations cause the cable to resonate, or sing, at frequencies of around 50–100 Hz depending on the specifics of the installation. Aeolian vibrations are most common in long spans of round cable installed at high tensions, and have the effect wearing away the cable sheath or fatiguing attachment hardware at the support points. Another wind-induced cyclic cable loading mechanism is galloping, which is characterized by larger amplitude and lower frequency movement. Galloping occurs at higher wind speeds on nonconcentric cable designs, such as "Figure-8" cables or round cables that have their shapes distorted by ice accumulations. Aerial cable galloping has the effect of overloading the attachment hardware, causing it to fail. Although these two phenomena may not affect the cable's optical performance to a significant extent, they are considerations for selecting hardware to properly support aerially installed cable. For this reason the selection of proper attachment hardware is a critical part of any aerial installation.

There are also a number of other potential hazards that must be accounted for when installing cables aerially because of the obvious exposure. These include shotgun damage, lighting strikes, gnawing rodents, and even the effects of high electric field potentials in installations where the cable is co-located near high tension power lines. Dielectric cables can minimize the effects of lightning or high field potentials, but do not render such cables impervious to their effects. Armoring can improve the survivability of cables to mechanical damage from gunshot and rodents, but adds considerable weight which can limit achievable span distances.

Indoor

The fact that most indoor cable is placed by hand in cable trays or pulled for very short distances into duct, conduit, or the like, makes the installation aspect somewhat less critical of a consideration for such designs. In some tall buildings the cable might be placed vertically for several stories, but must then be supported periodically along its length to prevent excessive stress from being placed on it.

9.2.2 Cable Components

Fiber Elements

Fibers in fiber optic cables are primarily offered in one of three basic forms, each with its own advantages and limitations as described below.

Individual Colored Fiber

Bare fiber is supplied by the manufacturer with diameter of about 245 μm. Most fiber is coated by the cable manufacturer with a thin layer of ink for the purpose of identification. Once colored, the fiber can be placed in a number of different cable designs for a variety of uses. The primary benefit of using loose individual fibers is the relative ease of manufacturing and the fact that the fibers can be packed relatively densely, especially for low fiber count applications. The main disadvantages to this design are that each fiber must be spliced individually and it is not recommended that fiber in this form be used to support the direct termination of fiber optic connectors. This can increase the cost of the materials, time, and space needed for installing these types of cable designs, and the differential costs can increase at higher fiber counts.

Ribbon Fiber

Ribbon cables employ groups of individual fibers locked in a tight planar array by the application of a matrix material that is similar to the coating placed on individual fibers (see Figure 9.26). The cured matrix then holds them parallel to one another, but generally enables the fibers to be physically separated into smaller groups by hand or with the use of specially designed ribbon-splitting tools. Ribbon fibers come in group sizes ranging from 2 to 36, but 12-fiber ribbons are currently the most commonly used in North America, especially in the outside plant. The advantages of ribbon cables are the greater fiber packing densities at high fiber counts, and the fact that all the fibers in two ribbons of the same type can be spliced to one another simultaneously in one operation. Current mass-fusion splicing equipment can support splicing up to 24 fibers at a time, resulting in significant cost savings for high fiber count installations. The cost benefit of splicing ribbon cable is not generally considered significant below fiber counts of about 72 for most applications, but low fiber count ribbons may still find uses where size restrictions are more of a consideration, such as in buried duct applications.

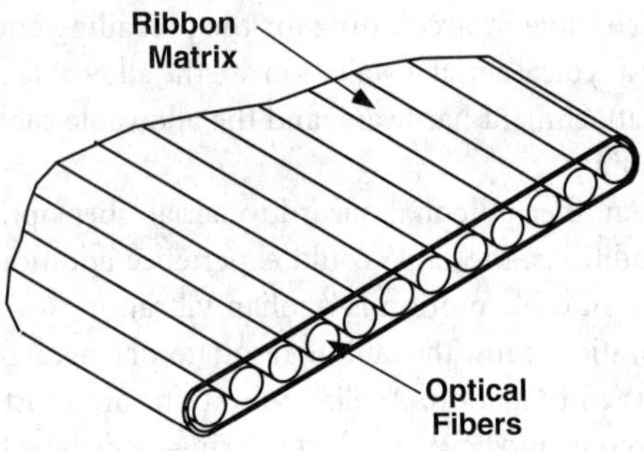

FIGURE 9.26 12-fiber ribbon. (Courtesy of Corning Cable Systems LLC and Corning® Inc.)

Buffered Fiber

For the vast majority of indoor cables, individual uncolored or colored optical fibers are coated with a buffer material to increase the diameter (see Figure 9.27). Although a number of sizes are available, 900 and 500 μm are the most common. The primary benefit of these fibers is that they are inherently more rugged, easier to handle, and can be directly terminated with standard field-installed optical fiber connectors. These features make them ideal for applications where fiber counts are low but which are termination intensive. Because of the added materials, buffered fiber cables are not well suited for high fiber count or long-distance applications due to the increased size, weight, and cost. To assist in the removal of the buffer layer from the fiber for termination or splicing purposes, these fibers may be coated with a low-friction slip layer applied to the coated fiber before the buffering process. These thin slip layers do not affect the performance of the fiber or the ability to strip the acrylate coating from the glass fiber itself.

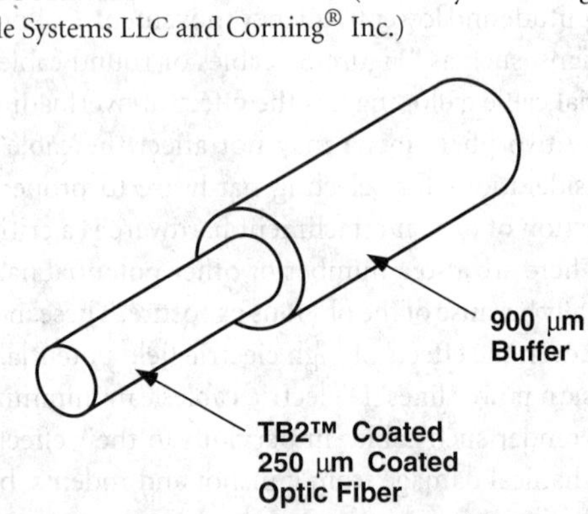

FIGURE 9.27 900 μm buffered fiber. (Courtesy of Corning Cable Systems LLC and Corning® Inc.)

Core Elements

Buffer Tubes and Subunits

The fundamental building block of any fiber optic cable is the buffer tube or subunit (see Figure 9.28). Buffer tubes are primarily associated with outdoor fiber optic cables and serve to house and organize the individual fiber elements. They typically contain a waterblocking material to protect the fiber from water ingress and to help isolate them from external stresses. The tube materials used must be flexible, yet provide sufficient hoop strength to ensure adequate mechanical resistance to crushing forces

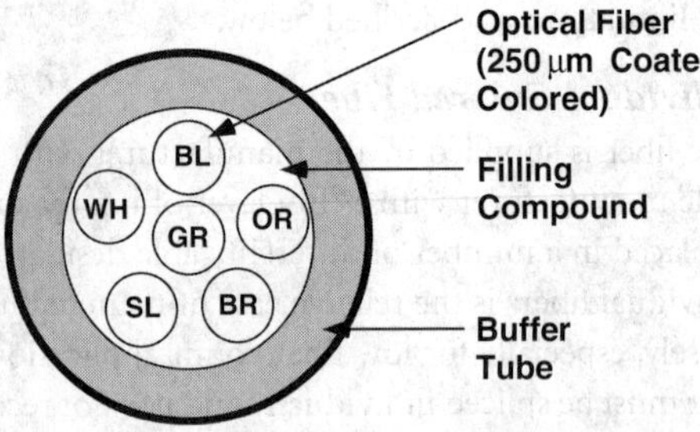

FIGURE 9.28 Cable buffer tubes (loose, ribbon, and bundled fiber). (Courtesy of Corning Cable Systems LLC and Corning® Inc.)

during installation and operation. Buffer tubes may contain just a few fibers and be relatively small in size, or may contain up to a thousand fibers or more in a single central tube. In addition to proving protection to the optical fibers, buffer tubes also serve to aid in the identification of individual fibers in multi-tube designs, and allow for groups of fibers to be protected when routed inside an optical closure or other optical hardware for termination.

In the case where the size of the buffer tube is small and several are located within the same cable core, multiple buffer tubes are stranded in a concentric, helical pattern that reverses direction periodically. The reason for this is to improve the performance of the cable under temperature and tensile load extremes, as well as in relatively tight bends, and to improve mid-span fiber accessibility. For cables with a large number of buffer tubes, the tubes may be arranged in two or more layers using the same type of stranding process for each.

Central tube designs are also fairly prevalent and typically support a smaller outer cable diameter for a given fiber count. In high fiber count ribbon applications, this can be a major benefit such as when the available ducting has a relatively small diameter. For central tube designs housing loose fibers, the identification of individual fibers is aided with the use of colored thread binders that also serve to segregate the fibers into groups, usually in multiples of twelve.

Most inside plant cables do not use buffer tubes, but may instead use subunits to organize the fibers within an overall outer cable sheath. A subunit serves many of the same purposes as a buffer tube except that it does not contain water blocking materials and it is usually made from materials that are chosen primarily for their fire-resistance properties. In some of these unitized designs the subunit itself may be structurally equivalent to a complete indoor cable. This allows the individual subunits to be broken away from the rest of the cable and routed independently, in order to deliver one or more fibers to where they are needed within a building. Such cables are usually called break-out, or fan-out, cables. The latter generally refers to cable in which each subunit contains only one fiber.

Not all cables require the use of buffer tubes in the traditional sense and some emerging designs may be tubeless to keep the overall cable size to a minimum. In these cases a single cable jacket may function as both a buffer tube and a jacket, with the fibers contained inside in a either loose, buffered, or ribbon form. However, such designs have seen only limited use and typically must rely on external structures such as trays, tubes, ducts, or microducts for additional protection.

Strength Members

The strength members of a cable are ultimately what enable it to survive the rigors of installation and the effects of large temperature extremes that might be experienced in various applications. Strength members themselves can take on a variety of forms and be placed in many different locations throughout the cable structure including centrally, in parallel inside the cable jacket, and applied as yarn rovings helically wound around the cable core or inner jacket. The most common strength members are glass-reinforced plastic (GRP) rods, rovings of aramid or fiberglass yarns, and steel wires that may be applied individually or as a unit of stranded steel wires. Even though the various other materials within the cable may contribute to the overall strength of the composite structure, discrete strength elements are usually needed for the purpose of meeting installation and in-service tensile performance requirements.

In stranded tube and stranded unit designs a central GRP or steel rod is typically used. The rod may be coated, called up-jacketing, for proper dimensional sizing to provide mechanical support to the fiber-bearing elements. In some indoor designs a group of yarns may also be used, upjacketed or not, to support the stranding of small units or individual 900 μm buffered fibers. The first layer of buffer tubes or subunits are then stranded around the central member and held in place by binding yarns. This central strength member serves a number of purposes, including acting as the primary tensile strength element, providing a means with which to anchor the cable to any hardware in which it is terminated, providing structural support to the stranded tubes or subunits also provide stiffness to the cable to prevent plastic elements from kinking during placement.

In central tube designs a buffer tube occupies the middle of the cable, so strength members must be placed elsewhere. To ensure adequate tensile strength, rigid GRP or steel rods, or the like, are usually bonded to the overall cable sheath to achieve the necessary tensile properties. These rigid elements cannot simply be placed loosely between the core and sheath as they would be prone to shift during the extreme bending or torsion forces experienced during installation, thereby potentially giving the cable unwieldy handing characteristics. These cable designs usually employ 2–6 rods placed on opposing sides of the cable, inside the sheath. As a result, the finished cable may exhibit preferential bend characteristics similar to those of noncircular cable types such as "Figure-8" designs.

Helically stranded tensile yarns can also provide the primary means of tensile strength for certain cable types, and can also be used to provide increased tensile strength for stranded or central tube designs through the use of dielectric yarns helically wrapped in one or more layers around the core. Tensile yarns are often used in lieu of rigid elements for indoor cables as such designs are required to be more flexible. The yarns, when used, can be held in place by a single binding yarn to aid in the manufacturing process. For low count indoor designs, the yarns may be applied in parallel to the core elements.

In some other designs the central strength element may also organize the fibers. In a slotted core or slotted rod type of cable, the central member is upjacketed with a thick layer of plastic incorporating grooves extruded in the outside. Buffer tubes, fiber bundles, or ribbons can then be placed in the grooves and covered with binders or tape to hold them in place. Other designs such as submarine or OPGW cables employ other materials as discussed later on.

Waterblocking

The effective blocking of water in cables employed in the outside plant is critical in preventing the ingress and migration of water throughout the cable. The presence of water in the cable can result in the crushing of fiber-bearing elements due to freezing effects, or can allow water to migrate into sensitive electronics or splice closures if not properly blocked. Even though optical fibers can generally survive for extended periods exposed to water alone, the long-term effects are not well understood and water may eventually cause the protective fiber coating to delaminate over long periods, especially if high temperatures are present as well.

Historically, cable waterblocking has been effectively achieved via the use of thixotropic gels. While very effective at blocking water, such materials are also useful in isolating fibers from external stresses by increasing the mechanical robustness of the cable. However, the use of such gels can present compatibility issues that affect cable material selection, and are a nuisance for installers terminating cables. They also can add significant weight to the cable itself resulting in shorter achievable installation distances for duct and aerial applications. Still, such designs have proven their mettle with hundreds of thousands of kilometers of such cable types successfully installed and continuing to operate error free for 20 years and beyond in direct buried, duct, and aerial applications.

Newer generation cables are employing gel-free, dry-core technologies to provide the waterblocking performance of the cable, should the protective jacket be compromised. These dry designs rely on the use of super absorbent polymers (SAPs), which absorb the water and swell on contact to keep the water from migrating further into the cable. They are generally applied to interstitial spaces in the core via the use of tapes, strings, and strength yarns. SAPs materials provide the cable with water blocking performance equivalent to the traditional gel designs, so they can be used in any normal outside plant environment. The primary benefit of such materials is that they are considerably less messy to work with than gels and can significantly reduce installation times. The performance of SAPs may be affected by exposure to moisture having a high ionic content, such as brackish water, so special tapes may be required in some cases. For completely gel-free designs, SAPs may be applied in powder, yarn, or other forms to the inside of the buffer tubes.

Other Core Components

Other elements not discussed above may be contained in cables designed for specific applications. These include binding yarns to hold buffer tubes or strength elements in place during manufacturing, filler rods to lend symmetry to certain stranded core designs, copper or coaxial wire for communications or other low-power applications, and additional tapes for improved moisture, shielding or flame resistance characteristics. Most cable designs also incorporate one or two special high tensile strength yarn cords, called ripcords, which can be used to split the outer jacket longitudinally for ready access to the core elements inside.

Sheath Elements & Options
Jacket Materials
In selecting a material for an outdoor cable jacket a manufacturer must choose one that is not only cost-wise, but also one having optimal performance characteristics for the intended environment. The characteristics considered crucial for a fiber optic cable jacket are toughness, tensile strength, heat and chemical resistance, flexibility, and long-term environmental stability. For the best combination of these factors, as well as cost, polyethylene is the material of choice for the vast majority of outside plant cable manufacturers. Various grades of polyethylene such as low, medium, and high density are available for applications requiring specific performance attributes.

Various additives can also be mixed in with the polyethylene resin by the material suppliers to enhance other material properties, as needed, to resist the demands of a specific type of environment. Carbon black (or furnace black) is the most common additive and is used to improve the jacket material's resistance to the effects of solar radiation. Adding carbon black also gives outside plant cables their characteristic black color. Other UV stabilizers in limited use allow for jacket colors other than black but may not be as effective.

Some specialized materials can be used to significantly improve the performance of the jacket in the presence of strong electrical fields. Track-resistant polyethylenes (TRPE) are used to inhibit cable jackets from eroding away when used in aerial applications that require the cable to be installed in areas of high electric field potentials. Tracking is an electrical phenomenon that occurs when a current induced on the surface of the cable jacket is interrupted, resulting in arcing. The current is established when water, from rain or fog, is mixed with contaminants in the air such as dust, pollution, salt spray, etc. The mixture coats the surface of a cable and sets up a conductive path in the presence of an electric field gradient. As the cable begins to dry, the path is interrupted and arcing occurs. The arcing creates grooves (tracking) in the cable jacket, which eventually deepen until the jacket wears through. Tracking is normally not an issue for cables installed near electrical systems with transmission voltages at or below 115 kV, but can occur at lower line voltages depending on the specifics of the installation environment.

Other materials such as alumina trihydrate and magnesium hydroxide can be added to polyethylene to provide some measure of fire resistance. The resultant materials are termed flame-retardant polyethylenes (FRPE). FRPE jacket materials are used in place of standard fire-resistant cable jackets when there are concerns that other compounds would release excessive combustion byproducts that are life-threatening, or which could damage sensitive electronics, if burned. Such materials do not have the same level of fire-resistance as the polyvinyl chloride (PVC) and polyvinylidene fluoride (PVDF) materials commonly used for indoor cable, so there are tradeoffs to consider when looking at the need for fire-resistance. Note that since jacket tracking is effectively caused by localized heating of the jacket material, some TRPEs and FRPEs may have similar material characteristics.

In cases where the polyethylene alone cannot be made to resist certain chemicals, such as hydrocarbon fuels, other materials can be extruded over the outer polyethylene jacket for an additional measure of protection for exposed cable. In these cases the added materials themselves may not be ideally suited for use as a stand-alone cable jacket due to certain material properties, cost, or other considerations. However, when extruded over a base polyethylene outer jacket the overall sheath combination results in composite structure that is resistant to a wide range of operating conditions and environmental concerns. Oversheaths consisting of nylon are available from some manufacturers and can be useful in heavy industrial environments where exposure to known aggressive chemicals is likely. Nylon also has been shown to improve the sheath's resistance to boring insects.

As stated before, the primary consideration for indoor cable is to ensure that the cable does not significantly contribute to the spread of flame and/or the generation of smoke in the event of a fire. Other attributes such as durability, a wide operating temperature range, flexibility, and cost are important as well, though not usually as critical as with outside plant applications. The materials most often used for indoor cables are PVC and PVDF which have excellent fire-resistant properties, but they are comparatively more expensive than those used for outdoor applications and may not perform well at very low temperatures.

Cable Sheaths and Armoring

Outside plant cables are categorized as either duct or armored. Armored cables are typically specified for direct-buried applications when the cable may be subject to attack by gnawing rodents, or if the conditions otherwise warrant the need for additional mechanical durability. Examples of the latter are when the cable is to be buried in rocky soil, or when exposure to gunshot is a concern. Metallic armoring may also be used to simplify the ability to locate buried cable through a process called toning. The major tradeoffs for employing cables with metallic armoring are that they are typically more expensive, and the metallic components will require bonding and grounding for safety everywhere the cable is terminated, or enters a building. By comparison, it is usually more expensive to install a single duct and to populate it with a cable versus just installing a single armored cable.

Several types of armors exist for use based on the needs of the application. Metallic armor tapes are usually corrugated before being formed concentrically around the cable core to allow for better flexibility in handling and installation. Depending on the manufacturing process, the armor may be sealed by bonding the polymer coating, if present, at the overlap. Most cable manufacturers use an electrolytic chrome-coated steel (ECCS) tape coated with a polymer layer to inhibit corrosion. ECCS armor provides a superior overall performance at a lower cost than other armor types. Stainless steel armor, which is sometimes used, is more corrosion resistant than the standard steel tapes, but is significantly more expensive and more difficult to form around the cable core. Copper clad steel tape may also be used as a moisture barrier, or in cables designed specifically for high temperature/steam-resistant applications. Another common moisture barrier option is laminated aluminum polyethylene (LAP).

A popular armoring choice for indoor cable designs is interlocking, or BX, armor. BX armor starts out as a flat strip of aluminum or steel metal which is roll formed to create a "hook" on each side of the tape. The tape is then helically wrapped around a finished cable with successive layers locking into previous layers. The result is a very rugged and highly flexible cable that can be used in applications where exposed cable may require an extra measure of security, or where excessive impact or compressive forces are possible. Because of the relatively high cost associated with their use, such armors are primarily relegated to indoor applications where shorts lengths of cable are exposed to potential damage, such as in a manufacturing plant where cables connecting equipment on the shop floor are exposed to mobile heavy equipment traffic.

For applications requiring all-dielectric cables, but for which the presence of protective armoring is also desired, some manufacturers offer dielectric armored cable designs. The armoring for these types of cables consists of an aramid tape or thick annular layer of aramid yarns wound around the cable core. The yarns may provide some protection against gnawing rodents and shotgun blasts, but these designs have only seen limited use. Protection against rodents in an all dielectric design can also be achieved by increasing the thickness of the standard outside cable jacket to increase the overall cable OD. In these cases the finished cable OD must usually be on the order of one inch or more, which may not be practical for cables with relatively small cores or low fiber counts, or in certain applications.

All-dielectric (duct) cables are used in applications where the additional protection afforded by metallic armoring is not needed or not wanted. Duct cable refers to all-dielectric cable designed primarily for use in a conduit, although such cable may be used in other applications as well. Some such uses are in aerial lashed installations, or in limited direct buried applications where the below-grade environment is more benign and rodent attacks are not a concern. All-dielectric cables may also be preferred in high lightning and high voltage areas in order to eliminate potential safety issues from induced currents.

Other materials may be incorporated into the overall cable sheath, such as shields, moisture barriers, and identification tapes. These materials see fairly limited use because of the minimal benefit/cost return they provide. Specific cable designs may also employ sheath constructions consisting of multiple jackets and layers of armor to improve the cable survivability in extremely harsh environments.

9.2.3   Cable Types

Combining the various attributes discussed to this point enables the construction of a wide range of cable types for various applications. The most common cable designs are described later on and the design of choice for a particular application depends on a number of factors. The primary considerations are fiber count, installation method, and the environment. Specific designs based on the basic constructions discussed (i.e., stranded loose-tube, central tube, and tight buffered), but which are specialized for a particular application, are discussed in Section 9.2.4.

Table 9.7 compares the major attributes of the various basic cable designs.

Stranded Loose Tube Cable

Stranded (buffer tube) loose tube cables are the most common design of fiber optic cable and were also the first ones in widespread use. The primary benefits of these designs are the large operating temperature window, the ability to access fiber within a short length of cable (midspan), and the relatively small size at moderate fiber counts when individual colored fibers are used. These designs can be easily adapted to almost every environment and application including direct buried, duct, aerial lashed, aerial self-supporting, and indoor. Common stranded loose tube cable types are illustrated in Fig 9.29.

Single Tube

The primary benefit of single tube designs is that they can have the smallest outer diameter size for a given fiber count. At counts above 12, the fibers must be grouped into ribbons or bundled together with colored threads to support the identification of individual fibers. Typical single tube cables are shown in Fig 9.30.

TABLE 9.7   General Comparison of Cable Design Attributes

	Individual Colored Fibers		Ribbon fiber		
Cable Attribute	Stranded Tube	Central Tube	Loose Tube	Central Tube	Buffered Fiber
Operating temperature range	Best	Good	Good	Good	Limited
High fiber count density (≥ 144)	Poor	Good	Good	Best	Poor
Moderate fiber count density (24 to 144)	Good	Good	Poor	Good	Poor
Low fiber count density (≤ 24)	Good	Best	Poor	Poor	Good
Midspan	Best	Good	Best	Good	NA
Splicing cost	Fair	Good	Best	Best	Poor
Fire resistance	Limited	Limited	Limited	Limited	Best
Direct termination	Fair	Fair	Fair	Fair	Best

Note: The intent of this table is to provide a simple comparison for various basic cable constructions. Special cable designs, including those employing exotic materials, can show an improvement over standard products in one or more categories

Buffered Fiber Cables

Tight buffered cables were the first generation of premises cables. The name 'tight buffered' is derived from the layer of thermoplastic or elastomeric material that is applied directly over the fiber coating as discussed earlier. The buffer coating on the fiber makes it easier to handle and supports direct termination with fiber optic connectors (Fig 9.31).

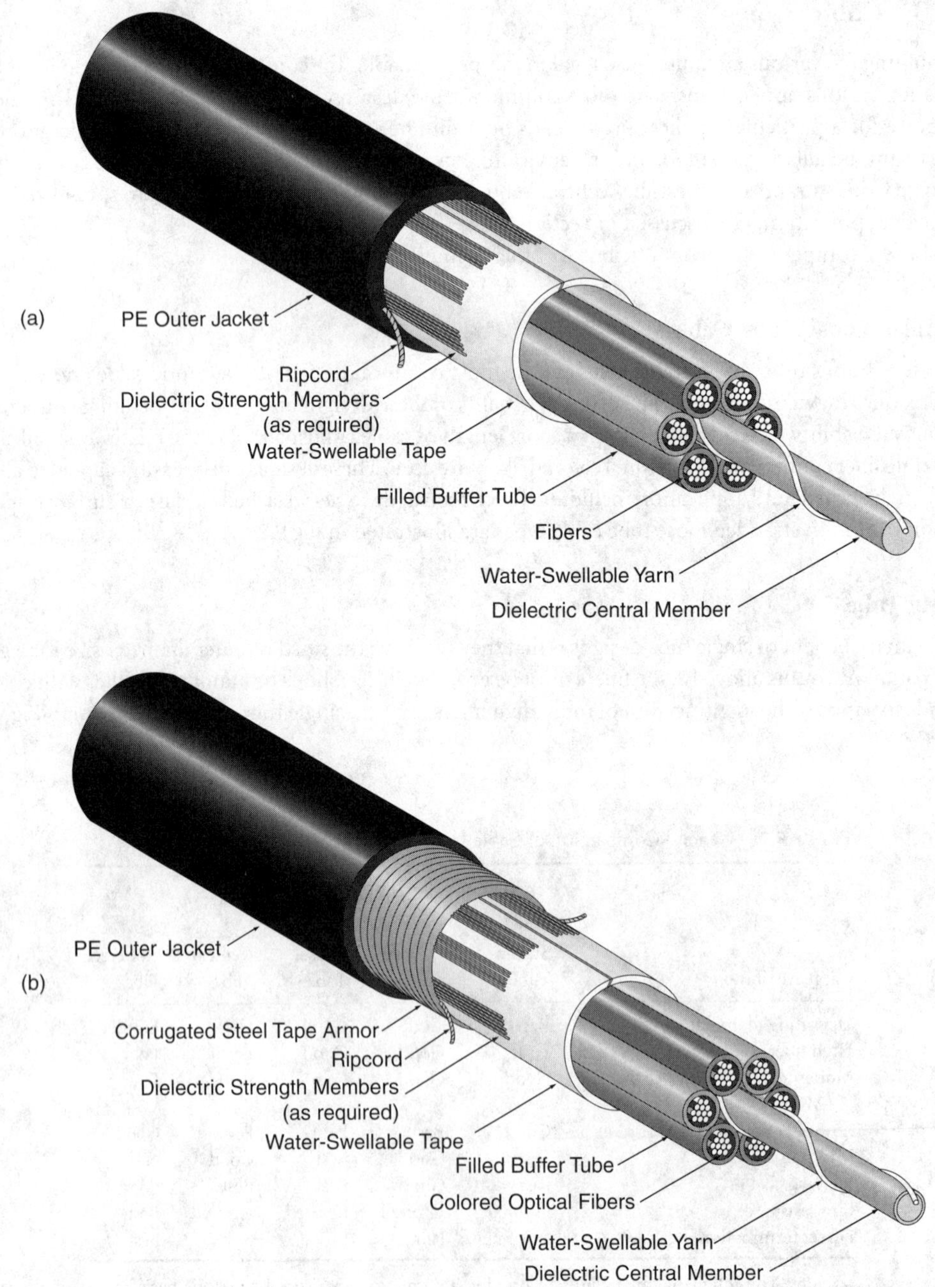

(a)
PE Outer Jacket
Ripcord
Dielectric Strength Members
(as required)
Water-Swellable Tape
Filled Buffer Tube
Fibers
Water-Swellable Yarn
Dielectric Central Member

(b)
PE Outer Jacket
Corrugated Steel Tape Armor
Ripcord
Dielectric Strength Members
(as required)
Water-Swellable Tape
Filled Buffer Tube
Colored Optical Fibers
Water-Swellable Yarn
Dielectric Central Member

FIGURE 9.29 Stranded loose tube cable examples; (a) ALTOS®, (b) ALTOS® Lite, (c) ALTOS® LST, (d) ALTOS® Ribbon. (Courtesy of Corning Cable Systems LLC and Corning® Inc.)

(c)

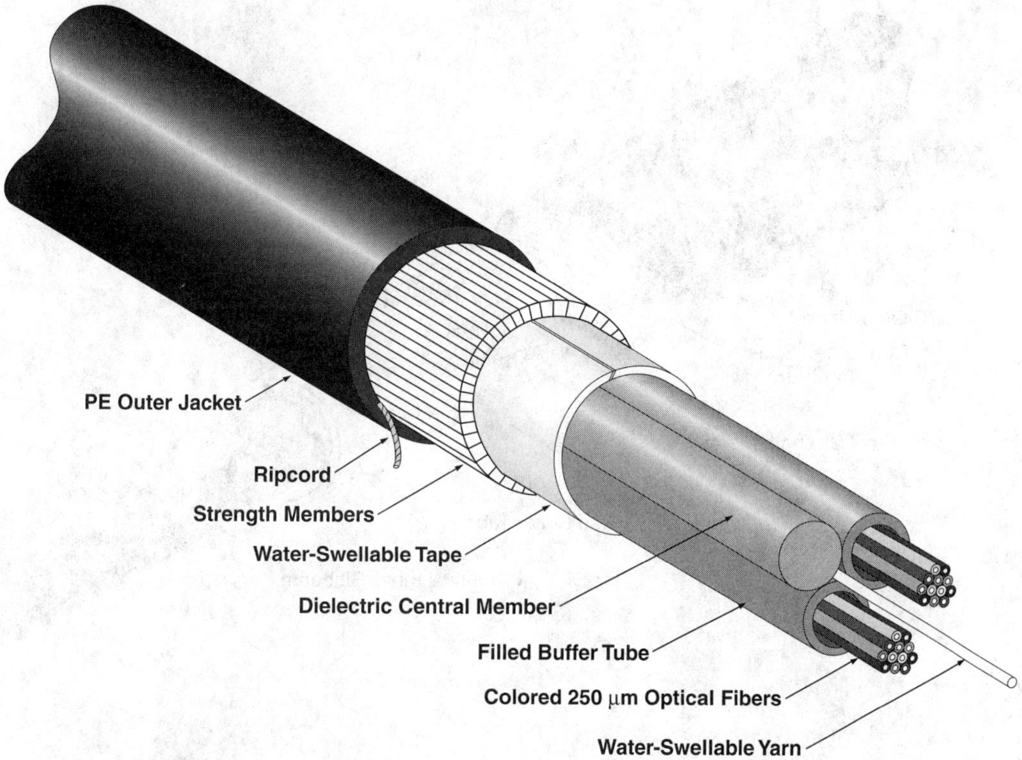

PE Outer Jacket

Ripcord

Strength Members

Water-Swellable Tape

Dielectric Central Member

Filled Buffer Tube

Colored 250 μm Optical Fibers

Water-Swellable Yarn

(d)

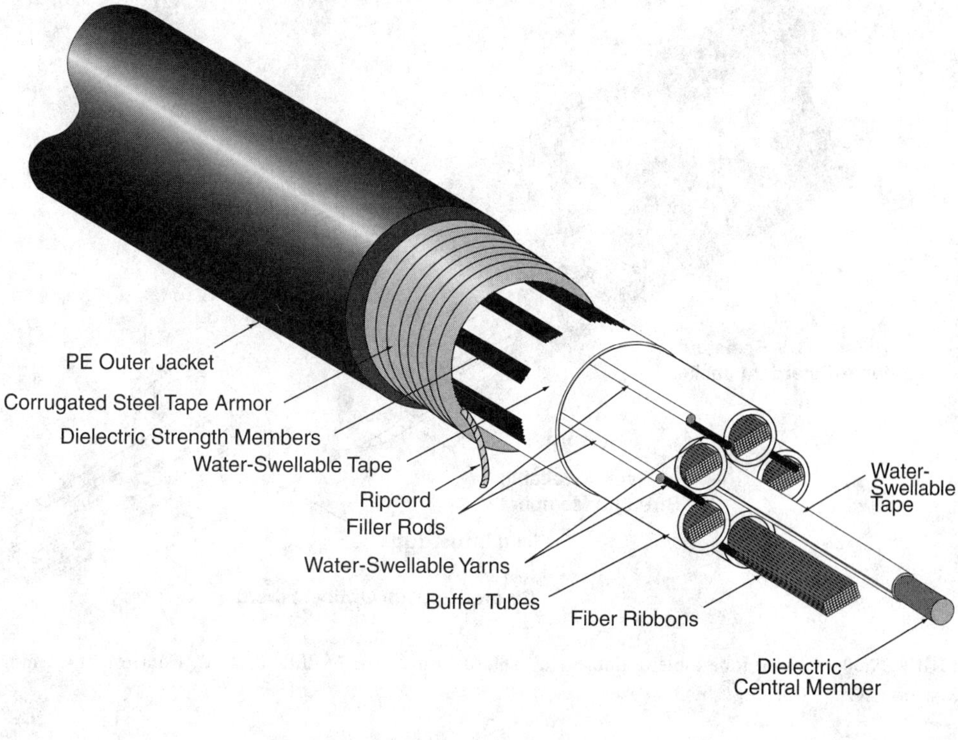

PE Outer Jacket

Corrugated Steel Tape Armor

Dielectric Strength Members

Water-Swellable Tape

Ripcord

Filler Rods

Water-Swellable Yarns

Buffer Tubes

Fiber Ribbons

Water-Swellable Tape

Dielectric Central Member

FIGURE 9.29   (*Continued*).

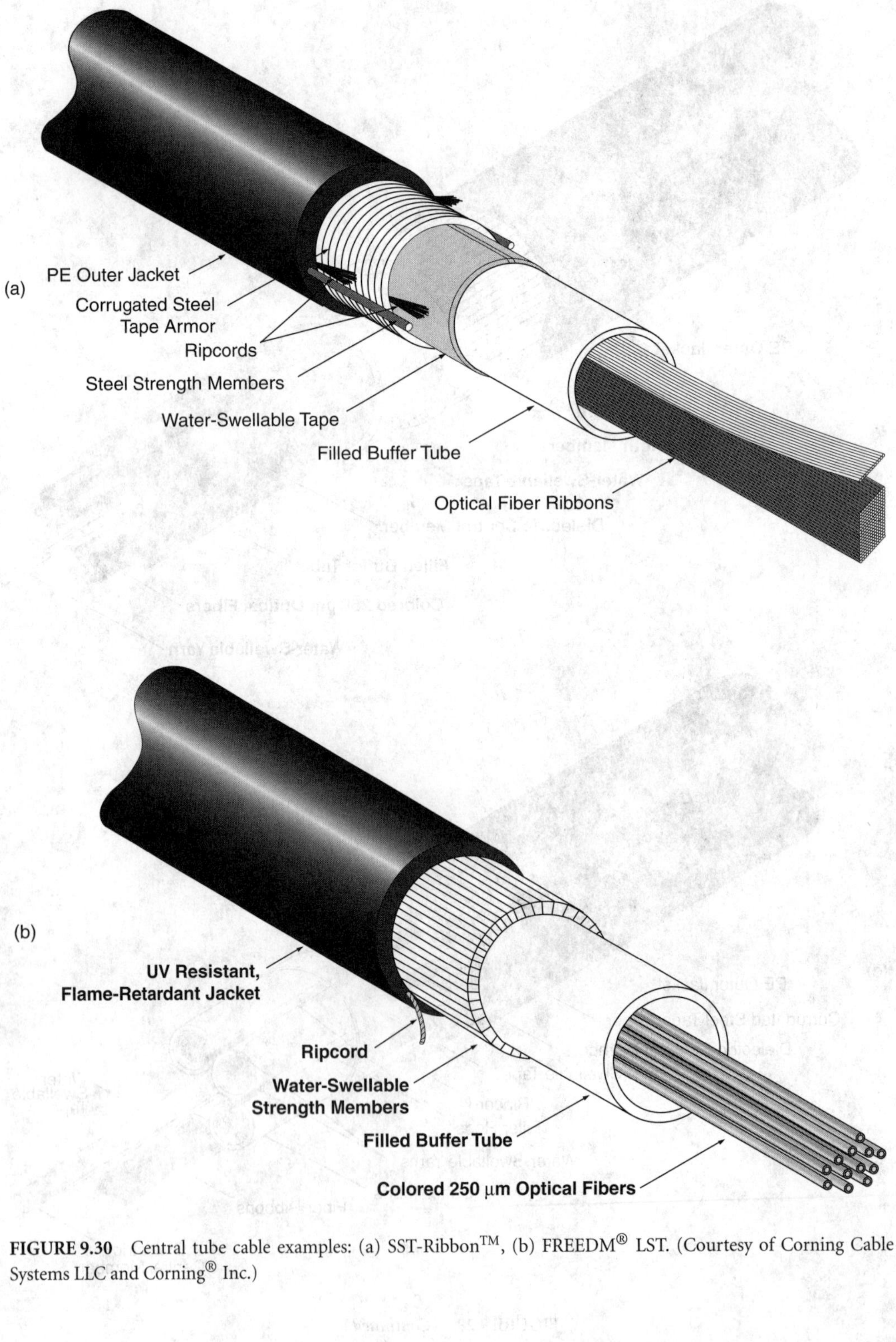

(a)

PE Outer Jacket

Corrugated Steel
Tape Armor

Ripcords

Steel Strength Members

Water-Swellable Tape

Filled Buffer Tube

Optical Fiber Ribbons

(b)

UV Resistant,
Flame-Retardant Jacket

Ripcord

Water-Swellable
Strength Members

Filled Buffer Tube

Colored 250 μm Optical Fibers

FIGURE 9.30 Central tube cable examples: (a) SST-Ribbon™, (b) FREEDM® LST. (Courtesy of Corning Cable Systems LLC and Corning® Inc.)

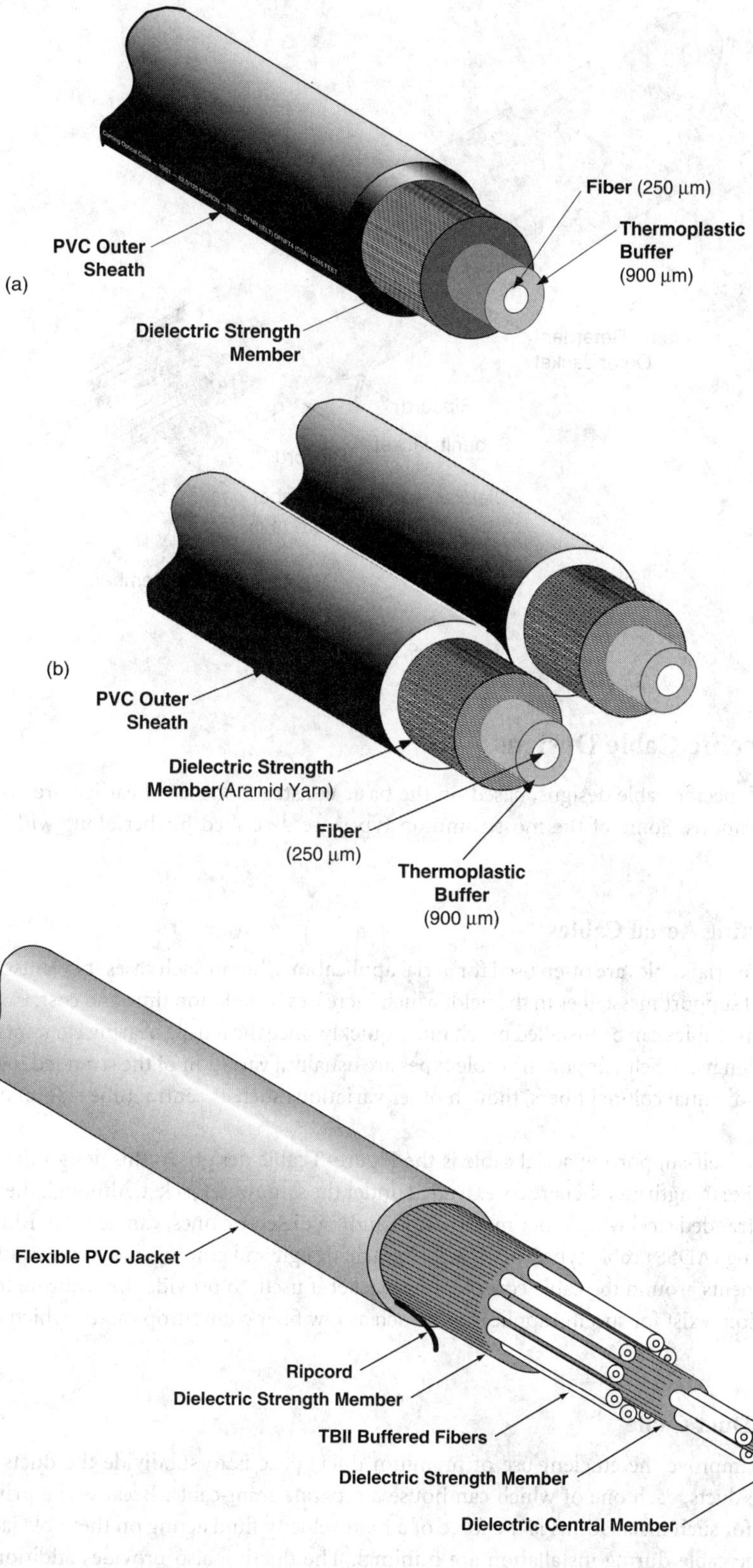

FIGURE 9.31 Fire-resistant buffered fiber cable examples: (a) SFC, (b) ZIP, (c) MIC®, (d) UMIC. (Courtesy of Corning Cable Systems LLC and Corning® Inc.)

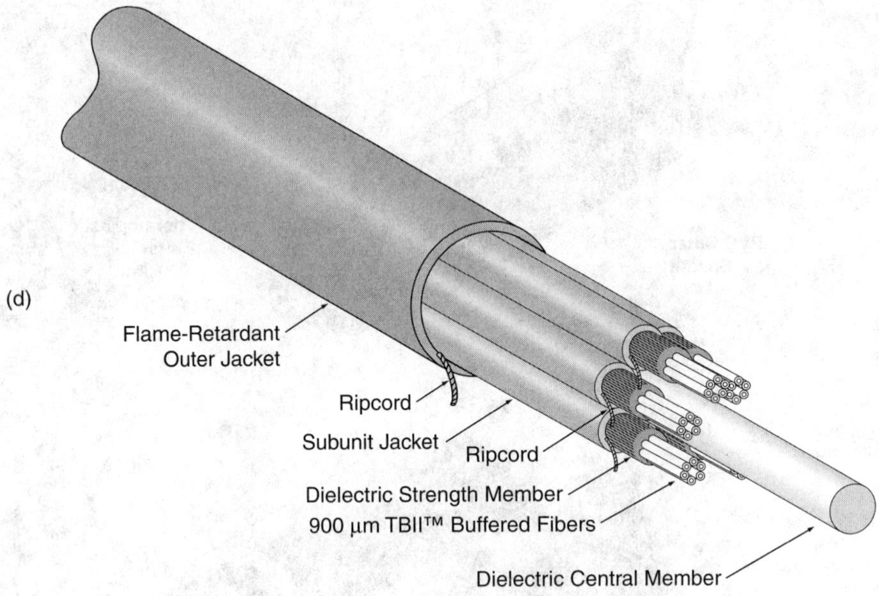

(d)

Flame-Retardant
Outer Jacket

Ripcord

Subunit Jacket
                    Ripcord

Dielectric Strength Member
900 µm TBII™ Buffered Fibers

Dielectric Central Member

FIGURE 9.31   (*Continued*).

9.2.4   Specific Cable Designs

A number of specific cable designs, based on the basic constructions listed earlier, are also available for certain applications. Some of the most common types are described further along with their potential uses.

Self-Supporting Aerial Cables

Standard terrestrial cables are often used for aerial applications, but in such cases they must be lashed to an existing aerial support messenger in the field, which increases installation time and cost. Conversely, aerial self-supporting cables can be installed much more quickly since the tensile bearing elements are integral to the cable design itself. Self-supporting cable types are usually a variation of the stranded loose tube design containing individual colored fibers, though other variations such as central tube ribbon designs may see limited use.

One type of self-supporting aerial cable is the Figure-8 cable design. In this design the cable core and separate tensile strength member are co-extruded under the same outer jacket. Although the tensile element is usually a stranded steel wire, other materials, including dielectric ones, can be used. Most all-dielectric self-supporting (ADSS) cable types are of a concentric design and employ one or more annular layers of strength elements around the cable core, or inner jacket if used, to provide the requisite tensile strength. Other variations exist for specific applications, such as low fiber count drop cables, which may use a GRP rods.

Miniature Duct Cable

One way to improve the efficient use of premium duct space is to subdivide the ducts using bundles of miniature ducts, each one of which can house a separate mini-cable. Because the primary source of locomotion for such installations is the force of a high velocity fluid acting on the cable jacket, the forces placed on the cable during installation are minimal. The ducting also provides additional mechanical protection for individual cables and isolates them from one another. For such applications, the size of the cable is more important than its ultimate mechanical durability, so resistance to tensile forces and other high mechanical loads, such as impact or compression, can be traded off somewhat to achieve an overall smaller cable cross section (Fig. 9.33).

Road Cable

To reduce the issues associated with installing cable in the last mile of all-optical networks, especially in densely populated metropolitan-type environments, alternative deployment methods are being developed.

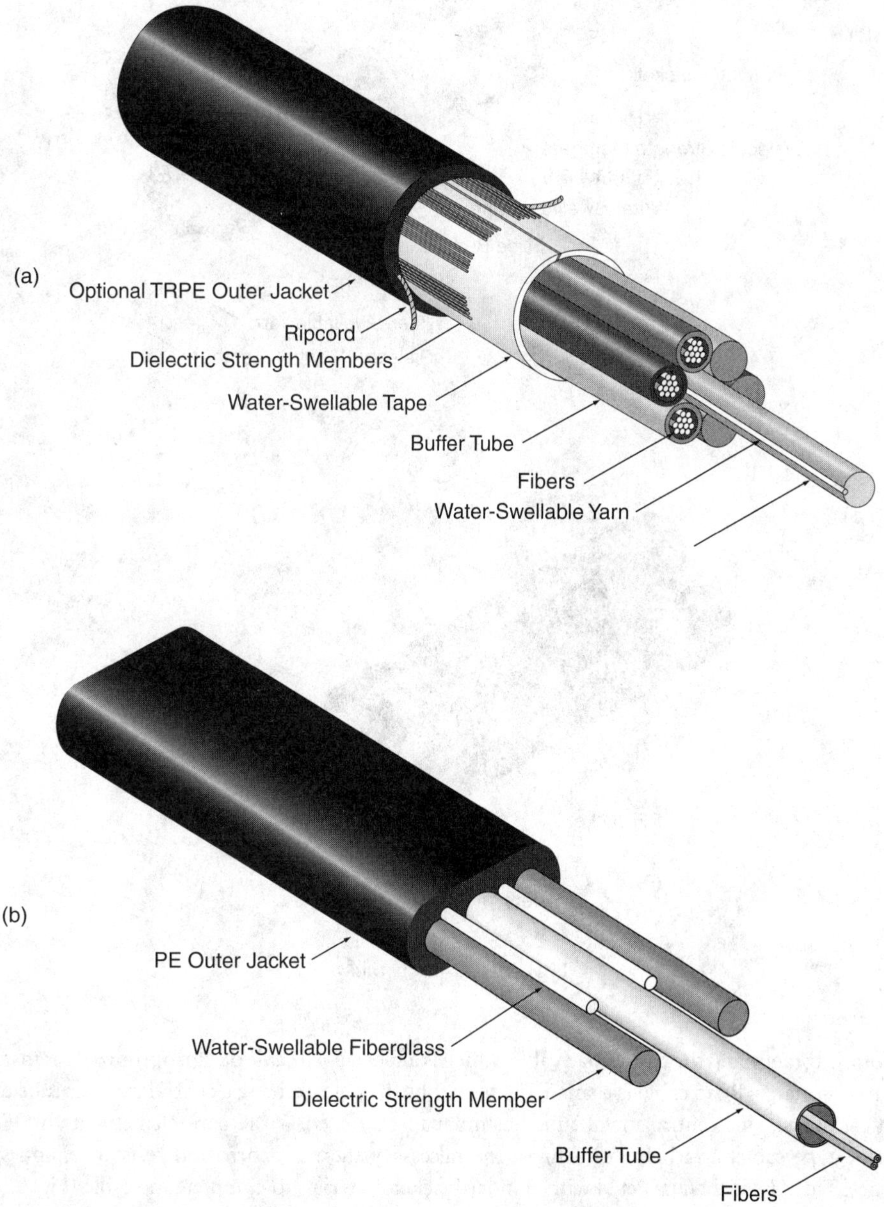

(a)
Optional TRPE Outer Jacket
Ripcord
Dielectric Strength Members
Water-Swellable Tape
Buffer Tube
Fibers
Water-Swellable Yarn

(b)
PE Outer Jacket
Water-Swellable Fiberglass
Dielectric Strength Member
Buffer Tube
Fibers

FIGURE 9.32 Aerial self-supporting loose tube cable examples: (a) Solo® short or medium, (b) SST (dielectric) Drop, (c) ALTOS® (d) SST Figure-8 Drop. (Courtesy of Corning Cable Systems LLC and Corning® Inc.)

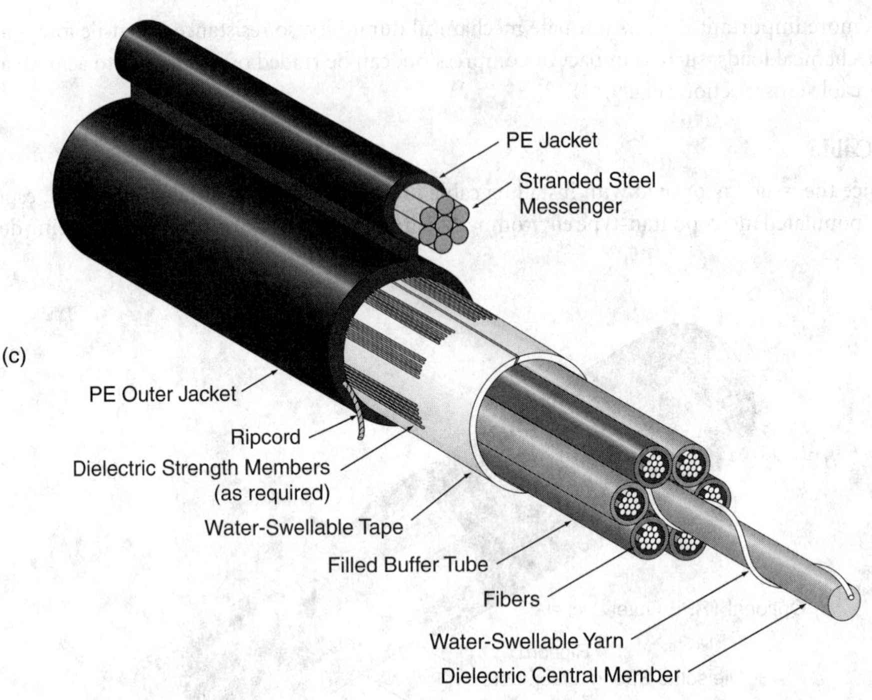

PE Jacket
Stranded Steel
Messenger

(c)

PE Outer Jacket
Ripcord
Dielectric Strength Members
(as required)
Water-Swellable Tape
Filled Buffer Tube
Fibers
Water-Swellable Yarn
Dielectric Central Member

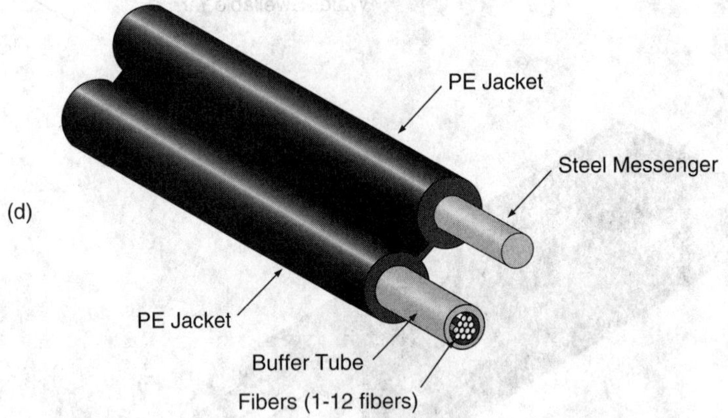

PE Jacket

Steel Messenger

(d)

PE Jacket

Buffer Tube
Fibers (1-12 fibers)

FIGURE 9.32   (*Continued*).

One promising method is the use of road cable, which is a specialized cable design optimized for installation within road surfaces. Road cable is a small, but very robust, cable utilizing a central metallic tube to house the optical fibers. This central metal tube is similar to that used as the core element of an OPGW or submarine type cable, and helps to achieve the necessary thermal, corrosion resistance, and strength properties. The tube is then jacketed with standard jacket materials to complete the cable (Fig. 9.34).

Ribbon Interconnect Cable

In addition to the standard buffered type fibers used in indoor systems, variations of loose tube and ribbon cable designs are also employed for certain applications. One design seeing widespread use is

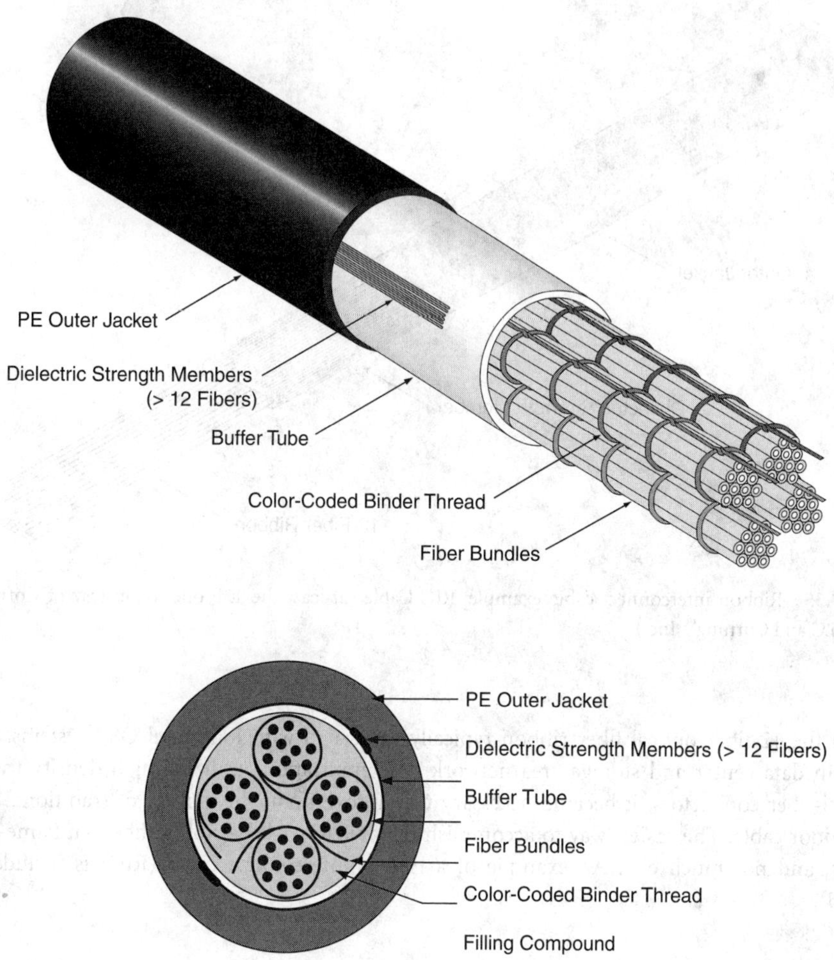

PE Outer Jacket

Dielectric Strength Members
(> 12 Fibers)

Buffer Tube

Color-Coded Binder Thread

Fiber Bundles

PE Outer Jacket

Dielectric Strength Members (> 12 Fibers)

Buffer Tube

Fiber Bundles

Color-Coded Binder Thread

Filling Compound

FIGURE 9.33 Microcable examples: SST Optimizer™. (Courtesy of Corning Cable Systems LLC and Corning® Inc.)

Outer Jacket

Copper Central Tube

Colored 250 µm Optical Fibers

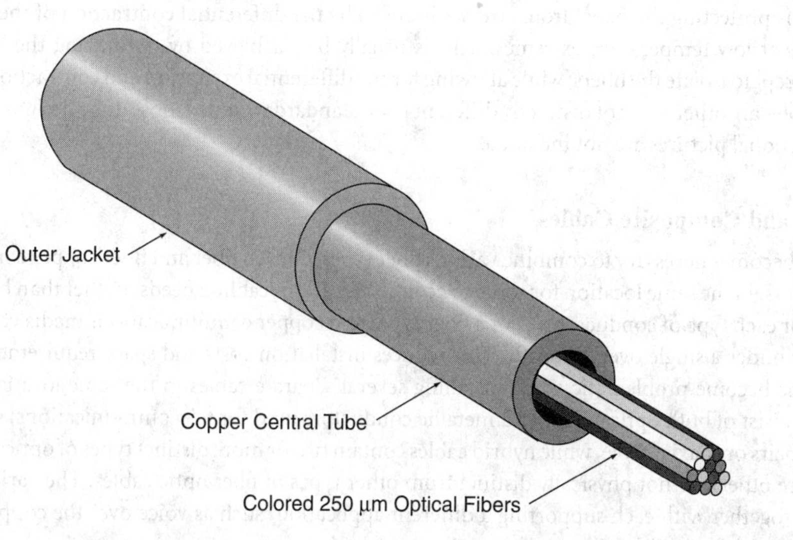

FIGURE 9.34 Road cable example: MCS® Road. (Courtesy of Corning Cable Systems LLC and Corning® Inc.)

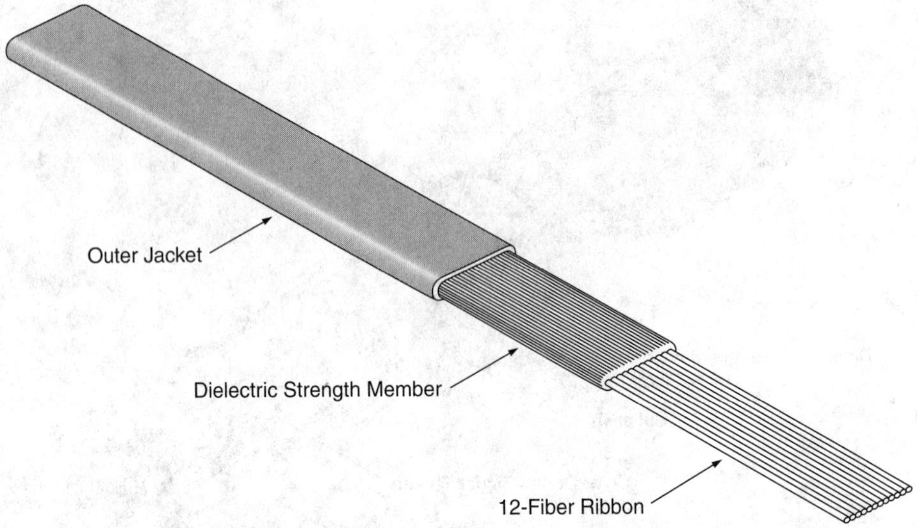

FIGURE 9.35   Ribbon interconnect cable example: RIC Cable (at least the 12F one. (Courtesy of Corning Cable Systems LLC and Corning® Inc.)

based on the 12-fiber optical fiber ribbon typically associated with long-haul OSP systems. With the increase in data center and storage area network type applications utilizing high density transceivers and multi-fiber connectors, it became necessary to improve upon the density of traditional low-fiber count indoor cable. The easiest way to accomplish this was to package ribbon fibers in flame-retardant materials, and not much else. An example of a ribbon interconnect cable (RIC) is included herein (Fig. 9.35).

Very Low Temperature Cable

To support the deployment of fiber networks in very cold environments (-55 to $-60°C$), modifications of standard cable designs are typically needed to improve their low temperature performance. The primary concern is protecting the fibers from stresses induced by the differential contraction of the various cable materials at low temperature extremes. This is usually best achieved by optimizing the basic stranded core concept to isolate the fibers, while allowing for the differential expansion and contraction forces. Since these cables are otherwise not distinctly different from standard stranded loose-tube designs representative cross-sectional pictures are not included.

Hybrid and Composite Cables

Often it becomes necessary to combine various fiber types, or even fiber and fiber copper communications conductors, in the same location for some telecommunications cabling needs. Rather than having separate cables for each type of conductor, various fiber types and copper communications media can be packaged together under a single overall sheath. This reduces installation costs and space requirements that could otherwise become problematic when installing several separate cables in the same location. Composite cables consist of both optical fiber and metallic conductors used for telecommunications, such as copper twisted pairs or coaxial cable, while hybrid cables contain two or more distinct types of optical fiber. Hybrid cables are otherwise not physically distinct from other types of fiber optic cables. The various media can be used together with each supporting a different application, such as voice over the copper and data or video over the fiber (Fig. 9.36).

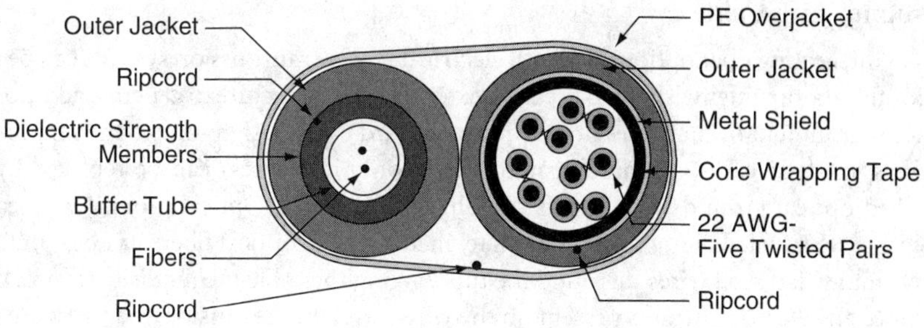

FIGURE 9.36   Composite cable example. (Courtesy of Corning Cable Systems LLC and Corning® Inc.)

Submarine Cable

Fiber optic cables connect every continent on earth (except Antarctica) in the form of submarine cables. These cables are specially designed for the rigors of a submarine environment such as high pressures, strong currents, biological attacks from sharks and other predators, and boating hazards such as anchors or trawler nets. The basis for the core cable is a hermetically sealed metallic tube that provides the primary protection for the fibers. The most common submarine cables incorporate one or more layers of steel rods around the core cable which are then flooded with an asphaltic compound for additional protection. This steel rod armoring increases cable strength, provides mechanical protection, and ensures the cable is sufficiently negatively buoyant such that it stays on the bottom of the ocean. In coastal applications, the cable may also be buried for additional security.

A variation of the submarine cable used for shallow water type applications on land uses a standard terrestrial cable design as the core cable instead of a steel tube. These cables are usually adequate for depths of a few hundred feet or less, and are typically much less costly than true transoceanic submarine cables (Fig. 9.37).

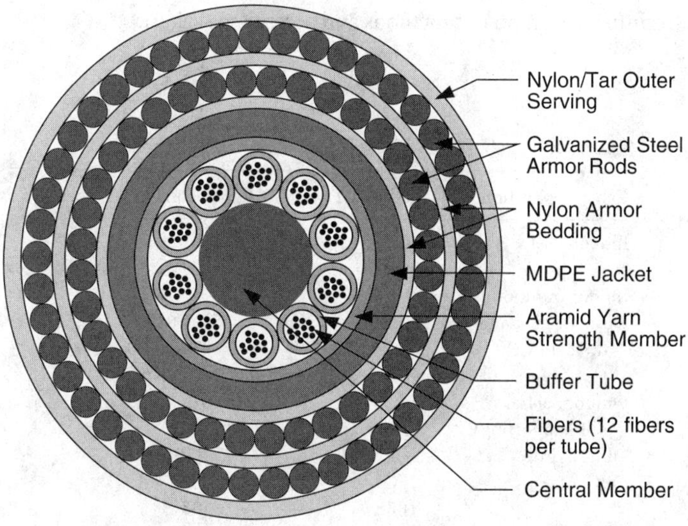

FIGURE 9.37   Submarine cable example. (Courtesy of Corning Cable Systems LLC and Corning® Inc.)

Optical Ground Wire Cable

The concept of integrating fiber optic cables with electrical power transmission systems can be attractive to some electric utility companies wishing to use their extensive infrastructure to generate additional revenue. However, the environment of such does not support standard cable designs due to the very long distances between towers. In these cases, standard ground wires (or shield wires) can be replaced by composite structures called optical ground wire, which contain optical fibers. Optical ground wire cable designs typically incorporate a sealed aluminum tube roll formed over the optical fibers or fiber unit. Aluminum alloy and aluminum clad steel wires surround the tube, giving the cable the requisite electrical and tensile properties. Since any electric currents present on the wire travel on the outside, the fibers are not affected by the presence of high electric field potentials unless overheated.

9.2.5 Fiber Optic Cable Standards and Specifications

Standards and specifications exist for fiber optic cables for the same reasons they exist for optical fiber. Key industry standards and specifications for fiber optic cable are addressed herein.

Fiber Optic Cable Standards Publications

Following are some of the key telecommunications industry fiber optic cable standards employed today for the various fiber optic cable types. For the sake of brevity, the standards are identified only by document number, the publishing organization, and the respective cable types or applications. Users of industry standards should purchase these from a reliable source to ensure they have the latest applicable version.

Cable Attribute Specifications

Each optical cable design normally undergoes a series of mechanical and environmental tests to ensure the cable performs as designed once deployed in the field. The tests are designed to simulate the installation, operating, and environmental conditions the cable will experience over its lifetime. Table 9.9 lists several tests typically performed on each cable design. In most tests, the acceptance criteria usually consist of attenuation measurements, visual inspections, or both.

As with optical fiber, various industry cable standards specify common performance attributes and each attribute has an associated test procedure, not listed here. Refer to the individual cable standards listed in Table 9.8 for detailed information on test methods and acceptance criteria.

TABLE 9.8 Fiber Optic Cable Detailed Industry Standards

Cable Application	Industry Forum	Document Number
Indoor	ICEA[a]	S-83-596
	IEC	60794-2 Series
Indoor-outdoor	ICEA	S-104-696
Outdoor duct and	ICEA	S-87-640
direct buried	IEC	60794-3-10
	RUS[b]	7CFR 1755.900
Outdoor aerial	IEC	60794-3-20
self-supporting	IEEE[c]	P1222
Optical drop	ICEA	S-110-717
OPGW	IEC	60794-4
	IEEE	1138

[a]ICEA is the Insulated Cable Engineers Association.
[b]RUS is the U.S. Department of Agriculture Rural Utility Service.
[c]IEEE is the Institute of Electrical and Electronics Engineers, Inc.

TABLE 9.9   Key Industry Specified Fiber Optic Cable Attributes

Attribute	Outdoor Aerial	Outdoor Buried or Duct	Indoor-Outdoor	Indoor
Temperature cycling	X	X	X	X
Aging	X	X	X	
Freezing	X	X	X	
Tensile strength	X	X	X	X
Impact resistance	X	X	X	X
Crush resistance	X	X	X	X
Cyclic flex	X	X	X	X
Twist	X	X	X	X
Bend (low and high temperature)	X	X	X	X
UV resistance	X	X	X	
Water penetration	X	X	X	
Flammability			X	X
Aeolian vibration	Aerial self-support only			
Galloping	Aerial self-support only			
Lightning	Conductive only	Conductive only	Conductive only	
Color coding	X	X	X	X
Identification and length marking	X	X	X	X
Packaging requirements	X	X	X	X
Optical attenuation	X	X	X	X

The following are the key specified performance attributes for the various fiber optic cable types. Other secondary parameters are also specified, while some are only characterized (i.e., recommended or typical).

Other Industry Cable Standards

Installation

The *National Electrical Code*®* (NEC®*) is a document issued every three years by the National Fire Protection Association®*. The document specifies the requirements for the safety of premises wire and cable installations, and also sets requirements to minimize the risk and spread of fire in commercial buildings and individual dwellings. Section 770 of the NEC defines the requirements for fiber optic cables. Although the NEC is advisory in nature, its content is generally adopted by local building authorities as Code.

The *National Electrical Safety Code*®** (NESC®**) addresses the safety of persons and equipment during the installation, operation, or maintenance of electric supply and communication lines, and the associated equipment. The NESC contains the basic guidance considered necessary for the safety of both the installers and the general public as a result of such installations. For example, the NESC contains information of the various environmental conditions that aerial cable plants may be exposed to in the different geographic regions of the U.S., which must be accounted for to ensure their survivability.

Testing

Each optical cable design normally undergoes a series of mechanical and environmental tests to ensure the cable performs as designed once deployed in the field. The tests are designed to simulate the installation, operating, and environmental conditions the cable will experience over its lifetime. Table 9.8 lists several applicable industry standards, while Table 9.9 lists tests typically performed on each cable design. Standard optical, mechanical, and environmental test procedures are defined in industry standard fiber optic test procedures (FOTPs). FOTPs are published as the TIA/EIA-455 series of documents. Each FOTP provides

---

* - Registered trademarks of the National Fire Protection Association, Inc.

** - Registered trademarks of the Institute of Electrical and Electronics Engineers, Inc.

TABLE 9.10   Fiber Optic Cable Color Code

Position No.[a,b,c]	Unit or Fiber Identifier		Position No.	Unit or Fiber Identifier	
1	1	Blue BL	7	7	Red RD
2	2	Orange OR	8	8	Black BK
3	3	Green GR	9	9	Yellow YL
4	4	Brown BR	10	10	Violet VI
5	5	Slate SL	11	11	Rose RS
6	6	White WH	12	12	Aqua AQ

[a]For positions 13–24, a stripe or dash is added.
[b]For positions 23–36, a double stripe or double dash is added.
[c]For ribbon, an appropriate unit identifier is printed on the ribbon matrix.

the information needed to conduct testing, including equipment and measurement specifications, and test set-ups, but they do not generally include acceptance criteria, which are instead listed in the various detailed industry product standards. The TIA also publishes a number of different standards that relate directly to certain applications, such TIA-568, "Commercial Building Telecommunications Cabling Standard." The TIA also develops optical fiber system level test procedures (OFSTPs) that are published as the TIA-526 series of documents.

Identification

The most widely referenced standard for identification by means of color coding is EIA/TIA-598, "Color Coding of Fiber Optic Cables." This document outlines the standard color coding and identification sequence used for fiber and fiber units, such as buffer tubes or binding threads. The twelve primary colors used and their sequence are listed in Table 9.10.

9.2.6   Optical Fiber Connectivity

Because of the distances involved or for reasons of complexity, fiber optic cables are almost never installed in continuous point-to-point links from transmitter to receiver. Optical fibers must be terminated at desired locations, spliced into long continuous spans, and connected to opto-electronic hardware or each other to support the construction of the passive optical system. The fibers are joined by one of two methods, splicing or connectorizing. In either case the termination must be protected within fiber optic hardware.

Splicing

Splicing is the joining of two separate fibers to create a continuous length of fiber. The fibers are aligned by one of several methods and then joined together physically and optically by one of two basic categories of splicing, mechanical or fusion.

In mechanical splicing, the ends of two separate fibers are physically held in place and aligned with the end faces of the fibers butted together. An index matching gel may occupy the space between the two fibers to minimize the amount of light reflected at the interface. Some light is reflected because the ends of the fibers are not in perfected alignment, so even though there is physical contact an air gap may exist if not for the gel. The internal components of the mechanical splice body must be precisely designed, manufactured, and assembled to ensure that the fiber cores are properly aligned, in order to minimize power loss due to a lateral core offset at the splice point. Mechanical splicing has relatively few applications as the performance is not as good as fusion splicing. It is primarily used in laboratories for fibers undergoing optical testing or in the emergency restoration of damaged sections of cable as a temporary fix until the system can be permanently repaired using fusion splicing techniques.

Fusion splicing is the permanent joining of two fibers by physically fusing them together to form a continuous length of glass fiber. In fusion splicing, the fibers are heated to their softening point of about 2000°C. The heat source is a pair of electrodes that create an electric arc to melt the ends of the optical fibers, which are then mated in the softened phase. After the electrical heat source is removed, the glass quickly cools and solidifies. Fusion splicing has significant benefits over mechanical splicing and is, therefore, used almost exclusively in outside plant, long-haul telecommunications systems to join fibers. Since no air gap is present in a completed fusion splice the amount of reflected light lost due to refractive index differences is inherently minimal, so index matching gel is not needed. Most importantly, fusion splicers can incorporate a variety of active alignment techniques to optimize the splice characteristics to improve the overall splice loss and to reduce reflected light, which can degrade the optical system's performance. Some fusion splicers can intentionally create precise high loss values, such as might be needed to attenuate signals that are too high for sensitive electronics in the receiver.

The primary methods of active fiber alignment employed by fusion splicers are profile and power through. In profile alignment systems (PAS), the splicer uses LEDs to light each fiber end from two sides and the profiles are then detected by cameras placed opposite the LEDs. The splicer then aligns the fiber core or cladding positions relative to each other, prior to initiating the fusion process. In power-through systems, or light injection/detection (LID®) systems, light is injected into one fiber and the power through the joint is continuously monitored by a receiver reading the other fiber. The splicer adjusts the alignment of the fibers in two or three axes to maximize the light coupled between them. The fuse time can be automatically optimized by the fusion splicer based on the prevailing environmental conditions to ensure that the minimum splice loss is achieved. In splicers where the power through is continuously monitored throughout, the splicing process is stopped when the loss is at its lowest level, as determined by a fitted curve representing the change in the power through value over time (Fig. 9.38).

Most splicers in use today can only splice one pair of fibers at a time, which can result in long installation times and increases the associated costs for high fiber count systems. For systems employing ribbon fiber cables, however, entire ribbons can be spliced together at once. Mass fusion splicers can simultaneously splice from 2 to 24 fibers, with 36 fiber splicers becoming available in the near future. Unlike with single fiber splicing, mass fusion splicers are not currently available with active alignment systems because of the inherent complexities involved in the simultaneous alignment and splicing of multiple fiber pairs. As a result, mass fusion splicers use metal or rigid polymer plates which have a series of parallel V-shaped grooves precisely machined into the top. The spacing of the V-grooves is roughly equal to the spacing

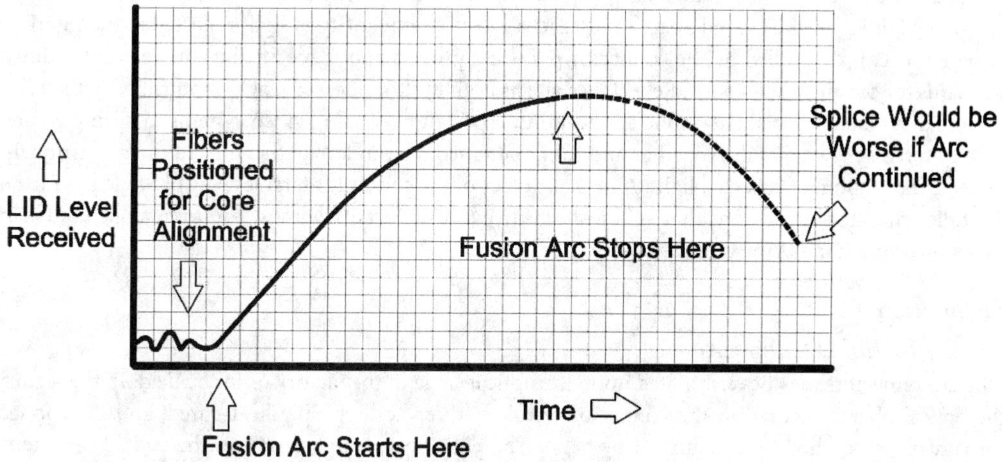

FIGURE 9.38 Splice time vs. optical loss curve. (Courtesy of Corning Cable Systems LLC and Corning® Inc.)

of the fibers in the ribbon matrix to simplify the fusion process and minimize alignment issues. Average splice loss values are slightly higher for mass fusion splices because the alignment is highly dependent on the cleanliness of the V-grooves and the glass geometries of the individual fibers. Also, special tooling is needed to strip and cleave the ribbonized fibers, increasing the start-up costs. Still, the benefits of mass fusion splicing for high fiber count systems are worth the added expense, as the overall cost to install a system can be significantly reduced due to the time saved in splicing alone.

Fiber Optic Connectors

Fiber optic connectors are devices that are installed (terminated) upon fiber ends to allow for the interconnection and remating of optical fibers. The ferrule of each connector houses the fiber and is the specific part of the connector aligned in a connector pair. When two connectors are mated, the fiber endfaces are mechanically aligned and held in physical contact to achieve the requisite **insertion loss** and reflectance performance. The connector also allows the fiber to be easily mated into the input and output ports of transceivers.

Fiber optic connectors are designed to be disconnected and reconnected easily by hand, which is an important feature in systems that are often reconfigured based on the needs of the time, such as in premises-type applications. They are available in a variety of designs and commonly employ a spring-loaded mechanism to establish and maintain the physical contact between the two connector end-faces. Most connectors also incorporate a keying scheme to ensure that the fiber end faces are consistently mated in the same position. While the majority of connectors are designed for only one fiber, some designs can house up to 24 optical fibers. The most common fiber optic connector designs in use are displayed in Fig. 9.39.

Connector Alignment and Mating

The alignment of the fibers in a connector pair is dependent upon the geometry of the optical fiber, the connector ferrules, and the sleeve in the adapter used to align the two connector ferrules. To aid in fiber alignment, some connector ferrules are tuned so as to align any offsets in the optical fiber position in the same direction with every connector, typically to the 0 or 180° relative position. This minimizes the light loss between two randomly mated connectors. The loss of optical power across a connector pair is called **insertion loss** and it is the most common parameter measured in evaluating the expected system performance.

Because index matching gels are not used to mate two connectors the joint may have an air gap that can create reflections. If large enough, these reflections can degrade system performance by a couple of methods. The reflected light can affect the operation of the transmitter, or can reduce the optical signal margin by adding to the overall system noise. As a result, the quantity of light reflected at the interface, or reflectance, is controlled by the connector polishing process and is specified in the relevant industry standards for certain applications. Reflectance affects both multimode and single-mode fiber systems, but is more of a concern for single-mode applications, especially in analog video systems. A spring is often used to force the two ferrules together and to maintain contact in a mated connector pair, which helps mitigate the amount of reflected light. Also the endface of the connector ferrules may be angled to further limit the reflectance in systems that are most sensitive to its effects. The most common angle used is 8°, which is common in CATV applications.

Installation

Factory Installed Connectors

Optical connectors may be terminated upon the optical cable in the factory or in the field. In the factory, the optical fiber is secured to the connector ferrule with epoxy, typically heat-cured, and the connector endface is polished by machine or by hand. The polishing process produces the endface geometry, (radius of curvature, apex offset, and fiber undercut) that is needed to ensure good optical performance (Fig 9.40).

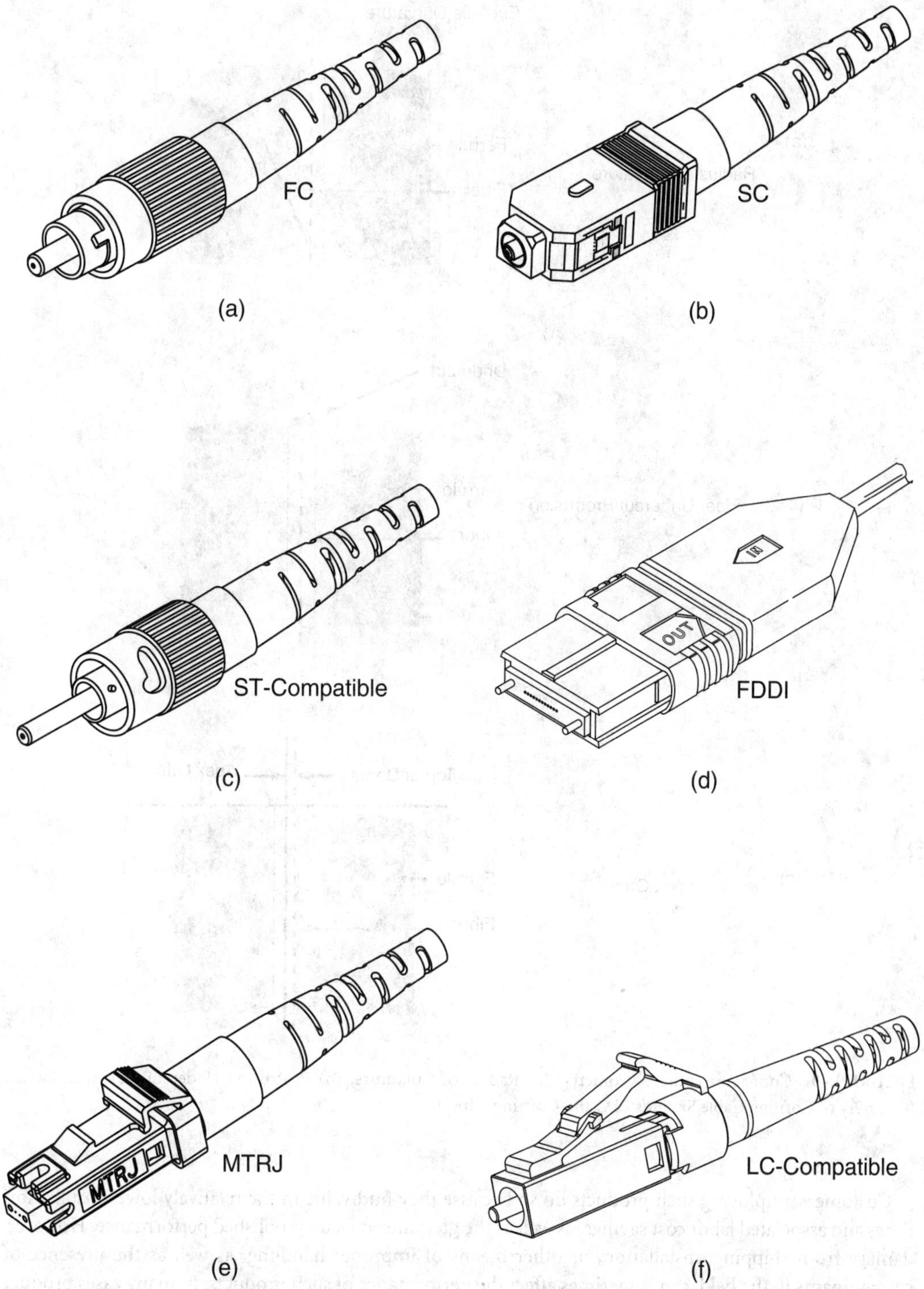

FIGURE 9.39 Common fiber optic connector types. (a) FC, (b) SC, (c) ST-compatiable, (d) FDDI, (e) MTRJ, (f) LC-compatible. (Courtesy of Corning Cable Systems LLC and Corning® Inc.)

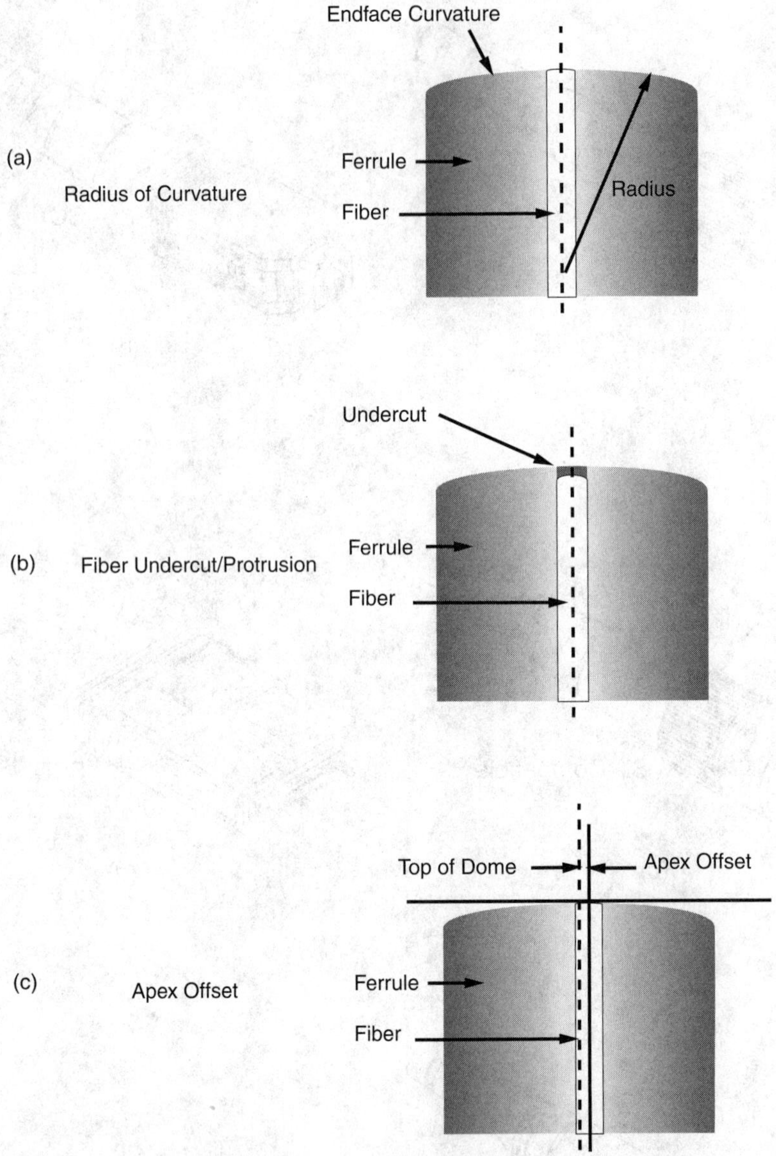

FIGURE 9.40 Connector endface geometry: (a) Radius of Curvature, (b) Protrusion/Undercut, (c) Apex Offset. (Courtesy of Corning Cable Systems LLC and Corning® Inc.)

Customers employing such products do so because they find value in the relatively lower installation times and associated labor cost savings, as well as the guaranteed factory-polished performance. However, damage from shipping, installation, or other means of improper handling, as well as the presence of contaminants in the field, can sometimes affect the performance of such products. In many cases product performance can be restored simply by cleaning. However, when cleaning is not enough, products can often be field-repaired depending on the exact nature of the issue.

To terminate optical cable in the field, a fiber optic pigtail may be used. A pigtail is typically a single- or multi-fiber cable that has been connectorized at the factory on one end only. The other end remains nonterminated and is spliced to the installed cable. This joint can be either a fusion splice or a mechanical splice. After splicing, the splice point is placed inside a protective splice tray, which is then housed in fiber optic hardware designed for indoor or outdoor use, as appropriate. Pigtail splicing yields factory quality performance while allowing connectors to be installed after the optical cable is placed.

Field Installed Connectors

For direct termination in the field connectors can be installed with an epoxy and polish procedure similar to that used in the factory. However, to speed up the termination process, a quick-cure anaerobic curing agent is often used. An adhesive and primer are mixed together to start the curing process and outside stimuli are not needed.

The as-installed performance of field installed connector products is largely dependent upon the skill of the craftsperson, as well as the conditions under which they are installed. In any event, the performance distribution of field terminated products is generally not as tight as products terminated by factory automated means, nor are the quality checks performed usually as detailed. Such product can be installed successfully in a wide range of applications but may not be appropriate in all cases. The success of such products also depends upon the system budget or specified acceptance criteria. Specifying field-terminated products at performance levels that challenge their capabilities can lead to significant scrap rates. Properly planning, specifying, ordering, and installing such products (i.e., following the manufacturers recommended procedures) can reduce scrap rates to low levels.

An even more advanced method of field termination utilizes no-epoxy, no-polish connectors. An optical fiber stub is cured into place in the fiber ferrule and polished at the factory. The field fiber is matched up to the fiber stub and a mechanical or fusion splice is made as appropriate inside the connector body. These connectors are typically very quick to install and require few consumables. Although single- and dual-fiber no-epoxy, no-polish connectors are the most common types used, twelve-fiber mechanical splice connectors are also available for use on ribbons.

Since fiber optic connectors must typically be terminated upon 900 μm buffered fibers, a fan-out kit with 900 μm tubing is used when directly connectorizing outside plant loose tube cables, where the fiber is normally only coated to 250 μm.

9.2.7 Chapter Summary

Optical fiber and fiber optical cable have dramatically changed the face of telecommunications over the last 20 years. As the industry matures the installation of optical cables and associated equipment will continue to reach further into businesses, schools, and homes, bringing a wider variety of services and improved quality of life to end users. In order to properly design, specify, order, install and grow an optical system, whether for a LAN, SAN, or a nationwide telecommunications network, knowing the basics of fiber and cable are critical in ensuring the completed system will support the user's telecommunications needs both presently and far into the future.

Defining Terms

Attenuation: The combination of optical energy losses from a light pulse traversing a length of optical fiber resulting from absorption and scattering, typically referred to as intrinsic attenuation, and bending losses, typically referred to as extrinsic attenuation, expressed in dB. However, attenuation is often used as a synonym for attenuation coefficient, expressed in dB/km.

Cladding: The layer of dielectric material that immediately surrounds the core of an optical fiber, the combination of which is fundamental to the ability of optical fiber to guide light at certain wavelengths. The cladding material for telecommunications grade optical fibers usually consists of a silica glass that cannot be physically separated from the core.

Core: The central region of an optical fiber consists of a silica-glass structure in which dopants are added to achieve the desired light transmission properties for a particular type of optical fiber. The distribution of dopants alters the index of refraction of the glass, which varies based on dopant concentration. The resulting refractive core properties are referred to as the index profile.

Insertion loss: The measured loss in optical power resulting from the scattering of light at the physical boundary between two optical waveguides (e.g., fibers). The light scattering is caused by the localized disruption in the refractive properties of the optical path due to the insertion of discrete components such as connectors, mechanical splices, couplers, etc. For fusion splices, this loss is typically referred to as "splice loss."

References (Sections 9.1 and 9.2)

Agrawal, G. P., *Nonlinear Fiber Optics*, 2nd ed., Academic Press, San Diego, California, 1995.

Hecht, J., *Understanding Fiber Optics*, 4th ed., Prentice-Hall, Columbus Ohio, 2002.

Mahlke, G. and Gössing, P., *Fiber Optic Cables*, 2nd ed., Siemens Aktiengesellschaft, Berlin, 1993.

Miller, S. E. and Chynoweth, A. G., *Optical Fiber Telecommunications*, Academic Press, Orlando, FL, 1979.

Miller, S. E. and Ivan P. K., *Optical Fiber Telecommunications II*, Academic Press, Orlando, FL, 1988.

Commercial Building Telecommunications Cabling Standard, ANSI/TIA/EIA-568-B, Telecommunication Industry Association, 2001.

EIA Standard Fiber Optic Terminology, EIA-440-A, Electronic Industries Association, 1989.

Optical Fibres - Part 2-10: Product Specifications - Sectional Specification for Category A1 Multimode Fibres, International Electrotechnical Commission, 2002.

Optical Fibres - Part 2-50: Product Specifications - Sectional Specification for Class B Single-Mode Fibres, International Electrotechnical Commission, 2002.

Recommendation G.651, Characteristics of a 50/125 μm multimode graded index optical fiber cable, International Telecommunications Union, 1998.

Recommendation G.652, Characteristics of a single-mode optical fiber cable, International Telecommunications Union, 2000.

Recommendation G.653, Characteristics of a dispersion-shifted single-mode optical fiber cable, International Telecommunications Union, 2000.

Recommendation G.655, Characteristics of a non-zero dispersion-shifted single-mode optical fiber cable, International Telecommunications Union, 2000.

Standard for Optical Fiber Premises Distribution Cable, ANSI/ICEA S-83-596-2001, Insulated Cable Engineers Association, 2001.

Standard for Indoor-Outdoor Optical Fiber Cable, ANSI/ICEA S-104-696-2001, Insulated Cable Engineers Association, 2001.

Standard for Outside Plant Communications Cable, ANSI/ICEA S-87-640-1999, Insulated Cable Engineers Association, 1999.

Standard for Optical Fiber Drop Cable, ICEA S-110-717-1999, Insulated Cable Engineers Association, 2003.

9.3 Optical Transmitters

Charles H. Cox, III

9.3.1 Introduction

The basic building block for any optical fiber system is the optical link, that is, all of the components necessary to take the desired electrical signal and modulate it onto an optical carrier, the fiber and other optical components to convey the optical signal from source to destination, and the components required to recover the desired electrical signal from the optical carrier. In this section the components that make up the optical transmitter will be reviewed.

There are primarily three wavelengths that are used in fiber optic links. All are in the near-infrared portion of the electromagnetic spectrum at wavelengths slightly longer than 0.65 μm, which is the red end of the visible spectrum. By far the dominant wavelengths in use at present are located in bands around 1.3 and 1.55 μm. The choice of these particular wavelength bands is based on the availability of certain desirable fiber properties, such as low attenuation (1.55 μm) or zero wavelength dispersion (1.3 μm). Early fiber optic links operated around 0.85 μm; this wavelength was chosen primarily because it corresponded to the availability of the first laser diodes. Although this wavelength offers neither of the fiber advantages of the longer wavelengths, it is still of interest because it offers the nearest term prospect for integration of gallium arsenide (GaAs) electronics with diode lasers, which at 0.85 μm are made in the GaAs material system.

Conceptually an optical transmitter could be based on modulating any one of the same properties of an RF carrier that are commonly modulated in an RF transmitter, that is, the optical carrier's amplitude, phase, or frequency. Although commonly referred to as amplitude modulation, strictly speaking AM of an optical carrier is *intensity modulation*, since in direct detection systems the square of the electric field amplitude is what is actually detected. We will concentrate on intensity modulation in this discussion.

Intensity modulation can be implemented using one of two general approaches (Cox, 1990). With *direct modulation, the electrical modulating signal is applied directly to the laser to change its intensity*. This implies that the modulating signal must be within the modulation bandwidth of the laser. As will be seen, only semiconductor diode lasers have sufficient bandwidth to be of practical interest for direct modulation. The alternative to direct modulation is *external*, or *indirect, modulation*. With *external modulation, the laser operates continuous wave (CW), and intensity modulation is impressed via a device that is typically external to the laser*. Since there is no modulation requirement on the laser for external modulation, this removes a major restriction on the choice of lasers that can be used for external modulation. Although both methods achieve the same end result, an intensity modulated optical carrier, there are a number of fundamental and implementational issues concerning these two approaches that gives each distinct advantages.

9.3.2 Direct Modulation

There are many books devoted entirely to various aspects of lasers in general and diode lasers in particular to which the reader can refer for detailed information (Thompson, 1980; Agrawal, 1986). However, it will facilitate the discussion in this section if we recall the two basic requirements for laser operation: the ability to produce stimulated emission and a cavity that is resonant at the stimulated emission wavelength. To produce *light* from a semiconductor is easy: forward biasing a semiconductor diode junction, which has been fabricated in a suitable material, results in the emission of *spontaneous* radiation. This is the basis for a light-emitting diode (LED). To produce laser light from a semiconductor requires a structure where an inverted population consisting of both electrons and holes can be set up, from which the spontaneous emission can trigger stimulated emissions. In a semiconductor the inverted population can be created and maintained by passing a DC current through the diode junction that has provisions to confine the electrons and holes.

The second criterion for laser operation is that these stimulated emissions must occur within an optical cavity that is resonant at the stimulated emission wavelength. The most common type of optically resonant cavity for diode lasers is one formed by two parallel, plane mirrors. Such a cavity is referred to as a Fabry-Perot after the original developers of this type of optical resonator. In a Fabry-Perot cavity, resonance exists for any wavelength, which after reflections off of both of the mirrors, is again in phase with the original wave. The wavelengths for which this criterion is satisfied are referred to as the Fabry-Perot modes of the cavity. It should be noted that if the range of energies over which stimulated emission can occur includes more than one Fabry-Perot mode, the laser can emit on several of those modes simultaneously.

Figure 9.41 shows a diode junction for generating the inverted population combined with a Fabry-Perot resonant cavity. When the junction is forward biased, one can obtain laser emission from such a structure. To get some of the light out of the cavity at least one, and often both, of the mirrors are only partially reflecting. It turns out to be easy to fabricate such mirrors in a single crystal material such as a semiconductor

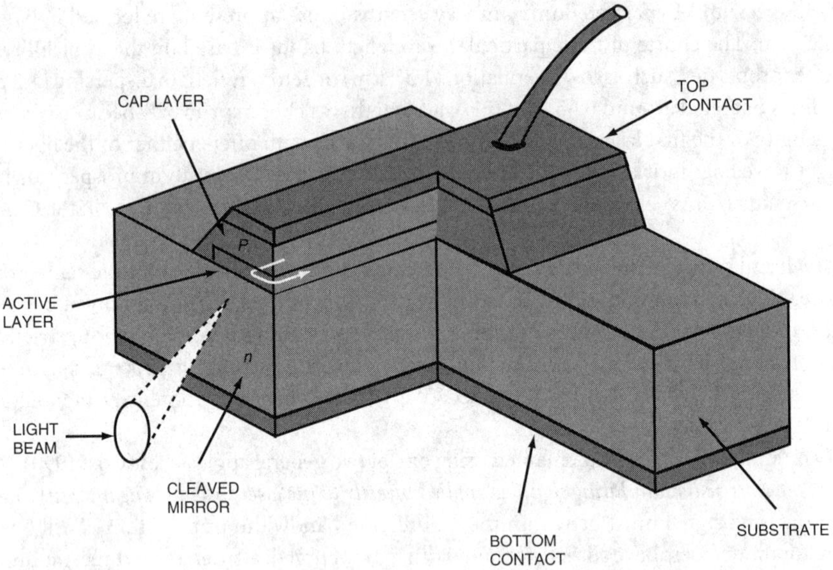

FIGURE 9.41   Semiconductor diode laser using a Fabry-Perot cavity.

since these materials crack or cleave along crystal planes, thereby insuring automatically that the resulting surfaces are optically smooth and parallel to each other. A convenient reflection magnitude is also easy to obtain: the Fresnel reflection at the crystal-to-air interface is about 32%, which is acceptable given the high gain of semiconductor lasers. To ensure a linear optical power vs. current function and a stable spatial optical mode, the light must be confined to a single-spatial-mode waveguide between the mirrors. Presently diode lasers for direct modulation are *index guided, which means that the light is confined by increasing the refractive index of a narrow stripe, so that the light is guided within the stripe via total internal reflection.* The increased index in the horizontal direction is via a compositional change in the semiconductor, whereas the vertical index increase is provided via semiconductor layers with a different refractive index.

A *Fabry-Perot diode laser* as described is by far the most common type as it is the laser that is in every CD player in the world. Optical powers from microwatts to tens of watts are commercially available and at a number of wavelengths of interest to fiber optics. However, as we shall see, the performance of this type of laser may be lacking in the more demanding applications. Part of the reason can be traced to the multiple Fabry-Perot modes. Hence, it would be desirable if it were possible to construct a cavity that is resonant at only one optical wavelength. The technique that has been adopted to improve the optical frequency selectivity of the cavity is to place a grating along the cavity, as shown in Fig. 9.42. The period of the grating is chosen such that there is only one resonant wavelength within the range where the semiconductor has significant stimulated emission. The balance between the magnitudes of the grating and facet reflections depends on the details of the laser design. In designs that require the grating reflections to dominate over the facet reflections, the facet reflections can be suppressed by antireflection coating the facets. *Lasers with internal grating are commonly referred to as distributed feedback (DFB) lasers*, since the reflections are distributed along the cavity. As will be seen, for a number of figures of merit DFBs out perform Fabry-Perots.

Direct Modulation Figures of Merit

Threshold Current

A typical plot of optical power P vs. laser current I for a diode laser is shown in Fig. 9.43. As the current is increased from zero, initially the optical output is dominated by spontaneous emission, and the output increases slowly with current. At low currents the probability of stimulated emission is low and so most of the spontaneous emission is absorbed by the semiconductor. Increasing the current increases the number

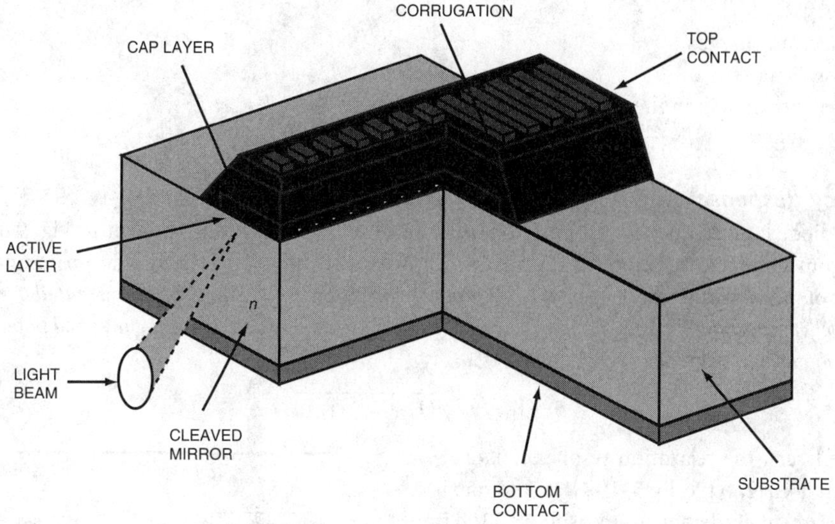

FIGURE 9.42   Semiconductor diode laser using a distributed feedback cavity.

of spontaneously generated photons, thereby increasing the number of stimulated photons. Eventually a current is reached where the number of stimulated photons equals the number of absorbed photons, which is defined as *transparency*. Above this current level stimulated emission begins to dominate the optical output, and the output increases at least 100 times more rapidly with laser current. *Extrapolating the straight line portion of the stimulated-emission dominated region down to where it intersects the current axis defines the threshold current.* This current is an important measure of laser performance because if it is too high, the heating caused

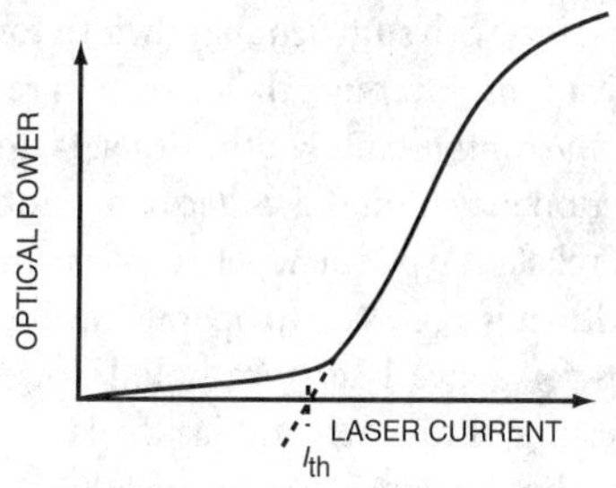

FIGURE 9.43   Plot of a representative optical power vs laser current for a diode laser.

by passing a large current through the junction will preclude operating the laser CW. Typical values of threshold current for present commercially available lasers are in the 5–50 mA range, with the lowest threshold current for an experimental device being around 0.5 mA (Lau, 1988). The threshold current increases with temperature which, usually necessitates some form of temperature control. This topic will be discussed in more detail subsequently.

Quantum and Slope Efficiency

Another laser figure of merit that comes from the P vs. I curve is the *slope efficiency* S_L. *This is simply the incremental slope of the P vs. I curve at a given laser bias current I_B,* and is defined as

$$S_L(\text{at } I = I_B) = \Delta P(I_B)/\Delta I_B$$

The higher the S_L is, the higher the efficiency in converting electrical modulation to optical modulation. Thus, the slope efficiency is an important parameter in link design. The slope efficiency is of practical utility because it can be easily measured on a packaged device without knowing any of the details of the laser's design or fabrication.

A closely related parameter is the *external differential quantum efficiency* η_L, which is the ratio of emitted photons to injected electrons. The quantum efficiency can be related to the slope efficiency as follows:

$$\eta_L = (e\lambda/hnc)S_L$$

where

e = electron charge

λ = wavelength

nc = velocity of light in the semiconductor

Frequency Response

A typical diode laser frequency response for small amplitude modulation is shown in Fig. 9.44. As can be seen from this plot, the modulation bandwidth extends from DC to the relaxation resonance. The frequency of the *relaxation resonance, f_{relax}, is proportional to the square root of bias current above threshold and inversely proportional to the geometric mean of the lifetimes of the carriers in the inverted population, τ_n, and the photons in the laser cavity, τ_p,*

$$f_{\text{relax}} \propto \sqrt{(I - I_{\text{th}})/\tau_p \tau_n}$$

For diode lasers, the relaxation resonance is usually at least 1 GHz, typically 5–10 GHz; the maximum experimentally demonstrated is about 30 GHz (Ralston, 1994; Morton, 1992). All other types of lasers have relaxation frequencies less than 1 MHz, and so with respect to this parameter diode lasers are the only lasers with sufficient bandwidth for practical applications. Although the relaxation resonance sets the *maximum* bandwidth, other factors such as distortion may limit the *usable* bandwidth to less than the relaxation frequency. To achieve maximum bandwidth, it is necessary to operate the diode laser at a bias far above laser threshold. However, the final decision as to where to bias the laser must be

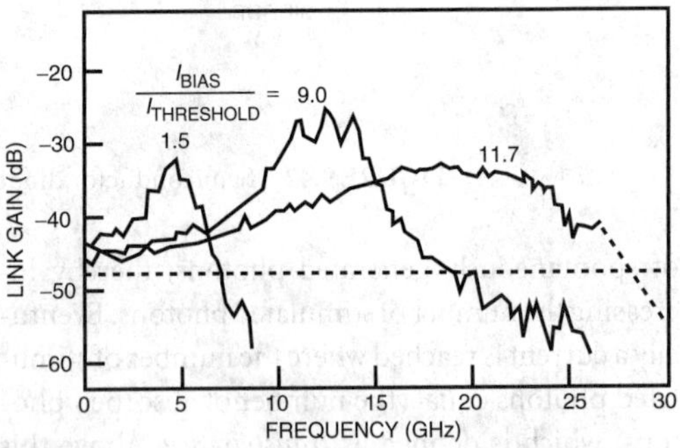

FIGURE 9.44 Plot of a typical frequency response for a high-frequency diode laser. (Lester, 1991)

tempered by the consideration of other laser parameters, such as the decrease in slope efficiency with increasing bias above threshold as shown in Fig. 9.43.

Relative Intensity Noise (RIN)

Recall that in laser operation stimulated emission dominates. But the spontaneous emission does not cease when stimulated dominates, thereby contributing one source of the small fluctuations in the laser intensity. In intensity modulated systems, these small fluctuations limit the minimum amplitude of the desired signal that can be used for modulation. The most commonly used *measure of the laser amplitude fluctuations is the relative intensity noise (RIN)* (Yammoto, 1983); it is defined as

$$\text{RIN} \propto \langle i^2 \rangle / \langle \bar{I} \rangle^2$$

where

i^2 = mean square photocurrent

$\bar{I}$ = average photocurrent.

Typical RIN values for Fabry-Perot lasers are from -120 to -140 dB Hz, whereas for DFB lasers values in the range from -150 to -165 dB Hz are common. Even with the lower RIN of DFBs, the link noise figure of directly modulated links is dominated by the RIN.

A typical plot of RIN vs. frequency for both Fabry-Perot and DFB lasers is shown in Fig. 9.45. The RIN frequency spectrum follows that of the small-signal frequency response with a peak in the RIN at the relaxation resonance. At low frequencies the RIN has a $1/f$ dependence. There is a broad, flat midband region, which ends with peaking in the RIN at the relaxation resonance. The RIN has its maximum value when the laser is biased near threshold. The RIN decreases as the square of increases in the laser bias above threshold, as can also be seen in Fig 9.45.

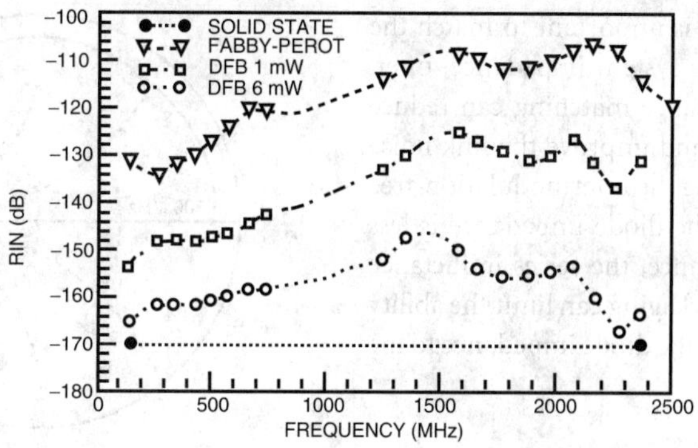

FIGURE 9.45 Plot of typical diode laser RIN vs. frequency with laser bias above threshold as a parameter.

Although it is common to think of the RIN as a single number, in some applications it is also important to know the RIN spectrum. For example, in applications where the laser output passes through a significant amount of a dispersive medium, such as a fiber, the effects of RIN increase if the laser has multiple modes. This is because in a multimode laser, an increase in the amplitude of one mode reduces the inverted population for all of the other modes in turn reducing their amplitude. Consequently, at different instants in time, the various modes have different amplitudes. Therefore, the resulting amplitude fluctuations for an *individual* mode in the presence of multiple modes can be much *greater* than the RIN of the combination of all the modes (Thompson, 1980, Sec. 7.7).

Frequency Chirp

The optical frequency of the mode or modes in a diode laser depends on the optical path length within the laser cavity. In turn the path length depends on the carrier density, which in turn depends in part on the amplitude of the modulating signal. The combination of all these effects is that the amplitude of the modulating signal changes the *frequency* as well as the intensity of the laser output. This incidental modulation of the optical frequency is referred to as *frequency chirp*. The change in laser frequency with laser drive current is positive, that is, an increase in drive current results in an increase in optical frequency. A typical value of chirp for a DFB is 1.2 GHz/mA. This is an important parameter in high-frequency, long-distance systems, since the chirp causes dispersion of the signal through the fiber, thereby limiting the distance-bandwidth capabilities of the link.

RF Impedance

Electrically a diode laser looks like, well, a diode, albeit a poor one. The doping profile results in a very low reverse breakdown voltage. This fact, combined with the laser's high speed, makes a diode laser particularly susceptible to damage from noise spikes and power supply glitches. When forward biased, the diode laser impedance can be modeled by the equivalent circuit shown in Fig. 9.46. At low frequencies the diode equivalent impedance is dominated by a small resistance, typically about 5 Ω. At higher frequencies the junction capacitance and bond wire inductance dominate the diode impedance. A typical Smith chart plot of a packaged diode laser impedance is shown in Fig. 9.47. As can be seen in this figure, the low-frequency impedance is dominated by the package bond wire inductance.

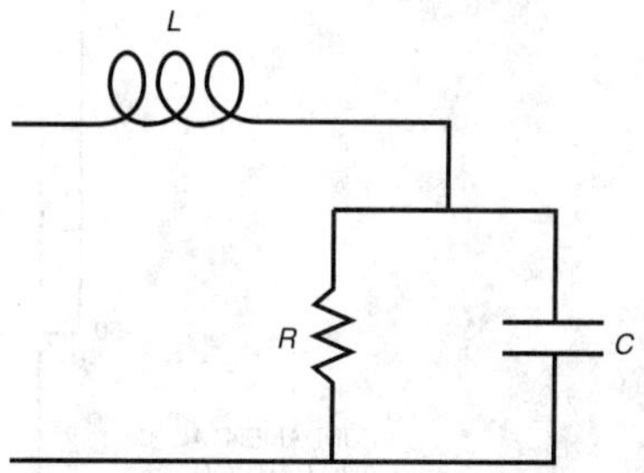

FIGURE 9.46 Schematic diagram of diode laser impedance model.

In many applications it is important to match the
diode impedance to the system impedance, often
50 or 75 Ω real. Impedance matching can reduce
the link RF-to-RF loss and improve the link noise
figure, thereby increasing the intermodulation-free
dynamic range. Since the diode impedance is less
than the system impedance, the series inductance
of the bond wire and packaging can limit the ability
to match the real part of the diode impedance to the
system impedance.

Linearity

Analog systems almost invariably have a linearity
requirement. The problem is that no realizable device is purely linear. To state this formally, the power

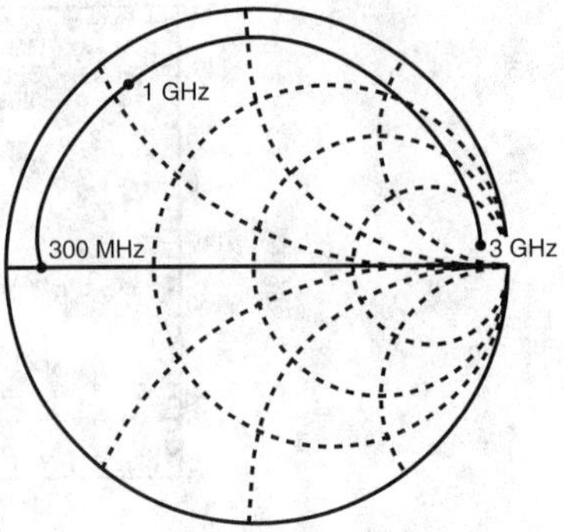

FIGURE 9.47   Smith chart plot of diode laser impedance
for laser bias at twice threshold current.

series expansion of any realizable device's transfer function has nonzero coefficients for at least some of
the terms other than the linear one. The effects of these nonlinear terms is clearly seen by examining the
output spectrum when two, closely spaced sine waves of frequencies f_i and f_j are input, as shown in
Fig. 9.48. The output spectrum consists partly of harmonics of the form mf_i and mf_j for each input sine
wave, where m is the order of the harmonic. For clarity only the second- and third-order harmonics are
shown in the figure: $2f_1$, $3f_1$, $2f_2$, and $3f_2$.

The output also contains intermodulation (IM) terms, which are of the form $mf_i \pm nf_j$, where $m + n$ is
the order of the intermodulation term. The figure also shows the two second-order and the four third-order
IM terms. Recall that for applications with a bandwidth of less than one octave, the third-order distortion
produces the dominant intermodulation signals that fall within the system pass band. Measures of third-
order distortion include the intermodulation-free dynamic range (IM-free DR) and the composite triple
beat (CTB). For broad-band, that is, multioctave, applications, both second- and third-order distortion
terms are important. Second-order measures of distortion include the IM-free DR, which this time contains
both second- and third-order spurious signals, and the composite second order (CSO). The IM-free DR is
used primarily by the radar and communication fields, where the presence of two strong signals could complicate the detection of a weaker signal. Consequently the IM-free DR is measured using two, closely spaced,
equal amplitude CW RF signals. The CTB and CSO are used primarily by the community antenna television

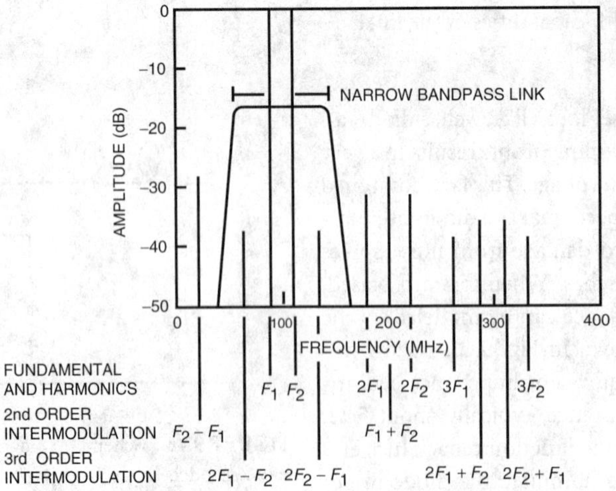

FIGURE 9.48   Representative output spectrum for a device with both second- and third-order distortion, when two
closely spaced sine waves are input.

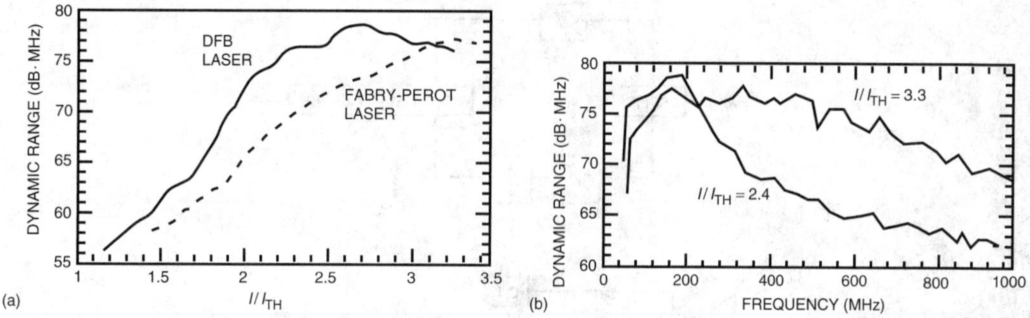

FIGURE 9.49 (a) IM-free DR for a Fabry-Perot, dashed curve, and DFB laser vs. laser bias at a fixed frequency; (b) IM-free DR for a Fabry-Perot vs. frequency at a fixed bias (Roussell, 1997).

(CATV) industry and are measured using a multi channel signal generator with the number and spacing of the channels chosen to equal the intended application. Since the measurement conditions for CTB and CSO are application specific, there is no exact conversion factor between IM-free DR and CSO/CTB. However, there are approximate techniques for converting between the two measures (Kim, 1989).

In an intensity modulated link the amplitude of the intermodulation terms depend primarily on the linearity of the modulation device, in this case the diode laser. *The IM-free DR is defined as the maximum difference between the noise floor and the fundamental output, which produces distortion terms of equal amplitude to the noise floor.* The noise floor in turn depends on the link bandwidth (BW), which varies by application. Consequently to make the IM-free DR measurements have general applicability, the results are often given in terms of a 1-Hz bandwidth. To use such results in a specific application simply requires scaling the 1-Hz data to the application bandwidth. The bandwidth scaling exponent depends on the order of the dominant distortion: for second-order distortion, the IM-free DR scales as $(BW)^{1/2}$ or the square root of bandwidth, for third-order distortion, the IM-free DR scales as $(BW)^{2/3}$.

The IM-free DR of diode lasers depends primarily on the laser type, the laser bias, the RF frequency, and the amount of the laser's own light that is reflected back into the laser. Experimental data for the first three parameters will be presented here. Figure 9.49(a) is a plot showing the IM-free DR of a typical Fabry-Perot and DFB lasers vs. laser bias at a fixed RF frequency. With the RF modulation amplitude fixed, the IM-free DR initially increases as the laser bias is increased above threshold. Once the bias is far enough above threshold to avoid clipping of the modulation waveform, the IM-free DR settles down to a value that is independent of bias, at least until the curvature of the P vs. I curve starts to become appreciable at high laser bias. Fixing the bias of a Fabry-Perot laser, Fig. 9.49(b) is a plot of IM-free DR vs. modulation frequency. Recall that the higher the bias, the higher the relaxation frequency. Consequently, the bias current for highest dynamic range increases as the operating frequency increases closer to the laser resonance frequency.

Linearization

Several important applications, such as the distribution of cable TV signals, require greater IM-free DR than is commonly available from the intrinsic performance of diode lasers. For these applications, the IM-free DR, or equivalently the CSO/CTB, can be increased through any one of a number of techniques. Although the details of these techniques are often proprietary to the company that developed them, the general approaches are well known.

A block diagram of *predistortion* (Bertelsmeier, 1984), the most common approach to linearization, is shown in Fig. 9.50(a). If the nonlinear transfer function, which is limiting the IM-free DR, is well known and stable, a circuit with the inverse of the nonlinearity can be inserted before the diode laser. The cascade of the two nonlinearities then has a more linear transfer function than the original single nonlinearity. However this is an open-loop approach, and so any changes in the offending nonlinearity will not trigger

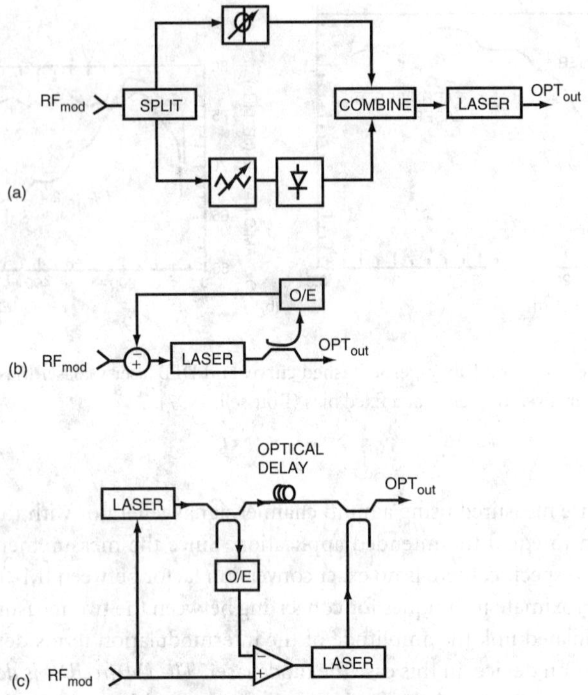

FIGURE 9.50 Block diagrams of linearization techniques: (a) predistortion, (b) feedback, (c) feed-forward.

a correction in the compensating nonlinearity. The results can range from less increase in IM-free DR to a degradation in IM-free DR! On the positive side, predistortion is simple to implement and does not limit the frequency response.

The *feedback* approach (Straus, 1978) to linearization, as shown in Fig. 9.50(b), can compensate for changes in the offending nonlinearity. However, feedback has its own limitations. Primary among them is the need to limit the phase delay around the feedback loop so that the loop is stable. Compounding this problem is the lack of feedback amplifiers with sufficient performance above about 150 MHz. Thus, to utilize feedback at high frequencies requires a substantial development effort covering feedback amplifier design and monolithic integration of this amplifier and the diode laser.

The third general category of linearization is *feedforward* (Hassin, 1993) and is shown in block diagram form in Fig. 9.50(c). This is probably the most complex to implement because it usually requires two diode lasers. This approach relies on the fact that lower amplitude modulation signals generate lower levels of distortion products. Thus, the distortion introduced by modulating the first laser is added back, but out of phase, to the second laser. However, because of the lower amplitude of the distortion signal used to modulate the second laser, the distortion by the second laser is much smaller.

The latter two techniques have in common that they can compensate for any order of nonlinearity, as opposed to predistortion, which is used primarily to cancel a particular order of nonlinearity, such as third order. Results for any of these techniques are not generally available because of the proprietary nature of the manufacturer's efforts. However it is known that, in general, these approaches have demonstrated about a 5-dB increase in the IM-free DR.

Direct Modulation Implementation Issues

Temperature Stabilization

Several diode laser parameters are temperature dependent, among them threshold current and slope efficiency. The threshold current increases with increasing temperature, eventually making lasing action impossible. The slope efficiency decreases with increasing temperature, and the leakage current around

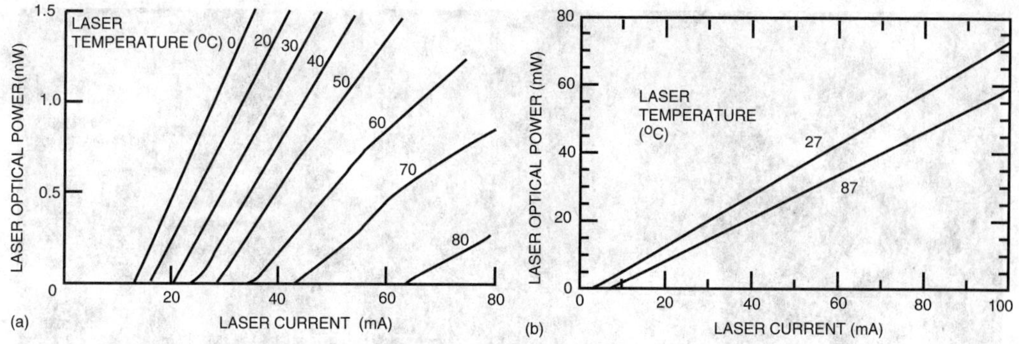

FIGURE 9.51 Plots of diode laser optical power vs. diode current with diode junction temperature as a parameter for: (a) a representative laser, (b) a laser designed for a low-temperature coefficient (Kazarinov, 1995).

the laser junction increases with temperature, which leads to a bending over of the laser's P vs. I curve. Figure 9.51(a) is a plot showing all of these effects. This trend is to first order independent of the laser cavity design, that is, Fabry-Perot or DFB.

The traditional way around the temperature dependence is to hold the laser at a constant junction temperature. This has led to diode laser packages which include thermoelectric coolers (TECs) and thermistors. By changing the direction of the current through the TEC, this device provides the capability to heat the junction when the ambient is too cold and cool the junction when the ambient is too high. When combined with the appropriate power supply and feedback controller, the diode temperature can be held constant over a range of 100°C of ambient change. However, the TEC can require up to 2 A at the temperature extremes, which may be a burden in some system designs.

A more recent trend is to design lasers whose performance parameters are less temperature dependent. One way to accomplish this is to change the laser's semiconductor layer structure. For example, a significantly lower temperature dependence is shown in Fig 9.51(b) which is a similar plot to Fig. 9.51(a) but for a laser in which the simple single active layer has been replaced by a multiple quantum well structure.

Fiber Coupling and Packaging

Virtually all diode lasers for communication applications needs to be coupled to a fiber. The desire to do the fiber coupling as efficiently as possible is hampered by the disparity between the wide, elliptical divergence of the laser beam and the narrow, circular acceptance angle of the fiber. Single-mode fibers, which are required for most applications because of their high distance-bandwidth product, have a numerical aperture of 0.12. Most lasers have an elliptical beam with a maximum numerical aperture of 0.5. The consequence of this numerical aperture mismatch is that the fiber coupling efficiency is typically around 10%. By lensing the end of the fiber, the fiber coupling can be increased to 70%. More radical approaches, include tapering the laser waveguide so that the laser produces a lower divergence, more circular beam.

Figure 9.52(a) is a photograph of a typical commercially available laser diode. The combination of the micrometer-scale tolerances and the rapid falloff in coupling efficiency with displacement, has forced most manufacturers to use active alignment for best results. The cost of active alignment adds to the laser cost, in some cases doubling it. Automated fiber coupling equipment is generally available.

The fiber coupling problem is further exacerbated in applications where an optical isolator, which will be discussed in more detail in the next section, must be inserted between the laser and the fiber. Isolators are typically most effective for collimated light, which means that circularizing and collimating optics need to be placed between the laser and the isolator. Even with all this paraphernalia, it is still possible to achieve fiber-coupled slope efficiencies of 0.2 W/A.

In many applications the concern is not the fiber coupling efficiency *per se* but the range of coupling efficiencies from unit to unit and/or over temperature. Unit-to-unit variation can eat into the design margin of the optical link, thereby forcing the user to design to the minimum fiber-coupled slope efficiency. Even in laser packages where the temperature of the laser is controlled, typically this does not include the fiber

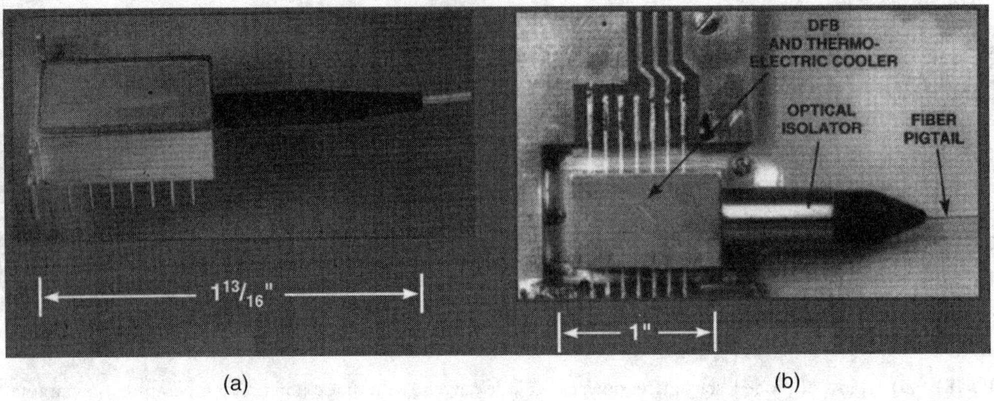

FIGURE 9.52 (a) Photograph of a typical commercially available fiber-coupled diode laser, (b) photograph of a fiber-coupled diode laser with an optical isolator.

attachment assembly. Figure 9.53 is a plot of the fiber-coupled slope efficiency vs. temperature for six lasers from three manufacturers. As can be seen, there is significant variation, both among manufacturers and among lasers from the same manufacturer. Thus, the link designer must also account for changes in fiber-coupled slope efficiency over temperature in the design.

Optical Isolation

Several of the laser's parameters, among them the laser's relative intensity noise and the laser's optical modes, are a strong function of the optical reflections of the laser's own light back into the laser cavity (Hirota, 1981). Thus, in high-performance applications, it is important to provide at least 30-dB, and often 60-dB, attenuation of the optical reflections. One simple way to provide the desired fiber-to-laser attenuation is to intentionally increase the coupling loss between the laser and the fiber. For example, by making the laser-to-fiber coupling -30 dB, optical reflections are attenuated by 60 dB, since they must traverse this loss twice. Increasing the coupling loss has the benefit of significantly simplifying the fiber coupling problem already discussed, thereby reducing the laser cost substantially. However, this approach is only useful in applications where there is substantial link margin since the reduced fiber coupling

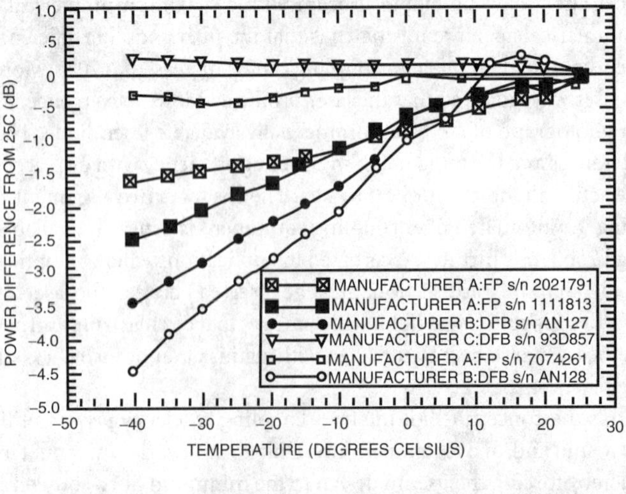

FIGURE 9.53 Plot of fiber-coupled slope efficiency vs. temperature for six diode lasers from three manufacturers.

attenuates the desired laser-to-fiber coupling as well. Continuing the example, 30 dB of fiber coupling loss results in 60 dB of RF-to-RF link loss.

An optical isolator provides a means of dramatically reducing the undesired fiber-to-laser coupling while only slightly decreasing the desired laser-to-fiber coupling. Typically an isolator introduces 1 dB of laser-to-fiber loss while providing 30 dB attenuation for fiber-to-laser reflections. Figure 9.52(b) is a photograph showing how an isolator is included in a fiber-coupled laser package. An isolator is clearly the superior solution technically, but results in increased laser package cost.

9.3.3 External Modulation

In contrast to direct modulation, external modulation requires two components: a laser that operates CW and a modulator that imposes the desired intensity modulation. Although slightly more complex, as will be seen, external modulation has several fundamental advantages over direct.

Modulators and Figures of Merit

Materials

The key material property for making a modulator is that a change in an electrical parameter must alter an optical property of the material. Other material properties that are also important are the optical loss and the stability, both thermal and optical. There are three classes of materials from which a modulator can be made: *inorganics*, *semiconductors*, and *nonlinear polymers*.

By far the most common modulator material is the inorganic lithium niobate, which is the benchmark material against which all newcomers are judged. In lithium niobate an applied electric field alters the optical index of refraction. The parameter that gives the optical index change per unit of applied electric field is the electro-optic coefficient, which is usually designated as r. The strongest electro-optic coefficient in lithium niobate is r_{33}. To use this coefficient requires an electrode configuration and modulator waveguide layout where the electric field and the optical polarization are along the z axis of the lithium niobate. A more meaningful measure of a material's electro-optic sensitivity is given by $n^3 r$; however; the justification of this figure of merit is beyond the scope of this discussion (Yariv, 1975).

As grown, electro-optic materials, such as lithium niobate, have an average electro-optic coefficient of zero, since all of the molecular dipole moments are randomly oriented. To align all of these dipole moments requires a process called poling: the material is heated above its Curie temperature, an electric field is applied, and then with the field still on, the material is cooled below its Curie temperature, thereby freezing the alignment of all the dipole moments. For wavelengths longer than about 1 μm, lithium niobate is quite stable over time and with optical power. For shorter wavelengths, some of the light is absorbed, which causes undesirable optical phase shifts. This later process is referred to as the *photorefractive effect*.

The most widely used method of forming low-loss waveguides in lithium niobate is by *diffusing in titanium* (Schmidt, 1974). Low-loss guides can also be formed by exchanging some of the lithium atoms with hydrogen, then annealing; this process is referred to as *annealed proton exchange* (Shah, 1975). The waveguide propagation losses are about the same for both processes.

Lithium niobate has probably succeeded not because it is the best with respect to sensitivity, loss, or stability, but because it has a reasonable combination of all three. However, lithium niobate is not without its drawbacks. Principal among them is the fact that it is not a semiconductor, and this makes integration with lasers, detectors, and electronics difficult at best. As was already mentioned, lithium niobate suffers from photorefractive effects at wavelengths less than about 1 μm. It is also difficult to process; etching, for example, is virtually impossible. These limitations have led to the investigation of alternate materials. Within the inorganic family there are other materials such as strontium barium niobate, which have a substantially higher electro-optic coefficient. But the waveguide optical loss is high. Further, higher electro-optic coefficient materials typically have Curie temperatures nearer room temperature, which makes thermal stability more of a problem.

Semiconductors are another class of materials that can exhibit the desired electro-optic effect. Although the electro-optic coefficient is lower in these materials, this deficiency can be partially offset by the higher *n* and the tighter optical waveguide confinement, which permits a tighter overlap between the electrical and optical fields. The tighter confinement, however, means higher fiber-to-waveguide coupling losses and somewhat higher waveguide losses. However, in direct bandgap semiconductors there is another mechanism, *electroabsorption* (EA), which can produce modulation. To appreciate the basis for this modulation approach recall that the longest wavelength absorbed by a semiconductor corresponds to the bandgap energy of the semiconductor. In some semiconductors structures, the effective bandgap energy is a function of the applied electric field. Consequently, if the wavelength to be modulated is chosen such that the semiconductor is almost transparent at zero applied electric field, then this wavelength is increasingly attenuated as the electric field is increased. This effect is greatly enhanced by using quantum wells, and in such semiconductor devices the modulation response can exceed that of lithium niobate. In principal semiconductor-based modulators have the advantage that they can be integrated with other electronic devices. However, this advantage has been only partially realized in practice largely because the fabrication parameters such as doping and layer composition can be quite different for optical and electronic devices.

Polymers are the third category of electro-optic materials. Based on nonlinear organic molecules, polymer modulators offer the potential for significant increase in electro-optic sensitivity. The basis for this hope lies in the considerable design flexibility afforded by polymers. In the inorganics and the semiconductors, the electro-optic properties are directly tied to the crystal structure; pick a crystal and you take whatever electro-optic properties come along with that crystal. With polymers, there exists considerably greater flexibility in designing a polymer that has the desired combination of properties. The myriad polymer structures combined with a plethora of dopants or chromophores offer an incredible array of possibilities. The belief is that somewhere within this array lies the utopian combination of modulator parameters.

Modulator Configurations

Early bulk modulators simply placed electrodes on a block of lithium niobate. The resulting electric field alters the phase of an optical wave passing through the lithium niobate. This configuration results in a relatively weak coupling between the electrical and optical fields. The low sensitivity of these modulators limits their application to sensing high electric fields such as those around power lines. Orders of magnitude greater sensitivity is possible by confining the light to an optical waveguide (Alferness, 1982). Electrodes placed along the waveguide produce much tighter coupling between the electrical and optical fields than is possible with bulk modulators.

As was mentioned in the Introduction, the most common optical modulation format presently in use is intensity modulation. And *the most common modulator design for accomplishing intensity modulation is based on a Mach-Zehnder interferometer.* The layout of a *Mach-Zehnder modulator (MZM)* (Martin, 1975), is shown in Fig. 9.54(a). The waveguides all propagate a single optical mode. The incoming CW light from the laser is typically split equally and fed into two ideally equal length arms, whose outputs recombine to feed the modulator output waveguide. With zero voltage on the modulator electrodes, the light in the two arms arrives at the output combiner in phase, as shown in the top half of Fig. 9.54(b). Thus, this bias condition results in maximum transmission through the modulator. When a voltage is applied to the electrodes, the resulting electric field, which is perpendicular to the waveguide, alters the refractive index in the two arms. Since the electro-optic effect is field-direction dependent, applying the voltage to the center electrode causes the index in one arm to increase and to decrease in the other arm. The effect of these index changes is the introduction of a relative phase shift between the light in the two arms when they recombine. The out-of-phase light attempts to excite higher order modes in the output waveguide, as indicated in the bottom half of Fig. 9.54(b). But since this guide is also single mode, all of these higher order modes are attenuated after propagating a sufficient distance. Thus, as the electrode voltage increases, the intensity in the output waveguide is decreases. For a sufficiently high voltage, the light in the two arms is 180° out of phase. All of the light is now in higher-order modes and none is in the fundamental mode. Consequently, no light propagates in the output waveguide. Further increasing the electrode voltage begins to bring the light in the two arms closer in phase until eventually they are completely in phase again. Thus the MZM transfer function is periodic in voltage, as shown by the solid curve in Fig. 9.55.

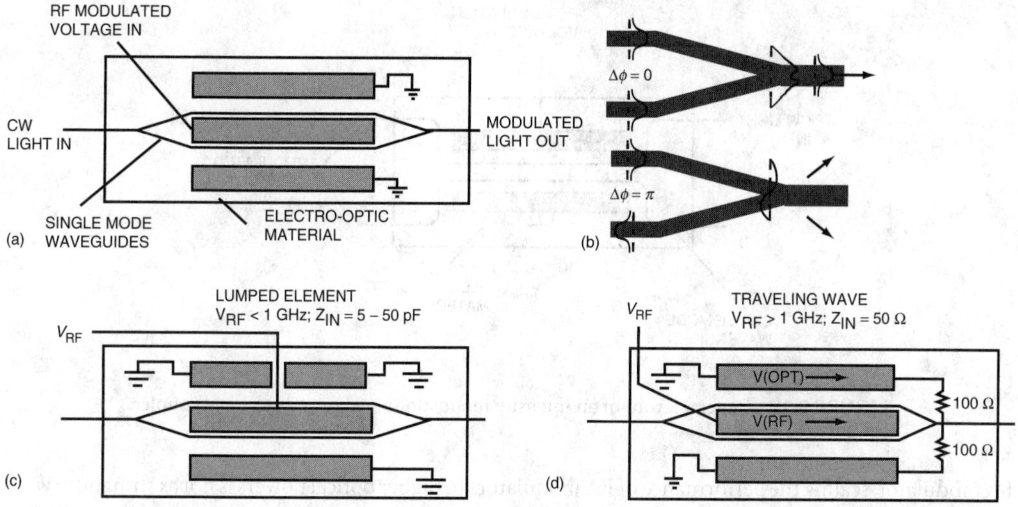

FIGURE 9.54 (a) General layout of an intensity modulator based on a Mach-Zehnder interferometer, (b) detail of output waveguides, (c) electrode configuration for lumped element modulators, (d) electrode configuration for traveling wave modulators.

A modulator similar to the MZM is based on a *directional coupler* (Papuchon, 1975). In a standard coupler, two waveguides are run parallel and close enough to each other that evanescent coupling can occur between them. The length of the coupling region is selected so that the desired degree of power is coupled from one guide to the other. To make a modulator from this passive coupler simply requires placing electrodes parallel to the waveguides, as shown in Fig. 9.56. The electric field from the electrodes alters the refractive index, which changes the coupling ratio, thereby altering the power between the two output guides. As can be seen from the dashed curve in Fig. 9.56, the directional coupler transfer function is not periodic and is more complex than the MZM.

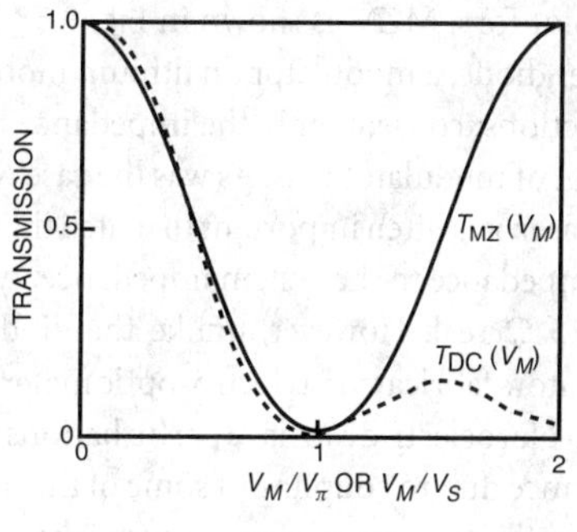

FIGURE 9.55 Modulation voltage to optical power transfer function for a MZM (solid line) and directional coupler modulator (dashed line).

Directional couplers theoretically offer about the same modulation response, bandwidth, and linearity as MZMs, but they are much more difficult to fabricate, especially if long electrode lengths are required. The directional coupler is favored over the MZM when a compact 2 × 2 switch is desired.

As was mentioned previously, semiconductors also offer the possibility of modulators based on the EA effect. The good news is that an *EA modulator* (Yu, 1995) has the potential for higher linearity, because it is not based on an interference of two phase modulated beams. The bad news is that there are several technical challenges to overcome. One is that the wavelength of the CW source must be matched to the absorption edge to within about 40 nm to get reasonable modulation sensitivity. Because the modulation depends on absorption, there is a tradeoff among insertion loss, drive voltage, and extinction. Another technical challenge in selecting the wavelength arises from the fact that the absorption change is a strong function of the wavelength separation between the material bandgap and the wavelength to be modulated. In turn this means that the modulation slope efficiency is temperature sensitive, since the material bandgap wavelength is temperatue dependent. A third technical challenge is the maximum optical power handling capacity of EA modulators. Since in the off state ideally all of the incident optical power is absorbed by the

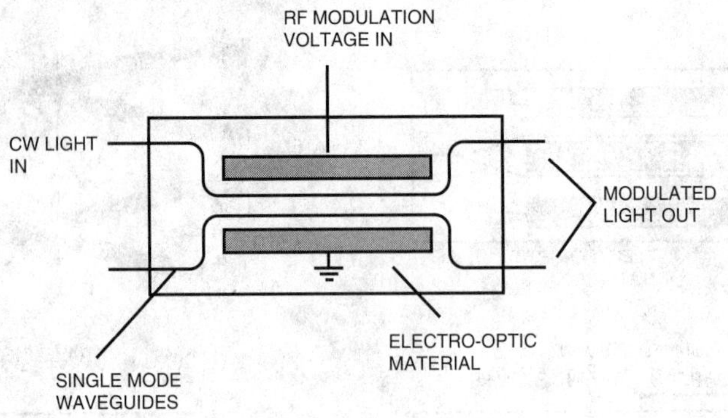

FIGURE 9.56 General layout of an intensity modulator based on a directional coupler.

EA modulator, scaling the performance of EA modulators to higher optical powers is not as straightforward as with interferometric modulators.

RF Impedance

For a low-frequency, Mach-Zehnder modulator, the impedance is dominated by the electrode capacitance and resistance as shown in Fig. 9.57. A Smith chart plot for a MZM is shown in Fig. 9.58. Unlike the laser diode, a modulator on lithium niobate has no junctions; consequently the impedance is independent of modulator bias. As was the case with the diode laser, it is often important to match the modulator impedance to the system impedance, typically 50 or 75 Ω real. However, unlike the diode laser, modulators fabricated in electro-optic materials that are also piezoelectric can have perturbations in their impedance due to coupling of some of the modulation signal into compressional waves. The coupling can be significant at frequencies where the modulator crystal is resonant. The dashed curve in Fig. 9.58 shows an acoustic resonance at 150 MHz. A number of techniques (Betts, Ray and Johnson, 1990) have been developed to suppress these acoustic modes to insignificant levels, as is evidenced by the solid curve in Fig. 9.56.

High-frequency modulators use traveling wave electrodes, for reasons described in the Frequency Response section. In this configuration the modulator impedance is ideally determined by the termination resistance at the end of the traveling wave electrodes. In practice the electrodes have loss that affects the impedance.

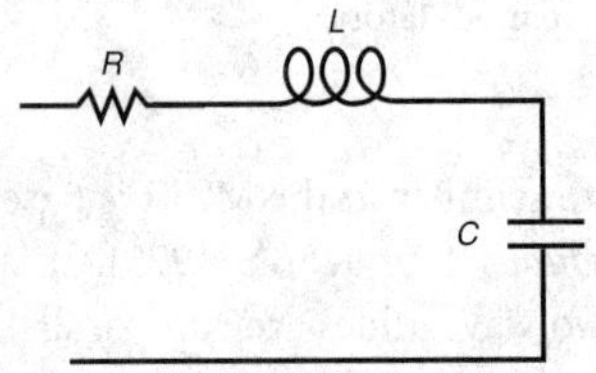

V_π or Switching Voltage

The voltage needed to change the modulator from on to off is an important figure of merit. In an MZM, this voltage introduces 180°, or π radians, of phase

FIGURE 9.57 Schematic diagram for a MZM with lumped element electrodes.

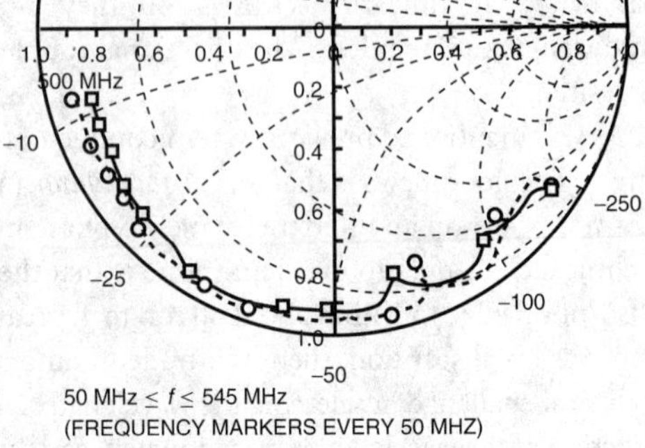

50 MHz $\leq f \leq$ 545 MHz
(FREQUENCY MARKERS EVERY 50 MHZ)

FIGURE 9.58 Smith chart plot of modulator impedance for lumped electrode MZM with (solid line) and without (dashed line) acoustic mode suppression.

shift; consequently it is referred to as V_π, as shown by the solid curve in Fig. 9.55. In a directional coupler modulator, which is often used for switching, this voltage is referred to as the switching voltage V_s, as shown by the dashed curve in Fig. 9.55. Although these definitions seem simple enough, there are several issues that require care to make sure the appropriate V_π is used in each case. For example, if the modulator includes separate electrodes for bias and modulation, these electrodes are typically of different lengths. Thus, the V_π used for bias control design is different from the V_π for link gain calculations. At DC, the voltage applied at the input connector is the voltage that appears across the modulator electrodes. However, as the frequency increases, the packaging and modulator impedance can result in a difference between these two voltages. Thus the DC V_π, in general, does not equal the RF V_π. Typical values of V_π for the modulating electrodes of commercially available lithium niobate modulators is 1–10 V, with the record low voltage of 0.3 V (Betts, 1989). The lower V_π correspond to lower frequency modulators. The commercially available GaAs MZM is a 20-GHz traveling wave device (see the next section for details about traveling wave modulators) with a V_π of 5.2 V (Walker, 1991), whereas a similar lithium niobate MZM would have a V_π of 3.7 V (Noguchi, 1993). MZMs in polymers with similar frequency response have V_π in the range 5–11 V (Wang, 1995). In analog links, the link gain depends on the inverse square of V_π or V_s, so even small decreases in this voltage can have a significant impact on link performance.

Frequency Response

The maximum frequency of the electro-optic effect in lithium niobate is probably in the hundreds of gigahertz. Thus, in practice the upper limit on the frequency response is set by electrical considerations. The key parameter that divides modulators into one of two types is the ratio of the modulation period of the maximum modulation frequency to the optical transit time past the electrodes. If the optical transit time is short relative to period of the maximum modulation frequency, then the modulator signal will essentially be constant during the optical transit time. In this case the electrodes can be treated as lumped elements, see Fig. 9.54(c), which means that the upper 3-dB frequency is simply determined by the RC rolloff due to the electrode capacitance and resistance as shown in Fig. 9.57. A typical frequency response of a low-frequency MZM is shown in Fig. 9.59(a).

At high frequencies where the period of the maximum modulation frequency is less than the optical transit time, lumped element electrodes would be ineffective because the modulation voltage would change during an optical transit time. The traditional method to achieve efficient modulation in this case is to make the electrode into a transmission line so that the electrical power copropagates with the optical power. Although this so-called velocity matched condition is easy to understand, it turns out to be difficult to implement. The reason for the difficulty is that in a straightforward design, the optical propagation velocity is about twice as fast as the microwave propagation velocity, which in turn is due to the much higher dielectric constant of lithium niobate at microwave frequencies ($\sim$100 GHz) as compared to the

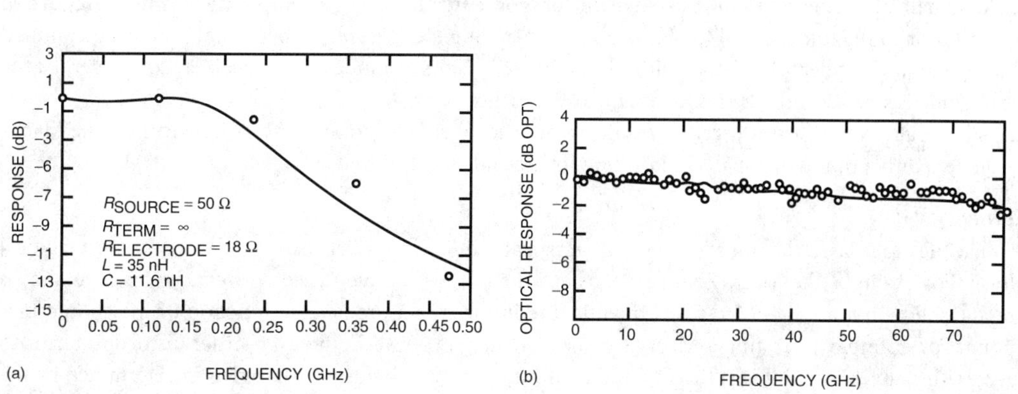

FIGURE 9.59 Frequency response of: (a) a lumped element modulator (Becker, 1984), (b) a velocity matched external modulator in lithium niobate (Noguchi, 1994).

dielectric constant at optical frequencies (~300 THz). Initial attempts at velocity matching were achieved by altering the electrode design (Alferness, 1984), but at a penalty of an increased V_π due to the smaller electrical-to-optical overlap.

Several groups (Noguchi, 1994; Gopalakrishnan, 1992; Dolfi, 1992) have demonstrated velocity matched modulators without significantly increased V_π. The frequency response shown in Fig. 9.59(b) has a V_π of 5 V and an upper 3-dB frequency of 75 GHz. It is important to note that bandwidth of modulator designs are commonly presented in terms of the frequency where the *optical* response is down by 3 dB. Since the electrical response of a link is the square of the optical response, the *electrical* response would be down 6 dB at the optical 3-dB bandwidth. The 3-dB electrical bandwidth is typically about half the optical bandwidth. Once a velocity match is achieved, it is theoretically possible to have an arbitrarily low V_π by simply making the electrodes longer. Unfortunately, the advantages of longer electrodes are limited in practice by the loss of the electrodes.

As the frequency goes up, it becomes increasingly difficult to couple the modulation signal to the electrodes. An elegant solution to this problem is to use radiative coupling in which the modulator electrode consists of dipole antennas. Coupling is accomplished when these dipoles are *illuminated* by the modulation signal from a waveguide. By segmenting the electrode and angling the modulator crystal with respect to the end of the waveguide, velocity matching can be achieved. Modulation at 94 GHz has been demonstrated using this technique (Bridges, 1990).

Optical Loss

There are three principal contributions to the optical loss. One is the loss of the straight waveguides. This loss is measured in decibels per centimeter and is typically 0.1–0.2 in lithium niobate, 0.5–1.0 in GaAs and 1 in polymers. The requirement for single-mode guides in most modulators constrains the tradeoff between optical loss and the magnitude of the refractive index step between the waveguide core and surrounding material. The larger the difference is in indices, the smaller the guide and the larger the waveguide loss, primarily due to Rayleigh scattering. In lithium niobate the waveguides are formed by doping, either titanium or protons, so that the index and hence the waveguide dimensions are essentially continuously adjustable. In semiconductor guides, the index delta is set by growing different materials such as gallium arsenide and aluminium gallium arsenide. Consequently, the index step is essentially set by the materials.

The need for low-loss straight waveguide drives modulator designers to use small index differences. Consequently, the light is weakly confined in the integrated waveguides and, therefore, only small bends (on the order of $1°$ in lithium niobate) can be made in straight waveguides without incurring substantial loss. However, even with small bends and careful fabrication, waveguide bends and branches are still a significant source of loss.

To be useful in systems virtually all modulators need to be coupled to optical fibers. The usual result of the waveguide loss-delta refractive index tradeoff discussed previously is that the waveguide optical mode field is smaller than the fiber's mode field diameter. Consequently, the fiber coupling loss is also a significant contributor to the total optical loss. The lowest fiber coupling losses have been achieved using confined core single-mode fibers attached to lithium niobate. In GaAs the index delta is larger, consequently, the waveguide is smaller and the fiber coupling loss is higher.

Adding all of these losses together results in fiber-to-fiber losses of commercial devices for the Mach-Zehnder configuration of 3–5 dB in lithium niobate and 10–12 dB in GaAs.

Linearity

Unlike the diode laser, the transfer function of many modulators is easily expressible in a simple analytical form. For example, the Mach-Zehnder transfer function is a simple raised cosine, $1 + \cos\phi$, as shown by the solid curve in Fig. 9.55. Consequently, the linearity is a predictable function of the chosen bias point. For example, operating a Mach-Zehnder at $V_\pi/2$ forces all of the even-order distortion terms to zero. Using this bias point in a broad-band application means that the IM-free DR is determined by the odd-order distortion, which is dominated by the third-order term. Recall that for narrow-band systems the second-order distortion terms fall outside the pass band and consequently can be filtered out. Thus, for such applications the bias point can be moved away from $V_\pi/2$ with no system consequence. One reason

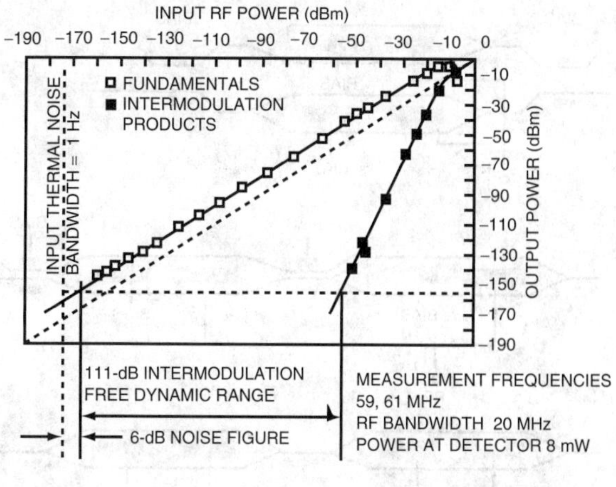

INPUT RF POWER (dBm)

FIGURE 9.60   Mach-Zehnder modulator IM-free DR with modulator biased at $V_\pi/2$.

for operating away from $V_\pi/2$ is to achieve a slightly lower noise figure (Farwell, 1993). For maximum IM-free DR in narrow-band applications, one would like to find a bias point where the third-order distortion is zero. In the Mach-Zehnder the only bias points where the third-order distortion is zero also force the fundamentals to zero, which means that these bias points are of limited interest. However, in the directional coupler modulator, bias points exist where either the second- or the third-order distortion is zero without the fundamental going to zero.

A typical plot of third-order IM-free DR for a Mach-Zehnder biased at $V_\pi/2$ is shown in Fig. 9.60. Although the data in this plot were taken at 60 MHz, they should be valid at any frequency. This is because for a modulator, unlike the diode laser, the shape of the electrical-to-optical transfer function is independent of RF frequency. This has been experimentally demonstrated up to 20 GHz (Betts, Cox and Ray, 1990). A directional coupler modulator, when biased where the second-order distortion is zero, has almost identical third-order IM-free DR.

Linearization

Although the IM-free DR for external modulators is high, it still may not be sufficient for some applications in CATV and antenna remoting. One way to increase the IM-free DR is to use any of the techniques described previously for the diode laser. However, because the transfer function of many modulators can be expressed analytically, this opens up an alternate class of all-optical linearization techniques. It is important to keep in mind a key distinction between these two classes of approaches. The *electronic* approaches generally target reducing the total distortion regardless of the order. Consequently, these approaches are most effective when the distortion has significant contributions at several orders. The *optical* approaches attempt to reduce a particular order of distortion, which makes these approaches most effective when the distortion is dominated by one order.

The optical approaches (Bridges, 1995) are based on connecting two or more modulators in series or parallel under the constraint that the transfer function of the combination is more linear under some bias condition than any of the individual transfer functions. The percentage bandwidth plays an important role in these all-optical linearization techniques; narrow-band applications, such as antenna remoting, require that the third-order distortion be reduced, but this can be done at the expense of *increasing* the second-order distortion. On the other hand wide-band applications, such as CATV, require that the third-order distortion must be reduced *without* increasing the second-order distortion. The upper limit on the IM-free DR improvement; regardless of the approach, is the *signal-to-noise dynamic range*, SNDR. This bound leaves substantial room for improvement; under typical operating conditions a standard Mach-Zehnder has an IM-free DR of 110 dB $Hz^{2/3}$ whereas the SNDR is 160 dB Hz.

Broad-band linearization of a Mach-Zehnder can be accomplished by connecting two Mach-Zehnders in either series (Skeie, 1991) or parallel (Korotky, 1990), as shown in simplified block diagram form in

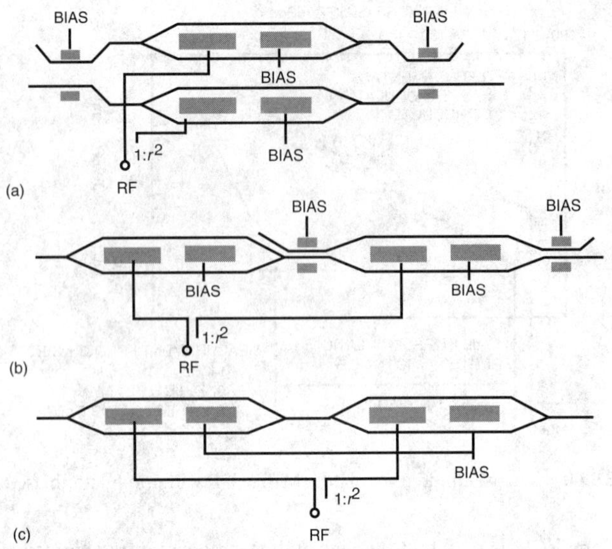

FIGURE 9.61　Three all-optical approaches to linearization of a Mach-Zehnder; broadband linearization can be accomplished via either: (a) parallel connection or, (b) series connection; (c) narrow-band linearization using a series connection.

Fig. 9.61. The bias control of either approach can be relatively complex, because the bias voltages for minimization of the second and third order are interdependent. For example, in the series approach, four bias voltages are required; one on each of the Mach-Zehnders and one on each of the directional couplers. However, IM-free DR improvements of 10–15 dB have been reported.

Recall that for a Mach-Zehnder there is no bias point where the thirds go to zero without the fundamentals also going to zero. Thus some form of linearization, for example, multiple modulators in the present discussion, is required for improved IM-free DR in narrow-band applications. However narrow-band linearization of a Mach-Zehnder is significantly simpler than broad-band linearization. Figure 9.61(c) shows a series connected Mach-Zehnder for narrow-band linearized performance (Betts, 1994). Although superficially similar to the broad-band modulator shown in Fig. 9.61(b), notice that the narrow-band configuration avoids the need for the two controllable directional couplers, thereby cutting in half the number of control voltages. Figure 9.62 is a plot of the IM-free DR for a narrow-band

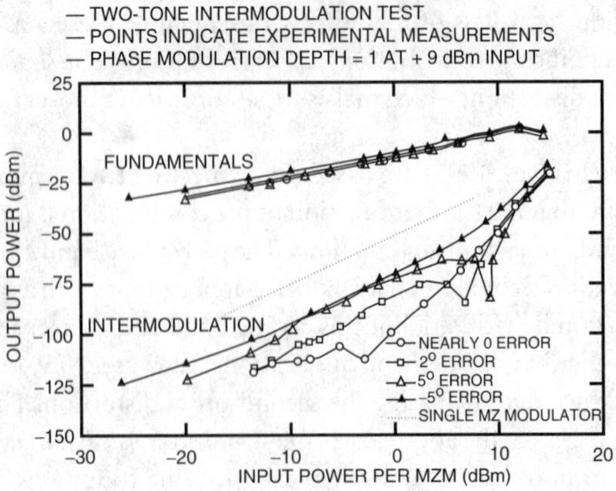

FIGURE 9.62　IM-free DR of a narrow-band, series-linearized Mach-Zehnder with bias control error as a parameter. For reference the IM-free DR of a standard Mach-Zehnder is also shown. (Betts, 1994).

series linearized Mach-Zehnder modulator. Note that with perfect bias control, the third-order distortion is completely canceled leaving only the fifth-order distortion, as evidenced in the plot by the distortion terms lying on a line with a slope of five. Under these conditions the IM-free DR of a linearized modulator scales as $(BW)^{4/5}$ rather than the $(BW)^{2/3}$ as a standard Mach-Zehnder does. It can be seen from the plot that with perfect bias control, 30 dB of improvement over a standard Mach-Zehnder is possible. However, the plot also shows the effects of bias voltage errors. In addition to the reduction in IM-free DR improvement, it is also important to note that the intermodulation signals no longer follow a single slope. This complicates the scaling of the results to bandwidths other than the measurement bandwidth, since no single power law is valid. Perhaps the best way to handle these cases is to measure and always specify the IM-free DR in the intended application bandwidth.

It is also possible to linearize a directional coupler modulator. As was mentioned previously, a directional coupler has bias points where either the second- or the third-order distortion theoretically go to zero. Thus for narrow-band applications the proper choice of bias point produces *linearized* performance. This converts the linearization problem to a bias control problem: the higher the required IM-free DR, the tighter the bias control needs to be. Consequently, linearization as is being discussed here is only required on directional couplers for wide-band applications, where both the second- and the third-order distortion must be suppressed. One way to achieve wide-band IM-free DR improvement in these modulators would be to move the second- and third-order nulls so they occur at the same voltage. It turns out that this is possible by adding an extra pair of electrodes to the directional coupler (Farwell, 1991). Only a DC bias (no RF modulation) is applied to this auxiliary electrode. This bias is adjusted in conjunction with the bias on the main RF electrodes to achieve minimum distortion of the second- and third-order distortion. Although it is theoretically possible to achieve linearization with only one additional pair of electrodes, in practice the control algorithm is significantly simpler if a third electrode pair is added.

Lasers for External Modulation and Figures of Merit

Types of Lasers

The source for a standard external modulation link operates CW; thus, there are no modulation requirements on the laser. This removes the primary constraint, which in the direct modulation case limited the laser choice to semiconductor lasers. Conceptually, then, any type of laser could serve as the source for an external modulation link. Early external modulation links used semiconductor diode lasers, but more recently the majority of external modulation links have been fed by solid-state lasers.

It is important to keep in mind that unlike the electronics field, the laser field uses semiconductor and solid state to reference two distinctly different types of lasers. Semiconductor lasers are almost exclusively excited or pumped electrically by passing an electric current through a diode junction. Typical semiconductor laser materials include GaAs for lasers emitting around 0.8 μm and indium phosphide-based materials for lasers emitting either at 1.3 or 1.55 μm. The exact emission wavelength can be set by choosing the composition of the active layer, for example, in GaAs lasers the wavelength can be changed from 0.87 to 0.64 μm by changing the concentration of aluminum from 0 to 40%. Solid-state lasers, the label for a class of lasers *not* including semiconductor lasers, are universally pumped optically. Common solid-state lasers used for fiber communication links include neodymium doped yttrium aluminum garnet, or Nd:YAG, which can be designed to emit at 1.06 or 1.3 μm, and erbium lasers which emit around 1.55 μm. Unlike their semiconductor counterparts, the emission wavelength of solid-state lasers is fixed by the energy levels in the host crystal. Consequently, in general the emission wavelength of solid state-lasers is much more strictly confined. The first solid-state lasers were pumped using gas discharge or flash lamps. This was relatively inefficient since only one of the flash lamp's myriad emission lines fell on the absorption wavelength of the solid state material. More recently semiconductor diode lasers have been designed that can emit all their energy at the solid-state absorption wavelength. Consequently, these diode-pumped solid-state lasers are more compact and more efficient (Koechner, 1995).

As was discussed in a previous section, external modulators are available that can handle almost 0.5 W of optical power, provided that this is at a wavelength where photorefractive effects are negligible. Further,

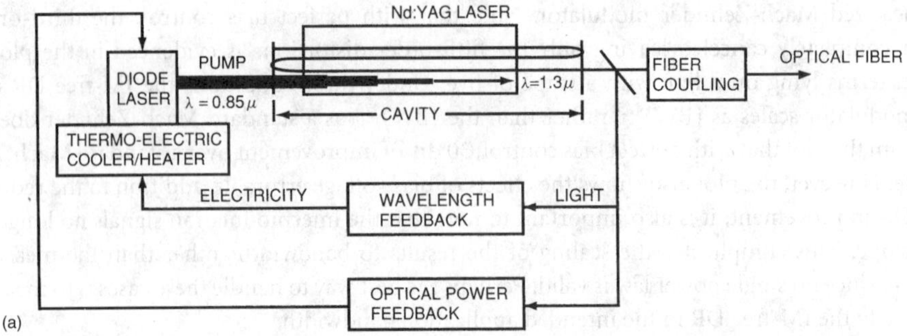

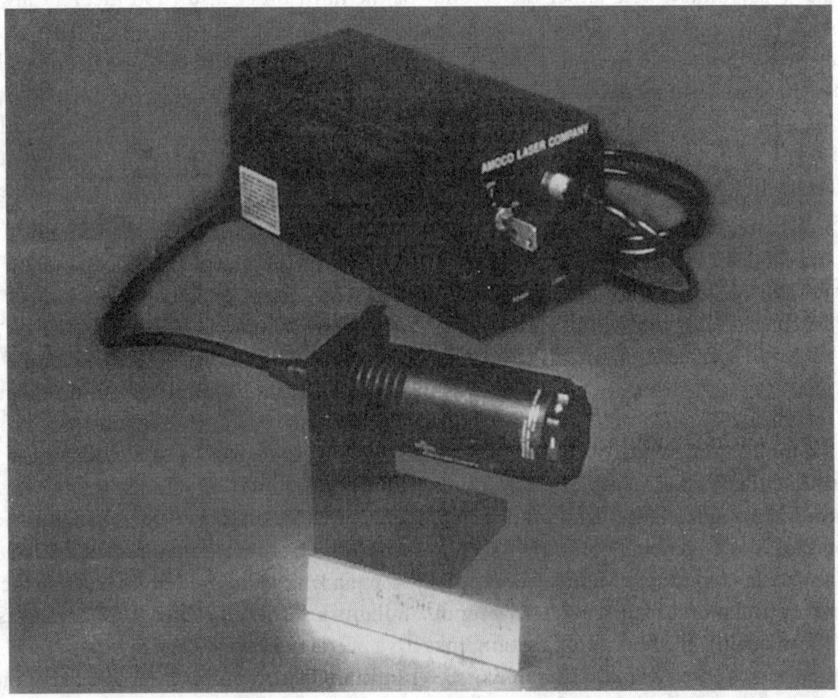

(a)

(b)

FIGURE 9.63 (a) Representative diagram, (b) photograph of a diode-pumped Nd:YAG laser.

the laser emission wavelength needs to be at one of the three common ones where single-mode optical fibers are available. A solid-state laser that meets both of these requirements is the Nd:YAG. Although the most common wavelength for these lasers is 1.06 μm, this is not a common fiber wavelength nor is it long enough for photo-refractive effects in lithium niobate to be negligible. Thus for external modulation applications a secondary emission wavelength of Nd:YAG at 1.319/1.338 μm is used. The strength of this longer wavelength is about half of the line at 1.06 μm, and so it is still possible to obtain sufficient optical power with reasonable efficiency.

A representative diagram of a diode-pumped solid-state laser is shown in Fig. 9.63(a). The solid-state laser cavity basically consists of the Nd:YAG rod and two mirrors. The mirror from which the desired output is taken is coated to transmit a portion of the 1.3-μm light and totally reflect the 0.8-μm pump light. The mirror at the other end of the cavity is coated to totally reflect the 1.3-μm light and antireflection coated to pass the 0.8-μm pump light. The nominal wavelength of the diode laser is chosen to correspond to the absorption wavelength of the Nd:YAG. To maintain laser operation, two feedback loops are generally required. One feedback loop keeps the pump emission wavelength at the same wavelength as the Nd:YAG absorption peak. Since the emission wavelength of diode lasers is temperature dependent, one way to do this is by using a feedback loop to change the diode laser temperature. The other feedback loop maintains a constant 1.3-μm optical power, typically via a feedback loop which controls the pump laser current. Since the 1.3-μm optical power depends on both the pump wavelength and the pump power, care must be taken in designing these feedback loops to ensure the desired operation. Figure 9.63(b) is a photograph of a commercially available diode-pumped solid-state laser.

The required optical power is also available from diode lasers. However, there are several drawbacks that limit the applicability of diode lasers at present. One is the wavelength: historically high-power diode lasers have only been available around 0.8 μm, which is well within the spectral range for photorefractive damage in lithium niobate. Another is that these high-power diode lasers achieve the power by forming linear arrays of individual diode lasers. As a result, these arrays have very low coupling efficiency to optical fibers. A third reason has to do with the high relative intensity noise of most diode lasers, a point that will be discussed in more detail in the RIN section. Diode laser designers have begun to address these shortcomings. For example, there is a diode laser that couples into a fiber 50 mW of 1.3-μm light from a single laser cavity, that is, not an array (Chen, 1995).

Spectrum and Output Power

Unless special laser cavity design features are included, the laser's spectral output contains emissions at more than one wavelength. Each of these emission wavelengths or lines corresponds to a different longitudinal mode of the laser cavity. Since it is the total optical power that is important in an intensity modulated link, and since it would be considerably more difficult to obtain the high optical powers in a single laser mode, all CW lasers for general purpose external modulation links are multimode. Care must be taken to distinguish the multiple *cavity* modes of a laser in this discussion from the multiple *propagation* modes of a multi-mode fiber. Thus, it is entirely possible to send the multiple laser cavity modes down the single propagation mode of a fiber.

For fiber applications, a Nd:YAG laser can be designed to operate on one of two lines near 1.3 μm: 1.319 or 1.338 μm. At either of these wavelengths, the laser will have multiple longitudinal modes. In Nd:YAG the total optical power is not spread uniformly among all of the lasing modes, but concentrated in two modes. To increase the number of modes generally requires changing the host crystal; for example, yttrium lanthanum fluoride- (YLF-) based lasers typically have their power distributed among seven modes.

The total optical power from a Nd:YAG laser operating at 1.3 μm can exceed 0.75 W. Largely because of their excellent beam quality, the fiber coupling efficiency for YAGs is typically 80%. Thus, over 0.5 W of fiber-coupled 1.3-μm light is now commercially available. Single-mode YAG lasers produce about a factor of 10 less power.

Relative Intensity Noise

Figure 9.64(a) is a plot of the RIN for a Nd:YAG laser. In comparing these data with the diode laser RIN shown in Fig. 9.45, it is clear that the RIN spectrum of solid-state lasers has the same basic shape as that for diode lasers. The most significant difference is that the relaxation frequency occurs at a much lower frequency. Since direct electrical modulation of the laser intensity is only feasible for frequencies below the relaxation oscillation, it is clear from this figure why solid-state lasers are rarely used for direct modulation. The low relaxation resonance also means that the operating frequencies of virtually all links are at frequencies above the relaxation peak. The RIN decreases as the square of the ratio of the operating frequency to the relaxation frequency. With relaxation frequencies around 100 kHz and operating frequencies of at least 10 MHz, the RIN of solid state is negligible in most applications.

In addition to the relaxation peak, there are peaks in the RIN spectrum at frequencies that correspond to the difference in frequency between the longitudinal modes. The typical mode spacing of early diode-pumped solid-state lasers was about 4 GHz. This puts the first intermodal RIN peak outside the

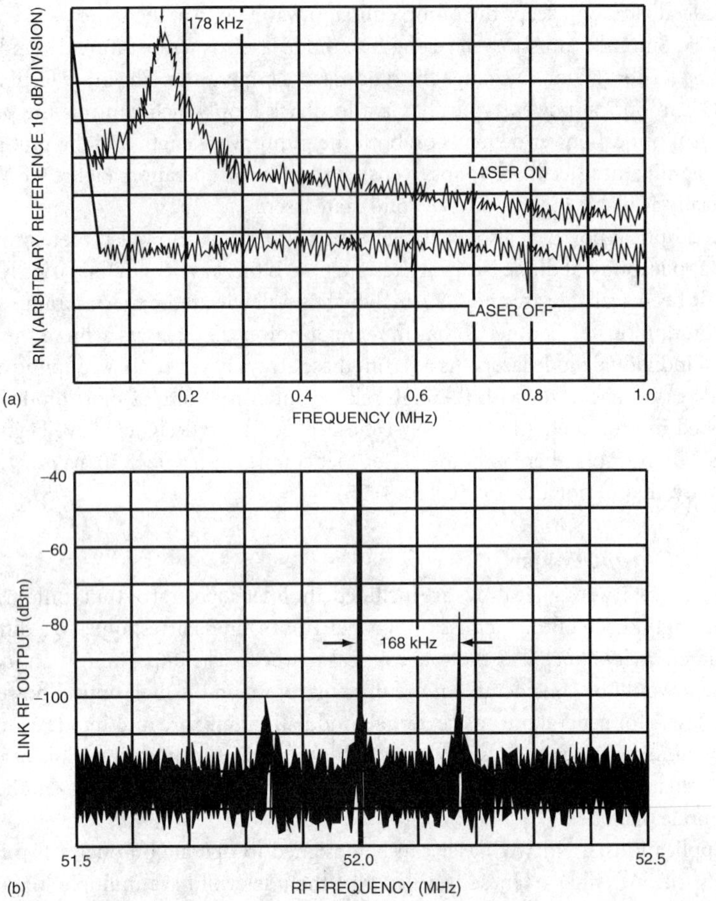

FIGURE 9.64   (a) Typical RIN spectrum for a Nd:YAG laser, (b) RIN spectrum from (a) appearing as sidebands around a single 52-MHz CW carrier.

bandwidth of many applications, such as CATV distribution. However, for wide-band data communication or high-frequency antenna remoting, it is desirable to have wider longitudinal mode spacing. To meet these needs, lasers with intermodal spacing of up to 40 GHz have been developed. Of course, the needs of these applications can also be met with single-mode lasers, albeit at a sacrifice in optical power.

The process of external intensity modulation translates the solid-state laser RIN spectrum up to the modulation carrier frequency resulting in RIN spectrum sidebands about each carrier of the modulation signal. To see an example of this process, consider the case where the modulation signal was a single 52-MHz CW carrier. Figure 9.64(b) shows this carrier with the solid state laser RIN from Fig. 9.64(a) appearing as side-bands. For some applications, these sidebands, especially the relaxation peak, can limit system performance. Fortunately this peak occurs at a low enough frequency that it is possible to use feedback to suppress the peak. The remainder of the solid-state laser RIN below the relaxation peak is due to the diode pump laser. Since the laser responds to modulation at frequencies below the relaxation peak, the low-frequency RIN at the output of the solid-state laser is simply a pass through of the RIN from the diode pump laser. As was seen previously, the diode laser has RIN that extends up into the gigahertz, but this higher frequency RIN is effectively filtered out by the solid-state laser's weak response to modulation above its relaxation frequency.

Implementation Issues

Maximum Optical Power (Cox, 1993)

As was previously mentioned, the RF performance of externally modulated links improves with increases in optical power. Thus, it is advantageous to operate such links at the maximum practical power. The highest optical power level in the link occurs in the fiber between the laser and the modulator. If the length of the fiber between these two components can be kept to less than a few hundred meters, then the maximum optical power will not be limited by stimulated Brillouin scattering in the optical fiber. The source end of a fiber CATV distribution network would fit this condition. In such cases the maximum optical power is limited by the modulator. In waveguide modulators, the power is probably ultimately limited by heating from the residual absorption of the guide and by nonlinear effects in the guide. In practice, the maximum power is limited by imperfections in the fiber attachment.

The power limit is wavelength and modulator configuration specific. The preceeding results are all for MZMs operating at 1.3 μm. Lithium niobate MZMs should be able to handle this same high power at longer wavelengths, although there does not appear to be experimental data to support this conjecture. At shorter wavelengths, that is, less than about 1 μm, photorefractive effects limit the usable maximum optical power. The power level for the onset of photorefractive effects depends on the waveguide formation process and the material. Consider the following results at 0.85 μm, a common diode laser wavelength. Titanium guides in lithium niobate show photorefractive damage at optical powers below 1 mW. However, as reported by Suchoski (1995), proton exchange guides in lithium niobate can support greater than 10 mW before the onset of photorefractive degradation, and proton exchange guides in lithium tantalate can support over 100 mW.

The optical power limits in other materials have not been as extensively studied. In GaAs, a MZM has been operated with 10 mW (Roussell, 1997), but the upper limit is unknown. It has been reported that more than 40 mW can propogate in an InGaAsP/InP waveguide (Yu, 1995). In this case also the upper limit is unknown. However, the upper limit for the EA modulator is expected to be lower than the MZM because when off, the EA modulator is absorbing all of the optical power, whereas the MZM disperses the optical power out of the guide and into the substrate when off.

Bias Control

To obtain the desired RF performance, primarily IM-free DR and noise figure, an external modulator needs to operate at a specific point along its transfer curve. A number of factors, among them temperature and photo-refractive effects, can shift the modulator operating point by at least the V_π voltage over time. Thus, to maintain a specific operating point typically requires some form of bias control circuit. There are lots of bias control techniques, and so the selection of the appropriate technique depends on application considerations and the particular bias point one is trying to maintain. A straightforward bias control approach to maintain bias points where either the second- or the third-order distortion is zero is to add a low frequency sine wave to the DC on the bias electrode. The sine wave perturbs or dithers the bias point, causing harmonics of the dither frequency to appear on the modulated optical signal. For ease of explanation, assume one desires to maintain a bias point where the second-order distortion is zero. The desired bias point is maintained by detecting the dither's second harmonic and using a feedback loop to adjust the modulator bias until the second harmonic is minimum. In an analogous manner the third harmonic can be used to maintain the bias point at a third harmonic null point.

Although a dither-based bias controller can maintain the bias point with a high degree of precision, the presence of a dither signal, however small, may be intolerable in some applications. There are several alternatives to a dither controller: however, perhaps the most elegant is to design and fabricate the modulator such that no controller is required. Such passively stabilized modulators have been demonstrated (Greenblatt, 1995). These approaches have reduced the bias drift by two orders of magnitude over an unstabilized modulator, which is sufficient for narrow-band applications using standard, that is, non linearized, modulators. To use passive bias in linearized applications or in broad-band applications will require another order of magnitude improvement in passive bias stability.

Polarization

In most electro-optic materials, the magnitude of the electro-optic coefficient depends on the polarization of the optical wave. For example, in lithium niobate the electro-optic coefficient is about three times larger for transverse electric (TE) light than it is for transverse magnetic (TM) light. Consequently, the V_π for TE light is about 3 times lower than for TM light. Since the link gain is proportional to $1/V_\pi^2$, variations in the polarization would cause up to a 10-dB change in link gain. Thus, in most applications it is essential that the polarization of the light at the modulator input be maintained over time. Since micro-bends in regular single-mode fiber cause random shifts in the state of polarization, it is generally unsuitable for use on modulator inputs. Therefore, virtually all modulator inputs have polarization maintaining (PM) single-mode fiber. Although this fiber is available from several manufacturers, at present it is at least a factor of 50 more expensive than regular single-mode fiber. PM fiber also requires special connectors and fiber fusing techniques. The costs and constraints of PM fiber have led to the investigation of alternatives.

Perhaps the most straightforward approach to eliminating the need for polarization would be to build a polarization independent modulator. For example, in lithium niobate, the next strongest electro-optic coefficient is r_{13}, which is about 1/3 as large as r_{33}. This coefficient can be used to build a polarization independent modulator since it applies to polarizations on both the x and y axes. Polarization independent modulators have been built, but their performance is substantially lower than the modulators that require polarized light (Spickermann, 1994).

Since it is almost never acceptable to trade RF performance for polarization independence, research has focused upon keeping the standard polarization sensitive modulators but eliminating the polarization maintaining fiber. One approach is to use a polarization controller at the modulator input to remove the changes in polarization that are introduced by standard single-mode fiber. By introducing controlled bends in a short piece of single-mode fiber placed just before the modulator, the polarization state can be restored to the required state. Most polarization controllers have a finite range and, consequently, must be reset periodically. Also packaging constraints at the modulator may preclude the use of such a controller.

Another approach to delivering polarized light to the modulator is the combination of a depolarized laser source, regular single-mode fiber and a polarizer at the modulator. The depolarized source (Burns, 1991) generates light of two orthogonal linear polarizations and launches these down a regular single-mode fiber. Microbends still alter the polarization states of each of these light waves, but on average the power in each linear polarization remains constant. At the modulator the desired linear polarization is extracted by passing the light through a polarizer. Alternatives for the polarizer include a length of *polarizing* fiber and proton exchange waveguides, which support only one polarization. Among the tradeoffs of the depolarized source approach is the loss of 3 dB of optical power as a result of the polarization process at the modulator.

Packaging and Fiber Attachment

These areas are a mixed bag relative to the corresponding areas with a diode laser. In several respects the tasks are easier with an external modulator. There is no junction, at least when the modulator is fabricated in either an inorganic or a polymer. Thus, there is no need to maintain a specific operating temperature, which eliminates the need for a TEC and thermistor. Also there is no feedback of optical power, which eliminates the rear facet photodiode commonly found in a diode laser package.

However, an external modulator is much larger than a diode laser and it has a relatively long aspect ratio: a typical modulator is a few millimeters wide and 60 mm long, whereas a typical diode laser chip is about 1 mm^2. To further complicate mounting a modulator, many of the inorganics including lithium niobate, have thermal expansion coefficients that are different for the three crystal directions. Despite these obstacles, several mounting techniques have been developed that permit the modulator to be used over 100°C temperature range and up to 15-Gs shock.

In virtually all cases the modulator has two fiber attachments, instead of the one the diode laser required. The fiber coupling loss in lithium niobate is low because the waveguide mode is reasonably closely matched to the fiber mode. However, active alignment is still required, which today is done manually. Automated fiber attachment equipment is in development to eliminate this labor intensive, and hence costly, step.

Figure 9.65 shows packaged MZMs from three commercial manufacturers.

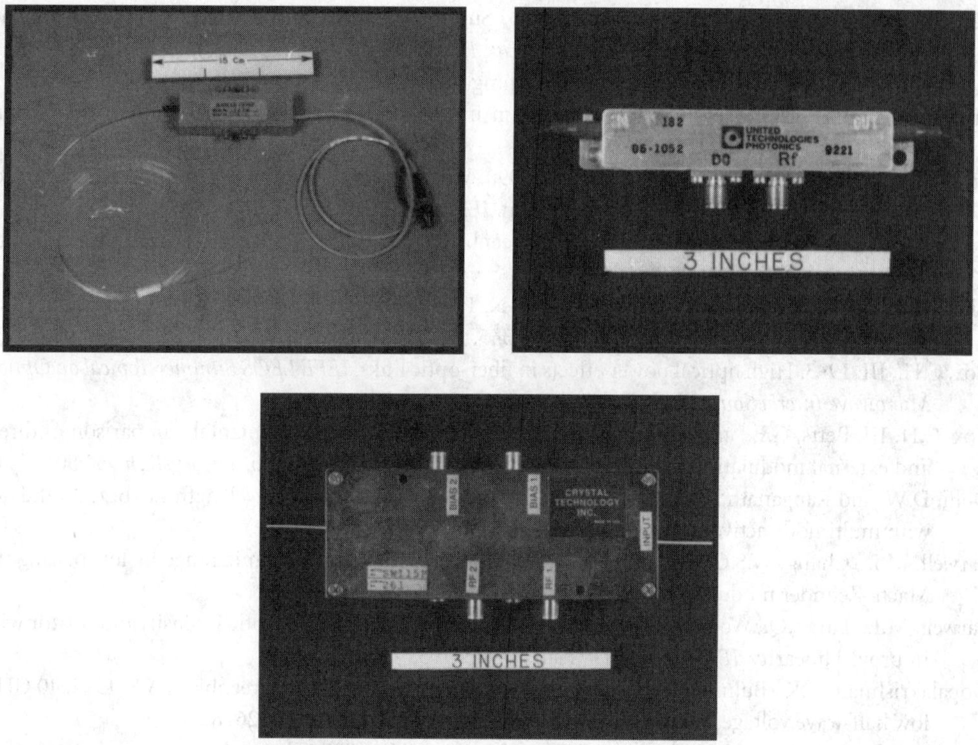

FIGURE 9.65 Photograph of lithium niobate MZMs from three commercial manufacturers.

Acknowledgments

The author would like to thank Gary Betts, Ed Ackerman, Roger Helkey of MIT Lincoln Labortory and Paul Yu of University of California at San Diego for reviewing drafts of the manuscript and offering numerous helpful suggestions. This work was sponsored by the Department of the Air Force under Contract F19628-95-C-0002. The views expressed in this section are those of the author and do not reflect the offical policy or position of the U.S. Government.

References

Agrawal, G.P. and Dutta, N.K. 1986. *Long-Wavelength Semiconductor Lasers*. Van Nostrand-Reinhold, New York.

Alferness, R.C. 1982. Waveguide electrooptic modulators. *IEEE Trans. Micro. Theory Tech*. MTT-30:1121–1137.

Alferness, R.C., Korotky, S.K., and Marcatili, E.A.J. 1984. Velocity-matching techniques for integrated optic traveling wave switch-modulators. *IEEE J. Quan. Elec*. QE-20:301–309.

Becker, R.A. 1984. Broad-band guided-wave electrooptic modulators. *IEEE J. Quan. Elec*. QE-20:723–727.

Bertelsmeier, M. and Zschunke, W. 1984. Linearization of broadband optical transmission systems by adaptive predistortion. *Frequenz* 38:206–212.

Betts, G.E. 1994. Linearized modulator for suboctave-bandpass optical analog links. *IEEE Trans. Micro. Theory Tech*. 42:2642–2649.

Betts, G.E., Cox, C.H. III, and Ray, K.G. 1990. 20 GHz optical analog link using an external modulator. *IEEE Phot. Tech. Lett*. 2:923–925.

Betts, G.E., Johnson, L.M., and Cox, C.H. 1989. High-sensitivity lumped-element bandpass modulators in LiNbO$_3$. *J. Lightwave Tech*. 7:2078–2083.

Betts, G.E., Ray, K.G., and Johnson, L.M. 1990. Suppression of acoustic effects in lithium niobate integrated-optical modulators. In *Technical Digest on Integrated Photonics Research*, Vol. 5, pp. 37–38. Optical Society of America, Washington, DC.

Bridges, W.B. and Schaffner, J.H. 1995. Distortion in linearized electrooptic modulators. *IEEE Trans. Micro. Theory Tech.* 43:2184–2197.

Bridges, W.B., Sheehy, F.T., and Schaffner, J.H. 1990. Wave-coupled LiNbO$_3$ electrooptic modulator for microwave and millimeter-wave modulation. *IEEE Phot. Tech. Lett.* 2:133–135.

Burns, W.K., Moeller, R.P., Bulmer, C.H., and Greenblatt, A.S. 1991. Depolarized source for fiber-optic applications. *Optics Lett.* 16:381–383.

Chen, T.R., Chen, P.C., Ungar, J., and Bar-Chaim, N. 1995. High-power operation of multiquantum well DFB lasers at 1.3 μm. *Elect. Lett.* 31:1344–1345.

Cox, C.H., III. 1993. High optical power effects in fiber-optic links. *IEEE LEOS Summer Topical on Optical Microwave Interactions*, Santa Barbara, CA., July 19–21, Paper T2.1.

Cox, C.H. III, Betts, G.E., and Johnson, L.M. 1990. An analytic and experimental comparison of direct and external modulation in analog fiber-optic links. *IEEE Trans. Micro. Theory Tech.*, 38:501–509.

Dolfi, D.W. and Ranganath, T.R. 1992. 50 GHz velocity-matched, broad wavelength LiNbO$_3$ modulator with multimode active section. *Elec. Lett.* 28:1197–1198.

Farwell, M.L., Chang, W.S.C., and Huber, D.R. 1993. Increased linear dynamic range by low biasing the Mach–Zehnder modulator. *IEEE Phot. Tech. Lett.* 3:779–782.

Farwell, M.L., Lin, Z.Q., Wooten, E., and Chang, W.S.C. 1991. An electro-optic intensity modulator with improved linearity. *IEEE Phot. Tech. Lett.* 3:7982–795.

Gopalakrishnan, G.K., Bulmer, C.H., Burns, W.K., Mcelhanon, R.W., and Greenblatt, A.S. 1992. 40 GHz, low half-wave voltage Ti:LiNbO$_3$ intensity modulator. *Elect. Lett.* 28:826–827.

Greenblatt, A.S., Bulmer, C.H., Moeller, R.P., and Burns, W.K. 1995. Thermal stability of bias point of packaged linear modulators in lithium niobate. *J. Lightwave Tech.* 13:2314–2319.

Hassin, D. and Vahldieck, R. 1993. Feedforward linearization of analog modulated laser diodes-theoretical analysis and experimental verification. *IEEE Trans. Micro. Theory Tech.* 41:2376–2382.

Hirota, O., Suematsu, Y., and Kwok, K.S. 1981. Properties of intensity noises of laser diodes due to reflected waves from single-mode optical fibers and its reduction. *IEEE J. Quantum Elec.* QE-17:1014–1020.

Kazarinov, R.F. and Belenky, G.L. 1995. Novel design of AlGaInAs-InP lasers operating at 1.3 μm. *IEEE J. Quan. Elec.* 31:423–426.

Kim., E.M., Tucker, M.E., and Cummings, S.L. 1989. Method for including CTBR, CSO and channel addition coefficient in multichannel AM fiber optic system models. *NCTA Technical Papers*, p. 238.

Koechner, W. 1995. *Solid-State Laser Engineering*, 4th ed. Springer Series in Optical Sciences. Springer-Verlag, New York.

Korotky, S.K. and Ridder, R.M. 1990. Dual parallel modulation schemes for low-distortion analog optical transmission. *IEEE J. Sele. Areas Commun.* 8:1377–1381.

Lau, K., Derry, P., and Yariv, A. 1988. Ultimate limit in low threshold quantum well GaAlAs Semiconductor lasers. *Appl. Phys. Lett.* 52:88–90.

Lester, L.F., Schaff, W.J., Song, X., and Eastman, L.F. 1991. Optical and RF characteristics of short-cavity-length multiquantum-well strained-layer lasers. *IEEE Phot. Tech. Lett.* 3:1049–1051.

Martin, W.E. 1975. A new waveguide switch/modulator for integrated optics. *Appl. Phys. Lett.* 26:552–553.

Morton, P.A. et al. 1992. 25 GHz bandwidth 1.55 μ GaInAsP p-doped strained multiquantum-well lasers. *Elec. Lett.* 28:2156–2157.

Noguchi, K., Mitomi, Kawano, K., and Yanagibashi, M. 1993. Highly-efficient 40 GHz bandwidth Ti:LiNbO$_3$ optical modulator employing ridge structure. *IEEE Pho. Tech. Lett.* 5:52–54.

Noguchi, K., Miyazawa, H., and Mitomi, O. 1994. 75 GHz broadband Ti:LiNbO$_3$ optical modulator with ridge structure. *Elec. Lett.* 30:949–951.

Papuchon, M. et al. 1975. Electrically switched optical directional coupler: COBRA. *Appl. Phys. Lett.* 27:289–291.

Ralston, J.D. et al. 1994. Low-bias-current direct modulation up to 33 GHz in InGaAs/GaAs/AlGaAs pseudomorphic MQW ridge-waveguide lasers. *IEEE Phot. Tech. Lett.* 6:1076–1079.

Roussell, H.V. and Helkey, R. 1997. High dynamic range optical links using Fabry-Perot lasers, to be published.

Schmidt, R.V. and Kaminouw, I.P. 1974. Metal diffused optical wave-guides in LiNbO$_3$. *Appl. Phys. Lett.* 25:458–460.

Seeds, A.J. 1992. Microwave opto-electronics. In the *IEEE Emerging Technology Series* (Video), 62 mins, released May.

Shah, M.L. 1975. Optical waveguides in LiNbO$_3$ by ion exchange techniques. *Appl. Phys. Lett.* 26:652–653.

Skeie, H. and Johnson, R.V. 1991. Linearization of electro-optic modulators by a cascade coupling of phase modulating electrodes. In *Integrated Optical Circuits*, SPIE Vol. 1583, pp. 153–164, SPIE.

Spickermann, R., Peters, M., and Dagli, N. 1994. Polarization independent GaAs/AlGaAs electro-optic modulator. *IEEE/LEOS Summer Topical Meeting on Integrated Optoelectronics*, July.

Straus, J. 1978. Linearized transmitters for analog fiber links. *Laser Focus* (Oct.):54–61.

Suchoski, P. 1995. High optical power at 0.85 μm in lithium niobate and lithium tantalate. (Personal communication).

Thompson, G.H.B. 1980. *Physics of Semiconductor Laser Devices*. Wiley, New York.

Walker, R.G. 1991. High-speed III-V semiconductor intensity modulators. *IEEE J. Quan. Elec.* 27:654–667.

Wang, W., Chen, D., Fetterman, H.R., Shi, Y., Steier, W.H., and Dalton, L.R. 1995. 40-GHz polymer electrooptic phase modulators. *IEEE Phot. Tech. Lett.* 7:638–640.

Yammoto, Y. 1983. AM and FM quantum noise in semiconductor lasers—Part I: Theoretical analysis. *IEEE J. Quan. Elec.* QE-19:34–46.

Yariv, A. 1975. *Quantum Electronics*, 2nd ed. Chap. 14. Wiley, New York.

Yu, P. 1995. Electro-absorption modulators. (Personal communication).

Further Information

Optical transmitters have been extensively studied in the past and are presently an active area of research. So although we have attempted to cover the basics in this section, there is much that had to be omitted and much that will be developed in the years following the publication of this book. To meet these needs, the reader is referred to the following journals and conferences:

Applied Physics Letters. American Institute of Physics, New York

Electronics Letters. IEE, London

Fiber and Integrated Optics. Taylor & Francis, Washington, DC

IEEE Journal of Quantum Electronics. Institute of Electrical and Electronics Engineers, New York

IEEE Photonics Technology Letters. Institute of Electrical and Electronics Engineers, New York

IEEE Transactions on Microwave Theory and Techniques. Institute of Electrical and Electronics Engineers, New York

Journal of Lightwave Technology. Institute of Electrical and Electronics Engineers, New York

Journal of Optical Communications. Fachverlag, Schiele & Schon, Berlin

Microwave and Optical Technology Letters. Wiley, New York

Optics Letters. Optical Society of America, New York

Conference on Lasers and Electro-optics (CLEO)

Integrated Photonic Research Conference

Optical Fiber Communications (OFC) Conference

IEEE MTT-S International Microwave Symposium

IEEE Lasers and Electro-Optics Society (LEOS) Annual Meeting

IEEE LEOS Summer Topical Meetings

Optical transmitters are rarely used as stand-alone components; more frequently they are used in conjunction with optical receivers as described in the next section to form a link for conveying an electrical signal over an optical fiber. For a discussion on how optical transmitters and receivers can be connected together to form a link, see, for example, Cox (1990) and Seeds (1992).

9.4   Optical Receivers

Paul Kit-Lai Yu

9.4.1   Introduction: Terminology in Optical Detection and Material Considerations

Semiconductor materials are commonly employed for optical detection mainly because the semiconductor detectors are usually small, require a low bias voltage, and can be easily assembled with the rest of the receiver. In semiconductors, photons interact with electrons and are absorbed in processes whereby the electrons make a transition from a lower to a higher energy state. These electrons are detected in an external circuit. The absorption is dependent on the photon energy and each material has its characteristic absorption spectrum. The absorption spectrum can be sensitive to electric field and temperature variations.

The absorptivity α_0, defined as the incremental decrease in optical intensity per unit length per unit incident power, is usually referred to as the *absorption coefficient* of the material. The photon energy $h\nu$ (where h is Planck's constant and ν is frequency) and wavelength λ are simply related by

$$\lambda = \frac{1.24}{h\nu} \tag{9.7}$$

where λ is measured in micrometer and $h\nu$ in electron volt.

The absorptivity α_0 describes the power decay of an optical beam as a function of its penetration into a given material. In general, a large α_0 is desirable in an optical detector for getting a high optical-to-electrical signal conversion efficiency. A small α_0 would require a large absorption region for sufficient absorption to take place. Too large an α_0, however, can result in a small penetration depth, and many of the high-density electrons and holes generated can be lost via near-surface recombination and are not detected.

An important performance parameter of an optical detector at a given wavelength is the **external quantum efficiency** η. It is defined as the ratio of the number of electrons generated to the number of incident photons before any photogain occurs. Here η takes into account the surface reflection loss and other losses in the detector. In general, η depends on the absorptivity of the materials and the dimensions of the absorption region. The photogain M can result from carrier injection in semiconductor materials, as in the case of photoconductive devices (see Sec. 9.4.3), or from impact ionization, as in the case of avalanche photodiodes (see Sec. 9.4.2).

The **responsivity** R, in amperes per watt, of a detector is the ratio of the output electrical response to the input optical power and can be expressed as

$$R = M\eta \frac{q}{h\nu} \tag{9.8}$$

where q is the electronic charge. For many applications, the optical power reaching the photodetector is small, and the responsivity remains unchanged as the optical power is varied slightly. The relation between R (or η) and wavelength is usually referred to as the **spectral response** characteristics. For applications where a large optical power level is incident on the detector, the responsivity can drop as a result of the carrier screening effect, as in the case of photodiode. Nonlinear response can also be a consequence of detection saturation.

The response speed of a photodetector can be very important for many analog and digital applications. The characteristics 3-dB **cutoff frequency** $f_{3\,dB}$ is inversely proportional to the response time of the detector. The $f_{3\,dB}$ can also be greatly reduced when detection saturation sets in.

The response time τ depends on the transit time of electrons and holes in the detector, the carrier diffusion process, the carrier multiplication process in the semiconductor, as well as the circuit time constant. Usually, τ can be determined from the $f_{3\,dB}$ obtained from a frequency modulation response measurement with a modulated optical carrier generated by direct modulation, or by external modulation of laser light, or simply by optical mixing of two laser sources whose emission wavelengths are temperature tuned. Alternatively, τ can be determined from the time-domain response of the detector when it is excited by optical pulses, such as those generated by mode-locked lasers, or those from a gain switched laser driven by a comb generator. Provided that the rise time and the fall time of the optical pulse are much shorter than those of the detector, τ can be estimated from both the rise time and fall time of the detector current pulse. For very high-speed detector (>40 GHz), a high-speed sampling oscilloscope is usually used to measure the total time-domain response, including the detector response time, the effect of the finite optical pulse width and the finite response time of the sampling circuit. The detector response can be de-embedded from the total time-domain response.

For detectors with photogain, an additional figure of merit, the *gain-bandwidth* (*MB*) product is defined. It is simply the product of the 3-dB bandwidth and the DC gain. In many cases, the *MB* product is closely related to device geometry and device parameters.

The quality of detected signals can be deteriorated by the presence of noise. The noise performance is a main concern in determining the overall system dynamic range for analog applications, and in determining the bit-error-rate for digital applications. Noise can arise from the transmitter, the electrical and optical amplifiers, and the receiver. For fiber-optic links, the common noise sources are the laser intensity noise; the Poison (shot) noise due to the quantized nature of photons and the finite response time of the detector; dark current noise due to generation and recombination of carriers in the detection region; and thermal (Johnson) noise due to the resistance of the external load of the detector. For far-infrared focal plane array application, the background radiation constitutes an additional noise source. Three commonly used measurements of noise performance are the signal-to-noise (S/N) ratio, **noise equivalent power (NEP)**, and **specific detectivity** (D^*) (Keyes, 1980). The S/N ratio is simply the ratio of detected electrical signal power to the noise power at the output of the detector (or receiver). The NEP of a detector is the optical power that must be incident on the detector so that S/N equals unity for a given wavelength, detector temperature, and resolution bandwidth. The specific detectivity is the inverse of the NEP normalized to the square root of the detector area A and the bandwidth B,

$$D^* = \frac{\sqrt{AB}}{\text{NEP}} \tag{9.9}$$

(although the noise equivalent power is normally used to characterize the sensitivity of a photodiode, the specific detectivity is commonly used even when the incident flux is larger than the detector).

For an optical detector, materials with good crystalinity and with high purity are usually preferred so as to minimize trappings and recombinations of photogenerated carriers at defect sites. In cases where heterostructure is used, close matching of crystal lattice constants of different materials ensures good interface quality. This is usually a safeguard against premature failure of detectors, as *interfacial states* or *traps* can cause gradual degradation of the dark current characteristics. In addition, two material-related issues are commonly encountered in detector designs: namely, the doping profile in the detection region of the device and the nature of the metal-semiconductor contact. The doping level affects the effective depletion region width and can be designed to fine tune the carrier transport. For instance, a low doping level of the depletion region is often desirable for photodiodes for fast carrier sweepout at low bias voltage, as well as for small device capacitance. The properties of the metal-semiconductor contact can affect carrier injection (and, thus, gain), the response speed, and the saturation optical power levels (and, thus, linearity) of the detector, as the contact itself can be viewed as a potential barrier.

For valence-band-to-conduction-band (or, simply, band-to-band) transition of electrons that involves the absorption of a photon, which is called the *intrinsic absorption process,* the photon energy is larger than that of the bandgap for large α_0 (Pancove, 1971). There are many ways to detect photons with an energy less than the bandgap energy. For instance, for far-infrared detection, such as those in an extrinsically doped silicon impurity band photoconductor, the photoabsorption occurs as the electron makes a transition from an impurity level within the bandgap to the nearby band. Free-carrier absorption can take place when the electron makes a transition within the same band while a photon is being absorbed. However, free-carrier absorption becomes significant only in the presence of large electron (or hole) density ($>10^{18}$ cm^{-3}). The absorption spectrum can change in the presence of an electric field. The Franz-Keldysh effect in bulk semiconductor and the quantum confined Stark effect in quantum wells exploits the electric field dependent shift of the absorption edge for below bandgap absorption (Pancove, 1971). These effects are currently used for various optical switches as well as for traveling-wave waveguide photodetector where a smaller α_0 can be counteracted with a longer interaction length (Kato et al., 1992). For semiconductors, the band-to-band absorption process, so far, results in the largest absorptivity simply due to the abundance in valance band and conduction band states available for optical transition.

Indirect bandgap semiconductors depend on a phonon or impurity assisted band-to-band transition process for absorption near the bandgap energy. This process has a smaller probability to occur when compared with transitions in the direct bandgap material, resulting in small absorptivity near band edge energy.

9.4.2　Photodiodes

Photodiodes typically have an absorption region where a strong electric field is applied so that the photo-generated carriers can be swept out to produce a current or voltage signal in an external circuit. Common types of photodiodes are pn junction photodiodes, p-intrinsic-n (PIN) photodiodes, Schottky junction photodiodes, and avalanche photodiodes (APDs).

A pn photodiode usually consists of a simple $p^+ n$ (or $n^+ p$) junction. For surface normal detection, the optical signal is incident at the top (or the bottom) surface. For some devices, such as optical switches or waveguide photodiodes, the optical beam can be incident parallel to the junction (Kato et al., 1992). In normal operation, the diode is reverse biased to drift apart the photogenerated electrons and holes collected at the depletion region. The external circuit load detects a signal current as the photogenerated carriers traverse the depletion region. The induced current stops when the carriers reach the bulk material outside the depletion region. The phenomenon gives rise to the transit time limit of the response speed. For a given velocity distribution $v(x)$ of the carriers and the thickness of the depletion region W, the carrier transit time T_{tr} can be estimated from

$$T_{\text{tr}} = \int_0^W \frac{dx}{v(x)} \tag{9.10}$$

Typically, for III–V semiconductors, each micrometer of the depletion region corresponds to about 10 ps of transit time when carriers are drifted at saturated velocity ($\sim 10^7$ cm/s).

The external quantum efficiency η of a top-illuminated photodiode with a thin p region is given by

$$\eta = (1-R_0)\left[e^{-\alpha_0 W_1}(1-e^{-\alpha_0 W}) + (e^{-W_0/L_n} - e^{-\alpha_0 W_1})\left(\frac{\alpha_0}{\alpha_0 + 1/L_n} \right) + e^{-\alpha_0(W_1+W)}\left(\frac{\alpha_0}{\alpha_0 + 1/L_p} \right) \right] \tag{9.11}$$

where R_0 denotes the reflectivity at the air-semiconductor interface, W_1 is the p region thickness, and L_n and L_p are diffusion lengths of electrons in the p region and holes in the n region, respectively. The first term inside the square bracket of Eq. (9.11) accounts for the photogenerated carriers within the depletion region; the second and third terms account for those in the p and n regions, respectively. For III–V photodiodes of this configuration, provided that the depletion region is sufficiently thick and the absorption coefficient there is in excess of 1000 cm^{-1}, the absorption in the n region can be ignored. In the case of heterojunction PIN photodiodes, for instance, where the p region consists of materials transparent to the incident photons, W_1 can be set to zero and the second and third terms can be ignored. As can be seen from Eq. (9.11), η can be enhanced by reducing R_0. One way to achieve this is by depositing an antireflection coating on the input facet of the photodiode. The quantum efficiency in some optimized heterojunction photodiodes can reach a value as high as 95% (Wake et al., 1987).

There is no internal photogain in photodiodes when carriers are swept out by a moderate strength electric field (10^3–10^4 V/cm). The responsivity of photodiodes can be obtained from Eq. (9.8) by setting M to unity. For instance, pn junction photodiodes made on silicon substrate with a resistivity less than

100 Ω-cm can have a depletion width of about 10 μm and a responsivity of ~ 0.4 A/W at 0.8–0.9-μm wavelength.

For homostructure pn junction photodiodes, it is difficult to optimize the quantum efficiency and the transit time at the same time due to the fact that the depletion region width depends on the applied electric field. This can be further complicated by the nonlinear relationship between the electron velocity and the electric field.

PIN Photodiodes

The PIN photodiode, consisting of a low-doped (either i or π) region sandwiched between p and n regions, can be a popular alternative to the simple pn photodiode. At a relatively low reverse bias, the intrinsic region becomes fully depleted, and the total depletion width, including those from the p and n regions, remains almost constant as the peak electric field at the depletion region is increased. A schematic diagram of a PIN photodiode with light incident from the left is shown in Fig. 9.66(a); its energy band diagram at reverse bias is depicted in Fig. 9.66(b). Typically, for a doping concentration of $\sim 10^{14}$–10^{15} cm^{-3} in the intrinsic region, a bias voltage of 5–10 V is sufficient to deplete several micrometers, and the electron velocity also reaches the saturation value. As mentioned earlier, however, transit time is not the only factor affecting the response speed of the PIN photodiode. Carrier diffusion and the resistance-capacitance (RC) time constant can also be important. For p and n regions less than one diffusion length in thickness, the response time due to diffusion alone is typically 1 ns/μm in p-type silicon and about 100 ps/μm in p-type III–V materials. The corresponding value for n-type III–V materials is several nanoseconds per micrometer due to the lower mobility of holes. The minimize the diffusion effect, very thin p and n regions can be employed; alternatively, a p or n region transparent to the incident light can be used (Wake et al., 1987).

For the 1.0–1.55-μm wavelength range suitable for optical fiber communication, germanium and a few III–V compound semiconductor alloys stand out as candidate materials for PIN photodiodes, primarily because of their large absorption coefficients. The absorption edge of germanium is near 1.6 μm at room temperature. Its α_0 is almost flat and is in the 10^4 cm^{-1} range over the 1.0–1.55-μm wavelength region (Ando et al., 1978). In comparison, silicon's absorption edge is near 1.1 μm. There are problems, however, in using germanium for making PIN photodiodes. Because of its small bandgap, the dark current arising from across the bandgap generation–recombination processes can degrade the signal-to-noise ratio.

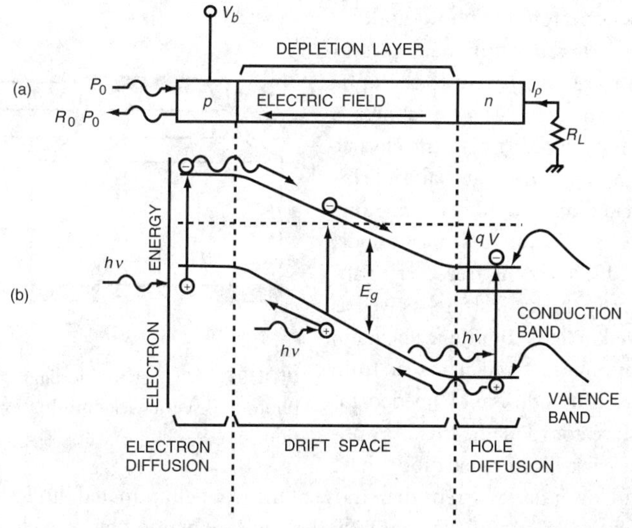

FIGURE 9.66 A PIN photodiode in operation: (a) structure: P_0 incident power, V_b bias voltage, R_0 reflectivity, I_p photocurrent, R_l load resistor, (b) energy band diagram at reverse bias, E_g bandgap energy.

This is further aggravtated by the surface recombination. So far, no satisfactory surface passivation technique has been found for germanium PIN photodiodes, and the surface leakage current tends to be very high and unstable, especially at high ambient temperature.

III–V compound semiconductors have largely replaced germanium as materials for fiber-optical compatible detectors. By properly selecting the material composition of III–V materials, they can be lattice matched to each other. For heterostructure PIN photodiodes, one can choose a material composition at the intrinsic region with a bandgap smaller than those of the p and n regions, while at the same time with an absorption edge wavelength sufficiently above the photon wavelength. The materials of particular interest are the ternary alloys $In_{0.53}Ga_{0.47}As$ and the quaternary $In_x Ga_{1-x}As_yP_{1-y}$ alloys lattice matched to InP. As is the case with germanium, unpassivated InGaAs surfaces also exhibit an unstable dark current. This problem, however, can be alleviated by passivating the surface with polyimide. PIN detectors of excellent performances in the 1–1.6-μm wavelength range have been demonstrated in the InGaAs/InP material system. To achieve a PIN photodiode with high yield and high reliability, planar fabrication process is often preferred; leakage current less than 25 pA has been achieved in some semiplanar photodiodes (Patillon et al., 1990).

For high-speed applications, both the junction area and the intrinsic layer thickness need to be small. An external resonant microcavity approach has been proposed to enhance η in such a situation (Dentai et al., 1987). In this approach the absorption region is placed inside a high Q cavity so that a large portion of the photons can be absorbed even with a small detection volume. An integral lens can be incorporated in the backside of the photodiode for focusing more photons to the detection region.

Two generic InGaAs/InP PIN photodiodes, top-illuminated and back-illuminated photodiodes, are commonly used. With liquid-phase epitaxy, a simple top-illuminated InGaAs/InP homojunction photodiode is usually made, due to the problem of *layer melt back* when the InP is grown on top of the InGaAs. With the advent of vapor-phase epitaxy employing high purity halide sources, and the metalorganic chemical vapor deposition using high purity hydride sources, one can readily made the InP-InGaAs-InP double heterostructure PIN photodiodes (Wake et al., 1987). Typically, a layer of $In_{0.53}Ga_{0.47}As$ with an absorption edge at 1.65 μm is grown on top of the InP material. The residual doping level of InGaAs layer is usually in the 5×10^{14}–$5 \times 10^{15} cm^{-3}$ range. A mesa is subsequently etched to provide electrical isolation and to define the junction area. Surface passivation materials (such as polyimide) and antireflection coating layer are then deposited. Dark currents below subnanoamperes have been observed for 50-μm-diam devices. Speeds higher than 20 GHz have been achieved. Lifetimes of 10^8 h have been observed. The disadvantage of this simple structure is that the contact area needed for forming the bonding pad introduced additional capacitance and dark current to the photodiodes.

Back illumination is possible for material systems whose substrates are transparent to the optical beam. In this case, the active-layer thickness is selected such that at the operating bias, the electric field punches through to n^+-InP material. In principle, a back-illuminated photodiode has the same basic features as the top-illuminated photodiode, although η can be reduced by the free-carrier absorption in the relatively thick substrate, as well as by the optics employed in light coupling. The main advantage of back illumination, as can be seen from Fig. 9.67, is that the front side is not needed for light coupling; therefore, for the same InGaAs optical detection area, the back-illuminated diodes have

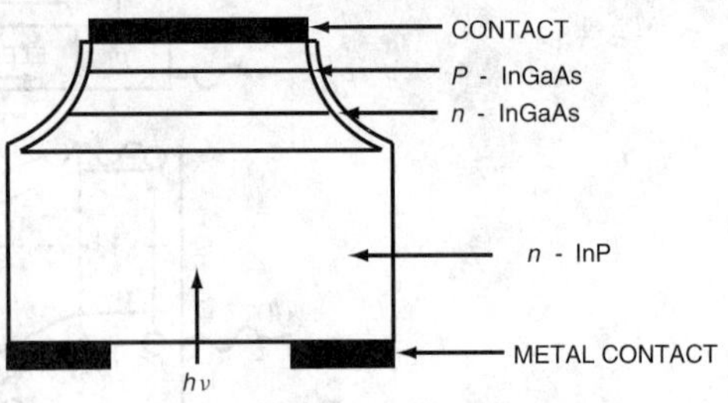

FIGURE 9.67 Schematic diagram of an InGaAs/InP PIN photodiode with back illumination and side wall passivation.

lower capacitance and lower dark current than those of the front-illuminated diodes. Optimal designs for broadband (>40 GHz) matching of PIN photodiodes can be obtained by considering the transit time, the geometric configuration including the diode area, and the transmission line characteristics together (Zhu et al., 1995).

Schottky Photodiodes

Schottky photodiodes can consist of an undoped semiconductor layer (usually n type) on bulk material and a metal layer deposited on top to form a Schottky barrier. An example is shown in Fig. 9.68. The metal is in the form of a very thin film ($\sim$100 Å) so that it is semitransparent to the incident radiation. Because of the strong electric field developed near the metal-semiconductor interface, photogenerated holes and electrons are separated and collected at the metal contact and in the bulk semiconductor materials, respectively. By increasing the reverse bias to these diodes, both the peak electric field and the width of the depletion region inside

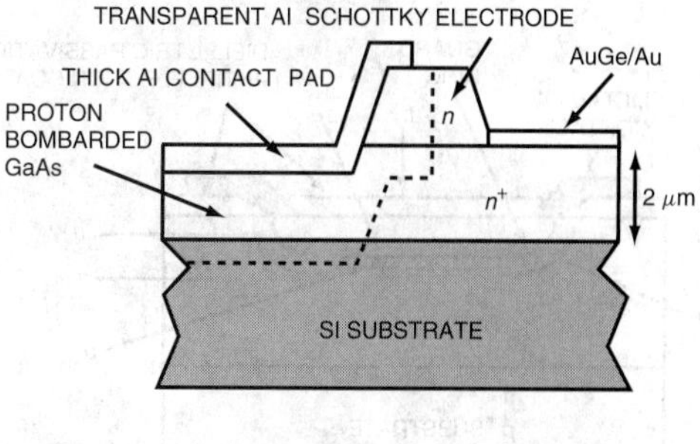

FIGURE 9.68 Schematic diagram of a GaAs Schottky photodiode.

the undoped layer can be increased and, thus, the responsivity can be increased. The complete absorption of optical signals within the high field region ensures high-speed performance, as absorption outside the high field region generates a diffusion current. Again, the undoped layer should not be too thick, as the photodiode response speed can then become transit time limited.

In comparison with PIN photodiodes, Schottky photodiodes are simpler in structure. Unfortunately, although good Schottky contacts are readily made on GaAs, for long wavelength InGaAsP/InP materials, stable Schottky barriers are difficult to achieve. One compromising approach to stabilize the barrier height is to employ a thin p^+-InGaAs layer between the metal and the detection layer (n-InGaAs). Recently, by placing a Au-Schottky contact on n-In$_{0.52}$Al$_{0.48}$As, which is lattice matched to InP, a wide spectral response range (0.4–1.6 μm) Schottky photodiode with a responsivity of 0.34 A/W at 0.8 μm and 0.42 A/W at 1.3 μm has been demonstrated (Hwang, Li, and Kao, 1991). Low-temperature molecular beam epitaxial (MBE) GaAs with much As precipitates for effective buried Schottky barriers has also been investigated for 1.3-μm detection.

For comparison, GaAs Schottky photodiodes with a 3-dB bandwidth larger than 100 GHz and operating at less than 4-V reverse bias have been reported (Wang and Bloom, 1983). This high speed is attained by restricting the photosensitive area to a very small mesa ($5 \times 5 \ \mu$m^2) to minimize capacitance and by using a semitransparent platinum Schottky contact such that good responsivity ($\sim$0.1 A/W) can be achieved with topside illumination.

Avalanche Photodiodes

Avalanche photodiodes are based on the *impact ionization effect*. When free carriers generated by photoabsorption are accelerated by a strong electric field, they gain sufficient energy to excite more electrons from the valence band to the conduction band, thus giving rise to an internal current gain (Emmons, 1967). Avalanche photodiodes usually require a higher bias voltage than PIN photodiodes to maintain the high electric field. Also, the current gain is not a linear function of the applied voltage and is quite sensitive to temperature variations.

Figure 9.69(a) shows a schematic diagram of a classic silicon avalanche photodetector that consists of an n^+ contact, a p-multiplying region, a π-drift region, and a p^+ contact. The corresponding electric field profile is depicted in Fig. 9.69(b). In operation, under the large reverse-bias voltage, the depletion region extends from the n^+ to the p^+ contact, with avalanche multiplication occurring in the high field region. When an electron-hole pair is photogenerated, electrons are injected into the multiplying region provided that the light is mostly absorbed to the right of the multiplication region shown in Fig. 9.69(b). Similarly, hole injection occurs for light absorbed to the left. Mixed injection of both carriers occurs for light absorbed inside the avalanche region. The doping profile and the position of the multiplication region need to be optimized for each material system to meet the receiver requirements. In general, it is important that carriers with a higher ionization coefficient are injected into the multiplying region in

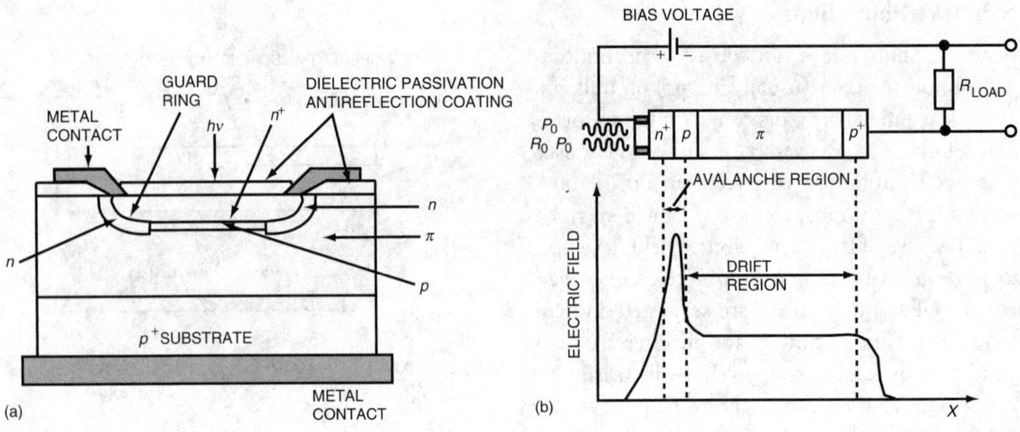

FIGURE 9.69 Schematic diagram of a silicon APD: (a) structure, (b) electric field distribution.

order to have the minimum excess noise due to the multiplication process. For instance, silicon APDs are generally designed so that electrons are the primary carriers that undergo multiplication since electrons in silicon have a larger ionization coefficient than holes. Consequently, both the hole injection or mixed injection are to be avoided. For low-noise silicon APDs, a long drift region is incorporated to ensure a large η and pure electron injection. The guard ring in the planar structures, shown in Fig. 9.69(a), prevents edge breakdown at the perimeter of the multiplying region. Reverse-bias voltage of up to 400 V is common for this kind of diode because of the large voltage drop across the long drift region.

The electric field required for impact ionization depends strongly on the bandgap of the material. The minimum energy needed for impact ionization, known as the *ionization threshold energy,* influences strongly the ionization rates (or coefficients) for electrons (α) and holes (β). These quantities are defined as the reciprocal of the average distance traveled by an electron or a hole, measured along the direction of the electric field, to create a secondary electron-hole pair. It should be noted that the threshold energy can be estimated by taking into account the conservation of carrier energy, crystal momentum, and group velocity during the ionization process. Avalanche multiplication is the manifestation of many consecutive occurrences of this process.

The multiplication is generally a three-body collision process and is, consequently, statistical in nature. As a result, it contributes a statistical noise component in excess of the shot noise already present in the diode. This excess noise has been studied for the case of arbitrary α/β ratios for both uniform and arbitrary electric field profiles (McIntyre, 1972). For the uniform electric field case, the excess noise factor for electrons F_n and holes F_p are given as

$$F_n = M_n \left[1 - (1 - k) \left(\frac{M_n - 1}{M_n} \right)^2 \right] \tag{9.12}$$

and

$$F_p = M_p \left[1 - \left(1 - \frac{1}{k} \right) \left(\frac{M_p - 1}{M_p} \right)^2 \right] \tag{9.13}$$

where $k = \beta/\alpha$ and M_n and M_p are the DC multiplication factors for pure electron and pure hole injection, respectively.

The DC multiplication factor M_n and M_p depend on α and β, as well as the width W_a of the avalanche region. For the case where $\beta = 0$, M_n can be expressed as

$$M_n = \frac{1}{1 - \int_0^{W_a} \alpha e^{-\alpha x} dx} \tag{9.14}$$

If an uniform field is assumed in the avalanche region, α becomes independent of position. M_n of Eq. (9.14) then increases exponentially with the number of ionizing carriers in the high field region, and no sharp breakdown will occur. Under these conditions, a current pulse is induced when the photogenerated electron starts to drift toward the multiplication region. The pulse increases in magnitude during the electron transit time through the high field region due to the electron impact ionization. It is reduced to zero once the last hole is swept outside the high field region. Neglecting the transit time inside the absorption region, the current pulse width is approximately the average of electron and hole transit times and decreases a little as the gain increases. The corresponding excess noise F (see Eq. (9.12) and Eq. (9.13)) is small because when M increases, statistical variation of the impact ionization process only causes a small fluctuation in the total number of carriers. Therefore, for low excess noise APDs, the ionization rate for one type of carrier should be much greater than that of the other, and the carrier with the larger ionization coefficient should initiate the multiplication process.

When both electron and hold ionize at the same rate (i.e., $\alpha = \beta$), APDs become quite noisy. In the case of a uniform electric field, the current gain can be expressed as

$$M_n = M_p = \frac{1}{1 - \alpha W_a} \tag{9.15}$$

which gives rise to a sharp breakdown ($M \to \infty$) situation as αW_a approaches unity. This corresponds to the situation when, on the average, each injected electron or hole produces one electron-hole pair during the transit through the high field region. The secondary electron and hole drift in opposite directions and gain energy in the avalanche region and generate further electron-hole pairs. This process persists for a long time when the gain is large. It is clear that for this situation with a large gain, there are fewer ionizing carriers in the high field region than the case when $\alpha \gg \beta$ (or $\beta \gg \alpha$). Thus, any statistical variation in the impact ionization process can produce a large fluctuation in the gain and cause considerable excess noise.

The gain can be a function of the signal frequency and can be approximated by

$$M(\omega) = \frac{M_0}{\left[1 + (\omega M_0 \tau_{\text{eff}})^2\right]^{1/2}} \tag{9.16}$$

where M_0 is the DC gain, ω is the angular frequency and τ_{eff} is approximately equal to $(\beta/\alpha)\tau_{\text{av}} \delta$, where δ is a parameter ranging from $\frac{1}{3}$ (at $\beta = \alpha$) to 2 (at $\beta \ll \alpha$) and τ_{av} is the average transit time. Figure 9.70 depicts the theoretical gain-bandwidth product as a function of M_0 for various α/β values.

For analog fiber-optic link applications, the noise floor at the receiver end is an important parameter in determining the overall link performance. For APD-based receiver, the noise floor can consist of the laser relative intensity noise, the receiver thermal noise, and the APD noise associated with the photocurrent. The APD noise power can be expressed as

$$i_{n,\text{APD}}^2 = 2qB[I_d + q\eta P_0/h\nu]M^2 F \tag{9.17}$$

where η is the quantum efficiency at unity gain, I_d is the dark current of the APD without multiplication, and F is the excess noise factor expressed in Eqs. (9.12) and (9.13). When the noise of the APD is dominant, furthur increase in M becomes less attractive for improving the link linear dynamic range performance.

Silicon APDs are used in the wavelength range of 0.4–1.0 μm because of the low dark current and a large α/β. The structure shown in Fig. 9.69(a) is employed in the case of silicon APDs to obtain carrier multiplication with very little excess noise. The high field $n^+ p$ junction serves as an electron-initiated

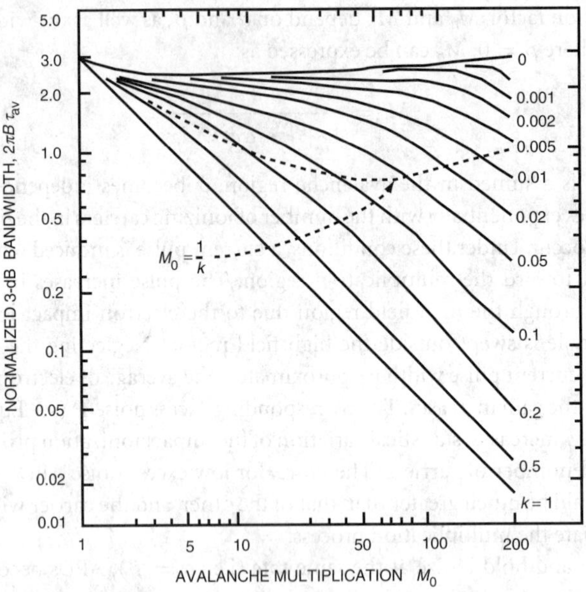

FIGURE 9.70   Normalized 3-dB bandwidth as a function of M_0 for several values of the ionization ratio k. For pure electron injection $k = \beta/\alpha$, for pure hole injection $k = \alpha/\beta$. Here, τ_{av} is the average of the electron and hole transit times in the avalanche region.

avalanche region and can be formed by diffusion or ion implantation. At bias below the breakdown of the $n^+ p$ junction, the π region is depleted. Beyond this point, the voltage drop across the wide π region increases with bias faster than that across the $n^+ p$, thus increasing the operating bias range for multiplication. The full depletion of the π region is essential for high response speed, as electrons generated by the absorption within the π region can then drift to the $n^+ p$ region with high velocity. Holes, on the other hand, drift opposite to the electrons and, thus, a pure electron current is injected into the multiplication region. (Although there are some hole injections due to the absorption in the n^+ region and the avalanche region, they are much smaller than in number than the electrons absorbed in the long drift region.) An excess noise factor of 4 at a gain of 100 has been achieved in such an APD. The response speed is limited mostly by the transit time across the π region. To increase the response speed without increasing the operating voltage, a built-in field in the π region can be profiled by grading the doping profile so as to enhance the carrier velocities. As for PIN photodiodes, an antireflective coating can be deposited at the back contact to increase η of a relatively narrow π region. Response times of 150–200 ps have been achieved.

For fiber-optic applications, germanium and III–V APDs have been used instead of silicon APDs because the spectral response of silicon cuts off near 1.1 μm. As can be estimated from Fig. 9.71, β/α ratio for germanium has a maximum value of 2. A relatively high excess noise factor ($F \approx 1/2M$), in addition to a relatively high dark current noise, is present in germanium APDs due to the small bandgap and surface leakage. Usually, for germanium APD, the usable gain is only around 10. Beyond this point, the noise power increases as M^3 and the signal to power ratio quickly degrades. By reducing the detection area of the APDs, dark current in the nanoampere range can be achieved.

III–V APDs are attractive alternatives to germanium APDs because the material band gap can be tailored to achieve both high absorption and low dark current. Early APD work in the ternary InGaAs and quanternary InGaAsP system have shown rather low usable gain because the difference in α and β in these materials is not as high as that in silicon. For homojunction InGaAs devices, the dark current increases exponentially with reverse bias as the bias approaches avalanche breakdown. This is attributed to the tunneling current effect, which is more profound in materials with small bandgap. To reduce the

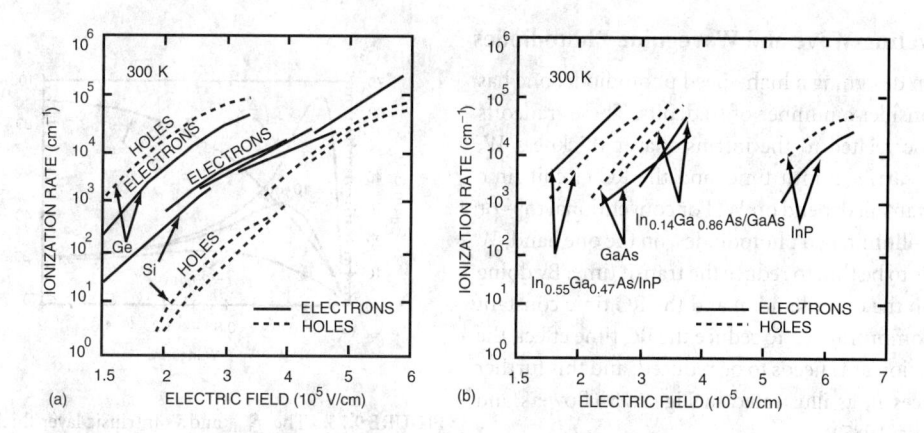

FIGURE 9.71 Experimental obtained ionization rates at 300 K: (a) silicon and germanium, (b) GaAs, InP, $In_{0.14}Ga_{0.86}As/GaAs$, $In_{0.53}Ga_{0.47}As/InP$.

tunneling current, a heterostructure that separates the absorption region from the multiplication region was proposed and is depicted in Fig. 9.72 (Susa et al., 1981). This is known as the *separate absorption and multiplication* (SAM) APD. With this arrangement, the low bandgap InGaAs material does not experience a high electric field, and thus the tunneling effect is reduced. The absorption in this structure takes place in the low bandgap material, which, like the silicon APDs, is low doped. The multiplication takes place in the wide bandgap *pn* junction, which usually consists of InP. Consequently, the noise of the SAM APDs reflects much of the ionization process in InP. Reasonably small unmultiplied dark currents are observed for APDs with an InGaAsP absorption region (nanoampere range) and InGaAs absorption region (tens of nanoamperes), whereas the gain is maintained in the range of 10–60. As in the case of silicon APDs (doubly zinc-diffused) guard ring structure can be used in a planar structure (see Fig. 9.69(a)) to prevent the edge breakdown at high field.

Trapping of holes at the InGaAs/InP heterojunction has been observed, which greatly degrades the speed of the APD. To overcome this problem, an additional layer of quaternary InGaAsP with intermediate energy bandgap (or layers of graded bandgap region) is incorporated. A relatively low-noise (as compared with germanium) APD can be achieved with proper layer thickness and doping profile. For instance, at a gain of 10, APDs with an excess noise figure of 5, a dark current of 20 nA, and a bandwidth in excess of 1 GHz have been demonstrated.

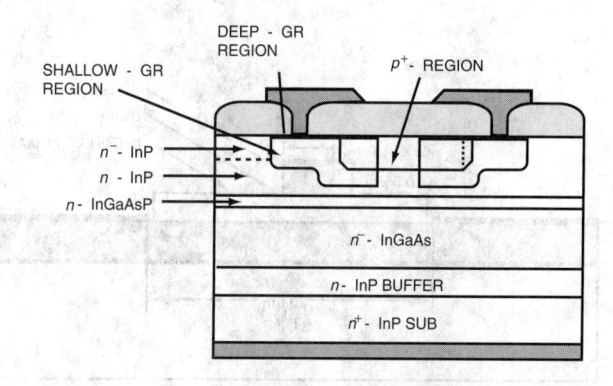

FIGURE 9.72 InGaAsP/InP APD with separated absorption and multiplication regions. The intermediate InGaAsP layer is for reducing the hole-trapping effect. GR stands for *guard ring*.

Traveling-Wave and Waveguide Photodiodes

When designing a high-speed photodiode, one has to consider a number of tradeoffs. These tradeoffs can be related to the intrinsic layer thickness W, as η, carrier transit time, and the RC circuit time constant all depend on W. For conventional top- or back-illuminated photodiodes, on the one hand, W needs to be thin to reduce the transit time. By doing so, on the other hand, η and the RC time constant are compromised. To reduce the RC time effect, the detection area needs to be reduced, and this further reduces η, as illustrated in Fig. 9.73 (Bowers and Burrus, 1987).

Waveguide photodiode relieves somewhat the tradeoff between the speed and η. This type of detector is generally side illuminated with light gui-

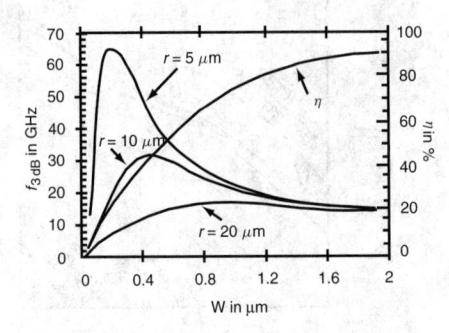

FIGURE 9.73 The $f_{3\,dB}$ and vs intrinsic layer thickness for PIN photodiode with surface normal illumination; r is the diode radius.

ded parallel to the intrinsic layer and incrementally absorbed. The optical waveguide structure for the waveguide is similar to that of the double heterostructure lasers. As example of a waveguide photodiode for 1.3–1.5-μm detection is shown in Fig. 9.74, where the mesa structure is used to confine the light in the lateral direction. Side illumination and optical waveguiding increases the interaction length of the light with the absorbing intrinsic layer. Efficiency η is limited by the coupling efficiency at the fiber to the waveguide interface and the scattering loss along the waveguide. When compared to the area photodiode, the waveguide photodiode usually has a smaller η since the cross section of the waveguide is not closely matched with that of the single-mode fiber. Because of the side illumination, however, the same value of η can be maintained as the total junction area is varied; therefore the waveguide detector has an improved bandwidth-efficiency performance when compared to a conventional detector. Efficiency η as high as 68% has been achieved for a waveguide photodiode with $f_{3\,dB}$ greater than 50 GHz (Kato et al., 1992). In contrast, as can be seen from Fig. 9.73, a conventional PIN photodiode with a η of 68% will be limted to a $f_{3\,dB}$ of about 30 GHz.

The waveguide photodiode whose cross section is shown in Fig. 9.74 consists of a rib-loaded PIN waveguide section integrated with a microwave coplanar transmission line on a semi-insulating InP substrate. The rib is 5 μm wide at the top and 20 μm long, the intrinsic InGaAs layer is 0.35 μm thick. For waveguide and photodiode, the $f_{3\,dB}$ is presently limited by the transit time across the intrinsic region and the RC time effect. Using a mushroom-mesa structure to reduce the RC time constant, a 110-GHz bandwidth waveguide photodiode with a η of 50% has been achieved.

A traveling-wave photodetector is a subset of the waveguide photodiode with the electrode designed to support traveling electrical waves so that the characteristic impedance is matched to that of the external

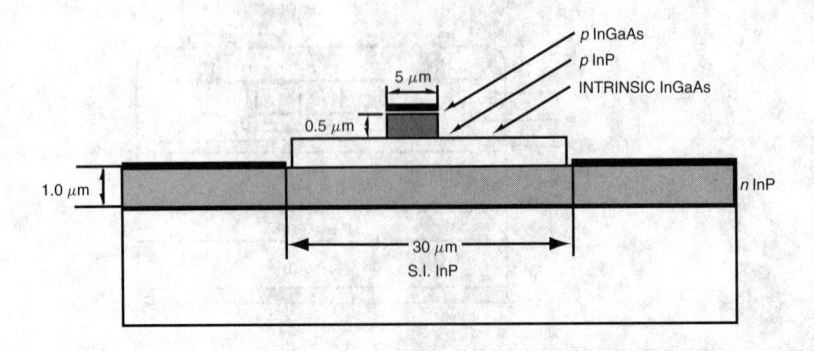

FIGURE 9.74   A cross-sectional view of an InGaAs/InP PIN waveguide photo-detector.

circuit (Giboney et al., 1995). The $f_{3\,\mathrm{dB}}$ is limited by the transit time across the intrinsic region and the velocity mismatch. The RC time constant is no longer a consideration for speed in the case of traveling wave photodiode, and the device can be viewed as a terminated section of a transmission line with a position-dependent photocurrent source distributed along its length, instead of being a lumped circuit element. The velocity-mismatch bandwidth limitation depends on α_0 and the mismatch between optical group velocity and microwave phase velocity. Bandwidth as high as 172 GHz has been reported for a GaAs/AlGaAs traveling-wave photodiode detecting in the 800–900-nm wavelength range (Giboney et al., 1995). The intrinsic GaAs region is 1 μm wide, 0.17 μm thick, and 7 μm long.

As noted earlier, the waveguide configuration poses a limitation to η due to the fiber coupling. The coupling can be improved through the use of lensed fibers, antireflection coating at the waveguide facet, or laterally tapered waveguide section.

9.4.3   Photoconductors

Photoconductive devices depend on the change in the electrical conductivity of the material upon irradiation. Although photoconductive effects in semiconductor materials have been studied for a long time, only recently are they employed for high-sensitivity, high-speed detection. In semiconductors, two types of photoconductors, extrinsic and intrinsic, are commonly used. For the intrinsic case, photoconduction is produced by the band-to-band absorption. The α_0 can be very large ($\sim 10^4$ cm^{-1}), in this case due to the large density of electronic states associated with the conduction and valence bands. For the extrinsic case, photons are absorbed at the impurity level and, consequently, free electrons are generated in the n-type semiconductor and free holes are generated in the p-type semiconductor. Extrinsic photoconduction is characterized by a lower sensitivity due to the smaller number of impurity states available for absorptive transition. These devices are usually operated at low temperatures (e.g., 77 K) to freeze out impurity carriers.

For the metal-semiconductor-metal (MSM) photoconductor shown in Fig. 9.75, a voltage is applied across the electrodes and light is illuminated from the top. The photosignal generates either a voltage change across a resistor in series with the photoconductor or a current change through the device. Frequently, the signal is detected as a change in voltage across a resistor matched to the dark resistance of the detector.

Performance Figure of Merit

The photoconductive gain M of photoconductors can be expressed as

$$M = \frac{\tau_n}{t_n} + \frac{\tau_p}{t_p} \tag{9.18}$$

where τ_n and τ_p are the lifetimes of the excess electrons and holes, respectively, and t_n and t_p are the corresponding transit times across the device. For cases where electron conduction is dominant and the response time of the device is τ, the gain-bandwidth product can be expressed as

$$MB = \frac{\tau_n}{t_n} \frac{1}{\tau} \tag{9.19}$$

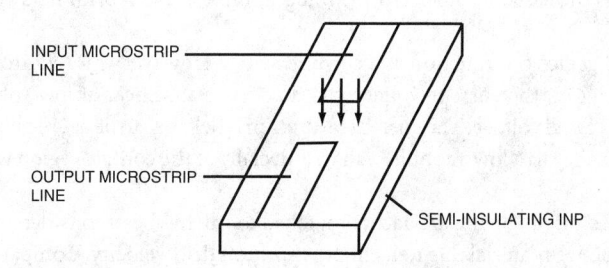

FIGURE 9.75   A MSM photoconductor fabricated in a microstrip-line configuration on semi-insulating InP.

For material with little carrier trapping, the response time is determined by carrier lifetime. As device dimensions get smaller, however, the response time can be altered by the nature of the metal contact and the method of biasing. With nonalloyed contacts to the MSM device, Schottky barriers can be formed at the metal–semiconductor interfaces, and the depletion region extends across the length of the photoconductor. In this case, the device operates more or less like a photodiode with τ_n replaced by t_n, and its MB value becomes $1/\tau$ (see Eq. (9.19)).

When a steady optical beam with power P_0 is incident on the photoconductor, the optical generation rate of electron-hole pairs, as a function of the distance x into the sample, is given by

$$g(x)dx = \frac{(1 - R_0)P_0}{h\nu}\alpha_0 e^{-\alpha_0 x}dx \tag{9.20}$$

from which η can be obtained as

$$\eta = (1 - R_0)(1 - e^{-\alpha_0 D}) \tag{9.21}$$

where D is the effective sample thickness. The steady-state excess electron and hole concentrations, Δn and Δp (as calculated from η, τ, and the device geometry), can be incorporated in the change of conductivity $\Delta\sigma$ given by

$$\Delta\sigma = q(\Delta n\mu_n + \Delta p\mu_p) \tag{9.22}$$

where μ_n and μ_p are the electron and hole mobilities, respectively.

For many signal-processing applications, the ratio of the on-resistance to the off-resistance of the photoconductive device is an important parameter, although the off-resistance is primarily limited by the material resistivity in the dark. The on-resistance depends on the carrier mobility and the excess carriers generated.

Because of the low surface recombination velocity of InP-related materials, planar photoconductive devices on iron-doped semi-insulating InP have been extensively studied. This material has a room temperature resistivity of 10^5–10^8 Ω-cm and an electron mobility of 1500–4500 cm^2/V-s depending on the crystal quality and iron concentration. The carrier lifetime ranges from less than 1 ps to 3 ns.

Array and Switch Applications

A typical photoconductor structure consists of either (1) a single narrow gap made on a metal stripe deposited on semi-insulating InP or (2) interdigitated MSM electrodes deposited on the substrate. Response times as short as 50 ps have been achieved in both devices. As noted earlier, the nature of metal contacts on semiconductors affects much of the device performance. For devices with nonalloyed Schottky contacts, a unity M is observed. However, the response time, especially the fall time, is shorter than that of devices with alloyed contacts. Alloyed contact devices show lower contact resistance and a higher photoconductive gain. Various contact materials (such as Au/Sn and Ni/Ge/Au), as well as alloying conditions (such as alloying temperature and duration), have been investigated. Compared to photodiodes, the planar photoconductor has advantages such as the ease of optoelectronic integration, especially with field effect transistor (FET) circuits.

A photoconductive device can function as a current switch. However, this requires a low on-resistance. For alloyed contact devices, the current-voltage characteristics are linear at low voltage, and the current tends to saturate at increased voltage. The current saturation is believed to be caused by electrons transferred from a high-mobility valley to a low-mobility valley, especially at the contact region where the electric field is relatively strong.

For 1.0–1.6-μm spectral range, InGaAs lattice matched to InP has been considered for photoconductive detectors. It is an attractive material due to its high saturation drift velocity. Compared to semi-insulating InP, however, semi-insulating InGaAs materials obtained so far tend to be less resistive ($\sim 10^5 \Omega$-cm) with a typical dark resistance in the range of kΩ, which compromises the device performance.

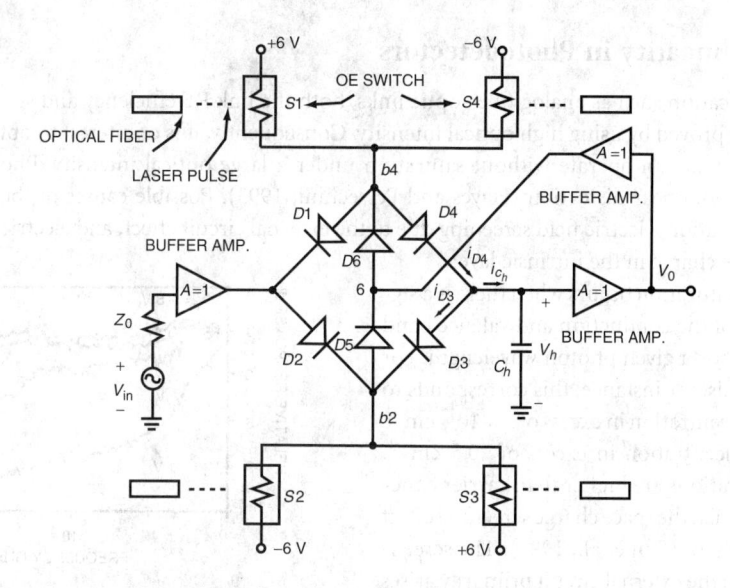

FIGURE 9.76 A diode-bridge type sample-and-hold circuit employing photoconductive switches.

Because of the fact that InP can be used to detect radiation ranges from near-IR to UV region with a good responsivity, it becomes very attractive for wide spectral response range detection. Together with its small surface recombination velocity, very fast detection (<90-ps rise time) has been achieved at the X ray wavelengths where $\alpha_0 \sim 10^3$ cm^{-1}. Recently, such devices have been considered for analog to digital (A/D) conversion where an analog electrical signal is time sampled and coded into a digital format. In this application, photoconductive devices operate as a time switch triggered by laser pulses and serve in the tracking of the analog signal voltage. When compared to electrical A/D, this can lead potentially to less time jitter as well as higher sampling rates. Up to 6-b resolution with a sampling rate greater than 1×10^9 samples per second has been predicted with the InP photoconductive switches.

For direct optoelectronic sample and hold, a photoconductor switch is placed between a signal source and a charging hold capacitor. Upon optical illumination, the switch is in the on-state. This scheme can be limited by the small hold capacitance required and by the time required to turn off the switch. To overcome the nonideal off-resistance of the photoconductive switches, it has been proposed and demonstrated that, by integrating the switches in a Wheatstone Bridge configuration, as shown in Fig. 9.76, when the laser pulses are fired accurately to steer the current for charging the hold capacitor, high-speed sample and hold can be achieved with high charging capacity and with low time jitter (Sun et al., 1991).

Another application of photoconductive device is in the generation of high-power, picosecond duration, electrical pulses. The photoconductor operates as a switching element that is connected via a low-impedance transmission line at the output side to a terminating load resistor. With the switch in the off-state, the input side of the line is charged to a high voltage. Upon laser excitation, the switch is turned on, and a high-power electrical pulse propagates down the transmission line. The on-impedance of the switching element needs to be much smaller than the line impedance so that a small fraction of the voltage (and, thus, power dissipation) drops across the switching element. The carrier lifetime determines the maximum time window when the switch is in the on-state. Silicon photoconductive switches have been used to generate gigawatt pulse power, and the duration of the electrical pulse is in the nanosecond range, depending on the length of the charging transmission line (Giorgi et al., 1988). Presently, III–V materials are inferior to silicon materials because of their relative shorter carrier lifetimes, which limit the duration of the electrical pulse. A photodiode switch has been investigated as an alternative to the photoconductive switch for such applications. Megawatt to gigawatt electrical pulses can potentially be generated (Giorgi et al., 1988). A detailed description of the switching mechanism is given in Sun et al. (1992).

9.4.4   Nonlinearity in Photodetectors

For many application such as analog fiber-optic links, both the link RF efficiency and spur-free dynamic range can be improved by using high optical intensity. Consequently, it is of interest to obtain high-speed photodetectors that can operate without saturation under a large optical intensity. Photodetector saturation causes nonlinear distortion (Hayes and Persechini, 1993). Possible causes of the saturation are absorption saturation, electric field screening due to the external circuit effect, and electric field screening due to the space charge in the intrinsic layer.

Absorption saturation occurs when there is a significant filling of the conduction and valence band states available for a given photon wavelength. For InGaAs materials, for instance, this corresponds to an electron concentration in excess of 2×10^{17} cm^{-3} and a hole concentration in excess of 10^{18} cm^{-3}. These concentrations are higher than carrier concentrations at which the space charge screening effect becomes important (Sun et al., 1992). The screening effect due to the external circuit primarily arises from the series resistance outside the junction region of the detector. For example, for a series resistance of 50 Ω, a photocurrent of 10 mA gives rise to a voltage reduction of 0.5 V across the intrinsic

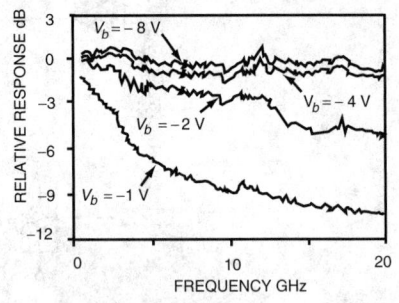

FIGURE 9.77   Waveguide photodiode frequency response with 10-mW optical power incident on the detector and at various detector bias voltages V_b.

region. This represents a small reduction when compared to typical bias (>5 V) across the detector.

Both photoconductor and photodiode operate under a high electric field to sweep out the free carriers; however, at a high flux of photogenerated carriers, the applied electric field can be screened by the electric field due to the electron-hole pairs. The nonlinear response can arise from a redistribution of the intrinsic region electric field, which is accompanied by a spatially dependent hole and electron velocity distributions (Williams, 1994). Consequently, the harmonics signal levels increase at high optical intensities. In the extreme case, even the fundamental RF signal response is affected. In principle, this screening effect can be reduced by subjecting the photodiode to a large reverse bias. Figure 9.77 shows the effect of bias to the frequency response of a waveguide photodetector under a high optical flux. Figure 9.78 shows the fundamental and second harmonics signal at −10 V as a function of the photocurrent. Additional nonlinear effect observed at high bias situation has been attributed to absorption at the highly doped contact layer of the photodiode (Williams, 1994).

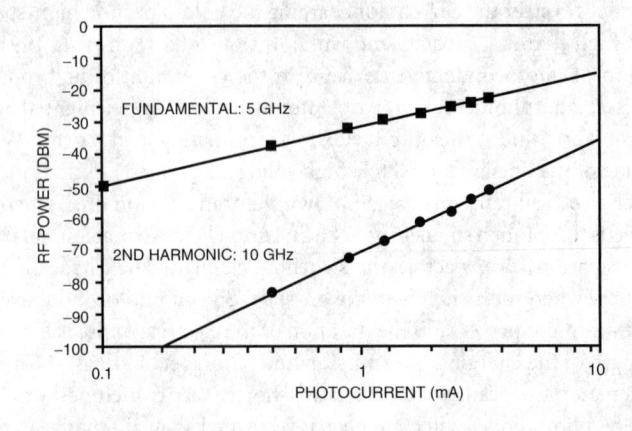

FIGURE 9.78   The fundamental (5 GHz) and second harmonic signal of a waveguide photodiode as a function of photocurrent. The fundamental signal is the beat signal between two frequency-stabilized YAG lasers.

Defining Terms

Cutoff frequency ($f_{3\,dB}$): The frequency at which the photodetector output decreases by 3 dB from the output at DC (or some low-frequency value).
External quantum efficiency (η): Number of electrons generated per incident photon.
Noise equivalent power (NEP): The amount of light equivalent to the noise level of the device.
Responsivity (R): The ratio of the output current to the input optical power, amperes per watt.
Specific detectivity (D^*): Inverse of the noise equivalent power normalized to the square root of the detector area and the resolution bandwidth.
Spectral response: The relation the responsivity (or external quantum efficiency) and wavelength.

Acknowledgment

The author is indebted to Dr. Charles Cox for helpful suggestions.

References

Ando, H., Kanbe, H., Kimura, T., Yamaoka, T., and Kaneda, T. 1978. Characteristics of germanium avalanche photodiodes in the wavelength region of 1–1.6 μm. *IEEE J. Quant. Electron.* QE-14(11):804–809.

Bowers, J.E. and Burrus, C.A., Jr. 1987. Ultrawide-band long-wavelength p-i-n photodetectors. *J. Lightwave Technol.* 5(10):1339–1350.

Dentai, A.G., Kuchibhotla, R., Campbell, J.C., Tsai, C., and Lei, C. 1991. High quantum efficiency, long wavelength InP/InGaAs microcavity photodiode. *Electron. Lett.* 27(23):2125–2127.

Emmons, R.B. 1967. Avalanche-photodiode frequency response. *J. Appl. Phys.* 38(9):3705–3714.

Giboney, K.S., Nagarajan, R.L., Reynolds, T.E., Allen, S.T., Mirin, R.P., Rockwell, J.W., and Bowers, J.E. 1995. Travelling-wave photodetectors with 172-GHz bandwidth and 76-GHz bandwidth-efficiency product. *IEEE Photonics Tech. Lett.* 7(4):412–414.

Giorgi, D., Yu, P.K.L., Long, J.R., Lew, V.D., Navapanich, T., and Zucker, O.S.F. 1988. Light-activated silicon diode switches. *J. Appl. Phys.* 63(3):930–933.

Hayes, R.R. and Persechini, D.L. 1993. Nonlinearity of p-i-n photodetectors. *IEEE Photonics Technol. Lett.* 5(1):70–72.

Hwang, K.C., Li, S.S., and Kao, Y.C. 1991. A novel high-speed dual wavelength InAlAs/InGaAs graded superlattice Schottky barrier photodiode for 0.8 and 1.3 μm detection. *Proceeding of SPIE* 1371:128–137.

Kato, K., Hata, S., Kawano, K., Yoshida, J., and Kozen, A. 1992. A high-efficiency 50 GHz InGaAs multimode waveguide photodetector. *IEEE J. Quant. Electron.* 28(12):2728–2735.

Keyes, R.J., ed. 1980. *Optical and Infrared Detectors, Topics in Applied Physics*, Vol. 19, Springer-Verlag, Berlin.

McIntyre, R.J. 1972. The distribution of gains in uniformly multiplying avalanche photodiodes: theory. *IEEE Trans. Electron. Devices* ED-19(6):702–713.

Pancove, J.I. 1971. *Optical Processes in Semiconductors*. Prentice-Hall, Englewood Cliffs, NJ.

Patillon, J.N., Andre, J.P., Chane, J.P., Gentner, J.L., Martin, B.G., and Martin, G.M. 1990. An optimized structure for InGaAs/InP photodiodes combining high-performance throughput and reliability. *Phillips Journal of Research* 44(5):465–480.

Sun, C.K., Wu, C.C., Chang, C.T., Yu, P.K.L., and McKnight, W.H. 1991. A bridge type optoelectronic sample and hold circuit. *IEEE J. Lightwave Technol.* 9(3):341–346.

Sun, C.K., Yu, P.K.L., Chang, C.T., and Albares, D.J. 1992. Simulation and interpretation of fast rise time light-activated p-i-n diode switches. *IEEE Trans. Electron Devices* 39(10):2240–2247.

Susa, N., Nakagome, H., Ando, H., and Kanbe, H. 1981. Characteristics in InGaAs/InP avalanche photodiodes with separated absorption and multiplication regions. *IEEE J. Quant. Electron.* 17(2):243–250.

Wake, D., Walling, R.H., Sargood, S.K., and Henning, I.D. 1987. $In_{0.53}Ga_{0.47}As$ PIN photodiode grown by MOVPE on a semi-insulating InP substrate for monolithic integration. *Electron. Lett.* 23(8):415–416.

Wang, S.Y. and Bloom, D.M. 1983. 100 GHz bandwidth planar GaAs Schottky photodiode. *Electron. Lett.* 19(14):554–555.

Williams, K.J. 1994. Nonlinear mechanisms in microwave photodetectors operated with high intrinsic region electric fields. *Appl. Phys. Lett.* 65(10):1219–1221.

Zhu, Z., Gillon, R., and Vorst, A.V. 1995. A new approach to broadband matching for p-i-n photodiodes. *Microwave and Optical Technol. Lett.* 8(1):8–13.

Further Information

Additional information on the topic of optical receivers can be obtained from the following sources:

Keyes, J., ed. 1980. *Optical and Infrared Detectors, Topics in Applied Physics,* Vol. 19. Springer-Verlag, Berlin.

Kingston, H. 1978. *Detection of Optical and Infrared Radiation.* Springer-Verlag, Berlin.

Sze, M. 1981. *Physics of Semiconductor Devices.* Wiley, New York.

Tsang, T., ed. 1995. *Photodetectors,* Semiconductors and Semimetals, Vol. 22, Part D. Academic Press, New York.

9.5   Optical System Design

David E. Rittenhouse

9.5.1   Introduction

Previously, optical fiber networks were designed to satisfy specific application(s) requirements, data, voice, or video. However, today the true benefits of optical fiber are being realized and used to design a structured cabling solution independent of specific applications—allowing for the effective use of fiber as a **universal transport system (UTS)**. With the establishment of standards for fiber type, performance, connectors, and components, the most significant design consideration now is connectivity, allowing the network to be flexible, manageable, and expandable. By complying with standards for the passive components, specifically, ANSI/TIA/EIA-568A(TIA,1995), the designer maximizes the prospects for future applications compatibility as systems change or new ones are implemented.

System Planning: Structured Cabling

Structured cabling will support both the applications of today as well as the future. UTS and optical fiber does just that, allowing today's systems to grow gracefully as requirements change without concern of obsolescence.

Fiber is the most attractive medium for structured cabling because of its ability to support the widest range of applications, at the fastest speeds, for the longest distance. Additionally, fiber has a number of intrinsic advantages beneficial to any application at any speed. Fiber is immune to electromagnetic interference (EMI) and radio frequency interference (RFI). Therefore, its signals cannot be corrupted by external interference.

Just as it is immune to EMI from outside sources, fiber produces no electronic emissions and, therefore, is not a concern of the Federal Communications Commission (FCC) or European emissions regulations. Crosstalk does not occur in fiber systems, and there are no shared sheath issues as with multipair UTP cables. Because all-dielectric cables can be used, grounding concerns can be eliminated and lightning effects dramatically reduced. Optical fibers are virtually impossible to tap, making it the most secure media type. Most importantly, optical bandwidth cannot be adversely affected by installation conditions. Compare this to the dramatic effects that an installer can have with crosstalk in a copper system.

A structured cabling solution should be designed and planned for in three distinct segments.

- **Interbuilding backbone**
- **Intrabuilding backbone**
- **Horizontal distribution**

Interbuilding Backbone

In campus environments, whether that be a university, industrial park, or military base, optical fiber is used extensively and sometimes exclusively in the interbuilding backbone. Carrying a complete spectrum of services, it is used for voice, data, and video applications. A cabling network is established from either a single hub or multiple hubs **main cross connect (MC)** in the facility to the **intermediate cross connects (IC)** in each building. The use of outside plant cables containing a combination of both 62.5-μm and single-mode fibers is the most common.

Intrabuilding Backbone

The intrabuilding backbone connects the intermediate cross connect in the building with each **telecommunications closet (TC)** and **horizontal crossconnect (HC)** within that building. An optical fiber intrabuilding backbone has emerged as the medium of choice for intrabuilding data backbone transmissions because of its ability to support multiple, high-speed networks in a much smaller cable without crosstalk concerns. Additionally, more users are using fiber to support voice and private branch exchange (PBX) applications in the intrabuilding backbone by placing small, remote PBX shelves at the telecommunications closet, eliminating the need for copper in the intrabuilding backbone altogether.

Horizontal Cabling

Fiber for horizontal applications is a reality because of the recent advances in the technology easing installation, reducing the overall system cost, and offering a system that is reliable, flexible, certifiable, and future proofed. Fiber horizontal systems can be implemented for today's network requirements, such as ethernet and token ring, and can easily accommodate growth to higher speed networks such as FDDI, ATM, fiber channel, and beyond. There is no concern that crosstalk, EMI/RFI, and FCC issues will be uncovered when frequencies are increased.

Whereas FTTD provides excellent benefits for the traditional 90-m, single-user cable network, the benefits of optical fiber—high bandwidth, low attenuation, and system operating margin—allow greater flexibility in cabling design. For open-systems furniture applications, the use of the multiuser outlet can be implemented to ensure the long-term survivability of the **horizontal cabling** network in the rapidly changing office area. Many users desire centralized hubs and switches in the building vs. the use of decentralized hubs and switches in each telecommunication closet. The increased distance capabilities of fiber and lack of crosstalk allows the user to implement centralized electronics with **centralized cabling**. This allows for simple administration, efficient use of electronic ports, and easy implementation of various network applications throughout the building.

Patch Cords and Equipment Cables

Although not part of the cabling infrastructure, patch cords and equipment cables are an important link in the network. Patch cords and equipment cables are factory-terminated assemblies, typically two fiber, used to connect equipment at cross connects and outlets or as cross-connect jumpers to connect two separate cabling segments.

Universal Transport System

The design philosophy in a universal transport system, as described in the following sections, will help the designer/user to plan and implement a backbone and horizontal network. This network will be application independent and will avoid recabling expenses. The communication network planner can be assured of maximum cable plant flexibility as well as minimum installation and administration costs by following the UTS recommendations. Thus, whether installing a complete fiber optic infrastructure or a simple **point-to-point link** for a specific application, the UTS end user receives a cabling system of maximum utility.

9.5.2 Network Design and Installation

One of the most challenging series of decisions a telecommunications manager makes is the proper design of an optical fiber cabling plant. Optical fiber cable, which has extremely high bandwidth, is a powerful telecommunications media that supports voice, data, video, and telemetry/sensor applications. However, the effectiveness of the media is greatly diminished if proper connectivity, which allows for flexibility, manageability, and versatility of the cable plant, is not designed into the system.

The traditional practice of installing cables or cabling systems dedicated to each new application is all but obsolete. Instead, designers and users alike are learning that proper planning of a structured cabling system can save time and money. For example, a cable that is installed for point-to-point links should also meet specifications for later upgrades as it becomes part of a much larger network. As a result, duplications of cable, **connecting hardware**, and installation labor can be avoided by anticipating future applications and providing additional fibers for unforeseen applications, as well as interfacing with local service providers.

UTS recommends a structured cabling design philosophy that uses a physical hierarchical star topology. The hierarchical star offers the user a communications transport system that can be installed to efficiently support all logical topologies (*star, bus, point-to-point, ring*).

The guidelines presented here follow the general recommendations set forth by the TIA/EIA-568A cabling standard (TIA,1995). This section will begin with an overview of the different logical topologies commonly used today. Following will be a discussion of the physical implementation of these logical topologies in the different premises environments (interbuilding backbones, intrabuilding backbones, and horizontal cabling).

A note regarding logical vs. physical topologies. The term logical topology refers to the method by which different nodes in a network communicate with one another. It involves protocols, access, and convention at the electronic circuit level. Physical topology simply refers to the physical arrangement of the cabling or cabling system by which nodes are attached (Fig. 9.79).

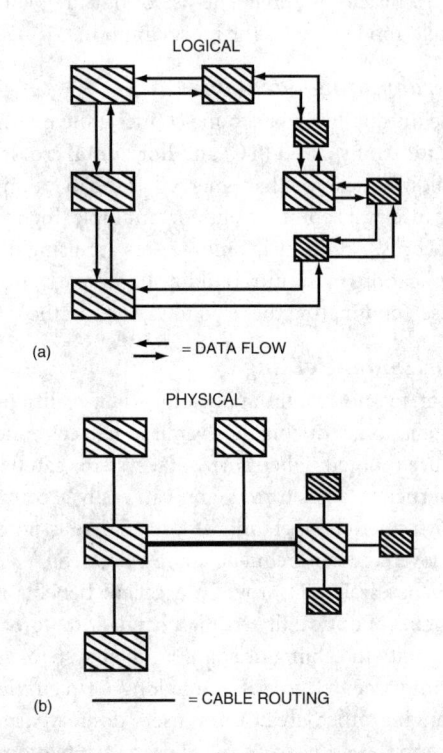

FIGURE 9.79 Networking topologies: (a) Logical, (b) Physical.

Logical Topology

To be universal for all possible applications, a structured cabling plan must support all logical topologies. These topologies define the electronic connection of the systems nodes. Fiber applications can support the following logical topologies:

- Point-to-point
- Star
- Ring
- Bus

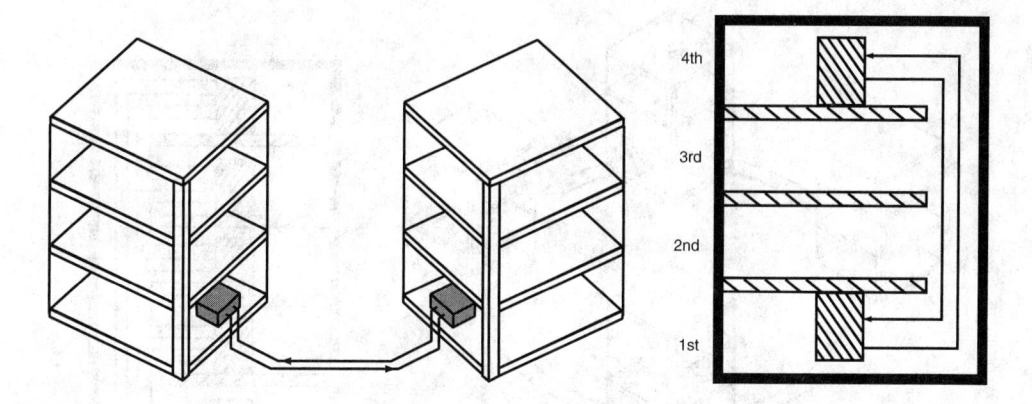

FIGURE 9.80   Logical point-to-point system.

Point-to-Point

Point-to-point logical topologies are still common in today's customer premises installations. Two nodes requiring direct communication are directly linked by the fibers, normally a fiber pair: one to transmit, one to receive (Fig. 9.80). Common point-to-point applications include:

- Fiber channel
- Terminal multiplexing
- Satellite up/down links
- ESCON
- Telemetry/sensor

Star

An extension of the point-to-point topology is the logical star topology. This is a collection of point-to-point topology links, all of which have a common node that is in control of the communications system (Fig. 9.81). Common star applications include:

- A switch, such as a PBX, ATM, or data switch
- A security video system with a central monitoring station
- An interactive video conference system serving more than 2 locations

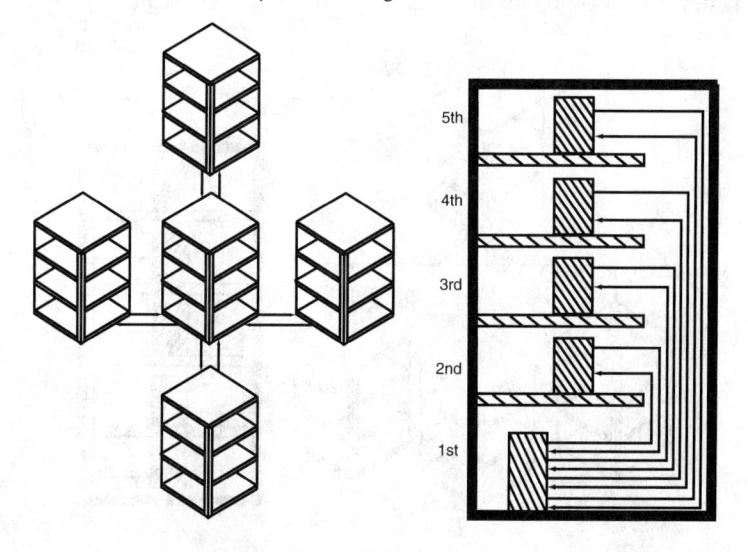

FIGURE 9.81   Logical star configuration.

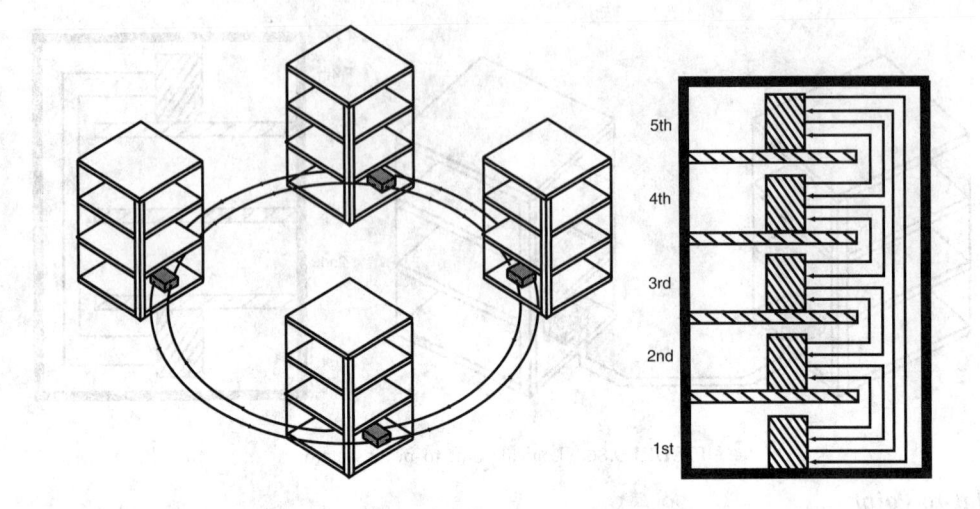

FIGURE 9.82   Logical counter-rotating ring system.

Ring

The logical ring topology, common in the data communications area, is supported by two primary standards: (1) the token ring (IEEE 802.5) and (2) FDDI (ANSI X3T9). In this topology, each node is connected to its adjacent nodes in a ring (Fig. 9.82). Examples of ring topologies are:

- FDDI
- Token ring

Nodes are connected in single or dual (counter-rotating) rings. With counter-rotating rings (most common), two rings transmit in opposite directions. If one node fails, one ring automatically loops back on the other, allowing the remaining nodes to function normally. This requires two fiber pairs per node rather than the one pair used in a simple ring. FDDI networks typically utilize a counter-rotating ring topology for the backbone and a single ring for the horizontal.

Logical Bus

The logical bus topology is also utilized by data communications and is supported by the IEEE 802.3 standards. All nodes share a common line (Fig. 9.83). Transmission occurs in both directions on the

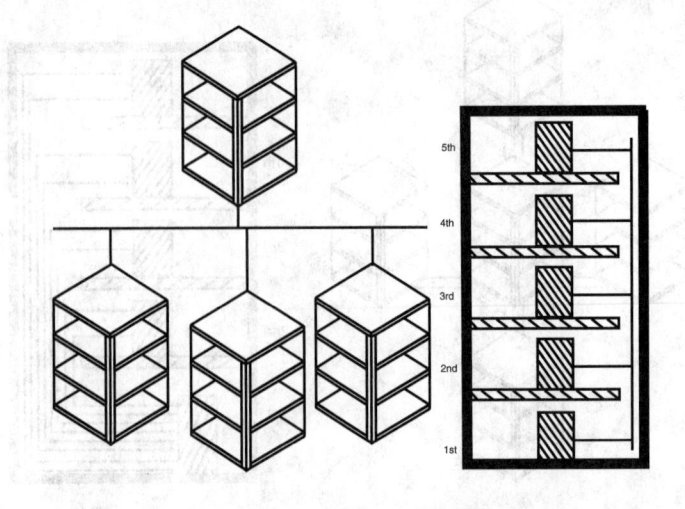

FIGURE 9.83   The logical bus topology.

common line rather than in one direction, as on a ring. When one node transmits, all of the other nodes receive the transmission at approximately the same time. The most popular systems requiring a bus topology are

- Ethernet
- Manufacturing automation protocol (MAP)
- Token bus

Physical Topologies, Logical Topology Implementation

All of the logical topologies noted earlier are easily implemented with a physical star cabling scheme, as recommended by the TIA/ EIA-568A cabling standard (TIA, 1995). Implementing logical point-to-point and star topologies on a physical star is straightforward.

Whereas the use of data networks that utilize bus or ring topology are predominant in the market, that is, ethernet, token ring, and FDDI, with the benefits of physical star cabling, electronic vendors and standards have developed solutions designed to interface with a star network. These applications are typically implemented with an intelligent hub or *concentrator*. This device in simple terms establishes the bus or ring in the backplane of the device and the connections are made from a central location(s). Therefore, from a physical connection these networks appear to be a star topology and are quite naturally best supported by a physical star cabling system.

Physical Stars Vs. Physical Rings

There are some situations where a physical ring topology would seem appropriate. But as can be seen in the following comparison of an optical fiber physical star topology to a physical ring topology, where advantages are indicated by a plus sign and disadvantages by a minus sign, a physical star topology is the best topology for supporting the varied requirements of a structured cabling system.

Physical Star

+ 1. It is flexible, supports all applications and topologies.
+ 2. It has acceptable connector loss with topology compatibility.
+ 3. Centralized fiber cross connects make administration and rearrangement easy.
+ 4. Existing outdoor ducts are often configured for the physical star, resulting in easy implementation.
+ 5. It easily facilitates the insertion of new buildings or stations into the network.
+ 6. It supported by TIA/EIA-568A cabling standard.
− 7. Cut cable will result in node failure unless redundant routing is used.
− 8. More fiber length is used than in a logical ring topology.

Physical Ring

− 1. It is less flexible.
− 2. It has unacceptable connector loss if star or bus logical topologies are required.
− 3. Connectors are located throughout system, making administration and rearrangement difficult.
− 4. Physical implementation can be difficult, often requiring new construction.
− 5. Insertion of new buildings or stations into a physical ring can cause disruptions to current applications or inefficient cable use.
− 6. Even with ring topology applications, the majority, if not all nodes, must be present and active.
+ 7. There is node survivability in event of cable cut if cable truly utilizes redundant paths.
+ 8. Less fiber lengths are used between nodes.

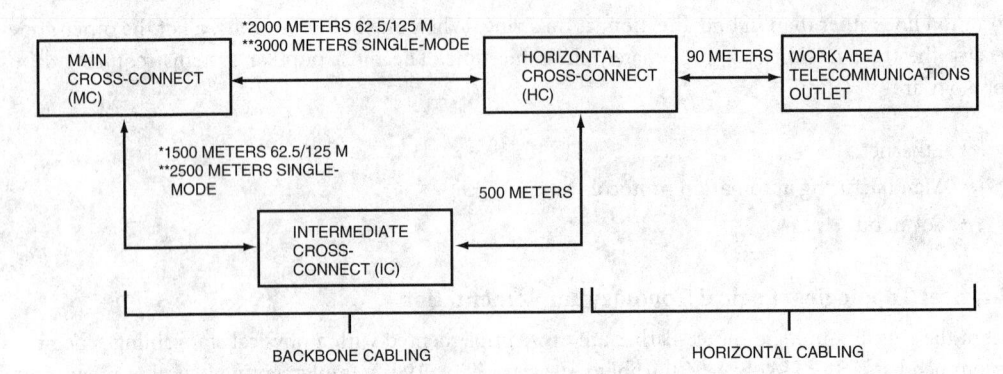

FIGURE 9.84   Commercial building telecommunications cabling standard (TIA/EIA-568A).

Redundancy and Zero Downtime

As protection against system downtime, many systems employ redundant electronic devices coupled to fibers in the same cable. The redundant system takes over immediately upon failure of the primary system. This protects against active device failure, although it does not help in the rare instance of a complete cable cut.

The ultimate protection against system downtime is achieved through redundant cable routing. Redundant fibers are placed in a second route to immediately take over in the event a cable is damaged. Redundant routing should be considered when zero downtime for the physical cable plant is required.

Physical Star Implementation

TIA/EIA-568A Cabling Standard

The more important recommendations of the TIA/ EIA-568A standard(TIA, 1995) as they relate to optical fiber and topologies are summarized as follows. The standard is based on a hierarchical star for the backbone and a single star for horizontal distribution (Fig. 9.84).

The rules for backbone cabling include a 2000-m for 62.5/125-μm (3000-m for single mode) maximum distance between the main cross connect and the **horizontal cross connect (HC)**, and a maximum of one intermediate cross connect between any HC and MC. The MC is allowed to provide connectivity to any number of ICs or HCs and the ICs are allowed to provide connectivity to any number of HCs.

The standard does not distinguish between interbuilding (outside) or intrabuilding (inside) backbones since these are determined by the facility size and campus layout. In most applications, however, one can envision the MC to IC being an interbuilding backbone link and the IC to HC being an intrabuilding backbone link (Fig. 9.85). The exceptions to this would be large high-rise buildings in which the backbone may be entirely inside the building or a campus containing small buildings with only one HC per building, thereby eliminating the requirement for an IC.

The horizontal cabling is specified to be a single star linking the horizontal cross-connect closet to the work area telecommunications outlet with a distance limitation of 90 m (Fig. 9.86). This distance limitation is not based on fiber capabilities but rather, it is based on copper distance limitations to support data requirements.

568SC Polarity

The simplex SC connector and **adapter** are keyed to ensure the same orientation of the connector upon mating. Upon installation of the duplex yoke, to make two simplex SC connectors into a duplex 568SC, the keying establishes the orientation of one fiber to the other (polarity) because the 568SC connector can only

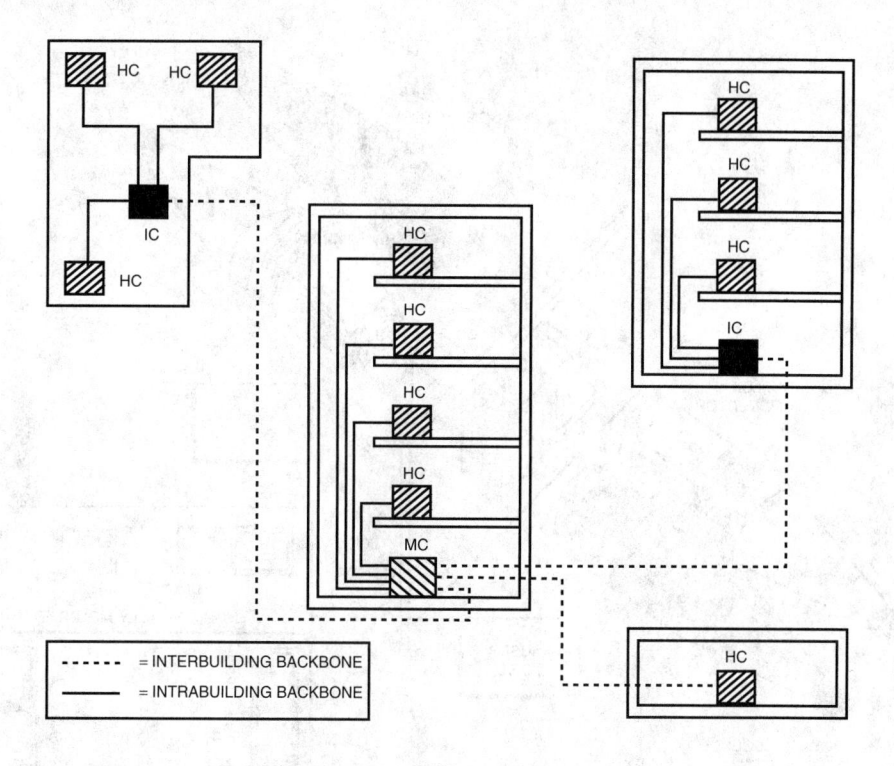

FIGURE 9.85   Backbone cabling system.

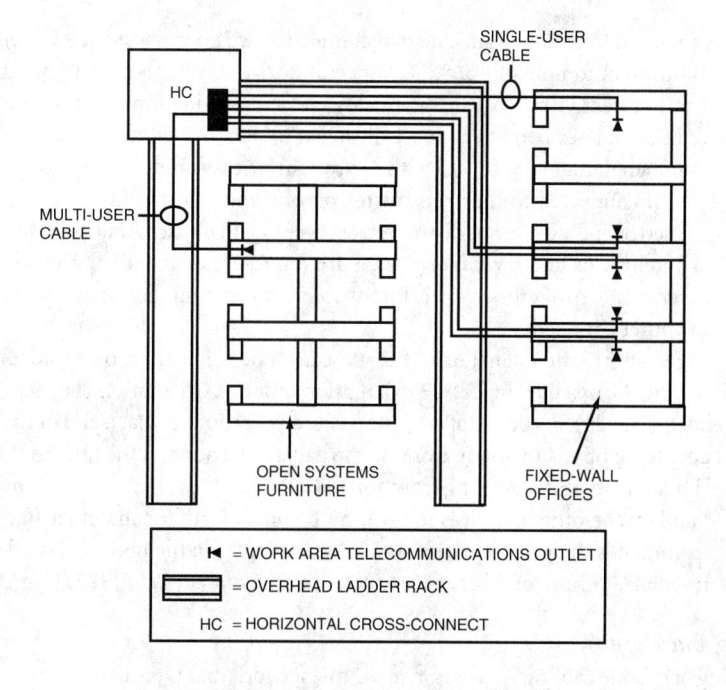

FIGURE 9.86   Horizontal cabling network.

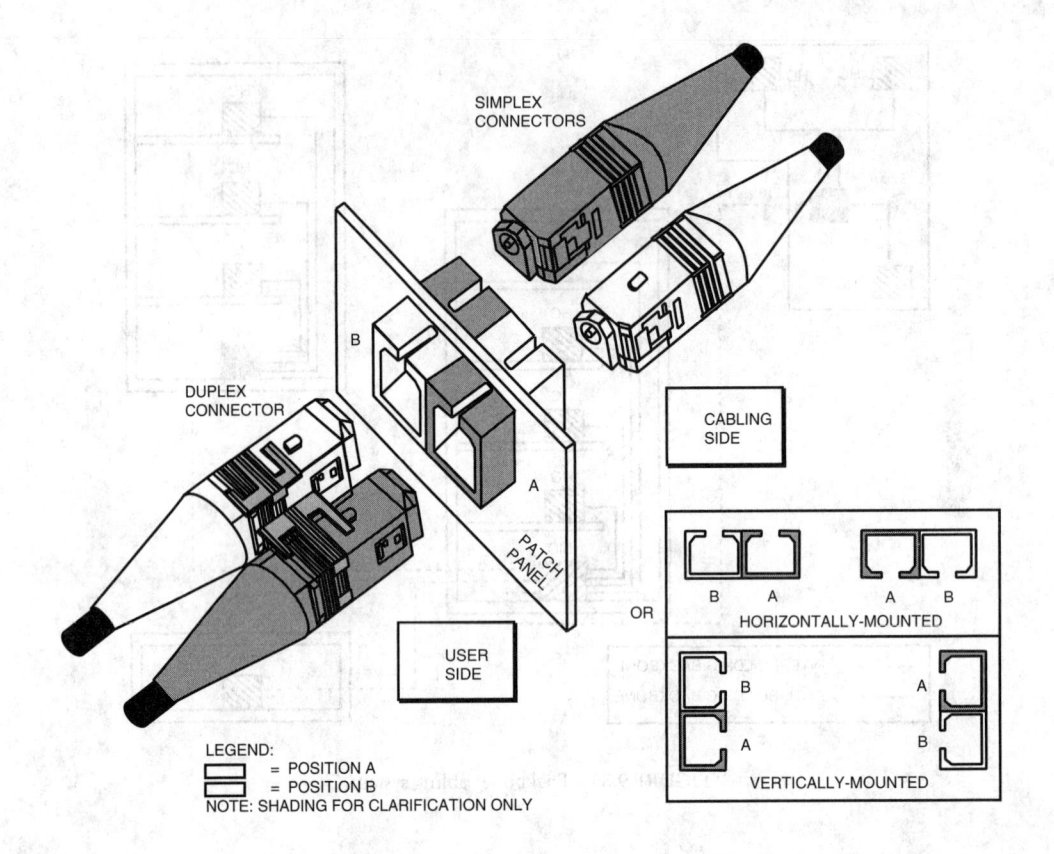

FIGURE 9.87   568SC connection configuration.

insert into the receptacle in one direction. The two connectors are now labeled as *position A* and *position B* (Fig. 9.87). It is important to note, the 568SC connector does not ensure correct polarity by itself.

To achieve transmit-to-receive polarity in an optical network, an odd number of crossovers must occur. A duplex fiber circuit comprises two fibers and is installed to pair an odd numbered fiber with the next consecutive even numbered fiber (1 with 2, 3 with 4). To achieve transmitter-to-receiver connectivity, the odd numbered fiber, if connected to the **transmitter** on one end, must connect to the receiver on the other end. The specified 568SC adapter performs a crossover in and of itself, that is, position A to position B. Therefore, if all patch cords and the cabling plant are duplexed, there will always be an odd number of crossovers. An even number of crossovers is impossible. An even number of crossovers would cause a transmitter to transmitter connection.

Polarity is managed at the patch panel or outlet. Based on the orientation of the adapter in the panel, the connector can either be installed in an A-B orientation or a B-A orientation (Fig. 9.87).

It is recommended that adapters be installed in opposite orientations on each end of the link (Fig. 9.88). By installing the adapter in opposite orientations, the installer still connects the fiber to the patch panel or outlet in a normal manner, that is, counting from top to bottom, left to right. On one end, however, fiber 1 is in position A and on the other end, it is in position B and so forth for the other fibers. Additionally, TIA/EIA-568A recommends the proper orientation of each adapter in the network based on whether it is at the main, intermediate, or horizontal cross connect, or at the outlet (TIA, 1995) (Fig. 9.89).

Interbuilding Backbone

The interbuilding backbone cabling is the segment of the network that typically presents the designer and user with the most options, especially in large networks. The interbuilding backbone is also, typically, the most constrained by physical considerations such as duct availability, right of way, and physical barriers. For this reason, this section will present a number of options for consideration.

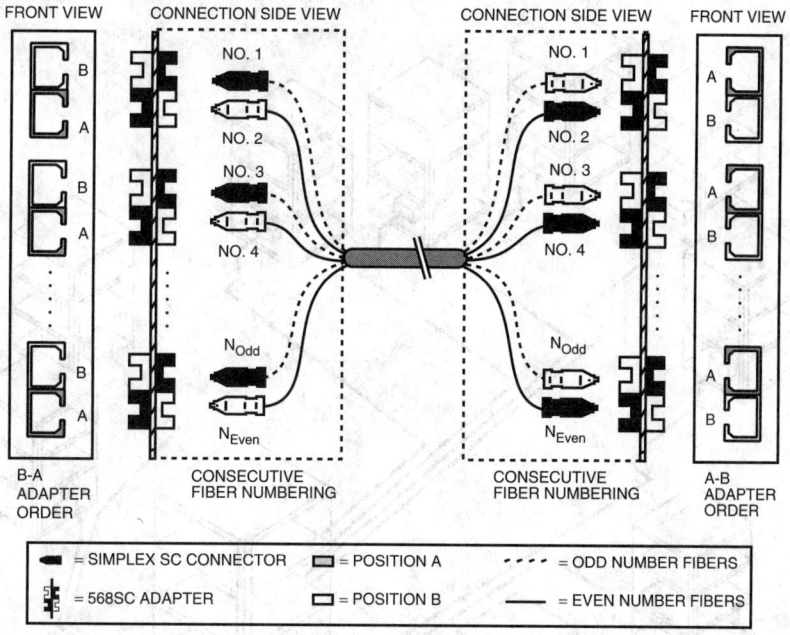

FIGURE 9.88   Crossover cabling with 568SC.

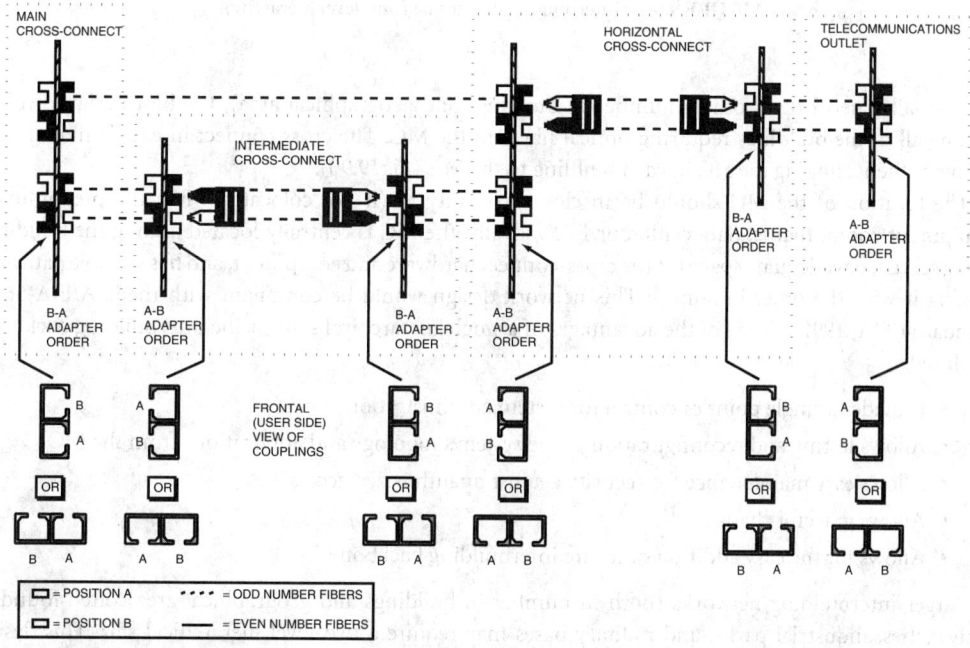

FIGURE 9.89   568SC adapter orientation in the network.

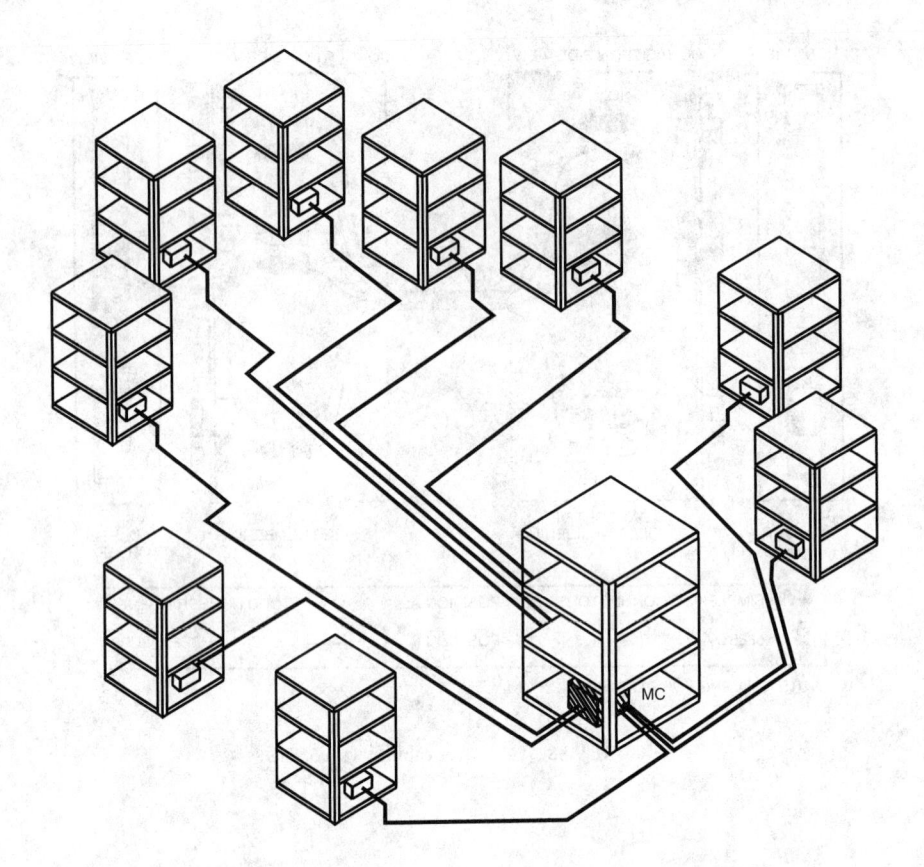

FIGURE 9.90 Interbuilding backbone (one-level hierarchy).

In smaller networks (both in number of buildings and geographical area), the best design involves linking all of the buildings requiring optical fibers to the MC. The cross connect in each building then becomes the IC linking the HC in each building to the MC (Fig. 9.90).

The location of the MC should be in close proximity to (if not colocated with) the predominant **equipment room**, that is, data center, or PBX. Ideally, the MC is centrally located among the buildings being served, has adequate space for the cross-connect hardware and equipment, and has suitable pathways linking it with the other buildings. This network design would be compliant with the TIA/EIA-568A standard (TIA, 1995). Some of the advantages of a single hierarchical star for the interbuilding backbone are listed:

- Provides a single point of control for system administration
- Allows testing and reconfiguration of the system's topology and applications from the MC
- Allows easy maintenance for security against unauthorized access
- Allows graceful change
- Allows for the easy addition of future interbuilding backbones

Larger interbuilding networks (both in number of buildings and geographical area) often found at universities, industrial parks, and military bases may require a two-level hierarchical star. This design provides an interbuilding backbone that does not link all of the buildings to the MC. Instead, it uses selected ICs to serve a number of buildings. The ICs are then linked to the MC. This option may be considered when the available pathways do not allow for all cables to be routed to an MC or when it is desirable to segment the network because of geographical or functional communication requirement groupings. In large networks, this often translates to a more effective use of electronics such as multiplexers,

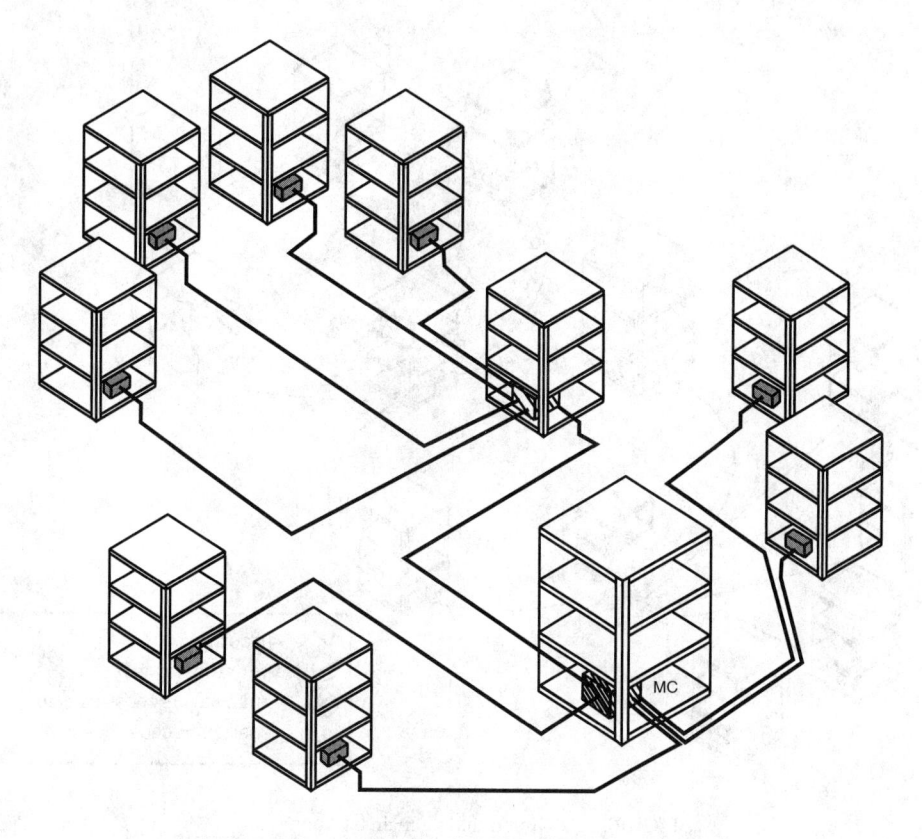

FIGURE 9.91   Interbuilding backbone (two-level hierarchy).

routers, or switches to better utilize the bandwidth capabilities of the fiber or to segment the network (Fig. 9.91).

It is recommended that no more than five ICs be used, unless unusual circumstances exist. If the number of interbuilding ICs is kept to a minimum, then the user can experience the benefits of segmenting the network without significantly sacrificing control, flexibility, or manageability.

It is strongly recommended that when such a hierarchical star for the interbuilding backbone is used, it be implemented by a physical star in all segments. This ensures that flexibility, versatility, and manageability are maintained. However, there are two main conditions where the user may consider a physical ring for linking the interbuilding ICs and MC. These are (1) that the existing conduit supports it or (2) that the primary (almost sole) purpose of the network is FDDI or token ring. It is seldom recommended to connect the outlying buildings in a physical ring (Fig. 9.92). The ideal design for a conduit system that provides a physical ring routing would dedicate X number of fibers of the cable to a ring and Y number of fibers to a star by expressing (not terminating) fibers through the ICs directly back to the MC (Fig. 9.93). This allows the end user the best of both worlds, although the design requires a more exacting knowledge of present and future communication requirements.

Intrabuilding Backbone

The intrabuilding backbone design between the building cross connect (MC or IC) and the HC is usually straightforward, though various options sometimes exist. A single hierarchical star design is recommended between the building cross connect and the HCs (Fig. 9.94). The only possible exception occurs in an extremely large building, such as a high rise, where a two-level hierarchical star may be considered (Fig. 9.95). The same options and decision processes apply here. Sometimes, building pathways link HC to HC in

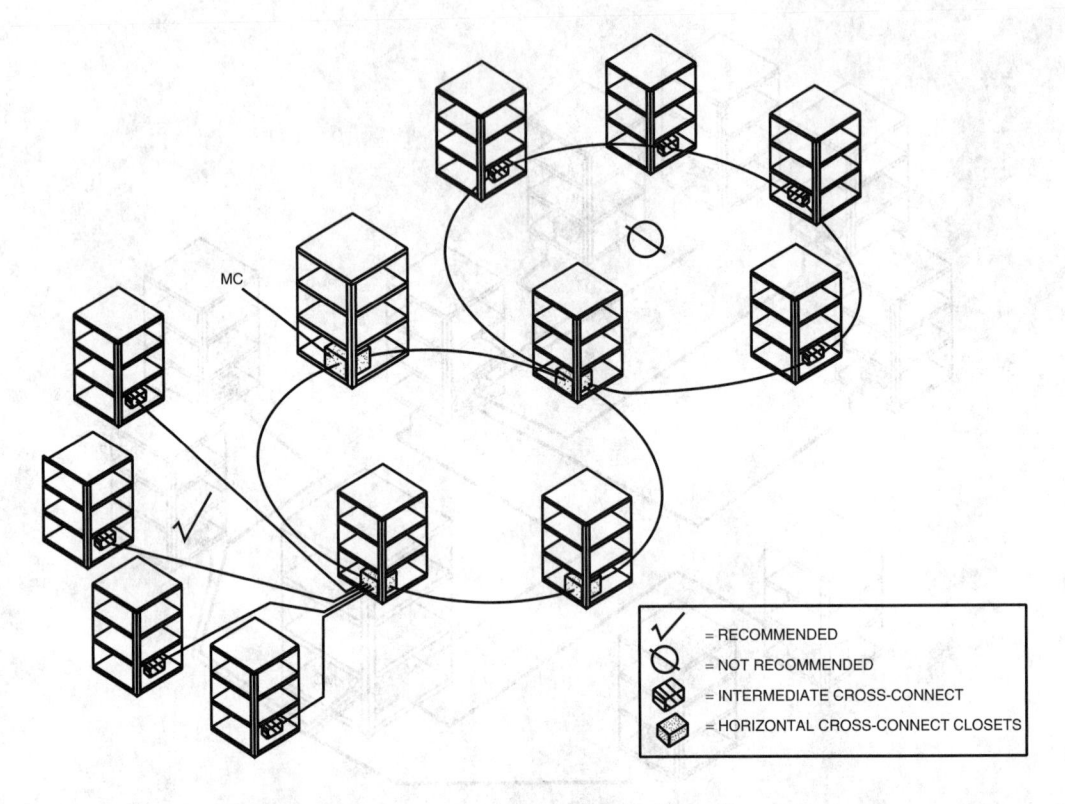

FIGURE 9.92   Interbuilding main backbone ring.

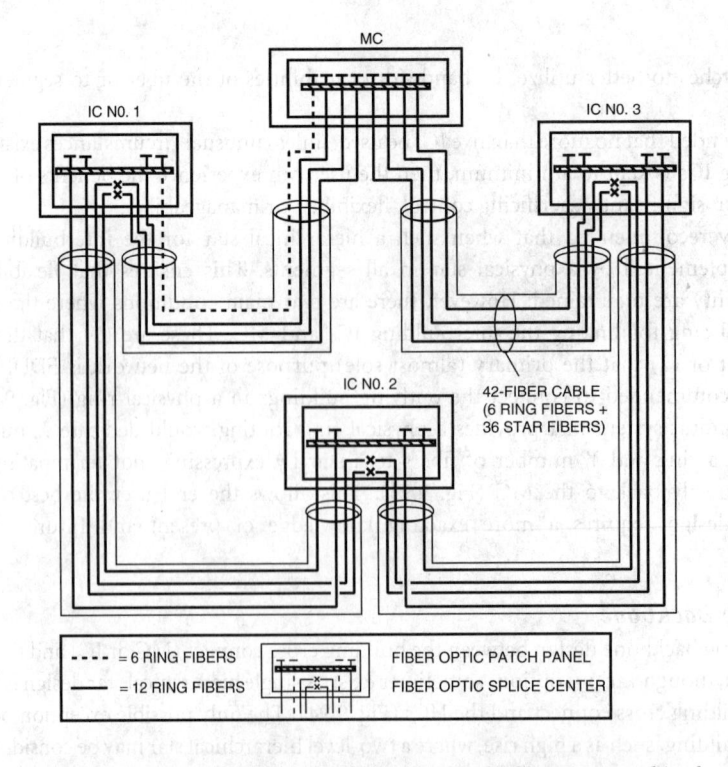

FIGURE 9.93   Main backbone ring and redundant backbone star combined.

addition to IC to HC, especially in buildings with multiple HCs per floor. Only in specific applications should a user design an HC-to-HC link. For example, this pathway might be of value in providing a redundant path between HC and IC, although direct connection between HCs should always be avoided (Fig. 9.96).

Horizontal Cabling

The benefits of fiber optics (high bandwidth, low attenuation, and system operating margin) allow greater flexibility in cabling design. Because of the distance limitations of copper and small operating margin, passing data over UTP requires compliance with stringent design rules. In addition to the traditional horizontal cabling network, optical fiber supports designs specifically addressing the use of open-systems furniture and/or centralized management.

The traditional network consists of an individual outlet for each user within 300 ft of the telecommunications closet. This network typically has data electronics equipment (hub, concentrator, or switch) located in each telecommunications closet within the building. Because electronics are located in each closet, this network is normally implemented with a small fiber-count backbone cable, 12- or 24-fiber count. This network uses a single cable per user in a physical star from the HC to the telecommunications outlet (Fig. 9.97).

For open-systems furniture applications employ a multiuser outlet network. The multiuser outlet calls for a high-fiber-count cable to be placed from the closet to an area in the open office where there is a fairly permanent structure, such as a wiring column or cabinet, in a grid type wiring scheme. At this structure, a multiuser outlet, capable of supporting 6–12 offices, is installed instead of individual outlets (Fig. 9.98). Fiber optic patch cords are then installed through the furniture raceways from the multiuser outlet to the office area. This allows the user to rearrange the furniture without disrupting or relocating the horizontal cabling. This type of network is supported in Annex G of TIA/EIA-568A, (TIA, 1995).

Centralized optical fiber cabling is intended as a cost-effective alternative to the optical horizontal cross connect when deploying 62.5 μm optical fiber cable in the horizontal in support of centralized electronics. Whereas category 5 UTP systems are limited to 100-m total length and require electronics in each closet for high-speed data systems,

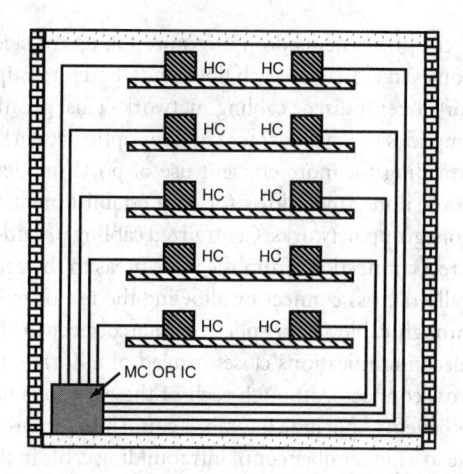

FIGURE 9.94 Interbuilding backbone (one-level hierarchy).

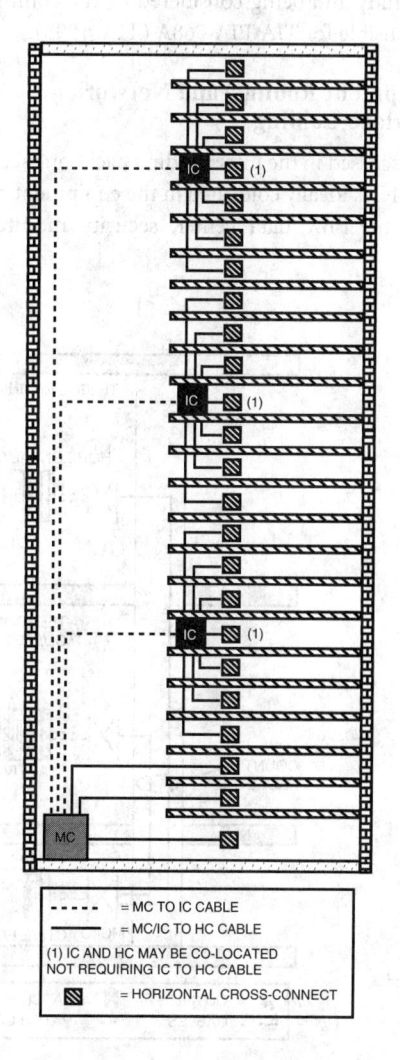

----- = MC TO IC CABLE
——— = MC/IC TO HC CABLE
(1) IC AND HC MAY BE CO-LOCATED
NOT REQUIRING IC TO HC CABLE
▨ = HORIZONTAL CROSS-CONNECT

FIGURE 9.95 Interbuilding backbone (two-level hierarchy).

fiber optic systems do not require the use of electronics in closets on each floor and, therefore, support a centralized cabling network. This greatly simplifies the management of fiber optic networks, providing for more efficient use of ports on electronic hubs, and allows for easy establishment of work group networks. Centralized cabling provides direct connections from the work areas to the centralized cross connect by allowing the use of pull-through cables, or a splice or interconnect in the telecommunications closet instead of a horizontal cross connect. Although each of these options has its benefits, splicing a low-fiber-count horizontal cable to a higher fiber count intrabuilding cable in the telecommunications closet is often the best choice (Fig. 9.99). This type of network is presently under study and being considered by the committee responsible for TIA/EIA-568A (TIA, 1995).

Equipment Room(s) and Network Interface Cabling

As discussed in the interbuilding backbone section, the MC is ideally colocated in the equipment room with the PBX, data center, security monitoring

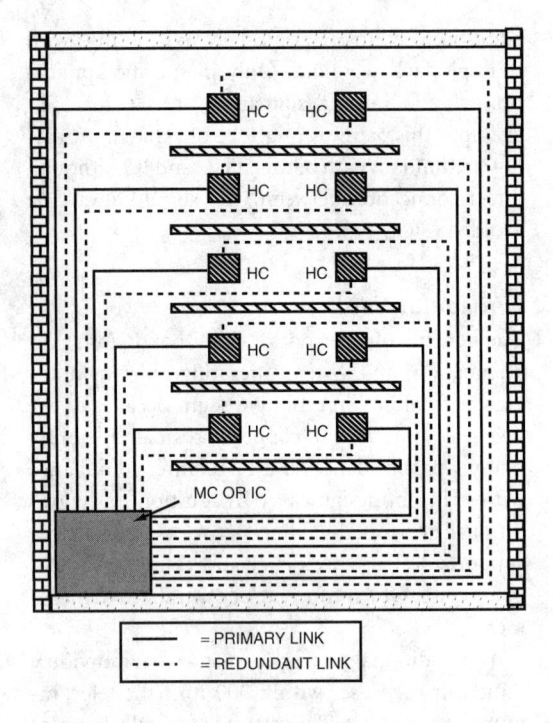

FIGURE 9.96   Intrabuilding backbone (redundant routing).

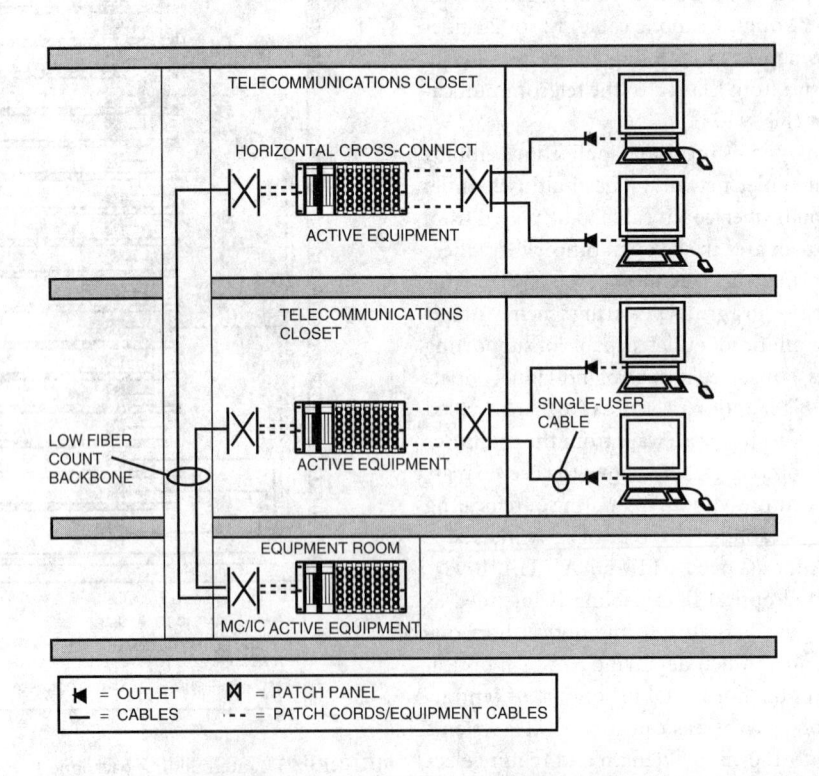

FIGURE 9.97   Traditional horizontal cabling system.

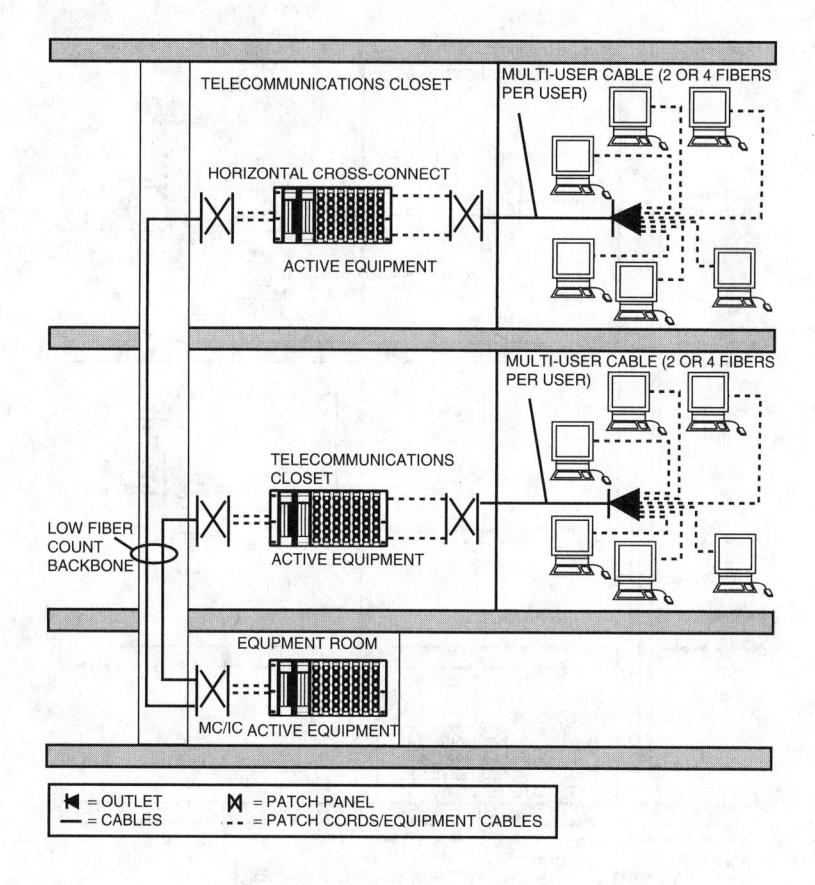

FIGURE 9.98 Multi-user outlet cabling.

equipment, or other active equipment being served. Although this is ideal, physical constraints sometimes make this impossible, such as when multiple equipment rooms are being served and are not centralized. The MC does not have to be colocated with any equipment room, and its location may be based entirely on geographic and physical constraints, that is, duct space and termination space. There should be only one primary MC. Connection to the equipment room(s) can then be provided either with fibers in separate sheaths or combined with the backbone fiber cables (Fig. 9.100).

For ultimate flexibility, manageability, and versatility of the fiber network all backbone cables and links to equipment room(s) should be terminated at the MC. Fibers can be cross connected to the required equipment room on an as-needed basis with the simple installation of a jumper. If there are link loss concerns, security concerns, or cost issues, selected fibers can be spliced through the MC to specific equipment rooms. Although this forfeits a degree of flexibility, as long as all of the cables come to the MC, the routing can be changed by altering the splice plan.

The same location considerations apply to the MC and network interface when contracted services are provided by a local common carrier, long distance company, or by-pass company. The designer/user should bear in mind that, when these services are provided via optical fibers, the fibers will almost certainly be single mode.

Multiple Sheaths and Splice Points

Multiple Cables

The physical design of the system should minimize splices whenever possible. The small size and long lengths of optical fiber cable allow, in most cases, the use of separate cable sheaths to serve each cross connect, closet, or outlet, providing conduit space allows. This avoids cable splices and results in fewer

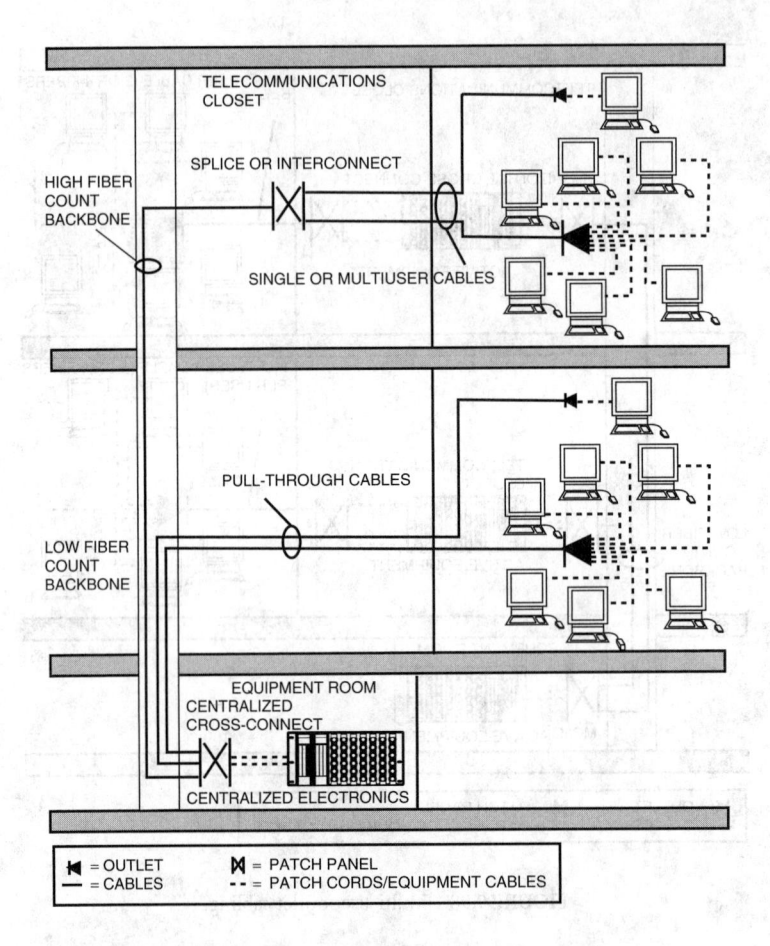

FIGURE 9.99   Centralized optical fiber cabling.

different fiber count cables, allowing for an easier installation, and typically avoiding minimum-order requirement issues. In short, the small incremental cost of additional sheaths usually offsets the cost of splicing different fiber count cables together (Fig. 9.101).

Consolidation Splice

When constrained duct or conduit space precludes the use of multiple sheaths, a splice should be used to consolidate the sheaths into one higher fiber count sheath. Combine as many cables as possible at a single splice point, since the incremental cost per additional fiber splice is less than the cost for splicing at different locations. It is important to analyze the entire system when planning splice points. For example, if a planned cross connect is near a manhole that is being considered for a splice point, it might make more sense to route the cables to the cross connect, so that the splice point is combined with the termination point. This can result in substantial labor cost savings (Fig. 9.102).

Splice Points

Splice point locations should be chosen only after considering the requirements for performing optical fiber splicing:

- To effectively perform a splice, the cable ends must reach a satisfactory work surface (preferably a vehicle or table that is clean and stable). The distance can be as much as 30 ft.
- The chosen location should have provisions for storing the slack cable after splicing is completed.

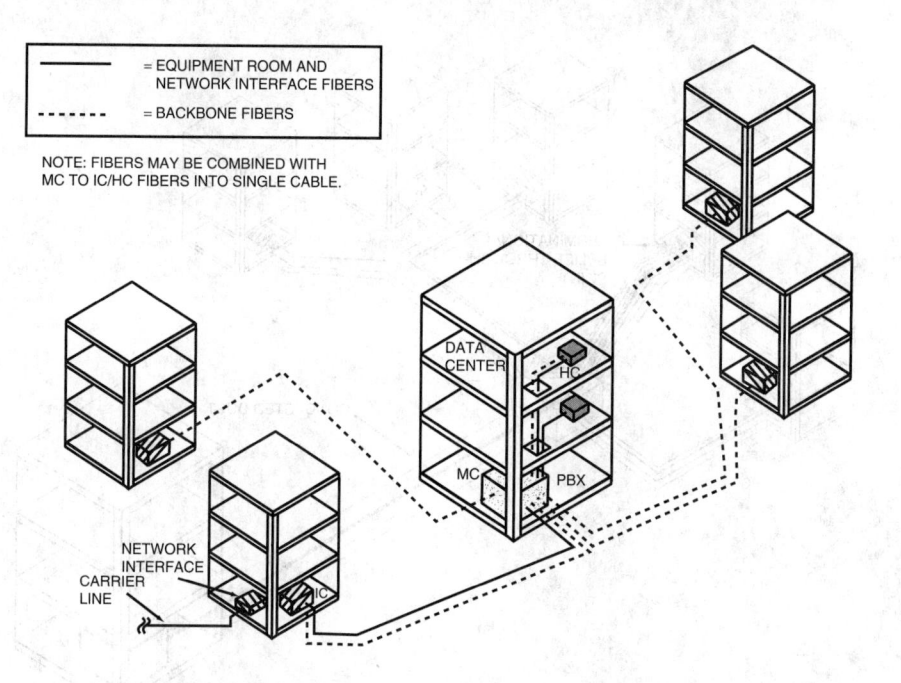

FIGURE 9.100   Equipment room(s) and network interface cabling.

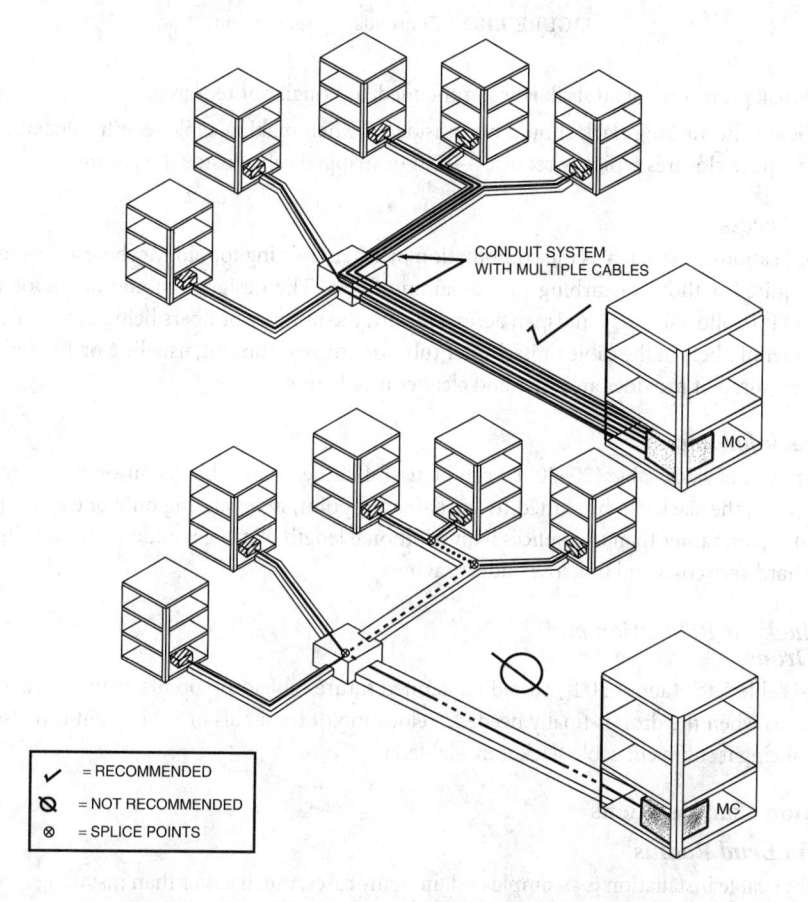

FIGURE 9.101   The use of multiple cables in wiring a facility.

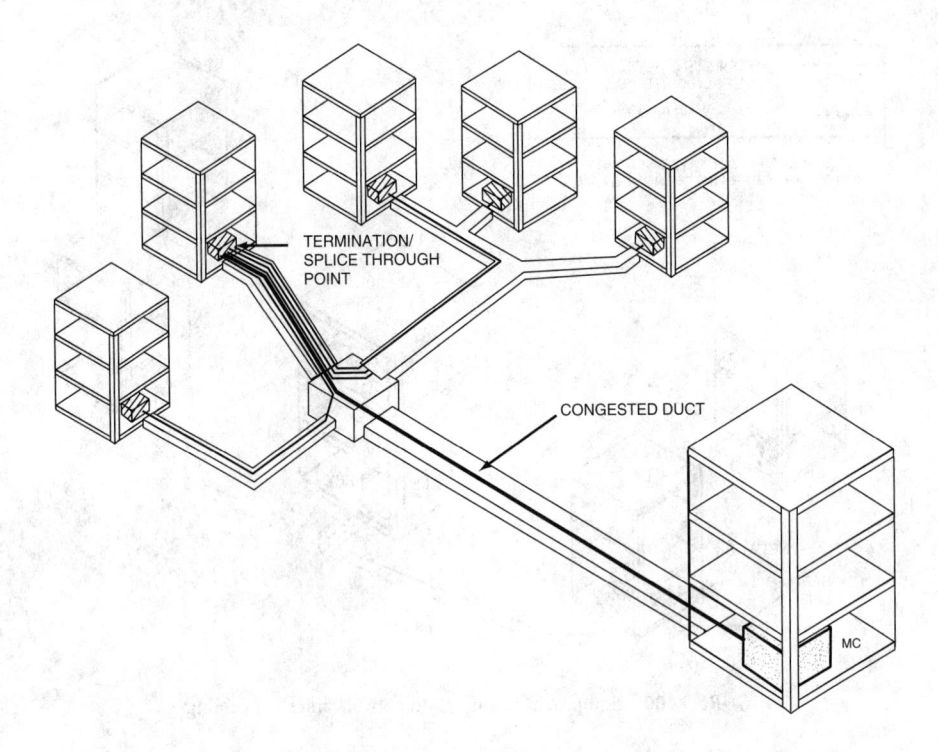

FIGURE 9.102   Consolidation splice point.

- Physical protection of all slack is recommended, although not required.
- Splicing and racking slack should be considered when making cable length calculations. Optical fiber splice closures typically require 8–10 ft of stripped cable inside the closure.

Midspan Access

Special applications exist, such as the combination of a star and ring topology, where access to individual fibers is required without disturbing the remaining fibers. The designer should allow for cable slack, normally 30 ft to allow for easy midspan access. Ideally, the number of fibers being accessed corresponds to the number of fibers in the cable units, buffer tube, or unitized subunit, usually 6 or 12 fibers. Although this is not required, it provides an easier and cleaner installation.

Cable Slack for Repair

A small amount of slack cable (20–30 ft) can be useful in the event cable repair or relocation is needed. If a cable is cut, the slack can be shifted to the damaged point, necessitating only one splice point in the permanent repair, rather than two splices if an additional length of cable is added. This results in reduced labor and hardware costs and link loss budget savings.

Cable Slack for Relocation and Future Drops

Additional cable slack (about 30 ft) stored at planned future cable drop points results in savings in labor and materials when the drop is finally needed. Relocation of terminals or cable plant can also take place without splicing if sufficient cable slack is available.

Installation Considerations

Minimum Bend Radius

Optical fiber cable installation is as simple and, in many cases, much easier than installing coaxial or UTP cable in the horizontal network. The most important factor in optical fiber cable installations is maintaining

| APPLICATION | FIBER COUNT | MINIMUM BEND RADIUS[b] | | | |
| | | LOADED | | UNLOADED | |
		cm	in	cm	in
INTERBUILDING	2–84	22.5	8.9	15.0	5.9
BACKBONE	86–216	25.0	9.9	20.0	7.9
INTRABUILDING	2–12	10.5	4.1	7.0	2.8
BACKBONE	14–24	15.9	6.3	10.6	4.2
	26–48	26.7	10.5	17.8	7.0
	48–72	30.4	12.0	20.3	8.0
	74–216	29.4	11.6	19.6	7.7
HORIZONTAL	2	6.6	2.6	4.4	1.7
CABLING	4	7.2	2.8	4.8	1.9

[a]SPECIFICATIONS ARE BASED ON REPRESENTATIVE CABLES FOR APPLICATIONS. CONSULT MANUFACTURER FOR SPECIFICATIONS OF SPECIFIC CABLES.
[b]FOR OPEN-SYSTEMS FURNITURE APPLICATIONS WHERE CABLES ARE PLACED, SIECOR'S 2- AND 4-FIBER HORIZONTAL CABLES HAVE A ONE-INCH BEND RADIUS. PLEASE CONTACT YOUR SIECOR REPRESENTATIVE FOR FURTHER INFORMATION.

FIGURE 9.103 Typical bend radius specifications.

the **cable's** minimum **bend radius**. Bending the cable tighter than the minimum bend radius may result in increased attenuation and broken fibers. If the elements of the cable, that is, buffer tube or sheath, are not damaged, when the bend is relaxed, the attenuation should return to normal. Cable manufacturers specify a minimum bend radii for cables under tension and for long-term installation (Fig. 9.103).

Maximum Tensile Rating

The cable's maximum tensile rating must not be exceeded during installation. This value is specified by the cable manufacturer (Fig. 9.104). Tension on the cable should be monitored when a mechanical pulling device is used. Hand pulls do not require monitoring. Circuitous pulls can be accomplished through the use of backfeeding or center-pull techniques. For indoor installations, pull boxes can be used to allow cable access for backfeeding at every third 90° bend. No residual tension should remain on the cable after installation except that which is due to the cable's own weight in a vertical rise.

| APPLICATION | FIBER COUNT | MAXIMUM TENSIL LOAD | | | |
| | | SHORT TERM | | LONG TERM | |
		N	lb	N	lb
INTERBUILDING	2–84	2700	608	600	135
BACKBONE	86–216	2700	608	600	135
INTRABUILDING	2–12	1800	404	600	135
BACKBONE	14–24	2700	608	1000	225
	26–48	5000	1124	2500	562
	48–72	5500	1236	3000	674
	74–216	2700	600	600	135
HORIZONTAL	2	750	169	200	45
CABLING	4	1100	247	440	99

[a]SPECIFICATIONS ARE BASED ON REPRESENTATIVE CABLES FOR APPLICATIONS. CONSULT MANUFACTURER FOR SPECIFIC INFORMATION.

FIGURE 9.104 Typical optical cable tensile ratings.

APPLICATION	FIBER COUNT	MAXIMUM VERTICAL RISE	
		m	ft
INTERBUILDING	2–84	194	640
BACKBONE	86–216	69	226
INTRABUILDING	2–12	500	1640
BACKBONE	14–24	500	1640
	26–48	500	1640
	48–72	500	1640
	74–216	69	226
HORIZONTAL	2	500	1640
CABLING	4	500	1640

a SPECIFICATIONS ARE BASED ON REPRESENTATIVE CABLES FOR APPLICATIONS. CONSULT MANUFACTURER FOR SPECIFICATIONS OF SPECIFIC CABLES.

FIGURE 9.105   Typical vertical rise values.

Maximum Vertical Rise

All optical fiber cables have a maximum vertical rise that is a function of the cable's weight and tensile strength (Fig. 9.105). This represents the maximum vertical distance the cable can be installed without intermediate support points. Some guidelines for vertical installations include:

- All vertical cable must be secured at the top of the run.
- The attachment point should be carefully chosen to comply with the cable's minimum bend radius while holding the cable securely.
- Long vertical cables should be secured when the maximum rise has been reached.

The use of a split mesh grip is recommended to secure the cable (Fig. 9.106). The optical fiber cables weight should be borne by the split mesh grip. **Riser** cables should be secured with cable ties where the cable is exposed to prevent unintentional movement. However, the cable's weight should be borne by the split mesh grip, not the cable ties.

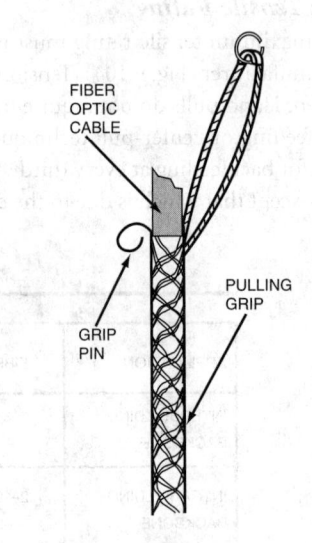

FIGURE 9.106   Vertical Rise Split Grip.

Pathways and Spaces

The consideration and design of proper cable pathways (i.e., conduit, cable trays, riser shafts, etc.) and termination spaces (i.e., main/intermediate cross connects, horizontal cross connects, and work area telecommunication outlets) are as important as the design of the cable network. Fortunately, a TIA/EIA standard addresses this area (TIA/EIA). Another useful document is the *Telecommunications Distribution Methods Manual* from the Building Industry Consulting Service International (BICSI). Listed next are just a few items to consider when designing or working with existing pathways and spaces.

Cable Protection

If future cable pulls in the same duct or conduit are a possibility, the use of innerduct to sectionalize the available duct space is recommended. Without this sectionalization, additional cable pulls can entangle an operating cable and could cause an interruption in service. Care should be exercised to ensure the

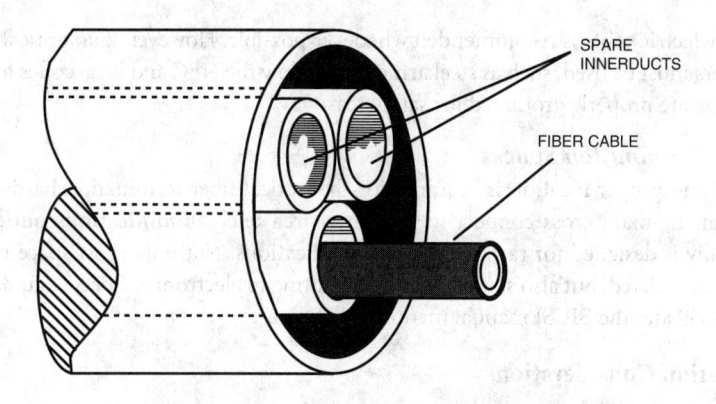

FIGURE 9.107   Proper duct utilization.

innerduct is installed as straight as possible, without twists that could increase the cable pulling tension (Fig. 9.107).

When the cable is installed in raceways, cable trays, or secured to other cables, consideration should be given to movement of the existing cables. Although optical fiber cable can be moved while in service without affecting fiber performance, it may warrant protection with conduit in places exposed to physical damage.

Duct Utilization

When pulling long lengths of cable through a duct or conduit, less than a 50% fill ratio by cross-sectional area is recommended (Fig. 9.108). For one cable, for example, this equates to a 0.71-in outside diameter cable in a 1-in inside diameter duct. Multiple cables can be pulled at once, as long as the tensile load is applied equally to all cables.

Fill ratios may dictate higher fiber counts in anticipation of future needs. One sheath can be more densely packed with fiber than multiple cable sheaths. In short, for customer premises applications, the cost of extra fibers is usually small when these extra fibers are not terminated until needed. For a difficult cable pull, extra fibers installed now but not terminated may be the most cost effective provision for the future.

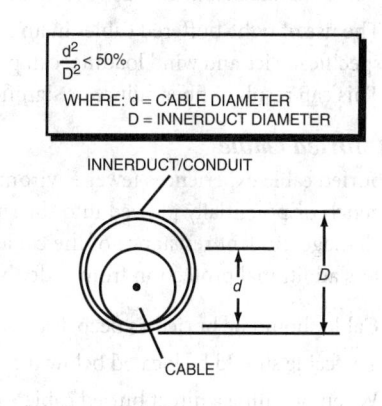

$$\frac{d^2}{D^2} < 50\%$$

WHERE: d = CABLE DIAMETER
D = INNERDUCT DIAMETER

FIGURE 9.108   Fill ratio guidelines.

Building and Fire Codes

Building and fire codes must be considered not only with intrabuilding telecommunications cabling but also most importantly when interbuilding telecommunications cabling enters the building. One of the most significant documents, while advisory in nature, is the **National Electrical Code** (**NEC**). The interbuilding backbone cable is usually not flame retardant. The NEC allows these cables to enter and be routed inside a building for 50 ft without conduit. This should be considered when planning the location of the main/intermediate cross connect in a building. If the unlisted cable must run more than 50 ft, then the designer/installer has two choices:

- To plan a splice point at the building entrance to transition from nonrated outdoor to indoor rated cable designs. The added loss of the splice is small and usually not significant in the link loss budget
- To enclose the outdoor cable in a proper flame-retardant conduit

The use of all-dielectric cable is recommended whenever possible. However, if an optical fiber cable with metallic members must be used, such as steel armoring, follow the NEC and local codes to ensure that the metallic members are properly grounded for safety reasons.

Optical Fiber Termination Spaces

Adequate space and proper location is required for the optical fiber termination hardware throughout a building, from the main cross connect to the **work area telecommunications outlet**. Optical fiber hardware typically is designed for rack and/or wall applications. Not only must space requirements for patch panels be considered, but also space for jumper routing to electronics. Specific guidance is provided in the TIA/EIA-569 and the BICSI manual mentioned earlier.

Other Installation Considerations

Although the typical pathway for the interbuilding backbone is a conduit/duct raceway, sometimes the designer may consider aerial or direct cable placement.

Aerial Cable

Aerial cable is exposed to many external forces. Wind and ice loading can put stresses on cables, and temperature changes are more extreme than in any other installation environment. Furthermore, aerial cables can be damaged by vehicular accidents, rodents, lightning, and even gunshots. Some guidelines follow:

- Aerial pole spans of less than 300 ft can be routinely engineered when a dedicated messenger wire is used.
- In some cases, optical fiber cable can be overlashed to the existing aerial plant.
- Cable is available in self-supporting designs, not requiring the use of a messenger.
- The use of tight-buffered cables in an aerial installation is not recommended. The external factors, specifically ice and wind loading, can place the fibers in tight-buffered cables under constant stress. This can result in fiber failure or significant increases in attenuation over time.

Direct Buried Cable

Direct buried cable experiences fewer environmental extremes than aerial cable. This cable is usually laid into a trench or potentially plowed into the soil. Primary dangers to underground cable are dig ups and rodent damage. Rodent resistance of the cable is provided by a coated steel tape armor. Installation in a duct offers additional protection from rodents, as well as dig ups. Some guidelines follow:

- Cable should be buried as deep as possible, at least 30 in underground, if possible.
- The cable should be located below the frost line to prevent damage due to frost heave.
- When planning a direct buried cable route, locate and avoid existing utilities.
- Plastic duct with an outside diameter greater than 1.5 in will also provide rodent resistance where armor cannot be used.
- A means of locating the cable, especially all-dielectric cable, should be provided.
- The use of tight-buffered cables in direct installations is not recommended. These cables are typically not waterblocked and are susceptible to water penetration and ensuing forces.

9.5.3 Fiber Count

The selection of the fiber count, or number of fibers used in the cable plant, is an extremely important decision that impacts both the current and future capabilities and cost of a communications network. The development and widespread use of fiber in all aspects of the network requires the designer to plan not only for the immediate needs but also for the evolution of future system requirements. Many of the isolated business or campus fiber links installed today will be integrated into the universal communication networks of tomorrow. Because these fiber systems will provide service for a number of different applications later, the number of fibers designed into the network today must be carefully considered.

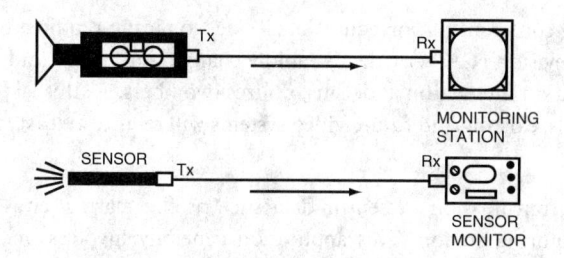

FIGURE 9.109   One-way (simplex) communications.

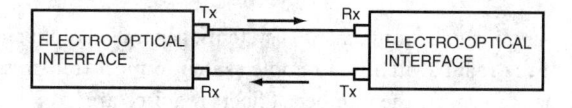

FIGURE 9.110   Two-way (duplex) communications.

The number of optical fibers required is dependent upon

- Intended end user application(s) requirements
- Level of electronics integration in the proposed network
- Future growth of the network
- Physical network topology

Application Requirements

Most of today's fiber optic system applications fall into three categories:

- *Simplex.* One fiber for transmit and receive (Fig. 9.109).
- *Duplex.* Two fibers, one for transmit and one for receive (Fig. 9.110).
- *Redundant duplex.* Four fibers, two for transmit and two for receive using separate duplex routing (Fig. 9.111).

Most of today's applications are duplex, utilizing two fibers to establish two-way communications—one fiber to transmit and the other fiber to receive. Note: It is possible to transmit and receive simultaneously using the same fiber. This is normally accomplished through the use of optical splitters inside the transceiver.

Video

Today's video applications employ a one- or two-fiber solution installed in either a point-to-point or physical star topology. The number of fibers depends on the sophistication of the proposed video system. Security video, for example, is a one-way, transmit type of application. However, certain systems require interaction at the camera for movement or voice communication and, thereby, require two fibers.

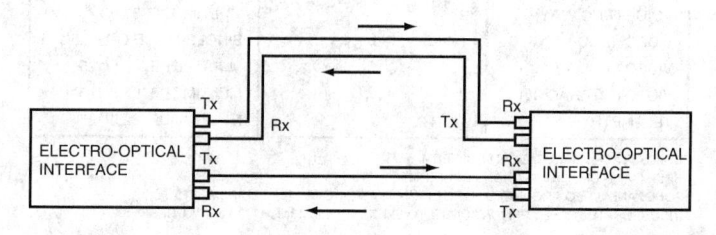

FIGURE 9.111   Redundant duplex communications.

Interactive video, video conferencing, or educational video typically requires two fibers. Broadcast or community antenna television (CATV) video is rapidly changing from the traditional one-fiber, source broadcast to receiver station interaction and, thus, utilizes two fibers. As this last form of video communication services expands, existing and future video systems will require at least two fibers.

Telemetry/Sensors

Telemetry involves the transmission of sensor data such as fire alarms, emissions, access, etc., to a distant or central monitoring station. This application generally involves one-way transmission and requires only one fiber. Certain types of telemetry, however, may be interactive. Correction signals, based on received data or signals, may be sent to the transmit locations. These applications would employ two fibers, one for transmit and one for receive.

Voice

Voice applications, which are normally multichannel systems, generally employ four fibers (two working and two protection) from the main switch to a remote switch to meet transmission requirements. The most significant factors that determine the number of fibers required are:

- Number of voice channels required between switch locations.
- Level of multiplexing to be employed. The same four 62.5-μm fibers that support a DS1/1.544 Mb/s (24 voice channels) network can be upgraded to support a DS3/ 44.736 Mb/s (672 voice channels) network or a DS3D/135 + Mb/s (2016 voice channels) network.

The determining factor is the cost of additional fibers vs. increased multiplexing equipment.

Data Communications

Today's data communications systems generally employ two fibers, however, four-fiber systems are also used. The determining factor in most situations will be the type of network application you are working with. Listed in Fig. 9.112 below are several current systems that most system designers are familiar with.

Electronic Integration

Electronics Versus Fiber Count

The diversity and depth of electronic integration in a system impacts the actual fiber count of the network. Simply put, you can use less fiber by increasing the electronics to collect or distribute the signals. In the same manner, you can reduce your dependence on electronics by adding fiber paths. The decision to be

NETWORK SYSTEM	NO. FIBERS REQUIRED TO SUPPORT NETWORK	FIBER TYPE, μm
ATM	2	SINGLE-MODE OR 62.5/125
10BASEF (ETHERNET)	2	MULTIMODE, 62.5/125
100BASEF (ETHERNET)	2	MULTIMODE, 62.5/125
TOKEN RING	4	MULTIMODE, 62.5/125
FDDI	4 (DAS[a]), 2 (SAS[b])	MULTIMODE, 62.5/125
FIBRE CHANNEL	2	MULTIMODE, 62.5/125
SONET[c]	4	SINGLE MODE
VOICE (TWO WAY)	2	MULTIMODE, 62.5/125
VIDEO[c] (BROADCAST)	1 OR 2	SINGLE-MODE OR 62.5/125
VIDEO SECURITY	1	MULTIMODE, 62.5/125
VIDEO (INTERACTIVE)	2	MULTIMODE, 62.5/125
TELEMETRY	1 OR 2	MULTIMODE, 62.5/125

[a]DUAL ATTACHMENT STATION FDDI TERM.
[b]SINGLE ATTACHMENT STATION FDDI TERM.
[c]SONET AND VIDEO ARE LISTED HERE DUE TO SYSTEM DESIGN REQUIREMENTS TO INCORPORATE THESE APPLICATIONS INTO EXISTING COMMUNICATION NETWORKS.

FIGURE 9.112 Network applications and fibers required.

made is to what level, if at all, bridges, routers, concentrators, or switches will be employed in the network. With the advent of FDDI and ATM, the designer/end user has an increased number of options. In the case of FDDI, this allows multiple networks such as ethernet and token ring to operate and includes voice channels. Add ATM, which accepts voice, data, or video over the same network, to this array and the designers options open up tremendously. The high bandwidth and distance capabilities of optical fiber allow the user the flexibility and ability to grow as requirements demand.

The decision to utilize electronics must be made by considering the cost of additional fibers vs. the cost of sophisticated electronics. It also involves the desirability of keeping networks separated vs. combined. Two networks are depicted in Fig. 9.113. The first network provides connectivity between locations by utilizing fibers. This has the added effect of isolating each **local area network (LAN)** in case of an electronics outage and, typically, providing the lowest electronics cost. The second network utilizes a router to reduce the fiber count. This concentrates all traffic into one system allowing for more efficient use of available fibers and easier management. Users often conclude that the most cost effective networks are those that require more fibers and less sophisticated electronics.

Future Growth

Single-Mode Fiber Planning

One decision that many system designers have been faced with is the inclusion of single mode fibers in the network. With the introduction of new high-speed networks, it has become more important to include single-mode into systems, especially the interbuilding backbone. TIA/ EIA-568A was modified to include single mode as a recognized media for backbone networks (TIA, 1995). A significant planning factor, which must be addressed, is that the design you install today will outlive all the electronics currently used. The standard fiber optic cable plant will have a typical useful lifetime of 20–30 years. This means that you are installing a network system that will be used for applications the designer cannot be aware of. Therefore, an efficient design will seek to include media that will handle any application. Single mode is certainly that media.

The use of hybrid cables containing single-mode and multimode fibers is recommended for interbuilding backbone networks. In a hybrid cable, single-mode and multimode fibers are placed in different buffer tubes or bundles within the same cable. This provides easy identification at splice points and allows termination into separate panels at cross-connect points. Additionally, use of hybrid cables reduces installation costs and duct/conduit congestion.

Backbone Spare Fiber Count Planning

Installing additional (spare) fibers for future use is recommended because of the versatility of fiber and almost certain growth of bandwidth requirements. Users should plan for growth in the backbone network, especially the interbuilding backbone. Extra fibers can be installed and left unterminated until needed. The cost of these unterminated spare fibers to the overall project is minimal. The major expense of cable installation is the labor associated with the actual cable placement and termination, especially in difficult installations with congested or inaccessible raceways. The incremental cost of installing extra fibers now will be much less than the cost if installing a new cable later.

The methods for deciding how many additional fibers to plan for are based on network analysis and percentage rules.

Network Analysis

Spare fiber counts can be determined by examining

- Future growth in existing areas
- Future growth from new construction or acquired areas
- Knowledge of proposed new network systems

For instance, if the designer/end user is aware that the organization will add 300 FDDI workstations within the next two years, then fibers required to network these added stations can be planned.

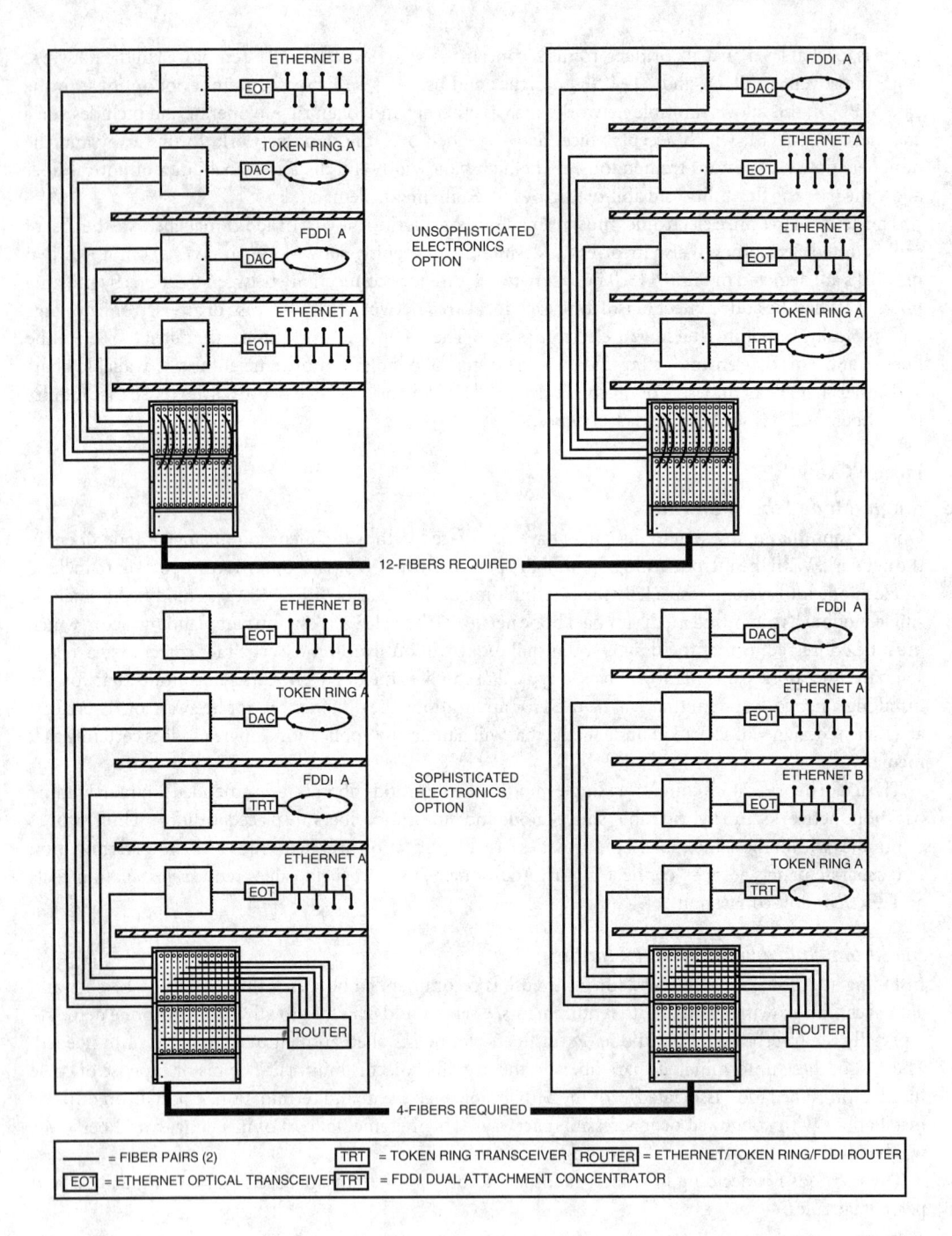

FIGURE 9.113 Electronics options.

Percentage Rules

Spare fiber counts can also be determined if limited future system information is available. The designer can utilize the 25, 50, or 100% rule.

- 25% Rule. If future network applications are defined and potential demand for unanticipated service is low, add 25–33% to the number of fibers currently required.

- 50% Rule. If the future network applications are defined and potential demand for unanticipated service is high, add 50% to the number of fibers currently required.
- 100% Rule. If the future network applications are uncertain and potential demand for unanticipated service is very high or if the installation of subsequent cable/fiber is extremely difficult, add 100% to the number of fibers currently required.

For example, a proposed system currently requires 24 fibers to service existing and defined future networks. If you employ the 25% rule then you would add 6 fibers to the cable run for a total of 30 fibers.

In the case of single-mode fibers, it is recommended that system designers look at adding either 25 or 50% to the number of fibers to ensure sufficient high-capacity media is available for future demands.

System designers should view the addition of spare fibers as a necessity that permits the system to accommodate the needs of the network as it expands and changes.

Backbone Standard Fiber Counts

Although a user can purchase a cable with almost any number of fibers, cable manufacturers, and cable distributors consider certain fiber counts standard and often stock these cables. Fiber counts in multiples of 6 and/12 are standard for backbone cabling and connecting hardware. For low-fiber counts (< 24), multiples of 6 are common (i.e., 6, 12, 18, and 24). For higher fiber counts (> 24), multiples of 12 are common (i.e., 36, 48, 60, 72, etc.). This should be considered when determining the fiber counts for your system.

Installation Considerations

An important aspect of overall system design is the physical construction of the existing system. Every system will have to be examined to determine if existing conduits, duct banks, and cable trays have sufficient space to handle the proposed cabling. This also includes future additions that may be required due to new service requirements. As previously discussed, the system designer must determine if there are sufficient fibers so that future cable installations are reduced or preferably eliminated.

Before a system design is complete and the fiber counts are determined, check to see if the physical facilities can accommodate additional cable. This includes proposed construction and any future additions that may be unseen.

Physical Network Topology

To better illustrate the approach to fiber count selection, we will review the requirements of an example system and propose a baseline fiber count for each of the three primary cable plant segments: interbuilding backbone, intrabuilding backbone, and horizontal cabling. The reader is advised that the solution to this example is one of many. Fiber counts depend on a number of variables that must be considered on a case-by-case basis.

Network Requirements

For the purpose of this discussion, this system will be

- A campus-type network that employs interbuilding and intrabuilding backbones plus horizontal cabling employed in accordance with UTS and EIA/TIA-568A recommendations.
- The system assumes a multiapplication end user environment consisting of voice, data, and video.
- The system allows for easy migration from separate LANs to an ATM or FDDI backbone network.
- As indicated in Fig. 9.114, the campus is configured in a physical star with a MC, ICs, HCs, and work area telecommunication outlets.

Interbuilding Backbone

Of the various segments of the fiber optic cable plant, the designer/user should spend the greatest effort in determining the proper fiber count for the interbuilding backbone. Not only is this the portion of

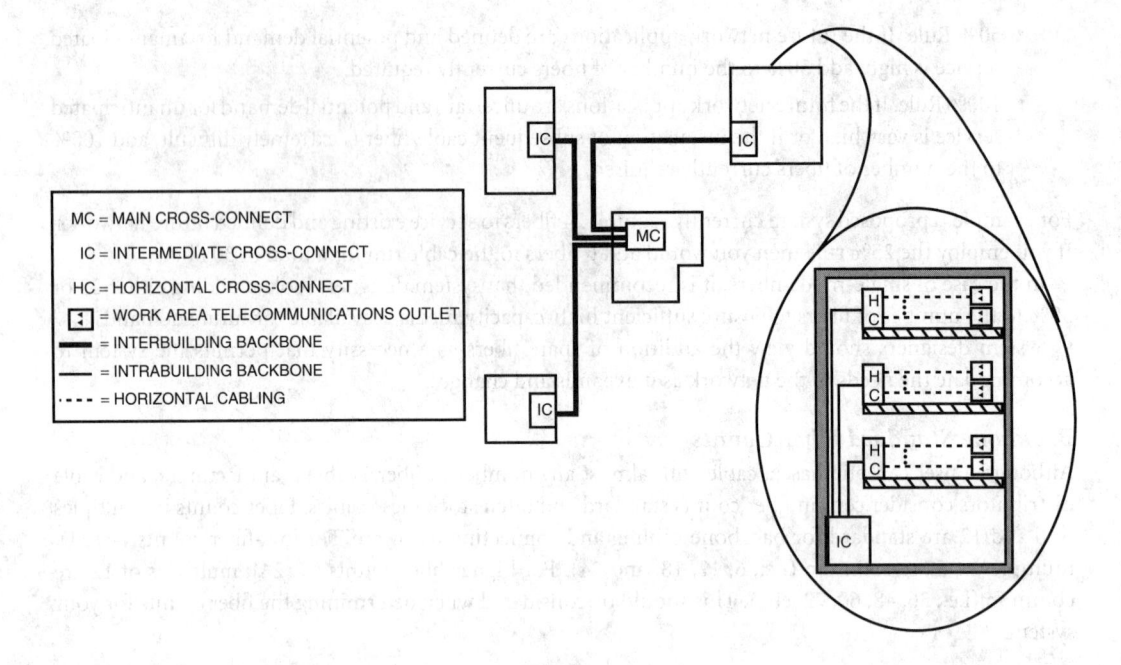

FIGURE 9.114 Example fiber network system.

the network that is most influenced by the decisions involving **multiplexing** and the use of bridges and routers, it is also the segment in which the end user must allow for adequate growth and flexibility because of the expense involved in placing interbuilding cables and constructing interbuilding pathways. Listed in Fig. 9.115 are the planning requirements for this example and the resulting fiber counts.

The user initially plans for the coexistence of ethernet, token ring, and FDDI vs. placing these applications onto one single FDDI backbone. This was based on a decision, initially, to use less sophisticated end electronics. This particular user decides to utilize a hybrid 60-fiber cable comprising 48 multimode (62.5-μm) fibers and 12 single-mode fibers. The additional 16 multimode fibers (50% rule) allow for

APPLICATION	NO. MULTIMODE FIBERS	NO. SINGLE-MODE FIBERS	COMMENTS
ATM	2	2	62.5 μm up to 155 Mb/s SM FOR 622 Mb/s
VIDEO (BROADCAST)	2	2	62.5 μm TO 4 CHANNELS SM FOR 4 CHANNELS
VIDEO (INTERACTIVE)	2		
FIBRE CHANNEL	6		3 CHANNELS
10BASEF (ETHERNET)	2		
TOKEN RING	4		
FDDI	4		
VOICE	4		DS3 RATE
SECURITY (INTERACTIVE)	4		
MULTIMODE SPARES (50% RULE)	16		
SINGLE-MODE SPARES (100% RULE)		4	
TOTAL REQUIREMENTS	48[a]	12[a]	

[a]MULTIMODE AND SINGLE-MODE FIBER COUNTS WERE INCREASED TO ACCOMMODATE STANDARD FIBER COUNTS IN OPTICAL CABLES.

FIGURE 9.115 Interbuilding backbone (example fiber count).

APPLICATION	NO. MULTIMODE FIBERS	NO. SINGLEMODE FIBERS	COMMENTS
ATM	2		62.5 μm up to 622 Mb/s
SECURITY VIDEO	2		REMOTE CONTROL-2F
10BASEF (ETHERNET)	2		
VOICE (REMOTE PBX SHELF)	4		
TOKEN RING	4		
FDDI	4		
MULTIMODE SPARES (50% RULE)	9		
TOTAL REQUIREMENTS	30[a]	SPECIAL REQUIREMENTS	

[a]MULTIMODE FIBER COUNT WAS INCREASED TO ACCOMMODATE STANDARD FIBER COUNTS IN OPTICAL CABLES.
SM = SINGLE-MODE

FIGURE 9.116 Intrabuilding backbone (decentralized network) for an example system.

future growth and the 12 single-mode fibers (100% rule) will be utilized for high data rate requirements. Additional growth can also be achieved by more effectively using the high bandwidth of fiber, such as moving from token ring to FDDI or implementing an ATM switch.

This particular example, although while not uncommon, did employ an extensive list of requirements. It is recommended that at least a 24–48 multimode fiber interbuilding backbone be considered, plus an additional 6–12 single-mode fibers. In large campus networks that may utilize an hierarchical star or ring interbuilding backbone, the designer/user should consider the fiber count requirements for the main backbone separate from the secondary backbone.

Intrabuilding Backbone

Traditionally data networks have been implemented with hubs or concentrators distributed to each telecommunications closet. This keeps the fiber count requirements relatively low in the intrabuilding backbone. This section analyzes such a network.

Although the determination of the proper fiber count in the intrabuilding backbone is typically less complicated than the interbuilding backbone, many of the same issues exist. These issues primarily concern establishing the list of requirements, both present and future, and determining the level of use of bridges and routers. Figure 9.116 indicates the example systems fiber count requirements for one of the intra-building backbone areas.

This intrabuilding backbone run comprises the fiber optic paths that support applications to the telecom-munications closet. This network supports the use of fiber-to-the-desk (FTTD) or copper but requires hubs or switches in the closet. This particular user decides to install a 30-fiber, 62.5-μm cable to each horizontal cross connect. The choice of 30 fibers included sufficient spares to handle expansion and future needs.

This particular example, although not uncommon, did employ an extensive list of requirements. It is recommended to utilize a minimum of a 24-fiber interbuilding backbone cable in support of distributed electronics.

In most cases, the installation of single mode in the intrabuilding backbone is not recommended unless special requirements exist. Cable placement costs for intrabuilding backbone are typically inexpensive when compared to the interbuilding backbone and horizontal wiring. Additionally, 62.5-μm multimode fiber can support extremely high data rates, in excess of 1 Gb/s, for the short distances (typically less than 300 m) found in the intrabuilding backbone.

Centralized Backbone Cabling

Although category 5 UTP systems are limited in total length and require electronics in the closet for high-speed data systems, fiber optic systems do not require the use of electronics in closets on each floor and, therefore, allow for a totally collapsed or centralized cabling network. This design allows the end user to place all of the data electronics in one closet in the building vs. multiple closets throughout the building, thus greatly simplifying the management of fiber optic networks and providing for more efficient use of ports on electronic hubs.

This design can be implemented in a number of ways: (1) pull-through cables from outlets or multiuser outlets to the single central closet in the building, (2) a splice between the horizontal cabling and a high-fiber-count backbone cable in the closet, or (3) a passive interconnect in each closet. Although each of these three options has its corresponding benefits, most users will choose option 2 because it allows for simpler cable placement and offers an acceptable degree of flexibility for adding users at a reasonable cost.

The intrabuilding backbone cable should be sized with sufficient spare capacity to permit additional horizontal circuits to be routed to the centralized cross connect without the need to pull additional intrabuilding backbone cables. The intrabuilding backbone fiber count should be sized to deliver present and future applications to the maximum work area density within the area served by the TC. If maximum density is not anticipated for a number of years, however, the user may elect to install a backbone based on present density allowing for additional cables in the future as the office area is fully occupied.

Horizontal Cabling

Fiber in the horizontal is a reality because of the recent advances in technology easing installation, reducing the overall system cost, and offering a system that is reliable, flexible, certifiable, and future proofed for the information era. A minimum of a two-fiber, 62.5-μm cable is recommended for data communications and a separate category 3 UTP cable for voice. In determining the fiber count of the horizontal data cables, there are two distinct network possibilities.

For hard-wired offices, a single-user cable will be installed to a single-user outlet. The fiber count of the single-user cable is based on the number of applications that user requires. Typically, it is a 2-fiber, 62.5-μm cable but sometimes a user has a requirement to run two separate networks requiring four fibers.

For open-systems furniture applications, the use of a multiuser cable and outlet can be implemented to ensure the long-term survivability of the horizontal cabling network in this rapidly changing office area. Multiuser cabling calls for a high-fiber count cable to be placed from the closet to an area in the open office where there is a fairly permanent structure, such as a wiring column or cabinet, in a grid type wiring scheme. At this structure, a multiuser outlet, capable of supporting 6–12 offices, is installed instead of individual outlets. Fiber optic patch cords are then installed through the furniture raceways from the multiuser outlet to the office area, allowing the user to rearrange the furniture without disrupting or relocating the horizontal cabling.

Defining Terms

Adapter: A mechanical media termination device designed to align and join fiber optic connectors. Often referred to as a coupling, bulkhead, or interconnect sleeve.

Backbone cabling: The portion of premises telecommunications cabling that provides connections between telecommunications closets, equipment rooms, and entrance facilities. The backbone cabling consists of the transmission media (optical fiber cable), main and intermediate cross connects, and terminations for the horizontal cross connect, equipment rooms, and entrance facilities. The backbone cabling can further be classified as interbuilding backbone (cabling between buildings), or intrabuilding backbone (cabling within a building).

Cable assembly: Optical fiber cable that has connectors installed on one or both ends. General use of these cable assemblies includes the interconnection of optical fiber cable systems and optoelectronic equipment. If connectors are attached to only one end of a cable, it is known as a pigtail. If connectors are attached to both ends, it is known as a jumper or patch cord.

Cable bend radius: Cable bend radius during installation infers that the cable is experiencing a tensile load. Free bend infers a smaller allowable bend radius since it is at a condition of no load.

Centralized cabling: A cabling topology used with centralized electronics connecting the optical horizontal cabling with intrabuilding backbone cabling passively in the telecommunications closet.

Composite cable: A cable containing both fiber and copper media per article 770 of the NEC.

Connecting hardware: A device used to terminate an optical fiber cable with connectors and adapters that provides an administration point for cross connecting between cabling segments or interconnecting to electronic equipment.

Entrance facility: An entrance to a building for both public and private network service cables including the entrance point at the building wall and continuing to the entrance room or space.

Equipment room: A centralized space for telecommunications equipment that serves the occupants of a building. An equipment room is considered distinct from a telecommunications closet because of the nature or complexity of the equipment.

Fiber bend radius: Radius a fiber can bend before the risk of breakage or increase in attenuation.

Horizontal cabling: That portion of the telecommunications cabling that provides connectivity between the horizontal cross connect and the work area telecommunications outlet. The horizontal cabling consists of transmission media, the outlet, the terminations of the horizontal cables, and horizontal cross connect.

Horizontal cross connect (HC): A cross connect of horizontal cabling to other cabling, for example, horizontal, backbone, equipment.

Hybrid cable: A fiber optic cable containing two or more different types of fiber, such as 62.5-μm multimode and single mode.

Interbuilding backbone: The portion of the backbone cabling between buildings. See backbone cabling.

Intermediate cross connect (IC): A secondary cross connect in the backbone cabling used to mechanically terminate and administer backbone cabling between the main cross connect and horizontal cross connect.

Intrabuilding backbone: The portion of the backbone cabling within a building. See backbone cabling.

Link: A telecommunications circuit between any two telecommunications devices, not including the equipment connector.

Local area network (LAN): A geographically limited communications network intended for the local transport of voice, data, and video. Often referred to as a customer premises network.

Main cross connect (MC): The centralized portion of the backbone cabling used to mechanically terminate and administer the backbone cabling, providing connectivity between equipment rooms, entrance facilities, horizontal cross connects, and intermediate cross connects.

Multifiber cable: An optical fiber cable that contains two or more fibers.

Multiplex: Combining two or more signals into a single bit stream that can be individually recovered.

Multiuser outlet: A telecommunications outlet used to serve more than one work area, typically in open-systems furniture applications.

National Electrical Code (NEC): Defines building flammability requirements for indoor cables. Note: Local codes take precedence but may refer to or require compliance to the NEC.

Plenum: An air-handling space such as that found above drop-ceiling tiles or in raised floors. Also, a fire-code rating for indoor cable.

Point-to-point: A connection established between two specific locations, as between two buildings.

Repeater: A device used to regenerate an optical signal to allow an increase in the system length.

Riser: Pathways for indoor cables that pass between floors. It is normally a vertical shaft or space. Also a fire-code rating for indoor cable.

Telecommunications closet (TC): An enclosed space for housing telecommunications equipment, cable terminations, and cross connects. The closet is the recognized cross connect between the backbone and horizontal cabling.

Transmitter: An electronic package used to convert an electrical information-carrying signal to a corresponding optical signal for transmission by fiber. The transmitter is usually a light emitting diode (LED) or laser diode.

Work area telecommunications outlet: A connecting device located in a work area at which the horizontal cabling terminates and provides connectivity for work-area patch cords.

References

BICSI. Building Industry Consultant Service International. *Telecommunications Distribution Manual.*

EIA. 1989. *EIA Standard Fiber Optic Terminology*. EIA-440-A, Electronic Industries Association.

ICEA. 1994. *Standard for Fiber Optic Outside Plant Communications Cable*. ANSI/ICEA S-87- 640-1992, Insulated Cable Engineers Association, Inc.

ICEA. 1994. *Standard for Fiber Optic Premises Distribution Cable*. ANSI/ICEA S-83-596-1994, Insulated Cable Engineers Association, Inc.

TIA. 1995. *Commercial Building Telecommunications Cabling Standard*. ANSI/TIA/EIA-568-A, Telecommunication Industry Association.

TIA/EIA. *Commercial Building Standard for Telecommunications Pathway and Spaces*. TIA/EIA-569, Telecommunication Industry Association.

10

Power Supplies and Regulation

Sadrul Ula
University of Wyoming

T. S. Kalkur
University of Colorado

Melissa S. Mattmuller
Purdue University

Robert J. Hofinger
Purdue University

Ashoka K.S. Bhat
University of Victoria

Badrul H. Chowdhury
University of Wyoming

Jerry C. Whitaker
Editor-in-Chief

Isidor Buchmann
Cadex Electronics, Inc.

10.1 Transformers

Sadrul Ula

10.1.1 Introduction

In the most basic definition, a transformer is a device that magnetically links two or more circuits for time-varying voltage and current. Magnetic coupling has a number of intrinsic advantages, including (1) DC isolation between the circuits, (2) the ability to match the voltage and current capability of a source to a load on the other side, (3) the ability to change the magnitude of the voltage and current from one side of the transformer to the other, and (4) the ability to change the phases of voltage and current from one side of the transformer to the other (three phase).

Transformers permeate nearly every facet of electrical engineering in one form or another. They are used in applications as diverse as power transmission and audio applications. Because the uses for transformers vary over such a wide range, and because the performance of a transformer depends on different critical design parameters for each application, it is essential that the design engineers have an understanding of transformer characteristics.

10.1.2 Electromagnetics of Transformers

In 1830 Faraday and Henry discovered that a voltage was induced in a coil of wire that was in the presence of a time-varying magnetic field. This relationship of the induced voltage to the magnetic flux is called Faraday's law, and is given in Eq. (10.1), where v is the induced voltage, ϕ is the magnetic flux, and N is the number of turns in the coil

$$v = N\frac{d\phi}{dt} \tag{10.1}$$

This fundamental relationship forms the basis for transformer operation. Take the case where two coils are set in close proximity, as illustrated in Fig. 10.1. A voltage, which is introduced on the first coil, produces magnetic flux. Part of this flux passes through, or links, the second coil, ϕ_{12}. As a result, a voltage is produced on the second coil. In the same way, a voltage on the second coil produces a flux, which links the first coil. The amount of flux that passes through both coils determines how tightly they are coupled.

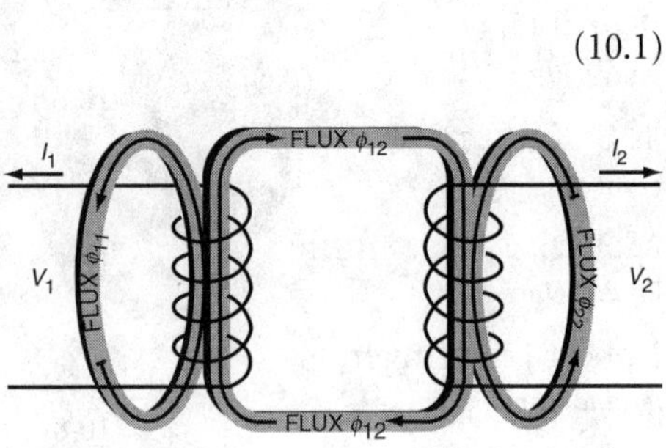

FIGURE 10.1 Linked coils.

Notice that not all of the flux that is produced by the first coil passes through the second coil. The nonlinking flux is called leakage flux, and produces an inductance in the coil, called leakage inductance. In Fig. 10.1, ϕ_{11} is the leakage flux of coil 1, and ϕ_{22} is the leakage flux of coil 2. In many transformers, a ring of magnetic material, called a core, is used to concentrate the flux, so that most of it passes through both coils and the leakage inductance is reduced.

In general, the coil into which power flows is called the *primary* coil, whereas the other coil is called the *secondary*.

Magnetic Cores

The ease with which a flux can be produced in a material is called its *permeability*, μ. More flux will be produced in a material with high permeability than in a one with a low permeability, given the same amount of current and the same number of turns in the coil. Core materials usually have a permeability thousands of times that of free space or air. The ratio of a material's permeability to the permeability of free space, called relative permeability, is often used. The actual permeability, which has units of webers per ampere-turn-meter, is found by multiplying the permeability of free space by the relative permeability.

The overall ability of a core to carry flux also depends on its size and shape, and its cross-sectional area. This is described by a quantity called permeance. The basic relationship of permeance to permeability in a core is defined in Eq. (10.2), where P is the permeance, μ is the permeability of the material, A is the cross-sectional area of the core, and l is the mean length of the flux path in the core. This equation assumes uniform flux distribution in the core and constant permeability inside the core. It does not take into account the variations in the length of the flux path from the inside of the core to the outside. The reciprocal of permeance is reluctance

$$P = \frac{\mu A}{l} = \frac{1}{R} \quad \text{or} \quad R = \frac{l}{\mu A} \tag{10.2}$$

Air Gaps

The technique of putting an air gap in the core is used to prevent saturation in some applications. Since the reluctance of the air gap is much larger than that of the core, the flux density in the core is greatly reduced. The formula for the reluctance of the air gap is the same as that for the core. In Eq. (10.3) l_a is the length of the air gap, μ_a is the permeability of air, and A_a is the cross-sectional area of the air gap. The cross-sectional area of the air gap is not necessarily the same as that of the core, due to the tendency of the flux to spread. This effect is called fringing, and it is more prevalent in applications where the air gap length is large compared to the cross-sectional area of the core. In most applications, fringing can be ignored.

$$R_a = \frac{l_a}{\mu_a A_a} \tag{10.3}$$

The total reluctance is the sum of the core reluctance and the air gap reluctance. This reluctance is used to calculate the flux in the core

$$R_{\text{total}} = R_a + R_c \tag{10.4}$$

$$\phi = \frac{N_i}{R_{\text{total}}} \tag{10.5}$$

Example Problem

A core made of iron and nickel has a relative permeability of 100,000, a path length of 30 cm, and a cross-sectional area of 9 cm². An air gap has been placed in the core, which is 5 mm deep. The total reluctance can be found by adding the reluctance of the air gap and the iron core. The core reluctance can be found using Eq. (10.2),

$$R_c = \frac{l_c}{\mu_c A_c} = \frac{l_c}{\mu_{Cr} \mu_{fs} A_c}$$

$$= \frac{3 \times 10^{-1}\,\text{m}}{(1 \times 10^5)(4\pi \times 10^{-7}\,\text{Wb/A} \cdot \text{turn} \cdot \text{m})(9 \times 10^{-4}\,\text{m}^2)} = 2.65 \times 10^3\,\frac{\text{A} \cdot \text{turns}}{\text{Wb}}$$

If the effects of fringing are neglected, the cross-sectional area of the air gap is the same as that of the core. The reluctance of the air gap can then be found using Eq. (10.2)

$$R_c = \frac{l_{\text{air}}}{\mu_{\text{air}} A_{\text{air}}} = \frac{l_{\text{air}}}{\mu_{\text{air}} \mu_{fs} A_c}$$

$$= \frac{5 \times 10^{-3} \text{ m}}{(1.0)(4\pi \times 10^{-7} \text{ Wb/A} \cdot \text{turn} \cdot \text{m})(9 \times 10^{-4} \text{ m}^2)} = 4.42 \times 10^6 \frac{\text{A} \cdot \text{turns}}{\text{Wb}}$$

The relationship between the flux passing through the coil and the current in the coil is given in Eq. (10.6), where N is the number of turns in the coil, i is the current, P is the permeance of the core material, and ϕ is the flux. The product of the current and the number of turns in the transformer is called the magnetomotive force (mmf). Notice the similarity of this equation to the one relating voltage and current in a circuit. The voltage is replaced by the mmf, impedance is replaced by the reluctance, and current is replaced by flux. The magnetizing force H is defined as the mmf divided by the length of the flux path for a continuous core.

$$\text{mmf} = N_i = R_\phi \tag{10.6}$$

$$H = \frac{\text{mmf}}{l} \tag{10.7}$$

Although the materials used to make transformer cores have high permeability, their ability to carry flux is not unlimited. The quotient of the flux divided by the area of the core is called the flux density. Equation (10.8) gives the formula for flux density. When the amount of flux density in the core gets too high, the permeability of the core material begins to decrease. This magnetic overloading of the core material is called saturation.

$$B = \frac{\phi}{A} \tag{10.8}$$

Mutual Inductance

The voltage of the second coil is related to the linking flux, ϕ_{12} by Faraday's law:

$$v_2 = \frac{d}{dt}(N_2 \phi_{12}) \tag{10.9}$$

The flux ϕ_{12} is related to the number of turns and the current as shown in Eq. (10.10) giving the relationship between the current in the first coil and the voltage in the second coil shown in Eq. (10.11). The quantity $N_1 N_2 P_{21}$ is called the mutual inductance since it relates the voltage on coil 2 to the differential of the current in coil 1. The mutual inductance between the coils can be increased by increasing the permeance of the core material which links them.

$$\phi_{12} = P_{21} N_1 i_1 \tag{10.10}$$

$$v_2 = N_1 N_2 P_{21} \frac{di_1}{dt} \tag{10.11}$$

Saturation, Hysteresis and Core Losses

Figure 10.2 shows the magnetization curve for a typical ferromagnetic material. Note that the curve follows two different paths, depending on whether the magnetizing force H is increasing or decreasing. This is called a hysteresis curve. It is caused by the fact that the magnetic particles in the core need to be rotated and realigned each time the polarity of the magnetizing force changes. This is why the magnetic force must be reversed to reduce the flux density to zero.

As the magnetizing force H increases, the flux density increases up to a point, then the curve flattens out. In this flattened region, the mmf can increase greatly, and only a small increase in the flux density

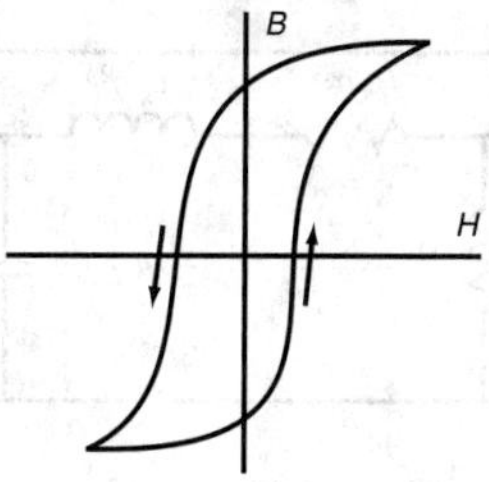

FIGURE 10.2 Typical magnetization curve.

will result. The core is said to be saturated in this region. The flattening of the curve indicates that the permeability has decreased from the value it had when there was only a little flux passing through the core.

10.1.3 The Ideal Transformer

Although no transformer is ideal in its characteristics, transformers approach their ideal characteristics in the operating range for which they were designed. The ideal transformer has no coil resistance and no core loses, so that it has no power loss. It also has no leakage inductance, since the permeability of the core is infinite, and the core material is able to carry an infinite amount of flux without saturating. Therefore, the mutual inductance is also infinite. The capacitance in an ideal transformer is negligible. The equations for an ideal transformer are given as follows:

$$v_1 i_1 = v_2 i_2 \tag{10.12}$$

$$\frac{v_1}{v_2} = \frac{N_1}{N_2} \tag{10.13}$$

$$\frac{i_1}{i_2} = \frac{N_2}{N_1} \tag{10.14}$$

$$\frac{Z_1}{Z_2} = \left(\frac{N_1}{N_2}\right)^2 \tag{10.15}$$

Equation (10.15) gives the effect of the transformer on an impedance on the secondary side. It is multiplied by the square of the turns ratio. The magnitude of the impedance as seen on the secondary side is called the reflected impedance.

Dot Convention

To eliminate ambiguity in the voltage and current polarity at the input and output of the transformer symbol, the dot convention is used. In circuit diagrams, a small dot is placed near one end of each coil, as shown in Fig. 10.3. The dot indicates a rise in voltage from the unmarked to the marked terminal on each coil. Under this convention, current

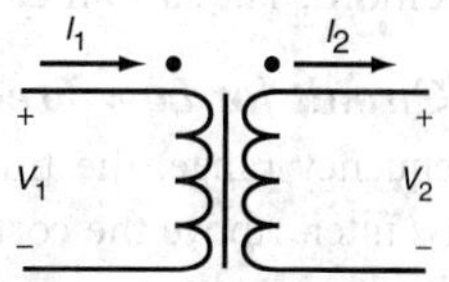

FIGURE 10.3 Dotted schematic symbol for a transformer.

into the dot on the primary side is labeled as having positive polarity, whereas current out of the dotted terminal on the other side is assigned positive polarity. This means that the power flow must be into the transformer on one side, and out of the transformer on the other side.

The side of the transformer into which power flows is called the primary; the other side is referred to as the secondary.

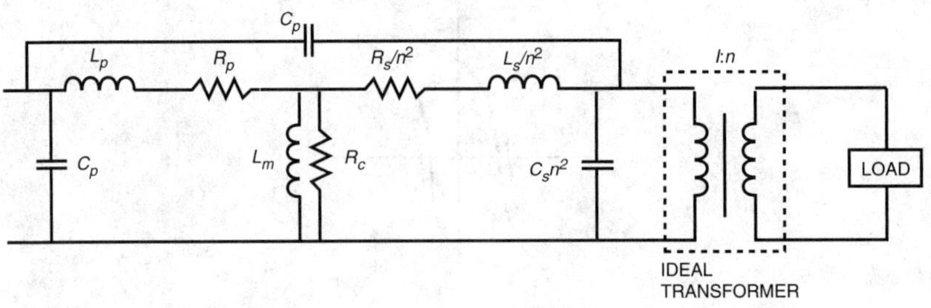

FIGURE 10.4   Equivalent circuit.

Transformer Equivalent Circuit

Equivalent circuits are often used to model the performance of transformers with greater accuracy. Although equivalent circuits are not exact replicas of real transformers, they are close enough to realize accurate results for most situations. The complete transformer equivalent circuit is shown in Fig. 10.4.

The leakage inductance of both coils has been modeled by an inductor in series with the load, since the current in the coils also produces leakage flux. These inductances are labeled L_p and L_s, respectively. Notice that the leakage inductance for the secondary side has been divided by the turns ratio n, squared because it was reflected to the primary side. Resistors labeled R_p and R_s have also been placed in series with the load to represent the resistance of the conductors used to wind the coils. Again, the secondary resistance has been divided by the square of the turns ratio, since it was reflected.

The mutual inductance is represented by shunt inductor L_m, since the magnetizing current is not coupled to the load. A resistor R_c is also placed in shunt, to represent the core loss due to hysteresis and eddy currents in the core.

The stray capacitances between turns of the coils are represented by a capacitor placed across the terminals of each coil. This capacitance is larger for coils with more turns. Although the capacitance is actually distributed, it is lumped for the equivalent circuit, in order to simplify the analysis. The capacitance from one coil to the other is represented by another capacitor placed in parallel with the leakage inductance and resistance. The transfer function of the complete equivalent circuit is given in Eq. (10.16),

$$\frac{V_{\text{out}}(j\omega)}{V_{\text{in}}(j\omega)}$$
$$= \frac{Z_L(Z_P Z_S + Z_M Z_C + Z_M Z_P + Z_M Z_S)}{Z_P Z_C Z_L + Z_P Z_C Z_M + Z_P Z_C Z_S + Z_P Z_S Z_L + Z_M Z_C Z_L + Z_M Z_C Z_S + Z_P Z_M Z_L + Z_M Z_S Z_L}$$

$$(10.16)$$

The transfer function of the complete equivalent circuit is quite complex. For this reason, the equivalent circuit is often broken up into several equivalent circuits, each of which is valid only for a certain set of operating conditions. The simplified circuits are then analyzed for their pertinent operating conditions.

Equivalent Circuit for Low Frequency

In the low-frequency range, the transformer acts like a high-pass filter. Above the corner frequency, the output voltage is nearly equal to the input voltage. Below the corner frequency the output voltage is substantially smaller than the input voltage, due to the presence of the shunt inductor L_m. The distributed capacitances and leakage inductances can be neglected. The equivalent circuit for low frequencies shown in Fig. 10.5, and the transfer functions for voltage and current are given in Eq. (10.17) and

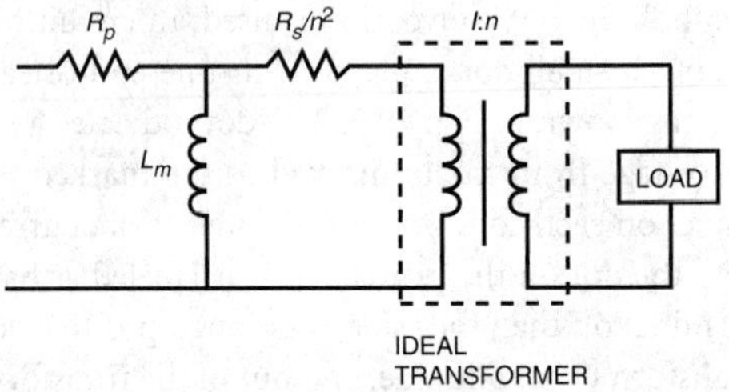

FIGURE 10.5   Low-frequency equivalent circuit.

Eq. (10.18), assuming the generator impedance is R_g. The impedance seen at the input of the transformer is given in Eq. (10.18), and the equation for the corner frequency of the transformer is given in Eq. (10.18) assuming a resistive load impedance.

$$\frac{V_{out}}{V_{in}} = \frac{j\omega L_m \cdot \dfrac{Z_l}{n^2}}{j\omega L_m\left(\dfrac{Z_l}{n^2} + \dfrac{R_s}{n^2}\right) + (R_p + R_g)\left(\dfrac{Z_l}{n^2} + \dfrac{R_s}{n^2} + j\omega L_m\right)} \tag{10.17}$$

$$\frac{I_{out}}{I_{in}} = \frac{j\omega L_m}{\dfrac{Z_l}{n^2} + \dfrac{R_s}{n^2} + j\omega L_m} \tag{10.18}$$

$$Z_{total} = \frac{j\omega L_m\left(\dfrac{Z_l}{n^2} + \dfrac{R_s}{n^2}\right) + (R_p + R_g)\left(\dfrac{Z_l}{n^2} + \dfrac{R_s}{n^2} + j\omega L_m\right)}{\dfrac{Z_l}{n^2} + \dfrac{R_s}{n^2} + j\omega L_m} \tag{10.19}$$

$$f_{corner,voltage} = \frac{2\pi L_m\left(\dfrac{R_l}{n^2} + \dfrac{R_s}{n^2} + R_p + R_g\right)}{(R_p + R_g)\left(\dfrac{R_l}{n^2} + \dfrac{R_s}{n^2}\right)} \tag{10.20}$$

Midband

This is the intended operating range for the transformer. The frequency is high enough that the mutual inductance can be neglected and low enough that the leakage inductance and distributed capacitance are still not a concern. For this reason, the only impedances of concern are the coil resistances and the core loss component. The resulting equivalent circuit is shown in Fig. 10.6, and the transfer functions for both voltage and current are given in Eq. (10.21) and Eq. (10.22). The formula for the input impedance is given by Eq. (10.23).

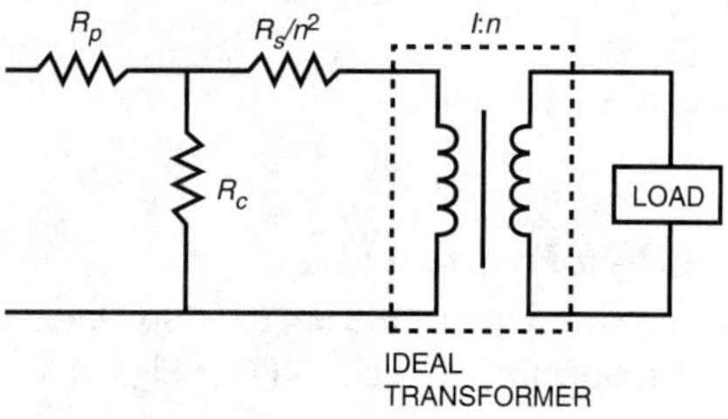

FIGURE 10.6 Midband equivalent circuit.

$$\frac{V_{out}}{V_{in}} = \frac{R_c \cdot \dfrac{Z_l}{n^2}}{R_c\left(\dfrac{Z_l}{n^2} + \dfrac{R_s}{n^2}\right) + (R_p + R_g)\left(\dfrac{Z_l}{n^2} + \dfrac{R_s}{n^2} + R_c\right)} \tag{10.21}$$

$$\frac{I_{out}}{I_{in}} = \frac{R_c}{\dfrac{Z_l}{n^2} + \dfrac{R_s}{n^2} + R_c} \tag{10.22}$$

$$Z_{total} = \frac{R_c\left(\dfrac{Z_l}{n^2} + \dfrac{R_s}{n^2}\right) + (R_p + R_g)\left(\dfrac{Z_l}{n^2} + \dfrac{R_s}{n^2} + R_c\right)}{\dfrac{Z_l}{n^2} + \dfrac{R_s}{n^2} + R_c} \tag{10.23}$$

High Frequency

At higher frequencies, the leakage inductance begins to dominate the transformer's response. The high-frequency equivalent circuit is shown in Fig. 10.7, and transfer functions for voltage and current are given in Eq. (10.23) and Eq. (10.24).

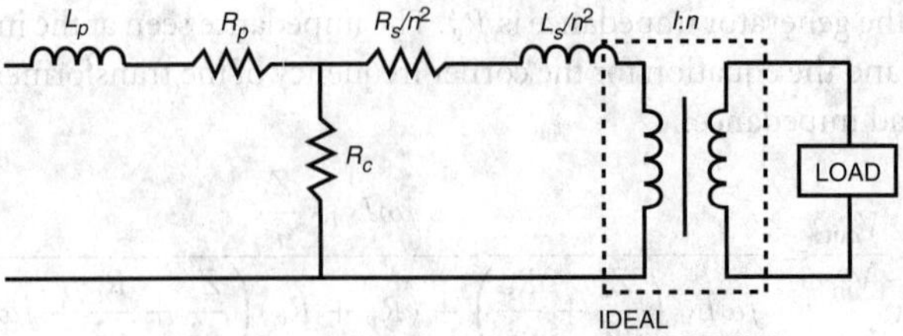

FIGURE 10.7 High-frequency model.

$$\frac{V_{\text{out}}}{V_{\text{in}}} = \frac{R_c \cdot \dfrac{Z_l}{n^2}}{R_c\left(\dfrac{Z_l}{n^2} + \dfrac{(R_s + j\omega L_s)}{n^2}\right) + (R_p + j\omega L_p + R_g)\left(\dfrac{Z_l}{n^2} + \dfrac{(R_s + j\omega L_s)}{n^2} + R_c\right)} \quad (10.24)$$

$$\frac{I_{\text{out}}}{I_{\text{in}}} = \frac{R_c}{\dfrac{Z_l}{n^2} + \left(\dfrac{j\omega L_s + R_s}{n^2} + R_c\right)} \quad (10.25)$$

$$Z_{\text{total}} = \frac{R_c\left(\dfrac{Z_l}{n^2} + \dfrac{(R_s + j\omega L_s)}{n^2}\right) + (R_p + j\omega L_p + R_g)\left(\dfrac{Z_l}{n^2} + \dfrac{(R_s + j\omega L_s)}{n^2} + R_c\right)}{\dfrac{Z_l}{n^2} + \dfrac{(R_s + j\omega L_s)}{n^2} + R_c} \quad (10.26)$$

Ultra High Frequency

At extremely high frequencies, the distributed capacitance of each coil begins to play a role, along with the leakage inductance of each coil (see Fig. 10.8).

$$\frac{V_{\text{out}}}{V_{\text{in}}} = \frac{Z_l}{Z_l + n^2 j\omega L_l(j\omega L_l C_s + 1)} \quad (10.27)$$

$$\frac{I_{\text{out}}}{I_{\text{in}}} = \frac{n^2}{j\omega C_p Z_l + n^2(1 - \omega^2 L_l C_p)(j\omega Z_l C_s + 1)} \quad (10.28)$$

$$Z_{\text{in}} = \frac{Z_l + n^2 j\omega L_l(Z_l j\omega C_s + 1)}{Z_l j\omega C_p + n^2 \omega^2 L_l C_p(Z_l j\omega C_s + 1) + n^2(Z_l j\omega C_s + 1)} \quad (10.29)$$

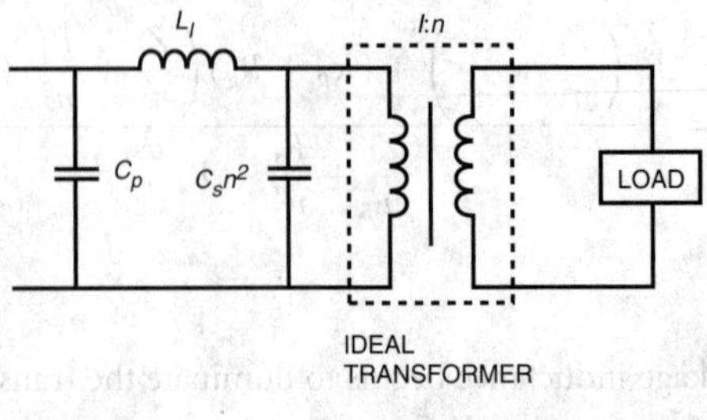

FIGURE 10.8 Ultra-high-frequency model.

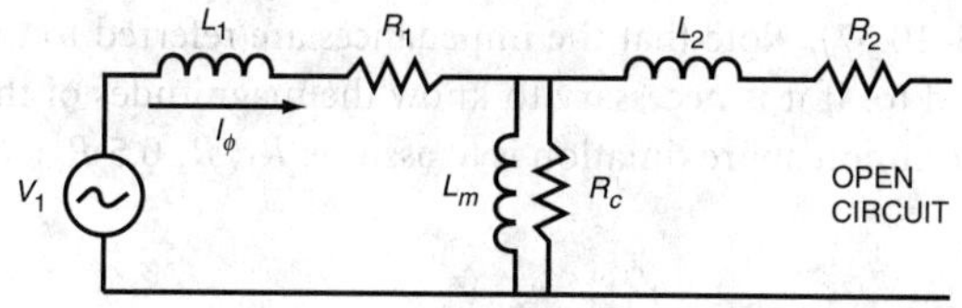

FIGURE 10.9 Connections for open-circuit test.

10.1.4 Testing to Determine Transformer Parameters

It is possible to quantify some of the important transformer parameters through testing. Two simple tests, the open-circuit and short-circuit tests, allow the user to identify the leakage inductance, the coil resistance, the core loss, and mutual inductance of the transformer. Since the characteristics of the transformer change with frequency, the tests should be done at the frequency (or frequencies) of interest.

The open-circuit test is done by leaving the terminals on one side of the transformer open, and applying the rated voltage to the coil on the other side. Figure 10.9 shows the equivalent circuit for this test, assuming the test is done at a midband frequency. Since the leakage inductance and core resistance are small compared to the mutual inductance and core loss resistance, they can be neglected for the open-circuit test.

The value of the exciting admittance is then given by Eq. (10.30), where I_ϕ and V_1 are phase values. The conductance component of the admittance is found by dividing the input power by the magnitude of the input voltage, as shown in Eq. (10.31). The susceptance component is then given by Eq. (10.32). The resistance and inductance, which account for the core loss, can then be found by taking the reciprocal of the conductance and susceptance, respectively. The values that are found are referred to the side of the transformer to which the voltage was applied.

$$Y_\phi = Y_{oc} = \frac{\hat{I}_\phi}{\hat{V}_1} \tag{10.30}$$

$$G_c = G_{oc} = \frac{P_1}{|V_1|^2} \tag{10.31}$$

$$B_m = B_{oc} = \sqrt{|Y_{oc}|^2 - G_{oc}^2} \tag{10.32}$$

$$R_{oc} = \frac{1}{G_{oc}} \tag{10.33}$$

$$L_m = \frac{1}{2\pi f B_m} \tag{10.34}$$

The short-circuit test is conducted by shorting one of the transformer's coils, and applying rated current to the other coil. Only a fraction of the rated voltage is needed to produce the rated current. The input current and power are then measured. Figure 10.10 shows a circuit diagram for this test.

Since the magnetization and core loss impedances are large compared to the leakage inductance and coil resistance, the input current is assumed to be the secondary current. Accordingly, the power input must go toward losses in the coil resistances. The sum of the leakage inductances and coil resistances can

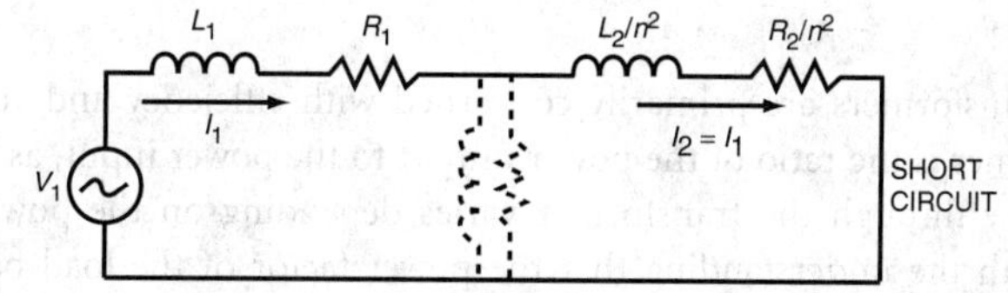

FIGURE 10.10 Short-circuit test setup.

be found using Eq. (10.35–10.37). Note that the impedances are referred to the side of the transformer that the voltage was applied to. If it is necessary to know the magnitudes of the primary and secondary impedances separately, a common approximation is to assume R_1, R_2 0.5 R_{eq}, and $X_{11} = X_{12} = 0.5X_{eq}$.

$$\hat{Z}_{eq} = \hat{Z}_{sc} = \frac{\hat{V}_{sc}}{\hat{I}_{sc}} \tag{10.35}$$

$$R_{eq} = R_{sc} = \frac{P_{sc}}{I_{sc}^2} \tag{10.36}$$

$$X_{eq} = R_{sc} = \sqrt{|Z_{sc}|^2 - R_{sc}^2} \tag{10.37}$$

Example Problem: Determining the Core Loss, Mutual Inductance, Leakage Inductance, and Coil Resistance

Open- and short-circuit tests were conducted on a 60-Hz, 500-W 120:6-V transformer. In the short-circuit test, the readings were 10 V, 4.16 A, and 24 W. The combined impedance of the coil resistance and the leakage inductance can be found using Eq. (10.35). The resistance and reactance of the transformer can then be found using Eq. (10.35). The resistance and reactance of the transformer can then be found using Eq. (10.36) and Eq. (10.37).

$$Z_{eq} = Z_{sc} = \frac{|\hat{V}_{sc}|}{|\hat{I}_{sc}|} = \frac{10.0\,\text{V}}{4.16\,\text{A}} = 2.40\,\Omega$$

$$R_{eq} = R_{sc} = \frac{P_{sc}}{I_{sc}^2} = \frac{24.0\,\text{W}}{(4.16\,\text{A})^2} = 1.39\,\Omega$$

$$X_{eq} = X_{sc} = \sqrt{|Z_{sc}|^2 - R_{sc}^2} = \sqrt{(2.40)^2 - (1.39)^2} = 1.96\,\Omega$$

The open-circuit test was done with the low-voltage side energized. The readings were 6.0 V, 1.2 A, and 0.8 W. The exciting admittance is then given by Eq. (10.30), and the conductance and susceptance components can be found using Eq. (10.31) and Eq. (10.32),

$$Y_\phi = Y_{oc} = \frac{I_\phi}{V_{oc}} = \frac{1.2\,\text{A}}{6.0\,\text{V}} = 0.21\,\Omega$$

$$G_c = G_{oc} = \frac{P_{oc}}{|V_{oc}|^2} - \frac{0.8\,\text{W}}{(6.0\,\text{V})^2} = 0.022\,\Omega$$

$$B_m = B_{oc} = \sqrt{|Y_{oc}|^2 - G_{oc}^2} = \sqrt{(0.2)^2 - (0.022)^2} = 0.198\,\Omega$$

$$R_{oc} = \frac{1}{G_{oc}} = \frac{\Omega}{0.022} = 45.4\,\Omega$$

$$L_m = \frac{1}{2\pi f B_m} = \frac{1}{2\pi(60.0)(0.198)} = 13.4\,\text{mH}$$

10.1.5 Applications

Power Transformers

Designers of power transformers are primarily concerned with efficiency and voltage regulation. The efficiency of a transformer is the ratio of the power output to the power input, as shown in Eq. (10.38). Because the power flow through the transformer varies depending on the power factor of the load, Eq. (10.38) is used with the understanding that the power factor of the load be unity. The losses in power transformers are due to the copper losses in the windings and the core losses. The copper losses vary with the square of the current; the core losses vary with the input voltage magnitude and frequency.

Because neither of these quantities depends on the power being consumed by the load, power transformers are rated by the volt-amperes (VA), which flows through them.

$$\%\text{Eff} = \frac{P_{\text{out}}}{P_{\text{in}}} \times 100 \qquad (10.38)$$

The regulation of a power transformer is a measure of the transformer's ability to maintain a constant output voltage for varying loads. The formula for regulation is given in Eq. (10.39). The primary voltage is held constant at the value required to produce rate voltage on the secondary at full load.

$$\text{Regulation}\% = \frac{V_{2oc} - V_{2fl}}{V_{2fl}} \times 100\% \qquad (10.39)$$

Wide-Band

Wide-band transformers are designed to be used over a frequency range, rather than at a single frequency. They are used in applications where waveforms that consist of a number of frequencies must be passed with a minimum of distortion. Some applications of wide-band transformers include audio, video, and sonar.

The power that is dissipated by the transformer in its midband with a resistive load is called the insertion loss. The formula for insertion loss is given in Eq. (10.40) where P_L is the power transmitted to the load and P_T is the power loss in the transformer due to winding resistance and core loss. Insertion loss is also found by taking the ratio of the load voltage with the actual transformer, E, over the load voltage with an ideal transformer, E'. This method is given in Eq. (10.41).

$$\text{Insertion Loss, dB} = 10 \ \log \frac{P_L}{P_L + P_T} \qquad (10.40)$$

$$\text{Insertion Loss, dB} = 20 \ \log \frac{E}{E'} \qquad (10.41)$$

Wide-band transformers are often used to match the load to the source for maximum power transfer and minimum reflection. A frequency used approximation is to select a turns ratio such that the generator resistance R_G is equal to the reflected load resistance, R_L/n^2. A more exact method is to set the generator resistance equal to the sum of the transformer coil resistances and the load, as shown in Eq. (10.42)

$$R_G = R_p + \frac{R_s}{n^2} + \frac{R_L}{n^2} \qquad (10.42)$$

Frequency Response of Wide-band Transformers

The equivalent circuit of Fig. 10.11(a) is used to analyze the low-frequency response of wide-band transformers. The critical element in the low-frequency range is the mutual inductance L_m. The corner frequency is given by Eq. (10.20). The midband response of the transformer can be found using the equivalent circuit of Fig. 10.6.

The high-frequency response of the wide-band transformer can be found using the equivalent circuit shown in Fig. 10.8. This circuit is somewhat complex, however, and a common practice is to include only

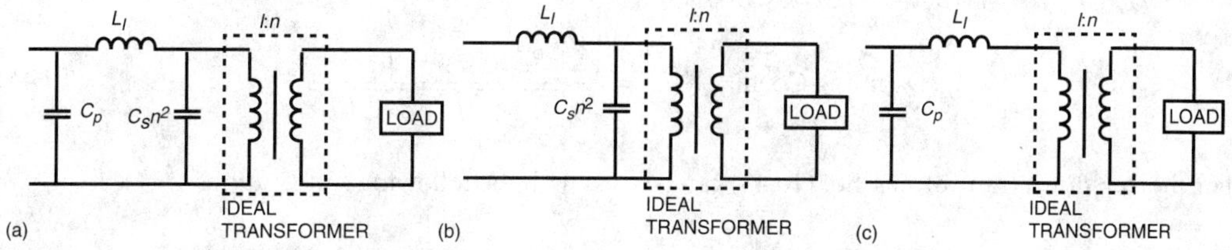

FIGURE 10.11 (a) Equivalent for 1:1, (b) equivalent for step up ($n > 1.0$), (c) equivalent for step down ($n < 1.0$).

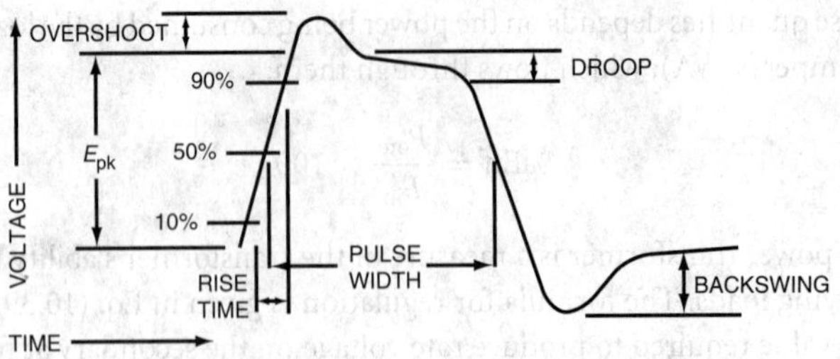

FIGURE 10.12   Pulse terminology.

one of the distributed capacitances, depending on whether the transformer is a step down, or step up, or 1:1 transformer. Assuming that one of the capacitances is dominant simplifies the equivalent circuit and makes the transfer function easier to analyze. The distributed capacitance will be larger in the coil with more turns, and the reflection of the larger capacitance to the other side of the transformer increases this effect. The equivalent circuits for 1:1, step up, and step down are given in Fig. 10.11(a), Fig. 10.11(b), and Fig. 10.11(c), respectively.

Pulse Transformers

Pulse transformers are used to transform video pulses. Although they are similar in design to wide-band transformers, they are designed with different specifications in mind. The shape of a typical pulse is shown in Fig. 10.12, which demonstrates how the important quantities in the pulse are specified.

Because of the complex nature of the pulse, it is often divided into three stages, which are analyzed separately. The rising edge of the pulse is analyzed using the high-frequency equivalent circuit for step up or step down wide-band transformers due to the high frequencies that are present. The top of the pulse is analyzed using the low-frequency equivalent circuit, since it is a DC voltage. The trailing edge is analyzed using the high-frequency model, with the assumption that the source is an open circuit.

Rising Edge

Because of the high frequencies present in the rising edge of the pulse, the high-frequency equivalent circuit is used to analyze it. For the step up transformer the ratio of the load voltage to the source voltage is given by Eq. (10.43). If $4c < b^2$, then the circuit is overdamped. If $4c > b^2$, then the circuit is underdamped, and the overshoot is larger. The damping factor is defined in Eq. (10.44). The rise time is less for lower damping factors, but the overshoot also increases for lower damping factors. A damping factor of 0.707 is optimum in the sense that it produces the shortest rise time with no overshoot.

$$\frac{V_L}{V_G} = \frac{R_L}{R_L + R_G}\left[1 - \exp\left(-\frac{b}{2}t\right)\left(\frac{b}{\sqrt{b^2 - 4c}}\sinh\frac{\sqrt{b^2 - 4c}}{2}t + \cosh\frac{\sqrt{b^2 - 4c}}{2}t\right)\right] \quad (10.43)$$

$$b = \frac{R_G}{L_L} + \frac{1}{R_L C_D} \qquad c = \frac{R_L + R_G}{L_L C_D R_L}$$

$$\text{Damping Coeff} = \frac{b}{2\sqrt{c}} = \frac{R_G R_L C_D + L_L}{2\sqrt{R_L L_L C_D (R_G + R_L)}} \quad (10.44)$$

For the step down transformer, Eq. (10.43) can be used with the following substitutions:

$$b = \frac{R_L}{L_L} + \frac{1}{R_G C_D} \qquad c = \frac{R_G + R_L}{L_L C_D R_G}$$

Top of Pulse

The top of the pulse is comparable to a DC input to the transformer. For this reason, the low-frequency equivalent circuit is used. As the pulse continues in the high state, the mutual inductance begins to conduct more current away from the load resistance. The ratio of the load voltage to the input voltage is given by Eq. (10.45). The negative exponential form of Eq. (10.45) indicates that the output voltage decays, or droops with time. The larger the mutual inductance is, the slower the rate of decay. It is necessary to know the soure resistance during the top of the pulse in order to accurately estimate the droop.

$$\frac{V_L}{V_G} = \frac{R_L}{R_L + R_G} \exp\left[\frac{R_L R_G}{L_M (R_L + R_G)} t\right] \tag{10.45}$$

Trailing Edge of Pulse

At the trailing edge of the pulse, the high-frequency model is used again. In this case, however, the source is assumed to be an open circuit. The distributed capacitance and the mutual inductance now produce a voltage on the load with stored energy. The equivalent circuit for the trailing edge is shown in Fig. 10.13, where R_T is the total shunt resistance, which include the core loss and the reflected load resistance.

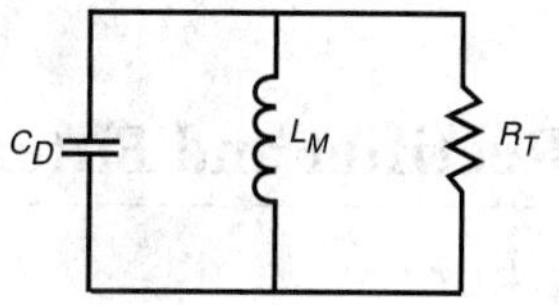

FIGURE 10.13 Equivalent circuit for trailing edge of pulse.

The response of the trailing-edge circuit is then given by Eq. (10.46), which relates the output voltage to the initial voltage at the time the pulse ends.

$$\frac{e(t)}{E_0} = \exp(-2\pi D T)\left(\cosh 2\pi T \sqrt{D^2 - 1} - \frac{D + F}{\sqrt{D^2 - 1}} \sinh 2\pi T \sqrt{1 - D^2}\right) \tag{10.46}$$

$$\text{Damping Factor} = D = \frac{1}{2 R_T \sqrt{L_M C_D}}$$

$$\text{Initial current factor} = F = \frac{\tau}{\sqrt{L_M C_D}}$$

$$\text{Normalized Time} = T = \frac{t}{2\pi \sqrt{L_M C_D}}$$

$$\tau = \text{Pulse Width}$$

References

Blume, L.F. et al. 1954. *Transformer Engineering*, 2nd ed. John Wiley & Sons, New York.

Fitzgerald, A.E., Kingsley, C., and Umans, S. 1990. *Electric Machinery*, 5th ed. McGraw-Hill, New York.

Gibbs, J.B. 1950. *Transformer Principles and Practice*, 2nd ed. McGraw-Hill, New York.

Grover, F.W. 1946. *Inductance Calculations*. Van Nostrand, U.S. Department of Commerce, Princeton, N.J.

Henney, K., ed. 1959. *Radio Engineering Handbook*, 5th ed. McGraw-Hill, New York.

Howe, J.G. et al. 1964. Final report for high power, high voltage, audio frequency transformer design manual. U.S. Navy Contract No. BSR 87721, Project Serial No. SR008-03-02, Task 9599 (Aug.).

Kajihara, H.H. 1956. Miniaturized audio transformer design for transistor aplications. *IRE Trans. Audio.* Jan.–Feb.

Landee, R.W. et al. 1957. *Electronic Designers Handbook*. McGraw-Hill, New York.

Lee R. 1955. *Electronic Transformers and Circuits*, 2nd ed. John Wiley & Sons, New York.

McPherson, George. 1981. *An Introduction to Electrical Machines and Transformers*. John Wiley & Sons, New York.

Ruthroff, C.L. 1959. Some broad-band transformers. *Proc. IRE.* 47(8)1337–1342.

Schlicke, H.M. 1961. *Dielectromagnetic Engineering*. Wiley, New York.

Slemon, G.R. 1966. *Magnetoelectric Devices*. Wiley, New York.

Further Information

The quarterly journal *IEEE Transactions on Power Delivery* covers the field of transformers mostly at power frequency ranges of 50 and 60 Hz.

The quarterly journal *IEEE Transactions on Magnetics* covers magnetic material aspects of transformers.

Information on transformers for high frequency applications is rather difficult to find. The following publications occasionally covers transformers for high frequency applications:

IEEE Circuits & Devices, the magazine of electronic and photonic systems.
Electronic Design, Penton Publishing Inc., Cleveland, Ohio.
IEE Proceedings Circuits, Devices and Systems (London, UK).
IEEE Transactions on Signals and Systems.

10.2   Rectifier and Filter Circuits

T. S. Kalkur

10.2.1   Rectifier Circuits

One of the most important building blocks in electronic power supplies is the **rectifier**. **Diodes** are used to form rectifiers in electronic power supplies.

Half-Wave Rectifier

The half wave rectifier circuit is shown in Fig. 10.14(a). When the input sinusoid voltage at A goes positive, the diode conducts, resulting in current in the load resistor R. When the input voltage goes negative, the diode is reverse biased and, hence, it does not conduct, resulting in negligle current across R. Therefore, the output voltage is given by

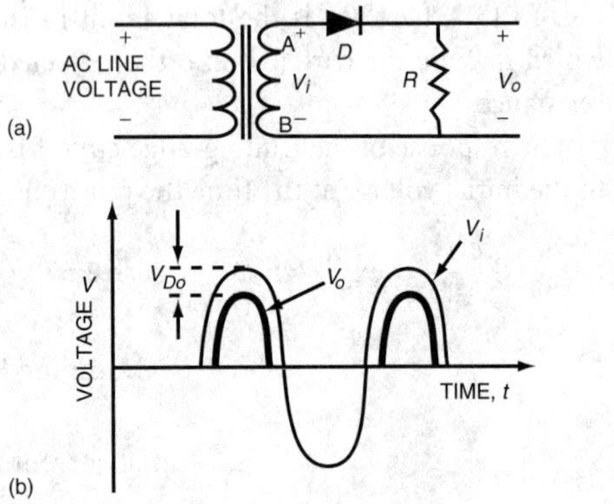

FIGURE 10.14   Half-wave rectifier: (a) circuit diagram, (b) input and output waveforms.

$$V_o = 0, \quad \text{if } V_i < V_{Do} \tag{10.47}$$

$$V_o = R/(R + r_D)V_i - V_{Do} R/(R + r_D) \tag{10.48}$$

where
$\quad V_i$ = the input voltage
$\quad r_D$ = diode resistance
$\quad V_{Do}$ = diode forward voltage $\simeq$ 0.7–0.8 V

Figure 10.15(b) shows the output voltage waveform for a sinusoidal input voltage. If $R \gg r_D$, then Eq. (10.48) simplifies to

$$V_o \simeq V_i - V_{Do} \tag{10.49}$$

The following aspects should be considered in designing the diode power supply:

- The current or power handling capability of the diode
- The *peak inverse voltage* (PIV) that the diode must be able to withstand without breakdown

When the input voltage goes negative, the diode is reverse biased, and the input voltage V_s appears across the diode.

The average output DC voltage, neglecting V_{Do} is given by

$$V_{DC} = \frac{1}{2\pi} \int_0^{2\pi} \frac{V_m R \sin wt \, d(wt)}{(R + r_D)} \tag{10.50}$$

$$= V_m R / \pi (R + r_D)$$

$$\cong V_m / \pi \quad \text{if } R \gg r_D \tag{10.51}$$

where V_m is the maximum value of V_o. The ripple factor of a half-wave rectifier is given by

$$r = [\pi^2/4 - 1]^{\frac{1}{2}} = 1.21 \tag{10.52}$$

Full-Wave Rectifier

The full-wave rectifier utilizes both halves of the input sinusoid; one possible implementation is shown in Fig. 10.15(a). The full-wave rectifier consists of two half-wave rectifiers connected to a load R. The secondary transformer winding is center tapped to provide two equal voltages V_i for each half-wave rectifier. When the node A is at positive polarity V_i with respect to node B, $D1$ will be forward biased and $D2$ will be reverse biased. Therefore, diode $D1$ will conduct and the current will flow through R back to the center tap of the transformer. When the node B is at positive polarity V_i with respect to node A, $D2$ will be forward biased and diode $D1$ will be reverse biased. The current conducted by $D2$ will flow through R and back to the center tap. The current through R is always in the same direction

(a)

(b)

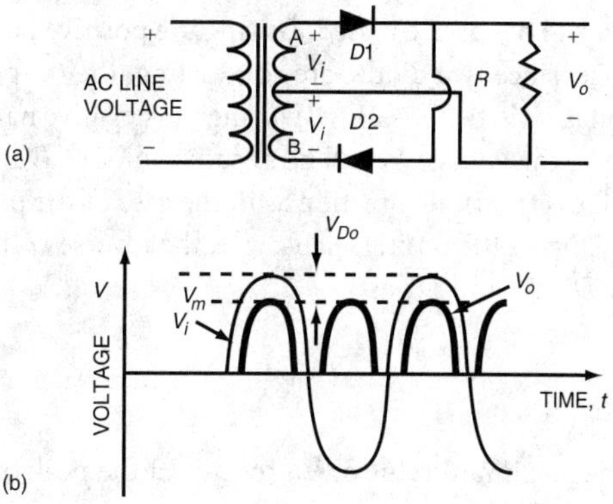

FIGURE 10.15 Full wave rectifier: (a) schematic diagram, (b) input and output waveforms.

giving rise to unipolar voltage V_o across R. The input and output voltage waveforms of the full-wave rectifier are shown in Fig. 10.15(b).

The output voltage of the full rectifier is given by

$$V_o = \left[\frac{R}{R_t + r_D + R}\right](V_s - V_{Do}) \tag{10.53}$$

where R_t is the resistance associated with the transformer. The DC value of the output voltage

$$V_{DC} \cong 2\left[\frac{R}{R + r_D + R_t}\right]\frac{V_m}{\pi} \tag{10.54}$$

$$\cong 2V_m/\pi, \quad \text{if } R \gg R + r_D + R_t \tag{10.55}$$

where V_m is the peak output voltage. Thus the full-wave rectifier produces double the output voltage of that of a half-wave rectifier. The ripple content of the full-wave rectifier is given by

$$r = (\pi^2/8 - 1)^{1/2} = 0.483 \tag{10.56}$$

The ripple factor for a full-wave rectifier is significantly less than that of a half-wave rectifier.

Another important factor to be considered in the design of full-wave rectifier is the peak inverse voltage of the diode. During the positive half-cycle, diode $D1$ is conducting and $D2$ is cut off. The voltage at the cathode of $D2$ will be at its maximum when V_o is at its peak value of $(V_i - V_{Do})$ and V_i at its peak value. Therefore, the peak inverse voltage

$$PIV = 2V_i - V_{Do} \tag{10.57}$$

is approximately twice that of the half-wave rectifier.

The Bridge Rectifier

The full-wave rectifier just discussed requires a center-tapped transformer. An alternative implementation of the full-wave rectifier, as shown in Fig. 10.16(a), is the bridge rectifier. This rectifier has four diodes arranged similar to Wheatstones bridge and does not require a center tapped transformer. Figure 10.16(b) shows the input and output voltage waveforms for the bridge rectifier. During the positive half-cycles of the input voltage, V_i is positive and the current is conducted through diode $D1$, resistor R, and diode $D2$. Meanwhile diodes $D3$ and $D4$ will be reverse biased. During the positive half-cycle, since two diodes are conducting, the output voltage will be $V_i - 2V_{Do}$. During the negative half-cycle, the voltage V_i will be negative, diodes $D3$ and $D4$ are forward biased, and the current through R follows the same direction as in the case of the positive half-cycle.

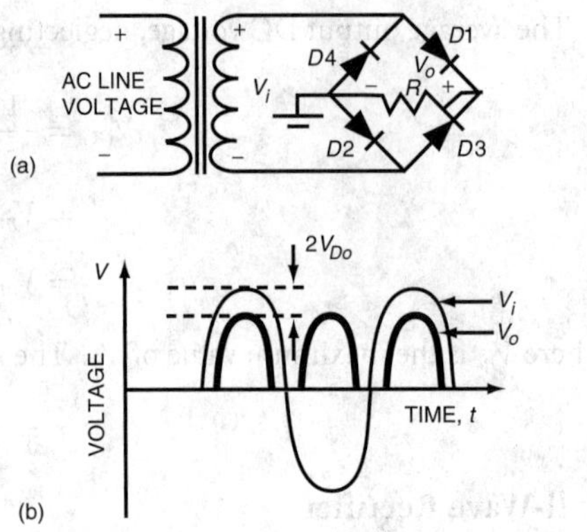

FIGURE 10.16   Bridge rectifier: (a) circuit schematic, (b) input and output voltage waveforms.

During the positive half-cycle, the reverse voltage across $D3$ can be determined from the loop formed by $D3$, R, and $D2$ as

$$V_{D3}(\text{reverse}) = V_o + V_{D2}(\text{ forward}) \tag{10.58}$$

The maximum value of V_{D3} occurs at the peak of V_o and is given by

$$\text{PIV} = V_i - 2V_{Do} + V_{Do}$$
$$= V_i - V_{Do} \tag{10.59}$$

The PIV is about half the value for the full-wave rectifier with a center tapped transformer, which is an advantage of the bridge rectifier.

Voltage Multipliers

The DC output voltage V_{DC} in rectifier circuits is limited by the peak value of the AC voltage applied V_i. Figure 10.17(a) shows the voltage doubler circuit composed of a clamp formed by $C1$ and $D1$ and a peak rectifier formed by $D2$ and $C2$. When exited by a sinusoid of amplitude V_p, the output of the clamping section reaches the negative peak value of $-2V_p$ as shown in Fig. 10.17(b). In response to this clamp section output voltage, the output of the peak rectifier reaches $-2V_p$. By connecting additional diode capacitor circuits it is possible to generate rectifier circuits that triple and quadruple the input voltage.

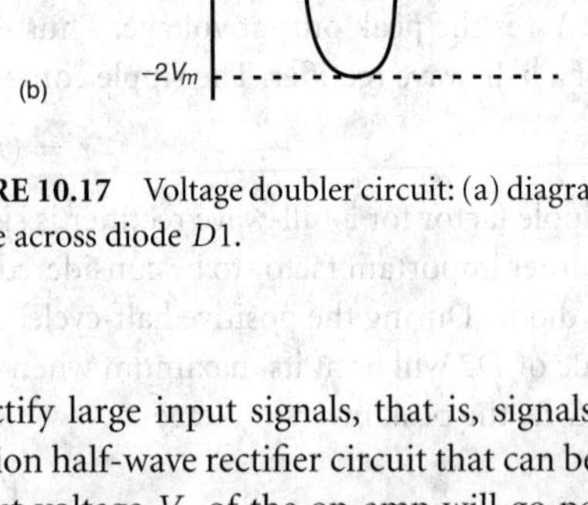

FIGURE 10.17   Voltage doubler circuit: (a) diagram, (b) voltage across diode $D1$.

Precision Half-Wave Rectifiers

The rectifier circuits discussed previously are useful to rectify large input signals, that is, signals with amplitude greater than 0.6 V. Figure 10.18(a) shows a precision half-wave rectifier circuit that can be used to rectify small input signals. If V_i goes positive, the output voltage V_a of the op-amp will go positive and the diode will conduct, thus establishing a closed feedback path between the op-amp output terminal

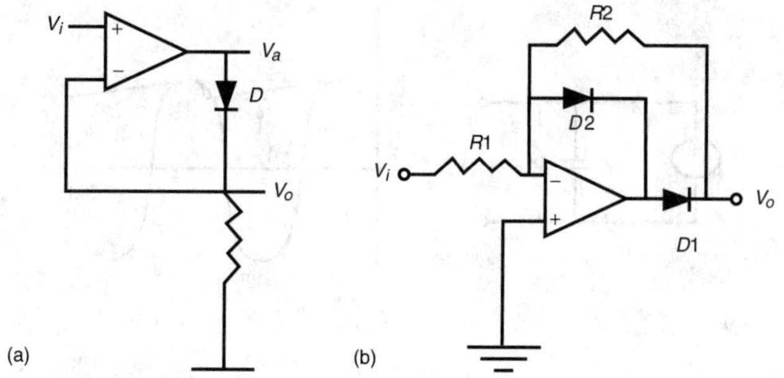

FIGURE 10.18   A precision rectifier: (a) circuit diagram, (b) an improved version of the precision half-wave rectifier.

and the negative input terminal. Because of the virtual ground, the ouput voltage $V_o = V_i$ for $V_i > 0$. Therefore, this circuit is useful to rectify very small signals.

When the input voltage goes negative, the op-amp output voltage V_a tends to follow and go negative. This reverse biases the diode with no current flowing through R. Therefore, the output voltage will be zero. The applied voltage V_i will be dropped across the two inputs of the op-amp. If V_i is more than a few volts, the op-amp might be damaged if input overvoltage protection is not provided. Another problem associated with this circuit is speed. When V_i goes negative, because of large reverse resistance of diode, the op-amp goes to saturation. Getting the op-amp out of saturation generally requires more time resulting in speed degradation.

An improved version of half-wave precision rectifier is shown in Fig. 10.18(b). When V_i goes positive, diode $D2$ conducts and closes the negative feedback loop around the op-amp. The diode $D1$ will be off, and no current flows through $R2$ resulting in zero output voltage. When the input voltage V_i goes negative, the output of the op-amp goes positive. This results in reverse bias for diode $D2$ and forward bias for diode $D1$. The output voltage V_o tracks the input voltage and by appropriately selecting $R1$ and $R2$, V_o can be equal to V_i.

In this circuit configuration, the feedback loop for the op-amp is always closed, thus avoiding the chances of the op-amp moving into saturation.

Precision Full-Wave Rectifier

Figure 10.19 shows the circuit diagram of the precision full-wave rectifier. When V_i is positive, V_F goes negative and V_E goes positive. Therefore, diode $D1$ will be in reverse bias and $D2$ will conduct because of forward bias, and the voltage $V_o = V_i$ for appropriate ratio of $R1$ and $R2$. When V_i goes negative, V_F goes postive and V_E goes negative. This results in diode $D2$ being cut off and $D1$ in conduction with $V_i = V_o$.

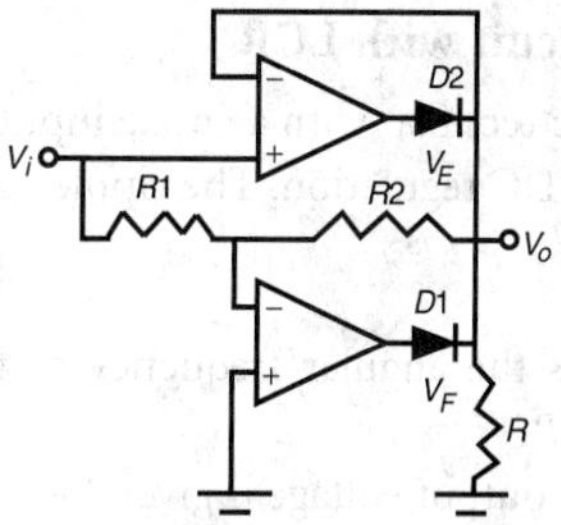

FIGURE 10.19   Precision full wave rectifier with a diode placed in the feedback loop.

Precision Bridge Rectifier

The precision bridge rectifier is shown in Fig. 10.20. When the input voltage V_i goes positive, diodes $D1$ and $D3$ are forward biased, the current flows through R, and the output voltage $V_o = V_i$. When the input voltage V_i goes negative, diodes $D2$ and $D4$ are forward biased and V_o follows V_i.

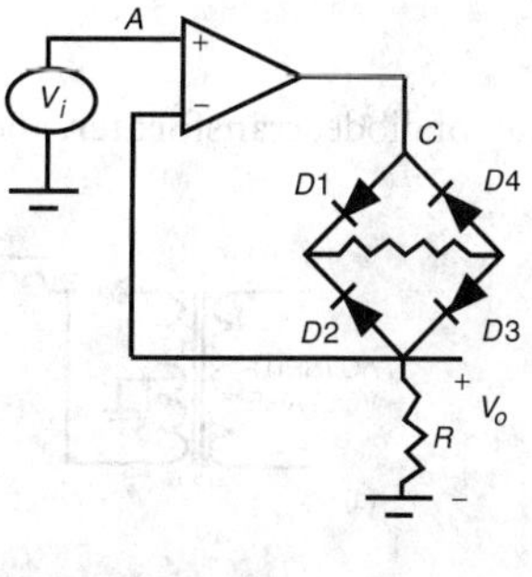

FIGURE 10.20   A precision bridge rectifier.

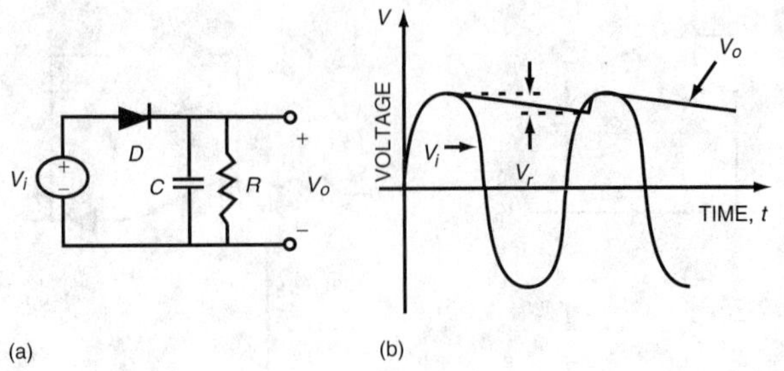

(a)   (b)

FIGURE 10.21   Half-wave rectifier and RC filter circuit: (a) schematic diagram, (b) input and output waveforms.

10.2.2   Filter Circuits

Resistor Capacitor (RC) Filters

Consider a diode half-wave rectifier with a filter capacitor C and a resistor R as shown in Fig. 10.21(a). When the input voltage V_i is positive, the diode conducts and charges the capacitor C. When the diode moves into cutoff, the capacitor discharges through the rest of the cycle until the time at which V_i exceeds the voltage at the capacitor. The output waveform for the rectifier with a capacitor is shown in Fig. 10.21(b). The **ripple voltage** V_r is given by

$$V_r \cong V_p/fCR \tag{10.60}$$

where V_p is the peak input voltage and f is the frequency of the input AC voltage. To decrease the ripple voltage, it is necessary to increase the value of RC. The average output voltage is

$$V_o \cong V_p - 0.5V_r \tag{10.61}$$

For full-wave rectifiers, the discharge period is half that of a half-wave rectifier. Therefore, the ripple voltage is given by

$$V_r = V_p/2fCR \tag{10.62}$$

Filter Circuit with LCR

A full wave rectifier with a choke input L type RC filter (as shown in Fig. 10.22(a)) improves the ripple factor and DC regulation. The ripple voltage across R and C for the first harmonic is given by

$$V_r = V_m/\left\{3(2)^{\frac{1}{2}}\pi\omega^2 LC\right\} \tag{10.63}$$

where ω is the angular frequency of the input waveform. The ripple voltage is independent of load resistance R.

The DC output voltage is given by

$$V_{DC} = V_m/\pi - R'I_{DC} \tag{10.64}$$

where I_{DC} is the direct DC current

$$R' = 1/6(2)^{\frac{1}{2}}\omega^2 LC \tag{10.65}$$

R' is the sum of diode, **transformer**, and inductor resistance.

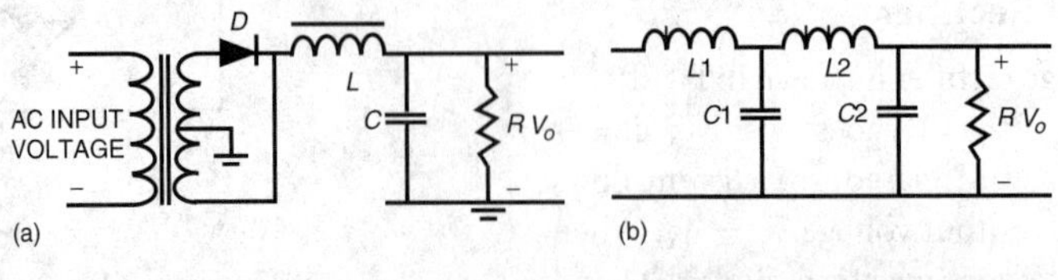

(a)   (b)

FIGURE 10.22   Filter circuits: (a) full wave rectifier with L-section filter, (b) multiple L-section filter.

Even without load, a resistance is added across the output, and this is called a *bleeder resistance.* The bleeder resistance performs the following functions:

- It bleeds the charge from the filter capacitor when the rectifier is turned off.
- It improves the regulation.
- It acts as a minimum load, preventing excessive voltage development in the filter inductance.

The ripple voltage is inversely proportional to inductance L and capacitance C. This discussion is valid when continuous current flows through the inductor. The current through the inductor is continuous if the inductance is greater than the critical inductance L_C, where L_C is greater than or equal to $R/(3\omega)$.

Filtering can be improved by cascading more energy storing elements L and C as shown in Fig. 10.22(b). The load voltage V_o is reduced due to the DC voltage drop across the inductor. This can be minimized by reducing the resistance associated with the inductors.

Defining Terms

Diode: A semiconductor device made up of a p and n type semiconductor junction.
Filter circuits: The circuits used to filter out AC ripple from DC.
Rectifier: A device that converts AC signals to DC.
Ripple voltage: The AC voltage associated with the DC in a rectifier circuit.
Transformer: A device used to step up or step down the AC voltage and to isolate mains supply from the rectifier circuit.

References

Alley, C.L. and Atwood, K.W. 1962. *Electronic Engineering.* Wiley, New York.
Burns, S.B. and Bond, P.R. 1987. *Principles of Electronic Circuits.* West, St. Paul, MN.
Gray, P.E. and Searle, C.L. 1969. *Electronic Principles.* Wiley, New York.
Horowitz, P. and Hill, W. 1989. *The Art of Electronics,* 2nd ed. Cambridge University Press, Cambridge, England, UK.
Millman, J. and Halkias, C.C. 1972. *Integrated Electronics: Analog and Digital Circuits and Systems.* McGraw-Hill, New York.
Sedra, A.S. and Smith, K.C. 1991. *Microelectronic Circuits,* 3rd ed. Saunders College Publishing, Philadelphia, PA.
Sze, S.M. 1985. *Semiconductor Devices, Physics and Technology.* Wiley, New York.

Further Information

Additional information on the topic of rectifiers and filter circuits is available from the following sources:

IEE Proceedings, Circuits, Devices and Systems. P.O. Box 96, Stenenage, SG1 2SD, U.K.
IEEE Transactions on Consumer Electronics, IEEE Publishing, Dept. 445, Hoss Lane, P.O. Box 1331, Piscataway, NJ 08855-1331.
International Journal of Electronics, Taylor Francis, Hampshire, GU34 1NW, U.K.

10.3   Voltage Regulation

Melissa S. Mattmuller

10.3.1   Introduction

An almost universal requirement of electronic systems is a regulated source of DC voltage. Whether the DC power originates with a battery or has been converted from widely available AC power, a voltage regulator is usually required to provide a steady DC voltage for a variety of load conditions. Figure 10.23 illustrates the place of voltage regulation in a typical power supply.

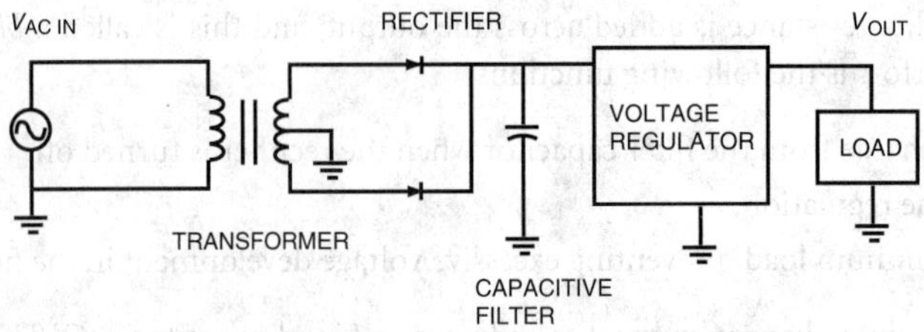

FIGURE 10.23   A typical power supply.

Although one may still find it useful to design and construct voltage regulators with discrete devices, today's precise but robust, bipolar monolithic voltage regulators make design a much easier task. They are available in increasing variety in both high- and low-power products. The low-power products include low voltage output regulators to drive low voltage (3–3.3 V) and very low voltage (sub 3.0 V) digital components.

10.3.2   Types of Voltage Regulators

Voltage regulators can be categorized as *discrete component circuits* or as *ICs*, as illustrated in Fig. 10.24. Examples of simple voltage regulators comprising discrete components open this chapter, but primarily IC regulators will be covered. Today a wide variety of IC regulators is available for almost any application or environment. IC voltage regulators are increasingly available for very low-power, high-power, and low-dropout applications. Also, many desirable features such as high precision and self-protection are common.

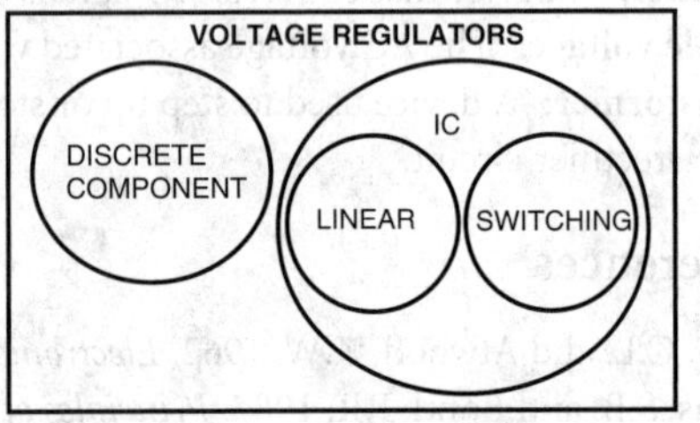

FIGURE 10.24   Voltage regulator groupings.

IC voltage regulators are usually divided into two categories, *linear* and *switching*. There are five basic categories of **linear** voltage **regulators**: positive adjustable, negative adjustable, fixed output, tracking, and floating. Any of these may be available with other attributes such as three-terminal packaging or **low dropout voltage** features. Some of the more popular linear IC regulators are summarized in Sec. 10.3.3 and references will guide the reader to more expansive literature.

Switching regulators have advanced significantly in the areas of size and weight reduction and improved efficiency, and they have the ability to perform voltage step-up, step-down, and voltage-inverting functions. Switching regulators are discussed in the Sec. 10.3.6.

10.3.3   Selecting a Voltage Regulator

In past years designing a power supply could become a tricky matter. Designers usually had to construct their own voltage regulators either of discrete components or using positive fixed regulators as building blocks. But with today's wide variety of IC voltage regulators it is likely that a single device may satisfy most design requirements. This is not just a matter of convenience, but it is important for reliability and maintainability reasons to keep the component count low. The following guidelines may offer some assistance in making the build/buy decision.

- If the load current is very low (50 mA or less) then either build, a simple zener diode may provide sufficient regulation, as shown in Fig. 10.25, or buy, for example, a 79 series in a TO-92 case.
- If the current load is less than one ampere then build. The two discrete voltage regulators (one op amp and the other transistor) discussed in this chapter are useful samples. Note that low-current IC regulators may be available to do the job just as easily.

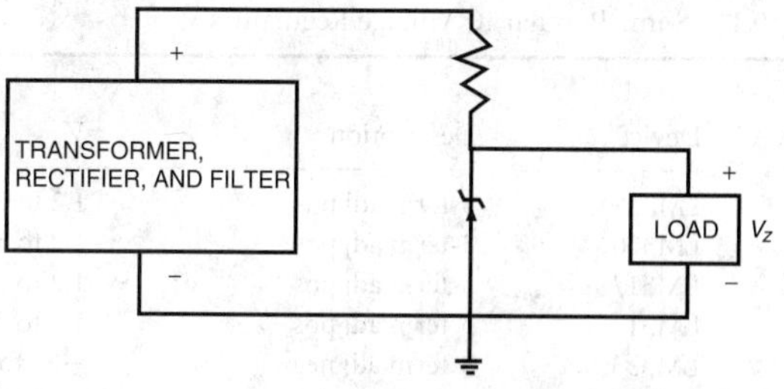

FIGURE 10.25 Simple zener regulator.

Or

- If a load current level from 50 mA to a few amperes is desired then buy. The designer may want to look for a linear IC regulator. As mentioned in the previous section, there are five basic categories of linear regulators: positive adjustable, negative adjustable, fixed output, tracking, and floating.

- If load currents of more than one ampere range are required then buy. The designer may want to investigate switching regulators, especially if the application requires high-current output at low voltages such as computer applications, or if efficiency, size, and weight are of primary concern. Recognize, however, that they are a source of high-frequency electrical noise. Switching power supplies are discussed extensively in Sec. 10.3.6.

Selecting a Linear IC Voltage Regulator

Generally regulators are chosen for use as their names would indicate: fixed positive regulators are used to generate the specified, rated positive voltage. Negative regulators, however, can be connected to provide positive voltages and positive fixed regulators can be utilized to make variable voltage regulators, or tracking regulators, or switching regulators, or even current regulators. The application notes contained in data books can be very useful for exploring the possible uses for the various regulators.

When the purpose of the regulated output is to provide $+/-$DC bias voltages as for an op amp, then a *dual, bipolar tracking regulator* is needed. In this way, not only will the positive and negative output voltages be symmetric but the midpoint of the signals will be at ground potential.

Many varieties of linear IC regulators are available with low dropout voltage requirements. The **dropout voltage** is the input-to-output voltage differential below which regulation is not provided. Thus, a low dropout voltage regulator may provide, for example, a regulated output voltage of 5 V for an input voltage of only 5.6 V. This is a considerable improvement from the earlier regulators requiring 2–3 V of headroom. Low dropout voltage regulators are available and are especially useful for battery powered systems where it is desired to extend the useful battery life. Two examples include automotive and laptop computer applications in which memory needs to be maintained for very low battery conditions.

One of the most popular IC regulator products is a *three-terminal regulator*. The output is either fixed or adjustable, and both categories are available in a variety of positive or negative output voltages and current ranges, as well as some with low dropout voltage requirements. Although it is not a difficult matter to make an adjustable voltage regulator from a fixed, three-terminal IC by providing external circuitry, three-terminal adjustable output regulators are readily available and easy to set.

Three-terminal regulators are frequently used as on-card regulators. That is, an unregulated power supply voltage is provided to each printed circuit board in a system, and a three-terminal voltage regulator provides the voltage for the circuits on that card, thereby eliminating coupling effects of a centralized power supply system.

The three-terminal voltage regulator is a popular choice for the following reasons:

- They are easy to use.
- They are robust with regard to their internal overcurrent and thermal protection.

TABLE 10.1 Some Popular IC Voltage Regulators

Output Current, A	Device	Description	V_{out}
5.0	LM338	3-term adj pos	1.2 to 32
3.0	LM350	3-term adj pos	1.2 to 32
1.5	LM317	3-term adj pos	1.2 to 37
0.5	LM317	3-term adj pos	1.2 to 37
3.0	LM333	3-term adj neg	−32 to −1.2
1.5	LM337	3-term adj neg	−32 to −1.2
3.0	LM323	3-term fxd pos	5
1.0	LM340	3-term fxd pos	5, 12, 15
3.0	LM345	3-term fxd neg	−5
1.5	LM320	3-term fxd neg	−5, −12, −15
1.5	LM330	Ldv[a] 0.6 V, 3-term fxd pos	5
0.5	LM2937	Ldv[a] 0.82 V	5, 8, 10, 12, 1, 5
0.5	MC33267	Ldv[a] 0.6 V or lower fxd pos	5 V
0.1	LM2951	Ldv[a] 0.38 V	5, or adj (1.24 to 29)
0.1	MC1468/15688	+/−15 V dual tracking	+/−15 V
0.15	MC1723	positive or negative adj	2 to 37

[a]Low dropout voltage.

- Fixed output regulators require no external circuitry (except, perhaps, two bypass capacitors) and the adjustable regulators may require as little as two external resistors to set the output.
- A wide variety of output currents and voltages are readily available at a low cost.

IC regulators are very versatile so regulator design can become a complicated matter and is considered to be beyond the scope of this work, but there are a few guidelines that can aid in the decision (Motorola, 1993).

- Determine which type of voltage regulator is best suited for the application of interest, based on the descriptions given previously. For example, negative, variable output, low dropout voltage, etc.
- Determine which products of this description meet or exceed the requirements for **line regulation**, **load regulation**, temperature coefficient of output voltage TC of V_O, and operating ambient temperature range.
- Select application circuits meeting the requirements for output current, output voltage, or any special features desired. From here, the designer may consider size, cost, and complexity.
- Because load current may be altered dramatically with the use of outboard, current-boosting transistors one need not be limited by specified output current. If possible the designer should choose a product whose output current gives plenty of margin, because output current performance will drop off as the temperature rises. This needs to be balanced with the fact that the rated maximum current may be delivered, if only for a time, in the event that the load is shorted.

Other selection criteria relate to individual device specifications. Product specifications and selection criteria are discussed in Sec. 10.3.6. Table 10.1 shows a small subset of the products currently available on the market. It is intended merely to illustrate the variety, not be a representative or complete market description.

10.3.4 Two Discrete Voltage Regulator Examples

Simple Op Amp Regulator

Figure 10.26 shows a stable, relatively simple voltage regulator, this one using a noninverting operational amplifier to provide the negative feedback control (Jacob, 1982, 80). The op amp buffers the voltage from the zener diode, placing a negligible load on the zener, allowing it to operate at a single, fixed point. The

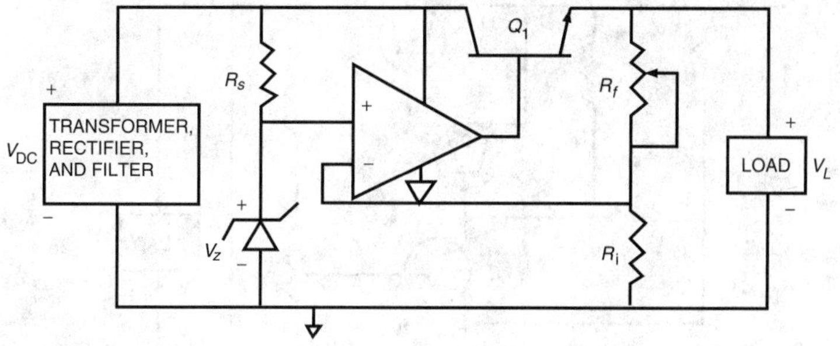

FIGURE 10.26   An op amp voltage regulator.

zener diode voltage then is very stable and the zener is not subjected to large swings in current and so one may select a low-power, highly precise reference diode.

The output voltage is altered by adjusting R_f

$$V_{\text{LOAD}} = V_z(1 + R_f/R_i)$$

The peak of the input voltage V_{DC} should be kept below the maximum supply voltage for the op amp (look for an op amp with a high supply voltage rating). Also, V_{DC} must always be at least 2 V above the output (load) voltage. This is necessary to keep the op amp out of saturation and to keep the zener diode in its breakdown region.

Notice that the minimum output voltage is V_z (when R_f is shorted), and so the zener voltage should be chosen to be smaller than the lowest output needed.

The op amp provides the base current that is then multiplied by the β of the **pass transistor** (so called because all of the load current is passed through it) to produce load current. The β of the transistor is determined by

$$\beta \geq \frac{I_{\text{load}}}{I_{\text{op amp max}}}$$

Also, the transistor must be able to dissipate the power produced by

$$P_Q = (V_{\text{DC}} - V_{\text{load}})I_{\text{load}}$$

where

V_{DC} = DC output voltage from the filter
V_{load} = load DC voltage
I_{load} = load DC current

A power metal oxide semiconductor field effect transistor (MOSFET) could have been used here. Be sure to carefully review the biasing and headroom requirements.

Second Discrete Voltage Regulator Example

The circuit of Fig. 10.27 shows how the highly stable zener voltage, or reference voltage, can be amplified using a negative-feedback amplifier to get a higher output voltage while retaining the temperature stability provided by the zener diode. A **Darlington pair** is used for the pass transistor for greater current gain. Here a voltage divider (R_1 and R_2, chosen to draw minimal current from the load) is used to feed a portion of the output voltage back to the base of Q_1 controlling its collector current. This control provides the regulation from the source. Recall that the input to this voltage regulator was an AC voltage source that has been rectified and filtered but probably had considerable **ripple**.

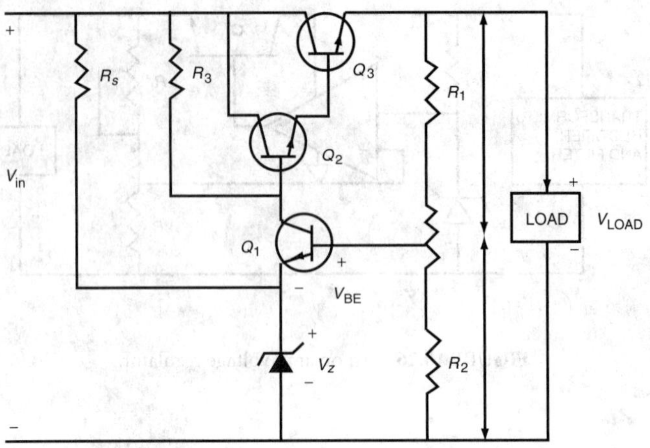

FIGURE 10.27 A discrete voltage regulator.

The closed-loop gain of this circuit A_{CL}, can be expressed as

$$A_{CL} = \frac{R_1}{R_2} + 1$$

and

$$V_{OUT} = A_{CL}(V_z + V_{BE})$$

Once the potentiometer is adjusted for the desired V_{OUT}, this output voltage remains constant despite changes in line voltage or load current.

The designer must check the power dissipated by the pass transistor (the Darlington transistor) to determine the heat sinking requirements.

Discrete Voltage Regulator with Current Limiting

Notice that the two voltage regulators discussed previously provide no short-circuit protection. In the event that the load terminals are shorted together, the pass transistor(s) and possibly other components could be destroyed. For this reason robust designs include some form of **current limiting**. A simple method of current limiting is illustrated in Fig. 10.28 but **fold-back current limiting** is more desirable because it reduces the power dissipated by reducing both the load current and voltage. This technique will be discussed in more detail in the next section.

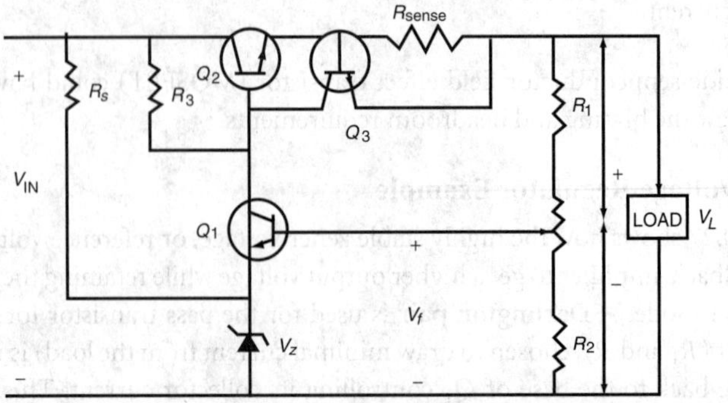

FIGURE 10.28 A voltage regulator with current limiting.

Notice the resistor marked R_{sense}. It is called a *current sensing* resistor and is very small in value. If the current through it is great enough, it will provide the base emitter of Q_3 with enough voltage to turn it on. Then Q_3 collector current will flow through R_3 and act to decrease the base voltage to Q_2. Since the transistor Q_2 is an emitter follower, decreasing its base voltage causes a decrease in the output voltage.

Now the maximum current that flows even for a shorted load I_{SL} is

$$I_{\text{SL}} = V_{\text{BE}}/R_{\text{sense}}$$

and the load for which regulation is lost can be estimated by

$$R_{\text{min}} = V_{\text{REG}}/I_{\text{SL}}$$

where V_{REG} is the regulated output voltage of the circuit.

10.3.5   IC Regulator Examples

Four examples are included here, a three-terminal standard application adjustable voltage regulator, a digital-to-analog controlled regulator, a current boosting configuration for a three-terminal positive regulator, and a positive or negative regulator with foldback current limiting.

Figure 10.29 shows a standard configuration for a positive or negative, three-terminal, adjustable voltage regulator.

The input capacitor (usually on the order of 0.1-μF disc or a 1-μF solid tantalum) is required if the regulator is more than 4 in. from the power supply filter, and the output capacitor is generally used to improve transient response performance, perhaps for stability (frequently 1-μF solid tantalum or 25-μF aluminum electrolytic).

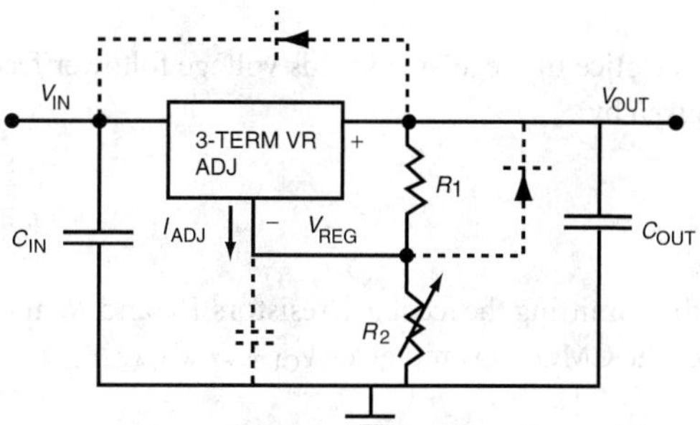

FIGURE 10.29   Configuration for an adjustable output three-terminal regulator.

Sometimes a bypass capacitor (10 μF) will be connected between the adjustment terminal and ground to improve **ripple rejection** and noise. Generally solid tantalum capacitors are used because of their low impedance even at high frequencies. If the bypass capacitor is used, then the addition of protection diodes for the bypass and output capacitors is a good idea. The details are left for the databooks.

The output voltage for the positive regulator can be expressed as

$$V_{\text{OUT}} = V_{\text{REG}} \times [(1 + R_2/R_1) + I_{\text{ADJ}} R_2]$$

and for the negative regulator

$$V_{\text{OUT}} = -V_{\text{REG}} \times [(1 + R_2/R_1) + I_{\text{ADJ}} R_2]$$

The $I_{\text{ADJ}} R_2$ term is frequently left out because $I_{\text{ADJ}} R_2$ is on the order of 10 mV.

Digital-to-Analog (D/A) Applications for Voltage Regulator

Figure 10.30 shows a different use of a three-terminal voltage regulator (Jacob, 1982). Here the three-terminal regulator has replaced the transistor of the previous circuits, and the D/A effectively replaces the zener diode. By using high quality but relatively inexpensive parts, the designer can acquire the robust power of a voltage regulator with the accuracy and resolution of a D/A converter.

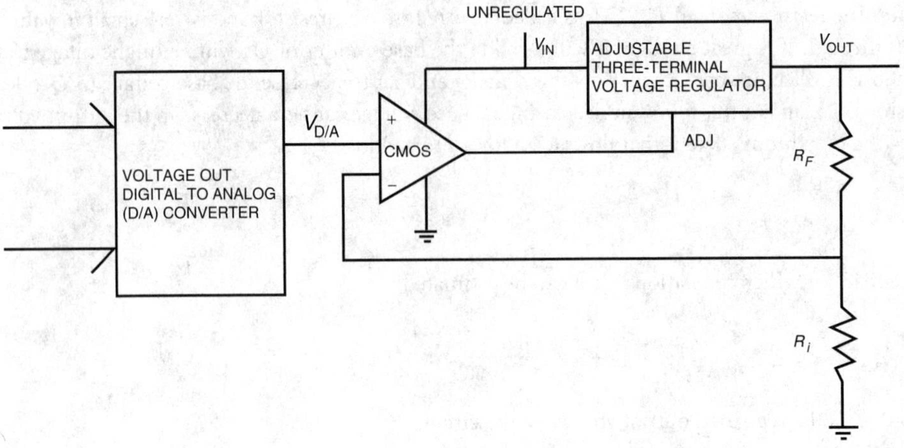

FIGURE 10.30   D/A controlled voltage regulator.

Notice the regulator in this voltage follower feedback loop. The closed-loop gain of this circuit is still given by

$$A_{\text{CL}} = 1 + \frac{R_F}{R_l}$$

thus omitting the feedback resistors R_F and R_l and feeding the regulated V_{OUT} back to the inverting input of the CMOS op amp gives $V_{\text{OUT}} = V_{\text{D/A}}$.

IC Three-Terminal Voltage Regulator Current Boosting

More current may be required of a power supply than is provided by the chosen IC regulator. Recall that the maximum power dissipated by the IC regulator (only available if appropriate heat sinking is provided) can be expressed as

$$P_{D\max} = (V_{\text{IN}} - V_{\text{REG}})I_{L\max}$$

and is called out in the device specification sheet. If a load is expected to draw close to the maximum, or if the heat sinking is in question, the regulator design should include a current boosting capability to prevent the IC device from entering thermal shutdown.

The output current capability may be increased through the use of a supplementary current boosting transistor. An example is shown in Fig. 10.31. This supplemental transistor will handle the current that the IC is not capable of passing. It is activated by the appropriate choice of a sensing resistor R_s. When the voltage across the sensing resistor is great enough to turn on the supplemental transistor, it acts to supply the additional current required of the load. This particular configuration provides output current to 5.0 A.

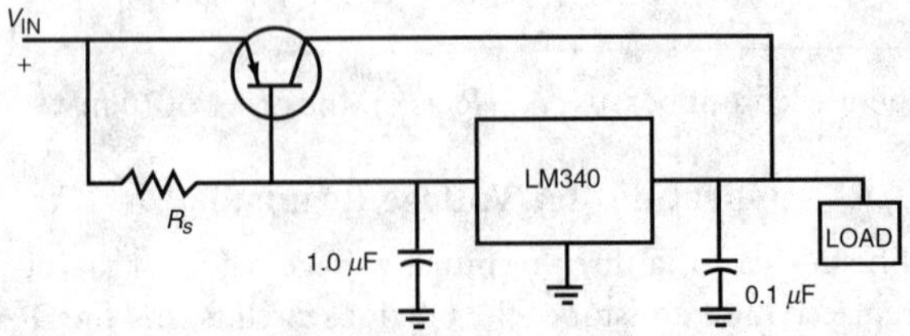

FIGURE 10.31   A three-terminal IC regulator with current boosting.

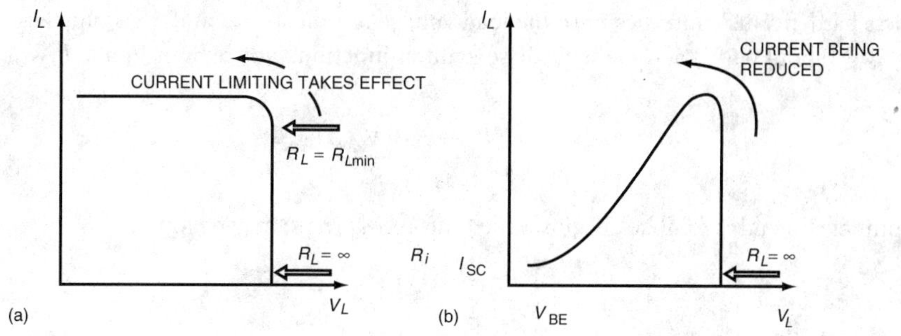

FIGURE 10.32 Current limiting schemes: (a) simple current limiting, (b) foldback current limiting.

IC Regulator Over Current Protection Schemes

Unlike discrete voltage regulators, IC voltage regulators routinely provide internal protection against overpower situations in the way of thermal shutdown. This means the device shuts off current from the internal pass transistor until it cools sufficiently and resets. If the load attempts to draw too much current, thermal shutdown would occur, the device would eventually cool, operation would resume with the uncorrected load, and the cycle would repeat.

Recall that with simple current limiting, discussed in Sec. 10.3.4, the current was held constant at some maximum value as the load resistance fell below R_{min} (the minimum value of load resistance for which regulation is provided). This is illustrated in Fig. 10.32(a). Although this offers some degree of protection for the pass transistor and diodes, it is still dissipating much power unnecessarily. This is especially undesirable because heat is usually responsible for variation and failure in voltage regulator circuits.

A preferable situation would be to design for foldback current limiting. This would prevent the high-power dissipation and cycling and simply cut back the load current and load voltage until the situation could be corrected. Figure 10.32(b) illustrates graphically the effect of foldback current limiting. Notice that foldback current limiting can actually bring the current down to a *fraction* of its maximum value.

Figure 10.33 shows a partial schematic of the 723 positive or negative adjustable voltage regulator, shaded. External circuitry to provide foldback current limiting is shown. Notice that as the load current rises it develops enough voltage across R_{SC} to turn on the current limit transistor Q_2. When Q_2 is activated it draws current from Q_1. Q_1 eventually turns off decreasing I_{LOAD} and in turn decreasing V_1 and V_{LOAD}. A difference in potential develops between the base and emitter of Q_2 making Q_2 conduct more; hence

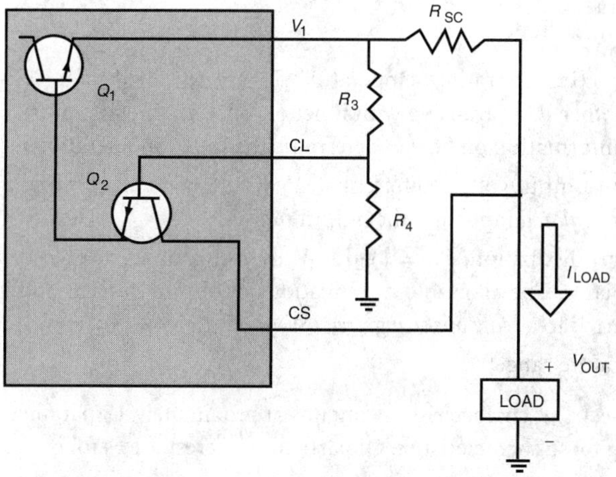

FIGURE 10.33 A 723 voltage regulator configured for foldback current limiting.

Q_1 conducts harder. This continues until the load voltage is almost zero and V_1 is only large enough to maintain V_{sense} (about 0.5 V) across the Q_2 base–emitter junction. Notice both V_1 and $I_{\text{LOAD}}(=I_{\text{SC}})$ have been lowered.

$$I_{\text{SC}} = \frac{(R_3 + R_4)(V_{\text{sense}})}{R_{\text{SC}} \times R_4}$$

and the point at which this foldback begins, called the *knee current*, may be found as

$$I_{\text{knee}} = \frac{(R_3 + R_4)V_{R4}}{R_{\text{SC}} \times R_4} - \frac{V_{\text{LOAD}}}{R_{\text{SC}}}$$

10.3.6 Reading the Voltage Regulator Specification Sheet

This section will be discussed in terms of an example, the LM317. First, it is important to realize that the industry part number or alphanumeric descriptor of the IC regulator is not unique to the manufacturer. The LM317, being a popular three-terminal adjustable regulator is found in several manufacturer's data books. It certainly could be the case, however, that a particular regulator is supplied by only one manufacturer. Some manufacturers provide a cross reference guide in the data book to aid in substitution or replacement decisions.

The regulator specification sheet starts with a general description of the part. Here the designer can quickly determine the applicability of the part to a particular need. Looking, for example, at the Motorola linear/interface ICs databook the following information in the general description of the LM317 can be obtained:

- The only available package is T-suffix Case 221A, a plastic body case with a tab mount.
- The upper limit on output current I_{OUT} is in excess of 1.5 A (assuming proper heat sinking; this is an implicit assumption throughout the specification sheets).
- Additional features are present such as internal thermal overload protection and the fact that short circuit current limiting is constant with temperature
- The adjustable range on output voltage is 1.2–37 V.
- Usually a standard application circuit is provided. This circuit shows exactly what resistors and capacitors are needed to use the regulator in its most basic configuration. This would be the circuit of choice if the part met the requirements of the designer without any kind of special features such as output current boosting or foldback current limiting. Frequently, these variations are detailed in a section on applications.

The maximum ratings include the following:

- Extremes in junction operating temperature T_J are specified as 0–125°C. (This specification sheet shows another IC is available with junction operating temperature range −40–125°C.) See Sec. 10.3.7 for information on how to determine the junction operating temperature.
- Maximum input-output voltage differential, $V_I - V_O$, is the maximum instantaneous difference between the *unregulated* input and the output pin,
- Maximum power dissipation, $P_D = (V_{\text{IN}} - V_{\text{OUT}})I_{\text{OUT}}$ cannot be exceeded without risking damage to the regulator. The allowable combinations of output current and voltage differentials are sometimes referred to as *safe operating area* (SOA).
- Storage temperature range.

Finally, a table of electrical characteristics contains approximately 15 parameters and lists underlying assumptions regarding these characteristics. Of particular interest are as follows:

- Percent line and load regulation, which indicate how well the regulator does its job given changes in line voltage and load current, respectively.

- Minimum and maximum load current. Note that maximum output current may never be achieved if proper heat sinking is not provided.
- Ripple rejection in decibels. This specification is for the line regulation, indicating the regulator's attenuation of the line voltage ripple. It is predicated on particular AC input signal levels and frequencies with a specified value of bypass capacitor.
- Thermal resistance for junction case θ_{JC} and thermal resistance for junction to ambient, θ_{JA}. (This assumes no heat sink is used). These two values are important for determining heat-sink requirements.

10.3.7 Voltage Regulator Packaging and Heat Sinking

Voltage regulators dissipate a significant amount of power as heat. This heat, generated at the IC junction, can, if not carried away, impair performance or destroy the device. (Remember performance specifications assume the use of proper heat sinking.) Data sheets indicates the maximum safe junction temperature (T_J) and the user must verify that it is not exceeded. The following relationships is used for this verification:

$$T_J = T_A + P_D\theta_{JA} \tag{10.66}$$

where

T_J = junction temperature, °C
T_A = ambient air temperature in close proximity to case, °C
P_D = power dissipated by device, W
θ_{JA} = thermal resistance from junction to ambient air, °C/W (see Fig. 10.34)

Note: if no heatsink is used this will come from the specification sheet; otherwise, it is determined as

$$\theta_{JA} = \theta_{JC} + \theta_{CS} + \theta_{SA} \tag{10.67}$$

where

θ_{JC} = thermal resistance from junction to case, °C/W
θ_{CS} = thermal resistance from case to heat sink, °C/W
θ_{SA} = thermal resistance from heat sink to ambient, °C/W

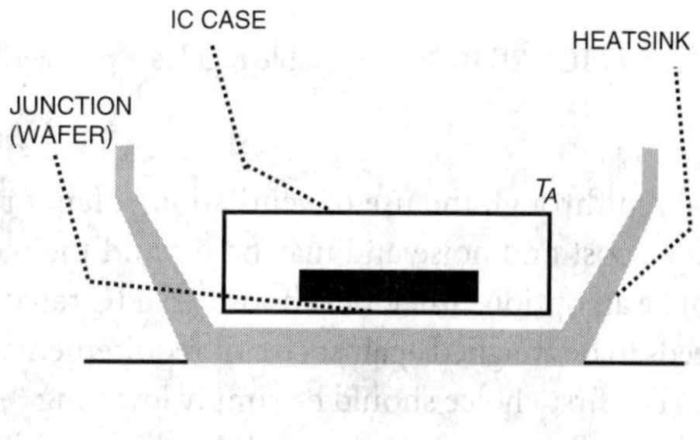

The device data sheets provide information regarding the following parameters: case options, see Fig. 10.35 for case types; maximum operating junction temperature $T_{J(max)}$ (°C); thermal resistance junction to ambient (no heat sink) θ_{JA}; and thermal resistance junction to case θ_{JC}.

To quickly check the junction temperature [Eq. (10.66)] use the θ_{JA} value from the data sheet and the calculated power dissipated for the circuit specific regulator application. (Note: the power calculation may be a simple matter, as in the case of a

FIGURE 10.34 Contributors to thermal resistance.

three-terminal voltage regulator [$(V_{IN} - V_{OUT}) \times I_{OUT}$]; or it may be more difficult, as in the case of a pulsed input or highly reactive loads. This is considered to be beyond the scope of this work.) If the maximum ambient temperature is not known, it is best to measure it. If that is not possible, an estimate may be made considering the operating environment. In open air, 50°C is not a bad number; but in a closed encasement with other electronics, 60°C may be more reasonable. Automotive and some plant environments are more severe (70–75°C).

If the calculated junction temperature exceeds the $T_{J(max)}$ of the data sheet, notice from Eq. (10.66) that it may be lowered by lowering the ambient temperature (T_A), lowering the power dissipated by the IC, or lowering the junction to ambient thermal resistance (θ_{JA}). Although a lower ambient temperature can be

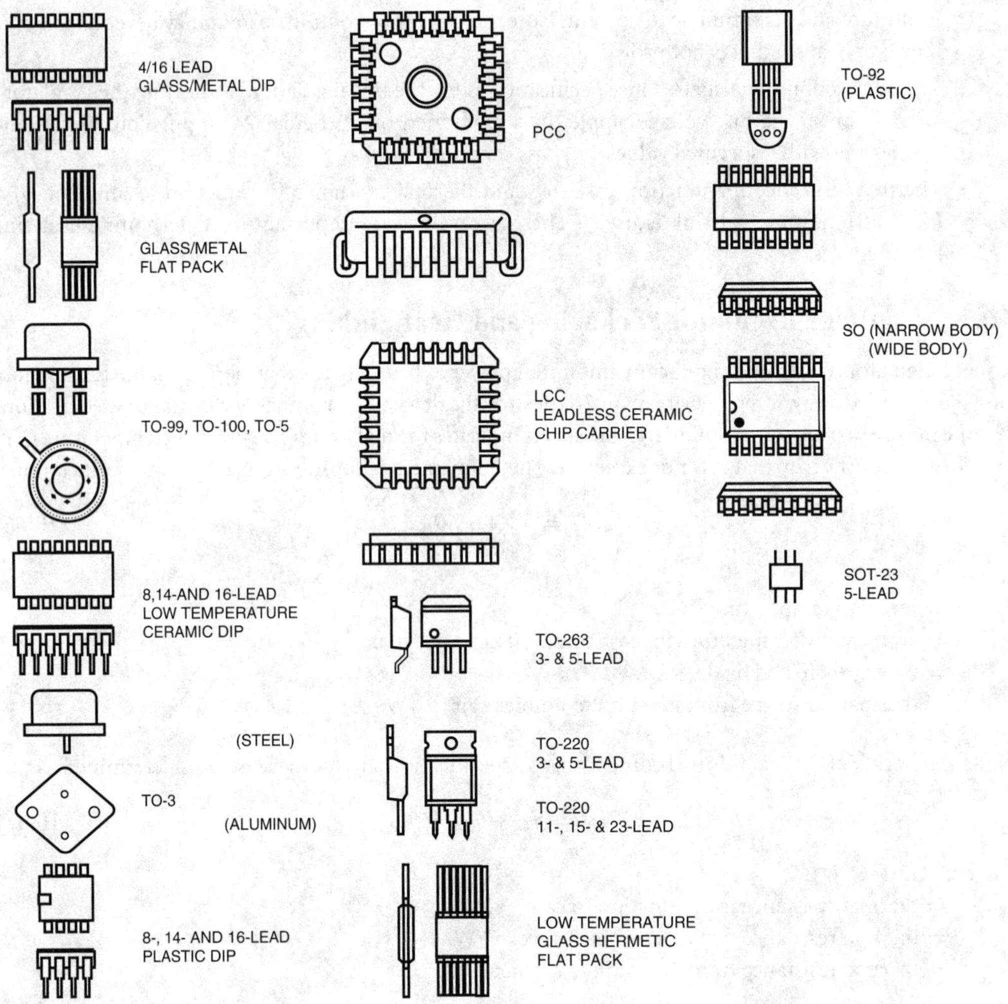

FIGURE 10.35   Available regulator package styles. (*Source:* Courtesy of National Semiconductor.)

realized through the use of ventilation, a fan, or other cooling techniques, these come at the expense of added cost and noise and may be beyond the scope of the designer. Lowering the operating power may not be an option either, but choosing an IC rated for a higher current may offer a lower θ_{JA}. This decision needs to be weighed against circuit requirements and the short-circuit protection offered by a lower rated IC. The first choice should be simply lowering θ_{JA} through use of a heat sink.

Once the need for a heat sink has been established, the designer can use Eqs. (10.66) and (10.67) to determine the maximum value of θ_{SA}. Equation (10.67) indicates that θ_{JA} is the sum of three thermal resistances: θ_{JC}, θ_{CS}, and θ_{SA}. The first, θ_{JC}, is provided in the device data sheet and estimates of θ_{CS} are provided in Table 10.2. Using these two values one may determine the necessary value of θ_{SA} and use this to select one of the commercially available heat sinks available. A sampling of θ_{SA} value is provided in Table 10.3.

A partial summary of heat sink θ_{SA} values is provided for a variety of manufacturers in Table 10.3.

10.3.8   Troubleshooting

Table 10.4 is a checklist that has been included courtesy of Motorola (1993).

TABLE 10.2 Case to Heat Sink Thermal Resistance, θ_{CS}, for Various Packages and Mounting Arrangements

	θ_{CS}			
	Metal-to-Metal[a]		Using an Insulator[a]	
Case	Dry, °C/W	With Heat-Sink Compound, °C/W	With Heat-Sink Compound, °C/W	Type
TO-240	0.5	0.1	0.36	3-mil mica
			0.28	anodized aluminum
TO-220	1.2	1.0	1.6	2-mil mica

[a]Typical values; heat-sink surface should be free of oxidation, paint, and anodization. (*Source:* Courtesy of Motorola.)

TABLE 10.3 Commercial Heat Sink Selection Guide

θ_{SA}, °C/W[a]	Manufacturer/Series or Part Number TO-204AA (TO-3)
0.3–1.0	Thermalloy—6441, 6443, 6450, 6470, 6560, 6660, 6690
1.0–3.0	Wakefield—641
	Thermalloy—6123, 6135, 6169, 6306, 6401, 6403, 6421, 6423, 6427, 6442, 6463, 6500
3.0–5.0	Wakefield—621, 623
	Thermalloy—6606, 6129, 6141, 6303
	IERC—HP
	Staver—V3-5-2
5.0–7.0	Wakefield—690
	Thermalloy—6002, 6003, 6004, 6005, 6052, 6053, 6054, 6176, 6301
	IERC—LB
	Staver—V3-5-2
7.0–10	Wakefield—672
	Thermalloy—6001, 6016, 6051, 6601
	IERC—LAμP
	Staver—V1-3, V1-5, V3-3, V3-5, V3-7
10–25	Thermalloy—6013, 6014, 6015, 6103, 6104, 6105, 6117
	TO-204AA (TO-5)
12–20	Wakefield—260
	Thermalloy—1101, 1103
20–30	Wakefield—209
	Thermalloy—1116, 1121, 1123, 1130, 1131, 1132, 2227, 3005
	IERC—LP
	Staver—F5-5
30–50	Wakefield—207
	Thermalloy—2212, 2215, 225, 2228, 2259, 2263, 2264
	Staver—F5-5, F6-5
	Wakefield—204, 205, 208
	Thermalloy—1115, 1129, 2205, 2207, 2209, 2210, 2211, 2226, 2230, 2257, 2260, 2262
	Staver—F1-5, F5-5
5.0–10	IERC—H P3 Series
	Staver—V3-7-225, V3-7-96
10–15	Thermalloy—6030, 6032, 6034
	Staver—V4-3-192, V-5-1
15–20	Thermalloy—6106
	TO-204AB
	Staver—V4-38, V6-2
20–30	Wakefield—295
	Thermalloy—6025, 6107
	TO-226AA (TO-92)

(*continued*)

TABLE 10.3 (*Continued*)

θ_{SA}, °C/W[a]	Manufacturer/Series or Part Number TO-204AA (TO-3)
46	Staver—F5-7A, F5-8
50	IERC—AUR
57	Staver—F5-7D
65	IERC—RU
72	Staver—F1-8, F2-7
80–90	Wakefield—292
85	Thermalloy—2224
20	Thermalloy—6007
30	Thermalloy—6010
32	Thermalloy—6011
34	Thermalloy—6012
45	IERC—LIC
60	Wakefield—650, 651

[a]All values are typical as given by the manufacturer or as determined from characteristic curves supplied by the manufacturer.

Staver Co., Inc., 41-51 N. Saxon Ave., Bay Shore, NY 11706.

IERC, 135 W. Mangolia Blvd., Burbank, CA 91502.

Thermalloy, P.O. Box 34829, 2021 W. Valley View Lane, Dallas, TX.

Wakefield Engin Ind: Wakefield, MA 01880. (*Source:* Courtesy of Motorola.)

TABLE 10.4 Voltage Regulator Troubleshooting Guidelines

Symptoms	Design Area to Check
Regulator oscillates:	Layout
	Compensation capacitor too small
	Input leads not bypassed
	External pass element parasitically oscillating
Loss of regulation at light loads:	Emitter–base resistor in PNP type boost configuration too large
	Absence of 1.0-mA minimum load
	Improper circuit configuration
Loss of regulation at heavy loads:	Input voltage or voltage differential too low
	External pass element gain too low
	Current limit too low
	Line resistance between sense points and load
	Inadequate heat sinking
IC regulator or pass element fails after warm-up or at high T_A:	Inadequate heat sinking
	Input voltage transient $V_{in(max)}$, V_{CEO}
Pass element fails during short circuit:	Insufficient pass element ratings SOA, $I_{C(max)}$
	Inadequate heat sinking
IC regulator fails during short circuit:	IC current or SOA capability exceeded
	Inadequate heat sinking
IC regulator fails during power-up:	Input voltage transient $V_{in(max)}$
	IC current or SOA capability exceeded as load (capacitor) is charged up
IC regulator fails during power-down:	Regulator reverse bias
Output voltage does not come up during power-up or after short circuit:	Out polarity reversal
	Load has latched up in some manner (usually seen with op amps, current sources, etc.)
Excessive 60 or 120 Hz output ripple:	Input supply filter capacitor ground loop

Source: Motorola. *Linear Interface ICs Device Data*, Vol. 1, 1993, pp. 3–609.

Defining Terms

Current limiting: A technique used to provide short-circuit protection. Typically a current sensing resistor is used to turn a transistor on and off that in turn is used to provide negative feedback to hold the output current at a high but safe value.

Darlington pair: A configuration of two transistors, the emitter of one feeding the base of the other, used to increase gain.

Dropout voltage: The input-output voltage differential at which the circuit ceases to regulate against further reductions in input voltage.

Foldback current limiting: A more sophisticated type of short-circuit protection in which the output current is cut dramatically back, not just held constant.

Line regulation: The percent change of output voltage for a change in input supply voltage.

Linear regulator: Voltage regulator using a transistor operating in the linear region.

Load regulation: The percent change of output voltage for a change in output current.

Low dropout voltage: Refers to a class of voltage regulators designed to operate at especially low input-output voltage differentials.

Pass transistor: A transistor used to boost the load current, so-called because all of the load current passes through it.

Quiescent current: That portion of the current not being supplied to the load.

Ripple: The fluctuation on the DC input signal to the voltage regulator, caused by the charging and discharging of the capacitor input filter.

Ripple rejection: A measure of how well the regulator attenuates the input ripple, usually specified in decibels.

Switching regulator: A voltage that operates by using a saturating switching transistor to chop an unregulated DC voltage. This chopped voltage is then filtered. The duty cycle of the switching transistor is varied to provide regulation of the output voltage.

Voltage regulator: That part of a power supply that accepts a filtered DC voltage and reduces, if not eliminates, the ripple, providing a continuous, smooth DC signal.

References

Jacob, J.M. 1982. *Applications and Design with Analog Integrated Circuits*, 2nd ed., pp. 80, 103–130. Regents/Prentice-Hall, Englewood Cliffs, NJ.

Malvino, A.P. 1993. *Electronic Principles*, pp. 842–879. Glencoe, Div. of Macmillan/McGraw-Hill, Westerville, OH.

Motorola. 1993. *Linear Interface ICs Device Data*, Vol. 1, pp. 3–609. Motorola Inc.

National Semiconductor. 1995. *National Power IC's Databook*, 1995 ed., pp. xxvii, 2–177.

Wurzburg, H.N.D. *Voltage Regulator Handbook Theory and Practice*. Motorola Inc.

Further Information

Data books provide the best source of information regarding voltage regulators and their use. The *Motorola Linear Interface ICs Device Data* Vol. 1 Addendum to Sec. 10.3.3, Power Supply Circuits is an excellent source of background information.

On-line information sources include:

The National Semiconductor BBS @ 408/739-1162 and
Circuit Cellar INK BBS @ 203/871-1988.

The trade journal *Power Conversion and Intelligent Motion (PCIM)* is another source of relevant information regarding power supplies.

10.4   Switching Power Supplies

Robert J. Hofinger

10.4.1   Introduction

Efficient conversion of electrical power is becoming a primary concern to electrical product manufacturers and to users alike. Switching supplies not only offer much higher efficiencies to the user but also greater flexibility to the product manufacturer. Recent advances in semiconductor, magnetic, and passive component technologies are making the switching power supply a popular choice in the power conversion arena today.

10.4.2   Linear Versus Switching Power Supplies

Historically, the linear **regulator** was the primary method of creating a fixed, regulated output voltage. It operated on the principal of inserting a power regulating transistor between the input voltage and the output voltage. By comparing the output voltage against a fixed reference voltage, a signal could be produced that would be proportional to the difference between these two inputs and that could then be used to control the power-regulating transistor. The power regulating transistor operated in a continuous conduction mode, dissipating large amounts of power at high-load currents, especially when the input-output voltage difference was large.

The power regulating transistor in a switching power supply operates in either the cutoff or saturated mode of operation. The switching regulator regulates by varying the on-off duty cycle of the power regulating transistor. Therefore, there is relatively little power dissipation in it. Also, because the switching regulator operates at a high frequency (much higher than the line frequency), the filtering elements used in the power supply can be made small, lightweight, low in cost, and very efficient.

Since the input-output voltage difference has very little effect on its operation and efficiency, very poorly filtered DC input voltage sources can be used, thereby eliminating the large and expensive line-frequency filtering elements.

Common Topologies

A **topology** is the arrangement of the power semiconductors, their magnetic storage elements, and the filtering elements used to make up a power supply. Each topology has its own merits and deficiencies. Each application requires that the designer review the needs and peculiarities of the system and choose the topology that most closely meets those requirements. Some of the factors that determine the suitability of a particular topology to those requirements are as follows:

1. Electrical isolation between the input and the output.
2. The ratio of the output voltage to the input voltage.
3. The magnitude of the voltages that are placed across the power semiconductors.
4. The magnitude of the currents that go through the power semiconductors.
5. A single output vs. multiple outputs.

Electrically isolated regulators are generally used in applications where the input comes from the AC mains, whereas nonisolated regulators are typically used for board-level regulation. The ratio of the output voltage to the input voltage determines whether one is to use a step-up or a step-down approach. There is even a topology that reverses the polarity of the output voltage. Multiple output voltages, which come from the addition of a transformer, can minimize the control circuitry. The control, or regulation of the power supply, however, can be related to only one output; therefore, the cross-regulation requirements should be reviewed to ensure the feasibility of this approach.

The input voltage variations and the output power requirements have an effect on the stresses that the regulator topology places on the power semiconductors. Minimizing these system stresses ensures a long product life (high Mean Time Between Failures—MTBF). At reduced DC input voltages and high output power, for example, large peak currents are experienced with a resultant high level of electromagnetic interference (EMI) radiation.

A few common topologies are as follows:

- Flyback
- Discontinuous
- Continuous
- Push-pull
- Half-bridge
- Full-bridge

Regulation

In the **pulse width modulated (PWM)** switching power supply, the output voltage is regulated by the ratio of either t_{on} to t_{off} or t_{on} to T, the switching period. The ratio of t_{on} to t_{off} can be varied in two ways. The switching period, $T = t_{on} + t_{off}$, may be kept constant and both t_{on} and t_{off} may be varied, or t_{on} may be kept constant and the switching period T varied in order to achieve the desired t_{off}. A constant switching period or frequency is generally preferred. Output ripple is generally easier to filter if the regulator is operated at a fixed frequency. Also, the noise generated by the switching operation is easier to tolerate if it is synchronous with some fundamental system frequency.

10.4.3   The Buck Regulator

The most elementary step-down regulator is the *buck regulator* whose topology can be seen in Fig. 10.36.

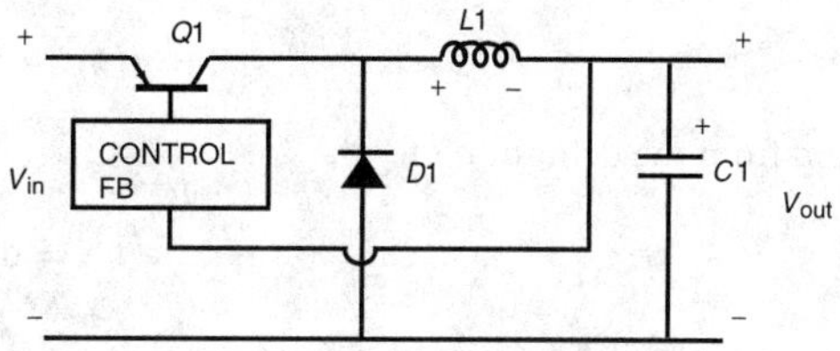

The buck regulator's operation can be seen as having two distinct time periods, which occur when the power switch $Q1$ is closed or is open. When the power switch is closed, referred to as the *on time* (denoted by t_{on}), the input voltage V_{in} is connected by the power switch $Q1$ to the left-hand or input side of the inductor. The output voltage V_{out}, which is held at a relatively constant voltage by the presence of the output capacitor $C1$, is connected to the right-hand or output side of inductor $L1$. Since the input side of the inductor is connected to V_{in}, the diode $D1$ is back biased. During the t_{on} period, there is a constant voltage connected across the inductor; hence, the inductor current ramps upward in a linear fashion, and energy is stored within the magnetic core of the inductor, see Fig. 10.37. The change in inductor current during this time is described by the relationship

$$\Delta i_{L1(on)} = ((V_{in} - V_{out}) * t_{on})/L1 \qquad (10.68)$$

During the time when the power switch $Q1$ is open, referred to as the *off time* (denoted as t_{off}), the energy stored within the inductor carries the load's energy requirements, and the current through the inductor starts to fall toward zero. Because of the negative

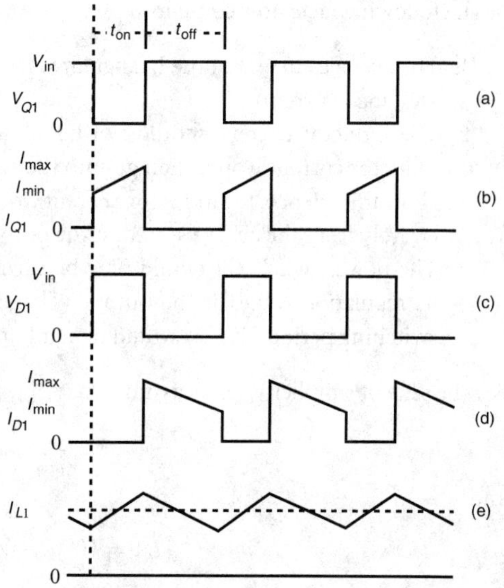

FIGURE 10.36   The buck regulator topology.

FIGURE 10.37   DC waveforms for a buck regulator.

slope of the current, the voltage across the inductor reverses polarity; hence, the input side of the inductor is clamped to one diode drop below ground by the diode $D1$, which is commonly called the *catch* or *free wheeling* diode. In this analysis, we will neglect the diode drop, that is, set it equal to zero. The inductor current now flows through the diode loop, thus maintaining its continuity and removing the stored energy from the inductor. The change in inductor current during this time is described by the relationship

$$\Delta i_{L1(\text{off})} = ((V_{\text{out}} - V_{\text{diode}}) * t_{\text{off}})/L1 \tag{10.69}$$

The off period ends when the power switch is turned on at a predetermined time, that is, a fixed frequency, or when the output voltage drops below a predefined design value and the power switch is turned on by the feedback control circuitry.

Regulation is accomplished by varying the on-to-off duty cycle of the power switch in order to keep $\Delta i_{L1(\text{on})} = \Delta i_{L1(\text{off})}$ (see Fig. 10.38). Combining the two previous equations, and disregarding the diode drop that becomes negligible, the relationship that describes the regulator's operation is

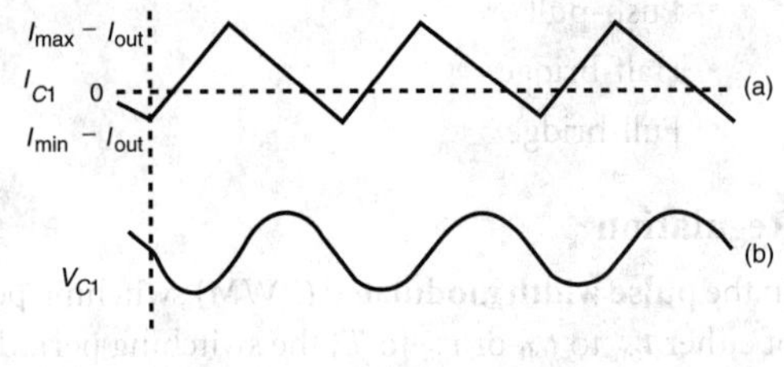

FIGURE 10.38   AC waveforms for a buck regulator.

$$V_{\text{out}} = V_{\text{in}} * t_{\text{on}}/(t_{\text{on}} + t_{\text{off}}) \tag{10.70}$$

Inductor $L1$

From Eq. (10.68)

$$L1 = ((V_{\text{in}} - V_{\text{out}}) * t_{\text{on}})/\Delta i_{L1(\text{on})} \tag{10.71}$$

and from the definition of

$$t_{\text{on}} = \text{duty cycle} * T$$

where the switching period $T = (t_{\text{on}} + t_{\text{off}})$

$$L1 = ((V_{\text{in}} - V_{\text{out}})/\Delta i_{L1(\text{on})}) * \text{duty cycle} * T \tag{10.72}$$

The following facts now come into play:

1. The average current of the triangular waveshape of the inductor current is equal to the average (or DC) load current.
2. The inductor current should not be allowed to fall to a zero value. If this were to happen, the inductor current would then go into the negative region. The forward biased diode $D1$ would not allow this. Hence, the inductor current would be held at zero. The load current would have to come entirely from the charge stored in the output capacitor $C1$ and the output voltage would drop.
3. The power switch $Q1$ would now be turned back on, due to the output voltage dropping below its regulation value, or the output voltage would go out of regulation until the end of the fixed switching period T. This would depend on the type of switching system used.

For the sake of simplicity, let us assume that $i_{(L1)}$ will not get to zero. Then

$$\Delta i_{L1(\text{on})} = 2 * i_{(\text{min})} \tag{10.73}$$

and

$$L1 = ((V_{\text{in}} - V_{\text{out}})/(2 * i_{(\text{min})})) * t_{\text{on}} \tag{10.74a}$$

or

$$L1 = ((V_{\text{in}} - V_{\text{out}})/(2 * i_{(\text{min})})) * \text{duty cycle} * T \tag{10.74b}$$

At maximum load current, the peak current should be limited so that the inductor will not saturate. A rule of thumb is to limit the peak inductor current to 1.25 times the maximum output load current. Then

$$\Delta i_{L1(on)} = 0.5\, i_{(max)} \tag{10.75}$$

and

$$L1 = ((V_{in} - V_{out})/(0.5 * i_{(max)})) * t_{on} \tag{10.76a}$$

or

$$L1 = 2 * ((V_{in} - V_{out})/i_{(max)}) * \text{duty cycle} * T \tag{10.76b}$$

This may take a few iterations to arrive at a comfortable compromise between $i_{(min)}$, $i_{(max)}$, duty cycle (or t_{on}), and the operating period T.

Capacitor $C1$

After the inductor $L1$ has been selected and the operating parameters have been decided, in accordance with Eq. (10.74) and Eq. (10.76), the next design is the selection of capacitor $C1$.

$C1$ is chosen to provide a maximum ripple voltage at the output node at the maximum load current. This ripple voltage is calculated from the average current per cycle that enters and leaves the capacitor. From

$$v_{(out)max} = \Delta Q / C1$$

where ΔQ is the area under the capacitor current triangle

$$\Delta Q = 1/2 * T/2 * \Delta i_{L1(on)}/2 \tag{10.77}$$

$$\Delta Q = (\Delta i_{L1(on)}/8) * T \tag{10.77a}$$

or where $T = 1/f$,

$$\Delta Q = \Delta i_{L1(on)}/(8 * f) \tag{10.77b}$$

therefore,

$$C1 = \Delta i_{L1(on)}/(8 * \Delta v_{(out)max} * f) \tag{10.78}$$

from Eq. (10.73)

$$C1_{min} = i_{min}/(4 * \Delta v_{(out)max} * f) \tag{10.79}$$

10.4.4   The Boost Regulator

The most elementary stepup regulator is the *boost regulator* whose topology can be seen in Fig. 10.39.

The boost regulator's operation can also be broken down into two distinct periods, which occur when the power switch $Q1$ is closed or is open. When the power switch $Q1$ is closed, which is referred to as the on time (denoted by t_{on}), the input voltage V_{in} is placed directly across inductor $L1$. This causes the current to ramp upward in a linear

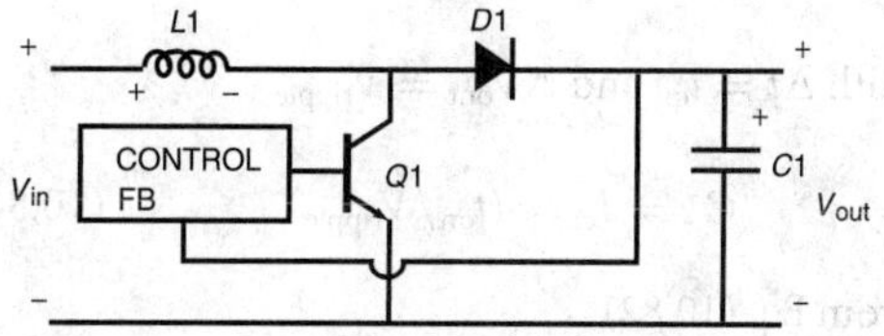

FIGURE 10.39   The boost regulator topology.

fashion from zero to some peak value and have energy stored within the magnetic core of the inductor proportional to this peak value, see Fig. 10.40. Since the right-hand end of the inductor and the anode of the diode $D1$ are effectively connected to ground, the diode is back biased, and no load current passes through the inductor during this period. The change in inductor current during this time is described by the relationship

$$\Delta i_{L1(on)} = (V_{in} * t_{on})/L1 \qquad (10.80)$$

When the power switch $Q1$ opens, which is referred to as the off time (denoted by t_{off}), the inductor voltage reverses polarity, and the output side (the right-hand end) flies back above the input voltage and is clamped by diode $D1$ at the output voltage. The current then begins to linearly ramp downward until the energy within the core is completely depl-

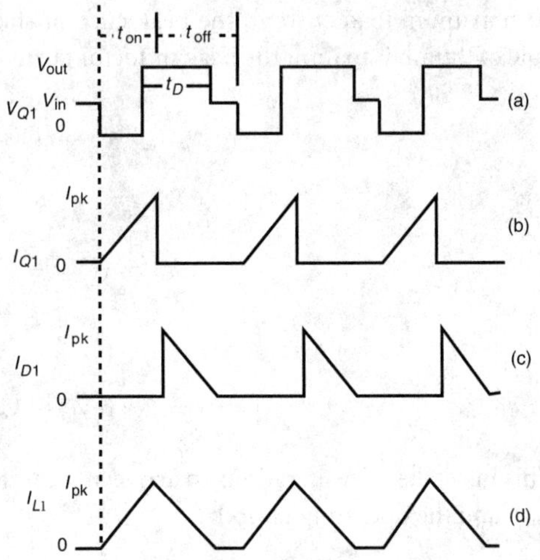

FIGURE 10.40   DC waveforms for a boost regulator.

eted. This is shown as time t_D. Hence, the output voltage V_{out} must be greater than the input voltage. The current waveform during the downward ramping period t_D, which is shown in Fig. 10.40 is described as

$$\Delta i_{L1(off)} = ((V_{out} - V_{in}) * t_D)/L1 \qquad (10.81)$$

Regulation is again accomplished by varying the on-to-off duty cycle of the power switch. The relationship that approximately describes its operation is

$$V_{out} = V_{in} * (1 + t_{on}/t_D) \qquad (10.82)$$

Note that during the on period, the energy required by the load is being supplied by the capacitor $C1$. Therefore, the amount of energy stored in the inductor $L1$ during the on period must be sufficient not only to carry the load during the off period (denoted by t_{off}), but to recharge the capacitor as well.

As with the buck regulator, the output voltage is varied by varying the ratio of the t_{on}/t_D times. Both t_{on} and t_D may be varied, as long as the sum of the t_{on} and t_D times is less than the switching period. This keeps the boost regulator in what is called the *discontinuous mode of operation*. The switching frequency is then constant, which as discussed earlier, is preferred from a system filtering standpoint.

Capacitor $C1$

During the time that the power switch $Q1$ is closed, the on time, the output capacitor $C1$ must supply the full load current (Fig. 10.41). While the capacitor is supplying the load current, its voltage will be decreasing at the rate of

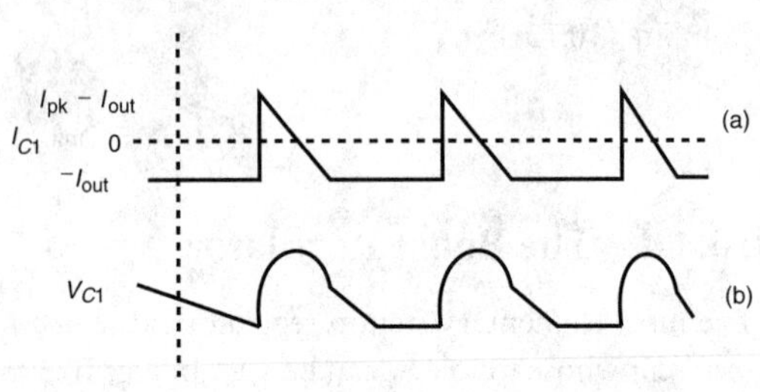

$$\Delta V_{out}/\Delta t = I_{out}/C1 \qquad (10.83)$$

with $\Delta t = t_{on}$ and $\Delta V_{out} = V_{ripple}$

$$C1 = I_{out} * (t_{on}/V_{ripple}) \qquad (10.84)$$

from Eq. (10.82)

FIGURE 10.41   AC waveforms for a boost regulator.

$$t_{off} = T * (V_{in}/V_{out})$$

and defining the switching period

$$T = t_{\text{on}} + t_{\text{off}} = 1/f$$

therefore

$$t_{\text{on}} = (V_{\text{out}} - V_{\text{in}})/(f * V_{\text{out}}) \tag{10.85}$$

substituting into Eq. (10.84), we get

$$C1 = I_{\text{out}} * ((V_{\text{out}} - V_{\text{in}})/(f * V_{\text{out}} * V_{\text{ripple}})) \tag{10.86}$$

Inductor $L1$

For the moment assume the system to be lossless; therefore

$$P_{\text{in}} = P_{\text{out}}$$

$$V_{\text{in}} * I_{\text{in(avg)}} = V_{\text{out}} * I_{\text{out(avg)}} \tag{10.87}$$

$$I_{\text{in(avg)}} = (V_{\text{out}}/V_{\text{in}}) * I_{\text{out(avg)}}$$

from Eq. (10.82)

$$(V_{\text{out}}/V_{\text{in}}) = 1 + t_{\text{on}}/t_{\text{off}} = T/t_{\text{off}} \tag{10.88}$$

therefore

$$I_{\text{in(avg)}} = I_{\text{out(avg)}} * T/t_{\text{off}} \tag{10.89}$$

The input current waveform (t_{on}) is a triangular waveform with the $I_{\text{in(avg)}} = 1/2\, I_{\text{pk}}$ since

$$V_{\text{in}} = L\, \Delta I_L/\Delta t \quad \text{or} \quad L = V_{\text{in}} \Delta t/\Delta I_L$$

To keep the I_L waveform continuous

$$I_{\text{in(avg)min}} = 1/2\, I_{\text{pk(min)}} \quad \text{or} \quad \Delta I_{L1} = 2 * I_{\text{in(avg)min}} \tag{10.90}$$

therefore

$$L1 = V_{\text{in}} * t_{\text{on}}/(2 * I_{\text{in(avg)min}}) \tag{10.91}$$

from Eq. (10.85)

$$t_{\text{on}} = (V_{\text{out}} - V_{\text{in}})/(f * V_{\text{out}})$$

therefore

$$L1 = 1/2 * (V_{\text{in}}/V_{\text{out}})^2 * ((V_{\text{out}} - V_{\text{in}})/(I_{\text{out}} * f)) \tag{10.92}$$

10.4.5 The Buck/Boost Regulator

The *buck/boost regulator* shown in the topology in Fig. 10.42 accepts a DC voltage input and provides a regulated output voltage of opposite polarity.

The output voltage of this regulator is the reverse polarity of the input voltage. When the power switch Q1 is closed, which is referred to as the on time (denoted by t_{on}), the input voltage V_{in} is connected across the inductor $L1$, and thus the current

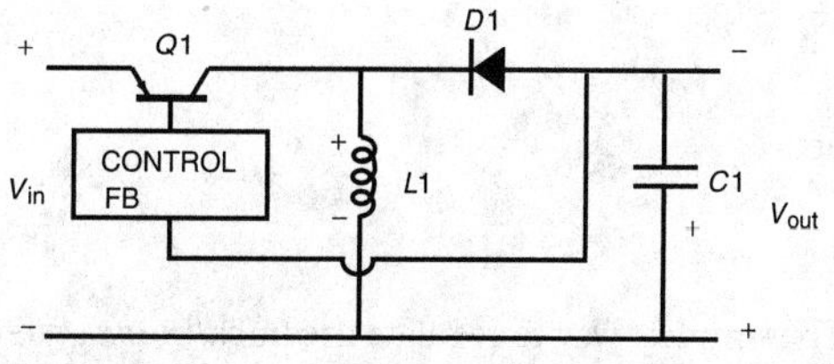

FIGURE 10.42 The buck/boost regulator topology.

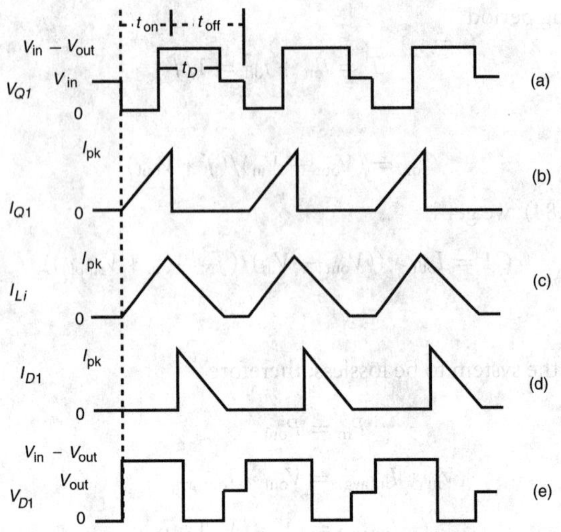

FIGURE 10.43   DC waveforms for a buck/boost regulator.

through the inductor ramps upward in a linear fashion, and energy is stored within the magnetic core of the inductor, see Fig. 10.43. The change in inductor current during this time is described by the relationship

$$\Delta I_{L1(on)} = (V_{in} * t_{on})/L1 \qquad (10.93)$$

The power switch $Q1$ is then opened, which is referred to as the off time (denoted by t_{off}) and the inductor voltage flies back above the input voltage and is clamped by the diode $D1$ at the output voltage V_{out}. The current then begins to linearly ramp downward for a time t_D, the time that the diode $D1$ conducts, until the energy within the magnetic core is completely depleted. Its waveform is shown in Fig. 10.43. The change in inductor current during this time is described by the relationship

$$\Delta I_{L1(off)} = (V_{out} * t_D)/L1 \qquad (10.94)$$

The relationship between the output voltage V_{out} and the input voltage V_{in} can be obtained by combining Eq. (10.93) and (10.94)

$$V_{out} = V_{in} * (t_{on}/t_D) \qquad (10.95)$$

Since the DC output current I_{out} is equal to the average current through the diode $D1$

$$I_{out} = I_{L1(pk)}/2 * (t_{on} + t_D)/T \qquad (10.96)$$

therefore

$$I_{L1(pk)} = 2 * I_{out} * T/(t_{on} + t_D)$$

From Eq. (10.96), we see that when I_{out} is a maximum, t_D is equal to t_{off}, or

$$T = t_{on} + t_D \qquad (10.97)$$

hence

$$I_{L1(pk)} = 2 * I_{out(max)} \qquad (10.98)$$

The simplest way to regulate the buck/boost system is to establish a fixed peak current $I_{L1(pk)}$, determined by $I_{out(max)}$, through the inductor, thereby allowing the on time t_{on} to vary inversely with changes in the input voltage and the sum of $t_{on} + t_D$ to vary directly with changes in the output current (see Fig. 10.44).

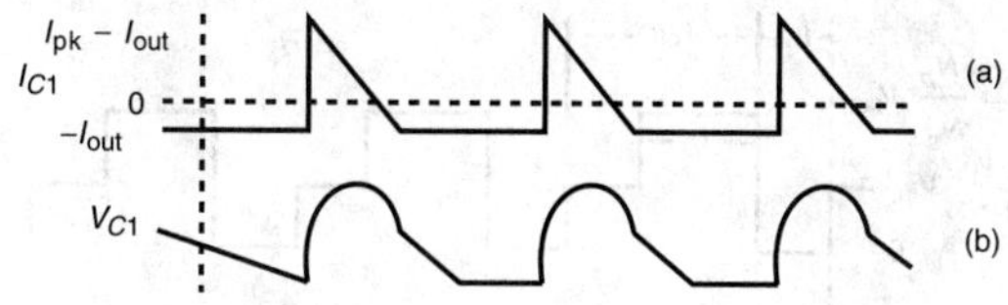

FIGURE 10.44 AC waveforms for a buck/boost regulator.

Inductor $L1$

From Eqs. (10.93), (10.94) and the relationship

$$\Delta I_{L1(on)} = I_{L1(pk)} = 2 * I_{out(max)} \tag{10.99}$$

when

$$t_{on} + t_D = T = 1/f$$
$$L1 = (V_{in} * V_{out})/(2 * I_{out(max)} * f) \tag{10.100}$$

Capacitor $C1$

During the time that the diode is not conducting, the output current has to come from the capacitor $C1$. From

$$\Delta v_{(out)max} = 1/C1 * I * \Delta t$$

where

$$\Delta t = T - t_D$$
$$I = I_{out(max)} \tag{10.101}$$
$$\Delta v_{(out)max} = V_{ripple}$$

$$V_{ripple} = 1/C1 * I_{out(max)} * (T - t_D)$$

or

$$C1 = [I_{out(max)} * (T - t_D)]/V_{ripple} \tag{10.102}$$

10.4.6 The Flyback Regulator

The topology of the *flyback regulator* can be seen in Fig. 10.45.

The operation of the flyback regulator, also called the *transformer coupled boost regulator*, can also be broken down into two distinct periods, where the power switch $Q1$ is closed or open. When the power switch is closed, which is referred to as the on time (denoted by t_{on}) the input voltage source is placed directly across the primary of the switching transformer. This causes the primary current to linearly ramp upward from zero and is described by

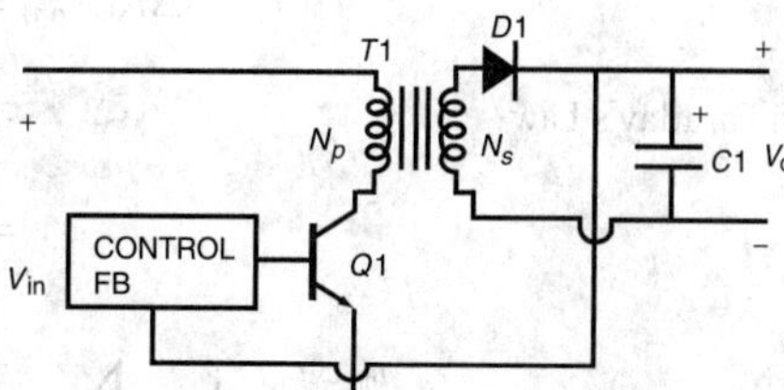

FIGURE 10.45 The flyback regulator topology.

$$\Delta i_{PRI(on)} = (V_{in} * t_{on})/L_{PRI} \tag{10.103}$$

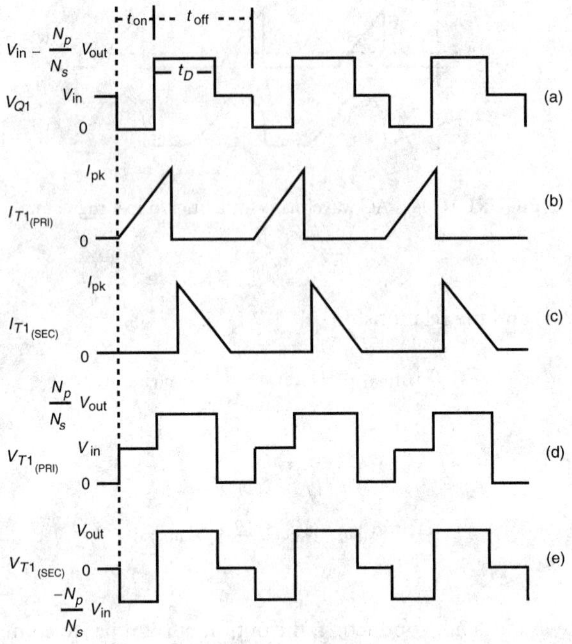

FIGURE 10.46   DC waveforms for a flyback regulator.

Because the polarity of the output voltage of the secondary winding of the switching transformer reverse biases the diode $D1$, no secondary current can flow. (On transformers are marked with a dot; the winding ends that are similarly marked are all positive at the same time, i.e., of the same polarity.) Therefore, all the energy during the on time is stored within the transformer primary inductance. The amount of energy stored during the t_{on} period is denoted by

$$W = 1/2 * L_{PRI} * I^2_{PRI(pk)} \tag{10.104}$$

The power switch $Q1$ then turns off and the secondary of the switching transformer flies back above the output voltage and is clamped by diode $D1$ at the output voltage. The energy stored within the transformer is then transferred to the secondary circuit. The secondary current then begins to linearly ramp downward until the energy within the transformer core is completely depleted. This is shown as t_D in Fig. 10.46 (see also Fig. 10.47) and is described as

$$\Delta i_{SEC(on)} = (V_{out} * t_D)/L_{SEC} \tag{10.105}$$

From Faraday's Law

$$v = N \, \Delta\phi/\Delta t$$

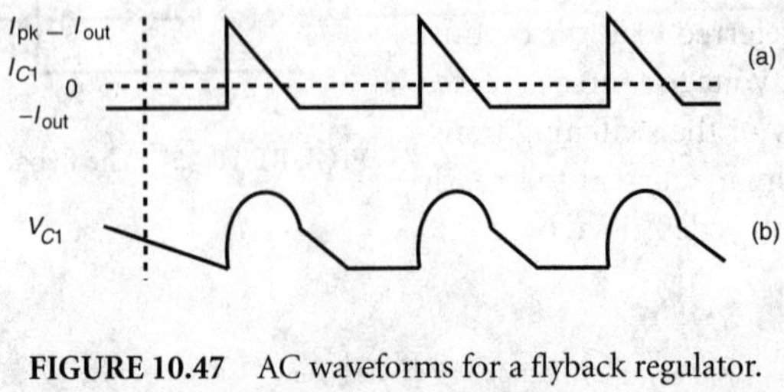

FIGURE 10.47   AC waveforms for a flyback regulator.

When $Q1$ is closed, t_{on}, the transformer flux builds up an amount $\Delta\phi = (V_{in} * t_{on})/N_P$. Now when $Q1$ is opened, the voltage across the winding reverses and causes $\Delta\phi/\Delta t$ to reverse. If the desired output voltage across N_S is V_{out}, then the transformer flux decreases an amount $\Delta\phi = (V_{out} * t_D)/N_S$.

In the steady state, the decrease in flux $\Delta\phi$ when $Q1$ is open must equal the increase in flux when $Q1$ is closed, for otherwise there would be a net change in flux per cycle and the flux would continue to build up to saturation. Hence

$$(V_{in} * t_{on})/N_P = (V_{out} * t_D)/N_S \tag{10.106}$$

$$V_{out} = V_{in} * (N_S/N_P) * (t_{on}/t_D) \tag{10.107}$$

Regulation is accomplished by varying the on-to-off duty cycle of the power switch. Care must be taken so that the voltage across the transistor $Q1$ when it is turned off does not exceed its maximum voltage rating. The maximum voltage that $Q1$ is subjected to is the input DC voltage plus the voltage coupled back from the secondary winding into the primary winding by the transformer turns ratio. The maximum voltage is

$$
\begin{aligned}
V_{Q1(max)} &= V_{in} + V_{out} * (N_P/N_S) \\
&= V_{in} + V_{in} * [(N_S/N_P) * (t_{on}/t_D) * (N_P/N_S)] \\
V_{Q1(max)} &= V_{in}(1 + (t_{on}/t_D))
\end{aligned}
\tag{10.108}
$$

Transformer $T1$

The switching transformer for the flyback regulator is in reality two inductors linked together by a common magnetic core. Therefore, in selecting the transformer, the dominating parameters are the primary and secondary inductances.

For the primary inductance, from Eq. (10.104)

$$W = 1/2 * L_{PRI} * I_{PRI(pk)}^2$$

since energy is equal to power × time

$$W = P_{out} * T = V_{out} * I_{out} * T$$

where the switching time

$$T = 1/f$$

therefore

$$L_{PRI} = (2 * V_{out} * I_{out})\big/\left(f * I_{PRI(pk)}^2\right) \tag{10.109}$$

and

$$L_{SEC} = (2 * V_{out} * I_{out})\big/\left(f * I_{SEC(pk)}^2\right) \tag{10.110}$$

Capacitor $C1$

The output capacitor $C1$ can be treated in a similar manner as a capacitive-input type filter. $C1$ is chosen so that it is large enough to supply the full load current for the entire half-cycle until the time when the half-wave rectified output rises high enough to forward bias diode $D1$. The secondary current then provides the current to the load as well as replenishing the charge lost by capacitor $C1$ when it supplied the full load current.

During the portion of the cycle when $C1$ is supplying load current, that current is considered to be constant. For a constant current out of a capacitor, its rate of change of voltage is given by the expression

$$\Delta V/\Delta T = I/C1 \tag{10.111}$$

since $\Delta T = T - t_{on}$

$$C1 = I * (T - t_{on})/\Delta V \qquad (10.112)$$

10.4.7 The Half-Forward Regulator

The topology for the elementary *half-forward regu-lator* can be seen in Fig. 10.48.

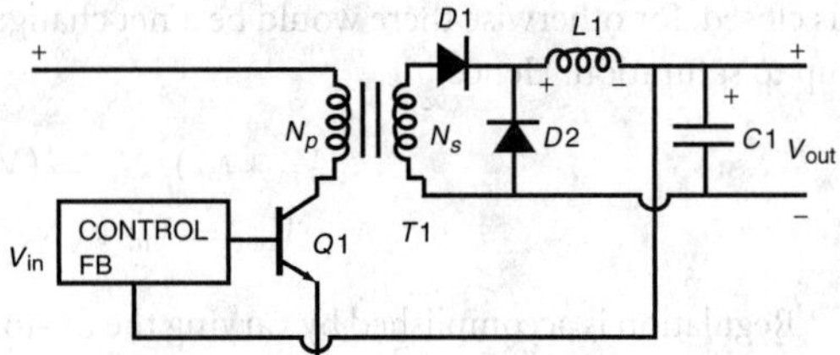

When the power switch $Q1$ is closed, which is referred to as the on time (denoted by t_{on}), the volt-age across the primary of transformer $T1$ is V_{in}, and thus the current through the transformer ramps up linearly. Its waveform during the on time, which is shown in Fig. 10.49(b) is described as

FIGURE 10.48 The half-forward regulator topology.

$$i_{PRI(on)} = ((V_{in}/L_{PRI}) * t) + I_{PRI(min)} \qquad (10.113)$$

During the on time, the secondary current is also ramping up linearly as

$$i_{SEC(on)} = (N_p/N_s) * i_{PRI(on)}$$

At the end of the on time, when the power switch $Q1$ opens, the transformer primary and the secondary currents have reached their maximums. The current through the inductor $L1$ is the same as the transformer secondary current, and energy is thus stored within the magnetic core of the inductor, see Fig. 10.49(c) and Fig. 10.50. The change in inductor current during this time is described by the relationship

$$\Delta i_{L1(on)} = ((V_{SEC} - V_{out}) * t_{on})/L1 \qquad (10.114)$$

where

$$V_{SEC} = (N_s/N_p) * V_{in}$$

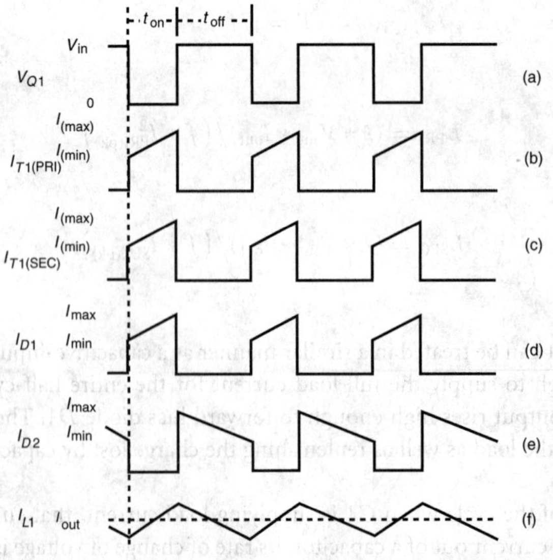

FIGURE 10.49 DC waveforms for a half-forward regulator.

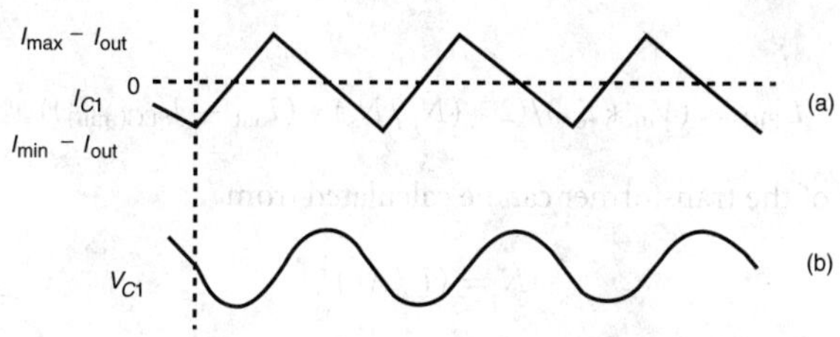

FIGURE 10.50 AC waveforms for a half-forward regulator.

hence

$$\Delta i_{L1(\text{on})} = (\{[(N_s/N_p) * V_{\text{in}}] - V_{\text{out}}\} * t_{\text{on}})/L1 \qquad (10.115)$$

During the time when the power switch $Q1$ is open, which is referred to as the off time (denoted as t_{off}), the energy stored within the inductor carries the load's energy requirements, and the current through the inductor starts to fall toward zero. Because of the negative slope of the current, the voltage across the inductor reverses polarity; hence, the input side of the inductor is clamped to one diode drop below ground by the diode $D2$, which is commonly called the catch or free-wheeling diode. The inductor current now flows through the diode loop, thus maintaining its continuity and removing the stored energy from the inductor. The change in inductor current during this time is described by the relationship

$$\Delta i_{L1(\text{off})} = (V_{\text{out}} * t_{\text{off}})/L1 \qquad (10.116)$$

The off period ends when the power switch is turned on at a predetermined time, that is, a fixed frequency, or when the output voltage drops below a predefined design value and the power switch is turned on by the feedback control circuitry.

Regulation is accomplished by varying the on-to-off duty cycle of the power switch in order to keep $\Delta i_{L1(\text{on})} = \Delta i_{L1(\text{off})}$. Combining the two previous equations, the relationship that describes the regulator's operation is

$$V_{\text{out}} = [(N_s/N_p) * V_{\text{in}}] * t_{\text{on}}/(t_{\text{on}} + t_{\text{off}}) \qquad (10.117)$$

Transformer $T1$

The switching transformer for the half-forward regulator transfers current, hence, power, in a linear fashion when the power switch is closed. Therefore, in selecting the transformer, the dominating parameters are again the primary and secondary inductances. During t_{on}

$$V_{\text{SEC}} = L_{\text{SEC}} * (\Delta I/\Delta T) \qquad \text{or} \qquad L_{\text{SEC}} = V_{\text{SEC}} * (\Delta T/\Delta I) \qquad (10.118)$$

where

$$\Delta T = t_{\text{on}}$$
$$\Delta I = I_{\text{SEC(max)}} - I_{\text{SEC(min)}}$$
$$V_{\text{SEC}} = N_s/N_p * V_{\text{in}}$$

since

$$I_{\text{out}} = (I_{\text{SEC(max)}} + I_{\text{SEC(min)}})/2$$
$$L_{\text{SEC}} = ((N_s/N_p) * V_{\text{in}} * t_{\text{on}})/(2 * (I_{\text{out}} - I_{\text{SEC(min)}})) \qquad (10.119)$$

in a similar fashion

$$L_{PRI} = (V_{in} * t_{on})/(2 * (N_p/N_s) * (I_{out} - I_{SEC(min)}))$$ (10.120)

The number of turns of the transformer can be calculated from

$$N = (L/A_L)^{\frac{1}{2}}$$ (10.121)

where

A_L = inductance index in milli-henry/turns^2
L = winding inductance
N = number of turns

Inductor $L1$

The power inductor $L1$ is chosen in the same manner as is done for the buck regulator, see Sec. 10.4.3. In this case, however, V_{in} is the peak secondary voltage V_{sec}. Since

$$V_{sec} = N_s/N_p * V_{in}$$
$$L1 = 2 * \{((N_s/N_p * V_{in}) - V_{out})/i_{(max)}\} * \text{duty cycle} * T$$ (10.122)

Capacitor $C1$

The output capacitor is determined in the same manner as is done for the buck regulator, see Sec. 10.4.3. From Eq. (10.79)

$$C1_{min} = i_{min}/(4 * \Delta v_{(out)max} * f)$$ (10.123)

10.4.8   The Push–Pull Regulator

The topology for the elementary *push-pull regulator* can be seen in Fig. 10.51.

In the operation of the push-pull regulator each power switch, $Q1$ and $Q2$, operate in the same way but 180° out of phase with each other. Each power switch has its own t_{on} and t_{off} time, see Fig. 10.52. During the time when either power switch is closed, the ramping transformer primary current and the resultant secondary current cause energy to be stored in inductor $L1$. There is also a dead time when both power switches are open. It is during this time that the energy stored in inductor $L1$ is moved to the output load and the current through the inductor starts to fall toward zero.

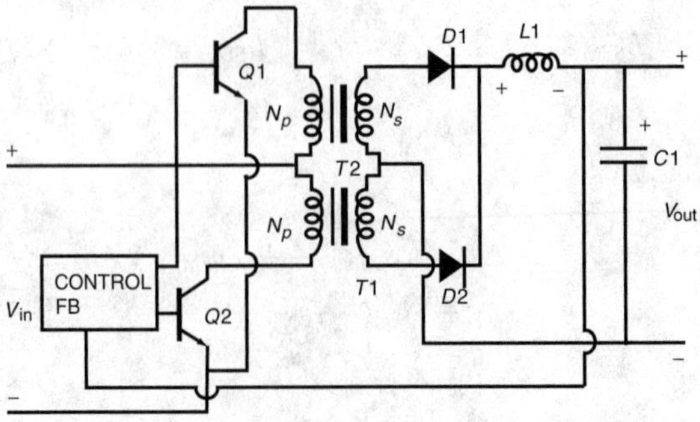

FIGURE 10.51   The push-pull regulator topology.

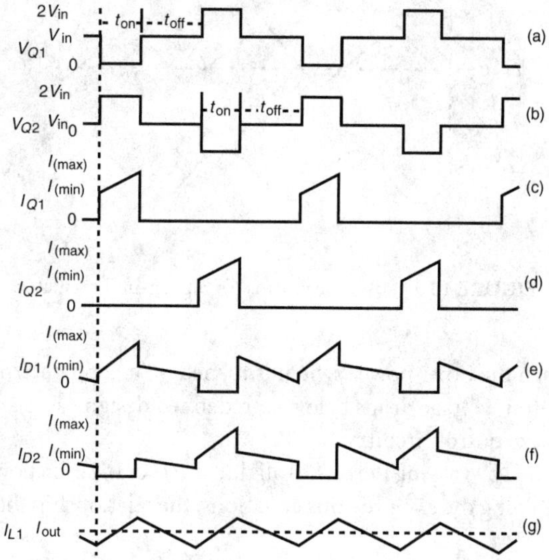

FIGURE 10.52 DC waveforms for a push-pull regulator.

Each half of the push-pull regulator thus resembles the half-forward regulator. The waveform of the primary current during the on time, which is shown in Fig. 10.52(c) and Fig. 10.52(d) is described as

$$i_{\text{PRI(on)}} = ((V_{\text{in}}/L_{\text{PRI}}) * t) + I_{\text{PRI(min)}} \tag{10.124}$$

During the on time, the secondary current is also ramping up linearly as

$$i_{\text{SEC(on)}} = (N_p/N_s) * i_{\text{PRI(on)}}$$

At the end of the on time, when the power switch opens, the transformer primary and secondary currents have reached their maximums. The current through the inductor $L1$ is the same as the transformer secondary current, and energy is thus stored within the magnetic core of the inductor, see Fig. 10.52(c) and Fig. 10.53. The change in inductor current during this time is described by the relationship

$$\Delta i_{L1(\text{on})} = ((V_{\text{SEC}} - V_{\text{out}}) * t_{\text{on}})/L1 \tag{10.125}$$

where

$$V_{\text{SEC}} = (N_s/N_p) * V_{\text{IN}}$$

hence

$$\Delta i_{L1(\text{on})} = (\{((N_s/N_p) * V_{\text{in}}) - V_{\text{out}}\} * t_{\text{on}})/L1 \tag{10.126}$$

During the time when both power switches are open, which is referred to as the off time (denoted as t_{off}), the energy stored within the inductor carries the load's energy requirements, and the current through the inductor starts to fall toward zero. Because of the negative slope of the current, the voltage across the inductor reverses polarity, and the input side of the inductor is clamped by its associated diode to one diode drop below ground. The inductor current now flows through the diode loop, thus maintaining its continuity and removing the stored energy from the inductor. The change in inductor current during this time is described by the relationship

$$\Delta i_{L1(\text{off})} = (V_{\text{out}} * t_{\text{off}})/L1 \tag{10.127}$$

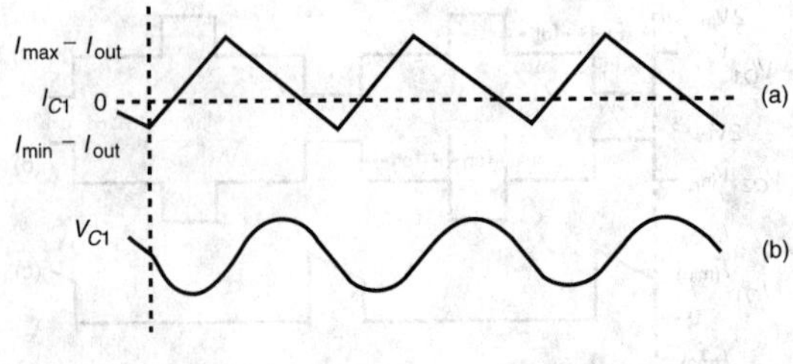

FIGURE 10.53   AC waveforms for a push-pull regulator.

The off period ends when the other power switch is turned on at a predetermined time, that is, a fixed frequency, or when the output voltage drops below a predefined design value and the other power switch is turned on by the feedback control circuitry.

Regulation is accomplished by varying the on-to-off duty cycle of the each power switch in order to keep $\Delta i_{L1(on)} = \Delta i_{L1(off)}$. Combining the two previous equations, the relationship that describes the regulator's operation is

$$V_{out} = ((N_s/N_p) * V_{in}) * t_{on}/(t_{on} + t_{off}) \tag{10.128}$$

where

$$t_{on} + t_{off} = \frac{1}{2} T$$

Note that in this configuration, each section of the push-pull regulator switches at twice the control switching frequency.

Transformer $T1$

The switching transformer for the push-pull regulator is similar in operation to that of the half-forward transformer, see Sec. 10.4.7. In this case, however, each half of the transformer carries the forward current only during its one-half cycle. From Eq. (10.119) and Eq. (10.120)

$$L_{SEC} = ((N_s/N_p) * V_{in} * t_{on})/(2 * (I_{out} - I_{SEC(min)})) \tag{10.129}$$

and

$$L_{PRI} = (V_{in} * t_{on})/(2 * (N_p/N_s) * (I_{out} - I_{SEC(min)})) \tag{10.130}$$

for each half of the center-tapped primary and secondary windings.

The number of turns of the transformer can be calculated from

$$N = (L/A_L)^{\frac{1}{2}} * 2 \tag{10.131}$$

where
A_L = inductance index in milli-henry/turns^2
L = winding inductance
N = number of turns (center tapped)

Inductor $L1$

The power inductor $L1$ is chosen in the same manner as is done for the buck regulator, see Sec. 10.4.3. In this case, however, V_{in} is the peak secondary voltage V_{sec}. Since

$$V_{sec} = N_s/N_p * V_{in}$$

$$L1 = 2 * \{[(N_s/N_p * V_{in}) - V_{out}]/i_{(max)}\} * \text{duty cycle} * T \tag{10.132}$$

Capacitor $C1$

The output capacitor is determined in the same manner as is done for the buck regulator, see Sec. 10.4.3. From Eq. (10.79)

$$C1_{min} = i_{min}/(4 * \Delta v_{(out)max} * f) \tag{10.133}$$

10.4.9 The Full-Bridge Regulator

The topology for the elementary *full-bridge regulator* can be seen in Fig. 10.54.

A disadvantage of the push-pull regulator of Fig. 10.51 is that when a power switch is not on, it must be able to withstand $2V_{in}$, that is, twice the DC input voltage. The full-bridge regulator of Fig. 10.54 reduces that requirement to V_{in}.

Each power switch combination, that is, Q1–Q3 and Q2–Q4, is on for a maximum of half the switching frequency. Hence, each power switch combination has its own t_{on} and t_{off} time, see Fig. 10.55. During the time when a power switch combination is closed, the ramping transformer current and its resultant secondary current cause energy to be stored in the inductor $L1$. The waveform of the primary current during the on time, which is shown in Fig. 10.55(c) and Fig. 10.55(d) (see also Fig. 10.56), is described as

$$i_{PRI(on)} = ((V_{in}/L_{PRI}) * t) + I_{PRI(min)} \tag{10.134}$$

During the on time, the secondary current is also ramping up linearly as

$$i_{SEC(on)} = (N_p/N_s) * i_{PRI(on)}$$

At the end of the on time, when the power switch opens, the transformer primary and the secondary currents have reached their maximums. The current through the inductor $L1$, which is the same as the transformer secondary current, starts to fall toward zero, and the energy stored in inductor $L1$ is moved to the output. The change in inductor current during this time is described by the relationship

$$\Delta i_{L1(on)} = ((V_{SEC} - V_{out}) * t_{on})/L1 \tag{10.135}$$

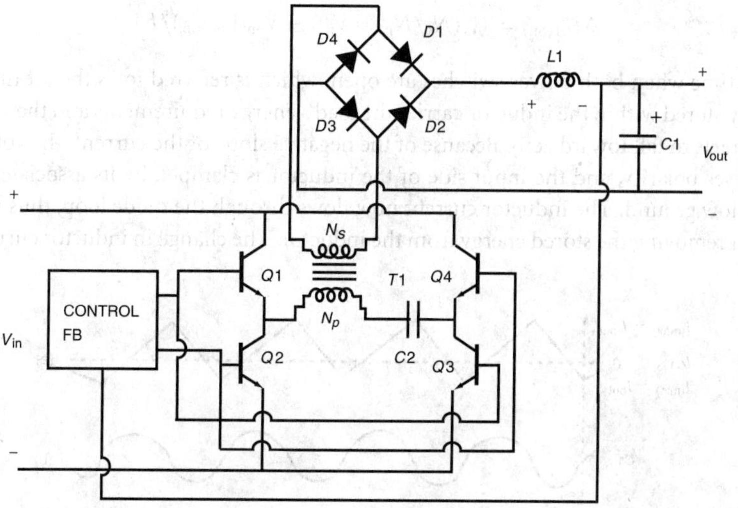

FIGURE 10.54 The full-bridge regulator topology.

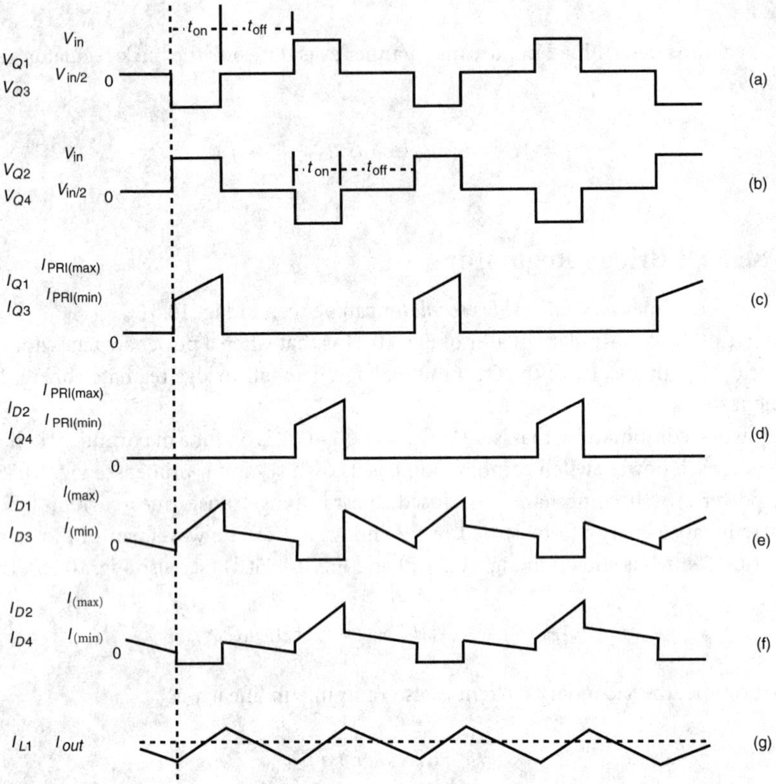

FIGURE 10.55   DC waveforms for a full-bridge regulator.

where

$$V_{\mathrm{SEC}} = (N_s/N_p) * V_{\mathrm{IN}}$$

hence,

$$\Delta i_{L1(\mathrm{on})} = (\{((N_s/N_p) * V_{\mathrm{in}}) - V_{\mathrm{out}}\} * t_{\mathrm{on}})/L1 \qquad (10.136)$$

During the time when both power switches are open, which is referred to as the off time (denoted as t_{off}), the energy stored within the inductor carries the load's energy requirements and the current through the inductor starts to fall toward zero. Because of the negative slope of the current, the voltage across the inductor reverses polarity, and the input side of the inductor is clamped by its associated diode to one diode drop below ground. The inductor current now flows through the diode loop, thus maintaining its continuity and removing the stored energy from the inductor. The change in inductor current during this

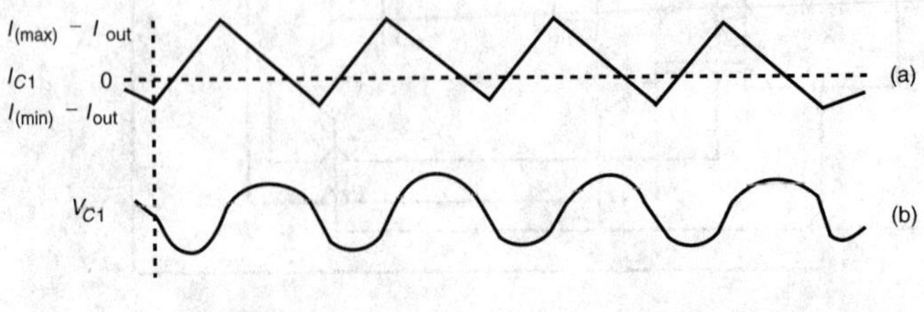

FIGURE 10.56   AC waveforms for a full-bridge regulator.

time is described by the relationship

$$\Delta i_{L1(\text{off})} = (V_{\text{out}} * t_{\text{off}})/L1 \tag{10.137}$$

The off period ends when the other power switch combination is turned on at a predetermined time, that is, a fixed frequency, or the output voltage drops below a predefined design value and the other power switch combination is turned on by the feedback control circuitry.

Regulation is accomplished by varying the on-to-off duty cycle of each power switch combination in order to keep $\Delta i_{L1(\text{on})} = \Delta i_{L1(\text{off})}$. Combining the two previous equations, the relationship that describes the regulator's operation is

$$V_{\text{out}} = ((N_s/N_p) * V_{\text{in}}) * t_{\text{on}}/(t_{\text{on}} + t_{\text{off}}) \tag{10.138}$$

where

$$t_{\text{on}} + t_{\text{off}} = \frac{1}{2}T$$

Note that in this configuration also, each section of the full-bridge regulator switches at twice the control switching frequency.

Transformer $T1$

The switching transformer for the push-pull regulator is similar in operation to that of the half-forward transformer, see Sec. 10.4.7. In this case, however, each half of the transformer carries the forward current only during its one-half cycle. From Eq. (10.119) and Eq. (10.120)

$$L_{\text{SEC}} = ((N_s/N_p) * V_{\text{in}} * t_{\text{on}})/(2 * (I_{\text{out}} - I_{\text{SEC(min)}})) \tag{10.139}$$

and

$$L_{\text{PRI}} = (V_{\text{in}} * t_{\text{on}})/(2 * (N_p/N_s) * (I_{\text{out}} - I_{\text{SEC(min)}})) \tag{10.140}$$

for each half of the center-tapped primary and secondary windings.

The number of turns of the transformer can be calculated from

$$N = (L/A_L)^{\frac{1}{2}} * 2 \tag{10.141}$$

where
  A_L = inductance index in milli-henry/turns^2
  L = winding inductance
  N = number of turns (center tapped)

Inductor $L1$

The power inductor $L1$ is chosen in the same manner as is done for the buck regulator, see Sec. 10.4.3. In this case, however, V_{in} is the peak secondary voltage V_{sec}. Since

$$V_{\text{sec}} = N_s/N_p * V_{\text{in}}$$

$$L1 = 2 * \{((N_s/N_p * V_{\text{in}}) - V_{\text{out}})/i_{(\text{max})}\} * \text{duty cycle} * T \tag{10.142}$$

Capacitor $C1$

The output capacitor is determined in the same manner as is done for the buck regulator, see Sec. 10.4.3. From Eq. (10.79)

$$C1_{\text{min}} = i_{\text{min}}/(4 * \Delta v_{(\text{out})\text{max}} * f) \tag{10.143}$$

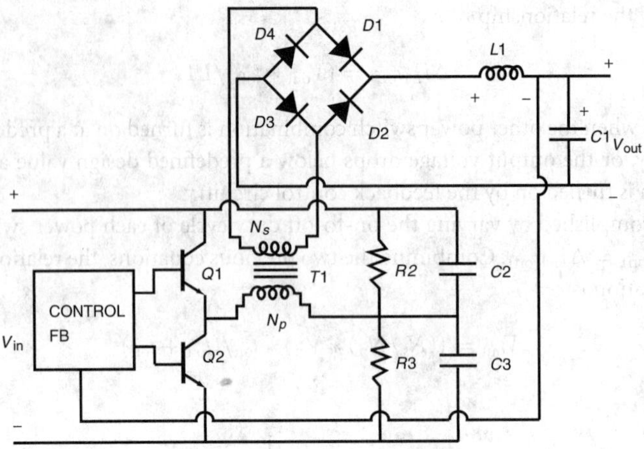

FIGURE 10.57   The half-bridge regulator topology.

10.4.10   The Half-Bridge Regulator

The topology for the elementary *half-bridge regulator* can be seen in Fig. 10.57.

The advantage of reducing the applied voltage across the invertor transistor from $2V_{in}$ to V_{in} is also offered by the half-bridge regulator of Fig. 10.58. The half-bridge regulator replaces two of the power switch transistors by two resistor/capacitor combinations that could be somewhat more economical. However, with the rapidly falling transistor prices and the fact that the two capacitors occupy a larger volume and footprint area than the two transistors, this advantage may be somewhat nullified.

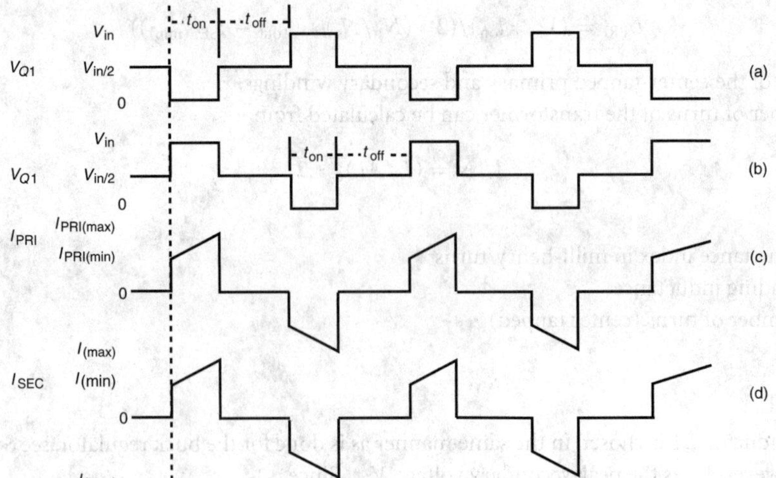

FIGURE 10.58   DC waveforms for a half-bridge regulator.

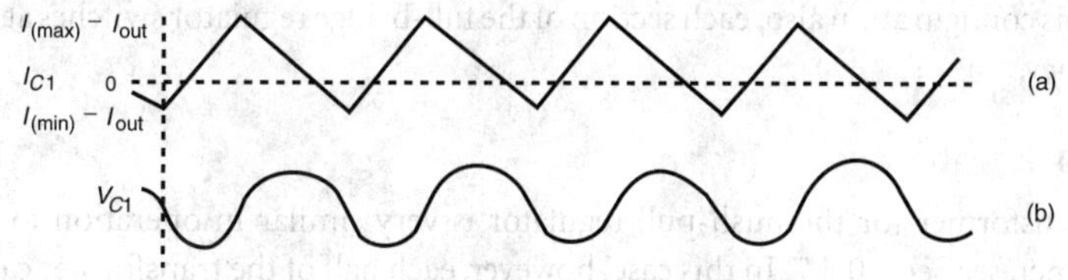

FIGURE 10.59   AC waveforms for a half-bridge regulator.

Like the full-bridge regulator, each power switch, that is, $Q1$ or $Q2$, is on for a maximum of half the switching frequency. Hence, each power switch has its own t_{on} and t_{off} time, see Fig. 10.58 and Fig. 10.59. During the time when a power switch is closed, the ramping transformer current and its resultant secondary current cause energy to be stored in inductor $L1$. The waveform of the primary current during the on time, which is shown in Fig. 10.58(c), is described as

$$i_{PRI(on)} = \{((V_{in}/2)/L_{PRI}) * t\} + I_{PRI(min)} \tag{10.144}$$

During the on time, the secondary current is also ramping up linearly as

$$i_{SEC(on)} = (N_p/N_s) * i_{PRI(on)}$$

At the end of the on time, when the power switch opens, the transformer primary and the secondary currents have reached their maximums. The current through inductor $L1$, which is the same as the transformer secondary current, starts to fall toward zero, and the energy stored in inductor $L1$ is transferred to the output. Assuming 1 V drops in the secondary circuit rectifier diodes, the change in inductor current during this time is described by the relationship

$$\Delta i_{L1(on)} = ((V_{SEC} - V_{out} - 2) * t_{on})/L1 \tag{10.145}$$

where

$$V_{SEC} = (N_s/N_p) * V_{in}$$

hence

$$\Delta i_{L1(on)} = (\{((N_s/N_p)V_{out}\} - * t_{on})/L1 \tag{10.146}$$

During the time when both power switches are open, which is referred to as the off time (denoted as t_{off}), the energy stored within the inductor carries the load's energy requirements, and the current through the inductor starts to fall toward zero. Because of the negative slope of the current, the voltage across the inductor reverses polarity; hence, the input side of the inductor is clamped to one diode drop below ground by its associated diode. The inductor current now flows through the diode loop, thus maintaining its continuity and removing the stored energy from the inductor. The change in inductor current during this time is described by the relationship

$$\Delta i_{L1(off)} = (V_{out} * t_{off})/L1 \tag{10.147}$$

The off period ends when the other power switch is turned on or the output voltage drops below a predefined design value and the other power switch is turned on by the feedback control circuitry.

Regulation is accomplished by varying the on-to-off duty cycle of the each power switch in order to keep $\Delta i_{L1(on)} = \Delta i_{L1(off)}$. Combining the two previous equations, the relationship that describes the regulator's operation is

$$V_{out} = ((N_s/N_p) * V_{in}) * t_{on}/(t_{on} + t_{off}) \tag{10.148}$$

where

$$t_{on} + t_{off} = \frac{1}{2}T$$

Note that, in this configuration also, each section of the full-bridge regulator switches at twice the control switching frequency.

Transformer $T1$

The switching transformer for the push-pull regulator is very similar in operation to that of the half-forward transformer, see Sec. 10.4.7. In this case, however, each half of the transformer carries the forward current only during its one-half cycle. From Eq. (10.119) and Eq. (10.120)

$$L_{\text{SEC}} = ((N_s/N_p) * V_{\text{in}} * t_{\text{on}})/(2 * (I_{\text{out}} - I_{\text{SEC(min)}})) \tag{10.149}$$

and

$$L_{\text{PRI}} = (V_{\text{in}} * t_{\text{on}})/(2 * (N_p/N_s) * (I_{\text{out}} - I_{\text{SEC(min)}})) \tag{10.150}$$

for each half of the center-tapped primary and secondary windings.

The number of turns of the transformer can be calculated from

$$N = (L/A_L)^{\frac{1}{2}} * 2 \tag{10.151}$$

where

A_L = Inductance index in milli-henry/turns^2
L = winding Inductance
N = number of turns (center tapped)

Inductor $L1$

The power inductor $L1$ is chosen in the same manner as is done for the buck regulator, see Sec. 10.4.3. In this case, however, V_{in} is the peak secondary voltage V_{sec}. Since

$$V_{\text{sec}} = N_s/N_p * V_{\text{in}}$$
$$L1 = 2 * \{[(N_s/N_p * V_{\text{in}}) - V_{\text{out}}]/i_{(\text{max})}\} * \text{duty cycle} * T \tag{10.152}$$

Capacitor $C1$

The output capacitor is determined in the same manner as is done for the Buck regulator, see Sec. 10.4.3. From Eq. (10.79)

$$C1_{\text{min}} = i_{\text{min}}/(4 * \Delta v_{(\text{out})\text{max}} * f) \tag{10.153}$$

Defining Terms

Flyback regulator: The technique by which the regulated load power is transferred during the *off* time of the switching device.

Forward regulator: The technique by which the regulated load power is transferred during the on time of the switching device.

Pulse width modulation (PWM): A modulation technique by which the width of a rectangular waveshape is varied in accordance with a particular error signal. Typically, the overall period is held constant, and the duty cycle is therefore varied.

Regulator: An electronic device that attempts to keep the output signal constant in spite of changes in the other system characteristics.

Topology: The overall design technique, or layout, by which an electronic device, such as a switching power supply regulator, operates.

References

Motorola. 1988. *A Designer's Guide for Switching Power Supply Circuits and Components.* Motorola Inc., SG79D Rev 5. UNITRODE Power Supply Design Seminar, UNITRODE Corp. (Call (603) 424-2410 or your local representative for specific dates and locations.)

Further Information

Motorola Corp., Technical Literature Center, P.O. Box 20912, Phoenix, AZ 85036.

PCIM Magazine, Intertec International Inc., 2472 Eastman Avenue Bldg 33, Ventura, CA 93003-5792, (805) 650–7070.
Proceedings of the International PCI Conference, Intertec International Inc., 2472 Eastman Avenue Bldg 33, Ventura, CA 93003-5792, (805) 650–7070.
Unitrode Corp., 7 Continental Blvd., Merrimack, NH 03054, (603) 424–2410.

10.5 Inverters

Ashoka K.S. Bhat

10.5.1 Introduction

The function of an **inverter** is to convert a DC input voltage to an AC output voltage. The output voltage and frequency may be fixed or variable depending on the application. Inverters find several applications, some of them are in variable speed AC motor drives, uninterruptible power supplies (UPS), mobile AC power supplies (operating on a battery), induction heating, etc. They are also used as the front stage in DC-to-DC converters.

Inverters can be broadly classified as: (1) **voltage-source inverters (VSI)**, and (2) **current-source inverters (CSI)**.

10.5.2 Voltage-Source Inverters

In the case of a VSI, input to the inverter is a constant DC voltage. This is usually obtained by connecting large capacitors at the input. There are basically three types of configurations:

- Half-bridge
- Full-bridge
- Push-pull

Half-Bridge

Figure 10.60(a) shows the half-bridge inverter circuit. A center tapped DC source is created by two large capacitors connected across the DC supply. The switch–diode pairs, $S1$, $D1$ and $S2$, $D2$, generate a square-wave voltage across the terminals A and B by connecting the sources $V1$ and $V2(V1 = V2 = V_s/2)$ alternately (Fig. 10.60(b)). The switches $S1$ and $S2$ are switching power devices (e.g., bipolar power transistors, metal oxide semiconductor field effect transistors (MOSFETs), insulated gate bipolar transistor (IGBT), silicon controlled rectifiers (SCRs), gate-turn-off thyristor (GTO)) (Fig. 10.60(d)). If SCRs are used as the switching devices, then a forced commutation circuit is needed unless natural or load commutation takes place. The diodes $D1$ and $D2$ carry the reactive current. The current flowing through the load depends on the nature of the load circuit. If the load is purely resistive (usually not possible due to wiring inductance), then the current and voltage waveforms are the same and the diodes $D1$, and $D2$ will never conduct. In the case of RLC type load (Fig. 10.60(c)), current waveshape depends on whether the load circuit is underdamped or overdamped. Sometimes, an inductor L_s and a capacitor C_s are added in series with R_L, to obtain load commutation or to shape the load current and (in turn, the switch currents)

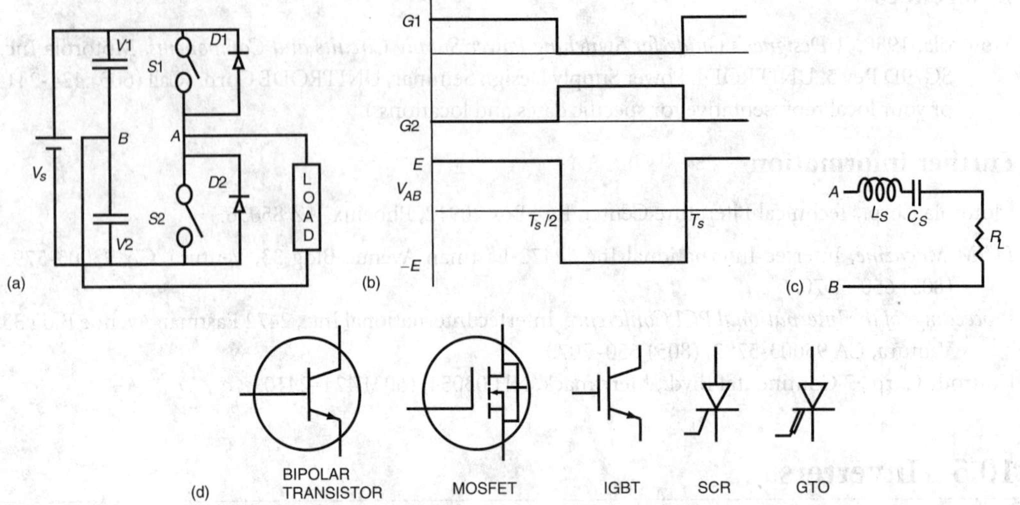

FIGURE 10.60 The half-bridge inverter configuration: (a) Basic circuit diagram, snubber circuits are not shown. $V1 = V2 = E = V_s/2$. (b) Gating pulses and the voltage v_{AB}. (c) RLC load circuit that can be connected across AB. (d) Some of the semiconductor switches that can be used in (a).

sinusoidally. An inverter which uses an RLC type load is usually referred to as a **resonant inverter** since the resonating nature of the RLC circuit is made use of in its operation. Two types of operation are possible.

Case 1: Below Resonance (Leading Power Factor (PF)) Mode of Operation

The inverter operates in this mode when the load is underdamped, and Fig. 10.61(a) shows the typical steady-state operating waveforms. The sequence of device conduction is also shown in the same figure. Assume that the switch $S1$ is carrying the load current. Since the load circuit is RLC and is underdamped, current is approximately sinusoidal. At $t = t_1$, current goes to zero naturally and tries to reverse. Switch $S1$ turns off and the diode $D1$ starts conducting, providing the path for the reverse current. At the end of the half-cycle, that is, at $t = T_s/2$ (where T_s is the period of the waveform equal to $1/f_s$ and f_s is the switching frequency), $S2$ is turned on. For an ideal switch-diode pair, $S2$ turns on immediately, applying a reverse voltage across $D1$. Therefore, $D1$ turns off, and $S2$ carries the load current. When the current

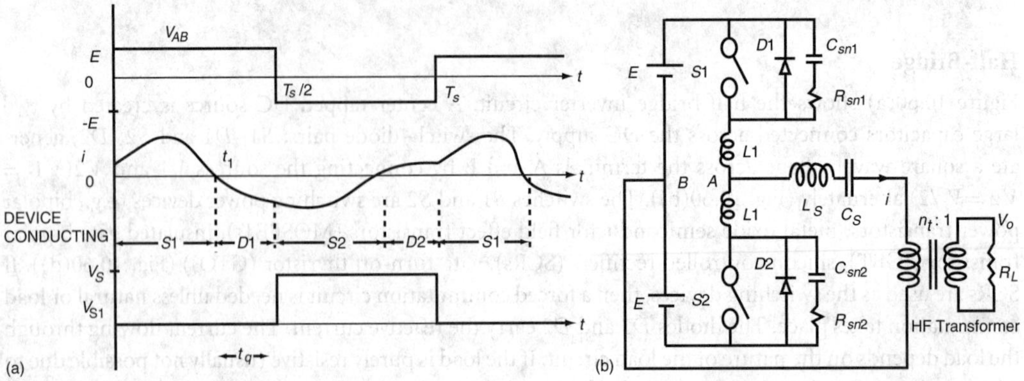

FIGURE 10.61 The half-bridge configuration: (a) operating waveforms for below resonance (or leading PF) mode of operation, (b) half-bridge configuration with the di/dt inductors ($L1, L1$) and snubber components. $C_{sn1} = C_{sn2} = C_{sn}$ and $R_{sn1} = R_{sn2} = R_{sn}$. High-frequency transformer is added for isolation and voltage scaling purpose.

through $S2$ goes to zero, $D2$ conducts and the cycle is complete when $S1$ is turned on at the beginning of the next cycle. The following important observations are to be made:

- During the conduction of $D1$ (or $D2$), a reverse voltage of a diode drop is applied across $S1$ (or $S2$). This time for which the diode conducts must be greater than the turn-off time of the switch for successful operation of the circuit.

- The switches turn off naturally, that is, load commutation takes place. This means thyristors can be employed in this type of operation provided the turn off time condition is satisfied, that is, $t_q > t_{off}$, where t_q is the diode conduction time and t_{off} is the turn off time of the thyristor used.

- Since the switches turn off with zero current, this operation is referred to as *zero current switching* (ZCS) and there are no turn off losses.

- The load current i is leading the voltage across AB, that is why this type of operation is referred to as leading PF mode of operation. In practice, usually the switching frequency is below the resonance frequency of the RLC circuit. Hence, it is also referred to as *operation below resonance*.

In the previous explanation, the turn-on time of the switch and the reverse recovery time of the diode were neglected. But in practice, this is not true, and due to the reverse recovery time of the diode, when a switch in the other arm is turned-on, the conducting diode will not turn off instantly. This means that there is a short interval during which the turned on switch and the reverse conducting diode short circuit the supply voltage. To take care of this, di/dt limiting inductors are connected in series with each switch (Fig. 10.61(b)). To reduce the size of these extra inductors and to reduce the losses, fast recovery diodes are used. If switches having slow recovery internal diodes (e.g., MOSFETs) are used, then the internal diodes must be bypassed by connecting external fast recovery diodes.

A snubber capacitor (C_{sn}) is connected across each switch in order to limit the rate of rise of voltage, dv/dt. When a switch is conducting, the capacitor across the other switch is charged to the supply voltage V_s. This capacitor gets discharged when the switch across it is turned on resulting in a large peak current through the switch. To limit this peak current, a resistor (R_{sn}) is connected in series with each snubber capacitor. These snubber resistors must be of proper power rating ($= C_{sn} V_s^2 f_s$). Losses in these snubber resistors increase if the switching frequency or the supply voltage increases.

Case 2: Above Resonance (Lagging PF) Mode of Operation

The gating signals and the steady-state waveforms for this mode of operation are shown in Fig. 10.62. This mode of operation occurs when the RLC type load is overdamped (Fig. 10.62(a)) or for an RL type load (Fig. 10.62(b)). Assume that $D1$ is conducting. The current through $D1$ goes to zero at t_1. The load current reverses after this. Since the gating signal is already applied to $S1$, current is transferred to $S1$. Because the antiparallel diode across $S1$ was conducting just prior to the turning on of $S1$, $S1$ turns on with *zero voltage switching* (ZVS). This means there are no turn on losses. The switch $S1$ is turned off at $T_s/2$ and the current through the load is transferred to $D2$. The next half-cycle is similar to the first half-cycle. When $S1$ or $S2$ is turned off, during the turn off time of the switch, snubber capacitor across the switch starts charging, whereas the snubber capacitor across the incoming diode starts discharging from the supply voltage. Since C_{sn1} and C_{sn2} are in series and directly across the supply, the sum of the voltages across these two capacitors must be equal to the supply voltage at any time. Therefore, when the voltage across the turned-off switch reaches V_s, the diode across the discharged capacitor starts conducting. The charging and discharging current of the capacitors flow through the load circuit. By properly choosing these snubber capacitors, the turn off losses can be minimized.

The following observations can be made:

- Lossy snubbers and di/dt inductors are not required, that is, only a capacitive snubber is used across the switch.

- Since there is enough time for the diodes to recover, internal diodes of MOSFETs, IGBTs and bipolar transistors can be used.

- Since the switches are turned off forcibly, a forced commutation circuit is required if SCRs are used as the switches.

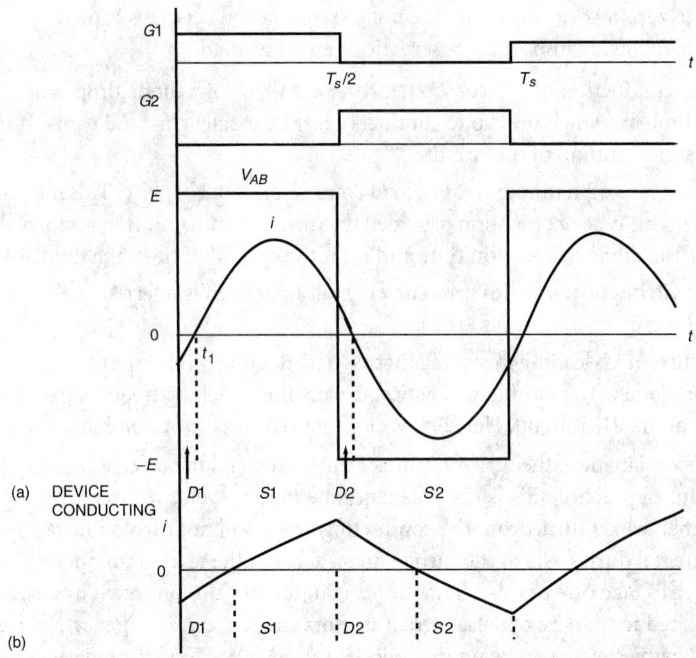

FIGURE 10.62 Operating waveforms of the half-bridge configuration, Fig. 1(a) for above resonance (or lagging PF) mode of operation: (a) RLC type overdamped load, (b) RL load.

When the load across AB is an RLC type, then the inverter is called a *series resonant inverter* or *series loaded resonant inverter*. In some applications, the commutating or resonating components are added to form the resonating tank. There are several other resonating tank circuit configurations possible. Other than the series resonant inverter, two more popular configurations are parallel loaded or parallel resonant inverter (Fig. 10.63(a)), and series-parallel (or LCC type) inverter (Fig. 10.63(b)). These resonant inverters are particularly useful in generating high-frequency voltage waveforms. They are used in induction heating, fluorescent lamp electronic ballasts, and other applications. The high-frequency transformer used in the circuits of Fig. 10.63 is for voltage scaling or **isolation**.

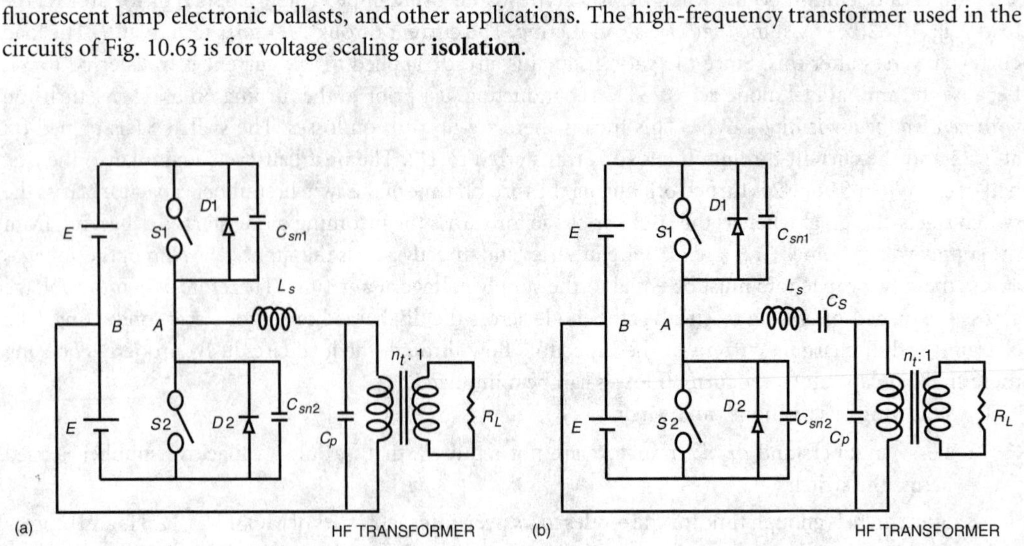

FIGURE 10.63 Inverter types: (a) parallel loaded or parallel resonant inverter, (b) series-parallel loaded (or LCC type) resonant inverter.

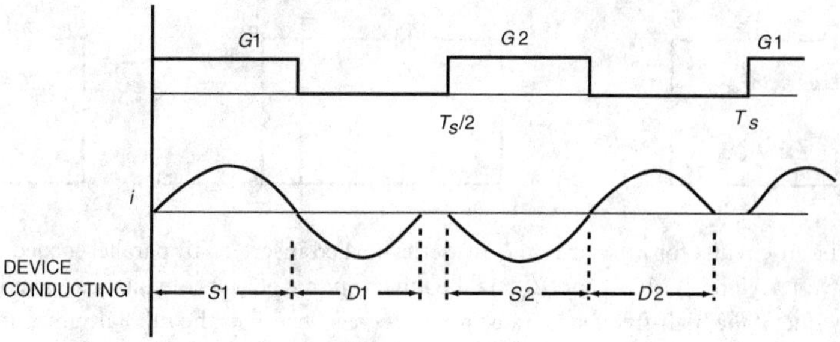

FIGURE 10.64   Operating waveforms for the discontinuous current mode of operation.

Power control in the case of resonant inverters is achieved either by variable frequency control or with fixed-frequency control. Fixed-frequency control is not applicable to the half-bridge configuration and, therefore, only variable frequency control method is explained here. This type of control can be used when the output frequency is not very important.

Variable Frequency Operation

The inverter operates near resonance frequency at full load and minimum input voltage. In this case the switching frequency is varied for power control. Two possibilities exist.

1. At full load and minimum input voltage, the switching frequency is slightly below resonance but enough turn-off time for the switches is provided. As the load current decreases or the input voltage increases, the inverter operating frequency is decreased below this value. The current can become discontinuous (Fig. 10.64) if the switching frequency is reduced below certain value. In addition to the disadvantages of below resonance operation mentioned earlier, this control method has an additional disadvantage, namely, the magnetics and input/output filters have to be designed for the lowest frequency of operation, increasing the inverter size.
2. At full load and minimum input voltage, the switching frequency is slightly above the resonance frequency. Power control is achieved by increasing the switching frequency above this value. This method has all of the advantages of above resonance operation explained earlier. SCRs cannot be used in this case since forced commutation or gate turn-off capability is required. Also, the switching frequency required may be very high at light load conditions, which increases the magnetic core losses and makes the control circuit design difficult.

Differential Equations (or Time-Domain Analysis) Approach

In this method, differential equations for the load circuit during different intervals are written, and they are solved to obtain the solutions for various state variables. This method is useful when transient response of the inverter has to be predicted. The steady-state solutions are obtained by matching the boundary conditions at the end of various intervals. This approach is complex when the order of the load circuit is higher.

Frequency-Domain Analysis

This approach is useful to analyze the inverter under steady-state conditions. The Fourier series expressions can be written taking into account all of the harmonics or using only the fundamental component in the simplified approximate analysis.

Fourier-Series Approach

In this method, the square-wave voltage across AB is represented by the Fourier series (note: $E = V_s/2$ for half-bridge and $\omega_s = 2\pi f_s$)

$$v_{AB} = \left(\frac{4E}{\pi}\right) \sum_{n=1,3...}^{\infty} \left[\frac{\sin(n\omega_s t)}{n}\right] \tag{10.154}$$

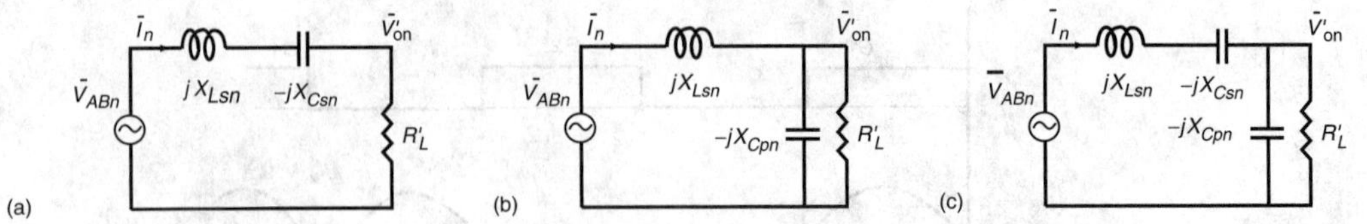

FIGURE 10.65 Phasor circuits for nth harmonic: (a) series-loaded inverter, (b) parallel-loaded inverter, (c) series–parallel resonant inverter. Note: (1) V_{ABn} and V'_{on} are the nth harmonic phasors of voltage v_{AB} and the output voltage referred to primary side of the high-frequency transformer, respectively; I_n is the nth harmonic phasor current at the output of the inverter. (2) $X_{Lsn} = n\omega_s L_s$, $X_{Csn} = 1/(n\omega_s C_s)$, and $X_{Cpn} = 1/(n\omega_s C_p)$ are the nth harmonic reactances of L_s, C_s, and C_p, respectively.

The nth harmonic phasor equivalent circuits across AB are shown in Fig. 10.65 for the three types of resonant inverters presented earlier. Using this circuit, the nth harmonic voltage and current equations can be written. Then the effect of all of the harmonics can be superposed to obtain the total response. In Fig. 10.65, the output voltage and the load resistor are referred to the primary side (using the turns ratio n_t of the transformer) and are denoted by $V'_{on}(= n_t V_{on})$ and $R'_L,(= n_t^2 R_L)$, respectively.

The current at the output of the inverter is

$$i = \left(\frac{4E}{\pi}\right) \sum_{n=1,3\ldots}^{\infty} \left[\frac{\sin(n\omega_s t - \phi_n)}{n Z_n}\right] \tag{10.155}$$

where Z_n is the nth harmonic impedance across the terminals A and B. For example, for a series loaded resonant inverter

$$Z_n = \left[R_L'^2 + \left\{n\omega_s L_s - 1/(n\omega_s C_s)\right\}^2\right]^{1/2} \tag{10.156}$$

and

$$\phi_n = \tan^{-1}[\{n\omega_s L_s - 1/(n\omega_s C_s)\}/R'_L] \tag{10.157}$$

The voltage and current through other components can be obtained in a similar way.

Approximate or Complex AC Circuit Analysis Approach
In this method, all of the harmonics except the fundamental are neglected. Although the method is an approximate approach, it can be used to design an inverter in a simple way. The fundamental voltage across AB is

$$v_{AB1} = (4E/\pi)(\sin(\omega_s t)) \tag{10.158}$$

Then the current at the output of the inverter is

$$i = (4E/(\pi Z_1))(\sin(\omega_s t - \phi_1)) \tag{10.159}$$

Depending on the value of ϕ_1, the inverter operates in lagging or leading PF mode. If ϕ_1 is negative, initial current (i at $t = 0$) is positive, and the inverter is operating in leading PF mode. If ϕ_1 is positive, initial current is negative with the inverter operating in the lagging PF mode, and forced commutation is necessary.

Using the approximate analysis, the gain of resonant inverters can be derived easily. For a series loaded resonant inverter (SLRI), considering only the fundamental component and neglecting all the harmonics (Fig. 10.65(a), with $n = 1$)

$$M(j\omega_s) = \frac{V'_{o1}}{V_{AB1}} = \frac{R'_L}{R'_L + j(w_s L_s - (1/w_s C_s))} \tag{10.160}$$

$$M = |M(j\omega_s)| = 1/\left(1 + (Q_s F - Q_s/F)^2\right)^{1/2} \tag{10.161}$$

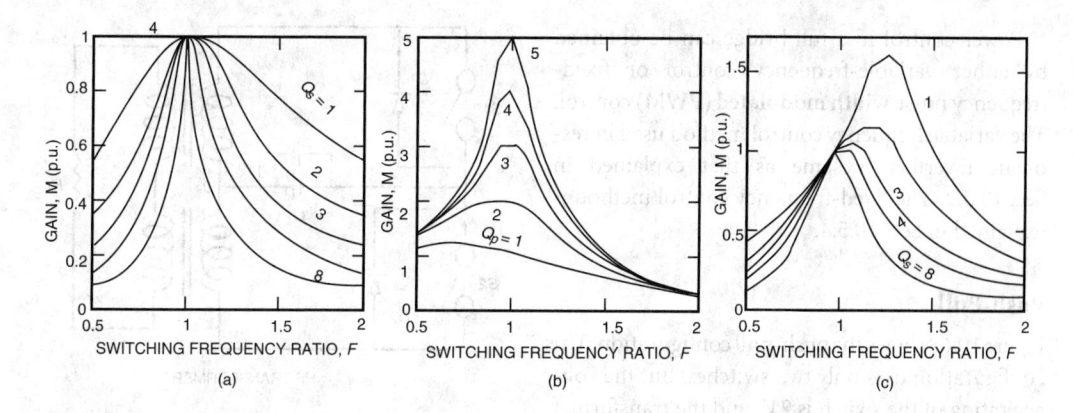

FIGURE 10.66 Inverter gain vs. switching frequency ratio, F: for (a) series-loaded inverter, (b) parallel-loaded inverter, (c) series–parallel inverter.

where $Q_s = (L_s/C_s)^{1/2}/R'_L$, $F = \omega_s/\omega_o$, $\omega_s = 2\pi f_s$ and $\omega_o = 1/(L_s C_s)^{1/2}$. Similarly, for a parallel loaded resonant inverter,

$$M = 1/\big((1 - F^2)^2 + (F/Q_p)^2\big)^{1/2} \qquad (10.162)$$

where $Q_p = R'_L/(L_s/C_p)^{1/2}$, $F = \omega_s/\omega_p$, and $\omega_p = 1/(L_s C_p)^{1/2}$, and for a series-parallel resonant inverter (Q_s and F same as those defined for SLRI)

$$M = 1/\Big(\big\{1 + (C_p/C_s)(1 - F^2)\big\}^2 + (Q_s F - Q_s/F)^2\Big)^{1/2} \qquad (10.163)$$

Figure 10.66 shows the plots of inverter gains vs. frequency ratio F obtained using the equations derived previously. It can be observed that the frequency variation required for load regulation is very wide for a series-loaded inverter, whereas it is very narrow for a parallel-loaded resonant inverter. But the inverter switch peak current decreases with the load current for a series-loaded inverter, whereas it does not decrease with the load current in the case of a parallel-loaded inverter. The series-parallel resonant inverter combines the advantages of series-loaded and parallel-loaded resonant inverters. Also, with a proper selection of C_s/C_p ratio, a near sine-wave output waveform which is almost load independent can be obtained.

Full-Bridge

Figure 10.67 shows the full-bridge inverter configuration. The basic operating principle is the same as the half-bridge inverter. However, the amplitude of the square-wave voltage across AB is V_s. Therefore, for the same power output, there is twice the output voltage, and the average current rating of the switches is half compared to the half-bridge configuration.

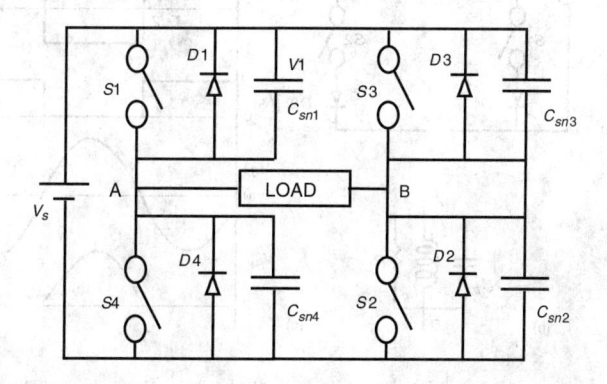

FIGURE 10.67 Basic circuit diagram of a full-bridge inverter.

Power control in a full bridge can be obtained by either variable-frequency control or fixed-frequency **pulse width modulated (PWM)** control. The variable frequency control method used in resonant inverters is same as that explained in Sec. 10.5.2. The fixed-frequency control method is explained in Sec. 10.5.4.

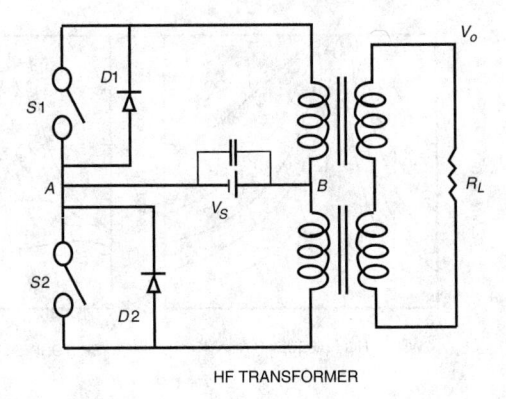

HF TRANSFORMER

FIGURE 10.68   Push–pull inverter configuration.

Push-Pull

Figure 10.68 shows the push-pull configuration. This configuration uses only two switches. But the voltage rating of the switch is $2V_s$ and the transformer size is larger (primary winding volt-ampere rating is $\sqrt{2}$ times the secondary winding or the load power) compared to the bridge configurations.

10.5.3   Current–Source Inverters (CSI)

This type of inverter is similar to the voltage-source type inverter. The input to the inverter is a current source, which is usually obtained by using a large inductor in series with the voltage source at the input. The voltage source at the input, in some cases (for example, in AC motor drives), is obtained by employing a phase controlled converter (for AC input) or a DC/DC chopper (for DC input). This allows the voltage to be controlled and, in turn, a controlled constant current source. The output of a CSI can be shorted, but cannot be open circuited. The peak current rating of the switches is equal to the DC current source and is lower compared to the VSI case. The disadvantages of a CSI are (1) slower dynamic response and (2) filters are required at the output to suppress the output voltage spikes.

In the case of a CSI (Fig. 10.69(a)), a DC current source is switched to obtain a square-wave current flowing at the output as shown in Fig. 10.69(c). The load used can be a parallel connection of R_L, L_p, and C_p, as shown in Fig. 10.69(b). This configuration is usually referred to as a *parallel resonant inverter*. The switches used in a CSI must be able to block the reverse voltage (similar to the capability of passing reverse

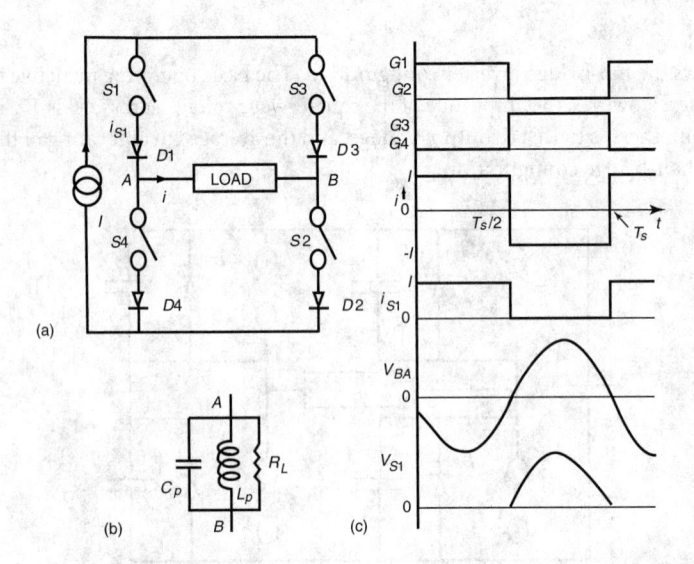

FIGURE 10.69   A current-source inverter: (a) circuit diagram, (b) load circuit used across A and B for parallel resonant inverter, (c) typical operating waveforms.

current for the switches of VSI). Therefore, diodes are connected in series with each switch, if transistors (MOSFETs, IGBTs, and bipolar) are used. The operating principle is similar to the voltage source case. Figure 10.69(c) illustrates operation corresponding to the lagging PF mode explained for a VSI.

The analysis approaches are similar to those explained for the VSI. For example, in the Fourier series approach, the equation for the square-wave current flowing in the output is given by

$$i = \left(\frac{4I}{\pi} \right) \sum_{n=1,3\ldots}^{\infty} \left[\frac{\sin(n\omega_s t)}{n} \right] \qquad (10.164)$$

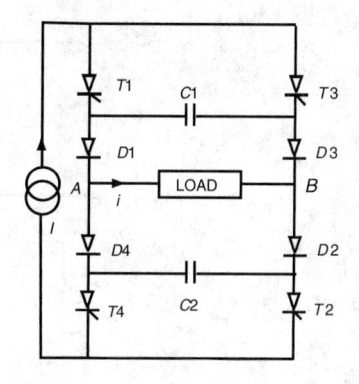

FIGURE 10.70   A single-phase current-source inverter using thyristors.

Figure 10.70 shows a single-phase CSI using thyristors. A three-phase version of this type of inverter is widely used in induction motor drives.

10.5.4   Voltage Control in Inverters

In most applications, output voltage of the inverter must be controlled. This may be to regulate the load voltage with variations in the load current or with variations in the DC (or AC if DC is derived from an AC supply) input voltage. In AC induction motor control applications, output voltage to frequency (V/f) must be kept constant. There are several ways to control the output voltage of an inverter. Some of them are listed next.

- Control the input DC voltage using a DC chopper.
- If the input is AC, use a phase controlled converter.
- Use an extra stage at the output to vary the output voltage.
- Control the inverter itself to vary the ratio of AC output voltage to the DC input voltage ratio.

The first three methods require an extra power stage. This causes additional losses and reduced efficiency. The fourth method involves the design of control electronics to properly vary the gating pulse pattern to obtain the desired output voltage control. This is called pulse width modulation (PWM). There are three methods of PWM possible:

- Single pulse width modulation.
- Multiple pulse width modulation.
- Sinusoidal pulse width modulation.

In many applications, the load voltage required must have low harmonic distortion. In some cases, a sine wave with a frequency equal to the utility line supply (60 Hz in North America) is required. In induction motor control applications, the output voltage, as well as frequency, must be controllable. In this case, it is possible to combine both the voltage control and sine wave generation in the inverter itself using sinusoidal PWM.

Single Pulse Width Modulation

Figure 10.71 shows the typical waveforms for this system. The gating signals G1 and G4 (or G3 and G2) are always 180° apart. The phase shift ϕ between the gating signals G1 and G3 is varied to get an output voltage of variable pulse width δ across the terminals A and B. When fixed-frequency control is used in a bridge converter, voltage waveform v_{AB} can be positive, negative, or zero. Zero voltage across AB is realized by turning on one switch and one diode on the positive side (or negative side), shorting the terminals A and B. The rms voltage across AB is a function of δ (or ϕ). When the inverter operates with minimum input voltage and maximum load power, $\delta = \pi$. As the load power decreases or the input voltage increases, output voltage

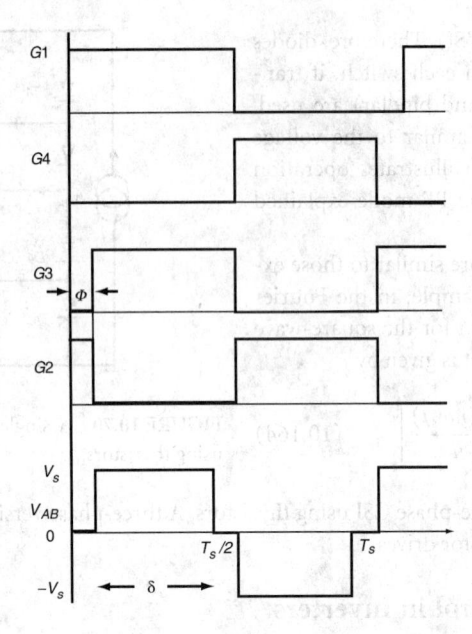

FIGURE 10.71 Typical operating waveforms of the bridge inverter when operating with fixed-frequency, phase-shift control.

is regulated by changing ϕ (in turn, δ). The same approach is used if voltage control is required across the load. The current through the load can be lagging or leading the voltage v_{AB}, depending on the load.

Fourier series analysis technique presented in Sec. 10.5.2 can be used for the steady-state analysis in this case also, with

$$v_{AB} = \left(\frac{4V_s}{\pi}\right) \sum_{n=1,3\ldots}^{\infty} \left[\frac{\sin(n\omega_s t)}{n}\right] \left[\sin\left(\frac{n\pi}{2}\right)\right] \left[\sin\left(\frac{n\delta}{2}\right)\right] \qquad (10.165)$$

The rms output voltage is given by

$$V_{ac\,rms} = V_s \sqrt{\delta/\pi} \qquad (10.166)$$

The amplitudes of output voltage harmonics are high in this control method. Filters can be added at the output to remove the harmonics and to generate the sine-wave of required frequency (such as 60 Hz), but the inverter then becomes bulky. This method can be used for power control in the case of resonant inverters also, and used in applications where the waveform harmonics may not be very important. This control method can further be applied to a CSI (Fig. 10.69). In the case of a CSI, zero current flow across the terminals A and B is obtained by turning on the switches of the same limb (i.e., S1 and S4 or S3 and S2).

Multiple Pulse Width Modulation

In this case, gating signals are generated by comparing a carrier signal (e.g., triangular wave) with a square wave to generate the output voltage v_{AB} as shown in Fig. 10.72. The number of pulses per half-cycle is $N = f_c/(2f)$, where f_c is carrier frequency and f is output frequency. If A_c and A_r are the amplitudes of the carrier and reference waveforms, respectively, then the modulation index is defined as

$$M = A_r/A_c \qquad (10.167)$$

M can be varied by varying A_r. This in turn varies δ^*, the range of variation being $0 < \delta^* < \pi/N$. This variation changes the output voltage from zero to the maximum value. This method reduces the output harmonics compared to single pulse PWM. When $\delta^* = \pi$, the output waveform is a square wave and has the highest distortion.

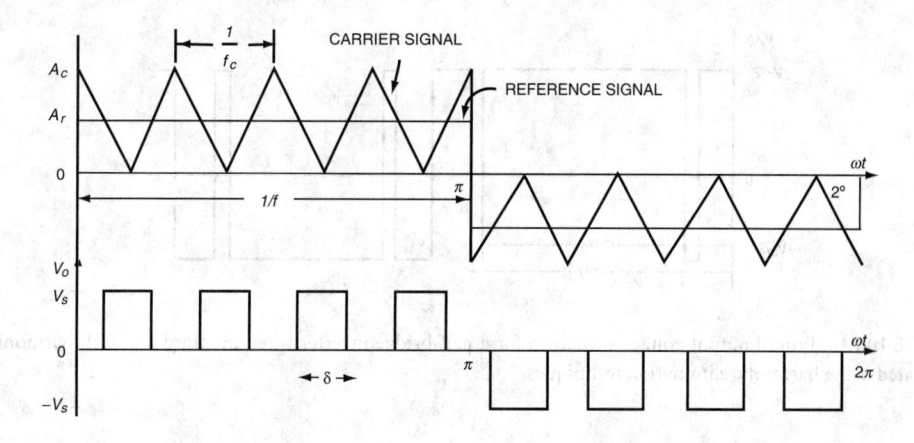

FIGURE 10.72   Waveforms to illustrate the multiple pulse width modulation technique.

Sinusoidal Pulse Width Modulation

When a sine wave voltage with low harmonic distortion is required across the load terminals, then sinusoidal PWM is used. As shown in Fig. 10.73, the gating signals are generated by comparing a triangular wave with the reference sine wave. Since the pulse widths of the output load voltage (shown for a full bridge) are modulated using a sine wave, the distortion is low and the harmonics of higher frequency can be easily filtered to get the sine wave voltage of required frequency. In this case also, similar to the multiple PWM, output voltage can be varied by changing the modulation index M. For $0 < M < 1$, lower order harmonics are eliminated. When a half-bridge configuration is used, then the output voltage can change only between $+V_s/2$ and $-V_s/2$.

10.5.5   Harmonic Reduction

The harmonics in the output voltage must be reduced to decrease the size of the filters used. If the amplitudes of lower order harmonics are reduced, then the size of the filter can be made smaller. This will also increase the speed of response of the inverter. The sinusoidal PWM technique discussed previously is a popular method. The disadvantages of this method are: (1) reduction of the fundamental component compared to a square-wave results in derating of the inverter by more than 17% and (2) increased switching losses

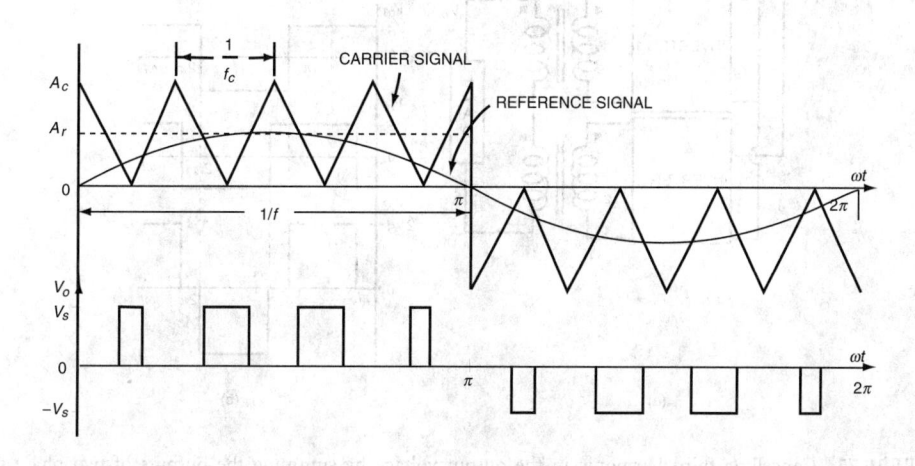

FIGURE 10.73   Waveforms to illustrate the sinusoidal pulse width modulation technique.

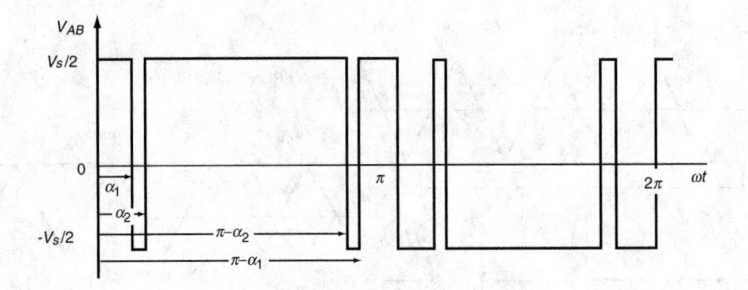

FIGURE 10.74 Typical output voltage waveform for a half-bridge inverter when the third and fifth harmonics are eliminated using harmonic elimination techniques.

are experienced because the switches are turned on and off several times in a cycle. The following two methods are also used for harmonic reduction of inverters.

1. **Harmonic elimination technique:** Selected harmonics can be eliminated by precalculated positioning of gating pulses. For example, for a half-bridge inverter with the output voltage, as shown in Fig. 10.74, third and fifth harmonics can be eliminated if $\alpha_1 = 23.62°$ and $\alpha_2 = 33.3°$. This method suffers from the disadvantages mentioned for sinusoidal PWM, but the number of switchings are less when only the first few lower order harmonics are eliminated.

2. **Phase-shifted inverters or harmonic cancellation technique:** In this case, the outputs of several inverters are phase shifted and combined at the output of the transformers to eliminate the low-order harmonics. For example, Fig. 10.75 shows the schematic to eliminate the third harmonic (in fact, all triple-n harmonics). The derating of each inverter in this case is about 14%. Using two such systems and phase shifting their outputs by 36°, a stepped wave can be generated to eliminate third and fifth harmonics. The same effect can be achieved by adding the outputs of two single pulse width modulated inverters generating the waveforms v_{o1} and v_{o2}.

Another disadvantage of these methods is that the voltage control has to be done at the input of the inverter (or by some other technique).

There are several other modulation techniques available (Rashid, 1993; Sen, 1987; Mohan, Undeland, and Robins, 1995). Among them is the hysteresis control technique.

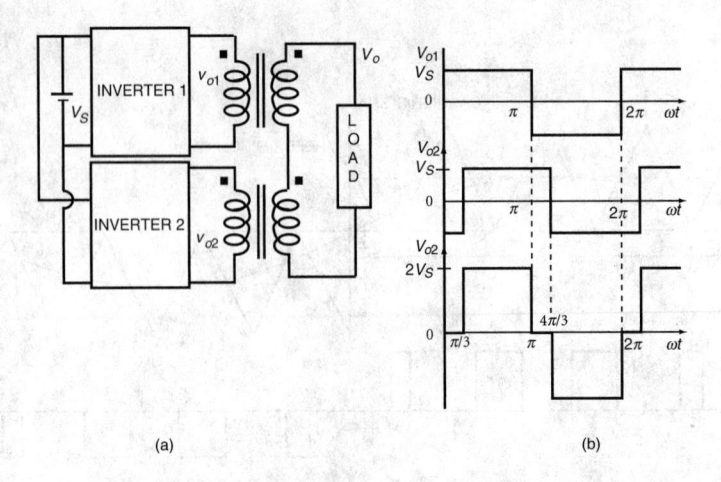

(a)   (b)

FIGURE 10.75 Cancelling third harmonic in the output voltage by summing the outputs of two phase-shifted inverters: (a) circuit diagram, (b) output voltage waveforms.

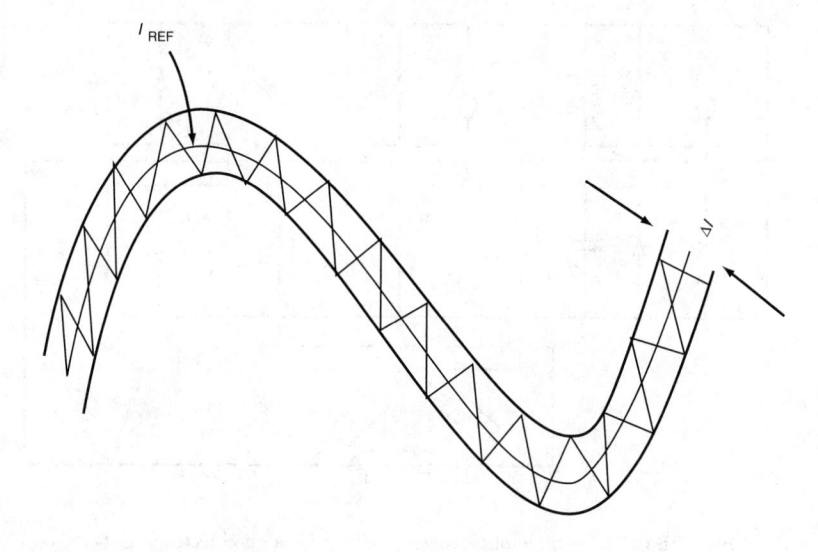

FIGURE 10.76   Hysteresis control technique.

Hysteresis Control

In this method, load current is sensed and compared with a reference sine wave. The inverter switches are controlled such that the output current does not exceed the set hysteresis limits $\pm \Delta I$ of the reference (Fig. 10.76). Although ΔI can be made very small to get almost a sinusoidal current, this will increase the switching frequency to a high value, which increases the switching losses. This method of control is popular in motor drive application.

Phase-Controlled Inverters

A phase-controlled inverter is a special type of inverter working on the utility line. It (shown in Fig. 10.77) can be used as an inverter when the firing angle α is in the range $\pi/2 < \alpha < \pi$, and the source is capable of supplying the energy to the AC source. This inverter requires the line voltage for commutation of the SCRs and, therefore, is referred to as a *line commutated inverter*. For a continuous current mode of operation,

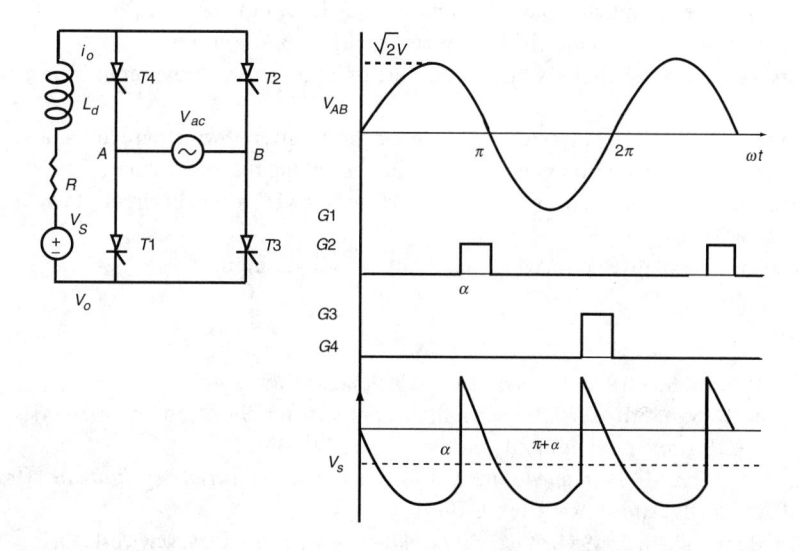

FIGURE 10.77   A single-phase phase-controlled inverter and typical operating waveforms for continuous current mode of operation.

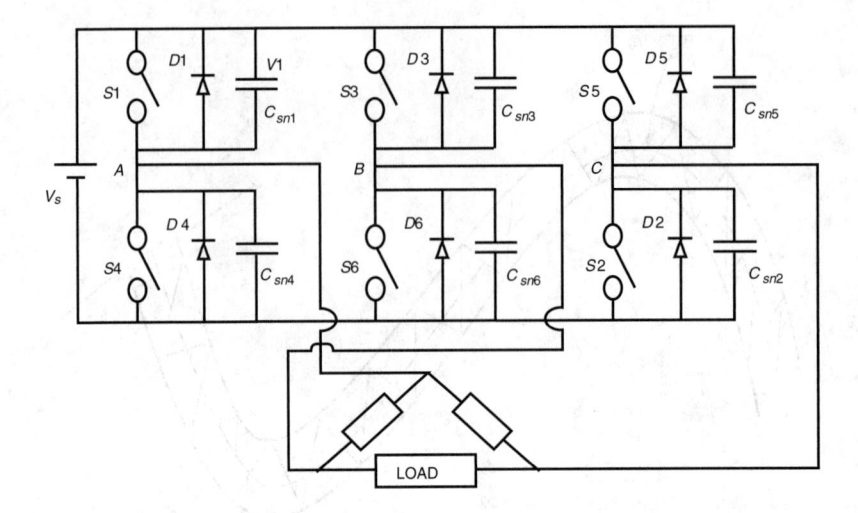

FIGURE 10.78   A three-phase inverter with the loads connected in delta.

the average output voltage is given by

$$V_o = (2\sqrt{2}V/\pi) \cos \alpha \qquad (10.168)$$

Since the average output current I_o is always positive and $V_o < 0$ for $\pi/2 < \alpha < \pi$, the output power is negative. Therefore, the converter operates in the inversion mode with the power being fed to the AC supply.

10.5.6   Three-Phase Inverters

For medium to high power applications, three-phase inverters are used. Figure 10.78 shows the basic circuit diagram of a three-phase VSI with the loads connected in delta. The loads can also be Y connected. The control of this inverter can be done by either 180° wide gating pulse control or 120° wide gating pulses.

Defining Terms

Current source inverters: Inverters having a constant DC current input.

Inverter: A power electronic circuit which converts DC input to AC output.

Isolated: A power electronic circuit which has ohmic isolation between the input source and the load circuit.

Pulse width modulated (PWM) inverters: A power electronic inverter which employs square-wave switching waveforms with varaition of pulse width for controlling the load voltage.

Resonant inverters: A power electronic inverter which employs LC resonant circuits to obtain sinusoidal switching waveforms.

Voltage source inverters: Inverters having a constant DC voltage input.

References

Bedford, F.E. and Hoft, R.G. 1972. *Inverter Circuits.* Wiley, New York.

Bhat, A.K.S. and Dewan, S.B. 1989. A generalized approach for the steady-state analysis of resonant inverters, *IEEE Trans. on Industry Applications* 35(2):326–338.

Brichant, F. 1984. *Force-Commutated Inverters—Design and Industrial Applications,* (translated by E. Griffin), North Oxford Academic, Oxford.

Dewan, S.B. and Straughen, A. 1984. *Power Semiconductor Circuits.* Wiley, New York.

Dubey, G.K., Doradla, S.R., Joshi, A., and Sinha, R.M.K. 1986. *Thyristorised Power Controllers.* Wiley Eastern Ltd., New Delhi.

Kassakian, J.G., Schlecht, M.F. , and Verghese, G.C. 1991. *Principles of Power Electronics*. Addison-Wesley, Reading, MA.

Mohan, N., Undeland, T.M., and Robins, W.R. 1995. *Power Electronics—Converters, Applications, and Design*. Wiley, New York.

Rashid, M.H. 1993. *Power Electronics—Circuits, Devices and Applications*. Prentice-Hall, Englewood Cliffs, NJ.

Sen, P.C. 1987. *Power Electronics*. Tata McGraw-Hill, New Delhi.

Vithayathil, J. 1995. *Power Electronics, Principles and Applications*. McGraw-Hill, New York.

Further Information

The following monthly magazines and conference records publish papers on the analysis, design and experimental aspects of inverter configurations and their applications:

IEEE Applied Power Electronics Conference Records
IEEE Industry Applications Conference Records
IEEE International Telecommunications Energy Conference Records
IEEE Power Electronics Specialists Conference Records
IEEE Transactions on Aerospace and Electronic Systems
IEEE Transactions on Industry Applications
IEEE Transactions on Industrial Electronics
IEEE Transactions on Power Electronics

For information and subscriptions or ordering any of these journals or conference records contact: IEEE Service Center, 445 Hoes Lane, P.O. Box 1331, Piscataway, NJ 08855-1331.

10.6   DC-to-DC Conversion

Ashoka K.S. Bhat

10.6.1   Introduction

DC-to-DC converters convert a DC input voltage to a variable or a required fixed DC output voltage. The output can be single or multiple. These converters are used in many applications including industrial, aerospace, telecommunications, and consumer products. Some of the requirements of DC-to-DC converters are small size, light weight, low cost, and high-power conversion efficiency. In addition to these, some applications require electrical isolation between the source and load. Some special applications require controlled direction of power flow.

10.6.2   Linear and Switch-Mode DC-to-DC Converters

DC-to-DC power converters can be mainly classified as: *linear* and *switch-mode* power converters.

In a linear DC-to-DC converter, the output voltage is regulated by dropping the extra input voltage across a series transistor. These regulators have very small output ripple, theoretically zero noise, large holdup time (typically, 1–2 ms), and fast response. The disadvantages of this type of converters are very low efficiency, no electrical isolation between the input and the output, larger volume and weight, and, in general, only single output possible. However, they are still used in very small regulated power supplies and in some special applications (e.g., magnet power supplies). Three terminal linear regulator integrated circuits (ICs) are readily available easy to use, and have built-in load short-circuit protection.

In a switch-mode DC-to-DC converter, power semiconductors are used as switches to connect or to remove the input DC supply voltage to the output. Since the switches operate in the on and off switching states, power conversion efficiency is very high, reducing the size and weight of heat sinks. With the availability of fast switching devices, high-frequency (HF) magnetics and capacitors, and high-speed control ICs, switching power converters have become very popular. They can be further classified as: *pulse width modulated (PWM)* converters and *soft-switched* converters.

10.6.3 Pulse Width Modulated Converters

In these converters, square-wave pulse width modulation is used to achieve voltage regulation. The **duty cycle** of the power semiconductor switch is varied to vary the average output voltage. The voltage (and current) waveform across the switch and at the output are square wave in nature, and they generally result in higher switching losses when the switching frequency is increased. Also, the switching stresses are high with the generation of large electromagnetic interference (EMI), which is difficult to filter. However, these converters are easy to control, well understood, and have wide load control range.

The Methods of Control

The PWM converters operate with fixed frequency, variable duty cycle. The three possible control methods (Severns and Bloom, 1988; Hnatek, 1981; Unitrode, 1984; Motorola, 1989; Philips, 1991) are briefly explained next:

1. *Direct duty cycle control:* This is the simplest control method. A fixed-frequency ramp is compared with the control voltage to obtain a variable duty cycle base drive signal for the transistor. Disadvantages of this method are that it provides no voltage feedforward to anticipate the effects of input voltage changes (slow response to sudden input changes, poor audio susceptibility, poor open-loop line regulation, requiring higher loop gain to achieve specifications) and it has poor dynamic response.
2. *Voltage feedforward control:* In this case the ramp amplitude varies in direct proportion to the input voltage. The open loop regulation is very good, and the problems in (1) are overcome.
3. *Current mode control:* In this method, as shown in Fig. 10.79, a second inner control loop compares the peak inductor current with the control voltage, which provides improved open-loop line regulation. All of the problems of the direct duty cycle control method (1) and (2) are corrected with this method. An additional advantage of this method is that the two pole second-order filter is reduced to a single pole (the filter capacitor) first-order filter resulting in simpler compensation networks. There are several current mode control ICs available in the market.

These control methods can be used in all of the PWM converter configurations. The converters can operate in either *continuous current mode* (CCM) or *discontinuous current mode* (DCM). If the current through the output inductor (e.g., refer to the circuit shown in Fig. 10.80(a)) never reaches zero, then the converter operates in CCM, otherwise DCM occurs.

PWM converters can be classified as: (1) single-ended and (2) double-ended converters. These converters may or may not have a HF transformer for isolation.

Nonisolated Single-Ended PWM Converters

The basic nonisolated single-ended converters are: (1) buck (step-down) (Fig. 10.80(a)), (2) boost (step-up) (Fig. 10.80(b)), (3) buck–boost (step down or up, also referred to as *flyback*) (Fig. 10.80(c)), and

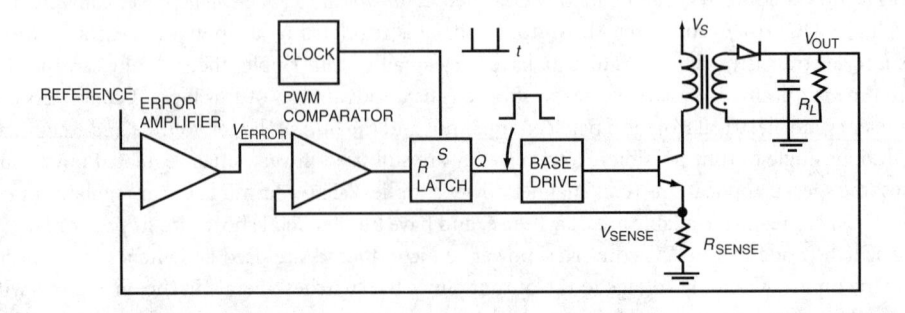

FIGURE 10.79   Current mode control technique illustrated for flyback converter.

TABLE 10.5 Output Voltage Expressions for
Single-Ended Converters (CCM Operation)[a]

Buck converter	$V_o = DV_{in}$
Boost converter	$V_o = V_{in}/(1 - D)$
Buck–boost converter	$V_o = DV_{in}/(1 - D)$
Cuk converter	$V_o = DV_{in}/(1 - D)$

[a]Note: duty ratio, $D = t_{on}/T_s$; V_{in} = DC input
voltage; V_o = output DC voltage.

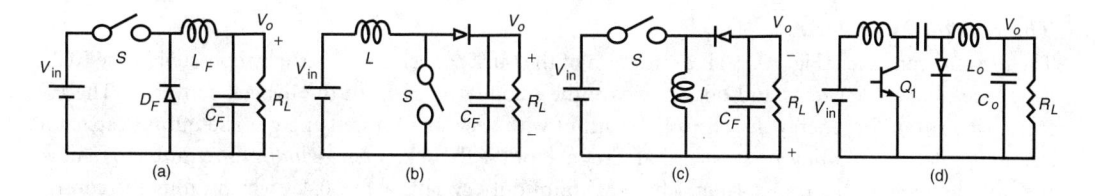

FIGURE 10.80 Converter types: (a) buck converter, (b) boost converter, (c) buck–boost converter, (d) nonisolated
Ćuk converter.

(4) Ćuk converter (Fig. 10.80(d)). The output voltage of a buck converter is always less than the input
voltage and the input current is pulsating. In a boost converter, the output voltage is greater than the
input voltage and the output current is pulsating. In a buck–boost converter, one can get an output
voltage less than or greater than the input voltage and both the input and output currents are pulsating.
The Ćuk converter provides the advantage of nonpulsating input-output current ripple requiring smaller
size external filters. The output voltage expression is the same as the buck–boost converter. Table 10.5
summarizes the expressions for the output voltage in these converters. There are many variations of
the basic nonisolated converters, and most of them use high-frequency transformer for ohmic isolation
between the input and the output.

Isolated Single-Ended Topologies

The Flyback Converter

The flyback converter (Fig. 10.81) is an isolated version of the buck–boost converter. In this converter,
when the transistor is on, energy is stored in the coupled inductor (not a transformer) and this energy is
transferred to the load when the switch is off. Flyback converters are used in the power range of 20–200 W.

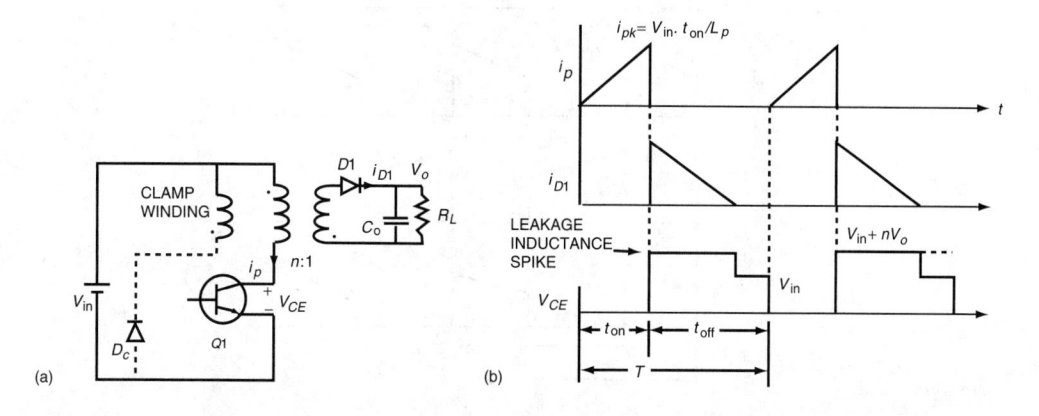

FIGURE 10.81 (a) Flyback converter. The clamp winding shown is optional and is used to clamp the transistor voltage
stress to $V_{in} + nV_o$. (b) Flyback converter waveforms without the clamp winding. The leakage inductance spikes vanish
with the clamp winding.

Some of the advantages of this converter are the following: (1) The leakage inductance is in series with the output diode when current is delivered to the output and, therefore, no filter inductor is required. (2) Cross regulation for multiple output converters is good. (3) It is ideally suited for high-voltage output applications. (4) It has the lowest cost.

Some of the disadvantages are the following: (1) Large output filter capacitors are required to smooth the pulsating output current. (2) Inductor size is large since air-gaps are to be provided. (3) Because of stability reasons, flyback converters are usually operated in the DCM, which results in increased losses. To avoid the stability problem, flyback converters are operated with current mode control explained in Sec. 10.6.3.

The Forward Converter

The forward converter (Fig. 10.82) is a variation of the buck converter. It is usually operated in the CCM to reduce the peak currents and does not have the stability problem of the flyback converter. The HF transformer transfers energy directly to the output with very small stored energy. The output capacitor size and peak current rating are smaller than they are for the flyback. A *reset winding* is required to remove the stored energy in the transformer. The maximum duty cycle is about 0.45, which limits the control range. This topology is used for power levels up to about 1 kW.

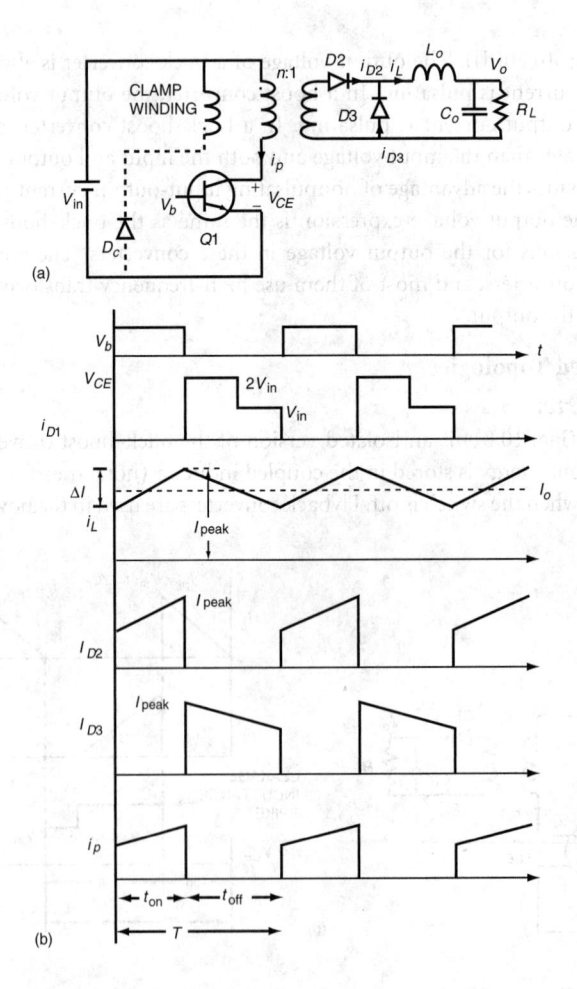

FIGURE 10.82 (a) Forward converter. The clamp winding shown is required for operation. (b) Forward converter waveforms.

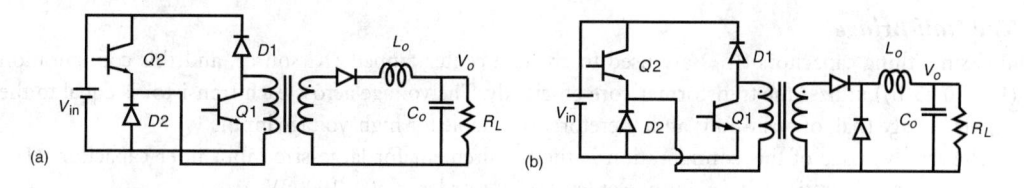

FIGURE 10.83   (a) Two transistor single-ended flyback converter, (b) two transistor single-ended forward converter.

The flyback and forward converters require the voltage rating of power transistors to be much higher than the supply voltage. The two transistor flyback and forward converters shown in Fig. 10.83 limit the voltage rating of transistors to the supply voltage.

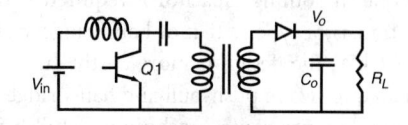

FIGURE 10.84   Sepic converter.

The Sepic converter shown in Fig. 10.84 is another isolated single-ended PWM converter.

Double-Ended PWM Converters

Usually, for power levels above 300 W, double-ended converters are used. Full-wave rectifiers are used in the double-ended converters and the ripple frequency of the output voltage is twice the switching frequency. Three important double-ended PWM converter configurations are push-pull (Fig. 10.85(a)), half-bridge (Fig. 10.85(b)) and full bridge (Fig. 10.85(c)).

The Push-Pull Converter

The duty ratio of each transistor in a push-pull converter (Fig. 10.85(a)) is less than 0.5. Some of the advantages are the transformer flux swings fully and the size of transformer is much smaller (typically half the size) than single-ended converters, and the output ripple is twice the switching frequency of transistors; therefore, smaller filters are needed.

Some of the disadvantages of this configuration are the transistors must block twice the supply voltage; flux symmetry imbalance can cause transformer saturation and special control circuitry is required to avoid this problem; and the use of a center-tap transformer requires extra copper resulting in a higher VA rating.

Current mode control (for the primary current) can be used to overcome the flux imbalance. This configuration is used in 100–500 W output range.

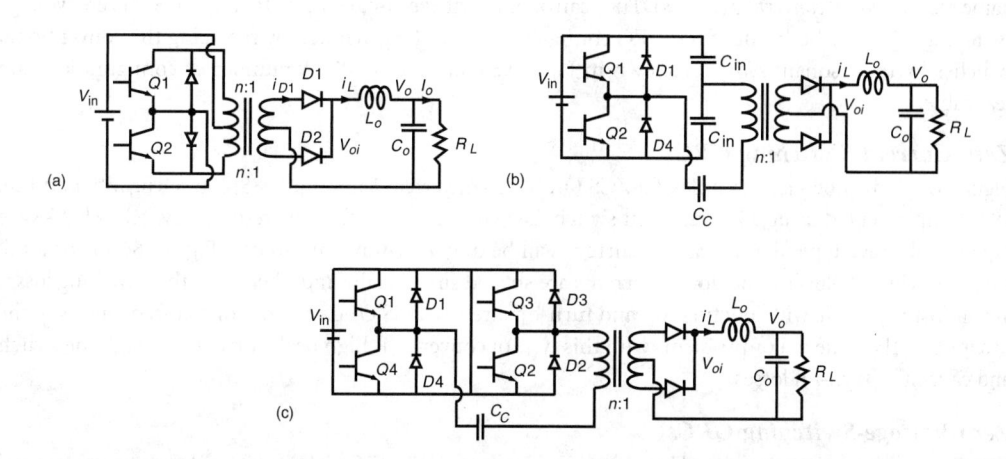

FIGURE 10.85   (a) Push–pull converter, (b) half-bridge converter, (c) full-bridge converter. Coupling capacitor C_c is used to avoid transformer saturation in (b) and (c).

The Half-Bridge

Two smoothing capacitors (C_{in}) are used to create a center tapped DC source, and this configuration (Fig. 10.85(b)) utilizes the transformer core efficiently. The voltage across each transistor is equal to the supply voltage (half of push-pull) and, therefore, is suitable for high-voltage inputs.

The disadvantage of this configuration is the requirement for large-size input filter capacitors. The half-bridge configuration is used for power levels of the order of 500–1000 W.

The Full Bridge

Only one smoothing capacitor is required at the input for this configuration (Fig. 10.85(c)) and for the same transistor type as that of half-bridge, output power can be doubled. Usually used for power levels above 1 kW, the design is more costly due to increased number of components (uses four transistors compared to two in push-pull and half-bridge converters.)

Using a proper control technique, a full-bridge converter can be operated in a *zero-voltage switching* (ZVS) mode (Dalal, 1990). This type of operation results in negligible switching losses. At reduced load currents, however, the ZVS property is lost. There has been much effort to overcome this problem, but the converter requires additional components to maintain the ZVS.

10.6.4   Soft-Switched Converters

Soft-switched converters can be classified as (a) **resonant converters**, and (b) **resonant** (zero-current or zero-voltage) **transition converters**. Similar to the PWM converters, there are two types in each case, namely, **single-ended** and **double-ended**.

Resonant DC-to-DC Converters

In the PWM converters explained in Sec. 10.6.3, LC (inductor and capacitor) resonating elements are added to generate resonant converter configurations. These converters generate sinusoidally varying voltage and/or current waveforms reducing the switching losses and the switch stresses, enabling the converter to operate at high switching frequencies resulting in reduced size, weight, and cost. Some other advantages of resonant converters are the leakage inductances of the HF transformer and the junction capacitances of semiconductors can be used profitably in the resonant circuit, and reduced EMI. The major disadvantage of resonant converters is increased peak current (or voltage).

Single-Ended Resonant Converters

Single-ended resonant converters generate quasisinusoidal voltage (or current) waveforms and, hence, the name *quasiresonant converters* (QRCs). They can operate with zero-current switching (ZCS) or zero-voltage switching (ZVS) or both. All of the QRC configurations can be generated by replacing the conventional switches by the resonant switches shown in Fig. 10.86 and Fig. 10.87. A number of configurations are realizable.

Zero-Current-Switching QRCs

Figure 10.86(d) shows an example of a ZCS QR buck converter (Kit Sum, 1988; Liu, Oruganti, and Lee, 1985) implemented using a ZC resonant switch. Depending on whether the resonant switch is half-wave type or full-wave type, the resonating current will be only half-wave sinusoidal (Fig. 10.86(e)) or a full sine wave (Fig. 10.86(f)). The device currents are shaped sinusoidally and, therefore, the switching losses are almost negligible with low turn-on and turn-off stresses. ZCS QRCs can operate at frequencies of the order of 2 MHz. The major problems with this type of converter is high peak currents through the switch and capacitive turn-on losses.

Zero-Voltage-Switching QRCs

ZVS QRCs (Kit Sum, 1988; Liu and Lee, 1986) are similar to ZCS QRCs. The auxiliary LC elements are used to shape the switching device's voltage waveform at off time in order to create a zero-voltage condition for the device to turn on. Figure 10.87(d) shows an example of ZVS QR boost converter implemented using ZV resonant switch. The circuit can operate in the half-wave mode (Fig. 10.87(e)) or in the full-wave mode

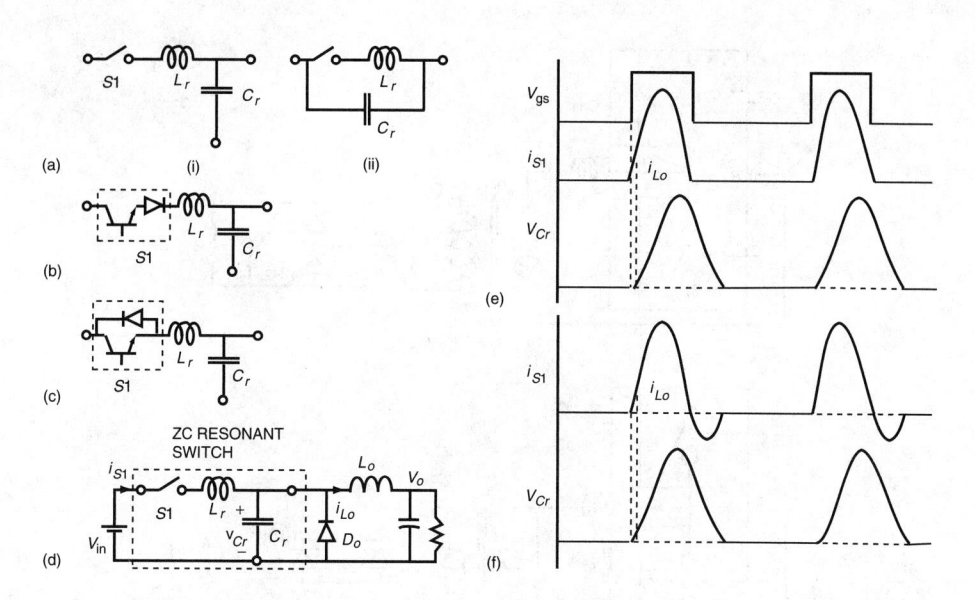

FIGURE 10.86 (a) Zero current resonant switch (i) L type (ii) M type. (b) half-wave configuration using L type ZC resonant switch. (c) full-wave configuration using L type ZC resonant switch. (d) implementation of ZCS QR buck converter using L type resonant switch. (e) operating waveforms for half-wave mode. (f) operating waveforms for full-wave mode.

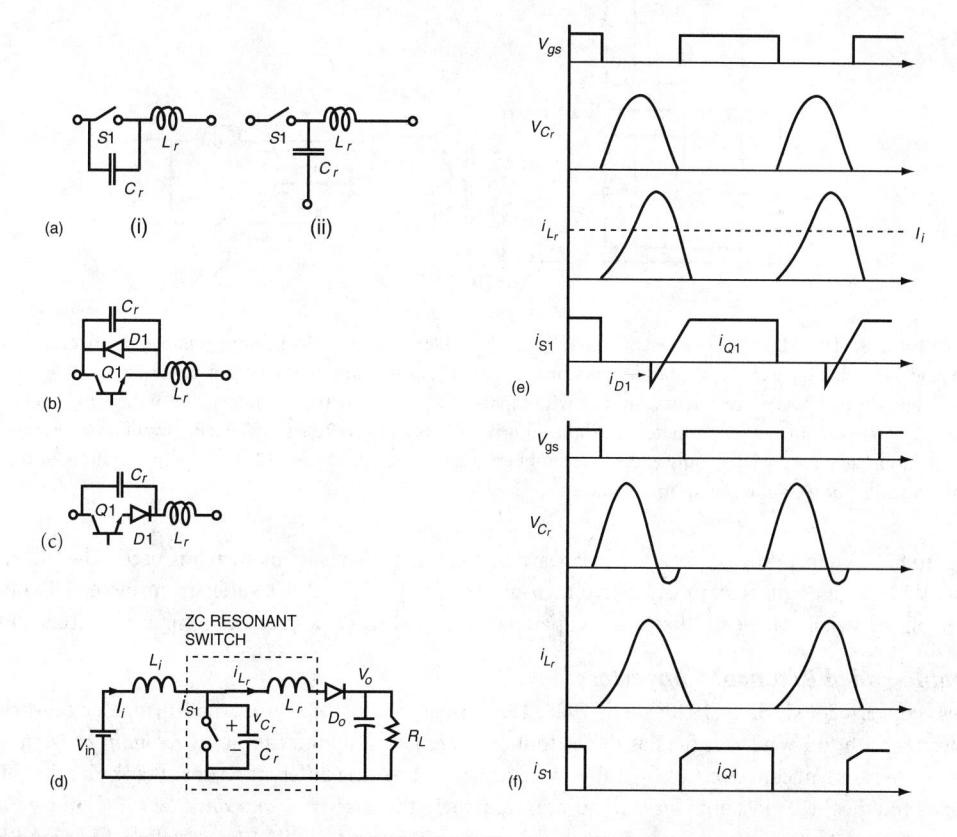

FIGURE 10.87 (a) Zero voltage resonant switches. (b) half-wave configuration using ZV resonant switch shown in Fig. 10.87(a)(i). (c) full-wave configuration using ZV resonant switch shown in Fig. 10.87(a)(i). (d) implementation of ZVS QR boost converter using resonant switch shown in Fig. 10.87(a)(i). (e) operating waveforms for half-wave mode. (f) operating waveforms for full-wave mode.

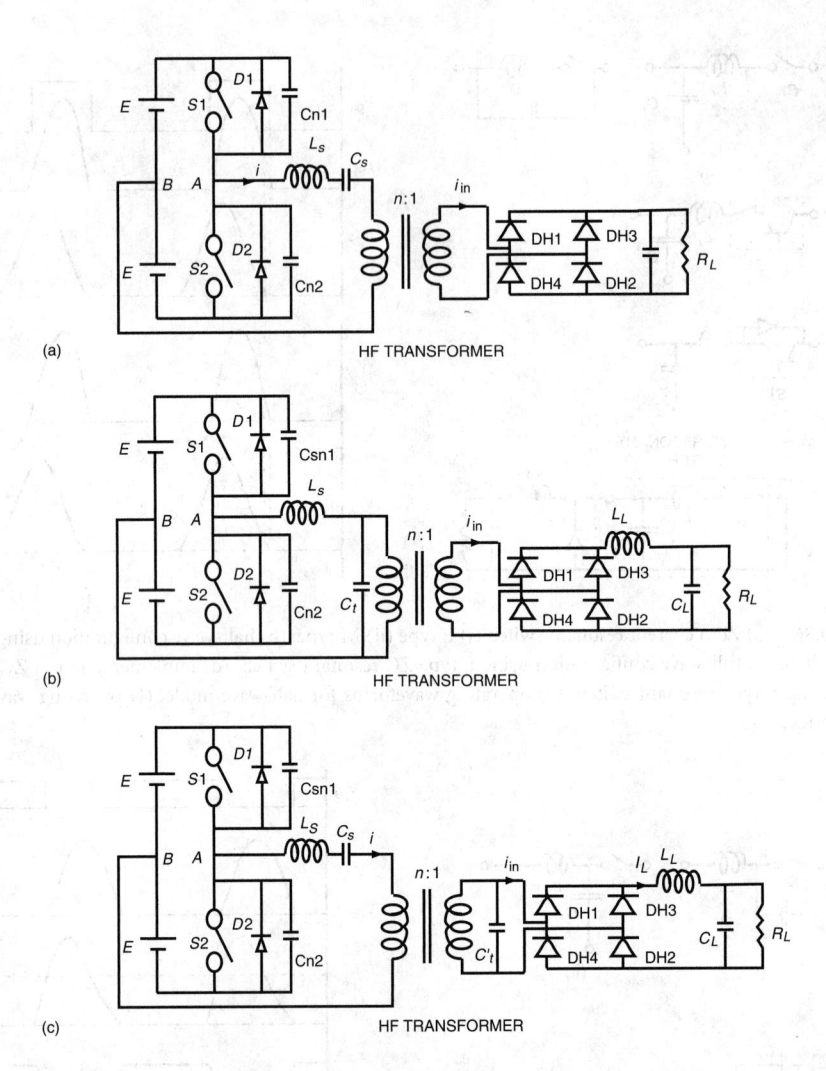

FIGURE 10.88 High-frequency resonant converter (half-bridge version) configurations suitable for operation above resonance. Cn1 and Cn2 are the snubber capacitors: (a) series resonant converter. (b) parallel resonant converter. (c) series-parallel (or LCC type) resonant converter with capacitor C_t placed on the secondary side of the HF transformer. [Note: (1) For operation below resonance, di/dt limiting inductors and RC (Resistor-Capacitor) snubbers are required. For operation above resonance, only capacitive snubbers are required as shown. (2) leakage inductances of the HF transformer can be part of resonant inductance.]

(Fig. 10.87(f)) depending on whether half-wave or full-wave ZV resonant switch is used. The full-wave mode ZVS circuit suffers from capacitive turn-on losses. The ZVS QRCs suffer from increased voltage stress on the switch. However, they can be operated at much higher frequencies compared to ZCS QRCs.

Double-Ended Resonant Converters

These converters (Kit Sum, 1988; Bhat, 1988, 1991; Steigerwald, 1988) use a half-bridge or full-bridge converter with full-wave rectifiers at the output, and are generally referred to as *resonant converters*. A number of resonant converter configurations are realizable by using different resonant tank circuits (Bhat, 1995b) and the three most popular configurations, namely, the *series resonant converter* (SRC), the *parallel resonant converter* (PRC) and the *series-parallel resonant converter* (SPRC) (also called LCC type PRC) are shown in Fig. 10.88(a) to Fig. 10.88(c). Load voltage regulation in such converters for input supply variations and load changes is achieved by either varying the switching frequency or using fixed-frequency (variable pulse width) control.

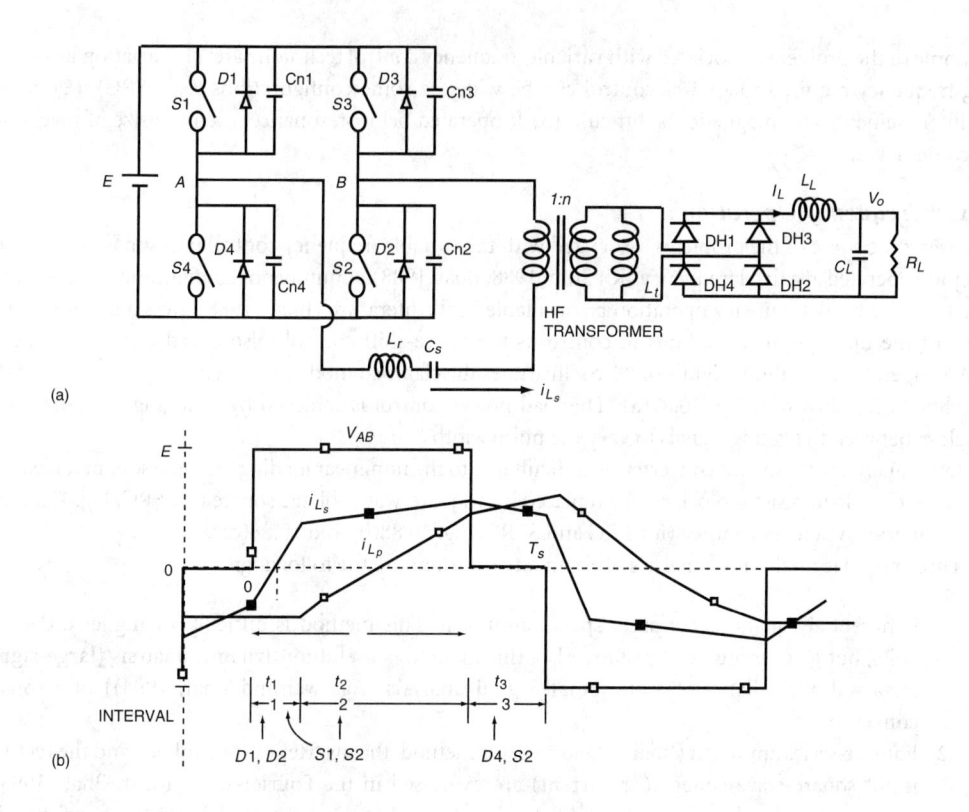

FIGURE 10.89 (a) Basic circuit diagram of a modified series resonant (LCL type) converter suitable for fixed-frequency operation with PWM (clamped-mode) control. This converter uses an inductive output filter. (b) waveforms to illustrate the operation of fixed-frequency PWM LCL type resonant converter working with a pulsewidth δ.

Series resonant converters (Fig. 10.88(a)) have high efficiency from full load to part load. Transformer saturation is avoided due to the series blocking resonating capacitor. The major problems with the SRC are (1) It requires a very wide change in switching frequency to regulate the load voltage and (2) the output filter capacitor must carry high ripple current (a major problem especially in low-output voltage, high-output current applications).

Parallel resonant converters (Fig. 10.88(b)) are suitable for low-output voltage, high-output current applications due to the use of filter inductance at the output with low ripple current requirements for the filter capacitor. The major disadvantage of the PRC is that the device currents do not decrease with the load current resulting in reduced efficiency at reduced load currents. The SPRC (Fig. 10.88(c)) takes the desirable features of SRC and PRC. It is possible to realize higher order resonant converters with improved characteristics and many of them are presented in Bhat (1991). Recently, modified series resonant converters (Bhat, 1994, 1995a) have been proposed. A modified SRC with an inductive output filter is shown in Fig. 10.89(a). The resonant inductor current is clamped approximately to the primary-side reflected load current. This configuration gives almost the same performance as the PWM converter but with reduced losses.

Variable Frequency Operation

Depending on whether the switching frequency is below the natural resonance frequency (ω_r) or above ω_r, the converter can operate in two different operating modes:

- Below resonance (leading PF) mode
- Above resonance (lagging PF) mode

These two modes of operation have been explained in Chapter 69. The operating principles, the advantages, and disadvantages are the same as those explained there.

Some of the problems associated with variable-frequency control technique are: (1) variation in switching frequency required for power control can be wide for some configurations (e.g., SRC), (2) design of filter elements and magnetics is difficult, (3) If operated below resonance, then the size of magnetics becomes large.

Fixed Frequency Operation

To overcome some of the problems associated with the variable frequency control of resonant converters, they are operated at a fixed frequency (Kit Sum, 1988; Bhat, 1988). A number of configurations and control methods for fixed-frequency operation are available in the literature (Bhat (1988) gives a list of papers). One of the most popular methods of control is the phase-shift control (also called clamped mode or PWM operation) method. Figure 10.89(b) illustrates the clamped mode fixed-frequency operation of the modified SRC shown in Fig. 10.89(a). The load power control is achieved by changing the phase-shift angle ϕ between the gating signals to vary the pulse width δ of v_{AB}.

Exact analysis of resonant converters is difficult due to the nonlinear loading on the resonant tanks. The rectifier-filter-load resistor block can be replaced by a square-wave voltage source (for SRC, Fig. 10.89(a)) or a square-wave current source (for PRC and SPRC, Fig. 10.88(b) and 10.88(c)).

Three popular methods of analysis of resonant converters are the following:

1. Differential equations (or **state-space**) approach: This method is difficult for higher order circuits, but gives more exact results. Also, this method is useful for dynamic analysis (**large-signal** (Agarwal and Bhat, 1995) and **small-signal analysis** (Agarwal and Bhat, 1994)) of resonant converters.

2. Fourier-series approach (Bhat, 1995b): In this method, the inverter output voltage and the rectifier input square-wave voltage (or current) are expressed in the Fourier-series form. Bhat (1995b) gives a generalized analysis of resonant converters using two-port model and the Fourier-series approach. The various waveforms for the converter can be predicted more accurately with this method.

3. Approximate analysis: Using fundamental components of the waveforms, an approximate analysis (Bhat, 1991; Steigerwald, 1988) using phasor circuit gives a reasonably good design approach. This analysis approach is illustrated for the SPRC in the next section.

Methods 2 and 3 give only steady-state analysis. The describing function method (Yang, Lee and Jovanovic, 1992) is an approximate but powerful method for the small signal analysis of resonant converters.

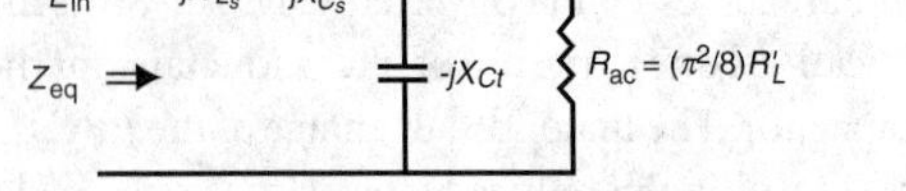

FIGURE 10.90 Phasor circuit model used for the analysis of the SPRC converter. $R_{ac} = (\pi^2/8)R'_L$ where $R'_L = n^2 R_L$. For voltage-source load [e.g., SRC of Fig. 10.88(a)] $R_{ac} = (8/\pi^2)R'_L$.

Approximate Analysis of SPRC

Using the fundamental waveforms, the rectifier-filter-load block can be replaced by an equivalent AC resistance R_{ac} and Fig. 10.90 shows the equivalent circuit at the output of the inverter. This phasor circuit is used for the analysis. All of the equations are normalized using the base quantities:

Base voltage, $V_B = E$ where $E = V_s/2$ for half-bridge and V_s for full-bridge.
Base impedance, $Z_B = R'_L = n^2 R_L$.
Base current, $I_B = V_B/Z_B$.

The converter gain (normalized output voltage in per unit (p.u.) referred to the primary-side) for an SPRC is given by (Bhat, 1988, 1991; Steigerwald, 1988)

$$M = \sin(\delta/2)/\left[(\pi^2/8)^2\left\{1 + (C_t/C_s)\left(1 - F_s^2\right)\right\}^2 + Q_s^2\left\{F_s - (1/F_s)\right\}^2\right]^{1/2} \quad \text{p.u.} \qquad (10.169)$$

where $\delta = \pi$ for variable-frequency operation,

$$Q_s = (L_s/C_s)^{1/2}/R'_L \tag{10.170}$$
$$F_s = f_s/f_r, \tag{10.171}$$

where f_s is the switching frequency, L_s includes leakage inductance of the HF transformer, and f_r is the series resonance frequency,

$$f_r = \omega_r/(2\pi) = 1/\left(2\pi(L_sC_s)^{1/2}\right) \tag{10.172}$$

The peak inverter output (same as the resonant inductor and the switch) current is given by

$$I_p = 4/(\pi|Z_{eq}|) \quad \text{p.u.} \tag{10.173}$$

where Z_{eq} is the equivalent impedance looking into the terminals AB given by

$$Z_{eq} = (B_1 + jB_2)/B_3 \quad \text{p.u.} \tag{10.174}$$

where

$$B_1 = (8/\pi^2)(C_s/C_t)^2(Q_s/F_s)^2 \tag{10.175}$$
$$B_2 = Q_s(F_s - (1/F_s))\left(1 + (8/\pi^2)^2(C_s/C_t)^2(Q_s/F_s)^2\right) - (C_s/C_t)(Q_s/F_s) \tag{10.176}$$
$$B_3 = 1 + (8/\pi^2)^2(C_s/C_t)^2(Q_s/F_s)^2 \tag{10.177}$$

The value of initial current I_0 is given by

$$I_0 = I_p \sin(-\phi) \quad \text{p.u.} \tag{10.178}$$

where

$$\phi = \tan^{-1}(B_2/B_1) \quad \text{rad} \tag{10.179}$$

If I_0 is negative, then the converter is operating in the lagging PF mode. The peak voltage across the capacitor C'_t (on the secondary side) is

$$V_{ctp} = (\pi/2)V_o \text{ V} \tag{10.180}$$

The peak voltage across C_s and the peak current through C'_t are given by

$$V_{csp} = (Q_s/F_s) \cdot I_p \quad \text{p.u.} \tag{10.181}$$
$$I_{ctp} = V_{ctp}/(X_{ctpu} \cdot R_L) \quad \text{A} \tag{10.182}$$

where

$$X_{ctpu} = (C_s/C_t)(Q_s/F_s) \quad \text{p.u.} \tag{10.183}$$

The plot of converter gain vs. the switching frequency ratio F_s can be obtained using Eq. (10.169). If the ratio C_s/C_t increases, then the converter takes the characteristics of SRC and the load voltage regulation requires very wide variation in the switching frequency. Lower values of C_s/C_t takes the characteristics of a PRC. Therefore, a compromised value of $C_s/C_t = 1$ is usually chosen in the design of such converters (Bhat, 1988, 1991; Steigerwald, 1988).

The converter gain and other equations just derived can be repeated for the SRC, PRC, and other resonant converters (Bhat, 1991).

Resonant Transition Converters

Major problem with the resonant converters is the high-current (or voltage) stress across the switches due to the generation of sinusoidal currents (and voltages). To overcome this problem, resonant transition converters have been proposed (Hua, Leu, and Lee, 1992; Hua et al., 1993). Basically, these converters use LC elements together with an auxiliary switch (the rating of which is much lower than the main switch) in parallel with the main switching device. Resonant transitions are used only during switching instants. These converters can be viewed as the switches employing active snubbers. There are two types of converters possible:

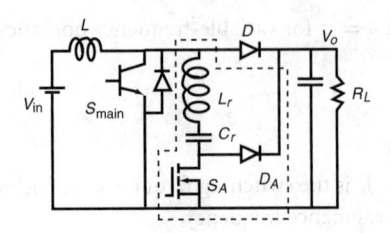

FIGURE 10.91 Zero-current-transition PWM boost converter. Elements shown in dashed lines are added to the original PWM boost converter shown in Fig. 10.2(b).

- *Zero-current-transition PWM converters* (Hua, Leu, and Lee, 1992): Figure 10.91 shows an example of zero-current transition boost converter. This approach is useful in high power converters when **insulated gate bipolar transistors (IGBTs)** or bipolar transistors are used as the switching devices.
- *Zero-voltage transition PWM converters* (Hua et al., 1993): Figure 10.92 shows an example of zero-voltage transition boost converter.

Some of the advantages of these converters are: fixed-frequency ZCS or ZVS PWM operation with characteristics closer to the PWM converters, and operation in ZCS or ZVS for both the switches and rectifier diodes for wide load and line voltage variations. The leakage inductance of the HF transformer, however, must be minimized in the isolated version of the converters. Also, extra components and drive circuits are required.

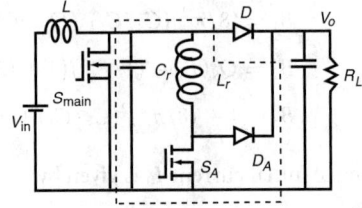

FIGURE 10.92 Zero-voltage-transition PWM boost converter. Elements shown in dashed lines are added to the original PWM boost converter shown in Fig. 10.2(b).

With the development of digital ICs operating on low voltage (of the order of 3 V) supplies, the use of MOSFETs as synchronous rectifiers with very low voltage drop ($\cong 0.2$ V) has become essential (Motorola, 1989) to increase the efficiency of the DC-to-DC converter.

Defining Terms

DC-to-DC converters: A power electronic circuit that converts DC input to DC output.

Duty cycle: Ratio of on time of the switch to the period of switching.

Insulated gate bipolar transistor (IGBT): A power semiconductor which can be turned on like a MOSFET with a collector to emitter conduction drop similar to a bipolar transistor.

Large signal analysis: Analysis used to study the behavior of a converter for large step changes in an input parameter (e.g., input supply voltage change, load change, etc.).

Resonant-transition-converter: A DC-to-DC converter that employs resonant transitions only during the instants of switching.

Small signal analysis: Analysis used to study the behavior of a converter for small changes in an input parameter (e.g., control voltage).

References

Agarwal, V. and Bhat, A.K.S. 1994. Small signal analysis of the LCC-type parallel resonant converter using discrete time domain modeling. *IEEE Power Electronics Specialists Conference Record*. Taipei, pp. 805–813 [revised version in 1995 in *IEEE Trans. on Industrial Electronics* 42(6):604–614].

Agarwal, V. and Bhat, A.K.S. 1995. Large signal analysis of the LCC-type parallel resonant converter using discrete time domain modeling. *IEEE Trans. on Power Electronics* 10(1):222–238.

Bhat, A.K.S. 1989. Fixed frequency PWM series-parallel resonant converter. *IEEE Industry Applications Conference Record*, San Diego, pp. 1115–1121 [revised version in 1992 in *IEEE Trans. on Industry Applications* 28(5)].

Bhat, A.K.S. 1991. A unified approach for the steady-state analysis of resonant converters. In *IEEE Trans. on Industrial Electronics* 38(4):251–259.

Bhat, A.K.S. 1994. Analysis and design of LCL-type resonant converter. *IEEE Trans. on Industrial Electronics* 41(1):118–124.

Bhat, A.K.S. 1995a. A fixed-frequency modified series-resonant converter: Analysis, design and experimental results. *IEEE Trans. on Power Electronics* 10(3):766–775.

Bhat, A.K.S. 1995b. A generalized steady-state analysis of resonant converters using two-port model and Fourier series approach. In *IEEE Applied Power Electronics Conference*, pp. 920–926. Dallas, TX.

Dalal, D.B. 1990. A 500 kHz multi-output converter with zero voltage switching. In *IEEE Applied Power Electronics Conference Record*, pp. 265–274, Los Angeles.

Hnatek, E.R. 1981. *Design of Solid-State Power Supplies*, 2nd ed. Van Nostrand Reinhold, New York.

Hua, G., Leu, C.S., and Lee, F.C. 1992. Novel zero-voltage-transition PWM converters. *IEEE PESC*. Taipei. pp. 55–61.

Hua, G., Yang, E.X., Jiang Y., and Lee, F.C. 1993. Novel zero-current-transition PWM converters. *IEEE PESC*. Taipei. pp. 538–543.

Kit Sum, K. 1988. Recent developments in resonant power conversion. Intertech Communications Inc., CA.

Liu, K.H. and Lee, F.C. 1986. Zero-voltage switching technique in DC/DC converters. In *IEEE Power Electronics Specialists Conference Record*, pp. 58–70, Vancouver.

Liu, K.H., Oruganti R., and Lee, F.C. 1985. Resonant switches—Topologies and characteristics. In *IEEE Power Electronics Specialists Conference Record*, pp. 106–116. Taipei.

Motorola. 1989. *Linear/Switchmode Voltage Regulator Handbook*. Arizona.

Philips. 1991. *Power Semiconductor Applications*. Philips Semiconductors, Netherlands.

Rashid, M.H. 1988. *Power Electronics: Circuits, Devices, and Applications*. Prentice-Hall, Englewood Cliffs, NJ.

Severns, R. and Bloom, G. 1988. *Modern Switching DC-to-DC Converters*. Van Nostrand, New York.

Steigerwald, R.L. 1988. A comparison of half-bridge resonant converter topologies. *IEEE Trans. on Power Electronics* PE-3(2):174–182.

Unitrode. 1984. *Unitrode Switching Regulated Power Supply Design Seminar Manual*. Unitrode Corporation, Lexington, KY.

Yang, E.X., Lee, F.C., and Jovanovic, M.M. 1992. Small-signal modeling of series and parallel resonant converters. *IEEE PESC*, pp. 785–792. Taipei.

Further Information

The following monthly magazines and conference records publish papers on the analysis, design and experimental aspects of DC-to-DC converter configurations and their applications:

IEEE Applied Power Electronics Conference Records
IEEE Industry Applications Conference Records
IEEE International Telecommunications Energy Conference Records
IEEE Transactions on Aerospace and Electronic Systems
IEEE Transactions on Industrial Electronics
IEEE Transactions on Industry Applications
IEEE Transactions on Power Electronics
IEEE Power Electronics Specialists Conference Records

For information and subscriptions or ordering any of the listed journals or conference records contact: IEEE Services Center, 445 Hoes Lane, P.O. Box 1331, Piscataway, NJ 08855-1331.

10.7   Power Distribution and Control

Badrul H. Chowdhury

10.7.1   Introduction

The power distribution system is that part of the power system which is concerned with delivering the power generated at remote generating stations to the consumers. Its place in the entire scheme of the power generation and transmission system is shown in Fig. 10.93. Transmission voltages can range up to 760 kV and subtransmission and distribution voltages are successively lower. Power from bulk sources is stepped down at substations for distribution to different points in a service area. Primary distribution feeders, originating at the substations, carry power at voltages such as 2.4, 4.16, and 13.8 kV or higher. Distributions transformers are used to step down from these primary voltage levels to 120 or 240 V for distribution over secondary lines to the individual customer's connection point. A generic scheme for distributing power is shown in Fig. 10.94. The primary system consists of the circuit between the distribution **substation** and the distribution **transformers**. A primary **feeder** includes a main feeder; usually a three-phase, four wire circuit; and lateral and sublateral branches tapped from the primary mains. The lateral circuits can be either single phase or three phase. **Fuses** and **reclosers** are freely used on feeders in order to minimize the portion of the circuit removed during faults.

The most common type of primary feeder is of the radial type, wherein the main feeder branches off into several lateral and sublateral branches. The current magnitude is the greatest in the circuit conductors that leave the substation and the smallest in the conductors farthest from the substation. Hence, if the same size of conductors are used in all circuits, then the capacity of those conductors which are far from the substation are underutilized. The reliability of service continuity of radial systems is low since the

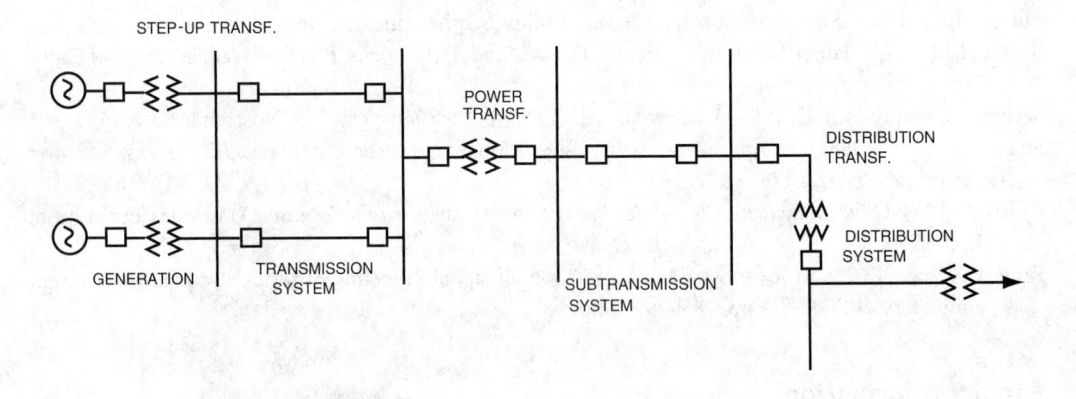

FIGURE 10.93   One-line diagram of a generation-transmission-distribution system.

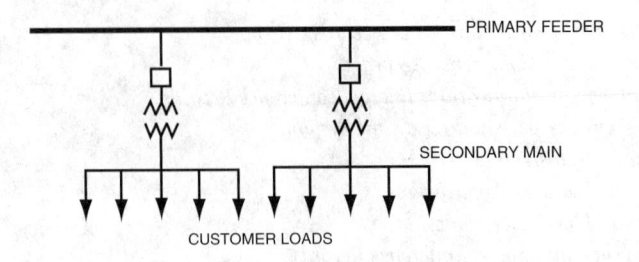

FIGURE 10.94   A generic primary distribution system.

occurrence of a fault at any location on the radial system causes power outage for all of the customers downstream on the feeder. If a higher service reliability is desired, the primary system could be designed as a loop where the feeder loops through the loads and returns to its origin, the substation. Thus, if a faulted section is isolated, the remaining sections can still be served, irrespective of their location. A specific design, called *primary networking*, is one where adjacent primary feeders emanating from different substations are interconnected together. This system has the capability of serving a load from several different directions, thus providing higher service reliability than the other designs.

The secondary system, as seen in Fig. 10.94, is that part of the distribution system that originates at the distribution transformers and ends at the consumer loads. Each secondary main may supply groups of customers. In some instances where greater service reliability is incorporated into the design, the secondary mains of several adjacent transformers may be connected together through a fuse or a recloser. This is referred to as *secondary banking*. If an even higher service reliability is desired in relation to secondary banking, the secondary mains in an area can all be connected together in a mesh or a network, similar to primary networking.

Single-Phase vs. Three-Phase Primary Systems

Single-phase circuits are mostly found in secondary systems and are typically used for supplying small motor and lighting loads. On the other hand, three-phase systems are used to supply larger industrial loads and are inherently more economical than single-phase systems. The following are the nominal voltage standards for most electric utilities serving residential and commercial customers in the U.S.:

- 120/240-V three-wire single-phase
- 120/240-V four-wire three-phase delta
- 208 Y/120-V four-wire three-phase wye
- 480 Y/277-V four-wire three-phase wye

The single-phase two-wire system, shown in Fig. 10.95(a), consists of a two-conductor circuit with constant voltage available between the conductors. The load is connected across the two lines. For safety reasons, one of the conductors can be grounded to form the neutral line. The single-phase, three-wire system, shown in Fig. 10.95(b), is the equivalent of two, two-wire systems combined so that a single wire serves as one wire of each of the two systems, the neutral. Hence, if one outer wire has a potential of 120 V above the neutral, the other outer wire will be 120 V below the neutral. The voltage between the two outer conductors will be 240 V.

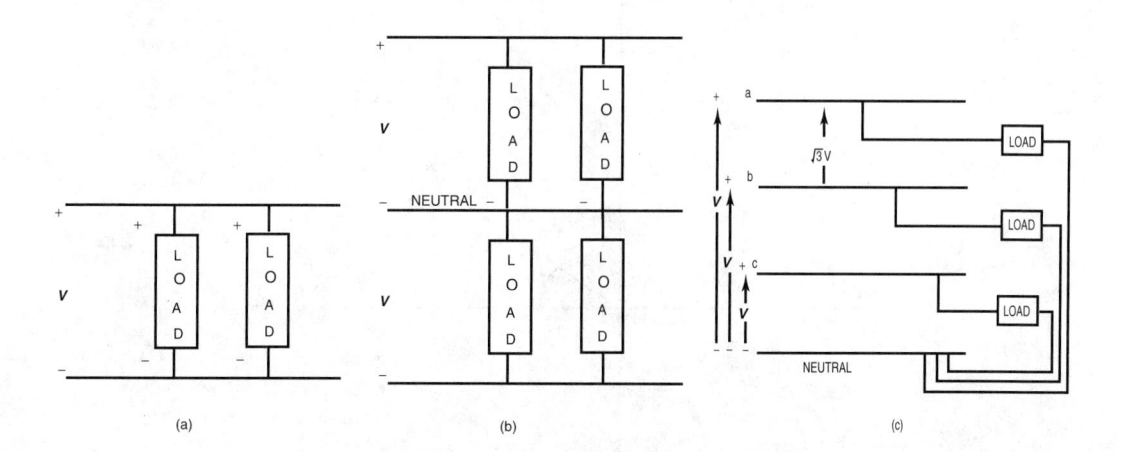

FIGURE 10.95 Distribution line arrangement.

In the three-phase, four-wire system, shown in Fig. 10.95(c), the voltage between each phase conductor and the neutral conductor is equal to 120 V. The individual phase voltages are also 120° out of phase with each other. If the positive sequence is followed, the a-phase leads the b-phase and the b-phase leads the c-phase by 120°. The voltages between the phases are $\sqrt{3}$ times the phase-to-neutral voltages of 120 V. Single-phase loads are connected between each phase and the neutral. If the loads in each phase can be balanced, then the currents in all three lines are also equal in magnitude with the phase angles being 120° out of phase. Under this condition, no current exists in the neutral conductor.

In the next section, the basics of complex algebra are introduced. Complex algebra is required whenever alternating current analysis is desired. Since most distribution systems are of the AC type, this tool is a necessity. In succeeding sections, distribution line equipment and substation equipment are discussed. Voltage analysis in power distribution systems are described at the end.

10.7.2　Basics of AC Analysis

Complex Numbers

A complex number is represented by a real part and an imaginary part. For example, in $A = a + jb$, A is the complex number; a is real part, sometimes written as Re(A); and b is the imaginary part of A, often written as Im(A). It is a convention to precede the imaginary component by the letter j (or i). This form of writing the real and imaginary components is called the *Cartesian* form and symbolizes the complex (or s) plane, wherein both the real and imaginary components can be indicated graphically. To illustrate this, consider the same complex number A when represented graphically as shown in Fig. 10.96. A second complex number B is also shown to illustrate the fact that the real and imaginary components can take on both positive and negative values. Figure 10.96 also shows an alternate form of representing complex numbers. When a complex number is represented by its magnitude and angle, for example, $A = r_A \underline{/\theta_A}$, it is called the *polar* representation.

To see the relationship between the Cartesian and the polar forms, the following equations can be used:

$$r_A = \sqrt{a^2 + b^2} \tag{10.184}$$

$$\theta_A = \tan^{-1} \frac{b}{a} \tag{10.185}$$

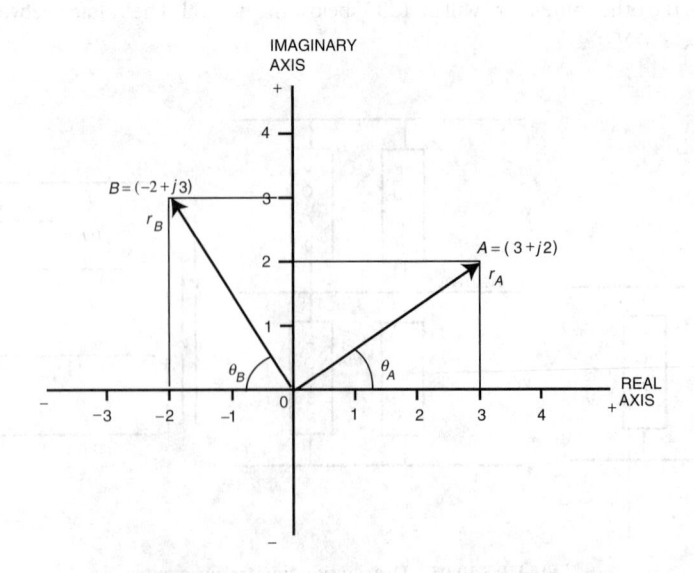

FIGURE 10.96　The s plane representing two complex numbers.

Conceptually, a better insight can be obtained by investigating the triangle shown in Fig. 10.97, and considering the trigonometric relationships. From this figure, one can see that

$$a = \text{Re}(A) = r_A \cos(\theta_A) \qquad (10.186)$$
$$b = \text{Im}(A) = r_A \sin(\theta_A) \qquad (10.187)$$

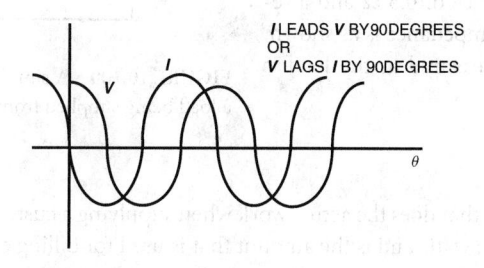

FIGURE 10.97 The relationship between Cartesian and polar forms.

Euler's Identity

The well-known Euler's identity is a convenient conversion of the polar and Cartesian forms into an exponential form, given by

$$\exp(j\theta) = \cos\theta + j\sin\theta \qquad (10.188)$$

Phasors

AC voltages and currents, appearing in distribution systems, can be represented by **phasors**, a concept useful in obtaining analytical solutions of 1-ϕ and 3-ϕ systems. A phasor is generally defined as a transform of sinusoidal functions from the time domain into the complex-number domain and given by the expression shown in Eq. (10.189) as

$$\boldsymbol{V} = V\exp(j\theta) = P\{V\cos(\omega t + \theta)\} = V\underline{/\theta} \qquad (10.189)$$

where $\boldsymbol{V}$ is the phasor, V is the magnitude of the phasor, and θ is the angle of the phasor. The convention used here is to use boldface symbols to symbolize phasor quantities. Graphically, in the time domain, the phasor $\boldsymbol{V}$ would be a simple sinusoidal wave shape as shown in Fig. 10.98. The concept of a phasor *leading* or *lagging* another phasor becomes very apparent from the figure. Phasor diagrams are also another effective medium of understanding the relationships between phasors. Figure 10.99 shows a phasor diagram for the phasors represented in Fig. 10.98. In this diagram, the convention of positive angles being read counterclockwise is used. The other alternative is certainly possible as well. It is quite apparent that a purely capacitive load could result in the phasors shown in Fig. 10.98 and Fig. 10.99.

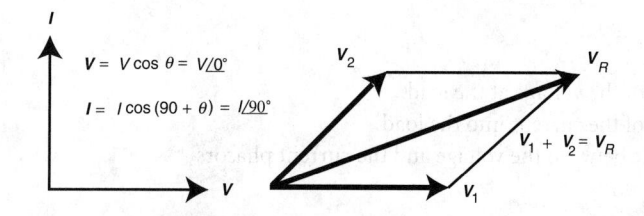

FIGURE 10.98 Wave forms representing leading/lagging phasors.

FIGURE 10.99 Phasor diagram showing phasor representation and phasor operation.

The Per Unit System

In the per unit system, basic quantities such as voltage, current, etc., are represented as certain percentages of base quantities. When so expressed, these per unit quantities do not need units, thereby making numerical analysis in power systems somewhat more easy to handle. Four quantities, namely, voltage, current, power, and impedance, encompass all variables required to solve a power system problem. Out of these, only two base quantities, corresponding to voltage (V_b) and power (S_b), are required to be defined. The other two base quantities can be derived from these two. Consider the following. Let

V_b = voltage base, kV
S_b = power base, MVA
I_b = current base, A
Z_b = impedance base, Ω

Then

$$Z_b = \frac{V_b^2}{S_b} \quad \Omega \tag{10.190}$$

$$I_b = \frac{V_b 10^3}{Z_b} \quad A \tag{10.191}$$

Power Calculations

Four different power definitions are encountered in power system calculations. These are (with their symbolic representation in parentheses): real power (P), reactive power (Q), complex power (S), and apparent power (T). Their individual definitions and relationships are discussed next with reference to Example 1.

Example 1

Figure 10.100 shows a single-phase distribution system with a single source supplying a load through a feeder which has a resistance of 0.5 Ω and a reactance of 1.2 Ω. The load impedance is as shown. Assume that the voltage measured at the load is $V_L = 10\underline{/0°}$.

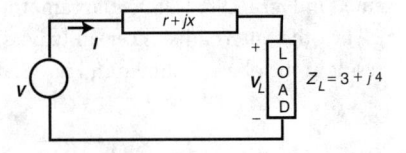

FIGURE 10.100   A simple single-phase circuit showing a load being supplied from a single source.

Real Power

This is the amount of power that does the actual work when supplying a customer's load; hence, the name **real power**. It is measured in watts and is the amount that is used for billing customers.

$$P = V_L I \cos\theta \quad W \tag{10.192}$$

where
V_L = rms value of the voltage at the load
I = rms value of the current into the load
θ = phase angle between the voltage and the current phasors

In Fig. 10.100, $V_L = 10$ V, $Z_L = 3 + j4 = 5\underline{/53.13°}$, $I = V_L/Z_L = 10\underline{/0°}/5\underline{/53.13°}$ A $= 2\underline{/-53.13°}$ A. Hence, $I = 2$ A; $\theta = -53.13°$. Therefore, from Eq. (10.192), $P = 12$ W.

Reactive Power

Reactive power is the amount of power that travels between the source and the load, without doing real work. It is the amount responsible for changing the power factor of the load. The reactive power is measured in var (or kvar).

$$Q = V_L I \sin\theta \quad \text{vars} \tag{10.193}$$

In Fig. 10.100, $Q = 16$ var.

Complex Power

When the real and reactive powers are expressed as the real and imaginary components, respectively, of a complex number expression, the latter form is called complex power, and the unit used is megavoltampere

$$S = P + jQ \quad \text{MVA} \tag{10.194}$$

Apparent Power

The magnitude of the complex power S is called the apparent power. Thus

$$T = |S| = \sqrt{P^2 + Q^2} \tag{10.195}$$

From Eqs. (10.192) and (10.193), one can see that

$$T = VI \tag{10.196}$$

In Fig. 10.100, $S = 20\underline{/53.13^\circ}$ MVA; $T = 20$ MVA.

Often the definition of the power terms can be best understood with the help of a power triangle as shown in Fig. 10.101. Note the relationship of this triangle with that of Fig. 10.97.

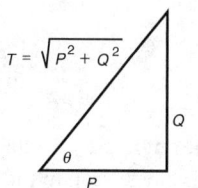

FIGURE 10.101 The power triangle.

Power Factor

The **power factor** (pf) of a load is defined as the cosine of the angle between the voltage across and the current through the load. Also, from Eq. (10.192) and Eq. (10.196), it is easy to see that the power factor is also the ratio of the real power and the apparent power. The power factor is further quantified by either a *leading* or a *lagging* label to represent capacitive or inductive loads, respectively. Thus

$$\text{pf} = \cos(\underline{/V} - \underline{/I}) = P/T \tag{10.197}$$

In Example 1, verify that the power factor of the load is 0.6 lagging.

Power Factor Correction

Often, the loads at industrial demand points can cause consumption of large reactive components, causing low power factors. It is easy to see from the power triangle in Fig. 10.101 that the power factor angle is directly related to the reactive power and that this angle can be made to go from lagging to leading by changing the amount of reactive power consumed at the load to reactive power injected into the line at the load. Capacitors are typically used in parallel with the loads in order to produce reactive power for suppressing the reactive power consumption. This reduction in reactive power consumed results in a reduced total current, thereby causing reduced power losses.

To illustrate power factor correction calculations, consider the same load of Example 1. It consumes $P = 12$ W of real power and $Q = 16$ var of reactive power to produce a power factor of 0.6 lagging. Assume that we want to improve the power factor to 0.9 lagging. Thus, we need to size a shunt capacitor to provide the necessary correction. From the power triangle, it is easy to that the new power factor is given by

$$\cos\left[\tan^{-1}\frac{Q_2}{P}\right] = 0.9 \tag{10.198}$$

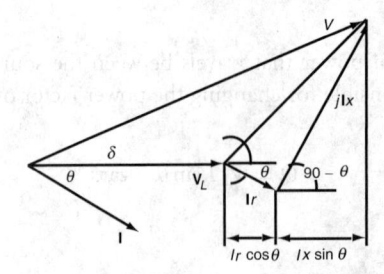

FIGURE 10.102 The phasor diagram for the circuit of Fig. 10.100.

where Q_2 represents new reactive power consumption at the load. If Q_c represents the amount of reactive power supplied by the capacitor, then

$$Q_c = Q - Q_2 \tag{10.199}$$

Voltage Regulation

The per cent voltage drop of a line with respect to the receiving-end voltage is termed the voltage regulation. It is defined as

$$\text{voltage regulation} = \frac{|V| - |V_L|}{|V_L|} \tag{10.200}$$

where $|V|$ is the sending-end voltage magnitude and $|V_L|$ is the receiving or load-end voltage magnitude. The higher the voltage drop in the line $|V| - |V_L|$, the higher is the regulation. The voltage at the source end can be found by the following relation:

$$V = V_L + I(r + jx) \tag{10.201}$$

Equation (10.201) can then be used to find the voltage drop in the line. However, an approximate relation can be found for the voltage drop by investigating the phasor diagram for the circuit shown in Fig. 10.100. The phasor diagram is shown in Fig. 10.102.

As seen in the figure, angle θ is the power factor angle. If the angle δ is small, then the approximate voltage drop is given by

$$\text{voltage drop} = |V| - |V_L| = Ir \cos \theta + Ix \sin \theta \tag{10.202}$$

For the problem of Example 1, Eq. (10.200) yields a voltage drop of 2.53 V whereas Eq. (10.202) yields a drop of 2.52 V. The difference in results from the two equations is, therefore, negligible.

Three-Phase Systems

If the 3-ϕ system is balanced, or close to being balanced, then a number of simplifying assumptions can be used to make the problem more tractable. For examples real power can be assumed to be three times the power consumed by a single phase.

Mathematically

$$P_{3\phi} = 3(VI \cos \theta) \quad \text{and} \quad Q_{3\phi} = 3(VI \sin \theta)$$

A significant advantage is achieved since only one phase of the balanced three-phase system can be used to determine voltages and currents in all three phases. This is illustrated by the following example.

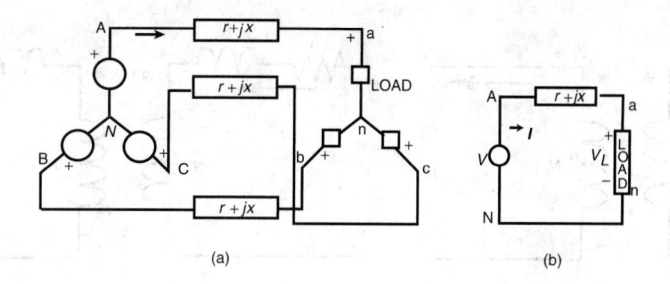

(a)   (b)

FIGURE 10.103 (a) A three-phase power system with wye-connected source and load. (b) the single-phase equivalent circuit.

Example 2

Consider the three-phase power system shown in Fig. 10.103(a). Assume the following data for the system:

- Phase voltages: $V_{AN} = 150\underline{/0°}$, $V_{BN} = 150\underline{/-120°}$, $V_{CN} = 150\underline{/120°}$.
- Loads connected in wye: $Z_{an} = Z_{bn} = Z_{cn} = 35 + j25\ \Omega = 43\underline{/35.54°}\ \Omega$.
- Transmission line: identical in each phase; $r = 5\ \Omega$, $x = 15\ \Omega$.
- Find the line currents and the voltages across the loads.

Since the system is fully balanced, we can use the single-phase equivalent circuit, shown in Fig. 10.103(b), for the analysis. Using Kirchhoff's circuit laws

$$I_{Aa} = V_{AN}/(r + jx + Z_{an}) = 2.65\underline{/-45°}\quad \text{A.}$$

Hence, $V_{an} = I_{Aa} Z_{an} = 114\underline{/-9.46°}$ V. Therefore, the desired results are

- Voltages:

$$V_{an} = 114\underline{/-9.46°}\quad V_{bn} = 114\underline{/-129.46°}\ \text{V}\quad V_{cn} = 114\underline{/110.54°}\ \text{V}$$

$$V_{ab} = 197.5\underline{/20.54°}\quad V_{bc} = 197.5\underline{/-99.46°}\ \text{V}\quad V_{ca} = 197.5\underline{/140.54°}\ \text{V}$$

- Currents:

$$I_{Aa} = 2.65\underline{/-45°}\ \text{A}\quad I_{Bb} = 2.65\underline{/-165°}\ \text{A}\quad I_{Cc} = 2.65\underline{/75°}\ \text{A}$$

The complete phasor diagram is shown in Fig. 10.104.

10.7.3 Distribution Line Equipment

Transformers

Transformers are the most common pieces of equipment found in all substations and distribution lines. Different types of transformers are used for varied purposes, from voltage level changes to phase angle regulation. Since the primary circuit of distribution systems are designed for high voltages in order to increase its load carrying capability, the voltage must be stepped down at the consumer points to ensure proper operation of all devices and equipment connected there. Thus, the most popular use for transformers is to change the voltage level from one side to another.

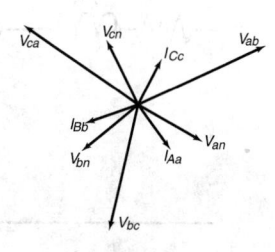

FIGURE 10.104 The complete phasor diagram for Example 2.

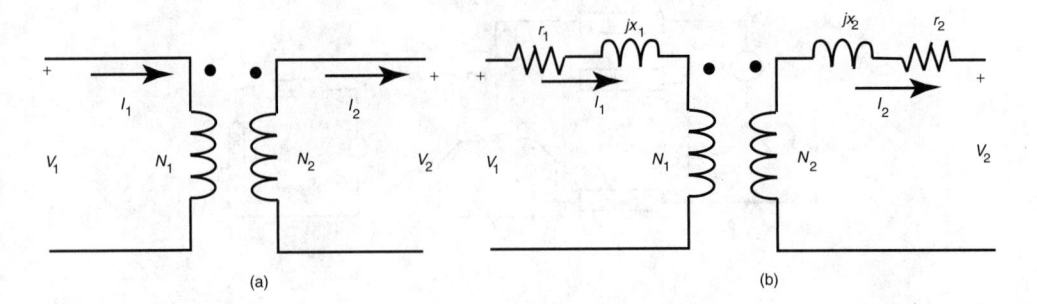

(a)   (b)

FIGURE 10.105   (a) Equivalent circuit of an ideal transformer. (b) approximate equivalent circuit of a nonideal transformer.

Transformers can be classified as *distribution* transformers and *power* transformers. The former type normally steps down voltages from primary voltage levels, such as 2400, 4160, or 13,800 V to 120 or 240 V. They are almost always located outdoors where they are hung from crossarms, mounted on poles directly, or placed on platforms. Power transformers are larger in size than distribution transformers and usually have auxiliary means for cooling. These transformers are typically installed at distribution substations for stepping down voltages from the subtransmission levels of 34.5 and 69 kV to primary distribution levels of up to 13.8 kV.

Single-Phase Transformers

A transformer consists of two (sometimes three) electrically isolated copper windings wound on an iron core. The choice of a ferromagnetic material for the core is to allow for high permeability so that the magnetic flux can be established easily. The core is usually laminated to reduce losses. Figure 10.105 shows a general representation of a single-phase transformer with the two windings on each side representing the primary and the secondary sides. In the figure subscript 1 denotes the primary quantities and subscript 2 signifies the same on the secondary side. For the ideal transformer of Fig. 10.105(a), the following relationship holds:

$$\frac{V_1}{V_2} = \frac{N_1}{N_2} = \frac{I_2}{I_1} = a \tag{10.203}$$

where N = the number of turns in the windings and a is the transformation ratio.

The nonideal characteristics of a transformer include core and winding losses, presence of leakage fluxes, and finite permeability of the core. Hence, the actual model should include the physical representations of these nonideal characteristics. This is shown in Fig. 10.105(b), where the shunt magnetization and core loss components have been ignored for the sake of convenience. Such approximations are common in transformer analysis and only cause trivial inaccuracies in computation. Figure 10.106 shows the equivalent circuit of Fig. 10.105(b) with the secondary-side impedance referred to the primary side.

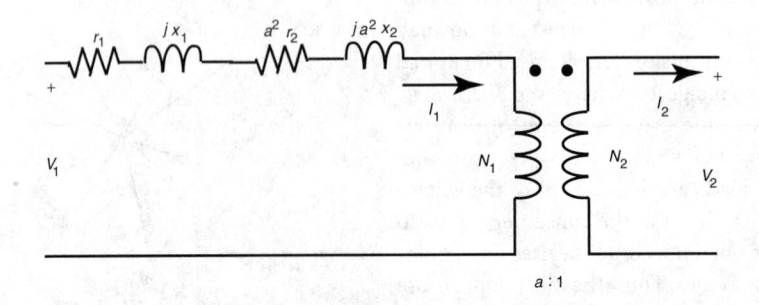

FIGURE 10.106   The approximate equivalent circuit as seen by the primary side.

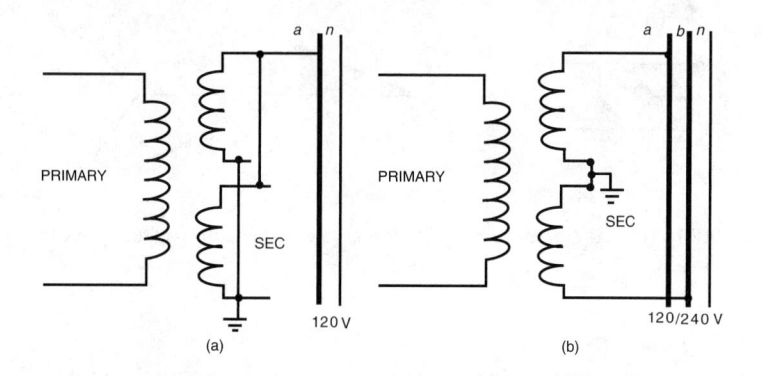

FIGURE 10.107 Single-phase distribution transformer connections.

Thus, the total impedance on the primary side is

$$Z_{eq} = \left(r_1 + a^2 r_2\right) + j\left(x_1 + a^2 x_2\right) \tag{10.204}$$

To find the voltage regulation of the transformer, one can use the following equations:

$$V_1 = a V_2 + I_1 Z_{eq} \tag{10.205}$$

where $I_1 = I_2/a$.

Single-phase distribution transformers typically have one high-voltage primary winding and two low-voltage secondary windings, which are rated at a nominal 120 V. The secondary coils may be connected in parallel to supply a two-wire 120-V circuit, shown in Fig. 10.107(a), or in series to supply a three-wire 120/240-V single circuit, shown in Fig. 10.107(b). As seen in the figures, one leg of the 120-V, two-wire system, and the middle leg of the 120/240-V, three-wire system is grounded to limit the voltage to ground on the secondary side.

In general, the 120/240-V three-wire connection is preferred since it has twice the load capacity of the 120-V system with only one-and-one-half times the amount of conductor. Each 120-V winding has one-half the total kilovoltampere rating of the transformer. The loads on the secondary side of the transformers are kept as nearly balanced as possible such that maximum transformer capacity can be utilized, as well as the neutral current kept to a minimum.

Three-Phase Transformers

A three-phase transformer bank can be easily created by using three single-phase transformers. The two sides of these three transformers can be either connected as a wye or a delta, thus allowing four possible types of connections: (1) wye-wye, (2) wye-delta, (3) delta-wye, and (4) delta-delta. These possibilities are shown in Fig. 10.108 to Fig. 10.112. As can be observed in the figures, the primary and secondary phase windings are drawn in parallel.

Another possible way of producing a three-phase transformer is to use a three-legged core with the primary and secondary windings of each phase placed on the three legs. Such a construction has an advantage over the bank construction in that a smaller amount of core material is used. However, the three-phase bank has higher versatility since it can operate even with one phase out of operation.

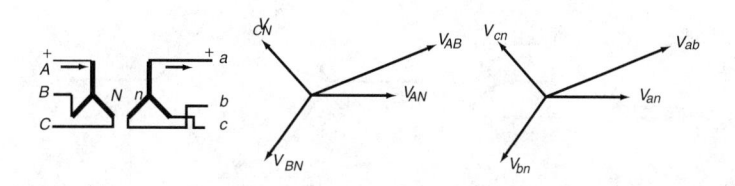

FIGURE 10.108 Y–Y transformer with 0° phase shift between primary and secondary.

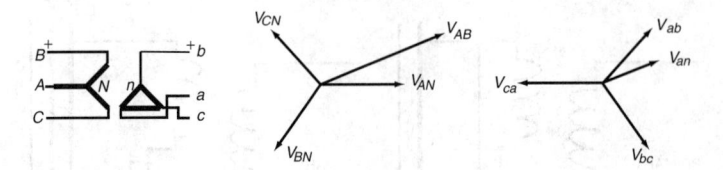

FIGURE 10.109   Y–Δ transformer with primary side lagging the secondary side by 30°.

The Wye–Wye Connection

With the wye–wye (Y–Y) connection, the secondary side is in phase with the primary circuit, and the ratio of primary to secondary voltage is the same as the ratio of turns in each of the phases. A possible connection is shown in Fig. 10.108.

Power distribution circuits supplied from a wye–wye bank often create series disturbances in communication circuits (e.g., telephone interference) in their immediate vicinity. However, one of the advantages of this connection is that when a system has changed delta to a four-wire wye to increase system capacity, existing transformers can be used.

The Wye–Delta Connection

In the Y–Δ connection, there is a 30° phase angle shift between the primary and secondary sides. The phase angle difference can be made either lagging or leading depending on the external connections of the transformer bank. The case with the primary side lagging is shown in Fig. 10.109, and the case with the primary side leading is shown in Fig. 10.110. The transformation ratio is $\sqrt{3}$ times the ratio of turns in each of the phases.

The Delta–Wye Connection

With this connection, the neutral of the secondary wye can be grounded and single-phase loads can be connected across the phase and the neutral conductors and three-phase loads can be connected across the phases. The phasor relationship between the primary and the secondary sides is shown in Fig. 10.111. The transformation ratio is $1/\sqrt{3}$ times the ratio of turns in each of the phases.

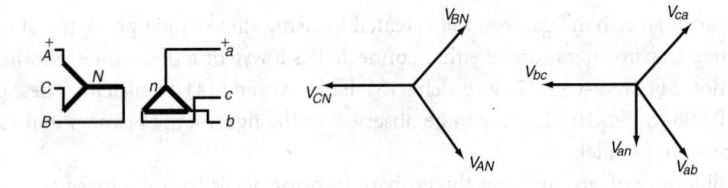

FIGURE 10.110   Y–Δ transformer with primary side leading the secondary side by 30°.

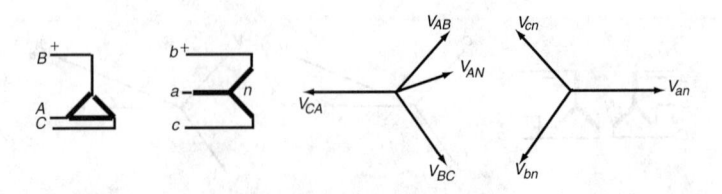

FIGURE 10.111   Δ–Y transformer with primary side leading the secondary side by 30°.

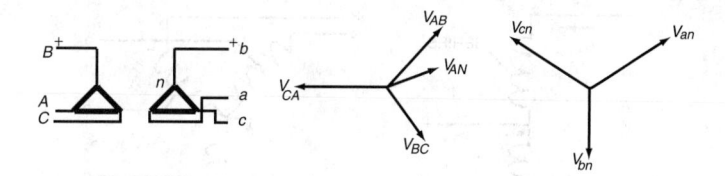

FIGURE 10.112 Δ–Δ transformer with 0° phase shift between primary and secondary sides.

The Delta–Delta Connection

This connection does not cause a phase shift between the primary and the secondary sides. The phasor relationship of this transformer is shown in Fig. 10.112. The transformation ratio is equal to the ratio of the turns in each of the phases. There is no problem from third-harmonic overvoltage or telephone interference since such disturbances get trapped in the delta and do not pass into the lines.

Other Three-Phase Transformer Connections

Open–Delta Connection

An advantage of the Δ–Δ connection is that if one of the single-phase transformers becomes damaged or removed for maintenance, the remaining two can be operated in the so-called open-delta connection. Since the currents in each of the two remaining transformers is the same as the line current, each transformer therefore carries $\sqrt{3}$ times the current it was carrying in the closed-delta connection. The open-delta bank continues to deliver three-phase currents and voltages in their correct phase relationship. To keep the transformers from being overloaded, however, it would be necessary to reduce the line currents by $1/\sqrt{3}$. Thus, the comparison of the three-phase capacity can be shown as follows:

$$S_{\text{closed}\Delta} = \sqrt{3}V_L I_L \tag{10.206}$$

$$S_{\text{open}\Delta} = \frac{\sqrt{3}V_L I_L}{\sqrt{3}} \tag{10.207}$$

$$\frac{S_{\text{open}\Delta}}{S_{\text{closed}\Delta}} = \frac{1}{\sqrt{3}} \tag{10.208}$$

Hence, even though the open-delta connection constitutes two-thirds or 66.67% of the original Δ–Δ connection, it can only supply 57.7% of the three transformers combined.

Scott or T Connection

The Scott or T connection is used when a two-phase (or a transformed three-phase) supply is needed from a three-phase system. In general, the T connection is used for getting a three-phase transformation, and the Scott connection is mainly used for obtaining a two-phase system. The two connection types are basically similar in design. Either connection requires two specially wound, single-phase transformers. The main transformer has a 50% tap on the primary winding, whereas the other transformer, called the *teaser* transformer, has an 86.6% tap. The main transformer is connected between two primary lines, whereas the teaser transformer is connected from the center tap of the main transformer to the third primary line. The secondary sides of the transformers are connected to serve a two-phase system. A Scott connection is shown in Fig. 10.113.

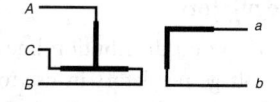

FIGURE 10.113 The T connection for three-phase to two-phase transformation.

Autotransformer

This is a specific type of transformer wherein the primary and secondary coils are connected in series with each other. The advantages of this connection are that it has lower losses, smaller excitation current

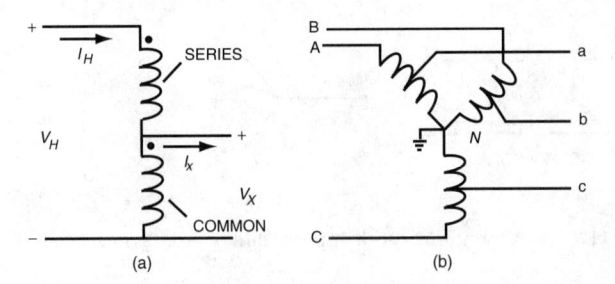

FIGURE 10.114   The autotransformer connection: (a) single phase (b) three phase.

requirement, weighs less, and is cheaper than the equivalent two-winding transformer. However, a specific disadvantage is the absence of electrical isolation between the primary and secondary sides. The single-phase autotransformer connection is shown in Fig. 10.114(a). The series and common windings are indicated in the figure. The corresponding three-phase connection is shown in Fig. 10.114(b).

Tap-Changing Transformers

Often, it becomes necessary to change the voltage in the transformer winding to allow for a varying input voltage on the transformer. Hence, the output voltage is held close to a constant level irrespective of the input voltage level. Changing taps is directly related to changing the transformation ratio. Two specific types of tap changers are available. These are (1) the no-load tap changer and (2) the under-load tap changer (ULTC). A transformer equipped with a no-load tap changer must always be disconnected from the circuit before the tap can be changed. On small distribution transformers, the covers must be removed, and an operating handle is used to make the tap change. For larger transformers, a handle is usually brought out through the cover, and the tap movement is brought about either manually by the use of a wheel or a motor. Under-load tap changers use the autotransformer principle and elaborate switching mechanisms. The ULTC is discussed in the section on voltage regulators.

Capacitors

Capacitors are used in distribution lines to keep line voltages at certain prespecified levels. Under heavy load conditions, the voltage has a tendency to sag. Capacitors can be used effectively in these situations to bring the voltage back up to the desired level. Capacitors also have an added capability of correcting the power factor at close to the load points.

A capacitor consists of two conducting metal plates separated by a dielectric insulating medium, such as an oil-impregnated paper. Capacitor sizes can vary from the 15–25 kvar range to the 200–300 kvar range. Capacitors can also be installed in banks with bank sizes ranging from 300 to 1800 kvar. Typically, provision is made for switching some of the units in the bank on and off as required to maintain a suitable voltage regulation. Distribution capacitors are mounted on crossarms or platforms.

Voltage Regulators

Voltage regulation on distribution lines can also be obtained by devices called voltage regulators. The majority of voltage regulators in use today are of the step type. These regulators can be used both in substations for bus voltage or individual feeder voltage regulation and pole mounted on overhead primary feeders.

The operating principle of the voltage regulator is somewhat similar to tap-changing transformers. Several taps are provided in the series winding. Most regulators are designed to correct the line voltage from 10% boost to 10% buck in 32 equal steps.

A voltage regulator also comes equipped with two other major components, namely, the tap-changing mechanism and the control mechanism. Line voltage sensing is provided through instrument transformers.

TABLE 10.6 Overhead Conductor Characteristics

Conductor	Type	Res, Ω/mi	Diam, in	GMR, ft	Ampere
1,000,000	AA	0.105	1.15	0.03680	698
556,500	ACSR	0.186	0.927	0.03110	730
500,000	AA	0.206	0.813	0.02600	483
336,400	ACSR	0.306	0.721	0.02440	530
4/0	ACSR	0.592	0.563	0.00814	340
2/0	AA	0.769	0.414	0.01250	230
1/0	ACSR	1.120	0.398	0.00446	230
1/0	CU	0.607	0.368	0.01113	310
#2	AA	1.540	0.292	0.00883	156
#2	ACSR	1.690	0.316	0.00418	180
#4	ACSR	2.550	0.257	0.00452	140
#10	CU	5.903	0.102	0.00331	80
#12	CU	9.375	0.081	0.00262	75
#14	CU	14.872	0.064	0.00208	20

The automatic voltage maintenance on the output of voltage regulators is achieved by dial settings (located on the control panel) of the adjustable resistance and reactance elements of a unit called the line drop compensator.

Overhead Conductors

Many different types of conductors are used on overhead distribution lines. They vary both in size and number, depending on the voltage level and the type of circuit. Copper, aluminum, and steel are the most commonly used material for overhead lines. Copper is used in three forms: hard drawn, medium-hard drawn, and soft drawn or annealed. Hard-drawn copper has the greatest strength and is used for circuits of relatively long spans (200 ft or more). However, its inflexibility makes it harder to work with. The soft-drawn variety is the weakest of the copper conductors. Its use is limited to short spans. The medium-hard-drawn copper conductor has found more widespread use in the medium range distribution circuits.

Steel wire is only about one-tenth as good a conductor as copper and, hence, is rarely used alone. However, it offers an economic advantage over the other types of conductors. Also, since steel wire is much stronger than copper, it permits longer spans and requires fewer supports.

Aluminum is only 60–80% as good a conductor as copper and only half as strong as copper. However, its property of lighter weight, as compared to copper and steel, and its relative advantage in transmitting AC power because of reduced skin effect, makes it suitable for overhead lines. Usually the aluminum wires are stranded on a core of steel wire to form what is termed as aluminum conductors steel reinforced (ACSR) conductor. The more strands in the ACSR conductor, the greater is its flexibility. Hence, today's larger conductors are all stranded and twisted in layers concentrically around a central steel wire.

Table 10.6 lists the characteristics of various conductors that are typically used on overhead distribution lines (IEEE, 1991).

Underground Cables

Underground construction of distribution lines is designed mostly in urban areas and is dictated by economics, congestion, and density of population. Although overhead lines have been ordinarily considered to be less expensive and easier to maintain, developments in underground cable and construction technology have narrowed the cost gap to the point where such systems are competitive in many urban and suburban residential installations.

The conductors used underground are different from overhead conductors in that they are insulated for their entire length and several of them may be combined under one protective sheath. The whole assembly is called an electric cable. These cables are either buried directly in the ground, or they may be installed in ducts buried in the ground.

The conductors in cables are usually made of copper or aluminum and are usually stranded. They are made of soft-drawn copper since they do not have to support any weight. Cables can be either single conductor or multiple conductor enclosed in a single sheath for economy.

Fuses and Fuse Cutouts

To prevent circuit overloads, short pieces of metal having low melting characteristics are installed in circuits. Fuses are designed to melt and disconnect the circuits they are placed on should the current in these circuits increase above the thermal ratings. Fuses designed to be used above 600 V are classified as *fuse cutouts*. Oil-filled cutouts are mainly used in underground installations and contain the fusible elements in an oil-filled tank. Expulsion-type cutouts are the most common type used on overhead primary feeders. In this cutout, the melting of the fusible element causes heating of a fiber fuse tube, which, in turn, produces deionizing gases to extinguish the arc. Expulsion type cutouts are classified as (1) open-fuse cutouts, (2) enclosed-fuse cutouts, and (3) open-link-fuse cutouts.

Automatic Reclosers

The recloser is another overcurrent device that automatically trips and recloses a preset number of times to clear or isolate faults. The concept of reclosing is derived from the fact that most faults are of the temporary type and can be cleared by de-energizing the circuit for a short period. Reclosers can be set for a number of operation sequences depending on the action desired. These typically include instantaneous trip and reclose operation followed by a sequence of time-delayed trip operations prior to lockout of the recloser. The minimum pick-up for most reclosers is typically set to trip instantaneously at two times the current rating.

An automatic line recloser is made of an interrupting chamber and the related contacts that operate in oil, a control mechanism to trigger tripping and reclosing, an operator integrator, and a lockout mechanism. An operating rod is actuated by a solenoid plunger that opens and closes the contacts in oil. Both single-phase and three-phase units are available.

Line Sectionalizer

The **line sectionalizer** is yet another overcurrent device, which is installed with backup circuit breakers or reclosers. It has coordination with the backup interrupting device and is designed to open after a preset number of tripping operations of the backup device. They are installed on poles or crossarms in overhead distribution systems. The standard continuous current rating for sectionalizers range from 10 to 600 A. Sectionalizers also come in both the single-phase and the three-phase variety.

Lightning Arrester

A lightning arrester is a device that protects expensive electrical apparatus from voltage surges caused by lightning. It provides a path over which the surge can pass to ground before it has the opportunity to pass through and damage the equipment.

A standard lightning arrester consists of an air gap in series with a resistive element. The resistive element is usually made of a material that allows a low-resistance path to the voltage surge, but presents a high-resistance path to the flow of line energy during normal operation. This material is known as the *valve* element. A common valve element material is silicon carbide. The voltage surge causes a spark that jumps across the air gap and passes through the resistive element to ground.

One type of arrester consists of a porcelain cylinder filled with some suitable material with electrodes at either end. The series gap assembly is usually at the top. In another type of arrester, called the expulsion type, there are two series gaps, one internal and one external. During a voltage surge, these gaps have to be bridged. Another gap formed between two internal electrodes and placed between two internal electrodes is assembled inside a fiber tube that functions as the valve element and serves to resist the flow of line energy when it comes through. This tube helps create nonconducting gases, which blow the arc formed through a vent, thus establishing a wall of resistance to the line energy.

10.7.4 Distribution Substation

Distribution substations serve as the source for primary distribution feeders. They receive bulk electric power at high voltages and reduce the voltage to distribution primary values. Also associated with a substation are provisions for protection from faults, for voltage regulation, and for data acquisition and monitoring.

The principle equipment generally installed in a distribution substation are power transformers, oil or air circuit breakers, voltage regulators, protective relays, air break and disconnect switches, surge arresters, measuring instruments, and in some instances, storage batteries and capacitors.

Some substations are entirely enclosed in buildings whereas others are built entirely in the open with all equipment enclosed in a metal-clad unit. The final design of the type of substation depends on economic factors, future load growth, and environmental, legal, and social issues.

Transformers have been discussed in an earlier section. Hence, the discussion in this section will be limited to **circuit breakers** and protection schemes.

Circuit Breakers

The function of circuit breakers is to protect a circuit from the harmful effects of a fault, in addition to energizing and de-energizing the same circuit during normal operation. Breakers are generally installed on both the incoming subtransmission lines and the outgoing primary feeders. These devices are designed to operate as quickly as possible (less than 10 cycles of the power frequency) to limit the impact of a fault on the expensive equipment installed at the substation. At the same time, the arc that forms between opening contacts also has to be quenched rapidly. Several schemes are available to extinguish the arc. The medium of oil is often used in this process. Some circuit breakers have no oil, but quench the arc by a blast of compressed air. These are called *air circuit breakers*. Yet another type encloses the contacts in a vacuum or a gas, such as sulfur hexafluoride, SF_6.

Air circuit breakers are typically used when fault currents are relatively small. These are characteristically simple, low cost and require little maintenance. The fault current flows through coils creating a magnetic field that tends to force the arc into ceramic chutes that stretch the arc, often with the aid of compressed air.

When the arc is extinguished through vacuum, the breaker is called a *vacuum circuit breaker*. Since a vacuum cannot theoretically sustain an arc, it can be an effective medium. However, owing to imperfections present in vacuum, a small arc of short duration can be produced. The construction of vacuum circuit breakers is simple, but the maintenance is more complex.

Breaker Schemes

The breaker schemes used at substations provide for varying degrees of reliability and maintainability on both the input and output sides. Each additional circuit breaker provides greater reliability and flexibility in maintaining the bus energized during fault or maintenance. However, the cost also increases with each circuit breaker. Hence, the selection of a particular substation scheme depends on safety, reliability, economy and simplicity. The most commonly used breaker schemes are (Gonen, 1986): (1) the single-bus, shown in Fig. 10.115(a), (2) the double-bus, double breaker, shown in Fig. 10.115(b), (3) the main-and-transfer bus, shown in Fig. 10.115(c), (4) the breaker-and-a-half, shown in Fig. 10.115(d), and (5) the ring bus, shown in Fig. 10.115(e).

Of these schemes, the single-bus scheme costs the least; however, it possesses a low reliability since the failure of the bus or any circuit breaker results in a total shutdown of the substation. The most expensive arrangement is the double-bus, double-breaker scheme. Each circuit is provided with two circuit breakers, and thus, any breaker can be taken out of service for maintenance without disruption of service at the

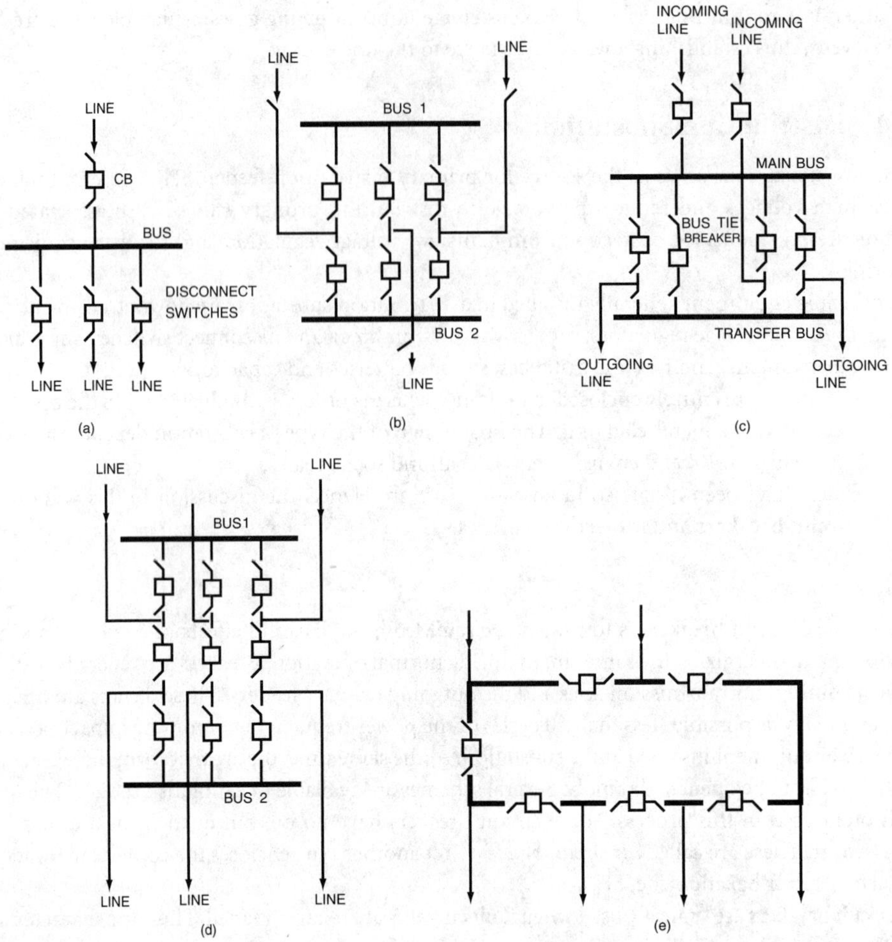

FIGURE 10.115   Substation bus and breaker arrangements.

substation. Also, feeder circuits can be connected to either bus. The main-and-transfer bus requires an extra breaker for the bus tie between the main and the auxiliary buses. The breaker-and-a-half scheme provides the most flexible operation with high reliability. The relaying and automatic reclosing, however, are somewhat complex.

10.7.5   Distribution System Analysis

A number of different mathematical analyses are regularly carried out by distribution planning engineers. Some of the more widely used analytical topics include

- Voltage analysis
- Power loss analysis
- Reactive power analysis
- Peak load analysis
- Motor start analysis
- Protection coordination

Only voltage analysis will be covered here.

Voltage Analysis

Distribution systems are designed to maintain service voltages within specified limits during normal and emergency conditions. Typical voltage limits are

- For service to residential customers, the voltage at the point of delivery shall not exceed 5% above or below the nominal voltage. This is equivalent to the band between 114 and 126 V for most utilities in the U.S.
- For service to commercial or industrial customers, the voltage at the point of delivery shall not exceed 7.5% above or below the nominal voltage.
- The maximum allowable voltage imbalance for a three-phase service should be 2.5%.

The goal of voltage analysis is to determine whether the voltages on different line sections stay within the specified limits under varying load conditions. Thus, voltage analysis facilitates the effective placement of capacitors, voltage regulators, and other voltage regulation devices on the distribution system. Load flow analysis is a computer-aided tool, which is typically used in this planning task. Load flows determine feeder voltages under steady state and at different load conditions.

Voltage analysis starts with a correct representation, or map, of the feeder circuits, starting at the substation. The map generally consists of details and electrical characteristics (e.g., kilovoltampere rating, impedance, etc.) of the conductors and cables on the system, substation and distribution transformers, series and shunt capacitors, voltage regulators, etc.

Before the analysis can begin, feeder loading must be known. Several different methods can be used for this task. If the utility maintains a database on each customer connected to a distribution transformer, it can use the billing data to determine the kilowatt hour supplied by each transformer for a given month. Methods can then be used to convert the kilowatt hour to a noncoincident peak kilovoltampere demand for all distribution transformers connected on the feeder. If this information is not available, the kilovoltampere rating of the transformer and a representative power factor can be used as the load. With the metered demand at the substation, the transformer loads can be allocated, for each phase, such that the allocated loads plus losses will equal the metered substation demand.

Accurately representing the load types or models is an important issue in voltage analysis. Several load models are available. These are

- Spot and distributed loads
- Wye and delta connected loads
- Constant power, constant current, constant impedance, or a combination

In addition, feeder models must also be applied before the analysis can begin. The feeder model can range from the most accurate model, which includes all three phases with exact phasing and spacings of the lines, to an approximate lumped model consisting of lumping the feeder load at the centroid of the feeder area. There are other possible models between the two extremes.

Impedance Calculation

An important step prior to starting voltage analysis is the determination of line impedance. Quite often, impedance tables are used for standard conductors. However, for greater accuracy, *Carson's equations* are recommended (Kersting, 1992).

A modification of Carson's equations is now used widely in calculating self-and mutual impedances in ohm per mile. The equations are

$$Z_{ii} = r_i + 0.0953 + j0.12134(\ell_n(1/GMR_i) + 7.93402) \tag{10.209}$$

$$Z_{ij} = 0.0953 + j0.12134(\ell_n(1/D_{ij}) + 7.93402) \tag{10.210}$$

where

Z_{ii} = self impedance of conductor i
Z_{ij} = mutual impedance between conductors i and j
r_i = resistance of conductor i
GMR_i = geometric mean radius of conductor i, ft
D_{ij} = distance between conductors i and j, ft

Equations (10.206) and (10.207) are used to compute a square $n \times n$ *impedance matrix*, where Z_{ii} are the diagonal terms and Z_{ij} are the symmetrical off-diagonal terms. Thus, for a four-wire overhead wye line section, it will result in a 4×4 matrix.

The impedance matrix needs to be reduced to a 3×3 *phase frame matrix* consisting of the self- and mutual equivalent impedances for the three phases. A standard method of reduction is the Kron reduction where the assumption is made that the line has a multigrounded neutral. Each element of the reduced matrix is determined by the equation

$$Z'_{ij} = Z_{ij} - Z_{in} Z_{nj} / Z_{nn} \qquad (10.211)$$

where the subscript n indicates neutral.

Often, the analysis of a feeder uses the positive and zero sequence impedances for the line sections. The 3×3 *sequence impedance matrix* can be obtained by the following transformation:

$$[Z_{012}] = [A]^{-1} [Z_{abc}][A] \qquad (10.212)$$

where

$[Z_{012}]$ = sequence impedance matrix
$[Z_{abc}]$ = phase frame impedance matrix
$[A]$ = symmetrical component transformation matrix

Analysis Methods

Two methods of voltage analysis are discussed briefly. These are

- K factors method
- Ladder network method

K Factors Method

This is an approximate method for finding out the voltage drop in a feeder section. In this method, a K_{drop} factor and a K_{rise} factor are computed for the primary conductors. The K_{drop} factor is defined as the per unit voltage drop per kilovoltampere-mile. A similar definition exists for the K_{rise} factor. The procedure to determine the K_{drop} factor is as follows:

Step 1. Assume a load of 1 kVA, 0.9 lagging power factor located 1 mi from the substation.
Step 2. Find the current by the following equation:

$$I = \frac{1\,kVA}{\sqrt{3} V_L\, kV} \underline{/-25.84^\circ} \quad A \qquad (10.213)$$

where V_L is the line-to-line voltage magnitude of the primary feeder.
Step 3. Determine the voltage drop from the substation to the load by

$$VD = re(Z^+ I) \quad V \qquad (10.214)$$

where Z^+ is the positive sequence impedance of the primary conductors in ohm per mile.
Step 4. Compute K_{drop} factor as

$$K_{drop} = (VD / V_{N,rated}) \quad \text{per unit drop/kVA-mi} \qquad (10.215)$$

where $V_{N,rated}$ is the rated line-to-neutral voltage.

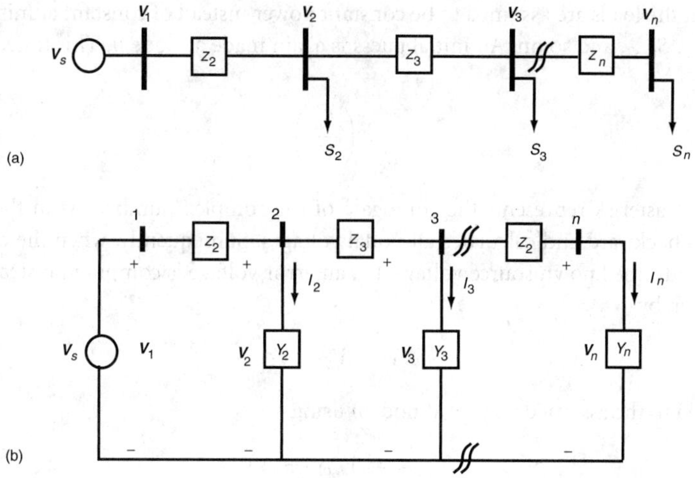

FIGURE 10.116 (a) A radial feeder, (b) its ladder representation.

Thus, if K_{drop} for a specific feeder has been determined to be 0.004%, then, for an actual load of $2500 + j1500$ kVA $= 2915.5\underline{/31°}$ at a distance of 0.5 mi from the substation, the per cent voltage drop is calculated by

$$VD = (K_{\text{drop}}) \quad (\text{kVA})(\text{length pu}) \tag{10.216}$$

On a 120-V base, this represents a voltage drop of

$$VD = (0.00004)(2915.5)(0.5)(120) = 7 \quad \text{V}$$

Thus, if the goal is to limit the voltage drop to within 5% of the nominal voltage of 120 V, then the drop represents a violation. This situation requires the use of voltage correction devices on the system.

Ladder Network Method of Voltage Analysis

This is an iterative technique used to solve linear electric networks of the ladder type. Since most radial distribution systems can be represented as ladder circuits, this method is effective in voltage analysis. An example of a distribution feeder and its equivalent ladder representation are shown in Fig. 10.116(a) and Fig. 10.116(b), respectively. It should be mentioned that Fig. 10.116(b) is a linear circuit since the loads are modeled as constant admittances. In such a linear circuit, the analysis starts with an initial guess of the voltage at node n. The current I_n is calculated as

$$I_n = V_n Y_n \tag{10.217}$$

With I_n known, the voltage drop in the line section can be calculated, and in turn, the preceding node $(n - 1)$ voltage can be found from Eq. (10.201). Consequently, the load current at node $n - 1$ can be found and the entire procedure repeats thereafter until one reaches node 1 where the source voltage V_1 is calculated. The calculated voltage V_1 is then compared with the known source voltage V_s and a correction factor is found by

$$\gamma = \frac{V_s}{V_1} \tag{10.218}$$

All calculated voltages are multiplied by this correction factor in order to reach the final solution.

In reality, a distribution system is nonlinear since loads are modeled as constant power. Hence, the exact ladder theory cannot be applied. A modification of the linear ladder network solution is used.

In this procedure, the loads are assumed to be constant power instead of constant admittances. Thus, the loads are S_n, S_{n-1}, S_{n-2}, and so on. An initial guess is again made at node n. This time, the current I_n is calculated as

$$I_n = \left(\frac{S_n}{V_n} \right)^* \tag{10.219}$$

where, superscript asterisk represents the conjugate of the complex number. As in the linear case, the analysis proceeds backward and calculates all nodal voltages. Subsequently, when the calculated voltage V_1 is compared with the known source voltage V_s, an error voltage is computed instead of a correction factor. This is given by

$$\Delta V = V_S - V_1 \tag{10.220}$$

The error is added to the assumed voltage at node n using

$$V_{n,\text{new}} = V_{n,\text{old}} + \Delta V \tag{10.221}$$

The entire set of calculations is then repeated by working backward from node n to the source. The procedure is iterated until the calculated voltage V_1 is within a prespecified tolerance of the known source voltage V_s.

Defining Terms

Circuit breaker: Devices that protect a circuit from the harmful effects of a fault, in addition to energizing and de-energizing the same circuit during normal operation.

Feeder: Lines that carry power from substations to distribution transformers.

Fuse: An overcurrent protection device that interrupts the line flow. Cannot be used repeatedly.

Line sectionalizer: An overcurrent protection device that interrupts the line flow. Used with backup protection devices.

Phasor: A form of representing sinusoids in the complex domain.

Power factor: Defined as the cosine of the angle between the voltage across and the current through the load.

Reactive power: The component of power that does no real work, but is responsible for changing the power factor.

Real power: The component of power that does the actual work.

Recloser: An overcurrent protection device that interrupts the line flow. Can be used repeatedly.

Substation: It serves as the source for primary distribution feeders. Receives bulk electric power at high voltages and reduces the voltage to distribution primary values.

Transformer: Equipment used to change the voltage level. Also used for phase angle regulation.

References

IEEE Working Group. 1996. Radial distribution test feeders. *IEEE Trans. on Power Systems* 6(3):975–985.

Gonen, T. 1986. *Electric Power Distribution System Engineering.* McGraw-Hill, New York.

Kersting, W.H. 1992. Distribution feeder analysis. In *Power Distribution Planning*, Chap. 6. Institute of Electrical and Electronics Engineers 92 EHO 361-6-PWR.

Further Information

Kersting, W.H. 1984. A method to teach the design and operation of a distribution system. *IEEE Trans. on Power Apparatus and Systems*, PAS-103(7):1945–1952.

Pansini, A.J. 1992. *Guide to Electrical Power Distribution Systems.* Prentice-Hall, Englewood Cliffs, NJ.

Pansini, A.J. 1991. *Power Transmission and Distribution.* The Fairmont Press, Lilburn, GA.

10.8   Power System Protection Alternatives

Jerry C. Whitaker

10.8.1   Introduction

Utility companies make a good-faith attempt to deliver clean, well-regulated power to their customers. Most disturbances on the AC line are beyond the control of the utility company. Large load changes imposed by customers on a random basis, PF correction switching, lightning, and accident-related system faults all combine to produce an environment in which tight control over AC power quality is difficult to maintain. Therefore, the responsibility for ensuring AC power quality must rest with the users of sensitive equipment.

The selection of a protection method for a given facility is as much an economic question as it is a technical one. A wide range of power-line conditioning and isolation equipment is available. A logical decision about how to proceed can be made only with accurate, documented data on the types of disturbances typically found on the AC power service to the facility. The protection equipment chosen must be matched to the problems that exist on the line. Using inexpensive basic protectors may not be much better than operating directly from the AC line. Conversely, the use of a sophisticated protector designed to shield the plant from every conceivable power disturbance may not be economically justifiable.

Purchasing transient-suppression equipment is only one element in the selection equation. Consider the costs associated with site preparation, installation, and maintenance. Also consider the operating efficiency of the system. Protection units that are placed in series with the load consume a certain amount of power and, therefore, generate heat. These considerations may not be significant, but they should be taken into account. Prepare a complete life-cycle cost analysis of the protection methods proposed. The study may reveal that the long-term operating expense of one system outweighs the lower purchase price of another.

The amount of money a facility manager is willing to spend on protection from utility company disturbances generally depends on the engineering budget and how much the plant has to lose. Spending $225,000 on system-wide protection for a highly computerized manufacturing center is easily justified. At smaller operations, justification may not be so easy.

The Key Tolerance Envelope

The susceptibility of electronic equipment to failure because of disturbances on the AC power line has been studied by many organizations. The benchmark study was conducted by the Naval Facilities Engineering Command (Washington, D.C.). The far-reaching program, directed from 1968 to 1978 by Lt. Thomas Key, identified three distinct categories of recurring disturbances on utility company power systems. As shown in Table 10.7, it is not the magnitude of the voltage, but the duration of the disturbance that determines the classification.

TABLE 10.7   Types of Voltage Disturbances Identified in the Key Report

Parameter	Type 1	Type 2	Type 3
Definition	Transient and oscillatory overvoltage	Momentary undervoltage or overvoltage	Power outage
Causes	Lightning, power network switching, operation of other loads	Power system faults, large load changes, utility company equipment malfunctions	Power system faults, unacceptable load changes, utility equipment malfunctions
Threshold[a]	200–400% of rated rms voltage or higher (peak instantaneous above or below rated rms)	Below 80–85% and above 110% of rated rms voltage	Below 80–85% of rated rms voltage
Duration	Transients 0.5–200 μs wide and oscillatory up to 16.7 ms at frequencies of 200 Hz to 5 kHz and higher	From 4–6 cycles, depending on the type of power system distribution equipment	From 2–60 s if correction is automatic; from 15 min to 4 h if manual

(After [1].)

[a] The approximate limits beyond which the disturbance is considered to be harmful to the load equipment.

TABLE 10.8   Causes of Power-Related Computer Failures, Northern Virginia, 1976

| Recorded Cause | Disturbance | | Number of Computer Failures |
	Undervoltage	Outage	
Wind and lightning	37	14	51
Utility equipment failure	8	0	8
Construction or traffic accident	8	2	10
Animals	5	1	6
Tree limbs	1	1	2
Unknown	21	2	23
Totals	80	20	100

(After [1].)

In the study, Key found that most data processing (DP) equipment failure caused by AC line disturbances occurred during bad weather, as shown in Table 10.8. According to a report on the findings, the incidence of thunderstorms in an area can be used to predict the number of failures. The type of power-transmission system used by the utility company was also found to affect the number of disturbances observed on power company lines (Table 10.9). For example, an analysis of utility system problems in Washington, D.C., Norfolk, VA, and Charleston, SC, demonstrated that underground power-distribution systems experienced one-third fewer failures than overhead lines in the same areas. Based on his research, Key developed the "recommended voltage tolerance envelope" shown in Fig. 10.117. The design goals illustrated are recommendations to computer manufacturers for implementation in new equipment.

Assessing the Lightning Hazard

As identified by Key in his Naval Facilities study, the extent of lightning activity in an area significantly affects the probability of equipment failure caused by transient activity. The threat of a lightning flash to a facility is determined, in large part, by the type of installation and its geographic location. The type and character of the lightning flash are also important factors.

The *Keraunic number* of a geographic location describes the likelihood of lightning activity in that area. Figure 10.118 shows the Isokeraunic map of the U.S., which estimates the number of lightning days per year across the country. On average, 30 storm days occur per year across the continental U.S. This number does not fully describe the lightning threat because many individual lightning flashes occur during a single storm.

The structure of a facility has a significant effect on the exposure of equipment to potential lightning damage. Higher structures tend to collect and even trigger localized lightning flashes. Because storm clouds tend to travel at specific heights above the earth, conductive structures in mountainous areas more readily attract lightning activity. The *plant exposure factor* is a function of the size of the facility and the Isokeraunic rating of the area. The larger the physical size of an installation, the more likely it is to be hit by

TABLE 10.9   Effects of Power-System Configuration on Incidence of Computer Failures

| Configuration | Number of Disturbances | | Recorded Failures |
	Undervoltage	Outage	
Overhead radial	12	6	18
Overhead spot network	22	1	23
Combined overhead (weighted[a])	16	4	20
Underground radial	6	4	10
Underground network	5	0	5
Combined underground (weighted[a])	5	2	7

(After [1].)

[a] The combined averages weighted based on the length of time monitored (30 to 53 months).

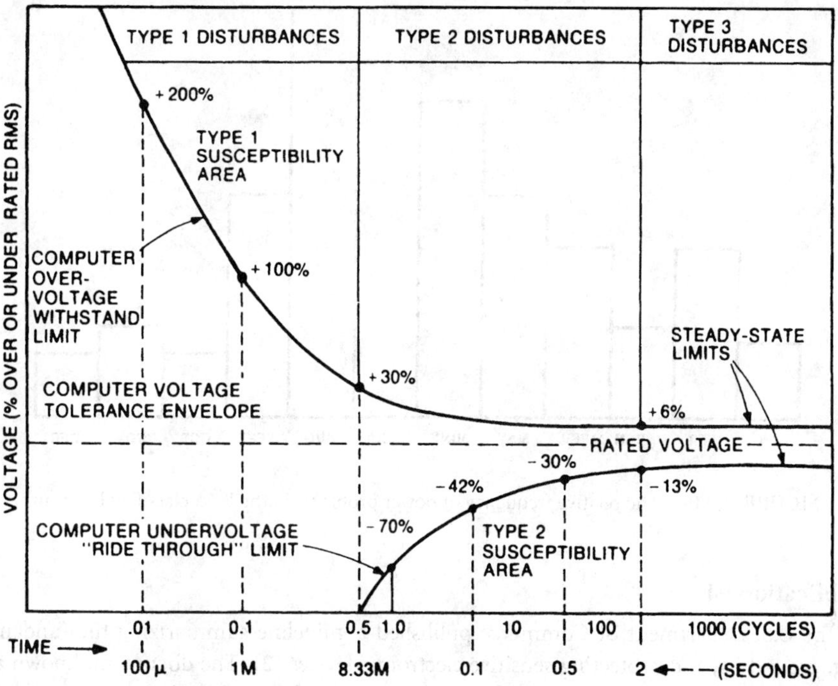

FIGURE 10.117 The recommended voltage tolerance envelope for computer equipment. This chart is based on pioneering work done by the Naval Facilities Engineering Command. The study identified how the magnitude *and* duration of a transient pulse must be considered in determining the damaging potential of a spike. The design goals illustrated in the chart are recommendations to computer manufacturers for implementation in new equipment. (After [1].)

lightning during a storm. The longer a transmission line (AC or RF), the more lightning flashes it is likely to receive.

The relative frequency of power problems is seasonal in nature. As shown in Fig. 10.119, most problems are noted during June, July, and August. These high problem rates primarily can be traced to increased thunderstorm activity.

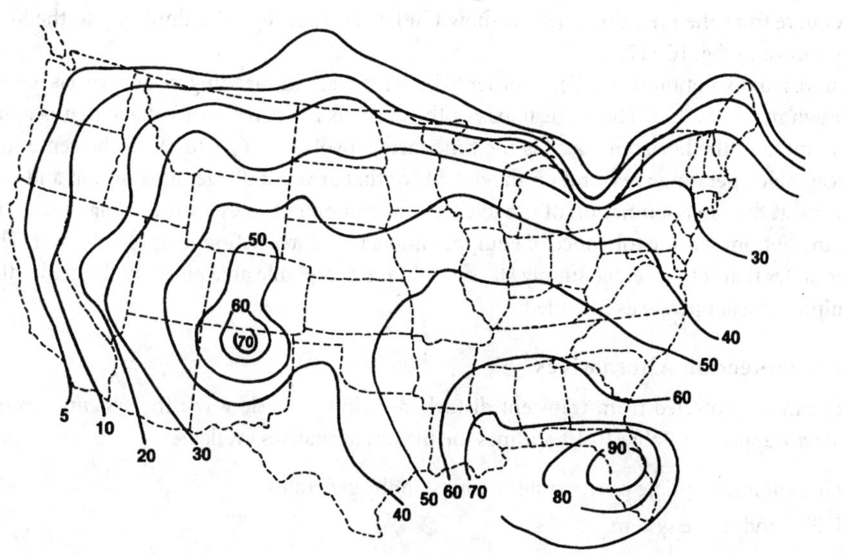

FIGURE 10.118 The Isokeraunic map of the U.S., showing the approximate number of lightning days per year.

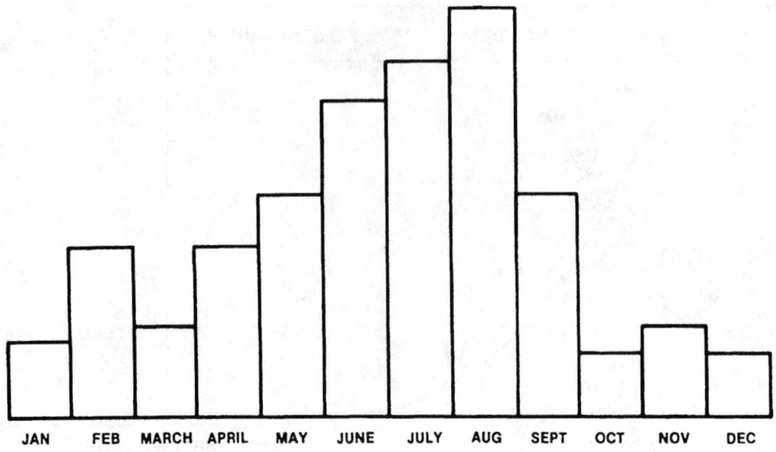

FIGURE 10.119　The relative frequency of power problems in the U.S., classified by month.

FIPS Publication 94

In 1983, the U.S. Department of Commerce published a guideline summarizing the fundamentals of powering, grounding, and protecting sensitive electronic devices [2]. The document, known as Federal Information Processing Standards (FIPS) Publication 94, was first reviewed by governmental agencies and then was sent to the Computer Business Equipment Manufacturers Association (CBEMA) for review. When the CBEMA group put its stamp of approval on the document, the data processing industry finally had an overarching guideline for power quality.

FIPS Pub. 94 was written to cover *automatic data processing equipment* (ADP), which at that time constituted the principal equipment that was experiencing difficulty running on normal utility-supplied power. Since then, IEEE standard P1100 was issued, which applies to all sensitive electronic equipment. FIPS Pub. 94 is a guideline intended to provide a cost/benefit course of action. As a result, it can be relied upon to give the best solution to typical problems that will be encountered, for the least amount of money.

In addition to approving the FIPS Pub. 94 document, the CBEMA group provided a curve that had been used as a guideline for their members in designing power supplies for modern electronic equipment. The CBEMA curve from the FIPS document is shown in Fig. 10.120. Note the similarity to the Key tolerance envelope shown in Fig. 10.117.

The curve is a susceptibility profile. In order to better explain its meaning, the curve has been simplified and redrawn in Fig. 10.130. The vertical axis of the graph is the percent of voltage that is applied to the power circuit, and the horizontal axis is the time factor involved (in μs to s). In the center of the chart is the acceptable operating area, and on the outside of that area is a danger area on top and bottom. The danger zone at the top is a function of the tolerance of equipment to excessive voltage levels. The danger zone on the bottom sets the tolerance of equipment to a loss or reduction in applied power. The CBEMA guideline states that if the voltage supply stays within the acceptable area given by the curve, the sensitive load equipment will operate as intended.

Transient Protection Alternatives

A facility can be protected from transient disturbances in two basic ways: the *systems* approach or the *discrete device* approach. Table 10.10 outlines the major alternatives available:

- UPS (uninterruptible power system) and standby generator
- UPS stand-alone system
- Secondary AC spot network
- Secondary selective AC network

Peak voltage and volt seconds expressed in percent of nominal or of slow average value are appropriate for less than 1/2 cycle

RMS voltage measurements of voltage envelope are appropriate for 1/2 cycle or longer

Short duration impulses are most frequently measured as maximum volts deviation from the power frequency sine wave, or as peak-to-peak volts measured through a high-pass filter or a coupling capacitor. It is sometimes expressed in percent of nominal peak or peak-to-peak voltage

Voltage breakdown concern area

Lack of stored energy in some manufacturer's equipment

Computer voltage

Tolerance envelope

Energy flow related problems

Noise and voltage breakdown problems

FIGURE 10.120 The CBEMA curve from FIPS Pub. 94. (From [2]. Used with permission.)

- Motor-generator set
- Shielded isolation transformer
- Suppressors, filters, and lightning arrestors
- Solid-state line-voltage regulator/filter

Table 10.11 lists the relative benefits of each protection method. Because each installation is unique, a thorough investigation of facility needs should be conducted before purchasing any equipment. The

TABLE 10.10 Types of Systemwide Protection Equipment Available to Facility Managers and the AC Line Abnormalities That Each Approach Can Handle

System	Type 1	Type 2	Type 3
UPS system and standby generator	All source transients; no load transients	All	All
UPS system	All source transients; no load transients	All	All outages shorter than the battery supply discharge time
Secondary spot network[a]	None	None	Most, depending on the type of outage
Secondary selective network[b]	None	Most	Most, depending on the type of outage
Motor-generator set	All source transients; no load transients	Most	Only brown-out conditions
Shielded isolation transformer	Most source transients; no load transients	None	None
Suppressors, filters, lightning arrestors	Most transients	None	None
Solid-state line voltage regulator/filter	Most source transients; no load transients	Some, depending on the response time of the system	Only brown-out conditions

(After [1] and [2].)

[a]Dual power feeder network.

[b]Dual power feeder network using a static (solid-state) transfer switch.

TABLE 10.11   Relative Merits of Systemwide Protection Equipment

System	Strong Points	Weak Points	Technical Profile
UPS system and standby generator	Full protection from power outage failures and transient disturbances; ideal for critical DP and life-safety loads	Hardware is expensive and may require special construction; electrically and mechanically complex; noise may be a problem; high annual maintenance costs	Efficiency 80–90%; typical high impedance presented to the load may be a consideration; frequency stability good; harmonic distortion determined by UPS system design
UPS system	Completely eliminates transient disturbances; eliminates surge and sag conditions; provides power outage protection up to the limits of the battery supply; ideal for critical load applications	Hardware is expensive; depending on battery supply requirements, special construction may be required; noise may be a problem; periodic maintenance required	Efficiency 80–90%; typical high impedance presented to the load may be a consideration; frequency stability good; harmonic content determined by inverter type
Secondary spot network[a]	Simple; inexpensive when available in a given area; protects against local power interruptions; no maintenance required by user	Not available in all locations; provides no protection from area-wide utility failures; provides no protection against transient disturbances or surge/sag conditions	Virtually no loss, 100% efficient; presents low impedance to the load; no effect on frequency or harmonic content
Secondary selective network[b]	Same as above; provides faster transfer from one utility line to the other	Same as above	Same as above
Motor-generator set	Electrically simple; reliable power source; provides up to 0.5 s power-fail ride-through in basic form; completely eliminates transient and surge/sag conditions	Mechanical system requires regular maintenance; noise may be a consideration; hardware is expensive; depending upon m-g set design, power-fail ride-through may be less than typically quoted by manufacturer	Efficiency 80–90%; typical high impedance presented to the load may be a consideration; frequency stability may be a consideration, especially during momentary power-fail conditions; low harmonic content
Shielded isolation transformer	Electrically simple; provides protection against most types of transients and noise; moderate hardware cost; no maintenance required	Provides no protection from brown-out or outage conditions	No significant loss, essentially 100% efficient; presents low impedance to the load; no effect on frequency stability; usually low harmonic content
Suppressors, filters, lightning arrestors	Components inexpensive; units can be staged to provide transient protection exactly where needed in a plant; no periodic maintenance required	No protection from Type 2 or 3 disturbances; transient protection only as good as the installation job	No loss, 100% efficient; some units subject to power-follow conditions; no effect on impedance presented to the load; no effect on frequency or harmonic content
Solid-state line voltage regulator/filter	Moderate hardware cost; uses a combination of technologies to provide transient suppression and voltage regulation; no periodic maintenance required	No protection against power outage conditions; slow response time may be experienced with some designs	Efficiency 92–98%; most units present low impedance to the load; usually no effect on frequency; harmonic distortion content may be a consideration

(After [1] and [2].)
[a] Dual power feeder network.
[b] Dual power feeder network using a static (solid state) transfer switch.

systems approach offers the advantages of protection engineered to a particular application and need, and (usually) high-level factory support during equipment design and installation. The systems approach also means higher costs for the end-user.

Specifying System-Protection Hardware

Developing specifications for systemwide power-conditioning/backup hardware requires careful analysis of various factors before a particular technology or a specific vendor is selected. Key factors in this process relate to the load hardware and load application. The electrical power required by a sensitive load may vary widely, depending on the configuration of the system. The principal factors that apply to system specification include the following:

- Power requirements, including voltage, current, power factor, harmonic content, and transformer configuration.
- Voltage-regulation requirements of the load.
- Frequency stability required by the load, and the maximum permissible *slew rate* (the rate of change of frequency per second).
- Effects of unbalanced loading.
- Overload and inrush current capacity.
- Bypass capability.
- Primary/standby path transfer time.
- Maximum standby power reserve time.
- System reliability and maintainability.
- Operating efficiency.

An accurate definition of *critical applications* will aid in the specification process for a given site. The potential for future expansion also must be considered in all plans.

Power requirements can be determined either by measuring the actual installed hardware or by checking the nameplate ratings. Most nameplate ratings include significant safety margins. Moreover, the load will normally include a *diversity factor*; all individual elements of the load will not necessarily be operating at the same time.

Every load has a limited tolerance to noise and harmonic distortion. Total harmonic distortion (THD) is a measure of the quality of the waveform applied to the load. It is calculated by taking the geometric sum of the harmonic voltages present in the waveform, and expressing that value as a percentage of the fundamental voltage. Critical DP loads typically can withstand 5% THD, where no single harmonic exceeds 3%. The power-conditioning system must provide this high-quality output waveform to the load, regardless of the level of noise and/or distortion present at the AC input terminals.

If a power-conditioning/standby system does not operate with high reliability, the results can often be disastrous. In addition to threats to health and safety, there is a danger of lost revenue or inventory, and hardware damage. Reliability must be considered from three different viewpoints:

- Reliability of utility AC power in the area.
- Impact of line-voltage disturbances on DP loads.
- Ability of the protection system to maintain reliable operation when subjected to expected and unexpected external disturbances.

The environment in which the power-conditioning system operates will have a significant effect on reliability. Extremes of temperature, altitude, humidity, and vibration can be encountered in various applications. Extreme conditions can precipitate premature component failure and unexpected system shutdown. Most power-protection equipment is rated for operation from 0 to 40°C. During a commercial power failure, however, the ambient temperature of the equipment room can easily exceed either value, depending on the exterior temperature. Operating temperature derating typically is required for altitudes in excess of 1000 ft.

TABLE 10.12   Power-Quality Attributes for Data Processing Hardware

Environmental Attribute	Typical Environment	Acceptable Limits for DP Systems	
		Normal	Critical
Line frequency	±0.1 to ±3%	±1%	±0.3%
Rate of frequency change	0.5 to 20 Hz/s	1.5 Hz/s	0.3 Hz/s
Over- and under-voltage	±5% to +6, −13.3%	+5% to −10%	±3%
Phase imbalance	2 to 10%	5% max	3% max
Tolerance to low power factor	0.85 to 0.6 lagging	0.8 lagging	less than 0.6 lagging, or 0.9 leading
Tolerance to high steady state peak current	1.3 to 1.6 peak, rms	1.0 to 2.5 peak, rms	Greater than 2.5 peak, rms
Harmonic voltages	0 to 20% total rms	10 to 20% total, 5 to 10% largest	5% max total, 3% largest
Voltage deviation from sine wave	5 to 50%	5 to 10%	3 to 5%
Voltage modulation	Negligible to 10%	3% max	1% max
Surge/sag conditions	+10%, −15%	+20%, −30%	+5%, −5%
Transient impulses	2 to 3 times nominal peak value (0 to 130% Vs)	Varies; 1.0 to 1.5 kV typical	Varies; 200 to 500 V typical
RFI/EMI normal and common modes	10 V up to 20 kHz, less at high freq.	Varies widely, 3 V is typical	Varies widely, 0.3 V typical
Ground currents	0 to 10 A plus impulse noise current	0.001 to 0.5 A or more	0.0035 A or less

(After [2].)

Table 10.12 lists key power-quality attributes that should be considered when assessing the need for power-conditioning hardware. The limits of acceptable power input are shown in graphical form in Fig. 10.121.

10.8.2   Motor-Generator Set

As the name implies, a motor-generator (m-g) set consists of a motor powered by the AC utility supply that is mechanically tied to a generator, which feeds the load (see Fig. 10.122.). Transients on the utility line will have no effect on the load when this arrangement is used. Adding a flywheel to the motor-to-generator shaft will protect against brief power dips (up to 0.5 s on many models). Figure 10.123 shows the construction of a typical m-g set. The attributes of an m-g include the following:

- An independently generated source of voltage can be regulated without interaction with line-voltage changes on the power source. Utility line changes of ±20% commonly can be held to within ±1% at the load.

- The rotational speed and inertial momentum of the rotating mass represents a substantial amount of stored rotational energy, preventing sudden changes in voltage output when the input is momentarily interrupted.

- The input and output windings are separated electrically, preventing transient disturbances from propagating from the utility company AC line to the load.

- Stable electrical characteristics for the load: (1) output voltage and frequency regulation, (2) ideal sine wave output, and (3) true 120° phase shift for three-phase models.

- Reduced problems relating to the power factor presented to the utility company power source.

The efficiency of a typical m-g ranges from 65 to 89%, depending on the size of the unit and the load. Motor-generator sets have been used widely to supply 415 Hz power to mainframe computers that require this frequency.

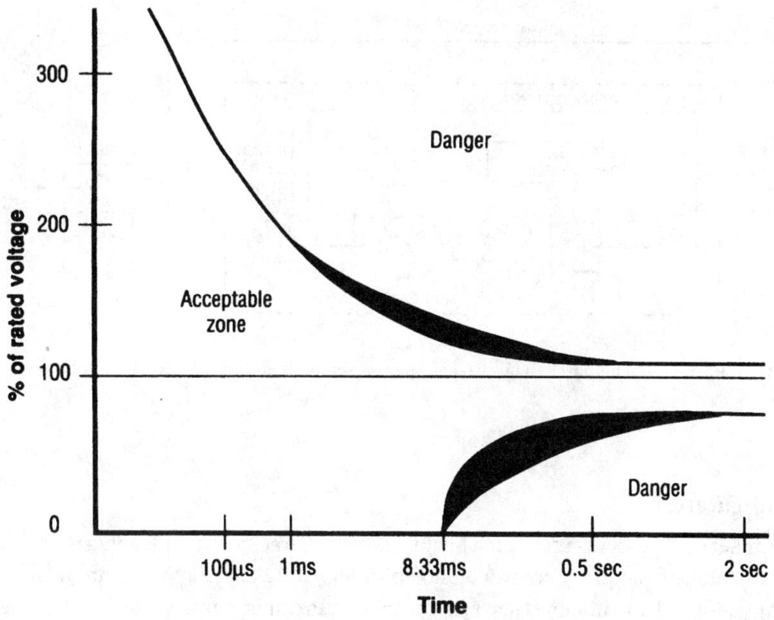

FIGURE 10.121 A simplified version of the CBEMA curve. Voltage levels outside the *acceptable zone* results in potential system shutdown and hardware and software loss. (From [2]. Used with permission.)

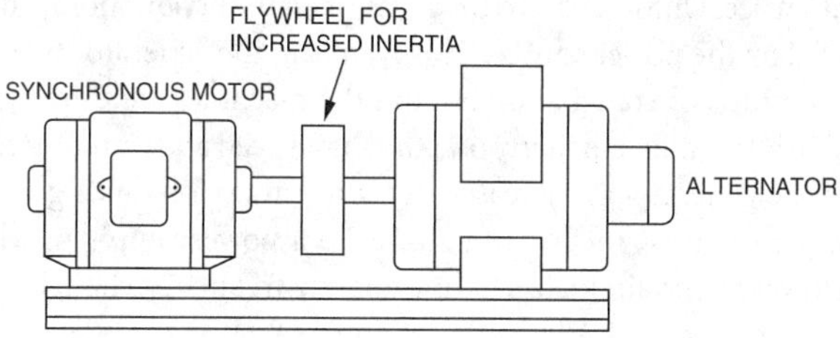

FIGURE 10.122 Two-machine motor-generator set with an optional flywheel to increase inertia and carry-through capability.

FIGURE 10.123 Construction of a typical m-g set. (Courtesy of Computer Power Protection.)

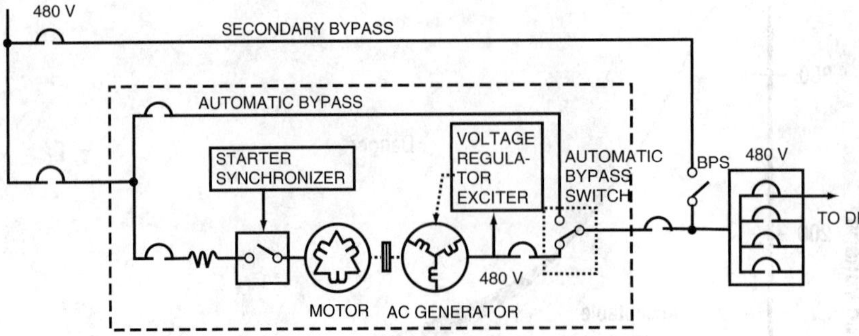

FIGURE 10.124 Simplified schematic diagram of an m-g set with automatic and secondary bypass capability. (After [2].)

System Configuration

There are a number of types of m-g sets, each having its own characteristics, advantages, and disadvantages. A simplified schematic diagram of an m-g is shown in Fig. 10.124. The type of motor that drives the set is an important design element. Direct-current motor drives can be controlled in speed independently of the frequency of the AC power source from which the DC is derived. Use of a DC motor, thereby, gives the m-g set the capability to produce power at the desired output frequency, regardless of variations in input frequency. The requirement for rectifier conversion hardware, control equipment, and commutator maintenance are drawbacks to this approach that must be considered.

The simplest and least expensive approach to rotary power conditioning involves the use of an induction motor as the mechanical source. Unfortunately, the rotor of an induction motor turns slightly slower than the rotating field produced by the power source. This results in the generator being unable to produce 60 Hz output power if the motor is operated at 60 Hz and the machines are directly coupled end-to-end at their shafts, or are belted in a 1:1 ratio. Furthermore, the shaft speed and output frequency of the generator decreases as the load on the generator is increased. This potential for varying output frequency may be acceptable where the m-g set is used solely as the input to a power supply in which the AC is rectified and converted to DC. However, certain loads cannot tolerate frequency changes greater than 1 Hz/s and frequency deviations of more than 0.5 Hz from the nominal 60 Hz value.

Low-slip induction motor-driven generators are available that can produce 59.7 Hz at full load, assuming 60 Hz input. During power interruptions, the output frequency will drop further, depending upon the length of the interruption. The capability of the induction motor to restart after a momentary power interruption is valuable. Various systems of variable-speed belts have been tried successfully. Magnetically controlled slipping clutches have been found to be largely unsatisfactory. Other approaches to make the induction motor drive the load at constant speed have produced mixed results.

Using a synchronous motor with direct coupling or a cogged 1:1 ratio belt drive guarantees that the output frequency will be equal to the motor input frequency. Although the synchronous motor is more expensive, it is more efficient and can be adjusted to provide a unity PF load to the AC source. The starting characteristics and the mechanical disturbance following a short line-voltage interruption depends, to a large extent, on motor design. Many synchronous motors that are not required to start under load have weak starting torque, and may use a *pony motor* to aid in starting. This approach is shown in Fig. 10.125. Those motors designed to start with a load have starting pole face windings that provide starting torque comparable to that of an induction motor. Such motors can be brought into synchronism while under load with proper selection of the motor and automatic starter system. Typical utility company AC interruptions are a minimum of six cycles (0.1 s). Depending upon the design and size of the flywheel used, the ride-through period can be as much as 0.5 s or more. The generator will continue to produce output power for a longer duration, but the frequency and rate of frequency change will most likely fall outside of the acceptable range of most DP loads after 0.5 s.

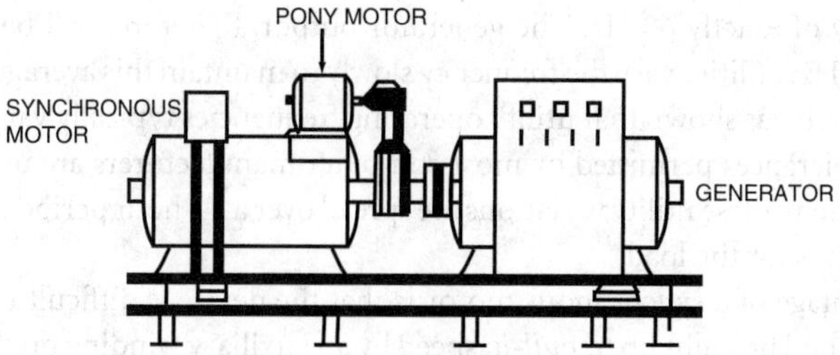

FIGURE 10.125 Use of a pony motor for an m-g set to aid in system starting and restarting. (After [2].)

If input power is interrupted and does not return before the output voltage and frequency begin to fall outside acceptable limits, the generator output controller can be programmed to disconnect the load. Before this event, a warning signal is sent to the DP control circuitry to warn of impending shutdown and to initiate an orderly interruption of active computer programs. This facilitates easy restart of the computer after the power interruption has passed.

It is important for users to accurately estimate the length of time that the m-g set will continue to deliver acceptable power without input to the motor from the utility company. This data facilitates accurate power-fail shutdown routines. It is also important to ensure that the m-g system can handle the return of power without operating overcurrent protection devices because of high inrush currents that may be required to accelerate and synchronize the motor with the line frequency. Protection against the latter problem requires proper programming of the synchronous motor controller to correctly disconnect and then reconnect the field current supply. It may be worthwhile to delay an impending shutdown for 100 ms or so. This would give the computer time to prepare for the event through an orderly interruption. It also would be useful if the computer were able to resume operation without shutdown, in case utility power returns within the ride-through period. Control signals from the m-g controller should be configured to identify these conditions and events to the DP system.

Generators typically used in m-g sets have substantially higher internal impedance than equivalent kVA-rated transformers. Because of this situation, m-g sets are sometimes supplied with an oversized generator that will be lightly loaded, coupled with a smaller motor that is adequate to drive the actual load. This approach reduces the initial cost of the system, decreases losses in the motor, and provides a lower operating impedance for the load.

Motor-generator sets can be configured for either horizontal installation, as shown previously, or for vertical installation, as illustrated in Fig. 10.126.

The most common utility supply voltage used to drive the input of an m-g set is 480 V. The generator output for systems rated at about 75 kVA or less is typically 208 Y/120 V. For larger DP systems, the most economical generator output is typically 480 V. A 480-208 Y/120 V three-phase isolating transformer usually is included to provide 208 Y/120 V power to the computer equipment.

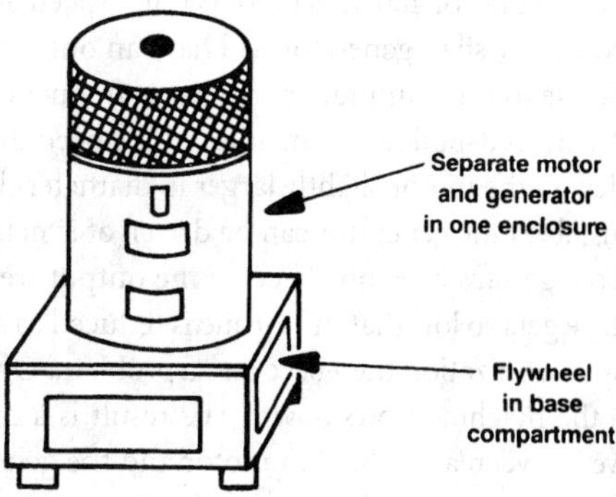

FIGURE 10.126 Vertical m-g set with an enclosed flywheel. (After [2].)

Motor-Design Considerations

Both synchronous and induction motors have been used successfully to drive m-g sets, and each has advantages and disadvantages. The major advantage of the synchronous motor is that while running normally, it is synchronized with the supply frequency. An 1800 rpm motor rotates at exactly 1800 rpm for a supply frequency of exactly 60 Hz. The generator output, therefore, will be exactly 60 Hz. Utility frequencies *average* 60 Hz; utilities vary the frequency slowly to maintain this average value under changing load conditions. Research has shown that utility operating frequencies typically vary from 58.7 to 60.7 Hz. Although frequency tolerances permitted by most computer manufacturers are usually given as ±0.5 Hz on a nominal 60 Hz system, these utility variations are spread over a 24-hour period or longer, and generally do not result in problems for the load.

The major disadvantage of a synchronous motor is that the device is difficult to start. A synchronous motor must be started and brought up to *pull-in* speed by an auxiliary winding on the armature, known as the *armortisseur* winding. The pull-in speed is the minimum speed (close to synchronous speed) at which the motor will pull into synchronization if excitation is applied to the field. The armortisseur winding is usually a squirrel-cage design, although it may be of the wound-rotor type in some cases. This winding allows the synchronous motor to start and come up to speed as an induction motor. When pull-in speed is achieved, automatic sensing equipment applies field excitation, and the motor locks in and runs as a synchronous machine. As discussed previously, some large synchronous motors are brought up to speed by an auxiliary pony motor.

The armortisseur winding can produce only limited torque, so synchronous motors usually are brought up to speed without a load. This requirement presents no problem for DP systems upon initial startup. However, in the event of a momentary power outage, problems can develop. When the utility AC fails, the synchronous motor must be disconnected from the input immediately, or it will act as a generator and feed power back into the line, thus rapidly depleting its stored (kinetic) rotational energy. During a power failure, the speed of the motor rapidly drops below the pull-in speed, and when the AC supply returns, the armortisseur winding must reaccelerate the motor under load until the field can be applied again. This requires a large winding and a sophisticated control system. When the speed of the m-g set is below synchronous operation, the generator output frequency may be too low for proper computer operation.

The induction motor has no startup problems, but it does have *slip*. To produce torque, the rotor must rotate at a slightly lower speed than the stator field. For a nominal 1800 rpm motor, the actual speed will be about 1750 rpm, varying slightly with the load and the applied input voltage. This represents a slip of about 2.8%. The generator, if driven directly or on a common shaft, will have an output frequency of about 58.3 Hz. This is below the minimum permissible operating frequency for most computer hardware. Special precision-built low-slip induction motors are available with a slip of approximately 0.5% at a nominal motor voltage of 480 V. With 0.5% slip, speed at full load will be about 1791 rpm, and the directly driven or common-shaft generator will have an output frequency of 59.7 Hz. This frequency is within tolerance, but close to the minimum permissible frequency.

A belt-and-pulley system adjustable-speed drive is a common solution to this problem. By making the pulley on the motor slightly larger in diameter than the pulley on the generator (with the actual diameters adjustable) the generator can be driven at synchronous speed.

Voltage sags have no effect on the output frequency of a synchronous motor-driven m-g set until the voltage gets so low that the torque is reduced to a point at which the machine pulls out of synchronization. Resynchronization then becomes a problem. On an induction motor, when the voltage sags, slip increases and the machine slows down. The result is a drop in generator output frequency. The adjustable-speed drive between an induction motor and the generator solves the problem for separate machines. If severe voltage sags are anticipated at a site, the system can be set so that nominal input voltage from the utility company produces a frequency of 60.5 Hz, 0.5 Hz on the high side of nominal frequency. Figure 10.127 charts frequency vs. motor voltage for three operating conditions:

- Slip compensation set high (curve A)
- Slip compensation set for 60 Hz (curve B)
- No slip compensation (curve C)

Through proper adjustment of slip compensation, considerable input-voltage margins can be achieved.

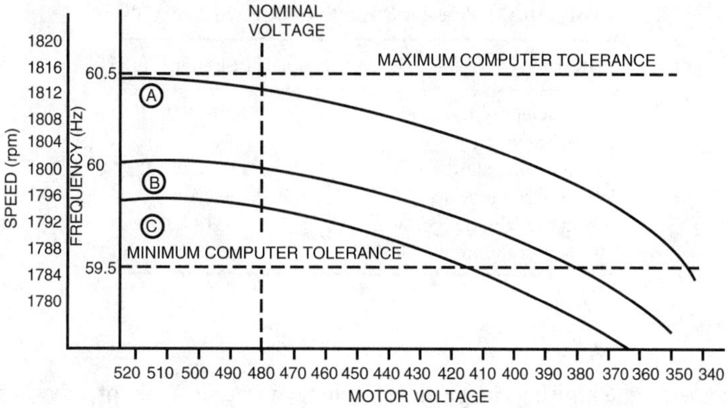

FIGURE 10.127 Generator output frequency vs. motor input voltage for an induction-motor-based m-g set.

Single-Shaft Systems

There are two basic m-g set machine mechanical designs used for DP applications: (1) separate motor-generator systems, and (2) single-shaft, single-housing units. Both designs can use either a synchronous or induction motor. In each case, there are advantages and disadvantages. The separate machine design (discussed previously) uses a motor driving a physically separate generator by means of a coupling shaft or pulley. In an effort to improve efficiency and reduce costs, manufacturers also have produced various types of single-shaft systems.

The basic concept of a single-shaft system is to combine the motor and generator elements into a single unit. A common stator eliminates a number of individual components, making the machine less expensive to produce and mechanically more efficient. The common-stator set substantially reduces mechanical energy losses associated with traditional m-g designs, and it improves system reliability as well. In one design, the stator is constructed so that alternate slots are wound with input and output windings. When it is fed with a three-phase supply, a rotating magnetic field is created, causing the DC-excited rotor to spin at a synchronous speed. By controlling the electrical characteristics of the rotor, control of the output at the secondary stator windings is accomplished.

Common-stator machines offer lower working impedance for the load than a comparable two-machine system. For example, a typical 400-kVA machine has approximately an 800-kVA frame size. The larger frame size yields a relatively low-impedance power source capable of clearing subcircuit fuses under fault conditions. The output of the unit typically can supply up to seven times the full-load current under fault conditions. Despite the increase in frame size, the set is smaller and lighter than comparable systems because of the reduced number of mechanical parts.

Flywheel Considerations

In an effort to achieve higher energy and power densities, m-g set designers have devoted considerable attention to the flywheel element itself. New composite materials and power electronics technologies have resulted in compact flywheel "batteries" capable of high linear velocity at the outside radius of the flywheel (*tip speed*) [3]. The rotational speed of the flywheel is important because the stored energy in a flywheel is proportional to the square of its rotational speed. Therefore, an obvious method for maximizing stored energy is to increase the speed of the flywheel. All practical designs, however, have a limiting speed, which is determined by the stresses developed within the wheel resulting from inertial loads. These loads are also proportional to the square of rotational speed. Flywheels built of composite materials weigh less, and hence develop lower inertial loads at a given speed. In addition, composites are often stronger than conventional engineering metals, such as steel. This combination of high strength and low weight enables extremely high tip speeds, relative to conventional wheels.

TABLE 10.13   Specific Strength of Selected Materials

Material	Specific Strength (in.^{3})
Graphite/epoxy	3,509,000
Boron/epoxy	2,740,000
Titanium and alloys	1,043,000
Wrought stainless steel	982,000
Wrought high-strength steel	931,000
7000 series aluminum alloys	892,000

(After [4].)

For a given geometry, the limiting *energy density* (energy per unit mass) of a flywheel is proportional to the ratio of material strength to weight density, otherwise known as the *specific strength*. Table 10.13 illustrates the advantage that composite materials offer in this respect.

Recent advances in composite materials technology may allow nearly an order of magnitude advantage in the specific strength of composites when compared to even the best common engineering metals. The result of this continuous research in composites has been flywheels capable of operation at rotational speeds in excess of 100,000 rpm, with tip speeds in excess of 1000 m/s.

These high speeds bring with them new challenges. The ultrahigh rotational speeds that are required to store significant kinetic energy in these systems virtually rule out the use of conventional mechanical bearings. Instead, most systems run on magnetic bearings. This relatively recent innovation uses magnetic forces to "levitate" a rotor, eliminating the frictional losses inherent in rolling element and fluid film bearings. Unfortunately, aerodynamic drag losses force most high-speed flywheels to operate in a partial vacuum, which complicates the task of dissipating the heat generated by ohmic losses in the bearing electromagnets and rotor. In addition, active magnetic bearings are inherently unstable, and require sophisticated control systems to maintain proper levitation.

The integrated generator of these systems is usually a rotating-field design, with the magnetic field supplied by rare-earth permanent magnets. Because the specific strength of these magnets is typically just fractions of that of the composite flywheel, they must spin at much lower tip speeds; in other words, they must be placed very near the hub of the flywheel. This compromises the power density of the generator. An alternative is to mount them closer to the outer radius of the wheel, but contain their inertial loads with the composite wheel itself. Obviously, this forces the designer to either derate the machine speed, or operate closer to the stress limits of the system.

Maintenance Considerations

Because m-g sets require some maintenance that necessitates shutdown, most systems provide bypass capability so the maintenance work can be performed without having to take the computer out of service. If the automatic bypass contactor, solid-state switch, and control hardware are in the same cabinet as other devices that also need to be de-energized for maintenance, a secondary bypass is recommended. After the automatic bypass path has been established, transfer switching to the secondary bypass can be enabled, taking the m-g set and its automatic bypass system out of the circuit completely. Some automatic bypass control arrangements are designed to transfer the load of the generator to the bypass route with minimum disturbance. This requires the generator output to be synchronized with the bypass power before closing the switch and opening the generator output breaker. However, with the load taken off the generator, bypass power no longer will be synchronized with it. Consequently, retransfer of the load back to the generator may occur with some disturbance. Adjustment for minimum disturbance in either direction requires a compromise in phase settings, or a means to shift the phase before and after the transfer.

The use of rotating field exciters has eliminated the need for slip rings in most m-g designs. Brush inspection and replacement, therefore, are no longer needed. However, as with any rotating machinery, bearings must be inspected and periodically replaced.

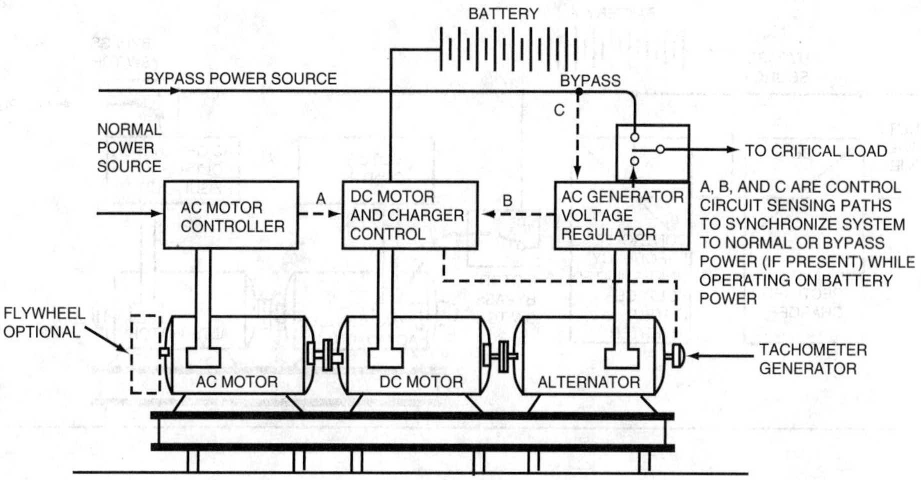

FIGURE 10.128 Uninterruptible m-g set with AC and DC motor drives. (After [2].)

Motor-Generator UPS

Critical DP applications that cannot tolerate even brief AC power interruptions can use the m-g set as the basis for an uninterruptible source of power through the addition of a battery-backed DC motor to the line-driven AC motor shaft. This concept is illustrated in Fig. 10.128. The AC motor normally supplies power to drive the system from the utility company line. The shafts of the three devices are all interconnected, as shown in the figure. When AC power is present, the DC motor serves as a generator to charge the battery bank. When line voltage is interrupted, the DC motor is powered by the batteries. Figure 10.129 shows a modified version of this basic m-g UPS using only a DC motor as the mechanical power source. This configuration eliminates the inefficiency involved in having two motors in the system. Power from the utility source is rectified to provide energy for the DC motor, plus power for charging the batteries. A complex control system to switch the AC motor off and the DC motor on in the event of a utility power failure is not needed in this design.

The m-g UPS also can be built around a synchronous AC motor, as illustrated in Fig. 10.130. Utility AC energy is rectified and used to drive an inverter, which provides a regulated frequency source to power the synchronous motor. The output from the DC-to-AC inverter need not be a well-formed sine wave, nor a

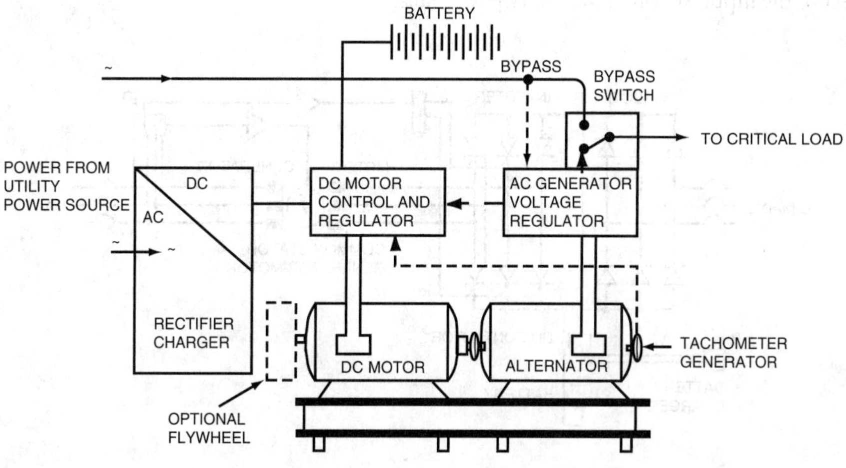

FIGURE 10.129 Uninterruptible m-g set using a single DC drive motor. (After [2].)

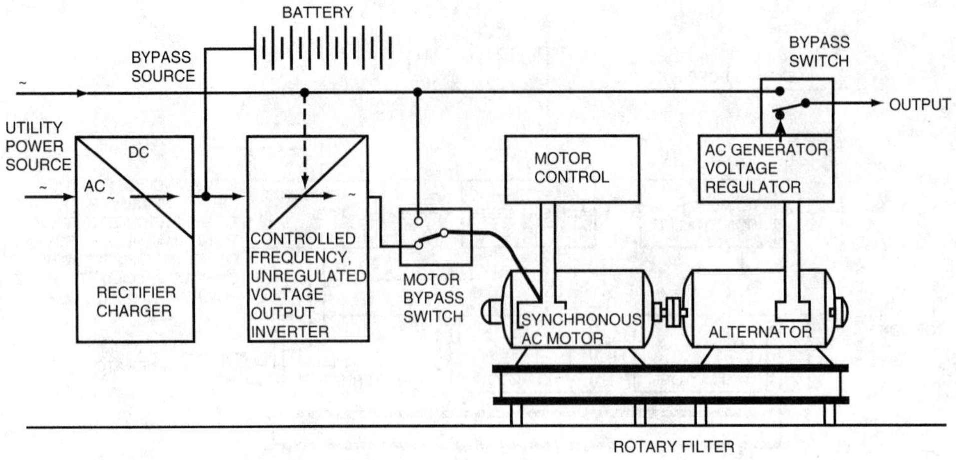

FIGURE 10.130 Uninterruptible m-g set using a synchronous AC motor. (After [2].)

well-regulated source. The output from the generator will provide a well-regulated sine wave for the load. The m-g set also can be operated in a bypass mode that eliminates the rectifier, batteries, and inverter from the current path, operating the synchronous motor directly from the AC line.

An m-g UPS set using a common stator machine is illustrated in Fig. 10.131. A feedback control circuit adjusts the firing angle of the inverter to compensate for changes in input power. This concept is taken a step further in the system shown in Fig. 10.132. A solid-state inverter bypass switch is added to improve efficiency. During normal operation, the bypass route is enabled, eliminating losses across the rectifier diodes. When the control circuitry senses a drop in utility voltage, the inverter is switched on, and the bypass switch is deactivated. A simplified installation diagram of the inverter/bypass system is shown in Fig. 10.133. Magnetic circuit breakers and chokes are included as shown. An isolation transformer is inserted between the utility input and the rectifier bank. The static inverter is an inherently simple design; commutation is achieved via the windings. Six thyristors are used. Under normal operating conditions, 95% of the AC power passes through the static switch; 5% passes through the inverter. This arrangement affords maximum efficiency, while keeping the battery bank charged and the rectifiers and inverter thyristors preheated. Preheating extends the life of the components by reducing the extent of thermal cycling that occurs when the load suddenly switches to the battery backup supply. The static switch allows for fast disconnect of the input AC when the utility power fails.

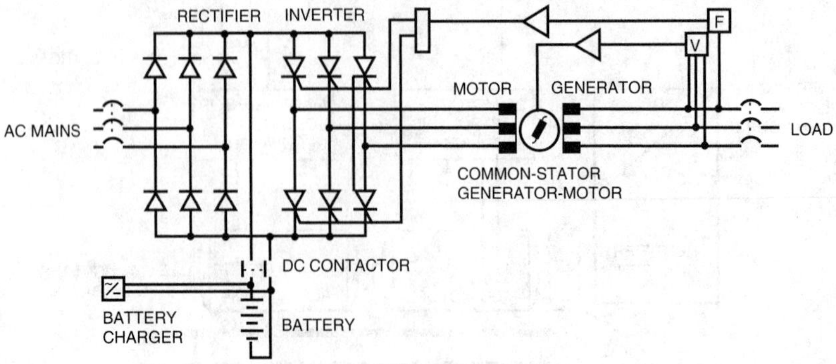

FIGURE 10.131 Common-stator UPS m-g set. The firing angle of the SCR inverters is determined by a feedback voltage from the generator output.

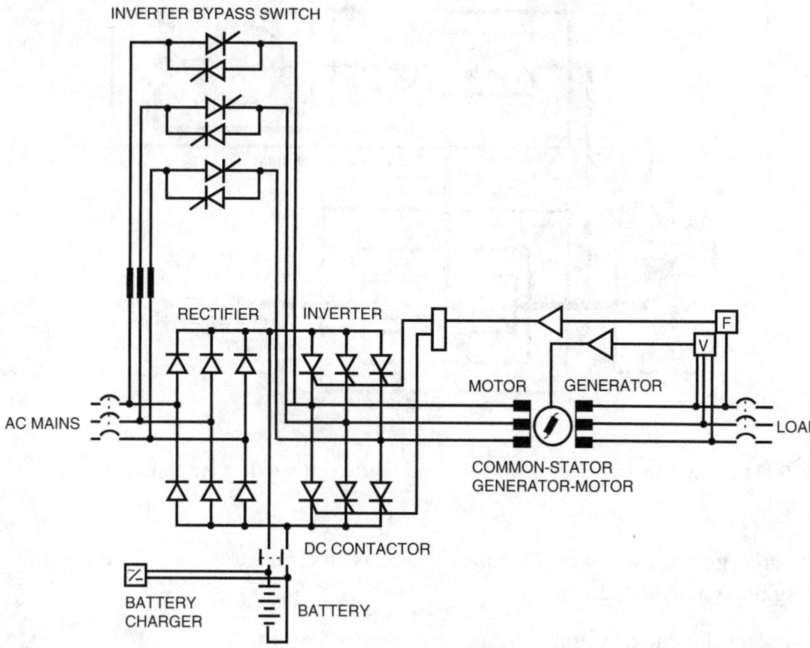

FIGURE 10.132   Common-stator UPS m-g set with a solid-state inverter bypass switch.

Kinetic Battery Storage System

As outlined previously, one of the parameters that limits the ride-through period of an m-g set is the speed decay of the flywheel/generator combination. As the flywheel slows down, the output frequency drops. This limits the useful ride-through period to 0.5 s or so. Figure 10.134 shows an ingenious modification of the classic power conditioning template that extends the potential ride-through considerably. As shown

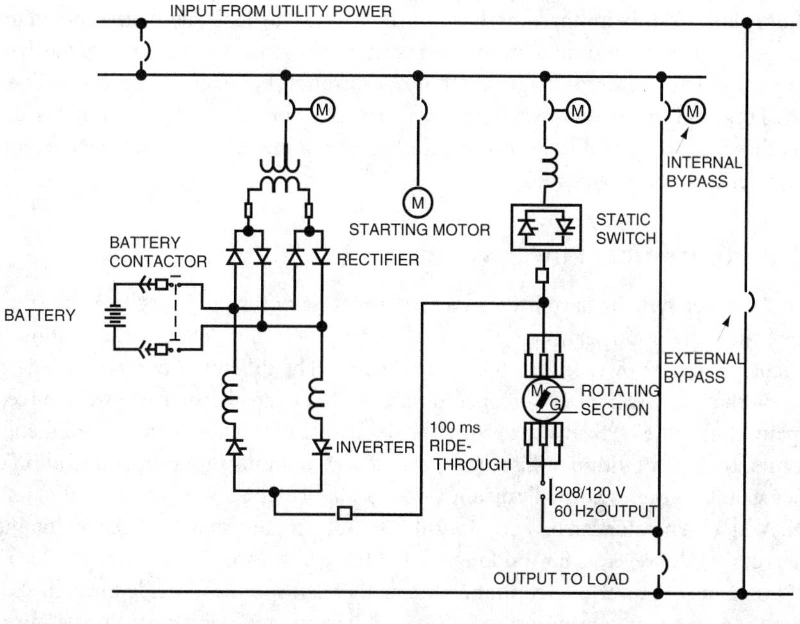

FIGURE 10.133   Power flow diagram of a full-featured UPS m-g set.

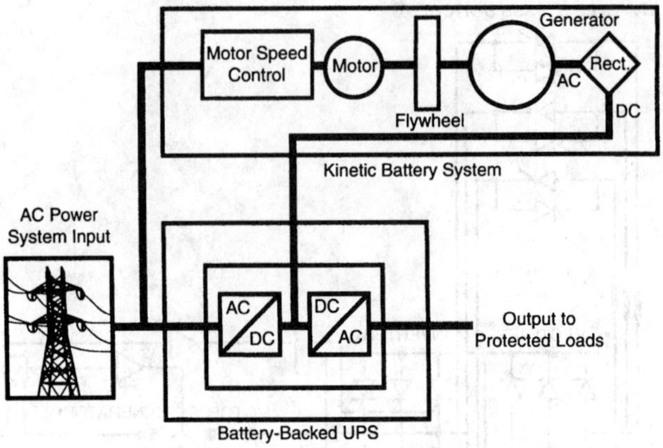

FIGURE 10.134   Functional block diagram of a kinetic battery m-g system for powering a UPS. (After [5].)

in the figure, an m-g set is used in a UPS-based system as an element of the DC supply grid. The major components of the system include

- Steel flywheel for energy storage
- Small drive motor, sized at 15 to 20% of the rated system output, to start the flywheel and maintain its normal operating speed
- Variable speed drive (VSD) to slowly ramp the flywheel up to speed and maintain it at the desired rpm
- Generator to convert the kinetic energy stored in the flywheel into electrical energy
- Diode bridge rectifier to convert the AC generator output to DC for use by the UPS bus, which continues to draw usable energy essentially independent of flywheel rpm.

Because the AC output of the generator is converted to DC, frequency is no longer a limiting factor, which allows the DC voltage output to be maintained over a much greater range of the flywheel's rpm envelope.

In operation, the small drive motor spins the flywheel while the variable speed drive maintains the proper motor speed [5]. Because the amount of stored kinetic energy increases by the square of the flywheel rpm, it is possible to greatly increase stored energy, and thus ride-through, by widening the flywheel's usable energy output range. These factors permit typical ride-through times from 10 s to several minutes, depending on loading and other operating conditions. Among the benefits of this approach include the reduced cycling of battery supplies and engine-generators systems.

10.8.3   Uninterruptible Power Systems

Uninterruptible power systems have become a virtual necessity for powering large or small computer systems where the application serves a critical need. Computers and data communications systems are no more reliable than the power from which they operate. The difference between UPS and *emergency standby power* is that the UPS is always in operation. It reverts to an alternative power source, such as the utility company, only if the UPS fails or needs to be deactivated for maintenance. Even then, the transfer of power occurs so quickly (within milliseconds) that it does not interrupt proper operation of the load.

Emergency standby power is normally off and does not start (manually or automatically) until the utility AC feed fails. A diesel generator can be started within 10 to 30 s if the system has been maintained properly. Such an interruption, however, is far too long for DP hardware. Most DP systems cannot ride through more than 8 to 22 ms of power interruption. Systems that can successfully ride through short-duration power breaks, as far as energy continuity is concerned, still may enter a fault condition because of electrical noise created by the disturbance.

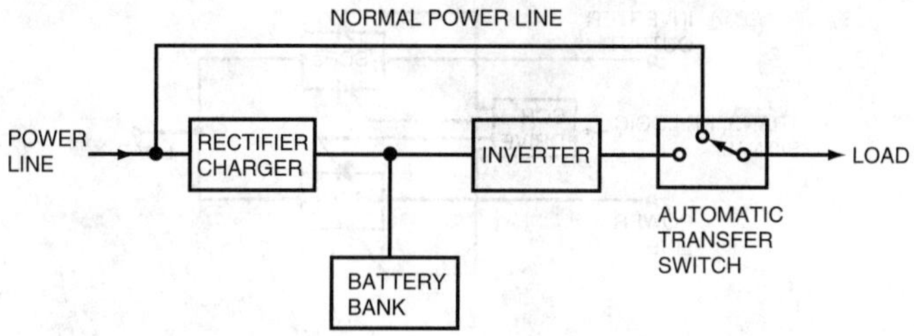

FIGURE 10.135   Forward-transfer UPS system.

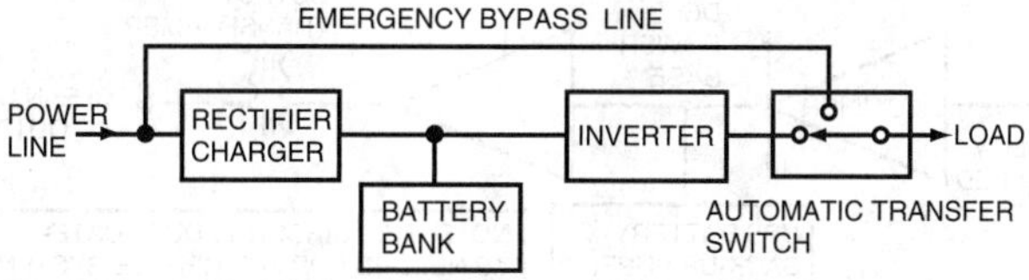

FIGURE 10.136   Reverse-transfer UPS system.

UPS hardware is available in a number of different configurations. All systems, however, are variations of two basic designs:

- *Forward-transfer mode:* The load normally is powered by the utility power line, and the inverter is idle. If a commercial power failure occurs, the inverter is started and the load is switched. This configuration is illustrated in Fig. 10.135. The primary drawback of this approach is the lack of load protection from power-line disturbances during normal (utility-powered) operation.
- *Reverse-transfer mode:* The load normally is powered by the inverter. In the event of an inverter failure, the load is switched directly to the utility line. This configuration is illustrated in Fig. 10.136. The reverse-transfer mode is, by far, the most popular type of UPS system in use for large-scale systems.

The type of load-transfer switch used in the UPS system is another critical design parameter. The continuity of AC power service to the load is determined by the type of switching circuit used. An electromechanical transfer switch, shown in Fig. 10.137, is limited to switch times of 20 to 50 ms. This time delay can cause sensitive load equipment to malfunction, and perhaps shut down. A control circuit actuates the relay when the sensed output voltage falls below a preset value, such as 94% of nominal. A static transfer switch, shown in Fig. 10.138, can sense a failure and switch the load in about 4 ms. Most

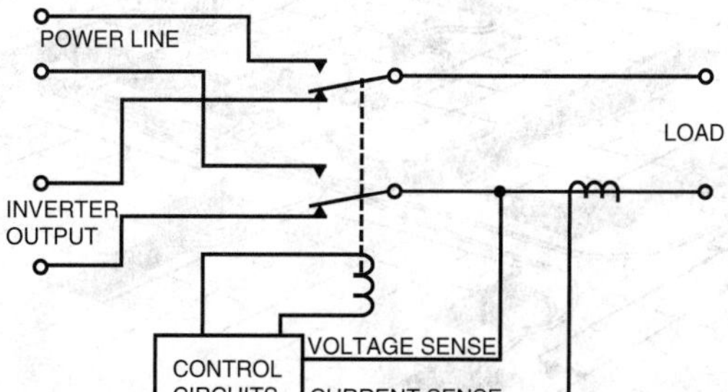

FIGURE 10.137   Electromechanical load-transfer switch.

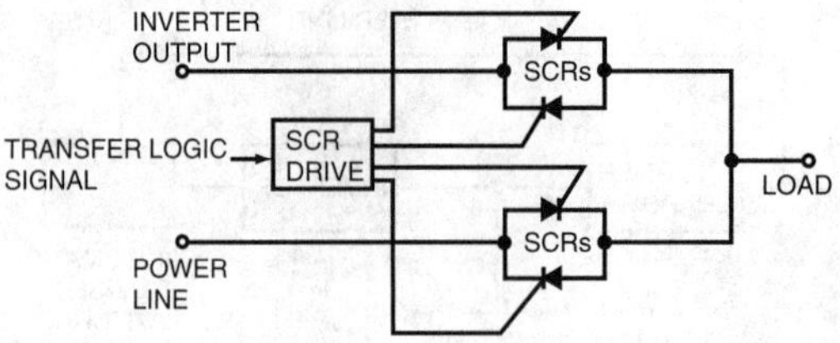

FIGURE 10.138   Static load-transfer switch.

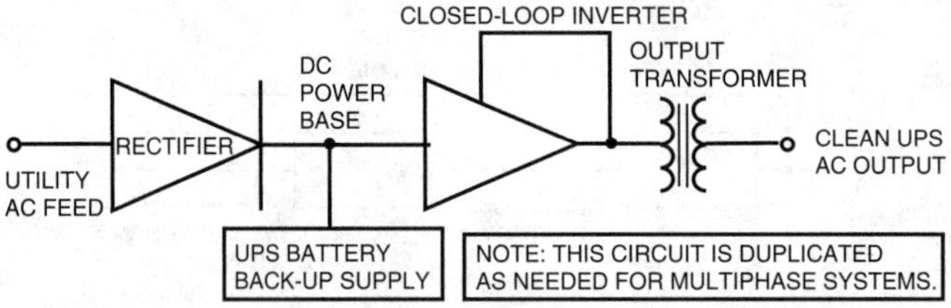

FIGURE 10.139   Block diagram of an uninterruptible power system using AC rectification to float the battery supply. A closed-loop inverter draws on this supply and delivers clean AC power to the protected load.

loads will ride through this short delay without any malfunction. To accomplish a smooth transition, the inverter output must be synchronized with the power line.

UPS Configuration

The basic uninterruptible power system is built around a battery-driven inverter, with the batteries recharged by the utility AC line. As shown in Fig. 10.139, AC from the utility feed is rectified and applied to recharge or *float* a bank of batteries. This DC power drives a single- or multiphase closed-loop inverter, which regulates output voltage and frequency. The output of the inverter is generally a sine wave, or pseudo sine wave (a stepped square wave). If the utility voltage drops or disappears, current is drawn from the batteries. When AC power is restored, the batteries are recharged. Many UPS systems incorporate a standby diesel generator that starts as soon as the utility company feed is interrupted. With this arrangement, the batteries are called upon to supply operating current for only 30 s or so, until the generator gets up to speed. A UPS system intended to power a computer center is illustrated in Fig. 10.140.

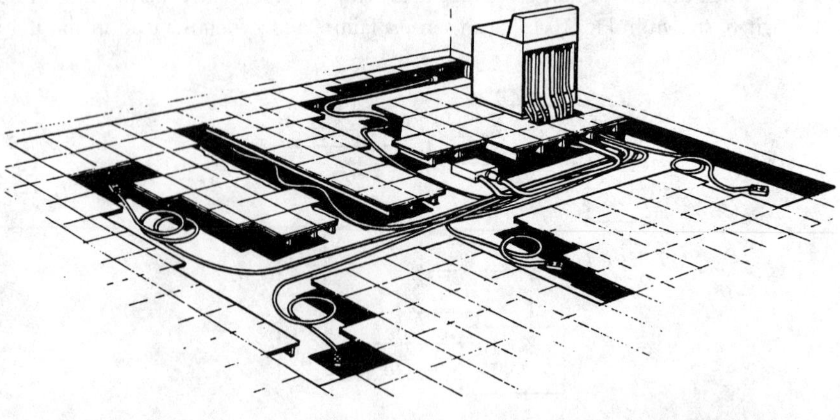

FIGURE 10.140   Installation details for a computer-room UPS power conditioner.

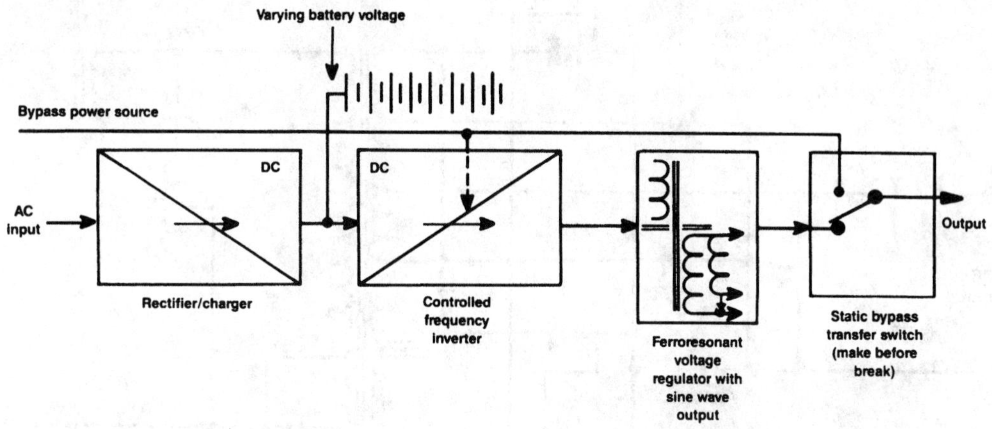

FIGURE 10.141   Simplified diagram of a static UPS system based on a ferroresonant transformer. (After [2].)

Power-Conversion Methods

Solid-state UPS systems that do not employ rotating machinery utilize one of several basic concepts. The design of an inverter is determined primarily by the operating power level. The most common circuit configurations include

- Ferroresonant inverter
- Delta magnetic inverter
- Inverter-fed L/C tank
- Quasi-square wave inverter
- Step wave inverter
- Pulse-width modulation inverter
- Phase modulation inverter

Ferroresonant Inverter

Illustrated in Fig. 10.141, the ferroresonant inverter is popular for low- to medium-power applications. A ferroresonant transformer can be driven with a distorted, unregulated input voltage and deliver a regulated, sinusoidal output when filtered properly. The ferroresonant transformer core is designed so that the secondary section is magnetically saturated at the desired output voltage. As a result, the output level remains relatively constant over a wide range of input voltages and loads. Capacitors, connected across the secondary, help drive the core into saturation and, in conjunction with inductive coupling, provide harmonic filtering. Regulated, sinusoidal three-phase output voltages are derived from two inverters operating into ferroresonant transformers in a *Scott-T* configuration. The basic ferroresonant inverter circuit, shown in Fig. 10.142, consists of an oscillator that controls SCR switches, which feed a ferroresonant transformer and harmonic filter. The saturated operating mode produces a regulated output voltage and inherent current limiting. Efficiency varies from 50 to 83%, depending on the load. Response time of the ferroresonant inverter is about 20 ms.

Although ferroresonant inverters are rugged, simple, and reliable, they do have disadvantages. First, such systems tend to be larger and heavier than electronically controlled inverters. Second, there is a phase shift between the inverter square wave and the output sine wave. This phase shift varies with load magnitude and power factor. When an unbalanced load is applied to a three-phase ferroresonant inverter, the output phases can shift from their normal 120° relationships. This results in a change in line-to-line voltages, even if individual line-to-neutral voltages are regulated perfectly. This voltage imbalance cannot be tolerated by some loads.

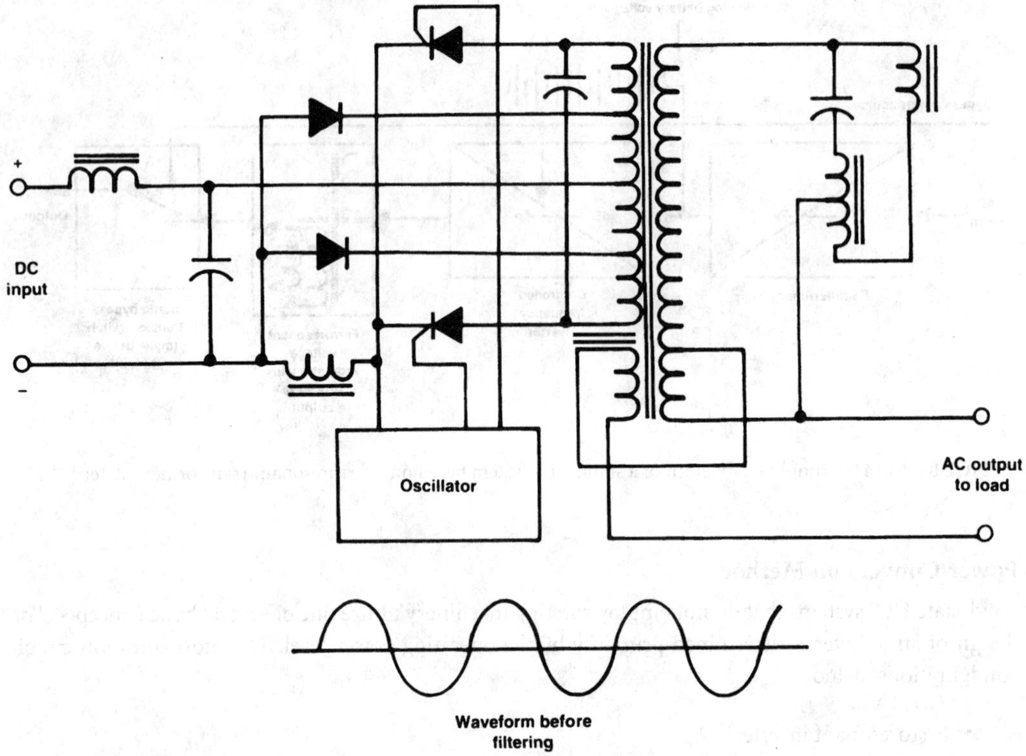

FIGURE 10.142   Simplified schematic diagram of a ferroresonant inverter. (After [2].)

Delta Magnetic Inverter

While most multiphase inverters are single-phase systems adapted to three-phase operation, the delta magnetic inverter is inherently a three-phase system. A simplified circuit diagram of a delta magnetic inverter is shown in Fig. 10.143. Inverter modules A1, B1, and C1 produce square wave outputs that are phase-shifted relative to each other by 120°. The waveforms are coupled to the primaries of transformer T1 through linear inductors. T1 is a conventional three-phase isolation transformer. The primaries of the device are connected in a delta configuration, reducing the third harmonic and all other harmonics that are odd-order multiples of the third. The secondaries of T1 are connected in a wye configuration to provide a four-wire three-phase output. Inductors L4–L9 form a network connected in a delta configuration to high-voltage taps on the secondary windings of T1. Inductors L4–L6 are single-winding saturating reactors, and L7–L9 are double-winding saturating reactors. Current drawn by this saturating reactor network is nearly sinusoidal and varies in magnitude, in a nonlinear manner, with voltage variations. For example, if an increase in load tended to pull the inverter output voltages down, the reduced voltage applied to the reactor network would result in a relatively large decrease in current drawn by the network. This, in turn, would decrease the voltage drop across inductors L1–L3 to keep the inverter output voltage at the proper level. The delta magnetic regulation technique is essentially a three-phase shunt regulator. Capacitors C1–C3 help drive the reactor network into saturation, as well as provide harmonic filtering in conjunction with the primary inductors.

Inverter-Fed L/C Tank

An inductor/capacitor tank is driven by a DC-to-AC inverter, as illustrated in Fig. 10.144. The tank circuit acts to reconstruct the sine wave at the output of the system. Regulation is accomplished by varying the capacitance or the inductance to control partial resonance and/or power factor. Some systems of this

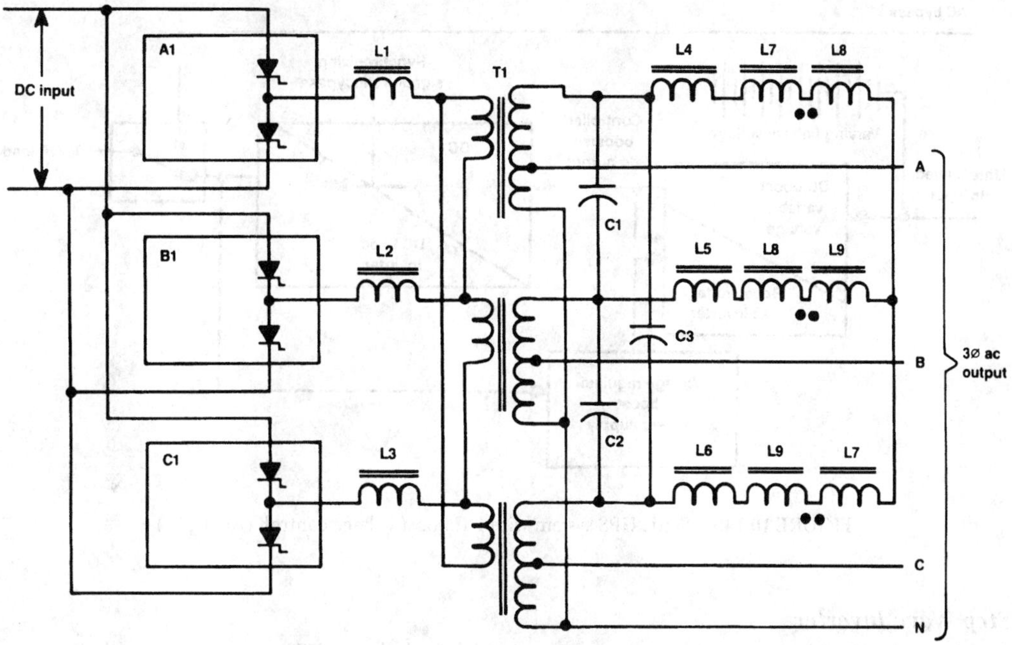

FIGURE 10.143 Simplified schematic diagram of the delta magnetic regulation system.

type use a saturable reactor in which electronic voltage-regulator circuits control the DC current in the reactor. Other systems, shown in Fig. 10.145, use a DC-to-variable-DC inverter/converter to control the UPS output through adjustment of the boost voltage. This feature permits compensation for changes in battery level.

Quasi-Square Wave Inverter

Shown in Fig. 10.146, the quasi-square wave inverter produces a variable-duty waveshape that must be filtered by tuned series and parallel inductive-capacitive networks to reduce harmonics and form a sinusoidal output. Because of the filter networks present in this design, the inverter responds slowly to load changes; response time in the range of 150 to 200 ms is common. Efficiency is about 80%. This type of inverter requires voltage-regulating and current-limiting networks, which increase circuit complexity and make it relatively expensive.

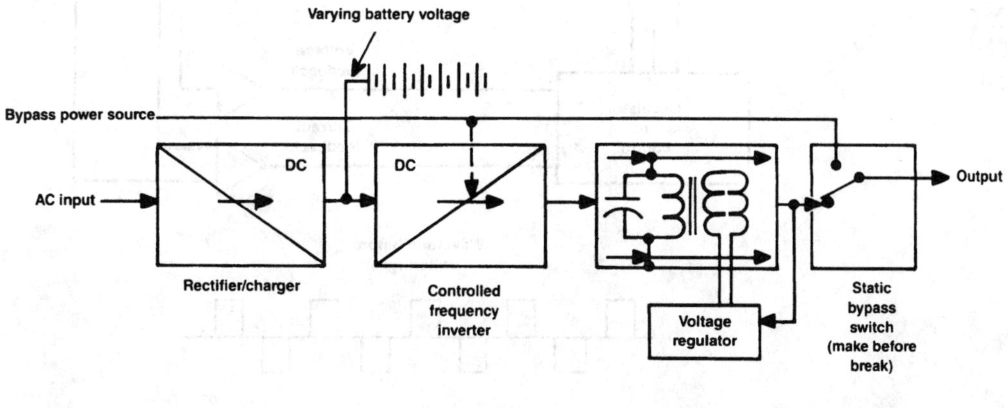

FIGURE 10.144 Static UPS system using saturable reactor voltage control. (After [2].)

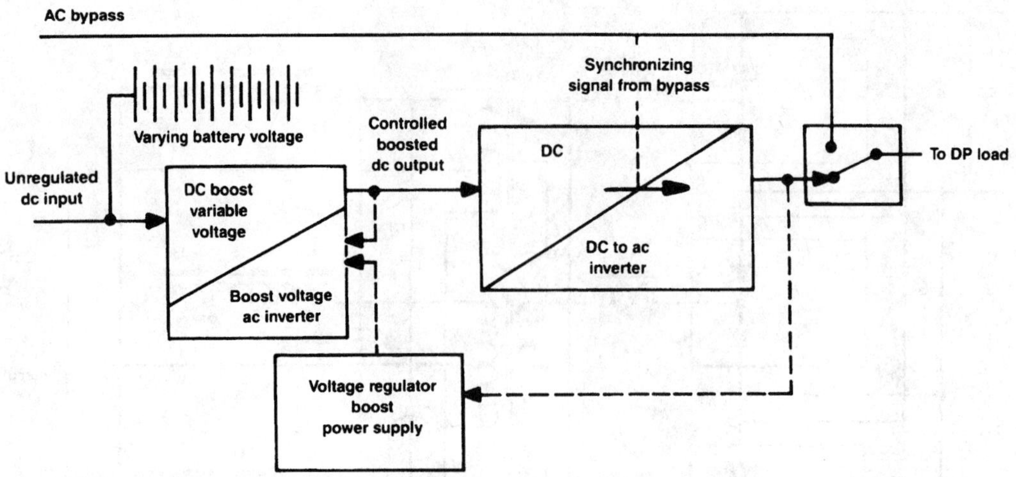

FIGURE 10.145   Static UPS system with DC boost voltage control. (After [2].)

Step Wave Inverter

A multistep inverter drives a combining transformer, which feeds the load. The general concept is illustrated in Fig. 10.147. The purity of the output sine wave is a function of the number of discrete steps produced by the inverter. Voltage regulation is achieved by a boost DC-to-DC power supply in series with the battery. Figure 10.148 and Figure 10.149 show two different implementations of single-phase units. Each system uses a number of individual inverter circuits, typically three or multiples of three. The inverters are

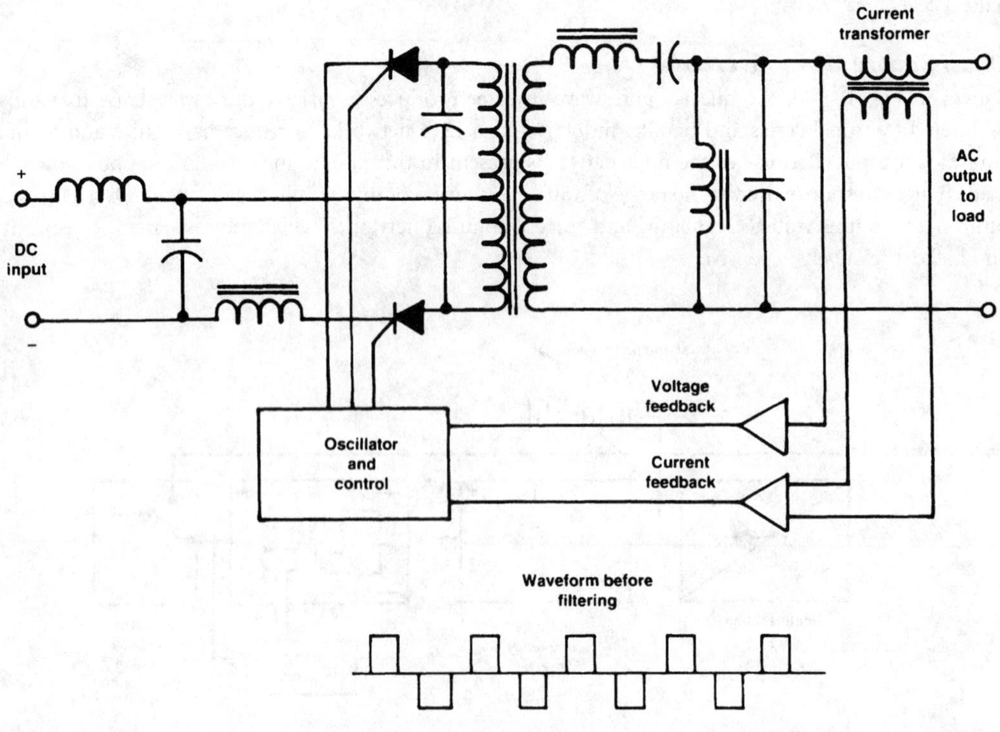

FIGURE 10.146   Simplified schematic diagram of a quasi-square wave inverter.

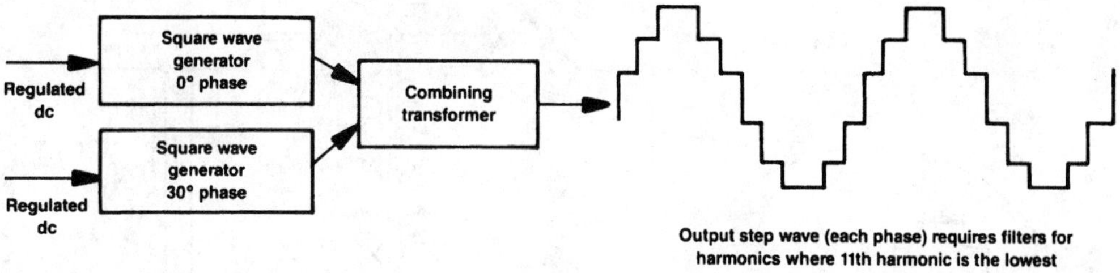

Output step wave (each phase) requires filters for harmonics where 11th harmonic is the lowest

FIGURE 10.147 Static UPS system using a stepped wave output. (After [2].)

controlled by a master oscillator; their outputs are added in a manner that reduces harmonics, producing a near-sinusoidal output. These types of systems require either a separate voltage regulator on the DC bus or a phase shifter. Response time is about 20 ms. Little waveform filtering is required, and efficiency can be as high as 85%. The step wave inverter, however, is complex and expensive. It is usually found in large three-phase UPS systems only.

Pulse-Width Modulation Inverter

Illustrated in Fig. 10.150, the pulse-width modulation (PWM) circuit incorporates two inverters, which regulate the output voltage by varying the pulse width. The output closely resembles a sine wave. Reduced filtering requirements result in good voltage-regulation characteristics. Response times close to 100 ms are typical. The extra inverter and feedback networks make the PWM inverter complex and expensive. Such systems usually are found at power levels greater than 50 kVA.

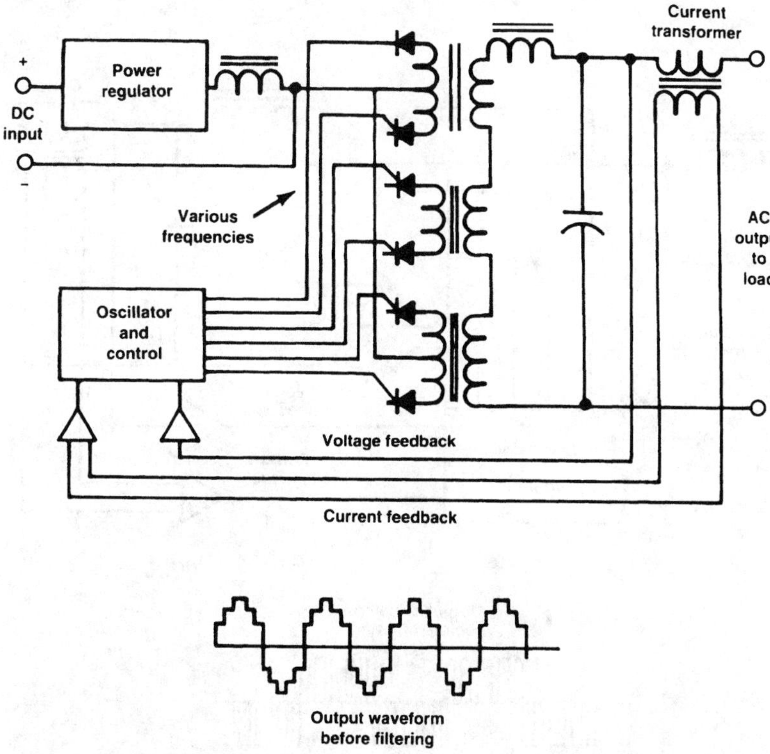

Output waveform before filtering

FIGURE 10.148 Step wave inverter schematic diagram.

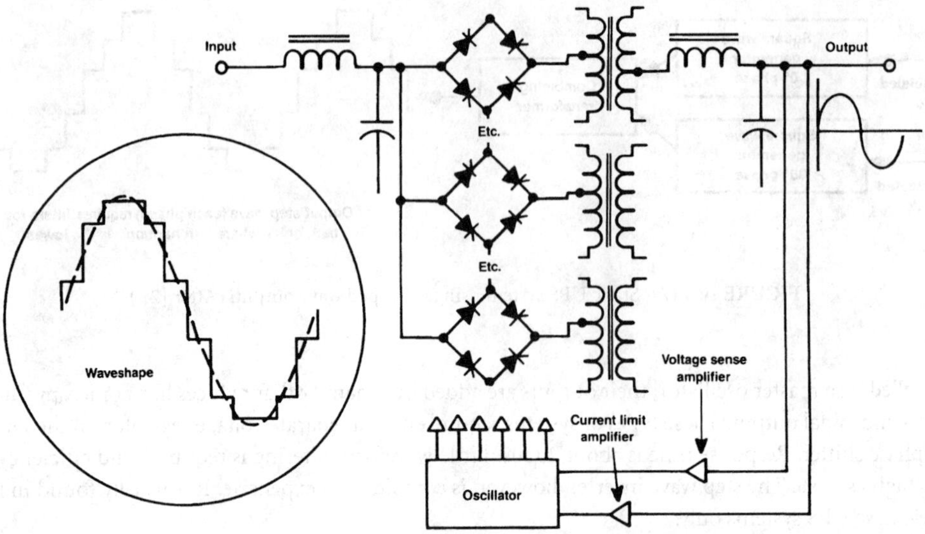

FIGURE 10.149   Step wave inverter and output waveform.

Phase Modulation Inverter

Illustrated in Fig. 10.151, this system uses DC-to-AC conversion through phase modulation of two square wave high-frequency signals to create an output waveform. The waveform then is filtered to remove the carrier signal and feed the load.

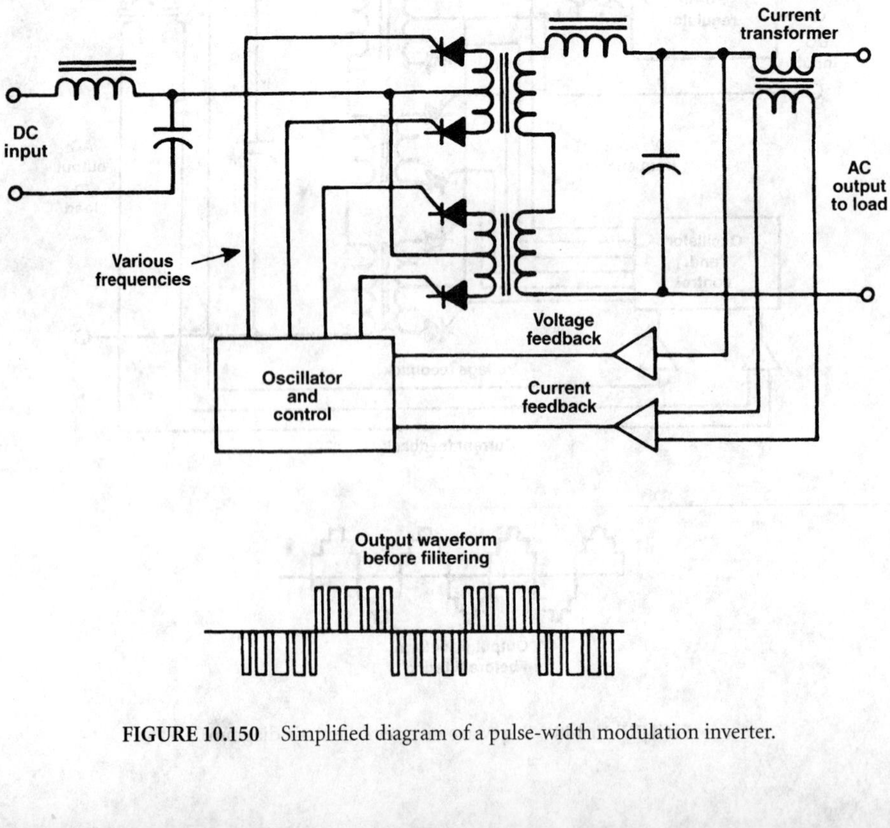

FIGURE 10.150   Simplified diagram of a pulse-width modulation inverter.

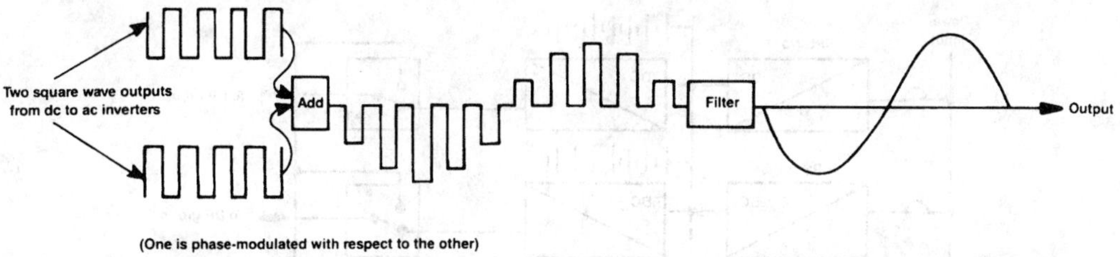

(One is phase-modulated with respect to the other)

FIGURE 10.151   Static UPS using a phase-demodulated carrier. (After [2].)

Redundant Operation

UPS systems can be configured as either a single, large power-conditioning/backup unit, or as several smaller systems arranged in a *parallel redundant* or *isolated redundant* mode. In the parallel redundant mode, the UPS outputs are connected together and share the total load. See Fig. 10.152. The power-output ratings of the individual UPS systems are selected to provide for operation of the entire load with any one UPS unit out of commission. In the event of expansion of the DP facility, additional UPS units can be added to carry the load. The parallel system provides the ability to cope with the failure of any single unit.

An isolated redundant system, illustrated in Fig. 10.153, divides the load among several UPS units. If one of the active systems fails, a static bypass switch will connect the affected load to a standby UPS system dedicated to that purpose. The isolated redundant system does not permit the unused capacity of one UPS unit to be utilized on a DP system that is loading another UPS to full capacity. The benefit of the isolated configuration is its immunity to systemwide failures.

Output Transfer Switch

Fault conditions, maintenance operations, and system reconfiguration require the load to be switched from one power source to another. This work is accomplished with an output transfer switch. As discussed previously, most UPS systems use electronic (static) switching. Electromechanical or motor-driven relays

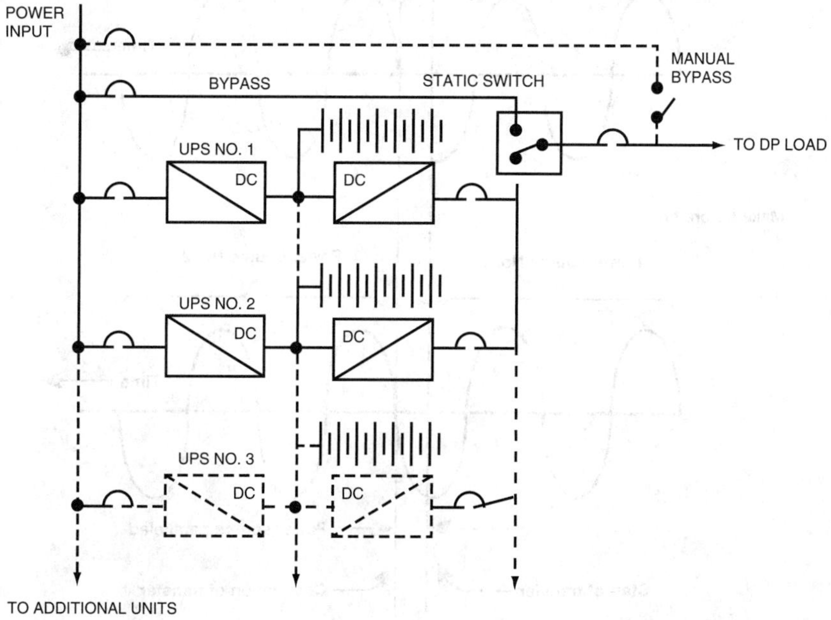

FIGURE 10.152   Configuration of a parallel redundant UPS system. (After [2].)

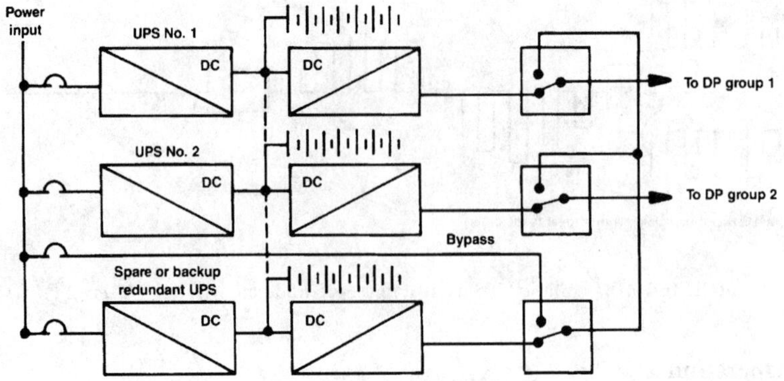

FIGURE 10.153   Configuration of an isolated redundant UPS system. (After [2].)

operate too slowly for most DP loads. Static transfer switches can be configured as either of the following:

- *Break-before-make:* Power output is interrupted before transfer is made to the new source.
- *Make-before-break:* The two power sources are overlapped briefly so as to prevent any interruption in AC power to the load.

Figure 10.154 illustrates each approach to load switching.

For critical-load applications, a make-before-break transfer is necessary. For the switchover to be accomplished with minimum disturbance to the load, both power sources must be synchronized. The UPS

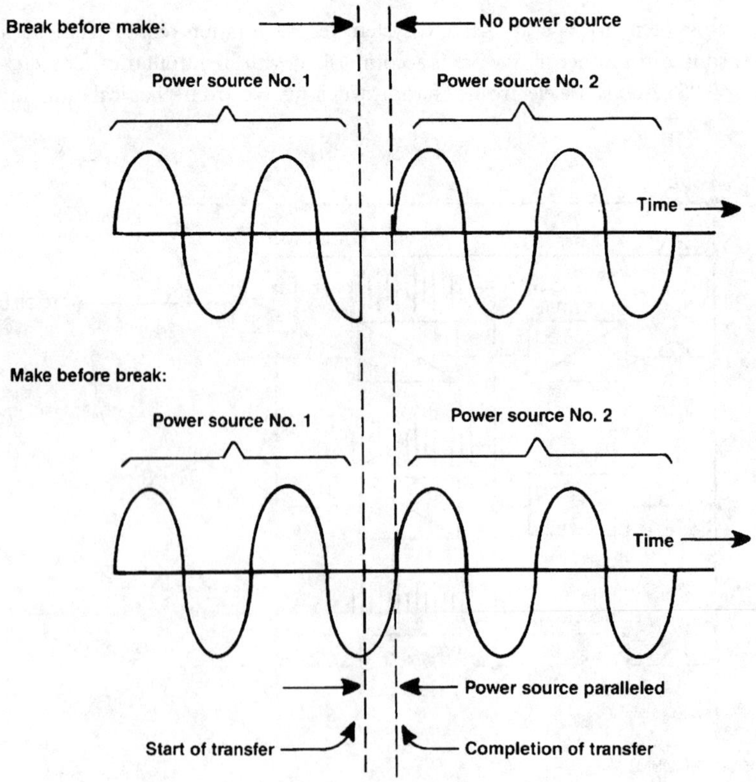

FIGURE 10.154   Static transfer switch modes.

system must, therefore, be capable of synchronizing to the utility AC power line (or other appropriate power source).

Battery Supply

UPS systems typically are supplied with sufficient battery capacity to carry a DP load for periods ranging from 5 min to 1 h. Long backup time periods usually are handled by a standby diesel generator. Batteries require special precautions. For large installations, they almost always are placed in a room dedicated to that purpose. Proper temperature control is important for long life and maximum discharge capacity.

Most rectifier/charger circuits operate in a *constant-current* mode during the initial charge period, and automatically switch to a *constant-voltage* mode near the end of the charge cycle. This provides maximum battery life consistent with rapid recharge. It also prevents excessive battery outgassing and water consumption. The charger provides a *float voltage level* for maintaining the normal battery charge, and sometimes a higher voltage to equalize certain devices.

Four battery types typically are found in UPS systems:

- *Semisealed lead calcium.* A gel-type electrolyte is used that does not require the addition of water. There is no outgassing or corrosion. This type of battery is used when the devices are integral to small UPS units, or when the batteries must be placed in occupied areas. The life span of a semisealed lead calcium battery, under ideal conditions, is about 5 years.
- *Conventional lead calcium.* The most common battery type for UPS installations, these units require watering and terminal cleaning about every 6 months. Expected lifetime ranges up to 20 years. Conventional lead-calcium batteries outgas hydrogen under charge conditions and must be located in a secure, ventilated area.
- *Lead-antimony.* The traditional lead-acid batteries, these devices are equivalent in performance to lead-calcium batteries. Maintenance is required every 3 months. Expected lifetime is about 10 years. To retain their capacity, lead-antimony batteries require a monthly equalizing charge.
- *Nickel-cadmium.* Advantages of the nickel-cadmium battery include small size and low weight for a given capacity. These devices offer excellent high- and low-temperature properties. Life expectancy is nearly that of a conventional lead-calcium battery. Nickel-cadmium batteries require a monthly equalizing charge, as well as periodic discharge cycles to retain their capacity. Nickel-cadmium batteries are the most expensive of the devices typically used for UPS applications.

10.8.4 Dedicated Protection Systems

A wide variety of power-protection technologies are available to solve specific problems at a facility. The method chosen depends upon a number of factors, not the least of which is cost. Although UPS units and m-g sets provide a high level of protection against AC line disturbances, the costs of such systems are high. Many times, adequate protection can be provided using other technologies at a fraction of the cost of a full-featured, facility-wide system. The applicable technologies include:

- Ferroresonant transformer
- Isolation transformer
- Tap-changing regulator
- Line conditioner

Ferroresonant Transformer

Ferroresonant transformers exhibit unique voltage-regulation characteristics that have proved valuable in a wide variety of applications. Voltage output is fixed by the size of the core, which saturates each half cycle, and by the turns ratio of the windings. Voltage output is determined at the time of manufacture and cannot be adjusted. The secondary circuit resonance depends upon capacitors, which work with the saturating inductance of the device to keep the resonance active. A single-phase ferroresonant transformer is shown in Fig. 10.155.

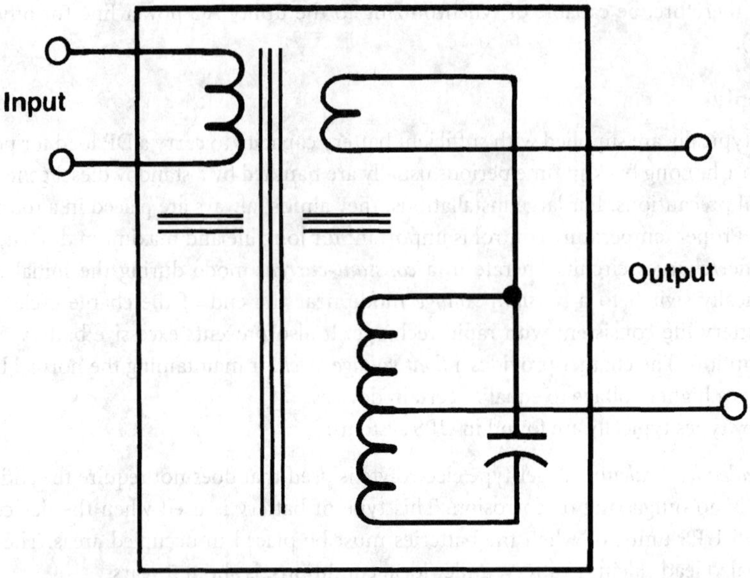

FIGURE 10.155   Basic design of a ferroresonant transformer. (After [2]).

Load currents tend to demagnetize the core, so output current is automatically limited. A ferroresonant transformer typically cannot deliver more than 125 to 150% of its full-load rated output without going into a current-limiting mode. Such transformers cannot support the normal starting loads of motors without a significant dip in output voltage.

Three-phase versions consisting of single-phase units connected in delta-delta, delta-wye, or wye-wye can be unstable when working into unbalanced loads. Increased stability for three-phase operation can be achieved with zigzag and other special winding configurations.

Shortcomings of the basic ferroresonant transformer have been overcome in advanced designs intended for low-power (2 kVA and below) voltage-regulator applications for DP equipment. Figure 10.156 illustrates one of the more common regulator designs. Two additional windings are included:

- *Compensating winding W_c*, which corrects for minor flux changes that occur after saturation has been reached.

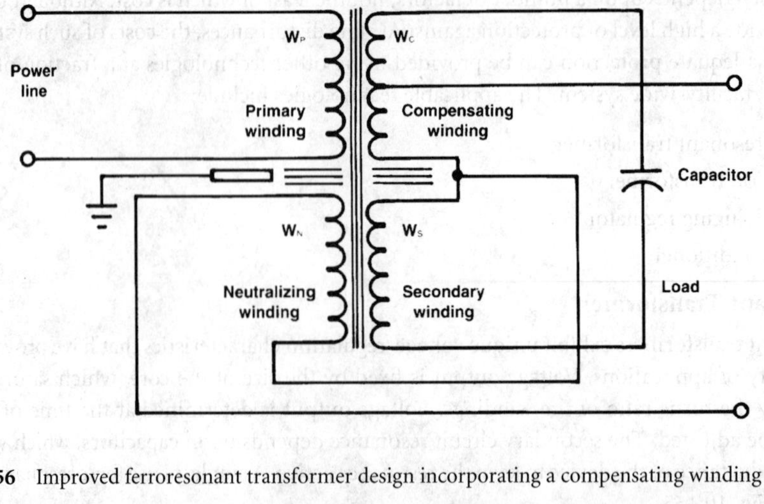

FIGURE 10.156   Improved ferroresonant transformer design incorporating a compensating winding and a neutralizing winding.

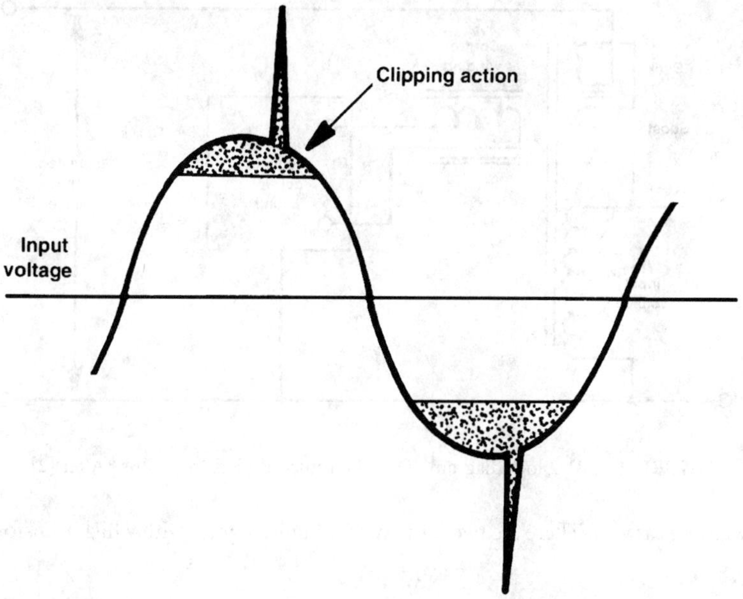

FIGURE 10.157 Clipping characteristics of a ferroresonant transformer.

- *Neutralizing winding* (W_n), which cancels out most of the harmonic content of the output voltage. Without some form of harmonic reduction, the basic ferroresonant transformer would be unsuitable for sensitive DP loads.

A unique characteristic of the ferroresonant regulator is its ability to reduce normal-mode impulses. Because the regulating capability of the device is based on driving the secondary winding into saturation, transients and noise bursts are clipped, as illustrated in Fig. 10.157.

The typical ferroresonant regulator has a response time of about 25 ms. Because of the tuned circuit at the output, the ferroresonant regulator is sensitive to frequency changes. A 1% frequency change will result (typically) in a 1.5% change in output voltage. Efficiency of the device is about 90% at full rated load, and efficiency declines as the load is reduced. The ferroresonant regulator is sensitive to leading and lagging power factors, and exhibits a relatively high output impedance.

Magnetic-Coupling-Controlled Voltage Regulator

An electronic magnetic regulator uses DC to adjust the output voltage through magnetic saturation of the core. The direct-current fed winding changes the saturation point of the transformer. This action, in turn, controls the AC flux paths through boost or buck coils to raise or lower the output voltage in response to an electronic circuit that monitors the output of the device. A block diagram of the magnetic-coupling-controlled voltage regulator is shown in Fig. 10.158. Changes in output voltage are smooth, although the typical response time of 5 to 10 cycles is too slow to prevent surge and sag conditions from reaching the load.

Figure 10.159 illustrates an electromechanical version of a saturable regulating transformer. Output voltage is controlled by varying the impedance of the saturable reactor winding in series with a step-up autotransformer. Response to sag and surge conditions is 5 to 10 cycles. Such devices usually exhibit high output impedance and are sensitive to lagging load power factor.

Isolation Transformer

Transients, as well as noise (RF and low-level spikes) can pass through transformers, not only by way of the magnetic lines of flux between the primary and the secondary, but through resistive and capacitive paths

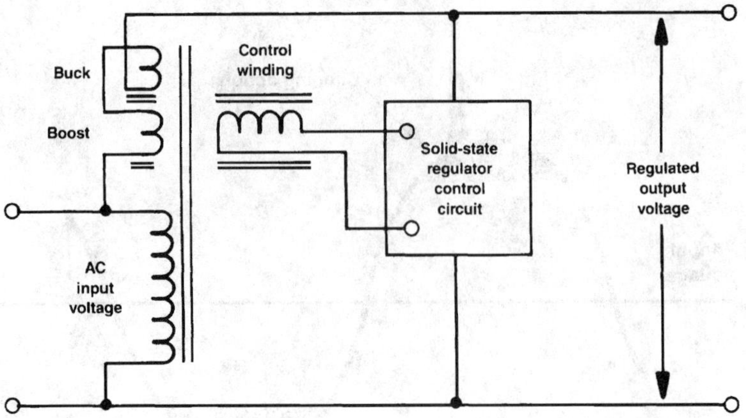

FIGURE 10.158   Block diagram of an electronic magnetic regulator. (After [2].)

between the windings as well. There are two basic types of noise signals with which transformer designers must cope:

- *Common-mode* noise. Unwanted signals in the form of voltages appearing between the local ground reference and each of the power conductors, including neutral and the equipment ground.
- Normal-mode noise. Unwanted signals in the form of voltages appearing in line-to-line and line-to-neutral signals.

Increasing the physical separation of the primary and secondary windings will reduce the resistive and capacitive coupling. However, it will also reduce the inductive coupling and decrease power transfer.

A better solution involves shielding the primary and secondary windings from each other to divert most of the primary noise current to ground. This leaves the inductive coupling basically unchanged. The concept can be carried a step further by placing the primary winding in a shielding box that shunts noise currents to ground and reduces the capacitive coupling between the windings.

One application of this technology is shown in Fig. 10.160, in which transformer noise decoupling is taken a step further by placing the primary and secondary windings in their own wrapped foil box shields. The windings are separated physically as much as possible for the particular power rating and are placed between Faraday shields. This gives the transformer high noise attenuation from the primary to the secondary, and from the secondary to the primary. Figure 10.161 illustrates the mechanisms involved. Capacitances between the windings, and between the windings and the frame, are broken into smaller

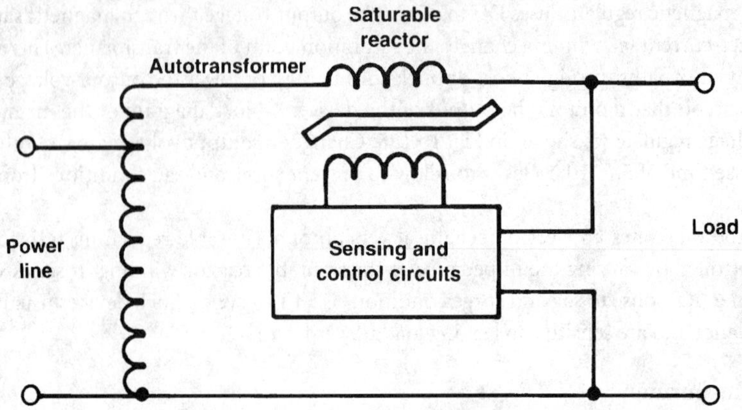

FIGURE 10.159   Autotransformer/saturable reactor voltage regulator.

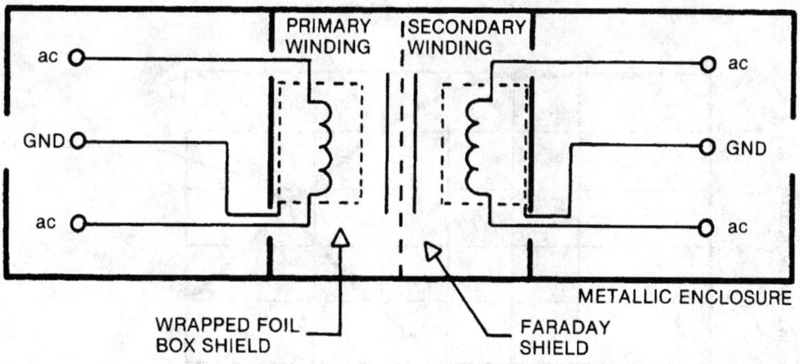

FIGURE 10.160 The shielding arrangement used in a high-performance isolation transformer. The design goal of this mechanical design is high common-mode and normal-mode noise attenuation.

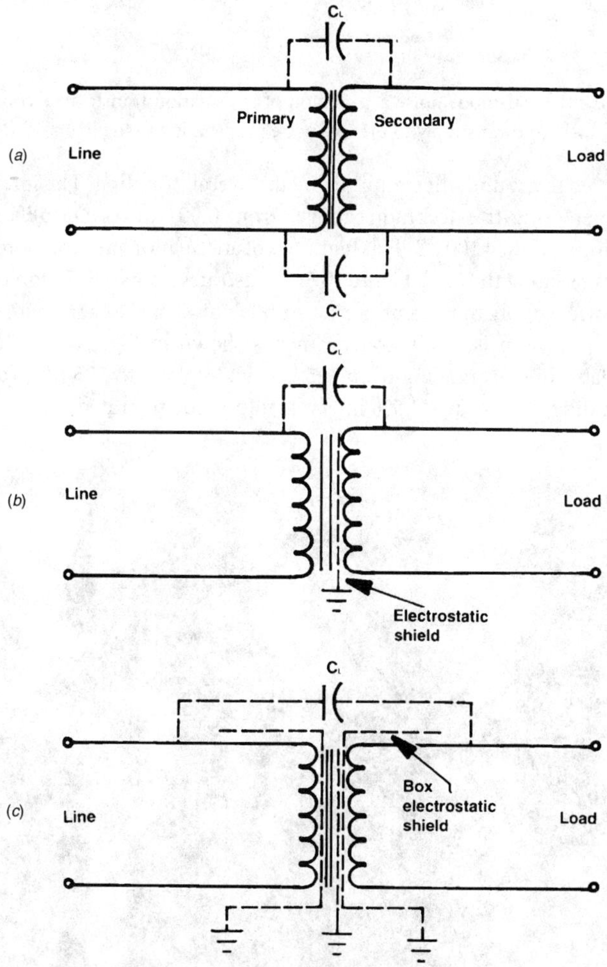

FIGURE 10.161 The elements involved in a noise-suppression isolation transformer: (*a*) conventional transformer with capacitive coupling as shown; (*b*) the addition of an electrostatic shield between the primary and the secondary; (*c*) transformer with electrostatic box shields surrounding the primary and secondary windings.

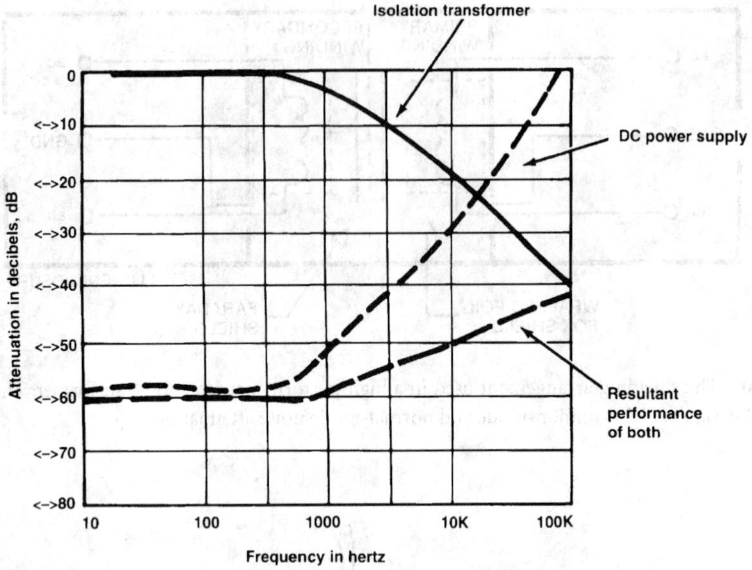

FIGURE 10.162   How the normal-mode noise attenuation of an isolation transformer combines with the filtering characteristics of the AC-to-DC power supply to prevent noise propagation to the load.

capacitances and shunted to ground, thus minimizing the overall coupling. The interwinding capacitance of a typical transformer using this technique ranges from 0.001 to 0.0005 pF. Common-mode noise attenuation is generally in excess of 100 dB. This high level of attenuation prevents common-mode impulses on the power line from reaching the load. Figure 10.162 illustrates how an isolation transformer combines with the AC-to-DC power supply to prevent normal-mode noise impulses from affecting the load.

A 7.5-kVA noise-suppression isolation transformer is shown in Fig. 10.163. High-quality isolation transformers are available in sizes ranging from 125 VA single-phase to 125 kVA (or more) three-phase. Usually, the input winding is tapped at 2.5% intervals to provide the rated output voltage, despite high

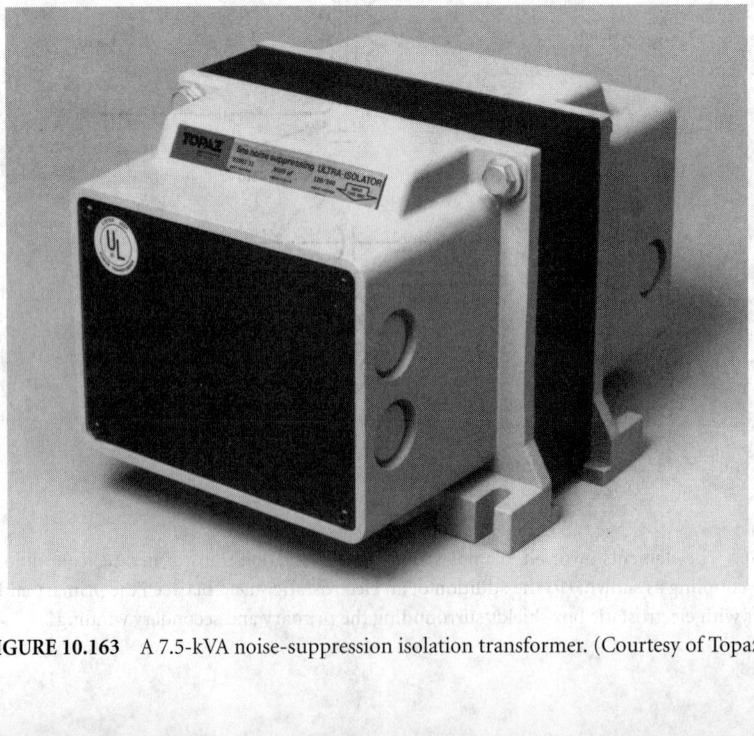

FIGURE 10.163   A 7.5-kVA noise-suppression isolation transformer. (Courtesy of Topaz.)

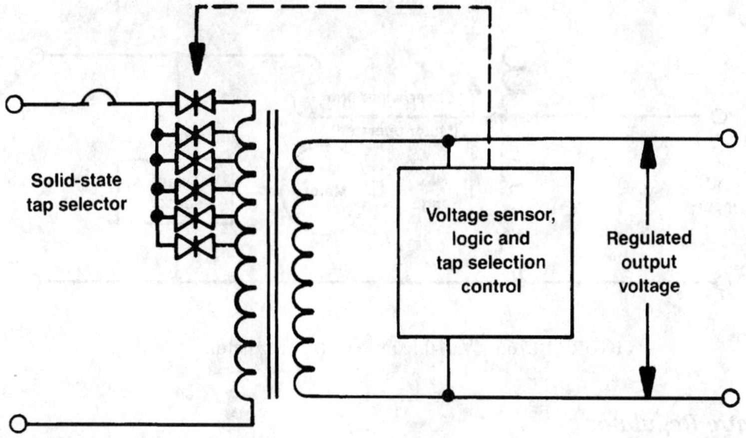

FIGURE 10.164 Simplified schematic diagram of a tap-changing voltage regulator. (After [2].)

or low average input voltages. The total tap adjustment range is typically from 5% above nominal to 10% below nominal. For three-phase devices, typical input-voltage nominal ratings are 600, 480, 240, and 208 V line-to-line for 15 kVA and larger.

Tap-Changing Regulator

The concept behind a tap-changing regulator is simple: adjust the transformer input voltage to compensate for AC line-voltage variations. A tap-changing regulator is shown in Fig. 10.164. Although simple in concept, the actual implementation of the system can become complex because of the timing waveforms and pulses necessary to control the SCR banks. Because of the rapid response time of SCRs, voltage adjustment can be made on a cycle-by-cycle basis in response to changes in both the utility input and the load. Tap steps are typically 2 to 3%. Such systems create no objectionable switching transients in unity power factor loads. For low-PF loads, however, small but observable transients can be generated in the output voltage at the moment the current is switched. This noise usually has no significant effect on DP loads. Other operating characteristics include:

- Low-internal impedance (similar to an equivalent transformer)
- High efficiency from full load to 25% or less
- Rapid response time (typically one to three cycles) to changes in the input AC voltage or load current
- Low acoustic noise level

An autotransformer version of the tap-changing regulator is shown in Fig. 10.165.

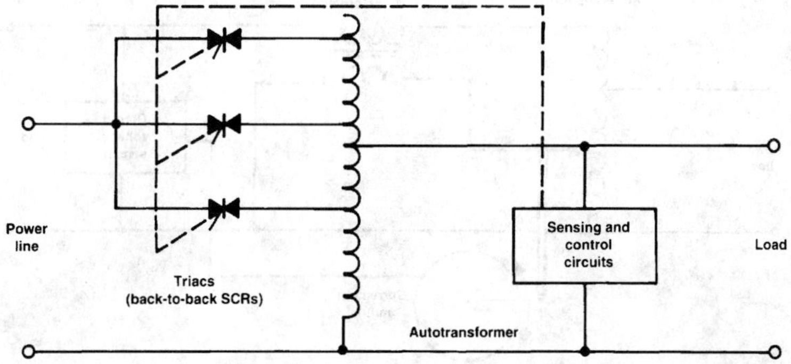

FIGURE 10.165 Tap-changing voltage regulator using an autotransformer as the power-control element.

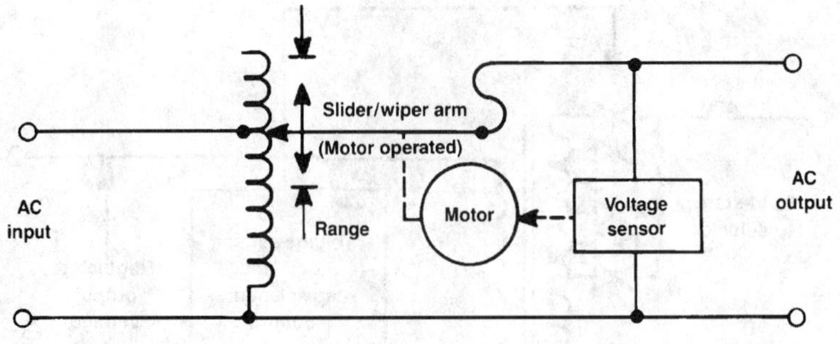

FIGURE 10.166   Variable ratio voltage regulator. (After [2].)

Variable Ratio Regulator

Functionally, the variable ratio regulator is a modified version of the tap-changer. Rather than adjusting output voltage in steps, the motor-driven regulator provides a continuously variable range of voltages. The basic concept is shown in Fig. 10.166. The system is slow, but generally reliable. It is excellent for keeping DP hardware input voltages within the optimum operating range. Motor-driven regulators usually are able to follow the steady rise and fall of line voltages that typically are experienced on utility company lines. Efficiency is normally good, approaching that of a good transformer. The internal impedance is low, making it possible to handle a sudden increase or decrease in load current without excessive undervoltages and/or overvoltages. Primary disadvantages of the variable ratio regulator are limited current ratings, determined by the moving brush assembly, and the need for periodic maintenance.

Variations on the basic design have been marketed, including the system illustrated in Fig. 10.167. A motor-driven brush moves across the exposed windings of an autotransformer, causing the series transformer to buck or boost voltage to the load. The correction motor is controlled by a voltage-sensing circuit at the output of the device.

The induction regulator, shown in Fig. 10.168, is still another variation on the variable ratio transformer. Rotation of the rotor in one direction or the other varies the magnetic coupling and raises or lowers the output voltage. Like the variable ratio transformer, the induction regulator is slow, but it has no brushes and requires little maintenance. The induction regulator has higher inductive reactance and is slightly less efficient than the variable ratio transformer.

Variable Voltage Transformer

Because of their application in voltage control systems, it is worthwhile to discuss the operation of variable voltage transformers in greater detail. A number of conventional transformers can have—to a limited

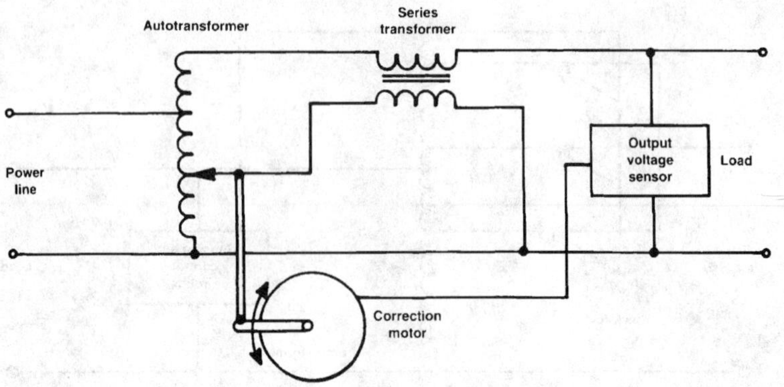

FIGURE 10.167   Motor-driven line-voltage regulator using an autotransformer with a buck/boost series transformer.

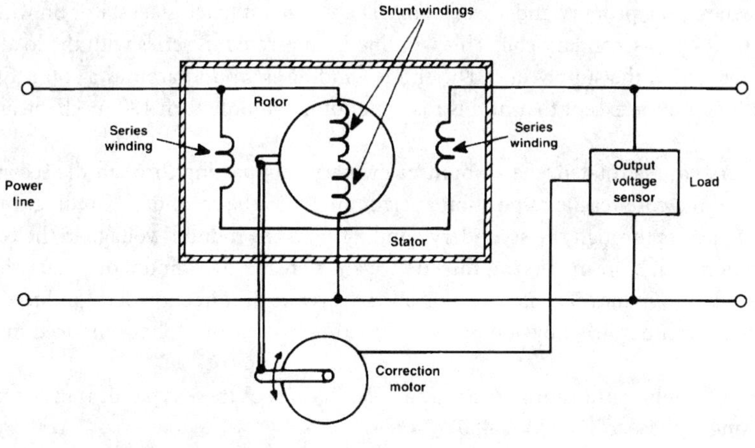

FIGURE 10.168   Rotary induction voltage regulator.

degree—their primary/secondary ratio changed through the use of taps located (usually) on the primary windings [6]. To obtain a greater degree of flexibility in changing the ratio between the primary and secondary coils, and thus allow greater secondary voltage changes, a variable voltage transformer is used. The amount of voltage change depends upon the basic construction of the device, which typically divides into one of two categories

- Brush type
- Induction type

Brush Type

To achieve a variable output voltage, one tap on the transformer secondary is fixed and the other tap is connected to a brush that slides along an uninsulated section of the transformer coil [6]. One way of accomplishing this objective is to have the coil wrapped around a toroidal shaped core. The voltage ratio relates to the location of the brush as it rides against the coil and depends upon where on the coil the brush is allowed to make contact. *Limited ratio* variable transformers are available, as well as full range units that can adjust the output voltage from 0 to about 120% of the incoming line voltage. When the output voltage exceeds the input voltage, it means that there are extra turns on the coil that extend beyond the windings that lie between the incoming power terminals (in effect, the unit becomes a step-up transformer).

Ratings start at less than 1 kVA for a 120 V, single-phase variable transformer. Basic units are wired in parallel and/or in series to obtain higher power capacity. Two units in parallel have double the current and kVA rating. The individual units are stacked on top of each other, bolted together, and operated from a common shaft that rotates the brushes. Operating in a configuration that combines parallel and series connections, with several units stacked together, this type of variable voltage transformer, at 480 V, can have a rating exceeding 200 kVA. Sometimes, a control mechanism is attached to the rotor that turns the brushes, allowing automatic adjustment of the output voltage.

An important characteristic of this type of transformer relates to brush contact and the amount of current flowing through the carbon brush. A rating based solely on output kVA can cause serious operational problems because, for a given kVA load, the current drawn depends on the output voltage. Because the output voltage is variable, a load of a given kVA value can draw a safe current at 100% voltage, but at 25% voltage, the current required to serve a load of the same kVA value would require four times as much current, which could cause the brush to overheat.

Induction Type

The induction type variable transformer does not use brushes [6]. The usual voltage change for these units is ±10%, but it can be greater. The device is essentially a variable-ratio autotransformer that uses two separate windings, a primary and a secondary. There is a laminated steel stator on which is wound a winding that serves as a secondary coil. This winding is connected in series with the load. The primary coil is connected across the supply line. The shunt winding is wound around a rotor. Construction is similar to that of a motor except that in this case, the rotor can only turn 180 mechanical and electrical degrees.

As the primary core is rotated, the amount of primary flux passing through the secondary winding is decreased until the core reaches a position at right angles to the secondary winding. In this position, no primary flux passes through the secondary windings and the induced voltage in the coil is zero. The continued rotation of the core in the same direction again increases the amount of flux passing through the secondary, but it is now in the opposite direction and so reverses the direction of the induced voltage. Thus, the output voltage can be varied by adding or subtracting from it the voltage induced in the secondary winding.

Both single and 3-phase transformers are available. Ratings of these types of transformers vary from 8 kVA, 120 V single-phase to 1500 kVA, 480V 3-phase.

Line Conditioner

A line conditioner combines the functions of an isolation transformer and a voltage regulator in one unit. The three basic types of line conditioners for DP applications are:

- *Linear amplifier correction system.* As illustrated in Fig. 10.169, correction circuitry compares the AC power output to a reference source, derived from a 60-Hz sine wave generator. A correction voltage is developed and applied to the secondary power winding to cancel noise and voltage fluctuations. A box shield around the primary winding provides common-mode impulse rejection (80 to 100 dB typical). The linear amplifier correction system is effective and fast, but the overall regulating range is limited.
- *Hybrid ferroresonant transformer.* As shown in Fig. 10.170, this system consists of a ferroresonant transformer constructed using the isolation techniques discussed in the section titled "Isolation Transformer." The box and Faraday shields around the primary and compensating windings give the transformer essentially the characteristics of a noise-attenuating device, while preserving the voltage-regulating characteristics of a ferroresonant transformer.

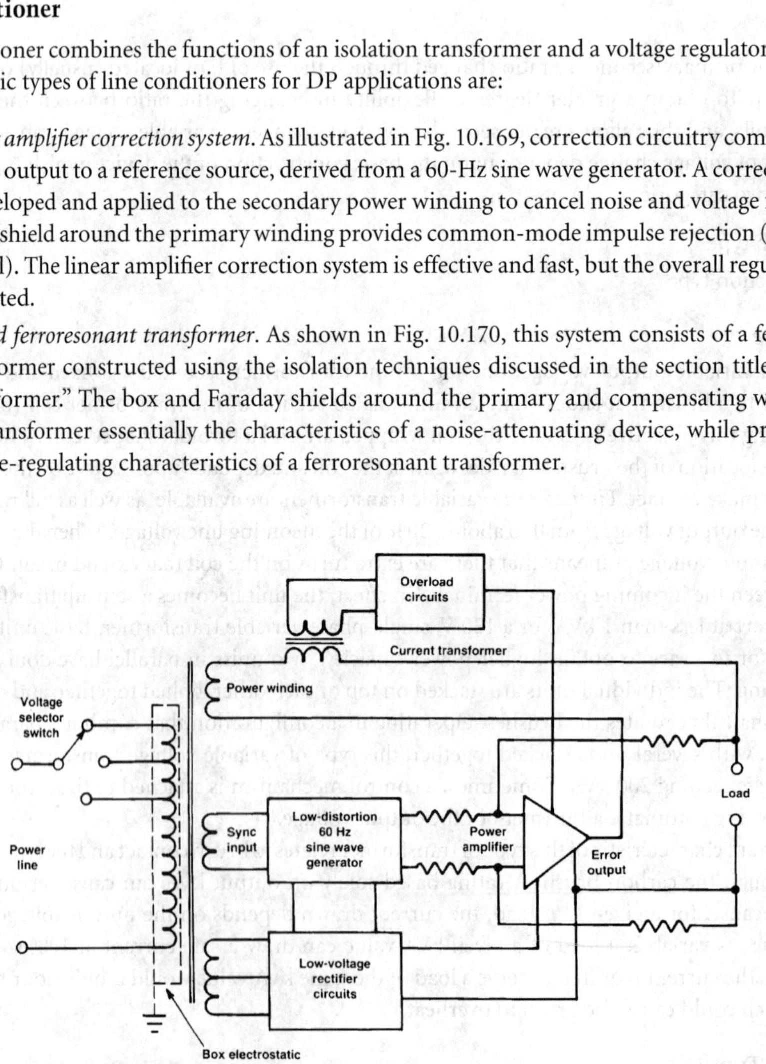

FIGURE 10.169   Power-line conditioner using linear amplifier correction.

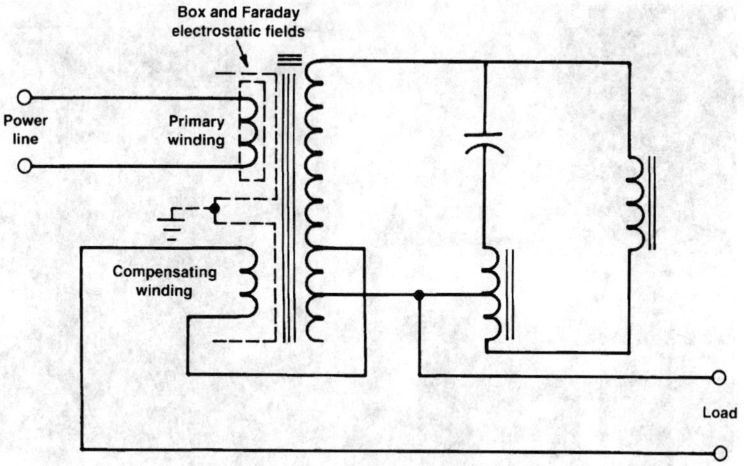

FIGURE 10.170   Line conditioner built around a shielded-winding ferroresonant transformer.

- *Electronic tap-changing, high-isolation transformer*. This system is built around a high-attenuation isolation transformer with a number of primary winding taps. SCR pairs control voltage input to each tap, as in a normal tap-changing regulator. Tap changing can also be applied to the secondary, as shown in Fig. 10.171. The electronic tap-changing, high-isolation transformer is an efficient design that effectively regulates voltage output and prevents noise propagation to the DP load.

Hybrid Transient Suppressor

A wide variety of AC power-conditioning systems are available, based on a combination of solid-state technologies. Most incorporate a combination of series and parallel elements to shunt transient energy. One such system is pictured in Fig. 10.172.

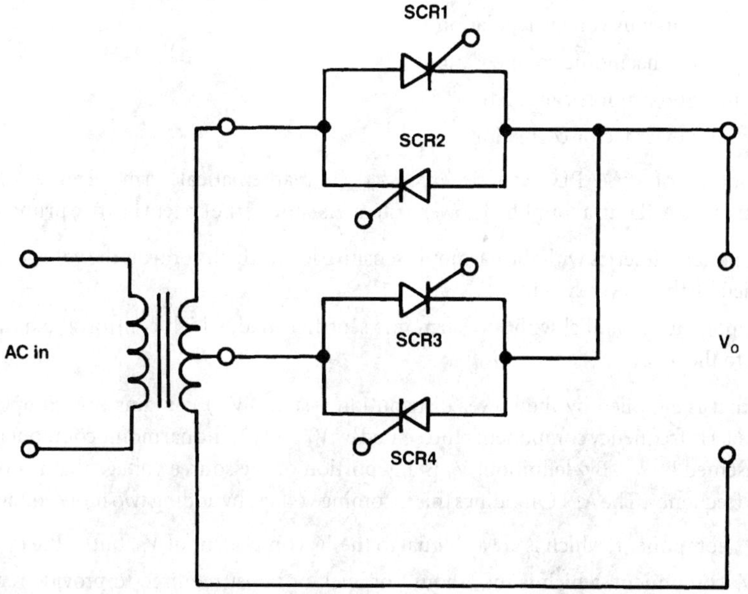

FIGURE 10.171   Secondary-side synchronous tap-changing transformer.

FIGURE 10.172   Interior view of a high-capacity hybrid power conditioner. (Courtesy of Control Concepts.)

Active Power Line Conditioner

The *active power line conditioner* (APLC) provides an adaptive solution to many power quality problems [7]. The major features of APLC include

- Near-instantaneous voltage regulation
- Source voltage harmonic compensation
- Load current harmonic cancellation
- Distortion power factor correction

The basic concept of the APLC can be expressed in mathematical terms. Figure 10.173 shows a block diagram of the APLC in a simplified power system. As shown, the system has two primary components

- A component in series with the harmonic sensitive load, which controls the voltage sine wave being supplied by the power system.
- A component in parallel with the harmonic sensitive load, which controls the harmonic current going to the load.

The voltage that is supplied by the power distribution system (V_{in}) contains two components: (1) the fundamental 60 Hz frequency component represented by V_f and (2) the harmonic component of the source voltage represented by V_h. By definition, V_h is any portion of the source voltage that is not of the 60 Hz fundamental frequency. The APLC modifies this incoming voltage by adding two more voltage components

- The V_h component, which is always equal to the V_h component of V_{in} but of the opposite polarity
- The V_r component, which is the amount of buck or boost required to provide regulation of the output voltage.

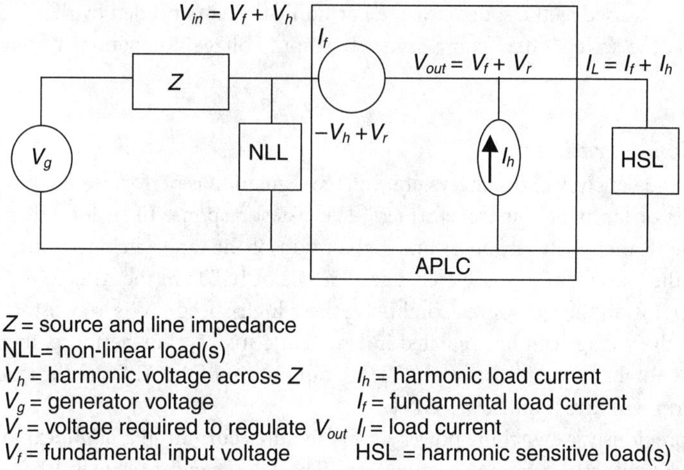

Z = source and line impedance
NLL= non-linear load(s)
V_h = harmonic voltage across Z I_h = harmonic load current
V_g = generator voltage I_f = fundamental load current
V_r = voltage required to regulate V_{out} I_l = load current
V_f = fundamental input voltage   HSL = harmonic sensitive load(s)

FIGURE 10.173   Basic conceptual diagram of the active power line conditioner. (After [7].)

Because the two V_h components cancel, the load receives a regulated sinusoidal voltage at the fundamental frequency with no harmonic components.

The load uses current (I_L) in a nonlinear fashion. The harmonic portion of the current I_h is—by definition—everything that is not the fundamental frequency component. The APLC becomes the source of I_h for the load circuit, leaving only the fundamental frequency portion (I_f) to be supplied by the power system. Because the harmonic portion of the load current is supplied locally by the APLC, the distortion power factor of the load is near unity.

A digital signal processor (DSP) is used to monitor the input and output parameters of the APLC. The DSP then makes any needed adjustment in the amount of voltage compensation and current injection by modifying the switching characteristics of the insulated-gate bipolar transistors (IGBTs) in the APLC.

A simplified schematic of the APLC power circuitry is given in Fig. 10.174. The DC bus (V_{dc}) is the energy storage device that supplies the power required for voltage compensation and current injection. The parallel "filter" serves two purposes. First, the IGBTs in this portion of the APLC keep the DC link capacitor

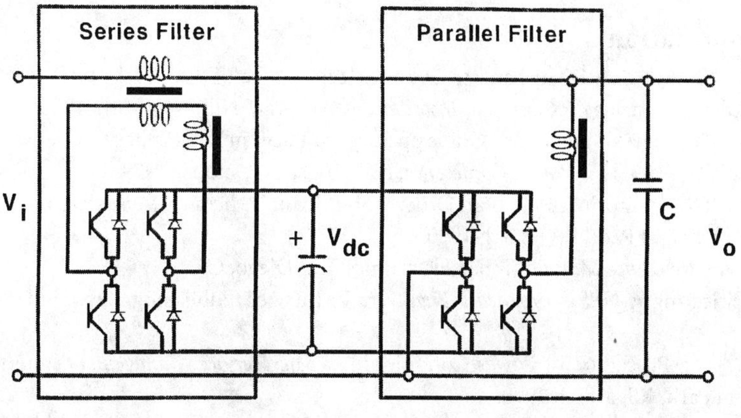

FIGURE 10.174   Series/parallel active power line conditioner power control circuit. (After [7].)

charged. Second, it is used as the source for the harmonic currents needed by the load. By using power from the DC link, the series "filter" compensates for input voltage fluctuations through the buck/boost transformer.

Application Considerations

In a conventional power control system system, voltage regulation is performed by using either ferroresonant transformers or tap switching transformers. The fastest response times for either of these solutions is about 1/2 cycle (8 ms). This response time is the result of the tap switching transformer's need for a zero crossing of the waveform to make a change. The use of IGBTs in the APLC allows full power to be switched at any point on the sine wave. Combining these high-speed power transistors with a fast digital signal processor, the voltage can be regulated in less than 1 ms [7]. Furthermore, the input voltage can exhibit up to 10% total harmonic distortion (THD), but output of the APLC to the load will typically see a voltage waveform with less than 1.0% THD [7].

As discussed previously, a switching power supply requires current in a nonlinear fashion. These load currents may have significant harmonic components. The APLC can cancel up to 100% load current THD at full load. The source current being supplied by the power distribution system will see a purely linear load with no more than 3% current THD [7].

References

Key, T. "The Effects of Power Disturbances on Computer Operation," IEEE Industrial and Commercial Power Systems Conference, Cincinnati, June 7, 1978.

Federal Information Processing Standards Publication No. 94, *Guideline on Electrical Power for ADP Installations*, U.S. Department of Commerce, National Bureau of Standards, Washington, D.C., 1983.

Plater, B. B., and J. A. Andrews, "Advances in Flywheel Energy Storage Systems," *Power Quality '97*, Intertec International, Ventura, CA, pp. 460–469, 1997.

Materials Selector 1987, *Materials Engineering*, Penton Publishing.

Weaver, E., J., "Dynamic Energy Storage System," *Power Quality Solutions/Alternative Energy*, Intertec International, Ventura, CA, pp. 373–380, 1996.

Berutti, A., *Practical Guide to Applying, Installing and Maintaining Transformers*, Robert B. Morgan (Ed.), Intertec Publishing, Overland Park, KS, p. 13, 1994.

Petrecca, R. M., "Mitigating the Effects of Harmonics and Other Distrubances with an Active Power Line Conditioner," *Power Quality Solutions*, Intertec International, Ventura, CA, 1997.

Further Information

"How to Correct Power Line Disturbances," Dranetz Technologies, Edison, NJ, 1985.

Lawrie, R., *Electrical Systems for Computer Installations*, McGraw-Hill, New York, 1988.

Martzloff, F. D., "The Development of a Guide on Surge Voltages in Low-Voltage AC Power Circuits," *14th Electrical/Electronics Insulation Conference*, IEEE, Boston, October 1979.

Newman, P., "UPS Monitoring: Key to an Orderly Shutdown," *Microservice Management*, Intertec Publishing, Overland Park, KS, March 1990.

Noise Suppression Reference Manual, Topaz Electronics, San Diego, CA.

Nowak, S., "Selecting a UPS," *Broadcast Engineering*, Intertec Publishing, Overland Park, KS, April 1990.

Pettinger, W., "The Procedure of Power Conditioning," *Microservice Management*, Intertec Publishing, Overland Park, KS, March 1990.

Smeltzer, D., "Getting Organized About Power," *Microservice Management*, Intertec Publishing, Overland Park, KS, March 1990.

10.9   Standby Power Systems

Jerry C. Whitaker

10.9.1   Introduction

When utility company power problems are discussed, most people immediately think of blackouts. The lights go out, and everything stops. With the facility down and in the dark, there is nothing to do but sit and wait until the utility company finds the problem and corrects it. This process generally takes only a few minutes. There are times, however, when it can take hours. In some remote locations, it can take even days.

Blackouts are, without a doubt, the most troublesome utility company problem that facility will have to deal with. Statistics show that power failures are, generally speaking, a rare occurrence in most areas of the country. They are also short in duration. Studies have shown that 50% of blackouts last 6 s or less, and 35% are less than 11 min long. These failure rates usually are not cause for concern to commercial users, except where computer-based operations, transportation control systems, medical facilities, and communications sites are concerned.

When continuity of operation is critical, redundancy must be carried throughout the system. The site never should depend on one critical path for AC power. For example, if the facility is fed by a single step-down transformer, a lightning flash or other catastrophic event could result in a transformer failure that would bring down the entire site. A replacement could take days or even weeks.

10.9.2   Blackout Effects

A facility that is down for even 5 min can suffer a significant loss of productivity or data that may take hours or days to rebuild. A blackout affecting a transportation or medical center could be life threatening. Coupled with this threat is the possibility of extended power-service loss due to severe storm conditions. Many broadcast and communications relay sites are located in remote, rural areas, or on mountaintops. Neither of these kinds of locations are well known for their power reliability. It is not uncommon in mountainous areas for utility company service to be out for days after a major storm. Few operators are willing to take such risks with their businesses. Most choose to install standby power systems at appropriate points in the equipment chain.

The cost of standby power for a facility can be substantial, and an examination of the possible alternatives should be conducted before any decision on equipment is made. Management must clearly define the direct and indirect costs and weigh them appropriately. Include the following items in the cost-vs.-risk analysis:

- Standby power-system equipment purchase and installation cost
- Exposure of the system to utility company power failure
- Alternative operating methods available to the facility
- Direct and indirect costs of lost uptime because of blackout conditions

A distinction must be made between **emergency** and **standby power sources**. Strictly speaking, emergency systems supply circuits legally designated as being essential for safety to life and property. Standby power systems are used to protect a facility against the loss of productivity resulting from a utility company power outage.

10.9.3   Standby Power Options

To ensure the continuity of AC power, many commercial/industrial facilities depend on either two separate utility services or one utility service plus on-site generation. Because of the growing complexity of electrical systems, attention must be given to power-supply reliability.

The engine-generator shown in Fig. 10.175 is the classic standby power system. An automatic transfer switch monitors the AC voltage coming from the utility company line for power failure conditions.

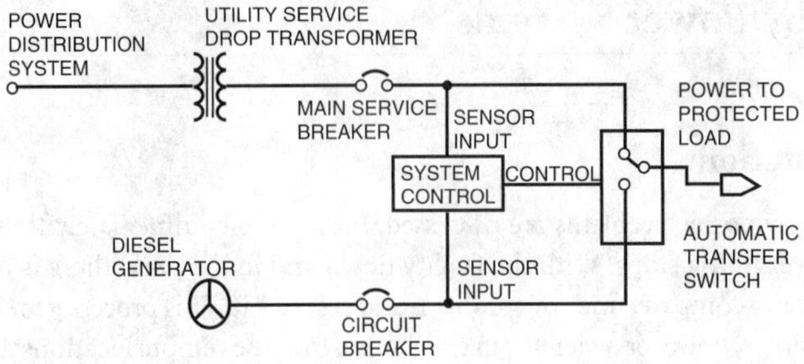

FIGURE 10.175   The classic standby power system using an engine-generator set. This system protects a facility from prolonged utility company power failures.

Upon detection of an outage for a predetermined period of time (generally 1–10 s), the standby generator is started; after the generator is up to speed, the load is transferred from the utility to the local generator. Upon return of the utility feed, the load is switched back after some predetermined safety time, and the generator is stopped. This basic type of system is used widely in industry and provides economical protection against prolonged power outages (5 min or more).

Dual Feeder System

In some areas, usually metropolitan centers, two utility company power drops can be brought into a facility as a means of providing a source of standby power. As shown in Fig. 10.176, two separate utility service drops—from separate power-distribution systems—are brought into the plant, and an automatic transfer switch changes the load to the backup line in the event of a mainline failure. The dual feeder system provides an advantage over the auxiliary diesel arrangement in that power transfer from main to standby can be made in a fraction of a second if a static transfer switch is used. Time delays are involved in the diesel generator system that limit its usefulness to power failures lasting more than several minutes.

The dual feeder system of protection is based on the assumption that each of the service drops brought into the facility is routed via different paths. This being the case, the likelihood of a failure on both power lines simultaneously is remote. The dual feeder system will not, however, protect against areawide power failures, which may occur from time to time.

The dual feeder system is limited primarily to urban areas. Rural or mountainous regions generally are not equipped for dual redundant utility company operation. Even in urban areas, the cost of bringing a second power line into a facility can be high, particularly if special lines must be installed for the feed. If two separate utility services are available at or near the site, redundant feeds generally will be less expensive than engine-driven generators of equivalent capacity.

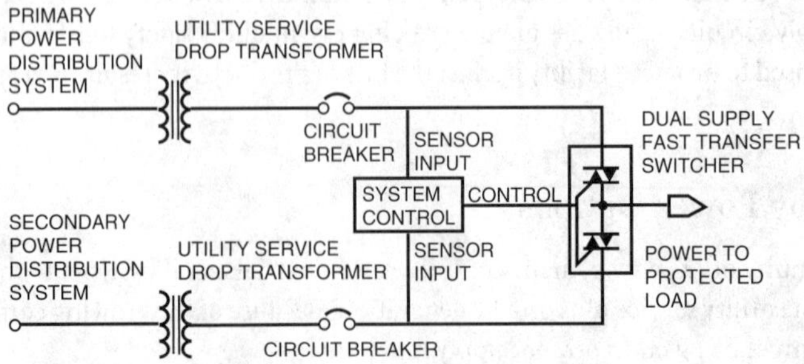

FIGURE 10.176   The dual utility feeder system of AC power loss protection. An automatic transfer switch changes the load from the main utility line to the standby line in the event of a power interruption.

Figure 10.177 illustrates a dual feeder system that utilizes both utility inputs simultaneously at the facility. Notice that during normal operation, both AC lines feed loads, and the tie circuit breaker is open. In the event of a loss of either line, the circuit-breaker switches reconfigure the load to place the entire facility on the single remaining AC feed. Switching is performed automatically; manual control is provided in the event of a planned shutdown on one of the lines.

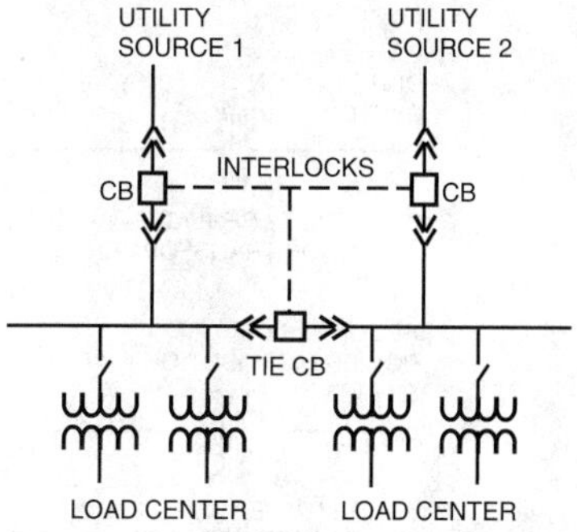

Peak Power Shaving

Figure 10.178 illustrates the use of a backup diesel generator for both standby power and *peak* **power shaving** applications. Commercial power customers often can realize substantial savings on utility company bills by reducing their energy demand during certain hours of the day. An automatic overlap transfer switch is used to change the load from the utility company system to the local diesel generator. The changeover is accomplished by a static transfer switch that does not disturb the operation of load equipment. This application of a standby generator can provide financial return to the facility, whether or not the unit is ever needed to carry the load through a commercial power failure.

FIGURE 10.177 A dual utility feeder system with interlocked circuit breakers.

Advanced System Protection

A more sophisticated power-control system is shown in Fig. 10.179, where a dual feeder supply is coupled with a **motor-generator (m-g) set** to provide clean, undisturbed AC power to the load. The m-g set will smooth over the transition from the main utility feed to the standby, often making a commercial power failure unnoticed by on-site personnel. An m-g typically will give up to $\frac{1}{2}$ s of power fail ride-through, more than enough to accomplish a transfer from one utility feed to the other. This standby power system is further refined in the application illustrated in Fig. 10.180, where a diesel generator has been added to the system. With the automatic overlap transfer switch shown at the generator output, this arrangement also can be used for peak demand power shaving.

Application Example

Figure 10.181 shows a simplified schematic diagram of a 220-kW **uninterruptible power system (UPS)** utilizing dual utility company feed lines, a 750-kVA gas engine generator, and five DC-driven

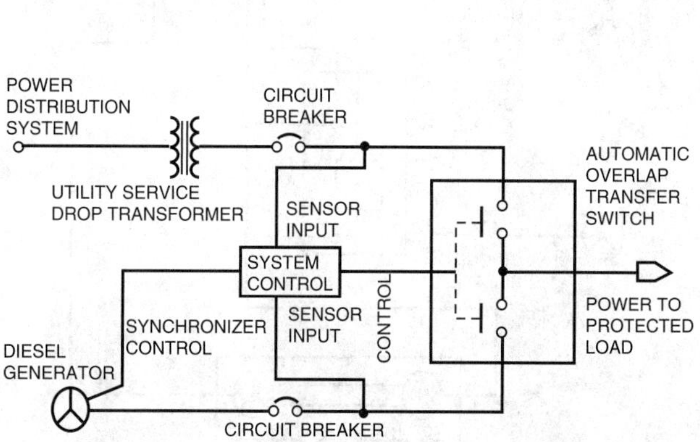

FIGURE 10.178 The use of a diesel generator for standby power and peak power shaving applications. The automatic overlap (*static*) transfer switch changes the load from the utility feed to the generator with sufficient speed that no disruption of normal operation is encountered.

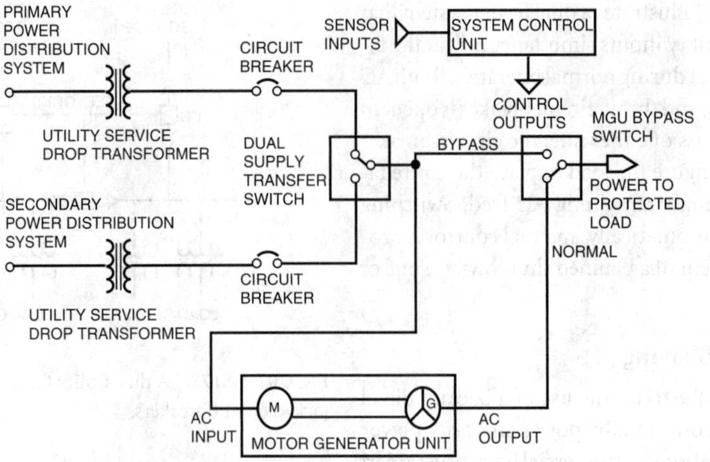

FIGURE 10.179 A dual feeder standby power system using a motor-generator set to provide power fail ride-through and transient-disturbance protection. Switching circuits allow the m-g set to be bypassed, if necessary.

motor-generator sets with a 20-min battery supply at full load. The five m-g sets operate in parallel. Each is rated for 100-kW output. Only three are needed to power the load, but four are on-line at any given time. The fifth machine provides redundancy in the event of a failure or for scheduled maintenance work. The batteries are always on-line under a slight charge across the 270-V DC bus. Two separate natural-gas lines, buried along different land routes, supply the gas engine. Local gas storage capacity also is provided.

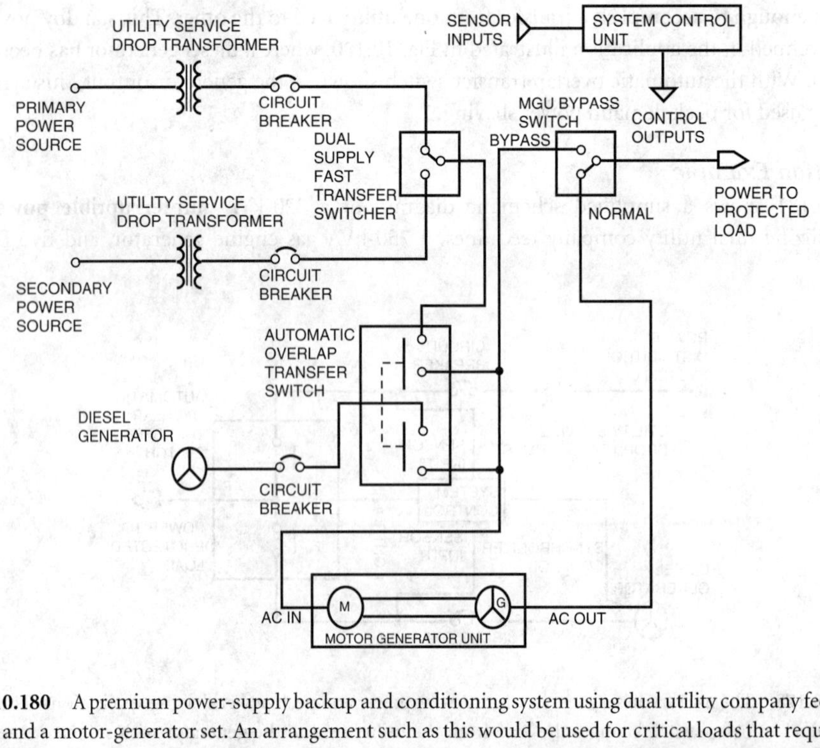

FIGURE 10.180 A premium power-supply backup and conditioning system using dual utility company feeds, a diesel generator, and a motor-generator set. An arrangement such as this would be used for critical loads that require a steady supply of clean AC.

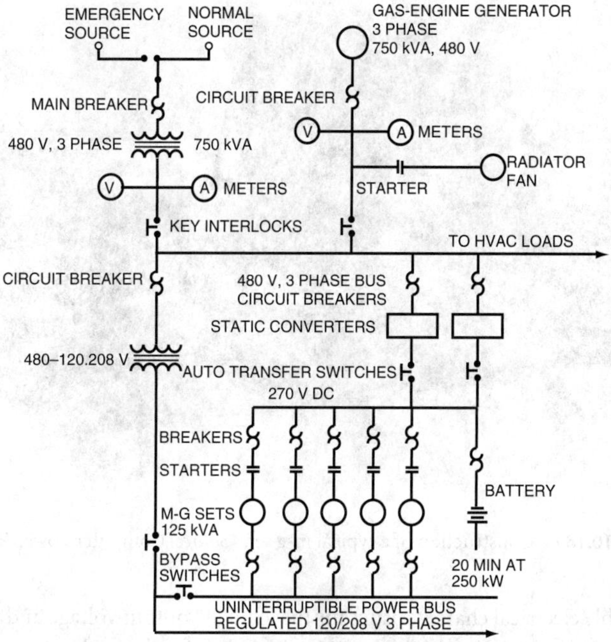

FIGURE 10.181 Simplified installation diagram of a high-reliability power system incorporating dual utility feeds, a standby gas-engine generator, and five battery-backed DC m-g sets.

Motor-Generator Set Operation

As the name implies, a motor-generator set consists of a motor powered by the AC utility supply that is mechanically tied to a generator, which feeds the load (see Fig. 10.182). Transients on the utility line will have no effect on the load when this arrangement is used. Adding a flywheel to the motor-to-generator shaft will protect against brief power dips. Figure 10.183 shows the construction of a typical m-g set. The attributes of an m-g include the following:

- An independently generated source of voltage can be regulated without interaction with line-voltage changes on the power source. Utility line changes of ±20% commonly can be held to within ±1% at the load.

- The rotational speed and inertial momentum of the rotating mass represents a substantial amount of stored rotational energy, preventing sudden changes in voltage output when the input is momentarily interrupted.

- The input and output windings are separated electrically, preventing transient disturbances from propagating from the utility company AC line to the load.

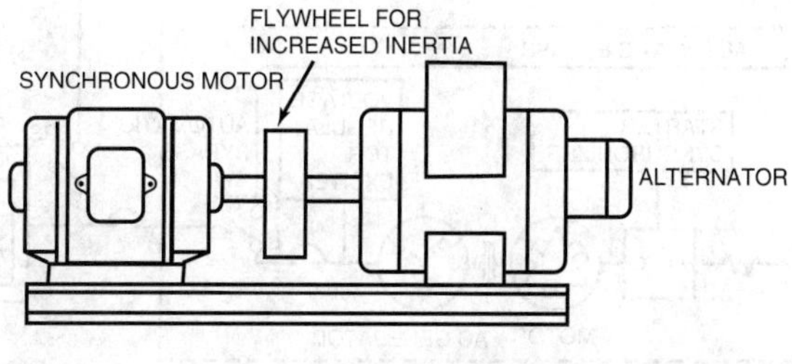

FIGURE 10.182 Two-machine motor-generator set with an optional flywheel to increase inertia and carry-through capability.

FIGURE 10.183   Construction of a typical m-g set. (*Source:* Computer Power Protection.)

- There are stable electrical characteristics for the load: (1) output voltage and frequency regulation, (2) ideal sine wave output, and (3) true 120° phase shift for three-phase models.
- There are reduced problems relating to the power factor presented to the utility company power source.

The efficiency of a typical m-g ranges from 65 to 89%, depending on the size of the unit and the load. Motor-generator sets have been used widely to supply 415-Hz power to mainframe computers that require this frequency.

System Configuration

There are a number of types of motor-generator sets, each having its own characteristics, advantages, and disadvantages. A simplified schematic diagram of an m-g is shown in Fig. 10.184. The type of motor that drives the set is an important design element. Direct-current motor drives can be controlled in speed independently of the frequency of the AC power source from which the DC is derived. Use of a DC motor, thereby, gives the m-g set the capability to produce power at the desired output frequency, regardless of variations in input frequency. The requirement for rectifier conversion hardware, control equipment, and commutator maintenance are drawbacks to this approach that must be considered.

The simplest and least expensive approach to rotary power conditioning involves the use of an induction motor as the mechanical source. Unfortunately, the rotor of an induction motor turns slower than the

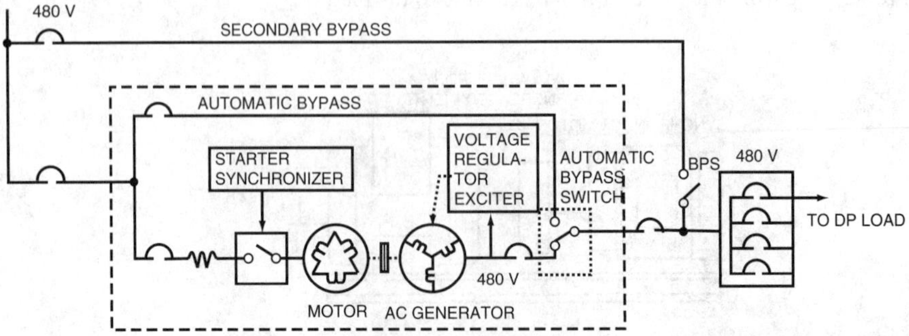

FIGURE 10.184   Simplified schematic diagram of an m-g set with automatic and secondary bypass capability.

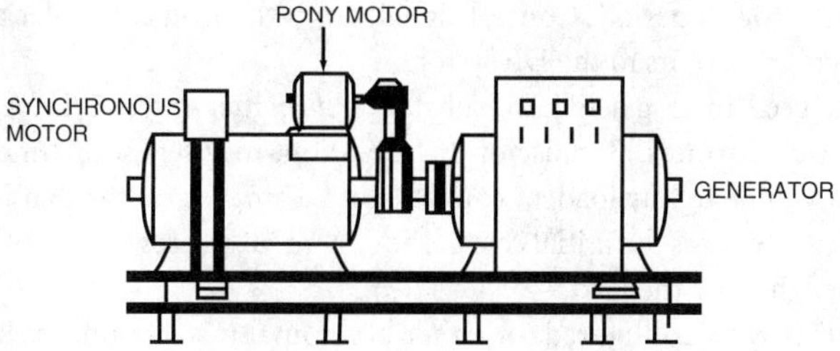

FIGURE 10.185   Use of a pony motor for an m-g set to aid in system starting and restarting.

rotating field produced by the power source. This results in the generator being unable to produce 60-Hz output power if the motor is operated at 60 Hz and the machines are directly coupled end-to-end at their shafts, or are belted in a 1:1 ratio. Furthermore, the shaft speed and output frequency of the generator decreases as the load on the generator is increased. This potential for varying output frequency may be acceptable where the m-g set is used solely as the input to a power supply in which the AC is rectified and converted to DC. However, certain loads cannot tolerate frequency changes greater than 1 Hz/s and frequency deviations of more than 0.5 Hz from the nominal 60-Hz value.

Low-slip induction motor-driven generators are available that can produce 59.7 Hz at full load, assuming 60-Hz input. During power interruptions, the output frequency will drop further depending on the length of the interruption. The capability of the induction motor to restart after a momentary power interruption is a valuable attribute. Various systems of variable-speed belts have been tried successfully. Magnetically controlled slipping clutches have been found to be generally unsatisfactory. Other approaches to make the induction motor drive the load at constant speed have produced mixed results.

Using a synchronous motor with direct coupling or a cogged 1:1 ratio belt drive guarantees that the output frequency will be equal to the motor input frequency. Although the synchronous motor is more expensive, it is more efficient and can be adjusted to provide a unity **power factor** (**PF**) load to the AC source. The starting characteristics and the mechanical disturbance following a brief line-voltage interruption depends, to a large extent, on motor design. Many synchronous motors that are not required to start under load have weak starting torque, and may use a *pony motor* to aid in starting. This approach is shown in Fig. 10.185. Those motors designed to start with a load have starting pole face windings that provide starting torque comparable to that of an induction motor. Such motors can be brought into synchronism while under load with proper selection of the motor and automatic starter system. Typical utility company AC interruptions are a minimum of six cycles (0.1 s). Depending on the design and size of the flywheel used, the ride-through period can be as much as 0.5 s or more. The generator will continue to produce output power for a longer duration, but the frequency and rate of frequency change will most likely fall outside of the acceptable range of most data processing (DP) loads after 0.5 s.

If input power is interrupted and does not return before the output voltage and frequency begin to fall outside acceptable limits, the generator output controller can be programmed to disconnect the load. Before this event, a warning signal is sent to the DP control circuitry to warn of impending shutdown and to initiate an orderly interruption of active computer programs. This facilitates easy restart of the computer after the power interruption has passed.

It is important for designers to accurately estimate the length of time that the m-g set will continue to deliver acceptable power without input to the motor from the utility company. These data facilitate accurate power-fail shutdown routines. It is also important to ensure that the m-g system can handle the return of power without operating overcurrent protection as a result of high inrush currents that may be required to accelerate and synchronize the motor with the line frequency. Protection against the latter problem requires proper programming of the synchronous motor controller to correctly disconnect and then reconnect the field current supply. It may be worthwhile to delay an impending shutdown for 100 ms or so. This would give the computer time to prepare for the event through an orderly interruption. It also would be useful if the computer were able to resume operation without shutdown, in case utility power returns within the ride-through period. Control signals from the m-g controller are often configured to identify these conditions and events to the DP system.

Generators typically used in m-g sets have substantially higher internal impedance than equivalent kilovolt-ampere-rated transformers. Because of this situation, m-g sets sometimes are supplied with an oversize generator that will be lightly loaded, coupled with a smaller motor that is adequate to drive the actual load. This approach reduces the initial cost of the system, decreases losses in the motor, and provides a lower operating impedance for the load.

Motor-generator sets may be configured for either horizontal installation, as shown previously, or for vertical installation.

The most common utility supply voltage used to drive the input of an m-g set is 480 V. The generator output for systems rated at about 75 kVA or less is typically 208 Y/120 V. For larger DP systems, the most economical generator output is typically 480 V. A 480-208 Y/120 V three-phase isolating transformer usually is included to provide 208 Y/120 V power to the computer equipment.

Motor-Design Considerations

Both synchronous and induction motors have been used successfully to drive m-g sets. Each has advantages and disadvantages. The major advantage of the synchronous motor is that while running normally, it is synchronized with the supply frequency. An 1800-r/min motor rotates at exactly 1800 r/min for a supply frequency of exactly 60 Hz. The generator output, therefore, will be exactly 60 Hz. Utility frequencies *average* 60 Hz; utilities vary the frequency slowly to maintain this average value under changing load conditions. Research has shown that utility operating frequencies typically vary from 58.7 to 60.7 Hz. Although frequency tolerances permitted by most computer manufacturers are usually given as ±0.5 Hz on a nominal 60-Hz system, these utility variations are spread over a 24-h period or longer, and generally do not result in problems for the load.

The major disadvantage of a synchronous motor is that the device is difficult to start. A synchronous motor must be started and brought up to **pull-in speed** by an auxiliary winding on the armature (the *armortisseur* winding). The pull-in speed is the minimum speed (close to synchronous speed) at which the motor will pull into synchronization if excitation is applied to the field. The armortisseur winding is usually a squirrel-cage design, although it may be of the wound-rotor type in some cases. This winding allows the synchronous motor to start and come up to speed as an induction motor. When pull-in speed is achieved, automatic sensing equipment applies field excitation, and the motor locks in and runs as a synchronous machine. As discussed previously, some large synchronous motors are brought up to speed by an auxiliary pony motor.

The armortisseur winding can produce only limited torque, so synchronous motors usually are brought up to speed without a load. This requirement presents no problem for DP systems upon initial startup. However, in the event of a momentary power outage, problems can develop. When the utility AC fails, the synchronous motor must be disconnected from the input immediately, or it will act as a generator and feed power back into the line, thus rapidly depleting its stored (kinetic) rotational energy. During a power failure, the speed of the motor rapidly drops below the pull-in speed, and when the AC supply returns, the armortisseur winding must reaccelerate the motor under load until the field can be applied again. This requires a large winding and a sophisticated control system. While the speed of the m-g set is below synchronous operation, the generator output frequency may be too low for proper computer operation.

The induction motor has no startup problems, but it does have *slip*. To produce torque, the rotor must rotate at slightly lower speed than the stator field. For a nominal 1800-r/min motor, the actual speed will be about 1750 r/min, varying slightly with the load and the applied input voltage. This represents a slip of about 2.8%. The generator, if driven directly or on a common shaft, will have an output frequency of approximately 58.3 Hz. This is below the minimum permissible operating frequency for most computer hardware. Special precision-built low-slip induction motors are available with a slip of approximately 0.5% at a nominal motor voltage of 480 V. With 0.5% slip, speed at full load will be about 1791 r/min, and the directly driven or common-shaft generator will have an output frequency of 59.7 Hz. This frequency is within tolerance, but close to the minimum permissible frequency.

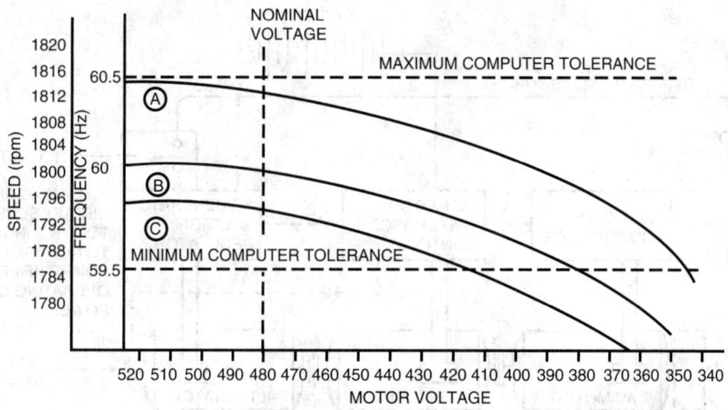

FIGURE 10.186 Generator output frequency vs. motor input voltage for an induction-motor-based m-g set.

A belt-and-pulley system adjustable-speed drive is a common solution to this problem. By making the pulley on the motor slightly larger in diameter than the pulley on the generator (with the actual diameters adjustable) the generator can be driven at synchronous speed.

Voltage sags have no effect on the output frequency of a synchronous motor-driven m-g set until the voltage gets so low that the torque is reduced to a point at which the machine pulls out of synchronization. Resynchronization then becomes a problem. On an induction motor, when the voltage sags, slip increases and the machine slows down. The result is a drop in generator output frequency. The adjustable-speed drive between an induction motor and the generator solves the problem for separate machines. If severe voltage sags are anticipated at a site, the system can be set so that nominal input voltage from the utility company produces a frequency of 60.5 Hz, 0.5 Hz on the high side of nominal frequency. Figure 10.186 charts frequency vs. motor voltage for three operating conditions:

- Slip compensation set high (curve *A*)
- Slip compensation set for 60 Hz (curve *B*)
- No slip compensation (curve *C*)

Through proper adjustment of slip compensation, considerable input-voltage margins can be achieved.

Single-Shaft Systems

There are two basic m-g set machine designs used for DP applications: (1) separate motor-generator systems and (2) single-shaft, single-housing units. Both designs can use either a synchronous or induction motor. In each case, there are advantages and disadvantages. The separate machine design (discussed previously) uses a motor driving a physically separate generator by means of a coupling shaft or pulley. In an effort to improve efficiency and reduce costs, manufacturers also have produced various types of single-shaft systems.

The basic concept of a single-shaft system is to combine the motor and generator elements into a single unit. A common stator eliminates a number of individual components, making the machine less expensive to produce and mechanically more efficient. The common-stator set substantially reduces mechanical energy losses associated with traditional m-g designs, and it improves system reliability as well. In one design, the stator is constructed so that alternate slots are wound with input and output windings. When it is fed with a three-phase supply, a rotating magnetic field is created, causing the DC-excited rotor to spin at a synchronous speed. By controlling the electrical characteristics of the rotor, control of the output at the secondary stator windings is accomplished.

Common-stator machines offer lower working impedance for the load than a comparable two-machine system. For example, a typical 400-kVA machine has approximately an 800-kVA frame size. The larger frame size yields a relatively low-impedance power source capable of clearing subcircuit fuses under fault

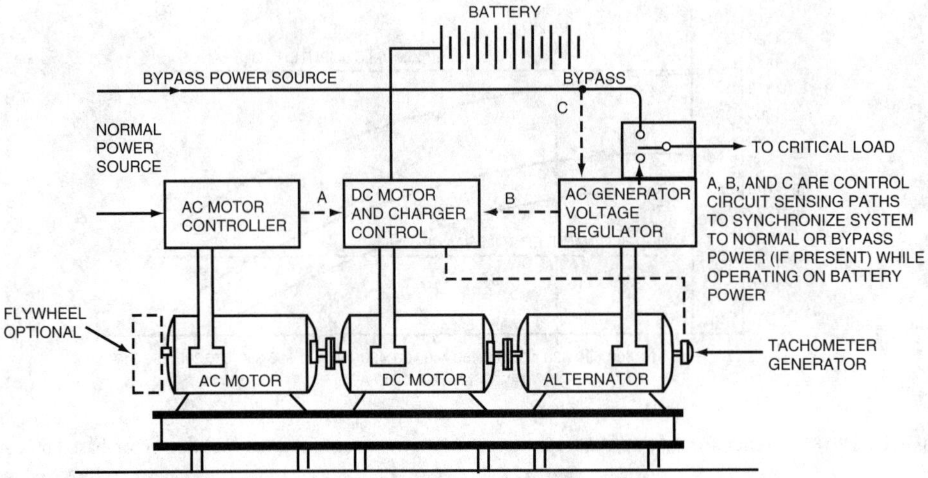

FIGURE 10.187 Uninterruptible m-g set with AC and DC motor drives.

conditions. The output of the unit typically can supply up to seven times the full-load current under fault conditions. Despite the increase in frame size, the set is smaller and lighter than comparable systems because of the reduced number of mechanical parts.

Motor-Generator UPS

Critical DP applications that cannot tolerate even brief AC power interruptions can use the m-g set as the basis for an uninterruptible source of power through the addition of a battery-backed DC motor to the line-driven AC motor shaft. This concept is illustrated in Fig. 10.187. The AC motor normally supplies power to drive the system from the utility company line. The shafts of the three devices all are interconnected, as shown in the figure. When AC power is present, the DC motor serves as a generator to charge the battery bank. When line voltage is interrupted, the DC motor is powered by the batteries. Figure 10.188 shows a modified version of this basic m-g UPS using only a DC motor as the mechanical power source. This configuration eliminates the inefficiency involved in having two motors in the system. Power from the utility source is rectified to provide energy for the DC motor, plus power for charging the

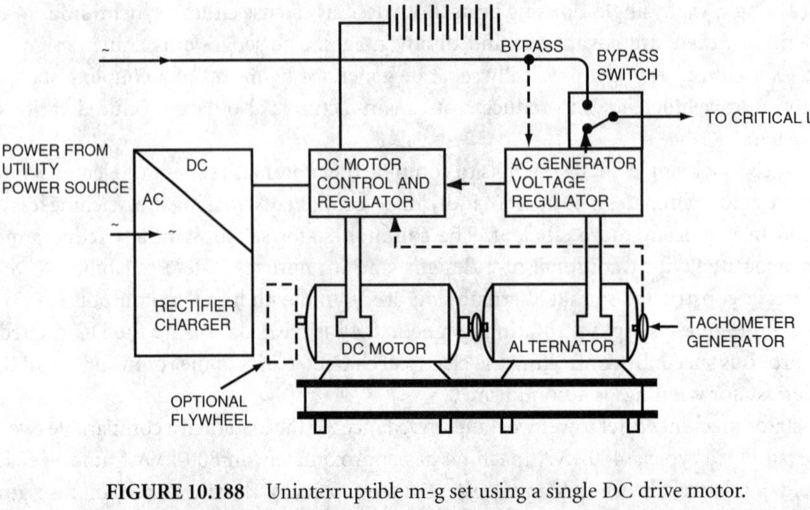

FIGURE 10.188 Uninterruptible m-g set using a single DC drive motor.

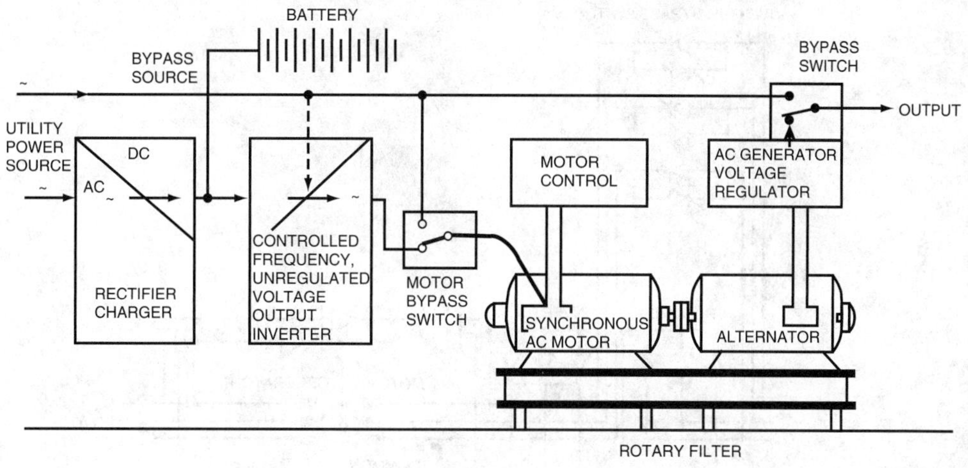

FIGURE 10.189 Uninterruptible m-g set using a synchronous AC motor.

batteries. A complex control system to switch the AC motor off and the DC motor on in the event of a utility power failure is not needed in this design.

The m-g UPS also can be built around a synchronous AC motor, as illustrated in Fig. 10.189. Utility AC energy is rectified and used to drive an inverter, which provides a regulated frequency source to power the synchronous motor. The output from the DC-to-AC inverter need not be a well-formed sine wave, nor a well-regulated source. The output from the generator will provide a well-regulated sine wave for the load. The m-g set also can be operated in a bypass mode that eliminates the rectifier, batteries, and inverter from the current path, operating the synchronous motor directly from the AC line.

An m-g UPS set using a common stator machine is illustrated in Fig. 10.190. A feedback control circuit adjusts the firing angle of the inverter to compensate for changes in input power. This concept is taken a step further in the system shown in Fig. 10.191. A solid-state inverter bypass switch is added to improve efficiency. During normal operation, the bypass route is enabled, eliminating losses across the rectifier diodes. When the control circuitry senses a drop in utility voltage, the inverter is switched on, and the bypass switch is deactivated. A simplified installation diagram of the inverter/bypass system is shown in Fig. 10.192. Magnetic circuit breakers and chokes are included as shown. An isolation transformer is inserted between the utility input and the rectifier bank. The static inverter is an inherently simple design; commutation is achieved via the windings. Six thyristors are used. Under normal operating conditions, 95% of the

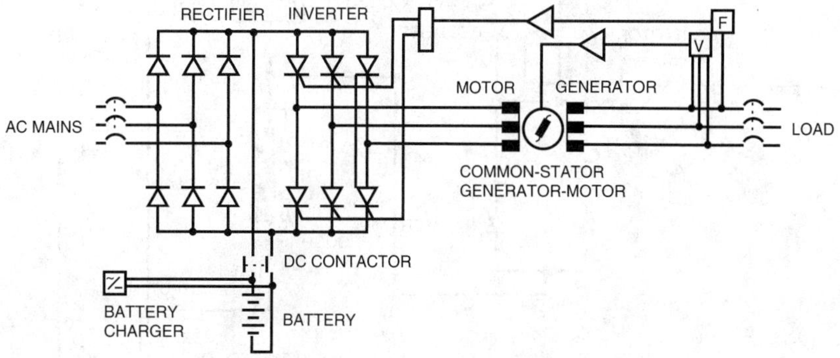

FIGURE 10.190 Common-stator UPS m-g set. The firing angle of the SCR inverters is determined by a feedback voltage from the generator output.

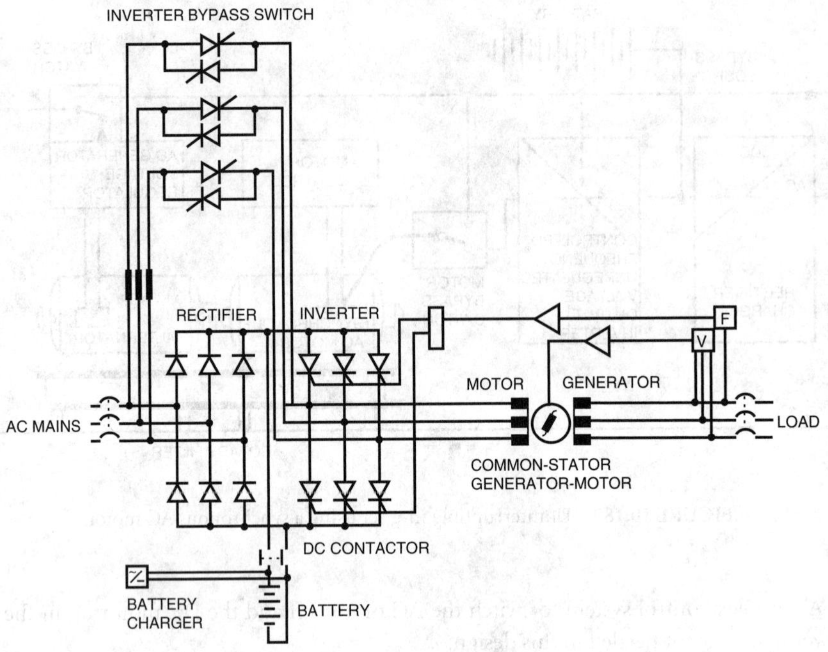

FIGURE 10.191   Common-stator UPS m-g set with a solid-state inverter bypass switch.

AC power passes through the static switch; 5% passes through the inverter. This arrangement affords maximum efficiency, while keeping the battery bank charged and the rectifiers and inverter thyristors preheated. Preheating extends the life of the components by reducing the extent of thermal cycling that occurs when the load suddenly switches to the battery backup supply. The static switch allows for fast disconnect of the input AC when the utility power fails.

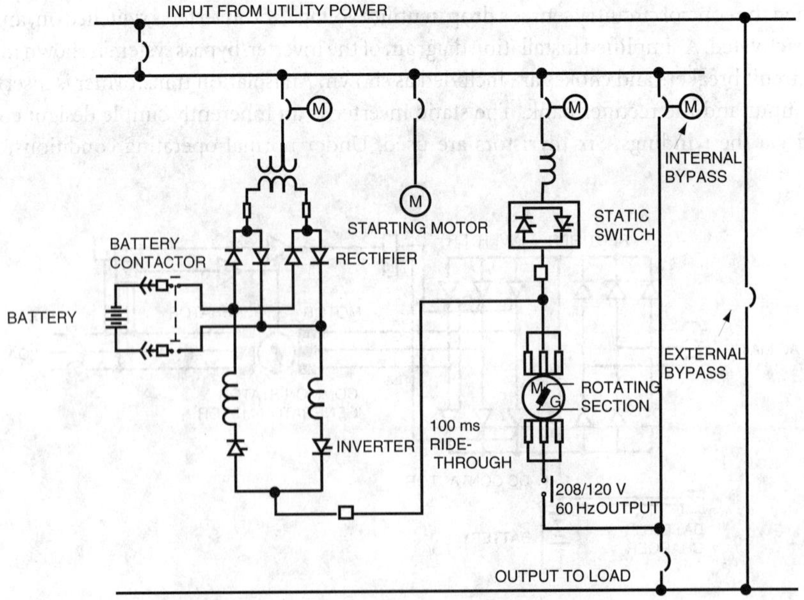

FIGURE 10.192   Power flow diagram of a full-featured UPS m-g set.

Maintenance Considerations

Because m-g sets require some maintenance that necessitates shutdown, most systems provide bypass capability so the maintenance work can be performed without having to take the computer out of service. If the automatic bypass contactor, solid-state switch, and control hardware are in the same cabinet as other devices that also need to be de-energized for maintenance, a secondary bypass is recommended. After the automatic bypass path has been established, transfer switching to the secondary bypass can be enabled, taking the m-g set and its automatic bypass system out of the circuit completely. Some automatic bypass control arrangements are designed to transfer the load of the generator to the bypass route with minimum disturbance. This requires the generator output to be synchronized with the bypass power before closing the switch and opening the generator output breaker. With the load taken off the generator, however, bypass power no longer is synchronized with it. Consequently, retransfer of the load back to the generator may occur with some disturbance. Adjustment for minimum disturbance in either direction requires a compromise in phase settings, or a means to shift the phase before and after the transfer.

The use of rotating field exciters has eliminated the need for slip rings in most m-g designs. Brush inspection and replacement, therefore, are no longer needed. However, as with any rotating machinery, bearings must be inspected and periodically replaced.

Generator System Options

Engine-generator sets are available for power levels ranging from less than 1 kVA to several thousand kilovolt-ampere or more. Machines also may be paralleled to provide greater capacity. Engine-generator sets typically are divided by the type of power plant used:

- *Diesel.* Advantages: rugged and dependable, low-fuel costs, low fire and/or explosion hazard. Disadvantages: somewhat more costly than other engines, heavier in smaller sizes.
- *Natural and liquefied-petroleum gas.* Advantages: quick starting after long shutdown periods, long life, low maintenance. Disadvantage: availability of natural gas during areawide power failure subject to question.
- *Gasoline.* Advantages: rapid starting, low initial cost. Disadvantages: greater hazard associated with storing and handling gasoline, generally shorter mean time between overhaul.
- *Gas turbine.* Advantages: smaller and lighter than piston engines of comparable horsepower, rooftop installations practical, rapid response to load changes. Disadvantages: longer time required to start and reach operating speed, sensitive to high input air temperature.

The type of power plant chosen usually is determined primarily by the environment in which the system will be operated and by the cost of ownership. For example, a standby generator located in an urban area office complex may be best suited to the use of an engine powered by natural gas, because of the problems inherent in storing large amounts of fuel. State or local building codes may place expensive restrictions on fuel-storage tanks and make the use of a gasoline- or diesel-powered engine impractical. The use of propane usually is restricted to rural areas. The availability of propane during periods of bad weather (when most power failures occur) also must be considered.

The generator rating for a standby power system should be chosen carefully and should take into consideration the anticipated future growth of the plant. It is good practice to install a standby power system rated for at least 50% greater output than the current peak facility load. This headroom gives a margin of safety for the standby equipment and allows for future expansion of the facility without overloading the system.

An engine-driven standby generator typically incorporates automatic starting controls, a battery charger, and automatic transfer switch. Control circuits monitor the utility supply and start the engine when there is a failure or a sustained voltage drop on the AC supply. The switch transfers the load as soon as the generator reaches operating voltage and frequency. Upon restoration of the utility supply, the switch returns the load after a preset delay and initiates engine shutdown. The automatic transfer switch must meet demanding requirements, including

- The ability to carry the full rated current continuously
- Ability to withstand fault currents without contact separation
- Ability to handle high inrush currents
- Ability to withstand repeated interruptions at full load without damage

The nature of most power outages requires a sophisticated monitoring system for the engine-generator set. Most power failures occur during periods of bad weather. Most standby generators are unattended. More often than not, the standby system starts, runs, and shuts down without any human intervention or supervision. For reliable operation, the monitoring system must check the status of the machine continually to ensure that all parameters are within normal limits. Time-delay periods usually are provided by the controller that require an outage to last from 5 to 10 s before the generator is started and the load is transferred. This prevents false starts that needlessly exercise the system. A time delay of 5–30 min usually is allowed between the restoration of utility power and return of the load. This delay permits the utility AC lines to stabilize before the load is reapplied.

The transfer of motor loads may require special consideration, depending on the size and type of motors used at a plant. If the residual voltage of the motor is out of phase with the power source to which the motor is being transferred, serious damage may result to the motor. Excessive current draw also may trip overcurrent protective devices. Motors above 50 hp with relatively high load inertia in relation to torque requirements, such as flywheels and fans, may require special controls. Restart time delays are a common solution.

Automatic starting and synchronizing controls are used for multiple-engine-generator installations. The output of two or three smaller units can be combined to feed the load. This capability offers additional protection for the facility in the event of a failure in any one machine. As the load at the facility increases, additional engine-generator systems can be installed on the standby power bus.

UPS System Options

An uninterruptible power system is an elegant solution to power outage concerns. The output of the UPS inverter may be a sine wave or pseudosine wave. The following design considerations are important:

- Power reserve capacity for future growth of the facility.
- Inverter current surge capability (if the system will be driving inductive loads, such as motors).
- Output voltage and frequency stability over time and with varying loads.
- Required battery supply voltage and current. Battery costs vary greatly, depending on the type of units needed.
- Type of UPS system (*forward-transfer* type or *reverse-transfer* type) required by the particular application. Some sensitive loads may not tolerate even brief interruptions of the AC power source.
- Inverter efficiency at typical load levels. Some inverters have good efficiency ratings when loaded at 90% of capacity, but poor efficiency when lightly loaded.
- Size and environmental requirements of the UPS system. High-power UPS equipment requires a large amount of space for the inverter/control equipment and batteries. Battery banks often require special ventilation and ambient temperature control.

Uninterruptible Power System Design

Uninterruptible power systems have become a virtual necessity for powering large or small computer systems where the application serves a critical need. Computers and data communications systems are no more reliable than the power from which they operate. The difference between UPS and emergency standby power is that the UPS is always in operation. It reverts to an alternative power source, such as the utility company, only if the UPS fails or needs to be deactivated for maintenance. Even then, the transfer of power occurs so quickly (within milliseconds) that it does not interrupt proper operation of the load.

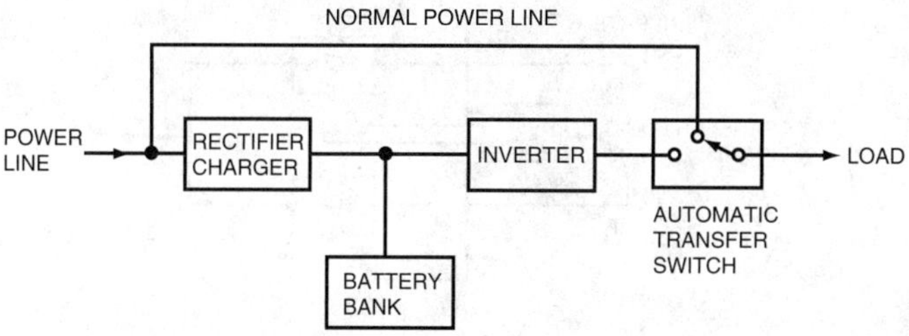

NORMAL POWER LINE

FIGURE 10.193   Forward-transfer UPS system.

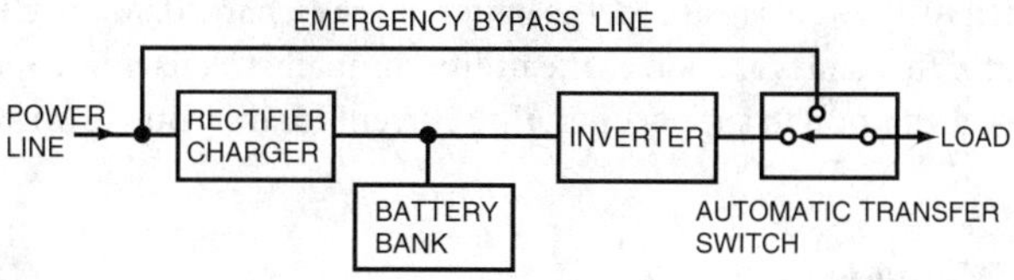

EMERGENCY BYPASS LINE

FIGURE 10.194   Reverse-transfer UPS system.

UPS hardware is available in a number of different configurations. All systems, however, are variations of two basic designs:

- **Forward-transfer mode.** The load normally is powered by the utility power line, and the inverter is idle. If a commercial power failure occurs, the inverter is started and the load is switched. This configuration is illustrated in Fig. 10.193. The primary drawback of this approach is the lack of load protection from power-line disturbances during normal (utility-powered) operation.

- **Reverse-transfer mode.** The load normally is powered by the inverter. In the event of an inverter failure, the load is switched directly to the utility line. This configuration is illustrated in Fig. 10.194. The reverse-transfer mode is, by far, the most popular type of UPS system in use.

The type of load-transfer switch used in the UPS system is a critical design parameter. The continuity of AC power service to the load is determined by the type of switching circuit used. An electromechanical transfer switch, shown in Fig. 10.195, is limited to switch times of 20–50 ms. This time delay may cause sensitive load equipment to malfunction, and perhaps shut down. A control circuit actuates the relay when the sensed output voltage falls below a preset value, such as 94% of nominal. A static transfer switch, shown in Fig. 10.196, can sense a failure and switch the load in about 4 ms. Most loads will ride through this short delay without any malfunction. To accomplish a smooth transition, the inverter output must be synchronized with the power line.

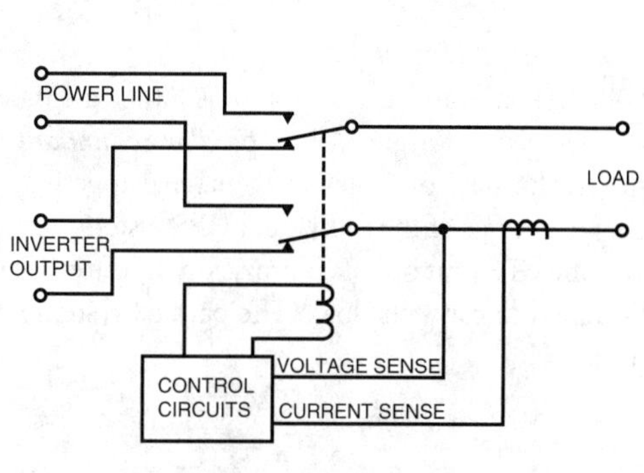

FIGURE 10.195   Electromechanical load-transfer switch.

UPS Configuration

The basic uninterruptible power system is built around a battery-driven inverter, with the batteries recharged by the utility AC line. As shown in Fig. 10.197, AC from the utility feed is rectified and applied to recharge or *float* a bank of batteries. This DC power drives a single- or multiphase closed-loop inverter, which regulates output voltage and frequency. The output of the inverter is generally a sine wave,

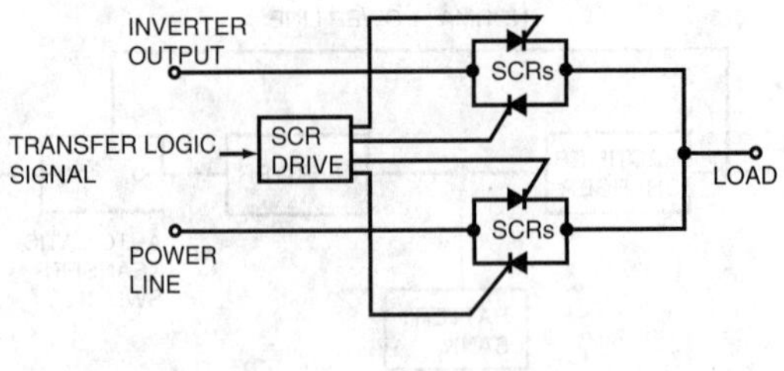

FIGURE 10.196   Static load-transfer switch.

or pseudosine wave (a stepped square wave). If the utility voltage drops or disappears, current is drawn from the batteries. When AC power is restored, the batteries are recharged. Many UPS systems incorporate a standby diesel generator that starts as soon as the utility company feed is interrupted. With this arrangement, the batteries are called upon to supply operating current for only 30 s or so, until the generator gets up to speed.

Power-Conversion Methods

Solid-state UPS systems that do not employ rotating machinery utilize one of several basic concepts. The design of an inverter is determined primarily by the operating power level. The most common circuit configurations include

- Ferroresonant inverter
- Delta magnetic inverter
- Inverter-fed L/C tank
- Quasisquare wave inverter
- Step wave inverter
- Pulse-width modulation inverter
- Phase modulation inverter

These and other types of inverters are discussed in Section 6.9.

Redundant Operation

UPS systems can be configured as either a single, large power-conditioning/backup unit, or as several smaller systems aranged in a *parallel redundant* or *isolated redundant* mode. In the parallel redundant mode, the UPS outputs are connected together and share the total load (see Fig. 10.198). The power-output ratings of the individual UPS systems are selected to provide for operation of the entire load with any one UPS unit out of commission. In the event of expansion of the DP facility, additional UPS units can be added to carry the load. The parallel system provides the ability to cope with the failure of any single unit.

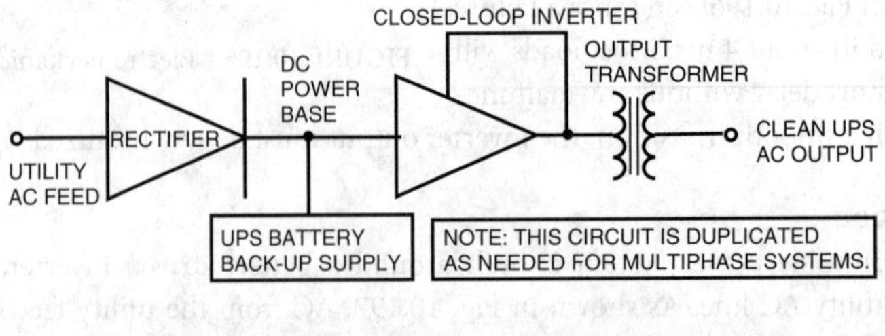

FIGURE 10.197   Block diagram of an uninterruptible power system using AC rectification to float the battery supply. A closed-loop inverter draws on this supply and delivers clean AC power to the protected load.

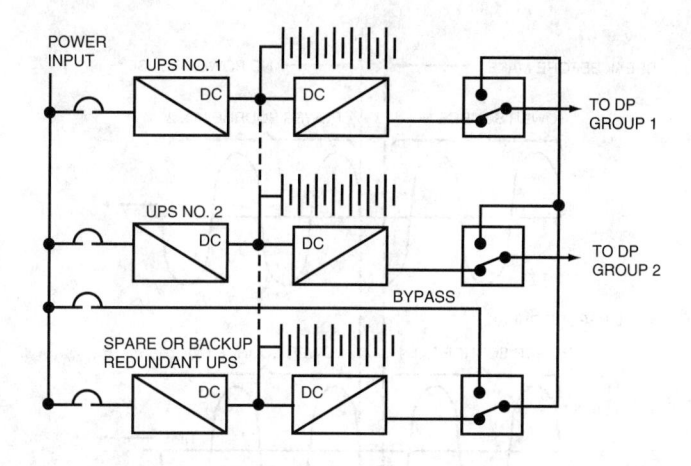

FIGURE 10.198 Configuration of a parallel redundant UPS system.

An isolated redundant system, illustrated in Fig. 10.199, divides the load among several UPS units. If one of the active systems fails, a static bypass switch will connect the affected load to a standby UPS system dedicated to that purpose. The isolated redundant system does not permit the unused capacity of one UPS unit to be utilized on a DP system that is loading another UPS to full capacity. The benefit of the isolated configuration is its immunity to system-wide failures.

Output Transfer Switch

Fault conditions, maintenance operations, and system reconfiguration require the load to be switched from one power source to another. This work is accomplished with an output transfer switch. As discussed previously, most UPS systems use electronic (static) switching. Electromechanical or motor-driven

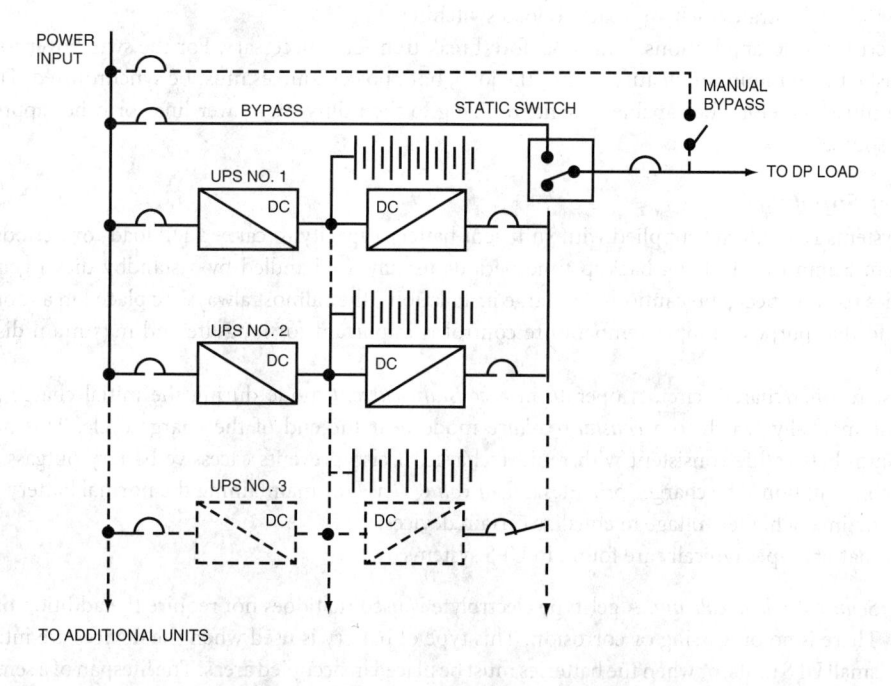

FIGURE 10.199 Configuration of an isolated redundant UPS system.

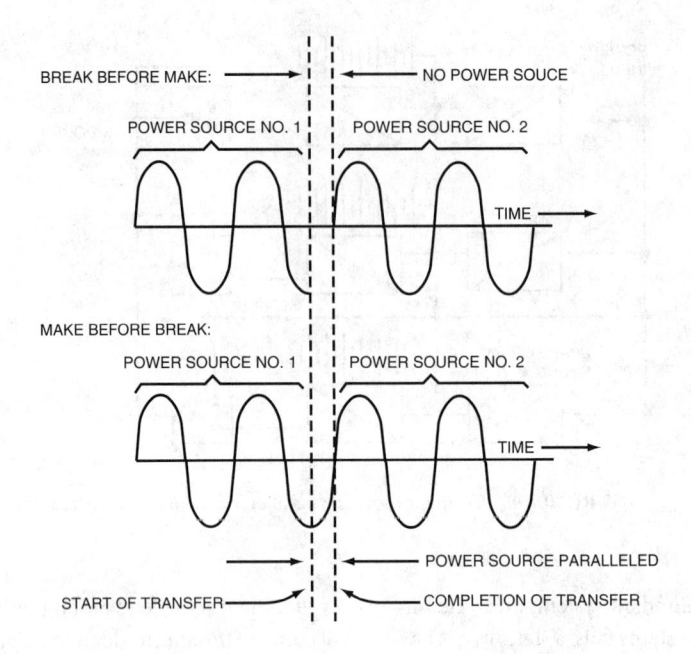

FIGURE 10.200   Static transfer switch modes.

relays operate too slowly for most DP loads. Static transfer switches can be configured as either of the following:

- *Break-before-make.* Power output is interrupted before transfer is made to the new source.
- *Make-before-break.* The two power sources are overlapped briefly so as to prevent any interruption in AC power to the load.

Figure 10.200 illustrates each approach to load switching.

For critical-load applications, a make-before-break transfer is necessary. For the switchover to be accomplished with minimum disturbance to the load, both power sources must be synchronized. The UPS system must, therefore, be capable of synchronizing to the utility AC power line (or other appropriate power source).

Battery Supply

UPS systems typically are supplied with sufficient battery capacity to carry a DP load for periods ranging from 5 min to 1 h. Long backup time periods usually are handled by a standby diesel generator. Batteries require special precautions. For large installations, they almost always are placed in a room dedicated to that purpose. Proper temperature control is important for long life and maximum discharge capacity.

Most rectifier/charger circuits operate in a *constant-current* mode during the initial charge period, and automatically switch to a *constant-voltage* mode near the end of the charge cycle. This provides maximum battery life consistent with rapid recharge. It also prevents excessive battery outgassing and water consumption. The charger provides a *float voltage level* for maintaining the normal battery charge, and sometimes a higher voltage to equalize certain devices.

Four battery types typically are found in UPS systems:

- *Semisealed lead calcium.* A gel-type electrolyte is used that does not require the addition of water. There is no outgassing or corrosion. This type of battery is used when the devices are integral to small UPS units, or when the batteries must be placed in occupied areas. The lifespan of a semisealed lead calcium battery, under ideal conditions, is about 5 years.

- *Conventional lead calcium.* The most common battery type for UPS installations, these units require watering and terminal cleaning about every 6 months. Expected lifetime ranges up to 20 years. Conventional lead-calcium batteries outgas hydrogen under charge conditions and must be located in a secure, ventilated area.
- *Lead-antimony.* The traditional lead-acid batteries, these devices are equivalent in performance to lead-calcium batteries. Maintenance is required every 3 months. Expected lifetime is about 10 years. To retain their capacity, lead-antimony batteries require a monthly equalizing charge.
- *Nickel-cadmium.* Advantages of the nickel-cadmium battery includes small size and low weight for a given capacity. These devices offer excellent high- and low-temperature properties. Life expectancy is nearly that of a conventional lead-calcium battery. Nickel-cadmium batteries require a monthly equalizing charge, as well as periodic discharge cycles to retain their capacity. Nickel-cadmium batteries are the most expensive of the devices typically used for UPS applications.

The operation and care of specific battery types is discussed in Section 10.11.

Standby Power-System Noise

Noise produced by backup power systems can be a serious problem if not addressed properly. Standby generators, motor-generator sets, and UPS systems produce noise that can disturb building occupants and irritate neighbors or landlords.

The noise associated with electrical generation usually is related to the drive mechanism, most commonly an internal combustion engine. The amplitude of the noise produced is directly related to the size of the engine-generator set. First consider whether noise reduction is a necessity. Many building owners have elected to tolerate the noise produced by a standby power generator because its use is limited to emergency situations. During a crisis, when the normal source of power is unavailable, most people will tolerate noise associated with a standby generator.

If the decision is made that building occupants can live with the noise of the generator, care must be taken in scheduling the required testing and exercising of the unit. Whether testing occurs monthly or weekly, it should be done on a regular schedule.

If it has been determined that the noise should be controlled, or at least minimized, the easiest way to achieve this objective is to physically separate the machine from occupied areas. This may be easier said than done. Because engine noise is predominantly low-frequency in character, walls and floor/ceiling construction used to contain the noise must be massive. Lightweight construction, even though it may involve several layers of resiliently mounted drywall, is ineffective in reducing low-frequency noise. Exhaust noise is a major component of engine noise but, fortunately, it is easier to control. When selecting an engine-generator set, select the highest quality exhaust muffler available. Such units often are identified as hospital-grade mufflers.

Engine-generator sets also produce significant vibration. The machine should be mounted securely to a slab-on-grade or an isolated basement floor, or it should be installed on vibration isolation mounts. Such mounts usually are specified by the manufacturer.

Because a UPS system or motor-generator set is a source of continuous power, it must run continuously. Noise must be adequately controlled. Physical separation is the easiest and most effective method of shielding occupied areas from noise. Enclosure of UPS equipment usually is required, but noise control is significantly easier than for an engine-generator because of the lower noise levels involved. Nevertheless, the low-frequency 120-Hz fundamental of a UPS system is difficult to contain adequately; massive constructions may be necessary. Vibration control also is required for most UPS and m-g gear.

Critical System Bus

Many facilities do not require the operation of all equipment during a power outage. Rather than use one large standby power system, key pieces of equipment can be protected with small, dedicated, uninterruptible power systems. Small UPS units are available with built-in battery supplies for computer systems and other

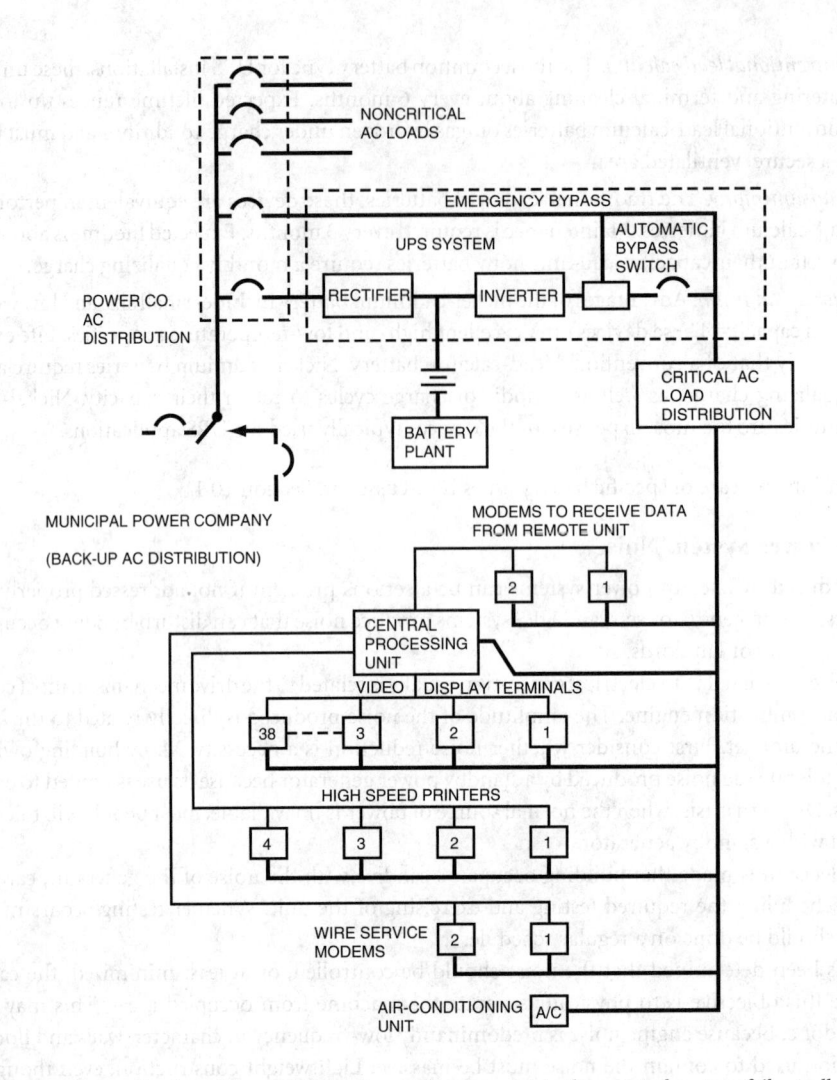

FIGURE 10.201   An application of the critical-load power bus concept. In the event of a power failure, all equipment necessary for continued operation is powered by the UPS equipment. Noncritical loads are dropped until commercial AC returns.

hardware. If cost prohibits the installation of a system-wide standby power supply (using generator or solid-state UPS technology), consider establishing a *critical load bus* that is connected to a UPS system or generator via an automatic transfer switch. This separate power supply is used to provide AC to critical loads, thus keeping the protected systems up and running. The concept is illustrated in Fig. 10.201. Unnecessary loads are dropped in the event of a power failure.

A standby system built on the critical load principle can be a cost-effective answer to the power-failure threat. The first step in implementing a critical load bus is to accurately determine the power requirements for the most important equipment. Typical power consumption figures can be found in most equipment instruction manuals. If the data is not listed or available from the manufacturer, it can be measured using a wattmeter.

When planning a critical load bus, be certain to identify accurately which loads are critical, and which can be dropped in the event of a commercial power failure. If air conditioning is interrupted but the computer equipment at a large DP center continues to run, temperatures will rise quickly to the point at which system components may be damaged or the hardware automatically shuts down. It may not be necessary to require cooling fans, chillers, and heat-exchange pumps to run without interruption. However, any outage should be less than 1–2 min in duration. Typically, liquid-cooled DP systems can withstand 1 min of cooling interruption. Air-cooled DP systems may tolerate 5–10 min of cooling interruption.

10.9.4 The Efficient Use of Energy

Utility company power bills are becoming an increasingly large part of the operating budgets of many facilities. To reduce the amount of money spent each month on electricity, engineers must understand the billing methods used by the utility. Saving energy is more complicated than simply turning off unnecessary loads. The amount of money that can be saved through a well-planned energy conservation effort is often substantial. Reductions of 20% are not uncommon, depending on the facility layout and extent of energy conservation efforts already under way. Regardless of any monetary savings that might be realized from a power-use-reduction program, the items discussed here should be considered for any well-run facility.

The rate structures of utility companies vary widely from one area of the country to another. Some generalizations can be made, however, with respect to the basic rate-determining factors. The four primary parameters used to determine a customer's bill are

- Energy usage
- Peak demand
- Load factor
- Power factor

These items often can be controlled, to some extent, by the customer.

Energy Usage

The kilowatthour (kWh) usage of a facility can be reduced by turning off loads such as heating and air conditioning systems, lights, and office equipment when they are not needed. The installation of timers, photocells, or sophisticated computer-controlled energy-management systems can make substantial reductions in facility kilowatthour demand each month. Common sense will dictate the conservation measures applicable to a particular situation. Obvious items include reducing the length of time high-power equipment is in operation, setting heating and cooling thermostats for reasonable levels, keeping office equipment turned off during the night, and avoiding excessive amounts of indoor or outdoor lighting.

Although energy conservation measures should be taken in every area of facility operation, the greatest savings generally can be found where the largest energy users are located. Transmitter plants, large machinery, and process drying equipment consume a large amount of power, and so particular attention should be given to such hardware. Consider the following:

- Use the waste heat from equipment at the site for other purposes, if practical. In the case of high-power RF generators, room heating can be accomplished with a logic-controlled power amplifier exhaust-air recycling system.
- Have a knowledgeable consultant plan the air conditioning and heating system at the facility for efficient operation.
- Check thermostat settings on a regular basis, and consider installing time-controlled thermostats.
- Inspect outdoor-lighting photocells regularly for proper operation.
- Examine carefully the efficiency of high-power equipment used at the facility. New designs may offer substantial savings in energy costs.

The efficiency of large-power loads, such as mainframe computers, transmitters, or industrial RF heaters, is an item of critical importance to energy conservation efforts. Most systems available today are significantly more efficient than their counterparts of just 10 years ago. Plant management often can find economic justification for updating or replacing an older system on the power savings alone. In virtually any facility, energy conservation can best be accomplished through careful selection of equipment, thoughtful system design, and conscientious maintenance practices.

Peak Demand

Conserving energy is a big part of the power bill reduction equation, but it is not the whole story. The *peak demand* of the customer load is an important criterion in the utility company's calculation of rate structures. The peak demand figure is a measure of the maximum load placed on the utility company system by a customer during a predetermined billing cycle. The measured quantities may be kilowatts, kilovolt-amperes, or both. Time intervals used for this measurement range from 15 to 60 min. Billing cycles may be annual or semiannual. Figure 10.202 shows an example of varying peak demand.

If a facility operated at basically the same power consumption level from one hour to the next and one day to the next, the utility company could predict accurately the demand of the load, then size its

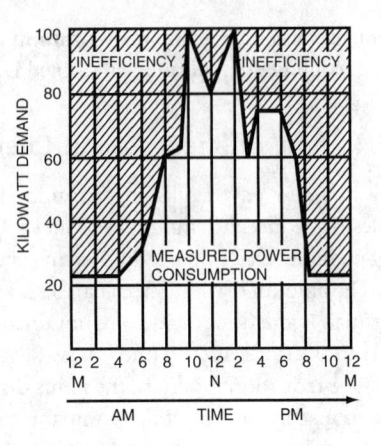

FIGURE 10.202   The charted power consumption of a facility not practicing energy-management techniques. Note the inefficiency that the utility company must absorb when faced with a load such as this.

equipment (including the allocation of energy reserves) for only the amount of power actually needed. For the example shown in the figure, however, the utility company must size its equipment (including allocated energy reserves) for the peak demand. The area between the peak demand and the actual usage is the margin of inefficiency that the customer forces upon the utility. The peak demand factor is a method used by utility companies to assess penalties for such operation, thereby encouraging the customer to approach a more efficient state of operation (from the utility's viewpoint).

Load shedding is a term used to describe the practice of trimming peak power demand to reduce high-demand penalties. The goal of load shedding is to schedule the operation of nonessential equipment so as to provide a more uniform power load from the utility company, and thereby a better kilowatthour rate. Nearly any operation has certain electric loads that can be rescheduled on a permanent basis or deferred as power demand increases during the day. Figure 10.203 illustrates the results of a load-shedding program. This more efficient operation has a lower overall peak demand and a higher average demand.

Peak demand reduction efforts can cover a wide range of possibilities. It would be unwise from an energy standpoint, for example, to test high-power standby equipment on a summer afternoon, when air conditioning units may be in full operation. Morning or evening hours would be a better choice, when

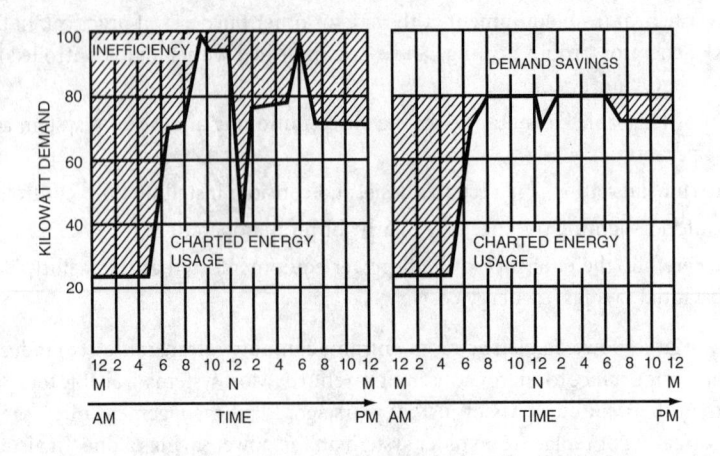

FIGURE 10.203   An example of the successful application of a load-shedding program. Energy usage has been spread more evenly throughout the day, resulting in reduced demand and, consequently, a better rate from the utility company.

the air conditioning is off and the demand of office equipment is reduced. Each operation is unique and requires an individual assessment of load-shedding options.

A computerized power-demand controller provides an effective method of managing peak demand. A controller can analyze the options available and switch loads as needed to maintain a relatively constant power demand from the utility company. Such systems are programmed to recognize which loads have priority and which loads are nonessential. Power demand then is automatically adjusted by the computer, based on the rate schedule of the utility company.

Load Factor

The load factor on an electric utility company bill is a product of the peak demand and energy usage. It usually is calculated and applied to the customer's bill each month. Reducing either the peak demand or energy usage levels, or both, will decrease this added cost factor. Reducing power factor penalties also will help to reduce load factor charges.

Power Factor

Power factor charges are the result of heavy inductive loading of the utility company system. A poor PF will result in excessive losses along utility company feeder lines because more current is required to supply a particular load with a low PF than would be demanded if the load had a PF close to unity. The power factor charge is a penalty that customers pay for the extra current needed to magnetize motors and other inductive loads. This magnetizing current does not show up on the service drop wattmeter. It is, instead, measured separately or prorated as an additional charge to the customer. The power factor penalty sometimes can be reduced through the addition of on-site **power factor correction** capacitors.

Power factor meters are available for measurement of a given load. It is usually less expensive in the long run, however, to hire a local electrical contractor to conduct a PF survey and recommend correction methods. Possible sources of PF problems include transmitters, blowers, air conditioners, heating equipment, and fluorescent and high-intensity discharge lighting-fixture ballasts.

Power Factor Fundamentals

Power factor is an important, but often misunderstood, characteristic of AC power systems. It is defined as the ratio of *true power* to *apparent power*, generally expressed as a percentage. (PF also may be expressed as a decimal value.) Reactive loads (inductive or capacitive) act on power systems to shift the current out of phase with the voltage. The cosine of the resulting angle between the current and voltage is the power factor.

A utility line that is feeding an inductive load (which is most often the case) is said to have a *lagging* power factor, whereas a line feeding a capacitive load has a *leading* power factor. See Fig. 10.204. A poor power factor will result in excessive losses along utility company feeder lines because more current is required to supply a given load with a low power factor than the same load with a power factor close to unity.

For example, a motor requiring 5 kW from the line is connected to the utility service entrance. If it has a power factor of 86%, the apparent power demanded by the load will be 5 kW divided by 86%, or more than 5.8 kW. The true power is 5 kW, and the apparent power is 5.8 kW. The same amount of work is being done by the motor, but the closer the power factor is to unity, the more efficient the system. To expand on this example, for a single-phase electric motor the actual power is the sum of several components, namely

- The work performed by the system, specifically lifting, moving, or otherwise controlling an object
- Heat developed by the power lost in the motor winding resistance
- Heat developed in the motor iron through eddy-current and hysteresis losses
- Frictional losses in the motor bearings
- Air friction losses in turning the motor rotor

All of these values are expressed in watts or kilowatts, and can be measured. They represent the actual power. The apparent power is determined by measuring the current and voltage with an ammeter and a

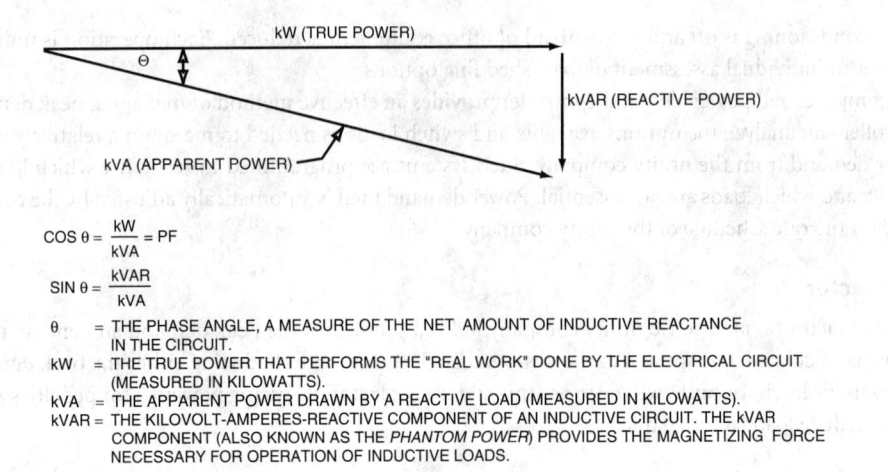

COS θ = $\dfrac{kW}{kVA}$ = PF

SIN θ = $\dfrac{kVAR}{kVA}$

θ = THE PHASE ANGLE, A MEASURE OF THE NET AMOUNT OF INDUCTIVE REACTANCE
     IN THE CIRCUIT.
kW = THE TRUE POWER THAT PERFORMS THE "REAL WORK" DONE BY THE ELECTRICAL CIRCUIT
     (MEASURED IN KILOWATTS).
kVA = THE APPARENT POWER DRAWN BY A REACTIVE LOAD (MEASURED IN KILOWATTS).
kVAR = THE KILOVOLT-AMPERES-REACTIVE COMPONENT OF AN INDUCTIVE CIRCUIT. THE kVAR
      COMPONENT (ALSO KNOWN AS THE *PHANTOM POWER*) PROVIDES THE MAGNETIZING FORCE
      NECESSARY FOR OPERATION OF INDUCTIVE LOADS.

FIGURE 10.204 The mathematical relationships of an inductive circuit as they apply to power factor (PF) measurements. Reducing the kVAR component of the circuit causes θ to diminish, improving the PF. When kW is equal to kVA, the phase angle is zero and the power factor is unity (100%).

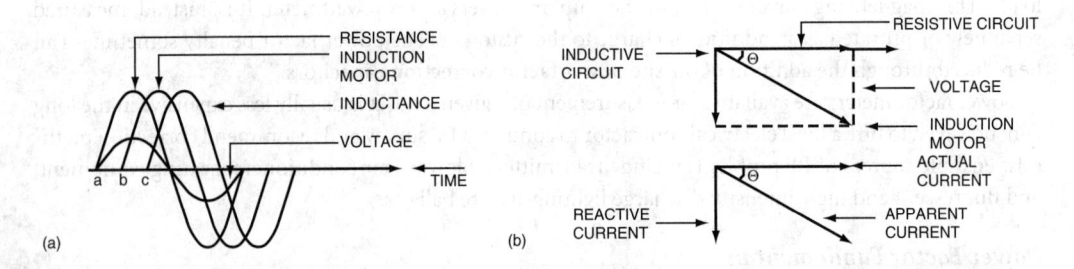

FIGURE 10.205 The effects of inductance in a circuit: (a) waveforms for resistive and inductive loads, (b) vector diagrams of inductive loads.

voltmeter, then calculating the product of the two. In the single-phase motor example, the apparent power thus obtained is greater than the actual power. The reason for this is the power factor.

The PF reflects the differences that exist between different types of loads. A soldering iron is a purely resistive load that converts current directly into heat. The current is called *actual current* because it contributes directly to the production of actual power. On the other hand, the single-phase electric motor represents a partially inductive load. The motor current consists of actual current that is converted into actual power, and a magnetizing current that is used to generate the magnetic field required for operation of the device. This magnetizing current corresponds to an exchange of energy between the power source and the motor, but it is not converted into actual power. This current is identified as the *reactive current* in the circuit.

As illustrated in Fig. 10.205, in a resistive circuit, the current is in phase with the voltage. In a purely inductive circuit, the current lags the voltage by 90°. This relationship can be represented graphically by vectors, as shown in the figure. For a circuit with both inductive and resistive components, as in the motor example, the two conditions exist simultaneously. The distribution between the actual power and the reactive power is illustrated in Fig. 10.206.

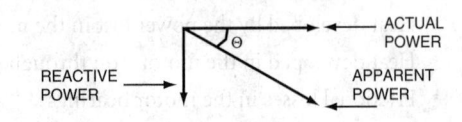

FIGURE 10.206 Vector diagram showing that apparent power is the vector sum of the actual and reactive power in a circuit.

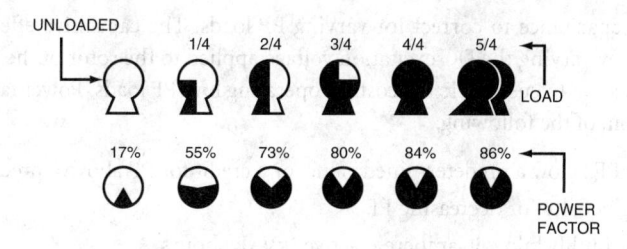

FIGURE 10.207   Changes in power factor for an induction motor as a function of motor loading.

The power factor, which has been defined previously as the ratio between actual power and apparent power, is also the cosine of the angle θ. The greater the angle θ becomes, the lower the power factor.

Determining the power factor for a given element of a power-distribution system is complicated by the wide variety of loads typically connected. Different loads present different PF components:

- Lighting. The PF for most incandescent lamps is unity. Fluorescent lamps usually have a low power factor; 50% is typical. Fluorescent lamps sometimes are supplied with compensation devices to correct for low power factor. Mercury vapor lamps have a low PF; 40–60% is typical. Again, such devices may be supplied with compensation devices.

- Electric motors. The PF of an induction motor varies with the load, as shown in Fig. 10.207. Unloaded or lightly loaded motors exhibit a low PF. Figure 10.208 illustrates the variation of PF and reactive power for varying loads on a three-phase induction motor. Synchronous motors provide good PF when their excitation is adjusted properly. Synchronous motors also can be overexcited to exhibit a leading power factor and, therefore, may be used to improve the power factor of a facility.

- Heating systems. Most heating systems used in ovens or dryers present a PF of close to unity.

- Welding equipment. Electric arc welders usually have a low PF; 60% is typical.

- Distribution transformers. The PF of a transformer varies considerably as a function of the load applied to the device, as well as the design of the transformer. An unloaded transformer would be highly inductive and, therefore, would exhibit a low PF.

To keep the power factor as close as possible to 100%, utility companies place capacitor banks at various locations in the distribution system, offsetting the inductive loading (lagging power factor) of most user equipment. The goal is to create an equal amount of leading PF in the system to match the lagging PF of the load. When balanced, the power factor is 100%. In practice, this is seldom attainable because loads are switched on and off at random times, but utilities routinely maintain an overall power factor of approximately 99%. To accomplish this, capacitor banks are switched automatically to compensate for changing load conditions.

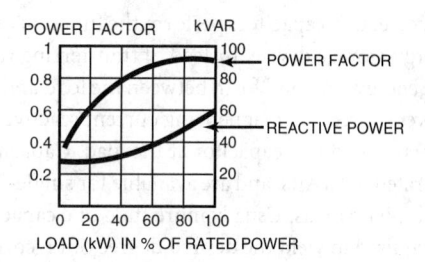

FIGURE 10.208   The relationship of reactive power and power factor to percentage of load for a 100-kW, three-phase induction motor.

Static capacitors are used commonly for power factor correction. These devices are similar to conventional oil-filled high-voltage capacitors. Operating voltages range from 230 V to 13.8 kV for pole-mounted applications. The PF correction capacitors are connected in parallel with the utility lines as close as practical to the low-PF loads. The primary disadvantage of static PF correction capacitors is that they cannot be adjusted for changing power factor conditions. Remotely operated relays may be used, however, to switch capacitor banks in and out of the circuit as required. Power factor correction also can be accomplished by using *synchronous capacitors* connected across utility lines. Synchronous capacitors may be adjusted to provide varying capacitance to correct for varying PF loads. The capacitive effect of a synchronous capacitor is changed by varying the DC excitation voltage applied to the rotor of the device.

Utilities usually pass on to customers the costs of operating low-PF loads. Power factor may be billed as one, or a combination, of the following:

- A penalty for PF below a predetermined value, or a credit for PF above a predetermined value
- An increasing penalty for decreasing PF
- A charge on monthly kilovolt-ampere reactive (kVAR) hours
- A straight charge for the maximum value of kilovoltampere used during the month

Aside from direct costs from utility companies for low-PF operation, the end-user experiences a number of indirect costs as well. When a facility operates with a low overall PF, the amount of useful electrical energy available inside the plant at the distribution transformer is reduced considerably because of the amount of reactive energy that the transformer(s) must carry. Figure 10.209 illustrates the reduction in available power from a distribution transformer when presented with varying PF loads. Figure 10.210 illustrates the increase in I^2R losses in feeder and branch circuits with varying PF loads. These conditions result in the need for oversized cables, transformers, switchgear, and protection circuits.

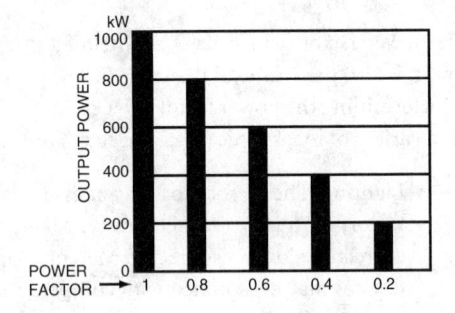

FIGURE 10.209   The effects of power factor on output of a transformer.

On-Site Power Factor Correction

The first step in correcting for low PF is to determine the current situation at a given facility. Clamp-on power factor meters are available for this purpose. Power factor can be improved in two ways:

- Reduce the amount of reactive energy by eliminating unloaded motors and transformers.
- Apply external compensation capacitors to correct the low-PF condition.

PF correction capacitors perform the function of an energy-storage device. Instead of transferring reactive energy back and forth between the load and the power source, the magnetizing current reactive energy is stored in a capacitor at the load. Capacitors are rated in kVARs, and are available for single- and multiphase loads. Usually more than one capacitor is required to yield the desired degree of PF correction. The capacitor rating required in a given application may be determined by using lookup tables provided by PF capacitor manufacturers. Installation options include

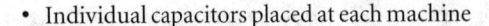

FIGURE 10.210   The relationship between power factor and percentage losses in system feeder and branch circuits.

- Individual capacitors placed at each machine
- A group or bank installation for an entire area of the plant
- A combination of the two approaches

When rectifier loads creating harmonic load current are the cause of a low-PF condition, the addition of PF correcting capacitors will not necessarily provide the desired improvement. The capacitors, in some cases, may actually raise the line current and fail to improve the power factor. Harmonic currents generally are most apparent in the neutral of three-phase circuits. Conductors supplying three-phase rectifiers using a neutral conductor require a neutral conductor that is as large as the phase conductors. A reduced neutral should not be permitted. When adding capacitors for PF correction, be careful to avoid any unwanted voltage resonances that might be excited by harmonic load currents.

If a delta/wye connected power transformer is installed between the power source and the load, the power factor at the transformer input generally will reflect the average PF of the loads on the secondary. This conclusion works on the assumption that the low PF is caused by inductive and capacitive reactances in the loads. However, if the load current is rich in harmonics from rectifiers and switching regulators, some of the harmonic currents will flow no farther toward the power source than the transformer delta winding. The third harmonic and multiples of three will flow in the delta winding and will be significantly reduced in amplitude. By this means, the transformer will provide some improvement in the PF of the total load.

An economic evaluation of the cost vs. benefits, plus a review of any mandatory utility company limits that must be observed for PF correction, will determine how much power factor correction, if any, may be advisable at a given facility. Correction to 85% will satisfy most requirements. No economic advantage is likely to result from correcting to 95% or greater. Overcorrecting a load by placing too many PF correction capacitors may reduce the power factor after reaching unity, and may cause uncontrollable overvoltages in low-kVA-capacity power sources.

PF correcting capacitors usually offer some benefits in absorbing line-voltage impulse-type noise spikes. However, if the capacitors are switched on and off, they will create significant impulses of their own. Switching may be accomplished with acceptably low disturbance through the use of soft-start or preinsertion resistors. Such resistors are connected momentarily in series with the capacitors. After a brief delay (0.5 s or less) the resistors are short circuited, connecting the capacitors directly across the line.

Installation of PF correction capacitors at a facility is a complicated process that requires a knowledgeable consultant and licensed electrician. The local utility company should be contacted before any effort is made to improve the PF of a facility.

Defining Terms

Critical system bus: A circuit in an AC power distribution system designated as feeding critical loads that must continue to operate in the event of a utility company power failure.

Emergency power source: A supply circuit or system designated as being essential for safety to life and property.

Forward-transfer mode: An operational state of an uninterruptible power system in which the load is normally powered by the utility line, and the inverter is idle.

Motor-generator set (m-g): A power processing machine consisting of a motor powered by the AC utility supply that is mechanically connected to a generator, which feeds the load.

Power factor: The ratio of *true power* to *apparent power*, generally expressed as a percentage (see the text).

Power factor correction: The practice of placing or connecting one or more devices (usually capacitors) onto an AC power circuit in order to improve its operating power factor.

Power shaving: The practice of a utility customer reducing its demand from the system during peak periods in order to enjoy a more attractive energy rate from the utility.

Pull-in speed: The minimum speed (close to the synchronous speed) at which a motor will pull into synchronization if excitation is applied to the field.

Reverse-transfer mode: An operational state of an uninterruptible power system in which the load is normally powered by the inverter. In the event of an inverter failure, the load is switched directly to the utility line.

Standby power source: A supply circuit or system used to protect a facility against the loss of productivity resulting from a utility company power outage.

Uninterruptible power system (UPS): A power processing device used to protect a sensitive load from
disturbances on the utility AC line. A UPS system typically consists of a battery bank that drives
an inverter to produce power for the load, line switching hardware, and supervisory control
circuitry.

References

Acard, C. 1981. Power factor measurement and correction. *Plant Electrical Systems*. Intertec, Overland
Park, KS, March.

Angevine, E. 1989. Controlling generator and UPS noise. *Broadcast Engineering*. Intertec, Overland Park,
KS, March.

Baietto, R. 1989. How to calculate the proper size of UPS devices? *Microservice Management*. Intertec,
Overland Park, KS, March.

DOC. 1983. *Guideline on Electrical Power for ADP Installations*. Federal Information Processing Standards
Pub. No. 94. U.S. Department of Commerce, National Bureau of Standards, Washington, DC.

Dranetz Technologies. 1985. How to Correct Power Line Disturbances. Edison, NJ.

Fardo, S. and Patrick, D. 1985. *Electrical Power Systems Technology*. Prentice-Hall, Englewood Cliffs, NJ.

Fink, D. and Christiansen, D., eds. 1989. *Electronics Engineers' Handbook*, 3rd ed. McGraw-Hill, New York.

Highnote, R.L. 1979. *The IFM Handbook of Practical Energy Management*. Inst. for Management, Old
Saybrook, CT.

Jay, F. 1984. *IEEE Standard Directory of Electrical and Electronics Terms*, 3rd ed. IEEE, New York.

Kramer, R.A. 1991. Analytic and operational consideration of electric system reliability. *Proceedings of the
Reliability and Maintainability Symposium*. IEEE, New York.

Lawrie, R. 1988. *Electrical Systems for Computer Installations*. McGraw-Hill, New York.

Newman, P. 1990. UPS monitoring: Key to an orderly shutdown. *Microservice Management*. Intertec,
Overland Park, KS, March.

Nowak, S. 1990. Selecting a UPS. *Broadcast Engineering*. Intertec, Overland Park, KS, April.

Pettinger, W. 1990. The procedure of power conditioning. *Microservice Management*. Intertec, Overland
Park, KS, March.

Smeltzer, D. 1990. Getting organized about power. *Microservice Management*. Intertec, Overland Park, KS,
March.

Smith, M. 1989. Planning for standby AC power. *Broadcast Engineering*. Intertec, Overland Park, KS,
March.

Whitaker, J.C. 1989. *AC Power Systems Handbook*, 2nd ed., CRC Press, Boca Raton, FL.

Further Information

Most manufacturers of emergency power supply equipment offer detailed applications booklets and bul-
letins that explain the available options. The following book is also recommended for an overview of AC
power quality considerations:

AC Power Systems Handbook, 2nd ed., written by J.C. Whitaker, CRC Press, Boca Raton, FL.

10.10 Facility Grounding

Jerry C. Whitaker

10.10.1 Introduction

The attention given to the design and installation of a facility ground system is a key element in the day-to-
day reliability of the plant. A well-designed and -installed ground network is invisible to the engineering
staff. A marginal ground system, however, will cause problems on a regular basis. Grounding schemes can

range from simple to complex, but any system serves three primary purposes:

- Provides for operator safety.
- Protects electronic equipment from damage caused by transient disturbances.
- Diverts stray radio frequency energy from sensitive audio, video, control, and computer equipment.

Most engineers view grounding mainly as a method to protect equipment from damage or malfunction. However, the most important element is operator safety. The 120 V or 208 V AC line current that powers most equipment can be dangerous, even deadly, if handled improperly. Grounding of equipment and structures provides protection against wiring errors or faults that could endanger human life.

Proper grounding is basic to protection against AC line disturbances. This applies if the source of the disturbance is lightning, power-system switching activities, or faults in the distribution network. Proper grounding is also a key element in preventing radio frequency interference in transmission or computer equipment. A facility with a poor ground system can experience RFI problems on a regular basis. Implementing an effective ground network is not an easy task. It requires planning, quality components, and skilled installers. It is not inexpensive. However, proper grounding is an investment that will pay dividends for the life of the facility.

Any ground system consists of two key elements: (1) the earth-to-grounding electrode interface outside the facility, and (2) the AC power and signal-wiring systems inside the facility.

Terms and Codes

A *facility* can be defined as something that is built, installed, or established to serve a particular purpose [1]. A facility is usually thought of as a single building or group of buildings. The National Electrical Code (NEC) uses the term *premises* to refer to a facility when it defines premises wiring as the interior and exterior (facility) wiring, such as power, lighting, control, and signal systems. Premises wiring includes the service and all permanent and temporary wiring between the service and the load equipment. Premises wiring does not include wiring internal to any load equipment.

The Need for Grounding

The Institute of Electrical and Electronic Engineers (IEEE) defines grounding as a conducting connection, whether intentional or accidental, by which an electric circuit or equipment is connected to the earth, or to some conducting body of relatively large extent that serves in place of the earth. It is used for establishing and maintaining the potential of the earth (or of the conducting body) or approximately that potential, on conductors connected to it, and for conducting ground current to and from the earth (or the conducting body) [2]. Based on this definition, the reasons for grounding can be identified as

- Personnel safety by limiting potentials between all non-current-carrying metal parts of an electrical distribution system.
- Personnel safety and control of electrostatic discharge (ESD) by limiting potentials between all non-current-carrying metal parts of an electrical distribution system and the earth.
- Fault isolation and equipment safety by providing a low-impedance fault return path to the power source to facilitate the operation of overcurrent devices during a ground fault.

The IEEE definition makes an important distinction between *ground* and *earth*. *Earth* refers to mother earth, and *ground* refers to the equipment grounding system, which includes equipment grounding conductors, metallic raceways, cable armor, enclosures, cabinets, frames, building steel, and all other non-current-carrying metal parts of the electrical distribution system.

There are other reasons for grounding not implicit in the IEEE definition. Overvoltage control has long been a benefit of proper power system grounding, and is described in IEEE Standard 142, also known as the *Green Book* [3]. With the increasing use of electronic computer systems, noise control has become associated with the subject of grounding, and is described in IEEE Standard 1100, *the Emerald Book* [4].

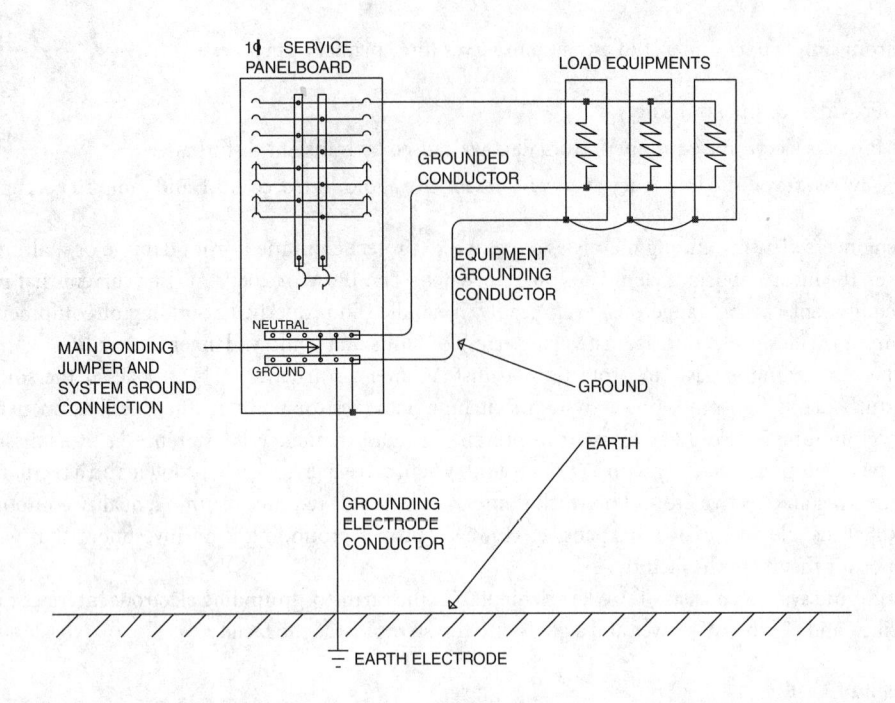

FIGURE 10.211   Equipment grounding and system grounding. (Courtesy of Intertec Publishing. After [5].)

Equipment Grounding

Personnel safety is achieved by interconnecting all non-current-carrying metal parts of an electrical distribution system, and then connecting the interconnected metal parts to the earth [5]. This process of interconnecting metal parts is called equipment grounding and is illustrated in Fig. 10.211, where the equipment grounding conductor is used to interconnect the metal enclosures. Equipment grounding ensures that there is no difference of potential, and thus no shock hazard between non-current-carrying metal parts anywhere in the electrical distribution system. Connecting the equipment grounding system to earth ensures that there is no difference of potential between the earth and the equipment grounding system. It also prevents static charge buildup.

System Grounding

System grounding, which is also illustrated in Fig. 10.211, is the process of intentionally connecting one of the current-carrying conductors of the electrical distribution system to ground [5]. The figure shows the neutral conductor intentionally connected to ground and the earth. This conductor is called the *grounded* conductor because it is intentionally grounded. The purpose of system grounding is overvoltage control and equipment safety through fault isolation. An ungrounded system is subject to serious overvoltages under conditions such as intermittent ground faults, resonant conditions, and contact with higher voltage systems. Fault isolation is achieved by providing a low-impedance return path from the load back to the source, which will ensure operation of overcurrent devices in the event of a ground fault. The system ground connection makes this possible by connecting the equipment grounding system to the low side of the voltage source. Methods of system grounding include solidly grounded, ungrounded, and impedance-grounded.

Solidly grounded means that an intentional zero-impedance connection is made between a current-carrying conductor and ground. The single-phase (1ϕ) system shown in Fig. 10.211 is solidly grounded. A solidly grounded, three-phase, four-wire, wye system is illustrated in Fig. 10.212 The neutral is connected directly to ground with no impedance installed in the neutral circuit. The NEC permits this connection to be made at the service entrance only [6]. The advantages of a solidly grounded wye system include

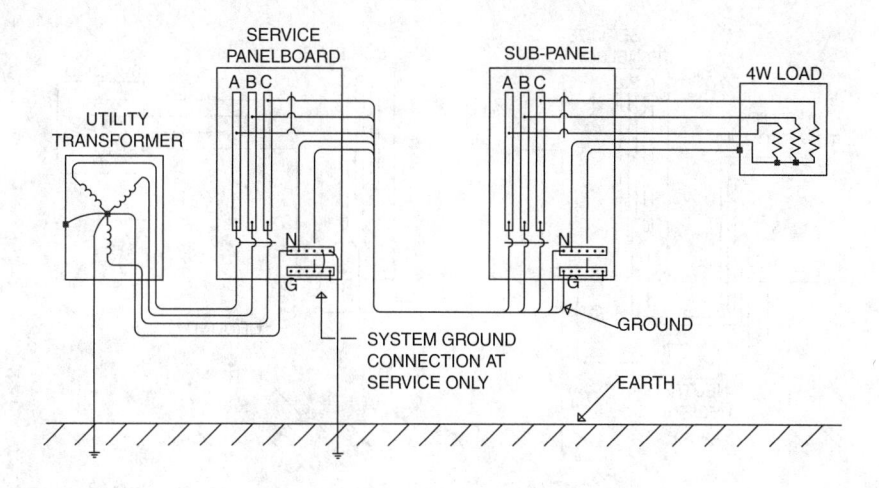

FIGURE 10.212   Solidly grounded wye power system. (Courtesy of Intertec Publishing. After [5].)

reduced magnitude of transient overvoltages, improved fault protection, and faster location of ground faults. There is one disadvantage of the solidly grounded wye system. For low-level arcing ground faults, the application of sensitive, properly coordinated, ground fault protection (GFP) devices is necessary to prevent equipment damage from arcing ground faults. The NEC requires arcing ground fault protection at 480Y/277V services, and a maximum sensitivity limit of 1200 A is permitted. Severe damage is less frequent at the lower voltage 208 V systems, where the arc may be self-extinguishing.

Ungrounded means that there is no intentional connection between a current-carrying conductor and ground. However, charging capacitance will create unintentional capacitive coupling from each phase to ground, making the system essentially a capacitance-grounded system. A three-phase, three-wire system from an ungrounded delta source is illustrated in Fig. 10.213. The most important advantage of an ungrounded system is that an accidental ground fault in one phase does not require immediate removal of power. This allows for continuity of service, which made the ungrounded delta system popular in the past. However, ungrounded systems have serious disadvantages. Because there is no fixed system ground point, it is difficult to locate the first ground fault and to sense the magnitude of fault current. As a result, the fault is often permitted to remain on the system for an extended period of time. If a second fault should

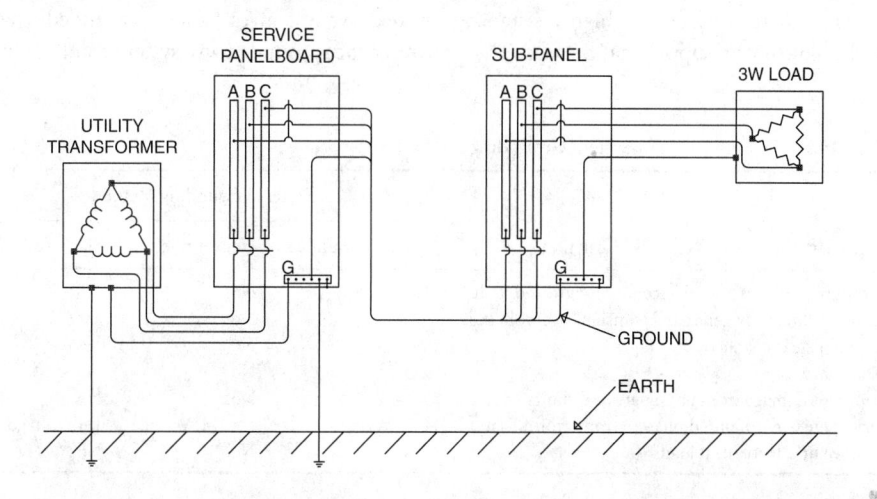

FIGURE 10.213   Ungrounded delta power system. (Courtesy of Intertec Publishing. After [5].)

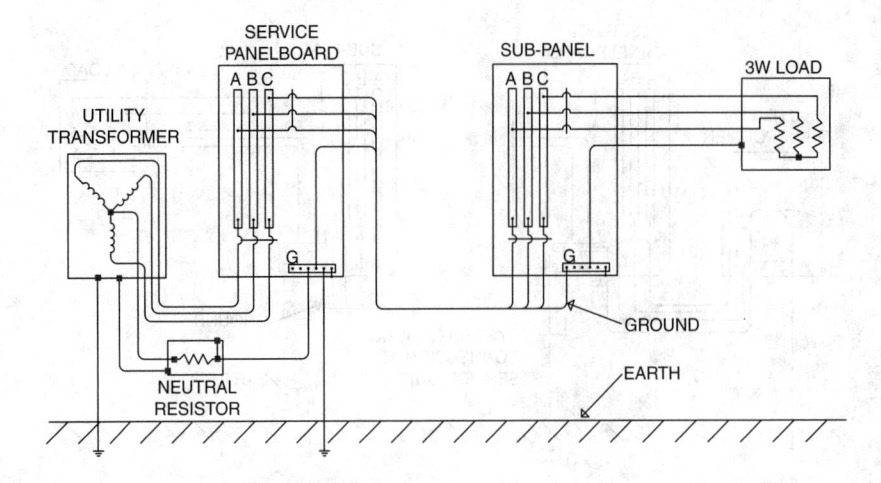

FIGURE 10.214   High-resistance grounded wye power system. (Courtesy of Intertec Publishing. After [5].)

occur before the first one is removed, and the second fault is on a different phase, the result will be a double line-to-ground fault causing serious arcing damage. Another problem with the ungrounded delta system is the occurrence of high transient overvoltages from phase-to-ground. Transient overvoltages can be caused by intermittent ground faults, with overvoltages capable of reaching a phase-to-ground voltage of from six to eight times the phase-to-neutral voltage. Sustained overvoltages can ultimately result in insulation failure and thus more ground faults.

Impedance-grounded means that an intentional impedance connection is made between a current-carrying conductor and ground. The high-resistance grounded wye system, illustrated in Fig. 10.214, is an alternative to solidly grounded and ungrounded systems. High-resistance grounding will limit ground fault current to a few amperes, thus removing the potential for arcing damage inherent in solidly grounded systems. The ground reference point is fixed, and relaying methods can locate first faults before damages from second faults occur. Internally generated transient overvoltages are reduced because the neutral to ground resistor dissipates any charge that may build up on the system charging capacitances.

Table 10.14 compares the three most common methods of system grounding. There is no one *best* system grounding method for all applications. In choosing among the various options, the designer must consider the requirements for safety, continuity of service, and cost. Generally, low-voltage systems should be operated solidly grounded. For applications involving continuous processes in industrial plants or where shutdown might create a hazard, a high-resistance grounded wye system, or a solidly grounded wye system with an alternate power supply, may be used. The high-resistance grounded wye system combines many of

TABLE 10.14   Comparison of System Grounding Methods

Characteristic Assuming No Fault Escalation	System Grounding Method		
	Solidly Grounded	Ungrounded	High Resistance
Operation of overcurrent device on first ground fault	Yes	No	No
Control of internally generated transient overvoltages	Yes	No	Yes
Control of steady-state overvoltages	Yes	No	Yes
Flash hazard	Yes	No	No
Equipment damage from arcing ground-faults	Yes	No	No
Overvoltage (on unfaulted phases) from ground-fault[a]	L-N Voltage	>> L-L Voltage	L-L Voltage
Can serve line-to-neutral loads	Yes	No	No

(After [3].)

[a] L = line, N = neutral

the advantages of the ungrounded delta system and the solidly grounded wye system. IEEE Standard 142 recommends that medium-voltage systems less than 15 kV be low-resistance grounded to limit ground fault damage yet permit sufficient current for detection and isolation of ground faults. Standard 142 also recommends that medium-voltage systems over 15 kV be solidly grounded. Solid grounding should include sensitive ground fault relaying in accordance with the NEC.

The Grounding Electrode

The process of connecting the grounding system to earth is called *earthing*, and consists of immersing a metal electrode or system of electrodes into the earth [5]. The conductor that connects the grounding system to earth is called the *grounding electrode conductor*. The function of the grounding electrode conductor is to keep the entire grounding system at earth potential (i.e., voltage equalization during lightning and other transients) rather than for conducting ground-fault current. Therefore, the NEC allows reduced sizing requirements for the grounding electrode conductor when connected to *made* electrodes.

The basic measure of effectiveness of an earth electrode system is called *earth electrode resistance*. Earth electrode resistance is the resistance, in ohms, between the point of connection and a distant point on the earth called *remote earth*. Remote earth, about 25 ft from the driven electrode, is the point where earth electrode resistance does not increase appreciably when this distance is increased. Earth electrode resistance consists of the sum of the resistance of the metal electrode (negligible) plus the contact resistance between the electrode and the soil (negligible) plus the soil resistance itself. Thus, for all practical purposes, earth electrode resistance equals the soil resistance. The soil resistance is nonlinear, with most of the earth resistance contained within several feet of the electrode. Furthermore, current flows only through the electrolyte portion of the soil, not the soil itself. Thus, soil resistance varies as the electrolyte content (moisture and salts) of the soil varies. Without electrolyte, soil resistance would be infinite.

Soil resistance is a function of soil resistivity. A 1 m^3 sample of soil with a resistivity ρ of 1 Ωm will present a resistance R of 1 Ω between opposite faces. A broad variation of soil resistivity occurs as a function of soil types, and soil resistivity can be estimated or measured directly. Soil resistivity is usually measured by injecting a known current into a given volume of soil and measuring the resulting voltage drop. When soil resistivity is known, the earth electrode resistance of any given configuration (single rod, multiple rods, or ground ring) can be determined by using standard equations developed by Sunde [6], Schwarz [7], and others.

Earth resistance values should be as low as practicable, but are a function of the application. The NEC approves the use of a single *made* electrode if the earth resistance does not exceed 25 Ω. IEEE Standard 1100 reports that the very low earth resistance values specified for computer systems in the past are not necessary. Methods of reducing earth resistance values include the use of multiple electrodes in parallel, the use of ground rings, increased ground rod lengths, installation of ground rods to the permanent water level, increased area of coverage of ground rings, and the use of concrete-encased electrodes, ground wells, and electrolytic electrodes.

Earth Electrode

Earth electrodes may be *made* electrodes, *natural* electrodes, or *special-purpose* electrodes [5]. Made electrodes include driven rods, buried conductors, ground mats, buried plates, and ground rings. The electrode selected is a function of the type of soil and the available depth. Driven electrodes are used where bedrock is 10 ft or more below the surface. Mats or buried conductors are used for lesser depths. Buried plates are not widely used because of the higher cost when compared to rods. Ground rings employ equally spaced driven electrodes interconnected with buried conductors. Ground rings are used around large buildings, around small unit substations, and in areas having high soil resistivity.

Natural electrodes include buried water pipe electrodes and concrete-encased electrodes. The NEC lists underground metal water piping, available on the premises and not less than 10 ft in length, as part of a preferred grounding electrode system. Because the use of plastic pipe in new water systems will impair the effectiveness of water pipe electrodes, the NEC requires that metal underground water piping be supplemented by an additional approved electrode. Concrete below ground level is a good electrical conductor. Thus, metal electrodes encased in such concrete will function as excellent grounding electrodes. The application of concrete-encased electrodes is covered in IEEE Standard 142.

10.10.2   Establishing an Earth Ground

The grounding electrode is the primary element of any ground system. The electrode can take many forms. In all cases, its purpose is to interface the electrode (a conductor) with the earth (a semiconductor). Grounding principles have been refined to a science. Still, however, many misconceptions exist regarding grounding. An understanding of proper grounding procedures begins with the basic earth-interface mechanism.

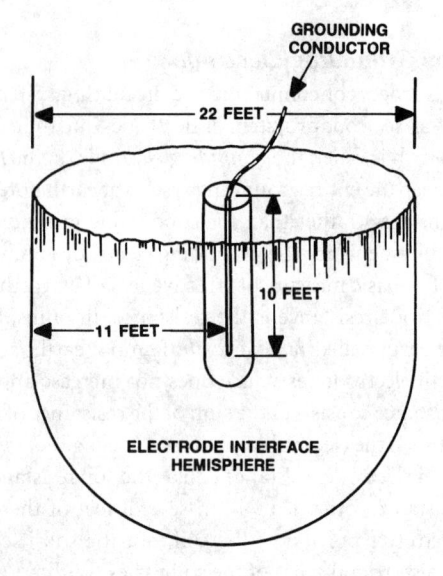

Grounding Interface

The grounding electrode (or ground rod) interacts with the earth to create a hemisphere-shaped volume, as illustrated in Fig. 10.215. The size of this volume is related to the size of the grounding electrode. The length of the electrode has a much greater effect than the diameter. Studies have demonstrated that the earth-to-electrode resistance from a driven ground rod increases exponentially with the distance from that rod. At a given point, the change becomes insignificant. It has been found that for maximum effectiveness of the earth-to-electrode interface, each ground rod requires a hemisphere-shaped volume with a diameter that is approximately 2.2 times the rod length [9].

FIGURE 10.215   The effective earth-interface hemisphere resulting from a single driven ground rod. The 90% effective area of the rod extends to a radius of approximately 1.1 times the length of the rod. (After [10].)

The constraints of economics and available real estate place practical limitations on the installation of a ground system. It is important, however, to keep the 2.2 rule in mind because it allows the facility design engineer to use the available resources to the best advantage. Figure 10.216 illustrates the effects of locating ground rods too close (less than 2.2 times the rod length). An overlap area is created that effectively wastes some of the earth-to-electrode capabilities of the two ground rods. Research has shown, for example, that two 10-ft ground rods driven only 1 ft apart provide about the same resistivity as a single 10-ft rod.

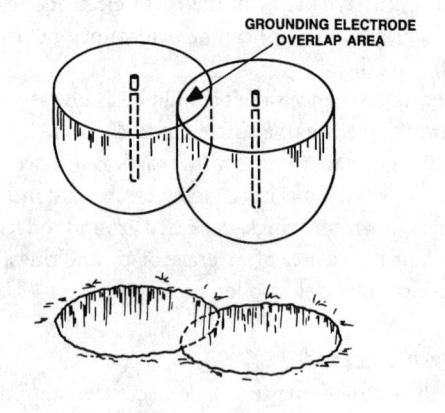

There are three schools of thought with regard to ground-rod length. The first approach states that extending ground-rod length beyond about 10 ft is

FIGURE 10.216   The effect of overlapping earth interface hemispheres by placing two ground rods at a spacing less than 2.2 times the length of either rod. The overlap area represents wasted earth-to-grounding electrode interface capability. (After [10].)

of little value for most types of soil. The reasoning behind this conclusion is presented in Fig. 10.217a, where ground resistance is plotted as a function of ground-rod length. Beyond 10 ft in length, a point of diminishing returns is reached.

The second school of thought concludes that optimum earth-to-electrode interface is achieved with long (40 ft or greater) rods, driven to penetrate the local water table. When planning this type of installation,

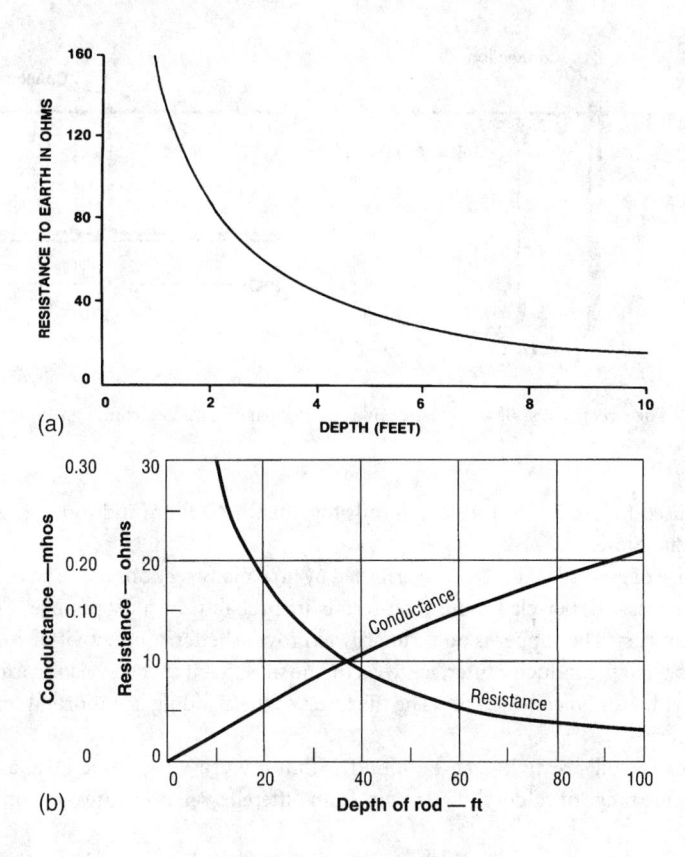

FIGURE 10.217 Charted grounding resistance as a function of ground-rod length: (a) data demonstrating that ground-rod length in excess of 10 ft produces diminishing returns (1-in.-diameter rod) [10], (b) data demonstrating that ground system performance continues to improve as depth increases. (Chart b courtesy of Intertec Publishing. After [12].)

consider the difficulty that may be encountered when attempting to drive long ground rods. The foregoing discussion assumes that the soil around the grounding electrode is reasonably uniform in composition. Depending upon the location, however, this assumption may not hold true.

The third school of thought concludes that the optimum ground rod installation is achieved by using the longest possible rod depth (length). Data to support this conclusion is given in Fig. 10.217b, which plots ground rod performance as a function of depth in soil of uniform resistivity. This curve does not take into account seasonal moisture content, changing chemical composition at different soil layers, and frozen soil conditions.

Given these three conflicting approaches, the benefits of hiring an experienced, licensed professional engineer to design the ground system can be readily appreciated.

Horizontal grounding electrodes provide essentially the same resistivity as an equivalent-length vertical electrode, given uniform soil conditions. As Fig. 10.218 demonstrates, the difference between a 10-ft vertical and a 10-ft horizontal ground rod is negligible (275 Ω vs. 250 Ω). This comparison includes the effects of the vertical connection element from the surface of the ground to the horizontal rod. Taken by itself, the horizontal ground rod provides an earth-interface resistivity of approximately 308 Ω when buried at a depth of 36 in.

Ground rods come in many sizes and lengths. The more popular sizes are $\frac{1}{2}$, $\frac{5}{8}$, $\frac{3}{4}$, and 1 in. The $\frac{1}{2}$-in. size is available in steel with stainless-clad, galvanized, or copper-clad rods. All-stainless-steel rods also are available. Ground rods can be purchased in unthreaded or threaded (sectional) lengths. The sectional sizes are typically $\frac{9}{16}$-in. or $\frac{1}{2}$-in. rolled threads. Couplers are made from the same materials as the rods.

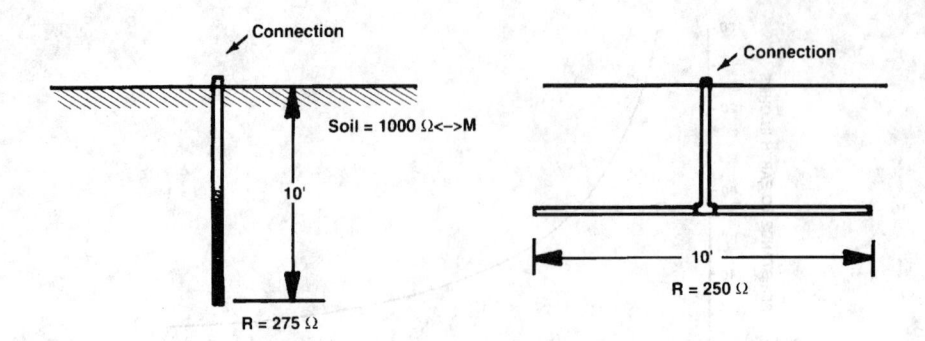

FIGURE 10.218   The effectiveness of vertical ground rods compared with horizontal ground rods. (After [10].)

Couplers can be used to join 8- or 10-ft length rods together. A 40-ft ground rod, for example, is driven one 10-ft section at a time.

The type and size of ground rod used is determined by how many sections are to be connected and how hard or rocky the soil is. Copper-clad ⅝-in. × 10-ft rods are probably the most popular. Copper cladding is designed to prevent rust. The copper is not primarily to provide better conductivity. Although the copper certainly provides a better conductor interface to earth, the steel that it covers is also an excellent conductor when compared with ground conductivity. The thickness of the cladding is important only insofar as rust protection is concerned.

Wide variations in soil resistivity can be found within a given geographic area, as documented in Table 10.15. The wide range of values shown results from differences in moisture content, mineral content, and temperature.

Temperature is a major concern in shallow grounding systems because it has a significant effect on soil resistivity [11]. During winter months, the ground system resistance can rise to unacceptable levels because of freezing of liquid water in the soil. The same shallow grounding system can also suffer from high resistance in the summer as moisture is evaporated from soil. It is advisable to determine the natural frost line and moisture profile for an area before attempting design of a ground system.

Figure 10.219 describes a four-point method for in-place measurement of soil resistivity. Four uniformly spaced probes are placed in a linear arrangement and connected to a ground resistance test meter. An alternating current (at a frequency other than 60 Hz) is passed between the two most distant probes, resulting in a potential difference between the center potential probes. The meter display in ohms of resistance can then be applied to determine the average soil resistivity in ohm-centimeters for the hemispherical area between the C1 and P2 probes.

Soil resistivity measurements should be repeated at a number of locations to establish a resistivity profile for the site. The depth of measurement can be controlled by varying the spacing between the probes. In no case should the probe length exceed 20% of the spacing between probes.

After the soil resistivity for a site is known, calculations can be made to determine the effectiveness of a variety of ground system configurations. Equations for several driven rod and radial cable configurations are given in [11], which, after the solid resistivity is known, can be used for the purpose of estimating total

TABLE 10.15   Typical Resistivity of Common Soil Types

Type of Soil Resistivity in Ω/cm	Average	Minimum	Maximum
Filled land, ashes, salt marsh	2,400	600	7,000
Top soils, loam	4,100	340	16,000
Hybrid soils	6,000	1,000	135,000
Sand and gravel	90,000	60,000	460,000

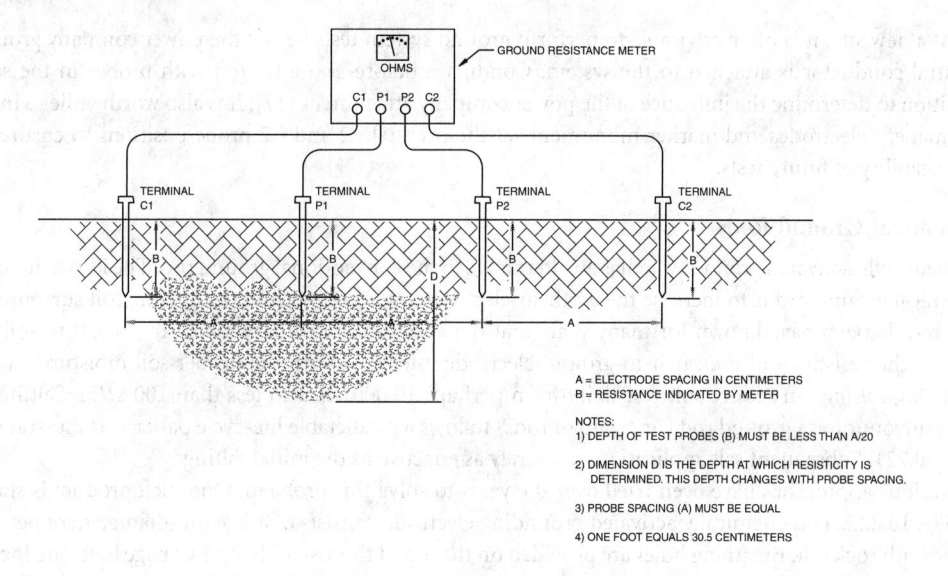

A = ELECTRODE SPACING IN CENTIMETERS
B = RESISTANCE INDICATED BY METER

NOTES:
1) DEPTH OF TEST PROBES (B) MUST BE LESS THAN A/20

2) DIMENSION D IS THE DEPTH AT WHICH RESISTICITY IS
   DETERMINED. THIS DEPTH CHANGES WITH PROBE SPACING.

3) PROBE SPACING (A) MUST BE EQUAL

4) ONE FOOT EQUALS 30.5 CENTIMETERS

FIGURE 10.219   The four-point method for soil resistivity measurement. (After [11].)

system resistance. Generally, driven rod systems are appropriate where soil resistivity continues to improve with depth or where temperature extremes indicate seasonal frozen or dry soil conditions. Figure 15.220 shows a typical soil resistivity map for the U.S.

Ground Electrode Testing

Testing of all ground electrodes before they are connected to form a complex network is a fairly simple process that is well described in the documentation included with all ground electrode meters. This instructional process, therefore, will not be described here. Also, the system as a whole should be tested after all interconnections are made, providing a benchmark for future tests.

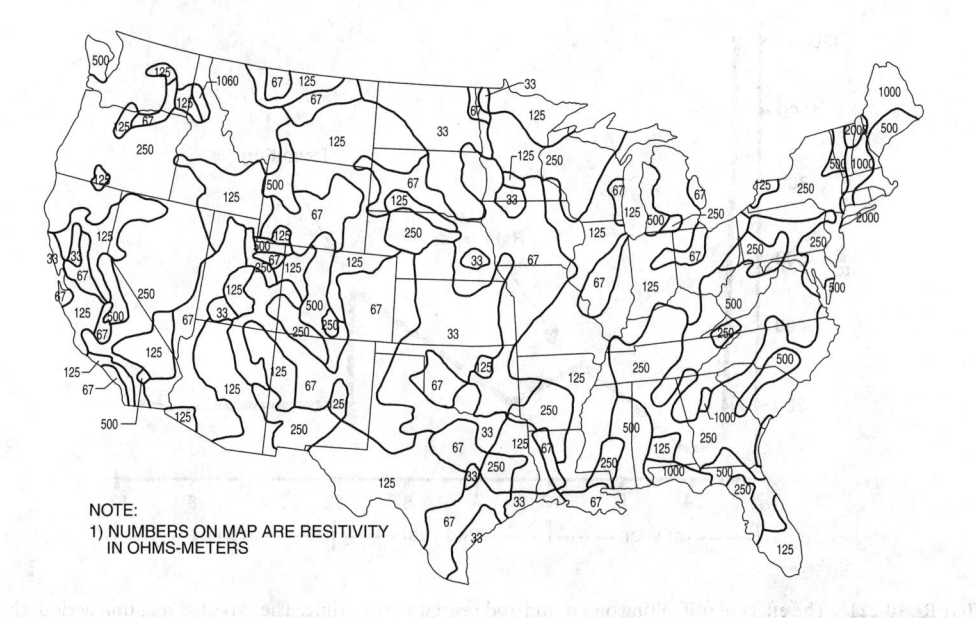

NOTE:
1) NUMBERS ON MAP ARE RESITIVITY
   IN OHMS-METERS

FIGURE 10.220   Typical soil resistivity map for the U.S. (After [11].)

At a new site, it is often advisable to perform ground system tests before the power company ground/ neutral conductor is attached to the system. Conduct a before-and-after test with probes in the same position to determine the influence of the power company attachment [11]. It is also worthwhile to install permanent electrodes and marker monuments at the original P2 and C2 probe positions to ensure the repeatability of future tests.

Chemical Ground Rods

A chemically activated ground system is an alternative to the conventional ground rod. The idea behind the chemical ground rod is to increase the earth-to-electrode interface by conditioning the soil surrounding the rod. Experts have known for many years that the addition of ordinary table salt (NaCl) to soil will reduce the resistivity of the earth-to-ground electrode interface. With the proper soil moisture level (4 to 12%), *salting* can reduce soil resistivity from perhaps 10,000 Ω/m to less than 100 Ω/m. Salting the area surrounding a ground rod (or group of rods) follows a predictable life-cycle pattern, as illustrated in Fig. 10.221. Subsequent salt applications are rarely as effective as the initial salting.

Various approaches have been tried over the years to solve this problem. One such product is shown in Fig. 10.222. This chemically activated grounding electrode consists of a 2-½-in.-diameter copper pipe filled with rock salt. Breathing holes are provided on the top of the assembly, and seepage holes are located at the bottom. The theory of operation is simple. Moisture is absorbed from the air (when available) and is then absorbed by the salt. This creates a solution that seeps out of the base of the device and conditions the soil in the immediate vicinity of the rod.

Another approach is shown in Fig. 10.223. This device incorporates a number of ports (holes) in the assembly. Moisture from the soil (and rain) is absorbed through the ports. The metallic salts subsequently absorb the moisture, forming a saturated solution that seeps out of the ports and into the earth-to-electrode hemisphere. Tests have shown that if the moisture content is within the required range, earth resistivity can be reduced by as much as 100:1. Figure 15.224 shows the measured performance of a typical chemical ground rod in three types of soil.

Implementations of chemical ground-rod systems vary depending on the application. Figure 10.225 illustrates a counterpoise ground consisting of multiple leaching apertures connected in a spoke fashion to a central hub. The system is serviceable in that additional salt compound can be added to the hub at

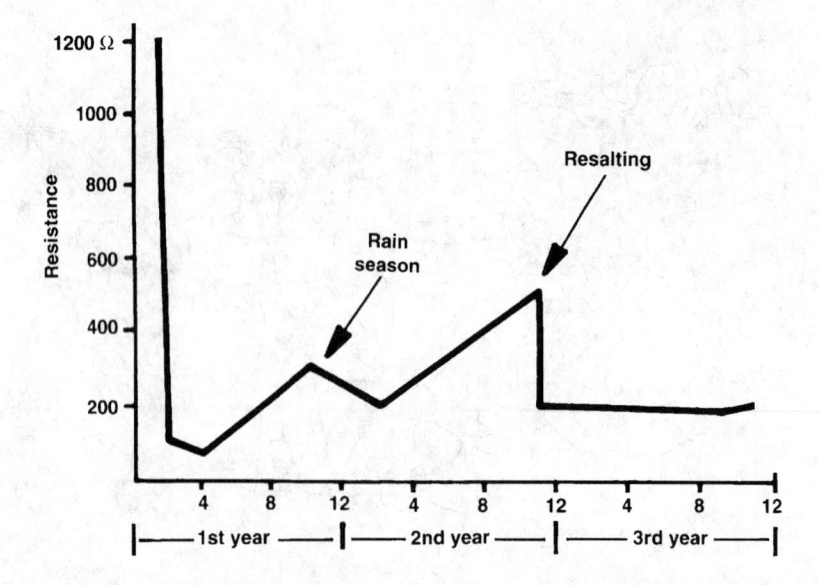

FIGURE 10.221   The effect of soil salting on ground-rod resistance with time. The expected resalting period, shown here as 2 years, varies depending on the local soil conditions and the amount of moisture present. (After [10].)

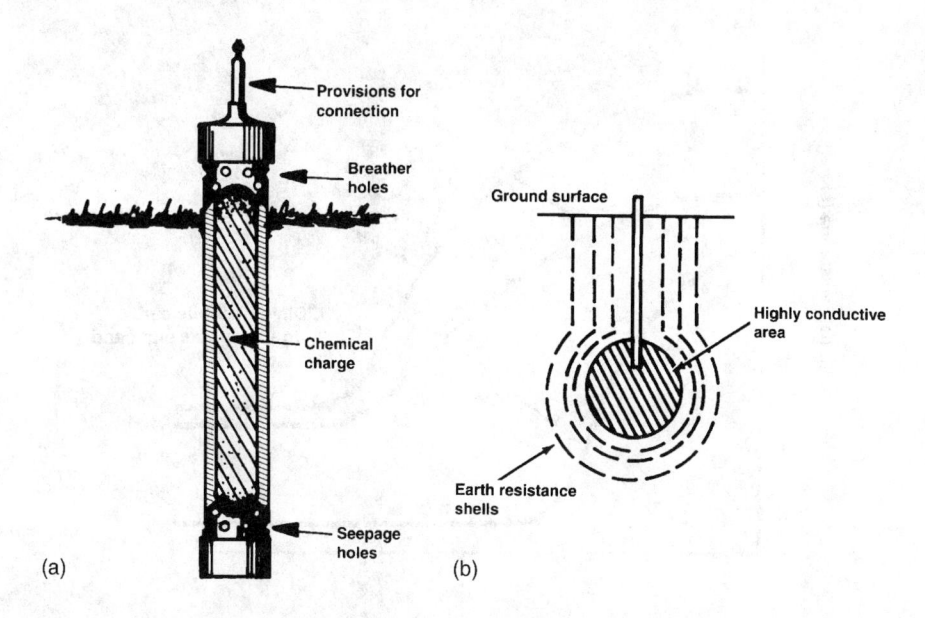

FIGURE 10.222 An air-breathing chemically activated ground rod: (a) breather holes at the top of the device permit moisture penetration into the chemical charge section of the rod; (b) a salt solution seeps out of the bottom of the unit to form a conductive shell. (After [10].)

required intervals to maintain the effectiveness of the ground. Figure 10.226 shows a counterpoise system made up of individual chemical ground rods interconnected with radial wires buried below the surface.

Ufer Ground System

Driving ground rods is not the only method of achieving a good earth-to-electrode interface [9]. The concept of the *Ufer* ground has gained interest because of its simplicity and effectiveness. The Ufer approach (named for its developer), however, must be designed into a new structure. It cannot be added on later. The Ufer ground takes advantage of the natural chemical- and water-retention properties of concrete to provide an earth ground. Concrete typically retains moisture for 15 to 30 days after a rain. The material has a ready supply of ions to conduct current because of its moisture-retention properties, mineral content, and inherent pH. The large mass of any concrete foundation provides a good interface to ground.

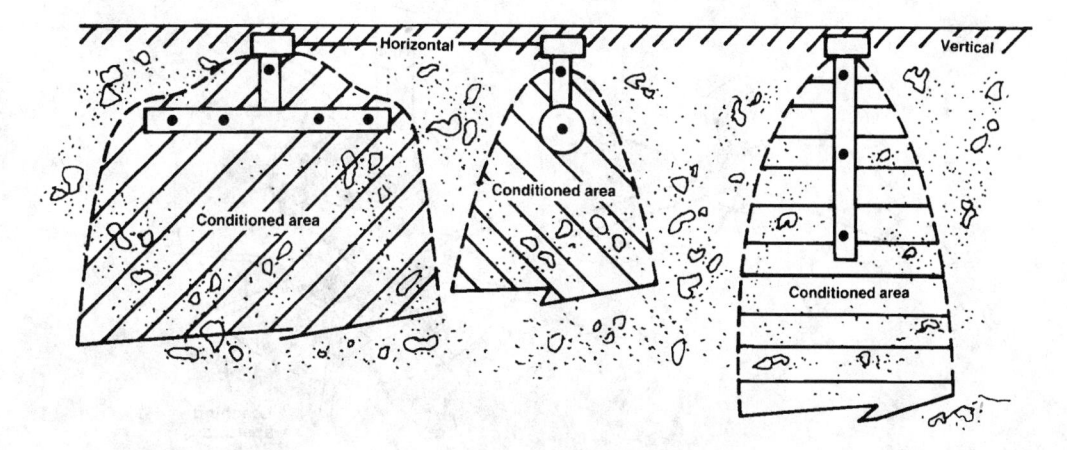

FIGURE 10.223 An alternative approach to the chemically activated ground rod. Multiple holes are provided on the ground-rod assembly to increase the effective earth-to-electrode interface. Note that chemical rods can be produced in a variety of configurations. (After [10].)

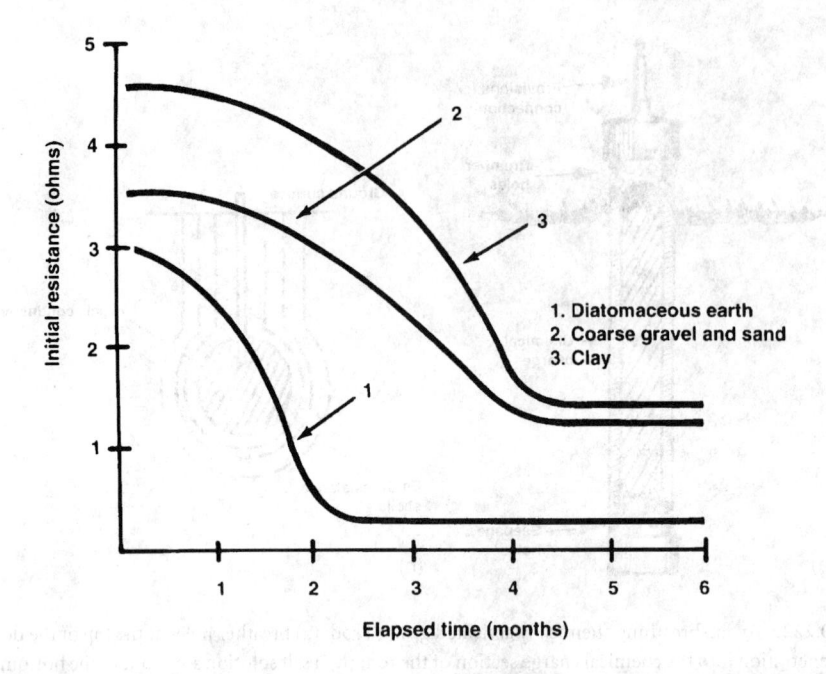

FIGURE 10.224   Measured performance of a chemical ground rod. (After [10].)

A Ufer system, in its simplest form, is made by routing a solid-copper wire (no. 4 gauge or larger) within the foundation footing forms before concrete is poured. Figure 10.227 shows one such installation. The length of the conductor run within the concrete is important. Typically a 20-ft run (10 ft in each direction) provides a 5 Ω ground in 1000 Ω/m soil.

As an alternative, steel reinforcement bars (rebar) can be welded together to provide a rigid, conductive structure. A ground lug is provided to tie equipment to the ground system in the foundation. The rebar must be welded, not tied, together. If it is only tied, the resulting poor connections between rods can result in arcing during a current surge. This can lead to deterioration of the concrete in the affected areas.

The design of a Ufer ground is not to be taken lightly. Improper installation can result in a ground system that is subject to problems. The grounding electrodes must be kept a minimum of 3 in. from the

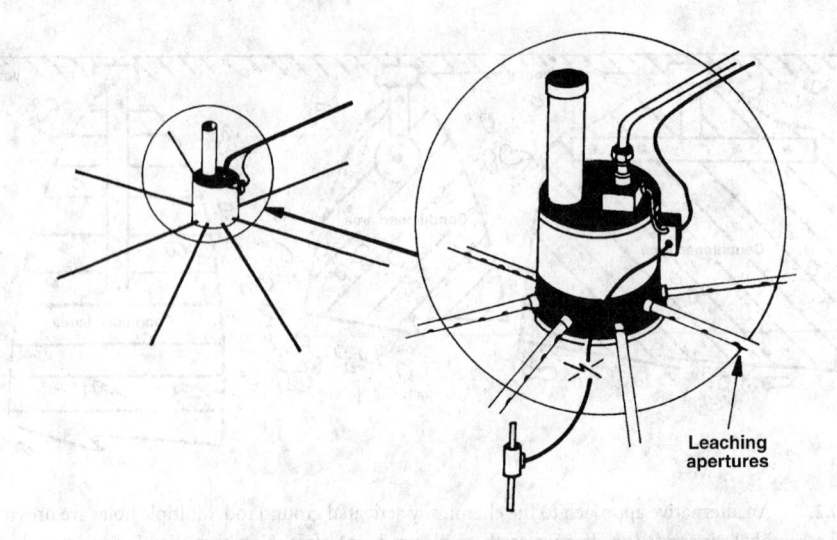

FIGURE 10.225   Hub and spoke counterpoise ground system. (After [10].)

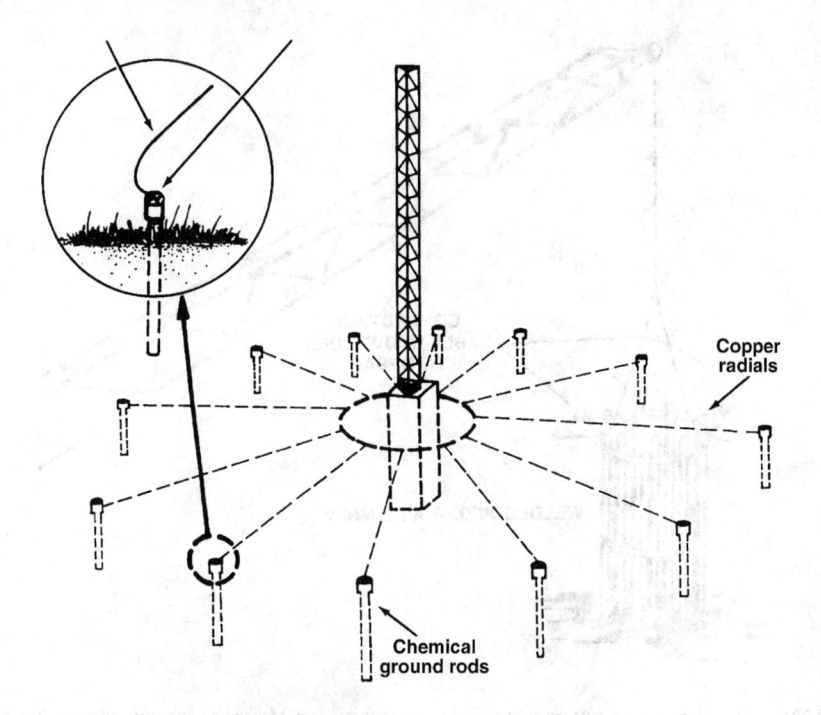

FIGURE 10.226 Tower grounding scheme using buried copper radials and chemical ground rods. (After [10].)

bottom and sides of the concrete to avoid the possibility of foundation damage during a large lightning surge. If an electrode is placed too near the edge of the concrete, a surge could turn the water inside the concrete to steam and break the foundation apart.

The Ufer approach also can be applied to guy-anchor points or a tower base, as illustrated in Fig. 10.228. Welded rebar or ground rods sledged in place after the rebar cage is in position can be used. By protruding

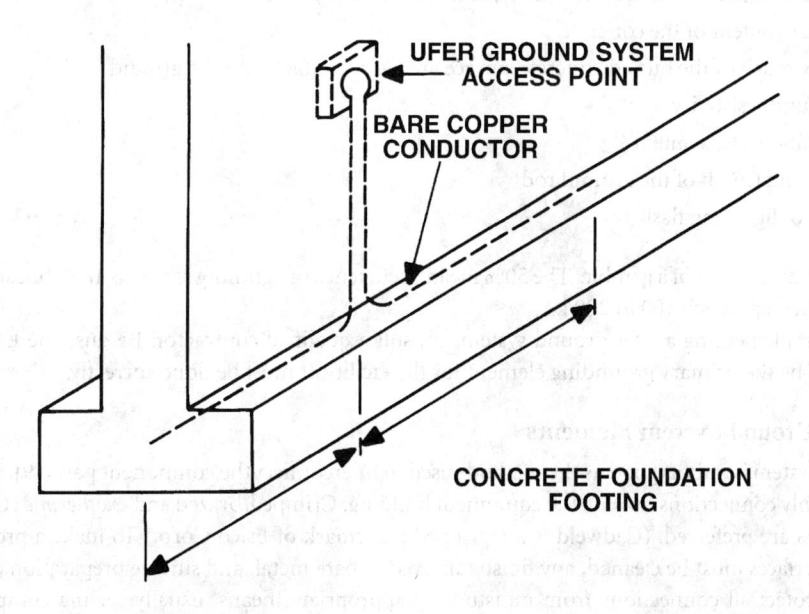

FIGURE 10.227 The basic concept of a Ufer ground system, which relies on the moisture-retentive properties of concrete to provide a large earth-to-electrode interface. Design of such a system is critical. Do not attempt to build a Ufer ground without the assistance of an experienced contractor. (After [9].)

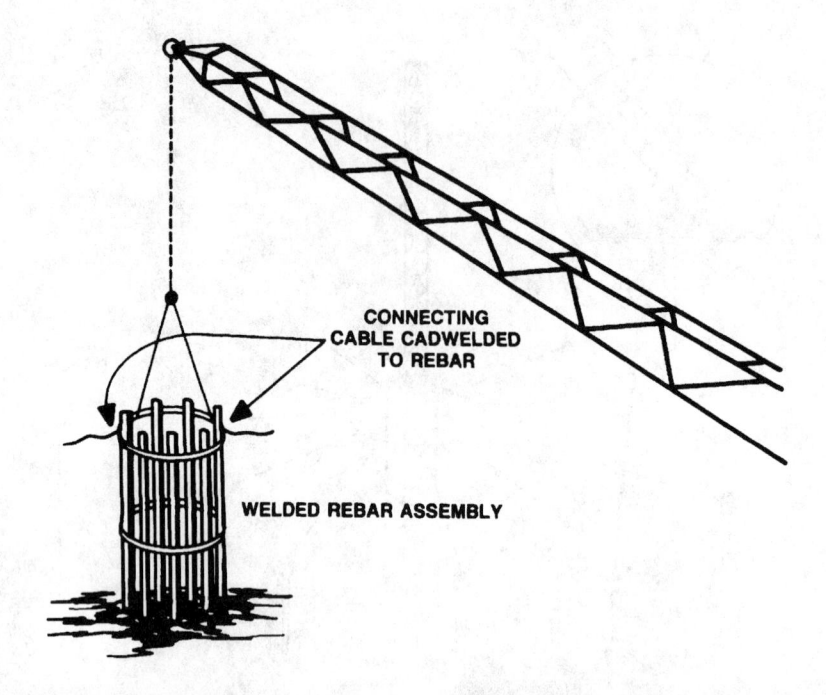

FIGURE 10.228   The Ufer ground system as applied to a transmission-tower base or guy-wire anchor point. When using this type of ground system, bond all rebar securely to prevent arcing in the presence of large surge currents. (After [9].)

below the bottom concrete surface, the ground rods add to the overall electrode length to help avoid thermal effects that can crack the concrete. The maximum length necessary to avoid breaking the concrete under a lightning discharge is determined by the following:

- Type of concrete (density, resistivity, and other factors)
- Water content of the concrete
- How much of the buried concrete surface area is in contact with the ground
- Ground resistivity
- Ground water content
- Size and length of the ground rod
- Size of lightning flash

The last variable is a bit of a gamble. The 50% mean occurrence of lightning strikes is 18 A, but superstrikes can occur that approach 100 to 200 kA.

Before implementing a Ufer ground system, consult a qualified contractor. Because the Ufer ground system will be the primary grounding element for the facility, it must be done correctly.

Bonding Ground-System Elements

A ground system is only as good as the methods used to interconnect the component parts [9]. Do not use soldered-only connections outside the equipment building. Crimped/brazed and *exothermic* (*Cadwelded*) connections are preferred. (Cadweld is a registered trademark of Erico Corp.) To make a proper bond, all metal surfaces must be cleaned, any finish removed to bare metal, and surface preparation compound applied. Protect all connections from moisture by appropriate means, usually sealing compound and heat-shrink tubing.

It is not uncommon for an untrained installer to use soft solder to connect the elements of a ground system. Such a system is doomed from the start. Soft-soldered connections cannot stand up to the acid

Cadweld metal

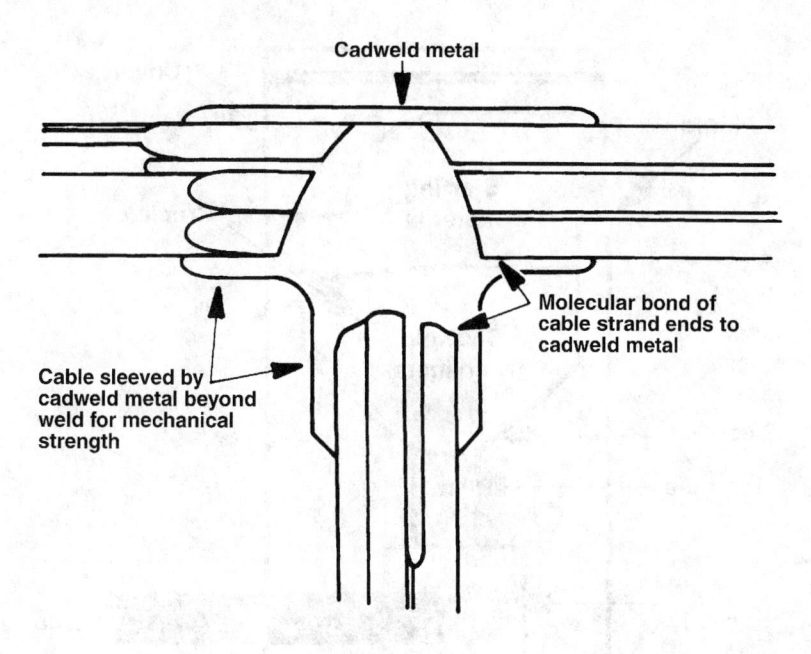

**Molecular bond of
cable strand ends to
cadweld metal**

**Cable sleeved by
cadweld metal beyond
weld for mechanical
strength**

FIGURE 10.229   The exothermic bonding process. (After [9].)

and mechanical stress imposed by the soil. The most common method of connecting the components of a ground system is silver soldering. The process requires the use of brazing equipment, which may be unfamiliar to many facility engineers. The process uses a high-temperature/high-conductivity solder to complete the bonding process. For most grounding systems, however, the best approach to bonding is the exothermic process.

Exothermic Bonding

Exothermic bonding is the preferred method of connecting the elements of a ground system [9]. Molten copper is used to melt connections together, forming a permanent bond. This process is particularly useful in joining dissimilar metals. In fact, if copper and galvanized cable must be joined, exothermic bonding is the only acceptable means. The completed connection will not loosen or corrode and will carry as much current as the cable connected to it. Figure 10.229 illustrates the bonding that results from the exothermic process.

   The bond is accomplished by dumping powdered metals (copper oxide and aluminum) from a container into a graphite crucible and igniting the material by means of a flint lighter. Reduction of the copper oxide by the aluminum produces molten copper and aluminum oxide slag. The molten copper flows over the conductors, bonding them together. The process is illustrated in Fig. 10.230. Figure 10.231 shows a

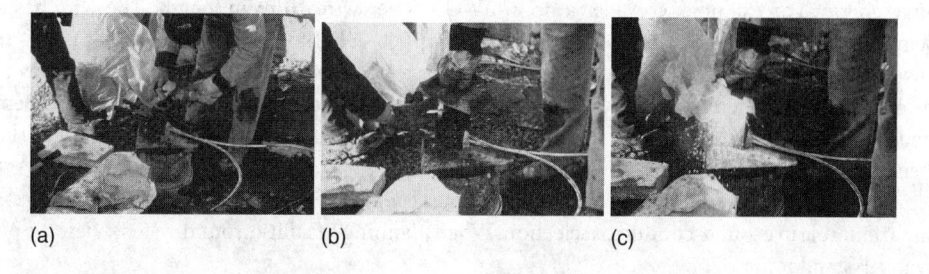

(a)   (b)   (c)

FIGURE 10.230   Exothermic bonding is the preferred method of joining the elements of a ground system. This photo sequence illustrates the procedure: (a) the powdered copper oxide and aluminum compound are added to the mold after the conductors have been mechanically joined; (b) final preparation of the bond before igniting; (c) the chemical reaction that bonds the materials together.

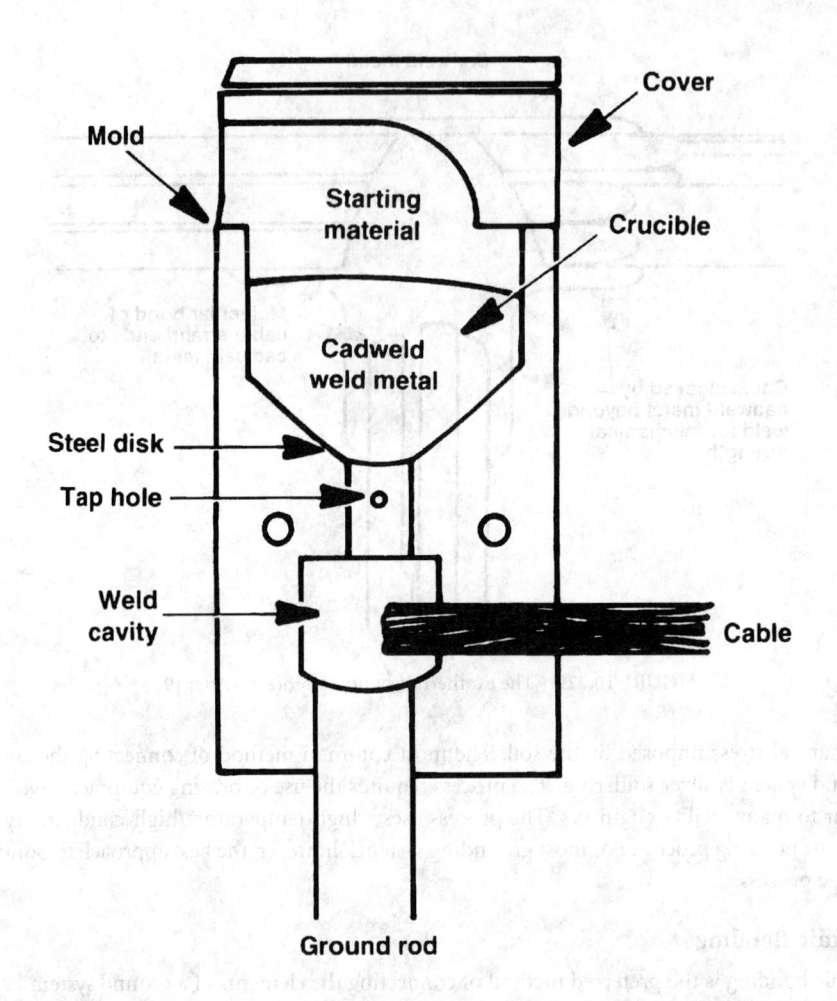

FIGURE 10.231   Typical exothermic bonding mold for connecting a cable to a ground rod. (After [9].)

typical mold. A variety of special-purpose molds are available to join different-size cables and copper strap. Figure 10.232 shows the bonding process for a copper-strap-to-ground-rod interface.

Ground-System Inductance

Conductors interconnecting sections or components of an earth ground system must be kept as short as possible to be effective [9]. The inductance of a conductor is a major factor in its characteristic impedance to surge energy. For example, consider a no. 6 AWG copper wire 10 m in length. The wire has a DC resistance of 0.013 Ω and an inductance of 10 μH. For a 1000-A lightning surge with a 1 μs rise time, the resistive voltage drop will be 13 V, but the reactive voltage drop will be 10 kV. Furthermore, any bends in the conductor will increase its inductance and further decrease the effectiveness of the wire. Bends in ground conductors should be gradual. A 90°-bend is electrically equivalent to a 1/4-turn coil. The sharper the bend, the greater the inductance.

Because of the fast rise time of most lightning discharges and power-line transients, the *skin effect* plays an important role in ground-conductor selection. When planning a facility ground system, view the project from an RF standpoint.

The effective resistance offered by a conductor to radio frequencies is considerably higher than the ohmic resistance measured with direct currents. This is because of an action known as the skin effect, which causes the currents to be concentrated in certain parts of the conductor and leaves the remainder of the cross-section to contribute little or nothing toward carrying the applied current.

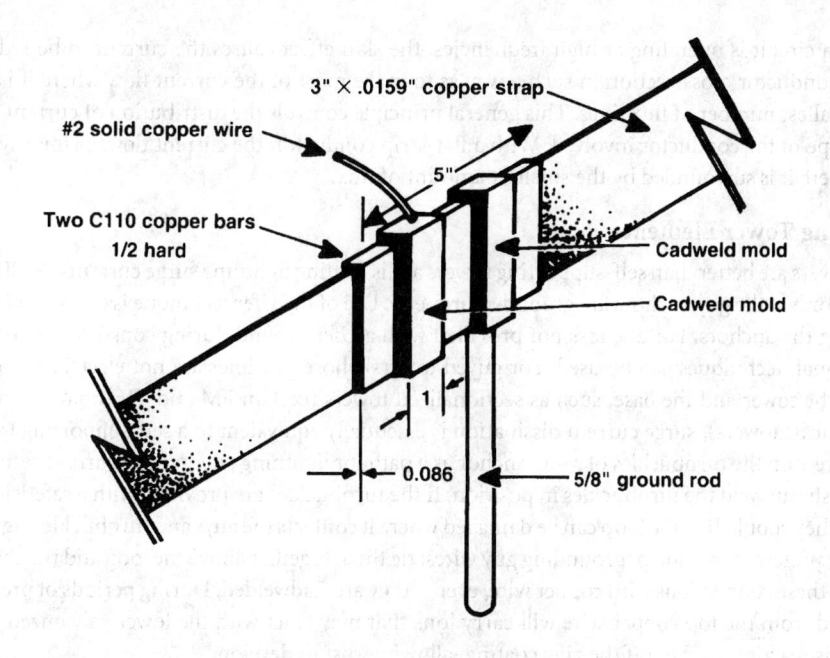

FIGURE 10.232 Cadweld mold for connecting a copper strap to a ground rod. (After [9].)

When a conductor carries an alternating current, a magnetic field is produced that surrounds the wire. This field continually is expanding and contracting as the AC current wave increases from zero to its maximum positive value and back to zero, then through its negative half-cycle. The changing magnetic lines of force cutting the conductor induce a voltage in the conductor in a direction that tends to retard the normal flow of current in the wire. This effect is more pronounced at the center of the conductor. Thus, current within the conductor tends to flow more easily toward the surface of the wire. The higher the frequency, or the faster the rise time of the applied waveform, the greater the tendency for current to flow at the surface. Figure 10.233 shows the distribution of current in a radial conductor.

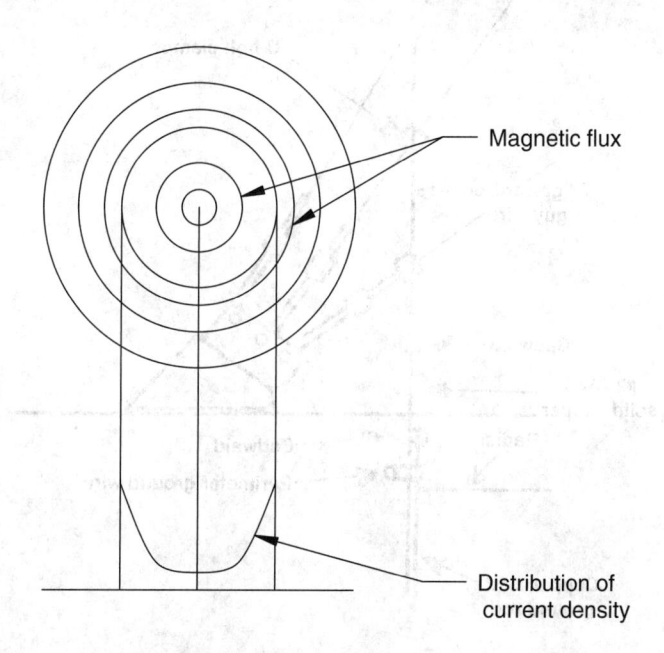

FIGURE 10.233 Skin effect on an isolated round conductor carrying a moderately high-frequency signal.

When a circuit is operating at high frequencies, the skin effect causes the current to be redistributed over the conductor cross-section in such a way as to make most of the current flow where it is encircled by the smallest number of flux lines. This general principle controls the distribution of current regardless of the shape of the conductor involved. With a flat-strip conductor, the current flows primarily along the edges, where it is surrounded by the smallest amount of flux.

Grounding Tower Elements

Guyed towers are better than self-supporting towers at dissipating lightning surge currents [9]. This is true, however, only if the guy anchors are grounded properly. Use of the Ufer technique is one way of effectively grounding the anchors. For anchors not provided with a Ufer ground during construction, other, more conventional, techniques can be used. For guyed towers whose guy lines are not electrically continuous between the tower and the base, such as sectionalized towers (certain FM and TV broadcast towers and AM broadcast towers), surge current dissipation is essentially equivalent to a self-supporting tower.

Never rely on the turnbuckles of a guy anchor as a path for lightning energy. The current resulting from a large flash can weld the turnbuckles in position. If the turnbuckles are provided with a safety loop of guy cable (as they should be), the loop can be damaged where it contacts the guys and turnbuckle. Figure 10.234 shows the preferred method of grounding guy wires: tie them together above the loop and turnbuckles. Do not make these connections with copper wire, even if they are Cadwelded. During periods of precipitation, water shed from the top copper wire will carry ions that may react with the lower galvanized (zinc) guy wires. This reaction washes off the zinc coating, allowing rust to develop.

The best way to make the connection is with all-galvanized materials. This includes the grounding wire, clamps, and ground rods. It may not be possible to use all galvanized materials because, at some point, a connection to copper conductors will be required. *Battery action* caused by the dissimilar metal junction may allow the zinc to act as a sacrificial anode. The zinc eventually will disappear into the soil, leaving a bare steel conductor that can fall victim to rust.

Ruling out an all-galvanized system, the next best scheme uses galvanized wire (guy-wire material) to tie the guy wires together. Just above the soil, Cadweld the galvanized wire to a copper conductor that

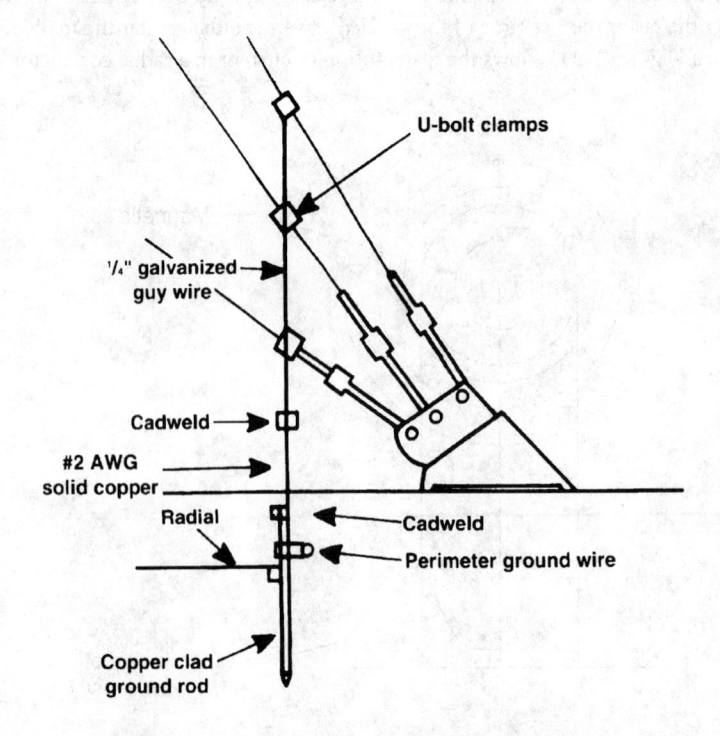

FIGURE 10.234   Recommended guy-anchor grounding procedure. (After [9].)

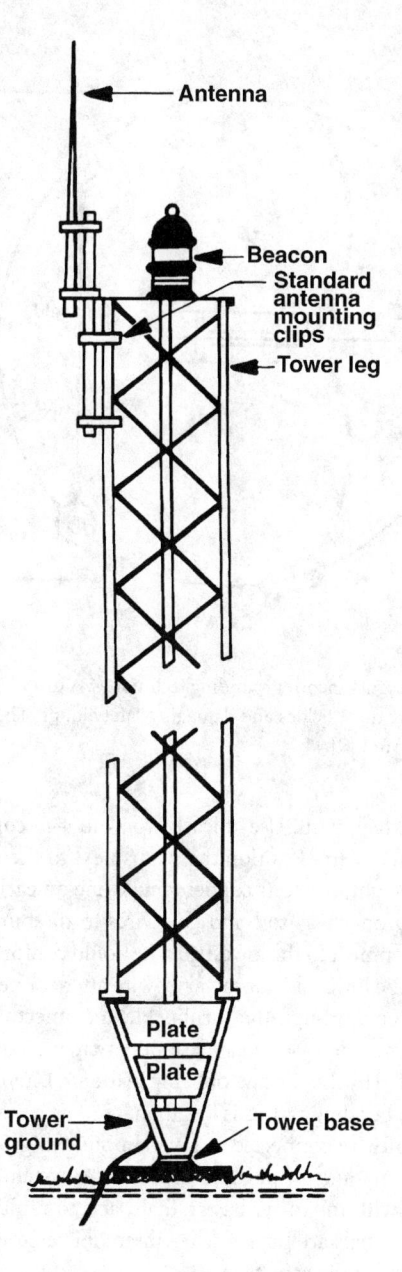

FIGURE 10.235   Ground-conductor dressing for the base of a guyed tower. (After [9].)

penetrates below grade to the perimeter ground system. The height above grade for the connection is determined by the local snowfall or flood level. The electric conductivity of snow, although low, can cause battery action from the copper through the snow to the zinc. The Cadwelded joint should be positioned above the usual snow or flood level.

Ground-Wire Dressing

Figure 10.235 illustrates the proper way to bond the tower base ground leads to the buried ground system [9]. Dress the leads close to the tower from the lowest practical structural element at the base. Keep the conductors as straight and short as possible. Avoid any sharp bends. Attach the ground wires to the tower only at one or more existing bolts (or holes). Do not drill any holes into the tower. Do not loosen any

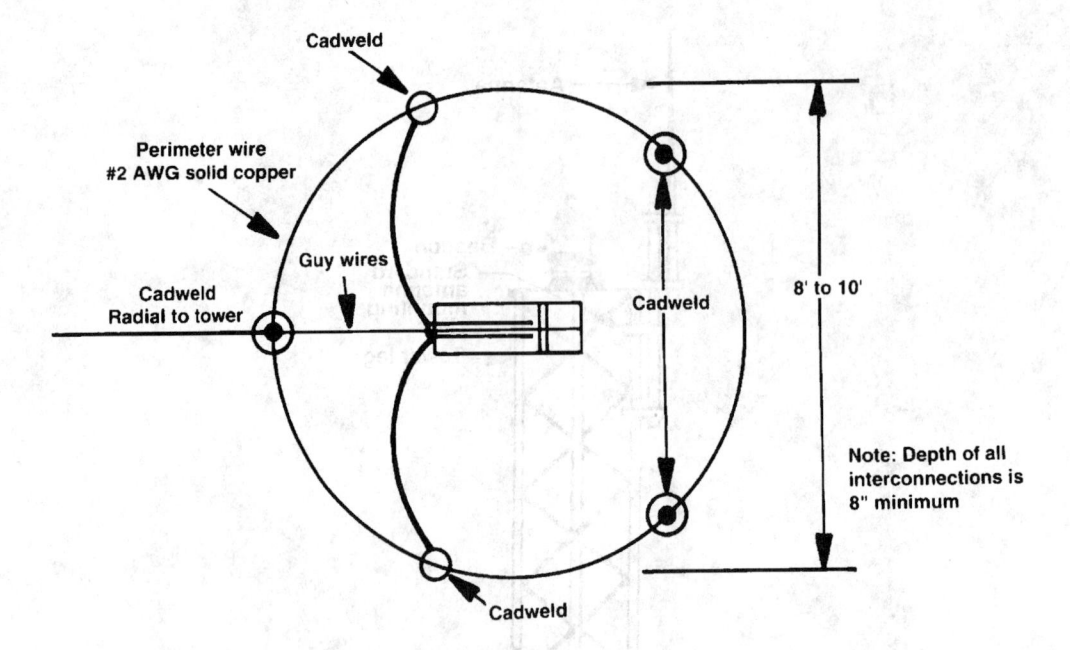

FIGURE 10.236 Top view of proper guy-anchor grounding techniques. A properly dressed and installed ground wire prevents surge currents from welding turnbuckles and damaging safety loops. The perimeter ground connects to the tower base by way of a radial wire. (After [9].)

bolts to make the ground-wire attachment. Use at least two 3- to 4-in copper straps between the base of the tower and the buried ground system. Position the straps next to the concrete pier of the tower base. For towers more than 200 ft in height, use four copper straps, one on each side of the pier.

Figure 10.236 illustrates the proper way to bond guy wires to the buried ground system. The lead is dressed straight down from the topmost to the lowest guy. It should conform as close to vertical as possible, and be dressed downward from the lower side of each guy wire after connecting to each wire (Fig. 10.234). To ensure that no arcing will occur through the turnbuckle, a connection from the anchor plate to the perimeter ground circle is recommended. No. 2 gauge copper wire is recommended. This helps minimize the unavoidable inductance created by the conductor being in the air. Interconnect leads that are suspended in air must be dressed so that no bending radius is less than 8 in.

A *perimeter* ground, a circle of wire connected at several points to ground rods driven into the earth, should be installed around each guy-anchor point. The perimeter system provides a good ground for the anchor, and when tied together with the tower base radials, acts to rapidly dissipate lightning energy in the event of a flash. Tower base radials are buried wires interconnected with the tower base ground that extend away from the center point of the structure.

The required depth of the perimeter ground and the radials depends upon soil conductivity. Generally speaking, however, about 8 in. below grade is sufficient. In soil with good conductivity, the perimeter wire may be as small as no. 10 gauge. Because no. 2 gauge is required for the segment of conductor suspended in air, it may be easier to use no. 2 throughout. An added advantage is that the same size Cadweld molds may be used for all bonds.

Facility Ground Interconnection

Any radial that comes within 2 ft of a conductive structure must be tied into the ground system [9]. Bury the interconnecting wire, if possible, and approach the radial at a 45° angle, pointing toward the expected surge origin (usually the tower). Cadweld the conductor to the radial and to the structure.

For large-base, self-supporting towers, the radials should be split among each leg pad, as shown in Fig. 10.237. The radials may be brought up out of the soil (in air) and each attached at spaced locations

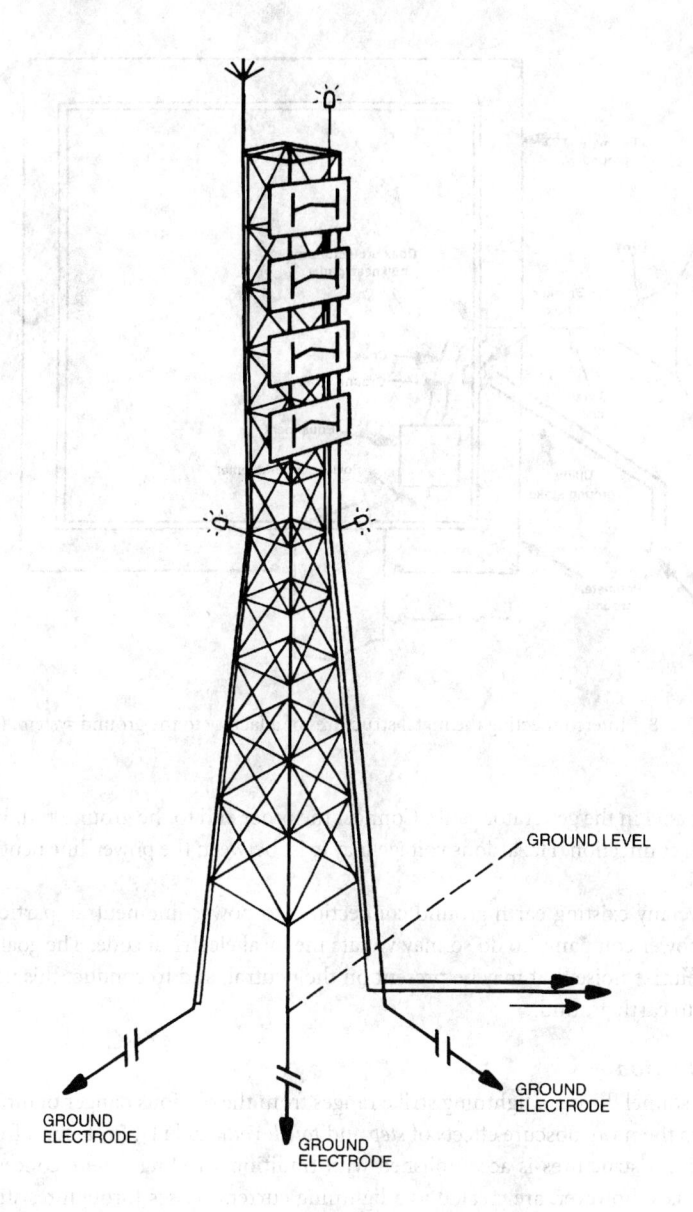

FIGURE 10.237   Interconnecting a self-supporting tower to the buried ground system.

around the foot pad. Some radials may have to be tied together first and then joined to the foot pad. Remember, if space between the radial lines can be maintained, less mutual inductance (coupling) will exist, and the system surge impedance will be lower.

It is desirable to have a continuous one-piece ring with rods around the equipment building. Connect this ring to one, and only one, of the tower radials, thus forming no radial loops. See Fig. 10.238. Bury the ring to the same depth as the radials to which it interconnects. Connect power-line neutral to the ring. *Warning*: Substantial current may flow when the power-line neutral is connected to the ring. Follow safety procedures when making this connection.

Install a ground rod (if the utility company has not installed one) immediately outside the generator or utility company vault, and connect this rod to the equipment building perimeter ground ring. Route a no. 1/0 insulated copper cable from the main power panel inside the generator vault to the ground rod outside the vault. Cut the cable to length, strip both ends, and tie one end to the power-line neutral at

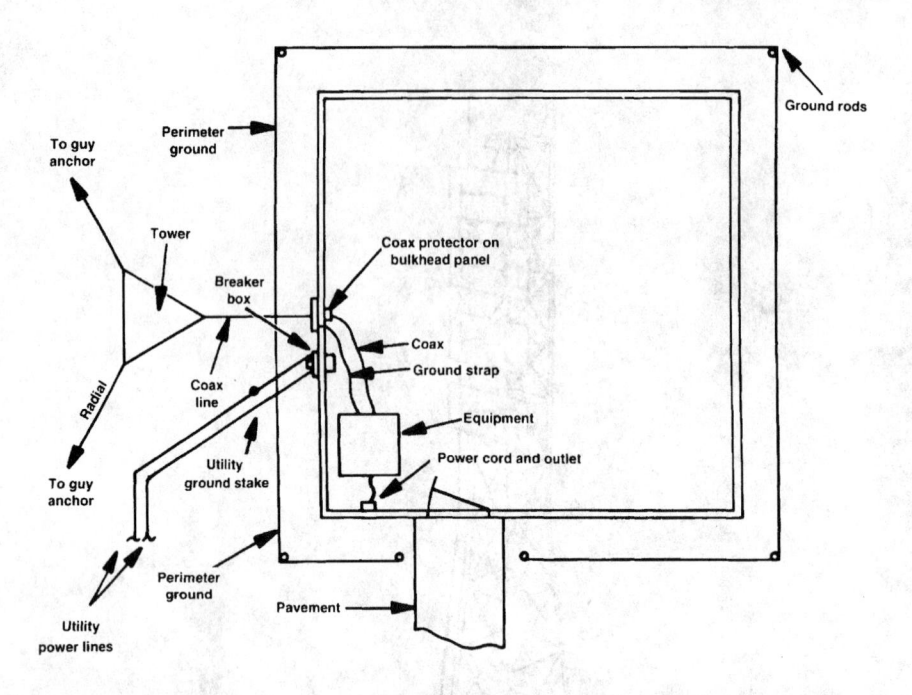

FIGURE 10.238   Interconnecting the metal structures of a facility to the ground system. (After [9].)

the main power panel in the generator vault. Connect the other end to the ground rod. *Warning:* Use care when making this connection. Hazardous voltage may exist between the power-line neutral and any point at earth potential.

Do not remove any existing earth ground connections to power-line neutral, particularly if they are installed by the power company. To do so may violate the local electrical code. The goal of this interconnection is to minimize noise that may be present on the neutral, and to conduct this noise as directly as possible outside to earth ground.

Personnel Protection

The threat to personnel during a lightning strike ranges from the obvious danger of direct contact with a lightning strike to the more obscure effects of *step* and *touch voltages* [11]. Protection from a direct strike when near or within structures is accomplished with traditional rolling sphere concept methods. Step and touch potentials, however, are created as a lightning current passes through resistive soil and other available paths as it dissipates into the earth. A person in contact with only one point of the gradient will simply rise and fall in potential with the gradient without injury. A person in contact with multiple points on the earth or objects at different potentials along the gradient, however, will become part of the current path and may sustain injury or death.

Figure 10.239 [13, 14] indicates a number of methods for protecting personnel from the direct and secondary effects of lightning. A typical tower/transmitter site is used as an example. A technician responding to a service problem during a thunderstorm would likely exit his or her vehicle outside the gate, unlock and open the gate, and move the vehicle into the inside yard. The technician would then leave the vehicle, and enter the building.

The threat of a direct lightning strike to the technician has been minimized by establishing a protective zone over the areas to be traversed. This zone is created by the tower and air terminals mounted atop light poles.

Step potentials are minimized through the use of a ground mat buried just below the surface of the area where the technician is expected to be outside the vehicle. Ground mats are commercially available, fabricated in a 6-in. × 6-in. square pattern using no. 8 AWG bare copper wire. Each intersection is welded,

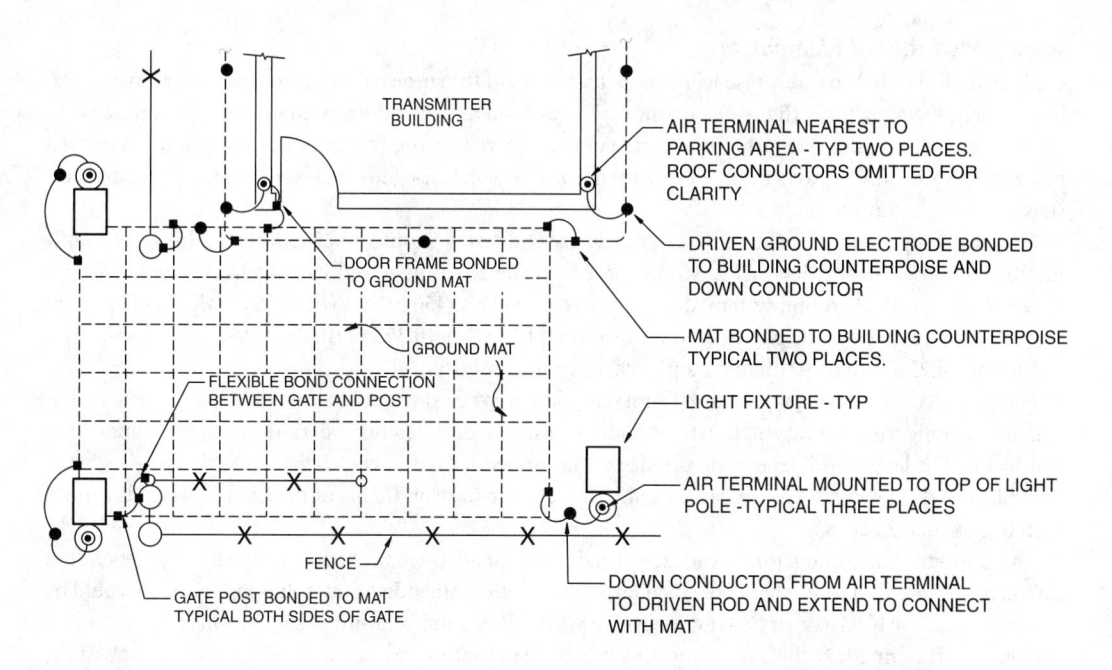

FIGURE 10.239   Protection methods for personnel at an exposed site. (After [11].)

creating, for all practical purposes, an *equipotential area* that short-circuits the step potential gradient in the area above the mat. The mat, as a whole, will rise and fall in potential as a result of the lightning current discharges; however, there will be very little difference in potential between the technician's feet. Mats should be covered with six in. of crushed stone or pavement.

The threat of dangerous touch potentials is minimized by bonding the following elements to the ground system:

- Personnel ground mat
- Fence at each side of the gate opening
- Door frame of the transmitter building
- Flexible bonding connection between the swing gate and its terminal post

Such bonding will ensure that the object being touched by the technician is at or near the same potential as his or her feet.

Bonding both sides of the gate opening to the mat helps to ensure that the technician and both sides of the gate are at approximately the same potential while the gate is being handled. The flexible bond between the gate and its support post can be accomplished using a commercially available kit or by Cadwelding a short length of flexible 2/0 AWG welding cable between the two elements.

Grounding on Bare Rock

A bare rock mountaintop location provides special challenges to the facility design engineer [9]. There is no soil, thus there are no ground rods. Radials are the only means to develop a ground system. Install a large number of radials, laid straight, but not too taut. The portions not in contact with the rock are in air and form an inductance that will choke the surge current. Because rock is not conductive when it is dry, keep the radials short. Only a test measurement will determine how short the radials should be. A conventional earth-resistance tester will tell only half the story (besides, ground rods cannot be placed in rock for such a measurement). A dynamic ground tester offers the only way to obtain the true surge impedance of the system.

Rock-Based Radial Elements

On bare rock, a radial counterpoise will conduct and spread the surge charge over a large area. In essence, it forms a leaky capacitor with the more conductive earth on or under the mountain [9]. The conductivity of the rock will be poor when dry, but quite good when wet. If the site experiences significant rainfall before a lightning flash, protection will be enhanced. The worst case, however, must be assumed: an early strike under dry conditions.

The surge impedance, measured by a dynamic ground tester, should be 25 ω or less. This upper-limit number is chosen so that less stress will be placed on the equipment and its surge protectors. With an 18-kA strike to a 25 Ω ground system, the entire system will rise 450 kV above the rest of the world at peak current. This voltage has the potential to jump almost 15.750 in. (0.35 in./10 kV at standard atmospheric conditions of 25°C, 30 in. of mercury and 50% relative humidity).

For nonsoil conditions, tower anchor points should have their own radial systems or be encapsulated in concrete. Configure the encapsulation to provide at least 3 in. of concrete on all sides around the embedded conductor. The length will depend on the size of the embedded conductor. Rebar should extend as far as possible into the concrete. The dynamic ground impedance measurements of the anchor grounds should each be less than 25 Ω .

The size of the bare conductor for each tower radial (or for an interconnecting wire) will vary, depending on soil conditions. On rock, a bare no. 1/0 or larger wire is recommended. Flat, solid-copper strap would be better, but may be blown or ripped if not covered with soil. If some amount of soil is available, no. 6 cable should be sufficient. Make the interconnecting radial wires continuous, and bury them as deep as possible; however, the first 6 to 10 in. will have the most benefit. Going below 18 in. will not be cost-effective, unless in a dry, sandy soil where the water table can be reached and ground-rod penetration is shallow. If only a small amount of soil exists, use it to cover the radials to the extent possible. It is more important to cover radials in the area near the tower than at greater distances. If, however, soil exists only at the outer distances and cannot be transported to the inner locations, use the soil to cover the outer portions of the radials.

If soil is present, install ground rods along the radial lengths. Spacing between ground rods is affected by the depth that each rod is driven; the shallower the rod, the closer the allowed spacing. Because the ultimate depth a rod can be driven cannot always be predicted by the first rod driven, use a maximum spacing of 15 ft when selecting a location for each additional rod. Drive rods at building corners first (within 24 in. but not closer than 6 in. to a concrete footer unless that footer has an encapsulated Ufer ground), then fill in the space between the corners with additional rods.

Drive the ground rods in place. Do not auger; set in place, then back-fill. The soil compactness is never as great on augured-hole rods when compared with driven rods. The only exception is when a hole is augured or blasted for a ground rod or rebar and then back-filled in concrete. Because concrete contains lime (alkali base) and is porous, it absorbs moisture readily, giving it up slowly. Electron carriers are almost always present, making the substance a good conductor.

If a Ufer ground is not being implemented, the radials may be Cadwelded to a subterranean ring, with the ring interconnected to the tower foot pad via a minimum of three no. 1/0 wires spaced at 120° angles and Cadwelded to the radial ring.

10.10.3 Transmission-System Grounding

Nature can devastate a communications site. Lightning can seriously damage an unprotected facility with little or no warning, leaving an expensive and time-consuming repair job. The first line of defense is proper grounding of the communications system.

Transmission Line

All coax and waveguide lines must include grounding kits for bonding the transmission line at the antenna [9]. On conductive structures, this can be accomplished by bonding the tail of the grounding kit to the structure itself. Remove all nonconductive paint and corrosion before attachment. Do not drill holes, and do not loosen any existing tower member bolts. Antenna clamping hardware can be used, or an all-stainless-steel hose clamp of the appropriate size can be substituted. The location of the tower-top ground is not as critical as the bottom grounding kit.

On nonconductive structures, a no. 1/0 or larger wire must be run down the tower [9]. Bond the transmission-line grounding kit to this down-run. Keep the wire as far away from all other conductive runs (aviation lights, coax, and waveguide) as possible. Separation of 2 ft is preferred; 18 in. is the minimum. If any other ground lines, conduit, or grounded metallic-structure members must be traversed that are closer than 18 in., they too must be grounded to the down-lead ground line to prevent flashover.

At the point where the coax or waveguide separates from the conductive structure (metal tower), a coax or waveguide grounding kit must be installed. Secure the connection to a large vertical structure member with a small number of joints. Attach to a structural member as low as possible on the tower.

Dress the grounding kit tails in a nearly straight 45° downward angle to the tower member. On nonconductive structures, a metal busbar must be used. Ground the bar to one or more of the vertical downconductors, and as close to ground level as possible.

Coaxial cables, lighting conduit, and other lines on the tower must be secured properly to the structure. Figure 10.240 illustrates several common attachment methods.

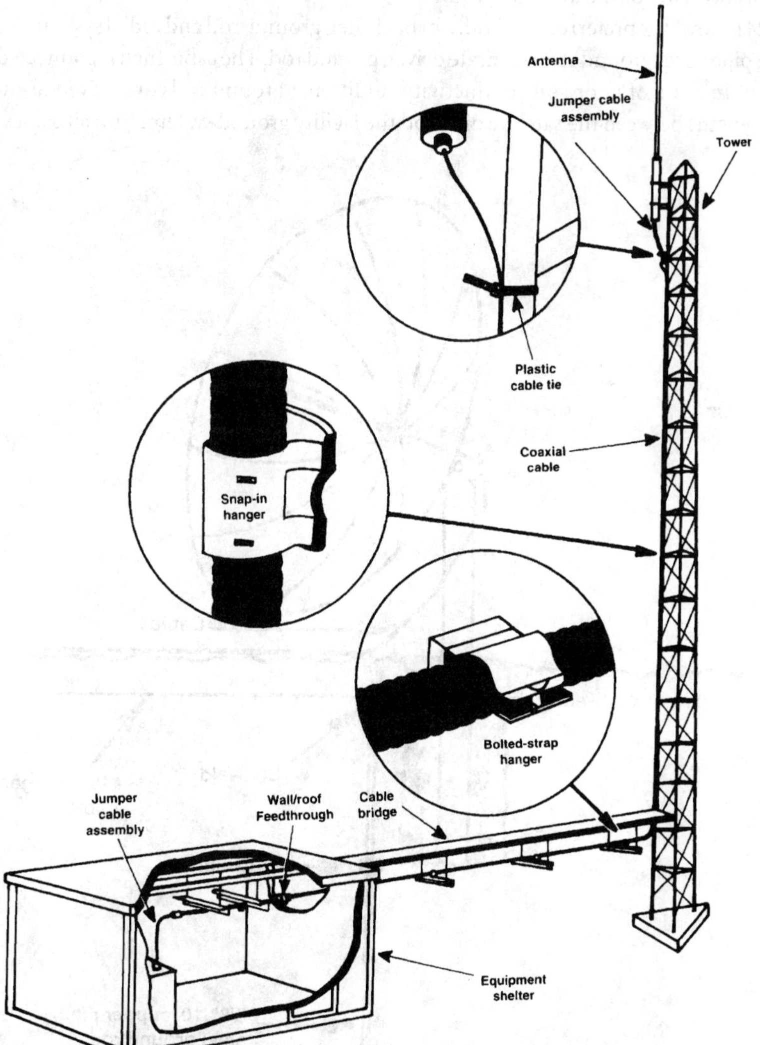

FIGURE 10.240 Transmission-line mounting and grounding procedures for a communications site.

Cable Considerations

Ground-strap connections must withstand weathering and maintain low electrical resistance between the grounded component and earth [9]. Corrosion impairs ground-strap performance. Braided-wire ground straps should not be used in outside installations. Through capillary action resembling that of a wick, the braid conveys water, which accelerates corrosion. Eventually, advancing corrosion erodes the ground-strap cable. Braid also can act as a duct that concentrates water at the bond point. This water speeds corrosion, which increases the electrical resistance of the bond. A jacketed, seven-strand copper wire strap (no. 6 to no. 2) is recommended for transmission-line grounding at the tower.

Satellite Antenna Grounding

Most satellite receiving/transmitting antenna piers are encapsulated in concrete [9]. Consideration should be given, therefore, to implementing a Ufer ground for the satellite dish. A 4-in.-diameter pipe, submerged 4 to 5 ft in an 18-in.-diameter (augered) hole will provide a good start for a Ufer-based ground system. It should be noted that an augered hole is preferred because digging and repacking the soil around the pier will create higher ground resistance. In areas of good soil conductivity (100 Ω/m or less), this basic Ufer system may be adequate for the antenna ground.

Figure 10.241 shows the preferred method: a hybrid Ufer/ground-rod and radial system. A cable connects the mounting pipe (Ufer ground) to a separate driven ground rod. The cable then is connected to the facility ground system. In areas of poor soil conductivity, additional ground rods are driven at increments (2.2 times the rod length) between the satellite dish and the facility ground system. Run all cables underground

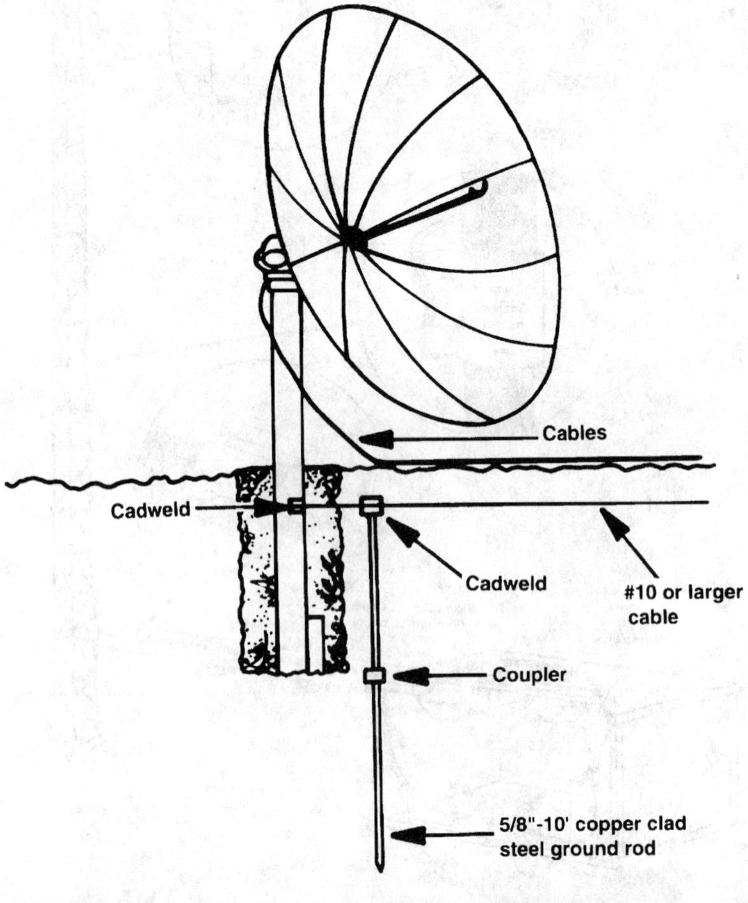

FIGURE 10.241   Grounding a satellite receiving antenna. (After [9].)

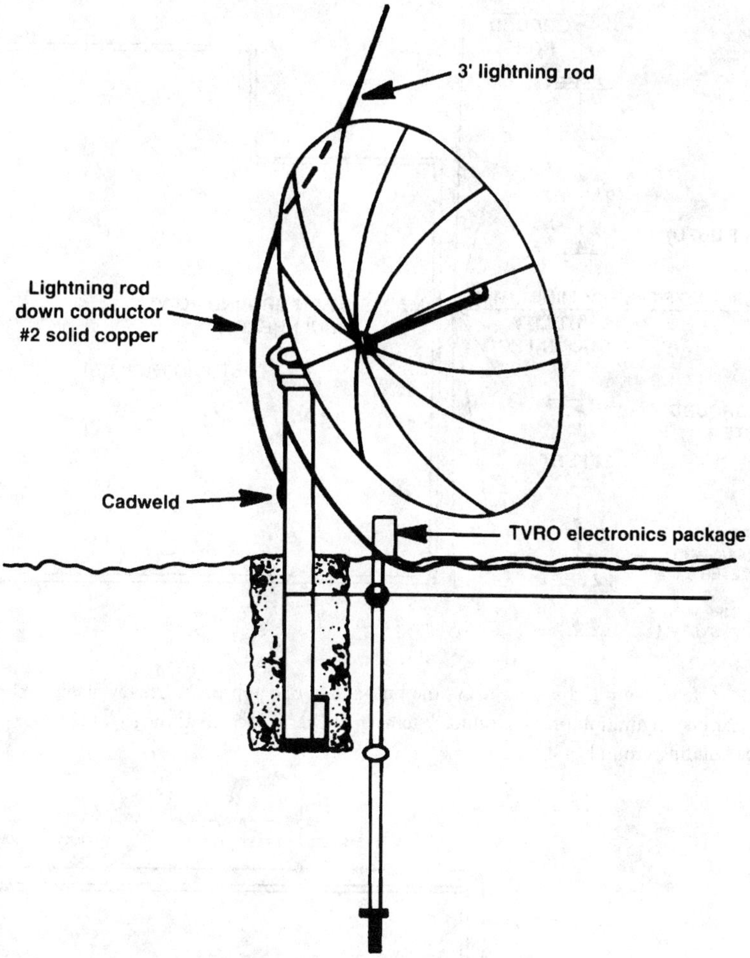

FIGURE 10.242 Addition of a lightning rod to a satellite antenna ground system. (After [9].)

for best performance. Make the interconnecting copper wire no. 10 size or larger; bury the wire at least 8 in. below finished grade. Figure 10.242 shows the addition of a lightning rod to the satellite dish.

10.10.4 Designing a Building Ground System

After the required grounding elements have been determined, they must be connected into a unified system [9]. Many different approaches can be taken, but the goal is the same: establish a low-resistance, low-inductance path to surge energy. Figure 10.243 shows a building ground system using a combination of ground rods and buried bare-copper radial wires. This design is appropriate when the building is large or located in an urban area. This approach also can be used when the facility is located in a highrise building that requires a separate ground system. Most newer office buildings have ground systems designed into them. If a comprehensive building ground system is provided, use it. For older structures (constructed of wood or brick), a separate ground system will be needed.

Figure 10.244 shows another approach in which a perimeter ground strap is buried around the building and ground rods are driven into the earth at regular intervals (2.2 times the rod length). The ground ring consists of a one-piece copper conductor that is bonded to each ground rod.

If a transmission or microwave tower is located at the site, connect the tower ground system to the main ground point via a copper strap. The width of the strap must be at least 1% of the length and, in any

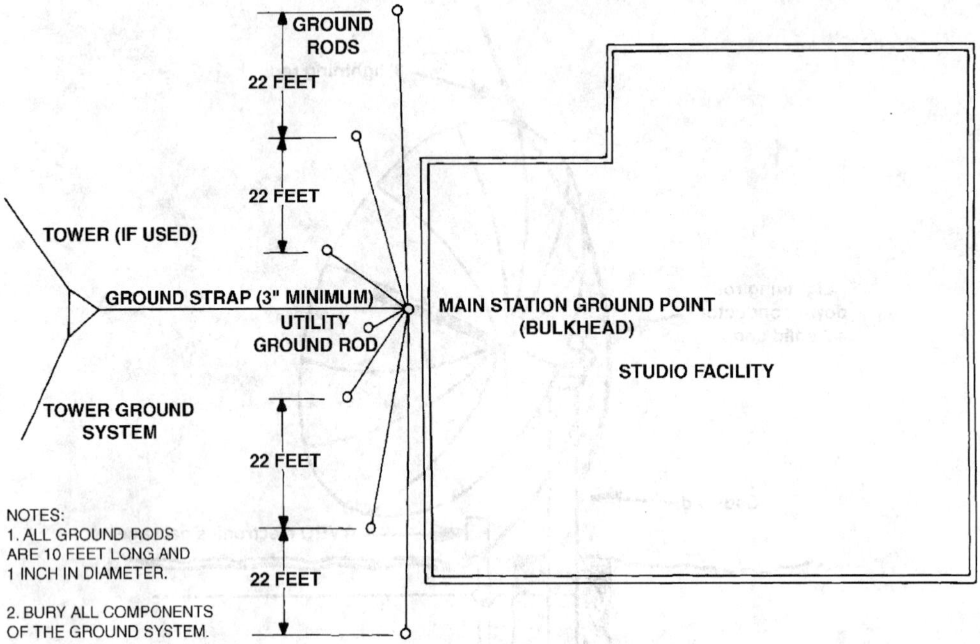

FIGURE 10.243 A facility ground system using the hub-and-spoke approach. The available real estate at the site will dictate the exact configuration of the ground system. If a tower is located at the site, the tower ground system is connected to the building ground as shown.

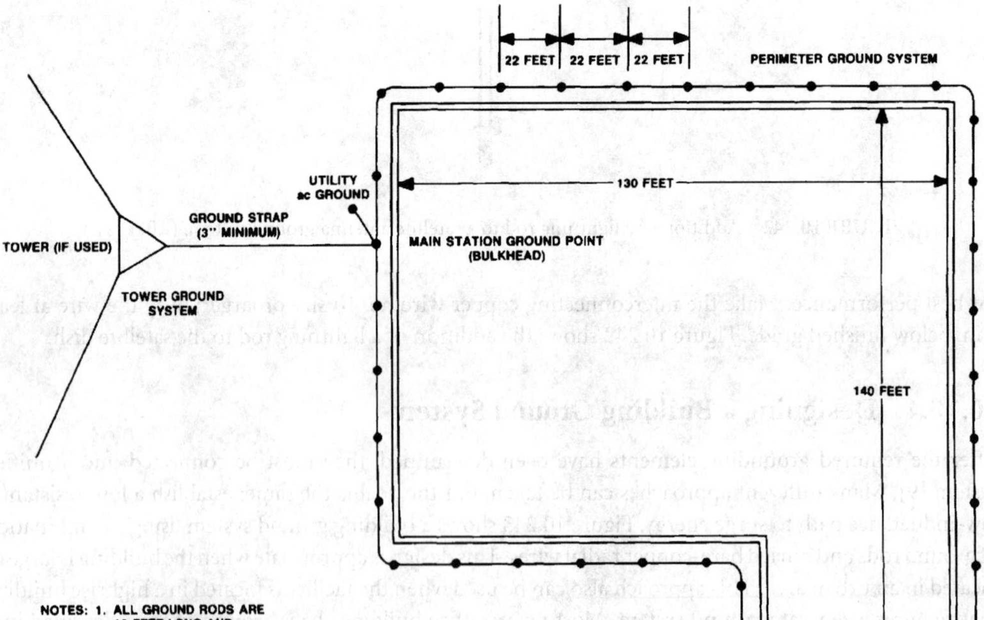

FIGURE 10.244 Facility ground using a perimeter ground-rod system. This approach works well for buildings with limited available real estate.

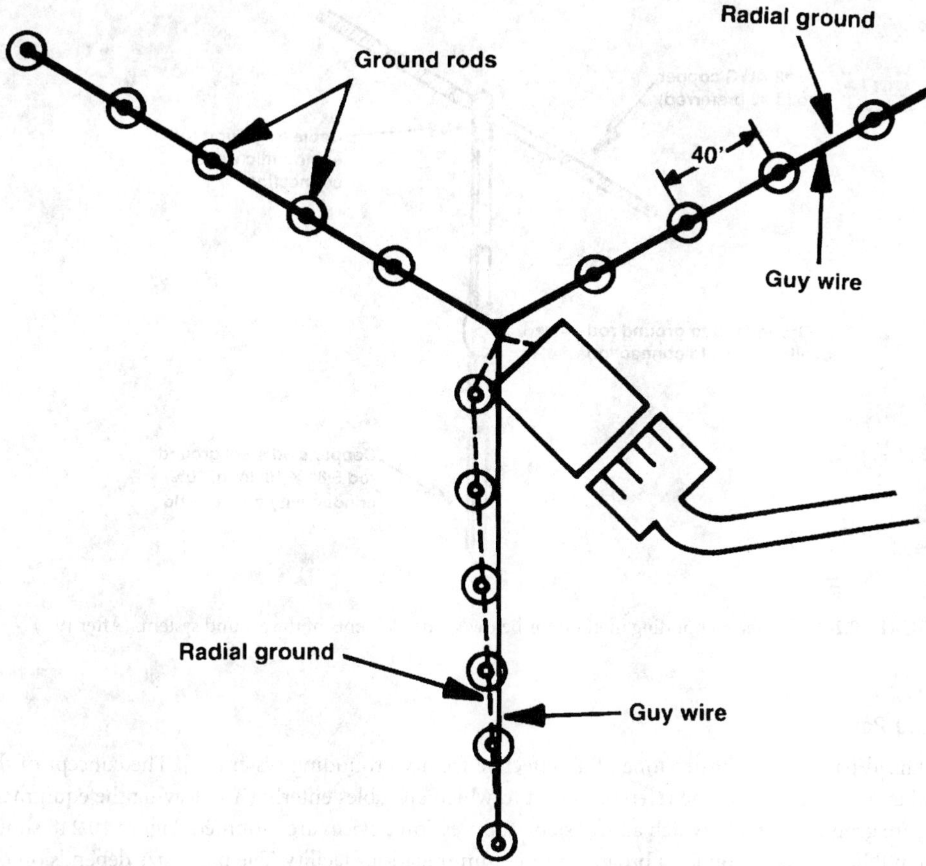

FIGURE 10.245 A typical guy-anchor and tower-radial grounding scheme. The radial ground is no. 6 copper wire. The ground rods are $^5/_8$ in. $\times 10$ ft. (After [9].)

event, not less than 3 in. wide. The building ground system is not a substitute for a tower ground system, no matter what the size of the tower. The two systems are treated as independent elements, except for the point at which they interconnect.

Connect the utility company power-system ground rod to the main facility ground point as required by the local electrical code. Do not consider the building ground system to be a substitute for the utility company ground rod. The utility rod is important for safety reasons and must not be disconnected or moved. Do not remove any existing earth ground connections to the power-line neutral connection. To do so may violate local electrical code.

Bury all elements of the ground system to reduce the inductance of the overall network. Do not make sharp turns or bends in the interconnecting wires. Straight, direct wiring practices will reduce the overall inductance of the system and increase its effectiveness in shunting fast-rise-time surges to earth. Figure 10.245 illustrates the interconnection of a tower and building ground system. In most areas, soil conductivity is high enough to permit rods to be connected with no. 6 bare-copper wire or larger. In areas of sandy soil, use copper strap. A wire buried in low-conductivity, sandy soil tends to be inductive and less effective in dealing with fast-rise-time current surges. As stated previously, make the width of the ground strap at least 1% of its overall length. Connect buried elements of the system as shown in Fig. 10.246.

For small installations with a low physical profile, a simplified grounding system can be implemented, as shown in Fig. 10.247. A grounding plate is buried below grade level, and a ground wire ties the plate to the microwave tower mounted on the building.

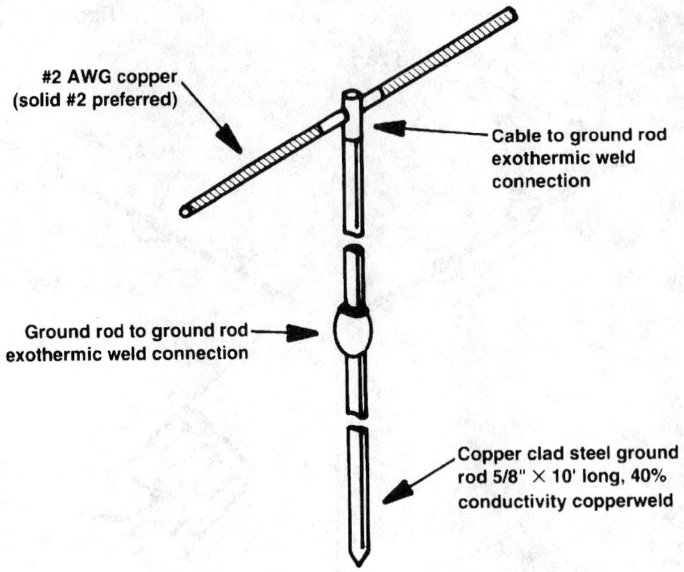

FIGURE 10.246   Preferred bonding method for below-grade elements of the ground system. (After [9].)

Bulkhead Panel

The bulkhead panel is the cornerstone of an effective facility grounding system [9]. The concept of the bulkhead is simple: establish one reference point to which all cables entering and leaving the equipment building are grounded and to which all transient-suppression devices are mounted. Figure 10.248 shows a typical bulkhead installation for a broadcast or communications facility. The panel size depends on the spacing, number, and dimensions of the coaxial lines, power cables, and other conduit entering or leaving the building.

To provide a weatherproof point for mounting transient-suppression devices, the bulkhead can be modified to accept a subpanel, as shown in Fig. 10.249. The subpanel is attached so that it protrudes through an opening in the wall and creates a secondary plate on which transient suppressors are mounted and grounded. A typical cable/suppressor-mounting arrangement for a communications site is shown in Fig. 10.250. To handle the currents that may be experienced during a lightning strike or large transient on the utility company AC line, the bottommost subpanel flange (which joins the subpanel to the main bulkhead) must have a total surface-contact area of at least 0.75 in.^{2} per transient suppressor.

Because the bulkhead panel will carry significant current during a lightning strike or AC line disturbance, it must be constructed of heavy material. The recommended material is 1/8-in. C110 (solid copper) 1/2-hard. This type of copper stock weighs nearly 5-1/2 lb/ft^2 and is rather expensive. Installing a bulkhead, while sometimes difficult and costly, will pay dividends for the life of the facility. Use 18-8 stainless-steel mounting hardware to secure the subpanel to the bulkhead.

Because the bulkhead panel establishes the central grounding point for all equipment within the building, it must be tied to a low-resistance (and low-inductance) perimeter ground system. The bulkhead establishes the *main facility ground point*, from which all grounds inside the building are referenced. A typical bulkhead installation for a small communications site is shown in Fig. 10.251.

Bulkhead Grounding

A properly installed bulkhead panel will exhibit lower impedance and resistance to ground than any other equipment or cable grounding point at the facility [9]. Waveguide and coax line grounding kits should be installed at the bulkhead panel as well as at the tower. Dress the kit tails downward at a straight 45° angle using 3/8-in. stainless-steel hardware to the panel. Position the stainless-steel lug at the tail end, flat

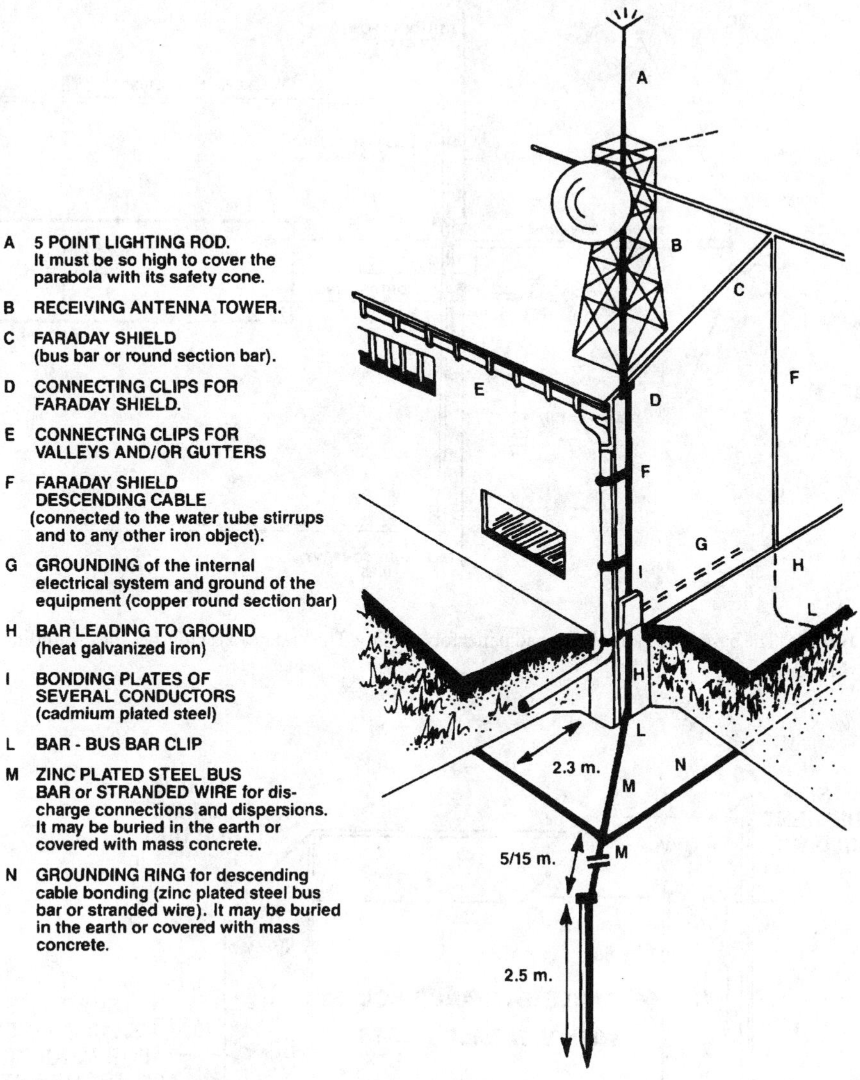

A **5 POINT LIGHTING ROD.**
It must be so high to cover the parabola with its safety cone.

B **RECEIVING ANTENNA TOWER.**

C **FARADAY SHIELD**
(bus bar or round section bar).

D **CONNECTING CLIPS FOR FARADAY SHIELD.**

E **CONNECTING CLIPS FOR VALLEYS AND/OR GUTTERS**

F **FARADAY SHIELD DESCENDING CABLE**
(connected to the water tube stirrups and to any other iron object).

G **GROUNDING** of the internal electrical system and ground of the equipment (copper round section bar)

H **BAR LEADING TO GROUND**
(heat galvanized iron)

I **BONDING PLATES OF SEVERAL CONDUCTORS**
(cadmium plated steel)

L **BAR - BUS BAR CLIP**

M **ZINC PLATED STEEL BUS BAR or STRANDED WIRE** for discharge connections and dispersions. It may be buried in the earth or covered with mass concrete.

N **GROUNDING RING** for descending cable bonding (zinc plated steel bus bar or stranded wire). It may be buried in the earth or covered with mass concrete.

FIGURE 10.247 Grounding a small microwave transmission tower.

against a cleaned spot on the panel. Joint compound will be needed for aluminum and is recommended for copper panels.

Because the bulkhead panel will be used as the central grounding point for all the equipment inside the building, the lower the inductance to the perimeter ground system, the better. The best arrangement is to simply extend the bulkhead panel down the outside of the building, below grade, to the perimeter ground system. This will give the lowest resistance and the smallest inductive voltage drop. This approach is illustrated in Fig. 10.252.

If cables are used to ground the bulkhead panel, secure the interconnection to the outside ground system along the bottom section of the panel. Use multiple no. 1/0 or larger copper wire or several solid-copper straps. If using strap, attach with stainless-steel hardware, and apply joint compound for aluminum bulkhead panels. Clamp and Cadweld, or silver-solder for copper/brass panels. If no. 1/0 or larger wire is used, employ crimp lug and stainless-steel hardware. Measure the DC resistance. It should be less than 0.01 Ω between the ground system and the panel. Repeat this measurement on an annual basis.

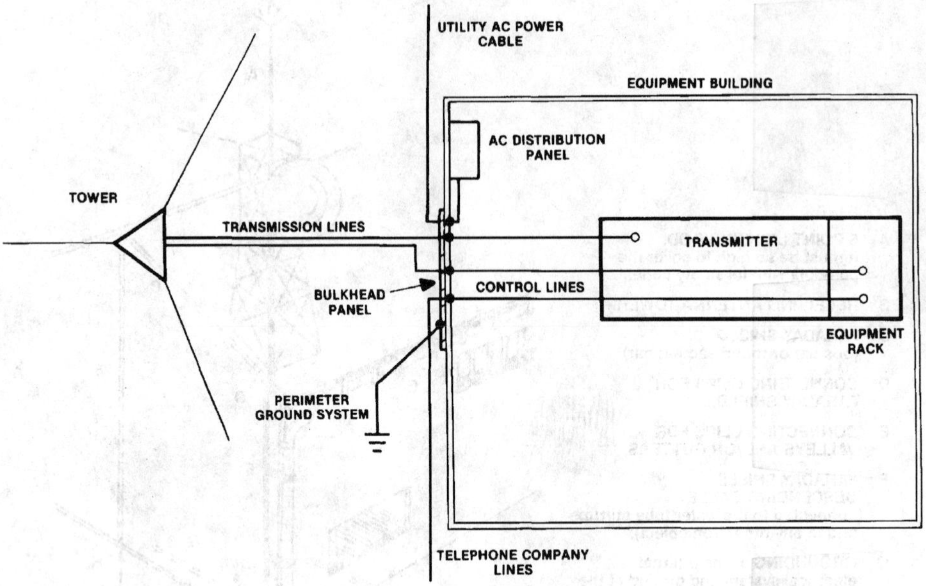

FIGURE 10.248 The basic design of a bulkhead panel for a facility. The bulkhead establishes the grounding reference point for the plant.

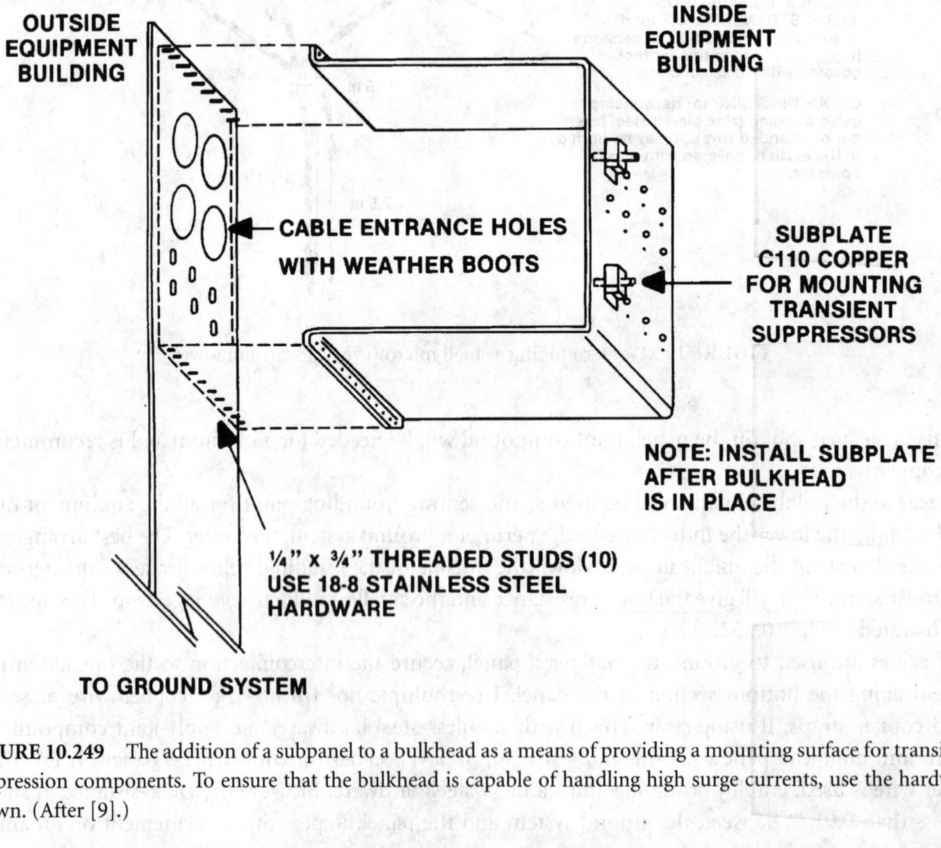

FIGURE 10.249 The addition of a subpanel to a bulkhead as a means of providing a mounting surface for transient-suppression components. To ensure that the bulkhead is capable of handling high surge currents, use the hardware shown. (After [9].)

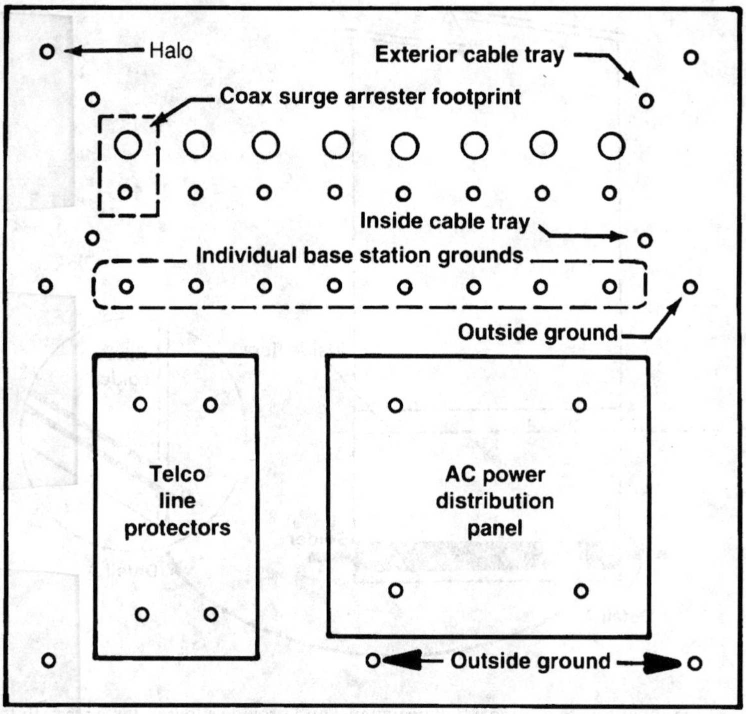

FIGURE 10.250 Mounting-hole layout for a communications site bulkhead subpanel.

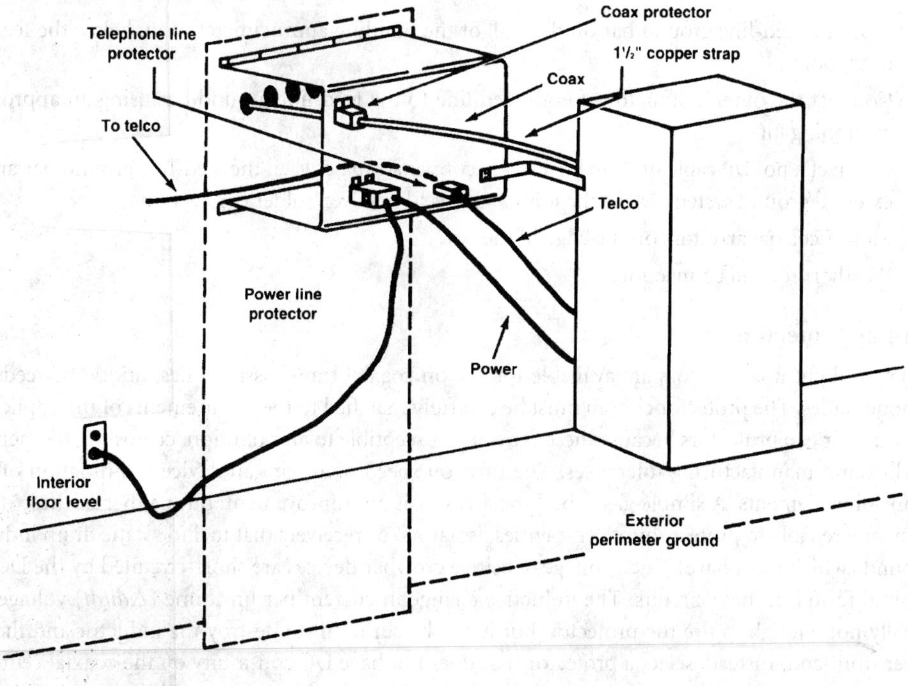

FIGURE 10.251 Bulkhead installation at a small communications site. (After [9].)

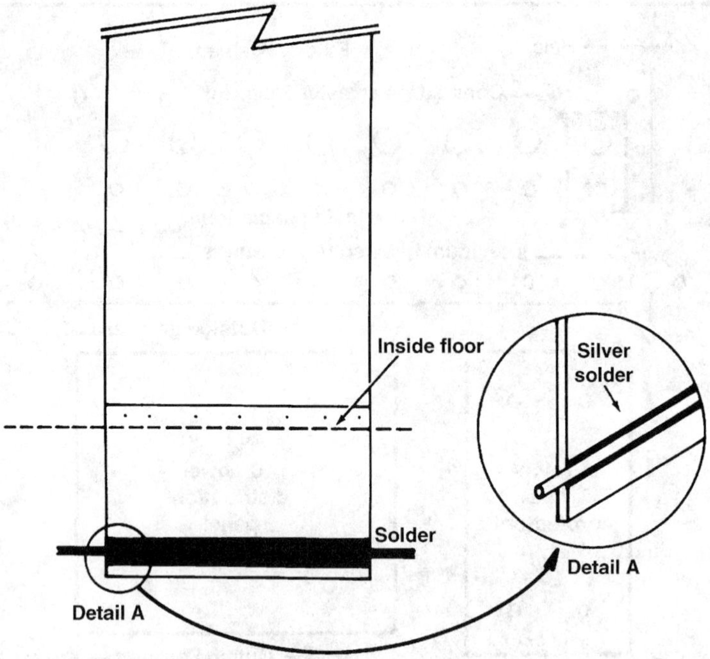

FIGURE 10.252   The proper way to ground a bulkhead panel and provide a low-inductance path for surge currents stripped from cables entering and leaving the facility. The panel extends along the building exterior to below grade. It is silver-soldered to a no. 2/0 copper wire that interconnects with the outside ground system. (After [9].)

If the antenna feed lines do not enter the equipment building via a bulkhead panel, treat them in the following manner:

- Mount a feed-line ground bar on the wall of the building approximately 4 in. below the feed-line entry point.
- Connect the outer conductor of each feed line to the feed-line ground bar using an appropriate grounding kit.
- Connect a no. 1/0 cable or 3- to 6-in.-wide copper strap between the feed-line ground bar and the external ground system. Make the joint a Cadweld or silver-solder connection.
- Mount coaxial arrestors on the edge of the bar.
- Weatherproof all connections.

Lightning Protectors

A variety of lightning arrestors are available for use on coaxial transmission lines, utility AC feeds, and telephone cables. The protector chosen must be carefully matched to the requirements of the application. Do not use air gap protectors because these types are susceptible to air pollution, corrosion, temperature, humidity, and manufacturing tolerances. The turn-on speed of an air gap device is a function of all of the foregoing elements. A simple gas-tube-type arrestor is an improvement, but neither of these devices will operate reliably to protect shunt-fed cavities, isolators, or receivers that include static drain inductors to ground (which most have). Such voltage-sensitive crowbar devices are short-circuited by the DC path to ground found in these circuits. The inductive change in current per unit time ($\Delta di/dt$) voltage drop is usually not enough to fire the protector, but it can be sufficient to destroy the inductor and then the receiver front end. Instead, select a protector that does not have DC continuity on the coaxial center pin from input to output. Such units have a series capacitor between the gas tube and the equipment center pin that will allow the voltage to build up so that the arrestor can fire properly.

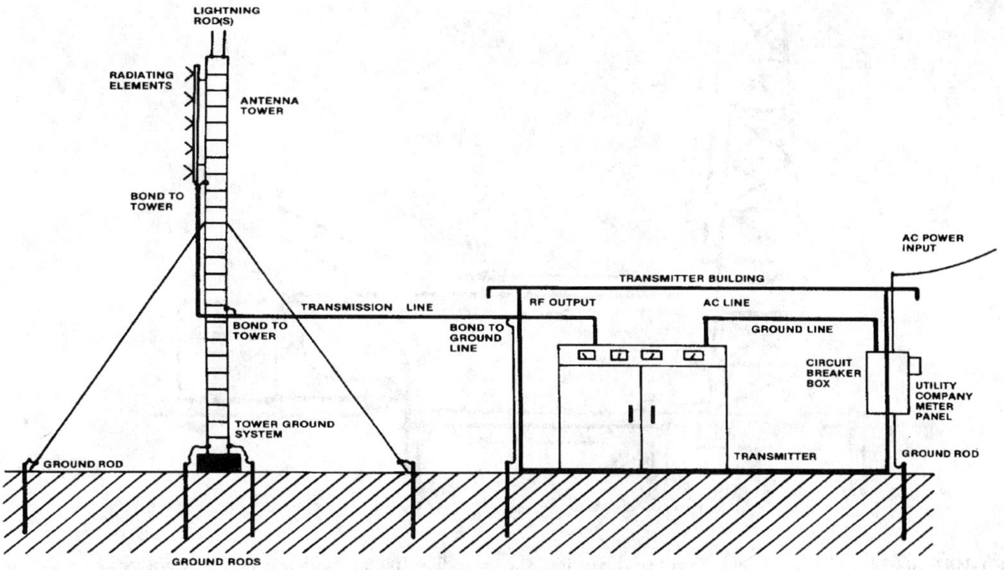

FIGURE 10.253 A common, but not ideal, grounding arrangement for a transmission facility using a grounded tower. A better configuration involves the use of a bulkhead panel through which all cables pass into and out of the equipment building.

Typical Installation

Figure 10.253 illustrates a common grounding arrangement for a remotely located grounded-tower (FM, TV, or microwave radio) transmitter plant. The tower and guy wires are grounded using 10-ft-long copper-clad ground rods. The antenna is bonded to the tower, and the transmission line is bonded to the tower at the point where it leaves the structure and begins the horizontal run into the transmitter building. Before entering the structure, the line is bonded to a ground rod through a connecting cable. The transmitter itself is grounded to the transmission line and to the AC power-distribution system ground. This, in turn, is bonded to a ground rod where the utility feed enters the building. The goal of this arrangement is to strip all incoming lines of damaging overvoltages before they enter the facility. One or more lightning rods are mounted at the top of the tower structure. The rods extend at least 10 ft above the highest part of the antenna assembly.

Such a grounding configuration, however, has built-in problems that can make it impossible to provide adequate transient protection to equipment at the site. Look again at the example. To equipment inside the transmitter building, two grounds actually exist: the utility company ground and the antenna ground. One ground will have a lower resistance to earth, and one will have a lower inductance in the connecting cable or copper strap from the equipment to the ground system.

Using the Fig. 10.253 example, assume that a transient overvoltage enters the utility company meter panel from the AC service line. The overvoltage is clamped by a protection device at the meter panel, and the current surge is directed to ground. But *which ground*, the utility ground or the antenna ground?

The utility ground surely will have a lower inductance to the current surge than the antenna ground, but the antenna probably will exhibit a lower resistance to ground than the utility side of the circuit. Therefore, the surge current will be divided between the two grounds, placing the transmission equipment in series with the surge suppressor and the antenna ground system. A transient of sufficient potential will damage the transmission equipment.

Transients generated on the antenna side because of a lightning discharge are no less troublesome. The tower is a conductor, and any conductor is also an inductor. A typical 150-ft self-supporting tower may exhibit as much as 40 μH inductance. During a fast-rise-time lightning strike, an instantaneous voltage drop of 360 kV between the top of the tower and the base is not unlikely. If the coax shield is bonded to

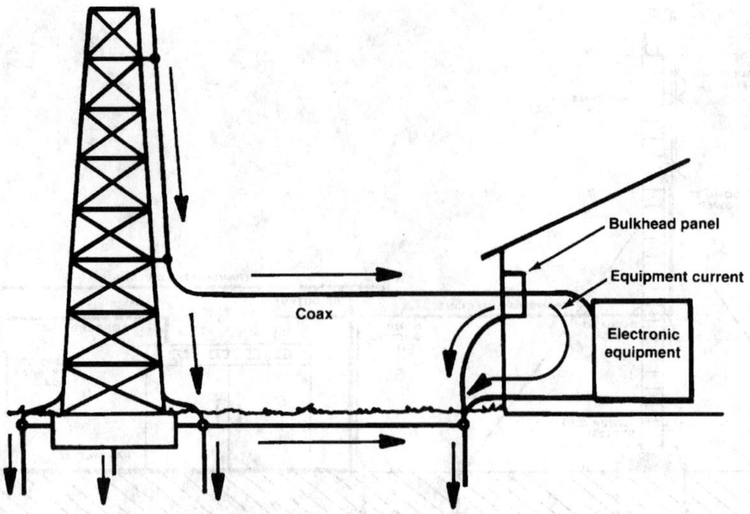

FIGURE 10.254 The equivalent circuit of the facility shown in Fig. 10.253. Note the discharge current path through the electronic equipment.

the tower 15 ft above the earth (as shown in the previous figure), 10% of the tower voltage drop (36 kV) will exist at that point during a flash. Figure 10.254 illustrates the mechanisms involved.

The only way to ensure that damaging voltages are stripped off all incoming cables (coax, AC power, and telephone lines) is to install a bulkhead entrance panel and tie all transient-suppression hardware to it. Configuring the system as shown in Fig. 10.255 strips away all transient voltages through the use of a single-point ground. The bulkhead panel is the ground reference for the facility. With such a design, secondary surge current paths do not exist, as illustrated in Fig. 10.256.

Protecting the building itself is another important element of lightning surge protection. Figure 10.257 shows a common system using multiple lightning rods. As specified by NFPA 78, the dimensions given in the figure are as follows:

- A = 50-ft maximum spacing between air terminals
- B = 150-ft maximum length of a coursing conductor permitted without connection to a main perimeter or downlead conductor
- C = 20- or 25-ft maximum spacing between air terminals along an edge

Checklist for Proper Grounding

A methodical approach is necessary in the design of a facility ground system. Consider the following points:

1. Install a bulkhead panel to provide mechanical support, electric grounding, and lightning protection for coaxial cables, power feeds, and telephone lines entering the equipment building.
2. Install an internal ground bus using no. 2 or larger solid-copper wire. (At transmission facilities, use copper strap that is at least 3 in. wide.) Form a *star* grounding system. At larger installations, form a *star-of-stars* configuration. Do not allow ground loops to exist in the internal ground bus. Connect the following items to the building internal ground system:

- Chassis racks and cabinets of all hardware
- All auxiliary equipment
- Battery charger
- Switchboard
- Conduit
- Metal raceway and cable tray

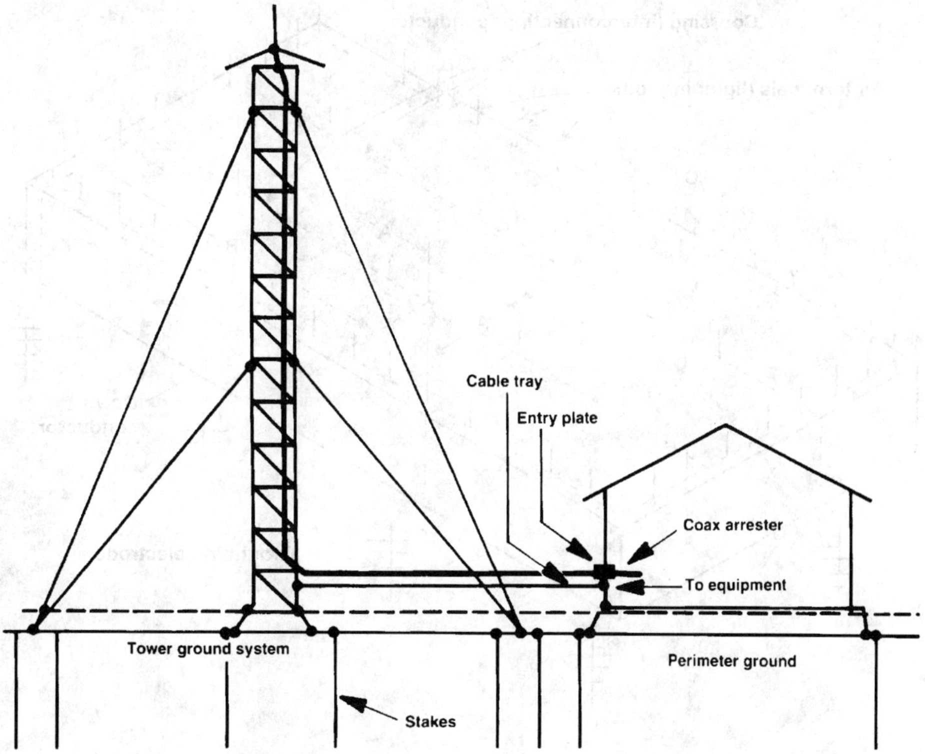

FIGURE 10.255 The preferred grounding arrangement for a transmission facility using a bulkhead panel. With this configuration, all damaging transient overvoltages are stripped off the coax, power, and telephone lines before they can enter the equipment building.

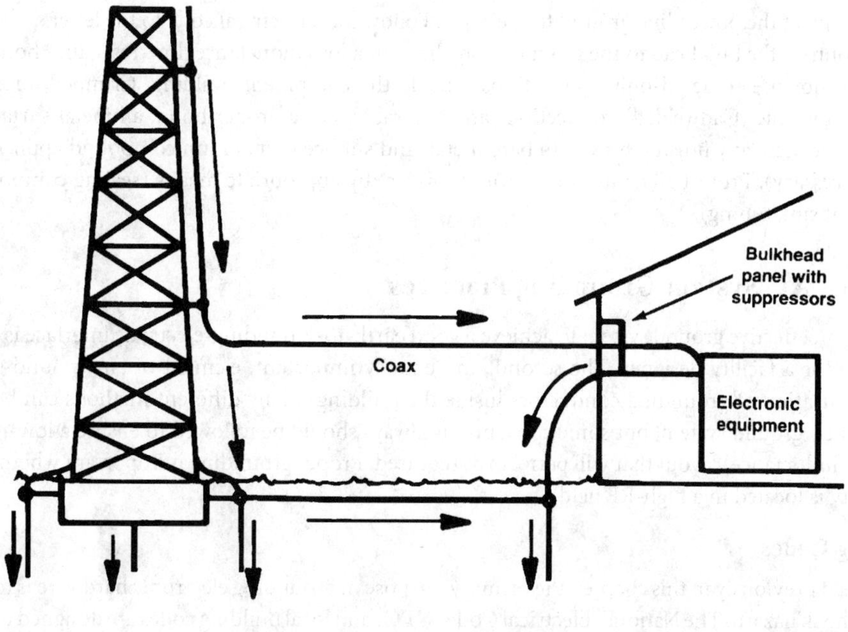

FIGURE 10.256 The equivalent circuit of the facility shown in Fig. 10.255. Discharge currents are prevented from entering the equipment building.

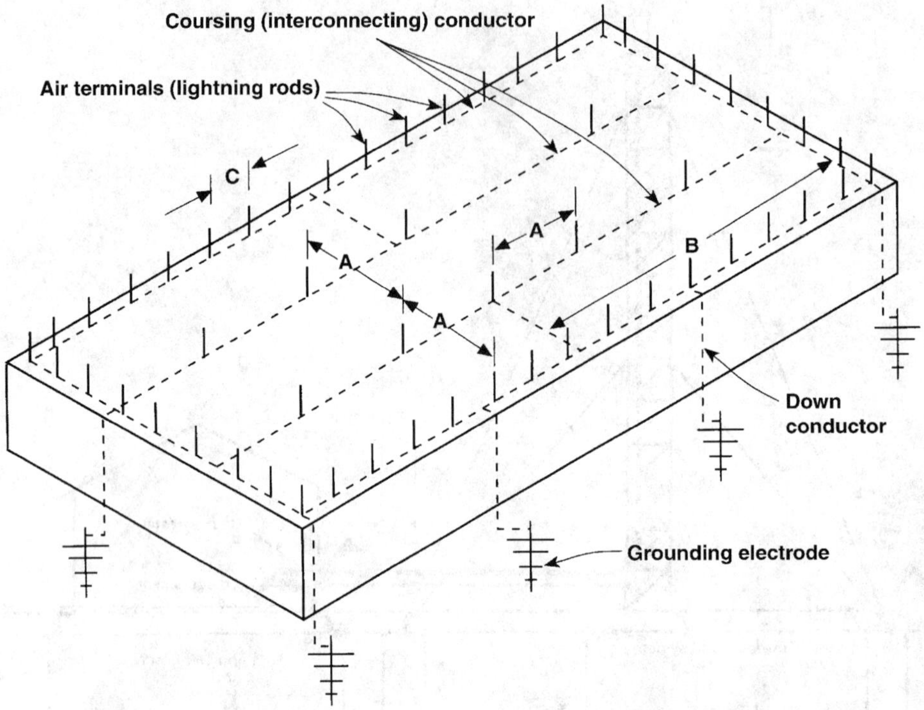

FIGURE 10.257   Conduction lightning-protection system for a large building. (After [12].)

3. Install a tower earth ground array by driving ground rods and laying radials as required to achieve a low earth ground impedance at the site.

4. Connect outside metal structures to the earth ground array (towers, metal fences, metal buildings, and guy-anchor points).

5. Connect the power-line ground to the array. Follow local electrical code to the letter.

6. Connect the bulkhead to the ground array through a low-inductance, low-resistance bond.

7. Do not use soldered-only connections outside the equipment building. Crimped, brazed, and exothermic (Cadwelded) connections are preferable. For a proper bond, all metal surfaces must be cleaned, any finish removed to bare metal, and surface preparation compound applied (where necessary). Protect all connections from moisture by appropriate means (sealing compound and heat sink tubing).

10.10.5   AC System Grounding Practices

Installing an effective ground system to achieve a good earth-to-grounding-electrode interface is only half the battle for a facility designer. The second, and equally important, element of any ground system is the configuration of grounding conductors inside the building. Many different methods can be used to implement a ground system, but some conventions always should be followed to ensure a low-resistance (and low-inductance) layout that will perform as required. Proper grounding is important whether or not the facility is located in a high-RF field.

Building Codes

As outlined previously in this chapter, the primary purpose of grounding electronic hardware is to prevent electric shock hazard. The National Electrical Code (NEC) and local building codes are designed to provide for the safety of the workplace. Local codes should always be followed. Occasionally, code sections are open to some interpretation. When in doubt, consult a field inspector. Codes constantly are being changed or

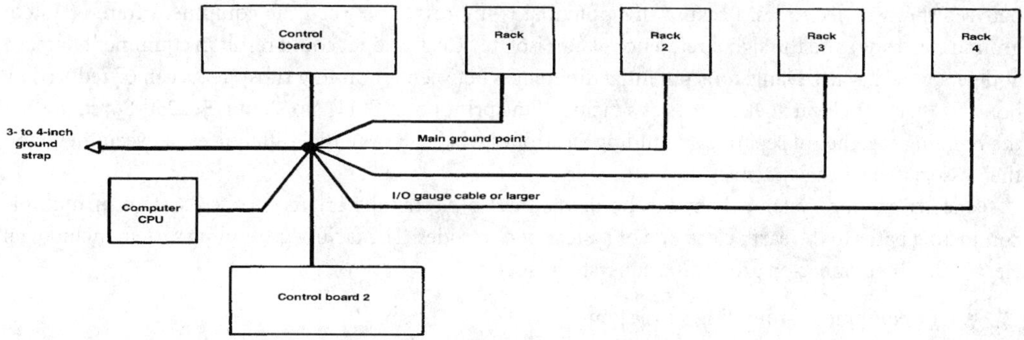

FIGURE 10.258 Typical facility grounding system. The *main facility ground point* is the reference from which all grounding is done at the plant. If a bulkhead entrance panel is used, it will function as the main ground point.

expanded because new situations arise that were not anticipated when the codes were written. Sometimes, an interpretation will depend upon whether the governing safety standard applies to building wiring or to a factory-assembled product to be installed in a building. Underwriters Laboratories (UL) and other qualified testing organizations examine products at the request and expense of manufacturers or purchasers, and list products if the examination reveals that the device or system presents no significant safety hazard when installed and used properly.

Municipal and county safety inspectors generally accept UL and other qualified testing laboratory certification listings as evidence that a product is safe to install. Without a listing, the end-user might not be able to obtain the necessary wiring permits and inspection sign-off. On-site wiring must conform with local wiring codes. Most codes are based on the NEC. Electrical codes specify wiring materials, wiring devices, circuit protection, and wiring methods.

Single-Point Ground

Single-point grounding is the basis of any properly designed facility ground network. Fault currents and noise should have only one path to the facility ground. Single-point grounds can be described as *star* systems in which radial elements circle out from a central hub. A star system is illustrated in Fig. 10.258. Note that all equipment grounds are connected to a *main ground point*, which is then tied to the facility ground system. Multiple ground systems of this type can be cascaded as needed to form a *star-of-stars* facility ground system. The key element in a single-point ground is that each piece of equipment has one ground reference. Fault energy and noise are then efficiently drained to the outside earth ground system. The single-point ground is basically an extension of the bulkhead panel discussed previously.

Isolated Grounding

Isolated grounding schemes, where the signal reference plane is *isolated* from equipment ground but connected to an *isolated* electrode in the earth, do not work, are unsafe, and violate the NEC [5]. It is thought by some people that the isolated earth connection is *clean* because there is no connection between it and the *dirty* system ground connection at the service entrance. The clean, isolated earth connection is also viewed (incorrectly) as a point where noise currents can flow into the earth and be dissipated. Kirchoff's current law teaches that any current flowing into the isolated ground must return to the source through another earth connection. Current cannot be dissipated. It must always return to its source. Even lightning current is not dissipated into the earth. It must have a return path (i.e., the electrostatic and electromagnetic fields that created the charge build-up and the lightning strike in the first place).

Consider what might happen if such a system were subjected to a lightning strike. Assume that a transient current of 2000 A flows into the earth and through an earth resistance of 5 Ω between the system ground electrode and the isolated electrode. A more realistic resistance might be even higher, perhaps 25 Ω; 2000 A flowing through 5 Ω results in a voltage drop or transient potential of 10,000 V between the two electrodes. Because this potential is impressed between the equipment frame (system ground electrode) and the signal reference plane (isolated electrode), it could result in equipment damage and personnel hazard. Dangerous potential differences between grounding subsystems can be reduced by bonding together all earth electrodes at a facility. Fine print note (FPN) No. 2, in NEC 250-86, states that the bonding together of separate grounding electrodes will limit potential differences between them and their associated wiring systems [6].

A facility ground system, then, can be defined as an electrically interconnected system of multiple conducting paths to the earth electrode or system of electrodes. The facility grounding system includes all electrically interconnected grounding subsystems such as

- The equipment grounding subsystem
- Signal reference subsystem
- Fault protection subsystem
- Lightning protection subsystem

Isolated ground (IG) receptacles, which are a version of single-point grounding, are permitted by the NEC. Proper application of IG receptacles is very important. They must be used with an insulated equipment grounding conductor, not a bare conductor. Also, only metallic conduit should be used.

Separately Derived Systems

A separately derived system is a premises wiring system that derives its power from generator, transformer, or converter windings and that has no direct electrical connection, including a solidly connected grounded circuit conductor, to supply conductors originating in another system [6]. Solidly grounded, wye-connected, isolation transformers used to supply power to computer room equipment are examples of separately derived systems [5]. Figure 10.259 illustrates the bonding and grounding requirements of separately derived systems. NEC 250-26 permits the bonding and grounding connections to be made at the source of the separately derived system or at the first disconnecting means. Other examples of separately derived systems include generators and UPS systems. Note that all earth electrodes are bonded together via the equipment grounding conductor system. This is consistent with the recommendations listed in NEC 250-86.

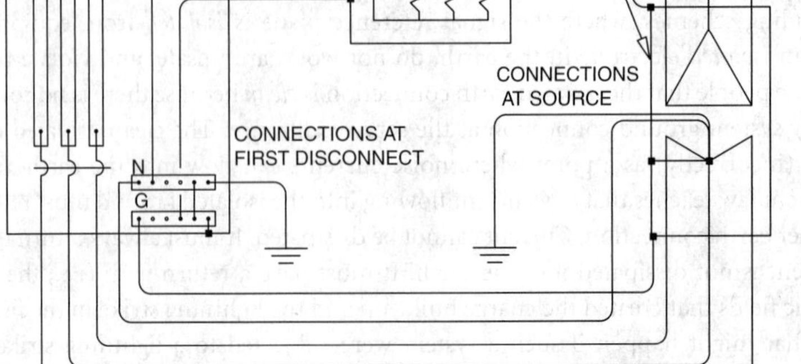

FIGURE 10.259 Basic scheme of the separately derived power system. (Courtesy of Intertec Publishing. After [5].)

Grounding Terminology

- *Grounded conductor.* A system or circuit conductor that is intentionally grounded [6].
- *Equipment grounding conductor.* The conductor used to connect the noncurrent-carrying metal parts of equipment, raceways, and other enclosures to the system grounded conductor, the grounding electrode conductor, or both, at the service equipment or at the source of a separately derived system [6].
- *Main bonding jumper.* The connection between the grounded circuit conductor and the equipment grounding conductor at the service [6].
- *Grounding electrode conductor.* The conductor used to connect the grounding electrode to the equipment grounding conductor, to the grounded conductor, or to both, of the circuit at the service equipment or at the source of a separately derived system [6].
- *Service.* The conductors and equipment for delivering energy from the electricity supply system to the wiring system of the premises served [6].
- *Service conductors.* The supply conductors that extend from the street main or from transformers to the service equipment of the premises supplied [6].
- *Service equipment.* The necessary equipment, usually consisting of a circuit breaker or switch and fuses, and their accessories, located near the point of entrance of supply conductors to a building or other structure, or an otherwise defined area, and intended to constitute the main control and means of cutoff of the supply [6].
- *Equipotential plane.* A mass of conducting material that offers a negligible impedance to current flow, thus producing zero volts (equipotential) between points on the plane.
- *Floating signal grounding.* A non-grounding system in which all electronic signal references are isolated from ground.
- *Single-point grounding.* A grounding system in which all electronic signal references are bonded together and grounded at a single point.
- *Multipoint grounding.* A grounding system in which all electronic signal references are grounded at multiple points.

Facility Ground System

Figure 10.260 illustrates a star grounding system as applied to an AC power-distribution transformer and circuit-breaker panel. Note that a central ground point is established for each section of the system: one in the transformer vault and one in the circuit-breaker box. The breaker ground ties to the transformer vault ground, which is connected to the building ground system. Figure 10.261 shows single-point grounding

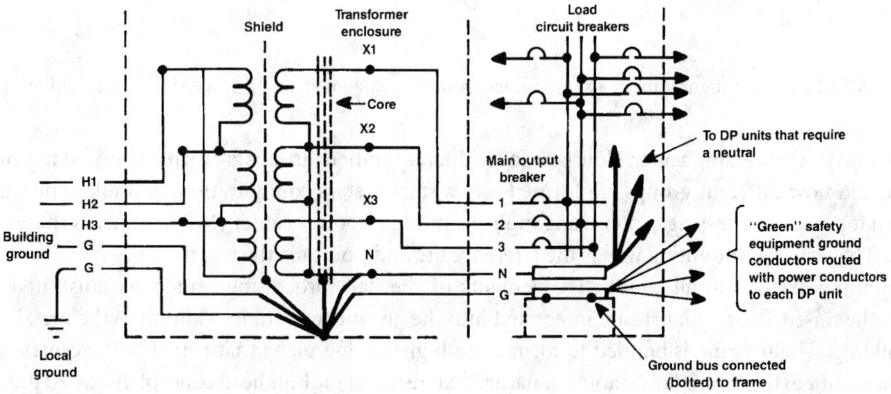

FIGURE 10.260 Single-point grounding applied to a power-distribution system. (After [15].)

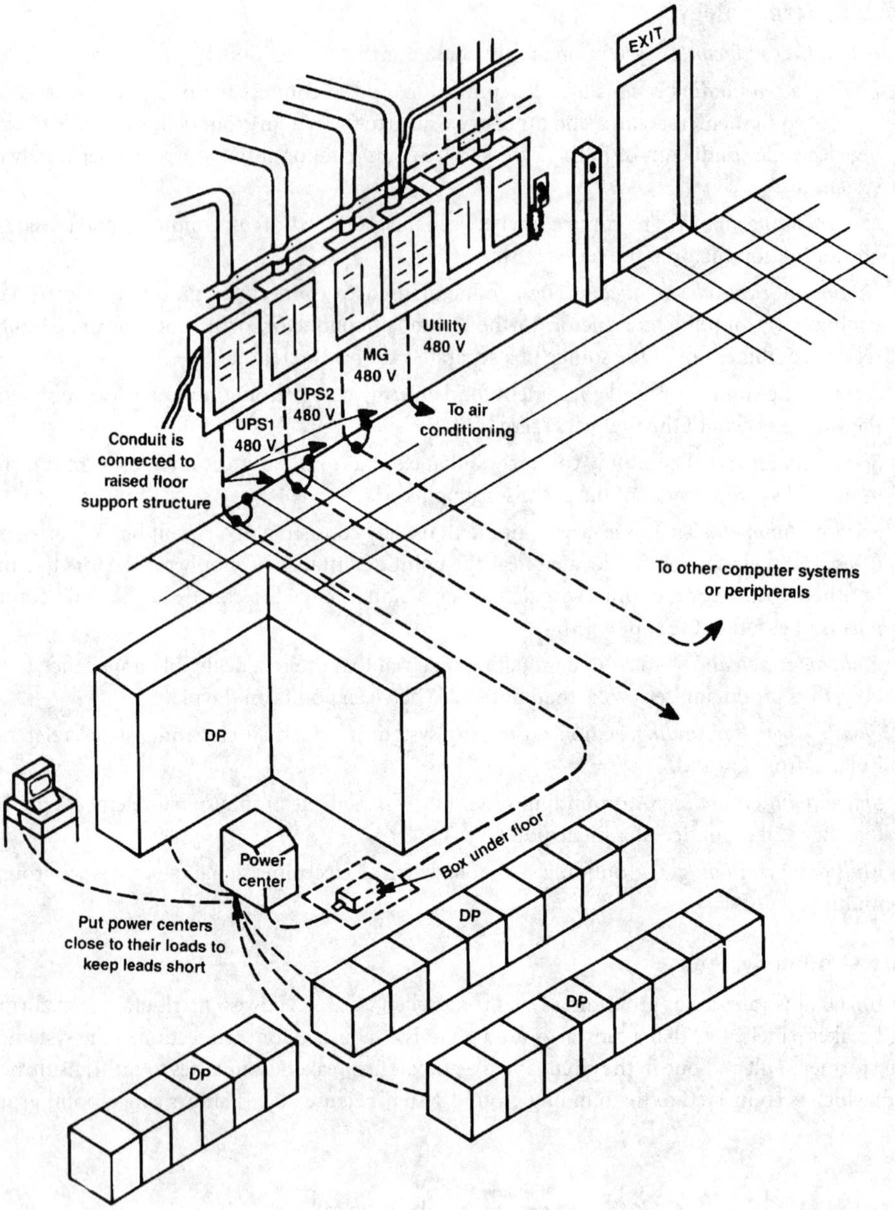

FIGURE 10.261   Configuration of a star-of-stars grounding system at a data processing facility. (After [15].)

applied to a data processing center. Note how individual equipment groups are formed into a star grounding system, and how different groups are formed into a star-of-stars configuration. A similar approach can be taken for a data processing center using multiple modular power center (MPC) units, as shown in Fig. 10.262. The terminal mounting wall is the reference ground point for the room.

Grounding extends to all conductive elements of the data processing center. As illustrated in Fig. 10.263, the raised-floor supports are integrated into the ground system, in addition to the metal work of the building. The flooring is bonded to form a mesh grounding plane that is tied to the central ground reference point of the room. This is another natural extension of the bulkhead concept discussed previously. Figure 10.264 shows connection detail for the raised-floor supports. Make sure to use a flooring system that can be bolted together to form secure electrical—as well as mechanical—connections.

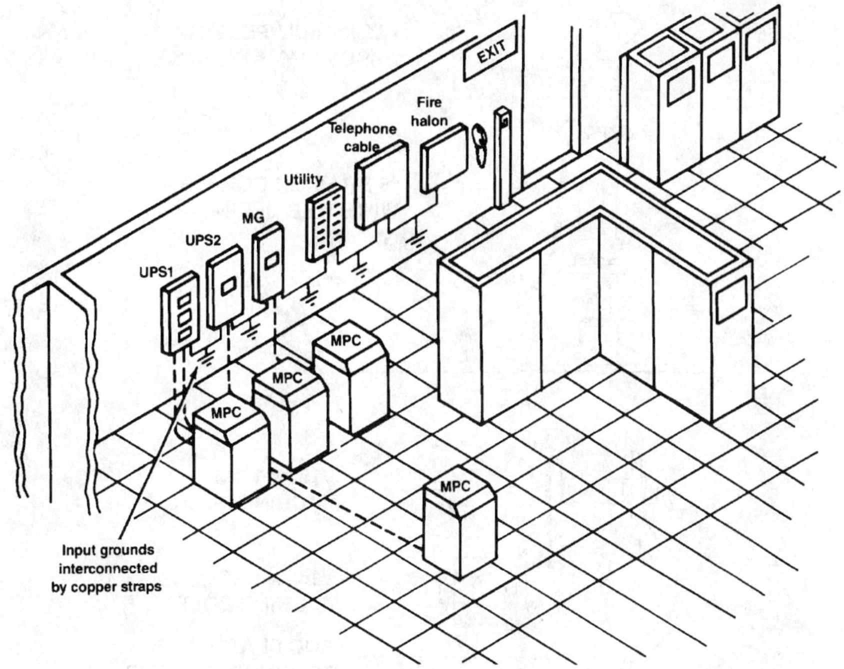

FIGURE 10.262 Establishing a star-based, single-point ground system using multiple modular power centers. (After [15].)

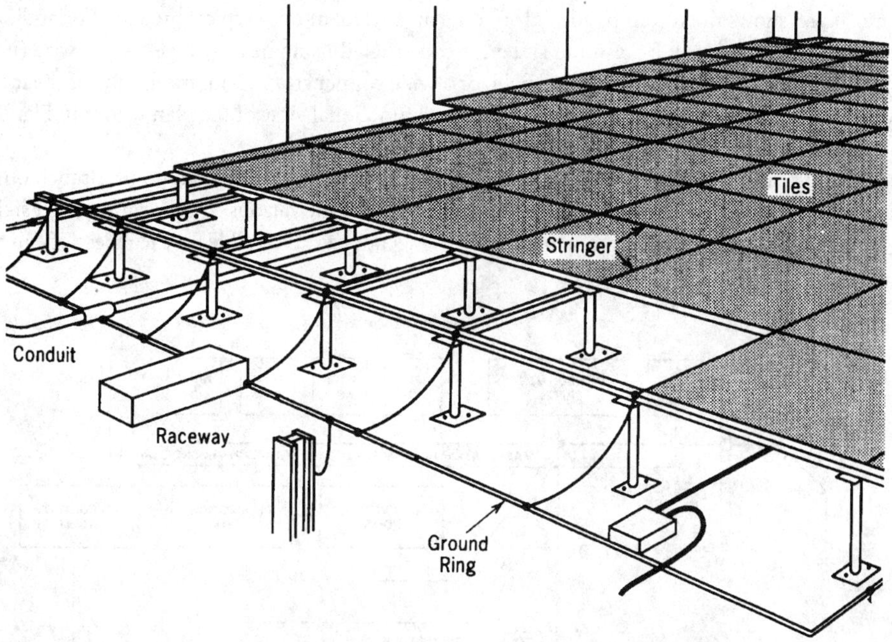

FIGURE 10.263 Grounding plan for a raised-floor support system.

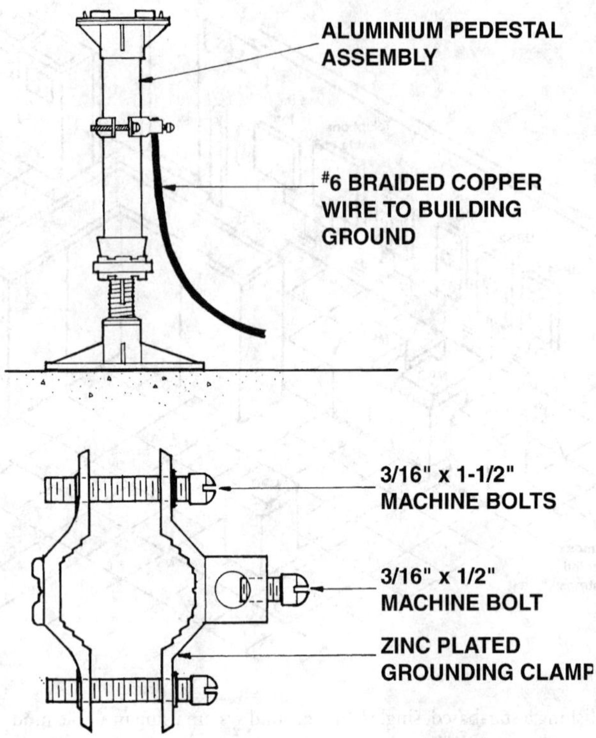

FIGURE 10.264   Bonding detail for the raised-floor support system.

Figure 10.265 shows the recommended grounding arrangement for a typical broadcast or audio/video production facility. The building ground system is constructed using heavy-gauge copper wire (no. 4 or larger) if the studio is not located in an RF field, or a wide copper strap (3-in. minimum) if the facility is located near an RF energy source. Figure 10.266 gives a functional view of the plan shown in Fig. 10.265. Note the bulkhead approach.

Run the strap or cable from the perimeter ground to the main facility ground point. Branch out from the main ground point to each major piece of equipment and to the various equipment rooms. Establish a local *ground point* in each room or group of racks. Use a separate ground cable for each piece of equipment

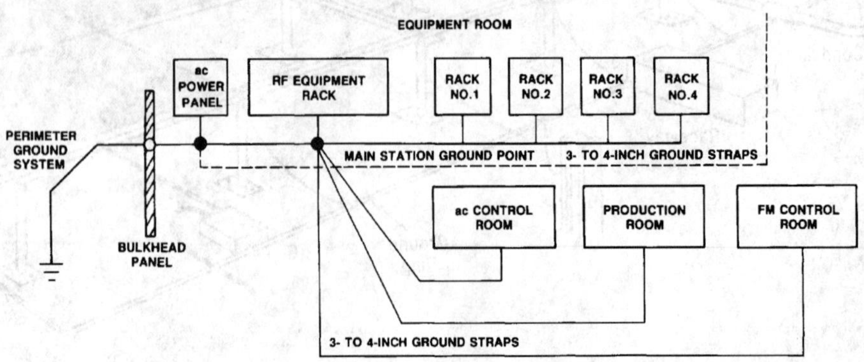

FIGURE 10.265   Typical grounding arrangement for individual equipment rooms at a communications facility. The ground strap from the main ground point establishes a *local ground point* in each room, to which all electronic equipment is bonded.

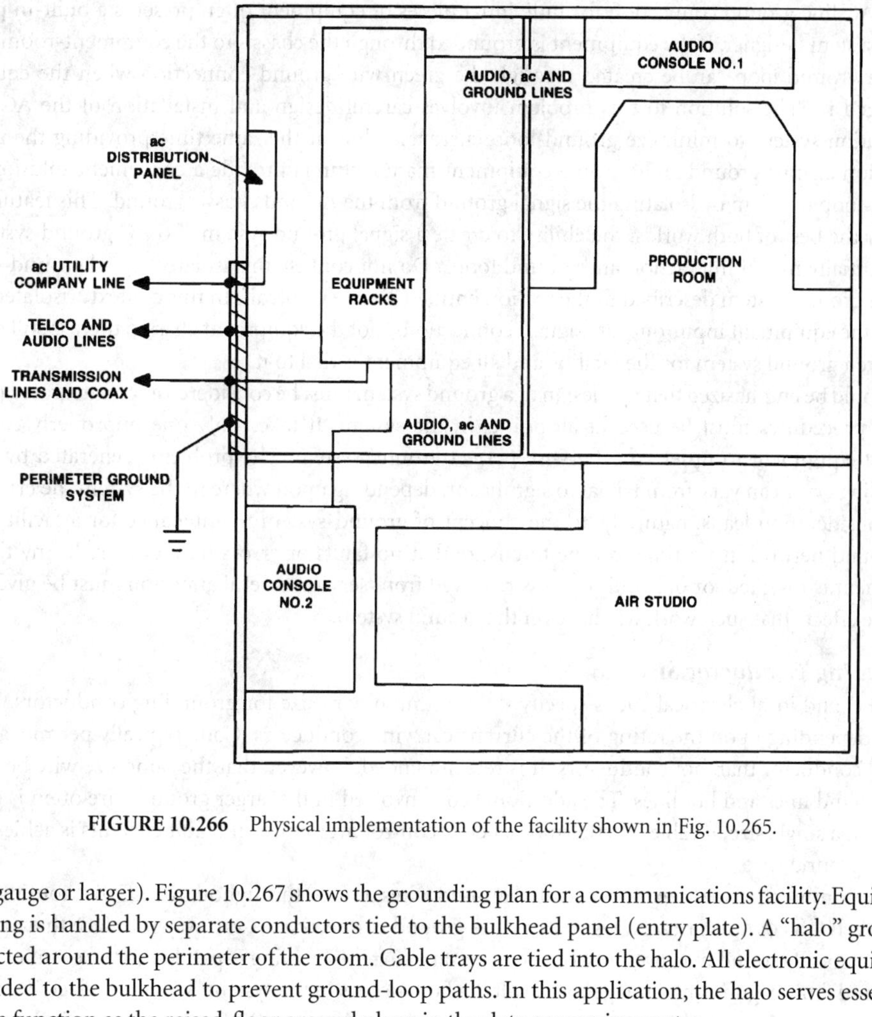

FIGURE 10.266 Physical implementation of the facility shown in Fig. 10.265.

(no. 12 gauge or larger). Figure 10.267 shows the grounding plan for a communications facility. Equipment grounding is handled by separate conductors tied to the bulkhead panel (entry plate). A "halo" ground is constructed around the perimeter of the room. Cable trays are tied into the halo. All electronic equipment is grounded to the bulkhead to prevent ground-loop paths. In this application, the halo serves essentially the same function as the raised-floor ground plane in the data processing center.

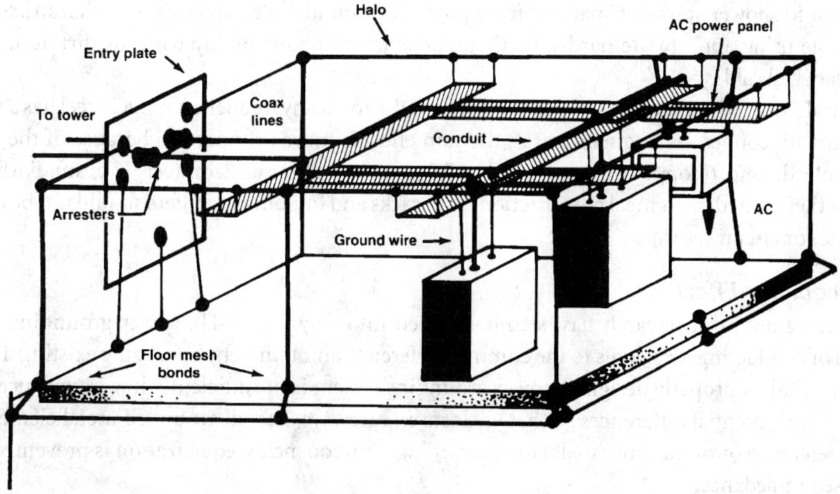

FIGURE 10.267 Bulkhead-based ground system including a grounding halo.

The AC line ground connection for individual pieces of equipment often presents a built-in problem for the system designer. If the equipment is grounded through the chassis to the equipment-room ground point, a ground loop can be created through the green-wire ground connection when the equipment is plugged in. The solution to this problem involves careful design and installation of the AC power-distribution system to minimize ground-loop currents, while at the same time providing the required protection against ground faults. Some equipment manufacturers provide a convenient solution to the ground-loop problem by isolating the signal ground from the AC and chassis ground. This feature offers the user the best of both worlds: the ability to create a signal ground system and AC ground system that are essentially free of interaction and ground loops. Do not confuse this isolated signal ground with the isolated ground system described in the section entitled "Noise Control." In this context, "isolated" refers only to the equipment input/output (signal) connections, not the equipment chassis; there is still only one integrated ground system for the facility, and all equipment is tied to it.

It should be emphasized that the design of a ground system must be considered as an integrated package. Proper procedures must be used at all points in the system. It takes only one improperly connected piece of equipment to upset an otherwise perfect ground system. The problems generated by a single grounding error can vary from trivial to significant, depending upon where in the system the error exists. This consideration leads, naturally, to the concept of ground-system maintenance for a facility. Check the ground network from time to time to ensure that no faults or errors have occurred. Any time new equipment is installed, or old equipment is removed from service, careful attention must be given to the possible effects that such work will have on the ground system.

Grounding Conductor Size

The NEC and local electrical codes specify the minimum wire size for grounding conductors. The size varies, depending upon the rating of the current-carrying conductors. Code typically permits a smaller ground conductor than hot conductors. It is recommended, however, that the same size wire be used for both ground lines and hot lines. The additional cost involved in the larger ground wire often is offset by the use of a single size of cable. Furthermore, better control over noise and fault currents is achieved with a larger ground wire.

It is recommended that separate insulated ground wires be used throughout the AC distribution system. Do not rely on conduit or raceways to carry the ground connection. A raceway interface that appears to be mechanically sound may not provide the necessary current-carrying capability in the event of a phase-to-ground fault. Significant damage can result if a fault occurs in the system. When the electrical integrity of a breaker panel, conduit run, or raceway junction is in doubt, fix it. Back up the mechanical connection with a separate ground conductor of the same size as the current-carrying conductors. Loose joints have been known to shower sprays of sparks during phase-to-ground faults, creating a fire hazard. Secure the ground cable using appropriate hardware. Clean attachment points of any paint or dirt accumulation. Properly label all cables.

Structural steel, compared with copper, is a poor conductor at any frequency. At DC, steel has a resistivity 10 times that of copper. As frequency rises, the skin effect is more pronounced because of the magnetic effects involved. A no. 6 copper wire can have less RF impedance than a 12-in steel "I" beam. Furthermore, because of their bolted, piecemeal construction, steel racks and building members should not be depended upon alone for circuit returns.

High-Frequency Effects

A significant amount of research has been conducted into why the 60-Hz power grounding system is incapable of conducting RF signals to the common reference point, thereby equalizing existing differences in potential [16]. A properly designed power grounding system is of sufficiently low impedance at 60 Hz to equalize any potential differences so that enclosures, raceways, and all grounded metal elements are at the same reference (ground) potential. However, at higher frequencies, equalization is prevented because of increased impedance.

The impedance of a conductor consists of three basic components: resistance, capacitive reactance, and inductive reactance. Although the inductance L (in Henrys) will be constant for a conductor of a given length and cross-sectional area, the inductive reactance X_L will vary according to the frequency f of the applied voltage as follows:

$$X_L = 2\pi f L \tag{10.222}$$

Therefore, at 60 Hz, the inductive reactance will be $377 \times L$; and at 30 MHz, it will be $(188.5 \times 10^6) \times L$. It is evident, then, that at 60 Hz the equipment grounding conductor is a short-circuit, but at 30 MHz, it is effectively an open circuit.

In addition to increased inductive reactance at higher frequencies, there is also stray capacitance and stray inductance between adjacent conductors and/or between conductors and adjacent grounded metal, as well as resonance effects. These factors combine to yield an increase in apparent conductor impedance at higher frequencies.

If conductors are connected in a mesh or grid to form a multitude of low-impedance loops in parallel, there will be little voltage difference between any two points on the grid at all frequencies from 60 Hz up to a frequency where the length of one side of the square represents about $\frac{1}{10}$ wavelength. A grid made up of 2-ft squares (such as a raised computer floor) will at any point provide an effective equipotential ground reference point for signals up to perhaps 30 MHz.

Power-Center Grounding

A modular power center, commonly found in computer-room installations, provides a comprehensive solution to AC power-distribution and ground-noise considerations. Such equipment is available from several manufacturers, with various options and features. A computer power-distribution center generally includes an isolation transformer designed for noise suppression, distribution circuit breakers, power-supply cables, and a status-monitoring unit. The system concept is shown in Fig. 10.268. Input power is fed to an isolation transformer with primary taps to match the AC voltage required at the facility. A bank of circuit breakers is included in the chassis, and individual preassembled and terminated cables supply AC power to the various loads. A status-monitoring circuit signals the operator if any condition is detected outside normal parameters.

The ground system is an important component of the MPC. A unified approach, designed to prevent noise or circulating currents, is taken to grounding for the entire facility. This results in a clean ground connection for all equipment on-line.

The use of a modular power center can eliminate the inconvenience associated with rigid conduit installations. Distribution systems also are expandable to meet future facility growth. If the plant is ever relocated, the power center can move with it. MPC units usually are expensive. However, considering the costs of installing circuit-breaker boxes, conduit, outlets, and other hardware on-site by a licensed electrician, the power-center approach may be economically viable. The use of a power center also will make it easier to design a standby power system for the facility. Many computer-based operations do not have a standby generator on site. Depending on the location of the facility, it might be difficult or even impossible to install a generator to provide standby power in the event of a utility company outage. However, by using the power-center approach to AC distribution for computer and other critical load equipment, an uninterruptible power system can be installed easily to power only the loads that are required to keep the facility operating. With a conventional power-distribution system—where all AC power to the building, or a floor of the building, is provided by a single large circuit-breaker panel—separating the critical loads from other nonessential loads (such as office equipment, lights, and heating equipment) can be an expensive detail.

Isolation Transformers

One important aspect of an MPC is the isolation transformer. The transformer serves to

- Attenuate transient disturbances on the AC supply lines.
- Provide voltage correction through primary-side taps.
- Permit the establishment of a local ground reference for the facility served.

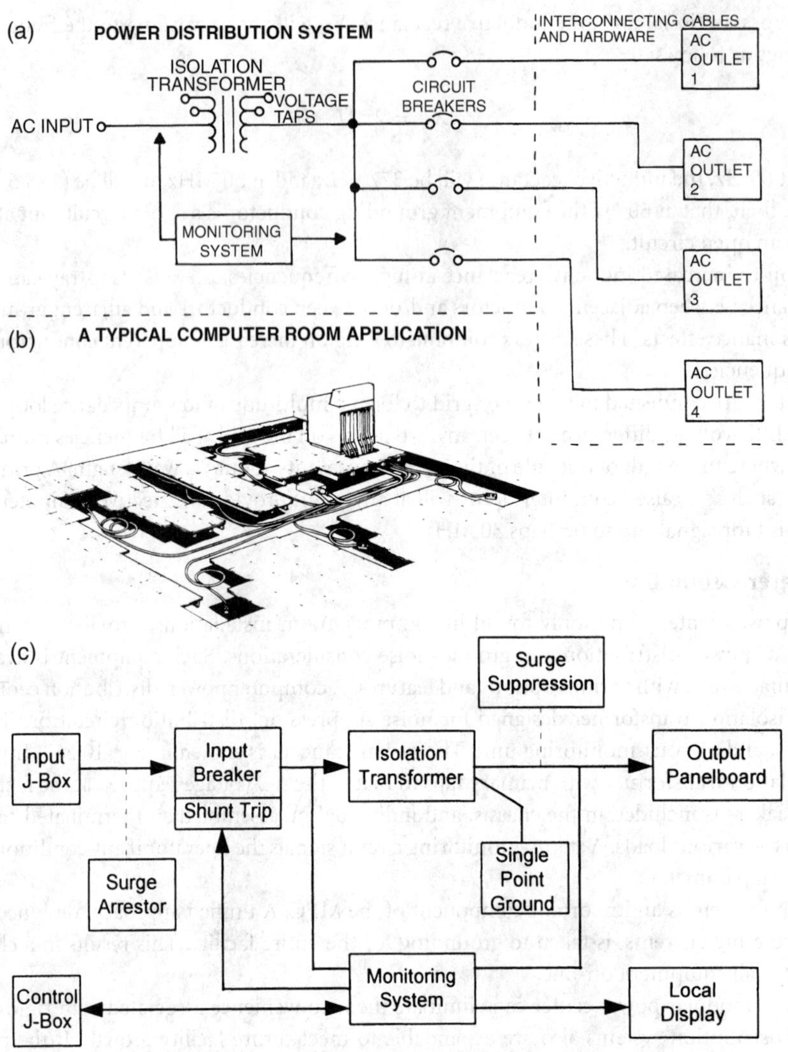

FIGURE 10.268 The basic concept of a computer-room modular power center: (*a*) basic line drawing of system, (*b*) typical physical implementation, (*c*) functional block diagram. Both single- and multiphase configurations are available. When ordering an MPC, the customer can specify cable lengths and terminations, making installation quick and easy. (After [17].)

Whether or not an MPC is installed at a facility, consideration should be given to the appropriate use of an isolation transformer near a sensitive load.

The AC power supply for many buildings originates from a transformer located in a basement utility room. In large buildings, the AC power for each floor can be supplied by transformers closer to the loads they serve. Most transformers are 208 Y/120 V three-phase. Many fluorescent lighting circuits operate at 277 V, supplied by a 408 Y/277 V transformer. Long feeder lines to DP systems and other sensitive loads raise the possibility of voltage fluctuations based on load demand and ground-loop-induced noise.

Figure 10.269 illustrates the preferred method of power distribution in a building. A separate dedicated isolation transformer is located near the DP equipment, providing good voltage regulation and permitting the establishment of an effective single-point star ground in the data processing center. Note that the power-distribution system voltage shown in the figure (480 V) is maintained until the DP step-down isolation transformer. Use of this higher voltage provides more efficient transfer of electricity throughout

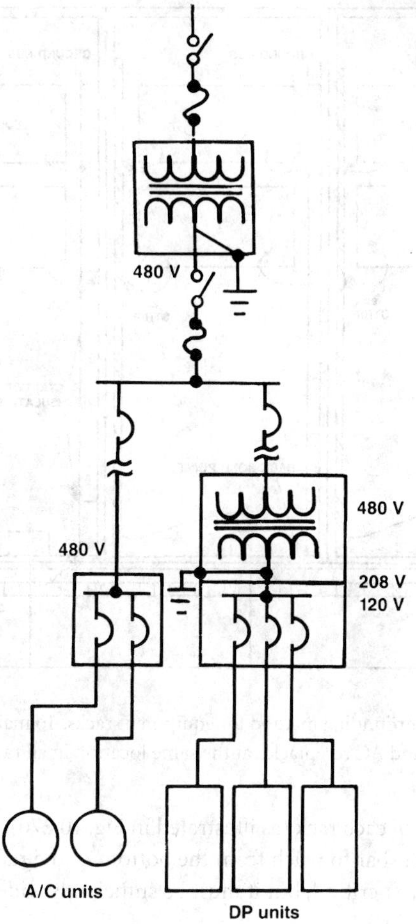

FIGURE 10.269   Preferred power-distribution configuration for a data processing site. (After [15].)

the plant. At 480 V, the line current is about 43% of the current in a 208 V system for the same conducted power.

There are a number of tradeoffs with regard to facility AC power system design. An experienced, licensed electrical contractor and/or registered professional engineer should be consulted during the early stages of any renovation or new construction project.

Grounding Equipment Racks

The installation and wiring of equipment racks must be planned carefully to avoid problems during day-to-day operations. Figure 10.270 shows the recommended approach. Bond adjacent racks together with ³⁄₈- to ½-in.-diameter bolts. Clean the contacting surfaces by sanding down to bare metal. Use lockwashers on both ends of the bolts. Bond racks together using at least six bolts per side (three bolts for each vertical rail). After securing the racks, repaint the connection points to prevent corrosion.

Run a ground strap from the *main facility ground point,* and bond the strap to the base of each rack. Spot-weld the strap to a convenient spot on one side of the rear portion of each rack. Secure the strap at the same location for each rack used. A mechanical connection between the rack and the ground strap can be made using bolts and lockwashers, if necessary. Be certain, however, to sand down to bare metal before making the ground connection. Because of the importance of the ground connection, it is recommended that each attachment be made with a combination of crimping and silver-solder. Keep the length of strap between adjacent bolted racks as short as possible by routing the strap directly under the racks.

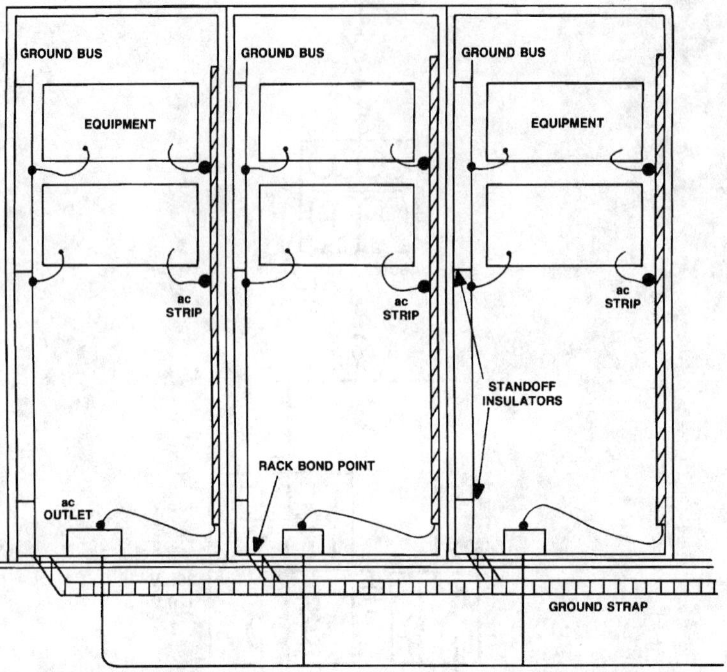

FIGURE 10.270 Recommended grounding method for equipment racks. To make assembly of multiple racks easier, position the ground connections and AC receptacles at the same location in all racks.

Install a vertical ground bus in each rack (as illustrated in Fig. 10.270). Use about 1-$\frac{1}{2}$-in.-wide, $\frac{1}{4}$-in.-thick copper busbar. Size the busbar to reach from the bottom of the rack to about 1 ft short of the top. The exact size of the busbar is not critical, but it must be sufficiently wide and rigid to permit the drilling of $\frac{1}{8}$-in. holes without deforming.

Mount the ground busbar to the rack using insulated standoffs. Porcelain standoffs commonly found in high-voltage equipment are useful for this purpose. Porcelain standoffs are readily available and reasonably priced. Attach the ground busbar to the rack at the point where the facility ground strap attaches to the rack. Silver-solder the busbar to the rack and strap at the same location in each rack used.

Install an orange-type isolated AC receptacle box at the bottom of each rack. The orange-type outlet isolates the green-wire power ground from the receptacle box. Use insulated standoffs to mount the AC outlet box to the rack. The goal of this arrangement is to keep the green-wire AC and facility system grounds separate from the AC distribution conduit and metal portions of the building structure. Try to route the power conduit and facility ground cable or strap via the same physical path. Keep metallic conduit and building structures insulated from the facility ground line, except at the bulkhead panel (main grounding point). Carefully check the local electrical code before proceeding.

Although the foregoing procedure is optimum from a signal-grounding standpoint, it should be pointed out that under a ground fault condition, performance of the system can be unpredictable if high currents are being drawn in the current-carrying conductors supplying the load. Vibration of AC circuit elements resulting from the magnetic field effects of high-current-carrying conductors is insignificant as long as all conductors are within the confines of a given raceway or conduit. A ground fault will place return current outside of the normal path. If sufficiently high currents are being conducted, the consequences can be devastating. Sneak currents from ground faults have been known to destroy wiring systems that were installed exactly to code. As always, consult an experienced electrical contractor.

Mount a vertical AC strip inside each rack to power the equipment. Insulate the power strip from the rack using porcelain standoffs or use a strip with an insulated (plastic) housing. Such a device is shown in Fig. 10.271. Power equipment from the strip using standard three-prong grounding AC plugs. Do not

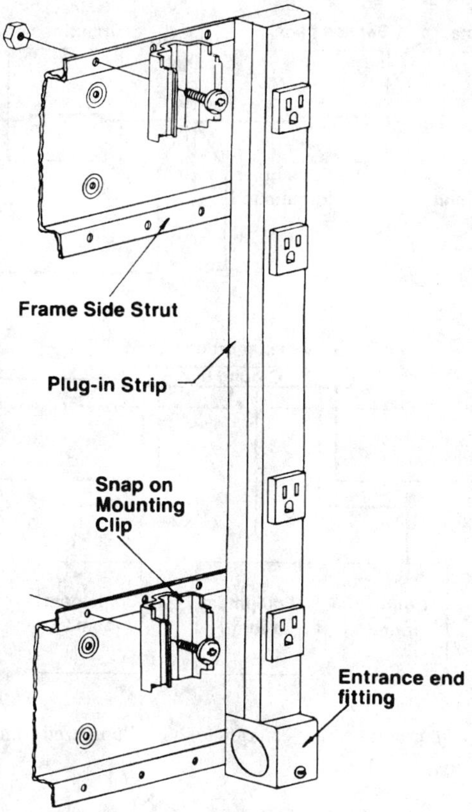

FIGURE 10.271 Rack mounting detail for an insulated-shell AC power strip.

defeat the safety ground connection. Equipment manufacturers use this ground to drain transient energy. Furthermore, defeating the safety ground will violate local electrical codes.

Mount equipment in the rack using normal metal mounting screws. If the location is in a high-RF field, clean the rack rails and equipment-panel connection points to ensure a good electrical bond. This is important because in a high-RF field, detection of RF energy can occur at the junctions between equipment chassis and the rack.

Connect a separate ground wire from each piece of equipment in the rack to the vertical ground busbar. Use no. 12 stranded copper wire (insulated) or larger. Connect the ground wire to the busbar by drilling a hole in the busbar at a convenient elevation near the equipment. Fit one end of the ground wire with an enclosed-hole solderless terminal connector (no. 10-sized hole or larger). Attach the ground wire to the busbar using no. 10 (or larger) hardware. Use an internal-tooth lockwasher between the busbar and the nut. Fit the other end of the ground wire with a terminal that will be accepted by the ground terminal at the equipment. If the equipment has an isolated *signal* ground terminal, tie it to the ground busbar as well.

Whenever servicing equipment in the rack, make certain to disconnect the AC power cord before removing the unit from the rack and/or disconnecting the rack ground wire. During any service work, make the first step removal of the AC power plug; when reassembling the unit, make the last step reinsertion of the plug. Do not take chances.

Figure 10.272 shows each of the grounding elements discussed in this section integrated into one diagram. This approach fulfills the requirements of personnel safety and equipment performance.

Follow similar grounding rules for simple one-rack equipment installations. Figure 10.273 illustrates the grounding method for a single open-frame equipment rack. The vertical ground bus is supported by insulators, and individual jumpers are connected from the ground rail to each chassis.

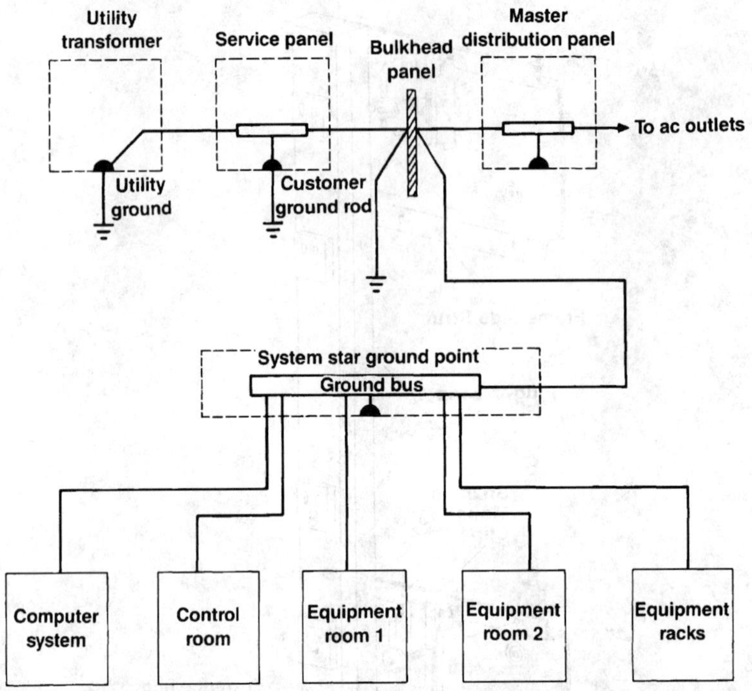

FIGURE 10.272   Equivalent ground circuit diagram for a medium-sized commercial/industrial facility.

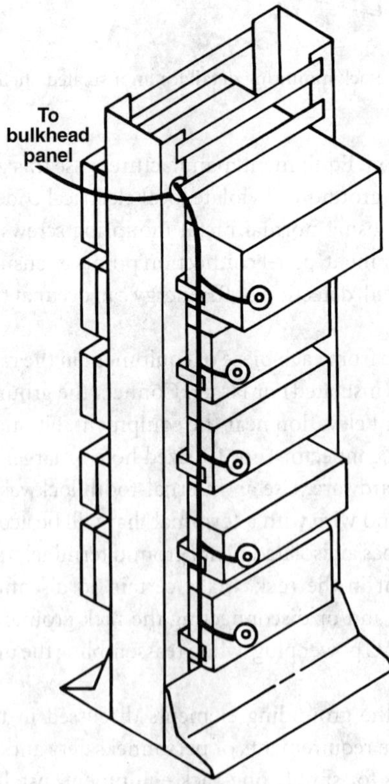

FIGURE 10.273   Ground bus for an open-frame equipment rack.

10.10.6 Grounding Signal-Carrying Cables

Proper ground-system installation is the key to minimizing noise currents on signal-carrying cables. Audio, video, and data lines are often subject to AC power noise currents and RFI. The longer the cable run, the more susceptible it is to disturbances. Unless care is taken in the layout and installation of such cables, unacceptable performance of the overall system can result.

Analyzing Noise Currents

Figure 10.274 shows a basic source and load connection. No grounds are present, and both the source and the load float. This is the optimum condition for equipment interconnection. Either the source or the load can be tied to ground with no problems, provided only one ground connection exists. *Unbalanced* systems are created when each piece of equipment has one of its connections tied to ground, as shown in Fig. 10.275. This condition occurs if the source and load equipment have unbalanced (single-ended) inputs and outputs. This type of equipment utilizes chassis ground (or common) for one of the conductors. Problems are compounded when the equipment is separated by a significant distance.

As shown in Fig. 10.275, a difference in ground potential causes current flow in the ground wire. This current develops a voltage across the wire resistance. The ground-noise voltage adds directly to the signal itself. Because the ground current is usually the result of leakage in power transformers and line filters, the 60-Hz signal gives rise to hum of one form or another. Reducing the wire resistance through a heavier ground conductor helps the situation, but cannot eliminate the problem.

By amplifying both the high side and the ground side of the source and subtracting the two to obtain a *difference signal*, it is possible to cancel the ground-loop noise. This is the basis of the *differential input* circuit, illustrated in Fig. 10.276. Unfortunately, problems still can exist with the unbalanced-source-to-balanced-load system. The reason centers on the impedance of the unbalanced source. One side of the line will have a slightly lower amplitude because of impedance differences in the output lines. By creating an output signal that is out of phase with the original, a balanced source can be created to eliminate this error. See Fig. 10.277.

IDEAL SOURCE AND LOAD

FIGURE 10.274 A basic source and load connection. No grounds are indicated, and both the source and the load float.

$$V_L = V_S + V_{GROUND}$$

FIGURE 10.275 An unbalanced system in which each piece of equipment has one of its connections tied to ground.

$$V_1 = V_S + V_{GROUND}$$
$$V_2 = V_{GROUND}$$
$$V_1 - V_2 = V_S$$

FIGURE 10.276 Ground-loop noise can be canceled by amplifying both the high side and the ground side of the source and subtracting the two signals.

$$v_1 - V_2 = 2V_S$$

FIGURE 10.277 A balanced source configuration where the inherent amplitude error of the system shown in Fig. 10.66 is eliminated.

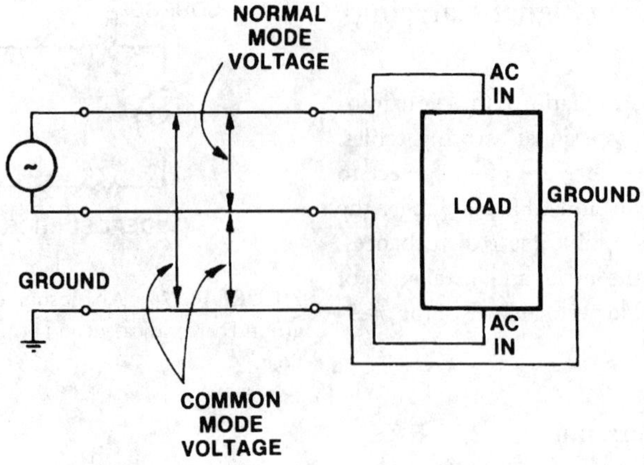

FIGURE 10.278   The principles of normal-mode and common-mode noise voltages as they apply to AC power circuits.

As an added benefit, for a given maximum output voltage from the source, the signal voltage is doubled over the unbalanced case.

Types of Noise

Two basic types of noise can appear on AC power, audio, video, and computer data lines within a facility: *normal mode* and *common mode*. Each type has a particular effect on sensitive load equipment. The normal-mode voltage is the potential difference that exists between pairs of power (or signal) conductors. This voltage also is referred to as the *transverse-mode* voltage. The common-mode voltage is a potential difference (usually noise) that appears between the power or signal conductors and the local ground reference. The differences between normal-mode and common-mode noise are illustrated in Fig. 10.278.

The common-mode noise voltage will change, depending upon what is used as the ground reference point. It is often possible to select a ground reference that has a minimum common-mode voltage with respect to the circuit of interest, particularly if the reference point and the load equipment are connected by a short conductor. Common-mode noise can be caused by electrostatic or electromagnetic induction.

In practice, a single common-mode or normal-mode noise voltage is rarely found. More often than not, load equipment will see both common-mode and normal-mode noise signals. In fact, unless the facility wiring system is unusually well-balanced, the noise signal of one mode will convert some of its energy to the other mode.

Common-mode and normal-mode noise disturbances typically are caused by momentary impulse voltage differences among parts of a distribution system that have differing ground potential references. If the sections of a system are interconnected by a signal path in which one or more of the conductors are grounded at each end, the ground offset voltage can create a current in the grounded signal conductor. If noise voltages of sufficient potential occur on signal-carrying lines, normal equipment operation can be disrupted. See Fig. 10.279.

Noise Control

Noise control is an important aspect of computer and electronic system design and maintenance [5]. The process of noise control through proper grounding techniques is more correctly called *referencing*. For this discussion, electronic systems can be viewed as a multiplicity of signal sources transmitting signals to a multiplicity of loads. Practically speaking, the impedance of the signal return path is never zero, and dedicated return paths for each source-load pair are not always practical. Packaged electronics systems typically incorporate a common signal reference plane that serves as a common return path for numerous source-load pairs. (See Fig. 10.280a.) The signal reference plane may be a large dedicated area on a circuit

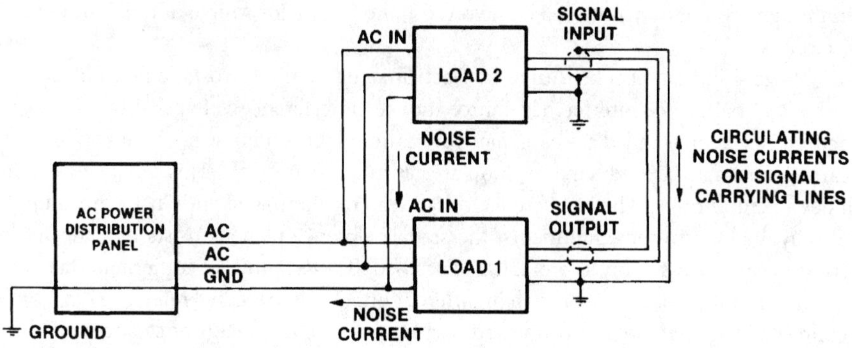

FIGURE 10.279 An illustration of how noise currents can circulate within a system because of the interconnection of various pieces of hardware.

board, the metal chassis or enclosure of the electronic equipment, or the metal frame or mounting rack that houses several different units. Ideally, the signal reference plane offers zero impedance to the signal current. Practically, however, the signal reference plane has a finite impedance. The practical result is illustrated in Fig. 10.280b, and is called *common-impedance* or *conductive coupling*.

Because a practical signal reference plane has a finite impedance, current flow in the plane will produce potential differences between various points on the plane. Source-load pairs referenced to the plane will, therefore, experience interference as a result. Z_R is common to both circuits referenced to the plane in Fig. 10.280b. Thus, I_1 and I_2 returning to their respective sources will produce interference voltages by flowing through Z_R. The total interference voltage drop seen across Z_R causes the source reference A to be at a different potential than the load reference B. This difference in potential is often called *ground voltage*

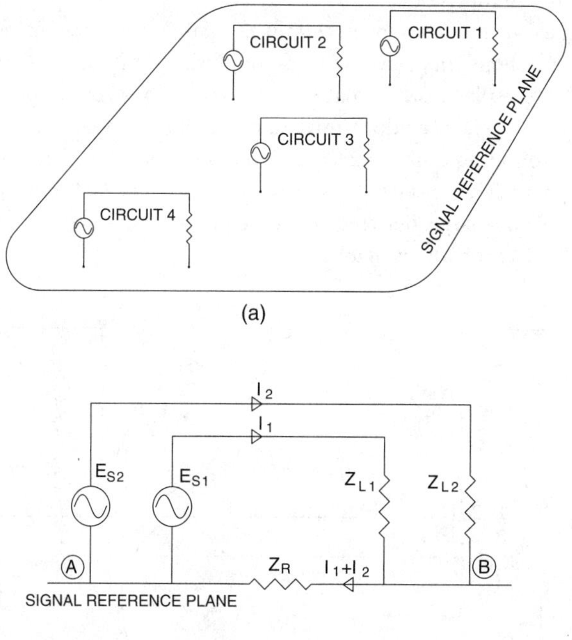

FIGURE 10.280 The equipotential plane: (a) individual signal sources, (b) net effect on the signal reference plane. (Courtesy of Intertec Publishing. After [5].)

shift (even though ground may not even be involved), and is a major source of noise and interference in electronic circuits.

Ground voltage shifts can also be caused by electromagnetic or electrostatic fields in close proximity to the source-load pairs. The interference source induces interference voltages into any closed loop by antenna action. This loop is called a *ground loop* (even though ground may not be involved). Interference voltages can be minimized by reducing the loop area as much as possible. This can be very difficult if the loop includes an entire room. The interference voltage can be eliminated entirely by breaking the loop.

Within individual electronic equipments, the signal reference plane consists of a metal plate or the metal enclosure or a rack assembly as previously discussed. Between units of equipment that are located in different rooms, on different floors, or even in different buildings, the signal reference planes of each unit must be connected together via interconnected wiring such as coax shields or separate conductors. This action, of course, increases the impedance between signal reference planes and makes noise control more difficult. Reducing noise caused by common-impedance coupling is a matter of reducing the impedance of the interconnected signal reference planes.

Regardless of the configuration encountered (i.e., circuit board, individual electronic equipment, or equipments remotely located within a facility or in separate buildings), the next question to be answered is: should the signal reference be connected to ground? Floating signal grounding, single-point grounding, and multipoint grounding are methods of accomplishing this signal-reference-to-ground connection.

IEEE Standard 1100 recommends multipoint grounding for most computers. For effective multipoint grounding, conductors between separate points desired to be at the same reference potential should be less than 0.1 wavelength ($\lambda/10$) of the interference or noise-producing signal. Longer conductor lengths will exhibit significant impedance because of the antenna effect and become ineffective as equalizing conductors between signal references. On printed circuit cards or within individual electronics equipments, short conductor lengths are usually easy to achieve. For remotely located equipments, however, short equalizing conductor lengths may be impossible. Equipments in relatively close proximity can be equalized by cutting bonding conductors to an odd multiple of $\lambda/2$ at the noise frequency. The distribution of standing waves on bonding conductors is illustrated in Fig. 10.281. The impedance, Z, is minimum for odd multiples of $\lambda/2$ but maximum for odd multiples of $\lambda/4$.

For remotely located electronics equipment within the same room, equalizing potential between respective signal reference systems may have to be accomplished with an equipotential plane. An ideal equipotential signal reference plane is one that has zero volts (thus zero impedance) between any two points on the plane. Because an *ideal* equipotential plane is not attainable, a *nominal* equipotential plane is accepted. Multipoint grounding connections are made to the plane, which ensures minimum ground voltage shift between signal reference systems connected to the plane. Collectively, the current flow in an equipotential plane can be quite large. Between any two equipments, however, the current flow should be low because of the many current paths available.

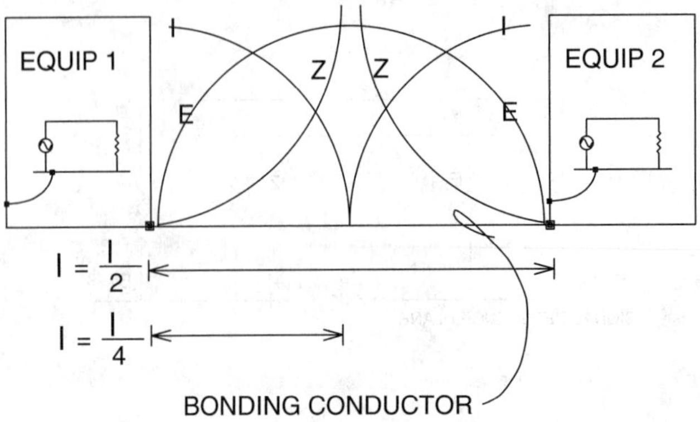

FIGURE 10.281   Standing waves on a bonding conductor resulting from the antenna effect. (After [5].)

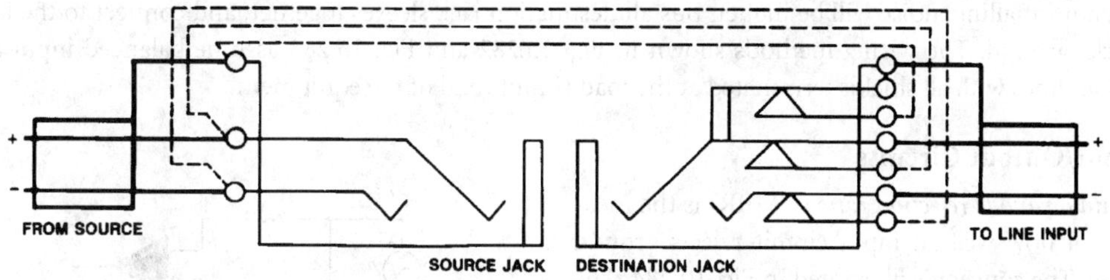

FIGURE 10.282 Patch-panel wiring for seven-terminal normalling jack fields. Use patch cords that connect ground (sleeve) at both ends.

Practical versions of an equipotential plane include the following:

- Bolted stringer system of a raised computer floor
- Flat copper strips bonded together at 2 ft centers
- Copper conductors bonded together at 2 ft centers
- Single or multiple, flat copper strips connected between equipment

Patch-Bay Grounding

Patch panels for audio, video, and data circuits require careful attention to planning to avoid built-in grounding problems. Because patch panels are designed to tie together separate pieces of equipment, often from remote areas of a facility, the opportunity exists for ground loops. The first rule of patch-bay design is to never use a patch bay to switch low-level (microphone) signals. If mic sources must be patched from one location to another, install a bank of mic-to-line amplifiers to raise the signal levels to 0 dBm before connection to the patch bay. Most video output levels are 1 V P-P, giving them a measure of noise immunity. Data levels are typically 5 V. Although these line-level signals are significantly above the noise floor, capacitive loading and series resistance in long cables can reduce voltage levels to a point that noise becomes a problem.

Newer-design patch panels permit switching of ground connections along with signal lines. Figure 10.282 illustrates the preferred method of connecting an audio patch panel into a system. Note that the source and destination jacks are *normalled* to establish ground signal continuity. When another signal is plugged into the destination jack, the ground from the new source is carried to the line input of the destination jack. With such an approach, jack cords that provide continuity between sleeve (ground) points are required.

If only older-style, conventional jacks are available, use the approach shown in Fig. 10.283. This configuration will prevent ground loops, but because destination shields are not carried back to the source

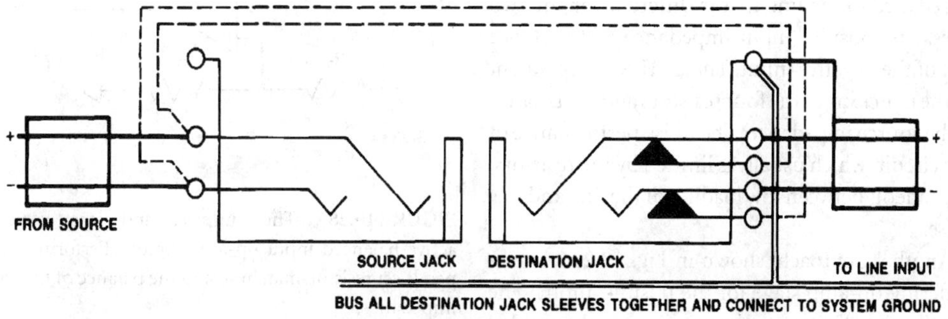

BUS ALL DESTINATION JACK SLEEVES TOGETHER AND CONNECT TO SYSTEM GROUND

FIGURE 10.283 Patch-panel wiring for conventional normalling jack fields. Use patch cords that connect ground (sleeve) at both ends.

when normalling, noise will be higher. Bus all destination jack sleeves together, and connect to the local (rack) ground. The wiring methods shown in Fig. 15.282 and Fig. 15.283 assume balanced input and output lines with all shields terminated at the load (input) end of the equipment.

Input/Output Circuits

Common-mode rejection ratio (CMRR) is the measure of how well an input circuit rejects ground noise. The concept is illustrated in Fig. 10.284. The input signal to a differential amplifier is applied between the plus and minus amplifier inputs. The stage will have a certain gain for this signal condition, called the *differential gain*. Because the ground-noise voltage appears on both the plus and minus inputs simultaneously, it is common to both inputs.

The amplifier subtracts the two inputs, giving only the difference between the voltages at the input terminals at the output of the stage. The gain under this condition should be zero, but in practice it is not. CMRR is the ratio of these two gains (the differential gain and the common-mode gain) in decibels. The larger the number, the better. For example, a 60-dB CMRR means that a ground signal common to the two inputs will have 60 dB less gain than the desired differential signal. If the ground noise is already 40 dB below the desired signal level, the output noise level will be 100 dB below the desired signal level. If, however, the noise is already part of the differential signal, the CMRR will do nothing to improve it.

$$\text{DIFFERENTIAL GAIN} = \frac{V_{OUT}}{V_{SOURCE}}$$

$$\text{COMMON MODE GAIN} = \frac{V_{OUT}}{V_{SOURCE}}$$

$$\text{CMRR (DB)} = 20 \, \text{LOG}_{10} \left(\frac{\text{DIFFERENTIAL GAIN}}{\text{COMMON MODE GAIN}} \right)$$

FIGURE 10.284 The concept of common-mode rejection ratio (CMRR) for an active-balanced input circuit: (a) differential gain measurement; (b) calculating CMRR.

Active-balanced I/O circuits are the basis of nearly all professional audio interconnections (except for speaker connections), many video signals, and increasing numbers of data lines. A wide variety of circuit designs have been devised for active-balanced inputs. All have the common goal of providing high CMRR and adequate gain for subsequent stages. All also are built around a few basic principles.

Figure 10.285 shows the simplest and least expensive approach, using a single operational amplifier (op-amp). For a unity gain stage, all the resistors are the same value. This circuit presents an input impedance to the line that is different for the two sides. The positive input impedance will be twice that of the negative input. The CMRR is dependent on the matching of the four resistors and the balance of the source impedance. The noise performance of this circuit, which usually is limited by the resistors, is a tradeoff between low loading of the line and low noise.

Another approach, shown in Fig. 10.286, uses a buffering op-amp stage for the positive input. The positive signal is inverted by the op-amp, then added

$$\text{GAIN} = \frac{R_2}{R_1}$$

FIGURE 10.285 The simplest and least expensive active-balanced input op-amp circuit. Performance depends on resistor-matching and the balance of the source impedance.

to the negative input of the second inverting amplifier stage. Any common-mode signal on the positive input (which has been inverted) will cancel when it is added to the negative input signal. Both inputs have the same impedance. Practical resistor-matching limits the CMRR to about 50 dB. With the addition of an adjustment potentiometer, it is possible to achieve 80 dB CMRR, but component aging will usually degrade this over time.

Adding a pair of buffer amplifiers before the summing stage results in an *instrumentation-grade* circuit, as shown in Fig. 10.287. The input impedance is increased substantially, and any source impedance effects are eliminated. More noise is introduced by the added op-amp, but the resistor noise usually can be decreased by reducing impedances, causing a net improvement in system noise.

Early active-balanced output circuits used the approach shown in Fig. 10.288. The signal is buffered to provide one phase of the balanced output. This signal then is inverted with another op-amp to provide the other phase of the output signal. The outputs are taken through two resistors, each of which is half of the desired source impedance. Because the load is driven from the two outputs, the maximum output voltage is double that of an unbalanced stage.

The circuit in Fig. 10.288 works reasonably well if the load is always balanced, but it suffers from two problems when the load is not balanced. If the negative output is shorted to ground by an unbalanced load connection, the first op-amp is likely to distort. This produces a distorted signal at the input to the other op-amp. Even if the circuit is arranged so that the second op-amp is grounded by an unbalanced load, the distorted output current probably will show up in the output from coupling through grounds or circuit-board traces. Equipment that uses this type of balanced stage often provides a second set of output connections that are wired to only one amplifier for unbalanced applications.

The second problem with the circuit in Fig. 10.288 is that the output does not *float*. If any voltage difference (such as power-line hum) exists between the local ground and the ground at the device receiving the signal, it will appear as an addition to the signal. The only ground-noise rejection will be from the CMRR of the input stage at the receive end.

The preferred basic output stage is the electronically balanced and floating design shown in Fig. 10.289. The circuit consists of two op-amps that are cross-coupled with positive and negative feedback. The output of each amplifier is dependent on the input signal and the signal present at the output

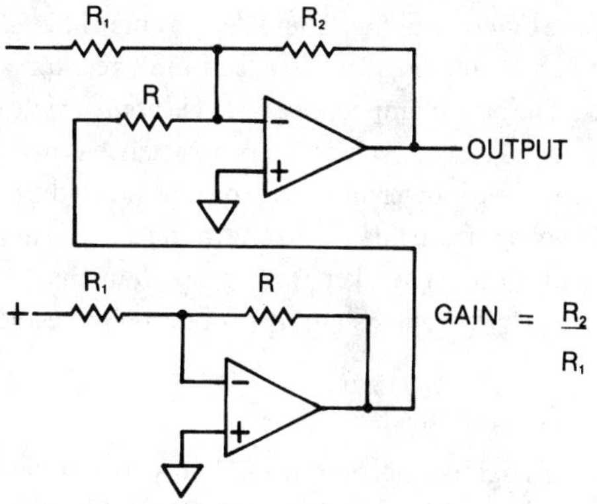

FIGURE 10.286 An active-balanced input circuit using two op-amps; one to invert the positive input terminal and the other to buffer the difference signal. Without adjustments, this circuit will provide about 50 dB CMRR.

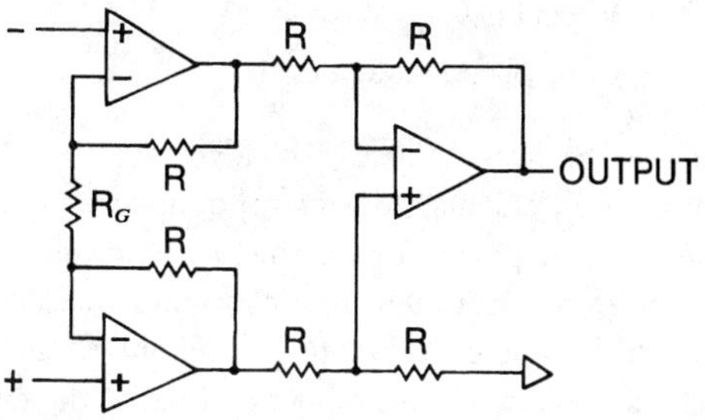

FIGURE 10.287 An active-balanced input circuit using three op-amps to form an instrumentation-grade circuit. The input signals are buffered, then applied to a differential amplifier.

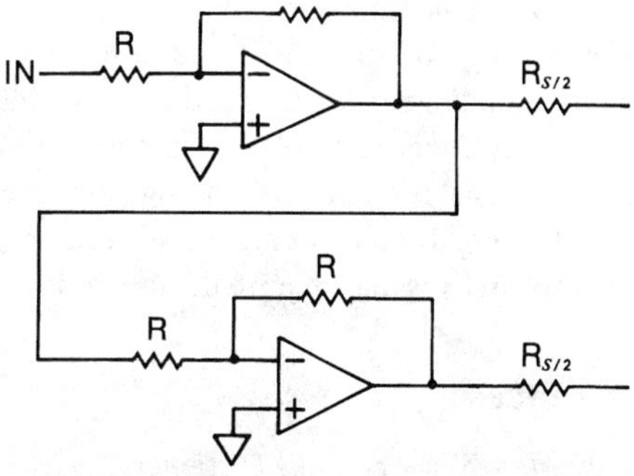

FIGURE 10.288 A basic active-balanced output circuit. This configuration works well when driving a single balanced load.

of the other amplifier. These designs may have gain or loss, depending on the selection of resistor values. The output impedance is set via appropriate selection of resistor values. Some resistance is needed from the output terminal to ground to keep the output voltage from floating to one of the power-supply rails. Care must be taken to properly compensate the devices. Otherwise, stability problems can result.

R_5 MAY BE OMITTED

FIGURE 10.289   An electronically balanced and floating output circuit. A stage such as this will perform well even when driving unbalanced loads.

Cable Routing

Good engineering practice dictates that different signal levels be grouped and separated from each other. It is common practice to separate cables into the following groups:

- AC power
- Speaker lines
- Line-level audio
- Microphone-level audio
- Video lines
- Control and data lines

Always use two-conductor shielded cable for all audio signal cables. This includes both balanced and unbalanced circuits, and microphone-level and line-level cables. On any audio cable connecting two pieces of equipment, tie the shield at one end only. Connect at the receiving end of signal transmission. On video coaxial cables running to outlet jacks mounted on plates, isolate the connector from the plate. The shield should connect to ground only at the equipment input/output or patch panel. For data cables, carefully follow the recommendations of the equipment manufacturer. The preferred interconnection method for long data cables is fiber optics, which eliminates ground-loop problems altogether.

Overcoming Ground-System Problems

Although the concept of equipment grounding seems rather basic, it can become a major headache if not done correctly. Even if all of the foregoing guidelines are followed to the letter, there is the possibility of ground loops and objectionable noise on audio, video, or data lines. The larger the physical size of the facility, the greater the potential for problems. An otherwise perfect ground system can be degraded by a single wiring error. An otherwise clean signal ground can be contaminated by a single piece of equipment experiencing a marginal ground fault condition.

If problems are experienced with a system, carefully examine all elements to track down the wiring error or offending load. Do not add AC line filters and/or signal line filters to correct a problem system. In a properly designed system, even one in a high-RF field, proper grounding and shielding techniques will permit reliable operation. Adding filters merely hides the problem. Instead, correct the problem at its source. In a highly complex system such as a data processing facility, the necessity to interconnect a large number of systems may require the use of fiber-optic transmitters and receivers.

References

1. *Webster's New Collegiate Dictionary*.
2. IEEE Standard 100, *Definitions of Electrical and Electronic Terms*, IEEE, New York.
3. IEEE Standard 142, "Recommended Practice for Grounding Industrial and Commercial Power Systems," IEEE, New York, 1982.

4. IEEE Standard 1100, "Recommended Practice for Powering and Grounding Sensitive Electronics Equipment," IEEE, New York, 1992.

5. DeWitt, William E., "Facility Grounding Practices," in *The Electronics Handbook,* Jerry C. Whitaker (Ed.), CRC Press, Boca Raton, FL, pp. 2218–2228, 1996.

6. NFPA Standard 70, "The National Electrical Code," National Fire Protection Association, Quincy, MA, 1993.

7. Sunde, E. D., *Earth Conduction Effects in Transmission Systems,* Van Nostrand Co., New York, 1949.

8. Schwarz, S. J., "Analytical Expression for Resistance of Grounding Systems," *AIEE Transactions,* 73, Part III-B, 1011–1016, 1954.

9. Block, Roger, "The Grounds for Lightning and EMP Protection," PolyPhaser Corporation, Gardnerville, NV, 1987.

10. Carpenter, Roy, B., "Improved Grounding Methods for Broadcasters," *Proceedings, SBE National Convention,* Society of Broadcast Engineers, Indianapolis, 1987.

11. Lobnitz, Edward A., "Lightning Protection for Tower Structures," in *NAB Engineering Handbook,* 9th ed., Jerry C. Whitaker (Ed.), National Association of Broadcasters, Washington, D.C., 1998.

12. DeDad, John A., (Ed.), "Basic Facility Requirements," in *Practical Guide to Power Distribution for Information Technology Equipment,* PRIMEDIA Intertec, Overland Park, KS, p. 24, 1997.

13. Military Handbook 419A, "Grounding, Bonding, and Shielding for Electronic Systems," U.S. Government Printing Office, Philadelphia, PA, December 1987.

14. IEEE 142 (Green Book), "Grounding Practices for Electrical Systems," IEEE, New York.

15. Federal Information Processing Standards Publication No. 94, *Guideline on Electrical Power for ADP Installations,* U.S. Department of Commerce, National Bureau of Standards, Washington, D.C., 1983.

16. DeDad, John A., (Ed.), "ITE Room Grounding and Transient Protection," in *Practical Guide to Power Distribution for Information Technology Equipment,* PRIMEDIA Intertec, Overland Park, KS, pp. 83–84, 1997.

17. Gruzs, Thomas M., "High Availability, Fault-Tolerant AC Power Distribution Systems for Critical Loads, *Proceedings,* Power Quality Solutions/Alternative Energy, Intertec International, Ventura, CA, pp. 22, September 1996.

Further Information

Benson, K. B., and Jerry C. Whitaker, *Television and Audio Handbook for Engineers and Technicians,* McGraw-Hill, New York, 1989.

Block, Roger, "How to Ground Guy Anchors and Install Bulkhead Panels," *Mobile Radio Technology,* PRIMEDIA Intertec, Overland Park, KS, February 1986.

Davis, Gary, and Ralph Jones, *Sound Reinforcement Handbook,* Yamaha Music Corporation, Hal Leonard Publishing, Milwaukee, WI, 1987.

Defense Civil Preparedness Agency, "EMP Protection for AM Radio Stations," Washington, D.C., TR-61-C, May 1972.

Fardo, S., and D. Patrick, *Electrical Power Systems Technology,* Prentice-Hall, Englewood Cliffs, NJ, 1985.

Hill, Mark, "Computer Power Protection," *Broadcast Engineering,* PRIMEDIA Intertec, Overland Park, KS, April 1987.

Lanphere, John, "Establishing a Clean Ground," *Sound & Video Contractor,* PRIMEDIA Intertec, Overland Park, KS, August 1987.

Lawrie, Robert, *Electrical Systems for Computer Installations,* McGraw-Hill, New York, 1988.

Little, Richard, "Surge Tolerance: How Does Your Site Rate?" *Mobile Radio Technology,* Intertec Publishing, Overland Park, KS, June 1988.

Midkiff, John, "Choosing the Right Coaxial Cable Hanger," *Mobile Radio Technology,* Intertec Publishing, Overland Park, KS, April 1988.

Mullinack, Howard G., "Grounding for Safety and Performance," *Broadcast Engineering,* PRIMEDIA In-
    tertec, Overland Park, KS, October 1986.

Schneider, John, "Surge Protection and Grounding Methods for AM Broadcast Transmitter
    Sites," *Proceedings, SBE National Convention,* Society of Broadcast Engineers, Indianapolis,
    1987.

Sullivan, Thomas, "How to Ground Coaxial Cable Feedlines," *Mobile Radio Technology,* PRIMEDIA In-
    tertec, Overland Park, KS, April 1988.

Technical Reports LEA-9-1, LEA-0-10, and LEA-1-8, Lightning Elimination Associates, Santa Fe Springs,
    CA.

10.11   Batteries

Isidor Buchmann

10.11.1   Introduction

Research has brought about a number of different battery chemistries, each offering distinct advantages over the others. Today's most common and promising chemistries available include:

- Nickel cadmium (NiCd). Used for portable radios, cellular phones, video cameras, power tools and some biomedical instruments. NiCds have a good load characteristics, are economically priced and are simple to use.
- Nickel metal hydride (NiMH). Used for cellular phones and laptop computers where high energy is of importance and cost is secondary.
- Sealed lead acid (SLA). Used for biomedical equipment, wheel chairs, UPS systems, and other heavier applications where energy-to-weight ratio is not critical and low battery cost is desirable.
- Lithium ion (Li-ion). Used for laptop computers and some video cameras, this chemistry will likely replace some NiCds for high energy-density applications.
- Lithium polymer (Li-polymer). This battery has the highest energy density and lowest **self-discharge** but its load characteristics only suit low current applications.
- Reusable alkaline. Used for light duty applications. Because of its low self-discharge, this battery is suitable for portable entertainment devices and other noncritical appliances that are used only occasionally.

10.11.2   Choice of Battery Chemistries

No single battery offers all of the answers; rather, each chemistry is based on a number of compromises. Table 10.16 compares the pros and cons of the six most common batteries: the sealed NiCd, NiMH, SLA, Li-ion, Li-polymer, and reusable alkaline.

It is interesting to observe that the NiCd has the shortest charge time, delivers the highest load current, and offers the lowest cost-per-cycle, but is most demanding on exercise requirements. For applications where high-energy density is critical, regular exercise is impractical, and cost is secondary, the NiMH is probably the best choice. Not without problems, NiMH batteries have a cycle life one-third that of NiCds. Furthermore, field use has revealed that the NiMH also needs some level of exercise to maximize service life, but to a lesser extent than the NiCd. In comparison, the SLA needs little or no maintenance but has a low-energy density.

Among rechargeable batteries, the NiCd has been around the longest (since 1950). It is also one of the best understood chemistries and has become a standard against which other batteries are compared. This section compares the different chemistries relative to the NiCd.

TABLE 10.16　Operating Characteristics of Common Battery Types

	NiCd	NiMH	SLA	Li-ion	Li-polymer	Reusable alkaline
Energy density, Wh/kg	50	75	30	100	175	80 (initial)
Cycle life[a] (typical)	1500	500	200–300	300–500	150	10 (to 65%)
Fast-charge time,[b] h	1 1/2	2–3	8–15	3–6	8–15	3–4
Self-discharge[c]	moderate[d]	high	low	low	very low	very low
Cell voltage,[e] (nominal) V	1.25	1.25	2	3.6	2.7	1.5
Load current,[f]	very high	moderate	low	high	low	very low
Exercise requirement[g]	/30 days	/90 days	/180 days	N/A	N/A	N/A
Battery cost[h]	low	moderate	very low	very high	high	very low
(estimated, ref. only)	(50.00)	(80.00)	(25.00)	(100.00)	(90.00)	(5.00)
Cost per cycle[i] ($)	0.04	0.16	0.10	0.25	0.60	0.50
In commercial use since	1950	1970	1970	1990	(1997)	1990

[a]Cycle life indicates the typical number of charge-discharge cycles before the capacity decreases from the nominal 100 to 80% (65% for the reusable alkaline).

[b]Fast-charge time is the time required to fully charge an empty battery.

[c]Self-discharge indicates the self-discharge rate when the battery is not in use.

[d]Moderate refers to 1–2% capacity-loss per day.

[e]Cell voltage multiplied by the number of cells provides the battery terminal voltage.

[f]Load current is the maximum recommended current the battery can provide. High refers to a discharge rate of 1C; very high is a current higher than 1 C. **C rate** is a unit by which charge and discharge times are scaled. If discharged at 1, a 1000 mAh battery provides a current of 1000 mA; if discharged at 0.5C, the current is 500 mA.

[g]**Exercise** requirement indicates the frequency the battery needs exercising to achieve maximum service life.

[h]Battery cost is the estimated commercial price of a commonly available battery.

[i]Cost-per-cycle indicates the operating cost derived by taking the average price of a commercial battery and dividing it by the cycle count.

The Nickel Cadmium (NiCd) Battery

Among the rechargeable batteries, the NiCd remains a popular choice for applications such as portable radios, video cameras, data loggers, and power tools. There is a renewed interest in this chemistry, and many equipment manufacturers are switching back to the NiCd after having recognized the limitations of other chemistries. Some of the distinct advantages of the NiCd over other battery chemistries are

- Fast and simple charge
- High number of charge–discharge cycles (if properly maintained, the NiCd provides several thousand cycles)
- Excellent load performance, even at cold temperatures
- Simple storage and transportation (the NiCd is accepted by most air freight companies)
- Easy to recharge after prolonged storage
- Forgiving if abused
- Economical price

The NiCd is a strong and reliable worker; hard labor poses no problem. Fast charge is preferable to slow charge and pulse-charge to DC charge. Improved performance is achieved by interspersing discharge pulses between charge pulses. Commonly referred to as *burp* or *reverse load* charge, this charge method promotes the recombination of gases generated during fast charge. The result is a cooler and more effective charge than with conventional DC chargers. The battery is stimulated while charging, a function that improves battery performance. Research has shown that reverse load adds 15% to the life of the NiCd battery.

The NiCd does not respond well to sitting in chargers for days and being used only occasionally for brief periods. In fact, the NiCd is the only battery type that performs best if periodically discharged to 1 V per cell (full discharge). All other chemistries react best with shallow discharges. So important is this periodic full discharge that, if omitted, the NiCd gradually loses performance due to crystalline formation, also referred to as **memory**.

The Nickel Metal-Hydride (NiMH) Battery

The NiMH battery has certain advantages over the NiCd. For example:

- The NimH provides 30% more capacity over a *standard* NiCd.
- The NiMH is affected by memory to a lesser extent than the NiCd. Periodic exercise cycles need to be done less often.
- Because of its low toxic metals content, the NiMH is labeled environmentally friendly.

In other aspects the NiMH lags behind the NiCd. For example:

- Number of cycles. The NiMH is rated for only, 500 charge-discharge cycles. It does not respond well to full discharging during each use; its longevity is in direct relation to the depth of discharge. (In comparison, the NiCd can accept several thousand full discharge-charge cycles. NiCd batteries for satellite applications are designed to last for 15–20 years and provide 70,000 cycles).
- Ease of fast charge. The NiMH battery requires a more complex algorithm for full-charge detection than the NiCd, if temperature sensing is not available. In addition, the NiMH cannot accept as quick a fast charge as the NiCd.
- Discharge current. The recommended discharge current of the NiMH is considerably less than that of the NiCd. For best results, manufacturers recommend a load current of $0.2C$ (one-fifth of the rated capacity). This shortcoming may not be critical for small DC loads, but in applications requiring high power or pulsed load, such as in transceivers, power tools, and GSM cellular telephones, the more rugged NiCd is the recommended choice.
- High self-discharge. Both NiMH and NiCd are affected by high self-discharge. The NiCd loses about 10% of its capacity within the first 24 h, after which the self-discharge settles to about 10% per month. For the NiMH the self-discharge is typically twice as much because the hydrogen atoms try to escape. Selecting materials that improve hydrogen bonding reduces the capacity of the battery. Research engineers are faced with a compromise between reasonably low self-discharge and high-energy density.
- Capacity. The NiMH delivers about 30% more capacity than a NiCd of the same size. The comparison is made with the *standard*, rather than *ultra-high capacity* NiCd. Some ultra-high capacity NiCd cells provide a capacity level similar to the NiMH.
- Price. The price of the NiMH is higher than that of the NiCd. Price may not be a major issue if the user requires high capacity and small size.

The NiMH is not new. In the 1970s, this battery chemistry was tested and consequently dropped because it was considered unsuitable for the applications intended. However, the modern NiMH has been improved and will likely maintain a strong market niche.

The Sealed Lead-Acid (SLA) Battery

Another commonly used chemistry is the sealed lead-acid battery. The *flooded* version is found in automobiles, but most portable equipment use the *sealed* version, also referred to as *gelcell* or *SLA*.

The SLA is commonly used when high power is required, weight is not critical, and cost must be kept low. The typical current range of the SLA is 2 to 30 Ah. Applications that fall into this category are wheel chairs, UPS units, and emergency lighting. Because of its minimal maintenance requirements and predictable storage, the SLA gained ground in the biomedical industry during the 1980s.

The SLA is not subject to memory. No harm is done by leaving the battery on float charge for a prolonged time. If removed from the charger, the SLA retains the charge four times longer than the NiCd. Based on the per ampere-hour cost, the SLA is the lowest priced battery, but in terms of cost-per-cycle, it is more expensive than the NiCd.

On the negative side, the SLA does not lend itself well to fast charging. Typical charge times are 8–16 h. The SLA must always be stored in a charged state because a discharged SLA will sulphate. If left discharged, a recharge is difficult or impossible.

Unlike the NiCd, a shallow discharge is preferred for SLA. A full discharge causes extra strain and reduces the number of times the battery can be recharged, similar to a mechanical device that wears down when being strained. In fact, each discharge–charge cycle robs the battery of a small amount of capacity. This wear-down characteristic also applies to other chemistries, including the NiMH and to a lesser extent the NiCd.

An additional difficulty with the SLA is a relatively low load current. This current is further reduced at cold temperatures. Compared to other rechargeables, the energy density is low, making the SLA unsuitable for devices demanding compact size and long play time. Because of its high lead content, the SLA is not environmentally friendly.

The Rechargeable Lithium Battery

The energy density of the rechargeable lithium is the highest among commercial batteries and the self-discharge is very low. There are several types of lithium batteries of interest; the most promising chemistries are the lithium-ion (Li-ion) and the lithium polymer (Li-polymer).

The lithium-ion provides an energy density that is twice that of the NiCd at comparable load currents. In addition, the Li-ion deep-discharges well. In fact, the Li-ion behaves similarly to the NiCd as far as charge and discharge characteristics are concerned.

Unlike other rechargeable lithium batteries, the Li-ion is stable and safe because no metallic lithium is used. However, certain precautions must still be taken. For reasons of safety and longevity, each cell must be equipped with a control circuit to limit the voltage peaks during charge and to prevent the voltage from dropping too low on discharge. In addition, the control circuit must limit the maximum charge and discharge current. With these precautions in place, the danger of metallic lithium formation is eliminated.

The lithium-polymer provides more than three times the energy density of the NiCd and features a very low self-discharge. As far as charge and discharge characteristics are concerned, the lithium polymer behaves more like the SLA battery. Charge time is 8–16 h. Shallow rather than deep discharge is preferred.

Disposal of all lithium-based batteries may cause some headaches. The concern is not their toxic metal content but the danger of explosion if moisture creeps into the cells when they corrode.

Reusable Alkaline

The idea of recharging alkaline batteries is not new. In fact, ordinary alkaline batteries have been recharged for years for applications such as flashlights in commercial aircraft. The difference between a *reusable* and a *standard* alkaline lies in the energy density and the number of times the battery can be recharged. Whereas the standard alkaline offers maximum energy density, the reusable alkaline provides a higher cycle count but compromises on the energy density.

One of the major drawbacks of the reusable alkaline is the limited number of times the battery can be cycled. The longevity is a direct function of the depth of discharge; the deeper the discharge, the fewer cycles the battery endures.

An additional limitation of the reusable alkaline is its low load current of 400 mA maximum (lower current provides better results). Although adequate for portable radios, CD players, tape recorders, and flashlights, 400 mA is insufficient to power many other devices.

Another negative aspect of the reusable alkaline is its relative high cost per cycle when compared to the NiCd. For many applications, this still represents a saving when compared to nonreusable alkalines.

To compare the operating cost between the standard and reusable alkaline, a study was done on flashlight batteries for hospital use. The low-intensity care unit achieved measurable savings by using the reusable alkaline, but the high-intensity unit did not attain the same saving; deeper and more frequent cycling of the batteries reduced their service life and offset any cost advantage.

10.11.3 Proper Charge Methods

The critical factor in terms of performance and longevity of all battery chemistries lies in the way the batteries are cared for. In this section, we examine the reasons why some batteries do not perform as expected and fall short of their life expectancy.

Chargers for the NiCd Battery

Commercial chargers are often not designed in the best interests of the battery. This is especially true of NiCd chargers that measure the charge state through temperature sensing. Temperature sensing is inaccurate because of the wide tolerances of the sensing thermistor and its positioning in respect to the cells. Ambient temperatures and exposure to the sun while charging also come into play. To assure full charge under all conditions, charger manufacturers tend to cook the batteries. Any prolonged temperature above 45°C (113°F) is harmful to the battery and should be avoided.

More advanced NiCd chargers use the delta temperature method for full-charge detection. Although still dependent on a thermistor, this type of charger is kinder on the batteries than those using the absolute battery temperature method.

With a microcontroller, more precise full-charge detection is achieved by monitoring the battery voltage and terminating the charge when a certain voltage signature occurs. Such a signature is the **negative delta V (NDV)**.

The NDV is ideal for *open-lead* NiCd chargers and battery analyzers because the batteries they service do not need a thermistor. The NDV has drawbacks, however; it only works well with matched NiCd packs. A battery with **mismatched cells** does not produce a sufficient negative slope to terminate the fast charge, and the battery may overheat due to overcharge. Chargers using the NDV must feature other charge-termination methods to provide safe charging under all conditions.

Mismatched cells are found in brand new batteries and those that have aged. Common reasons for poorly matched cells are: poor quality control at the cell manufacturing level and inadequate cell matching when packing the batteries. If not too far apart, the cells in a new pack adapt to each other after a few charge-discharge cycles in a manner similar to players in a winning sports team. In old packs, mismatched cells are irreconcilable; although still usable, the performance of these batteries is poor.

Chargers for the NiMH Battery

Unlike NiCd chargers, the chargers for NiMH batteries cannot utilize the NDV for full-charge detection because NiMH batteries do not generate an adequate voltage drop at full charge, even with perfectly matched cells. Reducing the delta to respond to a smaller voltage drop may terminate the fast charge halfway through the charge cycle. If temperature sensing is not available, an algorithm must be used that identifies a certain voltage signature that occurs at the full-charge state. Because of the complexity in circuitry and firmware required, most NiMH chargers are still based on temperature sensing.

The NiMH battery is sensitive to the amount of **trickle charge** applied to maintain full charge. In spite of its inherent high self-discharge, the trickle charge of the NiMH must be set lower than that of the NiCd. A trickle charge that is acceptable for the NiCd will overheat the NiMH and cause irreversible damage.

Chargers for the SLA Battery

The charge algorithm of the SLA differs from that of the NiCd and NiMH in that **voltage limit** rather than **current limit** is used. The charge time of the SLA is a slow 16 h. With higher charge currents and multistage charge methods, the charge time can be reduced to about 10 h. In no way can the SLA be charged as quickly as the NiCd.

A multistage charger applies three charge stages consisting of *constant-current charge, topping charge,* and *float charge* (see Fig. 10.290). During the constant-current stage, the battery charges to 70% in about 5 h, the remaining 30% is completed by the topping charge. The slow topping charge, lasting another 5 h, is essential for the well-being of the battery. If deprived, the SLA eventually loses the ability to accept a full charge and the performance of the battery is reduced. The third stage is the float charge, which compensates for the self-discharge after the battery has been fully charged. During the constant current charge, the SLA battery is charged at a high current, limited by the charger itself. Once the voltage limit is reached, the topping charge begins and the current starts to drop gradually. Full charge is reached when the current drops to a preset level or reaches a low-end plateau.

The proper setting of the cell voltage limit is critical and is related to the conditions under which the battery is charged. A typical voltage limit range is from 2.30 to 2.45 V. If a slow charge is acceptable, or if the

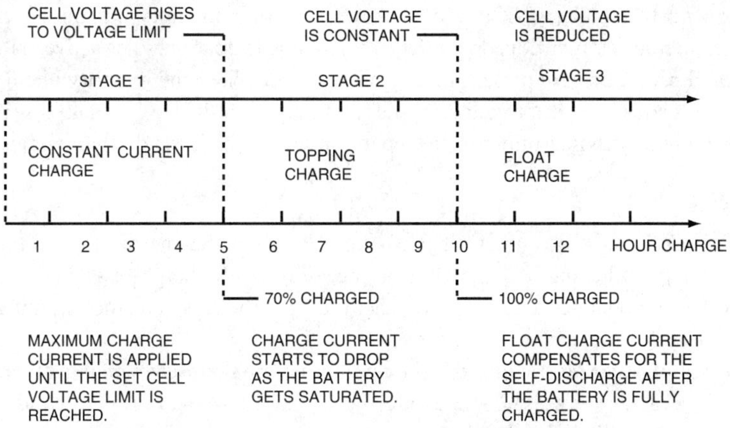

FIGURE 10.290 Charge stages of a SLA battery.

room temperature exceeds 30°C (86°F), the recommended voltage limit is 2.35 V /cell. If a faster charge is required and the room temperature remains below 30°C, 2.40 or 2.45 V/cell may be used. Table 10.17 compares the advantages and disadvantages of the different voltage settings.

10.11.4 Battery Maintenance

A common difficulty with battery-powered equipment is the progressive deterioration in reliability after the first year of service. This phenomenon is mostly due to premature aging of the battery, a reversible capacity loss that is induced by memory. This loss occurs gradually without the user's knowledge. Although fully charged, the battery eventually regresses to a point where it can hold less than half of its original capacity, resulting in unexpected down time.

Down time almost always occurs at critical moments. Under normal conditions, the battery holds enough power until recharged. During heavy activities and longer than expected duties, a marginal battery cannot provide the extra power needed and the equipment fails.

In many ways, a rechargeable battery exhibits human-like characteristics: it needs good nutrition, it prefers moderate room temperature and, in case of the NiCd battery, requires regular exercise to prevent the phenomenon called memory.

Memory: Myth or Fact?

There is some misconception about the word memory. Memory is commonly blamed for all battery failures known to man. Originally, the word *memory* was derived from *cyclic memory*, meaning that a NiCd battery can remember how much discharge was required on previous discharges. Improvements in battery technology have virtually eliminated this phenomenon.

TABLE 10.17   Recommended Charge Voltage Limit on SLA Battery

	2.30–2.35 V/cell	2.40–2.45 V/cell
Advantage	Maximum service life; battery remains cool on charge; battery may be charged at ambient temperature exceeding 30°C (86°F).	Faster charge times; higher and more consistent capacity readings; less subject to damage due to undercharge condition.
Disadvantage	Slow charge time; capacity readings may be low and in-consistent. Produces under-charge condition, which may cause sulphation and capacity loss if the battery is not periodically cycled.	Battery life may be reduced due to elevated battery temperature while charging. A hot battery may fail to reach the cell voltage limit, causing harmful overcharge.

The problem with the modern NiCd battery is not the cyclic memory but the effects of *crystalline formation*. (When memory is mentioned we refer to crystalline formation.) The active materials of a NiCd battery (nickel and cadmium) are present in crystalline form. When the memory phenomenon occurs, these crystals grow, forming spike or tree-like crystals that cause the NiCd to gradually lose performance. In advanced stages, these crystals may puncture the separator, causing high self-discharge or an electrical short.

Regular Exercise

Crystalline formation is most prominent if the battery is left in the charger for days, or if repeatedly recharged without a periodic full discharge. It is not necessary to discharge a NiCd before each charge. A full discharge to 1 V per cell once a month is sufficient to keep the crystal formation under control. Such a discharge/charge cycle is referred to as *exercise*.

If no exercise is applied for several months, the crystals ingrain themselves, making them more difficult to dissolve. In such a case, exercise is no longer effective in restoring a battery and **recondition** is required. Recondition is a slow, deep discharge that drains the cell to a voltage threshold below 1 V per cell. Laboratory tests have shown that a NiCd cell should be discharged to at least 0.6 V to effectively dissolve the more resistant crystalline buildup. When applying recondition, the current must be carefully controlled to prevent cell reversal.

In the following study, four batteries afflicted with various degree of memory are analyzed. The batteries are first fully charged, followed by a discharge to 1 V per cell. The resulting capacities are plotted on a scale of 0–120% in the first column of Fig. 10.291. Additional discharge-charge cycles are applied and the battery capacities are plotted in the subsequent columns. The dotted line represents exercise, the solid line recondition. Exercise and recondition cycles are applied manually at the discretion of the research technician. Battery A responded well to exercise alone. This result is typical of a battery that had been in service for only a few months or had been exercised periodically. Batteries B and C required recondition (solid line) to recover full performance. The capacity of the new battery was further enhanced after recondition was applied. When examined after six months of field use, the batteries still showed excellent capacity readings.

The importance of exercise and recondition are further emphasized by a study carried out by GTE Government Systems in Virginia, to determine what percentage of batteries needed replacing within the first year of use. One group of batteries received charge only, another group was exercised, and a third group received recondition. The batteries studied were used for portable radios on the aircraft carriers U.S.S. Eisenhower with 1500 batteries and U.S.S. George Washington with 600 batteries, and the destroyer U.S.S. Ponce with 500 batteries.

Table 10.18 shows that with charge only (charge-and-use), the annual percentage of batteries failure on the U.S.S. Eisenhower was 45%. By applying exercise, the failure was reduced to 15%. By far the best results were achieved with recondition; its failure rate dropped to 5%. The same results were achieved with the U.S.S. George Washington and the U.S.S. Ponce.

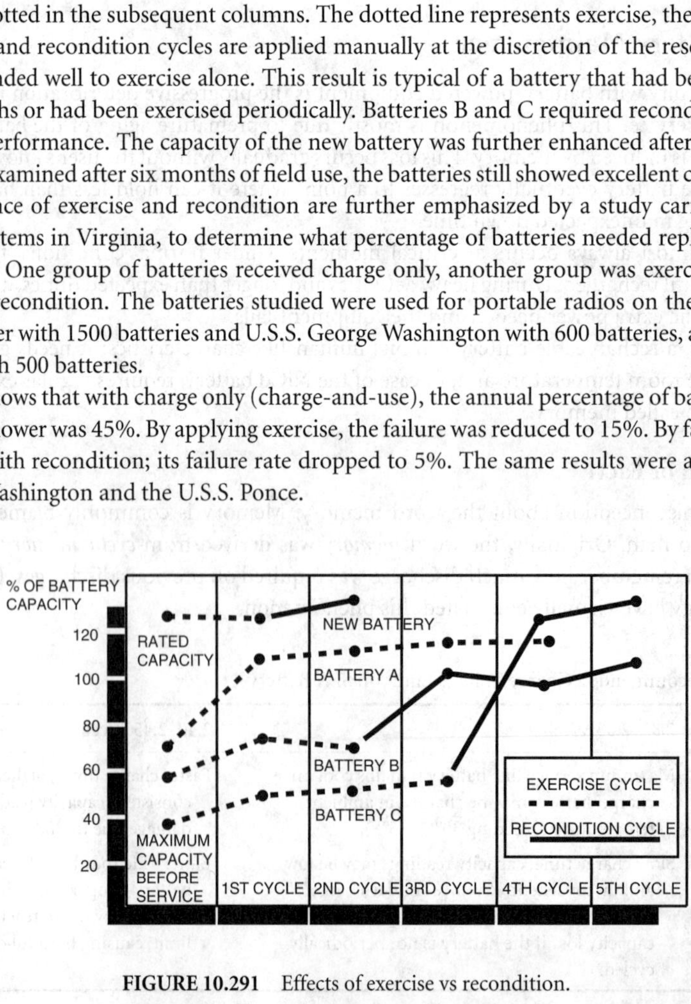

FIGURE 10.291   Effects of exercise vs recondition.

TABLE 10.18 Annual Percentage of Batteries Requiring Replacement on the U.S.S. Eisenhower, U.S.S. George Washington, and U.S.S. Ponce as a Function of Battery Maintenance

Maintenance Method	Annual Percentage of Batteries Requiring Replacement, %
Charge only (charge-and-use)	45
Exercise only (discharge to 1 V/cell)	15
Reconditioning (secondary deep discharge)	5

The report states further that a battery analyzer that features the exercise and recondition functions and costs $U.S. 2500 would pay for itself in less than one month on battery savings alone. No mention was made of the benefits of increased system reliability, an issue that is of equal or larger importance when the safety of human lives is at stake.

In respect to restoring weak batteries through exercise and recondition, not all batteries respond equally well; an older battery may show low and inconsistent capacity readings, or even get worse with each cycle applied. If this occurs, the battery should be replaced. This type of battery can be compared to a very old man to whom exercise is harmful. There are older NiCd batteries, however, that recover to near original capacity when serviced. Caution should be applied when "re-hiring" these older batteries because they may exhibit high self-discharge.

Self-Discharge

The NiCd and the NiMH exhibit a high self-discharge. If left on the shelf, a new NiCd loses about 10% of its capacity in 24 h. A problem arises if the battery self-discharges within a day, a phenomenon that is not uncommon, especially for an older battery. The cause of high self-discharge is a damaged separator, a thin, delicate insulator that isolates the positive and negative cell plates (see Fig. 10.292). Once damaged, the separator can no longer be repaired. External forces that harm the sensitive separator are: uncontrolled

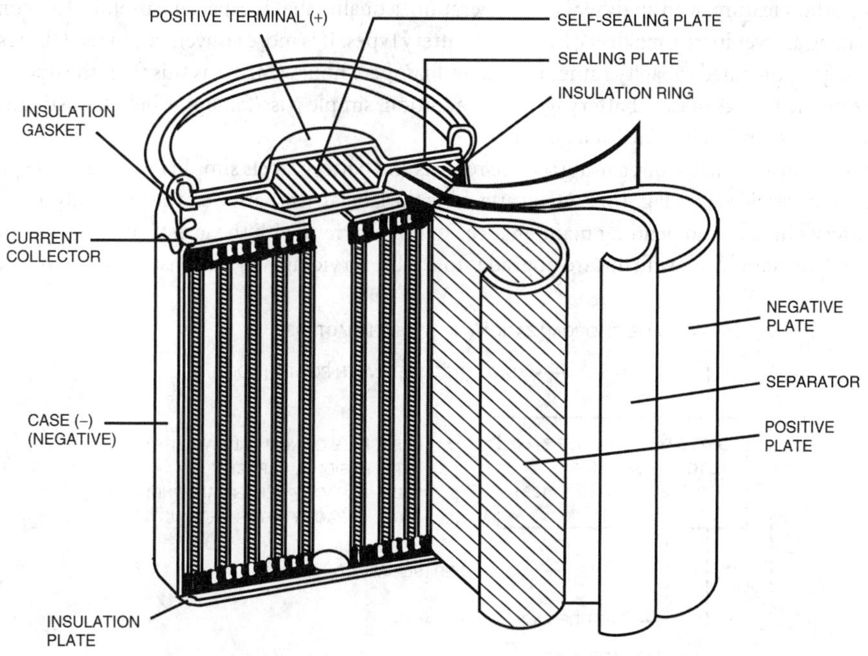

FIGURE 10.292 Drawing of a NiCd cell. The negative and positive plates are rolled together and placed in a metal cylinder. The positive plate is sintered and filled with nickel hydroxide. The negative plate is coated with cadmium active material. The two plates are isolated by a separator, which are moistened with the electrolyte.

crystalline formation due to lack of exercise, poorly designed chargers that boil the battery, and old age. In some instances, cell manufacturers use a separator material that is unsuitable.

The self-discharge of a battery can most conveniently be measured with a battery analyzer. A battery should be discarded if the resulting capacity loss over 24 h at a storage temperature of 22°C (70°F) is more than 30%. No amount of cycling will improve this condition.

Battery Analyzers

The technology of battery analyzers has advanced significantly. No longer are users limited to analog units with fixed charge/discharge currents and programs that only apply one or two discharge-charge cycles.

An advanced battery analyzer evaluates the condition of each battery and implements the appropriate service to restore its performance. A recondition cycle is applied automatically if a user-selected capacity level cannot be reached. Battery chemistry and voltage and current rates are user programmable. These parameters are stored in interchangeable cups and configure the analyzer to the correct function when installed.

Because the service is performed against preset parameters, the batteries are charged and discharged under true field conditions, a feature that provides accurate test results as well as fast service. Batteries with shorted, mismatched or **soft cells** are identified in minutes, their deficiencies displayed and, if necessary, the service halted. The derived battery capacities are organized into *residual* and *final* capacities. Problems, such as insufficient capacity reserve at the end of field use, are easily identified to allow necessary corrections.

To enable servicing all batteries, including those with known defects, an analyzer must feature multiple redundant charge-termination algorithms to ensure that the batteries are charged quickly and safely and without overheating. Without this provision, battery packs with shorted or mismatched cells may sustain damage due to overcharge. User-selectable programs address the different battery needs and include: a *prime* mode to prepare a new battery for field use, an *auto* mode to recondition batteries unable to reach a user-set target capacity, and a *custom* mode to allow the operator to set a sequence of unique cycle modes composed of charge, discharge, recondition, trickle charge, or any combination thereof, including rest periods and repeats.

An important feature of an analyzer is easy operation, a quality that is especially useful when confronted with servicing an ever increasing fleet of different battery types. It is more convenient to read the test results in percentage of the rated capacity rather than in milliampere hours (mAh) as this frees the operator from memorizing the ratings of each battery tested. In addition, simple pass/fail lights help to distinguish good batteries from unserviceable ones at a glance.

Analyzers that are capable of printing service reports and battery labels simplify the task of keeping track of the batteries serviced (see Fig. 10.293). By attaching the label to the batteries, the user is always informed of the battery's history and pending maintenance. To keep current with the latest battery developments, a battery analyzer should allow firmware upgrades to enable servicing new chemistries when introduced.

```
                    ┌─ NAME OF ORGANIZATION ( UP TO 15 CHARACTERS)
                    ┆
                         ┌─ DATE (RETAINED WHEN POWER IS OFF)
                         ┆
    ┌────────────────────┆────┐
    ┆ BATTERY SHOP    SEP 24/96 ┆   ┌─ CAPACITIES OF NOMINAL BATTERY RATING
    ┆ AUTO  17,  55,  117%      ┆   ┆  1ST — RESIDUAL CAPACITY
    ┆ [1] PASS:          #1234  ┆   ┆  2ND — CAPACITY BEFORE RECONDITION
    └────────────────────────────┘   └─ 3RD — CAPACITY AFTER RECONDITION
                         ┆
                         └─ ID NUMBER (SERIALIZED)
                    ┆
                    └─ TEST RESULT
              ┆
              └─ SELECTED STATION (1 OF 4)
    ┆
    └─ SELECTED PROGRAM
```

FIGURE 10.293   Sample battery label.

Advanced battery analyzers are capable of networking into a computer to form a workstation servicing 100 batteries or more simultaneously. User friendly software assists in collecting the battery test results for the database from which inventory, service reports, and graphs are generated. The software allows remote controlling of the analyzers through a central computer keyboard, including the downloading of firmware into the analyzer's flash memory to upgrade existing firmware, or to custom-program an analyzer for a special function. The battery identification number and parameters (chemistry, voltage, rating) are printed on bar code labels and attached to the batteries. By reading the bar code labels, the battery to be serviced is identified and the battery analyzer is automatically configured to the correct parameters for the battery intended.

The advancements in battery technology may not be as impressive as those achieved in microelectronics but with today's intelligent battery analyzers, the task of providing quality battery maintenance is simplified and the cost is lowered. By following the maintenance procedures recommended by the battery manufacturer, today's rechargeable batteries are capable of supplying a truly reliable and economical energy source.

10.11.5 Advancements in Battery Rapid-Testing

Portable batteries for cell phones, laptops, and cameras may be rapid-tested by applying a number of load pulses while observing the relationship between voltage and current. Ohm's Law is used to calculate the internal resistance. Comparing the readings against a table of values estimates the battery's state-of-health.

The load pulse method described above does not work well for larger batteries and AC conductance is commonly used. An AC voltage is applied to the battery terminals that floats as a ripple on top of the battery's DC voltage and charges and discharges the battery alternatively.

AC conductance has been incorporated into a number of hand-held testers to check batteries for vehicular and stationary batteries. To offer simple, compact and low-cost units, these testers load the battery with repetitive current pulses. Because these pulses are not voltage controlled, the thermal battery voltage[1] may be surpassed. The thermal voltage threshold of a lead-acid battery is 25 mV per cell. Exceeding this voltage is similar to over-driving an audio amplifier. Amplified noise and distortion is the result.

AC conductance provides accurate readings, provided the battery is fully charged, has rested or has been briefly discharged prior to taking the reading. AC conductance tends to become unreliable on low charge and sometimes fails a good battery. At other times, a faulty battery may pass as good. The sensitivity to the battery's state-of-charge is a common complaint by users. AC conductance works best in identifying batteries with definite deficiencies.

AC conductance is noninvasive, quick and the test instruments are relatively inexpensive. There are, however, some fundamental problems. Most commercial testers use only one frequency, which is commonly below 100 Hz. Multi-frequency systems would be more accurate but require complex data interpretation software and expensive hardware. In this section we focus on *electrochemical impedance spectroscopy* (EIS), a method that overcomes some of the shortcomings of AC conductance.

Electrochemical Impedance Spectroscopy (EIS)

EIS evaluates the electrochemical characteristics of a battery by applying an AC potential at varying frequencies and measuring the current response of the electrochemical cell. The frequency may vary from about 100 μHz to 100 kHz. 100 μHz is a very low frequency that takes more than two hours to complete one full cycle. In comparison, 100 kHz completes 100,000 cycles in one second.

Applying various frequencies can be envisioned as going through different layers of the battery and examining its characteristics at all levels. Similar to tuning the dial on a broadcast radio, in which individual

---

[1]Batteries are nonlinear systems. The equations, which govern the battery's response becomes linear below 25 mV/cell at 25°C. This voltage is called the battery thermal voltage.

stations offer various types of music, so too does
the battery provide different information at varying frequencies.

Battery resistance consists of pure Ohmic resistance, inductance, and capacitance. Figure 10.294
illustrates the classic Randles model.

Capacitance is responsible for the capacitor effect; and the inductance is accountable for the so-
called magnetic field, or coil effect. The voltage on
a capacitor lags behind the current. On a magnetic
coil, on the other hand, the current lags behind the
voltage. When applying a sine wave to a battery, the
reactive resistance produces a phase shift between
voltage and current. This information is used to evaluate the battery.

FIGURE 10.294   Randles model of a lead acid battery.
The overall battery resistance consists of pure Ohmic resistance, inductance, and capacitance. There are many other
models.

EIS has been used for a number of years to perform in-flight analysis of satellite batteries, as well as
estimating grid corrosion and water loss on aviation and stationary batteries. EIS gives the ability to observe
the kinetic reaction of the electrodes and analyze changes that occur in everyday battery usage. Increases
in impedance readings hint at minute intrusion of corrosion and other deficiencies. Impedance studies
using the EIS methods have been carried out on lead-acid, nickel-cadmium, nickel-metal-hydride and
lithium-ion batteries. Best results are obtained on a single cell.

One of the difficulties of EIS is data interpretation. It is easy to amass a large amount of data; making
practical use of it is more difficult. Analyzing the information is further complicated by the fact that the
readings are not universal and do not apply equally to all battery makes and types. Rather, each battery
type generates its own set of signatures. Without well-defined reference readings and software to interpret
the results, gathering information has little meaning for the ordinary person.

Modern technology can help by storing characteristic settings of a given battery type in the test instrument. Advanced digital signal processors are able to carry out millions of instructions per second.
Software translates the data into a single reading. EIS has the potential of becoming a viable alternative
to AC conductance in checking automotive, traction, and stationary batteries. Noteworthy advancements
are being made in this field.

Defining Terms

Cell mismatch: Cells within a battery pack that have different capacity levels.

C-rate: Unit by which charge and discharge times are scaled. At $1C$, the charge current equals the battery
rating, that is, a 1000-mAh battery is charged and discharged at 1000 mA.

Current-limiting chargers: A charger that keeps the charge current constant during the charge process
but allows the voltage to fluctuate.

Exercise: Commonly understood as one or more discharge cycles to 1 V per cell (NiCd and NiMH) with
subsequent recharge.

Memory: Reversible capacity loss found on NiCd and to a lesser extent on NiMH batteries. The modern
definition of memory refers to crystalline formation of the cell plates.

Negative delta V (NDV): The NDV is a drop in battery voltage after full charge is reached. During the
initial charge state the NiCd battery terminal voltage rises quickly, then levels off. After the 70%
charge level, the voltage rises again, peaks at full charge and drops slightly.

Recondition: Secondary discharge, which drains the battery of its remaining energy, during which the
crystalline formation (memory) is dissolved.

Self-discharge: Capacity loss due to internal leakage between the positive and negative cell plates.

Soft cell: A cell whose voltage rises well above the specifications during charging. This high voltage is
caused by high cell impedance.

Trickle charge: Maintenance charge to keep the battery in full-charge state after the charge cycle. Trickle charge compensates for self-discharge.

Voltage-limiting charger: A charger that limits the voltage after a preset voltage table is reached but allows the current to drop while maintaining the voltage limit.

References

Berndt, D. 1993. *Maintenance-Free Batteries: Lead-Acid, Nickel Cadmium Nickel-Hydride: A Handbook of Battery Technology*. Wiley, New York.

Buchmann, I. 1995. *Strengthening the Weakest Link: The Importance of Battery Maintenance*. Cadex Electronics, Burnaby, BC, Canada.

Chow, A.T. and Keyes, C.E. 1991. *Lift Truck Battery Charger Efficiency*. Ontario Hydro, Toronto, Canada.

Grant, J.C., ed. 1975. *Nickel-Cadmium Battery Application Engineering Handbook*, 2nd ed. General Electric Corp., Gainesville, FL.

Further Information

Recommended battery seminars are: the Power Conference, Annual Battery Conference, The International Seminar on Primary and Secondary Battery Technology and Application, and the Power Source Conference, *Mobile Communications* magazine frequently feature articles addressing new battery technologies. This information is commonly user-oriented and provides hints on how to get the most out of rechargeable batteries. Additional battery information can be found on the Web under NiCd, NiMH, Battery Analyzer, or Battery, in general.

11

Packaging Electronic Systems

Ravindranath Kollipara
LSI Logic Corporation

Vijai K. Tripathi
Oregon State University

Jerry E. Sergent
BBS PowerMod, Inc.

Glenn R. Blackwell
Purdue University

Donald White
Don White Consultants, Inc.

Zbigniew J. Staszak
Technical University of Gdansk

11.1 Printed Wiring Boards

Ravindranath Kollipara and Vijai K. Tripathi

11.1.1 Introduction

Printed wiring board (PWB) is, in general, a layered dielectric structure with internal and external wiring that allows electronic components to be mechanically supported and electrically connected internally to each other and to the outside circuits and systems. The components can be complex packaged very large-scale integrated (VLSI), RF, and other ICs with multiple I/Os or discrete surface mount active and passive components. PWBs are the most commonly used packaging medium for electronic circuits and systems. Electronic packaging has been defined as the design, fabrication, and testing process that transforms an electronic circuit into a manufactured assembly. The main functions of the packagings include signal distribution to electronic circuits that process and store information, power distribution, heat dissipation, and the protection of the circuits.

The type of assembly required depends on the electronic circuits, which may be discrete, integrated, or hybrid. Integrated circuits are normally packaged in plastic or ceramic packages, which are electrically connected to the outside I/O connections and power supplies with pins that require plated through holes (PTH) or pins and pads meant for surface mounting the package by use of the surface mount technology (SMT). These plastic or ceramic packages are connected to the PWBs by using the pins or the leads associated with the packages.

In addition to providing a framework for interconnecting components and packages, such as IC packages, PWBs can also provide a medium (home) for component design and placement such as inductors and capacitors. PWBs are, generally, a composite of organic and inorganic dielectric material with multiple layers. The interconnects or the wires in these layers are connected by **via** holes, which can be plated and are filled with metal to povide the electrical connections between the layers. In addition to the ground planes, and power planes used to distribute bias voltages to the ICs and other discrete components, the signal lines are distributed among various layers to provide the interconnections in an optimum manner. The properties of importance that need to be minimized for a good design are the signal delay, distortion, and crosstalk noise induced primarily by the electromagnetic coupling between the signal lines. Coupling through substrate material and ground bounce and switching noise due to imperfect power/ground planes also leads to degradation in signal quality.

In this chapter, the basic properties and applications of PWBs are introduced. These include the physical properties, the wiring board design, and the electrical properties. The reader is referred to other sources for detailed discussion on these topics.

11.1.2   Board Types, Materials, and Fabrication

The PWBs can be divided into four types of boards: (1) rigid boards, (2) flexible and rigid-flex boards, (3) metal-core boards, and (4) injection molded boards. The board that is most widely used is the rigid board. The boards can be further classified into single-sided, double-sided, or multilayer boards. The ever increasing packaging density and faster propagation speeds, which stem from the demand for high-performance systems, have forced the evolution of the boards from single-sided to double-sided to multilayer boards. On single-sided boards, all of the interconnections are on one side. Double-sided boards have connections on both sides of the board and allow wires to crossover each other without the need for jumpers. This was accomplished at first by Z-wires, then by eyelets, and at present by PTHs. The increased pin count of the ICs has increased the routing requirements, which led to multilayer boards. The necessity of a controlled impedance for the high-speed **traces**, the need for bypass capacitors, and the need for low inductance values for the power and ground distribution networks have made the requirement of power and ground planes a must in high-performance boards. These planes are possible only in multilayer boards. In multilayer boards, the PTHs can be buried (providing interconnection between inner layers), semiburied (providing interconnection from one of the two outer layers to one of the internal layers), or through vias (providing interconnection between the outer two layers).

The properties that must be considered when choosing PWB substrate materials are their mechanical, electrical, chemical, and thermal properties. Early PWBs consisted of copper-clad phenolic and paper laminate materials. The copper was patterned using resists and etched. Holes were punched in the laminate to plug the component leads, and the leads were soldered to the printed copper pattern. At present, the copper-clad laminates and the prepregs are made with a variety of different matrix resin systems and reinforcements (ASM, 1989). The most commonly used resin systems are fire resistant (FR-4) difunctional and polyfunctional epoxies. Their glass transition temperatures T_g range from 125 to 150°C. They have well-understood processability, good performance, and low price. Other resins include high-temperature, one-component epoxy, polyimide, cyanate esters, and polytetrafluoroethylene (PTFE) (trade name Teflon). Polyimide resins have high T_g, long-term thermal resistance, low coefficient of thermal expansion (CTE), long PTH life and high reliability, and are primarily used for high-performance multilayer boards with a large number of layers. Cyanate esters have low dielectric constants and high T_g, and are used in applications where increased signal speed and improved laminate-dimensional stability are needed. Teflon has the lowest dielectric constant, low dissipation factor, and excellent temperature stability but is difficult to process and has high cost. It is used mainly in high-performance applications where higher densities and transmission velocities are required.

The most commonly used reinforcement is continuous filament electrical (E-) glass. Other high-performance reinforcements include high strength (S-) glass, high modulus (D-) glass, and quartz (Seraphim et al., 1989). The chemical composition of these glasses determines the key properties of CTE and dielectric constant. As the level of silicon dioxide (SiO_2) in the glass increases, the CTE and the dielectric constant decrease. The substrates of most rigid boards are made from FR-4 epoxy resin-impregnated E-glass cloth. Rolls of glass cloth are coated with liquid resin (A-stage). Then the resin is partially cured to a semistable state (B-stage or prepreg). The rolls are cut into large sheets and several sheets are stacked to form the desired final thickness. If the laminates are to be copper clad, then copper foils form the outside of the stack. The stack is then laminated and cured irreversibly to form the final resin state (C-stage) (ASM, 1989).

The single-sided boards typically use phenolic or polyester resins with random mat glass or paper reinforcement. The double-sided boards are usually made of glass-reinforced epoxy. Most multilayer boards are also made of glass-reinforced epoxy. The internal circuits are made on single- or double-sided copper-clad laminates. The inner layers are stacked up with B-stage polymer sheets separating the layers. Rigid pins are used to establish layer-to-layer orientation. The B-stage prepreg melts during lamination and reflows. When it is cured, it glues the entire package into a rigid assembly (ASM, 1989). An alternative approach to pin-parallel composite building is a sequential buildup of the layers, which allows buried vias. Glass reinforced polyimide is the next most used multilayer substrate material due to its excellent handling strength and its higher temperature cycling capability. Other laminate materials include Teflon and various resin combinations of epoxy, polyimide, and cyanate esters with reinforcements. The dielectric thickness of the flat laminated sheets ranges from 0.1 to 3.18 mm (0.004 to 0.125 in.), with 1.5 mm (0.059 in) being the most commonly used for single- and double-sided boards. The inner layers of a multilayer board are thinner with a typical range of 0.13–0.75 mm (0.005–0.030 in.) (ASM, 1989). The number of layers could be 20 or more. The commonly used substrate board materials and their properties are listed in Table 11.1.

The most common conductive material used is copper. The substrate material can be purchased as copper-clad laminates or unclad laminates. The foil thickness of copper-clad laminates are normally expressed in ounces of copper per square foot. The available range is from 1/8 to 5 oz with 1-oz copper (0.036 mm or 0.0014 in.) being the most commonly used cladding. The process of removing copper in the unwanted regions from the copper-clad laminates is called subtractive technology. If copper or other metal is added on to the unclad laminates, the process is called additive technology. Other metals used during electroplating may include Sn, Pb, Ni, Au, and Pd. Selectively screened or stenciled pastes that contain silver or carbon are also used as conducting materials (ASM, 1989).

TABLE 11.1 Wiring Board Material Properties

Material	ε_r'	$\varepsilon_r''/\varepsilon_r'$	CTE ($\times 10^{-6}/°C$) x, y	z	T_g, °C
FR-4 epoxy-glass	4.0–5.0	0.02–0.03	16–20	60–90	125–135
Polyimide-glass	3.9–4.5	0.005–0.02	12–14	60	>260
Teflon	2.1	0.0004–0.0005	70–120		—
Benzocyclobutene	2.6	0.0004	35–60		>350
High-temperature one-component epoxy-glass	4.45–4.55	0.02–0.022			170–180
Cyanate ester-glass	3.5–3.9	0.003–0.007			240–250
Ceramic	~10.0	0.0005	6–7		—
Copper	—	—	17		—
Copper/Invar/Copper	—	—	3–6		—

Solder masks are heat- and chemical-resistant organic coatings that are applied to the PWB surfaces to limit solderability to the PTHs and surface-mount pads, provide a thermal and electrical insulation layer isolating adjacent circuitry and components, and protect the circuitry from mechanical and handling damage, dust, moisture, and other contaminants. The coating materials used are thermally cured epoxies or ultraviolet curable acrylates. Pigments or dyes are added for color. Green is the standard color. Fillers and additives are added to modify rheology and improve adhesion (ASM, 1989).

Flexible and rigid-flex boards are required in some applications. A flexible printed board has a random arrangement of printed conductors on a flexible insulating base with or without cover layers (ASM, 1989). Like the base material, conductor, adhesive, and cover layer materials also should be flexible. The boards can be single-sided, double-sided, or multilayered. However, multilayered boards tend to be too rigid and are prone to conductor damage. Rigid-flex boards are like multilayered boards with bonding and connections between layers confined to restricted areas of the wiring plane. Connections between the rigid laminated areas are provided by multiconductor layers sandwiched between thin base layers and are flexible. The metal claddings are made of copper foil, beryllium copper, aluminium, inconel, or conductive polymer thick films with copper foil being the most commonly used. Typical adhesives systems used include polyester, epoxy/modified epoxy, acrylic, phenolics, polyimide, and fluorocarbons. Laminates, which eliminate the use of adhesives by placing conductors directly on the insulator, are called adhesiveless laminates. Dielectric base materials include polyimide films, polyester films, aramids, reinforced composites, and fluorocarbons. The manufacturing steps are similar to those of the rigid boards. As insulating film or coating applied over the conductor side acts as a permanent protective cover. It protects the conductors from moisture, contamination, and damage and reduces stress on conductors during flexing. Pad access holes and registration holes are drilled or punched in an insulating film coated with an adhesive, and the film is aligned over the conductor pattern and laminated under heat and pressure. Often, the same base material is used as the insulating film. When coatings are used instead of films, they are screen printed onto the circuit, leaving pad areas exposed. The materials used are acrylated epoxy, acrylated polyurethane, and thiolenes, which are liquid polymers and are cured using ultraviolet (UV) radiation or infrared (IR) heating to form a permanent, thin, tough coating (ASM, 1989).

When selecting a board material, the thermal expansion properties of the material must be a consideration. If the components and the board do not have a closely matched CTE, then the electrical and/or mechanical connections may be broken and reliability of the board will suffer. The widely used epoxy-glass PWB material has a CTE that is larger than that of the encapsulating material (plastic or ceramic) of the components. When leaded devices in dual-in-line (DIP) package format are used, the mechanical forces generated by the mismatches in the CTEs are taken by the PTHs, which accommodate the leads and provide electrical and mechanical connection. When surface mount devices (SMDs) are packaged, the solder joint, which provides both the mechanical and electrical connection, can accommodate little stress without deformation. In such cases, the degree of component and board CTE mismatch and the thermal environment in which the board operates must be considered and if necessary, the board must be CTE tailored for the components to be packaged. The three typical approaches to CTE tailoring are constraining dielectric structures, constraining metal cores, and constraining low-CTE metal planes (ASM, 1989; Seraphim et al., 1989).

Constraining inorganic dielectric material systems include cofired ceramic printed boards and thick-film ceramic wiring boards. Cofired ceramic multilayer board technology uses multiple layers of green ceramic tape into which via holes are made and on which circuit metallization is printed. The multiple layers are aligned and fired at high temperatures to yield a much higher interconnection density capability and much better controlled electrical properties. The ceramic used contains more than 90% of alumina and has a much higher thermal conductivity than that of epoxy-glass making heat transfer efficient. The thick-film ceramic wiring boards are usually made of cofired alumina and are typically bonded to an aluminium thermal plane with a flexible thermally conducting adhesive when heat removal is required. Both organic and inorganic fiber-reinforcements can be used with the conventional PWB resin for CTE tailoring. The organic fibers include several types of aramid fibers, notably Kevlar 29 and 49 and Technora HM-50. The inorganic fiber most commonly used for CTE tailoring is quartz (ASM, 1989).

In the constraining metal core technology, the PWB material can be any one of the standard materials like epoxy-glass, polyimide-glass, or Teflon-based materials. The constraining core materials include metals, composite metals, and low-CTE fiber-resin combinations, which have a low-CTE material with sufficient strength to constrain the module. The two most commonly used constraining core materials are copper-invar-copper (CIC) and copper-molybdenum-copper (CMC). The PWB and the core are bonded with a rigid adhesive. In the constraining low-CTE metal plane technology, the ground and power planes in a standard multilayer board are replaced by an approximately 0.15-mm-thick CIC layers. Both epoxy and polyimide laminates have been used (ASM, 1989).

Molded boards are made with resins containing fillers to improve thermal and mechanical properties. The resins are molded into a die or cavity to form the desired shape, including three-dimensional features. The board is metalized using conventional seeding and plating techniques. Alternatively, three-dimensional prefabricated films can also be transfer molded and further processed to form structures with finer dimensions. With proper selection of filler materials and epoxy compounds or by molding metal cores, the molded boards can be CTE tailored with enhanced thermal dissipation properties (Seraphim et al., 1989).

The standard PWB manufacturing primarily involves five technologies (ASM, 1989).

1. Machining. This involves drilling, punching, and routing. The drill holes are required for PTHs. The smaller diameter holes cost more and the larger aspect ratios (board thickness-to-hole diameter ratios) resulting from small holes make plating difficult and less reliable. They should be limited to thinner boards or buried vias. The locational accuracy is also important, especially because of the smaller features (pad size or annular ring), which are less tolerant to misregistration. The registration is also complicated by substrate dimensional changes due to temperature and humidity fluctuations, and material relaxation and shifting during manufacturing operations. The newer technologies for drilling are laser and water-jet cutting. CO_2 laser radiation is absorbed by both glass and epoxy and can be used for cutting. The typical range of minimum hole diameters, maximum aspect ratios, and locational accuracies are given in Table 11.2.

2. Imaging. In this step the artwork pattern is transferred to the individual layers. Screen printing technology was used early on for creating patterns for print-and-etch circuits. It is still being used for simpler single- and double-sided boards because of its low capital investment requirements and high volume capability with low material costs. The limiting factors are the minimum line width and spacings that can be achieved with good yields. Multilayer boards and fine line circuits are processed using photoimaging. The photoimageable films are applied by flood screen printing, liquid roller coating, dip coating, spin coating, or roller laminating of dry films. Electrophoresis is also being used recently. Laminate dimensional instability contributes to the misregistration and should be controlled. Emerging photoimaging technologies are direct laser imaging and imaging by liquid crystal light valves. Again, Table 11.2 shows typical minimum trace widths and separations.

3. Laminating. It is used to make the multilayer boards and the base laminates that make up the single- and double-sided boards. Prepregs are sheets of glass cloth impregnated with B-stage epoxy resin and are used to bond multilayer boards. The types of pressing techniques used are hydraulic cold or hot press lamination with or without vacuum assist and vacuum autoclave nomination. With these

TABLE 11.2 Typical Limits of PWB Parameters

Parameter	Limit
Minimum trace width, mm	0.05–0.15
Minimum trace separation, mm	~0.25
Minimum PTH diameter, mm	0.2–0.85
Location accuracy of PTH, mm	0.015–0.05
Maximum aspect ratios	3.5–15.0
Maximum number of layers	20 or more
Maximum board thickness, mm	7.0 or more

techniques, especially with autoclave, layer thicknesses, and dielectric constants can be closely controlled. These features allow the fabrication of controlled-impedance multilayer boards of eight layers or more. The autoclave technique is also capable of laminating three-dimensional forms and is used to produce rigid-flex boards.

4. Plating. In this step, metal finishing is applied to the board to make the necessary electrical connections. Plating processes could be wet chemical processes (electroless or electrolytic plating) or dry plasma processes (sputtering and chemical vapor deposition). Electroless plating, which does not need an external current source, is the core of additive technology. It is used to metallize the resin and glass portions of a multilayer board's PTHs with high aspect ratios (>15:1) and the three-dimensional circuit paths of molded boards. On the other hand, electrolytic plating, which requires an external current source, is the method of choice for bulk metallization. The advantages of electrolytic plating over electroless plating include greater rate of plating, simpler and less expensive process, and the ability to deposit a broader variety of metals. The newer plasma processes can offer pure, uniform, thin foils of various metals with less than 1-mil line widths and spacings and have less environmental impacts.

5. Etching. It involves removal of metals and dielectrics and may include both wet and dry processes. Copper can be etched with cupric chloride or other isotropic etchants, which limit the practical feature sizes to more than two times the copper thickness. The uniformity of etching is also critical for fine line feature circuits. New anisotropic etching solutions are being developed to extend the fine line capability.

The typical range of minimum trace widths, trace separations, printed through hole diameters, and maximum aspect ratios are shown in Table 11.2. The board cost for minimum widths, spacings, and PTH diameters, and small tolerances for the pad placements is high and reliability may be less.

11.1.3   Design of Printed Wiring Boards

The design of PWBs has become a challenging task, especially when designing high-performance and high-density boards. The designed board has to meet signal-integrity requirements (crosstalk, simultaneously switching output (SSO) noise, delay, and reflections), **electromagnetic compatibility (EMC)** requirements (meeting **electromagnetic interference (EMI)** specifications and meeting minimum **susceptibility** requirements), thermal requirements (being able to handle the mismatches in the CTEs of various components over the temperature range, power dissipation, and heat flow), mechanical requirements (strength, rigidity/flexibility), material requirements, manufacturing requirements (ease of manufacturability, which may affect cost and reliability), testing requirements (ease of testability, incorporation of **test coupons**), and environmental requirements (humidity, dust). When designing the circuit, the circuit design engineer may not have been concerned with all of these requirements when verifying his circuit design by simulation. But the PWB designer, in addition to wiring the board, must ensure that these requirements are met so that the assembled board functions to specification. Computer-aided design (CAD) tools are indispensable for the design of all but simple PWBs.

The PWB design process is part of an overall system design process. As an example, for a digital system design, the sequence of events may roughly follow the order shown in Fig. 11.1 (ASM, 1989; Byers, 1991). The logic design and verification are performed by a circuit designer. The circuit designer and the signal-integrity and EMC experts should help the board designer in choosing proper parts, layout, and electrical rule development. This will reduce the number of times the printed circuit has to be reworked and speeds up the product's time to market. The **netlist** generation is important for both circuit simulation and PWB design programs. The netlist contains the nets (a complete set of physical interconnectivity of the various parts in the circuit) and the circuit devices. Netlist files are usually divided into the component listings and the pin listings.

Once the circuit is successfully simulated by the circuit designer, the PWB designer starts on the board design. The board designer should consult with the circuit designer to determine critical path circuits, line-length requirements, power and fusing requirements, bypassing schemes, and any required guards

```
        ┌──────────────┐
        │ LOGIC DESIGN │◄──────────────┐
        └──────┬───────┘               │
               │              ┌─────────────────────┐
               │              │ DESIGN VERIFICATION │
               │              │ BY CIRCUIT SIMULATION│
               ▼              └─────────┬───────────┘
        ┌──────────────────┐           │
        │ SCHEMATIC CAPTURE│───────────┘
        └──────┬───────────┘
               │
               ▼
        ┌──────────────────┐
        │ RULES DEVELOPMENT│
        └──────┬───────────┘
               │              ┌──────────────┐
               │              │ ADJUST RULES │
               │              └──────▲───────┘
               ▼                     │
   ┌────────────────────────────┐    │
   │ PCB DESIGN                 │    │
   │  SELECTION OF BOARD SIZE   │  ┌─────────────────────┐
   │   AND SHAPE                │  │ SIGNAL-INTEGRITY,EMI│
   │  PARTS LAYOUT ON THE BOARD │──│ AND THERMAL ANALYSIS│
   │  TRACE ROUTING             │  └─────────────────────┘
   │  DESIGN RULE CHECK         │
   └────────────┬───────────────┘
                │
                ▼
      ┌────────────────────┐
      │ ARTWORK GENERATION │
      └─────────┬──────────┘
                │
                ▼
      ┌─────────────────────────┐
      │ FABRICATION OF THE BOARD│
      └─────────┬───────────────┘
                │
                ▼
   ┌──────────────────────────────────┐
   │ TESTING AND DEBUGGING OF THE BOARD│
   └──────────────────────────────────┘
```

FIGURE 11.1 Overall design flow.

or shielding. First the size and shape of the board are chosen from the package count and the number of connectors and switches in the design netlist. Some designs may require a rigidly defined shape, size and defined connector placement, for example, PWB outlines for computer use. This may compromise the optimum layout of the circuit (Byers, 1991).

Next the netlist parts are chosen from the component library and placed on the board so that the interconnections form the simplest pattern. If a particular part is not available in the library, then it can be created by the PCB program's library editor or purchased separately. Packages containing multiple gates are given their gate assignments with an eye toward a simpler layout. There are software packages that automize and speed up the component placement process. However, they may not achieve complete component placement. Problems with signal path length and clock skewing may be encountered, and so part of the placement may have to be completed manually. Placement can also be done interactively (Byers, 1991).

Once the size and shape of the board is decided on, any areas on the board that cannot be used for parts are defined using fill areas that prevent parts placement within their perimeter. The proper parts placement should result in a cost-effective design and meet manufacturing standards. Critical lines may not allow vias in which case they must be routed only on the surface. Feedback loops must be kept short and equal line length requirements and shielding requirements, if any, should be met. Layout could be on a grid with components oriented in one direction to increase assembly speed, eliminate errors, and improve inspection. Devices should be placed parallel to the edges of the board (ASM, 1989; Byers, 1991).

The actual placement of a component may rely on the placement of a part before it. Usually, the connectors, user accessible switches and controls, and displays may have rigidly defined locations. Thus, these should be placed first. Next I/O interface chips, which are typically next to the connectors, are placed. Sometimes a particular group of components may be placed as a block in one area, for example, memory chips. Then the automatic placement program may be used to place the remaining components. The algorithms can be based on various criteria. Routability, the relative ease with which all of the traces can be completed, may be one criterion. Overall total connection length may also be used to find an optimal layout. Algorithms based on the requirements for distances, density, and number of crossings can be formulated to help with the layout. After the initial placement, a placement improvement program can be run that fine tunes the board layout by component swapping, logic gate swapping, or when allowed even when allowed even when pin swapping. Parts can also be edited manually. Some programs support block editing, macros and changing of package outlines. The PWB designer may check with the circuit design engineer after a good initial layout is completed to make sure that all of the requirements are met when the board is routed (Byers, 1991).

The last step in the PWB layout process is the placement of the interconnection traces. Most PCB programs have an autorouter software that does most, if not all, of the trace routing. The software uses the wire list netlist and the placement of the parts on the board from the previous step as inputs and decides which trace should go where based on an algorithm. Most routers are point-to-point routers. The most widely used autorouter algorithm is the Lee algorithm. The Lee router works based on the cost functions that change its routing parameters. Cost functions may be associated with direction of traces (to force all of the tracks on a particular layer of the board vertically, the cost of the horizontal tracks can be set high), maximum trace length, maximum number of vias, trace density, or other criteria. In this respect, Lee's algorithm is quite versatile. Lee routers have a high trace completion rate (typically 90% or better) but are slow. There are other autorouters that are faster than the Lee router but their completion rate may not be as high as the Lee router. These include Hightower router, pattern router, channel router, and gridless router (ASM, 1989; Byers, 1991). In many cases, fast routers can be run first to get the bulk of the work done in a short time and then the Lee router can be run to finish routing the remaining traces.

The autorouters may not foresee that placing one trace may block the path of another. There are some routers called clean-up routers that can get the offending traces out of the way by rip-up or shove-aside. The rip-up router works by removing the offending trace, completing the blocked trace, and proceeding to find a new path for the removed trace. The rip-up router may require user help in identifying the trace that is to be removed. Shove-aside routers work by moving traces when an uncompleted trace has a path to its destination but does not have enough room for the track. Manual routing and editing may be needed if the automatic trace routing is not 100% complete. Traces may be modified without deleting them (pushing and shoving, placing or deleting vias, or moving segments from one layer to another) and trace widths may be adjusted to accommodate a connection. For example, a short portion of a trace width may be narrowed so that it may pass between the IC pads without shorting them and is called necking down (Byers, 1991).

Once the parts layout and trace pattern are found to be optimal, the board's integrity is verified by subjecting the trace pattern to **design rule** check for digital systems. A CAD program checks to see that the tracks, vias, and pads have been placed according to the design rule set. The program also makes sure that all of the nodes in each net are connected and that there are no shorted or broken traces (Byers, 1991). Any extra pins are also listed. The final placement is checked for signal-integrity, EMI compliance, and thermal performance. If any problems arises, the rules are adjusted and the board layout and/or trace routing are modified to correct the problems.

Next a netlist file of the PWB artwork is generated, which contains the trace patterns of each board layer. All of the layers are aligned using placement holes or datum drawn on each layer. The artwork files are sent to a format photoplotter and photographic transparencies suitable for PWB manufacturing are produced. From the completed PWB design, solder masks, which apply an epoxy film on the PWB prior to flow soldering to prevent solder bridges from forming between adjacent traces, are also generated. Silkscreen mats could also be produced for board nomenclature. The PWB programs may also support drill file formats that numerically control robotic drilling machines used to locate and drill properly sized holes in the PWB (Byers, 1991).

Finally, the board is manufactured and tested. Then the board is ready for assembly and soldering. Finished boards are tested for system functionality and specifications.

11.1.4 PWB Interconnection Models

The principal electrical properties of the PWB material are relative dielectric constant, loss tangent, and dielectric strength. When designing high-speed and high-performance boards, the wave propagation along the traces should be well controlled. This is done by incorporating power and signal planes in the

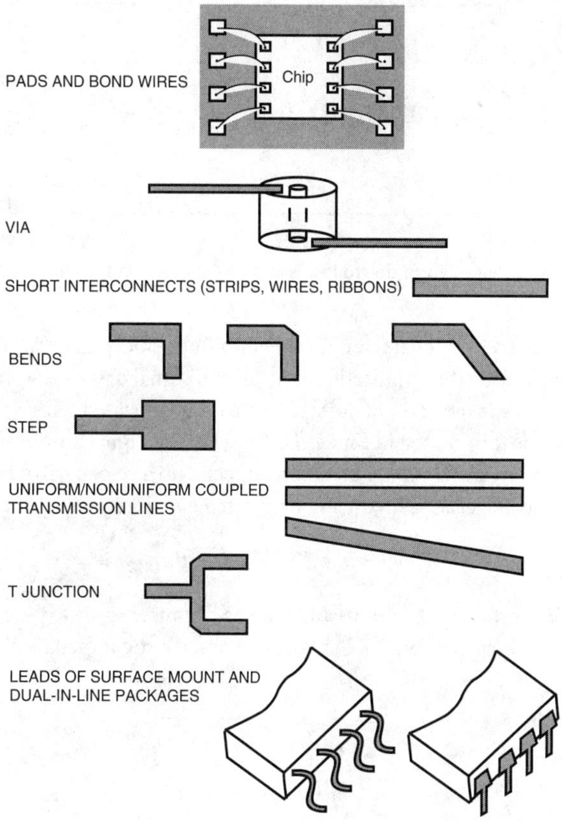

FIGURE 11.2 PWB interconnect examples.

boards, and utilizing various transmission line structures with well-defined properties. The electrical analog and digital signals are transferred by all of the PWB interconnects and the fidelity of these connections is dependent on the electrical properties of interconnects, as well as the types of circuits and signals. The interconnects can be characterized as electrically short and modeled as lumped elements in circuit simulators if the signal wavelength is large as compared to the interconnect length ($\lambda > 30 * l$). For digital signals the corresponding criterion can be expressed in terms of signal rise time being large as compared to the propagation delay (e.g., $T_r > 10\, T_d$). Propagation delay represents the time taken by the signal to travel from one end of the interconnect to the other end and is obviously equal to the interconnect length divided by the signal velocity, which in most cases corresponds to the velocity of light in the dielectric medium. Vias, bond wires, pads, short wires and traces, and bends in PWBs shown in Fig. 11.2 can be modeled as lumped elements, whereas, long traces must be modeled as distributed circuits or transmission lines.

The transmission lines are characterized by their characteristic impedances and propagation constants, which can also be expressed in terms of the associated distributed parameters R, L, G, and C per unit length of the lines (Magnuson, Alexander, and Tripathi, 1992). In general, the characteristic parameters are expressed as

$$\gamma \equiv \alpha + j\beta = \sqrt{(R + j\omega L)(G + j\omega C)}$$

$$Z_0 = \sqrt{\frac{R + j\omega L}{G + j\omega C}}$$

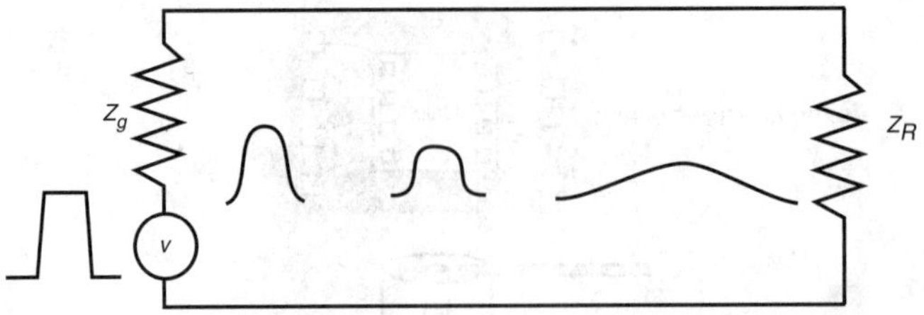

FIGURE 11.3   Signal degradation due to losses and dispersion as it travels along an interconnect.

Propagation constant $\gamma = \alpha + j\beta$ characterizes the amplitude and phase variation associated with an AC signal at a given frequency or the amplitude variation and signal delay associated with a digital signal. The characteristic impedance is the ratio of voltage to current associated with a wave and is equal to the impedance the lines must be terminated in for zero reflection. The signal amplitude, in general, decreases and it lags behind in phase as it travels along the interconnect with a velocity, in general, equal to the group velocity. For example, the voltage associated with a wave at a given frequency can be expressed as

$$V = V_0 e^{-\alpha z} \cos(\omega t - \beta z)$$

where V_0 is the amplitude at the input $(z = 0)$ and z is the distance.

For low loss and lossless lines the signal velocity and characteristic impedance can be expressed as

$$v = \frac{1}{\sqrt{LC}} = \frac{1}{\sqrt{\mu_0 \varepsilon_0 \varepsilon_{r\text{eff}}}}$$

$$Z_0 = \sqrt{\frac{L}{C}} = \frac{\sqrt{\mu_0 \varepsilon_0 \varepsilon_{r\text{eff}}}}{C}$$

The effective dielectric constant $\varepsilon_{r\text{eff}}$ is an important parameter used to represent the overall effect of the presence of different dielectric mediums surrounding the interconnect traces. For most traces $\varepsilon_{r\text{eff}}$ is either equal to or approximately equal to the ε_r, the relative dielectric constant of the board material, except for the traces on the top layer where $\varepsilon_{r\text{eff}}$ is a little more than the average of the two media.

The line losses are expressed in terms of the series resistance and shunt conductance per unit length (R and G) due to conductor and dielectric loss, respectively. The signals can be distorted or degraded due to these conductor and dielectric losses, as illustrated in Fig. 11.3 for a typical interconnect. The resistance is, in general, frequency dependent since the current distribution across the conductors is nonuniform and depends on frequency due to the skin effect. Because of this exclusion of current and flux from the inside of the conductors, resistance increases and inductance decreases with increasing frequency. In the high-frequency limit, the resistance and inductances can be estimated by assuming that the current is confined over the cross section one skin depth from the conductor surface. The skin depth is a measure of how far the fields and currents penetrate into a conductor and is given by

$$\delta = \sqrt{\frac{2}{\omega \mu \sigma}}$$

The conductor losses can be found by evaluating R per unit length and using the expression for γ or by using an incremental inductance rule, which leads to

$$\alpha_c = \frac{\beta \Delta Z_0}{2 Z_0}$$

where α_c is the attenuation constant due to conductor loss, Z_0 is the characteristic impedance, and ΔZ_0 is the change in characteristic impedance when all of the conductor walls are receded by an amount $\delta/2$. This expression can be readily implemented with the expression for Z_0. The substrate loss is accounted for by assigning the medium a conductivity σ, which is equal to $\omega\varepsilon_0\varepsilon_r''$. For many conductors buried in near homogeneous medium

$$G = C\frac{\sigma}{\varepsilon}$$

If the lines are not terminated in their characteristic impedances, there are reflections from the terminations. These can be expressed in terms of the ratio of reflected voltage or current to the incident voltage or current and are given as

$$\frac{V_{\text{reflected}}}{V_{\text{incident}}} = -\frac{I_{\text{reflected}}}{I_{\text{incident}}} = \frac{Z_R - Z_0}{Z_R + Z_0}$$

where Z_R is the termination impedance and Z_0 is the characteristic impedance of the line. For a perfect match, the lines must be terminated in their characteristic impedances. If the signal is reflected, the signal received by the receiver is different from that sent by the driver. That is, the effect of mismatch includes signal distortion, as illustrated in Fig. 11.4, as well as ringing and increase in crosstalk due to multiple reflections resulting in an increase in coupling of the signal to the passive lines.

The electromagnetic coupling between the interconnects is the factor that sets the upper limit to the number of tracks per channel or, in general, the interconnect density. The time-varying voltages and currents result in capacitive and inductive coupling between the interconnects. For longer interconnects, this coupling is distributed and modeled in terms of distributed self- and mutual line constants of the multiconductor transmission line systems. In general, this coupling results in both the near- and far-end crosstalk as illustrated in Fig. 11.5 for two coupled microstrips. Crosstalk increases noise margins and degrades signal quality. Crosstalk increases with longer trace coupling distances, smaller separation between traces, shorter pulse rise and fall times, larger magnitude currents or voltages being switched, and decreases with the use of adjacent power and ground planes or with power and ground traces interlaced between signal traces on the same layer.

The commonly used PWB transmission line structures are microstrip, embedded microstrip, stripline, and dual stripline, whose cross sections are shown in Fig. 11.6. The empirical CAD oriented expressions for transmission line parameters, and the models for wires, ribbons, and vias are given in Table 11.3.

The traces on the outside layers (microstrips) offer faster clock and logic signal speeds than the stripline traces. Hooking up components is also easier in microstrip structure than in stripline structure. Stripline offers better noise immunity for RF emissions than microstrip. The minimum spacing between the signal traces may be dictated by maximum crosstalk allowed rather than by process constraints. Stripline allows closer spacing of traces than microstrip for the same layer thickness. Lower characteristic impedance structures have smaller spacing between signal and ground planes. This makes the boards thinner allowing drilling of smaller diameter holes, which in turn allow higher circuit densities. Trace width and individual layer thickness tolerances of $\pm10\%$ are common. Tight tolerances ($\pm2\%$) can be specified, which would

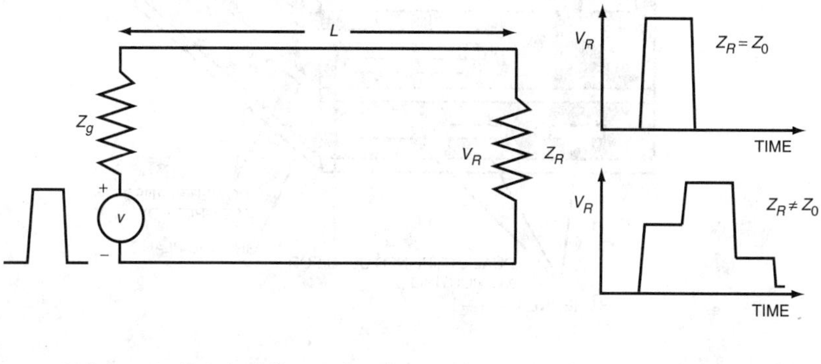

FIGURE 11.4 Voltage waveform is different when lines are not terminated in their characteristic impedance.

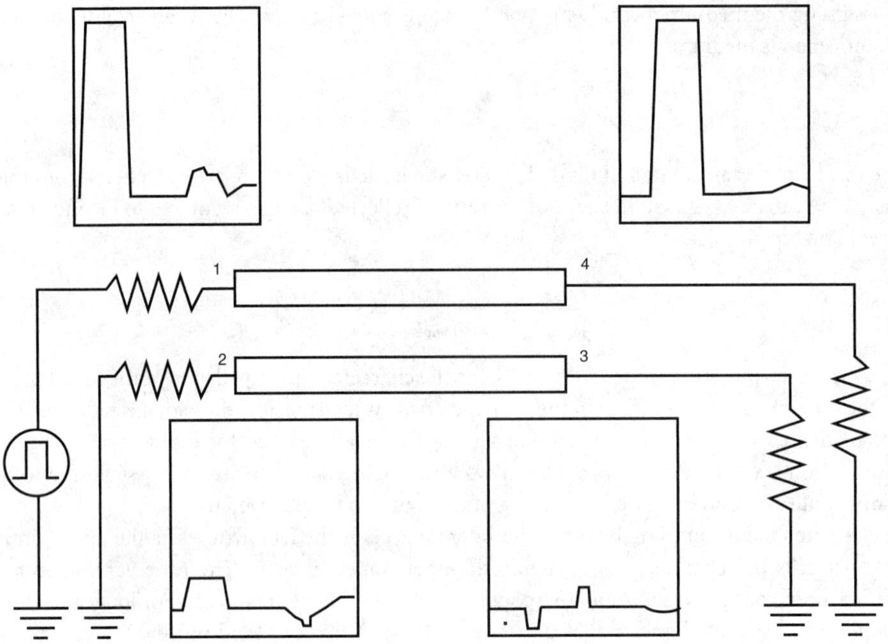

FIGURE 11.5 Example of near-end and far-end crosstalk signal.

result in higher board cost. Typical impedance tolerances are ±10%. New statistical techniques have been developed for designing the line structures of high-speed wiring boards (Mikazuki and Matsui, 1994).

The low dielectric constant materials improve the density of circuit interconnections in cases where the density is limited by crosstalk considerations rather than by process constraints. Benzocyclobutene is one of the low dielectric constant polymers with excellent electrical, thermal, and adhesion properties.

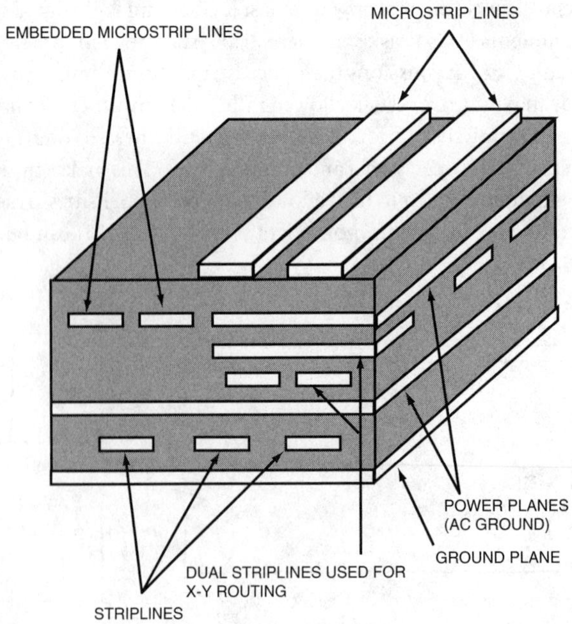

FIGURE 11.6 Example transmission line structures in PWBs.

TABLE 11.3   Interconnect Models

Interconnect type	Model

Round wire

$$L \cong 0.002l \left[\ell n \left(\frac{2l}{r} - 0.75 \right) \right], \mu H; l > r$$

l, r in cm.

Straight rectangular bar or ribbon

$$L \cong 0.002l \left[\ell n \frac{2l}{b+c} + 0.5 + 0.2235 \frac{b+c}{l} \right], \mu H$$

b, c, l in cm.

Via

$$L = 0.002l \left[\ell n \frac{2l}{r+W} - 1 + \xi \right], \mu H$$

where

$$\xi = 0.25 \left[\cos \left(\frac{r}{r+W} \frac{\pi}{2} \right) - 0.07 \sin \left(\frac{r}{r+W} \pi \right) \right]$$

l, r, W in cm.

Round wire over a ground plane and parallel round wires

$$Z_0 = \frac{120}{\sqrt{\varepsilon_r}} \cosh^{-1} \frac{d}{2r}, \text{ohm}$$

$$= \frac{120}{\sqrt{\varepsilon_r}} \ell n \frac{d}{r}, \text{ohm}; d \gg r$$

Microstrip

$$\varepsilon_{r\text{eff}} = \frac{\varepsilon_r + 1}{2} + \frac{\varepsilon_r - 1}{2} \left[1 + \frac{10h}{W_e} \right]^{-a.b}$$

where

$$\frac{W_e}{h} = \frac{W}{h} + \frac{1.25}{\pi} \frac{t}{h} \left[1 + \ell n \frac{2[h^4 + (2\pi W)^4]^{0.25}}{t} \right]$$

$$a = 1 + \frac{1}{49} \ell n \left\{ \frac{\left[\left(\frac{W_e}{h} \right)^4 + \left(\frac{W_e}{52h} \right)^2 \right]}{\left[\left(\frac{W_e}{h} \right)^4 + 0.432 \right]} \right\}$$

$$+ \frac{1}{18.7} \ell n \left[1 + \left(\frac{W_e}{18.1h} \right)^3 \right]$$

$$b = 0.564 \left[\frac{\varepsilon_r - 0.9}{\varepsilon_r + 3} \right]^{0.053}$$

$$Z_0 = \frac{60}{\sqrt{\varepsilon_{r\text{eff}}}} \ell n \left[\frac{F_1 h}{W_e} + \sqrt{1 + \left(\frac{2h}{W_e} \right)^2} \right]$$

with $F_1 = 6 + (2\pi - 6)e^{-\left(30.666 \frac{h}{W_e} \right)^{0.7528}}$

(continued)

TABLE 11.3   Interconnect Models (*Continued*)

Interconnect type	Model

Buried microstrip

Z_0 same expression as microstrip with

$$\varepsilon_{r\,\text{eff}} = \varepsilon_r \left[1 - e^{K \frac{h'}{h}} \right]$$

where

$$K = \ell n \left\{ \cfrac{1}{\left[1 - \cfrac{\varepsilon_{r\,\text{eff}}(h' = h)}{\varepsilon_r} \right]} \right\}$$

$[\varepsilon_{r\,\text{eff}}(h' = h)]$ is given by the microstrip formula

$$\varepsilon_{r\,\text{eff}} = \varepsilon_r$$

Stripline

$$Z_0 = \frac{30}{\sqrt{\varepsilon_{\text{reff}}}} \ell n \left[1 + \frac{4}{\pi W_n} \left(\frac{8}{\pi W_n} + \sqrt{\frac{8}{\pi W_n} + 6.27} \right) \right]$$

where

$$W_n = \frac{W}{b-t} + \frac{t_n}{\pi(1-t_n)}$$

$$\times \left\{ 1 - \frac{1}{2} \ell n \left[\left(\frac{t_n}{2 - t_n} \right)^2 + \left(\frac{0.0796 t_n}{1.1 t_n + \dfrac{W}{b}} \right)^m \right] \right\}$$

$$t_n = \frac{t}{b} \quad \text{and} \quad m = b \left(\frac{1 - t_n}{3 - t_n} \right)$$

Asymmetric and dual striplines

$$Z_0 = \frac{80Y}{0.918\sqrt{\varepsilon_r}} \ell n \left[\frac{3.8 h_1 + 1.9 t}{0.8 W + t} \right]$$

$$Y = \left[1.0636 + 0.33136 \frac{h_1}{h_2} - 1.9007 \left(\frac{h_1}{h_2} \right)^3 \right]$$

In addition, water absorption is lower by a factor of 15 compared to conventional polyimides. Teflon also has low dielectric constant and loss, which are stable over wide ranges of temperature, humidity, and frequency (ASM, 1989; Evans, 1994).

In surface mount technology, the components are soldered directly to the surface of a PWB, as opposed to through-hole mounting. This allows efficient use of board real estate, resulting in smaller boards and simpler assembly. Significant improvement in electrical performance is possible with the reduced package parasitics and the short interconnections. However, the reliability of the solder joint is a concern if there are CTE mismatches (Seraphim et al., 1989).

In addition to establishing an impedance-reference system for signal lines, the power and ground planes establish stable voltage levels for the circuits (Montrose, 1995). When large currents are switched, large voltage drops can be developed between the power supply and the components. The planes minimize the voltage drops by providing a very small resistance path and by supplying a larger capacitance and lower inductance contribution when two planes are closely spaced. Large decoupling capacitors are also added between the power and ground planes for increased voltage stability. High-performance and high-density boards require accurate computer simulations to determine the total electrical response of the components and complex PWB structures involving various transmission lines, vias, bends, and planes with vias.

11.1.5 Signal-Integrity and EMC Considerations

The term signal-integrity refers to the issues of timing and quality of the signal. The timing analysis is performed to ensure that the signal arrives at the destination within the specified time window. In addition, the signal shape should be such that it causes correct switching and avoids false switching or multicrossing of the threshold. The board features that influence the signal-integrity issues of timing and quality of signal are stub length, parallelism, vias, bends, planes, and termination impedances. Increasing clock speeds and signal edge rates make the signal-integrity analysis a must for the design of high speed PWBs. The signal-integrity tools detect and help correct problems associated with timing, crosstalk, ground bounce, dispersion, resonance, ringing, clock skew, and susceptibility (Maliniak, 1995). Some CAD tool vendors offer stand-alone tools that interface with simulation and layout tools. Other vendors integrate signal-integrity capabilities into the PWB design flow. Both prelayout and postlayout tools are available. Prelayout tools incorporate signal-integrity design rules up front and are used before physical design to investigate tradeoffs. Postlayout tools are used after board layout and provide a more in-depth analysis of the board.

The selection of the signal-integrity tools depends on the accuracy required and on the run times, in addition to the cost. Tools employing full-wave approach solve the Maxwell's field equations and provide accurate results. However, the run times are longer because they are computation intensive. Accurate modeling of radiation losses and complex three-dimensional structures including vias, bends, and tees can only be performed by full-wave analysis. Tools employing a hybrid approach are less accurate but run times are short. If the trace dimensions are smaller (roughly by a factor of 10) compared to the free space wavelength of the fields, then quasistatic methods can be used to compute the distributed circuit quantities like self- and mutual inductances and capacitances. These are then used in a circuit simulator like SPICE (*s*imulated *p*rogram for *i*ntegrated *c*ircuit *e*mphasis) to compute the time-domain response, from which signal delay and distortion are extracted (Maliniak, 1995).

The inputs for the signal-integrity tools are the device SPICE or behavioral models and the layout information including the thickness of the wires, the metal used, and the dielectric constant of the board. The input/output buffer information specification (IBIS) is a standard used to describe the analog behavior of a digital IC's I/O buffers. The IBIS models provide a nonproprietary description of board drivers and receivers that are accurate and faster to simulate. The models are based on DC I-V curves, AC transition data, and package parasitic and connection information. A SPICE-to-IBIS translation program is also available, which uses existing SPICE models to perform simulation tests and generate accurate IBIS models. An IBIS on-line repository, which maintains the IBIS models from some IC vendors, is also available. Other IC vendors provide the IBIS models under nondisclosure agreements.

The PWBs, when installed in their systems, should meet the North American and international EMC compliance requirements. EMC addresses both emissions, causing EMI and susceptibility, vulnerability to EMI. The designer needs to know the manner in which electromagnetic fields transfer from the circuit boards to the chassis and/or case structure. Many different layout design methodologies exist for EMC (Montrose, 1995). To model EMC, the full electromagnetic behavior of components and interconnects, not simple circuit models, must be considered. A simple EMC model has three elements: (1) a source of energy, (2) a receptor, and (3) a coupling path between the source and receptor. All three elements must be present for interference to exist. The emissions could be coupled by either radiation or by conduction. Emissions are **suppressed** by reducing noise source level and reducing propagation efficiency. Immunity is enhanced by increasing susceptor noise margin or reducing propagation efficiency. In general, the higher the frequency, the greater is the likelihood of a radiated coupling path and the lower the frequency, the greater is the likelihood of a conducted coupling path. The five major considerations in EMI analysis are frequency, amplitude, time, impedance, and dimensions (Montrose, 1995).

The use of power and ground planes embedded in the PWB with proper use of stripline and microstrip topology is one of the important methods of suppression of the RF energy internal to the board. Radiation may still occur from bond wires, lead frames, sockets, and interconnect cables. Minimizing lead inductance of PWB components reduces radiated emissions. Minimizing trace lengths of high-speed traces reduces radiation. On single- or double-sided boards, the power and ground traces should be adjacent to each other to minimize loop currents, and signal paths should parallel the power and ground traces. In multilayer boards, signal routing planes should be adjacent to solid image planes. If logic devices in multilayer boards have asymmetrical pull-up and pull-down current ratios, then an imbalance is created in the power and ground plane structure. The flux cancellation between RF currents that exist within the board traces, components, and circuits, in relation to a plane is not optimum when pull-up and pull-down current ratios are different and traces are routed adjacent to a power plane. In such cases, high-speed signal and clock traces should be routed adjacent to a ground plane for optimal EMI suppression (Montrose, 1995).

To minimize fringing between the power and ground planes and the resulting RF emissions, power planes should be physically smaller than the closest ground planes. Typical reduction size is 20 times the thickness between the power and the ground plane on each side. Any signal traces on the adjacent routing plane located over the absence of copper area must be rerouted. Selecting proper type of grounding is also critical for EMI suppression. The distance between any two neighboring ground point locations should not exceed 1/20 of the highest frequency wavelength generated on the board in order to minimize RF ground loops. Moats are acceptable in an image plane only if the adjacent signal routing layer does not have traces that cross the moated area. Partition the board into high-bandwidth, medium-bandwidth, and low-bandwidth areas, and isolate each section using partitions or moats. The peak power currents generated when logic gates switch states may inject high-frequency switching noise into the power planes. Fast edge rates generate greater amount of spectral bandwidth of RF energy. The slowest possible logic family that meets adequate timing margins should be selected to minimize EMI effects (Montrose, 1995).

Surface mount devices have lower lead inductance values and smaller loop areas due to smaller package size as compared to through hole mounted ones and, hence, have lower RF emissions. Decoupling capacitors remove RF energy generated on the power planes by high-frequency components and provide a localized source of DC power when simultaneous switching of output pins occur. Bypassing capacitors remove unwanted RF noise that couples component or cable common-mode EMI into susceptible areas. Proper values and proper placement on the board of decoupling and bypassing capacitors prevent RF energy transfer from one circuit to another and enhance EMC. The capacitance between the power and ground planes also acts as a decoupling capacitor (Montrose, 1995).

Clock circuits may generate significant EMI and should be designed with special attention. A localized ground plane and a shield may be required over the clock generation circuits. Clock traces should be considered critical and routed manually with characteristic terminations. Clock traces should be treated as transmission lines with good impedance control and with no vias or minimum number of vias in the traces to reduce reflections and ringing, and creation of RF common-mode currents. Ground traces should be placed around each clock trace if the board is single or double sided. This minimizes crosstalk and provides a return path for RF current. Shunt ground traces may also be provided for additional RF suppression. Bends should be avoided, if possible, in high-speed signal and clock traces and, if present, bend angles should not be smaller than 90° and they must be chamfered. Analog and input/output sections should be isolated from the digital section. PWBs must be protected from electrostatic discharge (ESD) that might enter at I/O signal and electrical connection points. For additional design techniques and guidelines for EMC compliance and a listing of North American and international EMI requirements and specifications, refer to Montrose (1995).

Signal-integrity and EMC considerations should be an integral part of printed circuit board (PCB) design. Signal integrity, EMI problems, and thermal performance should be addressed early in the development cycle using proper CAD tools before physical prototypes are produced. When boards are designed taking these considerations into account, the signal quality and signal-to-noise ratio are improved. Whenever possible, EMC problems should be solved at the PCB level than at the system level. Proper layout of PWB ensures EMC compliance at the level of cables and interconnects.

Defining Terms

Design rules: A set of electrical or mechanical rules that must be followed to ensure the successful manufacturing and functioning of the board. These may include minimum track widths and track spacings, track width required to carry a given current, maximum length of clock lines, and maximum allowable distance of coupling between a pair of signal lines.

Electromagnetic compatibility (EMC): The ability of a product to coexist in its intended electromagnetic environment without causing or suffering functional degradation or damage.

Electromagnetic interference (EMI): A process by which disruptive electromagnetic energy is transmitted from one electronic device to another via radiated or conducted paths or both.

Netlist: A file of component connections generated from a schematic. The file lists net names and the pins, which are a part of each net in the design.

Schematic: A drawing or set of drawings that shows an electrical circuit design.

Suppression: Designing a product to reduce or eliminate RF energy at the source without relying on a secondary method such as a metal housing.

Susceptibility: A relative measure of a device or system's propensity to be disrupted or damaged by EMI exposure.

Test coupon: Small pieces of board carrying a special pattern, made alongside a required board, which can be used for destructive testing.

Trace: A node-to-node connection, which consists of one or more tracks. A track is a metal line on the PWB. It has a start point, an end point, a width, and a layer.

Via: A hole through one or more layers on a PWB that does not have a component lead through it. It is used to make a connection from a track on one layer to a track on another layer.

References

ASM. 1989. *Electronic Materials Handbook*, Vol. 1, Packaging, Sec. 5, Printed Wiring Boards, pp. 505–629. ASM International, Materials Park, OH.

Beckert, B.A. 1993. Hot analysis tools for PCB design. *Comp. Aided Eng.* 12(1):44–49.

Byers, T.J. 1991. *Printed Circuit Board Design With Microcomputers*. Intertext, New York.

Evans, R. 1994. Effects of losses on signals in PWBs. *IEEE Trans. Comp. Pac., Man. Tech. Pt. B: Adv. Pac.* 17(2):217–222.

Magnuson, P.C, Alexander, G.C., and Tripathi, V.K. 1992. *Transmission Lines and Wave Propagation*, 3rd ed. CRC Press, Boca Raton, FL.

Maliniak, L. 1995. Signal analysis: A must for PCB design success. *Elec. Design* 43(19):69–82.

Mikazuki, T. and Matsui, N. 1994. Statistical design techniques for high speed circuit boards with correlated structure distributions. *IEEE Trans. Comp. Pac., Man. Tech.* 17(1):159–165.

Montrose, M.I. 1995. *Printed Circuit Board Design Techniques for EMC Compliance*. IEEE Press, New York.

Seraphim, D.P., Barr, D.E., Chen, W.T., Schmitt, G.P., and Tummala, R.R. 1989. Printed-circuit board packaging. In *Microelectronics Packaging Handbook*, eds. Rao R. Tummala and Eugene J. Rymaszewski, pp. 853–921. Van Nostrand-Reinhold, New York.

Simovich, S., Mehrotra, S., Franzon, P., and Steer, M. 1994. Delay and reflection noise macromodeling for signal integrity management of PCBs and MCMs. *IEEE Trans. Comp. Pac., Man. Tech. Pt. B: Adv. Pac.* 17(1):15–20.

Further Information

Electronic Packaging and Production Journal.

IEEE Transcations on Components, Packaging, and Manufacturing Technology—Part A.

IEEE Transcation on Components, Packaging, and Manufacturing Technology—Part B: Advanced Packaging.

Harper, C.A. 1991. *Electronic Packaging and Interconnection Handbook.*
PCB Design Conference, Miller Freeman Inc., 600 Harrison Street, San Francisco, CA 94107.
Printed Circuit Design journal.

Proceedings of the International Symposium on Electromagnetic Compatibility, sponsored by IEEE Transactions on EMC.

11.2 Hybrid Microelectronics Technology

Jerry E. Sergent

11.2.1 Introduction

Hybrid microelectronics technology is one branch of the electronics packaging discipline. As the name implies, the hybrid technology is an integration of two or more technologies, utilizing a film-deposition process to fabricate conductor, resistor, and dielectric patterns on a ceramic substrate for the purpose of mounting and interconnecting semiconductors and other devices as necessary to form an electronics circuit.

The method of deposition is what differentiates the hybrid circuit from other packaging technologies and may be one of two types: *thick film* or *thin film*. Other methods of metallizing a ceramic substrate, such as direct bond copper, active metal brazing, and plated copper, may also be considered to be in the hybrid family, but do not have a means for directly fabricating resistors and are not considered here. Semiconductor technology provides the **active components**, such as integrated circuits, transistors, and diodes. The **passive components**, such as resistors, capacitors, and inductors, may also be fabricated by thick- or thin-film methods or may be added as separate components.

The most common definition of a hybrid circuit is a circuit that contains two or more components, one of which must be active, which are mounted on a substrate that has been metallized by one of the film technologies. This definition is intended to eliminate single-chip packages, which contain only one active component, and also to eliminate such structures as resistor networks. A hybrid circuit may, therefore, be as simple as a diode-resistor network or as complicated as a multilayer circuit containing in excess of 100 integrated circuits.

The first so-called hybrid circuits were manufactured from polymer materials for use in radar sets in World War II and were little more than carbon-impregnated epoxies printed across printed circuit traces. The modern hybrid era began in the 1960s when the first commercial cermet thick-film materials became available. Originally developed as a means of reducing the size and weight of rocket payloads, hybrid circuits are now used in a variety of applications, including automotive, medical, telecommunication, and commercial, in addition to continued use in space and military applications. Today, the hybrid microelectronics industry is a multibillion dollar business and has spawned new technologies, such as the multichip module technology, which have extended the application of microelectronics into other areas.

11.2.2 Substrates for Hybrid Applications

The foundation for the hybrid microcircuit is the substrate, which provides the base on which the circuit is formed. The substrate is typically ceramic, a mixture of metal oxides and/or nitrides and glasses that are formulated to melt at a high temperature. The constituents are formed into powders, mixed with an organic binder, patterned into the desired shape, and fired at an elevated temperature. The result is a hard, brittle composition, usually in the configuration of a flat plate, suitable for metallization by one of the film processes. The ceramics used for hybrid applications must have certain mechanical, chemical, and electrical characteristics as described in Table 11.4.

The most common material is 96% alumina (Al_2O_3), with a typical composition of 96% Al_2O_3, 0.8% MgO, and 3.2% SiO_2. The magnesia and silica form a magnesium-alumino-silicate glass with the alumina

TABLE 11.4 Desirable Properties of Substrates for Hybrid Applications

- Electrical insulator
- Chemically and mechanically stable at processing temperatures
- Good thermal dissipation characteristics
- Low dielectric constant, especially at high frequencies
- Capable of being manufactured in large sizes and maintain good dimensional stability throughout processing
- Low cost
- Chemically inert to postfiring fabrication steps
- Chemically and physically reproducible to a high degree

when the substrate is manufactured at a firing temperature of 1600°C. The characteristics of alumina are presented in Table 11.5.

Other materials used in hybrid applications are described as follows:

- *Beryllium oxide, BeO, (Beryllia)*. Beryllia has excellent thermal properties (almost 10 times better than alumina at 250°C), and also has a lower **dielectric constant** (6.5).
- *Steatite*. Steatite has a lower concentration of alumina and is used primarily for low-cost thick-film applications.
- *Glass*. Borosilicate or quartz glass can be used where light transmission is required (e.g., displays). Because of the low melting point of these glasses, their use is restricted to the thin-film process and some low-temperature thick-film materials.
- *Enamel steel*. This is a glass or porcelain coating on a steel core. It has been in commercial development for some time and has found some limited acceptance in thick-film applications.
- *Silicon carbide, SiC*. SiC has a higher **thermal conductivity** than beryllia, but also has a high dielectric constant (40), which renders it unsuitable for high-speed applications.
- *Silicon*. Silicon has a very high thermal conductivity and matches the thermal expansion of semiconductors perfectly. Silicon is commonly used in thin-film applications because of the smooth surface.
- *Aluminum nitride (AlN)*. AlN substates have a thermal conductivity less than that of beryllia, but greater than alumina. At higher temperatures, however, the gap between the thermal conductivities

TABLE 11.5 Typical Properties of Alumina

Property	96% Alumina	99% Alumina
Electrical Properties		
Resistivity	10^{14} Ω-cm	10^{16} Ω-cm
Dielectric constant	9.0 at 1 MHz, 25°C	9.2 at 1 MHz, 25°C
	8.9 at 10 GHz, 25°C	9.0 at 10 GHz, 25°C
Dissipation factor	0.0003 at 1 MHz, 25°C	0.0001 at 1 MHz, 25°C
	0.0006 at 10 GHz, 25°C	0.0002 at 10 GHz, 25°C
Dielectric strength	16 V/mm at 60 Hz, 25°C	16 V/mm at 60 Hz, 25°C
Mechanical Properties		
Tensile strength	1760 kg/cm^2	1760 kg/cm^2
Density	3.92 g/cm^3	3.92 g/cm^3
Thermal conductivity	0.89 W/°C-in at 25°C	0.93 W/°C-in at 25°C
Coefficient of thermal expansion	6.4 ppm/°C	6.4 ppm/°C
Surface Properties		
Surface roughness, avg	0.50 mm CLA*	0.23 mm CLA*
Surface flatness, avg (Camber)	0.002 cm/cm	0.002 cm/cm

*Center line average.

of AlN and BeO narrows somewhat. In addition, the **temperature coefficient of expansion** (**TCE**) of aluminum nitride is close to that of silicon, which lowers the stress on the bond under conditions of temperature extremes. Certain of the copper technologies, such as plated copper or **direct-bonded copper** (**DBC**), may be used as a conductor system if the AlN surface is properly prepared.

11.2.3   Thick-Film Materials and Processes

Thick-film technology is an additive process that utilizes screen printing methods to apply conductive, resistive, and insulating films, initially in the form of a viscous paste, onto a ceramic substrate in the desired pattern. The films are subsequently dried and fired at an elevated temperature to activate the adhesion mechanism to the substrate.

There are three basic types of thick-film materials:

- Cermet thick-film materials, a combination of glass ceramic and metal (hence the name cermet). Cermet films are designed to be fired in the range 850–1000°C.
- **Refractory** thick-film materials, typically tungsten, molybdenum, and titanium, which also may be alloyed with each other in various combinations. These materials are designed to be cofired with ceramic substrates at temperatures ranging up to 1600°C and are postplated with nickel and gold to allow component mounting and wire bonding.
- Polymer thick films, a mixture of polymer materials with conductor, resistor, or insulating particles. These materials cure at temperatures ranging from 85–300°C.

This chapter deals primarily with cermet thick-film materials, as these are most directly associated with the hybrid technology.

Thick-Film Materials

A cermet thick-film paste has four major ingredients: an active element, an adhesion element, an organic binder, and a solvent or thinner. The combination of the organic binder and thinner are often referred to as the vehicle, since it acts as the transport mechanism of the active and adhesion elements to the substrate. When these constituents are mixed together and milled for a period of time, the result is a thick, viscous mixture suitable for screen printing.

Active Element

The nature of the material that makes up the active element determines the electrical properties of the fired film. If the active material is a metal, the fired film will be a conductor; if it is a conductive metal oxide, it will be a resistor; and, if it is an insulator, it will be a dielectric. The active metal is in powder form ranging from 1 to 10 μm, with a mean diameter of about 5 μm.

Adhesion Element

The adhesion element provides the adhesion mechanism of the film to the substrate. There are two primary constituents used to bond the film to the substrate. One adhesion element is a glass, or frit, with a relatively low melting point. The glass melts during firing, reacts with the glass in the substrate, and flows into the irregularities on the substrate surface to provide the adhesion. In addition, the glass flows around the active material particles, holding them in contact with each other to promote sintering and to provide a series of three-dimensional continuous paths from one end of the film to the other.

A second class of conductor materials utilizes metal oxides to provide the adhesion. In this case, a pure metal is placed in the paste and reacts with oxygen atoms on the surface of the substrate to form an oxide. The primary adhesion mechanism of the active particles is sintering, both to the oxide and to each other, which takes place during firing. Conductors of this type offer improved adhesion and have a pure metal surface for added bondability, solderability, and conductivity. Conductors of this type are referred to as fritless materials, oxide-bonded materials, or molecular-bonded materials.

A third class of conductor materials utilizes both reactive oxides and glasses. These materials, referred to as mixed-bonded systems, incorporate the advantages of both technologies and are the most frequently used conductor materials.

Organic Binder

The organic binder is a **thixotropic** or **pseudoplastic fluid** and serves two purposes: it acts as a vehicle to hold the active and adhesion elements in suspension until the film is fired, and it gives the paste the proper fluid characteristics for screen printing. The organic binder is usually referred to as the nonvolatile organic since it does not readily evaporate, but begins to burn off at about 350°C. The binder must oxidize cleanly during firing, with no residual carbon, which could contaminate the film. Typical materials used in this application are ethyl cellulose and various acrylics.

For nitrogen-fireable films, where the firing atmosphere can contain only a few ppm of oxygen, the organic vehicle must decompose and thermally depolymerize, departing as a highly volatile organic vapor in the nitrogen blanket provided as the firing atmosphere.

Solvent or Thinner

The organic binder in its usual form is too thick to permit screen printing, which necessitates the use of a solvent or thinner. The thinner is somewhat more volatile than the binder, evaporating rapidly above 70–100°C. Typical materials used for this application are terpineol, butyl carbitol, and certain of the complex alcohols into which the nonvolatile phase can dissolve.

The ingredients of the thick-film paste are mixed together in proper proportions and milled on a three-roll mill for a sufficient period of time to ensure that they are thoroughly mixed and that no agglomeration exists.

Thick-Film Conductor Materials

Most thick-film conductor materials are noble metals or combinations of noble metals, which are capable of withstanding the high processing temperatures in an oxidizing atmosphere. The exception is copper, which must be fired in a pure nitrogen atmosphere.

Thick-film conductors must perform a variety of functions:

- Provide a path for electrical conduction
- Provide a means for component attach
- Provide a means for terminating thick-film resistors

The selection of a conductor material depends on the technical, reliability, and cost requirements of a particular application. Figure 11.7 illustrates the process compatibility of the various types of conductors.

| | | PROCESS | | | |
		AU WIRE BONDING	AL WIRE BONDING	EUTECTIC BONDING	SN/PB SOLDER	EPOXY BONDING
MATERIAL	AU	Y	N	Y	N	Y
	PD/AU	N	Y	N	Y	Y
	PT/AU	N	Y	N	Y	Y
	AG	Y	N	N	Y	Y
	PD/AG	N	Y	N	Y	Y
	PT/AG	N	Y	N	Y	Y
	PT/PD/AG	N	Y	N	Y	Y
	CU	N	Y	N	Y	N

FIGURE 11.7 Process compatibility of thick-film conductors.

Gold

Gold (Au) is used in applications where a high degree of reliability is required, such as military and space applications, and where eutectic die bonding is necessary. Gold is also used where gold ball bonding is required or desired. Gold has a strong tendency to form intermetallic compounds with other metals used in the electronic assembly process, especially tin (Sn) and aluminum (Al), and the characteristics of these alloys may be detrimental to reliability. Therefore, when used in conjunction with tin or aluminum, gold must frequently be alloyed with other noble metals, such as platinum (Pt) or palladium (Pd) to prevent AuSn or AuAl alloys from forming. It is common to use a PtAu alloy when PbSn solder is to be used, and to use a PdAu alloy when Al wire bonding is required to minimize intermetallic compound formation.

Silver

Silver (Ag) shares many of the properties of gold in that it alloys readily with tin and leaches rapidly into molten Sn/Pb solders in the pure state. Pure silver conductors must have special bonding materials in the adhesion mechanism or must be alloyed with Pd and/or Pt in order to be used with Sn/Pb solder. Silver is also susceptible to **migration** when moisture is present. Although most metals share this characteristic to a greater or lesser degree, Ag is highly susceptible due to its low ionization potential.

Alloys of Silver with Platinum and/or Palladium

Alloying Ag with Pd and/or Pt slows down both the leaching rate and the migration rate, making it practical to use these alloys for soldering. These are used in the vast majority of commercial applications, making them by far the most commonly used conductor materials. Until recently, silver-bearing materials have been rarely used in multilayer applications due to their tendency to react with and diffuse into dielectric materials, causing short circuits and weakened voltage handling capability. Advancements in these materials have improved this property. A disadvantage of Pd/Pt/Ag alloys is that the electrical resistance is significantly increased. In some cases, pure Ag is nickel (Ni) plated to increase leach resistance.

Copper

Originally developed as a low-cost substitute for gold, copper is now being selected when solderability, leach resistance, and low resistivity are required. These properties are particularly attractive for power hybrid circuits. The low resistivity allows the copper conductor traces to handle higher currents with a lower voltage drop, and the solderability allows power devices to be soldered directly to the metallization for better thermal transfer.

Thick-Film Resistor Material

Thick-film resistors are formed by adding metal oxide particles to glass particles and firing the mixture at a temperature/time combination sufficient to melt the glass and to sinter the oxide particles together. The resulting structure consists of a series of three-dimensional chains of metal oxide particles embedded in a glass matrix. The higher the metal oxide-to-glass ratio, the lower is the resistivity and vice versa.

Referring to Fig. 11.8, the electrical resistance of a material in the shape of a rectangular solid is given by the classic formula

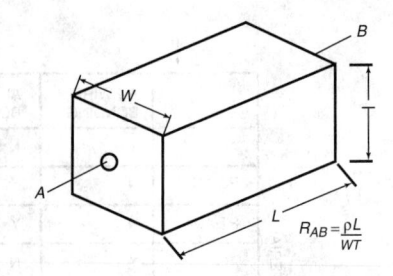

FIGURE 11.8 Resistance of a rectangular solid.

$$R = \frac{\rho_B L}{WT} \tag{11.1}$$

where

R = electrical resistance, Ω
ρ_B = bulk resistivity of the material, ohms-length
L = length of the sample in the appropriate units
W = width of the sample in the appropriate units
T = thickness of the sample in the appropriate units

A bulk property of a material is one that is independent of the dimensions of the sample. When the length and width of the sample are much greater than the thickness, a more convenient unit to use is the sheet resistance, which is equal to the bulk resistivity divided by the thickness,

$$\rho_s = \frac{\rho_B}{T} \tag{11.2}$$

where ρ_s is the sheet resistance in ohms/square/unit thickness.

The sheet resistance, unlike the bulk resistivity, is a function of the dimensions of the sample. Thus, a sample of a material with twice the thickness as another sample will have half the sheet resistivity, although the bulk resistivity is the same. In terms of the sheet resistivity, the electrical resistivity is given by

$$R = \frac{\rho_s L}{W} \tag{11.3}$$

If the length is equal to the width (the sample is a square), the electrical resistance is the same as the sheet resistivity independent of the actual dimensions of the sample. This is the basis of the units of sheet resistivity, ohms/square/unit thickness. For thick-film resistors, the standard adopted for unit thickness is 0.001 in. or 25 mm of *dried* thickness. The specific units for thick-film resistors are $\Omega/\square/0.001$ (read ohms per square per mil) of dried thickness. For convenience, the units are generally referred to as, simply, $\Omega/\square$.

A group of thick-film materials with identical chemistries that are blendable is referred to as a *family* and will generally have a range of values from 10 $\Omega/\square$ to 1 M$\Omega/\square$ in decade values, although intermediate values are available as well. There are both high and low limits to the amount of material that may be added. As more and more material is added, a point is reached where there is not enough glass to maintain the structural integrity of the film. A practical lower limit of sheet resistivity of resistors formed in this manner is about 10 $\Omega/\square$. Resistors with a sheet resistivity below this value must have a different chemistry and often are not blendable with the regular family of materials. At the other extreme, as less and less material is added, a point is reached where there are not enough particles to form continuous chains, and the sheet resistance rises very abruptly. Within most resistor families, the practical upper limit is about 2 M$\Omega/\square$. Resistor materials are available to about 20 M$\Omega/\square$, but the chemical composition of these materials is not amenable to blending with lower value resistors.

The active phase for resistor formulation is the most complex of all thick-film formulations due to the large number of electrical and performance characteristics required. The most common active material used in air-fireable resistor systems is ruthenium, which can appear as RuO_2 (ruthenium dioxide) or as $BiRu_2O_7$ (bismuth ruthenate).

Properties of Thick-Film Resistors

The conduction mechanisms in thick-film resistors are very complex and have not yet been well defined. Mechanisms with metallic, semiconductor, and insulator properties have all been identified. High ohmic value resistors tend to have more of the properties associated with semiconductors, whereas low ohmic values tend to have more of the properties associated with conductors:

- High ohmic value resistors tend to have a more negative *temperature coefficient of resistance* (*TCR*) than low ohmic value resistors. This is not always the case in commercially available systems due to the presence of TCR modifiers, but always holds true in pure metal oxide–glass systems.
- High ohmic value resistors exhibit substantially more current noise than low ohmic value resistors.

- High ohmic value resistors are more susceptible to high voltage pulses and static discharge than low ohmic value resistors. Some resistor families can be reduced in value by more than an order of magnitude when exposed to an *electrostatic discharge (ESD)* charge of only moderate value.

Temperature Coefficient of Resistance

All materials exhibit a change in resistance with material, either positive or negative, and many are nonlinear to a high degree. By definition, the TCR at a given temperature is the slope of the resistance-temperature curve at that temperature. The TCR of thick-film resistors is generally linearized over one or more intervals of temperature, typically in the range from -55 to $+125°C$, as shown in Eq. (11.4) as

$$\text{TCR} = \frac{R(T_2) - R(T_1)}{R(T_1)(T_2 - T_1)} \times 10^6 \qquad (11.4)$$

where

$\quad$ TCR $=$ temperature coefficient of resistance, ppm/°C
$\quad R(T_2) =$ resistance at temperature T_2
$\quad R(T_1) =$ resistance at temperature T_1
$\qquad T_1 =$ temperature at which $R(T_1)$ is measured, reference temperature
$\qquad T_2 =$ temperature at which $R(T_2)$ is measured

A commercial resistor paste carefully balances the metallic, nonmetallic, and semiconducting fractions to obtain a TCR as close to zero as possible. This is not a simple task, and the hot TCR may be quite different from the cold TCR. Paste manufacturers give both the hot and cold values when describing a resistor paste. Although this does not fully define the curve, it is adequate for most design work.

It is important to note that the TCR of most materials is not linear, and the single point measurement is at best an approximation. The only completely accurate method of describing the temperature characteristics of a material is to examine the actual graph of temperature vs. resistance. The TCR for a material may be positive or negative. By convention, if the resistance increases with increasing temperature, the TCR is positive. Likewise, if the resistance decreases with increasing temperature, the TCR is negative.

Voltage Coefficient of Resistance (VCR)

The expression for the VCR is similar in form to the TCR and may be represented by Eq. (11.5) as

$$\text{VCR} = \frac{R(V_2) - R(V_1)}{R(V_1)(V_2 - V_1)} \qquad (11.5)$$

where

$\quad R(V_1) =$ resistance at V_1
$\quad R(V_2) =$ resistance at V_2
$\qquad V_1 =$ voltage at which $R(V_1)$ is measured
$\qquad V_2 =$ voltage at which $R(V_2)$ is measured

Because of the semiconducting nature of resistor pastes, the VCR is always negative. As V_2 is increased, the resistance decreases. Also, because higher resistor decade values contain more glass and oxide constituents, and are more semiconducting, higher paste values tend to have more negative VCRs than lower values.

The VCR is also dependent on resistor length. The voltage effect on a resistor is a gradient; it is the volt/mil rather than the absolute voltage that causes resistor shift. Therefore, long resistors show less voltage shift than short resistors, for similar compositions and voltage stress.

Resistor Noise

Resistor noise is an effective method of measuring the quality of the resistor and its termination. On a practical level, noise is measured according to MIL-STD-202 on a Quantech Noise Meter. The resistor current noise is compared to the noise of a standard low noise resistor and reported as a **noise index** in decibel.

The noise index is expressed as microvolt/volt/frequency decade. The noise measured by MIL-STD-202 is 1/f noise, and the measurement assumes this is the only noise present. However, there is thermal or white noise present in all materials that is not frequency dependent and adds to the 1/f noise. Measurements are taken at low frequencies to minimize the effects of thermal noise. The noise index of low-value resistors is lower than high-value resistors because the low-value resistors have more metal and more free electrons. Noise also decreases with increasing resistor area (actually resistor volume).

High-Temperature Drift

The high-temperature drift characteristic of a resistor is an important material attribute, as it affects the long-term performance of the circuit. A standard test condition is 125°C for 1000 h at normal room humidity. A more aggressive test would be 150°C for 1000 h or 175°C for 40 h. A trimmed thick-film resistor is expected to remain within 1% of the original value after testing under these conditions.

Power Handling Capability

Drift due to high power is usually due to internal resistor heating. It is different from thermal aging in that the heat is generated at the point-to-point metal contacts within the resistor film. When a resistor is subjected to heat from an external source, the whole body is heated to the test temperature. Under power, local heating can result in a much higher temperature. Because lower value resistors have more metal and, therefore, many more contacts, low-value resistors tend to drift less than higher value resistors under similar loads.

Resistor pastes are generally rated at 50 W/in^2 of active resistor area. This is a conservative figure, however, and the resistor can be rated at 100 or even 200 W/in^2. A burn-in process will substantially reduce subsequent drift since most of the drift occurs in the first hours of load.

Process Considerations

The process windows for printing and firing thick-film resistors are extremely critical in terms of both temperature control and atmosphere control. Small variations in temperature or time at temperature can cause significant changes in the mean value and distribution of values. In general, the higher the ohmic value of the resistor, the more dramatic is the change. As a rule, high ohmic values tend to decrease as temperature and/or time is increased, whereas very low values ($<100\ \Omega/\square$) may tend to increase.

Thick-film resistors are very sensitive to the firing atmosphere. For resistor systems used with air fireable conductors, it is critical to have a strong oxidizing atmosphere in the firing zone of the furnace. In a neutral or reducing atmosphere, the metallic oxides that comprise the active material reduces to pure metal at the temperatures used to fire the resistors, dropping resistor values by more than an order of magnitude. Again, high ohmic value resistors are more sensitive than low ohmic value. Atmospheric contaminants, such as vapors from hydrocarbons or halogenated hydrocarbons, breaks down at firing temperatures, creating a strong reducing atmosphere. For example, one of the breakdown components of hydrocarbons is carbon monoxide, one of the strongest reducing agents known. Concentrations of fluorinated hydrocarbons in a firing furnace of only a few ppm can drop the value of a 100 kΩ resistor to below 10 kΩ.

Properties of Thick-Film Dielectric Materials

Multilayer Dielectric Materials

Thick-film dielectric materials are used primarily as insulators between conductors, either as simple crossovers or in complex multilayer structures. Small openings, or vias, may be left in the dielectric layers so that adjacent conductor layers may interconnect. In complex structures, as many as several hundred vias per layer may be required. In this manner, complex interconnection structures may be created. Although the majority of thick-film circuits can be fabricated with only three layers of metallization, others may require several more. If more than three layers are required the yield begins dropping dramatically with a corresponding increase in cost.

Dielectric materials used in this application must be of the devitrifying or recrystallizable type. These materials in paste form are a mixture of glasses that melt at a relatively low temperature. During firing, when they are in the liquid state, they blend together to form a uniform composition with a higher melting point than the firing temperature. Consequently, on subsequent firings they remain in the solid state, which maintains a stable foundation for firing sequential layers. By contrast, vitreous glasses always melt at the same temperature and would be unacceptable for layers either to sink and short to conductor layers underneath, or to swim and form an open circuit. Additionally, secondary loading of ceramic particles is used to enhance devitrification and to modify the TCE.

Dielectric materials have two conflicting requirements in that they must form a continuous film to eliminate short circuits between layers while, at the same time, they must maintain openings as small as 0.010-in. In general, dielectric materials must be printed and fired twice per layer to eliminate pinholes and prevent short circuits between layers.

The TCE of thick-film dielectric materials must be as close as possible to that of the substrate to avoid excessive bowing, or warpage, of the substrate after several layers. Excessive bowing can cause severe problems with subsequent processing, especially where the substrate must be held down with a vacuum or where it must be mounted on a heated stage. In addition, the stresses created by the bowing can cause the dielectric material to crack, especially when it is sealed within a package. Thick-film material manufacturers have addressed this problem by developing dielectric materials that have an almost exact TCE match with alumina substrates. Where a serious mismatch exists, matching layers of dielectric must be printed on the bottom of the substrate to minimize bowing, which obviously increases the cost.

Dielectric materials with higher dielectric constants are also available for manufacturing thick-film capacitors. These generally have a higher loss tangent than chip capacitors and utilize a great deal of space. Although the initial tolerance is not good, thick-film capacitors can be trimmed to a high degree of accuracy.

Overglaze Materials

Dielectric overglaze materials are vitreous glasses designed to fire at a relatively low temperature, usually around 550°C. They are designed to provide mechanical protection to the circuit, to prevent contaminants and water from spanning the area between conductors, to create solder dams, and to improve the stability of thick-film resistors after trimming.

When soldering a device with many leads, it is imperative that the volume of solder under each lead be the same. A well-designed overglaze pattern can prevent the solder from wetting other circuit areas and flowing away from the pad, keeping the solder volume constant. In addition, overglaze can help to prevent solder bridging between conductors.

Overglaze material has long been used to stabilize thick-film resistors after laser trim. In this application, a green or brown pigment is added to enhance the passage of a yttrium–aluminum–garnet (YAG) laser beam. Colors toward the shorter wavelength end of the spectrum, such as blue, tend to reflect a portion of the YAG laser beam and reduce the average power level at the resistor. There is some debate as to the effectiveness of overglaze in enhancing the resistor stability, particularly with high ohmic values. Several studies have shown that, although overglaze is undoubtedly helpful at lower values, it can actually increase the resistor drift of high-value resistors by a significant amount.

Processing Thick-Film Circuits

Screen Printing

The thick-film fabrication process begins with the generation of artwork from the drawings of the individual layers created during the design stage. The artwork represents the exact pattern to be deposited on a 1:1 scale. A stainless-steel wire mesh screen ranging from 80 to 400 mesh count is coated with a photosensitive material and exposed using the artwork for the layer to be printed. The selection of the mesh count depends on the linewidth to be printed. For 0.010 in lines and spaces, a 325 mesh is adequate. For lines down to 0.005 in., a 400 mesh screen is required, and for coarse materials such as solder paste, an 80 mesh screen is needed. The unexposed photoresistive material is washed away leaving openings in the screen where paste is to be deposited. The screen is mounted into a screen printer that has, at a minimum, provisions for controlling squeegee pressure, squeegee speed, and the snapoff distance. Most thick-film materials are printed by the off-contact process in which the screen is separated from the substrate by a distance called the snapoff. As the squeegee is moved, it stretches the screen by a small amount, creating tension in the wire mesh. As the squeegee passes, the tension in the screen causes the mesh to snap back, leaving the paste on the screen. The snapoff distance is probably the most critical parameter in the entire screen printing process. If it is too small, the paste will tend to be retained in the screen mesh, resulting in an incomplete print; if it is too large, the screen will be stretched too far, resulting in a loss of tension. A typical snapoff distance for an 8×10 in. screen is 0.025 in.

Drying

Thick-film pastes are dried prior to firing to evaporate the volatile solvents from the printed films. If the volatile solvents are allowed to enter the firing furnace, flash evaporation may occur, leaving pits or craters in the film. In addition, the byproducts of these materials may result in reduction of the oxides that comprise the fired film. Most solvents in thick-film pastes have boiling points in the range of 180–250°C. Because of the high surface area/volume of deposited films, drying at 80–160°C for a period of 10–30 min is adequate to remove most of the solvents from wet films.

Firing

Belt furnaces having independently controlled heated zones through which a belt travels at a constant speed are commonly used for firing thick films. By adjusting the zone temperature and the belt speed, a variety of time vs. temperature profiles can be achieved. The furnace must also have provisions for atmosphere control during the firing process to prevent reduction (or, in the case of copper, oxidation) of the constituents.

Figure 11.9 illustrates two typical thick-film circuits.

11.2.4   Thin-Film Technology

Thin-film technology, in contrast to thick-film technology, is a subtractive technology in that the entire substrate is coated with several layers of material and the unwanted material is etched away in a succession of photoetching processes. The use of photolithographic processes to form the patterns enables much narrower and more well-defined lines than can be formed by the thick-film process. This feature promotes the use the thin-film technology for high-density and high-frequency applications.

Thin-film circuits typically consist of three layers of material deposited on a substrate. The bottom layer serves two purposes: it is the resistor material and also provides the adhesion to the substrate. The adhesion mechanism of the film to the substrate is an oxide layer, which forms at the interface between the film and the substrate. The bottom layer must, therefore, be a material that oxidizes readily. The middle layer acts as an interface between the resistor layer and the conductor layer, either by improving the adhesion of the conductor or by preventing diffusion of the resistor material into the conductor. The top layer acts as the conductor layer.

Deposition Technology

The term thin film refers more to the manner in which the film is deposited onto the substrate, as opposed to the actual thickness of the film. Thin films are typically deposited by a vacuum deposition technique or by electroplating.

Sputtering

Sputtering is the principal method by which thin films are applied to substrates. In the most basic form of sputtering, a current is established in a conducting plasma formed by striking an arc in a partial vacuum of approximately 10-μ pressure with a potential applied. The gas used to establish the plasma is typically an inert gas, such as argon, that does not react with the target material. The substrate and a target material are situated in the plasma with the substrate at ground potential and the target at a high potential, which may be AC or DC. The high potential attracts the gas ions in the plasma to the point where they collide with the

FIGURE 11.9 Thick-film circuits: Hermetic (above) and nonhermetic.

target with sufficient kinetic energy to dislodge microscopically sized particles with enough residual kinetic energy to travel the distance to the substrate and adhere. This process is referred to as *triode sputtering*.

Ordinary triode sputtering is a very slow process, requiring hours to produce useable films. By utilizing magnets at strategic points, the plasma can be concentrated in the vicinity of the target, greatly speeding up the deposition process. The potential that is applied to the target is typically RF energy at a frequency of approximately 13 MHz, which may be generated by a conventional electronic oscillator or by a magnetron. The magnetron is capable of generating considerably more power with a correspondingly higher deposition rate.

By adding small amounts of other gases, such as oxygen and nitrogen, to the argon, it is possible to form oxides and nitrides of certain target materials on the substrate. It is this technique, called *reactive sputtering*, which is used to form tantalum nitride, a common resistor material.

Evaporation

The evaporation of a material into the surrounding area occurs when the vapor pressure of the material exceeds the ambient pressure and can take place from either the solid state or the liquid state. In the thin film process, the material to be evaporated is placed in the vicinity of the substrate and heated until the vapor pressure of the material is considerably above the ambient pressure. The evaporation rate is directly proportional to the difference between the vapor pressure of the material and the ambient pressure and is highly dependent on the temperature of the material.

Evaporation must take place in a relatively high vacuum ($<10^{-6}$ torr) for three reasons:

1. To lower the vapor pressure required to produce an acceptable evaporation rate, thereby lowering the required temperature required to evaporate the material.
2. To increase the mean free path of the evaporated particles by reducing the scattering due to gas molecules in the chamber. As a further result, the particles tend to travel in more of a straight line, improving the uniformity of the deposition.
3. To remove atmospheric contaminants and components, such as oxygen and nitrogen, which tend to react with the evaporated film.

At 10^{-7} torr, a vapor pressure of 10^{-2} torr is required to produce an acceptable evaporation rate. For most metals, this temperature is in excess of 1000°C. The refractory metals, or metals with a high melting point such as tungsten, titanium, or molybdenum, are frequently used as carriers, or *boats*, to hold other metals during the evaporation process. To prevent reactions with the metals being evaporated, the boats may be coated with alumina or other ceramic materials.

In general, the kinetic energy of the evaporated particles is substantially less than that of sputtered particles. This requires that the substrate be heated to about 300°C to promote the growth of the oxide adhesion interface. This may be accomplished by direct heating of the substrate mounting platform or by radiant infrared heating.

There are several techniques by which evaporation can be accomplished. The two most common of these are resistance heating and electron-beam (E-beam) heating.

Evaporation by resistance heating usually takes place from a boat made with a refractory metal, a ceramic crucible wrapped with a wire heater, or a wire filament coated with the evaporant. A current is passed through the element, and the generated heat heats the evaporant. It is somewhat difficult to monitor the temperature of the melt by optical means due to the propensity of the evaporant to coat the inside of the chamber, and control must be done by empirical means.

The E-beam evaporation method takes advantage of the fact that a stream of electrons accelerated by an electric field tend to travel in a circle when entering a magnetic field. This phenomenon is utilized to direct a high-energy stream of electrons onto an evaporant source. The kinetic energy of the electrons is converted into heat when they strike the evaporant. E-beam evaporation is somewhat more controllable since the resistance of the boat is not a factor, and the variables controlling the energy of the electrons are easier to measure and control. In addition, the heat is more localized and intense, making it possible to evaporate metals with higher 10^{-2} torr temperatures and lessening the reaction between the evaporant and the boat.

Comparison Between Sputtering and Evaporation

The adhesion of a sputtered film is superior to that of an evaporated film, and is enhanced by presputtering the substrate surface by random bombardment of argon ions prior to applying the potential to the target. This process removes several atomic layers of the substrate surface, creating a large number of broken oxygen bonds and promoting the formation of the oxide interface layer. The oxide formation is further enhanced by the residual heating of the substrate that results from the transfer of the kinetic energy of the sputtered particles to the substrate when they collide.

It is difficult to evaporate alloys such as NiCr due to the difference between the 10^{-2} torr temperatures. The element with the lower temperature tends to evaporate somewhat faster, causing the composition of the evaporated film to be different from the composition of the alloy. To achieve a particular film composition, the composition of the melt must contain a higher portion of the material with the higher 10^{-2} torr temperature and the temperature of the melt must be tightly controlled. By contrast, the composition of a sputtered film is identical to that of the target.

Evaporation is limited to the metals with lower melting points. Refractory metals and ceramics are virtually impossible to deposit by evaporation.

Reactive deposition of nitrides and oxides is very difficult to control.

Electroplating

Electroplating is accomplished by applying a potential between the substrate and the anode, which are suspended in a conductive solution of the material to be plated. The plating rate is a function of the potential and the concentration of the solution. In this manner, most metals can be plated to a metal surface.

This is considerably more economical and results in much less target usage. For added savings, some companies apply photoresist to the substrate and electroplate gold only where actually required by the pattern.

Photolithographic Processes

In the photolithographic process, the substrate is coated with a photosensitive material that is exposed through a pattern formed on a glass plate. Ultraviolet light, X rays, and electron beams may all be used to expose the film. The photoresist may be of the positive or negative type, with the positive type being prevalent due to its inherently higher resistance to the etchant materials. The unwanted material that is not protected by the photoresist may be removed by wet, or chemical etching, or by dry, or sputter etching, in which the unwanted material is removed by ion bombardment. In essence, the exposed material acts as a sputtering target. The photoresist, being more compliant, simply absorbs the ions and protects the material underneath.

Thin-Film Materials

Virtually any inorganic material may be deposited by the sputtering process, although RF sputtering is necessary to deposit dielectric materials or metals, such as aluminum, that oxidize readily. Organic materials are difficult to sputter because they tend to release absorbed gases in a high vacuum, which interferes with the sputtering process. A wide variety of substrate materials are also available, but these in general must contain or be coated with an oxygen compound to permit adhesion of the film.

Thin-Film Resistors

Materials used for thin-film resistors must perform a dual role in that they must also provide the adhesion to the substrate, which narrows the choice to those materials that form oxides. The resistor film begins forming as single points on the substrate in the vicinity of substrate faults or other irregularities, which might have an excess of broken oxygen bonds. The points expand into islands, which in turn join to form continuous films. The regions where the islands meet are called *grain boundaries* and are a source of collisions for the electrons. The more grain boundaries present, the more negative is the TCR. Unlike thick-film resistors, however, the boundaries do not contribute to the noise level. Further, laser trimming does not create microcracks in the glass-free structure, and the inherent mechanisms for resistor drift are not present in thin films. As a result, thin-film resistors have better stability, noise, and TCR characteristics than thick-film resistors.

The most common types of resistor material are nichrome (NiCr) and tantalum nitride (TaN). Although NiCr has excellent stability and TCR characteristics, it is susceptible to corrosion by moisture if not passivated by sputtered quartz or by evaporated silicon monoxide (SiO). TaN, on the other hand, may be passivated by simply baking in air for a few minutes. This feature has resulted in the increased use of TaN at the expense of NiCr, especially in military programs. The stability of passivated TaN is comparable to that of passivated NiCr, but the TCR is not as good unless annealed for several hours in a vacuum to minimize the effect of the grain boundaries. Both NiCr and TaN have a relatively low maximum sheet resistivity on alumina, about 400 $\Omega/\square$ for NiCr and 200 $\Omega/\square$ for TaN. This requires lengthy and complex patterns to achieve a high value of resistance, resulting in a large area and the potential for low yield.

The TaN process is the most commonly used due to its inherently high stability. In this process N_2 is introduced into the argon gas during the sputtering process, forming TaN by reacting with pure Ta atoms on the surface of the substrate. By heating the film in air at about 425°C for 10 min, a film of TaO is formed over the TaN, which is virtually impervious to further O_2 diffusion at moderately high temperatures, which helps to maintain the composition of the TaN film and to stabilize the value of the resistor. TaO is

FIGURE 11.10   A thin-film circuit.

essentially a dielectric and, during the stabilization of the film, the resistor value is increased. The amount of increase for a given time and temperature is dependent on the thickness and sheet resistivity of the film. Films with a lower sheet resistivity increase proportionally less than those with a higher sheet resistivity. The resistance increases as the film is heated longer, making it possible to control the sheet resistivity to a reasonable accuracy on a substrate-by-substrate basis.

Other materials, based on chromium and rhenium alloys, are also available. These materials have a higher sheet resistivity than NiCr and TaN and offer the same degree of stability.

Barrier Materials

When Au is used as the conductor material, a barrier material between the Au and the resistor is required. When gold is deposited directly on NiCr, the Cr has a tendency to diffuse through the Au to the surface, which interferes with both wire bonding and eutectic die bonding. To alleviate this problem, a thin layer of pure Ni is deposited over the NiCr, which also improves the solderability of the surface considerably. The adhesion of Au to TaN is very poor. To provide the necessary adhesion, a thin layer of 90Ti/10W may be used between the Au and the TaN.

Conductor Materials

Gold is the most common conductor material used in thin-film hybrid circuits because of the ease of wire and die bonding and its high resistance to tarnish and corrosion. Aluminum and copper are also used in some applications. A thin-film circuit is illustrated in Fig. 11.10.

11.2.5   Resistor Trimming

One of the advantages of the hybrid technology over other **packaging** technologies is the ability to adjust the value of resistors to a more precise value by a process called *trimming*. By removing a portion of the resistor with a laser or by abrasion, as shown in Fig. 11.11, the value can be increased from the as-processed state to a predetermined value. The laser trimming process can be highly automated and can trim a resistor to a tolerance of better than 1% in less than a second.

More than any other development, laser trimming has contributed to the rapid growth of the hybrid microelectronics industry over the past few years. Once it became possible to trim resistors to precision

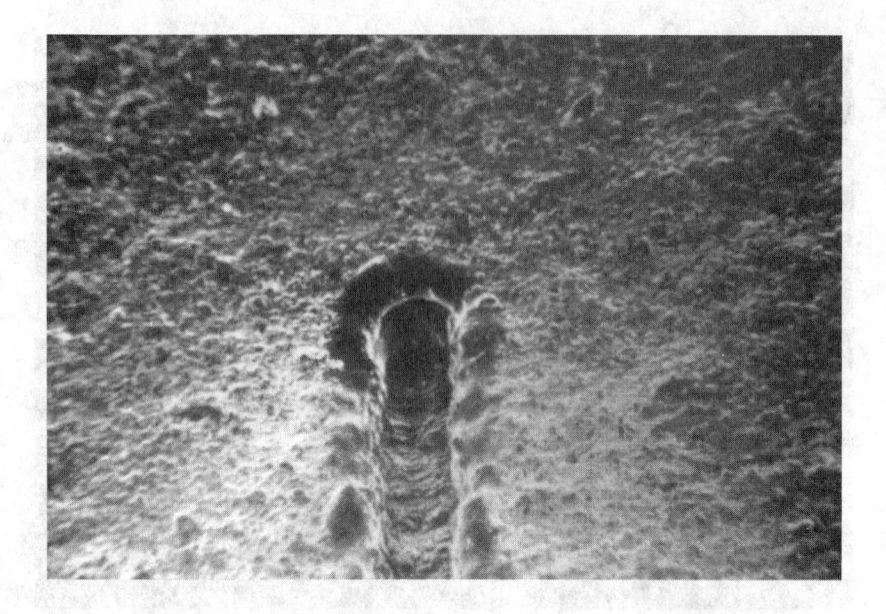

FIGURE 11.11   A laser-trimmed thick-film resistor.

values economically, hybrid circuits became cost competitive with other forms of packaging and the technology expanded rapidly.

Lasers used to trim thick-film resistors are generally a YAG crystal, which has been doped with neodymimum. YAG lasers are of lower power and are capable of trimming film resistors without causing serious damage if the parameters are adjusted properly. Even with the YAG laser, it is necessary to lower the maximum power and spread out the beam to avoid overtrimming and to increase the trimming speed. This is accomplished by a technique called *Q-switching*, which decreases the peak power and widens the pulse width such that the overall average power is the same. One of the effects of this technique is that the first and last pulses of a given trim sequence have substantially greater energy than the intervening ones. The first pulse does not create a problem, but the last pulse occurs within the body of the resistor, penetrating deeper into the substrate and creating the potential for resistor drift by creating minute cracks (microcracks) in the resistor, which emanate from the point of trim termination.

The microcracks occur as a result of the large thermal gradient, which exists between the area of the last pulse and the remainder of the resistor. Over the life of the resistor, the cracks will propagate through the body of the resistor until they reach a termination point, such as an active particle or even another microcrack. The propagation distance can be up to several mils in a high-value resistor with a high concentration of glass. The microcracks cause an increase in the value of the resistor and also increase the noise by increasing the current density in the vicinity of the trim. With proper design of the resistor, and with proper selection of the cut mode, the amount of drift can be held to less than 1%.

The propagation of the microcracks can be accelerated by the application of heat, which also has the effect of stabilizing the resistor. Where high-precision resistors are required, the resistors may be trimmed slightly low, exposed to heat for a period of time, and retrimmed to a value in a noncritical area.

The amount of drift depends on the amount of propagation distance relative to the distance from the termination point of the trim to the far side of the resistor. This can be minimized by either making the resistor larger or by limiting the distance that the laser penetrates into the resistor. As a rule of thumb, the value of the resistor should not be increased by more than a factor of two to minimize drift.

In the abrasive trimming process, a fine stream of sand propelled by compressed air is directed at the resistor, abrading away a portion of the resistor and increasing the value. Trimming resistors to a high precision is more difficult with abrasive trimming due to the size of the particle stream, although tolerances of 1% can be achieved with proper setup and slow trim speeds. It has also proven difficult to automate

TABLE 11.6 Comparison of Thick-Film and Thin-Film Circuits

Thick Film	Thin Film
Multilayer structures are much simpler and more economical to fabricate than with the thin-film technology.	The lines and spaces are much smaller than can be attained by the thick-film process. In a production environment, the thin-film process can produce lines and spaces 0.001 in. in width, whereas the thick film process is limited to 0.01 in.
The range of resistor values is wider. Thin-film designers are usually limited to a single value of sheet resistivity from which to design all of the resistors in the circuit.	The line definition available by the thin-film process is considerably better than that of the thick-film process. Consequently, the thin-film process operates much better at high frequencies.
For a given application, the thick-film process is usually less expensive.	The electrical properties of thin-film resistors are substantially better than thick-film resistors in terms of noise, stability, and precision.

the abrasive trimming process to a high degree, and it remains a relatively slow process. In addition, substantially larger resistors are required.

Despite these shortcomings, abrasive trimming plays an important role in the hybrid industry. The cost of an abrasive trimmer is substantially less than that of a laser trimmer, and the setup time is much less. In a developmental mode where only a few prototype units are required, abrasive trimming is generally more economical and faster than laser trimming.

In terms of performance, abrasively trimmed resistors are more stable than laser-trimmed resistors and generate less noise because no microcracks are generated during the abrasive trimming process. In the power hybrid industry, resistors that carry high currents or that dissipate high power are frequently abrasively trimmed by trimming a groove in the middle of the resistor. This technique minimizes current crowding and further enhances the stability.

11.2.6   Comparison of Thick-Film and Thin-Film Technologies

Most circuits can be fabricated with either the thick- or the thin-film technology. The ultimate result of both processes is an interconnection pattern on a ceramic substrate with integrated resistors, and, to a degree, they can be considered to be competing technologies. Given a particular set of requirements, however, the choice between technologies is usually quite distinct. Table 11.6 summarizes the particular advantages of each technology.

11.2.7   The Hybrid Assembly Process

The assembly of a hybrid circuit involves mechanically mounting the components on the substrate and electrically connecting the components to the proper substrate traces.

In the most fundamental configuration, the **semiconductor die** is taken directly from the wafer without further processing (generally referred to as a bare die). In this process, the die is mechanically mounted to the substrate with epoxy, solder, or by direct eutectic bonding of the silicon to the substrate metallization, and the electrical connections are made by bonding small wires from the bonding pads to the appropriate conductor. This is referred to as chip-and-wire technology, illustrated in Fig. 11.10.

Other approaches require subsequent processing of the die at the wafer level. The most common configurations are *ultrasonic bonding* (Fig. 11.12), *tape automated bonding* (*TAB*), shown in Fig. 11.13, and the so-called *flip chip* or *bumped chip* approach, depicted in Fig. 11.14. These are discussed later.

The Chip-and-Wire Process

Bare semiconductor die may be mounted with epoxy, solder, or by direct eutectic bonding. The most common method of both active and passive component attachment is with epoxy, both conductive and nonconductive. This method offers a number of advantages over other techniques, including ease of repair,

FIGURE 11.12   Ultrasonic bonding of heavy aluminum wire.

reliability, productivity, and low process temperature. Most epoxies have a filler added or electrical and/or thermal conductivity. The most common material used for conductive epoxies is silver, which provides both. Other conductive filler materials include gold, palladium–silver for reduced silver migration, and copper plated with a surface coating of tin. Nonconductive filler materials include aluminum oxide and magnesium oxide for improved thermal conductivity. For epoxy attach, it is preferred that the bottom surface of the die be gold. Solder attachment of a die to a substrate is commonly used in the power hybrid industry. For this application, a coating of Ti/Ni/Ag alloy on the bottom is favored to gold. Since many

FIGURE 11.13   Tape automated bonding technique.

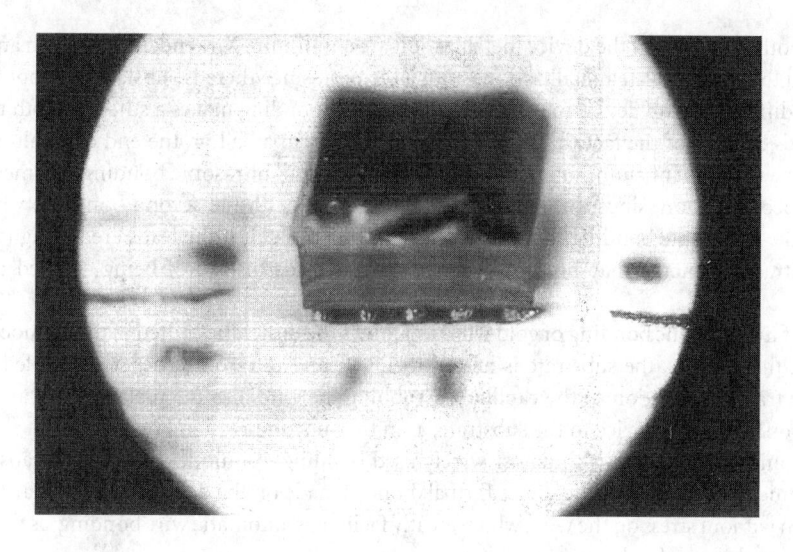

FIGURE 11.14 Flip chip bonding technique.

power dice are, by necessity, physically large, a compliant solder, such as one of the PbIn alloys is preferred. Eutectic bonding refers to the direct bonding of a silicon device to gold metallization, which takes place at 370°C with a combination of 94% gold and 6% silicon by weight. Gold is the only metal to which the eutectic temperature in combination with silicon is sufficiently low to be practical.

Wire bonding is used to make the electrical connections from the aluminum contacts to the substrate metallization or to a lead frame, from other components, such as chip resistors or chip capacitors, to substrate metallization, from package terminals to the substrate metallization, or from one point on the substrate metallization to another.

There are two basic methods of *wire bonding, thermocompression wire bonding* and *ultrasonic wire bonding*. Thermocompression wire bonding, as the name implies, utilizes a combination of heat and pressure to form an intermetallic bond between the wire and a metal surface. In pure thermocompression bonding, a gold wire is passed through a hollow capillary and a ball formed on the end by means of an electrical arc. The substrate is heated to about 300°C, and the ball is forced into contact with the bonding pad on the device with sufficient force to cause the two metals to bond. The capillary is then moved to the bond site on the substrate, feeding the wire as it goes, and as the wire is bonded to the substrate by the same process, except that the bond is in the form of a stitch, as opposed to the ball on the device. The wire is then clamped and broken at the stitch by pulling and another ball formed as described. Thermocompression bonding is rarely used for the following reasons:

- The high substrate temperature precludes the use of epoxy for device mounting.
- The temperature required for the bond is above the threshold temperature for gold-aluminum intermetallic compound formation. The diffusion rate for aluminum into gold is much greater than for gold into aluminum. The aluminum contact on a silicon device is very thin, and, when it diffuses into the gold, voids, called Kirkendall voids, are created in the bond area, increasing the electrical resistance of the bond and decreasing the mechanical strength.
- The thermocompression bonding action does not effectively remove trace surface contaminants that interfere with the bonding process.

The ultrasonic bonding process uses ultrasonic energy to vibrate the wire (primarily aluminum wire) against the surface to blend the lattices together. Localized heating at the bond interface caused by the scrubbing action, aided by the oxide on the aluminum wire, assists in forming the bond. The substrate itself is not heated. Intermetallic compound formation is not as critical, as with the thermocompression bonding process, as both the wire and the device metallization are aluminum. Kirkendall voiding on an aluminum wire bonded to gold substrate metallization is not as critical, since there is substantially more aluminum available to diffuse than on device metallization. Ultrasonic bonding makes a stitch on both the first and second bonds because of the fact that it is very difficult to form a ball on the end of an aluminum wire because of the tendency of aluminum to oxidize. For this reason, ultrasonic bonding is somewhat slower than thermocompression, since the capillary must be aligned with the second bond site when the first bond is made. Ultrasonic bonding to package leads may be difficult if the leads are not tightly clamped, since the ultrasonic energy may be propagated down the leads instead of being coupled to the bond site.

The use of thermosonic bonding of gold wire overcomes the difficulties noted with thermocompression bonding. In this process, the substrate is heated to 150°C and ultrasonic energy is coupled to the wire through the transducer action of the capillary, scrubbing the wire into the metal surface and forming a ball-stitch bond from the device to the substrate, as in thermocompression bonding.

Thermosonic gold bonding is the most widely used bonding technique, primarily because it is faster than ultrasonic aluminum bonding. Once the ball bond is made on the device, the wire may be moved in any direction without stress on the wire, which greatly facilitates automatic wire bonding, as the movement need only be in the x and y directions.

By contrast, before the first ultrasonic stitch bond is made on the device, the circuit must be oriented so that the wire moves toward the second bond site only in the direction of the stitch. This necessitates rotational movement, which not only complicates the design of the bonder, but increases the bonding time as well.

The wire size is dependent on the amount of current that the wire is to carry, the size of the bonding pads, and the throughput requirements. For applications where current and bonding size is not critical, wire 0.001 in. in diameter is the most commonly used size. Although 0.0007-in. wire is less expensive, it is difficult to thread through a capillary without frequent breaking and consequent line stoppages.

For high-volume applications, gold is the preferred method due to bonding speed. Once a ball bond is made, the wire can be dressed in any direction. Stitch bonding, on the other hand, requires that the second bond be lined up with the first prior to bonding, which necessitates rotation of the holding platform with a corresponding loss in bonding rate.

Figure 11.12 illustrates aluminum ultrasonic bonding of heavy wire, whereas thermosonic bonding is illustrated in Fig. 11.9 and Fig. 11.10.

Devices in die form have certain advantages over packaged devices in hybrid circuit applications, including

- Availability: Devices in the chip form are more readily available than other types and require no further processing.
- **Thermal management:** Devices in the chip form are mounted in intimate contact with the substrate, which maximizes the contact area and improves the heat flow out of the circuit.
- Size: Except for the flip-chip approach, the chip-and-wire approach utilizes the least substrate area.
- Cost: Since they require no special processing, devices in the chip form are less expensive than packaged devices.

Coupled with these advantages, however, are at least two disadvantages

- Fragility: Devices in the chip form are susceptible to mechanical handling, static discharge, and corrosion as a result of contamination. In addition, the wire bonds require only a few grams of force in the normal direction to fail.
- Testability: This is perhaps the most serious problem with devices in the die form. It is difficult to perform functional testing at high speeds and at temperature extremes as a result of the difficulties in probing. Testing at cold temperatures is virtually prohibitive as a result of moisture condensation. The net result when several devices must be utilized in a single hybrid is lower yields at first

electrical test. If the yield of a single chip is 0.98, then the initial yield that can be expected in a hybrid circuit with 10 devices is $0.98^{10} = 0.817$. This necessitates added troubleshooting and repair time with a corresponding detrimental effect on both cost and reliability.

Tape Automated Bonding and Flip-Chip Bonding

The tape automated bonding process was developed in the early 1970s, and has since been adopted by several companies as a method of chip packaging. There are two basic approaches to the TAB process, the *bumped-chip process* and the *bumped-tape process*.

In one form of the bumped-chip process, the wafer is passivated with silicon nitride and windows are opened up over the bonding pads by photoetching. The wafer is then coated with thin layers of titanium, palladium, and gold applied in succession by sputtering. The titanium and palladium act as barrier layers to prevent the formation of intermetallic compounds between the gold and the aluminum. Using a dry film photoresist with a thickness of about 0.0015 in., windows are again opened over the bonding pads and gold is electroplated to a thickness of about 0.001 in. at these points. The photoresist is removed, and the wafer is successively dipped in gold, palladium, and titanium etches, leaving only the gold bumps over the bonding pads. To prepare the tape, a copper foil, is laminated to a 35-mm polyimide film with holes punched at intervals. A lead frame, which is designed to align with the bumps on the chip and to give stress relief after bonding to the substrate, is etched in the copper foil, and the ends of the leads are gold plated. The remainder of the leads may be either tin plated or gold plated, depending on whether the leads are intended to be attached to the substrate by soldering or by thermocompression bonding. The wafer is mounted to a ceramic block with a low-temperature wax, and the die is separated by sawing through completely. The leads are attached to the bumps by thermocompression bonding (usually referred to as the inner-lead bonding process) with a heated beryllia **thermode** whose tip size is the same as that of the die. During the bonding process, the wax melts allowing the chip to be removed from the ceramic block. The inner-lead bonding process may be highly automated by using pattern recognition systems to locate the chip with respect to the lead frame. A TAB assembly mounted to the substrate is shown in Fig. 11.13.

The bumped-tape process is somewhat less expensive than the bumped-chip process, and the technology is not as complex. Early efforts at bumped-tape bonding utilized the same gold-plated leads as in the bumped-chip process and proved somewhat unreliable due to the direct gold-aluminum contact. More recently, the use of sputtered aluminum on the lead frame has eliminated this problem, and has further reduced the processing temperature by utilizing ultrasonic bonding to attach the leads. Low-cost devices may use only a strip of copper or aluminum, whereas devices intended for pretesting use a multilayer copper-polyimide structure. The so-called area array TAB, developed by Honeywell, utilizes a multilayer structure, which can form connections within the periphery of the chip, greatly increasing the lead density without increasing the size of the chip and simplifying the interconnection structure on the chip itself.

TAB technology has other advantages over chip-and-wire technology in the areas of high frequency and high-lead density. TAB may be used on pads about one-third the size required by gold thermosonic bonding, and the larger size of the TAB leads lowers the series inductance, increasing the overall operating frequency. These factors contribute to the use of TAB in the very high-speed integrated circuit (VHSIC) program, which utilizes chips with in excess of 400 bonds operating at high speeds. In addition, the devices are in direct contact with the substrate, easing the problem of removing heat from the device. This is a critical feature, since high-speed operation is usually accompanied by increased power dissipation.

Flip-chip technology, as shown in Fig. 11.14, is similar to TAB technology in that successive metal layers are deposited on the wafer, ending up with solder-plated bumps over the device contacts. One possible configuration utilizes an alloy of nickel and aluminum as an interface to the aluminum bonding pads. A thin film of pure nickel is plated over the Ni/Al, followed by copper and solder. The copper is plated to a thickness of about 0.0005 in., and the solder is plated to about 0.003 in. The solder is then reflowed to form a hemispherical bump. The devices are then mounted to the substrate face down by reflow solder methods. During reflow, the face of the device is prevented from contacting the substrate metallization by the copper bump. This process is sometimes referred to as the *controlled collapse* process.

Testing of flip-chip devices is not as convenient compared to TAB devices, since the solder bumps are not as amenable to making ohmic contact as the TAB leads. Testing is generally accomplished by lining the devices over a **bed of nails**, which is interfaced to a testing system. At high speeds, this does not realistically model the configuration that the device actually sees on a substrate, and erroneous data may result.

Flip-chip technology is the most space efficient of all of the packaging technologies, since no room outside the boundaries of the chip is required. Further, the contacts may be placed at any point on the chip, reducing the net area required and simplifying the interconnection pattern.

Surface mount techniques have also been successfully used to assembly hybrid circuits. The metallized ceramic substrate is simply substituted for the conventional printed circuit board and the process of screen printing solder, component placement, and reflow solder is identical.

Defining Terms

Active component: An electronic component that can alter the waveshape of an electronic signal, such as a diode, a transistor, or an integrated circuit. By contrast, ideal resistors, capacitors, and inductors leave the waveform intact and are referred to as passive components.

Bed of nails: A method of interfacing a component or electronic circuit to another assembly, consisting of inverted, spring-loaded probes located at predetermined points.

Dielectric constant: A measurement of the ability of a material to hold electric charge.

Direct bond copper (DBC): A method of bonding copper to a substrate, which blends a portion of the copper with a portion of the substrate at the interface.

Migration: Transport of metal ions from a positive surface to a negative surface in the presence of moisture and electric potential.

Noise index: A measurement of the amount of noise generated in a component as a result of random movement of carriers from one energy state to another.

Packaging: The technology of converting an electronic circuit from a schematic or prototype to a finished form.

Passive components: Components that, in the ideal form, cannot alter the waveshape of an electronic signal.

Plasma: A gas that has been ionized to produce a conducting medium.

Pseudoplastic fluid: A fluid in which the viscosity is nonlinear, with the rate of change lessening as the pressure is increased.

Refractory metal In this context, a metal with a high melting point.

Semiconductor die: A semiconductor device in the unpackaged state.

Temperature coefficient of expansion (TCE): A measurement of the dimensional change of a material as a result of temperature change.

Thermal conductivity: A measurement of the ability of a material to conduct heat.

Thermal management: The technology of removing heat from the point of generation in an electronics circuit.

Thermode: A device used to transmit heat from one surface to another to facilitate lead or wire bonding.

Thixotropic fluid: A fluid in which the viscosity is nonlinear, with the rate of change increasing as the pressure is increased.

Vapor pressure: A measurement of the tendency of a material to evaporate. A material with a low vapor pressure will evaporate more readily.

References

Elshabini-Riad, A. 1996. *Handbook of Thin Film Technology.* McGraw-Hill, New York.

Hablanian, M. 1990. *High-Vacuum Technology.* Marcel Dekker, New York.

Harper, C. 1996. *Handbook of Electronic Packaging.* McGraw-Hill, New York.

Sergent, J. E. and Harper, C. 1995. *Handbook of Hybrid Microelectronics*, 2nd ed. McGraw-Hill, New York.

Further Information

For further information on hybrid circuits see the following:

ISHM Modular Series on Microelectronics.

Tummala, R.R. and Rymaszewski, E. 1989. *Microelectronics Packaging Handbook*. Van Nostrand Reinhold, New York.

Licari, J.J. 1995. *Multichip Module Design, Fabrication, and Testing*. McGraw-Hill, New York.

Proceedings of ISHM Symposia, 1967–1995. *Journal of Hybrid Microelectronics*, published by ISHM.

11.3 Surface Mount Technology

Glenn R. Blackwell

11.3.1 Introduction

This chapter is intended for the practising engineer who is familiar with standard **through-hole** (insertion mount) circuit board design, manufacturing, and test, but who now needs to learn more about the realm of surface mount technology (SMT). Numerous references will be given, which will also be of help to the engineer who may be somewhat familiar with SMT, but needs more in-depth information in specific areas. The reader with little knowledge about SMT is referred to a basic introductory article (Mims, 1987) and to a journal series which covered design of an SMT memory board (Leibson, 1987).

11.3.2 Definition

Surface mount technology is a collection of scientific and engineering methods needed to design, build, and test products made with electronic components, which mount to the surface of the printed circuit board without holes for leads (Higgins, 1991) as illustrated in Fig. 11.15. This definition notes both the breadth of topics necessary to understand SMT, as well as that the successful implementation of SMT will require the use of *concurrent engineering* (Classon, 1993; Shina, 1991). Concurrent engineering means that a team of design, manufacturing, test, and marketing people will concern themselves with board layout, parts and parts placement issues, soldering, cleaning, test, rework, and packaging before any product

Molded Dip, SOIC, micro-SMD - 8 Pin

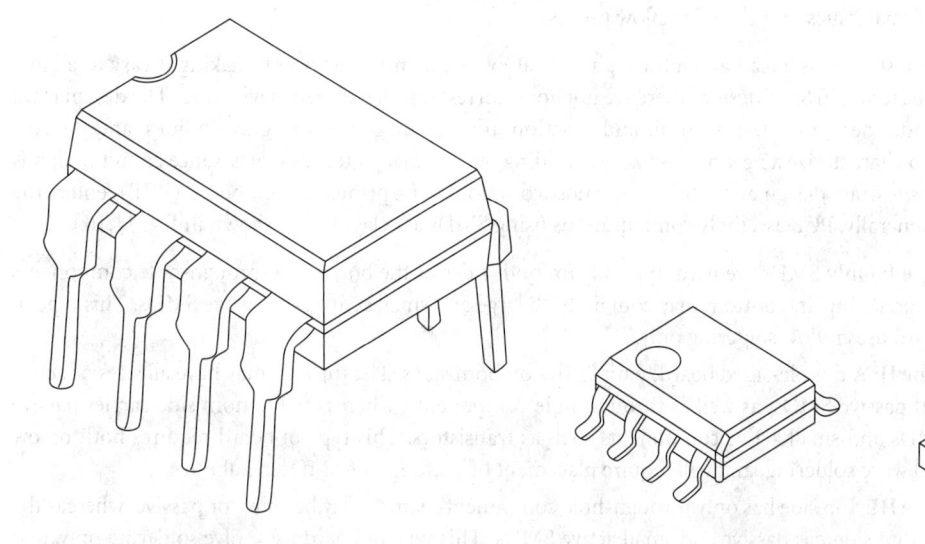

FIGURE 11.15 Comparison of 8-lead DIP and SMT components (National Semiconductor, with permission).

is manufactured. The careful control of all of these issues improves both yield and reliability of the final product. In fact, SMT cannot be reasonably implemented without the use of concurrent engineering and the principles contained in *design* for *manufacturability (DFM)* and *design for testability (DFT)* and, therefore, any facility that has not embraced these principles should do so if SMT is to be successfully implemented.

Considerations in the Implementation of SMT

The main reasons to consider implementation of SMT include the following:

- Reduction in circuit board size
- Reduction in circuit board weight
- Reduction in number of layers in the circuit board
- Reduction in trace lengths on the circuit board

Note that not all of these reductions may occur in any given product redesign from **through-hole technology** (THT) to SMT. Reduction in circuit board size can vary from 40 to 60% (TI, 1984). By itself, this reduction presents many advantages in packaging possibilities. Camcorders and cell phones are only possible through the use of SMT. The reduction in weight means that circuit boards and components are less susceptible to vibration problems. Reduction in the number of layers in the circuit board means the bare board will be significantly less expensive to build. Reduction in trace lengths means that high-frequency signals will have fewer problems or that the board will be able to operate at higher frequencies than a through-hole board with longer trace/lead lengths. Of course, there are some disadvantages using SMT. During the assembly of a through-hole board, either the component leads go through the holes or they do not, and the component placement machines can typically detect the difference in force involved and signal for help. During SMT board assembly, the placement machine does not have such direct feedback, and accuracy of final soldered placement becomes a stochastic (probability-based) process, dependent on such items as the following:

- Component pad design
- Accuracy of the PCB artwork and fabrication, which affects the accuracy of pad location
- Accuracy of solder paste deposition location and deposition volume
- Accuracy of adhesive deposition location and volume if adhesive is used
- Accuracy of placement machine vision system(s)
- Variations in component sizes from the assumed sizes
- Thermal issues in the solder reflow process.

In THT test there is a lead and a through-hole at every potential test point, making it easy to align a bed-of-nails tester. In SMT designs there are not holes corresponding to every device lead. The design team must consider design for test, form fit and function, time-to-market, existing capabilities, and the cost and time to characterize a new process when deciding on a change of technologies. Once circuit design is complete, substrate design and fabrication, most commonly of a printed circuit board (PCB), enters the process. Generally, PCB assembly configurations using SMDs are classified as shown in Fig. 11.16.

- **Type I**: Only SMDs are used, typically on both sides of the board. No through-hole components are used. Top and bottom may contain both large and small active and passive SMDs. This type of board uses reflow soldering only.
- **Type II**: A double-sided board, with SMDs on both sides. The top side may have all sizes of active and passive SMDs, as well as through-hole components, whereas the bottom side carries passive SMDs and small active components such as transistors. This type of board requires both reflow and wave soldering, and will require placement of bottom-side SMDs in adhesive.
- **Type III**: Top side has only through-hole components, which may be active or passive, whereas the bottom side has passive and small active SMDs. This type of board uses wave soldering only, and also requires placement of the bottom-side SMDs in adhesive.

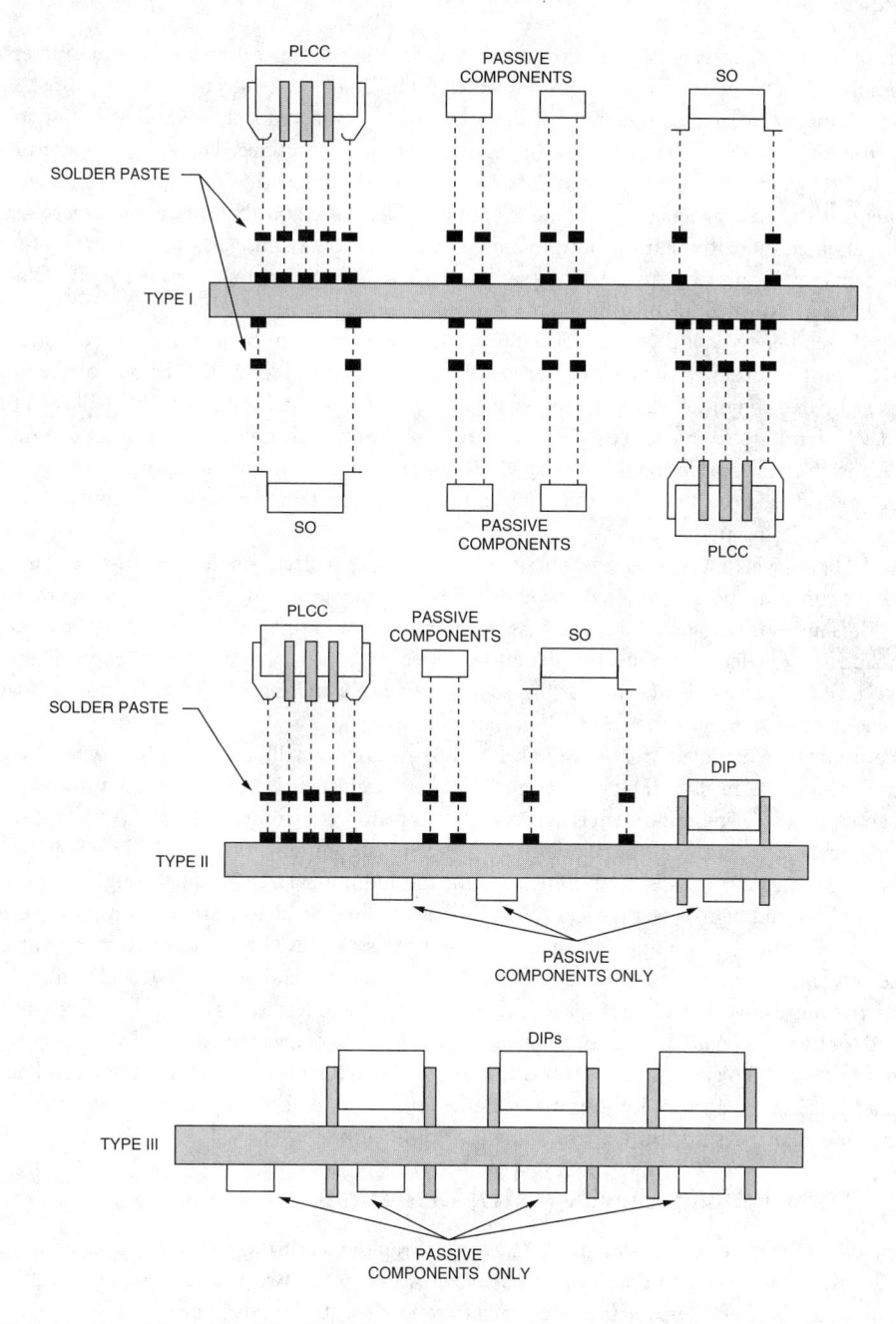

FIGURE 11.16 Type I, II, and III SMT circuit boards. (Source: Intel. 2000. Packaging Handbook. Intel Corp., Santa Clara, CA. With permission.)

A type I bare board will first have solder paste applied to the component pads on the board. Once solder paste has been deposited, active and passive parts are placed in the paste. For prototype and low-volume lines this can be done with manually guided X-Y tables using vacuum needles to hold the components, whereas in medium and high-volume lines automated placement equipment is used. This equipment will pick parts from reels, sticks, or trays, then place the components at the appropriate pad locations on the board, hence the term *pick and place* equipment.

After all parts are placed in the solder paste, the entire assembly enters a reflow oven to raise the temperature of the assembly high enough to reflow the solder paste and create acceptable solder joints at the component lead/pad transistions. Reflow ovens most commonly use convection and IR heat sources to heat the assembly above the point of solder liquidus, which for 63/37 tin-lead eutectic solder is 183° C. Because of the much higher thermal conductivity of the solder paste compared to the IC body, reflow soldering temperatures are reached at the leads/pads before the IC chip itself reaches damaging temperatures. The board is inverted and the process repeated. It is good practice to reflow the "bottom" side first with small components, then upon inversion of the board apply solder paste and then place the larger parts. It is generally accepted that although the bottom-side paste will reflow again, parts smaller than 24-lead ICs will maintain their location due to the strong surface tension of the bottom-side molten Sn-Pb solder. This will have to be investigated further with changes to lead-free solder.

If mixed-technology type II is being produced, the board will then be inverted, an adhesive will be dispensed at the centroid of each SMD, parts will be placed, the adhesive will be cured, the assembly will be rerighted, through-hole components will be mounted, and the circuit assembly will then be wave soldered, which will create acceptable solder joints for both the through-hole components and bottom-side SMDs. A design team considering the use of SMT IC components on bottom-side must verify with the IC manufacturer's data sheets that wave-soldering of the ICs is acceptable. Intel, for example, recommends against wave-soldering any of its ICs.

A type III board will first be inverted, adhesive dispensed, SMDs placed on the bottom-side of the board, the adhesive cured, the board rerighted, through-hole components placed, and the entire assembly wave-soldered. It is imperative to note that only passive components and small active SMDs can be successfully bottom-side wave-soldered without considerable experience on the part of the design team and the board assembly facility. It must also be noted that successful wave soldering of SMDs requires a dual-wave machine with one turbulent wave and one laminar wave.

It is common for a manufacturer of through-hole boards to convert first to a type II or type III substrate design before going to an all-SMD type I design. This is especially true if amortization of through-hole insertion and wave-soldering equipment is necessary. Many factors contribute to the reality that most boards are mixed-technology type II or type III boards. Although most components are available in SMT packages, through-hole connectors are still commonly used for the additional strength the through-hole soldering process provides, and high-power devices such as three-terminal regulators are still commonly through-hole due to off-board heat-sinking demands. Both of these issues have been addressed by manufacturers and solutions are available that allow type I boards with connectors and power devices (Holmes, 1993).

Again, it is imperative that all members of the design, build, and test teams be involved from the design stage. Today's complex board designs mean that it is entirely possible to exceed the ability to adequately test a board if test is not designed-in, or to robustly manufacture a board if in-line inspections and handling are not adequately considered. Robustness of both test and manufacturing are only assured with full involvement of all parties to overall board design and production.

11.3.3 Surface Mount Device (SMD) Definitions

The new user of surface mount designs (SMDs) must rapidly learn the packaging sizes and types for SMDs. Resistors, capacitors, and most other passive devices come in two-terminal packages, as shown in Fig. 11.17, which have end-terminations designed to rest on substrate pads/lands.

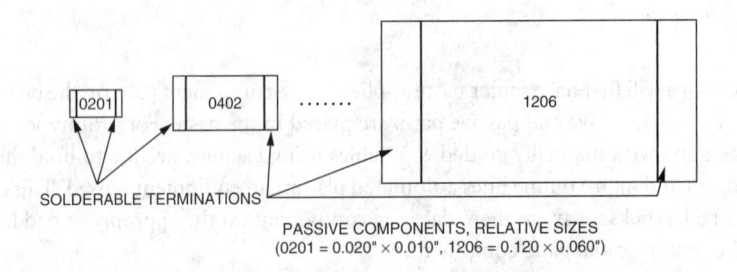

PASSIVE COMPONENTS, RELATIVE SIZES
(0201 = 0.020" × 0.010", 1206 = 0.120 × 0.060")

FIGURE 11.17 Example of passive relative component sizes (not to scale).

Basic SMD ICs come in a wide variety of packages, from 8-lead *small outline packages* (SOLs) to 208-lead quad flat packs (QFPs). Examples of SO20 and QFP128 packages are shown in Fig. 11.18.

Other common commercial packages currently include *plastic leaded chip carriers (PLCCs)*, which may as *bumpered quad flat packs* (BQFPs) as shown in Fig. 11.18, *tape automated bonding* (TAB), *ball grid array* (BGA), and other newer technologies, and the IC possibilities become overwhelming. Additionally, bare die may be placed on the substrate with either the bond pads up and wire-bonding to the substrate (*chip on board, or COB*) or they may be "flipped" with their bond pads down (*flip chip*), with solderable bumps attached to the die bond pads. Figure 11.19 shows a flip chip example with adhesive underfill which helps prevent cracking of the solder joints. Beker and Adams discuss flip chip issues in more detail.

As another example of a foray into newer technologies, Achong and Chen discuss BGA land pattern and assembly issues. The reader is referred to

FIGURE 11.18   SO20 and PQFP128 packages.

the standards of the Institute for Interconnecting and Packaging Electronic Circuits (IPC) to find the latest package standards.

Each IC manufacturer's data books will have packaging information for their products. The engineer should be familiar with the term *lead pitch*, which means the center-to-center distance between IC leads. All DIP ICs have a lead pitch of 0.1″ (= 100 mils), while the pitch on SMT ICs may be in thousandths of an inch, also known as mils, or may be in millimeters. Common pitches are 0.050 in. (50-mil pitch), 0.025 in. (25-mil pitch), frequently called fine pitch; and 0.020 in. and smaller, frequently called ultrafine pitch. Metric equivalents are 1.27, 0.635, and 0.508 mm, and smaller. Conversions from metric to inches are easily approximated if one remembers that 1 mm is approximately 40 mils. It is important to note that virtually all new ICs are designed to metric standards. Common pitch values include 1, 0.8, 0.65 and 0.5 mm.

11.3.4 Substrate Design Guidelines

As noted previously, substrate (typically PCB) design has an effect not only on board/component layout, but also on the actual manufacturing process. Incorrect land design or layout can negatively affect the placement process, the solder process, the test process, or any combination of the three. Substrate design must take into account the mix of surface mount devices that are available for use in manufacturing.

The considerations which will be noted here as part of the design process are neither all encompassing, nor in sufficient detail for a true SMT novice to adequately deal with all of the issues involved in the process.

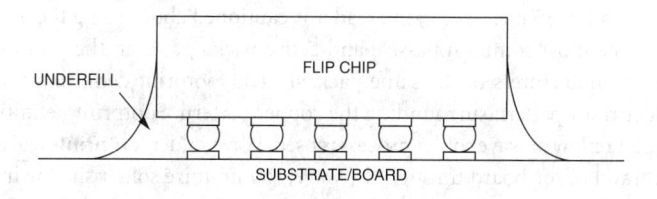

FIGURE 11.19   Flip chip with underfill.

They are intended to guide an engineer through the process, allowing access to more detailed information, as necessary. General references are noted at the end of this chapter, and specific references will be noted as applicable. In addition, conferences such as the APEX, National Electronics Production and Productivity Conference (NEPCON) and Surface Mount International (SMI) are invaluable sources of information for both the beginner and the experienced SMT engineer. It should be noted that although these guidelines are noted as steps, they are not necessarily in an absolute order, and several back-and-forth iterations among the steps may be required to result in a final satisfactory process and product. After the circuit design (schematic capture) and analysis, step one in the process is to determine whether all SMDs will be used in the final design making of a type I board, or whether a mix of SMDs and through hole parts will be used, leading to a type II or type III board. This is a decision that will be governed by some or all of the following considerations:

- Current parts stock
- Existence of current through hole (TH) placement and/or wave solder equipment
- Amortization of current TH placement and solder equipment
- Existence of reflow soldering equipment
- Cost of new reflow soldering equipment
- Desired size of the final product
- Panelization of smaller type I boards
- Thermal issues related to high-power circuit sections on the board

It is desirable to segment the board into areas based on function: RF, low power, high power, etc., using all SMDs where appropriate, and mixed-technology components as needed. Power portions of the circuit may point to the use of through-hole components if air flow is the primary heat-dissipation method, although circuit board materials are available that will sink substantial amounts of heat. Using one solder technique (reflow or wave) can be desirable from a processing standpoint, and may outweigh other considerations. Another primary concern is minimization of EMI/RFI. Many references address these issues XXX.

Step two in the SMT process is to define all of the footprints of the SMDs under consideration for use in the design. The footprint is the copper pattern, commonly made up of pads or lands, on the circuit board upon which the SMD will be placed. The *Intel Packaging Handbook* says, "The design of land patterns is very critical, because it determines solder joint strength and thus the reliability of solder joints, and also impacts solder defects, cleanability, testability, and repair or rework. In other words, the very producibility or success of SMT is dependent upon the land pattern design. The lack of standardization of surface mount packages has compounded the problem of standardizing the land pattern. There are a variety of package types offered by the industry, and the variations in a given package type can be numerous. For example, for the small outline package (SOP), there are not only two lead types (gull-wing and J-lead), but there are multiple body types such as narrow wide and thin (as well as a variety of pitch dimensions *gb*). In addition, the tolerance on components varies significantly, adding to the manufacturing problems for SMC users."

Footprint examples are shown in Fig. 11.20, and footprint recommendations are available from IC manufacturers and in the appropriate data books. They are also available in various ECAD packages used for the design process, as well as in the several references that include an overview of the SMT process (Hollomon, 1995; Prasad, 1997). However, the reader is cautioned about using the general references for anything other than the most common passive and active packages. Even the position of pin 1 may be different among IC manufacturers of the same package. The footprint definition may also include the position of the solder resist pattern surrounding the copper pattern. Footprint definition sizing will vary depending on whether reflow or wave solder process is used. Wave solder footprints will require recognition of the direction of travel of the board through the wave, to minimize solder shadowing in the final fillet, as well as to meet requirements for solder thieves. The copper footprint must allow for the formation of an appropriate, inspectable solder fillet.

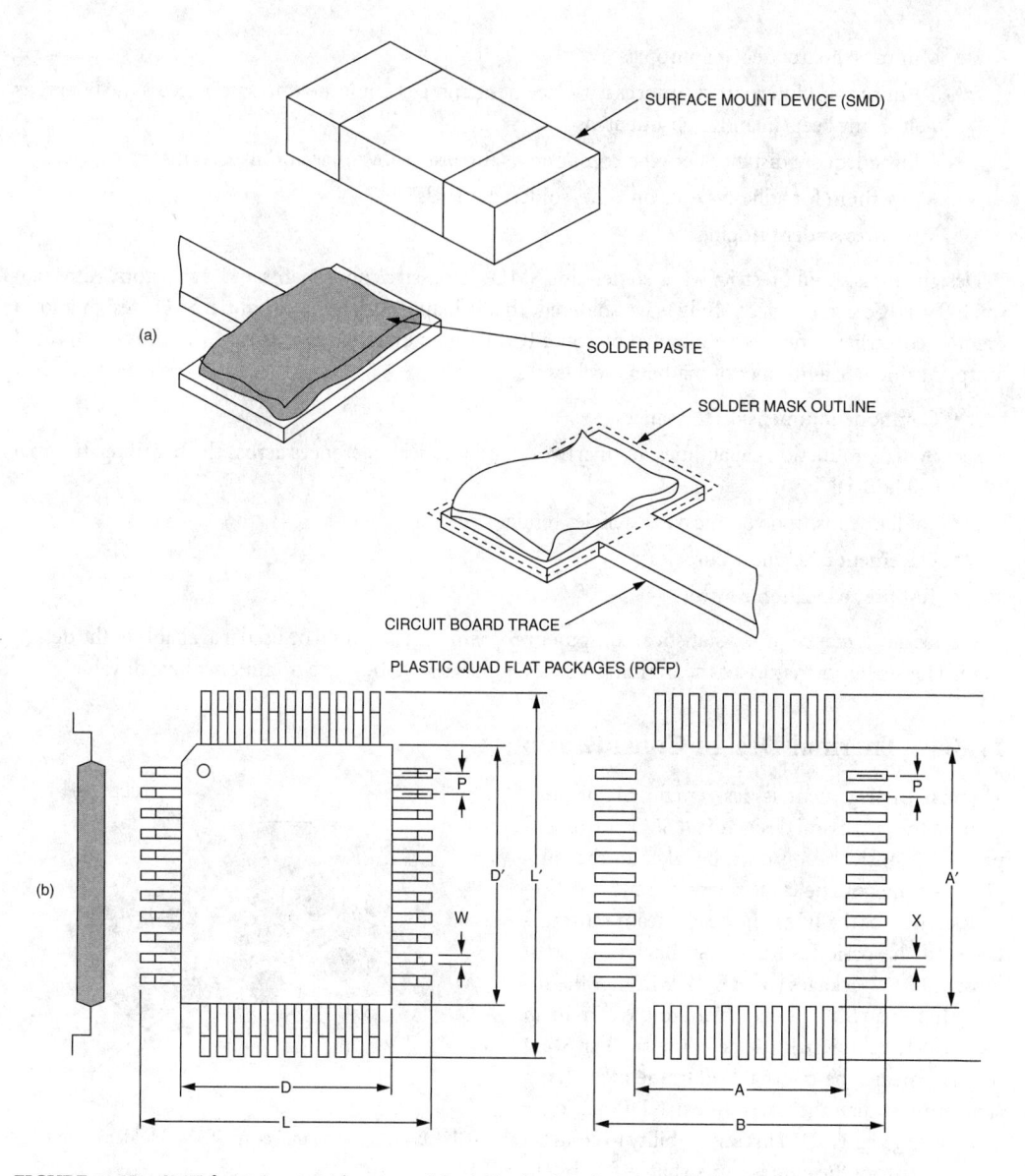

FIGURE 11.20 SMT footprint considerations: (a) passive land and component (b) QFP component and footprint (Source: National Semiconductor, With permission.)

If done as part of the electronic design automation (EDA) process using appropriate electronic computer-aided design (ECAD) software, the software will automatically assign copper directions to each component footprint, as well as appropriate coordinates and dimensions. These may need adjustment based on considerations related to wave soldering, test points, RF, power and EMI/RFI issues, and board production limitations. Allowing the software to select 5-mil traces when the board production facility to be used can only reliably do 8-mil traces would be inappropriate. Likewise, the solder mask patterns must be governed by the production capabilities.

Final footprint and trace decisions will

- Allow for optimal solder fillet formation
- Minimize necessary trace and footprint area
- Allow for adequate test points

- Minimize board area, if appropriate
- Set minimum interpart clearances for placement and test equipment to safely access the board, as well as any keep-in and keep-out areas
- Allow adequate distance between components forpost reflow operator inspections
- Allow room for adhesive dots on wave-soldered boards
- Minimize solder bridging

Design teams should restrict wave-solder-side SMDs to passive components and transistors. Although small SMT ICs can be successfully wave soldered, this is inappropriate for an initial SMT design and as mentioned earlier is not recommended by some IC manufacturers. Decisions that will provide optimal footprints include a number of mathematical issues:

- Component dimension tolerances
- Board production capabilities, both artwork and physical tolerances across the board relative to a 0-0 fiducial
- Solder deposition volume consistencies refillet sizes
- Placement machine accuracies
- Test probe location control

These decisions may require a statistical computer program, which should be used if available to the design team. The stochastic nature of the overall process suggests a statistical programmer will be of value.

11.3.5   Thermal Design Considerations

Thermal management issues remain major concerns in the successful design of an SMT board and product. Consideration must be taken of the varibles affecting both board temperature and junction temperature of the IC.-1The design team must understand the basic heat transfer characteristics of most SMT IC packages (ASME, 1993). Since the silicon chip of an SMD is equivalent to the chip in an identical-function DIP package, the smaller SMD package means the internal lead frame metal has a smaller mass than the lead frame in a DIP package, as shown in Fig. 11.21. This lesser ability to conduct heat away from the chip is somewhat offset by the

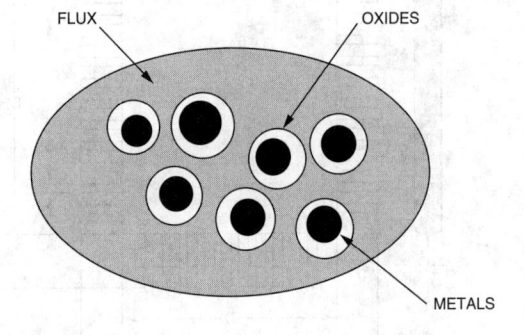

FIGURE 11.21   The make-up of SMT solder paste.

leadframe of many SMDs being constructed of copper, which has a lower thermal resistance than the Kovar and Alloy 42 materials commonly used for DIP packages. However, with less metal and shorter lead lengths to transfer heat to ambient air, more heat is typically transferred to the circuit board itself. Several board thermal analysis software packages are available, and are highly recommended for boards that are expected to develop high thermal gradients (Flotherm). XXX Since all electronics components generate heat in use, and elevated temperatures negatively affect the reliability and failure rate of semiconductors, it is important that heat generated by SMDs be removed as efficiently as possible. In most electronic devices and assemblies, heat is removed primarily by some combination of conduction and convection, although some radiation effects are present.

The design team needs to have expertise with the variables related to thermal transfer:

- Junction temperature: T_j
- Thermal resistances: $\Theta_{jc}, \Theta_{ca}, \Theta_{cs}, \Theta_{sa}$
- Power dissipation: P_D
- Thermal characteristics of substrate material

SMT packages have been developed to maximize heat transfer to the substrate. These include various analog and digital ICs with integral heat spreaders and various power transistor packages. Note that all of these devices are designed primarily for processing with the solder paste process, and some specifically recommend against their use with wave-solder applications. In the conduction process, heat is transferred from one element to another by direct physical contact between the elements. Ideally the material to which heat is being transferred should not be adversely affected by the transfer. As an example, the **glass transition temperature** T_g of FR-4 is 125°C. Heat transferred to the board has little or no detrimental effect as long as the board temperature stays at least 50°C below T_g. Good heat sink material exhibits high thermal conductivity, which is not a characteristic of fiberglass. Therefore, the traces and other copper features must be depended on to provide the thermal transfer path (Choi, Kim, and Ortega, 1994). Conductive heat transfer is also used in the transfer of heat from THT IC packages to heat sinks, which also requires use of thermal grease to fill all air gaps between the package and the flat surface of the sink.

The discussion of lead properties of course does not apply to leadless devices such as *leadless ceramic chip carriers* (LCCCs). Design teams using these and similar packages must understand the better heat transfer properties of the alumina used in ceramic packages and must match **coefficients of thermal expansion (CTEs or TCEs)** between the LCCC and the substrate since there are no leads to bend and absorb mismatches of expansion. Use of ceramic devices may lead the design team to consider the use of Low Temperature Co-fired Ceramic (LTCC) substrates and assembly technologies XXX .

Since the heat transfer properties of the system depend on substrate material properties, it is necessary to understand several of the characteristics of the most common substrate material, FR-4 fiberglass. The glass transistion temperature has already been noted, and board designers must also understand that multilayer FR-4 boards do not expand identically in the X, Y, and Z directions as temperature increases. Plate-through-holes will constrain z-axis expansion in their immediate board areas, whereas non-through-hole areas will expand further in the z axis, particularly as the temperature approaches and exceeds T_g (Lee et al., 1984). This unequal expansion can cause delamination of layers and plating fracture. If the design team knows that there will be a need for higher abilities to dissipate heat and/or needs for higher glass transition temperatures and lower coefficients of thermal expansion than FR-4 possesses, many other materials are available, examples of which are shown in Table 11.7.

Note in the table that copper-clad Invar has both variable T_g, and variable thermal conductivity depending on the volume mix of copper and Invar in the substrate. Copper has a high TCE, and Invar has a low TCE, and so the TCE increases with the thickness of the copper layers. In addition to heat transfer considerations, board material decisions must also be based on the expected stress and humidity in the application.

Convective heat transfer involves transfer due to the motion of molecules, typically airflow over a heat sink. Convective heat transfer, like conductive, depends on the relative temperatures of the two media involved. It also depends on the velocity of air flow over the boundary layer of the heat sink. Convective heat transfer is primarily affected when forced air flow is provided across a substrate and when convection effects are maximized through the use of heat sinks. The rules that designers are familiar with, when designing THT heat-sink device designs, also apply to SMT designs.

The design team must consider whether passive conduction and convection will be adequate to cool a populated substrate or whether forced-air cooling or liquid cooling will be needed. Passive conductive

TABLE 11.7 Thermal Properties of Common PCB Materials Glass TCE-Thermal Substrate Transition, Coefficient of Thermal Moisture

Material	Temperature, T_g, °C	X-Y Expansion, PPM/°C	Conductivity, W/M °C	Absorption, %
FR-4 epoxy glass	125	13-18	0.16	0.10
Polymide glass	250	12-16	0.35	0.35
Copper-clad Invar	Depends on resin	5-7	160XY-15-20Z	NA
Poly aramid fiber	250	3-8	0.15	1.65
Alumina/ceramic	NA	5-7	20-45	NA

cooling is enhanced with thermal layers in the substrate, such as the previously mentioned copper/Invar. There will also be designs that rely on the traditional through-hole device with heat sink to maximize heat transfer. An example of this would be the typical three-terminal voltage regulator mounted on a heat sink or directly to a metal chassis for heat conduction, for which standard calculations apply (Motorola, 1993).

Many specific examples of heat transfer may need to be considered in board design and, of course, most examples involve both conductive and convective transfer. For example, the air gap between the bottom of a standard SMD and the board affects the thermal resistance from the case to ambient, Θ_{ca}. A wider gap will result in a higher resistance, due to poorer convective transfer, whereas filling the gap with a thermal-conductive epoxy will lower the resistance by increasing conductive heat transfer. Thermal-modeling software is the best way to deal with these types of issues, due to the rigorous application of computational fluid dynamics (CFD) (Lee, 1994).

11.3.6 Adhesives

In the surface mount assembly process, type II and type III boards will always require adhesive to mount the SMDs for passage through the solder wave. This is apparent when one envisions components on the bottom side of the substrate with no through hole leads to hold them in place. Adhesives will stay in place after the soldering process and throughout the life of the substrate and the product, since there is no convenient means for adhesive removal once the solder process is complete. Additionally, adhesives can be used to enhance both thermal and electrical conduction between device features and board features. This means use of an adhesive must consider a number of both physical and chemical characteristics:

- Electrical conductivity
- Thermal conductivity
- Thermal cofficient of expansion similar to the substrate and the components
- Stable in both storage and after application, prior to curing
- Stable physical drop shape, retains drop height and fills z-axis distance between the board and the bottom of the component. Thixotropic with no adhesive migration.
- Noncorrosive to substrate and component materials
- Nonconductive electrically
- Chemically inert to flux, solder, and cleaning materials used in the process
- Curable as appropriate to the process: UV, oven, or air cure
- Removable for rework and repair
- Once cured, unaffected by temperatures in the solder process

Adhesive can be applied by stenciling/screening techniques similar to solder paste stencil/screen application, by pin-transfer techniques, and by syringe deposition. Stencil and pin-transfer techniques are suitable for high-volume production lines with few product changes over time. Syringe deposition, an X-Y table riding over the board with a volumetric pump and syringe tip, is more suitable for lines with a varying product mix, prototype lines, and low-volume lines where the open containers of adhesive necessary in pin-transfer and screen techniques are avoided. Newer syringe systems are capable of handling high-volume lines. See Fig. 11.22 for methods of adhesive deposition.

If type II or type III assemblies are used and thermal transfer between components and the substrate is a concern, the design team must consider thermally conductive adhesives.

Regardless of the type of assembly, the type of adhesive used, or the curing technique used, adhesive volume and height must be carefully controlled.

Detailed information on the appropriate volumes, application, and use of adhesives in electronic manufacturing is beyond the scope of this chapter. The reader wishing further information is referred to Prasad.

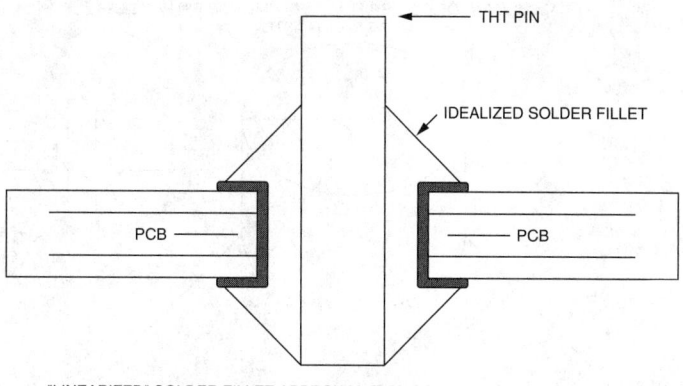

"LINEARIZED" SOLDER FILLET APPROXIMATION FOR THE TRUNCATED CONE
SOLDER PASTE CALCULATION.

FIGURE 11.22   Idealized cone approximation of solder fillet.

11.3.7   Solder Paste and Joint Formation

Standard SMT Paste Technology

Solder joint formation is the culmination of the entire process. Regardless of the quality of the design, or any other single portion of the process, if high-quality reliable solder joints are not formed, the final product is not reliable. It is at this point that PPM levels take on their finest meaning. For a medium-size substrate (nominal 6 × 8 in.), with a medium density of components, a typical mix of active and passive parts on the topside and only passive and 3- or 4-terminal active parts on bottomside, there may be in excess of 1000 solder joints/board. If solder joints are manufactured at the 3 sigma level (99.73% good joints, or 0.27% defect rate, or 2700 defects/1 milllion joints) there will be 2.7 defects per board! At the 6 sigma level, of 3.4 PPM, there will be a defect on 1 board out of every 294 boards produced. If your anticipated production level is 1000 units/day, you will have 3.4 rejects based solely on solder joint problems, not counting other sources of defects.

Solder paste may be deposited by syringe, or by screen or stencil printing techniques. Stencil techniques are best for high-volume/speed production although they do require a specific stencil for each board design. Syringe and screen techniques may be used for high-volume lines and are also suited to mixed-product lines where only small volumes of a given board design are to have solder paste deposited. Syringe deposition is the only solder paste technique that can be used on boards that already have some components mounted. It is also well suited for prototype lines and for any use requires only software changes to develop a different deposition pattern (see Fig. 11.23). Solder joint defects have many possible origins:

- Poor or inconsistent solder paste quality
- Inappropriate solder pad design/shape/size/trace connections
- Substrate artwork or production problems: for example, mismatch of copper and mask, warped substrate
- Solder paste deposition problems: for example, wrong volume or location
- Component lead problems: for example, poor coplanarity or poor tinning of leads
- Placement errors: for example, part rotation or X-Y offsets
- Reflow profile: for example, preheat ramp too fast or too slow; wrong temperatures created on substrate
- Board handling problems: for example, boards get jostled prior to reflow

Once again, a complete discussion of all of the potential problems that can affect solder joint formation is beyond the scope of this chapter. Many references are available which address the issues. An excellent

CALCULATION OF SOLDER FILLET VOLUME FOR PIP PROCESS
(PIP = PIN IN PASTE)

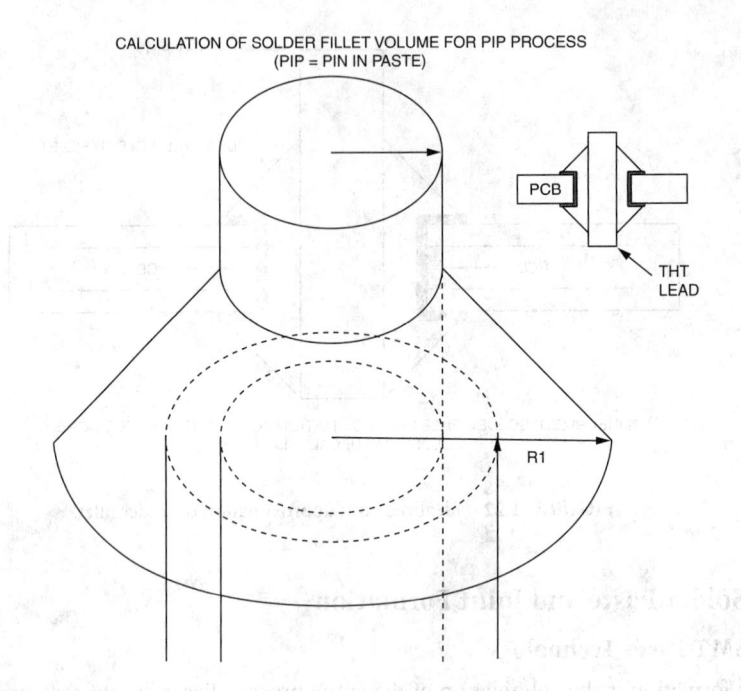

R1 = RADIUS OF SOLDER PAD ON BOARD
R2 = RADIUS OF THT LEAD
H = HEIGHT OF SOLDER FILLET = R1-R2, ASSUMES 45 DEGREE FILLET ANGLE

ASSUMPTIONS: THAT THE SOLDER FILLET IS CORRECTLY FORMED WITH NO VOIDS
              FOR SINGLE-SIDED BOARDS, THERE IS NO SOLDER FILL OF THE THROUGH-HOLE IN THE PCB
              FOR DOUBLE-SIDED BOARDS, THE TOP AND BOTTOM FILLETS ARE IDENTICAL
              FOR DOUBLE-SIDED BOARDS, THERE IS COMPLETE SOLDER FILL OF THE THROUGH-HOLE IN THE PCB
              THROUGH-HOLE SOLDER FILL VOLUME IS = THE VOLUME OF THE HOLE (PI R2² T) MINUS THE VOLUME OF
                  THE LEAD IN THE HOLE (PI R2² T), WHERE HOLE LENGTH T IS = PCB THICKNESS
              VOLUME OF THE TRUNCATED CONE THAT DEFINES THE TOP AND BOTTOM FILLET IS:
                  $V_{cone} = 1/3\ PIH\ (R1^2 + R1R2 + R2^2)$

FIGURE 11.23 Calculation measurements for truncated cone approximation of solder fillet.

overview of solder joint formation theory is found in Lau (1991). Update information about this and all SMT topics is available each year at conferences, such as SMI and NEPCON.

Although commonly used solder paste for both THT and SMT production contains 63-37 eutectic tin-lead solder, other metal formulations are available, including 96-4 tin-silver (*silver solder*). The fluxes available are also similar, with typical choices being made between RMA, water-soluble, and no-clean fluxes. The correct decision rests as much on the choice of flux as it does on the proper metal mixture. A solder paste supplier can best advise on solder pastes for specific needs. Many studies are in process to determine a no-lead replacement for lead-based solder in commercial electronic assemblies. The design should investigate the current status of these studies as well as the status of no-lead legislation as part of the decision making process.

To better understand solder joint formation, one must understand the make-up of solder paste used for SMT soldering. The solder paste consists of microscopic balls of solder, most commonly tin-lead with the accompanying oxide film, flux, and activator and thickener solvents as shown in Fig. 11.24.

Regardless of the supplier, frequent solder paste tests are advisable, especially if the solder is stored for prolonged periods before use. At a minimum, viscosity, percent metal, and solder sphere formation should be tested (Capillo, 1990). Solder sphere formation is particularly important since acceptable particle sizes will vary depending on the pitch of the smallest pitch part to be used, and the consistency of solder sphere formation will affect the quality of the final solder joint. Round solder spheres have the smallest surface area for a given volume, and therefore will have the least amount of oxide formation. Uneven distribution

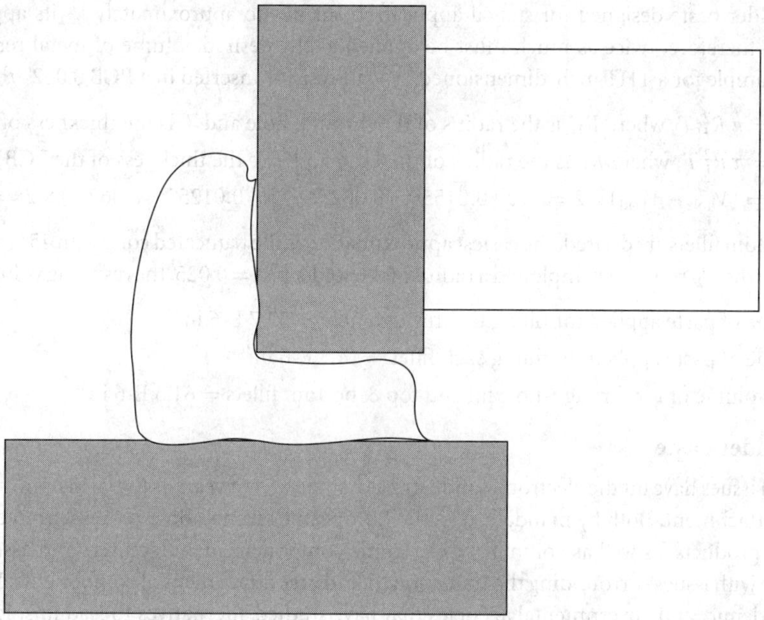

PASSIVE COMPONENT PLACED INTO SOLDER PASTE.
(EXAGGERATED)

FIGURE 11.24 Part placed into solder paste with a passive part.

of sphere sizes within a given paste can lead to uneven heating during the reflow process, with the result that the unwanted solderballs will be expelled from the overall paste mass at a given pad/lead site. Fine-pitch paste has smaller ball sizes and, consequently, more surface area on which oxides can form.

It should be noted at this point that there are three distinctly different solder balls referred to in this chapter and in publications discussing SMT. The solder sphere test refers to the ability of a volume of solder to form a ball shape due to its inherent surface tension when reflowed (melted). This ball formation is dependent on minimum oxides on the microscopic metal balls that make up the paste, the second type of solder ball. It is also dependent on the ability of the flux to reduce the oxides that are present, as well as the ramp up of temperature during the preheat and drying phases of the reflow oven profile. Too steep a time/temperature slope can cause rapid escape of entrapped volatile solvents, resulting in expulsion of small amounts of metal, which will form undesirable solder balls of the third type, that is, small metal balls scattered around the solder joint(s) on the substrate itself rather than on the tinned metal of the joint.

Pin in Paste Technology

Pin in paste (PIP) technology, also know as alternate assembly and reflow technology (AART) by Universal Instruments and others, allow the use of THT parts with solder paste and reflow soldering. Paste is deposited onto the board at through-hole locations, the leads of a THT part are placed through the paste, and at the conclusion of the placement process the board and components are reflowed. The two major concerns with the use of the PIP process are the deposition of an adequate volume of solder paste and the use of THT components whose body can withstand reflow soldering temperatures. In the wave solder process, only component leads were raised to soldering temperatures, but in the reflow process the entire part must withstand the process temperatures.

The solder paste volume calculation consists of two parts, (1) solder fill for the through-hole and (2) solder for fillet inspection. The second part may or may not be necessary, depending on whether the solder joints will undergo visual inspection or not. Lack of necessity for fillets reduces the amount of solder paste that must be deposited. This makes the deposition easier. The solder fill calculation is straightforward: (volume of the through hole) − (volume of the part lead). Consideration must then be given that solder paste designed for stencil application shrinks to approximately $\frac{1}{2}$ its applied-volume after reflow—therefore, twice as much must be applied as the desired volume of metal remaining after reflow. An example for a THT inch-dimensioned $\frac{1}{2}$ watt resistor inserted in a PCB 0.062″ thick would be

- $V_{\text{hole}} = \pi R_H^2 T$, where R_H is the radius of the through hole and T is the thickness of the PCB.
- $V_{\text{lead}} = \pi R_L^2 T$, where R_L is the radius of the Lead and T is the thickness of the PCB.
- $V_{\text{paste}} = (V_{\text{hole}} - V_{\text{lead}}) \times 2 = ((\pi * 0.0155''^2 * 0.062'') - (\pi * 0.0125''^2 * 0.062'')) \times 2 = 32.7\,\text{E-6 in}^3$.

If top and bottom fillets are desired, the easiest approximation is the truncated cone with 45° sides, shown in Fig. 11.22. For the $\frac{1}{2}$ resistor example and a radius of the solder pad = 0.025, the resulting volumes would be

- Volume of paste applied for filling the through hole = 32.7 E-6 in^3
- Volume of paste applied for filling each fillet = 14.3 E-6 in^3
- Total volume of for through-hole fill and top & bottom fillets = 61.3 E-6 in^3

No Lead Solder Paste

Few technical issues have hit the electronics industry as hard in recent years as the switch from Sn-Pb solder to lead-free attachment. Both by mandate, e.g., the European Commission requirement to eliminate lead in consumer products, as well as voluntarily, electronic component manufacturers and assemblers must come to grips with issues surrounding the transition to lead-free attachment of components. While various industry, academic, and governmental agencies that have studied alternatives to lead in solder now seem to agree on SnAgCu as the preferred alternative, as discussed by Bradley, Handwerker, and Sohn, there is still much to be done to create a minimal-problem transition.

Lau and Liu discuss the issues surrounding the various formulations of SnAgCu solder. Laske and Jensen discuss implementation issues with SnAgCu. They note that although Bi-containing alloys have the advantage of low melting temperature, Pb contamination is a very real concern for the next decade, from existing components, board finishes, etc., and Bi alloys do not tolerate lead contamination. Although they do not feel it is a major issue, Lau and Liu note that the elevated melting temperature of 217°C will bring with it major process control concerns. Laske and Jensen do present a lead-free process implementation used by one Motorola facility.

If the elevated temperatures or other issues make one feel that the SnAgCu alloys may not be appropriate for their facility, Dreezen and Luyckx present information on lead free assembly with conductive adhesives.

11.3.8 Parts Inspection and Placement

Briefly, all parts must be inspected prior to use. Functional parts' testing should be performed on the same basis as for through-hole devices. Each manufacturer of electronic assemblies is familiar with the various processes used on through-hole parts, and similar processes must be in place on SMDs. Problems with solderability of leads and lead **planarity** are two items which can lead to the largest number of defects in the finished product. Solderability is even more important with SMDs than with through-hole parts, since all electrical and mechanical strength rests within the solder joint, there being no hole-with-lead to add mechanical strength.

Lead coplanarity is defined as follows. If a multilead part, for example, an IC, is placed on a planar surface, lack of coplanarity exists if the solderable part of any lead does not touch that surface. Coplanarity requirements vary depending on the pitch of the component leads and their shape, but generally out-of-plane measurements should not exceed 4 mils (0.004 in.) for 50-mil pitch devices, and 2 mils for 25-mil pitch devices.

All SMDs undergo thermal shocking during the soldering process, and particularly if the SMDs are to be wave soldered (type I or type II boards), which means they will be immersed in the molten solder wave for 2-4 s. Therefore, all plastic-packaged parts must be controlled for moisture content. If the parts have not been stored in a low-humidity environment (<25% relative humidity) the absorbed moisture will expand during the solder process and crack the package, a phenomenon known as pop corning, since the crack is accompanied by a loud pop, and the package expands due to the expansion of moisture, just like popcorn, expands.

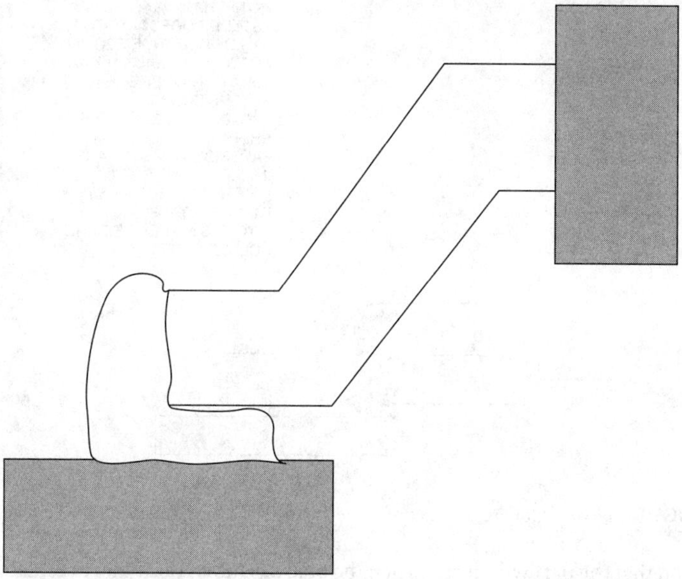

GULL WING LEAD PLACED INTO SOLDER PASTE,
(EXAGGERATED)

FIGURE 11.25   Part placed into solder paste with an active part.

Parts Placement

Proper parts placement not only places the parts within an acceptable window relative to the solder pad pattern on the substrate, the placement machine will apply enough downward pressure on the part to force it halfway into the solder paste (see Fig. 11.25 and Fig. 11.26).

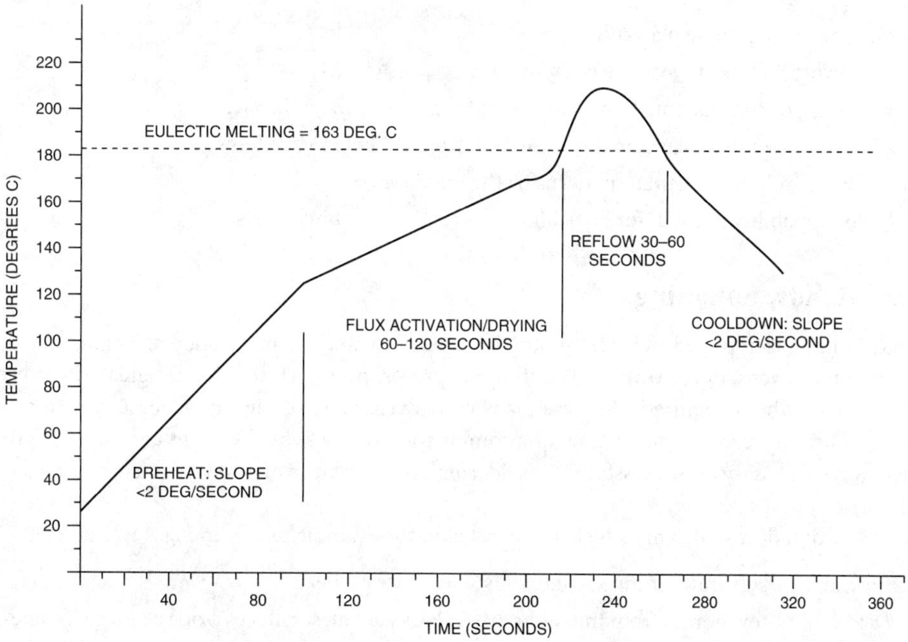

FIGURE 11.26   Typical thermal profile for SMT reflow soldering (types I or II assemblies). (Source: Cox, R.N. 1992. Reflow Technology Handbook. Research Inc., Minneapolis, MN. With permission.)

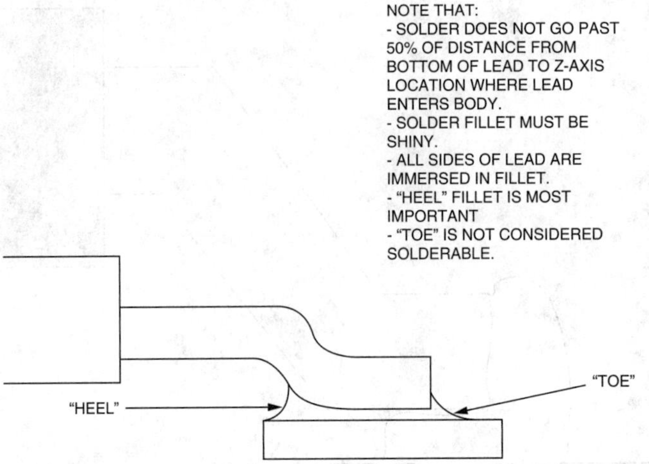

NOTE THAT:
- SOLDER DOES NOT GO PAST
50% OF DISTANCE FROM
BOTTOM OF LEAD TO Z-AXIS
LOCATION WHERE LEAD
ENTERS BODY.
- SOLDER FILLET MUST BE
SHINY.
- ALL SIDES OF LEAD ARE
IMMERSED IN FILLET.
- "HEEL" FILLET IS MOST
IMPORTANT
- "TOE" IS NOT CONSIDERED
SOLDERABLE.

FIGURE 11.27   General inspection criteria for J-lead and gull wing components.

This assures both that the part will sit still when the board is moved and that acceptable planarity offsets among the leads will still result in an acceptable solder joint. Parts placement may be done manually for prototype or low-volume operations, although the author suggests the use of guided X-Y tables with vacuum part pickup for even the smallest operation. Manual placement of SMDs does not lend itself to repeatable work. For medium- and high-volume work, a multitude of machines are available. See Fig. 11.27 for the four general categories of automated placement equipment.

One good source for manufacturer's information on placement machines and most other equipment used in the various SMT production and testing phases is the annual *Directory of Suppliers to the Electronics Manufacturing Industry* published by *Electronic Packaging and Production* (see Further Information section). Among the elements to consider in the selection of placement equipment, whether fully automated or X-Y-vacuum assist tables, are

- Volume of parts to be placed/hour
- Conveyorized links to existing equipment
- Packaging of components to be handled, tubes, reels, trays, bulk, etc.
- Ability to download placement information from CAD/CAM systems
- Ability to modify placement patterns by the operator
- Vision capability needed, for board fiducials and/or fine-pitch parts

11.3.9   Reflow Soldering

Once SMDs have been placed in solder paste, the assembly will be reflow soldered. This can be done in either batch-type ovens, or conveyorized continuous-process ovens. The choice depends primarily on the board throughput/hour required. Whereas many early ovens were of the vapor phase type, most ovens today use IR heating, convection heating, or a combination of the two. The ovens are zoned to provide a thermal profile necessary for successful SMD soldering. An example of an oven thermal profile is shown in Fig. 11.26.

The phases of reflow soldering, which are reflected in the example profile in Fig. 11.26 are

- *Preheat:* The substrate, components, and solder paste preheat.
- *Dry:* Solvents evaporate from the solder paste. Flux activates, reduces oxides, and evaporates. Both low- and high-mass components have enough soak time to reach temperature equilibrium.
- *Reflow:* The solder paste temperature exceeds the liquidus point and reflows, wetting both the component leads and the board pads. Surface tension effects occur, minimizing wetted volume.

- *Cooling:* The solder paste cools below the liquidus point, forming acceptable (shiny and appropriate volume) solder joints.

The setting of the reflow profile is not trivial. It will vary depending on whether the flux is RMA, water soluble, or no clean, and it will vary depending on both the mix of low- and high-thermal mass components, and on how those components are laid out on the board. See Cox's *Reflow Technology Handbook* for additional information. The profile should exceed the liquidus temperature of the solder paste by 20–25°C. Although final setting of the profile will depend on actual quality of the solder joints formed in the oven, initial profile setting should rely heavily on information from the solder paste vendor, as well as the oven manufacturer. Remember that the profile shown in Fig. 11.26 is the profile to be developed on the substrate, and the actual control settings in various stages of the oven itself may be considerably different, depending on the thermal inertia of the product in the oven and the heating characteristics of the particular oven being used.

Defects as a result of poor profiling include

- Component thermal shock
- Solder splatter
- Solder balls formation
- Dewetted solder
- Cold or dull solder joints

Note in Fig. 11.27 that the "toe" fillet on the gull-wing package is generally considered an unreliable indicator of a good solder joint. During the manufacturing process this end of the lead may be cut after tinning, leaving bare copper exposed at the end of the lead.

It should be noted that many other problems may contribute to defective solder joint formation. One example would be placement misalignment, which contributes to the formation of solder bridges. Other problems that may contribute to defective solder joints include poor solder mask adhesion and unequal solder land areas at opposite ends of passive parts, which creates unequal moments as the paste liquifies and develops surface tension. Wrong solder paste volumes, whether too much or too little, will create defects, as will board shake in placement machines and coplanarity problems in IC components. Many of these problems should be covered and compensated for during the design process and the qualification of SMT production equipment. Final analysis of the process is performed based on the quality of the solder joints formed in the reflow process. Whatever criteria may have been followed during the overall process, solder joint quality is the final determining factor of the correctness of the various process steps. As noted earlier, the quality level of solder joint production is a major factor in successful board assembly. A primary criteria is the indication of wetting at the junction of the reflowed solder and the part termination.

The discussion in section 11.3.7.3 noted the higher melting point of the SnAgCu lead-free solder. As Apell notes, "Due to the higher melting points of lead-free solder formulations, the profiling requirements will change. . . . peak temperature and time above liquidus must be achieved without overheating the assembly or components."

Presentation of criteria for all the various SMD package types and all of the possible solder joint problems is beyond the scope of this chapter. The reader is directed to Hollomon (1995), Hwang (1989), Lau (1991), Klein-Wassink (1989), and Prasad (1997) for an in-depth discussion on these issues.

11.3.10   Prototype Systems

Systems for all aspects of SMT assembly are available to support low-volume/prototype needs. These systems will typically have manual solder-paste deposition and parts placement systems, with these functions being assisted for the user. Syringe solder paste deposition may be as simple as a manual medical-type syringe dispenser, which must be guided and squeezed freehand. More sophisticated systems will have the syringe mounted on an *X-Y* arm to carry the weight of the syringe, and will apply air pressure to the top of the syringe with a foot-pedal control, freeing the operator's arm to guide the syringe to the proper location on the substrate and perform the negative z-axis manuever, which will bring the syringe tip into the proper location and height above the substrate. Dispensing is then accomplished by a timed air-pressure burst applied to the top of the syringe under foot-pedal control. Paste volume is likewise determined by trial and error with the time/pressure relation, and depends on the type and manufacturer of paste being dispensed.

Parts placement may be as simple as tweezers and progress to hand-held vacuum probes to allow easier handling of the components. As mentioned previously, X-Y arm/tables are available that have vacuum-pick nozzles to allow the operator to pick a part from a tray, reel, or stick, and move the part over the correct location on the substrate. The part is then moved down into the solder paste, the vacuum is turned off manually or automatically, and the nozzle is raised away from the substrate.

Soldering of prototype/low-volume boards may be done by contact soldering of each component, by a manually guided hot-air tool, or in a small batch or conveyorized oven. Each step up in soldering sophistication is, of course, accompanied by a increase in the investment required.

For manufacturers with large prototype requirements, it is possible to set up an entire line that would involve virtually no hardware changes from one board to another using the following steps:

- ECAD design and analysis produce Gerber files.
- CNC circuit board mill takes Gerber files and mills out two-sided boards.
- Software translation package generates solder-pad centroid information.
- Syringes older paste deposition system takes translated Gerber file and dispenses appropriate amount at each pad centroid.
- Software translation package generates part centroid information.
- Parts placement equipment places parts based on translated part centroid information.
- Assembly is reflow soldered.

The only manual process in this system is adjustment of the reflow profile based on the results of soldering an assembly. The last step in the process would be to test the finished prototype board. This system could also be used for very small volume production runs, and all components as described are available. With a change from milled boards to etched boards, the system can be used as a flexible assembly system.

Defining Terms

Coefficient of thermal expansion (CTE or TCE): A measure of the ratio between the measure of a material and its expansion as temperature increases. May be different in X, Y, and Z axes. Expressed in parts per million per degree Celsius, it is a measure that allows comparison of materials that are to be joined.

Coplanarity: A simplified definition of *planarity*, which is difficult to measure. Noncoplanarity is the distance between the highest and lowest leads and is easily measured by placing the IC on a flat surface, such as a glass plate. The lowest leads will then rest on the plate, and the measured difference to the lead highest above the plate is the measurement of noncoplanarity

Glass transition temperature (T_g): Below T_g, a polymer substance, such as fiberglass, is relatively linear in its expansion/contraction due to temperature changes. Above T_g, the expansion rate increases dramatically and becomes nonlinear. The polymer will also lose its stability, that is, an FR-4 board droops above T_g.

Gull wing: An SMD lead shape as shown in Fig. 11.18 for SOPs and QFPs, and in Fig. 11.26. So called because it looks like a gulls wing in flight.

Plastic leaded chip carrier (PLCC): Shown in Fig. 11.18, it is a mature SMT IC package and is the only package that has the leads bent back under the IC itself.

Planarity: Lying in the same plane. A plane is defined by the exit of the leads from the body of the IC. A second plane is defined at the average of the lowest point all leads are below the first plane. Nonplanarity is the maximum variation in mils or millimeters of any lead of an SMD from the lowest point plane.

Quad flat pack: Any flat pack IC package that has leads on all four sides.

Through hole: Also a *plate through hole* (PTH). A hole in a substrate that extends completely through a substrate, is solder plated, and is intended for a component lead.

Through-hole technology: The technology of using leaded components which require holes through the substrate for their mounting (insertion) and soldering.

Via: A hole in the substrate intended only to establish electrical connectivity, and not intended to hold a component lead.

References

ASME. 1993. Conduction heat transfer measurements for an array of surface mounted heated components. In *Proceedings of the 1993 ASME Annual Meeting*, Vol. 263, pp. 69–78. Am. Soc. of Mech. Engr., Heat Transfer Div.

Achong, C. and Chen, A. BGA land pattern and assembly issues. *Printed Circuit Design and Mnaufacture*, v21 #2, Feb. 2004.

Apell, M.C. Considerations for lead-free reflow soldering. *Circuits Assembly*, v15 #3, March 2004.

Becker, K-F, Adams, T. Optimizing processes and materials for flip chip reliability. *Circuits Assembly*, v.15, #2, Feb. 2004.

Bradley, E., Handwerker, C., and Sohn, J.E. NEMI Report: A single lead-free alloy is recommended. *Surface Mount Technology Magazine*, v.17 #1, Jan. 2003.

Capillo, C. 1990. *Surface Mount Technology, Materials, Processes and Equipment*, Chap. 3, 7, 8. McGraw-Hill, New York.

Choi, C.Y., Kim, S.J., and Ortega, A. 1994. Effects of substrate conductivity on convective cooling of electronic components. *Journal of Electronic Packaging* 116(3):198–205.

Classon, F. 1993. *Surface Mount Technology for Concurrent Engineering and Manufacturing*. McGraw-Hill, New York.

Cox, N.R. 1992. *Reflow Technology Handbook*. Research, Inc., Minneapolis, MN.

Dreezen, G. and Luyckx, G. Lead free assembly: conductive adhesives offer a real alternative. *Global SMT & Packaging*, v.3 #7, Sept. 2003.

Flotherm. Advanced thermal analysis of packaged electronic systems. Flomerics, Inc., Westborough, MA.

Higgins, C. 1991. Signetics Corp. presentation, Nov.

Hollomon, J.K., Jr. 1995. *Surface Mount Technology for PC Board Design*. Prompt, Indianapolis, IN.

Holmes, J.G. 1993. Surface mount solution for power devices. *Surface Mount Technology* 7(9):18–20.

Hwang, J.S. 1989. *Solder Paste in Electronics Packaging*. Van Nostrand Reinhold, New York.

Intel. 2000. *Packaging Handbook*. Intel Corp., Santa Clara, CA.

Klein-Wassink, R.J. 1989. *Soldering in Electronics*. Electrochemical Publishing Co., Ltd.

Lasky, R. and Jensen, T. Best practices in implementing Pb-free assembly. *Global SMT & Packaging*, v3 #8, Nov. 2003.

Lau, J.H., ed. 1991. *Solder Joint Reliability*. Van Nostrand Reinhold, New York.

Lau, J.H. and Liu, K. Global trends in lead-free soldering, part I. *Advanced Packaging*, Jan. 2004.

Lau, J.H. and Liu, K. Global trends in lead-free soldering, part II. *Advanced Packaging*, Feb. 2004.

Lea, C. 1988. *A Scientific Guide to SMT*. Electrochemical Publishing Co., Ltd.

Lee, L.C. et al. 1984. Micromechanics of multilayer printed circuit board. *IBM Journal of Research and Development* 28(6).

Lee, T.Y. 1994. Application of a CFD tool for system-level thermal simulation. *IEEE Trans. Components, Packaging, and Manufacturing Tech.* Part A 17(4):564–571.

Leibson, S.H. 1987. The promise of surface mount technology. *EDN Magazine* 5-28-87, pp. 165–174, five parts through 7-3-87.

Marcoux, P.P. 1992. *Fine Pitch Surface Mount Technology*. Van Nostrand Reinhold, New York.

Mims, F.M., III. 1987. Surface mount technology: an introduction to the packaging revolution. *Radio-Electronics* (11):58–90.

Motorola. 1993. *Linear/Interface IC Device Databook*, Vol. 1, Sec. 3 Addendum. Motorola, Inc.

Philips. 1991. *Signetics Surface Mount Process and Application Notes*. Philips Semiconductor, Sunnyvale CA.

Prasad, R.P. 1997. *Surface Mount Technology Principles and Practice*, 2nd ed. Van Nostrand Reinhold, New York.

Rowland, R. 1993. *Applied Surface Mount Assembly*. Van Nostrand Reinhold, New York.

Shina, S.G. 1991. *Concurrent Engineering and Design for Manufacture of Electronic Products*. Van Nostrand Reinhold, New York.

Texas Instruments. 1984. *How to Use Surface Mount Technology*. Texas Instruments, Dallas, TX.

Vicari, D. Pin-in-Paste, or Alternative Assembly and Reflow Technology. Universal Instruments white paper, at http://www.uic.com/wcms/WCMS.nsf/index/White_Papers_3.html#pinpaste

Further Information

Specific journal references are available on any aspect of SMT. A search of the COMPENDEX Engineering Index 1987-present will show over 1500 references specifically to SMT topics.

Education/Training: a partial list of organizations which specialize in education and training directly related to issues in surface mount technology:

Electronic Manufacturing Productivity Facility. 714 North Senate Ave., Indianapolis, IN 46202-3112. 317-226-5607

SMT Plus, Inc. 5403-F Scotts Valley Drive, Scotts Valley, CA 95066; 408-438-6116

Surface Mount Technology Association (SMTA), 5200 Wilson Rd, Ste 100, Edina, MN 55424-1338. 612-920-7682.

Institute for Interconnecting and Packaging Electronic Circuits (IPC), 7380 N, Lincoln Ave., Lincolnwood, Il 60646-1705, 708-677-2850.

Conferences directly related to SMT:

Surface Mount International (SMI). Sponsored by SMTA.

National Electronics Packaging and Production Conference (NEPCON). Coordinated by Reed Exhibition Co., P.O. Box 5060, Des Plaines, IL 60017-5060. 708-299-9311.

Journals covering various aspects of SMT:

Advanced Packaging, published bimonthly, P.O. Box 159, Libertyville, IL 60048-0159.

Advances in Electronic Packaging, Transactions of the ASME, American Society of Mechanical Engineers.

Circuits Assembly, published monthly, 2000 Powers Ferry Center, Suite 450, Marietta, GA 30067.

Electronic Packaging & Production, published monthly, 1350 E. Touhy Ave., Des Plaines, IL 60018-3358. Also annual

Directory of Suppliers to the Electronics Manufacturing Industry Cahners, 8773 South Ridgeline Blvd., Highlands Ranch, CO 80126-2329.

IEEE Transactions on Components, Packaging and Manufacturing Technology, Institute for Electrical and Electronic Engineers, 345 East 47th St., New York, NY 10017.

Journal of Electronic Packaging, Transactions of the ASME, American Society of Mechanical Engineers.

Journal of Surface Mount Technology, published quarterly by the Surface Mount Technology Assn., 5200 Willson Rd., Suite 200, Edina, MN 55424-1338.

Printed Circuit Design, published monthly, 2000 Powers Ferry Center, Suite 450, Marietta, GA 30067.

Surface Mount Technology, published monthly, P.O. Box 159, Libertyville, IL 60048-0159.

11.4   Shielding and EMI Considerations

Donald White

11.4.1   Introduction

Two of many requirements of electronic-system packaging are to ensure that the corresponding equipments are in compliance with applicable national and international **electromagnetic compatibility (EMC)**

regulations for emission and immunity control. In the U.S., this refers to FCC, Rules and Regulations, Parts 15.B on emission, and internationally to CISPR 22 and IEC-1000-4 (formerly, IEC 801 family) for emission and immunity control.

Emission and immunity (susceptibility) control each have two parts: conducted (on hard wire) and radiated (radio-wave coupling). This section deals only with the latter on **electromagnetic interference (EMI) radiated emission** and susceptibility control.

In general, reciprocity applies for EMI and its control. When a shield is added to reduce radiated emission, it also reduces radiated immunity. Quantitatively speaking, however, reciprocity does not apply since the bilateral conditions do not exist, viz., the distance from the source to the shield does not equal the distance from the shield to the victim. Thus, generally, expect different shielding effectiveness values for each direction.

11.4.2 Cables and Connectors

One principal method of dealing with EMI is to shield the source, the victim, or both, so that only intentional radiation will couple between them. The problem, then, is to control unintentional radiation coupling into or out of victim circuits, their PCBs, their housings, and equipment interconnecting cables.

Generalizing, the greatest cause of EMI failure is from interconnecting cables. Cables are often referred to as *the antenna farm* since they serve to unintentionally pickup or radiate emissions to or from the outside world thereby causing EMI failures. However, cables are not the subject of this section. Connectors are loosely referred to as devices to stop outside **conducted emissions** from getting in or vice versa.

11.4.3 Shielding

If the system is designed correctly in the first place, little shielding may be needed to clean up the remaining EMI. One classical example is the design and layout of printed circuit boards (PCBs). Multilayer boards radiate at levels 20–35 dB below levels from double-sided, single-layer boards. Since logic interconnect trace radiation is the principal PCB radiator, it follows that multilayer boards are cost effective vs. having to shield the outer box housing.

Shielding Effectiveness

Figure 11.28 shows the concept of shielding and its measure of effectiveness, called **shielding effectiveness (SE)**. The hostile source in the illustration is at the left and the area to be protected is on the right. The metal barrier in the middle is shown in an edge-elevation view.

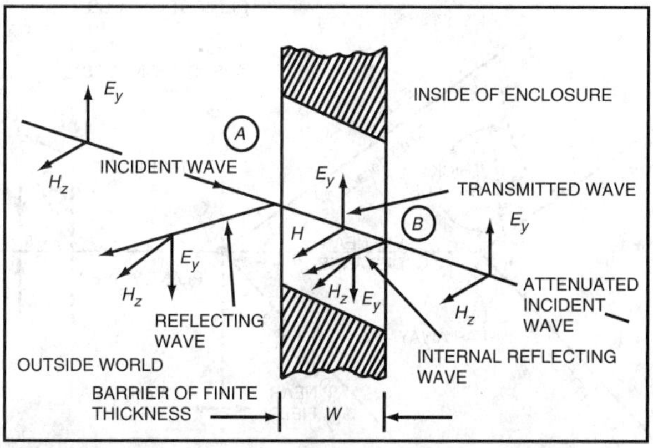

SHIELDING EFFECTIVENESS RATIO = REFLECTION LOSS × ABSORPTION LOSS
$SE_{dB} = R_{dB} + A_{dB}$ FOR ONLY FAR FIELD: $SE_{dB} = 20 \log_{10} (E_{y\,before}/ E_{y\,after})$
$= 20 \log_{10} (H_{z\,before}/ H_{z\,after})$

FIGURE 11.28 Shielding effectiveness analysis.

The oncoming electromagnetic wave (note the orthogonal orientation of the electric and magnetic fields and the direction of propagation: $E \times H =$ Poynting's vector) is approaching the metal barrier. The wave impedance ($Z_w = E/H$) is 377 Ω in the **far field** (distance $> \lambda/2\pi$). Z_w is greater in the **near field** (distance $< \lambda/2\pi$) for high-impedance waves or E fields and less than 377 Ω in the near field for low-impedance waves or H fields. The wave impedance is related to the originating circuit impedance at the frequency in question.

The wavelength λ, in meters, used in the previous paragraph is related to the frequency in megahertz.

$$\lambda_m = 300/f_{\mathrm{MHz}} \tag{11.6}$$

Since the wave impedance and metal barrier impedance Z_b (usually orders of magnitude lower) are not matched, there is a large reflection loss RL_{dB} established at the air-metal interface

$$RL_{\mathrm{dB}} = 20\log_{10}(Z_w/(4 * Z_b)) \tag{11.7}$$

where Z_w is the wave impedance equal to 377 Ω in the far field.

Figure 11.28 shows that what is not reflected at the metal barrier interface continues to propagate inside where the electromagnetic energy is converted into heat. This process is called *absorption loss* AL_{dB} defined as

$$AL_{\mathrm{dB}} = 8.6t/\delta \tag{11.8}$$

$$\delta = 1/\sqrt{\pi f \sigma \mu} \tag{11.9}$$

where

  t = metal thickness in any units
  δ = skin depth of metal in same units
  f = frequency, Hz
  σ = absolute conductivity of the metal, mho
  μ = absolute permeability of the metal, H/m

The shielding effectiveness SE_{dB} is the combination of both the reflection and absorption losses

$$SE_{\mathrm{dB}} = RL_{\mathrm{dB}} + AL_{\mathrm{dB}} \tag{11.10}$$

Figure 11.29 shows the general characteristics of shielding effectiveness. As mentioned before, the interface between near and far fields occur at the distance $r = \lambda/2\pi$. The electric field is shown decreasing

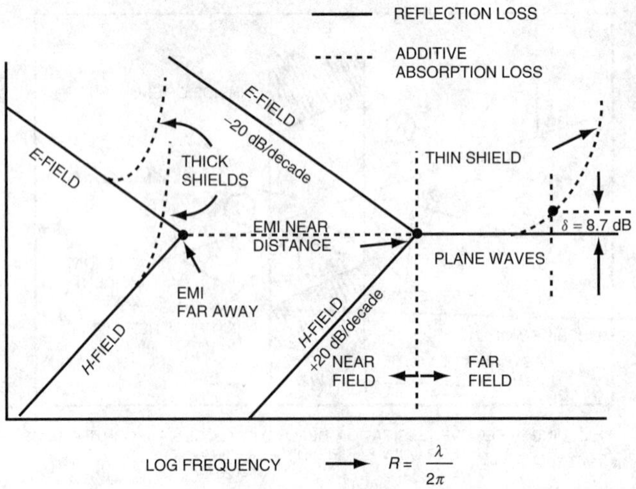

FIGURE 11.29 The general characteristics of shielding effectiveness.

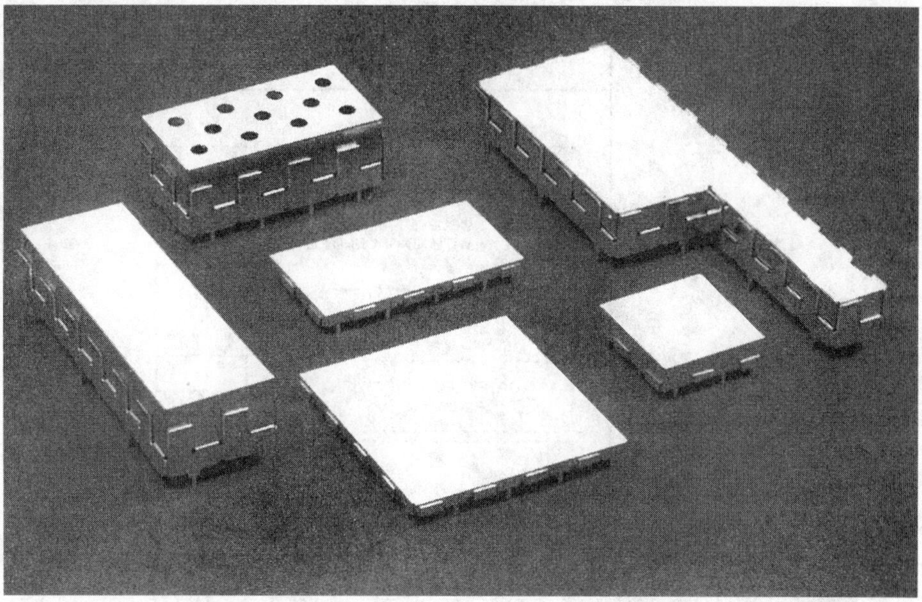

FIGURE 11.30   PCB component shielding. (*Source:* emf-emi control^R, Inc. Copyright 1995.)

at −20 dB/decade with increasing frequency while the magnetic field increases at +20 dB/decade. The two fields meet at the far-field interface, called *plane waves* at higher frequencies.

Figure 11.29 shows the absorption loss (dotted line) appearing when the metal thickness is approaching a fraction of one skin depth (adds 8.6 dB attenuation).

Different Levels of Shielding

Shielding may be performed at the component level, PCB or backplane level, card-cage level, box level, or entire system level (such as a shielded room). The latter is excluded from this discussion.

Component Shielding

Figure 11.30 shows a shield at the component level used on a clock on a PCB. The shield is grounded to the signal return, V_0, or board ground traces, as most conveniently located.

The argument for shielding at the component level is that after using a multilayer board only the components can significantly radiate if further radiation reduction is needed.

Tests indicate that shielding effectiveness of component shields at radio frequencies above 10 MHz may vary from 10 to 25 dB over radiation from no shields.

Printed-Circuit Board Shielding

The reason that multilayer PCBs radiate at much lower levels than double-sided, single-layer boards is that the trace return is directly under the trace in the V_{cc} or V_0 planes, forming a microstrip line. This is similar to the technique of using a phantom ground plane on single-layer boards as shown in Fig. 11.31. The trace height is now small compared to a trace that must wander over one or more rows as in single-layer boards.

The remaining traces left to radiate in multilayer boards are the top traces (for example, north-south) and the bottom traces on the solder side (east-west). Many military spec PCBs shield both the top and bottom layers, but industrial boards generally do not do this because of the extra costs. They also make maintenance of the PCBs nearly impossible.

To shield the top side, the board would have to be broached to clear the surface-mount components. The bottom side shield can simply be a single-sided board (foil on the outside) brought up to the solder tabs. The board can be secured in place and bonded to the PCB (at least at the four corners). For greater shielding (see **aperture leakage**), additional bonding is needed.

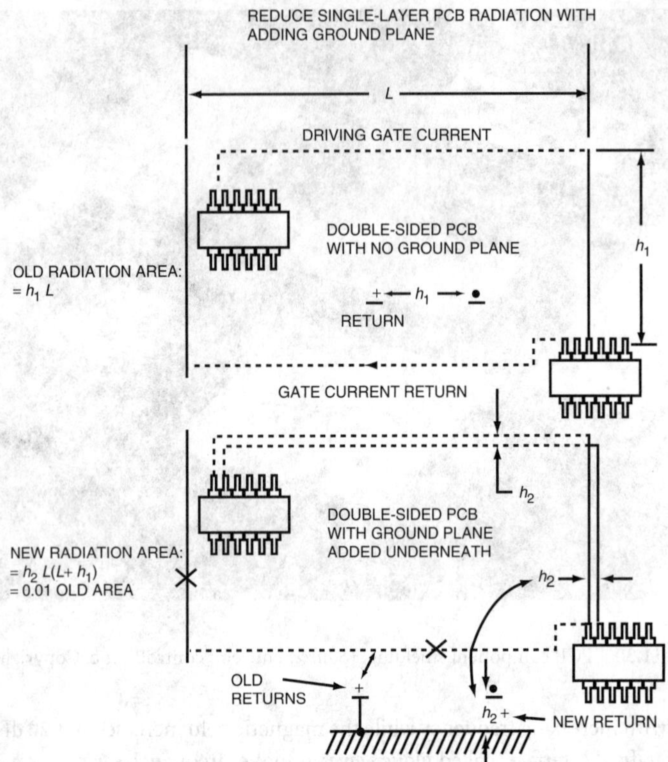

REDUCE SINGLE-LAYER PCB RADIATION WITH
ADDING GROUND PLANE

FIGURE 11.31　Using a phantom-ground plane in single-layer boards to reduce radiation.

Card-Cage Shielding

Card-cage shielding (metalizing all six sides of a plastic card cage) can add additional shielding effectiveness if the cards are single-layer boards. If multilayer boards, expect no shielding improvement since the trace-to-ground dimensions are much less than those of the card cage.

Shielding at the Box or Housing Level

Classically, the most effective level to shield is at the box-housing level. The attitude often is what radiated EMI you do not stop at the lower levels can be accomplished at the box level by adding conductive coatings and aperture-leakage control.

How much shielding is required at the box level to comply with FCC Parts 15.B or CISPR 22, Class B (residential locations)? Naturally, this depends on the complexity and number of PCBs, the clock frequency, the logic family, etc. At the risk of generalizing, however, if little EMI control is exercised at the PCB level, roughly 40 dB of shielding effectiveness will be required for the entire box over the frequency range of roughly 50–500 MHz.

To achieve a 40-dB overall box shielding, a deposited conductive coating would have to provide at least 50 dB. The coating impedance Z_b in ohms/square is determined from Eq. (11.7)

$$SE_{dB} = 20 \log_{10}(Z_w/(4 * Z_b))$$

where Z_w is the wave impedance equal to 377 Ω in the far field. For $SE = 50$ dB, solving the above equation for Z_b

$$Z_b = Z_w/(4 * 10^{SE}/20)$$
$$94/10^{2.5} = 0.3 \ \Omega/\text{sq.}$$

(11.11)

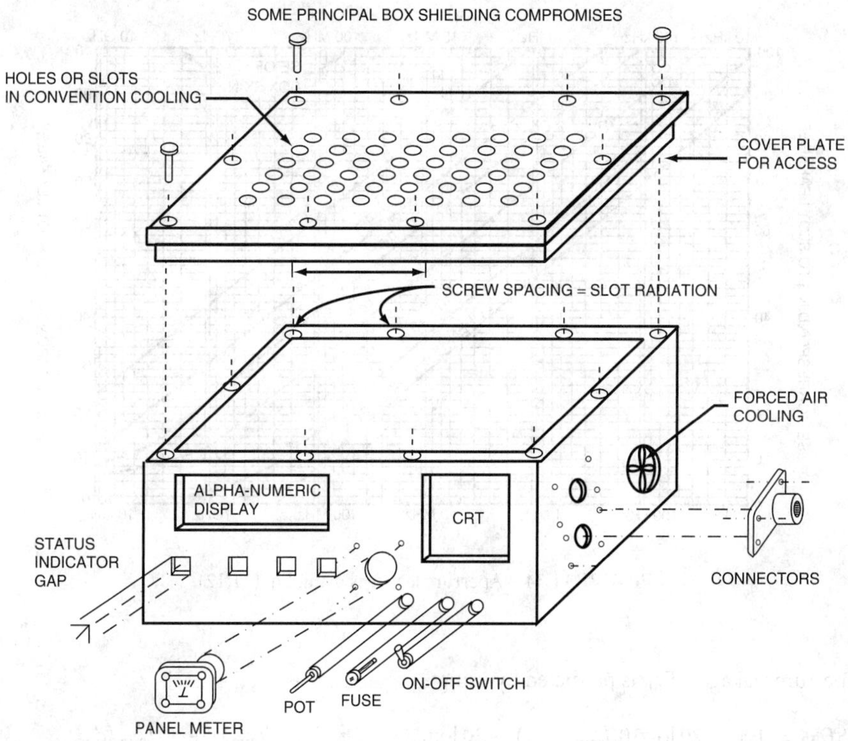

FIGURE 11.32 Principal aperture leakages in boxes.

This means that the conductive coating service house needs to provide the equivalent of a conductive coating of about 0.3 Ω/sq.

The electroless process is one of the principal forms of conductive coatings in use today. This process coats both inside and outside the plastic housing. One mil (24 μm) of copper is more than enough to achieve this objective as long as it is uniformly or homogeneously deposited.

Sometimes, a flashing of nickel is used over an aluminum or copper shield to provide abrasion protection to scratches and normal abuse. The flashing may be of the order of 1 μm. Nickel does not compromise the aluminum or copper shielding. It simply adds to the overall shielding effectiveness.

11.4.4 Aperture-Leakage Control

As it develops, selecting and achieving a given shield-ing effectiveness of the conductive coating is the easy part of the box or housing shielding design. Far more difficult is controlling the many different aperture leakages depicted in Fig. 11.32. Therein it is seen that there are large apertures, such as needed for windows for cathode ray tube (CRT) and light emitting diode (LED) displays, holes for convection cooling, and slots for mating panel members.

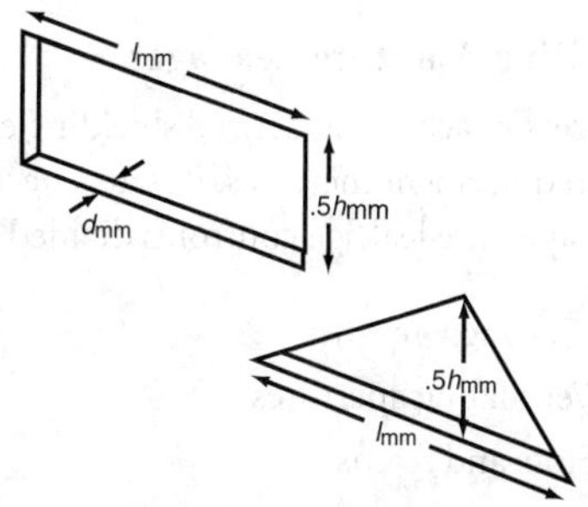

To predict aperture leakages, the geometry is first defined in Fig. 11.33. The longest aperture dimen-sion is l_{mm}, given in millimeters. The shortest aper-ture dimension is h_{mm}, given in millimeters. The thickness of the metal or conductive coating is d_{mm}, also given in millimeters.

FIGURE 11.33 Aperture dimensions used in mathematic models.

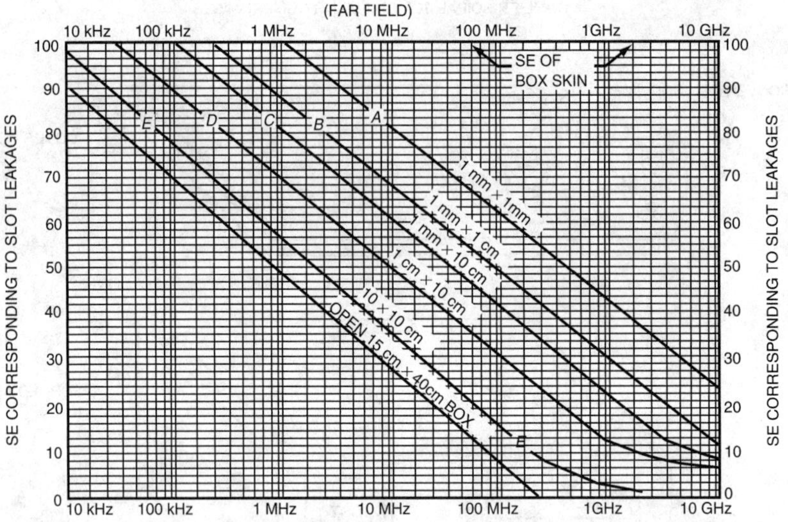

FIGURE 11.34   Aperture leakages using Eq. (11.12).

The aperture leakage SE_{dB} is predicted by using

$$SE_{dB} = 100 - 20 \log 10(l_{mm} f_{MHz}) + 20 \log_{10}(1 + ln_e(l_{mm}/h_{mm})) + 30 l_{mm}/d_{mm} \qquad (11.12)$$

with all terms having been previously defined.

To gain a better mental picture of the shielding effectiveness of different apertures, Eq. (11.12) is plotted in Fig. 11.34.

Illustrative Example 1

Even small scratches can destroy the performance of conductive coatings. For example, suppose that a single (inside) conductive coating has been used to achieve 50 dB of shielding effectiveness in the 50 MHz–1 GHz spectrum. Further, suppose that a screwdriver accidentally developed a 4-in. (102-mm) × 10-mil (0.25-mm) scratch. From Eq. (11.12), the resulting shielding effectiveness to the seventh harmonic of a 90-MHz onboard clock is

$$SE_{dB} = 100 - 20 \log_{10}(102 * 630) + 20 \log_{10}(1 + ln_e(102/0.25))$$
$$= 100 - 96 + 17 = 21dB$$

Controlling Aperture Leakages

The success in achieving desired shielding effectiveness, then, is to control all major aperture leakages. As the SE requirement increases, it is also necessary to control minor aperture leakages as well.

Most aperture leakage control is divided into four sections

- Viewing windows
- Ventilating apertures
- Slots and seams
- All other apertures

Viewing Apertures

The first of the four aperture-leakage categories is viewing aperture windows. This applies to apertures such as CRTs, LED displays, and panel meters.

There are two ways to shore up leaky windows: (1) knitted wire mesh or screens and (2) thin-conductive films. In both cases, the strategy is to block the RF while passing the optical viewing with minimum attenuation.

Knitted-Wire Meshes and Screens

The amount of RF blocking (aperture leakage) from a knitted-wire mesh or screen is

$$SE_{dB} = 20\log_{10}(\lambda/2/D)$$
$$= 103 - (f_{MHz}D_{mm})dB$$

(11.13)

where D_{mm} is the screen mesh wire separation in millimeter.

Figure 11.35 shows one method for mounting screens and wire mesh. Mesh also comes in the form of metalized textiles with wire-to-wire separations as small as 40 μm, hence the wires are not even discernible, but appear more like an antiglare screen.

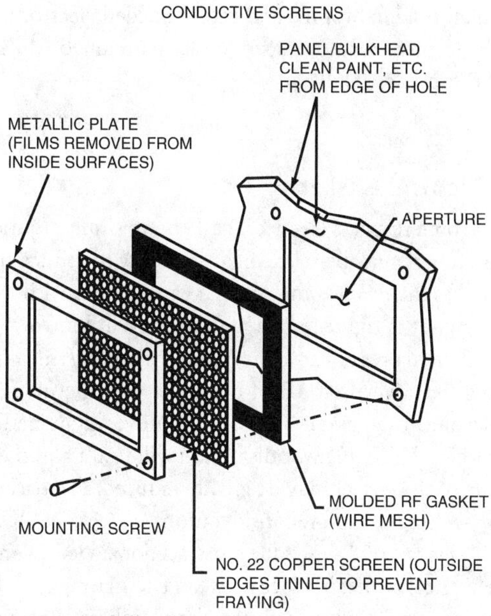

Thin Films

Another approach to providing RF shielding is to make a thin-film shield, which provides a surface impedance while passing most of the optical light through the material. Operating on the reflection-loss principle, typical shielding effectiveness of about 30 dB may be obtained (from Eq. (11.7), film impedances correspond to about 3 Ω/square). When the impedance is lowered to gain greater SE, optical blockage begins to exceed 50%. Figure 11.36 shows the performance of typical thin films.

As also applicable to screens and meshes, thin films must be bonded to the aperture opening 360° around the periphery. Otherwise, the film is compromised and SE suffers significantly.

Ventilating Holes

Convection cooling requires many small holes or slots at the top cover or upper panels of an electronic enclosure. If the array of holes is tight with relatively little separation (the subtends is less than 10% of a wavelength), the aperture leakage corresponds to that of a single hole defined in Eq. (11.12).

Additional attenuation is achievable by making the depth of the hole comparable to or exceeding the dimension of the hole or slot. This, then, generates the waveguide-beyond-cutoff effect in which the last term of Eq. (11.12) now applies as well.

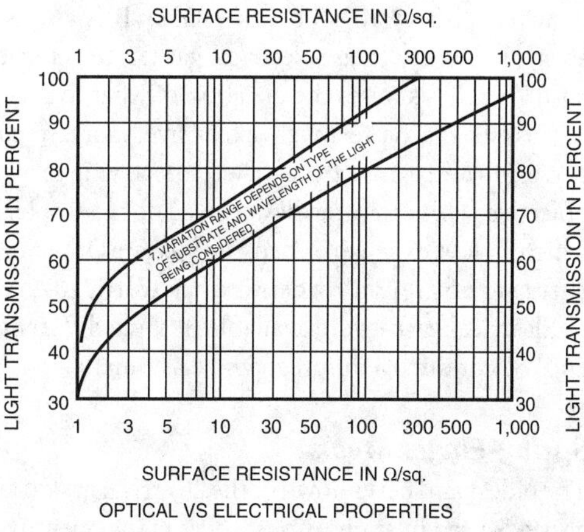

Illustrative Example 2

Suppose an egg-crate array of square holes is placed in the top of a monitor for convection cooling. The squares are 4 mm on a side and 5 mm deep. Determine the shielding effectiveness of this configuration at 1.7 GHz.

FIGURE 11.35 Method of mounting wire screen over an aperture.

FIGURE 11.36 Effectiveness of thin-films shields.

From Eq. (11.12)

$$SE_{dB} = 100 - 20\log_{10}(4 * 1700) + 20\log_{10}(1 + ln_e(4/4)) + 30 * 5/4$$
$$= 100 - 77 + 0 + 38 = 61 \text{ dB}$$

This is only possible with the electroless process of conductive coating of a plastic enclosure since the metal must be plated through the hole as well for the waveguide-beyond-cuttoff concept to work.

Commercial versions of this process are shown in Fig. 11.37 for the hexagonal honeycomb used in all shielded rooms and most shielded electronic enclosures. Air flow is smoothly maintained while *SE* exceeds 100 dB.

FIGURE 11.37 Hexagonal honeycomb shields.

Electrical Gaskets

In some respects electrical gaskets are similar to mechanical gaskets. First, the tolerances of mating metal members may be relaxed considerably to reduce the cost of the fabrication. A one penny rubber washer makes the leakless mating of a garden hose with its bib a snap. The gasket does not work, however, until the joint is tightened enough to prevent blowby. The electrical equivalent of this is the joint must have sufficient pressure to achieve an *electrical seal* (meaning the advertised *SE* of the gasket is achievable). Second, the gasket should not be over-tightened. Otherwise *coldflow* or distortion will result and the life of the gasket may be greatly reduced. Electrical gaskets have similar considerations.

Figure 11.38 shows the cross-sectional view of an electrical gasket in place between two mating metal members. The gap between the members is to be filled with a metal bonding material (the gasket), which provides a low-impedance path in the gap between the members, thereby avoiding a slot radiator discussed in earlier sections.

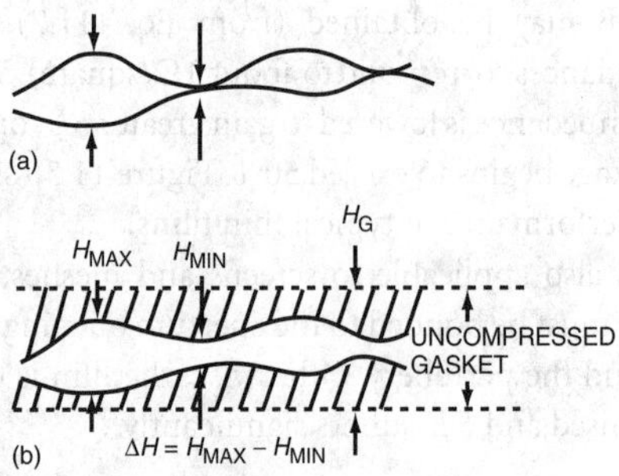

Usually, it is not necessary to use electrical gaskets to achieve *SE* of 40 dB below 1 GHz. It is almost

FIGURE 11.38 Electrical gasket filling the gap between mating metal parts: (a) joint unevenness $= \Delta H$ just touching, (b) compressed gasket in place.

always necessary to use electrical gaskets to achieve 60-dB *SE*. This raises the question of what to do to achieve between 40 and 60 dB overall *SE*?

Gaskets can be relatively expensive, increase cost of installation and maintenance and are subject to corrosion and aging effects. Therefore, trying to avoid their use in the 40–60 dB range becomes a motivated objective. This can often be achieved, for example, by replacing a cover plate with an inverted shallow pan, which is less expensive than the gasket. Other mechanical manipulations to force the electromagnetic energy to go around corners is often worth 10–20 dB more in equivalent *SE*.

Electrical gaskets are available in three different forms: (1) spring-finger stock, (2) wire-mesh gaskets, and (3) conductive impregnated elastomers.

Spring-Finger Stock

The oldest and best known of the electrical gasket family is the spring-finger stock shown in Fig. 11.39. This gasket is used in applications where many openings and closings of the mating metal parts are anticipated. Examples include shielded room doors, shielded electronic enclosures, and cover pans.

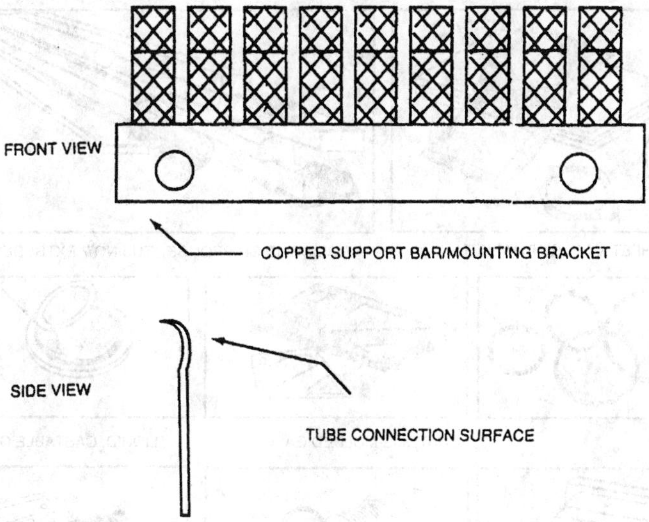

FIGURE 11.39   Spring finger gaskets.

Spring-finger gaskets may be applied by using adhesive backing, sweat soldering into place, riveting, or eyeleting.

Wire-Mesh Gaskets

Another popular electrical gasket is the multilayer, knitted-wire mesh. Where significant aperture space needs to be absorbed, the gasket may have a center core of spongy material such as neophrem or rubber. Figure 11.40 shows a number of methods of mounting wire-mesh gaskets.

Conductive-Impregnated Elastomers

A third type of gasket is the conductive-impregnated elastomers. In this configuration, the conductive materials are metal flakes or powder dispersed throughout the carrier. This gasket and some of its applications are shown in Fig. 11.41.

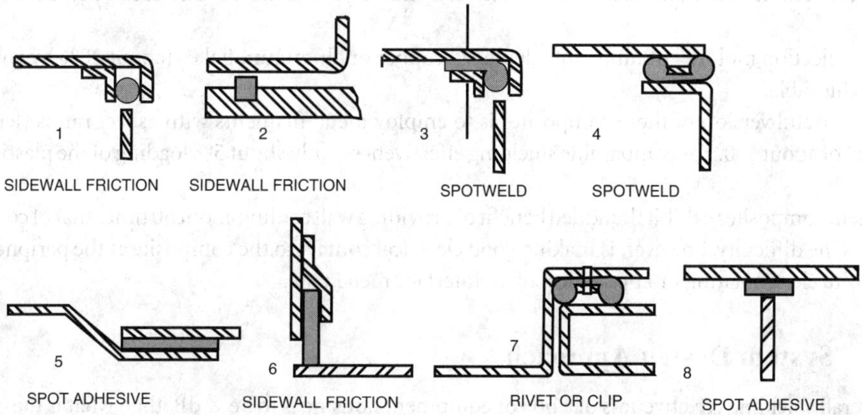

FIGURE 11.40   Knitted-wire mesh mountings.

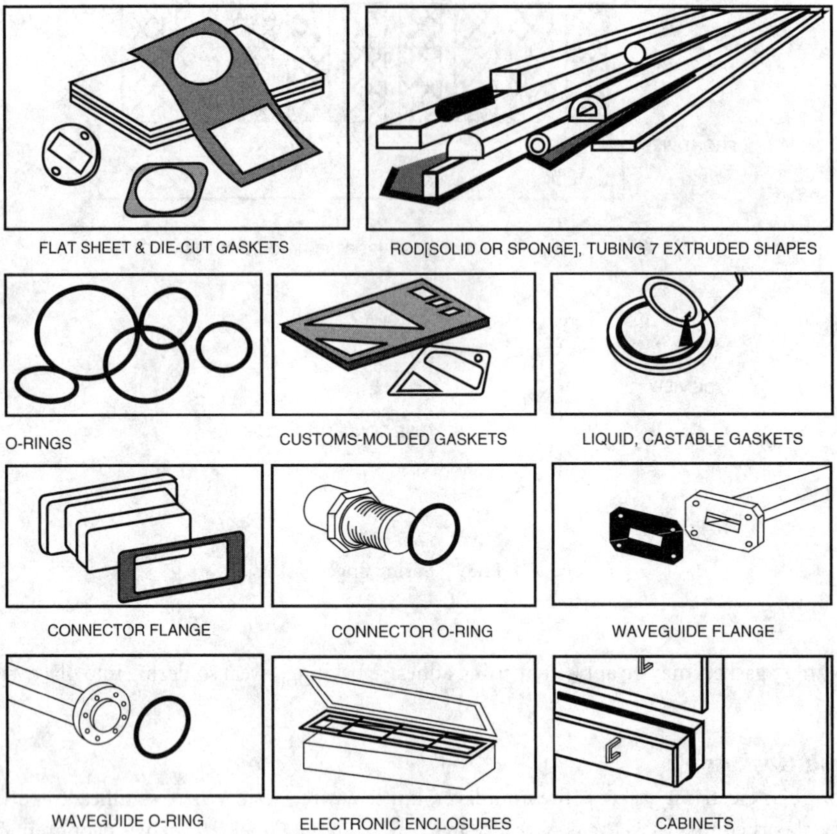

FIGURE 11.41   Conductive elastomeric gaskets.

Shielding Composites

Another material for achieving shielding is called composites. Here the plastic material is impregnated with flakes or powder of a metal such as copper or aluminum. Early versions were not particularly successful since it was difficult to homogeneously deploy the filler without developing leakage spots or clumps of the metal flakes. This lead to making the composites into multilayers of sheet material to be formed around a mold.

Today, injection molding of uniformly deployed copper or aluminum flakes (about 20% by volume) is readily achievable.

A more useful version of these composites is to employ metal filaments with aspect ratios (length-to-diameter) of about 500. For comparable shielding effectiveness, only about 5% loading of the plastic carrier is needed.

Filament composites exhibit the added benefit of providing wall cooling about 50 times that of conductive coatings. One difficulty, however, is making good electrical contact to the composite at the periphery. This may require electroplating or application of an interface metal tape.

11.4.5   System Design Approach

If the overall shielding effectiveness of a box or equipment housing is to be X dB, then what is the objective for the design of each of the leaky apertures? And, what must be the shielding effectiveness of the conductive coating of the housing?

Leaky apertures combine like resistors in parallel. The overall SE of the enclosure is poorer than any one aperture:

$$SE_{\text{overall}} = 20\log_{10}\left[\sum \text{antilog } SE_N\right] \tag{11.14}$$

where SE_N is the SE of the Nth aperture.

The required SE_N may be approximated by assuming that each aperture leakage must meet the same shielding amount, hence

$$SE_N = SE_{\text{overall}} + 20\log_{10} N \tag{11.15}$$

Illustrative Example 3

A box housing has three separate aperture leakages: the CRT, convection cooling holes, and seams at mating conductive coating parts. Determine the aperture leakage requirements of each to obtain an overall SE of 40 dB.

From Eq. (11.14)

$$SE_{\text{each}} = 40\,\text{dB} + 20\log_{10}(3) = 50\,\text{dB}$$

To fine tune this value, the aperture leakage can be relaxed in one area if tightened in another. This is often desired where a relaxation can avoid an expensive additive such as an electrical gasket. In this example, it may be desired to relax the seam gasket to produce 45 dB, if the other two leakages are tightened to 60 dB.

Finally, the basic SE of the conductive coating must exceed the SE of each aperture leakage (or the most stringent of the aperture leakages). A margin of 10 dB is suggested since 6 dB will produce a 4-dB degradation at the aperture, where as 10 dB will allow a 3-dB degradation. Thus, in example 3, the SE of the base coating would have to be 60 dB to permit controlling three apertures to 50 dB to result in an overall SE of 40 dB.

Defining Terms

Aperture leakage: The compromise in shielding effectiveness resulting from holes, slits, slots, and the like used for windows, cooling openings, joints, and components.

Broadband EMI: Electrical disturbance covering many octaves or decades in the frequency spectrum or greater than receiver bandwidth.

Common mode: As applied to two or more wires, all currents flowing therein which are in-phase.

Conducted emission (CE): The potential EMI which is generated inside the equipment and is carried out of the equipment over I/O lines, control leads, or power mains.

Coupling path: The conducted or radiated path by which interfering energy gets from a source to a victim.

Crosstalk: The value expressed in dB as a ratio of coupled voltage on a victim cable to the voltage on a nearby culprit cable.

Differential mode: On a wire pair when the voltages or currents are of opposite polarity.

Electromagnetic compatibilty (EMC): Operation of equipments and systems in their installed environments which cause no unfavorable response to or by other equipments and systems in same.

Electromagnetic environmental effects (E3): A broad umbrella term used to cover EMC, EMI, RFI, electromagnetic pulse (EMP), electrostatic discharge (ESD), radiation hazards (RADHAZ), lightning, and the like.

Electromagnetic interfrence (EMI): When electrical disturbance from a natural phenomena or an electrical/electronic device of system causes an undesired response to another.

Electrostatic discharge (ESD): Fast risetime, intensive discharges from humans, clothing, furniture, and other charged dielectric sources.

Far field: In radiated-field vernacular, EMI point source distance grater than about 1/6 wavelength.

Ferrites: Powdered magnetic material in the form of beads, rods, and blocks used to absorb EMI on wires and cables.

Field strength: The radiated voltage or current per meter corresponding to electric or magnetic fields.

Narrowband EMI: Interference whose emission bandwidth is less than the bandwidth of EMI measuring receiver or spectrum analyzer.

Near field: In radiated-field vernacular, EMI point source distance less than about 1/6 wavelength.

Noise-immunity level: The voltage threshold in digital logic families above which a logic zero may be sensed as a one and vice versa.

Normal mode: The same as differential mode, emission or susceptibility from coupling to/from wire pairs.

Radiated emission (RE): The potential EMI that radiates from escape coupling paths such as cables, leaky apertures, or inadequately shielded housings.

Radiated susceptibility: Undesired potential EMI that is radiated into the equipment or system from hostile outside sources.

Radio frequency interference (RFI): Exists when either the transmitter or receiver is carrier operated, causing unintentional responses to or from other equipment or systems.

Shielding effectiveness (*SE*): Ratio of field strengths before and after erecting a shield. *SE* consist of absorption and reflection losses.

Transients: Short surges of conducted or radiated emission during the period of changing load conditions in an electrical or electronic system or from other sources.

References

EEC. Predicting, analyzing, and fixing shielding effectiveness and aperture leakage control. EEC Software Program #3800.

Kakad, A. 1995. ePTFE gaskets: A competitive alternate. *emf-emi control Journal* (Jan.–Feb.):11.

Mardiguian, M. *Electromagnetic Shielding*, Vol. 3, EMC handbook series. EEC Press.

Squires, C. 1994. Component aperture allocation for overall box shielding effectiveness, part 1, strategy. *emf-emi control Journal* (Jan.–Feb.):18.

Squires, C. 1994. Component aperture leakage allocation for overall box shielding effectiveness, part 2, design. *emf-emi control Journal* (March–April):18.

Squires, C. 1994. EMF containment or avoidance. *emf-emi control Journal* (July–Aug.):27.

Squires, C. 1995. The many facets of shielding. *emf-emi control Journal* (March–April):11.

Squires, C. 1995. The ABCs of shielding. *emf-emi control Journal* (March–April):28.

Squires, C. 1995. Predicting and controlling PCB radiation. *emf-emi control Journal* (Sep.–Oct.):17.

Stephens, E. 1994. Cable shield outside the telco plant. *emf-emi control Journal* (Jan.–Feb.):18.

Vitale, L. 1995. Magnetic shielding of offices and apartments. *emf-emi control Journal* (March–April):12.

White, D. 1994. Building EM advances from shielding to power conditioning. *emf-emi control Journal* (Jan.–Feb.):17.

White, D. 1994. RF vs. low frequency magnetic shielding. *emf-emi control Journal* (Jan.–Feb.).

White, D.1994. Shielding to RF/microwave radiation exposure. *emf-emi control Journal*.

White, D. 1994. What to expect from a magnetic shielded room. *emf-emi control Journal*.

White, D. 1995. How much cable shielding is required. *emf-emi control Journal* (March–April):16.

White, W. Fire lookout tower personnel exposed to RF radiation. *emf-emi control Journal*.

Further Information

The Institute of Electrical and Electronics Engineers (IEEE) conducts an annual symposium on EMC. Copies of their Convention Record are available from IEEE, Service Center, 445 Hoes Lane, Piscataway, NJ 08855.

11.5 Heat Management

Zbigniew J. Staszak

11.5.1 Introduction

Thermal *cooling/heat* **management** is one of four major functions provided and maintained by a packaging structure or system, with the other three being: (1) mechanical support, (2) electrical interconnection for power and signal distribution, and (3) environmental protection, all aimed at effective transfer of the semiconductor chip performance to the system. Heat management can involve a significant portion of the total packaging design effort and should result in providing efficient heat transfer paths at all packaging levels, that is, level 1 packaging (chip(s), carrier—both single-chip and multichip modules), level 2 (the module substrate—card), level 3 (board, board-to-board interconnect structures), and level 4 (box, rack, or cabinet housing the complete system), to an ultimate **heat sink** outside, while maintaining internal device temperature at acceptable levels in order to control thermal effects on circuits and system performance.

The evolution of chip technology and of packaging technology is strongly interdependent. Very large-scale integrated (VLSI) packaging and interconnect technology are driven primarily by improvements in chip and module technologies, yield, and reliability.

Packaging performance limits are thus being pushed for high-speed (subnanosecond edge speeds), high-power ($\sim$50–100 W/cm^2) chips with reliable operation and low cost. However, there is a tradeoff between power and delay; the highest speed demands that gates be operated at high power. Though packaging solutions are dominated by applications in digital processing for computer and aerospace (avionics) applications, analog and mixed analog/digital, power conversion, and microwave applications cannot be overlooked. The net result of current chip and substrate characteristics is that leading-edge electrical and thermal requirements that have to be incorporated in packaging can be met with some difficulty and only with carefully derived *electrical design rules* (Davidson and Katopis, 1989) and complex thermal management techniques (ISHM, 1984).

The drive toward high functional density, extremely high speed, and highly reliable operation is constrained by available **thermal design** and *heat removal techniques*. A clear understanding of thermal control technologies, as well as the thermal potential and limits of advanced concepts, is critical. Consequently, the choice of thermal control technology and the particular decisions made in the course of evolving the thermal packaging design often have far-reaching effects on both the reliability and cost of the electronic system or assembly (Nakayama, 1986; Antonetti, Oktay, and Simons, 1989).

Temperature changes encountered during device fabrication, storage, and normal operation are often large enough to place limits on device characterization and lifetime. Temperature related effects, inherent in device physics, lead to performance limitations, for example, propagation delays, degradation of noise margins, decrease in electrical isolation provided by reverse-biased junctions in an IC, and other effects. **Thermally** enhanced **failures**, such as oxide wearout, fracturing, package delamination, wire bond breakage, deformation of metallization on the chip, and cracks and voids in the chip, substrate, die bond, and solder joints lead to reliability limitations (Jeannotte, Goldmann, and Howard, 1989).

The dependence of the **failure rate** on temperature is a complex relationship involving, among other things material properties and environmental conditions, and in a general form can be described by the Arrhenius equation depicted in Fig. 11.42.

FIGURE 11.42 Normalized failure rate $\lambda(T)/\lambda(T_R)$ as a function of temperature showing an exponential dependence of failure rate with temperature; for example, for $E_A = 0.5$ eV an increase of 10°C in temperature almost doubles, whereas an increase of 20°C more than triples the failure rate at about room temperature.

This relation is well suited for device-related functional failures; however, for thermally induced structural failures it has to be applied with care since these failures depend both on localized temperature as well as the history of fabrication and assembly, thermal cycling for test purposes (thermal shock testing), and powering up during operation (operational thermal stress), thus being more complicated in nature.

$$\lambda(T) = \lambda(T_R) \exp\left[\frac{E_A}{k}\left(\frac{1}{T_R} - \frac{1}{T}\right)\right]$$

where

$\lambda(T)$ = failure rate (FIT) at temperature T, K

$\lambda(T_R)$ = failure rate (FIT) at temperature T_R, K

E_A = activation energy, eV, typical values for integrated
circuits are 0.3–1.6 eV

k = Boltzman's constant, 8.616×10^{-5} eV/K

In the definition of $\lambda(T)$, FIT is defined as one failure per billion device hours (10^{-9}/h) or 0.0001%/1000 h. Improvements in thermal parameters and in removable heat densities require enhancements in chip technology, package design, packaging materials compositions, assembly procedures, and cooling techniques (systems). These must be addressed early in the design process to do the following:

- Reduce the rise in temperature and control the variation in the device operating temperature across all the devices, components, and packaging levels in the system.
- Reduce deformations and stresses produced by temperature changes during the fabrication process and normal operation.
- Reduce variations in electrical characteristics caused by thermal stresses.

11.5.2 Heat Transfer Fundamentals

In the analysis of construction of VLSI-based chips and packaging structures, all modes of heat transfer must be taken into consideration, with natural and forced air/liquid convection playing the main role in the cooling process of such systems. The temperature distribution problem may be calculated by applying the (energy) conservation law and equations describing heat conduction, convection, and radiation (and, if required, phase change). Initial conditions comprise the initial temperature or its distribution, whereas boundary conditions include adiabatic (no exchange with the surrounding medium, i.e., surface isolated, no heat flow across it), isothermal (constant temperature), or/and miscellaneous (i.e., exchange with external bodies, adjacent layers, or surrounding medium). Material physical parameters, and thermal conductivity, **specific heat**, thermal coefficient of expansion, and **heat transfer coefficients**, can be functions of temperature.

Basic Heat Flow Relations, Data for Heat Transfer Modes

Thermal transport in a solid (or in a stagnant fluid: gas or liquid) occurs by *conduction*, and is described in terms of the *Fourier equation* here expressed in a differential form as

$$q = -k\nabla T(x, y, z)$$

where

q = heat flux (power density) at any point x, y, z, W/m^2

k = thermal conductivity of the material of conducting medium (W/m-degree), here assumed
to be independent of x, y, z (although it may be a function of temperature)

T = temperature, °C, K

In the one-dimensional case, and for the transfer area A(m) of heat flow path length L (m) and thermal conductivity k not varying over the heat path, the temperature difference ΔT(°C, K) resulting from the conduction of heat $Q(\text{W})$, normal to the transfer area, can be expressed in terms of a *conduction* **thermal resistance** θ (degree/W). This is done by analogy to electrical current flow in a conductor, where heat flow $Q(\text{W})$ is analogous to electric current $I(\text{A})$, and temperature $T(^\circ\text{C}, \text{K})$ to voltage $V(\text{V})$, thus making thermal resistance θ analogous to electrical resistance $R(\Omega)$ and thermal conductivity k (W/m-degree) analogous to electrical conductivity, $\sigma(1/\Omega - \text{m})$:

$$\theta = \frac{\Delta T}{Q} = \frac{L}{kA}$$

Expanding for multilayer (n layer) composite and rectilinear structure

$$\theta = \sum_{i=1}^{n} \frac{\Delta l_i}{k_i A_i}$$

where
$\quad \Delta l_i$ = thickness of the ith layer, m
$\quad k_i$ = thermal conductivity of the material of the ith layer, W/m degree
$\quad A_i$ = cross-sectional area for heat flux of the ith layer, m^2

In semiconductor packages, however, the heat flow is not constrained to be one dimensional because it also spreads laterally. A commonly used estimate is to assume a 45° heat spreading area model, treating the flow as one dimensional but using an effective area A_{eff} that is the arithmetic mean of the areas at the top and bottom of each of the individual layers Δl_i of the flow path. Assuming the heat generating source to be square, and noting that with each successive layer A_{eff} is increased with respect to the cross-sectional area A_i for heat flow at the top of each layer, the thermal (spreading) resistance θ_{sp} is expressed as follows:

$$\theta_{\text{sp}} = \frac{\Delta l_i}{k A_{\text{eff}}} = \frac{\Delta l_i}{k A_i \left(1 + \frac{2\Delta l_i}{\sqrt{A_i}}\right)}$$

On the other hand, if the heat generating region can be considered much smaller than the solid to which heat is spreading, then the semi-infinite heat sink case approach can be employed. If the **heat flux** is applied through a region of radius R, then either $\theta_{\text{sp}} = 1/\pi kR$ for uniform heat flux and the maximum temperature occurring at the center of the region, or $\theta_{\text{sp}} = 1/4kR$ for uniform temperature over the region of the heat source (Carstaw and Jaeger, 1967).

The preceding relations describe only static heat flow. In some applications, however, for example, switching, it is necessary to take into account transient effects. When heat flows into a material volume $V(\text{m}^3)$ causing a temperature rise, thermal energy is stored there, and if the heat flow is finite, the time required to effect the temperature change is also finite, which is analogous to an electrical circuit having a capacitance that must be charged in order for a voltage to occur. Thus the required power/heat flow Q to cause the temperature ΔT in time Δt is given as follows:

$$Q = \rho c_p V \frac{\Delta T}{\Delta t} = C_\theta \frac{\Delta T}{\Delta t}$$

where
$\quad C_\theta$ = thermal capacitance, W-s/degree
$\quad \rho$ = density of the medium, g/m^3
$\quad c_p$ = specific heat of the medium, W-s/g-degree

Again, we can make use of electrical analogy, noting that thermal capacitance C_θ is analogous to electrical capacitance $C(\text{F})$.

A rigorous treatment of multidimensional heat flow leads to a time-dependent heat flow equation in a conducting medium, which in Cartesian coordinates, and for $Q_V(\text{W/m}^3)$ being the internal heat

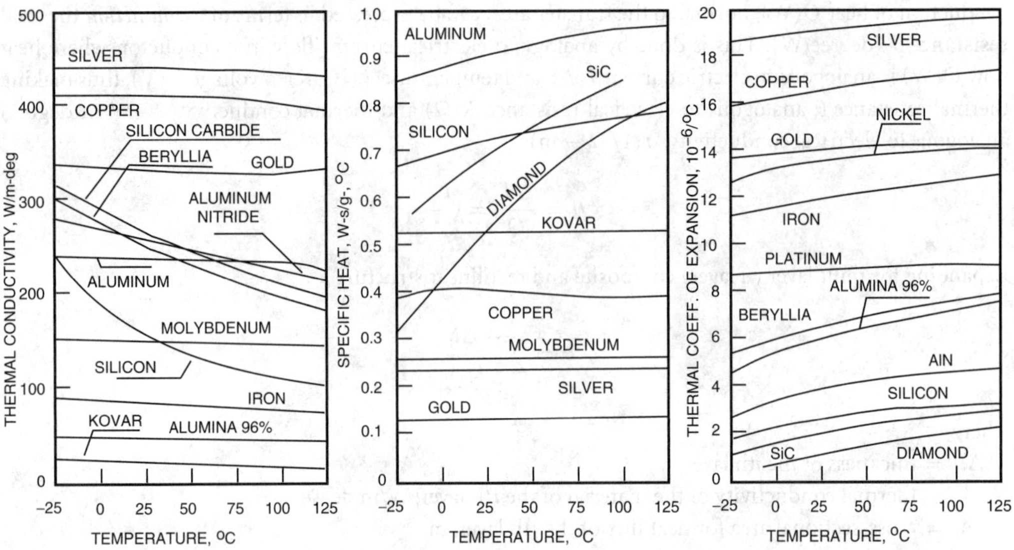

FIGURE 11.43 Temperature dependence of thermal conductivity k, specific heat c_p, and coefficient of thermal expansion (CTE) β of selected packaging materials.

source/generation, is expressed in the form of

$$k\nabla^2 T(x,y,z,t) = -Q_V(x,y,z,t) + \rho c_p \frac{\partial T(x,y,z,t)}{\partial t}$$

An excellent treatment of analytical solutions of heat transfer problems has been given by Carslaw and Jaeger (1967). Although analytical methods provide results for relatively simple geometries and idealized boundary/initial conditions, some of them are useful (Newell, 1975; Kennedy, 1960). However, thermal analysis of complex geometries requires multidimensional numerical computer modeling limited only by the capabilities of computers and realistic CPU times. In these solutions, the designer is normally interested in the behavior of device/circuit/package over a wide range of operating conditions including temperature dependence of material parameters, finite dimensions and geometric complexity of individual layers, nonuniformity of thermal flux generated within the active regions, and related factors.

Figure 11.43 displays temperature dependence of thermal material parameters of selected packaging materials, whereas Table 11.8 summarizes values of parameters of insulator, conductor, and semiconductor materials, as well as gases and liquids, needed for thermal calculations, all given at room temperature. Note the inclusion of the thermal coefficient of expansion $\beta(^\circ C^{-1}, K^{-1})$, which shows the expansion and contraction ΔL of an unrestrained material of the original length L_o while heated and cooled according to the following equation:

$$\Delta L = \beta L_o (\Delta T)$$

Convection heat flow, which involves heat transfer between a moving fluid and a surface, and *radiation heat flow*, where energy is transferred by electromagnetic waves from a surface at a finite temperature, with or without the presence of an intervening medium, can be accounted for in terms of heat transfer coefficients $h(W/m^2\text{-degree})$. The values of the heat transfer coefficients depend on the local transport phenomena occurring on or near the package/structure surface. Only for simple geometric configurations can these values be analytically obtained. Little generalized heat transfer data is available for VLSI type conditions, making it imperative to create the ability to translate real-life designs to idealized conditions (e.g., through correlation studies). Extensive use of empirical relations in determining heat transfer correlations is made through the use of dimensional analysis in which useful design correlations relate the transfer coefficients to geometrical/flow conditions (Furkay, 1984).

TABLE 11.8 Selected Physical and Thermal Parameters of Some of the Materials Used in VLSI Packaging Applications (at Room Temperature, $T = 27°C$, 300 K)[a]

Material Type	Density, ρ, g/cm^3	Thermal Conductivity, k, W/m-°C	Specific Heat, c_p, W-s/g-°C	Thermal Coeff. of Expansion, β, 10^{-6}/°C
Insulator Materials:				
Aluminum nitride	3.25	100–270	0.8	4
Alumina 96%	3.7	30	0.85	6
Beryllia	2.86	260–300	1.02–1.12	6.5
Diamond (IIa)	3.5	2000	0.52	1
Glass-ceramics	2.5	5	0.75	4–8
Quartz (fused)	2.2	1.46	0.67–0.74	0.54
Silicon carbide	3.2	90–260	0.69–0.71	2.2
Conductor Materials:				
Aluminum	2.7	230	0.91	23
Beryllium	1.85	180	1.825	12
Copper	8.93	397	0.39	16.5
Gold	19.4	317	0.13	14.2
Iron	7.86	74	0.45	11.8
Kovar	7.7	17.3	0.52	5.2
Molybdenum	10.2	146	0.25	5.2
Nickel	8.9	88	0.45	13.3
Platinum	21.45	71.4	0.134	9
Silver	10.5	428	0.234	18.9
Semiconductor Materials (lightly doped):				
GaAs	5.32	50	0.322	5.9
Silicon	2.33	150	0.714	2.6
Gases:				
Air	0.00122	0.0255	1.004	3.4×10^3
Nitrogen	0.00125	0.025	1.04	10^2
Oxygen	0.00143	0.026	0.912	10^2
Liquids:				
FC-72	1.68	0.058	1.045	1600
Freon	1.53	0.073	0.97	2700
Water	0.996	0.613	4.18	270

[a]Approximate values, depending on exact composition of the material. (*Source:* Compiled based in part on: Touloukian, Y.S. and Ho, C.Y. 1979. *Master Index to Materials and Properties.* Plenum Publishing, New York.)

For *convection*, both free-air and forced-air (gas) or -liquid convection have to be considered, and both flow regimes must be treated: **laminar flow** and **turbulent flow**. The detailed nature of convection flow is heavily dependent on the geometry of the thermal duct, or whatever confines the fluid flow, and it is nonlinear. What is sought here are crude estimates, however, just barely acceptable for determining whether a problem exists in a given packaging situation, using as a model the relation of *Newton's law of cooling* for convection heat flow Q_c (W)

$$Q_c = h_c A_S (T_S - T_A)$$

where

h_c = average convective heat transfer coefficient, W/m^2-degree
A_S = cross-sectional area for heat flow through the surface, m^2
T_S = temperature of the surface, °C, K
T_A = ambient/fluid temperature, °C, K

For *forced convection* cooling applications, the designer can relate the temperature rise in the coolant temperature $\Delta T_{\text{coolant}}$ (°C, K), within an enclosure/heat exchanger containing subsystem(s) that obstruct the fluid flow, to a volumetric flow rate $G(\text{m}^3/\text{s})$ or fluid velocity $v(\text{m/s})$ as

$$\Delta T_{\text{coolant}} = T_{\text{coolant-out}} - T_{\text{coolant-in}} = \frac{Q_{\text{flow}}}{\rho G c_p} = \frac{Q_{\text{flow}}}{\rho v A c_p} = \frac{Q_{\text{flow}}}{\dot{m} c_p}$$

where

$T_{\text{coolant-out/in}}$ = the outlet/inlet coolant temperatures, respectively, °C, K

$\quad\quad Q_{\text{flow}}$ = total heat flow/dissipation of all components within the enclosure upstream of the component of interest, W

$\quad\quad\quad \dot{m}$ = mass flow rate of the fluid, g/s

Assuming a fixed temperature difference, the **convective heat transfer** may be increased either by obtaining a greater heat transfer coefficient h_c or by increasing the surface area. The heat transfer coefficient may be increased by increasing the fluid velocity, changing the coolant fluid, or utilizing *nucleate boiling*, a form of immersion cooling.

For nucleate boiling, which is a liquid-to-vapor phase change at a heated surface, increased heat transfer rates are the results of the formation and subsequent collapsing of bubbles in the coolant adjacent to the heated surface. The bulk of the coolant is maintained below the boiling temperature of the coolant, while the heated surface remains slightly above the boiling temperature. The **boiling heat transfer** rate Q_b can be approximated by a relation of the following form:

$$Q_b = C_{\text{sf}} A_S (T_S - T_{\text{sat}})^n = h_b A_S (T_S - T_{\text{sat}})$$

where

C_{sf} = constant, a function of the surface/fluid combination, W/m²-K^n (Rohsenow and Harnett, 1973)

T_{sat} = temperature of the boiling point (saturation) of the liquid, °C, K

$\quad n$ = coefficient, usual value of 3

$\quad h_b$ = boiling heat transfer coefficient, $C_{\text{sf}}(T_S - T_{\text{sat}})^{n-1}$, W/m²-degree

Increased heat transfer of surface area in contact with the coolant is accomplished by the use of *extended surfaces*, plates or pin fins, giving the heat transfer rate Q_f by a fin or fin structure as

$$Q_f = h_c A \eta (T_b - T_f)$$

where

A = full wetted area of the extended surfaces, m²

η = fin efficiency

T_b = temperature of the fin base, °C, K

T_f = temperature of the fluid coolant, °C, K

Fin efficiency η ranges from 0 to 1; for a straight fin $\eta = \tanh mL/mL$, where $m = \sqrt{2h_c/k\delta}$, $L = $ the fin length (m), and $\delta = $ the fin thickness (m) (Kern and Kraus, 1972).

Formulas for heat convection coefficients h_c can be found from available empirical correlation and/or theoretical relations and are expressed in terms of dimensional analysis with the dimensionless parameters: Nusselt number Nu, Rayleigh number Ra, Grashof number Gr, Prandtl number Pr, and Reynolds number Re, which are defined as follows:

$$Nu = \frac{h_c L_{\text{ch}}}{k}, \quad\quad Ra = Gr\,Pr, \quad\quad Gr = \frac{g\beta\rho^2}{\mu^2}L_{\text{ch}}^3\Delta T, \quad\quad Pr = \frac{\mu c_p}{k}, \quad\quad Re = v L_{\text{ch}} \frac{\rho}{\mu}$$

where

L_{ch} = characteristic length parameter, m
g = gravitational constant, 9.81 m/s^2
μ = fluid dynamic viscosity, g/m-s
$\Delta T = T_S - T_A$, degree

Examples of such expressions for selected cases used in VLSI packaging conditions are presented next. Convection heat transfer coefficients h_c averaged over the plate characteristic length are written in terms of correlations of an average value of Nu vs. Ra, Re, and Pr.

1. For natural (air) convection over external flat horizontal and vertical platelike surfaces,

$$Nu = C(Ra)^n$$

$$h_c = (k/L_{ch})Nu = (k/L_{ch})C(Ra)^n = C'(L_{ch})^{3n-1}\Delta T^n$$

where

C, n = constants depending on the surface orientation (and geometry in general), and the value of the Rayleigh number, see Table 11.9.
$C' = kC[(g\beta\rho^2/\mu^2)Pr]^n$

Most standard applications in electronic equipment, including packaging structures, appear to fall within the laminar flow region (see Fig. 11.44). Ellison (1987) has found that the preceding expressions for natural air convection cooling are satisfactory for cabinet surfaces; however, they significantly underpredict the heat transfer from small surfaces. By curve fitting the empirical data under laminar conditions, the following formula for natural convection to air for small devices encountered in the electronics industry was found:

$$h_c = 0.83 f(\Delta T/L_{ch})^n (\text{W/m}^2\text{-degree})$$

where $f = 1.22$ and $n = 0.35$ for vertical plate, $f = 1.00$ and $n = 0.33$ for horizontal plate facing upward, that is, upper surface $T_S > T_A$, or lower surface $T_S < T_A$, and $f = 0.50$ and $n = 0.33$ for horizontal plate facing downward, that is, lower surface $T_S > T_A$, or upper surface $T_S < T_A$.

TABLE 11.9 Constants for Average Nusselt Numbers for Natural Convection[a] and Simplified Equations for Average Heat Transfer Coefficients h_c (W/m^2-degree) for Natural Convection to Air over External Flat Surfaces (at Atmospheric Pressure)[b]

Configuration	L_{ch}	Flow Regime	C	n	h_c	h_c for Natural Convection Air Cooling $C'(27°C)$	$C'(75°C)$
Vertical plate	H						
		$10^4 < Ra < 10^9$ laminar	0.59	0.25	$C'(\Delta T/L_{ch})^{0.25}$	1.51	1.45
		$10^9 < Ra < 10^{12}$ turbulent	0.13	0.33	$C'(\Delta T)^{0.33}$	1.44	1.31
Horizontal plate (heated side up)	$WL/[2(W+L)]$						
		$10^4 < Ra < 10^7$ laminar	0.54	0.25	$C'(\Delta T/L_{ch})^{0.25}$	1.38	1.32
		$10^7 < Ra < 10^{10}$ turbulent	0.15	0.33	$C'(\Delta T)^{0.33}$	1.66	1.52
(heated side down)							
		$10^5 < Ra < 10^{10}$	0.27	0.25	$C'(\Delta T/L_{ch})^{0.25}$	0.69	0.66

where H, L, and W are height, length, and width of the plate, respectively

[a]Compilation based on two sources. (*Sources*: McAdams, W.H. 1954. *Heat Transmission*. McGraw-Hill, New York, and Kraus, A.D. and Bar-Cohen, A. 1983. *Thermal Analysis and Control of Electronic Equipment*. Hemisphere, New York.)

[b]Physical properties of air and their temperature dependence, with units converted to metric system, are depicted in Fig. 11.44. (*Source*: Keith, F. 1973. *Heat Transfer*. Harper & Row, New York.)

FIGURE 11.44 Temperature dependence of physical properties of air: density ρ, thermal conductivity k, dynamic viscosity μ, and CTE β. Note that specific heat c_p is almost constant for given temperatures, and as such is not displayed here.

2. For forced (air) convection (cooling) over external flat plates

$$Nu = C(Re)^m(Pr)^n$$

$$h_c = (k/L_{ch})C(Re)^m(Pr)^n = C''(L_{ch})^{m-1}v^m$$

where

C, m, n = constants depending on the geometry and the Reynolds number, see Table 11.10
$\quad C'' = kC(\rho/\mu)^m(Pr)^n$
$\quad L_{ch}$ = length of plate in the direction of fluid flow (characteristic length)

Experimental verification of the these relations, as applied to the complexity of geometry encountered in electronic equipment and packaging systems, can be found in literature.

Ellison (1987) has determined that for *laminar forced air flow*, better agreement between experimental and calculated results is obtained if a correlation factor f, which depends predominantly on air velocity, is used, in which case the heat convection coefficient h_c becomes

$$h_c = f(k/L_{ch})Nu$$

Though the correlation factor f was determined experimentally for laminar flow over 2.54×2.54 cm ceramic substrates producing the following values: $f = 1.46$ for $v = 1$ m/s, $f = 1.56 - v = 2$ m/s, $f = 1.6 - v = 2.5$ m/s, $f = 1.7 - v = 5$ m/s, $f = 1.78 - v = 6$ m/s, $f = 1.9 - v = 8$ m/s, $f = 2.0 - v = 10$ m/s, and we expect the correlation factor to be somewhat different for other materials and other plate sizes, the quoted values are useful for purposes of estimation.

TABLE 11.10 Constants for Average Nusselt Numbers for Forced Convection[a], and Simplified Equations for Average Heat Transfer Coefficients h_c (W/m²-degree) for Forced Convection Air Cooling over External Flat Surfaces and at Most Practical Fluid Speeds (at Atmospheric Pressure)[b]

Flow Regime		C	m	n	h_c	h_c for Forced Convection Air Cooling	
						$C''(27°C)$	$C''(75°C)$
$Re < 2 \times 10^5$	laminar	0.664	0.5	0.333	$C''(v/L_{ch})^{0.5}$	3.87	3.86
$Re > 3 \times 10^5$	turbulent	0.036	0.8	0.333	$C''v^{0.8}/(L_{ch})^{0.2}$	5.77	5.34

[a](*Compilation based on four sources:* Keith, F. 1973. *Heat Transfer.* Harper & Row, New York. Rohsenow, W.M. and Choi, H. 1961. *Heat, Mass, and Momentum Transfer.* Prentice–Hall, Englewood Cliffs, NJ. Kraus, A.D. and Bar-Cohen, A. 1983. *Thermal Analysis and Control of Electronic Equipment.* Hemisphere, New York. Moffat, R.J. and Ortega, A. 1988. Direct air cooling in electronic systems. In *Advances in Thermal Modeling of Electronic Components and Systems,* ed. A. Bar-Cohen and A.D. Kraus, Vol. 1, pp. 129–282. Hemisphere, New York.)

[b]Physical properties of air and their temperature dependence, with units converted to metric system, are depicted in Fig. 11.44 (*Source:* Keith, F. 1973. *Heat Transfer.* Harper & Row, New York.)

Buller and Kilburn (1981) performed experiments determining the heat transfer coefficients for laminar flow forced air cooling for integrated circuit packages mounted on printed wiring boards (thus for conditions differing from that of a flat plate), and correlated h_c with the air speed through use of the Colburn J factor, a dimensionless number, in the form of

$$h_c = J\rho c_p v (Pr)^{-2/3} = 0.387(k/L_{ch})Re^{0.54}Pr^{0.333}$$

where

$J = 0.387^*(Re)^{-0.46}$

L_{ch} = redefined characteristic length, allows to account for the three-dimensional nature of the package, $[(A_F/C_F)(A_T/L)]^{0.5}$

W, H = width and height of the frontal area, respectively

$A_F = W, H$, frontal area normal to air flow

$C_F = 2(W + H)$, frontal circumference

$A_T = 2H(W + L) + (W + L)$, total wetted surface area exposed to flow

L = length in the direction of flow.

Finally, Hannemann, Fox, and Mahalingham (1991) experimentally determined average heat transfer coefficient h_c for finned heat sink of the fin length L(m), for forced convection in air under laminar conditions at moderate temperatures and presented it in the form of

$$h_c = 4.37(v/L)^{0.5}(\text{W/m}^2\text{-degree})$$

Radiation heat flow between two surfaces or between a surface and its surroundings is governed by the *Stefan–Boltzman equation* providing the (nonlinear) **radiation heat transfer** in the form of

$$Q_r = \sigma A F_T \left(T_1^4 - T_2^4\right) = h_r A(T_1 - T_2)$$

where

Q_r = radiation heat flow, W

σ = Stefan-Boltzman constant, 5.67×10^{-8} W/m^2-K^4

A = effective area of the emitting surface, m^2

F_T = exchange radiation factor describing the effect of geometry and surface properties

T_1 = absolute temperature of the external/emitting surface, K

T_2 = absolute temperature of the ambient/target, K

h_r = average **radiative heat transfer** coefficient, $\sigma F_T(T_1^4 - T_2^4)/(T_1 - T_2)$, W/m^2-degree

For two-surface radiation exchange between plates, F_T is given by

$$F_T = \cfrac{1}{\cfrac{1 - \varepsilon_1}{\varepsilon_1} + \cfrac{1 - \varepsilon_2}{\varepsilon_2}\cfrac{A_1}{A_2} + \cfrac{1}{F_{1-2}}}$$

where

ε_1 = emissivity of material 1

ε_2 = emissivity of material 2

A_1 = radiation area of material 1

A_2 = radiation area of material 2

F_{1-2} = geometric view factor (Ellison, 1987; Siegal and Howell, 1981)

The emissivities of common packaging materials are given in Table 11.11. Several approximations are useful in dealing with radiation:

- For a surface that is smaller compared to a surface by which it is totally enclosed (e.g., by a room or cabinet), $\varepsilon_2 = 1$ (no surface reflection) and $F_{1-2} = 1$, then $F_T = \varepsilon_1$.

TABLE 11.11 Emissivities of Some of Materials Used in Electronic Packaging

Aluminum, polished	0.039–0.057
Copper, polished	0.023–0.052
Stainless steel, polished	0.074
Steel, oxidized	0.80
Iron, polished oxidized	0.14–0.38
	0.31–0.61
Porcelain, glazed	0.92
Paint, flat black lacquer	0.96–0.98
Quartz, rough, fused	0.93–0.075

- For most packaging applications in a human environment, $T_1 \approx T_2$, then $h_r = 4\sigma F_T T_1^3$.
- For space applications, where the target is space with T_2 approaching 0 K, $Q_r = \sigma F_T T_1^4$.

For purposes of physical reasoning, convection and radiation heat flow can be viewed as being represented by (nonlinear) thermal resistances, *convective θ_c* and *radiational θ_r*, respectively,

$$\theta_{c,r} = \frac{1}{h A_S}$$

where h is either the convective or radiative heat transfer coefficients, h_c or h_r, respectively, the total heat transfer, both convective and radiative, $h = h_c + h_r$. Note that for nucleate boiling ηh_c should be used.

11.5.3 Study of Thermal Effects in Packaging Systems

The *thermal evaluation* of solid-state devices and integrated circuits (ICs), and VLSI-based packaging takes two forms: *theoretical analysis* and *experimental characterization*. Theoretical analysis utilizes various approaches: from simple to complex closed-form analytical solutions and numerical analysis techniques, or a combination of both. Experimental characterization of the device/chip junction/surface temperature(s) of packaged/unpackaged structures takes both direct, infrared microradiometry, or liquid crystals and thermographic phosphorous, or, to a lesser extent, thermocouples, and indirect (parametric) electrical measurements.

Thermal Resistance

A *figure-of-merit* of thermal performance, thermal resistance is a measure of the ability of its mechanical structure (material, package, and external connections) to provide heat removal from the active region and is used to calculate the device temperature for a given mounting arrangement and operating conditions to ensure that a maximum safe temperature is not exceeded [Baxter 1977]. It is defined as

$$\theta_{JR} = \frac{T_J - T_R}{P_D}$$

where

$\quad \theta_{JR}$ = thermal resistance between the junction or any other circuit element/active region that generates heat and the reference point, °C/W, K/W

$\quad T_J, T_R$ = temperature of the junction and the reference point, respectively, °C

$\quad P_D$ = power dissipation in the device, W

The conditions under which the device is thermally characterized have to be clearly described. In fact, thermal resistance is made up of constant terms that are related to device and package materials and geometry in series with a number of variable terms. These terms are related to the heat paths from the package boundary to some point in the system that serves as the reference temperature and are determined by the method of mounting, the printed-wiring-board (PWB) if used, other heat generating components on the board or in the vicinity, air flow patterns, and related considerations. For discrete devices, thermal parameters—**junction temperature**, thermal resistance, *thermal time constants*—are well defined since the region of heat dissipation and measurement is well defined. For integrated circuits, however, heat is generated at multiple points (resistors, diodes, or transistors) at or near the chip surface resulting in a nonuniform temperature distribution. Thus θ_{JR} can be misleading unless temperature uniformity is assured or thermal resistance is defined with respect to the hottest point on the surface of the chip. A similar situation exists with a multichip package, where separate thermal resistances should be defined for each die. As a result, the data on spatial temperature distribution are necessary, as well as the need to standardize the definition of a suitable reference temperature and the configuration of the test arrangement, including the means of mounting the device(s) under test.

In thermal evaluation of microelectronic packages/modules, the *overall (junction-to-coolant)* thermal resistance θ_{tot}, including forced convection and characterizing the requirements to cool a chip within a single- or multichip module, can be described in terms of the following:

- **Internal thermal resistance** θ_{int}, largely bulk conduction from the circuits to the case/package surface, thus also containing thermal spreading and contact resistances

$$\theta_{int} = \frac{T_{chip} - T_{case/pckg}}{P_{chip}}$$

Thermal contact resistance R_c, occurring at material interface, is dependent on a number of factors including the surface roughness, the properties of the interfacing materials, and the normal stress at the interface. The formula(s) relating the *interfacial normal pressure* P to R_c have been given (Yovanovich and Antonetti, 1988) as

$$R_c = 1/C[P/H]^{0.95}$$

where C is a constant related to the surface texture, and the conductivities of the adjoining materials, k_1 and k_2, and H is the microhardness of the softer surface.

- External thermal resistance θ_{ext}, primarily convective, from the case/package surface to the coolant fluid (gas or liquid)

$$\theta_{ext} = \frac{T_{case/pckg} - T_{coolant-out}}{P_m}$$

- Flow (thermal) resistance θ_{flow}, referring to the transfer of heat from the fluid coolant to the ultimate heat sink, thus associated with the heating of the fluid as it absorbs energy passing through the heat exchanger

$$\theta_{flow} = \frac{T_{coolant-out} - T_{coolant-in}}{Q_{flow}}$$

thus expressing the device junction/component temperature as follows:

$$T_j = \Delta T_{j\text{-}chip} + P_{chip}\theta_{int} + P_m\theta_{ext} + Q_{flow}\theta_{flow} + T_{coolant-in}$$

where

$\quad T_{chip} =$ the chip temperature

$\quad T_{case/pckg} =$ the case/package temperature (at a defined location)

$\quad P_{chip}, P_m =$ the power dissipated within the chip and the module, respectively

$\quad \Delta T_{j\text{-}chip} =$ the junction to chip temperature rise (can be negligible if power level are low)

Note that in the literature, the total thermal resistance θ_{tot}, usually under natural convection cooling conditions, is also referred to as θ_{j-a} (junction–ambient), θ_{int} as θ_{j-c} (junction–case), and θ_{ext} as θ_{c-a} (case–ambient).

Thermal Modeling/Simulation

The problem of theoretical thermal description of solid-state devices, circuits, packages, and assemblies requires the solution of heat transfer equations with appropriate initial and boundary conditions and is solved with the level of sophistication ranging from approximate analytical through full numerical of increased sophistication and complexity. To fully theoretically describe the thermal state of a solid-state/VLSI device and, in particular, the technology requirements for large-area packaged chips, thus ensuring optimum design and performance of VLSI-based packaging structures, thermal modeling/simulation has to be coupled with thermal/mechanical modeling/simulation leading to analysis of thermal stress and tensions, material fatigue, reliability studies, **thermal mismatch**, etc., and to designs both low thermal resistance and low thermal stress coexist.

Development of a suitable calculational model is basically a trial- and error process. In problems involving complex geometries, it is desirable to limit the calculation to the smallest possible region, by eliminating areas weakly connected to the region of interest, or model only part of the structure if geometric symmetry, symmetrical loading, symmetric power/temperature distribution, isotropic material properties, and other factors can be applied. In addition, it is often possible to simplify the geometry considerably by replacing complex shapes with simpler shapes (e.g., of equivalent volume and resistance to heat flow). All of these reduce the physical structure to a numerically (or/and analytically) tractable model. Also, decoupling of the component level problem from the substrate problem, as the chip and the package can be assumed to respond thermally as separate components, might be a viable concept in evaluation of temperature fields. This is true if thermal time constants differ by orders of magnitude. Then the superposition can be used to obtain a temperature response of the total package.

The accuracy with which any given problem can be solved depends on how well the problem can be modeled, and in the case of *finite-difference* (with electrical equivalents) or *finite-element models*, which are the most common approaches to thermal (and thermal–mechanical) analysis of packaging systems, on the fineness of the spatial subdivisions and control parameters used in the computations. To improve computational efficiency and provide accurate results in local areas of complexity, some type of *grid continuation strategy* should be used. A straightforward use of uniform discretization meshes seems inappropriate, since a mesh sufficiently fine to achieve the desired accuracy in small regions, where the solution is rough, introduces many unnecessary unknowns in regions where the solution is smooth, thus unnecessarily increasing storage and CPU time requirements. A more sophisticated approach involves starting with coarse grids with relatively few unknowns, inexpensive computational costs, and low accuracy, and using locally finer grids on which accurate solutions may be computed. Coarse grids are used to obtain good starting iterates for the finer meshes and can also be used in the creation of the finer meshes by indicating regions where refinement is necessary.

Electrical network finite-difference models for study of various phenomena occurring in solid-state devices, circuits, and systems have been widely reported in literature (Ellison 1987; Fukuoka and Ishizuka, 1984; Riemer, 1990). One of the advantages of such a technique is a simple physical interpretation of the phenomena in question in terms of electrical signals and parameters existing in the network/circuit model (see Fig. 11.45). For all but very simple cases, the equivalent circuits are sufficiently complex that computer solution is required. It is important to note, however, that once the equivalent circuit is established the analysis can readily be accomplished by existing network analysis programs, such as SPICE (SPICE2G User's Manual).

Finite element models (FEMs) have been gaining recognition as tools to carry out analyses of VLSI packages due to the versatility of FEM procedures for the study of a variety of electronic packaging problems, including wafer manufacturing, chip packaging, connectors, board-level simulation, and system-level simulation. Thermal/mechanical design of VLSI packages (Miyake, Suzuki, and Yamamoto, 1985) requires an effective method for the study of specific regions of complexity (e.g., sharp corners, material discontinuity, thin material layers, and voids). As fully refined models are too costly, especially in complex two- and three-dimentional configurations, *local analysis procedures* (Simon et al., 1988) can be used for the problems of this kind. The local analysis approach to finite element analysis allows a very refined mesh to be introduced in an overall coarse FEM so that geometric and/or material complexities can be accurately modeled.

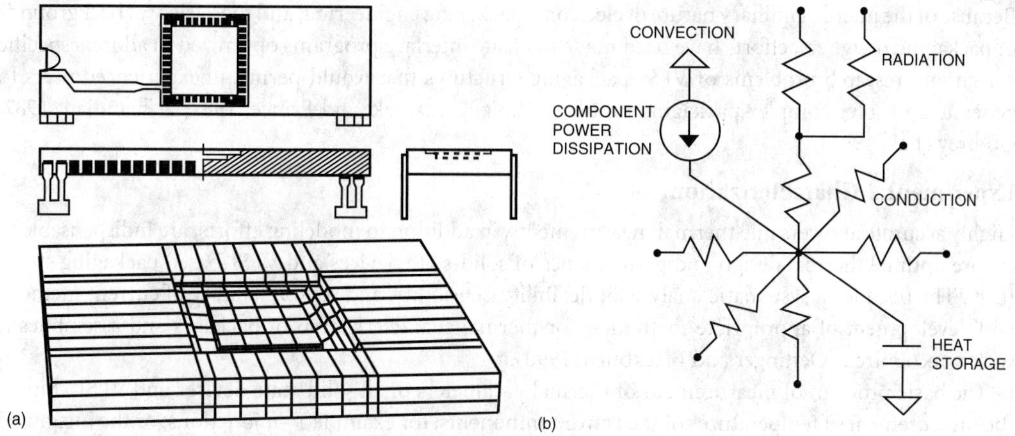

(a)   (b)

FIGURE 11.45 (a) Views of a ceramic package and its general mesh in three-dimensional discretization of this rectangular type packaging structure. (b) a lumped RC thermal model of a unit cell/volume with (nonlinear) thermal resistances representing conduction, convection, and radiation; capacitances account for time-dependent thermal phenomena, and current sources represent heat generation/dissipation properties (thermal current). Nodal points, with voltages (and voltage sources) corresponding to temperatures, can be located either in the center or at corners of unit cells. Each node may be assigned an initial (given) temperature, heat generation/dissipation rate, and modes of heat transport in relation to other nodes. Voltage controlled current sources could be used to replace resistances representing convection and radiation.

This method allows efficient connection of two bordering lines with differing nodal densities in order to form an effective transition from coarse to refined meshes. Linear (or) arbitrary constraints are imposed on the bordering lines in order to connect the lines and reduce the number of independent freedoms in the model. Refined mesh regions can be nested one within another as multilevel refined meshes to achieve greater accuracy in a single FEM, as illustrated in Fig. 11.46.

Several analytical and numerical computer codes varying widely in complexity and flexibility are available for use in the thermal modeling of VLSI packaging. The programs fall into the two following categories:

1. *Thermal characterization* including, analytical programs such as TXYZ (Albers, 1984) and TAMS (Ellison, 1987), as well as numerical analysis programs based on a finite-difference (with electrical equivalents) approach such as TRUMP (Edwards, 1969), TNETFA (Ellison, 1987), SINDA (COSMIC, 1982b).

2. *General nonlinear codes*, such as ANSYS (Swanson) ADINAT (Adina), NASTRAN (COSMIC, 1982a), NISA (COSMIC, 1982c), and GIFTS (Casa) (all based on finite-element approach), to name only a few.

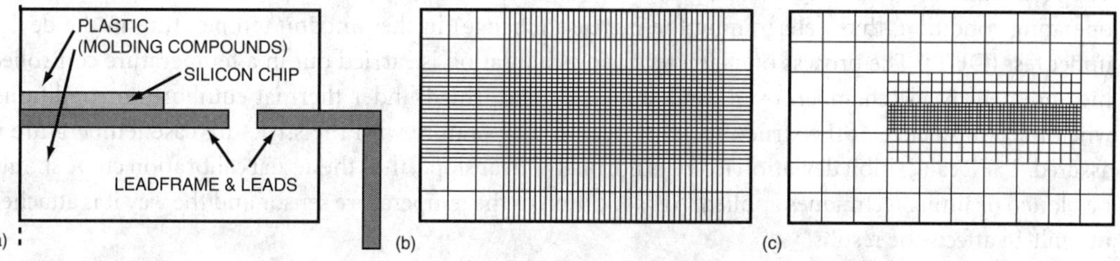

(a)   (b)   (c)

FIGURE 11.46 Example of grid continuation strategy used in discretization of a packaging structure: (a) a two-dimensional cross-section of a plastic package, (b) fine refinement in the area of interest, that is, where gradients of the investigated parameter(s), thermal or thermal-mechanical, are expected to be most significant, (c) a two-dimensional two-level refinement.

Because of the interdisciplinary nature of electronic packaging engineering, and the variety of backgrounds of packaging designers, efforts have been made to create interface programs optimized to address specific design and research problems of VLSI-packaging structures that would permit unexperienced users to generate and solve complex packaging problems (Shiang, Staszak, and Prince, 1987; Nalbandian, 1987; Godfrey et al., 1993).

Experimental Characterization

Highly accurate and versatile thermal measurements, in addition to modeling efforts, are indispensable to ensure optimal thermal design and performance of solid-state devices and VLSI-based packaging structures. The need for a systematic analysis of flexibility, sensibility, and reproducibility of current methods and development of appropriate method(s) for thermal characterization of packages and assemblies is widely recognized (Oettinger and Blackburn, 1990).

The basic problem of measurement of thermal parameters of all solid-state devices and VLSI-chips is the measurement of temperatures of the active components, for example $p-n$ junctions, or the integrated circuit chip surface temperature. Nonelectrical techniques, which can be used to determine the operating temperature of structures, involve the use of infrared microradiometry, liquid crystals, and other tools, and require that the surface of the operating device chip is directly accessible. The electrical techniques for measuring the temperature of semiconductor chips can be performed on fully packaged devices, and use a *temperature sensitive electrical parameter* (TSEP) of the device, which can be characterized/calibrated with respect to temperature and subsequently used as the temperature indicator.

Methodology for experimental evaluation of transient and steady-state thermal characteristics of solid-state and VLSI-based devices and packaging structures is dependent on the device/structure normal operating conditions. In addition, validity of the data obtained has to be unchallengeable; each thermal measurement is a record of a unique set of parameters, so as to assure repeatability, everything must be the same each time, otherwise the data gathered will have little relevance. Also, the choice of thermal environmental control and mounting arrangement, still air, fluid bath, temperature controlled heat sink, or wind tunnel, must be addressed.

In general, the idea of thermal measurements of all devices is to measure the active region temperature rise $\Delta T(t)$ over the reference (case, or **ambient**) **temperature** T_R, caused by the power P_D being dissipated there, for example, during the bias pulse width. Expanding the time scale, steady-state values can be obtained. Employing electrical methods in thermal measurements, the TSEP that is to be used should be consistent with the applied heating method, whereas the measurements have to be repeatable and meaningful. Most commonly used TSEPs are $p-n$ junction forward voltage or emitter-base voltage (bipolar devices), threshold voltage or intrinsic drain source diode voltage (metal oxide semiconductors (MOSs)). There are two phases in the experimental evaluation process:

- *Calibration* of the TSEP (no power or negligibly low power dissipated in the device) characterizing the variation of TSEP with temperature.
- *Actual measurement* (device self-heated).

The TSEP is monitored/recorded during or immediately after the removal of a **power** (heating) **dissipation** operating condition, thus determining the increase (change) in the junction temperature in the device under test (DUT). The process of characterization/calibration is carried out in a temperature controlled measurement head (chamber, oven) and should be performed under thermal equilibrium conditions, when the equivalence of the structure active region temperature with the structure case temperature is assured. TSEP can exhibit deviation from its normally linear slope, thus the actual calibration curve should be plotted or fitting techniques applied. Also, location of the temperature sensor and the way it is attached or built-in affects the results.

Figure 11.47 illustrates the idea of the emitter-only-switching method (Oettinger and Blackburn, 1990), which is a popular and fairly accurate approach for thermal characterization of transistors. Its principle can also be used in thermal measurement of other device/circuits as long as there is an accessible $p-n$ junction that can be used as a TSEP device. This method is intended both for steady-state and transient measurements. The circuit consists of two current sources of different values, heating current i_H and

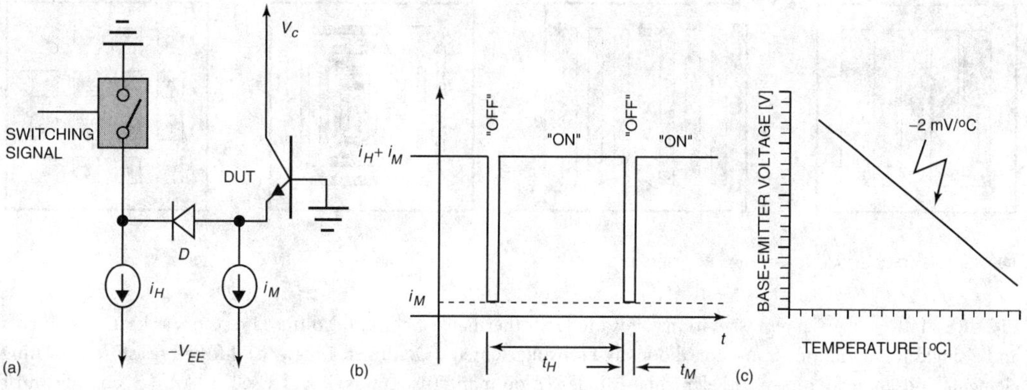

FIGURE 11.47 (a) Schematic of measurement circuit using the emitter-switching-only method for measuring temperature of an npn bipolar transistor, (b) heating and measurement currents waveforms; heating time t_H, measurement time t_M, (c) calibration plot of the TSEP: base-emitter voltage vs. temperature.

measurement current i_M, and a diode D, and a switch. When the switch is open, the current flowing through the device-under-test is equal to $i_H + i_M$ (power dissipation phase). When the switch is closed, the diode is reverse biased, and the current of the DUT is equal to i_M (measurement phase); i_M should be small enough to keep power dissipation in the DUT during this phase negligible.

An approach that has been found effective in experimental evaluation of packaged IC chips employs the use of specially designed test structures that simulate the materials and processing of real chips, but not the logical functions. These look like product chips in fabrication and in assembly and give (electrical) readouts of conditions inside the package (Oettinger, 1984). Any fabrication process is possible that provides high fabrication yield. Such devices may be mounted to a wide variety of test vehicles by many means. In general, they can be used to measure various package parameters—thermal, mechanical, and electrical—and may be thought of as a process monitor since the obtained data reflect the die attach influence or lead frame material influence, or any number of factors and combination of factors.

For thermal measurements, the use of these structures overcomes inherent difficulties in measuring temperatures of active integrated chips. Many IC chips have some type of accessible TSEP element (p–n junction), parasitic, input protection, base-emitter junction, output clamp, or isolation diode (substrate diode). Such TSEP devices, as in discrete devices, have to be switched from the operating (heating) mode to measurement (sensing) mode, resulting in unavoidable delays and temperature drops due to cooling processes. Also, the region of direct electrical access to the TSEP does not always agree with the region of highest temperature.

In thermal test chips, heating and sensing elements might be (or are) separated. This eliminates the necessity of electrical switching, and allows measurements to be taken with the DUT under continuous biasing. Also, heating and sensing devices can be placed on the chip in a prescribed fashion, and access to temperature sensing devices can be provided over all of the chip area resulting in variable heat dissipation and sensed temperature distribution. This provides thermal information from all areas of the chip, including the ones of greatest **thermal gradients**. In addition, arrays of test chips can be built up from a standard chip size. It is desirable to use a die of the same or nearly same size as the production or proposed die for thermal testing and to generate a family of test dies containing the same test configuration on different die sizes. The particular design of the (thermal) test chip depends on the projected application, for example, chip die attachment, or void study, or package characterization. Examples of pictorial representations of thermal test chips are given in Fig. 11.48.

More sophisticated, multisensor test chips contain test structures for the evaluation of thermal resistance, electrical performance, corrosion, and thermomechanical stress effects of packaging technologies (Mathuna, 1992). However, multisensor chips have to be driven by computerized test equipment (Alderman, Tustaniwskyi, and Usell, 1986; Boudreaux et al., 1991) in order to power the chip in a manner that

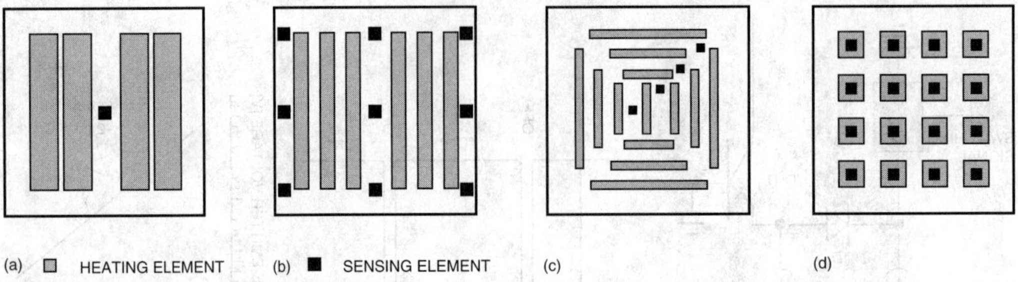

(a) ▢ HEATING ELEMENT   (b) ■ SENSING ELEMENT   (c)   (d)

FIGURE 11.48 Pictorial representations of examples of thermal test chips: (a) diffused resistors as heating elements and a single *p–n* junction (transistor, diode) as a sensing element (Oettinger, 1984), (b) diffused resistors as heating elements and an array of sensing elements—diodes (Alderman, Tustaniwskyi, and Usell, 1986), (c) polysilicon or implant resistor network as heating elements and diodes as sensing elements (Boudreaux et al., 1991), (d) an array of heaters and sensors; each heater/sensor arrangement consists of two (bipolar) transistors having common collector and base terminals, but separate emitters for heating and sensing (Staszak et al., 1987).

permits spatially uniform/nonuniform power dissipation and to sense the test chip temperature at multiple space points, all as a function of time, and to record temperatures associated with separate thermal time constants of various parts of the tested structure, ranging from tens and hundreds of nanoseconds on the transistor level, tens and hundreds of microseconds for the chip itself, tens to hundreds of microseconds for the substrate, and seconds and minutes for the package.

11.5.4   Heat Removal/Cooling Techniques in Design of Packaging Systems

Solutions for adequate cooling schemes and optimum thermal design require an effective combination of expertise in electrical, thermal, mechanical, and materials effect, as well as in industrial and manufacturing engineering. Optimization of packaging is a complex issue: compromises between electrical functions, an optimal heat flow and heat related phenomena, and economical manufacturability have to be made, leading to packaging structures that preserve the performance of semiconductors with minimum package delay to the system. Figure 11.49 shows the operating ranges of a variety of heat removal/cooling techniques.

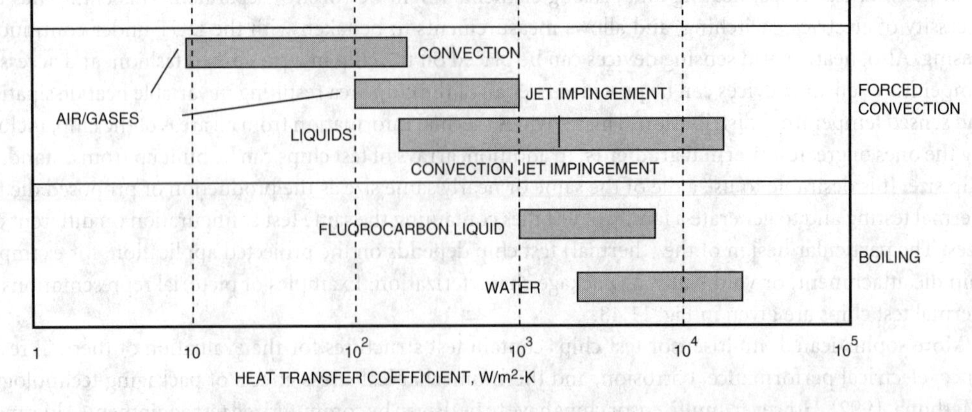

FIGURE 11.49 Heat transfer coefficient for various heat removal/cooling techniques.

The data are presented in terms of heat transfer coefficients that correlate dissipated power or heat flux with a corresponding increase in temperature.

Current single-chip packages, dual in line package (DIP) (cerdip), pin grid array (PGA) (ceramic, TAB, short pin, pad-grid array), and flat packages (fine chip ceramic, quad flat package (QFP)), surface-mounted onto cards or board and enhanced by heat sinks, are capable of dissipating not more than a few watts per square centimeter, thus are cooled by natural and forced-air cooling. Multichip packaging, however, demands power densities beyond the capabilities provided by air cooling. The state-of-the-art of multichip packaging, multilayer ceramic (cofired multilayer thick-film ceramic substrate, glass-ceramics/copper substrate, polyimide-ceramics substrate), thin-film copper/aluminum-polyimide substrates, chip on board with TAB, and silicon substrate (silicon-on-silicon), require that in order to comply with the preferential increase in temperature not exceeding, for example, 50°C, sophisticated cooling techniques are indispensable. Traditional liquid forced cooling that employs a cold plate or heat exchanger, which physically contacts the device to be cooled, is good for power densities below 20 W/cm². For greater power densities, a possible approach is a water jet impingement to generate a fluid interface with a high rate of heat transfer (Mahalingham, 1985). In most severe cases, forced liquid cooling (and boiling) is often the only practical cooling approach: either indirect water cooling (microchannel cooling (Tuckerman and Pease, 1981)) or immersion cooling (Yokouchi, Kamehara, and Niwa, 1987) (in dielectric liquids like fluorocarbon). Cryogenics operation of CMOS devices in liquid nitrogen is a viable option (Clark et al., 1992).

A major factor in the cooling technology is the quality of the substrate, which also affects the electrical performance. All of the substrate materials (e.g., alumina/polyimide, epoxy/kevlar, epoxy-glass, beryllia, alumina, aluminum nitride, glass-ceramics, and diamond or diamondlike films (Tzeng et al., 1991)) have the clear disadvantage of a relative dielectric constant significantly greater than unity. This results in degraded chip-to-chip propagation times and in enhanced line-to-line capacitive coupling. Moreover, since chip integration will continue to require more power per chip and will use large-area chips, the substrate must provide for a superior thermal path for heat removal from the chip by employing high **thermal conductivity** materials as well as assuring thermal expansion compatibility with the chip. Aluminum nitride and silicon carbide ceramic with good thermal conductivity ($k = $ 100–200 W/m-°C and 260 W/m-°C, respectively) and good thermal expansion match (**cofficient of thermal expansion (CTE)** $=$ 4×10^{-6}/°C and 2.6×10^{-6}/°C, respectively) to the silicon substrate (CTE $= 2.6 \times 10^{-6}$/°C), copper–invar–copper ($k = 188$W/m-°C and thermal expansion made to approximate that of silicon), and diamond for substrates and heat spreaders ($k = $ 1000–2000 W/m-°C) are prime examples of the evolving trends. However, in cases where external cooling is provided not through the substrate base but by means of heat sinks bonded on the back side of the chip or package, high conductivity substrates are not required, and materials such as alumina ($k = 30$ W/m-°C and CTE $= 6 \times 10^{-6}$/°C) or glass-ceramics ($k = 5$ W/m−°C and a tailorable CTE $= 4-8 \times 10^{-6}$/°C, matching that required by silicon or GaAs, CTE $= 5.9 \times 10^{-6}$/°C), with low dielectric constants for low propagation delays, are utilized. Designers should also remember thermal enhancement techniques for minimizing the thermal contact resistance (including greases, metallic foils and screens, composite materials, and surface treatment (Fletcher, 1990)).

The evolution of packaging and interconnects over the last decade comprises solutions ranging from air-cooled single-chip packages on a printed wiring board, custom, or semicustom chips and new materials and technologies, that is, from surface mount packages through silicon-on-silicon (also known as wafer scale integration) and chip-on-board packaging techniques, to perfected multichip module technology with modules operating in liquid-cooled systems.

11.5.5 Concluding Remarks

Active research areas in heat management of electronic packaging include all levels of research from the chip through the complete system. The trend toward increasing packing densities, large-area chips, and increased power densities, imply further serious heat removal and heat-related considerations. These problems can only be partly handled by brute-force techniques, coupled with evolutionary improvements in existing packaging technologies. However, requirements to satisfy the demands of high performance functional density, speed, power/heat dissipation, and levels of customization give rise to the need for exploring new approaches to the problem and for refining and better understanding existing approaches.

Techniques employing forced air and liquid convection cooling through various advanced schemes of cooling systems, including direct cooling of high-power dissipation chips (such as immersion cooling that allow also for uniformity of heat transfer coefficients and coolant temperature throughout the system), very fine coolant channels yielding an extremely low thermal resistance, and nucleate boiling heat transfer as well as cryogenic operation of CMOS devices, demand subsequent innovations in the format of packages and associated components/structures in order to accommodate such technologies.

Moreover, the ability to verify packaging designs, and to optimize materials and geometries before hardware is available, requires further developments of models and empirical tools for thermal design of packaging structures and systems. Theoretical models have to be judiciously combined with experimentally determined test structures for thermal (and thermal stress) measurements, creating conditions for equivalence of theoretical and empirical models. This in turn will allow generation of extensive heat transfer data for a variety of packaging structures under various conditions, thus translating real-world designs into modeling conditions.

Nomenclature

A = area, m^2

A_{eff} = effective cross-sectional area for heat flow, m^2

A_S = cross-sectional area for heat flow through the surface, m^2

A_i = cross-sectional area of the material of the ith layer, m^2

C = capacitance, F; constant

C_θ = thermal capacitance, W-s/degree

C_{sf} = constant, a function of the surface/fluid combination in the boiling heat transfer, W/m^2-K^n

c_p = specific heat, W-s/g-degree

E_A = activation energy, eV

F_T = total radiation factor; exchange factor

F_{1-2} = radiation geometric view factor

f = constant

G = volumetric flow rate, m^3/s

Gr = Grashof number $(g\beta\rho^2/\mu^2)(L_{ch})^3 \Delta T$, dimensionless

g = gravitational constant, m/s^2

H = height (of plate), m

h = heat transfer coefficient, W/m^2-degree

h_b = boiling heat transfer coefficient, W/m^2-degree

h_c = average convective heat transfer coefficient, W/m^2-degree

h_r = average radiative heat transfer coefficient, W/m^2-degree

I, i = current, A

i_M = measurement current, A

i_H = heating currents, A

J = Colburn factor, dimensionless

k = thermal conductivity, W/m^2-degree

k_i = thermal conductivity of the material of the ith layer, W/m^2-degree

k = Boltzman's constant, 8.616×10^{-5} eV/K

L = length (of plate, fin), m

L_{ch} = characteristic length parameter, m

m = mass flow, g/s; constant, $\sqrt{hc/k\delta}$

n = constant; coefficient

Nu = Nusselt number, dimensionless

P_{chip} = power dissipated within the chip, W

P_D	= power dissipation, W
P_m	= power dissipated within the module, W
Pr	= Prandtl number, dimensionless, $\mu c_p / k$
Q	= heat flow, W
Q_b	= boiling heat transfer rate, W
Q_c	= convection heat flow, W
Q_f	= heat transfer rate by a fin/fin structure, W
Q_{flow}	= total heat flow/dissipation of all components within the enclosure upstream of the component of interest, W
Q_r	= radiation heat flow, W
Q_V	= volumetric heat source generation, W/m³
q	= heat flux, W/m²
R	= resistance, Ω; radius, m
R_c	= thermal contact resistance, °C/W, K/W
Ra	= Raleigh number, $Gr\,Pr$, dimensionless
Re	= Reynolds number, $v L_{\text{ch}}(\rho/\mu)$, dimensionless
T	= temperature, °C, K
T_A	= ambient temperature, °C, K
T_b	= temperature of the fin base, °C, K
$T_{\text{case/pckg}}$	= temperature of the case/package at a defined location, °C, K
T_{chip}	= temperature of the chip, °C, K
$T_{\text{coolant-in/out}}$	= temperature of the coolant inlet/outlet, °C, K
T_f	= temperature of the fluid coolant, °C, K
T_J	= temperature of the junction, °C, K
T_R	= reference temperature, °C, K
T_S	= temperature of the surface, °C, K
T_{sat}	= temperature of the boiling point saturation of the liquid, °C, K
t	= time, s
t_M	= measurement time, s
t_H	= heating time, s
V	= voltage, V; volume, m³
v	= fluid velocity, m/s
W	= width (of plate), m
β	= coefficient of thermal expansion, 1/°C
Δl_i	= thickness of the ith layer, m
ΔT	= temperature difference, °C, K
$\Delta T_{j\text{-chip}}$	= temperature rise between the junction and the chip, °C, K
Δt	= time increment, s
δ	= fin thickenss, m
ε	= emissivity
η	= fin efficiency
θ	= thermal resistance, °C/W, K/W
δ	= fin thickness, m
$\theta_{\text{ext/int}}$	= external/internal thermal resistance, ° C/W, K/W
θ_{JR}	= thermal resistance between the junction or any other circuit element/active region that generate heat and the reference point, °C/W, K/W
θ_{flow}	= flow (thermal) resistance, °C/W, K/W
θ_{sp}	= thermal spreading resistance, °C/W, K/W
λ	= failure rate,
μ	= dynamic viscosity, g/m-s
ρ	= density, g/m³
σ	= electrical conductivity, $1/\Omega$ m; Stefan–Boltzman constant, 5.673×10^{-8} W/m²-K⁴

Defining Terms[1]

Ambient temperature: Temperature of atmosphere in intimate contact with the electrical part/device.

Boiling heat transfer: A liquid-to-vapor phase change at a heated surface, categorized as either *pool boiling* (occurring in a stagnant liquid) or *flow boiling*.

Coefficient of thermal expansion (CTE): The ratio of the change in length dimensions to the change in temperature per unit starting length.

Conduction heat transfer: Thermal transmission to heat energy from a hotter region to a cooler region in a conducting medium.

Convection heat transfer: Transmission of thermal energy from a hotter to a cooler region through a moving medium (gas or liquid.)

External thermal resistance: A term used to represent thermal resistance from a convenient point on the outside surface of an electronic package to an ambient reference point.

Failure, thermal: The temporary or permanent impairment of device or system functions caused by thermal disturbance or damage.

Failure rate: The rate at which devices from a given population can be expected or were found to fall as a function of time; expressed in FITs, where FIT is defined as one failure per billion device hours (10^{-9}/h or 0.0001% /1000 h of operation).

Flow regime, laminar: A constant and directional flow of fluid across a clean workbench. The flow is usually parallel to the surface of the bench.

Flow regime, turbulent: Flow where fluid particles are disturbed and fluctuate.

Heat flux: The outward flow of heat energy from a heat source across or through a surface.

Heat sink: The supporting member to which electronic components, or their substrate, or their package bottom is attached. This is usually a heat conductive metal with the ability to rapidly transmit heat from the generating source (component).

Heat transfer coefficient: Thermal parameter that encompasses all of the complex effects occurring in the convection heat transfer mode, including the properties of the fluid gas/liquid, the nature of the fluid motion, and the geometry of the structure.

Internal thermal resistance: A term used to represent thermal resistance from the junction or any heat generating element of a device, inside an electronic package, to a convenient point on the outside surface of the package.

Junction temperature: The temperature of the region of transistor between the *p* and *n* type semiconductor material in a transistor or diode element.

Mean time to failure (MTTF): A term used to express the reliability level. It is the arithmetic average of the lengths of time to failure registered for parts or devices of the same type, operated as a group under identical conditions. The reciprocal of the failure rate.

Power dissipation: The dispersion of the heat generated from a film circuit when a current flows through it.

Radiation heat transfer: The combined process of emission, transmission, and absorption of thermal energy between bodies separated by empty space.

Specific heat: The quantity of heat required to raise the temperature of 1 g of a substance 1°C.

Temperature cycling: An environmental test where the (film) circuit is subjected to several temperature changes from a low temperature to a high temperature over a period of time.

Thermal conductivity: The rate with which a material is capable of transferring a given amount of heat through itself.

Thermal design: The schematic heat flow path for power dissipation from within a (film) circuit to a heat sink.

---

[1] Based in part on *Hybrid Microcircuit Design Guide* published by ISHM and IPC, and Tummala, R.R. and Rymaszewski, E.J. 1989. *Microelectronics Packaging Handbook*. Van Nostrand Reinhold, New York.

Thermal gradient: The plot of temperature variances across the surface or the bulk thickness of a material being heated.

Thermal management or control: The process or processes by which the temperature of a specified component or system is maintained at the desired level.

Thermal mismatch: Differences of coefficients of thermal expansion of materials that are bonded together.

Thermal network: Representation of a thermal space by a collection of conveniently divided smaller parts, each representing the thermal property of its own part and connected to others in a prescribed manner so as not to violate the thermal property of the total system.

Thermal resistance: A thermal characteristic of a heat flow path, establishing the temperature drop required to transport heat across the specified segment or surface; analogous to electrical resistance.

References

Adina. *ADINAT*, User's Manual. Adina Engineering, Inc., Watertown, MA.

Albers, J. 1984. *TXYZ: A Program for Semiconductor IC Thermal Analysis*. NBS Special Pub. 400-76, Washington, DC.

Alderman, J., Tustaniwskyi, J., and Usell, R. 1986. Die attach evaluation using test chips containing localized temperature measurement diodes. *IEEE Trans. Comp., Hybrids, and Manuf. Technol.* 9(4):410–415.

Antonetti, V.W., Oktay, S., and Simons, R.E. 1989. Heat transfer in electronic packages. In *Microelectronics Packaging Handbook*, eds. R.R. Tummala and E.J. Rymaszewski, pp. 167–223. Van Nostrand Reinhold, New York.

Bar-Cohen, A. 1987. Thermal management of air- and liquid-cooled multichip modules. *IEEE Trans. Comp., Hybrids, and Manuf. Technol.* 10(2):159–175.

Bar-Cohen, A. 1993. Thermal management of electronics. In *The Engineering Handbook*, ed. R.C. Dorf, pp. 784–797. CRC Press, Boca Raton, FL.

Baxter, G.K. 1977. A recommendation of thermal measurement techniques for IC chips and packages. In *Proceedings of the 15th IEEE Annual Reliability Physics Symposium*, pp. 204–211, Las Vegas, NV. IEEE Inc.

Blodgett, A.J. 1983. Microelectronic packaging. *Scientific American* 249(1):86–96.

Boudreaux, P.J., Conner, Z., Culhane, A., and Leyendecker, A.J. 1991. Thermal benefits of diamond inserts and diamond-coated substrates to IC packages. In *1991 Gov't Microcircuit Applications Conf. Digest of Papers*, pp. 251–256.

Buller, M.L. and Kilburn, R.F. 1981. Evaluation of surface heat transfer coefficients for electronic modular packages. In *Heat Transfer in Electronic Equipment, HTD*, Vol. 20, pp. 25–28. ASME, New York.

Carlson, D.M., Sullivan, D.C., Bach, R.E., and Resnick, D.R. 1989. The ETA 10 liquid-cooled supercomputer system. *IEEE Trans. Electron Devices* 36(8):1404–1413.

Carslaw, H.S. and Jaeger, J.C. 1967. *Conduction of Heat in Solids*. Oxford University Press, Oxford, England, UK.

Casa. *GIFTS*, User's Reference Manual. Casa Gifts, Inc., Tucson, AZ.

Clark, W.F., El-Kareh, B., Pires, R.G., Ticomb, S.L., and Anderson, R.L. 1992. Low temperature CMOS—A brief review. *IEEE Trans. Comp., Hybrids, and Manuf. Technol.* 15(3):397–404.

COSMIC. 1982a. *NASTRAN Thermal Analyzer*, COSMIC Program Abstracts No. GSC-12162. University of Georgia, Athens, GA.

COSMIC. 1982b. *SINDA—Systems Improved Numerical Differencing Analyzer*, COSMIC Prog. Abst. No. GSC-12671. University of Georgia, Athens, GA.

COSMIC. 1982c. *NISA: Numerically Integrated Elements for Systems Analysis*, COSMIC Prog. Abst. University of Georgia, Athens, GA.

Davidson, E.E. and Katopis, G.A. 1989. Package electrical design. In *Microelectronics Packaging Handbook*, ed. R.R. Tummala and E.J. Rymaszewski, pp. 111–166. Van Nostrand Reinhold, New York.

Edwards, A.L. 1969. *TRUMP: A Computer Program for Transient and Steady State Temperature Distribution in Multidimensional Systems; User's Manual*, Rev. II. University of California, LRL, Livermore, CA.

Ellison, G.N. 1987. *Thermal Computations for Electronic Equipment*. Van Nostrand Reinhold, New York.

Fletcher, L.S. 1990. A review of thermal enhancement techniques for electronic systems. *IEEE Trans. Comp., Hybrids, and Manuf. Technol.* 13(4):1012–1021.

Fukuoka, Y. and Ishizuka, M. 1984. Transient temperature rise for multichip packages. In *Thermal Management Concepts in Microelectronic Packaging*, pp. 313–334. ISHM Technical Monograph Series 6984-003, Silver Spring, MD.

Furkay, S.S. 1984. Convective heat transfer in electronic equipment: An overview. In *Thermal Management Concepts in Microelectronic Packaging*, pp. 153–179. ISHM Technical Monograph Series 6984-003, Silver Spring, MD.

Godfrey, W.M., Tagavi, K.A., Cremers, C.J., and Menguc, M.P. 1993. Interactive thermal modeling of electronic circuit boards. *IEEE Trans. Comp., Hybrids, Manuf. Technol.* CHMT-16(8)978–985.

Hagge, J.K. 1992. State-of-the-art multichip modules for avionics. *IEEE Trans. Comp., Hybrids, and Manuf. Technol.* 15(1):29–41.

Hannemann, R., Fox, L.R., and Mahalingham, M. 1991. Thermal design for microelectronic components. In *Cooling Techniques for Components*, ed. W. Aung, pp. 245–276. Hemisphere, New York.

ISHM. 1984. *Thermal Management Concepts in Microelectronic Packaging*, ISHM Technical Monograph Series 6984-003. International Society for Hybrid Microelectronics, Silver Spring, MD.

Jeannotte, D.A., Goldmann, L.S., and Howard, R.T. 1989. Package reliability. In *Microelectronics Packaging Handbook*, ed. R.R. Tummala and E.J. Rymaszewski, pp. 225–359. Van Nostrand Reinhold, New York.

Keith, F. 1973. *Heat Transfer*. Harper & Row, New York.

Kennedy, D.P. 1960. Spreading resistance in cylindrical semiconductor devices. *J. App. Phys.* 31(8):1490–1497.

Kern, D.Q. and Kraus, A.D. 1972. *Extended Surface Heat Transfer*. McGraw-Hill, New York.

Krus, A.D. and Bar-Cohex, A. 1983. *Thermal analysis and Control of Electronic Equipment*. Hemisphere, New York.

Mahalingham, M. 1985. Thermal management in emiconductor device packaging. *Proceedings of the IEEE* 73(9):1396–1404.

Mathuna, S.C.O. 1992. Development of analysis of an automated test system for the thermal characterization of IC packaging technologies. *IEEE Trans. Comp., Hybrids, and Manuf. Technol.* 15(5):615–624.

McAdams, W.H. 1954. *Heat Transmission*. McGraw-Hill, New York.

Miyake, K., Suzuki, H., and Yamamoto, S. 1985. Heat transfer and thermal stress analysis of plastic-encapsulated ICs. *IEEE Trans. Reliability* R-34(5):402–409.

Moffat, R.J. and Ortega, A. 1988. Direct air cooling in electronic systems. In *Advances in Thermal Modeling of Electronic Components and Systems*, ed. A. Bar-Cohen and A.D. Kraus, Vol. 1, pp. 129–282. Hemisphere, New York.

Murano, H. and T. Watari, 1992. Packaging technology for the NEC SX-3 supercomputers. *IEEE Trans. Comp., Hybrids, and Manuf. Technol.* 15(4):401–417.

Nakayama, W. 1986. Thermal management of electronic equipment: A review of technology and research topics. *Appl. Mech. Rev.* 39(12):1847–1868.

Nalbandian, R. 1987. Automatic thermal and dynamic analysis of electronic printed circuit boards by ANSYS finite element program. In *Proceeding of the confrence on Integrating Design and Analysis*, pp. 11.16–11.33, Newport Beach, CA, March 31–April 3.

Newell, W.E. 1975. Transient thermal analysis of solid-state power devices—making a dreaded process easy. *IEEE Trans. Industry Application* IA-12(4):405–420.

Oettinger, F.F. 1984. Thermal evaluation of VLSI packages using test chips—A critical review. *Solid State Tech.* (Feb):169–179.

Oettinger, F.F. and Blackburn, D.L. 1990. *Semiconductor Measurement Technology: Thermal Resistance Measurements*. NIST Special Pub. 400-86, Washington, DC.

Ohsaki, T. 1991. Electronic packaging in the 1990s: The perspective from Asia. *IEEE Trans. Comp., Hybrids, and Manuf. Technol.* 14(2):254–261.

Riemer, D.E. 1990. Thermal-Stress Analysis with Electrical Equivalents. *IEEE Trans. Comp., Hybrids, Manuf. Technol.* 13(1):194–199.

Rohsenow, W.M. and Choi, H. 1961. *Heat, Mass, and Momentum Transfer*. Prentice-Hall, Englwood, NJ.

Rohsenow, W.M. and Hartnett, J.P. 1973. *Handbook of Heat Transfer*. McGraw-Hill, New York.

Shiang, J.-J., Staszak, Z.J., and Prince, J.L. 1987. APTMC: An interface program for use with ANSYS for thermal and thermally induced stress modeling simulation of VLSI packaging. In *Proceedings of the Conference in Integration Design and Analysis*, pp. 11.55–11.62, Newport Beach, CA, March 31–April 3.

Siegal, R. and Howell, J.R. 1981. *Thermal Radiation Heat Transfer*. Hemisphere, New York.

Simon, B.R., Staszak, Z.J., Prince, J.L., Yuan, Y., and Umaretiya, J.R. 1988. Improved finite element models of thermal/mechanical effects in VLSI packaging. In *Proceedings of the Eighth International Electronic Packaging Conference*, pp. 3–19, Dallas, TX.

SPICE. *SPICE2G User's Manual*. University of California, Berkeley, CA.

Staszak, Z.J., Prince, J.L., Cooke, B.J., and Shope, D.A. 1987. Design and performance of a system for VLSI packaging thermal modeling and characterization. *IEEE Trans. Comp., Hybrids, and Manuf. Technol.* 10(4):628–636.

Swanson. *ANSYS Engineering Analysis System*, User's Manual. Swanson Engineering Analysis System, Houston, PA.

Touloukian, Y.S. and Ho, C.Y. 1979. *Master Index to Materials and Properties*. Plenum Publishing, New York.

Tuckerman, D.P. and Pease, F. 1981. High performance heat sinking for VLSI. *IEEE Electron Device Lett.* EDL-2(5):126–129.

Tummala, R.R. 1991. Electronic packaging in the 1990s: The perspective from America. *IEEE Trans. Comp., Hybrids, and Manuf. Technol.* 14(2):262–271.

Tummala, R.R. and Ahmed, S. 1992. Overview of packaging of the IBM enterprise system/9000 based on the glass-ceramic copper/thin film thermal conduction module. *IEEE Trans. Comp., Hybrids, and Manuf. Technol.* 15(4):426–431.

Tzeng, Y., Yoshikawa, M., Murakawa, M., and Feldman, A. 1991. *Application of Diamond Films and Related Materials*. Elsevier Science, Amsterdam.

Watari T. and H. Murano, 1995. Packaging technology for the NEC SX supercomputers. *IEEE Trans. Comp., Hybrids, and Manuf. Technol.* 8(4):462–467.

Wessely, H., Fritz, O., Hor, M., Klimke, P., Koschnick, W., and Schmidt, K.-K. 1991. Electronic packaging in the 1990s: The perspective from Europe. *IEEE Trans. Comp., Hybrids, and Manuf. Technol.* 14(2):272–284.

Yokouchi, K., Kamehara, N., and Niwa, K. 1987. Immersion Cooling for High-Density Packaging. *IEEE Trans. Comp., Hybrids, and Manuf. Technol.* 10(4):643–646.

Yovanovich, M.M. and Antonetti, V.M. 1988. Application of thermal contact resistance theory to electronic packages. In *Advances in Thermal Modeling of Electronic Components and Systems*, Vol. 1, eds. A. Bar-Cohen and A.D. Kraus, pp. 79–128. Hemisphere, New York.

Further Information

Additional information on the topic of heat management and heat related phenomena in electronic packaging systems is available from the following sources:

Journals:

IEEE Transactions on Components, Packaging, and Manufacturing Technology, published by the Institute of Electrical and Electronics Engineers, Inc., New York (formerly: *IEEE Transactions on Components, Hybrids, and Manufacturing Technology*).

International Journal for Hybrid Microelectronics, published by the International Society for Hybrid Microelectronics, Silver Spring, MD.

Journal of Electronic Packaging, published by the American Society of Mechanical Engineers, New York.

Conferences:

IEEE Semiconductor Thermal and Temperature Measurement Symposium (SEMI-THERM), organized annually by IEEE Inc.

Intersociety Conference on Thermal Phenomena in Electronic Systems (I-THERM), organized biannually in the U.S. by IEEE Inc, ASME, and other engineering societies.

Electronics Components and Technology Conference (formerly: Electronic Components Conference), organized annually by IEEE Inc.

International Electronic Packaging Conference, organized annually by the International Electronic Packaging Society.

European Hybrid Microelectronics Conference, organized by the International Society for Hybrid Microelectronics.

In addition to the references, the following books are recommended:

Aung, W. ed. 1991. *Cooling Techniques for Computers.* Hemisphere, New York.

Dean, D.J. 1985. *Thermal Design of Electronic Circuit Boards and Packages.* Electrochemical Publications, Ayr, Scotland.

Grigull, U. and Sandner, H. 1984. *Heat Conduction*, International Series in Heat and Mass Transfer. Springer-Verlag, Berlin.

Seely, J.H. and Chu, R.C. 1972. *Heat Transfer in Microelectronic Equipment.* Marcel Dekker, New York.

Sloan, J.L. 1985. *Design and Packaging of Electronic Equipment.* Van Nostrand Reinhold, New York.

Steinberg, D.S. 1991. *Cooling Techniques for Electronic Equipment.* Wiley, New York.

12

Communication Principles

Leon W. Couch, II
University of Florida

Robert Kubichek
University of Wyoming

Ken Seymour
AT&T Wireless Services

Rodger E. Ziemer
University of Colorado

Kurt L. Kosbar
University of Missouri

William H. Tranter
*Virginia Polytechnic Institute
and State University*

Oktay Alkin
Southern Illinois University

Fred Wylie
Audio Processing Technology, Ltd.

William J.J. Roberts
Atlantic Coast Technologies, Inc.

Yariv Ephraim
George Mason University

Gopal Lakhani
Texas Tech University

12.1   Intelligence Coding[1]

Leon W. Couch, II

12.1.1   Introduction

The purpose of a communication system is to transmit message intelligence of a source to a sink (user). The exact waveform present at the sink is *unknown* until after it is received; otherwise, no intelligence would be transmitted and there would be no need for the communication system. More intelligence is communicated to the sink when the user at the sink is "surprised" by the message that was transmitted, that is, the transmission of intelligence implies the communication of messages that are not known ahead of time (*a priori*). On the other hand, in energy or power systems it is known *a priori* what the received waveform will be, and it is the energy imparted by the received waveform that is the desired quantity.

Noise limits our ability to communicate. If there were no noise, we could communicate messages electronically to the outer limits of the universe using an infinitely small amount of power. This has been obvious on an intuitive basis since the early days of radio. The theory that describes noise and the effect of noise on the transmission of information, however, was not developed until the 1940s by such persons as S.O. Rice (1944) and C.E. Shannon (1948).

Communication systems may be described by the block diagram shown in Fig. 12.1. Regardless of the particular applications, all communications systems involve three main subsystems: the *transmitter*, the *channel*, and the *receiver*. The message from the source is represented by the intelligence input waveform $m(t)$. The message delivered to the sink is denoted by $\tilde{m}(t)$. The tilde indicates that the message received may not be the same as that transmitted; that is, the message at the sink $\tilde{m}(t)$ may be corrupted by noise in the channel or there may be other impairments in the system, such as undesired filtering or undesired nonlinearities. The message information may be in analog or digital form, depending on the particular system, and it may represent audio, video, or some other type of intelligence. In multiplexed systems there may be multiple input and output message sources and sinks. The spectra (or frequencies) of $m(t)$ and $\tilde{m}(t)$ are concentrated about $f = 0$; consequently, they are said to be *baseband signals*.

The signal processing block at the transmitter conditions the source for more efficient transmission. For example, in a digital system the processor might consist of a microprocessor that provides redundancy reduction of the source. It may also provide channel encoding so that error detection and correction can be used at the signal processing unit at the receive site to reduce errors caused by noise in the channel. In an analog communication system the signal processor may be nothing more than an analog low-pass filter. In a hybrid system, the signal processor might take samples of an analog input and digitize them to provide a pulse code modulation (PCM) *digital word* for transmission over the system. The signal at the output of the transmitter signal processor is also a baseband signal since it has frequencies concentrated near $f = 0$.

---

[1] Material in this chapter is adapted from Couch, L.W., II. 1995. *Modern Communication Systems*, Chaps. 1 and 5. Prentice-Hall, Englewood Cliffs, NJ. Used with permission.

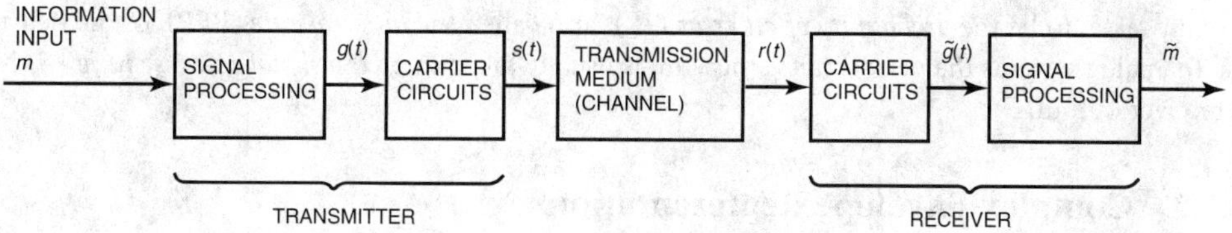

FIGURE 12.1 A basic communication system. (*Source*: Couch, L.W., II. 1995. *Modern Communication Systems*. Prentice-Hall, Englewood Cliffs, NJ. With permission.)

The transmitter carrier circuits convert the processed baseband signal into a frequency band that is appropriate for propagation over the transmission medium of the channel. For example, if the channel consists of a twisted-pair telephone line, the spectrum of $s(t)$ will be in the audio range of, for example, 300–3700 Hz, but if the channel consists of a fiber optic link, the spectrum of $s(t)$ will be at light frequencies. If the channel propagates baseband signals, no carrier circuits are necessary, and $s(t)$ can be just the output of the processing circuits at the transmitter. Carrier circuits are needed when the channel can only propagate frequencies located in a band around f_c where $f_c \gg 0$. (The c subscript denotes carrier frequency.) In this case $s(t)$ is said to be a *bandpass* signal since it is designed to have frequencies located in a band about f_c. For example, an amplitude modulated (AM) broadcasting station with an assigned frequency of 850 kHz has a carrier frequency of $f_c = 850$ kHz. The mapping of the baseband input information waveform $m(t)$ into the bandpass signal $s(t)$ is called **modulation** ($m(t)$ is the audio signal in AM broadcasting). Any bandpass signal has the form

$$s(t) = R(t)\cos[\omega_c t + \theta(t)] \tag{12.1}$$

where $\omega_c = 2\pi f_c$. If $R(t) = 1$ and $\theta(t) = 0$, $s(t)$ would be a pure sinusoid of frequency $f = f_c$ with *zero* bandwidth. In the modulation process provided by the carrier circuits, the baseband input waveform $m(t)$ causes $R(t)$ and/or $\theta(t)$ to change as a function of $m(t)$. These fluctuations in $R(t)$ and $\theta(t)$ cause $s(t)$ to have a *finite* bandwidth that depends on the characteristics of $m(t)$ and on the mapping functions used to generate $R(t)$ and $\theta(t)$.

Channels may be classified into two categories: *hardwire* and *softwire*. Some examples of hardwire channels are twisted-pair telephone lines, coaxial cables, waveguides, and fiber optic cables. Some typical softwire channels are air, vacuum, and seawater. It should be realized that the general principles of digital and analog modulation apply to all types of channels, although channel characteristics may impose constraints that favor a particular type of signaling. In general, the channel medium attenuates the signal so the noise level of the channel and/or the noise introduced by an imperfect receiver cause the delivered information $\tilde{m}(t)$ to be deteriorated from that of the source. The channel noise may arise from natural electrical disturbances (e.g., lightning) or from man-made sources such as high-voltage transmission lines, ignition systems of cars, or even switching circuits of a nearby digital computer. The channel may contain active amplifying devices, such as repeaters in telephone systems or satellite transponders in space communication systems. These devices, of course, are necessary to help keep the signal above the noise level. In addition, the channel may provide *multiple paths* between its input and output that have different time delays and attenuation characteristics. Even worse, these characteristics may vary with time. This variation produces signal fading at the channel output. You have probably observed this type of fading when listening to distant shortwave stations.

The receiver takes the corrupted signal at the channel output and converts it to a baseband signal that can be handled by the receiver baseband processor. The baseband processor "cleans up" this signal and delivers an estimate of the source information $\tilde{m}(t)$ to the communication system output.

The goal of the engineer is to design communication systems so that the information is transmitted to the sink with as little deterioration as possible while satisfying design constraints such as allowable transmitted energy, allowable signal bandwidth, and cost. In digital systems the measure of deterioration is usually taken to be the *probability of bit error* (P_e), also called the *bit error rate* (*BER*), of the delivered data. In analog systems the performance measure is usually taken to be the *signal-to-noise ratio* (S/N) at the receiver output.

12.1.2 Complex Envelope Representation

All **bandpass waveforms**, whether they arise from a modulated signal, interfering signals, or noise, may be represented in a convenient form by using an appropriate complex envelope. Here, $v(t)$ will be used to denote the bandpass waveform canonically; specifically, $v(t)$ can represent the signal when $s(t) = v(t)$, the noise when $n(t) = v(t)$, the filtered signal plus noise at the channel output when $r(t) = v(t)$, or any other type of bandpass waveform. Any bandpass waveform can be represented by (Couch, 1995)

$$v(t) = \text{Re}\{g(t)e^{j\omega_c t}\} \tag{12.2a}$$

when Re{·} denotes the real part of {·}, $g(t)$ is called the **complex envelope** of $v(t)$, and f_c is the associated carrier frequency (hertz) where $\omega_c = 2\pi f_c$. Furthermore, two other equivalent representations are

$$v(t) = R(t)\cos[\omega_c t + \theta(t)] \tag{12.2b}$$

and

$$v(t) = x(t)\cos\omega_c t - y(t)\sin\omega_c t \tag{12.2c}$$

where

$$g(t) = x(t) + jy(t) = |g(t)|e^{j\underline{/g(t)}} = R(t)e^{j\theta(t)} \tag{12.3}$$

$$x(t) = \text{Re}\{g(t)\} = R(t)\cos\theta(t) \tag{12.4a}$$

$$y(t) = \text{Im}\{g(t)\} = R(t)\sin\theta(t) \tag{12.4b}$$

$$R(t) = |g(t)| = \sqrt{x^2(t) + y^2(t)} \tag{12.5a}$$

$$\theta(t) = \underline{/g(t)} = \tan^{-1}\left(\frac{y(t)}{x(t)}\right) \tag{12.5b}$$

The waveforms $g(t)$, $x(t)$, $y(t)$, $R(t)$, and $\theta(t)$ are all **baseband waveforms**, and, except for $g(t)$, they are all real waveforms. Thus, Eq. (12.2) is a low-pass-to-bandpass transformation. $R(t)$ is called the **real envelope** of $v(t)$. Here, $x(t)$ is said to be the *in phase modulation* associated with $v(t)$, and $y(t)$ is said to be the *quadrature modulation* associated with $v(t)$. The polar form of $g(t)$, represented by $R(t)$ and $\theta(t)$, is given by Eq. (12.3) where the identities between Cartesian and polar coordinates are given by Eq. (12.4) and Eq. (12.5). $R(t)$ and $\theta(t)$ are real waveforms and, in addition, $R(t)$ is always nonnegative. $R(t)$ is said to be the *amplitude modulation* (AM) on $v(t)$, and $\theta(t)$ is said to be the *phase modulation* (PM) on $v(t)$. In general, bandpass noise includes both AM, $R(t)$, and PM, $\theta(t)$, noise components.

The usefulness of the complex envelope representation for bandpass waveforms cannot be overemphasized. In modern communication systems, the bandpass signal is often partitioned into two channels, one for $x(t)$ called the I (in-phase) channel and one for $y(t)$ called the Q (quadrature-phase) channel.

12.1.3 Representation of Modulated Signals

Modulation is the process of encoding the source intelligence $m(t)$ (modulating signal) into a bandpass signal $s(t)$ (**modulated signal**). Consequently, the modulated signal is just a special application of the bandpass representation. The modulated signal is given by

$$s(t) = \text{Re}\{g(t)e^{j\omega_c t}\} \tag{12.6}$$

TABLE 12.1(a) Complex Envelope Functions for Various Types of Modulation[a]

Type of Modulation	Mapping Functions $g[m]$	Corresponding Quadrature Modulation	
		$x(t)$	$y(t)$
AM	$A_c[1 + m(t)]$	$A_c[1 + m(t)]$	0
DSB-SC	$A_c m(t)$	$A_c m(t)$	0
PM	$A_c \exp\{jD_p m(t)\}$	$A_c \cos[D_p m(t)]$	$A_c \sin[D_p m(t)]$
FM	$A_c \exp\left\{ jD_f \int_{-\infty}^{t} m(\sigma)\,d\sigma \right\}$	$A_c \cos\left[D_f \int_{-\infty}^{t} m(\sigma)\,d\sigma \right]$	$A_c \sin\left[D_f \int_{-\infty}^{t} m(\sigma)\,d\sigma \right]$
SSB-AM-SC[b]	$A_c[m(t) \pm j\hat{m}(t)]$	$A_c m(t)$	$\pm A_c \hat{m}(t)$
SSB-PM[b]	$A_c \exp\{jD_p[m(t) \pm j\hat{m}(t)]\}$	$A_c e^{\mp D_p \hat{m}(t)} \cos[D_p m(t)]$	$A_c e^{\mp D_p \hat{m}(t)} \sin[D_p m(t)]$
SSB-FM[b]	$A_c \exp\left\{ jD_f \int_{-\infty}^{t}[m(\sigma) \pm j\hat{m}(\sigma)]\,d\sigma \right\}$	$A_c \exp\left\{ \mp D_f \int_{-\infty}^{t}\hat{m}(\sigma)\,d\sigma \right\}$ $\times \cos\left[D_f \int_{-\infty}^{t} m(\sigma)\,d\sigma \right]$	$A_c \exp\left\{ \mp D_f \int_{-\infty}^{t}\hat{m}(\sigma)\,d\sigma \right\}$ $\times \sin\left[D_f \int_{-\infty}^{t} m(\sigma)\,d\sigma \right]$
SSB-EV[b]	$A_c \exp\{\ln[1 + m(t)] \pm j\widehat{\ln}[1 + m(t)]\}$	$A_c[1 + m(t)]$ $\times \cos\{\widehat{\ln}[1 + m(t)]\}$	$\pm A_c[1 + m(t)]$ $\times \sin\{\widehat{\ln}[1 + m(t)]\}$
SSB-SQ[b]	$A_c \exp[(1/2)\{\ln[1 + m(t)]$ $\pm j\widehat{\ln}[1 + m(t)]\}]$	$A_c \sqrt{1 + m(t)}$ $\times \cos\{\frac{1}{2}\widehat{\ln}[1 + m(t)]\}$	$\pm A_c \sqrt{1 + m(t)}$ $\times \sin\{\frac{1}{2}\widehat{\ln}[1 + m(t)]\}$
QM	$A_c m_1(t) + jm_2(t)$	$A_c m_1(t)$	$A_c m_2(t)$

[a] $A_c > 0$ is a constant that sets the power level of the signal as evaluated by use of (12.13); L, linear; NL, nonlinear; $[\hat{\cdot}]$ is the Hilbert transform (i.e., $-90°$ phase-shifted version) of $[\cdot]$.

[b] Use upper signs for upper sideband signals and lower signs for lower sideband signals.

(*Source:* Couch, L.W., II. 1995. *Modern Communication Systems*, pp. 234–235. Prentice-Hall, Englewood Cliffs, NJ. With permission.)

where $\omega = 2\pi f_c$. Here, f_c is the carrier frequency. The complex envelope $g(t)$ is a function of the modulating signal $m(t)$. That is

$$g(t) = g[m(t)] \tag{12.7}$$

Table 12.1 gives the big picture for the modulating problem. Examples of the mapping function $g[m]$ are given for amplitude modulation, double-sideband suppressed carrier (DSB-SC), phase modulation, frequency modulation (FM), single-sideband AM suppressed carrier (SSB-AM-SC), single-sideband PM (SSB-PM), single-sideband FM (SSB-FM), single-sideband envelope detectable (SSB-EV), single-sideband square-law detectable (SSB-SQ), and quadrature modulation (QM). Digitally modulated bandpass signals are obtained when $m(t)$ is a digital baseband signal, for example, the output of a transistor transistor logic (TTL) circuit.

Obviously, it is possible to use other $g(m)$ functions that are not listed in Table 12.1. The question is: Are they useful? The $g(m)$ functions are desired that are easy to implement and that will give desirable spectral properties. Furthermore, in the receiver the inverse function $m(g)$ is required. The inverse should be single valued over the range used and should be easily implemented. The mapping should suppress as much noise as possible so that $m(t)$ can be recovered with little corruption.

TABLE 12.1(b) Complex Envelope Functions for Various Types of Modulation[a]

Type of Modulation	Corresponding Amplitude and Phase Modulation		Linearity	Remarks
	$R(t)$	$\theta(t)$		
AM	$A_c\lvert 1 + m(t)\rvert$	$\begin{cases} 0, & m(t) > -1 \\ 180°, & m(t) < -1 \end{cases}$	L[c]	$m(t) > -1$ required for envelope detection
DSB-SC	$A_c\lvert m(t)\rvert$	$\begin{cases} 0, & m(t) > 0 \\ 180°, & m(t) < 0 \end{cases}$	L	Coherent detection required
PM	A_c	$D_p m(t)$	NL	D_p is the phase deviation constant (rad/V)
FM	A_c	$D_f \displaystyle\int_{-\infty}^{t} m(\sigma)\,d\sigma$	NL	D_f is the frequency deviation constant (rad/V-s)
SSB-AM-SC[b]	$A_c\sqrt{[m(t)]^2 + [\hat{m}(t)]^2}$	$\tan^{-1}[\pm\hat{m}(t)/m(t)]$	L	Coherent detection required
SSB-PM[b]	$A_c e^{\mp D_p \hat{m}(t)}$	$D_p m(t)$	NL	
SSB-FM[b]	$A_c \exp\left\{ \mp D_f \displaystyle\int_{-\infty}^{t} \hat{m}(\sigma)\,d\sigma \right\}$	$D_f \displaystyle\int_{-\infty}^{t} m(\sigma)\,d\sigma$	NL	
SSB-EV[b]	$A_c[1 + m(t)]$	$\pm \widehat{\ln}\,[1 + m(t)]$	NL	$m(t) > -1$ is required so that the $\ln(\cdot)$ will have a real value
SSB-SQ[b]	$A_c\sqrt{1 + m(t)}$	$\pm\frac{1}{2}\widehat{\ln}\,[1 + m(t)]$	NL	$m(t) > -1$ is required so that $\ln(\cdot)$ will have a real value
QM	$A_c\sqrt{m_1^2(t) + m_2^2(t)}$	$\tan^{-1}[m_2(t)/m_1(t)]$	L	Used in NTSC color television; requires coherent detection

[a] $A_c > 0$ is a constant that sets the power level of the signal as evaluated by use of (12.13); L, linear; NL, nonlinear; $[\hat{\cdot}]$ is the Hilbert transform (i.e., $-90°$ phase-shifted version) of $[\cdot]$.

[b] Use upper signs for upper sideband signals and lower signs for lower sideband signals.

[c] In the strict sense, AM signals are not linear because the carrier term does not satisfy the linearity (superposition) condition.

(*Source:* Couch, L.W., II. 1995. *Modern Communications Systems*, pp. 234–235. Prentice-Hall, Englewood Cliffs, NJ. With permission.)

12.1.4 Spectrum and Power of Bandpass Signals

The spectrum of a bandpass signal is directly related to the spectrum of its complex envelope. If a bandpass voltage (or current) waveform is represented by

$$v(t) = \mathrm{Re}\{g(t)e^{j\omega_c t}\} \tag{12.8}$$

then the **voltage** (or **current**) **spectrum** for this bandpass waveform is

$$V(f) = \frac{1}{2}(G(f - f_c) + G^*(-f - f_c)) \tag{12.9}$$

and the **power spectral density (PSD)** of the waveform is

$$\wp_v(f) = \frac{1}{4}(\wp_g(f - f_c) + \wp_g(-f - f_c)) \tag{12.10}$$

where $G(f) = \Im[g(t)]$ is the **Fourier transform** of $g(t)$ and $\wp_g(f)$ is the PSD of $g(t)$. Starting with Eq. (12.8), Eq. (12.9) can be proven as follows:

$$v(t) = \mathrm{Re}\{g(t)e^{j\omega_c t}\} = \frac{1}{2}g(t)e^{j\omega_c t} + \frac{1}{2}g^*(t)e^{-j\omega_c t}$$

Thus,

$$V(f) = \Im(v(t)) = \frac{1}{2}\Im(g(t)e^{j\omega_c t}) + \frac{1}{2}\Im(g^*(t)e^{-j\omega_c t}) \tag{12.11}$$

If we use $\Im(g^*(t)) = G^*(-f)$ and the frequency translation property of Fourier transforms, this equation becomes

$$V(f) = \frac{1}{2}\{G(f - f_c) + G^*(-(f + f_c))\} \tag{12.12}$$

which reduces to Eq. (12.9). In a similar way Eq. (12.10) can also be proved (Couch, 1995, p. 236).

The total average normalized power of a bandpass waveform, $v(t)$, is (Couch, 1995, p. 237)

$$P_v = \langle v^2(t)\rangle = \int_{-\infty}^{\infty} \wp_v(f)\,df = \frac{1}{2}\langle |g(t)|^2\rangle \tag{12.13}$$

where normalized implies that the load is equivalent to 1 Ω. Here, $\langle[\cdot]\rangle$ denotes the average value of $[\cdot]$, which is defined by

$$\langle[\cdot]\rangle = \lim_{T\to\infty}\frac{1}{T}\int_{-T/2}^{T/2}[\cdot]\,dt \tag{12.14a}$$

If $[\cdot]$ is periodic, this average may be evaluated by integrating over one period

$$\langle[\cdot]\rangle = \frac{1}{T_0}\int_{a}^{a+T_0}[\cdot]\,dt \quad \text{periodic case} \tag{12.14b}$$

where T_0 is the period of $[\cdot]$ and a is any convenient real constant. It is also realized that

$$\langle v^2(t)\rangle = V_{\text{rms}}^2 \tag{12.15}$$

where V_{rms} is the rms value of $v(t)$.

Another type of power rating, called the *peak envelope power* (PEP) is useful for transmitter specifications. PEP is the average power that would be obtained if $|g(t)|$ were to be held constant at its peak value. This is equivalent to evaluating the average power of an unmodulated RF sinusoid that has a peak value of $A_P = \max[v(t)]$. The normalized PEP is given by

$$P_{\text{PEP}} = \frac{1}{2}[\max|g(t)|]^2 \tag{12.16}$$

A proof of this theorem follows by applying the definition of PEP to Eq. (12.13). The PEP is useful for specifying the power capability of AM, SSB, and television transmitters.

Example: Spectrum and Power of an AM Signal

Evaluate the magnitude spectrum for an amplitude modulated signal. From Table 12.1, the complex envelope of an AM signal is

$$g(t) = A_c[1 + m(t)]$$

so that the spectrum of the complex envelope is

$$G(f) = A_c\delta(f) + A_c M(f) \tag{12.17}$$

where $\delta(f)$ is a Dirac delta function and $M(f)$ is the spectrum of the modulating signal $m(t)$. Using Eq. (12.6), we obtain the AM signal waveform

$$s(t) = A_c[1 + m(t)]\cos\omega_c t$$

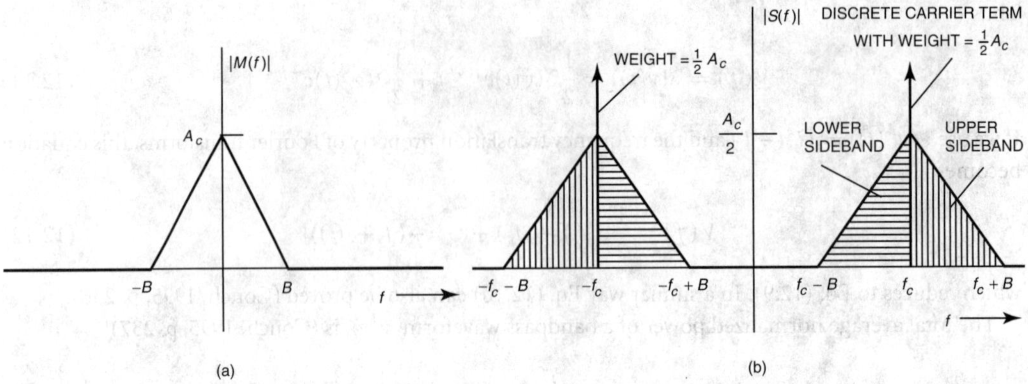

FIGURE 12.2 Spectrum of an AM signal: (a) magnitude spectrum of modulation, (b) magnitude spectrum of AM signal. (*Source:* Couch, L.W., II. 1995. *Modern Communication Systems*, p. 239. Prentice-Hall, Englewood Cliffs, NJ. With permission.)

and, using Eq. (12.9), the AM spectrum

$$S(f) = \frac{1}{2} A_c [\delta(f - f_c) + M(f - f_c) + \delta(f + f_c) + M(f + f_c)] \tag{12.18}$$

where, since $m(t)$ is real, $M^*(f) = M(-f)$ and $\delta(f) = \delta(-f)$ (the delta function is defined to be even). Suppose that the magnitude spectrum of the modulation happens to be a triangular function, as shown in Fig. 12.2(a). This spectrum might arise from an analog audio source where the bass frequencies are emphasized. The resulting AM spectrum, using Eq. (12.18), is shown in Fig. 12.2(b). Since $G(f - f_c)$ and $G^*(-f - f_c)$ do not overlap, the magnitude spectrum is

$$|S(f)| = \begin{cases} \frac{1}{2} A_c \delta(f - f_c) + \frac{1}{2} A_c |M(f - f_c)|, & f > 0 \\ \frac{1}{2} A_c \delta(f + f_c) + \frac{1}{2} A_c |M(-f - f_c)|, & f < 0 \end{cases} \tag{12.19}$$

The 1 in $g(t) = A_c(1 + m(t))$ causes delta functions to occur in the spectrum at $f = \pm f_c$ where f_c is the assigned carrier frequency.

Using Eq.(12.13), we obtain the total average normalized signal power

$$P_s = \frac{1}{2} A_c^2 \langle |1 + m(t)|^2 \rangle = \frac{1}{2} A_c^2 \langle 1 + 2m(t) + m^2(t) \rangle$$

$$= \frac{1}{2} A_c^2 (1 + 2\langle m(t) \rangle + \langle m^2(t) \rangle)$$

If we assume that the DC value of the modulation is zero, then $\langle m(t) \rangle = 0$, and the average normalized signal power becomes

$$P_s = \frac{1}{2} A_c^2 [1 + P_m] \tag{12.20}$$

where $P_m = \langle m^2(t) \rangle$ is the normalized power in the modulation $m(t)$, $\frac{1}{2} A_c^2$ is the carrier power, and $\frac{1}{2} A_c^2 P_m$ is the power in the sidebands of $s(t)$.

Specific numerical results depend on the modulation waveforms used and on the value of A_c. For example, assume $A_c = 100$ and that the modulating signal is a 1-kHz sine-wave test tone given by $m(t) = 0.8 \sin(2\pi 1000t)$. Then $P_m = \langle m^2(t) \rangle = M_{\text{rms}}^2 = (0.8/\sqrt{2})^2 = 0.320$.

Using Eq. (12.20), the **normalized average** signal **power** is

$$P_s = \frac{1}{2}(100)^2 [1 + 0.320] = 6600 \text{ W} \tag{12.21}$$

Furthermore, if $s(t)$ is a voltage waveform that appears across a 50-Ω resistive load R_L, then the **actual** (i.e., unnormalized) **average power** dissipated in the load for this AM signal is

$$(P_s)_{\text{actual}} = \frac{(S_{\text{rms}})^2}{R_L} = \frac{P_s}{R_L} = \frac{6600}{50} = 132 \text{ W} \tag{12.22}$$

The actual average carrier power is $\frac{1}{2}A_c^2/R_L = 100$ W and the actual average sideband power is $\frac{1}{2}A_c^2 P_m/R_L = 32$ W. Using Eq. (12.16), the actual PEP is

$$(P_{\text{PEP}})_{\text{actual}} = \frac{A_c^2}{2R_L}[1 + \max m(t)]^2 = \frac{(100)^2}{2(50)}(1 + (0.8))^2 = 324 \text{ W} \tag{12.23}$$

12.1.5 Bandpass Filtering and the Equivalent Low-Pass Filter

If a bandpass signal $v_1(t)$ is filtered by a bandpass filter, then the filtered output signal is

$$v_2(t) = v_1(t) * h(t) \tag{12.24}$$

where the asterisk denotes the convolution operation and $h(t)$ is the impulse response of the bandpass filter. In the frequency domain this becomes

$$V_2(f) = H(f)V_1(f) \tag{12.25}$$

where $H(f)$ is the transfer function of the bandpass filter. In general, the mathematics involved with bandpass filter problems are more difficult than those for low-pass filter problems.

A shortcut technique will be given for modeling a bandpass filter by using an equivalent low-pass filter that has a complex valued impulse response. This is shown in Fig. 12.3. Here, $v_1(t)$ and $v_2(t)$ are the input and output bandpass waveforms with the corresponding complex envelopes $g_1(t)$ and $g_2(t)$. The impulse response of the bandpass filter, $h(t)$, can also be represented by its corresponding complex envelope $k(t)$. In addition, as shown in Fig. 12.3(a), the frequency domain description $H(f)$ can be expressed in terms of $K(f)$ with the help of Eq. (12.8) and Eq. (12.9). Figure 12.3(b) shows a typical bandpass frequency response characteristic $|H(f)|$. The complex envelopes for the input, output, and impulse response of a

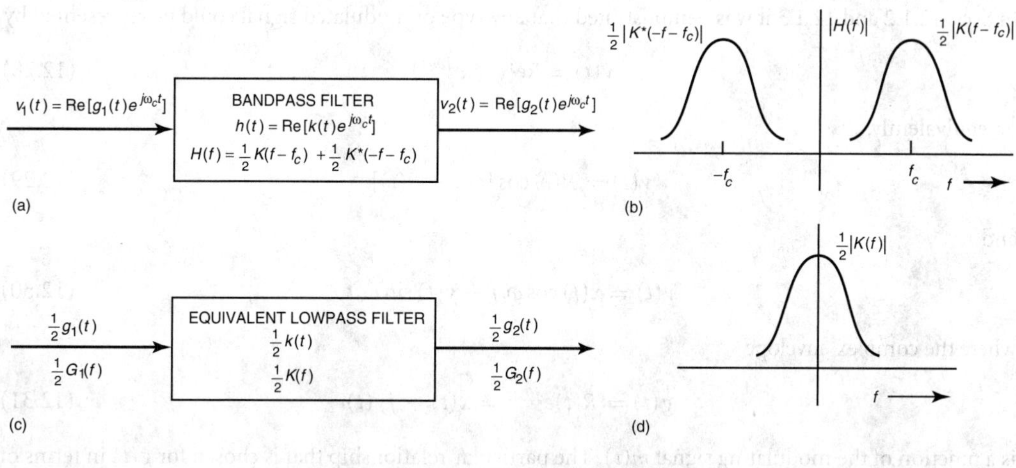

FIGURE 12.3 Bandpass filtering: (a) bandpass filter, (b) typical bandpass filter frequency response, (c) equivalent (complex impulse response) lowpass filter, (d) typical equivalent lowpass filter frequency response. (*Source*: Couch, L.W., II. 1995. *Modern Communication Systems*, p. 240. Prentice-Hall, Englewood Cliffs, NJ. With permission.)

bandpass filter are related by (Couch, 1995, p. 240)

$$\frac{1}{2}g_2(t) = \frac{1}{2}g_1(t) * \frac{1}{2}k(t) \tag{12.26}$$

where $g_1(t)$ is the complex envelope of the input and $k(t)$ is the complex envelope of the impulse response. It also follows that

$$\frac{1}{2}G_2(f) = \frac{1}{2}G_1(f)\frac{1}{2}K(f) \tag{12.27}$$

These equations indicate that any bandpass filter systems may be described and analyzed by using an *equivalent low-pass filter* as shown in Fig. 12.3(c). A typical equivalent low-pass frequency response characteristic is shown in Fig. 12.3(d). Since equations for equivalent low-pass filters are usually much less complicated than those for bandpass filters, the equivalent low-pass filter system model is very useful. This is the basis for computer programs that simulate bandpass communication systems by using equivalent low-pass models.

It is realized that the linear bandpass filter can produce variations in the phase modulation at the output, $\theta_2(t)$, where $\theta_2(t) = /g_2(t)$, as a function of the amplitude modulation on the input complex envelope $R_1(t)$, where $R_1(t) = |g_1(t)|$. This is called *AM-to-PM conversion*. Similarly, the filter can also cause variations on the AM at the output, $R_2(t)$, because of the PM on the input, $\theta_1(t)$. This is called *PM-to-AM conversion*.

Since $h(t)$ represents a linear filter, $g_2(t)$ will be a linear filtered version of $g_1(t)$; however, $\theta_2(t)$ and $R_2(t)$, the PM and AM components of $g_2(t)$, will be a *nonlinear* filtered version of $g_1(t)$ since $\theta_2(t)$ and $R_2(t)$ are nonlinear functions of $g_2(t)$. The analysis of the nonlinear distortion is very complicated. Although many analysis techniques have been published in the literature, none has been entirely satisfactory. Panter (1965) gives a three-chapter summary of some of these techniques, and a classical paper is also recommended (Bedrosian and Rice, 1968). Furthermore, nonlinearities that occur in a practical system will also cause nonlinear distortion and AM-to-PM conversion effects. Nonlinear effects can be analyzed by several techniques, including power-series analysis (Couch, 1995). If a nonlinear effect in a bandpass system is to be analyzed, a Fourier series technique that uses the Chebyshev transform has been found to be useful (Spilker, 1977).

12.1.6 Generalized Transmitters

Transmitters generate the modulated signal at the carrier frequency f_c from the modulating signal $m(t)$. In Secs. 12.1.2 and 12.1.3 it was demonstrated that any type of modulated signal could be represented by

$$v(t) = \text{Re}\{g(t)e^{j\omega_c t}\} \tag{12.28}$$

or, equivalently,

$$v(t) = R(t)\cos[\omega_c t + \theta(t)] \tag{12.29}$$

and

$$v(t) = x(t)\cos\omega_c t - y(t)\sin\omega_c t \tag{12.30}$$

where the complex envelope

$$g(t) = R(t)e^{j\theta(t)} = x(t) + jy(t) \tag{12.31}$$

is a function of the modulating signal $m(t)$. The particular relationship that is chosen for $g(t)$ in terms of $m(t)$ defines the type of modulation that is used, such as FM, SSB, or FM (see Table 12.1). A generalized approach may be taken to obtain universal transmitter models that may be reduced to those used for a particular modulation type. We will also see that there are equivalent models that correspond to different

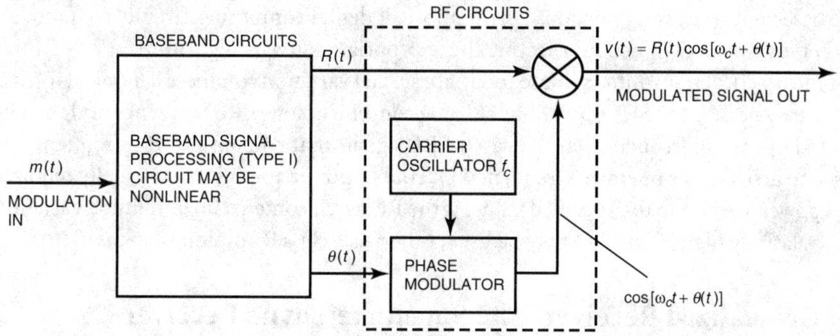

FIGURE 12.4 Generalized transmitter using the AM–PM generation technique. (*Source:* Couch, L.W., II. 1995. *Modern Communication Systems*, p. 282. Prentice-Hall, Englewood Cliffs, NJ. With permission.)

circuit configurations, yet they may be used to produce the same type of modulated signal at their outputs. It is up to the designer to select an implementation method that will maximize performance, yet minimize cost based on the state of the art in circuit development. There are two canonical forms for the generalized transmitter, as indicated by Eq. (12.29) and Eq. (12.30). Equation (12.29) describes an AM–PM type circuit as shown in Fig. 12.4. The baseband signal processing circuit generates $R(t)$ and $\theta(t)$ from $m(t)$. The R and θ are functions of the modulating signal $m(t)$ as given in Table 12.1(b) for the particular modulation type desired. The signal processing may be implemented either by using nonlinear analog circuits or a computer that incorporates the R and θ algorithms under software program control. In the implementation using a computer, one analog-to-digital converter (ADC) will be needed at the input and two digital-to-analog converters (DACs) will be needed at the output. The remainder of the AM–PM canonical form requires radio frequency (RF) circuits, as indicated in the figure.

Figure 12.5 illustrates the second canonical form for the generalized transmitter. This uses in-phase and quadrature-phase (IQ) processing. Similarly, the formulas relating $x(t)$ and $y(t)$ to $m(t)$ are shown in Table 12.1(a), and the baseband signal processing may be implemented by using either analog hardware or digital hardware with software. The remainder of the canonical form utilizes RF circuits as indicated.

Once again, it is stressed that any type of signal modulation, AM, FM, SSB, quadrature phase shift keying (QPSK), etc., may be generated by using either of these two canonical forms. Both of these forms conveniently separate baseband processing from RF processing. Digital techniques are especially useful to

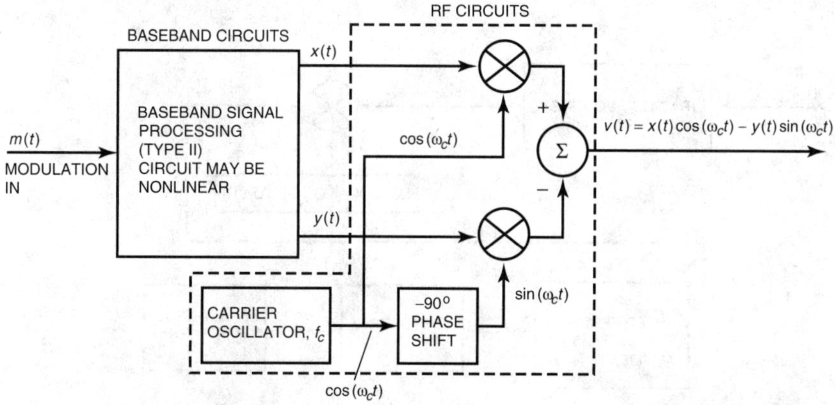

FIGURE 12.5 Generalized transmitter using the quadrature generation technique. (*Source:* Couch, L.W., II. 1995. *Modern Communication Systems*, p. 282. Prentice-Hall, Englewood Cliffs, NJ. With permission.)

realize the baseband processing portion. Furthermore, if digital computing circuits are used, any desired modulation type can be obtained by selecting the appropriate software algorithm.

Most of the practical transmitters in use today are special variations on these canonical forms. Practical transmitters may perform the RF operations at some convenient lower RF frequency and then up-convert to the desired operating frequency. In the case of RF signals that contain no AM, frequency multipliers may be used to arrive at the operating frequency. Of course, power amplifiers are usually required to bring the output power level up to the specified value. If the RF signal contains no amplitude variations, class C amplifiers (relatively high efficiency) may be used; otherwise, class B amplifiers are used.

12.1.7 Generalized Receiver: The Superheterodyne Receiver

The receiver has the job of extracting the source information from the received modulated signal that has been corrupted by noise. Often it is desired that the receiver output be a replica of the modulating signal that was present at the transmitter input. There are two main classes of receivers: the *tuned radio frequency* (TRF) receiver and the *superheterodyne* receiver.

The TRF receiver consists of a number of cascaded high-gain RF bandpass stages that are tuned to the carrier frequency f_c, followed by an appropriate detector circuit (envelope detector, product detector, FM detector, etc.). The TRF is not very popular because it is difficult to design tunable RF stages so that the desired station can be selected and yet have a narrow bandwidth so that adjacent channel stations are rejected. In addition, it is difficult to obtain high gain at radio frequencies and to have sufficiently small stray coupling between the output and input of the RF amplifying chain so that the chain will not become an oscillator at f_c.

Most receivers employ the superheterodyne receiving technique as shown in Fig. 12.6. (*Dual-conversion* superheterodyne receivers can also be built. Here a second mixer and a second IF stage follow the first IF stage shown in Fig. 12.6.) This technique consists of either down converting or up converting the input signal to some convenient frequency band, called the *intermediate frequency* (IF) band, and then extracting the information (or modulation) by using the appropriate detector. This basic receiver structure is used for the reception of all types of bandpass signals, such as television, FM, AM, satellite, and radar signals. The RF amplifier has a bandpass characteristic that passes the desired signal and provides amplification to override additional noise that is generated in the mixer stage. The RF filter characteristic also provides some rejection of adjacent channel signals and noise, but the main adjacent channel rejection is accomplished by the IF filter.

The IF filter is a bandpass filter that selects either the up-conversion or down-conversion component (whichever is chosen by the receiver designer). When up conversion is selected, the complex envelope of the IF (bandpass) filter output is the same as the complex envelope for the RF input, except for the effect of

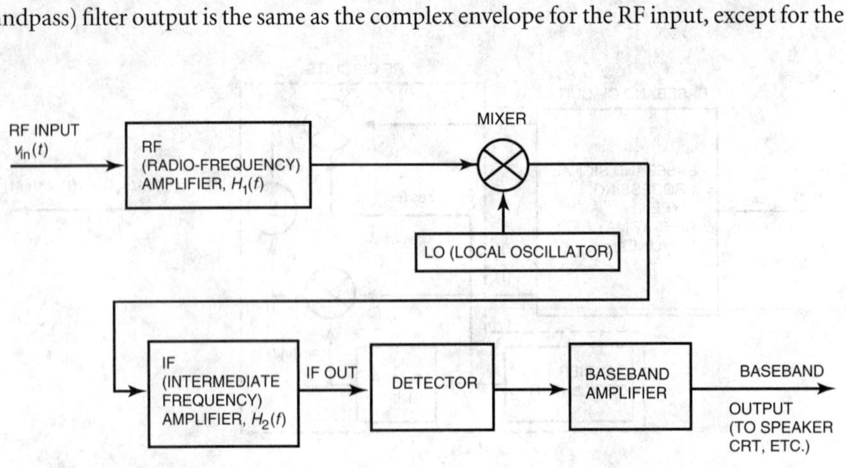

FIGURE 12.6 A superheterodyne receiver. (*Source:* Couch, L.W., II. 1995. *Modern Communication Systems*, p. 283. Prentice-Hall, Englewood Cliffs, NJ. With permission.)

RF filtering $H_1(f)$ and IF filtering $H_2(f)$. However, if down conversion is used with $f_{LO} > f_c$, the complex envelope at the IF output will be the conjugate of that for the RF input. This means that the sidebands of the IF output will be inverted (i.e., the upper sideband on the RF input will become the lower sideband, etc., on the IF output). If $f_{LO} < f_c$, the sidebands are not inverted. In some special applications the local oscillator (LO) frequency is selected to be the carrier frequency ($f_{LO} = f_c$), and then the superheterodyne receiver becomes a direct-conversion receiver also called a *homodyne* or a *synchrodyne* receiver. In this case the IF filter is replaced by a low-pass filter so that the mixer-LPF combination becomes a product detector and the detector stage of Fig. 12.6 is deleted. Thus, the direct-conversion receiver is the same as a TRF receiver with product detection.

The center frequency selected for the IF amplifier is chosen based on the following three considerations:

- The frequency should be such that a stable high-gain amplifier can be economically attained.
- The frequency needs to be low enough so that with practical circuit elements in the IF filters, values of Q can be attained that will provide a steep attenuation characteristic outside the bandwidth of the IF signal. This decreases the noise and minimizes the interference from adjacent channels.
- The frequency needs to be high enough so that the receiver image response can be made acceptably small.

The *image response* is the reception of an unwanted signal located at the image frequency due to insufficient attenuation of the image signal by the RF amplifier filter. The image is best illustrated by an example.

Example: AM Broadcast Superheterodyne Receiver

Assume that an AM broadcast band radio is tuned to receive a station at 850 kHz and that the LO frequency is on the high side of the carrier frequency. If the IF frequency is 455 kHz, the LO frequency will be $850 + 455 = 1305$ kHz (see Fig. 12.7). Furthermore, assume that other signals are present at the RF input of the radio and, particularly, that there is a signal at 1760 kHz; this signal will be down converted by the mixer to $1760 - 1305 = 455$ kHz. That is, the undesired (1760-kHz) signal will be translated to 455 kHz and will be added at the mixer output to the desired (850-kHz) signal, which was also down converted to 455 kHz. This *undesired* signal that has been converted to the IF band is called the *image signal*. If the gain of the RF amplifier is down by, say, 25 dB at 1760 kHz as compared to 850 kHz, and if the undesired signal is 25 dB stronger at the receiver input than the desired signal, both signals will have the same level when translated to the IF. In this case the undesired signal will definitely interfere with the desired signal in the detection process.

For down converters (i.e., $f_{IF} = |f_c - f_{LO}|$) the image frequency is

$$f_{image} = \begin{cases} f_c + 2f_{IF}, & \text{if } f_{LO} > f_c \quad \text{(high-side injection)} \\ f_c - 2f_{IF}, & \text{if } f_{LO} < f_c \quad \text{(low-side injection)} \end{cases} \qquad (12.32a)$$

FIGURE 12.7 Spectra of signals and transfer function of an RF amplifier in a superheterodyne receiver. (*Source:* Couch, L.W., II. 1995. *Modern Communication Systems*, p. 285. Prentice-Hall, Englewood Cliffs, NJ. With permission.)

TABLE 12.2   Some Popular IF Frequencies in the U.S.

IF Frequency	Application
262.5 kHz	AM broadcast radios (in automobiles)
455 kHz	AM broadcast radio
10.7 MHz	FM broadcast radio
21.4 MHz	FM two-way radio
30 MHz	Radar receivers
43.75 MHz (video carrier)	TV sets
60 MHz	Radar receivers
70 MHz	Satellite receivers

Source: Couch, L.W., II. 1995. *Modern Communication Systems*, p. 285. Prentice-Hall, Englewood Cliffs, NJ. With permission.

where f_c is the desired RF frequency, f_{IF} is the IF frequency, and f_{LO} is the local oscillator frequency. For up converters (i.e., $f_{IF} = f_c + f_{LO}$) the image frequency is

$$f_{image} = f_c + 2f_{LO} \tag{12.32b}$$

From Fig. 12.7 it is seen that the image response will usually be reduced if the IF frequency is increased, since f_{image} will occur farther away from the main peak (or lobe) of the RF filter characteristic $|H_1(f)|$.

We also realize that other spurious responses (in addition to the image response) will occur in practical mixer circuits. These must also be taken into account in good receiver design.

Table 12.2 illustrates some typical IF frequencies that have become *de facto* standards. For the intended application, the IF frequency is low enough that the IF filter will provide good adjacent channel signal rejection when circuit elements with realizable Q are used; yet the IF frequency is large enough to provide adequate image-signal rejection by the RF amplifier filter.

The type of detector selected for use in the superheterodyne receiver depends on the intended application. For example, a product detector may be used in a phase shift keyed (PSK) digital system, and an envelope detector is used in AM broadcast receivers. If the complex envelope $g(t)$ is desired for generalized signal detection or for optimum reception in digital systems, the $x(t)$ and $y(t)$ quadrature components, where $x(t) + jy(t) = g(t)$, may be obtained by using quadrature product detectors, as illustrated in Fig. 12.8. Here, $x(t)$ and $y(t)$ could be fed into a signal processor to extract the modulation information. Disregarding the effects of noise, the signal processor could recover $m(t)$ from $x(t)$ and $y(t)$ (and, consequently, demodulate the IF signal) by using the inverse of the complex envelope generation functions given in Table 12.1(a).

FIGURE 12.8   IQ (in-phase and quadrature-phase) detector. (*Source:* Couch, L.W., II. 1995. *Modern Communication Systems*, p. 286. Prentice-Hall, Englewood Cliffs, NJ. With permission.)

The superheterodyne receiver has many advantages and some disadvantages. The main advantage is that extraordinarily high gain can be obtained without instability (self-oscillation). The stray coupling between the output of the receiver and the input does not cause oscillation because the gain is obtained in disjoint frequency bands: RF, IF, and baseband. The receiver is easily tunable to another frequency by changing the frequency of the LO signal (which may be supplied by a frequency synthesizer) and by tuning the bandpass of the RF amplifier to the desired frequency. Furthermore, high-Q elements, which are needed to produce steep filter skirts for adjacent channel rejections, are needed only in the fixed tuned IF amplifier. The main disadvantage of the superheterodyne receiver is the response to spurious signals that will occur if one is not careful with the design.

A discussion of receivers would not be complete without considering some of the causes of *interference*. Often the receiver owner thinks that a certain signal, such as an amateur radio signal, is causing the difficulty. This may or may not be the case. The origin of the interference may be at any of the three following locations:

- At the interfering signal *source*, the transmitter may generate out-of-band signal components (such as harmonics) that fall in the band of the desired signal.

- At the *receiver* itself, the front end may overload or produce spurious responses. Front-end overload occurs when the RF and/or mixer stage of the receiver is driven into the nonlinear range by the interfering signal and the nonlinearity causes cross modulation on the desired signal at the output of the receiver RF amplifier.

- In the *channel*, a nonlinearity in the transmission medium may cause undesired signal components in the band of the desired signal.

For more discussion of receiver design and examples of practical receiver circuits, the reader is referred to the *ARRL Handbook* (ARRL, 1994).

Defining Terms

Actual average power: For voltage waveform impressed across a resistive load R_L, the actual average power is V_{rms}^2/R_L watts. For a current waveform flowing through a resistive load R_L, the actual average power is $I_{rms}^2 R_L$ watts.

Bandpass waveform: The spectrum of the waveform is nonzero for frequencies in some band concentrated about a frequency $f_c \gg 0$; f_c is called the carrier frequency.

Baseband waveform: The spectrum of the waveform is nonzero for frequencies near $f = 0$.

Complex envelope: The function $g(t)$ of a bandpass waveform $v(t)$ where the bandpass waveform is described by $v(t) = \text{Re}\{g(t)e^{j\omega_c t}\}$.

Current spectrum: The Fourier transform of a current waveform.

Fourier transform: If $w(t)$ is a waveform, then the Fourier transform of $w(t)$ is

$$W(f) = \Im[w(t)] = \int_{-\infty}^{\infty} w(t)e^{-j2\pi f t}\, dt$$

where f has units of hertz.

Modulated signal: The bandpass signal $s(t) = \text{Re}\{g(t)e^{j\omega_c t}\}$ where fluctuations of $g(t)$ are caused by an intelligence source such as audio, video, or data.

Modulation: The intelligence or information source, $m(t)$, that causes fluctuations in a bandpass signal.

Normalized average power: The average power dissipated in a 1-Ω resistive load. The normalized average power is the average of the square of the voltage or current waveform.

Power spectral density (PSD): A function of frequency $\wp_w(f)$ associated with a waveform $w(t)$, such that the area under the PSD over a frequency band gives the normalized average power for that waveform corresponding to that band of frequencies [Couch 1995, p. 62].

Real envelope: The function $R(t) = |g(t)|$ of a bandpass waveform $v(t)$ where the bandpass waveform is described by

$$v(t) = \text{Re}\{g(t)e^{j\omega_c t}\} = R(t)\cos[\omega_c t + \theta(t)]$$

Voltage spectrum: The Fourier transform of a voltage waveform.

References

ARRL. 1994. *The ARRL Handbook for Radio Amateurs*, 72nd ed., ed. R. Schetgen. American Radio Relay League, Newington, CT.

Bedrosian, E.B. and Rice, S.O. 1968. Distortion and crosstalk of linearly filtered angle-modulated signals. *Proc. of the IEEE* 56(Jan.):2–13.

Couch, L.W., II. 1995. *Modern Communication Systems*. Prentice-Hall, Englewood Cliffs, NJ.

Panter, P.F. 1965. *Modulation, Noise and Spectral Analysis*. McGraw-Hill, New York.

Rappaport, T.S. 1989. Characteristics of UHF multipath radio channels in factory buildings. *IEEE Trans. on Antenna and Propagation* 37(Aug.):1058–1069.

Rice, S.O. 1954. Mathematical analysis of random noise, In *Selected Papers on Noise and Stochastic Processes*, ed. N. Wax. Dover, New York. (Reprinted from *Bell Sys. Tech. J.* 23(July 1944):282–333 and 24(Jan. 1945):46–156.)

Shannon, C.E. 1948. A mathematical theory and communication. *Bell Sys. Tech. J.* 27(July):379–423 and 27(Oct.):623–656.

Spilker, J.J. 1977. *Digital Communication by Satellite*. Prentice-Hall, Englewood Cliffs, NJ.

Further Information

In addition, the following books are recommended:

Couch, L.W., II. 1993. *Digital and Analog Communication Systems*, 4th ed. Macmillan, New York, now Prentice-Hall, Englewood Cliffs, NJ.

Peterson, R.L., Ziemer, R.E., and Borth, D.E. 1995. *Introduction to Spread Spectrum Communications*. Prentice-Hall, Englewood Cliffs, NJ.

Proakis, J.G. 1995. *Digital Communications*, 3rd ed. McGraw-Hill, New York.

12.2   Amplitude Modulation

Robert Kubichek

12.2.1   Amplitude Modulation

There are two basic types of communication systems, *baseband* systems and *passband* systems. In baseband systems, the signal is transmitted without modifying the frequency content. A simple intercom is an example of this approach. Here, a microphone senses the input or **message signal**, and injects the resulting signal $m(t)$ into a cable or *channel*. At the receiver, the signal is filtered to remove noise, amplified and reproduced into sound using a speaker. In passband communication systems, the message signal is *modulated* by translating its spectrum to a new frequency location called the *carrier frequency*, f_c. **Modulation** offers numerous advantages over baseband communication including

- Maximize efficiency: Signals can be modulated into regions of the spectrum where there is lower noise and interference, or better propagation characteristics.
- Frequency-domain multiplexing (FDM): Multiple signals can be modulated into nonoverlapping frequency bands and transmitted simultaneously over the same channel. In commercial broadcasting, for example, this allows AM and FM signals to be multiplexed onto the same radio frequency channel.

- Physical considerations: The physical size of antennas and other electronic components tend to decrease with increasing frequency. This makes it feasible to build smaller radio receivers, transmitters, and antennas when higher frequencies are used.

Frequency Domain Concepts

To understand the basic concepts of modulation, we first review time- and frequency-domain representations of signals. The information present in $m(t)$ can be completely specified by a complex function of speech amplitude vs. frequency, $M(f)$, obtained using the *Fourier transform*. Since $M(f)$ contains all of the information in $m(t)$, it is possible to go back and forth between $m(t)$ and $M(f)$ using the forward and inverse Fourier transforms.

Forward transform:

$$M(f) = \int_{-\infty}^{\infty} m(t) e^{-j2\pi ft} \, dt \qquad (12.33)$$

Inverse transform:

$$m(t) = \int_{-\infty}^{\infty} M(f) e^{j2\pi ft} \, df \qquad (12.34)$$

Although we usually think of positive-valued frequencies, Eq. (12.33) shows that negative frequency values are also mathematically valid. We can get a better feel for this idea by considering the signal $\sin 2\pi f_c t$, which is a pure tone with frequency f_c Hz. The relationship $\sin 2\pi f_c t = -\sin 2\pi (-f_c)t$ shows that a pure tone at $-f_c$ Hz has the same magnitude as the positive frequency pure tone, but is different in phase by $180°$.

Modulation Theorem

The *modulation theorem* states that when any signal is multiplied by a sine-wave signal of frequency f_c, the resulting signal has a spectrum similar to the original, but translated out to frequency $\pm f_c$. That is, consider a signal $x(t)$ produced by multiplying $m(t)$ against a tone signal

$$x(t) = m(t) \cos (2\pi f_c t) \qquad (12.35)$$

The frequency-domain result is

$$X(f) = 0.5\{M(f - f_c) + M(f + f_c)\} \qquad (12.36)$$

where $M(f - f_c)$ and $M(f + f_c)$ represent the entire message spectrum translated to two new locations, the carrier frequency f_c and the negative carrier frequency $-f_c$, respectively. This multiplication process is called *mixing* or *heterodyning*. Figure 12.9 shows an example of the modulation theorem where $M(f)$ represents the baseband spectrum and $X(f)$ is the modulated spectrum. A second example in Fig. 12.10 shows the passband signal $x(t)$ from Fig. 12.9 being modulated to produce components at $2f_c$, 0, and $-2f_c$ Hz. This approach can be used in a receiver to *demodulate* the passband signal and reproduce the

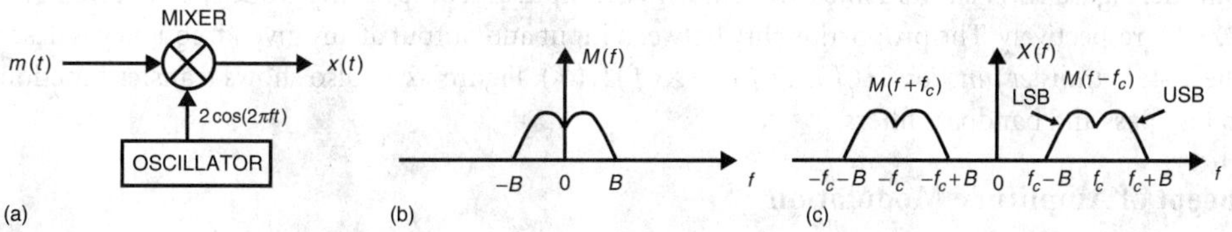

FIGURE 12.9 Example of the modulation theorem: (a) the modulator, (b) the input message spectrum, (c) the modulated spectrum composed of upper and lower sidebands (USB and LSB).

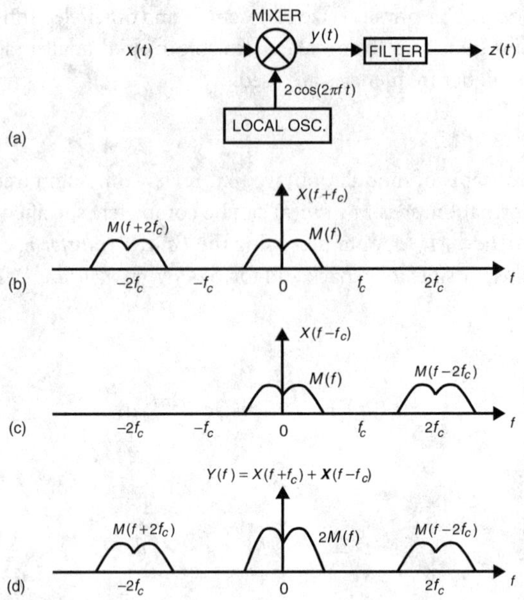

FIGURE 12.10 Example of modulation theorem used in demodulation: (a) the demodulator, (b) and (c) the two components of the output signal, (d) the combined demodulator output spectrum before filtering.

original baseband signal. This is accomplished simply by removing the spectral components at $\pm 2 f_c$ using a filter.

The spectrum $M(f)$ is a complex function that can be represented either by real and imaginary or by magnitude and phase components $|M(f)|$ and $\phi(f)$, respectively. When $m(t)$ is real valued, the magnitude spectrum is an even function ($|M(f)| = |M(-f)|$) and phase is an odd function ($\phi(f) = -\phi(-f)$) of frequency. Most basic communication concepts can be illustrated using only the magnitude spectrum $|M(f)|$; this will be the approach taken in this chapter.

Bandwidth: Baseband versus Passband

Bandwidth is defined as the range of positive frequencies occupied by a signal. Thus, for the baseband signal shown in Fig. 12.9(b), the bandwidth equals the highest frequency present, B Hz. By comparison, the modulated signal shown in Fig. 12.9(c) has a bandwidth of $2B$ Hz, which occupies double the bandwidth of the baseband signal and, therefore, represents poor spectral efficiency.

Linear Systems

Low-pass and *bandpass* filters (denoted LPF and BPF, respectively) are fundamental components in amplitude modulation systems. These filters, along with the communication channel itself, are often modeled as *linear systems*, which implies that system output amplitude is strictly proportional to the system input amplitude. Figure 12.11 shows a linear system with the input and output amplitude spectra given by $X(f)$ and $Y(f)$, respectively. The proportionality between input and output at any given frequency is described by the system *transfer function* $H(f)$: $Y(f) = X(f)H(f)$. Figure 12.11 also shows transfer functions for both low-pass and bandpass filters.

Concept of Amplitude Modulation

We can write Eq. (12.33) in more general terms

$$w(t) = A(t) \cos \left[2\pi f_c t + \theta(t) \right]$$

(12.37)

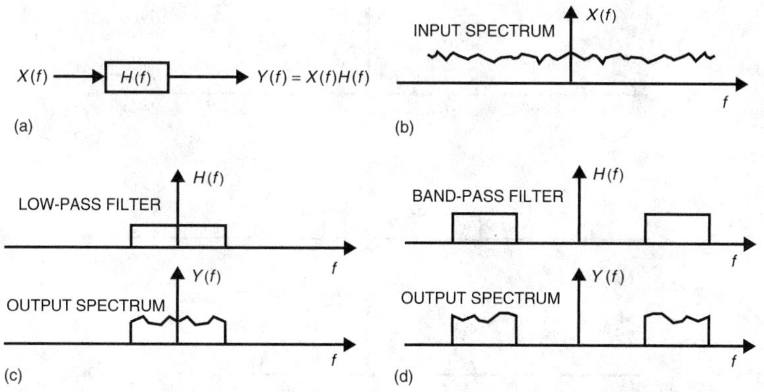

FIGURE 12.11 Example of low-pass and bandpass linear filters: (a) the linear system, (b) the input signal spectrum, (c) low-pass filter and output spectrum, (d) a bandpass filter and output spectrum.

where both the amplitude and phase are allowed to vary as functions of time. If the message signal $m(t)$ affects only the amplitude $A(t)$, the resulting signal is termed **amplitude modulation**. This generic category includes the well known commercial broadcast amplitude modulation (AM) system, as well as **double sideband (DSB), single sideband (SSB)**, and **vestigial sideband (VSB)** transmission. When the message signal is transmitted by modulating the phase component $\theta(t)$, the result is called *angle modulation*. This includes *frequency modulation* (FM) and *phase modulation* (PM) techniques to be covered in Section 12.3.

Double Sideband-Suppressed Carrier (DSB-SC)

A transmission system directly implementing Eq. (12.35) is shown in Fig. 12.9(a) and is called double-sideband (DSB) modulation. The two lobes observed in the magnitude spectrum in Fig. 12.9(c) are called the *upper sideband* (USB) and *lower sideband* (LSB). The USB and LSB are mirror images of each other (but with opposite phases) and each contains sufficient information to reconstruct the message signal $m(t)$.

A transmitter requires an *oscillator* to generate the **carrier** *signal* $c(t) = \cos(2\pi f_c t)$ and a *mixer* to multiply $m(t)$ and $c(t)$.

An alternative modulation system is shown in Fig. 12.12. This technique uses an electronic switch rather than a carrier and a mixer. To see why this works, consider that switching $m(t)$ on and off is the same as multiplying $m(t)$ with a square wave $s(t)$ having frequency f_c; $x(t) = m(t)s(t)$. Using Fourier series, $s(t)$ can be written as a trigonometric expansion:

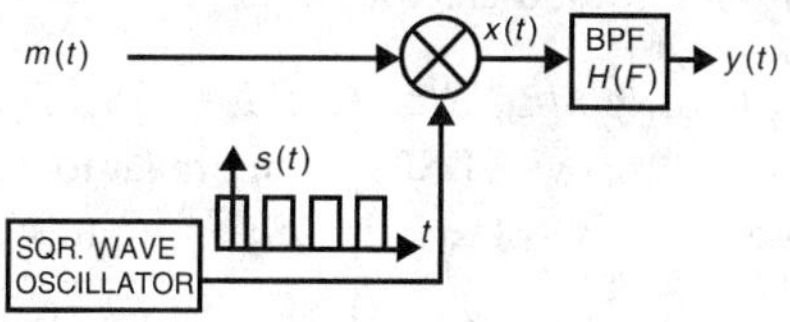

FIGURE 12.12 DSB-SC modulation using a switching circuit.

$$s(t) = 0.5 + a_1 \cos 2\pi f_0 t + a_3 \cos 2\pi 3 f_0 t + a_5 \cos 2\pi 5 f_0 t + \cdots$$

where a_i represent the ith Fourier series coefficient. The switching output is then given by

$$x(t) = 0.5m(t) + a_1 m(t) \cos 2\pi f_0 t + a_3 m(t) \cos 2\pi 3 f_0 t + a_5 m(t) \cos 2\pi 5 f_0 t + \cdots$$

Applying the modulation theorem to each term, we see that the expansion represents an infinite sum of DSB signals with passbands centered at 0, $\pm f_0$, $\pm 3 f_0$, $\pm 5 f_0$, etc. A bandpass filter selects the desired DSB signal at $\pm f_c$ and attenuates the remaining undesired harmonics as shown in Fig. 12.13.

Figure 12.10 shows a simple design for a receiver (or **detector** or *demodulator*). The receiver requires a local oscillator to generate $2\cos(2\pi f_c t)$, a mixer, and a low-pass filter at the mixer output to remove the

FIGURE 12.13 The switched signal spectrum, $X(f)$, can be filtered to create a DSB signal: (a) the switched signal spectrum, (b) the bandpass filter response, (c) the DSB output spectrum.

components at $\pm 2 f_c$. Receiver operation is described in Table 12.3 in both time- and frequency-domain equations.

Ideally, the receiver oscillator, or *local* oscillator, produces a carrier signal in phase with the transmitter oscillator, that is, $C_{\text{trans}}(t) = C_{\text{rcvr}}(t) = \cos(2\pi f_c t)$; this is called *coherent detection* or *product detection*. Keeping both oscillators perfectly in phase, however, can be quite difficult. To find out what happens if the transmitter carrier $C_{\text{trans}}(t)$ and receiver carrier $C_{\text{rcvr}}(t)$ have different phases, we can write $C_{\text{trans}}(t) = \cos(2\pi f_c t)$ and $C_{\text{rcvr}}(t) = \cos(2\pi f_c t + \varphi)$, where φ represents phase error. An analysis similar to Table 12.3 gives the result $z(t) = m(t) \cos(\varphi)$. This shows that any nonzero phase errors will decrease the receiver output amplitude by an amount $\cos(\varphi)$, the worst case occurring when $\varphi = \pi/2$, which produces zero output.

To minimize phase error effects, DSB systems often employ *phase-locked loops* in the receivers to lock the local oscillator into phase synchony with the received signal. In another approach, a pilot signal is transmitted simultaneously with the DSB signal but on a different frequency. The pilot is typically a sine wave in phase with $C_{\text{trans}}(t)$ and chosen to be harmonically related to f_c. It is used at the receiver to generate an in-phase local oscillator signal.

Double Sideband Plus Carrier (DSB+C)

Figure 12.14 shows a DSB system similar to that in Fig. 12.9 except for the constant DC level added to $m(t)$ prior to the mixer stage. The mixer output signal is, thus

$$x(t) = [A + m(t)] \cos(2\pi f_c t) = A \cos(2\pi f_c t) + m(t) \cos(2\pi f_c t) \tag{12.38}$$

This consists of a simple DSB signal (the second term) plus a carrier of amplitude A (the first term). The carrier concentrates all of its energy at frequency f_c and shows up in the magnitude spectrum as a spike. This is

TABLE 12.3 Demodulation Equations

Description	Time Domain	Frequency Domain
Received signal	$x(t) = m(t) \cos(2\pi f_c t)$	$X(f) = 0.5\{M(f + f_c) + M(f - f_c)\}$
Mixer output	$y(t) = 2x(t) \cos(2\pi f_c t)$	$Y(f) = 2^*0.5\{X(f + f_c) + X(f - f_c)\}$
	$= 2m(t) \cos(2\pi f_c t) \cos(2\pi f_c t)$	$= 0.5M(f + 2f_c) + 0.5M(f + f_c - f_c)$
		$+ 0.5M(f - f_c + f_c) + 0.5M(f - 2f_c)$
Some algebra: (use the trig identity: $\cos^2 A = 0.5 + 0.5 \cos(2A)$	$y(t) = m(t) + m(t) \cos(4\pi f_c t)$	$Y(f) = 0.5M(f + 2f_c) + M(f) + 0.5M(f - 2f_c)$
Low-pass filter output	$z(t) = m(t)$	$Z(f) = M(f)$

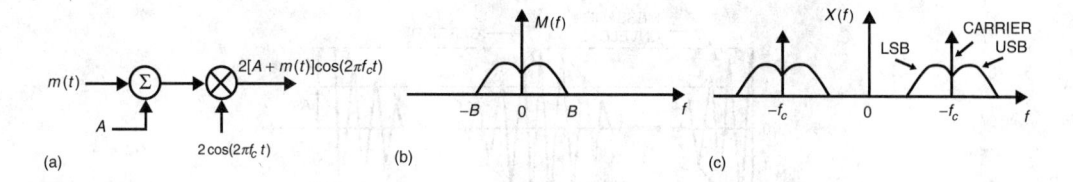

FIGURE 12.14 Insertion of carrier producing DSB+C signal showing (a) DSB+C modulator, (b) the message spectrum, (c) the DSB+C spectrum.

the classic AM transmission scheme and is also known as DSB plus carrier (DSB+C). The system described in the last section is often called DSB-suppressed carrier or DSB-SC to differentiate it from DSB+C.

Figure 12.15 illustrates the DSB-SC and DSB+C waveforms. In DSB+C, the message signal is contained in the positive (or negative) *envelope* and the carrier is confined within the envelope boundary. No simple relationship between $m(t)$ and the envelope is evident in DSB-SC waveform.

The amount of signal impressed on the carrier is measured by a *modulation index*, defined as

$$m = -\min[m(t)]/A$$

which is sometimes stated as a percentage. Figure 12.16, for example, shows AM signals with 100, 50, and 120% modulation indices. The 120% modulation case shows that the envelope containing $m(t)$ is distorted, making it impossible to recover $m(t)$ during demodulation. This is called *overmodulation*.

The modulation index is also indicative of the power efficiency of the AM transmission. From Eq. (12.38) it is clear that the term due to the carrier conveys no useful information, whereas the sideband term contains all of the message signal content. Efficiency can be defined as the ratio of energy containing useful information to the total transmitted energy, and can be written in terms of the modulation index:

$$E = m^2/(2 + m^2) \times 100\%$$

The highest efficiency is at 100% ($m = 1$) modulation and results in only 33% efficiency. Lower modulation values result in even lower efficiencies. By contrast, suppressed-carrier DSB systems are 100% efficient since they waste no energy in a carrier; this is a key disadvantage of AM systems.

The DSB+C signal can be generated without a mixing circuit by using a nonlinear device (NLD) such as a diode, as in Fig. 12.17. For example, suppose the input x and output y of an NLD can be modeled as

$$y = a_1 x + a_2 x^2$$

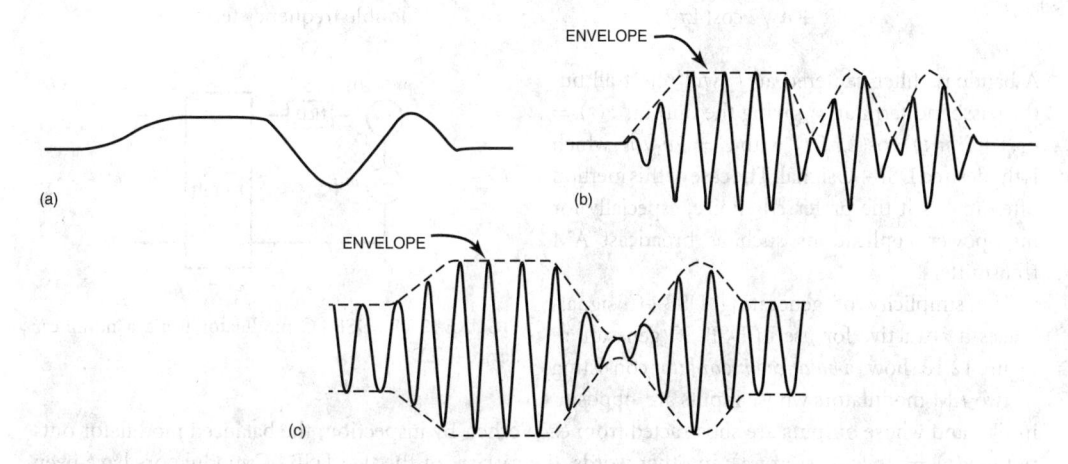

FIGURE 12.15 The envelope of the DSB-SC signal (b) cannot be used to reconstruct the message signal (a). In contrast, the message is preserved in the DSB+C envelope (c).

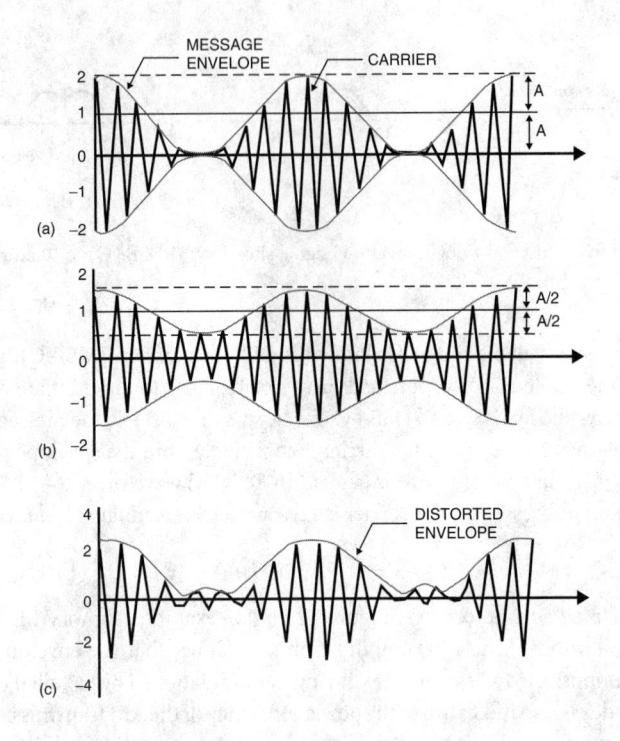

FIGURE 12.16   DSB+C signal with (a) 100%, (b) 50%, (c) 120% modulation index.

Noting that the NLD input is $(\cos 2\pi f_c t + m(t))$, the output is given by

$$y(t) = a_1(\cos(2\pi f_c t) + m(t)) + a_2(\cos(2\pi f_c t) + m(t))^2$$

which can be simplified to give

$$
\begin{aligned}
y(t) = &\; a_1 m(t) + a_2 m^2(t) + a_2/2 && \text{baseband terms}\\
&+ a_1 \cos(2\pi f_c t) + 2a_2 m(t)\cos(2\pi f_c t) && \text{passband terms}\\
&+ a_2/2 \cos(4\pi f_c t) && \text{double frequency term}
\end{aligned}
$$

A bandpass filter centered at f_c will reject all but the passband term at f_c giving the outputs $z(t) = a_1(1 + Cm(t))\cos(2\pi f_c t)$, with $C = 2a_2/a_1$, which is the desired DSB+C signal. The ease of this method often makes it the preferred choice, especially for high-power applications such as broadcast AM transmitters.

    The simplicity of generating DSB+C signals makes it attractive for use in DSB-SC generation. Figure 12.18 shows a *balanced modulator* consisting of two AM modulators whose inputs are opposite

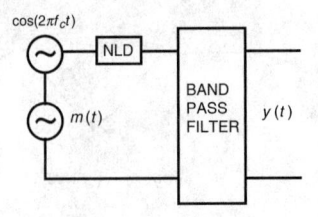

FIGURE 12.17   DSB+C modulator using a nonlinear device (NLD).

in sign and whose outputs are substracted from each other. By inspection, the balanced modulator output is $y(t) = 2m(t)\cos(2\pi f_c t)$; in other words, the carriers of the two DSB+C modulators have been suppressed to yield a DSB-SC signal. The balanced modulator can be thought of as a multiplier or mixer circuit operating on $m(t)$ and $\cos(2\pi f_c t)$.

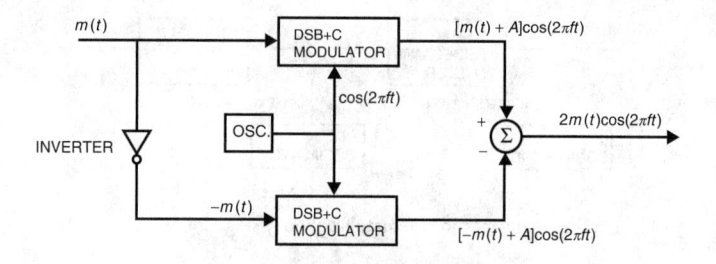

FIGURE 12.18 Balanced modulator to generate DSB-SC signals.

The DSB+C signal can be demodulated coherently, as described for DSB-SC signals. The real benefit of using DSB+C systems, however, is the ability to receive using *envelope detection* methods.

These receivers do not require coherent local oscillators and are much cheaper and easier to built than coherent receivers. Envelope detection is illustrated by the simple systems shown in Fig. 12.19. In Fig. 12.19(a), a diode clips the incoming signal to isolate the positive envelope of the signal. A resistor-capacitor (RC) stage is used to low-pass filter the diode output to remove most of the carrier component. The resulting signal is capacitively coupled to remove the unwanted DC component and then amplified for output through earphones or a speaker. Figure 12.19(b) shows a more general design called *rectifier detection*. A potential drawback with these techniques occurs if $m(t)$ contains significant low-frequency information: this energy will be attenuated by the capacitive coupling.

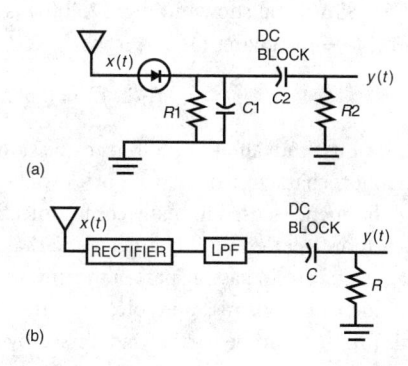

FIGURE 12.19 DSB+C detection using (a) envelope detection, (b) rectifier detection.

Single-Sideband

Double-sideband methods transmit redundant information since the baseband signal is duplicated in both the upper and lower sidebands. Consequently, the required bandwidth is twice the needed amount. In contrast, a single-sideband signal contains only the USB or LSB, thus reducing bandwidth by one-half. This is the primary advantage of SSB communication.

The most common technique for generating SSB signals is by applying a sharp bandpass filter to the DSB-SC signal, as shown in Fig. 12.20. This sideband filter passes only the desired sideband and suppresses the other sideband before the signal is amplified and transmitted.

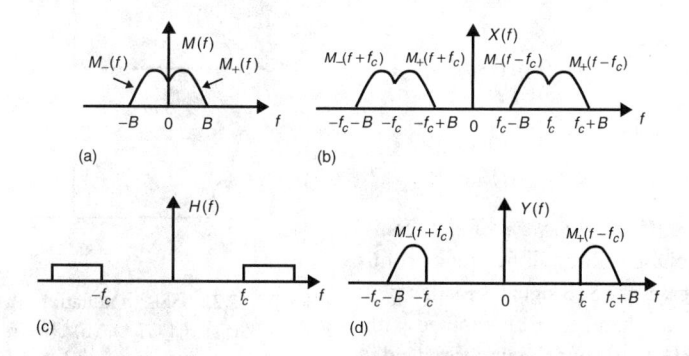

FIGURE 12.20 USB single sideband generation using a sharp BPF to remove the LSB: (a) shows the message spectrum, (b) the DSB-SC spectrum, (c) the sideband filter response, (d) the resulting USB SSB spectrum.

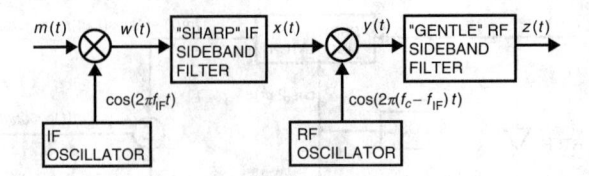

FIGURE 12.21   SSB generation using an intermediate frequency stage followed by a second modulator stage.

To help understand SSB operation, we note that the baseband signal can be expressed as the sum of a negative frequency part $M_-(f)$ and a positive frequency part $M_+(f)$: $M(f) = M_-(f) + M_+(f)$, as shown in Fig. 12.20. Similarly, the passband DSB-SC signal can be written as

$$X(f) = 0.5\{M_-(f + f_c) + M_+(f + f_c) + M_-(f - f_c) + M_+(f - f_c)\}$$

The USB-SSB signal shown in Fig. 12.20(d) is created by suppressing the lower sidebands, $M_+(f + f_c)$ and $M_-(f - f_c)$, giving

$$Y_{USB}(f) = 0.5\{M_-(f + f_c) + M_+(f - f_c)\}$$

A significant disadvantage of SSB transmission results from the need for sideband filters with extremely sharp rolloff characteristics. Such "brick wall" response is difficult to achieve in the real world, especially at high frequencies used in many communications systems. A typical bandpass filter (BPF) response has a much more gentle rolloff characteristic that could result in attenuating the low-frequency part of the message signal, or in passing part of the unwanted sideband, thus causing signal distortion. Consequently, single-sideband transmission typically suffers from poor low-frequency response.

This problem can be partially addressed by creating a DSB-SC signal at a low intermediate frequency f_{IF}, where $f_{IF} \ll f_c$, as shown in Fig. 12.21. A sharp sideband filter (which is much easier to implement at low frequency) is used to create a SSB-SC signal at frequency f_{IF}. This signal is modulated up to the desired carrier frequency where there are again two sidebands present. Since these are separated by a distance of $2f_{IF}$, the undesired sideband can be removed using an easily implemented BPF with gentle rolloff response as in Fig. 12.22.

An alternative method for SSB generation not requiring sideband filters is called the *phase-shifting method*. Mathematical analysis shows that the time-domain representations of the USB and LSB signals can be written as

$$y_{USB}(t) = m(t)\cos(2\pi f_c t) - m_H(t)\sin(2\pi f_c t)$$

and

$$y_{LSB}(t) = m(t)\cos(2\pi f_c t) + m_H(t)\sin(2\pi f_c t)$$

where $m_H(t)$ is the *Hilbert transform* of $m(t)$ computed as a $-90°$ phase shift of all frequency components. In both cases the SSB signal is made up of a DSB signal (the left-hand term) combined with a second term that cancels out the undesired sideband. The right-hand term is equal in magnitude to the DSB term, but has one sideband exactly $180°$

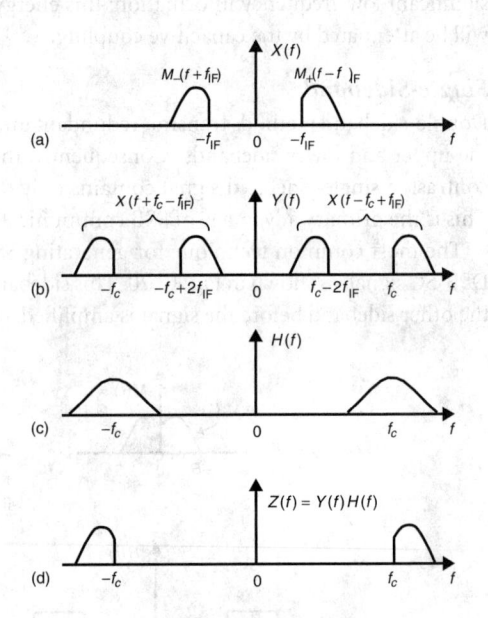

FIGURE 12.22   SSB modulation: (a) intermediate frequency SSB signal $X(f)$, (b) after modulation to passband, (c) gentle rolloff sideband filter response, (d) removal of unwanted sideband to produce the final SSB signal $Z(f)$.

out of phase. Thus, one of the sidebands is canceled when this term is added to or subtracted from the DSB component. Figure 12.23 shows a phase-shifting USB modulator implementing these equations. The primary disadvantage of this approach is the need for precision wide-band phase-shift networks that can be difficult to implement.

Although SSB generation is rather complex, detection can be accomplished using a coherent demodulation system identical to that used for DSB-SC signals (Fig. 12.10(a)). For example, when a USB SSB signal $Y_{USB}(f)$ is mixed with the local oscillator signal $2\cos(2\pi f_c t)$, the modulation theorem gives $Z(f) = 2*0.5\{Y_{USB}(f + f_c) + Y_{USB}(f - f_c)\}$, or

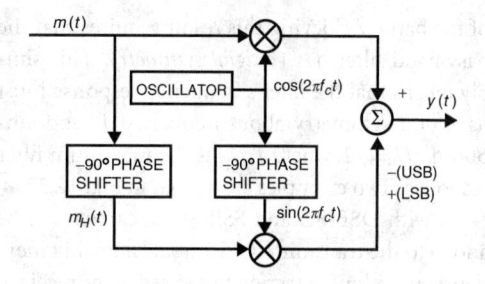

FIGURE 12.23 SSB modulator using the phase-shifting method.

$$Z(f) = 0.5\{M_-(f + f_c + f_c) + M_+(f - f_c + f_c) + M_-(f + f_c - f_c) + M_+(f - f_c - f_c)\}$$
$$= 0.5M_-(f + 2f_c) + 0.5M_+(f - 2f_c) + M_+(f) + M_-(f)$$
$$= 0.5M_-(f + 2f_c) + 0.5M_+(f - 2f_c) + M(f)$$

The result is the desired baseband signal plus unwanted harmonics at $\pm 2f_c$, which are removed by the receiver LPF.

As with the DSB receiver, achieving coherent detection requires a local oscillator in-phase with the transmitter oscillator. This can be done by transmitting a pilot signal, using a phase-locked loop (PLL) in the receiver circuit, or by simply tuning the local oscillator for the best output signal.

Alternatively, a carrier can be inserted into the SSB signal, which can then be demodulated using simple envelope detection. This SSB+C approach has the advantages of both low bandwidth and low-complexity receiver design, but is inefficient in power use due to the added carrier and usually performs poorly in the presence of noise.

Vestigial Sideband (VSB)

The chief disadvantage of SSB is poor low-frequency response resulting from the difficulty of realizing perfect sideband filters or, for the phase-shifting method, perfect wide-band 90° phase shifters. On the other hand, double sideband modulation methods have good low-frequency response, but suffer from excessive bandwidth. **Vestigial sideband (VSB)** modulation is a compromise between these two techniques, offering excellent low frequency response along with reasonably low bandwidth.

A VSB signal can be generated using a sideband filter similar to that used in SSB transmission. For VSB, however, a controlled portion of the rejected sideband is retained. Figure 12.24 shows a comparison of SSB and VSB sideband filters and the corresponding output signals. Note that low-frequency energy in the demodulated baseband signal is significantly attenuated by the SSB sideband filter. In contrast, no distortion is apparent in the demodulated VSB signal. As shown in the figure, this is because the negative and positive frequency components (labeled A and B) add constructively at baseband to perfectly reconstruct the low-frequency portion

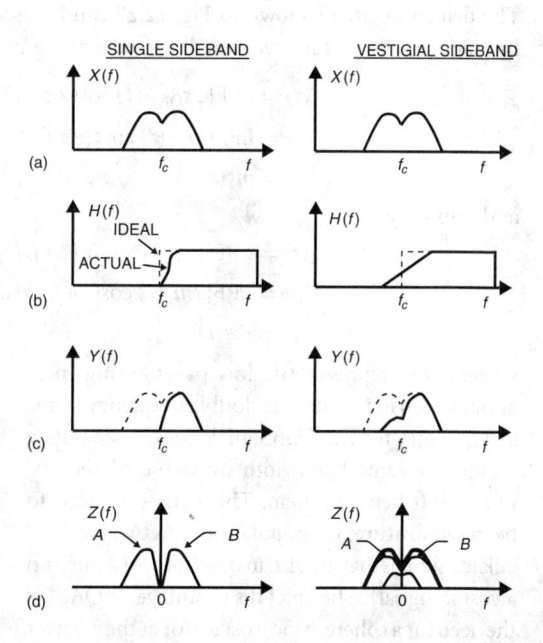

FIGURE 12.24 Comparison of SSB and VSB generation: (a) DSB-SC spectrum, (b) SSB and VSB sideband filter response, (c) spectra of filtered signal, (d) reconstruction of baseband signal.

of the band. Achieving this result requires that the pass-band filter has *vestigial symmetry*. This simply means that the filter's frequency response must have odd symmetry about frequency f_c and amplitude $H_{max}/2$, where H_{max} is the maximum filter response. Two examples are shown in Fig. 12.25.

As with DSB-SC and SSB-SC, a carrier can be added to the transmitted VSB signal allowing inexpensive envelope detection to be used in the receiver. This scheme is used in commercial broadcast television and makes it possible to mass produce relatively inexpensive high-quality receivers.

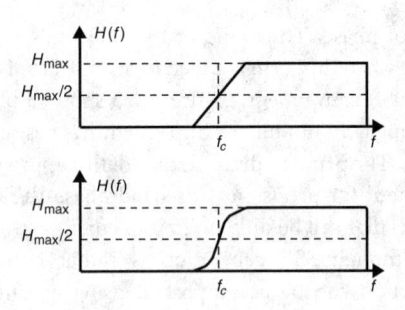

FIGURE 12.25 Two examples of sideband filter responses having vestigial symmetry.

Quadrature Amplitude Modulation (QAM)

Single-sideband transmission makes very efficient use of the spectrum; for example, two SSB signals can be transmitted within the bandwidth normally required for a single DSB signal. However, DSB signals can achieve the same efficiency by means of *quadrature amplitude modulation* (QAM), which permits two DSB signals to be transmitted and received simultaneously using the same carrier frequency. Suppose we want to transmit two message signals, $m_1(t)$ and $m_2(t)$. The QAM DSB modulator shown schematically in Fig. 12.26 can be represented mathematically as

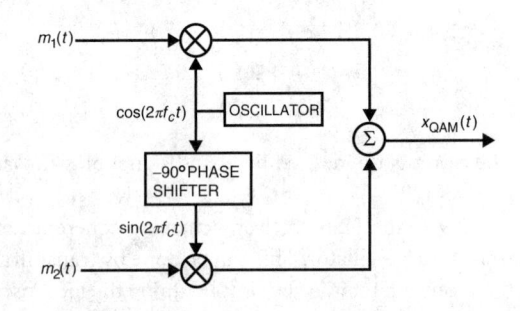

FIGURE 12.26 QAM modulator.

$$x(t) = m_1(t)\cos(2\pi f_c t) + m_2(t)\sin(2\pi f_c t)$$

The detection circuit shown in Fig. 12.27 can be used to receive the QAM signal. To show that the two message signals are fully recovered, we see that the two outputs are

$$\begin{aligned}
y_1(t) &= \text{LPF}\{2x_{QAM}(t)\cos(2\pi f_c t)\} \\
&= \text{LPF}\{2m_1(t)\cos^2(2\pi f_c t) + 2m_2(t)\sin(2\pi f_c t)\cos(2\pi f_c t)\} \\
&= m_1(t)
\end{aligned}$$

and, similarly

$$\begin{aligned}
y_2(t) &= \text{LPF}\{2x_{QAM}(t)\sin(2\pi f_c t)\} \\
&= \text{LPF}\{2m_1(t)\cos(2\pi f_c t)\sin(2\pi f_c t) + 2m_2(t)\sin^2(2\pi f_c t)\} \\
&= m_2(t)
\end{aligned}$$

where LPF$\{\cdot\}$ represents the low-pass filtering operation, which eliminates the double frequency terms at $2f_c$. Thus, the two DSB signals coexist separately within the same bandwidth by virtue of the 90° phase shift between them. The signals are said to be in **quadrature**. Demodulation uses two local oscillator signals that are also in quadrature (a sine and a cosine signal). The chief disadvantage of QAM is the need for a coherent local oscillator at the receiver exactly in-phase with the transmitter oscillator signal. Slight errors in phase or frequency can cause both loss of signal and interference between the two signals (called *cochannel interference* or *crosstalk*).

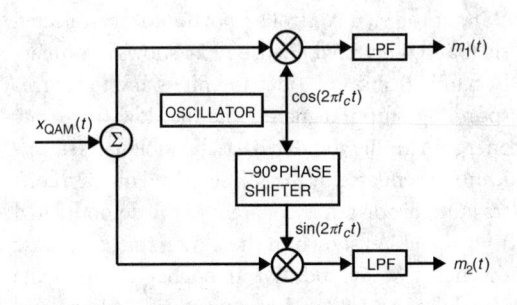

FIGURE 12.27 QAM demodulator.

Noise Effects

Real-world communication channels adversely affect the received signal due to noise and attenuation. A communication receiver designed to operate in the presence of noise is similar to the designs shown but with an additional bandpass filter inserted immediately before the receiver input (see Fig. 12.28). The filter removes noise and interference energy outside the passband.

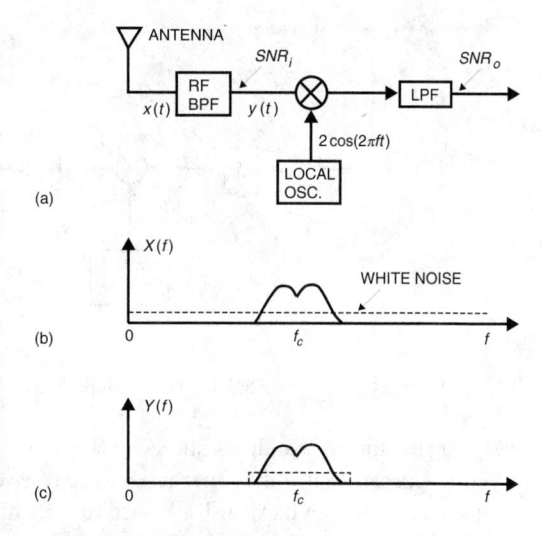

FIGURE 12.28 Receiver noise considerations: (a) demodulator using RF bandpass filter showing where SNR_i and SNR_o are measured, (b) received signal spectrum, (c) bandpass output spectrum.

The *input signal-to-noise ratio* (SNR_i) is defined as the ratio of received signal power to noise power at the band-pass filter output: $SNR_i = S_i/NB$, where S_i is the received signal power, N is the spectral noise power in watts per hertz, and B is the bandwidth of the baseband signal $m(t)$. An *output signal-to-noise ratio* (SNR_o) is also defined and is measured at the receiver output. This latter measure is indicative of the receiver's output quality since it measures the noise content of the output signal. Another useful measure of performance is the ratio of output to input SNR values, i.e., $P = SNR_o/SNR_i$. Values of P equal to 1.0 or greater indicate good performance, whereas smaller values of P imply poorer system performance. Table 12.4 shows P values for each amplitude modulation method. A few points are worth discussing in greater detail. First, these results show that for a fixed level of transmitted power, the suppressed-carrier systems all produce about the same level of noise and signal power at the receiver output. In particular, we note that SSB-SC requires the same transmitter power as DSB-SC, even though DSB-SC uses both sidebands whereas SSB uses only one. The reason for this is that none of the DSB-SC energy is wasted: the receiver utilizes both sidebands to reproduce the output signal. The remaining systems, DSB+C, SSB+C, and VSB+C all require much more transmitter power to achieve the same output SNR as the suppressed carrier methods, assuming envelope detection is being used. Another fact not shown in the table is that envelope detection systems exhibit a *threshold effect*. That is, when the input signal SNR_i drops below a certain value, the output SNR begins to decline at a much more rapid rate. For DSB+C signals, this threshold occurs when received SNR is below approximately 10 dB, which is much lower level than the 30 dB or so needed for good reception. For most practical applications, the threshold effect has a minimal effect on amplitude modulation systems.

Superheterodyne Receivers

Early receiver designs used a tunable local oscillator (LO) to select the desired station and a bandpass filter to reject stations at nearby frequencies, out-of-band noise, and interference. The filter was *ganged* with the local oscillator so that its center frequency tracked the oscillator frequency (Fig. 12.29). However,

TABLE 12.4 Performance of Modulation Techniques in Noise: Performance Ratio P

Modulation Technique	Suppressed Carrier	Carrier Present
DSB	1	$P_m^2/[A^2 + P_m^2]$ using envelope detection P_m is average power in $m(t)$
SSB	1	Much less than DSB+C
VSB	≈ 1	Much less than DSB+C
QAM	1	Not applicable

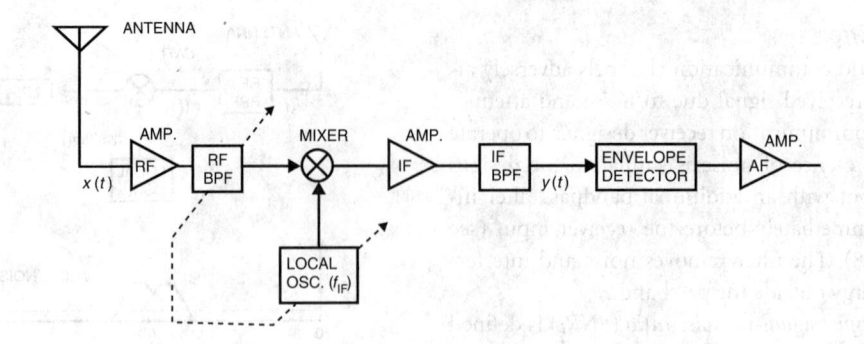

FIGURE 12.29   Superheterodyne AM (DSB+C) receiver.

designing the tunable bandpass filter with sharp cut-off response is difficult. A cheaper and much more effective approach called the **superheterodyne** receiver is shown in Fig. 12.30. The front-end *RF section* of this receiver uses an inexpensive ganged tunable filter with a gentle rolloff response to eliminate most noise and interference. It is standard practice for the LO to be tuned to 455 kHz above the desired station (hence, the name super heterodyning) for voice reception. (Automobile receiver designs often use an IF of 262.5 kHz.) This translates the desired station to a fixed *intermediate frequency* (IF) of 455 kHz. The IF section contains a sharp-rolloff bandpass filter; this is practical because of the fixed center frequency f_{IF} and because f_{IF} is relatively low. The IF BPF filter provides the needed *selectivity* to reject adjacent stations that were not removed by the RF bandpass filter. An envelope detector follows the IF stage to produce the audio output signal.

   Although an RF bandpass filter with sharp rolloff is not required in this technique, there is a limit on how broad its response can be. Consider a station located at a frequency $2f_i = 910$ kHz above the desired frequency. If this station is not removed by the RF section, it will be moved to the IF frequency on top of the desired station and cause unacceptable interference, as shown in Fig. 12.29. The

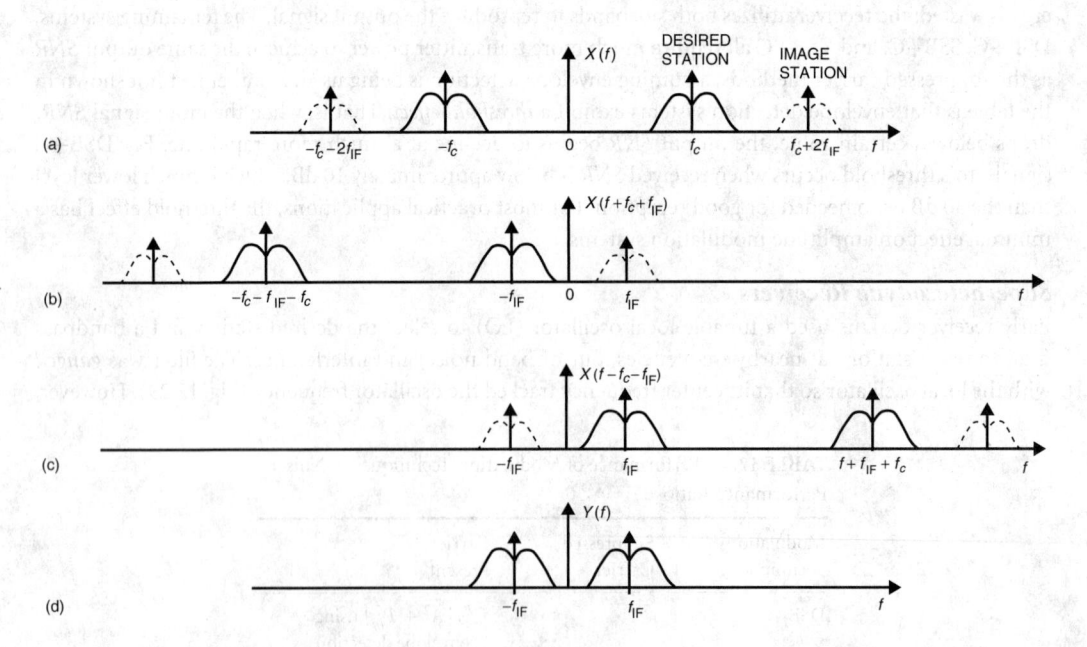

FIGURE 12.30   Superheterodyne receiver operation: (a) the desired signal and undesired image spectra (shows as a dotted line), (b) and (c) the two components of the mixer output, (d) the IF bandpass output given by $Y(f) = \text{BPF}(X(f + f_c + f_{IF}) + X(f - f_c - f_{IF}))$. Notice that the image signal must be removed by the RF stage or it will interfere with the desired signal in the IF stage.

TABLE 12.5   Comparison of Amplitude Modulation Techniques

Modulation Scheme	Advantages	Disadvantages	Comments
DSB-SC	Good power efficiency. Good low-frequency response.	More difficult to generate than DSB+C. Detection requires coherent local oscillator, pilot, or phase-locked loop (PLL). Poor spectrum efficiency.	
DSB+C (AM)	Easier to generate than DSB-SC, especially at high-power levels. Inexpensive receivers using envelope detection.	Poor power efficiency. Poor Spectrum efficiency. Poor low-frequency response. Exhibits threshold effect in noise.	Used in commercial AM.
SSB-SC	Excellent spectrum efficiency.	Complex transmitter design. Complex receiver design (same as DSB-SC). Poor low-frequency response.	Used in military communication systems, and to multiplex multiple phone calls onto long-haul microwave links.
SSB+C	Good spectrum efficiency. Low receiver complexity.	Poor power efficiency. Complex transmitters. Poor low-frequency response. Poor noise performance.	
VSB-SC	Good spectrum efficiency. Excellent low-frequency response. Transmitter easier to build than for SSB.	Complex receivers (same as DSB-SC).	
VSB+C	Good spectrum efficiency. Good low-frequency response. Inexpensive receivers using envelope detection.	Poor power efficiency. Poor performance in noise.	Used in commercial TV
QAM	Good low-frequency response. Good spectrum efficiency.	Complex receivers. Sensitive to frequency and phase errors.	Two SSB signals may be preferable.

unwanted signal is called an *image* and must be removed at the receiver's front end for acceptable receiver performance.

Superheterodyning is also used in TV, FM, and other receiver designs where the use of an intermediate frequency permits effective IF filtering strategies. The relative merits of the modulation schemes discussed in this Section are summarized in Table 12.5.

Defining Terms

Amplitude modulation: Linear modulation schemes that imprint the message signal onto the amplitude component of a carrier. In contrast, *angle modulation* methods operate by modulating the phase of the carrier. Amplitude modulation schemes include DSB-SC, DSB+C, SSB-SC, SSB+C, VSB-C, VSB+C, and QAM, as well as numerous digital communications methods.

Carrier: A sinusoidal signal of frequency f_c (the *carrier frequency*). It is mixed with a *message* signal to produce a modulated transmitter output signal centered at f_c. In *suppressed carrier* systems, no carrier component is present in the spectrum, whereas in added-carrier systems, such as broadcast AM, the carrier contains a significant amount of power

Detection: The process of converting the transmitted passband signal back into a baseband message signal. *Envelope detection* operates by extracting the envelope of the received signal to reconstruct the message signal, whereas *product* or *coherent detection* uses a local oscillator in phase with the transmitter oscillator along with a mixing circuit. Also known as *demodulation*.

Double sideband (DSB): An amplitude modulation transmission scheme where both sidebands are transmitted.

Message signal: The input audio signal or program that is desired to be transmitted. This can be speech, music, or a digital signal.

Modulation: The process of multiplying, or *mixing*, a message signal with a carrier signal. This causes the message spectrum to be translated out to the positive and negative carrier frequencies. In other words, it transforms a *baseband* signal (the message) into a *passband* signal (the transmitter output).

Quadrature: The condition when two signals having the same carrier frequency exhibit a 90° phase difference.

Sideband: A component of an amplitude modulated waveform. A modulated signal has a spectrum that is symmetric about frequency f_c, such that the upper-half is a mirror image of the lower-half. These are termed the *upper sideband* and *lower sideband*; each sideband contains an entire copy of the program information.

Single sideband (SSB): A transmission scheme using only one sideband to convey information.

Superheterodyne: A receiver design that mixes the input signal with a carrier tuned to f_{IF} Hz above the desired station. This translates it to an *intermediate frequency* f_{IF}, where efficient fixed frequency filters can be utilized. *Image stations* located at $f_c + 2f_{IF}$ must be removed by the RF section to prevent interference.

Vestigial sideband (VSB): A transmission scheme using one sideband plus a carefully controlled portion of the other sideband. This system provides excellent fidelity at all frequencies and are still spectrally efficient.

References

Carlson, A.B. 1986. *Communication Systems—An Introduction to Signals and Noise in Electrical Engineering*, 3rd ed. McGraw-Hill, New York.

Couch, L.W. II. 1995. *Modern Communication Systems—Principles and Applications*. Prentice-Hall, Englewood Cliffs, NJ.

Haykin, S. 1994. *Communication Systems*, 3rd ed. Wiley, New York.

Haykin, S. 1989. *An Introduction to Analog and Digital Communications*. Wiley, New York.

Lathi, B.P. 1989. *Modern Digital and Analog Communication Systems*, 2nd ed. Holt, Rinehart, and Wieston, Orlando, FL.

Proakis, J.G. and Salehi, M. 1994. *Communication Systems Engineering*. Prentice-Hall, Englewood Cliffs, NJ.

Schwartz, M. 1990. *Information Transmission, Modulation, and Noise*, 4th ed. McGraw-Hill, New York.

Shanmugam, K.S. 1979. *Digital and Analog Communication Systems*. Wiley, New York.

Ziemer, R.E. and Tranter, W.H. 1995. *Principles of Communications—Systems, Modulation, and Noise*, 4th ed. John Wiley and Sons, New York.

Further Information

Crutchfield, E.B., ed. c1985. *NAB Engineering Handbook*. Washington, D.C. Dept. of Science and Technology, National Association of Broadcasters.

Fink, D.G. and Christiansen, D., eds. 1989. *Electronics Engineers' Handbook*, 3rd ed. McGraw-Hill, New York.

IEEE Journal On Selected Areas In Communications, Institute of Electrical and Electronics Engineers.

IEEE Transactions on Communication, Institute of Electrical and Electronics Engineers.

IEEE Transactions On Vehicular Technology, Institute of Electrical and Electronics Engineers.

IRE Transactions on Vehicular Communications, Institute of Radio Engineers. Professional Group on Vehicular Communications.

Safford, E.L., Jr. 1971. *A Guide To Radio & TV Broadcast Engineering Practice*, 1st ed. Tab Books, Blue Ridge Summit, PA.

Wharton, W., Metcalfe, S., and Platts, G.C. 1992. *Broadcast Transmission Engineering Practice*. Focal Press, Boston, MA.

12.3 Frequency Modulation

Ken Seymour

12.3.1 Introduction

Frequency modulation (FM) is one of the most widely used modes of modulation. From applications in commercial broadcasting, television audio, cordless phones, to cellular and mobile communications, FM is indeed both a reliable and important form of modulation. The brainchild of Edwin H. Armstrong, FM was first demonstrated in December 1933 (Lewis, 1991, p. 256) as a solution to eliminate the static and noise problems that plagued AM communications.

In amplitude modulation (AM), interference, such as static, lightning, and manmade noise, cause the amplitude of an RF signal to vary widely. This is because these noises are predominately amplitude modulated signals in composition. The noise is added and superimposed on the transmitted AM signal carrying the desired intelligence. This increases the overall amplitude of the signal as shown in Fig. 12.31. These added variations are then demodulated at the receiver, and the noise is passed onto the audio section, where they are reproduced as clicks, pops, and various other objectionable noises. The problems associated with amplitude modulation are overcome in frequency modulation. FM receivers are designed to reduce the amplitude variations of an incoming signal. This is done without effecting the frequency modulated waveform that contains the desired intelligence.

12.3.2 The Modulated FM Carrier

The composition of the unmodulated AM carrier and FM carrier are identical. That is, the RF carrier is a sinusoidal waveform operating at a specific period or frequency. As the frequency is increased, the period of the waveform decreases as more cycles are completed per second. When an AM carrier is modulated, the amplitude of the carrier is effected. When the FM carrier is modulated, the frequency of the carrier varies by an amount that is proportional to amplitude of the modulating waveform; this occurs at a rate that is determined by the modulation frequency.

To better understand how the modulating signal affects a carrier that is being frequency modulated, it is best to use illustrative examples. Figure 12.32 graphically illustrates what happens when a carrier is frequency modulated. Figure 12.32(a) shows one period of an audio signal that will be used to frequency modulate a carrier. The AC signal is positive for 180° and swings negative for the remaining 180° to complete one 360° cycle.

Figure 12.32(b) illustrates the effect of how the audio signal (a) affects the carrier. At time $t = <0$, we see the RF carrier operating at a specific frequency. This is sometimes referred to as *center* or *resting* frequency. At time $t = 0$, the modulating signal (a) is applied to the RF carrier. As the amplitude of signal

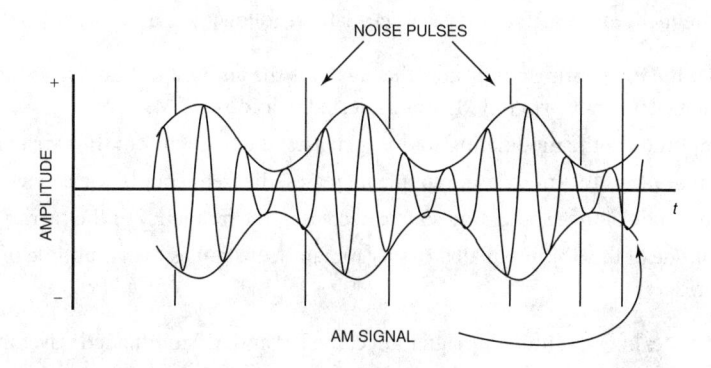

FIGURE 12.31 Typical AM signal with induced noise.

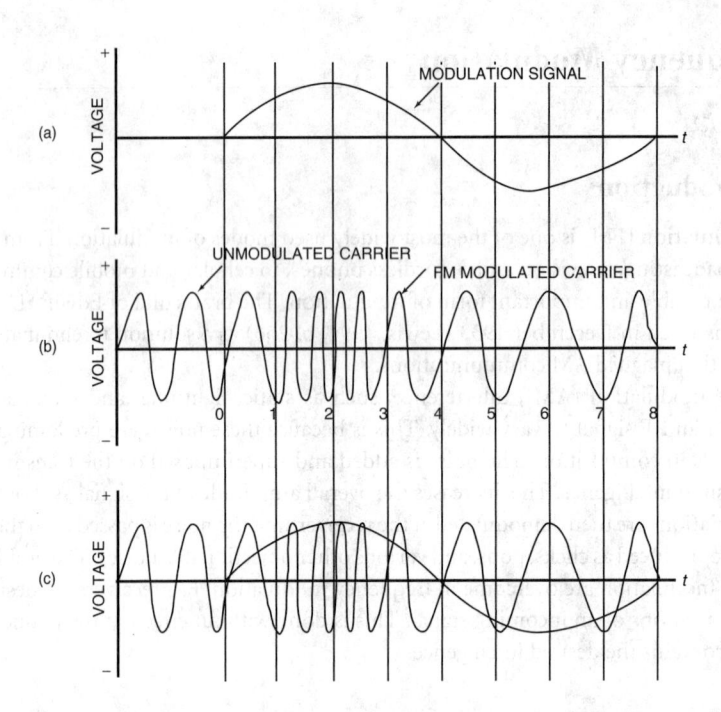

FIGURE 12.32   Frequency modulating an RF carrier.

(a) swings positive, the frequency of the RF carrier also begins to change. At time $t = 1$, the frequency of the modulated RF carrier has increased proportionally, resulting in a greater number of cycles occurring in a given interval of time. At time $t = 2$, the amplitude of the modulating signal (a) reaches its maximum. At the same time, the RF carrier has increased to its maximum frequency. At time $t = 3$, the modulating waveform (a) begins to decrease in amplitude and the carrier frequency (b) also begins to decrease. At time $t = 4$, the modulating frequency returns to zero and the carrier frequency returns to the resting frequency.

For the last 180° portion of the modulation cycle, the amplitude of the modulating signal goes negative. At time $t = 6$, the modulating signal (a) has decreased to its maximum negative value and the frequency of the modulated carrier (b) also reaches its minimum. In fact, it has decreased to a frequency below that of the unmodulated carrier. At time $t = 7$, the modulation amplitude begins its return journey back to zero. Then, at $t = 8$, the carrier frequency returns to the resting frequency. Figure 12.32(c) shows waveforms (a) and (b) superimposed on each other to better illustrate the waveform relationships.

When an FM signal is received, it is the amount of frequency shift that is produced in the modulated waveform that determines the audio intensity or volume that is heard on the speaker of the receiver. To summarize, the frequency modulated carrier observes these following characteristics:

- The higher the modulating amplitude, the greater is the amount of frequency shift away from the resting frequency. This form of FM is also referred to as **direct FM**.
- As the amplitude of the modulating source increases, the frequency of the carrier increases.
- As the amplitude of the modulating source increases, the frequency of the carrier increases.
- As the amplitude of the modulating source decreases, the frequency of the carrier decreases.
- The amplitude of the FM modulated carrier remains constant as the amplitude of the modulating source varies.

Table 12.6 illustrates how a modulating signal affects an FM and AM modulated waveform. For example, as the amplitude of the modulating signal increases, the overall frequency swing on FM increases. With AM, the amplitude of the carrier increases.

TABLE 12.6 Summary of Modulation Effects on AM vs. FM

| | Modulating Amplitude | |
	Increases	Decreases
FM	Frequency swing increases	Frequency swing decreases
AM	Carrier level increases	Carrier level decreases

12.3.3   Frequency Deviation

As we saw in the previous section, the amplitude of the modulating signal plays an important part in the overall characteristic of the carrier frequency. The peak difference between the modulated carrier and the frequency of the carrier is known as the **frequency deviation** (Code, 1993). The peak difference between the minimum and maximum frequency values is known as the *frequency (or carrier) swing* (Code, 1993). This can be defined as

$$\Delta f = fpc - fc$$

where
Δf = frequency deviation
fpc = peak frequency of the modulated carrier (minimum or maximum)
fc = frequency of the carrier (unmodulated)

Example 12.1

A commercial FM broadcast station operates on a frequency of 97.1 MHz. On a modulation peak, the frequency increases to 97.13 MHz. Determine the (1) frequency deviation and (2) the frequency swing. The solution is as follows:

1. Applying the deviation equation: $fpc = 97.13$ and $fc = 97.10$. Substituting terms: $\Delta f = 97.13 - 97.10$. This results in a frequency deviation of $+30$ kHz.
2. The frequency swing would be the peak difference of the minimum and maximum frequency deviation or $(2)(\Delta f) = (2)(30 kHz) = 60$ kHz.

It is important to remember, that there are two cases of frequency deviation. In the preceding example, the frequency was increased as a result of the modulating signal. Likewise in FM, when a carrier is modulated, the carrier frequency also goes negative. Therefore, the swing is defined as $(2)(\Delta f)$. The FCC places limits on the amount of deviation that a frequency can swing. With commercial FM stations, the frequency deviation is $+/-75$ kHz or 150 kHz total frequency swing. Depending on the FM communication service, the FCC imposes different frequency deviation requirements.

12.3.4   Percent of Modulation in FM

In amplitude modulation, 100% modulation is defined as the point where the amplitude of the RF carrier rises to twice the normal amplitude at its maximum, and drops to zero at its minimum. Anything greater than 100% modulation, in AM, causes distortion to the modulated wave.

With FM, it is the amount of frequency deviation that determines the degree of modulation. The frequency deviation that corresponds to 100% modulation is an arbitrary value defined by the FCC or other appropriate licensing authority as related to the FM service in use. For commercial FM broadcasting, 100% modulation is reached when the carrier frequency deviation reaches 75 kHz. In television, 100% modulation is set at a frequency deviation of $+/-25$ kHz by the FCC.

If the carrier is modulated above 100%, distortion and spurious sidebands are not produced as in AM. To avoid interference between adjacent stations, the FCC has set the channel spacing for commercial FM at 200 kHz. This gives ample guard band for stations operating at 100% modulation, which accounts for a +/−75 kHz frequency deviation or a total frequency swing of 150 kHz.

12.3.5   Modulation Index

In FM, the **modulation index** is often used more frequently than **percentage** of **modulation**. Modulation index is defined as the ratio of the frequency deviation to the frequency of the modulating signal (National, 1985, p. 3.3-65). The term has no units and is expressed as a decimal. For a constant frequency deviation, the modulation index drops as the frequency of the modulating signal increases. During the transmission of an FM signal, the modulation index varies as the modulation frequency varies. As we shall see later, this relationship is important for determining the bandwidth requirements of an FM signal

$$mf = \Delta f / fs$$

where
  mf = modulation index
  Δf = frequency deviation
  fs = frequency of the modulating signal

Example 12.2

A 97.1-MHz carrier frequency is modulated by a 10-kHz audio signal source. This produces a frequency deviation of +/−40 kHz. Determine the modulation index. The solution is

$$mf = 40/10$$
$$= 4$$

12.3.6   Bandwidth and Sidebands Produced by FM

Frequency modulation differs from amplitude modulation in that the modulated wave consists of the carrier frequency and numerous sideband components that are generated for each modulating frequency. Recall that AM consists of a carrier and a upper and lower sideband. The bandwidth of an AM signal is determined by the highest frequency of the modulating signal. For example, a carrier is modulated by a audio signal which contains frequencies up to 4000 Hz. The AM bandwidth would therefore be: $(2)(4000) = 8000$ Hz.

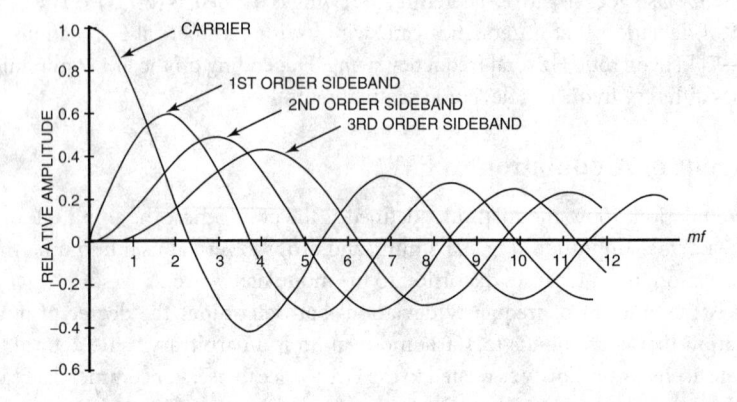

FIGURE 12.33   Relationship of carrier and sideband amplitudes to modulation index mf (Bessel functions).

TABLE 12.7 Bessel Functions: Values of Carrier and Sideband Amplitudes

Modul. Index, mf	Carrier, fc	Amplitude of Sidebands											
		$f1$	$f2$	$f3$	$f4$	$f5$	$f6$	$f7$	$f8$	$f9$	$f10$	$f11$	$f12$
0.00	1.00	—	—										
0.25	0.98	0.12	0.01										
0.50	0.94	0.24	0.03	—									
1.00	0.77	0.44	0.11	0.02	—								
1.50	0.51	0.56	0.23	0.06	0.01	—							
2.00	0.22	0.58	0.35	0.13	0.03	0.01							
2.50	−0.05	0.50	0.45	0.22	0.07	0.02	—						
3.00	−0.26	0.34	0.49	0.31	0.13	0.04	0.01	—					
4.00	−0.40	−0.07	0.36	0.43	0.28	0.13	0.05	0.02	—	—			
5.00	−0.18	−0.33	0.05	0.36	0.39	0.26	0.13	0.05	0.02	0.01	—		
6.00	0.15	−0.28	−0.24	0.11	0.36	0.36	0.25	0.13	0.06	0.02	0.01	—	
7.00	0.30	0.00	−0.30	−0.17	0.16	0.35	0.34	0.23	0.13	0.06	0.02	0.01	—
8.00	0.17	0.23	−0.11	−0.29	−0.10	0.19	0.34	0.32	0.22	0.13	0.06	0.03	0.01

In FM, the amplitude of modulating signal is the primary factor in determining the amount of bandwidth. This was illustrated in Fig. 12.32 where the difference in the amplitude of the audio modulating signal produced a difference frequency change, or deviation. This deviation could shift the frequency of the carrier by 75 kHz or more, depending on the amplitude of the modulating signal.

The sidebands generated in FM are spaced on both sides of the carrier at frequency intervals equal to the modulating frequency and its multiples. To better understand all of the components of FM sidebands, we will analyze a FM modulated carrier using *Bessel functions* and *spectral diagrams*. Figure 12.33 illustrates the mathematical relationship of the Bessel function, the FM waveform theoretically contains an infinite number of side frequencies on both sides of the carrier. The side frequencies are spaced at intervals that correspond to the modulating frequency, which can be represented as

$$fc +/-fs +/-2fs +/-3fs +/-4fs \cdots \infty$$

Notice that the Bessel functions illustrated in Fig. 12.33 resemble damped sine waves. The Bessel curves enable us to understand the component of the FM sidebands. The number of sidebands depends totally on the modulation index (mf). Table 12.7 is a tabular representation of the Bessel curve illustrated in Fig. 12.33. It shows the amplitudes of the sidebands for audio modulating harmonics where they decrease to a value close to the amplitude of the unmodulated carrier. The following example illustrates how to determine FM side bands using the Bessel tables.

Example 12.3

Analyze and draw a spectral diagram of: (1) an AM carrier modulated 100% by a 12.5-kHz audio source and (2) an FM carrier modulated by an 12.5-kHz audio source. Assume a frequency deviation of +/−75 kHz. The solution is as follows:

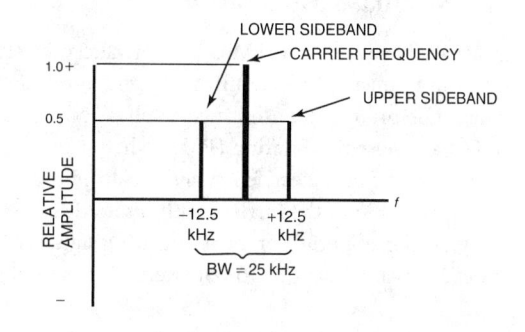

FIGURE 12.34 Example 12.3(1) AM spectral diagram.

1. The spectral diagram is easy to illustrate as shown in Fig. 12.34. Notice that there are two significant sidebands, upper and lower. The total bandwidth is 25 kHz.
2. Next, we need to determine the modulation index. This is defined as

$$mf = \Delta f/fs$$

Substituting terms, $mf = 75,000/12,500 = 6$.

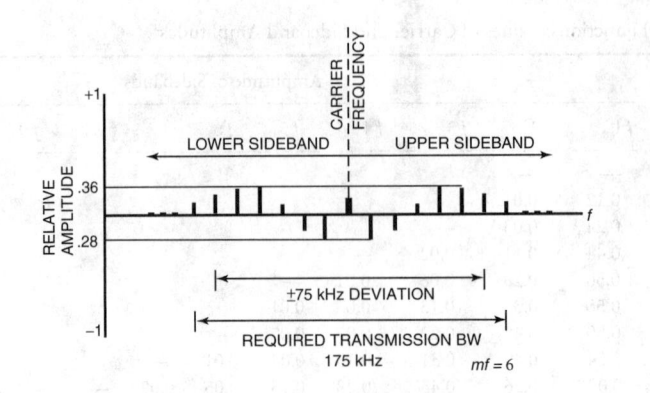

FIGURE 12.35   Example 12.3(2) FM spectral diagram.

Referring to Table 12.7, with a modulation index of 6, we see that there are 10 harmonics plus the carrier. The amplitudes of the sideband harmonics are taken from the table and are illustrated in the spectral diagram shown in Fig. 12.35.

There also may be upper and lower sideband frequencies that extend beyond the allowable deviation in an FM signal, in this example beyond the $+/-75$-kHz limit. The total number of sideband frequencies that are produced is different for each different value of modulation index. The greater the modulation index is, the greater the number of sideband components.

12.3.7   Narrow-Band versus Wide-Band FM

The *narrow-band* (NBFM) and *wide-band* (WBFM) terms refer to the amount of frequency deviation (Δf) that is present for a specific transmitted FM signal. This directly correlates to the amount of spectral bandwidth that the transmission occupies. NBFM is typically used for communications services that occupy less spectral bandwidth. This service is used for two-way voice communications, amateurs, and by governmental agencies. The FCC limits the frequency deviation to less than $+/-15$ kHz for these services. WBFM on the other hand, typically uses a frequency deviation greater than $+/-15$ kHz.

12.3.8   Phase Modulation

So far, we have seen how FM can be produced by shifting the frequency of the carrier above and below a resting frequency as determined by the amplitude of the modulation signal. Frequency modulation can also be produced by shifting the phase of the carrier relative to an arbitrary reference point. This is known as **phase modulation (PM)**. When a carrier is phase modulated, the input signal is designed to alter the phase of the carrier. When the amplitude of the modulating signal swings positive, the greater is the phase shift of the carrier results as it advances or leads in phase. This results in a greater frequency swing. As the amplitude of the modulating signal goes negative, the carrier will lag in phase. This method of frequency modulation is often referred to as *indirect FM*.

12.3.9   FM Transmission Principles

The basic transmission principles used today for frequency modulation falls into two categories. One type, *direct FM*, is the modulation process where the frequency of the transmitter oscillator varies in accordance with the amplitude of the modulating signal. The other type, *indirect FM*, obtains a frequency modulated waveform by phase modulating the carrier.

The method of modulating an FM transmitter vs modulating an AM transmitter differs significantly. With AM, modulation generally takes place in a higher level stage of the transmitter at the final RF power amplifier stages. With FM, modulation takes place in a much lower stage of the transmitter, usually the

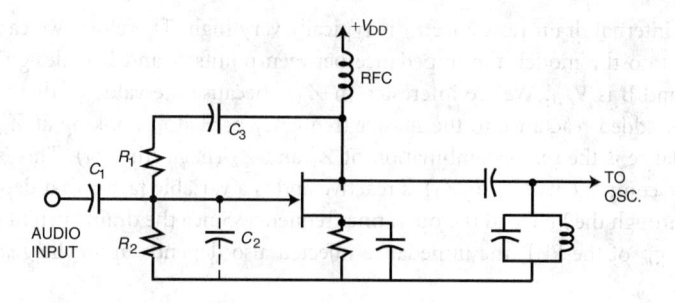

FIGURE 12.36 Simplified reactance modulator.

master oscillator itself. The lower level modulation of FM is generally performed before any frequency multiplication occurs to minimize increasing phase shift and frequency deviation.

To frequency modulate a carrier directly, an active device is used in conjunction with the input signal to produce a variable reactance (capacitive or inductive) across its output. If the variable reactance is placed across a tuned tank circuit, the effective capacitance or inductance of the tank will change. This, in turn, will change the resonant frequency of the circuit. This methodology was primarily used to generate FM in the early days of FM broadcast service. Today, *voltage controlled oscillators* are commonly used, and circuits built around variactor diodes are found.

Direct FM Modulators

Over the years, many types of circuits have been used to produce direct FM. In each case, a reactance device is used to shunt capacitive or inductive reactance across an oscillator. The value of capacitive or inductive reactance is made to vary as the amplitude of the modulating signal varies. Since the reactive load is placed across an oscillator tuned circuit, the frequency of the oscillator will therefore shift by a predetermined amount, thereby creating an FM signal. A Typical example of a reactance modulator is shown in Fig. 12.36. The circuit uses a field effect transistor (FET) where the modulating signal is applies to the modulator through C_1. The actual components that effect the overall reactance consist of R_1 and C_2. Typically, the value of C_2 is small as this is the input capacitance to the FET. This may only be a few picofarads. However, this capacitance will generally be much larger by a significant amount due to the *Miller effect*. Capacitor C_3 has no significant effect on the reactance of the modulator. It is strictly a blocking capacitor, which keeps DC from effecting the gate bias of the FET.

To further understand the performance of the reactance modulator, an equivalent circuit of Fig. 12.36 is represented in Fig. 12.37. The FET is represented as a current source $g_m V_g$, with the internal drain resistance r_d. The impedances Z_1 and Z_2 are a combination of resistance and capacitive reactance, which are designed to provide a $90°$, phase shift. This will be evident as the analysis of the reactance FET proceeds.

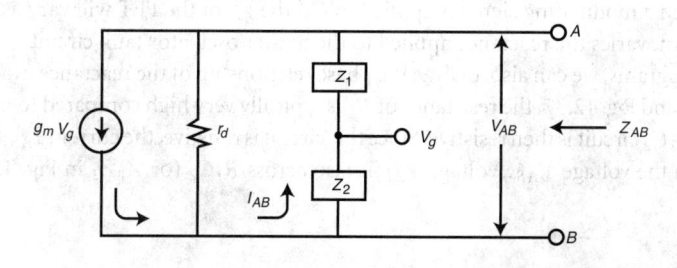

FIGURE 12.37 Equivalent circuit of the reactance modulator.

In an FET, the internal drain resistance r_d is typically very high. Therefore, we can neglect it in our analysis. Looking into the model, the impedance between points A and B is designated Z_{AB} and the voltage across A and B is V_{AB}. We are interested in Z_{AB}, because the value of this impedance will indicate the value of added reactance to the master frequency oscillator. Looking at Z_{AB}, it appears that the output impedance is the series combination of Z_1 and Z_2 (neglecting r_d). This is not exactly true, because one of the components (Z_1 or Z_2) is reactive and is a variable factor that depends on the drain current flowing through the FET and the operating frequency. Since the drain current depends upon the transconductance g_m of the FET, the impedance injected also depends upon the g_m. This is shown as follows.

The impedance is defined as

$$Z_{AB} = V_{AB}/I_{AB} \tag{12.39}$$

By definition, we also know that the circuit current is (r_d is neglected)

$$I_{AB} = g_m V_g \tag{12.40}$$

Substituting terms, Eq. (12.40) into Eq. (12.39)

$$Z_{AB} = V_{AB}/g_m V_g \tag{12.41}$$

Referring to Fig. 12.37, V_g is derived using the voltage divider principle

$$V_g = V_{AB}(Z_2/(Z_1 + Z_2)) \tag{12.42}$$

Substituting terms, Eq. (12.42) into Eq. (12.41)

$$Z_{AB} = \frac{V_{AB}}{g_m V_{AB}(Z_2/(Z_1 + Z_2))} \tag{12.43}$$

$$Z_{AB} = ((Z_1 + Z_2)/Z_1))/(g_m) \tag{12.44}$$

$$Z_{AB} = (1/g_m)((Z_1 + Z_2)/Z_2) \tag{12.45}$$

$$Z_{AB} = (1/g_m)(1 + Z_1/Z_2) \tag{12.46}$$

$$Z_{AB} = (1/g_m) + (1/g_m)(Z_1/Z_2) \tag{12.47}$$

Equation (12.47) represents the impedance seen when looking into the reactive FET circuit, at points A and B. Since the unit for transconductance is given in mho, the term $1/g_m$ will therefore be resistive in ohms. The equation then states that the impedance across points A and B consists of a *resistance in series with a reactance*. In this case, the reactance Z_1 is R_1 (purely resistive) and Z_2 is C_2. As Eq. (12.47) also illustrates, the transconductance value of the FET is a key term and plays an important role in determining the overall added reactance. Since the transconductance of the FET is dependent on the gate voltage,[2] it is apparent that when a modulating signal is applied to C_1, the g_m of the FET will vary as the audio voltage varies. This, in turn, varies the reactance applied to the master oscillator tank circuit.

Using vector diagrams, we can also analyze the phase relationship of the reactance modulator. Referring back to Fig. 12.36 and Fig. 12.37, the resistance of R_1 is typically very high compared to the capacitive reactance of C_2. The R_1C_2 circuit is then resistive. Since this circuit is resistive, the current I_{AB} that flows through it is in phase with the voltage V_{AB}. Voltage V_{AB} is also across R_1C_2 (or Z_1Z_2 in Fig. 12.37). This is true

---

[2] $g_m = \left(\frac{I_d}{V_{gs}}\right)\Big|V_{DS}$

because current and voltage tends to be in phase in a resistive network. However, voltage V_{C2}, which is across C_2, is out of phase with I_{AB}. This is because the voltage that is across a capacitor lags behind its current by 90°. This is illustrated in the vector diagram of Fig. 12.38.

By design, we also see that V_{C2} is also the voltage applied to the gate of the FET, V_g. Since the drain current variations in the FET are a direct result of variations in gate voltage, the drain current I_d is in phase with the gate voltage V_g. This is shown in Fig. 12.38 with I_d next to V_{C_2}. As previously shown, we can put V_{AB} next to I_{AB} since they are in phase. The vector diagram shows that I_d, the drain current of the FET, is 90° behind V_{AB}. Notice that V_{AB} is also across the oscillator tuned circuit. Thus, we now have a circuit where the drain current lags the voltage across the oscillator tank by 90°. AC analysis has proven that the current through an inductance lags behind the voltage across the inductor by 90°. The voltage, V_{AB} therefore behaves the same as inductance. Again, as this inductance varies, so to will the total impedance of the oscillator tank circuit, which in turn results in FM.

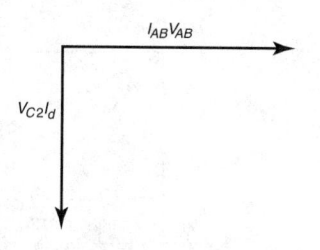

FIGURE 12.38   Vector diagram of reactance modulator producing FM. Note: $V_g = V_{C_2}$

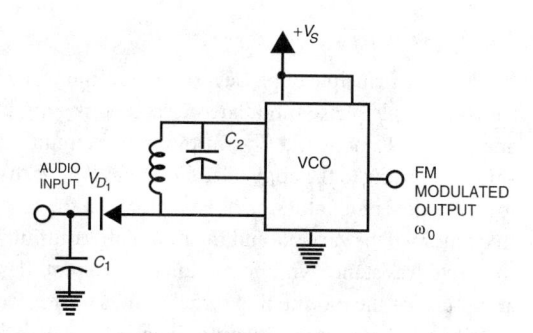

FIGURE 12.39   Voltage controlled direct-FM modulator.

Voltage Controlled Oscillator (VCO) Direct-FM Modulators

Another one of the more common direct-FM modulation techniques in use today uses an analog voltage controlled oscillator (VCO) in a phase locked loop (PLL) arrangement. This is shown in Fig. 12.39. In this configuration, a VCO produces a desired carrier frequency, which is in turn modulated by applying the audio signal to the VCO input via a variactor diode. A variactor diode is generally used to vary the capacitance of an circuit. Therefore, oscillator tank the variactor behaves as a variable capacitor whose capacitance varies as the input voltage across it varies. As the input capacitance of the VCO is changed by the variactor, the output frequency of the VCO is shifted, which produces a direct-FM modulated signal.

Indirect-FM Modulators

Thus far we have only discussed the direct method of producing FM and just briefly mentioned the indirect method. Historically, the indirect method, or phase modulation, was originally developed by Armstrong in the early 1930s and results published in 1936 (Armstrong, 1936). His was the first method to provide a practical system for producing an FM signal as many of the early FM transmitters used his method of modulation.

Simply stated, in phase modulation, the phase of the carrier signal deviates away from its resting position as modulation is applied. This is accomplished by passing a fixed RF carrier through a time-delay network, which makes the carrier change in phase. If the time-delay network is made to vary in accordance with the amplitude of an input signal, the delay network will change the phase of the carrier in accordance with the applied audio. The resultant output from the time-delay network will then be a phase modulated signal. The center frequency of the carrier is typically produced by using a stable oscillator circuit, such as a crystal, that is resonate at some frequency lower than that of the final desired output frequency. For this reason, there are typically many stages of frequency multiplication that are used following the phase modulator.

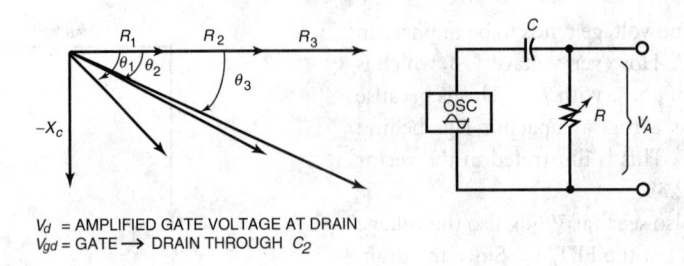

V_d = AMPLIFIED GATE VOLTAGE AT DRAIN
V_{gd} = GATE $\longrightarrow$ DRAIN THROUGH C_2

FIGURE 12.40 Phase modulator principle and vector diagram.

The basic principle of phase modulation can easily be illustrated in the following example. Figure 12.40 shows a simple phase modulated circuit consisting of a crystal oscillator (providing a center frequency) and a series RC network. The output of this circuit is taken across the variable resistor R whose resistance varies according to the applied audio signal. The vector diagram illustrates the impedance of the RC circuit for three different values of resistance. With the resistance variable, the resultant phase of the oscillator frequency will be variable and the modulated output voltage V_{AB} varies in phase. Resistance R_2 represents the circuit resistance with no modulation present. Resistance R_1 represents the circuit resistance when the amplitude of the modulating signal swings positive, and resistance R_3 represents the circuit resistance as the amplitude of the modulating signal swings negative. The component $-X_C$ is the capacitive reactance of C.

When the modulating signal is applied to the circuit, it alternately causes the resistance to decrease R_1 and to increase R_3. The phase of the current through the RC circuit varies as long as the phase angle of the impedance varies. As we have shown, as R varies, the resultant phase angle θ varies from θ_1 to θ_3. The output voltage V_{AB} also follows the phase change in the current and impedance, and the result is a phase modulated signal.

In a real application, the variable resistance R in the example can be replaced with an active device such as an FET. For an FET to function properly, it must act like a variable resistance. This happens in the FET, for example, as the dynamic drain resistance is placed in parallel with the load resistor R_4 as illustrated in Fig. 12.41. The capacitive reactance of C_2 determines the amount of phase shift that occurs with the overall variable resistance of the FET. This example is one of the most simple types of phase modulators.

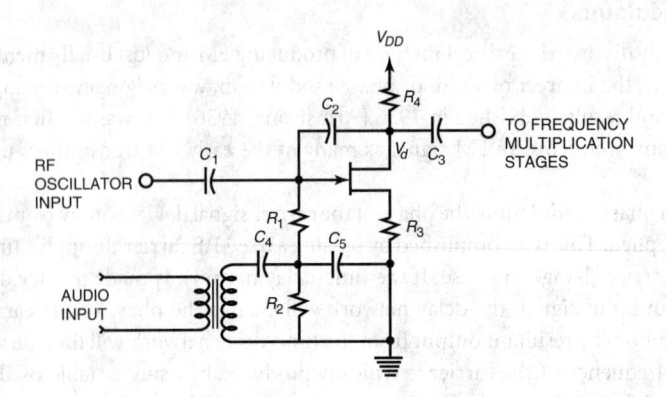

FIGURE 12.41 Simple FET phase modulator.

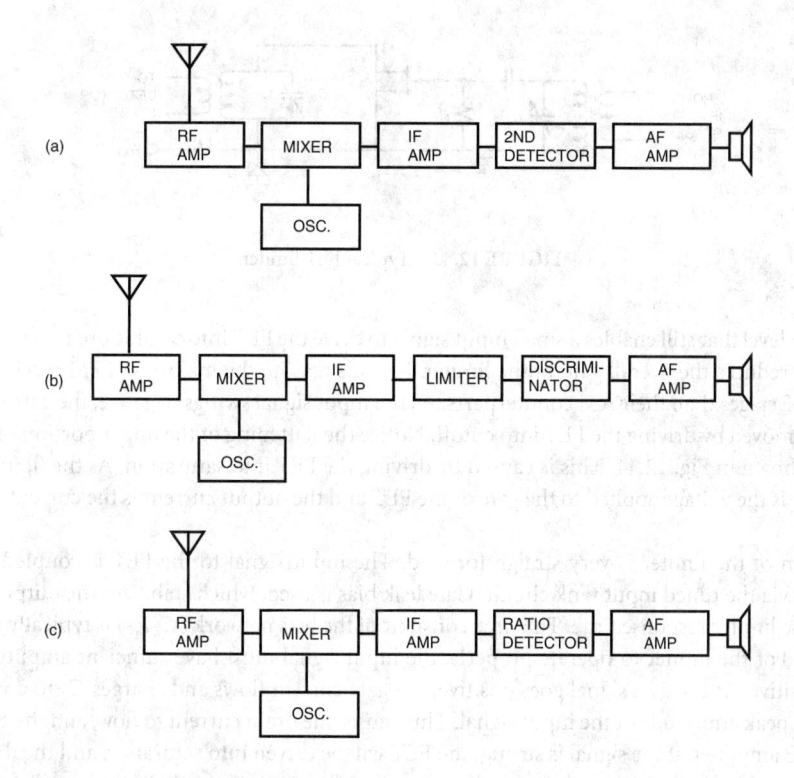

FIGURE 12.42 Functional block diagrams of typical AM and FM receivers: (a) AM receiver, (b) FM receiver using a Foster–Seeley discriminator and limiter, (c) FM receiver using a ratio detector.

12.3.10   FM Reception Principles

Regardless of frequency, the receivers for FM communications are similar to those for AM. The functional layout for both types of receivers are similar in that a superheterodyne circuit is used. Both receiver types contain RF amplification, mixing, a local oscillator, IF amplification, detection, and amplification. There are, however, a few important design differences. Retrieval of the intelligence of an FM signal requires a slightly different detector circuit and some form of signal limiting. FM requires a detector that is designed to discriminate between a positive and negative frequency deviation. In addition, any variation in the amplitude of the carrier represents undesirable noise. This is removed by a **limiter** stage before demodulation occurs. A functional comparison between an AM receiver and two types of FM receivers is illustrated in Fig. 12.42.

As is apparent in the illustration, the major circuit differences between the AM and FM receiver consist of the limiter, **discriminator**, and **ratio-detector** stages. There are other minor differences, but these will not be discussed in detail since they are not unique to FM receiver design. These include differences in the receiver tuning and RF amplification stages and final amplification. The tuning range and RF amplification stages are different only in the design of the received frequency.

Limiters

The limiter stage of an FM receiver is basically an IF amplifier that is designed to saturate and clip off undesired AM and noise components from a signal prior to FM detection. With proper design, the output of the limiter will have a constant amplitude as the output signal is applied to the following detector stages. Figure 12.43 shows a limiter circuit designed with an FET.

When the noise levels on the positive peaks exceed a specified design level, limiting action is produced by driving the FET out of the active region into saturation. To help ensure saturation, the supply voltage is

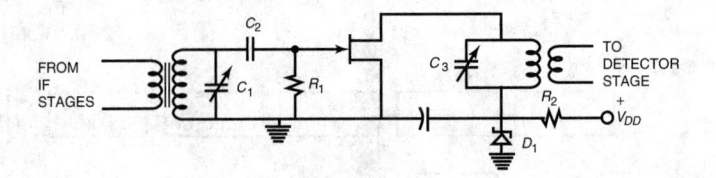

FIGURE 12.43   Typical FET limiter.

reduced to a level that still enables a small input signal to drive the FET into saturation. Lowering the drain voltage also reduces the overall gain of the limiter. For this reason, the majority of FM receivers typically have more IF stages than their AM counterparts. As the input signal swings negative, the extreme negative peaks are removed by driving the FET into cutoff. Notice the flattening of the upper portion of the output waveform shown in Fig. 12.44. This is caused by driving the FET into saturation. As the figure shows, an input signal is the voltage applied to the gate of the FET and the output current is the current through the drain.

Operation of the limiter is very straightforward. The input signal to the FET is coupled to the gate through C_2 via the tuned input tank circuit. Gate leak bias is used, which stabilizes the output signal and improves the limiter response time. The time constant of the bias network $(C_2 R_1)$ is typically in the range of 1–15 μs. For the limiter to operate properly, the input signal must have sufficient amplitude to drive the gate positive. As the gate signal goes positive, the gate current flows and charges C_2 to a value almost equal to the peak amplitude of the input signal. This causes the drain current to flow, and the FET acts like a normal RF amplifier. If the signal is strong, the FET will be driven into saturation and the drain current reaches its peak level. As the signal drops, the capacitor discharges through R_1 and produces a negative voltage, which reduces the drain current. The drain voltage is reduced to a level below that of other stages in the receiver by R_2 and zener D_1.

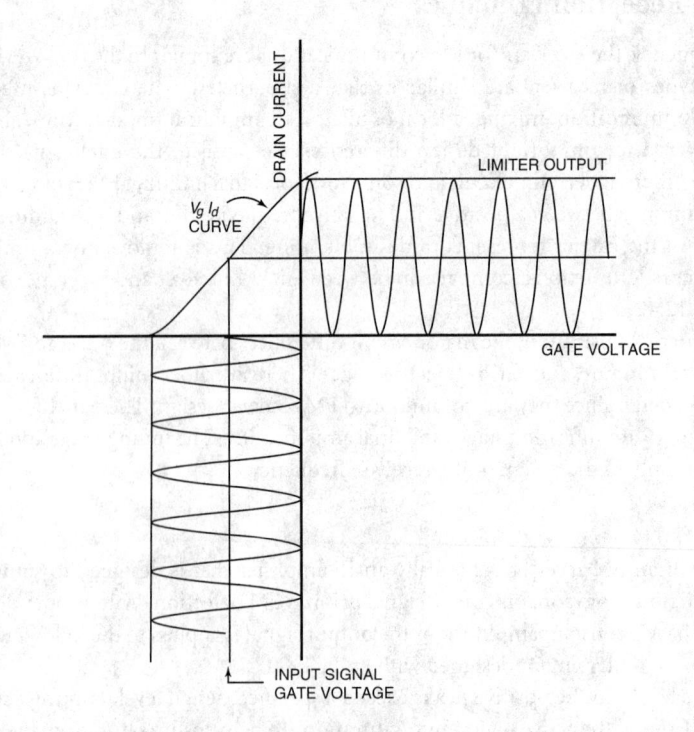

FIGURE 12.44   Dynamic transfer characteristics of an FM limiter (I_d vs. V_g).

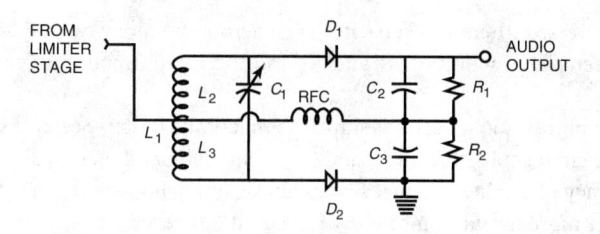

FIGURE 12.45 A simplified Foster–Seeley discriminator.

FM Detectors

The two most popular classic detector circuits for FM are the Foster–Seeley discriminator and the ratio detector. Other types of FM detection used today include phase-locked loop circuitry. The basic function of the discriminator is to convert the frequency swings in the FM signal back into amplitude variations for further audio processing. The discriminator is, therefore, susceptible to both amplitude and frequency variations. For this reason, the Foster–Seeley detector is always preceded by a limiter stage. The ratio detector on the other hand, acts like a limiter, and so there is no need for the use of a separate limiter stage.

Discriminators and Ratio Detectors

Perhaps the most frequently used FM detector is the Foster–Seeley discriminator. The design is simple and its operational characteristics are less critical than other types. A typical Foster–Seeley descriminator circuit is illustrated in Fig. 12.45. The tuned circuit L_1C_1 is tuned to the center resting frequency of the IF stage. With no modulation present, the voltage developed across the center tap of L_1 (L_2 and L_3) is out of phase with the voltage across L_1. When this occurs, the current through D_1 and D_2 are equal, and the voltage developed across R_1 and R_2 will be equal and opposite in polarity. The output voltage from the circuit will, therefore, be zero.

When modulation occurs, the frequency varies above and below the carrier resting frequency. The tuned tank circuit, L_1C_1, becomes inductive or capacitive, depending on which way the signal shifts from the resting frequency. As the tuned circuit turns reactive, the phase angle changes between voltage and current. The resulting current flow is different flowing through D_1 from that flowing through D_2. As a rectified voltage is developed across R_1 and R_2, the difference between these two voltages will no longer be zero. This is, then, the reproduced output voltage that is produced as the modulation frequency varies.

The ratio detector is another commonly used FM detector. Notice that there are some close similarities between this and the Foster–Seeley discriminator. The major difference, as illustrated in Fig. 12.46, is that

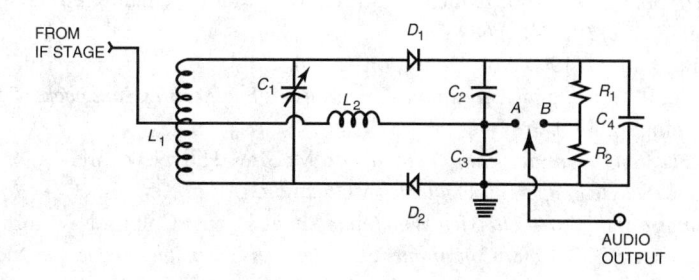

FIGURE 12.46 A simplified ratio detector.

one of the diodes is reversed, there is a capacitor (C_4) across the other two capacitors, and the output voltage is taken between the junction of R_1 and R_2 and the taped junction of capacitors C_2 and C_3 at terminals A and B.

The basic operation of the ratio detector is similar to that of the Foster–Seeley discriminator. The input is coupled to the tuned circuit consisting of L_1 and C_1. At the resting frequency, diodes D_1 and D_2 conduct equally. As the frequency of the input signal swings above and below the resting frequency, either diode D_1 or D_2 will conduct more heavily. Since one of the diodes is reversed, current also flows through the entire circuit path of D_2 through the time constant of $C_4 R_1 R_2$, through D_1 and back into the top of L_1. After several RF cycles, C_4 charges to the peak value of the voltage across L_2. Variations in amplitude of the incoming signal have little effect on the charging of C_4, and the voltage across the capacitor remains fairly constant (due to a long time constant). This results in a practically constant voltage across $R_1 + R_2$ and $C_2 + C_3$. This, then, results in a constant output voltage. Since the output voltage was not effected by amplitude variations, the need for limiting is reduced.

With no modulation applied to the circuit, both diodes conduct equally and C_4 charges up. This results in a constant output voltage of zero at terminals AB. When the frequency changes as a result of modulation on the carrier, diode D_1 or D_2 will conduct more heavily. This makes the voltage ratio charged on C_2 and C_3 unequal, which gives the name ratio detector. The voltage at terminals AB is, therefore, the output signal of the demodulated FM signal.

Defining Terms

Direct FM: Frequency modulation produced by changing the frequency of a carrier as a result of applying a modulating signal.

Discriminator: A detector used in an FM receiver to demodulate the FM signal.

Frequency deviation: The peak difference between modulated wave and the carrier frequency.

Frequency modulation: A system of modulation where the instantaneous radio frequency varies in proportion to the instantaneous amplitude of the modulating signal and the instantaneous radio frequency is independent of the frequency of the modulating signal.

Limiter: A stage of an FM receive that is designed to saturate and clip off undesired AM and noise components from a signal prior to FM detection.

Modulation index: The ratio of the frequency deviation to the frequency of the modulating signal.

Percentage modulation: The ratio of the actual frequency deviation to the frequency deviation defined as 100% modulation, expressed in percentage. For FM broadcast stations, a frequency deviation of $+/-75$ kHz is defined as 100% modulation.

Phase modulation: Frequency modulation produced by shifting the phase of the carrier relative to an arbitrary reference point. This method of frequency modulation is often referred to as indirect FM.

Ratio detector: A detector used in an FM receiver to demodulate the FM signal.

References

Armstrong, E.H. 1936. A method of reducing disturbances in radio signaling by a system of frequency modulation. *Proc. of IRE* 24(5):689–740.

Code of Federal Regulations. 1993. Vol. 47, Section 73.310.

Crutchfield, E.B., ed. 1985. *National Association of Broadcasters Engineering Handbook*, 17th ed., pp. 3.3-63– 3.3-68. Washington, D.C.

Inglis, A.F. 1988. *Electronic Communications Handbook*. McGraw-Hill, New York.

Klapper, J., ed. 1970. *Selected Papers on Frequency Modulation*. Dover, New York.

Lewis, T. 1991. *Empire of the Air—The Men Who Made Radio*. Harper Collins, New York.

Rohde, L.U. and Bucher, T.T. 1988. *Communications Receivers Principles & Design*. McGraw-Hill, New York.

12.4 Pulse Modulation

Rodger E. Ziemer

12.4.1 Introduction

Pulse modulation is important for many applications including telephone call transmission, compact disks for music, airline passenger communication systems, and digital control systems, among others. The reasons for the use of pulse modulation even in cases where the information is analog in form are many. For example, in telephone call transmission, the overloading of cable trays by copper wire in the late 1960s in part led to the development of the T carrier system, which employs **time-division multiplexing** (TDM). Thus, several separate telephone calls can be carried by the same transmission line. Airline passenger communication systems were similarly developed based on TDM to save in weight of the transmission lines required to address several functions at each passenger's seat, including the attendant calling function and several channels of entertainment media. In the late 1980s compact disks and the associated recording and playback technology were developed to provide high-fidelity music with little danger of degradation due to damage of the medium (i.e., the compact disk). Control systems in aircraft, particularly high-performance military aircraft, rely on the servomechanism commands and responses being conveyed by transmission lines to the various control points with these commands being represented in digital format. Spacecraft similarly utilize baseband digital pulse modulation to distribute commands and acquire data from the various sensors and systems on board.

12.4.2 The Sampling Theorem

The basis for pulse modulation applications is representing analog signals as properly spaced samples. The theoretical justification for this is Shannon's **sampling theorem** (Ziemer and Tranter, 2002), which may be stated succinctly as follows:

> *Lowpass Uniform Sampling Theorem:* A signal with no frequency components above W Hz can be uniquely represented by uniformly spaced samples taken at intervals spaced no greater than $1/2W$ sec apart.

Various forms of this sampling theorem are often used, including a bandpass uniform sampling version and quadrature sampling for bandpass signals.

The proof of the low-pass uniform sampling theorem (hereafter referred to simply as the sampling theorem) is based on the following Fourier transform theorems (Ziemer and Tranter, 2002):

> *Multiplication Theorem:* Given two signals, $x(t)$ and $y(t)$, with Fourier transforms $X(f)$ and $Y(f)$, where f is the frequency in Hz, the Fourier transform of their product is the convolution of their Fourier transforms:

$$x(t)y(t) \leftrightarrow X(f) * Y(f) = \int_{-\infty}^{\infty} X(\lambda)Y(f - \lambda)\, d\lambda \qquad (12.48)$$

where the double-headed arrow denotes a Fourier transform pair.

> *Fourier Transform of the Ideal Sampling Waveform:* The Fourier transform of a doubly infinite train of uniformly spaced impulses is itself a doubly-infinite train of uniformly spaced impulses, or

$$\sum_{n=-\infty}^{\infty} \delta(t - nT_s) \leftrightarrow f_s \sum_{n=-\infty}^{\infty} \delta(f - nf_s), \quad f_s = \frac{1}{T_s} \qquad (12.49)$$

Consider a low-pass signal $x(t)$ with Fourier transform $X(f)$ bandlimited so that none of its spectral components lie above W Hz. Ideal impulse sampling of $x(t)$ gives

$$x_\delta(t) = x(t) \sum_{n=-\infty}^{\infty} \delta(t - nT_s) = \sum_{n=-\infty}^{\infty} x(nT_s)\delta(t - nT_s) \tag{12.50}$$

Use of Eq. (12.48) and Eq. (12.49) to obtain the Fourier transform of $x_\delta(t)$ gives

$$X_\delta(f) = X(f) * f_s \sum_{n=-\infty}^{\infty} \delta(f - nf_s) = f_s \sum_{n=-\infty}^{\infty} X(f - nf_s) \tag{12.51}$$

where $\delta(f - nf_s) * X(f) = X(f - nf_s)$ has been used and the asterisk denotes convolution. Equation (12.51) is sketched in Fig. 12.47(b) for an assumed $X(f)$ shown in Fig. 12.47(a). It is seen that if the sampling frequency f_s is greater than $2W$, then the baseband portion of the sampled signal spectrum can be separated from the translated spectra centered around nonzero multiples of f_s, and the result is an undistorted version of the original signal spectrum $X(f)$. If the signal to be sampled, $x(t)$, is bandlimited but $f_s < 2W$, the translates of $X(f)$ centered at $f = \pm f_s$ will overlap with the component centered at $f = 0$. The resulting error is referred to as **aliasing**. If the signal to be sampled, $x(t)$, is not strictly bandlimited, then the sampling frequency cannot be chosen large enough to prevent overlap of the translates making up the sampled signal spectrum. Perfect distortionless recovery is impossible and

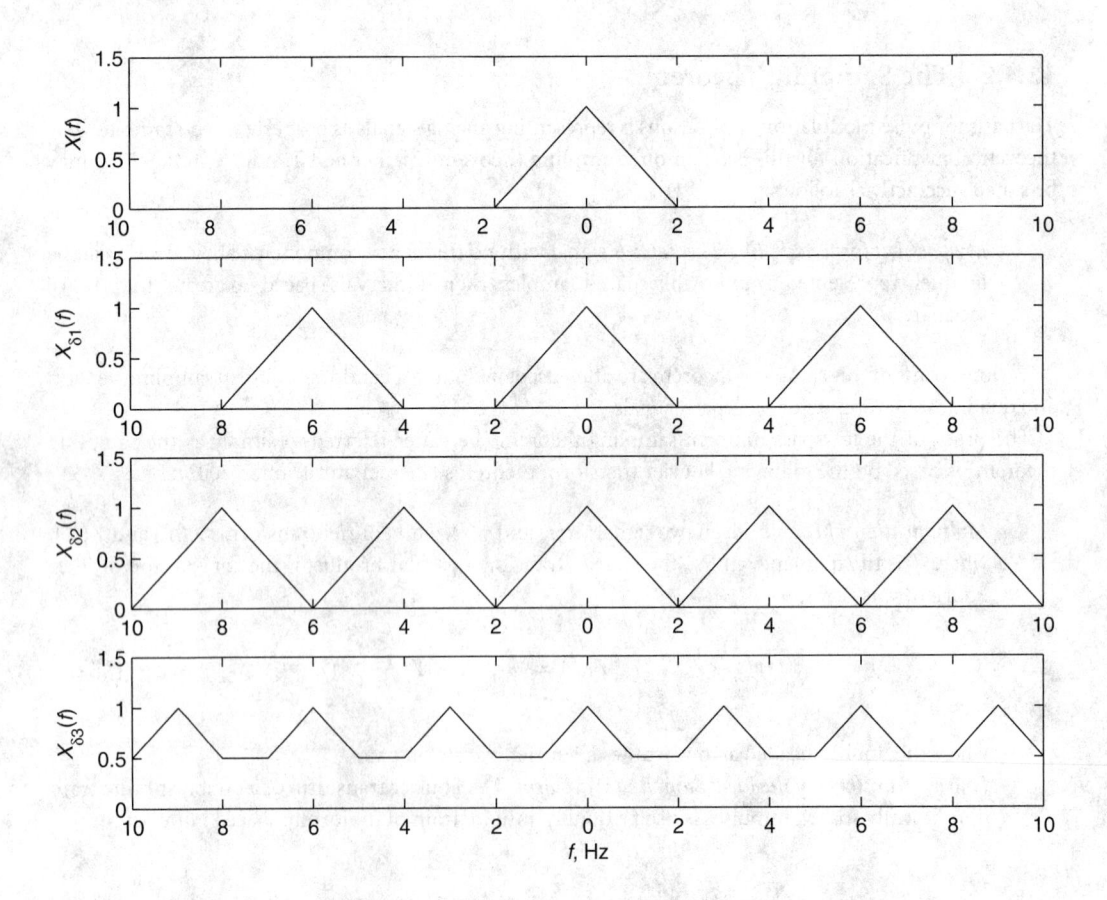

FIGURE 12.47   Spectra for sampling of a strictly bandlimited signal: (a) spectrum of signal to be sampled, (b) sampled signal spectrum for $f_s > 2W$, (c) sampled signal spectrum for $f_s = 2W$ (d) sampled signal spectrum for $f_s < 2W$. (All maximum spectral levels scaled to unity.)

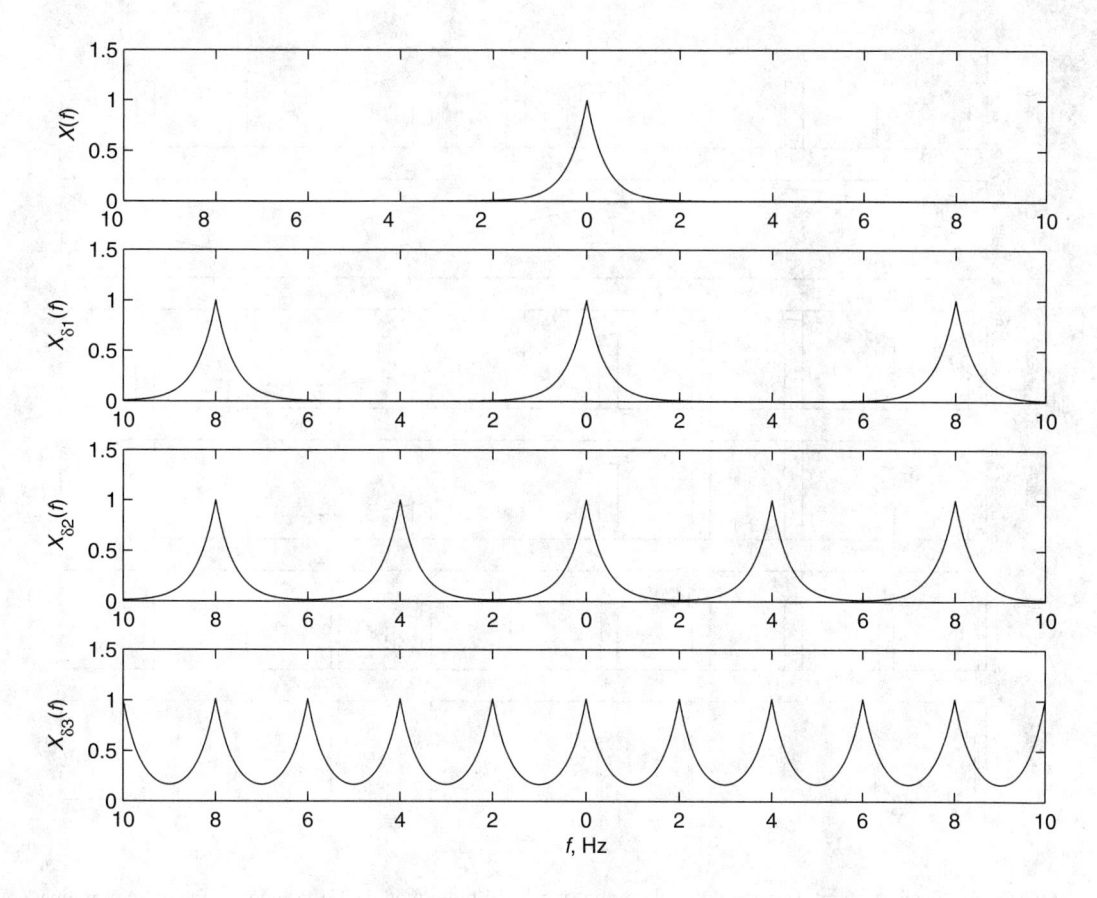

FIGURE 12.48 Spectra for sampling a signal with nonbandlimited spectrum: (a) spectrum of signal to be sampled, (b) sampled signal spectrum for very large f_s (minimal aliasing), (c) sampled signal spectrum for f_s moderately large (some aliasing), (d) sampled signal spectrum for f_s too small (significant aliasing).

distortion results due to the filtering operation used for signal recovery, as does aliasing. The aliasing due to a nonbandlimited signal spectrum is illustrated in Fig. 12.48.

Clearly, sampling by means of impulses is impossible from a practical standpoint. A more practical waveform is the train of square pulses illustrated in Fig. 12.49(a). The sampling operation can be done in two ways with such a pulse train. The first, referred to as **natural sampling**, consists of multiplication of the signal to be sampled, $x(t)$ as shown in Fig. 12.49(b), by a pulse train, such as that shown in Fig. 12.49(a), as illustrated in Fig. 12.49(c). The second, called **flat top sampling**, is accomplished by sampling the waveform $x(t)$ at the instants nT_s and holding that value for the duration of the pulses making up the pulse train as illustrated in Fig. 12.49(d). For the former, the difference in the spectrum from ideal sampling is provided by the transform pair

$$p_\tau(t) = \begin{Bmatrix} 1, & 0 \le t \le \tau \\ 0, & \text{otherwise} \end{Bmatrix} \leftrightarrow \tau \operatorname{sinc}(\tau f) e^{-j\pi\tau f} \tag{12.52}$$

where τ is the pulse width and $\operatorname{sinc}(u) = \sin(\pi u)/(\pi u)$. The ideal sampling waveform (Eq. (12.49)) can be converted to the natural sampling waveform, shown in Fig. 12.49(a), by convolving $p_\tau(t)$ with Eq. (12.49). The Fourier transform of the natural sampled waveform is, therefore, the product of the transform in Eq. (12.49) and the transform in Eq. (12.52), and the overall effect on the spectrum is to multiply each term of Eq. (12.51) by $\tau \operatorname{sinc}(n\tau f_s)e^{-jn\pi\tau f_s}$. Thus, the spectrum is not distorted, but each term in the spectrum is multiplied by the weighting function $\tau \operatorname{sinc}(n\tau f_s)$. On the other hand, flat top sampling can

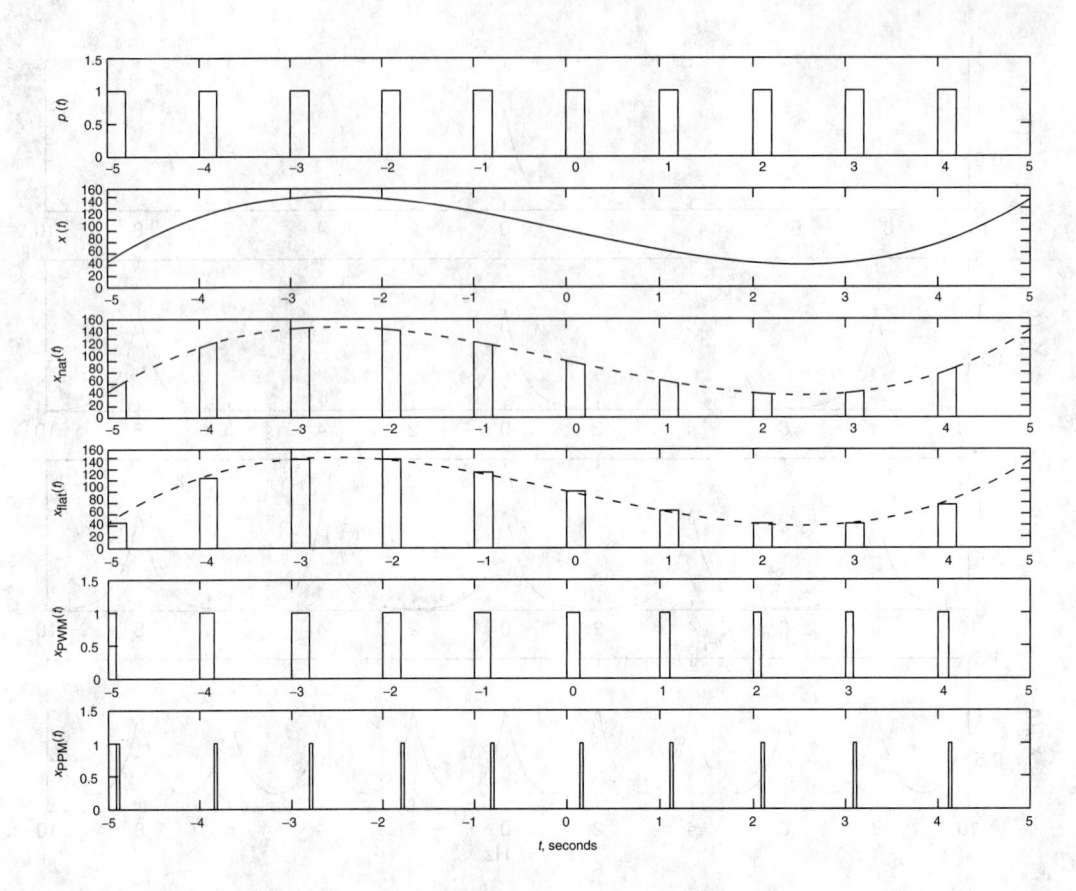

FIGURE 12.49 llustration of practical sampling of a signal and pulse modulation techniques: (a) pulse sampling function, (b) signal to be sampled, (c) natural sampling, (d) flat top sampling or pulse-amplitude modulation, (e) pulse-width modulation, (f) pulse-position modulation.

be represented as

$$x_{\text{flat}}(t) = \sum_{n=-\infty}^{\infty} x(nT_s) p_\tau(t - nT_s)$$

$$= p_\tau(t) * \sum_{n=-\infty}^{\infty} x(nT_s)\delta(t - nT_s) \qquad (12.53)$$

$$= p_\tau(t) * x_\delta(t)$$

In analogy to Eq. (12.48), the Fourier transform of the convolution of two signals is the product of their respective Fourier transforms. Thus, the spectrum of Eq. (12.53) is

$$X_{\text{flat}}(f) = \tau f_s \text{sinc}(\tau f) \sum_{n=-\infty}^{\infty} X(f - nf_s) \qquad (12.54)$$

It is seen that in the case of flat top sampling, the sampled signal spectrum is a distorted version of $X(f)$ because of the factor $\tau f_s \text{sinc}(\tau f)$. If $\tau \ll T_s$, this distortion is small and the original signal can be recovered almost exactly. If not, the factor $\tau f_s \text{sinc}(\tau f)$ must be compensated for by an inverse filter having a frequency response of the form $[\text{sinc}(\tau f)]^{-1}$ before recovery of $x(t)$ takes place.

Some additional remarks with respect to Fig. 12.49 relate to other ways of representing a sampled signal as shown in Fig. 12.49(e) Fig. 12.49(f). In Fig. 12.49(e), the samples are represented as pulses the width of which vary in proportion to the sample values, while in Fig. 12.49(f) they are represented in terms of pulse delays from the points of initial sampling in proportion to the sample values. The former is called pulse width modulation (PWM) and the latter is called pulse position modulation (PPM). Pulse width modulation is often used to control the speed of DC moters used in various variable-speed drive applications.

12.4.3 Analog-to-Digital Conversion

Although analog samples can be transmitted from the source to destination, and historically this was done, it is now more usual to digitize the samples of an analog signal and transmit them in digital form. To do so, the additional steps of quantization and encoding are required first. This process is called **analog-to-digital (A/D) conversion**. This quantization is usually accomplished in terms of a binary code for each sample. Two methods for such representation will be discussed here. First, the base two representation for any set of positive integers is of the form

$$x(nT_s) = b_{m-1} \times 2^{m-1} + \cdots + b_2 \times 2^2 + b_1 \times 2^1 + b_0 \times 2^0 \qquad (12.55)$$

where m is the number of bits b_i in the representation, also called the **wordlength**. As an example, consider the sample values of Fig. 12.49, which are given in Table 12.8, where the sample values are rounded to the nearest integer value. A disadvantage of the binary code is that several bits can change for a one-unit change in the sample value. For example, the **binary representation** of 15_{10} is 01111 whereas the binary representation of 16_{10} is 10000. An alternative representation, where a one decimal-unit change in sample magnitude results in a single digit change in the binary representation, is the **Gray code**, which can be found from the binary code by means of the following algorithm:

$$\begin{aligned} g_0 &= b_0 \\ g_i &= b_i \oplus b_{i-1}, \quad i > 0 \end{aligned} \qquad (12.56)$$

The Gray code representation of the samples shown in Fig. 12.47 is given in the fourth column of Table 12.8.

Representation of the sample values by a binary code, although straightforward and intuitively satisfying, is not always the best course to follow. For example, large values are represented by the same number of bits as smaller values, and this may not be the best road to follow since smaller values may be more probable than the larger values. Two methods for representing more probable values of the samples may be employed: (1) nonuniform **quantization** and (2) compression of the signal to be sampled and quantized followed by subsequent expansion at the receiver, referred to as **companding**. The objective is to compress the signal before sampling and quantization such that the dynamic range is increased with very little loss in

TABLE 12.8   Sample Values and Binary Representations for the Waveform of Fig. 12.49(b).

Time, nT_s	Sample Value (rounded)	Binary Code	Gray Code
−5	45	00101101	01110111
−4	124	01111100	10000100
−3	157	10011101	10100111
−2	157	10011101	10100111
−1	134	10000110	10001010
0	100	01100100	10101100
1	66	01000010	11000110
2	43	00101011	01111101
3	43	00101011	01111101
4	76	01001100	11010100
5	155	10011011	10101101

terms of signal to quantization noise power ratio. One such compressor characteristic is called the μ-**law compressor**, which is described by the input-output characteristic (Couch, 1990)

$$y(x) = V \operatorname{sgn}(x) \frac{\ln\left[1 + \frac{\mu|x|}{V}\right]}{\ln(1 + \mu)} \qquad (12.57)$$

where V is the maximum input signal amplitude, μ is a parameter, and

$$\operatorname{sgn}(x) = \begin{cases} 1, & x > 0 \\ 0, & x = 0 \\ -1, & x < 0 \end{cases} \qquad (12.58)$$

The value $\mu = 255$ is used in the telephone system in the U.S. At the receiving end, the samples are put through a nonlinearity which is the inverse of Eq. (12.57).

Companders are an attempt at minimizing the error in representing a waveform in terms of quantized samples based on a fixed wordlength. Another approach at minimizing the average number of bits per sample is to use a variable length code. The optimum variable wordlength code for independent samples is called a **Huffman code** after its inventor (Blahut, 1990). It is an algorithm for obtaining the minimum average number of bits per sample based on the probability distribution of the samples, with the lower probability samples being assigned the most number of bits and the most probable sample values being assigned the least number of bits. The process of encoding samples at a source is called **source encoding** and can be viewed as removing redundancy from the source output. Another type of encoding, called **channel encoding** or *error correction encoding*, adds redundancy in terms of extra bits appended to each encoded sample so that errors can be corrected at the reception point.

12.4.4   Baseband Digital Pulse Modulation

After the samples have been quantized, they are transmitted through a channel, received, and converted back to their approximate original form. The reason for the modifier "approximate" is that they will invariably suffer some degradation from noise and channel-induced distortion. This will be explored shortly. It is first important to point out that transmission does not necessarily mean to a remote location. For example, storage of data on a computer hard disk and retrieval may happen in the same physical location.

The question at hand is in what format, hereafter called modulation, should the quantized samples be placed for faithful transmission through the channel and subsequent reproduction? The answer depends on the channel. In the case of magnetic recording media, low-frequency response of the medium and recording and pickup heads is poor, so a type of modulation must be used that minimizes the low-frequency content of the signal being recorded. A myriad of formats is possible. We will discuss only a few. In Fig. 12.50, several possible formats are shown.

The first is called **nonreturn-to-zero** (NRZ) polar because the waveform does not return to zero during each signaling interval, but switches from $+V$ to $-V$, or vice versa at the end of each signaling interval (NRZ unipolar uses the levels V and 0). NRZ-mark (also called differentially encoded) represents a data 1 by no change of level and a data 0 by a change of level as shown in Fig. 12.50(b). **Unipolar return-to-zero** (URZ) format, shown in Fig. 12.50(b), returns to zero in each signaling interval. Since bandwidth is inversely proportional to pulse duration, it is apparent that URZ requires twice the bandwidth that NRZ does. Another point about URZ is that it has a nonzero DC component, whereas this is usually not the case for NRZ unless there are more 1s than 0s present in the data or vice versa. An advantage of URZ over NRZ is that a pulse transition is guaranteed in each signaling interval whereas this is not the case for NRZ. Thus, in cases where there are long strings of 1s or 0s, it may be difficult to synchronize the receiver to the starting and stopping times of each pulse in the case of NRZ. A very important modulation format from synchronization considerations is NRZ-mark, also known as **differential encoding**, where an initial reference bit is chosen and a subsequent 1 is encoded as no change from the reference and a 0 is encoded as a change. After the initial reference bit, the current bit serves as a reference for the next bit, etc.

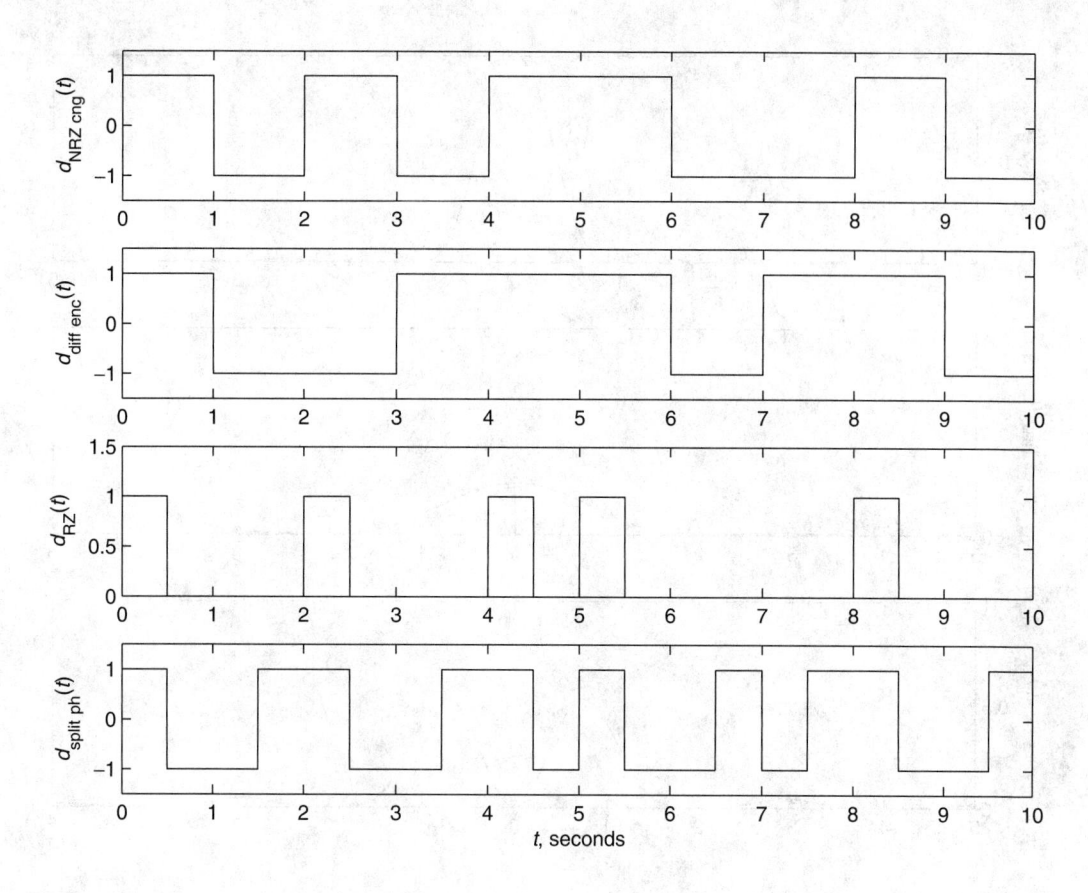

FIGURE 12.50 Various baseband modulation formats: (a) nonreturn-to-zero change, (b) differentially encoded NRZ-mark), (c) return-to-zero, (d) split phase.

An example of this modulation format is shown in Fig. 12.50(b). Another baseband data modulation format that guarantees a transition in each signaling interval and does not have a DC component is known as Manchester, biphase, or **split phase**, which is illustrated in Fig. 12.50(d). It is produced by doing a logical OR of the data clock with an NRZ-formatted signal. The result is a + to − transition for a logic 1, and a − to + zero crossing for a logic 0. Several other data formats have been proposed and employed in the past, but we will consider only these.

An important consideration of any data format is its **bandwidth occupancy**. It can be shown (Ziemer and Tranter, 2002) that NRZ-polar has the power spectral density [$\text{sinc}(z) = \sin(\pi z)/(\pi z)$]

$$S_{\text{NRZ}}(f) = V^2 T_b \, \text{sinc}^2(T_b f) \tag{12.59}$$

whereas URZ format has the power spectrum

$$S_{\text{RZ}}(f) = \frac{V^2 T_b}{16} \text{sinc}^2\left(\frac{T_b f}{2}\right) + \frac{V^2}{16} \sum_{\substack{m=-\infty \\ m \,\text{odd}}}^{\infty} \frac{4}{(\pi m)^2} \delta\left(f - \frac{m}{T_b}\right) + \frac{V^2}{16}\delta(f) \tag{12.60}$$

where the delta function at $f = 0$ reflects the nonzero DC level of the unipolar return-to-zero format. On the other hand, differentially encoded (NRZ-mark) and split phase are formats with zero DC level. The power spectral density of the former is the same as for NRZ, and that of the latter is

$$S_{\text{SP}}(f) = V^2 T_b \text{sinc}^2\left(\frac{T_b f}{2}\right) \sin^2\left(\frac{\pi T_b f}{2}\right) \tag{12.61}$$

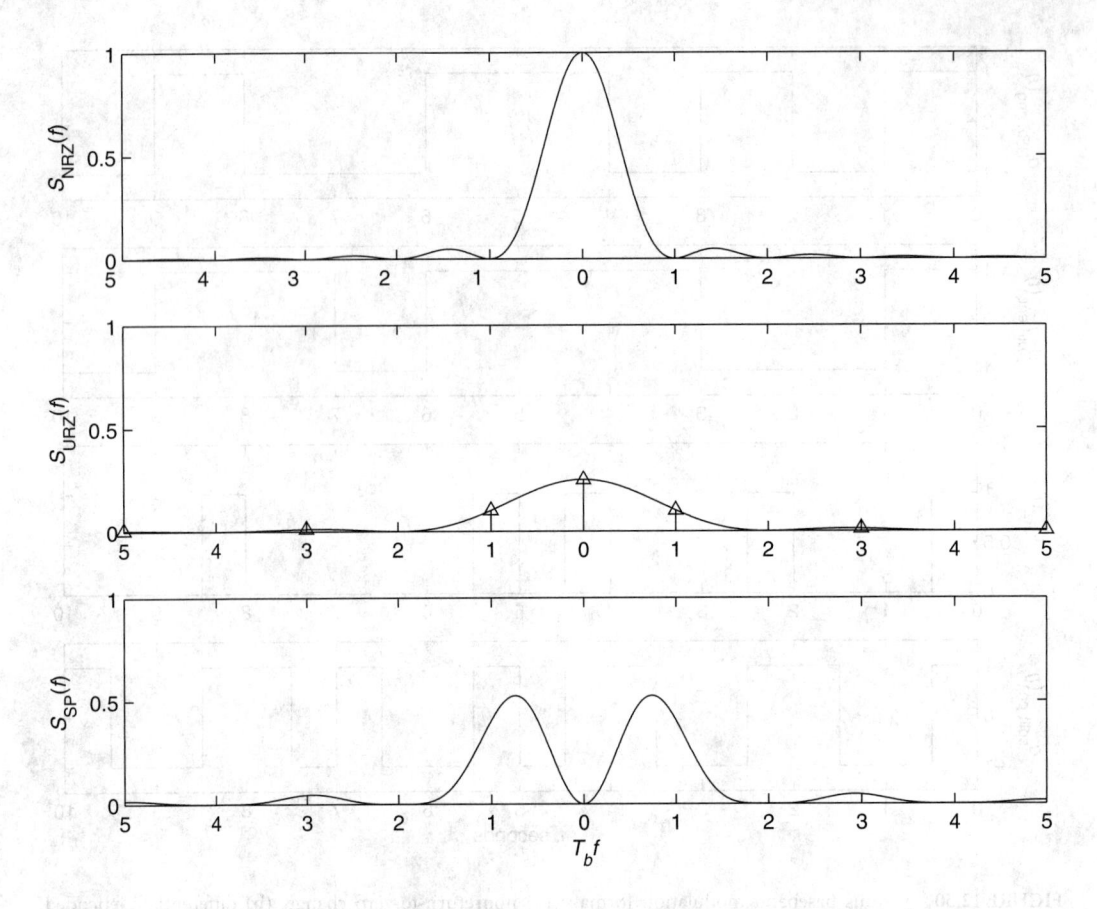

FIGURE 12.51 Power spectra for baseband modulation formats: (a) nonreturn-to-zero, (b) unipolar return-to-zero, (c) split phase. (Amplitudes of pulse trains scaled to have unit energy.)

The total average power in each case is obtained by integrating the power spectrum over $-\infty < f < \infty$. Spectra for NRZ, URZ, and split phase are plotted in Fig. 12.51, where it is noted that nonreturn-to-zero occupies half the bandwidth of URZ and split phase.

Of the four modulation formats discussed, NRZ and URZ assume zero memory between pulses, and differentially encoded and split phase have memory imposed between pulses. Split phase has zero power density at $f = 0$ with the result that its bandwidth is double that of nonreturn-to-zero. In a sense, the zero power density at $f = 0$ is obtained in the case of split phase by imposing a particular type of memory between pulses. More general memory structures are used between pulses for applications such as magnetic recording. These can be classified as **line codes**. It is beyond the scope of this chapter to go into this subject here. A simple example is provided by assuming a square pulse function of width T_b for each bit, but with successive pulse multipliers related by $a_k = A_k - A_{k-1}$ where $A_k = \pm 1$ represent the bit value in signaling interval k. Thus the multiplier for pulse k can assume the values $2 (A_k = 1$ and $A_{k-1} = -1)$, $0 (A_k = 1$ and $A_{k-1} = 1)$, or $-2 (A_k = -1$ and $A_{k-1} = 1)$. The power spectral density of this pulse modulation format can be shown to be (Ziemer and Tranter, 2002)

$$S_{DC}(f) = 4V^2 T_b \, \text{sinc}^2(T_b f) \sin^2(\pi T_b f) \tag{12.62}$$

Such pulse modulation is referred to as *dicode*. If the original bit stream is precoded with differential encoding, it is referred to as *duobinary*, a modulation scheme that was invented by Lender (1966).

Before leaving the subject of pulse modulation formats, we discuss one more principle involved in choosing pulse shapes with bandwidth occupancy in mind, known as **Nyquist's pulse-shaping criterion**. The idea is to find pulse shapes $p(t)$, with bandlimited spectra, that have the zero **intersymbol-interference** property, given by

$$p(nT_b) = \begin{cases} 1, & n = 0 \\ 0, & n \neq 0 \end{cases} \tag{12.63}$$

The condition (12.63) says that if the ouput of the transmitter/channel/receiver filter cascade is of the form

$$x_{\text{rec}}(t) = V \sum_{k=-\infty}^{\infty} a_k p(t - kT_b) \tag{12.64}$$

where a_k represent the bit values (± 1), then sampling at intervals of T_b ensures that a given pulse sample is not influenced by preceeding or following pulses. This is referred to as zero intersymbol interference. Nyquist's pulse shaping criterion states that Eq. (12.64) holds if the Fourier transform of $p(t)$, $P(f)$, satisfies

$$\sum_{k=-\infty}^{\infty} P\left(f + \frac{k}{T_b}\right) = T_b, \quad |f| \leq \frac{1}{2T_b} \tag{12.65}$$

One pulse family that has this property is that family having **raised-cosine spectra**. This pulse shape family is given by

$$p(t) = \frac{\cos 2\pi\beta t}{1 - (4\beta t)^2} \text{sinc}(t/T_b) \tag{12.66}$$

which has the spectrum

$$P(f) = \begin{cases} T_b, & |f| \leq \frac{1}{2T_b} - \beta \\ \frac{T_b}{2}\left[1 + \cos\left(\frac{\pi(|f| - 1/(2T_b) + \beta)}{2\beta}\right)\right], & \frac{1}{2T_b} - \beta < |f| \leq \frac{1}{2T_b} + \beta \\ 0, & |f| > \frac{1}{2T_b} + \beta \end{cases} \tag{12.67}$$

The parameter β determines the bandwidth of the pulse spectrum and its rate of decrease to zero. This pulse shape family and the corresponding spectra are shown in Fig. 12.52.

12.4.5 Detection of Pulse Modulation Formats

With no dependency between signaling pulses, the optimum detection scheme for a digitally modulated pulse train in additive white Gaussian noise (AWGN) is to pass the received pulse train plus noise through a filter matched to the basic signaling pulse, sample the output at the time of the peak output signal, and compare this sample with a suitably chosen threshold. A **matched filter** has impulse response proportional to the time reverse of the signal to which it is matched. It can be shown that the matched filtering and sampling of a signal is equivalent to correlating the received signal plus noise with a replica of the basic pulse shape, sampling, and threshold comparison. These two basic detector structures are shown in Fig. 12.53(a) and Fig 12.53(b). For a rectangular pulse shape, the **correlation receiver** is equivalent to integrating the pulse being detected over its width, sampling at the end of the pulse, and comparing with a threshold. This receiver structure, called an **integrate-and-dump detector**, is shown in Fig. 12.53(c). NRZ pulse modulation fits this description, and it can be shown that the average probability of making an error in detecting NRZ using this procedure is given by

$$P_E = Q\left(\sqrt{\frac{2E_b}{N_0}}\right), \quad \text{NRZ} \tag{12.68}$$

where

$$Q(x) = \int_x^\infty \frac{\exp(-u^2/2)}{\sqrt{2\pi}} \, du \tag{12.69}$$

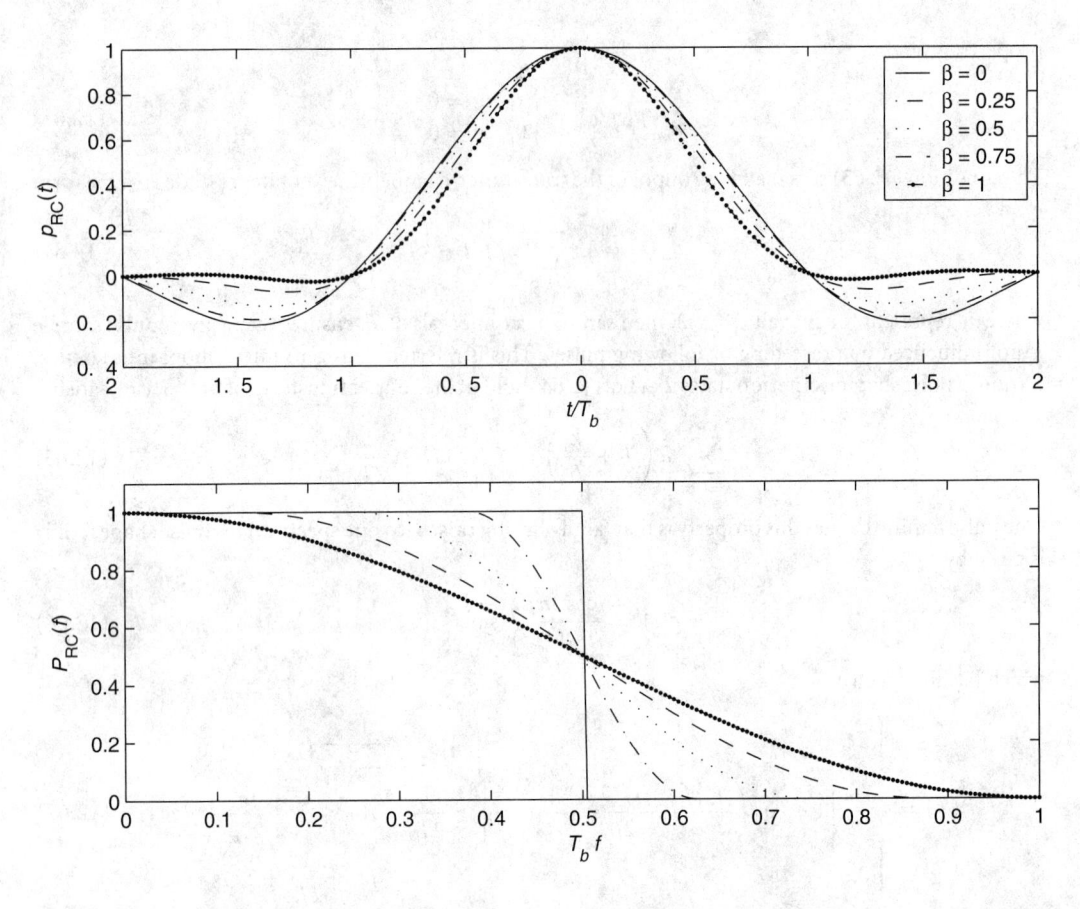

FIGURE 12.52 The raised cosine family of (a) pulse shapes and (b) spectra.

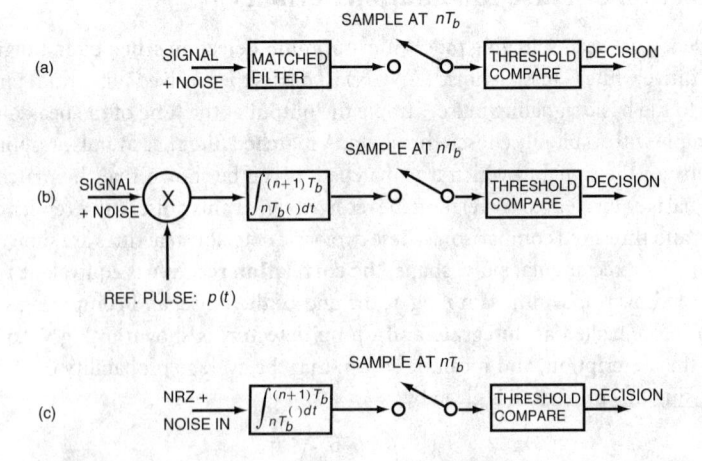

FIGURE 12.53 Detectors for pulse modulated signals: (a) matched filter, (b) correlation, (c) integrate-and-dump for square-pulse signals.

is the so-called Q-function, N_0 is the one-sided noise power spectrum level, and $E_b = V^2 T_b$ is the energy in the pulse.

Since URZ has half the pulse width of NRZ, we scale the amplitude by $2^{1/2}$ relative to NRZ in order to do a comparison on an equal energy basis. However, since the level zero is used to represent a logic 0 (as opposed to $-V$ for NRZ), its performance is a factor of two worse than that of NRZ, with the result for the **probability of error** being

$$P_E = Q\left(\sqrt{\frac{E_b}{N_0}}\right), \quad \text{URZ} \tag{12.70}$$

where E_b is now the average bit energy. For differential encoding, essentially two bit errors result each time there is a bit error (the present bit serves as a reference for a succeeding bit) so that for a given E_b/N_0 the probability of error is double that of NRZ. This amounts to about a 0.8-dB shift on the E_b/N_0 axis in a plot of P_E vs. E_b/N_0. Finally, detection of split phase is accomplished by the same detector as for NRZ with a multiplication by the clock preceeding the detector. The probability of error is the same as for NRZ. The P_E results for these pulse modulation formats are plotted in Fig. 12.54 as a function of E_b/N_0.

12.4.6 Analog Pulse Modulation

The concentration in this section has been on digital pulse modulation methods. Analog pulse modulation, whereby some attribute of a pulse is made to vary in a one-to-one correspondence with the signal samples,

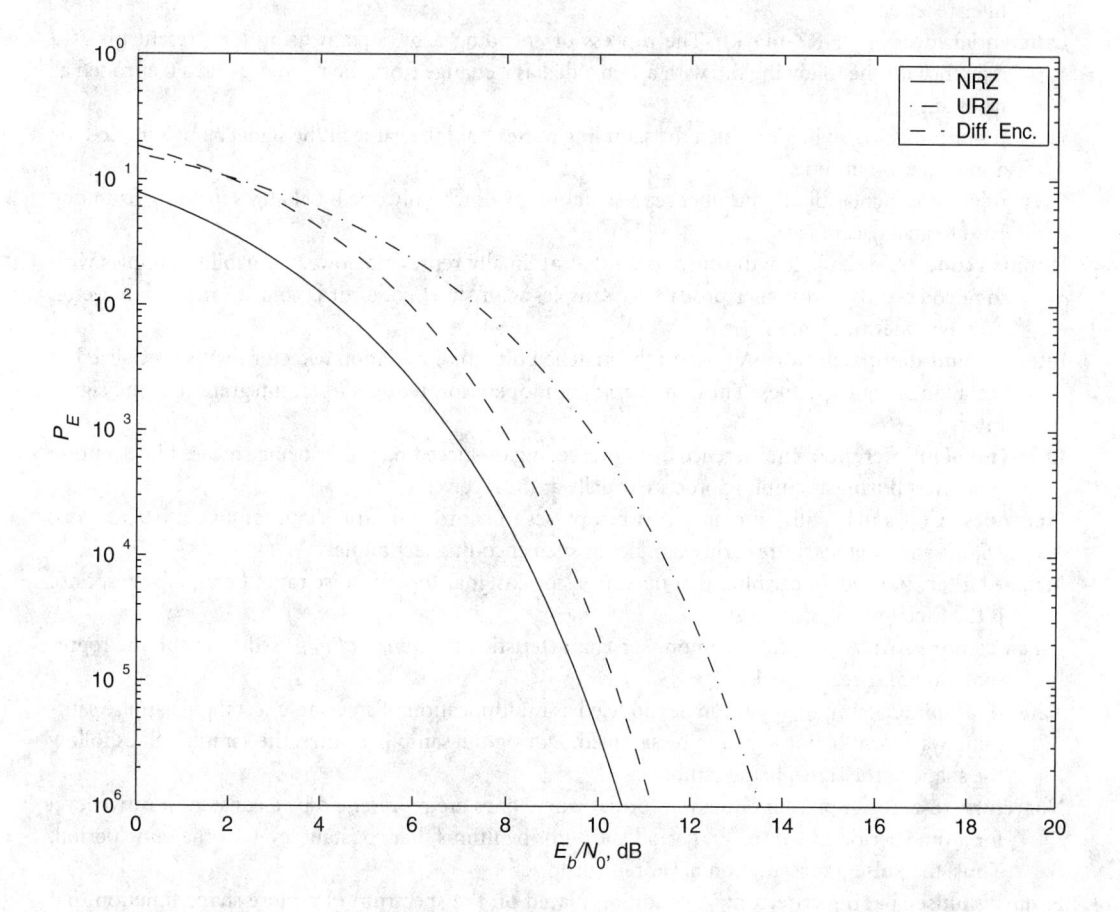

FIGURE 12.54 Probability of error curves for NRZ, URZ, and differentially encoded data formats.

is not as important now as it once was. The main types of analog pulse modulation are pulse amplitude, pulse width, and pulse position modulation. (Ziemer and Tranter, 2002).

Defining Terms

Aliasing: Distortion arising from the representation of a signal by its samples due to too low a sampling rate relative to the signal bandwidth.

Analog-to-digital (A/D) conversion: The process of approximating an analog sample in terms of a finite-precision number, usually in binary form.

Bandwidth occupancy: The amount of bandwidth in Hz occupied by a signal. Quite often, an approximate measure must be used because the total bandwidth extent of typical signals is infinite.

Binary representation: Representation of sample values in terms of a binary number.

Channel encoding: The process of appending redundant bits onto the digital information sent through a channel with the goal of being able to correct errors at the receiver.

Companding: The process of compressing a signal in amplitude before A/D conversion and then expanding it after digital-to-analog conversion. The purpose is to get more resolution for small and moderate sample values and less at large amplitudes. Without quantization, companding would result in no waveform distortion.

Correlation receiver: A receiver structure that correlates received signal plus noise with replicas of the possible received signal shapes. It is completely equivalent in terms of performance to the matched filter receiver.

Differential encoding (NRZ-mark): The process of encoding a bit stream using the present bit as a reference for the following bit, with a 1 encoded as a change from the reference and a 0 encoded as no change.

Flat top sampling: Sampling in which the sampling pulses hold the value of the signal being sampled for a short period of time.

Gray code: A particular binary number representation in which only one bit changes in going from one level to an adjacent level.

Huffman code: A variable-length source code that optimally represents lower probability samples with long codewords and higher probability samples with short codewords so as to minimize overall average codeword length.

Integrate-and-dump detector: A form of the matched filter or correlation detector that is specialized to rectangular signal shapes. Thus, the correlation operation reduces to an integration of the signal interval.

Intersymbol interference: Interference from preceeding or suceeding pulses being smeared into a pulse of interest during a sampling process, usually at the receiver.

Line codes: Codes that utilize memory between pulses to control spectral shape, among other reasons. *Dicode* and *duobinary* are early examples of such encoding techniques.

Matched filter: A fixed, linear filter that maximizes peak signal-to-rms noise ratio. Its impulse response is the time reverse of the signal.

μ-Law compressor: A particular compressor characteristic (see *companding*) used in telephone representation of speech signals.

Natural sampling: Sampling that can be modeled as multiplication of a periodic rectangular-pulse sampling waveform by the signal to be sampled. During the sampling times, the sample values follow the shape of the signal being sampled.

Nonreturn-to-zero: A pulse modulation format where ones are represented by a constant positive level for a time period called the bit period and zeros by minus that constant level for the same period. Thus, the pulse representation never returns to zero.

Nyquist's pulse-shaping criterion: A condition placed on the spectrum of a pulse shape function that guarantees zero intersymbol interference when samples are taken at a proper sampling rate.

Probability of error: A measure of performance for a digital receiver. Over a long string of received symbols (signals), it is approximately the ratio of the number of symbols received erroneously divided by the total number.

Pulse modulation: A process whereby information is impressed on a pulse train carrier for transmission through a channel.

Quantization: One of the steps in A/D conversion whereby samples assuming a continuum of values are approximated by finite-precision values.

Raised-cosine spectra: The spectrum of one family of pulses that obeys Nyquist's pulse-shaping criterion.

Sampling Theorem: One of several theorems, the most common of which is called the low-pass sampling theorem and says that a low-pass bandlimited signal of bandwidth W Hz may be represented in terms of samples taken periodically at a minimum rate of $2W$ per second.

Source code: Any coding technique to represent the output of a source, usually with the objective of minimizing the average wordlength.

Split phase: A pulse modulating format that amounts to nonreturn-to-zero multiplied by the data clock, which assumes ± 1 values (also called Manchester or biphase).

Time-division multiplexing: Interlacing pulse samples from different sources in time so that they can be transmitted through a common channel.

Unipolar return-to-zero: A pulse modulation format wherein ones are represented by a positive pulse level during the first-half of the bit period and zero during the last-half and zeros are represented by a zero level.

Wordlength: As pertaining to A/D conversion, the number of digits, usually bits, used to represent a sample.

Bibliography

Blahut, R. E. 1990. *Digital Transmission of Information*. Addison-Wesley, Reading, MA.

Couch, L. W., II. 2001. *Digital and Analog Communication Systems*, 6th ed. Prentice-Hall, Upper Saddle River, NJ.

Haykin, S. 2000. *Communication Systems*, 4th ed. Wiley, New York.

Lender, A. 1966. Correlative level encoding for binary data transmission. *IEEE Spectrum* (Feb.):104–115.

Proakis, J. G. 2001. *Digital Communications*, 4th ed. McGraw-Hill, New York.

Ziemer, R. E. and Peterson, R. L. 2000. *Introduction to Digital Communication*, 2nd ed. Prentice-Hall, Upper Saddle River, NJ.

Ziemer, R.E. and Tranter, W.H. 2002. *Principles of Communications: Systems, Modulation, and Noise*, 5th ed. Wiley, New York.

There are many books written at the senior/first-year graduate level that discuss the material in this article in addition to those given in the references. Examples include the following.

Sklar, Bernard. 2001. *Digital Communications: Fundamentals and Applications*, 2nd ed. Prentice-Hall, Upper Saddle River, NJ.

Gibson, J.D. 1993. *Principles of Digital and Analog Communications*, 2nd ed. Prentice-Hall, Upper Saddle River, NJ.

Pulse modulation is important in the digital representation of speech and video waveforms. Information on encoding methods for such signals may be found in the book:

Jayant, N.S. and Noll, P. 1984. *Digital Coding of Waveforms: Principles and Applications to Speech and Video*. Prentice-Hall, Upper Saddle River, NJ.

A special communications applications area where pulse modulation is used to represent speech waveforms is wireless communications. A reference book for this extensive area is the following.

Mark, J. W. and Zhuang, W. 2003. *Wireless Communications and Networking*. Prentice-Hall, Upper Saddle River, NJ.

12.5 Digital Modulation

Rodger E. Ziemer

12.5.1 Introduction

Digital modulation is necessary before digital data can be transmitted through a bandpass channel. Examples of such channels are microwave line of sight, satellite, optical fiber, and cellular mobile radio. Modulation is the process of varying some attribute of a carrier waveform, such as amplitude, phase, or frequency, in accordance with the message to be transmitted. In the case of digital modulation, the message sequence is a stream of digits, quite often binary valued. In the simplest case, this parameter variation is on a symbol-by-symbol basis (zero memory), and the carrier parameters that can be varied are amplitude (**amplitude-shift keying (ASK)**), phase (**phase-shift keying (PSK)**), or frequency (**frequency-shift keying (FSK)**). So-called higher order modulation schemes impose memory over several symbol periods. Modulation techniques can be classified as **binary** or **M-ary** depending on whether one of two possible **signals** or $M > 2$ signals per signaling interval can be sent. If the latter, and if the source digits are binary, it is clear that several bits must be grouped together in order to make up an M-ary word. Another classification for digital modulation techniques is **coherent** vs. **noncoherent**, depending on whether a reference carrier at the receiver coherent with the received carrier is necessary for demodulation (coherent) or not (noncoherent).

12.5.2 Detection of Binary Signals in Additive White Gaussian Noise

Binary, Coherent Modulation Schemes

The simplest possible digital communications system is one which transmits a sequence of binary symbols represented for convenience by $\{0, 1\}$ from a transmitter to a receiver over a channel that degrades the transmitted signal with **additive white Gaussian noise (AWGN)** of two-sided spectral density $N_0/2$. The transmitted binary symbols are associated with two signaling waveforms, denoted as $s_1(t)$ and $s_2(t)$, defined to exist over the time interval $(0, T_b)$. (The symbol T_b will be used to denote **bit period**. Later, T_s will be used to denote **symbol period** when signaling schemes that select from more than two possible transmitted signals are discussed). One of these signals is transmitted each T_b seconds so that the information transmission rate is $R_b = 1/T_b$ binary symbols (bits) per second (b/s). During signaling interval k, the transmitter associates a symbol, such as a 1, with $s_1(t - kT_b)$ and the other symbol, a 0, with $s_2(t - kT_b)$. The receiver is assumed to have perfect knowledge of both $s_1(t)$ and $s_2(t)$, including the precise time at which they could be received and the probability that they were transmitted, assumed to be equally likely, but the receiver does not know which signal was, in fact, transmitted. During each T_b-signaling interval, the receiver observes the signaling waveform contaminated in AWGN and processes this information so as to minimize the probability of making an error.

It can be shown (Ziemer and Tranter, 2002; Wozencraft and Jacobs, 1965; Blahut, 1990; Proakis, 2002) that the minimum probability of error is achieved when the receiver guesses the transmitted signal to be that signal which, given the received signal plus noise waveform, was most likely to have been transmitted. Such a receiver is called a **maximum-likelihood receiver**. For equally likely binary symbols transmitted in AWGN, it can be shown (Ziemer and Tranter, 2002) that the minimum **probability of error** is

$$P_E = Q[\sqrt{z(1 - R_{12})}] \tag{12.71}$$

where

$$Q(x) = \int_x^\infty \frac{e^{-u^2/2}}{\sqrt{2\pi}} du \tag{12.72}$$

is the Q function. The quantities z and R_{12} are defined as

$$z = \frac{E_1 + E_2}{2N_0} = \frac{E_b}{N_0} \tag{12.73}$$

and

$$R_{12} = \frac{\sqrt{E_1 E_2}}{E_b} \rho_{12} \tag{12.74}$$

in which E_i, $i = 1, 2$, is the energy of signal i defined as

$$E_i = \int_0^{T_b} |s_i(t)|^2 \, dt \tag{12.75}$$

and $E_b = (E_1 + E_2)/2$ is the average signal energy. The parameter ρ_{12} is the normalized **correlation coefficient** between signals, which is given by

$$\rho_{12} = \frac{1}{\sqrt{E_1 E_2}} \int_0^{T_b} s_1(t) s_2(t) \, dt \tag{12.76}$$

If $R_{12} = 0$, the signaling scheme is said to be **orthogonal**, whereas if $R_{12} = -1$, the signaling scheme is said to be **antipodal**. An example of the former is coherent, binary FSK and an example of the latter is binary PSK. The probability of error is shown in Fig. 12.55 as a function of $z = E_b/N_0$ in decibels for these two cases.

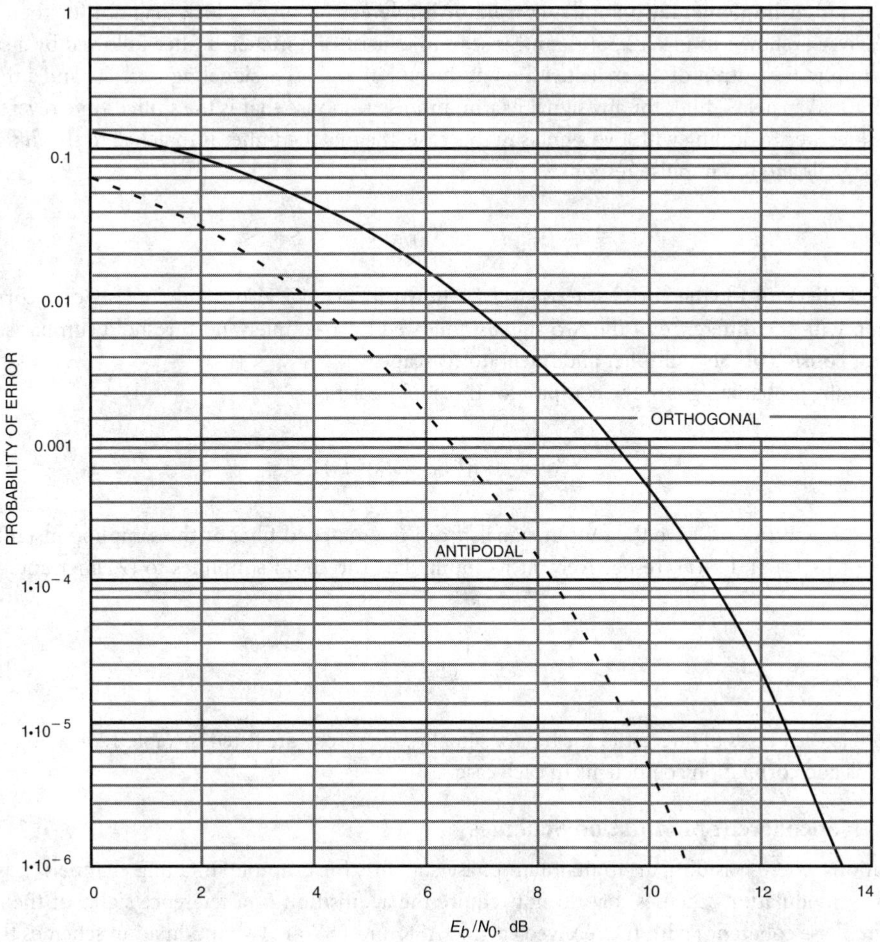

FIGURE 12.55 Probabilty of error for orthogonal (solid curve) and antipodal (dashed curve) signaling.

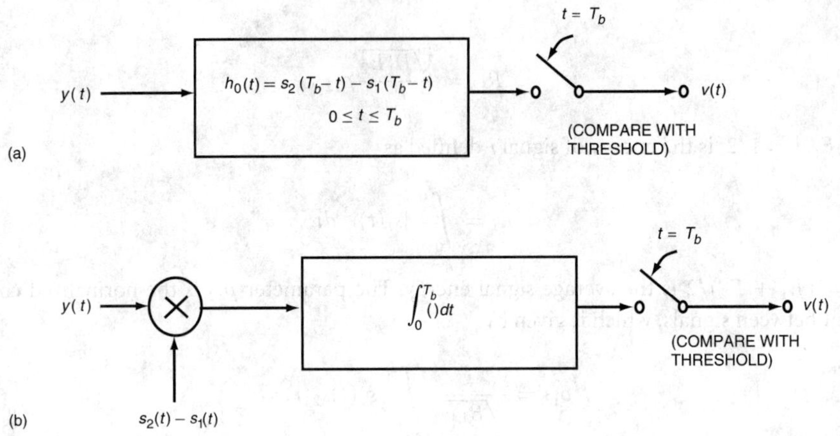

FIGURE 12.56 Implementations of the minimum probability of error receiver for binary signal reception: (a) matched filter, (b) correlator.

The receiver for these binary signaling schemes can have one of two equivalently performing structures—a **matched filter implementation** and a **correlator implementation**. A block diagram for the matched filter receiver is shown in Fig. 12.56(a), and is seen to consist of a matched filter followed by a sampler, which samples the output of the matched filter at the end of each T_b-s signaling interval, and a threshold comparator. A matched filter for any signal has an impulse response that is the shifted time reverse of the signal. Since we are dealing with two signals in this case, the matched filter is matched to the *difference* of the two signals and has impulse response

$$h(t) = s_2(T_b - t) - s_1(T_b - t), \quad 0 \le t \le T_b \tag{12.77}$$

A block diagram for the correlator receiver is shown in Fig. 12.56(b), and consists of a correlation operation with the difference of the two signals, followed by a sampler and threshold comparison. The correlator consists of the multiplier and integrator cascade.

For equally probable signals, the comparator threshold is set at

$$k = \frac{1}{2}(s_{o1}(T_b) + s_{o2}(T_b)) \tag{12.78}$$

where $s_{o1}(T_b)$ and $s_{o2}(T_b)$ are the output signals from the matched filter at the sampling instant corresponding to $s_1(t)$ and $s_2(t)$, respectively, at its input. The threshold simplifies to (Ziemer and Tranter, 2002)

$$k = \frac{1}{2}(E_2 - E_1) \tag{12.79}$$

Several special cases of interest in the binary signaling hierarchy are listed in Table 12.9, along with the thresholds and correlation coefficients in each case.

Binary, Noncoherent Modulation Schemes

In situations where it is difficult to maintain phase stability, for example in fading channels, it is useful to employ modulation schemes that do not require the acquisition of a reference signal at the receiver that is in phase coherence with the received carrier. ASK and FSK are two modulation schemes that lend themselves well to noncoherent detection. Receivers for detection of ASK and FSK noncoherently are shown in Fig. 12.57.

TABLE 12.9 Characteristics of Coherent Binary Digital Modulation Schemes

Name	Signal	Threshold Eq (12.78)	Signal Corr. Coeff., R_{12}
Antipodal baseband[a] signaling	$s_{1,2}(t) = \pm\sqrt{\dfrac{E_b}{T_b}}, \quad 0 \le t \le T_b$	0	-1
Amplitude-shift keying	$\left.\begin{aligned} &s_1(t) = 0 \\ &s_2(t) = \sqrt{\dfrac{4E_b}{T_b}}\cos(\omega_c t) \end{aligned}\right\} \quad 0 \le t \le T_t$	$E_2/2$	0
(Binary) phase-shift keying (PSK)	$s_{1,2}(t) = \sqrt{\dfrac{2E_b}{T_b}}\sin(\omega_c t \pm \cos^{-1} m), \quad 0 \le t \le T_b$ where $\lvert m \rvert \le 1$ is the modulation index	0	$2m^2 - 1$
Biphase-shift keying (BPSK)	$s_{1,2}(t) = \pm\sqrt{\dfrac{2E_b}{T_b}}\cos(\omega_c t), \quad 0 \le t \le T_b$	0	-1
Frequency-shift keying (FSK)	$\left.\begin{aligned} &s_1(t) = \sqrt{2E_b/T_b}\,\cos(\omega_c t) \\ &s_2(t) = \sqrt{2E_b/T_b}\,\cos[(\omega_c + \Delta\omega)t] \end{aligned}\right\} \quad 0 \le t \le T_b$ $\Delta\omega = \pi \times \text{integer}/T_b$	0	0

[a]In all cases, E_b is the average signal energy per bit, E_1 is the energy of signal 1, and E_2 is the energy of signal 2. All signalling schemes except antipodal baseband are referred to as coherent, because the carrier phase must be known at the receiver to implement the matched or correlator detector.

For noncoherent reception of binary ASK, the error probability for large signal-to-noise ratios is well approximated by (Ziemer and Tranter, 2002)

$$P_E \cong \frac{1}{2}e^{-z/2}, \quad z \gg 1 \text{ (noncoherent ASK)} \tag{12.80}$$

where $z = E_b/N_0$ as before. For noncoherent detection of binary FSK, the probability of error is exactly given by (Ziemer and Tranter, 2002)

$$P_E = \frac{1}{2}e^{-z/2} \quad \text{(noncoherent FSK)} \tag{12.81}$$

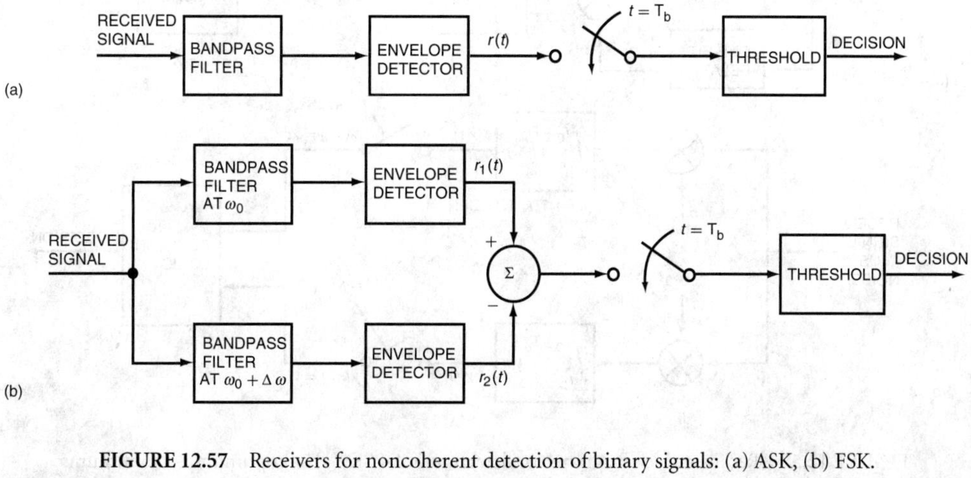

FIGURE 12.57 Receivers for noncoherent detection of binary signals: (a) ASK, (b) FSK.

TABLE 12.10 An Example Illustrating the Differential Encoding Process

Message sequence:		1	0	0	1	1	1	0
Encoded sequence:	1	1	0	1	1	1	1	0
Transmitted phase radians:	0	0	π	0	0	0	0	π

Thus, both perform the same for large signal-to-noise ratios. To compare this with coherent detection of FSK, the asymptotic approximation for the Q function given by

$$Q(x) = \frac{e^{-x^2/2}}{\sqrt{2\pi}\, x}, \quad x \gg 1 \tag{12.82}$$

is employed. Application of this to Eq. (12.71) with $R_{12} = 0$ gives

$$P_E = \frac{e^{-z/2}}{\sqrt{2\pi z}}, \quad z \gg 1 \text{ (coherent FSK)} \tag{12.83}$$

Since the dominant behavior comes through the exponent in Eq. (12.83) it follows that coherent and noncoherent FSK have very nearly the same error probability performance at large signal-to-noise ratios, with coherent FSK slightly better due to the $z^{-1/2}$ in the denominator of Eq. (12.83).

There is one other binary modulation scheme which is, in a sense, noncoherent. It is **differentially coherent PSK (DPSK)**, in which the phase of the preceding bit interval is used as a reference for the current bit interval. This technique depends on the channel being stable enough so that phase changes due to channel perturbations from a given bit interval to the succeeding one are inconsequential. It also depends on there being a known phase relationship from one bit interval to the next. This is ensured by differentially encoding the bits before phase modulation at the transmitter. **Differential encoding** is illustrated in Table 12.10. An arbitrary reference bit is chosen to start the process off. In Table 12.10 a 1 has been chosen. For each bit of the encoded sequence, the present bit is used as a reference for the following bit in the sequence. A 0 in the message sequence is encoded as a transition from the state of the reference bit to the opposite state in the encoded message sequence; a 1 is encoded as no change of state. Using these rules, it is seen that the encoded sequence shown in Table 12.10 results.

Block diagrams of two possible receiver structures for DPSK are shown in Fig. 12.58. The first is suboptimum, but relatively simple to implement. The second is the optimum receiver for DPSK in AWGN.

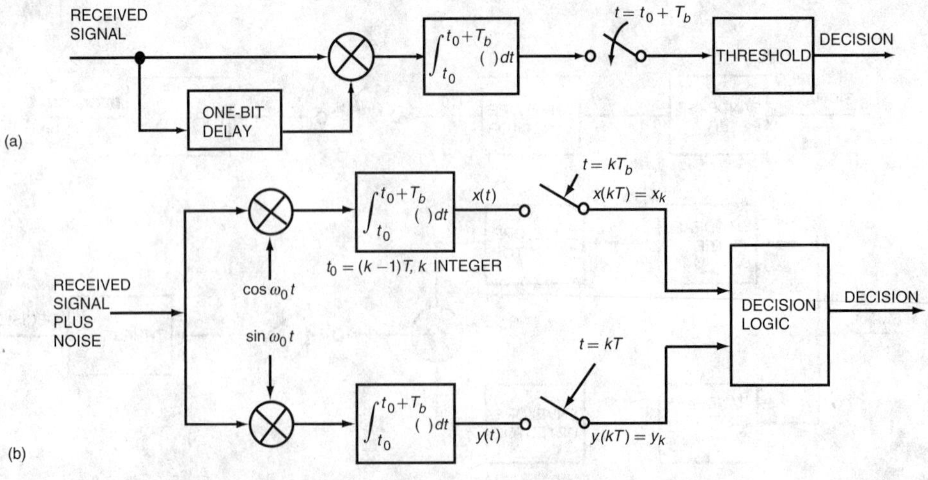

FIGURE 12.58 Receiver block diagrams for detection of DPSK: (a) suboptimum, (b) optimum.

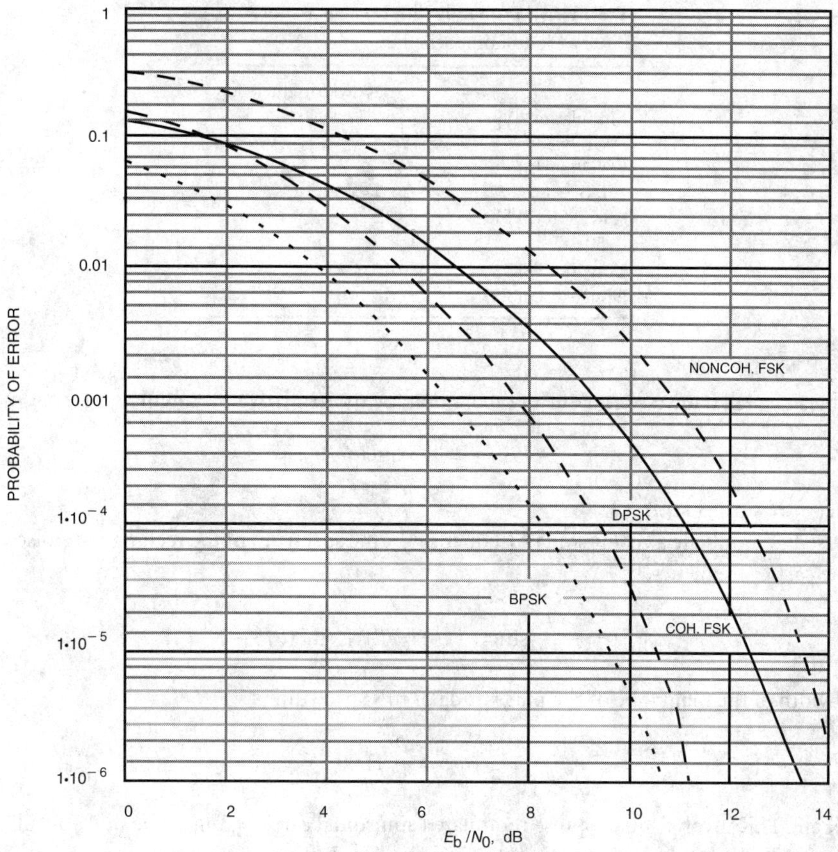

FIGURE 12.59   Comparison of error probabilities for BPSK, DPSK, coherent FSK, and noncoherent FSK.

Its probability of error performance can be shown to be (Ziemer and Tranter, 2002)

$$P_E = \frac{1}{2}e^{-z} \text{ (DPSK)} \tag{12.84}$$

This can be compared with BPSK by again making use of the asymptotic approximation for the Q function given by Eq. (12.82) in Eq. (12.71) with $R_{12} = -1$ to give the following approximate result for BPSK for large signal-to-noise ratios:

$$P_E = \frac{e^{-z}}{2\sqrt{\pi z}}, \quad z \gg 1 \text{ (BPSK)} \tag{12.85}$$

The exponential behavior for DPSK and BPSK is the same for large z; BPSK is slightly better due to the factor $z^{-1/2}$ in Eq. (12.85). Error probabilities for BPSK, DPSK, coherent binary FSK, and noncoherent binary FSK are compared in Fig. 12.59. It is seen that less than 1 dB of degradation results in going from coherent to noncoherent detection at $P_E = 10^{-6}$.

Bandwidth Efficiency

The error probability as a function of E_b/N_0 for a given modulation scheme tells only half of the story, and is often referred to as a measure of its **power efficiency**. Also important is its **bandwidth effficiency**, defined to be the ratio of the bandwidth required to accept a given data rate divided by the data rate.

TABLE 12.11   Bandwidth Efficiencies for
Binary Modulation Schemes

Modulation Type	Bandwidth Efficiency, b/s/Hz
Rectangular pulse baseband	1
ASK, PSK, BPSK, DPSK	0.5
Coherent FSK	0.4
Noncoherent FSK	0.25

For example, it is well known from Fourier theory that the spectrum of a rectangular pulse of duration T is

$$S(f) = AT\,\text{sinc}(fT) \tag{12.86}$$

where $\text{sinc}(x) = \sin(\pi x)/(\pi x)$.

When used to modulate a cosinusoid of frequency f_c, the spectrum of the rectangular pulse is centered around the carrier frequency, f_c

$$S_m(f) = \frac{AT}{2}\{\text{sinc}((f - f_c)T) + \text{sinc}((f + f_c)T)\} \tag{12.87}$$

The bandwidth of the main lobe of the magnitude of this spectrum is

$$B_{\text{RF}} = \frac{2}{T}\,\text{Hz} \tag{12.88}$$

Since ASK and PSK involve square-pulse modulated sinusoidal carriers, this is the null-to-null main lobe bandwidth for these modulation schemes with T replaced by T_b. For coherent FSK, note that the minimum frequency spacing between cosinusoidal bursts at frequencies f_c and $f_c + \Delta f$ is $1/2T_b$ Hz in order to maintain orthogonality of the two signals. The first null of the sinc function at frequency f_c must have $1/T_b$ Hz below it, and the one at frequency $f_c + \Delta f$ must have $1/T_b$ Hz above it, giving a total bandwidth for FSK of

$$B_{\text{FSK}} = \frac{2.5}{T_b}\,\text{Hz} \tag{12.89}$$

For noncoherent FSK, the frequency spacing between tones should be $2/T_b$ to allow separation at the receiver by filtering. Allowing the additional $1/T_b$ on either side due to the sinc-function spectrum for each tone, the total bandwidth required for noncoherent FSK is

$$B_{\text{NFSK}} = \frac{4}{T_b}\,\text{Hz} \tag{12.90}$$

Since $1/T_b$ is the data rate R_b in b/s, the bandwidth efficiencies, defined as the b/s per hertz of bandwidth, of the various modulation schemes considered are as given in Table 12.11.

12.5.3   Detection of *M*-ary Signals in Additive White Gaussian Noise

Signal Detection in Geometric Terms

It is useful to view digital data transmission in geometric terms for several reasons. First, it provides a general framework that makes the analysis of several types of digital data transmission methods easier, particularly *M*-ary systems. Second, it provides an insight into the digital data transmission problem that allows one to see intuitively the power-bandwidth tradeoffs possible. Third, it suggests ways to improve upon standard modulation schemes. The mathematical basis for the geometric approach is known as signal space (Hilbert space in mathematical literature) theory. The first book to use this approach in the U.S. was Wozencraft and Jacobs (1965), which was based, in part, on earlier work by the Russian Kotelnikov (1960). An early paper in the literature approaching signal detection from a geometric standpoint is Arthurs and Dym (1962). An overview of signal space concepts is given next.

The Gram---Schmidt Procedure

Given a finite set of signals, denoted $s_i(t)$, $i = 1, 2, \ldots, M$ for $0 \leq t \leq T_s$, it is possible to find an orthonormal basis set in terms of which all signals in the set can be represented. (Because more than two possible signals can be sent during each signaling interval, the parameter T_s will now be used to denote the signaling, or *symbol, interval*). The procedure is known as the **Gram–Schmidt procedure**, and is easy to describe once some notation is defined. (For the most part, signals will be real. However, it is sometimes convenient to represent signals as phasors or complex exponentials). The **scalar product** of two signals, u and v, defined over the interval $(0, T_s)$, is defined as

$$(u, v) = \int_0^{T_s} u(t) v^*(t) dt \tag{12.91}$$

and the **norm** of a signal is defined as

$$\|u\| = \sqrt{(u, u)} \tag{12.92}$$

In terms of this notation, the Gram–Schmidt procedure is as follows:

1. Set $v_1(t) = s_1(t)$ and define the first orthonormal basis function as

$$\phi_1(t) = \frac{v_1(t)}{\|v_1\|} \tag{12.93}$$

2. Set $v_2(t) = s_2(t) - (s_2, \phi_1)\phi_1(t)$ and let the second orthonormal basis function be

$$\phi_2(t) = \frac{v_2(t)}{\|v_2\|} \tag{12.94}$$

3. Set $v_3(t) = s_3(t) - (s_3, \phi_2)\phi_2(t) - (s_3, \phi_1)\phi_1(t)$ and let the next orthonormal basis function be

$$\phi_3(t) = \frac{v_3(t)}{\|v_3\|} \tag{12.95}$$

4. Continue until all signals have been used. If one or more of the steps yield $v_j(t)$ for which $\|v_j(t)\| = 0$, omit these from consideration so that a set of $K \leq M$ orthonormal functions is obtained. This is called a *basis set*.

Using the orthonormal basis set thus obtained, an arbitrary signal in the original set of signals can be represented as

$$s_j(t) = \sum_{i=1}^{K} S_{ij} \phi_i(t), \quad j = 1, 2, \ldots, M \tag{12.96}$$

where

$$S_{ij} = (s_j, \phi_i) = \int_0^{T_s} S_j(t) \phi_i^*(t) dt \tag{12.97}$$

With this procedure, any signal of the set can be represented as a point in a signal space (the coordinates of $s_j(t)$ are $S_{1j}, S_{2j}, \ldots, S_{Kj}$). The representation of the signal in this space will be referred to as the signal vector. Thus, the signal detection problem can be viewed geometrically. This is discussed in the following subsection.

Geometric View of Signal Detection

Given a set of M signals as discussed, an M-ary digital communication system selects one of them with equal likelihood each contiguous T_s-s interval and sends it through a channel in which AWGN of two-sided spectral density $N_0/2$ is added. Letting the noise be represented by $n(t)$ and supposing that the jth signal is transmitted, the coordinates of the noisy received signal, here called the components of the data vector, are

$$Z_i = S_{ij} + N_i, \quad i = 1, 2, \ldots, K \text{ (signal } j \text{ transmitted)} \tag{12.98}$$

where

$$N_i = (n, \phi_i) = \int_0^{T_s} n(t)\phi_i^*(t)dt \tag{12.99}$$

Schematic diagrams of two receiver front ends that can be used to compute the coordinates of the data vector are shown in Fig. 12.60. The first is called the **correlator implementation**, and the second is called the **matched filter implementation**. Because the noise components are linear transformations of a Gaussian random process, they are also Gaussian, and can be shown to have zero means and covariances

$$\text{cov}(N_i, N_j) = E\{N_i N_j\} = \frac{N_0}{2}\delta_{ij} \tag{12.100}$$

where δ_{ij} is the Kronecker delta, which is zero for the indices equal and zero otherwise. Consequently, the signal coordinates (12.96) are Gaussian with means S_{ij}, zero covariances, and variances of $N_0/2$. Thus,

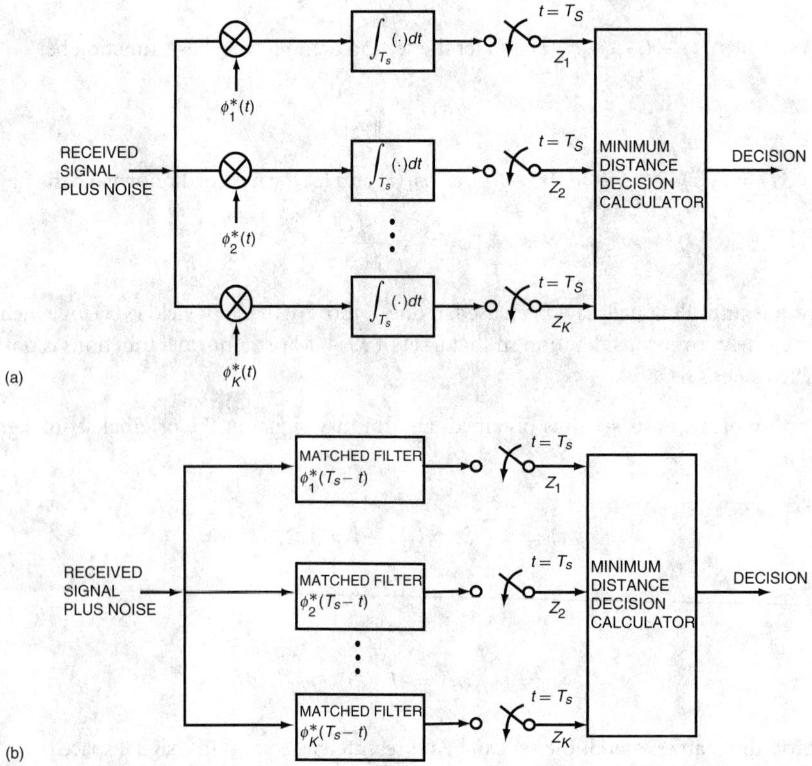

FIGURE 12.60 Receiver configurations for computing data vector components: (a) correlator realization, (b) matched filter realization.

given signal $s_j(t)$ was sent, the joint conditional probability density function of the received data vector components is

$$p[z_1, z_2, \ldots, z_k \mid s_j(t)] = (\pi N_0)^{-K/2} \exp\left\{-\frac{1}{N_0} \sum_{i=1}^{K} (z_i - s_{ij})^2\right\}, \quad j = 1, 2, \ldots, M \qquad (12.101)$$

A reasonable strategy for deciding on the signal that was sent is to choose the most likely, that is, if each signal is transmitted with equal probability, maximize the conditional probability density function (12.101) by choosing the appropriate signal vector, which can also be shown equivalent to minimizing the average probability of error (Wozencraft and Jacobs, 1965). Given the form of Eq. (12.101), this is accomplished by minimizing its exponent. Minimizing the exponent is equivalent to minimizing the sum of the squares of the differences between the components of the received data vector and those of the signal vector, that is, choosing the signal point that is closest in Euclidian distance to the received data point. This is illustrated in the following specific examples.

M-ary Phase-Shift Keying (MPSK)
Consider MPSK for which the signal set is

$$s_i(t) = \sqrt{\frac{2E_s}{T_s}} \cos\left[2\pi\left(f_c t + \frac{i-1}{M}\right)\right], \quad 0 \le t \le T_s; i = 1, 2, \ldots, M \qquad (12.102)$$

where E_s is the signal energy, T_s is the signal duration, and f_c is the carrier frequency in hertz. An $M = 4$ MPSK system is referred to as **quadriphase-shift keying (QPSK)**. The Gram–Schmidt procedure could be used to find the orthonormal basis set for expanding this signal set, but it is easier to expand Eq. (12.101) using trigonometric identities as

$$s_i(t) = \sqrt{E_s}\left[\cos\left(\frac{2\pi(i-1)}{M}\right)\sqrt{\frac{2}{T_s}}\cos\omega_c t - \sin\left(\frac{2\pi(i-1)}{M}\right)\sqrt{\frac{2}{T_s}}\sin\omega_c t\right]$$

$$= \sqrt{E_s}\left[\cos\left(\frac{2\pi(i-1)}{M}\right)\phi_1(t) - \sin\left(\frac{2\pi(i-1)}{M}\right)\phi_2(t)\right] \qquad (12.103)$$

$$0 \le t \le T_s; i = 1, 2, \ldots, M$$

where $\omega_c = 2\pi f_c$, and it follows that

$$\phi_1(t) = \sqrt{\frac{2}{T_s}}\cos\omega_c t, \quad 0 \le t \le T_s \quad \text{and} \quad \phi_2(t) = \sqrt{\frac{2}{T_s}}\sin\omega_c t, 0 \le t \le T_s \qquad (12.104)$$

Note that in this case $K = 2 \le M$. A signal space diagram is shown in Fig. 12.61(a) (only the ith signal point is shown). The best decision strategy, as discussed, chooses the signal point in signal space closest in **Euclidian distance** to the received data point. This is accomplished in Fig. 12.61(a) by dividing the signal space up into pie-shaped decision regions associated with each signal point. If the received data point lands in a given region, the decision is made that the corresponding signal was transmitted. The probability of error is the probability that, given a certain signal was transmitted, the noise causes the data vector to land outside of the corresponding decision region. The **probability of *symbol* error** P_S can be upper and lower bounded by (Peterson, Ziemer, and Borth, 1995)

$$Q\left(\sqrt{\frac{2E_s}{N_0}}\sin\frac{\pi}{M}\right) \le P_S \le 2Q\left(\sqrt{\frac{2E_s}{N_0}}\sin\frac{\pi}{M}\right) \qquad (12.105)$$

which is obtained by considering two half-planes above and below the wedge corresponding to the ith signal as shown in Fig. 12.61(b) and Fig. 12.61(c). The upper bound is very tight for M moderately large, a fact which follows by comparing the areas of the plane with one wedge excluded in Fig. 12.61(a) with the area of the two half-planes in Fig. 12.61(b) and Fig. 12.61(c). Probability of error curves will be shown later.

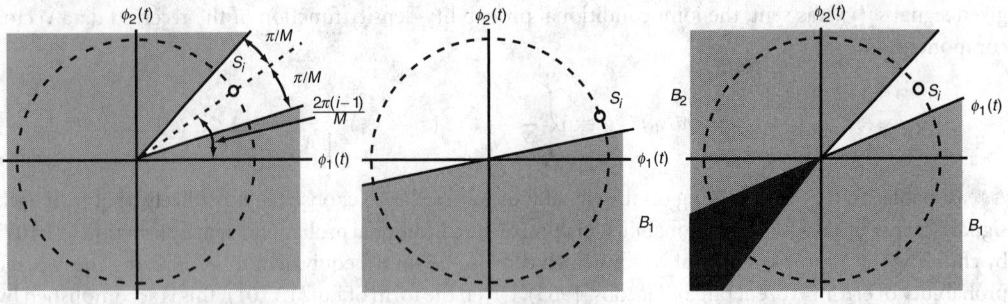

(a) DECISION REGION (UNSHADED) FOR iTH TRANSMITTED PHASE (b) INTEGRATION REGION FOR COMPUTING LOWER BOUND ON $P_b(\varepsilon)$ (c) INTEGRATION REGION FOR COMPUTING UPPER BOUND ON $P_b(\varepsilon)$

FIGURE 12.61 Signal space diagrams for MPSK: (a) diagram showing typical transmitted signal point with decision region; (b) and (c) half-planes for bounding the error probability. The probability of error is greater than the probability of the received data vector falling into one half-plane, but less than the probability of it falling into either one of both half-planes.

Coherent M-ary Frequency-Shift Keying (CMFSK)

The signal set for this modulation scheme is given by

$$s_i(t) = \sqrt{\frac{2E_s}{T_s}} \cos\{2\pi[f_c + (i-1)\Delta f)t], \quad 0 \le t \le T_s, i = 1, 2, \ldots, M \qquad (12.106)$$

where f_c is the lowest tone frequency and Δf is the frequency spacing between tones, both in hertz. The orthonormal basis function set for this modulation scheme is

$$s_i(t) = \sqrt{\frac{2}{T_s}} \cos\{2\pi[f_c + (i-1)\Delta f)t]\}, \quad 0 \le t \le T_s, i = 1, 2, \ldots, M \qquad (12.107)$$

Note that the signal space is M dimensional as opposed to two dimensional for the case of MPSK. A signal space diagram for $M = 3$ is shown in Fig. 12.62. For moderately large M, the symbol error probability can be tightly upper bounded by (Peterson, Ziemer, and Borth, 1995)

$$P_s \le (M-1)Q\left(\sqrt{\frac{E_s}{N_0}}\right) \qquad (12.108)$$

Curves showing bit error probability as a function of signal-to-noise ratio will be presented later.

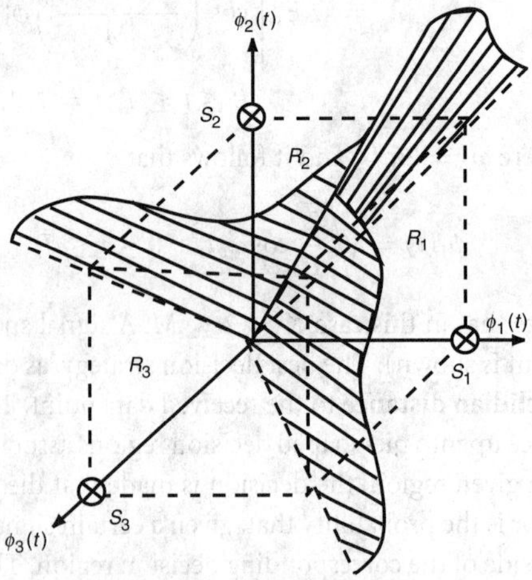

FIGURE 12.62 Signal space showing the possible transmitted signal points and decision boundaries for coherent 3-ary FSK

M-ary Quadrature-Amplitude-Shift Keying (MQASK)

The M-ary **quadrature-amplitude-shift keying** (MQASK) modulation scheme uses the two-dimensional space of MPSK, but with multiple amplitudes. Many such two-dimensional configurations have been considered, but only a simple rectangular grid of signal points will be considered here. The signal set can be expressed as

$$s_i(t) = \sqrt{\frac{2}{T_s}}(A_i \cos\omega_c t + B_i \sin\omega_c t), \quad 0 \le t \le T_s \qquad (12.109)$$

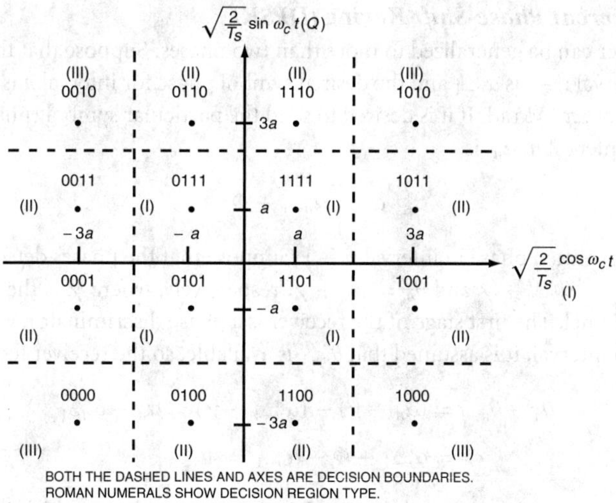

FIGURE 12.63 Signal space diagram for 16-ary QASK.

where ω_c is the carrier frequency in radians per second, and A_i and B_i are amplitudes taking on the values

$$A_i, B_i, = \pm a, \pm 3a, \dots, \pm \sqrt{M}a \tag{12.110}$$

where M is assumed to be a power of 4. The parameter a can be related to the average signal energy by

$$a = \sqrt{\frac{3E_s}{2(M-1)}} \tag{12.111}$$

A signal space diagram is shown in Fig. 12.63 for $M = 16$ with optimum partitioning of the decision regions. Each signal point is labeled with a Roman numeral I, II, or III. For the type I regions, the probability of correct reception is

$$P(C \mid \mathrm{I}) = \left[1 - 2Q\left(\sqrt{\frac{2a^2}{N_0}} \right) \right]^2 \tag{12.112}$$

For the type II regions, the probability of correct reception is

$$P(C \mid \mathrm{II}) = \left[1 - 2Q\left(\sqrt{\frac{2a^2}{N_0}} \right) \right] \left[1 - Q\left(\sqrt{\frac{2a^2}{N_0}} \right) \right] \tag{12.113}$$

and for the type III regions, it is

$$P(C \mid \mathrm{III}) = \left[1 - Q\left(\sqrt{\frac{2a^2}{N_0}} \right) \right]^2 \tag{12.114}$$

In terms of these probabilities, the probability of symbol error is (Peterson, Ziemer, and Both, 1995)

$$P_s = 1 - \frac{1}{M} \left[(\sqrt{M} - 2)^2 P(C \mid \mathrm{I}) + 4(\sqrt{M} - 2) P(C \mid \mathrm{II}) + 4 P(C \mid \mathrm{III}) \right] \tag{12.115}$$

Bit error probability plots will be provided later.

Differentially Coherent Phase-Shift Keying (DPSK)

DPSK discussed earlier can be generalized to more than two phases. Suppose that the transmitted carrier phase for symbol interval $n-1$ is α_{n-1} and the desired symbol phase for interval n is β_n, which is assumed to take on a multiple of $2\pi/M$ rad. If it is desired to send the particular symbol (phase) $\beta_n = \Phi$, then the *transmitted phase* at interval n, α_n, is

$$\alpha_n = \alpha_{n-1} + \Phi \tag{12.116}$$

where α_{n-1} is the transmitted phase in interval $n - 1$. Suppose that the phases detected corresponding to α_{n-1} and α_n are $\theta_{n-1} = \alpha_{n-1} + \gamma$ and $\theta_n = \alpha_n + \gamma$, respectively, where γ is the unknown phase shift introduced by the channel. The first stage of the receiver is a phase discriminator, which detects θ_n. From the previous decision interval, it is assumed that θ_{n-1} is available, so the receiver forms the difference

$$\theta_n - \theta_{n-1} = (\alpha_n + \gamma) - (\alpha_{n-1} + \gamma) = \alpha_n - \alpha_{n-1}$$
$$= \alpha_{n-1} + \Phi - \alpha_{n-1} = \Phi \tag{12.117}$$

where no noise is assumed. In the presence of noise, the receiver must decide which $2\pi/M$ region $\theta_n - \theta_{n-1}$ falls into; hopefully, in this example, this is the region centered on Φ. Thus, a correct decision will be made at the receiver when

$$\Phi - \frac{\pi}{M} < \theta_n - \theta_{n-1} \le \Phi + \frac{\pi}{M} \tag{12.118}$$

A receiver block diagram implementing this decision strategy is shown in Fig. 12.64. The error probability can be bounded by (Prabhu, 1982)

$$P_1 \le P_s \le 2P_1 \tag{12.119}$$

where P_1 is upper and lower bounded by

$$P_1 \le \frac{\pi}{2} \frac{\cos\left(\frac{\pi}{2M}\right)}{\sqrt{\cos(\pi/M)}} Q\left[2\sqrt{\frac{E_s}{N_0}} \sin\left(\frac{\pi}{2M}\right)\right] \tag{12.120}$$

$$P_1 \ge \frac{1}{2} \frac{\cos\left(\frac{\pi}{2M}\right)}{\sqrt{\cos(\pi/M)}} \left[1 - 2Q\left(\pi\sqrt{\frac{E_s}{N_0}} \sin\left(\frac{\pi}{M}\right)\right)\right] Q\left[2\sqrt{\frac{E_s}{N_0}} \sin\left(\frac{\pi}{2M}\right)\right]$$

Noncoherent M-ary FSK

The signal set for noncoherent MFSK can be expressed as

$$s_i(t) = \sqrt{\frac{2E_s}{T_s}} \cos(\omega_i t + \alpha), \quad 0 \le t \le T_s; i = 1, 2, \ldots, M \tag{12.121}$$

where ω_i is the radian frequency of the ith signal, E_s is the symbol energy, T_s is the symbol duration, and α is the unknown phase, which is modeled as a uniformly distributed random variable in $(0, 2\pi)$. The

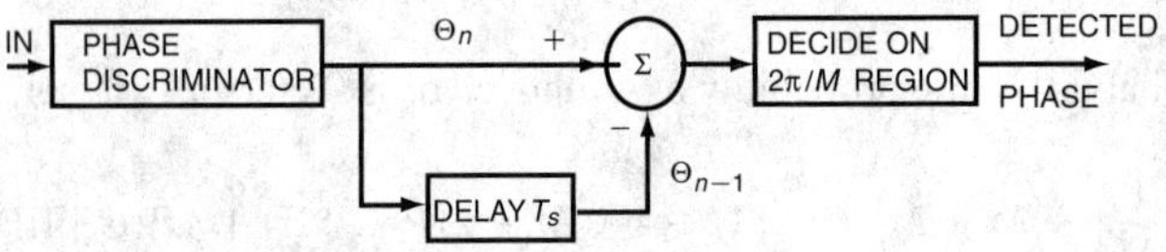

FIGURE 12.64 Block diagram for an M-ary differential PSK receiver.

signal space is $2M$ dimensional and can be defined by the basis functions

$$\left.\begin{aligned}\phi_{xi}(t)&=\sqrt{\frac{2}{T_s}}\,\cos\omega_i t\\[2mm]\phi_{yi}(t)&=\sqrt{\frac{2}{T_s}}\,\sin\omega_i t\end{aligned}\right\},\quad 0\le t\le T_s; i=1,2,\ldots,M \tag{12.122}$$

A fairly lengthy derivation (Peterson, Ziemer, and Borth, 1995) results in the symbol error probability expression

$$P_s=\sum_{k=1}^{M-1}\binom{M-1}{k}\frac{(-1)^k}{k+1}\exp\left(-\frac{k}{k+1}\frac{E_s}{N_0}\right) \tag{12.123}$$

which reduces to the result for binary noncoherent FSK for $M=2$. The optimum receiver is an extension of Fig. 12.57 with a parallel set of in-phase and quadrature filters and envelope detectors for each possible transmitted signal.

12.5.4 Comparison of Modulation Schemes

Bandwidth Efficiency

There are several ways to compare bandwidth of M-ary digital modulation schemes. One way is to compute out-of-band power as a function of bandwidth of an ideal brick wall filter. This requires integration of the power spectrum of the various modulation schemes being compared. The basis for bandwidth comparison used here will be the bandwidth required for the mainlobe of the signal spectrum, which makes for somewhat simpler computation without undue loss of accuracy. For example, replacing T by the symbol duration, T_s, in Eq. (12.87), the radio-frequency bandwidth of the mainlobe of a modulation scheme, which uses a single modulated frequency to transmit the information, such as M-ary PSK, M-ary DPSK, or QASK, is

$$B_{\text{RF}}=\frac{2}{T_s}=2R_s\text{ Hz} \tag{12.124}$$

where R_s is the symbol rate. However, for an M-ary modulation scheme, the symbol duration is related to the bit duration by

$$T_s=T_b\log_2 M=\frac{\log_2 M}{R_b} \tag{12.125}$$

Thus, the bandwidth in terms of bit rate for such modulation schemes is

$$B_{\text{RF}}=\frac{2R_b}{\log_2 M}\text{ (MPSK, MDPSK, MQASK)} \tag{12.126}$$

Now the ratio of bit rate to required bandwidth is called the bandwidth efficiency of a modulation scheme. In the case at hand, the bandwidth efficiency is

$$\text{bandwidth efficiency}=\frac{R_b}{B_{\text{RF}}}=0.5\log_2 M\text{ (MPSK, MDPSK, MQASK)} \tag{12.127}$$

For schemes using multiple frequencies to transmit the information such as M-ary FSK, a more general approach is used. Each symbol is represented by a different frequency. For coherent M-ary FSK, the minimum separation per frequency required to maintain othogonality is $1/2T_s$ Hz. The two outside frequencies use $1/T_s$ Hz for the half of the mainlobe on the left and right of the composite spectrum. The $M-2$ interior frequencies require a minimum separation of $1/2T_s$ Hz (there are actually $M-1$ interior slots) for a total RF bandwidth of

$$B_{RF} = \frac{1}{T_s} + \frac{M-1}{2T_s} + \frac{1}{T_s} = \frac{M+3}{2T_s} \text{ (coherent MFSK)} \tag{12.128}$$

Substitution of Eq. (12.125) gives a bandwidth efficiency of

$$\text{bandwidth efficiency} = \frac{R_b}{B_{RF}} = \frac{2\log_2 M}{M+3} \text{ (coherent MFSK)} \tag{12.129}$$

For noncoherent MFSK, the minimum separation of the frequencies used to represent the symbols is taken as $2/T_s$ Hz for a total RF bandwidth of

$$B_{RF} = \frac{1}{T_s} + \frac{2(M-1)}{T_s} + \frac{1}{T_s} = \frac{2M}{T_s} \text{ (noncoherent MFSK)} \tag{12.130}$$

with a bandwidth efficiency of

$$\text{bandwidth efficiency} = \frac{R_b}{B_{RF}} = \frac{\log_2 M}{2M} \text{ (noncoherent MFSK)} \tag{12.131}$$

A comparison of Eq. (12.127) with Eq. (12.129) and (12.131) shows that the bandwidth efficiency of MPSK and MQASK *increases* with M whereas the bandwidth efficiency of MFSK *decreases* with M. This can be attributed to the dimensionality of the signal space staying constant with M for the former modulation schemes, whereas it increases with M for the latter. A comparison of bandwidth efficiencies for the various modulation schemes considered in this chapter is given in Table 12.12.

Power Efficiency

The power efficiency of a modulation method is indicated by the value of E_b/N_0 required to yield a desired bit error probability, such as 10^{-6}. This being the case, it is necessary to convert from symbol error probability to bit error probability and from E_s/N_0 to E_b/N_0 for M-ary modulation schemes. The latter is straightforward since the difference between symbol energy and bit energy is symbol duration vs. bit duration, which are related by

$$T_s = T_b \log_2 M \tag{12.132}$$

giving

$$\frac{E_b}{N_0} = \frac{1}{\log_2 M} \frac{E_s}{N_0} \tag{12.133}$$

Conversion between symbol error probability and bit error probability is somewhat more complicated. First, considering two-dimensional modulation schemes such as MPSK and MQASK, it is assumed that

TABLE 12.12 Comparison of Bandwidth Efficiencies for Various Modulation Methods

M	MPSK, MDPSK, MQASK[a]	Coherent MFSK	Noncoherent MFSK
2	0.5	0.400	0.250
4	1.0	0.571	0.250
8	1.5	0.545	0.188
16	2.0	0.421	0.125
32	2.5	0.286	0.078
64	3.0	0.179	0.047

[a] For MQASK, only values of M of a power of four are applicable.

the most probable errors are those in favor of adjacent signal points and that encoding is used that results in a single bit change in going from one signal point to an adjacent one (i.e., Gray encoding). Since there are $\log_2 M$ bits per symbol, the result is that bit error probability is approximately related to symbol error probability by

$$P_s = \frac{P_b}{\log_2 M} \tag{12.134}$$

Finally, consider MFSK for which each symbol occupies a separate dimension in the signal space. Thus, all symbol errors are equally probable, which means that each symbol error occurs with probability $P_s/(M-1)$. Suppose that for a given symbol error k bits are in error. There are

$$\binom{\log_2 M}{k}$$

ways that this can happen, since each symbol represents $\log_2 M$ bits. This gives the average number of bit errors per symbol error as

$$\text{Ave. no. of bit errors per symbol error} = \sum_{k=1}^{\log_2 M} k \binom{\log_2 M}{k} \frac{P_s}{M-1} = \frac{M \log_2 M}{2(M-1)} \tag{12.135}$$

Since there are a total of $\log_2 M$ bits per symbol, the average bit error probability in terms of symbol error probability is

$$P_b = \frac{M}{2(M-1)} P_s \tag{12.136}$$

The various modulation schemes considered in this section are compared on the basis of bit error probability versus E_b/N_0 in Fig. 12.65 to Fig. 12.69.

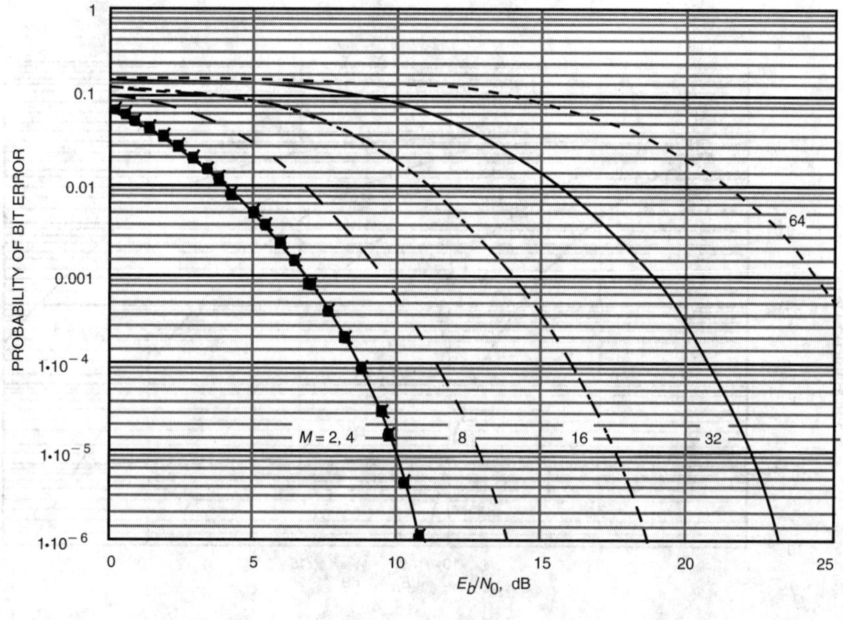

FIGURE 12.65 Bit error probability vs. E_b/N_0 for coherent M-ary PSK.

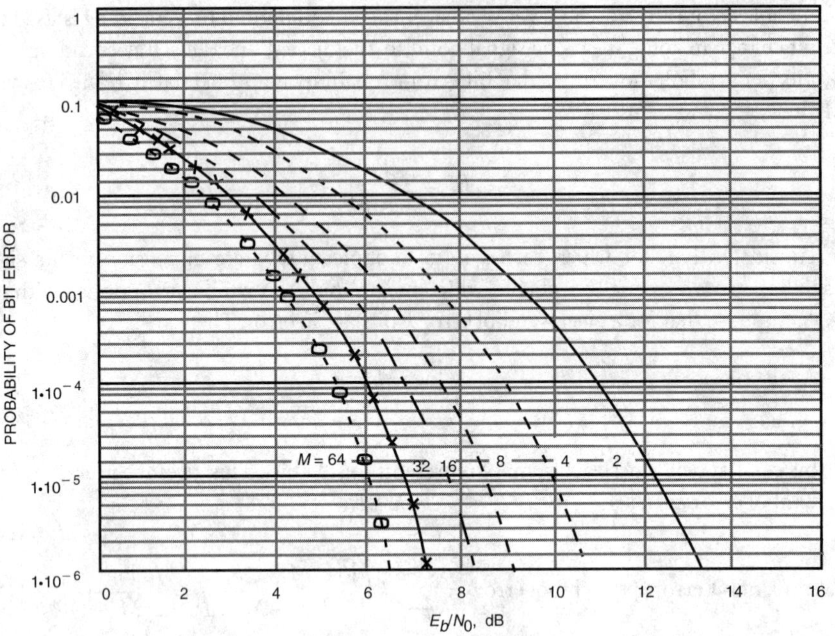

FIGURE 12.66   Bit error probability vs. E_b/N_0 for coherent M-ary FSK.

Other Important Types of Digital Modulation Schemes

There are many other important digital modulation formats that have not been discussed here. In this section, several of these will be mentioned with appropriate references given so that the reader will have a place to start to learn more about them.

A modulation scheme related to QPSK is **offset QPSK** (OQPSK) (Ziemer and Tranter, 2002). This modulation format is produced by allowing only $\pm 90°$ phase changes in a QPSK format. Furthermore,

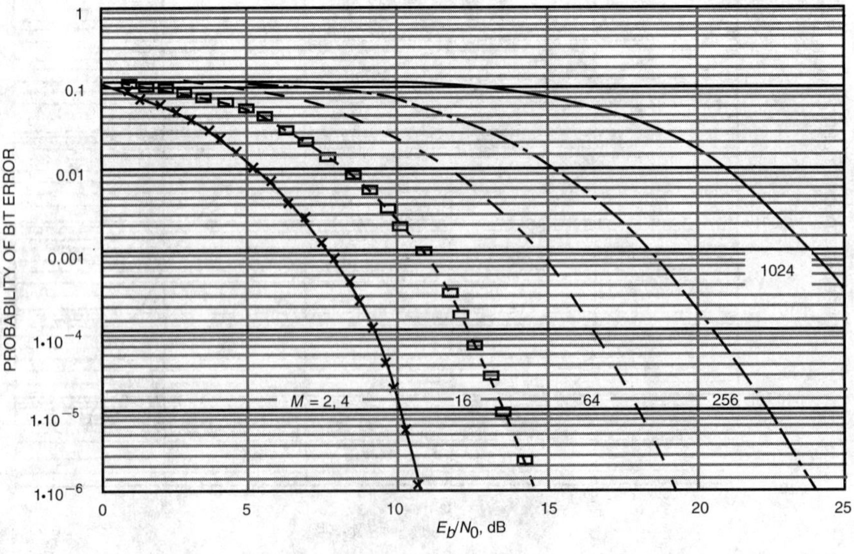

FIGURE 12.67   Bit error probability vs. E_b/N_0 for coherent M-ary QASK.

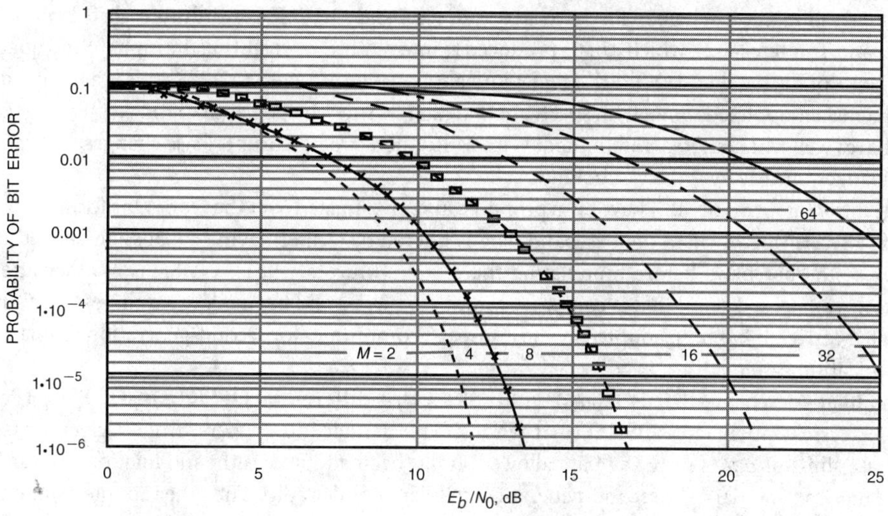

FIGURE 12.68 Bit error probability vs. E_b/N_0 for M-ary DPSK.

the phase changes can take place at multiples of a half-symbol interval, or a bit period. The reason for limiting phase changes to $\pm 90°$ is to prevent the large envelope deviations that occur when QPSK is filtered to limit sidelobe power, and then regrowth of the sidelobes after amplitude limiting is used to produce a constant-envelope signal. This is typically encountered in satellite communications where, due to power efficiency considerations, hard limiting repeaters are used in the satellite communications system.

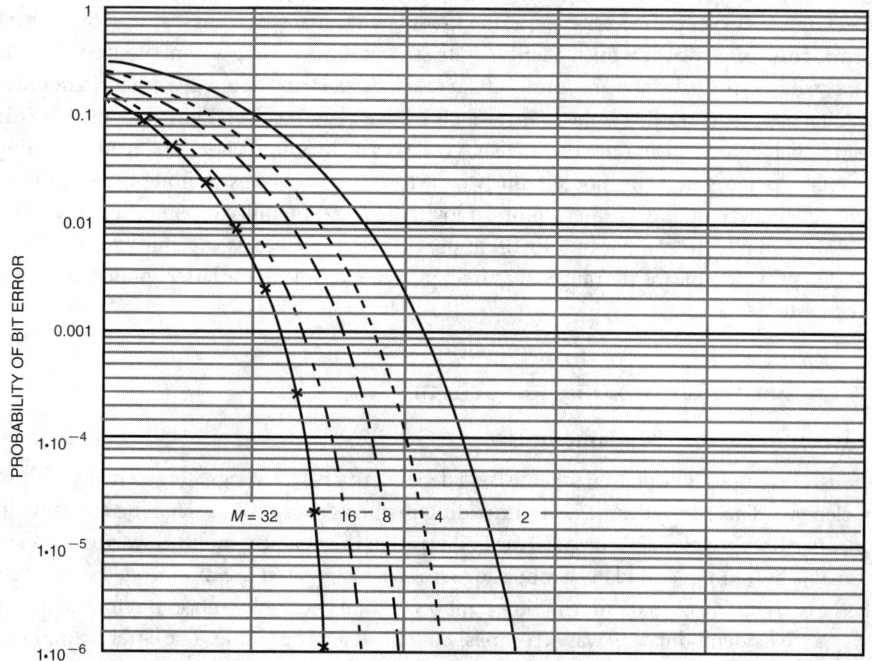

FIGURE 12.69 Bit error probability vs. E_b/N_0 for noncoherent M-ary FSK.

Another modulation scheme closely related to QPSK and OQPSK is **minimum shift keying (MSK)** (Ziemer and Tranter, 2002), which can be produced from OQPSK by weighting the inphase and quadrature components of the baseband OQPSK signal with half-sinusoids. As with OQPSK, this is a further attempt at producing a modulated signal with spectrum that has less sidelobe power than BPSK, QPSK, or OQPSK and behaves well when filtered and limited. With MSK, there are no abrupt phase changes, but rather, the phase changes linearly over the symbol interval.

Many different forms of MSK have been proposed and investigated over the years. One form is **Gaussian-filtered MSK (GMSK)**, where the baseband data is filtered by a filter having a Gaussian-shaped transfer function at the baseband before modulating the carrier (Murota, 1981). GMSK is another attempt at producing a signal spectrum that is compact around the carrier and having low envelope deviation of the modulated carrier. It has been adopted by the European community for second-generation cellular-mobile communication systems.

A modulation scheme that is related to 8-PSK is $\pi/4$-**differential QPSK** ($\pi/4$-DQPSK) (Peterson, Ziemer). It is essentially an 8-PSK format with differential encoding where, from a given phase state, only phase shifts of $\pm\pi/4$ or $\pm3\pi/4$ are allowed. It has been adopted as the modulation format for one North American standard for second-generation cellular-mobile radio. The other adopted standard uses direct-sequence spread-spectrum modulation.

A whole host of modulation schemes may be grouped under the heading **continuous-phase modulation (CPM)** (Sklar, 1988). These modulation formats employ continuous phase trajectories over one or more symbols to get from one phase to the next in response to input data changes. MSK and GMSK are special cases of CPM. CPM schemes are employed in an attempt to simultaneously improve power and bandwidth efficiency.

A class of modulation that is important for military and cellular radio communications is spread spectrum (Peterson, Ziemer, and Borth, 1995). Spread spectrum modulation is any modulation format that utilizes a transmission bandwidth much wider than that required to transmit the message signal itself, independent of the message signal bandwidth. It can be categorized several ways. The two most common types of spread spectrum modulation are **direct-sequence spread spectrum (DSSS)** and **frequency-hop spread spectrum (FHSS)**. In the former type, the modulated signal spectrum is spread by multiplication with a pseudonoise binary code, which changes state several times during a symbol interval. In the latter type, the spectrum is spread by hopping the modulated signal spectrum about in a pseudorandom manner. The code that is used for spreading at the transmitter is known at the receiver. Thus, it is possible to despread the received signal at the receiver once the receiver's code is synchronized with the spreading coding on the received signal. Spread spectrum modulation is used for several reasons. Among these are to hide the modulated signal from an enemy interceptor, to lower the susceptability of jamming by an unfriendly or unintentional source, to combat multipath, to provide multiple access capability for the modulation scheme, and to provide a means for range measurement. An example of the latter application is the **global positioning system (GPS)**.

12.5.5 Higher Order Modulation Schemes

Error Correction Coding Fundamentals

The previously discussed modulation schemes had dependency between signaling elements only over one signaling element. There are advantages to providing memory over several signaling elements from the standpoint of error correction. Historically, the first way that this was accomplished was to encode the data (usually binary) by adding redundant symbols for error correction and use the encoded symbol stream to modulate the carrier. The ratio of information symbols to total encoded symbols is referred to as the **code rate**. At the receiver, demodulation was accomplished followed by decoding. The latter operation allowed some channel-induced errors to be corrected. Since redundant symbols are added at the transmitter, it is necessary to use a larger bandwidth by a factor of one over the code rate than if no encoding is employed in order to keep the same information rate through the channel with coding as without coding. Thus, the signal-to-noise ratio is smaller by a factor of the code rate with coding than without coding, and the raw bit error probability is higher. Since the code can correct errors, however, this compensates partially for the higher raw bit error probability to a degree. If the code is powerful enough, the overall bit error probability through the channel is lower with coding than without. Usually, this depends on the signal-to-noise ratio: for small values of the signal-to-noise ratio, the compensation by coding is not enough, and the overall bit errror probability is higher with coding than without; for sufficiently large values of the signal-to-noise ratio, the error correction capability of the code more than makes up for the higher raw bit error probability, and a lower overall error probability is obtained. The **coding gain** is defined as the ratio of the signal-to-noise ratios without and with coding, or if they are expressed in decibels, the difference between the signal-to-noise ratios without and with coding, at a given bit error probability.

There are two widely used types of coding: **block** and **convolutional**. We will discuss these assuming that binary data is used. For block coding, the information bits are taken a block at a time (k of them) and correction bits are added to the block to make an overall block of length n. The ratio k/n is the code rate in this case. The trick is to introduce dependencies between the bits in the blocks of length n so that error correction can be accomplished at the receiver. The field of block coding dates back to work by Shannon in the late 1940s, and many powerful block codes are now available with efficient decoding techniques to make them implementable. Examples are Hamming, Bose–Chaudhuri–Hocquenghem (BCH), and Golay codes for binary codes; a nonbinary family of codes that is very powerful is the Reed–Solomon codes, which are used in compact disc audio technology. Depending on the code rate and the code used, block codes can provide coding gains from less than 1 to 10 or more dB at a probability of bit error of 10^{-6}.

A second class of error correction codes are convolutional codes. A simple convolutional encoder is shown in block diagram form in Fig. 12.70. It consists of a shift register and two or more modulo-2 adders to combine the input bit and bits residing in the shift register. It is assumed that a clock synchronizes the shifting of the bits into and down the shift register. When a new bit is shifted into the first stage of the shift register, all former bits in the shift register ripple to the right. In this case, the raw data is fed into the shift register bit-by-bit and the outputs of the modulo-2 adders are sampled in turn after each bit is fed in. Since two encoded bits are obtained for each input bit, this is a rate-$\frac{1}{2}$ convolutional encoder. If there were three modulo-2 adders for the bits in the shift register, a rate-$\frac{1}{3}$ code would result. Code rates greater than $\frac{1}{2}$ can be achieved by deleting bits periodically at the output of the encoder. This process is called **puncturing**. At the decoder, arbitrary symbols are inserted at each puncturing time (the period of the puncturing is known at the decoder, but the values of the punctured symbols are, of course, not known) and the decoder treats the arbitrarily inserted bit as a possible error. Convolutional codes have been in existence since the 1960s. A very efficient decoding method for convolutional codes is called the **Viterbi algorithm**, which was invented by Andrew Viterbi in the late 1960s. We will not present this algorithm in detail here, but the reader is referred to the references (Peterson, Ziemer, and Borth, 1995). The performance of convolutional

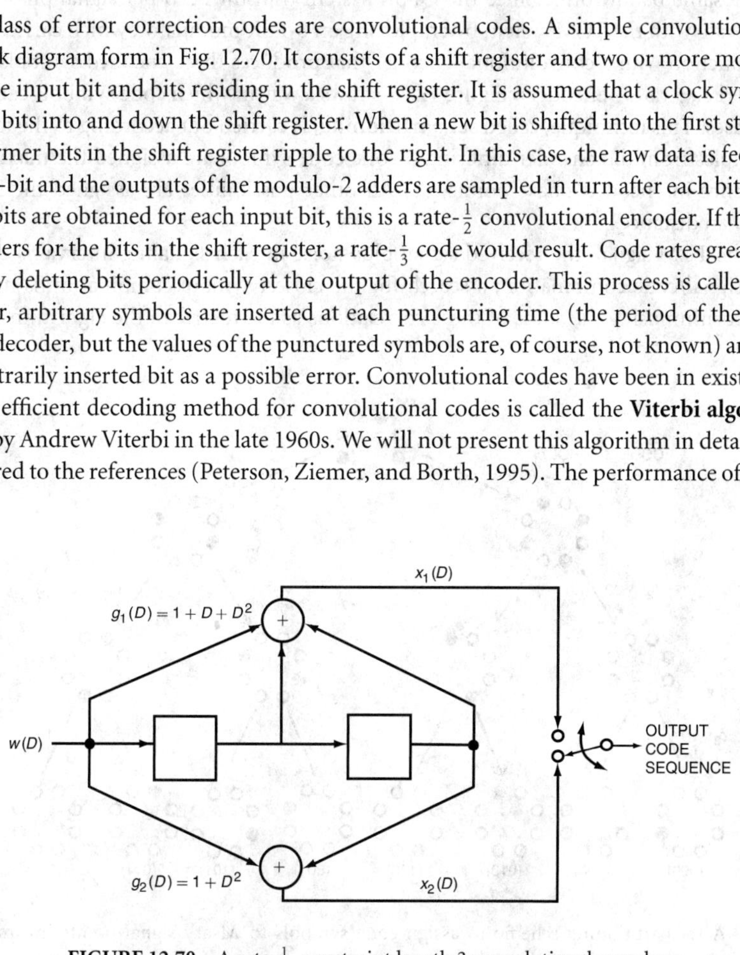

FIGURE 12.70 A rate-$\frac{1}{2}$, constraint length 3 convolutional encoder.

codes depends on the number of stages in the shift register (the number of stages plus 1 is called the **constraint length** of the code), the code rate, and whether hard or soft decisions are fed into the decoder. A **hard decision** is a definite 1–0 decision at the output of the detector; a **soft decision** is a quantized version of the detector output before a hard decision is made. Coding gains for convolutional codes range from a few decibels to 6 or 7 dB at a probability of error of 10^{-6} for soft decisions, constraint lengths of 8 or 9, and rates of $\frac{1}{2}$ or $\frac{1}{3}$. The higher the rate of a punctured code (i.e., the closer the rate is to 1), the less coding gain yielded by puncturing. Code rates close to 1 can be achieved by methods other than puncturing.

In closing, we note that coding used in conjunction with modulation always expands the required transmission bandwidth by the inverse of the code rate, assuming the overall bit rate is kept constant. In other words, power efficiency goes up but bandwidth efficiency goes down with use of a well-designed code.

Trellis-Coded Modulation

In attempting to simultaneously conserve power and bandwidth, Ungerboeck (1987) in the 1970s began to look at ways to combine coding and modulation. His solution was to use coding in conjunction with M-ary modulation to increase the minimum Euclidian distance between those signal points that are most likely to be confused without increasing the average power or bandwidth over an uncoded scheme transmitting the same number of bits per second. The technique is called **trellis-coded modulation (TCM)**. An example will illustrate the procedure.

In the example, we wish to compare a TCM system and a QPSK system operating at the same data rates and, hence, the same bandwidths. Since the QPSK system transmits 2 b per signal phase, we can keep that same data rate with the TCM system by employing an 8-PSK modulator, which carries 3 b per signal phase, in conjunction with a convolutional encoder that produces three encoded symbols for every two input data bits (i.e., a rate $\frac{2}{3}$ encoder). Figure 12.71 shows a way in which code symbols can be associated with signal phases. This technique, called **set partitioning**, is carried out according to the following rules: (1) All code triples from the encoder differing by one code symbol are assigned the maximum possible Euclidian distance in encoded phase points; (2) All other possibilities are assigned the next to largest possible Euclidian distance.

To decode the TCM signal, the received signal plus noise in each signaling interval is correlated with each possible signal/code combination; a search for the closest symbol combination is made by means of a Viterbi algorithm using the sums of these correlations as the metric in the algorithm (i.e., Euclidian

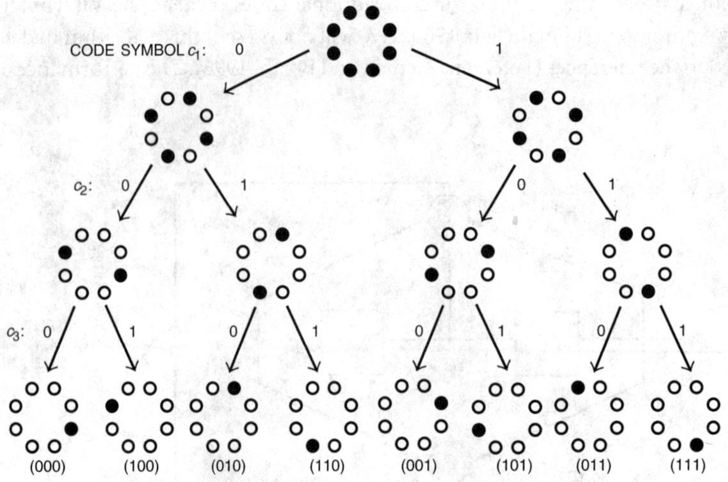

FIGURE 12.71 A set partitioning scheme to assign code symbols to M-ary signal points in order to maximize Euclidian distance between signal points.

distance). With the level of this chapter, the reader is encouraged to consult the references for the details of the decoding procedure (Ziemer and Tranter, 2002; Blahut, 1990; Proakis, 1995). Using this procedure, it can be shown that a coding gain results in going from the QPSK system to the combined coded/8-PSK system at *no increase in bandwidth*. In fact, for the system considered here, Ungerboeck has shown that an asymptotic (i.e., in the limit of small bit error probabilities) coding gain in excess of 3.5 dB results. More powerful codes and larger phase shifts result in asympotic coding gains of over 5.5 dB. Viterbi et al. (1989) has discussed the use of rate $\frac{1}{2}$, constraint length 7 convolutional codes, for which VLSI circuit implementations are available at a low cost, to provide various coding gains using puncturing and MPSK.

Defining Terms

Additive white Gaussian noise (AWGN): Noise that is additive to the signal having a constant power spectrum over all frequencies and Gaussian amplitude distribution.

Amplitude-shift keying (ASK): A signaling scheme whereby the digital data modulates the amplitude of a carrier.

Antipodal signaling: A signaling scheme which represents the digital data as plus (logic 1) or minus (logic 0) a basic pulse shape.

Bandwidth efficiency: The ratio of bit rate to bandwidth occupied for a digital modulation scheme. Technically, it is dimensionless, but it is usually given the dimensions of bits/second/hertz.

Binary signaling: Any signaling scheme where the number of possible signals sent during any given signaling interval is two.

Bit period: The symbol duration for a binary signaling scheme.

Block code: Any encoding scheme that encodes the information symbols block-by-block by adding a fixed number of error correction symbols to a fixed block length of information symbols.

Code rate: The ratio of the number of information bits into to encoded symbols out of a coder corresponding to those input bits.

Coding gain: The amount of improvement in E_b/N_0 at a specified probability of bit error provided by a coding scheme.

Coherent modulation: Any modulation technique that requires a reference at the receiver which is phase coherent with the received modulated carrier signal for demodulation.

Constraint length: The number of shift register stages plus one for a convolutional encoder. It is also the span of information symbols that determines the error correction symbols for a convolutional code.

Continuous-phase modulation (CPM): Any phase modulation scheme where the phase changes continuously over one or more symbol intervals in response to a change of state of the input data.

Convolutional code: Any encoding scheme that encodes a sliding window of information symbols by means of a shift register and two or more modulo-2 adders for the bits in the shift register that are sampled to produce the encoded output.

Correlation coefficient: A measure of the similarity between two signals normalized to a maximum absolute value of 1. If the correlation coefficient of two signals is zero, they are called orthogonal; if -1, they are called antipodal.

Correlator implementation: A receiver implementation that multiplies the incoming signal plus noise during any given signaling interval by a replica of each possible signal transmitted, integrates over the signaling interval, and uses these data for determining the transmitted signal.

Differential encoding: A coding scheme that compares the present differentially encoded bit with the next data bit; if a 1, the next encoded bit is the same as the reference, whereas if a 0, the next encoded bit is opposite the reference.

Differentially coherent PSK (DPSK): The result of using a differentially encoded bit stream to phase-shift key a carrier.

Digital modulation: The variation of some attribute of a carrier, such as amplitude, phase, or frequency, in a one-to-one correspondence with a message taking on a discrete set of values.

Direct-sequence spread-spectrum (DSSS): A modulation scheme employing a transmission bandwidth much wider than that required for the message signal, where the additional spectral spreading is achieved by multiplication with a binary pseudonoise code having many state changes per data symbol.

Euclidian distance: The length of the vector that is the difference of two other vectors in signal space; forms the Euclidian distance between the points defined by these vectors.

Frequency-hop spread-spectrum (FHSS): A modulation scheme employing a transmission bandwidth much wider than that required for the message signal, where the additional spectral spreading is achieved by hopping the modulated signal spectrum pseudorandomly in frequency.

Frequency-shift keying (FSK): A signaling scheme whereby the digital data modulates the frequency of a carrier.

Gaussian MSK (GMSK): MSK in which the baseband signal has been filtered with a filter having a Gaussian-shaped amplitude response before carrier modulation.

Global positioning system: A system consisting of 24 satellites spaced around the Earth with precise timing sources and emitting DSSS signals, at least four of which can be acquired by a receiver simultaneously in order to determine the position of the platform on which the receiver is located.

Gram–Schmidt procedure: An algorithm for obtaining a set of orthonormal functions from a set of signals.

Hard decision: The process of making a definite 1–0 decision at the output of a detector, usually in conjunction with coding. Thus information is discarded before the decoding process.

M-ary signaling: Any signaling scheme where the number of possible signals sent during any given signaling interval is M; binary signaling is a special case with $M = 2$. MPSK uses M phases of a sinusoidal carrier.

Matched filter implementation: A receiver implementation that has a parallel bank of matched filters as the first stage, one matched to each possible transmitted signal. A matched filter has an impulse response that is the time reverse of the signal to which it is matched.

Maximum-likelihood receiver: A receiver that bases its decisions on the probabilities of the received waveform given each possible transmitted signal.

Minimum-shift keying (MSK): OQPSK with the in-phase and quadrature baseband signal components weighted by half-sinusoids. The phase changes linearly over a bit interval.

Noncoherent modulation: Any modulation scheme not requiring a reference at the receiver in phase coherence with the received modulated signal in order to perform demodulation.

Norm: A function of a signal that is analogous to its length.

Offset quadriphase-shift keying (OQPSK): A QPSK scheme where phase shifts can be only $\pm\pi/2$ and can occur at half-symbol intervals.

Orthogonal signaling: Any signaling scheme that uses a signal set that have zero correlation coefficient between any pair of signals.

Phase-shift keying (PSK): A signaling scheme whereby the digital data modulates the phase of a carrier.

$\pi/4$-differential quadriphase-shift keyed signaling ($\pi/4$-DQPSK): A differentially coherent phase-shift keyed modulation scheme in which one of eight possible phases is transmitted each signaling interval, and in which the possible phase changes from the current phase is limited to $\pm\pi/4$ and $\pm 3\pi/4$.

Power efficiency: The energy-per-bit over noise power spectral density (E_b/N_0) required to provide a given probability of bit error for a digital modulation scheme.

Probability of (bit) (symbol) error: The average number of symbol errors over a long string (ideally infinite) of transmitted signals for any digital modulation scheme. If expressed in terms of the number of bits per symbol, the probability of bit error results.

Puncturing: A method to increase the rate of a code, usually convolutional, by periodically deleting symbols from the encoded output. At the decoder, arbitrary symbols are inserted in the places the deleted symbols would have occupied, and the decoder corrects a large majority of these if necessary.

Quadrature-amplitude-shift keying (QASK): A digital modulation scheme in which both the phase and amplitude of the carrier takes on a set of values in one-to-one correspondence with the digital data to be transmitted.

Quadriphase-shift keying (QPSK): An MPSK (see M-ary signaling) system that utilizes four signals in the signal set distinguished by four phases 90° apart.

Scalar product: A function of two signals that is analogous to their inner or dot product in ordinary geometric terms.

Set partitioning: A procedure for implementing TCM encoding rules in order to maximize Euclidian distance between signal points.

Soft decisions: The use of detector outputs quantized to more than two levels (i.e., a 1 or a 0) to feed into a decoder with the intention of improving the coding gain over what is achievable with hard decisions.

Symbol period: The duration of the signaling element for an M-ary modulation scheme.

Trellis-coded modulation (TCM): The process of using convolutional codes of suitable rates in conjunction with M-ary modulation to achieve coding gain without bandwidth expansion.

Viterbi algorithm: An efficient algorithm for decoding convolutional codes, among other applications, that implements maximum likelihood estimation.

References

Arthurs, A. and Dim, H. 1962. On the optimum detection of digital signals in the presence of white Gaussian noise—A geometric interpretation and a study of three basic data transmission systems. *IRE Trans. on Comm. Systems* CS-10(Dec.).

Blahut, R.E. 1990. *Digital Transformation of Information.* Addison-Wesley, Reading, MA.

Kotelnikov, V.A. 1960. *The Theory of Optimum Noise Immunity.* Dover, New York.

Murota, K. 1981. GMSK modulation for digital mobile radio telephony. *IEEE Trans. on Comm.* COM-29(July): 1044–1050.

Peterson, R.L., Ziemer, R.E., and Borth, D.E. 1995. *Introduction to Spread Spectrum Communications.* Prentice-Hall, Englewood Cliffs, NJ.

Prabhu, V.K. 1982. Error rate bounds for differential PSK. *IEEE Trans. on Comm.* COM-30(Dec.):2547–2550.

Proakis, J.G. 2002. *Digital Communications,* 4th ed. McGraw-Hill, New York.

Sklar, B. 1988. *Digital Communications: Fundamentals and Applications.* Prentice-Hall, Englewood Cliffs, NJ.

Ungerboeck, G. 1987. Trellis-coded modulation with redundant signal sets, part I: Introduction and trellis-coded modulation with redundant signal sets, part II: State of the art. *IEEE Comm. Mag.* 25(Feb.):5–11 and 12–21.

Viterbi, A.J., Wlolf, J.K., Zehavi, E., and Padovani, R. 1989. A pragmatic approach to trellis-coded modulation. *IEEE Comm. Mag.* 27(July):11–19.

Wozencraft, J.M. and Jacobs, I.M. 1965. *Principles of Communications Engineering.* Wiley, New York (out of print, but available from Waveland Press, Prospect Heights, IL).

Ziemer, R.E. and Tranter, W.H. 2002. *Principles of Communications: Systems, Modulation, and Noise,* 5th ed. Wiley, New York.

Further Information

The references were chosen to back up statements made, provide additional information in certain areas, or provide historical background. Some additional sources that give simpler treatments in many respects than the references are the following:

Roden, M.S. 1991. *Analog and Digital Communication Systems,* 3rd ed. Prentice-Hall, Englewood Cliffs, NJ.

Peebles, P.Z., Jr. 1987. *Digital Communication Systems,* Prentice-Hall, Englewood Cliffs, NJ.

Schwartz, M. 1990. *Information Transmission, Modulation, and Noise,* 4th ed. McGraw-Hill, New York.

Gibson, J.D. 1989. *Principles of Digital and Analog Communications.* Wiley, New York.

12.6   Spread Spectrum Systems

Kurt L. Kosbar and William H. Tranter

12.6.1   Why Use Spread Spectrum?

Radio communication systems are designed to move the information contained in a low-frequency message waveform through a channel that only passes high-frequency signals. High-frequency signals allow for channel assignment using frequency division multiplexing and allow signals to be radiated using antennas of reasonable size. The process referred to as *modulation* translates a low-frequency message signal to a high-frequency modulated signal. Modulation may be analog or digital. Examples of analog modulation are amplitude modulation (AM), frequency modulation (FM), and a variety of other formats such as phase modulation (PM), double sideband (DSB), single sideband(SSB), and VSB. A similar set of modulation formats were developed for digital data messages (Sklar, 1988) and examples include ASK, FSK, BPSK, QPSK, and MSK. All of these modulation techniques fall into a general class that we will call **narrow-band (NB) modulation** techniques since the result of the modulation process is a NB bandpass signal. If an analog message has a bandwidth of W Hz, or a digital source has a data rate of W bits/second, the output of a NB transmitter (modulator) will span a bandwidth of αW Hz. The exact bandwidth will depend on the choice of modulation and various system parameters, but in most cases of practical interest the transmitted bandwidth will be within an order of magnitude (a factor of 10) of W Hz. In other words, α will be less than 10 for most applications.

Since NB signals have a limited bandwidth, multiple users can share a band of the electromagnetic spectrum by using frequency division multiple access (FDMA). In this format, the available spectral space is divided into nonoverlapping segments, or channels, and each user is given exclusive use of one channel. The narrower the bandwidth of the transmitted signals, the more channels and users a given bandwidth can support. The architects of NB FDMA systems typically make a number of assumptions about the operating environment, including the following:

- Aside from the intended transmitter, there will be no other devices producing high-power signals within the allocated channel, that is, interference is negligible.
- Electronic devices in close proximity to the transmitter will not be adversely affected by the radiated signal.
- There is no need to conceal the existence of the transmitter.
- It should be easy to receive and demodulate the signal.
- The channel model will be fairly simple, such as an attenuation and a small amount of additive noise.
- Reflections or multipath problems will be insignificant.

These assumptions were quite reasonable during the early development of commercial FDMA systems since these systems used a relatively small number of high-power transmitters at fixed locations. Regulatory agencies such as the FCC carefully controlled the radiation patterns, power levels, location and frequency assignments of the transmitters to avoid interference.

Military users could not afford to use the same set of assumptions. An enemy may intentionally generate interfering signals to disrupt (jam) the communication link, or secretly intercept and demodulate the radiated signal (eavesdrop). Sometimes the transmitted message is of little interest, but detecting the presence and location of a transmitted signal may be very useful. More than one weapon system has been developed to locate and destroy a source of electromagnetic (EM) radiation. These difficulties lead designers to search for a modulation technique that would be very difficult to jam, easy for friendly users to demodulate, difficult for unfriendly users to demodulate, and difficult for enemies to detect the presence of a transmitter. The result of this research was a completely new class of modulation techniques now know as **spread spectrum (SS)** communications. As the name suggests, signals generated by SS modulators and transmitters cover a much wider frequency range than NB signals. The

TABLE 12.13   Similarity of Military and Commercial Problems

Military Problem	Commercial Problem
Hostile Jammers	• Electronic equipment that unintentionally radiates interfering signals.
	• New communication services must share bands with existing NB transmitters
Jammers that intentionally create multipath	• Multipath in indoor and urban communication environments
Need to conceal the existence of the transmitted signal	• Need to generate signals that will not interfere with existing communication links
	• Need to generate signals that are unlikely to cause EMI
Prevent unauthorized demodulation	• Prevent unauthorized demodulation
New multiple access format	• New multiple access format

radiated bandwidth may be many orders of magnitude greater than the message bandwidth or message data rate.

Commercial communication system designers initially ignored SS techniques. Spread spectrum systems were considered exotic, expensive, and necessary only when hostile parties were attempting to jam or intercept the communication signal. This attitude toward SS techniques has changed within the past few decades for a number of reasons as outlined in Sec. 12.6.7 to follow. Although military and commercial users have different motivations, the technical problems faced by these systems are very similar. Table 12.13 summarizes these similarities.

12.6.2   Jamming and Eavesdropping

Figure 12.72 is a high level block diagram that we will use throughout this section. The goal is to transmit data from the source, through the friendly transmitter and receiver to the authorized end user. We assume the designer has complete control over the friendly transmitter and receiver. The designer has little or no control over the design and operation of the two related systems that we call the jammer and the unauthorized receiver. In a military environment these devices may be operated by hostile parties. For commercial systems, the *jammer* is any piece of equipment that radiates EM signals, whereas the unauthorized receiver is any device that is susceptible to electromagnetic interference (EMI). If the friendly receiver link is operational even when the jammer is producing a high-power signal, we call the friendly system **jam resistant**. If the friendly receiver can demodulate the signal, but it is very difficult (or unlikely) that the

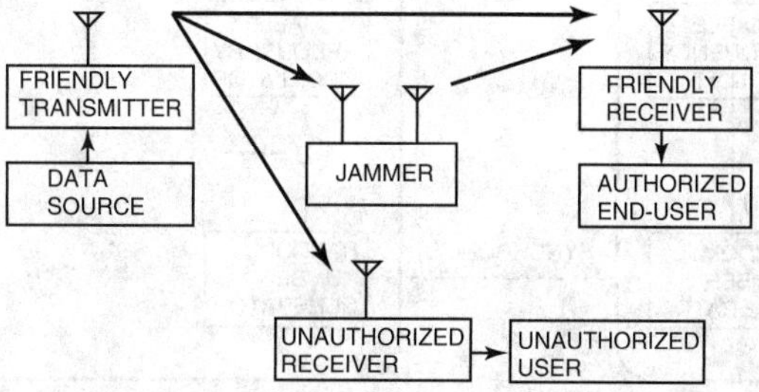

FIGURE 12.72   Communication system threats.

unauthorized receiver can do the same, we say the friendly system has a **low probability of intercept (LPI)**. Spread spectrum systems are both jam resistant and exhibit an LPI.

Jam resistance and LPI come at a cost. Spread spectrum systems are typically more complex, require greater bandwidth, and have lower message bandwidths and lower message data rates than NB systems. In spite of these drawbacks SS systems are attractive for some applications. Spread spectrum techniques provide the designer with a set of design tradeoffs that are unavailable with conventional NB systems.

The remainder of this section will discuss two fundamental SS modulation techniques: **frequency hopping (FHSS)** and **direct sequence (DSSS)**. After introducing the basic architecture of these SS systems, their performance in the face of interfering signals will be addressed. Designers accustomed to working with NB systems will notice an unfamiliar device in all SS systems. A pair of SS transmitters and receivers must synchronously generate pseudorandom waveforms called *spreading signals*.

12.6.3 Basic Spread Spectrum Techniques

Frequency Hopping

One of the simplest ways to disrupt a NB communication link is for a jammer to produce a high-power sinusoid near the friendly carrier frequency. The friendly system can obviously avoid this threat by switching to a new carrier frequency. If the new carrier is sufficiently far away from the jamming tone, the IF filters in the friendly receiver will remove the interfering signal. The selection of a new carrier frequency may be difficult for a variety of reasons. A truly hostile and intelligent jammer may monitor the friendly signal and intentionally shift the jamming frequency to match it. In a more benign setting, the friendly transmitter may be unaware of which frequencies are being jammed at the friendly receiver. A reverse link from the friendly receiver to the friendly transmitter would solve this problem, but increases system complexity. Even if the friendly link is not being jammed, the friendly transmitter will typically be unaware if the unauthorized receiver is able to demodulate the radiated signal.

Some of these problems can be overcome by quickly switching, or hopping, the friendly carrier frequency over a wide band. This is the fundamental concept behind frequency-hopped spread spectrum systems, as shown in Fig. 12.73. A number of excellent texts discuss this topic such as Sklar (1988); Peterson, Ziemer, and Borth (1995); and Simon et al. (1985). The system design is very similar to conventional NB systems. The only modification to the transmitter is that the conventional fixed frequency oscillator is replaced by a variable frequency oscillator that can change frequency very quickly under electronic control. In Fig. 12.73 the message source and variable frequency oscillator are fed into a DSB modulator (a simple multiplier), although any form of NB modulation can be used. The oscillator frequency used at any instant in time is

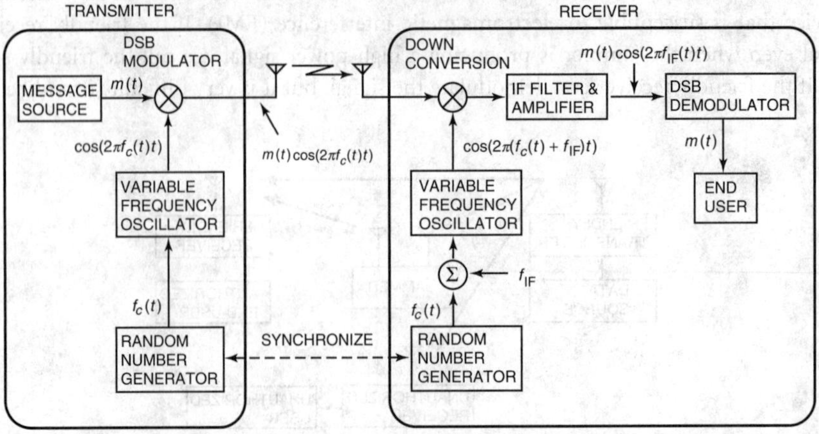

FIGURE 12.73 Frequency hopped spread spectrum system.

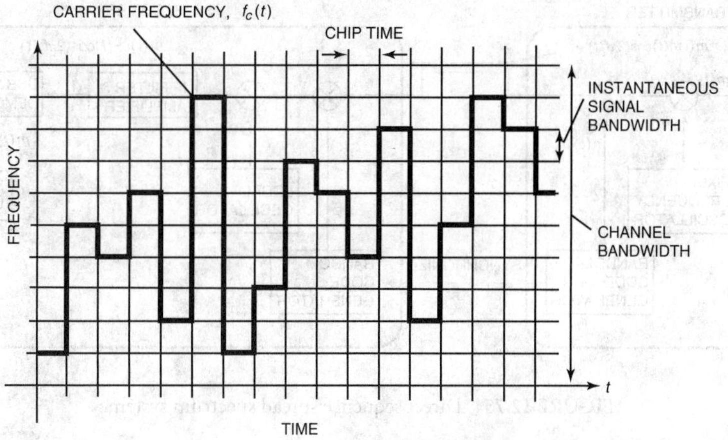

FIGURE 12.74 Sample hopping pattern for an FHSS system.

controlled by a pseudorandom number (PN) generator. A sample output of the PN generator output is shown in Fig. 12.74. Typically, the PN generator remains at each frequency for a fixed time interval called the *chip time*. At the end of the chip time, the transmitter will rapidly change to a new carrier frequency, selected at random by the PN generator, over the channel bandwidth. The transmitter and receiver use identical and synchronized PN generators. This ensures that the output of the receiver's downconverter is always at the IF frequency, regardless of the carrier frequency. There are two classes of FHSS systems based on the chip time and the symbol time of the message source. In *fast hopping systems* the chip time is less than the symbol time, whereas the converse is true in a *slow hopping system*.

Frequency hopping accomplishes one of the major goals of an SS system: it makes it difficult to jam the transmitted signal over an extended time. If the hopping rate is sufficiently rapid, this system can also be immune to some multipath problems. However, FHSS signals are fairly easy to detect and demodulate. A spectrum analyzer or bank of energy detectors tuned to different frequency bands can detect the presence of an FHSS signal. Once the carrier frequency is located, a conventional receiver can demodulate the signal. There will be a processing delay from the time the friendly transmitter changes frequencies until the time the hostile receiver detects the change, but a delay line can store the signal long enough to permit demodulation. Frequency hopping is also only moderately successful at eliminating EMI. If the unauthorized receiver is an electronic device sensitive to a particular frequency range, it may see bursts of energy in the frequency band. In many cases short bursts of interference may be just as damaging as interference from a constant source.

Direct Sequence

Frequency hopped transmitters are easy to detect because at any given time they are radiating large amounts of energy in a limited bandwidth. For a system to have an LPI, it must continuously spread the transmitted power over a wide-frequency band. This will reduce the power spectral density (PSD) level and make the transmitted signal resemble background noise. An unauthorized or unintended receiver will only detect a slight increase in the noise floor, due to the presence of the signal, and will find it extremely difficult to demodulate the signal. Instantaneous frequency spreading can be accomplished by using the DSSS system shown in Fig. 12.75. The message is modulated using a conventional NB modulator at a fixed carrier frequency. Prior to transmission, the modulated signal is multiplied by a high-rate PN sequence. The chip time of the sequence is much less than the message symbol time, as shown in Fig. 12.76. Arbitrarily complex PN sequences can be used, but most systems of practical interest select the bilevel sequence ($c(t) = +1$ or -1). The DSSS receiver can strip the PN spreading code from the signal by multiplying the received signal by a locally produced replica of $c(t)$. When the code is a bilevel sequence, the output of the despreading

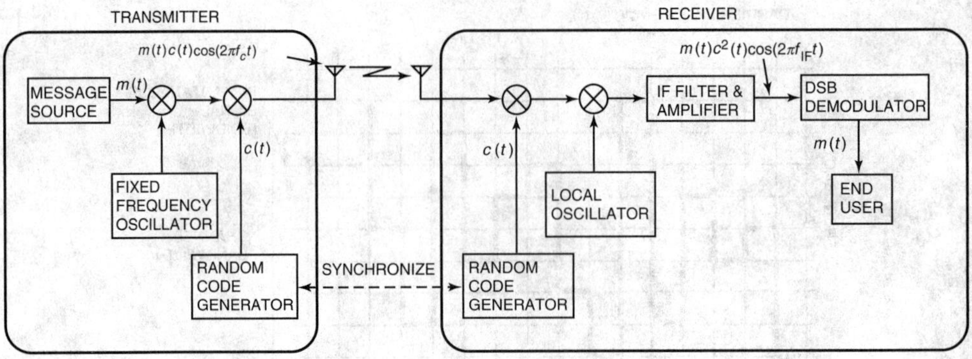

FIGURE 12.75 Direct sequence spread spectrum system.

multiplier is then

$$c^2(t)m(t)\cos(\omega_c t) = m(t)\cos(\omega_c t)$$

This despread signal can be demodulated by a conventional NB receiver.

The LPI characteristics of DSSS systems can be observed in the spectra of Fig. 12.76. Since the chip time of the PN generator is much less than the symbol time of the message source, the PN spectrum is much wider than the NB signal. From basic Fourier transform theory, we know that multiplication in time is equivalent to convolution in the frequency domain. Since the NB signal spectra is narrow compared to the spreading code, it will act almost the same as an impulse function. The result of the convolution will essentially be the PN spectrum shifted to the carrier frequency. This very wide bandwidth signal will resemble thermal noise over than bandwidth of the receiver.

12.6.4 Performance in Jamming Environments

Spread spectrum techniques are often used to improve the system performance when the received signal contains both the desired and interfering waveforms. The sensitivity of systems to a jamming waveform will depend on the modulation format and on the characteristics of the interference.

The jamming of a communication link is often viewed as a game. The transmitter can deliver a fixed amount of power S to the receiver, and the end user specifies a maximum allowable bit error rate (BER).

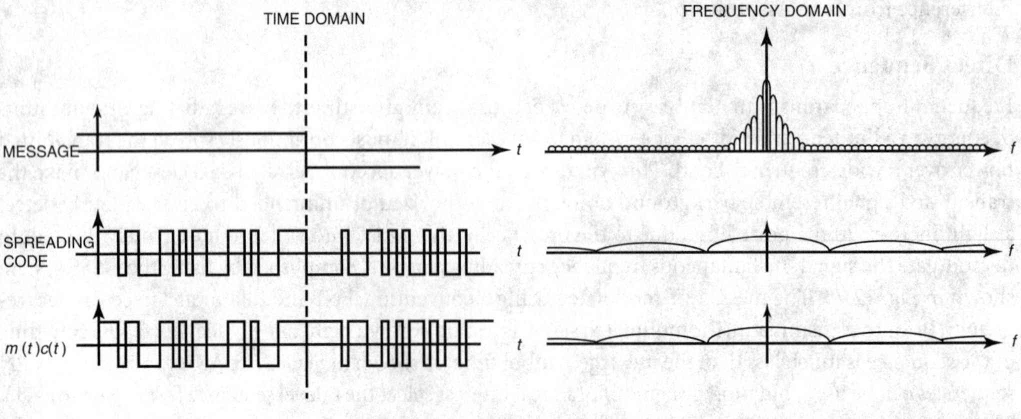

FIGURE 12.76 Direct sequence signals.

The jammer can deliver a signal with J watts of power to the friendly receiver and will try to force the BER above the acceptable level. An important parameter in these systems is the *jammer-to-signal ratio (JSR)*, J/S, which is often greater than unity. The jammer can always win this game by making J sufficiently large. Spread spectrum techniques can only increase the threshold where the jammer will eventually win the game. The change in the JSR level required for the jammer to win the game is called the *system processing gain*, and may be many orders of magnitude.

The jammer can use a wide variety of signals, including white Gaussian noise, bandlimited white Gaussian noise, pulsed noise, continuous tone, multiple tones, or a time-delayed copy of the transmitted signal (intentional multipath). A formidable adversary will be adept at observing the friendly signal and altering the jamming waveform to maximize the damage. The SS system designer should consider many jamming waveforms, to help minimize the chance the jammer will find one that is particularly effective. In a commercial environment, an electronic device may inadvertently act like a jammer by radiating one of these signals. Designers often analyze commercial systems in the same way as military systems because Murphy's law guarantees that an unintentional jammer will be just as effective as the most capable enemy jammer.

The Jammer Must Not Know the Spreading Code

A jammer can do a tremendous amount of damage if he knows the spreading format, carrier frequency, symbol rate, chip rate, and the spreading code used by the friendly system. Given all of this information, the jammer will be able to disrupt or eavesdrop on the SS link as effectively as a simple NB link. Designers and users of secure communication systems find it very difficult to keep the jammer from learning many of these system parameters. A conservative designer will assume the jammer has access to a complete description of the transmitter architecture. Somewhat surprisingly, a SS system can still be made secure by prohibiting the jammer from learning how to predict the output of the PN spreading code generator in real time. Even if the jammer knows the architecture of the PN generator, it is often impossible to predict the exact output without knowing the initial state of the generator (this requires a PN generator that is more complex than the simple *binary shift register (BSR)* generators discussed earlier). In the discussion that follows, we will assume the jammer knows all of the critical parameters of the friendly system *except* the exact sequence produced by the PN spreading code generator.

Additive White Gaussian Noise (AWGN) Jamming

One of the simplest ways to degrade the performance of any communication system is to raise the noise floor from N_0 to $N_0 + \Delta$ W/Hz over the entire band of the transmitted signal. This will lower the SNR at the receiver and degrade the performance of any communication system. For a NB system, the jamming power required to effect a change of Δ W/Hz over a band of W Hz is ΔW W. In a SS system the transmitted bandwidth is $Y \gg W$ Hz, so to create the same change in the noise floor, the jammer must now produce Y/W times as much energy.

Continuous Wave (CW) Jammers

Another simple jamming signal is a sinusoidal tone, or continuous wave (CW) signal. This type of jammer produces a high-power sinusoid or other very narrow bandwidth waveform. Conventional NB transmitters often look like CW jammers to a SS signal. First consider the effect of a CW jammer on a DSSS system as shown in Fig. 12.77. The despreading mixer in the receiver collapses the DSSS signal back to a NB signal for processing by the IF filter. This mixer has the exact opposite effect on the jamming tone. It will multiply the CW signal by a high-rate PN code, effectively spreading it over a very wide bandwidth. Some of the jamming signal will still fall within the IF bandwidth, but the majority will fall well outside the passband of the IF filters, significantly decreasing the JSR at the demodulator.

The **processing gain** is the improvement in the JSR at the demodulator output compared to the JSR at the receiver antenna input. Nearly all of the friendly signal power passes through the IF filter, which has a bandwidth of approximately W Hz. Since the spreading code has a bandwidth of Y Hz, the jamming signal power will be spread over Y Hz. This means that the JSR at the IF output will be approximately

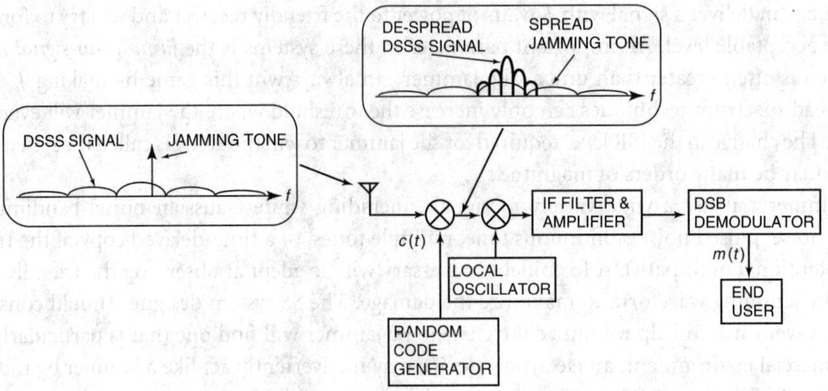

FIGURE 12.77   Response of a DSSS receiver to a jamming tone.

Y/W times less than the JSR at the antenna. There is a second, more subtle, benefit to DSSS receivers. Even though the jammer produced a continuous tone, the demodulator will see the interference as bandlimited white noise because of the multiplication by the spreading code. Most demodulators are designed under the assumption that the interference is white Gaussian noise. Nearly all jamming signals will look like white noise by the time they reach the demodulator output in these systems.

An FHSS system response to a CW jammer is slightly different and is summarized in Fig. 12.78. As the carrier frequency is hopped over a discrete set of frequencies, different frequency bands will be passed by the receiver's IF filter (recall that the local oscillator at the receiver will hop in the same pattern as the transmitter's oscillator). The shaded regions on the time/frequency graph represent the bands passed by the IF filter. During the majority of the hops the jamming tone (represented by the heavy line) will not be passed by the IF filter. When the jammer is lucky enough to hit the right frequency, all of the jammer energy is passed through the IF filter to the demodulator. During these chips, the data or message produced by the demodulator may be in error. There are many well-known error detection and correction techniques for dealing with errors, including the use of interleavers and error correcting codes. These devices are often used in FHSS systems.

To measure the jam resistance of a FHSS system, assume that the IF filter has a bandwidth of W Hz and the range of frequencies available to the system is Y Hz. If the jammer does not know the hopping pattern, the probability the signal will pass through the IF filter is W/Y, the reciprocal of the processing gain. The jammer will obviously be more successful if the jammer can determine the hopping pattern and make the corresponding change to the jamming frequency.

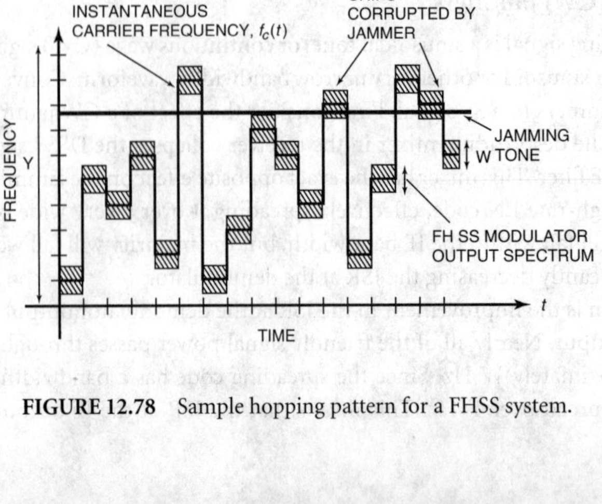

FIGURE 12.78   Sample hopping pattern for a FHSS system.

There are other techniques that receivers can use to combat CW jammers, particularly if the jammer uses a constant frequency. A narrow band exciser can be inserted before the despreading mixer. This device will actively search for powerful, narrow-band interference signals, then form a notch filter to remove the interfering signal. This will also remove a segment of the friendly signal, but may still improve overall system performance. A truly hostile jammer can defeat the exciser by hopping the jamming carrier frequency. These devices are much more useful in benign environments, such as when new commercial systems must share a band with existing NB transmitters.

Partial-Band Jammers

There are several jamming strategies between the two extremes of AWGN and CW jamming. Two possible alternatives are to send a collection of tones at different frequencies or transmit bandlimited AWGN that covers only a fraction of the friendly signal bandwidth. We will refer to both of these techniques as *partial-band jamming*. The advantage to partial-band jamming is most easily seen with FHSS systems. If a jammer can generate twice the power required to disable a NB demodulator, it would be foolish for the jammer to concentrate all of this power into a single tone. The jammer will be more successful by generating two lower power tones at widely differing frequencies. This will double the chance that the FHSS system will see an interfering signal and still guarantee the system will be disabled when it does see one of these tones. Depending on the JSR, the jammer may wish to generate more than two tones. For a given JSR, there will be an optimal number of tones the jammer should produce. If the jammer exceeds this optimal number, the power in each tone will become too small to effectively jam the friendly signal. There is a similar tradeoff with the bandwidth of a bandlimited AWGN jamming signal. It is well known that for a given JSR there is an optimal bandwidth that the jammer should use to increase the average BER to the extent possible.

There is typically little reason to use partial band jamming against DSSS systems. By spreading the jamming signal over a wider bandwidth, the jammer lowers the PSD of the jamming signal within the passband of the IF filter.

Pulse Jammers

In some cases the jammer has both an average power and peak power constraint. If the peak power limit is less stringent than the average power limit, the jammer may wish to deliver short bursts of interference rather than a constant signal. This type of jamming is also common in commercial systems, such as is found in computer local area networks, where each device may transmit a short burst of information and then remain silent for an extended time.

The exact analysis of pulse jammers can be rather difficult, and the analysis is dependent on a variety of parameters. Pulse jammers have the advantage that it is difficult for the friendly receiver to use an adaptive device to remove the jamming signal. Besides degrading demodulator performance, pulse jammers can disrupt a variety of support subsystems in the receiver. For example, pulse jammers can disrupt automatic gain control (AGC) systems. Most AGC subsystems use low-pass filters to find the average signal power. This means that when a pulse jammer is first turned on it can saturate the AGC amplifiers, although immediately after the end of the pulse the amplifier gains may be so low that the signal is lost due to quantizing noise or some other nonlinear distortion effect. Other devices sensitive to pulse jammers include carrier, symbol, and **code synchronization** devices.

Repeat-Back (Multipath) Jammers

Rather than generating a completely new waveform, some jammers prefer to intentionally introduce a multipath component. These jammers are reflective devices, where the friendly signal is received and then reflected or retransmitted in the direction of the receiver. In military systems the repeat-back jammer may be an active device; however, in commercial applications it is typically known as multipath interference. Multipath is a significant problem in urban areas where signals can reflect off buildings and in indoor wireless communication systems where high carrier frequencies make a host of structures and equipment effective reflectors of EM radiation.

Direct sequence SS systems can combat this type of interference effectively because the autocorrelation function of the spreading code is typically very small for time delays greater than one chip time. This means that the delayed signal will not be despread at the receiver, but will simply appear as another noise source at the despreader output. To demonstrate this effect, assume that the received signal consists of the direct path signal and a reflected signal of equal strength but delayed by τ s. Therefore

$$r(t) = m(t)\cos(2\pi f_c t)c(t) + m(t)\cos(2\pi f_c(t-\tau))c(t-\tau)$$

and the output of the despreading mixer is

$$m(t)\cos(2\pi f_c t)c^2(t) + m(t)\cos(2\pi f_c(t-\tau))c(t-\tau)c(t)$$
$$= m(t)\cos(2\pi f_c t) + m(t)\cos(2\pi f_c(t-\tau))c(t-\tau)c(t)$$

The second term will typically not despread because $c(t-\tau)c(t)$ will be a function of time. If τ is greater than a chip time, this product will resemble a PN sequence and be largely removed by the IF filter. This property will make DSSS receivers fairly insensitive to multipath interference as long as the differential delay between the main path and the multipath components is greater than a chip time. There will be some synchronization problems with these receivers, since it is important that the local PN reference is aligned to the strongest signal (usually the direct path) and not fooled into becoming synchronized with one of the multipath components.

Slow FHSS systems are sensitive to multipath jamming. The problem can sometimes be avoided by switching to a fast FHSS format. In a hostile environment, one can typically assume there is some minimum time difference between the main path and any multipath signal. This minimum time delay will be determined by calculating the difference in distance of the transmitter–jammer–receiver path over the transmitter–receiver direct path. If the chip time is less than the minimum time difference, the multipath jamming signal will always be at least one chip time behind the correct frequency. A judicious choice of hopping patterns will ensure that a jammer that is at least one chip time behind will seldom, if ever, be at a frequency that can cause interference. Of course, if the jammer is very close to the transmitter, it may be impossible to hop sufficiently fast. This may be the case for many indoor and urban communication systems. In these cases, the friendly users must simply use other NB techniques for dealing with the multipath problem, and FHSS will provide little assistance in overcoming the problem.

12.6.5   Random Number Generators

As we have seen, both FHSS and DSSS transmitters and receivers use random number generators, a device not typically found in NB systems. Frequency hopped SS systems use generators that produce integers, whereas DSSS systems typically use binary generators. Despite this difference, the underlying structure of the generators is very similar. The structure of the maximal length binary shift register generator is shown in Fig. 12.79.

The BSR generator contains a series of D type flip-flops (F/F) all cycled by a common clock. The number of F/F in the shift register will vary depending on the application, but is typically between 10 and 100. The BSR generator also contains a series of modulo-two adders, or exclusive-OR gates. Figure 12.79 shows one XOR gate for each F/F, with a switch between the gate and the F/F. The designer of the generator must select the number of F/F in the BSR and which of the switches will be closed. Open switches will feed a logic 0 to the XOR gate, meaning that the gate is not needed. The designer (or user) must also select an appropriate clock frequency and the initial state of the BSR.

In many applications we would like the PN generator to produce a sequence that is unpredictable and never repeats. The BSR generator can come close to, but does not achieve, this ideal goal. The output of a sample BSR generator is shown in Fig 12.80. Notice that the output seems to toggle randomly, but after 63 clock cycles the pattern repeats. Since BSR generators are finite state machines with no inputs, their outputs must be periodic. If the generator has N F/F, and each F/F can take on one of two states, the generator has only 2^N unique states. After no more than 2^N clock cycles, the generator must return to a

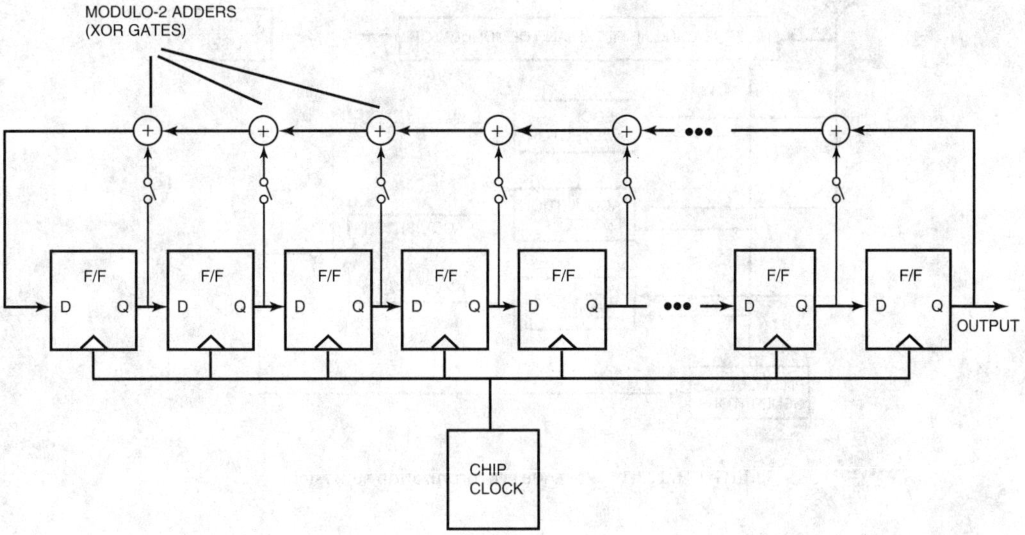

MODULO-2 ADDERS
(XOR GATES)

FIGURE 12.79 Binary shift register PN sequence generator.

state that it has already occupied, at which time the generator enters a loop it can never exit. The period of the generator will be determined by both N and the settings of the feedback switches. By carefully selecting the switch positions, the period of the output can reach a maximum of $2^N - 1$ clock cycles. Tables of optimal switch settings have been tabulated for various BSR lengths up to 89. Note that a $N = 89$ BSR generator clocked at 1 GHz will have a period of approximately 10^{10} years.

When the generator is first turned on the F/F must be set to predetermined states. For a BSR this may be any set of values agreed upon by the transmitter and receiver, except the all 0 state. A BSR initiated to all 0s will stay in that state forever. This set of initial values, along with the real-world time at which the generator is set to this state, is known as the PN generator **key**. Two identical SS systems (same carrier frequency, same modulation format) can share a common channel by simply using different keys. This format is called *code division multiple access (CDMA)* and is becoming popular in a variety of new communication systems.

The BSR generators discussed previously have one serious flaw: they are not secure. An intelligent enemy can determine the generator key and architecture after observing $2N$ output symbols. Secure communication systems must use more elaborate generators. Although the internal mechanism of secure generators vary, they will still be finite state machines that must be initialized by a key and have an output that is ultimately periodic.

Recall that FHSS systems require generators that produce integers. This is easily accomplished by increasing the clock rate of the BSR generator. An 8 state output can be produced by increasing the BSR

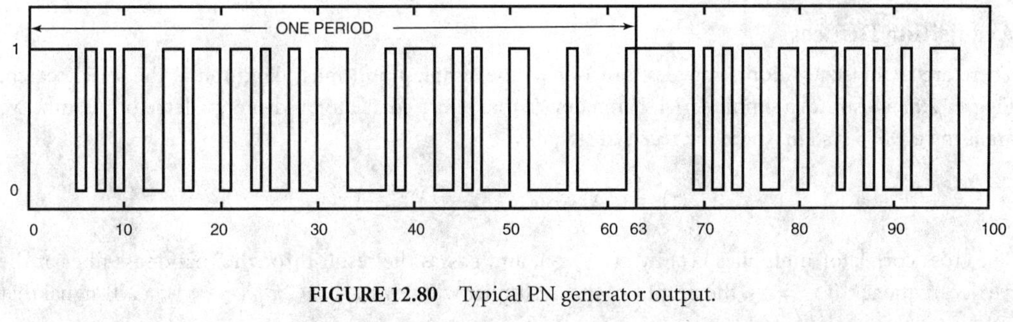

FIGURE 12.80 Typical PN generator output.

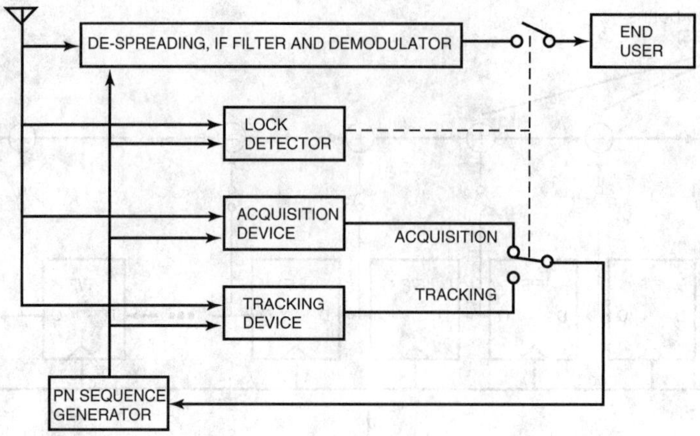

FIGURE 12.81 PN code synchronization subsystem.

clock by a factor of 3 and using the output of the last three F/F to generate a 3 bit number. This same strategy can be used to generate arbitrarily large integers.

12.6.6 Code Synchronization

All communication systems require synchronization in some form. In most cases the receiver needs a fairly accurate measure of the carrier frequency, and in digital communication systems the receiver often performs carrier phase synchronization and symbol synchronization. Besides these traditional synchronization tasks, SS systems must also synchronize the PN generators used at the transmitter and receiver since it is impossible for the receiver to demodulate the message without proper PN code synchronization. Pseudonoise code synchronization concerns may limit how rapidly a SS system be turned on and may also limit the minimum SNR at which the link can be maintained.

Code synchronization is divided into two separate tasks: initial timing acquisition and code tracking. Figure 12.81 is a high-level block diagram of a code synchronization system. The lock detector determines if the subsystem is in the acquisition or tracking mode. If the lock detector thinks the difference between the received and local PN codes exceeds a chip time, it will connect the acquisition device to the PN generator and inform the end user that the demodulated data is not valid. The acquisition device will reduce the timing error to approximately one chip time or less. At this point the lock detector will turn control of the PN generator over to the tracking device. This device further reduces the timing error and maintains it to within a small fraction of a chip time. The lock detector will constantly monitor the synchronization state. If at any time the timing error becomes large, the lock detector will switch control back to the acquisition device, causing the subsystem to reacquire the signal. Figure 12.81 is a conceptual block diagram only. The hardware that implements these functions is often shared between the different blocks.

Acquisition Devices

There are many acquisition strategies, but one of the simplest and most common is the serial search. Figure 12.82 illustrates a simple serial search device that uses a correlator and energy detector. Assume we are using a DSSS system where the received signal is

$$r(t) = m(t)\cos(2\pi f_c(t - \tau))c(t - \tau_1)$$

The cross correlator multiplies $r(t)$ by $c(t - \tau_2)$, and passes the result through a bandpass filter at the carrier frequency. If $\tau_1 = \tau_2$, the output of the multiplier will be $m(t)\cos(2\pi f_c(t - \tau))$, a NB signal that

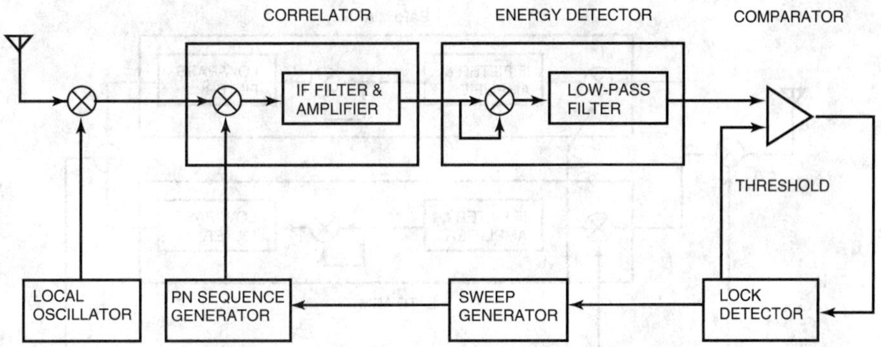

FIGURE 12.82 Serial search acquisition device.

will pass through the IF filter. This will generate a large signal at the IF filter output, causing the energy detector to produce a large output voltage. If there is a large difference between τ_1 and τ_2, the multiplier output will still have a PN code multiplying the NB signal. Very little of this wide-band signal will pass through the IF filter, and so the energy detector output will be quite small.

In sequential search acquisition, the PN generator is initiated with a key that is as close to the correct value as possible. The sweep generator forces the PN generator to sweep a wide range of timing offsets. Eventually one of these offsets will be within a chip time of the true offset, producing a significant output of the energy detector. The lock detector will sense the large energy detector output, halt the sweep generator, and turn control over to a tracking device. There are a variety of search patterns the sweep generator can use when hunting for the correct timing offset. Some of the more common sweep patterns are shown in Fig. 12.83.

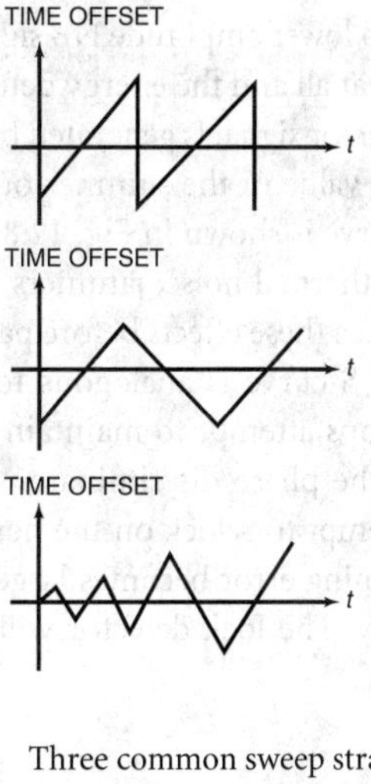

FIGURE 12.83 Three common sweep strategies.

Tracking Devices

Once the acquisition device has placed the PN generator to within a chip time of the correct value, control is turned over to a code tracking device. *Delay-locked loops (DLL)* are frequently used for code tracking and are closely related the phase-locked loops (PLL) widely used in NB systems. The DLL can be used in a variety of configurations to track PN codes in FHSS and DSSS systems. The only DLL discussed in this section will be the noncoherent early/late DLL for DSSS code tracking. Figure 12.84 shows the basic components of the DLL are the PN sequence generator, two cross correlation arms, a loop filter, and a voltage controlled clock. These devices are analogous to the VCO, phase detector, and loop filter used in a conventional PLL.

The PN code generator in a DLL must be modified to produce both an early and late output. If the on-time output of the DLL is $c(t)$, then the early output will be $c(t + \Delta/2)$ and the late output will be $c(t - \Delta/2)$. The time difference Δ is under the control of the designer and is typically on the order of 0.5 to 1.0 chip times. Figure 12.85 shows the output of the early and late correlator arms for different timing errors. When the received PN code aligns with one of the locally generated references the received signal will be perfectly despread. This will produce a high-power, NB signal that will pass through the IF filter and be seen by the energy detector. As the timing error increases, the signal will be less perfectly despread,

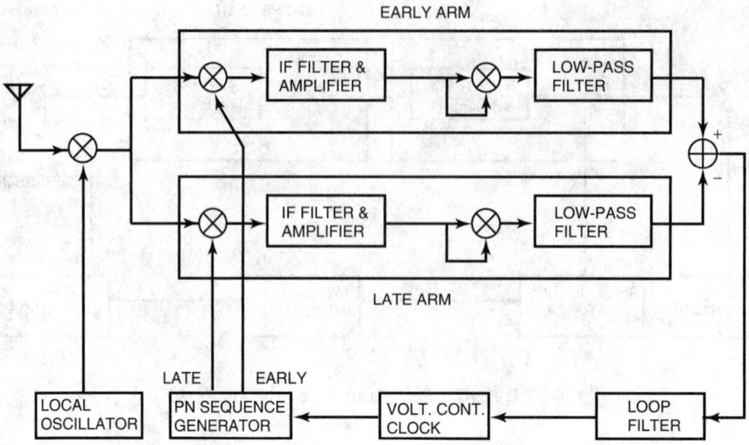

FIGURE 12.84   Noncoherent early/late DLL.

resulting in a lower amplitude NB signal. When the timing error exceeds one chip time, the signal will not be despread at all and the energy detector output will be negligible.

A useful error signal is generated by subtracting the output of the early and late cross-correlation arms. The average value of the summer output as a function of the timing error is called the *loop S-curve*. A sample S-curve is shown in Fig. 12.85. The signal produced by the summer will have a number of noise terms from thermal noise, jammers, and the so-called self-noise from the PN code. The loop filter will need to reduce these effects before passing the signal on to the voltage controlled clock.

The DLL S-curve is analogous to the sinusoidal phase detector characteristic of the PLL. In both cases the loops attempt to maintain a small timing error, keeping the error in the linear region of the S-curve of the phase discriminator characteristic. If the PLL error becomes large, the loop will cycle slip and attempt to relock on the next stable lock point. A large DLL timing error is more catastrophic. Once the timing error becomes large, the error term becomes negligible and the timing error will wander aimlessly. The lock detector will need to detect this large timing error and initiate a reacquisition cycle.

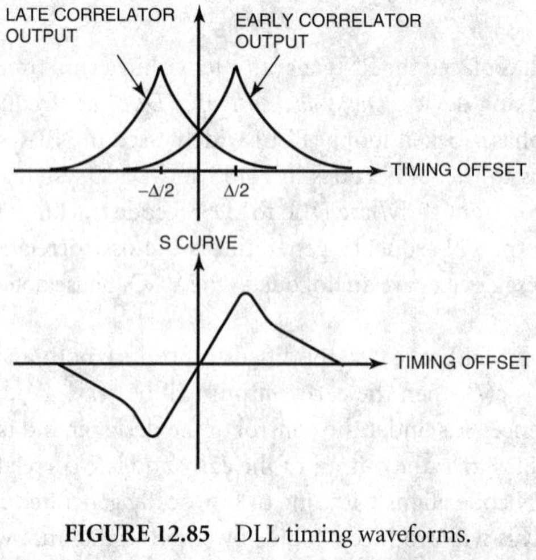

FIGURE 12.85   DLL timing waveforms.

12.6.7 Applications of Spread Spectrum Systems

An excellent tutorial paper on the theory of spread spectrum was published in 1992 (Pickholtz, Schilling, and Milstein). In this paper six characteristics of spread spectrum systems were identified. These were as follows:

- Antijamming
- Anti-interference
- Low probability of intercept
- Multiple user random access communications with selective addressing capability
- High-resolution ranging
- Accurate universal timing

Over the past decade spread spectrum systems have matured significantly. The operating characteristics of these systems are becoming better understood. In addition, dramatic advances in the technology for implementing SS systems allows them to be fabricated using a small number of integrated circuits. In addition, current SS systems provide much better performance, higher carrier frequencies, and higher data rates than those of just a few years ago. Although the basic list of characteristics remain unchanged, there has been a shift from military to commercial applications (Schilling et al., 1995).

The previous pages discussed in some detail the antijamming, anti-interference, and low probability of intercept characteristics of spread spectrum systems. In this section we look at several applications, especially those arising from the multiple access characteristics.

Wireless Local Area Networks (LANs)

One important application of spread spectrum communication systems is in the implementation of wireless local area networks (LANs). The attraction of such networks is obvious. A computer with access to a wireless LAN is completely untethered and a hard (wired) connection to a network is unnecessary. Wireless LANs are clearly advantageous in a situation in which computers are highly mobile (roaming), such as with laptop computers being moved from office to office or from an office to a conference room. Each computer carries its own RF transmitter and receiver, and such a system becomes very practical if licensing requirements can be avoided.

For systems such as wireless LANs to operate without a FCC site license, the industrial, scientific, and medical (ISM) frequency bands were created. There are two such bands, 902–928 MHz and 2.400–2.484 GHz. The FCC requires that transmission on these bands must use either frequency hopping or direct sequence spread spectrum transmission. The more restrictive IEEE 802.11 standard, however, limits FHSS to the 2.400–2.484 GHz band, so that the lower frequency band is reserved for DSSS systems. FCC requirements also dictate certain parameters that influence the hopping pattern for FHSS systems. For example, the FCC dictates that in the 2.4-GHz frequency band, a FHSS transmitter must not spend more than 0.4 s in any one frequency slot every 30 s and that the transmitter must hop through at least 75 frequency slots in the frequency hopping pattern (Rigney, 1995).

Defining Terms

Code synchronization: Spread spectrum transmitters and receivers must synchronously generate identical pseudorandom sequences. Code synchronization is the act of bringing these two generators into alignment.

Direct sequence: A spread spectrum format where a narrow-band signal is multiplied by a high rate pseudonoise code prior to transmission.

Frequency hopping: A spread spectrum format where the frequency of the transmitted signal is rapidly changed in a pseudorandom pattern.

Jam resistant systems: Systems that can reliably transmit information even when interfering signals are present. The power in the interfering signal may be significantly greater than the power of the transmitted signal.

Key: A piece of data the describes the initial state of a pseudorandom generator. It can be difficult or impossible to determine the output of the generator if one does not know the key.

Low probability of intercept: The transmitted signal in some spread spectrum systems resembles background noise. It is difficult for unauthorized users to detect and demodulate these signals.

Narrow–band modulation: Any one of a variety of modulation techniques where the bandwidth of the transmitted signal is comparable with the bandwidth (or data rate) of the message. Common examples include AM, FM, PM, ASK, FSK, BPSK, MSK, and QPSK.

Processing gain: Jammers must use more power to disrupt a spread spectrum link than to disrupt a narrow-band link. The amount the jammer must increase power is known as the processing gain of the spread spectrum system.

Pseudorandom noise: A deterministic, but noiselike, waveform typically produced by a digital pseudorandom number generator. Two pseudorandom generators will generate identical waveforms if they are initialized with the same seed and clocked synchronously.

Spread spectrum modulation: A modulation format where the bandwidth of the transmitted signal is typically orders of magnitude wider than the bandwidth (or data rate) of the message. Common examples include frequency hopping (FHSS) and direct sequence (DSSS).

References

Boyle, P. 1995. Wireless LANs: No strings attached. *PC Mag.* (Jan. 10).

Peterson, R.L., Ziemer, R.E., and Borth, D.E. 1995. *Introduction to Spread Spectrum Communications.* Prentice-Hall, Englewood Cliffs, NJ.

Pickholtz, R.L., Schilling, D.L., and Milstein, L.B. 1992. Theory of spread-spectrum communications—A tutorial. *IEEE Trans. on Comm.* COM-30(5).

Rigney, S. 1995. Spread the signal. *PC Mag.* (Jan. 10).

Schilling, D.L., Milstein, L.B., Pickholtz, R.L., Kullback, M., and Miller, F. 1995. Spread spectrum for commercial applications. *IEEE Comm. Mag.* (April).

Simon, M.K., Omura, J.K., Scholtz, R.A., and Levitt, B.K. 1985. *Spread Spectrum Communications* Vol. I, II, and III. Computer Science Press, Rockville, MD.

Sklar, B. 1988. *Digital Communications, Fundamentals and Applications.* Prentice-Hall, Englewood Cliffs, NJ.

Viterbi, A.J. 1995a. *CDMA: Principles of Spread Spectrum Communication.* Addison-Wesley, Reading, MA.

Viterbi, A.J. 1995b. Multiuser detection for CDMA systems. *IEEE Personal Comm.* (April).

Further Information

Additional information on spread spectrum communications is availalbe in the following texts and technical journals:

IEEE Transaction on Communications, Vols. Com-25, 28, and 30; special issues on spread spectrum and military communications.

IEEE Journal on Selected Areas in Communications, Vol. SAC-2, SAC-3, 8, 10; special issues on spread spectrum, mobile radio and military communications.

IEEE Transactions on Vehicular Technology, Vol. 40, May 1991; special issue on mobile radio.

Peterson, R.L., Ziemer, R.E., and Borth, D.E. 1995. *Introduction to Spread Spectrum Communications.* Prentice-Hall, Englewood Cliffs, NJ.

Viterbi, A.J. 1995. Multiuser detection for CDMA systems. *IEEE Personal Communications*, April.

12.7 Digital Coding Schemes

Oktay Alkin

12.7.1 Introduction

Two fundamental issues in assessing the performance of a digital communication system are:

- The reliability of the system in terms of accurately transmitting information from one point to another
- The rate at which information can be transmitted with an acceptable level of reliability

In an ideal communication system, information would be transmitted at an infinite rate with an infinite level of reliability. In reality, however, there are fundamental limitations affecting the performance of any communication system. No physical system is capable of responding to changes in an instantaneous manner, and the range of frequencies that a system could reliably handle is limited, thus leading to the concept of *bandwidth*. In addition, *random noise* affects any signal being transmitted through any communication medium. Finite bandwidth and additive random noise are two fundamental limitations that prevent us from achieving an infinite rate of transmission with infinite reliability, and a compromise must be found. What makes the problem even more challenging is the fact that the reliability and the rate of information transmission usually work against each other, that is, for a given system, a higher rate of transmission means a lower degree of reliability and vice versa. To affect the balance in our favor, we need to improve the efficiency and the robustness of the communication system. **Source coding** and **channel coding** are the means for accomplishing this task, and will be discussed in the following subsections. First, some background in **information theory** needs to be established.

Since we are concerned about transmitting information from one point to another, a mathematical means for measuring information is needed. Consider an information source that produces symbols from an alphabet

$$A = \{a_0, a_1, \ldots, a_{M-1}\} \tag{12.137}$$

with associated probabilities

$$\text{Prob}\,\{a_k\} = p_k, \qquad k = 0, \ldots, M-1 \tag{12.138}$$

subject to the constraints $p_k \geq 0$ for all k, and $\sum_{k=0}^{M-1} p_k = 1$. We will assume that the source generates symbols from the alphabet given in Eq. (12.138) at a rate of r symbols per second. We will further assume that subsequent symbols generated are statistically independent of each other. Such an information source is referred to as a *discrete memoryless source*. An example might be a digital computer or some kind of a data acquisition device.

If we were to observe the output of this information source for one symbol interval, our level of surprise after observing the symbol would depend on the probability of that symbol being generated. A symbol with low probability would cause a bigger surprise than a symbol with higher probability. Furthermore, if any symbol in the source alphabet has a probability of 1, observing that symbol would cause no surprise at all, and no information would be gained. Thus, it would intuitively make sense to associate the *amount of information* gained by observing a symbol with the *degree of surprise* caused by that observation. We will define the amount of information gained by observing the symbol a_k at the source output as

$$
\begin{aligned}
I(a_k) &= \log_2\left(\frac{1}{p_k}\right) \\
&= -\log_2(p_k)
\end{aligned}
\tag{12.139}
$$

expressed in *bits*. [Note in this context we are using the term *bit* as a unit of information. This should not be confused with the common use of this term in computer terminology as a *binary digit*. One *binary digit* does not necessarily contain one *bit of information*.]

Example 1

Consider a source that generates symbols from the alphabet $A = \{a_0, a_1\}$ with probabilities $p_0 = 0.7$ and $p_1 = 0.3$. If we observe the symbol a_0 at the source output, the information gained by this observation is

$$I(a_0) = -\log_2(0.7) = 0.5146 \text{ bits}$$

Similarly, the information gained after observing the symbol a_1 would be

$$I(a_1) = -\log_2(0.3) = 1.7370 \text{ bits}$$

To assess the efficiency of a source, we need to know the average amount of information expected to be gained from the observation of one source symbol. The **entropy** of a source is defined as

$$H(A) = E\{I(a_k)\}$$

$$= -\sum_{k=0}^{M-1} p_k \log_2(p_k)$$

(12.140)

and represents the expected amount of information per symbol. Note that entropy is a general characteristic of the source, and represents the expected average amount of information before an observation. Once the symbol is observed, the amount of information gained depends on the symbol, and can be computed as in example 1.

Example 2

Consider again the binary source in example 1. The entropy of the source is

$$H(A) = -0.7 \log_2(0.7) - 0.2 \log_2(0.2)$$

$$= 0.8813 \text{ bits/symbol}$$

Once an observation is made, the amount of information gained is either 0.5146 bits or 1.7370 bits depending on the symbol observed. If a large number of observations are made, however, the amount of information per symbol is expected to average around 0.8813 bits/symbol.

As the previous example illustrates, the entropy of a source is a function of only the symbol probabilities. For a binary source $p_1 = 1 - p_0$, and the entropy can be written as

$$H(A) = -p_0 \log_2(p_0) - (1 - p_0) \log_2(1 - p_0)$$

(12.141)

A graph of $H(A)$ as a function of p_0 is given in Fig. 12.86. Note that the entropy is maximum when $p_0 = p_1 = 0.5$, that is, when the uncertainty of the symbol to be generated is maximum. If an information source generates symbols from an alphabet of M distinct symbols, it can be shown that the entropy is maximum if all symbols of the alphabet are equally likely, that is

$$p_0 = p_1 = \cdots = p_{M-1} = \frac{1}{M}$$

(12.142)

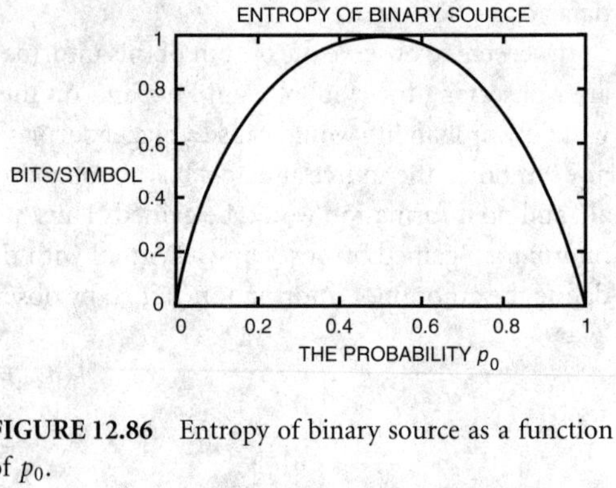

FIGURE 12.86   Entropy of binary source as a function of p_0.

Example 3

Consider a discrete memoryless source that generates symbols from the alphabet

$$A = \{a_0, a_1, a_2, a_3\}$$

(12.143)

with corresponding probabilities

$$P(a_0) = 0.50$$
$$P(a_1) = 0.30$$
$$P(a_2) = 0.15$$
$$P(a_3) = 0.05$$

If the output of the source is to be transmitted through a binary channel, it needs to be *encoded* in binary form. Let us assume that the following encoding scheme is used:

$$a_0 \Rightarrow 00$$
$$a_1 \Rightarrow 01$$
$$a_2 \Rightarrow 10$$
$$a_3 \Rightarrow 11$$

It can easily be shown that the entropy of this source is 1.6477 bits/symbol. All symbols are coded into 2-bits codewords.

Source Coding Theorem

Given a discrete memoryless source with entropy $H(\mathcal{A})$ and an average codeword length of L, it is possible to encode the output of the source in such a way that the resulting average codeword length can be made as close to the lower bound

$$L \geq H(\mathcal{A}) \tag{12.144}$$

as desired. Note that the theorem establishes a lower bound on L, and claims that it could be approached, however, we are not told how to accomplish this. Methods of source encoding will be the subject of the next subsection.

Example 4

Consider again the source in example 3. The average codeword length is significantly greater than the entropy of the source. Note that the symbol a_0 occurs with a higher probability than the other symbols, hence it would be wise to assign a shorter codeword to a_0 perhaps at the expense of longer codewords to some of the other symbols. Consider the alternative encoding scheme

$$a_0 \Rightarrow 0$$
$$a_1 \Rightarrow 10$$
$$a_2 \Rightarrow 110$$
$$a_3 \Rightarrow 111$$

It can be shown that the average codeword length for this case is 1.7 bits/symbol, significantly closer to the source entropy compared to the encoding scheme of example 3. Based on the source coding theorem, however, there is still room for further improvement, perhaps by using another coding scheme.

12.7.2 Source Coding

Examples 3 and 4 illustrate an important concept: The most intuitive coding scheme may not always be the most efficient. In example 3, each symbol was assigned a codeword of length 2. If N symbols are transmitted, the number of codewords to be transmitted would be $2N$. In example 4, symbols were assigned codewords of different lengths resulting in an average codeword length of 1.7. For large N, this would translate to about $1.7N$ codewords for the transmission of the same information. Thus, the coding scheme of example 4 allows us to get closer to the boundary set by the source coding theorem.

Most information sources generate signals that contain redundancies. For example, consider a page of text. If we want to code this text in binary so that it can be transmitted over a digital channel, we might opt for a scheme that assigns 7 bits to each symbol on the page including space and punctuation characters. (One example of such a coding scheme is American Standard Code for Information Interchange (ASCII).) If the page is set up to hold 60 rows with 80 characters in each line, transmitting a page of text would require transmitting 4,800 characters, or equivalently, 33,600 binary digits. It is very unlikely, however, that the page will contain 33,600 bits of information. Another example is a picture that is made up of pixels each of which represent one of 256 grayness levels. If we use a fixed coding scheme that assigns 8 binary digits to each pixel, a 100 by 100 picture of random patterns and a 100 by 100 picture that consists of only white pixels would both be coded into the same number of binary digits, although the latter has significantly less information than the former.

One simple method of source encoding is the *Huffman coding* technique, which is based on the idea of assigning a codeword to each symbol of the source alphabet such that the length of each codeword is approximately equal to the amount of information conveyed by that symbol. As a result, symbols with lower probabilities get longer codewords. Huffman coding is achieved through the following process:

- List source symbols in descending order of probabilities.
- Assign a binary-0 and a binary-1, respectively, to the last two symbols in the list.
- Combine the last two symbols in the list into a new symbol with its probability equal to the sum of two symbol probabilities.
- Reorder the list, and continue in this manner until only one symbol is left.
- Trace the binary assignments in reverse order to obtain the codeword for each symbol.

The following example should help clarify this procedure.

Example 5

Consider a discrete memoryless source that generates symbols from the alphabet

$$\mathcal{A} = \{a_0, a_1, a_2, a_3, a_4, a_5, a_6, a_7\} \tag{12.145}$$

with corresponding probabilities

$$P(a_0) = 0.46$$
$$P(a_1) = 0.18$$
$$P(a_2) = 0.12$$
$$P(a_3) = 0.08$$
$$P(a_4) = 0.07$$
$$P(a_5) = 0.04$$
$$P(a_6) = 0.03$$
$$P(a_7) = 0.02$$

Figure 12.87 illustrates the Huffman coding procedure for this source. Obtained codewords are

$$a_0 \Rightarrow 1$$
$$a_1 \Rightarrow 000$$
$$a_2 \Rightarrow 011$$
$$a_3 \Rightarrow 0010$$
$$a_4 \Rightarrow 0011$$
$$a_5 \Rightarrow 0101$$
$$a_6 \Rightarrow 01000$$
$$a_7 \Rightarrow 01001$$

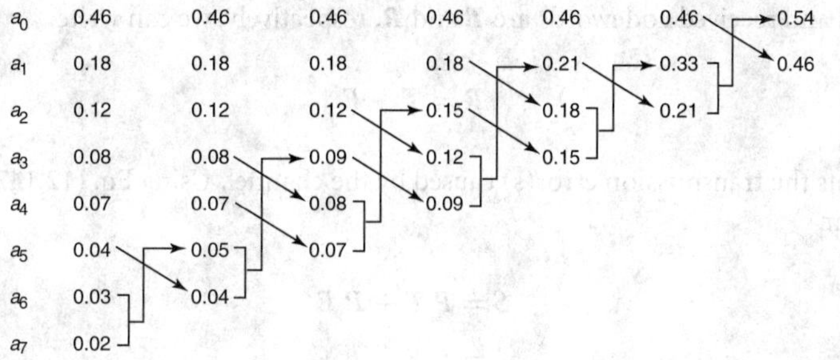

FIGURE 12.87   Huffman coding algorithm.

A tree diagram for decoding a coded sequence of symbols is shown in Fig. 12.88.

It can easily be verified that the entropy of the source under consideration is 2.3382 bits/symbol, and the average codeword length using Huffman coding is 2.37 bits/symbol.

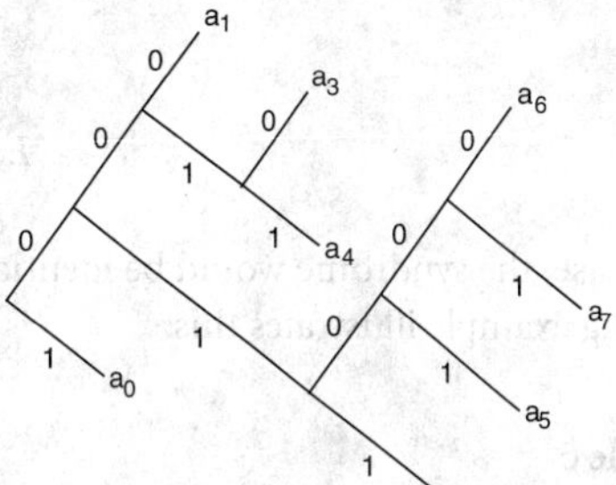

FIGURE 12.88   Huffman code tree for Example 5.

12.7.3   Forward Error Correction Coding

In the previous section, we discussed the need for removing redundancies from the message signal in order to increase efficiency in transmission. From an efficiency point of view, the ideal scenario would be to obtain an average wordlength that is numerically equal to the entropy of the source. From a practical perspective, however, this would make it impossible to detect or correct errors that may occur during transmission. Some redundancy must be added to the signal in a controlled manner to facilitate detection and correction of transmission errors. This is referred to as channel encoding.

A variety of techniques exists for detection and correction of errors. In this section, we will consider *linear block codes*. The term **block code** refers to the fact that the message signal is partitioned into blocks (codewords) of fixed length.

Let a codeword be represented as a vector in the form

$$T = (m_1 \, m_2 \, \cdots \, m_k \, c_1 \, c_2 \, \cdots \, c_{n-k})^T$$

where the leftmost k bits $m_1, m_2, \ldots, m_k$ are the message bits, and the following $n - k$ bits $c_1, c_2, \ldots, c_{n-k}$ are the parity bits added by the channel encoder. Parity bits are selected such that

$$h_{i,1}m_1 + h_{i,2}m_2 + \cdots + h_{i,k}m_k + c_i = 0 \qquad i = 0, \ldots, n - k \qquad (12.146)$$

where additions are done in modulo-2. In matrix form, Eq. (12.146) can be written as

$$P\,T = 0 \qquad (12.147)$$

where P is the *parity-check matrix* defined as

$$P = [H \; I_{n-k}]$$

$$= \begin{bmatrix} h_{1,1} & h_{1,2} & \cdots & h_{1,k} & 1 & 0 & \cdots & 0 \\ h_{2,1} & h_{2,2} & \cdots & h_{2,k} & 0 & 1 & \cdots & 0 \\ \cdots & & & & & & & \\ \cdots & & & & & & & \\ h_{n-k,1} & h_{n-k,2} & \cdots & h_{n-k,k} & 0 & 0 & \cdots & 1 \end{bmatrix} \qquad (12.148)$$

If the transmitted and received codewords are T and R, respectively, we can write

$$R = T + E \qquad (12.149)$$

where E represents the transmission error(s) caused by the channel. Using Eq. (12.147) and Eq. (12.149) together, we obtain

$$S = PT + PE$$
$$= PE \qquad (12.150)$$

The vector S is called the *syndrome*. In the absence of any bit errors, the syndrome should be zero.

Now, let us assume that the vector E has a binary 1 in the ith position indicating a bit error, that is

$$E = (0\,0\,\ldots,\,1\,\ldots\,0)$$

In this case, the syndrome would be identical to the ith column of the matrix parity-check matrix P. The following example illustrates this.

Example 6
The parity-check matrix for a particular code is

$$P = \begin{bmatrix} 1 & 1 & 1 & 1 & 1 & 0 & 0 & 0 \\ 1 & 0 & 1 & 0 & 0 & 1 & 0 & 0 \\ 1 & 1 & 1 & 0 & 0 & 0 & 1 & 0 \\ 1 & 0 & 0 & 1 & 0 & 0 & 0 & 1 \end{bmatrix}$$

If the bitstream 10011000 is received, the syndrome would be

$$S = \begin{bmatrix} 1 & 1 & 1 & 1 & 1 & 0 & 0 & 0 \\ 1 & 0 & 1 & 0 & 0 & 1 & 0 & 0 \\ 1 & 1 & 1 & 0 & 0 & 0 & 1 & 0 \\ 1 & 0 & 0 & 1 & 0 & 0 & 0 & 1 \end{bmatrix} \cdot \begin{bmatrix} 1 \\ 0 \\ 0 \\ 1 \\ 1 \\ 0 \\ 0 \\ 0 \end{bmatrix} = \begin{bmatrix} 1 \\ 1 \\ 1 \\ 0 \end{bmatrix}$$

This is clearly identical to the fourth column of the parity-check matrix, and indicates a possible bit error in the fourth bit position. We can not be absolutely sure since this scheme only detects single bit errors. In the case of multiple bit errors, the syndrome would be the sum of corresponding columns of the parity-check matrix. For example, the situation may also indicate that bits 2 and 6 are in error. On the other hand, in most communication systems, the probability of having multiple bit errors within one codeword is significantly smaller than the probability of a single bit error.

The set of equations in Eq. (12.146) can also be written in the alternative form

$$h_{i,1}m_1 + h_{i,2}m_2 + \cdots + h_{i,k}m_k = c_i \qquad i = 0,\ldots,n-k \qquad (12.151)$$

or in matrix form as

$$
\begin{bmatrix}
h_{1,1} & h_{i,2} & \cdots & h_{1,k} \\
h_{2,1} & h_{2,2} & \cdots & h_{2,k} \\
\cdots & & & \\
\cdots & & & \\
h_{n-k,1} & h_{n-k,2} & \cdots & h_{n-k,k}
\end{bmatrix}
\cdot
\begin{bmatrix}
m_1 \\
m_2 \\
\cdots \\
m_k
\end{bmatrix}
=
\begin{bmatrix}
c_1 \\
c_2 \\
\cdots \\
\cdots \\
c_{n-k}
\end{bmatrix}
\tag{12.152}
$$

Based on Eq. (12.152), a *generator matrix* can be constructed as

$$
G = \begin{bmatrix} I_{n-k} H \end{bmatrix}
$$

$$
=
\begin{bmatrix}
1 & 0 & \cdots & 0 \\
0 & 1 & \cdots & 0 \\
\cdots & & & \\
\cdots & & & \\
0 & 0 & \cdots & 1 \\
h_{1,1} & h_{i,2} & \cdots & h_{1,k} \\
h_{2,1} & h_{2,2} & \cdots & h_{2,k} \\
\cdots & & & \\
\cdots & & & \\
h_{n-k,1} & h_{n-k,2} & \cdots & h_{n-k,k}
\end{bmatrix}
\tag{12.153}
$$

so that codewords can be obtained from message bit sequences as

$$
T = G M
\tag{12.154}
$$

Defining Terms

Block codes: Error detecting codes that are obtained by appending special bits to data bits.

Channel coding: Adding redundancy into information in a controlled manner to facilitate detection and possible correction of errors that may occur during transmission.

Entropy: Average information gained per symbol by observing the output of an information source.

Error correcting codes: Codes that are added to data for locating erroneous bits and correcting them at the receiver end.

Information theory: The mathematical theory that deals with quantifying and statistically analyzing information.

Source coding: Encoding the output of an information source to remove redundancy.

References

Hamming, R.W. 1950. Error detecting and error correcting codes. *Bell Syst. Tech. J.* 29(April):147–160.

Hamming, R.W. 1980. *Coding and Information Theory*. Prentice-Hall, Englewood Cliffs, NJ.

Haykin, S. 1988. *Digital Communications*. Wiley, New York.

Lin S. and Costello, D.J., Jr., 1983. *Error Control Coding: Fundamentals and Applications*. Prentice-Hall, Englewood Cliffs, NJ.

12.8 Audio Compression Techniques

Fred Wylie

12.8.1 Introduction

The transmission and storage of digital signals present difficulties to professional audio users due to the large and expensive circuit bandwidth and computer storage space required to cope with such a signal. The CD and DAT formats, real-time 16-bit linear *pulse code modulation* (**PCM**) processes, each transfers data at 1.411 Mb/s for stereo and require circuit bandwidth of at least 700 kHz to avoid distorting the digital signal during transmission. In practice, additional overhead bits are added to the signal for channel coding, synchronization, and error correction, and this increases the bandwidth demands yet again. The commonly quoted bandwidth figure for distribution and transmission circuits is 1.5 MHz. Compare this with the requirement for two separate 20-kHz circuits to distribute the same stereo audio in its analog format. The mathematics quickly demonstrate that it would take around 630 megabytes of computer storage space to hold the contents of a single 60-min stereo CD or DAT tape.

The general result is that it is too costly and complex to fully implement a 16-bit linear PCM infrastructure, except over very short distances and within studio areas.

12.8.2 Digital Audio Data Compression

Significant developments in high speed digital signal processing (**DSP**) hardware have resulted in a number of alternative real-time, or in reality, near instantaneous digital compression coding systems offering sizeable reductions in data rates. These rates require significantly lower bandwidths for transmission and occupy less computer storage space and yet subjectively still deliver high quality audio.

Applications

Digital audio data compression systems are currently in use throughout the world with many applications. For broadcast telecommunications this includes RF studio transmitter links (**STL**) and other fixed digital cable links and satellite-based outside broadcast links; all of which can be adapted by selecting different data rates for either speech or high-quality music. New eras of broadcasting unfold with the introduction of digital audio broadcasting (**DAB**) and high-definition television (**HDTV**), both of which will use audio data reduction techniques.

Data compression can be used to increase the capacity of numerous audio storage products such as hard and magneto-optical (MO) disk-based digital audio recorders and work stations. A modest compression ratio of 4:1 enables a 1-h stereo CD to be stored on 160 megabyte (**MB**) of disk space and to store some 30 s of full bandwidth stereo on a 1.44 MB, 31/2-in floppy diskette. Cinema audiences are now being offered virtual audio splendor with audio surround systems incorporating data compression.

Definition of Real-Time Data Compression

Real-time data compression is the reduction of the bit rate of linear PCM with a minimum of processing delay and imperceptible loss of quality, either from signal distortion or injected noise.

Analog audio transmission requires fixed input and output bandwidths, and this implies that in a real-time compression system the quality, bandwidth, and distortion/noise level of both the original and the decoded output sound should not be subjectively different, thus giving the appearance of a lossless and real-time process.

Without exception all real-time bit rate reduction systems can be referred to as *lossy*. In other words the digital audio signal at the output is not identical to the input signal PCM data stream. However, some *adaptive differential pulse code modulation* (**ADPCM**) compression algorithms are to all intents and purposes lossless by losing as little as 2% of the original signal and others are in the *adaptive pulse code modulation* (**APCM**) or transform family of algorithms, which remove around 80% of the original signal.

A common application of data compression, which illustrates lossless compression, is the fax machine. This requires a variable bandwidth and some statistical knowledge of the incoming data. A blank page is processed much more quickly than a complicated page, the feed speed dynamically changing as the text detail changes. This adjusts the bandwidth of the scanner input to maintain a fixed 9600 Bd (9.6 kb/s) rate into the telecom line.

Redundancy and Irrelevancy

A complex linear PCM signal contains a great deal of information, some of which the human ear cannot hear and can, therefore, be deemed to be irrelevant. The same signal, depending on its complexity, contains information that is highly predictable and, therefore, can be made redundant.

Redundancy, measurable and quantifiable, can be removed in the coder and replaced again in the decoder, a process often referred to as statistical compression. On the other hand, irrelevancy, referred to as perceptual coding, once removed from the signal cannot be replaced and is irretrievably lost. This is entirely a subjective process with each proprietary algorithm using a different psychoacoustic model.

Critically perceived signals, such as pure tones, are high in redundancy and low in irrelevancy. They compress quite easily, almost totally a statistical compression process. Conversely, noncritically perceived signals, such as complex audio or noisy signals, are low in redundancy but high in irrelevancy. These compress easily in the perceptual coder but with the total loss of all of the irrelevancy content.

Human Auditory System

The sensitivity of the human ear is biased toward the lower end of the audible frequency spectrum, around 3 kHz. At 50 Hz, the bottom end of the spectrum, and 17 kHz, at the top end, the sensitivity of the ear is down by approximately 50 dB relative to that at 3 kHz (see Fig. 12.89). Additionally, there are very few audio signals, music or speech based, that carry fundamental frequencies above 4 kHz. Taking advantage of these characteristics of the ear, the structure of audible sounds and the redundancy content of the PCM signal is the basis used by the designers of the ADPCM or *predictive* range of compression algorithms.

Another well-known feature of the hearing process is that loud sounds will mask out quieter sounds at a similar or near frequency. This compares with the action of an automatic gain control, turning the gain down when subjected to loud sounds, thus making quieter sounds less likely to be heard. For example

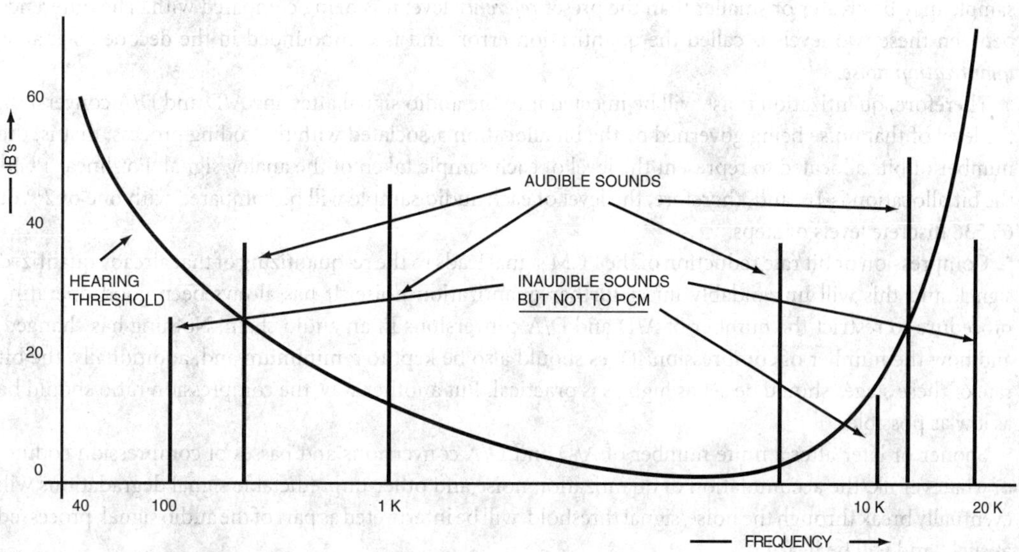

FIGURE 12.89 Frequency response of the human ear.

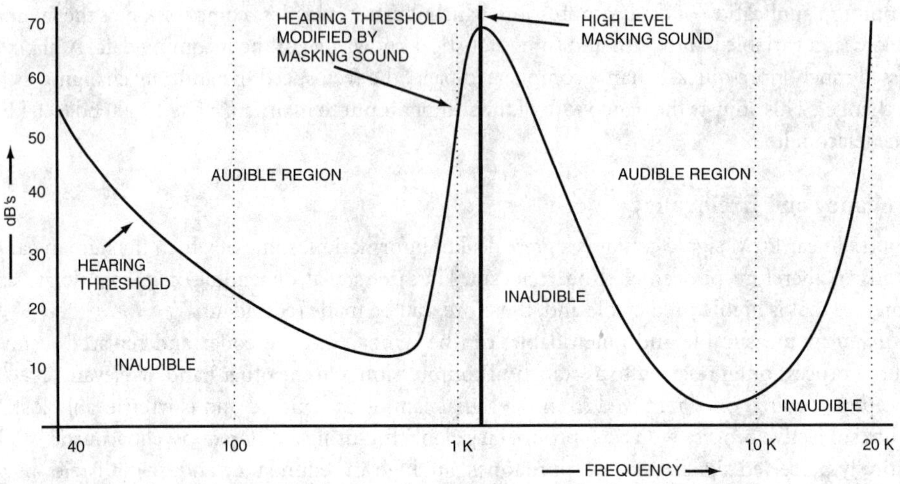

FIGURE 12.90 Masking effect of high-level sound.

(see Fig. 12.90), with a 1 kHz tone level of 70 dB it would require levels of greater than 40 dB at 750 Hz and 2 kHz for those frequencies to be heard. The ear also exercises a degree of temporal masking, being exceptionally tolerant of sharp transient sounds.

It is by mimicking these additional psychoacoustic features of the human ear and identifying the irrelevancy content of the PCM signal that the APCM, or *transform* range of low bit rate algorithms operate, adopting the principle that if the ear is unable to hear the sound then there is no point in transmitting it in the first place.

Quantization

Quantization is the process of converting an analog signal to its representative digital format or, as in the case with compression, the requantizing of an already converted signal. It is the limiting of a finite level measurement of a signal sample to a specific preset integer value. This means that the *actual* level of the sample may be greater or smaller than the preset *reference* level it is being compared with. The difference between these two levels is called the quantization error, and is compounded in the decoded signal as *quantization noise*.

Therefore, quantization noise will be injected into the audio signal after any A/D and D/A conversion, the level of that noise being governed by the bit allocation associated with the coding process, that is, the number of bits allocated to represent the level of each sample taken of the analog signal. For linear PCM the bit allocation is 16, and, therefore, the level of each audio sample will be compared with one of 2^{16} or 65,536 discrete levels or steps.

Compression or bit rate reduction of the PCM signal leads to the re-quantizing of that already quantized signal, and this will unavoidably inject further quantization noise. It has always been good operating procedure to restrict the number of A/D and D/A conversions in an audio chain. Nothing has changed, and now the number of compression stages should also be kept to a minimum and, additionally, the bit rate of these stages should be set as high as is practical. Put another way, the compression ratio should be as low as possible.

Sooner or later after a finite number of A/D and D/A conversions and passes of compression coding, of whatever ilk, the accumulation of quantization noise and other unpredictable signal degradations will eventually break through the noise/signal threshold, will be interpreted as part of the audio signal, processed as such, and will be heard.

Therefore, when considering using compression it is crucial to first select a compression coder–decoder (**codec**) best suited for the specific purpose, anticipating the possibility or requirement for any further compression. Second, retain as far as possible the signal in the PCM format between any further stages of compression, signal routing, or distribution. The latter may not be very practical, and additional D/A and A/D conversions between stages may be unavoidable.

Sampling Frequency and Bit Rate

The bit rate of a digital signal is defined by

$$\text{sampling frequency} \times \text{bit resolution} \times \text{no. of audio channels}$$

The rules regarding the selection of a sampling frequency are based on Nyquist's theorem, which states that the rate must be at least twice the highest audible frequency required at the analog output of the decoder. In particular, this ensures that the lower sideband of the sampling frequency does not encroach into the wanted base band audio. Objectionable and audible aliasing effects would occur if the two bands should overlap. In practice, the sampling rate is set slightly above twice the highest audible frequency, which makes the filter designs less complex and less expensive.

In the case of a stereo CD, with the audio signal having been sampled at 44.1 kHz, produces audio bandwidths of approximately 20 kHz for each channel. The resulting audio bit rate is 44.1 kHz $\times$ 16 $\times$ 2 = 1.411 M b/s. Similarly, an FM stereo broadcaster would generate a PCM signal of 1.024 Mb/s (32 kHz $\times$ 16 $\times$ 2) to deliver 15 kHz stereo audio.

Why 44.1-kHz Sampling

The unique 44.1-kHz sampling frequency used in the CD is historical. The early digital stereo audio recordings were made on VHS video recorders using the Sony F1 format. The equipment would only record a video signal with line sync signals; therefore, the digital audio signal had to be reformatted to resemble a video signal. Three 16-bit PCM words were carried on each active line of the video signal, whether NTSC or PAL, and quite remarkably, even with these differing line and picture standards a common sampling frequency of 44.1 kHz produced 20-kHz stereo audio on both systems.

Compression Ratio

Compression algorithms come with differing and often selectable bit rates related to a combination of sampling frequency and compressed bits per PCM sample (compression ratio). For example, a compression ratio of 6:1 can be translated as 2.66 b/PCM sample and with a sampling frequency of 48 kHz produces a bit rate of (2.66 $\times$ 48 kHz) = 128 kb/s per mono channel capable of delivering greater than 20-kHz audio bandwidth.

The same FM stereo broadcaster distributing a linear PCM signal at 1.024-Mb/s bit rate would have the option of using a compression coder offering a modest 4:1 compression ratio and the same sampling frequency of 32 kHz. The coding process would generate a new reduced output bit rate of 256 kb/s (for stereo), and this when subsequently decoded would produce the same 15-kHz bandwidth audio.

Higher compression ratios up to 18:1 can reduce the PCM bit rate even further. This may be satisfactory for the final delivery of audio or the capture of a single shot audio but if further compression is anticipated, even at lower compression ratios, then this more aggressive compression may render the signal more vulnerable to noise and distortion breakthrough and in the case of a stereo signal may, alter the stereo image, depending on the program content.

The misuse of compression can lead to difficulties, particularly with tandem or multiple passes, and bandwidths or noise floors defined by an earlier stage of aggressive compression cannot be subsequently improved by later less aggressive stages.

12.8.3   Processing and Propagation Delay

Compression coding is not a wholly real time but a near instantaneous process and, with any of todays popular algorithms, will introduce processing delay. This delaying of the output audio with reference to the input can range from a few milliseconds in some codecs to tens of milliseconds in others. This can be a crucial factor in the selection of a codec for a two-way interview or interactive applications where delays in excess of 20 ms can be problematical. In a one-way audio capture or delivery application the delay is of no consequence.

Propagation delay is unavoidable and is due to the length of the satellite, extremely long terrestrial circuit paths, or mixes of both. A 1000-km totally digital telecommunication path in full duplex mode has a propagation delay of 3 ms in each direction, and this is comparable with two people in a conversation while standing approximately 1 m apart. Whether in a short or long haul two-way hook-up, it is clear that the introduction of a compression codec with a long processing delay will have a dramatic effect.

History of Data Compression

It was in the early 1960s that data compression was first used by two Consultative Committees for International Telegraph and Telephone (**CCITT**) G711 compliant systems: A law in Europe and μ law in the U.S. The sampling frequency was 8 kHz and the then 12-bit PCM format was reduced to 8 b, delivering an audio bandwidth of 4 kHz at a bit rate of 64 kb/s, ideal for telephone speech. Eventually, 64 kb/s was adopted as the bit rate for a single ISDN B, basic rate telecommunication channel.

The unique NICAM system was developed by the British Broadcasting Corporation (BBC) during the 1970s to make more efficient use of its leased transmitter and interstudio communications network, which was carrying linear PCM signals and distributing channels of audio in mono and stereo. NICAM reduces a 32-kHz sampled, 14-bit linear signal down to 10 b, a modest compression. The technique analyzes 1-ms blocks of 32 PCM samples and compresses them into one of five ranges, depending on the amplitude of the largest sample in the block and signaled by a 3-bit scale factor word. Some of the most significant and least significant bits (**MSBs** or LSBs) of each digital word are discarded depending on which range the block is allocated. Nicam 3, with 338 kb/s allocated to each 15-kHz audio channel enables 24 audio channels to be distributed in the same carrier bandwidth where previously only 13 linear channels could be accommodated, an 80% increase.

12.8.4 Modern Compression Techniques

Common compression formats are based on

- *Predictive* or ADPCM time-domain coding
- *Transform* or APCM frequency-domain coding.

It is in their approach to dealing with the redundancy and irrelevancy of the PCM signal that these techniques differ.

Subband coding variations of both of these techniques figure highly within the audio industry. The present range of popular algorithms process the PCM signal in narrow discrete frequency bands, splitting the frequency range of the signal into 2, 4, 32, 256, 576, or 1024 subbands, dependent on the system being used. The bit allocation is then either dynamically adapted or fixed, for each subband. In doing so the frequency-domain redundancies within the audio signals can be exploited, allowing for a reduction in the coded bit rate, compared to PCM, for a given signal fidelity. Since the signal energies in each subband are unequal at any instant in time this will also lead to spectral redundancies. By altering the short-term coding resolution in each band, according to the energy of the subband signal the quantization noise can be reduced across all bands. This process compares favorably with the noise characteristics of a PCM coder performing at the same overall bit rate.

On its own, subband coding incorporating PCM in each band is capable of providing an improvement in performance compared with that of full band PCM coding, both being fed with the same complex, constant level input signal. This improvement or gain is defined as the ratio of the variations in quantization errors generated in each case while both are operating at the same transmission data rate. The gain increases as the number of subbands increases and varies with the extent of the complexity of the input signal. Figure 12.91 illustrates the variation of subband gain with the number of subbands for four essentially stationary, but differing, complex audio signals.

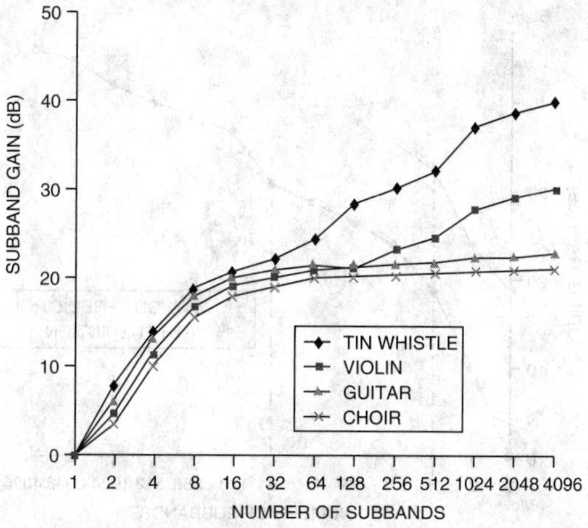

FIGURE 12.91 Variation of subband gain with the number of subbands.

In the practical implementations of compression codecs a number of factors tend to limit the number of subbands employed. First, the level variation of normal audio signals leads to an averaging of the energy across bands and a reduction in the coding gain. The coding or processing delay and computational complexity are two further problems that must also be addressed.

The key issue in the analysis of a subband framework is in determining the likely improvement associated with high-quality audio and in also determining the relationships between that gain, the number of subbands, and the response of the filter bank used to create those subbands.

12.8.5   Subband ADPCM Coding

G722 was the first commercial quality compression algorithm in this category and was approved by the CCITT in the mid-1970s. Although originally designed to compress 14-bit PCM sampled at 16 kHz, it is still very much in use for 16-bit PCM, 7-kHz bandwidth, speech only applications. The algorithm processes the PCM signal in 2 subbands with 6-bit resolution for the 50 Hz–4 kHz subband, and 2 b for the other 4–7 kHz subband.

Apart from subband coding the other key components of the subband ADPCM algorithm, which collectively achieve compression, are

- Linear prediction
- Adaptive quantization

Linear Prediction

Any reduction in potential gain from the use of just four subbands can be compensated for by linear prediction, which aims to remove spectral redundancies remaining in the subbands. With the inclusion of linear prediction, the total performance of the coder can easily match the subband gain associated with the use of many more subbands, without resorting to an increase in filter complexity or coding delay.

Redundancy is removed by subtracting a predicted signal, derived from coder lookup tables, from the incoming subband signal. This action creates a difference signal, which is then requantized. If the prediction is accurate enough, then the magnitude of the difference or error signal will be quite small, much less than the original subband signal. This allows the bit allocation to be effectively reduced. The degree to which the subband signal is attenuated is termed the *prediction gain* and represents an improvement in the coded S/N ratio compared with linear PCM.

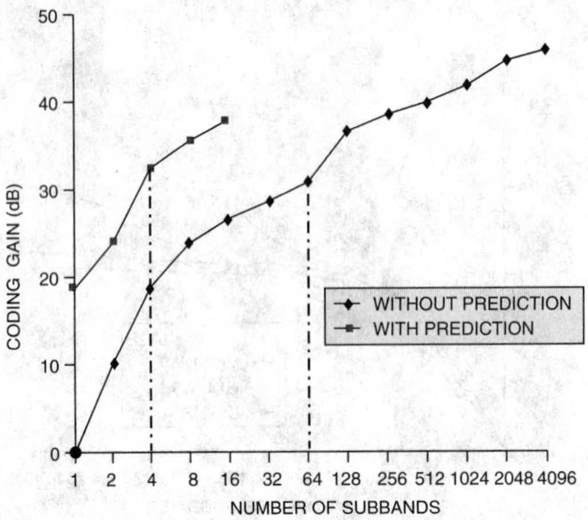

FIGURE 12.92   Subband gain with and without prediction.

In the decoder, a signal, identical to that predicted and removed in the coder and similarly derived from lookup tables, is added back again to the now decoded error or difference signal. This enables the reconstruction of the original linear PCM signal with a minimal loss of information.

The success of linear predictive coding is highly dependent on the predictability or correlation of the subband signals. Prediction gain is reduced in the higher subbands and for most musical instruments the gain is mainly in the first two bands. For the trombone signal the 4-band coder with prediction is seen to give comparable gain to a 64-band coder without prediction. (See Fig. 12.92.)

Adaptive Quantization

Audio exhibits relatively slowly varying energy fluctuations with respect to time, and adaptive quantization exploits this feature by dynamically adjusting the quantizer step size to match the level of the incoming signal. This is a continous process and provides an almost constant and optimal signal/quantization noise ratio across the operating range of the quantizer.

Time-domain algorithms implicitly model the hearing process and indirectly exploit a degree of irrelevancy by accepting that the human ear is less sensitive at higher frequencies. This is achieved in the subband derivative by allocating more bits to the lower frequency bands. This is the only application of psychoacoustics exercised in ADPCM. All of the information contained in the PCM signal is processed, audible or not; no attempt is made to remove irrelevant information.

12.8.6   Subband APCM Coding

The APCM or transform processor acts in a similar fashion to the automatic gain control in the ear, continually adjusting in response to the dynamics of the incoming audio signal at allfrequencies.

Transform coding takes a time block of signal and analyzes it for frequency, energy, and identification of the irrelevancy content. Again, to exploit the spectral response of the ear, the frequency spectrum of the signal is divided into a number of subbands and the most important criteria is coded with a bias toward the more sensitive low frequencies. At the same time, using psychoacoustic masking techniques, those frequencies that it is assumed will be masked by the ear are also identified and removed. Therefore, the data generated describe the frequency content and the energy level at those frequencies, with more bits being allocated to the higher energy frequencies than to those with lower energy.

The larger the time block of signal being analyzed the better the frequency resolution will be and the greater the amount of irrelevancy identified. The penalty, however, is an increase in coding delay and a decrease in temporal resolution. A balance has been struck with advances in perceptual coding techniques and psychoacoustic modeling leading to increased efficiency.

This hybrid arrangment of working with time-domain subbands and simultaneously carrying out a spectral analysis can be achieved by using a dynamic bit allocation process for each subband.

Additionally, some of these systems exploit the significant redundancy between stereo channels by a technique known as joint stereo coding. Once the common information between left and right channels of a stereo signal has been identified it is only coded once, thus reducing the bit rate demands yet again.

Each of the subbands has its own defined masking threshold. The output data from each of the filtered subbands is requantized with just enough bit resolution to maintain adequate headroom between the quantization noise and the masking threshold for each band. In more complex coders, for example, ISO/MPEG layer 3, any spare bit capacity is utilized by those subbands with the greater need for increased masking threshold separation.

It is the maintenance of these signal to masking threshold ratios that is crucial if further compression is contemplated in any post production or onward transmission process. In extremis, one of the limits for data compression is reached when the quantization noise rises above the masking threshold and is detected as part of the wanted audio, processed as such, and is then heard. Out of limits harmonic and phase distortion are other limiting factors.

Defining Terms

ADPCM: Adaptive differential pulse code modulation.
APCM: Adaptive pulse code modulation.
ASPEC: Adaptive spectral perceptual entropy coding.
ATC: Adaptive transform coding.
ATRAC: Adaptive transform acoustic coding.
BER: Bit error rate.
CCITT: Consultative committee for international telephone and telegraph.
CODEC: Coder-decoder.
CRC: Cyclic redundancy code.
DAB: Digital audio broadcasting.
DCC: Digital compact cassette.
DSP: Digital signal processor.
EBU: European broadcasting union.
FFT: Fast fourier transform.
HDTV: High-definition television.
ISDN: Integrated services digital network.
ISO: International Standards Organisation.
ITU: International Telecommunications Union.
MB: Megabyte.
MD: MiniDisc.
MDCT: Modified discrete cosine transform.
MO: Magneto-optical.
MPAC: Multichannel perceptual audio coding.
MPEG: Moving Pictures Experts Group.
MSB: Most significant bit.
MUSICAM: Masking pattern universal subband integrated coding and multiplexing.
NICAM: Near instantaneously companded audio multiplex.
PAC: Perceptual audio coding.
PASC: Precision adaptive subband coding.
PCM: Pulse code modulation.
PLCC: Plastic leadless chip carrier.
PSTN: Public switched telephone network.
QMF: Quadrature mirror filter.
RAM: Random access memory.

ROM: Read only memory.
SCFSI: Scale factor select information.
SEDAT: Spectrum efficient digital audio technology.
SMR: Signal-to-mask ratio.
STL: Studio to transmitter link.

References

Brandenburg, K. and Stoll, G. 1992. ISO-MPEG-1 audio: A generic standard for coding of high quality
    digital audio. In *92nd AES Convention*. Revised 1994. *J. Aud. Eng. Soc.* 42(10).
Smyth, S. 1992. Digital audio data compression. *Broadcast Engineering* (Feb.):52–60.
Stallings, W. 1992. *ISDN and Broadband ISDN*, 2nd ed. Macmillan, New York.

12.9   Aural Noise Reduction Systems

William J.J. Roberts and Yariv Ephraim

12.9.1   Introduction

There are many examples of voice and aural systems that have to battle with the detrimental effects of
noise. Pay phones are often located in noisy environments, such as airports and near busy streets. Air–
ground communications suffer from both loud cockpit noise and noise introduced by the communications
channel. Long-distance communications suffer primarily from the noise arising in the channel. Cellular
telephone systems have to contend with noise introduced by traffic, wind, and other people talking.
Hearing aid users must struggle with acoustic distortions due to reverberations in a room and ambient
noise. The performance of speech recognition systems suffers greatly as noise is introduced into the
system. Researchers have studied ways to reduce the effects of noise since the 1940s. In this presentation,
we characterize the different types of noise, and review some of the techniques that are used to reduce the
deleterious effects of noise. The techniques discussed will generally be those that have been useful to the
problem of noise reduction of speech signals, but they may be applied to the noise reduction of other types
of signals. The advantages and disadvantages of each technique will be discussed, and the assumptions,
upon which each technique relies, will be stated. Some of the techniques require special circumstances
such as two or more sensors, or the ability to gain access to the signal prior to its corruption by noise, and
these circumstances will be listed where appropriate. Because of the limited number of references allowed
in this publication, the references given will generally be of a tutorial nature and will not necessarily reflect
the originators of a particular approach.

12.9.2   Noise Sources

Signals that interfere with a desired signal are referred to as *noise*. Noise may be deterministic in nature
or it may be random in nature. An example of **deterministic noise** is the 50- or 60-Hz line hum due
to the coupling between an audio system and any nearby power lines. **Random noise** includes thermal
noise, due to the random thermal motion of charge carriers, and shot noise, due to the discrete nature of
charge particles. Often noise can look like the clean signal itself. An example of this is a second competing
speaker in a speech recognition system. This is a particularly odious type of noise as it is statistically
similar to the clean signal itself. Typical noise sources include cafeterias, traffic, music, competitive speech,
reverberations, and communication channels.

    Noise can also be related in some manner to the clean signal. It may be independent of the clean signal
or it may depend statistically on the clean signal. The noise can be additive, as in the case of a noisy
communications channel, it may be multiplicative, as in the case of speckle in radar systems, or it may
have some other relationship to the signal, such as convolutional reverberation noise.

The amount of noise in a signal can be characterized by the **signal-to-noise ratio (SNR)**, which is the ratio of the power of the signal to the power of the noise. The SNR is usually expressed in decibels.

Thermal noise is very well modeled by the Gaussian distribution implying that a histogram plot of thermal noise values looks somewhat like that of the normal or Gaussian distribution. This is a very important property as it permits this type of noise to be treated as a mathematical entity allowing the development of closed-form solutions. As with thermal noise there is a mathematical distribution that is useful in modeling speech. It is the **hidden Markov model (HMM)** and it is also mathematically tractable and is employed in many speech systems, see Rabiner (1989).

In this chapter, noise generally will be assumed to be uncorrelated with, and additive to, the signal. This assumption is not particularly restrictive and holds for many situations of interest. Generally the noise will also be assumed to be **white**. This is also not a restrictive assumption, since if the noise is not white it can always be made so with an appropriate linear transformation, see Van Trees (1968).

12.9.3 Performance Measures

Noise reduction systems are designed by optimizing a performance measure; however, in the case of speech, the most perceptually meaningful distortion measure is not known. Furthermore, measures correlated with speech perception are not easily quantified in mathematical terms. Hence, mathematically tractable measures are often used in designing noise reduction systems, and actual performance is evaluated on human listeners. Two aspects of the noisy speech signal that attempts are often made to improve are speech quality and intelligibility. Quality is a subjective measure, which is expressed in terms of how pleasant the signal sounds, or how much effort is required by the listeners in order to understand the message. Intelligibility, which is an objective criterion, measures the amount of information that a listener can extract from the signal. A given signal may be of high quality and low intelligibility or vice versa. The aspect of the signal to which we wish to devote the most attention depends on the application. High quality is generally preferred for telephone systems where intelligibility is already high. For an air–ground communications system, where mistakes may be disastrous, quality may be sacrificed for an improvement in intelligibility. The goal of speech enhancement for listening purposes may be expressed as that of maximum quality improvement without significant decreases in speech intelligibility.

The most mathematically tractable distortion measure is the *mean square error* (MSE), originally proposed by Gauss in the late 1700s. Its utility for speech processing, however, has been questioned (Lim and Oppenheim, 1979). For example, changing the short-term phase of a speech signal may results in a signal that sounds very similar to the original but the MSE between the two signals could be very large.

A modification of the MSE criterion which allows some masking of the input noise has been found to be useful. Here the system is designed so that the spectrum of the residual noise is similar to that of the speech. This is illustrated in Fig. 12.93 for a typical spectrum of a vowel. This figure shows the shaped spectrum of the residual noise, as well as the spectrum of white noise with the same power. A system that results in this shaped noise provided higher SNR in the perceptually important spectral regions such as in the vicinity of the second and

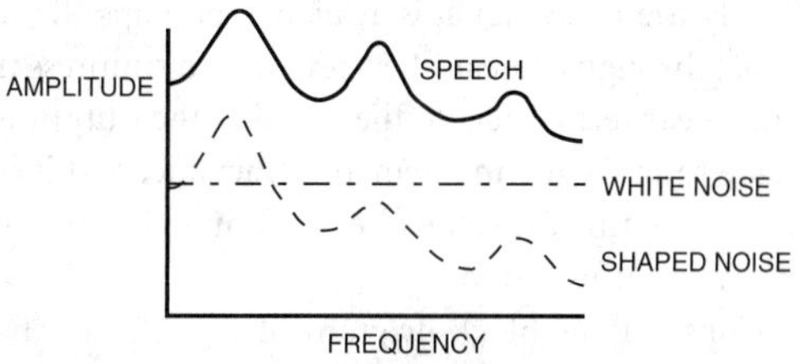

FIGURE 12.93 Shaped noise spectra.

third formants, that is, the peaks in the spectral envelope of the signal. This allows spectral components of the noise in these regions to be fully or partially masked by the corresponding spectral components of the speech signal. This interesting phenomenon is explained by studies of the human aural system that suggest that humans may be more tolerant of a particular frequency of noise when there is a large-signal component near that frequency and correspondingly less tolerant of noise in frequency bands where the signal energy is low. The noise near the large-signal components is effectively masked by those components. Significant progress has been made in understanding and utilizing noise masking in designing speech processing systems (Jayant, Johnston, and Safranek, 1993).

12.9.4 Configurations

The applied noise reduction technique for a particular situation depends on the physical configuration of the system. For instance, the clean signal may be available for preprocessing prior to its degradation by noise. Access to a reference noise source, recorded by a second microphone, may be available for cancellation of the main noise source. In many other situations, we are required to suppress noise from a noisy signal where neither reference to the noise nor access to the clean signal is possible. This is the most difficult situation and has been the subject of intensive research. To illustrate these concepts, consider a pilot communicating with a control tower. The control tower may be relatively noise free and thus the signal from the air-traffic controller may be preprocessed prior to being transmitted to the pilot. In the noisy cockpit, there will be a microphone to record the pilot's speech and if a second microphone is located such that it receives cockpit noise only, noise cancellation techniques are possible. Back in the control tower, however, only the noisy signal from the pilot is available for processing. Although all of these configurations are addressed in this chapter, the main thrust is toward the single microphone situation.

In some situations we may have training data allowing us to learn some statistical aspects of the speech signal. In this case, the learned statistics may be used in designing the noise reduction systems.

12.9.5 Accomplishments

The problem of noise reduction in speech has been an active area of research for over three decades, and numerous solutions with various degrees of success have been developed, see Lim and Oppenheim (1979), Ephraim (1992), and the references therein. Currently there are no noise reduction systems that achieve improvements in both quality and intelligibility for people with normal hearing and where intelligibility is measured by the DRT using nonsense syllables. However, many of the approaches to be described here yield significant improvements in quality with only some sacrifice in intelligibility. The *spectral subtraction* approach is perhaps the most successful as it is conceptually simple, easy to implement, and yields good results. Most techniques will result in quality improvement at high SNR; however, the problem of enhancing speech at low SNR remains unsolved. Noise reduction for speech coding and speech recognition has met with more success, and algorithms that improve both quality and intelligibility have been developed, see Juang (1991) and the references therein.

12.9.6 Techniques

Linear Filtering

If only the noisy signal is available, perhaps the most straightforward approach is to pass the degraded signal through a linear filter designed to suppress the noise without significantly affecting the clean signal. The linear term refers to the fact that the output is a linear function of the input. The linear constrain is made to make the mathematics tractable, and it often provides effective filters. If the system is operated in discrete time, and the filter is of finite impulse response (FIR), then the filter can be implemented as a matrix multiplication.

Consider the block diagram of Fig. 12.94. The output from the linear filter is used to provide an estimate of the desired response or clean signal. The estimation error is the difference between the estimate and the actual desired response.

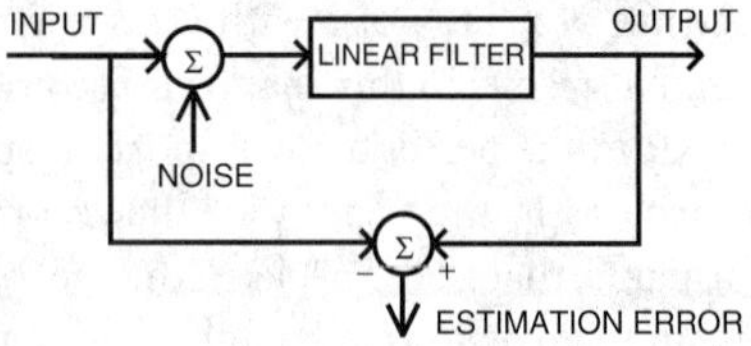

Some considerations that are necessary in the design of the filter include whether the filter is finite duration impulse response or infinite duration impulse response (IIR). Also the choice of optimality

FIGURE 12.94 The linear filtering problem.

criterion must be addressed. The design of such filters is in the domain of optimal filter design, the theory of which was pioneered by Norbet Wiener and Andrei Kolmologrove in the 1940s and extended by Kalman and others in the 1960s. Wiener's solution for the optimal linear estimate of a signal in the minimum MSE sense requires the solution of an integral equation known as the *Wiener–Hopf equation*. The resulting linear filter is optimal in the minimum MSE sense for the case of Gaussian signals and additive Gaussian noise. Kalman's solution requires description of speech in terms of a dynamic system and is also optimal in the minimum MSE sense. In both cases a priori information about the signal and noise, such as the spectra and the parameters of the dynamical system are required. Implementation of these classical solutions will require estimation from either the noisy data or from *training data*. Furthermore, these estimates will need to be periodically updated to accommodate the fact that speech is not strictly stationary.

Lim and Oppenheim (1979) question the effectiveness of Wiener filtering for speech due to the reliance on the mean squared error criterion. In Ephraim and Van Trees (1995), linear estimators that minimize signal distortion subject to a permissible noise energy or spectrum were designed. These estimators allow noise masking and have performed better than plain Wiener filtering techniques. They constitute Wiener filters with an adjustable noise spectrum or energy and were implemented using signal and noise statistics estimated from the noisy signal. The implementation of this approach using the signal subspace technique will be discussed in the following.

Spectral Subtraction

Spectral subtraction (Lim and Oppenheim, 1979) is currently the standard filtering approach for speech enhancement as it is simple, intuitive, easy to implement, and provides reasonable results. This approach operates on a frame by frame basis where each frame consists of 20–30 ms of speech samples. In the spectral subtraction method, the discrete Fourier transform (DFT) of the signal vector is computed, and those spectral components whose variance is smaller than that of the noise are nulled. The surviving spectral components are modified by a gain function. The resulting set of nulled and modified spectral components constitute the spectral components of the enhanced signal. The inverse DFT is applied to get the enhanced signal estimate. A block diagram of the spectral subtraction process is shown in Fig. 12.95.

The choice of gain function is motivated by different perceptual criteria. An often used criterion is the minimum MSE between spectral components of the clean signal and its estimate. The result is a Wiener gain function that constitutes the ratio of the variances of the clean and noisy spectral components. Since the Wiener gain factor was proved to be too mild, families of related gain factors have been used (Lim and Oppenheim, 1979), for instance

$$\alpha(k) = \left(\frac{|Z(k)|^a - b E\{V(k)^a\}}{|Z(k)|^a} \right)^c$$

Here $Z(k)$ and $V(k)$ denote the kth spectral components of the noisy signal and noise process, respectively, and $a > 0, b > 0, c > 0$. The Wiener gain function is obtained when $a = 2, b = 1, c = 1$.

The spectral subtraction approach is very popular as it is easy to implement, makes minimal assumptions about the signal and noise, and does not require training data. It results in reasonably clear enhanced signals; however, a major drawback is the residual noise has annoying tonal characteristics referred to as *musical noise*. This effect is due to the random nulling of the spectral components of the enhanced signal.

Many variations of spectral subtraction algorithms have been developed, see Lim and Oppenheim (1979), Ephraim (1992), and references therein. A notable approach was proposed by McAulay and Malpass (1980) who modified the basic estimator to take into account the fact that speech is not always present in

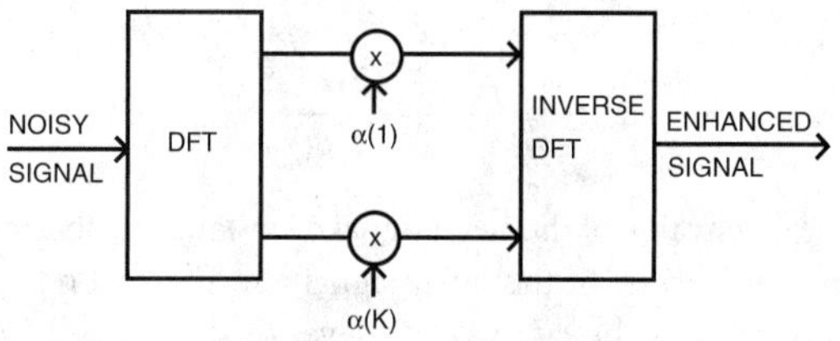

FIGURE 12.95 Block diagram of the spectral subtraction process.

the noisy observations. A weighted combination of the estimate for the clean signal given the noisy signal and the clean signal given that the noisy signal is noise only is applied. Since the optimal estimate of the clean signal given that the clean signal is absent is zero, the resulting estimator is simply the product of the estimator for the clean signal given that the signal is present and the posterior probability of the signal being present given the noisy signal. Significant improvement of the spectral subtraction estimator was achieved using this approach.

Signal Subspace Approaches

Another possible approach for speech enhancement is to utilize the existence of the **signal subspace** of speech signals (Ephraim and Van Trees, 1995). The signal subspace technique is motivated by the observation that sample clean speech covariance matrices have some eigenvalues that are practically zero. The implication of this fact is that K-dimensional speech vectors are confined to a *subspace* of the K-dimensional Euclidean space. The white noise covariance, however, has equal eigenvalues and, hence, noise vectors occupy the entire K dimension Euclidean space. Thus the vector space of the noisy signal can be decomposed into a signal plus noise subspace and a noise only subspace. Speech enhancement can be accomplished by removing the **noise subspace** and masking the remaining noise in the signal subspace.

The eigenvalues of the noisy speech are given by the addition of the eigenvalues of the clean speech and the eigenvalues of the noise. Thus a typical plot of the ordered noisy signal eigenvalues would look like Fig. 12.96.

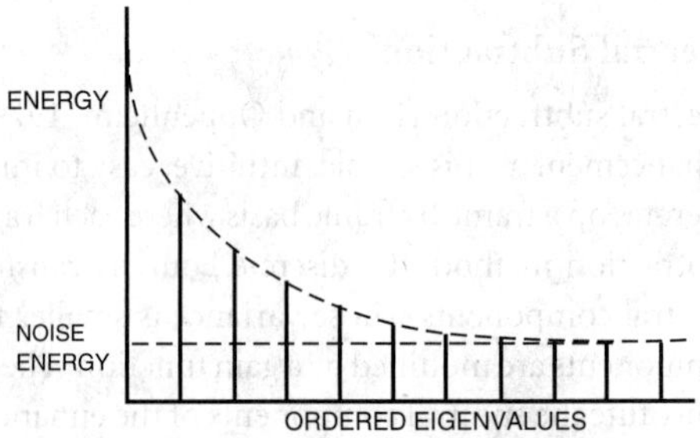

FIGURE 12.96   Ordered noisy eigenvalues.

It can be shown that the eigenvectors corresponding to eigenvalues that are above the noise energy span the signal subspace. Thus, in theory, the signal subspace can be identified from eigendecomposition of the covariance matrix of the noisy signal using the noise variance. In practice these second-order statistics are not available, and they must be estimated from the noisy data. Hence, more sophisticated approaches must be used to identify the signal subspace, see Ephraim and Van Trees (1995).

If the noisy signal is projected onto the signal subspace, a least squares (LS) estimate is obtained. The projection can be accomplished by applying the *Karhunen–Loeve transform* (KLT) (Van Trees, 1968) to the noisy signal and nulling transform components obtained from eigenvectors in the noise subspace.

It is possible to further reduce the noise, at the expense of some signal distortion, by modifying the projected signal. If the linear estimate that minimizes signal distortion for a given spectrum of residual noise is used, then an estimator as shown in Fig. 12.97 results. Here $\alpha(k)$ represents the gain function applied to the kth spectral component. An example of a useful $\alpha(k)$ is given by

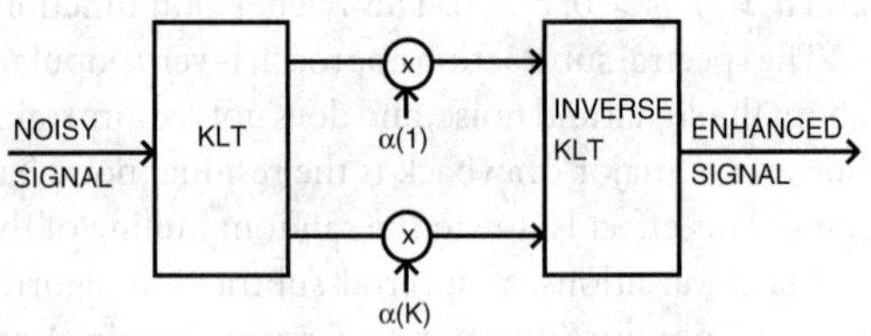

FIGURE 12.97   Block diagram of gain function modified KLT components.

$$\alpha(k) = \left(\frac{\lambda_y(k)}{\lambda_y(k) + \sigma_w^2}\right)^{\gamma/2}$$

where $\lambda_y(k)$ is the kth sample eigenvalue of the clean signal covariance as obtained by subtracting the noise variance σ_w^2 from the sample eigenvalue of the noisy signal covariance, and $\gamma \geq 1$ is an experimentally determined constant that determines the suppression level of the noise as well as the resulting signal distortion. If M denotes the signal subspace dimension then $\alpha(k) = 0$ for $k > M$. More aggressive noise suppression can be obtained by using $\alpha(k) = \exp\{-\gamma\sigma_w^2/\lambda_y(k)\}$.

A performance evaluation of the subspace technique appears in Ephraim and Van Trees (1995) where it was found that a majority of listeners preferred speech that had been processed with this techniques to speech process by spectral subtraction.

The signal subspace approach and the spectral subtraction approach are closely related (see Fig. 12.95 and Fig. 12.97). They mainly differ in the transform applied to the noisy signal. Spectral subtraction uses the DFT and the subspace approach uses the KLT. The conditions under which the DFT is equivalent to the KLT are known (Gray, 1977) and occur when the noisy signal covariance **matrix** is **circulant**. For this case spectral subtraction is a signal subspace approach and is optimal in the same sense as used in designing the filter of Fig. 12.97. If the speech in a given vector has a **Toeplitz** covariance **matrix**, that is, this vector may be a portion of a **stationary process**, then it can be shown that spectral subtraction estimates asymptotically coincides with the linear estimator of Fig. 12.97 if both use the Wiener gain function. In this case, spectral subtraction is an asymptotically optimal linear estimator in the minimum MSE sense.

The KLT depends on the second-order statistics of the noisy signal and is known to be computationally intensive. The DFT is computationally much less intensive and is a signal independent transformation.

The signal subspace has been the subject of intensive research in the statistical literature where approximations to the KLT, in the form of wavelet transforms, are being investigated. Under certain conditions, the wavelet transform was shown to be optimal in the minimax sense, see Donoho (1995).

Model-Based Approaches

Model-based speech enhancement utilizes a statistical model of the signal and noise to optimize a performance measure. For example, minimization of the MSE can be performed by implementing the conditional mean estimator. The models are estimated using training data. Empirical distributions or parametric models can be used. Using the sample distributions of the clean signal and noise to implement the conditional mean estimator is conceptually simple and can be shown to be asymptotically optimal (Porter and Boll, 1984). Since the conditional mean is a function of the noisy signal, however, the sample average must be calculated for each newly observed vector of the noisy signal.

Based on both theoretical knowledge and experimental observations, parametric models of the source and noise can be chosen. The model parameters are then estimated from training data. This method has some important advantages:

- If the model chosen accurately reflects the true underlying model for the data, then the chosen model will represent any sample function of the source and not just those in the training set. The sample densities of the training data only represents samples that are in the training set.
- Generally the chosen model has only a small number of parameters to estimate and, thus, the estimates can be expected to be reliable.
- If the chosen model is mathematically tractable, we can find closed-form expressions for the desired estimators.

The most popular statistical model for speech signals is the hidden Markov model, a tutorial on which can be found in Rabiner (1989). This model assumes that each speech vector is generated from a finite sequence of states, each of which is characterized by a Gaussian process with a given mean and covariance. Transitions from one state to another are made in a Markovian manner. Iterative algorithms are know for model training. This model has been used in Ephraim (1992) in designing speech enhancement systems. The major disadvantages of this technique are that its performance depends on statistical similarity between the training and test data and its implementation is computationally intensive.

Another example of a model based system is the one proposed by Ephraim and Malah (1984) in which the spectral components of speech and noise in a given vector are modeled as statistically independent Gaussian random variables. The variance of a given spectral component of the speech signal varies from frame to frame and is estimated from the noisy signal. Under these assumptions, the minimum MSE estimate of the short-time spectral amplitude is derived. This approach outperforms spectral subtraction as it provides enhanced signals with similar clarity but with drastically reduced musical noise.

Note that in these two techniques, speech is not assumed to be Gaussian but rather a mixture of Gaussian processes. In the HMM system, each mixture component has a fixed mean and covariance, whereas in the minimum MSE short-time spectral amplitude estimation system, the mean and variance are estimated from the noisy data.

Model based approaches made significant progress in the speech recognition and enhancement problems in the 1980s. A comprehensive treatment of this subject can be found in the tutorial paper by Ephraim (1992).

Signal Preprocessing

In some situations, such as when speech is generated in a clean environment and then transmitted to listeners in a noisy environment, the signal can be preprocessed to increase its robustness to noise (Lim and Oppenheim, 1979). The technique in Niederjohn and Grotelueschen (1976) is to first apply high-pass filtering to the clean signal, normalize the amplitude of the filtered signal so that it remains constant within a certain range, and finally restore the original power to the clean signal. These operations emphasize spectral and temporal components of the speech signal that are known to be important for understanding speech, such as consonants and spectral component in the vicinity of the second and third formants. This approach achieves significant improvement in the intelligibility of the noisy signal.

Adaptive Noise Cancellation

Adaptive noise canceling makes use of an auxiliary or reference input in addition to the noisy signal from the primary sensor. If a reference for the noise is desired, the auxiliary microphone is placed at a position in the noise field where the signal is weak or undetectable. This scheme operates on the basis of two main assumptions:

1. The signal and noise at the input to the primary sensor are uncorrelated.
2. The noise input to the reference sensor is correlated with the noise component at the primary input.

The reference signal is filtered to obtain an estimate of the noise, which is then subtracted from the primary sensor signal. As a result, the noise in the noisy signal is attenuated or eliminated by cancellation. The filtering of the reference signal is done adaptively allowing the filter parameters to be adjusted automatically according the dynamics of the noise. The filter design may make little use of a priori knowledge of the signal. The filter parameters are adjusted based on the overall output of the noise canceler. Ordinarily, it is inadvisable to subtract noise from a received signal because of the possibility of a disastrous increase in the average power of the output noise. However, in circumstances where adaptive noise cancellation is appropriate, and proper provisions are made, levels of noise reduction may be achieved that are not attainable by other means (Widrow et al., 1975). An adaptive noise cancellation is illustrated in Fig. 12.98.

The most common filter adaptation algorithm is the *least-mean-square* (LMS) algorithm. The LMS algorithm adjusts the filter parameters in the opposite direction to the gradient of the squared amplitude

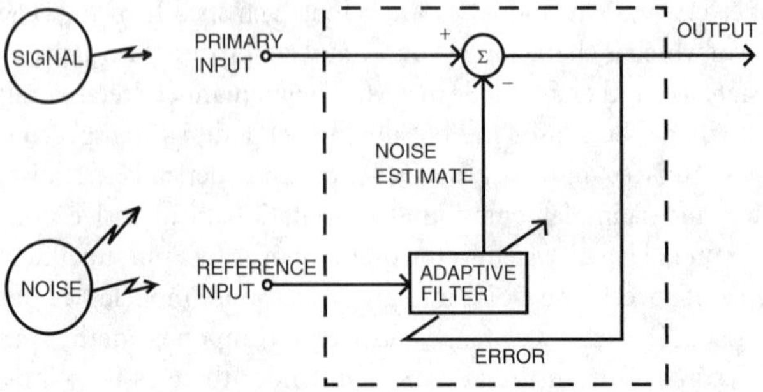

FIGURE 12.98   The adaptive noise cancellation concept.

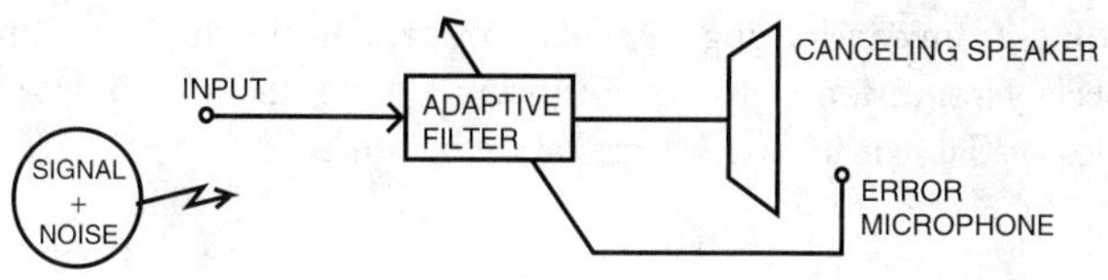

FIGURE 12.99   Active noise cancellation scheme.

of the error signal. The LMS algorithm is simple and yields good performance when an appropriate noise reference exists.

In some applications the noise reference may be created from the main noisy signal. For example, if the noise is white, a delayed version of the noisy signal will decorrelate the noise and leave the signal in both inputs correlated. In this case, the signal may be predicted and subtracted from the primary input. The predicted signal is used as the signal estimate. The major limitation is its slow rate of convergence which adversely affects its usefulness in some applications of speech signals.

The earliest work on adaptive noise cancelers can be traced to the mid 1960s to Kelly of Bell Telephone Laboratories, who first proposed the use of an adaptive filter for echo cancellation of speech signals. Kelly's contribution is cited in the book by Haykin (1991). Widrow et al. (1975) extended the concept and built a device to cancel 60-Hz interference at the output of an electrocardiograph amplifier and recorder. Lim and Oppenheim (1979) discuss the use of adaptive noise cancellation on speech and find that, at some SNR, the technique actually decreases the intelligibility of the speech signal.

Active Noise Cancellation

In active noise cancellation, the noisy input signal is adaptively filtered to produce a signal that is supplied to a canceling microphone. The output from the canceling microphone destructively interferes with the noise (Oppenheim et al., 1994). A block diagram appears in Fig. 12.99.

The purpose of the error microphone is to measure the unwanted noise field after destructive interference. The adaptive filter takes this measurement and produces an estimate of the noise process, which is then used to adjust the filter parameters so as to improve the cancellation effect. In general, there may be a multiple error microphones producing an improved noise estimate; however, Oppenheim et al. (1994) presents a scheme using only a single input sensor. The approach is evaluated using both computer generated noise and with recordings of aircraft noise.

Comb Filtering

Comb filtering capitalizes on the periodicity of voiced speech signals. It uses **pitch** information to design a comb filter that passes the harmonics of the signal while rejecting noise between these harmonics. This and other related techniques are discussed in Lim and Oppenheim (1979). The main drawback with this approach is that accurate estimates of the voiced speech pitch are required and such estimates are difficult to obtain from noisy speech.

12.9.7   Comments

Some of the major approaches for noise reduction from speech signals were reviewed. Emphasis of the physical circumstances where each is applicable and the theoretical assumptions upon which each is based were considered. Aural noise reduction systems have been an active area of research for many decades. The theory of Weiner and Kolmologrove was advanced in the 1940s and has been applied for many different speech models since then. However, the utility of the MSE criterion upon which these methods are based has been questioned for speech. In addition, these methods require knowledge of the spectra of speech which, due to the fact that speech is not strictly stationary, are difficult to obtain. Thus approaches, which subscribe a parametric model to the speech signal, have arisen. The MSE criterion has also been applied in the spectral domain to yield the successful spectral subtraction technique. The theoretical justification for spectral subtraction comes from considering the signal subspace of speech signals. Improved performance can be obtained in special circumstances where, for instance, separate noise only signals are available, or where the clean signal is available prior to its degradation by noise.

Defining Terms

Circulant matrix: A Toeplitz matrix in which every row beginning from the second is obtained from the proceeding row through a cyclic shift.
Deterministic noise: Noise that is of known form but with one or more unknown parameters.
Hidden Markov model (HMM): A statistical model consisting of several subsources or states controlled by a Markovian process.
Noise subspace: A complimentary subspace to the signal subspace that contains noise only.
Pitch period: Fundamental period of periodic portions of speech.
Random noise: Noise that fluctuates in an unpredictable manner.
Signal subspace: A subspace of the Euclidean space R^K that contains both signal information and noise.
Signal-to-noise ratio (SNR): Ratio of the signal power to the noise power, usually expressed in decibels.
Stationary process: A process whose probability law does not change with time.
Toeplitz matrix: A matrix with all diagonal values equal.
White noise: Noise exhibiting a flat spectrum.

References

Donoho, D.L. 1995. Denoising by soft thresholding. *IEEE Trans. Information Theory* 41(3):613–627.
Ephraim, Y. 1992. Statistical model based speech enhancement systems. *Proc. IEEE* 80(10):1526–1555.
Ephraim, Y. and Malah, D. 1984. Speech enhancement using a minimum mean-square error short time spectral amplitude estimator. *IEEE Trans. Acoustics, Speech, and Signal Processing* 32(6):1109–1121.
Ephraim, Y. and Van Trees, H.L. 1995. A signal subspace approach for speech enhancement. *IEEE Trans. on Speech and Audio Processing* 3(4):251–266.
Gray, R.M. 1977. Toeplitz and circulant matrices II. Stanford Electron. Lab. Tech. Rept. 6504-1, April. Stanford Univ., Stanford, CA.
Haykin, S.S. 1991. *Adaptive Filter Theory.* Prentice-Hall, Engelwood Cliffs, NJ.
Jayant, N., Johnston, J., and Safranek, R. 1993. Signal compression based on models of human perception. *Proc. IEEE*, 81(10):1385–1422.
Juang, B.H. 1991. Speech recognition in adverse environments. *Computer Speech and Language* 5:275–294.
Lim, J.S. and Oppenheim, A.V. 1979. Enhancement and bandwidth compression of noisy speech. *Proc. IEEE* 67(Dec.):1586–1604.
McAulay, R.J. and Malpass, M.L. 1980. Speech enhancement using a soft-decision noise suppression filter. *IEEE Trans. Acoustics, Speech, and Signal Processing* 28(April):137–145.
Niederjohn, R.J. and Grotelueschen, J.H. 1976. The enhancement of speech intelligibility in high noise levels by high-pass filtering followed by rapid amplitude compression. *IEEE Trans. Acoustics, Speech and Signal Processing* 24(Aug.):202–207.
Oppenheim, A.V., Weinstein, E., Zangi, K.C., Feder, M., and Gauger, D. 1994. Single-sensor active noise cancellation. *IEEE Trans. Speech and Audio Processing* 2(2):285–290.
Porter, J.E. and Boll, S.F. 1984. Optimal estimators for spectral restoration of speech. *Proc. Int. Conf. Acoust., Speech, Signal Processing* March 18A.2.1–18A.2.4.
Rabiner, L.R. 1989. A tutorial on hidden Markov models and selected applications in speech recognition. *Proc. IEEE* 77(Feb.):257–286.
Van Trees, H.L. 1968. *Detection, Estimation and Modulation Theory* (Part I). Wiley, New York.
Widrow, B. Ulover, J.R., McCool, J.M., Kaunitz, J., Williams, C.S., Hearn, R.H., Zeidler, J.R., Dong, E., and Goodin, R.C. 1975. Adaptive noise cancellations: Principles and applications. *Proc. IEEE* 63(Dec.):1692–1716.

Further Information

A further treatment of noise reduction applied to speech can be found in:

Boll, S.F. 1992. Speech enhancement in the 1980s: Noise suppression with pattern matching. In *Advances in Speech Processing*, eds. S. Furui and M.M. Sonhdi. Marcel Dekker, New York.

Makhoul, J., Crystal, T.H., Green, D.M., Hogan, D., McAulay, R.J., Pisoni, D.B., Sorkin, R.D., and Stockham, T.G. 1989. *Removal of Noise from Noisy-Degraded Speech Signals*. National Academy Press, Washington DC.

Further information about speech processing can be obtained in the journal *IEEE Transactions on Speech and Audio Processing*, and the annual *Proceedings of the IEEE International Conference on Acoustics, Speech, and Signal Processing* (ICASSP).

12.10 Video Compression Techniques

Gopal Lakhani

12.10.1 Demand for Video Compression

Virtually all applications of video and visual communication deal with enormous amount of data. Because of the volume of data, compression is an integral part of all video applications. A compression system reduces the volume of data by exploiting **spatial** and **temporal redundancies** and by eliminating the data, which cannot be displayed suitably by the associated display or imaging devices. Its main objective is to retain as little data as possible, but sufficient to reproduce the original images without causing much distortion.

A compression system consists of the following components:

- *Digitization, sampling, and segmentation:* Conversion of analog signals on a regular grid of picture elements; division of video input first into frames and then into blocks.
- *Redundancy reduction:* Decorrelation of data into fewer useful data using some invertible transformation techniques.
- *Entropy reduction:* Representation of digital data using fewer bits by dropping lesser significant information. This component causes distortion and it is the main contributor in *lossy* compression.
- *Entropy coding:* Assignment of codewords (bit strings) of shorter length to more likely image symbols and codewords of longer length to lesser likely symbols in order to minimize the average number of bits needed to code an image.

This section presents the core compression techniques of widely known video standards. The structure of this compression system is shown in Fig. 12.100.

12.10.2 Video Image Representation

For the purpose of image processing, digitized video can be treated as a sequence of frames (*images*) with each frame represented as an array of picture elements (*pixels*). For color video, a pixel is represented by three primary components: red (R), green (G), and blue (B). For effective coding, the three color components are converted to another coordinate system, called YUV, where Y denotes the luminance (brightness), and U and V, called the *chrominance*, denote the strength and vividness of the color. This conversion is described as

$$Y = 0.3R + 0.6G + 0.1B$$
$$U = B - Y$$
$$V = R - Y$$

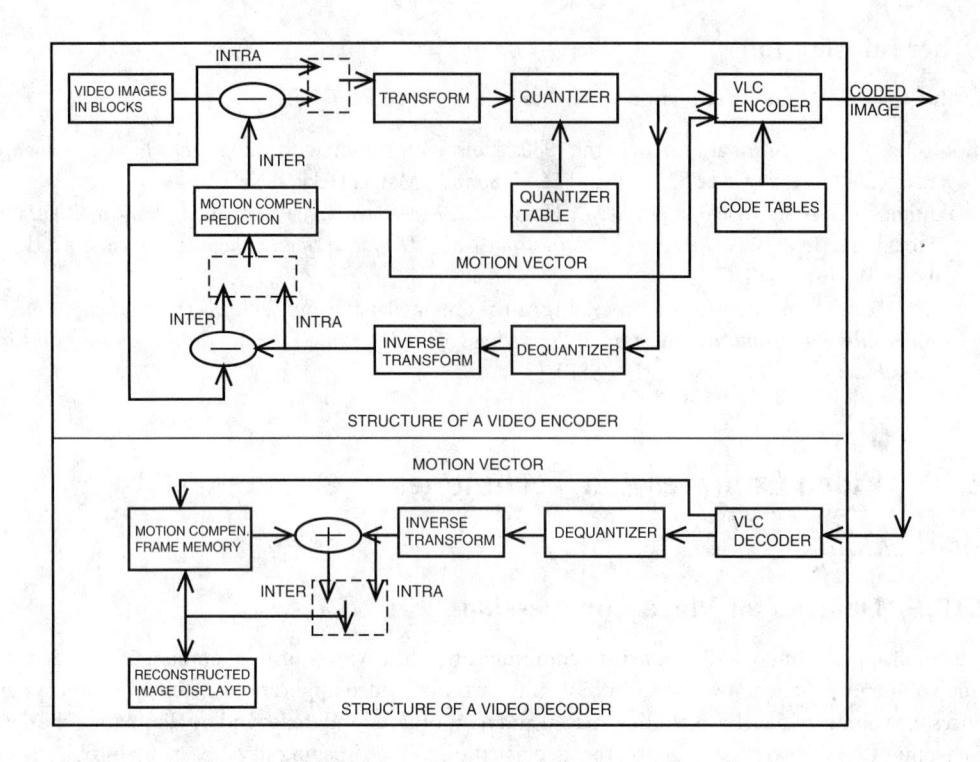

FIGURE 12.100   Structure of a typical coder and decoder.

A color coordinate system, followed in many compression standards, is *YCrCb*; the *Cr* and *Cb* components are obtained as follows:

$$Cr = (V/1.6) + 0.5$$
$$Cb = (U/2) + 0.5$$

There is a higher order of redundancy in the *Cr* and *Cb* components. A popular method is to subsample the chrominance components 2:1 along each dimension, resulting in 4 luminance values, but only one *Cr* and one *Cb* value for each 2 × 2 array of the image. The luminance component is divided into four components, labeled as *Y*1, *Y*2, *Y*3, and *Y*4. The subsampled *Cr* and *Cb* data is obtained by averaging the chrominance values of the four pixels in the array. Figure 12.101 shows this subsampling pattern. Thus, a region of 16 × 16 pixels of the original 24-b frame can be represented by four *Y* and one of each *Cr* and *Cb* blocks, each expressed as a 8 × 8 array of 8-b integers. This group of the six blocks is called the *minimum coded unit* (MCU) since these are stored and decoded together for efficient computing. A 352 pixels per

FIGURE 12.101   2:1 Subsampling of chrominance pixels.

line by 288 line frame can be segmented into 396 (22 × 18) MCUs. Note that this subsampling alone reduces the total size of the original data by one-half.

12.10.3 Scalar Quantization of Image Data

Scalar quantization is the process of mapping the input to a small number of output levels. Since the small number of levels can be coded using shorter codes, the code size of the image is reduced. The quantization and dequantization is performed according to the following equations:

$$\text{Quantization:} \quad \hat{X} = \text{round}\left(\frac{X}{Q}\right)$$

$$\text{Dequantization:} \quad \bar{X} = \hat{X} * Q$$

In these equations, X denotes the input and Q is a quantizer step size. The coder codes $\hat{X}$ as a representation of X. The difference, $X - \bar{X}$ is the loss due to quantization. A careful selection of quantizer steps is very important since it results in high compression, but at the same time it also introduces small losses in the reconstructed image.

12.10.4 Intraframe Compression Techniques

Before a frame is quantized, it is desirable to eliminate redundant data in the frame. There are many techniques for reducing the spatial redundancy, but by far, **DPCM** and **discrete cosine transformation-** (**DCT-**) based coding are the most widely used techniques for intraframe compression. (Rabbani and Jones (1991) discuss many image compression techniques. See Further Information.)

Differential Pulse Code Modulation (DPCM)

This compression method encodes the difference of the video signals (pulses). It can be applied for both the **intraframe** coding, which exploits the spatial redundancy, and the **interframe** coding, which exploits the temporal redundancy. In the intramode, the difference is formed between the values of two neighboring pixels of the same frame, and in the intermode, it is between the value of the same pixel on two consecutive frames. In either mode, the value of the target pixel is predicted using the reconstructed values of the previously coded neighboring pixels and this value is subtracted from the original value to form the differential image. The differential image is then quantized and encoded. Figure 12.102 presents the structure of a DPCM **codec**. Its computation is described in the following for intraframe coding.

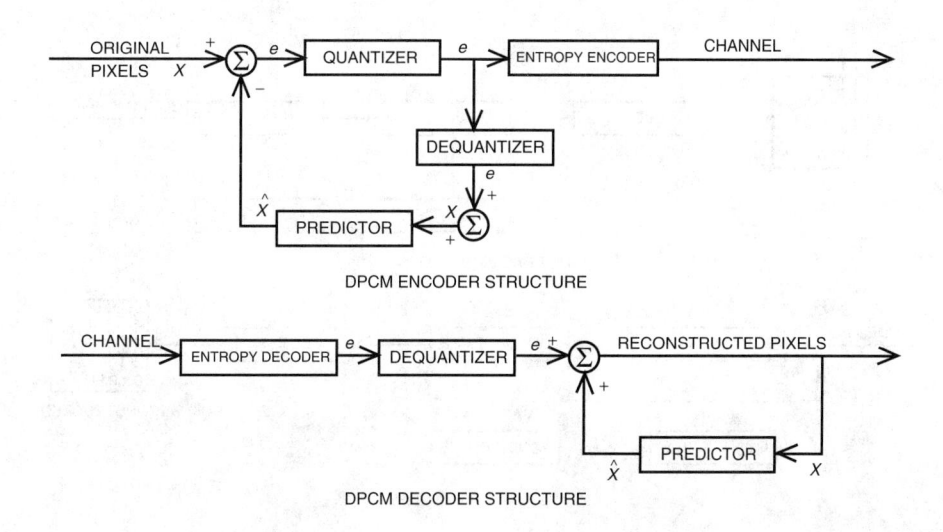

FIGURE 12.102 DPCM system block diagram.

In Figure 12.102, X denotes the actual value of the target pixel and $\hat{X}$ denotes the predicted value that the decoder would compute using some prechosen criterion. Here, e denotes the difference, $X - \hat{X}$, and $\bar{e}$ denotes its quantized value, which when dequantized is denoted by $\tilde{e}$. In the coded image, X is represented by $\tilde{e}$. During decompression, $\tilde{e}$ is decoded, and the reconstructed value $\tilde{X}$ of X is obtained by adding $\tilde{e}$ to $\hat{X}$. $\tilde{X}$ is used to display X, which the coder also uses for all future use of X.

Effectiveness of DPCM depends on the prediction method. For intraframe coding, a common method is to use the values of three previously coded pixels A, B, and C, where A is immediate to the left of X, C is above X and B is above A. A simple predictor is $\hat{X} = A - B + C$, and a predictor based on some statistical derivation is $\hat{X} = 0.90A - 0.81B + 0.90C$.

DCT-Based Transform Coding

The main advantage of DCT is that it *decorrelates* the pixels (i.e., converts statistically dependent pixel values into independent coefficients) very efficiently, and in the process, it packs the signal energy of the image block onto a small number of coefficients. Another big advantage of using DCT is that a number of its fast implementations are available. Rao and Yip (1990) provide in-depth study of DCT for image compression. A block diagram of a DCT-based codec is shown in Fig. 12.103.

Let B denote the DCT matrix and B^T be its transpose. For 8×8 size image blocks, the matrix of B is defined by:

$$B(i, j) = 0.5C \cos\{2i\pi(2j + 1)/16\}$$

where row and column indices, i and j vary from 0 to 7, respectively, and

$$C = \begin{cases} 1/\sqrt{2} & \text{if } i = 0 \\ 1 & \text{otherwise} \end{cases}$$

Note that DCT is an orthogonal transformation and, therefore, $B^{-1} = B^T$.

Let f denote a 8×8 image block and F be its transformed block. Then f and F can be obtained from each other by applying the forward DCT (FDCT) and inverse DCT (IDCT), respectively, which are defined as

- FDCT: $F = BfB^T$
- IDCT: $f = B^TFB$

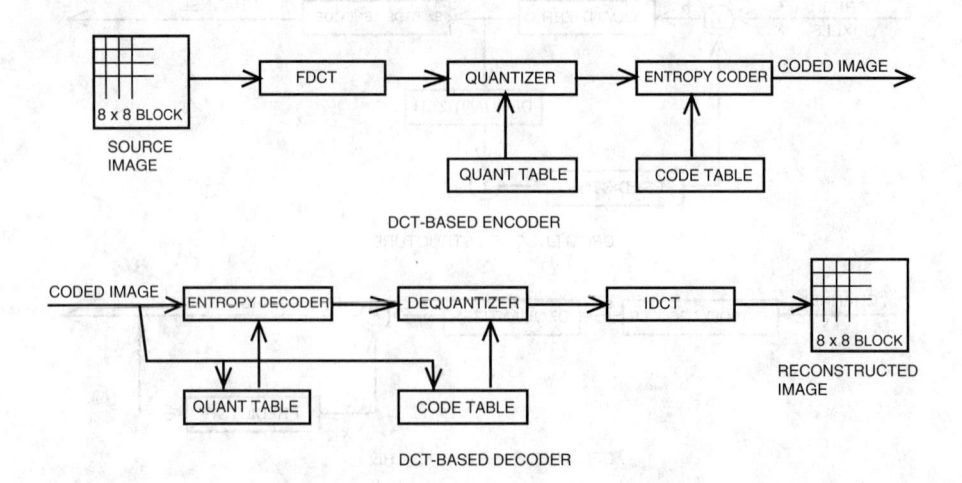

FIGURE 12.103 A block diagram of DCT-based image compression system.

To demonstrate effectiveness of DCT, Fig. 12.104 presents a complete example. It considers a luminance block f taken from a real image, the corresponding F, the block $\mathcal{F}$ of DCT coefficients of f obtained by quantizing the elements of F by 16 uniformly, and the reconstructed image $\hat{f}$ of f.

Note that all but five elements of F are very small, which signifies that DCT packs the signal energy onto a few coefficients and decorrelates the pixel values efficiently.

The element $F(0,0)$ is called the *DC coefficient*, since it denotes the weighted average of the pixel values in the block, and the remaining elements are called *AC coefficients*, since each coefficient is a difference of weighted sum of pixels in various subblocks.

To encode only the nonzero elements of $\mathcal{F}$, a commonly used method is to scan the matrix in a zig-zag order and produce a list of pairs, each containing a run-length of zeros and the next nonzero DCT coefficient. The procedure is called the run-length encoder. For $\mathcal{F}$, the following list will be produced.

$$f = \begin{bmatrix} 85 & 86 & 87 & 89 & 91 & 92 & 94 & 94 \\ 96 & 96 & 96 & 95 & 95 & 94 & 94 & 94 \\ 113 & 111 & 109 & 105 & 101 & 98 & 95 & 94 \\ 126 & 124 & 119 & 114 & 108 & 102 & 98 & 96 \\ 130 & 128 & 124 & 118 & 112 & 107 & 102 & 100 \\ 125 & 124 & 121 & 117 & 114 & 110 & 108 & 106 \\ 115 & 115 & 114 & 114 & 113 & 113 & 113 & 112 \\ 107 & 108 & 109 & 111 & 113 & 114 & 116 & 116 \end{bmatrix}$$

$$F = \begin{bmatrix} 854 & 31 & 0 & 0 & 0 & 0 & -1 & 0 \\ -63 & 0 & 0 & 0 & 0 & 0 & 0 & 0 \\ -38 & -44 & 0 & 0 & 0 & 0 & -1 & 0 \\ 0 & 0 & 0 & 0 & 0 & 0 & 0 & 0 \\ 0 & 0 & 0 & 0 & 0 & 0 & 0 & 0 \\ 0 & 0 & 0 & 0 & 0 & 0 & -1 & 0 \\ 0 & 0 & 0 & 0 & 0 & 0 & 0 & 0 \\ 0 & 0 & 0 & 0 & 0 & 1 & 0 \end{bmatrix}$$

$$\mathcal{F} = \begin{bmatrix} 53 & 2 & 0 & 0 & 0 & 0 & 0 & 0 \\ -4 & 0 & 0 & 0 & 0 & 0 & 0 & 0 \\ -2 & -3 & 0 & 0 & 0 & 0 & 0 & 0 \\ 0 & 0 & 0 & 0 & 0 & 0 & 0 & 0 \\ 0 & 0 & 0 & 0 & 0 & 0 & 0 & 0 \\ 0 & 0 & 0 & 0 & 0 & 0 & 0 & 0 \\ 0 & 0 & 0 & 0 & 0 & 0 & 0 & 0 \\ 0 & 0 & 0 & 0 & 0 & 0 & 0 & 0 \end{bmatrix}$$

$$\hat{f} = \begin{bmatrix} 84 & 85 & 87 & 89 & 91 & 93 & 94 & 94 \\ 95 & 95 & 95 & 95 & 94 & 94 & 94 & 93 \\ 112 & 110 & 108 & 104 & 100 & 96 & 93 & 92 \\ 125 & 123 & 118 & 112 & 106 & 100 & 95 & 93 \\ 130 & 127 & 123 & 117 & 110 & 104 & 100 & 97 \\ 125 & 123 & 120 & 116 & 112 & 109 & 106 & 104 \\ 114 & 114 & 114 & 113 & 113 & 113 & 112 & 112 \\ 107 & 107 & 109 & 111 & 113 & 115 & 116 & 117 \end{bmatrix}$$

FIGURE 12.104 An illustration of DCT computation.

$$(53)(0,2)(0,-4)(0,-2)(4,-3)(0,0)$$

$(0,0)$ is added to the list to denote the end of block (EOB).

This list is finally encoded in the form of bit strings using an entropy encoder, such as the Huffman or arithmetic coder, which have been widely used in data compression (Amsterdam, 1986).

The process of decompression is completely the reverse of the compression process. The coded bit strings are first decoded, and the dequantization is performed by multiplying each element by the chosen quantizer step. The resulting block is then inverse DCT transformed to obtain the reconstructed block. The difference e of f and $\hat{f}$ denotes the loss due to compression. In the compression literature, the loss is expressed as the *root mean square error* (RMSE), and it is defined as

$$\text{RMSE} = \frac{1}{64}\sqrt{\sum_{i=0}^{7}\sum_{j=0}^{7} e^2(i,j)}$$

For f in Fig. 12.104, the RMSE is 0.047.

12.10.5 Interframe Coding

A simple method for interframe coding is to compute the difference of a frame with the previous frame and then apply the DCT-based method to code the difference image. This method exploits temporal redundancy, but is not very efficient, because pixels of difference images are poorly correlated and DCT does not compress uncorrelated data efficiently. To support this statement, results of a compression experiment are given in Table 12.14. A gray-level video sequence of 40 frames was compressed using an H.261 codec (Turletti, 1993). Table 12.14 gives percentage of nonzero elements at every stage of the compression. Here, 75% of the frames were interframe coded, and an overall compression ratio of 4.7:1 was achieved, which is by no standard adequate for video compression. For this reason, hybrid methods are used for interframe

TABLE 12.14   Compression Ratio for Intra- and Interframe Coding

Coding Mode	Nonzero Pixels	Nonzero DCT Coeff.	Nonzero Quant. Coeff.	Compression Ratio, bpp
Intra	99.99	25.44	10.50	1.32
Inter	92.68	53.93	18.90	1.85

Source: Adapted from Ayyagari, V. 1994. A bit plane coding algorithm for still and video compression. Master's Thesis, Texas Tech Univ., Lubbock, TX.

coding (Naveen and Wood, 1994). The common theme of these methods is to consider motion in the video and code information pertaining to displacement of image blocks.

12.10.6   Motion Compensation

For the motion-compensated interframe coding, the target frame is divided into nonoverlapping, fixed-size blocks, and then each block is compared with blocks of the same size of some reference frame to find its best match. To limit the search, a small neighborhood is selected in the reference frame, and then a search is performed by stepwise translation of the target block. To reduce mathematical complexity, a simple block matching criterion, such as the mean of the absolute difference of pixels, is used to find a best matched block. The position of the best matched block determines the displacement of the target block, and its location is denoted by a (motion) vector. Block matching is computationally expensive and, therefore, a number of heuristics have been developed (Pirsh, Demassieux, and Gehrke, 1995; Naveen and Wood, 1994). A simple method, known as one-at-a-time search (OTS), is shown in Fig. 12.105. First the target block is moved along one direction and the best match is found, and then it is moved along the perpendicular direction to find the best match in that direction. Figure 12.106 shows the target frame in terms of the best match blocks in the reference frame. This much detail of the motion-compensated interframe codec is now sufficient to follow the diagram of Fig. 12.100.

12.10.7   H.261: A Videoconference Coding Standard

H.261 is a standard to describe the data organization of a low-bit rate visual communication over telephone lines. It is also known as $p \times 64$ kb/s codec, since it identifies the data rate of the communication, where p is between 1 and 30. It uses DCT and **variable length coding (VLC)** techniques for intraframe coding and an optional block-based motion compensated coding for interframe coding. The frame previous to the target frame is used for **motion compensation (MC)** prediction. The range of the motion vector is ±15 pixels with integer values, and the size of blocks for matching is 16×16.

H.261 describes the organization of the compressed bit stream, and not how it should be produced. The coded data is arranged in a hierarchical structure consisting of four layers, each with its own header and a number of data blocks. The input picture is partitioned into 8 lines by 8 pixels per line image blocks. The lowest layer containing the quantized transformed coefficients of image blocks is called the block layer.

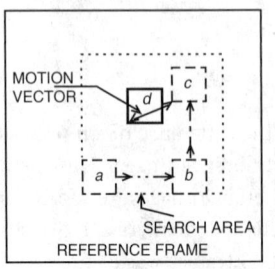

a: INITIAL POSITION OF TARGET BLOCK DURING THE SEARCH

b: BEST MATCH ALONG x-AXIS

c: BEST MATCH ALONG y-AXIS GIVEN b

d: LOCATION OF TARGET BLOCK IN THE TARGET FRAME

OTS SEARCH FOR THE BEST MATCHED BLOCK

FIGURE 12.105   A simplified search of a best matched block.

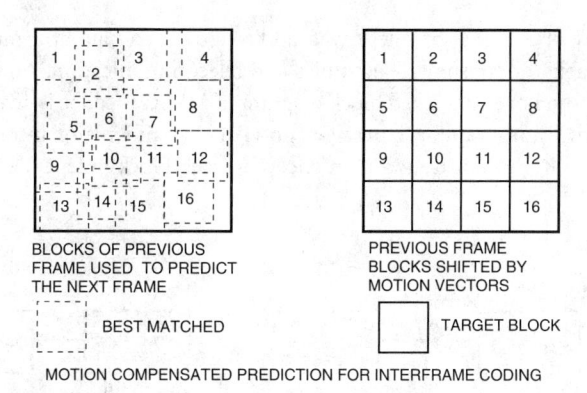

BLOCKS OF PREVIOUS
FRAME USED TO PREDICT
THE NEXT FRAME

PREVIOUS FRAME
BLOCKS SHIFTED BY
MOTION VECTORS

BEST MATCHED

TARGET BLOCK

MOTION COMPENSATED PREDICTION FOR INTERFRAME CODING

FIGURE 12.106 Motion compensated prediction.

The so-called *macroblock* (MB) contains four luminance Y and two chrominance Cr and Cb blocks. MBs pertaining to three rows of image blocks form a group of blocks (GOB). There are two picture size formats: *common intermediate format* (CIF) and quarter-CIF (QCIF). Figure 12.107 gives arrangement of various types of blocks for a CIF picture.

As expected, headers contain different information about the coded picture such as the location of blocks in the picture, method of coding, quantizer step size, and motion vectors. There are four types of coding

- Intra: transformed-quantized representation of the original input.
- Inter: the difference pixels with the previous frame; zero-motion vectors are not coded.
- Inter with MC: predicted residual with respect to the previous frame; the displacement (nonzero motion vector) is coded as part of the MB header; no specific method is recommended for block matching.
- Inter MC with filtering: the residual filtered by a predefined filter to reduce block artifacts, which occur routinely at low-bit rate coding.

12.10.8 MPEG: Standards for Audio and Video Compression

The Moving Photographic Expert Group (MPEG) set up international standards for compression of digital audio and video transmission at various bit rates. The first version, MPEG-1, was developed primarily for coding of video for digital storage media, for example, CD-ROM, at the rates of 1–1.5 Mb/s. The goal

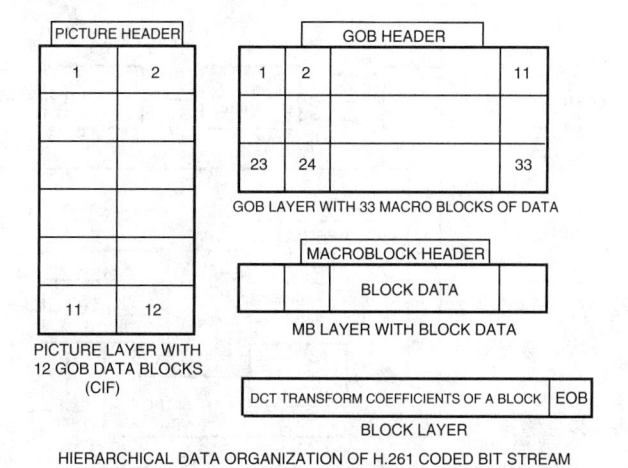

PICTURE LAYER WITH
12 GOB DATA BLOCKS
(CIF)

GOB LAYER WITH 33 MACRO BLOCKS OF DATA

MB LAYER WITH BLOCK DATA

BLOCK LAYER

HIERARCHICAL DATA ORGANIZATION OF H.261 CODED BIT STREAM

FIGURE 12.107 Organization of H.261 coded data.

of MPEG-2 was to describe coding of interleaved audio and video data at a much higher rate, such as 10 Mb/s, and is suitable for consumer electronics and telecommunication. An overview of the video coding of MPEG-1 is given next. A detailed block diagram of MPEG-1 codec is given in Fig. 12.108.

MPEG uses the **Joint Photographic Expert Group (JPEG)** standard for intraframe coding by first dividing each frame into 8×8 blocks, and then encoding each block independently compressed using

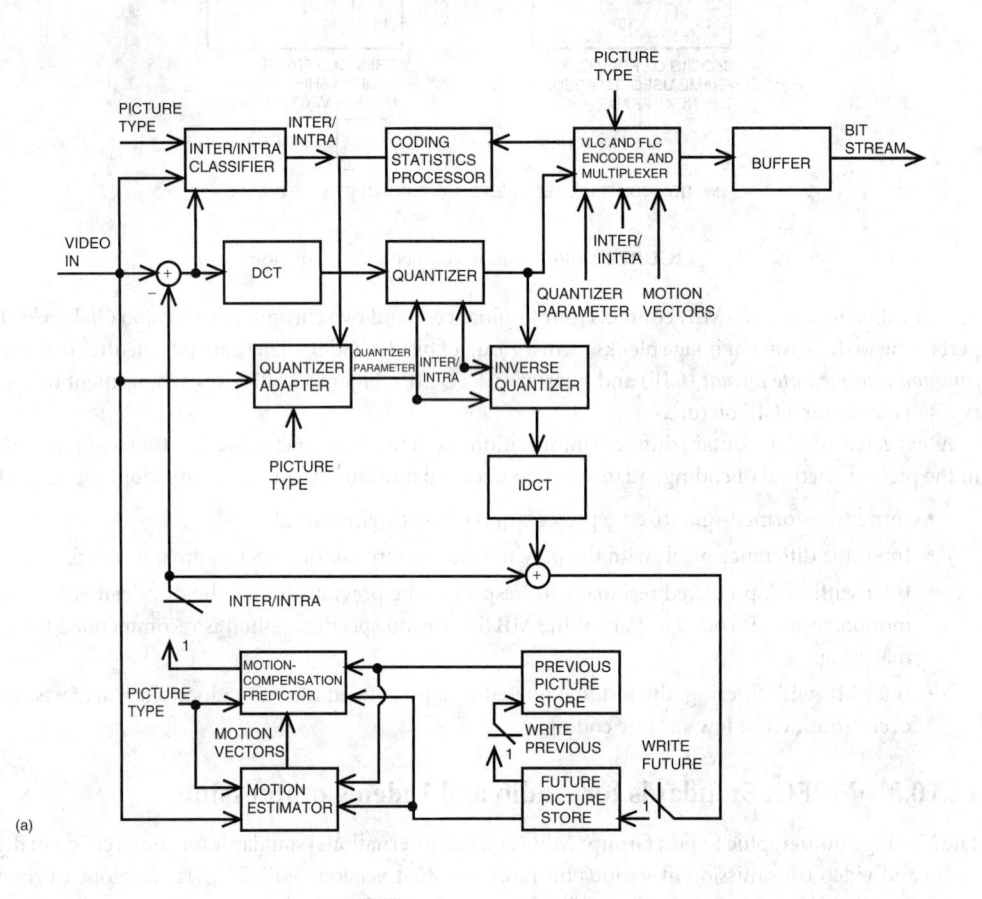

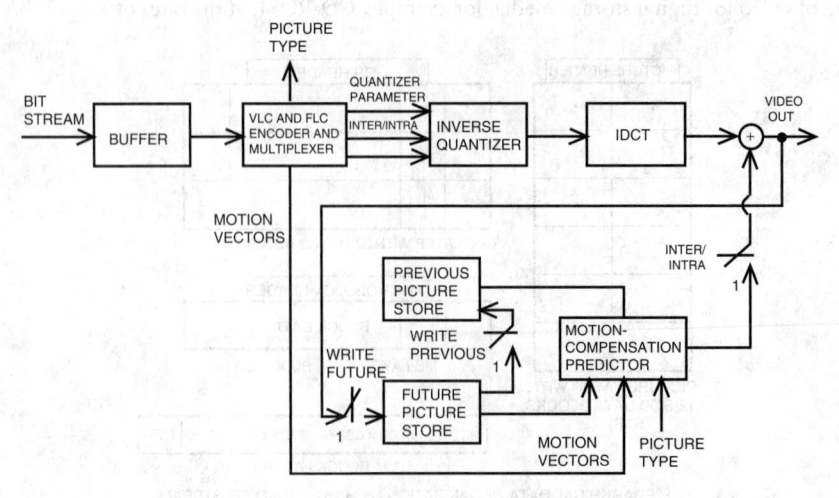

FIGURE 12.108 A typical MPEG-1 codec. (*Source:* Adapted from Arvind, R. et al. 1993. Images and video coding standards, p. 86. *AT&T Technical J.*)

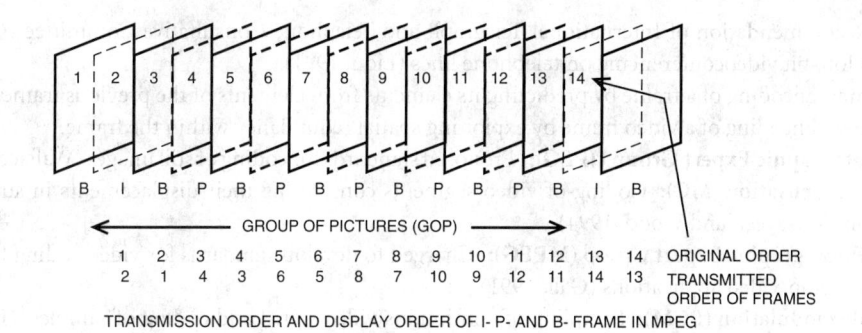

FIGURE 12.109   Bidirectional prediction in MPEG coding.

the DCT-based method illustrated in Fig. 12.103. Interframe coding is based on MC prediction, but its implementation is far more complex than the one used in H.261. The main difference between the two standards is that MPEG allows bidirectional, temporal prediction. As usual, a block matching algorithm is used to find the best matched block, which can belong to either the past frame (i.e., forward prediction) or the future frame (i.e., backward prediction). In fact, the best matched block can be the average of two blocks, one from the previous and the other from the next frame of the target frame (i.e., interpolation). In any case, the placement of the best matched block(s) is used to determine the motion vector(s); blocks predicted on the basis of interpolation have two motion vectors. Frames that are bidirectionally predicted are never used themselves as reference frames.

Figure 12.109 shows a group of pictures of 14 frames with two different orderings. Pictures marked I are intraframe coded. A P-picture is predicted using the most recently encoded P or I picture in the sequence. A macroblock in a P picture can be coded using either the intraframe or the forward predicted method. A B-picture macroblock can be predicted using either or both of the previous or the next I and/or P pictures. To meet this requirement, the transmission order and display order of frames are different. The two orders are also shown in Fig. 12.109.

Like H.261, the MPEG coded bit stream is divided into several layers. The outermost layer, called *video sequence*, contains the basic global information, such as the size of frames, bit rate, and frame rate. GOP layer contains information on fast search and random access of the video data. A GOP can be of arbitrary length. The picture layer contains a coded frame. Its header contains the type I, P, B, and the position of the frame in the GOP. The next three layers are the *slice*, *macroblock*, and the *block layer*, which basically correspond to the GOB, MB, and block layers of H.261.

The size of the luminance and the chrominance components of the input picture recommended by the MPEG-1 standard are 360×240 and 180×120, respectively, and the recommended frame rate for the coding of the NTSC-compatible television signal is 29.97 frames/s. MPEG uses blocks of size 16×16 for block matching, and the search is limited to an area of 31×31.

Roughly speaking, the coding algorithms and prediction methods of MPEG-1 and MPEG-2 are very much the same, but MPEG-2 contains more features, such as interlaced video and scalability. In the interlaced video, a frame may be divided into fields such as one containing the odd numbered lines and other containing the even numbered lines of the picture. A block can be coded either in the frame mode, as in MPEG-1, or in the field mode to exploit the correlation between the two fields. Scalability allows coding for multiple-grade and multiple-bandwidth service.

Pirsch, Demassieux, and Gehrke (1995) provide numerous references on MPEG and relevant topics for further studies.

Defining Terms

Codec: Coder–decoder programs of a compression system.
Discrete cosine transformation (DCT): Decorrelates image data efficiently [Rao and Yip 1990].
DPCM: Predictive coding of differences of samples.

H.261: Recommendation of International Telegraph and Telephone Consultative Committee (CCITT) on low-bit videoconferencing on telephone lines (Liou, 1991).

Interframe: Encoding of a frame by predicting its elements from elements of the previous frame.

Intraframe: Encoding of a video frame by exploiting spatial redundancy within the frame.

Joint Photographic Expert Group (JPEG): Proposed standards for coding of still images (Wallace, 1991).

Motion compensation (MC): Coding of video segments considering their displacements in successive frames (Naveen and Wood, 1994).

Moving Photographic Expert Group (MPEG): Charged to develop standards for video coding for various audio/video applications (Gall, 1991).

Pulse code modulation (PCM): Sampling of continuous video signal and coding of samples. (Rabbani and Jones discuss basics of image techniques. See Further Information.)

Quantization: Dropping of the less significant bits of image values to gain higher compression.

Redundancies: (1) Spatial—correlation within a still image or a video frame data, (2) spectral—correlation between different color components of image elements, (3) temporal—correlation between successive frames of a video file.

Variable length coding (VLC): An entropy coding method (Amsterdam, 1986).

References

Arvind, R., Cash, G.L., Duttweller, D.L., Hang, H.-M., Haskell, B.G., and Puri, A. 1993. Images and video coding standards. *AT&T Technical J.,* 67–88.

Liou, M. 1991. Overview of the $p \times 64$ kbits/s video coding standard. *Comm. of ACM* 34(4):59–63.

Gall, D.L. 1991. MPEG: A video compression standard for multimedia applications. *Comm. of ACM* 34(4): 46–58.

Ayyagari, V. 1994. A bit plane coding algorithm for still and video compression. Master's Thesis, Texas Tech Univ., Department of Computer Science, Lubbock, TX.

Pirsch, P., Demassieux, N., and Gehrke, W. 1995. VLSI architecture for video compression—A survey. *Proc. IEEE* 83(2):220–246.

Wallace, G.K. 1991. The JPEG still picture compression. *Comm. of ACM* 34(4):30–62.

Amsterdam, J. 1986. Data compression with Huffman coding. *BYTE* 11(5):99–108.

Rao, K.R. and Yip, P. 1990. *Discrete Cosine Transform.* Academic, New York.

Naveen, T. and Wood, J.W. 1994. Motion compensation multiresolution transmission of high definitiona video. *IEEE Trans. Circuits and Systems for Video Technology* 4(1):29–40.

Turletti, T. 1993. H.261 software codec for videoconferencing over the Internet. (Report). Institut National de Recherche en Informatique et an Automatique, 2004 routedes Lucioles, B.P. 93, 06902 Sophia-Antipolis, France.

Further Information

This chapter provided only a basic description of techniques used in video compression standards H.261 and MPEG. There are a number of periodicals that publish articles dealing with video compression. A few noted ones follow:

IEEE Transactions on Circuits and Systems for Video Technology
IEEE Transactions on Communications
Multimedia, a new journal by ACM and Springer

In addition, the Society of Photo-Optical Instrumentation Engineers (SPIE), P.O. Box 10, Bellingham, WA 98227 has published numerous proceedings of conferences on relevant subject areas.

The following books are recommended as preparatory material:

Pennebaker, W. and Mitchell, J.L. 1993. *JPEG: Still Image Data Compression Standard.* Van Nostrand Reinhold, New York.

Rabbani, M. and Joans, P.W. 1991. *Digital Image Compression Techniques*, Vol. TT7. SPIE Press, Bellingham, WA.

13

Electromagnetic Radiation

Pingjuan L. Werner
Pennsylvania State University

Anthony J. Ferraro
Pennsylvania State University

Douglas H. Werner
Pennsylvania State University

Gerhard J. Straub
Hammett and Edison, Inc.

Jerry C. Whitaker
Editor-in-Chief

Robert A. Surette
Shively Labs

13.1 Antenna Principles

Pingjuan L. Werner, Anthony J. Ferraro, and Douglas H. Werner

13.1.1 Antenna Types

Antennas find extensive utilization in communication systems ranging from cellular telephone, television, radio, radar, and numerous other applications. An antenna is designed to launch an electromagnetic signal with desired characteristics; this could be direction of radiation, area of coverage, strength of emission, beamwidth, sidelobe levels, etc. Any metal structure carrying a time varying electrical current will radiate electromagnetic waves as dictated by the well established Maxwell equations. Antennas however are purposefully designed to radiate with certain specified characteristics.

Antenna types are numerous; generally a metallic structure such as wires or metal surfaces serve as the radiator. Simple configurations like a straight length of wire excited from a signal source at the center, for instance, is called a **dipole**. Dipoles can be grouped to form what is known as an *antenna array* for enhancing some characteristic like power gain. A wire curved into a closed circuit, like a circle, forms a simple **loop antenna**. On the other hand, certain useful antennas do not bear any resemblance to the wire types. For instance, the common satellite television receiving antenna is actually a paraboloidal reflector collecting and focusing the received electromagnetic signal to a feed antenna. Such an antenna is called an

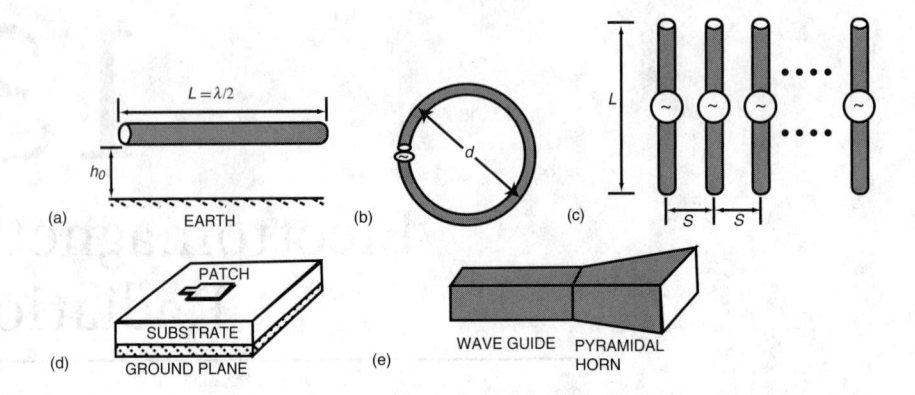

FIGURE 13.1   Examples of different types of antennas: (a) dipole, (b) loop, (c) array, (d) patch, (e) pyramidal horn.

aperture. Antennas can either receive, or transmit, or be designed to do both. The *laws of reciprocity* allow the receiving characteristics to be defined from the transmitting characteristics.

Figure 13.1 depicts several common types of antennas. In Fig. 13.1(a), a half-wave dipole is shown oriented parallel to and above the Earth. The antenna length is adjusted to be near one-half of the emitted wavelength of the electromagnetic signal. If the frequency is changed, then the structure is no longer one half-wave in length, and all of the characteristics will change. One of the most important characteristics is the impedance that the center of the antenna, called the *feed-point*, presents to the signal source. A circular loop antenna is shown in Fig. 13.1(b). Like the dipole, the characteristics of the loop depend on whether the loop circumference is small or large compared to the signal wavelength. Figure 13.1(c) is an array of dipoles that are parallel to each other and spaced a fixed distance between individual elements. A typical *patch antenna* is shown in Fig. 13.1(d). The patch antenna is constructed from a rectangular sheet of metal on a substrate. The electrical properties of the substrate influence the characteristics of this antenna. Figure 13.1(e) is a *pyramidal electromagnetic horn* connected to a waveguide for delivering the excitation to the mouth of the horn. This is a type of aperture antenna in which a knowledge of the electromagnetic fields across the face of the mouth are used to calculate the electromagnetic fields radiated by the horn.

13.1.2   Antenna Bandwidth

Antennas can find use in systems that require narrow or large bandwidths depending on the intended application. Bandwidth is a measure of the frequency range over which a parameter, such as impedance, remains within a given tolerance. Dipoles, for example, by their nature are very narrow band.

For narrow-band antennas, the percent bandwidth can be written as

$$\frac{(f_U - f_L)}{f_C} \times 100 \tag{13.1}$$

where

f_L = lowest useable frequency
f_U = highest useable frequency
f_C = center design frequency

In the case of a broadband antenna it is more convenient to express bandwidth as

$$\frac{f_U}{f_L} \tag{13.2}$$

One can arbitrarily define an antenna to be *broadband* if the impedance, for instance, does not change significantly over one octave ($f_U/f_L = 2$).

The design of a broadband antenna relies in part on the concept of a **frequency-independent** antenna. This is an idealized concept, but understanding of the theory can lead to practical applications. Broadband antennas are of the helical, biconical, spiral, and **log-periodic** types. Frequency independent antenna concepts are discussed later in this chapter. Some newer concepts employing the idea of fractals are also discussed for a new class of wideband antennas.

Narrow-band antennas can be made to operate over several frequency bands by adding resonant circuits in series with the antenna wire. Such traps allow a dipole to be used at several spot frequencies, but the dipole still has a narrow band around the central operating frequency in each band. Another technique for increasing the bandwidth of narrow-band antennas is to add parasitic elements, such as is done in the case of the open-sleeve antenna (Hall, 1992).

13.1.3 Antenna Parameters

Current Distribution

To calculate the characteristics of an antenna, the distribution of current on the elements must be known in advance. From the current distribution, the vector potential can be computed, and ultimately the electric field E and magnetic field H can be found. With this information, the radiation pattern, polarization, **directivity**, and other parameters can be described.

In the case of an aperture such as that shown in Fig. 13.1(e), the primary currents are flowing on the wave-guide and horn surfaces and are too difficult to find. In this case, the standard practice is to introduce what are known as *equivalent currents*. For the rectangular aperture at the mouth of the horn, equivalent electric and magnetic currents can be found from the equivalence principle so that the radiation from the equivalents is the same as from the primary currents. In any event, the current must be known before proceeding with the analysis of an antenna. Several techniques are available for determining the current distribution including transmission line approximations, the **method of moments (MoM)** and measurements.

The transmission line approximation looks at a center fed dipole, as in Fig. 13.1(a), as an open-circuited transmission line that has been opened up or fanned out. This interpretation suggests a sinusoidal current distribution with current nodes at the ends as shown in Fig. 13.2. The current distribution can be written as

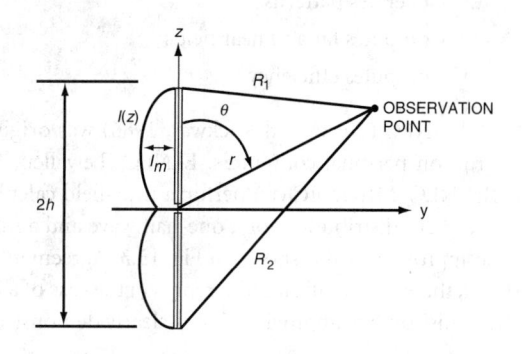

FIGURE 13.2 Transmission line approximation for the current distribution of a center fed dipole.

$$I(z) = I_m \sin \beta(h - z) \quad z > 0$$

$$I(z) = I_m \sin \beta(h + z) \quad z < 0 \tag{13.3}$$

where

$\beta = $ propagation constant
$h = $ half-length of the center fed dipole
$z = $ distance measured along the dipole with the center as the origin
$I_m = $ maximum current that occurs along the wire.

Exact closed-form expressions for the fields may be derived based on the assumed sinusoidal current distribution of Eq. (13.3). The z component of the free space electric field intensity E is given by

$$E_z = -j30 I_m \left(\frac{e^{-j\beta R_1}}{R_1} + \frac{e^{-j\beta R_2}}{R_2} - 2\cos(\beta h)\frac{e^{-j\beta r}}{r} \right) \tag{13.4}$$

where R_1, R_2, and r are as indicated in Fig. 13.2. The far-field radiation pattern is of primary interest for most practical antenna applications in communications. The far-field approximation is valid for very large values of r. In particular, using the sinusoidal current distribution given in Eq. (13.3), it can be shown that the corresponding far-zone θ component of the electric field is

$$E_\theta = \frac{60 I_m}{r} \left[\frac{\cos(\beta h) - \cos(\beta h \cos \theta)}{\sin \theta} \right] e^{-j\beta r} \tag{13.5}$$

For preliminary investigations the sinusoidal distribution is useful. For more accurate analysis, however, the MoM is required. The MoM solves an integral equation for the current distribution required to satisfy the boundary conditions on the surface of the antenna wires. There are several user friendly software packages available for the analysis of complex antenna geometries.

The **Numerical Electromagnetics Code** (NEC-2) (Burke and Poggio, 1981) is an antenna simulation program for modeling wire-based antennas. NEC has many features that include

- Ability to handle wire arcs
- Coordinate transformations
- Generates cylindrical structures
- Scales all of the units
- Surface patches
- Many different ground types
- Networks integrated into the wires
- Generates patterns
- Computes far and near fields
- Computes efficiency

MININEC (Logan and Rockway, 1986) was originally written in order to allow antenna simulations to run on personal computers. ELNEC (Lewallen, 1993) was developed as a user friendly interface to MININEC. MININEC can perform near-field calculations, whereas ELNEC does not. A comparison of the current distribution for a one-half wave and a full-wave dipole using the MoM and the transmission line approximation is shown in Fig. 13.3. Agreement is quite good for the half-wave dipole; on the other hand, there are significant differences in the case of a full-wave dipole. Hence, the limited usefulness of the transmission line approximation is clearly demonstrated in Fig. 13.3(b).

Polarization

Antennas create a state of electromagnetic polarization generally described as *linear, circular* or *elliptical*. Figure 13.4(a) depicts a linear polarized electric field (vertical polarization in this case). This would result from one of the dipoles shown in Fig. 13.1(c). The polarization of the wave is related to the orientation of the antenna wire. The electric field in Fig. 13.4(a) oscillates up and down at the angular frequency ω.

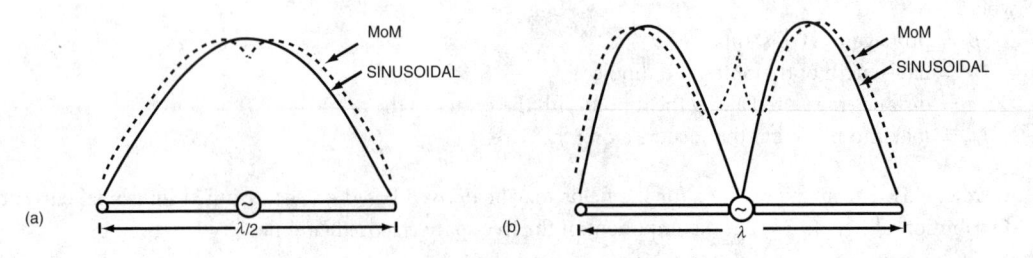

FIGURE 13.3 Comparison of current distributions using the transmission line approximation (sinusoidal) and method of moments: (a) half-wave dipole, (b) full wave dipole.

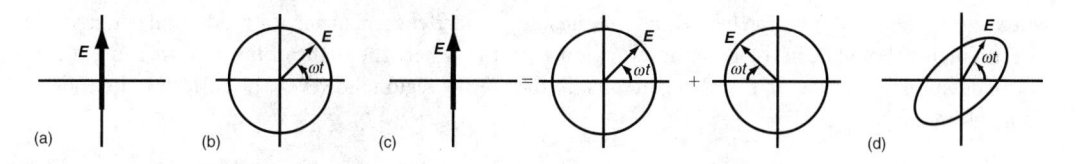

FIGURE 13.4 Various wave polarization states: (a) vertical polarization, (b) right-hand circular polarization, (c) representing linear polarization as the superposition of two circular polarized states, (d) right-handed elliptical polarization.

Circular polarization is shown in Fig. 13.4(b); here the electric vector stays constant in amplitude but rotates at an angular frequency ω. If the wave is approaching the observer, then the rotation direction shown in Fig. 13.4(b) is termed right handed.

A linear polarized wave can be decomposed into two oppositely rotating circular polarized waves having equal magnitudes as shown in Fig. 13.4(c). A circular polarized wave can be generated from two linear polarized waves using two orthogonal dipoles with equal amplitude excitation but 90° electrical phase shift between the excitations. The sense of rotation is controlled by selection of either a leading or lagging 90° phase shift. In general, the polarization of the wave can be elliptical as shown in Fig. 13.4(d). Elliptical polarized waves can be decomposed into two oppositely rotating circular waves of unequal magnitudes and specific electrical phase shift between the circular components.

In a communication system, there can be a polarization mismatch between the transmitting and receiving polarizations. If the transmitted wave is linear polarized, say in the vertical direction, and the receiving antenna is linear polarized in the horizontal direction then, theoretically, there is no far-field received signal at the receiver. If the transmitted wave is linear polarized and the receiving antenna is circular polarized then there is a 3-dB loss in received signal since the linear wave decomposes into two equal amplitude waves as shown in Fig. 13.4(c). This might not be an undesirable situation since the direction of the linear polarized signal could change with time due to propagation environment changes or due to motion of the source (a satellite mounted antenna). In this case, the received signal will be constant in amplitude if received on a circular polarized antenna with a 3-dB signal loss. Polarization mismatch requirements will dictate the type of receiving and transmitting polarizations required in the design of the system.

Fields

The regions around an electrically small radiator contain three types of field terms, that is, electrostatic field terms (fields that decrease as $1/r^3$ where r is the distance from the center of the radiator), induction field terms (fields that decrease as $1/r^2$), and the radiation field or **Fraunhofer region** term (fields that decrease as $1/r$). The electrostatic and induction field terms do not contribute to the radiated power and are responsible for the reactive component to the input impedance of the radiator. The complete electric field expression of a very short ideal dipole is given by

$$\mathbf{E} = j\omega\mu\frac{I_0 l}{4\pi}\left[1 + \frac{1}{j\beta r} + \frac{1}{(j\beta r)^2}\right]\frac{e^{-j\beta r}}{r}\sin\theta\,\boldsymbol{\theta} + j\omega\mu\frac{I_0 l}{2\pi}\left[\frac{1}{j\beta r} + \frac{1}{(j\beta r)^2}\right]\frac{e^{-j\beta r}}{r}\cos\theta\,\mathbf{r} \qquad (13.6)$$

where

l = length of the electrically short radiator
I_0 = current in the ideal dipole
μ = magnetic permeability of the medium
$\boldsymbol{\theta}, \mathbf{r}$ = unit vectors in the theta and radial directions

For larger radiators the **Fresnel** and Fraunhofer regions are important. The boundary between the two regions has been arbitrarily taken to be

$$\frac{2L^2}{\lambda} \qquad (13.7)$$

where L is the largest dimension of the antenna and λ is the wavelength. For accurate radiation patterns the measurements should be in the Fraunhofer region or farther where the shape of the pattern is essentially independent of the distance r from the antenna. In the Fresnel region, however, the pattern is dependent upon distance r.

Radiation Patterns

The radiation pattern of an antenna is a curve plotted in either Cartesian or polar coordinates from which the electric field intensity or a quantity proportional to the electric field can be depicted in various polar directions. The radiation pattern is a three-dimensional plot, but for simplicity it is usually plotted as a two dimensional cut along the three-dimensional display. Conventionally, two important cuts are the E-plane and H-plane plots. In the case of a dipole such as that shown in Fig. 13.2, the electric field is in the theta direction, such that it is parallel to any plane that contains the dipole. A plot of the intensity in that plane is called the E-plane radiation pattern. The magnetic field intensity is phi directed and encircles the dipole. The plane that has the dipole element perpendicular to it and also has the magnetic field intensity parallel to it yields the H-plane radiation pattern.

Figure 13.5(a) illustrates both the E- and H-plane patterns for a one-half wavelength dipole oriented along the z coordinate axis. The ELNEC program was used to compute both the current distribution and the radiation pattern. The current distribution, which is graphed using the wire as the axis, is close to what the transmission line approximation would give. The E-plane is shown for the observer in the xz plane. Radiation patterns presume that the observer moves around the antenna while in the far-field region, keeping a constant distance from the antenna. This is a free space pattern with no ground system included. It clearly shows the direction of maximum radiation and that this antenna has a rather broad pattern. Also displayed are null directions where the fields ideally go to zero. Plotted is a quantity proportional to the field intensity, or one could also choose an option to use the decibel scale where usually the maximum is normalized to 0 dB. The H-plane plot for the half-wave dipole is shown in Fig. 13.5(b). Because of symmetry, the dipole is expected to have uniform field intensity as the observer moves around the antenna in the H-plane or specifically the xy plane. For a longer dipole of one full wavelength, Fig. 13.5(c) displays the E-plane radiation pattern. Clearly the main beam of field intensity is narrower than the previous example, and the current distribution takes on a different form having a minimum of current at the center feed point. This case, which has already been considered in Fig. 13.3(b), is one in which the transmission

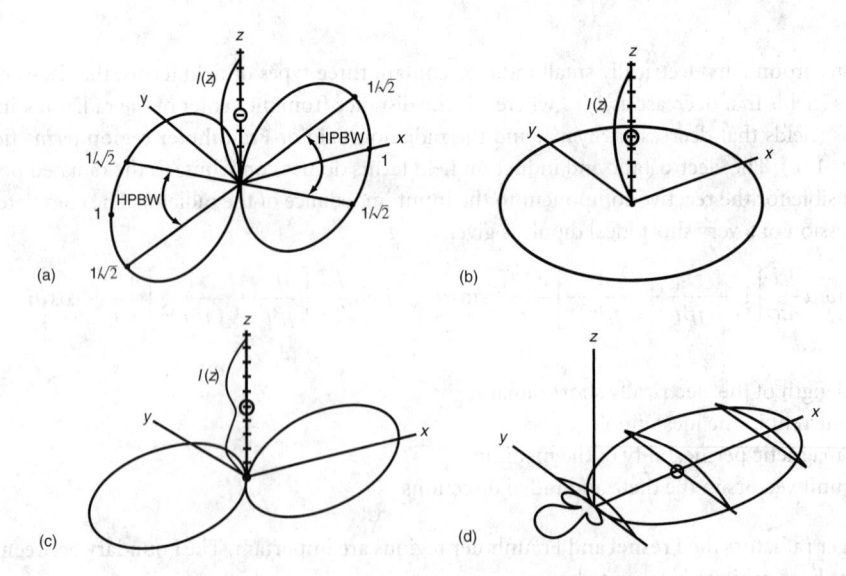

FIGURE 13.5 (a) E-plane plot of the radiation pattern of a half-wave dipole with half power beamwidth shown, (b) H-plane plot of the radiation pattern of a half-wave dipole, (c) E-plane plot of the radiation pattern of a one-wavelength dipole, (d) E-plane plot of the radiation pattern of a three element Yagi.

line approximation would incorrectly predict a null at the feed point. However, using the transmission line current distribution would not show any significant difference in radiation pattern, but would create a sizeable error in the computation of impedance at the center point. A more complex pattern is shown in Fig. 13.5(d), which results from a three element Yagi array. In this type of antenna a central element is fed from a source and two other elements are unfed (parasitic), but due to the mutual coupling with the fed element there is a current induced in the other two. The fed element is close to a half-wavelength long, whereas the element on the right is slightly longer and the one on the left is slightly shorter. Proper pruning of the lengths and spacings results in an E-plane pattern that shows the signal is now predominantly directed to the right and that there is a smaller back lobe in the opposite direction. The front-to-back ratio defines the ratio of the field to the right referenced to that directed to the left. The front-to-back ratio is usually measured in decibels. Also evident in Fig. 13.5(d) are smaller sidelobes in the back direction.

A radiation pattern provides good visualization of the distribution of the field intensity in different compass directions. More accurate values could be tabulated if desired. A measure of the broadness of the main beam of energy is called the half-power-beamwidth (HPBW). In the linear plots this is the point where the field diminishes to $1/\sqrt{2}$ of the maximum as illustrated in Fig. 13.5(a).

For an antenna that is not straight or one that is not aligned along the z axis as in the previous example, there can be two components of fields, one that is phi and one that is theta directed. In this case patterns can be found for either component or for the magnitude of the total field. Sometimes a pattern is displayed as the power pattern, in which case the points on the plot are proportional to the square of the field pattern values.

Pattern Multiplication

Pattern multiplication is a procedure that allows the determination of the radiation pattern of an array of identical radiators, spaced equally along a baseline. The method permits the rapid visualization of the effect of changing the spacings, the orientation of the radiators, the type of radiators, and the magnitude and phase shift of excitation to the radiators. Central to the method is to first determine the radiation pattern of an array of **isotropic elements** having the same number, spacings and excitations as the original array. Isotropic elements are fictitious elements that radiate equal intensity in all directions. With information of the isotropic array pattern and knowledge of the element pattern one can construct the resultant pattern of the original array graphically. Use of computers can give the pattern directly, but the multiplication technique does allow better insight into the performance of the array and permits visualization of array parameter changes.

Figure 13.6 illustrates the geometry for an array of N isotropic elements arranged along the z axis, having spacings d with equal excitation magnitudes I and electrical phase shift α between pairs of elements. It can be shown that the radiation pattern for this array may be expressed as

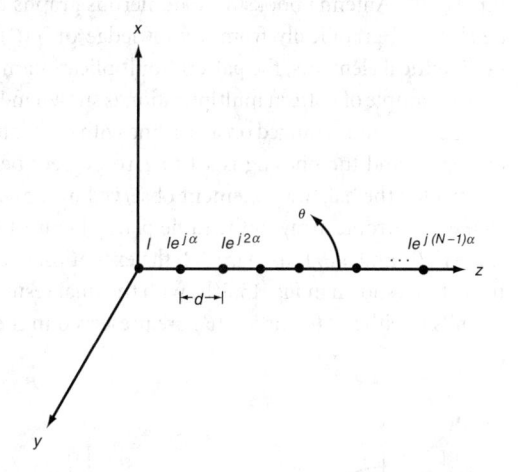

FIGURE 13.6 Geometry for an array of N isotropic elements having spacing d, equal excitation magnitudes I, and phase shift α between pairs of elements.

$$AF(\theta) = \frac{1}{N} \frac{\sin\left(\frac{N\psi}{2}\right)}{\sin\left(\frac{\psi}{2}\right)} \tag{13.8}$$

where

$$\psi = \beta d \cos\theta + \alpha \tag{13.9}$$

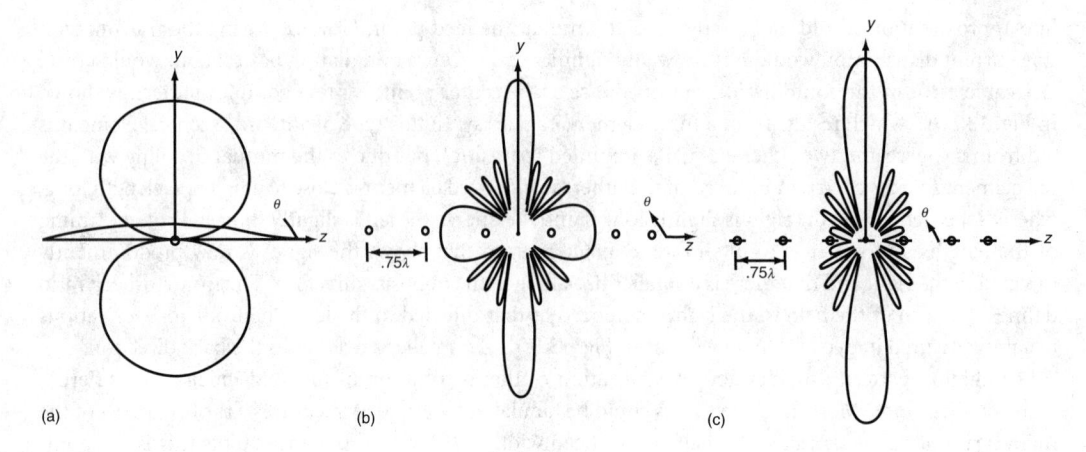

FIGURE 13.7 Example of Pattern multiplication: (a) pattern of a dipole element, (b) pattern of a six-element isotropic array, spacing $d = 0.75\lambda$ and $\alpha = 0°$, (c) result of multiplications of (a) and (b).

If the isotropic elements are replaced by elements having radiation pattern function $f(\theta)$, then it can be shown that the resultant pattern of the array is given by

$$g(\theta) = f(\theta) \times AF(\theta) \qquad (13.10)$$

Modification of the element type or orientation requires the updated $f(\theta)$ in order to graph the resultant pattern $g(\theta)$. On the other hand, changes in the geometry of the isotropic array requires a new function for $AF(\theta)$. Antenna books have numerous graphs of $AF(\theta)$ and $f(\theta)$ for different situations; $g(\theta)$ can be sketched graphically from a knowledge of $f(\theta)$ and $AF(\theta)$. For the case of unequal spacings and nonidentical elements, the pattern multiplication method is not applicable.

An example of pattern multiplication is shown in Fig. 13.7. Here a six element array of half-wave dipoles are collinear and arranged on a baseline with center to center spacings of 0.75 wavelengths. The excitations are equal, and the phasing is set to zero degrees between the individual elements. Figure 13.7(a) is the pattern for the half-wave element observed in the yz-plane, whereas Fig. 13.7(b) is the pattern of the six element isotropic array in the same plane. Figure 13.7(c) shows the resultant from the multiplication of the $AF(\theta)$ and $f(\theta)$. In Fig. 13.8, the excitation between elements is changed to 180°, and this results in the pattern shown in Fig. 13.8(b) with the final result of multiplication shown in Fig. 13.8(c). The location of nulls in either $f(\theta)$ or $AF(\theta)$ are preserved in the final pattern, as these figures confirm.

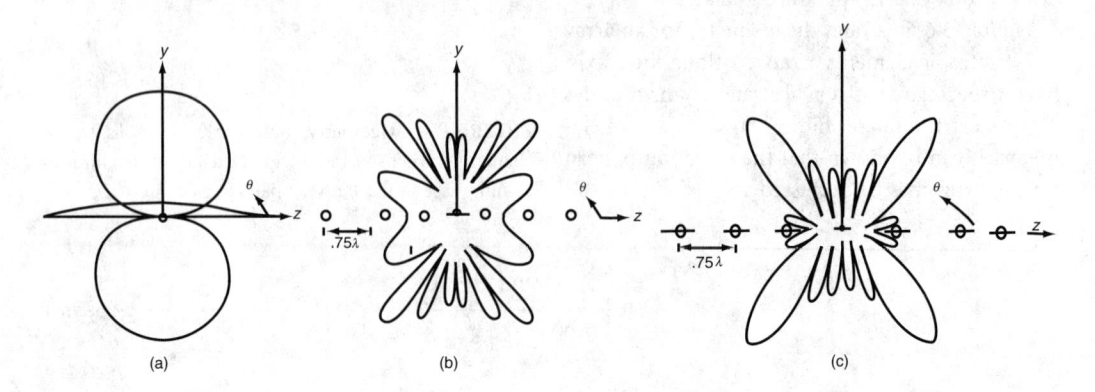

FIGURE 13.8 A second example of pattern multiplication: (a) same as Fig. 13.7(a), (b) pattern of six-element isotropic array with spacing $d = 0.75\lambda$ but α changed to 180°, (c) result of multiplications of (a) and (b).

TABLE 13.1   Radiation Resistance for Various Elements

Element Type	Element Length, λ	Radiation Resistance, Ω
Short dipole in free space	$L < 0.1$	$R_{\mathrm{rad}} = 20\pi^2 (L/\lambda)^2$
Short monopole on a perfectly conducting ground plane	$L < 0.05$	$R_{\mathrm{rad}} = 40\pi^2 (L/\lambda)^2$
1/2 wavelength dipole in free space	$L = 0.5$	$R_{\mathrm{rad}} = 73$
3/2 wavelength dipole in free space	$L = 1.5$	$R_{\mathrm{rad}} = 105.5$

Impedance

The input impedance of an antenna plays an important role in the matching of the source to the antenna. Knowledge of the impedance over the operating bandwidth is of concern. The *real* part of the impedance is primarily due to the radiation resistance, and in part due to the ohmic loss of the conductors. The radiation resistance is the equivalent resistance, which if connected to the source in place of the antenna absorbs the same power as radiated by the antenna. Impedance can be determined a number of ways. Use of the method of moments gives the most definitive results subject to modeling limitations. Method of moments software was discussed in the first part of Sec. 13.1.3.

From the transmission line approximation for the current distribution, the far electric fields can be computed. From this the Poynting vector, (watts per square meter), can be integrated over a large spherical surface to find the total radiated power. The radiation resistance at the feed point is then found from this radiated power and the known current at the feed point. The near fields dictate the *imaginary* part of the impedance. Some typical values of radiation resistance for various elements, which have small diameter, are given in Table 13.1. The section on dipole characteristics gives the input impedance of dipoles of various lengths computed from the latest moment method formulation described in Sec. 13.1.4.

Mutual Impedance

Several elements collected in an array have interactions among themselves. Since in an array environment the elements could be closely spaced in units of wavelengths, the interaction between them can result in significant mutual coupling measured in terms of the mutual impedance. It is convenient to treat the antenna array as an N-port network in order to determine mutual impedances. Hence, for an antenna array the mutual impedances can be computed from knowledge of the total fields acting upon an element.

The port equations for an N-element array with element excitation V_i and element current I_j can be written as

$$
\begin{aligned}
V_1 &= I_1 Z_{11} + I_2 Z_{12} + \cdots + I_N Z_{1N} \\
V_2 &= I_1 Z_{21} + I_2 Z_{22} + \cdots + I_N Z_{2N} \\
&\vdots \\
V_N &= I_1 Z_{N1} + I_2 Z_{N2} + \cdots + I_N Z_{NN}
\end{aligned}
\tag{13.11}
$$

Where Z_{ii} is the self-impedance of the ith element and Z_{ij} is the mutual impedance between the ith and jth elements. These lead to expressions for the driving point or input impedance viewed by each source connected to the elements. Thus, from these port equations one obtains

$$
\begin{aligned}
Z_{D1} &= \frac{V_1}{I_1} = Z_{11} + \left(\frac{I_2}{I_1}\right) Z_{12} + \cdots + \left(\frac{I_N}{I_1}\right) Z_{1N} \\
Z_{D2} &= \frac{V_2}{I_2} = \left(\frac{I_1}{I_2}\right) Z_{21} + Z_{22} + \cdots + \left(\frac{I_N}{I_2}\right) Z_{2N} \\
&\vdots \\
Z_{DN} &= \frac{V_N}{I_N} = \left(\frac{I_1}{I_N}\right) Z_{N1} + \left(\frac{I_2}{I_N}\right) Z_{N2} + \cdots + Z_{NN}
\end{aligned}
\tag{13.12}
$$

Consider the simple case of a two-element array of half-wave dipoles separated by a half-wavelength with elements parallel and side by side. The driving point impedances for each element reduces to

$$Z_{D1} = Z_{11} + \left(\frac{I_2}{I_1}\right) Z_{12}$$

$$Z_{D2} = Z_{22} + \left(\frac{I_1}{I_2}\right) Z_{21}$$

(13.13)

The known impedance for a thin half-wave dipole is $73 + j42.5$ and available tables (Hickman and Tillman, 1961) for the mutual impedance give a value of $-12.5 - j30$. At this point knowledge of the complex current ratios is required to complete the analysis. Several interesting cases are $I_1/I_2 = 1$ and $I_1/I_2 = -1$. The mutual impedance tables referred to assumed that the array elements had sinusoidal current distributions. Each case gives the following results:

- If $I_1/I_2 = 1$, then $Z_{D1} = Z_{D2} = 60.5 + j12.5 \,\Omega$.
- If $I_1/I_2 = -1$, then $Z_{D1} = Z_{D2} = 85.5 + j72.5 \,\Omega$.

The moment method gives the driving point impedance directly without any need for separate calculations for the self- and mutual terms. The same problem repeated with the ELNEC code yields for a wire radius of 1.0 mm the following results:

- If $I_1/I_2 = 1$, then $Z_{D1} = Z_{D2} = 60.08 + j9.38 \,\Omega$.
- If $I_1/I_2 = -1$, then $Z_{D1} = Z_{D2} = 97.43 + j70.66 \,\Omega$.

For the case of a two-element Yagi where only one element is fed and the other obtains excitation from only the mutual coupling, the driving point impedance is derived as

$$Z_{D1} = Z_{11} - Z_{12}^2 / Z_{22}$$

since V_2 and, hence, Z_{D2} are zero. The expression shows that the subtraction of the two terms can result in a low input impedance, which can restrict bandwidth and lower efficiency. Numerical results from ELNEC for the Yagi with two parallel and side-by-side half-wavelength dipoles spaced 0.1 and 0.05 wavelengths are as follows:

- If spacing is 0.1λ, then $Z_{D1} = 22.29 + j59.27 \,\Omega$.
- If spacing is 0.05λ, then $Z_{D1} = 5.50 + j31.15 \,\Omega$.

Generally, if the dipoles in an array are not parallel or collinear, the computation for self- and mutual impedances is more difficult, and using the method of moments code is a more suitable approach.

Directivity, Directive Gain, and Power Gain

The **directivity** of an antenna provides a measure of performance compared to a reference antenna. The reference is either the hypothetical isotropic element or the practical half-wave dipole. Specifically, if the antenna under test and the reference are supplied with the same input power, then the directivity would simply compare the ratio of the power densities of the former with the latter. The directivity is used to express the performance in the direction of maximum power density from the antenna under test. It is presumed that all of the power supplied to the antenna is radiated, that is, it is 100% efficient. The *directive gain* is similar to directivity except it can be used for any direction and not necessarily the direction of maximum radiation. The power gain takes into account the efficiency of the antenna system since all of the power supplied is not radiated. This can include the losses in the transmission line feed configuration.

Figure 13.9 shows the concept for defining the directive gain and directivity. The antenna under test and the isotropic radiator are supplied with the same power W_R, which is 100% radiated. The power density P is measured from both antennas at the same distance R. The directive gain $D(\theta, \phi)$ is the ratio

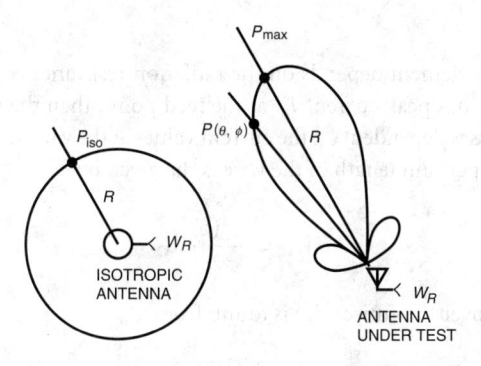

FIGURE 13.9 Comparison of power densities between an antenna and an isotropic reference. W_R is the power radiated, $P(\theta, \phi)$ is the power density, P_{max} is the maximum power density and R is the range at which measurements are made.

$P(\theta, \phi)/P_{iso}$. The quantity P_{iso} is the power density from the isotropic antenna that is radiating W_R watts. If P is measured in the direction of maximum, P_{max}, then the ratio is called the directivity, otherwise it is known as directive gain. From the three-dimensional radiation pattern $f(\theta, \phi)$, the directive gain and directivity can be computed from the following:

$$D(\theta, \phi) = \frac{4\pi f^2(\theta, \phi)}{\int_0^{2\pi} \int_0^{\pi} f^2(\theta, \phi) \sin\theta \, d\theta \, d\phi} \tag{13.14}$$

when $f^2(\theta, \phi) = 1$, that is, in the direction of P_{max}

$$D = \frac{4\pi}{\int_0^{2\pi} \int_0^{\pi} f^2(\theta, \phi) \sin\theta \, d\theta \, d\phi} \tag{13.15}$$

where $f^2(\theta, \phi)$ is the power pattern, which is the square of the field pattern.

Table 13.2 gives the values of directivity for some common antenna types. The results are referenced to both the isotropic and half-wave dipole.

The *power gain* G is derived in a similar manner, except the total power input to the antenna is used in place of the radiated power; thus, the relationship between G and D is

$$G = (\text{eff}) \times D \tag{13.16}$$

where eff is the antenna efficiency.

TABLE 13.2   Directivity for Various Antenna Types

Antenna Type	Directivity, dB, Isotropic Reference	Directivity, dB, Half-Wave Dipole Reference
Isotropic	0	−2.15
Free space half-wave dipole	2.15	0
Short dipole	1.76	−0.39
Free space full-wave dipole (from ELNEC)	3.92	1.77
Three element Yagi of Fig. 13.5(d) (from ELNEC)	9.52	7.37

Efficiency

The efficiency of an antenna element depends on the radiation resistance R_{rad} and the ohmic resistance R_Ω. If a dipole, for instance, has peak current I_{in} at the feed point, then the radiated power is $I_{in}^2/2R_{rad}$, whereas the ohmic power loss is dependent on the current values at the various positions along the antenna wire. The ohmic power loss per unit length of the wire is the given by

$$dW_\Omega = \frac{I^2(z)}{2}R_\Omega \tag{13.17}$$

The total ohmic power lost over the entire wire is found from

$$W_\Omega = \int \frac{I^2(z)}{2}R_\Omega dz \tag{13.18}$$

The calculation of radiation resistance has been discussed in an earlier subsection. Here an approximate value of R_Ω will be found. This information may be used to find the efficiency from the expression

$$\text{eff} = \frac{W_{rad}}{W_{rad} + W_\Omega} \tag{13.19}$$

Figure 13.10(a) depicts the geometry used for calculating the surface impedance of a flat conductor 1×1 m but very deep. The **surface impedance** (Stutzman and Thiele, 1981) for this case is shown for a good conductor to be

$$Z_s = \sqrt{\frac{j\omega\mu_0}{\sigma}} \tag{13.20}$$

where

σ = conductivity of the wire
μ_0 = permeability of free space
ω = angular frequency of the antenna source

The real part of Z_s gives the surface resistance R_s to be

$$R_s = \sqrt{\frac{\omega\mu_0}{2\sigma}} \tag{13.21}$$

In the case of a cylindrical wire, as shown in Fig. 13.10(b), the small section with dimensions Δz and Δs can approximate the flat conductor of Fig. 13.10(a). At high frequencies, due to skin effect, the current will be confined to the surface and so a deep section of conductor is not required. Under these circumstances,

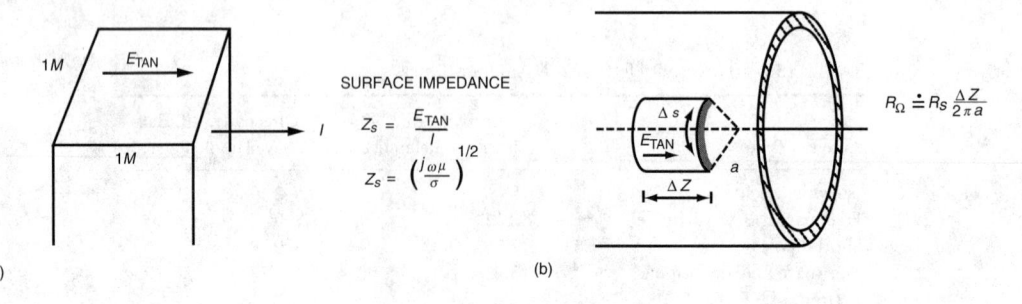

(a)   (b)

FIGURE 13.10 (a) Surface impedance for a flat thick conductor, (b) application of the surface impedance for finding the ohmic resistance of a cylindrical conductor at high frequencies.

unfolding the cylindrical surface will approximate a flat conductor of dimensions $2\pi a$ and Δz. This results in an approximate ohmic resistance of

$$R_\Omega = R_s \frac{\Delta z}{2\pi a} \qquad (13.22)$$

From a knowledge of the current distribution discussed earlier, the total ohmic power loss can be computed. A less strenuous method of computing the efficiency is with a method of moments code. Since wire conductivity can be inputted into the code, it is possible to evaluate the directivity (assuming perfectly conducting wires) and then repeat with wire conductivity included. From a knowledge of D and G, efficiency can be inferred.

13.1.4   Antenna Characteristics

Antennas over ground can be modeled by several methods. The simplest is to assume a perfectly conducting Earth. This would be an excellent approximation for some situations, such as an antenna above sea water. For this case the method of images can be used. If Earth parameters do not warrant the perfectly conducting approximation, then the Fresnel reflection coefficients can be used to compute the ground reflected wave, which is superimposed with the direct wave. For antennas very close to Earth the **Sommerfeld/Norton solution** is used. This method uses the exact solution for fields in the presence of Earth. The NEC code discussed earlier uses this method as an option, although computations require more time to complete.

Effect of the Earth

Perfectly Conducting Ground Plane

If the Earth can be approximated as perfectly conducting, then the method of images can be applied and the solution is straightforward. Figure 13.11 shows a horizontal and vertical dipole elevated above a perfectly conducting and infinitely large ground plane. For the horizontal case, it can be shown that there is an image that can account for the correct amplitude and phase of the reflected wave. The image has the same current distribution as the antenna above ground but is 180° out of phase. The image is the same distance below the ground plane interface as the antenna is above the ground plane interface. The image and original antenna acting together in free space properly accounts for the reflected wave from the ground plane. The combination of the direct and ground reflected waves can be done by using the dipole field equation given earlier. For the vertical dipole, the image has the same current distribution as the original antenna and is in phase. The distance requirements of the image are indicated in Fig. 13.11.

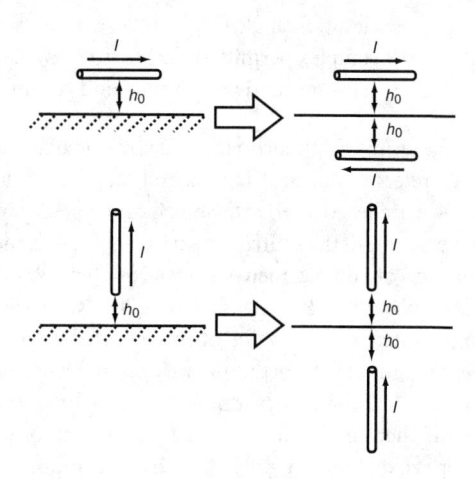

FIGURE 13.11   (a) Horizontal dipole above perfectly conducting ground plane and the image equivalent, (b) vertical dipole above perfectly conducting ground plane and the image equivalent.

The driving point impedance of the dipole above Earth is easily calculated from the self- and mutual impedances already discussed. This gives for the case of a single dipole above a perfectly conducting Earth

$$Z_{D1} = \frac{V_1}{I_1} = Z_{11} + \left(\frac{I_2}{I_1}\right) Z_{12} \qquad (13.23)$$

for the horizontal dipole

$$I_1/I_2 = -1$$

therefore

$$Z_{D1} = Z_{11} - Z_{12}$$

for the vertical dipole

$$I_1/I_2 = 1 \tag{13.24}$$

therefore

$$Z_{D1} = Z_{11} + Z_{12} \tag{13.25}$$

Imperfectly Conducting Ground Plane

For the case of a ground defined by the electrical permittivity and conductivity, the reflection coefficient can be computed (Kraus and Carver, 1973). The reflection coefficient defines the amplitude of the reflected wave relative to the direct wave so that the two fields can be added to determine the far-field radiation pattern. The reflection coefficients for horizontal and vertical polarizations are given respectively by

$$\rho_H = \frac{\sin \alpha - \sqrt{\varepsilon_r - \cos^2 \alpha}}{\sin \alpha + \sqrt{\varepsilon_r - \cos^2 \alpha}} \tag{13.26}$$

$$\rho_V = \frac{\varepsilon_r \sin \alpha - \sqrt{\varepsilon_r - \cos^2 \alpha}}{\varepsilon_r \sin \alpha + \sqrt{\varepsilon_r - \cos^2 \alpha}} \tag{13.27}$$

where

α = elevation angle for the reflected ray (see Fig. 13.12)
ε_r = complex permittivity, $\varepsilon_r' - j\sigma/\omega\epsilon_0$
ε_r', σ = actual ground permittivity and conductivity parameters of the Earth, respectively.

The terminology *horizontal* and *vertical* polarizations refers to the electric field and magnetic field being parallel to the Earth interface, respectively. For example, if the source shown in Fig. 13.12 is the end view of a dipole, then the far electric field is perpendicular to the figure and, consequently, parallel to the interface. The value of the reflected electric field is computed from the quantity ρ_H. If the source in Fig. 13.12 is a vertical dipole in the plane of the figure, then the far magnetic field is parallel to the interface or the far electric field has a component perpendicular to the interface; hence, the ρ_V is used to determine the reflected field.

The resultant far field from a short horizontal dipole of length ds perpendicular to the page would be expressed by

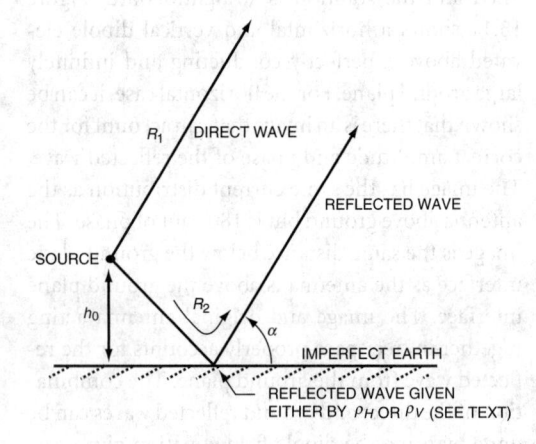

FIGURE 13.12   Direct and reflected waves for an antenna above an imperfect Earth.

$$E = j30\beta I_0 ds \left(\frac{e^{-j\beta R_1}}{R_1} + \rho_H \frac{e^{-j\beta R_2}}{R_2} \right) \tag{13.28}$$

For a short vertical dipole of length ds parallel to the page, the resultant far field would be expressed by

$$E = j30\beta I_0 ds \cos \alpha \left(\frac{e^{-j\beta R_1}}{R_1} - \rho_V \frac{e^{-j\beta R_2}}{R_2} \right) \tag{13.29}$$

Several examples (Jordan and Balmain, 1968) of the influence of an imperfect ground on an antenna are available in the literature.

The Sommerfeld/Norton Method

The NEC codes provide for several options of modeling a structure over ground. For the perfectly conducting ground the code generates an image as discussed earlier. For a finitely conducting ground plane the code generates an image modified by the Fresnel reflection coefficients as discussed previously. The alternate method uses the Sommerfeld/Norton exact solution for the fields in the presence of Earth and is claimed to be accurate for wire antennas close to the ground. However, computational times are considerably longer. NEC also includes the option to model a radial-wire ground screen as an approximation and a two medium ground approximation, such as a cliff overlooking a different medium.

Irregular Terrain

The assumption of a flat ground plane is not always realistic because terrain can be very irregular. Some recent work (Young, 1994) has developed a practical computer code for the simulation of antenna patterns over three-diimensional terrain using the **uniform geometrical theory of diffraction (UTD)**. The software is capable of generating a three-dimensional terrain model from existing terrain databases. Using existing diffraction programs, patterns can be obtained. This code is useful for siting antennas and determining optimum locations for high frequency (HF) communications. Various two- and three-dimensional generic shape profiles were analyzed and tabulated as a quick reference guide in selecting an antenna site.

Dipole Characteristics

Consider a center-fed dipole antenna of length $2L$ and diameter $2a$ oriented along the z axis, as illustrated in Fig. 13.13. If it is assumed that the current on the surface of the wire dipole is circumferentially invariant, then it has been shown that the exact form of the current distribution $I(z')$ on the dipole may be determined by solving the following integral equation (Pocklington, 1897):

$$\int_{-L/2}^{L/2} \left[\frac{\partial^2}{\partial z^2} + \beta^2 \right] K(z - z') I(z') dz' = -j4\pi\omega\varepsilon E_z^i(z) \tag{13.30}$$

where E_z^i represents the field incident or impressed on the surface of the wire by a source and

$$K(z - z') = \frac{1}{2\pi} \int_0^{2\pi} \frac{e^{-j\beta R'}}{R'} d\phi' \tag{13.31}$$

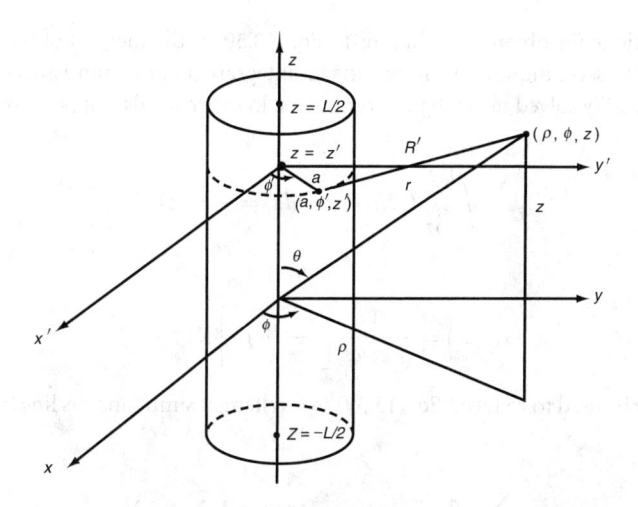

FIGURE 13.13 Dipole antenna geometry.

is the cylindrical wire kernel, in which

$$R' = \sqrt{(z - z')^2 + \rho^2 + a^2 - 2\rho a \cos(\phi - \phi')} \qquad (13.32)$$

This is known as Pocklington's integral equation in which the unknown $I(z')$ appears under the integral.

For modeling of thin-wire antennas, that is, wires with radii $a \lesssim 0.01\lambda$, the reduced kernel approximation of Eq. (13.31) is often used. This approximation is

$$K(z - z') \approx K_0(z - z') = \frac{e^{-j\beta r_0}}{r_0} \qquad \text{for} \qquad \rho \geq a \qquad (13.33)$$

where

$$r_0 = \sqrt{(z - z')^2 + \rho^2} \qquad (13.34)$$

It should be noted that Eq. (13.33) is exact for a vanishingly thin wire ($a = 0$) and becomes approximate as soon as the wire radius begins to increase ($a > 0$). On the other hand, for moderately thick wires with radii in the range $0.01\lambda \lesssim a \lesssim 0.1\lambda$, the complete kernel expression of Eq. (13.31) must be considered in order to achieve an accurate solution. Several techniques for evaluating Eq. (13.31) have been discussed in Werner (1995). These techniques include numerical integration schemes, as well as analytical methods. Among the more useful techniques is a recently found exact representation of Eq. (13.31) given by (Werner, 1993, 1995)

$$K(z - z') = -\frac{e^{-j\beta R}}{R} \sum_{n=0}^{\infty} \sum_{k=0}^{2n} A_{nk} \frac{(\beta^2 \rho a)^{2n}}{(\beta R)^{2n+k}} \qquad (13.35)$$

where

$$R = \sqrt{(z - z')^2 + \rho^2 + a^2} \qquad (13.36)$$

and the coefficients A_{nk} may be determined through the following set of recursions:

$$A_{nk} = \frac{(2n + 1 - k)(2n + k)}{j(2k)} A_{nk-1}, \quad n > 0 \quad \text{and} \quad k > 0 \qquad (13.37)$$

$$A_{n0} = \begin{cases} -1, & n = 0 \\ -\dfrac{1}{(2n)^2} A_{n-10}, & n > 0 \end{cases} \qquad (13.38)$$

A common technique for obtaining solutions to Eq. (13.30) is the method of moments (Harrington, 1968). The beauty of this technique is that it converts the integral equation to an associated matrix equation, which can then be readily solved using digital computers. In other words, suppose we express Eq. (13.30) in the form

$$\int_{-L/2}^{L/2} I(z')G(z, z')dz' = -E_z^i(z) \qquad (13.39)$$

where

$$G(z, z') = \frac{1}{j4\pi\omega\varepsilon} \left[\frac{\partial^2}{\partial z^2} + \beta^2 \right] K(z - z') \qquad (13.40)$$

Then the MoM may be used to reduced Eq. (13.39) to a system of simultaneous linear algebraic equations of the form

$$\sum_{n=1}^{N} Z_{mn} I_n = V_m \qquad \text{for } m = 1, 2, \ldots, N \qquad (13.41)$$

which may be written in compact matrix notation as

$$[Z_{mn}][I_n] = [V_m] \qquad (13.42)$$

where $[Z_{mn}]$, $[I_n]$, and $[V_m]$ are known as the impedance, the current and voltage matrices, respectively. Hence, the problem has been reduced from solving for the unknown current distribution $I(z')$ in Eq. (13.30) to solving for the I_n in Eq. (13.42).

There are several powerful antenna analysis codes available that are based on a MoM formulation. These codes are used by antenna engineers to aid in the design of complex antenna systems where it is important to be able to predict the interaction of antennas with each other and their environment. Some of the more popular antenna modeling codes, which have already been briefly discussed include NEC, ELNEC, and MININEC.

The MoM formulations used in these codes are essentially valid for the analysis of electrically thin wires since they are based on the reduced kernel approximation of Eq. (13.33). A thin-wire antenna analysis capability is sufficient for many practical applications. However, there are some situations that may require the analysis of moderately thick-wire antennas, especially at higher frequencies ($100\,\text{MHz} \le f \le 1\,\text{GHz}$). For these cases, a MoM code has been developed at the Pennsylvania State University, Applied Research Laboratory, which is capable of accurately analyzing moderately thick- as well as thin-wire antennas. This code uses a robust treatment of the cylindrical wire kernel Eq. (13.31) in its analysis of thicker wires. Figure 13.14 graphically illustrates the improvements in input impedance predictions for moderately thick wires when using a robust treatment of the kernel as compared to a strictly reduced kernel thin-wire treatment. The code predictions as well as measured values of input resistance and reactance for various moderately thick monopoles are plotted in Fig. 13.14(a) and Fig.13.14(b), respectively.

Figure 13.15 contains design curves of impedance as a function of length for a center fed dipole in free space. The data for these curves was generated using the Pennsylvania State University, MoM code. The MoM may also be used to determine radiation patterns and directivity of a particular antenna. The radiation patterns for several popular dipole antennas, including the half-wave and full-wave dipole, have already been discussed in Sec. 13.1.3 with their corresponding directivities listed in Table 13.2.

Loop Antennas

A thin wire bent into a closed contour is called a loop antenna. The terminals of a loop antenna are formed by a small discontinuity or gap in the conducting wire. Loops have been used in such diverse applications as radio receiving antennas, direction finding, magnetic field-strength probes, as well as array elements.

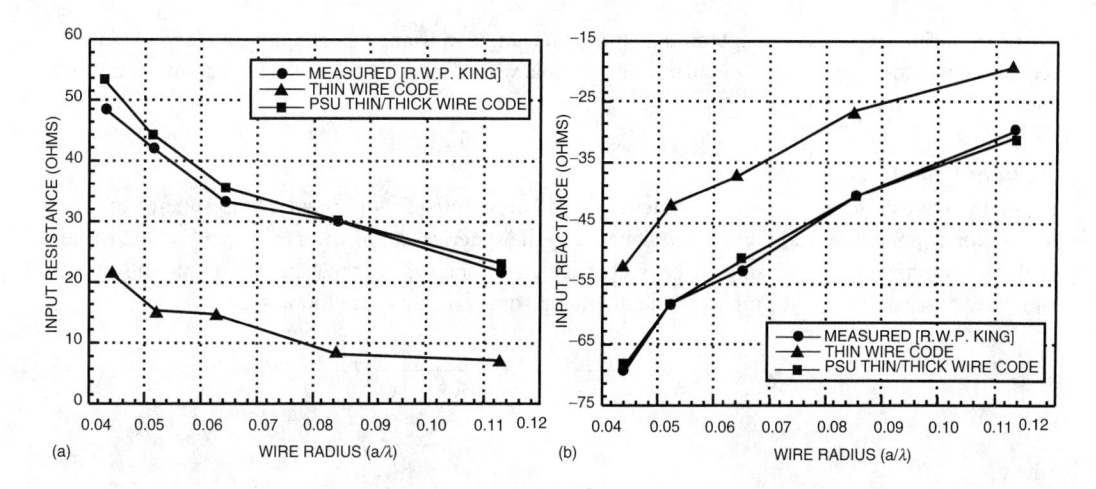

FIGURE 13.14 Code predictions for the input impedance of a 0.4λ monopole compared to measurements: (a) input resistance, (b) input reactance.

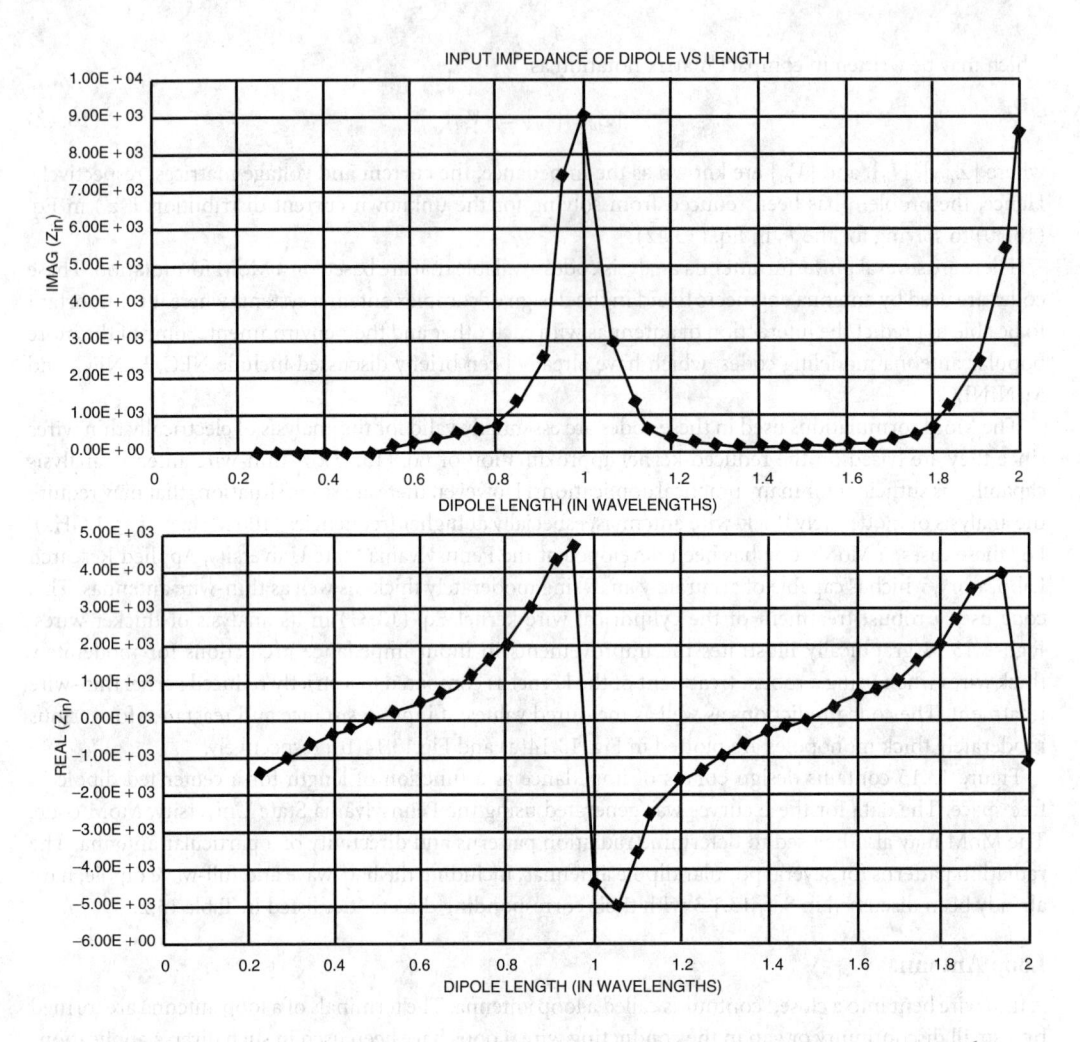

FIGURE 13.15   Design curves of input impedance vs. dipole length for a fixed radius of $a = 1 \times 10^{-6}\lambda$.

There are various types of loop antennas including rectangular, triangular, rhombic, and circular. In this section we will focus on the thin circular loop antenna, which is one of the most popular and commonly used configurations.

Radiated Fields

Figure 13.16 illustrates the coordinate system that will be adopted for the circular loop antenna of radius a. The source point and field point are designated by the spherical coordinates $(r' = a, \theta' = 90°, \phi')$ and (r, θ, ϕ), respectively. Hence, using the geometry depicted in Fig. 13.16 it may be shown that the distance from the source point on the loop to the field point at some arbitrary location in space is

$$R' = \sqrt{R^2 - 2ar\,\sin\theta\,\cos(\phi - \phi')} \qquad (13.43)$$

where

$$R = \sqrt{r^2 + a^2} \qquad (13.44)$$

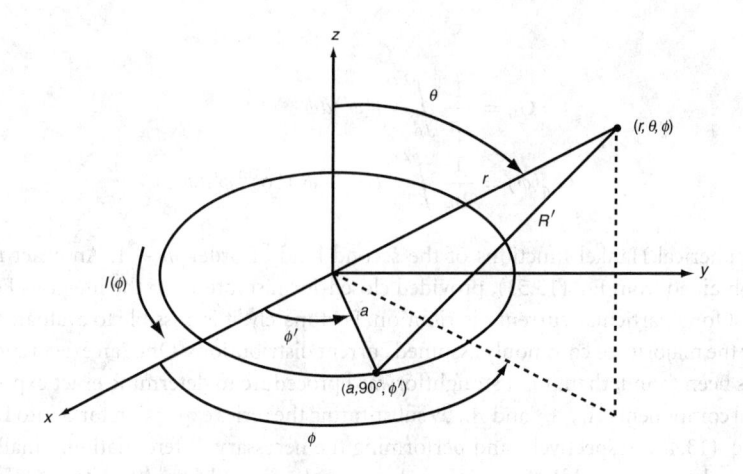

FIGURE 13.16   Circular loop antenna geometry.

The vector potential for the circular loop with a ϕ-directed current $I(\phi)$ may be expressed in the general form (Balanis, 1982)

$$\boldsymbol{A}(r,\theta,\phi) = A_r(r,\theta,\phi)\boldsymbol{r} + A_\theta(r,\theta,\phi)\boldsymbol{\theta} + A_\phi(r,\theta,\phi)\boldsymbol{\phi} \tag{13.45}$$

where

$$A_r(r,\theta,\phi) = \frac{\mu a \sin\theta}{4\pi} \int_0^{2\pi} I(\phi')\sin(\phi-\phi')\frac{e^{-j\beta R'}}{R'}d\phi' \tag{13.46}$$

$$A_\theta(r,\theta,\phi) = \frac{\mu a \cos\theta}{4\pi} \int_0^{2\pi} I(\phi')\sin(\phi-\phi')\frac{e^{-j\beta R'}}{R'}d\phi' \tag{13.47}$$

$$A_\phi(r,\theta,\phi) = \frac{\mu a}{4\pi} \int_0^{2\pi} I(\phi')\cos(\phi-\phi')\frac{e^{-j\beta R'}}{R'}d\phi' \tag{13.48}$$

A general method for evaluating the vector potential integrals defined in Eqs. (13.46–13.48) has been developed in (Werner, 1996). This exact integration technique is based on the fact that Eqs. (13.46–13.48) may be expressed in the following way:

$$A_r(r,\theta,\phi) = -\frac{\mu}{j2\beta r}\frac{d}{d\phi}\Im(r,\theta,\phi) \tag{13.49}$$

$$A_\theta(r,\theta,\phi) = -\frac{\mu}{j2\beta r \tan\theta}\frac{d}{d\phi}\Im(r,\theta,\phi) \tag{13.50}$$

$$A_\phi(r,\theta,\phi) = \frac{\mu}{j2\beta r \cos\theta}\frac{d}{d\phi}\Im(r,\theta,\phi) \tag{13.51}$$

where the integral

$$\Im(r,\theta,\phi) = \frac{1}{2\pi}\int_0^{2\pi} I(\phi')e^{-j\beta R'}d\phi' \tag{13.52}$$

is common to all three components of the vector potential. This integral may be expressed in the form of an infinite series given by

$$\Im(r,\theta,\phi) = G_0 e^{-j\beta R} + \sum_{m=1}^{\infty} G_m(\phi)\frac{(\beta^2 ar\sin\theta)^m}{m!}\frac{h_{m-1}^2(\beta R)}{(\beta R)^{m-1}} \tag{13.53}$$

where

$$G_0 = \frac{1}{2\pi} \int_0^{2\pi} I(\phi')d\phi' \tag{13.54}$$

$$G_m(\phi) = \frac{1}{2\pi} \int_0^{2\pi} I(\phi + \phi') \cos^m \phi' d\phi' \tag{13.55}$$

and $h_{m-1}^{(2)}$ are spherical Hankel functions of the second kind of order $m - 1$. An exact representation for $\Im$ may be obtained from Eq. (13.53), provided closed-form solutions to the integrals Eq. (13.54) and Eq. (13.55) exist for a particular current distribution. Fortunately, it is possible to evaluate these integrals analytically for the majority of commonly assumed current distributions. Once an exact representation for a specific $\Im$ has been found, then it is a straightforward procedure to determine exact expressions for the vector potential components A_r, A_θ, and A_ϕ by substituting the series expansion for $\Im$ into Eq. (13.46), Eq. (13.47), and Eq. (13.48), respectively, and performing the necessary differentiation. Finally, expressions for the electric and magnetic fields of the circular loop may be derived from its vector potential by making use of the relationships

$$H = \frac{1}{\mu}(\nabla \times A) \tag{13.56}$$

$$E = \frac{1}{j\omega\varepsilon} \nabla \times H = \frac{1}{j\omega\mu\varepsilon}[\nabla(\nabla \cdot A) + \beta^2 A] \tag{13.57}$$

Scalar equations for the three components of the magnetic field, which are in terms of the vector potential components, follow directly from Eq. (13.56). These equations are given by

$$H_r(r,\theta,\phi) = \frac{1}{\mu r \sin\theta}\left[\frac{\partial}{\partial\theta}(\sin\theta\, A_\phi) - \frac{\partial}{\partial\phi}A_\theta\right] \tag{13.58}$$

$$H_\theta(r,\theta,\phi) = \frac{1}{\mu r}\left[\frac{1}{\sin\theta}\frac{\partial}{\partial\phi}A_r - \frac{\partial}{\partial r}(r\, A_\phi)\right] \tag{13.59}$$

$$H_\phi(r,\theta,\phi) = \frac{1}{\mu r}\left[\frac{\partial}{\partial r}(r\, A_\theta) - \frac{\partial}{\partial\theta}A_r\right] \tag{13.60}$$

Likewise, the vector form of the electric field Eq. (13.57) may be separated into three scalar components, which results in

$$E_r(r,\theta,\phi) = \frac{\eta}{j\beta r \sin\theta}\left[\frac{\partial}{\partial\theta}(\sin\theta\, H_\phi) - \frac{\partial}{\partial\phi}H_\theta\right] \tag{13.61}$$

$$E_\theta(r,\theta,\phi) = \frac{\eta}{j\beta r}\left[\frac{1}{\sin\theta}\frac{\partial}{\partial\phi}H_r - \frac{\partial}{\partial r}(r\, H_\phi)\right] \tag{13.62}$$

$$E_\phi(r,\theta,\phi) = \frac{\eta}{j\beta r}\left[\frac{\partial}{\partial r}(r\, H_\theta) - \frac{\partial}{\partial\theta}H_r\right] \tag{13.63}$$

where

$$\eta = \sqrt{\frac{\mu}{\varepsilon}} \tag{13.64}$$

is the characteristic impedance of the medium.

Electrically Small Circular Loop

Circular loop antennas in which the radius of the loop is less than about 0.04λ may be considered electrically small. Electrically small-loop antennas are not usually used in transmitting applications because of their very low-radiation resistance, making them inefficient radiators. On the other hand, small circular loops are frequently employed in receiving applications. The standard approximation for the current distribution on an electrically small circular loop is uniform, that is, $I(\phi) = I_0$ where I_0 is a constant (Balanis, 1982).

The integral form of the vector potential for a uniform current loop is well known and may be expressed as

$$\mathbf{A} = A_\phi(r,\theta)\phi \tag{13.65}$$

where

$$A_\phi(r,\theta) = \frac{a\mu I_0}{2}\frac{1}{\pi}\int_0^\pi \cos\phi' \frac{e^{-j\beta R'}}{R'}d\phi' \tag{13.66}$$

and R' of Eq. (13.43) reduces to

$$R' = \sqrt{R^2 - 2ar\sin\theta\cos\phi'} \tag{13.67}$$

The procedure outlined in the Radiated Fields subsection may be followed to find an analytical expression for the uniform current vector potential integral of Eq. (13.66). This results in the following exact series representation for the near-zone of the loop:

$$A_\phi(r,\theta) = \frac{\beta a \mu I_0}{2j}\sum_{m=1}^\infty \frac{[(\beta^2 ar\sin\theta)/2]^{2m-1}}{m!(m-1)!}\frac{h_{2m-1}^{(2)}(\beta R)}{(\beta R)^{2m-1}} \tag{13.68}$$

The corresponding exact representations of the magnetic and electric field components may be obtained from Eqs. (13.58–13.60) and (13.61–13.63), respectively, by making use of Eq. (13.68). These field expressions are given by

$$H_r(r,\theta) = \frac{\beta(\beta a)^2 I_0 \cos\theta}{2j}\sum_{m=1}^\infty \frac{[(\beta^2 ar\sin\theta)/2]^{2m-2}}{[(m-1)!]^2}\frac{h_{2m-1}^{(2)}(\beta R)}{(\beta R)^{2m-1}} \tag{13.69}$$

$$H_\theta(r,\theta) = \frac{\beta(\beta a)^2 I_0 \sin\theta}{2j}\left[\frac{(\beta r)^2}{2}\sum_{m=1}^\infty \frac{[(\beta^2 ar\sin\theta)/2]^{2m-2}}{m!(m-1)!}\frac{h_{2m}^{(2)}(\beta R)}{(\beta R)^{2m}}\right.$$
$$\left. -\sum_{m=1}^\infty \frac{[(\beta^2 ar\sin\theta)/2]^{2m-2}}{[(m-1)!]^2}\frac{h_{2m-1}^{(2)}(\beta R)}{(\beta R)^{2m-1}}\right] \tag{13.70}$$

$$H_\phi = 0 \tag{13.71}$$

$$E_r = E_\theta = 0 \tag{13.72}$$

$$E_\phi(r,\theta) = -\frac{\eta\beta(\beta a)I_0}{2}\sum_{m=1}^\infty \frac{[(\beta^2 ar\sin\theta)/2]^{2m-1}}{m!(m-1)!}\frac{h_{2m-1}^{(2)}(\beta R)}{(\beta R)^{2m-1}} \tag{13.73}$$

Small-loop approximations to the fields may be obtained from Eq. (13.69), Eq. (13.70), and Eq. (13.73) by retaining only the first term ($m=1$) in each series and recognizing that $R \to r$ as $a \to 0$. Under these conditions, we find that

$$H_r(r,\theta) \approx \frac{j\beta a^2 I_0 \cos\theta}{2r^2}\left[1+\frac{1}{j\beta r}\right]e^{-j\beta r} \tag{13.74}$$

$$H_\theta(r,\theta) \approx -\frac{(\beta a)^2 I_0 \sin\theta}{4r}\left[1+\frac{1}{j\beta r}-\frac{1}{(\beta r)^2}\right]e^{-j\beta r} \tag{13.75}$$

$$E_\phi(r,\theta) \approx \frac{(\beta a)^2 \eta I_0 \sin\theta}{4r}\left[1+\frac{1}{j\beta r}\right]e^{-j\beta r} \tag{13.76}$$

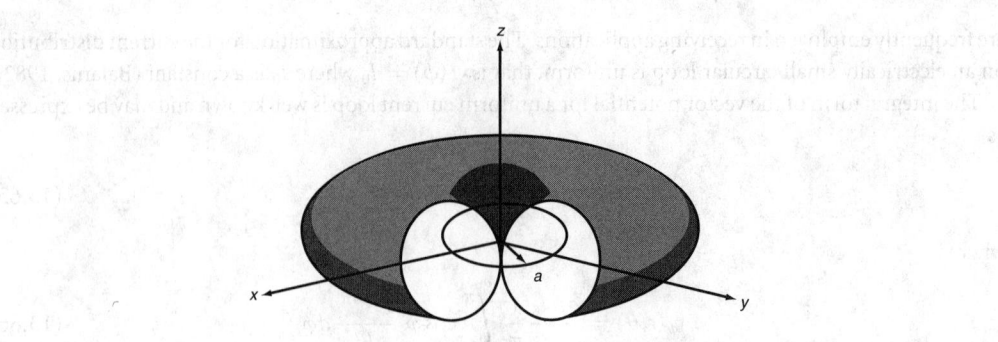

FIGURE 13.17 Three-dimensional radiation pattern produced by an electrically small-loop antenna with a uniform current distribution.

In the far zone of the loop, that is, when $\beta r \gg 1$, expressions (13.74)–(13.76) may be used to show that the fields reduce to the simple form given by

$$H_r \approx H_\phi = E_r = E_\theta = 0 \tag{13.77}$$

$$H_\theta \approx -\frac{(\beta a)^2 I_0 e^{-j\beta r}}{4r} \sin\theta \tag{13.78}$$

$$E_\phi \approx \eta \frac{(\beta a)^2 I_0 e^{-j\beta r}}{4r} \sin\theta \tag{13.79}$$

Hence, the far-field radiation pattern of an electrically small-loop antenna has a $\sin\theta$ variation as demonstrated by Eq. (13.78) and Eq. (13.79). Figure 13.17 illustrates the three-dimensional radiation pattern that would be produced by a small loop. Finally, by using Eq. (13.15) with $f(\theta, \phi) = \sin\theta$, it follows that the directivity of a small loop is

$$D = \frac{3}{2} \quad \text{or} \quad 1.76\,\text{dB} \tag{13.80}$$

which is the same as that of an infinitesimal or short dipole (see Table 13.2). The corresponding radiation resistance of a small circular loop with circumference $C = 2\pi a$ is given by the formula (Balanis, 1982)

$$R_{\text{rad}} = 20\pi^2 \left(\frac{C}{\lambda}\right)^4 \Omega \tag{13.81}$$

Circular Loops with Nonuniformly Distributed Current

The uniform current analysis that was presented earlier in this section is valid for electrically small loops. However, as the radius of the loop increases, the current is no longer uniform as it begins to vary along the circumference of the loop. One common assumption for the variation of the loop current under these conditions is a consinusoidal distribution of the form (Balanis, 1982; Lindsey, 1960)

$$I(\phi) = I_n \cos(n\phi); \quad n = 0, 1, 2, \ldots \tag{13.82}$$

This assumption may be generalized to include loops having an arbitrary current distribution represented by the following Fourier cosine series (Werner, 1996; Storer, 1956):

$$I(\phi) = \sum_{n=0}^{\infty} I_n \cos(n\phi) \tag{13.83}$$

This form of the current distribution is very general and may be applied to the analysis of the radiation characteristics of loop antennas for a wide variety of circumstances (King and Smith, 1981; King, 1969).

Exact expressions for the vector potential and electromagnetic fields may be derived from the current distributions Eq. (13.82) and Eq. (13.83) by following the procedure outlined earlier in this section (Werner, 1996). The resulting field expressions, although useful, are complicated and beyond the scope of this chapter.

The actual form of the current distribution on a particular loop antenna may be determined by performing a method of moments analysis, as discussed in two earlier sections. It is a straightforward procedure to find the input impedance, radiation pattern, and gain of a loop once the current distribution is known. Method of moments generated plots of the input impedance vs. loop radius may be found in Fig. 13.6 of Harrington (1968).

Yagi–Uda Arrays

A **Yagi–Uda** (or Yagi) antenna is a parasitic linear array that contains one driven element and one or more parasitic elements. A parasitic element is called a *reflector* when the radiation pattern enhancement is from the parasitic to the driver, or it is referred to as a *director* when the radiation enhancement is from the driver to the parasitic. A reflector and director can be obtained by properly adjusting the length of the parasitic. Usually, a reflector is cut longer than the driver and a director is cut shorter than the driver, so that the main beam is enhanced in the same direction by both reflector and director. Numerous curves (Hall, 1992) are available to show the effect of element spacings and lengths on input resistance, input reactance, front-to-back ratio, gain, and radiation patterns. These features are for two, three, and four element Yagis. Generally a Yagi having closely spaced elements has a low-radiation resistance and suffers from having a narrow bandwidth of 3–5% of the center frequency.

The numerical electromagnetic codes described in the first subsection of Sec. 13.1.3 are suitable for analyzing Yagi arrays. However, these codes can only be used for thin-wire analysis (wire radius less than 0.01λ). They do have limitations on the radius of the elements and are considered to be thin-wire codes. There are many applications where thick wires are used to construct antennas. The same wire radius for relatively lower frequencies are no longer a thin wire at a much higher frequency. Also, thick wires, such as large diameter aluminum or copper tubing, are often used to increase antenna radiation resistance, lower power loss, as well as to increase mechanical stability. In any of these cases, the thin wire codes in the first subsection of Sec. 13.1.3 can no longer provide accurate results. The electromagnetic code developed at the Pennsylvania State University, discussed in the second subsection of Sec. 13.1.4, has improved the moment method to allow very thick conductors to be modeled (wire radius from 0.01 to 0.1λ) and, consequently, can give improved accuracy in the design of a Yagi. Figure 13.18 shows the geometry for a six-element Yagi array where a thicker wire was used for the driven element.

The maximum gain of a Yagi can be improved by increasing the number of directors. However, there is no significant improvement by having more than one reflector (Stutzman and Thiele, 1981). The gain or the front-to-back ratio (front-to-back ratio was discussed in an earlier section) can be controlled by adjusting the spacing between elements and the length of the elements.

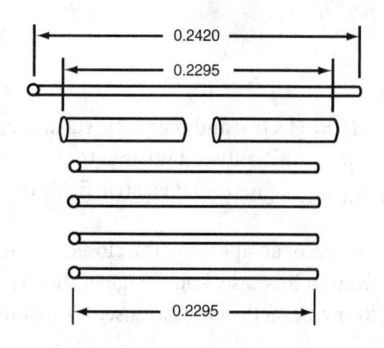

FIGURE 13.18 Geometry for a six-element Yagi array with a thick-wire driven element.

Input impedance, which is measured at the feeding point of the driver, is affected by the reflector and directors due to the mutual coupling effect. The amount of effect on resistance and reactance on the driven element by the parasitic depends on the spacing and the length of the elements. A good front-to-back ratio gives maximum forward signal and minimum rearward signal. Normally, the best front-to-back ratio can not be obtained without sacrificing the maximum gain. An optimum design, which gives a maximum front-to-back ratio with a small sacrifice in gain, is obtainable by proper adjustment of the spacing and length of Yagi elements.

TABLE 13.3 Characteristics of Equally Spaced Yagi–Uda Antennas (conductor radii = 0.0015λ)

No. of Elements	Spacing, λ	Reflector, λ	*Driver*, λ	Directors, λ	Gain, dB	Z_{in}, Ω
3	0.25	0.479	0.453	0.451	9.5	$22.79 + j7.62$
4	0.25	0.486	0.463	0.456	11.03	$11.94 + j13.76$
5	0.25	0.477	0.451	0.442	11.03	$45.87 + j4.07$
6	0.25	0.484	0.459	0.446	12.18	$23.04 + j10.26$
7	0.25	0.477	0.454	0.434	12.06	$53.31 + j2.86$

Tables 13.3 and 13.4 give design data for thin-wire and thick-wire antennas, respectively. Yagis having 3–7 elements provide the designer with the various options for gain and input impedance. (All calculations were made with the Pennsylvania State University method of moments code discussed earlier). Note the dramatic increase in input impedance for the case of a six-element Yagi array that was achieved by using a thicker driven element.

13.1.5 Apertures

Equivalence Theorem

The calculation of the characteristics of a wire antenna started from the knowledge of the current distribution. In the case of an aperture antenna, shown in Fig. 13.1(e), there is no easy solution to finding the distribution of current. In this case the currents that flow on the inner surfaces of the waveguide and the horn are the primary currents responsible for radiation. Problems of this type are more easily solved using the *equivalence theorem*. The theorem states that if the electric and magnetic fields are known over a closed surface, then this surface can be replaced by a distribution of magnetic and electric surface currents. Maintaining the inner volume of the closed surface as source free, then the two surface currents can reproduce the fields exterior to the closed surface. This equivalence leads to powerful methods for analyzing apertures. The electric and magnetic surface currents are given by

$$J_s = n \times H_A$$
$$M_s = -n \times E_A$$

(13.84)

where
$$J_s, M_s = \text{electric and magnetic surface currents}$$
$$n = \text{unit outward normal to the closed surface}$$
$$H_A, E_A = \text{magnetic and electric fields over the closed surface}$$

In the case of an aperture, the closed surface would include the open aperture, such as the mouth of the horn. Even in this case, knowledge of the distribution of the electric and magnetic fields over the aperture is not known exactly. One can use, for instance, waveguide theory to estimate the field distribution over

TABLE 13.4 Characteristics of Equally Spaced Yagi–Uda Antennas (driven element radius = .015λ, all other conductor radii = .0015λ)

No. of Elements	Spacing, λ	Reflector, λ	*Driver*, λ	Directors, λ	Gain, dB	Z_{in}, Ω
3	0.25	0.479	0.453	0.451	9.54	$35.71 + j65.16$
4	0.25	0.486	0.463	0.456	11.08	$35.70 + j65.16$
5	0.25	0.477	0.451	0.442	11.04	$17.92 + j59.84$
6	0.25	0.484	0.459	0.446	12.20	$70.34 + j56.04$
7	0.25	0.477	0.454	0.434	12.07	$34.84 + j59.43$

the aperture. Or in some instances a good engineer-ing guess can be made. With these approximations in mind, the equivalence theorem still remains a good technique for calculating the radiation characteristics of aperture structures.

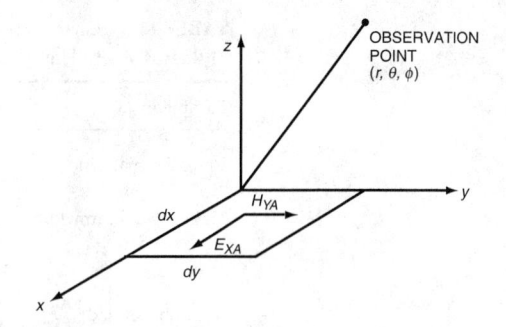

FIGURE 13.19 Geometry of an elementary aperture: a Huygen's source.

Huygens' Sources

Figure 13.19 gives the geometry for an elementary aperture of dimensions $dx \times dy$ having a uniform electric field E_{XA} distribution across it. Visualizing this distribution as part of a wavefront, then the magnetic field distribution H_{YA} is computed based on the electric field and the intrinsic impedance, η_0, of free space; that is, $H_{YA} = E_{XA}/\eta_0$. Applying the equivalence theorem, it can be shown that the far field from this elementary aperture is

$$E_\theta = \frac{jE_{XA}dxdye^{-j\beta r}}{2\lambda r}\cos\phi(1 + \cos\theta) \tag{13.85}$$

$$E_\phi = \frac{-jE_{XA}dxdye^{-j\beta r}}{2\lambda r}\sin\phi(1 + \cos\theta) \tag{13.86}$$

The application of these results to a larger aperture is found from Huygens' principle. This states that each point on a wavefront sends out secondary waves that can be combined to form a new wavefront. Thus by knowing the distribution of amplitude and phase of the fields across the aperture under examination, the superposition of the Huygens' source fields across the aperture can determine the remote field on a wavefront elsewhere.

Practical Apertures

Figure 13.20 shows several practical aperture antennas. Figure 13.20(a) to Figure 13.20(d) pertain to horn types, whereas Fig. 13.20(e) is the paraboloidal antenna, which is popular with satellite television receiving systems for home use. Shown in Fig. 13.20(a) to Fig. 13.20(c) is a TE_{10} electric field distribution. The terminology TE_{10} implies a transverse electric field, as shown, with one-half of a cosine variation in one direction and no variation in the other direction, hence the subscript 10. In Fig. 13.20(a), it is called a

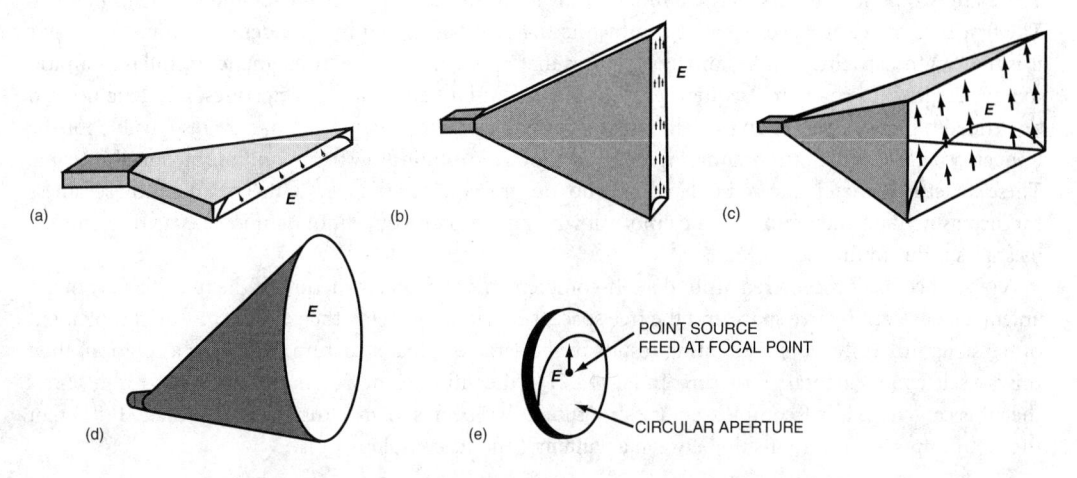

FIGURE 13.20 Some practical aperture antennas: (a) sectoral H-plane, (b) sectoral E-plane, (c) pyramidal, (d) conical, (e) paraboloid.

TABLE 13.5 Characteristics of a Circular Aperture
of Radius *a* with Uniform Excitation

Parameter	Result
Radiation pattern	$\dfrac{2J_1(\beta a \sin\theta)}{\beta a \sin\theta}$
Half-power beamwidth, rad	$1.02\dfrac{\lambda}{2a}$
Directivity	$\dfrac{4\pi}{\lambda^2}(\pi a^2)$
Side lobe level, dB	-17.6

sectorial *H*-plane because the horn is flared out in the *H*-field direction; whereas in Fig. 13.20(b), the horn is a sectorial *E*-plane because the horn is flared out in the *E*-field direction. Flaring out in both directions gives the pyramidal horn shown in Fig. 13.20(c). Universal radiation patterns and directivity curves (Stutzmam and Thiele, 1981) provide design data.

The paraboloidal antenna derives its name from the paraboloidal shape of the reflecting metallic surface. A source is placed at the feed point, like a horn, and the reflected wave illuminates the circular aperture. For the simplistic case of a uniform field distribution over the circular aperture, the useful parameters for this case are given in Table 13.5.

The quantity J_1 is the Bessel function of the first kind of order unity. Although a uniform distribution is not realistic, it shows the general characteristics of a paraboloidal reflector antenna. For more accuracy, it is usually assumed that there is a parabolic taper (Stutzman and Thiele, 1981) to the field distribution with the maximum being at the center and tapering to smaller values at the edge of the circular aperture. The value of the field at the edges is a function of the beamwidth of the feed antenna.

13.1.6 Wideband Antennas

Antennas can be designed to work over several octaves in frequency. Dipoles or arrays of dipoles as discussed in earlier sections are very frequency dependent. This section discusses some of the principles that in the ideal sense lead to frequency-independent characteristics. Generally, the ideal structure is infinite in extent; hence, it is not practical. However, truncating these ideal structures to finite dimensions can lead to wide-band operation.

Frequency-Independent Principles

There are two basic concepts that are inherent in the conception of a frequency-independent antenna. The first is to arrive at a geometry of conductors that can be defined independent of a dimension. For it is the lengths, spacings, and diameters of wires that appear in the electromagnetic formulas as a ratio to wavelength, thus changing frequency results in nonconstant antenna characteristics. If a structure can be defined only by angles, then one eliminates the characteristic dimensions that are responsible for the frequency dependance of an antenna. Figure 13.21 shows the infinite bow tie and infinite biconical antenna. These structures when truncated in length are no longer frequency independent. They do find usefulness for increasing the bandwidth over a cylindrical wire dipole. This concept of defining a structure entirely by angles is due to Rumsey (1966).

A second concept is centered around a self-complementary planar structure. If the metallic part of the antenna is replaced by free space and the free space is replaced by metal, then this forms the complement of the structure. If the complementing results in the same original structure except for a rotation, then this is a self-complementary structure. In Fig. 13.21, if the infinite bow tie has α equal to 90° for instance, then this can create a self-complementary structure. It has been shown (Stutzman and Thiele, 1981) that the input impedance relationship between an antenna and its complement is

$$Z_A(f) \times Z_{AC}(f) = \frac{Z_0^2}{4} \tag{13.87}$$

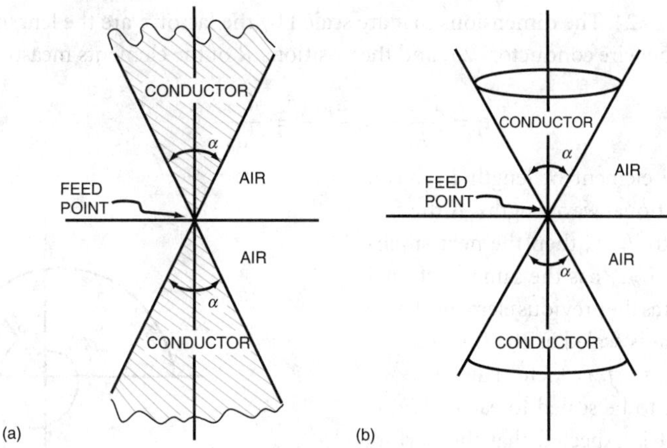

FIGURE 13.21 Two antenna shapes defined by an angle α: (a) bow tie, (b) biconical.

where $Z_A(f)$ and $Z_{AC}(f)$ are the input impedances for the antenna and its complement and $Z_0 = 120\pi$ is the characteristic impedance of free space. For a self-complementary structure $Z_A = Z_{AC}$, since it is exactly the original structure, so that

$$Z_A(f) = Z_{AC}(f) = 60\pi \quad \Omega \tag{13.88}$$

From this one infers that the input impedance is independent of frequency and equals 60π Ω.

The spiral shape has been used to develop a series of wide-band antennas. Figure 13.22 shows two arms of an antenna constructed from conductors with edges forming a spiral shape. The equation of a single spiral as shown in Fig. 13.23 is

$$\rho = \rho_0 e^{a\phi} \tag{13.89}$$

where

ρ_0 = radius when $\phi = 0$

a = constant that controls the rate of expansion of the spiral

This is an infinite spiral, it continues into the origin as well as out to infinity. It is completely described by an angle and not by any characteristic dimensions. The parameters ρ_0 and a control the specific characteristics of the spiral. The shape in Fig. 13.22 is shown with truncation near the origin. In practice, the feed is connected to the two conductors near the origin. The antenna in Fig. 13.22 can be made self-complementary and, if infinite in extent, will be frequency independent as already discussed. Truncation at the feed point leads to an upper use-

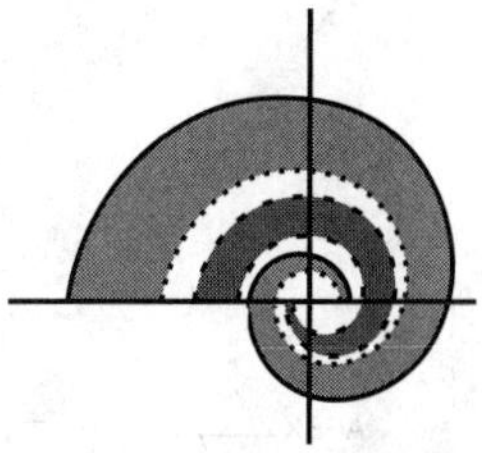

FIGURE 13.22 Spirals for forming the two arms of a self-complementary antenna.

ful frequency limit, whereas truncation of the outer region leads to a lower frequency limit. Wide-band operation is achieved by controlling the dimensions of the truncated regions.

Log-Periodic Antennas

The most practical form of a wide-band antenna originates from the log-periodic dipole array (LPDA) (Stutzman and Thiele, 1981). The property of a spiral like that shown in Fig. 13.23 is that the ratio of distances from the origin of adjacent arms is a fixed quantity τ. This idea is carried over into the design of the

LPDA shown in Fig. 13.24. The dimensions that are scaled by the factor τ are the lengths of the elements L, the diameters of the wire conductors $2a$, and the positions R of the elements measured from the apex

$$\tau = \frac{R_{n+1}}{R_n} = \frac{L_{n+1}}{L_n} = \frac{a_{n+1}}{a_n} \qquad (13.90)$$

At frequency f_n, the element of length L_n has a certain length in fractional wavelengths. If the frequency is increased to f_n/τ, then the next smallest element of length L_{n+1} has the same fractional length in wavelengths as the previous element. Therefore, the array in total is scaled as the frequency is increased to f_n/τ, f_n/τ^2, f_n/τ^3, etc. This relationship allows the array to be scaled to each of these frequencies so that it is expected that the performance of the array at these frequencies should be identical. This would be absolutely true if the LPDA were infinite in both directions toward and away from the origin. Because the relationship of the frequencies is τ, the performance is repeatable in log f, hence the name log periodic. A useful design curve for finding the gain of a LPDA is given in Figs. 6–30 of Stutzman and Thiele (1981). Note the optimum gain is defined at the point where the parameter τ is the smallest, thus leading to a smaller number of required elements. In practice, the lowest frequency of operation is where the longest element is approximately a half-wave long and the upper frequency of operation is where the shortest element is approximately a half-wave long,

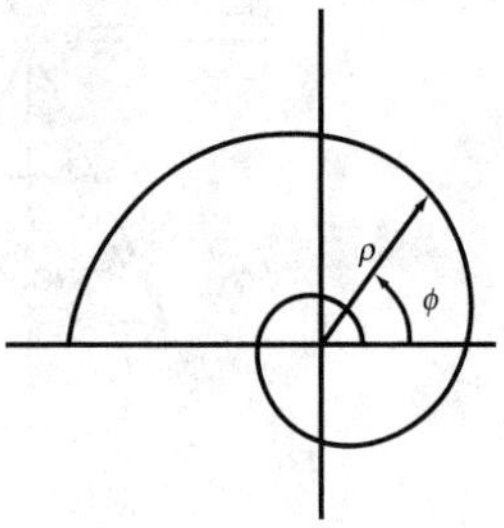

FIGURE 13.23   Basic spiral shape.

$$L_1 \simeq \frac{\lambda_{\text{lower}}}{2} \qquad (13.91)$$

and

$$L_N \simeq \frac{\lambda_{\text{upper}}}{2} \qquad (13.92)$$

The mainbeam of the LPDA is away from the apex, that is, toward the shortest elements. A few extra elements are added onto each end for approximately complying with the idea that the structure should be infinite in the direction of the traveling wave.

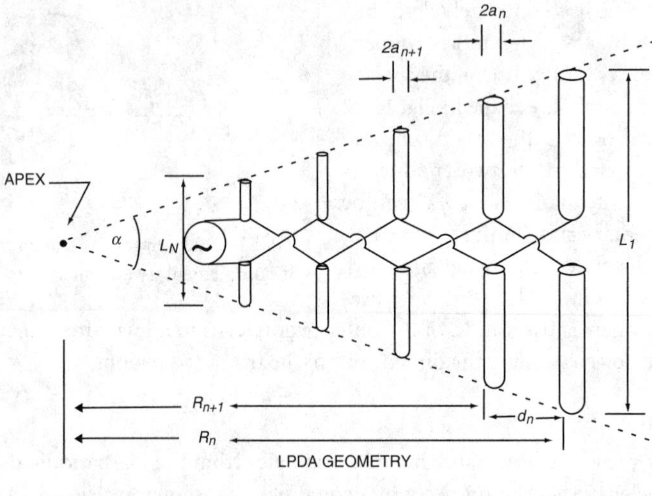

FIGURE 13.24   LPDA geometry.

Frequency Independent Phased Arrays

If the LPDA is part of an array, for example, of equally spaced elements, such as those discussed in Sec. 13.1.3 (Pattern Multiplication subsection), then a characteristic dimension is introduced (the spacings between the LPDAs) and the wide-band operation can be seriously affected. A concept for maintaining log-periodic behavior and still arraying the elements is found in the three-dimensional frequency-independent phased array (FIPA) (Breakall, 1992). In this array, elements in the log periodic, which make up this array, are positioned such that each wire's height above a ground plane is some constant (i.e., 0.25λ) in wavelengths at each wire's resonant frequency. At the prescribed height in wavelengths above the ground plane, each wire's center is also at a constant spacing distance in wavelengths (i.e., 0.60λ) from adjacent wire centers. Scaling of the antenna elements in a log-periodic sense causes the physical wire lengths, heights, and interwire spacings to get progressively smaller as one moves toward the feed end of the element. The scaling factor is the same τ factor as used in the classical log-periodic design.

Figure 13.25 shows an overhead view of a 4 × 4 circularly polarized model of the three-dimensional FIPA. The dark highlighted wires depict the planar nature of the array with the excitation frequency creating large currents at the resonant element frequency. The figure is shown for operation at a midfrequency. The 4 × 4 planar array maintains fixed lengths, spacings, and heights in wavelengths as the frequency is changed. Therefore the gain, impedance and radiation pattern remain fairly constant making this a respectable wide-band array. The feed wires shown in this figure are a single wire for simplicity.

The three-dimensional FIPA antenna described overcomes traditional phased-array limitations by using arrays of log-periodic elements.

FIGURE 13.25 A 4 × 4 3D-FIPA (top view) shown excited at the midfrequency.

Self-Similar Fractal Antennas

It has been recognized that one of the fundamental properties of frequency-independent antennas is their ability to retain the same shape under certain scaling transformations. It has been demonstrated that this property of self-similarity is also shared by many fractals (Mandelbrot, 1983). This commonality has led to the notion that fractal geometric principles be used to provide a natural extension to the traditional approaches for classification, analysis, and design of frequency-independent antennas (D.H. Werner and P.L.Werner). This theory allows the classical interpretation of frequency-independent antennas to be generalized to include the radiation from structures that are not only self-similar in the smooth or discrete sense but also in the rough sense.

The standard approach of categorizing frequency-independent antennas is to consider them as being constructed from a multiplicity of adjoining cells. Each cell is identical to the previous cell except for a scaling factor. This periodicity may be characterized in terms of the constant tau (Lo and Lee, 1988)

$$\tau = D_n / D_{n+1} \tag{13.93}$$

where D_n represents the dimension of the nth cell and D_{n+1} represents the dimension of the next adjoining cell. In view of the recent introduction of fractal geometry, however, a natural extension to this classical definition of broadband antennas is to consider them as being composed of a sequence of self-similar cells with a similarity factor of τ. For instance, scaling the logarithmic spiral defined in Eq. (13.89) by a factor of τ yields the same spiral rotated by a constant angle $\ell n(\tau)/a$. This suggests that the logarithmic spiral is self-similar, with a similarity factor $\tau = e^{2\pi|a|}$. If rotations are disregarded, however, then the logarithmic spiral can be considered self-similar for any real scaling factor τ. The self-similarity arguments developed for the log spiral are easily generalized to apply to the conical log spiral as well.

Another example of a broadband antenna in common use is the log-periodic dipole antenna discussed earlier in this section. Log-periodic dipole antennas exhibit self-similarity in their geometrical structure at discrete frequencies, as illustrated by Fig. 13.26. The lengths and spacings of adjacent dipole elements are scaled by the same similarity factor τ. Figure 13.26 also demonstrates how the log-periodic dipole antenna may be constructed through the use of an initiator and an associated generator.

The scaling properties of log-periodic dipole antennas have also been exploited in the design of a three-dimensional frequency-independent phased array (3D-FIPA) presented earlier in this section. The 3D-FIPA concept can be thought of as having many separate $N \times N$ vertically stacked self-similar planar dipole subarrays. The lengths, heights, and horizontal spacings of the dipoles in each consecutive subarray are physically scaled by a similarity factor of τ. This has the desired effect of ensuring that the dimensions of each dipole subarray remain electrically invariant. Other concepts for broadband arrays of log periodics

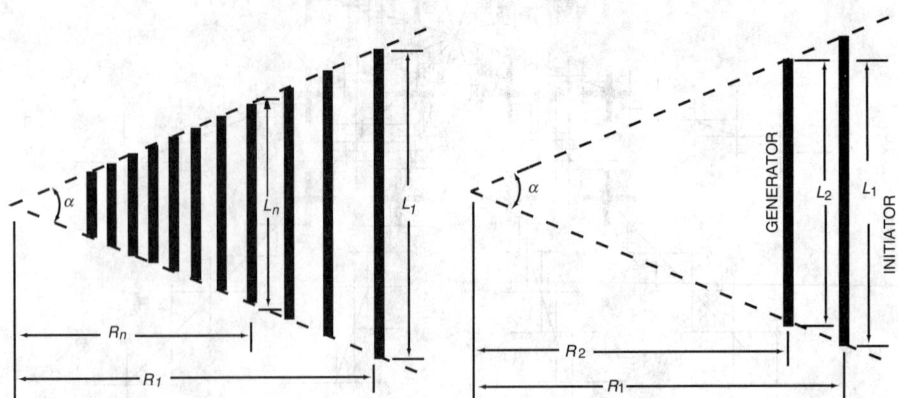

FIGURE 13.26 Fractal geometric description of a log-periodic dipole antenna with initiator and generator.

are discussed in Duhammel and Berry (1958), Mei and Moberg (1965), and Johnson and Jasik (1984), which also take advantage of self-similarity in their designs.

A special class of nonuniformly but symmetrically spaced linear arrays, known as Weierstrass arrays, which possess a self-similar geometrical structure, have been investigated in D.H. Werner and P.L. Werner (1996, 1995) and Werner (1994). It can be shown that the array factor of a Weierstrass array satisfies the self-similarity relation

$$f(\gamma v) = \gamma^{(2-D)} f(v) \tag{13.94}$$

when

$$v = \cos\theta - \cos\theta_0 \tag{13.95}$$

and

$$\gamma = \tau^k \quad \text{for } k = 0, 1, 2, \ldots \tag{13.96}$$

provided $0 < \tau < 1$ and $1 < D < 2$. The parameters τ and D represent the similarity factor and fractal dimension, respectively. The box-counting definition of fractal dimension for a given fractal F was adopted in this case. In other words

$$D = \lim_{\delta \to 0} \frac{\ln N_\delta(F)}{\ln(1/\delta)} \tag{13.97}$$

where N_δ represents the smallest number of sets of diameter at most δ required to cover the fractal F (Falconer, 1988). The connection between the box-counting definition of fractal dimension and the intuitive Euclidean concept of dimension is discussed in Voss (1988) and Jaggard (1990).

The self-similarity in the geometrical structure and radiation pattern of infinite Weierstrass arrays suggest that they may be used as multiband arrays that maintain the same radiation characteristics at an infinite number of frequencies. For example, this multiband performance may be achieved by selecting a sequence of discrete frequencies that satisfy the relationship

$$\frac{f_{k+1}}{f_k} = \tau \quad \text{for } k = 0, 1, \ldots \tag{13.98}$$

By making use of the multiband properties of Eqs. (13.94), (13.96) and (13.98), it may easily be demonstrated that the directive gain obeys the relationship

$$G(\gamma v) = \frac{|f(\gamma v)|^2}{\dfrac{1}{2} \displaystyle\int_{-1-u_0}^{1-u_0} |f(\gamma v)|^2 dv} = \frac{|f(v)|^2}{\dfrac{1}{2} \displaystyle\int_{-1-u_0}^{1-u_0} |f(v)|^2 dv} = G(v) \tag{13.99}$$

where $u_0 = \cos\theta_0$. This suggests that the directive gain of an infinite Weierstrass array is a log periodic function of frequency with a log period of τ, that is

$$G(w) = G(w + kT) \quad \text{for } k = 0, 1, \ldots \tag{13.100}$$

where

$$w = \log v \tag{13.101}$$

$$T = \log \tau \tag{13.102}$$

Defining Terms

Dipole antenna: A straight length of wire excited by a signal source at its center.

Directivity: Provides a measure of performance compared to a reference antenna, such as the isotropic element.

Fraunhofer region: The region that the field from an antenna is a good representation of the far-field expected values. The pattern is essentially independent of the distance from the antenna.

Frequency independent: The term used to define an antenna that has some parameter, such as impedance, independent of frequency. This is an idealization but the concepts can lead to broadband operation.

Fresnel region: In this region the pattern is dependent on the distance from the antenna and does not represent the far-field expected values.

Isotropic source: A fictitious antenna element that radiates equal intensity in all directions. This source is usually the reference antenna element to which gain or directivity of an actual antenna is compared to.

Log periodic: A particular type of broadband antenna in which an antenna parameter, such as impedance, is essentially constant at multiple frequencies that are related logarithmically.

Loop antenna: A type of antenna that is constructed by bending a thin wire into a closed contour.

Method of moments: A numerical method for transforming an integral equation to an associated matrix equation, which can then be readily solved using a computer.

Numerical electromagnetic code: Various antenna software that compute the parameters of an antenna method of moments formulation. NEC, MININEC, and ELNEC are names of some of the popular codes for modeling antennas.

Sommerfeld/Norton solution: A technique that provides the exact solution for fields of wires in the presence of a ground. It is used in antenna modeling software such as NEC for greater accuracy when wires are close to ground.

Surface impedance: The ratio of the tangential electric field on a conductor divided by the resulting linear current density that flows in the conductor.

Uniform theory of diffraction (UTD): Concept allows diffracted fields to be computed for various types of incident waves, such as spherical, plane, and cylindrical.

Yagi–Uda: A Form of antenna array in which there is only one driven element. The other elements derive the current excitations from the mutual coupling between elements. Element spacing must be a small fraction of a wavelength in order to have sufficient current excitation.

References

Balanis, C.A. 1982. *Antenna Theory, Analysis, and Design*. Harper & Row, New York.

Breakall, J.K. 1992. Introduction to the three-dimensional frequency-independent phased array (3D-FIPA), A new class of phased array design. *IEEE Antennas and Propagation Society International Symposium Digest*, Vol. III, pp. 1414–1417, July, Chicago IL.

Burke, G.J. and Poggio, A.J. 1981. Numerical electromagnetic code (NEC), User's Guide. Lawrence Livermore Laboratory, Livermore, CA.

DuHammel, R.H. and Berry, D.G. 1958. Logarithmically periodic antenna arrays. *Wescon Convention Record*, Pt. 1, pp. 161–174.

Falconer, K. 1988. *Fractal Geometry*. Wiley, New York.

Hall, G. 1992. *The ARRL Antenna Book*. The American Radio Relay League, Newington, CT.

Harrington, R.F. 1968. *Field Computation by Moment Methods*. Macmillian, New York.

Hickman, C.E. and Tillman, J.D. 1961. The mutual impedance between identical, parallel dipoles. Bulletin No. 25, Engineering Experiment Station, Univ. of Tennessee, Knoxville, TN.

Jaggard, D.L. 1990. On fractal electrodynamics. In *Recent Advances in Electromagnetic Theory*. Springer-Verlag, New York.

Johnson R.C. and Jasik, H. 1984. *Antenna Engineering Handbook*. McGraw-Hill, New York.

Jordan, E.C. and Balmain, K.G. 1968. *Electromagnetic Waves and Radiating Systems*, 2nd ed. Prentice-Hall, Englewood Cliffs, NJ.

King, R.W.P. 1969. The loop antenna for transmission and reception. In *Antenna Theory*, Pt. I, McGraw-Hill, New York.

King, R.W.P. and Smith, G.S. 1981. *Antennas in Matter: Fundamentals, Theory and Applications.* MIT Press, Cambridge, MA.

Kraus, J.D. and Carver, K.R. 1973. *Electromagnetics,* 2nd ed. McGraw-Hill, New York.

Lewallen, R. 1993. ELNEC antenna analysis program. P. O. Box 6658, Beaverton, OR.

Lindsay, J.E., Jr. 1960. A circular loop antenna with nonuniform current distribution. *IRE Trans. Antennas Propagat.* AP-8(4):439–441.

Lo, Y.T. and Lee, S.W. 1988. *Antenna Handbook.* Van Nostrand Reinhold, New York.

Logan, J.C. and Rockway, J.W. 1986. The new MININEC (version 3): A mini-numerical electromagnetic code. Tech. Doc. 938, Naval Ocean System Center, San Diego, CA.

Mandelbrot, B.B. 1983. *The Fractal Geometry of Nature.* W.H. Freeman, New York.

Mei, K.K., Moberg, M.W., Rumsey, V. H., and Yeh, Y.S. 1965. Directive frequency independent arrays. *IEEE Trans. on Antennas and Propagation* AP-13(5):807–809.

Pocklington, H.C. 1897. Electrical oscillations in wire. *Cambridge Philosophical Soc. Proc.* London, England, 9:324–332.

Rumsey, V. 1966. *Frequency Independent Antennas.* Academic Press, New York.

Storer, J.E. 1956. Impedance of thin-wire loop antennas. *AIEE Trans.* Part I. Communication and Electronics 75(Nov.):606–619.

Stutzman, W.L. and Thiele, G.A. 1981. *Antenna Theory and Design,* 2nd ed. Wiley, New York.

Voss, R.F. 1988. Fractals in nature: From characterization to simulation. In *The Science of Fractal Images.* Springer-Verlag, New York.

Werner, D.H. 1993. An exact formulation for the vector potential of a cylindrical antenna with uniformly distributed current and arbitrary radius. *IEEE Trans. on Antennas and Propagation* 41(8):1009–1018.

Werner, D.H. 1994. Fractal radiators. In *Proceedings of IEEE Dual-Use Technologies & Applications Conference,* Vol. I, pp. 478–482. 4th Annual IEEE Mohawk Valley Section, SUNY Inst. of Technology at Utica/Rome, New York.

Werner, D.H. 1995. Analytical and numerical methods for evaluating the electromagnetic field integrals associated with current-carrying wire antennas. In *Advanced Electromagnetism: Foundations, Theory and Applications.* World Scientific, Ltd.

Werner, D.H. 1996. An exact integration procedure for vector potentials of thin circular loop antennas. *IEEE Trans. on Antennas and Propagation* 44(2):157–165.

Werner, D.H. and Werner, P.L. 1995. On the synthesis of fractal radiation patterns. *Radio Science* 30(1): 29–45.

Werner, D.H. and Werner, P.L. 1996. Frequency independent features of self-similar fractal antennas. *IEEE Trans. on Antennas and Propagation Society International Symposium Digest,* July, Baltimore, MD.

Young, J.S. 1994. Simulation of antenna patterns over 3-dimensional irregular terrain using the uniform geometrical theory of diffraction (UTD) [development of the PAINT system]. Ph.D. thesis in Electrical Engineering, Pennsylvania State Univ., University Park, PA.

Further Information

Additional information on Antenna Technology is available from the following sources:

The IEEE Antennas and Propagation Society (AP-S) publishes a monthly journal, *IEEE Transactions on Antennas and Propogation,* and holds an annual conference. IEEE Headquarters is located in Piscataway, NJ.

International Unions of Radio Scinece (URSI) holds an annual meeting in conjunction with AP-S in the summer and publishes abstracts of conference papers. URSI is headquartered in Brussels, Belgium.

Radio Science is a bimonthly journal, which is published by the American Geophysical Union and cosponsored by International Union of Radio Science (URSI). The American Geophysical Union is headquartered in Washington, D.C.

13.2   Radio Wave Propagation

Gerhard J. Straub

13.2.1   Introduction

From the sparks of the beginning of radio to the present day congested radio frequency environment, the understanding of radio wave propagation plays a vital role in any communications system. With the ever increasing demand to communicate farther with less power and with less interference comes the demand to know how to design a reliable radio frequency propagation path and how to evaluate the potential for system outages and interference. Without a firm grasp of radio wave propagation principles, the system engineer may be forced to specify equipment with performance in excess of that necessary so as to feel secure that the planned communications system will perform as expected. Use of higher than necessary power to establish the desired communications link may result in better path reliability; however, there is the increased risk of interference to others and, in many instances, a violation of applicable rules and regulations. The basics of radio wave propagation are the same whether one is designing a system for operation at 100 kHz or 1000 MHz, but each frequency range has its own advantages, disadvantages, and peculiarities that must be understood if optimum use of the electromagnetic spectrum is to be achieved.

13.2.2   Radio Wave Basics

To visualize a radio wave, consider the image of a sine wave being traced across the screen of an oscilloscope. As the image is traced, it sweeps across the screen at a specified rate, constantly changing amplitude and phase with relation to its starting point at the left side of the screen. Consider the left side of the screen to be the antenna, the horizontal axis to be distance instead of time, and the sweep speed to be the speed of light, or at least very close to the speed of light, and the propagation of the radio wave is visualized. To be correct, the traveling, or propagating, radio wave is really a wavefront, as it

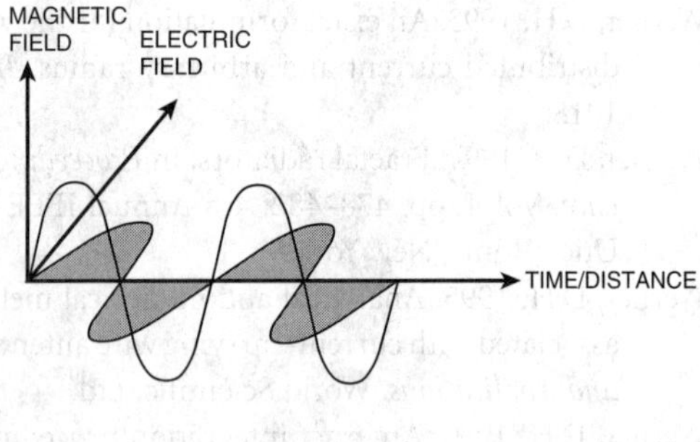

FIGURE 13.27   Propagation of wavefront.

comprises an electric field component and an orthogonal magnetic field component as shown in Fig. 13.27. The distance between wave crests is defined as the **wavelength** and is calculated by

$$\lambda = \frac{c}{f}$$

where
   λ = wavelength, m
   c = the speed of light, approximately 2.998×10^8 m/s
   f = frequency, Hz

At any point in space far away from the antenna, on the order of 10 wavelengths or 10 times the aperture of the antenna to avoid near-field effects, the electric and magnetic fields will be orthogonal and remain constant in amplitude and phase in relation to any other point in space. The polarization of the radio wave is defined by the polarization of the electric field, horizontal if parallel to the Earth's surface and vertical if perpendicular to it. Typically, polarization can be determined by the orientation of the antenna radiating elements.

An **isotropic antenna** is one that radiates equally in all directions. To state this another way, it has a gain of unity. If this isotropic antenna is located in an absolute vacuum and excited with a given amount of power at some frequency, as time progresses the radiated power must be equally distributed along the surface of an ever expanding sphere surrounding the isotropic antenna as in Fig. 13.28. The power density at any point on the surface of this imaginary sphere is simply the radiated power divided by the surface area of the sphere, or,

$$P_d = \frac{P_t}{4\pi D^2}$$

where

P_d = power density, W/m^2
D = distance from antenna, m
P_t = radiated power, W

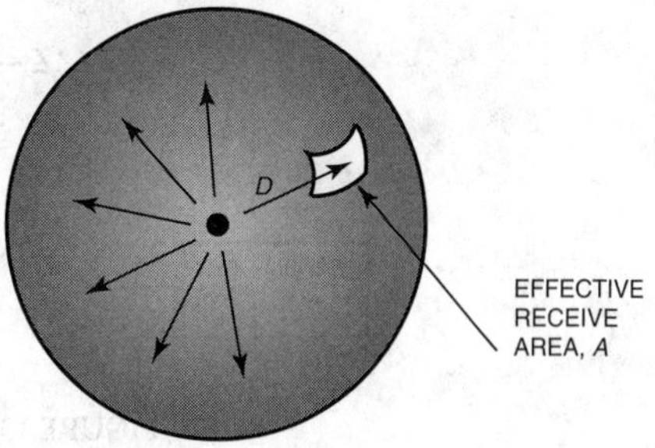

FIGURE 13.28 Power density distribution for an isotropic radiator.

Since power and voltage, in this case power density and electric field strength, are related by impedance, it is possible to determine the electric field strength as a function of distance given that the impedance of free space is taken to be approximately 377 Ω

$$E = \sqrt{Z P_d} = 5.48 \frac{\sqrt{P_t}}{D}$$

where E is the electric field strength in volts per meter. Converting to units of kilowatts for power, the equation becomes

$$E = 173 \frac{\sqrt{P_{t(kW)}}}{D} \text{ V/m}$$

which is the form in which the equation is usually seen. Since a half-wave dipole has a gain of 2.15 dB over that of an isotropic radiator (dBi), the equation for the electric field strength from a half-wave dipole is

$$E = 222 \frac{\sqrt{P_{t(kW)}}}{D} \text{ V/m}$$

From these equations it is evident that, for a given radiated power, the electric field strength decreases linearly with the distance from the antenna and power density decreases as the square of the distance from the antenna.

13.2.3 Free Space Path Loss

A typical problem in the design of a radio frequency communications system requires the calculation of the power available at the output terminals of the receive antenna. Although the gain or loss characteristics of the equipment at the receiver and transmitter sites can be ascertained from manufacturer's data, the effective loss between the two antennas must be stated in a way that allows for the characterization of the transmission path between the antennas. The ratio of the power radiated by the transmit antenna to the power available at the receive antenna is known as the *path loss* and is usually expressed in decibels. The minimum loss on any given path occurs between two antennas when there are no intervening obstructions and no ground losses. In such a case when the receive and transmit antennas are isotropic, the path loss is known as **free space path loss**.

If the transmission path is between isotropic antennas, then the power received by the receive antenna is the power density at the receive antenna multiplied by the effective area of the antenna and is expressed as

$$P_r = \frac{P_t A}{4\pi D^2}$$

where A is the effective area of the receive antenna in square meters.

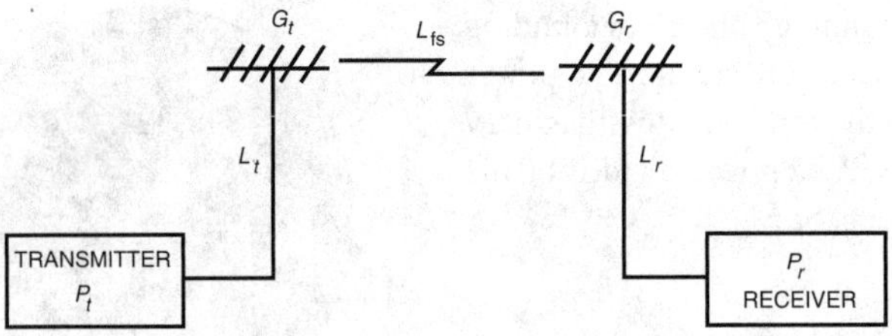

FIGURE 13.29   Path loss variables.

The effective area of an isotropic antenna is defined as $\lambda^2/4\pi$. Note that an isotropic antenna is not a point source, but has a defined area; this is often a misunderstood concept. As a result, the received power is

$$P_r = \frac{P_t}{4\pi D^2} \cdot \frac{\lambda^2}{4\pi} = P_t \left(\frac{\lambda}{4\pi D}\right)^2$$

The term $(\lambda/4\pi D)^2$ is the free space path loss. Expressed in decibels with appropriate constants included for consistency of units, the resulting equation for free space path loss, written in terms of frequency, becomes

$$L_{fs} = 32.5 + 20\log D + 20\log f$$

where
  D = distance, km
  f = frequency, MHz

The equation for the received power along a path with no obstacles and long enough to be free from any near-field antenna effects, such as that in Fig. 13.29, then becomes

$$P_r = P_t - L_t + G_t - L_{fs} + G_r - L_r$$

where
  P_r = received power, dB
  P_t = transmitted power, dB
  L_t = transmission line loss, dB
  G_t = gain of transmit antenna referenced to an isotropic antenna, dBi
  L_{fs} = free space path loss, dB
  G_r = gain of receive antenna, dBi
  L_r = line loss of receiver downlead, dB

It should be pointed out that the only frequency-dependent term in the equation for free space path loss occurs in the expression for the power received by an isotropic antenna. This is a function of the antenna area and, as stated previously, the area of an isotropic radiator is defined in terms of wavelength. As a result, the calculated field strength at a given distance from sources with equal radiated powers but on frequencies separated by one octave will be identical, but the free space path loss equation will show 6-dB additional loss for the higher frequency path. To view this another way, for the two paths to have the same calculated loss, the antennas for both paths must have equal effective areas. An antenna with a constant area has higher gain at higher frequencies. As a result, to achieve the same total path loss over these two paths, the higher frequency path requires a higher gain antenna, but the required effective areas of the antennas for the two paths are equal. The most important concept to remember is that the resultant field strength and power density at a given distance for a given radiated power are the same regardless of frequency, as long as the path approximates a free space path, but that the free space path loss increases by 6 dB for a doubling of frequency or distance.

The representation of the radio wave path in Fig. 13.29 and the previous discussion have only considered a direct path between the receiver and transmitter. In reality, there are two major modes of propagation: the **skywave** and the **groundwave**. The skywave refers to propagation via the ionosphere, which consists of several layers of ionized particles in the Earth's atmosphere from approximately 50 to several hundred kilometers in altitude. Some frequencies will be reflected by the ionosphere resulting in potentially long-distance propagation. This propagation mode is discussed in detail in a later section.

The other major mode of propagation is known as groundwave propagation. Groundwave propagation itself consists of two components, the space wave and the surface wave. The space wave also has two components known as the direct path and the reflected path. The direct path is the commonly depicted line-of-sight path that has been previously discussed and is represented in Fig. 13.29. The reflected path is that path that ends at the receiver by way of reflection from the ground or some other object. Note that there may be multiple reflected paths. The surface wave is that portion of the wavefront that interacts with and travels along the surface of the Earth. The surface wave, which will be discussed in more detail in the next section, is commonly incorrectly called the groundwave.

13.2.4   Reflection, Refraction, and Diffraction

As with light, the direction of propagation of a radio wave may be changed by reflection, **refraction**, or **diffraction**. Like light from a mirror, the propagation of a radio wave may be abruptly changed by reflection from a smooth surface. Smooth in this case is a relative term since the surface must be smooth in terms of wavelength. As a result, although a surface may not appear smooth to the eye of an observer, that is at optical frequencies, the surface may be very smooth and serve as an excellent reflector at the frequencies of interest. Reflection from a perfectly conducting surface results in no energy loss and a complete phase reversal at the reflecting surface. In reality some energy will be lost in the process, as there is no such perfectly conducting surface to be found, but large metal objects or bodies of water may come very close. Perfect reflection results in equal angles of arrival and departure of the direct and reflected wavefronts relative to the reflecting surface.

Assume for a moment that an RF propagation path has been established parallel to and at some height above a perfectly conducting surface. The receive antenna will intercept the transmitted energy by two separate paths: the direct path and the reflected path from the conducting surface. There will be a path height such that the total reflected path length will be 180° longer than the direct path. At this height, the direct and reflected fields will add to result in a 6-dB increase in field strength, since there is a 180° change in phase at the point of reflection. The path height at the point of reflection under these conditions is known as the radius of the first **Fresnel zone** (Fig. 13.30). As the path height at the point of reflection is increased farther, a point will be reached where the direct and reflected paths will be equal in terms of phase. At this point, due to the phase reversal at the reflection point, the fields from the direct and reflected paths will cancel. The height of the path above the reflection point under these conditions is known as the radius of the second Fresnel zone. These conditions repeat themselves as the path height is increased with every odd Fresnel zone radius resulting in a field strength increase and every even Fresnel zone radius resulting in field strength cancellation, or at least a very significant reduction. The Fresnel zone radius increases with increasing distance from the transmitter reaching a maximum at the path midpoint and then decreases with decreasing distance to the receiver. This is a three-dimensional phenomenon, since reflections may occur from a surface on any side of the path. To be more precise, the first Fresnel zone is defined as the locus of all points from which a reflected path will have a pathlength one-half wavelength greater in length. These points form an ellipsoid with the transmitter and receiver antennas as focal points.

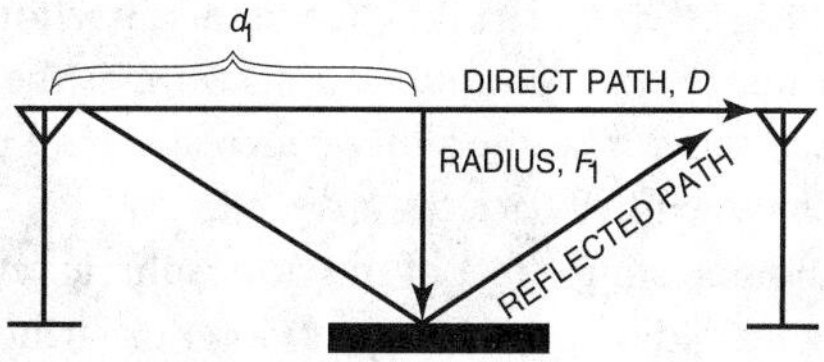

FIGURE 13.30   First Fresnel zone: Reflected path length equals direct path length plus 180°.

Although it is not necessarily incorrect, and many times in fact useful, to visualize the path of the radio frequency energy as a ray between the transmitting and receiving antennas, the path actually has dimensions that are important to the path designer. It is generally accepted that an RF path must have at least 0.6 Fresnel zone radius clearance to any obstruction to be considered an unobstructed or free space path for which the free space path loss equation is directly applicable. A formula for calculating the approximate first Fresnel zone radius at any point along the path is

$$F_1 = \frac{\sqrt{(D - d_1)d_1\lambda}}{D}$$

where

F_1 = radius of the first Fresnel zone
d_1 = distance from the transmitter to the point of interest

Assuming a 50-km path the radius of the first Fresnel zone at the path midpoint is approximately 61 m at 1 GHz, 194 m at 100 MHz, and 612 m at 10 MHz. From these examples, it is apparent that free space path loss conditions do not often exist for lower frequency systems. If there is less than 0.6 Fresnel zone radius clearance on a given path, then it is presumed that diffraction effects must be considered.

Diffraction may occur on any path that does not meet the requirements for free space propagation. When an obstacle is placed in the path of an electromagnetic wavefront, some energy may be bent around the obstacle such that areas expected to be completely shadowed from the source of the energy may actually receive a signal. This diffraction is really an interference effect between the radiated energy from the source and currents induced in the surface of the obstruction. Any time a path is subject to diffraction effects, losses greater than those predicted by free space path equations will occur. The minimum additional loss occurs when the path is partially obstructed by a sharp edge. This is known as *knife-edge diffraction* and paths may actually be designed to rely on this propagation mode. Where a path is at grazing or near grazing incidence over a gently rolling surface, such as a smooth hill or the surface of the Earth, losses can be very high and may render the path unusable. Several methods for calculating diffraction losses have been developed, including those of the National Institute of Standards and Technology compiled in a long standing reference for non-line-of-sight propagation loss calculation methods (Rice et al., 1967). It should be understood that although propagation paths using a mode such as knife-edge diffraction are sometimes said to have obstacle gain, the total path loss will never be less than that of a free space path of the same length.

Refraction is a change of direction of the wavefront due to a change of the refractive index of the medium of propagation. As a wavefront, either at light or radio frequencies, passes through the interface between two mediums with different refractive indices, as shown in Fig. 13.31, the velocity of propagation of the portion of the wavefront entering the medium with the smaller refractive index will be increased relative to that in the denser medium. As the entire wavefront enters the new medium, the

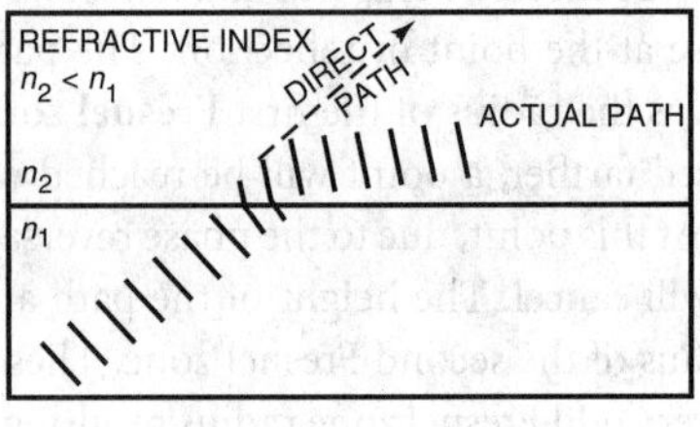

FIGURE 13.31   Effect of refraction: The wavefront is bent toward the denser medium.

resulting effect will be to change the direction of propagation of the wavefront. The medium through which terrestrial communications travel is the Earth's atmosphere, which decreases in density, as well as refractive index, as the altitude increases. As a result, as the wavefront increases in altitude, it undergoes a constant bending back toward the Earth.

An interesting effect of this atmospheric refraction is that, depending on the actual conditions at the time, the radio horizon may be closer or farther away than the true horizon. For an antenna aimed at the horizon, with no intervening terrain or other obstructions, the radiated energy, using a ray analogy, travels toward the horizon. At the horizon, as the Earth's surface curves away, the path of propagation begins to increase in altitude. As the altitude increases, the density of the atmosphere decreases and the propagation path is gradually bent back toward the surface of the Earth, effectively traveling beyond the horizon and somewhat following the curvature of the Earth. As a result, the effective communications range, even for line-of-sight type links, is somewhat beyond the horizon due to the refraction. Stated another way, the distance to the radio horizon, or the **effective Earth radius**, has been increased.

Under typical atmospheric conditions, the effective Earth radius used in propagation planning is 4/3 of the actual Earth radius. Assuming 4/3 Earth radius, the distance to the radio horizon derived from equations in White (1975) can be calculated by

$$d = 3.56\sqrt{Kh}$$

where
d = distance to the radio horizon, km
h = height of the antenna above ground, m
K = effective Earth radius multiplier, usually 4/3

By rearranging this equation, the amount of Earth bulge at any point along a path can be calculated by

$$h = \frac{(D - d_1)d_1}{12.69K}$$

where h is the height of the Earth bulge in meters and D and d_1 are expressed in kilometers.

To aid in planning line-of-sight propagation paths without performing calculations, many times the desired path is plotted on what is known as 4/3 Earth paper as shown in Fig. 13.32. This graph paper is curved with height on the vertical axis and distance on the horizontal axis. The curve of the coordinate system represents the effective curvature of the Earth's surface. The height of the path endpoints as well as the height of significant intervening terrain or manmade obstructions are plotted at the appropriate distances to determine whether or not the path is obstructed. As stated previously, however, clearance of the direct path of transmission based on ray theory is not sufficient, as the radius of the first Fresnel zone must also be considered.

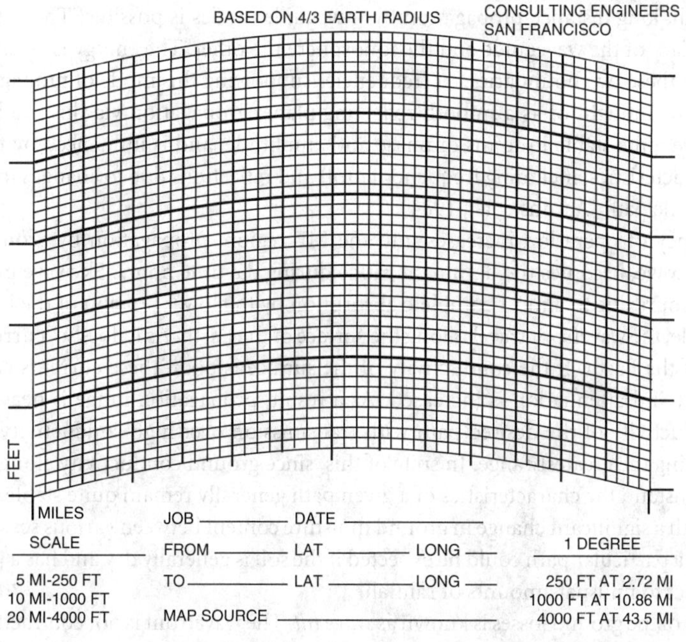

FIGURE 13.32 4/3 Earth paper for path design.

Although the preceding discussion applies generally to all frequencies, each portion of the electromagnetic spectrum has unique qualities that make it suitable for specific tasks. Line-of-sight type links are normally established on VHF and higher frequencies, whereas the lower frequencies are normally suitable for longer distance communications. A brief discussion of the propagation characteristics of various frequency ranges follows.

13.2.5   Very Low Frequency (VLF), Low Frequency (LF), and Medium Frequency (MF) Propagation

This portion of the frequency spectrum includes frequencies below 3 MHz and has been the mainstay of communications systems since the beginning of wireless communications. Use of higher frequencies has only been a relatively recent occurrence. This portion of the spectrum today is primarily used for relatively low data rate, long distance communications and long range navigation. The Omega and LORAN navigation systems operate in this frequency range. Also found in this portion of the spectrum is the familiar AM or medium wave broadcast band. Frequencies in this range are characterized by very stable communications over a relatively large distance. A major disadvantage to this portion of the spectrum is that due to the very long wavelength, greater than 100 m at the high frequency end of this range and measured in terms of kilometers in the lower portion of this range, efficient antennas or antennas with considerable directional characteristics are very difficult to construct. As a result, communications systems operating in this frequency range typically require acres of antennas and transmitters with very high power. Atmospheric and man-made noise can also cause significant disruption to communications on these frequencies.

At these frequencies, propagation is primarily by surface wave and waveguide modes. The direct and reflected waves of the space wave essentially cancel leaving only the surface wave component. Again, due to the proximity of the ground, at least in typical installations, horizontally polarized antennas experience very high losses and so most antennas in this frequency range are vertically polarized. At the very low end of the frequency range, significant surface penetration is possible so that communications with subsurface facilities, submarines for instance, are possible.

When the distance from the surface of the Earth to the lower boundary of the ionosphere is on the order of one wavelength, long distance propagation by waveguide modes is possible. The surface of the Earth forms one boundary of the waveguide and the ionosphere the other. The energy is essentially continually reflected between these two boundaries. At frequencies supporting this mode of propagation, the surface of the Earth can be thought of as a smooth reflecting plane. Propagation via this mode results in characteristics that are quite stable in terms of phase and amplitude, and is influenced by typical waveguide propagation characteristics, including frequency cutoff and reflections due to changes in properties of the waveguide along the path (Aarons et al., 1984).

As the frequency of operation increases into the MF region, propagation by groundwave, or more correctly surface wave, becomes the dominant mode during daytime hours. Skywave propagation is rare during daylight hours due to almost complete absorption by the lower portion of the ionosphere. In this propagation mode, the wavefront travels along the surface of the Earth. As it does so, currents are generated on the surface of the Earth. If the conductivity of the surface is poor, these currents cause energy to be dissipated as heat and attenuation is increased. As a result, propagation range is heavily dependent on the ground conductivity of the desired path with transmission over high conductivity surfaces, such as sea water, providing the greatest range. In spite of this, since ground conductivity along a particular path remains fairly constant, the characteristics of a given path generally remain quite stable. The exception to this is in areas with a significant change in ground moisture content between various seasons. For example, a large change in a particular path could be expected if the soil is generally dry and has a poor conductivity and then is subject to unusual amounts of rainfall.

Another effect of the ground losses is known as *wave tilt*. The wavefront is not confined to the immediate surface of the Earth but extends significantly upward. The ground losses tend to reduce the velocity of travel of the portion of the wavefront nearest the ground. As a result, the electric field vector, instead of remaining

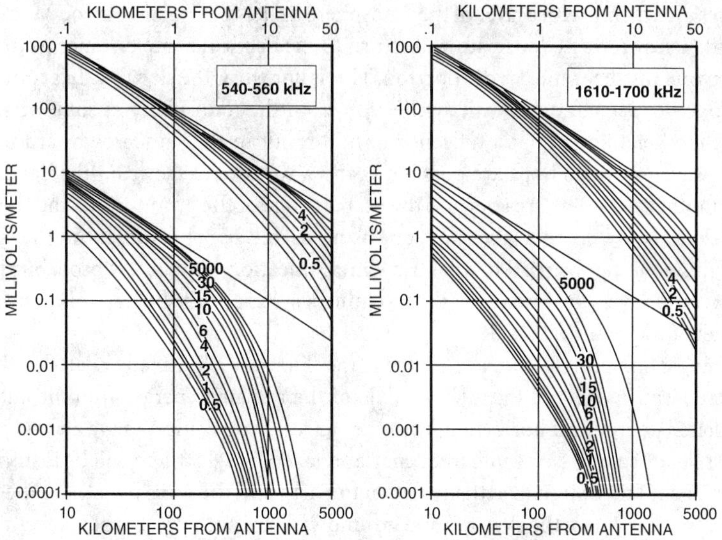

FIGURE 13.33 Ground wave field strength; electric field strength vs. distance for various soil conductivities for the band edges of the AM broadcast band. Conductivities are in millisiemens per meter.

vertical, begins to lean in the direction of travel. This wave tilt becomes more pronounced with increasing frequency, and attenuation increases with increasing wave tilt, as the wavefront alters its polarization from vertical with no horizontal component to having a significant horizontally polarized component, because of the effect of the Earth's short circuiting condition on the horizontally polarized component. This effect and the effect of ground conductivity can be seen in Fig. 13.33 showing FCC propagation curves for the high- and low-frequency limits of the AM broadcasting band for various soil conductivities.

After sunset, when the D layer of the ionosphere (the source of MF skywave attenuation) disappears (see next section), skywave propagation is possible. With this propagation mode, coverage beyond normal surface wave range is possible as the energy directed upward is reflected back to Earth at great distances. This particular mode of propagation is treated in more detail in the next section on HF propagation. It should be noted that not all skywave propagation is beneficial. Energy radiated at high angles can return to Earth within the groundwave coverage area of a station. If the groundwave and skywave field strengths are approximately equal, then severe fading can occur, effectively reducing the groundwave range of that station during daytime hours. Also, skywave interference inside the normal groundwave service area can be experienced from other stations removed by such a distance that groundwave interference is not a concern.

13.2.6 HF Propagation

The HF portion of the radio spectrum, from about 3 to 30 MHz, has recently experienced renewed interest. Systems operating in this frequency range are capable of worldwide communications without the use of satellites or other types of relay stations. This capability can be enjoyed with relatively low-power levels and unsophisticated antennas with consequently relatively low cost.

Beyond horizon propagation on HF circuits is made possible by reflection from the ionosphere. Actually, there are several specific layers that form this region of ionized particles created primarily by solar radiation and occurring from approximately 50 to 350 km above the surface of the Earth. Because of the dependency on solar radiation, the ionization density of these layers, and the resulting propagation characteristics, are highly dependent on not only solar activity variations, but seasonal and diurnal variations as well. As a result, any communications system operating in this region must vary frequency and times of operation for the expected conditions.

HF radio waves interacting with a layer of the ionosphere undergo refraction in the direction of the Earth. If the wave is sufficiently refracted, its path will return to the Earth at some distance from the transmitter, normally well beyond the theoretical radio horizon. This is known as the skywave. It is common to visualize this phenomenon with ray concepts with a reflection from the ionosphere at some virtual height. This virtual height is the height at which a reflection would result in the same downward path as traversed by the refracted wave. Multiple hops are possible as the wave is reflected from the Earth back up to the ionosphere for another pass. The area beyond the usable range of the groundwave and the point of return to Earth of the skywave is known as the skip zone as no usable signal is present in this area. Knowledge of the extent of this zone can be used to establish communications with a low probability of intercept by unwanted receivers. Conversely, inadvertently designing a system with the desired receiver located in the skip zone will yield disappointing results.

The range of single hop propagation and extent of the skip zone are directly related to the virtual height of the ionosphere layer in use and the takeoff angle of the radiated energy from the antenna, the angle between the radiated energy and horizontal, with the longest propagation range and skip zone occurring with the smallest takeoff angle. For some frequency and takeoff angle, there will be insufficient refraction for the wave to be bent far enough to actually return to Earth and the wave penetrates the ionosphere and is lost into space. In areas where the skywave and groundwave are both present at relatively equivalent field strength, severe distortion and fading can occur as the two waves alternatively cancel and reinforce each other. This is a rather annoying problem for some AM broadcasters, who depend primarily on groundwave coverage, and some work has been done on antiskywave antennas to eliminate, or at least significantly reduce, radiation upward toward the ionosphere at high enough angles to return to Earth within the groundwave coverage area. Figure 13.34 demonstrates the interaction of the virtual height, takeoff angle, and communications range.

The virtual height of an ionospheric layer can be measured with a device known as an ionosonde. This device radiates a swept frequency signal directly toward the zenith and times the return echo much like a conventional radar system. Half the round-trip travel time multiplied by the speed of light yields the virtual height of the ionosphere layer. The frequency at which no echo is returned is known as the critical frequency for that layer.

A very useful HF propagation mode is known as *near vertical incidence skywave* (NVIS) propagation. As the name implies, this mode utilizes high-angle radiation to provide relatively short range propagation, perhaps on the order of 0 to several hundred kilometers. Energy radiated toward the zenith, below the

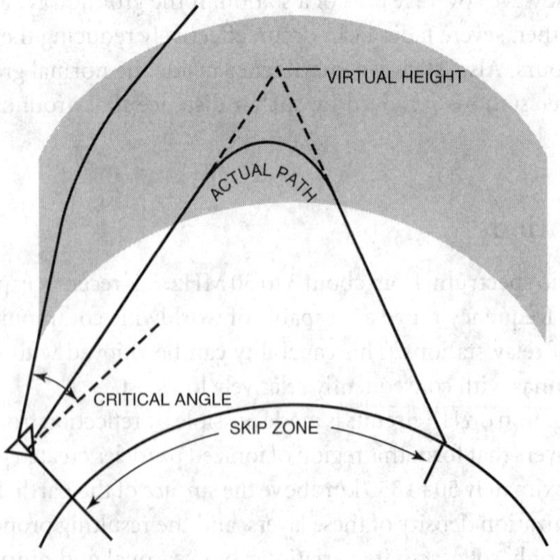

FIGURE 13.34 The mechanism of skywave propagation.

critical frequency of the pertinent ionospheric layer, is reflected down around the transmitter. As a result, frequencies used for NVIS propagation are usually in the lower-half of the HF spectrum. Note that directional characteristics of the transmitting antenna are not important, or necessarily desirable, as the returned energy from the ionosphere forms an essentially omnidirectional radiation pattern around the transmitter.

This form of propagation has found followers in the military and in agencies charged with providing emergency communications. Since the transmitted energy is reaching the receiver from such a high arrival angle, it is essentially unaffected by terrain, foliage, or other terrestrial attenuators, resulting in a relatively uniform field strength throughout the coverage area. Using this propagation mode it is possible to communicate from deep canyons or over local hills and mountains without the use of relays or repeaters that would be required if line-of-sight type propagation modes were being utilized in areas shielded by terrain features. From a security standpoint, this propagation mode is useful as it is extremely difficult to use direction finding techniques to locate the transmitter of a signal that has a very large angle of arrival. For NVIS propagation, it is desirable to have the majority of the radiation aimed at high angles with almost no radiation at the horizon. Specifically, it is important not to have groundwave and skywave propagation paths of similar loss characteristics simultaneously to any given receiver site or signal cancellation or severe fading and distortion could occur. As a result, NVIS antennas are typically horizontally polarized antennas mounted less than one quarter wavelength above ground. Because of the high angle radiation, the resultant antenna pattern and coverage area is essentially omnidirectional.

There are three major layers of the ionosphere that are of importance in HF propagation. The lowest layer of concern is known as the D layer and is only present during daylight hours. Of all of the important ionospheric layers, the D layer is located in the region of highest atmospheric density. Because of this, recombination of ionized particles occurs rather rapidly and constant solar energy is required to sustain ionization in this layer. As the sun sets, this layer recombines and disappears. This layer is primarily responsible for almost complete absorption of frequencies below about 4 MHz. As energy from the electromagnetic wave sets electrons in motion, there is a high probability that the energy will be absorbed in a collision with a neutral particle. More precisely, the electromagnetic energy has been changed to useless, at least for radio wave propagation purposes, kinetic energy. Note that as the angle of incidence of the wave entering the layer is increased, which also means that the takeoff angle from the antenna is increased, the absorption is decreased. This occurs because of the shorter distance traveled in the ionized layer. However, as the angle of incidence is decreased, so is the usable communications range.

The next higher layer of the ionosphere at an altitude of approximately 115 km is the E layer. This is generally considered to be the lowest useful region of the ionosphere. Still low enough so that recombination occurs quickly, this layer is also only present during daylight hours, forming a useful ionized density around midday and disappearing after sunset.

The most useful ionosphere layer for over the horizon propagation is known as the F layer. This layer is located in the least dense portion of the atmosphere at an altitude of approximately 300 km at night. During daylight hours, this layer essentially splits into two separate layers known as the F1 and F2 layers at altitudes of approximately 200 and 300 km, respectively. This layer remains ionized throughout the night with minimum ionization density just before sunrise. Figure 13.35 shows the relative positions of these layers.

In characterizing the propagation characteristics between two geographical locations at a given time of day and time of year with a given level of solar activity, two frequencies are of considerable importance. These are the *maximum usable frequency* (MUF) and the *lowest usable frequency* (LUF). Propagation between the two points must occur on a frequency between these two limits. Lower frequencies require a lower electron density for sufficient refraction for usable service and are consequently reflected from lower ionospheric levels. Also, absorption in the lower levels increases with decreasing frequency. As a result, the LUF is typically determined by the absorption in the lower regions of the ionosphere. The MUF is usually determined by the peak electron density in the ionosphere, which occurs in the F layer, as higher electron density is required for sufficient refraction. The optimum operating frequency, known as the *frequence optimum de travail* (FOT), is typically 75–80% of the MUF where the refracting layer electron density is sufficient for reflection and the absorption in lower ionospheric layers is low. Note that frequencies refracted back to Earth by the F layer experience absorption as they pass through the D and E layers twice,

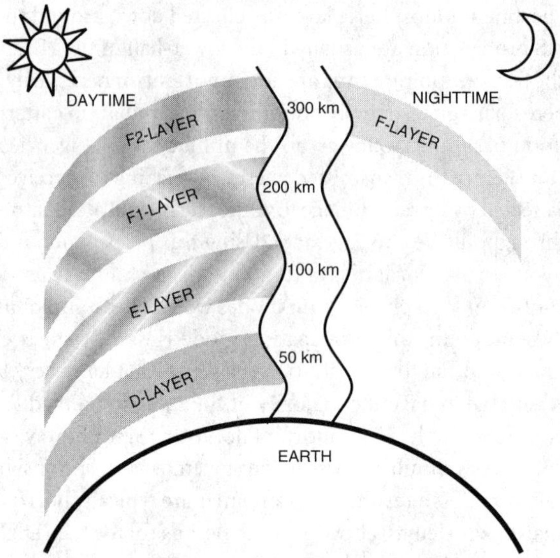

FIGURE 13.35 The ionosphere.

assuming daytime propagation. As a result, the MUF usually gradually increases to its highest point in the afternoon and then gradually decreases, reaching a minimum during the hours of darkness.

Although HF prediction programs are available that will compute the MUF, LUF, and FOT based on available solar data, HF propagation is subject to some unpredictability, making the choice of frequency and time for a particular propagation path not a precise science. As with line-of-sight-type propagation modes on higher frequencies, HF propagation paths are subject to multipath interference. Ionospheric conditions may support reflection from more than one layer or multiple hop propagation from energy radiated at a higher angle from the transmitter. This type of multipath interference may result in deep fading as the phase of the fields from the two paths alternatively cancel and reinforce each other. Proper system design and choice of frequency can help to reduce this type of interference.

Other HF propagation anomalies are difficult to predict and occur without warning. Many of these anomalies are the result of solar flares. A shortwave fade (SWF) is a sudden and complete absorption of HF radio waves in the D region of the ionosphere. It occurs on the sunlit side of the Earth and is caused by ultraviolet and X-ray emissions from a solar flare. The fade occurs approximately 8 min after the solar event and can last from a few minutes to a few hours. Protons entering the ionosphere near the magnetic poles can cause a complete loss of HF propagation in these regions. This effect may occur several hours after the flare. Ionospheric storms are another potential effect that can drastically alter the expected MUF and may occur one to two days after the solar event. Sporadic ionization in the E layer, known as sporadic E, may occur at any time and is not necessarily solar related. This effect can isolate the F layer, altering the path characteristics such that communication is disrupted or perhaps enhanced. Even VHF frequencies may be affected by sporadic E conditions causing a significant increase in effective range.

13.2.7 VHF and UHF Propagation

VHF and UHF frequencies, considered to be 30–3000 MHz, are used primarily for line-of-sight communications. Aircraft, public service, government, Amateur, and business communications as well as FM and TV broadcasting have allocations in this frequency range. These frequencies are typically used for local area and point-to-point communications. Usable propagation range is usually limited to the radio horizon. At these frequencies, antenna heights that result in sufficient path clearance for free space propagation to be applicable are practical. Ionospheric skywave propagation at these frequencies, at least the higher portion of the band, is essentially nonexistent with the exception of occasional sporadic E propagation.

Transmit antennas are located as high above ground as practical in order to have the greatest range to the radio horizon. Communications paths requiring propagation to areas beyond line of sight from the transmitter, either because the receiver is located over the radio horizon or is obstructed by terrain or manmade obstacles, are routinely constructed using repeaters located on mountain tops or tall buildings.

Since they are essentially line-of-sight paths, propagation at VHF and UHF is relatively consistent, although, there are several potential disturbances to reliable propagation at these frequencies. Wavelengths in this frequency range vary from 0.1 to 10 m. At these wavelengths, many natural and manmade objects appear smooth enough to exhibit good reflective properties. As a result, the potential for radiated energy to arrive at the receiver from the direct path to the transmitter as well as various reflected paths is quite high. If the energy from the reflected path and from the direct path reach the receiver in phase, the strength of the received signal is enhanced. If, on the other hand, the energy from the reflected path arrives out of phase with that from the direct path, then there will be complete signal cancellation. This phenomenon is known as *multipath interference* and can be apparent as picket-fence-type noise on a VHF signal, such as that from an FM broadcast station while in a moving vehicle. The picket fencing occurs with an interval related to the wavelength of the operating frequency as the receiver passes through standing waves of destructive interference between the direct and reflected signal. If stopped at a fixed location, moving only a few inches usually can make a significant difference in the strength of the received signal. For analog video signals, such multipath problems may cause "ghosting" in pictures as the reflected signal arrives slightly time delayed from the direct signal.

Although VHF and UHF frequencies are used primarily for line-of-sight paths, propagation beyond the radio horizon is not only possible, but sometimes depended on. Such propagation can also unexpectedly cause disruption to established communications paths and interference to paths in use far beyond the radio horizon. Tropospheric **ducting** is one such phenomena that occurs mostly in warm marine climates. Tropospheric ducting occurs because of a large and rapid change in the atmospheric index of refraction and can result in propagation distances as large as 4000 km (Hutchinson, 1985). This type of propagation can occur in two different ways, but both are typically caused by temperature inversions where atmospheric temperature increases with altitude. In one type of tropospheric ducting, a well-defined boundary between warm and cool air, and the resulting rapid change in refractive index, bends the traveling wave significantly and essentially traps it between the inversion layer and the surface of the Earth, resulting in greatly enhanced communications range. For this to occur, however, the wave must have a very shallow arrival angle at the inversion layer.

The second type of tropospheric ducting occurs between two layers of air with temperature inversions. In this case the wave is trapped between the two layers and propagates much like a wave in a waveguide as shown in Fig. 13.36. Also, as in a waveguide, the dimensions of the atmospheric duct result in some frequency selectivity for this mode of propagation. For this mode of propagation to be useful, both the transmitter and receiver antennas must be located within the duct. Such ducting can disrupt an existing

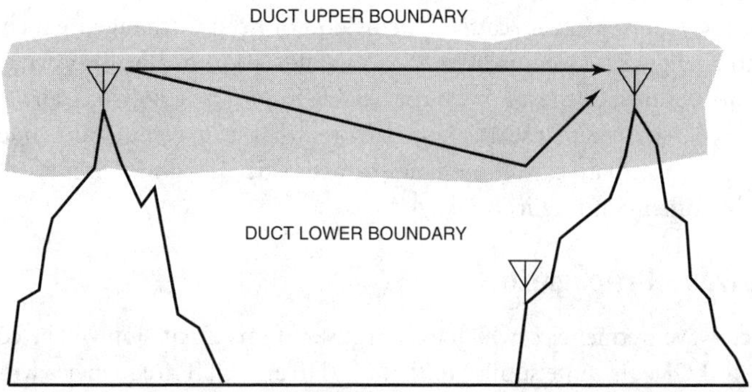

DUCT UPPER BOUNDARY

DUCT LOWER BOUNDARY

FIGURE 13.36 The mechanism of atmospheric ducting; The propagation path is limited by the boundaries of the duct. Communications with lower receive antenna not possible.

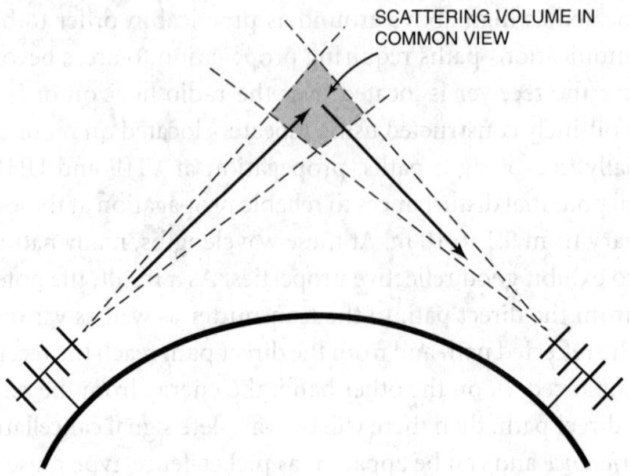

FIGURE 13.37   The principle of tropospheric scattering.

communications system, especially if highly directive antennas are used, by trapping the transmitted wave inside a duct with the receive antenna located either above or below the duct. This problem can occur with aircraft flying within a duct trying to communicate with a receiver on the ground well below the duct. At higher frequencies, such ducting can cause radar systems to miss close targets or to detect targets at ranges that are theoretically beyond detection.

Another method of propagation beyond the radio horizon is known as tropospheric scatter or *troposcatter* and is shown in Fig. 13.37. In this mode of propagation, energy is reflected and refracted by small changes in the atmospheric index of refraction and by various particulate matter in the troposphere, such as dust. Communications range is limited by the height of the scattering region as both stations must have a line of sight to it. The amount of energy scattered depends on the density of the scattering particles and the power impinging on that area, and so systems regularly using this mode are typically designed with high transmit power levels.

Another mode of propagation that extends the range of VHF systems is known as meteor scatter or *meteor burst* propagation. In this propagation mode, the traveling wave is reflected from the ionized trail of a meteor passing through the Earth's atmosphere. These ionized trails occur at an altitude of approximately 100 km, resulting in a single-hop communications range of approximately 1500–2000 km. Frequencies in the range of 40–60 MHz are most effective for meteor scatter work, although higher frequencies are sometimes used. While meteor paths are sporadic, they are sufficient in number that digital systems with acceptable throughput can be established. Because of the orbit of most meteors and the Earth around the sun, the number of meteors and, hence, meteor scatter performance is typically highest in the early morning hours (Weitzen and Ralston, 1988). There are seasonal variations as well. Remote sensing systems have been established around this propagation mode. Meteor scatter systems have some distinct advantages besides over the horizon propagation. Meteor scatter systems do not need to change frequency over the course of the day and with changing seasons as HF systems would. Additionally, in order for communication to occur between two stations, both stations must be able to "see" the meteor trail at a proper angle for reflection to occur. This makes meteor scatter systems somewhat secure from unintended reception as well as providing some level of antijam capability. This characteristic also allows simplified frequency sharing or reuse as opposed to other communications systems.

13.2.8   Microwave Propagation

Propagation at microwave frequencies, which for purposes of this discussion will be considered anything above approximately 3 GHz, is quite similar to that at VHF and UHF frequencies with the addition of a few peculiarities. Because of the very small wavelength at these frequencies, obtaining appropriate Fresnel zone clearance is usually relatively easy. However, also because of the very small wavelength, energy in this frequency range may be readily absorbed by atmospheric gases or scattered by particulate matter in the atmosphere. Vegetation and foliage become obstructions at these frequencies and, if a path is through an area of deciduous trees, the performance between the various seasons can be substantially different. For this reason, propagation paths at microwave frequencies should be completely clear of any type of obstruction, often requiring visual inspection of the path.

Rain attenuation may be significant on paths operating at frequencies above approximately 10 GHz. The actual amount of attenuation is a function of many variables including the size and shape of the drops and the instantaneous intensity of the rainfall as well as the frequency of operation. For a frequency of 13 GHz, rainfall intensity of 1 in/h can result in over 1 dB/km of additional attenuation (White, 1975). The amount of attenuation increases with rainfall intensity and frequency. Significantly less attenuation is caused by snow and fog as the water density is considerably less. Absorption by atmospheric gases, oxygen to be specific, may be an additional consideration for paths operating in the tens of gegahertz range. As a result, if a path is being planned in areas subject to periods of locally intense rainfall, this additional attenuation must be accounted for in the design of the path fade margin. Other causes of atmospheric absorption are usually secondary considerations for most typical paths.

At these frequencies, diversity receiving systems are often used to combat various forms of multipath fading. Although a path may perform flawlessly most of the time, it may be subject to occasional deep fades. These fades can be very brief or can last for minutes or even hours. The fading is primarily caused by multipath effects. The fading conditions are not constant, however, as changes in the refractive index of the atmosphere effectively change the point of reflection. These fades can also be caused by changes on the ground between the receiver and transmitter. Indeed, these multipath effects can occur with no true reflective surface, but rather from atmospheric effects themselves, such as the already discussed temperature inversion. To minimize these effects, either frequency or space diversity are used.

Diversity systems are based on the premise that fades are a function of frequency and distance from the point of reflection, since the direct and reflected signals must have a phase difference of 180°. Using another frequency, preferably far removed from the primary frequency, usually ensures that both paths will not have simultaneous fades. The drawback to this system is a that it is an inefficient use of spectrum. Alternatively, space diversity can be used. With a space diversity system, two receive systems are used with antennas spaced many wavelengths apart, usually vertically. The desired distance is a function of path length and wavelength. With such spaced antennas, the point of reflection will not be the same at both heights so that while one antenna is in a deep fade, the other is not. It is important to note that a diversity system will not improve reliability due to obstructions, such as foliage, insufficient path clearance, or other such nonfrequency specific problems.

At microwave frequencies, non-line-of-sight paths can be established with the use of passive repeaters as well as active repeaters. At the millimeter wavelengths involved, passive reflectors can be constructed with sufficient gain to make their use practical. The most common passive repeater is a large, flat, billboard-looking reflector that acts exactly like an optical mirror. As with an active repeater, the passive repeater must have line of sight to both the receiver and transmitter. A single passive reflector can not act as a relay when all three components of the link are in-line or are nearly in-line. In such a case, two closely spaced passive reflectors are required.

Communications between Earth and space are also conducted in this frequency range as well as at lower frequencies. The primary concern with space communications is to choose a frequency high enough that will not be refracted back to Earth by the ionosphere and low enough so that absorption by atmospheric gases is not significant. Because of the lower powers and great distances involved, high-gain antennas are typically required and rain and other causes of atmospheric absorption are important considerations. Further complicating space communications is the RF noise associated with thermal radiation from celestial objects, notably the sun. This interference can completely destroy a path when the sun is directly in line with the space borne transmitter.

Defining Terms

Diffraction: Change in direction of propagating energy around an object caused by interference between the radiated energy and induced currents in the object.

Ducting: A change in typical propagation conditions caused by anomalous atmospheric conditions which result in a waveguide effect.

Effective Earth radius: The assumed Earth radius required to result in a distance to the horizon equivalent to the radio horizon.

Free space path loss: The amount of attenuation of RF energy on an unobstructed path between isotropic antennas.

Frequence optimum de travail (FOT): The optimum working frequency.

Fresnel zone: A locus of points along a path where a reflection results in a change of overall path length by $n \times 180°$.

Groundwave: That portion of radiated radio frequency energy that is not influenced by the ionosphere.

Isotropic antenna: An antenna with a gain of unity.

Refraction: Change in direction of propagating radio energy caused by a change in the refractive index, or density, of a medium.

Skywave: That portion of radiated radio frequency energy that propagates skyward and interacts with the ionosphere.

Wavelength: The distance traveled by a radio frequency wavefront in one cycle.

References

Aarons, J., ed. 1984. Ionospheric radio wave propagation. *Handbook of Geophysics and Space Environments,* Chap. 10, 1983 Revision. Environmental Research Paper 879. Air Force Geophys. Lab, Hanscom AFB, MA.

Al'pert, Y.L., 1974. *Radio Wave Propagation and the Ionosphere.* Consultants Bureau, Plenum, New York.

Bekefi, G. and Barrett, A.H. 1977. *Electromagnetic Vibrations, Waves and Radiation.* MIT Press, Cambridge, MA.

Bothias, L. 1987. *Radio Wave Propagation.* McGraw-Hill, New York.

Department of the Army. 1953. *Antennas and Radio Propagation.* U.S. Government Printing Office, Washington, DC.

Hutchinson, C.L., ed. 1985. *The ARRL Handbook for the Radio Amateur.* The American Radio Relay League, Newington, CT.

Rice, P.L., Longley, A.G., Norton, K.A., and Barsis, A.P. 1967. *Transmission Loss Predictions for Tropospheric Communication Circuits.* NBS Technical Note 101. U.S. Department of Commerce, Washington, DC.

Shibuya, Sh. 1987. *A Basic Atlas of Radio-Wave Propagation.* Wiley, New York.

Straw, R.D., ed. 1994. *The ARRL Antenna Book.* American Radio Relay League, Newington, CT.

Watt, A.D. 1967. *VLF Radio Engineering.* Pergamon. New York.

Weitzen, J.A. and Ralston, W.T. 1988. Meteor scatter: an overview. *IEEE Trans. Antennas Propagation* 36(12):1813–1819.

White, R.F., ed. 1975. *Engineering Considerations for Microwave Communications Systems.* GTE Lenkurt, San Carlos, CA.

Further Information

Additional information on the topic of radio wave propagation is available from the following sources:

The American Radio Relay League, Newington, Connecticut, publishes *QST,* a monthly magazine with HF propagation predictions and occasional articles regarding radio wave propagation. The organization also publishes books, aimed primarily at amateur radio operators, that discuss radio wave propagation.

The IEEE Antennas and Propagation Society, New York, NY, publishes *Transactions on Antennas and Propagation* on a monthly basis.

Near real-time information concerning HF propagation is broadcast by National Institute of Standard and Technology radio stations WWV and WWVH.

13.3 Practical Antenna Systems

Jerry C. Whitaker

13.3.1 Introduction

Transmission is accomplished by the emission of coherent electromagnetic waves in free space from one or more radiating elements that are excited by RF currents. Although, by definition, the radiated energy is composed of mutually dependent magnetic and electric vector fields, it is conventional practice to measure and specify radiation characteristics in terms of the electric field only.

The purpose of an antenna is to efficiently radiate the power supplied to it by the transmitter. A simple antenna, consisting of a single vertical element over a ground plane can do this job quite well at low to medium frequencies. Antenna systems may also be required to concentrate the radiated power in a given direction and minimize radiation in the direction of other stations sharing the same or adjacent frequencies. To achieve such directionality may require a complicated antenna system that incorporates a number of individual elements or towers and matching networks.

As the operating frequency increases into VHF and above, the short wavelengths permit the design of specialized antennas that offer high directivity and gain.

Operating Characteristics

Wavelength is the distance traveled by one cycle of a radiated electric signal. The frequency of the signal is the number of cycles per second. It follows that the frequency is inversely proportional to the wavelength. Both wavelength and frequency are related to the speed of light. Conversion between the two parameters can be accomplished with the formula

$$c = f \times \lambda$$

where c is the speed of light, f is the operating frequency, and λ is the wavelength. The velocity of electric signals in air is essentially the same as that of light in free space (2.9983×10^{10} cm/sec).

The *electrical length* of a radiating element is the most basic parameter of an antenna

$$H = \frac{H_t \times F_o}{2733}$$

where H is the length of the radiating element in electrical degrees, H_t is the length of the radiating element in feet, and F_o is frequency of operation in kHz. Where the radiating element is measured in meters

$$H = \frac{H_t \times F_o}{833.23}$$

The *radiation resistance* of an antenna is defined by

$$R = \frac{P}{I^2}$$

where R is the radiation resistance, P is the power delivered to the antenna, and I is the driving current at the antenna base.

Antenna Bandwidth

Bandwidth is a general classification of the frequencies over which an antenna is effective. This parameter requires specification of acceptable tolerances relating to the uniformity of response over the intended operating band.

Strictly speaking, *antenna bandwidth* is the difference in frequency between two points at which the power output of the transmitter drops to one-half the midrange value. The points are called *half-power points*. A half-power point is equal to a VSWR of 5.83:1, or the point at which the voltage response drops to 0.7071 of the midrange value. In most communications systems, a VSWR of less than 1.2:1 within the occupied bandwidth of the radiated signal is preferable.

Antenna bandwidth depends on the radiating element impedance and the rate at which the reactance of the antenna changes with frequency.

Bandwidth and RF coupling go hand in hand, regardless of the method used to excite the antenna. All elements between the transmitter output circuit and the antenna must be analyzed, first by themselves, and then as part of the overall system bandwidth. In any transmission system, the *composite bandwidth*, not just the bandwidths of individual components, is of primary concern.

Polarization

Polarization is the angle of the radiated electric field vector in the direction of maximum radiation. Antennas may be designed to provide horizontal, vertical, or circular polarization. Horizontal or vertical polarization is determined by the orientation of the radiating element with respect to Earth. If the plane of the radiated field is parallel to the ground, the signal is said to be *horizontally polarized*. If it is at right angles to the ground, it is said to be *vertically polarized*. When the receiving antenna is located in the same plane as the transmitting antenna, the received signal strength will be maximum.

Circular polarization (CP) of the transmitted signal results when equal electrical fields in the vertical and horizontal planes of radiation are out-of-phase by 90° and are rotating a full 360° in one wavelength of the operating frequency. The rotation can be clockwise or counterclockwise, depending on the antenna design. This continuously rotating field gives CP good signal penetration capabilities because it can be received efficiently by an antenna of any random orientation. Figure 13.38 illustrates the principles of circular polarization.

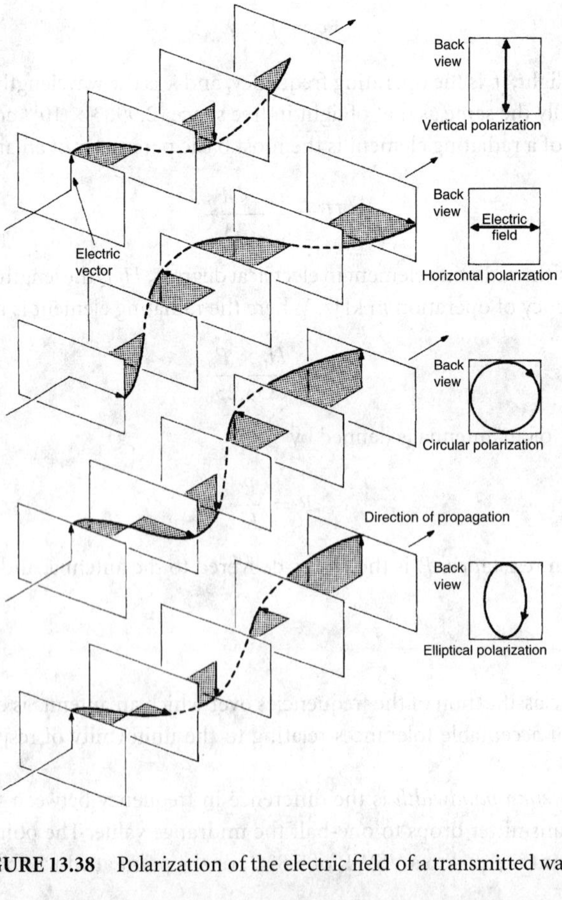

FIGURE 13.38　Polarization of the electric field of a transmitted wave.

Antenna Beamwidth

Beamwidth in the plane of the antenna is the angular width of the directivity pattern where the power level of the received signal is down by 50% (3 dB) from the maximum signal in the desired direction of reception.

Antenna Gain

Directivity and gain are measures of how well energy is concentrated in a given direction. Directivity is the ratio of power density in a given direction to the power density that would be produced if the energy were radiated isotropically. The reference can be linearly or circularly polarized. Directivity is usually given in dBi (decibels above isotropic).

Gain is the field intensity produced in a given direction by a fixed input power to the antenna, referenced to a dipole. It is frequently used as a figure of merit. It is closely related to directivity, which in turn is dependent upon the radiation pattern. High values of gain are usually obtained with a reduction in beamwidth.

An antenna is typically configured to exhibit "gain" by narrowing the beamwidth of the radiated signal to concentrate energy toward the intended coverage area. The actual amount of energy being radiated is the same with a unity-gain antenna or a high-gain antenna, but the useful energy (commonly referred to as the *effective radiated power*, or ERP) can be increased significantly.

Electrical *beam tilt* can also be designed into a high-gain antenna. A conventional antenna typically radiates more than half of its energy above the horizon. This energy is lost for practical purposes in most applications. Electrical beam tilt, caused by delaying the RF current to the lower elements of a multi-element antenna, can be used to provide more useful power to the service area.

Pattern optimization is another method that can be used to maximize radiation to the intended service area. The characteristics of the transmitting antenna are affected, sometimes greatly, by the presence of the supporting tower, if side-mounted, or by nearby tall obstructions (such as another transmitting tower) if top-mounted. Antenna manufacturers use various methods to reduce pattern distortions. These generally involve changing the orientation of the radiators with respect to the tower and adding parasitic elements.

Space Regions

Insofar as the transmitting antenna is concerned, space is divided into three regions:

- *Reactive near-field region.* This region is the area of space immediately surrounding the antenna in which the reactive components predominate. The size of the region varies, depending on the antenna design. For most antennas, the reactive near-field region extends 2 wavelengths or less from the radiating elements.
- *Radiating near-field region.* This region is characterized by the predictable distribution of the radiating field. In the near-field region, the relative angular distribution of the field is dependent on the distance from the antenna.
- *Far-field region.* This region is characterized by the independence of the relative angular distribution of the field with varying distance. The pattern is essentially independent of distance.

Impedance Matching

Most practical antennas require some form of impedance matching between the transmission line and the radiating elements. The implementation of a matching network can take on many forms, depending on the operating frequency and output power.

The *negative sign* convention is generally used in impedance matching analysis. That is, if a network delays or retards a signal by $\theta°$, the phase shift across the network is said to be "minus θ degrees." For example, a 1/4-wave length of transmission line, if properly terminated, has a phase shift of $-90°$. Thus, a *lagging* or low-pass network has a negative phase shift, and a *leading* or high-pass network has a positive phase shift. There are three basic network types that can be used for impedance matching: L, *pi*, and *tee*.

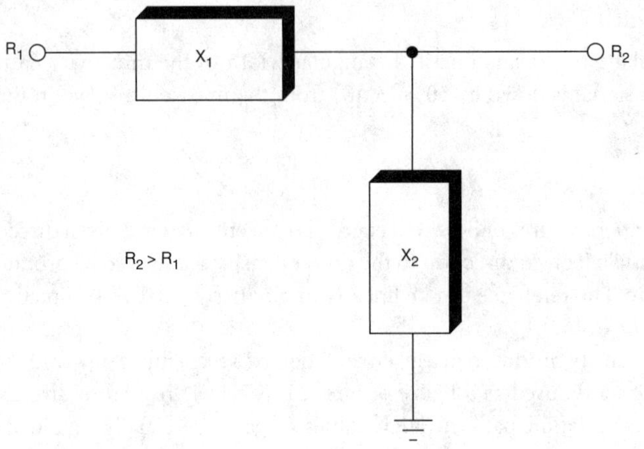

Equation 1:

$$Q = \sqrt{\frac{R_2}{R_1} - 1} = \left|\frac{X_1}{R_1}\right| = \left|\frac{R_2}{X_2}\right|$$

Equation 2:

$$X_2 = \frac{\pm R_2}{Q}$$

Equation 3:

$$X_1 = \frac{-R_1 R_2}{X_2}$$

Equation 4:

$$\Theta = TAN^{-1}\left(\frac{R_2}{X_2}\right)$$

Where

R_1 = L network input resistance (ohms)
R_2 = L network output resistance (ohms)
X_1 = Series leg reactance (ohms)
X_2 = Shunt leg reactance (ohms)
Q = Loaded Q of the L network

FIGURE 13.39   L network parameters.

L Network

The L network is shown in Fig. 13.39 The loaded Q of the network is determined from Equation 1. Equation 2 defines the shunt leg reactance, which is negative (capacitive) when θ is negative, and positive (inductive) when θ is positive. The series leg reactance is found using Equation 3, the phase shift via Equation 4, and the currents and voltages via Ohm's law. Note that R_2 (the resistance on the shunt leg side of the L network) must always be greater than R_1. An L network cannot be used to match equal resistances, or to adjust phase independently of resistance.

Tee Network

The tee network is shown in Fig. 13.40. This configuration can be used to match unequal resistances. The tee network has the added feature that phase shift is independent of the resistance transformation ratio. A tee network can be considered simply as two L networks back-to-back. Note that there are two loaded Qs associated with a tee network: an input Q and an output Q. To gauge the bandwidth of the tee network, the lower value of Q must be ignored. Note that the Q of a tee network increases with increasing phase shift.

Equations 5 through 14 describe the tee network. It is a simple matter to find the input and output currents via Ohm's law, and the shunt leg current can be found via the Cosine law (Equation 12). Note that this current increases with increasing phase shift. Equation 13 describes the mid-point resistance of a tee network, which is always higher than R_1 or R_2. Equation 14 is useful when designing a *phantom tee network*; that is, where X_2 is made up only of the antenna reactance and there is no physical component in place of X_2. Keep in mind that a tee network is considered as having a lagging or negative phase shift

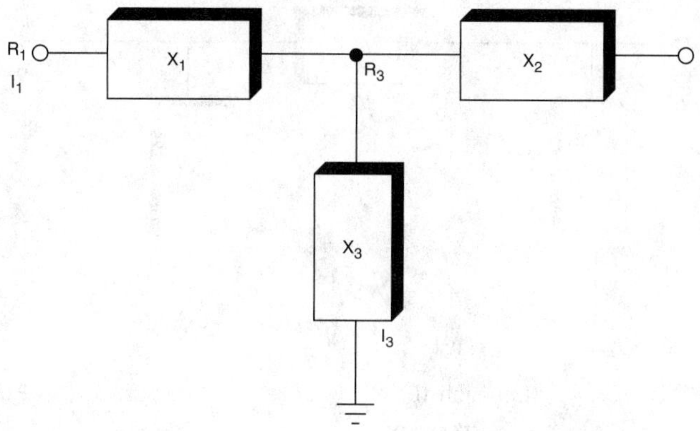

Equation 5:

$$X_3 = \frac{\sqrt{R_1 R_2}}{SIN\,(\Theta)}$$

Equation 6:

$$X_1 = \frac{R_1}{TAN\,(\Theta)} - X_3$$

Equation 7:

$$X_2 = \frac{R_2}{TAN\,(\Theta)} - X_3$$

Equation 8:

$$Q_1 = \left| \frac{X_1}{R_1} \right|$$

Equation 9:

$$Q_2 = \left| \frac{X_2}{R_2} \right|$$

Equation 10:

$$I_1 = \sqrt{\frac{P}{R_1}}$$

Equation 11:

$$I_2 = \sqrt{\frac{P}{R_2}}$$

Equation 12:

$$I_3 = \sqrt{I_1^2 + I_2^2 = 2(I_1)(I_2)COS\,(\Theta)}$$

Equation 13:

$$R_3 = (Q_2^2 + 1)R_2$$

Equation 14:

$$\Theta = TAN^{-1}\left(\frac{X_1}{R_1}\right) \pm TAN^{-1}\left(\frac{X_2}{R_2}\right)$$

Where

R_1 = Tee network input resistance (ohms)
R_2 = Tee network output resistance (ohms)
I_1 = Tee network input current (amps)
I_2 = Tee network output current (amps)
I_3 = Shunt element current (amps)
X_1 = Network input element reactance (ohms)
X_2 = Network output element reactance (ohms)
X_3 = Network shunt element reactance (ohms)
P = Input power (watts)
Q_1 = Input loaded Q
Q_2 = Output loaded Q
R_3 = Midpoint resistance of the network (ohms)

FIGURE 13.40 Tee network parameters.

when the shunt leg is capacitive (X_3 negative), and vice versa. The input and output arms can go either negative or positive, depending on the resistance transformation ratio and desired phase shift.

Pi Network

The pi network is shown in Fig. 13.41. It can also be considered as two L networks back-to-back and, therefore, the same comments about overall loaded Q apply. Note that susceptances have been used in Equations 15 through 19 instead of reactances in order to simplify calculations. The same conventions regarding tee network currents apply to pi network voltages (Equations 20, 21, and 22). The mid-point resistance of a pi network is always less than R_1 or R_2. A pi network is considered as having a negative or lagging phase shift when Y_3 is positive, and vice versa.

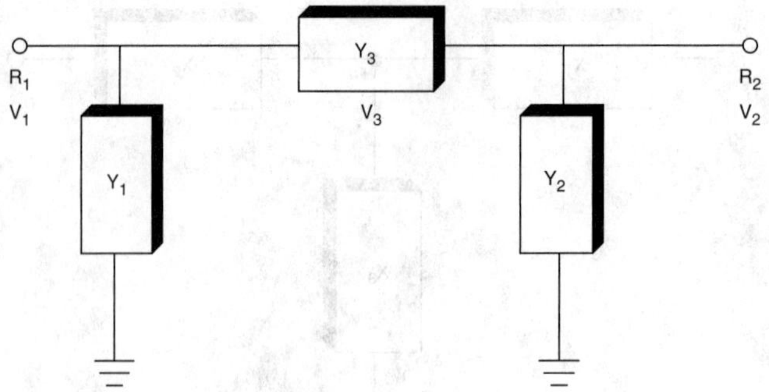

Equation 15:

$$Y_3 = \frac{1}{-\text{SIN}\,(\Theta)\sqrt{R_1 R_2}}$$

Equation 16:

$$Y_1 = \frac{\text{TAN}\,(\Theta)}{R_1 - Y_3}$$

Equation 17:

$$Y_2 = \frac{\text{TAN}\,(\Theta)}{R_2 - Y_3}$$

Equation 18:

$$Q_1 = |R_1 Y_1|$$

Equation 19:

$$Q_2 = |R_2 Y_2|$$

Equation 20:

$$V_1 = \sqrt{R_1 P}$$

Equation 21:

$$V_2 = \sqrt{R_2 P}$$

Equation 22:

$$V_3 \sqrt{V_1^2 + V_2^2 - 2(V_1)(V_2)\,\text{COS}\,(\Theta)}$$

Equation 23:

$$R_3 = \frac{Q_2^2 + 1}{R_2}$$

Where

R_1 = Pi network input resistance (ohms)
R_2 = Output resistance (ohms)
V_1 = Input voltage (volts)
V_2 = Output voltage (volts)
V_3 = Voltage across series element (volts)
P = Power input to pi network (watts)
Y_1 = Input shunt element susceptance (mhos)
Y_2 = Output shunt element susceptance (mhos)
Y_3 = Series element susceptance (mhos)
Q_1 = Input loaded Q
Q_2 = Output loaded Q

FIGURE 13.41   Pi network parameters.

Line Stretcher

A *line stretcher* makes a transmission line look longer or shorter in order to produce sideband impedance symmetry at the transmitter PA (see Fig. 13.42). This is done to reduce audio distortion in an envelope detector, the kind of detector that most AM receivers employ. Symmetry is defined as equal sideband resistances, and equal—but opposite sign—sideband reactances.

There are two possible points of symmetry, each 90° from the other. One produces sideband resistances greater than the carrier resistance, and the other produces the opposite effect. One side will create a pre-emphasis effect, and the other a de-emphasis effect.

Depending on the Q of the transmitter output network, one point of symmetry may yield lower sideband VSWR at the PA than the other. This results from the Q of the output network opposing the Q of the antenna in one direction, but aiding the antenna Q in the other direction.

13.3.2   Antenna Types

The *dipole antenna* is simplest of all antennas, and the building block of most other designs. The dipole consists of two in-line rods or wires with a total length equal to 1/2-wave at the operating frequency. Figure 13.43 shows the typical configuration, with two 1/4-wave elements connected to a transmission line.

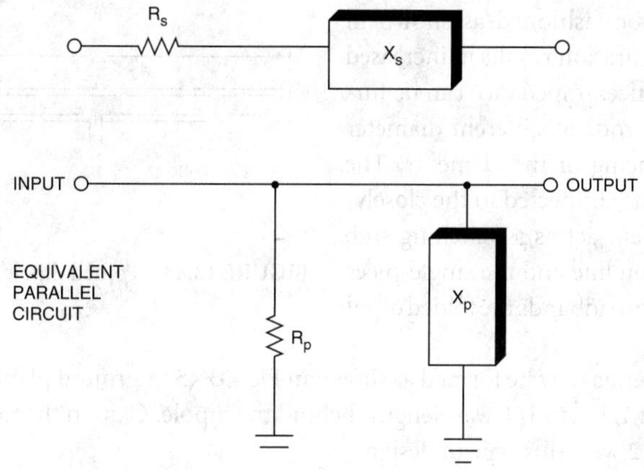

Equation 25:

$$R_p = \frac{R_{S^2} + X_{S^2}}{R_S}$$

Equation 26:

$$X_p = \frac{R_{S^2} = X_{S^2}}{X_S}$$

Where

R_S = Series configuration resistance (ohms)
R_p = Parallel configuration resistance (ohms)
X_S = Series reactance (ohms)
X_p = Parallel reactance (ohms)

FIGURE 13.42   Line stretcher configuration.

The radiation resistance of a dipole is on the order of 73 Ω. The bandwidth of the antenna may be increased by increasing the diameter of the elements, or by using cones or cylinders rather than wires or rods, as shown in the figure. Such modifications also increase the impedance of the antenna.

The dipole can be straight (in-line) or bent into a V-shape. The impedance of the V-dipole is a function of the V angle. Changing the angle effectively tunes the antenna. The vertical radiation pattern of the V-dipole antenna is similar to the straight dipole for angles of 120° or less.

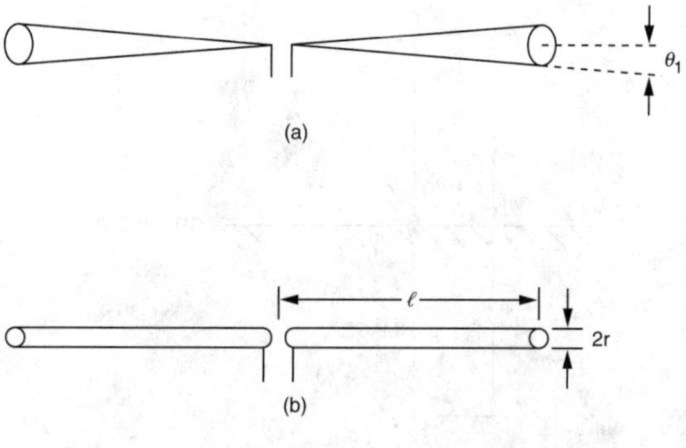

FIGURE 13.43   Half-wave dipole antenna: (a) conical dipole, and (b) conventional dipole.

A *folded dipole* can be fashioned as shown in Fig. 13.44. Such a configuration results in increased bandwidth and impedance. Impedance can be further increased by using rods of different diameter and by varying spacing of the elements. The 1/4-wave dipole elements connected to the closely-coupled 1/2-wave element act as a matching stub between the transmission line and the single-piece 1/2-wave element. This broadbands the folded dipole antenna by a factor of 2.

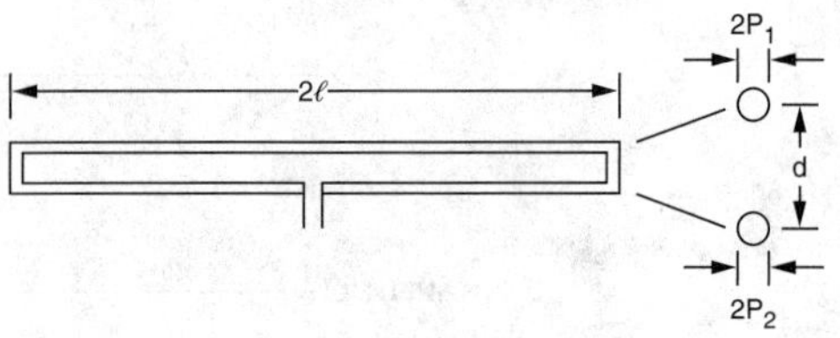

FIGURE 13.44   Folded dipole antenna.

A *corner-reflector* antenna may be formed as shown in Fig. 13.45. A ground plane or flat reflecting sheet is placed at a distance of 1/16- to 1/4-wavelengths behind the dipole. Gain in the forward direction can be increased by a factor of 2 with this type of design.

Quarter-Wave Monopole

A conductor placed above a ground plane forms an image in the ground plane such that the resulting pattern is a composite of the *real antenna* and the *image antenna* (see Fig. 13.46). The operating impedance is one-half of the impedance of the antenna and its image when fed as a physical antenna in free space. An example will help illustrate this concept. A 1/4-wave monopole mounted on an infinite ground plane has an impedance equal to one-half the free-space impedance of a 1/4-wave dipole. It follows, then, that the theoretical characteristic resistance of a 1/4-wave monopole with an infinite ground plane is 37 Ω.

For a real-world antenna, an infinite ground plane is neither possible nor required. An antenna mounted on a ground plane that is 2 to 3 times the operating wavelength has about the same impedance as a similar antenna mounted on an infinite ground plane.

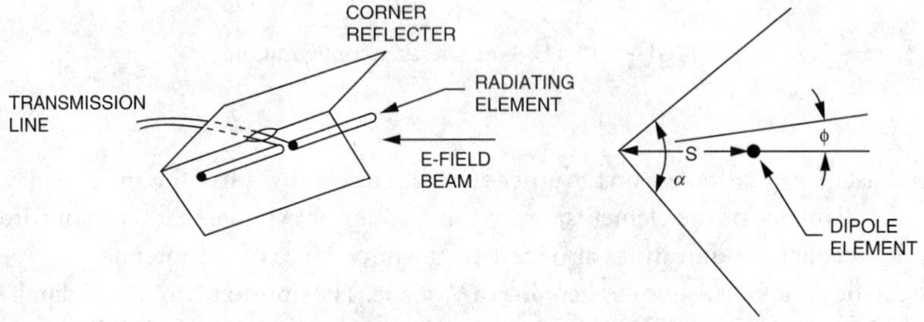

FIGURE 13.45   Corner-reflector antenna.

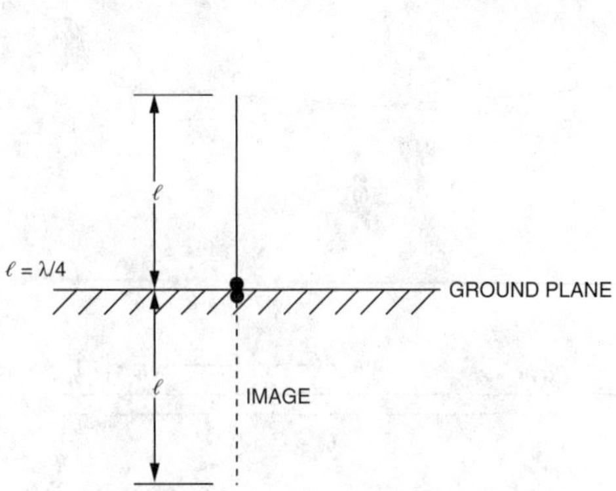

FIGURE 13.46   Vertical monopole mounted above a ground plane.

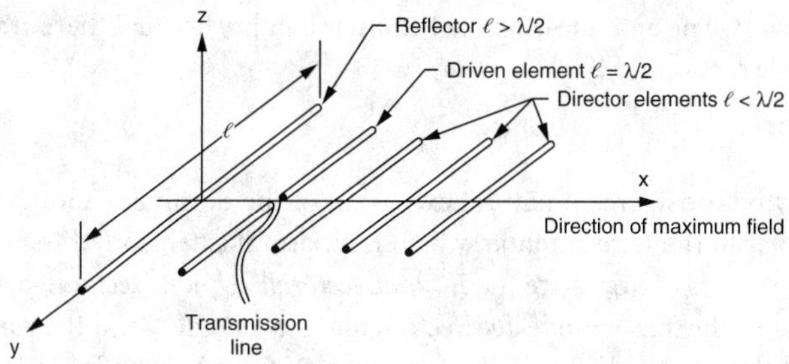

FIGURE 13.47   The Yagi-Uda array.

Log-Periodic Antenna

The log-periodic antenna can take on a number of forms. Typical designs include the *conical log spiral*, *log-periodic V*, and *log-periodic dipole*. The most common of these antennas is the log-periodic dipole. The antenna can be fed either by using alternating connections to a balanced line, or by a coaxial line running through one of the feeders from front to back. In theory, the log-periodic antenna can be designed to operate over many octaves. In practice, however, the upper frequency is limited by the precision required in constructing the small elements, feed lines, and support structure of the antenna.

Yagi-Uda Antenna

The Yagi-Uda is an *end-fire array* consisting typically of a single driven dipole with a *reflector dipole* behind the driven element, and one or more *parasitic director elements* in front (see Fig. 13.47). Common designs use from one to seven director elements. As the number of elements is increased, directivity increases. Bandwidth, however, decreases as the number of elements is increased. Arrays of more than four director elements are typically narrowband.

The radiating element is $^1/_2$-wavelength at the center of the band covered. The single reflector element is slightly longer, and the director elements are slightly shorter, all spaced approximately 1/4-wavelength from each other.

Table 13.6 demonstrates how the number of elements determines the gain and beamwidth of a Yagi-Uda antenna.

Waveguide Antenna

The waveguide antenna consists of a dominant-mode-fed waveguide opening onto a conducting ground plane. Designs may be based on rectangular, circular, or coaxial waveguide (also called an *annular slot*). The slot antenna is simplicity itself. A number of holes of a given dimension are placed at intervals along a section of waveguide. The radiation characteristics of the antenna are determined by the size, location,

TABLE 13.6   Typical Characteristics of Single-Channel Yagi-Uda Arrays

No. of Elements	Gain (dB)	Beam Width (°)
2	3–4	65
3	6–8	55
4	7–10	50
5	9–11	45
9	12–14	37
15	14–16	30

Source: Benson, B., Ed., *Television Engineering Handbook*, McGraw-Hill, New York, 1986, 388.

and orientation of the slots. The antenna offers optimum reliability because there are no discrete elements, except for the waveguide section itself.

Horn Antenna

The horn antenna may be considered a natural extension of the dominant-mode waveguide feeding the horn in a manner similar to the wire antenna, which is a natural extension of the two-wire transmission line. The most common types of horns are the *E-plane sectoral, H-plane sectoral, pyramidal horn* (formed by expanding the walls of the $TE_{0,1}$-mode-fed rectangular waveguide), and the *conical horn* (formed by expanding the walls of the $TE_{1,1}$-mode-fed circular waveguide). Dielectric-loaded waveguides and horns offer improved pattern performance over unloaded horns. Ridged and tapered horn designs improve the bandwidth characteristics. Horn antennas are available in single and dual polarized configurations.

Reflector Antenna

The reflector antenna is formed by mounting a radiating feed antenna above a reflecting ground plane. The most basic form of reflector is the loop or dipole spaced over a finite ground plane. This concept is the basis for the parabolic or spherical reflector antenna. The *parabolic reflector antenna* may be fed directly or through the use of a subreflector in the focal region of the parabola. In this approach, the subreflector is illuminated from the parabolic surface. The chief disadvantage of this design is the aperture blockage of the subreflector, which restricts its use to large-aperture antennas. The operation of a parabolic or spherical reflector antenna is typically described using physical optics.

Parabolic reflector antennas are usually illuminated by a flared-horn antenna with a flare angle of less than 18°. A rectangular horn with a flare angle less than 18° has approximately the same aperture field as the dominant-mode rectangular waveguide feeding the horn.

Spiral Antenna

The bandwidth limitations of an antenna are based on the natural change in the critical dimensions of the radiating elements caused by differing wavelengths. The spiral antenna overcomes this limitation because the radiating elements are specified only in angles. A two-arm equiangular spiral is shown in Fig. 13.48. This common design gives wideband performance. Circular polarization is inherent in the antenna. Rotation of the pattern corresponds to the direction of the spiral arms. The gain of a spiral antenna is typically slightly higher than a dipole.

The basic spiral antenna radiates on both sides of the arms. Unidirectional radiation is achieved through the addition of a reflector or cavity.

FIGURE 13.48   Two-arm equiangular spiral antenna.

Array Antenna

The term "array antenna" covers a wide variety of physical structures. The most common configuration is the *planar array antenna*, which consists of a number of radiating elements regularly spaced on a rectangular or triangular lattice. The *linear array antenna*, where the radiating elements are placed in a single line, is also common. The pattern of the array is the product of the element pattern and the array configuration. Large array antennas may consist of 20 or more radiating elements.

Correct phasing of the radiating elements is the key to operation of the system. The radiating pattern of the structure, including direction, can be controlled through proper adjustment of the relative phase of the elements.

13.3.3   Antenna Applications

An analysis of the applications of antennas for commercial and industrial use is beyond the scope of this chapter. It is instructive, however, to examine three of the most obvious antenna applications: AM and FM radio, and television. These applications illustrate antenna technology as it applies to frequencies ranging from the lower end of the MF band to the upper reaches of UHF.

AM Broadcast Antenna Systems

Vertical polarization of the transmitted signal is used for AM broadcast stations because of its superior groundwave propagation, and because of the simple antenna designs that it affords. The Federal Communications Commission (FCC) and licensing authorities in other countries have established classifications of AM stations with specified power levels and hours of operation. Protection requirements set forth by the FCC specify that some AM stations (in the U.S.) reduce their transmitter power at sunset, and return to full power at sunrise. This method of operation is based on the propagation characteristics of AM band frequencies. AM signals propagate further at nighttime than during the day.

The different day/night operating powers are designed to provide each AM station with a specified coverage area that is free from interference. Theory rarely translates into practice insofar as coverage is concerned, however, because of the increased interference that all AM stations suffer at nighttime.

The tower visible at any AM radio station transmitter site is only half of the antenna system. The second element is a buried ground system. Current on a tower does not simply disappear; rather, it returns to Earth through the capacitance between the Earth and the tower. Ground losses are greatly reduced if the tower has a radial copper ground system. A typical single-tower ground system is made up of 120 radial ground wires, each 140 electrical degrees long (at the operating frequency), equally spaced out from the tower base. This is often augmented with an additional 120 interspersed radials 50 ft long.

Directional AM Antenna Design

When a nondirectional antenna with a given power does not radiate enough energy to serve the station's primary service area, or radiates too much energy toward other radio stations on the same or adjacent frequencies, it is necessary to employ a directional antenna system. Rules set out by the FCC and regulatory agencies in other countries specify the protection requirements to be provided by various classes of stations, for both daytime and nighttime hours. These limits tend to define the shape and size of the most desirable antenna pattern.

A directional antenna functions by carefully controlling the amplitude and phase of the RF currents fed to each tower in the system. The directional pattern is a function of the number and spacing of the towers (vertical radiators), and the relative phase and magnitude of their currents. The number of towers in a directional AM array can range from two to six, or even more in a complex system. One tower is defined as the *reference tower*. The amplitude and phase of the other towers are measured relative to this reference.

A complex network of power splitting, phasing, and antenna coupling elements is required to make a directional system work. Figure 13.49 shows a block diagram of a basic two-tower array. A power divider

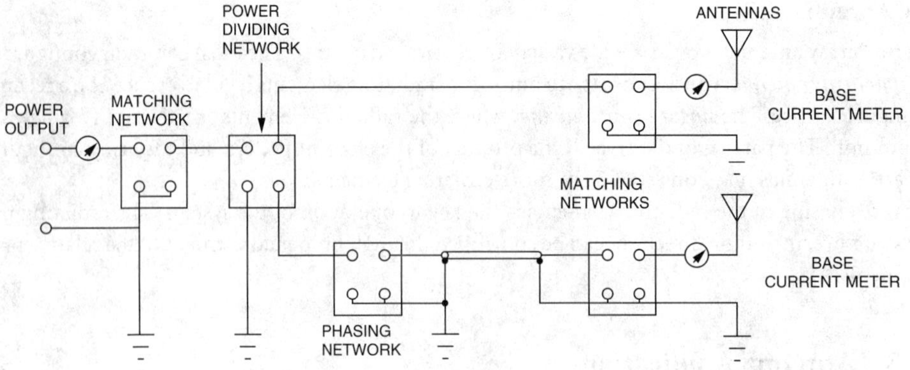

FIGURE 13.49   Block diagram of an AM directional antenna feeder system for a two-tower array.

network controls the relative current amplitude in each tower. A phasing network provides control of the phase of each tower current, relative to the reference tower. Matching networks at the base of each tower couple the transmission line impedance to the base operating impedance of the radiating towers.

In practice, the system shown in the figure would not consist of individual elements. Instead, the matching network, power dividing network, and phasing network would all usually be combined into a single unit, referred to as the *phasor*.

Antenna Pattern Design

The pattern of any AM directional antenna system (array) is determined by a number of factors, including

- Electrical parameters (phase relationship and current ratio for each tower)
- Height of each tower
- Position of each tower with respect to the other towers (particularly with respect to the reference tower)

A *directional array* consists of two or more towers arranged in a specific manner on a plot of land. Figure 13.50 shows a typical three-tower array, as well as the pattern such an array would produce. This is an *in-line array*, meaning that all the elements (towers) are in line with one another. Notice that the *major lobe* is centered on the same line as the line of towers, and that the *pattern nulls* (*minima*) are positioned symmetrically about the line of towers, protecting co-channel stations A and B at true bearings of 315° and 45°, respectively.

Figure 13.51 shows the same array, except that it has been rotated by 10°. Notice that the pattern shape is not changed, but the position of the major lobe and the nulls follow the line of towers. Also notice that the nulls are no longer pointed at the stations to be protected.

If this directional antenna system were constructed on a gigantic turntable, the pattern could be rotated without affecting the shape. But, to accomplish the required protections and to have the major lobe(s) oriented in the right direction, there is only one correct position. In most cases, the position of the towers will be specified with respect to a single reference tower. The location of the other towers will be given in the form of a distance and bearing from that reference. Occasionally, a reference point, usually the center of the array, will be used for a geographic coordinate point.

Bearing

The bearing or azimuth of the towers from the reference tower or point is specified clockwise in degrees from *true north*. The distinction between true and *magnetic north* is vital. The magnetic North Pole is not at the true or geographic North Pole. (In fact, it is in the vicinity of 74° north, 101° west, in the islands of northern Canada.) The difference between magnetic and true bearings is called variation or *magnetic declination*. Declination, a term generally used by surveyors, varies for different locations. It is

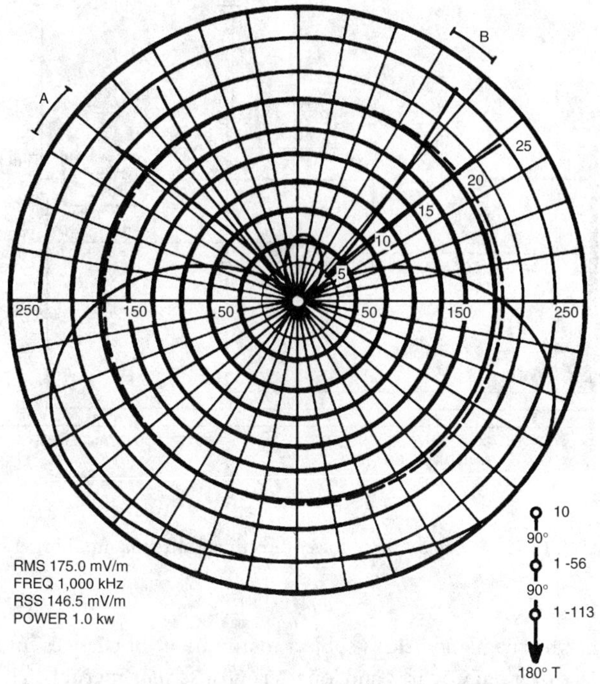
B
A
25
20
15
10
5
250
150
50
50
150
250
RMS 175.0 mV/m
FREQ 1,000 kHz
RSS 146.5 mV/m
POWER 1.0 kw
10
90°
1 -56
90°
1 -113
180° T

FIGURE 13.50 Radiation pattern generated with a three-tower in-line directional array using the electrical parameters and orientation shown.

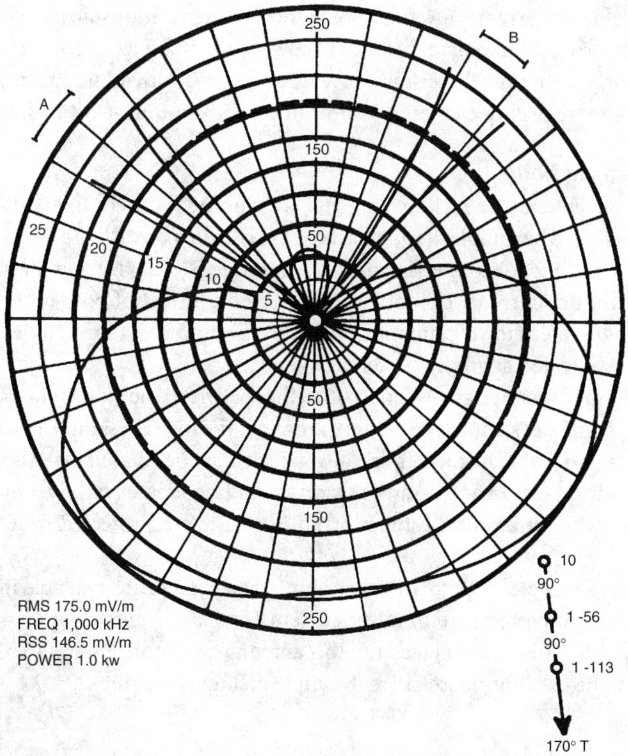
250
B
A
150
25
20
15
10
5
50
50
150
250
RMS 175.0 mV/m
FREQ 1,000 kHz
RSS 146.5 mV/m
POWER 1.0 kw
10
90°
1 -56
90°
1 -113
170° T

FIGURE 13.51 Radiation pattern produced when the array of Fig. 13.50 is rotated to a new orientation.

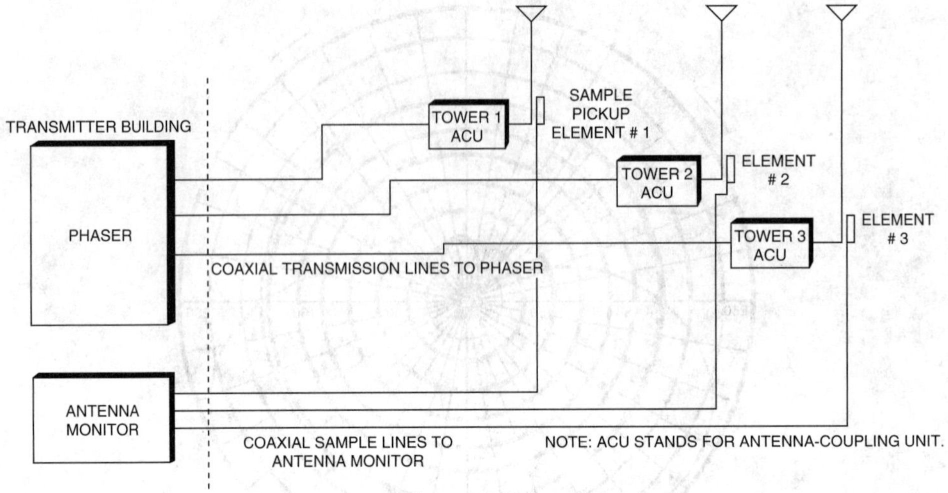

FIGURE 13.52 A typical three-tower directional antenna monitoring system.

not a constant. The Earth's magnetic field is subject to a number of changes in intensity and direction. These changes take place over daily, yearly, and long-term (or *secular*) periods. The secular changes result in a relatively constant increase or decrease in declination over a period of many years.

Antenna Monitoring System

Monitoring the operation of an AM directional antenna basically involves measuring the power into the system, the relative value of currents into the towers, their phase relationships, and the levels of radiated signal at certain monitoring points some distance from the antenna. Figure 13.52 shows a block diagram of a typical monitoring system for a three-tower array. For systems with additional towers, the basic layout is extended by adding more pickup elements, sample lines, and a monitor with additional inputs.

Phase/Current Sample Loop

Two types of phase/current sample pickup elements are commonly used: the *sample loop* and *torroidal current transformer* (TCT). The sample loop consists of a single turn unshielded loop of rigid construction, with a fixed gap at the open end for connection of the sample line. The device must be mounted on the tower near the point of maximum current. The loop can be used on towers of both uniform and nonuniform cross-section. It must operate at tower potential, except for towers of less than 130 electrical degrees height, where the loop can be operated at ground potential.

When the sample loop is operated at tower potential, the coax from the loop to the base of the tower is also at tower potential. To bring the sample line across the base of the tower, a sample line isolation coil is used.

A shielded torroidal current transformer can also be used as the phase/current pickup element. Such devices offer several advantages over the sample loop, including greater stability and reliability. Because they are located inside the tuning unit cabinet or house, TCTs are protected from wind, rain, ice, and vandalism.

Unlike the rigid, fixed sample loop, torroidal current transformers are available in several sensitivities, ranging from 0.25 to 1.0 V per ampere of tower current. Tower currents of up to 40 A can be handled, providing a more usable range of voltages for the antenna monitor. Figure 13.53 shows the various arrangements that can be used for phase/current sample pickup elements.

Sample Lines

The selection and installation of the sampling lines for a directional monitoring system are important factors in the ultimate accuracy of the overall array.

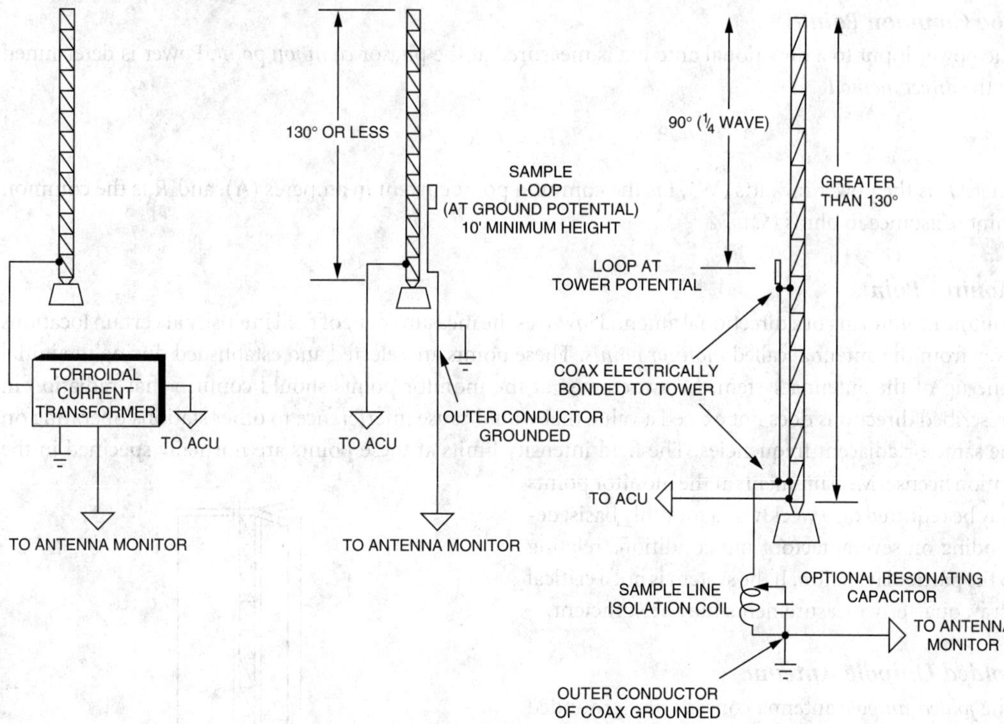

FIGURE 13.53 Three possible circuit configurations for phase sample pickup.

With *critical arrays* (antennas requiring operation within tight limits specified in the station license), all sample lines must be of equal electrical length and installed in such a manner that corresponding lengths of all lines are exposed to equal environmental conditions.

While sample lines may be run above ground on supports (if protected and properly grounded), the most desirable arrangement is direct burial using jacketed cable. Burial of sample line cable is almost a standard practice because proper burial offers good protection against physical damage and a more stable temperature environment.

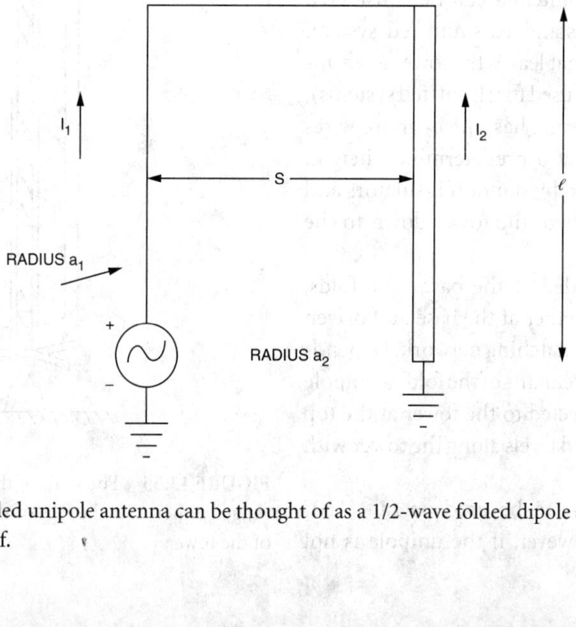

FIGURE 13.54 The folded unipole antenna can be thought of as a 1/2-wave folded dipole antenna perpendicular to the ground and cut in half.

The Common Point

The power input to a directional antenna is measured at the phasor *common point*. Power is determined by the *direct method*

$$P = I^2 R$$

where P is the power in watts (W), I is the common point current in amperes (A), and R is the common point resistance in ohms (Ω).

Monitor Points

Routine monitoring of a directional antenna involves the measurement of field intensity at certain locations away from the antenna, called *monitor points*. These points are selected and established during the initial tune-up of the antenna system. Measurements at the monitor points should confirm that radiation in prescribed directions does not exceed a value that would cause interference to other stations operating on the same or adjacent frequencies. The field intensity limits at these points are normally specified in the station license. Measurements at the monitor points may be required on a weekly or a monthly basis, depending on several factors and conditions relating to the particular station. If the system is not a critical array, quarterly measurements may be sufficient.

Folded Unipole Antenna

The *folded unipole* antenna consists of a grounded vertical structure with one or more conductors folded back parallel to the side of the structure. It can be visualized as a half-wave folded dipole perpendicular to the ground and cut in half (see Fig. 13.54). This design makes it possible to provide a wide range of resonant radiation resistances by varying the ratio of the diameter of the folded-back conductors in relation to the tower. Top loading can also be used to broaden the antenna bandwidth. A side view of the folded unipole is shown in Fig. 13.55.

The folded unipole antenna can be considered a modification of the standard shunt-fed system. Instead of a slant wire that leaves the tower at an approximate 45° angle (as used for shunt-fed systems), the folded unipole antenna has one or more wires attached to the tower at a predetermined height. The wires are supported by standoff insulators and run parallel to the sides of the tower down to the base.

The tower is grounded at the base. The folds, or wires, are joined together at the base and driven through an impedance matching network. Depending on the tuning requirements of the folded unipole, the wires may be connected to the tower at the top and/or at predetermined levels along the tower with shorting stubs.

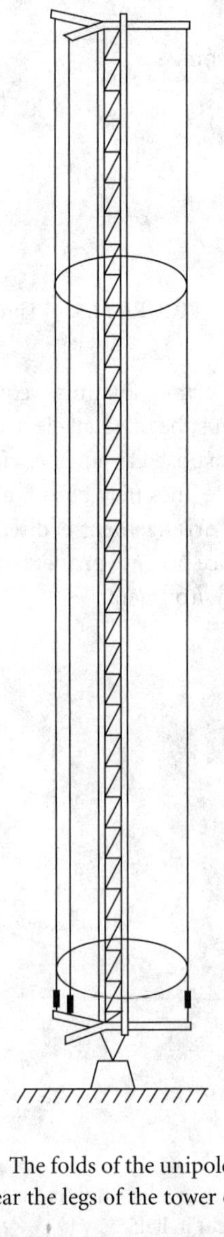

The folded unipole can be used on tall (130° or greater) towers. However, if the unipole is not

FIGURE 13.55 The folds of the unipole antenna are arranged either near the legs of the tower or near the faces of the tower.

divided into two parts, the overall efficiency (unattenuated field intensity) will be considerably lower than the normally expected field for the electrical height of the tower.

FM Broadcast Antenna Systems

The propagation characteristics of VHF FM radio are much different than for MF AM. There is essentially no difference between day and night FM propagation. FM stations have relatively uniform day and night service areas with the same operating power.

A wide variety of antennas is available for use in the FM broadcast band. Nearly all employ circular polarization. Although antenna designs differ from one manufacturer to another, generalizations can be made that apply to most units.

Antenna Types

There are three basic classes of FM broadcast transmitting antennas in use today: *ring stub* and *twisted ring*, *shunt-* and *series-fed slanted dipole*, and *multi-arm short helix*. While each design is unique, all have the following items in common:

- The antennas are designed for side-mounting to a steel tower or pole.
- Radiating elements are shunted across a common rigid coaxial transmission line.
- Elements are placed along the rigid line every one wavelength.
- Antennas with one to seven bays are fed from the bottom of the coaxial transmission line.
- Antennas with more than seven bays are fed from the center of the array to provide more predictable performance in the field.
- Antennas generally include a means of tuning out reactances after the antenna has been installed through the adjustment of variable capacitive or inductive elements at the feed point.

Figure 13.56 shows a shunt-fed slanted dipole antenna that consists of two half-wave dipoles offset 90°. The two sets of dipoles are rotated 22.5° (from their normal plane) and are *delta-matched* to provide a 50-Ω impedance at the radiator input flange. The lengths of all four dipole arms can be matched to

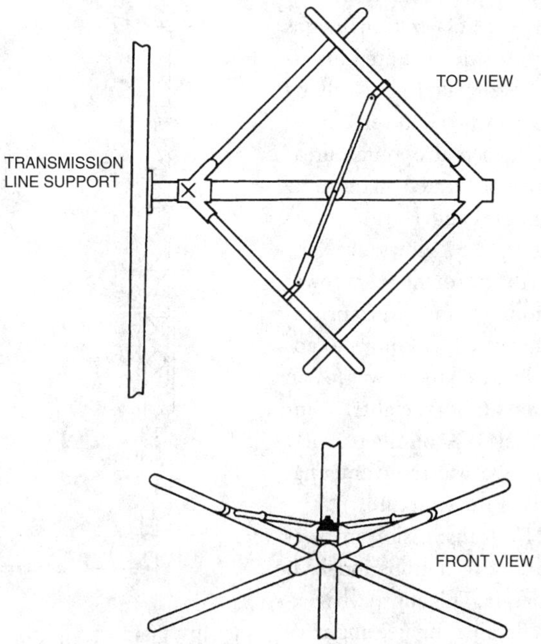

FIGURE 13.56 Mechanical configuration of one bay of a circularly polarized FM transmitting antenna.

TABLE 13.7 Various Combinations of Transmitter Power and
Antenna Gain that will Produce 100-kW ERP for an FM Station

No. Bays	Antenna Gain	Transmitter Power (kW)
3	1.5888	66.3
4	2.1332	49.3
5	2.7154	38.8
6	3.3028	31.8
7	3.8935	27.0
8	4.4872	23.5
10	5.6800	18.5
12	6.8781	15.3

resonance by mechanical adjustment of the end fittings. Shunt-feeding (when properly adjusted) provides equal currents in all four arms.

Wide-band *panel antennas* are a fourth common type of antenna used for FM broadcasting. Panel designs share some of the characteristics listed previously, but are intended primarily for specialized installations in which two or more stations will use the antenna simultaneously. Panel antennas are larger and more complex than other FM antennas, but offer the possibility for shared tower space among several stations and custom coverage patterns that would be difficult or even impossible with more common designs.

The ideal combination of antenna gain and transmitter power for a particular installation involves the analysis of a number of parameters. As shown in Table 13.7, a variety of pairings can be made to achieve the same ERP.

Television Antenna Systems

Television broadcasting uses horizontal polarization for the majority of installations worldwide. More recently, interest in the advantages of circular polarization has resulted in an increase in this form of transmission, particularly for VHF channels.

Both horizontal and circular polarization designs are suitable for tower-top or side-mounted installations. The latter option is dictated primarily by the existence of a previously installed tower-top antenna. On the other hand, in metropolitan areas where several antennas must be located on the same structure, either a stacking or candelabra-type arrangement is feasible. Figure 13.57 shows an example of antenna stacking on the top of the Sears Tower in Chicago, where numerous TV and FM transmitting antennas are located. Figure 13.58 shows a candelabra installation atop the Mt. Sutro tower in San Francisco. The Sutro tower supports eight TV antennas on its uppermost level. A number of FM transmitting antennas and two-way radio antennas are located on lower levels of the structure.

Another approach to TV transmission involves combining the RF outputs of two stations and feeding a single wide-band antenna. This approach is expensive and requires considerable engineering analysis to produce a combiner system that will not degrade the performance of either transmission

Intertower spacing = 100 ft

FIGURE 13.57 Twin-tower antenna array atop the Sears Tower in Chicago. Note how antennas have been stacked to overcome space restrictions.

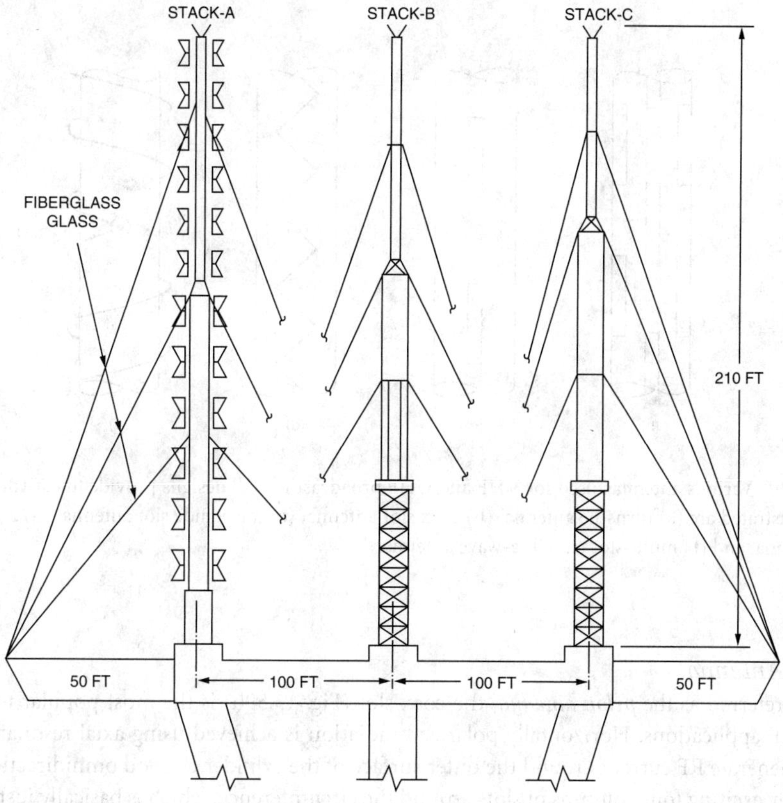

FIGURE 13.58 Installation of TV transmitting antennas on the candelabra structure at the top level of the Mt. Sutro tower in San Francisco. This installation makes extensive use of antenna stacking.

system. In the Mt. Sutro example (Fig. 13.58), it can be seen that two stations (channels 4 and 5) are combined into a single transmitting antenna.

Top-Mounted Antenna Types

The typical television broadcast antenna is a broadband radiator operating over a bandwidth of several megahertz with an efficiency of over 95%. Reflections from the antenna and transmission line back to the transmitter must be kept small enough to introduce negligible picture degradation. Furthermore, the gain and pattern characteristics of the antenna must be designed to achieve the desired coverage within acceptable tolerances. Tower-top, pole-type antennas designed to meet these parameters can be classified into two categories: *resonant dipoles* and *multi-wavelength traveling-wave elements*.

The primary considerations in the design of a top-mounted antenna are the achievement of uniform omnidirectional azimuth fields and minimum windloading. A number of different approaches have been tried successfully. Figure 13.59 illustrates the basic mechanical design of the most common antennas.

Turnstile Antenna

The turnstile is the earliest and most popular resonant antenna for VHF broadcasting. The antenna is made up of four *batwing*-shaped elements mounted on a vertical pole in a manner resembling a turnstile. The four batwings are, in effect, two dipoles fed in quadrature phase. The azimuth-field pattern is a function of the diameter of the support mast. The pattern is usually within 10 to 15% of a true circle.

The turnstile antenna is made up of several layers, usually six layers for channels 2 through 6 and twelve layers for channels 7 through 13. The turnstile is not suitable for side-mounting, except for standby applications in which coverage degradation can be tolerated.

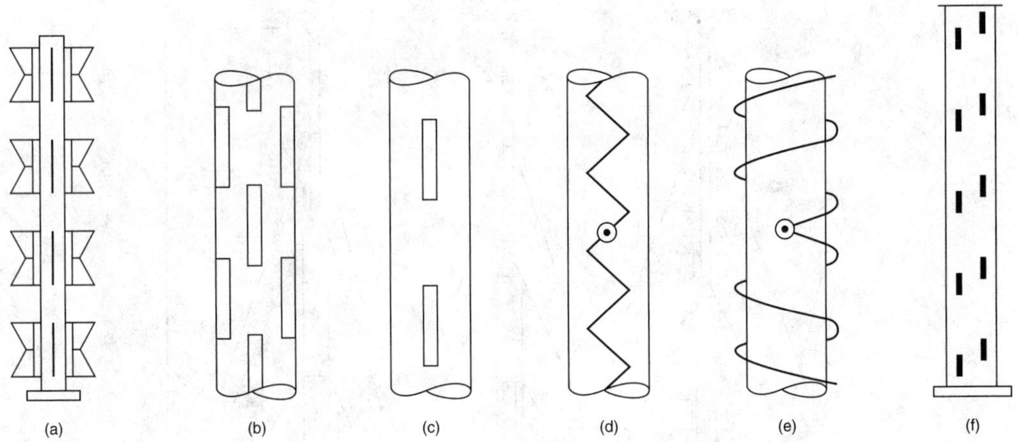

FIGURE 13.59 Various antennas used for VHF and UHF broadcasting. All designs provide linear (horizontal) polarization. Illustrated are (a) turnstile antenna, (b) coax slot antenna, (c) waveguide slot antenna, (d) zigzag antenna, (e) helix antenna, and (f) multi-slot traveling-wave antenna.

Coax Slot Antenna

Commonly referred as the *pylon antenna*, the coax slot (Fig. 13.59b) is the most popular top-mounted unit for UHF applications. Horizontally polarized radiation is achieved using axial resonant slots on a cylinder to generate RF current around the outer surface of the cylinder. A good omnidirectional pattern is achieved by exciting four columns of slots around the circumference, which is basically just a section of rigid coaxial transmission line.

The slots along the pole are spaced approximately one wavelength per layer, and a suitable number of layers are used to achieve the desired gain. Typical gains range from 20 to 40. By varying the number of slots around the periphery of the cylinder, directional azimuth patterns can be achieved.

Waveguide Slot Antenna

The UHF waveguide slot (Fig. 13.59c) is a variation on the coax slot antenna. The antenna is simply a section of waveguide with slots cut into the sides. The physics behind the design is long and complicated. However, the end result is the simplest of all antennas. This is a desirable feature in field applications because simple designs translate into long-term reliability.

Zigzag Antenna

The zigzag is a panel array design that utilizes a conductor routed up the sides of a three- or four-sided panel antenna in a "zigzag" manner (see Fig. 13.59d). With this design, the vertical current component along the zigzag conductor is mostly canceled out, and the antenna can effectively be considered an array of dipoles. With several such panels mounted around a polygonal periphery, the required azimuth pattern can be shaped by proper selection of feed currents to the various elements.

Helix Antenna

A variation on the zigzag, the helix antenna (Fig. 13.59e) accomplishes basically the same goal using a different mechanical approach. Note the center feed point shown in the figure.

VHF Multi-slot Antenna

Mechanically similar to the coax slot antenna, the VHF multi-slot antenna (Fig. 13.59f) consists of an array of axial slots on the outer surface of a coaxial transmission line. The slots are excited by a traveling wave inside the slotted line. The azimuth pattern is typically within 5% of omnidirectional. The antenna is generally about 15 wavelengths long.

Circularly Polarized Antennas

Circular polarization holds the promise of improved penetration into difficult coverage areas. There are a number of points to weigh in the decision to use a circularly polarized (CP) antenna, not the least of which is that a station must double its transmitter power in order to maintain the same ERP if it installs a CP antenna. This assumes equal vertical and horizontal components.

It is possible, and sometimes desirable, to operate with *elliptical polarization*, in which the horizontal and vertical components are not equal. Furthermore, the azimuth patterns for each polarization can be customized to provide the most efficient service area coverage.

Three major antenna types have been developed for CP television applications: the *normal mode helix*, *various panel antenna designs*, and the *interlaced traveling wave array*.

Normal Mode Helix

The normal mode helix consists of a supporting tube with helical radiators mounted around the tube on insulators. The antenna is broken into sub-arrays, each powered by a divider network. The antenna is called the "normal mode helix" because radiation occurs normal to the axis of the helix, or perpendicular to the support tube. The antenna provides an omnidirectional pattern.

Panel Antennas

The basic horizontally polarized panel antenna can be modified to produce circular polarization through the addition of vertically polarized radiators. Panel designs offer broad bandwidth and a wide choice of radiation patterns. By selecting the appropriate number of panels located around the tower, and the proper phase and amplitude distribution to the panels, a number of azimuth patterns can be realized. The primary drawback to the panel is the power distribution network required to feed the individual radiating elements.

Interlaced Traveling Wave Array

As the name implies, the radiating elements of this antenna are interlaced along an array, into which power is supplied. The energy input at the bottom of the antenna is extracted by the radiating elements as it moves toward the top. The antenna consists of a cylindrical tube that supports the radiating elements. The elements are slots for the horizontally polarized component and dipoles for the vertically polarized component. The radiating elements couple RF directly off the main input line; thus, a power dividing network is not required.

Side-Mounted Antenna Types

Television antennas designed for mounting on the faces of a tower must meet the same basic requirements as a top-mounted antenna—wide bandwidth, high efficiency, predictable coverage pattern, high gain, and low windloading—plus the additional challenge that the antenna must work in a less-than-ideal environment. Given the choice, no broadcaster would elect to place its transmitting antenna on the side of a tower instead of at the top. There are, however, a number of ways to solve the problems presented by side-mounting.

Butterfly Antenna

The butterfly is essentially a batwing panel developed from the turnstile radiator. The butterfly is one of the most popular panel antennas used for tower-face applications. It is suitable for the entire range of VHF applications. A number of variations on the basic batwing theme have been produced, including modifying the shape of the turnstile-type wings to rhombus or diamond shapes. Another version utilizes multiple dipoles in front of a reflecting panel.

For CP applications, two crossed dipoles or a pair of horizontal and vertical dipoles are used. A variety of cavity-backed crossed-dipole radiators are also utilized for CP transmission.

The azimuth pattern of each panel antenna is unidirectional, and three or four such panels are mounted on the sides of a triangular or square tower to achieve an omnidirectional pattern. The panels can be fed in-phase, with each one centered on the face of the tower, or fed in rotating phase with the proper mechanical offset. In the latter case, the input impedance match is considerably better.

Directionalization of the azimuth pattern is realized by proper distribution of the feed currents to individual panels in the same layer. Stacking layers provides gains comparable with top-mounted antennas.

The main drawbacks of panel antennas include (1) high windload, (2) complex feed system inside the antenna, and (3) restrictions on the size of the tower face, which determine to a large extent the omnidirectional pattern of the antenna.

UHF Side-Mounted Antennas

Utilization of panel antennas in a manner similar to those for VHF applications is not always possible at UHF installations. The high gains required for UHF broadcasting (in the range of 20 to 40, compared with gains of 6 to 12 for VHF) require far more panels with an associated complex feed system.

The zigzag panel antenna has been used for special omnidirectional and directional applications. For custom directional patterns, such as a cardioid shape, the pylon antenna can be side-mounted on one of the tower legs. Many stations, in fact, simply side-mount a pylon-type antenna on the leg of the tower that faces the greatest concentration of viewers. It is understood that viewers located behind the tower will receive a poorer signal; however, given the location of most TV transmitting towers—usually on the outskirts of their licensed city or on a mountaintop—this practice is often acceptable.

Broadband Antennas

Radiation of multiple channels from a single antenna requires the antenna to be broadband in both pattern and impedance (VSWR) characteristics. As a result, a broadband TV antenna represents a significant departure from the narrowband, single-channel pole antennas commonly used for VHF and UHF. The typical single-channel UHF antenna uses a series feed to the individual radiating elements, while a broadband antenna has a branch feed arrangement. The two feed configurations are shown in Fig. 13.60.

- At the design frequency, the series feed provides co-phased currents to its radiating elements. As the frequency varies, the electrical length of the series line feed changes such that the radiating elements are no longer in-phase outside of the designed channel. This electrical length change causes significant beam tilt out of band, and an input VSWR that varies widely with frequency.

- In contrast, the branch feed configuration employs feed lines that are nominally of equal length. Therefore the phase relationships of the radiating elements are maintained over a wide span of frequencies. This provides vertical patterns with stable beam tilt, a requirement for multi-channel applications.

- The basic building block of the multi-channel antenna is the broadband panel radiator. The individual radiating elements within a panel are fed by a branch feeder system that provides the panel with a single input cable connection. These panels are then stacked vertically and arranged around a supporting spine or existing tower to produce the desired vertical and horizontal radiation patterns.

Bandwidth

The ability to combine multiple channels in a single transmission system depends on the bandwidth capabilities of the antenna and waveguide or coax. The antenna must have the necessary bandwidth in both pattern and impedance (VSWR). It is possible to design an antenna system for low-power applications using coaxial transmission line that provides whole-band capability. For high-power systems, waveguide bandwidth sets the limits of channel separation.

Antenna pattern performance is not a significant limiting factor. As frequency increases, the horizontal pattern circularity deteriorates, but this effect is generally acceptable. Also, the electrical aperture increases with frequency, which narrows the vertical pattern beamwidth. If a high-gain antenna were used over a wide bandwidth, the increase in electrical aperture might make the vertical pattern beamwidth unacceptably narrow. This, however, is usually not a problem because of the channel limits set by the waveguide.

Horizontal Pattern

Because of the physical design of a broadband panel antenna, the cross-section is larger than the typical narrowband pole antenna. Therefore, as the operating frequencies approach the high end of the UHF band, the *circularity* (average circle to minimum or maximum ratio) of an omnidirectional broadband antenna generally deteriorates.

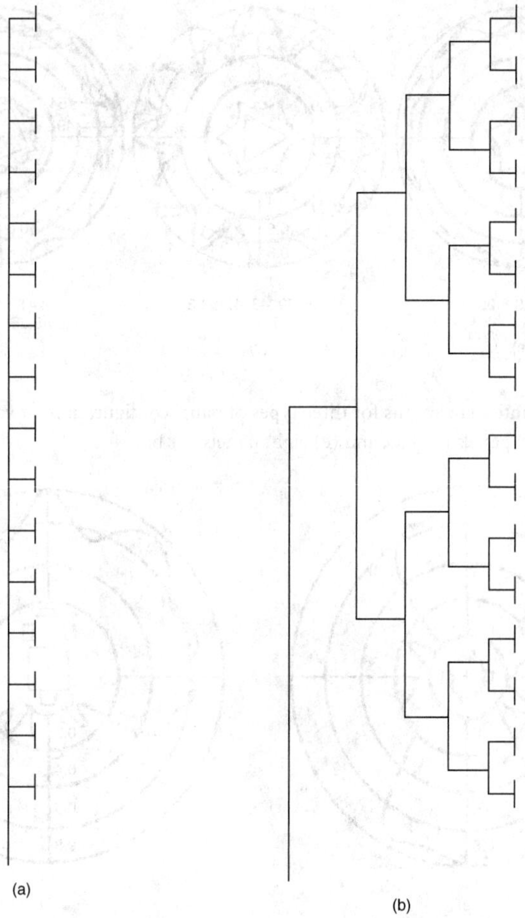

FIGURE 13.60 Antenna feed configurations: (a) series feed, and (b) branch feed.

Improved circularity is possible by arranging additional panels around the supporting structure. Typical installations have used five, six, and eight panels per bay. These are illustrated in Fig. 13.61 along with measured patterns at different operating channels. These approaches are often required for power handling considerations, especially when three or four transmitting channels are involved.

The flexibility of the panel antenna allows directional patterns of unlimited variety. Two of the more common applications are shown in Fig. 13.62. The peanut and cardioid types are often constructed on square support spines (as indicated). A cardioid pattern may also be produced by side-mounting on a triangular tower. Different horizontal radiation patterns for each channel may also be provided, as indicated in Fig. 13.63. This is accomplished by changing the power and/or phase to some of the panels in the antenna with frequency.

Most of these antenna configurations are also possible using a circularly polarized panel. If desired, the panel can be adjusted for elliptical polarization with the vertical elements receiving less than 50% of the power. Using a circularly polarized panel will reduce the horizontally polarized ERP by half (assuming the same transmitter power).

13.3.4 Phased-Array Antenna Systems

Phased-array antennas are steered by tilting the phase front independently in two orthogonal directions called the array coordinates. Scanning in either *array coordinate* causes the beam to move along a cone

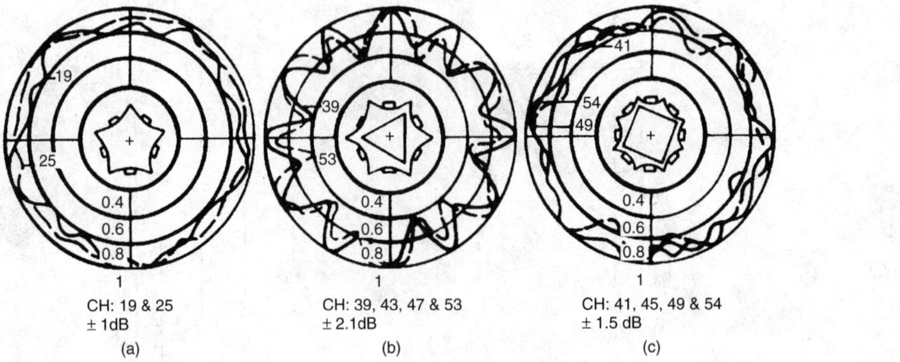

FIGURE 13.61 Measured antenna patterns for three types of panel configurations at various operating frequencies: (a) five panels per bay, (b) six panels per bay, and (c) eight panels per bay.

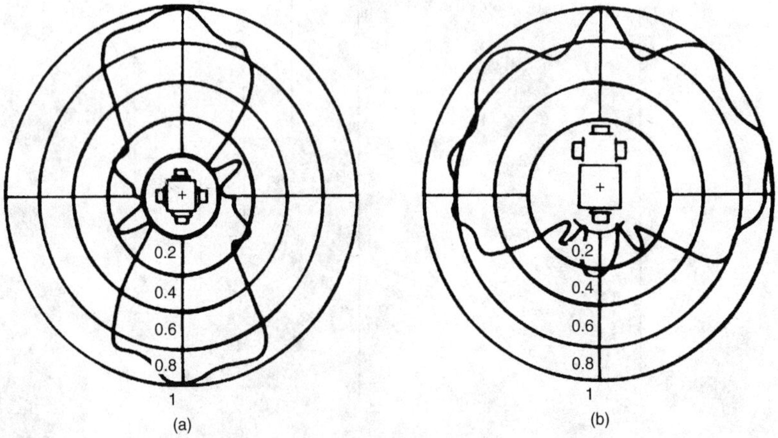

FIGURE 13.62 Common directional antenna patterns: (a) peanut, and (b) cardioid.

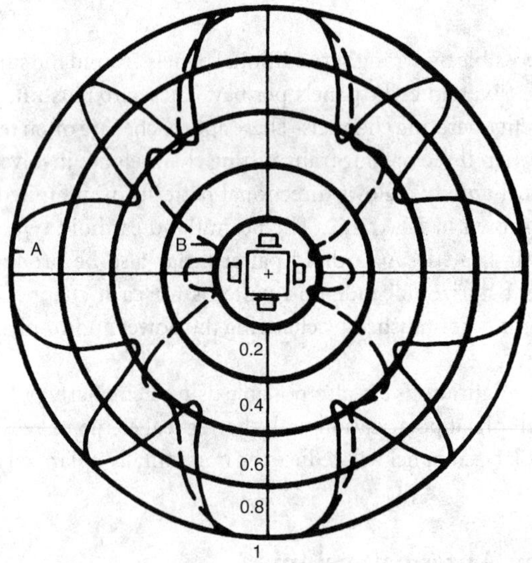

FIGURE 13.63 Use of a single antenna to produce two different rediation patterns, omnidirectional (trace "A") and peanut (trace "B").

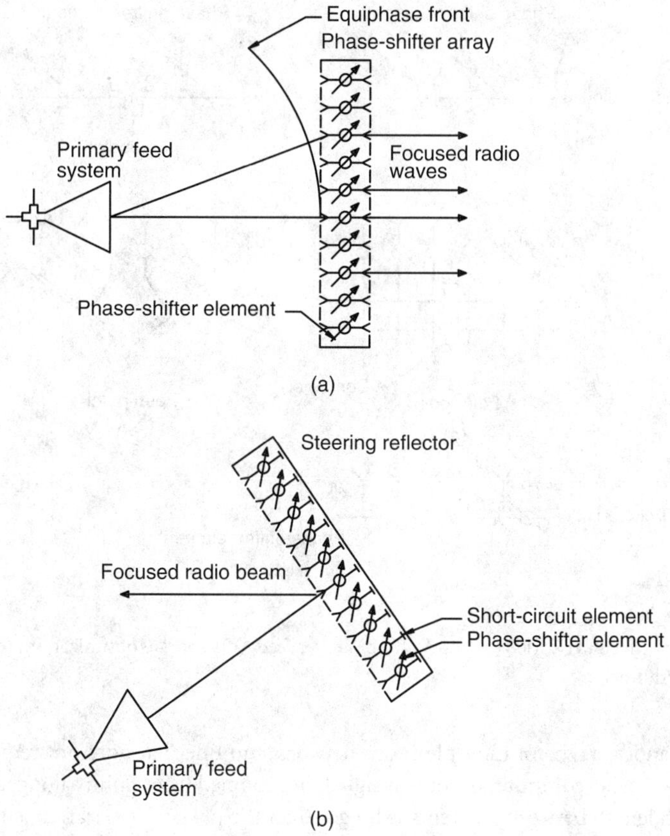

FIGURE 13.64 Optical antenna feed systems: (a) lens, and (b) reflector.

the center of which is at the center of the array. As the beam is steered away from the array normal, the projected aperture in the beam's direction varies, causing the beamwidth to vary proportionately.

Arrays can be classified as either *active* or *passive*. Active arrays contain duplexers and amplifiers behind every element or group of elements of the array. Passive arrays are driven from a single feed point. Active arrays are capable of high-power operation. Both passive and active arrays must divide the signal from a single transmission line among all the elements of the system. This can be accomplished through one of the following methods:

- *Optical feed*: a single feed, usually a monopulse horn, is used to illuminate the array with a spherical phase front (illustrated in Fig. 13.64). Power collected by the rear elements of the array is transmitted through the phase shifters to produce a planar front and steer the array. The energy can then be radiated from the other side of the array, or reflected and reradiated through the collecting elements. In the latter case, the array acts as a steerable reflector.

- *Corporate feed*: a system utilizing a series-feed network (Fig. 13.65) or parallel-feed network (Fig. 13.66). Both designs use transmission-line components to divide the signal among the elements. Phase shifters can be located at the elements or within the dividing network. Both the series- and parallel-feed systems have several variations, as shown in the figures.

- *Multiple-beam network*: a system capable of forming simultaneous beams with a given array. The Butler matrix, shown in Fig. 13.67, is one such technique. It connects the N elements of a linear array to N feed points corresponding to N beam outputs. The phase shifter is one of the most critical components of the system. It produces controllable phase shift over the operating band of the array. Digital and analog phase shifters have been developed using both ferrites and pin diodes.

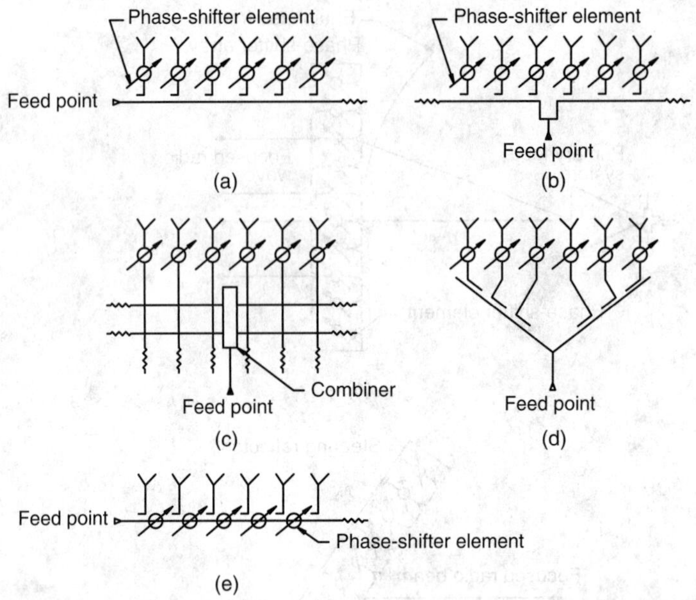

FIGURE 13.65 Series-feed networks: (a) end feed, (b) center feed, (c) separate optimization, (d) equal path length, and (e) series phase shifters.

Frequency scan is another type of multiple-beam network, but one that does not require phase shifters, dividers, or beam-steering computers. Element signals are coupled from points along a transmission line. The electrical path length between elements is longer than the physical separation, so a small frequency change will cause a phase change between elements that is large enough to steer the beam. This technique can be applied only to one array coordinate. If a two-dimensional array is required, phase shifters are normally used to scan the other coordinate.

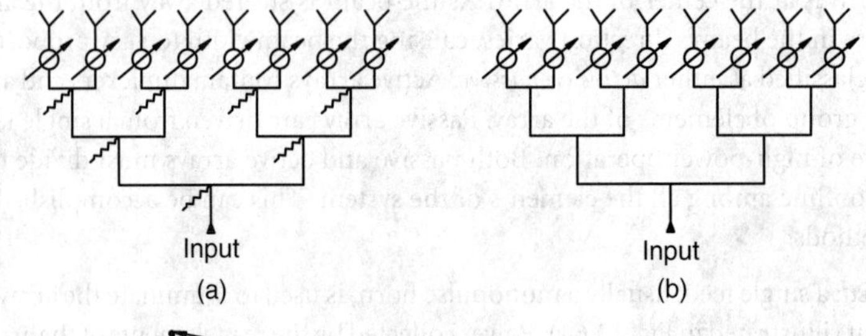

FIGURE 13.66 Types of parallel-feed networks: (a) matched corporate feed, (b) reactive corporate feed, (c) reactive stripline, and (d) multiple reactive divider.

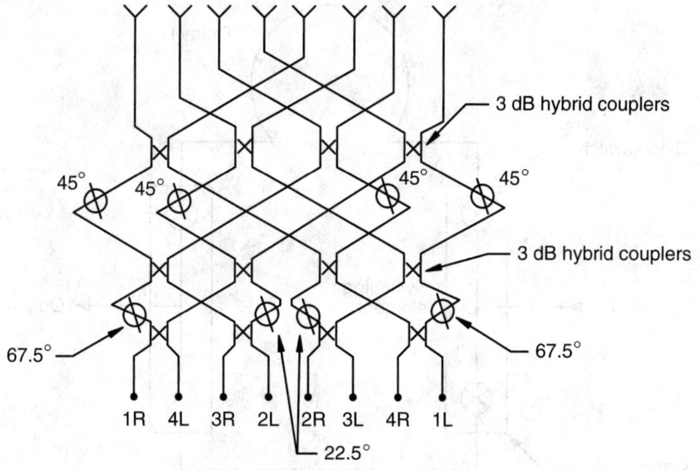

FIGURE 13.67   The Butler beam-forming network.

Phase-Shift Devices

The design of a phase shifter must meet two primary criteria

- Low transmission loss
- High power-handling capability

The Reggia-Spencer phase shifter meets both requirements. The device, illustrated in Fig. 13.68, consists of a ferrite rod mounted inside a waveguide that delays the RF signal passing through the waveguide, permitting the array to be steered. The amount of phase shift can be controlled by the current in the solenoid, because of the effect a magnetic field has on the permeability of the ferrite. This design is a

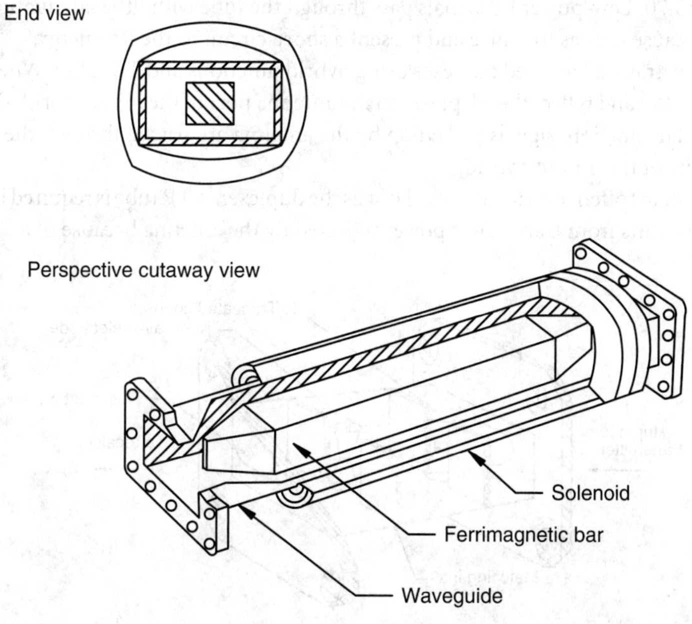

FIGURE 13.68   Basic concept of a Reggia-Spencer phase shifter.

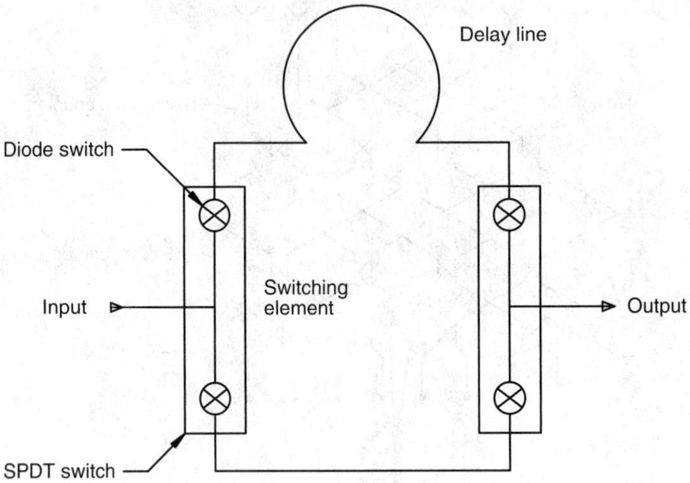

FIGURE 13.69   Switched-line phase shifter using pin diodes.

reciprocal phase shifter, meaning that the device exhibits the same phase shift for signals passing in either direction (forward or reverse). Nonreciprocal phase shifters are also available, where phase-shift polarity reverses with the direction of propagation.

Phase shifters may also be developed using pin diodes in transmission line networks. One configuration, shown in Fig. 13.69, uses diodes as switches to change the signal path length of the network. A second type uses pin diodes as switches to connect reactive loads across a transmission line. When equal loads are connected with 1/4-wave separation, a pure phase shift results.

Radar System Duplexer

The duplexer is an essential component of any radar system. The switching elements used in a duplexer include gas tubes, ferrite circulators, and pin diodes. Gas tubes are the simplest. A typical gas-filled TR tube is shown in Fig. 13.70. Low-power RF signals pass through the tube with little attenuation. Higher-power signals, however, cause the gas to ionize and present a short-circuit to the RF energy.

Figure 13.71 illustrates a balanced duplexer using hybrid junctions and TR tubes. When the transmitter is on, the TR tubes fire and reflect the RF power to the antenna port of the input hybrid. During the receive portion of the radar function, signals picked up by the antenna are passed through the TR tubes and on to the receiver port of the output hybrid.

Newer radar systems often use a ferrite circulator as the duplexer. A TR tube is required in the receiver line to protect input circuits from transmitter power reflected by the antenna because of an imperfect match.

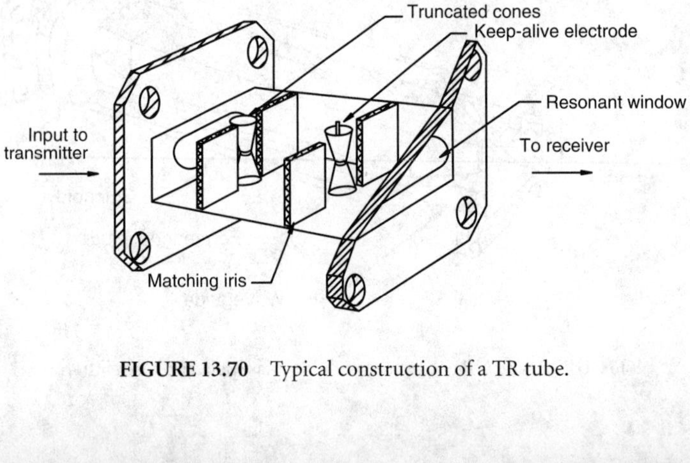

FIGURE 13.70   Typical construction of a TR tube.

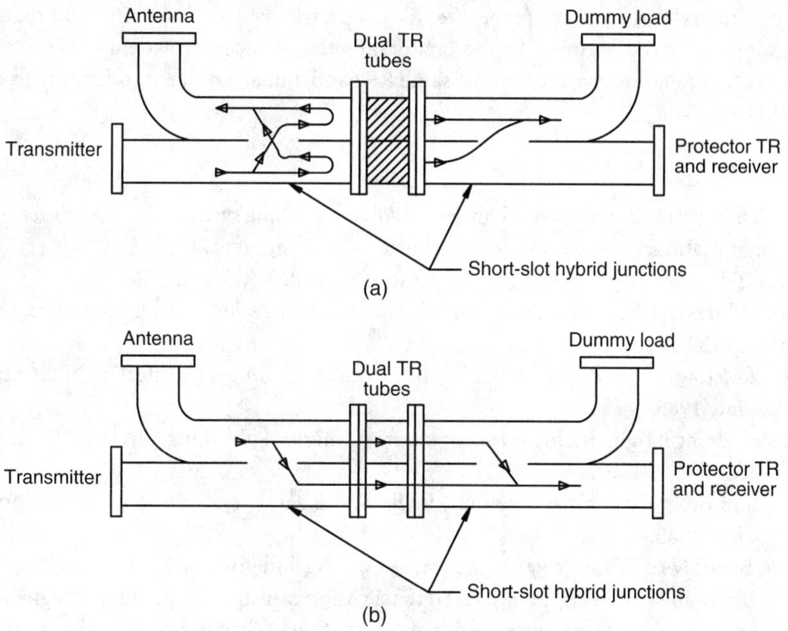

FIGURE 13.71 Balanced duplexer circuit using dual TR tubes and two short-slot hybrid junctions: (a) transmit mode, and (b) receive mode.

A four-port circulator generally is used with a load between the transmitter and receiver ports so that power reflected by the TR tube is properly terminated.

Pin diode switches have also been used in duplexers to perform the protective switching function of TR tubes. Pin diodes are more easily applied in coaxial circuitry, and at lower microwave frequencies. Multiple diodes are used when a single diode cannot withstand the expected power.

Microwave filters are sometimes used in the transmit path of a radar system to suppress spurious radiation, or in the receive signal path to suppress spurious interference.

Harmonic filters commonly are used in the transmission chain to absorb harmonic energy output by the system, preventing it from being radiated or reflected back from the antenna. Figure 13.72 shows a filter in which harmonic energy is coupled out through holes in the walls of the waveguide to matched loads.

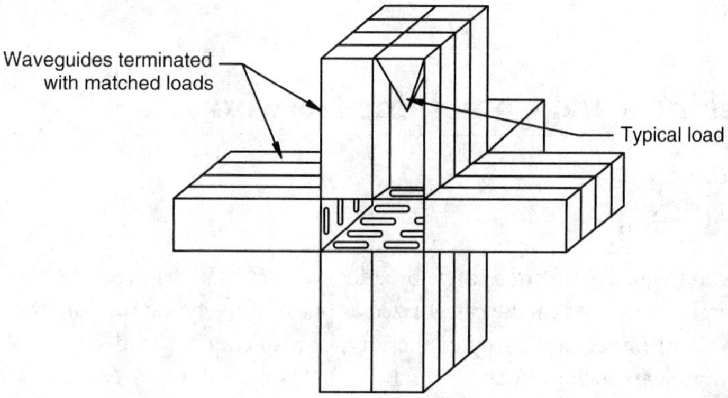

FIGURE 13.72 Construction of a dissipative waveguide filter.

Narrow-band filters in the receive path, often called *preselectors*, are built using mechanically tuned cavity resonators or electrically tuned TIG resonators. Preselectors can provide up to 80 dB suppression of signals from other radar transmitters in the same RF band, but at a different operating frequency.

References

Anders, M.B. A case for the use of multi-channel broadband antenna systems, *NAB Engineering Conference Proceedings*, National Association of Broadcasters, Washington, D.C., 1985.

Benson, B., Ed. *Television Engineering Handbook*, McGraw-Hill, New York, 1986.

Benson, B. and Whitaker, J. Ed. *Television and Audio Handbook for Technicians and Engineers*, McGraw-Hill, New York, 1989.

Bingeman, G. AM tower impedance matching, in *Broadcast Engineering*, Intertec Publishing, Overland Park, KS, July 1985.

Bixby, J. AM DAs: doing it right, in *Broadcast Engineering*, Intertec Publishing, Overland Park, KS, February 1984.

Chick, E. B. Monitoring directional antennas, in *Broadcast Engineering*, Intertec Publishing, Overland Park, KS, July 1985.

Crutchfield, E. B. Ed. *NAB Engineering Handbook*, 7th ed. Washington, D.C., 1985.

Dienes, G. Circularly and elliptically polarized UHF television transmitting antenna design, in *Proceedings of the NAB Engineering Conference*, National Association of Broadcasters, Washington, D.C., 1988.

Fink, D. and Christiansen, D. Eds. *Electronics Engineers' Handbook*, 3rd ed. McGraw-Hill, New York, 1989.

Jordan, E. C. Ed. *Reference Data for Engineers: Radio, Electronics, Computer and Communications*, 7th ed. Howard W. Sams Company, Indianapolis, IN, 1985.

Howard, G. P. The Howard AM Sideband Response Method, in *Radio World*, Falls Church, VA, August 1979.

Mullaney, J. H. The *Consulting Radio Engineer's Notebook*, Mullaney Engineering, Gaithersburg, MD, 1985.

Mullaney, J. H. The Folded Unipole Antenna, *Broadcast Engineering*, Intertec Publishing, Overland Park, KS, July 1986.

Mullaney, J. H. The Folded Unipole Antenna for AM Broadcast, *Broadcast Engineering*, Intertec Publishing, Overland Park, KS, January 1960.

Mullaney, J. H. and Howard, G. P. SBNET: A Fortran Program for Analyzing Sideband Response and Design of Matching Networks, Mullaney Engineering, Gaithersburg, MD, 1970.

Mayberry, E. and Stenberg, J. UHF Multi-Channel Antenna Systems, *Broadcast Engineering*, Intertec Publishing, Overland Park, KS, March 1989.

Raines, J. K. Folded Unipole Studies, *Think Book Series*, Multronics, 1968–1969.

Raines, J. K. Unipol: A Fortran Program for Designing Folded Unipole Antennas, Mullaney Engineering, Gaithersburg, MD, 1970.

13.4 Combiners and Combining Networks

Robert A. Surette

13.4.1 Introduction

Transmitting several frequencies from a single broadband antenna system requires the use of combining networks. In general, a combiner can be categorized as one of three types: runout, branched (star point), and balanced (constant impedance). Any of these types may employ band-reject (notch) or bandpass filters. Finally, components such as isolators can be added to a combiner network tailor the basic system to a specific application. This section discusses the use of filters, isolators, and other components in FM, UHF-TV, VHF-TV, and land mobile communication combiner networks.

13.4.2   Why Combiners Are Used

As populations migrate from rural areas to the suburbs, it has become more desirable to construct large transmission facilities that can reach these heavily populated areas from central locations. Of course, these prime locations have become more valuable, and so it makes sense to use each to its fullest potential. This can best be done by sharing a transmitter site and a common antenna among several users. To accomplish this, the industry uses combiners of various types and sizes.

For example, in San Francisco, Toronto, and Montreal, huge towers have been built to consolidate as many broadcasting facilities as possible, including VHF-TV, UHF-TV, FM, and land mobile communications services. This approach has proven very effective, not only using real estate economically, but spreading the tower costs over many users. In fact, the CN tower carries virtually all of the services in the Toronto area. These installations, of course, employ combiner networks to prevent interference among services.

Federal Communications Commission (FCC) Isolation Requirements

Why not just merge two or more signals together in a tee and send them to a common antenna? Because unless they are stopped by some means, the signals will backfeed up each other's input lines and into the transmitters where spurious intermodulation products (*spurs*) will be generated within the final cavities of the transmitters. These spurs will then be transmitted up to the antenna and broadcast. FCC rules specify that spurs must be 80 dB below the carrier frequency, so care must be taken to prevent them.

To keep a transmitter from generating spurs, it must be isolated from unwanted signals by approximately 40 dB. Therefore, the basic design criterion for a combiner is to provide 40 dB of isolation, plus a 6–10 dB safety margin. The closer the spacing of the frequencies between transmitters, the more filtering that is required to attain the 46–50 dB of isolation. The farther the spacing, the less filtering that is required. Modern FM combiners provide isolation of 75–80 dB, and along with transmitter losses of 13 dB, ensure that the 80-dB FCC spur rule is met.

Group Delay Equalization

The complex signals that make up today's broadcasts require extensive use of the full channel bandwidth. Any filter system has a tendency to throw the edges of the channel out of phase with the fundamental frequency, inducing crosstalk among the subcarriers. This is called *group delay* or *group delay difference*. Modern transmitter/combiner systems include group delay equalization devices, which compensate for this phase distortion. In FM, this is normally done at the combiner input, using high-power components. A television channel, with its many subcarriers, requires similar equalization, but here this is commonly achieved using low-level equipment within the transmitter. In contrast, a land mobile channel uses such a narrow bandwidth that group delay typically is not an issue. With these principles in mind, the type of combiner and the components used will depend on the number of channels to be combined and their frequency spacing.

13.4.3   History of FM Combiners in the U.S.

Dr. Andrew Alford designed several of the early combiner systems in the U.S. A major installation was the multichannel system on the Empire State Building in New York in the mid-1950s. This system used pairs of hybrids with a phase-delaying cavity in one leg of each hybrid pair. In the mid-1960s, for the John Hancock Building in Chicago, he designed a runout combiner system, again using quadrature hybrids, but the phase matching was accomplished with coaxial *runout* lines instead of cavities, to combine 10 stations into one broadband antenna. Both of these early combiners have since been replaced by balanced bandpass systems.

Many manufacturers now design and build combining systems. Just outside Houston, Texas, the Senior Road project is a 10 channel system. This early 1980s installation was the last of the large notch-filter balanced combiners in the U.S. In Miami, Florida, the first high-powered, expandable balanced combiner system using bandpass filtering and group delay equalization was installed in the mid 1980s.

13.4.4   Components of Combiners

Tee and Star Junctions

A tee junction, shown in Fig. 13.73, is a coaxial component that provides for two RF signals to flow into a common path; a star junction is a tee with more than two input paths. This basic coaxial component is one of the building blocks of a *branched combiner*.

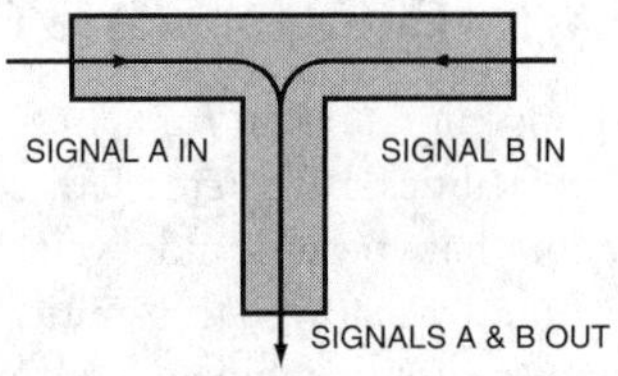

SIGNAL A IN   SIGNAL B IN

SIGNALS A & B OUT

FIGURE 13.73   Tee junction schematic diagram.

Filters

Filters sort RF frequencies, attenuating some while allowing others to pass readily. Depending on the filter design, a narrow bandwidth may be attenuated (*band-reject type*), or only a narrow bandwidth may be passed (*bandpass type*).

Bandpass Filters

Figure 13.74 shows the basic configuration of a bandpass filter cavity. The cavity is tuned to a resonant frequency by adjusting the length of the tuning probe. When RF energy is applied to the input coupling loop, the loop inductively couples the energy into the cavity. Energy is transferred through the cavity and is inductively coupled to the output coupling loop. As Fig. 13.76, curve A indicates, the transfer of energy is maximized at the resonant frequency of the cavity. A cavity that is larger overall is more efficient in that it will provide a lower insertion loss at the desired frequency at comparable isolation of other frequencies.

The placement, size, and rotation of the coupling loops are chosen to give a range of coupling that corresponds with a range of attenuation required for a given application. Figure 13.75 is an impedance diagram for a single bandpass cavity, and Fig. 13.76 shows its frequency response. As the loops are adjusted away from full coupling (curves A) to less coupling (curves B) and still less coupling (curves C), the impedance curve (Fig. 13.75) moves farther from the center of the chart for a given frequency change, and the isolation increases (Fig. 13.76). The increase in insertion loss is represented by the movement away from the center of the chart at f_0 for both the impedance curve (Fig. 13.75) and the response curve (Fig. 13.76).

In most cases, a single bandpass cavity will not yield enough isolation or impedance band width for modern transmitters. To increase the isolation and improve impedance bandwidth, a second cavity may be added to the first, as shown in Fig. 13.77.

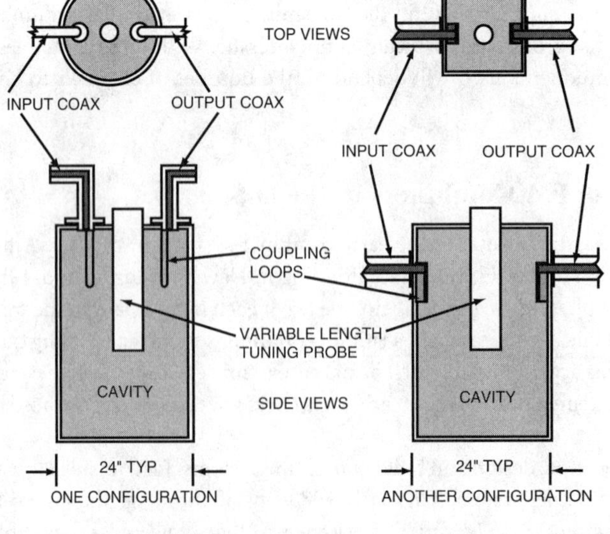

TOP VIEWS

INPUT COAX   OUTPUT COAX

INPUT COAX   OUTPUT COAX

COUPLING LOOPS

VARIABLE LENGTH TUNING PROBE

CAVITY   SIDE VIEWS   CAVITY

24" TYP   24" TYP

ONE CONFIGURATION   ANOTHER CONFIGURATION

FIGURE 13.74   Typical bandpass filter cavity design.

When two identical cavities are coupled 1/2-wave apart, the impedances superimpose themselves as shown in Fig. 13.78, and when the results are added together mathematically, their much-improved combined impedance bandwidth is as shown in Fig. 13.79. Curve A of Fig. 13.80 shows the frequency response of the two-cavity filter.

When still more isolation is required, more cavities can be added. Figure 13.81 is an impedance diagram for a three-cavity system; Fig. 13.82 for a four-cavity system. Curves B and C of Fig. 13.80 show the frequency responses of a three-cavity system and a four-cavity system, respectively. A four cavity bandpass filter is about as large a filter system as is needed for most high-isolation applications.

It is important to realize that the optimum spacing of the cavities is 1/2-wave, and that as the frequency increases the electrical wavelength decreases. Therefore, the length of the intercavity coax must be shortened. At the top end of the FM band, the 24-in-diam cavities used for high-power applications are difficult to link together, because the large cavities themselves approach 1/4 electrical wavelength in radius. As a result, when the intercavity coax is added the mechanical spacing is longer than 1/2-wavelength. In this case, the coupling loops can be manipulated to compensate for the extra length, so that the impedance bandwidth of the cavities is maintained.

This problem can also be prevented by building the cavities contiguous to each other and coupling them through an iris in the wall between them, as shown in Fig. 13.83. The electrical 1/2-wave spacing is maintained by the coupled fields between the cavities.

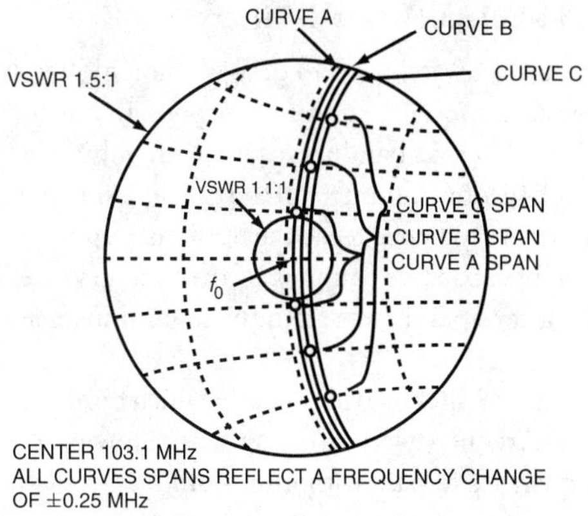

CENTER 103.1 MHz
ALL CURVES SPANS REFLECT A FREQUENCY CHANGE OF ±0.25 MHz

FIGURE 13.75 Impedance diagram for a single cavity, showing the effects of loop adjustment.

CENTER 99.5 0 MHz   SPAN 4.00 MHz

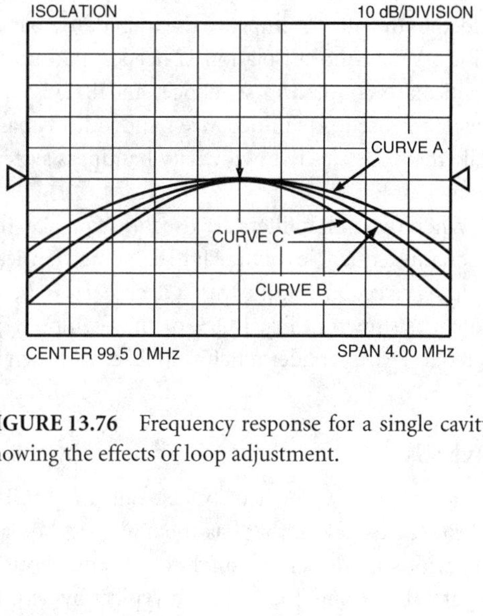

FIGURE 13.76 Frequency response for a single cavity, showing the effects of loop adjustment.

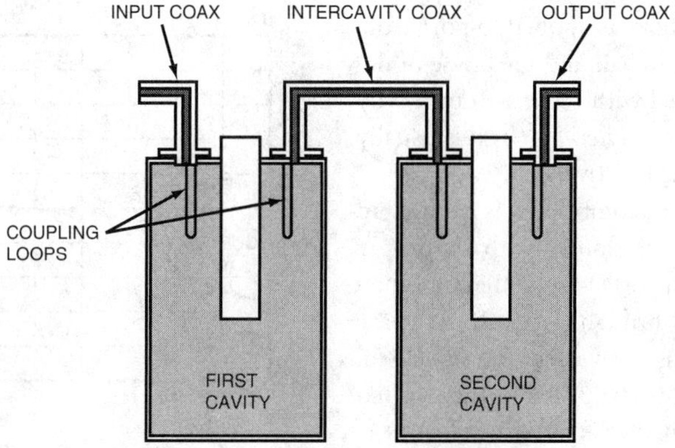

FIGURE 13.77 Loop-coupled cavities.

Band-Reject or Notch Filters

There are several ways to design a notch filter, but they all accomplish the same purpose. In one form, a notch filter is a bandpass cavity with only an input coupling loop, mounted off the transmission line by means of a matched tee. Other designs employ some form of capacitive coupling of the tuned frequency into a cavity and away from the main transmission line.

A notch filter provides a path, which removes the tuned frequency from the system, allowing other frequencies to pass with minimum loss, instead of passing the RF energy through the cavity as described in the last section. The response curve of a single-cavity notch filter is shown in Fig. 13.84; its impedance plot is basically the same as that of a bandpass cavity (Fig. 13.74). When a single notch cavity is used, an impedance matching network is added to the filter to improve the impedance bandwidth. When more isolation is needed two notch cavities are coupled in sequence, and if 1/2-wave spacing is used, the impedance bandwidth is basically the same as for a two-cavity bandpass system (shown in Fig. 13.79).

When two notch filters are used in sequence, the resonant frequencies of the cavities can be identical, yielding a response curve with a deep narrow-band notch, as shown in Fig. 13.85, or they can be staggered to give a broader notch response, as shown in Fig. 13.86.

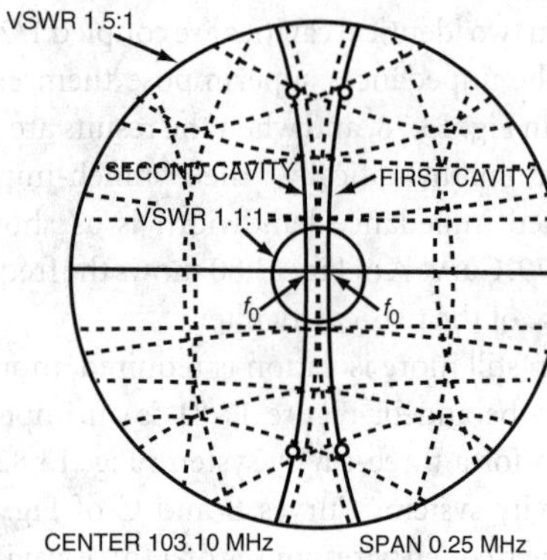

Hybrids

A *quadrature hybrid* (usually just called *hybrid*) is a broadband device that has the ability to operate in various modes either singly or simultaneously. Figures 13.87 and 13.88 show a typical hybrid. In Fig. 13.87, the hybrid is acting as a power and phase splitter. When an RF signal is applied to port 1 the hybrid splits the signal in half and the phase of the port-4 output is delayed with respect to port 3 by 90°. Port 2 is called the *isolated port* and is usually terminated with a 50-Ω resistive load.

Use of a hybrid for combining is shown in Fig. 13.88. If two equal RF signals, with the proper phase, are introduced at ports 1 and 2, the combined signal exits the hybrid through port 4. If the phase of the two input signals is reversed, the signal will exit the hybrid through port 3. Again the isolated port is usually terminated with a resistive load.

The hybrid's third mode of operation is called the *reflected mode* and is shown in Fig. 13.89. When

FIGURE 13.78 Impedance diagram showing the superimposition of two identical cavities $\lambda/2$ apart.

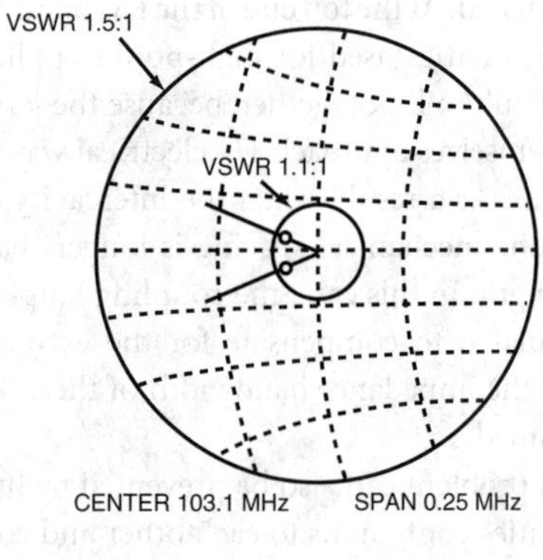

FIGURE 13.79 Combined impedance diagram for a two-cavity system.

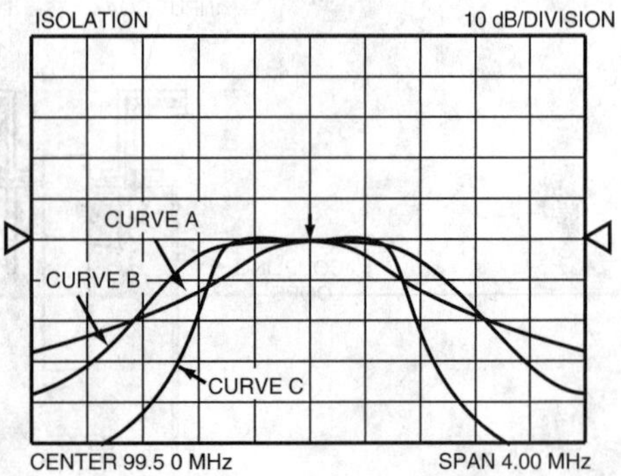

FIGURE 13.80 Typical frequency response diagrams: (a) Two-, (b) three-, (c) four-cavity bandpass filters.

two identical devices with high impedance to the signal frequency, such as bandpass or band-reject cavities, are attached to ports 3 and 4 of the hybrid, the signal entering at port 1 is reflected and exits the hybrid through port 2.

Two hybrids can be connected together to form what some engineers call a *hybrid ring*. As shown in Fig. 13.90, when an RF signal is applied to port 1, the energy flows diagonally across and exits port 4 with little loss; ports 2 and 3 are isolated by a factor of 35 dB. If the two hybrids are tuned for a specific frequency this isolation can be optimized to 40 dB.

The multiple operational modes of the quadature hybrid make it the backbone of the *balanced* and *runout* combiners, which will be described later in this chapter.

Resistive Loads

Resistive loads, often called *dummy* loads, are used in many applications and can be manufactured in many sizes and impedances depending on the requirements. In a dummy load, incoming power is absorbed and converted to heat. The heat must then be dissipated to the surrounding air, so that the power rating of a dummy load is determined by the size of the resistor and the amount of heat that can be dissipated before the resistor overheats and fails. If enough resistors can be chained together with enough cooling, they can dissipate a substantial amount of RF energy. Most coaxial systems use either 50- or 75-Ω dummy loads, depending on the base impedance of the system.

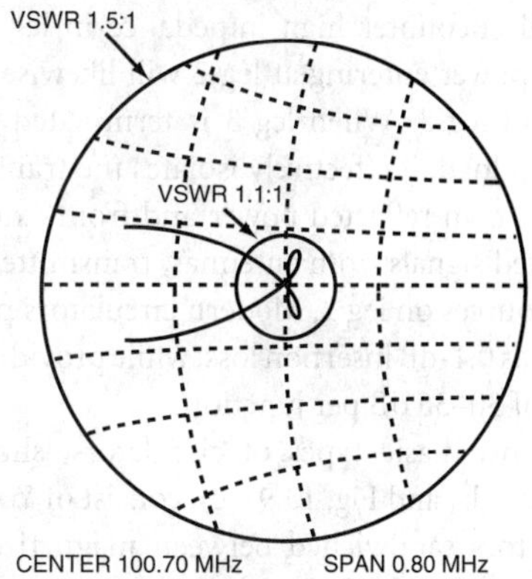

Isolators and Circulators

FIGURE 13.81 Impedance diagram for a three-cavity bandpass filter.

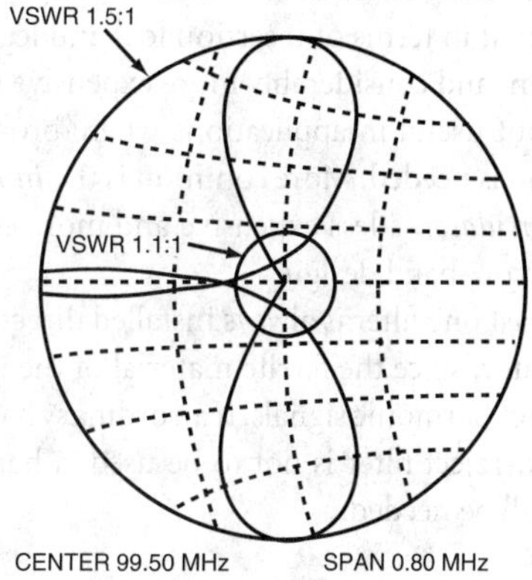

FIGURE 13.82 Impedance diagram for a four-cavity bandpass filter.

An isolator provides nonreciprocal transmission of a desired RF signal. The working component of an isolator is a *circulator*, a three-port device that will conduct a signal between its ports in a circular mode, as shown in Fig. 13.91(a). Power from the transmitter, introduced at leg 1, will flow with low loss to leg 2,

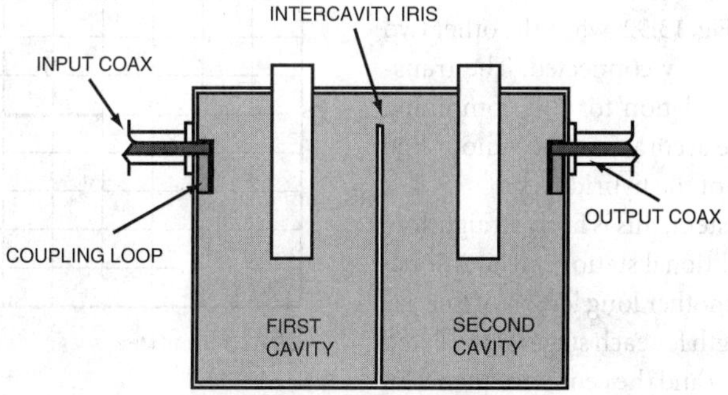

FIGURE 13.83 Iris-coupled cavities.

but will encounter high impedance in leg 3. Reflected power entering at leg 2 will likewise be diverted to leg 3. When leg 3 is terminated with a matched load, it effectively isolates the transmitter from its own reflected power and from externally generated signals from antennas, transmitters, and other sources on leg 2. Modern circulators provide as little as 0.1-dB insertion loss, while providing isolation of 30–35 dB per junction.

The two basic types of circulators, shown in Fig. 13.91(b) and Fig. 13.91(c), consist of Y-shaped conductors sandwiched between magnetized ferrite discs. The final shape, dimensions, and type of material varies according to frequency of operation, power handling requirements, and the method of coupling. The distributed constant circulator is the older design; it is a broadband device, not quite as efficient in terms of insertion loss and leg-to-leg isolation, and considerably more expensive to produce, but useful in applications where broadband isolation is needed. More common is the *lump constant circulator*, a less expensive and more efficient but narrow-band design.

At least one filter is always installed directly after an isolator, since the ferrite material of the isolator generates harmonic signals. If an ordinary bandpass or band-reject filter is not to be used, a harmonic filter will be needed.

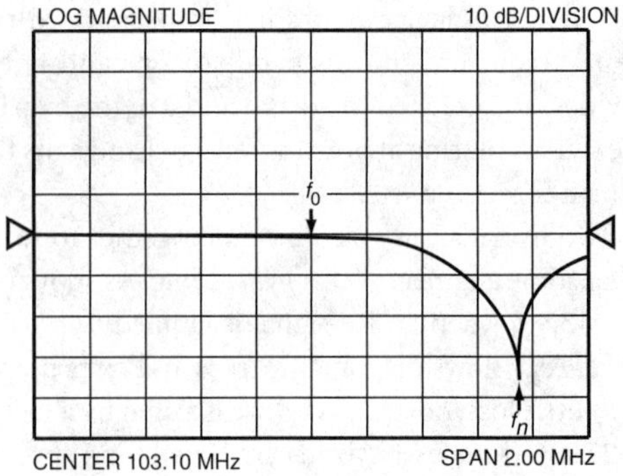

13.4.5 Types of Combiners

Runout Combiners

In the early days of FM combining systems, the easiest way known to combine two frequencies was to use a hybrid ring where one leg contained an extended path equaling an even multiple of both electrical lengths. This was accomplished by adding a long length of coaxial transmission line (the runout) between the two hybrids to connect two of the internal ports, as shown in Fig. 13.92, while the other two internal ports were directly connected. The transmitter-to-transmitter isolation for this combining system depends on the accuracy of the runout line length and the tuning of the hybrid ring.

For a two-station system, this is fairly straightforward. But for each additional station, an additional combining stage and another long length of line are necessary. The line length for each stage is calculated from the new frequency and the center frequency of the preceding stage. The major disadvantage of this method is that the runout lines tend to be hundreds

FIGURE 13.84 Typical frequency response for a single cavity band-reject filter.

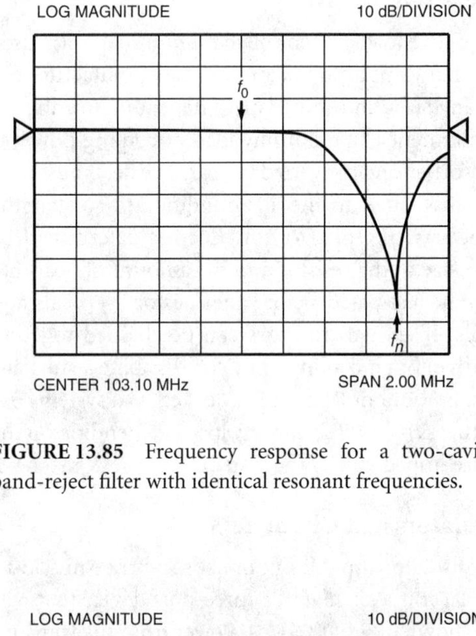

FIGURE 13.85 Frequency response for a two-cavity band-reject filter with identical resonant frequencies.

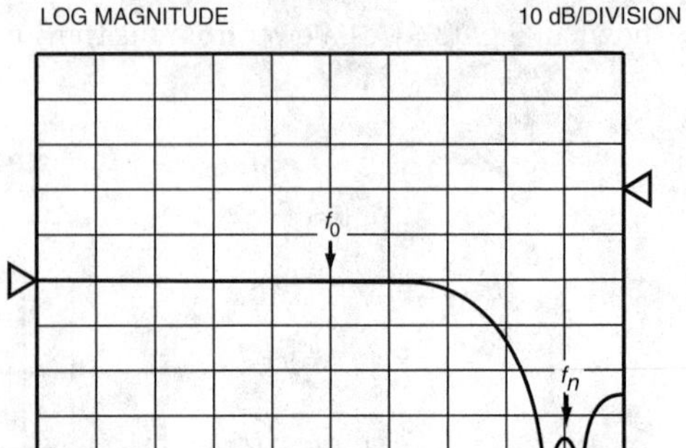

FIGURE 13.86 Frequency response for a two-cavity band-reject filter with staggered resonant frequencies.

of feet long and are difficult to store and protect. Also, this type of combiner is narrow band, because it is line-length dependent. Finally, the isolation is not high enough for modern transmitters. Therefore, notch filters have had to be added to retrofit existing systems.

A variation of the FM runout combiner used a filter in place of the long runout line to provide the same electrical delay. However, this modification did not substantially improve the electrical performance of the system, and notch filtering has also been added to retrofit these systems.

Runout combiners are no longer used for high-powered FM broadcast systems, but they are commonly used for low-powered UHF TV combiners, where shorter wavelengths and lower power allow the unit and its runout line to fit in a standard rack mount.

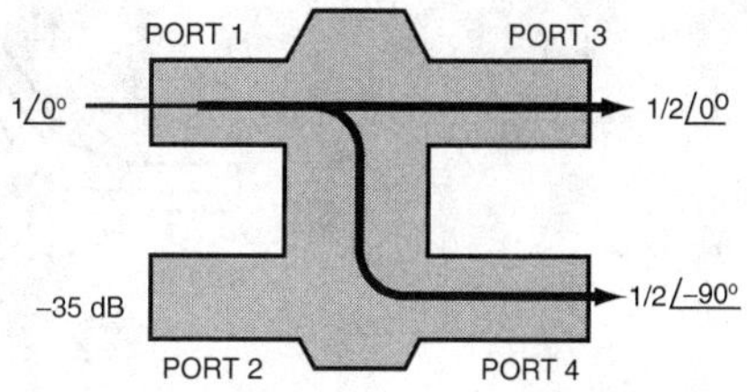

Branched or Star Point Combiners

A branched combiner is a simple combination of a tee junction and the required number of filters to ensure a sufficient amount of isolation. For example, an FM branched combiner containing a three-cavity bandpass filter in series with a two-cavity band-reject filter may be used to combine two close-spaced (0.8 MHz) frequencies. This close spacing requires maximum filtration.

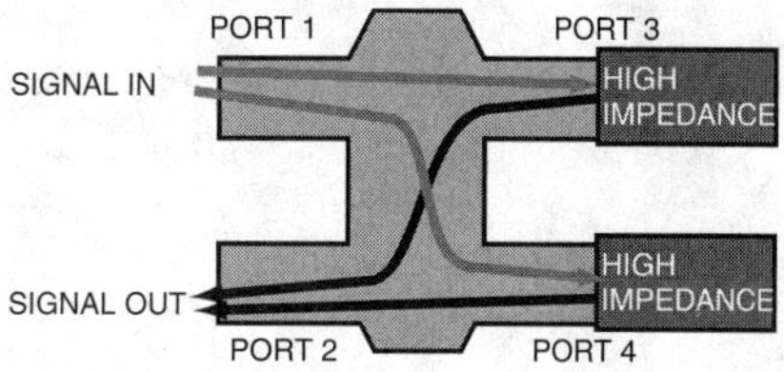

Refer again to Fig. 13.80, curve B (response curve for a three-cavity bandpass filter) and Fig. 13.86 (response curve for a two-cavity band-reject filter). When these filters are added in series, the resulting

FIGURE 13.87 A hybrid used as a signal splitter.

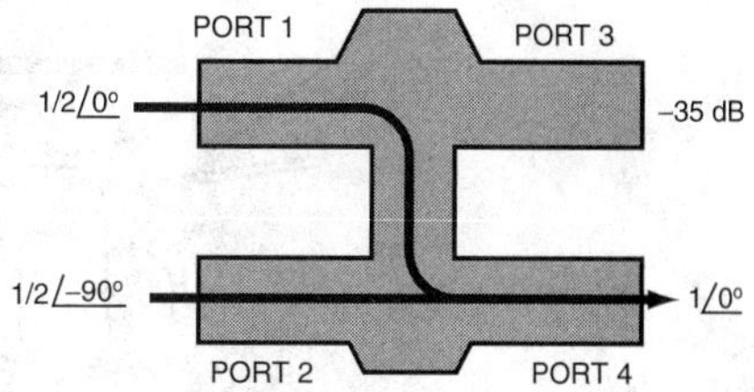

FIGURE 13.88 A hybrid used as a signal combiner.

FIGURE 13.89 A hybrid used as a signal reflector.

response curve is shown in Fig. 13.93. Note that the insertion loss for the pass frequency f_1 is only about 0.25 dB and the isolation at the reject frequency f_2 is greater than 50 dB across the channel. The impedance plot, Fig. 13.94, is the combination of impedance plots, Fig. 13.81 and Fig. 13.79, for the same filter combination. The filter system is tuned for one of the channels in a two-station branched combiner, and forms one leg of the combiner. The second leg is a similar combination of filters tuned for the other frequency, and its response plot resembles a mirror-image of Fig. 13.93. The basic schematic, Fig. 13.95, shows the combiner configuration. TX1 and TX2 are the signals from transmitters 1 and 2 as they enter the combiner. The length of the coaxial line between each set of filters and the tee junction is tuned to

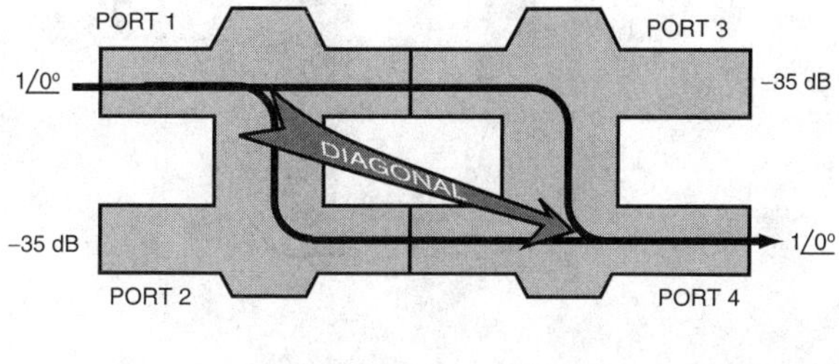

FIGURE 13.90 A hybrid ring, showing low-loss diagonal energy flow.

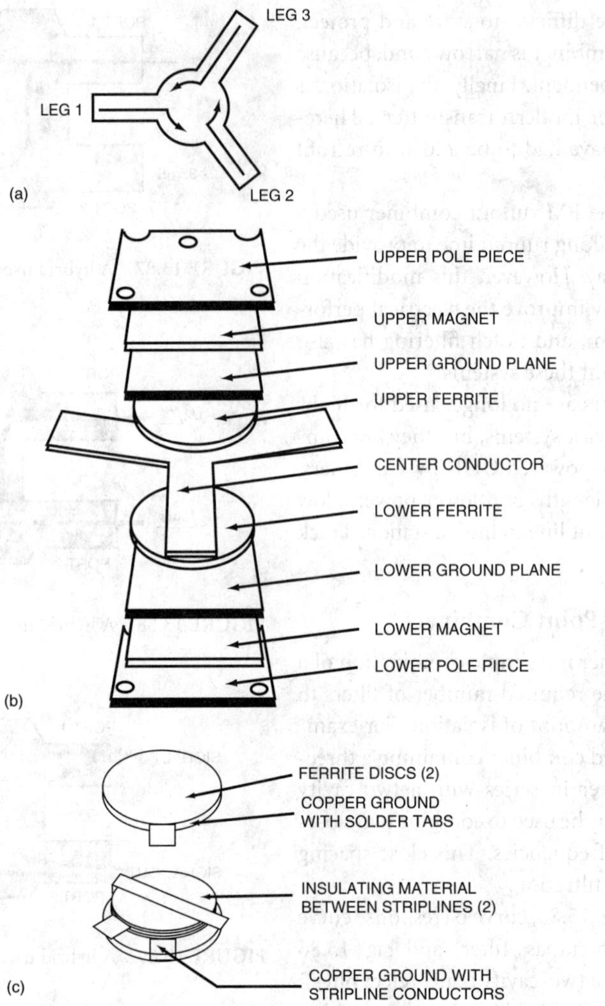

FIGURE 13.91 Types of circulators: (a) circulator schematic with power flow arrows, (b) distributed constant circulator, (c) lump constant circulator.

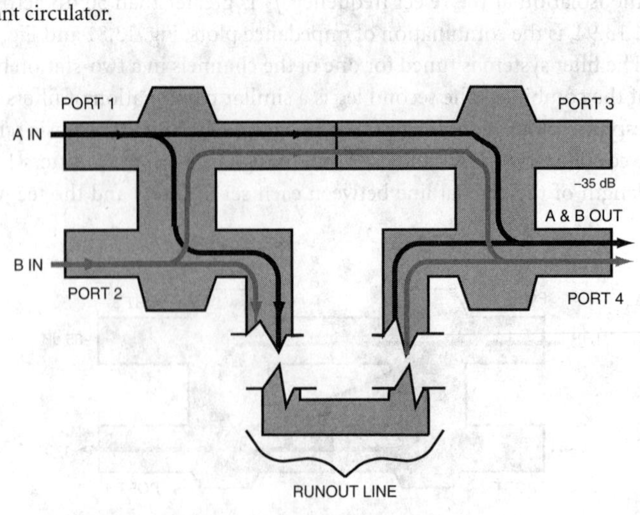

FIGURE 13.92 Runout combiner schematic diagram.

provide a high impedance to the other frequency, so that the power flow of each signal is through its own filter, out of the tee junction, and up to the antenna.

A branched combiner is efficient for a two-station installation, but as more stations are tied into the common tee junction, not only does the tee junction become impractically large, but the mechanics of tuning the interconnecting coax become more complex. Above about four or five stations, a balanced combiner becomes more practical and cost effective.

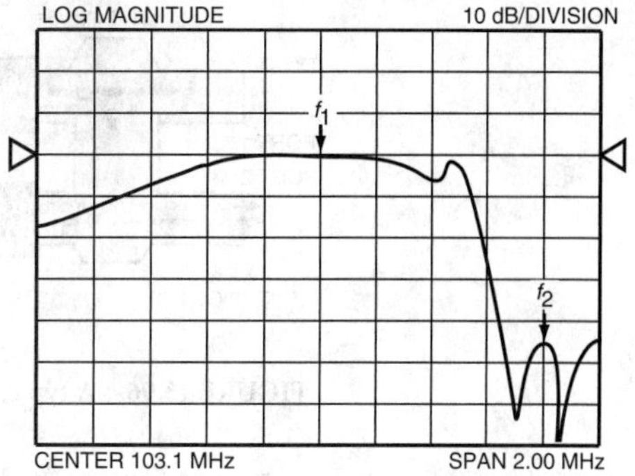

Balanced Combiners

The balanced combiner, like the runout combiner, uses a hybrid ring as the basis for its operation. Each leg of the ring contains an identical set of either bandpass or band-reject filters, hence the term balanced.

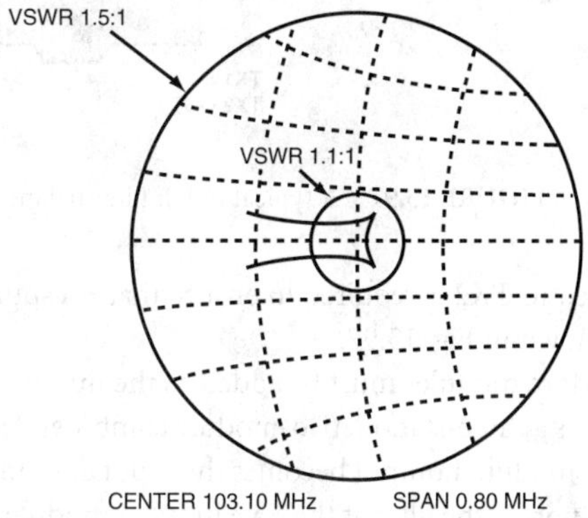

Notch Filter Balanced Combiners

In the notch-filter balanced combiner (Fig. 13.96), both notch filters within the hybrid ring are tuned to reflect TX1's frequency, which enters the combiner (black arrows) at the narrow-band input port, port 1. That signal is reflected by the filters, and exits at the broadband output port, port 2. TX2 (gray arrows) enters the broadband input port of the module, port 3, passing through in the diagonal mode shown in Fig. 13.90, with minimal loss in the reject cavities. The isolation of TX1 through the module to TX2 is a function of both the hybrid ring isolation of 35 dB and the depth of the notch cavities, which can approach 35 dB, as shown in Fig. 13.86. The isolation of TX2 to TX1 is only that of the hybrid ring, about 35 dB. Therefore, additional filtering

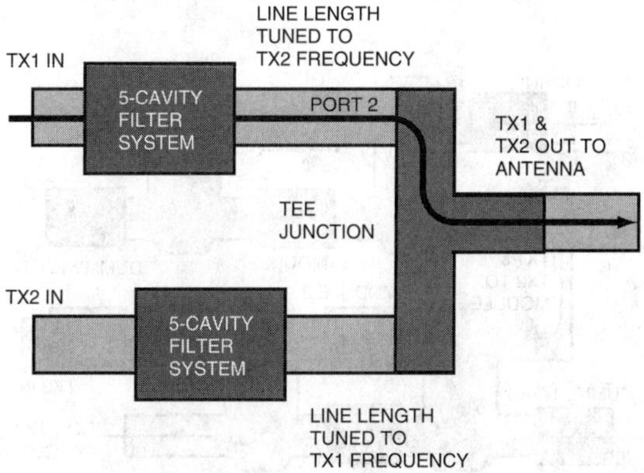

FIGURE 13.93 Response curve for a filter system consisting of a three-cavity bandpass filter and a two-cavity band-reject filter in series.

FIGURE 13.94 Impedance diagram for a filter system consisting of a three-cavity bandpass filter and a two-cavity band-reject filter in series.

FIGURE 13.95 A branched combining system.

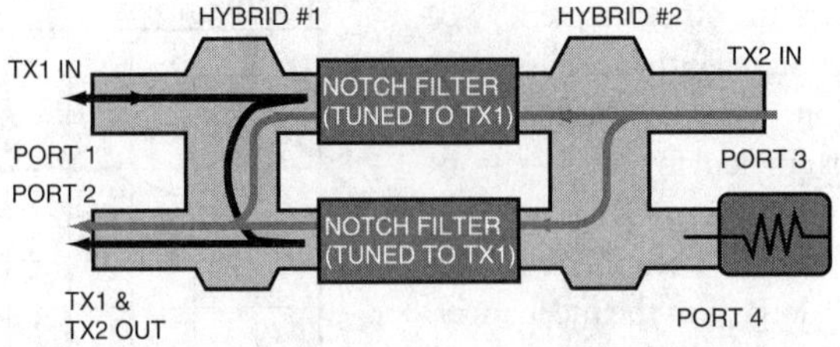

FIGURE 13.96   A typical notch-filter balanced combiner.

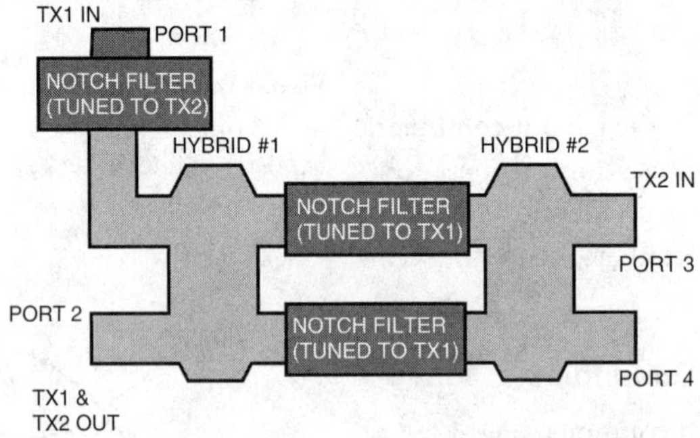

FIGURE 13.97   A typical notch filter balanced combiner with an added notch filter at the TX1 input.

tuned to TX2 is required to ensure that no spurs are generated within the transmitters. This added filter is shown in Fig. 13.97.

More modules must be added as the number of stations entering the system is increased. Figure 13.98 shows a three-station, two-module combiner. Note the filter in the TX1 input line, as described previously. The module 1 output becomes the input at module 2 port 1. Since both TX1 and TX2 are entering module 2 at port 1, the filter at the TX3 input at module 2s port 3 must now reject both frequencies, TX1 and TX2. This can be done with two reject filters or a bandpass filter tuned to TX3.

A further extension of this pattern is shown in Fig. 13.99, showing a five-station, four-module combiner. Note that in this example, each input filter is a bandpass tuned to the frequency of that input. If reject filters were to be used at the various inputs, each input would have to filter all of the frequencies previously introduced. Therefore, port 3 of module 2 would have to contain two filters; port 1 of module 3, three filters; and port 3 of module 4, four filters. This complicates the system. This situation is avoided by the use of a combination of internal band-reject and external bandpass filters, as shown.

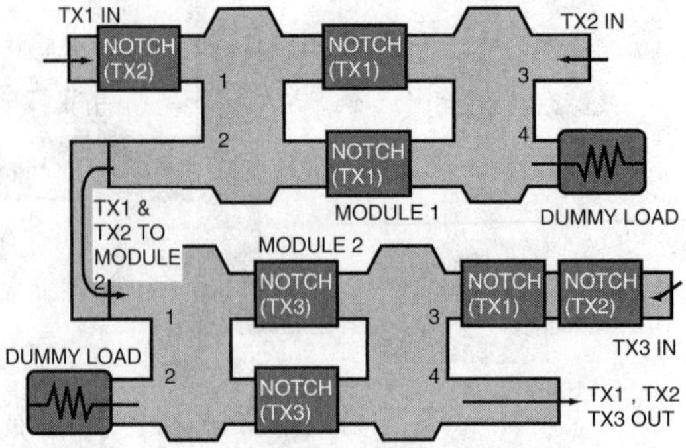

FIGURE 13.98   A typical three-station two-module notch-filter combiner.

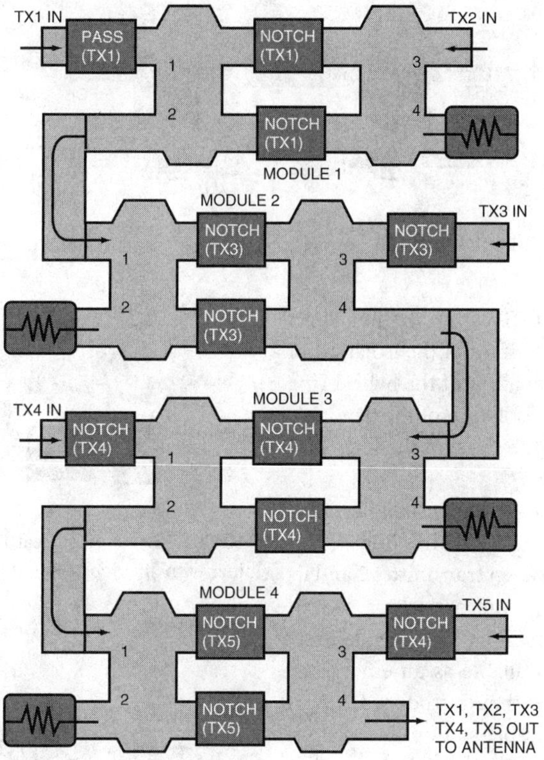

FIGURE 13.99   A five-station, four-module balanced combiner system.

In some cases, instead of having a station located at port 3 of module 1, that port is terminated in a 50-Ω load and can be used as an emergency input for any station in the system. By providing an extra port in this way, a damaged module can be bypassed. Because of the particular configuration of that port, as long as the external filter at port 1 of module 1 is a bandpass filter, no input filtering is necessary.

A disadvantage of using internal notch filters is that if the two filters in a module are not identically tuned, an imbalance occurs within the hybrid ring, reducing the isolation to a point where a spur can be generated within a transmitter. Once a spur has been generated, there are no filters within the system to reject that spur, since the filters are tuned only to the expected frequencies. Therefore, the spur is broadcast.

A second disadvantage of using internal notch filters is that since each module in turn has to conduct the accumulated power of all of the previous modules, for a high-powered system each module must be larger than the previous one, and the power rating of the system can handle is limited by the size of the final module. Third, because of their electrical characteristics, notch-filter combiners tend to be narrow-band in nature, especially when the frequencies combined are close together. Because of these and other limitations of notch-filter systems, they are seldom used. Modern FM combiners use bandpass filters.

Bandpass Filter Balanced Combiners

In a bandpass filter balanced combiner system, the filters used within the hybrid ring are bandpass instead of notch filters. The basic layout is the same as that of a notch combiner. The difference in the power flow is shown in Fig. 13.100, as compared to Fig. 13.96. In the notch system, the filters are used to reflect signal TX1 entering the combiner's narrow-band input port (port 1). In the bandpass system the narrow-band signal TX1 also enters port 1, but passes through the hybrid ring bandpass filters and out the broadband output port (port 4), while the broadband input signal TX2, entering at port 3, is reflected by the filters and exits at port 4.

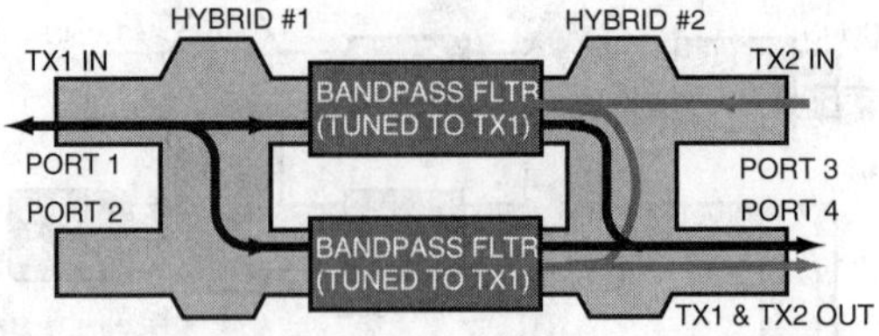

FIGURE 13.100   A bandpass-filter balanced combiner.

The isolation of TX1 to TX2 is only the hybrid ring isolation of about 35 dB, and the isolation of TX2 to TX 1 is the combination of the hybrid ring isolation (35 dB) and the isolation of the bandpass filter, which depends on the slope of the filter's response curve.

For a single-module two-station installation, since the isolation of the TX1 to TX2 path is only 35 dB, a filter must be added between transmitter 2 and its input port (Fig. 13.101). Alternatively, a second module may be added, and port 3 of module 1 terminated in 50 Ω (and available as an emergency input port). Signal TX2 is then introduced at port 1 of module 2. A multiple-station installation (Fig. 13.102) is an extension of the latter configuration, where each frequency has its own module.

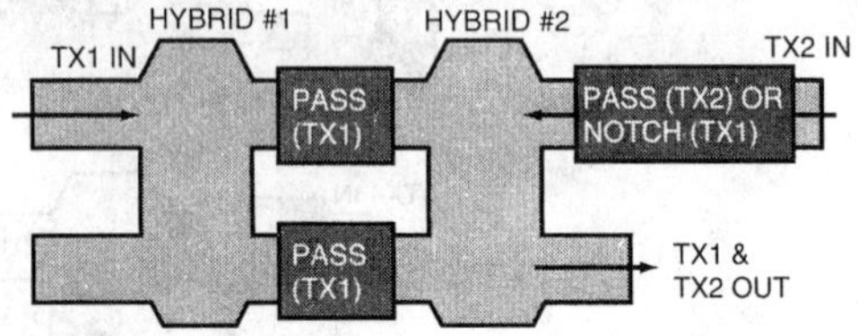

Unlike notch systems, in a bandpass system the accumulated power entering each module flows only through the output hybrid, so the power-handling capacity of the system is limited only by the size of the output hybrids and interconnecting transmission line, not the entire module.

When two FM broadcast stations are 1.2 MHz apart or closer, the bandpass filter's response curve will not provide quite enough isolation, allowing a small amount of signal interaction. This interaction narrows the group delay curve of the module which is farthest from the antenna, as shown in Fig. 13.103. This phase distortion can be compensated for by adding a *group delay equalizer*, as shown in Fig. 13.104. This is a hybrid containing two deadended bandpass filters, which selectively delay the carrier frequency relative to the sidebands, yielding a net group delay curve, which approaches a square wave, as shown in Fig. 13.105. Group delay equalization can be achieved with high-power components at the input to the module or with low-power components at the output of the exciter of the transmitter.

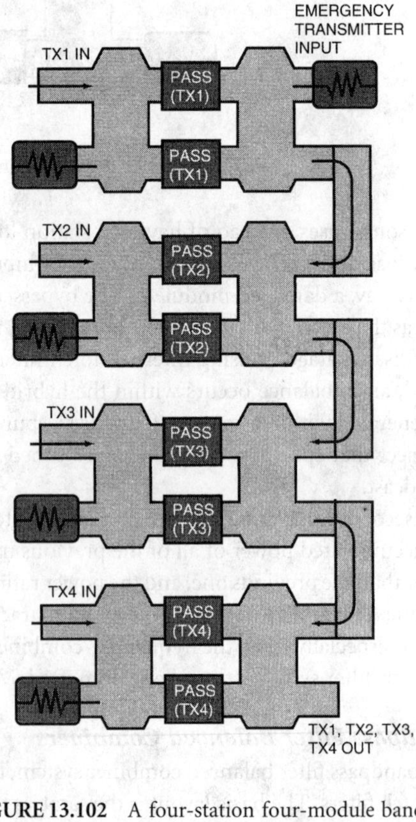

FIGURE 13.101   A typical bandpass-filter balanced combiner with an added filter at the TX2 input.

FIGURE 13.102   A four-station four-module bandpass filter balanced combiner.

Television Combiners

The first VHF-TV combiners were balanced combiners (Fig. 13.106), which used one notch cavity per leg of the hybrid ring. The aural or audio portion of the TV channel used the narrow-band port of the combiner

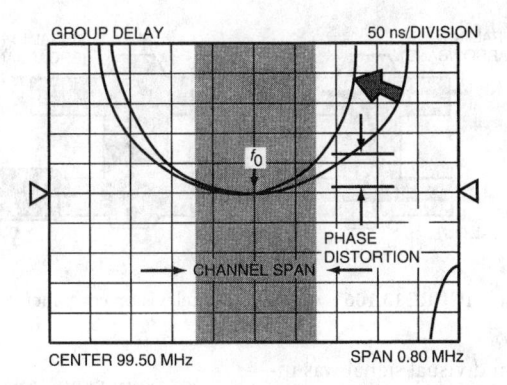

FIGURE 13.103 Asymmetrical phase distortion in the module farthest from the antenna.

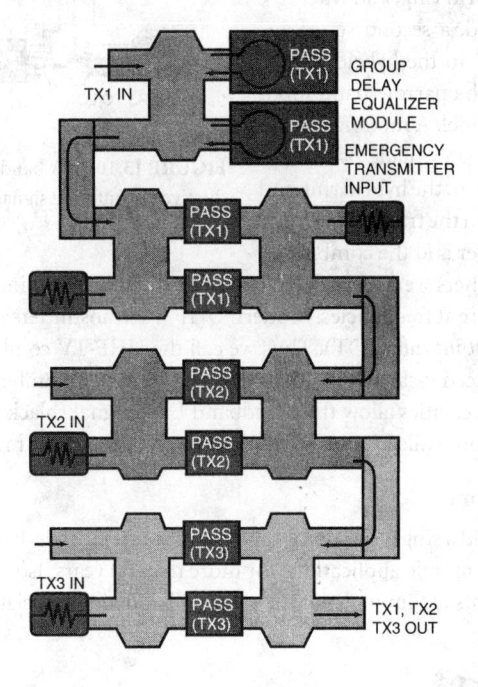

FIGURE 13.104 A bandpass-filter balanced combiner with a group delay equalizer.

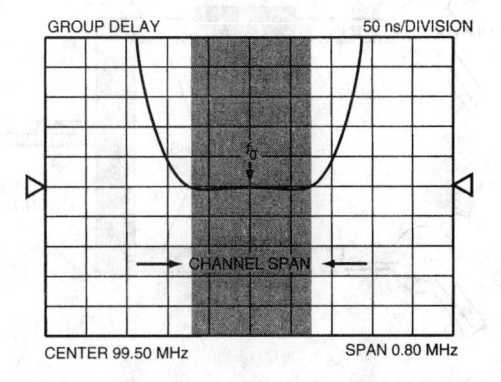

FIGURE 13.105 Group delay for a bandpass filter module with a group delay equalizer.

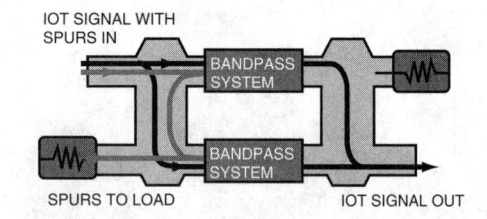

FIGURE 13.106 A notch-filter television combiner.

module, while the broadband visual signal was introduced at the broadband port of the module.

With the advent of color TV a second set of notch cavities was added to the hybrid ring, and when TV stereo sound was introduced, a second set of aural notch cavities was added to the hybrid ring to increase the bandwidth on the narrow-band input port for the stereo aural channel. A modification to the basic TV combiner was later made by removing the color notch cavities from the hybrid ring and placing a single notch cavity in the transmission line between the visual transmitter and the combiner.

FIGURE 13.107 A bandpass-filter combiner for an inductive output tube signal.

For years UHF-TV combiners were basically the same as VHF-TV combiners, because the transmitters were the same design at different frequencies. Modern UHF-TV transmitters use a new type of amplifying device called an *inductive output tube* (IOT). Thus we call the UHF-TV combiner an IOT combiner. The IOT combiner is still a balanced system (Fig. 13.107), but instead of a number of specifically tuned notch cavities, a series of bandpass cavities allow the broadband UHF signal (black arrows) to pass through the combiner with little distortion while keeping spurs (gray arrows) to a minimum.

Combiners Using Isolators

Combiners using isolators and a single-bandpass cavity in place of multiple-bandpass cavities in each input line have been in use in land mobile applications for more than 30 years. Isolators lend themselves readily where the power requirements are limited and the channel bandwidth is relatively narrow. Figure 13.108

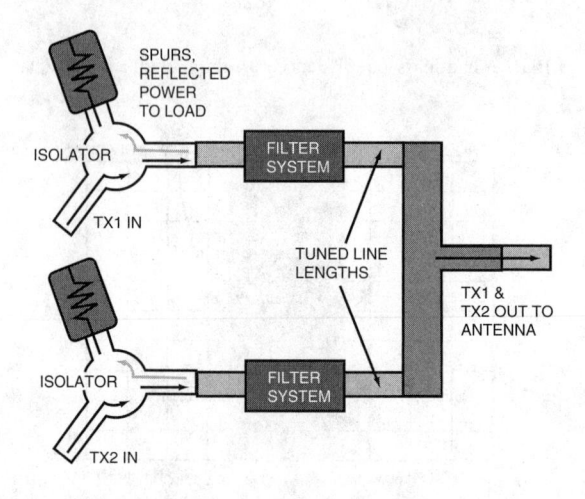

FIGURE 13.108 A branched combiner with isolators.

shows the schematic of a two-station branched combiner using an isolator and a single-filter cavity in each leg instead of a multiple-cavity filter as shown in Fig. 13.109.

Figure shows a *star-junction combiner*, earlier referred to as a star point. This is an extension of the branched combiner concept to multiple-channel combining, made possible because of the relatively small size of the components and the low-power requirements. Combiners up to 20 inputs have been built successfully using this method, provided that sufficient channel-to-channel spacing is available.

Figure 13.110 shows a line-segmented or line-insertion combiner, where the various transmitter inputs are introduced sequentially. The line lengths between the cavities and the tee junctions are critical, each about 1/2-wavelength long, adjusted for effects of cavity loop effective length, effective length of the tee connections, and velocity factor of the cable type employed.

The branched (or star-junction) type and the line-segmented type combiner are comparable in electrical performance. The choice of configuration is generally dictated by mechanical layouts, frequencies of transmitters to be combined, and cable characteristics. Occasionally, combinations of the two types are found, in which groups of star-junction combined transmitters are coupled to other groups by branches or line segments. In all cases, however, cable lengths and cavity loop adjustments are critical.

Figure 13.111 shows a two-station hybrid combiner, where two transmitters are merged in a single hybrid. Since there are no filters between the isolators and the hybrid, harmonic filters are added to remove the harmonics generated by the isolators. More stations may be added by adding additional combiner stages, in effect pyramiding the combiners. This configuration is used where channel-to-channel frequency spacing is very small (≈ 25 kHz),

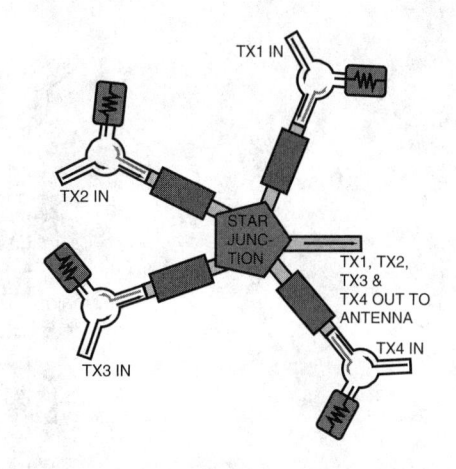

FIGURE 13.109   A branched combiner with isolators.

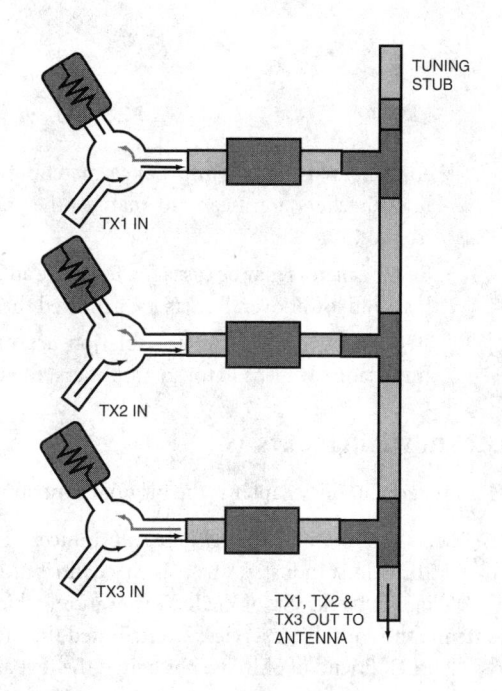

FIGURE 13.110   A line-segmented style combiner.

and since half the transmitter power is always lost in the load, this design is not for applications where high losses would be a problem.

Isolator combiners work well in land mobile applications because the RF power requirements are only a few hundred watts, compared to the broadcast industry, where many kilowatts of power are the standard.

Other caveats in the use of isolators include the following:

- It is critical that the impedance at all ports of the circulator, including the dummy load, be matched precisely. Poor matches result in poor isolation.
- Characteristics of transmission line and antennas influence isolator performance. The power rating of the load termination at the last isolator stage before feeding the antenna must be sized to handle the maximum, worst-case reflected power level.

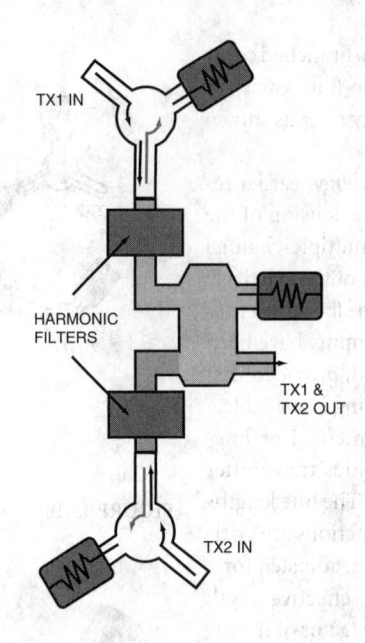

FIGURE 13.111   A hybrid isolator combiner.

- An isolator is not an impedance matching device. Proper cable lengths and the possible use of line matchers are means to match other devices to the characteristic design impedance of an isolator.
- Single isolators may be cascaded to secure added isolation. Better overall performance, less insertion loss, and lower overall costs are achieved through using multiple junction isolators.
- Benefits other than system isolation are provided by isolators. These include the firm source impedance of the isolator to amplifiers, filters, duplexers, and other devices.

Acknowledgments

The material in this chapter is the original contribution of the authors.

Robert A. Surette of Shively Labs, a producer of combiners, antennas, and other passive products for the FM broadcast industry, was primary contributor and oversaw the compilation of the chapter.

William F. Lieske, Sr. of EMR Corporation, which specializes in intermodulation control devices and antenna site consulting services, contributed the material on isolators and land mobile combiners.

Albert G. Friend of Shively Labs edited the text and created the illustrations.

Further Information

NAB Engineering Handbook, 8th ed., chap. 2.4, Diplexers, Combiners, and Filters.

Microwave Filters, Impedance-Matching Networks, and Coupling Structure. 1980. Artech House Inc.
Lieske, W. 1984. *Intermod Control.* Wiesner Publishing.
Liao, S. 1980. *Microwave Devices and Circuits.* Prentice-Hall, Englewood Cliffs, NJ.

14

Information Recording and Storage

David Stafford
Quantegy, Inc.

Jerry C. Whitaker
Editor-in-Chief

Praveen Asthana
IBM Corporation

Fabrizio Pollara
California Institute of Technology

Jerome R. Breitenbach
California Polytechnic State University

14.1 Magnetic Tape

David Stafford

14.1.1 Introduction

Since its earliest use magnetic tape has been constantly evolving: paper backing has given way to polyester **base films**, chemical technology has advanced allowing higher performing particles to be used. It is true to say that many of the advances in recording formats would not have been possible if the associated advances in magnetic tape technologies had not also taken place.

This evolution has resulted in tape being a great deal more than just, as some people say, rust on plastic. Magnetic tape is an integral part of the recording process, as important a part as the recorder on which it is running. Many users tend to forget that, although tape is a relatively inexpensive item, its value, after the recording is finished, can be immense.

14.1.2 Trends in Magnetic Media

Although magnetic media, and their uses, have changed across many market segments since their inception, there is, perhaps, no segment that more typifies the changes that have taken place than the experience of the professional videotape market. Consequently, while acknowledging the significant changes that have been experienced in other segments of the magnetic tape market, the author chooses to concentrate mainly on the professional videotape market.

Applications

Just as videotape formats, and the magnetic tape used with them, have become more complex and demanding, so have the end user applications. In the early days of magnetic tape usage, simple recording and reproduction were the basic uses of the technology. Although this is still true, today's applications demand much more from magnetic tape than did their predecessors. Still framing, slow motion control, stop frame animation, high end graphics compiling, multiple pass performance (particularly in commercial and program insertion systems), and a host of other innovative applications demand a great deal from magnetic tape. In fact, magnetic media is now such an integral part of the production process many users tend to forget its fragile nature. The developments in videotape format technology have demanded that today's magnetic media must be stronger, more **durable**, of much higher quality and surface integrity, and utilize highly sophisticated chemicals and formulations compared with their predecessors. These factors present the magnetic media manufacturers with a complex set of design criteria. Many times one particular design criteria may contradict another, and the design of a particular product becomes a situation of balancing all requirements and producing the best possible product.

14.1.3 Manufacture of Magnetic Tape

The manufacture of magnetic tape, or *recording media* as it has more recently been named, is a complex process involving a wide variety of sciences and manufacturing disciplines. The process itself is essentially linear in concept, that is, one portion of the process follows its predecessor and cannot be started until its predecessor is compete. Clearly, the manufacturing process varies somewhat from manufacturer to manufacturer, and the intricacies of the process are closely guarded secrets. Consequently, the description of the process that follows is a generic one, and it is intended to give an overview of the process rather than an in-depth discussion of any one portion of the entire process. The flow chart shown in Fig. 14.1 graphically represents the entire manufacturing process.

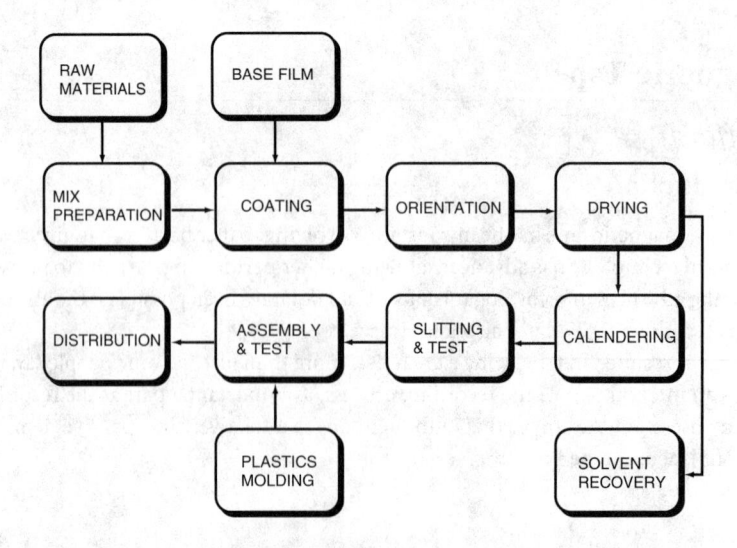

FIGURE 14.1 Process flow digram for the manufacture of magnetic tape.

Research and Development

The first step in the manufacture of a recording media product actually occurs well before the coating commences: research and development. The various chemists involved in the development effort determine the appropriate master formulation for a tape product of given performance parameters. Once this has been achieved, the formulation is then passed onto the manufacturing side for full-scale production.

This research and development phase may take several months, or even years, before a suitable formulation is developed. During this time, many trials, or experiments are conducted. Each of these trials examines a different chemical component and how its performance will affect the performance of the final product.

Typically, these trials are performed on coating lines especially developed for this purpose. They are usually 5–10 in wide (i.e., considerably narrower than their production counterparts) and capable of much shorter roll lengths. The use of these research and development lines enables the development effort to proceed in an environment conducive to rapid feedback and test data.

Mix Constituents

The formulation contains several components, each performing a specific function. Although the types of constituents may differ from one formulation to another, essentially all recording media coatings contain some or all of the following.

Magnetic Particles

As discussed earlier, the type of magnetic particles used has changed dramatically over the history of magnetic recording. Early audio tape products used ferric oxide (FeO_2) with coercivities in the 300–350 Oersted range. Trending toward higher energy particles, manufacturers then moved to gamma-ferric oxide, cobalt doped gamma-ferric oxide, metal particles, and barium-ferrite particles.

The trend is clearly toward metal particle products exhibiting high coercivities (1500–2200 Oe) and high retentivity performance. Generally, magnetic particles can be described as acicular in shape (somewhat like a sewing needle). The smaller size of metal particles enables more particles to be packed into a given surface area on the tape and allows much higher data packing densities.

Binders

In a very crude analogy, binders can be likened to glue. In this sense they provide the cross linking necessary to glue all of the mix ingredients together (*cohesion*) and to ensure the coating does not shed or flake off the base film (**adhesion**). Binders also add considerably to the stiffness of the finished tape product. Stiffness is required so that the tape wraps the **helical scanner** and transport assemblies properly and head-to-tape intimacy is maintained. Stiffness is often measured as coating modulus.

The term binder is a generic term covering a wide range of polymeric material such as polyurethane and vinyl chloride. A trend is toward the use of funtionalized binder systems. These binders more effectively provide the cross linking necessary compared with their nonfunctionalized counterparts. Moreover, functionalized binder systems allow the formulating chemist to use fewer solvents and dispersants, thus simplifying the formulation chemistry. An added advantage is that as formulations become less complex in terms of the number of ingredients, the effect of random variations in raw chemicals is minimized. This helps to ensure a more consistent product.

Plasticizers

Many of the guides in tape transport require wrap angles of greater than $90°$. This is particularly true in videotape transports. In some formats the tape must be moved around the scanner at the helix angle; consequently, some of these guides force tape in an other than vertical plane. This type of physical requirement places great demands on a tape product. There is a clear need for the tape to be flexible yet have relatively high **coating modulus**. Plasticizers perform this function. Without plasticizers, tape would be somewhat brittle and unable to survive the torture test of a transport. To be fair, however, equipment manufacturers design their transports so as to place the minimum possible physical requirements on the tape.

Head Cleaning Agents

Typically, head cleaning agents are added to formulations. Usually, this is in the form of alumina particles of a specific size. The purpose of these particles is to ensure the heads of the machine remain clean. In addition, these cleaning agents also keep the remaining portions of the tape path free from dirt and contamination.

There are a number of questions that are raised when discussing the amount of head cleaning agent. These pertain to head life and transport maintenance. Regarding head life, it is frequently stated that the alumina added to a tape formulation is responsible for premature head deterioration. This is not necessarily the case. The amount of head cleaning agent added is extremely important. If too much alumina is added certainly the abrasivity of the tape is higher and head life deterioration becomes an issue. If too little alumina is added the tape may not be able to keep the transport clean enough to ensure good head-to-tape intimacy and a poor quality recording results. Either way, there are many other factors that influence the abrasivity of a given tape. The type of particles, binder systems, and other formulation ingredients all play a role in determining a particular tape's abrasivity.

Regarding transport maintenance, the head cleaning agents added to a tape's formulation are no substitute for good engineering practices in the form of routine cleaning and transport maintenance. Routine cleaning and alignment checking are part of any good preventative maintenance regime and will help to reduce the incidence of dropouts and machine-induced physical damage.

Solvents

All of the mix ingredients are added together to form a *slurry*. Part of the formulation will include some solvents. These are used to dissolve other mix chemicals such as binders during the milling process. In addition, these solvents add to the formulation's rheological properties and effect the manner in which the finished mix flows. This directly impacts the manner in which the particular formulation performs on the coating line.

These solvents are not required in the final tape product. Accordingly, they are typically evaporated off the tape during the drying process. Many companies collect the solvent laden air and route it to some form of solvent recovery. Here the solvents are recovered using a distillation process. This is an environmental friendly process and typically yields solvents of better quality than those originally purchased from suppliers.

Dispersants

Because of their magnetic nature, particles tend to clump together in their raw form. This situation is not desirable as it would tend to result in some areas on the tape surface highly saturated with particles, whereas other areas may be entirely void of magnetic material. Dispersants are added to the formulation to aid in dispersing the particles evenly throughout the mix. The **dispersion** process is completed in mix preparation and can be achieved through a variety of methods. These methods are discussed in the subsequent section on mix preparations.

Fungicides

Surprisingly, fungus growth on the surface of tape is not uncommon. However, this situation is almost entirely confined to storage or operating environments with very high humidity. Fungicides are added to help prevent this phenomenon. Despite manufacturers' best efforts, however, poor storage environments are commonplace. Any reputable manufacturer will be able to provide information on the proper storage and handling procedures for their magnetic tape products.

Antistatic Agents

One of the causes of dropouts (i.e., the momentary loss of head-to-tape intimacy) is static electricity buildup. The buildup causes dust and dirt particles to attach themselves to the tape surface and cause a lack of head-to-tape intimacy. Antistatic agents are, therefore, added to tape formulations, particularly backcoat formulations. In most cases, carbon black is used for this purpose.

Mix Preparation

Once all of the formulation ingredients have been combined, it is necessary to *mill* the slurry so that two conditions are met. First, the magnetic particles, and other mix ingredients, must be thoroughly dispersed throughout the slurry. Second, all ingredients must be dissolved so that no particles larger than a given size remain. This process is achieved in a milling process, which can be performed in a variety of ways. Regardless of the method chosen, however, it is normal to filter the mix after milling to ensure no foreign particles were introduced during that process.

At the end of the milling process, the slurry is ready for coating and is a relatively thick, viscous liquid with a consistency somewhat like house paint.

Ball or Pebble Mills

Ball mills are the oldest of the milling technologies. Simply put, the ball mill is a horizontally oriented cylindrical rotating vessel into which the formulation is poured. The vessel is then rotated for a predetermined amount of time. The time is determined during the formulation development process. Inside the vessel is the milling media. In this case the media can either be steel balls or smooth pebbles. The media in the mill act on the formulation ingredients causing them to be fully dispersed throughout the slurry.

Sand Mills

Sand mills are a more recent technology and operate on a flow system rather than the static batch system of the ball mills. Sand mills are actually banks of milling vessels vertically oriented. The sand mills themselves do not rotate; rather, the mix is pumped through each vessel from bottom to top. As the mix is pumped through the vessel it must pass through a series of rotating disks and milling media. The media in this case is typically zirconium silicate. The number of vessels through which the mix must pass is determined during the formulation development process.

High Solids Mixers

The most recent advances in milling technology are high solids mixers, the most common of which, double planetary mixers, can be likened to an industrial dough mixer in a bakery. High solids mixers can operate in either a batch mode, like ball mills, or a continuous flow mode, like sand mills. As the name implies, this type of technology requires a much more viscous mix than do ball or sand mills. Consequently, this technology provides an efficient and effective milling procedure.

Base Film

As mentioned earlier, many of the base films in common use are *polyethylene terephthalate* films, usually known simply as polyester or PET films. Base films are made by a handful of manufacturers around the world. Films come in a variety of widths, lengths, thicknesses, and tensile properties. Different magnetic tape products call for different tensile properties.

Base films are becoming thinner with newer tape formats. This presents a unique challenge to the tape manufacturer. It is essential that tensile properties are maintained despite the move toward thinner films. Base film derives a lot of its stiffness from its thickness. Obviously, the thinner the film, the less stiff it may be.

Base film suppliers are constantly developing newer products to meet the changing needs of the tape manufacturer. *Polyethylene naphthalate* (PEN) and *polyaramide* base films are the next generation of films. As a general rule of thumb, both PEN and polyaramide films provide greater tensile strength than the same thickness PET film. This fact makes both PEN and polyaramide films attractive candidates for large-scale use in the future.

Coating

Coating can be achieved by a variety of methods with each method being desirable for certain types of coatings.

Knife Coating

Knife coating is probably the oldest type of coating method employed in the manufacture of magnetic tape. As Fig. 14.2 indicates, the wet mix is fed onto the base film and then past the knife blade. The blade height (with respect to the film) determines how much mix is left on the film. Knife coating is particularly good for relatively thick coatings such as those found on analog professional audio mastering tapes.

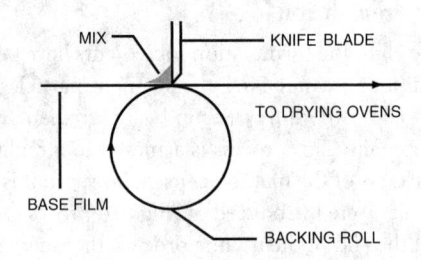

FIGURE 14.2   The knife coating process.

Reverse Roll

Reverse roll coating is somewhat newer and more complex than knife coating and, consequently, is capable of coating much thinner coatings. As indicated in Fig. 14.3, in this method, the mix is picked up from a holding vessel by the pick up roller and transferred to the applicator roll. A metering roll then meters the amount of mix on the applicator roll. The amount of mix remaining on the pick up roll is determined by the distance between the two rolls. The mix is then transferred to the base film.

Reverse roll coating was the mainstay of the tape industry for many years and is still in common use. Reverse roll coating is better suited to thinner coatings (compared to knife coating) such as those found on some older professional videotape formats.

Gravure Coating

Gravure coating comes to the tape industry from the printing industry. As Fig. 14.4 shows, the mix is picked up from a mix vessel and applied to the gravure roll. The gravure roll is pitted with cells, small holes of given diameter and depth. The diameter and depth of the cell, as well as the viscosity of the mix, together determine how much mix is coated onto the base film. Gravure coating is an ideal method for very thin coatings such as consumer VHS and the backcoats found on many professional video tapes.

More Recent Methods

Two other methods, multiple layer and evaporation, are in production use today. Multiple layer coating is a proprietary method capable of coating, as the name implies, multiple layers on a single base film. The layers can be coated at the same time and be of different formulations. The different layers, then, can serve different purposes in the recording process. Nonmagnetic layers are also possible.

Metal Evaporation (ME)

Metal evaporation tape (MET) is a relative newcomer to the tape industry and is generally restricted to newer videotape formats. As Fig. 14.5 indicates, the base film is passed over a cooled drum in a vacuum

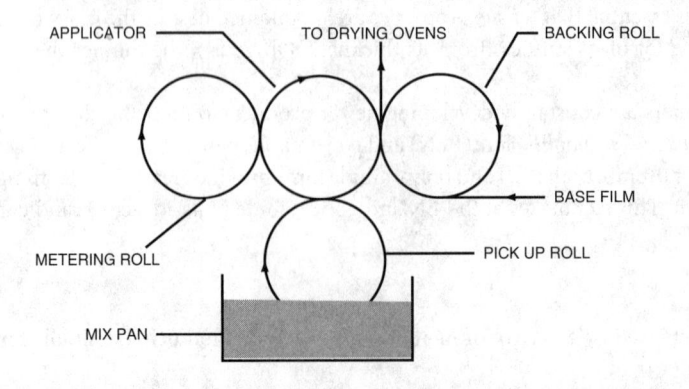

FIGURE 14.3   Reverse roll coating.

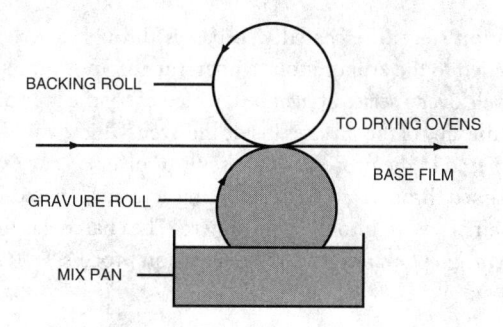

FIGURE 14.4 The gravure coating process.

chamber. A pure metal rod is bombarded with electrons, and a metal vapor is produced. The vapor rises in the chamber and, by virtue of the cooler film surface, adheres to the base film. The metal evaporation technique can coat extremely thin magnetic layers. In fact, ME is capable of coatings that are measured in terms of angstroms rather than micrometers or microinches.

Orientation

After leaving the coating area, the web, as it is now called, passes through the **orientation process**. This process physically moves the individual particles into a specific direction. The direction desired is always identical to the direction of head travel across the tape, or as near to this direction as possible. In the case of the original videotape format quadruplex, develop by Ampex Corporation in 1956, the head traveled in a lateral or transverse direction. The particles in quad tape were therefore **oriented** in the lateral, or transverse, direction. Orienting particles increases electrical output and decreases noise introduced into the recorded or replayed signal by the tape itself. Since the introduction of helical scan technology, particles must be oriented in roughly a longitudinal direction.

Drying, Solvent Evaporation, and Recovery

As noted earlier, the mix formulation contains a number of different solvents. These solvents are, of course, undesirable in a finished product and, therefore, must be removed. The solvents are removed during the drying process.

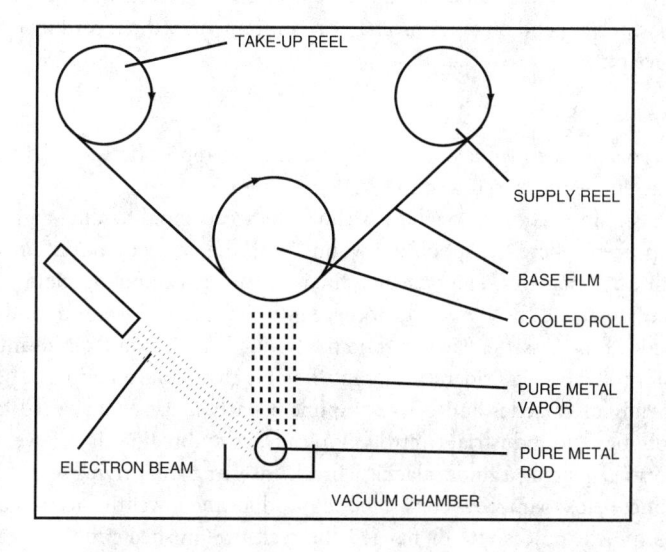

FIGURE 14.5 The metal evaporation process.

After leaving the orientation stage, the coated web passes through a series of ovens. Each oven is a different temperature (matched to the evaporation temperature of one of the solvents in use). Within the ovens, the web typically travels over a series of *plenums*, which carry the tape on a cushion of hot air. This prevents the tape from touching any of the surfaces inside the oven, which would damage the coated surface.

The solvent laden air gathered in the process is then sent to solvent recovery. This is a relatively simple process whereby the air is passed through a vessel containing activated charcoal. The charcoal absorbs the evaporated solvent and the air is passed into the atmosphere. The charcoal is then flushed with steam and the solvents recovered in a *cracking tower* as part of a distillation process.

Calendering

After passing through the ovens, the coated web then passes through a series of calenders. Calendering may either be performed in-line, that is, in a continuous process, or off-line, which involves a separate set of calender stacks to those found on the coating line itself.

A calender stack consists of two rolls located side by side. The meeting point of the two rolls is called a *nip*. One of the rolls provides pressure and the other heat. The calendering process can be roughly likened to ironing. During the process, which may be multiple numbers of nips, the tape surface is smoothed. Smoother surfaces tend to give higher electrical outputs than rougher ones.

Take Up

If the calendering process is performed in-line, the last task in the coating line is to wind the coated and calendered web onto a hub ready for the slitting process. This station acts in the very same manner as the take up reel on a reel-to-reel audio recorder.

Slitting

Of course, a jumbo roll of tape is useless to the end user. It must be slit and packaged into a usable form. Slitting takes the jumbo roll and cuts it into strands of the appropriate width for the given format. Each strand is then wound onto a hub and is then known as a *pancake*. All slitters operate in a similar manner. The web passes through a series of blades, known as *arbors*. There are two major parts to the arbor. The first is the male cutting blade, which is typically positioned above the web surface. The second is the female guides, which are typically positioned below the surface.

The male and female portions of the arbor operate together to form the slitting process. Edge weave, so-called *country laning*, and edge quality are all critical factors that are controllable during this process. As noted earlier tape formats are becoming more demanding on the integrity of the magnetic surface. As well as the tape surface itself, edge quality becomes a major issue. Defects in edges are greatly magnified in smaller formats, which depend on every mechanical component of the recording process being within a very small tolerance range.

Assembly

Having now mixed, coated, calendered, and slit the tape into the appropriate lengths and widths, the final task is to load and assemble the cassettes or to add flanges to reels.

In cassette formats manufacturers now have a choice between spool loading and *in-cassette* loading. Spool loading is a process where the spools are wound with the correct amount of tape prior to being placed into the unfinished cassettes. The cassette is then assembled around the spools. In-cassette loading assembles the cassette first, and then winds the footage into the cassette by cutting and splicing the leader material from outside of the cassette. Each method has both advantages and disadvantages and depends, to a great extent, on production considerations, most notably that of automation.

Large-scale assembly of cassettes and their component parts is largely achieved through the use of relatively simple robotics and industrial controls. Typically, assembly lines have several stations where a certain task is performed: placing a guide, placing a break, inserting and driving screws into final position, etc. The use of robotics helps to improve consistency of product and lower the overall cost of manufacture.

Having now assembled the cassette or flanged the reel, the finished product is then placed in the appropriate packaging, labeled, and boxed ready for use.

Defining Terms

Adhesion: A condition that describes the degree to which a formulation is affixed to the associated base film.

Base film: A generic name for different types of polyester films that are used to support the formulation(s).

Binder: Material used to "bind" all the formulation ingredients together in a cohesive manner. Also used to ensure good adhesion to the base film.

Cobalt doped oxide: Typically gamma ferric oxide particles that have been doped, or treated, with cobalt in order to achieve a higher coercivity.

Coercivity: A measurement of the magnetic characteristics of a given particle. Technically, coercivity refers to the amount of magnetic energy required to bring a saturated magnetic particle to a zero, or demagnetized, state.

Component: A system whereby the components of a video signal remain in their component form, i.e., one luminance signal and two color difference signals. This method is used in order to maximize the luminance and chrominance bandwidth.

Composite: A system whereby the components of a video signal are combined into one signal.

Dispersion: Typically the degree to which a magnetic particle has been distributed through a formulation.

Durability: The ability of a piece of magnetic recording tape, or media, to perform under multiple shuttle, play or still frame conditions.

Helical scan: A method of video recorder and format design where the recording, replay, and flying erase heads are mounted on a rotating cylinder that lies at the helix angle.

High-energy tape: A generic term for magnetic recording media of higher than approximately 500 Oe. This family of products could use a variety of magnetic particles including cobalt dope gamma ferric oxide, BaFe, or metal particles.

Orientation: A process during which the magnetic particles are physically moved and aligned in a certain direction. This process is performed before the coated surface becomes dry and hard.

Orientation direction: The direction in which the magnetic particles are oriented during the manufacturing process, usually described with reference to the linear motion of the tape on the transport. For example in the Quadruplex format, the tape motion is linear but the orientation direction is traverse, or lateral. This is because the orientation direction matches the direction of head travel across the tape surface.

Packing density: A method of describing the amount of information that can be recorded on a given piece of magnetic tape of given size.

Quadruplex: The first practical videotape recording format. The format uses a series of four heads (hence quad.) mounted on the head drum. The head drum rotates transversely with respect to the direction of tape travel.

Retentivity: The amount of magnetic energy retained by a given particle when that particle has been raised to a saturated level.

References

Ampex Corporation, Magnetic Tape Division. 1988. *A Complete Guide to Media & Formats* T50-02.

Jorgensen, F. 1988. *The Complete Handbook of Magnetic Recording*, 3rd ed. McGraw-Hill, New York.

Mee, C.D. and Daniel, E.D. 1987. *Magnetic Recording Handbook, Technology & Applications*. McGraw-Hill, New York.

Further Information

Jorgensen, F. 1988. Magnetic Recording. *The Complete Handbook of Magnetic Recording*, 3rd ed. McGraw-Hill, New York.

Stafford, D. 1994. Technology Trends. Continuing Developments in Magnetic Tape Technology. *Broadcast Engineering*, Oct.

Stafford, D. 1995. VTR Cleaning and Maintenance. The Care & Feeding of VTRs & Videotape. *Television Broadcast*, Sept.

14.2   Data Storage Systems

Jerry C. Whitaker

14.2.1   Introduction

The astounding volume of data being transmitted between systems today has created an obvious need for data management. As a result, more servers—whether they are PCs, UNIX workstations, minicomputers, or supercomputers—have assumed the role of information providers and managers. The number of networked or connectable systems is increasing by leaps and bounds as well, thanks to the widespread adoption of the client-server computing model. Hard disk storage plays an important role in enabling improvements to networked systems, because the vast and growing ocean of data needs to reside somewhere. It also has to be readily accessible, placing a demand upon storage system manufacturers to not only provide high-capacity products, but products that can access data as fast as possible and to as many people at the same time as possible. Such storage also needs to be secure, placing an importance on reliability features that best ensure data will never be lost or otherwise rendered inaccessible to network system users.

14.2.2   Redundant Arrays of Independent Disks (RAID) Systems

The common solution to providing access to many gigabytes of data to users fast and reliably has been to assemble a number of drives together in a gang or array of disks, known as **redundant arrays of independent disks (RAID)** subsystems. Simple RAID subsystems are basically a cluster of up to five or six disk drives assembled in a cabinet that are all connected to a single controller board. The RAID controller orchestrates read and write activities in the same way a controller for a single disk drive does, and treats the array as if it were in fact a single or **virtual drive**. RAID management software that resides in the host system provides the means to manage data to be stored on the RAID subsystem. A typical RAID configuration is illustrated in Fig. 14.6.

RAID Elements

Despite its multidrive configuration, the individual disk drives of a RAID subsystem remain hidden from users; the subsystem itself is the virtual drive, though it can be infinitely large. The phantom virtual drive is created at a lower level within the host operating system through the RAID management software. Not only does the software set up the system to address the RAID unit as if it were a single drive, it allows the subsystem to be configured in ways that best suit the general needs of the host system.

RAID subsystems can be optimized for performance, highest capacity, fault tolerance, or a combination of these attributes. Different **RAID levels** have been defined and standardized in accordance with these

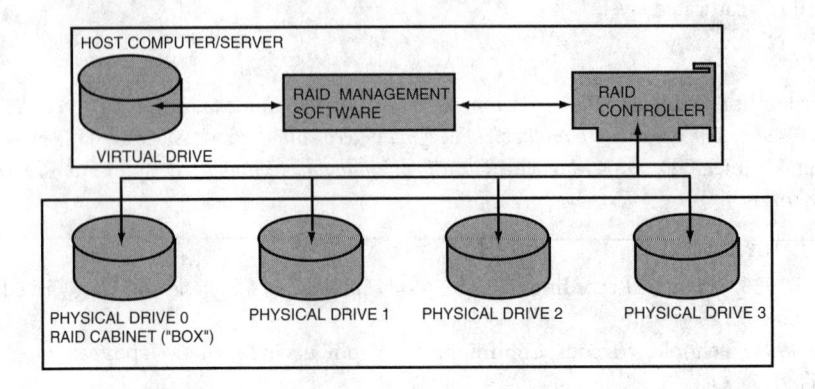

FIGURE 14.6   A typical RAID configuration. (*Source:* Adapted from Heyn, T. 1995. The RAID Advantage. Seagate Technology Paper, Seagate, Scotts Valley, CA.)

general optimization parameters. There are six such standardized levels RAID, called RAID 0, 1, 2, 3, 4, and 5, depending on performance, redundancy and other attributes required by the host system.

The RAID controller board is the hardware element that serves as the backbone for the array of disks; it not only relays the input/output (I/O) commands to specific drives in the array, but provides the physical link to each of the independent drives so that they may easily be removed or replaced. The controller also serves to monitor the integrity of each drive in the array to anticipate the need to move data should it be placed in jeopardy by a faulty or failing disk drive (a feature known as *fault tolerance*).

RAID Levels

The RAID 0–5 standards offer users and system administrators a host of configuration options. These options permit the arrays to be tailored to their application environments. Each of the various configurations focus on maximizing the abilities of an array in one or more of the following areas:

- Capacity
- Data availability
- Performance
- Fault tolerance

RAID Level 0

An array configured to RAID level 0 is an array optimized for performance, but at the expense of fault tolerance or data integrity. RAID level 0 is achieved through a method known as **striping**. The collection of drives (*virtual drive*) in a RAID level 0 array has data laid down in such a way that it is organized in stripes across the multiple drives. A typical array can contain any number of stripes, usually in multiples of the number of drives present in the array. Take, as an example, a four-drive array configured with 12 stripes (four stripes of designated space per drive). Stripes 0, 1, 2, and 3 would be located on corresponding hard drives 0, 1, 2, and 3. Stripe 4, however, appears on a segment of drive 0 in a different location than stripe 0; stripes 5–7 appear accordingly on drives 1, 2, and 3. The remaining four stripes are allocated in the same even fashion across the same drives such that data would be organized in the manner depicted in Fig. 14.7. Practically any number of stripes can be created on a given RAID subsystem for any number of drives: 200 stripes on two disk drives is just as feasible as 50 stripes across 50 hard drives. Most RAID subsystems, however, tend to have between 3 and 10 stripes.

The reason RAID 0 is a performance-enhancing configuration is that striping enables the array to access data from multiple drives at the same time. In other words, because the data is spread out across a number of drives in the array, it can be accessed faster because it is not bottled up on a single drive. This is especially

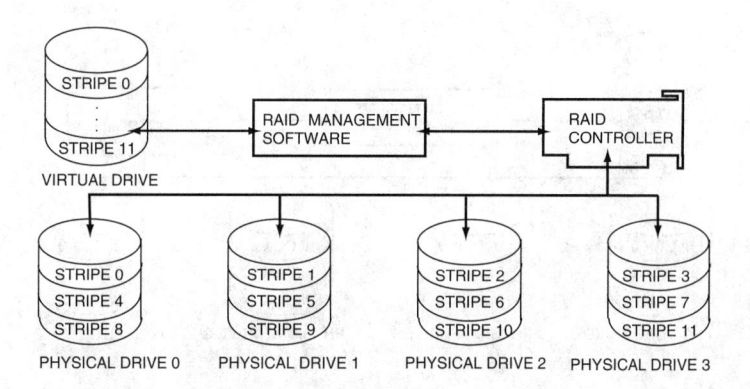

FIGURE 14.7 In a RAID level 0 configuration, a virtual drive comprises several stripes of information. Each consecutive stripe is located on the next drive in the chain, evenly distributed over the number of drives in the array. (*Source:* Adapted from Heyn, T. 1995. The RAID Advantage. Seagate Technology Paper, Seagate, Scotts Valley, CA.)

beneficial for retrieving a very large file, because it can be spread out effectively across multiple drives and accessed as if it were the size of any of the fragments it is organized into on the data stripes.

The downside to RAID level 0 configurations is that it sacrifices fault tolerance, raising the risk of data loss because no room is made available to store redundant data. If one of the drives in the RAID 0 fails for any reason, there is no way of retrieving the lost data, as can be done in other RAID implementations.

RAID Level 1

The RAID level 1 configuration employs what is known as *disk mirroring*, which is done to ensure data reliability or a high degree of fault tolerance. RAID 1 also enhances read performance, but the improved performance and fault tolerance come at the expense of available capacity in the drives used. In a RAID level 1 configuration, the RAID management software instructs the subsystem's controller to store data redundantly across a number of the drives (mirrored set) in the array. In other words, the same data is copied and stored on different disks to ensure that, should a drive fail, the data is available somewhere else within the array. In fact, all but one of the drives in a mirrored set could fail and the data stored to the RAID 1 subsystem would remain intact. A RAID level 1 configuration can consist of multiple mirrored sets, whereby each mirrored set can be a different capacity. Usually the drives making up a mirrored set are of the same capacity. If drives within a mirrored set are of different capacities, the capacity of a mirrored set within the RAID 1 subsystem is limited to the capacity of the smallest capacity drive in the set, hence the sacrifice of available capacity across multiple drives.

The read performance gain can be realized if the redundant data is distributed evenly on all of the drives of a mirrored set within the subsystem. The number of read requests and total wait state times both drop significantly, inversely proportional to the number of hard drives in the RAID, in fact. To illustrate, suppose three read requests are made to the RAID level 1 subsystem (see Fig. 14.8). The first request looks for data in the first block of the virtual drive; the second request goes to block 0, and the third seeks from block 2. The host-resident RAID management software can assign each read request to an individual drive. Each request is then sent to the various drives, and now—rather than having to handle the flow of each data stream one at a time—the controller can send three data streams almost simultaneously, which in turn reduces system overhead.

RAID Level 2

RAID level 2 is rarely used in commercial applications, but is another means of ensuring data is protected in the event drives in the subsystem incur problems or otherwise fail. This level builds fault tolerance around Hamming error correction code (ECC), which is often used in modems and solid-state memory devices as a means of maintaining data integrity. ECC tabulates the numerical values of data stored on specific blocks in the virtual drive using a formula that yields a checksum. The checksum is then appended to the end of the data block for verification of data integrity when needed.

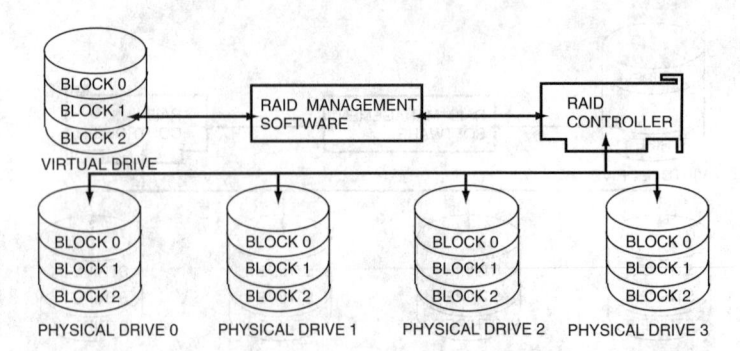

FIGURE 14.8   A RAID level 1 subsystem provides high data reliability by replicating (*mirroring*) data between physical hard drives. In addition, I/O performance is boosted as the RAID management software allocates simultaneous read requests between several drives. (*Source:* Adapted from Heyn, T. 1995. The RAID Advantage. Seagate Technology Paper, Seagate, Scotts Valley, CA.)

TABLE 14.1 Example of Error Correction Coding

Assume the phrase being stored is **HELLOTHERE**. The checksum is computed for every 10 bytes of data.

Data being stored:	H	E	L	L	O	T	H	E	R	E
Numerical representation:	72	69	76	76	79	84	72	69	82	69
Checksum formula:	$x1$	$x2$	$x3$	$x4$	$x5$	$x6$	$x7$	$x8$	$x9$	$x10$
Multiplied out:	72	138	228	304	395	504	504	414	738	690
[Check]sum of all values:	72	+138	+228	+304	+395	+504	+504	+414	+738	+690
										=3987

Thus, the data is stored on the drive as: 72 69 76 76 79 84 72 69 82 69 3987

As the data is read back from the drive, the same calculations with the data segment are made. The newly computed checksum is compared against the previously stored checksum, thus verifying data integrity.

Source: Adapted from Heyn, T. 1995. The RAID Advantage. Seagate Technology Paper, Seagate, Scotts Valley, CA.

As data is read back from the drive, ECC tabulations are again computed, and specific data block checksums are read and compared against the most recent tabulations. If the numbers match, the data is intact; if there is a discrepancy, the lost data can be recalculated using the first or earlier checksum as a reference point, as illustrated in Table 14.1.

This form of ECC is actually different from the ECC technologies employed within the drives themselves. The topological formats for storing data in a RAID level 2 array is somewhat limited, however, compared to the capabilities of other RAID implementations, which is why it is not commonly used in commercial applications.

RAID Level 3

This RAID level is essentially an adaptation of RAID level 0 that sacrifices some capacity, for the same number of drives, but achieves a high level of data integrity or fault tolerance. It takes advantage of RAID level 0 data striping methods, except that data is striped across all but one of the drives in the array. This drive is used to store parity information for maintenance of data integrity across all drives in the subsystem. The parity drive itself is divided into stripes, and each parity drive stripe is used to store parity information for the corresponding data stripes dispersed throughout the array. This method achieves high data transfer performance by reading from or writing to all of the drives in parallel or simultaneously but retains the means to reconstruct data if a given drive fails, maintaining data integrity for the system. This concept is illustrated in Fig. 14.9. RAID level 3 is an excellent configuration for moving very large sequential files in a timely manner.

The stripes of parity information stored on the dedicated drive are calculated using the Exclusive OR function. By using Exclusive OR with a series of data stripes in the RAID, any lost data can easily be recovered. Should a drive in the array fail, the missing information can be determined in a manner similar to solving for a single variable in an equation.

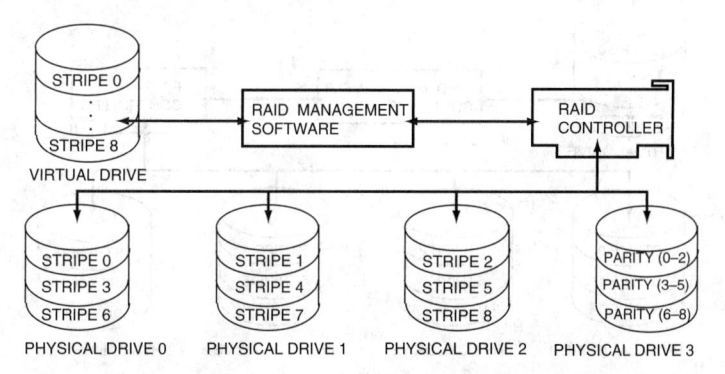

FIGURE 14.9 A RAID level 3 configuration is similar to a RAID level 0 in its utilization of data stripes dispersed over a series of hard drives to store data. In addition to these data stripes, a specific drive is configured to hold parity information for the purpose of maintaining data integrity throughout the RAID subsystem. (*Source:* Adapted from Heyn, T. 1995. The RAID Advantage. Seagate Technology Paper, Seagate, Scotts Valley, CA.)

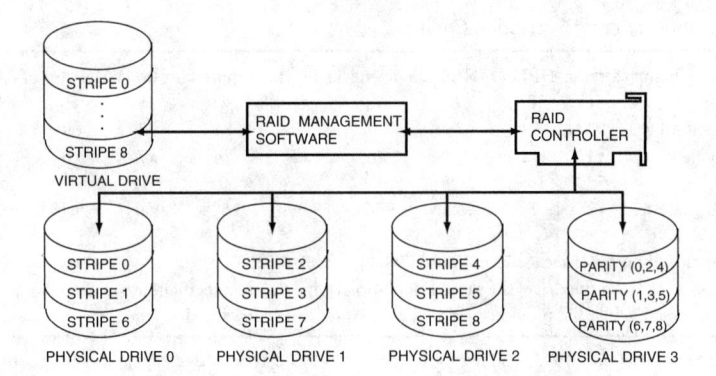

FIGURE 14.10 RAID level 4 builds on RAID level 3 technology by configuring parity stripes to store data stripes in a nonconsecutive fashion. This enables independent disk management, ideal for multiple-read-intensive environments. (*Source:* Adapted from Heyn, T. 1995. The RAID Advantage. Seagate Technology Paper, Seagate, Scotts Valley, CA.)

RAID Level 4

This level of RAID is similar in concept to RAID level 3, but emphasizes performance for different applications, e.g., database files vs large sequential files. Another difference between the two is that RAID level 4 has a larger stripe depth, usually of two blocks, which allows the RAID management software to operate the disks more independently than RAID level 3 (which controls the disks in unison). This essentially replaces the high data throughput capability of RAID level 3 with faster data access in read-intensive applications. (See Fig. 14.10.)

A shortcoming of RAID level 4 is rooted in an inherent bottleneck on the parity drive. As data is written to the array, the parity encoding scheme tends to be more tedious in write activities than with other RAID topologies. This more or less relegates RAID level 4 to read-intensive applications with little need for similar write performance. As a consequence, like level 3, level 4 does not see much common use in commercial applications.

RAID Level 5

This is the last of the most common RAID levels in use, and is probably the most frequently implemented. RAID level 5 minimizes the write bottlenecks of RAID level 4 by distributing parity stripes over a series of hard drives. In so doing it provides relief to the concentration of write activity on a single drive, which in turn enhances overall system performance. (See Fig. 14.11.)

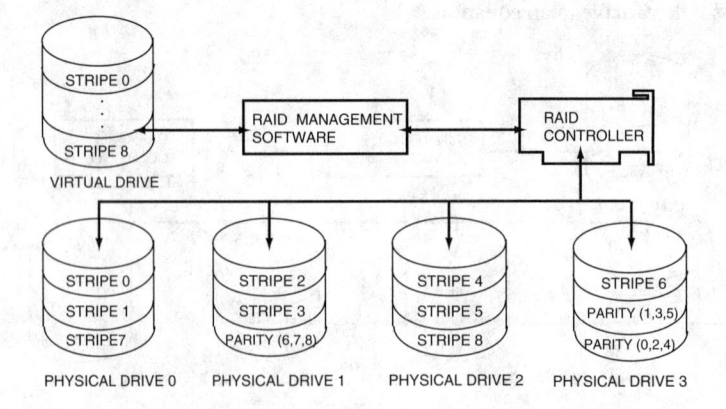

FIGURE 14.11 RAID level 5 overcomes the RAID level 4 write bottleneck by distributing parity stripes over two or more drives within the system. This better allocates write activity over the RAID drive members, thus enhancing system performance. (*Source:* Adapted from Heyn, T. 1995. The RAID Advantage. Seagate Technology Paper, Seagate, Scotts Valley, CA.)

TABLE 14.2 Summary of RAID Level Properties

RAID Level	Capacity	Data Availability	Data Throughput	Data Integrity
0	High	Read/write high	High I/O transfer rate	
1		Read/write high		Mirrored
2	High		High I/O transfer rate	ECC
3	High		High I/O transfer rate	Parity
4	High	Read high		Parity
5	High	Read/write high		Parity
6		Read/write high		Double parity
10		Read/write high	High I/O transfer rate	Mirrored
53			High I/O transfer rate	Parity

Source: Adapted from Heyn, T. 1995. The RAID Advantage. Seagate Technology Paper, Seagate, Scotts Valley, CA.

The way RAID level 5 reduces parity write bottlenecks is relatively simple. Instead of allowing any one drive in the array to assume the risk of a bottleneck, all of the drives in the array assume write activity responsibilities. This distribution frees up the concentration on a single drive, improving overall subsystem throughput. The RAID level 5 parity encoding scheme is the same as levels 3 and 4, and maintains the system's ability to recover lost data should a single drive fail. This can happen as long as no parity stripe on an individual drive stores the information of a data stripe on the same drive. In other words, the parity information for any data stripe must always be located on a drive other than the one on which the data resides.

Other RAID Levels

Other, less common RAID levels have been developed as custom solutions by independent vendors. Those levels include

- RAID level 6, which emphasizes ultrahigh data integrity
- RAID level 10, which focuses on high I/O performance and very high data integrity
- RAID level 53, which combines RAID levels 0 and 3 for uniform read and write performance

Custom RAID Systems

Perhaps the greatest advantage of RAID technology is the sheer number of possible adaptations available to users and systems designers. RAID offers the ability to customize an array subsystem to the requirements of its environment and the applications demanded of it. The inherent variety of configuration options provides several ways in which to satisfy specific application requirements, as detailed in Table 14.2. Customization, however, does not stop with a RAID level. Drive models, capacities, and performance levels have to be factored in as well as what connectivity options are available.

Defining Terms

Fiber channel-arbitrated loop (FC-AL): A high-speed interface protocol permitting high data transfer rates, large numbers of attached devices, and long-distance runs to remote devices using a combination of fiber optic and copper components.

Redundant arrays of independent disks (RAID): A configuration of hard disk drives and supporting software for mass storage whose primary properties are high capacity, high speed, and reliability.

RAID level: A standardized configuration of RAID elements, the purpose of which is to achieve a given objective, such as highest reliability or greatest speed.

Single connector attachment (SCA): A cabling convention for certain interface standards that simplifies the interconnection of devices by combining data and power signals as a common, standardized port.

Small computer systems interface (SCSI): A cabling and software protocol used to interface multiple devices to a computer system. These devices may be internal and/or external to the computer itself. Several variations of SCSI exist.

Striping: A hard disk storage organization technique whereby data is stored on multiple physical devices so as to increase write and read speed.

Thermal calibration: A housekeeping function of a hard disk drive used to maintain proper alignment of the head with the disk surface.

Virtual drive: An operational state for a computing device whereby memory is organized to achieve a specific objective that usually does not exist as an actual physical entity. For example, RAM in the computer may be made to appear as a physical drive, or two or more hard disks can be made to appear as a single physical drive.

References

Anderson, D. 1995. Fiber channel-arbitrated loop: The preferred path to higher I/O performance, flexibility in design. Seagate Technology Paper No MN-24, Seagate, Scotts Valley, CA.

Heyn, T. 1995. The RAID Advantage. Seagate Technology Paper, Seagate, Scotts Valley, CA.

Tyson, H. 1995. Barracuda and elite: Disk drive storage for professional audio/video. Seagate Technology Paper No. SV-25, Seagate, Scotts Valley, CA.

Further Information

The technology behind designing and manufacturing hard disk drives is beyond the scope of this chapter, and because most of the applied use of disks involves treating the drive as an operational component (black box, if you will), the best source of current information is usually drive manufacturers. Technical application notes and detailed product specifications are typically available at little or no cost.

14.3   Optical Storage Systems

Praveen Asthana

14.3.1   Introduction

Recordable optical disk drive technology provides a well-matched solution to the increasing demands for removable storage. An optical disk drive provides, in a sense, infinite storage capabilities: Extra storage space is easily acquired by using additional media cartridges (which are relatively inexpensive). Such cost effective storage capabilities are welcome in storage-intensive modern computer applications such as desktop publishing, computer aided design/computer aided manufacturing (CAD/CAM), or multimedia authoring.

14.3.2   The Optical Head

The purposes of the optical head are to transmit the laser beam to the optical disk, focus the laser beam to a diffraction limited spot, and to transmit readout signal information from the optical disk to the data and servo-detectors.

The laser diode is a key component in optical storage, whether the recording technology is magneto-optic, ablative WORM, or phase change. Early generations of optical drives used infrared lasers emitting in the 780 nm or 830 nm wavelengths. Later generation of drives use red laser wavelengths emitting at around 690 nm. The lasers are typically rated to have a maximum continuous output power in the 40-mW range and are index guided to ensure good wavefront quality.

In a laser diode, light is emitted from the *facets*, which are the cleaved ends of the waveguide region of an index guided laser. The facet dimensions are small enough (order of a few micrometers) for diffraction to take place as light is emitted from the facet. As a result the output beam has a significant divergence angle. In many commercial laser diodes, the width of the facet (i.e., the dimensions parallel to the pn-junction plane) is much larger than the height (or the direction perpendicular to the junction plane), which causes the divergence angles to be unequal in the directions parallel and perpendicular to the laser junction. The spatial profile of the laser beam some distance from the laser is thus elliptical.

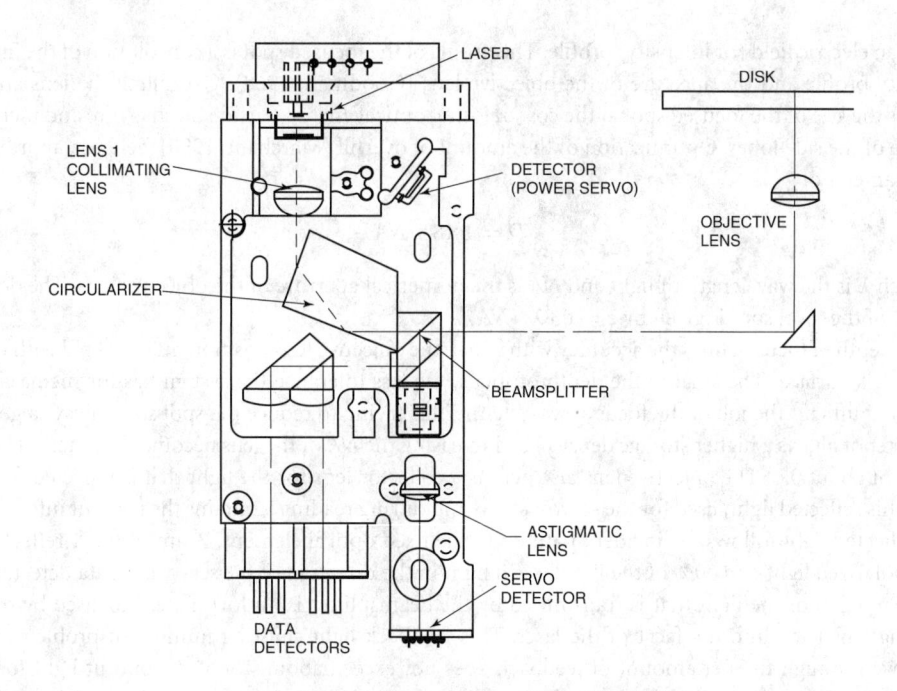

FIGURE 14.12 Drawing of the optical head in a typical magneto-optic disk drive. Shown is a split-optics design, which consists of a fixed set of components (the laser, detectors, and polarizing optics) and a movable system consisting of a beam bender and an objective lens that focuses the light on the disk. (*Source:* Asthana, P. 1994. Laser Focus World, Jan., p. 75. Penwell Publishing Co., Nashua, N.H. Used by permission.)

The basic layout of the optical head of a magneto-optic drive is shown in Fig. 14.12. The laser is mounted in a heat sink designed to achieve an athermal response. The output from the laser diode is collimated by lens 1. A prismlike optical element called a circularizer is then used to reduce the ellipticity of the laser beam. The beam then passes through a polarizing beamsplitter, which reflects part (30%) of the beam toward a detector and transmits the rest toward the disk.

The output of the laser is linearly polarized in the direction parallel to the junction (which will be referred to as P-polarization). The ratio of the intensity of the P-polarization component of the emitted light to the ratio of the intensity in the S-polarization component is > 25:1. This polarizing beam splitter is designed to transmit 70% of the P-polarized light and 100% of the S-polarized light. The light that is reflected is incident on a light detector, which is part of a power servoloop designed to keep the laser at a constant power. Without a power servoloop, the laser power will fluctuate with time as the laser junction heats up, which can adversely affect the read performance.

The beam transmitted by the beam splitter travels to a turning (90°) mirror, called a *beam bender*, which is mounted on a movable actuator. During track seeking operations, this actuator can move radially across the disk. The beam reflected by the turning mirror is incident on an objective lens (also mounted on the actuator), which focuses the light on the disk. This type of optical head design in which the laser, the detectors, and most of the optical components are stationary while the objective lens and beambender are movable is called a *split optics design*. In early optical drive designs, the entire optical head was mounted on a actuator and moved during seeking operations. This led to slow **seek times** (~200 ms) because of the mass on the actuator. A split optics design, which is possible because coherent light can be made highly collimated, lowers the mass on the actuator and thus allows much faster seek times.

The size of the focal spot formed by the objective lens on the disk depends on the *numerical aperture* (NA) of the lens and the amount of *overfill* of the lens (i.e., the amount with which the diameter of the incident collimated beam exceeds the aperture of the objective lens). The numerical aperture is given by $NA = n \sin \theta_{max}$, in which n is the refractive index of the lens and θ_{max} is the incidence angle of a light ray focused through the margin of the lens. The beam of light incident on the objective lens usually has a

Gaussian electric field (or intensity) profile. The profile of the focused spot is a convolution of the incident intensity profile and the aperture of the objective lens (Goodman, 1968). Overfilling the lens aperture reduces the size of the focused spot at the cost of losing optical energy outside the aperture and increasing the size of the side lobes. Optimization of the amount of overfill (Marchant, 1990) yields an approximate spot diameter of

$$D = 1.18\lambda/NA \tag{14.1}$$

in which λ is the wavelength of light and NA is the numerical aperture of the objective lens. The depth of focus z of the focal spot is given by $z = 0.8\lambda/(NA)^2$.

The depth of focus defines the accuracy with which the objective lens position must be held with respect to the disk surface. The smaller the depth of focus, the less tolerance the system has for media tilt and the more difficult the job of the focus servosystem. Thus trying to reduce the spot size (always a goal as a smaller spot allows a higher storage density) by increasing the NA of the lens becomes impractical beyond an NA of about 0.6. The objective lens also acts as a collector lens for the light that is reflected from the disk. This reflected light, used for the servosystems and during reading, contains the readout information. The reflected light follows the incident path up to the fixed optical element. Beamsplitter 1 reflects all of the S polarized light and 30% of the P polarized light in the direction of the servo and data detectors.

The portion of the light that is transmitted by the beamsplitter is, unfortunately, focused by the collimating lens back into the facet of the laser. This feedback light causes a number of problems in the laser, even though the net amount of feedback does not exceed about 7% of the output light for most magneto-optic media. Optical feedback affects the laser by causing the laser to mode hop randomly, which results in a random amplitude fluctuation in the output. This amplitude noise can be a serious problem and so techniques must be used to control the laser noise (such as injection of high-frequency current) or HFM (Arimoto, 1986). Increasing the HFM current in general can decrease the amount of noise, but as a practical matter, the injection current cannot be made arbitrarily large as it may then violate limits on allowable radiation from computer accessories. Optical feedback also decreases the threshold and increases the slope of the power-current (or PI) curve of the laser, but these effects are not really a problem.

The light reflected by beamsplitter 1 is further split by beamsplitter 2 into servo and data components. The light reflected by the beamsplitter is incident on the data detectors. For magneto-optic read back, two detectors are used in a technique known as *differential detection* (i.e., the difference in the signals incident on the two detectors is taken). The light transmitted through beamsplitter 2 is incident onto a special multielement servodetector and is used to generate the servosignals. The mechanism by which these signals are generated are the subject of the following discussion on the **servosystem**.

The Servosystem

The servosystem is what enables the focused laser spot to be positioned with accuracy onto any of the tracks on the disk and ensure that it can be moved to any track on the disk as required. The high track densities (18,000 tracks/in) on optical disks require that the laser spot position be controlled to within a fraction of a micrometer. To be able to move across the entire disk surface requires a large actuator, but such an actuator will be too massive to quickly respond to the rapid changes in the track position (due to run-out in the disk) as the disk spins. Therefore, a compound actuator consisting of a coarse actuator and a fine actuator is used to control the radial position of the laser beam on the disk. The fine actuator, which has a very low mass, can change the spot position rapidly over a limited range. The coarse actuator has a slower response, but has a much wider range of motion and is used for long seek operations. Optical disks have a continuous spiral groove (as in a phonograph record) to provide information on the relative track location.

In addition to **tracking** and seeking, the laser spot in an optical drive must be kept in perfect focus on the disk regardless of the motion of the disk (there can be quite a lot of vertical motion if the disk has tilt or is slightly warped). To do this, the objective lens must be constantly adjusted to correct for the axial motion of the disk surface as the media spins. The lens position is controlled by a focus servomechanism.

Figure 14.13 shows a block diagram of an optical drive servocontrol system (combined tracking and focusing). The return beam contains information on the focus and position of the spot, which is processed

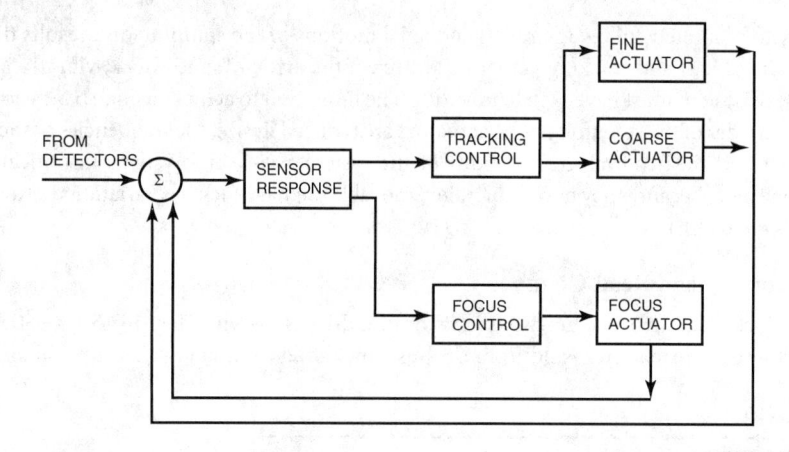

FIGURE 14.13 Block diagram of the servocontrol system for an optical disk drive.

by the servo detectors. The feedback signals derived from the detectors allow the system to maintain control of the beam.

The focus control system requires a feedback signal that accurately indicates the degree and direction of focus error (Braat and Bouwhuis, 1978; Earman, 1982). To generate a focus error signal, an astigmatic lens is used to focus onto a quadrant detector, a portion of the light reflected from the disk. In perfect focus, the focal spot is equally distributed on the four elements of the quad detector (as shown in Fig. 14.14). However, if the lens is not in focus, the focal spot on the detector is elliptical because of the optical properties of the astigmatic lens. The unequal distribution of light on the detector quadrants generates a *focus error signal* (FES). This signal is normalized with respect to light level to make it independent of laser power and disk reflectivity. The focus actuator typically consists of an objective lens positioned by a small linear voice coil motor. The coils are preferably mounted with the lens to reduce moving mass, while the permanent magnets are stationary. The lens can be supported by either a bobbin on a sliding pin or elastic flexures. The critical factors in the design are range of motion, acceleration, freedom from resonances, and thermal considerations.

Once the spot is focused on the active surface, it must find and maintain position along the desired track. This is the role of the tracking servo. The same quadrant detector that is used to generate the focus signal can be used to generate the *tracking error signal* (TES). The beam returning from the disk contains first-order diffraction components; their intensity depends on the position of the spot on the tracks and varies the light falling along one axis of the quadrant detector. The TES is the normalized difference of the current from the two halves of the detector, and it peaks when the spot passes over the cliff between a land and groove (Mansuripur, 1987; Braat and Bouwhuis, 1978).

As a matter of terminology, the **seek time** usually refers to the actual move time of the actuator. The time to get to data, however, is called the **access time** which includes the latency of the spinning disk in addition to the seek time. For example, a drive spinning a disk at 3600 rpm has a latency of $(0.5 \times (60/3600))$ s or 8 ms. Thus, a 3600-rpm drive with a seek time of 30 ms will have an access time of 38 ms. The standard way to measure seek time is 1/3 of the full stroke of the actuator (i.e., the time it takes to cover 1/3 of the disk). This is a historical artifact from the times of early hard disk drives.

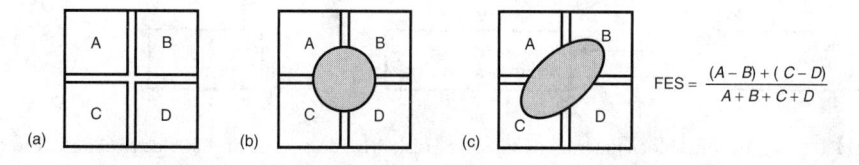

FIGURE 14.14 Focus control system: (a) the quad detector, (b) the spot of light focused on the quad by the astigmatic lens (circular implies objective lens is in focus), and (c) the spot of light on the quad when the objective lens is out of focus.

The ability to accurately follow the radial and axial motions of the spinning disk results directly from the quality of the focus and tracking actuators. To reject the errors due to shock, vibration, and media runout, the servosystem must have high bandwidth. The limitation to achieving high bandwidth is usually the resonance modes of the actuator. As the actuators are reduced in size, the frequencies of the resonances become higher and the achievable bandwidth of the system rises. Servosystems in optical drives face additional challenges because they have to handle removable media, which has variations between different disks (such as media tilt).

Optical Recording and Read Channel

A schematic block diagram of the functions in an optical drive is shown in Fig. 14.15. The SCSI controller handles the flow of information to and from the host (including commands). The optical disk controller

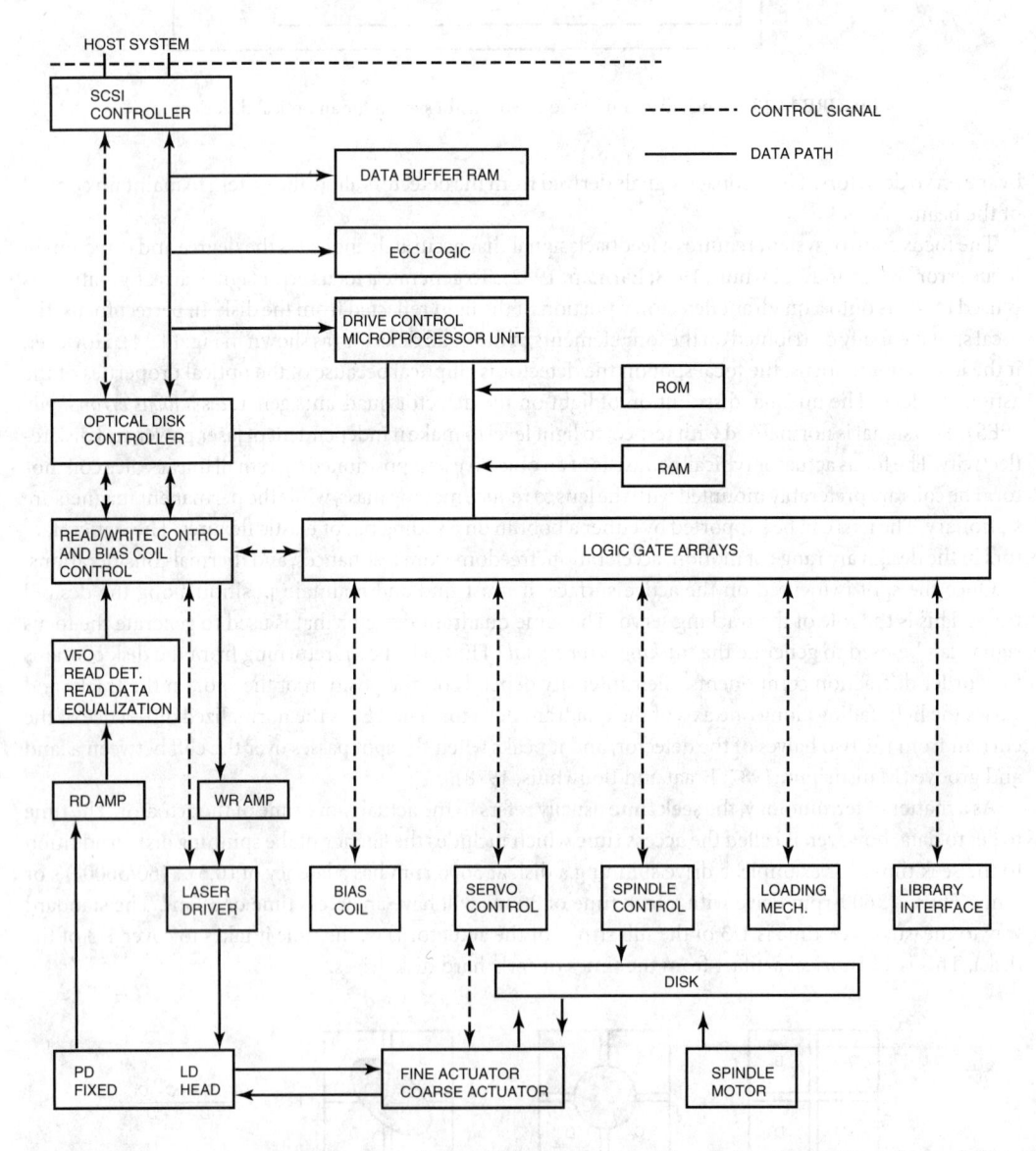

FIGURE 14.15 A functional block diagram of an optical disk drive describing the key electrical functions and interfaces. Some of the abbreviations are: VFO is the variable field oscillator, which is used to synchronize the data, RD amp is the read signal amplifier, WR amp is the write signal amplifier, PD fixed is the photodetextor assembly, and LD head is the laser diode.

is a key controller of the data path. It interprets the commands from the SCSI controller and channels data appropriately through the buffer random access memory (RAM) to the write channel, or from the read channel to the output. The drive control microprocessor unit controls, through the logic gate arrays, all the functions of the optical drive including the servo control, spindle motor, actuators, laser driver, etc.

Data that are input to the drive over the SCSI for recording are first broken up into fixed **block sizes** (of, for example, 512 kilobyte or 1024 kilobytes length) and then stored in the data buffer RAM. Magneto-optic, phase-change, and WORM drives can be classified as *fixed block architecture* technologies in which data blocks are recorded much like in hard drives (Marchant, 1990). Blocks of data can be placed anywhere on the disk in any sequence. The current CD recordable drive is, on the other hand, an example of a non-fixed block architecture (because its roots are in CD audio). In common CD-R drives, input data are recorded sequentially (like a tape player) and can be of any continuous length.

Error correction and control (**ECC**) bytes are added to each block of data. Optical drives use Reed–Solomon codes, which are able to reduce the error rate from 1E-5 to about 1E-13 (Golomb, 1986). After the addition of ECC information, the data are encoded with run length limited (RLL) modulation codes (Tarzaiski, 1983; Treves and Bloomberg, 1986) in order to increase efficiency and improve detection. Special characters are inserted in the bit stream such as a synchronization character to tell where the data begins. All in all, the overhead required to store customer data is about 20%.

In most current optical drives, recording is based on **pulse position modulation** (**PPM**) techniques. In basic PPM recording, the presence of a mark signifies a one and the absence signifies a zero bit. Thus, a 1001 bit sequence would be recorded as mark-space-space-mark. In **pulse width modulation** (**PWM**) recording, the edges of the mark represent the one bits (Sukeda et al., 1987). Thus a 1001 bit sequence would be recorded as a single mark of a certain size. A longer bit sequence such as 100001 would be represented by a longer mark, hence the term pulse *width* modulation. The use of pulse width modulation allows a higher linear density of recording than pulse position modulation.

A comparison of PWM recording with the current technique of PPM is shown in Fig. 14.16. It is more difficult to implement PWM technology (a more complicated channel is required), and PWM writing has greater sensitivity to thermal effects. Thus implementing PWM is a challenging task for drive vendors.

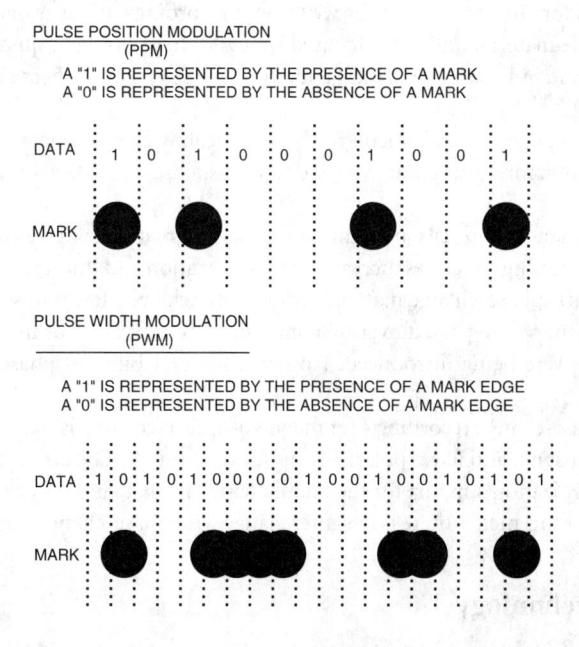

FIGURE 14.16 Schematic showing the mark spacing for PPM and PWM recording. The PPM recording is used in most current optical drives. The PWM recording increases capacity by as much as 50% and will be used in almost all forthcoming writable optical drives. PWM recording, however, requires much tighter tolerances than PPM recording.

On readback, the light reflects from the disk and travels to a photodetector (or two detectors in the case of magneto-optic drives). For WORM, phase-change, and CD-R disks, the signal is intensity modulated and thus can be converted to current from the detectors and amplified prior to processing. The WORM, phase-change, and CD-R disks have a high contrast ratio between mark and no-mark on the disk and thus provide a good signal-to-noise ratio. In magneto-optic drives, the read signal reflected off the disk is not intensity modulated but *polarization* modulated. Thus polarization optics and the technique of differential detection must be used to convert the *polarization* modulation to intensity modulation (as is discussed later in the section on magneto-optic recording).

To extract the 1s and 0s from the noisy analog signal derived from the photodetectors, optical drives use a number of techniques, such as equalization, which boosts the high frequencies and thus provides greater discrimination between spots. Using an analog-to-digital converter, the analog data signal is converted into channel bits. The channel bits are converted back into customer data bytes using basically the reverse of the encoding process. The data is clocked into the decoder, which removes the modulation code. The remaining special characters are removed from the data, which is then fed into the ECC alignment buffer to correct any errors (up to 40 bytes long). Once data has been read from the disk, it is stored in a RAM buffer and then output to whatever readout device is hooked by SCSI to the drive. With this basic understanding of how data is recorded on a spinning disk, we can turn to the specific items such as recording physics that delineate the various recording technologies.

Phase-Change Recording

Phase-change recording takes advantage of the fact that certain materials can exist in multiple *metastable* (i.e. normally stable) crystalline phases, each of which have differing optical properties (such as reflectivity). Thermal energy (as supplied by the focused beam of a high-power laser) above some threshold can be used to switch from one metastable state to another (Ovshinsky, 1970; Takenaga et al., 1983). Energy below the switching threshold should have no effect. In this way a low-power focused spot can be used to read out the recorded information without affecting it.

To achieve this kind of multiple metastable states, phase-change materials typically are a mixture of several elements such as germanium, tellurium, and antimony ($Ge_2Sb_2Te_5$). Phase-change materials are available that are suited for either rewritable or write-once recordings. In an erasable material, recording is affected by melting the material under the focused spot and then cooling it quickly enough to freeze it in an amorphous phase. Rapid cooling is critical and, thus, the design of the heat sinking capability of the material is important.

Erasure of the phase-change material is achieved by an annealing process, that is, heating the material to just below the melting point for a long enough period to recrystallise the material and erase any amorphous marks.

The fact that phase-change materials are a mixture of several materials makes recyclability difficult to achieve. The melting/annealing processes increase phase segregation and thus reduce the number of cycles that can be achieved. Early phase-change materials could only achieve a few thousand cycles, which is one of the primary reasons they were passed over by many companies in favor of magneto-optical products when rewritable drives were being introduced. However, the cyclability of phase-change materials has increased substantially.

The advantage of phase-change recording over magneto-optic recording is that a simpler head design is possible (since no bias magnet and fewer polarizing optics are needed). The disadvantages include the fact that there is less standardization support for the phase-change format, and fewer companies produce the drives. For the consumer this means there is less interchange with phase-change than with magneto-optic.

14.3.3 Worm Technology

The earliest writable optical drives were, in fact, WORM drives, and although the rewritable drive is more popular for daily storage needs, the WORM technology has a clear place in data storage because it allows permanent archiving capability.

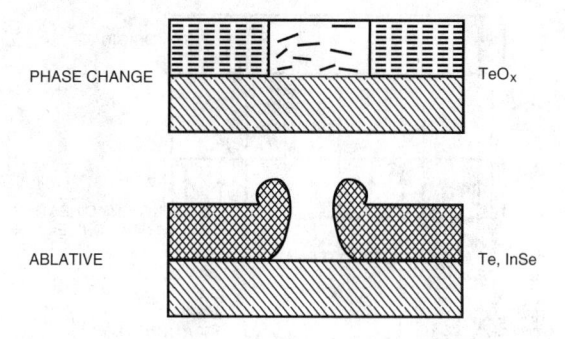

FIGURE 14.17   Permanent WORM recording provides the highest level of data security available in a removable storage device. In ablative WORM recording, marks are physically burned into the material. In phase-change worm, the recording process is a physical change in the material, which results in a change in reflectivity.

There are a number of different types of write-once technologies that are found in commercial products. Ablative WORM disks consist of tellurium-based alloys. Writing of data is accomplished by using a high-powered laser to burn a hole in the material (Kivits et al., 1982). A second type of WORM material is what is known as textured material, such as in a moth's eye pattern. The actual material is usually a platinum film. Writing is accomplished by melting the textured film to a smooth film and thus changing the reflectivity. Phase-change technology provides a third type of WORM technology using materials such as tellurium oxide. In the writing process, amorphous (dark) material is converted to crystalline (light) material by application of the heat from a focused laser beam (Wrobel et al., 1982). The change cannot be reversed. A comparison of phase-change and ablative WORM recording is shown schematically in Fig. 14.17.

14.3.4   Magneto-Optic Technology

In a magneto-optic drive, data recording is achieved through a thermomagnetic process (Mayer, 1958), also known as *Curie point writing* as it relies on the threshold properties of the Curie temperature of magnetic materials. In this process, the energy within the focused optical spot heats the recording material past its Curie point (about 200°C), a threshold above which the magnetic domains of the material are susceptible to moderate (about 300 G) external magnetic fields. Application of an external magnetic field is used to set the state of the magnetization vector (which represents the polarization of the magnetic domains) in the heated region to either up (a one bit) or down (a zero bit). When the material is cooled to below the Curie point this orientation of the magnetic domains is fixed. An illustration of the magneto-optic recording process is given in Fig. 14.18. This recording cycle has been shown to be highly repeatable (>1 million cycles) in any given region without degradation of the material. This is an important aspect if the material is to be claimed as fully rewritable.

In any practical recording process, it is necessary to have a sharp threshold for recording. This ensures the stability of the recorded information both to environmental conditions as well as during readout. Thermo-magnetic recording is an extremely stable process. Unless heated to high temperatures (>100°C), the magnetic domains in magneto-optic recording films are not affected by fields under several kilogauss in strength (in comparison, the information stored on a magnetic floppy is affected by magnetic fields as low as a 100 G). The coercivity of a magneto-optic material remains high until very close to the Curie temperature. Near the Curie temperature (about 200°C), the coercivity rapidly drops by two or three orders of magnitude as the magnetic domain structure becomes disordered.

Readout of the recorded information can safely be achieved with a laser beam of about 2-mW power at the disk, a power level which is high enough to provide good signal strength at the detectors, but low enough not to affect the recorded information because any media heating from it is far below the Curie threshold. During readout, the magnetic state of the recorded bits of information is sensed through the *polar-Kerr effect* by a low-power linearly polarized readout beam. In this effect the plane of polarization of the light beam is rotated slightly (0.5°) by the magnetic vector. The direction of rotation, which defines

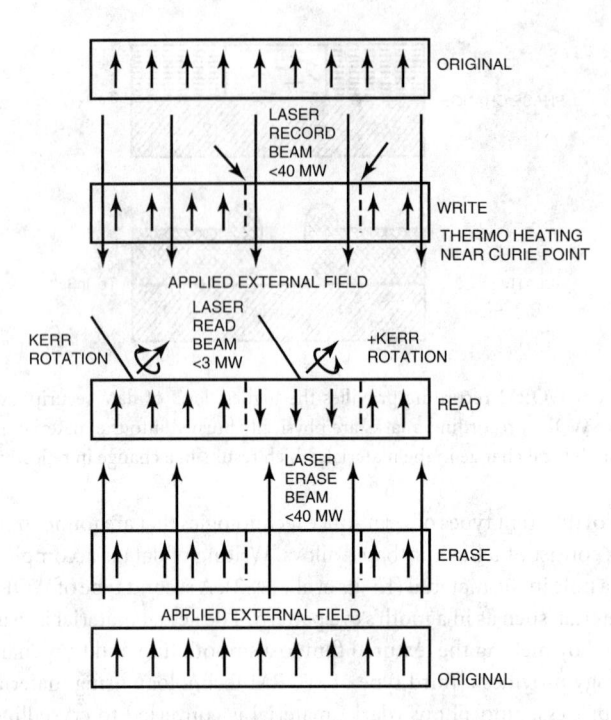

FIGURE 14.18   Schematic showing the recording, readout, and erasure processes of a magneto-optic material.

whether the bit is a one or a zero, is detected by the readout detectors and channel. Although the tiny amount of Kerr rotation results in a very small amount of signal modulation riding on a large DC bias, the technique of differential detection permits acceptable signal-to-noise ratio (SNR) to be achieved.

The output signal in an MO recording system is the signal from the light falling on one detector minus the signal from the light falling on the other detector. By placing a polarizing beamsplitter at 45° to the incident polarization, the two data detectors get the signals (Mansuripur, 1982)

$$
\begin{aligned}
d_1 &= \frac{I_0}{2}(\cos\theta_k/2 - \sin(\theta_k/2))^2 \approx \frac{I_0}{2}(1 - \theta_k) \\
d_2 &= \frac{I_0}{2}(\cos\theta_k/2 + \sin(\theta_k/2))^2 \approx \frac{I_0}{2}(1 + \theta_k)
\end{aligned}
\tag{14.2}
$$

in which d_1 and d_2 refer to the detector signals, I_0 is the incident intensity, and $\theta_k/2$, the rotation angle, is assumed to be small. The readout signal is taken as

$$
s = (d_1 - d_2)/(d_1 + d_2)
\tag{14.3}
$$

As can be seen, this signal does not contain intensity noise from either the laser or from reflectivity variations of the disk. But the signal is very sensitive to polarization noise, which may be introduced by polarization sensitive diffraction, by substrate birefringence effects, by inhomogeneities in the magnetooptic films, or by other polarization sensitive components.

Early MO recording media, such as manganese bismuth (MnBi) thin films (Chen et al., 1968), were generally crystalline in nature. The magnetic domains followed the crystalline boundaries and thus were irregular in shape (Marchant, 1990). The crystalline nature of the films caused optical scattering of the readout signal, and the irregular domains led to noise in the recorded signal. The combination degraded the SNR sufficiently to make polycrystalline magneto-optic media impractical.

The discovery in 1976 of magneto-optic materials based on the rare earth/transition metal (RE/TM) alloys (Choudhari et al., 1976) provided a practical material system for rewritable magneto-optic recording. These materials were amorphous and thus allowed acceptable signal-to-noise ratio to be obtained. Most commercial magneto-optic films today are based on terbium iron cobalt (TbFeCO).

14.3.5 Compact Disk-Recordable (CD-R)

The writable version of the popular CD-R disk looks very much like a stamped CD-ROM disk, and can be played in most CD-ROM players.

One of the reasons why CD-audio and CD-ROM have become so successful is the strict, uniform standards that all of the manufacturers of these products have adhered to. The standards were drawn up by Philips and Sony and are described by the colors of the books that they were first printed in. The standards describing CD-audio are found in the Red Book, those describing CD-ROM are found in the Yellow book, and those describing CD-R are found in the Orange Book. These books describe the physical attributes that the disks must meet (such as reflectivity, track pitch, etc.), as well as they layout of recorded data (Bouwhuis et al., 1985).

In prerecorded CD-ROM disks, the information is stamped as low reflectivity pits on a high reflectivity background. The disk has a reflectivity of 70% (achieved by using an aluminum layer), whereas the pits have a reflectivity of 30% (these reflectivity specifications have been defined in the Red Book). The CD-ROM drive uses this difference in reflectivity to sense the information stamped on the disk. To be compatible with CD-ROM readers, a CD-R disk must also use this reflectivity difference when recording data. To accomplish this, a CD-R disk is coated with an organic polymer that can change its local reflectivity permanently upon sufficient heating by a laser spot. The structure of a CD-R disk is shown in Fig. 14.19.

When the organic dye polymer is locally heated by the focused spot of a laser beam, polymeric bonds are broken or altered resulting in a change in the complex refractive index within the region. This refractive index change results in a change in the material reflectivity. There are a half-dozen organic dye polymers that are commercially being used. Two examples are phthalocyanine and poly-methane cyanine.

Like the CD-ROM drives, the CD-R drives have relatively low performance (when compared with optical or hard drives). The seek times are on the order of a few hundred milliseconds, whereas the maximum data rate for a 4X speed drive is about 600 kilobytes/s. The seek time is slow because the CD-R drives spin the disks in *constant linear velocity* (CLV) mode as defined in the Red Book standards. Constant linear velocity means that the disk rotation speed varies with the radius at which the read head is positioned in such a way as to ensure that the linear velocity is constant with radius.

Pure data devices such as hard disks and optical WORM drives, however, can handle constant angular velocity (CAV) operation in which the linear velocity and data rate increase with radial position. The disadvantage of CLV operation is that it results in a very slow access time. When a seek of the optical head is performed, the motor speed has to be adjusted according to the radial position of the head. This takes time and thus lengthens the seek operation to 250–300 ms. In contrast, constant angular velocity devices like optical WORM disks have seek times on the order of 40 ms.

The roots of CD-R are found in audio CD, and thus some of the parameters, recording formats, and performance features are based on the Red Book standard (which has tended to be a handicap). The CD format (Bouwhuis et al., 1985) is not well suited for random access block oriented recording. The CD format leads to a sequential recording system, much like a tape recorder.

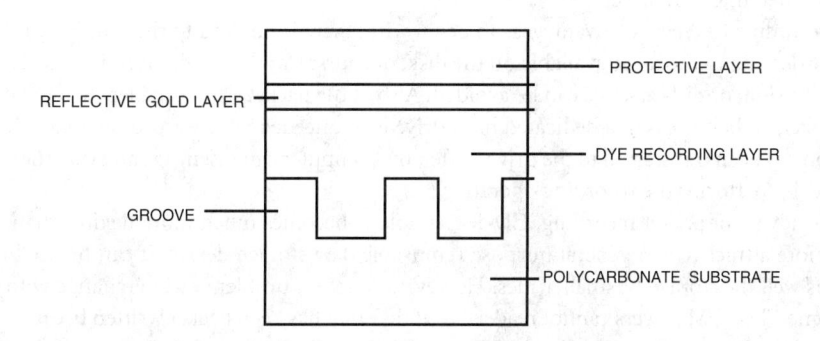

FIGURE 14.19 The basic structure of a CD-R disk.

Recording Modes of CD-R

To understand the attributes and limitations of CD-R, it is important to understand the various recording modes that it can operate in. For fixed block architecture devices, the question of recording modes never comes up as there is only one mode, but in CD-R, there are four modes (Erlanger, 1994).

The four recording methods in CD-R drives are

- Disk-at-once (or single session)
- track-at-once
- Multisession
- Incremental packet recording

In disk-at-once recording, one recording session is allowed on the disk, whether it fills up the whole disk or just a fraction of the disk. The data area in a single session disk consists of a lead-in track, the data field, and a lead out track. The lead-in track contains information such as the table of contents (TOC). The lead-in and lead-out are particularly important for interchange with CD-ROM drives. In single session writing, once the lead-in and lead-out areas are written, the disk is considered finalized and further recording (even if there are blank areas on the disk) cannot take place. After the disk is finalized, it can be played back on a CD-ROM player (which needs the lead-in and lead-out tracks present just to read the disk).

Having just the capability of recording a single session can be a quite a limitation for obvious reasons, and so the concept of *multisession* recording was introduced. An early proponent of multisession recording was Kodak, which wanted multisession capability for its photo-CD products. In multisession recording, each session is recorded with its own lead-in and lead-out areas. Multisession recorded disks can be played back in CD-ROM drives that are marked multisession compatible (assuming that each session on the disk has been finalized with lead-in and lead-out areas). Unfortunately, the lead-in and lead-out areas for each session take up lots of overhead (about 15 megabytes). With this kind of overhead, the ultimate maximum number of sessions that can be recorded on a 650 megabyte disk is 45 sessions.

Rather than do multisession recording, the user may choose track-at-once recording. In this type of recording, a number of tracks (which could represent distinct instances of writing) could be written within each session. The maximum number of tracks that can be written on the whole disk is 99. However, the disk or session must be finalized before it can be read on a CD-ROM drive.

Because of the way input data are encoded and spread out, it is imperative to maintain a constant stream of information when recording. If there is an interruption in the data stream, it affects the whole file being recorded (not just a sector as in MO or WORM drives). If the interruption is long enough, it will cause a blank region on the disk and will usually lead to the disk being rendered useless.

Many of these problems or inconveniences can be alleviated through a recording method called *packet recording*. In packet recording, the input data are broken up into packets of specified size (for example 128 kilo-bytes or 1 megabyte). Each packet consists of a link block, four run-in blocks, the data area, and two run-out blocks. The run-in and run-out blocks help delineate packets and allow some room for stitching, that is, provide some space for overlap if perfect synching is not achieved when recording an adjacent packet in a different CD-R drive.

Packet recording has several advantages. To begin with, there is no limit to the number of packets that can be recorded (up to the space available on the disk, of course), and so limitations imposed by track-at-once, multisession, or disk-at-once can be avoided. Also, if the packet size is smaller than the drive buffer size (as is likely to be the case), a dedicated hard drive is not needed while recording. Once the packet of information has been transferred to the drive buffer, the computer can disengage and do other tasks while the CD-R drive performs the recording operation.

With the advent of packet recording, CD-R technology becomes much more flexible than in the past and thus more attractive as a general purpose removable data storage device. It can be used for backup purposes as well the storage of smaller files. However, there is a problem of interchange with CD-ROM players. Some CD-ROM players cannot read a CD-R disk that has been packet written because they post a hard error when they encounter the link block at the beginning of each packet. Packet written CD-R disks can be read on CD-R drives and on CD-ROM drives that are packet enabled.

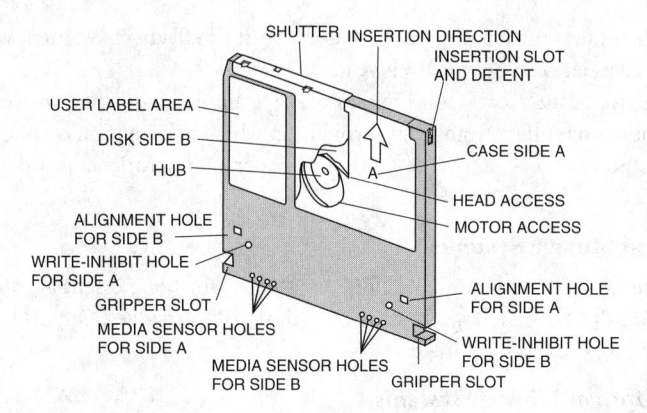

FIGURE 14.20 Schematic showing a 5.25-in disk cartridge. The various slots in the cartridge can be sensed by the drive and thus provide information about the disk to the drive.

14.3.6 Optical Disk Systems

Disks

An important component of the optical storage system is, of course, the media cartridges. In fact, the greatest attraction of optical storage is that the storage media is removable and easily transported, much like a floppy disk. Most of the writable media that is available comes in cartridges of either 3.5- or 5.25-in. form factors. An example of the 5.25-in. cartridge is shown in Fig. 14.20. The cartridge has some sensory holes to allow its media type to be easily recognized by the drive. The tracks on the media are arranged in a spiral fashion. The first block of data is recorded on the innermost track (i.e., near the center). Recordable CD media is cartridgeless and is played in CD-R/CD-ROM drives either using a caddy or through tray loading (like audio CD players).

Cartridged media has been designed to have a long shelf and archival life (without special and expensive environmental control) and to be robust enough to survive the rigors of robotic jukeboxes and customer handling. Magneto-optic and WORM media is extremely stable and so data can be left on the media with great confidence (Okino, 1987). A chart showing projected lifetimes (after accelerated aging) of IBM WORM media is given in Fig. 14.21, which indicates that the lifetimes for 97.5% of the media surfaces for shelf and archival use are projected to exceed 36 and 510 years, respectively, for storage at 30°C/80%

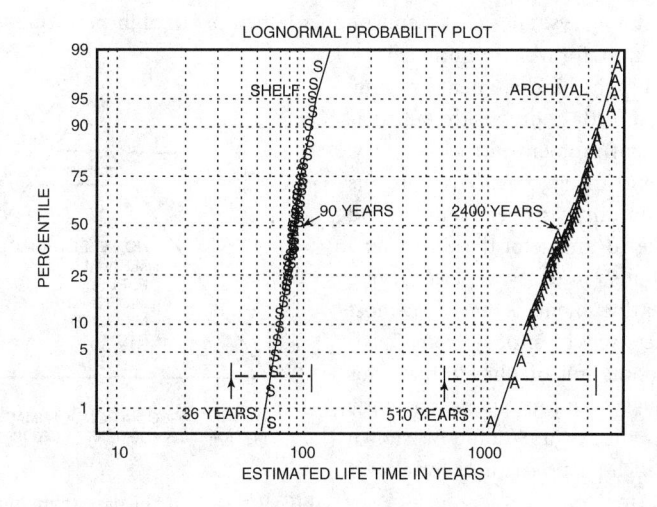

FIGURE 14.21 Estimated lifetime of an IBM WORM disk. Lognormal plot of lifetimes for shelf and archival use. (*Source:* Wong, J.S. et al., 1993. Life expectancy of IBM write-once media. IBM White Paper, IBM Publications.)

relative humidity. Shelf life is the length of time that data can be effectively written, whereas archival life is the length of time that data can be effectively read.

Optical media is perhaps the most stable digital storage technology: magnetic drives are prone to head crashes, tape deteriorates or suffers from print through (in which information on one layer is transferred to another layer in the tightly wound tape coil), and paper or microfiche also deteriorate with time (Rothenberg, 1995).

Automated Optical Storage Systems

Optical drives can be extended to automated storage systems, which are essentially jukeboxes consisting of one or more optical drives and a large number of optical disk cartridges. Optical libraries can provide on-line, direct-access, high-capacity storage.

Applications of Optical Library Systems

Optical library systems fit well into a large computer-based environment, such as client-server systems, peer-to-peer local area networks, or mainframe-based systems. In such environments, there is a distinct storage hierarchy based on cost/access trade-off of various types of data. This hierarchy is shown schematically in Fig. 14.22 as a pyramid. The highest section of the pyramid contains the highest performance and highest cost type of memory. The most inexpensive (on a cost/megabyte) basis and lowest performance is that of tape. Optical libraries are an

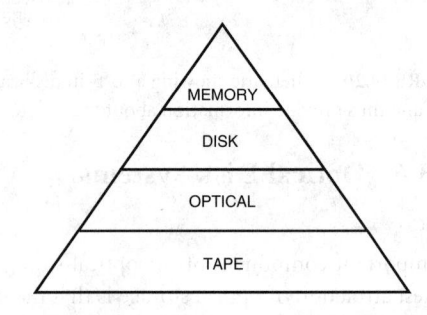

FIGURE 14.22   The storage hierarchy.

important part of this segment because they provide storage with performance capabilities approaching that of magnetic, but at a cost approaching that of tape.

An optical library contains a cartridge transport mechanism called an autochanger. The autochanger moves optical cartridges between an input/output slot (through which cartridges can be inserted into the library), the drives (where the cartridges are read or written), and the cartridge storage cells.

Sophisticated storage systems provide a hierarchical storage management capability in which data are automatically migrated between the various layers of the pyramid depending on the access needs of the data. For example, a telephone company can keep current billing information on magnetic storage, whereas billing information older than a month can be stored in optical libraries, and information older than six months can be stored on tape. It does not make sense to keep data that the system does not frequently access on expensive storage systems such as semiconductor memory or high-performance magnetic drives.

One of the applications for which automated optical storage is ideally suited is document imaging. Document imaging directly addresses the substantial paper management problem in society today. The way we store and manage paper documents has not changed significantly in over a century, and largely does not take advantage of the widespread availability of computers in the workplace. Trying to retrieve old paper documents is one of the most inefficient of business activities today (some businesses have stockrooms full of filing cabinets or document boxes). The way information is currently stored can be represented by a pyramid (as shown in Fig. 14.23), in which paper accounts for almost all of the document storage.

An application aimed squarely at re-engineering the way we handle and store documents is

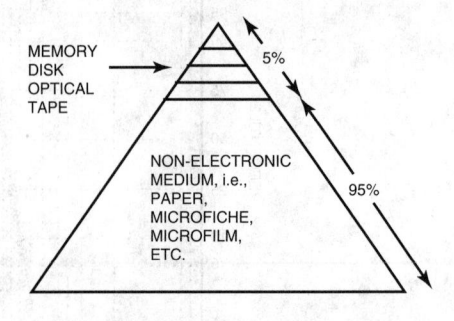

SOURCE: ASSOCIATION FOR IMAGE AND INFORMATION MANAGEMENT (AIIM), 1990

FIGURE 14.23   The paper pyramid. Most of the world's documents are in the form of paper, only 5% are in electronic form.

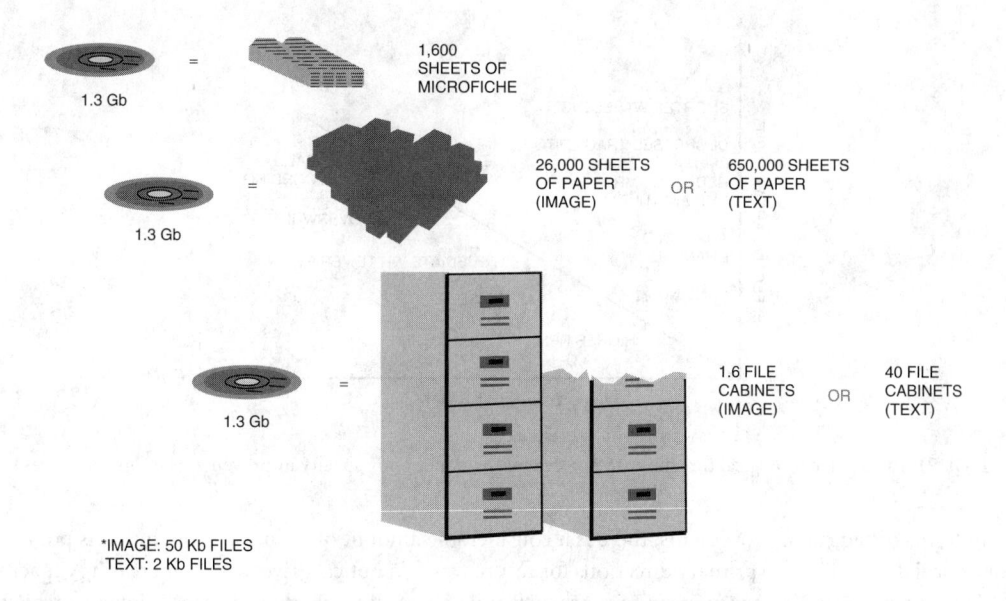

FIGURE 14.24 Equivalent storage capacity of a single 1.3 gigabyte optical disk and microfiche, paper, and filing cabinets.

document imaging. In document imaging, documents are scanned onto the computer and stored as computer readable documents. Storing documents in computer format has many advantages, including rapid access to the information. Recall of the stored documents is as easy as typing in a few key search words. Document imaging, however, is very storage intensive and, thus, a high-capacity, low-cost storage technology is required, ideally with fast random access. Optical storage is now recognized as the optimal storage technology in document imaging. Optical libraries can accommodate the huge capacities required to store large numbers of document images (text and/or photos), and optical drives can provide random access to any of the stored images. Figure 14.24 illustrates the equivalent storage capacity of a single 1.3 gigabyte optical disk.

Future Technology and Improvements in Optical Storage

Optical drives and libraries, like any high-technology products, will see two types of improvement processes. The first is an incremental improvement process in which quality and functionality will continuously improve. The second is a more dramatic improvement process in which disk capacity and drive performance will improve in distinct steps every few years.

For the library, the improvements will be in the speed of the automated picker mechanism as well as the implementation of more sophisticated algorithms for data and cartridge management. Most of the improvements in the library systems will arise out of improvements in the drives themselves.

For the optical drive, the incremental improvement process will concentrate on the four core technology elements: the laser, the media, the recording channel, and the optomechanics. The laser will see continual improvements in its beam quality (reduction of wavefront aberrations and astigmatism), lifetime (which will allow drive makers to offer longer warranty periods), and power (which will allow disks to spin faster). The media will see improvements in substrates (reduction of tilt and birefringence), active layers (improved sensitivity), and passivation (increased lifetimes). The optics and actuator system will see improvements in the servosystem to allow finer positioning, reductions in noise, smaller optical components, and lighter actuators for faster seek operations. The recording channel and electronics will see an increase in the level of electronics integration, a reduction in electronic noise, the use of lower power electronics, and better signal processing and ECC to improve data reliability. One of the paramount directions for improvement will be in the continuous reduction of cost of the drives and media. This step is necessary to increase the penetration of optical drives in the marketplace.

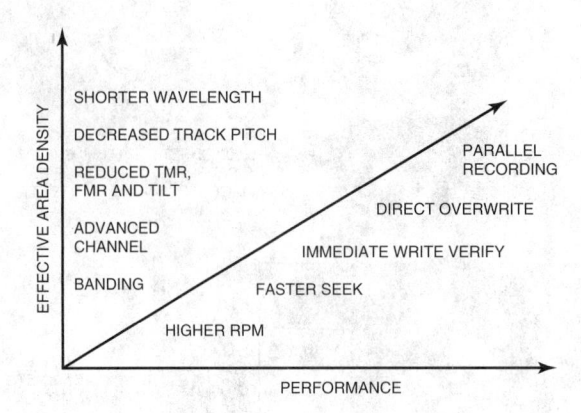

FIGURE 14.25   Technological directions to achieve performance and capacity improvements in optical drives.

In terms of radical improvements, there is a considerable amount of technical growth that is possible for optical drives. The two primary directions for future work on optical drives are (1) increasing capacity and (2) improving performance specifications (such as data rate and seek time). The techniques to achieve these are shown schematically in Fig. 14.25.

The performance improvements will include spinning the disk at much higher revolutions per minute to get higher data rates, radically improving the seek times for faster access to data, and implementing direct overwrite and immediate verify techniques to reduce the number of passes required when recording. Finally, the use of parallel recording (simultaneous recording or readback of more than one track), either through the use of multielement lasers or through special optics, will improve the data rates significantly.

In optical disk products, higher capacities can only be achieved by higher areal densities since the size of the disk cannot increase from presently established standard sizes. There are a number of techniques that are being considered for improving the storage capacity of optical drives. These include

- Use of shorter wavelength lasers or *superresolution* techniques to achieve smaller spot sizes
- Putting the data tracks closer together for greater areal density
- Reducing the sensitivity to *focus misregistration* (FMR) (i.e., misfocus) and *tracking misregistration* (TMR) (i.e., when the focused spot is not centered on the track)
- Reducing the sensitivity to media tilt

Finally, improvements in the read channel such as the use of partial-response-maximum-likelihood (PRML) will enable marks in high-density recording to be detected in the presence of noise.

Acknowledgments

The author is grateful to two IBM Tucson colleagues: Blair Finkelstein for discussions and suggestions on the read channel section and Alan Fennema for writing some of the paragraphs in the servo section.

Defining Terms

Compact disk erasable/compact disk recordable (CD-E/CD-R): These define the writable versions of the compact disk. CD-E is a rewritable media and CD-R is a write-once media.

Data block size: The data that is to be recorded on an optical disk is formatted into minimum block sizes. The standard block size, which defines a single sector of information, is 512 byte. Many DOS/windows programs expect this block size. If the block size can be made larger, storage usage becomes more efficient. The next jump in block size is 1024 byte, often used in Unix applications.

Device driver: This is a piece of software that enables the host computer to talk to the optical drive. Without this piece of software, you cannot attach an optical drive to a PC and expect it to work. As optical drives grow in popularity, the device driver will be incorporated in the operating system itself. For example, the device driver for CD-ROM drives is already embodied in current operating systems.

Error correction and control (ECC): These are codes (patterns of data bits) that are added to raw data bits to enable detection and correction of errors. There are many types of error correcting codes. Compact disk drives, for example, use cross interleaved Reed–Solomon codes.

Magneto-optical media: An optical recording material in which marks are recorded using a thermomagnetic process. That is, the material is heated until the magnetic domains can be changed by the application of a modes magnetic field. This material is rewritable.

Optical jukebox: This is very similar to a traditional jukebox in concept. A large number of disks are contained in a jukebox and can be accessed at random for reading or writing.

Phase-change media: An optical recording material consisting of an alloy that has two metastable phases with different optical properties. Phase-change media can be rewritable or write-once.

Pulse-position modulation (PPM): A recording technique in which a mark on the disk signifies a binary 1 and its absence signifies a binary 0. A 1001 bit sequence is mark-space-space-mark.

Pulse-width modulation (PWM): A recording technique in which the edges of the mark represent the ones and the length of the mark represents the number of zeros. Thus, a 1001 sequence is represented by one mark.

Seek time/access time: The two terms are often used interchangeably, which is incorrect. The seek time is, by convention, defined as the length of time taken to seek across one-third of the full stroke of the actuator (which is from the inner data band to the outer data band on the disk). The access time is the seek time plus some latency for settling of the actuator and for the disk to spin around appropriately. The access time really states how quickly you can get to data.

Servosystem: The mechanism, which through feedback and control, keeps the laser beam on track and in focus on the disk—no easy task on a disk spinning at 4000 rpm.

Tracking: The means by which the optical stylus (focused laser beam) is kept in the center of the data tracks on an optical disk.

References

Arimoto, A. et al. 1986. Optimum conditions for high frequency noise reduction method in optical videodisk players. *Appl. Opt.* 25(9):1.

Asthana, P. 1994. A long road to overnight success. *IEEE Spectrum* 31(10):60.

Bouwhuis, G., Braat, J., Huijser, A., Pasman, J., van Rosmalem, G., and Schouhamer Immink, K. 1985. *Principles of Optical Disk Systems*. Adam Hilger, Bristol, England, UK.

Braat, J. and Bouwhuis, G. 1978. Position sensing in video disk read-out. *Appl. Opt.* 17:2013.

Chen, D., Ready, J., and Bernal, G. 1968. MnBi thin films: Physical properties and memory applications. *J. Appl. Phys.* 39:3916.

Choudhari, P., Cuomo, J., Gambino, R., and McGuire, T. 1976. U.S. Patent #3,949,387.

Earman, A. 1982. Optical focus servo for optical disk mass data storage system application. *SPIE Proceedings* 329:89.

Erlanger, L. 1994. Roll your own CD. *PC Mag.* (May 17):155.

Golomb, S. 1986. Optical disk error correction. *BYTE* (May):203.

Goodman, J. 1968. *Introduction to Fourier Optics*. McGraw-Hill, San Francisco.

Inoue, A. and Muramatsu, E. 1994. Wavelength dependency of CD-R. *Proceedings of the Optical Data Storage Conference*. Optical Society of America, May, p. 6.

Kivits, P., de Bont, R., Jacobs, B., and Zalm, P. 1982. The hole formation process in tellerium layers for optical data storage. *Thin Solid Films* 87:215.

Mansuripur, M., Connell, G., and Goodman, J.W. 1982. Signal and noise in magneto-optical readout. *J. Appl. Phys.* 53:4485.

Mansuripur, M. 1987. Analysis of astigmatic focusing and push-pull tracking error signals in magneto-optical disk systems. *Appl. Opt.* 26:3981.

Marchant, A. 1990. *Optical Recording*. Addison-Wesley, Reading, MA.

Mayer, L. 1958. Curie point writing on magnetic films. *J. Appl. Phys.* 29:1003.

Okino, Y. 1987. Reliability test of write-once optical disk. *Japanese J. Appl. Phys.* 26.

Ovshinsky, S. 1970. Method and apparatus for storing and retrieving information. U.S. Patent #3,530,441.

Rothenberg, J. 1995. Ensuring the longevity of digital documents. *Sci. Am.* (Jan.).

Sukeda, H., Ojima, M., Takahashi, M., and Maeda, T. 1987. High density magneto-optic disk using highly controlled pit-edge recording. *Japanese J. Appl. Phys.* 26:243.

Takenaga, M. et al. 1983. New optical erasable medium using tellurium suboxide thin film. *SPIE Proc.* 420:173.

Tarzaiski, R. 1983. Selection of 3f (1,7) code for improving packaging density on optical disk recorders. *SPIE Proc.* 421:113.

Treves, D. and Bloomberg, D. 1986. Signal, noise, and codes in optical memories. *Optical Eng.* 25:881.

Wrobel, J., Marchant, A., and Howe, D. 1982. Laser marking of thin organic films. *Appl. Phys. Lett.* 40:928.

Further Information

There are a number of excellent books that provide an overview of optical disk systems. A classic is *Optical Recording* by Alan Marchant (Addison-Wesley, Reading, MA, 1990), which provides an overview of the various types of recording as well as the basic functioning of an optical drive. A more detailed study of optical disk drives and their opto-mechanical aspects is provided in *Principles of Optical Disc Systems* by G. Bouwhuis, J. Braat, A. Huijser, J. Pasman, G. van Rosmalen, and K. Schouhamer Immink (Adam Hilger Ltd., Bristol, England, 1985). An extensive study of magneto-optical recording is presented in *The Physical Properties of Magneto-Optical Recording* by Masud Mansuripur (Cambridge University Press, London, 1994).

For recent developments in the field of optical storage, the reader is advised to attend the meetings of the International Symposium on Optical Memory (ISOM) or the Optical Data Storage (ODS) Conferences (held under the auspices of the IEEE or the Optical Society of America).

For information on optical storage systems and their applications, a good trade journal is the *Computer Technology Review. Imaging Magazine* usually has a number of good articles on the applications of optical libraries to document imaging, as well as periodic reviews of commercial optical products.

14.4   Error Correction

Fabrizio Pollara

14.4.1   Background

Digital data transmitted over channels with impairments such as noise, distortions, and fading is inevitably delivered to the user with some errors. A similar situation occurs when digital data is stored on devices such as magnetic or other media that contain imperfections. The rate at which errors occur is a very important design criterion for digital communication links and for data storage. Usually, it must be kept smaller than a given preassigned value, which depends on the application. Error correction techniques based on the addition of **redundancy** to the original message can be used to control the error rate. The most obvious, albeit not efficient, example is the repetition of the same message. In this chapter we illustrate several efficient error correction techniques.

In 1948 Claude Shannon revolutionized communication technology with the publication of his classic paper in which he showed that the most efficient solution for reliable communications necessarily involved *coding* (i.e., introducing a controlled amount of redundancy) the selected message, before transmission over the channel. Shannon did not say exactly how this coding should be done: he only proved mathematically that efficient coding schemes must exist. Since 1948, a whole generation of later researchers has validated Shannon's work by devising explicit and practical coding schemes, which are now part of nearly every modern digital communication system. Satellite communication systems, high-performance military systems, computer communication networks, high-speed modems, and compact-disk and magnetic recording and playback systems, all rely heavily on sophisticated coding schemes to enhance their performance.

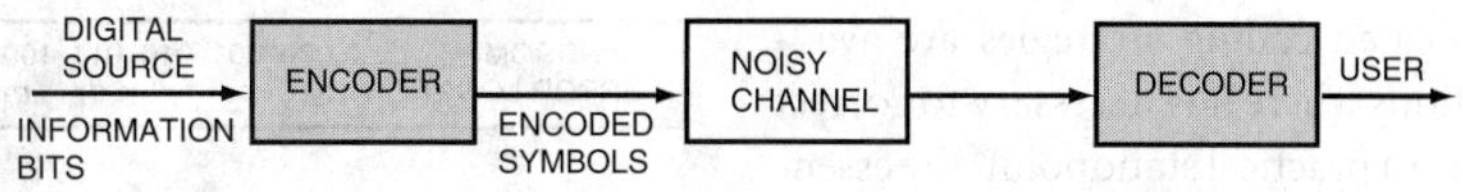

FIGURE 14.26   Coded communication system.

Shannon showed that to every communication channel we can associate a **channel capacity** C, and that there exist error correcting codes such that information can be transmitted across a noisy channel at rates less than C, with arbitrarily low bit error rate. In fact, an important implication of Shannon's theory is that it is wasteful to build a channel that is too good; it is more economical to use an appropriate code. Until Shannon's discovery, it was believed that the only way to overcome channel noise was to use more powerful transmitters or build larger antennas.

A communication system connects a data source to a data user through a channel. Microwave links, coaxial cables, telephone circuits, and even magnetic tapes are examples of channels. Coded digital communication systems (or data storage systems) include a device that creates appropriate redundant bits (the encoder) and a corresponding device (the decoder) that uses this redundancy to correct some of the errors introduced by the channel, as shown in Fig. 14.26. The encoder takes a block of information bits and produces a larger block of channel symbols, which is called a **codeword**. The channel in Fig. 14.26 contains all of the devices necessary to transmit digital information on a physical medium, including: a modulator to convert each symbol of the codeword into a corresponding analog symbol from a finite set of possible analog symbols (waveforms), and a demodulator to convert each received channel output signal into one of the codeword symbols. Each demodulated symbol is a best estimate of the transmitted symbol, but the demodulator makes some errors because of channel noise.

14.4.2   Introduction

The basic ideas of efficient error correction are illustrated in the following example of a *Hamming code*, a single-error-correcting code belonging to the first class of practical codes invented by Richard Hamming in 1948.

> Example of single-error-correcting code: Hamming code. The data to be transmitted consists of blocks of four bits denoted by $\{d_1, d_2, d_3, d_4\}$. Three redundant bits $\{p_1, p_2, p_3\}$, the parity-check bits, are computed from the bits in the data block and are appended to this block by the *encoder*. The parity-check p_1 is set to the value 0 if the number of 1s among the bits d_2, d_3, and d_4 is even and is set to 1 otherwise. Similarly, p_2 is set to 0 or 1 by the parity, that is, the even- or oddness, of the number of 1s among d_1, d_3, and d_4. Finally, p_3 is determined by the parity of d_1, d_2, and d_4. For example, if the data bits are $\{1011\}$, then the parity bits will be $\{010\}$. The combination of the data bits and the parity bits (seven bits, in this example) is sent over the channel or stored in a memory device and is called the codeword.
>
> When the codeword is recovered, possibly with some bit errors, the decoder checks to see if the three parity-check rules are still satisfied. The decoders computes a 0 for legitimate parity checks and a 1 for incorrect parity checks, thus generating three bits that are called the **syndrome**. For example, if $\{1001010\}$ is received, the first syndrome bit is set to 1, because the received data bits $d_2 = 0, d_3 = 0$ and $d_4 = 1$ have odd parity but the first parity-check bit is 0. Similarly, the second and third syndrome bits are set to 1 and 0, to produce the syndrome $\{110\}$. This syndrome is then used to find the location of the error using the one-to-one correspondence between syndromes and error locations illustrated in Fig. 14.27, for this example. In our example, the syndrome is $\{110\}$ which indicates that an error occurred at location d_3, the third data bit. The decoder will then change $\{1001010\}$ into $\{1011010\}$ and deliver to the user the corrected data $\{1011\}$. This strategy will successfully locate any single error in a block of four data bits encoded with three additional parity bits. If more than one error occurs this method will fail.

More sophisticated coding strategies are available for applications where it is necessary to correct more errors. From a practical standpoint, the essential limitation of all coding and decoding schemes proposed to date has not been a lack of good codes but the complexity (and cost) of the decoder. For

SYNDROME	000	001	010	011	100	101	110	111
ERROR LOCATION	–	p_3	p_2	d_1	p_1	d_2	d_3	d_4

FIGURE 14.27 Syndrome table.

this reason, efforts have been directed toward the design of coding and decoding schemes that could be easily implemented.

The Benefits of Error Correction Coding

The amount of redundancy inserted by the code is usually expressed in terms of the **code rate** R. This rate is the ratio of the number of information bits k in a block to the total number of transmitted symbols n in a codeword ($n = k+$ number of redundant symbols, i.e., $n > k$, or, equivalently, $R = k/n < 1$).

Let us assume that the we want to transmit data at a rate of R_b information bits per second. The redundancy of the code forces us to increase the transmitted symbol rate R_s to $R_s = R_b/R > R_b$. If the transmitted energy per information bit E_b is fixed, then the received energy per transmitted symbol is reduced from E_b to $E_s = R \times E_b$. If we did not perform any special decoding operations, the bit error rate (BER) would actually be worse than its original value obtained by using no coding at all! However, if we use the redundancy to correct some of the errors and the code was well selected, the BER at the output of the decoder is better than that at the output of the demodulator of the original uncoded system. This improvement in BER is obtained with the same E_b and R_b of the uncoded system. This explains how coding can improve BER at given E_b and R_b. Alternatively, the effect of coding can be seen as a reduction of required E_b at given BER and R_b, or as an increase in R_b (throughput) at given BER and E_b. The ratio by which the throughput is increased, or the transmitter power can be decreased, is called the *coding gain* of a given coding system.

14.4.3 The Development of Error Correction Codes

The Hamming codes were disappointingly weak compared with the far stronger codes promised by Shannon's theoretical results. The major advance came when Bose and Ray-Chaudhuri (1960) and Hocquenghem (1959) found a large class of multiple-error-correcting codes (the BCH codes), and Reed and Solomon (1960) found a related class of codes for nonbinary channels. All of these codes were based on algebraic properties of finite fields (**Galois fields**).

A completely different attack to the construction of efficient codes, more probabilistic in flavor, was the discovery in the late 1950s of **convolutional codes**, based on the properties of binary sequences generated by shift registers. These codes gained widespread popularity when an efficient decoding algorithm was discovered in 1967. The use of an algebraic **block code** cascaded with a convolutional code (**concatenated codes**) provided the most powerful codes available until the early 1990s.

The major progress in the 1970s was the discovery of family of codes that are good asymptotically, that is, when the block size of the encoded message becomes arbitrarily large. More efficient and asymptotically good codes, based on algebraic-geometric properties, were discovered in the late 1980s.

Codes that are very efficient for transmission on channels with limited bandwidth were discovered in the early 1980s, by combining the coding operation with an appropriate choice of the modulation scheme, that is, the set of waveforms that are actually used to transmit the encoded symbols. These codes have a major practical significance in the design of virtually any commercial data communication system.

More recently, the development of a new class of codes (**turbo codes**) promises to fill the last gap toward achieving the ultimate limits set by Shannon's theory.

14.4.4 Code Families

Error correction codes can be divided into two different classes: block codes and convolutional codes. The encoder for a block code breaks the continuous sequence of information bits into k-bit sections or blocks, and then operates on these blocks independently according to the particular code used. With each possible information block is associated an n-tuple of channel symbols, where $n > k$. The quantity n is referred to as the *code length* or *block length*. Codes can be defined on larger than binary alphabets, where each information sample is not just a bit but a symbol taken from an alphabet with q entries.

The other type of codes, called convolutional codes, operate on the information sequence without breaking it up into independent blocks. Rather, the encoder for a convolutional code processes the information continuously and associates each long (perhaps semi-infinite) information sequence with a code sequence containing more symbols.

Block codes and convolutional codes have similar error-correcting capabilities and the same fundamental limitations. In particular, Shannon's fundamental theorems hold for both types of codes.

Which Codes Are Good Codes?

Block codes are judged by three parameters: the block length n, the information length k, and the minimum distance d. The minimum distance is a measure of the amount of difference between the two most similar codewords.

The **Hamming distance** $d(x, y)$ between two q-ary sequences x and y of length n is the number of places in which they differ. The *minimum distance* of a code is the Hamming distance of the pair of codewords with smallest Hamming distance. More refined comparisons among codes make use of the **weight distribution** of a code, which is the list of the Hamming distances of each codeword from a given reference codeword. We will see that for **linear codes** the choice of the reference is irrelevant.

Suppose that a codeword is transmitted, and a single error is made by the channel. Then the received word is at a Hamming distance of 1 from the transmitted codeword. If the distance to every other codeword is larger than 1, then the decoder will properly correct the error if it assumes that the closest codeword to the received word was actually transmitted. In general, if t errors occur and if the distance from the received word to every other codeword is larger than d, then the decoder will properly correct the errors if it assumes that the closest codeword to the received word was actually transmitted. This always occurs if $d \geq 2t + 1$, as illustrated geometrically in Fig. 14.28, where each sphere of radius t consist of all n-tuples within distance t of codeword C, which we think as being the center of the sphere. Nonintersecting spheres of radius t can be drawn about each of the codewords. A received word in a sphere is decoded as the codeword at the center of that sphere. If t or fewer errors occur, then the received word is always in the proper sphere, and

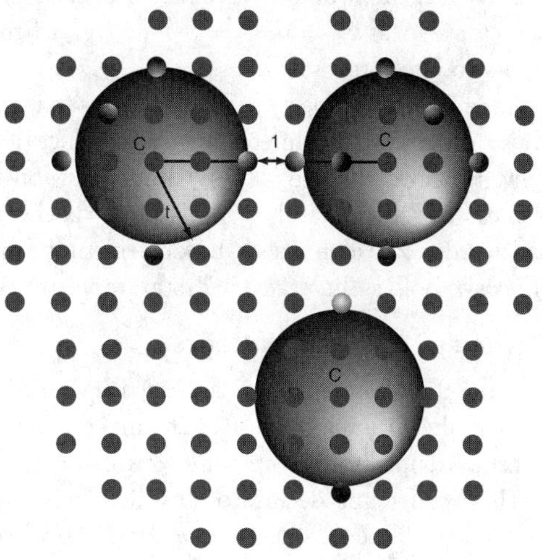

FIGURE 14.28 Geometric representation of a t-error correcting code ($t = 2$).

the decoding is correct. Some received words that have more than t errors will be in a decoding sphere about another codeword and, hence, will be decoded incorrectly. Other received words that have more than t errors will lie in the interstitial space between decoding spheres. This case can be handled in one of the two following methods: (1) An *incomplete decoder* decodes only those received words lying in a decoding sphere about a codeword. Other received words have more than the allowable number of errors and are declared by the decoder to be unrecognizable (uncorrectable error patterns). (2) A *complete decoder* decodes every received word (whether inside or outside a sphere) by selecting the closest codeword. When more than t errors occur, a complete decoder will often decode incorrectly but occasionally will find the correct codeword. A complete decoder is used when it is preferable to have a best guess of the message than to have no estimate at all.

14.4.5   Linear Block Codes

Block encoding can be regarded as a table-lookup operation, where each of the $M = q^k$ (q is the alphabet size of information symbols) codewords $x_1, x_2, \ldots, x_m, \ldots, x_M$ is stored in an n-stage register of a memory bank, and whenever message $u_m = (u_{m1}, \ldots, u_{mk})$ is to be transmitted, the corresponding signal vector $x_m = (x_{m1}, \ldots, x_{mn})$ is read from memory and used as the output of the encoder. We are primarily concerned with binary alphabets, that is, $q = 2$; thus, initially, we take the entries u_{mi} from $\{0, 1\}$ for all m, i.

The linear coding operation for binary data can be represented by

$$x_{m1} = u_{m1}g_{11} \oplus u_{m2}g_{21} \oplus \cdots \oplus u_{mk}g_{k1}$$
$$x_{m2} = u_{m1}g_{12} \oplus u_{m2}g_{22} \oplus \cdots \oplus u_{mk}g_{k2}$$
$$\vdots \qquad\qquad\qquad\qquad\qquad (14.4)$$
$$x_{mn} = u_{m1}g_{1n} \oplus u_{m2}g_{2n} \oplus \cdots \oplus u_{mk}g_{kn}$$

where $\oplus$ denotes modulo-2 addition, and $g_{ij} \in \{0, 1\}$ for all i, j. The term $u_{mi}g_{ij}$ is an ordinary multiplication, so that u_{mi} enters into the particular combination for x_{mj} if and only if $g_{in} = 1$. The matrix

$$G = \begin{bmatrix} g_{11}g_{12} \cdots g_{1n} \\ g_{21}g_{22} \cdots g_{2n} \\ \vdots \\ g_{k1}g_{k2} \cdots g_{kn} \end{bmatrix} = \begin{bmatrix} g_1 \\ g_2 \\ \vdots \\ g_k \end{bmatrix} \qquad (14.5)$$

is called the **generator matrix** of the linear code and g_i are its row vectors. Thus Eq. (14.4) can be expressed in vector form as $x_m = u_m G = \sum_{i=1}^{k} u_{mi}g_i$, where $\sum$ means modulo-2 addition and both u_m and x_m are binary row vectors.

An important property of linear codes is that the modulo-2 termwise sum of two codewords x_m and x_l is also a codeword. An interesting consequence of this property is that the set of Hamming distances from a given codeword to the $(M - 1)$ other codewords is the same for all codewords.

It can be shown that the generator matrix G can always be rearranged so that the first k entries of each codeword x_m are identical to the k entries of the message u_m and the remaining $n - k$ entries are the parity check symbols, as illustrated in our example of a Hamming code. In this case the code is called **systematic**.

Decoding Linear Block Codes

Suppose the message $u = u_1, \ldots, u_k$ is encoded into the codeword $x = x_1, \ldots, x_n$, which is then sent through the channel. Because of channel noise, the received vector $y = y_1, \ldots, y_n$, may be different from x. Let us define the error vector $e = y \oplus x = (e_1, \ldots, e_n)$.

The decoder must decide from y which message u or which codeword x was transmitted. Since $x = y \oplus e$, it is enough if the decoder finds e. The decoder can never be certain of what the true e was. Its strategy, therefore, will be to choose the most likely error vector e, given that y was received. Provided the codewords are all equally likely, this strategy is optimum in the sense that it minimizes the probability of the decoder making a mistake and is called **maximum likelihood decoding**.

We now illustrate a table lookup technique for decoding systematic linear codes, which is a generalization of our example on the Hamming code. Define the $n \times (n - k)$ matrix H^T by

$$H^T = \begin{bmatrix} g_{1,k+1} \cdots g_{1n} \\ g_{2,k+1} \cdots g_{2n} \\ \vdots \\ g_{k,k+1} \cdots g_{kn} \\ 1\,0\,0 \cdots\cdots 0 \\ 0\,1\,0 \cdots\cdots 0 \\ \vdots \\ 0\,0\,0 \cdots\cdots 1 \end{bmatrix} \qquad (14.6)$$

The matrix H is called the *parity-check matrix*. Any code vector multiplied by H^T yields the null vector $xH^T = 0$. Now consider postmultiplying any received vector y by H^T. The resulting $(n-k)$-dimensional binary vector is called the syndrome of the received vector and is given by $s = yH^T$. The decoding procedure is outlined as follows:

- Step 1. Initially, prior to decoding, for each of the 2^{n-k} possible syndromes s, store the corresponding minimum weight vector e satisfying $eH^T = s$. A table of 2^{n-k} n-bit entries will result.
- Step 2. From the n-dimensional received vector y, generate one $(n-k)$-dimensional syndrome s by the linear operations $s = yH^T$; this requires an n-stage register and $n - k$ modulo-2 adders.
- Step 3. Do a table lookup in the table of step 1 to obtain $\hat{e} = e$ corresponding to the syndrome s found from step 2.
- Step 4. Obtain the most likely code vector by the operation $x_m = y \oplus \hat{e}$. The first k symbols are the data symbols.

Examples of Linear Codes: Reed—Solomon Codes

After the discovery of Hamming codes, the development of the theory of **cyclic codes** provided the first large family of efficient error correcting codes (BCH codes). The central concept of cyclic codes is based on mathematical abstractions called *ideals* in the theory of finite fields. Cyclic codes are important practically because they can be implemented using high-speed shift-register-based encoders and decoders, at gigabit-per-second data rates. Within the family of cyclic codes there are specific classes of codes that are extremely powerful: Golay codes, BCH, and Reed–Solomon codes. Cyclic codes are defined in terms of a generator polynomial, which is multiplied by the data polynomial to obtain the codeword polynomial. These polynomials are defined in a finite field, and shift register circuits are used to perform the required polynomial multiplications and divisions. A popular family of codes called *cyclic redundancy check codes* (CRC) was derived from cyclic codes by *shortening* (reducing the block length) of some cyclic codes. CRC codes are primarily used for **error detection**, that is, to signal the presence of errors in the received message. Error detection, which is generally easier to obtain than error correction, is used in applications where it is extremely important to get the correct message or otherwise flag it as unreliable and discard it or ask for *retransmission*, according to a prescribed protocol as used in automatic-repeat-request (ARQ) systems. Reed–Muller codes (1954) are not used in practice any longer, but were the first codes to enable correction of multiple errors. They have fast decoding algorithms, but their rates are typically low. Reed–Muller codes are, however, very important ingredients for the construction of modern more complex codes.

Reed–Solomon (RS) codes, which can be viewed as a nonbinary extension of BCH codes, are a particularly interesting and useful class of linear block codes. The block length n of an RS code is $q - 1$, with q being the alphabet size of the symbols. It can be seen that these codes are useful only for large alphabet sizes. RS codes with k information symbols and block length n have a minimum distance $d = n - k + 1$, which is a property shared by all codes belonging to the class of *maximum distance separable* codes. RS codes are particularly effective on channels that produce errors in clusters, which are called *bursty* channels. Using symbols with $q = 2^m$ for some m, the block length is $n = 2^m - 1$. For an arbitrarily chosen odd minimum distance d, the number of information symbols is $k = n - d + 1$ and any combination of $t = (d - 1)/2 = (n - k)/2$ errors can be corrected. If we represent each letter in a codeword by m binary digits, then we can obtain a binary code with km information bits and block length nm bits. Any noise sequence that alters at most t of these n binary m-tuples can be corrected, and thus the code can correct all bursts of length $m(t - 1) + 1$ or less, and many combinations of multiple shorter bursts.

14.4.6 Convolutional Codes

These codes can be generated by a linear shift-register circuit that performs a convolution operation on the information sequence. Convolutional codes were successfully decoded by **sequential decoding** algorithms in the late 1950s. A much simpler decoding algorithm—the **Viterbi algorithm**—was developed in 1967 and made low-complexity convolutional codes very popular. Unfortunately, this algorithm is impractical for stronger convolutional codes.

A rate $R = 1/n$ convolutional encoder is a linear finite-state machine with one input, n outputs, and a K-stage shift register, where $K = m+1$ is called the constraint length of the code and m is the memory of the finite-state machine. An encoder with $R = 1/2$ and $K = 3$ is shown in Fig. 14.29. Such an encoder has 2^m possible states and is described by the taps of the shift register that are used to form each output by modulo-2 additions. The set of taps of each output is described by a polynomial with binary coefficients, where a coefficient 1 indicates a tap and a coefficient 0 indicates no tap. The example in Fig. 14.29 shows

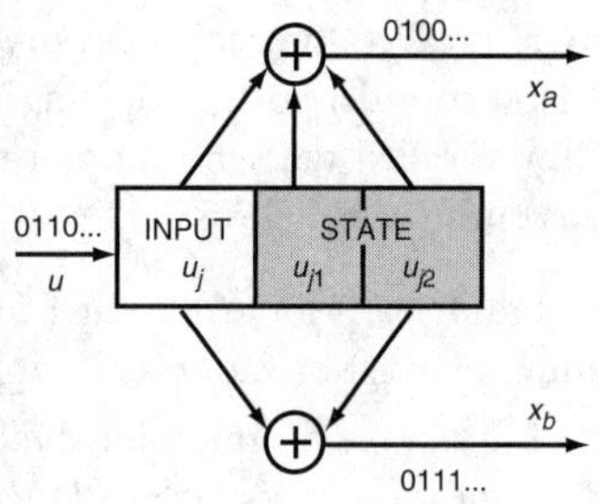

FIGURE 14.29 Example of convolutional encoder ($K = 3$).

how each input bit produces two output symbols, based on the present input and on two previous inputs that are stored in the shaded register cells after having been shifted in at previous ticks of the clock.

A finite-state machine such as that shown in Fig. 14.29 can be described by a *state diagram*, as shown in Fig. 14.30(a). The state transitions are indicated by directed edges labeled by $u_j/x_{0j}x_{1j}$, that is, the input and the two outputs associated with a transition from a state $u_{j-1}u_{j-2}$ to state u_ju_{j-1}. The **trellis diagram** shown in Fig. 14.30(b) gives an equivalent description of the state transition of this finite-state machine stressing the evolution in time. Here a path traversing states from left to right corresponds to the evolution with time of the encoder. As for block codes, the properties of convolutional codes depend on the Hamming distances of possible output symbol stream, which correspond the codewords of the code. Good codes have large minimum distances, where the minimum distance can be determined by finding the minimum output weight (number of ones) of any path that starts in state 00 and end in the same state after some digression. The choice of a reference state is irrelevant due to the linearity of these codes. The trellis diagram shown in Fig. 14.30(b) was a crucial element in the discovery of an efficient method for decoding these codes: the Viterbi algorithm (1967), from the name of its inventor. This algorithm is an efficient method to compare the distance of a received symbol stream to that of all of the possible paths on the trellis and to select that which has the least Hamming distance. The efficiency of the algorithm is given by the fact that at each state (a node in the trellis) the accumulated distances of paths coming into the state can be compared, and only the path of least distance can be preserved. The other paths can be discarded forever without loss of optimality. This considerably reduces the amount of operations necessary to compute distances to all possible paths. Once this pruning of the trellis has been done, the surviving paths are the only candidates to be at minimum distance from the received sequence. If the trellis is terminated at some time, there will be only one survivor into state 00. Otherwise, as in common practice, if the sequence is semi-infinite, the survivor into the state with minimum accumulated distance will be selected. Once a surviving path is selected, the corresponding sequence of decoded bits can be read following the only

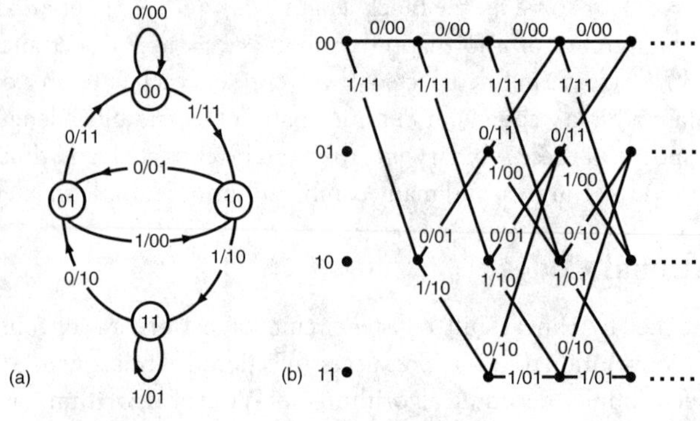

FIGURE 14.30 Examples of state and trellis diagrams.

possible backward path along the survivor. It has been shown (Forney, 1973) that the Viterbi algorithm implements, in fact, maximum-likelihood decoding, that is, the optimum decoding strategy.

Sequential decoding is another method used earlier to decode convolutional codes: its complexity does not increase exponentially with the memory *m* of the code, but its performance can only approximate that of a true maximum-likelihood decoder, as the Viterbi decoder. Sequential decoding is a procedure for systematically searching through a *code tree* and using received information as a guide, with the objective of eventually tracing out the path representing the actually transmitted information sequence. The two best known sequential decoding algorithms are the *stack algorithm* and the *Fano algorithm*. Sequential decoding is an example of incomplete decoding which yields two types of decoding failure. The first, called an *undetected error*, occurs when the decoder accepts a number of wrong hypotheses and moves ahead anyway. The second, called a *buffer overflow*, occurs when the number of computations permitted per block is exceeded. In this case, the frame cannot be decoded, and is considered an erased or *deleted frame*. Convolutional codes, which are sequentially decoded, typically have a large enough constraint length so that the undetected error probability of the decoder is negligible compared to the probability that a block cannot be successfully decoded in the time allowed.

For applications requiring large coding gain and very low bit error probabilities the idea of cascading two codes has been widely used and called concatenated coding. The most powerful combination in use is that of an outer Reed–Solomon code with an inner convolutional code. The decoding is done in two stages, which makes it intrinsically suboptimal, but its performance is a good approximation of the optimum decoder. Since the inner decoder produces errors clustered in bursts it is usually necessary to use a technique called **interleaving** to scatter such errors and ease the job of the outer decoder. Typically, the inner Viterbi decoder corrects enough errors so that a high-rate outer code can reduce the error probability to the desired level.

Turbo Codes

Coding theorists have traditionally attacked the problem of constructing powerful codes by developing codes with a lot of structure, which lends to feasible decoders. However, coding theory suggests that codes chosen at random should be pretty good if their block size is large enough. The challenge to find practical decoders for almost random, large codes had not been seriously considered until recently.

In 1993 a new class of concatenated codes called turbo codes, was introduced. These codes, which can be viewed as convolutional or block codes, can achieve near-Shannon-limit error correction performance with reasonable decoding complexity.

A turbo encoder is a combination of two simple convolutional encoders. For a block of *k* information bits, each constituent code generates a set of parity bits. The turbo code consists of the information bits and both sets of parity, as shown in Fig. 14.31. The key innovation is an interleaver *P*, which permutes the original *k* information bits before encoding the second code. If the interleaver is well chosen, information blocks that correspond to error-prone codewords in one code will correspond to error-resistant codewords in the other code. The resulting code performs much like Shannon's random codes, which can approach optimum performance at the price of a prohibitively complex decoder. Turbo decoding uses two simple decoders matched to the constituent codes. Each decoder sends likelihood estimates of the decoded bits to the other decoder, and uses the corresponding estimates from the other decoder as *a priori* likelihoods. The turbo decoder iterates between the outputs of the two constituent decoders until reaching satisfactory convergence.

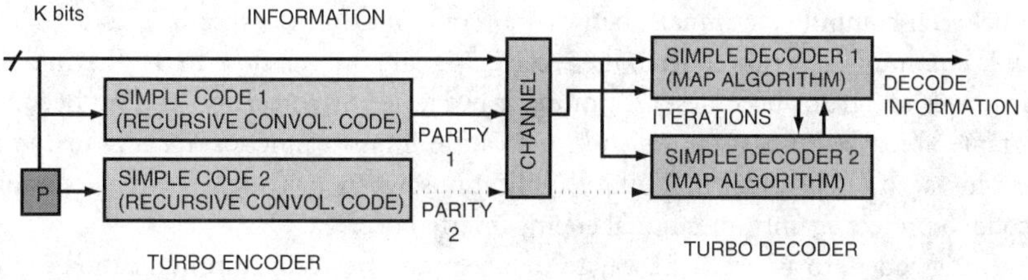

FIGURE 14.31 Example of turbo encoder/decoder.

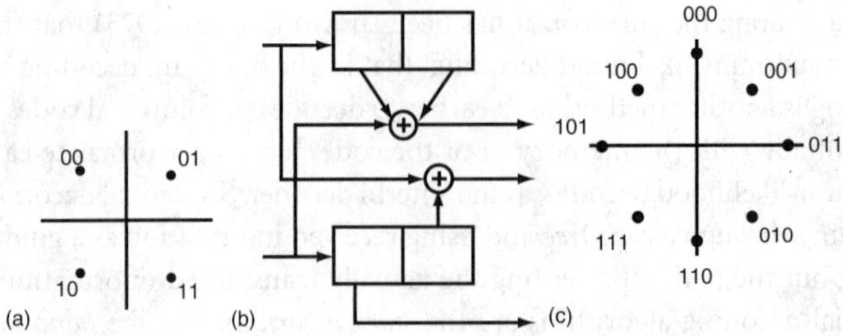

FIGURE 14.32   Example of trellis-coded modulation.

Turbo codes outperform even the most powerful codes known to date, but more importantly they are simpler to decode. To achieve their phenomenal performance, turbo codes require the use of large interleavers, but not much larger than those used by current concatenated codes.

14.4.7   Trellis Coded Modulation

Various efforts to improve the signal-to-noise ratio in channels with additive noise and *intersymbol inter- ference*, which is characteristic of severely bandlimited channels (e.g., the telephone channel), led to the original development of *set partitioning* techniques that culminated in Ungerboeck's development (1982) of **trelliscoded modulation** (TCM).

Ungerboeck showed how large Hamming distance between differing data sequences does not neces- sarily imply large Euclidean distance between modulated data sequences, unless the assignment of coded signals to modulated signals is cleverly made. He proposed a new method of set partitioning, which not only provided a method for the assignment of coded signals to channel signals, but also provided a simple formula for the lowerbound of the Euclidean distance between modulated data sequences. Figure 14.32(b) and Figure 14.32(c) illustrate an example of encoder and modulated signal assignment for trellis-coded modulation, which we will compare to the popular uncoded quadrature phase shift keying (QPSK) scheme of Fig. 14.32(a).

The trellis code (Fig. 14.32(b)) replaces the QPSK signal constellation with an 8-ary PSK signal constel- lation so as to introduce eight points in the complex plane that represents phase modulation. Instead of using the eight points to increase the data rate by modulating three bits at a time into a channel symbol, the trellis code modulates only two bits of data at a time into a three-bit channel symbol. The three code bits define one of eight possibilities that are labeled in the phase diagram in Fig. 14.32(c). Every pair of incoming data bits is mapped into three code bits by the trellis encoder in order to create a waveform. The transmitted trellis-coded waveform has the same data rate as a QPSK waveform and uses the same bandwidth, but the power requirement is reduced by 2.5 times.

TCM can also be used to increase the transmitted data rate at a fixed transmitted power. Much more complicated signal constellations are required for this purpose.

14.4.8   Applications

Error correcting codes have been applied to a variety of communication systems. Digital data are commonly transmitted between computer terminals, between aircraft, and from spacecraft. Codes can be used to achieve reliable communication even when the received signal power is close to the thermal noise power. As the radio waves spectrum becomes ever more crowded, error-correction coding becomes an even more important subject because it allows communication links to function reliably in the presence of interference. This is particularly true in military applications, where it is often essential to employ an error correcting code to protect against intentional enemy interference.

Error correcting codes are an excellent way to reduce power needs in communications systems where power is scarce, as in communication relay satellites because weak received messages can be recovered correctly with the aid of the code.

Data flow within computer systems usually is intolerant of even very low error rates, because a single error can destroy a computer program, and error correcting codes are becoming important in these applications. Bits can be packed more tightly into dense computer memories, by using simple Hamming codes. Codes are also widely used to protect data in digital tapes and disks.

Communication is also important within complex systems, where a large data flow may exist between subsystems, as in digital switching systems, and in digital radar signal-processing systems. These internal data might be transferred either by hardwired lines or by more sophisticated time-shared data bus systems. In either case, error correcting techniques are becoming important to ensure proper performance.

Defining Terms

Block codes: Codes that take an information block of length k and produce a codeword of length $n > k$ (n is the block length or code length).

Burst error correction: An error correction method that is effective against errors occurring in clusters, as opposed to random errors.

Channel capacity: The maximum information rate that can be transmitted across a channel, with arbitrarily low bit error rate.

Code rate: The ratio of the number of information bits k in a block to the total number of transmitted symbols n in a codeword ($n = k +$ no. of redundant symbols).

Codeword: The sequence of symbols obtained by adding redundant symbols to the original message.

Concatenated codes: Codes based on cascading two codes separated by an interleaver.

Convolutional codes: Codes generated by a linear shift-register circuit that performs a convolution operation on the information sequence.

Cyclic codes: Codes defined in terms of a generator polynomial, which is multiplied by the data polynomial to obtain the codeword polynomial. Polynomials are defined in a finite field, and shift register circuits are used to perform the required polynomial multiplications and divisions; examples: Golay codes, BCH, and Reed–Solomon codes.

Error detection: The mechanism by which a code can detect the presence of errors in a codeword. Error patterns that cannot be detected are called undetected errors.

Galois field: A Galois field (or finite field) is a set of objects or elements on which two operations $+$ and $\cdot$ are defined, obeying some specific rules (i.e., operations are commutative and distributive). Example: The integers $\{0, 1, \ldots, p - 1\}$ modulo p form a finite field if p is prime.

Generator matrix: The $k \times n$ matrix that produces a codeword (of length n) of a block code, by multiplication by the information block (of length k).

Hamming distance: The number of places in which two sequences of equal length differ. The minimum distance of a code is the smallest Hamming distance between any pair of codewords.

Interleaving: A device that scrambles the symbols from several codewords so that symbols from a given codeword are well separated.

Linear codes: Codes such that the linear combination of any set of codewords is a codeword.

Maximum-likelihood decoding: A decoding procedure that maximizes the probability of a received sequence given any of the possible codewords. If codewords are equally likely, this decoding method yields the minimum possible error probability.

Redundancy: Additional symbols (parity check symbols) appended to the original message to allow for error correction.

Sequential decoding: A procedure for systematically searching through a code tree with the objective of approximating the solution for the path at minimum distance from the received sequence. Examples: Fano algorithm, stack algorithm.

Syndrome: A vector computed by the decoder to find the location of correctable errors.

Systematic codes: Codes whose first k symbols of each codeword are identical to the information block. All codes can be reordered into systematic form.

Trellis coded modulation: A combined method for joint coding and modulation based on the design of convolutional codes matched to the modulation signal set in such a way to maximize the Euclidean distance between modulated sequence.

Trellis diagram: A graph used to represent the evolution in time of a finite-state machine by defining states as vertices and possible state transitions as edges.

Turbo codes: Codes generated by the parallel concatenation of two (or more) simple convolutional encoders, separated by interleavers.

Viterbi algorithm: An efficient method for decoding convolutional codes based on finding the path on a trellis diagram that is at minimum distance from the received sequence.

Weight distribution: The list of the Hamming distances of each codeword from a given reference codeword.

References

Berlekamp, E.R. 1968. *Algebraic Coding Theory*. McGraw-Hill, New York.

Berlekamp, R.R., ed. 1974. *Key Papers in the Development of Coding Theory*. IEEE Press, New York.

Blahut, R. 1983. *Theory and Practice of Error Control Codes*. Addison-Wesley, Reading, MA.

Bose, R.C. and Ray-Chaudhuri, D.K. 1960. On a class of error correcting binary group codes. *Info. and Control* 3:68–79.

Clark, G.C. and Cain, J.B. 1981. *Error-Correction Coding for Digital Communication*. Plenum, New York.

Fire, P. 1959. A class of Multiple-Error-Correcting Binary Codes for Non-independent Errors. Sylvania Report No. RSL-E-2, Sylvania Electronic Defense Laboratory, Reconnaissance Systems Division, Mountain View, Calif., March.

Forney, G.D. 1973. The Viterbi algorithm. *Proc. IEEE* 61:268–278.

Forney, G.D. 1967. *Concatenated Codes*. MIT Press, Cambridge, MA.

Hamming, R.W. 1980. *Coding and Information Theory*. Prentice-Hall, Englewood Cliffs, NJ.

Hocquenghem, A. 1959. Codes correcteurs d'erreurs. *Chiffres* 2:147–156.

Lidl, R. and Niederreiter, H. 1983. *Finite Fields*. Addison-Wesley, Reading, MA.

MacWilliams, F.J. and Sloane, N.J. 1977. *The Theory of Error-Correcting Codes*. North-Holland, Amsterdam.

Muller, D.E. 1954. Application of Boolean algebra to switching circuit design and to error detection. *IEEE Trans. Computers* 3:6–12.

Reed, I.S. 1954. A class of multiple-error-correcting codes and the decoding scheme. *IEEE Trans. Info. Theory* 4:38–49.

Reed, I.S. and Solomon, G. 1960. Polynomial codes over certain finite fields. *J. SIAM* 8:300–304.

Schouhamer Immink, K.A. 1989. Coding techniques for the noisy magnetic recording channel—A state-of-the-art report. *IEEE Trans. Comm.* 37(5):413–419.

Schouhamer Immink, K.A. 1991. *Coding Techniques for Digital Recorders*. Prentice-Hall, Englewood Cliffs, NJ.

Shannon, C.E. 1948. A mathematical theory of communication. *Bell Sys. Tech. J.* 27:379:423, 623:656.

Ungerboeck, G. 1982. Channel coding with multilevel/phase signals. *IEEE Trans. Inf. Theory* (Jan.).

Viterbi, A.J. 1967. Error bounds for convolutional codes and an asymptotically optimum decoding algorithm. *IEEE Transactions on Information Theory* IT-13: 260–269.

Viterbi, A.J. and Omura, J.K. 1979. *Principles of Digital Communication and Coding*. McGraw-Hill, New York.

Further Information

Most of the coding systems discussed in this chapter are too complex for a complete and detailed explanation. For in-depth treatment of general concepts in coding theory it is suggested to consult Clark and Cain (1981), Berlekamp (1974), Berlekamp (1968), MacWilliams and Sloane (1977), Blahut (1983), Viterbi and Omura (1979), and Hamming (1980). These references are a representative cross section of the literature on the subject of error correcting codes.

On the subject of coding for magnetic recording more details can be found in Schouhamer Immink (1989). The compact disk coding problem is discussed in Schouhamer Immink (1991). The origins of concatenated codes are treated in Forney (1967). The important mathematical tools of finite fields are treated in Lidl and Niederreiter (1983).

14.5 Data Compression

Jerome R. Breitenbach

14.5.1 Introduction

The field of data compression is concerned with devising practical and widely applicable methods for efficiently representing digital information. The primary reason for compressing information is to cause it to consume less space on the medium on which it is stored. In addition, though, if the information is to be transmitted to another location, then its prior compression will make it less taxing on the transmission medium; in contemporary parlance, data compression conserves "bandwidth." Moreover, if the amount of compression is significant, and the compression and decompression algorithms are sufficiently fast, then the time required for the overall communication process—compression, transmission, and decompression— can also be less, because the time used for compression and decompression will be more than offset by the speedier transmission of the compressed data.

This section presents an overview of the both theory and practice of data compression. The discussion is quite general, as it applies to any kind of digital information regardless of what the digits represent (text, audio, video, etc.). Thus, although such additional knowledge about a data source can be quite useful toward its compression (for example, some voice compression techniques are based on the physical structure of the human vocal tract), no use will be made here of any special character of the source.

14.5.2 The Data Compression Problem

The diagram in Fig. 14.33 displays the typical setup for *data compression*, also known as *source coding*. First, an *information source* randomly produces a stream of digital information to be stored, which is then later retrieved from storage for presentation to an *information sink*. Being digital, the output of the source is a sequence $(u_1, u_2, \ldots)$ of digits; for simplicity, we will focus on *binary* data, where the digits are *bits* taking on the two possible values 0 and 1. The *storage medium* is also assumed to be binary, and noncorrupting as well; that is, any sequence of 0s and 1s stored on the medium can be retrieved exactly. (Binary sources and storage media are by far the most prevalent. Nevertheless, everything in this chapter can be reformulated to apply without the binary assumption.) Thus, for the sake of our analyses, the storage medium is nothing more than a direct line from its input to its output.

If we were to apply the output of the source directly to the storage medium, and then present the output of the medium directly to the sink, then we would obtain the basic desired result—flawless storage and retrieval of the source output. Our objective, however, is also to minimize the number of bits stored on the medium in order to obtain this result; therefore, pre- and postprocessing are placed around it.

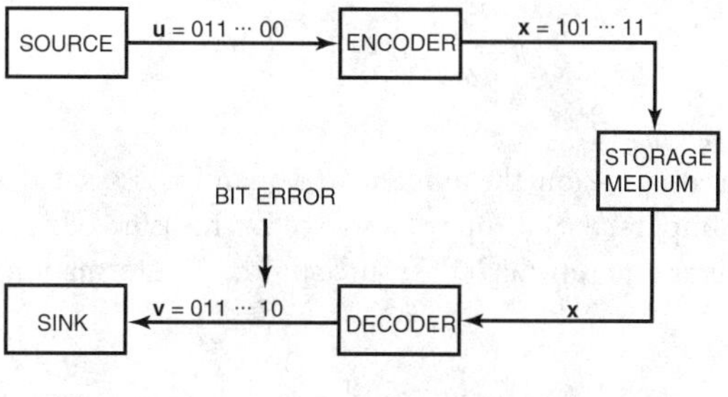

FIGURE 14.33 Basic setup for data compression.

Namely, an *encoder* converts the bit stream from the source into another bit stream applied to the medium, with the conversion rule being called a *code*. (In the context of data compression, the use of a code has nothing to do with encryption; it is not the intention of the encoder to *hide* the information (though the codes we shall discuss can be used for that purpose).) Correspondingly, a *decoder* converts the stored bit stream $(x_1, x_2, \ldots)$ into that presented to the sink. Ideally, of course, the total processing is such that the sink bit stream $(v_1, v_2, \ldots)$ is identical to that of the source. If the encoder is such that, for an appropriate decoder, this is indeed the case for every possible source sequence, then the code is called *lossless* or *noiseless*; otherwise, it is *lossy*.

As shown in the figure, one kind of code is the *block code*. Here, the encoder waits for the source to emit a *block* $\mathbf{u} = (u_1, \ldots, u_k)$ of k bits, for some fixed $k \geq 1$; then, according to the rule of the code, it converts this block into a corresponding *codeword* $\mathbf{x} = (x_1, \ldots, x_n)$ of $n \geq 1$ bits. At the output of the medium, the decoder converts $\mathbf{x}$ into another k-bit block $\mathbf{v} = (v_1, \ldots, v_k)$, hopefully the same as $\mathbf{u}$, to present to the sink. (Thus, the encoder is essentially a lookup table specifying the $\mathbf{x}$ corresponding to each possible $\mathbf{u}$, as is the decoder.) If the value of n is the same for all possible blocks $\mathbf{u}$, then we have a *fixed-length code* of *length* n; whereas if n is allowed to vary, then we have a *variable-length code*. The main question to be addressed in the sequel is: For a given performance criterion (e.g., an upper bound on the probability of decoder error), how many bits n, on average, are required to store the k source bits?

The *rate* R of a block code is defined to be the average number of bits stored per source bit; that is

$$R \triangleq E\{n\}/k \quad \text{bits/source bit}$$

In general, the expectation operation $E\{\cdot\}$ is required because, for a variable-length code, the code length n is a function of the random source block $\mathbf{u}$ and, thus, is random too; however, for a fixed-length code, the definition simplifies to

$$R = n/k \quad \text{bits/source bit}$$

Clearly, R provides a measure of how much compression is achieved by the code; the smaller R is, the better. For the case of no encoding/decoding already mentioned, we have $R = 1$; thus our intention is to achieve some $R < 1$. An essentially equivalent question to the one just posed is: For a given performance criterion, how small can we make R?

14.5.3 Data Compression via Fixed-Length Codes

The compression of a given stream of information is made possible by its source being statistically biased. First, consider the simplest type of nontrivial source, the *discrete memoryless source (DMS)*. A DMS is discrete in that its output is digital, as previously described; it is memoryless in that the output digit u_i at any given time i is statistically independent of the output digits u_j ($j \neq i$) at all other times. Also, it is assumed that the u_i ($i = 1, 2, \ldots$) are all taken from the same finite set of possible values, as well as each having the same probability distribution over these values. In the binary case, then, a DMS is completely characterized by the single probability

$$p = \Pr(u_i = 1) \quad \text{(all } i)$$

since $1 - p = \Pr(u_i = 0)$ (all i).

For the purposes of data compression, the fundamental quantity associated with an information source is its *entropy*. As the name implies, the entropy of a source is a measure of its randomness; but, as we will see, it also expresses the average rate at which the source produces information. In general, the entropy H of a source is defined as

$$H \triangleq \lim_{k \to \infty} \frac{1}{k} E \left\{ \log_2 \frac{1}{P(\mathbf{u})} \right\} \quad \text{bits/source bit} \tag{14.7}$$

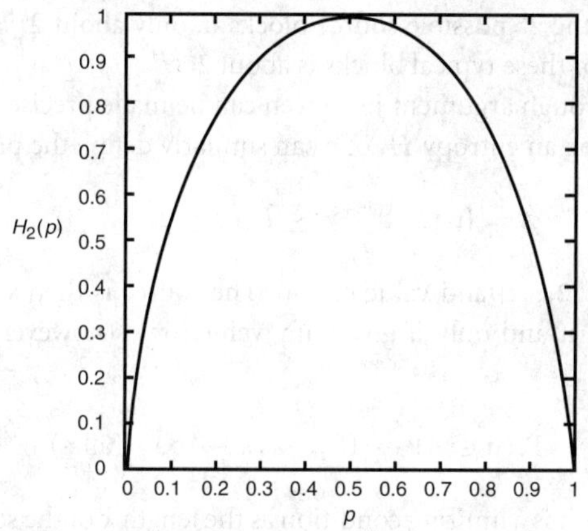

FIGURE 14.34 The binary entropy function.

(assuming the limit exists), where $P(\mathbf{u})$ is the probability distribution of $\mathbf{u}$, and thus

$$E\left\{\log_2 \frac{1}{P(\mathbf{u})}\right\} = \sum_{\mathbf{u}} P(\mathbf{u}) \log_2 \frac{1}{P(\mathbf{u})}$$

(the entropy of k-bit source block $\mathbf{u}$). For a binary DMS, H depends only on the probability p, and is given by the *binary entropy function*

$$H_2(p) \overset{\triangle}{=} -p\log_2 p - (1-p)\log_2(1-p) \quad (0 < p < 1) \tag{14.8}$$

with $H_2(0) \overset{\triangle}{=} 0$ and $H_2(1) \overset{\triangle}{=} 0$ (thereby making the function continuous). As shown by Fig. 14.34, the binary entropy function is nonnegative and symmetric about $p = \frac{1}{2}$, at which point it takes on its maximum value of 1.

To understand the fundamental significance of a source's entropy, recall that it follows from the law of large numbers that if k is large, then for a binary DMS it is highly probable that about kp bits of the k-bit source block $\mathbf{u}$ are 1s; that is, defining the set

$$A = \{\mathbf{u} \mid \mathbf{u} \text{ has approximately } kp \text{ 1s}\}$$

we have

$$\Pr(\mathbf{u} \in A) \approx 1 \quad \text{for } k \text{ large} \tag{14.9}$$

Also, each of the blocks in A has about the same probability of occurring

$$P(\mathbf{u}) \approx p^{kp}(1-p)^{k-kp} = 2^{-kH_2(p)} \quad (\mathbf{u} \in A)$$

by Eq. (14.14); thus, since $H = H_2(p)$, we have

$$P(\mathbf{u}) \approx 2^{-kH} \quad (\mathbf{u} \in A) \tag{14.10}$$

Therefore, combining Eq.(14.15) and Eq.(14.16), we obtain

$$1 \approx \Pr(\mathbf{u} \in A) = \sum_{\mathbf{u} \in A} P(\mathbf{u}) \approx |A| \cdot 2^{-kH} \quad (k \text{ large})$$

where $|A|$ is the size of the set A; hence

$$|A| \approx 2^{kH} \quad (k \text{ large})$$

That is, if k is large, then of the 2^k possible source blocks $\mathbf{u}$, only about 2^{kH} of them are likely to occur; and, the probability of each of these typical blocks is about 2^{-kH}.

Without much work, the rough argument just given can be made precise and rigorous. In general, for any information source having an entropy H, one can similarly define the precise source-block set

$$A = \{\mathbf{u} \mid 2^{-k(H+\epsilon)} \le P(\mathbf{u}) \le 2^{-k(H-\epsilon)}\}$$

for each block length $k \in \{1, 2, \ldots\}$ and value $\epsilon > 0$. The source is then said to possess the *asymptotic equipartition property (AEP)* if and only if given any value for ϵ (however small), we have Eq. (14.9); that is

$$\Pr(\mathbf{u} \in A) \to 1 \quad \text{as } k \to \infty \quad \text{(all } \epsilon)$$

The AEP is asymptotic in that it is a limiting condition as the length k of the source block becomes infinite; and, by virtue of Eq. (14.9) and Eq. (14.10), the elements of the set A provide a near equipartition of the total probability 1.

Thus, we have seen that every binary DMS possesses the AEP; in fact, so does every DMS. An information source need not be memoryless, though, to possess the AEP; every *stationary ergodic source*—that is, a discrete source for which the output is a **stationary ergodic random process**—possesses the AEP (and, therefore, an entropy too). Even some nonstationary sources possess the AEP. For example, if the source sequence $(u_1, u_2, \ldots)$ is a time-invariant **Markov chain**, then its entropy is given by the formula

$$H = \sum_{u_i} \bar{P}(u_i) \sum_{u_{i+1}} P(u_{i+1} \mid u_i) \log \frac{1}{P(u_{i+1} \mid u_i)}$$

(where the choice of $i \in \{1, 2, \ldots\}$ is arbitrary), in which $\bar{P}(\cdot)$ is the average distribution given by

$$\bar{P}(u) = \lim_{k \to \infty} \frac{1}{k} \sum_{i=1}^{k} \Pr(u_i = u) \quad \text{(all } u)$$

and, if the chain is also indecomposable, then the source possesses the AEP.

The importance of the AEP for data compression is shown by the *source coding theorem for fixed-length codes*. Defining the probability P_e of decoder error as

$$P_e = \Pr(\mathbf{v} \ne \mathbf{u})$$

the theorem states that for any source having an entropy H and possessing the AEP, if we are given values $\delta > 0$ and $\delta' > 0$, then there exists a fixed-length code of rate R and a corresponding decoder such that

$$R \le H + \delta \quad \text{and} \quad P_e \le \delta'$$

On the other hand, the *converse* to the theorem states that this result no longer holds if we replace the first condition with $R \le H - \delta$. Roughly put, the theorem says that using a fixed-length code on a source possessing the AEP, the error probability P_e can be made arbitrarily small for any code rate R that is greater than (perhaps arbitrarily close to) the source entropy H; whereas the converse says that P_e cannot be made arbitrarily small when R is any less than H. Put even more roughly, the entropy of a source possessing the AEP is the minimum rate at which reliable fixed-length encoding can be attained.

The proof of the theorem is quite simple; one merely allots storage space only to the approximately 2^{kH} source blocks in the typical set A, which requires approximately kH bits. A decoder error then occurs only when the source emits an atypical block $\mathbf{u}$, the probability P_e of which can be made arbitrarily small by making the code length k sufficiently large. Of course, such a compression scheme is lossy if $H < 1$; however, it is *nearly* lossless if P_e is made vanishingly small (which can be guaranteed by so choosing δ'). As an example, for a binary DMS with $p = 0.1$, we find from Eq. (14.8) that $H \approx 0.469$ bit/source bit;

therefore, if k is allowed to be large, then less than half as many storage bits are required to reliably store the k bits of **u**.

14.5.4   Data Compression via Variable-Length Codes

Suppose we do not require the codewords to all have the same length; then, this creates a decoding difficulty not present before. Namely, it is common in Fig. 14.33 that, instead of merely encoding a single source block of length k, this same encoding step is repeated several times; that is, for some fixed value k, the source sequence $(u_1, u_2, \ldots)$ is parsed into a succession of k-bit source blocks

$$(u_1, \ldots, u_k), \qquad (u_{k+1}, \ldots, u_{2k}), \qquad \cdots$$

each of which is encoded according to the rule of the code to produce a corresponding succession of codewords having respective lengths $n_1, n_2, \ldots$,

$$(x_1, \ldots, x_{n_1}), \qquad (x_{n_1+1}, \ldots, x_{n_1+n_2}), \qquad \cdots$$

which when concatenated form the resulting code sequence $(x_1, x_2, \ldots)$. In the case of a fixed-length code (where $n_1 = n_2 = \cdots = n$), the decoder can easily parse the code sequence back into n-bit blocks, and then decode each block individually; however, when n is allowed to vary, it is not always possible to determine the proper parsing. For example, consider the following simple code with $k = 1$:

u	x
0	0
1	00

Here, even though each source block **u** is assigned a unique codeword **x**, the two source sequences 01 and 10 both produce the same code sequence 000.

In light of this problem, we require all of our variable-length codes to be *uniquely decodable*; that is, no two finite-length source sequences can produce the same code sequence, and thus it is possible to decode any finite-length code sequence that the encoder might emit. A particularly convenient kind of uniquely decodable code is a *prefix-free* code, which is defined as any block code satisfying the *prefix condition* whereby no codeword is a prefix of any other. To decode a code sequence $(x_1, x_2, \ldots)$ generated by a prefix-free code, the decoder simply determines the smallest value of n for which $(x_1, \ldots, x_n)$ is a codeword, and decodes it; then, this process is applied, repeatedly, to the remainder of the code sequence.

Not every uniquely decodable code, though, is prefix free; for example, the reader might enjoy precisely describing a decoder for the following code:

u	x
0	0
1	01

Given any uniquely decodable code, however, there always exists a prefix-free code having the same number of codewords, with the same respective lengths. In fact, it can be shown that given any M positive integers $n_1, n_2, \ldots, n_M$, there exists a uniquely decodable code, or prefix-free code, with M binary codewords of lengths $n_1, n_2, \ldots, n_M$, if and only if these numbers satisfy the *Kraft inequality*:

$$\sum_{i=1}^{M} 2^{-n_i} \leq 1$$

Using the Kraft inequality, one can prove the *source coding theorem for variable-length codes* (often called the *noiseless coding theorem*): For any source having an entropy H, if we are given a value $\delta > 0$, then there exists a uniquely decodable variable-length code of rate

$$R \le H + \delta$$

for which every source block is assigned a unique codeword. The *converse* to this theorem, though, states that this result no longer holds if we replace the previous condition with $R \le H - \delta$. Thus, once again, the entropy of a source provides the minimum rate at which it can be reliably encoded. Comparing this result with the theorem for fixed-length codes, we see that the AEP is not required here; moreover, uniquely decodable codes are truly lossless, as compared with the only near-lossless performance of fixed-length codes. The amount of compression that will be obtained by a fixed-length code is known from the outset, however, and can be as close as desired to H bits/source bit. Whereas for a variable-length code, such performance can only be obtained *on average*; that is, it is still possible for the source to emit a particular sequence that will not be compressed so well (and might even be expanded!). In addition, if the codewords are transmitted over some noisy medium before they are decoded, then a variable-length code will be more vulnerable to any errors in the transmission, because these can destroy the uniquely decodable property of the code.

14.5.5 Huffman Coding

We now present some well-established encoding methods that achieve what the aforementioned source coding theorems promise. Continuing our focus on binary-input/binary-output coding, the matter is straightforward for fixed-length codes. Here, given a source, source-block length k, and code length $n \le k$, one merely assigns the 2^n possible codewords $\mathbf{x}$ to the 2^n highest probability source blocks $\mathbf{u}$ one-to-one; then, the remaining $2^k - 2^n$ source blocks are mapped into these same codewords arbitrarily (thus, if $n < k$, some get reused). The best corresponding decoder will then err only when the source emits a block other than one of the 2^n most probable source blocks. Clearly, this encoding/decoding scheme is optimal in that no other scheme meeting the same constraints can yield a smaller probability P_e of decoder error.

For variable-length codes, however, the coding problem is complicated by the need to have a uniquely decodable code. Also, the performance criterion is different; namely, given a source and block length k, a code is considered optimal if and only if it attains the minimum average length $E\{n\}$ over all uniquely decodable codes, thus minimizing the rate R. Obviously, we should assign shorter codewords to higher probability source blocks, and the codeword lengths must satisfy the Kraft inequality. To obtain the actual codewords of an optimal code, one can use the following encoding algorithm due to David A. Huffman, which in addition yields a *prefix-free* code.

We will explain the Huffman algorithm by a simple example. Assuming a binary DMS with $p = 0.1$, and taking $k = 2$, the first two columns of Table 14.3 show the $2^k = 4$ possible source blocks $\mathbf{u}$ and their corresponding probabilities $P(\mathbf{u})$. The algorithm proceeds by first forming a *code tree*, of which the source-block probabilities are the *leaves*. Choosing any smallest two of these probabilities (0.09 and 0.01, the last two probabilities, were chosen, though the second and fourth would have done just as well; whenever the Huffman algorithm allows for more than selection, the choice is immaterial as far as the resulting value of $E\{n\}$ is concerned) we sum them into a new probability (0.10), and indicate this

TABLE 14.3 Huffman Coding Example

u	P(u)	Code Tree	x
00	0.81		0
01	0.09		10
10	0.09		110
11	0.01		111

action on the tree by drawing *branches* from the first two probabilities to a single point corresponding to the new one. This reduction step is then performed repeatedly on the remaining probabilities (thus, the second step applies to 0.81, 0.09, and 0.10), until eventually leading to the total probability of 1 at the tree's *root*. Having completed the tree, the two bits 0 and 1 are assigned (in any order) to each pair of branches that join together. Finally, the codeword **x** assigned to each **u** is obtained by tracing the path through the tree *from its root* back to the appropriate leaf, reading off the bits encountered on the way. In our example, the average codeword length turns out to be $E\{n\} = 1.29$ bit, thereby giving a code rate of $R = 0.645$ bit/source bit. Note that by the source entropy $H \approx 0.469$ bit/source bit given earlier, we have

$$H \le R < H + 1/k$$

This is always the result when the binary Huffman encoding algorithm is applied to a DMS (even if nonbinary); in particular, this fact implies that $R \to H$ as $k \to \infty$. (Indeed, applying the algorithm to the same source with $k = 3$ yields a code for which $R \approx 0.533$ bit/source bit—closer to H.) For an arbitrary source, the algorithm yields

$$E\left\{\log_2 \frac{1}{P(\mathbf{u})}\right\} \le R < E\left\{\log_2 \frac{1}{P(\mathbf{u})}\right\} + 1/k$$

where, by Eq. (14.7), the expectations can be made arbitrarily close to H (if it exists) by taking k sufficiently large; so, again, $R \to H$ as $k \to \infty$ (which by the source coding theorem and converse for fixed-length codes must be the case, since for each k the code is optimal).

For example, the Huffman encoding algorithm may be incorporated into facsimile (fax) machines to speed up the telephone transmission of printed documents. First, each page of the document is sampled along horizontal lines while converting the scanned black and white pixels into 0s and 1s. Then, since the alternating black and white segments of these lines usually comprise many consecutive pixels each (especially the white segments), a simple lossless encoding method called *run-length coding* is applied. Here, the binary sequence for each line is represented by a sequence of numbers telling how many 1s begin the sequence, followed by how many 0s, and then how many more 1s, etc.; for example, 00001110 would become (0, 4, 3, 1). Finally, each of these number sequences is Huffman encoded (according to the presumed statistics of an average document) and transmitted. Of course, the inverses of these operations are applied, in reverse order, at the receiver. The net improvement in transmission speed is by about a factor of 7.

14.5.6 Arithmetic Coding

Ignoring the fact that codewords must have integer lengths, then for a given source and block length k, the average length $E\{n\}$ of a block code constrained by the Kraft inequality is minimized if and only if each source block **u** is assigned a codeword of length $n(\mathbf{u}) = -\log_2 P(\mathbf{u})$. Thus, a reasonable strategy for designing a low-rate uniquely decodable code is to try to obtain $n(\mathbf{u}) \approx -\log_2 P(\mathbf{u})$ for all **u**. The technique of *arithmetic coding*, developed by several researchers, provides a clever way doing this. It is so called because it associates codewords with real numbers; therefore, typical data-processing operations can be expressed as equivalent arithmetic operations (e.g., shifting a bit sequence one position to the left is equivalent to multiplying the corresponding binary number by 2).

Recall that given any infinite binary expansion

$$0.x_1 x_2 \cdots \qquad (x_1, x_2, \ldots, \in \{0, 1\}) \tag{14.11}$$

the *n*-bit truncation

$$0.x_1 x_2 \cdots x_n \tag{14.12}$$

approximates the given number to within an amount somewhere from 0 to less than 2^{-n}; therefore, given any $\delta > 0$, if we take n to be the integer for which

$$-\log_2 \delta \le n < -\log_2 \delta + 1$$

then the n-bit truncation approximates the given number to within an accuracy of δ. Similarly, one can show that given any subinterval $[a, b)$ of the unit interval $[0, 1)$, there exists an n-bit number (Eq. (14.12)) for which

$$n < -\log_2(b - a) + 2$$

such that the number and *all* of its extensions (Eq. (14.11)) lie within $[a, b)$. (For any real numbers α and β, $[\alpha, \beta)$ is the set of all real numbers r such that $\alpha \le r < \beta$.) Accordingly, in arithmetic coding each k-bit source block $\mathbf{u}$ is assigned a subinterval $[a_\mathbf{u}, b_\mathbf{u})$ of $[0, 1)$, where $b_\mathbf{u} - a_\mathbf{u} = P(\mathbf{u})$, such that the subintervals for different source blocks do not overlap; this is possible since

$$\sum_\mathbf{u} P(\mathbf{u}) = 1$$

Then, the codeword $\mathbf{x} = (x_1, \dots, x_n)$ corresponding to each $\mathbf{u}$ is formed from the bits of the shortest binary number (Eq. (14.12)) lying within $[a_\mathbf{u}, b_\mathbf{u})$ for which all of its extensions (Eq. (14.11)) also lie within $[a_\mathbf{u}, b_\mathbf{u})$; or, if there is more than one such number (as happens in the example below), then the smallest one (say) is selected. It follows that we obtain a prefix-free code for which the codeword assigned to each source block $\mathbf{u}$ has a length $n(\mathbf{u}) < -\log_2 P(\mathbf{u}) + 2$. Thus, although such a code need not be optimal (as we will see), its rate R approaches the source entropy H as $k \to \infty$.

A convenient way to uniquely assign subintervals $[a_\mathbf{u}, b_\mathbf{u})$ of $[0, 1)$ to source blocks $\mathbf{u}$ as stipulated above is to utilize the cumulative distribution function

$$F(\mathbf{u}') \overset{\triangle}{=} \Pr(\mathbf{u} < \mathbf{u}') = \sum_{\mathbf{u} < \mathbf{u}'} P(\mathbf{u}) \quad \text{(all source blocks } \mathbf{u}')$$

where $\mathbf{u} = (u_1, \dots, u_k)$ is considered to be less than $\mathbf{u}' = (u_1', \dots, u_k')$ if and only if the corresponding binary number $0.u_1 u_2 \cdots u_k$ is less than the binary number $0.u_1' u_2' \cdots u_k'$; namely, we let

$$[a_\mathbf{u}, b_\mathbf{u}) = [F(\mathbf{u}), F(\mathbf{u}) + P(\mathbf{u})) \quad \text{(all } u) \tag{14.13}$$

Proceeding this way, the arithmetic encoding of a binary DMS with $p = 0.1$ produces the results shown in Table 14.4 when $k = 2$. (Regarding the codeword $\mathbf{x}$ for $\mathbf{u} = 10$, note that the 5-bit number 0.11110 and all of its extensions lie within the subinterval $[a_\mathbf{u}, b_\mathbf{u})$, whereas no shorter binary number has this property; however, the smallest 5-bit number having this property is 0.11101.) In Table 14.4 the average code-length $E\{n\} = 1.69$ bit, making the rate $R = 0.845$ bit/source bit. Comparing this with the optimum code obtained earlier via Huffman encoding for this same source and k, we see that an arithmetic code need not be optimal; in particular, it can be wasteful for low-probability source blocks (the last codeword could be shortened by 3 bit without upsetting the prefix condition).

Nevertheless, arithmetic coding has an implementation advantage in that it can be performed *recursively* as bits enter the encoder; because, using Eq. (14.13), the subinterval $[a_{\mathbf{u},i}, b_{\mathbf{u},i})$ assigned to the i-bit source block $(u_1, \dots, u_i)$ gradually shrinks as i increases, until we finally obtain $[a_\mathbf{u}, b_\mathbf{u}) = [a_{\mathbf{u},k}, b_{\mathbf{u},k})$ when $i = k$.

TABLE 14.4   Arithmetic Coding Example

$\mathbf{u}$	$P(\mathbf{u})$	$\approx -\log_2 P(\mathbf{u})$	$[a_\mathbf{u}, b_\mathbf{u})$	$\mathbf{x}$
00	0.81	0.3	$[0, 0.81)$	0
01	0.09	3.5	$[0.81, 0.9)$	1101
10	0.09	3.5	$[0.9, 0.99)$	11101
11	0.01	6.6	$[0.99, 1)$	1111111

TABLE 14.5 Arithmetic Coding Done Recursively

i	u_i	$[a_{\mathbf{u},i}, b_{\mathbf{u},i})$
1	0	$[0, 0.9)$
2	1	$[0.81, 0.9)$
3	1	$[0.891, 0.9)$
4	0	$[0.891, 0.8991)$

Specifically, for a binary DMS with probability p, we can define the initial values of $a_{\mathbf{u},0} = 0$ and $b_{\mathbf{u},0} = 1$, and then for each $i = 1, \ldots, k$ let

$$a_{\mathbf{u},i} = a_{\mathbf{u},i-1} + u_i(1 - p)(b_{\mathbf{u},i-1} - a_{\mathbf{u},i-1}) \tag{14.14a}$$

and

$$b_{\mathbf{u},i} = b_{\mathbf{u},i-1} - (1 - u_i)p(b_{\mathbf{u},i-1} - a_{\mathbf{u},i-1}) \tag{14.14b}$$

as determined by the bits u_i of the emitted source block $\mathbf{u} = (u_1, \ldots, u_k)$. For example, if $p = 0.1$ and $\mathbf{u} = 0110$ ($k = 4$), we obtain the values shown in Table 14.5 (notice the agreement between $[a_{\mathbf{u},2}, b_{\mathbf{u},2})$ in the table and the subinterval $[a_{\mathbf{u}}, b_{\mathbf{u}})$ given in Table 14.4 for $\mathbf{u} = 01$). Therefore, $[a_{\mathbf{u}}, b_{\mathbf{u}}) = [0.891, 0.8991)$, from which the appropriate codeword $\mathbf{x}$ can be obtained. In particular, unlike Huffman encoding, which typically must wait for much (if not all) of the source block $\mathbf{u}$ before it can it can begin to emit the corresponding codeword, here after just the first two source bits we can determine that the encoder output must begin with 11, since only then can the eventual binary number $0.x_1 x_2 \cdots x_n$ lie within the intermediate subinterval $[a_{\mathbf{u},2}, b_{\mathbf{u},2})$ and, thus, within the final subinterval $[a_{\mathbf{u}}, b_{\mathbf{u}})$.

Similarly, the corresponding decoder can also be designed to operate recursively as it comes to know the codeword $(x_1, \ldots, x_n)$ bit-by-bit; for after the first $j \leq n$ code bits are known, the decoder can emit as many as possible of its output bits v_i ($i = 1, \ldots, k$) such that each v_i is that value of u_i in Eq. (14.14) for which the binary number $0.x_1 x_2 \cdots x_j$ and all of its extensions lie within the subinterval $[a_{\mathbf{u},i}, b_{\mathbf{u},i})$. Hence, arithmetic coding has an implementation advantage over Huffman coding when the block length k is large. Moreover, should one decide to change k, then a new Huffman code would need to be calculated from scratch, whereas the encoder and decoder for arithmetic coding could be easily modified.

14.5.7 Universal Codes

In contrast to the preceding, suppose we have a binary DMS for which the probability distribution of the output is unknown; that is, we do not know the value of p and, thus, are unable to apply either the Huffman or arithmetic encoding algorithms. We now show that one can nevertheless design an *adaptive* lossless encoding/decoding scheme such that, as the block-length $k \to \infty$, the code rate R approaches the source entropy H (whatever it happens to be). That is, there exists a *universal code* for all sources of this type. The following is but one way to proceed.

First, for the given value of k, the encoder observes the k-bit source block $\mathbf{u}$ emitted by the source, from which it forms an estimate $\hat{p}$ of p

$$\hat{p} = \langle \mathbf{u} \rangle / k \tag{14.15}$$

where $\langle \mathbf{u} \rangle$ is the number of 1s in $\mathbf{u}$. Then, the encoder performs the arithmetic encoding algorithm under the assumption that $\hat{p}$ is, indeed, the value of p; let the resulting codeword be $\hat{\mathbf{x}}$. Finally, to enable the decoder to reproduce $\mathbf{u}$ from $\hat{\mathbf{x}}$, the encoder essentially stores *both* $\hat{p}$ and $\hat{\mathbf{x}}$; specifically, its output is the composite codeword

$$\mathbf{x} = \langle \mathbf{u} \rangle_b \circ \hat{\mathbf{x}},$$

where $\langle \mathbf{u} \rangle_b$ is an N-bit binary representation of $\langle \mathbf{u} \rangle$ for some fixed number N and $\circ$ indicates concatenation. Clearly, since both k and N are known to the decoder, it can reverse these steps to determine $\mathbf{u}$ from $\mathbf{x}$; moreover, since the arithmetic code for each $\hat{p}$ is prefix free, the code is uniquely decodable.

To show that the performance claimed is indeed obtained, recall from the previous section that for each source block $\mathbf{u}$, the length $\hat{n}(\mathbf{u})$ of the intermediate codeword $\hat{\mathbf{x}}$ satisfies the bound

$$\hat{n}(\mathbf{u}) < -\log_2 \hat{P}(\mathbf{u}) + 2 \quad \text{(all } \mathbf{u}\text{)}$$

where $\hat{P}(\mathbf{u})$ is the probability distribution for $\mathbf{u}$ under the assumption that $\hat{p}$ is the value of p,

$$\hat{P}(\mathbf{u}) = \hat{p}^{\langle \mathbf{u} \rangle}(1 - \hat{p})^{k - \langle \mathbf{u} \rangle} \quad \text{(all } \mathbf{u}\text{)}$$

Combining these facts, and using Eq. (14.8) and Eq. (14.15), we obtain

$$\hat{n}(\mathbf{u}) < k H_2(\hat{p}) + 2 \quad \text{(all } \mathbf{u}\text{)}$$

Also note that the estimate $\hat{p}$ in Eq.(14.15) is just the relative frequency of 1s in $\mathbf{u}$; thus, by the law of large numbers, we have

$$\Pr(\hat{p} \approx p) \approx 1 \quad \text{for } k \text{ large}$$

Therefore

$$E\{H_2(\hat{p})\} \approx H_2(p) = H \quad (k \text{ large})$$

More precisely, a rigorous version of this argument would show that for some $\delta > 0$, we have

$$|E\{H_2(\hat{p})\} - H| \le \delta$$

where $\delta \to 0$ as $k \to \infty$. Finally, since the possible values of $\langle \mathbf{u} \rangle$ are $0, 1, \ldots, k$, it is sufficient for the storage of $\langle \mathbf{u} \rangle_b$ that N be the largest integer for which $N \le \log_2 k + 1$. Hence, altogether, the average length of the composite codeword $\mathbf{x}$ is

$$E\{n\} = E\{\hat{n}\} + N < E\{k H_2(\hat{p}) + 2\} + (\log_2 k + 1) \le k(H + \delta) + \log_2 k + 3$$

Therefore, dividing by k, we obtain

$$R < H + \delta + (\log_2 k + 3)/k \to H \quad \text{as } k \to \infty$$

and from this it follows that R (which by the converse to the variable-length source coding theorem cannot be less that H) must approach H as $k \to \infty$.

14.5.8 Lempel–Ziv Coding

Of all of the source codes presently in use, the most important are the following two, commonly abbreviated *LZ77* and *LZ78*, which were published by Abraham Lempel and Jacob Ziv in 1977 and 1978. Though distinct (it is incorrect to speak generally of *the* Lempel–Ziv encoding algorithm), these two lossless codes are structurally similar in that they both perform data compression by storing a sequence of *pointers* that cite *phrases*, which are previous segments of the source-sequence, within a variable *dictionary* maintained at both the encoder and the decoder. The idea behind such a scheme is that any source that is not completely random (e.g., a binary DMS with an entropy of less than 1 bit/source bit) will often duplicate some of its past behavior; when a duplication is sufficiently long, it is more efficient to store a reference (pointer) to a previous occurrence rather than the duplication itself. The widespread use of these codes is due to their speed, as well as their being universal for all stationary ergodic sources (which need not be memoryless, as was assumed in the previous section).

The algorithm LZ77 operates by passing the source block $\mathbf{u} = (u_1, \ldots, u_k)$ through a sliding *window* of a fixed width w, initially filled with 0s. Assuming that, for some $i \geq 0$, the first i source bits $u_1, \ldots, u_i$ have already been encoded, then the window contains the last w of these bits; that is, defining $u_j = 0\,(j \leq 0)$ in order to conveniently express the initial 0s, the window is the w-bit block $(u_{i-w+1}, \ldots, u_i)$. In a practical implementation, the window would be followed by a finite buffer containing the next several yet to be encoded source bits; the dictionary would be comprised of all of the source blocks that start somewhere within the window and end somewhere before the last bit of the buffer. However, to streamline the explanation that follows, the buffer is omitted; thus, the dictionary consists of every finite source block starting within the window.

The algorithm matches as much of the remainder $(u_{i+1}, \ldots, u_k)$ of $\mathbf{u}$ as possible to a phrase in the dictionary. Specifically, using a pointer $p \in \{1, \ldots, w\}$ to tell where (in terms of how many bits before u_{i+1}) in the window the phrase begins, along with the length $l \in \{1, \ldots, k - i - 1\}$ of the phrase, the encoder chooses the values of p and l to obtain a match

$$(u_{i+1-p}, \ldots, u_{i+l-p}) = (u_{i+1}, \ldots, u_{i+l})$$

(where the block on the left is from the dictionary). In addition, p and l are chosen to make l as large as possible; or, if no such match can be obtained, then the encoder sets $p = l = 0$. The first source bit u_{i+l+1} not included in the matching block is now annexed onto it to yield the extended match

$$(u_{i+1}, \ldots, u_{i+l+1})$$

of length $l + 1$; hence, for the case of no match, the extended match is just the single bit u_{i+1}.

All of the pertinent information about the extended match is contained in the ordered triple

$$(\text{starting point, length, last bit}) = (p, l + 1, u_{i+l+1})$$

because, using just this information and the previously decoded bits $u_1, \ldots, u_i$, a decoder that mirrors the encoding process (i.e., keeps track of the window) can determine the present extended match. Therefore, using an efficient uniquely decodable code, the triple is encoded into a binary codeword that is stored as the next part of the overall codeword $\mathbf{x} = (x_1, \ldots, x_n)$. Finally, the encoder slides the window down the source block $\mathbf{u}$ to make the last bit of the extended match the new last bit of the window. Repeating this process, parsing $\mathbf{u}$ into a succession of extended matches and encoding the corresponding triples, the entire source block is eventually encoded.

As an example of applying LZ77, suppose we wish to encode the source block

$$\mathbf{u} = 101010100110 \tag{14.16}$$

with a window width $w = 5$. Using an overbar to indicate the window, an up-arrow to indicate the pointer, and boldface type to indicate the extended match, the algorithm sequentially produces triples as follows:

$$\overline{00000}\mathbf{1}01010100110 \rightarrow (0, 1, 1)$$
$$\overline{00000}\mathbf{10101}0100110 \rightarrow (2, 8, 0)$$
$$00000101010\overline{10}0110 \rightarrow (3, 2, 1)$$
$$0000010101001\overline{00110} \rightarrow (0, 1, 0)$$

thus, the resulting codeword is

$$\mathbf{x} = (0, 1, 1)_b \circ (2, 8, 0)_b \circ (3, 2, 1)_b \circ (0, 1, 0)_b$$

where the subscript b indicates the binary encoding of the triple. Since the first source bit in this example is 1, no match is possible at the start. The second step shows that a phrase and its match can overlap, and even be longer than window; nevertheless, the reader (acting as the decoder) should be able to calculate the second extended match from the corresponding triple and window alone. The third step shows that more than one value of the pointer is sometimes possible (viz., we could have taken $p = 5$). Finally, since only one unencoded source bit is left for the last extended match, no match is performed, and the bit becomes the extended match.

In LZ78, the source block **u** is again parsed into segments; however, there is no window. Instead, as each segment is identified, it becomes a new phrase in an ever growing dictionary (initially empty) by which the encoder performs the parsing. Namely, after the first $i \geq 0$ source bits $u_1, \ldots, u_i$ have been encoded, the encoder determines the largest length $l \in \{1, \ldots, k - i - 1\}$ for which the subsequent block $(u_{i+1}, \ldots, u_{i+l})$ matches a phrase in dictionary; then, the first source bit u_{i+l+1} not included in the matching block is annexed onto it to form the extended match

$$(u_{i+1}, \ldots, u_{i+l+1})$$

which is expressed as the order-pair

$$\text{(phrase number, last bit)} = (p, u_{i+l+1}) \tag{14.17}$$

Here, the pointer p gives the number (in order of occurrence) of the phrase in the dictionary which provided the match; or, if no match is possible, then $p = 0$ and the extended match is just u_{i+1}. (Of course, a practical encoder has a finite memory and, therefore, must place an upper limit on p.) Using an efficient uniquely decodable code, the pair is encoded into a binary codeword that is stored as the next part of the overall codeword $\mathbf{x} = (x_1, \ldots, x_n)$; finally, the extended block is added to the dictionary as a new phrase. This process is repeated until **u** is exhausted.

Applying LZ78 to the source-block in Eq. (14.16), we obtain the parsing

$$\mathbf{u} = 1 \circ 0 \circ 0 \circ 10 \circ 101 \circ 00 \circ 11 \circ 0$$

Because, for instance, when the fourth match is attempted, the dictionary is $\{1, 0, 10, 101\}$ and the unencoded portion of **u** is 00110; therefore, the longest possible match is with the second phrase 0 (i.e., $p = 2$), thereby making the extended match 00 (the next phrase). The final codeword is then

$$\mathbf{x} = (0,1)_b \circ (0,0)_b \circ (1,0)_b \circ (3,1)_b \circ (2,0)_b \circ (1,1)_b \circ (0,0)_b$$

As before, it happened that only one unencoded source bit was left for the last extended match, and so no match was performed.

Because the **u** given in Eq. (14.16) is rather short, these runs of LZ77 and LZ78 do not display the compression to be expected with longer source blocks. For another example, then, consider the encoding of the all-0 sequence $\mathbf{u} = 00 \cdots$; here, LZ78 gives

$$\mathbf{u} = 0 \circ 00 \circ 000 \circ 0000 \circ 00000 \circ 000000 \circ \cdots \quad \Rightarrow$$
$$\mathbf{x} = (0,0)_b \circ (1,0)_b \circ (2,0)_b \circ (3,0)_b \circ (4,0)_b \circ (5,0)_b \circ \cdots \tag{14.18}$$

Also, suppose that each pair from Eq.(14.17) is encoded as

$$(p, u_{i+l+1})_b = (N\text{-bit binary representation of } p) \circ u_{i+l+1}$$

where N is the largest integer for which

$$N \leq \log_2[(\text{number of phrases currently in dictionary}) + 1] + 1$$

then (as the reader can verify), the general encoding from **u** to **x** is uniquely decodable, whereas for the present example we have

$$\mathbf{x} = 00 \circ 010 \circ 100 \circ 1100 \circ 1000 \circ 1010 \circ \cdots = 00010100110010001010 \cdots \tag{14.19}$$

Comparing Eq. (14.19) with Eq. (14.18), we see that $\mathbf{x}$ eventually grows slower than $\mathbf{u}$; indeed, for each $j \in \{1, 2, \ldots\}$, the jth phrase of $\mathbf{u}$ has length j, whereas for the respective addition to $\mathbf{x}$ we have

$$(\text{length of } j\text{th encoded pair}) \leq \log_2 j + 2$$

It follows that even though the dictionary and bit allocation N grow indefinitely, we obtain $n/k \to 0$ as $k \to \infty$; this is as expected, since we are essentially compressing the output of a binary DMS with $p = 0$, for which the entropy is 0 bit/source bit.

14.5.9 Rate-Distortion Theory

All of our discussion thus far has been directed toward coding that achieves lossless or near-lossless performance at rates R that are above, but perhaps arbitrarily close to, the source entropy H. What if we attempt to achieve even greater compression—that is, encode at some $R < H$? Then, according to the coding theorem converses already discussed, such performance is no longer possible. Instead, the best we can hope for is that the sink sequence $(v_1, v_2, \ldots)$ will replicate the source-sequence $(u_1, u_2, \ldots)$ with a minimal amount of *distortion*. The study of how well this can be done is called *rate-distortion theory*.

To express the fidelity with which a particular k-bit sink block $\mathbf{v} = (v_1, \ldots, v_k)$ represents a given k-bit source block $\mathbf{u} = (u_1, \ldots, u_k)$, we first introduce a digitwise *distortion measure*; this is any function $d(u, v)$ of $u, v \in \{0, 1\}$ for which

$$d(u, v) \geq 0 \quad (\text{all } u, v)$$

Then, respectively averaging over all bits of the two blocks, we obtain the blockwise distortion

$$d(\mathbf{u}, \mathbf{v}) \triangleq \frac{1}{k} \sum_{i=1}^{k} d(u_i, v_i) \quad (\text{all } \mathbf{u}, \mathbf{v})$$

Finally, averaging over the statistics of source, we obtain the *average distortion* $\bar{d}$ between $\mathbf{u}$ and $\mathbf{v}$,

$$\bar{d} \triangleq E\{d(\mathbf{u}, \mathbf{v})\} = \sum_{\mathbf{u}} P(\mathbf{u})[\mathbf{u}, \mathbf{v}(\mathbf{u})]$$

where we have explicitly indicated on the right the functional dependence of $\mathbf{v}$ on $\mathbf{u}$, as determined by the encoder and decoder. Just what distortion measure is chosen depends on the application; for example, a common choice is

$$d(u, v) = \begin{cases} 0, & u = v \\ 1, & u \neq v \end{cases} \quad (\text{all } u, v) \tag{14.20}$$

for which

$$d(\mathbf{u}, \mathbf{v}) = \frac{\text{number of bits in which } \mathbf{u} \text{ and } \mathbf{v} \text{ differ}}{k} \quad (\text{all } \mathbf{u}, \mathbf{v})$$

thereby making $\bar{d}$ the bit-error rate (BER) produced by the processing of the encoder/decoder. (Having focused on binary sources, the possible distortion measures $d(\cdot, \cdot)$ are rather limited. If instead, for example, the source emitted a sequence of *bytes*, then one might choose a distortion measure that more heavily weights an error in the most significant bit.)

To proceed further, we define the *mutual information* $I(\mathbf{u}; \mathbf{v})$ between any two random blocks $\mathbf{u}$ and $\mathbf{v}$, with joint probability-distribution $P(\mathbf{u}, \mathbf{v})$ and individual probability-distributions $P(\mathbf{u})$ and $P(\mathbf{v})$, by

$$I(\mathbf{u}; \mathbf{v}) \triangleq E\left\{ \log_2 \frac{P(\mathbf{u}, \mathbf{v})}{P(\mathbf{u})P(\mathbf{v})} \right\} = \sum_{\mathbf{u}, \mathbf{v}} P(\mathbf{u}, \mathbf{v}) \log_2 \frac{P(\mathbf{u}, \mathbf{v})}{P(\mathbf{u})P(\mathbf{v})}$$

As the name suggests, $I(\mathbf{u}; \mathbf{v})$ provides a measure of the information (entropy, randomness) that $\mathbf{u}$ and $\mathbf{v}$ have in common. Note that once the distribution $P(\mathbf{u})$ and the conditional probability distribution $P(\mathbf{v} \mid \mathbf{u})$ are specified, then the distributions $P(\mathbf{u}, \mathbf{v})$ and $P(\mathbf{v})$ follow; hence, $I(\mathbf{u}; \mathbf{v})$ can be thought of as a function of $P(\mathbf{u})$ and $P(\mathbf{v} \mid \mathbf{u})$. In the present situation, $P(\mathbf{u})$ is given by the source, whereas

$$P(\mathbf{v} \mid \mathbf{u}) = \begin{cases} 1, & \mathbf{v} = \mathbf{v}(\mathbf{u}) \\ 0, & \text{otherwise} \end{cases} \quad (\text{all } \mathbf{u}, \mathbf{v})$$

since $\mathbf{v}$ is a deterministic function of $\mathbf{u}$. However, fixing $P(\mathbf{u})$, suppose we let $P(\mathbf{v} \mid \mathbf{u})$ vary over *all* possible conditional probability distributions for $\mathbf{v}$ given $\mathbf{u}$; then, the average distortion between $\mathbf{u}$ and $\mathbf{v}$ would be calculated as

$$\bar{d} = \sum_{\mathbf{u}, \mathbf{v}} P(\mathbf{u}) P(\mathbf{v} \mid \mathbf{u}) d(\mathbf{u}, \mathbf{v})$$

From here, the *rate-distortion function* $R(d)$ for the given source and distortion measure would be defined as

$$R(d) \triangleq \lim_{k \to \infty} \frac{1}{k} \min_{P(\mathbf{v} \mid \mathbf{u}): \bar{d} \le d} I(\mathbf{u}; \mathbf{v}) \quad \text{bits/source bit} \tag{14.21}$$

(for those values $d \ge 0$ for which the limit exists), where the minimum is performed over all conditional distributions $P(\mathbf{v} \mid \mathbf{u})$ that result in an average distortion $\bar{d} \le d$.

The significance of the rate-distortion function stems from the *source coding theorem with a fidelity criterion*, which states that for any stationary ergodic source and accompanying distortion measure, if we are given a distortion d for which $R(d)$ is defined, and a value $\delta > 0$, then there exists a fixed-length code of rate

$$R \le R(d) + \delta$$

and a corresponding decoder, which together yield an average distortion

$$\bar{d} \le d + \delta$$

However, the *converse* to this theorem states that we must always have $R \ge R(\bar{d})$. Observing that it follows immediately from Eq. (14.21) that $R(d)$ is nonincreasing vs. d (as one would expect, since increasing the allowable distortion should not require a higher encoding rate) the theorem and converse roughly say that $R(d)$ is the minimum rate R at which source coding can attain a given distortion d.

As a simple example, consider a binary DMS with $p = \frac{1}{2}$ (which has an entropy of $H = 1$ bit/source bit), and take the distortion measure given in Eq. (14.20). It can then be shown that the resulting rate-distortion

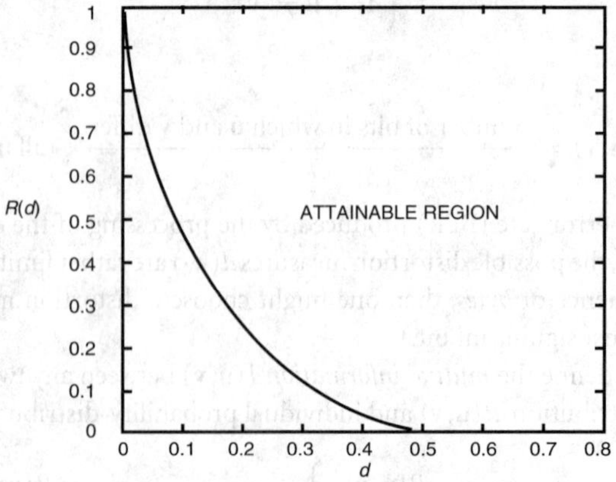

FIGURE 14.35   Example of a rate-distortion function.

function $R(d)$ is defined for all $d \geq 0$, and

$$R(d) = \begin{cases} 1 - H_2(d), & 0 \leq d < \frac{1}{2} \\ 0, & d \geq \frac{1}{2} \end{cases}$$

As shown in Fig. 14.35, we have $R(d) \leq H$ (all d), as one would expect from the fixed-length source coding theorem. At the lower extreme, the minimum value of $R(d) = 0$ is first obtained at $d = \frac{1}{2}$, which is also what one would expect, because no storage is necessary to achieve $\bar{d} = \text{BER} = \frac{1}{2}$. Also, for instance, since $R(0.12) \approx 0.471$ bit/source bit, it follows that a BER of 12% can be obtained by encoding at a rate $R = \frac{1}{2}$ bit/source bit, which is much better than the best BER (25%) attainable by simply discarding every other bit.

Defining Terms

Ergodic random processs: A random process that, with probability 1, never locks into a mode that is unrepresentative of its overall statistics.

Markov chain: A discrete random process having a memory of 1 time unit; that is, if the output of the process is the sequence $(u_1, u_2, \ldots)$, then

$$P(u_{i+1} \mid u_1, \ldots, u_i) = P(u_{i+1} \mid u_i) \qquad (i = 1, 2, \ldots)$$

The chain is *time invariant* if and only if these conditional probabilities are the same for all i; it is then *indecomposable* if and only if there is an output value u such that for all output values u', we have $\Pr(u_{i+j} = u \mid u_i = u') > 0$ (where i is arbitrary) for some $j \in \{1, 2, \ldots\}$.

Stationary random process: A random process for which the statistics do not vary with time.

References

Bell, T.C., Cleary, J.G., and Witten, I.C. 1990. *Text Compression*. Prentice-Hall, Englewood Cliffs, NJ.

Cover, T.M. and Thomas, J.A. 1991. *Elements of Information Theory*. Wiley, New York.

Gallager, R.G. 1968. *Information Theory and Reliable Communication*. Wiley, New York.

Lucky, R.W. 1989. *Silicon Dreams*. St. Martin's Press, New York.

Mansuripur, M. 1987. *Introduction to Information Theory*. Prentice-Hall, Englewood Cliffs, NJ.

Pierce, J.R. 1980. *An Introduction to Information Theory: Symbols, Signals and Noise*, 2nd, revised ed. Dover, New York.

Shannon, C.E. and Weaver, W. 1949. *The Mathematical Theory of Communication*. University of Illinois Press, Chicago, IL.

Wyner, A.D. and Ziv, J. 1994. The sliding-window Lempel–Ziv algorithm is asymptotically optimal. *Proc. of the IEEE* 82(June):872–877.

Further Information

The theoretical foundation of data compression comprises the source coding portion of *information theory*. The book by Pierce provides an elementary introduction to this field, although a more advanced, yet accessible, introduction can be found in the concise volume by Mansuripur; for a complete treatment, the reader is referred to the works by Cover and Thomas, Gallager, and others. Also worth mentioning is the still interesting book by Shannon and Weaver, in which the original paper on information theory by Claude Shannon appears, preceded by an introduction from the second author. The popular book by Lucky discusses data compression in the context of text, speech, and video compression. As for specific implementation details of the various coding methods, as well as historical perspective, the work by Bell, Cleary, and Witten is recommended.

Current research on data compression can be found in the related journals and conferences of the Institute of Electrical and Electronics Engineers (IEEE) and the Association for Computing Machinery (ACM); in particular, see the *IEEE Transactions on Information Theory* and the *Journal of the ACM*.

15

Wired Communications Systems

Tsong-Ho Wu
Transtech Networks, Inc.

Brent Allen
Bell Northern Research

Rodger E. Ziemer
University of Colorado

15.1 Network Switching Concepts

Tsong-Ho Wu

15.1.1 Introduction

Switching is the control or routing of signals in either physical or virtual circuit form to transmit data between specific points in a network. A switched network is a communication network that has switching facilities or centers to perform the function of message switching, packet switching, or circuit switching. The primary purpose of the network switching function is to maximize the transmission system efficiency by sharing the capacity of the transmission system among communications circuits.

In general, there are two major types of network switching techniques: packet switching and circuit switching. The message switching system may be viewed as a special case of the packet switching system. The traditional circuit switching concept is designed to handle and transport stream-type traffic, such as voice and video. A circuit-switched connection is set up with a fixed bandwidth for the duration of a connection, which provides fixed throughput and constant delay. In other words, network resources are dedicated in circuit-switched systems. In contrast, packet switching is primarily designed to efficiently carry data communications traffic. Its characteristics are access with buffering, statistical multiplexing, and variable throughput and delay, which is a consequence of the dynamic sharing of communication resources

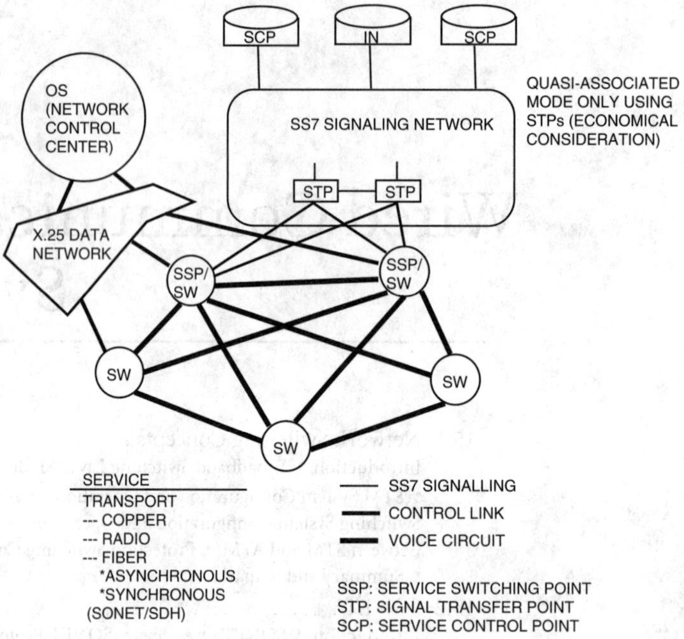

FIGURE 15.1   Today's telecommunications network infrastructure.

to improve utilization. Today, it is used exclusively for data applications. Examples of circuit-switched networks include telephony networks and private line networks. Examples of packet-switched networks include X.25 (with variable packet sizes) and **asynchronous transfer mode (ATM)** networks (with the fixed-cell size).

Telecommunications services are primarily supported by three separate switched networks using two major switching concepts. As depicted in Fig. 15.1, the circuit-switched network supports voice transport, which includes copper, radio, and/or fiber transmission systems using asynchronous or synchronous optical network (SONET) transmission technology. The signaling system needed to support switched voice services is supported by a specialized (delay-sensitive) packet-switched SS7 network. The network supporting the access to the operations systems from the service transport system is a packet-switched X.25 network.

In the future telecommunication network infrastructure, both the narrowband services of today and emerging broadband services will be supported by an integrated B-integrated services digital network (B-ISDN). ATM has been adopted by the International Telecommunications Union (ITU-T) as the core technology to support B-ISDN transport. ATM is a high-speed, integrated multiplexing and switching technology that transmits information using fixed-length cells in a connection-oriented manner. Physical interfaces for the user-network interface (UNI) of 155.52 Mb/s and 622.08 Mb/s provide integrated support for high-speed information transfer and various communication modes, such as circuit and packet modes, and constant/variable/burst bit-rate communications.

In the ATM network, services, control, signaling, and operations and maintenance (OAM) messages are carried by the same physical network, but they are logically separated from each other by using different virtual channels (VCs) or virtual paths (VPs) with different quality of service (QoS) requirements.

ATM is growing and maturing rapidly. It has been implemented today in products by many equipment suppliers and deployed by customers who anticipate such business advantages as:

- Enabling high-bandwidth applications including desktop video, digital libraries, and real-time image transfer

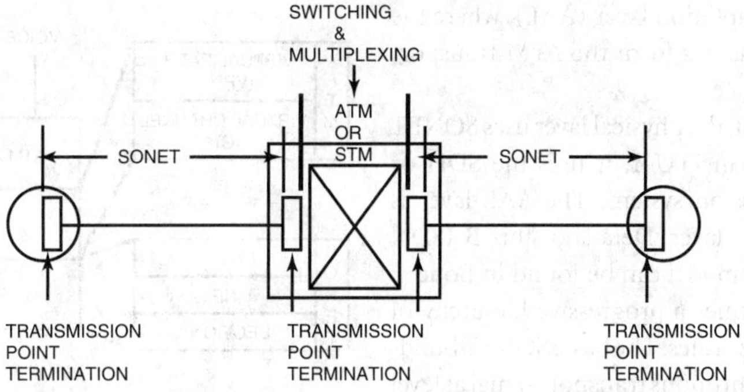

FIGURE 15.2 Relationship between SONET transmission, STM, and ATM switching.

- Coexistence of different types of traffic on a single network platform to reduce both the transport and operations costs
- Long-term network scalability and architectural stability

In addition, ATM has been used in both local and wide area networks. It can support a variety of high layer protocols and will cope with future network speeds of gigabits per second.

In North America standards, as specified by the American National Standards Institute (ANSI) T1S1 committee, ATM cells are transported over the SONET system as the physical layer. (Note that in ITU-T standards, there are two options for the transmission standards: synchronous digital hierarchy (SDH) and cell-based transmission. SDH is functionally equivalent to SONET, although some of the rates are different and the use of some overhead and multiplexing schemes may also be different. For the cell-based transmission, ATM cells will not be accommodated by the SDH frame within the transmission link.) SONET is the North America's standard synchronous transmission network technology for optical signal transmission (Ballart and Ching, 1989). In this multilayer network environment, the switching of the service connection may be implemented at the SONET layer using synchronous transfer mode (STM) or at the ATM layer using ATM. STM is a time division multiplexing and switching technology that uses a similar switching concept as traditional circuit switching systems.

Figure 15.2 depicts the relationship between SONET transmission, STM, and ATM switching. The end-to-end ATM connection is established and transported through a set of transmission links that are terminated at nodes for processing and switching (called transfer mode). The nodal transfer mode may be existing STM or emerging ATM, depending on types of services and QoS being supported. STM switching technology has been widely deployed in today's telecommunications networks, whereas ATM is expected to be widely deployed in the near future. Without loss of generality, we will assume in this section that STM is operated on the SONET system.

To help understand how the broadband SONET/ATM network is used to support high-speed data and video services (discussed subsequently), this section focuses on reviewing switching concepts used in the SONET layer (STM switching) and the ATM layer (ATM switching). The switching concept is based on the characteristics of the signal switching format (e.g., time slot or virtual connection). Thus, this section will first review signal switching formats and their bandwidth relationship in a multilayer SONET/ATM network environment. The operations of STM and ATM switching concepts will then be explained. The protection switching concept for SONET/ATM networks will also be discussed.

15.1.2 Broadband Switching Layer Model and Formats

Multilayer Transport Model

Figure 15.3 shows the B-ISDN signal transport protocol reference model defined in CCITT Rec. I.321. [The International Telegraph and Telephone Consultative Committee (CCITT) has been renamed the International Telecommunication Union (ITU) since 1994.] Three layers are defined in the model: physical,

ATM, and ATM adaptation layer (AAL), where the physical and ATM layers form the ATM transport platform.

In North America, the physical layer uses SONET standards, whereas in ITU-T, it uses the SDH or cell-based transmission system. The AAL layer is a service-dependent layer. Details of this B-ISDN transport reference model can be found in Boudec [1992]. SONET defines a progressive hierarchy of optical signal and line rates. The basic SONET building block is the synchronous transport signal at level 1 (STS-1) signal operating at 51.840 Mb/s. Higher rate signals (STS-N) are multiples (N) of this basic

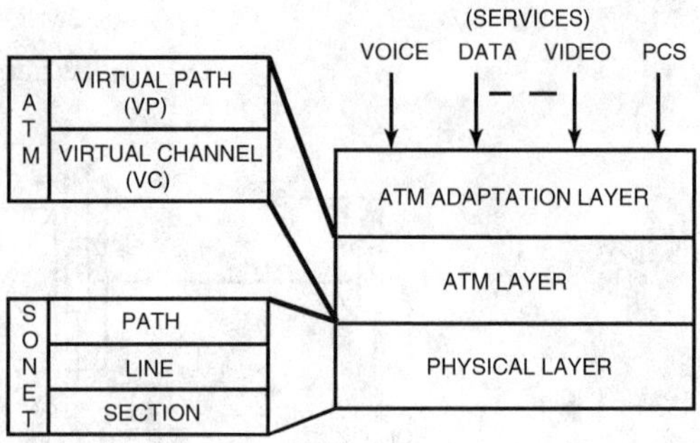

FIGURE 15.3 B-ISDN reference model.

rate. Values of N currently used in standards are 1, 3, 12, 24, and 48, with 192 likely in the near future. Broadband services requiring more than one STS-1 payload capacity are transported by concatenated STS-1s. The function of the SONET layer is to carry ATM cells or non-ATM frames (e.g., STS-1) in a high-speed and transparent manner, as well as to provide very fast protection switching capability to ATM cells or non-ATM frames whenever needed. SONET offers opportunities to implement new network architectures in a cost-effective manner (Wu, 1992) and has been widely deployed by local exchange carriers (LECs) since 1991.

The major function of the ATM layer is to provide fast multiplexing and routing for data transfer based on the header information. The ATM layer is divided into two sublayers: virtual path and virtual channel. A VC is a generic term used to describe a unidirectional communication capability for the transport of ATM cells. The VC identifies an end-to-end connection that can be set via either provisioning (permanent VC) or near real-time call setup (switched VC). Either way the bandwidth capacity of that connection is determined in real-time depending on traffic to be transported over that connection.

The VP accommodates a set of different VCs, all with the same terminations (source and destination). The VPs are managed by network systems, whereas VCs can be managed, end-to-end, by users with ATM terminals. For example, a business user may be provisioned with a VP to another user location to provide the equivalent of leased circuits, and another VP to the serving central office for switched services. Each VP may include several VCs for wide area network (WAN), MAN, private branch exchange (PBX), and video conference traffic.

The VC and VP are identified through virtual channel identifier (VCI) and virtual path identifier (VPI), respectively. For large networks, VCIs and VPIs are assigned on a per link basis (i.e., local significance). Thus, translation of VPI (or VCI) is necessary at intermediate ATM switches for an end-to-end VP (or VC). The switching operation for ATM cells will be discussed in Sec. 15.1.4.

Figure 15.4 shows the relationship among the virtual channel, virtual path, SONET's STS path, and SONET link. The SONET physical link (e.g., an optical carrier level-48 (OC-48) optical system) provides a bit stream to carry digital signals. This bit stream may include one of multiple digital paths, such as SONET STS-3c, STS-12c, or STS-48c. Each digital path may carry ATM cells via multiple VPs, each having multiple VCs. The switching method used for SONET's STS paths is STM using a hierarchical **time slot interchange** (**TSI**) concept, whereas the switching method for VPs/VCs uses a nonhierarchical ATM switching concept. In addition, STM performs network rerouting through physical network reconfiguration, whereas ATM performs network rerouting using logical network reconfiguration through the update of the routing table.

Physical Path vs. Virtual Path

The switching principles used for SONET's STS Paths (STM) and ATM VPs/VCs (ATM) are completely different due to different characteristics of corresponding path structures. The STS path uses a physical path structure, whereas the VP/VC uses a logical path structure, as depicted in Fig. 15.5. The physical path concept used in the SONET STM system has a hierarchical structure with a fixed capacity for each physical path. For example, the VT1.5 and STS-1 have the capacity of 1.728 and 51.84 Mb/s, respectively. To transport the optical signals over fiber, VT1.5s are multiplexed to a STS-1 and then to STS-12, STS-48

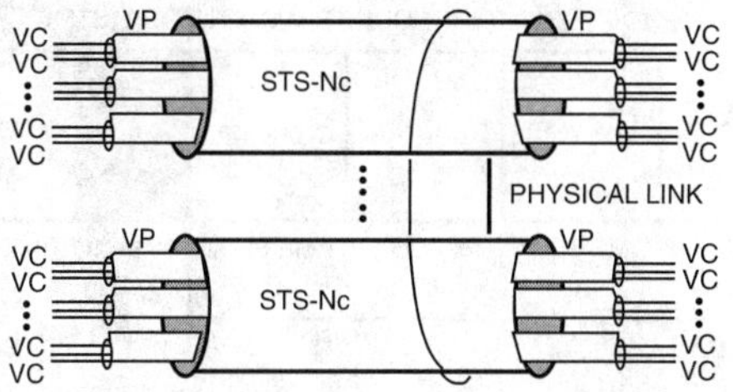

SWITCHING TECHNOLOGY	SONET/STM	ATM
UNITS	VT/STS-1/STS-Nc	ATM CELLS IN VCs/VPs
METHOD	TIME SLOT INTERCHANGE (TSI)	SELF-ROUTING WITH ROUTING TABLE
SIGNAL ROUTING PATH	HIERARCHICAL	NONHIERARCHICAL
NETWORK REROUTING CHARACTERISTIC	PHYSICAL NETWORK RECONFIGURATION	LOGICAL NETWORK RECONFIGURATION

FIGURE 15.4   Relationship among the VC, VP, digital path and physical link.

with other multiplexed streams for optical transport. Thus, a SONET transport node may equip a variety of switching equipment needed for each hierarchy of the signals. In contrast, the VP transport system is physically nonhierarchical (see Fig. 15.5(b)) and its capacity can be varied, which ranges from 0 (for protection) up to the line rate or STS-Nc, depending on applications. This nonhierarchical multiplexing structure provides a natural grooming characteristic, which may simplify the nodal system design. Detailed discussions between STM and ATM VP path structures and their implications on network designs can be found in Sato, Ueda, and Yoshikai (1991).

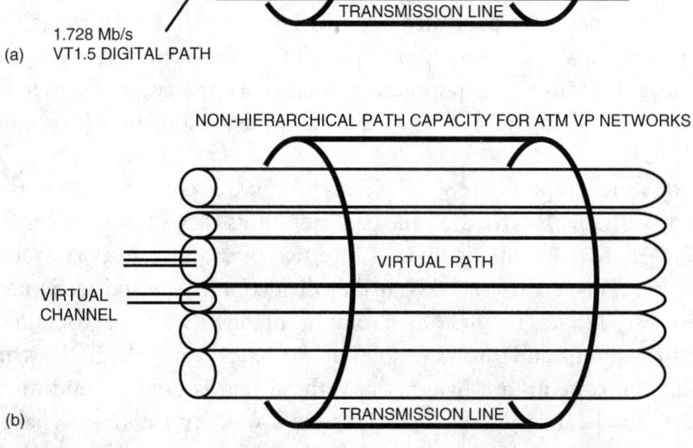

FIGURE 15.5   Physical path vs. virtual path: (a) physical path structure, (b) virtual path structure.

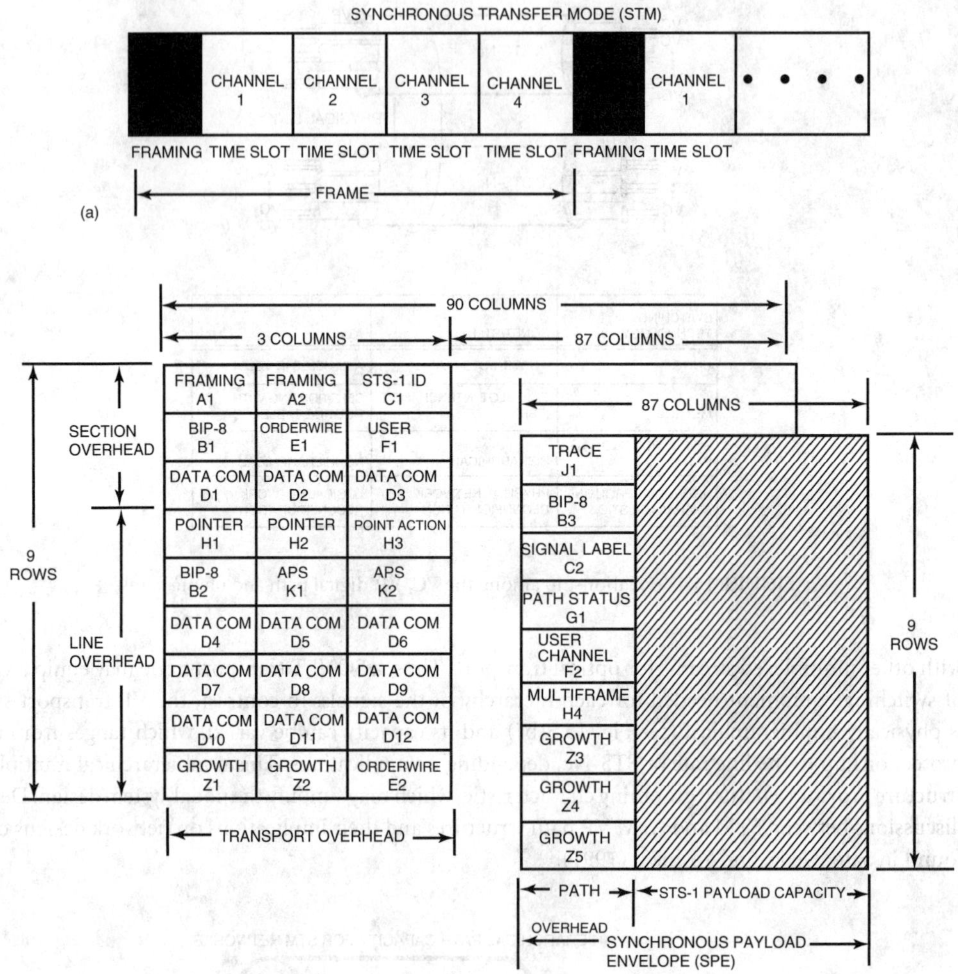

FIGURE 15.6 STM concept and SONET STS-1 example: (a) STM (framing) concept, (b) an example of SONET STS-1 frame.

Channel Formats for STM and ATM

In SONET/STM, data channels are represented by time slots and identified through the relative time slot location within the frame, as depicted in Fig. 15.6(a). Thus, STM switching is performed through time slot interchange. One example of STM equipment is the digital cross-connect system (DCS), which cross connects VT1.5s (1.728 Mb/s) and/or STS-1s (51.84 Mb/s), depending on the type of DCSs (wideband DCS or broadband DCS).

Figure 15.6(b) shows an example of the SONET STS-1 frame that corresponds to a frame in Fig. 15.6(a). In this example, each VT1.5 path carried on the payload of this STS-1 frame represents a channel (time slot) in Fig. 15.6(a). The STS-1 frame is divided into two portions: transport overhead and information payload, where the transport overhead is further divided into line and section overheads. The STS frame is transmitted every 125 μs. The first three columns of the STS-1 frame contain transport overhead and the remaining 87 columns and 9 rows (a total of 783 bytes) carry the STS-1 synchronous payload envelope (SPE). The SPE contains a 9-byte path overhead that is used for end-to-end service performance monitoring. Optical carrier level-1 (OC-1) is the lowest level optical signal used at equipment and network interfaces. An OC-1 is obtained from a STS-1 after scrambling and electrical-to-optical conversion.

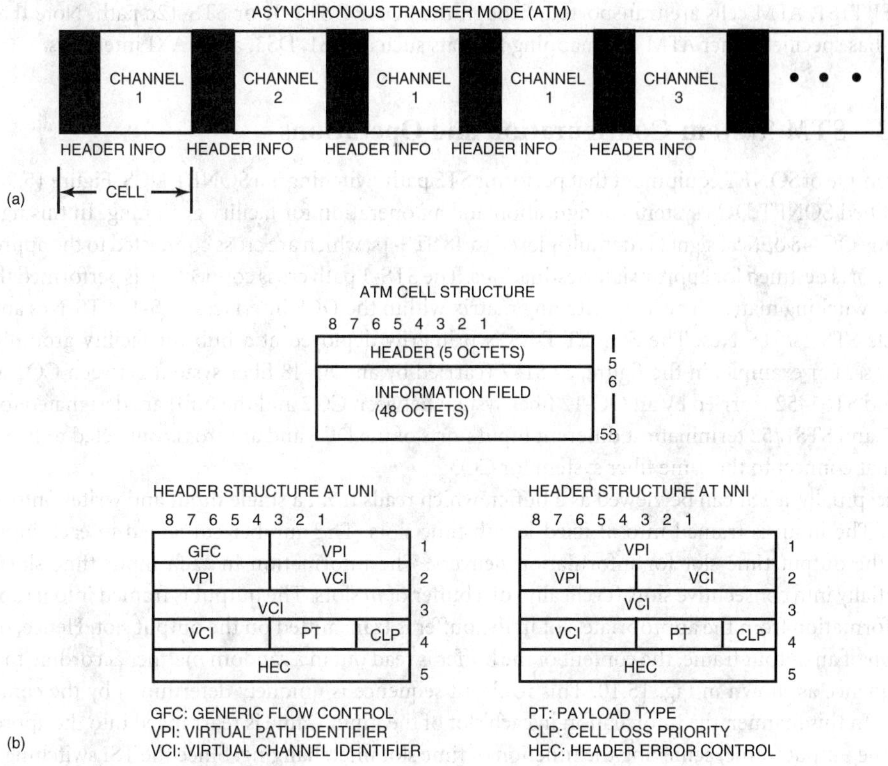

ASYNCHRONOUS TRANSFER MODE (ATM)

| CHANNEL 1 | CHANNEL 2 | CHANNEL 1 | CHANNEL 1 | CHANNEL 3 | • • • |

HEADER INFO   HEADER INFO   HEADER INFO   HEADER INFO   HEADER INFO

(a) |← CELL →|

ATM CELL STRUCTURE

8 7 6 5 4 3 2 1

| HEADER (5 OCTETS) | 5 |
| INFORMATION FIELD (48 OCTETS) | 6 ... 53 |

HEADER STRUCTURE AT UNI

8 7 6 5 4 3 2 1

GFC	VPI	1
VPI	VCI	2
VCI		3
VCI	PT	CLP	4
HEC		5

HEADER STRUCTURE AT NNI

8 7 6 5 4 3 2 1

VPI		1
VPI	VCI	2
VCI		3
VCI	PT	CLP	4
HEC		5

GFC: GENERIC FLOW CONTROL
VPI: VIRTUAL PATH IDENTIFIER
VCI: VIRTUAL CHANNEL IDENTIFIER

PT: PAYLOAD TYPE
CLP: CELL LOSS PRIORITY
HEC: HEADER ERROR CONTROL

(b)

FIGURE 15.7 ATM concept and cell formats: (a) ATM (fixed cell) concept, (b) ATM cell format.

In contrast, ATM channels are represented by a set of fixed-size cells (see Fig. 15.7(a)) and are identified through the channel indicator in the cell header (see Fig. 15.7(b)). Thus, ATM switching is performed on a cell-by-cell basis based on routing information in the cell header.

The ATM cell has two parts: the header and the payload. The header has 5 bytes and the payload has 48 bytes. As mentioned earlier, the major function of the ATM layer is to provide fast multiplexing and routing for data transfer based on information included in the 5-byte header. This 5-byte header includes information not only for routing (i.e., VPI and VCI fields), but also fields used to (1) indicate the type of information contained in the cell payload (e.g., user data or network operations information), (2) assist in controlling the flow of traffic at the UNI, (3) establish priority for the cell, and (4) facilitate header error control and cell delineation functions. One key feature of this ATM layer is that ATM cells can be independently labeled and transmitted on demand. This allows facility bandwidth to be allocated as needed, without fixed hierarchical channel rates required for STM networks. Since ATM cells may be sent either periodically or randomly (i.e., in bursts), both constant and variable bit-rate services are supported at a broad range of bit rates. The connections supported at the VP layer are either permanent or semipermanent connections that do not require call control, real-time bandwidth management, and processing capabilities. The connections at the VC layer may be permanent, semipermanent, or switched virtual connections (SVCs). SVCs require a signaling system to support its establishment, tear-down, and capacity management.

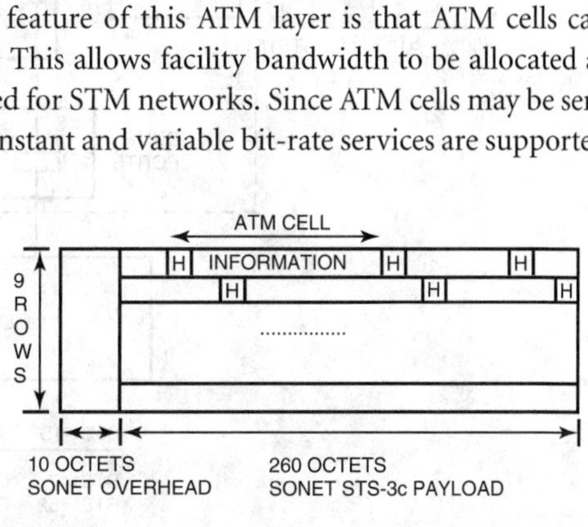

Figure 15.8 shows an example of ATM cell mapping in SONET payloads for high-speed transport.

ATM CELL

	H	INFORMATION	H		H	
		H			H		H
		....				

9 ROWS

|←10 OCTETS→|←————260 OCTETS————→|

10 OCTETS
SONET OVERHEAD

260 OCTETS
SONET STS-3c PAYLOAD

FIGURE 15.8 ATM cell mapping on SONET payloads.

In ANSI T1S1, ATM cells are transported through the SONET STS-3c or STS-12c path. Note that ATM Forum has specified other ATM cell mapping formats such as DS1, DS3, and TAXI interfaces.

15.1.3 STM System Configuration and Operations

One example of SONET equipment that performs STS path switching is a SONET DCS. Figure 15.9 depicts a simplified SONET DCS system configuration and an operation for facility grooming. In this figure, an incoming OC-48 optical signal is demultiplexed to 48 STS-1s, which are cross connected to the appropriate output ports destined for appropriate destinations. The STS-1 path cross connection is performed through the TSI switching matrix. The TSI switching matrix within the DCS interfaces STS-1s/STS-Ncs and cross connects STS-1s/STS-Ncs. The SONET DCS is primarily deployed at a hub for facility grooming and test access. For example, in the figure, STS1#7 (carried by an OC-48 fiber system between CO1 and the hub) and STS1#52 (carried by an OC-12 fiber system between CO2 and the hub) are designated for CO3. STS1#7 and STS1#52 terminate at different input ports of the DCS and are cross connected to two output ports that connect to the same fiber system for CO3.

Conceptually, a TSI can be viewed as a buffer, which reads from a single input and writes onto a single output. The input is framed into m fixed-length time slots. The number contained in each input time slot is the output time slot for information delivery. The information in each input time slot is read sequentially into consecutive slots (cyclically) of a buffer of m slots. The output is framed into n time slots, and information from the appropriate slot in the buffer is transmitted on the output slot. Hence, over the duration of an output frame, the content of the buffer is read out in a random manner according to a read-out sequence as shown in Fig. 15.10. This read-out sequence is uniquely determined by the connection pattern. In this manner, the information in each slot of the input frame is rearranged into the appropriate slot in the output frame, achieving the function of time slot interchanging. Since the TSI switching matrix performs physical STS path switching from one input port to one output port, TSI rerouting involves physical switch reconfiguration.

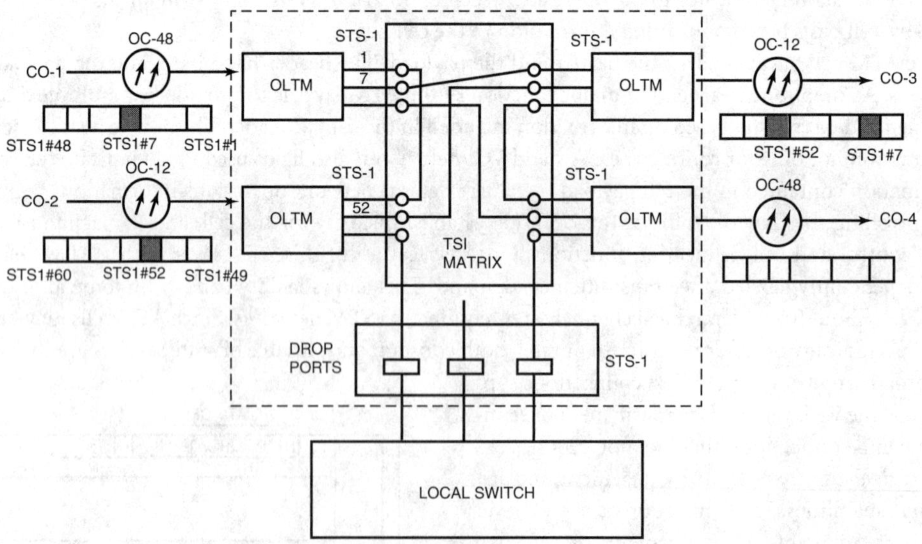

FIGURE 15.9 An example of SONET/STM system configuration and operation.

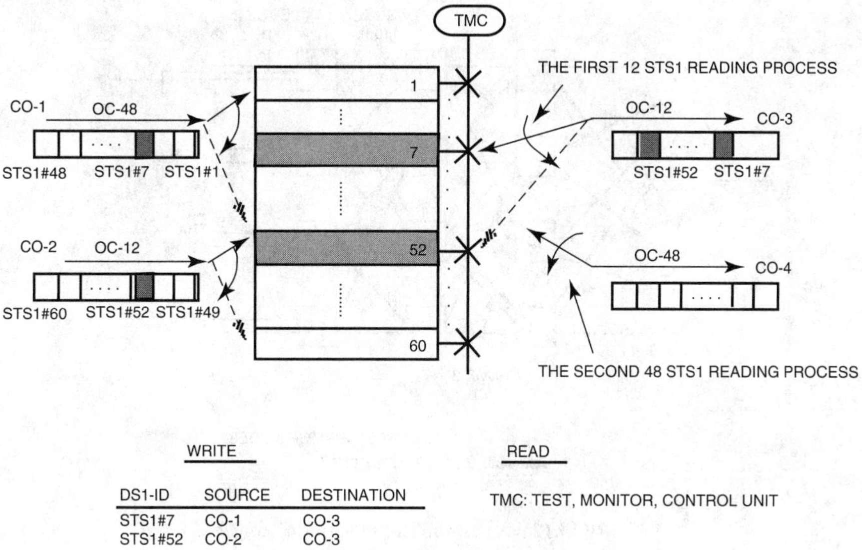

FIGURE 15.10   TSI switching concept.

15.1.4   ATM Cell Switching System Configuration and Operations

Figure 15.11 illustrates a simplified ATM cell switching system configuration and a switching principle. The ATM cell entering input port 3 of the switch arrives with VPI = 9 and VCI = 4 (for example, the first ATM cell in the figure). The call processor has been alerted through the routing table that the cell must leave the ATM switch with VPI = 4 and VCI = 5 on output port 4. The call processor directs the virtual channel identifier converter to remove the 9 VPI and 4 VCI and replaces them with 4 and 5, respectively. If the switch is a large multistage switch, the call processor further directs the virtual channel identifier converter to create a tag that travels with the cell, identifying the internal multistage routing within the

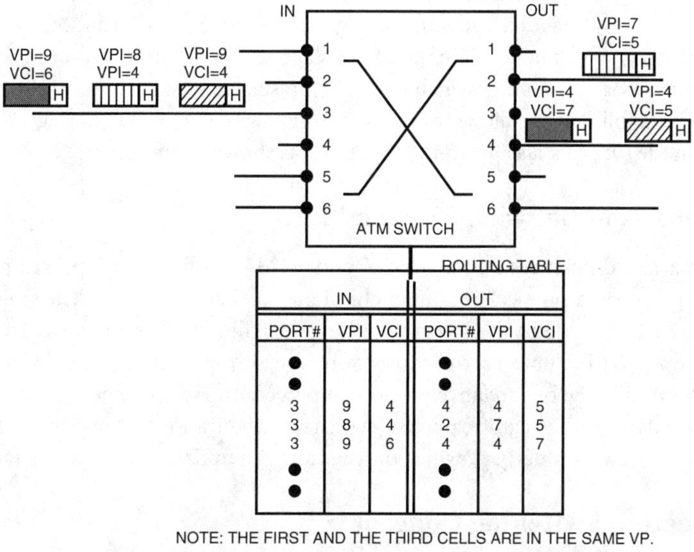

FIGURE 15.11   An example of ATM cell switching principle.

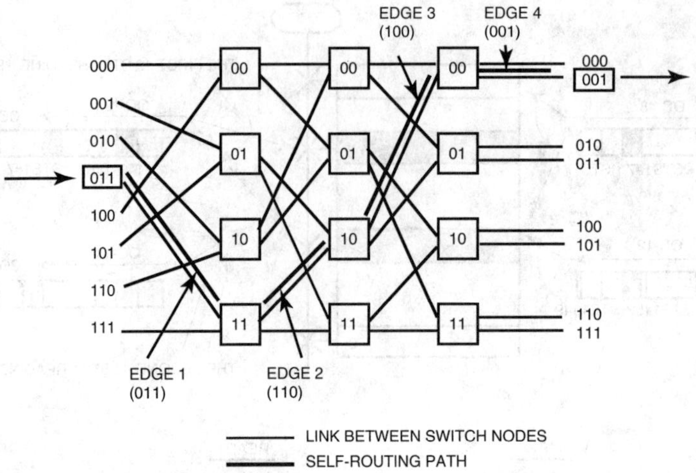

FIGURE 15.12   An example of self-routing concept.

ATM switch matrix. Note that ATM cells belonging to the same VP must have the same VPIs in the input port (e.g., VPI = 9 in input port 3) or the output port (e.g., VPI = 4 in output port 4).

The ATM switch implemented for large telecommunications networks usually uses a **self-routing** switching matrix. (Note: ATM switches designed for enterprise networks may not use the self-routing switching matrix.) An ATM cell is self-routing if the header of the cell contains all of the information needed to route through a switching network. In the following, we use a simple example to explain the self-routing concept, as depicted in Fig. 15.12. The example switching network is a shuffle exchange network, which has $N = 2^n$ inputs and 2^n outputs, interconnected by n stages of 2^{n-1} binary switch nodes. In this example, $n = 3$ and $N = 8$. We assume that the ATM cell is to be routed from input port 4 (i.e., code 011) to output port 2 (i.e., code 001). When the cell arrives input port 4 (011), the call processor will check with VPI/VCI routing table, identifies the destination output 2 (001), and it generates a routing label which combines the source and destination addresses: 011001. Through this code, the system identifies edges in each stage where each edge is represented by three consecutive digits from the left side to the right of the routing label. In this case, 4 edges are identified: edge 1: 011, edge 2: 110, edge 3: 100, and edge 4: 001. Each edge (ijk) connects switch node (ij) in the $M - 1$ stage and switch node (jk) in stage M, if this edge is edge M. For example, edge 2 connects switch node (11) in stage 1 to switch node (10) in stage 2. Thus, a self-routing path is established as follows: Input port (011)-switch node (11) of stage 1-switch node (10) of stage 2-switch node (00) of stage 3-output port (001), as shown in the figure.

15.1.5   Comparison Between STM and ATM

Table 15.1 summarizes differences between STM and ATM switching concepts. In general, the STM system switches signals on a physical and hierarchical basis, whereas ATM switches signals on a logical and nonhierarchical basis, due to their corresponding switching path structures. Thus, ATM may be more flexible than its STM counterpart in terms of bandwidth on demand, bandwidth allocation, and transmission system efficiency, but a relatively more complex control system is needed. Since STM physically switches the signals, its rerouting requires physical switch reconfiguration that may make network rerouting much slower than its ATM counterpart, which only requires an update of the routing table.

15.1.6   Protection Switching Concept

The function of **protection switching** is to reroute new and existing connections around the failure area if a network failure occurs. The basic requirement for network protection switching is to restore services as fast

TABLE 15.1 Comparison Between STM and ATM

System Parameters	STM	ATM
Switching unit	Time slot (STS paths)	Fixed-size cell (VPs/VCs)
Path structure	Physical and hierarchical	Logical and non-hierarchical
Number of allowable path (STS, VP) capacities	Limited	More
Switching system complexity	Simpler	Complex
Network nodal complexity	Complex (hierarchical)	Simpler (non-hierarchical)
Transmission system efficiency (on average)	Lower	Higher
Method of rerouting	Physical network reconfiguration	Logical network reconfiguration
Bandwidth on demand	More difficult	Easier

as possible in a cost-effective manner. ATM technology is more flexible and efficient than STM technology in terms of bandwidth allocation and transmission efficiency. However, these advantages come at an expense of requiring a much more complex control systems, especially when network restoration is an important consideration in system requirements. Unlike STM transport, the service quality in ATM networks is affected by both network component failure and network congestion (because resources are not dedicated).

In addition to fast restoration times, other possible network restoration objectives include the maximal restoration ratio, and the minimal spare capacity and overhead associated with network restoration. A network restoration system may be designed to optimize tradeoffs of these objectives or may be designed for some specific objective (such as minimum protection switching time), depending on the network planning strategy and service requirements being supported.

For SONET/ATM networks, network restoration can be performed at the physical (SONET) or ATM layer. For physical layer protection, service restoration can be performed at the STS-3Nc level (where $N = 1$, 4, or 16) or at the physical transmission link level. For ATM layer protection, network restoration can be performed at the VC or VP (or group of VPs) level. Service restoration at the VC level is typically slower and more expensive than that performed at the VP level due to greater VC-based ATM network complexity. The restoration system complexity at the VP level is similar to that at the STS-1/STS-3c level. Determining which layer (SONET, VP, or VC) is an appropriate restoration layer depends on the SONET/ATM network architecture and other factors such as costs and QoS.

The major differences between STM and ATM network restoration technologies are in the restoration unit (e.g., STS-1 or VP) and the network reconfiguration technology. SONET networks restore traffic on physical paths (e.g., STS-3c) and/or links (e.g., SONET links) through the physical network reconfiguration using time slot interchange technology. In contrast, ATM networks restore traffic carried on VPs/VCs through the logical network reconfiguration by modifying the VPI/VCI routing table.

SONET protection switching systems, such as automatic protection switching systems (APSs) and SONET self-healing rings, have been widely deployed in LECs. The impacts of emerging ATM technology on network restoration (compared to SONET network restoration) may be significant due to following factors:

- Separation of capacity allocation and physical route assignment for VPs/VCs that would reduce required spare capacity (e.g., the capacity of protection VPs/VCs can be zero in normal conditions)
- More OAM bandwidth with allocation on demand that would reduce delays for restoration message exchanges, and the detection of system degradation (i.e., soft failure) in the ATM network much more quickly than its SONET counterpart
- Nonhierarchical path multiplexing (for VPs) that would simplify the survivable network design and help reduce intranode processing delays and required spare capacity.

Table 15.2 summarizes the relative comparison between STM and ATM network protection technologies. In general, ATM layer protection may require less spare capacity than SONET layer protection at the

TABLE 15.2   Comparison of SONET and ATM Layer Protection Technologies

Protection Layer	ATM Layer	SONET Layer (STM)
Protection protocol complexity	Moderate	Simple (APS/rings) moderate (DCS mesh)
Restoration time	Moderate	Fast (APS/rings) slow (DCS mesh)
Spare capacity needed	Least	Moderate
Equipment needed	Less	More
Network management systems/OSs	Under development	Already developed
Equipment availability	Available soon	Available now
Protection targets	Node and link	Node and link

expense of a slower restoration time and a more complex control system. How best to use a combination of these two layer protection schemes in the same network is a challenging task for network planners and engineers who are seeking cost-effective solutions for SONET/ATM network protection. More details of network protection schemes and operations for SONET/ATM networks can be found in Wu (1992, 1995); Sato, Ueda, and Yoshikai (1991); and Wu et al. (1994).

15.1.7   Summary and Remarks

We have reviewed the basic switching concepts that are used in emerging SONET/ATM broadband transport networks. Implementations of such switching concepts may be different between enterprise networks and public telecommunications networks due to traditionally different QoS perspectives and operations experiences from computer and telecommunications industries. Such differences may result in two sets of ATM switching network infrastructures that could coexist for a period of time. An open question is how and when they may be merged as an integrated ATM network infrastructure. This challenge may require the change of traditional QoS perspectives, definitions, and requirements from both the computer and telecommunications industries.

Defining Terms

Asynchronous transfer mode (ATM): An international standard high-speed, integrated multiplexing and switching technology that transmits information using fixed length cells in a connection-oriented manner.

Protection switching: A function to reroute new and existing connections around the failure area if a network failure occurs.

Self-routing: An ATM cell is self-routing if the header of the cell contains all of the information needed to route through a switching network.

Switching: Control or routing of signals in either physical or virtual circuit form to transmit data between specific points in a network.

Synchronous optical network (SONET): A North American standard synchronous transmission network technology for optical signal transmission.

Synchronous transfer mode (STM): A time division multiplexing and switching technology that uses a similar switching concept as traditional circuit switching systems.

Time slot interchange (TSI): A technique to exchange the location of time slots within a frame.

References

Ballart, R. and Ching, Y.-C. 1989. SONET: Now it's the standard optical network. *IEEE Comm. Mag.* (March):8–15.

Boudec, J.-Y.L. 1992. The asynchronous transfer mode: A tutorial. *Computer Networks and ISDN Systems* 24:279–309.

Sato, K., Ueda, H., and Yoshikai, N. 1991. The role of virtual path crossconnection. *IEEE Mag. Lightwave Telecomm. Sys.* 2(3):44–54.

Wu, T.-H. 1992. *Fiber Network Service Survivability.* Artech House, May.

Wu, T.-H. 1995. Emerging technologies for fiber network survivability. *IEEE Comm. Mag.* (Feb.).

Wu, T.-H., McDonald, J.C., Sato, K., and Flanagan, T.P., eds. 1994. Integrity of public telecommunications networks. *IEEE J. Selected Areas in Comm.* (Jan.).

Further Information

Additional information on the topic of network switching concept is available from the following sources:

IEEE Communications Magazine, a monthly periodical dealing with broadband communication networking technology. The magazine is published by IEEE.

IEEE Journal on Selected Areas in Communications, a monthly periodical dealing with current research and development on broadband communication networking technology. The magazine is published by IEEE.

15.2   SONET

Brent Allen

15.2.1   Introduction

Synchronous optical network (SONET) refers to an American National Standards Institute (ANSI) standard for fiber optic telecommunications interfaces. It was developed in the late 1980s and telecommunications equipment providing the interface continues to be deployed throughout the North American network. Closely related to SONET is the *synchronous digital hierarchy* (SDH) interface, which is an international fiber optic telecommunications interface standard prepared by the International Telecommunications Union (ITU), an agency of the United Nations. Although SONET and SDH are documented in different standards, the two interfaces are compatible. SONET can be considered a subset of SDH, in the sense that it selects among various options allowed in SDH.

As illustrated in Fig. 15.13, the public telephony network consists of users with telephones, connected to a switching system over an **access** network. To allow connections between users that are not connected to the same switching system, the switching systems are all connected to each other over the **trunk** network. Although access networks can be localized to the area surrounding the switching system, trunk networks span cities, countries, and even the globe.

Today, the majority of connectivity in the trunk network, and much of the connectivity in the access network, is provided using fiber optic transmission. This is referred to as the transport infrastructure. This infrastructure also provides direct digital connections, called private lines, between businesses for private voice, data, and video networks. In fact, the amount of **bandwidth** allocated for private lines rivals that allocated for regular voice.

Although the design and deployment of fiber optic telecommunications systems began in the late 1970s and early 1980s, these systems featured proprietary optical interfaces (Fig. 15.14). The only standard interfaces at that time were the DS1 (1.544-Mb/s interface consisting of 24 digitally encoded voice channels), which was and continues to be the main trunk interface to voice switching systems, and the DS3 (44.736-Mb/s interface consisting of 28 DS1s), which was the primary interface to interoffice transmission systems, including radio and fiber. Both the DS1 and DS3 are electrical interfaces and as such are limited in terms of reach, essentially restricted to connections between equipment within a building.

The creation of a network on the scale of the telecommunications network requires interface standards. Such networks require equipment developed by different manufacturers (and owned by different network providers) to be connected, and this requires an interface specification. A standard results when the equipment manufacturers, as well as network providers and users, come together to negotiate the interface specification.

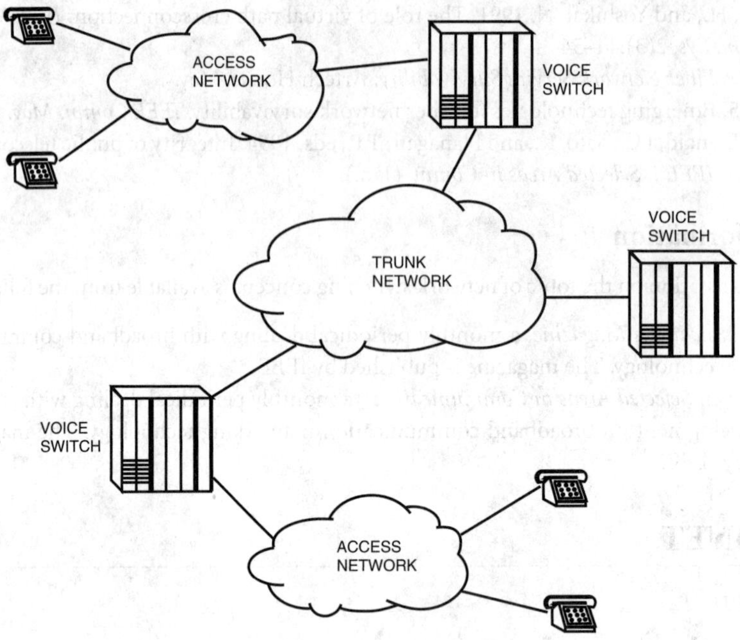

FIGURE 15.13 Telephony network architecture.

Since pre-SONET fiber systems featured proprietary optical interfaces, their application was restricted to the point-to-point transport of DS1s and DS3s. The development of the SONET standard is now resulting in the deployment of all optical networks, the benefits of which include very low noise floors, high reliability and survivability, flexibility, and almost unlimited capacity.

15.2.2   SONET Overview

The structural elements of SONET, and the way they are combined to create a complete SONET signal, are shown in Fig. 15.15. A complete, optically transmitted SONET signal is called an OC-N, where OC is optical carrier and N may be 1, 3, 12, 24, 48, or 192. The OC-N is the optical equivalent of the STS-N where STS is synchronous transport signal. The STS-N or OC-N consists of a **multiplex** of N STS-1s.

Carried within the STS-1 is a *synchronous payload envelope* (SPE). In SONET, all **payloads** are transported within an SPE, and SONET provides a variety of SPEs to accommodate the different bandwidth requirements of various payloads. The STS-1 SPE has a payload capacity of approximately 50 Mb/s and therefore can carry one DS3 (44.736 Mb/s). SONET provides a mechanism, called *concatenation*, whereby the capacity of several STS-1s can be combined to carry a larger SPE and, therefore, a larger payload. For example, M STS-1s can be concatenated to carry an STS-Mc SPE, which will have a payload capacity of approximately $M \times 50$ Mb/s.

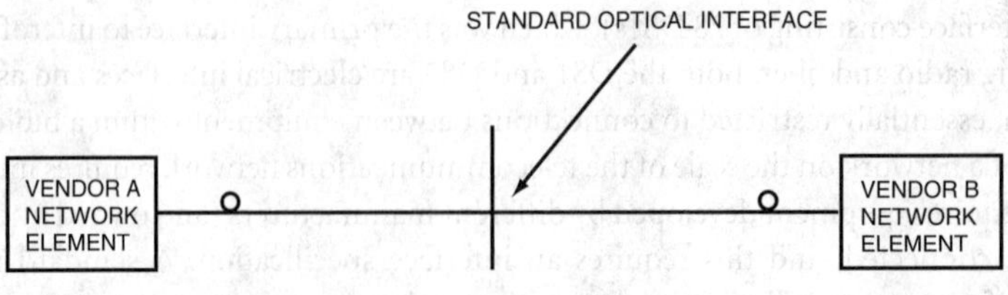

FIGURE 15.14 Multivendor midspan meet.

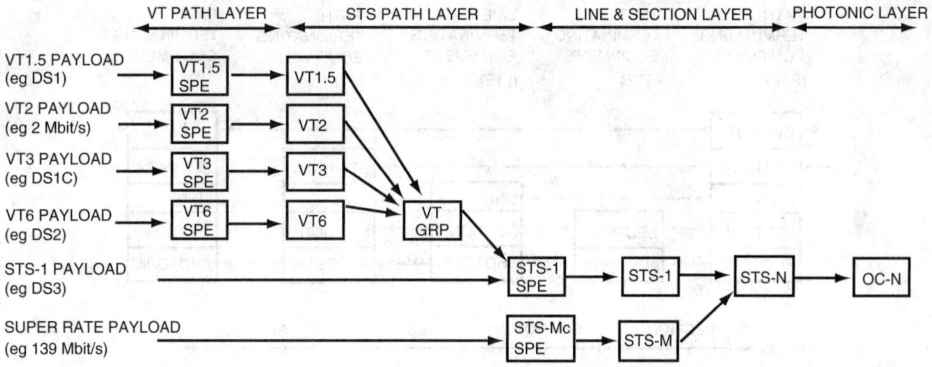

FIGURE 15.15   SONET protocol structure.

The STS-1 SPE can be subdivided into *virtual tributaries* (VTs). These VTs contain VT SPEs in order to carry smaller payloads, such as the DS1 (1.544 Mb/s). There are four VT sizes, which offer different payload capacities ranging from approximately 1.5 to 6 Mb/s. Like-VTs are combined into VT groups, which are then multiplexed into the STS-1 SPE.

Layered Architecture

The SONET protocol is defined in layers to highlight the interface functionality that is required on a particular network element. The layers are illustrated in Fig. 15.16. The interface functionality of a particular network element is closely associated with the overall function of the element, and the functionality associated with all of the layers may not be required.

The *photonic layer* defines the physical parameters of the optical link, such as optical pulse shape, transmit power, wavelength, etc.

The *section layer* provides basic functions that are required before the information contained within the SONET signal can be extracted and processed. Framing is the process of finding the alignment of the constituent bits and bytes within a serial bit stream so that the information contained can be located. Scrambling is the process of pseudorandomizing the outgoing bitstream. A certain density of 0-1 and 1-0

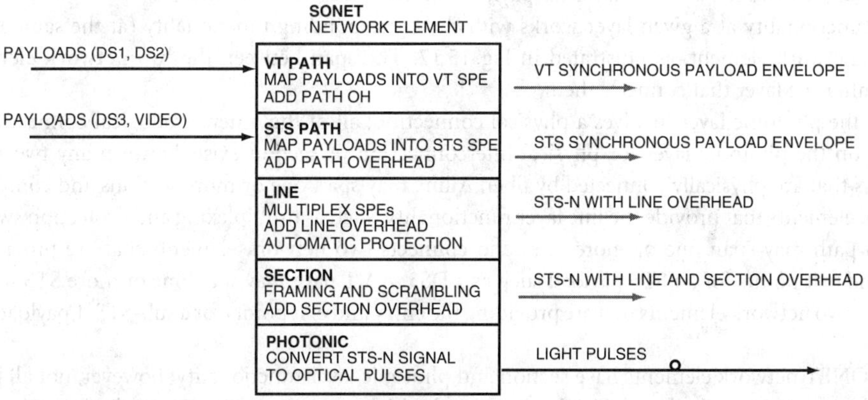

FIGURE 15.16   SONET layers as applied to network element functionality. (*Source:* Adapted from ANSI 1991. *Digital Hierarchy—Optical Interface Rates and Formats Specification (SONET)*, ANSI T1.105-1991, Fig. 1, p. 38. American National Standards Institute, Washington, DC. With permission from the Alliance for Telecommunications Industry Solutions.)

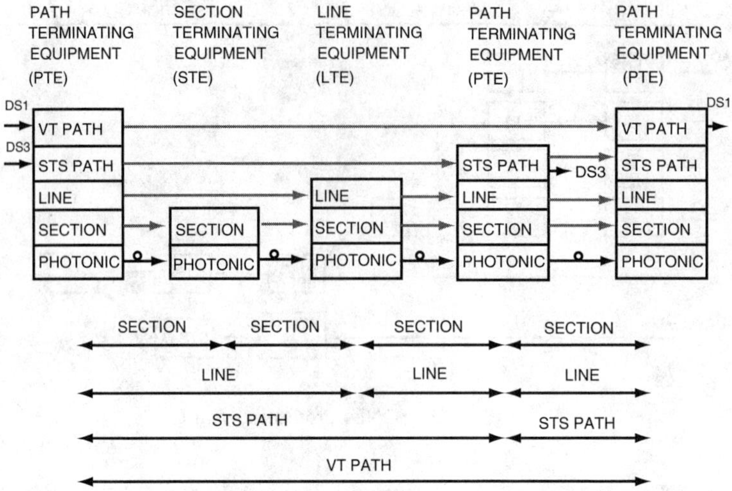

FIGURE 15.17 SONET layers as applied to network connectivity. (*Source*: Adapted from ANSI. 1991. *Digital Hierarchy—Optical Interface Rates and Formats Specification (SONET)*, ANSI T1.105-1991, Fig. 28, p. 58. American National Standards Institute, Washington, DC. With permission from the Alliance for Telecommunications Industry Solutions.)

transitions is required by the receiver in order to be able to derive a clock with which to receive the digital signal. **Overhead** is added for the purpose of monitoring the error performance of the interface at the section layer and communicating with a centralized network management system.

The *line layer* provides the *multiplexing* of lower rate SONET structures up to the final line rate. Overhead is added for the purpose of monitoring the error performance of the interface at the line layer, communicating with the network element on the other end of the line to coordinate protection switching, and communicating with a centralized network management system to report situations such as equipment failure or fiber breaks.

The *path layers* provide the mapping of payloads into SONET so that they can be transported across a SONET network. Overhead is added to the payload to allow the integrity of the path layer connection to be verified. There are two granularities of path layer; the STS path layer for payloads on the order of 50 Mb/s or higher, and the VT path layer for payloads on the order of 1.5–6 Mb/s.

The functionality at a given layer works with the corresponding functionality (at the same layer) in another network element, as illustrated in Fig. 15.17. The span between the two **network elements** is named after the layer that connects them.

Only the photonic layer involves a physical connection; all of the others are logical connections that depend on the photonic layer for physical interconnection. A section exists between any two network elements that are physically connected by fiber. A line may span one or more sections and connects two network elements that provide the line layer functionality, such as multiplexing and protection switching. An STS path may span one or more lines and connects two network elements that are providing the entry and exit points for an STS payload such as a DS3. A VT path may span one or more STS paths and connects two network elements that are providing the entry and exit points for a sub-STS-1 payload such as a DS1.

All SONET network elements have section and photonic layer functionality; however, not all have the higher layers. A network element is classified by the highest layer supported on the interface. Thus a network element with path layer functionality is referred to as *path terminating equipment* (PTE), either VT PTE or STS PTE. Similarly, a network element with line layer functionality but no higher, is *line terminating equipment* (LTE). Finally, a network element with section layer functionality but no higher is *section terminating equipment* (STE).

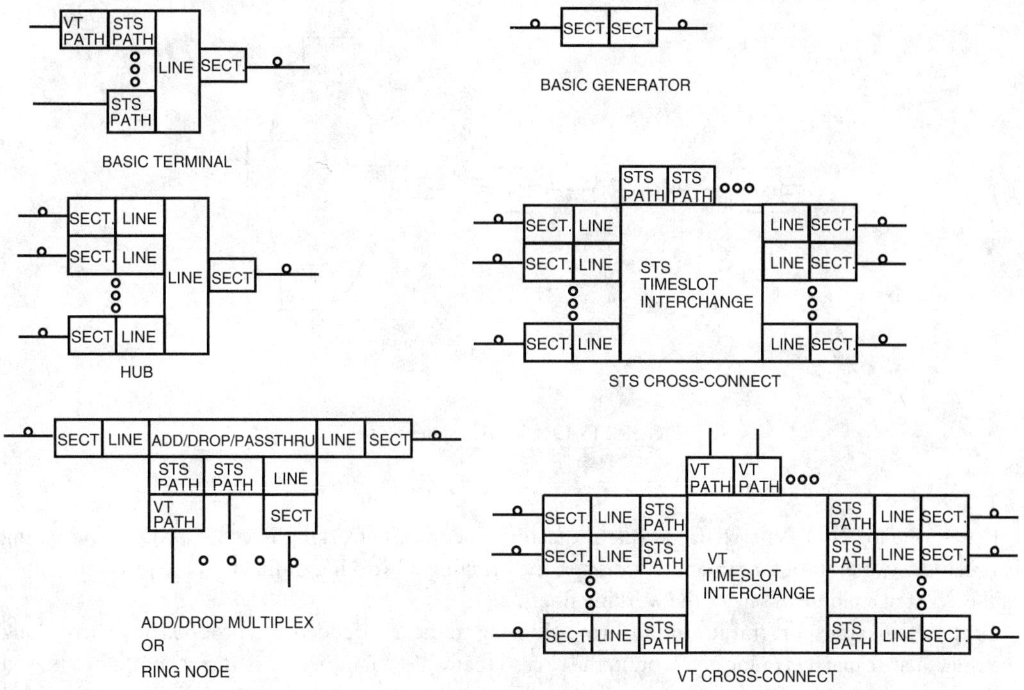

FIGURE 15.18 SONET network elements.

Network Elements

The functionality associated with the various SONET layers can be combined in a variety of ways to create **network elements**. The key network elements (NEs) are illustrated in Fig. 15.18.

The *basic terminal* interfaces a number of non-SONET payloads, maps those signals into the appropriate SONET path layer, and multiplexes them onto a SONET optical signal.

The *regenerator* increases the distance allowed between adjacent line terminals by regenerating the optical signal.

The *hub* is similar to the terminal, except that the tributary interfaces are also SONET signals. For example, a hub might multiplex four OC-3s into an OC-12. It is common for basic terminals to also have hub capabilities.

An *add/drop multiplexer* (ADM) adds and drops signals from a SONET signal. Any traffic that is not added/dropped can pass through directly. The added/dropped signals may be non-SONET signals, in which case the appropriate path terminations are required, or they may be SONET signals.

SONET *cross connects* provide path layer interconnect between a large number of SONET or non-SONET interfaces. The STS cross connect provides STS path layer connections, and the VT cross connect provides VT path layer connections. The connections can be changed via an operations interface.

System Configurations

SONET network elements are combined to create systems that are classified based on the mechanism used to provide **traffic** protection and **survivability**. The two main types of protection provided in SONET are *linear* and *ring* as shown in Fig. 15.19.

Linear systems transport traffic along a single route, which may consist of one or more working fibers. A **standby or protection fiber** exists along route to provide a backup path for 1 or more of the **working fibers**. A one plus one (1 + 1) system has one protection fiber and one working fiber. The traffic is permanently bridged onto both fibers so that the receiving end can autonomously choose the fiber that is operating

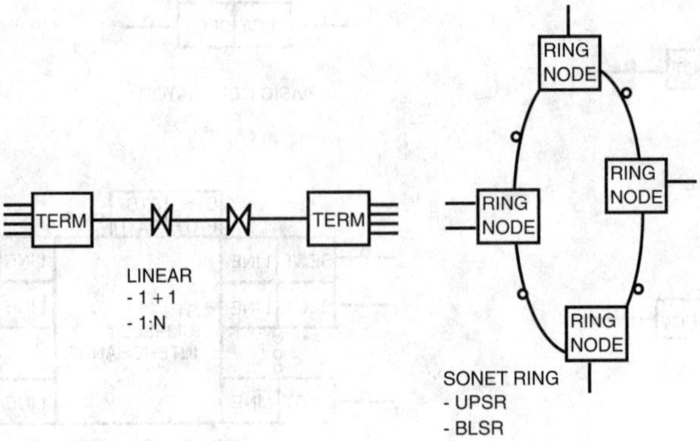

FIGURE 15.19   SONET system types.

better. A one for N $(1:N)$ system has one protection fiber and N working fibers. A protocol operating between the two ends of the system coordinates the bridging of working traffic onto the protection fiber in the event of a failure of one of the working fibers.

Ring systems transport traffic around a ring allowing traffic to be added and dropped anywhere along the ring. Spare capacity is allocated around the ring so that when a failure occurs at any point, the affected traffic can be restored using the spare capacity. In a *unidirectional path switched ring* (UPSR), traffic added to the ring is bridged onto both directions such that the drop node can autonomously select the better path. In a *bidirectional line switched ring* (BLSR), the traffic affected by a failure at any point on the ring is rerouted the other way around the ring. A protocol operating around the ring coordinates this action.

Networks

SONET networks are created by combining network elements and systems. Hubs, ADMs, and cross-connects are used to interconnect linear systems to each other and to rings. The elements and systems used depend on the application and topology of the network and the degree of survivability required.

In the example shown in Fig. 15.20, the ring may be part of an access network collecting voice traffic to be connected to the voice switch and private line traffic to be connected to the interoffice transport network. The VT cross connect consolidates traffic having the same destination from the switch and the

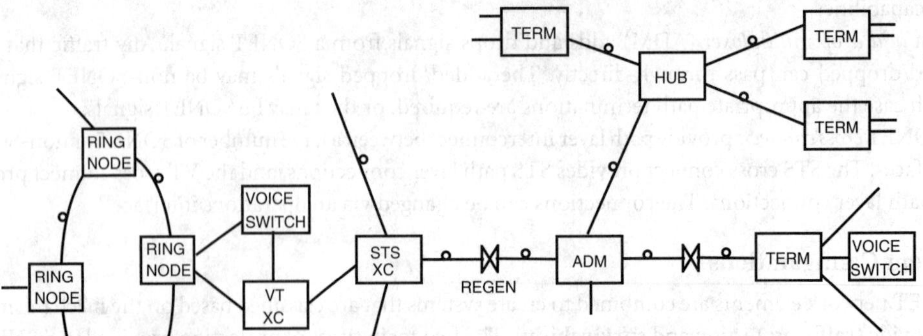

FIGURE 15.20   Example of SONET network.

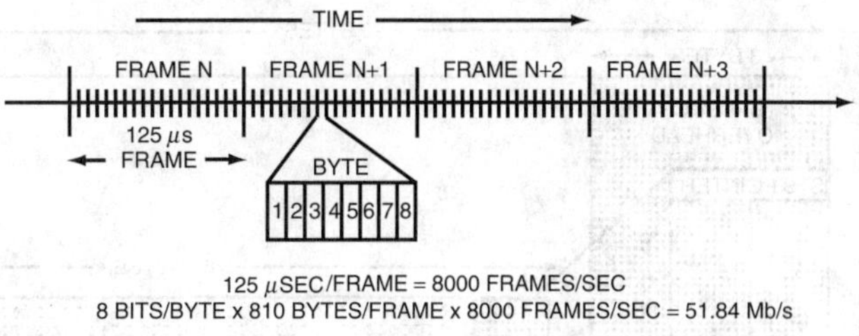

FIGURE 15.21 Serial transmission over SONET.

access ring to fill up STS-1s before they are put onto the interoffice network. The STS cross connect routes the STS-1 to the appropriate destination. The ADM and hub split traffic off of the main route to the end destinations.

15.2.3 SONET Protocol

A SONET signal is transmitted as a continuous serial bit stream as illustrated in Fig. 15.21. The bit stream is organized into bytes, and the bytes are grouped in 125-μs blocks called *frames*. At the basic STS-1 rate, there are 810 bytes per frame, which results in a basic rate of 51.84 Mb/s. The rate of an OC-N is exactly N times the basic rate.

The 810 bytes of an STS-1 frame are most conveniently represented as a matrix consisting of 9 rows and 90 columns as shown in Fig. 15.22. The intersection of a row and a column is one byte. The order of transmission is from left to right and from top to bottom.

The first three columns carry transport **overhead**. Transport overhead consists of section overhead and line overhead and is allocated to support section and line layer functionality. It is considered overhead because it is not available to carry payload. The remaining 87 columns of the STS-1 frame carry the STS-1 synchronous payload envelope as illustrated in Fig. 15.23.

The STS-1 SPE has its own frame consisting of 9 rows and 87 columns. The first column of the SPE is path overhead, which is allocated to support path layer functionality. The remainder of the SPE is available for payload. Although the SPE fits within the STS-1 SPE capacity, the first byte of the SPE does not necessarily occupy the first byte position within the SPE capacity. Thus, the SPE is represented as being offset from the STS-1 frame.

The STS *payload pointer* is a line layer function that locates the start of the SPE frame within the SPE capacity. The pointer overhead bytes carry a binary number, which is an offset, in bytes, into the SPE capacity to the beginning of the next SPE frame.

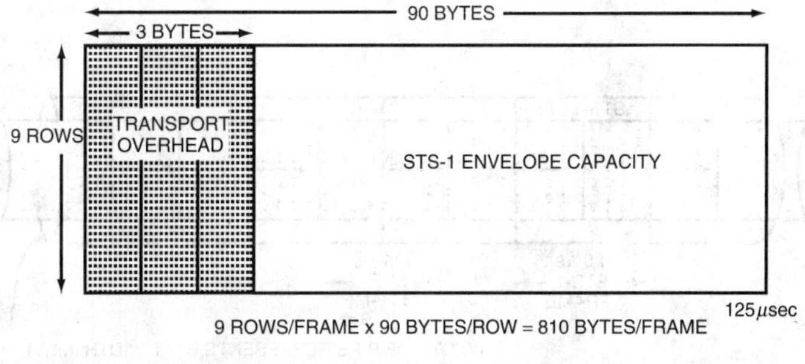

FIGURE 15.22 STS-1 frame structure. (*Source*: Adapted from ANSI. 1991. *Digital Hierarchy—Optical Interface Rates and Formats Specification (SONET)*, ANSI T1.105-1991, Fig. 7, p. 42. American National Standards Institute, Washington, DC. With permission from the Alliance for Telecommunications Industry Solutions.)

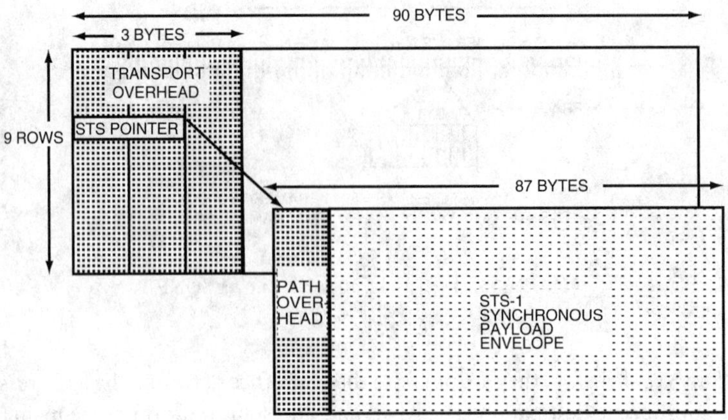

FIGURE 15.23 STS-1 frame with synchronous payload envelope. (*Source:* Adapted from ANSI. 1991. *Digital Hierarchy—Optical Interface Rates and Formats Specification (SONET),* ANSI T1.105-1991, Fig. 8, p. 42. American National Standards Institute, Washington, DC. With permission from the Alliance for Telecommunications Industry Solutions.)

It is useful to think of the STS-1 as being a pipe of digital information, as illustrated in Fig. 15.24. The diameter of the pipe represents the bandwidth flowing through the pipe. The length of the pipe represents time which can be marked off in 125-μs segments representing one frame of transmission. Within the STS-1 pipe is another, slightly smaller pipe, which is the STS-1 SPE. The difference in diameter between the STS-1 pipe and the SPE pipe represents the transport overhead. The pipe diagram helps visualize how the SPE frame alignment can be offset with respect to the STS-1 frame and how the STS-1 pointer indicates the offset to the beginning of the SPE frame.

Multiplexing

An STS-N or OC-N is created by byte-interleaving N frame-aligned STS-1s as shown in Fig. 15.25. Since the STS-1s are frame aligned, the transport overheads from each individual STS-1 will come out as a block occupying the first $3 \times N$ columns of the STS-N, effectively becoming the transport overhead for the STS-N.

Only certain rates (values of N) are defined as standard rates and have physical (photonic) interfaces specified. The currently defined values of N and their corresponding bit rates are shown in Table 15.3. The currently defined SDH rates are also shown.

For SONET, the rates start at the basic 51.84-Mb/s OC-1. SDH has a similar hierarchy but it begins at a basic rate of 155.52 Mb/s, which is called the synchronous transport module—level 1 (STM-1).

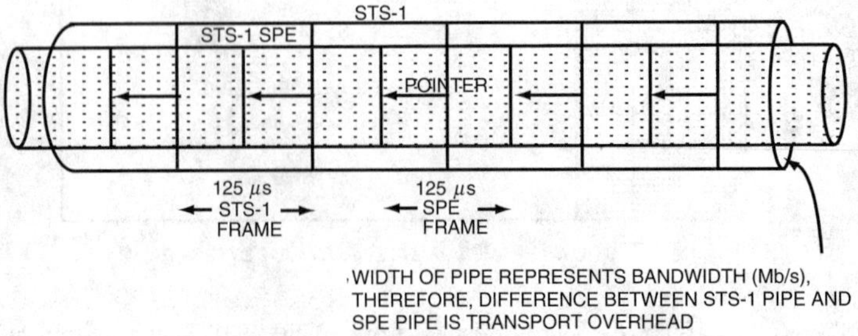

FIGURE 15.24 Pipe representation of STS-1 with SPE.

TABLE 15.3 SONET/SDH Bit Rates

SONET	SDH	Rate, Mb/s
OC-1	N/A	51.84
OC-3	STM-1	155.52
OC-12	STM-4	622.08
OC-24	N/A	1244.16
OC-48	STM-16	2488.32
OC-192	STM-64	9953.28

The pipe analogy can be extended to show an STS-N or OC-N carrying N STS-1 SPEs as illustrated in Fig. 15.26. Note that the STS-N becomes a single pipe, although it still contains a separate pointer for each STS-1 SPE.

Concatenation

SONET provides the ability, called *concatenation*, to combine the SPE capacities of N STS-1s to carry a single STS-Nc SPE. This STS-Nc SPE has approximately N times the payload capacity of an STS-1 SPE.

Using the pipe analogy as shown in Fig. 15.27, the STS-Nc SPE looks like a single pipe within the STS-N, however it has N times the diameter (bit rate) of the STS-1 SPE. Note also that there is a single pointer for the STS-Nc SPE.

Payload Pointer Mechanism

The pointer mechanism serves two key purposes: the first is SPE phase identification, and the second is SPE frequency justification. The need for SPE phase identification is illustrated using Fig. 15.28.

Consider a SONET hub that is multiplexing the SPEs from three separate OC-1s into a single OC-3. Although we could assume that the arriving SPEs are aligned within the corresponding STS-1 SPE

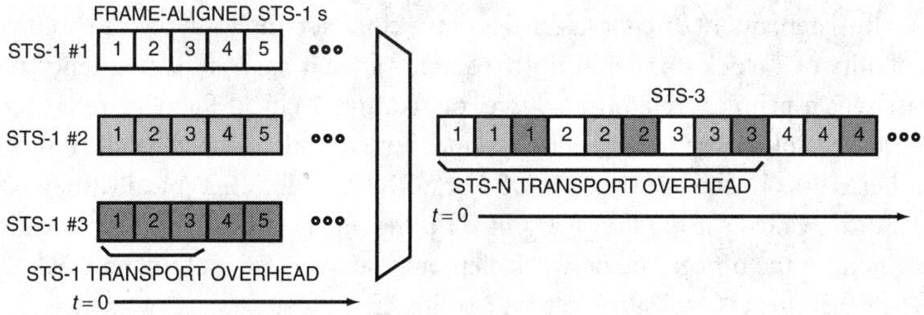

FIGURE 15.25 Example of STS-N byte interleaving.

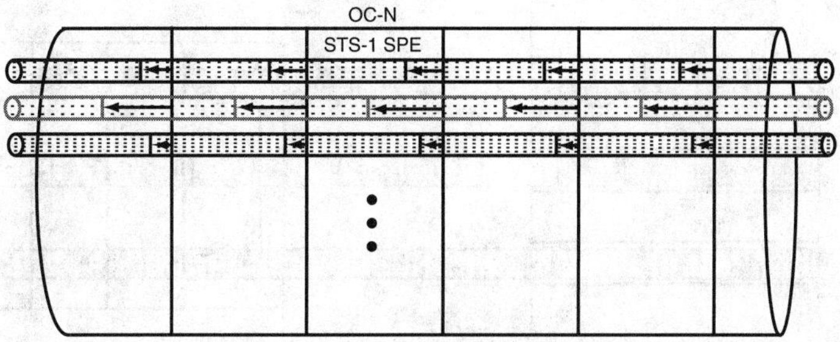

FIGURE 15.26 Pipe diagram for OC-N/STS-N.

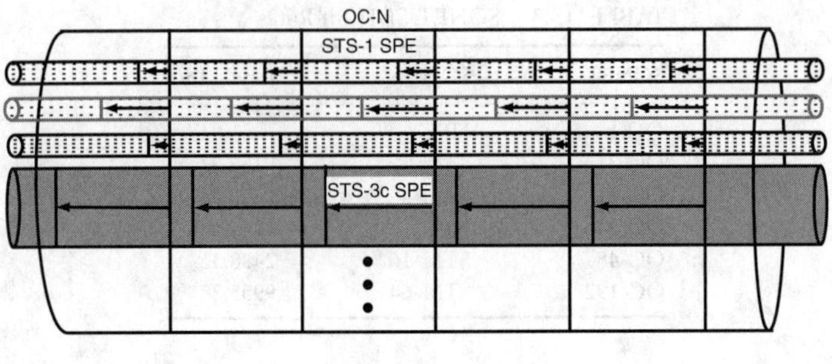

FIGURE 15.27   Pipe diagram including concatenated STS-1s.

capacities, we cannot assume that the arriving STS-1 frames are aligned since they have arrived from different places.

SONET requires that the STS-1s within an STS-N be frame aligned so that the transport overhead from each STS-1s can be combined and used as the transport overhead for the STS-N. One way to align the incoming STS-1s would be to use a 125-μs buffer to delay the early arrivals to align with the latest arrival. However, 125-μs delay in every transport network element results in an unacceptably high additional propagation delay for delay sensitive services such as voice. Furthermore, if any of the incoming OC-1s is not perfectly synchronous with respect to the hub, then the buffer will periodically overflow or underflow, causing corruption of the payload. The SONET solution is to decouple the phase alignment of the SPE from that of the STS-1 and use a pointer to indicate the offset. Thus buffers are not required and the SPE encounters minimal delay through the hub.

The second purpose of pointers is SPE frequency justification. This is the process of accommodating an offset in the SPE frame rate relative to the frame rate of the STS-1. Of course, in a perfectly synchronous network in which the timing distribution never fails, such a capability would not be required. However, reality is less than perfect.

Timing in the telecommunications network is distributed through a hierarchy of clocks. The position of a clock is the hierarchy is dependent on characteristics of the clock such as accuracy and holdover capabilities. Holdover is the ability of a clock to maintain is frequency after it has lost its frequency reference. At the top of the hierarchy is a primary reference source, or stratum 1 clock. Primary reference sources often lock to a satellite-based source, or they may be a standalone atomic clock. The clock from this source is distributed to other clocks in the network via DS1 or SONET facilities. Typically, there is one clock per network provider office called the *building integrated timing supply* (BITS) and that clock is distributed to each network element in the office. The network element itself will have a built-in clock that locks to the BITS reference. The NE clock is typically stratum 3 or lower.

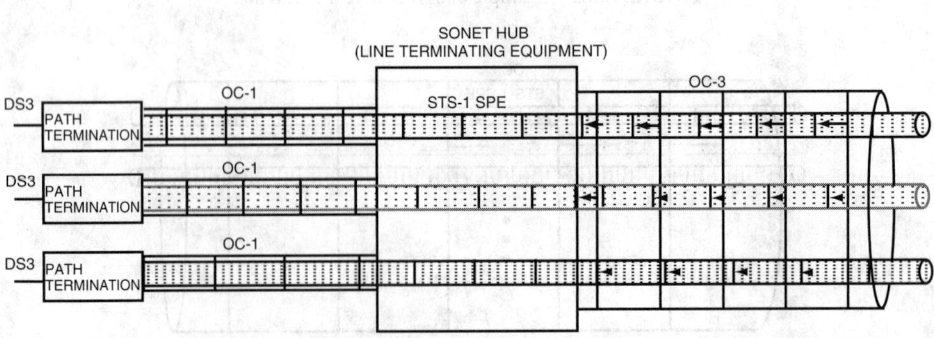

FIGURE 15.28   Pipe diagram illustrating need for pointer mechanism.

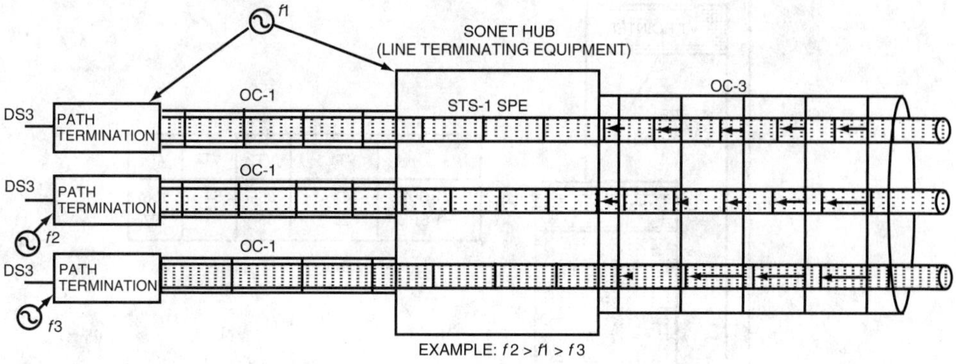

FIGURE 15.29 Pipe diagram illustrating need for frequency justification.

In this architecture, there are a number of reasons why the overall network may not be perfectly synchronous. First, atomic clocks are extremely accurate, but no two atomic clocks will be precisely the same. Second, faults in the timing distribution can cause portions of the network to be temporarily **asynchronous** with respect to the rest. Finally, although the network may be exactly synchronous over the long term, there can be shorter term phase and frequency variations that can be quite large.

The impact of imperfect network synchronization on SONET facilities is illustrated in Fig. 15.29. The hub and the first OC-1 are synchronized to the same reference, $f1$, whereas the second and third OC-1s are each locked to different references. The result is that the incoming OC-1 frames, although containing exactly the same number of bytes, have different lengths when measured in time. The SPEs from these STS-1s must be transferred to STS-1s that are frame aligned and exactly synchronous; a requirement for synchronous multiplexing. As the figure illustrates, the phase relationship between the STS-1 and the SPEs from the second and third OC-1s will be constantly changing because of the difference in frame rates. The rate of change in this phase relationship is highly exaggerated in the figure. Typically, even under worst-case conditions, the offset will remain constant for some time and then change by one byte in one direction or the other. This event is called an increment operation or a decrement operation, depending of whether the pointer offset increases or decreases.

Virtual Tributaries

The STS-1 SPE allows the bandwidth of the fiber transport network to be managed in 50-Mb/s chunks. For example, an STS cross connect can provide the flexible interconnection of STS-1 SPEs between a large number of OC-N interfaces. In many parts of the network, however, the STS-1 granularity is too coarse to allow efficient bandwidth management.

The STS-1 SPE can be subdivided into virtual tributaries (VTs) to allow bandwidth management at finer granularities ranging from approximately 1.5 up to 6 Mb/s. There are four sizes of VT defined, the VT1.5, VT2, VT3, and VT6. These sizes correspond roughly to rates of 1.7, 2, 3, and 6 Mb/s, respectively. The sizes are all multiples of 9 bytes so that they will occupy an integer number of columns in the 9 row structured STS-1 SPE frame.

As illustrated in Fig. 15.30, the VT consists of a overhead part and an SPE capacity part, just as the STS-1 consists of the transport overhead and the STS-1 SPE capacity. The overhead part of the VT only contains one function, the VT payload pointer. The VT SPE capacity carries the VT SPE, which has its own frame structure, the alignment of which may be offset with respect to the VT. The value of the offset is indicated by the VT pointer.

The pipe analogy can be extended to describe the transport of virtual tributaries. In Fig. 15.31, the pipe corresponding to the STS-1 is not shown. The outer pipe is the STS-1 SPE. The VT pointers effectively become part of the STS-1 SPE since they are in a fixed position. The VT pointers identify the alignment of

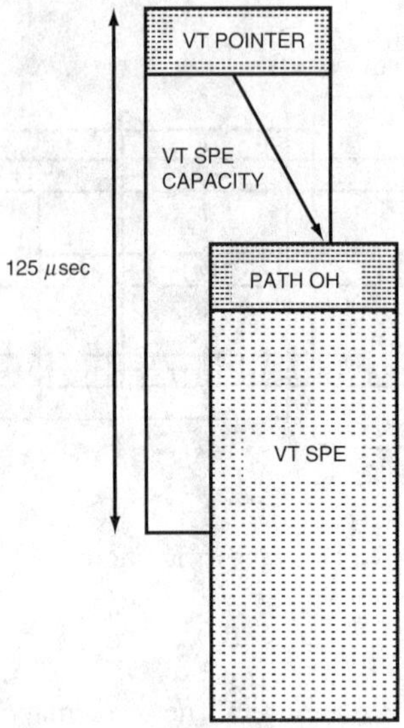

	VT1.5	VT2	VT3	VT6
VT SIZE	27	36	54	108
VT SPE	26	35	53	107

FIGURE 15.30 Virtual tributary structure.

the VT SPEs within the STS-1 SPE. Note that the VT pointer only occurs once every four 125-μs frames, effectively creating a VT superframe that spans four consecutive frames of the STS-1 SPE.

15.2.4 SONET Automatic Protection Switching

Automatic protection switching (APS) refers to the ability of SONET network elements to detect a failure and transfer the affected traffic to another line.

Linear Protection

The most basic protection system is a linear $1 + 1$ system, illustrated in Fig. 15.32. The term *linear* differentiates from ring systems and the $1 + 1$ indicates that there is 1 working fiber and 1 standby fiber and that the traffic in both directions is permanently bridged onto both the working and the standby fiber.

When one end of the system detects that the side from which it is currently selecting the traffic fails, it automatically switches to select the traffic from the other side after first verifying that the switch will restore the traffic and that there are no other reasons why the switch cannot be made. A linear $1 + 1$ system can be made survivable by diversely routing the protection channel as is shown in the figure. In this way, a fiber cut on the working channel can still be restored on the protection channel.

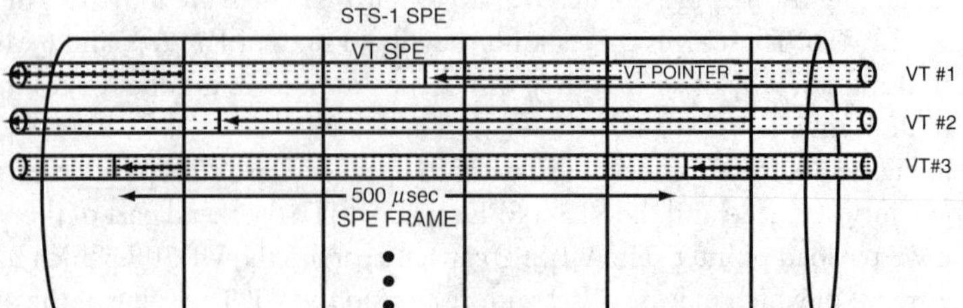

FIGURE 15.31 Pipe diagram for VTs within STS-1 SPE.

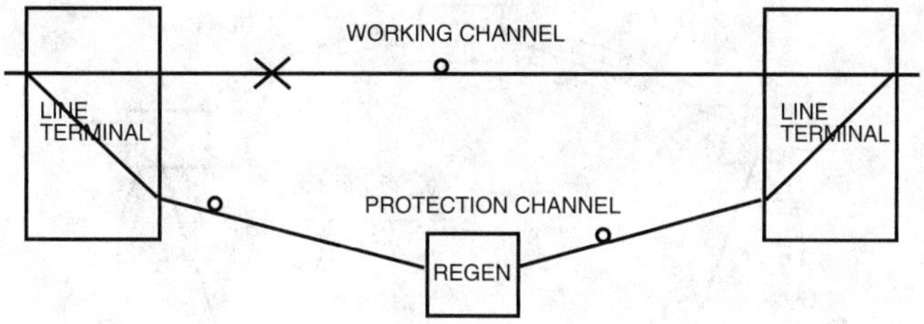

FIGURE 15.32 $1 + 1$ linear protection.

A linear 1:N (1 for N) protection system has one protection fiber providing a restoration path for up to N working fibers, as shown in Fig. 15.33. Such a system is not survivable since it is not practical to diversely route all of the fibers. Therefore, a fiber cut, even if it does not affect the protection channel, is likely to affect more than one working channel, and the protection channel can only restore one working channel at a time.

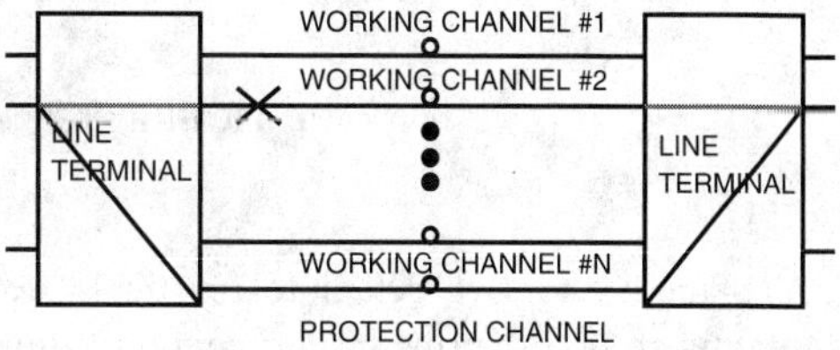

FIGURE 15.33 1:N linear protection.

A 1:N system involves more complicated signaling between the two ends of the system to coordinate the switch. This is because the failed channel must first be bridged onto the protection channel before the receiving end can do the switch, and the end that must do the bridge may not even be directly aware that there is a failure.

Ring-Based Protection

Ring-based APS provides an increased level of survivability for SONET networks by allowing as much traffic as possible to be restored, even in the event of a cable cut or a node failure. In a ring, a number of network elements are connected in a closed loop. In the event of a disruption at any point on the ring, the affected traffic can be restored by rerouting it the other way around the ring.

The two main types of ring are the *unidirectional path switched ring* (UPSR) and the *bidirectional line switched ring* (BLSR).

A 5-node UPSR is illustrated in Fig. 15.34. The ring consists of two fibers on each span, transmitting in opposite directions, between adjacent nodes. Traffic entering the ring at any node is transmitted in both directions around the ring until it reaches its exit point. Each node on the ring is line terminating, such that the connection across the ring is a path layer connection. At the exit point, the path layer integrity information is examined for each of the two paths and a selection made.

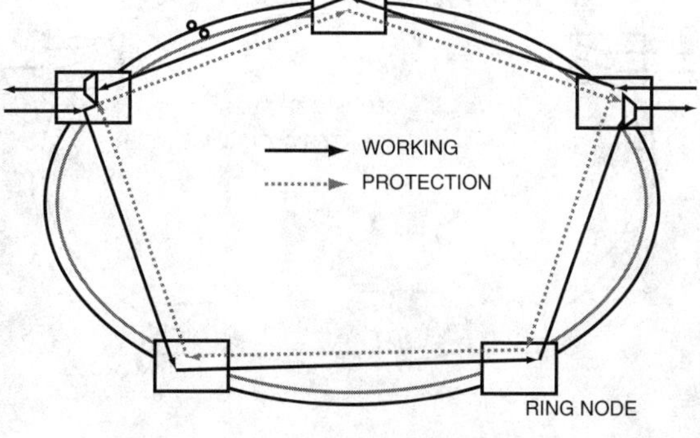

FIGURE 15.34 Unidirectional path switch ring.

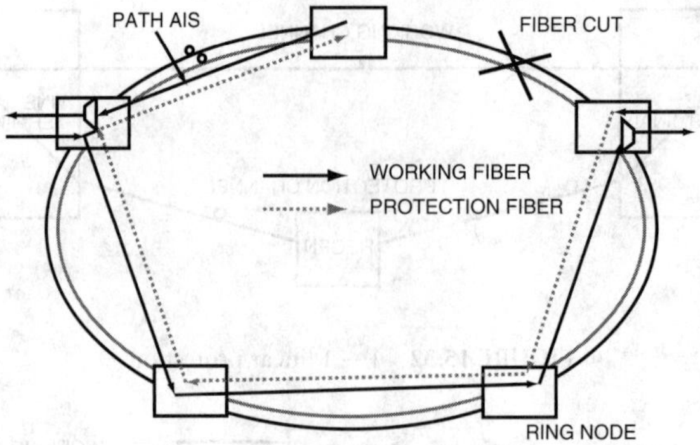

FIGURE 15.35   UPSR failure example.

All of the selectors will default to a particular direction, such that under nonfailure conditions, all traffic is rotating around the ring in the same direction, hence the name unidirectional. Since the selection is made on a per path basis, using path layer integrity information, the ring is called *path switched*.

Figure 15.35 illustrates a cable cut on the UPSR. The channel traveling from left to right is unaffected by the cut so the selector at the right does not change. However, the channel traveling from right to left is affected. The top node will detect the line failure and send an *path alarm indication signal* (AIS) on downstream to the node on the left. Upon detecting path AIS, the selector at the left will switch to the traffic traveling in the clockwise direction, restoring the bidirectional connection. Note that no signaling is required to coordinate any of the switching.

In a BLSR, as illustrated in Fig. 15.36, the two directions of a single connection occupy the same segment of the ring, that is, both directions of the ring are used even under nonfailure conditions, hence the name bidirectional.

There are two types of BLSR, the two-fiber BLSR and the four-fiber BLSR. In the two-fiber BLSR, there are two fibers, transmitting in opposite directions, between adjacent nodes. In the four-fiber BLSR, there are four fibers between adjacent nodes, two in one direction and two in the other. The four-fiber BLSR supports more traffic on the ring and allows $1 + 1$ span switching between adjacent nodes in addition to ring switching.

Note that the BLSR can support more than one connection on the ring using the same channels, as long as the connections do not overlap on any spans. This is not true of the UPSR, which uses an entire channel in both directions all the way around the ring for each and every connection. Therefore, in general, the

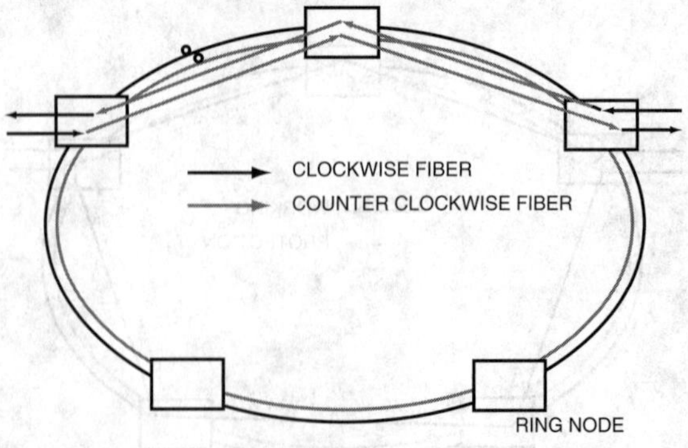

FIGURE 15.36   Bidirectional line switched ring.

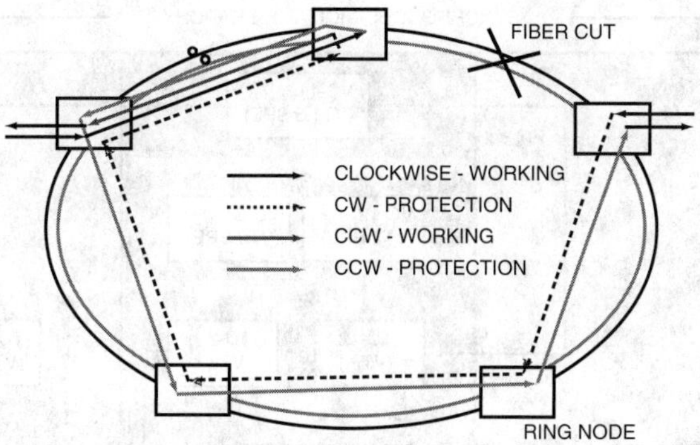

FIGURE 15.37   BLSR failure example.

BLSR can support many more connections than a UPSR, even if they are both operating at the same line rate.

In the BLSR, half of the capacity in each direction is assigned to working traffic and half is assigned to protection. If it is a two-fiber ring, then half of the STS-1s on each fiber are assigned to working and half to protection. If it is a four-fiber ring, then two fibers, one in each direction, are assigned to working and the other two are assigned to protection.

When a failure occurs and a ring switch is required, the nodes adjacent to the failure and on either side of the failure will reroute (loopback) all of the working traffic that was crossing the failed span onto the protection capacity around the ring. This is illustrated in Fig. 15.37. Since the decision to switch is made by the nodes adjacent to the failure using line layer integrity information, the ring is called *line switched*. Signaling is required to coordinate the switch between the nodes adjacent to the failure. Bytes within the line overhead are used to carry the protocol.

15.2.5   SONET/SDH Relationship

SDH has a similar structure although it uses a different terminology and has some options that are not allowed in SONET. The structure of SDH reflects two main viewpoints: one from North America and one from the European community, as represented by the European Telecommunications Standards Institute (ETSI). The North American (SONET) viewpoint is that the optical network should be managed at the 50-Mb/s granularity (AU-3), whereas the ETSI viewpoint is that the network should be managed at the 150-Mb/s granularity (AU-4).

To address the ETSI view, SDH allows VTs and STS-1 SPEs to be carried in an STS-3c SPE, whereas in SONET they can only be carried within an STS-1. There is also one other difference. SDH allows the VT1.5 SPE to be carried within a VT2. This is not allowed in SONET. However, in SDH it means that a network that can be managed at the VT2 level only will be able to carry both VT1.5 payloads as well as VT2 payloads.

Figure 15.38 shows the structure of SDH and highlights those components that are part of the SONET option, those that are part of the ETSI option, and those that are common. SDH terminology is shown, with the equivalent SONET terminology given in parentheses. Note that there are two boundaries within the common part, which cannot be crossed without choosing either the ETSI option or the SONET option. The first is that one option must be chosen to decide whether to carry the VT1.5 SPE within a VT2 or a VT1.5. The second is that one option must be chosen to decide whether to map the VT group into an STS-3c SPE or an STS-1 SPE. There is complete commonality in systems that map payloads directly into the STS-3c SPE.

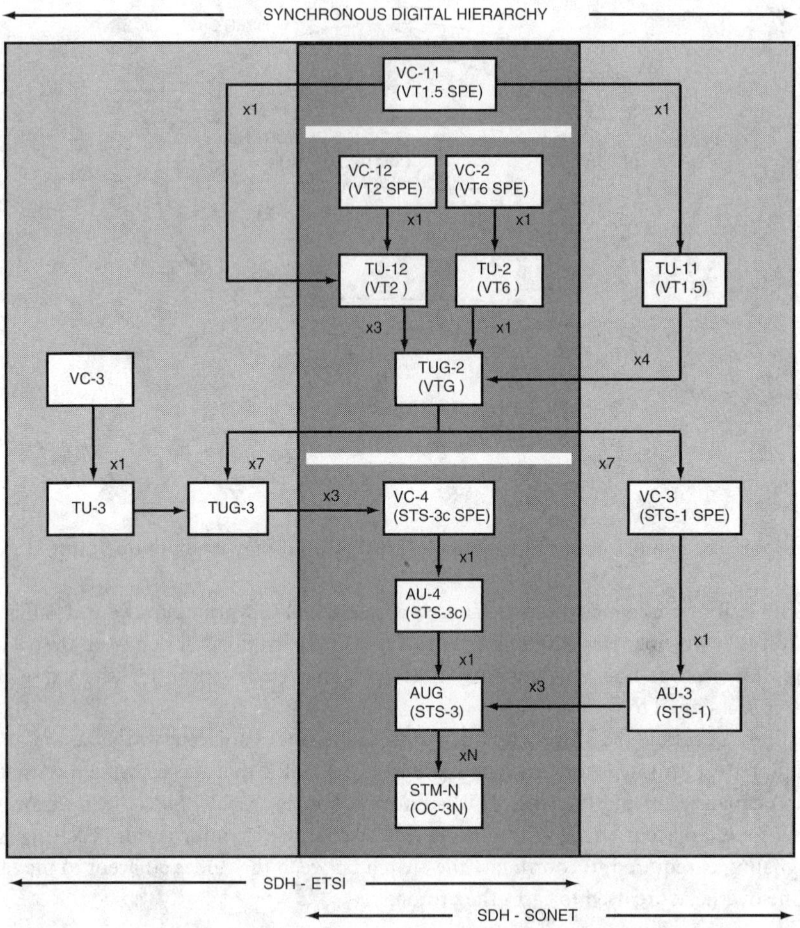

FIGURE 15.38 SONET/SDH relationship.

Defining Terms

Access: The portion of the network that connects end user terminals, such as telephones, to the service element, such as a digital switch.

Asynchronous: A network or system is asynchronous if the digital transmission clock rates are not referenced to a common timing source.

Bandwidth: The speed with which digital information is being transported, usually measured in millions of bits per second (Mb/s).

Multiplex: The process of combining a number of lower speed digital signals onto one higher speed digital signal.

Network element: A network element is any device in the telecommunications network that handles end user traffic.

Overhead Bytes within the SONET frame allocated to operations, administration, maintenance, and providing functions. They do not carry payload.

Payload: An end user signal that is transported and switched by the network.

Standby or protection fiber: Fibers that are available to restore traffic that has been affected by a failure elsewhere in the system or network, are considered standby or protection fibers.

Survivable: A network is considered survivable if it can restore traffic connectivity in the event of a complete fiber cable cut, or network element failure.

Synchronous: A network or system is synchronous if all digital transmission clock rates are referenced, directly or indirectly, to a common timing source.

Trunk: A trunk is a connection between service elements, such as digital switches.

Traffic: End user information that is transported and switched by the network.

Working fiber: Fibers that carry live traffic under normal (non-failure) conditions are considered working fibers.

References

ANSI. 1991. *Digital Hierarchy—Optical Interface Rates and Formats Specification (SONET)*, ANSI T1.105-1991. American National Standards Institute, Washington, DC.

Further Information

Additional information on the topic of SONET and SDH is available from the following sources:

ETS DE/TM-01015, *Generic Functional Requirements for SDH Transmission Equipment*, European Telecommunications Standards Institute (ETSI).

ITU-T G.709, *Synchronous Multiplexing Structure*, International Telecommunication Union.

TR-NWT-000253, *Synchronous Optical Network (SONET) Transport Systems: Common Generic Criteria*, Bellcore.

15.3 Facsimile Systems

Rodger E. Ziemer

15.3.1 Introduction

Facsimile combines copying with data transmission to produce an image of a *subject copy* at another location, either nearby or distant. Although the Latin phrase *facsimile* means to make similar, since 1815 the compressed phrase *facsimile* has been taken to mean exact copy of a transmission (Quinn, 1989). The image of the subject copy is referred to as a *facsimile,* or *record,* copy. Often the abbreviated reference "fax" is used in place of the longer term facsimile.

Facsimile was invented by Alexander Bain in 1842; Bain's system used a synchronized pendulum arrangement to send a facsimile of dot patterns and record them on electrosensitive paper. Over the years, much technological development has taken place to make facsimile a practical and affordable document transmission process. An equally important role in the wide acceptance of facsimile for image transmission has been the adoption of standards by the International Telephone and Telegraph Union (ITU-T), formerly the Consultative Committee on International Telephone and Telegraph (CCITT). The advent of a nationwide dial telephone network in the 1960s provided impetus for the rebirth of facsimile after television put the damper on early facsimile use. Group 2 fax machines, which appeared in the mid-1970s, were capable of transmitting a page within a couple of minutes. These machines, based on analog transmission methods, were developed by Graphic Sciences and 3M. The group 3 fax machines, developed in the mid-1970s by the Japanese, are based on digital transmission technology and are capable of transmitting a page in 20 s or less. They can automatically switch to an analog mode to communicate with the older groups 1 and 2 fax machines. Group 4 fax units offer the highest resolution at the fastest rates, but rely on digital telephone lines (Quinn, 1989). Group 3 facsimile will be featured in the remainder of this article. Group 3 facsimile refers to apparatus that is capable of transmitting an 8.5 × 11 in page over telephone-type circuits in 1 min or less. Detailed standards for group 3 equipment may be found in Recommendation T.4 of ITU-T, Vol. VII [ITU].

Facsimile transmission involves the separate processes of *scanning, encoding, modulation, transmission, demodulation, decoding,* and *recording* (Stamps, 1982). Each of these will be described in detail in this section.

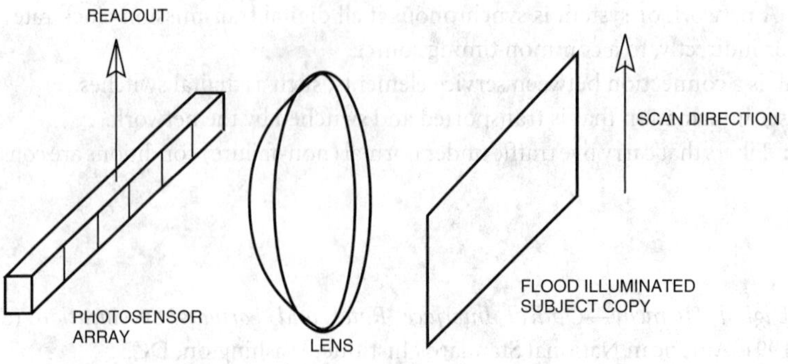

FIGURE 15.39 Arrangement for scanning by means of a linear photosensitive array.

15.3.2 Scanning

Before transmission of the facsimile signal, the **subject copy** must be scanned. This involves the sensing of the diffuse reflectance of light from the elemental areas making up the subject copy. For ITU-T group 3 high-resolution facsimile, these elemental areas are rectangles 1/208 in. wide by 1/196 in. high. The signal corresponding to an elemental area is called a **pixel** which stands for *picture element*. For pixels that can assume only one of two possible states (i.e., white on black or vice versa), the term used is a **pel**. Various arrangements of illuminating sources, light sensing transducers, and mechanical **scanning** methods can be employed (Stamps, 1982). For more than six sweeps per second across the subject copy, electronic scanning utilizing a cathode-ray tube or photosensitive arrays, or laser sources with polygon mirrors is utilized. A photosensitive array arrangement for scanning a flood-illuminated subject copy is illustrated in Fig. 15.39. This is the most often encountered scanning mechanism for modern facsimile scanners, and the sensors are typically silicon photosensitive devices. Two photosensor arrays in common use are photodiode arrays and charge-coupled-device linear image sensors. For digital facsimile, the array is typically composed of 1728 sensors in a row 1.02 in long with the optics designed so that an 8.5-in subject copy can be scanned.

15.3.3 Encoding

The output of the photosensor array for one scan or row of the subject copy consists of 1728 pels (1s or 0s) since group 3 facsimile recognizes only black or white. Typically, facsimile subject copy is 85% white. The data from scanning the subject copy is reduced through **run-length encoding**. In the encoding process, it is assumed that a white pel (0) always occurs first. A white run is the number of 0s until the first 1 is encountered (the run length is zero if the first pel is a 1); after a white run, a black run must follow with length equal to the number of 1s until the first 0 is encountered. All possible run lengths of white and black are then encoded into a binary code using a modified Huffman encoding technique (Jayant and Noll, 1984). On an average, fewer binary symbols are needed to encode run lengths of the subject copy than if the binary values of the pels themselves were transmitted. For group 3 facsimile, compression is optionally extended to the vertical dimension through employment of a relative element address designate (READ) code.

Run lengths from 0 to 63 are encoded by *terminating codes*, and run lengths in equal multiples of 64 from 64 to 1728 are encoded by *makeup codes*. Thus any run length up to 1728 can be described by a makeup code plus an appropriate terminating code. Additional makeup codewords are available for equipment that accommodates wider paper while maintaining the same resolution (Stamps, 1982). Tables of modified Huffman run length terminating and makeup codes are given in Stamps (1982).

15.3.4 Modulation and Transmission

Transmission of the encoded facsimile signal makes use of signaling techniques based on ITU-T recommendations V.27 (standard) and V.29 (optional addition). The former utilizes 8-phase modulation at 4800 b/s, and the latter employs 16-QAM quadrature amplitude modulation (QAM) at 9600 b/s with adaptive, linear equalization (a process that combats effects of intersymbol interference). A facsimile telephone call consists of a five phases, labeled A – E (Stamps, 1982). In phase A, the telephone call is placed, with a training sequence sent consisting of signals to establish carrier detection, automatic gain control, timing synchronization, and adjust equalizer tap settings; phase B consists of the called station responding with a confirmation to receive (CFR) signal. The response is a 300 b/s binary coded frequency-shift keyed signal (1 = 1650 ± 6 Hz and 0 = 1850 ± 6 Hz), except for the equalizer training sequence, which is at the fast data rate of the digital modem. In phase C the encoded facsimile image is transmitted. Phase D consists of the end-of-transmission signal consisting of six consecutive end of lines (EOLs), with receipt required from the receiver. If no more images are to be sent or received, phase E (going on-hook) is affected at both terminals.

Demodulation and Decoding

Demodulation consists of the inverse of the modulation process. Standard techniques are used to demodulate the phase modulated or QAM signals. Also included in the demodulation process is equalization. The decoding process converts the run-length-encoded information to a series of 1s and 0s corresponding to the black and white pels of the image. The demodulated and decoded signal is then used to control the recording of the image.

15.3.5 Recording

Recording of the image at the receiver to produce the **record copy** is affected by applying electricity, heat, light, ink jet, or pressure to a recording medium (Stamps, 1982). Xerography or ink jet techniques can be used to record on plain paper. Other recording means using electricity, heat, or pressure require specially coated papers. Marking transducers are used to apply the image to the recording medium. For more description of recording processes, see Stamps (1982).

Whereas character-oriented text is readily transmitted between personal computers by means of **teletex** or computer mail, facsimile transmission in conjunction with personal computers extends this capability to images (Hayashi and Motegi, 1989).

15.3.6 Group 4 Facsimile

As mentioned previously, group 4 facsimile apparatus is used mainly on public data networks of the circuit-switched, packet switched, or integrated services digital network (ISDN) varieties. Group 4 facsimile machines are subdivided into the following three classes (Yasuda et al., 1985):

- Class 1 with the minimum requirement that such equipment can send and receive documents containing facsimile-encoded information.
- Class 2, which in addition to the class 1 capabilities, must be able to receive **mixed-mode documents**.
- Class 3, which in addition to class 1 and class 2 capabilities, must be able to generate and send **mixed-mode documents**.

An additional feature of the specifications for group 4 facsimile is that the resolution is equal in the horizontal and vertical directions. Standard resolution for class 1 is 200 pels per 25.4 mm, and that for classes 2 and 3 is 200 and 300 pels per 25.4 mm. When operating as a mixed-mode terminal, a receiving density of 240 pels per 25.4 mm is required, which is optional for all three classes. Bit rates range from 2.4 to 48 kb/s with 64 kb/s for ISDN.

Defining Terms

Facsimile: The process of making an exact copy of a document through scanning of the subject copy, electronic transmission of the resultant signals modulated by the subject copy, and making a record copy at a remote location.

Mixed-mode documents: Document containing both character and facsimile information within a page. Such documents can be handled by group 4 facsimile very efficiently. Group 3 facsimile treats each document as an image to be transmitted pixel by pixel. Mixed-mode documents are subdivided into increasingly smaller parts, such as pages, frames, and blocks. A block is a rectangular area that can contain only one category (character or facsimile information).

Pel: A picture element which has been encoded as black or white, with no gray scale in between.

Pixel: A picture element of a subject or record copy that is represented in shades of gray.

Record copy: The copy of the document made at the receiving end of a facsimile system.

Run length encoding: The assignment of a codeword to each possible run of 0s (white pel sequence) or run of 1s (black pel sequence in a scan of the subject copy).

Scanning: The process of scanning the subject copy in a facsimile transmission from left to right and from top to bottom.

Subject copy: The document that is scanned and transmitted in a facsimile system.

References

Hayashi, K. and Motegi, C. Personal computer image communications using facsimile. *IEEE Journal on Selected Areas in Communications* 7(Feb.):276–282.

International Telephone and Telegraph Consultative Committee, Blue Book, Vol. VII.3.

Jayant, N.S. and Noll, P. 1984. *Digital Coding of Waveforms,* Chap. 10. Prentice-Hall, Englewood Cliffs, NJ.

Quinn, G.V. 1989. *The FAX Handbook* Tab Books. Blue Ridge Summit, PA.

Stamps, G.M. 1982. Facsimile systems. In *Electronic Engineers' Handbook,* ed. D. G. Fink, pp. 20-87–20-107. McGraw-Hill, New York.

Yasuda, Y., Yamazaki, Y., Kamae, T., and Kobayashi, K. 1985. Advances in FAX. *IEEE Proc.* 73(April):706–730.

Further Information

Cannon, D.L. and Luecke, G. 1980. *Understanding Communications Systems,* Chap. 9. Texas Instruments Inc., Dallas, TX.

Chamzas, C. and Duttweiler, D.L. 1989. Encoding facsimile images for packet-switched networks. *IEEE J. on Selected Areas in Comm.* 7(June):857–864.

Costigan, D.M. 1978. *Electronic Delivery of Documents and Graphics.* Van Nostrand Reinhold, New York.

Held, G. 1987. *Data Compression: Techniques and Applications: Hardware and Software Considerations,* 2nd ed. Wiley, New York.

Kawabe, M., Kato, T., and Sato, T. Group 3 error-free facsimile terminal for analog cellular networks. *IEEE Trans. on Veh. Tech.* 45(Feb.):64–73.

McComb, G. 1991. *Troubleshooting and Repairing Fax Machines.* Tab Books, Blue Ridge Summit, PA.

McConnell, K., Bodson, D., and Schaphorst, R. 1992. *FAX: Digital Facsimile Technology and Applications.* Artech House, Boston, MA.

16
Wireless Communications Systems

Dennis F. Doelitzsch
3-D Communications Corporation

Stanley Salek
Hammett and Edison, Inc.

Almon H. Clegg
CCi

Roy W. Rising
ABC-TV

Curtis J. Chan
Chan and Associates

Jerry C. Whitaker
Editor-in-Chief

Michel D. Yacoub
University of Campinas

Harry E. Young
MTA-EMCI

Daniel F. DiFonzo
Planar Communications Corporation

16.1   Radio Broadcasting

Dennis F. Doelitzsch

16.1.1   Standard Broadcasting (Amplitude Modulation)

Standard broadcasting refers to the transmission of voice and music received by the general public in the 535–1705 kHz frequency band. Amplitude modulation is used to provide service ranging from that needed for small communities to higher powered broadcast stations needed for larger regional areas. The **primary service area** is defined as the area in which the ground or surface-wave signal is not subject to objectionable interference or objectionable fading. The **secondary service area** refers to an area serviced by **skywaves** and not subject to objectionable interference. **Intermittent service area** refers to an area receiving service from either a surface wave or a skywave but beyond the primary service area and subject to some interference and fading.

Frequency Allocations

The carrier frequencies for standard broadcasting in the U.S. (referred to internationally as medium wave broadcasting) are designated in the Federal Communications Commission (FCC) Rules and Regulations (FCC, 1993). A total of 117 carrier frequencies are allocated from 540 to 1700 kHz in 10-kHz intervals. Each carrier frequency is required by the FCC rules to deviate no more than $+/-20$ Hz from the allocated frequency, to minimize heterodyning from two or more interfering stations.

Modulation Techniques

Double-sideband full-carrier modulation, commonly called amplitude modulation (AM), is used in standard broadcasting for sound transmission. Basically, the amplitude of an analog radio frequency (RF) carrier is controlled by an analog audio frequency (AF) modulating signal. The resulting RF waveform consists of a carrier wave plus two additional signals that are the result of the modulation process: an upper-sideband signal, which is equal in frequency to the carrier plus the audio modulating frequency, and a lower-sideband signal, which equals the carrier frequency minus the audio modulating frequency.

Figure 16.1 shows a fully amplitude modulated signal. When the peak value of the modulated carrier wave (produced by a positive excursion of the audio modulating signal) is twice the value of the unmodulated carrier and the minimum value of the RF carrier reaches zero (produced by a negative excursion of the modulating signal), 100% modulation occurs. If the transmitter is well designed, with sufficient headroom, it is possible to exceed 100% positive modulation with an asymmetrical waveform, but the negative modulation may never exceed 100% (the RF carrier at zero or completely cut off) without causing distortion of the audio signal. The power contained in the audio sidebands created in the process of modulating the carrier 100% is equal to 25% of the carrier power in both the upper and lower sidebands. As a result, the overall power transmitted during periods of 100% modulation is 150% of the unmodulated carrier power.

The overall bandwidth is equal to twice the highest audio modulating frequency. Although full fidelity is possible with amplitude modulation, the FCC requires standard broadcast stations to limit the fidelity, thus restricting occupied bandwidth of the transmitted signal. Typical modulation frequencies for voice and music range from 50 Hz to 10 kHz. Each channel is generally thought of as 10 kHz in width, and thus the frequency band is designated from 535 to 1705 kHz; however, when the modulation frequency exceeds 5 kHz, the radio frequency bandwidth of the channel exceeds 10 kHz and adjacent channel interference may occur. To improve the high-frequency performance of transmission and to compensate for the high-frequency rolloff of many consumer receivers, FCC rules require that stations boost the high-frequency amplitude of transmitted audio using pre-emphasis techniques and to use *brick wall* filters to eliminate any modulation frequencies greater than 10 kHz.

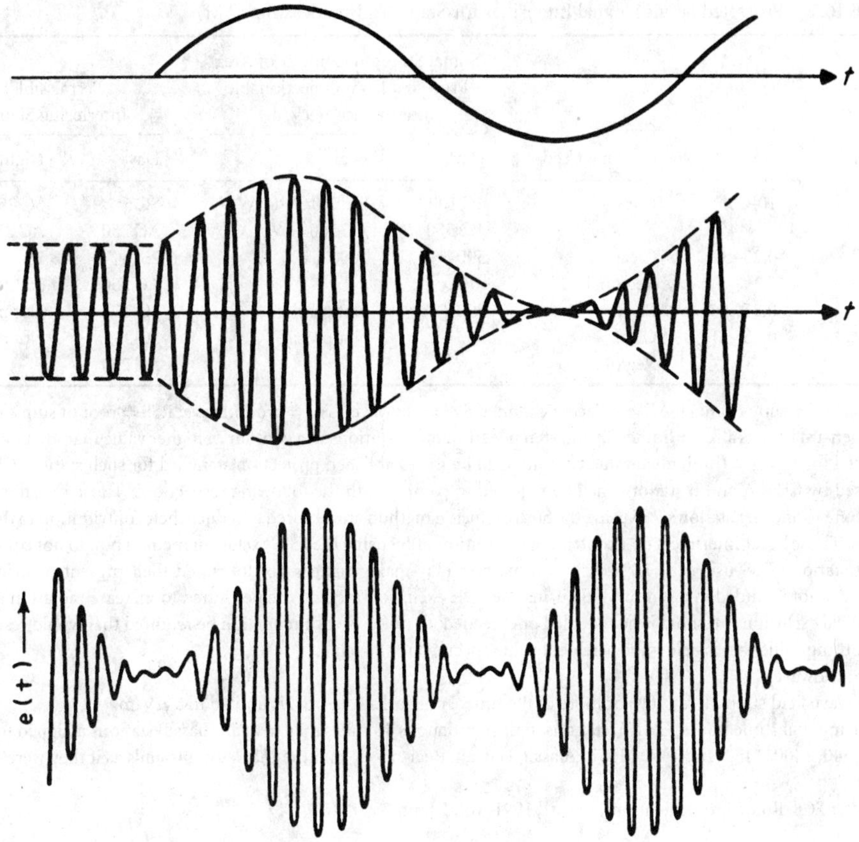

FIGURE 16.1   100% modulation of AM carrier wave.

Channel and Station Classifications

In standard broadcast (AM), stations in the U.S. are classified by the FCC according to their operating power, protection from interference, and hours of operation. A Class A station operates with 10–50 kW of power servicing a large area with primary, secondary, and intermittent coverage and is protected from interference both day and night. These stations are called *clear channel* stations because the channel is cleared of nighttime interference over a major portion of the country. Class B stations operate full time with transmitter powers of 0.25–50 kW and are designed to render primary service only over a principal center of population and contiguous rural area. Whereas nearly all Class A stations operate with 50 kW, most Class B stations must restrict their power to 5 kW or less to avoid interfering with other stations. Class B stations operating in the 1605–1705 kHz band are restricted to a power level of 10 kW daytime and 1 kW nighttime. Class C stations operate on six designated channels (1230, 1240, 1340, 1400, 1450, and 1490) with a maximum power of 1 kW or less full time and render primarily local service to smaller communities. Class D stations operate on Class A or B frequencies with Class B transmitter powers during daytime, but nighttime operation, if permitted at all, must be at low power (less than 0.25 kW) with no protection from interference.

Although Class A stations cover large areas at night (approximately a 1000 km radius), the nighttime coverage of Class B, C, and D stations is limited by interference from other stations, electrical devices, and atmospheric conditions to a relatively small area. Class C stations, for example, have an interference-free nighttime coverage radius of approximately 8–16 km. As a result there may be large differences in the area

TABLE 16.1   Protected Service Signal Intensities for Standard Broadcasting (AM)

0 Class of Station	Power (kW)	Class of Channel Used	Signal Strength Contour of Area Protected from Objectionable Interference[a] (μV/m)		Permissible Interfering Signal	
			Day[b]	Night	Day[b]	Night[c]
A	10–50	Clear	SC 100 AC 500	SC 500 50% SW AC 500 GW	SC 5 AC 250	SC 25 AC 250
B	0.25–50	Clear Regional	500	2000[b]	25 AC 250	25 250
C	0.25–1	Local	500	Not precise[d]	SC 25	Not precise
D	0.25–50	Clear Regional	500	Not precise	SC 25 AC 250	Not precise

Note: SC = same channel; AC = adjacent channel; SW = skyway; GW = groundwave; RSS = root of sum squares.

[a] When a station is already limited by interference from other stations to a contour of higher value than that normally protected for its class, this higher-value contour shall be the established protection standard for such station. Changes proposed by Class A and B stations shall be required to comply with the following restrictions. Those interferers that contribute to another station's RSS using the 50% exclusion method are required to reduce their contribution to that RSS by 10%. Those lesser interferers that contribute to a station's RSS using the 25% exclusion method but do not contribute to that station's RSS using the 50% exclusion method may make changes not to exceed their present contribution. Interferers not included in a station's RSS using the 25% exclusion method are permitted to increase radiation as long as the 25% exclusion threshold is not equaled or exceeded. In no case will a reduction be required that would result in a contributing value that is below the pertinent value specified in the table.

[b] Groundwave.

[c] Skywave field strength for 10% or more of the time. For Alaska, Class SC is limited to 5 μV/m.

[d] During nighttime hours, Class C stations in the contiguous 48 states may treat all Class B stations assigned to 1230, 1240, 1340, 1400, 1450, and 1490 kHz in Alaska, Hawaii, Puerto Rico and the U.S. Virgin Islands as if they were class C stations.

Source: FCC Rules and Regulations, revised 1991; vol. III, pt. 73.182(a).

that the station covers daytime vs. nighttime. With over 4900 AM stations licensed for operation by the FCC, interference, both day and night, is a factor that significantly limits the service stations may provide. In the absence of interference, a daytime signal strength of 2 mV/m is required for reception in populated towns and cities, whereas a signal of 0.5 mV/m is generally acceptable in rural areas without large amounts of man made interference. Secondary nighttime service is provided in areas receiving a 0.5-mV/m signal 50% or more of the time without objectionable interference. Table 16.1 indicates the interference contour overlap limits. However, it should be noted that these limits apply to new stations and modifications to existing stations. Nearly every station on the air was allocated prior to the implementation of these rules with interference criteria that were less restrictive.

Field Strength

The field strength produced by a standard broadcast station is a key factor in determining the primary and secondary service areas and interference limitations of possible future radio stations. The FCC and others have conducted significant testing to determine what levels of interference are acceptable from adjacent and co-channel stations and have used that data in determining how close stations may be located to one another. If the transmitted power, antenna radiation characteristics, and ground conductivity are known, the extent of coverage (or interference) in a given direction for a particular station can be calculated with a high degree of accuracy. These calculations form the basis of the FCC's station allocation system and are based on field intensities with the unit of measure as volts per meter (the voltage induced in an antenna 1 m in length). The protected service contours and permissible interference contours for standard broadcast stations are shown in Table 16.1. This table, along with a knowledge of the field strength of existing broadcast stations, may be used in determining the potential for establishing new standard broadcast stations.

Propagation

One of the major factors in the determination of field strength is the propagation characteristic, described as the change in electric field intensity with an increase in distance from the broadcast station antenna. This variation depends on a number of factors including frequency, distance, surface dielectric constant, surface loss tangent, polarization, local topography, and time of day. Generally speaking, surface-wave propagation occurs over shorter ranges both during day and night periods. Skywave propagation in the AM broadcast band permits longer ranges and occurs during night periods, and thus some stations must either reduce power or cease to operate at night to avoid causing interference. Propagation curves in the broadcast industry are frequently referred to a reference level of 100 mV/m at 1 km; however a more general expression of surface-wave propagation may be obtained by using the Bremmer series (Bremmer, 1949). A typical surface-wave propagation curve for electric field strength as a function of distance is shown in Fig. 16.2 for an operating frequency of 770–810 kHz. The ground conductivity varies from 0.1 to 5000 mS/m, and the ground relative dielectric constant is 15.

Skywave propagation is much less predictable than surface-wave propagation, because it necessarily involves some fading and less predictable field intensities and is most appropriately described in terms of statistics or the percentage of time a particular field strength level is found. Figure 16.3 shows skywave propagation of a 100 mV/m inverse field strength at a distance of 1 km for midpoint path latitudes of 35–50°.

Transmitters

Standards that cover AM broadcast transmitters are given in the Electronic Industry Association (EIA) Standard (EIA, 1948). Parameters and methods for measurement include the following:

- Carrier output rating
- Carrier power output capability
- Carrier frequency range
- Carrier frequency stability
- Carrier shift
- Carrier noise level
- Magnitude of RF harmonics
- Normal load impedance
- Transmitter output circuit adjustment facilities
- RF and audio interface definitions
- Modulation capability
- Audio input level for 100% modulation
- Audio frequency response
- Audio frequency harmonic distortion
- Rated power supply
- Power supply variation
- Operating temperature characteristics
- AC power input

Standard AM broadcast transmitters range in power output from 5 W up to 50 kW units. Modern transmitters utilize low-voltage, high-current semiconductor devices to generate the RF power. However, in a high-power transmitter, it may take hundreds of transistors to generate the rated power. These transistors are combined into modules, the outputs of which are combined together. This design approach has the added benefit that in the event of a module failure, the transmitter continues to operate, but at slightly reduced power. A block diagram of a 10 KW solid state AM transmitter is shown in Fig. 16.4.

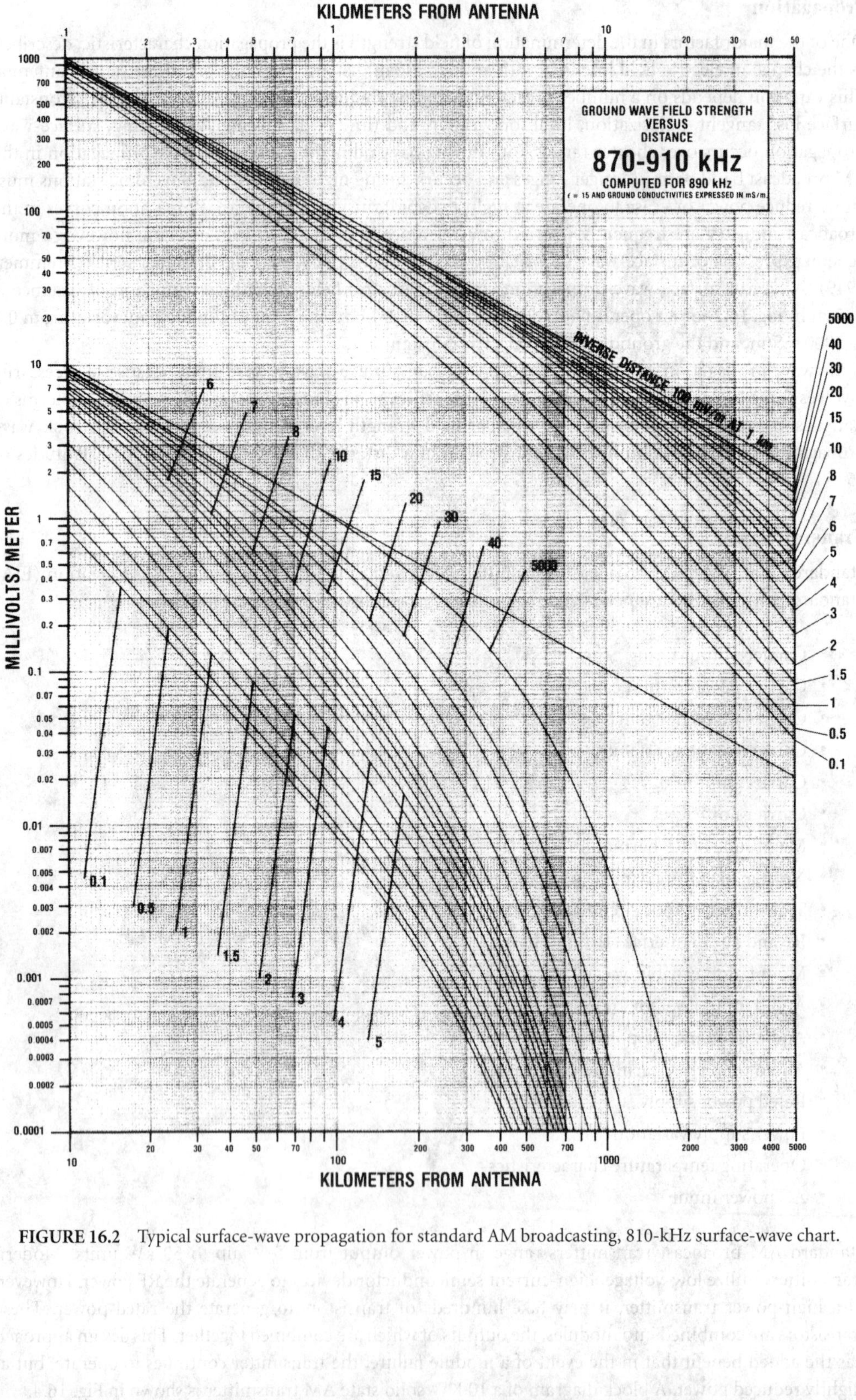

FIGURE 16.2 Typical surface-wave propagation for standard AM broadcasting, 810-kHz surface-wave chart.

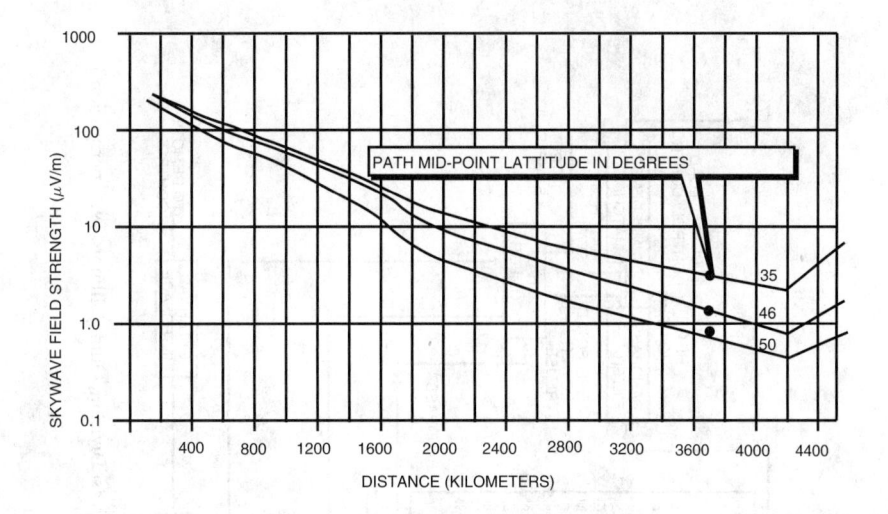

FIGURE 16.3   Skywave propagation formula for standard AM broadcasting. (*Source:* FCC. 1993. FCC Rules and Regulations, revised 1993; vol. III, pt. 73.190(b).)

Antenna Systems

The antenna system for a standard AM broadcast station typically consists of four parts:

1. Vertical radiator. This is a tower, which may be self-supporting or guyed. The tower may be base insulated (a large insulator is installed at the base to electrically isolate the tower) or electrically grounded. An insulated tower is series fed at the base. A grounded tower is shunt or slant wire fed at a point on the tower that presents a usable radiation resistance. If the tower is guyed, the guy wires have insulators inserted in them at intervals of 1/8 wavelength or less.
2. Ground system. The ground system usually consists of 120 or more quarter-wavelength radials of #10 softdrawn bare copper wire buried a few inches underground. The ground system is very important, even though it is hidden under the ground's surface, since it determines the efficiency and stability of the system. The ground system can be improved further by interspersing more radials or installing an 8-m^2 ground screen under the base of the tower.
3. Antenna tuning unit (ATU). The tower's complex impedance at the feed point must be *matched* to the characteristic impedance of the transmission line so that maximum transfer of power may occur. Elements of inductive and capacitive reactance (coils and capacitors) are combined into a T or L network to accomplish this task. The ATU will also contain provisions for measuring base RF antenna current.
4. Transmission line. The transmission line conducts the RF signal from the transmitter to the ATU. It is generally a coaxial cable of 50-Ω characteristic impedance, which is buried several feet beneath the ground's surface to protect it from damage.

Typical heights for AM broadcast towers range from 50 to 150 m. The surface-wave radiation efficiency of an antenna (the signal strength at 0° elevation) is determined by the radiation characteristic of the tower. The efficiency increases as the height of the tower increases up to a maximum at approximately 5/8 wavelength. In addition, when the surface-wave efficiency increases, the radiation at angles above the horizon decreases, thus minimizing nighttime skywave interference to other stations. However, 5/8 wavelength antennas, especially at the lower standard broadcast band frequencies, are more than 150 m tall and become cost prohibitive and difficult to construct. As a result, 1/4 wavelength antennas are the generally accepted standard for AM broadcast stations.

When the circular radiated signal pattern must be modified to prevent interference to other stations or to provide better service in a particular direction, additional antenna systems may be combined in a phased array to produce the desired field intensity contours. For example, if a station power increase would cause

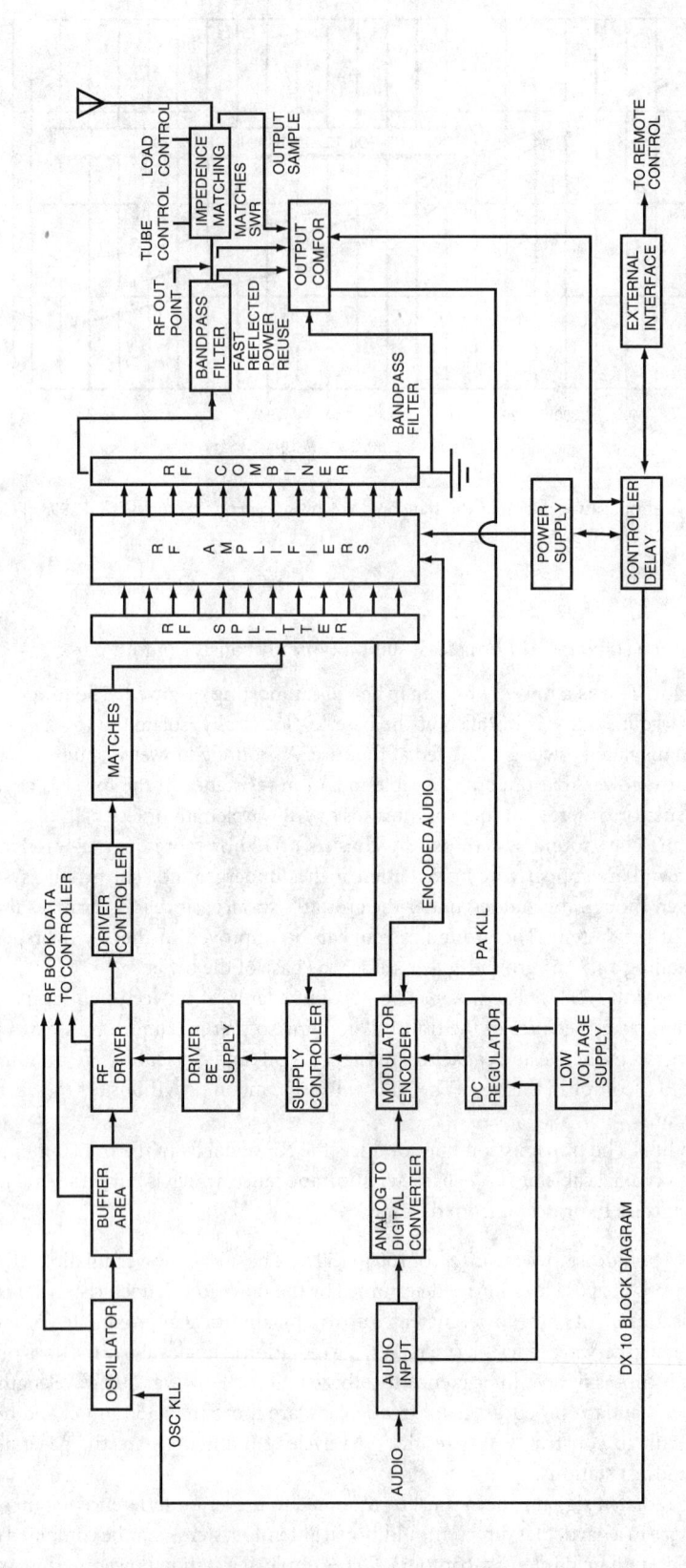

FIGURE 16.4   Block diagram and picture of typical 10-KW Solid State AM transmitter. (*Source:* Harris Corporation, Broadcast Division, Quincy, Illinois. With permission.)

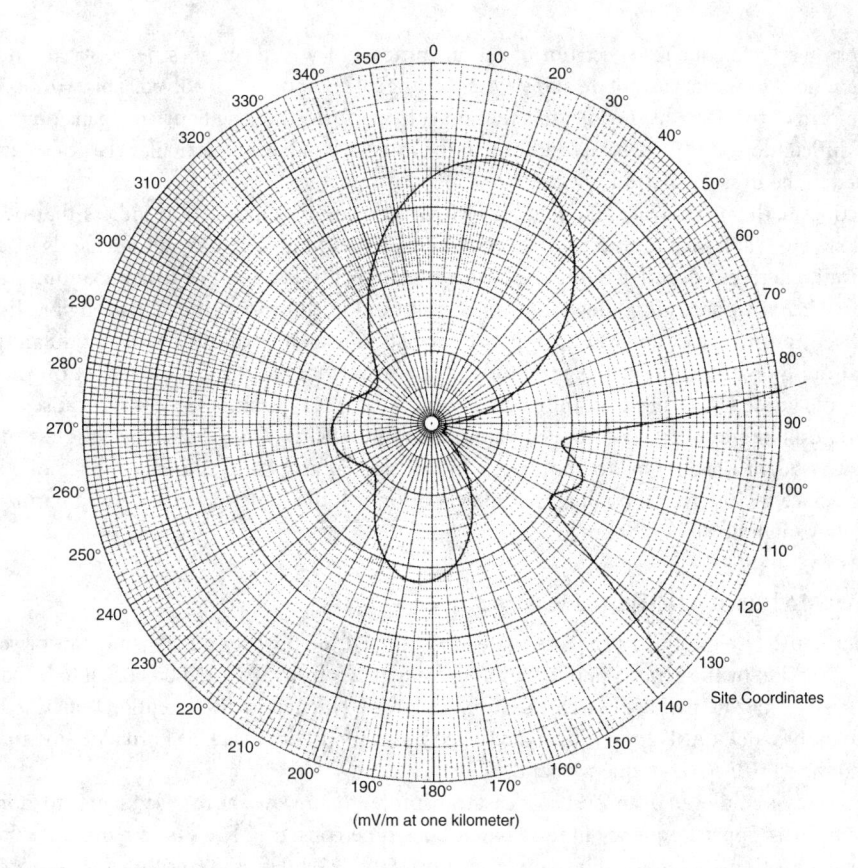

(mV/m at one kilometer)

FIGURE 16.5 Directional AM antenna pattern for three element array. (*Source:* Independent Broadcast Consultants, Trumansburg, NY. Reprinted with permission of du Treil, Lundin & Rackly, Consulting Engineers.)

interference with existing stations, a directional array could be designed that would tailor the coverage to protect the existing stations while allowing increases in radiation in other directions. The protection requirements can generally be met with arrays consisting of 4 towers or less, but complex arrays have been constructed consisting of 12 or more towers to meet stringent requirements for particular locations. The directional pattern in Fig. 16.5 shows the type of shaping that may be required to meet interference requirements.

16.1.2 Frequency Modulation

Frequency modulation utilizes the audio modulating signal to vary the *frequency* of the RF carrier. The greater the amplitude of the modulating frequency, the greater the frequency deviation from the center carrier frequency. The rate of the frequency variation is a direct function of the frequency of the audio modulating signal. In FM modulation, multiple pairs of sidebands are produced. The actual number of sidebands that make up the modulated wave is determined by the modulation index (MI) of the system. The modulation index is a function of the frequency deviation of the system and the applied modulating signal.

$$MI = \text{frequency deviation/modulating frequency}$$

As the MI increases there are more sidebands produced. As the modulating frequency increases for a given maximum deviation, there will be a smaller number of sidebands spaced at wider intervals. Unlike amplitude modulation, which has a percentage of modulation directly proportional to the carrier power, the percentage of modulation in FM is generally referenced to the maximum allowable occupied bandwidth set by regulation. For example, FM broadcast stations are required to restrict frequency deviation of $+/-75$ kHz from the main carrier. This is referred to as 100% modulation for FM broadcast stations.

To determine the frequency spectrum of a transmitted FM waveform, it is necessary to compute a Fourier series or Fourier expansion to show the actual signal components involved. This work is difficult for a waveform of this type, as the integrals that must be performed in the Fourier expansion or Fourier series are difficult to solve. The actual result is that the integral produces a particular class of solution that is identified as the **Bessel function**.

Supporting mathematics will show that an FM signal using the modulation indices that occur in a broadcast system will have a multitude of sidebands. To achieve no distortion, all sidebands would have to be transmitted, received, and demodulated. However, in reality, the $+/-$ 75-kHz maximum deviation, which is used in FM broadcasting, allows the transmission and reception of audio with negligible distortion.

The power transmitted by an FM transmitter is virtually constant, regardless of the modulating signal. Additional noise or distortion of the amplitude of the waveform has virtually no effect on the quality of the received audio. Thus, FM transmitters may utilize Class C type amplifiers, which cause amplitude distortion and allow greater powers to be generated, but are inherently more efficient than the less amplitude distorting Class A or Class B types of amplifiers. In addition, atmospheric and man-made noise has very little effect, since the receiver clips all amplitude variations off the signal prior to demodulation. This is why FM allows high fidelity with very low noise.

Frequency Modulation (FM) Broadcasting

The monophonic system of FM broadcasting was developed to allow sound transmission of voice and music for reception by the general public for audio frequencies from 50 to 15,000 Hz, all to be contained within a $+/-$ 75-kHz RF bandwidth. This technique provided higher fidelity reception than was available with standard broadcast AM along with less received noise and interference. FM broadcasting in the U.S. is allocated the 88–108 MHz frequency band.

Pre-emphasis is employed in an FM broadcast transmitter to improve the received signal-to-noise ratio. The pre-emphasis upper-frequency limit is based on a time constant of 75 μs as required by the FCC for FM broadcast transmitters. Audio frequencies from 50 to 2120 Hz are transmitted with normal FM; whereas audio frequencies from 2120 Hz to 15 kHz are emphasized with a larger modulation index. There is significant signal-to-noise improvement at the receiver, which is equipped with a matching **de-emphasis** circuit.

FM Broadcast Modulation Techniques

FM stereo was developed in 1961 to provide transmission capability for a left- and right-stereo audio signal. Stereophonic transmission is accomplished by adding the left- and right-channel stereo information together in the baseband signal. In addition, a left-minus-right channel is added and frequency multiplexed on a subcarrier of 38 kHz using double sideband suppressed carrier modulation. An unmodulated 19-kHz subcarrier is derived from the 38-kHz subcarrier to provide a synchronous demodulation reference for the stereophonic receiver. In the receiver, a synchronous detector at 38 kHz recovers the left-minus-right channel information, which is then combined with the left-plus-right channel information in sum and difference combiners to produce the original left-channel and right-channel signals. The system works well and, because the baseband mono signal is a combination of both left and right audio, provides full compatibility with monophonic FM receivers.

Subsidiary Communications Authorization (SCA)

In addition to its main channel and 38-kHz subcarrier stereo broadcasting, an FM station may utilize additional subcarriers for private commercial use. These SCAs may be used in a variety of ways, such as paging, data transmission, specialized foreign language programs, radio reading services, utility load management, and background music. An FM stereo station may utilize multiplex subcarriers within the range of 53–99 kHz with up to 20% total SCA modulation of the main carrier using any form of modulation, including techniques for the transmission of digital information. The only requirement is that the station does not exceed its occupied bandwidth limitations or cause interference to itself.

TABLE 16.2 FM Station Classifications, Powers, and Antenna Heights

Station Class	Maximum ERP	HATT, m(ft)	Distance, km
A	6 kW (7.8 dBk)	100(328)	28
B1	25 kW (14.0 dBk)	100(328)	39
B	50 kW (17.0 dBk)	150(492)	52
C3	25 kW (14.0 dBk)	100(328)	39
C2	50 kW (17.0 dBk)	150(492)	52
C1	100 kW (20.0 dBk)	299(981)	72
C	100 kW (20.0 dBk)	600(1968)	92

Source: FCC Rules and Regulations, revised 1991; vol. III, pt. 73.211(b)(1).

Frequency Allocations

The 100 carrier frequencies for FM broadcast range from 88.1 to 107.9 MHz and are equally spaced every 200 kHz. The channels from 88.1 to 91.9 MHz are reserved for educational and noncommercial broadcasting, and those from 92.1 to 107.9 MHz for commercial broadcasting. Each channel has a 200-kHz bandwidth. The maximum frequency swing under normal conditions is $+/- 75$ kHz. Stations operating with an SCA may, under certain conditions, exceed this level, but in no case may exceed a frequency swing of $+/- 82.5$ kHz. The carrier center frequency is required to be maintained within $+/- 2000$ Hz. The frequencies used for FM broadcasting limit the coverage to essentially line-of-sight distances. As a result, consistent FM coverage is limited to a maximum receivable range of a few hundred kilometers depending on the antenna height above average terrain (HAAT) and **effective radiated power** (ERP). The actual coverage area for a given station can be reliably predicted after the power and the antenna height are known. Either increasing the power or raising the antenna will increase the coverage area.

Station Classifications

In FM broadcast, stations are classified by the FCC according to their maximum allowable ERP and the transmitting antenna HAAT in their service area. Class A stations provide primary service to a radius of about 28 km with 6000 W of ERP at a maximum HAAT of 100 m. The most powerful class, Class C, operates with maximums of 100,000 W of ERP and heights up to 600 m with a primary coverage radius of over 92 km. The powers and heights above average terrain for all classes are shown in Table 16.2. All classes may operate at antenna heights above those specified but must reduce the ERP accordingly. Stations may not exceed the maximum power specified, even if antenna height is reduced. The classification of the station determines the allowable distance to other cochannel and adjacent channel stations.

Field Strength and Propagation

The field strength produced by an FM broadcast station depends on the ERP, antenna HAAT, local terrain, tropospheric scattering conditions, and other factors. From a statistical point of view, however, an estimate of the field intensity may be obtained from Fig. 16.6. A factor in the determination of new licenses for FM broadcast is the separation between allocated cochannel and adjacent channel stations, the class of station, and the antenna heights. The spacings are given in Table 16.3. The primary coverage of all classes (except B and B1, which are 0.5 and 0.7 mV/m, respectively) is the 1.0-m V/m contour. The distance to the primary contour, as well as to the *city grade* or 3.16-mV/m contour may be estimated using Fig. 16.5. Although FM broadcast propagation is generally thought of as line of sight, larger ERPs along with the effects of diffraction, refraction, and tropospheric scatter allow coverage slightly greater than line of sight.

Transmitters

FM broadcast transmitters typically range in power output from 10 W to 50 kW and have performance standards similar to those specified above for AM broadcast transmitters. Digital technology is used extensively in the processing and exciter stages, allowing for precision modulation control, whereas solid-state devices are predominately used for RF drivers and amplifiers up to several kilowatts. High-power

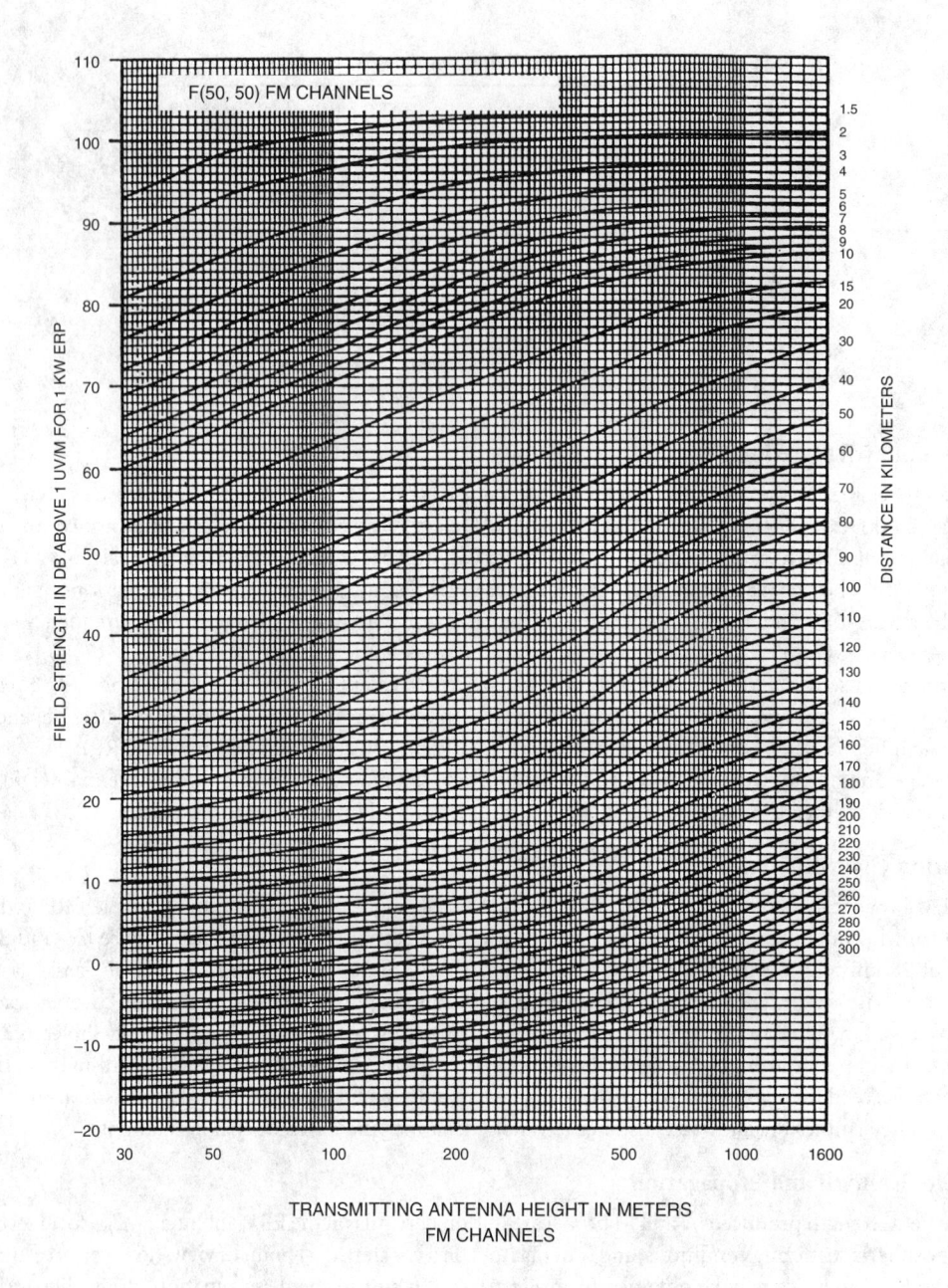

FIGURE 16.6   Propagation curves for FM broadcasting. (*Source:* FCC 1993. FCC Rules and Regulations, revised 1993; vol III, pt. 73.333.)

transmitters still rely mainly on tubes in the final stages for the generation of RF power. A block diagram and picture of a typical high-power FM transmitter are shown in Fig. 16.7.

Antenna Systems

FM broadcast antenna systems are required to have a horizontally polarized component. Most antenna systems, however, are circularly polarized, having both horizontal and vertical components. The antenna

TABLE 16.3 Distance Separation Requirements for FM Stations.

Station Class Relation	Minimum Distance Separation Requirements, km (mi)			
	Co-Channel	200 kHz	400/600 kHz	10.6/10.8 MHz
A to A	115(71)	72(45)	31(19)	10(6)
A to B1	143(89)	96(60)	48(30)	12(7)
A to B	178(111)	113(70)	69(43)	15(9)
A to C3	142(88)	89(55)	42(26)	12(7)
A to C2	166(103)	106(66)	55(34)	15(9)
A to C1	200(124)	133(83)	75(47)	22(14)
A to C	226(140)	165(103)	95(59)	29(18)
B1 to B1	175(109)	114(71)	50(31)	14(9)
B1 to B	211(131)	145(90)	71(44)	17(11)
B1 to C3	175(109)	114(71)	50(31)	14(9)
B1 to C2	200(124)	134(83)	56(35)	17(11)
B1 to C1	233(145)	161(100)	77(48)	24(15)
B1 to C	259(161)	193(120)	105(65)	31(19)
B to B	241(150)	169(105)	74(46)	20(12)
B to C3	211(131)	145(90)	71(44)	17(11)
B to C2	211(131)	145(90)	71(44)	17(11)
B to C1	270(168)	195(121)	79(49)	27(17)
B to C	274(170)	217(135)	105(65)	35(22)
C3 to C3	153(95)	99(62)	43(27)	14(9)
C3 to C2	177(110)	117(7)	56(35)	17(11)
C3 to C1	211(131)	144(90)	76(47)	24(15)
C3 to C	237(147)	176(109)	96(60)	31(19)
C2 to C2	190(118)	130(81)	58(36)	20(12)
C2 to C1	224(139)	158(98)	79(49)	27(17)
C2 to C	237(147)	176(109)	96(60)	31(19)
C1 to C1	245(152)	177(110)	82(51)	34(21)
C1 to C	270(168)	209(130)	105(65)	35(22)
C to C	290(180)	241(150)	105(65)	48(30)

Source: FCC Rules and Regulations, revised 1991; vol. III, pt. 73.207.

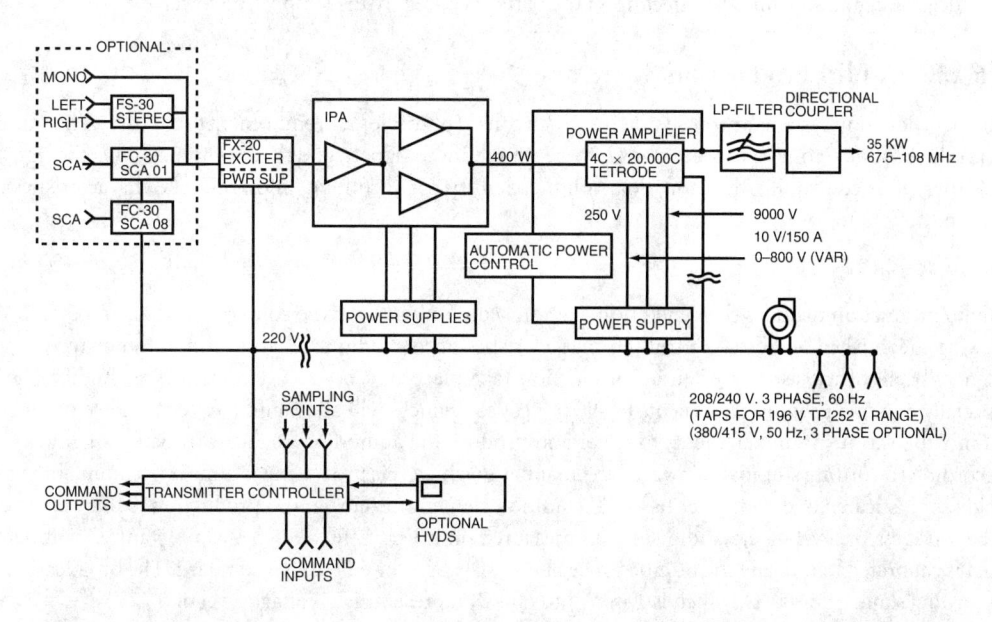

FIGURE 16.7 Block diagram of typical FM transmitter. (*Source:* Broadcast Electronics, Quincy, IL. With permission.)

TABLE 16.4 Power and Antenna Gain Combinations to Achieve 50- kW ERP

Transmitter Power, kW	No. Bays	Antenna Gain, dB	Feedline Efficiency, %	Effective Radiated Power, kW
30.13	4	3.29	77.8	50.0
13.58	8	6.52	82.0	50.0
10.19	12	8.37	71.3	50.0

system, which usually consists of several individual radiating bays, is fed as a phased array mounted on a tower, has a radiation characteristic that concentrates the transmitted energy in the horizontal plane toward the population to be served, minimizing the radiation out into space and down toward the ground. Thus, the ERP toward the horizon is increased with power gains up to 10 dB. That means that a 5-kW transmitter coupled to an antenna system with a 10-dB gain would have an ERP of 50 kW.

The RF power is transferred to the antenna via a transmission line. At FM broadcast frequencies and powers, coaxial transmission line is used, ranging in size from 7/8 in outside diameter to 3 in. The dielectric (the material between the inner and outer conductors in coaxial transmission lines) is foam or for air types, either dry air or nitrogen. These coax cables are not 100% efficient, and some power is lost as heat in the line. The larger diameter air type cables have the least loss. The combination of transmitter power output (TPO), feedline loss, and antenna gain determine the actual ERP of the station. For a given ERP, a station may install a large, multibay antenna with high gain and relatively low TPO or use a low-gain antenna and a high TPO. The choice is determined by a number of factors, including service area, terrain in the area, and economics.

The radiation pattern of a nondirectional FM broadcast antenna system is designed to radiate equally well in all directions. However, when mounted on a tower as a support structure, the structure itself distorts the uniformity of the pattern, especially on towers with larger cross-sectional dimensions of over 36 in. It is possible to optimize the omnidirectional pattern using additional parasitic elements to offset the effects of the tower and achieve a more circular pattern. The use of such elements also allows an intentional directionalization of the pattern to avoid interference with other stations or to meet spacing requirements. In order to meet FCC requirements, a directional FM antenna system must meet strict standards. Figure 16.8 is a plot of the horizontal and vertical components of a typical nondirectional circularly polarized FM broadcast antenna showing the effect upon the pattern caused by the supporting tower.

16.1.3 Audio Production Systems

Audio production systems are those systems which consist of studios and equipment to capture, generate, or reproduce sounds, convert those sounds into electronic audio signals, and then combine or *mix* the signals together while controlling the individual volume, density, tonal content, and other like characteristics of each of the sounds or signals.

Studios

Audio production includes the generation of audio, combining the audio sources together, and then processing the finished audio to control the characteristics of the audio prior to transmission or recording. Generally, all of the aspects of audio production take place in a broadcast or recording studio. These specially equipped rooms are as acoustically perfect as finances will allow and have elaborate equipment to mix the various sound elements together and produce the audio for broadcast (broadcast studios) or recording (recording studios). However, the manner in which this processing is done is becoming increasingly sophisticated and computer based. A common element in most audio production is the VU meter. The loudness or level of the audio signal is measured on a VU meter, which is simply an AC voltmeter that is calibrated to indicate the number of decibels (dB) above or below a set standard. The broadcast and recording standard is 0 VU, which is 1 mW into 600 Ω, representing a voltage level of 0.775 V.

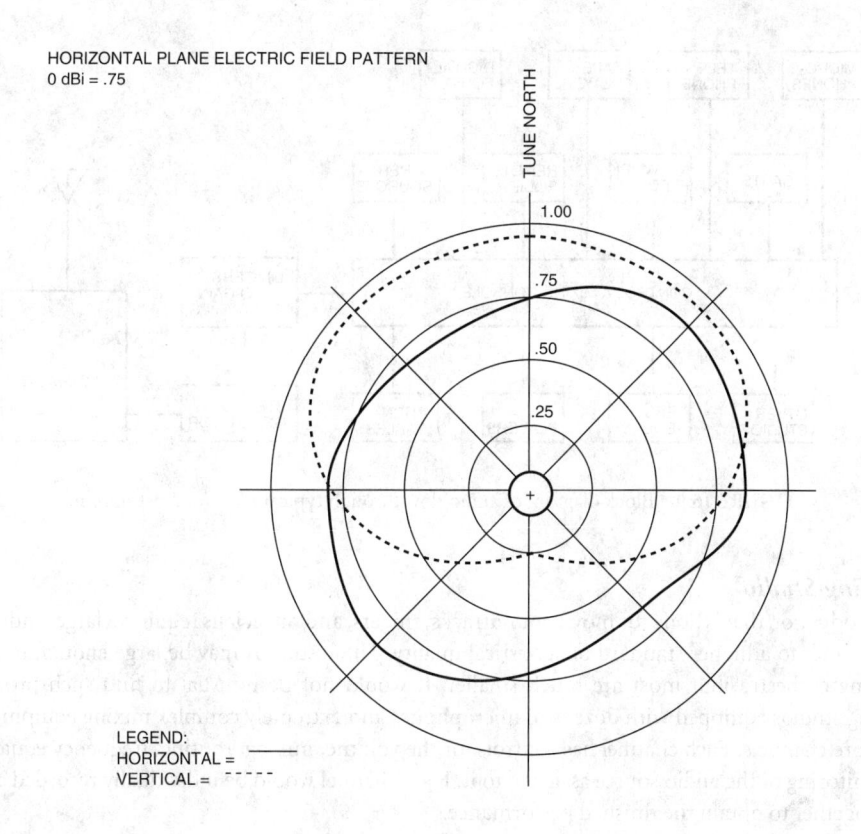

FIGURE 16.8 Typical nondirectional 92.5-MHz FM antenna characteristics showing the effect of the tower structure on the horizontal and vertical radiation patterns. (*Source:* Electronics Research, Inc. Chandler, IN.)

Audio production systems have rapidly replaced analog with digital as the preferred method for production, storage, and transmission. Analog signals have limited bandwidth and are subject to increasing noise and degradation with multiple generations, unlike digital signals, which are virtually unlimited as to bandwidth and offer signal to noise ratios of over 100 dB.

The basic elements of studio production are

- Audio source equipment such as microphones, musical instruments, CD players, reel-to-reel tape machines, digital tape machines, cassette players, and so forth, virtually any source that can generate or reproduce audio signals
- Audio mixing equipment such as an audio mixing console (either analog or digital)
- Audio processing equipment such as equalizers, limiters, compressors, and digital effects synthesizers and processors
- Audio storage equipment such as digital hard disk storage, reel-to-reel tape equipment, digital tape machines, cassette recorders, minidisk, and CD recorders.

Computer-based systems are becoming common that input a variety of sources, mix, process, and store, all within the computer environment, eliminating much of the equipment found in conventional studios.

The three most common types of audio production studios are: 1) recording studios used for mastering music and movie sound track recordings, 2) on-air studios used for assembling the audio elements of a radio show, and 3) production studios at broadcast stations used to record commercials and other audio elements for later broadcast.

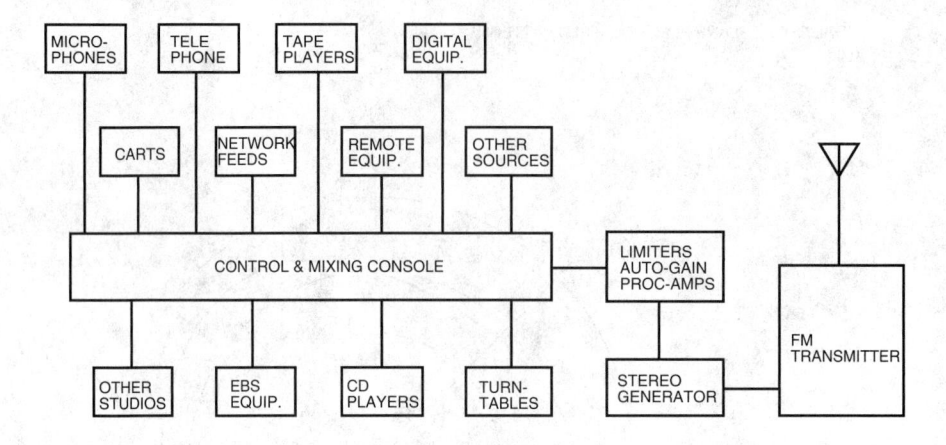

FIGURE 16.9   Block diagram of audio flow through typical FM broadcast station.

Recording Studio

Audio production that is done for movie soundtracks, singers, and musicians requires a large studio, which has been built to a higher standard of acoustical quality. These studios may be large enough to record a symphony orchestra, but most are much smaller. It would not be unusual to find such professional recording studios equipped with dozens of microphones and extremely complex mixing equipment with 64 or more channels. Each channel has controls for the volume, mixing, routing, frequency equalization, and monitoring of the audio source assigned to it. Each channel would be individually recorded and later mixed together to obtain the finished performance.

On-Air Studio

An on-air studio differs from the recording and production studios in that studios of this sort are primarily concerned with controlling the many different audio sources that are used for programming the station. A typical on-air studio may have as many as 30 or more different sources of audio to control, including network audio, other studios, and so forth. Generally, only two or three microphones are available for use and the acoustical properties of the studio are tailored for voice work. Figure 16.9 shows a block diagram of the signal flow and audio source equipment in a typical FM radio station on-air studio.

Production Studio

A production studio in a radio station is used to prepare various audio elements for later broadcast. For example, the production of a typical radio commercial would utilize many production elements. An announcer's voice may be converted into audio signals by a microphone, then equalized, limited, and enhanced by processing equipment, mixed at a preset volume in a mixing console, and finally recorded onto a single track of a multitrack digital editing workstation. Similarly, sound effects (either studio generated or from a sound effects audio library) may be originated, processed, mixed, and recorded onto other tracks. A group of singers and an orchestra may be used, following the same process to be placed on additional tracks. An audio editor would manipulate the multitrack digital editor to produce a perfectly edited, finished commercial for use at one or more radio stations.

Defining Terms

Bessel function: A mathematical integration that produces a set of curves, which in graph form can be used to determine the amplitude of FM sidebands as a result of the modulations process.

Digital audio: Audio signals that have been converted from a continuous range of analog signal levels and frequencies into a digital code.

Effective radiated power: The radiated power output from an antenna in a specified direction, including the effects of transmitter power, transmission line losses, and antenna power gain.

Intermittent service area: The surface-wave service area of a standard broadcast station beyond the primary service area and subject to some interference and fading.

Pre-emphasis–de-emphasis: The process of increasing the relative signal level of some audio frequencies during radio broadcasting (pre-emphasis) and decreasing the level during reception (de-emphasis) to reduce noise and improve quality.

Primary service area: The surface-wave service area of a standard broadcast station that is not subject to objectionable interference or objectionable fading.

Secondary service area: The service area of a standard broadcast station served by a skywave signal only, not subject to objectionable interference and in which the signal is subject to intermittent variations in signal strength.

Skywave: Refers to the ionospheric reflection of a standard broadcast station signal during nighttime hours allowing signals to be received hundreds of miles beyond the radio horizon.

Surface wave: Refers to the propagation of a standard broadcast station signal, which closely follows the surface of the Earth.

References

EIA. 1948. Standard TR-101A, Electrical Performance Standards for AM Broadcast Transmitters. Electronic Industries Association.

FCC. 1993. Federal Communications Commission, Rules and Regulations. Vol. III, Pts. 73 and 74, revised.

Further Information

National Association of Broadcasters Engineering Handbook, 9th ed. The National Association of Broadcasters, Washington, DC, 1996.

Broadcast Engineering Magazine. Primedia, 9800 Metcalf, Overland Park, KS 66212-2215.

16.2   Digital Audio Broadcasting

Stanley Salek and Almon H. Clegg

16.2.1   Introduction

Digital audio broadcasting (DAB) is the term applied to a few independently developed emerging technologies that promise to provide consumers with a new and better aural broadcast system. DAB offers dramatically better reception over existing terrestrial analog AM and FM broadcasts by providing improved audio quality and by integrating technology to provide resistance to interference in stationary and mobile/portable reception environments. Additionally, the availability of a digital data stream direct to consumers opens the prospect of providing extra services to augment basic sound delivery.

As of this writing, DAB transmission and reception systems are available from at least five system developers. Two of the systems, from Sirius Satellite Radio and XM Satellite Radio, transmit via satellite, using ground-based augmentation transmitters as needed, while the system from the Eureka 147/DAB consortium can be deployed as a satellite, terrestrial, or hybrid satellite/terrestrial system. The remaining systems, from iBiquity Digital Corporation and the Digital Radio Mondiale Consortium, are designed for terrestrial signal delivery, operating in established broadcasting frequency bands. This section provides a general overview of the technical aspects of DAB systems, as well as a summary of three example transmission systems.

16.2.2   The Need for DAB

In the years since the early 1980s, the consumer marketplace has undergone a great shift toward digital electronic technology. The explosion of personal computer use has led to greater demands for information, including full multimedia integration. Over the same time period, compact disc (CD) digital audio technology overtook long-playing records and analog audio cassettes as the consumer audio playback medium of choice. Similar digital transcription methods and effects also have been incorporated into commonly available audio and video equipment, including digital video disc (DVD) and hard disk/flash memory-based recorders and players. Additionally, the presently ongoing transition to high definition television broadcasting systems has included the incorporation of fully digital methods for video and audio transmission. Because of these market pressures, the radio broadcast industry determined that the existing analog methods of broadcasting needed updating to keep pace with the expectations of the advancing audio marketplace.

Along with providing significantly enhanced audio quality, DAB systems have been developed to overcome the technical deficiencies of existing AM and FM analog broadcasting systems. The foremost problem of current broadcast technology, as perceived by the industry, is its susceptibility to interference. AM medium-wave broadcasts, operating in the 530- to 1700-kHz frequency range, are prone to disruption by fluorescent lighting and by power system distribution networks, as well as by numerous other unintentional radiators, including computer and telephone systems. Additionally, natural effects, such as nighttime skywave propagation, interference between stations, and nearby lightning discharges, cause irritating service disruption to AM reception. FM broadcast transmissions in the 88- to 108-MHz band are much more resistant to these types of interference. However, multipath propagation and abrupt signal fading, especially found in urban and mountainous areas containing a large number of signal reflectors and shadowers (e.g., buildings and terrain), can seriously degrade FM reception, particularly to mobile receivers.

16.2.3   DAB System Design Goals

DAB systems are designed with several technical goals in mind. The first goal is to create a service that delivers "compact disc quality" stereo sound for broadcast to consumers. The second is to overcome the interference problems of current AM and FM broadcasts, especially under portable and mobile reception conditions. Third, DAB must be spectrally efficient, in that total bandwidth should be no greater than that currently used for corresponding analog broadcasts. Fourth, the DAB system should provide space in its data stream to allow for the addition of ancillary services, such as textual program information display, software downloading, or subscription data services. Finally, to foster rapid consumer acceptance, DAB receivers must not be overly cumbersome, complex, or expensive.

In addition to these goals, desired features include reduced RF transmission power requirements (when compared to AM and FM broadcast stations having the same signal coverage), a mechanism to seamlessly fill-in coverage areas that are shadowed from the principal transmitted signal, and the ability to easily integrate DAB receivers into personal, home, and automotive sound systems.

16.2.4   Historical Background

DAB development work began in Europe in 1986, with the initial goal to provide high-quality audio services to consumers directly by satellite. Companion terrestrial systems were developed to evaluate the technology being considered, as well as to provide fill-in service in small areas where the satellite signals were shadowed. A consortium of European technical organizations known as Eureka-147/DAB demonstrated the first working terrestrial DAB system in Geneva, Switzerland, in September 1988. Subsequent terrestrial demonstrations of the system followed in Canada in the summer of 1990, and in the U.S. in 1991 (See *http://www.worlddab.org/eureka.aspx*).

VHF and UHF transmission frequencies between 200 and 900 MHz were used for the demonstrations with satisfactory results. Because most VHF and UHF frequency bands suitable for DAB are already in use (or in use by/reserved for digital television and other services), an additional Canadian study in 1991 evaluated frequencies near 1500 MHz (L-band) for use as a potential worldwide DAB allocation. This study concluded that L-band frequencies would support a DAB system such as Eureka-147/DAB, while continuing to meet the overall system design goals.

During the scheduled 1992 World Administrative Radio Conference (WARC-92), frequency allocations for many different radio systems were debated. As a result of WARC-92, a worldwide L-band standard allocation at 1452 to 1492 MHz was designated for both satellite and terrestrial digital radio broadcasting. However, because of existing government and military uses in L-band, the U.S. was excluded from the standard. Instead, an S-band allocation at 2310 to 2360 MHz was substituted. Additionally, nations including Japan, China, and Russia opted for an extra S-band allocation in the 2535- to 2655-MHz frequency range.

During the same time period, most DAB system development work in the U.S. shifted from out-band (i.e., UHF, L-band, and S-band) to in-band, because of uncertainty as to the suitability of using S-band frequencies for terrestrial broadcasting. In-band terrestrial systems would merge DAB services with existing medium wave AM and VHF FM broadcasts, using novel adjacent- and co-channel modulating schemes. Between 1992 and 2000, competing system proponents demonstrated proprietary methods of extracting a compatible digital RF signal from co-channel analog AM and FM broadcast transmissions.

Standardization activities for DAB systems were started in the early 1990s by the Consumer Electronics Group of the Electronic Industries Association (EIA), now known as the Consumer Electronics Association (CEA). In 1996, field testing of some systems was conducted by EIA in the San Francisco, California, Bay Area. Standardization and field testing activities for IBOC AM and FM systems was begun shortly thereafter by the National Radio Systems Committee, a joint committee of the National Association of Broadcasters (NAB) and the CEA. In October 2002, the FCC conditionally approved deployment of proprietary analog/digital in-band on-channel (IBOC) AM and FM technology developed by system proponent iBiquity Digital Corporation (See *http://www.ibiquity.com*). As of mid-2004, IBOC standards activities are ongoing, with a permanent Federal Communications Commission (FCC) authorization for use of the systems by all U.S. broadcasters anticipated in the near future.

In 1998, a worldwide consortium known as Digital Radio Mondiale (DRM) was formed, with the goal of producing "near FM quality" digital broadcasting in AM broadcast bands below 30 MHz (See *http//www.drm.org*). While sharing some technical similarity to the iBiquity AM IBOC system, the non-proprietary DRM system has been primarily demonstrated using all-digital transmission, without a modulated co-channel analog AM component. In 2001, DRM received International Telecommunications Union (ITU) approval, and a technical standard was published. As of mid-2004, a number of international broadcasters have been conducting over-air DRM system testing.

The inception of S-band satellite broadcasting in the U.S. began in 1997, when the FCC auctioned two 12.5 MHz-wide frequency bands to private operators. Commercial Satellite DAB commenced in late 2001 and 2002 when systems licensees XM Satellite Radio and Sirius Satellite Radio began broadcasting to subscribers. Both systems offer numerous commercial-free music channels and other content (See *http://www.xmsatelliteradio.com* and *http://www.sirius.com*).

16.2.5 Technical Overview of DAB

Regardless of the actual signal delivery system used, all DAB systems share a common overall topology. Figure 16.10 presents a block diagram of a typical DAB transmission system.

To maintain the highest possible audio quality, program material must originate from digital sources, such as CD players and digital audio recorders, or digital audio feeds from network sources. Analog sources, such as microphones, are converted to a digital audio data stream using an analog-to-digital (A/D) converter, prior to switching or summation with the other digital sources.

The linear digital audio data stream from the studio is then applied to the input of a source encoder. The purpose of this device is to reduce the required bandwidth of the audio information, helping to produce a spectrally efficient RF broadcast signal. For example, 16-bit linear digital audio sampled at 48 kHz (the standard professional rate) requires a data stream of 1.536 megabits/s to transmit a stereo program in a serial format. This output represents a bandwidth of approximately 1.5 MHz, much greater than that used by an equivalent analog audio modulating signal (Smyth, 1992). Source encoders can reduce the data rate by factors of 8:1 or more, yielding a much more efficient modulating signal.

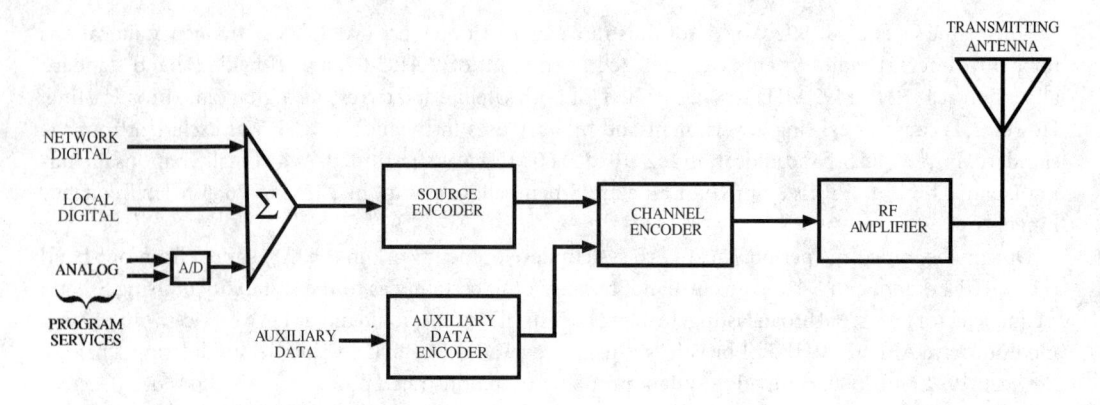

FIGURE 16.10   An example of DAB transmission system. (*Source:* Hammett & Edison, Inc., Consulting Engineers.)

Following the source encoder, the resulting serial digital signal is applied to the input of the channel encoder, a device that modulates the transmitted RF wave with the reduced-rate audio information. Auxiliary serial data, such as program information and receiver control functions, also can be input to the channel encoder (or combined within the source encoder) for simultaneous transmission.

The channel encoder uses sophisticated modulating techniques to accomplish the goals of interference cancellation and high spectral efficiency. Methods of interference cancellation include expansion of time and frequency diversity of the transmitted information, as well as the inclusion of error correction codes in the data stream. Time diversity involves transmitting the same information multiple times by using a predetermined time interval. Frequency diversity, such as that produced by spread-spectrum, multiple-carrier, or frequency-hopping systems, provides the means to transmit identical data on several different frequencies within the bandwidth of the system. At the receiver, real-time mathematical processes are used to locate the required data on a known frequency at a known time. If the initial information is found to be unusable because of signal interference, the receiver simply uses the same data found on another frequency or at another time, producing seamless demodulation.

Spectral efficiency is a function of the modulation system used. Among the modulation formats that have been evaluated for DAB transmission are QPSK, M-ary QAM, and MSK (Springer, 1992). Using these and other formats, digital transmission systems that use no more spectrum than their analog counterparts have been designed.

The RF output signal of the channel encoder is amplified to the appropriate power level for transmission. Because the carrier-to-noise (C/N) ratio of the modulated waveform is not generally so critical as that required for analog communications systems, relatively low transmission power often can be used. Depending on the sophistication of the data recovery circuits contained in the DAB receiver, the use of C/N ratios as low as 6 dB are possible without causing a degradation to the received signal.

DAB reception is largely the inverse of the transmission process, with the inclusion of sophisticated error correction circuits. Fig. 16.11 shows a typical DAB receiver.

DAB reception begins in a similar manner as is used in virtually all radio receivers. A receiving antenna feeds an appropriate stage of RF selectivity and amplification, from which a sample of the coded DAB signal is derived. This signal then drives a channel decoder, which reconstructs the audio and auxiliary data streams. To accomplish this task, the channel decoder must demodulate and de-interleave the data contained on the RF carrier and then apply appropriate computational and statistical error correction functions.

The source decoder converts the reduced bit-rate audio stream back to pseudolinear at the original sampling rate. The decoder computationally expands the mathematically reduced data and fills the gaps left from the extraction of irrelevant audio information with averaged code or other masking data. The output of the source decoder feeds audio digital-to-analog (D/A) converters, and the resulting analog stereo audio signal is amplified for the listener.

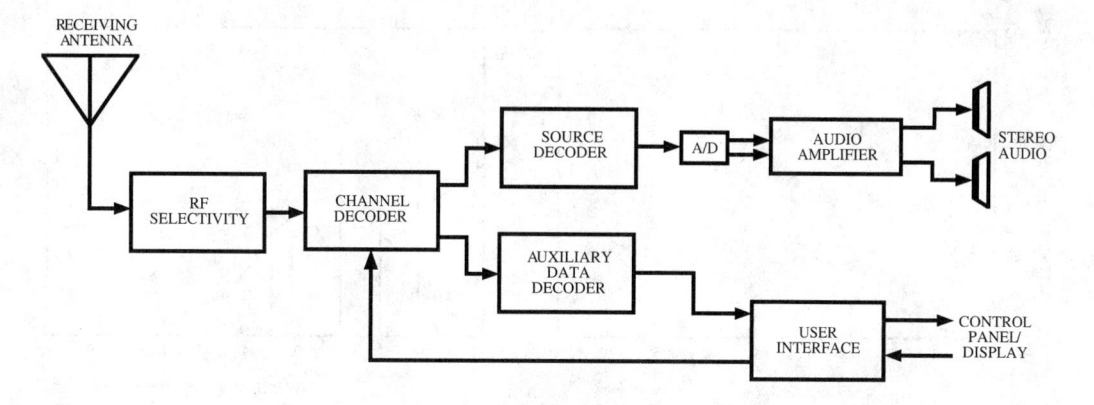

FIGURE 16.11   An example of DAB receiver. (*Source:* Hammett & Edison, Inc., Consulting Engineers.)

In addition to audio extraction, DAB receivers are generally capable of decoding auxiliary data. These data can be used in conjunction with the user interface to control receiver functions, or for a completely separate purpose. A typical user interface contains a data display screen in addition to the usual receiver tuning and audio controls. This data screen can be used to obtain information about the programming, news reports, sports scores, advertising, or any other useful data sent by the station or an originating network. Also, external interfaces could be used to provide a software link to personal computer systems or other digital information and productivity devices.

16.2.6   Audio Compression and Source Encoding

The development of digital audio encoding started with research into pulse-code modulation (PCM) in the late 1930s and evolved, shortly thereafter to include work on the principles of digital PCM coding. Linear predictive coding (LPC) and adaptive delta pulse-code modulation (ADPCM) algorithms had evolved in the early 1970s and later were adopted into standards such as C.721 (published by the CCITT) and CD-I (Compact Disc-Interactive). At the same time, algorithms were being invented for use with phoneme-based speech coding. Phonetic coding, a first-generation "model-based" speech-coding algorithm, was mainly implemented for low bit-rate speech and text-to-speech applications. These classes of algorithms for speech further evolved to include both code excited linear predictive (CELP) and vector selectable excited linear predictive (VSELP) algorithms by the mid-1980s. In the late 1980s, these classes of algorithms were also shown to be useful for high-quality audio music coding. They were put to commercial use from the late 1970s to the latter part of the 1980s.

Subband coders evolved from the early work on quadrature mirror filters in the mid-1970s and continued with polyphase filter-based schemes in the mid-1980s. Hybrid algorithms employing both subband and ADPCM coding were developed in the latter part of the 1970s and standardized (e.g., CCITT G.722) in the mid- to late 1980s. Adaptive transform coders for audio evolved in the mid-1980s from speech coding work done in the late 1970s. By employing psychoacoustic noise-masking properties of the human ear, perceptual encoding evolved from early work of the 1970s and where high-quality speech coders were employed. Music quality bit-rate reduction schemes such as motion picture expert group (MPEG), precision adaptive subband coding (PASC), and adaptive transform acoustic coding (ATRAC) have been developed. Further refinements to the technology will focus attention on novel approaches such as wavelet-based coding and the use of entropy coding schemes.

Audio coding for digital broadcasting in the early days employed one of the many perceptual encoding schemes previously mentioned or some variation thereof. Fundamentally, they depend on two basic psychoacoustic phenomena: (1) the threshold of human hearing, and (2) masking of nearby frequency components. In the early days of hearing research, Harvey Fletcher, a researcher at Bell Laboratories, measured the hearing of many human beings and published the well-known Fletcher-Munson threshold-of-hearing

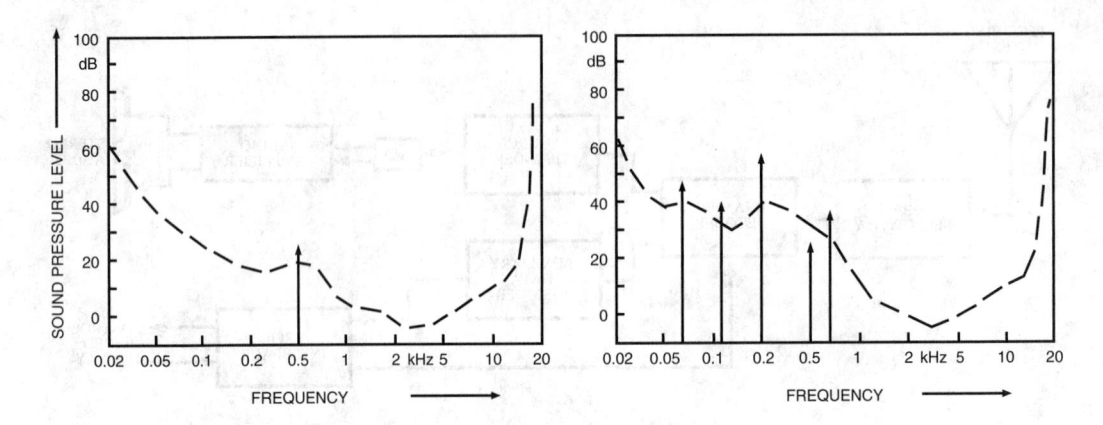

FIGURE 16.12   An example of the masking effect. Based on the hearing threshold of the human ear (dashed line), a 500-Hz sinusoidal acoustic waveform, shown at A on the left graph, is easily audible at relatively low levels. However, it can be masked by adding nearby higher-amplitude components, as shown on the right. (*Source:* CCi.)

chart. Basically it states that, depending on the frequency, audio sounds below certain levels cannot be heard by the human ear. Further, the masking effect, simply stated, is when two frequencies are very close to each other and one is higher in acoustic level than the other, the weaker of the two is masked and will not be heard. These two principles allow for as much as 80% of the data representing a musical signal to be discarded.

Figure 16.12 shows how introduction of frequency components affects the ear's threshold of hearing vs. frequency. Figure 16.13 shows how the revised envelope of audibility results in the elimination of components that would not be heard.

The electronic implementation of these algorithms employs a digital filter that breaks the audio spectrum into many subbands, and various coefficient elements are built into the program to decide when it is permissible to remove one or more of the signal components. The details of how the bands are divided and how the coefficients are determined are usually proprietary to the individual system developers. Standardization groups and system proponents have spent many worker-hours of evaluation attempting to determine the most accurate coding system for given compression ratios or data transfer rates.

During recent years, the algorithms used for encoding music signals have improved greatly, even though the fundamental methods remain as outlined above. Experience in the art of producing both hardware and software encoders and the frequent use of listening test panels has brought about great improvements. "Tweaking" of the overall technology has improved sound quality while keeping the compression and transfer rates as needed for the broadcast service to which it is applied. This optimization has benefited the prospects for high quality sound in digital transmission and broadcasting systems.

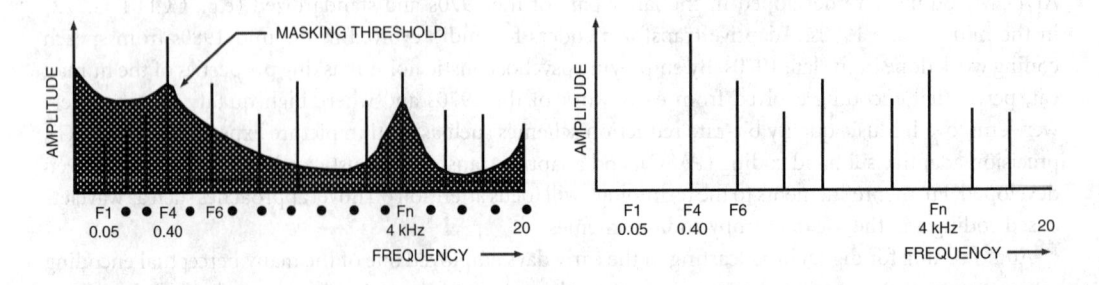

FIGURE 16.13   Source encoders use an empirically derived masking threshold to determine which audio components can be discarded (left). As shown on the right, only the audio components with amplitudes above the masking threshold are retained. (*Source:* CCi.)

Some examples of coding schemes that have been considered for use in broadcasting channels are: perceptual audio coder (PAC) that extends the idea of perceptual coding to stereo pairs. It uses both L/R (left/right) and M/S (sum/difference matrixing to abate noise generation) coding, switched in both frequency and time in a signal dependent fashion. The adaptive transform acoustic coder (ATRAC) has been developed for portable digital audio, but because of its robust coding scheme and availability of commercial integrated circuits, it also lends itself to broadcast use. This coder uses hybrid frequency mapping employing a signal splitting scheme that breaks the band into three sub-bands (0–5.5, 5.5–11, and 11–22 kHz) followed by a suitable dynamically windowed modified discrete cosine transform (MDCT). Advanced audio coding (AAC) is a high-quality perceptual audio coding technology appropriate for many broadcast and electronic music-distribution applications. Coding efficiency is superior to that of MP3, providing higher quality audio at lower transmitted bit rates. Developed and standardized as an ISO/IEC specification by industry leaders, AAC is supported by a growing number of hardware and software manufacturers as the logical successor to MP3 for consumer and computer products, as well as for broadcasting (See *www.aac-audio.com* and *http://www.iis.fraunhofer.de/amm/techinf/aac/* and *http://www.mp3prozone.com*).

Broadcast studios must have available in their facilities the necessary transforms to convert the many digital formats to their broadcast algorithm of choice. Hence, within the broadcast community, expert knowledge of all types and generations of audio compression formats is necessary. If an audio signal passes through several successive compression or bit reduction algorithms, there may be unexpected audible artifacts produced. It is the responsibility of the broadcast engineers to learn the compatibility issues and adjust the signal processing methods to make the necessary conversions with minimum degradation of audio quality. Consideration must be given to the format of the source audio program content and its legacy and compatibility with the coder being used for the final transmission or download. In the case of Internet streaming or downloading, the source coding used may sound best when listened to on a particular software decoder. Hence, there are many choices in the marketplace, each claiming certain improvements over its competitors.

An interesting modern source coding example is CT-aacPlus audio compression. CT-aacPlus is the combination of AAC, a highly efficient global standard based on the collaborative work of several experts on perceptual audio encoding, and additional coding from Coding Technologies' Spectral Band Replication (SBR^{TM}) technology. SBR creates additional bitrate efficiency and utilizes subjective improvements worked out by the German Fraunhofer Institute, the inventor of MP3 (See *http://www.codingtechnolgies.com* and *http://www.xmradio.com*). With this combination of AAC and SBR, CT-aacPlus has been tested by audio experts and found to provide certain audio improvements. In a double-blind listening test, AAC alone was 33% more efficient when compared to previous generations of competing algorithms. Double-blind listening tests have established that the CT-aacPlus combination is as much as 30% more efficient than AAC. As a result of these tests, CT-aacPlus has been adopted by the DRM consortium and accepted by the MPEG standardization group as the "reference model" for the upcoming versions of MPEG-4 (a standard for a new form of audio and video compression) (See *http://www.ebu.ch/trev_291-dietz.pdf*).

16.2.7 System Example: Eureka-147/DAB

Eureka-147/DAB was the first fully developed DAB system that has demonstrated a capability to meet virtually all the described system goals. Developed by a European consortium, it is an out-band system in that its design is based on the use of a frequency spectrum outside the AM and FM radio broadcast bands. Out-band operation is required because the system packs up to 16 stereophonic broadcast channels (plus auxiliary data) into one contiguous band of frequencies, which can occupy a total bandwidth of several megahertz. Thus, overall efficiency is maintained, with 16 digital program channels occupying about the same total bandwidth as 16 equivalent analog FM broadcast channels. System developers have promoted Eureka-147/DAB for satellite transmission, as well as for terrestrial applications in locations that have a suitable block of unused spectrum in the L-band frequency range or below.

The ISO/MPEG-2 source encoding/decoding system has been defined for use with the Eureka-147/DAB system. Originally developed by IRT (Institut für Rundfunktechnik) in Germany as MUSICAM (Masking pattern-adapted Universal Subband Integrated Coding And Multiplexing), the system works by dividing the original digital audio source into 32 subbands. As with the source encoders described earlier, each of the bands is digitally processed to remove redundant information and sounds that are not perceptible to the human ear. Using this technique, the original audio, sampled at a rate of 768 kilobits/sec per channel, is reduced to as little as 96 kilobits/sec per channel, representing a compression ratio of 8:1.

The Eureka-147/DAB channel encoder operates by combining the transmitted program channels into a large number of adjacent narrowband RF carriers, which are each modulated using QPSK and grouped in a way that maximizes spectrum efficiency known as orthogonal frequency-division multiplex (OFDM). The information to be transmitted is distributed among the RF carriers and is also time-interleaved to reduce the effects of selective fading. A guard interval is inserted between blocks of transmitted data to improve system resistance to intersymbol interference caused by multipath propagation. Convolutional coding is used in conjunction with a viterbi maximum-likelihood decoding algorithm at the receiver to make constructive use of echoed signals and to correct random errors (Alard and Lassalle, 1988).

RF power levels of just a few tens of watts per program channel have been used, providing a relatively wide coverage area, depending on the height of the transmitting antenna above surrounding terrain. This low power level is possible because the system can operate at a C/N ratio of less than 10 dB, as opposed to the more than 30 dB that is required for high-fidelity demodulation of analog FM broadcasts.

Another demonstrated capability of the system is its ability to use "gap filler" transmitters to augment signal coverage in shadowed areas. A gap filler is simply a system that directly receives the DAB signal at an unobstructed location, provides RF amplification, and retransmits the signal, on the same channel, into the shadowed area. Because the system can make constructive use of signal reflections (within a time window defined by the guard interval and other factors), the demodulated signal is uninterrupted on a mobile receiver when it travels between an area served by the main signal into the service area of the gap filler.

16.2.8   System Example: iBiquity FM IBOC

As disclosed in recent public FCC proceedings, the iBiquity FM IBOC DAB system provides a novel proprietary method for the transition from analog FM broadcasting to a hybrid digital/analog technology, and finally to an all-digital system, using existing broadcast spectrum in the 88–108 MHz FM broadcast band. Source coding uses a proprietary method known as high definition coder (HDC), which like CT-aacPlus has its roots in AAC. The present implementation operates at 96 kilobits/sec for both stereo channels, representing a compression ratio of 16:1, or even greater if the stream is split to allow both audio and data transmission or transmission of two or more separate but simultaneously delivered audio programs.

As shown in Figure 16.14, groupings of primary low level OFDM carriers are added to the FM analog signal, both below and above the host signal. Optional extended carriers also may be added, although their presence is not required for basic system operation. In addition to the FM analog host carrier frequency, reference subcarriers (used by the receiver/decoder for synchronization) are included with the IBOC carriers. As illustrated, the injection level of the individual carriers is quite low, and is easily contained within the spectral emission mask specified by the U.S. FCC Rules.

Because the ultimate system goal is for a transition to an all-digital broadcast system, the methodology and spectrum loading has been predefined to be compatible with receivers marketed to operate using the hybrid system. Figure 16.15 shows the spectral content of the defined all-digital system. Note that the power of the primary carrier groupings, previously adjacent to the analog FM signal, have been increased, and groupings of secondary carriers have been added in the location of the former FM analog signal spectrum. The higher level OFDM carriers are maintained in an approximate 100 to 200 kHz offset from the carrier frequency so that interference to or from co-channel and adjacent-channel stations is not degraded subsequent to the all-digital broadcasting transition.

For the hybrid (analog compatible) system, the IBOC DAB signal can be combined with the analog FM signal by way of a number of potential methods. Three of the more common combining methods

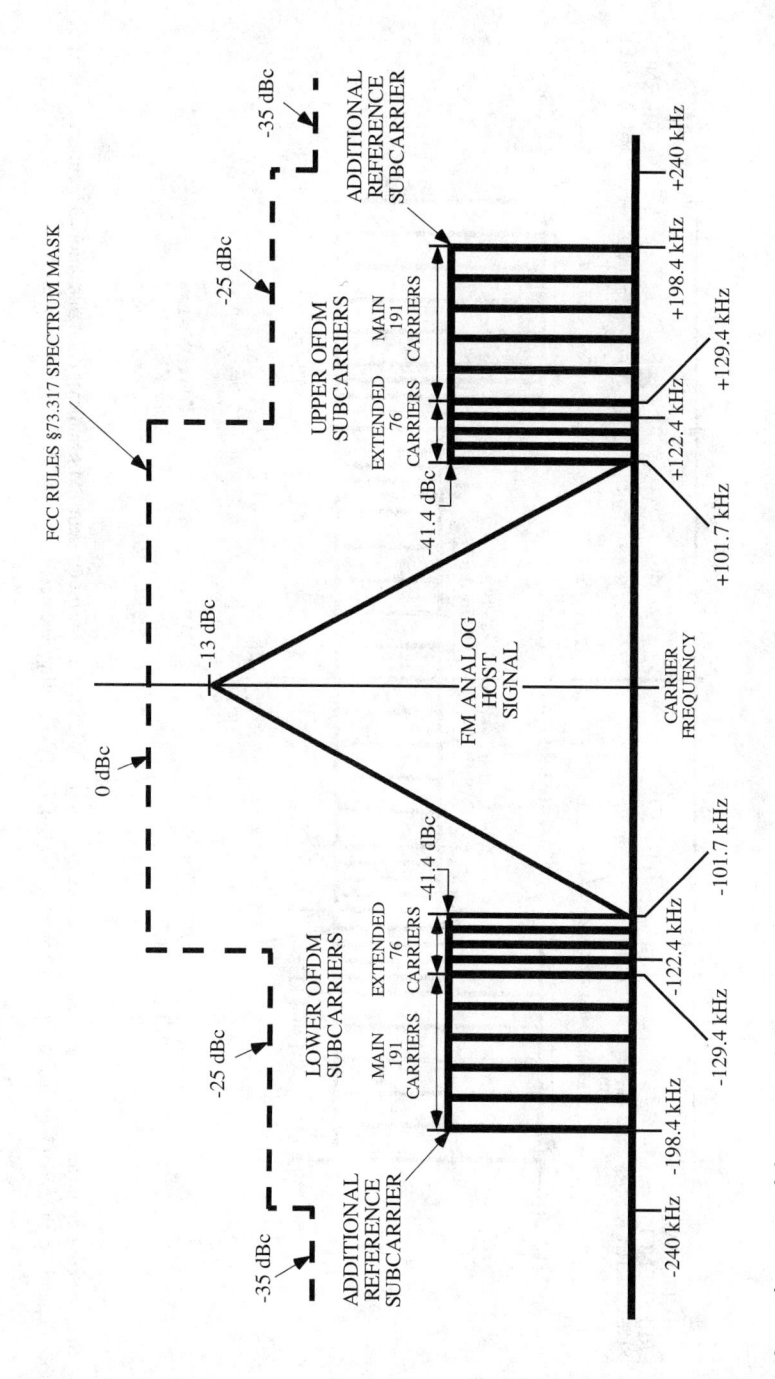

FIGURE 16.14 Spectral content of the iBiquity hybrid FM analog/IBOC DAB system (dBc = decibels relative to carrier level). (*Source:* Hammett & Edison, Inc., Consulting Engineers.)

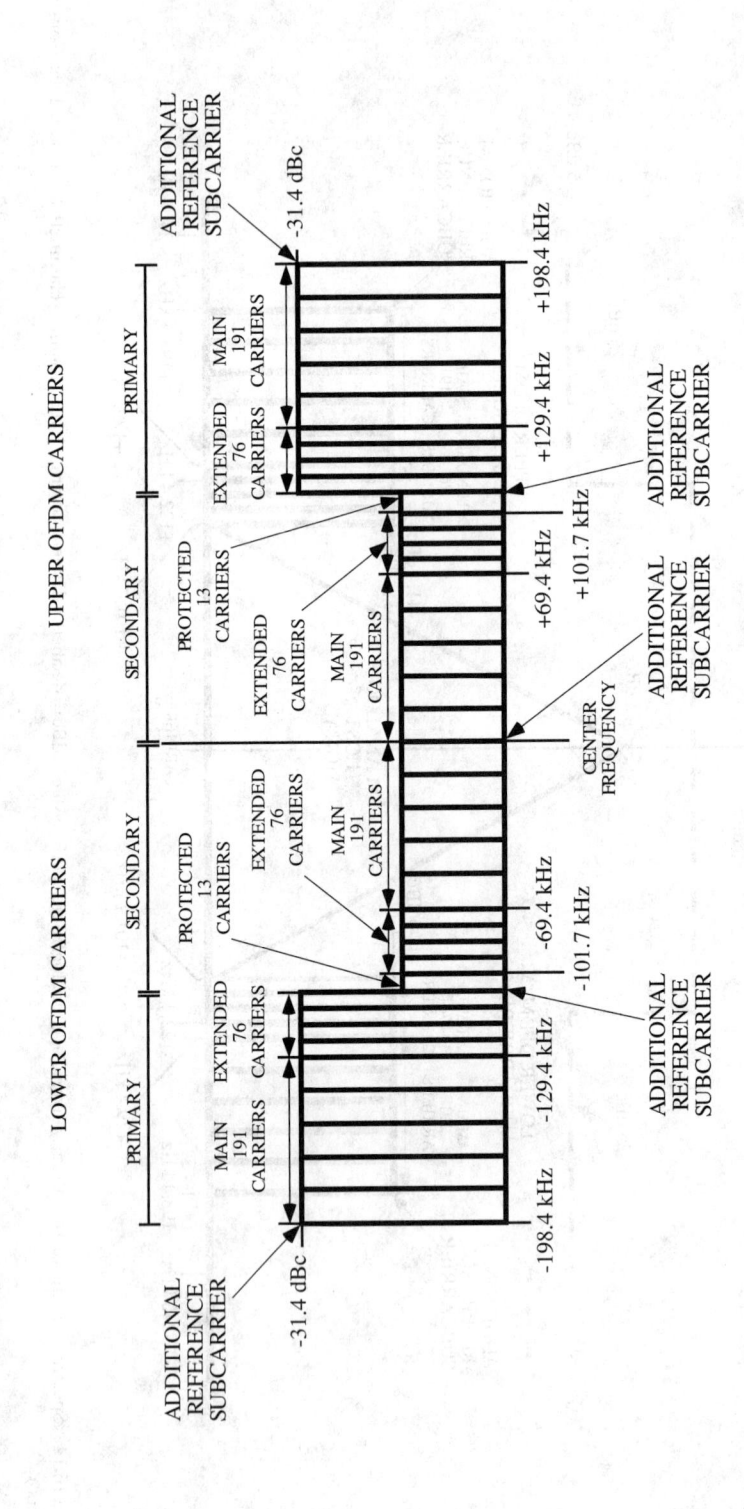

FIGURE 16.15 Spectral content of the iBiquity all-digital IBOC DAB system intended for use in the present FM broadcast band allocation. The injection level of the secondary carriers is variable, depending on carrier loading. (*Source:* Hammett & Edison, Inc., Consulting Engineers.)

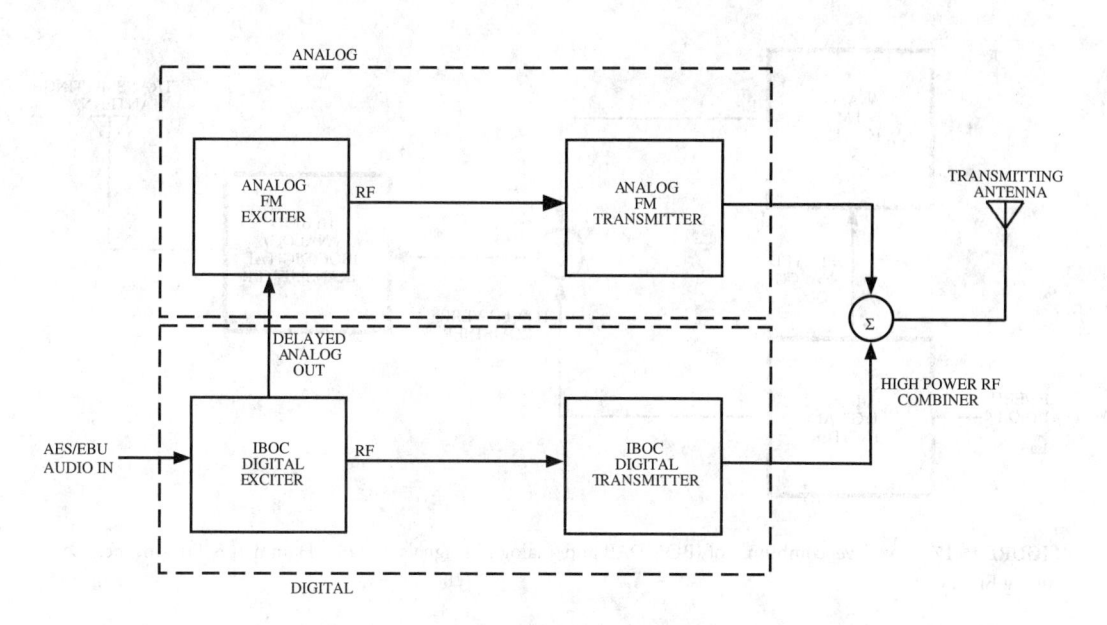

FIGURE 16.16   High level combining of IBOC DAB and analog FM signals. (*Source:* Hammett & Edison, Inc., Consulting Engineers.)

are illustrated in the following block diagrams. In all cases, the FM transmitter analog audio is delayed by several seconds to achieve time alignment with the recovered digital audio signal in the receiver. Thus, the system can automatically and seamlessly return to analog audio demodulation in locations where digital demodulation becomes impaired.

Figure 16.16 depicts the high level combining method, by which full-power analog and digital transmission systems operate in parallel, and the two signals are combined just prior to the input of the transmitting antenna system. This method has the benefit of maintaining high linearity of the IBOC DAB signal, since it never passes through the same active stages as the analog FM signal. However, the combiner must be carefully designed to minimize losses, particularly to the analog FM signal, which for many FM stations operates at power levels of 10 kilowatts or more. Thus, in order to increase the analog combining efficiency, the digital combining efficiency must be decreased, requiring much higher DAB power than is actually transmitted, often on the order of 10 dB higher. For example, to achieve 100 watts of DAB power at the input to the antenna system, 1000 watts of DAB power must be generated. This method is most popular with moderate power FM stations, since the amount of DAB power wasted in the combining process remains reasonable.

One alternative to high level combining is low level combining, shown in Figure 16.17. This method combines the low power output signals of the analog FM and IBOC exciter units and applies the composite signal to the transmitter. This method requires exceptional linearity in the transmitter RF amplifier stages to prevent distortion of the IBOC carriers. It is generally suitable for transmitters operating at an FM analog output power of 10 kilowatts or less.

A third method of combining analog FM and IBOC carriers is to employ separate transmitting antennas, as depicted in Figure 16.18. Field testing has shown this method to be acceptable, provided that the transmitting antennas have similar radiation patterns, are located close to each other, and are installed at comparable radiation center heights. This method is of significant interest to stations that operate with very high analog FM transmitter power of generally more than 30 kilowatts, where the high level combining method is impractical due to efficiency issues and low level combining is not possible due to the prohibitive cost of producing a high power linear transmitter to operate at VHF frequencies. The use of separate antennas is also of interest to many other FM broadcasters, since the method allows the existing analog FM facility to be left in place.

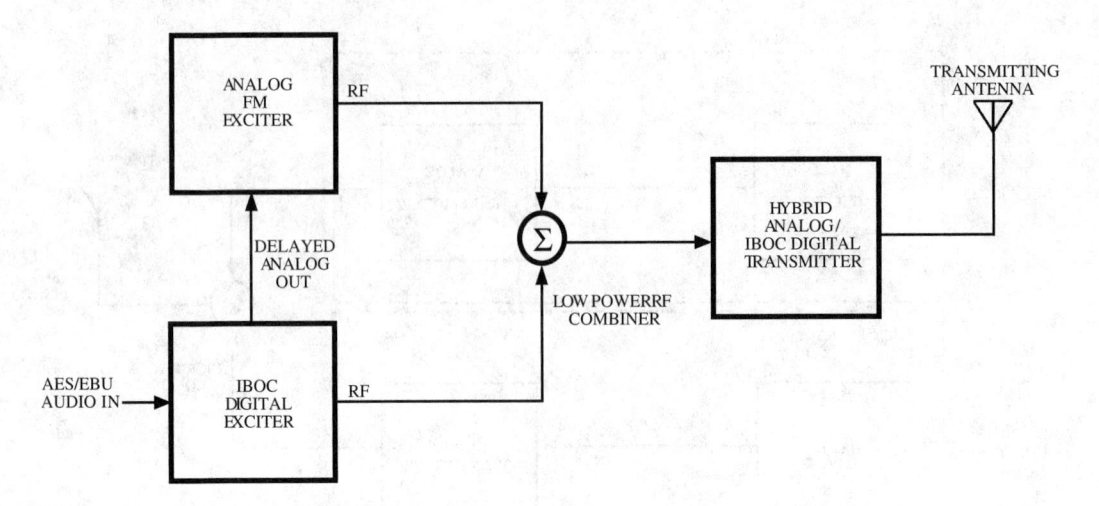

FIGURE 16.17   Low level combining of IBOC DAB and analog FM signals. (*Source:* Hammett & Edison, Inc., Consulting Engineers.)

16.2.9   System Example: iBiquity AM IBOC

As with the FM IBOC system, the proprietary iBiquity AM IBOC transmission system is capable of operating in two modes, hybrid and all-digital, with only the hybrid mode exhibiting interoperability with existing analog broadcasting. As with the FM IBOC system, source coding employs HDC, operating at 36 kilobits/sec for both stereo channels, representing a compression ratio of almost 43:1. A 400 bit/sec data stream is reserved for receiver program identification, but the data path can be expanded further at

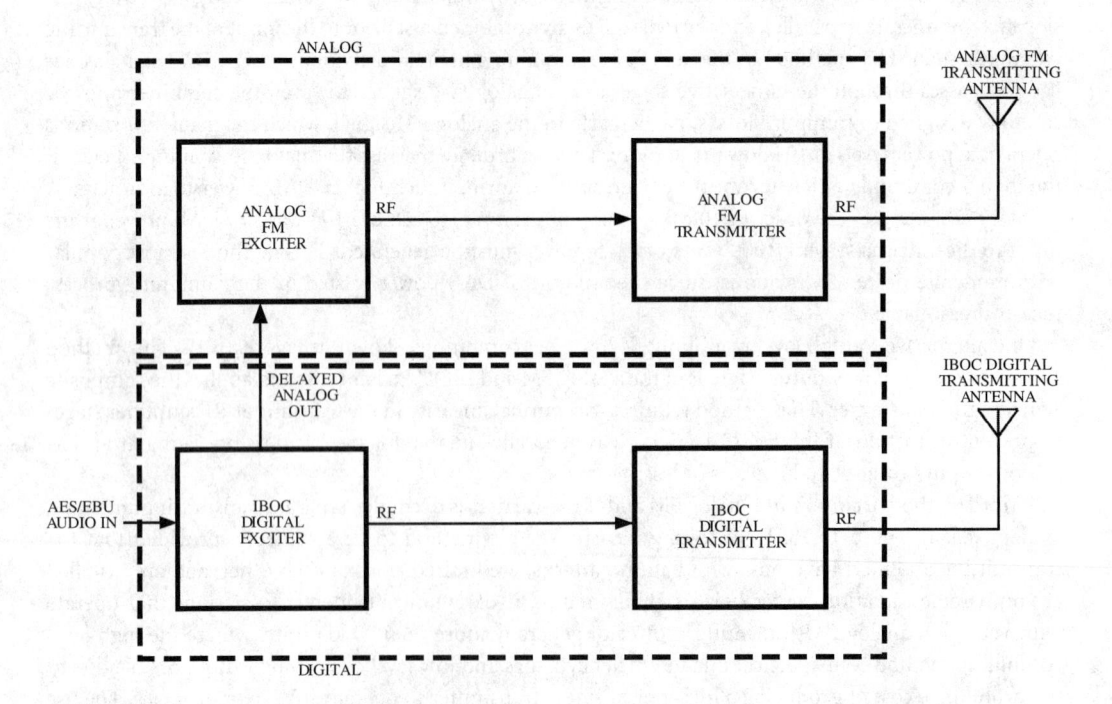

FIGURE 16.18   Combining of IBOC DAB and analog FM signals using separate transmitting antennas. (*Source:* Hammett & Edison, Inc., Consulting Engineers.)

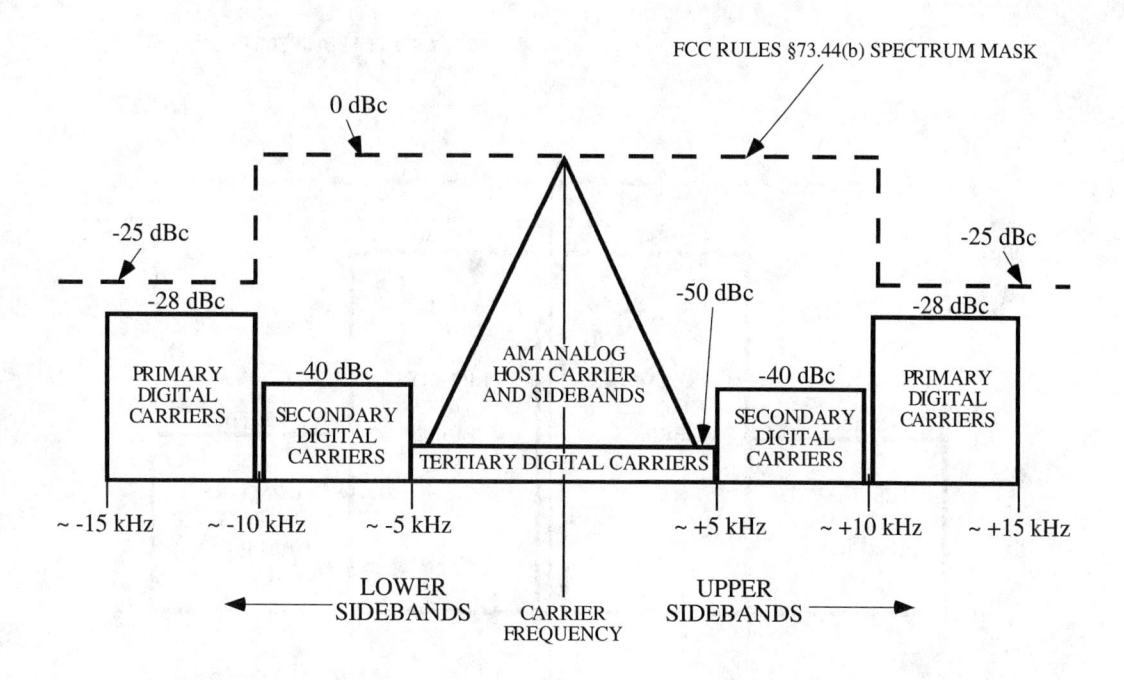

FIGURE 16.19 Spectral content of the iBiquity hybrid AM analog/IBOC DAB system. (*Source:* Hammett & Edison, Inc., Consulting Engineers.)

the expense of using a higher audio compression ratio. According to the developers, HDC is capable of producing acceptable audio quality at bit rates as low as 20 kilobits/sec.

Accompanying Figure 16.19 provides a drawing of the spectrum occupancy of the hybrid AM IBOC transmission system. Similar to the related FM IBOC system, transmissions are carried by a series of OFDM carriers, separated into primary, secondary, and tertiary groupings. The primary digital carrier groupings are offset ±10 to ±15 kHz from the carrier frequency, while the secondary digital carrier groupings are offset ±5 to ±10 kHz of the carrier frequency. The tertiary carriers are located within ±5 kHz of the carrier frequency, sharing same spectrum as the AM analog host sidebands. As shown by the dashed line, this arrangement of digital carriers fits within the U.S. FCC Rules spectrum mask.

In general, the system requires that the analog audio bandwidth be limited to 5 kHz, to protect the secondary digital carriers. An alternate mode of the system allows up to 7.5 kHz analog audio bandwidth, which generates some interference to the secondary digital carriers. Because the digital sidebands carry duplicative information that is inverted about the carrier frequency, recovery of the lower portion of the lower secondary digital subcarriers, along with the upper portion of the upper secondary digital subcarriers, still results in full recovery of the secondary digital information. The drawback is that recovery of both secondary sidebands is required for use of 7.5 kHz analog audio bandwidth, while recovery of just one secondary sideband is required for use of 5 kHz analog audio bandwidth. Thus, IBOC digital service area for 7.5 kHz analog audio bandwidth may be impaired in the presence of adjacent-channel interfering signals.

Accompanying Figure 16.20 provides a representation of the spectrum occupancy for the defined all-digital IBOC AM system. As with the FM IBOC system, it is anticipated that a transition to this system would take place after most stations have implemented the hybrid IBOC system and when market indicators have shown IBOC receiver penetration exceeding a defined threshold.

Note that the unmodulated carrier signal is retained in the all-digital system, but the primary digital carrier groupings now immediately surround the carrier, replacing the hybrid system tertiary digital carriers. Defined secondary and tertiary carriers occupy the former positions of the hybrid system secondary digital carrier groupings, and no information is transmitted beyond ±10 kHz of the center carrier frequency.

FCC RULES § 73.44(b) SPECTRUM MASK

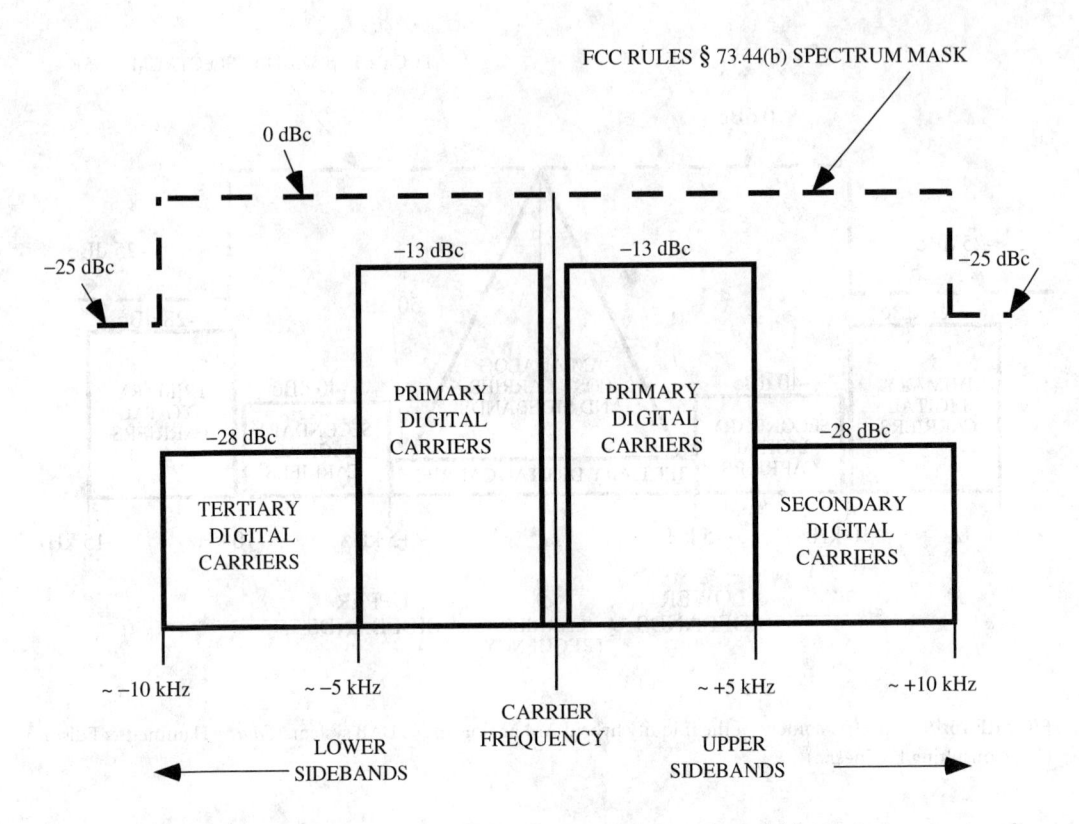

FIGURE 16.20 Spectral content of the iBiquity all-digital IBOC DAB system intended for use in the present medium wave AM broadcast band allocation. (*Source:* Hammett & Edison, Inc., Consulting Engineers.)

As a result of its design, it is anticipated that the eventual migration to all-digital transmission will enhance station digital coverage over the hybrid system and exhibit improved performance on transmitting antenna systems exhibiting narrow bandwidth.

Figure 16.21 provides a block diagram of a typical AM IBOC installation. The conversion of an analog station requires the addition of an exciter, an auxiliary service unit (ASU), and appropriate audio processing.

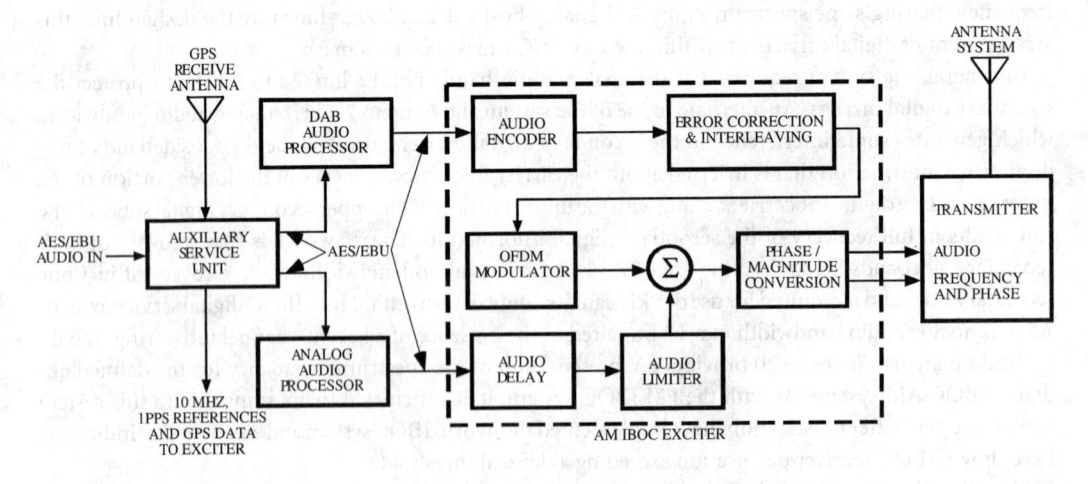

FIGURE 16.21 Combining of IBOC DAB and analog AM signals. (*Source:* Hammett & Edison, Inc., Consulting Engineers.)

The auxiliary service unit (ASU) accepts AES/EBU digital audio as input, handles sampling rate conversions, and feeds the digital audio inputs of separate AM and IBOC audio processors. The ASU also contains a GPS-synchronized clock to generate the carrier reference and other synchronizing signals used by the exciter. The DAB audio processor controls and levels audio of the IBOC digital transmission, while the analog audio processor controls analog audio. The exciter contains a delay circuit for the analog audio path so that digital-to-analog transitions in receivers are time-aligned, similar to the FM IBOC system.

The exciter DAB path encodes the source audio, introduces error correction and interleaving overhead, and generates the coded OFDM carriers. The DAB and analog paths are then summed and converted to magnitude/phase signals that feed the broadcast transmitter. The exciter output signals feed the transmitter analog audio input and frequency-determining stages of the transmitter. The hardware topology of an all-digital conversion is the same as that shown, except that the analog audio processor and related exciter circuitry is not used.

Defining Terms

Channel encoder: A device that converts source-encoded digital information into an analog RF signal for transmission. The type of modulation used depends on the particular digital audio broadcasting (DAB) system, although most modulation techniques employ methods by which the transmitted signal can be made more resistant to frequency-selective signal fading and multipath distortion effects.

In-band on-channel (IBOC): A method of combining digital audio broadcasting signals, intended for reception by radio listeners, with the signals of existing analog broadcast services operating in the medium wave AM and VHF FM broadcast bands.

Gap filler: A low-power transmitter that boosts the strength of transmitted DAB RF signals in areas which normally would be shadowed due to terrain obstruction. Gap fillers can operate on the same frequency as DAB transmissions or on alternate channels that can be located by DAB receivers using automatic switching.

Source encoder: A device that substantially reduces the data rate of linearly digitized audio signals by taking advantage of the psychoacoustic properties of human hearing, eliminating redundant and subjectively irrelevant information from the output signal. Transform source encoders work entirely within the frequency domain, while time-domain source encoders work primarily in the time domain. Source decoders reverse the process, using various masking techniques to simulate the properties of the original linear data.

References

Alard, M. and Lassalle, R. Principles of modulation and channel coding for digital broadcasting for mobile receivers. In *Advanced Digital Techniques for UHF Satellite Sound Broadcasting* (collected papers), European Broadcasting Union, pp. 47–69, 1988.

Bruno, R. *Digital Audio and Video Compression, Present and Future*, presented to the Delphi Club, Tokyo, Japan, July 1992.

Chouinard, G. and Conway, F. Broadcasting systems concepts for digital sound. In *Proceedings of the 45th Annual Broadcast Engineering Conference*, National Association of Broadcasters, 1991, pp. 257–266.

Conway, F. Voyer, R. Edwards, S. and Tyrie, D. Initial experimentation with DAB in Canada. In *Proceedings of the 45th Annual Broadcast Engineering Conference*, National Association of Broadcasters, 1991, pp. 281–290.

Kuh, S. and Wang, J. Communications systems engineering for digital audio broadcast. In *Proceedings of the 45th Annual Broadcast Engineering Conference*, National Association of Broadcasters, 1991, pp. 267–272.

Moose, P.H. and Wozencraft, J.M. Modulation and coding for DAB using multi-frequency modulation. In *Proceedings of the 45th Annual Broadcast Engineering Conference*, National Association of Broadcasters, 1991, pp. 405–410.

Smyth, S. Digital audio data compression. *Broadcast Engineering Magazine*, pp. 52–60, Feb. 1992.

Springer, K.D. *Interference between FM and Digital M-PSK Signals in the FM Band*, National Association of Broadcasters, 1992.

First Report and Order to Mass Media Docket No. 99-325, *Digital Audio Broadcasting Systems and Their Impact on the Terrestrial Radio Broadcast Service*, Federal Communications Commission, Washington, D.C., 2002.

Further Information

The National Association of Broadcasters (See *http://www.nab.org*) publishes periodic reports on the technical, regulatory, and political status of DAB in the U.S. Additionally, their Broadcast Engineering Conference proceedings published since 1990 contain a substantial amount of information on emerging DAB technologies.

IEEE Transactions on Broadcasting, typically published quarterly by the Institute of Electrical and Electronics Engineers, Inc., periodically includes papers on digital broadcasting (See *http://www.ieee.org*).

Additionally, the periodic newspaper publication *Radio World* (See *http://www.rwonline.com*) provides continuous coverage of DAB technology, including proponent announcements, system descriptions, field test reports, and broadcast industry reactions.

16.3   Audio Interconnection

Roy W. Rising

16.3.1   Introduction

System interfacing problems have existed since the first audio engineer tried to wire two pieces of equipment together. Devices that work flawlessly alone can fail miserably when connected. Initially, transformers were required to correct for differences in ground potentials between pieces of equipment and to eliminate electrical noise picked up in cabling. Transformers also interfaced the internal unbalanced high-impedance characteristics of active devices with balanced transmission lines having much lower impedances. Transformers added significant cost, weight, and sometimes distortion to the equipment.

As the performance of electronic equipment improved, transformers kept pace with reduced distortion and broadened frequency response. Their physical size and weight, however, made them inconvenient for use in the ever shrinking chassis' of transistorized equipment. As solid-state technology became standard, designers sought ways to eliminate transformers. The result was electronically (active) balanced inputs and outputs (I/O).

The move away from transformers introduced several unexpected problems. Installation practices that provide acceptable performance with transformer-isolated equipment may not work for active balanced I/O equipment. Active balanced circuits can be less forgiving of wiring faults and short circuits; some output stages will self-destruct if shorted. Avoidance of ground loops is more important with an active balanced system than with transformers. Active balanced input circuits can provide excellent noise rejection under ideal conditions. In the real world, however, their noise rejection can deteriorate rapidly. Transformers remain the best choice for the toughest situations.

16.3.2 Balanced and Unbalanced Systems

Nearly all professional audio systems use balanced inputs and outputs because of the noise immunity they afford. Figure 16.22 shows a basic source and load connection. No grounds are present, and both the source and load float. This is the optimum condition for equipment interconnection.

Either the source or the load may be tied to ground with no problems, provided only one ground connection exists. Unbalanced systems are created when each piece of equipment has one of its connections tied to ground, as shown in Fig. 16.23. This condition occurs if the source and load equipment have unbalanced (single-ended) inputs and outputs. This type of equipment uses chassis ground for one of the audio conductors. This is a common practice in consumer audio products. Equipment with unbalanced inputs and outputs is cheaper to build but when various pieces of audio gear are tied together, trouble often results. These problems are compounded when the equipment is separated by a significant distance.

As shown in Fig. 16.23, a difference in ground potential causes current flow in the ground wire. This current develops a voltage across the wire resistance, and the ground noise voltage adds directly to the signal itself. Because the ground current usually results from leakage in power transformers and line filters, the 60-Hz signals give rise to hum. Reducing the wire resistance through a heavier ground wire helps reduce the hum, but cannot eliminate it completely.

By amplifying both the high side and the ground side of the source and subtracting the two to obtain a *difference signal*, it is possible to cancel the ground loop noise (see Fig. 16.24). This is the basis of the **differential input** circuit. Unfortunately, problems still can exist with the unbalanced source to balanced load system shown in the drawing. The reason centers around the impedance of the unbalanced source. One side of the line will have a slightly lower amplitude because of impedance differences in the output lines.

By creating an output signal that is out of phase with the original, a balanced source can be created to eliminate this error (see Fig. 16.25). As an added benefit, for a given maximum output voltage from the source, the signal voltage is doubled over the unbalanced case.

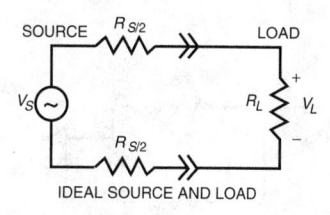

FIGURE 16.22 A basic source and load connection. No grounds are indicated and both the source and the load float.

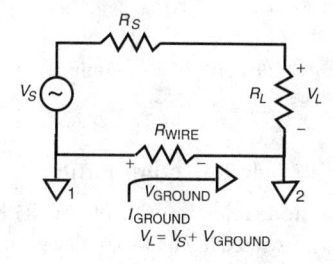

FIGURE 16.23 An unbalanced system in which each piece of equipment has one of its connections tied to ground.

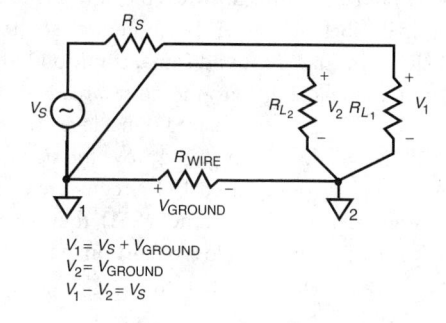

$$V_1 = V_S + V_{GROUND}$$
$$V_2 = V_{GROUND}$$
$$V_1 - V_2 = V_S$$

FIGURE 16.24 Illustration of how ground loop noise can be canceled by amplifying both the high side and the ground side of the source and subtracting the two signals.

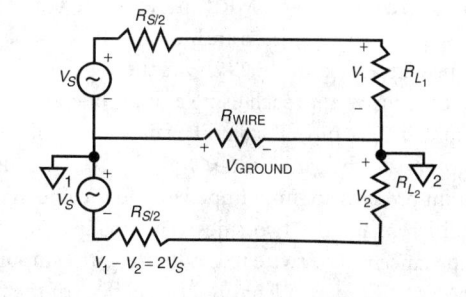

$$V_1 - V_2 = 2V_S$$

FIGURE 16.25 A balanced source configuration where the inherent amplitude error of the system shown in Fig. 16.24 is eliminated.

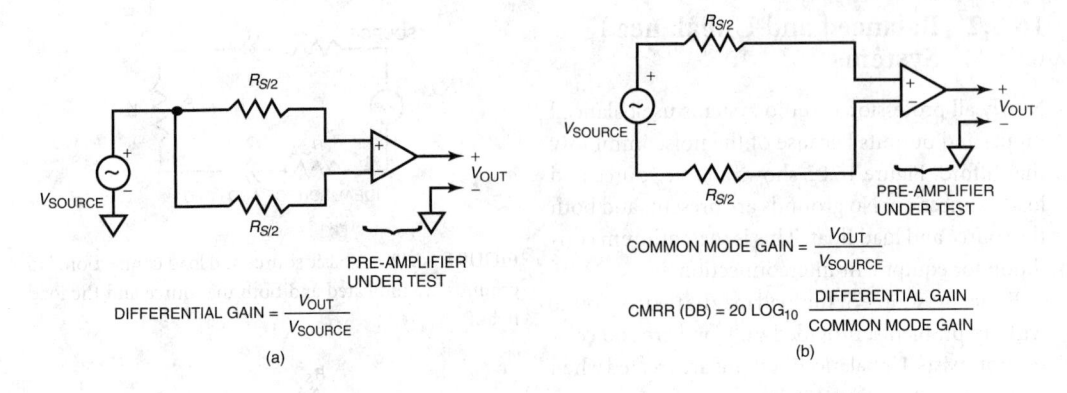

FIGURE 16.26   Common mode rejection ratio for an active balanced input circuit.

Common Mode Rejection Ratio

Common mode rejection ratio (CMRR) is the measure of how well an input rejects ground noise. The concept is illustrated in Fig. 16.26. The input signal to a differential amplifier is applied between the plus and minus amplifier inputs. The stage will have a certain gain for this signal condition, called the **differential gain**. Because the ground noise voltage appears on both the plus and minus inputs simultaneously, it is common to both inputs. The amplifier subtracts the two inputs, giving only the difference between the voltages at the input terminals at the output of the stage. The gain under this condition should be zero, but in practice it is not. CMRR is the ratio of these two gains (the differential gain and the **common mode gain**) in decibel. The larger the number, the better is the performance of the circuit. For example, a 60-dB CMRR means that a ground signal common to the two inputs will have 60 dB less gain than the desired differential signal. If the ground noise already is 40 dB below the desired signal level, the output noise level will be 100 dB below the desired signal level. If, however, the noise is already part of the differential signal, the CMRR will do nothing to improve the signal.

Although active balanced I/O circuits are not as effective as transformers in rejecting ground noise and radio frequency interference (RFI), they usually are adequate if well designed. Because of their advantages in cost, weight, and low distortion, active balanced I/O circuits generally are the best choice for all but the most demanding environments.

Active Balanced Input Circuits

A wide variety of circuit designs have been devised for active balanced inputs. All have the common goal of providing high CMRR and adequate gain for subsequent stages. All also are built around a few basic principles. Figure 16.27 shows the simplest and least expensive approach using a single operational amplifier (op-amp). For a unity gain stage all of the resistors are the same value, typically $10\,k\Omega$. This circuit presents an input impedance to the line that is different for the two sides. The positive input impedance will be twice that of the negative input.

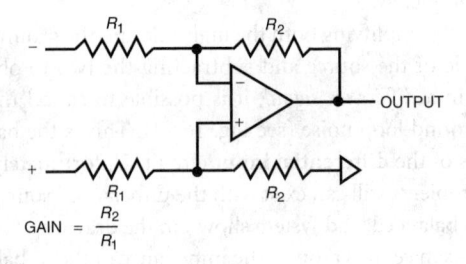

FIGURE 16.27   The simplest and least expensive active balanced input op-amp circuit. Performance depends on resistor matching and the balance of the source impedance.

The CMRR is dependent on the matching of the four resistors and the balance of the source impedance. The noise performance of this circuit, which usually is limited by the resistors, is a tradeoff between low loading of the line and low noise.

Another approach, shown in Fig. 16.28, uses a buffering op-amp stage for the positive input. The positive signal is inverted by the op-amp and then added to the negative input of the second inverting amplifier stage. Any common mode signal on the positive input (which has been inverted) will cancel when it is added to the negative input signal. Both inputs have the same impedance. The matching of resistors limits the CMRR to about 50 dB. With the addition of an adjustment pot, it is possible to achieve 80-dB CMRR, but component aging will degrade this value over time.

Adding a pair of buffer amplifiers before the summing stage results in an instrumentation grade circuit, as shown in Fig. 16.29. The input impedance is increased substantially, and floating source signal imbalance effects are eliminated. Additional noise is introduced by the added op-amp, but the resistor noise usually can be decreased by reducing impedances, causing a net improvement in system noise.

Both instrumentation and operational amplifiers are available as single-package devices. These incorporate laser-trimmed resistors that are critically matched for optimum performance. Fixed and adjustable gain units are offered with gains up to 1000 (60 dB), equivalent input noise of -128 dB, and total harmonic distortion (THD) of 0.0001% at 20 kHz.

CMRR performance for active balanced input circuits is highly dependent on the source balance. The ideal figure of 120 dB is achieved with a shorted input or perfectly balanced source. Source imbalances of 0.2–20 Ω are not uncommon. Figure 16.30 compares these effects at 60 Hz for transformer vs. active balanced inputs. At high frequencies, capacitance tolerance between the conductors and shield in typical wiring is $\pm$5% resulting in similar CMRR deterioration.

Another potential problem with active balanced inputs is their maximum input voltage limit. While easily handling the signal voltage, common-mode voltages in broadcasting environments can drive the sum of the two into clipping. Precisely matched input attenuation can help this, but makeup gain increases the circuit's noise. For installations near broadcast transmitters, the common mode range should be $\pm$50 V from DC through the significant harmonics of the transmitter. Transformer inputs easily withstand this and more.

Active Balanced Output Circuits

The classic active balanced output circuit used an approach similar to that shown in Fig. 16.31. The signal is buffered to provide one phase of the

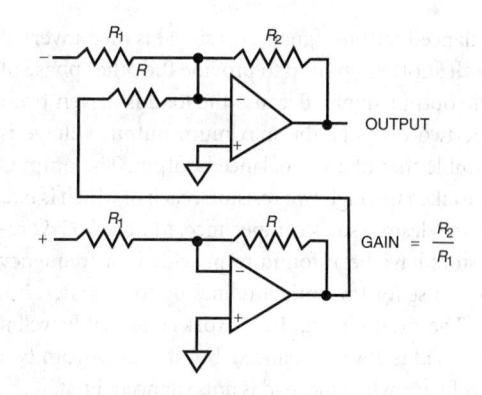

FIGURE 16.28 An active balanced input circuit using two op-amps, one to invert the positive input terminal and the other to buffer the difference signal. Without adjustments, this circuit will provide about 50-dB CMRR.

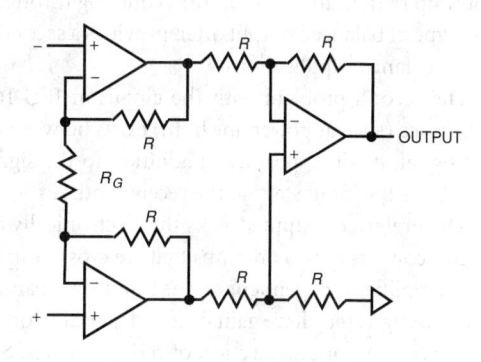

FIGURE 16.29 An active balanced input circuit using three op-amps to form an instrumentation-grade circuit. The input signals are buffered and then applied to a differential amplifier.

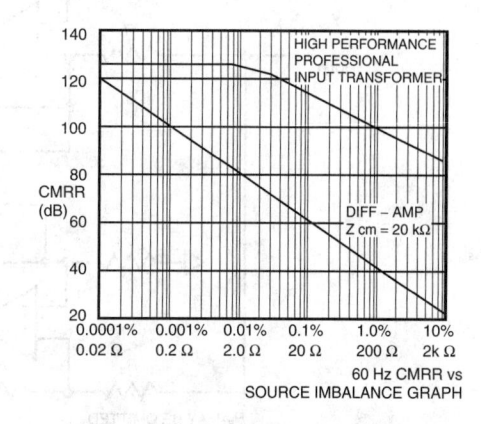

FIGURE 16.30 CMRR as a function of source imbalance. (*Source:* Pacific Research & Engineering Corp., Carlsbad, CA. With Permission.)

balanced output signal. This signal is then inverted with another op-amp to provide the other phase of the output signal. Because the load is driven from the two outputs, the maximum output voltage is double that of an unbalanced output. The outputs are taken through two resistors, each of which is half of the desired source impedance. Matched 33-Ω resistors have been found to provide best frequency response for transmission lines up to 2000 ft.

The circuit in Fig. 16.31 works reasonably well if the load is always balanced, but it suffers from two problems when the load is not balanced. First, if the negative output is shorted to ground by an unbalanced load connection, the first op-amp is likely to distort. This produces a distorted signal at the input to the other op-amp. Even if the circuit is arranged

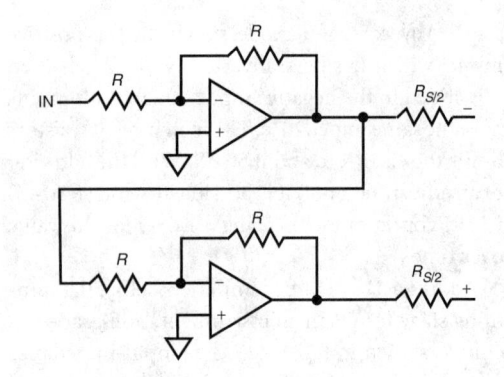

FIGURE 16.31   A basic active balanced output circuit. This configuration works well when driving a single balanced load.

so that the second op-amp is grounded by an unbalanced load, the distorted output current probably will show up in the audio output from coupling through grounds or circuit board traces. Equipment that uses this type of balanced output often provides a second set of output jacks that are wired to only one amplifier for unbalanced applications.

The second problem with the circuit in Fig. 16.31 is that the output does not float. If any voltage difference (such as power line hum) exists between the local ground and the ground at the device receiving the signal, it will appear as an addition to the signal. The only ground-noise rejection will be from the CMRR of the input stage at the receive end.

The preferred output stage is the electronically balanced and floating design shown in Fig. 16.32. The circuit consists of two op-amps that are cross-coupled with positive and negative feedback. The output of each amplifier is dependent on the input signal and the signal present at the output of the other amplifier. These designs may have gain or loss depending on the selection of resistor values. The output impedance is set via appropriate selection of resistor values. Some resistance is needed from the output terminal to ground to keep the output voltage from floating to one of the power supply rails. Care also must be taken to properly compensate the devices or stability problems may result.

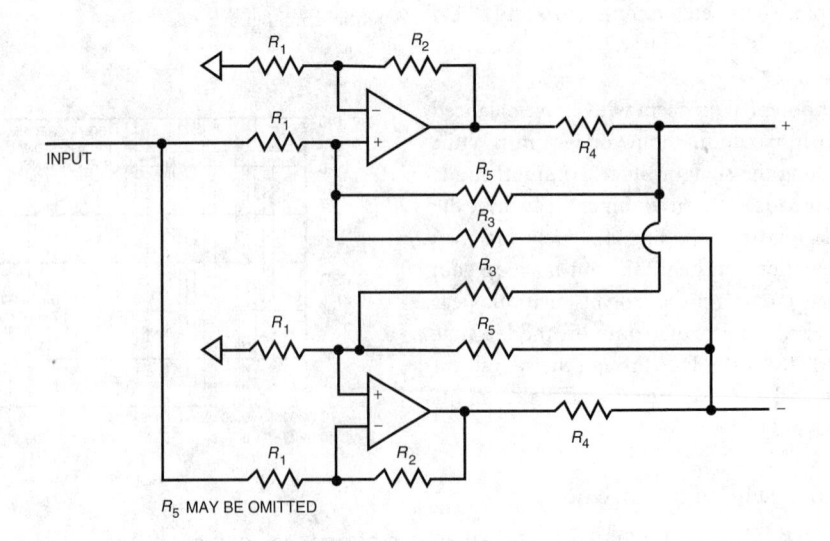

FIGURE 16.32   An electronically balanced and floating output circuit. A stage such as this will perform well even when driving unbalanced loads.

16.3.3 Interconnecting Audio Equipment

Susceptibility to RFI is a common problem with active-balanced circuits. Strong radio signals often can be rectified by nonlinearities in the input operational amplifiers or transistors. Although wideband, low-distortion circuits are less prone to this problem, they are not immune to it. Therefore, signals outside the range of the active circuits must be filtered before they are inadvertently amplified or demodulated.

To reduce this problem, most manufacturers add small series resistors and capacitors to ground at the input terminals. Inductors also may be added, but they are susceptible to external magnetic fields. If package shielding is inadequate, the inductors may, in fact, pick up as much noise as they are supposed to filter out. Another RFI defense consists of placing donut-shaped ferrite beads on the input leads of affected stages.

Successful equipment interfacing is a function of how well the I/O circuits were designed and the care taken by the equipment installer. Mixing of balanced and unbalanced inputs and outputs can be done if care is taken in planning where signals go and how wiring is to be performed. It is important to remember that the ground potential of one device may not be the same as the ground potential of another.

Wiring Guidelines

When it comes to interconnecting audio systems, theory and practice often go in divergent directions. A system design may look elegant on paper, yet turn out to be a nightmare when all of the hardware is installed and turned on. Strict adherence to some basic wiring conventions will help prevent equipment interfacing problems:

- The microphone is the starting point for most audio work. Microphones are considered to be floating sources. Use only shielded twisted-pair cable for all microphone wiring. Connect the microphone shield only to the microphone case and input chassis (see Fig. 16.33). If a grounded-shell XLR connector is used, the shell pin may be connected to the ground (shield) wire. The shell ground provides complete RFI shielding at the point where two cables join. This allows users to serially connect several microphone cables and maintain complete shielding. Be certain, however, that the XLR shells are not allowed to come in contact with any grounded metallic structures or fixtures. For this reason, most microphone cables are wired with the shells not grounded. Special microphone cables containing two twisted pairs have proven effective for reducing induced hum pickup in problem locations. This cable is also useful for stereo microphones.
- For audio equipment other than microphones, use only two-conductor shielded cable and tie the shield of each cable to the signal ground point at the respective destination equipment chassis. Do not connect the source end of the shield. Do not use the shield of an audio cable to provide safety (AC) grounding under any circumstances. Instead, build a separate ground system to tie the equipment together.

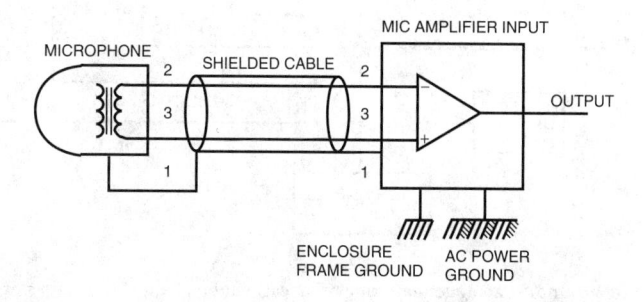

FIGURE 16.33 Interconnection arrangement for a floating source such as a microphone. Note that the microphone shield is bonded to the microphone case. It should also be bonded to the microphone preamplifier chassis.

Equipment Grounding

It is a common misconception that the only way for an audio system to be free from hum and buzz is to secure a good Earth ground. While this certainly will help, it is only one part of the picture. The term *grounding* when applied to audio often refers loosely to the interconnection of individual chassis with the Earth. It is better, however, to view grounding as an entire system. One ground system mistake can seriously impair the performance of an otherwise perfect arrangement. Grounding schemes can range from simple to complex, but any system serves three primary purposes:

- It provides for operator safety.
- It protects electronic equipment from damage caused by transient disturbances.
- It diverts stray radio frequency energy away from sensitive audio equipment.

Many engineers view grounding mainly as a method to protect equipment from damage or malfunction. However, the most important element is operator safety. The 120-V AC line current that powers most audio equipment can be dangerous, even deadly, if improperly handled. Grounding of equipment and structures provides protection against wiring errors or faults that could endanger human life.

Several different approaches can be taken to designing a studio or facility ground system. But some conventions should always be followed to ensure a low resistance (and low inductance) ground that will perform as required. Proper grounding is important whether or not the facility is located in an RF field.

Figure 16.34 shows the recommended grounding arrangement for a typical broadcast facility. The configuration is referred to as a **star ground system** because of the way the various pieces of equipment are connected in a radial pattern around the central ground point. For larger systems, a star of stars system is used, as illustrated in Fig. 16.35.

Distribution Amplifiers

Often it is necessary in an audio system to connect a single source to several different loads. Although this can be accomplished simply by tying the inputs in parallel, it may be a risky solution. The parallel arrangement is simple but leaves the audio system vulnerable to problems caused by changing levels, ground loops, and wiring faults. With a parallel hookup, shorting one of the input lines will result in a complete loss of signal to all of the other inputs.

Distribution amplifier (DA) systems have been developed to solve these problems. DAs are available in stand alone and modular configurations. Stand-alone DAs are appropriate if only a small number of

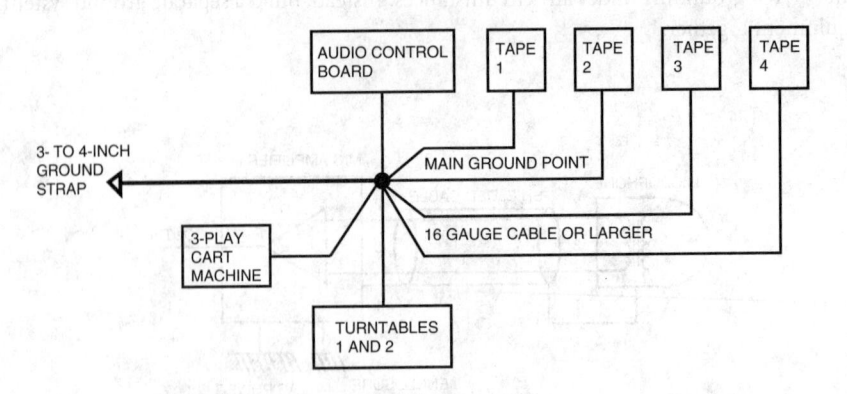

FIGURE 16.34 A typical grounding arrangement for a broadcast studio facility. The *main station ground point* is the reference from which all grounding is done at the facility.

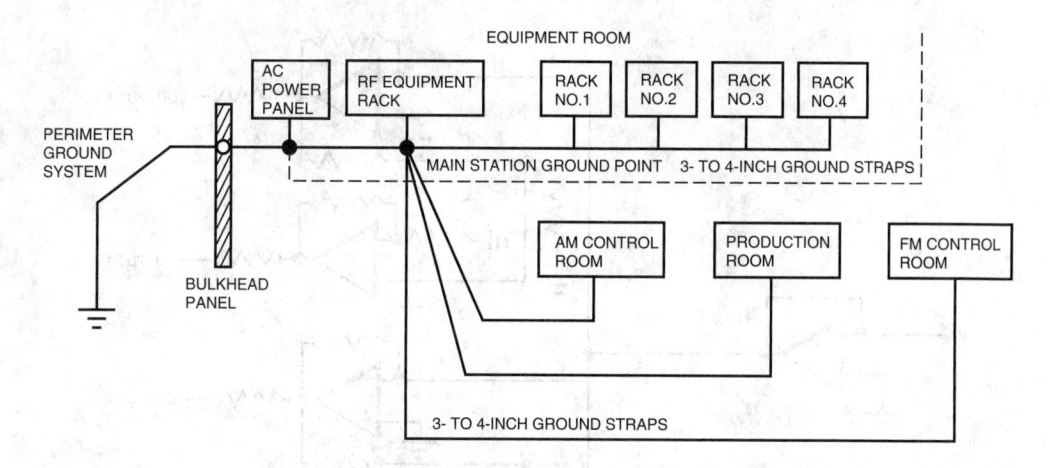

FIGURE 16.35 A typical grounding arrangement for individual equipment rooms at a broadcast facility. The ground strap from the station ground point establishes a *local ground point* in each room, to which all source and control equipment is bonded.

sources need to be distributed. Modular DAs are better suited to multisource distribution. Generally, the modular approach is best if more than five channels must be distributed and they all can be fed conveniently from the same physical location. DAs are available with either fixed or adjustable output levels. If all of the devices being fed require approximately the same input level, then a fixed output DA with an overall gain control will be adequate. If, on the other hand, the expected destination levels vary, individual output adjustments are recommended.

DAs use transformer and active balanced inputs. Active balanced outputs may use one of two basic approaches: a resistive divider network or individual driver amplifiers. Figure 16.36 shows the resistive

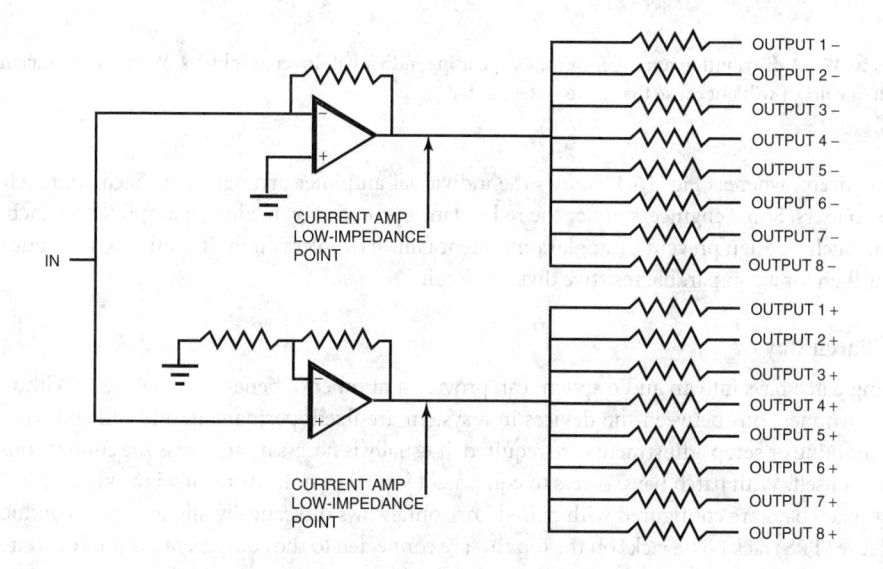

FIGURE 16.36 A distribution amplifier output stage using resistive dividers. This system relies on a single high-current driver stage to feed all outputs.

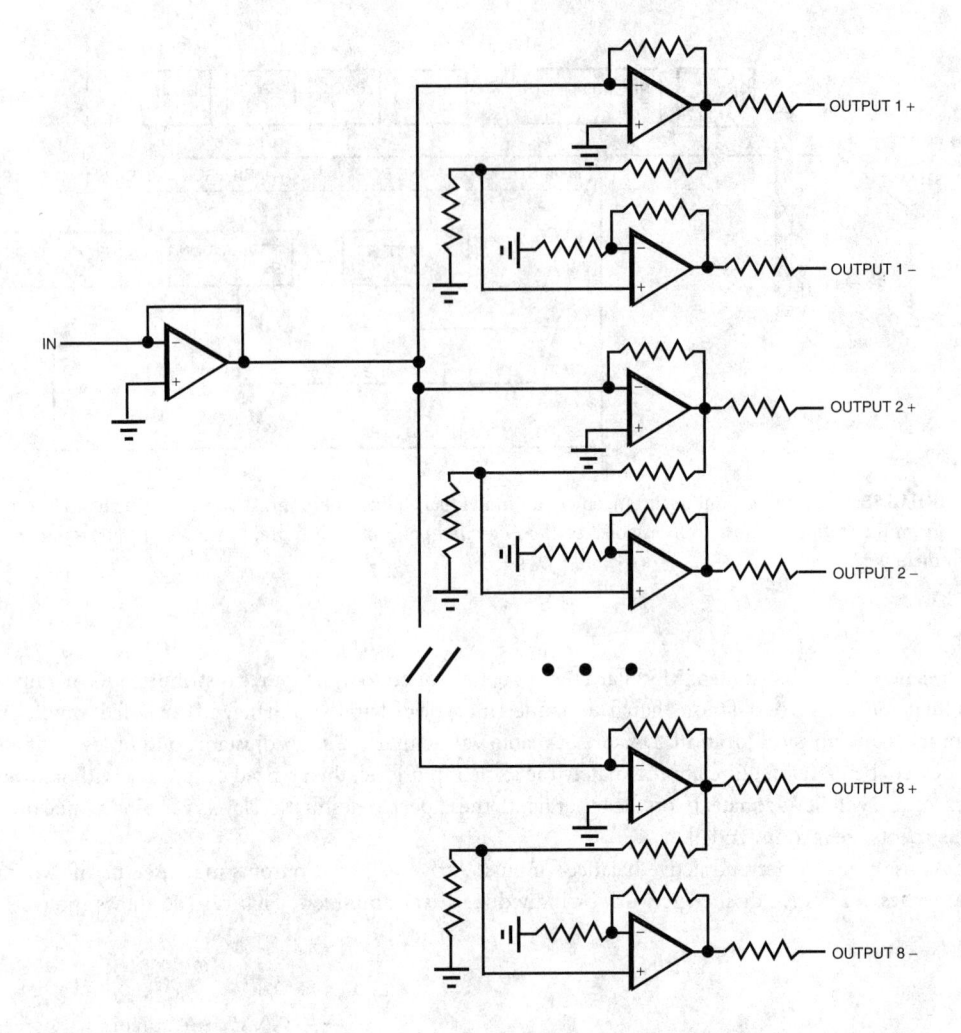

FIGURE 16.37 A distribution amplifier output stage using individual driver amplifiers. With this arrangement, the failure of one driver will not cause the entire system to fail.

divider concept, whereas Fig. 16.37 shows the individual amplifier arrangement. Each approach has its benefits to users. Some engineers prefer the redundancy provided by having an amplifier for each output terminal. Such a design prevents a single component failure from crashing the entire system, but the cost is higher than for a comparable resistive divider circuit.

Audio Patch Bays

Designing patch bays into an audio system can provide a number of benefits to the users. Without patch bays, interconnections between the devices in a system are fixed, permanent, and difficult to access. If troubleshooting or setup adjustments are required, it usually is necessary to access the connections on the equipment itself. With patch bays, access to equipment inputs or outputs readily is available.

Most patch bays are configured with paired horizontal rows of vertically aligned three-conductor **tip, ring, sleeve (TRS)** jacks. The jacks on the top row are connected to the equipment outputs or sources. The bottom row of jacks are connected to the inputs or destinations. Although similar in appearance, PJ-051 patch plugs are not the same as $\frac{1}{4}$-in. stereo phone plugs. The latter has a larger tip diameter and will damage the patch bay jacks.

If the signal path changes frequently, patch cords are used to complete the circuit between the desired jacks in the patch bay. Where parts of the system's circuit path are relatively permanent, switching jacks can be used. These jacks incorporate TRS terminals that are an integral part of the contacts that mate with the patch cord plug. When a pair of jacks is **normaled** (connected) to each other, the signal is routed through the pair; see Fig. 16.38. When a patch cord is inserted in either of the two normaled jacks, the circuit path is broken and the signal is routed through the cord. A more flexible approach, made possible by newer source and destination equipment (which does not require termination, typically 600 Ω), is the *half-normaled* arrangement shown in Fig. 16.39. The source jack is wired to monitor the source without breaking its path to the destination jack's normaling contacts. The destination jack is wired as before. The insertion of a plug breaks the path and introduces a new source.

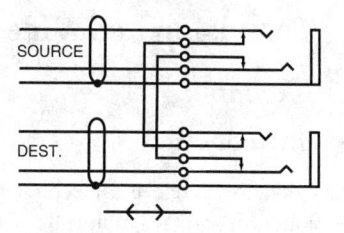

FIGURE 16.38 Schematic and symbol for *full-normaled* jacks.

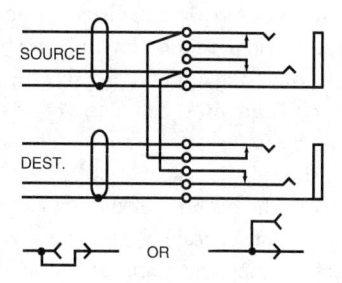

FIGURE 16.39 Schematic and symbols for *half-normaled* jacks.

Defining Terms

Common mode gain: The gain of a differential amplifier fed by an unbalanced output signal (one referenced to ground).

Common mode rejection ratio (CMRR): A measure of how well a balanced input circuit rejects ground noise; it is the ratio of the differential gain and the common mode gain in decibels.

Differential gain: The gain of a differential amplifier fed by a balanced output signal.

Differential input: An amplifier input in which the signal is applied two complementary terminals that are not referenced to ground.

Normaled: The condition when, by design, two circuits are connected when a plug is absent from a patch panel point. The circuit is interrupted with the plug inserted, resulting in a new signal path.

Star ground system: A ground network constructed around a single, reference ground point so as to represent a star pattern.

Tip, ring, sleeve (TRS): The connection nomenclature for a three-conductor plug of the type used in patch bays.

Reference

Tremaine, H.M. 1969. Installation techniques. In *Audio Cyclopedia*, 2nd ed., p. 1565. Howard W. Sams & Co., Indianapolis, IN.

Further Information

Numerous publications are available from the Audio Engineering Society (New York, NY), which serves the professional audio industry. AES also sponsors yearly trade shows and technical seminars that address various areas of audio technology.

Davis, G. and Jones, R. 1987. *Sound Reinforcement Handbook*. Yamaha Music Corp., Buena Park, CA.
Whitaker, J.C. 2000. *The Communications Facility Design Handbook*. CRC Press, Boca Raton, FL.
Whitaker, J.C. 2003. *Master Handbook of Audio Production*. McGraw-Hill, New York.

16.4   Television and Video Production Systems

Curtis J. Chan

16.4.1   Introduction

Significant advances in television technology have been made since the early days of broadcasting. Present day television incorporates a multitude of diverse technological improvements in digital signal processing, tape and disk media recording, optical and electronic imaging, and distribution and transmission technologies. In its simplest form, the technology of television is based on the conversion of light rays from still or moving scenes and pictures into electronic signals for transmission, manipulation, or storage, and subsequent reconversion into visual images on a viewing or display device. A similar function is provided in the production of motion picture film; however, where film records the brightness variations of a complete scene on a single frame in a short exposure no longer than a fraction of a second, the elements of a television picture must be **scanned** one segment at a time. In the television system, a scene is dissected into a frame composed of a mosaic of picture elements (**pixels**). A picture element is the smallest area of a television picture capable of being delineated by an electrical signal passed through the system or part thereof. The number of picture elements or pixels in the completed picture, and their geometric characteristics of vertical height and horizontal width, provide information on the total amount of detail that the **raster** can display and on the sharpness of that detail. A pixel is defined as the smallest area of a television image that can be transmitted within the parameters of the system. This process is accomplished by

- Analyzing the image with a photoelectric device in a sequence of horizontal scans from the top to the bottom of the image to produce an electrical signal in which the brightness and color values of the individual picture elements are represented as voltage levels of a video waveform
- Transmitting the values of the picture elements in sequence as voltage levels of a video signal
- Reproducing the image of the original scene in a video signal display of parallel scanning lines on a viewing screen

Conceptually, the processing of a television signal starts with a camera whose lens focuses an image of a scene onto a photoelectric (charge-coupled device (CCD)) imager, which converts the optical signal into an electrical signal. The photosensitive surface of the imager is scanned with an electron beam, converting the represented optical image into an electrical signal in which variations in brightness are converted into their electrical voltage equivalent. The signal from the camera is then input to a processing amplifier that adds horizontal and vertical synchronizing pulses to time the picture tube scanning in synchronism with the camera. The processing amplifier also readies the signal for transmission where it is amplified and used to modulate a radio frequency (RF) carrier, using amplitude modulation (AM) for broadcasting.

The end user's receiver then receives the transmitted signal through a bandwidth-limited transmission channel, which consists of any one or combination of coaxial cable, microwave, satellite, fiber optic, or broadcast transmitter. In the television receiver, the modulated RF carrier is amplified and detected to restructure the video signal and synchronizing pulses. These signals are then used to drive the display device, usually a cathode-ray tube (CRT) or other display device, which reproduces an image of the original scene.

In the same way, the information content of a color picture produced by a compatible color television system can be described by the following:

- The information content consists of a combination of monochrome and color information since it is brought to the picture by the monochrome signal and by an auxiliary coloring signal called the chrominance signal.
- The information content consists of a combination of luminance and chromaticity information since it describes the distribution of luminance levels and chromaticity values in the picture.
- The information also consists of red, green, and blue primaries.

The information content of a video signal can be broken down into basic component definitions consisting of (1) luminance level, (2) monochrome level, (3) chromaticity value, (4) raster coordinates, (5) point image, (6) line image, and (7) picture element. Various recording, delivery, and playback systems can be developed through the use of the following.

The *luminance level* at a given point in a picture is the ratio of the luminance at that point to the maximum luminance occurring anywhere in the picture during an interval in time. Its value is between zero and unity. The reciprocal of the lowest luminance level in the scene is called the *contrast level* of that scene and the reciprocal of the luminance range is called the *contrast range*.

The *monochrome level* at a given point in a color television image is the luminance level in the monochrome picture, which appears in the absence of the chrominance signal.

The *chromaticity value* at a given point in a color television picture is the quantitative specification of the chromaticity at that point in terms of a suitable set of chromaticity coordinates.

The *raster coordinates* of a given point in a picture are a pair of numbers specifying the position of that point with respect to a suitable set of reference axes. The horizontal and vertical axes may be chosen along the upper edge and left-hand edge of the raster, respectively, and a distance equal to the raster height may be designated as unity along both axes. Therefore, raster coordinates of a point in a raster with a 4/3 aspect ratio would range between zero and unity for the vertical coordinate and between zero and 1.33 for the horizontal one.

The *point image* is the pattern of light formed on a television display in response to light received by a television camera from a single object point in its field of view. The position of any given point image on a display raster is determined by the position of the corresponding object point together with whatever geometric distortions may exist in the raster. Its position is also impacted by the line structure in the scanning raster unless its size is large in comparison with the distance between adjacent scanning lines, as happens when the object point lies beyond the camera field of focus.

The *line image* is the pattern of light formed on a television display in response to light received by a television camera from a uniformly illuminated ideally narrow line in the field of view. The position of any given line image on a reproducing raster is determined by the position of the corresponding object line in a manner analogous to that for a point image. A line image oriented perpendicular to the scanning path may be called a *transverse* line image, and one parallel to this path may be called a *longitudinal* line image.

As described earlier, a *picture element* is the smallest area of a television picture capable of being delineated by an electrical signal passed through the system or part thereof. The number of picture elements (pixels) in a complete picture and their geometric characteristics of vertical height and horizontal width provide information on the total amount of detail that the raster can display and on the sharpness of that detail, respectively.

Limiting Horizontal and Vertical Resolution

The image detail perceived by the eye is determined by the resolution capability of the reproducing system or the number of basic elements that can be reproduced. The overall *limiting resolution* of a television picture is expressed as the maximum number of alternate black and white lines that can be distinguished in a dimension equal to the height of the picture where both the black and white lines are counted. The detail of a scanned television picture is defined in terms of both **horizontal (H)** and vertical resolution.

The vertical resolution is determined by the number of horizontal scanning lines, which is the total number of horizontal black and white lines that can be distinguished or resolved in the picture height. Similarly, the horizontal resolution, the number of vertical lines that can be distinguished in a dimension equal to the picture height, is twice the number of cycles of the highest frequency in the video bandwidth in the time interval required for the scanning beam to traverse this dimension.

The *horizontal limiting resolution* of a television system is usually determined by its electrical bandwidth. Assuming that the aperture response (a universal criterion for specifying picture definition and other aspects of imaging system performance) of the system components is adequate to make the black-and-white line pair (two television lines) visible, the equation for horizontal resolution R_H is, therefore

$$R_H = 2C_H B_W / A_R N_L F_R$$

TABLE 16.5 Horizontal Resolution of the NTSC and PAL Broadcast Systems

	NTSC	PAL
B_W, MHz	4.2	5.0
N_L	525	625
C_H	0.85	0.80
Aspect ratio A_R	1.33 (4:3)	1.33 (4:3)
F_R, frames/s	29.97	25
R_H, lines	340	409

where

R_H = horizontal resolution

C_H = fraction of time in each scanning line devoted to the transmission of picture information after subtracting the time required for horizontal blanking

B_W = system bandwidth

A_R = aspect ratio

N_L = total number of scanning lines per frame

F_R = frame rate per second

For comparison, the horizontal resolution R_H of two broadcast systems discussed later are compared in Table 16.5. They are the National Television Systems Committee (NTSC) broadcast systems (U.S. and Japan) and phase alternate line (PAL) broadcast systems (Europe).

The *vertical limiting resolution* of a television system is the number of horizontal lines that can be distinguished in a dimension equal to the picture height and is limited by the number of active (visible) scanning lines. However, the actual value is lower because the vertical detail in the image is randomly located with respect to the scanning lines. The ratio of the vertical resolution to the number of scanning lines is called the *Kell factor* (K_F). It is not a precise number because it is subjective and depends on the width of the scanning lines. From 1934 to 1940, the Kell factor was given a value from 0.53 to 0.85. However, through earlier tests, K_F has been given an approximate value of 0.7.

The vertical resolution R_V is given by

$$R_V = C_V K_F N_L$$

where

R_V = vertical resolution

C_V = fraction of scanning lines that are visible after subtracting the lines that are removed by vertical blanking

K_F = Kell factor

N_L = total number of scanning lines

Table 16.6 compares the vertical resolutions of the NTSC and PAL broadcast systems.

TABLE 16.6 Vertical Resolution of the NTSC and PAL Broadcast Systems

	NTSC	PAL
C_V	0.92	0.92
N_L	525	625
K_F	0.7	0.7
R_V (lines)	343	408

Picture Elements (Pixels)

A television raster can be described as a grid of picture elements or pixels. The height of each pixel is the spacing between scanning lines divided by the Kell factor, and its width is the space on a scanning line occupied by one-half the cycle of the highest frequency in the transmission bandwidth. The number of pixels, which is equal to the product of the bandwidth and the fraction of time occupied by visible scanning lines, is a rough indicator of the information content of the image. The number of pixels (N_P) in a given image is

$$N_P = 2CK_F B_W/R_F$$

where

C = fraction of time occupied by active scanning after subtracting vertical and horizontal blanking periods

K_F = Kell factor

B_W = bandwidth of the system, Hz

R_F = frame rate, frames/s

For normal broadcast systems as described previously, the approximate number of pixels equals 148,000–150,000 for NTSC, 210,000 for PAL and 250,000 for sequential color with memory (SECAM).

Bandwidth Requirements

The bandwidth required by a television signal is one-half the number of pixels transmitted per second. The bandwidth B_W is given by the equation

$$B_W = 0.8F_R N_L R_H$$

where

B_W = system bandwidth

F_R = number of frames or complete pictures transmitted each second

N_L = number of scanning lines

R_H = horizontal resolution

The numerical constant of 0.8 is derived by the following:

$$B_W = \text{(cycles per frame)} (F_R)$$
$$\text{Cycles per frame} = (N_L)(\text{cycles per line})$$
$$\text{Cycles per line} = (0.5)(\text{aspect ratio})(R_H)/0.84 = 0.8R_H$$

The factor 0.5 is the ratio of the number of cycles to the number of lines resolved. The aspect ratio in the equation is required because the horizontal resolution is measured in lines per picture height. The factor 0.84 is the fraction of the horizontal scanning interval that is devoted to signal transmission where the remainder is utilized for horizontal blanking. Choosing the optimum combination of B_W, F_R, N_L, and R_H for NTSC involved a series of tradeoffs. However, it can be said that the spectrum requirements of present day television systems required a wide bandwidth to resolve the fine detail of the picture while maintaining a high enough picture repetition rate to avoid objectionable flicker. As explained in the following text, the use of *interlace scanning* was devised to effectively reduce image flicker.

Scanning Lines and Fields

The image pattern of electrical charges on a camera tube target, corresponding to the brightness levels of a scene, are converted to a video signal in a sequential order of picture elements in the scanning process (see Fig 16.40). At the end of each horizontal line sweep, the video signal is blanked while the beam returns rapidly to the left side of the scene to start scanning the next line. This process continues until the image has been scanned from top to bottom to complete one field scan. That is why the television system employs a uniform linear scanning system. In other words, the camera sees the scene at a uniform rate and the receiver reconstructs the picture at the same rate as the camera scans it.

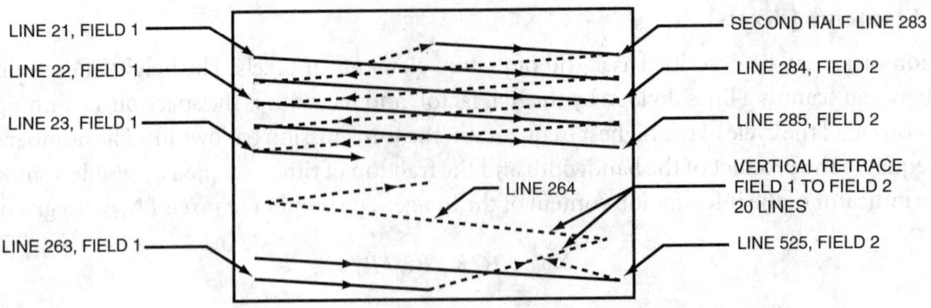

FIGURE 16.40   The interlace scanning pattern (raster) of the television image. (*Source:* Electronics Industries Association.)

After completion of this first field scan, at the midpoint of the last line, the beam again is blanked as it returns to the top center of the target where the process is repeated to provide a second field scan. The spot size of the beam as it impinges upon the target must be fine enough to leave unscanned areas between lines for the second scan. The pattern of scanning lines covering the area of the target, or the screen of a picture display, is called a raster. The ratio of the width of the raster and the resulting image is known as the picture ratio or more correctly, *aspect ratio.*

In general, countries with a 60-Hz power line standard use 525 horizontal lines laid down in sequence to form one complete picture, or frame, and there are 30 frames reproduced in every second. Countries with a 50-Hz power line standard, in general, produce a 625-line picture frame with 25 frames reproduced every second.

Interlaced Scanning

While 30 frames presented in rapid sequence will present an illusion of motion, flicker will be quite noticeable, particularly at the levels of illumination used on the television display device. Therefore, a repetition rate of 60 is more desirable to reduce the perception of flicker. However, signal bandwidths present a problem. Thus, to increase the rate, **interlaced** scanning is used. In so doing, the television frame is then divided into two parts, or **fields**, with each field presenting half of the information.

The net effect of the two fields is that the image repetition rate is increased by scanning all of the odd numbered lines and then scanning all of the even numbered lines. Then 60 images are presented to the eye without having to increase the system bandwidth. Because of the half-line offset for the start of the beam return to the top of the raster and for the start of the second field, the lines of the second field lie in between the lines of the first field. Thus, the lines of the two are interlaced. The two interlaced fields constitute a single television **frame**. Therefore, the main purpose of interlace scanning is to effectively reduce image flicker.

Reproduction of the camera image on a CRT is accomplished by a reversing of the process, with the scanning beam modulated in intensity by the video signal applied to an element of the electron gun. This control voltage to the CRT varies the brightness of each picture element on the phosphor screen. As a result of the retentivity of human vision, the eye perceives the rapidly repeated sequential display of scanning lines and rasters on the receiver display device as a continuous image.

Blanking of the scanning beam during the return trace is provided for in the video signal by a blacker-than-black pulse waveform. In addition, in most receivers and monitors another blanking pulse is generated from the horizontal and vertical scanning circuits and applied to the CRT electron gun to ensure a black screen during scanning retrace.

In conclusion, the interlaced scanning format, standardized for monochrome and compatible color, was chosen primarily for two partially related and equally important reasons:

- To eliminate viewer perception of the intermittent presentation of images, known as flicker
- To reduce video bandwidth requirements for an acceptable flicker threshold level

Perception of flicker is dependent primarily on two conditions:

- The brightness level of an image
- The relative area of an image in a picture

The 30-Hz transmission rate for a full 525-line television frame is comparable to the highly successful 24-frame-per-second rate of motion-picture film. However, at the higher brightness levels produced on television screens, if all 483 lines (525 less blanking) of a television image were to be presented sequentially as single frames, viewers would observe a disturbing flicker in picture areas of high brightness. For a comparison, motion-picture theaters on average produce a screen brightness of 10–25 fL, whereas a direct-view CRT may have a highlight brightness of 50–80 fL.

Through the use of interlaced scanning, single field images with one-half the vertical resolution capability of the 525-line system are provided at the high flicker-perception threshold rate of 60 Hz. Higher resolution of the full 490 lines (525 less vertical blanking) of vertical detail is provided at the lower flicker-perception threshold rate of 30 Hz. The result is a relatively flickerless picture display at a screen brightness of well over 50–75 fL, more than double that of motion-picture film projection. Both 60-Hz fields and 30-Hz frames have the same horizontal resolution capability.

The second advantage of interlaced scanning, compared to progressive scanning, is a reduction in video bandwidth for an equivalent flicker threshold level. Progressive scanning of 525 lines would have to be completed in 1/60 sec to achieve an equivalent level of flicker perception. This would require a line scan to be completed in half the time of an interlaced scan. The bandwidth then would double for an equivalent number of pixels per line.

The standards adopted by the Federal Communications Commission (FCC) for monochrome television in the U.S. specified a system of 525 lines per frame, transmitted at a frame rate of 30 Hz, with each frame composed of two interlaced fields of horizontal lines. Initially in the development of television transmission standards, the 60-Hz power line waveform was chosen as a convenient reference for vertical scan. Furthermore, in the event of coupling of power line hum into the video signal or scanning/deflection circuits, the visible effects would be stationary and less objectionable than moving **hum bars** or distortion of horizontal-scanning geometry. In the United Kingdom and much of Europe, the 50-Hz interlaced system was chosen for many of the same reasons. With improvements in television receivers, the power line reference was replaced with a stable crystal oscillator.

The existing 525-line monochrome standards were retained for color in the recommendations of the NTSC for compatible color television in the early 1950s. The NTSC system, adopted in 1953 by the FCC, specifies a scanning system of 525 horizontal lines per frame, with each frame consisting of two interlaced fields of 262.5 lines at a field rate of 59.94 Hz; 42 of the 525 lines in each frame are blanked as black picture signals and reserved for transmission of the vertical scanning synchronizing signal. This results in 483 visible lines of picture information.

Synchronizing Video Signals

In monochrome television transmission, two basic synchronizing signals are provided to control the timing of picture-scanning deflection: horizontal sync pulses at the line rate and vertical sync pulses at the field rate in the form of an interval of wide horizontal sync pulses at the field rate. Included in the interval are equalizing pulses at twice the line rate to preserve interlace in each frame between the even and odd fields (offset by a half-line).

In color transmissions, a third synchronizing signal is added during horizontal scan blanking to provide a frequency and phase reference for color signal encoding circuits in cameras and decoding circuits in receivers. These synchronizing and reference signals are combined with the picture video signal to form a composite video waveform.

The scanning and color-decoding circuits in receivers must follow the frequency and phase of the synchronizing signals to produce a stable and geometrically accurate image of the proper color **hue** and **saturation**. Any change in timing of successive vertical scans can impair the interlace of the even and odd fields in a frame. Small errors in horizontal scan timing of lines in a field can result in a loss of resolution in vertical line structures. Periodic errors over several lines that may be out of the range of the horizontal scan automatic frequency control circuit in the receiver will be evident as jagged vertical lines.

16.4.2 Television Industry Standards

There are three primary analog color transmission standards in use today:

- *NTSC* is used in the U.S., Canada, Central America, most of South America, and Japan. In addition, NTSC is used in various countries or possessions heavily influenced by the U.S.
- *PAL* is used in England, most countries and possessions influenced by the British Commonwealth, many western European countries, and China. Variation exists in PAL systems.
- *SECAM* is used in France, countries and possessions influenced by France, and the former USSR, including former East Germany, and other areas influenced by Russia.

The three standards are incompatible for a variety of reasons.

Television transmitters in the U.S. operate in three frequency bands:

- Low-band VHF, channels 2–6
- High-band VHF, channels 7–13
- UHF, channels 14–83 (UHF channels 70–83 currently are assigned to mobile radio services)

Table 16.7 shows the frequency allocations for channels 2–83. Because of the wide variety of operating parameters for television stations outside the U.S., this section will focus primarily on TV transmission as it relates to the U.S.

TABLE 16.7   Frequency Bands for Television Broadcasting in the U.S.

Channel Designation	Frequency Band, MHz	Channel Designation	Frequency Band, MHz	Channel Designation	Frequency Band, MHz
2	54–60	29	560–566	56	722–728
3	60–66	30	566–572	57	728–734
4	66–72	31	572–578	58	734–740
5	76–82	32	578–584	59	740–746
6	82–88	33	584–590	60	746–752
7	174–180	34	590–596	61	752–758
8	180–186	35	596–602	62	758–764
9	186–192	36	602–608	63	764–770
10	192–198	37	608–614	64	770–776
11	198–204	38	614–620	65	776–782
12	204–210	39	620–626	66	782–788
13	210–216	40	626–632	67	788–794
14	470–476	41	632–638	68	794–800
15	476–482	42	638–644	69	800–806
16	482–488	43	644–650	70	806–812
17	488–494	44	650–656	71	812–818
18	494–500	45	656–662	72	818–824
19	500–506	46	662–668	73	824–830
20	506–512	47	668–674	74	830–836
21	512–518	48	674–680	75	836–842
22	518–524	49	680–686	76	842–848
23	524–530	50	686–692	77	848–854
24	530–536	51	692–698	78	854–860
25	536–542	52	698–704	79	860–866
26	542–548	53	704–710	80	866–872
27	548–554	54	710–716	81	872–878
28	554–560	55	716–722	82	878–884
				83	884–890

Maximum power output limits are specified by the FCC for each type of service. The maximum **effective radiated power (ERP)** for low-band VHF is 100 kW, for high-band VHF it is 316 kW, and for UHF it is 5 MW. The ERP of a station is a function of transmitter power output (TPO) and antenna gain. ERP is determined by multiplying these two quantities together, and subtracting transmission line loss.

The second major factor that affects the coverage area of a TV station is antenna height, known in the broadcast industry as *height above average terrain* (HAAT). HAAT takes into consideration the effects of the geography in the vicinity of the transmitting tower. The maximum HAAT permitted by the FCC for a low- or high-band VHF station is 1000 ft (305 m) east of the Mississippi River, and 2000 ft (610 m) west of the Mississippi. UHF stations are permitted to operate with a maximum HAAT of 2000 ft (610 m) anywhere in the U.S. (including Alaska and Hawaii).

The ratio of visual output power to **aural** output power can vary from one installation to another; however, the aural is typically operated at between 10 and 20% of the visual power. This difference is the result of the reception characteristics of the two signals. Much greater signal strength is required at the consumer's receiver to recover the visual portion of the transmission than the aural portion. The aural power output is intended to be sufficient for good reception at the fringe of the station's coverage area, but not beyond. It is of no use for a consumer to be able to receive a TV station's audio signal, but not the video.

Two classifications of **low power TV (LPTV)** stations have been established by the FCC to meet certain community needs. They are

- **Translator**. A low-power system that rebroadcasts the signal of another station on a different channel. Translators are designed to provide fill-in coverage for a station that cannot reach a particular community because of the local terrain. Translators operating in the VHF band are limited to 100-W power output (ERP), and UHF translators are limited to 1 kW.
- **Low-Power Television.** A service recently established by the FCC designed to meet the special needs of particular communities. LPTV stations operating on VHF frequencies are limited to 100-W ERP and UHF stations are limited to 1 kW. LPTV stations originate their own programming and can be assigned by the FCC to any channel, as long as sufficient protection against interference to a full-power station is afforded.

The **composite video** waveform is shown in Fig. 16.41. The actual radiated signal is inverted, with modulation extending from the synchronizing pulses at maximum carrier level (100%) to reference picture white at 7.5%. Because an increase in the amplitude of the radiated signal corresponds to a decrease in picture brightness, the polarity of modulation is termed negative.

Composite Video

The term composite is used to denote a video signal that contains: picture luminance and chrominance information and timing information for synchronization of scanning and color signal processing circuits (blanking, sync, and color burst). The television waveform is a complex combination of pulses and sine waves conveying seven different types of information necessary to satisfactorily reconstruct a television picture. The seven components, when properly combined into one continuous electrical wave, form a television *composite video waveform* or simply, composite video. The seven components of the composite video signal are

- Horizontal line synchronizing pulses
- Color sync (burst)
- Setup (black) level
- Picture elements
- Color hue (tint)
- Color saturation (vividness)
- Field synchronizing pulses

Referring to Fig. 16.42, nearly one-third of the total peak-to-peak voltage amplitude range of the composite video waveform is used for synchronizing information (horizontal, color burst, field synchronizing pulses).

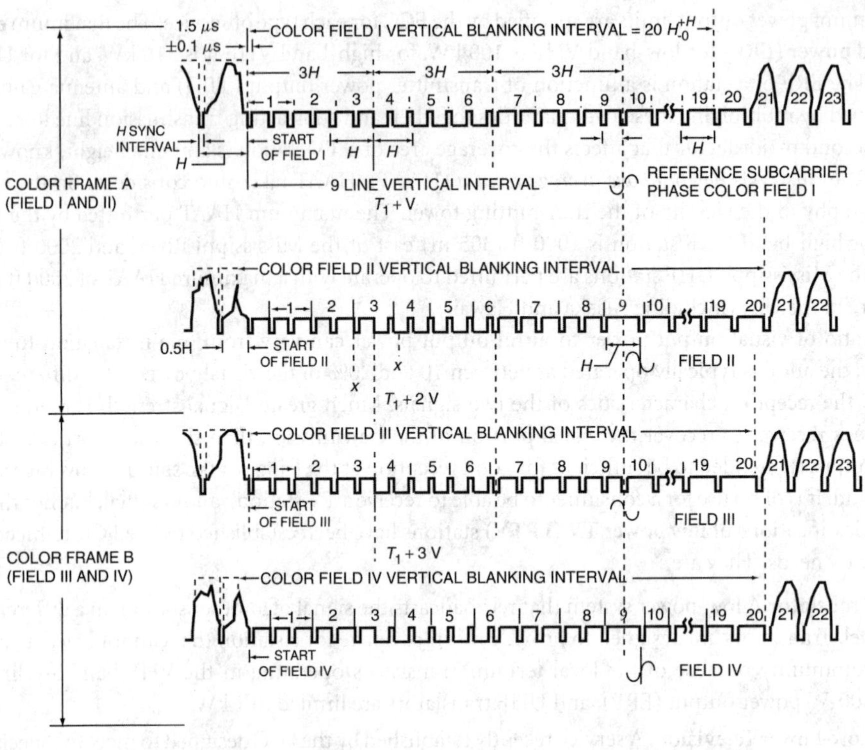

FIGURE 16.41 The principle components of the NTSC color system. (*Source*: Electronics Industries Association.)

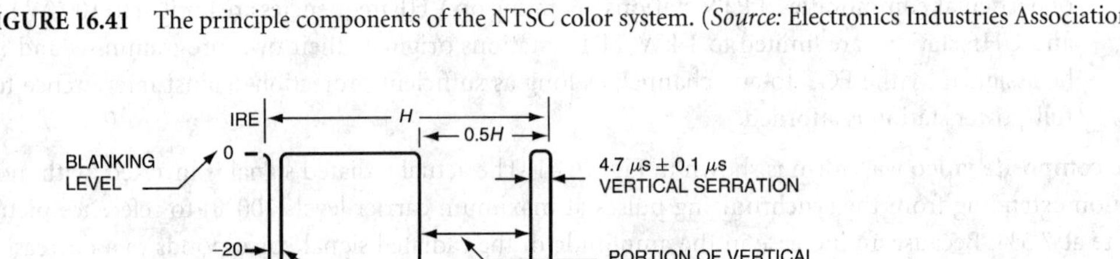

FIGURE 16.42 Sync pulse widths for the NTSC color system. (*Source*: Electronics Industries Association.)

If the remaining two-thirds of the composite video is removed (**setup** level, picture elements, color hue, and saturation), only a series of pulses will remain. The series of pulses are called *composite sync.* This may be further defined as follows:

Line, or *horizontal sync,* occurs once every 63.5 μs, or at approximately a 15-kHz rate and is used to synchronize the movement of the scanning electron beam from left to right in the receiver to that in the camera.

Field, or *vertical sync,* occurs once every 16.6 ms, or at approximately a 60-Hz rate and is used to synchronize the movement of the scanning electron beam from bottom to top of the picture between subsequent fields of information.

Color sync, or *burst,* a 3.58-MHz occurs one per line for a specific duration and at a specific time and is used to ensure that the color information displayed by the receiver is identical to that picked up by the camera.

The nominal video and sync levels (in IRE units) are listed in Table 16.8

The Horizontal Blanking Interval

As shown in Fig. 16.42, the *horizontal blanking interval* includes *front porch, leading edge of sync, horizontal sync, breezeway, color burst,* and *back porch.* The total time duration of the interval is approximately 11.1 μs. The definitions of the various components are described next.

Front Porch

The primary function of the front porch period is to blank the television receiver just before the start of horizontal retrace, the time during which the electron beam is returned to the left side of the raster to begin tracing the next picture line. In the days of quadruplex videotape recorders, the reproduce heads were switched during this time and the switch time was called *front porch switch, switch time pulse* or *blanking switch pulse.*

Horizontal Sync

The leading edge of the pulse synchronizes horizontal line sweeps of the receiver and camera. In videotape recorders, this edge is used to identify electronically the start of the blanking interval.

Back Porch

In the television receiver, the back porch interval provides picture blanking until the start of the active portion of the line. In some videotape recorders, the interval is used to re-establish the DC level of the signal after detection.

Color Sync or Burst

The color burst signal is normally included only during color transmission and is the color sync portion of the horizontal blanking interval. It provides a fixed reference of constant phase and amplitude to the receiver to enable it to decode correctly the varying phase and amplitude of the color information on the line that follows.

There are a number of other timing signals in the television system that are either derived from horizontal rate information or have a definite relationship to it. They are the following:

TABLE 16.8 Video and Sync Levels in IRE Units

Signal Level	IRE Level
Reference white	100
Blanking level width measurement	20
Color burst sine wave peak	+20 to −20
Reference black	7.5
Blanking	0
Sync pulse width measurement	−20
Sync level	−40

Horizontal Drive
This is a pulse approximately 6.35 μs long whose leading edge coincides with the leading edge of horizontal sync.

Burst Flag or Burst Key
This is a pulse approximately 2.3 μs long whose leading edge is delayed approximately 5 μs, from the leading edge of sync. It is used to gate in the correct amount of color subcarrier on the back porch of the composite video signal.

Color Subcarrier
To minimize the visual effects of the color subcarrier signal on black and white (B/W) televisions, it was decided that the line pattern should be stationary and cross the picture at 45° from the horizontal. This was achieved by making the subcarrier frequency an odd multiple of half the horizontal line frequency. In that case, alternate lines in the same field have the subcarrier 180° out of phase or the line ends at half a subcarrier cycle. Because of the odd number of lines in the raster, the interlaced field has a $\frac{1}{4}$ cycle shift. One of the results of this system is that it takes two frames (four fields) for the color signal to get back to where it started with regard to sync. Thus, there are four fields in the complete color picture. Also, the beat between the color signal and the sound carrier was considered. To make this beat stationary, it was decided to alter the subcarrier and the scanning frequencies. The final value for the subcarrier signal was decided to be 3.579545 MHz and horizontal sync is 2/455 times this value or 15,734.3 Hz, changing vertical sync to 59.94 Hz. The color subcarrier of 3.579545 MHz is an odd multiple of half-line frequency to produce minimum sound and picture interference on B/W receivers and minimum sound interference on color televisions.

Initial calculations indicated that a frequency 455 times half-line rate would be ideal.

$$F_{sc} = (15{,}750/2)(455) = 3.583125 \text{ MHz}$$

However, this frequency when beat with the sound carrier of 4.5 MHz produced an even harmonic of the line rate resulting in visible picture artifacts as a function of audio content. Instead, the horizontal frequency was changed from 15,750 to 15,734.3 Hz (actually 15,734.264 Hz) and the vertical line rate was changed from 60 to 59.94 Hz. Taking the 455th harmonic of half the new line frequency resulted in

$$F_{sc} = (15{,}734.264)(455)/2 = 3.579545 \text{ MHz}.$$

Thus, a continuous RF signal of exactly 3.579545 MHz ± 5 Hz is used to define phase and amplitude of the coloring information in the video waveform. The relationship between color subcarrier and the prime timing signals are

Horizontal rate = (2) (3.579545 MHz)/455 = 15,734.3 Hz
Horizontal period = 63.55 μs
Vertical rate = H Rate/262.5 = 15,734.3/262.5 = 59.94 Hz
Vertical period = 16.68 ms
Frame rate = H Rate/525 = 15,734.3/525 = 29.97 Hz
Frame period = 33.36 ms = 1/29.97 Hz

Composite Blanking
This is a signal that contains both vertical and horizontal rate timing pulses. The pulse width for horizontal blanking is 11.1 μs in duration and 21 lines in duration for vertical blanking. The vertical blanking period is approximately 21 lines in duration for NTSC. The most important area of the vertical interval is the nine-line period comprising the pre-equalizing pulses, the serrated vertical sync pulse, and the postequalizing pulses. The balance of the interval is used in transmission systems for the insertion of active test signal information and other resetting or calibration signals for videotape recording.

Pre-equalizing Pulses

The function of the six pre-equalizing pulses present at the start of the vertical blanking interval is to ensure that the television picture maintains proper interlace. There are six of these pulses, each approximately 2.4 μs wide, occurring during a period of three television lines, and so they are said to occur at *twice line rate* or *half-line period*, a frequency of approximately 31.5 kHz. The way in which these pulses maintain interlace is determined by the fact that although there are always six pulses present at the start of the vertical interval, the timing relationship of these six pulses differ on succeeding intervals.

Vertical Sync

The vertical sync pulse portion of the vertical blanking interval is called the serrated pulse, since it is broken into six equal parts, or serrations, each approximately 4.5 μs wide. The vertical sync has a much greater average negative value than the equalizing pulse period. This large negative-to-positive value ratio is widely used in television equipment to develop a single pulse, which provides the vertical time reference for the system.

Post-equalizing Pulses

These six pulses, which are identical to the pre-equalizing pulses, have no current use in modern day television systems.

Back Porch

The balance of the vertical blanking interval from the end of the nine-line synchronizing period to the start of the first active scanning line is called the back porch and contains the correct number of horizontal sync pulses and line intervals to complete the interval. In a color signal, burst is also present in its proper relationship to horizontal sync. The back porch period is also used for the insertion of special test and reference signals to check the quality of television transmission and recording devices.

Field and Frame Identification

All modern day videotape recorders, editing systems and other processing devices place great reliance on some means of identifying a television frame or field in the reproduced picture so that the playback may be electronically compared to the station synchronizing signals. Each television picture is formed from two interlaced fields, which are then called a frame. There is one complete vertical blanking interval associated with each field. Although each blanking interval contains the same type of information, they are different from each other by the following.

If we consider the vertical interval to be the start of the field, and compare the intervals for fields one and two, six differences become apparent:

1. The start of field one is preceded by a full line of video; the start of field two is preceded by one-half line of video.
2. The leading edge of vertical sync of field one is time coincident with the leading edges of horizontal sync pulses; for field two, it is half-way between.
3. The first active scanning line in field one starts at the top left corner raster; in field two, it starts at the top center.
4. The last active scanning line in field one ends at the bottom center of the raster; in field two, it ends at the bottom right corner.
5. The scanning lines for field one are numbered beginning with the first equalizing pulse, in field two with the second equalizing pulse.
6. The first horizontal sync pulse in the interval for field one occurs one-half line after the last post-equalizing pulse; in field two, it occurs a full line after the last postequalizing pulse.

Color Signal Encoding

To facilitate an orderly introduction of color television broadcasting in the U.S. and other countries with existing monochrome services, it was essential that the new transmissions be compatible. In other words, color pictures would provide acceptable quality on unmodified monochrome receivers. In addition, because of the limited availability of RF spectrum, another related requirement was the need to fit approximately 2-MHz bandwidth of color information into the 4.2-MHz video bandwidth of the existing 6-MHz broadcasting channels with little or no modification of existing transmitters. This is accomplished using the band-sharing color signal system developed by the NTSC and by taking advantage of the fundamental characteristics of the eye regarding color sensitivity and resolution.

The video-signal spectrum generated by scanning an image consists of energy concentrated near harmonics of the 15,734-Hz line scanning frequency. Additional lower amplitude sideband components exist at multiples of 60 Hz (the field scan frequency) from each line scan harmonic. Almost no energy exists halfway between the line scan harmonics, that is, at odd harmonics of one-half line frequency. Thus, these blank spaces in the spectrum are available for the transmission of a signal for carrying color information and its sideband. In addition, a signal modulated with color information injected at this frequency is of relatively low visibility in the reproduced image because the odd harmonics are of opposite phase on successive scanning lines and in successive frames, requiring four fields to repeat. Furthermore, the visibility of the color video signal is reduced further by the use of a subcarrier frequency near the cutoff of the video bandpass.

FIGURE 16.43 Vectorscope representation for vector and chroma amplitude relationships in the NTSC system. (*Source:* Electronics Industries Association.)

In the NTSC system, the red, green, blue (RGB) outputs from the camera are fed to an encoder whose functions are the following and can be seen in Fig. 16.43.

The first function is to derive the brightness or luminance signal E_Y by suitable combinations of E_R, E_G, and E_B. The luminance signal is derived from components of the three primary colors, red, green, and blue in the proportions for *reference white* E_Y as follows:

$$E_Y = 0.299E_R + 0.587E_G + 0.114E_B$$

Rounding off becomes

$$E_Y = 0.30E_R + 0.59E_G + 0.11E_B$$

These transmitted values equal unity for white and thus result in the reproduction of colors on monochrome receivers at the proper luminance level. This is known as the constant-luminance principle. *Luminance*, therefore, is the portion of the signal that will be used by B/W receivers to display the brightness information of the color camera signal and consists of the proportions of the three camera pickup tubes already stated. It will be combined with the decoded color difference signals in the color receiver to give the original RGB signals that existed at the color camera outputs before encoding.

The second function of the encoder is to derive the I (in phase) color difference signal E_I by suitable combinations of E_R, E_G, and E_B in the proportion

$$E_I = 0.60E_R - 0.28E_G - 0.32E_B$$

and to pass this signal through a 1.5-MHz low-pass filter to the I channel input of a balanced modulator.

The third encoder function is to derive the Q (quadrature) color difference signal E_Q by suitable combinations of E_R, E_G, and E_B in the proportion

$$E_Q = 0.21E_R - 0.52E_G - 0.31E_B$$

and to pass this signal through a 0.5-MHz low-pass filter to the Q channel input of a balanced modulator.

The fourth encoder function is to use the bandlimited I and Q signals to modulate a suppressed subcarrier balanced modulator, the output phase of which represents hue and the output amplitude of which represents saturation. This is the chrominance signal and consists of sideband information generated by a suppressed subcarrier modulator, which is quadrature modulated by a wide-band color difference signal (I) and a narrow-band color difference signal (Q). The instantaneous phase of the chroma signal defines hue; the instantaneous amplitude of chroma defines saturation. The chroma signal cancels out in a B/W receiver because of the half line offset between subcarrier and line rate causing the chroma phase to reverse itself at the frame rate. The chroma signal in a color receiver is decoded to its original (R-Y) and (B-Y) components; the missing (G-Y) is developed by combining portions of R-Y and B-Y and the three color difference signals (R-Y), (B-Y) and (G-Y) are combined with the luminance signal $+Y$ to give R, G, and B signals to drive the individual guns of the CRT.

The fifth function is to combine the luminance (E_Y) signal with the chrominance signal.

The last encoder function is to add sync and color burst to the combined luminance and chrominance signal. Color burst consists of 8–11 cycles of the 3.579545-MHz subcarrier inserted in the back porch of horizontal blanking at an amplitude of 40-**IRE** divisions peak-to-peak where IRE is a unit named for the Institute of Radio Engineers. Its function is to phase lock a complementary crystal oscillator in the color receiver to the encoder to permit correct resultant hue on the color receiver after decoding. The final composite color video output of the encoder consists of the combination of luminance, chrominance, sync, and color burst.

In conclusion, the color signal consists of two chrominance components, $I(E_I)$ and $Q(E_Q)$, transmitted as amplitude modulated sidebands of two 3.579545-MHz subcarriers in quadrature (differing in phase by 90°). The subcarriers are suppressed, leaving only the sidebands in the color signal. Suppression of the carriers permits demodulation of the color signal as two separate color signals in a receiver by reinsertion of a carrier of the phase corresponding to the desired color signal. This system for recovery of the color signals is called *synchronous demodulation*.

Chrominance Bandwidth Requirements

Extensive research done by NTSC members in the late 1940s and early 1950s proved that the color acuity of human vision was at maximum for certain colors and minimum for others. By placing colored chips in a random pattern and having individuals with normal color vision identify the colors and then repeating the test by rearranging the chips and doubling the distance each time, these studies showed that in the region of yellow-orange the eye could distinguish color differences when the detail was quite small, whereas in the region of green-cyan, small objects were perceived as shades of gray.

On the basis of these tests, it was decided to shift the R-Y and B-Y signals 33° counterclockwise to place the shifted R-Y signal on the axis of maximum acuity and the shifted B-Y signal on the axis of minimum acuity. The bandwidth of the shifted R-Y signal was limited to 1.5 MHz by a low-pass filter and this bandlimited shifted R-Y signal was referred to as the I signal (in phase). The bandwidth of the shifted B-Y signal was limited to 0.5 MHz by a low-pass filter and this bandlimited shifted B-Y signal was referred to as the Q signal (quadrature).

As both signals were shifted the same amount, their original 90° relationship was maintained and the previous analysis of R-Y and B-Y to generate saturation and hue information still applied. Therefore, the values of E_I and E_Q as stated previously with the reduction factors are

$$E_I = 0.60E_R - 0.28E_G - 0.32E_B$$
$$E_Q = 0.21E_R - 0.52E_G - 0.31E_B$$

The vectorial in Fig. 16.43 shows the location of the primary and complementary colors with respect to the I and Q axis for 100% saturated corrected signals.

Color-Signal Decoding

Each of the two chroma signal carriers (E_I and E_Q) can be recovered individually by means of **synchronous detection** (using quadrature synchronous detectors), followed by suitable filtering. The E_I and

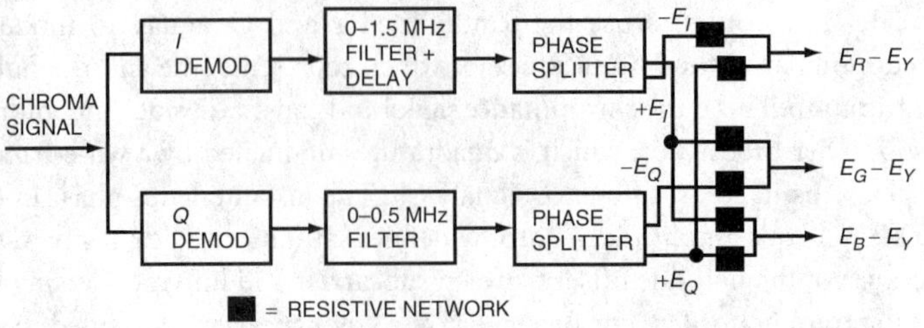

FIGURE 16.44   Matrix for obtaining color difference signals E_I and E_Q signals.

E_Q components can be used in a simple matrix to derive the needed R-Y, B-Y, and G-Y signals. The equations that form the matrix are

$$E_B - E_Y = 1.7E_Q - 1.10E_I$$

$$E_R - E_Y = 0.62E_Q + 0.96E_I$$

$$E_G - E_Y = -0.65E_Q - 0.28E_I$$

In performing the matrix operation, the E_I and E_Q signals are time coincident by adding delay in the wide-band E_I channel. Following such delay equalization, the preceding matrix operation might be performed by using a pair of phase splitters, to obtain plus and minus E_I and E_Q signals, and then combining these signals through suitable resistive networks (see Fig. 16.44). A set of quadrature demodulators may be used to obtain directly the desired color difference signals R-Y and B-Y. If needed, a third demodulator can be employed to obtain the remaining color difference signal G-Y. Primary color signals of the form E_R, E_G, and E_B are required to operate a three gun color display in a television receiver. These signals may be obtained by combining the respective color difference signals with the luminance signal. This combining function may occur in external adding circuits, so that the primary color signals E_R, E_G, and E_B are applied, respectively, to one control electrode of each gun of the display. These signals can also be combined in the picture tube by applying a negative polarity luminance signal $-E_Y$ to each of the cathodes and by applying the appropriate color difference signals $E_R - E_Y$, $E_B - E_Y$, and $E_G - E_Y$ to the respective control grids of the picture tube electron guns. The former operation can be then called *RGB operation* and the latter *color-difference operation*.

16.4.3   Transmission Equipment

Television transmitters are classified in terms of their operating band, power level, final-tube type, translators, and cooling method. The transmitter is divided into two basic subsystems: The *visual* section accepts the video input, amplitude modulates a radio frequency (RF) carrier, and amplifies the signal to feed the antenna system. The *aural* section accepts the audio input, frequency modulates a separate RF carrier and amplifies the signal to feed the antenna system. In the U.S., the ratio of aural power permitted by FCC Rules and Regulations may be between 10 and 20% of the peak signal power. The visual and aural signals are combined to feed a single radiating system.

The differences in transmitter design between low VHF, high VHF, and UHF are mainly in the up-converter and the output power level. Low VHF are channels 2–6, and high VHF are channels 7–13. UHF are channels 14–83, except for channels 70–83, presently assigned to mobile radio service. Present-day practice is to use an exciter operating at or near the intermediate frequency (IF) followed by an up-converter, which translates the IF signal to the desired channel frequency. In most cases, distortion corrections can be done at IF, although some are done at baseband and RF (see Table 16.9).

TABLE 16.9   Power and Height Limitations

Band	Channels	ERP, kW	Height, ft (m)	Zone
Low VHF	2–6	100	1000 (305)	1
			2000 (610)	2, 3
High VHF	7–13	316	1000 (305)	1
			2000 (610)	2, 3
UHF	14–69	5000	2000 (610)	1, 2, 3

Transmitter Classification

- *Power Output.* Output power refers to the peak power of the visual transmitter.
- *Modulation Level.* Transmitters classified by level are of the IF type. Some of the reasons for using IF modulation are ease of correcting distortions, ease in vestigial sideband shaping, and economic advantages to the manufacturer. This is because only one type of exciter is required for the VHF band, and only one basic design is required for the UHF band.
- *Final-Stage Type.* Solid-state devices generally are used for VHF transmitters and for lower level UHF transmitters, and **klystrons** are used for UHF above about 10 kW. The choice is dictated by the device that will best operate at the required frequency, factoring in initial cost, operating cost per hour, and the cost of the auxiliary equipment.
- *Cooling.* Cooling can be by water, air, vapor, or a combination of these. Again, operating frequency, operating costs, device life, noise, and efficiency determine the final selection.
- *Translators.* Low-power stations, or translators, rebroadcast another station's programs on another frequency, usually in the grade B service area of the mother station. They do not originate their own programs and their power output is limited to 100-W VHF and 100-W UHF.
- *Low-Power Stations.* These are translators, except that they originate programs and can be assigned to any channel meeting the FCC rules. The power limitations are 100-W VHF and 1000-W UHF with the stipulation that they do not interfere with other television stations even if it requires reducing their power output to a few watts.

Transmitter Design Considerations

Each manufacturer has a particular philosophy with regard to the design and construction of a broadcast TV transmitter. Some generalizations can, however, be made with respect to basic system design.

When the power output of a TV transmitter is discussed, the visual section is the primary consideration. Output power refers to the peak power of the visual section of the transmitter (peak of sync). The FCC-licensed effective radiated power is equal to the transmitter power output minus feedline losses times the power gain of the antenna.

A low-band VHF station can achieve its maximum 100-kW power output through a wide range of transmitter and antenna combinations. A 35-kW transmitter coupled with a gain of 4 antenna would work, as would a 10-kW transmitter feeding an antenna with a gain of 12. Reasonable parings for a high-band VHF station would range from a transmitter with a power output of 50 kW feeding an antenna with a gain of 8, to a 30-kW transmitter connected to a gain of 12 antenna. These combinations assume reasonable feedline losses. To reach the exact power level, minor adjustments are made to the power output of the transmitter, usually by a front panel power trim control.

UHF stations, to achieve their maximum licensed power output, are faced with installing a very high-power transmitter. Typical pairings include a transmitter rated for 220 kW and an antenna with a gain of 25, or a 110-kW transmitter and a gain of 50 antenna. In the latter case, the antenna could pose a significant problem. UHF antennas with gains in the region of 50 are possible, but not advisable for most installations because of the coverage problems that can result. High-gain antennas have a narrow vertical radiation pattern that can reduce a station's coverage in areas near the transmitter site.

At first examination, it might seem reasonable and economical to achieve licensed ERP using the lowest transmitter power output possible and highest antenna gain. Other factors, however, come into play that make the most obvious solution not always the best solution. Factors that limit the use of high-gain antennas include

- The effects of high-gain designs on coverage area and signal penetration
- Limitations on antenna size because of tower restrictions, such as available vertical space, weight, and windloading
- The cost of the antenna

The amount of output power required of a transmitter will have a fundamental effect on system design. Power levels dictate whether the unit will be of solid-state or vacuum tube design; whether air, water, or vapor cooling must be used; the type of power supply required; the sophistication of the high-voltage control and supervisory circuitry; and many other parameters.

Elements of the Transmitter

The design of a television transmitter may differ widely from manufacturer to manufacturer depending upon the frequency band and power rating. All transmitters, however, must include the following components:

1. Aural pre-emphasis to pre-emphasize the high-frequency components in accordance with FCC requirements.
2. Processing amplifiers to correct deficiencies such as improper sync level, in the input visual signal and to provide DC restoration.
3. Visual and aural carrier generators with the frequency tolerances specified by the FCC.
4. Visual and aural modulators.
5. Filters to remove the lower sidebands of the modulated visual carriers and to remove spurious frequency components from both carriers.
6. Linearity correction devices in the visual chain to compensate for incidental phase modulation and differential gain. Ideally, these corrections should be made in both the video and RF domains.
7. Intermediate and final power amplifiers.

Figure 16.45 shows the audio, video, and RF paths for a typical television transmitter. Referring to Fig. 16.45, video is first input to a video processing amplifier and equalizer unit within the video processing block. The processing amplifier performs the following operations: (1) video input amplifications, (2) receiver delay equalization, (3) video clamping and hum cancellation, (4) phase-delay compensation, (5) white clipping, and (6) sync regeneration.

The processed video is converted to amplitude modulation of an IF frequency, either on a carrier generated by a crystal oscillator or by a frequency synthesizer. The modulated IF is band-shaped in a vestigial sideband filter using a surface-acoustic-wave (SAW) filter. Envelope delay correction is not required for the SAW filter because of the uniform delay characteristics of the device. Envelope-delay compensation may, however, be needed for other parts of the transmitter. The SAW filter provides many benefits to transmitter designers and operators because it requires no adjustments and is stable with respect to temperature and time. Envelope-delay compensation, however, might be required for other parts of the transmitter.

The modulated visual IF signal is processed for distortion corrections and subsequent RF transmission in the following stages: (1) power amplifier linearity correction, (2) power amplifier group delay correction, (3) driver amplifier linearity correction, (4) incidental phase modulation correction, (5) RF up-conversion, and (6) channel bandpass filtering.

The IF power amplifier and the final power amplifiers then raise the power level to the station license requirement. An RF sample, obtained from a directional coupler installed at the output of the transmitter,

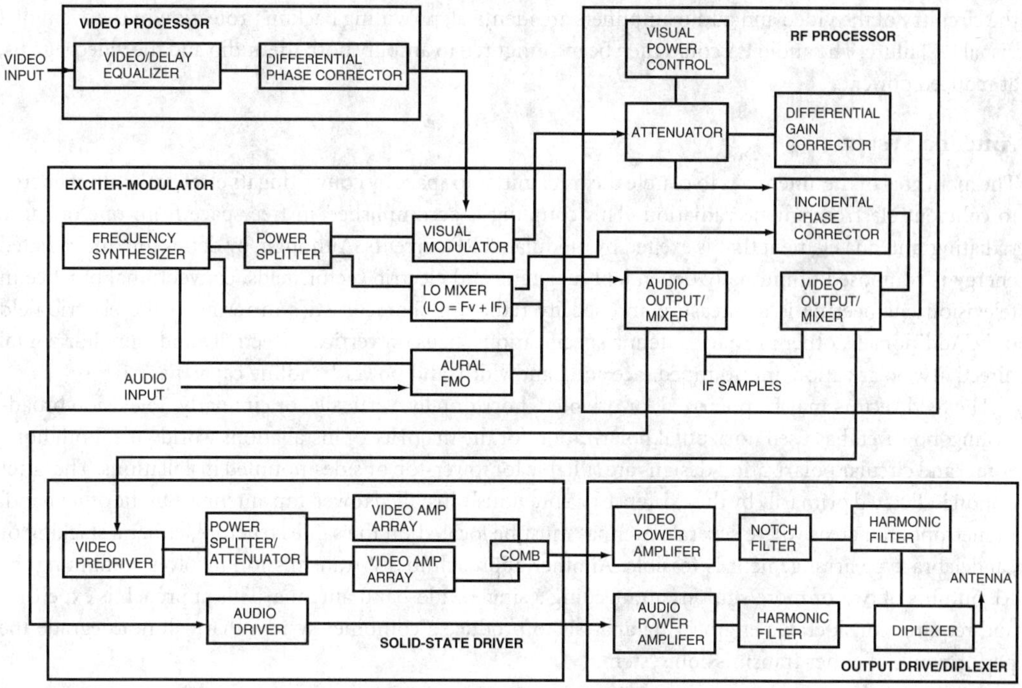

FIGURE 16.45 Simplified block diagram of a VHF television transmitter.

is used in an automatic power-level control circuit located in the RF processor section. A color-notch filter is used to provide the necessary additional attenuation of the lower sideband at 3.58 MHz, and a harmonic filter is used to attenuate out-of-band radiation as required by the FCC. Additional sample couplers provide signals for the following monitoring facilities: (1) voltage standing wave ratio (VSWR) metering, (2) VSWR protection, (3) RF for the demodulator, and (4) RF for the spectrum analyzer and sideband units.

Concurrently with the processing of video, audio is input and pre-emphasized to the 75-μs standard and used to frequency modulate the IF carrier generated in the oscillator section. The modulated signal is then up-converted to the final RF carrier frequency and passed to the intermediate power amplifier (IPA) and power amplifier (PA) stages. The PA raises the output energy of the transmitter to the desired RF operating level. Harmonic filters are employed to attenuate out-of-band radiation of the audio and video signals to ensure compliance with FCC requirements. Filter designs vary depending on the manufacturer, however, most are of coaxial construction utilizing L and C components housed within a pre-packaged assembly. Stubfilters are also used, typically adjusted to provide maximum attenuation at the second harmonic of the operating frequency of the video and the audio carriers.

In addition, a color notch filter is required at the output of the transmitter because imperfect linearity of the IPA and PA stages introduces unwanted modulation products. The filtered video and audio outputs are fed to a hybrid diplexer where the two signals are combined to feed the antenna. For installations that require dual-antenna feedlines, a hybrid combiner with quadrature-phased outputs is used. Depending on the design and operating power, the color notch filter, aural and visual harmonic filters, and diplexer may be combined into a single mechanical unit.

The transmitter block diagram of Fig. 16.45 also shows separate video and audio PA stages. This configuration is normally used for high-power transmitters. Low-power designs often use a combined mode in which the audio and video signals are added prior to the PA. This approach offers a simplified system, but at the cost of additional precorrection of the input video signal. PA stages often are configured so that the circuitry of the video and audio amplifiers are identical, providing backup protection in the event of a visual PA failure. The audio PA could then be reconfigured to amplify both the audio and the video signals, at reduced power.

Antenna System

The main goal of the antenna is to couple the transmitter to space by converting its electrical output energy to coherent electromagnetic radiation. This coupling is accomplished in free space from one or more radiating-antenna element that is excited by modulated RF currents. Although, by definition, the radiated energy is composed of mutually dependent magnetic and electric vector fields, conventional practice in television engineering is to measure and specify radiation characteristics in terms of the electric field only. Additionally, other primary antenna specifications focus on vertical directivity and gain, horizontal directivity, polarization, input impedance and bandwidth, and power handling capacity.

The field vectors may be polarized, or oriented, horizontally, vertically, or circularly. Television broadcasting, however, has used horizontal polarization for the majority of installations worldwide. Both horizontal and circular polarization designs are suitable for tower-top or side-mounted installations. The latter option is dictated primarily by the existence of a previously installed tower-top antenna. On the other hand, in metropolitan areas where several antennas must be located on the same structure, either a stacking or candelabra-type arrangement is feasible. Another approach to TV transmission involves combining the RF outputs of two or more stations and feeding a single wide-band antenna. This approach is expensive and requires considerable engineering analysis to produce a combiner system that will not degrade the performance of either transmission system.

Vertical Directivity and Gain

Transmitting antennas increase the signal strength on the Earth's surface by minimizing the power that is radiated into space and concentrating it at angles toward or just below the horizon. This property, the antenna's *directivity*, is indicated by a radiation pattern, a plot of the radiation intensity at different elevations.

Radiation patterns can best be described by the Fig. 16.46, which depicts the vertical radiation patterns for three representative antennas. These are a low-band low-gain VHF antenna, a high-band medium-gain VHF antenna, and a high-gain UHF antenna.

The points on the pattern where the field strength is 70% of the maximum are known as the *half-power points*, and the angular separation of these points is the *beam width*. The gain of the antenna is increased by narrowing the beam width to concentrate the radiated power in the desired directions, and the product of the gain and beam width tends to be constant.

The *aperture*, an imaginary cylinder in space surrounding the radiating elements, is used to calculate the directivity of the antenna. The cylindrical aperture is illuminated by the radiating elements, and for

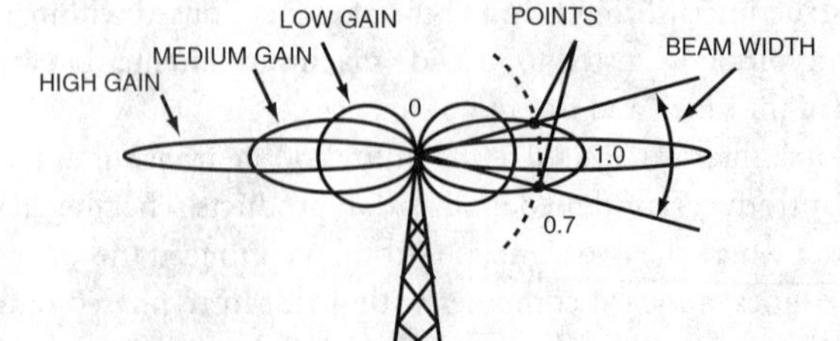

FIGURE 16.46   Vertical antenna radiation patterns.

analytic purposes, each point on the aperture can be considered to be a source of radiant energy. The antenna's directivity is based on the dimensions of the aperture and by the amplitude and phase of the radiant energy at each point on its surface. In the most basic assumption, all points on the aperture are illuminated by a field of the same phase and equal amplitude. The beam width, the angular separation between the half-power points at which the radiation intensity is 70% of that at the beam maximum, is inversely proportional to the ratio of the aperture height (h) to the wavelength (λ). Assuming that the illumination of the aperture is uniform, the beam width can be calculated as

$$\text{beam width} = 48(\lambda/h)^\circ$$

This equation does not apply if the illumination of the aperture is not uniform. However, in general, it can be said that the longer the aperture relative to the wavelength, the taller the antenna and the narrower the radiated beam of energy.

Antenna's Horizontal Directivity

The FCC Rules make a distinction between antennas that are intentionally designed to be directional and those that are directional by virtue of unavoidable imperfections in design. As a result with the FCC's bad experiences with AM directional antennas, which were overused to increase the number of stations on each frequency, the FCC ruled that directional antennas cannot be used in television to reduce the spacing between stations. Also, the FCC ruled that the maximum ERP from directional antennas cannot exceed the maximum allowed for non-directional antennas. The major purpose of directional antennas is to avoid wasting energy over sparsely populated areas and to provide the maximum allowed ERP to densely populated areas with lower total power.

Antenna Bandwidth, VSWR, and Echoes

One of the early challenges of antenna design was to achieve the proper input impedance over the bandwidth of a single channel because antenna components are frequency sensitive. Since that time, modern antenna design has made it possible to achieve good response over multiple channels.

The indicator of antenna bandwidth is the range of frequencies at which its input impedance approximates the characteristic impedance of the transmission line. This is measured by the VSWR of the transmission line feeding the antenna, and a number of 1.05 or less is desirable from 750 kHz below to 5.45 MHz above the visual carrier. The VSWR at the color subcarrier 3.58 MHz above the carrier is very important. A low VSWR is not an absolute guarantee of a satisfactory match between antenna and transmission line, and reflections from the antenna may cause visible echoes and ghosts, even though the VSWR is within limits. Also, the visibility of the echoes depends on their amplitude and their separation from the primary signal. For example, echoes having an amplitude of 1.5% or more will be visible for a transmission line length of 1000 ft. For a 500 ft length, visibility begins for echoes having an amplitude of 3% or greater.

16.4.4 Television Reception

The broadcast channels in the U.S. are 6 MHz wide for transmission on conventional 525-line standards. The minimum signal level at which a television receiver will provide usable pictures and sound is called the *sensitivity level*. The FCC has set up two standard signal level classifications, grades A and B, for the purpose of licensing television stations and allocating coverage areas. Grade A refers to urban areas relatively near the transmitting tower; Grade B use ranges from suburban to rural, and other fringe areas a number of miles from the transmitting antenna.

Many sizes and form factors of receivers are manufactured. Portable personal types include pocket-sized or hand-held models with picture sizes of 2–4 in diagonal for monochrome and 4–6 in for color. Large screen sizes are available in monochrome where low cost and lightweight are prime requirements. However, except where portability is important, the majority of television program viewing is in color. The 19- and 27-in sizes dominate the market.

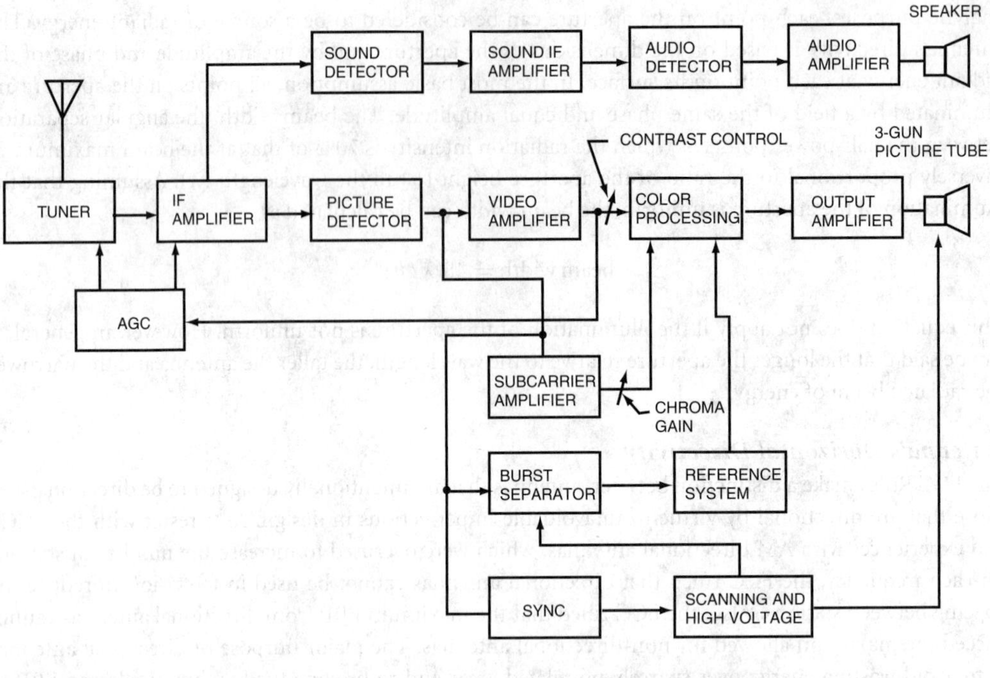

FIGURE 16.47 Simplified schematic block diagram of a color television receiver.

Television receiver functions may be broken down into several interconnected blocks. With the increasing use of very large-scale integrated circuits (VLSI), the isolation of functions has become less obvious in the design and service of television receivers. The typical functional configuration of a receiver using a tri-gun picture tube is shown in Fig. 16.47.

Display Systems

Color video displays may be classified under the following categories:

- Direct view CRT
- Large screen display, optically projected from a CRT
- Large screen display, projected from a modulated light beam
- Large area display of individually driven light-emitting CRTs or incandescent picture elements
- Flat panel matrix of transmissive or reflective picture elements
- Flat panel matrix of light-emitting picture elements

Cathode Ray Tube Display Designs

The direct-view CRT is the dominant display device in television. The attributes offered by CRTs include the following:

- High brightness
- High resolution
- Excellent gray scale reproduction
- Low cost compared to other types of displays

From the standpoint of television receiver manufacturing simplicity and low cost, packaging of the display device as a single component is attractive. The tube itself is composed of only three basic parts: an electron gun, an envelope, and a shadow-mask phosphor screen. The luminance efficiency of the electron optical

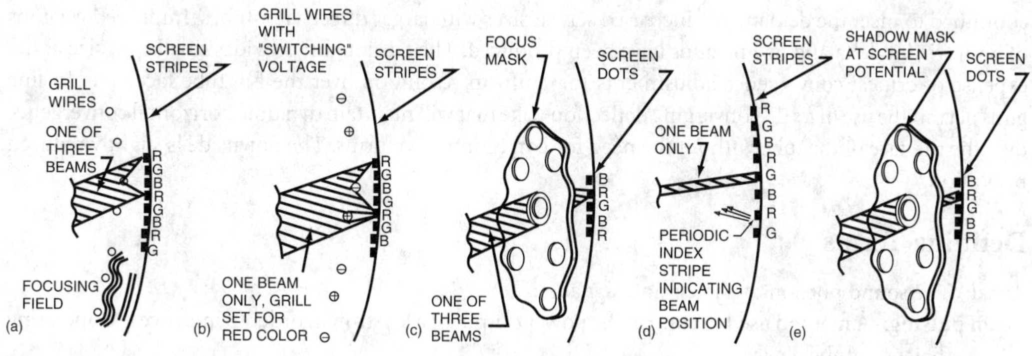

FIGURE 16.48 Color selection schemes (plan view): (a) three-beam focus grill, wires at 30–40% of screen voltage, (b) one-beam focus grill, wires at 30–40% of screen voltage, with 300–500 V color switching between alternate wires, (c) focus mask, three beams, mask at 30–40% of screen voltage, (d) index scheme, electron beam smaller than phosphor stripe, periodic index signal (secondary electrons or light) indicating beam position, (e) shadow mask at screen voltage, no focusing, three beams, can consist of holes and dots as shown, or slots and stripes. Note: in(a), (c), and (e) two additional beams approach at other angles to excite the other two colors. (*Source: Television Engineering Handbook.* 1986. K. Blair Benson, ed. McGraw-Hill, New York.)

system and the phosphor screen is high. A peak beam current of under 1 μA in a 25-in tube will produce a highlight brightness of up to 100 fL. The major drawback is the power required to drive the horizontal sweep circuit and the high accelerating voltage necessary for the electron beam. This requirement is partially offset through generation of the screen potential and other lower voltages by rectification of the scanning flyback voltage.

In retrospect, many CRT designs were researched during the early days of television. Figure 16.48 depicts five of the more important approaches. They can be categorized as follows.

Figure 16.48(a) shows a three-beam grill focusing design, which offers very high brightness because of minimal mask intercept. Figure 16.48(b) shows a one-beam grill focusing design, which offers less brightness but with the simplicity of a single beam operation. However, both designs suffer in their execution because secondary electrons generated at the grill and screen are all attracted without color selection to the high voltage of the screen where they reduce the contrast of the display. Screening process is another inherent problem with grill focusing designs due to the absence of an optical shadowing mask. Screening requires the use of printing masters and interchangeable parts, which presents a daunting task. For these reasons, focus-grill tubes are no longer built.

In Fig. 16.48(c), a large-hole focus mask, with the mask at a fraction of the screen potential, will focus the beamlets to permit increased mask transmission and screen brightness, but at the expense of reduced contrast. Figure 16.48(d) depicts an *index* color selection topology. This scheme makes it possible to monitor the position of a single beam and control which screen color is being excited at any instant through a periodic index signal generated by the beam, from either secondary electrons or from a fast response light-emitting phosphor. Again, two major drawbacks can be deduced from these early designs. First, high speed complex circuitry was needed to monitor the position of the beam and properly control it. Second, the beam itself must be no wider than a phosphor stripe, even when deflected over the full area of the picture tube.

In recent years, the shadow mask design as depicted in Fig. 16.48(e) has become a dominant player because of its improvements in brightness and contrast over previous designs, but without having to contend with the complex design issues of previous designs.

As consumer demands drive manufacturers to produce larger picture sizes, the weight and depth of the CRT and the higher power and voltage requirements become serious limitations. These are reflected in sharply increasing receiver costs. To withstand the atmospheric pressures on the evacuated glass envelope, CRT weight increases exponentially with the viewable diagonal. Nevertheless, manufacturers have continued to meet the demand for increased screen sizes with larger direct-view tubes. Improved versions of both tri-dot delta and in-line guns have been produced. The tri-dot gun provides small spot size at the expense of critical convergence adjustments for uniform resolution over the full tube faceplate. In-line guns permit the use of a self-converging deflection yoke that will maintain dynamic horizontal convergence over the full face of the tube without the need for correction waveforms. The downside is slightly reduced resolution.

Defining Terms

Aural: The sound portion of a television signal.

Beam pulsing: A method used to control the power output of a klystron in order to improve the operating efficiency of the device.

Blanking: The portion of a television signal that is used to blank the screen during the horizontal and vertical retrace periods.

Composite video: A single video signal that contains luminance, color, and synchronization information. NTSC, PAL, and SECAM are all examples of composite video formats.

Effective radiated power: The power supplied to an antenna multiplied by the relative gain of the antenna in a given direction.

Equalizing pulses: In an encoded video signal, a series of 2X line frequency pulses occurring during vertical blanking, before and after the vertical synchronizing pulse. Different numbers of equalizing pulses are inserted into different fields to ensure that each field begins and ends at the right time to produce proper interlace. The 2X line rate also serves to maintain horizontal synchronization during vertical blanking.

Field: One of the two (or more) equal parts of information into which a frame is divided in interlace video scanning. In the NTSC system, the information for one picture is divided into two fields. Each field contains one half of the lines required to produce the entire picture. Adjacent lines in the picture are contained in alternate fields.

Frame: The information required for one complete picture in an interlaced video system. For the NTSC system, there are two fields per frame.

H (horizontal): In television signals, H may refer to any of the following: the horizontal period or rate, horizontal line of video information, or horizontal sync pulse.

Hue: One of the characteristics that distinguishes one color from another. Hue defines color on the basis of its position in the spectrum (red, blue, green, yellow, etc.). Hue is one of the three characteristics of television color. Hue is often referred to as *tint*. In NTSC and PAL video signals, the hue information at any particular point in the picture is conveyed by the corresponding instantaneous phase of the active video subcarrier.

Hum bars: Horizontal black and white bars that extend over the entire TV picture and usually drift slowly through it. Hum bars are caused by an interfering power line frequency or one of its harmonics.

Interlaced: A shortened version of *interlaced scanning* (also called *line interlace*). Interlaced scanning is a system of video scanning whereby the odd- and even numbered lines of a picture are transmitted consecutively as two separate interleaved fields.

IRE: A unit equal to 1/140 of the peak-to-peak amplitude of a video signal, which is typically 1 V. The 0 IRE point is at blanking level, with the sync tip at −40 IRE and white extending to +100 IRE. IRE stands for *Institute of Radio Engineers*, an organization proceeding the IEEE, which defined the unit.

Klystron: An amplifier device for UHF and microwave signals based on velocity modulation of an electron beam. The beam is directed through an input cavity, where the input RF signal polarity initializes a *bunching effect* on electrons in the beam. The bunching effect excites subsequent cavities, which increase the bunching through an energy flywheel concept. Finally, the beam passes an output cavity that couples the amplified signal to the load (antenna system). The beam falls onto a collector element that forms the return path for the current and dissipates the heat resulting from electron beam bombardment.

Low power TV (LPTV): A television service authorized by the FCC to serve specific confined areas. An LPTV station may typically radiate between 100 and 1000 W of power, covering a geographic radius of 10 to 15 m.

Pixel: The smallest distinguishable and resolvable area in a video image. A pixel is a single point on the screen. The word pixel is derived from *picture element*.

Raster: A predetermined pattern of scanning the screen of a CRT. Raster may also refer to the illuminated area produced by scanning lines on a CRT when no video is present.

Saturation: The intensity of the colors in the active picture; the voltage levels of the colors. Saturation relates to the degree by which the eye perceives a color as departing from a gray or white scale of the same brightness. A 100% saturated color does not contain any white; adding white reduces saturation. In NTSC and PAL video signals, the color saturation at any particular instant in the picture is conveyed by the corresponding instantaneous amplitude of the active video subcarrier.

Scan: One sweep of the target area in a camera tube, or of the screen in a picture tube.

Setup: A video term relating to the specified base of an active picture signal. In NTSC, the active picture signal is placed 7.5 IRE units above blanking (0 IRE). Setup is the separation in level between the *video blanking* and *reference black* levels.

Synchronous detection: A demodulation process in which the original signal is recovered by multiplying the modulated signal by the output of a synchronous oscillator locked to the carrier.

Translator: An unattended television or FM broadcast repeater that receives a distant signal and retransmits the picture and/or audio locally on another channel.

Vectorscope: An oscilloscope-type device used to display the color parameters of a video signal. A vectorscope decodes color information into R-Y and B-Y components, which are then used to drive the X and Y axis of the scope. The total lack of color in a video signal is displayed as a dot in the center of the vectorscope. The angle, distance around the circle, magnitude, and distance away from the center indicate the phase and amplitude of the color signal.

References

Benson, K.B. and Whitaker, J. 1990. *Television and Audio Handbook for Technicians and Engineers*. McGraw-Hill, New York.

Benson, K.B. and Whitaker, J., ed. 1991. *Television Engineering Handbook*, rev. ed. McGraw-Hill, New York.

Inglis, A.F. 1993. *Video Engineering*. McGraw-Hill, New York.

Whitaker, J. 1991. *Maintaining Electronic Systems*. CRC Press, Boca Raton, FL.

Whitaker, J. 1991. *Radio Frequency Transmission Systems: Design and Operation*. McGraw-Hill, New York.

Further Information

Additional information on the topic of television system technology is available from the following sources:

Broadcast Engineering magazine, a monthly periodical dealing with television and radio technology. The magazine, published in Overland Park, KS, is free to qualified subscribers.

The Society of Motion Picture and Television Engineers, which publishes a monthly journal, and holds an annual conference and convention in the fall. The SMPTE is headquartered in White Plains, NY.

The National Association of Broadcasters, which holds an annual engineering conference and trade show in the spring. The NAB is headquartered in Washington, D.C.

In addition, the following books are recommended:

NAB. 1996. *National Association of Broadcasters Engineering Handbook*, 9th ed. National Association of Broadcasters, Washington, DC.

Whitaker, J.C. and Benson, K.B., eds. 2003. *Standard Handbook of Video and Television Engineering*, 4th ed. McGraw-Hill, New York, N.Y.

16.5 ATSC Video, Audio, and PSIP Transmission[1]

Jerry C. Whitaker

16.5.1 Overview of the ATSC Digital Television System

The digital television (DTV) standard has ushered in a new era in television broadcasting. The impact of DTV is more significant than simply moving from an analog system to a digital system. Rather, DTV permits a level of flexibility wholly unattainable with analog broadcasting. An important element of this flexibility is the ability to expand system functions by building upon the technical foundations specified in ATSC standards such as the ATSC digital television standard (A/53) [1] and the digital audio compression (AC-3) standard (A/52) [2].

With NTSC, and its PAL and SECAM counterparts, the video, audio, and some limited data information are conveyed by modulating an RF carrier in such a way that a receiver of relatively simple design can decode and reassemble the various elements of the signal to produce a program consisting of video and audio, and perhaps related data (e.g., closed captioning). As such, a complete program is transmitted by the broadcaster that is essentially in finished form. In the DTV system, however, additional levels of processing are required after the receiver demodulates the RF signal. The receiver processes the digital bit stream extracted from the received signal to yield a collection of program elements (video, audio, and/or data) that match the service(s) that the consumer has selected. This selection is made using system and service information that is also transmitted. Audio and video are delivered in digitally compressed form and must be decoded for presentation. Audio may be monophonic, stereo, or multi-channel. Data may supplement the main video/audio program (e.g., closed captioning, descriptive text, or commentary) or it may be a stand-alone service (e.g., a stock or news ticker).

The nature of the DTV system is such that it is possible to provide new features that build upon the infrastructure within the broadcast plant and the receiver. One of the major enabling developments of digital television, in fact, is the integration of significant processing power in the receiving device itself. Historically, in the design of any broadcast system—be it radio or television—the goal has always been to concentrate technical sophistication (when needed) at the transmission end and thereby facilitate simpler receivers. Because there are far more receivers than transmitters, this approach has obvious business advantages. While this trend continues to be true, the complexity of the transmitted bit stream and compression of the audio and video components require a significant amount of processing power in the receiver, which is practical because of the enormous advancements made in computing technology. Once a receiver reaches a certain level of sophistication (and market success) additional processing power is essentially "free."

DTV System Overview

The DTV standard describes a system designed to transmit high quality video and audio and ancillary data over a single 6 MHz channel. The system can deliver about 19 Mbps in a 6 MHz terrestrial broadcasting channel and about 38 Mbps in a 6 MHz cable television channel. This means that encoding high-definition (HD) video sources at 885^2 Mbps (highest rate progressive input) or 994^3 Mbps (highest rate interlaced picture input) requires a bit rate reduction by about a factor of 50 (when the overhead numbers are added, the rates become closer). To achieve this bit rate reduction, the system is designed to be efficient in utilizing available channel capacity by exploiting complex video and audio compression technology.

---

[1]This section is based on: ATSC, "Guide to the Digital Television Standard, Doc. A/54A, Advanced Television Systems Committee, Washington, D.C., 2003.

[2]$720 \times 1280 \times 60 \times 8 \times 2 = 884.736$ Mbps (the 2 represents the 4:2:2 or color subsampling, an RGB with full bandwidth would be a 3)

[3]$1080 \times 1920 \times 29.97 \times 8 \times 2 = 994.333$ Mbps (the 2 represents the 4:2:2 or color subsampling, an RGB with full bandwidth would be a 3)

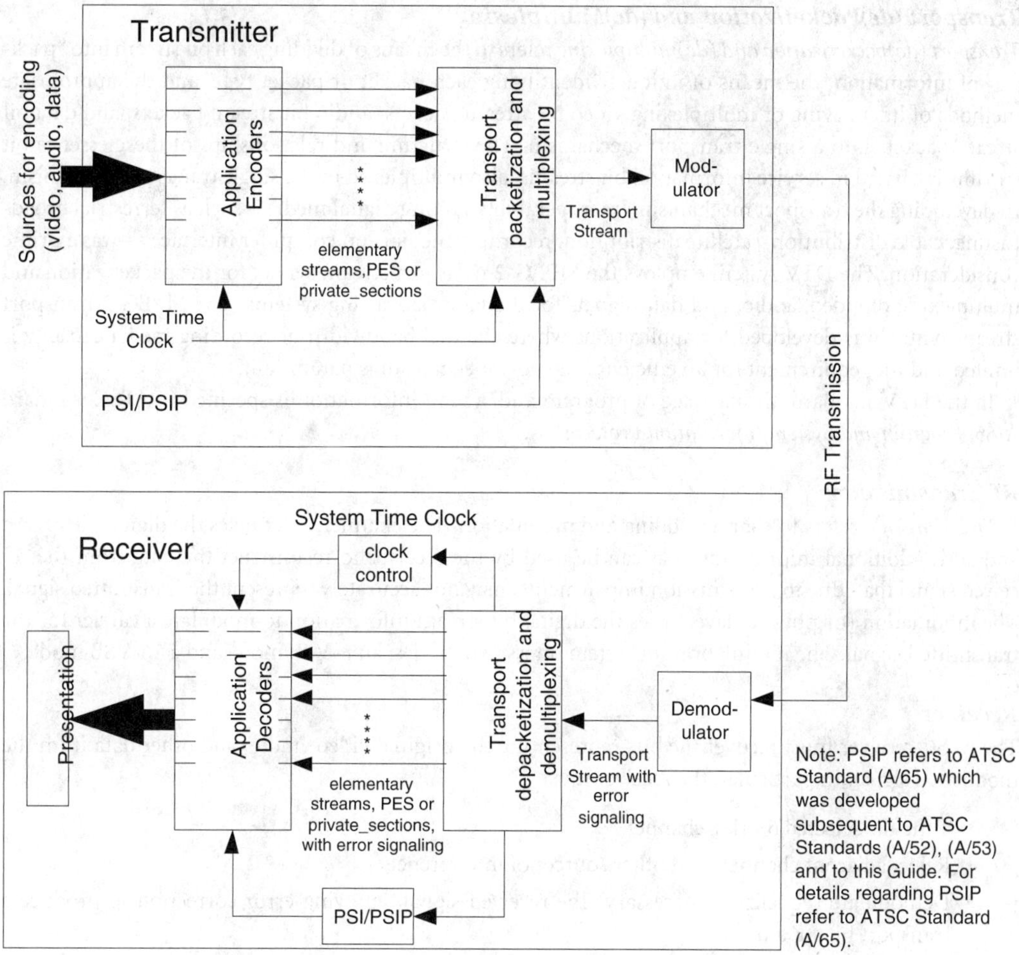

FIGURE 16.49   Block diagram of functionality in a transmitter/receiver pair.

The compression scheme used for DTV service optimizes the throughput of the transmission channel by representing the video, audio, and data sources with as few bits as possible while preserving the level of quality required for the given application.

The RF/transmission subsystems described in the DTV standard are designed specifically for terrestrial and cable applications. The structure is such that the video, audio, and service multiplex/transport subsystems are useful in other applications as well.

System Block Diagram

A basic block diagram representation of the DTV system is shown in Figure 16.49. According to this model, the digital television system consists of four major elements, three within the broadcast plant plus the receiver.

Application Encoders/Decoders

The *application encoders/decoders* as shown in Figure 16.49, refer to the bit rate reduction methods (also known as data compression) appropriate for application to the video, audio, and ancillary digital data streams. The purpose of compression is to minimize the number of bits needed to represent the audio and video information. The DTV system employs the MPEG-2 video stream syntax for the coding of video and the ATSC standard *digital audio compression* (*AC-3*) for the coding of audio [2].

Transport (de)Packetization and (de)Multiplexing

Transport (de)packetization and (de)multiplexing refers to the means of dividing each bit stream into "packets" of information, the means of uniquely identifying each packet or packet type, and the appropriate methods of interleaving or multiplexing video bit stream packets, audio bit stream packets, and data bit stream packets into a single transport mechanism. The structure and relationships of these essence bit streams is carried in service information bit streams, also multiplexed in the single transport mechanism. In developing the transport mechanism, interoperability among digital media—such as terrestrial broadcasting, cable distribution, satellite distribution, recording media, and computer interfaces—was a prime consideration. The DTV system employs the MPEG-2 transport stream syntax for the packetization and multiplexing of video, audio, and data signals for digital broadcasting systems. The MPEG-2 transport stream syntax was developed for applications where channel bandwidth or recording media capacity is limited and the requirement for an efficient transport mechanism is paramount.

In the DTV standard, the carriage of program and system information is specified in ATSC standard A/65, *Program and System Information Protocol* [3].

RF Transmission

RF Transmission refers to channel coding and modulation. The channel coder takes the digital bit stream and adds additional information that can be used by the receiver to reconstruct the data from the received signal that, due to transmission impairments, may not accurately represent the transmitted signal. The modulation (or physical layer) uses the digital bit stream information to modulate a carrier for the transmitted signal. The modulation subsystem offers two modes: an 8-VSB mode and a 16-VSB mode.

Receiver

The ATSC receiver must recover the bits representing the original video, audio, and other data from the modulated signal. In particular, the receiver must

- Tune the selected 6 MHz channel
- Reject adjacent channels and other sources of interference
- Demodulate (equalize as necessary) the received signal, applying error correction to produce a transport bit stream
- Identify the elements of the bit stream using a transport layer processor
- Select each desired element and send it to its appropriate processor
- Decode and synchronize each element
- Present the programming

16.5.2 Video System

The MPEG-2 specification is organized into a system of profiles and levels, so that applications can ensure interoperability by using equipment and processing that adhere to a common set of coding tools and parameters [4]. The DTV standard is based on the MPEG-2 main profile. The main profile includes three types of frames (I-frames, P-frames, and B-frames), and an organization of luminance and chrominance samples (designated 4:2:0) within the frame. The main profile does not include a scalable algorithm, where scalability implies that a subset of the compressed data can be decoded without decoding the entire data stream. The high level includes formats with up to 1152 active lines and up to 1920 samples per active line, and for the Main Profile is limited to a compressed data rate of no more than 80 Mbps. The parameters specified by the DTV standard represent specific choices within these constraints.

Compatibility with MPEG-2

The DTV video compression system does not include algorithmic elements that fall outside the specifications for MPEG-2 main profile. Thus, video decoders that conform to the MPEG-2 MP@HL can be expected to decode bit streams produced in accordance with the DTV standard. Note that it is not

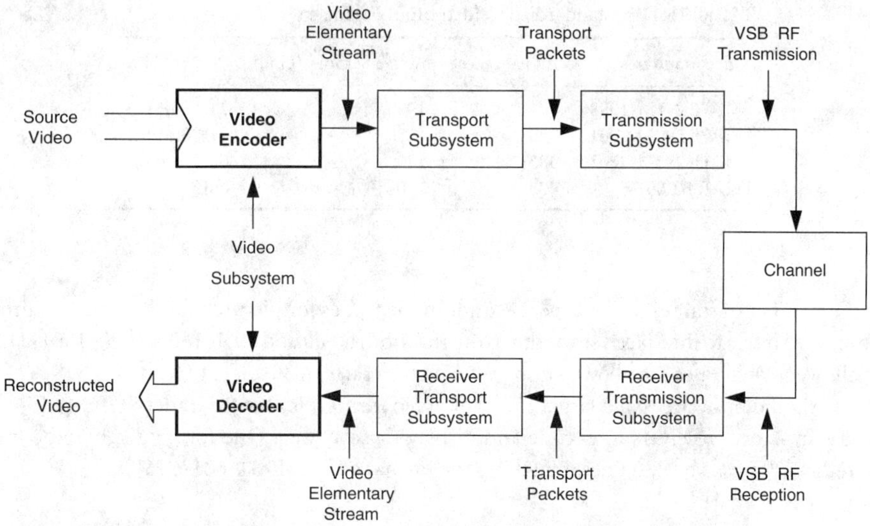

FIGURE 16.50   Video coding in relation to the ATSC DTV system.

necessarily the case that all video decoders which are based on the DTV standard will be able to properly decode all video bit streams that comply to MPEG-2 MP@HL.

Overview of Video Compression

The DTV video compression system takes in an analog video source signal and outputs a compressed digital signal that contains information that can be decoded to produce an approximate version of the original image sequence. The goal is for the reconstructed approximation to be imperceptibly different from the original for most viewers, for most images, for most of the time. In order to approach such fidelity, the algorithms are flexible, allowing for frequent adaptive changes in the algorithm depending on scene content, history of the processing, estimates of image complexity, and perceptibility of distortions introduced by the compression.

Figure 16.50 shows the overall flow of signals in the ATSC DTV system. Note that analog signals presented to the system are digitized and sent to the encoder for compression, and the compressed data then are transmitted over a communications channel. On being received, the compressed signal is decompressed in the decoder, and reconstructed for display (following error-correction, as necessary).

Video Preprocessing

Video preprocessing converts the analog input signals to digital samples in the form needed for subsequent compression. The analog input signals are typically composite for standard-definition signals or components consisting of luminance (Y) and chrominance (Cb and Cr) signals.

Video Compression Formats

Table 16.10 lists the compression formats allowed in the DTV standard.

In Table 16.10, "vertical lines" refers to the number of active lines in the picture. "Pixels" refers to the number of pixels during the active line. "Aspect ratio" refers to the picture aspect ratio. "Picture rate"

TABLE 16.10   DTV Compression Formats

Vertical Lines	Pixels	Aspect Ratio	Picture Rate
1080	1920	16:9	60I, 30P, 24P
720	1280	16:9	60P, 30P, 24P
480	704	16:9 and 4:3	60P, 60I, 30P, 24P
480	640	4:3	60P, 60I, 30P, 24P

TABLE 16.11 Standardized Video Input Formats

Video Standard	Active Lines	Active Samples/Line	Picture Rate
SMPTE 274M-1998	1080	1920	24P, 30P, 60I
SMPTE 296M-2001	720	1280	24P, 30P, 60P
SMPTE 293M-2003	483	720	60P
ITU-R BT.601-5	483	720	60I

refers to the number of frames or fields per second. In the values for picture rate, "P" refers to progressive scanning and "I" refers to interlaced scanning. Note that both 60.00 and 59.94 ($60 \times 1000/1001$) Hz picture rates are allowed. Dual rates are allowed also at the picture rates of 30 and 24 Hz.

Designers should be aware that a larger range of video formats is allowed under SCTE 43 [5], and that consumers may expect receivers to decode and display these as well. One format likely to be frequently encountered is 720 pixels by 480 lines, matching the format of ITU-R BT. 601-5 [6].

Possible Video Inputs

While not required by the DTV standard, there are certain digital television production standards, shown in Table 16.11, that define video formats which relate to compression formats specified by the standard.

The compression formats may be derived from one or more appropriate video input formats. It may be anticipated that additional video production standards will be developed in the future that extend the number of possible input formats.

Sampling Rates

For the 1080-line format, with 1125 total lines per frame and 2200 total samples per line, the sampling frequency will be 74.25 MHz for the 30.00 frames per second (fps) frame rate. For the 720-line format, with 750 total lines per frame and 1650 total samples per line, the sampling frequency will be 74.25 MHz for the 60.00 fps frame rate. For the 480-line format using 704 pixels, with 525 total lines per frame and 858 total samples per line, the sampling frequency will be 13.5 MHz for the 59.94 Hz field rate. Note that both 59.94 fps and 60.00 fps are acceptable as frame or field rates for the system.

For both the 1080- and 720-line formats, other frame rates, specifically 23.976, 24.00, 29.97 and 30.00 fps rates are acceptable as input to the system. The sample frequency will be either 74.25 MHz (for 24.00 and 30.00 fps) or 74.25/1.001 MHz for the other rates. The number of total samples per line is the same for either of the paired picture rates. (See SMPTE 274M [7] and SMPTE 296M [8]).

The six frame rates noted here are the only allowed frame rates for the DTV standard. In this discussion, any references to 24 fps include both 23.976 and 24.00 fps, references to 30 fps include both 29.97 and 30.00 fps, and references to 60 fps include both 59.94 and 60.00 fps.

For the 480-line format, there may be 704 or 640 pixels in the active line. The interlaced formats are based on ITU-R B.T 601-5 [6]; the progressive formats are based on SMPTE 294M [9].

If the input is based on ITU-R BT.601-5 or SMPTE 294M, it will have 483 active lines with 720 pixels in the active line. Only 480 of the 483 active lines are used for encoding. Only 704 of the 720 pixels are used for encoding; the first eight and the last eight are dropped. The 480-line, 640 pixel picture format is not related to any current video production format. It does, however, correspond to the IBM VGA graphics format and may be used with ITU-R BT.601-5 sources by using appropriate resampling techniques.

Colorimetry

For the purposes of the DTV standard, "colorimetry" means the combination of color primaries, transfer characteristics, and matrix coefficients. Video inputs conforming to SMPTE 274M and SMPTE 296M have the same colorimetry. Video inputs corresponding to ITU-R BT.601-5 should have SMPTE 170M colorimetry [10].

Active Format Description (AFD)

The DTV standard makes provisions for conveying *active format description* (AFD) data [1]. The term "active format" in this context refers to that portion of the coded video frame containing "useful information." For example, when 16:9 aspect ratio material is coded in a 4:3 format (such as 480i), letterboxing may be used to avoid cropping the left and right edges of the widescreen image. The black horizontal bars at the top and bottom of the screen contain no useful information, and in this case the AFD data would indicate 16:9 video carried inside the 4:3 rectangle. The AFD data solve a troublesome problem in the transition from conventional 4:3 display devices to widescreen 16:9 displays, and also address the variety of aspect ratios that have been used over the years by the motion picture industry to produce feature films.

There are, of course, a number of different types of video displays in common usage—ranging from 4:3 CRTs to widescreen projection devices and flat-panel displays of various design. Each of these devices may have varying abilities to process incoming video. In terms of input interfaces, these displays may likewise support a range of input signal formats—from composite analog video to IEEE 1394.

Possible video source devices include cable, satellite, or terrestrial broadcast set-top (or integrated receiver-decoder) boxes, media players (such as DVDs), analog or digital VHS tape players, and personal video recorders.

Although choice is good, this wide range of consumer options presented two problems to be solved:

- First, no standard method had been agreed upon to communicate to the display device the "active area" of the video signal. Such a method would be able, for example to signal that the 4:3 signal contains within it a letterboxed 16:9 video image.
- Second, no standard method had been agreed upon to communicate to the display device, for all interface types, that a given image is intended for 16:9 display.

The AFD solves these problems and, in the process, provides the following benefits:

- Active area signaling allows the display device to process the incoming signal to make the highest-resolution and most accurate picture possible. Furthermore, the display can take advantage of the knowledge that certain areas of video are currently unused and can implement algorithms that reduce the effects of uneven screen aging.
- Aspect ratio signaling allows the display device to produce the best image possible. In some scenarios, lack of a signaling method translates to restrictions in the ability of the source device to deliver certain otherwise desirable output formats.

Active Area Signaling

A consumer device such as a cable or satellite set-top box cannot reliably determine the active area of video on its own. Even though certain lines at the top and bottom of the screen may be black for periods of time, the situation could change without warning. The only sure way to know active area is for the service provider to include these data at the time of video compression and to embed it into the video stream.

Figure 16.51 shows 4:3- and 16:9-coded images with various possible active areas. The group on the left is either coded explicitly in the MPEG-2 video syntax as 4:3, or the uncompressed signal provided in NTSC timing and aspect ratio information (if present) indicates 4:3. The group on the right are coded explicitly in the MPEG-2 video syntax as 16:9, provided with NTSC timing and an aspect ratio signal indicating 16:9, or provided uncompressed with 16:9 timing across the interface.

As can be seen in the figure, a pillar-boxed display results when a 4:3 active area is displayed within a 16:9 area, and a letterboxed display results when a 16:9 active area is displayed within a 4:3 area. It is also apparent that double-boxing can also occur, for example when 4:3 material is delivered within a 16:9 letterbox to a 4:3 display. Or, when 16:9 material is delivered within a 4:3 pillar-box to a 16:9 display.

For the straight letter- or pillar-box cases, if the display is aware of the active area it may take steps to mitigate the effects of uneven screen aging. Such steps could, for example, involve using gray instead of black bars. Some amount of linear or nonlinear stretching and/or zooming may be done as well using the knowledge that video outside the active area can safely be discarded.

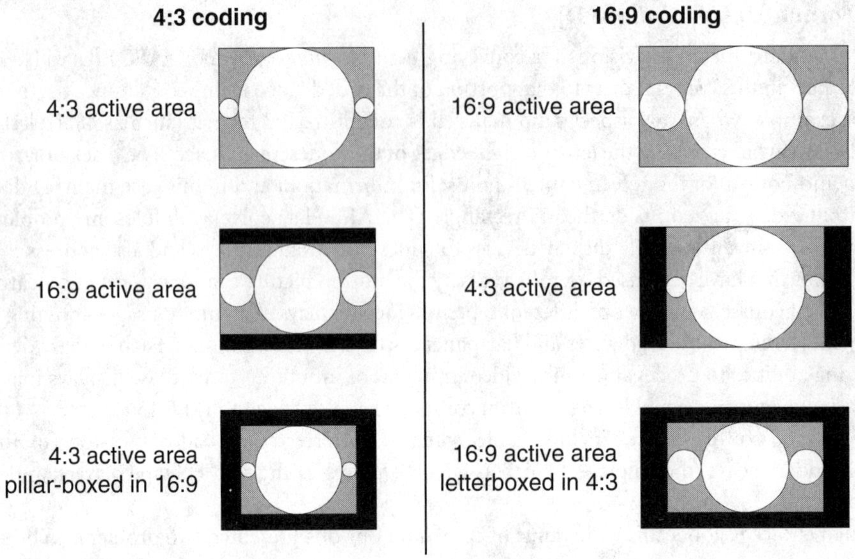

4:3 coding   ## 16:9 coding

4:3 active area   16:9 active area

16:9 active area   4:3 active area

4:3 active area
pillar-boxed in 16:9   16:9 active area
                                                 letterboxed in 4:3

FIGURE 16.51   Coding and active area.

16.5.3   Audio System

As illustrated in Figure 16.52, the audio subsystem comprises the audio encoding/decoding function and resides between the audio inputs/outputs and the transport subsystem. The audio encoder(s) is (are) responsible for generating the audio elementary stream(s), which are encoded representations of the baseband audio input signals. The flexibility of the transport system allows multiple audio elementary

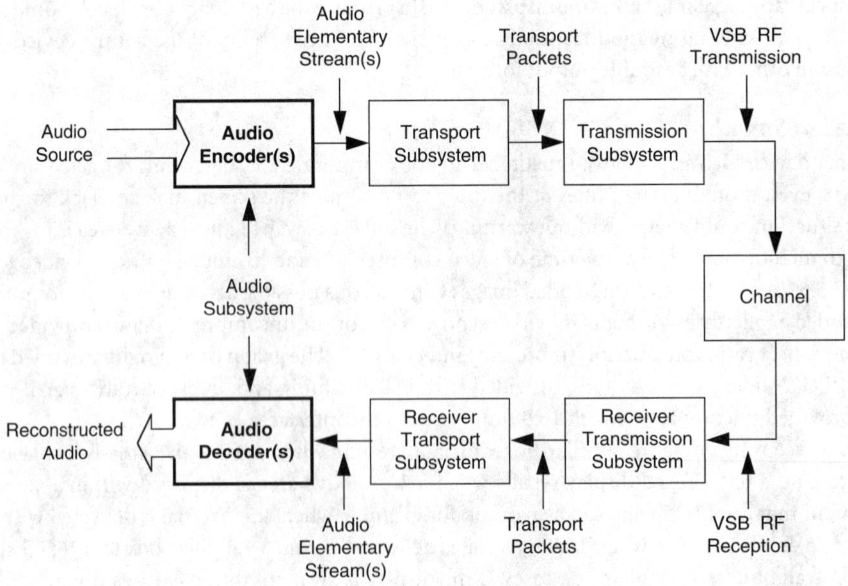

FIGURE 16.52   Audio subsystem within the DTV system.

streams to be delivered to the receiver. At the receiver, the transport subsystem is responsible for selecting which audio streams(s) to deliver to the audio subsystem. The audio subsystem is responsible for decoding the audio elementary stream(s) back into baseband audio.

An audio program source is encoded by a digital television audio encoder. The output of the audio encoder is a string of bits that represent the audio source, and is referred to as an *audio elementary stream*. The transport subsystem packetizes the audio data into PES packets, which are then further packetized into transport packets. The transmission subsystem converts the transport packets into a modulated RF signal for transmission to the receiver. At the receiver, the signal is demodulated by the receiver transmission subsystem. The receiver transport subsystem converts the received audio packets back into an audio elementary stream, which is decoded by the digital television audio decoder. The partitioning shown is conceptual, and practical implementations may differ. For example, the transport processing may be broken into two blocks; one to perform PES packetization, and the second to perform transport packetization. Or, some of the transport functionality may be included in either the audio coder or the transmission subsystem.

Audio Encoder Interface

The audio system accepts baseband audio inputs with up to six audio channels per audio program bit stream. The channelization is consistent with ITU-R recommendation BS-775, "Multi-channel stereophonic sound system with and without accompanying picture." The six audio channels are: left, center, right, left surround, right surround, and low frequency enhancement (LFE). Multiple audio elementary bit streams may be conveyed by the transport system.

The bandwidth of the LFE channel is limited to 120 Hz. The bandwidth of the other (main) channels is limited to 20 kHz. Low frequency response may extend to DC, but is more typically limited to approximately 3 Hz (−3 dB) by a DC blocking high-pass filter. Audio coding efficiency (and thus audio quality) is improved by removing DC offset from audio signals before they are encoded.

Input Source Signal Specification

Audio signals that are input to the audio system may be in analog or digital form. Audio signals should have any DC offset removed before being encoded. If the audio encoder does not include a DC blocking high-pass filter, the audio signals should be high-pass filtered before being applied to the audio encoder. For analog input signals, the input connector and signal level are not specified. Conventional broadcast practice may be followed. One commonly used input connector is the 3-pin XLR female (the incoming audio cable uses the male connector) with pin 1 ground, pin 2 hot or positive, and pin 3 neutral or negative.

For digital input signals, the input connector and signal format are not specified. Commonly used formats such as the AES 3-1992 two-channel interface may be used [11]. When multiple two-channel inputs are used, the preferred channel assignment is

Pair 1:   Left, Right
Pair 2:   Center, LFE
Pair 3:   Left Surround, Right Surround

Sampling Frequency

The system conveys digital audio sampled at a frequency of 48 kHz, locked to the 27 MHz system clock. If analog signal are employed, the A/D converters should sample at 48 kHz. If digital inputs are employed, the input sampling rate should be 48 kHz, or the audio encoder should contain sampling rate converters that convert the sampling rate to 48 kHz. The sampling rate at the input to the audio encoder must be locked to the video clock for proper operation of the audio subsystem.

In general, input signals should be quantized to at least 16-bit resolution. The audio compression system can convey audio signals with up to 24-bit resolution.

Summary of Service Types

The following service types are defined in the digital audio compression (AC-3) standard [2] and in the ATSC digital television standard [1]:

- **Complete main audio service (CM):** This is the normal mode of operation. All elements of a complete audio program are present. The audio program may be any number of channels from 1 to 5.1[4].

- **Main audio service, music and effects (ME):** All elements of an audio program are present except for dialogue. This audio program may contain from 1 to 5.1 channels. Dialogue may be provided by a D associated service (that may be simultaneously decoded and added to form a complete program).

- **Associated service: visually impaired (VI):** This is typically a single-channel service, intended to convey a narrative description of the picture content for use by the visually impaired, and intended to be decoded along with the main audio service. The VI service also may be provided as a complete mix of all program elements, in which case it may use any number of channels (up to 5.1).

- **Associated service: hearing impaired (HI):** This is typically a single-channel service, intended to convey dialogue that has been processed for increased intelligibility for the hearing impaired, and intended to be decoded along with the main audio service. The HI service also may be provided as a complete mix of all program elements, in which case it may use any number of channels (up to 5.1).

- **Associated service: dialogue (D):** This service conveys dialogue intended to be mixed into a main audio service (ME) that does not contain dialogue.

- **Associated service: commentary (C):** This service typically conveys a single-channel of commentary intended to be optionally decoded along with the main audio service. This commentary channel differs from a dialogue service, in that it contains optional instead of necessary program content. The C service also may be provided as a complete mix of all program elements, in which case it may use any number of channels (up to 5.1).

- **Associated service: emergency message (E):** This is a single-channel service, which is given priority in reproduction. If this service type appears in the transport multiplex, it is routed to the audio decoder. If the audio decoder receives this service type, it will decode and reproduce the E channel while muting the main service.

- **Associated service: voice-over (VO):** This is a single-channel service intended to be decoded and added into the center loudspeaker channel.

Multi-Lingual Services

Each audio bit stream may be in any language. In order to provide audio services in multiple languages a number of main audio services may be provided, each in a different language. This is the (artistically) preferred method, because it allows unrestricted placement of dialogue along with the dialogue reverberation. The disadvantage of this method is that as much as 384 kbps is needed to provide a full 5.1-channel service for each language. One way to reduce the required bit-rate is to reduce the number of audio channels provided for languages with a limited audience. For instance, alternate language versions could be provided in 2-channel stereo with a bit-rate of 128 kbps or, a mono version can be supplied at a bit-rate of approximately 64–96 kbps.

Another way to offer service in multiple languages is to provide a main multi-channel audio service (ME) that does not contain dialogue. Multiple single-channel dialogue associated services (D) can then be provided, each at a bit-rate of approximately 64–96 kbps. Formation of a complete audio program requires that the appropriate language D service be simultaneously decoded and mixed into the ME service. This method allows a large number of languages to be efficiently provided, but at the expense of artistic limitations. The single-channel of dialogue would be mixed into the center reproduction channel,

---

[4]5.1 channel sound refers to a service providing the following signals: right, center, left, right surround, left surround, and low frequency effects (LEF).

TABLE 16.12 Typical Audio Bit Rate

Type of Service	Number of Channels	Typical Bit Rates
CM, ME, or associated audio service containing all necessary program elements	5	384–448 kbps
CM, ME, or associated audio service containing all necessary program elements	4	320–384 kbps
CM, ME, or associated audio service containing all necessary program elements	3	192–320 kbps
CM, ME, or associated audio service containing all necessary program elements	2	128–256 kbps
VI, narrative only	1	64–128 kbps
HI, narrative only	1	64–96 kbps
D	1	64–128 kbps
D	2	96–192 kbps
C, commentary only	1	64–128 kbps
E	1	64–128 kbps
VO	1	64–128 kbps

and could not be panned. Also, reverberation would be confined to the center channel, which is not optimum. Nevertheless, for some types of programming (sports, etc.) this method is attractive due to the savings in bit rate it offers. Some receivers may not have the capability to simultaneously decode an ME and a D service.

Stereo (two-channel) service without artistic limitation can be provided in multiple languages with added efficiency by transmitting a stereo ME main service along with stereo D services. The D and appropriate language ME services are simply combined in the receiver into a complete stereo program. Dialogue may be panned, and reverberation may be placed included in both channels. A stereo ME service can be sent with high quality at 192 kbps, while the stereo D services (voice only) can make use of lower bit-rates, such as 128 or 96 kbps per language. Some receivers may not have the capability to simultaneously decode an ME and a D service.

Note that during those times when dialogue is not present, the D services can be momentarily removed, and their data capacity used for other purposes.

Audio Bit Rates

The information in Table 16.12 provides a general guideline as to the audio bit rates that are expected to be most useful. For main services, the use of the LFE channel is optional and will not affect the indicated data rates.

The audio decoder input buffer size (and thus part of the decoder cost) is determined by the maximum bit rate that must be decoded. The syntax of the AC-3 standard supports bit rates ranging from a minimum of 32 kbps up to a maximum of 640 kbps per individual elementary bit stream. The bit rate utilized in the DTV system is restricted in order to reduce the size of the input buffer in the audio decoder, and thus the receiver cost. Receivers can be expected to support the decoding of a main audio service, or an associated audio service that is a complete service (containing all necessary program elements), at a bit rate up to and including 448 kbps. Transmissions may contain main audio services, or associated audio services that are complete services (containing all necessary program elements), encoded at a bit rate up to and including 448 kbps. Transmissions may contain single-channel associated audio services intended to be simultaneously decoded along with a main service encoded at a bit rate up to and including 128 kbps. Transmissions may contain dual-channel dialogue associated services intended to be simultaneously decoded along with a main service encoded at a bit rate up to and including 192 kbps. Transmissions have a further limitation that the combined bit rate of a main and an associated service that are intended to be simultaneously reproduced is less than or equal to 576 kbps.

16.5.4   DTV Transport

The transport subsystem employs the fixed-length transport stream packetization approach defined in ISO/IEC13818-1 [12], which is usually referred to as the MPEG-2 systems standard. This approach is well-suited to the needs of terrestrial broadcast and cable television transmission of digital television. The use of relatively short, fixed-length packets matches well with the needs and techniques for error protection in both terrestrial broadcast and cable television distribution environments.

The ATSC DTV transport may carry a number of television programs. The MPEG-2 term "program" corresponds to an individual digital TV channel or data service, where each program is composed of a number of MPEG-2 program elements (i.e., related video, audio, and data streams). The MPEG-2 systems standard support for multiple channels or services within a single, multiplexed bit stream enables the deployment of practical, bandwidth-efficient digital broadcasting systems. It also provides great flexibility to accommodate the initial needs of the service to multiplex video, audio, and data while providing a well-defined path to add additional services in the future in a fully backward-compatible manner. By basing the transport subsystem on MPEG-2, maximum interoperability with other media and standards is maintained.

Referring to Figure 16.49, the transport subsystem resides between the application (e.g., audio or video) encoding and decoding functions and the transmission subsystem. At its lowest layer, the encoder transport subsystem is responsible for formatting the encoded bits and multiplexing the different components of the program for transmission. At the receiver, it is responsible for recovering the bit streams for the individual application decoders and for the corresponding error signaling. The transport subsystem also incorporates other higher-level functionality related to identification of applications and synchronization of the receiver.

PSIP

The *Program and system information protocol* (PSIP) is data that are transmitted along with a station's DTV signal that tells DTV receivers important information about the station and what is being broadcast. Described in ATSC Standard A/65 [3], the most important function of PSIP is to provide a method for DTV receivers to identify a DTV station and to determine how a receiver can tune to it. PSIP identifies both the DTV channel and the associated NTSC (analog) channel. It helps maintain the current channel branding because DTV receivers will electronically associate the two channels, making it easy for viewers to tune to the DTV station even if they do not know the RF channel number.

In addition to identifying the channel number, PSIP tells the receiver whether multiple program channels are being broadcast and, if so, how to find them. It identifies, for example, whether the programs are closed-captioned, conveys V-chip information, and if data are associated with the program.

PSIP Structure

PSIP is a collection of tables, each of which describes elements of typical digital television services [13]. Figure 16.53 shows the primary components and the notation used to describe them. The packets of the base tables are all labeled with a base *packet identifier* (PID). The base tables are

- System time table (STT)
- Rating region table (RRT)
- Master guide table (MGT)
- Virtual channel table (VCT)

The event information tables (EITs) are a second set of tables, whose packet identifiers are defined in the MGT. The extended text tables (ETTs) are a third set of tables, and similarly, their PIDs are defined in the MGT.

The system time table is a small data structure that fits in one transport stream packet and serves as a reference for time-of-day functions. Receivers can use this table to manage various operations and scheduled events, as well as display time-of-day.

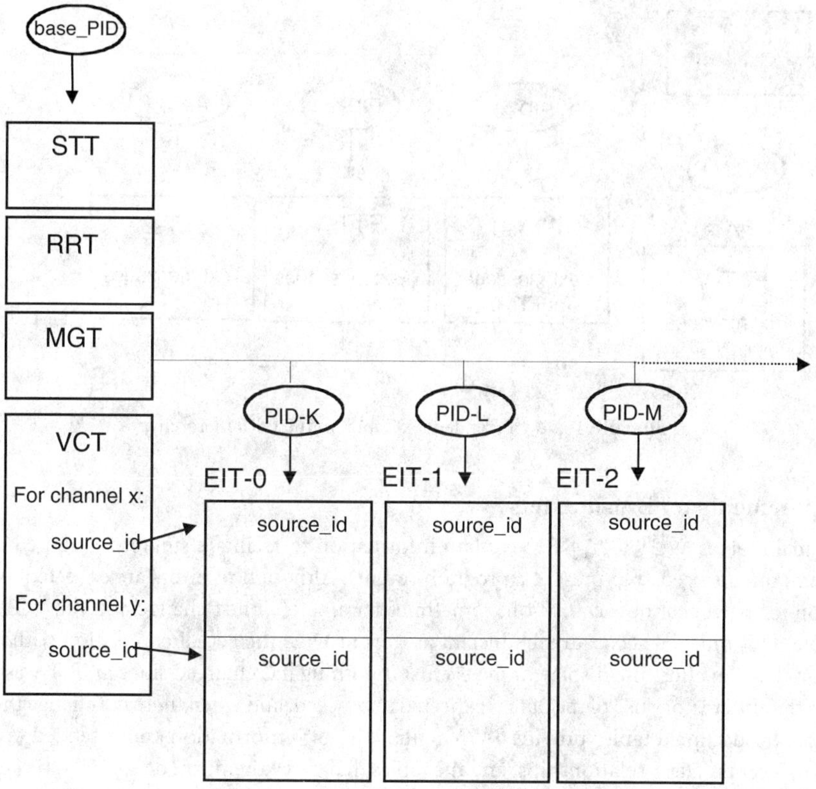

FIGURE 16.53 Overall structure of the PSIP tables.

The rating region table has been designed to transmit the rating system in use for each country using the ratings. In the U.S., this is incorrectly but frequently referred to as the "V-chip" system; the proper title is television parental guidelines (TVPG). Provisions have been made for multi-country systems.

The master guide table provides indexing information for the other tables that comprise the PSIP Standard. It also defines table sizes necessary for memory allocation during decoding, defines version numbers to identify those tables that need to be updated, and generates the packet identifiers that label the tables.

The virtual channel table, also referred to as the Terrestrial VCT (TVCT), contains a list of all the channels that are or will be on-line, plus their attributes. Among the attributes given are the channel name and channel number.

There are up to 128 event information tables, EIT-0 through EIT-127, each of which describes the events or television programs for a time interval of 3 h. Because the maximum number of EITs is 128, up to 16 days of programming may be advertised in advance. At minimum, the first four EITs must always be present in every transport stream, and 24 are recommended. Each EIT-k may have multiple instances, one for each virtual channel in the VCT.

As illustrated in Figure 16.54, there may be multiple extended text tables, one or more channel ETT sections describing the virtual channels in the VCT, and an ETT-k for each EIT-k, describing the events in the EIT-k. These are all listed in the MGT. An ETT-k contains a table instance for each event in the associated EIT-k. As the name implies, the purpose of the ETT is to carry text messages. For example, for channels in the VCT, the messages can describe channel information, cost, coming attractions, and other related data. Similarly, for an event such as a movie listed in the EIT, the typical message would be a short paragraph that describes the movie itself. Extended text tables are optional in the ATSC system.

FIGURE 16.54   Extended text tables in the PSIP hierarchy.

PSIP Requirements for Broadcasters

The three main tables (VCT, EIT, STT) contain information to facilitate suitably equipped receivers to find the components needed to present a program (event). Although receivers are expected to use stored information to speed channel acquisition, sometimes parameters must change and the VCT and EIT-0 are the tables that must be accurate each instant as they provide the actual connection path and critical information that can affect the display of the events. If nothing has changed since an EIT was sent for an event, then the anticipatory use of the data is expected to proceed, and when there is a change the new parts would be used. Additional tables provide TV parental advisory information and extended text messages about certain events. These relationships, and the tables that carry them, are designed to be kept with the DTV signal when it is carried by a cable system.

The Basics

There are certain "must have" items and "must do" rules of operation. If the PSIP elements are missing or wrong, there may be severe consequences, which vary depending on the design of receiver. The following are key elements that must be set and/or checked by each station:

- **Transport stream identification** (TSID). The preassigned TSIDs must be set correctly in all three locations (PAT, VCT common information, and virtual channel-specific information).
- **System time table** (SST). The STT time should be checked daily and locked to house time. Ideally, the STT should be inserted into the TS within a frame before each seconds-count increment of the house time with the to-be-valid value.
- **Short channel name**. This is a seven-character name that can be set to any desired name indicating the virtual channel name. For example, a station's call letters followed by SD1, SD2, SD3, and SD4

TABLE 16.13   Mandatory PSIP Table Suggested Repetition Rates

PSIP Table	Transmission Cycle
MGT	Once every 150 ms
TVCT	Once every 400 ms
EIT-0	Once every 0.5 sec
EIT-1	Once every 3 sec
EIT-2 and EIT-3	Once every minute
STT	Once every second
RRT (not required in some areas[a])	Once every minute

[a]The U.S. is one such area.

TABLE 16.14 Suggested Repetition Rates for Optional PSIP Tables

	PSIP Table	Transmission Cycle
	A/65A specifies the following repetition rates for DCC per specified conditions.	
DCCT	DCC request in progress	150 msec
	2 seconds prior to DCC request	400 msec
	No DCC	n/a
DCCSCT		Once per hour
ETT		Once every minute
EIT-4 and higher		Once every minute
DET		A later version of this Recommended Practice will address data services.

to indicated various SDTV virtual channels or anything else to represent the station's identity (e.g., WNABSD1, KNABSD2, WNAB-HD, KIDS, etc.).

- **Major channel**. The previously assigned, paired NTSC channel is the major channel number.
- **Service type**. The service type selects DTV, NTSC, audio only, data, etc., and must be set as operating modes require.
- **Modulation mode**. A code for the RF modulation of the virtual channel.
- **Source ID**. The source ID is a number that associates virtual channels to events on those channels. It typically is automatically updated by PSIP equipment or updated from an outside vendor. Proper operation of this feature should be confirmed.
- **Service location descriptor** (SLD). Contains the MPEG references to the contents of each component of the programs plus a language code for audio. The PID values for the components identified here and in the PMT must be the same for the elements of an event/program. Some deployed systems require separate manual setup, but PID values assigned to a VC should seldom change.

The maximum cycle time/repetition rate of the tables should be set or confirmed to conform with the suggested guidelines given in Table 16.13 for mandatory PSIP tables and Table 16.14 for optional PSIP tables.

It is recommended that broadcasters send populated EITs covering at least three days. The primary cycle time guidelines are illustrated in Figure 16.55.

The recommended table cycle times given in this section result in a minimal demand on overall system bandwidth. Considering the importance of the information that these PSIP tables provide to the receiver, the bandwidth penalty is trivial.

FIGURE 16.55 Recommended PSIP table cycle times.

Chan	Name	6:00 PM	6:30 PM	7:00 PM	7:30 PM	8:00 PM	8:30 PM
6-0	XYZ	City Scene		Travel Log		Movie: Speed II	
6-1	XYZ	City Scene		Travel Log		Movie: Speed II (HD)	
6-2	XYZ	Movie: Star Trek6 The Voyage Home				Tune 6-1 for Movie: Speed II (HD)	
6-3	LNC	Local News		Airport Info			

FIGURE 16.56   Example electronic program guide.

Program Guide

Support for an electronic program guide (EPG) or interactive program guide (IPG) is another important function enabled by the PSIP standard. The concept is to provide a way for viewers to find out "what's on" directly from their television sets, similar to the *guide channels* that are typically available for cable and satellite broadcast services. In a terrestrial broadcast environment, there is no single authority that determines what programs are on all the channels, so each broadcaster needs to include this type of information within the broadcast stream. Viewers' receivers will be able to scan all the available channels and create a program guide channel from the consolidated information. The value of this type of guide information is high, and it will continue to increase in the DTV environment. Viewers will have the ability to choose not only what channel to watch, but also be able to select from multiple options within a broadcast. Examples include selecting from a set of alternative audio tracks in different languages, or choosing one of several SDTV programs shown at the same time on different virtual channels.

The more viewers rely on these guides, the more critical the accuracy and detail of the information they contain will become. Figure 16.56 shows what a typical electronic program guide might look like.

Building the On-Screen Display or EPG

The EPG has the dual functionality of announcing future programs and providing critical information about the current program. It contains program names and planned broadcast times as well as other information about an event. The data can be combined to build a receiver on-screen display such as that shown in Figure 16.57.

16.5.5   RF Transmission

The ATSC VSB system offers two basic operational modes: a terrestrial broadcast mode (8-VSB) and a high data rate mode (16-VSB) intended for cable applications. Both modes provide a pilot, segment syncs, and a training sequence (as part of data field sync) for acquisition and operation. The two system modes can use the same carrier recovery, demodulation, sync recovery, and clock recovery circuits. Adaptive equalization for the two modes can use the same equalizer structure with some differences in the decision feedback and adaptation of the filter coefficients. Furthermore, both modes use the same Reed-Solomon (RS) code and circuitry for *forward error correction* (FEC). The terrestrial broadcast mode is optimized for maximum service area and provides a data payload of approximately 19.4 Mbps in a 6 MHz channel. The high data rate mode, which provides twice the data rate at the cost of reduced robustness for channel degradations such as noise and multipath, provides a data payload of 38.8 Mbps in a single 6 MHz channel.

In order to maximize service area, the terrestrial broadcast mode incorporates trellis coding, with added precoding that allows the data to be decoded after passing through a receiver comb filter, used selectively to suppress analog co-channel interference. The high-data-rate mode is designed to work in a cable environment, which is less severe than that of the terrestrial system. It is transmitted in the form

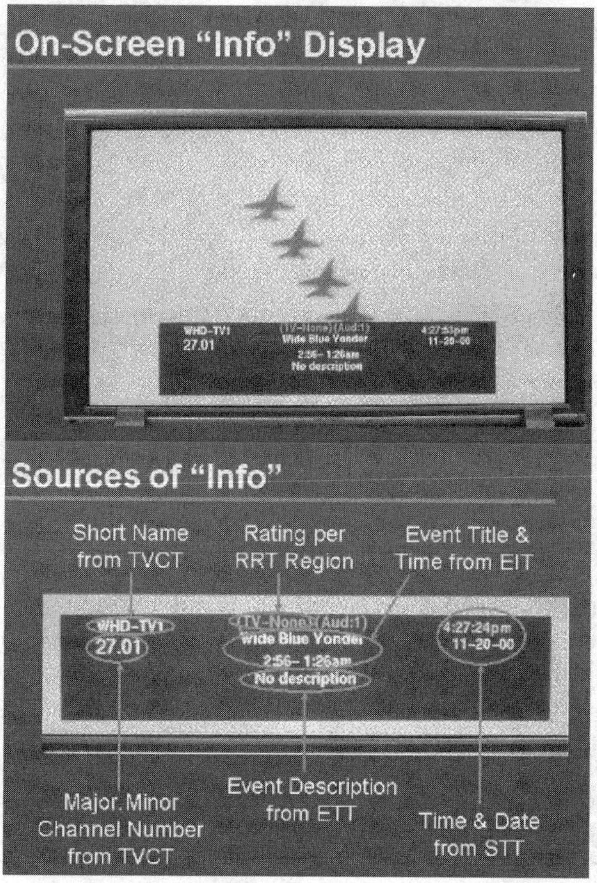

FIGURE 16.57 Illustration of how the various PSIP tables could combine to produce the on-screen display at the receiver.

of more data levels (bits/symbol). No trellis coding or precoding for an analog broadcast interference rejection (comb) filter is employed in this mode.

VSB transmission with a raised-cosine roll-off at both the lower edge (pilot carrier side) and upper edge (Nyquist slope at 5.38 MHz above carrier) permits equalizing just the in-phase (I) channel signal with a sampling rate as low as the symbol (baud) rate. The raised-cosine shape is obtained from concatenating a root-raised cosine in the transmitter with the same shape in the receiver. Although energy in the vestigial sideband and in the upper band edge extends beyond the Nyquist limit frequencies, the demodulation and sampling process aliases this energy into the baseband to suppress *intersymbol interference* (ISI) and thereby avoid distortion. With the carrier frequency located at the −6 dB point on the carrier-side raised-cosine roll-off, energy in the vestigial sideband folds around zero frequency during demodulation to make the baseband DTV signal exhibit a flat amplitude response at lower frequencies, thereby suppressing low-frequency ISI. Then, during digitization by synchronous sampling of the demodulated I signal, the Nyquist slope through 5.38 MHz suppresses the remnant higher-frequency ISI. With ISI due to aliasing thus eliminated, equalization of linear distortions can be done using a single A/D converter sampling at the symbol rate of 10.76 Msamples/s and a real-only (not complex) equalizer also operating at the symbol rate. In this simple case, equalization of the signal beyond the −6 dB points in the raised-cosine roll-offs at channel edges is dependent on the in-band equalization and cannot be set independently. A complex equalizer does not have this limitation, nor does a fractional equalizer sampling at a rate sufficiently above the symbol rate.

TABLE 16.15 Parameters for VSB Transmission Modes

Parameter	Terrestrial Mode	High Data Rate Mode
Channel bandwidth	6 MHz	6 MHz
Excess bandwidth	11.5%	11.5%
Symbol rate	10.76 Msymbols/s	10.76 Msymbols/s
Bits per symbol	3	4
Trellis FEC	2/3 rate	None
Reed-Solomon FEC	T = 10 (207, 187)	T = 10 (207, 187)
Segment length	832 symbols	832 symbols
Segment sync	4 symbols per segment	4 symbols per segment
Frame sync	1 per 313 segments	1 per 313 segments
Payload data rate	19.39 Mbps	38.78 Mbps
Analog co-channel rejection	Analog rejection filter in receiver	N/A
Pilot power contribution	0.3 dB	0.3 dB
C/N threshold	14.9 dB	28.3 dB

The 8-VSB signal is designed to minimize interference and RF channel allocation problems. The VSB signal is designed to minimize peak-energy-to-average-energy ratio, thereby minimizing interference into other signals, especially adjacent and "taboo" channels. To counter the man-made noise that often accompanies over-the-air broadcast signals, the VSB system includes an interleaver that allows correction of an isolated single burst of noise up to 190 microseconds in length by the (207,187) RS FEC circuitry, which locates as well as corrects up to 10 byte errors per data segment. This was done to allow VHF channels, which are often substantially affected by man-made noise, to be used for DTV broadcasting. If soft-decision techniques are used in the trellis decoder preceding the RS circuitry, the location of errors can be flagged, and twice as many byte errors per data segment can be corrected, allowing correction of an isolated burst of up to 380 microseconds in length.

The parameters for the two VSB transmission modes are shown in Table 16.15.

Bit Rate Delivered to a Transport Decoder by the Transmission Subsystem

As outlined previously, all data in the ATSC DTV system is transported in MPEG-2 transport packets. The useful data rate is the amount of MPEG-2 transport data carried end-to-end including MPEG-2 packet headers and sync bytes. The exact symbol rate of the transmission subsystem is given by

$$\frac{4.5}{286} \times 684 = 10.7 \ldots \text{million symbols/second (megabaud)} \qquad (16.1)$$

The symbol rate must be locked in frequency to the transport rate.

The numbers in the formula for the ATSC symbol rate in 6 MHz systems are related to NTSC scanning and color frequencies. Because of this relationship, the symbol clock can be used as a basis for generating an NTSC color subcarrier for analog output from a set top box. The repetition rates of data segments and data frames are deliberately chosen not to have an integer relationship to NTSC or PAL scanning rates, to insure that there is no discernible pattern in co-channel interference.

The particular numbers used are

- 4.5 MHz = the center frequency of the audio carrier offset in NTSC. This number was traditionally used in NTSC literature to derive the color subcarrier frequency and scanning rates. In modern equipment, this may start with a precision 10 MHz reference, which is then multiplied by 9/20.
- 4.5 MHz/286 = the horizontal scan rate of NTSC, 15734.2657 + ... Hz (note that the color subcarrier is 455/2 times this, or 3579545 +5/11 Hz).
- 684: this multiplier gives a symbol rate for an efficient use of bandwidth in 6 MHz. It requires a filter with Nyquist roll-off that is a fairly sharp cutoff (11% excess bandwidth), which is still realizable with a reasonable surface acoustic wave (SAW) filter or digital filter.

In the terrestrial broadcast mode, channel symbols carry three bits/symbol of trellis-coded data. The trellis code rate is 2/3, providing 2 bits/symbol of gross payload. Therefore the gross payload is

$$10.76\ldots \times 2 = 21.52\ldots \text{Mbps (megabits/second)} \qquad (16.2)$$

To find the net payload delivered to a decoder it is necessary to adjust (16.2) for the overhead of the *data segment sync, data field sync,* and Reed-Solomon FEC.

To get the net bit rate for an MPEG-2 stream carried by the system (and supplied to an MPEG transport decoder), it is first noted that the MPEG sync bytes are removed from the data stream input to the 8-VSB transmitter and replaced with segment sync, and later reconstituted at the receiver. For throughput of MPEG packets (the only allowed transport mechanism), segment sync is simply equivalent to transmitting the MPEG sync byte, and does not reduce the net data rate. The net bit rate of an MPEG-2 stream carried by the system and delivered to the transport decoder is accordingly reduced by the data field sync (one segment of every 313) and the Reed-Solomon coding (20 bytes of every 208):

$$21.52\ldots \text{Mbps} \times \frac{312}{313} \times \frac{188}{208} = 19.39\ldots \text{Mbps} \qquad (16.3)$$

The net bit rate supplied to the transport decoder for the high data rate mode is

$$19.39\ldots \text{Mbps} \times 2 = 38.78\ldots \text{Mbps} \qquad (16.4)$$

Performance Characteristics of Terrestrial Broadcast Mode

The terrestrial 8-VSB system can operate in a signal-to-additive-white-Gaussian-noise (S/N) environment of 14.9 dB. The 8-VSB segment error probability curve including 4-state trellis decoding and (207,187) Reed-Solomon decoding in Figure 16.58 shows a segment error probability of 1.93×10^{-4}. This is equivalent to 2.5 segment errors/second, which was established by measurement as the *threshold of visibility* (TOV) of errors in the prototype equipment [14]. Particular product designs may achieve somewhat better performance for subjective TOV by means of error masking.

The *cumulative distribution function* (CDF) of the peak-to-average power ratio, as measured on a low power transmitted signal with no nonlinearities, is plotted in Figure 16.59. The plot shows that 99.9% of the time the transient peak power is within 6.3 dB of the average power.

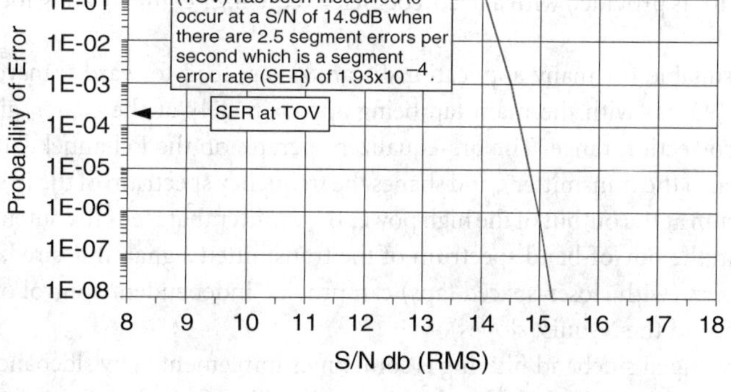

Note: Threshold of visibility (TOV) has been measured to occur at a S/N of 14.9dB when there are 2.5 segment errors per second which is a segment error rate (SER) of 1.93×10^{-4}.

SER at TOV

FIGURE 16.58 Segment error probability, 8-VSB with 4 state trellis decoding, RS (207, 187).

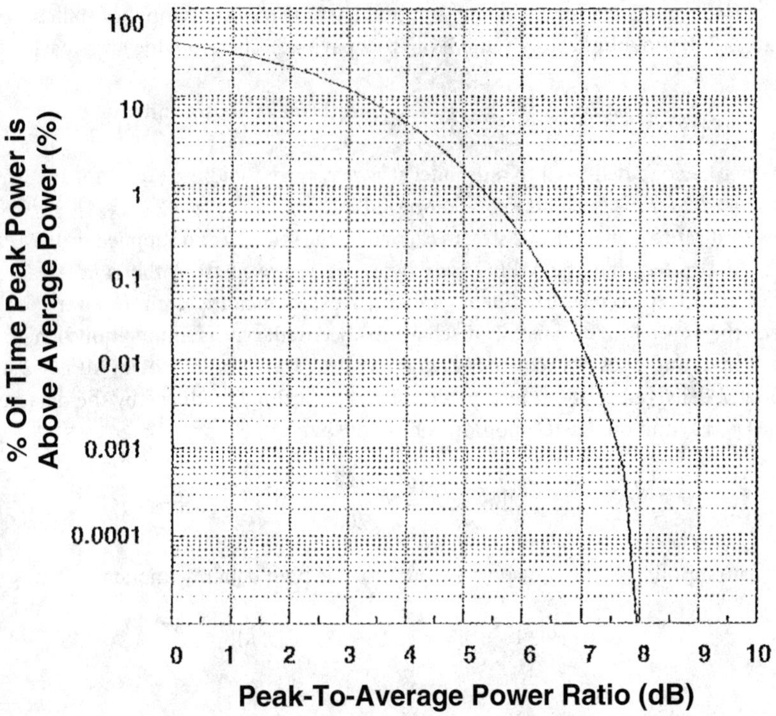

FIGURE 16.59 Cumulative distribution function of 8-VSB peak-to-average power ratio (in ideal linear system).

Transmitter Signal Processing

A pre-equalizer filter is recommended for use in over-the-air broadcasts where the high power transmitter may have significant in-band ripple or significant roll off at band edges. Pre-equalization is typically required in order to compensate the high-order filter used to meet a stringent out-of-band emission mask, such as the U.S. FCC required mask [15]. This linear distortion can be measured by an equalizer in a reference demodulator ("ideal" receiver) employed at the transmitter site. A directional coupler, which is recommended to be located at the sending end of the antenna feed transmission line, supplies the reference demodulator a small sample of the antenna signal feed. The equalizer tap weights of the reference demodulator are transferred to the transmitter pre-equalizer for pre-correction of transmitter linear distortion. This is a one-time procedure of measurement and transmitter pre-equalizer adjustment. Alternatively, the transmitter pre-equalizer can be made continuously adaptive. In this arrangement, the reference demodulator is provided with a fixed-coefficient equalizer compensating for its own deficiencies in ideal response.

A pre-equalizer suitable for many applications is an 80 tap, feed-forward transversal filter. The taps are symbol-spaced (93 ns) with the main tap being approximately at the center, giving approximately ± 3.7 microsecond correction range. The pre-equalizer operates on the I channel data signal (there is no Q channel data signal in the transmitter), and shapes the frequency spectrum of the IF signal so that there is a flat in-band spectrum at the output of the high power transmitter that feeds the antenna for transmission. There is no effect on the out-of-band spectrum of the transmitted signal. If desired, complex equalizers or fractional equalizers (with closer-spaced taps) can provide independent control of the outer portions of the spectrum (beyond the Nyquist slopes).

The transmitter vestigial sideband filtering is sometimes implemented by sideband cancellation, using the phasing method. In this method, the baseband data signal is supplied to digital filtering that generates in-phase and quadrature-phase digital modulation signals for application to respective D/A converters.

This filtering process provides the root raised cosine Nyquist filtering and provides compensation for the $(\sin x)/x$ frequency responses of the D/A converters, as well. The baseband signals are converted to analog form. The in-phase signal modulates the amplitude of the IF carrier at zero degrees phase, while the quadrature signal modulates a 90° shifted version of the carrier. The amplitude-modulated quadrature IF carriers are added to create the vestigial sideband IF signal, canceling the unwanted sideband and increasing the desired sideband by 6 dB.

Upconverter and RF Carrier Frequency Offsets

Modern analog TV transmitters use a two-step modulation process. The first step usually is modulation of the data onto an IF carrier, which is the same frequency for all channels, followed by translation to the desired RF channel. The digital 8-VSB transmitter applies this same two-step modulation process. The RF upconverter translates the filtered flat IF data signal spectrum to the desired RF channel. For the same approximate coverage as an analog transmitter (at the same frequency), the average power of the DTV signal is on the order of 12 dB less than the analog peak sync power (when operating on the same frequency).

The nominal frequency of the RF upconverter oscillator in DTV terrestrial broadcasts will typically be the same as that used for analog transmitters (except for offsets required in particular situations).

Note that all examples in this section relate to a 6 MHz DTV system. Values may be modified easily for other channel widths.

Nominal DTV Pilot Carrier Frequency
The nominal DTV pilot carrier frequency is determined by fitting the DTV spectrum symmetrically into the RF channel. This is obtained by taking the bandwidth of the DTV signal—5,381.1189 kHz (the Nyquist frequency difference or one-half the symbol clock frequency of 10,762.2378 kHz)—and centering it in the 6 MHz TV channel. Subtracting 5,381.1189 kHz from 6,000 kHz leaves 618.881119 kHz. Half of that is 309.440559 kHz, precisely the standard pilot offset above the lower channel edge. For example, on channel 45 (656–662 MHz), the nominal pilot frequency is 656.309440559 MHz.

Requirements for Offsets
There are two categories of requirements for pilot frequency offsets:

1. Offsets to protect lower adjacent channel analog broadcasts, mandated by FCC rules in the U.S., and which override other offset considerations.
2. Recommended offsets for other considerations such as co-channel interference between DTV stations or between DTV and analog stations.

Upper DTV Channel into Lower Analog Channel
This is the overriding case mandated by the FCC rules in the U.S.—precision offset with a lower adjacent analog station, full service or low power television (LPTV).

The FCC Rules, Section 73.622(g)(1), states that

> "DTV stations operating on a channel allotment designated with a "c" in paragraph (b) of this section must maintain the pilot carrier frequency of the DTV signal 5.082138 MHz above the visual carrier frequency of any analog TV broadcast station that operates on the lower adjacent channel and is located within 88 km. This frequency difference must be maintained within a tolerance of ±3 Hz."

This precise offset is necessary to reduce the color beat and high-frequency luminance beat created by the DTV pilot carrier in some receivers tuned to the lower adjacent analog channel. The tight tolerance assures that the beat will be visually cancelled, since it will be out of phase on successive video frames.

Note that the frequency is expressed with respect to the lower adjacent analog video carrier, rather than the nominal channel edge. This is because the beat frequency depends on this relationship, and therefore the DTV pilot frequency must track any offsets in the analog video carrier frequency. The offset in the

FCC rules is related to the particular horizontal scanning rate of NTSC, and can easily be modified for PAL. The offset O_f was obtained from

$$O_f = 455 \times \left(\frac{F_h}{2}\right) + 191 \times \left(\frac{F_h}{2}\right) - 29.97 = 5,082,138 \, \text{Hz} \qquad (16.5)$$

Where F_h = NTSC horizontal scanning frequency = 15,734.264 Hz.

The equation indicates that the offset with respect to the lower adjacent chroma is an odd multiple (191) of one-half the line rate to eliminate the color beat. However, this choice leaves the possibility of a luma beat. The offset is additionally adjusted by one-half the analog field rate to eliminate the luma beat. While satisfying the exact adjacent channel criteria, this offset is also as close as possible to optimal comb filtering of the analog co-channel in the digital receiver. Note additionally that offsets are to higher frequencies rather than lower, to avoid any possibility of encroaching on the lower adjacent sound. (It also reduces the likelihood of the automatic fine tuning (AFT) in the analog receiver experiencing lock-out because the signal energy including the pilot is moved further from the analog receiver bandpass.)

Other Offset Cases

The FCC rules do not consider other interference cases where offsets help. The offset for protecting lower-adjacent analog signals takes precedence. If that offset is not required, other offsets can minimize interference to co-channel analog or DTV signals.

In co-channel cases, DTV interference into analog TV appears noise-like. The pilot carrier is low on the Nyquist slope of the IF filter in the analog receiver, so no discernable beat is generated. In this case, offsets to protect the analog channel are not required. Offsets are useful, however, to reduce co-channel interference from analog TV into DTV. The performance of the analog rejection filter and clock recovery in the DTV receiver will be improved if the DTV carrier is 911.944 kHz below the NTSC visual carrier. In other words, in the case of a 6 MHz NTSC system, if the analog TV station is not offset, the DTV pilot carrier frequency will be 338.0556 kHz above the lower channel edge instead of the nominal 309.44056 kHz. As before, if the NTSC station is operating with a ±10 kHz offset, the DTV frequency will have to be adjusted in the same direction. The formula for calculating this offset is

$$F_{\text{pilot}} = F_{\text{vis(n)}} - 70.5 \times F_{\text{seg}} = 338.0556 \, \text{Hz (for no NTSC analog offset)} \qquad (16.6)$$

Where
F_{pilot} = DTV pilot frequency above lower channel edge
$F_{\text{vis(n)}}$ = NTSC visual carrier frequency above lower channel edge
  = 1250 kHz for no NTSC offset (as shown)
  = 1240 kHz for minus offset
  = 1260 kHz for plus offset
F_{seg} = ATSC data segment rate; = symbol clock frequency/832 = 12,935.381971 Hz

The factor of 70.5 is chosen to provide the best overall comb filtering of analog color TV co-channel interference. The use of a value equal to an integer +0.5 results in co-channel analog TV interference being out-of-phase on successive data segment syncs.

Note that in this case the frequency tolerance is plus or minus one kHz. More precision is not required. Also note that a different data segment rate would be used for calculating offsets for 7 or 8 MHz systems.

Co-channel DTV into DTV

If two DTV stations share the same channel, interference between the two stations can be reduced if the pilot is offset by one and a half times the data segment rate. This ensures that the frame and segment syncs of the interfering signal will each alternate polarity and be averaged out in the receiver tuned to the desired signal.

TABLE 16.16 DTV Pilot Carrier Frequencies for Two Stations (Normal offset above lower channel edge: 309.440559 kHz)

	DTV Pilot Carrier Frequency Above Lower Channel Edge			
Channel Relationship	NTSC Station Zero Offset	NTSC Station +10 kHz Offset	NTSC Station −10 kHz Offset	DTV Station No Offset
DTV with lower adjacent NTSC	332.138 kHz ±3 Hz	342.138 kHz ±3 Hz	322.138 kHz ±3 Hz	
DTV co-channel with NTSC	338.056 kHz ±1 kHz	348.056 kHz ±1 kHz	328.056 kHz ±1 kHz	
DTV co-channel with DTV	+19.403 kHz above DTV	+19.403 kHz above DTV	+19.403 kHz above DTV	328.8436 kHz ±10 Hz

The formula for this offset is

$$F_{offset} = 1.5 \times F_{seg} = 19.4031 \text{ kHz} \tag{16.7}$$

Where

F_{offset} = offset to be added to one of the two DTV carriers
F_{seg} = 12,935.381971 Hz (as defined previously)

This results in a pilot carrier 328.84363 kHz above the lower band edge, provided neither DTV station has any other offset.

Use of the factor 1.5 results in the best co-channel rejection, as determined experimentally with the prototype equipment. The use of an integer +0.5 results in co-channel interference alternating phase on successive segment syncs.

Summary: DTV Frequency

Table 16.16 summarizes the various pilot carrier offsets for different interference situations in a 6 MHz system (NTSC environment). Note that if more than two stations are involved the number of potential frequencies will increase. For example, if one DTV station operates at an offset because of a lower-adjacent-channel NTSC station, a co-channel DTV station may have to adjust its frequency to maintain a 19.403 kHz pilot offset. If the NTSC analog station operates at an offset of plus or minus 10 kHz, both DTV stations should compensate for that. Cooperation between stations will be essential in order to reduce interference.

Frequency Tolerances

The tightest specification is for a DTV station with a lower adjacent NTSC analog station. If both NTSC and DTV stations are at the same location, they may simply be locked to the same reference. The co-located DTV station carrier should be 5.082138 MHz above the NTSC visual carrier (22.697 kHz above the normal pilot frequency). The co-channel DTV station should set its carrier 19.403 kHz above the co-located DTV carrier.

If there is interference with a co-channel DTV station, the analog station is expected to be stable within 10 Hz of its assigned frequency.

While it is possible to lock the frequency of the DTV station to the relevant NTSC station, this may not be the best option if the two stations are not at the same location. It will likely be easier to maintain the frequency of each station within the necessary tolerances. Where co-channel interference is a problem, that will be the only option.

In cases where no type of interference is expected, a pilot carrier-frequency tolerance of ±1 kHz is acceptable, but in all cases, good practice is to use a tighter tolerance if practicable.

Defining Terms

Anchor frame: In MPEG-2, a video frame that is used for prediction. I-frames and P-frames are generally used as anchor frames, but B-frames are never anchor frames.

Asynchronous transfer mode (ATM): A digital signal protocol for efficient transport of both constant-rate and bursty information in broadband digital networks. The ATM digital stream consists of fixed-length packets called "cells," each containing 53 8-bit bytes—a 5-byte header and a 48-byte information payload.

Bidirectional pictures (or **B-pictures** or **B-frames**): In MPEG-2, pictures that use both future and past pictures as a reference. This technique is termed *bidirectional prediction*. B-pictures provide the most compression. B-pictures do not propagate coding errors as they are never used as a reference.

Bit rate: The rate at which the compressed bit stream is delivered from the channel to the input of a decoder.

Block: A block is an 8-by-8 array of pel values or DCT coefficients representing luminance or chrominance information.

Byte-aligned: A bit in a coded bit stream is byte-aligned if its position is a multiple of 8-bits from the first bit in the stream.

Compression: Reduction in the number of bits used to represent an item of data.

Constant bit rate: A mode of operation in a digital system where the bit rate is constant from start to finish of the compressed bit stream.

Conventional definition television (CDTV): This term is used to signify the *analog* NTSC television system as defined in ITU-R Recommendation 470. See also *standard definition television*.

CRC: The cyclic redundancy check to verify the correctness of data.

D-frame: A frame coded according to an MPEG-1 mode that uses DC coefficients only.

Data element: An item of data as represented before encoding and after decoding.

Decoded stream: The decoded reconstruction of a compressed bit stream.

Decoder: An embodiment of a decoding process.

Digital storage media (DSM): A digital storage or transmission device or system.

Discrete cosine transform: A mathematical transform that can be perfectly undone and which is useful in image compression.

Elementary stream (ES): A generic term for one of the coded video, coded audio, or other coded bit streams in a DTV system.

Encoder: An embodiment of an encoding process.

Entitlement control message (ECM): Entitlement control messages are private conditional access information that specify control words and possibly other stream-specific, scrambling, and/or control parameters.

Entitlement management message (EMM): Entitlement management messages are private conditional access information that specify the authorization level or the services of specific decoders. They may be addressed to single decoders or groups of decoders.

Entropy coding: Variable length lossless coding of the digital representation of a signal to reduce redundancy.

Entry point: Refers to a point in a coded bit stream after which a decoder can become properly initialized and commence syntactically correct decoding. The first transmitted picture after an entry point is either an I-picture or a P-picture. If the first transmitted picture is not an I-picture, the decoder may produce one or more pictures during acquisition.

Event: A collection of elementary streams with a common time base, an associated start time, and an associated end time.

Field: For an interlaced video signal, a "field" is the assembly of alternate lines of a frame. Therefore, an interlaced frame is composed of two fields, a top field and a bottom field.

Frame: A frame contains lines of spatial information of a video signal. For progressive video, these lines contain samples starting from one time instant and continuing through successive lines to the bottom of the frame. For interlaced video, a frame consists of two fields, a top field and a bottom field. One of these fields will commence one field later than the other.

Group of pictures (GOP): In MPEG-2, a group of pictures consists of one or more pictures in sequence.

High-definition television (HDTV): A system that provides significantly improved picture quality with more visible detail, a wide screen format (16:9 aspect ratio), and may be accompanied by digital surround-sound capability.

High level: A range of allowed picture parameters defined by the MPEG-2 video coding specification that corresponds to high-definition television.

Huffman coding: A type of source coding that uses codes of different lengths to represent symbols which have unequal likelihood of occurrence.

Intra coded pictures (or **I-pictures** or **I-frames**): MPEG-2 pictures that are coded using information present only in the picture itself and not depending on information from other pictures. I-pictures provide a mechanism for random access into the compressed video data. I-pictures employ transform coding of the pel blocks and provide only moderate compression.

Level: A range of allowed picture parameters and combinations of picture parameters in the DTV system.

Main level: A range of allowed picture parameters defined by the MPEG-2 video coding specification with maximum resolution equivalent to ITU-R Recommendation 601.

Main profile: A subset of the syntax of the MPEG-2 video coding specification that is supported over a large range of applications.

Motion vector: A pair of numbers that represent the vertical and horizontal displacement of a region of a reference picture for MPEG-2 prediction.

MPEG: Refers to standards developed by the ISO/IEC JTC1/SC29 WG11, *Moving Picture Experts Group.* MPEG may also refer to the Group itself.

MPEG-1: Refers to ISO/IEC standards 11172-1 (Systems), 11172-2 (Video), 11172-3 (Audio), 11172-4 (Compliance Testing), and 11172-5 (Technical Report).

MPEG-2: Refers to ISO/IEC standards 13818-1 (Systems), 13818-2 (Video), 13818-3 (Audio), 13818-4 (Compliance).

Packet: A packet consists of a header followed by a number of contiguous bytes from an elementary data stream. It is a layer in the system coding syntax of the DTV system.

Payload: Payload refers to the bytes that follow the header byte in a packet. The transport stream packet header and adaptation fields are not payload.

PES packet: The data structure used to carry elementary stream data. It consists of a packet header followed by PES packet payload.

PES stream: A PES stream consists of PES packets, all of whose payloads consist of data from a single elementary stream, and all of which have the same stream ID number.

Picture: Source, coded, or reconstructed image data. A source or reconstructed picture consists of three rectangular matrices representing the luminance and two chrominance signals.

Pixel: "Picture element" or "pel." A pixel is a digital sample of the color intensity values of a picture at a single point.

Predicted pictures or **P-pictures** or **P-frames:** MPEG-2 pictures that are coded with respect to the nearest *previous* I or P-picture. This technique is termed *forward prediction*. P-pictures provide more compression than I-pictures and serve as a reference for future P-pictures or B-pictures. P-pictures can propagate coding errors when P-pictures (or B-pictures) are predicted from prior P-pictures where the prediction is flawed.

Profile: A defined subset of the syntax specified in the MPEG-2 video coding specification.

Program clock reference (PCR): A time stamp in the transport stream from which decoder timing is derived.

Program specific information (PSI): Normative data that are necessary for the demultiplexing of transport streams and the successful regeneration of programs.

Program: A program is a collection of program elements. Program elements may be elementary streams. Program elements need not have any defined time base; those that do have a common time base and are intended for synchronized presentation.

Quantizer: A processing step that intentionally reduces the precision of DCT coefficients.

Random access: The process of beginning to read and decode the coded bit stream at an arbitrary point.

Scrambling: The alteration of the characteristics of a video, audio, or coded data stream in order to prevent unauthorized reception of the information in a clear form. This alteration is a specified process under the control of a conditional access system.

Source stream: A single, nonmultiplexed stream of samples before compression coding.

Splicing: The concatenation performed on the system level or two different elementary streams.

Standard definition television (SDTV): This term is used to signify a *digital* television system in which the quality is approximately equivalent to that of NTSC. This equivalent quality may be achieved from pictures sourced at the 4:2:2 level of ITU-R Recommendation 601 and subjected to processing as part of the bit rate compression. The results should be such that when judged across a representative sample of program material, subjective equivalence with NTSC is achieved. Also called *standard digital television*.

Start codes: 32-bit codes embedded in the coded bit stream that are unique. They are used for several purposes including identifying some of the layers in the coding syntax.

STD input buffer: A first-in, first-out buffer at the input of a system target decoder for storage of compressed data from elementary streams before decoding.

System clock reference (SCR): A time stamp in the program stream from which decoder timing is derived.

System header: The system header is a data structure that carries information summarizing the system characteristics of the DTV multiplexed bit stream.

System target decoder (STD): A hypothetical reference model of a decoding process used to describe the semantics of the DTV multiplexed bit stream.

Time-stamp: A term that indicates the time of a specific action such as the arrival of a byte or the presentation of a presentation unit.

TOV: Threshold of visibility, used to judge the quality of an analog or digital video signal.

Variable bit rate: Operation in a digital system where the bit rate varies with time during the decoding of a compressed bit stream.

Video buffering verifier (VBV): A hypothetical decoder that is conceptually connected to the output of an encoder. Its purpose is to provide a constraint on the variability of the data rate that an encoder can produce.

8 VSB: Vestigial sideband modulation with 8 discrete amplitude levels.

16 VSB: Vestigial sideband modulation with 16 discrete amplitude levels.

References

1. ATSC Standard A/53B. 2001. ATSC Digital Television Standard. Advanced Television Systems Committee, Washington, D.C.
2. ATSC Standard A/52A. 2001. Digital Audio Compression (AC-3). Advanced Television Systems Committee, Washington, D.C.
3. ATSC Standard A/65B. 2003. Program and System Information Protocol. Advanced Television Systems Committee, Washington, D.C.
4. ISO/IEC IS 13818-2. International Standard (1996), MPEG-2 Video.
5. ANSI/SCTE 43. Digital Video Systems Characteristics Standard for Cable Television. Society of Cable Telecommunications Engineers, Exton, PA.
6. ITU-R BT.601-5. Encoding parameters of digital television for studios.
7. SMPTE 274M 1998. Standard for Television—1920 × 1080 Scanning and Analog and Parallel Digital Interfaces for Multiple Picture Rates. Society of Motion Picture and Television Engineers, White Plains, N.Y.

8. SMPTE 296M. 2001. Standard for Television—1280 × 720 Progressive Image Sample Structure, Analog and Digital Representation and Analog Interface. Society of Motion Picture and Television Engineers, White Plains, N.Y.

9. SMPTE 294M. 2001. 720 × 483 Active Line at 59.94-Hz Progressive Scan Production—Bit-Serial Interfaces. Society of Motion Picture and Television Engineers, White Plains, N.Y.

10. SMPTE 170M. 1999. Standard for Television—Composite Analog Video Signal, NTSC for Studio Applications. Society of Motion Picture and Television Engineers, White Plains, N.Y.

11. AES3. 1997. Serial Transmission Format for Two-Channel Linearly Represented Digital Audio Data. Audio Engineering Society, New York, N.Y.

12. ISO/IEC IS 13818-1.2000 (E). International Standard, Information technology–Generic coding of moving pictures and associated audio information: Systems.

13. ATSC Recommended Practice A/69. 2002. Program and System Information Protocol Implementation Guidelines for Broadcasters. Advanced Television Systems Committee, Washington, D.C.

14. ATSC Recommended Practice A/54A. 2003. Guide to the Use of the Digital Television Standard. Advanced Television Systems Committee, Washington, D.C.

15. FCC. 1998. Memorandum Opinion and Order on Reconsideration of the Sixth Report and Order. Federal Communications Commission, Washington, D.C.

16.6 Propagation Considerations for Mobile Radio Systems

Michel D. Yacoub

16.6.1 Introduction

Mobile communications systems have been driven to use high frequencies due to the congestion at the low portion of the frequency spectrum, where the available bandwidth was not adequate to satisfy the great demand for mobile services. Dealing with high frequencies, however, usually leads to intricate problems, which are severely aggravated by the mobility of the users. Any small-sized object may work as a scatterer, and the multiple reflections provoke multipath propagation. As a consequence, the received signal fades rapidly. With the users moving from one place to another, large variations on the mean signal level are experienced due to different environments. In addition to this, cochannel and adjacent channel interference and a number of different types of noise may influence the transmission quality. Other kinds of phenomena, such as Doppler shift, random frequency modulation, etc., are also present in the signal.

16.6.2 Basic Propagation Models

The propagation of energy in a mobile radio environment is strongly influenced by several factors, including natural and artificial relief, propagation frequency, antenna heights, and others. A precise characterization of the signal variability in this environment constitutes a hard task. Deterministic methods, such as those described by the *free-space, plane earth,* and *knife-edge diffraction* models, are restricted to very simple applications. They are useful, however, in providing the basic propagation mechanisms.

In radio propagation studies one measure of interest is the **path loss** l, defined as the ratio between the transmitted power w_t and the received power w_r. It is frequently expressed in decibel as

$$L = 10 \log l = 10 \log \frac{w_t}{w_r} \tag{16.8}$$

In the next subsections we describe the three mentioned propagation models with the aim of determining their respective path losses.

Free-Space Model

A free-space condition represents an ideal situation, characterized by a line-of-sight propagation between transmitter and receiver, in which the radio propagation phenomenon is basically affected by the distance d between transmitter and receiver and by the wavelength λ of the propagated wave. In such a condition, the ratio between the received power w_r and the transmitted power w_t is given by the Friis free-space transmission formula

$$\frac{w_r}{w_t} = G_t G_r \left(\frac{\lambda}{4\pi d}\right)^2 \tag{16.9}$$

where G_t and G_r are the transmitting and receiving antenna gain, respectively.

Using the relation $c = \lambda f$, where c is the wave propagation velocity assumed to be 3×10^8 m/s and f is the frequency, for isotropic and lossless antennas ($G_t = G_r = 1$) the path loss expressed in decibel is

$$L = 20 \log f + 20 \log d + 32.44 \tag{16.10}$$

In Eq. (16.10) f is given in megahertz and d in kilometers.

Plane Earth Model

When two elevated antennas within line of sight of each other are mounted over a flat terrain, the received signal will be the resultant of the vector sum of the waves traveling through a direct path and those reaching the antenna through indirect paths. The indirect signals comprise mostly the reflected and surface waves. Reflection effects must be considered whenever the dimension of the surface obstructing the signal is large compared to its wavelength. Surface waves are only effective at a few wavelengths above the ground and are usually neglected for mobile radio communications purposes. Accordingly, the ratio between received and transmitted powers can be written as

$$\frac{w_r}{w_t} = G_t G_r \left(\frac{\lambda}{4\pi d}\right)^2 |1 + \rho e^{j\Delta\varphi}|^2 \tag{16.11}$$

where $\Delta\varphi$ is the phase difference between direct and indirect paths and ρ is the reflection coefficient. For grazing angles or frequencies above 100 MHz and incidence angles less than $10°$, ρ tends to -1. Writing $\Delta\varphi$ as a function of the wavelength λ and taking the geometry shown in Fig. 16.60 into account we obtain

$$\Delta\varphi = \frac{2\pi d}{\lambda} \left\{ \left[\left(\frac{h_t + h_r}{d}\right)^2 + 1\right]^{1/2} - \left[\left(\frac{h_t - h_r}{d}\right)^2 + 1\right]^{1/2} \right\} \tag{16.12}$$

Using the approximations $(1 + x)^{1/2} \cong 1 + x/2$, for small x, and $\sin(\Delta\varphi/2) \cong \Delta\varphi/2$, we have

$$\frac{w_r}{w_t} = G_t G_r \left(\frac{h_t h_r}{d^2}\right)^2 \tag{16.13}$$

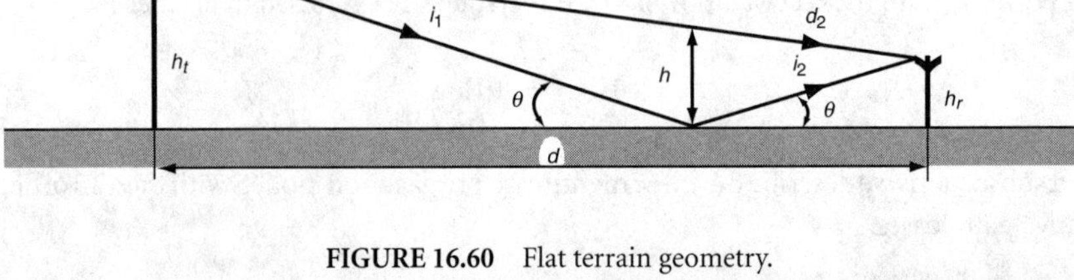

FIGURE 16.60   Flat terrain geometry.

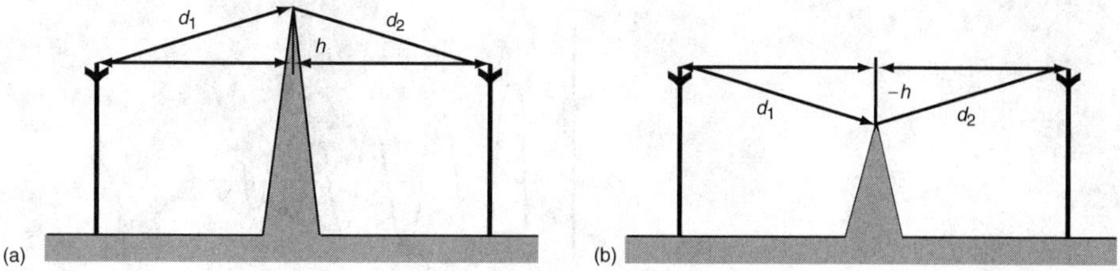

FIGURE 16.61 Knife-edge diffraction geometry.

Assuming isotropic and lossless antennas ($G_t = G_r = 1$), the path loss is given by

$$L = 40 \log d - 20 \log(h_t h_r) \tag{16.14}$$

Knife-Edge Diffraction Model

If an obstruction exists that intercepts a radio wave propagated by the transmitter, the receiver may be reached by diffracted rays. In other words, even for situations in which receivers are placed in shadowed regions, radio communications can still be provided. The diffracted rays comprise the vector sum of the secondary waves resulting from the obstructing surface. Although the diffraction effects are usually more relevant for the obstructed conditions, they can also be sensed in a line-of-sight situation. This is illustrated in Fig. 16.61(a) and Fig. 16.61(b), respectively.

For a single and sharp obstruction, which characterizes the knife-edge condition, the diffracted field E can be calculated as a function of the radiated field E_0 using the classical diffraction theory, in which case

$$\frac{E}{E_0} = \frac{(1+j)}{2} \int_x^\infty \exp\left(-j\frac{\pi}{2}t^2\right) dt = F \exp(j\Delta\varphi) \tag{16.15}$$

where F is the diffraction coefficient, $\Delta\varphi$ is the phase difference between the direct path and the indirect path, and x is the Fresnel–Kirchoff diffraction parameter given by

$$x = -h\sqrt{\frac{2(d_1 + d_2)}{\lambda d_1 d_2}} \tag{16.16}$$

In Eq. (16.16) λ is the wavelength and h, d_1, and d_2 are the dimensions shown in Fig. 16.61.

Therefore, the loss due to diffraction is given by

$$L = 10 \log \left|\frac{E}{E_0}\right|^2 = 20 \log F \tag{16.17}$$

Figure 16.62 shows the diffraction loss as a function of the parameter x.

Some Considerations on the Basic Propagation Models

Free-Space Model

Note that, from Eq. (16.10), in free-space conditions the path loss rate due to the distance is 20 dB/decade or 6 dB/octave. Moreover, although it has been said that the free-space model describes an ideal propagation condition, it can indeed be accurately used in satellite communication systems and short line-of-sight radio links.

Plane Earth Model

Note that, from Eq. (16.14), the path loss rate due to the distance is 40 dB/decade or 12 dB/octave. In the same way, the loss rate due to the antenna heights is −20 dB/decade or −6 dB/octave. In other words, a gain is obtained with the increase of the antenna height. A flat terrain condition can be assumed for

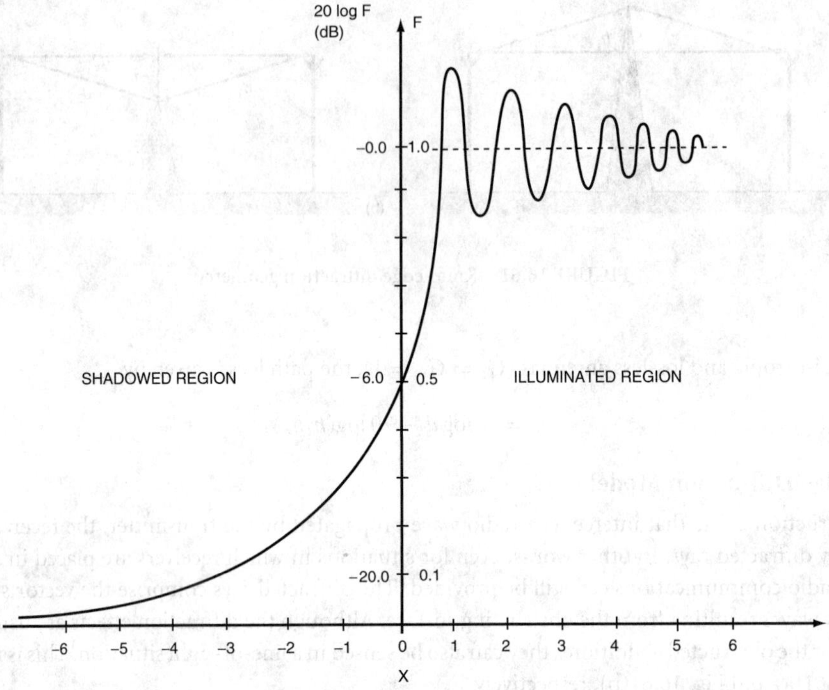

FIGURE 16.62   Knife-edge diffraction loss.

distances less than a few tens of kilometers, in which case the Earth curvature is negligible. Moreover, in general, for mobile radio applications, because the antennas' heights are small compared to the distance separating the transmitter and receiver, the grazing angle assumption is also reasonable. Equation (16.13) is a valid approximation for situations in which the loss due to reflection exceeds that of the free space, this occurring when $d > 12h_t h_r/\lambda$. Because, in practice, a perfectly flat terrain condition does not exist, a measure of smoothness S is utilized. This measure is given by the Rayleigh criterion as a function of the standard deviation σ of the mean heights of the terrain irregularities, the incidence angle θ of the ray, and the wavelength λ, such that

$$S = \frac{4\pi\sigma\sin\theta}{\lambda} \qquad (16.18)$$

A surface is considered to be smooth if $S < 0.1$, whereas it is considered as rough if $S > 10$.

Knife-Edge Diffraction Model

An element positioned at any given point between transmitter and receiver is said to produce diffraction effects at the receiver if its height h obstructs the Fresnel zones in the concerned radio path. It is recognized that the first Fresnel zone bounds the volume contributing significantly to wave propagation. Therefore, an obstruction should be considered as a diffracting element if

$$h > \sqrt{\frac{\lambda d_1 d_2}{(d_1 + d_2)}} \qquad (16.19)$$

The simple knife-edge diffraction model applies to single obstructing elements only. In practical situations a radio path will certainly encounter several obstructions, and the exact computation of the total diffraction loss is cumbersome. Many approximate solutions are available, and their accuracy varies according to the application. The Bullington method (Bullington, 1947) replaces the obstructions by an equivalent edge found at the intersection point between the line connecting transmitter-first-obstruction-from-transmitter and that connecting receiver-first-obstruction-from-receiver. The Deygout method (Deygout, 1966) selects a dominant obstruction and calculates the corresponding loss. The losses due to the remaining knife edges are determined with respect to this dominant edge.

16.6.3 Field Strength Prediction Models

Generally speaking, the best known field strength prediction models currently available arose as a combination of some of the previously described analytical models with data obtained in field measurements. The resulting empirical models include several adjustment parameters, most of them lacking theoretical explanations, and their applicability is restricted to situations in which the experiments were performed. Predictions for situations other than the original ones must be exercised with care and are, usually, not recommended, unless further adjustments can be performed with field measurements.

Okumura Model

The Okumura model (Okumura et al., 1968) is based on field measurements taken in the Tokyo area. It provides an initial excess path loss estimate $A(f, d)$ with respect to the free-space loss L_F, as a function of the operating frequency f, ranging from 100 to 3000 MHz, and of the distance d, ranging from 1 to 100 km. This initial estimate has been obtained for a quasismooth terrain and for a base station antenna height h_{tb} of 200 m and a mobile station antenna height h_{rb} of 3 m. Correction factors are then introduced to account for conditions other than the initial ones. In particular, a gain $G(\text{area}, f)$ as a function of the frequency f can be encountered for *suburban area*, *quasiopen area*, and *open area*. Moreover, gains $G(h_t)$ and $G(h_r)$ due to base station and mobile station antenna heights, h_t and h_r, respectively, must be computed. Accordingly, the general path loss formulation L, in decibels, of the Okumura model is

$$L = L_F + A(f, d) - G(\text{area}, f) - G(h_t) - G(h_r) \tag{16.20}$$

Apart from the free-space loss, which can be calculated in analytical form, all of the other parameters in Eq. (16.20) are provided in graph form. The plots show that there is a gain ranging from 5 to 13 dB as the frequency goes from 100 to 3000 MHz due to a change from urban to suburban area. In the same way, the gain is reasonably constant and approximately equal to 12 dB due to a change from suburban to quasiopen area, whereas this is approximately 6 dB for a change from quasiopen to open area. As far as the antenna heights are concerned, the gain is 20 dB/decade for the base station antenna and 20 dB/decade for the mobile station antenna height if $3 < h_r < 10$ m, but 10 dB/decade if $h_r < 3$ m.

The Okumura model is probably one of the most complete prediction models and has become a comparison standard for other models.

Hata Model

Because the Okumura model is totally based on information provided by empirical curves, its automated version can become cumbersome. Hata (1980) has proposed an empirical formulation that fits the Okumura curves within a useful range of application. Within such a range the Hata model gives predictions almost indistinguishable from those provided by Okumura. The general path loss L of the Hata formulation, in decibels, for distances d, in kilometers, and base and mobile antenna heights h_t and h_r, in meters, within the range

$$150 \leq f \leq 1000 \, \text{MHz}$$
$$30 \leq h_t \leq 200 \, \text{m}$$
$$1 \leq h_r \leq 10 \, \text{m}$$
$$1 \leq d \leq 20 \, \text{km}$$

is as follows:

Urban area

$$L_{\text{urban}} = 69.55 + 26.16 \log f - 13.82 \log h_t - A(h_r) + (44.9 - 6.55 \log h_t) \log d \qquad (16.21)$$

where $A(h_r)$ is a correction factor computed according to the size of the city.

For a small or medium-sized city and $1 \leq h_r \leq 10$ m

$$A(h_r) = (1.1 \log f - 0.7)h_r - (1.56 \log f - 0.8) \qquad (16.22)$$

For a large city and $f \leq 200$ MHz

$$A(h_r) = 8.29 \log^2(1.54h_r) - 1.1 \qquad (16.23)$$

For a large city and $f \geq 400$ MHz

$$A(h_r) = 3.2 \log^2(11.75h_r) - 4.97 \qquad (16.24)$$

Suburban area

$$L = L_{\text{urban}} - 2 \log^2(f/28) - 5.4 \qquad (16.25)$$

Rural (quasiopen) area

$$L = L_{\text{urban}} - 4.78 \log^2(f) + 18.33 \log f - 35.94 \qquad (16.26)$$

Rural area (open area)

$$L = L_{\text{urban}} - 4.78 \log^2(f) + 18.33 \log f - 40.94 \qquad (16.27)$$

COST 231-Hata Model

This model (ETR 103 1995) has been developed by the European Cooperation in the Field of Scientific and Technical Research (Euro-Cost), COST 231 subgroup on propagation models. It has been derived from Okumura's propagation curves in the upper frequency band. It is, therefore, an extension of the Hata model for frequencies f in the range $1500 \leq f \leq 2000$ MHz. The application of this model is restricted to large and small cells.

The parameters considered here are the same as those defined previously in the subsection on the Hata model. The basic loss for rural areas are the those expressed in Eq. (16.26) and Eq. (16.27). The path loss is as follows:

Urban area

$$L = 46.3 + 33.9 \log f - 13.82 \log h_t - A(h_r) + (44.9 - 6.55 \log h_r) \log d \qquad (16.28)$$

where $A(h_r)$ is a correction factor computed according to the urban environment.

For medium-sized and suburban centers with moderate tree density

$$A(h_r) = (1.1 \log f - 0.7)h_r - (1.58 \log f - 0.8) \qquad (16.29)$$

For metropolitan centers

$$A(h_r) = (1.1 \log f - 0.7)h_r - (1.58 \log f + 2.2) \qquad (16.30)$$

Lee Model

The Lee model (Lee, 1986) is based on the well-accepted assumption that the ratio between the powers received at two different distances from the base station is proportional to the ratio of these distances to the power of the path loss coefficient. Accordingly, the model is completely determined if the received power at a given distance and the path loss coefficient for the required environment are known. These two parameters are easily obtained, for a given set of basic initial conditions, by carrying out field measurements. Or, alternatively, by using data already available in the literature. Correction factors are then included in order to account for conditions other than the basic initial ones. There are four correction factors, and they are related to the base station and mobile station antenna heights in addition to their respective gains.

This constitutes a simpler version of the model, known as the *area-to-area mode*. The model is further enhanced by incorporating the loss due to diffraction L_k, or reflection L_r, as applicable. In such a situation it assumes a *point-to-point mode*, in which case the terrain profile information is taken into account for a more accurate prediction. Assuming a basic path loss L_b for the initial conditions h_{tb} and h_{rb} (representing the base and mobile antenna heights) and G_{tb}, and G_{rb} (their respective gains), the total path loss L at a distance d and for the new base and mobile antenna heights h_t and h_r and gains G_t and G_r, respectively, is given by

$$L = L_b + 10\alpha \log\left(\frac{d}{d_b}\right) - C + m(oL_k - \bar{o}L_r) \qquad (16.31)$$

where L, L_b, L_k, and L_r are given in decibels, α is the path loss coefficient, m is a binary parameter denoting a point-to-point mode ($m = 1$) or an area-to-area mode ($m = 0$), o is a binary parameter denoting obstructive path ($o = 1$) or nonobstructive path ($o = 0$) (the bar implies the NOT binary operation), d_b is an initial distance, and

$$C = 10\log\left[\left(\frac{h_t}{h_{tb}}\right)^2 \left(\frac{h_r}{h_{rb}}\right)^x \left(\frac{G_t}{G_{tb}}\right)\left(\frac{G_r}{G_{rb}}\right)\right]$$

is the correction factor, with $x = 2$ for $h_r > 10$ m or $x = 1$ for $h_r < 10$.

Some values of L_b, for a transmission power of 10 W (40 dB(1 mW)) and for the basic initial conditions of $d_b = 1$ mi (1.6 km), $h_{tb} = 100$ ft (30.48 m), $h_{rb} = 10$ ft (3 m), $G_{tb} = 4$ (6 dB) with respect to $\lambda/2$ dipole, and $G_{rb} = 1$ (0 dB) also with respect to $\lambda/2$ dipole, are given in Table 16.17. Also shown are the values of α for the respective environment.

The loss due to diffraction may be computed by any of the methods available. On the other hand, the loss due to reflection is determined as

$$L_r = 20\log\left(\frac{h_{te}}{h_t}\right) \qquad (16.32)$$

where h_{te} is the effective base station antenna height, representing the imaginary height the base station antenna should have in order to touch the ground extended from the reflection point that is closer to the mobile station, as shown in Fig. 16.63. Accordingly, this can either represent a real loss ($h_{te} < ht$) or a gain ($h_{te} > h_t$).

TABLE 16.17   One-Mile Intercept Path Losses and
Path Loss Coefficients for Different Environments

Environment	L_b, dB	α
Free space	85	2.0
Open area	89	4.35
Suburban area	101.7	3.84
Urban area		
Newark	104	4.31
Philadelphia	110	3.68
Tokyo	124	3.05
New York	117	4.8

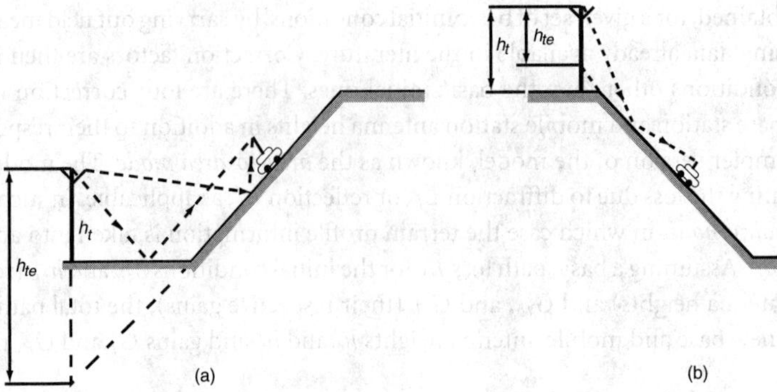

FIGURE 16.63 Effective antenna height (Lee model).

COST 231–Walfish–Ikegami Model

The COST 231 subgroup on propagation models (ETR 103 1995) has developed a prediction method in which street canyon propagation and radio paths obstructed by buildings are treated separately. It is based on measurements performed in the city of Stockholm, for the line-of-sight case, and on a combination of the Walfish and Bertoni (1988) and Ikegami et al. (1984) models, for the obstructed path situation. The model is applicable to large and small as well as micro cells for frequencies f, base station antenna height h_t, mobile station antenna height h_r, and distances d, within the following range:

$$800 \leq f \leq 2000 \, \text{MHz}$$
$$4 \leq h_t \leq 50 \, \text{m}$$
$$1 \leq h_r \leq 3 \, \text{m}$$
$$0.02 \leq d \leq 5 \, \text{km}$$

The Line-of-Sight (Street Canyon) Case

The path loss is given by

$$L = 42.6 + 26 \log d + 20 \log f \tag{16.33}$$

where $d \geq 0.020$ is given in kilometers, and f in megahertz.

The Obstructed Case

In the obstructed case the model also considers as parameters (1) the mean height of the buildings h_B, (2) the road width r_w, (3) the mean distance between the buildings d_B, and (4) the road orientation with respect to the direct radio path φ, with h_B, r_w, and d_B in meters and φ in degree. The total loss L, in decibels, is composed by three terms, all in decibels; namely, the free-space loss L_F, the roof top-to-street diffraction and scatter loss L_D, and the multiscreen loss L_S, as

$$L = L_F + L_D + L_S \tag{16.34}$$

The free-space loss is given by Eq. 16.10. The other losses are detailed next. The roof top-to-street diffraction and scatter loss is

$$L_D = -16.9 - 10 \log r_w + 10 \log f + 20 \log(h_B - h_r) + L_\varphi \tag{16.35}$$

where

$$L_\varphi = -10 + 0.354\varphi \qquad 0 \leq \varphi < 35°$$
$$L_\varphi = 2.5 + 0.075(\varphi - 35) \qquad 35° \leq \varphi < 55°$$
$$L_\varphi = 4.0 - 0.114(\varphi - 55) \qquad 55° \leq \varphi < 90°$$

If Eq. (16.35) gives a negative value, then L_D is assumed to be equal to zero. The multiscreen diffraction loss is

$$L_S = -18 \log(1 + h_t - h_B) - 9 \log d_B + k_a + k_d \log d + k_f \log f \qquad (16.36)$$

where the first term after the equal sign is assumed to be zero if $h_t - h_B \leq 0$ and

$$
\begin{aligned}
k_a &= 54 & h_t - h_B &> 0 \\
k_a &= 54 - 0.8(h_t - h_B) & d &\geq 0.5 \text{ and } h_t - h_B \leq 0 \\
k_a &= 54 - 0.8(h_t - h_B)\frac{d}{0.5} & d &< 0.5 \text{ and } h_t - h_B \leq 0 \\
k_d &= 18 & h_t - h_B &> 0 \\
k_d &= 18 - 15\left[\frac{(h_t - h_B)}{h_B}\right] & h_t - h_B &\leq 0 \\
k_f &= -4 + 0.7\left(\frac{f}{925} - 1\right) & & \text{for medium-sized cities and suburban centers with moderate} \\
& & & \text{tree densities} \\
k_f &= -4 + 1.5\left(\frac{f}{925} - 1\right) & & \text{for metropolitan centers}
\end{aligned}
$$

In case detailed data on the structure of buildings and roads are not available, the following default values are recommended: $d_B = 20, \ldots, 50$ m, $r_w = d_B/2$, $h_B = 3 \times$ (number of floors) + roof height, in meters, and $\varphi = 90°$. The roof height is assumed 3 m for pitched roofs, or 0 m for flat roofs.

16.6.4 Statistical Distributions of the Received Envelope

The field prediction models give the mean value of the received power as a function of parameters that more strongly influence the propagation of the mobile radio signal. However, a countless number of variables exist that deviate the instantaneous signal from its mean value. For instance, different obstructions may attenuate, diffract, reflect, or scatter the signal differently, and an infinitude of obstacles are encountered by the signal on its way from transmitter to receiver. Accordingly, because it is not viable or practicable to take all of the phenomena influencing the mobile radio propagation into account the received signal is treated on a statistical basis.

The random variation in the signal level characterizes the fading phenomenon. There are two types of fading: (1) long-term or **slow fading** and (2) short-term or **fast fading**. The slow fading is related to the variability of the area mean signal level due to shadowing. The fast fading is associated with the variability of the local mean signal level due to multipath propagation.

In this section we describe the several distributions that better characterize the variability of received envelope. All of these distributions have proved to be satisfactory fits to data obtained by field measurements.

Lognormal Distribution

The lognormal distribution describes the long-term variation of the received envelope. The corresponding probability density function is given by

$$p(R) = \frac{1}{\sqrt{2\pi}\sigma_R} \exp\left[-\frac{1}{2}\left(\frac{R - M_R}{\sigma_R}\right)^2\right] \qquad (16.37)$$

where R is the signal envelope (the local mean), M_R is the area mean, and σ_R is the standard deviation, usually in the range 5–12 dB, all given in decibels.

Rayleigh Distribution

The Rayleigh distribution describes the short-term variation of the received envelope in a multipath propagation environment where the scattered waves are assumed to be uniformly distributed around the receiver. The corresponding probability density function is given by

$$p(r) = \frac{r}{\sigma_r^2} \exp\left(-\frac{r^2}{2\sigma_r^2}\right) \tag{16.38}$$

where r is the signal envelope and $\sqrt{\pi/2}\sigma_r$ is the local mean.

Rice Distribution

The Rice distribution describes the short-term variation of the received envelope in a multipath environment where a line of sight is also present. The corresponding probability density function is given by

$$p(r) = \frac{r}{\sigma_r^2} \exp\left(-\frac{r^2 + a^2}{2\sigma_r^2}\right) I_0\left(\frac{ar}{\sigma_r^2}\right) \tag{16.39}$$

where r is the signal envelope, $a^2/2$ is the power of the direct wave, σ_r^2 is the power of the scattered waves, and

$$I_0\left(\frac{ar}{\sigma_r^2}\right) = \frac{1}{2\pi} \int_0^{2\pi} \exp\left(\frac{ar\cos\theta}{\sigma_r^2}\right) d\theta$$

is the modified zeroth-order Bessel function. Note that if $a = 0$ in Eq. (16.39) the Rice distribution degenerates into the Rayleigh distribution.

Nakagami Distribution

The Nakagami distribution describes the short-term variation of the received envelope in a multipath environment where the scattered waves are assumed to reach the receiver in clusters. The corresponding probability density function is given by

$$p(r) = \frac{2}{\Gamma(m)} \left(\frac{m}{\Omega}\right)^m r^{2m-1} \exp\left(-\frac{m}{\Omega}r^2\right) \tag{16.40}$$

where r is the signal envelope, $\Omega = E(r^2)$ is the mean square value, $m = \Omega^2/\text{Var}[r^2] \geq 1/2$ is the fading factor, and $\Gamma(m) = \int_0^\infty x^{m-1} \exp(-x)dx$ is the gamma function.

The Nakagami distribution reduces to the Rayleigh distribution for $m = 1$. Moreover, the Nakagami and the Rice distributions are approximately coincident for the following relation:

$$k = \frac{\sqrt{m^2 - m}}{m - \sqrt{m^2 - m}} \quad m > 1$$

where $k = a^2/2\sigma_r^2$ is the Rician factor.

Combined Distributions

The overall variability of the signal, in which slow fading and fast fading are combined, can be described through several distributions. The basic idea used to obtain these distributions is to assume the mean value of the fast fading distributions to be lognormally distributed. Therefore, the overall density is calculated as

$$p(r) = \int_{-\infty}^{\infty} p(r|R)p(R)dR \tag{16.41}$$

where $p(r|R)$ is one of the fast fading densities given by Eqs. (16.38), (16.39), or (16.40), $p(R)$ is the lognormal density given by Eq. (16.37), and R is the mean value of the corresponding densities given in a logarithmic form. The respective densities obtained are named shadowed-Rayleigh density, shadowed-Rice density, or shadowed-Nakagami density. In particular, the shadowed-Rayleigh density is also known as the Suzuki density (Suzuki, 1977). Unfortunately, it does not seem to be possible to put these densities in a closed-form expression. Accordingly, they appear in their integral form and are computed by numerical methods.

16.6.5   Radio Coverage

System specifications require that an adequate reception must be accomplished anywhere within the mobile radio service area; 100% coverage could be achieved if the phenomena affecting the radio propagation in a mobile radio environment were deterministic. The received signal is a random variable, however, and the problem of radio coverage must be treated on a statistical basis, such that the outcomes of the prediction calculations should be interpreted as random events occurring with a given probability. Therefore, coverage requirements are specified as the proportion of locations where the received signal must be greater than a certain threshold considered to be satisfactory.

Suppose that at a given distance from the base station the *mean signal strength* is considered to be known. We want to determine the service area radius such that the mobiles experience a received signal above a certain threshold with a stipulated probability. The mean signal strength can be determined either by any one of the prediction models quoted previously or by field measurements. In the same way, the corresponding statistics are those mentioned before. In the calculations that follow we assume a lognormal environment, in which case only the large-scale variations are accounted for. The other environments can be analyzed in a like manner (Yacoub, 1993; Leonardo and Yacoub, 1993a, 1993b), although this may not be of interest if some sort of diversity is implemented because, in this case, the effects of the small-scale variations are minimized.

Propagation Model

Define m_w and k as the mean powers at distances x and x_0, respectively, such that

$$m_w = k \left(\frac{x}{x_0} \right)^{-\alpha} \tag{16.42}$$

where α is the path loss coefficient. Expressed in decibels, $M_w = 10 \log m_w$, $K = 10 \log k$ and

$$M_w = K - 10\alpha \log \left(\frac{x}{x_0} \right) \tag{16.43}$$

Define the received power as $w = r^2/2$, where r is the received signal envelope. Let $p(W)$ be the probability density function of the power W, where $W = 10 \log w$. In a lognormal environment r has a lognormal distribution and

$$p(W) = \frac{1}{\sqrt{2\pi}\,\sigma_w} \exp \left(-\frac{(W - M_w)^2}{2\sigma_w^2} \right) \tag{16.44}$$

where M_W is the mean and σ_w is the standard deviation, all given in decibels. Define w_T and $W_T = 10 \log w_T$ as the threshold above which the received signal is considered to be satisfactory. The probability that the received signal is below this threshold is its *probability distribution function, $P(W_T)$*, such that

$$P(W_T) = \int_{-\infty}^{W_T} p(W) dW = \frac{1}{2} + \frac{1}{2} \text{erf} \left[\frac{(W_T - M_W)^2}{2\sigma_w^2} \right] \tag{16.45}$$

where $\text{erf}(\cdot)$ is the error function, defined as

$$\text{erf}(y) = \frac{2}{\sqrt{\pi}} \int_0^y \exp(-t^2) dt \tag{16.46}$$

Base Station Coverage

The problem of estimating the service area can be approached in two ways. In the first approach, we may wish to determine the proportion β of locations at x_0 where the received signal power w is above the threshold power w_T. In the second approach we may estimate the proportion μ of the circular area defined by x_0 where the signal is above this threshold. In the first case this proportion is averaged over the perimeter of the circumference (cell border), whereas in the second approach the average is over the circular area (cell area).

The proportion β equals the probability that the signal at x_0 is greater than this threshold. Hence

$$\beta = \text{prob}(W \geq W_T) = 1 - P(W_T) \tag{16.47}$$

Using Eq. (16.43) and Eq. (16.45) in Eq. (16.47) we obtain

$$\beta = \frac{1}{2} - \frac{1}{2}\text{erf}\left[\frac{W_T - K + 10\alpha \log(x/x_0)}{\sqrt{2}\sigma_w}\right] \tag{16.48}$$

This probability is plotted in Fig. 16.64, for $x = x_0$ (cell border).

Let $\text{prob}(W \geq W_T)$ be the probability of the received power W being above the threshold W_T within an infinitesimal area dS. Accordingly, the proportion μ of locations within the circular area S experiencing such a condition is

$$\mu = \frac{1}{S} \int_S [1 - P(W_T)]ds \tag{16.49}$$

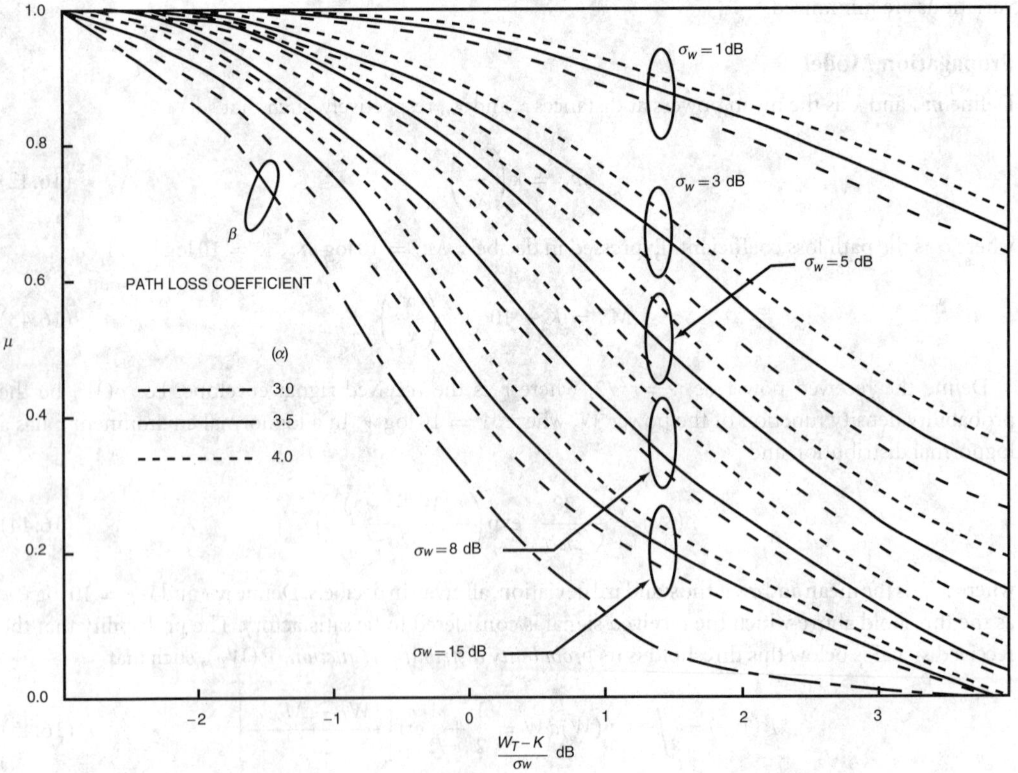

FIGURE 16.64 Proportion of locations where the received signal is above a given threshold (the dashdot line corresponds to the β approach and the other lines to the μ approach).

where $S = \pi x_0^2$ and $dS = x\, dx\, d\theta$. Note that $0 \le x \le x_0$ and $0 \le \theta \le 2\pi$. Therefore, solving for $d\theta$, we obtain

$$\mu = 2 \int_0^1 u\beta\, du \tag{16.50}$$

where $u = x/x_0$ is the normalized distance.

With Eq. (16.48) in Eq. (16.50) we have the result

$$\mu = 0.5\left\{ 1 + \mathrm{erf}(a) + \exp\left(\frac{2ab+1}{b^2}\right)\left[1 - \mathrm{erf}\left(\frac{ab+1}{b}\right)\right]\right\} \tag{16.51}$$

where $\alpha = (K - W_T)/\sqrt{2}\sigma_w$ and $b = 10\alpha \log(e)/\sqrt{2}\sigma_w$.

These probabilities are plotted in Fig. 16.64 for different values of standard deviation and path loss coefficients.

Application Examples

From the theory that has been developed it can be seen that the parameters affecting the probabilities β and μ for cell coverage are the path loss coefficient α, the standard deviation σ_w, the required threshold W_T, and a certain power level K measured or estimated at a given distance x_0 from the base station.

The applications that follow are illustrated for two different standard deviations: $\sigma_w = 5$ dB and $\sigma_w = 8$ dB. We assume the path loss coefficient to be $\alpha = 4$ (40 dB/decade), the mobile station receiver sensitivity to be -116 dB (1 mW), and the power level estimated at a given distance from the base station as being that at the cell border, $K = -102$ dB (1 mW). The receiver is considered to operate with a SINAD of 12 dB for the specified sensitivity. Assuming that cochannel interference levels are negligible and given that a signal-to-noise ratio S/N of 18 dB is required, the threshold W_T will be -116 dB (1 mW) $+ (18 - 12)$ dB (1 mW) $= -110$ dB (1 mW).

Moreover, three cases will be explored, as follows:

Case 1: We want to estimate the probabilities β and μ that the received signal exceeds the given threshold, first, at the border of the cell (β approach) and, second, within the area delimited by the cell radius (μ approach).

Case 2: It may be interesting to estimate the cell radius x_0 such that the received signal be above the given threshold with a given probability (say 90%), first, at the perimeter of the cell and, second, within the cell area. This problem implies the calculation of the mean signal strength K at the distance x_0 (the new cell border) of the base station. Given K and given that at a distance x_0 (the former cell radius) the mean signal strength M_w is known (note that in this case $M_w = -102$ dB(1 mW)) the ratio x_0/x can be estimated.

Case 3: To fulfill the coverage requirement, rather than calculating the new cell radius, as in case 2, a signal strength at a given distance can be estimated such that a proportion of the locations at this distance (β approach) or within the area delimited by this distance (μ approach) will experience a received signal above the required threshold. This corresponds to calculating the value of the parameter K already carried out in case 2 for the various situations.

The calculation procedures are now detailed for $\sigma_w = 5$ dB. Results are also shown for $\sigma_w = 8$ dB.

Case 1

Using the given parameters we obtain $(W_T - K)/\sigma_w = -1.6$. With this value in Fig. 16.64 we obtain Table 16.18.

TABLE 16.18 Probability That the Received Signal Exceeds -116 dB (1 mW) for $S/N = 18$ dB Given That at the Cell Border the Mean Signal Power is -102 dB (1 mW)

Standard Deviation, dB	β Approach (border coverage), %	μ Approach (area coverage), %
5	97	100
8	84	95

TABLE 16.19 Normalized Radius of a Cell Where the Received Signal Power is Above -116 dB (1 mW) with 90% Probability for $S/N = 18$ dB, Given That at a Reference Distance from the Base Station (the Cell Border) the Received Mean Signal Power is -102 dB (1 mW)

Standard Deviation, dB	β Approach (border coverage)	μ Approach (area coverage)
5	1.10	1.38
8	0.88	1.27

Note, from Table 16.18, that the signal at the cell border exceeds the receiver sensitivity with 97% probability for $\sigma_w = 5$ dB and with 84% probability for $\sigma_w = 8$ dB. If, on the other hand, we are interested in the area coverage rather than in the border coverage, then these figures change to 100% and 95%, respectively.

Case 2

From Fig. 16.64, with $\beta = 90\%$ we find $(W_T - K)/\sigma_w = -1.26$. Therefore, $K = -103.7$ dB(1 mW). Because $M_w - K = -10\alpha \log(x/x_0)$, then $x_0/x = 1.10$. Again, from Fig. 16.64, with $\mu = 90\%$ we find $(W_T - K)/\sigma_w = -0.48$, yielding $K = -107.6$ dB (1mW). Because $M_w - K = -10\alpha \log(x/x_0)$, then $x_0/x = 1.38$. These results are summarized in Table 16.19.

Note, from Table 16.19, that in order to satisfy the 90% requirement at the cell border the cell radius can be increased by 10% for $\sigma_w = 5$ dB. If, on the other hand, for the same standard deviation the 90% requirement is to be satisfied within the cell area rather than at the cell border, a substantial gain in power is achieved. In this case the cell radius can be increased by a factor of 1.38. For $\sigma_w = 8$ dB and 90% coverage at the cell border the cell radius should be reduced to 88% of the original radius. For area coverage an increase of 27% of the cell radius is still possible.

Case 3

The values of the mean signal power K are taken from case 2 and shown in Table 16.20.

TABLE 16.20 Signal Power at the Cell Border such that 90% of the Locations Will Experience a Received Signal Above -116 dB (1 mW) for $S/N = 18$ dB

Standard Deviation, dB	β Approach, dB (1 mW) (border coverage)	μ Approach, dB (1 mW) (area coverage)
5	-103.7	-107.6
8	-99.8	-106.2

16.6.6   Summary and Conclusions

A precise characterization of the signal variability in the mobile radio environment constitutes a hard task. Deterministic methods, such as those described by the free-space, plane earth, and knife-edge diffraction models, are restricted to very simple applications.

Generally speaking, the best known field strength prediction models currently used arose as a combination of the deterministic methods with data obtained in field measurements. They include several adjustment parameters, most of them lacking theoretical explanations, and their applicability is restricted to situations in which the experiments were performed. Predictions for situations other than the original ones must be exercised with care.

Because it is not viable or practicable to take all of the phenomena influencing the mobile radio propagation into account the received signal is treated on a statistical basis. Several distributions exist that well characterize the variability of the received envelope and they have proved to have a satisfactory fit to data obtained by field measurements.

Hence, the problem of radio coverage must be treated on a statistical basis, such that the outcomes of the prediction calculations should be interpreted as random events occurring with a given probability. Coverage requirements are specified as the proportion of locations where the received signal must be greater than a certain threshold considered to be satisfactory.

Defining Terms

Fast fading (also short-term fading): Fading associated with the variability in the local mean signal level due to multipath propagation.

Large cells and small cells: Cells having their base station antennas installed above the roof tops. Large and small cells differ in maximum range, which is 1–3 km for the latter.

Microcells: Cells having their base station antennas installed below the roof tops. Microcells have a maximum range of 0.5–1 km.

Path loss: The ratio between the transmitted power and the received power.

Slow fading (also long-term fading): Fading related to the variability of the area mean signal level due to shadowing.

References

Bullington, K. 1947. Radio propagation at frequencies above 30 megacycles. *Proc. IRE* 35(10):1122–1136.

Deygout J. 1966. Multiple knife-edge diffraction of microwaves. *IEEE Trans. Antennas and Propagation* AP-14(4):480–489.

ETR 103. 1995. European digital cellular telecommunication system (phase 2); Radio network planning aspects (GSM 03.30), ETSI Technical Report, ETR 103, Feb.

Hata, M. 1980. Empirical formula for propagation loss in land-mobile radio services. *IEEE Trans. Vehicular Tech.* VT-29(3):317–325.

Ikegami, F., Yoshida, S., Takeuchi, T., and Umeriha, M. 1984. Propagation factors controlling mean field strength on urban streets. *IEEE Trans. Antennas and Propagation* AP-32(8):822–829.

Lee, W.C.Y. 1986. *Mobile Communications Design Fundamentals*. Howard W. Sams, Indianapolis, IN.

Leonardo, E.J. and Yacoub, M.D. 1993. (Micro)Cell coverage area using statistical methods. In *Proceedings of the IEEE Global Telecommunications Conference*, GLOBECOM'93 pp. 1227–1231. (Houston, TX), Dec.

Leonardo, E.J. and Yacoub, M.D. 1993. A statistical approach for cell coverage area in land mobile radio systems. *Proc. 7th IEE Conf. on Mobile and Personal Comm.* pp. 16–20. (Brighton, UK), Dec.

Okumura, Y., Ohmori, E., Kawano, T., and Fukuda, K. 1968. Field strength and its variability in VHF and UHF land mobile service. *Rev. Elec. Comm. Lab* 16 (Sept.–Oct.):825–873.

Suzuki, H. 1977. A statistical model for urban radio propagation. *IEEE Trans. Comm.* 25(7):673–680.

Walfish, J. and Bertoni, H.L. 1988. Theoretical model of UHF propagation in urban environments. *IEEE Trans. Antennas and Propagation* AP-36(12):1788–1796.

Yacoub, M.D. 1993. *Foundations of Mobile Radio Engineering*. CRC Press, Boca Raton, FL.

Further Information

The fundamentals of mobile radio engineering in connection with many practical examples and applications as well as an overview of the main topics involved can be found in Yacoub, M.D. 1993. *Foundations of Mobile Radio Engineering*. CRC Press, Boca Raton, FL.

Several periodicals cover this subject in depth. Some are as follows:

IEEE Journal on Selected Areas in Communications
IEEE Personal Communication Magazine
IEEE Transactions on Antennas and Propagation
IEEE Transactions on Communications
IEEE Transactions on Vehicular Technology

The Vehicular Technology Society and the Communications Society are very active in promoting conferences in the area.

16.7 Cellular Radio

Harry E. Young

16.7.1 Cellular Service Evolution

In 1946, two-way mobile service was introduced in St. Louis. The service had hardly begun when the inherent disadvantages of the two-way mobile service were recognized. The major disadvantage, of course, was the tremendous physical separation required for cochannel interference reasons. In 1947, Bell Telephone Laboratories began exploring a concept that would reuse frequencies by utilizing small cells with low-powered mobile units. These cells could be linked together using a computer to permit mobility while simultaneously greatly increasing the number of subscribers that could be served by the system. However, the necessary computer technology did not become available until the mid-1960s when electronic switching systems were developed.

In spite of the fact that the cellular concept was developed in the U.S., other countries introduced cellular as a commercial service before the U.S. This was largely due to the fact that it took almost 13 years from the time cellular service was technically possible until the first commercial cellular systems were actually activated in 1983. This delay was essentially due to the regulatory process in the U.S., and a study by AT&T later concluded that the U.S. economy lost about $86 billion due to this regulatory delay. Consequently, commercial cellular service first became available in Japan in 1979, quickly followed by systems in the Scandinavian countries in 1980 and 1981. Many other countries have since implemented cellular service and it is now available in many countries throughout the world.

There is no worldwide compatibility for cellular service, however. The system in the U.S. is based on the original AT&T design, **advanced mobile phone service (AMPS)**. AMPS is used in approximately 70 other countries in the world and is the most widely used standard at this time. Some of the other more common standards used for cellular service include **nordic mobile telephone (NMT)**, total access communications service (TACS), and **global system for mobile (GSM)** communications.

The original NMT was designed for rather small systems of 20–30,000 subscribers and provides 180 channels, each of which utilizes 25 or 30 kHz of bandwidth in the 450-MHz band. A later version of NMT provides greater capacity in the 900-MHz band. NMT 900 provides 1000 channels, each using 25-kHz of bandwidth, or 2000 channels of 12.5 kHz each. NMT, in one form or another, is used in about 30 countries in the world.

TACS is used in Europe, the United Kingdom, and about two dozen other countries. A specialized form of TACS is used in Japan and is called Japan Total Access Communications System (**JTACS**). TACS provides 1320 channels of 25 kHz each.

Because different cellular standards were used by various countries in Europe, a single standard was required to provide roaming capability. Thus, the GSM system has evolved as an all-digital service that is used in Europe and many other countries. GSM is designed to work in the 900-MHz band and provides eight digital time slots for every 200-kHz channel.

Not only do the various standards use different protocols, they operate on different frequencies so that worldwide compatibility is not possible between the different systems.

16.7.2   Regulation and Competition

To promote competition, the **Federal Communications Commission (FCC)** issued two cellular licenses for each defined geographical area. These geographical areas are known as a **metropolitan statistical area (MSA)** and **rural service area (RSA)**. One of these licenses was awarded to a subsidiary of a company that provides landline telephone service in the given geographical area and is known as the *wireline* licensee, or B band. The other was issued to a company not associated with the landline **local exchange carrier (LEC)** in that area and is called the *nonwireline* licensee, or A band. Because of mergers and acquisitions in recent years, this distinction has become confusing because wireline companies often operate as nonwireline carriers outside of the franchised areas of their landline LEC.

A cellular licensee may serve all or just a portion of a designated MSA or RSA. Usually most of a MSA is served, although large portions of a RSA may be left unserved. The actual service area is defined by the cellular carrier and is called the **cellular geographic service area (CGSA)**. Debate continues as to whether the FCC's effort to promote competition has been successful because there is often little difference in the prices charged by the wireline and nonwireline licensees or in the quality of service. Nonetheless, the overall price of cellular service has continued to drop and the number of subscribers has increased dramatically since service began in 1983. As of January 1996, there were approximately 34 million cellular subscribers in the U.S. and it is expected that there will be almost 65 million by the end of 2005.

16.7.3   Cellular Radio Spectrum Requirements

Each cellular licensee in a MSA or RSA is authorized by the FCC to use 25 MHz of spectrum within the 800-MHz range. Originally, the FCC allocated 20 MHz to each carrier, which was obtained by deleting UHF television channels 70–83 because they were lightly used. An additional 5 MHz per carrier was authorized in 1985 due to growing capacity problems in some areas. This 25 MHz of spectrum is divided for use by base and mobile stations within each system so that different frequencies are used for base-to-mobile and mobile-to-base. This type of operation is called *full duplex* and permits simultaneous conversation or transmission. It must be remembered that two channels are required for each conversation in order to operate on a full-duplex basis.

The AMPS cellular systems in the U.S. are based on an analog technology that employs frequency modulation (FM). With this technology, each channel occupies 30 kHz of bandwidth yielding a total of 832 channels, or 416 channel pairs, for each cellular operator. Of these 416 channel pairs, 21 are dedicated to control functions for call setup and handoff purposes. This leaves 395 channel pairs available for voice services.

Existing cellular systems use **frequency division multiple access (FDMA)** multiplexing, which divides the total band allocated to a cellular operator (25 MHz) into discrete channels. Since each channel has a bandwidth of 30 kHz, the system has a total of 832 channels available. Each conversation requires the use of two pairs of frequencies, and so there are 416 pairs of frequencies available to each operator, each of which can potentially be assigned to a cellular user at any given time. Mobile equipment that utilizes FDMA is less complex than that used with other multiplexing techniques and is generally less expensive. Because each channel requires the use of a separate transmitter and receiver, however, FDMA needs considerably more equipment at the nonmobile, or *base station* site. FDMA can be used with analog or digital transmission systems.

Introduction of Digital Cellular

Initially, the development of digital cellular was the result of major concerns by the cellular carriers about their ability to meet a rapidly growing subscriber base. With growth rates of over 40%, it did not appear that existing analog technology would be able to cope with the projected numbers of subscribers within a few years.

In an effort to hasten the development of digital cellular, the Cellular Telecommunications Industry Association (CTIA) issued a **user performance requirements (UPR)** document in 1988 that outlined the service requirements. Key elements of the UPR were

- A tenfold increase in capacity
- Dual mode during an undefined transition period
- New features (such as calling number identification and improved security)
- Equipment to be available by 1991
- Standards for high quality

A request was made to the **Telecommunications Industry Association (TIA)**, which is a standards body, to develop a digital specification based on the UPR. In 1991, the TIA released Interim Standard 54 (IS-54), which uses **time division multiple access (TDMA)** technology. In 1993 the TIA also approved IS-95, a digital standard based on **code division multiple access (CDMA)** technology. Debate continues within TIA over a third technology, **extended time division multiple access (E-TDMA)**, which has been promoted by Hughes Network Systems.

With TDMA, each radio channel is divided into time slots. Each individual conversation is converted to a digital signal, which is then assigned to one of these time slots. The number of time slots per radio channel can vary, because it is a function of the system design. There are at least two time slots per channel, and usually more, which means that TDMA has the potential of serving several times more customers with the same amount of bandwidth as FDMA techniques.

TDMA is a more complex system than FDMA because the voice has to be digitized, or coded, then stored in a buffer for assignment to a vacant time slot, and finally transmitted. Consequently, signal transmission is not continuous and the transmission rate must be several times greater than the coding rate. Also, because more information is contained within the same bandwidth, TDMA equipment must have more sophisticated techniques of equalizing the received signal in order to maintain signal quality. TDMA offers improved handoffs between cells because of a mobile-assisted feature called **mobile assisted handoff (MAHO)**, improved sound quality, greatly improved security since the transmission is digitally encoded, and the possibility of new features such as receipt of caller identification. These advantages are somewhat offset by higher power consumption in the set, more weight and complexity to the set, and a higher unit cost of the set.

CDMA was developed by the military to prevent communications channels from being detected, and then jammed, by the enemy. The signal intelligence (speech) is converted to a digital signal, which is then mixed with a randomlike code. The total signal, speech plus the randomlike code, is then transmitted over a wide band of frequencies through *spread spectrum* techniques. It is difficult for an outside party to detect the radio signal and intelligence when it is spread over a wide bandwidth in a randomlike manner. There is the further potential for multiple radio systems to share the same spectrum with only limited interference.

Unlike FDMA and TDMA, the spread spectrum transmission used by CDMA requires channels that have a relatively immense bandwidth. For example, the channels in the IS-95 digital cellular standard require 1.25 MHz of bandwidth but other CDMA schemes require an even greater amount of bandwidth. However, it is theorized that CDMA could accommodate perhaps 10–20 times as many users for the same amount of total bandwidth as FDMA can. Until now, CDMA has been used for specialized purposes with a small numbers of users. A major question to be answered is whether one service, with tens of thousands of users employing CDMA devices, can function and not interfere with other wireless services.

CDMA provides the ability to keep an existing cell connection while the call is transferred to an adjacent cell. This feature, called soft handoff, is not available with TDMA or FDMA multiplexing. CDMA also allows the same frequencies to be used in adjacent cells, thus somewhat simplifying the system design. It also has the same disadvantages of TDMA in terms of sets that are more complex, require more power, and cost more. As of this writing production models of dual-mode CDMA and analog sets have just become available.

A finite limit for an analog cellular system is almost impossible to determine. Some additional capacities have been achieved by improving the throughput of the cellular switches. Other increases have resulted from increasingly expensive network solutions such as cell-splitting and the deployment of microcells that contain some call-processing abilities. In some instances, subscribers may simply be added to the system, which results in a reduction of service quality because the ability of the subscriber to access the network without being blocked is reduced. There is no magic number that determines when the capacity limits of the analog cellular system have been exceeded. There are simply too many variables, including market size, geography, usage patterns, and network design, to set a finite limit. The largest market in the U.S., Los Angeles, however, supported over 750,000 subscribers in 1993 without the implementation of digital. This implies that most markets in the country could support similar levels of subscribers without deploying digital cellular.

In spite of the fact that system capacity can be met using analog technology, increasing capacity by converting to digital can still be a more economical solution. The number of subscribers that may be served by a given amount of spectrum can be increased by a factor of 3 to almost 20, depending on the type of technology that is used. Capital costs are about $1000 per subscriber for an analog-only system. Comparable capital costs for a TDMA-based system may be as low as $247 per subscriber, whereas those for a CDMA-based system are predicted to be $109 per subscriber. Moreover, digital also permits the carrier to offer new subscriber features, increase battery life for a portable unit, improve sound quality, provide increased voice security, and improve handoffs. CDMA technology, in particular, should help enhance the data capabilities of the cellular systems.

16.7.4　Basic Cellular Architecture

Instead of having a single large, powerful transmitter to cover a geographical area, cellular places transceivers in numerous small cells within an area. The transceivers are controlled by a central processor, or switch, so that the subscriber can move between cells and still maintain service. This provides the frequency reuse capability that gives cellular the potential for far greater capacity than early two-way mobile systems.

The central processor, or switch, is also referred to as the **mobile switching center (MSC)**. The MSC was formerly referred to as the **mobile telephone switching office (MTSO)**. It provides the switching and radio control functions. Links extend from the MSC to the **cell sites** and also provide connections to the **public switched telephone network (PSTN)**.

Figure 16.65 illustrates a typical cellular system. In this diagram, the cell arrangements are shown in a hexagonal pattern. Cellular systems are engineered using this hexagonal arrangement in order to determine the frequency reuse pattern. This method is used to ensure enough separation exists between any two cells using the same frequencies so that interference does not occur. Typically, the system may be designed using a 4-cell, 7-cell, or 12-cell reuse pattern. If the world were flat and there were no obstructions, the radio waves could be confined to these hexagonal cells, but in reality, the radio signals can, at best, only approximate the hexagonal shape. Special measures are needed for areas that are difficult to serve with normal antennas, like tunnels. Still, the use of hexagonal cells is a useful engineering tool.

Cell Sites

Early cellular systems used cell sites that typically covered a radius of 10 miles or more. As traffic increased, or better coverage was needed in certain areas, the size of the cell sites decreased. Presently, the average cell

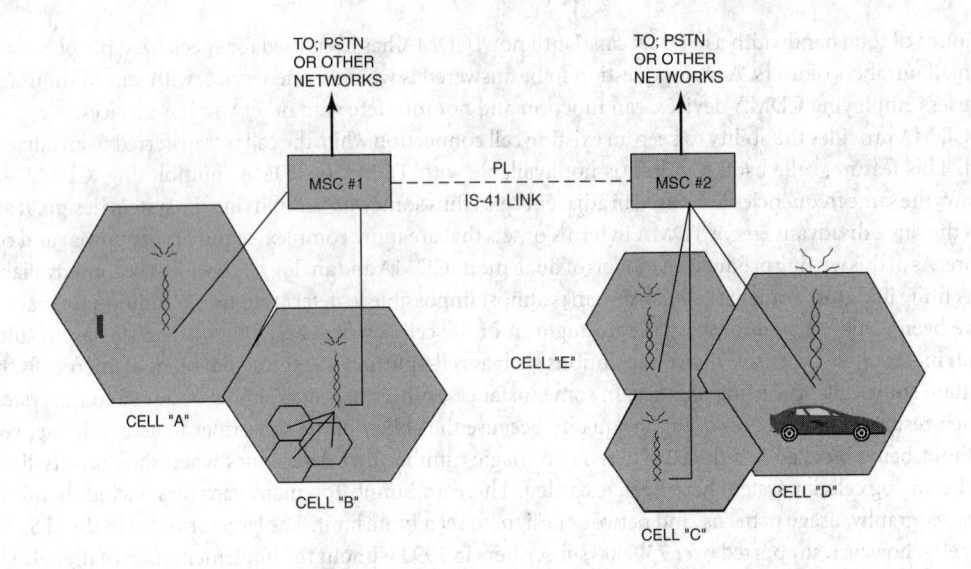

FIGURE 16.65   Configuration of a typical cellular system.

site has a radius of about 3–5 mi but this is also diminishing. The cells can range in size from a radius of less than 1 to 25 mi or more with the range being determined by the transmitted signal power and height of the antenna.

As traffic grows in a cellular system, additional channels and cells are added until all of the available spectrum is in service. Metropolitan areas may use all of the spectrum but RSAs rarely need a full complement of channels.

The number of transceivers in a given cell depends on the traffic that the cell is expected to handle. The maximum number of radios in any given cell site, however, depends on the reuse pattern used in the system design. This can be determined by dividing the total number of channels by the reuse pattern. For example, if the system is designed for a reuse pattern of $N = 4$, and the maximum number of voice channel pairs is 395, then the maximum number of channels in any given cell cannot exceed 98.

Definitions

Figure 16.65 displays several microcells that are distributed between two separate cellular systems. There are larger cells, such as cell site A and cell site B, as well as smaller cells, such as cell site C. In addition, there are very small cells within cell site B.

In any discussion of wireless systems employing a cellular-type architecture, the terms *cell site*, *macrocell*, *microcell*, and even *picocell*, are often used. A precise definition of each term is not possible because even standards bodies, such as the Telecommunications Industry Association, or industry associations such as CTIA or the Personal Communications Industry Association (PCIA) have not officially defined each term. Macrocells, microcells, and picocells are all considered to be cell sites in the cellular network architecture and the following definition does exist for a cell site.

A cell site is a transmitter/receiver location, operated by the cellular carrier (or **wireless carrier**) through which radio links are established between the cellular system and the mobile units. The area reliably served by a given cell site is referred to as a cell.

The other terms generally relate to the size of the cell site, and this is where the ambiguity of terms occurs. Determining the precise term to use based on the radio frequency (RF) coverage area of the cell is not of critical importance because the term cell site can always be used. To add confusion to matters, the term *base station* is often considered to be synonymous with the term cell site. Nonetheless, in spite of the lack of formal definition for the other terms, the following general definitions are usually accepted by most engineers or technicians in the industry:

- Macrocell: A cell site having an RF coverage area diameter of greater than 2000 ft.
- Microcell: A cell site having an RF coverage area diameter of greater than 400 ft but less than 2000 ft.
- Picocell: A cell site having an RF coverage area diameter of less than 400 ft.

In Fig. 16.65, all of the cell sites, except for the microcells, would probably qualify as macrocells. The cell sites within cell site B fit the definition for microcells.

Purpose of Microcells

Microcells are deployed in a network for a number of reasons, such as

- Increase system capacity: One of the methods used to increase capacity in a cellular system is to split the existing cells into smaller cells. Sometimes a large cell, or macrocell, is divided into several smaller cells. This, of course, increases the frequency reuse capability within the network to provide additional capacity.
- Improve signal reception: In many large buildings, penetration of the RF signal from an outdoor cell site location becomes very difficult. A similar problem exists with certain outdoor locations, such as tunnels or bridges. The use of microcells located within the building or tunnel provides a more direct route for the RF signal, especially those from lower powered portable units.
- Provide inconspicuous coverage: Cellular users expect their service to work in any location within their expected service area. Often, they are also among the more affluent residents of a community and are the first to protest any proposal to construct a cellular tower in their neighborhood. The use of a microcell provides a relatively inconspicuous means of correcting a service problem without constructing a large tower.
- Reduce handoff switching: In traditional cellular systems, most handoffs between cells are controlled by the MSC. Increasing the number of cells, as well as decreasing the size of the cells themselves, increases the load requirements of the MSC. Some microcells do not need to have handoffs controlled by the MSC, which can also help increase the overall system capacity.
- Provide temporary capacity: In some situations, such as sporting events or traffic congestion, additional capacity is needed in a certain location but only on a temporary basis. One method of achieving this capacity is to borrow capacity from an adjacent cell. This can be accomplished with a microcell connected to a donor cell.

Cellular Power Levels

By design, cellular systems operate at a lower signal power level than traditional two-way mobile systems. Base stations have a maximum power limitation of 100 W *effective radiated power* (ERP) except for those located in rural service areas, which can transmit as high as 500 W ERP. Mobile units designed for use in cars are capable of transmitting at 3 W ERP whereas portable, hand-held, units have a maximum power of 600 mW. Cellular is often perceived as a high-powered wireless service for vehicular traffic and sometimes viewed as not suitable for low-powered applications. These power levels, though, represents maximum levels. The actual levels for mobiles are automatically adjusted by instructions from the base stations. Mobiles can transmit using power levels as low as 7 mW. Base station levels are set according to system design parameters but are often much lower than the maximum of 100 W.

Number of Subscribers Served

The capacity of any wireless system is greatly affected by the number of voice channels that is available to serve subscribers. The number of subscribers that can utilize any given voice channel is a function of the traffic characteristics of the subscribers. These characteristics are the number of calls generated, as well as the length of each call. There are traffic tables available that can predict the number of channels required to serve a given load of traffic, measured in the quantity *erlang*. One erlang is equal to the continuous occupancy of one channel.

Cellular systems are engineered to provide a B.01 grade of service, that is, 1 of every 100 call attempts may be blocked. Using this grade of service, a system can typically accommodate 20–30 subscribers for each voice channel in a set. For an analog system, only one subscriber can use a voice channel at any one time, but multiple subscribers can share a single channel since they do not all use the channel at the same time.

Cell Splitting

To minimize interference, a certain distance must be maintained between cells using the same frequencies. However, this distance can be reduced without disturbing the cell reuse pattern. As the size of the cells are reduced, the same frequencies can be utilized in more cells, which in turn means more subscribers can be accommodated on the system. Particularly in congested areas, the cellular operator often splits an existing cell into two or more smaller cells. New transceivers are placed and the power of the transmitters are reduced in order to confine the signals to the newly created cells. For example, a cell that originally had a radius of 8 mi could be split into four cells with each new cell having a 2 mi radius. For the existing analog systems, cell splitting is an effective way to increase system capacity, although some practical limitations are reached. Suitable locations for cell sites becomes more difficult and the processing load on the switch rapidly increases because handoffs are more frequent.

Cell Site to MSC Links

Several methods may be used to connect cell sites of any type back to an MSC. The links may use two-wire or four-wire analog circuits, digital, fiber, or microwave facilities. The choice often depends on the type of microcell used, the facilities available in a given area, and of course, cost.

Two-wire analog facilities using copper wires are often used for microcells having small capacities, such as stand-alone applications. These systems do not need to be connected to the MSC or switch, but do require a normal line-side connection to the LEC's switch for access to the PSTN. This connection employs a standard, two-wire circuit, which usually uses copper wire that is part of the LEC's local loop. The copper wire is normally 22–26 gauge wire and the loss of the loop may be as high as 8 dB at 1000 Hz. Four-wire analog circuits are not as common now as they were when cellular was first introduced in the U.S. This is because the LECs have modernized their facilities in the last decade so that digital capabilities are more widely available. There are still some instances where digital is not available, or the voice channel requirements are very small, that a four-wire analog link is used.

Digital links are quite common between cell sites and the MSC or switching center. These links may be several miles in length or contained within a building. Generally, a **line** at the DS1 rate (1.544 Mb/s) is chosen for these applications, which requires some care in facility selection for proper performance. For intrabuilding applications, 22-gauge shielded wiring is the preferred choice in order to reduce the possibility of interference. Nonshielded wire in other sizes will work but are more susceptible to interference. Digital links at the DS3 rate (45 Mb/s) are also used for configurations requiring large numbers of voice channels.

Fiber links are a common way of connecting microcells to a donor cell. Because of its wide bandwidth capabilities, the fiber facility can transport the entire cellular spectrum if necessary. A number of these links utilize an analog signal that modulates the entire RF signal with an optical signal for transmission. Other equipment is available that digitizes the signal before it enters the fiber facility. This equipment is a little more tolerant of signal degradation than the analog variety. Both types can usually use fiber facilities that are over 40 km in length without requiring an optical repeater.

Microwave links from the MSC to cell site locations are quite common in cellular systems for normal macrocells. Microcells, because of their size and location, are generally more difficult to serve with the 2- or 6-GHz microwave systems normally employed in these situations. As an alternative, some companies have used 23- or 38-GHz links to connect microcells. Because of the high frequency, a very small antenna may be used for the microwave. However, the range between the donor cell and microcell is very limited and is also subject to fading.

Mobile Switching Center

Only digital switches are now used for cellular service. This has been the case practically since the introduction of cellular service in the U.S. although there were a few switches initially that were based on an analog, electronic switch design. Digital switching, however, does not necessarily mean digital radio transmission in the cellular system. It simply refers to the manner in which calls are processed by the switch.

Like many landline switches, the MSCs are *stored program controlled* (SPC) devices whose functions are controlled by software instructions. The systems are usually full availability, nonblocking switches, meaning any input port can be connected to any output port.

In addition to completing connections from the radio portion of the system to the landline network, or other wireless networks, the MSC also performs the radio controller function in an AMPS-type cellular network. Because of this feature, it has been estimated that an MSC takes 4–10 times the processing of a landline switch for an equivalent number of subscribers.

In most digital switches, the number of ports are configured in multiples of 24 in order to accommodate a **DS1**-rate input or output channel. Not all of the channels may be activated in any single DS1. As shown in Fig. 16.65, some of the ports are connected to the cell sites by the MSC–cell site links, others are connected to the PSTN or other networks, whereas still others provide links between MSCs themselves.

Cellular switches may be connected to the PSTN using several different configurations. The initial connection offered to cellular carriers was a trunk-side connection to an end office and is known as a Type 1 connection. Later, as a result of an FCC order, cellular carriers were able to obtain connections from a tandem office, which are known as Type 2A. There are also other variants available, such as high-usage **trunks** to specific **end offices** (Type 2B), trunks to 911 tandems (Type 2C), trunks to operator services tandems (Type 2D), and 56 kb/s circuits used for **signaling system 7 (SS7)** applications (Type S).

Other functions that are often part of the MSC are the **home location register (HLR)**, *visitor location register* (VLR), and the **authentication center (AC)**.

The HLR function is a major component of the seamless network so often discussed. It is often an inherent part of most switches, although it can also be located externally from the switch. The HLR stores registration and service profiles of the subscribers of that particular system.

VLRs are used to store information regarding subscribers who have roamed into another area. It also can be located externally from the MSC but is generally an inherent part of the MSC too. The AC is a new feature that is intended to provide additional authentication information. These data change with each call so the AC must be queried for each call in order to continue call processing. Like the HLR and VLR, the AC function can also be separated from the MSC. The MSC contains software and hardware for recording call detail information. This information is normally transmitted to a billing center over voice-grade data links, but magnetic tapes are also used for this function. The billing system itself is separate from the MSC and is often a service provided by a vendor to the cellular carrier.

Operational systems allow personnel to perform maintenance, change subscriber data, change software within the switch, and provide alarms.

IS-41 Standard

Initially the protocols for messages between MSCs were proprietary to each vendor. Unless this was changed, this fact would have greatly inhibited the development of the seamless network the cellular industry was seeking. As a result, the TIA developed a standard called *interim standard number 41* (IS-41).

IS-41 allows MSCs manufactured by different vendors to exchange information regarding subscribers. Using Fig. 16.65, this allows MSC 2 to query the HLR of MSC 1 when a subscriber from MSC 1 enters the service area of MSC 2.

The major functional entities of the IS-41 standard are the *mobile station*, which is also known as the *cellular subscriber station* (CSS), the *base station* (BS), MSC, HLR, VLR, and AC. An initial version of IS-41, called IS-41 revision 0, was approved in late 1987. It provided handoffs between different systems and user validation but the validation process was done off line. Revision 0 only used X.25 protocol and the

transaction capabilities application part (TCAP) messages were based on the Consultative Committee on International Telephony and Telegraphy (CCITT) standard.

In 1990, IS-41, revision A was approved. It is not compatible with revision 0 because it used TCAP messages based on the American National Standards Institute (ANSI), instead of the CCITT standard. Revision A also added the use of the SS7 protocol as well as X.25, but the SS7 links had to be point-to-point between the respective MSCs. Moreover, revision A also added the ability to transfer features such as call waiting, three-way calling, and call forwarding.

Revision B of IS-41 was approved in 1991. It added the ability to minimize the paths between MSCs and added *global title translation* (GTT) capability so that direct links were not required between MSCs. GTT allows routing of SS7 messages on an SS7 network such as the North American Cellular Network (NACN). It also added support for dual-mode sets using TDMA multiplexing.

The latest version of IS-41, revision C, was approved in 1995. It adds support for the AC, short messaging (which is like alphanumeric paging), support for dual-mode CDMA sets, and the *data message handler* (DMH). The DMH is equipment that will allow real-time billing, which should help with fraud management.

Digital Synchronization

When digital switches and/or digital facilities are interconnected, some means of synchronizing the clock rates must be used in order to avoid errors. The synchronization is obtained from the bit stream of a digital signal, such as a DS1 rate (1.544 Mb/s) channel, that is used to connect the two networks. An agreement must be reached between the LEC and the cellular carrier regarding the primary source of timing.

Glare Resolution

When switches are interconnected by two-way trunk-side connections, there is always the possibility that the trunk will be simultaneously seized by both switches. This results in a condition called *glare* and is resolved by the switches containing instructions known as a *glare bit*. The glare bit indicates which switch is to yield first in the event of a glare condition. The normal convention is for the lower ranked office to yield first and so in the LEC network, end offices normally yield to tandem offices and tandem offices yield to **interexchange carrier (IC)** offices. It does not really matter which office yields first as long as it is understood when the connection is made what the convention will be. Therefore, in cases where the cellular carrier is interconnected with the LEC, an understanding regarding glare resolution must be reached as part of the interconnection agreement so that the glare bit may be properly set in both switches.

Numbering Requirements

Each cellular set used in North America has a telephone number encoded in the set. This number conforms to the **North American numbering plan (NANP)** and serves as the **mobile identification number (MIN)**. If the telephone number is changed for any reason, the set must be physically altered to accommodate the new MIN.

Cellular carriers are entitled to telephone number assignments just like any LEC. They often have entire **NXX** codes (blocks of 10,000 numbers) dedicated to them for cellular service. This is particularly true when the cellular carrier has a tandem (Type 2) interconnection with the LEC. However, entire codes can also be used with other types of interconnection as well.

In addition to the MIN, the sets have an **electronic serial number (ESN)** and **system identification (SID)** number encoded in them. The ESN is assigned by the manufacturer of the set, whereas the SID is assigned by the FCC for sets used in the U.S. A unique SID was assigned for each wireline and nonwireline carrier in each MSA and RSA, but cellular carriers have consolidated many of these SIDs. For example, the Washington and Baltimore systems originally had unique SIDs but the wireline carrier serving those areas has combined them into a single SID.

16.7.5   Call Processing

To process calls in a cellular system, the location and identity of the mobile user must be known by the MSC. This is true regardless of whether the mobile unit is operating in its home system or has traveled to a distant location and is considered to be roaming. To accomplish this task, the mobile unit transmits certain information each time the power is turned on, and again each time a call is originated. Specifically, this information consists of the mobile unit's MIN, ESN, and SID. In many cases, the operation of the set is the same for the user whether the mobile unit is used in its home system or not. This was not always the case, and there are still areas where different operations are required of the user but the seamlessness provided by IS-41 and SS7 networks is quickly eliminating these differences.

Technically, there are differences in the manner in which calls are processed between a user in a home system and one who is roaming. The basic call processing operations are described in the following paragraphs but in a seamless network, these processes are transparent to the user.

Call Processing-Home System

Dialing from a cellular phone is different from a landline telephone. Instead of receiving dial tone and then dialing the desired number, the cellular user first determines if service is available by verifying the presence of a service available indicator. This indicator is displayed when the mobile unit can receive a signal from one of the control channels in the cellular system. The mobile unit knows whether it is in its home area or not because the base station periodically broadcasts its identity, including the SID. The mobile unit compares this SID with its own and if they are the same, the mobile unit is operating in its home area. If the two SIDs do not match, the roam light is illuminated on the mobile unit to indicate a roaming condition exists.

Mobile-Land Calls-Home System

Assuming service is available, the mobile user enters the digits of the party being called and presses the send button on the mobile unit. The mobile unit transmits these digits to the base station using the control channel. This information may be received by more than one cell site, and the MSC will determine which cell site should initially handle the call. After ensuring the availability of a voice channel to that cell site and a trunk connection to the PSTN, the MSC seizes a trunk, Type 1, 2A, or 2B, depending on the connection used and call type, and outpulses the required number of digits to the LEC office. When the call is answered, the MSC connects the mobile to the voice channel so that conversation can begin.

Land-Mobile Calls-Home System

To reach a mobile subscriber in a cellular system, the landline customer dials the 7- or 10-digit mobile number, depending on the local dialing plan. The call is directed to either the end office providing a Type 1 or Type 2B connection, or the tandem office if a Type 2A connection is employed. In either case, the 7- or 10-digit number of the mobile subscriber is outpulsed by the LEC office to the MSC. If the MSC has not recorded the presence of the mobile unit when the power was first turned on, the location of the mobile subscriber is determined by broadcasting a paging code that includes the mobile subscriber's MIN to all cell sites via a separate control channel. The cell site with the strongest signal strength from the responding sites is selected and a voice channel is assigned by the MSC. A code is then outpulsed over the signaling channel, which causes the mobile unit to initiate a ringing signal. When the mobile subscriber answers, the voice channel is connected so that conversation can begin.

Cellular Roaming Call Processing

Once a mobile unit moves outside of its home area (the point where the SIDs no longer match), the unit is usually considered to be a roamer and is then served by a *visited* system. The following paragraphs describe the basic call processing techniques used for these types of calls and assume the use of autoregistration features used in seamless networks such as those provided by the North American Cellular Network.

Mobile-Land Calls-Visited System

Assume a cellular subscriber in Seattle travels to Miami and is part of the NACN. Once the set is turned on in Miami, the Miami system recognizes the MIN being broadcast from the mobile unit is from Seattle. The Miami system then queries its VLR to see if the user is already registered. If the user has not been previously registered there will be no information resident in the VLR. That being the case, the Miami system sends a TCAP message to Seattle requesting the service profile of the roaming subscriber. The information returned from Seattle validates the user and provides a profile of the services used by the subscriber in Seattle. In addition, the Seattle system requests a Miami *temporary local directory number* (TLDN) so that calls for the Seattle subscriber can be routed to Miami over the PSTN. When the subscriber makes a mobile-land call in Miami, the call is then processed just as in a home system because the Miami system has validated the user and, using IS-41, will usually provide the same features the subscriber has available in Seattle. The validation process may occur in real-time or off line. If off line, calls will be permitted if a roaming agreement exists between the two carriers even if the validation response has not yet been received by the Miami system. In essence, the Seattle-based subscriber does not have to do anything special to utilize the mobile phone in Miami.

Calls made by the Seattle subscriber in Miami are recorded and, through a clearinghouse function that usually involves a third-party vendor, the Seattle system eventually receives the data so that a bill can be rendered to the customer. The Miami system is ultimately reimbursed for these calls as part of this clearinghouse function.

Land-Mobile Calls-Visited System

Continuing with the example of the Seattle-based cellular subscriber, calls made to the subscriber are routed first to the Seattle MSC. This happens regardless of whether the call is made from a landline phone in Seattle or a landline phone in Miami. Upon receiving the call, the Seattle MSC queries its HLR and determines that the subscriber is currently in Miami. Using the TLDN provided by the Miami system, the call is forwarded to Miami where it is received by the Miami MSC. The Miami MSC translates the TLDN into the Seattle subscribers normal MIN and uses a broadcast page over all control channels in Miami to locate the subscriber. If the Seattle subscriber's mobile unit is activated, a response is made and the call is processed just like any other land-mobile call at that point.

Call Handoffs

All cellular systems perform call handoffs in which a call in progress is transferred from one cell to another. Many of these handoffs occur within the same cellular system or cells controlled by the same MSC. Others take place between cells controlled by different MSCs, and the MSCs may belong to different cellular systems.

In the analog mode of an AMPS cellular system, the handoffs are initiated by the base station and controlled by the MSC. For example, if mobile unit in cell A in Fig. 16.65 were to move close to cell B, the base station equipment in cell A would signal the MSC that the signal quality was diminishing. The MSC would initiate orders to adjacent cell sites to make signal measurements and report the results. Once the results had been received and analyzed by the MSC, the MSC would make a determination as to which cell site would receive the handoff. In this example, a message would be sent to the mobile unit, over the current voice channel, to change channels. This momentary use of the voice channel for signaling purposes is called blank and burst because the voice signal is blanked for a brief instant while a burst of signaling information is transmitted. The mobile unit changes channels as directed and the call is now transferred to the assigned channel in cell B. There is also a brief period, up to 400 ms, when no signal is received while the connection from the first cell is dropped and the connection to the new cell is being established.

Digital cellular, as previously explained, uses mobile assisted handoff to begin the handoff process. With MAHO, the mobile unit may make measurements on as many as 12 channels from adjacent cell sites. On command, it will report these results to the base station. Based on those results, the MSC may order a handoff to another channel controlled by an adjacent base station. The advantage of this method is a more accurate evaluation of the signal actually being received by the mobile unit. As previously explained,

TDMA still experiences a brief period of the mobile not receiving a signal while the handoff is occurring. Because of its ability to do soft handoffs, CDMA systems do not lose the signal because the connection is maintained with the old cell until the connection to the new cell is established.

Handoffs between cells controlled by different MSCs also take place, and IS-41 messages are used between the MSCs to control the process. The mechanics are similar to the handoffs described earlier except that when there is an adjacent system IS-41 messages must be sent to the adjacent MSC requesting the measurements (and results) and then initiating the handoff. The messages are sent over signaling links while the voice channels are part of dedicated lines between the MSCs. With IS-41, revision B, the capability exists to minimize the number of connections when more than two MSCs are involved in any one call.

Call Processing-Other Types of Calls

There are other types of calls that are processed in a cellular system. These can originate from home or visiting subscribers but are processed by the home system. They include mobile-mobile, emergency, and operator-assisted calls.

Mobile-Mobile Calls

Mobile subscribers can also reach other mobile subscribers, including those that operate on other cellular systems. If the call is to a mobile subscriber in the same cellular system, the MSC sets up the call, via the control channels, between the cell sites serving the two mobile subscribers. If the call is to a mobile subscriber in another system that is not directly interconnected with the originating system, the call is normally processed like the mobile- to-land call already described. This type of call would occur when a mobile subscriber of a wireline system places a call to a mobile subscriber served by a nonwireline system, or vice versa.

Emergency Services Calls

In most cases, the cellular carrier must make arrangements with the local public safety agency that is responsible for handling emergency calls before they begin forwarding such traffic to the LEC. Normally, the mobile user dials 911, but those digits may be translated into a seven digit number by the MSC in order to complete the call.

A major problem with calls from a mobile unit is the fact that the precise location may not be known, and the number of the mobile unit is not forwarded to the emergency answering location. Because of existing technical limitations, it is possible to provide either a general geographical location or identify the mobile unit, but not both. Typically, the general geographical location is used, and it is identified by associating a 7-digit telephone number with a specific sector of the antenna that receives the initial call. The MSC is able to discern which cell, and which sector, receives the initial call and will use this information to translate the 911 digits to the proper 7-digit number that is forwarded to the LEC for completion to the emergency answering location.

Operator Services Traffic

Traffic destined for operator services provided by the LEC is usually routed by the MSC over a Type 1 connection or over special direct trunks (Type 2D) to a LEC operator services tandem. The direct connection to the operator services tandem provides greater flexibility because the actual identification of the mobile unit placing the call is provided with this connection, whereas the Type 1 connection does not receive this information from the MSC. For example, a Type 2D will allow the LEC to provide *directory assistance call completion* (DACC) service whereby a call to directory assistance can be completed to the requested number by the LEC operator.

Because of heavy fraud activity, cellular carriers sometime restrict mobile users to placing long distance calls by using either a calling card, collect, or third-number billing options. This is possible because of a service called *selective class of call screening* (SCCS) that is offered to the cellular carriers by the LECs.

The LECs are not the only entities with operator services. Many ICs provide their own operators and these may be accessed via direct connections from the MSC to the IC or by using Type 1 or Type 2A trunks in the LEC network to reach the IC network.

Defining Terms

Advanced Mobile Phone Service (AMPS): An analog air interface standard for cellular service developed in the U.S. by AT&T Bell Laboratories. It is also known as the EIA/TIA-553 standard.

Authentication center (AC): An entity that manages encryption information for a wireless carrier. It may be associated with the home location register or it may be a separate function.

Cell site: A transmitter/receiver location, operated by the cellular carrier, through which radio links are established between the cellular system and mobile units. The area reliably served as a given cell site is referred to as a cell.

Cellular geographic service area (CGSA): The geographic area served by the cellular system within which a CMC is authorized to provide service.

Cellular system: A high-capacity land mobile radio system in which an assigned frequency spectrum is divided into discrete channels that are assigned in groups to cells covering the cellular geographic service area. The discrete cells can be reused in different cells within the same service area. Calls are handed off automatically from a channel in one cell to a different channel in an adjacent cell as the mobile unit moves across cell boundaries.

Code division multiple access (CDMA): A multiplexing technique where the signal intelligence is mixed with a randomlike code and then spread over a wide range of frequencies.

DS0: The 64-kb/s zero-level signal in the digital transmission hierarchy.

DS1: The 1.544 Mb/s first-level signal in the digital transmission hierarchy. In the time-division multiplexing hierarchy of the telephone network, DS1 is the initial level of multiplexing. Traditionally, 24 DS0 channels have been multiplexed up to the DS1 rate with each DS0 channel carrying the digital representation of an analog voice channel.

DS3: The 45 Mb/s third-level signal in the digital transmission hierarchy. It contains 28 DS1 signals or the equivalent of 672 analog voice channels.

Electronic serial number (ESN): Information electronically encoded by the manufacturer in a cellular set for identification purposes.

End office (EO): A LEC switching system within a LATA or market area where customer station loops are terminated for purposes of interconnection to each other and to trunks.

Federal Communications Commission (FCC): The federal agency empowered by law to regulate all interstate and foreign radio and wire communications services in the U.S. including radio, television, facsimile, telegraph, mobile, and telephone systems. The agency was created under the Communications Act of 1934.

Frequency division multiple access (FDMA): A multiplexing method whereby each channel is assigned a specific frequency for its use.

Global System for Mobile (GSM) Communications: A digital cellular standard developed in Europe.

Home location register (HLR): The location register to which end user identity information is assigned for record purposes.

Interexchange carrier (IC): A common carrier that provides services to the public between local exchanges on an intra or interLATA basis in compliance with local or Federal regulatory requirements and that is not an end user of the services provided.

Line: A central office connection that usually connects the switching system to customer equipment.

Local exchange carrier (LEC): A company that provides intraLATA telecommunications within a franchised territory.

Metropolitan statistical area (MSA): Sometimes known as standard metropolitan statistical areas (SMSAs), MSAs are areas based on counties as defined by the U.S. Census Bureau that are cities of 50,000 or more population and the surrounding counties. Using data from the 1980 census, the FCC allocated two cellular licenses in each of the 305 MSAs in the U.S.

Mobile identification number (MIN): A 10-digit number that conforms to the North American numbering plan that is used for identification purposes in wireless sets (such as a cellular).

Mobile switching center (MSC): A cellular switching system that is used to terminate mobile stations for purposes of interconnection to each other and to trunks interfacing with the public switched network.

Mobile telephone switching office (MTSO): A synchronous, but obsolete, term for mobile switching center.

North American numbering plan (NANP): A 10-digit number address used for direct dialing within world zone 1 which includes the U.S., Canada, Puerto Rico, the Virgin Islands, Jamaica, and Bermuda.

NPA code: A unique 3-digit code in the N 0/1 X series that identifies a numbering plan area. A NPA is the first three digits of the 10-digit destination number for all inter-NPA calls within the NANP area.

NXX code: A code normally used as a central office code. It may also be used as an NPA code or special NPA code.

Signaling System 7 (SS7): An internationally standardized, general purpose CCS system.

System identification (SID): An identity assigned by the FCC for cellular systems used to identify a specific cellular system. Initially, these numbers were unique for each provider and each MSA or RSA. Some systems have since been combined and so MSAs and RSAs are no longer necessarily unique.

Telecommunications Industry Association (TIA): A standards body accredited by ANSI that addresses telecommunications equipment and systems.

Time division multiple access (TDMA): A multiplexing technique in which individual conversations are converted to a digital format and assigned to a specific time slot in a channel. There are usually two or more time slots per channel.

Transaction capabilities application part (TCAP): In the SS7 protocol, the application layer of the transaction capability protocol.

Trunk: In a network, a communication path connecting two switching systems.

Wireless carrier (WC): A generic term used to describe entities providing wireless service. Such entities include, but are not limited to, cellular carriers, radio common carriers, and private carriers.

References

Boucher, N. 1992. *The Cellular Radio Handbook*. Quantum, Mendocino, CA.

Calhoun, G. 1988. *Digital Cellular Radio*. Artech House, Norwood, MA.

Calhoun, G. 1992. *Wireless Access and the Local Telephone Network*. Artech House, Norwood, MA.

Lee, W.C.Y. 1982. *Mobile Communications Engineering*. McGraw-Hill, New York.

Walker, J. ed. 1990. *Mobile Information Systems*. Artech House, Norwood, MA.

Young, H.E. 1996. *Wireless Basics*, 2nd ed. Intertec, Chicago.

Bellcore Technical Reference TR-NPL-000145, *Compatibility Information For Interconnection Of A Wireless Services Provider And A Local Exchange Carrier's Network*. Bell Communications Research, Livingston, NJ, 1993.

Bellcore Technical Reference TR-EOP-000352, *Cellular Mobile Carrier Interconnection Transmission Plan*. Bell Communications Research, Livingston, NJ, 1986.

Bellcore Special Report SR-TSV-002275, *BOC Notes On The LEC Network-1994*. Bell Communications Research, Livingston, NJ, 1994.

Digital Cellular: Economics And Comparative Technologies. Economic and Management Consultants International, Washington, DC, 1993.

TIA Interim Standard #41 (IS-41), Cellular Radiotelecommunications Intersystem Operations, Revision B, December, 1991. Telecommunications Industry Association, Washington, DC, 1991.

16.8   Satellite Communications

Daniel F. DiFonzo

16.8.1   Introduction

The impact of satellites on world communications since commercial operations began in the mid-1960s is such that we now take for granted many services that were not available a few decades ago: worldwide TV, reliable communications with ships and aircraft, wide-area data networks, communications to remote areas, direct TV broadcast to homes, position determination, and Earth observation (weather and mapping). New and proposed satellite services include global personal communications to hand-held portable telephones and broadband voice, video, and data to and from small user terminals at customer premises around the world.

Satellites function as line-of-sight microwave relays in orbits high above the Earth which can see large areas of the Earth's surface. This unique feature ensures satellites' continued growth even as fiber optic cables capture a larger market share of high-density point-to-point traffic. Satellites provide cost-effective access for areas with low (thin-route) communications traffic because Earth terminals can be installed in locations where the high investment cost of terrestrial facilities might not be warranted. Satellites are particularly well suited to wide-area coverage for broadcasting, mobile communications, and point-to-multipoint communications.

16.8.2   Satellite Applications

Figure 16.66 depicts several kinds of satellite links and orbits. The geostationary Earth orbit (GEO) is in the equatorial plane at an altitude of 35,786 km with a period of one sidereal day (23 h 56 min 4.09 sec). GEO satellites appear to be almost stationary from the ground (subject to small perturbations) and the

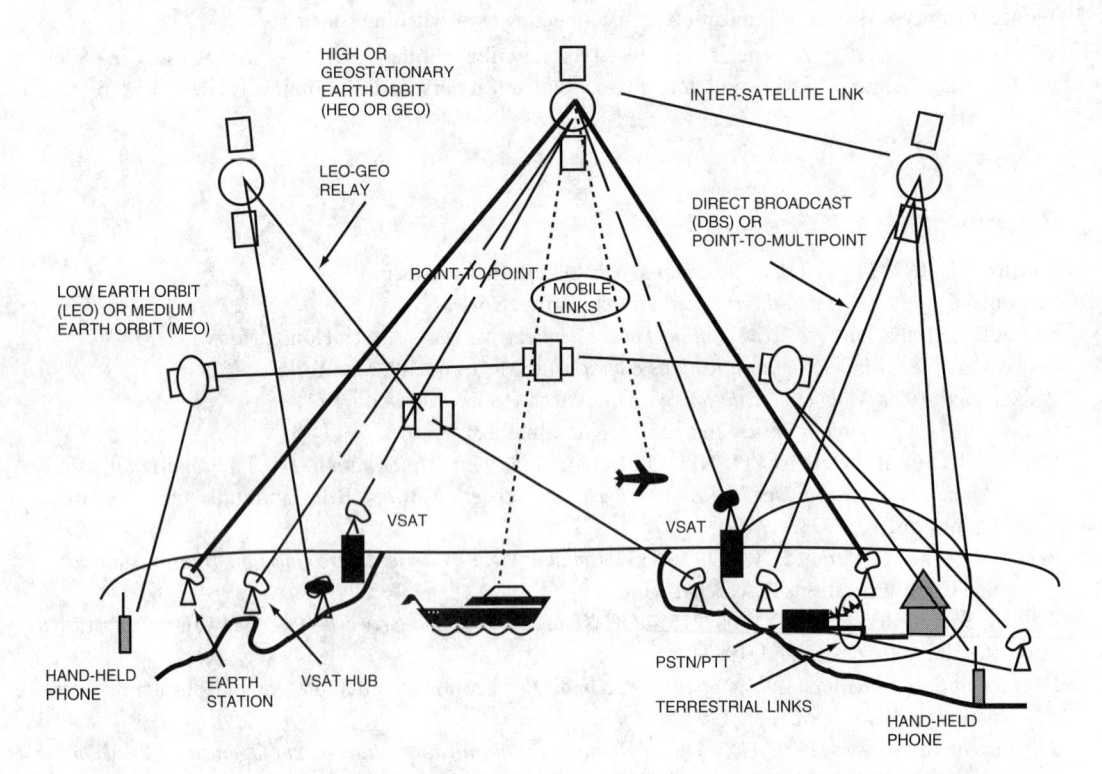

FIGURE 16.66   Several types of satellite links: illustrated are point-to-point, point-to-multipoint, VSAT, direct broadcast, mobile, personal communications, and intersatellite links.

Earth antennas pointing to these satellites may need only limited or no tracking capability. The orbits for which the highest altitude (apogee) is at or greater than GEO are sometimes referred to as high Earth orbits (HEO). Low Earth orbits (LEO) typically range from a few hundred km to about 2000 km. Medium Earth orbits (MEO) are at intermediate altitudes. Circular MEO orbits are sometimes called intermediate circular orbits (ICO). Several organizations have proposed MEO systems at an altitude of about 10,400 km for global personal communications at frequencies designated for mobile satellite services (MSS).

LEO systems for voice communications are called *big LEOs*. Constellations of so-called *little LEOs* operating below 1 GHz and having only limited capacity have been proposed for low data rate nonvoice services, such as paging and store and forward data for remote location and monitoring, for example, for freight containers and remote vehicles and personnel (Kiesling, 1996).

Initially, satellites were used primarily for point-to-point traffic in the GEO fixed satellite service (FSS), for example, for telephony across the oceans and for point-to-multipoint TV distribution to cable *head end* stations. Large Earth station antennas with high-gain narrow beams and high uplink powers were needed to compensate for limited satellite power. This type of system, exemplified by the early global network of the International Telecommunications Satellite Organization (INTELSAT), used Standard-A Earth antennas with 30 m diameters. Since then, many other satellite organizations have been formed around the world to provide international, regional, and domestic services (Rees, 1990).

As satellites have grown in power and sophistication, the average size of the Earth terminals has been reduced. High-gain satellite antennas and relatively high-power satellite transmitters have led to *very small aperture* Earth *terminals* (VSAT) with diameters of less than 2 m, modest powers of less than 10 W (Gagliardi, 1991) and to smaller *ultrasmall aperture terminals* (USAT) diameters typically less than 1 m. As depicted in Fig. 16.66, VSAT terminals may be placed atop urban office buildings, permitting private networks of hundreds or thousands of terminals, which bypass terrestrial lines. VSATs are usually incorporated into *star* networks where the small terminals communicate through the satellite with a larger *hub* terminal. The hub retransmits through the satellite to another small terminal. Such links require two *hops* with attendant time delays. With high-gain satellite antennas and relatively narrowband digital signals, direct single-hop *mesh* interconnections of VSATs may be used.

16.8.3   Satellite Functions

The traditional function of a satellite is that of a bent pipe quasilinear repeater in space. As shown in Fig. 16.67, *uplink* signals from Earth terminals directed at the satellite are received by the satellite's antennas, amplified, translated to a different *downlink* frequency band, channelized into *transponder channels,* further amplified to relatively high power, and retransmitted toward the Earth. Transponder channels are generally rather broad (e.g., bandwidths from 24 MHz to more than 100 MHz) and each may contain many individual or user channels.

The functional diagram in Fig. 16.67 is appropriate to a satellite using frequency-division duplex (FDD), which refers to the fact that the satellites use separate frequency bands for the uplink and downlink and where both links operate simultaneously. This diagram also illustrates a particular *multiple access* technique, known as frequency-division multiple access (FDMA), which has been prevalent in mature satellite systems.

Multiple access, to be discussed later, allows many different user signals to utilize the satellite's resources of power and bandwidth without interfering with each other. Multiple access systems segregate users by frequency division (FDMA) where each user is assigned a specific frequency channel, space-division multiple access (SDMA) by reusing the same frequencies on multiple spatially isolated beams, time-division multiple access (TDMA) where each user signal occupies an entire allocated frequency band but for only part of the time, polarization-division (PD) where frequencies may be reused on spatially overlapping but orthogonally polarized beams, and code-division multiple access (CDMA) where different users occupy the same bandwidth but use spread spectrum signals that contain orthogonal signaling codes (Sklar, 1988; Richharia, 1995).

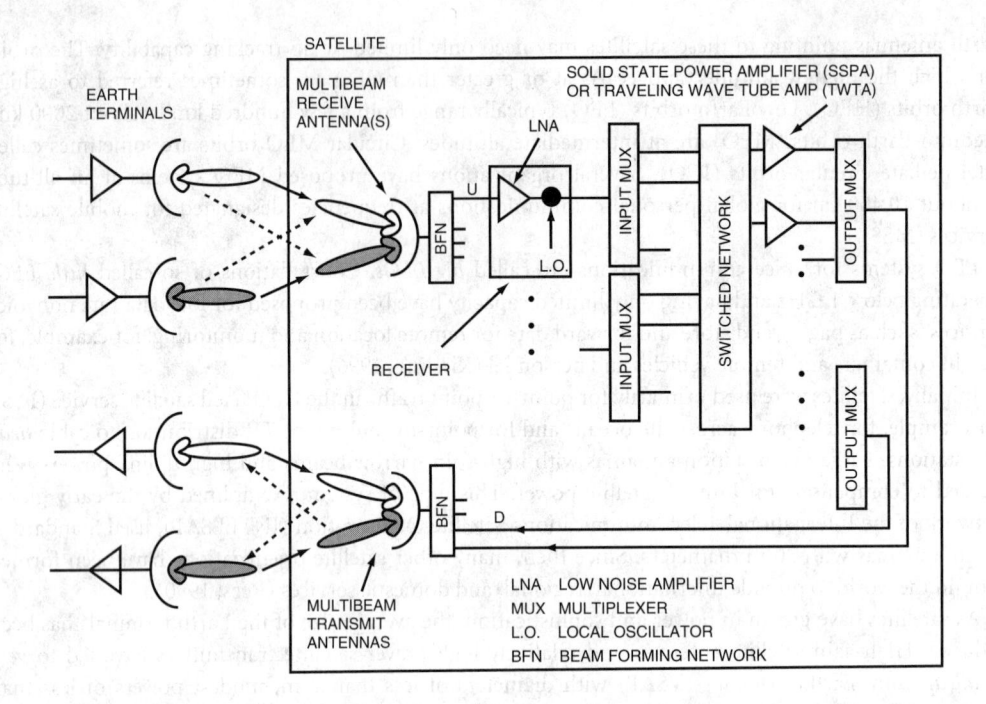

FIGURE 16.67 A satellite repeater receives uplink signals (U), translates them to a downlink frequency band (D), channelizes, amplifies to high power, and retransmits to Earth. Multiple beams allow reuse of the available band. Interference (dashed lines) can limit performance. Downconversion may also occur after the input multiplexers and several intermediate frequencies, and downconversions may be used.

Frequency modulation (FM) has been the most widely used modulation. However, advances in digital voice and video compression have led to the widespread use of digital modulation methods such as quadrature phase shift keying (QPSK) and quadrature amplitude modulation (QAM) (Schwartz, 1990; Sklar, 1988).

Newer satellite architectures incorporate digital modulations and onboard demodulation of the uplink signals to baseband bits, subsequent switching and assignment of the baseband signals to an appropriate downlink antenna beam, and remodulation of the clean baseband signals prior to downlink transmission. These *regenerative repeaters* or *onboard processors* permit flexible routing of the user signals and can improve the overall communications link by separating the uplink noise from that of the downlink. The baseband signals may be those of individual users or they may represent frequency-division multiplexed (FDM) or time-division multiplexed (TDM) signals from many users.

High-power *direct broadcast satellites* (DBS) or *direct-to-home* (DTH) satellites operate at Ku-band. In the U.S., satellites operating in the broadcast satellite service (BSS) with downlink frequencies of 12.2–12.7 GHz, deliver TV directly to home receivers having parabolic dish antennas as small as 46 cm (18 in) in diameter. DBS with digital modulation and compressed video is providing more than 150 National Television Systems Committee (NTSC) TV channels from a single orbital location having an allocation of 32 transponder channels, each with 24-MHz bandwidth.

Mobile satellite services operating at L-band around 1.6 GHz have revolutionized communications with ships and aircraft, which would normally be out of reliable communications range of terrestrial radio signals. The International Maritime Satellite Organization (INMARSAT) operates the dominant system of this type.

Links between LEO satellites (or the NASA Shuttle) and GEO satellites are used for data relay, for example, via the NASA tracking and data relay satellite system (TDRSS). Some systems will use intersatellite links (ISL) to improve the interconnectivity of a wide-area network. ISL systems would typically operate at frequencies such as 23 GHz, 60 GHz, or even use optical links.

16.8.4 Satellite Orbits and Pointing Angles

Reliable communication to and from a satellite requires a knowledge of its position and velocity relative to a location on the Earth. Details of the relevant astrodynamics formulas for satellite orbits are given in Griffin and French (1991) and Morgan and Gordon (1989). Launch vehicles needed to deliver the satellites to their intended orbits are described in Isakowitz (1994).

A satellite, having mass m in orbit around the Earth, having mass M_e, traverses an elliptical path such that the centrifugal force due to its acceleration is balanced by the Earth's gravitational attraction, leading to the equation of motion for two bodies:

$$\frac{d^2 r}{dt^2} + \frac{\mu}{r^3} r = 0 \tag{16.52}$$

where r is the radius vector joining the Earth's center and the satellite and $\mu = G(m + M_e) \approx G M_e = 398,600.5 \, \text{km}^3/\text{sec}^2$ is the product of the gravitational constant and the mass of the Earth. Because $m \ll M_e$, the center of rotation of the two bodies may be taken as the Earth's center, which is at one of the focal points of the orbit ellipse.

Figure 16.68 depicts the orbital elements for a geocentric right-handed coordinate system where the x axis points to the first point of Aries, that is, the fixed position against the stars where the sun's apparent path around the Earth crosses the Earth's equatorial plane while traveling from the southern hemisphere toward the northern hemisphere at the vernal equinox. The z axis points to the north and the y axis is in the equatorial plane and points to the winter solstice. The elements shown are: longitude or right ascension of the ascending node Ω measured in the equatorial plane; the orbit's inclination angle i relative to the equatorial plane; the ellipse semimajor axis length a; the ellipse eccentricity e; the argument (angle) of perigee ω, measured in the orbit plane from the ascending node to the satellite's closest approach to the Earth; and the true anomaly (angle) in the orbit plane from the perigee to the satellite v.

The mean anomaly M is the angle from perigee that would be traversed by a satellite moving at its mean angular velocity n. Given an initial value M_0, usually taken as 0 for a particular epoch (time) at perigee, the mean anomaly at time t is $M = M_o + n(t - t_0)$, where $n = \mu^{1/2}/a^{3/2}$. The eccentric anomaly E may then be found from Kepler's transcendental equation $M = E - e \sin E$, which must be solved numerically by, for example, guessing an initial value for E and using a root finding method. For small eccentricities, the series approximation $E \approx M + e \sin M + (e^2/2) \sin 2M + (e^3/8)(3 \sin 3M - \sin M)$ yields good accuracy

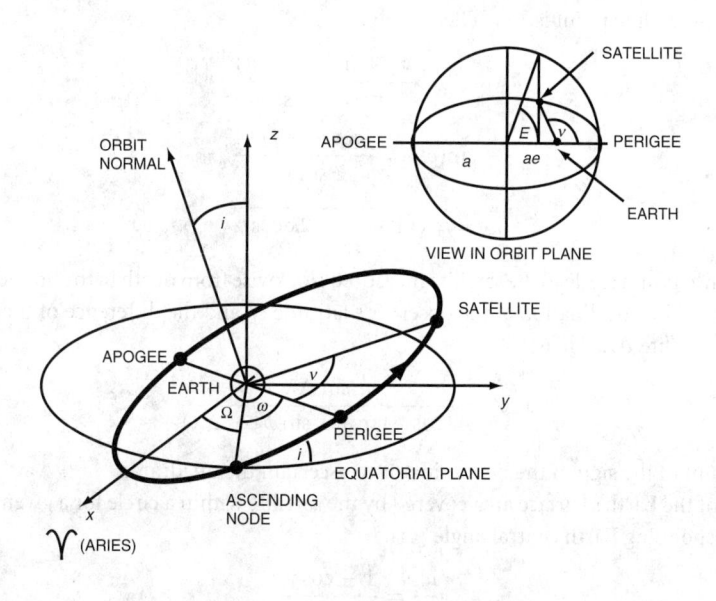

FIGURE 16.68 Orbital elements.

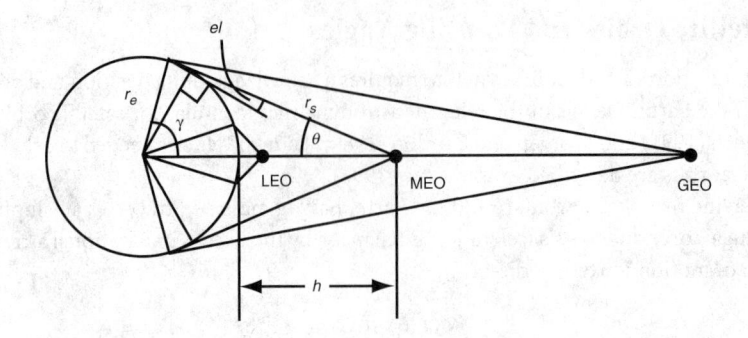

FIGURE 16.69 Geometry for a satellite in the plane defined by the satellite, the center of the Earth, and a point on the Earth's surface. The elevation angle *el* is the angle from the local horizon to the satellite. Shown to approximate scale are satellites at LEO, MEO (or ICO), and GEO.

(Morgan and Gordon, 1989, p. 806). Other useful quantities include the orbit radius r, the period P of the orbit (i.e., for $n(t - t_0) = 2\pi$), the velocity V, and the radial velocity V_v

$$r = a(1 - e \cos E) \tag{16.53}$$

$$P = 2\pi \sqrt{a^3/\mu} \tag{16.54}$$

$$V = \mu\left(\frac{2}{r} - \frac{1}{a}\right) \tag{16.55}$$

$$V_v = \frac{e(\mu a)^{1/2} \sin E}{a(1 - e \cos E)} \tag{16.56}$$

Figure 16.69 depicts quantities useful for communications links in the plane formed by the satellite, a point on the Earth's surface and the Earth's center. Shown to approximate scale for comparison are satellites at altitudes representing LEO, MEO (or ICO), and GEO orbits.

For a satellite at altitude h and Earth radius $r_e = 6378.14$ km, the slant range r_s, elevation angle to the satellite from the local horizon el, and the satellite's nadir angle θ are related by simple spherical trigonometry formulas. Note that $\theta + el + \gamma = 90°$, where γ is the Earth central angle and the ground range from the subsatellite point is γr_e. Then

$$k = \frac{r_e + h}{r_e} = \frac{\cos(el)}{\sin \theta} \tag{16.57}$$

$$\tan(el) = \frac{(\cos \gamma - 1/k)}{\sin \gamma} \tag{16.58}$$

$$r_s = r_e \sqrt{1 + k^2 - 2k \cos \gamma} \tag{16.59}$$

The Earth station azimuth angle to the satellite measured clockwise from north in the horizon plane is given in terms of the satellite's declination δ, the observer's latitude ϕ, and the difference of the east longitudes of observer and satellite $\Delta\lambda$. Then

$$\tan A = \frac{\sin \Delta\lambda}{(\cos \phi \tan \delta - \sin \phi \cos \Delta\lambda)} \tag{16.60}$$

taking due account of the sign of the denominator to ascertain the quadrant.

The fraction of the Earth's surface area covered by the satellite within a circle for a given elevation angle el and the corresponding Earth central angle γ is

$$\frac{a_c}{a_e} = \frac{1 - \cos \gamma}{2} \tag{16.61}$$

16.8.5   Communications Link

Figure 16.70 illustrates the elements of the RF link between a satellite and Earth terminals. The overall link performance is determined by computing the link equation for the uplink and downlink separately and then combining the results along with interference and intermodulation effects.

The uplink received carrier-to-noise ratio at the satellite is

$$(C/N)_u = EIRP_u + (G_{su} - 10\log T_{su}) + 228.6$$
$$-10\log\left(4\pi r_{su}^2\right) - 10\log\left(4\pi/\lambda_u^2\right) - A_u + \Gamma_u - B \tag{16.62}$$

where each term is a quantity expressed in decibel, for example, P (dBW) $= 10\log(p)$ (W). The downlink equation has identical form with appropriate downlink quantities substituted in Eq. 16.62. The relevant quantities (without subscripts) are described next.

The carrier-to-noise ratio (C/N) is the decibel *difference* between carrier and noise powers expressed in decibel relative to one watt (dBW), that is, (C/N) dBW $= 10\log c - 10\log n$. $EIRP = P + G$ dBW is the equivalent isotropically radiated power (EIRP) where the antenna gain G in decibels relative to isotropic (dBi) is the antenna gain *in the direction of the link*, that is, it is not necessarily the antenna's peak gain. The received noise power is $N = 10\log(kTb)$ where $k = 1.38 \times 10^{-23}$ J/K is

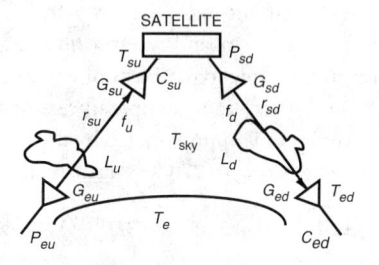

FIGURE 16.70   Quantities for a satellite RF link: $P =$ transmit power (dBW), $G =$ antenna gain (dBi), $C =$ received carrier power (dBW), $T =$ noise temperature (K), $L =$ dissipative loss (dB), $r =$ slant range (m), $f =$ frequency (Hz), $u =$ uplink, $d =$ downlink, $e =$ Earth, $s =$ satellite.

Boltzmann's constant, $10\log(k) = -228.6$ dBW/K/Hz, T is the system noise temperature in kelvins and $B = 10\log(\text{bandwidth (Hz)})$ is the bandwidth in decibel hertz. Then, $(G - 10\log T)$ dB/K is a figure of merit for the satellite or Earth receiving system. It is usually written as G/T and read as "gee over tee." The antenna gain and the noise temperature must be defined at the same reference point, for example, at the receivers input port.

The spreading factor $4\pi r_s^2$ is independent of frequency and depends *only* on the slant range distance r_{su} or r_{sd}. The gain of a 1-m^2 antenna is $10\log(4\pi/\lambda^2)$, where the wavelength $\lambda = c/f$, where f is the frequency in hertz, and $c = 2.9979 \times 10^8$ m/sec is the velocity of light. The decibel sum of the spreading factor and the gain of a 1-m^2 antenna is the frequency-dependent path loss. A is the signal attenuation due to dissipative losses in the propagation medium.

Γ is the polarization mismatch factor in decibel between the incident wave and the receive antenna, that is, $\Gamma = 10\log(\zeta)$ where $0 \leq \zeta \leq 1$. For nearly circularly polarized systems, having opposite senses of polarization, the decibel value of the polarization coupling term can be approximated as

$$\Gamma \approx 10\log\left(R_w^2 + R_A^2 + 2R_w R_A \cos\delta\right) - 24.8 \tag{16.63}$$

where R_w and R_A are the axial ratios in decibel of the wave and antenna, respectively, and δ is the difference between the tilt angles of the two polarization ellipses. For ideal antennas and no *depolarization* due to antenna imperfections or precipitation (Gagliardi, 1991; Pratt and Bostian, 1986; Sciulli, 1986) the coupling would be zero or $-\infty$ dB. The copolarized coupling factor would normally be ≈ 1, corresponding to $\Gamma = 0$ dB.

16.8.6   System Noise Temperature and G/T

The system noise temperature T incorporates contributions to the noise power radiated into the receiving antenna from the sky, ground, and galaxy, as well as the noise temperature due to circuit and propagation losses, and the noise figure of the receiver. The clear sky antenna temperature for an Earth station depends on the elevation angle since the antenna's sidelobes will receive a small fraction of the thermal noise power

radiated by the Earth, which has a noise tempera-
ture $T_{earth} \approx 290$ K. At 11 GHz, the clear sky *antenna*
noise temperature, T_{aclear}, ranges from 5 to 10 K at
zenith ($el = 90°$) to more than 50 K at $el = 5°$
(Pratt and Bostian, 1986).

As shown in Fig. 16.71, the system noise temperature is developed from the standard formula for
the equivalent temperature of tandem elements including the antenna in clear sky, propagation (rain)
loss of A dB, circuit losses between the aperture
and receiver of L_c dB, and receiver noise figure of
F dB (corresponding to receiver noise temperature
T_r K). The system noise temperature referred to
the receiver input is approximated by the following

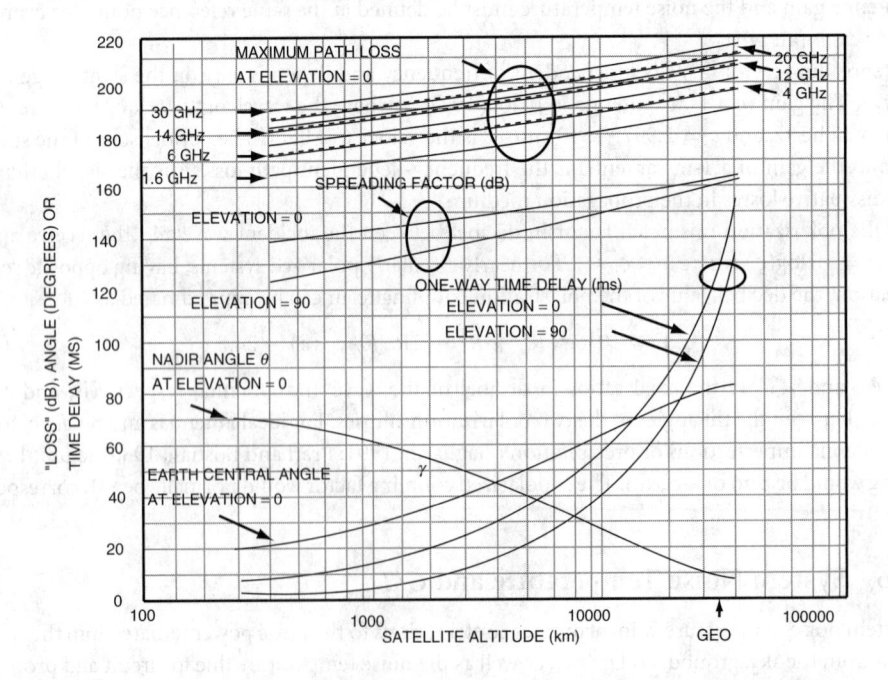

FIGURE 16.71 Tandem connection of antenna, loss elements such as waveguide and receiver front end. The noise
temperature depends on the reference plane but the G/T
is the same for both points shown.

equation where $T_{rain} \approx 270$ K is a reasonable approximation for the physical temperature of the rain:

$$T_{ed} = \left(T_{aclear} + 270(1 - 10^{-A/10})\right)/10^{L_c/10} + 290(1 - 10^{-L_c/10}) + 290(10^{F/10} - 1) \quad \text{K} \quad (16.64)$$

The system noise temperature is defined at a specific reference point such as the input to the receiver.
However, G/T is *independent* of the reference point when G correctly accounts for circuit losses. The
satellite's noise temperature is generally higher than an Earth terminal's under clear sky conditions because
the satellite antenna *sees* a warm Earth temperature of ≈ 150–300 K, depending on the proportion of clouds,
oceans, and land in the satellite antenna's beam, whereas a directive Earth antenna generally sees cold sky
and the sidelobes generally receive only a small fraction of noise power from the warm Earth. Furthermore,
a satellite receiving system generally has a higher noise temperature due to circuit losses in the beam forming
networks, protection circuitry, and extra components for redundancy.

Figure 16.72 illustrates the link loss factors, maximum nadir angle θ, Earth central angle γ, and earth-
space time delay as a function of satellite altitude. The path losses are shown for several satellite frequencies

FIGURE 16.72 Satellite link losses, spreading factors, maximum nadir angle θ-max, Earth central angle γ, and
one-way Earth–space time delay vs. satellite altitude, h km.

in use. The variation in path loss and Earth central angle is substantial. For example, L-band LEO personal communications systems to low cost handheld telephones with low gain (e.g., $G_u \approx 0$–3 dBi) need less link power than for MEO or GEO. On the other hand, more satellites are needed to provide full coverage since LEO satellites see much smaller fractions of the Earth.

The design for a constellation of satellites to serve communications needs, includes choices for the number of satellites, their orbital parameters, the satellite G/T and EIRP, etc. These mission analysis and design topics involve trades of many factors such as total communications capacity, link margins, space segment and Earth segment costs, reliability, interconnectivity, availability and cost of launch vehicles, mission lifetime, and system operations (Wertz and Larson, 1991).

16.8.7 Digital Links

For digital modulation systems, bit error rate (BER) is related to the dimensionless ratio (decibel difference) of energy per bit, E_b dB J to the total noise power density $N_o = 10 \log(kT)$ dB J (Sklar, 1988). For thermal noise

$$
\begin{aligned}
(E_b/N_o) &= (C/N) + B - R \\
&= (C/N_o) - R \text{ dB}
\end{aligned}
\tag{16.65}
$$

where $R = 10 \log(\text{transmission bit rate in bit per second})$, B is the bandwidth (decibel hertz), and (C/N_o) is the *carrier-to-thermal noise density ratio*, that is, C/N normalized to unit bandwidth. Curves relating the communications performance measure of BER vs. E_b/N_o for different modulations may be found in Sklar (1988). The link equation may then be expressed in terms of (E_b/N_o) and transmission data rate R, without explicit reference to the bandwidth

$$
(E_b/N_o) = EIRP + (G/T) + 228.6 - \left\{ 10 \log \left(4\pi r_s^2 \right) + 10 \log \left(\frac{4\pi}{\lambda^2} \right) \right\} - A + \Gamma - R \text{ dB} \tag{16.66}
$$

where the appropriate quantities are substituted depending on whether the uplink or downlink is being considered.

16.8.8 Interference and Total C/N and $E_b/(N_o + I_o)$

The C/N or $E_b/(N_o + I_o)$ for a complete transponder link combines the contributions of the uplink, downlink, and also the power sum of all interference signals due, for example, to intermodulation products generated in the output stages of the amplifiers, external interference from other systems, and intrasystem interference from reusing the same frequency band on spatially isolated or dual-polarized antenna beams to increase communications capacity. For most applications the total interference power may be taken as the power sum of interfering signals as long as they are not correlated with the desired carrier. The values for the interfering signals due to effects outside the satellite, for example, **frequency reuse** cross-polarization, multiple beam interferers, and other systems, must be obtained by carefully constructing the link equation for each case, taking into account the antenna gains for each polarization and beam direction of concern.

For an interference power i, and carrier power c (both in watt), the interference ratio c/i must be combined with the uplink and downlink c/n values to yield the total c/n. Here, the ratios are written in lower case to indicate they are *numerical power ratios* of quantities expressed in watts.

$$
\left(\frac{c}{n} \right)_{total} = \frac{1}{\left(\frac{c}{n} \right)_u^{-1} + \left(\frac{c}{n} \right)_d^{-1} + \left(\frac{c}{n} \right)_i^{-1}} \tag{16.67}
$$

Equation 16.67 applies to a bent pipe satellite. If onboard signal regeneration is used for digital transmission the uplink signal is demodulated and a *clean* set of baseband bits is remodulated. This has the effect of separating the accumulation of uplink and downlink noise contributions by causing the uplink noise to be effectively modulated onto the downlink carrier with the desired signal (Gagliardi, 1991). In that case, only the uplink or the downlink term in the denominator of Eq. (16.67) would be used as appropriate. Remodulation is also useful for intersatellite links. In each case, a savings in power or antenna size may be obtained at the expense of circuit and processing complexity.

The degradation to a digital link from interference follows a form similar to that of Eq. (16.67) but in terms of e_b/n_o, where the lower case quantities refer to numerical ratios. For a link that is subject to a *given* additive white noiselike interference power expressed as a ratio of desired signal power to interference power c/i, and assuming digital modulation with m bits per symbol,

$$\frac{e_b}{i_o} = \frac{1}{m}\frac{c}{i}$$ (16.68)

The ratio of energy per bit to total thermal noise plus interference power density is

$$\left(\frac{e_b}{n_o + i_o}\right) = \frac{1}{\left(\dfrac{e_b}{n_o}\right)^{-1} + \left(\dfrac{e_b}{i_o}\right)^{-1}}$$ (16.69)

For a system employing frequency reuse via dual polarizations, the polarization coupling factor Γ between a wave and antenna determines the interference power. The C/I due to polarization is the ratio of desired (copolarized) receive power and undesired (cross polarized) powers and is called the **polarization isolation**.

16.8.9 Some Particular Orbits

A *geosynchronous* orbit has a period that is a multiple of the Earth's rotation period, but it is not necessarily circular, and it may be inclined. Therefore, a *geostationary* Earth orbit (GEO) is a special case of a geosynchronous orbit where $e = 0$, $i = 0$, $k = (r_e + h)/r_e = 6.61$, and $h = 35,786$ km. When $el = 0$ the maximum nadir angle $\theta = 8.7°$, the maximum slant range is 41,680 km, and, from Eq. (16.58), $\gamma = 81.3°$. Therefore, a GEO satellite cannot see the Earth above 81.3° latitude.

Molniya and Tundra orbits are critically inclined ($i = 63.4°$) highly inclined elliptical orbits (HIEO) to cause the satellite's subsatellite ground trace to dwell at apogee at the same place each day. Such orbits whose subsatellite paths trace a repetitive loop (LOOPUS) allow several satellites to be phased to offer quasistationary satellite service at high latitudes. For full Earth coverage from a constellation of LEO satellites, circular polar constellations (Adams and Rider, 1987) and constellations of orbit planes with different inclinations, for example, Walker orbits (Walker, 1977) have received attention.

Table 16.21 compares the geometry, coverage, and some parameters relevant to the communications links for typical LEO, MEO (or ICO), and GEO systems. Reference should be made to Fig. 16.69 for the geometry and to the given equations for geometrical and link parameters.

16.8.10 Access and Modulation

Satellites act as central relay nodes, which are visible to a large number of users who must efficiently use the limited power and bandwidth resources. For detailed discussions of access issues see Gagliardi (1991), Pritchard et al (1993), Miya (1985), and Shimbo (1988). A brief summary of issues specific to satellite systems is now given.

Frequency-division multiple access has been the most prevalent access for satellite systems until recently. Individual users being assigned a particular frequency band may communicate at any time. Satellite filters subdivide a broad frequency band into a number of *transponder channels,* for example, the 500-MHz uplink FSS band from 5.925 to 6.425 GHz may be divided into 12 transponder channels of 36-MHz bandwidth plus guard bands. This limits the interference among adjacent channels in the corresponding downlink band of 3.7–4.2 GHz.

FDMA implies that several individual carriers coexist in the transmit amplifiers. To operate the amplifiers in a quasilinear region relative to their saturated output power to limit intermodulation products, the amplifiers, must be operated in a **backed off** condition. For example, to limit third-order intermodulation

TABLE 16.21 Comparison of Orbit and Link Parameters for LEO, MEO, and GEO for the Particular Case of Circular Orbits (Eccentricity $e = 0$) and for Elevation Angle $el = 10°$.

Earth radius r_e, km	6378.14		
Earth surface area a_e, km^2	511.2×10^6		
Elevation angle $el°$	10		
Orbit	LEO	MEO/ICO	GEO
Example system	Iridium$^®$	ICO-P	INTELSAT
Inclination $i°$	86.4	±45	0
Altitude h, km	780	10,400	35786
Semimajor axis radius a, km	7159	16,778	42164
Orbit period, min	100.5	360.5	1436.1
$(r_e + h)/r_e$	1.1222	2.6305	6.6107
Earth central angle $\gamma°$	18.658	58.015	71.433
Nadir angle $\theta°$	61.3	22	8.6
Nadir spread factor:			
$\quad 10\log(4\pi h^2)$, dB m^2	128.8	151.3	162.1
Slant range r_s, km	2325	14,450	40,586
Earth-space time delay, ms	2.6	51.8	139.1
Maximum spread factor:			
$\quad 10\log(4\pi r_s^2)$, dB m^2	138.3	154.2	163.2
$20\log(r_s/h)$, dB	9.5	2.9	1.1
Ground coverage area, km^2	13.433×10^6	120.2×10^6	174.2×10^6
Fraction of Earth area	0.026	0.235	0.34

power for two carriers in a conventional traveling wave tube (TWT) amplifier to ≈ -20 dB relative to the carrier, its input power must be reduced (*input backoff*) by about 10 dB relative to the power that would drive it to saturation. The output power of the carriers is reduced by about 4–5 dB (*output backoff*). Amplifiers with fixed bias levels will consume power even if no carrier is present. Therefore, DC-to-RF efficiency degrades as the operating point is backed off. For amplifiers with many carriers, the intermodulation products have a noiselike spectrum and the noise power ratio is a good measure of multicarrier performance.

When reusing the available frequency spectrum by multiple spatially isolated beams (SDMA), interference can result if the sidelobes of one beam receives or transmits substantial energy in the direction of the other beams. Two beams that point in the same direction may reuse frequencies provided that they are orthogonally polarized, for example, vertical and horizontal linear polarizations or right- and left-hand circular polarizations. Typical values of sidelobe or polarization isolation among beams reusing the same frequency bands are from 27 to 35 dB.

Time-division multiple access users (TDMA) share a common frequency band and are each assigned a unique time slot for their digital transmissions. At any instant the DC—RF efficiency is high because there is only one carrier in the transmit amplifier, which may be operated near saturation. A drawback is the system complexity required to synchronize widely dispersed users in order to avoid intersymbol interference caused by more than one signal appearing in a given time slot. Also, the total transmission rate in a TDMA satellite channel must be essentially the sum of the users' rates, including overhead bits such as for framing, synchronization and clock recovery, and source coding. Earth terminal hardware costs for TDMA have been higher than for FDMA. Nevertheless, TDMA systems are gaining acceptance for some applications as their costs decrease.

Code-division multiple access (CDMA) modulates each carrier with a unique pseudorandom code, usually by means of either a direct sequence or frequency hopping spread spectrum modulation. As the CDMA users occupy the same frequency band at the same time, the aggregate signal in the satellite amplifier is noiselike. Individual signals are extracted at the receiver by correlation processes. CDMA tolerates noiselike interference but does not tolerate large deviations from average loading conditions. One or more very strong carriers could violate the noiselike interference condition and generate strong intermodulation signals.

User access is via assignments of a frequency, time slot, or code. Fixed assigned channels allow a user unlimited access. However, this may result in poor utilization efficiency for the satellite resources and may imply higher user costs (analogous to a leased terrestrial line). Other assignment schemes include *demand assigned multiple access* (DAMA) and *random access* (e.g., for the Aloha concept). DAMA systems require the user to first send a channel request over a common control channel. The network controller (at another Earth station) seeks an empty channel and instructs the sending unit to tune to it (either in frequency or time slot). A link is maintained for the call duration and then released to the system for other users to request. Random access is economical for lightly used burst traffic such as data. It relies on random time of arrival of data packets and protocols are in place for repeat requests in the event of collisions. (See Gagliardi (1991)).

In practice, combinations of multiplexing and access techniques may be used. A broadband may be channelized or *frequency-division multiplexed* and FDMA may be used in each subband, for example, FDM/FDMA.

16.8.11 Frequency Allocations

Table 16.22 contains a partial list of frequency allocations for satellite communications. The World Administrative Radio Conference, WARC-92 allocated L-band frequencies for LEO personal communications services and for LEO small satellite data relay. The World Radiocommunication Conference, WRC-95,

TABLE 16.22 Partial List of Satellite Frequency Allocations, Frequencies in Gigahertz

Band	Uplink	Downlink	Satellite Service
VHF		0.137–0.138	Mobile
VHF	0.3120–0.315	0.387–0.390	Mobile
L-Band		1.492–1.525	Mobile
	1.610–1.6138		Mobile, Radio Astronomy
	1.613.8–1.6265	1.6138–1.6265	Mobile LEO
	1.6265–1.6605	1.525–1.545	Mobile
		1.575	Global Positioning System
		1.227	GPS
S-Band	1.980–2.010	2.170–2.200	MSS. Available Jan. 1, 2000
	(1.980–1.990)		(Available in U.S. in 2005)
	2.110–2.120	2.290–2.300	Deep-space research
		2.4835–2.500	Mobile
C-Band	5.85–7.075	3.4–4.2	Fixed (FSS)
	7.250–7.300	4.5–4.8	FSS
X-Band	7.9–8.4	7.25–7.75	FSS
Ku-Band	12.75–13.25	10.7–12.2	FSS
	14.0–14.8	12.2–12.7	Direct Broadcast (BSS) (U.S.)
Ka-Band		17.3–17.7	FSS (BSS in U.S.)
		22.55–23.55	Intersatellite
		24.45–24.75	Intersatellite
		25.25–27.5	Intersatellite
	27–31	17–21	FSS
Q	42.5–43.5, 47.2–50.2	37.5–40.5	FSS, MSS
	50.4–51.4		Fixed
		40.5–42.5	Broadcast Satellite
V	54.24–58.2-		Intersatellite
	59–64		Intersatellite

Sources: Final Acts of the World Administrative Radio Conference (WARC-92), Malaga–Torremolinos, 1992; 1995 World Radiocommunication Conference (WRC-95). Also, see Gagliardi, R.M. 1991. *Satellite Communications,* van Nostrand Reinhold, New York. Note that allocations are not always global and may differ from region to region in all or subsets of the allocated bands.

allocated S-band frequencies for mobile satellite services. Most of the other bands have been in force for years.

16.8.12 Satellite Subsystems

The major satellite subsystems are described in, for example, Griffin and French (1991). They are: propulsion, power, antenna, communications repeater, structures, thermal, **attitude** determination and control, telemetry, tracking, and command.

The satellite *antennas* typically are offset-fed paraboloids. Typical sizes are constrained by launch vehicles and have ranged from less than 1 m to more than 5 m for some applications. The Intelsat 6 satellite used a 3.2-m antenna at 4 GHz. Ku-band satellites may use a diameter > 2 m (i.e., $D > 80\lambda$). Multiple feeds in the focal region each produce a narrow *component beam* whose beamwidth is $\approx 65\lambda/D$ and whose directions are established by the displacement of the feed from the focal point. These beams are combined to produce a shaped beam with relatively high gain over a geographical region. Multiple beams are also used to reuse frequencies on the satellite. Figure 16.67 suggests that a satellite may have several beams for frequency reuse. In that case, the carriers occupying the same frequencies must be isolated from each other by either polarization orthogonality or antenna sidelobe suppression. As long as the sidelobes of one beam do not radiate strongly in the direction of another, both may use the same frequency band and increase the satellite's capacity.

The *repeaters* include the following main elements (see Fig. 16.67): A low noise amplifier (LNA) amplifies the received signal and establishes the uplink noise. The G/T of the satellite receiver includes the effect of losses in the satellite antenna, the noise figure of the LNA, and the noise temperature of the Earth seen from space (from 150 to 290 K depending on the percentage of the beam area over oceans and clouds). In a conventional repeater, the overall frequency band is downconverted by a local oscillator (LO) and mixer from the uplink band to the downlink band. It is channelized by an input multiplexer into a number (e.g., 12) of transponder channels. These channelized signals each are amplified by a separate high-power amplifier. Typically, a traveling wave tube amplifier (TWTA) is used with powers from a few watts to > 200 W for a DBS. Solid-state amplifiers can provide more than 15 W at C- and Ku-bands.

The *attitude determination and control system* (ADCS) must maintain the proper angular orientation of the satellite in its orbit in order to keep the antennas pointed to the Earth and the solar arrays aimed toward the sun (for example). The two prevalent stabilization methods are spin stabilization and body stabilization. In the former, the satellite body spins and the angular momentum maintains a gyroscopic stiffness. The latter uses momentum wheels to keep the spacecraft body orientation fixed. Components of this subsystem include the momentum wheels, torquers (which interact with the Earth's magnetic field), gyros, sun and Earth sensors, and thrusters to maintain orientation.

The *telemetry tracking and command* (TT&C) subsystem receives data from the ground and enables functions on the satellite to be activated by appropriate codes transmitted from the ground. This system operates with low data rates and requires omnidirectional antennas to maintain ground contact in the event the satellite loses its orientation.

The *power* subsystem comprises batteries and a solar array. The solar array must provide enough power to drive the communications electronics as well as the housekeeping functions and it must also have enough capacity to charge the batteries, which power the satellite during eclipse, that is, when it is shadowed and receives no power from the sun (Richharia 1995, 39). Typical battery technology uses nickel–hydrogen cells, which can provide a power density of more than 50 W-h/kg. Silicon solar cells can yield more than 170 W/m^2 at a satellite's beginning of life (BOL). Gallium arsenide solar cells (GaAs) yield more than 210 W/m^2. However, they are more expensive than silicon cells.

The space environment including radiation, thermal, and debris issues are described in Wertz and Larson (1991), and Griffin and French (1991). The structure must support all of the functional components and withstand the rigors of the launch environment. The thermal subsystem must control the radiation of heat to maintain a required operating temperature for critical electronics (Gilmore, 1994).

Defining Terms

Attitude: The angular orientation of a satellite in its orbit, characterized by roll (R), pitch (P), and yaw (Y). The roll axis points in the direction of flight, the yaw axis points toward the Earth's center, and pitch axis is perpendicular to the orbit plane such that $R \times P \longrightarrow Y$. For a GEO satellite, roll motion causes north–south beam pointing errors, pitch motion causes east–west pointing errors, and yaw causes a rotation about the subsatellite axis.

Backoff: Amplifiers are not linear devices when operated near saturation. To reduce intermodulation products for multiple carriers, the drive signal is reduced or backed off. Input backoff is the decibel difference between the input power required for saturation and that employed. Output backoff refers to the reduction in output power relative to saturation.

Beam and polarization isolation: Frequency reuse allocates the same bands to several independent satellite transponder channels. The only way these signals can be kept separate is to isolate the antenna response for one reuse channel in the direction or polarization of another. The beam isolation is the coupling factor for each interfering path and is always measured at the receiving site, that is, the satellite for the uplink and the Earth terminal for the downlink.

Bus: The satellite bus is the ensemble of all of the subsystems that support the antennas and payload electronics. It includes subsystems for electrical power, attitude control, thermal control, TT&C, and structures.

Frequency reuse: A way to increase the effective bandwidth of a satellite system when available spectrum is limited. Dual polarizations and multiple beams pointing to different Earth regions may utilize the same frequencies as long as, for example, the gain of one beam or polarization in the directions of the other beams or polarization (and vice versa) is low enough. Isolations of 27–35 dB are typical for reuse systems.

References

Adams, W.S. and Rider, L. 1987. Circular polar constellations providing continuous single or multiple coverage above a specified latitude. *J. Astro. Sci.* 35(2):155–192.

Björnstrom, G. 1993. Digital payloads: enhanced performance through signal processing. *ESA Journal* Vol. 17, pp. 1–29.

Briskman, R.D. 1996. Satellite radio technology. In *16th International Communications Satellite Systems Conference*, Feb. 25–29, pp. 821–825. AIAA, Washington, DC.

Chobotov, V.A. 1991. *Orbital Mechanics,* 2nd ed. AIAA, Washington, DC.

De Gaudenzi, S., Giannetti, F., and Luise, M. 1996. Advances in satellite CDMA transmission for mobile and personal communications. *Proc. IEEE* 84(1):18–39.

Feher, K. 1983. *Digital Communications: Satellite/Earth Station Engineering.* Prentice-Hall, Englewood Cliffs, NJ.

Feher, K. 1987. *Advanced Digital Communications.* Prentice-Hall, Englewood Cliffs, NJ.

Gagliardi, R.M. 1991. *Satellite Communications.* Van Nostrand Reinhold, New York.

Gedney, R. 1996. Considerations for satellites providing NII/GII integrated services using ACTS results. In *16th International Communications Satellite Conference*, AIAA Paper 96-1027-CP, pp. 344–353, Feb. 25–29. AIAA, Washington, DC.

Gilmore, D.G., ed. 1994. *Satellite Thermal Control Handbook.* Aerospace Corp. Press, El Segundo, CA.

Gordon, G. and Morgan, W. 1993. *Principles of Communication Satellites.* Wiley, New York.

Griffin, M.D. and French, J.R. 1991. *Space Vehicle Design.* AIAA, Washington, DC.

Helm, N.R. and Edelson, B.I. 1996. Satellites in the NII and GII. In *16th International Satellite Systems Conference*, Feb. 25–29, pp. 138–148. AIAA, Washington, DC.

Ippolito, L.J. 1989. *Propagation Effects Handbook for Satellite Systems Design.* NASA Ref. Pub. 1082(04).

Isakowitz, S.J. 1994. *International Reference Guide to Space Launch Systems,* 2nd ed. AIAA, Washington, DC.

Kiesling, J.D. 1996. Little LEOs, an important satellite service. In *16th International Communications Satellite Systems Conference,* Feb. 25–29, pp. 918–928. AIAA, Washington, DC.

Long, M. *World Satellite Almanac.* Howard W. Sams, Indianapolis, IN.

Long, M. *The 1991 World Satellite Annual.* MLE, Inc. Winter Beach, FL.

Miya, K., ed. 1985. *Satellite Communications Technology.* KDD Engineering and Consulting, Inc., Tokyo, Japan.

Morgan, W.L and Gordon, G.D. 1989. *Communications Satellite Handbook.* Wiley, New York.

Pocha, J.J. 1987. *An introduction to Mission Design for Geostationary Satellites.* D. Reidel, Dordrecht, The Netherlands.

Pratt, T. and Bostian, C.W. 1986. *Satellite Communications.* Wiley, New York.

Pritchard, W.L. et al. 1993. *Satellite Communications Systems Engineering.* 2nd ed. Prentice-Hall, Englewood Cliffs, NJ.

Rees, D. 1990. *Satellite Communications, The First Quarter Century of Service.* Wiley, New York.

Richharia, M. 1995. *Satellite Communications Systems, Design Principles.* McGraw-Hill, New York.

Roddy, D. 1996. *Satellite Communications,* 2nd ed. McGraw-Hill, New York.

Schwartz, E. 1990. *Information Transmission, Modulation, and Noise.* McGraw-Hill, New York.

Scott, A. 1993. *Understanding Microwaves.* Wiley, New York.

Shimbo, O. 1988. *Transmission Analysis in Communications Systems,* Vols. 1 and 2. Computer Science Press, New York.

Sklar, B. 1988. *Digital Communications.* Prentice-Hall, Englewood Cliffs, NJ.

Teledisic Corp. 1995. FCC filing.

Tuck, E.F. et al. 1994. The calling SM network: A global wireless communication system. *Int. J. Sat. Commun.* 12(1):45–61.

Walker, J.G. 1977. Continuous whole-Earth coverage by circular orbit satellite patterns. Tech. Rept. 77044, Royal Aircraft Establishment, Farnborough, Hants, U.K.

Wertz, J.R., ed. 1978. *Spacecraft Attitude Determination and Control.* D. Reidel, Dordrecht, The Netherlands.

Wertz, J.R. and Larsen W.J., eds. 1991. *Space Mission Analysis and Design.* Kluwer Academic, Dordrecht, The Netherlands.

Further Information

For a brief history of satellite communications see *Satellite Communications: The First Quarter Century of Service,* by D. Reese, Wiley, 1990.

Many of the organizations mentioned can be accessed via the Internet. Several examples include: NASA (http://nasa.gov); International Telecommunications Union (ITU) (http://itu.ch); INTELSAT (http://www.intelsat.int:8080); Inmarsat (http://www.worldserver.pipex.com/inmarsat/index.htm); and FCC (http://www.fcc.gov/).

17
Radar and Radionavigation

James M. Howell
Raytheon Corporation

Melvin L. Belcher, Jr.
Georgia Tech Research Institute

James A. Scheer
Georgia Tech Research Institute

Benjamin B. Peterson
U.S. Coast Guard Academy

Sanjay K. Mehta
Naval Undersea Warfare Center

Clifford G. Carter
Naval Undersea Warfare Center

Bernard E. McTaggart
Naval Undersea Warfare Center

Robert D. Hayes
RDH Incorporated

17.1   Radar Principles

James M. Howell

17.1.1   Introduction

Radar is an acronym for radio detection and ranging as these were primary functions during the early use of radar. Radars can also measure other target properties such as range rate (Doppler), angular location, amplitude statistics, and polarization scattering matrix. In its simplest form, a radar propagates a pulse from an antenna to a target. The target reflects the pulse in many directions with some of the energy backscattered toward the radar. The radar return is received by the radar and subjected to processing to allow its detection. Since the pulse travels at approximately the speed of light, the distance to the target can be determined based on the round trip time delay. Reflections from undesired targets are known as *clutter* and often include terrain, rain, man-made objects, etc. Usually, the radar will have a narrow beam so that the angular location of the target (i.e., azimuth and elevation) can also be determined by some technique such as locating the centroid of the target returns as the beam scans across the target or by comparing the signals received simultaneously or sequentially by different antenna patterns or overlapped beams. The radial velocity of the target can be determined by differencing the range measurements. Since the range measurements may not be very accurate, better range rate accuracy can be obtained by coherently measuring the Doppler frequency; that is, phase change from pulse-to-pulse in a given range cell. At microwave frequencies, the wavelength is quite small and, hence, small changes in range are readily detected. Generally, frequency is measured by using a **pulse Doppler** filter bank, pulse pair processing, or a CW frequency discriminator. Coherently measuring the frequency is also a good way for filtering moving targets from stationary or slowly moving clutter.

Radar parameters vary with the type of radar. Typically, the radar transmitted pulse width is 1 to 100 μs with a pulse repetition frequency (prf) of 200 Hz –10 KHz. If the antenna is a mechanically rotated reflector, it generally rotates $360°$ in azimuth at about 12–15 r/min. If the two-way time delay is T, the range to the target is $R = 0.5T \times c$, where $c = 3 \times 10^8$ m/s is the velocity of light.

17.1.2 Applications

Radars can be classified by frequency band, use, or platform, for example, ground based, shipborne, airborne, or spaceborne. Radars generally operate in the microwave regime although HF over-the-horizon (OTH) radars such as JINDALEE, OTHB, and ROTHR use similar principles in bouncing signals from the ionosphere to achieve long-range coverage. Radars are often denoted by the letter band of operation, for example, L-band (1–2 GHz), S-band (2–4 GHz), C-band (4–8 GHz), and X-band (8–12 GHz). Some classifications of radar are based on propagation mode (e.g., monostatic, bistatic, OTH, underground) or on scan method (mechanical, electronic, multibeam). Other classifications of radar are based on the waveform and processing, for example, pulse Doppler (PD), continuous wave (CW), FM/CW, synthetic aperture radar (SAR) or impulse (wideband video). Radars are often classified by their use: weather radar, police speed detection, navigation, precision approach radar, airport surveillance and air route surveillance, radio astronomy, fire control and weapon direction, terrain mapping and avoidance, missile fuzing, missile seeker, foliage penetration, subsurface or ground penetrating, acquisition, orbital debris, range instrumentation, imaging (e.g., SAR/ISAR), etc.

Search (or surveillance) radars are concerned with detection of targets out to long range and low-elevation angles to allow adequate warning on pop-up low-flying targets (e.g., sea skimmers). Since the search radar is more concerned with detection (i.e., presence or absence of targets) and can accommodate cruder accuracy in estimating target parameters such as azimuth angle, elevation angle, and range, search radars tend to have poorer range and angle accuracy than tracking radars. The frequency tends to be lower than track radars since RF power and antenna aperture are less expensive and frequency stability is better. Broad beams (e.g., *fan beam*) allow faster search of the volume. To first order, the radar search performance is driven by the *power-aperture* product (PA) to search the volume with a given *probability of detection* (PD) in a specified frame time. PA actually varies slightly in that to maintain a fixed false alarm rate per scan, more beam positions offer more opportunities for false alarms and, hence, the detection threshold must be raised, which increases the power to achieve the specified PD. With a phased-array antenna (i.e., electronically scanned beam), the probability of false alarm can be optimized by setting a high false alarm in the search beam and using a verify beam with higher threshold to confirm whether a search detection was an actual target or just a false alarm. The lower threshold in search allows less search power with some fraction of beams requiring the extra verify beams. The net effect on total required transmit power can be a reduction using this optimization technique. Search radars tend to use a fan beam or stacked receive beams to reduce the number of beam positions allowing more time in the beam for coherent processing to reduce clutter. Fill pulses are sometimes used to allow good clutter cancellation on second- or higher time-around clutter returns.

Track radars tend to operate at higher frequency and have better accuracy, that is, narrower beams and high range resolution. Simple radars track a single target with an early–late range tracker, Doppler speed gate, and conical scan or sequential lobing. More advanced angle trackers use monopulse or *conical scan on receive only* (COSRO) to deny inverse modulation by repeater jammers. The multifunction phased-array radar can be programmed to conduct searches with track beams assigned to individual detected targets. The tracks are maintained in track files. If time occupancy becomes a problem, the track pulses can be machine gunned out at the targets in range order, and on receive they are gathered in one after the other since the track window on each target is quite small. In mechanically rotated systems, track is often a part of search, for example, *track-while-scan* (TWS). A plot extractor clusters the primitive returns in range Doppler angle from a given target to produce a single plot. The plots are associated with the track files using scan-to-scan correlation gates. The number of targets that can be handled in a TWS system is limited by data processing rather than track power.

17.1.3   Radar Configuration and Subsystems

A typical radar consists of a transmitter, *receiver–exciter* (REX), antenna, microwave components, signal processor, and data processor. Transmitters use solid-state or tubes (hollow state) as RF amplifiers. The solid-state power amplifier uses bipolar transistors at the lower frequencies, occasionally impact avalanche transit times (IMPATTs) at higher frequencies, and usually GaAs field effect transistors (FETs) at the higher frequencies. Solid-state transmitters have advantages of low voltage, high reliability due to graceful failure of multiple combined output stages, and maintainability, which can often be performed with the radar on-line. Since solid-state devices are peak power limited, long pulses are used to provide the required energy for detection with pulse compression on receive to provide the required range resolution. The long pulse sets the minimum range since the receiver generally cannot be receiving while the transmitter is transmitting. A short pulse at a different frequency is often transmitted before the long pulse and is used to provide coverage in the region blanked by the long pulse. A different frequency prevents longer range returns from the short pulse being detected when receiving the long pulse. Tube transmitters use either a traveling wave tube (TWT), magnetron, crossfield amplifier (CFA), klystron, or twystron. Good design of the transmitter includes, low spurious signals, low droop and timing jitter, etc.

The antenna is usually a reflector, lens, phased array, or some hybrid combination (e.g., limited scan applications). The antenna can be either rotated mechanically or scanned electronically (inertialess beam agility) using phase shifters and/or time-delay units. Cost considerations favor the low-cost reflector except in those cases where an array is essential, for example, rapidly switching between hundreds of targets. The cost of hundreds of reflectors would be even more expensive. Multiple beam lenses (e.g., R2R, RKR, **Rotman**, Ruze) are usually scanned with RF switches or *variable power divider* (VPD) networks. The antenna pattern is generally reciprocal on transmit and receive. The antenna pattern is created on transmit by radiation coming from different portions of the aperture with phase differences depending on direction. Mathematically, the pattern is a Fourier transform of the aperture illumination. Amplitude tapering (or space tapering/thinning the active elements of a phased array) lowers the sidelobes but increases the beamwidth. Common phased-array sum tapers include Taylor (either standard with $1/\mu$ sidelobe envelope or generalized with $1/\mu^n$ falloff), Tchebysheff, convolved Taylor (i.e., $\sin x/x$ convolved with Taylor to provide rapid falloff outside a given window), and Hamming weighting (i.e., optimum cosine-on-pedestal). Difference tapers include Bayliss, transmuted Taylor (i.e., linear weighting of Taylor sum with sidelobes about 10 dB higher), etc. Shaped antenna beam systems use **Woodward** and/or pattern-zero **synthesis** techniques. Phase-only synthesis can also be used to generate shaped beams. This approach uses the principle of stationary phase to scan different parts of the aperture toward different angles (i.e., similar to shaped reflector synthesis or the design of nonlinear frequency modulation (NLFM) waveforms). Phase synthesis can also be used to produce a notch in the elevation pattern near the horizon to minimize the effects of standoff jamming or low-angle multipath. Phase synthesis of planar arrays having uniform amplitude can also use checker-board synthesis where the phase of adjacent elements is adjusted to create the proper in-phase signal, and the sign of the quadrature signals are rotated successively from element-to-element. The in-phase signals add coherently producing the desired beam, whereas the quadrature signals add quasirandomly creating a sidelobe hashlevel that decreases with the number of elements. Performance is equivalent to using attenuators to produce an amplitude taper across the uniform amplitude aperture with the exception that the lost energy is in the sidelobes rather than terminated in attenuators. The antenna beamwidth of an antenna is inversely proportional to both the aperture dimension and the RF frequency. Many antennas are rotated mechanically or scanned electronically, or use a hybrid combination of mechanical scan in azimuth and electronic scan in elevation. Phased arrays use diode or ferrite phase shifters for scanning the beam. The diode devices switch in nanoseconds, are reciprocal, and have small sensitivity to temperature. The ferrite devices switch in 10–100 μs and can be reciprocal or nonreciprocal as well as analog or digital. Ferrites have lower loss at the higher frequencies and can handle higher power levels. Flux drive is used for accurate phase setting. Frequency scan is sometimes used with slotted waveguide arrays for low cost, but it often limits frequency agility and makes each angle associate with a given frequency, which makes a jammer more effective against the radar.

The receiver–exciter (REX) provides precision timing and the high-quality frequency source, generates the waveform (digital synthesizer; waveform stored in programmable read-only memories (PROMs) and read out through a D/A converter), and provides the upconversion, amplification, and filtering needed for a drive signal to the transmitter. It provides timing signals for other subsystems and may include a test target generator for calibrating system performance. It also provides the downconversion on receive. The receiver should have good linearity over a wide dynamic range to prevent intermods and should have good image rejection (through the use of double downconversion). The front-end filtering (selectivity) provides rejection of interference and other undesired signals. The receiver can provide electronic counter-countermeasure (ECCM) features such as a Q-channel, sidelobe blanker, coherent sidelobe canceller loops, guard band blanking, Dicke-fix/log-fast time constant (FTC), etc. The receiver often provides sensitivity time control (STC) to prevent saturation at short ranges by clutter and undesired targets. Since RF STC at longer ranges can affect low sidelobe pulse compression waveforms, the STC can be performed digitally via threshold setting after pulse compression to prevent saturation by large objects. In a surveillance radar, such as an airport surveillance radar (ASR) or air route surveillance radar (ARSR), the antenna pattern should have a thumb at high-elevation angles to accomodate high-altitude detection when STC or a min-range short pulse is used. Pulse compression is sometimes accomplished in the receiver using analog **surface acoustic wave (SAW)** lines (e.g., **reflective array compressor (RAC)** or **impedance control (IMCON)** lines). The receiver local oscillators (LOs) should be stable, switchable from dwell-to-dwell for frequency diversity (decorrelate rain, chaff, and/or targets) or frequency jump burst (FJB) processing to allow wide-band waveforms to be passed through narrow-band receivers. A frequency swept local oscillator is needed for stretch processing, which enables wide-band waveforms to be passed through lower sampling rate A/D converters with high-range resolution over a narrow range window. The REX provides the analog anti-aliasing filter prior to A/D sampling. The receiver uses either direct sampling of the IF signal or coherent product detectors. An example of direct sampling is having the IF frequency at 4 times the sampling frequency. I and Q samples (i.e., in-phase and quadrature) are constructed by subtracting every other in-phase sample and every other quadrature sample. A Hilbert transform can be used to align the quadrature samples in time with the in-phase samples. Direct sampling avoids DC offsets from the A/D converter, and unlike the use of product detectors does not have to worry about amplitude and phase balance of the coherent product detectors. The video gain is set so that the thermal noise is set at one or two least significant bits (LSBs) of the A/D converter to provide a compromise between added noise due to A/D converter quantization, coherent extraction of signals below noise, and preservation of large dynamic range. Automatic gain control (AGC) can be used to protect the A/D converter from saturation. The A/D converter should have a low spurious level and the sample rate should be sufficient to reduce range straddling loss. If the sampling is at least Nyquist, the samples can be interpolated later for higher accuracy to minimize the peak sampling loss. The A/D converter often provides the interface to the signal processor.

Microwave components (IKEE) often include a circulator or gas transmit–receive (TR) switch (duplexer) or diplexer, RF filtering, pilot pulse injection, forward and backward couplers for monitoring power and voltage standing wave ratio (VSWR), rotary joint and associated slip rings, pressured waveguide for high power, arc detectors, etc.

The signal processor provides the digital processing for pulse compression, Doppler filtering, clutter map, constant false alarm rate (CFAR), and detection thresholding. In a *multifunction array radar* (MFAR), the signal processor may also provide scheduling (adaptive or template) and generate beam steering commands. For a wider repertoire of waveforms, modern radars provide pulse compression in the signal processor using the fast Fourier transform (FFT). Linear frequency modulated (LFM) waveforms are often used because the compressed waveform is invariant with respect to Doppler frequency (other than the range-Doppler coupling due to the associated range-Doppler ambiguity) and allows a common compressor to be used for all Doppler filters. To prevent degradation due to Doppler sensitivity of phase-coded waveforms, a different pulse compressor may be required for each Doppler filter, which greatly increases processing throughput. Inverse filters can be used to compensate for the Fresnel ripples of low time-bandwidth waveforms or to compensate for system errors. Uncompensated correlated errors can lead to large range sidelobes (i.e., paired echoes). LFM waveforms in track are stretch processed to allow digital pulse compression of wide-band signals using relatively low sample rate A/D converters over a relatively small range window. The receive LFM signal is deramped to a CW tone, which is amplitude weighted in time and fast Fourier transformed to yield a compressed pulse with low range sidelobes. Since the weighting is applied in the time domain it may not be properly aligned with the uncompressed pulse, and range sidelobes will degrade. A better approach is to apply the weighting in the frequency domain so that it is independent of target position within the range window. This can be accomplished over a larger range window by using 3 cascaded FFTs rather than a single FFT. If the LFM bandwidth is not large, stretch processing is not needed and the pulse can be passed through A/D converter and compressed via an FFT, followed by phase conjugation and spectral weighting for reducing range sidelobes, followed by an inverse FFT. This allows all-range processing in search rather than pulse compression over only a small track window.

The data processor is a general or special purpose computer(s). The data processor processes the detected signals and includes plot extraction (i.e., reducing multiple returns from a single target (range-Doppler-angle) during a scan to a single report), associate a target return with a given track file, provides prediction/filtering to maintain track files with track intiation, drop, and promotion logic. The data processor can also correlate (data fusion) the target return with data from other primary surveillance radar (PSR) or secondary surveillance radar (SSR) sensors. An *extended Kalman filter* (EKF) is usually used for tracking since it is recursive and includes an optimal averaging of the measurements and predictions using a builtin plant (physics of motion) model. A generalized Kalman tracker can be used to combine the data from multiple sensors. Other approaches include acceleration detection or adaptive bandwidth based on innovations alarm, and interacting muliple models (IMM). Since EKF requires accurate tuning and optimal selection of coordinates (**RVCC, RUV**, etc.), simpler radars often use polynomial filters (alpha-beta-gamma or ghk) with fixed gains. Tracking filters require a compromise between smoothing out noise vs. responding to fast changes (i.e., minimize dynamic lag). An adaptive bandwidth helps in this regard.

Radar Range Equation

The signal-to-noise ratio (SNR) for a single pulse from a monostatic pulsed radar in free space is given as

$$2E/No = SNR = \frac{P_t \tau G_T}{4\pi R^2}(A)\left(\frac{\sigma}{4\pi R^2}\right)\left(\frac{1}{L}\right)\left(\frac{1}{kTs}\right)$$

where Pt is the peak power, τ is the uncompressed pulsewidth, Gt is the transmit antenna gain, A is the effective receive aperture, σ is the radar cross section (RCS) of the target, R is the range, $L > 1$ represents system losses, k is Boltzmann's constant, and Ts is the system noise temperature given by

$$Ts = Ta + 290 \times (Lr\,NF - 1)$$

where Ta is antenna temperature, Lr is receive loss, and NF is the noise figure.

Since the receive antenna gain is given by

$$Gr = \frac{4\pi A}{\lambda^2}$$

SNR can be written as

$$SNR = \frac{P_t \tau G_T G_R \lambda^2}{(4\pi)^3 R^4 L k T s}$$

As written, SNR is proportional to λ^2. Expressed in other combinations of antenna aperture or gain, SNR varies inversely, directly, or independently of frequency. The units used in the radar range equation should be consistent. The system losses include transmit loss, signal processing losses (A/D converter quantization, range straddling, Doppler weighting and straddling, filter mismatch, constant false alarm rate (CFAR)), beamshape loss, scan loss, lens loss due to atmospheric divergence, atmospheric and rain attenuation, frequency squint over the bandwidth, etc. Range eclipsing and blind velocities associated with a given prf are assessed more accurately by their effect on probability of detection rather than treated simply as a loss. The signal and thermal noise should be calculated at the same point within the receiver chain with the system noise referenced to that location. For coherent integration and N pulses received in a coherent processing interval (CPI), the SNR can be increased by a factor of N with any additional losses placed in L. The product $P_t \tau$ accounts for the pulse energy and can be replaced in a pulse Doppler radar by $P_{ave} \times T_c$ where P_{ave} is the average power and Tc is the coherent integration time over the CPI.

In a search radar with a given prf, the number of pulses to be integrated in a beam depends on the search time over the volume and number of beam positions in that volume. Hence, N decreases linearly with aperture A since the beam area is inversely proportional to A. Since SNR depends on $GtGr$, radar sensitivity has an A^2 dependence. In search, however, the SNR has only a power-aperture dependence since N is inversely proportional to A. This would not be true, however, if the elevation coverage was quite small near the horizon and a large vertical beamwidth antenna were used. Doubling the vertical aperture would still provide the proper elevation coverage, but would increase the SNR in each beam by 6 dB since the horizontal beamwidth did not change. This would allow a 6-dB decrease in transmit power for the same performance rather than a 3-dB reduction as expected from the PA relationship. For a bistatic radar there are two ranges, Gt and Gr are different, and the bistatic radar cross section is used. To evaluate clutter-to-noise ratio (CNR), σ is replaced by the clutter coefficient and L is adjusted to show those losses particular to clutter (most signal processing losses, beam squint or beam shape loss are relevant only to the target). To include ground lobing effects, SNR is multiplied by a pattern factor F^4 where

$$F^4 = \left| 1 + D\rho \frac{E(\theta_R)}{E(\theta_d)} e^{j\frac{2\pi}{\lambda}\Delta r} \right|^4$$

where D is the spherical divergence factor and ρ factors into a diffuse term, shadowing term, and Fresnel specular reflection terms given by

$$\rho = \rho_{SHAD}\rho_s\rho_d$$

where the specular Fresnel reflection coefficient for horizontal and vertical polarizations are given by

$$\rho_h = \frac{\sin \chi - \sqrt{\varepsilon}}{\sin \chi + \sqrt{\varepsilon}}$$

$$\rho_v = \frac{\sqrt{\varepsilon} \sin \chi - 1}{\sqrt{\varepsilon} \sin \chi + 1}$$

$$\varepsilon = \varepsilon_r - j60\sigma\lambda$$

$$\rho_d = e^{-\left(\frac{4\pi\sigma_h \sin \chi}{\lambda}\right)^2}$$

where χ is the grazing angle and σ_h is the RMS height of the rough surface.

To reduce ground lobing effects in search, one desires a tall vertical aperture with a sharp cutoff near the horizon (e.g., generated by Orchard synthesis). With ground lobing there is always a null near the horizon, which degrades the detection of low flyers. Ground lobing extends the coverage at elevation angles corresponding to lobing peaks and reduces detection in the nulls. The target can have fades as it passes through the lobes of a two-dimensional surveillance radar. Height estimates can be crudely defined based on the aircraft velocity and the time between fades.

Various techniques can be used in search to improve detection performance. For example, if the target is large (e.g., a ship) and the pulse width is much shorter than the target, the return from a given transmitted pulse can be noncoherently integrated across the ship's length yielding essentially the same effect as if a number of pulses had been transmitted. Another approach is to perform detection on both the sum and difference receive beams. The thermal noise will be different in these channels providing independent looks. In most cases the target will be detected in the sum channel, but when the target is farther from the beam axis, detection in either the Azimuth (AZ) or Elevation (EL) difference channel may be more likely. The benefit in sensitivity might offset the higher cost associated with more processing throughput, particularly as computation become less expensive. Another approach is *track-before-detect* (TBD) where various trajectory threads are integrated noncoherently from scan-to-scan to allow detection of weaker targets. Many possible trajectories increase the processing. Detection on a given thread yields the track following detection.

In track, the angle accuracy is proportional to beamwidth $/\sqrt{\text{SNR}}$. One desires to size the radar system to achieve a given angle accuracy. The square of angle accuracy is proportional to beamwidth squared divided by SNR. Since beamwidth squared is proportional to aperture area, the angle accuracy is inversely proportional to PA^3. If a radar (e.g., phased array) performs both search and track, the radar can be optimized in cost by recognizing that the transmit RF power costs so much per watt and the aperture costs so much per square foot (or per element). One desires a given PA in search and a given PA^3 in track (i.e., two performance constraints and two radar parameter variables). Simple differentiating gives the optimum power and aperture sizes to minimize cost. Bucket curves can be constructed to show cost sensitivity for departing from the optimum design.

Rain Clutter

The reflectivity of rain, snow, fog, wind blown sand, etc. is given by η where η has units such as square meter per cubic meter. Here η depends on the distribution of the particles (e.g., Marshall–Palmer). The radar resolution cell is given by

$$V_c = c_1(\Delta r)(R\theta_{\text{AZ}})(R\theta_{\text{EL}})$$

where the constant c_1 depends on the effective pulse width and beamwidth (e.g., 1-way vs. 2-way beamwidths, shapes). The average clutter cross section is given by $\sigma_c = \eta V_c$. The clutter amplitude will actually fluctuate according to a Rayleigh distribution (central limit theorem and large number of rain drops) or, perhaps, K distributed with a large number of degrees of freedom. The CNR is similar to SNR with σ replaced by σ_c, and losses L and pattern factor F^4 adjusted appropriately. Since the clutter is distributed, it does not experience the same losses as a point target. The rain clutter will also experience the attenuation associated with rain. If the rain does not completely fill the elevation beam, the rain height should be used in place of $R\theta_{\text{EL}}$ with a possible reduction in antenna gain since the rain may only be in the lower portion of the beam at longer ranges. A more rigorous approach is to integrate the rain in elevation with the 2-way pattern, varying attenuation, and rain rate with elevation angle. Rain rate generally decreases in a Gaussian fashion with height.

Surveillance antennas are generally designed with circular polarization (CP), which will reject the returns from a symmetrical (single bounce) object such as a rain drop. Since the antenna pattern is not perfectly polarized in all directions, the antenna rain rejection performance can be stated in terms of an *integrated cancellation ratio* (ICR), which is obtained by integrating the rain through the *X*POL pattern over all angles and comparing that to the amount of rain received if the polarization were matched to the rain return. Practical values of antenna ICR are about 20 dB. Even if the antenna polarization were perfect, ICR is limited in practice by the increasingly nonspherical shape of the rain drops with rain rate, splashes from the ground that produce resonant dipolelike returns that are appreciably larger than the Mie scattering of the drops, and ground reflections at angles steeper than the Brewster angle, which tends to reverse the sense of CP so that it will not cancel after directly returning from a rain drop. A sharp cutoff antenna pattern (i.e., vertical aperture) will minimize the latter two limitations.

Land or Sea Clutter

The average cross section of terrain or ocean is given by σ_o. The clutter will have both a spatial and temporal distribution. Spatially the clutter tends to have a distribution with long tails such as Weibull, K distributed, or log-normal (particularly at low grazing angles, high-range resolution, higher sea states, and horizontal polarization). Temporally, the distribution is often Rayleigh. A common value of $\sigma_o = \gamma \sin \chi$ where γ is a constant depending on frequency and polarization, whereas χ is the grazing angle. The resolution cell area is, generally,

$$A_c = c_2 \frac{(\Delta r)(R\theta_{AZ})}{\cos \chi}$$

but can be different if the elevation beamwidth intercepts less ground than the pulse width. The clutter cross section is given by $\sigma_c = \sigma_o A_c$. The clutter tends to be smaller for horizontal polarization and this polarization is often used by shipboard surveillance radars.

Radar Range Equation with Jamming

Jamming will often dominate over the receiver thermal noise and, hence, one is interested in the signal-to-jamming power ratio (S/J). This can be obtained from the previous SNR equation by replacing the $No = kTs$ term by

$$J = \frac{\text{EIRP}\left(\dfrac{1}{B_j}\right)}{4\pi R j^2}\left(\frac{A_j}{L_j}\right)$$

where EIRP is the effective isotropic radiated power, i.e., product of jammer power and antenna gain toward the radar; B_j is the jammer bandwidth; R_j is the jammer range; A_j is the effective radar aperture in the direction of the jammer; and L_j are losses on the jammer. For a self-screening jammer, the target and jammer ranges are the same. Since $R_j = R$, S/J varies as $1/R^2$ rather than $1/R^4$ for thermal noise.

Radar Propagation

When the radar height is h and a spherical 4/3 effective Earth's radius is assumed, the range to the horizon is given by $d = 1.23\sqrt{h}$ where d is in nautical miles and h is in feet. The troposphere causes refractive bending according to Snell's law since the air density decreases with altitude bending the wave downward. This can be approximated in plotting coverage on a range–height (R–H) diagram by drawing rays as straight lines but using a 4/3 Earth radius. Anomalous propagation corresponds to a value larger than 4/3.

Ducting conditions can occur over land with temperature inversions over water causing energy to be trapped in the duct at elevation angles below a given critical angle. This causes strong clutter returns to arrive from ranges beyond the horizon since there is only energy spreading in the horizontal direction. These strong returns can appear at closer ranges (second time around) on a plan position indicator (PPI).

Plots of atmospheric attenuation can be found in many books. These curves are for a radar at the surface. Similar curves can be constructed for airborne radars by integrating the attenuation along the line of sight for an elevated radar. In general, atmospheric attenuation increases with frequency, lower radar heights, and lower elevation angles. There are atmospheric loss windows at 35, 94, and 140 GHz. An additional atmospheric loss is the lens loss due to the dispersive bending; that is, the bending is different at the upper and lower 3-dB points on the pattern spreading the energy out and reducing gain.

Attenuation also occurs with propagation through fog, dust, foliage, and missile plumes. At microwave frequencies, foliage attenuation is very severe, for example, about 1 dB/m. The loss is less for horizontal polarization (less tree trunk scatter) at lower frequencies. At long distances, the least loss propagation path through jungle is upward with diffraction along the canopy and diffraction back down into the foliage.

The attenuation through missile plumes can be severe and increases with engine thrust (more ionization) and metallic particles in the exhaust.

Rain and other particulate matter also cause attenuation through either scattering or absorption. The rain attenuation is given by $\alpha = ar^b$ where r is the rain rate in millimeter per hour. Rain can be both distributed (strataform) and in convective cells. The diameter of the rain cells decrease with rain rate. Rainfall is generally classified as drizzle (0.25 mm/h), light (1.0 mm/h), moderate (4 mm/h), heavy (16 mm/h), and excessive (>40 mm/h).

The ionosphere produces Faraday rotation, which can cause polarization loss for a linearly polarized signal and can break the return waveform into two waveforms due to right-handed circular polarization (RHCP) and left-handed circular polarization (LHCP) having different propagation velocities. The time delay is frequency dependent and, hence, the ionosphere can change the ramp on an LFM waveform unless compensated.

Antenna temperature is a contribution to the radar system temperature and depends on where the antenna pattern is pointed. A narrow beam elevated high above the Earth will have a low temperature, whereas a broad elevation beamwidth illuminating much of the ground and low elevation angles will have a higher temperature. On transmit, energy is transmitted in various directions as defined by the antenna pattern, that is, toward ground, various regions of sky including stars, sun, etc., coupled into the radome as losses or into nearby structures. By reciprocity, all of these terminations are at various temperatures that radiate inband energy, which returns to the receive antenna via the same paths that occur during transmit. Hence, the antenna temperature is obtained by integrating the environmental temperatures through the receive pattern. The temperature contribution from the sky into a narrow beam can be approximated by

$$ Tan\,t = 290 \left[1 - \frac{1}{\sqrt{L}} \right] $$

where L is the two-way atmospheric loss through the entire atmosphere at a given elevation angle.

The presence of the Earth creates lobing as the ground reflection interferes constructively or destructively with the direct signal. Ground roughness reduces the extent of the lobing, however. The elevation angle for the first lobe peak is $E = 0.25\lambda/h$ where h is the radar height. Other lobes occur at odd multiples of this angle, and nulls occur at even multiples. At the lower microwave frequencies, there is some diffraction over mountains, hills, buildings, and the Earth's horizon. These can be approximated by simple knife edge diffraction or diffraction formulas for a cylinder.

Diffuse scattering from rough ground comes primarily from a region called the glistening surface. The larger is the RMS slope of the surface, the larger is the region. At optical frequencies, the phenomena can be seen when the sun or moon rises over a slightly rough body of water.

17.1.4 Targets and RCS

The RCS of a target is the area assumed to intercept incident radiation on the target, which, when scattered isotropically, yields the actual power density received at the radar. A complex target is made up of many scatterers and, hence, the backscatter RCS varies with angular direction due to the constructive and destructive interference from the various scatterers. Similar to an antenna pattern, the lobes change more rapidly with angle when the target size is increased as measured in wavelengths. Changing frequency will cause the lobes to move when the frequency change is enough to make the phase difference along the target change by $180°$. Frequency agility is used to make the RCS change so that if it happens to be in a null at one frequency it will probably be at a peak on another frequency. The bistatic RCS varies with angle for a given illumination angle of the target. When the angle between the incident and reflected angles is small, the bistatic cross section is approximately the backscatter RCS at the bisector angle since the scatter phasing is equivalent.

When a re-entry vehicle (RV) enters the atmosphere, an ionized sheath begins to form in front of the RV nose. An RCS dip can occur when the ionized layer is sufficiently thick that reflections from the nose and front of the layer cancel. This will occur at a given altitude depending on the speed and entry angle of the RV and the wavelength of the radar.

Radar targets can be either fluctuating or nonfluctuating (steady). Particular models are Swerling targets SW0, SW1, SW2, SW3, and SW4. SW0 is a steady target. SW1 and SW2 have a probability density function (pdf) given by

$$p(\sigma) = \frac{1}{\sigma_{\text{ave}}} e^{-\sigma/\sigma_{\text{ave}}}$$

whereas the pdf for SW3 and SW4 are given by

$$p(\sigma) = \frac{4\sigma}{\sigma_{\text{ave}}} e^{-\frac{2\sigma}{\sigma_{\text{ave}}}}$$

SW1 and SW2 are chi square with $m = 1$, whereas SW3 and SW4 are chi square with $m = 2$. SW1 and SW2 are slow and fast fluctuating Rayleigh targets where *slow* means that the target RCS is essentially constant during the dwell but changes from scan-to-scan and *fast* means that the RCS fluctuates on a pulse-to-pulse basis. For a single pulse, SW1 and SW2 are the same and detection performance can be represented by

$$PD = PFA^{\frac{1}{1+SNR}}$$

Note that if SNR is zero, probability of detection (PD) is simply the probability of false alarm (PFA) as expected. When SNR approaches infinity, PD approaches unity as expected. Frequency agility will convert an SW1 target into an SW2 target. SW3 and SW4 targets are slow and fast fluctuating targets whose RCS consists of several small scatterers plus a dominant scatter. A fluctuating target can be compared to a steady target when they have the same mean value of RCS. At higher PDs (e.g., PD > 33%), an increased RCS fluctuation does not help raise the PD since it is quite high anyway, but a decreased RCS fluctuation can greatly lower the PD. When averaged over the fluctuation values, the result is a net fluctuation loss that can be quite large (e.g., 8 dB at PD = 0.9 and PFA = 10^{-6} for SW1 vs. SW0; i.e., SNR of 13 dB vs. 21 dB). The fluctuation loss for SW3 and SW4 is about half (e.g., 4 dB vs. 8 dB at PD = 0.9) that of SW1 and SW2 at all PDs above 33%, which is the same as 2 pulse frequency diversity with SW1 and SW2. At low PDs (e.g., PD < 33%), the steady target has very low performance, but an increased fluctuation can greatly improve performance and a decreased RCS fluctuation has little damage. The result for low PDs is that fluctuation is actually a gain rather than a loss.

The complex target provides not only amplitude scintillation but also provides angle glint in that the phase center associated with backscatter from a target can wander all over and even off the target. If two dominant scatterers produce an out of phase but equal amplitude scattering, the pattern scattered back to the tracking radar has a field variation with a large phase gradient. If the tracker attempts to align with the phase front, there will be a large angle error. In effect, monopulse sum and difference patterns are interchanged and the ratio inverted. This is the strategy of the cross-eyed repeater jammer trying to defeat a monopulse system. A ground bounce jammer attempts to strongly illuminate the ground so that a missile will home onto the ground image instead of the target.

The RCS of a sphere is πa^2 and is independent of direction. The frequency region where the circumference of the sphere is less than one wavelength is defined as the Rayleigh region, and the RCS is proportional to λ^{-4} making the individual return from small scatterers such as a raindrop quite low. The region where the circumference of the sphere is much greater than a wavelength is the optical region. The region in between is marked by oscillations about the geometrical RCS and is termed the *resonance region*. A resonant dipole has an RCS given by $0.866\lambda^2$. The RCS of a flat plate is

$$\sigma = G \times A = \frac{4\pi A^2}{\lambda^2}$$

A typical small aircraft has a nose-on RCS of about 1 m^2 (comparable to that of a person). Larger aircraft have RCS from 10 to 100 m^2 depending on size and aspect angle. Ships can have RCS of several 1000 m^2 depending on displacement. At the other end of the RCS spectra are birds and insects having very small individual RCS but often detected in the aggregate. Simple expressions are available for the geometrical optics RCS of simple targets such as a cylinder, dihedral, trihedral, cone, etc.

RCS patterns of radar targets can be obtained via the method of moments when the target is electrically small. The resonances of the mutual impedance matrix depend on the target type and are independent of angle of incidence making them potentially useful as angle-independent discriminants. For larger targets, physical diffraction can be used to compute the scattered pattern based on integrating the currents induced on the skin of the target via physical optics. Scattering can also be modeled using finite difference-time domain (FDTD), which is a discrete time-space representation of Maxwell's equations with increments chosen to satisfy the Courant stability criterion. For large targets, geometrical optics and geometrical theory of diffraction (GTD) are used to determine scattering from the contributing scatterers.

Stealth requires that scattering be minimized (e.g., absorbing or smoothly varying surfaces) or be redirected. External stores are avoided as are GTD edges and gaps. The fuselage is usually blended with the wings along curved contours. A metallic resonant radome can be used to hide large RCS objects such as antennas, and thin metalized foil can be used over the windshield with little visible attenuation. The backscatter from jet engine inlets can be analayzed using shooting and bouncing rays within the engine duct to establish the reflected field at the engine aperture. The RCS of fixed wing (aircraft) and rotary wing (helicopters) are often measured on scaled models. Propeller modulation can be seen in the Doppler spectrum and can be useful for classification or identification. There is, however, an ambiguity in determining the blade rotation unless the number of blades is known. HF or VHF frequencies are sometimes proposed for antistealth radars in that they resonate the target (target is like a resonant dipole) and shape has little effect on the RCS. In addition, RF absorber material is less effective at the lower frequencies.

Detection

Detection is a form of hypothesis testing where one asks: Is there a target present or not in this resolution cell? Generally, detection is based on exceeding a threshold and there are two types of errors (i.e., misdetection where the target was present, but the signal plus noise did not exceed the threshold, and false alarms where the target was not present but the noise alone exceeded the threshold). The PFA depends only on the threshold level. The PD depends on both the threshold and the SNR. Raising the threshold reduces the PFA but also reduces PD for a given SNR. Increasing the SNR, however, can restore the PD. The effect of threshold on PFA and PD can be plotted as a receiver operating characteristic (ROC) curve. The Neyman–Pearson criterion requires that for a given PFA a test is used that minimizes the missed-detection probability. The optimal test is a likelihood ratio compared to a threshold.

Cumulative probability of detection gives the probability that a target approaching the radar will be detected at least once by a given range. P_{cum} increases rapidly due to both the SNR increasing as the target approaches and the multiple opportunities of detection. PD over n scans is given by: $P_{cum} = 1 - (1 - P1) \times (1 - P2) \times \times \times (1 - Pn)$ where Pi is the probability of detection on the ith scan.

Binary integration can be used to prevent detections of nth-time-around targets. Each dwell has a different prf, and a detection is performed on each dwell based on a first threshold level. If the number of detections is m or greater out of n, then the target is declared to be present. Here, m is a second threshold and should be properly selected for optimum performance. Second time around targets and asynchronous pulsed interference will appear in different range cells and will most likely be rejected by the binary integration. The probability of detection with binary integration is given by

$$PD = \sum_{k=m}^{n} \binom{n}{m} P^k (1 - P)^{k-1}$$

where P is the probability of detection based on a single dwell. PFA can also be evaluated by replacing P with the PFA of a single dwell. Generally, $m = 0.5n$ is used to optimize performance.

Another type of detection is sequential detection where lower and upper thresholds are used. The likelihood function is computed based on all previous returns and is compared to the two thresholds.

If the likelihood function exceeds the upper threshold, a target present decision is declared. If it is less than the lower threshold, a no target decision is made. If the likelihood function is between the two thresholds, the decision is deferred and another return is included in the likelihood function and the process is repeated. The process can be forced to a decision if none has been made after N tries. The standard detection technique is merely a special case whereby the two thresholds approach each other. Hence, optimizing the two thresholds might improve some benefit over the single threshold. Another sequential detection approach uses a low threshold in search with detections confirmed by a verify pulse with higher threshold. The search-verify thresholds are optimized for lowest required power. The benefit of this approach is diminished when there are a large number of range cells at a given beam position.

Fluctuation loss is the increase in SNR to achieve the same PD and PFA for a fluctuating target as for a nonfluctuating target. This loss increases with PD, number of pulses noncoherently integrated, and correlation coefficient associated with the returns. As the correlation coefficient ρ approaches unity, the loss greatly increases. Even a small change of ρ from unity to 0.9 can reduce the fluctuation loss from 8 to 1.8 dB for a PD of 0.9. With frequency agility, the target backscatter is decorrelated, and the fluctuation loss in decibel is divided by Ne, the equivalent number of independent pulses. With two alternating frequencies and a SWI target, the returns are highly correlated and probabilities of detection of 1 out of 2, 2 out of 4, and 3 out of 6 would be the same if the antenna were not moving across the target.

There are several commonly used fluctuating models, which are related or approximate each other as limiting cases. These include chi square, log-normal, Rayleigh, Weibull, and Ricean probability density functions (pdf). Marcum derived the pdf for detection of a constant target with video integration in Gaussian (i.e., Rayleigh) noise in terms of the Q-function. Swerling considered 4 special cases of the chi-square distribution ($K = 1, N, 2$, and $2N$) where K is the fluctuation parameter relating to the number of degrees of freedom. His results are actually general in that non-integer values of K can be used. The Weinstock fluctuation uses $0.4 < K < 0.7$. The log-normal pdf is Gaussian when the RCS is in decibels relative to a square meter (dBsm), which emphasizes the large and small values via the tails. The Weibull distribution is defined in terms of a slope parameter beta. A Weibull with beta $= 1$ reduces to Rayleigh ($K = 1$). Ricean is made up of a constant and fluctuating component with ratio given by S. As S approaches zero, the Ricean approaches Rayleigh and as S approaches infinity, it approaches the nonfluctuating case.

Estimation

Detection performance depends only on the energy in the waveform and not the type of waveform. Estimation of certain target parameters, however, is dependent on waveform parameters such as bandwidth, time duration, etc. Tracking is a particular type of estimation applied to a moving object. An optimal estimator might be unbiased, provide the best LMS estimate among all linear estimators, based on maximum likelihood or Bayesian algorithms, and come close to satisfying the Cramer-Rao bound on estimation performance. The Cramer-Rao inequality bound on accuracy is given by

$$E[(\hat{\theta} - \theta)^2] = \left\{ \frac{\frac{dm(\theta)}{d\theta}}{\int \int \left[\frac{\partial \log[p(x,\theta)]}{\partial \theta} \right]^2 p(x;\theta)dx} \right\}$$

where θ is the value, $\hat{\theta}$ is the estimate, and m is the mean. For an unbiased estimator, the expected value is the true value and the numerator is unity.

The signal has parameters θ, which are to be estimated, but the signal is also masked by additive noise, that is, the observation is $x(t) = As(t|\theta) + n(t)$. We assume $p(x|\theta)$ is known, but desire to estimate the most likely estimate of θ based on the observations, that is $\hat{\theta}(x)$. The maximum likelihood estimator (MLE) maximizes the *a posteriori* probability density function $p(\theta|x)$, which is related to the known *a priori* pdf $p(x|\theta)$ via Bayes' rule. Rather than use $p(\theta|x)$, one uses $\log(p(x|\theta))$ since it is a monotonic function yielding the same best estimate but simplifies the arithmetic. In the case of Gaussian noise, the MLE reduces to the problem of estimation by weighted LMS.

Accuracy and Resolution

Accuracy relates to a measurement or prediction being close to the true value of target parameters. Precision relates to the fineness of the measurements, which may not be very accurate, but could be quite precise. Target parameters for which accuracy is important include range, angle, Doppler, and amplitude. Accuracy varies as a function of range. At long range, thermal noise effects tend to dominate. At intermediate ranges, accuracy is dominated by the instrumentation errors (relatively constant vs. range). At short ranges, angle glint effects can dominate since the angular extent of the target increases inversely with range. The accuracy of a given measurement due to thermal noise is given by $\sigma = K/\sqrt{\text{SNR}}$ where K has the same dimensions as the measurement, but is also inversely proportional to the effective width in the other domain of a Fourier transform (FT) pair (i.e., range or time has frequency or bandwidth as its FT pair). Hence, the K for a range measurement is inversely proportional to bandwidth and the K for a Doppler frequency measurement is inversely proportional to time extent of the waveform, that is, takes a long time to discern small differences in frequency. Since an antenna pattern is the FT of its aperture distribution, the K for angle accuracy is inversely proportional to effective aperture width.

Resolution pertains to the question: Is there one target present or many? If two targets are resolved in range (i.e., well separated compared to the compressed pulsewidth), there will be two distinct returns. As the targets get closer together, the returns begin to merge such that it is difficult to tell if there is one or two since the thermal noise tends to distort the combination. The presence of a dip between them yielding two peaks will depend on the relative phases of the two pulses. Typical resolution algorithms include the classical inflection or dip approach, as well as template matching algorithms that look for differences compared to the known response of a single point target. Multipath and thermal noise will affect the probability of correctly resolving two targets in range when two targets are present as well as the probability of false splits (i.e., claiming that two targets are present when only one is actually present). Similar algorithms are used for resolution in angle when the beam scans past the target. If frequency diversity is used on different prfs, this will cause the amplitude to fluctuate as the beam scans past the target making it even more difficult to determine whether there is one or two targets present. With a monopulse radar, one can examine the imaginary part of the complex monopulse ratio to determine if more than one target is present. A single target creates a quadrature value of zero, and multiple targets can create a nonzero value.

Clutter

Radar clutter includes ground, sea, aurora, discretes, cultured structures, birds and insects, vehicles that are not of interest (e.g., automobiles are moving clutter for an airport surveillance radar and are often excised via a highway map), etc. At an altitude where the temperature is approximately 0°C, relatively large snowflakes begin to be covered with melting water producing large radar returns called bright band.

An airborne early warning (AEW) used in a look-down overland mode will encounter both mainlobe and sidelobe clutter. Modern designs use time-averaged-clutter coherent airborne radar (TACCAR), displaced phase center antenna (DPCA), or pulse Doppler techniques. Pulse Doppler approaches use a medium (ambiguities in both range and Doppler) or high prf (ambiguities in range) to provide sufficient Doppler space to detect targets overland.

A clutter fence is sometimes placed in front of a ground-based radar to reduce ground clutter with some loss on low-angle targets. The fence edge can be curved or serrated to minimize diffraction effects that will allow some of the ground clutter to propagate over the fence.

CFAR

There are many types of constant false alarm rate approaches that have been implemented. Early approaches used log-FTC or limiting to reduce clutter residue. Feedback was sometimes used for adjusting the detection threshold by counting the alarms per scan or simply adjusting by the operator, based on viewing a PPI display. More recent digital implementations use range (or range and Doppler) cell averaging (e.g., log-cell averaging (CA) CFAR) with leading and trailing windows used to estimate the clutter level. Greatest-of logic is used to prevent clutter breakthrough on clutter edges and smallest-of logic is sometimes used to prevent adjacent targets or discretes from suppressing actual targets. Normally, a few guard cells are used around the center test cell to prevent a broad mainlobe or high near-in range sidelobes from contributing to the threshold. Sometimes a few largest returns are excised from the window to prevent large discretes from masking targets. To handle non-Rayleigh clutter distributions having large tails, a two-parameter threshold can be calculated. Increasing the number of samples in the window makes the estimate of the threshold better, reducing CFAR loss; but when the window is much larger than the spatial correlation of the clutter, the estimate may not be representative of the clutter in the test cell. More recently, ordered statistics approaches are being used; that is, test cell is compared to a threshold, which might be the 75% largest sample in the range window. This makes the approach somewhat insensitive to the tails of the distribution.

In many radars, a clutter map provides an adaptive threshold for the zero Doppler cell to allow super-clutter visibility as a target flies through clutter regions. A clutter map for a radar on a moving platform is somewhat difficult in that the measurements in polar coordinates must be slid along with the platform motion for averaging, but the resolution cell sizes are different.

Basically, a clutter map represents a temporal estimate of the clutter within a cell. A cell averaging CFAR represents a spatial average around a cell to estimate the clutter level within that cell.

Moving Target Indicator (MTI)

An MTI radar sends out several uniformly spaced pulses (sometimes staggered to minimize Doppler blindness) and generally uses a binomial canceller (or a generalized finite impulse response (FIR) filter or feedback filter) to cancel clutter. For stationary ground clutter, the phase difference from pulse-to-pulse is zero, and the weights are selected so that DC and very low frequency signals cancel. Moving targets will pass through the filter without canceling unless their velocity produces a pulse-to-pulse phase difference that is some multiple of $360°$. This condition is known as a blind velocity and is given by $v = 0.5n\lambda(\text{prf})$. The performance of an MTI radar in canceling clutter is defined by the improvement factor given by

$$I = (S/C)\text{out}/(S/C)\text{in} = G(f) \times CA = Vo \times SCV$$

where $G(f)$ is the integration gain at a given target Doppler, CA is the clutter attenuation, Vo is the visibility factor (i.e., signal-to-clutter ratio (SCR) for a given detection performance), and SCV is the subclutter visibility (i.e., amount by which clutter can exceed the signal and still be detected).

Most modern surveillance radars use a *moving target detector* (MTD) consisting of an MTI canceler followed by a Doppler filter bank or simply FIR filters with super clutter visibility and clutter maps derived from the zero Doppler filter.

In the past, a limiter was used to provide CFAR performance in strong clutter. Limiting (a nonlinearity) spreads clutter into the passband of the canceler reducing the improvement factor. Birds traveling with the wind at 40–50 kn can also be in the passband. They are usually seen around dawn and dusk as flocks move between roosting areas and fields. Sensitivity time control (STC) is the best defense against birds.

Tracking

Tracking involves both data association and the process of filtering, smoothing, or predicting. Data association involves determining the origin of the measurements (i.e., determine whether a return is a false alarm, clutter, or a valid target and assess which returns go with which tracks or is this the first return from a given target). Given that the return is properly associated, an algorithm is needed to include this latest measurement in a manner that will improve the estimate of the next expected position of the target. Early trackers, such as the alpha-beta-gamma filter, used precomputed fixed gains that were sometimes changed based on maneuver detection. They were simple to code and required small amounts of memory and throughput. As tracking advanced, radars began to use the EKF. Many filter states were used in ballistic missile defense (BMD) and early warning trackers. More modern tracking approaches use nonuniformly scheduled pulses, Kalman filtering of multiple sensors, nonlinear filters, *interacting multiple model* (IMM), **joint probabilistic data association (JPDA)**, and **multiple hypothesis tracking (MHT)**.

Decision-directed techniques, such as MHT, can result in a growing memory that must be pruned as possibilities are deemed unlikely. As target density or clutter increases, many false tracks can initiate but over time it becomes obvious which are actual and which are bogus. The Hough transform can be used to track/detect straight line trajectories or those generalized for curvature.

Phased arrays provide flexibility for minimizing the energy to track targets with a given accuracy or impact point prediction (IPP). Options include revisit interval, dwell time, and beamwidth spoiling.

A tracking radar can use a high prf that avoids range blindness on the tracked target while providing ample Doppler space free of clutter. If a search radar is ambiguous in range, a different prf must be used on each dwell to resolve range ambiguities. If a target is within the unambiguous range interval, the range cell where the detection occurs does not change. If the target is beyond the range ambiguity distance, the range cell number changes due to the range foldover. The Chinese remainder theorem can be used to unravel the true range based on several ambiguous range measurements. The range estimate, however, can be grossly in error if an unambiguous range cell number is off by a single range cell due to measurement noise. Other approaches of resolving range ambiguities avoid this problem. For example, the entire instrumented range can be laid out for each dwell with a return placed at all corresponding ambiguous range cells. By summing the dwells in each range cell, the one with the highest count will be the true range since they all occur in this cell. To prevent errors due to a slight range error, one can sum both the range cell and its adjacent neighboring cells over all dwells. This will ensure that slight misses will be correctly counted in the summation. Methods that resolve range ambiguities for a point target are not very effective for a weather radar where the target is distributed. Multiple targets can produce ghosts when unraveling unambiguous ranges.

Classification

Many instrumentation and early warning/BMD radars perform object classification based on radar signature measurements, for example, sorting reentry vehicle (RV) vs. decoy. This is usually obtained through deceleration of the body by the atmosphere, wake effects (mean and spread), microdynamic motion (nosetip precession), polarization, range profile, inverse synthetic aperture radar (ISAR) imaging, radar cross-section (rcs) statistics, etc.

Typical estimators include Bayesian approaches. The K factor describes the ability to resolve two types of objects, that is, the separation of their probability density functions normalized to the spread of the density function. Many lightweight traffic decoys (e.g., balloons) can be placed on a postboost vehicle (PBV) by replacing an RV, but the ability of the lightweight decoy to penetrate the defense is less than that of a heavier replica decoy. Munkers algorithm can be used for optimally assigning objects seen on one sensor to those seen by another sensor, that is, handover or target object mapping (TOM).

Much work has been performed in the past in identifying or classifying battlefield vehicles (e.g., truck, jeep, tank) based on high-range resolution measurements. *A priori* measured range profiles at various angles can be stored to be matched against by an unknown object. Sometimes features are extracted from the data such as spacing between largest spikes, order of magnitude of spikes, etc. Some of the work has involved the use of neural networks.

Imaging

The simplest imaging radars use high range resolution with Doppler processing, that is, FFTs within the range cells. For a rotating object, the Doppler frequency increases with distance from the axis of rotation and hence, maps cross range into Doppler to produce two-dimensional ISAR images. Range walk will limit Doppler resolution since it determines how many pulses can be processed in the Doppler filter. The cross-range resolution is related to the angle through which the target rotates during the coherent processing. ISAR is similar to the conventional noncoherent tomography (radon transform, back projection) used in X-ray processing. Since it is only the relative motion between radar and target that is important, the turning object in ISAR is equivalent to a stationary target and a synthetic circular SAR, that is, aircraft flying a circle about the target. More advanced ISAR imaging radars use polar processing to avoid the range walk problem. The most advanced imaging radars use extended coherent processing where an image is created by coherently overlaying images for several complete rotations of the object. *Maximum entropy method* (MEM) techniques can be used to extend the bandwidth to provide sharper images for a given actual RF bandwidth.

Airborne synthetic imaging radars (i.e., conventional SAR) use a small aperture on a moving platform. By storing the pulses and coherently combining them, a large synthetic array can be constructed that is focused at all ranges. The effective synthetic pattern is actually a 2-way pattern and the cross-range resolution at every range is about the same as the size of the physical antenna on the aircraft. At each range, the phases from a scatterer produces a quadratic runout (i.e., LFM) that varies with range. Each range cell is match filtered yielding a pulse compression in the azimuth direction. Since Doppler frequency is mapping into cross range, moving objects such as a train create a range-Doppler coupling and may image off the tracks. Stereoscopic imaging can be performed by using SAR mapping from two aircraft or multiple displaced apertures on the same aircraft. The phase difference in a common pixel for the two apertures will provide height data within the pixel.

Ambiguity Function and Waveform Design

Whereas detection depends only on waveform energy, measurement accuracy and resolution also depend on waveform shape. As target parameters (e.g., angular location of two targets) approach each other there is an associated ambiguity function that was originally defined by Woodward. For an optimal detector, the optimal filter is the matched filter given by $H(f) = S^*(f)$. It can be shown by Schwarz's inequality that this filter provides the highest peak SNR of all filters. It filters white noise with little effect on filtering the signal. The phase conjugation of the filter's spectrum compensates for the wide-band phase of the signal spectrum allowing pulse compression. The impulse response of the matched filter is simply a time reversed version of the signal so that convolution becomes correlation. The ambiguity function is a surface that expresses the filter output for all possible Doppler offsets of the signal; that is

$$\chi(t, f_d) = \frac{1}{2\pi} \int S^*(\omega)S(\omega + 2\pi f_d)e^{j\omega t}d\omega = \int s(\tau + t)e^{j2\pi f_d \tau}s(\tau)\,d\tau$$

The ambiguity function (AF) is generally defined as $|\chi|^2$. The energy under the AF is constant for all waveforms suggesting that it can be shaped but not eliminated. Wide-band signals can have a thumbtack mainlobe but must have rather high sidelobes since very little of the energy is in the mainlobe. Pulse trains will have periodic clumps of ambiguity. Range-Doppler (i.e., time-frequency Fourier transform pair) ambiguity is ultimately related to the Hseienberg uncertainty principle of simultaneously measuring time and frequency. The value of the AF along its principle plane ($tau = 0$ or $f = 0$) is related to the Fourier transform of the square of the magnitude of the waveform or spectrum. Tracking low RCS targets in a dense multiple target environment (e.g., RVs in the presence of a fragmented tank), requires low integrated range sidelobes (e.g., NLFM or weighted uniform burst (WUB)) having a good Q function where the Q function is the integration of the AF along range for a given Doppler offset.

Various waveforms have been designed for pulse compression. These include LFM with weighting, NLFM, and phase-coded waveforms. The use of phase-coded waveforms prevents a range gate pulloff repeater jammer from defeating a leading-edge tracker by taking advantage of range-Doppler coupling in a waveform such as LFM. A common phase-coded waveform is the Barker code. The Barker code is a biphase code with autocorrelation function having unity sidelobes. There are no known codes with more than 13 elements. The known codes are $N = 2, 3, 4, 5, 7, 11$, and 13. There are no known codes for $N < 12,000$ as verified by exhaustive search. Barker codes can be combined or nested. Frank codes are another type of phase-coded waveform. The Frank code consists of N^2 phases derived from an $N \times N$ phase matrix where the phases are given by $phi = 2 \times \pi \times p \times m/N$ where $m = 0, 1, \ldots, N-1$ represents columns and $p = 0, 1, \ldots, N-1$ represents the rows. The phases are similar to that of a Butler matrix or FFT. When cascaded the phase is similar to a stepped LFM and Doppler sensitivity is similar to LFM. Frank codes have about 10 dB lower sidelobes than the best biphase codes and have better Doppler tolerance. Another set of codes are the Kretschmer polyphase codes (P1, P2, P3, P4), which are related to Frank codes and stepped LFM. The P1 code is identical to Frank code for zero Doppler. P1 and P2 codes are more tolerant of receiver bandlimiting prior to pulse compression (e.g., antialiasing filter prior to A/D). Quadriphase codes are another phase-coded waveform. The quadriphase codes are derived from biphase codes such as the combined Barker code. The subpulses are overlapped with cosine rise/fall times providing a constant envelope and good spectrum rolloff to minimize electromagnetic interference (EMI). Complementary codes are another family. There exist waveforms whose autocorrelation functions are complementary in that when added together the side-lobes completely cancel for zero Doppler. Two approaches for using them consist of (1) transmit 2 wave-forms separated in time, receive first pulse with its matched filter, switch receivers to receive the second pulse with its matched filter, delay the pulses so that they are aligned yielding a compressed waveform with zero sidelobes; and (2) separate the two transmit pulses and build a matched filter for the combination, which will cause near-in sidelobes to completely cancel but far-out sidelobes do not cancel since one of the pulses is going through the matched filter portion designed for the other pulse. Another wide-band waveform makes use of Costas codes where the waveform is frequency hopped pseudo-randomly during the pulse.

Weather radars frequently make use of pulse pairs. The returns are first processed through a clutter filter to remove ground clutter; then the phase difference of successive pulses is calculated (e.g. Rummler pairs). Averaging the phase differences yields an estimate of the mean Doppler frequency and averaging the deviation above the mean provides an estimate of the variance of the weather spectrum. These parameters are often used for low-level wind shear detection since clear air microbursts are an aircraft landing hazard. Another waveform for measuring the Doppler frequency of a target in the presence of extended clutter at another Doppler frequency (e.g., target in chaff) is to use a Steudel quad waveform, which is a burst waveform consisting of a combination of 2-pulse MTI and pulse pairs. The first two pulses in the burst are processed through a single delay canceler providing an output signal where the clutter is canceled but the signal passes through. The next two pulses also pass through the canceler. Finally, the phase difference is taken between the outputs in the classical pulse pair fashion. The Steudel quad waveform has been proposed in BMD applications where a fast moving target is to be measured in the presence of lower velocity wake or pancaked chaff and debris.

Antennas

Antennas match the transmitter to free space and couple received signals from free space back into the radar. Hence, the antenna should have a good match over frequency, polarization purity, high gain, low sidelobes, etc.

A phased array consists of discrete radiators. The array is usually linear or planar, but can be cylindrical, spherical, or any conformal shape. Array design must consider mutual coupling, scan blindness effects, grating lobes, polarization purity, bandwidth, etc. Impedance match is a function of scan due to mutual coupling. The imbedded element pattern is given by

$$E(\theta) = \frac{4\pi A}{\lambda^2} \times \cos(\theta) \times (1 - \Gamma^2)$$

where Γ is the active reflection coefficient and is a summation of the coupling coefficients with phase shifter phasing. A is the unit cell area of a single element.

The ideal element pattern varies simply as $\cos(\theta)$ since Γ is matched to zero at all scan angles within real space. On receive, this ideal element's cell area collects all of the power incident on it from any angle without reflections. In a thinned array, dummy elements are generally used to ensure that the active element patterns are identical. In a thinned array, some elements can experience a reflection coefficient greater than unity at some scan angles since the reflection is made up of coupled power from other elements. On average, however, the reflection coefficient cannot exceed unity. In fact, the sum of the power radiated, dumped in dummy elements, and reflected back into the feed will add up to equal the total incident power. This is valid for any assumed coupling coefficients. The embedded element pattern in an array of open-ended waveguide radiators can be analyzed using modal analysis where the waveguide modes are matched to the free-space Floquet modes at the aperture–free-space interface. Performance in a large array can be evaluated using the periodic cell associated with an infinite array. Evaluation of element patterns in a finite array generally uses the Poison summation or method-of-moment (element-by-element) techniques. The potential to form a scan blindness (i.e., complete reflection) can occur when the phase of the mutual coupling coefficient is approximately linear with distance and the mainbeam is scanned to an angle where a grating lobe begins to form at endfire. At this scan angle, the phase shifters are set so that all elements on one side of a given element will couple in-phase into that element producing a large mismatch; that is, the phase lags of the coupling are offset by the phase advance of the phase shifter setting. Since the coupling is generally via a slow wave across the aperture, the scan angle creating the problem will occur slightly before the onset of the grating lobe. Ultralow sidelobe antennas (ULSA) have peak sidelobes below 40 dB and RMS sidelobes of below 55–60 dB.

Reflectors are often used in search radars and tracking radars. These are usually prime focus or employ a subreflector (Cassegrain or Gregorian). Design emphasis includes feed-strut blockage, offset geometries to avoid blockage, and attention to feed designs having low spillover. Subreflectors can be shaped for high gain or low sidelobes and can be made transparent to blockage using a twist reflector for the main reflector. Multibeam horns can be used for adaptive nulling, pattern shaping, or producing low sidelobes. The horns used with reflectors are generally sectorial, pyramidal, or conical. Multimodes, choke rings, corrugations, or dielectric lined walls can be used in feeds for pattern shaping. Search radars often use a doubly curved reflector with dual horns to create both a high beam and low beam. The signal is transmitted on the low beam for high gain at low elevation. Returns are initially received on the low beam to avoid near-in ground clutter and then switched to the high beam when ground clutter is reduced due to Earth's curvature.

A limited scan antenna achieves high gain and narrow beamwidth by using a low-cost large reflector and adds some limited beam scan with an array feed. One desires to design the antenna to achieve maximum gain over a given scan sector. On receive from a given angle, the array feed excitation should match that of the incident field on the array to achieve maximum gain, that is, matched filter with amplitude match and phase conjugation. If only phase shifters are available, performance may be slightly suboptimal. Design consists of shaping the main reflector surface, optimally locating the array and defining its size, and spacing the elements as far apart as possible for economy. Stangel's theorem is a powerful bound for assessing the optimal gain-scan product when combining different types of antennas (e.g., dishes combined with lenses and low-gain radiators) or array elements. In a given direction, each antenna has a different available gain. The optimal feed excitation for exciting this group of antennas is one where the amplitudes match the available gain in that direction and the phases are set for coherent addition of their patterns in that direction. If a given element has zero gain in a particular direction, then it is not excited when scanning in that direction. This approach provides maximum gain over the sector, but not necessarily low sidelobes. Phased arrays have generally been optimized for a given face, and the faces are usually oriented for good performance. True optimization, however, requires simultaneous excitation of multiple faces, which also provides a high degree of angular resolution, which can be useful for mainlobe nulling of jammers closely spaced near targets. When there are multiple array faces (e.g., 3), some element gain is not used effectively; for example, at an angle midway between two adjacent faces only one face is typically used throwing away 3 dB of available aperture. Tilting the array faces back by only 15° to favor long-range targets near the horizon causes almost half of the available element gain to be wasted since it is below the horizon. The dome antenna (i.e., a passive spherical lens fed by a zenith facing planar array) attempts to optimally redistribute the gain of each upward looking array element. The lens is designed to redistribute the element pattern; that is, take the high available gain near zenith and distribute it more toward the horizon. Since there is much more angular space near the horizon, however, the improvement is not drastic. For low sidelobes, all of the element patterns viewed through the lens should be similar, which requires that the lens be oversized. Hence, cost increases as sidelobe performance is increased. Several other peculiarities must be addressed with the design of the dome antenna; that is, AZ and EL difference patterns interchange and the polarization can rotate with azimuth scan direction. Most tradeoffs considering the added losses and costs associated with achieving good sum and difference patterns have shown that the dome has only a small advantage relative to an optimized set of planar faces.

17.1.5 ECCM

Jammers are typically barrage noise or repeater jammers. The former try to prevent all radar detections whereas the latter attempt to inject false targets to overload processing or attempt to pull trackers off the target. A standoff jammer attempts to protect a penetrating aircraft by increasing the level of noise in the radar's receiver. In such an environment, the radar should be designed with electronic counter countermeasures. These can include adaptive receive antennas (e.g., adaptive array or sidelobe canceler), polarization cancelers (defeated easily by jammer using independent jamming on horizontal and vertical polarizations), sidelobe blankers to prevent false pulses through the sidelobes, frequency and prf agility to make life more difficult for the repeater jammer, low probability of intercept (LPI) waveforms, spread spectrum waveforms that will decorrelate CW jammers, spoofer waveform with a false frequency on the leading edge of the pulse to defeat set-on repeaters or a spoofer antenna having an EIRP that covers the sidelobes of the main antenna and masks the transmitted pulses in those directions, receiver uses CFAR/Dicke-fix, guard band blanking, excision of impulsive noise in time domain, and excision of narrow-band jammers via the frequency domain, etc. In stressing cases, the radar can employ burnthrough (i.e., long dwells with noncoherent integration of pulses). Bistatic radars can also be used to avoid jamming. For example, a standoff (*sanctuary*) transmitter can be used with forward-based netted receive-only sensors [avoid antiradiation missiles (ARMs) and responsive jammers] to located targets via multilateration. Ultralow sidelobe antennas can be complemented with remote ARM decoy transmitters that cover the radar's sidelobes.

Adaptive antennas include both adaptive arrays and sidelobe cancelers. The adaptive array includes a number of low-gain elements whereas the sidelobe canceler has a large main antenna and one or more low-gain auxiliary elements having sufficient gain margin to avoid carryover noise degradation. The processing algorithms are either analog (e.g., Applebaum or Widrow LMS feedback) that can compensate for nonlinearities or are digital (sample matrix inversion or various eigenvector approaches including Gram–Schmidt and singular valved decomposition (SVD)). Systolic configurations have been implemented for increased speed using Givens rotations or Householder (conventional and hyperbolic) transformations. In a sidelobe canceller (SLC) the jamming signal is received in the sidelobe of the main antenna as well as in the low-gain auxiliary element. By weighting the auxiliary signal to match that of the main antenna and setting the phase difference to $180°$, the auxiliary signal can be added to the main channel yielding cancellation of the jammer. The weighting is determined adaptively since the main antenna is usually rotating. Target returns in the mainbeam are not canceled because they have much higher gain than their associated return in the auxiliary antenna. Since they are pulsed vs. the jammer being continuous, target returns have little effect in setting the adaptive weight. Since the closed-loop gain of an analog canceler is proportional to jamming level, the weights will converge faster on larger jammers creating an *eigenvalue spread*. To prevent the loop from becoming unstable, receiver gains must be set for a given convergence time on the largest expected jammer. Putting limiters or AGC in the loops will minimize the eigenspread on settling time. The performance of jammer cancellation depends on the nulling bandwidth since the antenna pattern is frequency sensitive and the receivers may not track over the bandwidth (i.e., weights at one edge of the band may not yield good nulling at the other end of the band). Broader bandwidth nulling is achieved through more advanced space-time processing; that is, channelize the spectrum into subbands that are more easily nulled or, equivalently, use adaptive tapped delay lines in each element to provide equalization of the bandpasses; that is, the adaptive filter for each element is frequency sensitive and can provide the proper weight at each frequency within the band. A Frost constraint can be included in digital implementations to maintain beamwidth, monopulse slope, etc., of the adapted patterns. If the jammers are closely spaced, mainlobe nulling may be required. Nulling the jammer will cause some undesired nulling of the target as the jammer-target angular separation decreases. This is limited by the aperture resolution. Difference patterns can be used as auxiliary elements with the sum beam. The adaptation will place nulls in the mainlobe of the sum pattern. They are actually more like conical scan where a difference pattern is added to a sum pattern to move the beam over. The mainbeam squints such that the jammer is placed in the null on the side of the mainbeam. Better angular resolution can be achieved by nulling with two separated array faces. The adaptive pattern can now have sharp nulls that cancel jammers with minimal target loss since the angular resolution is set by the much wider interferometric baseline.

Defining Terms

Conical scan: Angle estimate based on rotating an offset beam about the boresight axis.
Impedance control (IMCON): Bulk wave pulse compressor.

Joint probabilistic data association (JPDA): Updates a track with weighted sum of the various returns where the weights are proportional to the likelihood of being the correct association.

Monopulse: Angle estimate based on a single pulse using the ratio of the difference (odd) pattern to sum (even) pattern.

Multiple hypothesis tracking (MHT): Makes multiple hypotheses for track extension prior to a hard decision; generally uses all available data from the past.

Pulse Doppler: Radar processing of a coherent train of pulses usually with uniform spacing. Older versions implemented by analog filters; modern versions use fast Fourier transform or discrete Fourier transform processing.

Reflective array compressor (RAC): Pulse compressor built with reflecting grid of nonuniformly spaced grooves or reflectors.

Rotman lens: Microwave lens that forms multiple beams with the peaks fixed in space vs. frequency.

RUV coordinates: Natural phased-array tracking coordinates aligned with range and direction cosines.

RVCC coordinates: Radar tracking coordinates aligned with LOS, target velocity, and orthogonal to the other two.

Sensitivity time control (STC): Variation of attenuation (e.g., RF or IF) with range to minimize the returns from small objects at shorter ranges.

Surface acoustic wave (SAW): Pulse compressor built on an acoustic surface usually with interdigital transducers and/or reflective grooves.

Woodward synthesis: Constructing a signal or antenna pattern based on overlapping $\sin x/x$ subsignals.

Reference

Marshall, J.S. and Palmer, W.M. 1948. The distribution of raindrops with size. *J. Meteorol.* 5:165–166.

Further Information

Following IEEE Transactions: AES, Propagation of Antennas, Signals Tri-Service Radar Symposium

Aviation Week
Jane's Radars
Microwave Journal
National Radar Symposium
Radio Science

17.2 Radar System Implementation

Melvin L. Belcher, Jr. and James A. Scheer

17.2.1 Radar System Design, Evaluation, and Assessment

Although there is no standard cookbook procedure to radar system design or analysis, there is a systematic approach, based on understanding the problem and all related issues, and using a base of experience to apply engineering solutions to the problem. Understanding the problem includes understanding the mission requirements, understanding the environment, understanding the limitations imposed by the platform, and understanding the technology being applied or at the disposal of the designer. Using the understanding of these elements, the design can be developed, or an analysis of a given design approach can be performed. Figure 17.1 depicts a methodology for the design of a radar system, highlighting some of the significant driving elements.

The performance requirements given by the user are usually in terms of the probability that the sensor will perform a given task (e.g. P_d for probability of detection) and the probability that a false action will take place (e.g., P_{fa} for probability of false alarm on noise). It is not easy to express these parameters (P_d and P_{fa}) in terms of the radar characteristics, and so usually an intermediate calculation is performed.

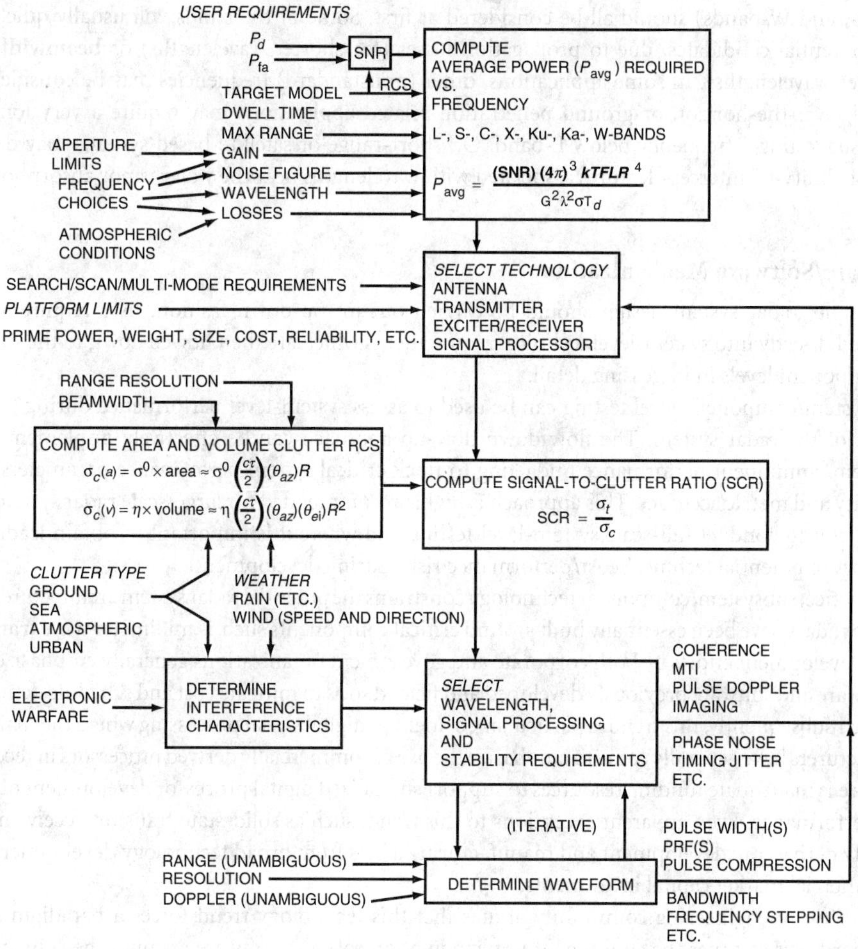

FIGURE 17.1 Methodology for the design of a radar system.

In the case of the examples of detection and false alarm, the signal-to-noise ratio (SNR) is the intermediate measure of performance. This is easily calculated from the radar parameters, and then that value (SNR) is related to P_d and P_{fa} through subsequent analysis.

Usually there is a set of implementation related requirements that is given by the user, based on the platform on which the system is to be implemented. These may be size, weight, and power limits, as is the case with a missile-borne seeker or airborne intercepter system, or in the case of a remotely located land-based system, the requirements may not be size and weight related as much as reliability and maintainability related.

A key element to the proper design of a system is the understanding of the phenomenology associated with the radar characteristics of the target, clutter, and atmospheric propagation environment. Since these are usually time varying values, the distributions and spectral characteristics are important, both for the target and interference signals.

Almost always there will be a limit to the size of the aperture that can be employed. In a limited-space installation, the size limits are easily determined. In a land-based system, exposure to weather and effects of wind may be what determine the maximum size of an aperture. It is not always appropriate to use the largest aperture that is allowable, but it usually is.

It is not always easy to determine the best radar frequency band to use, unless a parametric study is performed to determine performance as a function of frequency. The typical radar bands (L-, S-, C-, X-,

Ku-, Ka-, and W-bands) should all be considered at first. Some of the bands will usually quickly drop out as potential candidates, due to propagation losses (for shorter wavelengths) or **beamwidth** issues (at longer wavelengths). In some applications, other (nonstandard) frequencies may be considered. For example, over-the-horizon or ground penetration related applications may require a very long wavelength, suggesting a frequency below L-band. Or, short-range or satellite-based systems may desire the low-probability-of-intercept features associated with wavelengths near the water vapor absorption line at 60 GHz.

Hardware/Software Mechanization

In principle, radar system design should be carried out in top-down fashion: user requirements are translated directly into system-level technical requirements which are then flowed down to the subsystem and component levels in increasing detail.

Subsystem/component-level testing can be used to assess system-level performance during the development of the radar system. The flow-down/flow-up approach can be effectively implemented using subsystem/component performance budgeting to track critical system performance parameters such as sensitivity and metric accuracy. This approach is highly recommended for large-scale radar systems where it is difficult to conduct full-scale system-level testing and where it is important to obtain leading-edge indicators of potential technical/cost/performance risks during development.

In practice, subsystem/component technology constrains the potential radar system architecture. Special-purpose radars have been essentially built around critical components such as millimeter wave **transmitter** tubes. Development efforts by both corporate and government organizations generally emphasize the use of hardware and software previously developed and fielded so as to mitigate cost and schedule demands. As described subsequently, this trend is particularly evident in digital signal processing where the major radar manufacturers have generally come to emphasize the use of commercially derived processors in recognition that there is inadequate funding resources to support specialized digital processor development of comparable performance. Even apparent exceptions to this trend, such as solid-state transmit/receive modules, generally derive core development and manufacturing assets from broad technology development efforts or commercial market capital investments.

The software engineering community argues that this technology trend forces a paradigm shift "in which hardware assumes the component implementation role and software assumes the systems implementation role" (Chittister and Haimes, 1994). In radar development with the emphasis on using existing hardware technology, software engineers are responsible for programming general purpose computers and generating embedded software so as to support the required real-time control or data collection functions. However, radar system engineering skills provide the core mechanism for deriving a radar system design that meets user requirements within the constraints of available software and hardware technology. Radar hardware engineering imposes unique requirements including wideband signal synthesis and processing and very wide **dynamic range.** Although requirements vary markedly with implementation, radar hardware architecture is remarkably consistent as described in the following section. In contrast, radar software engineering seems characterized by a great deal of variation depending on such particulars as digital platform, operating system, and higher order language. Independent of these considerations or the associated computer aided software development tools, it is essential to maintain a rigorous system engineering process to correctly specify software functions. Given that hardware design generally precedes software development, it is imperative for the software specification process to reflect in-depth understanding of radar hardware capabilities and requirements.

17.2.2 Radar Functions

Application Requirements—Mission Requirements

System implementation should be based on the user's requirements, not on the historical background and experience of the developer. Often, a contractor is selected based on success at previous radar programs, but the favorite solution may not be the appropriate solution. The designer must thoroughly understand the mission requirements insofar as radar functions, target engagement scenario, and target and atmospheric characteristics are concerned.

A list of the most common mission related radar functions is given as follows:

Radar Functions. For a multimode phased array system, the radar resources must be divided among several functions. Long-range, high-altitude target defense requires a set of radar parameters that are somewhat diverse from those required for low-angle shorter range target defense. The long-range functions often consume nearly 100% of the radar resources, and so a separate system is often required for the other (short-range) threat.

Search/Surveillance/Detection. The search mode is usually the most demanding on the radar resources. Search parameters are usually specified in terms of the solid angle to be searched, the maximum range (sometimes as a function of elevation angle), and the revisit time allowed, for a given probability of detection and false alarm rate. The search mode usually requires that the direction of the antenna beam be changed to provide angular coverage. In general, the wider the beam, the less time it will take to search a given angular sector, out to some range. Also, it is usually the case that the larger the power/aperture product, regardless of wavelength, the faster the search can be conducted. Given a search volume, determined by the angular coverage and maximum range for each angular segment, for a given target radar cross section (RCS) characteristic, and revisit time, the so-called power-aperture product can be determined.

Acquisition. Acquisition refers to the process of declaring a detection to be a target and initiating the track mode on that target. It involves revisiting the detection and verifying its existence as a target of interest, and developing a preliminary set of tracking parameters on it. The ultimate function of the system is to track targets of interest. With a track-while-scan system, or a phased array radar in a track-while-search mode, it is not a trivial matter to determine when a target track should be initiated, particularly in the cases in which more than one target is in a given sector. Several samples of the target ensemble must be obtained before the target association process can be reliably conducted, in which case a track file can be initiated.

Track. Once a detection has been declared to be a target of interest, a (so-called) track file is established, which is a set of values defining its position in space (and Doppler) and its rate of change of these parameters with time. The longer a target is tracked, the better the estimate of its position, and velocity. Some systems are required to track multiple targets; 10 per face of a phased array radar or of a track-while-scan radar, is a typical number. The track function uses a small percentage of the radar resources. Typically, 1% of the time can be devoted to tracking a particular target. If there are 10 targets to be tracked, then this would consume only about 10% of the radar resources.

Target Identification. In a dense wartime environment, avoidance of fratricide has become an important consideration, making target recognition an additional function required of many fire control systems. This function can be performed cooperatively, by use of beacons and other identification-friend-or-foe (IFF) systems. However, sometimes these systems are not fully effective, and so noncooperative techniques must be considered. One technique involves imaging of the target, requiring a wide instantaneous **bandwidth** (to perform range profile of the target) and *inverse synthetic aperture radar* (ISAR) process, in which target motion is used to develop a cross-range image.

Fire Control. Once a target is identified by the search and track radar, often some form of engagement is required. This may be in the form of a missile or gun. The radar must provide some cueing and/or navigation information to the fire unit.

Illumination. Classically, this function is performed by a separate radar (illuminator) system, because a continuous wave (CW) signal has been required. Engagement of multiple targets has required cooperative timing of missile launch so that no two targets need be illuminated at a given time. Modification of the semiactive mode of defensive missiles can allow for a pulsed illuminator, which allows for multiple target illumination by a single system.

Damage Assessment. After a target of interest is engaged, sometimes the radar must determine the damage to that target. This requires a capability to image the target, providing a **high-resolution** description of the remains. The same features that are used to identify or classify the target are appropriate for the damage assessment task. The same techniques that are used for target recognition can be used to assess the damage caused by a defensive system. Imaging of the area around the target, and a microscopic look at the Doppler signature of that area can provide a measure of the damage.

Scenario

Range, Altitude, Search Volume. The volume that has to be searched determines the architecture of the array. A sector search (no more that $\pm 60°$ in azimuth) is achievable with a single array face, whereas more than that requires either mechanically scanning the array face or implementing more than one face.

Revisit Time. The revisit time required determines how well the radar can search the required search volume. Basically, the required power-aperture product can be determined from the target RCS characteristics, search volume, and revisit time parameters.

Target Density and Distribution in Angle (Sector). Given the multimode capability of a phase array system, the target density and angular distribution determines the scheduling parameters of the system. Multiple target tracks do not usually drive the system resources as much as the search function.

Target Characteristics

RCS and Fluctuation Model. The RCS and fluctuation statistics of the target are required to determine the detection performance of a given system. The probability density function of the (pdf) target plus noise plus clutter signal is compared to the pdf of the noise plus clutter to determine the receiver operating curves.

Speed. The speed of travel of a target determines the Doppler characteristics, as seen by the radar. This is important to determine the clutter suppression that is achieved in the target Doppler domain.

Height. The height of the target over the land or sea clutter determines the ability of the antenna system (beam characteristics) to separate some of the clutter from the received signal, at minimal loss of target signal. The remaining clutter must be removed by waveform and signal processing related approaches.

Moving Target Indication (MTI) and Stationary Target Indication (STI)

In a broad sense, a radar target is defined as the item the radar operator wants to detect, and other things that the radar may see are considered to be clutter. For a weather radar system, the target is the rain, hail, sleet, wind shear cell, etc., that may be detectable by the radar. In the case of a surface-to-air radar, one designed to detect aircraft, the weather is considered to be clutter.

This discussion relates to a system that is designed to detect hard, man-made targets, such as a vehicle, in the presence of ground clutter. This is generally the most taxing situation, particularly for the case in which the target of interest is not moving. It is often the case that the clutter signal is larger than the target signal, so the detection problem is one of determining what is target and what is clutter. This is often called *discrimination*, rather than detection, since the target is indeed detected, but not discriminated from the clutter.

There are two radar modes designed to detect such a target. The *moving target indication* (MTI) mode takes advantage of an identifying characteristic of a moving target, associated with the Doppler frequency offset imposed by the target motion. Since a moving target is at a different distance from the radar on a sample-to-sample (pulse-to-pulse) basis, the phase of the received signal is different on successive pulses, whereas the phase of the clutter signal is not. A coherent radar is designed to measure phase and can, therefore, detect a moving target in the presence of clutter, within the stability-imposed limits of the radar.

If the target of interest is not moving, the detection problem is considerably harder, requiring a *stationary target indication* (STI) technique. The radar must be designed to exploit some characteristic of the target that is different from that of the clutter. The basic techniques involve reducing the pulse width (range resolution) and beamwidth so that the range/azimuth resolution cell matches that of the expected target. This minimizes the clutter reflection, without losing any target signal. In this case, the target-to-clutter ratio is about optimum, which may not be good enough for reliable detection. Some other characteristic of the target must be exploited. Identification of the individual target scatterers by imaging (in range and/or cross range) sometimes provides discrimination capability. If not, then sometimes the polarimetric properties (*polarization scattering matrix* (PSM)) of the target are measured. At the extreme, in state-of-the-art systems, a combination of imaging, with resolution less than 1 ft (0.3-m) in range and cross range, and measuring the PSM of each major scatterer is combined for target discrimination and classification applications.

17.2.3   Radar Subsystem Implementation

Figure 17.2 is an overall block diagram of a radar system, not including the **signal processor** and beyond. The system depicted in the figure is a coherent system in that it will measure the phase of the received signal (vector) in addition to the amplitude. This capability lies in the ability to maintain a stable phase reference, by means of the stable local **oscillator** (STALO), coherent oscillator (COHO), and intermediate frequency (IF) oscillator. These signals are generated in the exciter. Once the RF signal is developed, by mixing the stable signals with each other, the RF signal is amplified and applied to the antenna system. The received signal is down-converted to IF and synchronously detected to provide the baseband signal for processing, normally consisting of spectral analysis (filtering and Fourier transform processing) and data processing such as *constant false alarm rate* (CFAR) detection, and tracking, etc.

The state-of-the-art in component development for the major system components is discussed here, but first a few words concerning one of the more basic radar performance metrics, signal-to-noise ratio, and the performance of components as related to that metric. Since the radar range equation has a factor of R^4 in the denominator, the system SNR must improve by a factor of 16 relative to some baseline value, to improve the single-look detection range by a factor of 2. A 1-dB (25%) change in radar SNR performance results in a range performance increase of only about 6%; a 0.5-dB change results in about 3% increase in range.

For example, noise figure values on the order of 2–2.5 dB are readily achievable at most microwave frequencies from L-band through X-band. By tighter selection process, and narrow-band tuning, noise

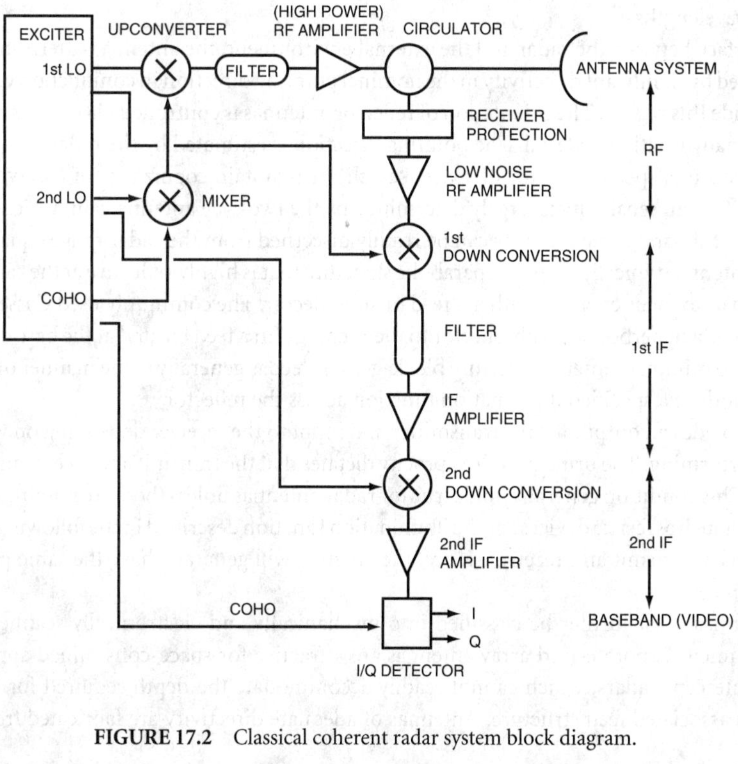

FIGURE 17.2   Classical coherent radar system block diagram.

figure values of 1–1.5 dB might be achievable, but at considerable cost; with field degradation, these values may revert back to the more common 2–2.5 dB values, all for a 6% improvement in range. If it is important to achieve a little more range performance, it is often easier to cool the receiver. Reducing the temperature from 290 to 230 K has the same effect as reducing the noise figure by 1 dB, or increasing the transmit power by 1 dB.

As subtle as they are, these improvements are only true if the system range performance is noise limited, not clutter limited. In many modern applications, the range performance is not noise limited, but rather it is clutter limited. For example, air-to-surface, surface-to-surface, and ground penetration applications often result in a poor signal-to-clutter ratio. Therefore, it is not always practical to strive for the utmost in power or noise figure performance, if the system performance is clutter limited.

Seldom is a system required to perform its detection process in a single pulse. The system is usually allowed to process an ensemble of pulses to perform detection. In this case, it is the average power and dwell time that are important, not the peak power and instantaneous bandwidth. The system designer should choose a transmitter with sufficient average power capability, select a pulse repetition frequency (PRF) based on Doppler sampling and unambiguous range considerations, and select a pulse width that operates the transmitter at its maximum duty cycle. If that pulse width is too wide for the desired range resolution, then the pulse should be coded, using phase coding or linear FM (as examples). The receiver must then use matched filtering (**pulse compression**) to obtain the resolution desired.

Antenna Subsystem

The antenna is the transitional device between the transmitter/**receiver** and free space. Although such a general definition can be fulfilled by any number of radiating structures, most modern radar antennas can be classified as either reflectors or arrays. (However, for applications requiring only modest directivity, a horn or other aperture antenna may be utilized.) Maximization of system performance as a function of cost generally mandates that a substantial fraction of the total cost be devoted to the antenna. The antenna may consist of 20% of the total system cost for relatively undemanding applications to well over 50% for solid-state array systems. Large reflector antennas can be produced relatively cheaply though it may be difficult to maintain stringent tolerances associated with low-**sidelobe** performance for a reflector that spans many wavelengths.

As the interface between the radar and the external environment, the antenna pattern should generally be characterized by significant directivity in the nominal pointing direction in conjunction with suppressed response outside this region. The orientation of reflector antennas is controlled via some manner of one or two dimensionally rotating pedestal. The pointing direction is estimated by the radar controller to either probe for targets in a specified sector of space (search) or maintain cognizance of a previously detected target (track). System sensitivity is largely determined by the two-way antenna gain, which is the product of the transmit gain and receive gain as can be readily discerned from the radar range equation.

Reflector antennas typically employ a parabolic structure that is highly reflective at the carrier frequency of the radar so as to focus energy at either a feed or subreflector. The commonly used Cassegrain reflector antenna employs a hyperboloidal subreflector to focus energy on a feed located at the base of the parabolic reflector so as to minimize antenna aperture blockage. The feed is generally some manner of horn antenna designed to produce a specified illumination function across the reflector.

The feed provides an output for the transmitter and input to the receiver depending on which function the radar is performing. The principle of reciprocity dictates that the transmit and receive antenna patterns are identical. This condition generally holds true for radar antennas unless there are nonlinear components in the antenna (such as ferrite devices) or the illumination function described in the following is deliberately changed between transmit and receive. Hence, the antenna will generally have the same polarization on transmit and receive.

Array antennas must further be classified into mechanically and electronically scanned deigns. Mechanically scanned corporate-feed array antennas are attractive for space-constrained applications such as airborne intercept radars, which cannot readily accommodate the depth required for a reflector antenna with its associated feed structure. Antennas of adequate directivity are fabricated from an array of individual elements that can be formed with individual wire elements or fabricated as slots in a waveguide among other techniques. The elements are fed using a constrained beamformer typically consisting of waveguide or coaxial cabling.

In the case of an electronically scanned antenna, the relative phase of each individual element is controlled so that the pointing direction is determined by the element-to-element phase differential. Given an element spacing of d and a desired scan direction of θ_0 from the antenna boresight (defined as the direction orthogonal to the aperture plane), transmitted/received signals will be in phase if the interelement phase shift is $(2\pi d/\lambda)\sin(\theta_0)$. Phased arrays implement electronic beam scanning by controlling an individually phase shifter for each element. Diode or ferrite devices are used to implement the commanded phase shift. Phase shifters can potentially switch phase sets in less than 1 μs but the computation and control process necessary to command the phase shifter setting typically imposes beam switching times on the order of 5–25 μs. This architecture can provide flexible one- or two-dimensional electronic beam scanning at the cost of a relatively complex element-level phase control subsystem.

The feed structure cost can be significantly lowered by illuminating the back face of the array with a feed horn rather than employing a corporate feed. This space-fed phased array sacrifices compactness and some degree of illumination function control. Feed scatter and aperture spillover degrades antenna sidelobe suppression.

Active arrays drive each radiating element or small groups of radiating elements with dedicated high-power amplifiers (HPAs) on transmit and low-noise amplifiers (LNAs) on receive. Modern active arrays package a HPA and LNA along with a phase shifter, precision attenuator, and control circuitry into a solid-state module that typically possesses an integrated radiating element. A radar antenna may be constructed from a few hundred such modules for short-range search applications up to several to tens of thousands for demanding long-range/high-resolution applications such as missile defense.

The beam pointing direction of frequency scanned arrays is determined by the carrier frequency of the radar. This architecture does not require element-level control as a fixed feed system is used that produces an element-to-element phase shift inversely proportional to wavelength. A serpentine feed feeding the elements in series is commonly used for this purpose. This technique affords one-dimensional electronic beam steering at relatively low cost but constrains waveform bandwidth and is susceptible to electronic countermeasurs (ECMs) as the radar's scanning can be estimated from measurement of its carrier frequency.

The fundamental pattern and resolution characteristics of antennas are summarized in Fig. 17.3. Antenna response and performance is generally characterized by its far-field pattern, which is independent of range (REF). The antenna near-field pattern is a function of range and is equivalent to the Fresnel region

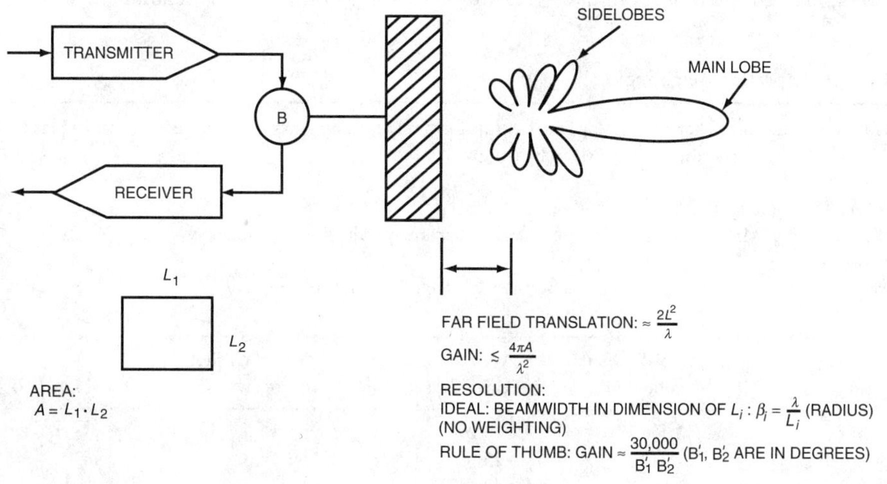

FIGURE 17.3 Aperture antenna characteristics.

encountered in optics, whereas the far field corresponds to the Fraunhofer region. The transition from near field to far field is generally defined as $2D^2/\lambda$ where D is the maximum extent of the antenna. In practice, the peak response regions are well formed at a range of D^2/λ but the antenna pattern nulls are not well defined until substantially greater ranges. In general, all radar operations will be associated with targets in the far field.

Gain is the product of both the directivity and the efficiency of the antenna. The corresponding directive gain is the ratio of the maximum radiation intensity produced by the antenna to that produced by an isotropic antenna given the same source. The efficiency of the antenna is degraded by a number of factors but the loss imposed by intentional tapering to suppress the antenna pattern outside the desired region is often the dominate source in radar applications. Major unintentional **loss** sources in the antenna and associated feed structure that degrade antenna efficiency include mismatch and ohmic losses. In addition, feed blockage is often a significant source of loss in reflector antennas.

As defined here and used in the radar range equation, gain is strictly due to passive components and does not include active component (amplifier) gain associated with the transmitter subsystem. This distinction is important in estimating the performance of solid-state active arrays.

The far-field antenna pattern is generally characterized in terms its **main lobe** and sidelobes. The main lobe extent denotes the far-field angular resolution of antenna along a given dimension. This value is approached by λ/L where λ is the radar carrier wavelength and L is the extent of the antenna in that dimension as illustrated in Fig. 17.3. As can be readily deduced, the directivity of the antenna is inversely proportional to the solid-angle extent of the main lobe. Radar antenna gain generally lies in the region of 25–45 dB inversely corresponding approximately to angular resolutions on the order of 1–10°.

Low sidelobe response is required to reject interference from clutter, electromagnetic countermeasures, and unintended electromagnetic interference that are resolved from the target. Peak sidelobe magnitude in most radar systems is at least 25 dB below the peak of the main lobe. In applications requiring detection of relatively small targets in the presence of strong clutter or active interference sources, the sidelobe level must be significantly lower. Given that the transmit and receive antenna patterns are identical, the composite rejection of sidelobe interference in decibels will be twice the nominal sidelobe level for clutter but active interference rejection is dependent solely on the receive pattern sidelobes.

Sidelobe suppression is highly dependent on the antenna illumination function, which characterizes the power distribution across the face of the antenna. The antenna illumination function and the resulting far-field pattern constitute a Fourier transform. Hence, illumination tapering can be used to suppress the sidelobe response of the antenna by implementing weighting techniques such as those delineated in Table 17.1. The Fourier transform of a smoothly tapering function such as a cosine-on-a-pedestal results in sidelobes suppressed below the nominal -13.3-dB peak value associated with the $\sin(x)/x$ far-field voltage

TABLE 17.1 Aperture Weighting

- Radar functions often characterized by Fourier transform relationship antenna aperture illumination, far field:
  Linear frequency modulation (LFM) pulse compression
  Doppler filtering
  SAR/ISAR imaging
- Corresponding MF of uniform weighted signal aperture produces high sidelobes/sidebands [REC(t) $\rightarrow$ $\sin(\pi f)/\pi f$]
- Weighting suppresses sidelobes at cost of decreasing effective signal aperture

Weighting	−3-dB Main Lobe Width	Peak Sidelobe, dB	Sidelobe Falloff, dB/octave	S/N Loss, dB
Uniform	0.866/W	−13.2	6	0
Taylor (25 dB, $\bar{n} = 3$)	1.049	−25	~6	0.46
Taylor (40 dB, $\bar{n} = 6$)	1.25/W	−40	~6	1.18
Hamming	1.33/W	−42.8	6	1.38

- Signal aperture (e.g., time duration for case of Doppler filtering)

that would otherwise characterize linear or rectangular aperture antenna's response along the principal plane. The cost of this enhanced rejection of sidelobe interference is broadened main lobe extent (degraded angular resolution) and loss in S/N.

Antenna sidelobe suppression is ultimately limited by the electrical and mechanical tolerances that can be maintained during manufacturing and operation. In the case of reflector antennas, feed blockage will further degrade the attainable degree of sidelobe suppression. Typical average sidelobe suppression relative to isotropic may be on the order of -3 dBi though specialized antennas with substantially lower sidelobe levels have been developed at the penalty of developing and maintaining tighter amplitude/phase/displacement tolerances.

One can interpret an antenna pattern as the Fourier transform of the sum of the desired antenna illumination function and an illumination error function that represents deviation from that function. The worst case consists of periodic errors across the antenna face that imposes a spatially coherent illumination error function that manifests as discrete sidelobes of significant magnitude. In modern antennas designed for good sidelobe suppression, the illumination error function can often be characterized by randomly disturbed amplitude, phase, and mechanical displacement errors resulting in randomly distributed sidelobes. Wang (1992) has provided useful expressions to estimate the resulting error sidelobe level for array antennas with such randomly distributed errors as well as random element failures. For example, adapting his expression yields the root-mean-square (rms) sidelobe magnitude in units of dBi as

$$\sim 10 \cdot \log[\pi \cdot [[1 + [10^{(\frac{a}{20})} - 1]^2] \cdot e^{[\theta^2 + (\frac{2 \cdot \pi}{\lambda} \cdot \sigma)^2]} - P_{go}]]$$

where:

a = rms element-level amplitude error (dB)

θ = rms element-level phase error (radians)

σ = three-dimensional rms displacement error of element (m), and

P_{go} = probability of give element being functional.

The cross-range resolution afforded by a given antenna can be closely approximated using simple trigonometric identities as the product of the angular resolution and the range to the target. It is necessary to estimate a given target's position with significantly better **accuracy** than the nominal resolution to support precision tracking applications such as fire control and air traffic control. Most modern radar systems employ a monopulse beamformer for that purpose. In the three-dimensional measurement case, the receive antenna feed simultaneously produces a sum beam, azimuth-difference, and elevation-difference beams which requires three receiver channels but is much more efficient at estimating target position to a given accuracy than other available techniques. Monopulse operation can be conceived as essentially comparing the amplitude of a return received in two independent offset lobes. The voltage measured at the output of the two difference channels is a function of the target's angular position relative to the nose of the beam. Monopulse **accuracy** degrades as the target approaches the edge of the main lobe and the difference-beam slope becomes increasingly nonlinear.

Composite tracking accuracy is generally dominated by antenna performance. Accuracy is determined by instrumentation and propagation error sources as well as the noise-limited precision. Instrumentation errors include antenna illumination errors as well as amplitude and phase imbalance among the monopulse receiver channels. Propagation errors are generally dominated by multipath and tropospheric refraction. Both error sources decrease in severity as elevation angle increases.

As will be described subsequently, tracking accuracy can be improved substantially by estimating the target position and velocity from multiple measurements. Errors that are independent from measurement to measurement, such as induced by thermal noise, are efficiently smoothed by track filtering. However, performance is sometimes limited in practice by systematic residual errors that are highly correlated over multiple measurements and are not significantly smoothed by track filtering. These errors include coordinate registration error associated with translating radar measurement coordinates into other coordinate systems used by the track filter and the end user.

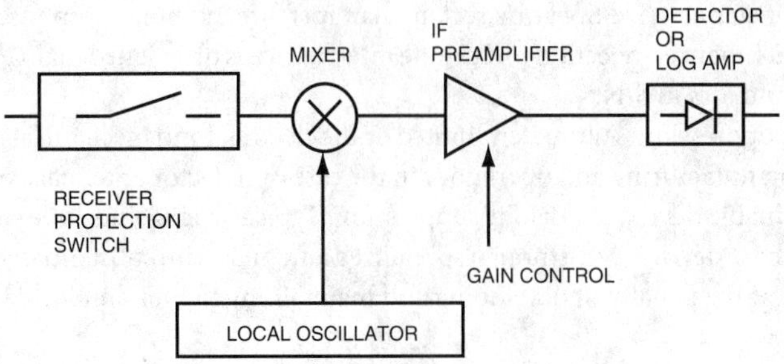

FIGURE 17.4   Simple superhetrodyne receiver block diagram.

Receiver Subsystem

Figure 17.2 showed the overall block diagram of a basic coherent radar system. The exciter develops all of the reference oscillator signals and the timing signals. The up converter stages generate the transmitted RF from the product of mixing the various reference oscillator signals. This signal is amplified to the desired transmit power level, and applied to the antenna system. The signal that comes back into the antenna, from the target scene, is directed to the receiver subsystem, which converts the RF back down to intermediate frequencies, which are convenient for amplifying and filtering the signal, and then to baseband (video). The in-phase/quadrature phase (I/Q) detector develops the in-phase and quadrature phase components of the received vector.

The receiver can be one of a variety of configurations. Complexity varies from a simple detector to a dual conversion superhetrodyne system. Figure 17.4 shows the block diagram of a relatively simple noncoherent (amplitude only) receiver. The RF signal is converted down to an intermediate frequency (IF) by means of a mixer, using a free running oscillator as the LO drive. The IF signal is amplified to an appropriate power level, and detected using a linear, square law, or log detector circuit. The log amplifier used in many of the latest noncoherent radar system provides the best dynamic range of the amplitude detectors.

Figure 17.5 is the block diagram of a typical log amplifier used in a non-coherent radar system receiver. It is configured as a series of limiting IF amplifiers, whose outputs are detected and added (summed) in a summing network/video amplifier circuit. For very low signal levels, none of the amplifier stages is saturated, and the circuit acts as a linear amplifier. As the signal level increases, successive stages of the circuit begin to saturate (beginning with the last stage) so that the marginal gain is decreased with increased signal level. The saturation characteristics are designed such that the output signal is proportional to the log of the input signal, producing the input–output characteristics shown in Fig. 17.6.

Figure 17.6 shows the output voltage as a function of input signal level, for a typical (hypothetical) log amplifier. At very low input signal levels, the curve is exponential when plotted on a log axis, but at some nominal input level the IF amplifiers begin to saturate, contributing less to the output, and the output

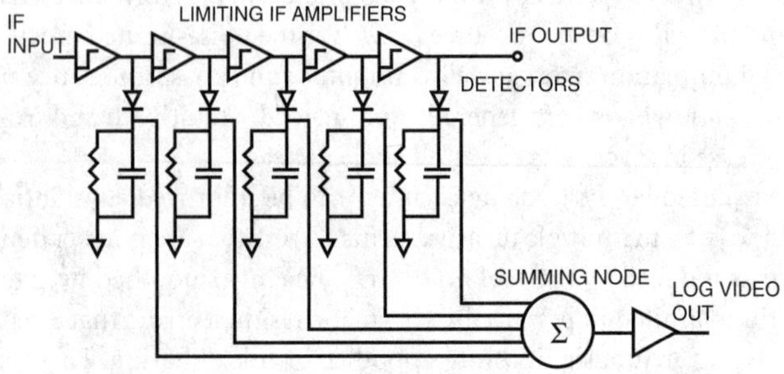

FIGURE 17.5   Log amplifier block diagram.

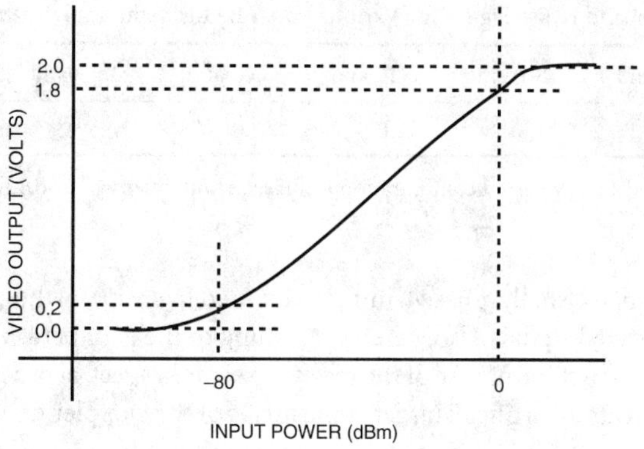

FIGURE 17.6   Typical log amplifier input–output characteristics.

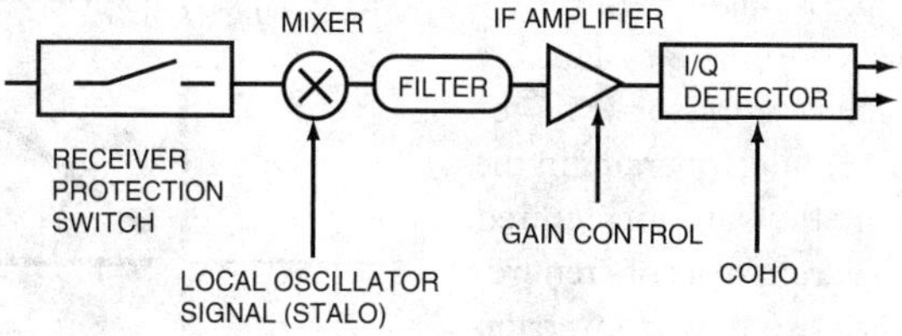

FIGURE 17.7   Coherent radar receiver block diagram.

voltage becomes a linear function of the input power, up to the point at which the first stage saturates, at which time there is no increase in output signal level. An 80-dB dynamic range is usually achievable with this circuit.

Figure 17.7 shows the block diagram of a relatively simple coherent receiver. The received RF signal is down converted to IF in a mixer, in much the same way as the superhetrodyne, with the additional feature that the local oscillator is a sample of the LO used in the transmitter upconverter process. In this way, the phase integrity of the received signal (relative to that of the transmitted signal) is maintained. Measurement of the received signal vector (amplitude and phase) is commonly performed by means of an I/Q detector, which produces two orthogonal components of the vector, the *in-phase* and *quadrature-phase* components. The reference signal for this phase sensitive detection is a sample of the coherent oscillator signal that is used in the transmitter upconversion process.

Figure 17.8 is a block diagram of a typical I/Q detector network. The received down-converted IF signal is split and applied to the input to two double balanced mixers, whose reference (LO) inputs are quadrature

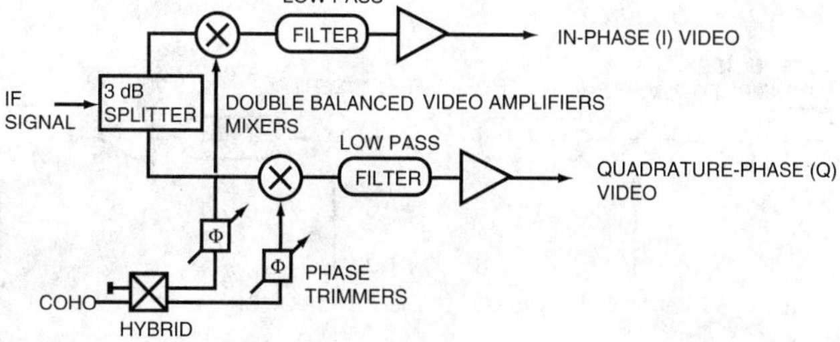

FIGURE 17.8   Coherent radar I/Q detector network block diagram.

TABLE 17.2 Double Sideband Noise Figure for Various Radar Bands, Achievable with Low-Noise RF Amplifiers

Radar band	L-band	S-band	C-band	X-band	Ku-band	Ka-band	W-band
Noise figure, dB	1.5	1.8	2.0	3.0	3.5	4.0	5.0[a]

[a]No low-noise amps available, preferred configuration is mixer/preamplifier with 3-dB additional loss.

samples of the COHO signal. Usually, phase trimmers are used to ensure orthogonality between the I and Q components of the detected signal. Also, careful matching of the gain between the I and Q channels is required to properly reconstruct the nature of the received vector in spectral processing. Nonorthogonality and gain imbalance will create an artificial target at the image of the Doppler frequency of the actual target, presenting a false target.

Figure 17.9 depicts the relationships between the orthogonal components of the received vector (I and Q) to the original vector, which has an amplitude and a phase. The expressions for tranforming from one to the other coordinate system are shown. Fortunately, the signal processor (MTI or fast Fourier transform (FFT)) often operates in the I/Q domain, so the transformations are not required.

Figure 17.10 is a block diagram of a coherent receiver configuration that has two downconversion stages preceding the I/Q detector network. This configuration is used with stepped frequency systems, since the second IF signal into the I/Q network can be of a constant frequency, while the RF is stepping in frequency on a pulse-to-pulse or dwell-to-dwell time frame.

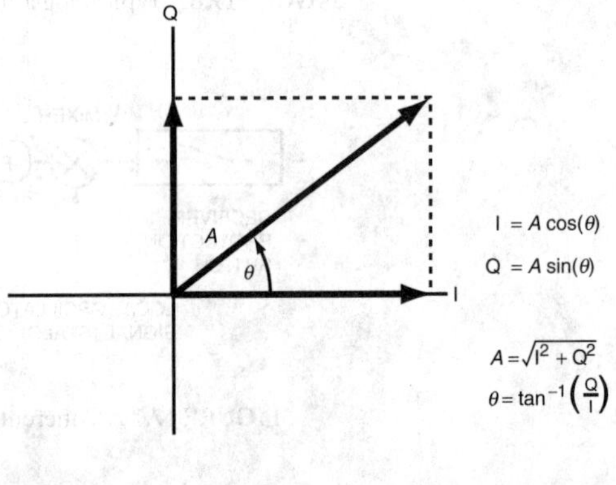

$$I = A\cos(\theta)$$
$$Q = A\sin(\theta)$$
$$A = \sqrt{I^2 + Q^2}$$
$$\theta = \tan^{-1}\left(\frac{Q}{I}\right)$$

FIGURE 17.9 Relationship between polar coordinate system variables (A, q) and rectangular coordinate system variables (I, Q).

Noise Figure

The sensitivity of the receiver depends on the thermal noise inherent at the input to the first stage of gain. The original system configurations did not have low noise RF amplifiers ahead of the mixer (downconverter) stage. In this case, the noise figure was the mixer loss plus the noise figure of the IF amplifier that followed. In modern systems, at frequencies up through Ka-band, the noise figure is established at a low noise RF amplifier preceding the mixer stage. Table 17.2 lists the double sideband noise figure obtainable using modern catalog-listed narrow-band (usually about 10%) low noise RF amplifiers. This does not include losses in the system prior to the amplifier, and does not include losses associated with built-in limiters if so configured. Since there is no low noise RF amplifier available at W- band, the figure given represents that achievable with a mixer/preamplifier combination.

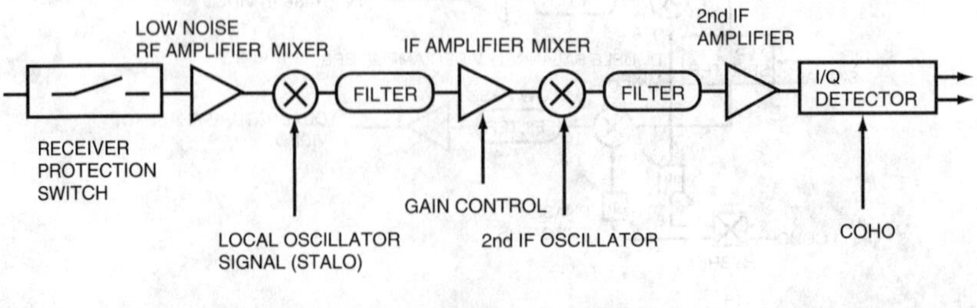

FIGURE 17.10 Dual conversion coherent receiver block diagram.

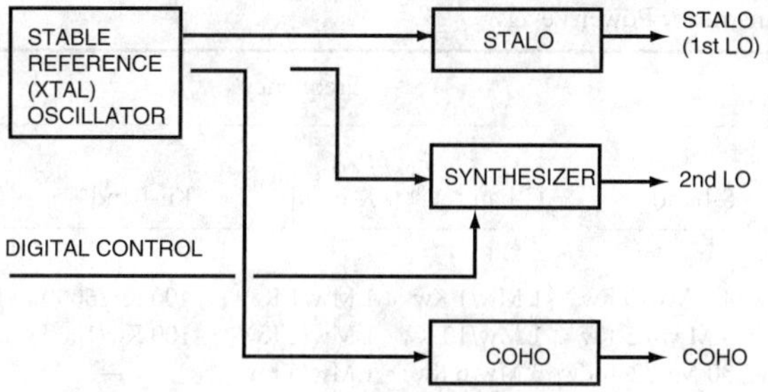

FIGURE 17.11 Coherent radar exciter subsystem block diagram.

In this case, there is an additional (approximate 3 dB) signal-to-noise ratio loss due to the fact that both noise sidebands are mixed to IF, while the signal appears in only one RF sideband.

Exciter Subsystem

One of the most stressing requirements for many radar systems is the requirement to detect a small (low RCS) target in the presence of a high clutter background. Detection of a cruise missile, near the surface of the ground or water, is a prime example. The most commonly applied technique involves performing a spectral analysis of the signal return from each range/azimuth cell, and determining whether the signal has a Doppler frequency offset indicative of the presence of a moving target.

The system exciter, which to a large extent is integrated with the receiver down-converter subsystem, develops the required IF and RF reference signals for the transmitter and receiver. Many of the components of the system **instabilities,** which lead to a limit to the clutter suppression performance, are derived in the exciter subsystem: timing jitter and oscillator phase noise being the two dominant contributors.

Figure 17.11 is a block diagram of a typical exciter subsystem that produces the coherent reference signals and stable oscillator signals used in the processes described previously. Typically three signals are required, a coherent oscillator signal, an *intermediate frequency local oscillator (IFLO)* signal, and a stable local oscillator signal, though there can be more (or fewer) as the system requirements dictate. The IFLO signal is often the one that is stepped in frequency to produce a stepped frequency RF signal to transmit. Usually all of the reference oscillators are phase locked or derived from a stable crystal oscillator as shown. Often the PRF and microtiming and control signals are also derived from this clock for system synchronization.

Transmitter Subsystem

To a large measure the performance of a radar system is determined by the parameters of the transmitter subsystem. Some of the primary factors to consider in transmitter selection and design are

- Frequency
- Bandwidth (instantaneous and tunable)
- Peak power
- Average power
- Pulse length
- Pulse repetition frequency
- Efficiency
- Coherence requirements
- Stability (amplitude and phase)
- Spurious outputs
- Cost
- Weight, volume, environmental sensitivity
- Reliability, maintainability

TABLE 17.3   Typical Transmitter Power Levels

Device	L-band	S-band	C-band	X-band P_{pk}/P_{avg}	Ku-band	Ka-band	W-band
Thermionic							
• Magnetron	25 Mw/60 Kw	4.7 Mw/5 Kw	1 Mw/1 Kw	1 Mw/1 Kw	400 Kw/600 W	150 Kw/75 W	2 Kw/1 W
• CFA	18 Mw/36 Kw	3 Mw/4.5 Kw	1 Mw/10 Kw	1 Mw/1 Kw	100 Kw/100 W	—	—
• Klystron	40 Mw/40 Kw	30 Mw/240 Kw	3 Mw/6 Kw	1 Mw/6 Kw	—	—	—
• EIA/EIO	—	—	—	—	—	1 Kw/10 W	1 Kw/10 W
• TWT	700 Kw/—	125 Kw/750 W	4.5 Mw/9 Kw	150 Kw/300 W	100 Kw/500 W	200 W/20 W	100 W/10 W
Solid State							
• GUNN	—	—	—	—	—	1 W/100 mw	1 W/20 mw
• I MPATT	—	—	—	—	—	10 W/150 mw	5 W/50 mw
• Transistor	200 W/50 W	200 W/50 W	100 W/25 W	20 W/5 W	—	—	—

The two major catagories of transmitters are coherent and noncoherent. A noncoherent radar system has the ability to measure the amplitude of the received signal, but has no ability to measure the phase (or at least the phase has no significance). A coherent system, on the other hand, can measure the phase of the received vector as well as its amplitude. Given this distinction, there is no need for the noncoherent transmitter to have particularly stable or predictable phase characteristics. The coherent radar transmitter must have a high degree of phase integrity.

The power requirements for the transmitter depend on the required detection range for a given target RCS, and other system parameters such as antenna size, receiver sensitivity, and dwell time. Short-range systems, or those with a large aperture and reasonable detection range requirements, may have the luxury of being able to employ solid-state technology for the transmitter. In this case, the device may be a transistor or diode, which provides a high degree of reliability. In general, solid-state devices can develop on the order of 100 mW of average power (at millimeter wave frequencies) to tens of watts of average power at microwave frequencies. Missile systems, close-in security systems, and automatic systems (including police speed timing radars) are examples of solid-state transmitters.

Longer range systems need power associated with thermionic (tube) type devices, such as magnetrons, cross-field amplifiers, traveling wave tube amplifiers (TWTAs), or klystron type devices. In general, tube type transmitters can develop up to hundreds of kilowatts of average power at microwave frequencies, though the most commonly used devices generate tens of kilowatts. At millimeter wave frequencies, the power is more modest, with tens of watts being a typical average power limit. Peak power levels can range from 100 W at millimeter wave frequencies to megawatts at microwave frequencies. Table 17.3 lists some typical power levels for different devices, at different frequencies. These are typical values, and do not represent the maximum levels achievable. In many cases, the table presents what has been done as opposed to "what could be done," where the difference is based on the needs of a particular program with sufficient resources and need to pay for the development of a particular device. Also, the absence of an entry in the table does not necessarily mean that the technology does not or could not exist.

The selected frequency of the system depends on the application, available space, and the results of the trade-off analysis described in Sec. 17.2.2. Usually, long-range surveillance systems are best designed to operate at the lower microwave bands, such as L-band (1 GHz), S-band (3 GHz), or C-band (6 GHz). Airborne interceptors are usually at 10 or 16 GHz (X- or Ku-band) and missile seeker radars are at millimeter wave bands (35 or 95 GHz). The longer wavelength systems experience less atmospheric attenuation, and generally more power, less system losses, and lower noise figure is available at lower frequencies. The airborne systems would suffer from the relatively wide beamwidth associated with lower frequencies, given the limited space available for the aperture. Yet, millimeter wave systems for this application would suffer from excessive atmospheric loss and lower available power. The missile seekers have a limited required range of operation and are driven more by beamwidth issues, particularly in the air-to-ground applications in which the target is in a severe clutter environment.

The bandwidth of the system determines the range resolution (δR). The wider the bandwidth, the finer the range resolution.

$$\delta R = \frac{\alpha c}{2B}$$

where

α = factor (about 1.3) determined by the processing weighting function
c = speed of electromagnetic energy (3×10^8 m/s)
B = bandwidth of the system, Hz

Using conventional signal processing, resulting in an α of about 1.3, about 650-MHz bandwidth is required for 1-ft (0.3-m) range resolution. In most cases this is adequate for target discrimination (in clutter) and in some cases it is sufficient for target classification and recognition.

The corresponding range-rate measurement resolution is

$$\delta_R = \frac{\lambda}{2} \frac{\alpha}{T}$$

where T is the coherent signal duration.

The peak power determines the single pulse signal to noise ratio. Though in some systems, during the search mode, it is desired that a detection is made on a the basis of a single pulse, more often than not, there is more than one pulse available with which to make a detection decision. In this case, the average power capability of the transmitter is important.

The length of the transmitted pulse, coupled with the PRF, determines the average power developed by the transmitter. The bandwidth of the pulse determines the instantaneous range resolution. For a monochromatic pulse, the bandwidth is approximated by the reciprocal of the pulse width. Therefore, range resolution and detection range are conflicting requirements. In most modern systems, however, the pulse is coded, such that it contains high-resolution range information, without sacrificing average power.

The pulse repetition frequency of the system is the rate at which pulses are transmitted. If range ambiguity is undesired, then the maximum PRF (minimum pulse repetition interval (PRI)) is determined by the time delay to the farthest target of interest. If Doppler sampling satisfying the Nyquist sampling criterion is desired, then the PRF needs to be at least as high as the highest Doppler frequency expected, based on the fastest target of interest. (I and Q detection is equivalent to doubling the sample rate.) Sign ambiguity issues require twice that rate. Often, there is a conflict between the minimum and maximum PRF. In this case, there are bound to be ambiguities, in range and/or Doppler. These ambiguities can be resolved by staggering the interpulse period on a dwell-to-dwell basis, and analyzing the response. Any target that does not change apparent Doppler frequency or range when the PRI is somewhat changed is not in an ambiguity region. A target that does change apparent Doppler or range is in an ambiguity region. The amount of apparent change in the measured parameter relative to the amount of PRI change is an indication of which ambiguity region the target is in.

Reasons for employing a low PRF are elimination of range ambiguities, lower data rates for data acquisition, easier power supply development, etc. Higher PRF is desired for Doppler sampling requirements and average power and number-of-pulses-integrated considerations.

The unambiguous range R_u to a target, for a given PRF is found from

$$R_u = c/2\text{PRF}$$

and so the maximum PRF is

$$\text{PRF}_{\max} = c/2R_u$$

which is not wavelength dependent.

If the Doppler frequency characteristics of a target are to be determined, then the sampling rate (PRF) must be sufficient to satisfy the Nyquist sampling criterion, two samples per Doppler cycle. In this case, the minimum PRF is found from

$$\text{PRF}_{\min} = \frac{4V_r}{\lambda}$$

where V_r is the radial component of velocity.

As the wavelength becomes shorter, it is increasingly difficult to avoid a range or Doppler ambiguity, or both.

Digital Processing

Digital processing functions for radar applications can generally be divided into signal processing and data processing functions. Signal processing is essentially characterized by single-measurement processing operations such as pulse compression, Doppler filtering, and detection thresholding that exist over time scales of milliseconds to tens of milliseconds. These operations are characterized by relatively simple algorithm structures and generally do not involve extensive iteration or branching. These operations can be efficiently implemented on specialized high-throughput processors to accommodate the relatively wide processing bandwidths associated with radar signal processing. The signal processor must generally operate in close synchronization with the other radar subsystems to ensure that it does not bottleneck system-level capability.

In contrast, data processing functions are generally implemented on a general purpose computer, which often operates asynchronously from the other radar subsystems. Data processing algorithms are characterized by extensive data-dependent flexibility and operate over time scales of tens to hundreds of seconds. Typical data processing functions include top-level radar control, track filtering, and control and display support. In radars with modest throughput requirements, signal processing and data processing functions may be readily integrated in a single processor.

Signal processing functions can further be divided into matched filtering approximation and information extraction. The radar signal processor first attempts to maximize S/N subject to both unintentional and intentional mismatch terms. Unintentional mismatch is due to ohmic and nonohmic losses imposed by the hardware and software constraints. Pulse compression consists of pulse-level matched filtering, whereas Doppler filtering approximates a matched filter to a coherent sequence of pulses.

Signal Processing

As noted previously, typical radar bandwidths can range from 1 MHz to significantly in excess of 100 MHz. These bandwidths have historically stressed signal processing technology and so a number of analog technologies have been investigated for pulse compression and related functions. Current implementation trends strongly favor digital processing with analog processing largely limited to the receiver/exciter applications previously described. The principal advantages of digital processing over analogy technologies, such as surface acoustic wave devices, are flexibility, long-duration signal processing, large dynamic range, and low perceived development risk. Accordingly, this subsection will focus on defining the parameters necessary to size a digital signal processor for radar applications.

As will be evident in the following subsections, a number of radar signal processing functions are implemented via the fast fourier, transform. The FFT is a computationally efficient implementation of a discrete Fourier transform. A number of high-throughput digital engines possess optimized hardware/software to perform FFTs.

Vendor claims of signal processing throughput, typically given in terms of complex operations per second or FFT butterflies per second, should be treated skeptically. Achieving peak throughput performance is often dependent on repetitively processing a given FFT size. In contrast, radar applications involving support of multiple waveforms and functions mandate an interleaved sequence of heterogeneous FFT sizes, which may result in peak performance significantly less than indicated by vendor literature. Instruction and data transfer operations may impose additional delays. Input-to-output processor latency for a typical mixture of radar application software tasks is generally required to reliably evaluate a candidate signal processor without imposing unreasonable performance margin requirements.

Pulse Compression

Pulse compression consists of matched filtering waveforms designed to provide the requisite range/Doppler resolution and sensitivity. Matched filtering provides appropriate signal processing for the detection and amplitude estimation of signals embedded in Gaussian noise independent of the statistical criterion utilized in the detection processor. Matched filtering maximizes S/N so that it also determines the achieved lower bound on measurement precision. The matched filter, $h_m(t)$, to any baseband-referenced radar waveform is given simply by

$$s(t) = h_m^*(-t)$$

where $*$ denotes the complex conjugate, or

$$S(f) = H^*(f)$$

in the frequency domain where the upper case letters denote the Fourier transform and f represents frequency.

Pulse compression requires a separate matched filter be supported for each waveform used by the radar system. In practice, this filter is implemented via fast convolution as illustrated in Fig. 17.12. From Fourier transform theory, frequency-domain multiplication is equivalent to time-domain convolution. Hence, receiver digitized output data is input to a FFT, multiplied by the Fourier transform of the matched filter response, and then passed through an inverse FFT (IFFT) to output time-domain data. The matched filter reference functions transforms are generally computed off-line and stored in memory to support real-time processing. Fast convolution significantly reduces the number of operations required compared to time-domain direct convolution of returns and the appropriate matched filter function.

Most programmable multiple function radar signal processors are optimized for performing FFTs. The FFT is ubiquitous in radar signal processing as it is used to support the fundamental operations of pulse compression and Doppler filtering. However, it is important for the signal processor to accommodate multiple size FFTs in a multiple tasking software environment necessary to satisfy the radar functional needs.

FFT sizing for fast convolution is illustrated in Fig. 17.13. Since a FFT implements a discrete Fourier transform, frequency-domain multiplication equates to circular convolution in the time domain. The desired

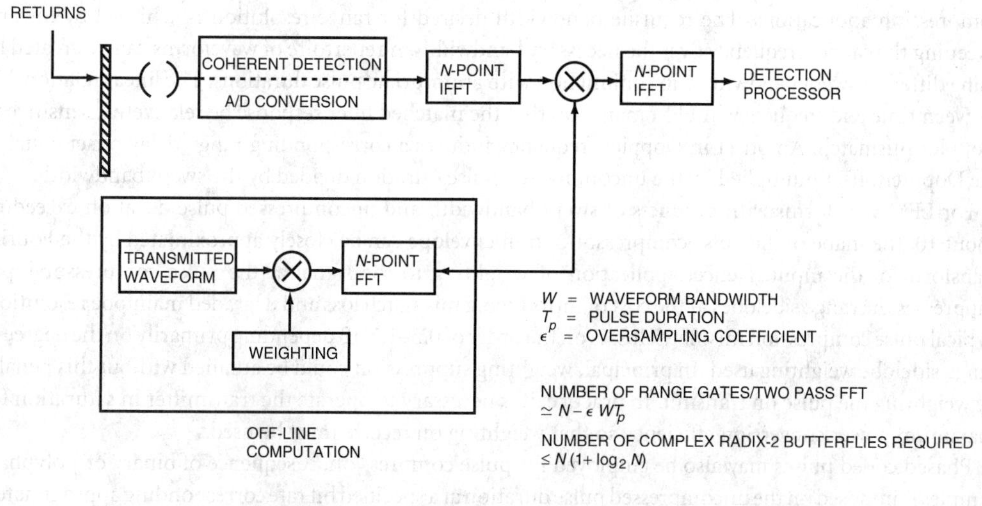

FIGURE 17.12 Pulse compression via fast convolution.

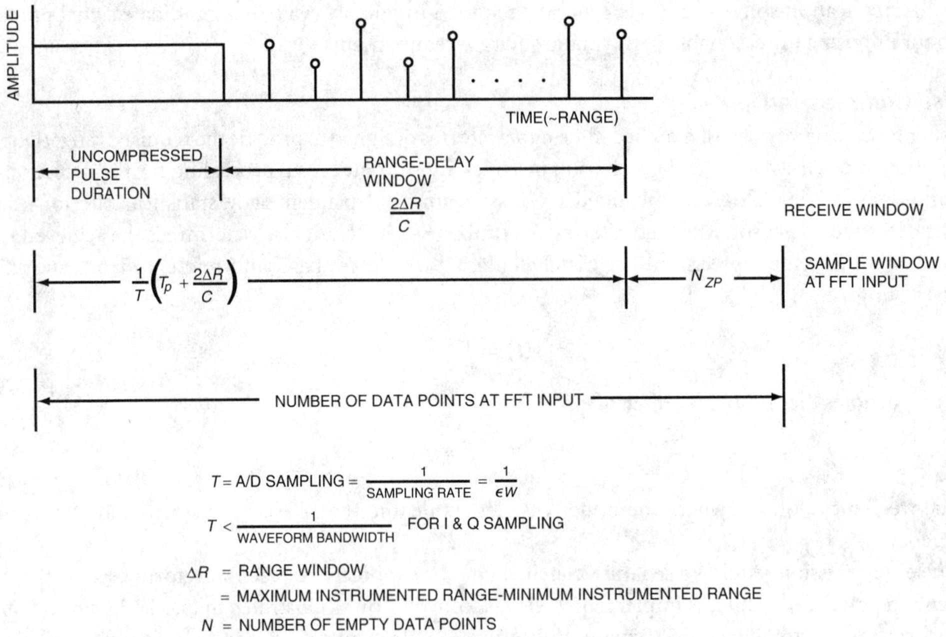

FIGURE 17.13   Fast convolution: FFT sizing.

time-domain linear convolution is achieved by increasing the size of the transform to accommodate the specified range window as illustrated. If the range window corresponding to the range interval over which target returns must be processed correspond to M range resolution cells, then the fast convolution transform size must be at least M points larger than the product of the sampling rate and pulse duration. For example, assuming a waveform bandwidth and complex sampling rate of 1 MHz and a pulse duration of 10 μs would mandate an FFT size of 64 to achieve a range window of about 2-km extent. An FFT size of 512 points would be required to achieve a range window of some 69-km duration as would be typically required for search.

It is often desirable to tradeoff somewhat degraded mainlobe resolution and associated S/N mismatch loss for reduced range sidelobes or Doppler sidebands via windowing, as previously described in the context of antenna illumination tapering. Linear frequency modulation (LFM) is frequently used for pulse compression applications. The requisite bandwidth desired for range resolution is achieved by linearly sweeping the carrier frequency over the necessary bandwidth. A repertoire of waveforms can be created by using different swept bandwidths in conjunction with appropriate pulse durations. The linear relationship between time and frequency in LFM mandates that the matched filter response be relatively insensitive to Doppler mismatch. An offset in Doppler frequency induces a corresponding range-delay offset equal to the Doppler offset multiplied by the uncompressed pulse duration divided by the swept bandwidth.

For LFM waveforms with products of swept bandwidth and uncompressed pulse duration exceeding about 10, the shape of the pulse compressor output envelope can be closely approximated by the Fourier transform of the input. Hence, application of weighting to the signal at the pulse compressor input suppresses the range sidelobes at the cost of introducing mismatch loss and degraded mainlobe resolution. Typical pulse compression mismatch loss is on the order of 0.5–1.2 dB depending primarily on the degree of range sidelobe weighting used. In principal, weighting suppression could be attained without this penalty by weighting the pulse on transmit. In practice, it is necessary to operate the transmitter in saturation for maximum power conversion efficiency so that weighting on receive must be used.

Phased coded pulses may also be employed for pulse compression. A sequence of binary or polyphase samples is imposed on the uncompressed pulse duration at a specified bit rate corresponding approximately to the effective waveform bandwidth. For the commonly used linear recursive sequence (LRS) waveforms, a binary sequence composed of antipodal phase states is used. The corresponding Doppler resolution is on the order of the inverse of the uncompressed pulse duration. The rms range-Doppler sidelobe level of match-filtered LRS waveforms relative to the peak response is given approximately by $10 \log(N)$ where N is the number of bits in the code sequence. For large N, the mainlobe-sidelobe range-Doppler response approximates a thumbtack shape.

Detection Processing

Automatic detection processing may be characterized as the decision-driven filtering operations that reduce the relatively wideband data stream of range-gate/Doppler-filter outputs to a series of target-return reports that can be accommodated by the radar data processor. Detection processors are generally designed to provide constant false alarm performance. The processor adjusts the detection threshold as a function of the local interference estimate based on the design false-alarm probability and the CFAR algorithm. CFAR thresholding prevents data processor and/or operator performance from being degraded with excessive return reports due to the combined effects of thermal noise, clutter returns, and other interference sources.

Detection-decision processing consists of a binary decision at the output of each range-gate/Doppler-filter output between two hypotheses: target-absent and target-plus-interference-present. This decision is generally implemented by comparing the cell output magnitude x_i, with a threshold value X_t, such that the presence of a target is declared if $x_i > X_t$.

Most operational CFAR processors employ some variation of cell-averaging CFAR (CA-CFAR). CA-CFAR provides an unbiased minimum-variance statistical estimate of homogeneous interference within the reference window. Interference is estimated by taking the mean across a local area termed the reference window, which is centered around the test cell. The CFAR window may be implemented across range, Doppler, angle, time, or some combination depending on the radar operational characteristics and design. An azimuth rotating radar may be designed to maintain a map of estimated local clutter and other interference sources estimated over multiple scans (time) as a function of range and azimuth. This mean interference is multiplied by a constant computed to provide a designated average probability of false alarm.

CA-CFAR performance is degraded by the presence of strong interfering signals within the CFAR window. The test cell is not included in the local interference estimation. A strong target return present in the reference window causes the local interference level to be overestimated, which elevates the detection threshold. The summation also omits one or more cells on each side of the test cell. These so-called guard cells prevent a cell straddling target return from contaminating the interference estimate.

CFAR loss is the incremental signal-to-interference (S/I) ratio required by a given CFAR threshold estimation scheme to achieve a specified detection probability over that required with full knowledge of the background interference. CFAR loss is effectively the S/I margin required to accommodate the error imposed by the local interference-estimation technique for an assumed interference PDF. The CFAR loss for a Swerling 1 target is plotted in Fig. 17.14 as a function of the number of independent reference cells for various false alarm probabilities. Note that CFAR loss can be kept below 1 dB at reasonable false-alarm probabilities by choosing a reference window affording in excess of 32 samples.

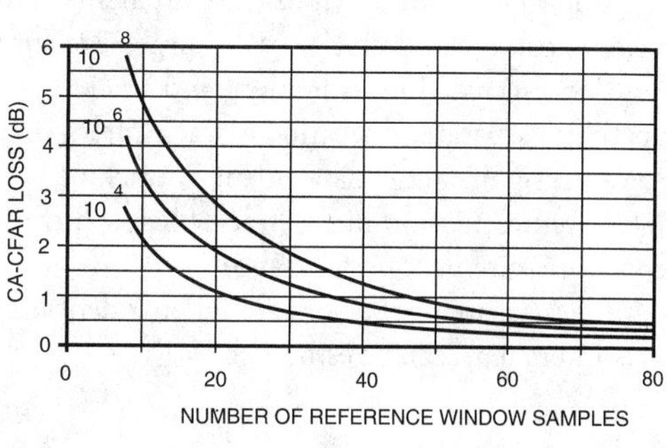

FIGURE 17.14   Swerling 1 CA-CFAR loss.

Alternative CFAR techniques are widely used to mitigate target masking effects at the cost of some increase in homogeneous CFAR loss. In general, these techniques are based on some manner of nonlinear processing the reference window samples prior to estimating the local interference level. Logarithmic rectification of the receiver output, rank ordering of reference window samples, and rejection of samples exceeding a given magnitude are common techniques for this purpose.

Data Processing

Radar Automation

Whereas operator effectiveness has been greatly enhanced by advances in controls and displays technology, radar system operation and control are increasingly automated. Modern workstation technology has allowed development of high-resolution color monitors capable of supporting three-dimensional perspective displays tailored to specific operational environment such as air traffic control. In a modern surveillance radar system, the **data processor** routinely initiates, maintains, and terminates tracks, drastically reducing the operator workload. Critical functions such as flightpath monitoring and projection coupled with multiple-source target identification monitoring affords the radar user the critical information needed to make decisions without overwhelming the operator with data.

To support applications with either very short timelines (such as point defense against low-flying missiles) or high-density multiple-target tracking, the radar control process must be automated as the requisite decision rate exceeds that which can be accommodated by an operator. Human operator intervention can be essential in coping with target and environmental parameters that are outside the nominal bounds implicit in the radar software.

Modern radars typically employ a general-purpose computer for these automation tasks. A real-time operating system is generally required to assure that the data processor performs critical functions within the latency constraints imposed by measurement rates and radar control requirements. A dedicated controller generally serves as the synchronizing interface from the data processor to the other subsystems such as the antenna steering unit, receiver, and transmitter. The implementation of this controller varies from a smart data bus to a dedicated digital processor of substantial capability depending on the design and sophistication of the radar. Multiple-function radar systems that interleave operations present the most challenging data processing requirements because of the associated need for deterministic run-time guarantees to assure that latency constraints are met across all combinations of conditions as well as the need to manage radar hardware resources such as duty cycle, occupancy, and beam pointing direction.

Target Tracking

Tracking essentially consists of maintaining cognizance of the position and course of a given target. This task is typically implemented using a tracking loop such as illustrated in Fig. 17.15. This tracking loop is implemented as a recursive digital filter in most modern tracking radars. The assumed equations of motion vary from a simple polynomial equation to high-order nonlinear models of the target flight characteristics. The track filter compares the update measurement with the value predicted by the track filter based on data smoothed from previous measurements. The weighted difference is used to update the estimated current position and velocity of the target. Tracking techniques such as Kalman filtering adapt this weight based on estimated measurement and track quality. Depending on target characteristics and processor capability, separate independent track filters may be implemented for each measurement dimension. However, significant enhancement in tracking accuracy can be used by employing a coupled coordinate system representation in the track filter, which better represents target maneuvering characteristics or sensor measurement performance.

If the measurement errors are independent look-to-look, the random error, after smoothing over the track interval σ_s, can be estimated as

$$\sigma_s = \sigma_m \cdot \sqrt{\frac{1}{N_m} \cdot \left(1 + \frac{3 \cdot (N_m - 1)}{N_m + 1}\right)}$$

where
  σ_m = standard deviation of the measurement error
  N_m = the number of measurements smoothed by the track filter

It is necessary to predict target motion over short periods for measurement updating and over long periods to satisfy user needs such as fire control or collision avoidance. The resulting prediction error for

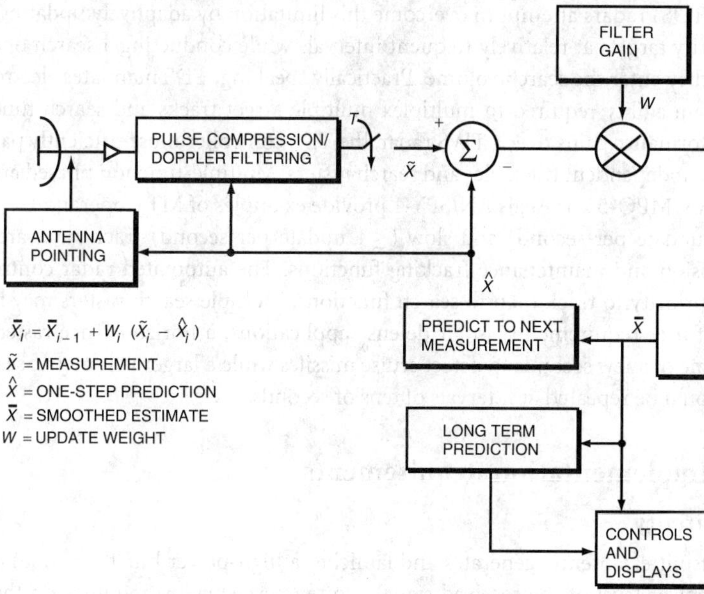

FIGURE 17.15 Tracking loop implementation.

the first-order polynomial filter given a prediction period of T_p beyond the end of the tracking interval is

$$\sigma_p = \sigma_m \cdot \sqrt{\frac{1}{N_m} \cdot \left(1 + \frac{12 \cdot \left(T_p + \frac{(N_m - 1)T}{2}\right)^2}{T^2 \cdot (N_m - 1)(N_m + 1)}\right)}$$

where T is the interval between measurements. This expression effectively presupposes that the target equations of motion are known by the track filter designer. Obviously this is not the case against such targets as maneuvering aircraft but is indicative of the performance that can be achieved tracking targets on ballistic trajectories. Maneuvering targets can be modeled as possessing random acceleration of a given magnitude and time constant so as to tune the track filter and estimate associated smoothing/prediction performance.

Target tracking can essentially be broken down into two classes of problems: single target tracking (STT) and multiple target tracking (MTT). STT applications are characterized by applications where the radar seeks to implement a measurement rate significantly higher than the maneuver bandwidth of the target. Targets are presupposed to be separated by multiple resolution cells. Microwave-band range instrumentation radars employing mechanically scanned antennas typify this type of design. Such radars operate at a few hundred pulses per second and typically use simple first-order polynomial tracking filters to keep the antenna oriented at the target and keep the target return positioned in range and Doppler acceptance gates.

In applications such as air surveillance or air defense, there are multiple targets distributed throughout the spatial volume of interest to the user. Targets may be closely spaced so that association of measurements to track files can be problematic. Radar resources must be shared among the multiple targets resulting in two different approaches. Track-while-scan (TWS) radars illuminate a given angular sector by repeatedly scanning a fixed illumination with a typical revisit period on the order of 10 sec. Tracks are generated by correlating measurements across multiple scans. TWS radar performance is constrained from achieving highly accurate tracking by the associated constraints on the update rate. In the event of densely spaced targets, the track filtering process may misassociate returns with the wrong track file so as to further degrade tracking accuracy.

Track-during-scan (TDS) radars attempt to overcome this limitation by adaptively updating the measurement of high-priority targets at relatively frequent intervals while conducting a search operation to acquire new targets as they enter the search volume. Practically speaking, TDS mandates electronic scanning to achieve the beam agility required to multiplex multiple target tracks and search functions. To achieve substantial performance gains over a TWS radar, the TDS controller must efficiently partition resources among multiple independent track files and search rasters. Multiple-function phased array radars such as the PATRIOT AN/MPQ-53 or Aegis AN/SPY-1 provide examples of MTT operation.

Typically, fast (≥ 1 update per second) and slow (≤ 1 update per second) track rates are available corresponding to precision and maintenance tracking functions. The automated radar control process generally gives higher priority to track than to search functions. Multiple search rasters may be used to facilitate satisfaction of user requirements. In air defense applications, a horizon fence raster could be deploy with a revisit time of a few seconds to detect cruise missiles while a larger volumetric search raster for aircraft detection could be repeated at intervals of tens of seconds.

17.2.4   Driving Implementation Requirements

Noise-Limited Sensitivity

Though the radar transmitter/antenna generates and launches a high-power burst (or tone) of electromagnetic energy, the power level of the received signal is often very small, sometimes on the order of -100 dBm (10^{-10} mW, or 10^{-13} W). The noise that exists in a receiver, due to the random motion of the electrons in the circuits, is sometimes greater than this level, so there is a challenge in detecting a target signal in the presence of the thermal noise.

The noise power can be predicted with a high degree of accuracy from

$$P_n = kTBF$$

where
  k = Boltzmann's constant
  T = standard temperature
  B = receiver bandwidth
  F = the noise factor

The signal power is easily computed; however there are some variables with uncertainty, in particular, the losses.

The radar range equation was introduced in Sec. 17.1 In practice, estimation of the composite system loss factor and the system noise temperature can be very demanding.

The composite system loss factor is generally tabulated from operational losses, propagation loss, transmit losses, and nonohmic receive losses. Operational losses are generally a consequence of target positional uncertainty and include beamshape loss and range-gate/Doppler-filter straddle. Propagation loss is imposed by multipath cancellation and atmospheric absorption. Transmit loss is due to ohmic heating and mismatch between the transmitter output and the antenna radiating structure.

Nonohmic receive loss is imposed by mismatch between RF and IF components as well as signal processing effects such as quantization noise, range sidelobe and Doppler filter weighting, and detection thresholding. Quantization noise is due to the finite number of bits in analog-to-digital conversion process. This source of loss is typically on the order of 0.1–0.3 dB.

Noise temperature is a more direct way of estimating the system noise floor than estimating the composite noise figure directly for low noise receiver chains. The **system noise temperature** incorporates ohmic receive loss since such components effectively raise the noise floor as well as attenuating the signal.

The process of estimating dynamic range and system noise temperate is illustrated in Fig. 17.16. The process is delineated in terms of an active array but can be readily adapted to other radar configurations. To estimate the performance of a conventional passive array, G_2 and NF_2 would be set to unity in magnitude. The case of $N = 1$ corresponds to a conventional mechanically scanned antenna. This particular case was

RADIATOR	LNA	N:1 COMBINER	AMPLIFIER	
$G_{1(s)}$	G_2	$G_{3(s)} = N$	G_4	NF_5
L_1	NF_2	L_3	NF_4	
	TOI_2		TOI_4	

SPURIOUS FREE DYNAMIC RANGE

$$DR = \left(\frac{Q_3}{N_0}\right)^{\frac{2}{3}}$$

N_0 = NOISE POWER AT THE OUTPUT OF THE DEVICE

$$N_0 = kTB$$

WHERE:

K = BOLTZMAN'S CONSTANT (1.38 × 10-23 Ws/deg K)
T = NOISE TEMPERATURE
B = SIGNAL BANDWIDTH,
Q_3 = THIRD-ORDER INTERCEPT AT THE OUTPUT OF THE DEVICE

$$\frac{1}{Q_3} = \sum_{i=1}^{N-1} \left[\frac{1}{Q_{3(i)} \cdot \prod_{j=i+1}^{N} G_j} \right] + \frac{1}{Q_{3(N)}}$$

WHERE:

$Q_{3(i)}$ = THIRD-ORDER INTERCEPT FOR THE i th DEVICE
G_j = GAIN OF THE j th DEVICE

SYSTEM NOISE TEMPERATURE

$$T_s = T_a + T_{e(1)} + \sum_{i=2}^{N} \left[\frac{T_{e(i)}}{\prod_{j=1}^{i-1} G_j} \right]$$

WHERE:
T_a = ANTENNA TEMPERATURE
$T_{e(i)}$ = EFFECTIVE NOISE TEMPERATURE FOR THE i th DEVICE
G_j = GAIN OF THE j th DEVICE

ACTIVE COMPONENTS

$$T_e = (\overline{NF} - 1) T_0$$

WHERE:
$\overline{NF}$ = NOISE FIGURE FOR THE AMPLIFIER
T_0 = 290 K (BY IEEE DEFINITION)

$$T_e = T_t \, L_{N\Omega} \cdot (L_\Omega - 1)$$

PASSIVE COMPONENTS

WHERE:
T_t = THERMAL TEMPERATURE OF THE DEVICE,
L_Ω = OHMIC LOSS FOR THE DEVICE
$L_{N\Omega}$ = NONOHMIC LOSS FACTOR FOR THE DEVICE

FIGURE 17.16 Computation of dynamic range and noise temperature. (*Source:* Robert Howard, Georgia Tech. Research Institute, Atlanta, GA.)

chosen since it illustrates the difference between spatial gain and amplifier gain in computing dynamic range; the former contributes directly to system dynamic range since the process of element combination effectively splits up the signal path.

Interference Environment

The radar system is designed to detect, track, and somehow treat targets of interest, in a nonideal environment. There are two major catagories of environmental effects that limit the radar performance, propagation, and interference. Propagation effects influence the classical $1/R^4$ term in the radar equation by either attenuating the signal or enhancing it. Interference takes the form of noise, clutter (man-made and natural), and intention or unintentional signal jamming.

In addition to the mission requirements, a comprehensive understanding of the environmental conditions is required. The following sections describe some of the major factors.

Ducting

Surface-based ducting (over land and sea) and evaporation ducting (over sea) affect the propagation of electromagnetic energy in that the signal is bent toward the Earth (usually) to increase the power density at a given range, extending the distance over which targets and clutter will be detected. Though this effect increases target signal, which may improve noise-limited detection performance, it also increases clutter signal, exacerbating a clutter limited performance limit.

Multipath

Multipath is a condition in which if a target is near a reflecting terrain of sea surface, the direct signal constructively or destructively interferes with the indirect signal, creating areas of enhanced or reduced signal, respectively. This condition varies with many parameters, such as sea state, ground moisture content, depression angle, wavelength, etc. Prediction of detection performance is highly dependent on accurate modeling of the multipath environment.

Clutter

In the past, for benign clutter conditions, radar detection performance has been limited by the presence of noise generated in the receiver. Low RCS targets flying near the Earth's surface create a problem in detection due to the interference of the clutter signal. So-called clutter limited performance analysis depends on a high-fidelity model of the clutter reflection characteristics.

- Sea clutter. The reflectivity (backscatter and forward scatter) from the sea depends on the sea state and wind conditions, as well as the relative angle between the look direction and the wind direction. The reflectivity is a strong function of wavelength, depression angle, and polarization of the EM wave.
- Land clutter. Land clutter reflectivity is also dependent on the wavelength, depression angle, and polarization of the EM wave. The type of land clutter (grass, desert, trees, urban, etc.) is a strong influence on the reflectivity as well.
- Propagation. Refraction of the RF signal across standard atmospheric conditions (apart from the ducting effect) causes the propagation of the signal to depart from straight line, line of sight. The effect varies for different standard conditions.
- Losses. Atmospheric losses result from water vapor and oxygen as well as other elements in the atmosphere. Though at most microwave frequencies, up to Ku-band, these effects are negligible, except at very long ranges, the additional effects of humidity, and worse, rain, sleet, and hail, are not.

Clutter Rejection

The system signal-to-interference ratio is determined by a variety of factors, in addition to the thermal noise. Therefore, it is more complete to say that the interference is noise plus clutter plus interference (such as jamming).

In many cases, the detection performance of the system is not limited by thermal ($kTBF$) noise but rather by clutter. The RCS of a clutter cell is estimated to be the average reflectivity of the clutter in that cell (σ^0) times the illuminated area, defined by the pulse width and the beamwidth,

$$\sigma_c = \sigma^0 R\theta c\tau/2$$

This is the signal with which the target must compete.

The clutter signal is complicated by the effects of range folding, particularly with a medium or high PRF system. A far-away target may be competing with close-in clutter in the case of the ambiguous range effect. An example of the effect is presented here. For $\sigma^0 = -20$ dB (0.01), $\theta^\circ = 1^\circ$ and $\tau = 1$ μs, the effective clutter RCS at a range of 1 km is about 25.5 m^2, or 14 dBm^2. A small target, (such as a 0-dBm^2 target) will be hidden by the clutter signal. If, as is often the case, the system is ambiguous in range, then there may exist a condition in which the target at some far range appears in the same range cell as close-in clutter. In this case, the close-in clutter will experience a $1/R^4$ loss associated with a small R (say, 1000 m) and the target will experience a $1/R^4$ loss associated with a large R (say, 100 km). In this case, for the ranges given, the target is further reduced by a factor of 80 dB! The total clutter signal with which the target signal has to compete consists of two components, the clutter that is in the same cell as the target (at 50 km) and the

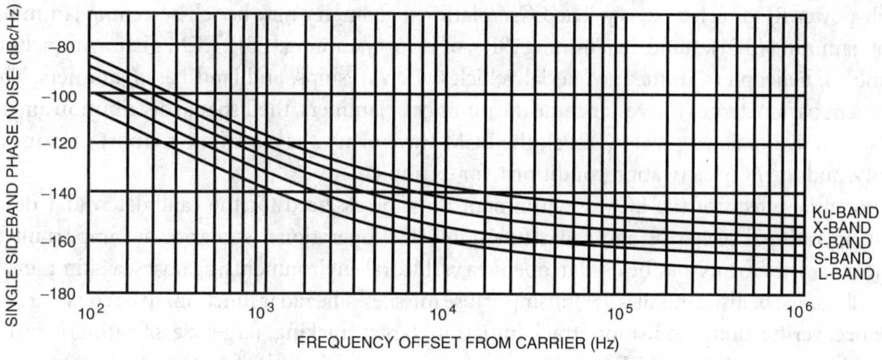

FIGURE 17.17 Phase noise characteristics (typical) for ultrastable sources at various frequencies.

clutter at 1 km. The clutter at 1 km has been calculated to be about 14 dBm^2, but its effect is increased by 68 dB because of the "close-in effect" to an equivalent RCS of 82 dBm^2, and the clutter at 50 km is 31 dBm^2. It is easy to see that the clutter may have to be supressed by 100 dB or more to be able to reliably detect the target.

Doppler processing offers a means to supress the clutter signal. Since the target is moving, it has a nonzero Doppler frequency offset, and the clutter has zero offset. A high-pass filter will reject the clutter signal and pass the target signal. As an alternative, an FFT process will separate the clutter signal from the target signal, and will also provide target velocity information.

An MTI filter is designed to remove the clutter signal, because it is nearly at DC (having no Doppler frequency offset), except that portion of the signal that makes it through the passband. The optimum filter shape depends on expected clutter spectrum.

Pulse Doppler (PD) processing generally provides better clutter rejection because the filters are narrower and farther away from the centroid of the clutter spectrum (in the frequency domain). The thermal noise is also reduced (relative to an MTI filter) because of the narrower bandwidth.

Unfortunately, in practice, the spectral analysis processing described previously does not work as well as desired. System instabilities reduce effectiveness of these processes by spreading the Doppler frequency of clutter, which is expected to be in the zero Doppler bin, into all of the other Doppler bins. Though a major portion of the clutter signal is in the first few bins, remembering that we want to reject the clutter by 100 dB, only a very small portion of the clutter is allowed in the Doppler bin of the target.

There are several primary contributors to system instability. The most prevalent is oscillator phase noise. Others include timing jitter, amplifier additive noise, A/D quantization, and I/Q detector effects.

Oscillator phase noise is one of the major contributors to this effect. Figure 17.17 shows the phase noise that can be expected from oscillators at a variety of common radar bands. The spectral spreading shown, modified by sampling effects and range ambiguity effects, is a measure of the subclutter visibility that can be achieved.

Electronic Countermeasures (ECM)

The ECM environment in which a radar system must operate is complicated by several factors. First, improvements in electronic warfare (EW) technology such as digital RF memory (DRFM) and solid-state amplifiers are making certain ECM techniques more feasible and effective. Second, EW technology and export systems are widely available to emerging nations as well as the rest of the world on the international market. Third, the performance demands placed on the area-defense and self-defense systems allow very little margin for ECM-induced degradation.

The ECM environment will potentially consist of active and passive techniques that attempt to achieve denial or deception against the radar in its various modes. Examples of possible ECM techniques include barrage noise, responsive spot noise, repeater noise, range false targets, angle false targets, Doppler false targets, range track error, angle track error, Doppler track error, active and passive decoys, chaff, and high-power RF directed energy. The ECM platform geometry may be self-screening jamming (SSJ), stand-off jamming (SOJ), stand-in jamming (SIJ), or escort jamming (ESJ). ECM platforms include fighters, bombers, helicopters, unmanned aerial vehicles (UAVs), ships, and land-based jammers. The ECM environment parameters of interest include the number of jammers, their spatial distribution and motion, the ECM effective radiated power (ERPs), the ECM waveforms, the electronic support measures (ESM) sensitivity, and the ECM activation conditions, for example.

The potential effects of the ECM environment must be assessed for the candidate radar designs for different operational scenarios and radar functions. The operational scenarios are important because different radar waveforms may be used in open sea vs. littoral environments, against sea-skimmers vs. high divers, and against ballistic missiles vs. antiship cruise missiles. The radar functions to be considered include surveillance, verification, acquisition, track initiation, target tracking, target classification/identification, and target designation functions. For each operational scenario, the radar functions that are at risk of being degraded by the specified ECM environments must be identified. The extent of the degradation of each function must be assessed by characterizing effects such as: reduced probability of detection, increased probability of false alarm, delay in acquisition, delay in track initiation, false target acquisition, false data association, track error, increased occupancy requirements, increased integration time requirements, and track file overload.

Electronic counter-countermeasures (ECCM) requirements must be based on the assessment of ECM degradations for each radar function. These ECCM requirements may potentially impact hardware design, software design, or both. Examples of the ECCM techniques and radar system design features that may be considered include: adaptive digital beam forming, sidelobe cancellation, sidelobe blanking, polarization diversity, polarization purity, high-resolution range, high-resolution Doppler, superresolution, mainbeam nulling, mainbeam cancellation, frequency agility, waveform agility, passive environmental sensing, spoof waveforms, sidelobe reduction, target identification, interleaved antenna beam waveforms, range-Doppler correlation, angle gating, monopulse-in-search, burnthrough, and passive tracking. The ECCM requirements for each candidate radar design may differ based on bandwidth, peak power, and duty cycle associated with the respective radar technologies.

Defining Terms

Accuracy: The amount of error in the estimated value of a signal parameter.

Bandwidth: The range of frequencies over which a system instantaneously operates, i.e., passes a signal, usually taken as the difference between the upper -3 dB point in the response and the lower -3 dB point.

Beamwidth: The range of angles over which the antenna system has instantaneous response, usually taken as the difference between the -3 dB points in the one-way antenna pattern.

Data processor: That portion of a radar system designed to operate on the narrow-band signal out of the signal processor, e.g., thresholding, Kalman filtering, tracking, etc. Dwell-to-dwell processing is performed in the data processor.

Dynamic range: The ratio between the maximum signal power within the linear operating regime of the radar receiver and the minimum detectable signal.

Instability: Random variations in component parameters leading to undesired amplitude, phase, or frequency variations in the received signal.

Loss: Ohmic loss in S/N is due to signal dissipation across a resistive load imposed by a component, whereas while nonohmic loss is due to impedance mismatch so that the reflectance coefficient between adjacent components is greater than 0. In a radar system, loss can also occur as a less-than-ideal gain in the signal processing function.

Mainlobe: A signal response within the region (in time, angle, and/or frequency) within which the radar is attempting to maximize its sensitivity.

Oscillator: Device for producing a precise sinusoidal reference signal.

Pulse compression: The coding and processing of a pulse of long duration to one of short time duration and relatively fine range resolution while providing S/N commensurate with the long pulse duration.

Receiver: That portion of a radar system allocated the task of amplifying, filtering, and detecting the energy reflected from a target and captured by the antenna subsystem.

Resolution: The degree to which two signals of nearly equal value can be separated in some dimension, such as range, angle, and/or frequency

Sidelobes: An undesired signal response in time, frequency, or angle outside the extent where the radar is attempting to maximize sensitivity

Signal processor: The portion of a radar system that operates on the wideband (video) signal by employing some algorithm for a particular effect, e.g., integration, high-pass filtering, and fast Fourier transform (FFT) processing. Pulse-to-pulse and single-dwell processing is performed in the signal processor.

System noise temperature: The equivalent temperature of a passive device yielding the same thermal noise power per unit bandwidth as the system under test.

Transmitter: That portion of a radar system allocated the responsibility for generating the radio frequency (RF) signal that is launched in the desired direction.

References

Balanis, C.A. 1982. *Antenna Theory and Design.* Harper and Row, New York.

Chittister, A.C. and Haimes, Y. 1994. Assessment and management of software technical risk. *IEEE Trans.* SMC-24(2):187–202.

Minkoff, J. 1992. *Signals, Noise and Active Sensors.* Wiley, New York.

Nathanson, F.E. 1991. *Radar Design Principles: Signal Processing and the Environment,* 2nd ed. McGraw-Hill, New York.

Nessmith, J.T. and Patton, W.T. 1984. Tracking Antennas. In *Antenna Engineering Handbook,* eds. R.C. Johnson and H. Jasik, 2nd ed. McGraw-Hill, New York.

Wang, H.S.C. 1992. Performance of phased-array antennas with mechanical errors. *IEEE Trans.* (2):535–545.

Further Information

Coherent Radar Performance Estimation, J.A. Scheer and J.L. Kurtz, eds. Artech House, Norwood, MA, 1993.

Principles of Modern Radar, J.L. Eaves and E.K. Reedy, eds. Van Nostrand Reinhold, New York, 1987.

Radar Handbook, 2nd ed., M.I. Skolnik, ed. McGraw-Hill, New York, 1990.

Radar Transmitters, G.W. Ewell, McGraw-Hill, New York, 1981.

17.3 Electronic Navigation Systems

Benjamin B. Peterson

17.3.1 Introduction

In an attempt to treat electronic navigation systems in a single chapter, one must either treat lightly or completely ignore many aspects of the subject. Before going into detail on some specific topics, I would like to point out what is discussed and what is not in this chapter and why, and where interested readers can go for missing topics or more details. The chapter begins with fairly brief descriptions of the U.S. Department of Defense satellite navigation system, NAVSTAR Global Positioning system (GPS), its Russian counterpart, Global Navigation Satellite System, (GLONASS), and LORAN C. For civil use of GPS and GLONASS there are numerous existing and proposed augmentations to improve accuracy and integrity which will be mentioned briefly. These include maritime Differential GPS, the FAA Wide Area Augmentation System (WAAS), and Local Area Augmentation System (LAAS) for aviation. The second half of the chapter looks in detail at how position is determined from the measurements from these systems, the relationships between geometry, and the statistics of position and time errors. It also considers how redundant information may be optimally used and how information from multiple systems can be integrated.

GPS and its augmentations were chosen because they already are and will continue to be the dominant radionavigation technologies well into the next century. GLONASS, could potentially become significant, by itself, and potentially integrated with GPS. LORAN C is included while other systems were not for a variety of reasons. The combination of its long history and military significance dating back to World War II, and its large user base, has resulted in a wealth of research effort and literature. Because the basic principle of LORAN C, that is, the measurement of relative times of arrival of signals from precisely synchronized transmitters, is the same as that of GPS, it is both interesting and instructive to analyze the solution for position of both systems in a unified way. While the future of LORAN C in the U.S. is uncertain (DoD/DoT, 1999), it is expanding in Europe and Asia and may remain significant on a worldwide basis for many years. Since LORAN C has repeatable accuracy adequate for many applications, because it is more resistant to jamming, has failure modes independent of those of GPS, and because its low frequency signals penetrate locations like urban canyons and forests better than those of GPS, analysis of its integration with GPS is felt useful and is included.

Of other major systems, VHF omnidirectional range (VOR), instrument landing system (ILS), distance measuring equipment (DME), and microwave landing system (MLS) are relatively shorter range aviation systems that are tentatively planned to be phased out in favor of GPS/WAAS/LAAS beginning in 2008. (DoD/DoT, 1999). Due to space and because their principles of operation differ from GPS they are not included. The U.S. Institute of Navigation in its 50th Anniversary issue of Navigation published several excellent historical overviews. In particular, see Enge (1995) for sections on VOR, DME, ILS, and MLS, and Parkinson (1995) for the history of GPS.

Recently, there has been a significant and simultaneous reduction in the printing of U.S. Government documents and increased distribution of this information via the Internet. Details on Internet and other sources of additional information are included near the end.

17.3.2 The Global Positioning System (NAVSTAR GPS)

GPS is a U.S. Department of Defense (DoD) developed, worldwide, satellite-based radionavigation system that will be the DoD's primary radionavigation system well into the 21st century. GPS full operational capability (FOC) was declared on July 17, 1995 by the Secretary of Defense and meant that 24 operational satellites (Block II/IIA) were functioning in their assigned orbits and the constellation had successfully completed testing for operational military functionality.

GPS provides two levels of service—a standard positioning service (SPS) and a precise positioning service (PPS). SPS is a positioning and timing service, which is available to all GPS users on a continuous, worldwide basis. SPS is provided on the GPS L1 frequency, which contains a coarse acquisition (C/A) code and a navigation data message. The current official specifications state that SPS provides, on a daily basis, the capability to obtain horizontal positioning accuracy within 100 m (95% probability) and 300 m (99.99% probability), vertical positioning accuracy within 140 m (95% probability), and timing accuracy within 340 nanosec (95% probability). For most of the life of GPS, the civil accuracy was maintained at approximately these levels through the use of selective availability (SA) or the intentional degradation of accuracy via the dithering of satellite clocks. In his Presidential Decision Directive in 1996, President Clinton committed to terminating SA by 2006. On May 1, 2000, in a White House press release (White House, 2000) the termination of SA was announced and a few hours later it was turned off. Work is starting on a new civil GPS signal specification with revised accuracy levels. Very preliminary data, as this are begin written in late May 2000, indicate 95% horizontal accuracy of approximately 7 m will be possible under some conditions. When SA was on, it was the dominant error term; accuracy for all receivers with a clear view of the sky was easily predictable and independent of other factors. Now, it is expected that terms such as multipath and ionospheric delay, which vary greatly with location, time, and receiver and antenna technology will dominate, and exact prediction of error statistics will be more difficult.

The GPS L1 frequency also contains a precision (P) code that is not a part of the SPS. PPS is a highly accurate military positioning, velocity, and timing service which is available on a continuous, worldwide basis to users authorized by the DoD. PPS is the data transmitted on GPS L1 and L2 frequencies. PPS is designed primarily for U.S. military use and is denied to unauthorized users by the use of cryptography. Officially, P-code capable military user equipment provides a predictable positioning accuracy of at least 22 m (2 drms) horizontally and 27.7 m (2 sigma) vertically, and timing/time interval accuracy within 90 nanosec (95% probability).

The GPS satellites transmit on two L-band frequencies: L1 = 1575.42 MHz and L2 = 1227.6 MHz. Three pseudo-random noise (PRN) ranging codes are in use giving the transmitted signal direct sequence, spread spectrum attributes. The coarse/acquisition (C/A) code has a 1.023 MHz chip rate, a period of one millisecond (msec), and is used by civil users for ranging and by military users to acquire the P-code. Bipolar-phase shift key (BPSK) modulation is utilized. The transmitted PRN code sequence is actually the Modulo-2 addition of a 50Hz navigation message and the C/A code. The SPS receiver demodulates the received code from the L1 carrier, and detects the differences between the transmitted and the receiver-generated code. The SPS receiver uses an exclusive or truth table, to reconstruct the navigation data, based upon the detected differences in the two codes. Ward (1994–1995) contains an excellent description of how receivers acquire and track the transmitted PRN code sequence.

The precision (P) code has a 10.23 MHz rate, a period of seven days, and is the principle navigation ranging code for military users. The Y-code is used in place of the P-code whenever the antispoofing (A-S) mode of operation is activated. *Antispoofing* (A-S) guards against fake transmissions of satellite data by encrypting the P-code to form the Y-code. A-S was exercised intermittently through 1993 and implemented on January 31, 1994. The C/A code is available on the L1 frequency and the P-code is available on both L1 and L2. The various satellites all transmit on the same frequencies, L1 and L2, but with individual code assignments.

Each satellite transmits a navigation message containing its orbital elements, clock behavior, system time and status messages. In addition, an almanac is provided that gives the approximate data for each active satellite. This allows the user set to find all satellites once the first has been acquired. Tables 17.4 and 17.5 include examples of ephemeris and almanac information for the same satellite at approximately the same time.

The nominal GPS constellation is composed of 24 satellites in six orbital planes, (four satellites in each plane). The satellites operate in circular 20,200-km altitude (26,570 km radius) orbits at an inclination angle of 55° and with a 12-h period. The position is therefore the same at the same sidereal time each day, that is, the satellites appear four minutes earlier each day.

The GPS control segment consists of five monitor stations (Hawaii, Kwajalein, Ascension Island, Diego Garcia, Colorado Springs), three ground antennas, (Ascension Island, Diego Garcia, Kwajalein), and a master control station (MCS) located at Falcon AFB in Colorado. The monitor stations passively track all satellites in view, accumulating ranging data. This information is processed at the MCS to determine

TABLE 17.4 Ephemeris Information

Satellite Ephemeris Status for PRN07
Ephemeris Reference Time = 14:00:00 02/10/1995, Week Number = 0821
All Navigation Data is Good, All Signals OK
Code on L2 Channel = Reserved, L2 P Code Data = On
Fit Interval = 4 hours, Group Delay = 1.396984e-09 s
Clock Correction Time = 14:00:00 02/10/1995, af0 = 7.093204e-04 s
af1 = 4.547474e-13 s/s, af2 = 0.000000e + 00 s/s2
Semi-Major Axis = 26560.679 km, Eccentricity = 0.007590
Mean Anom = −23.119829, Delta n = 2.655361e-07, Perigee = −144.510162
Inclination = 55.258496, Inclination Dot = 1.100943e-08
Right Ascension = −10.880005, Right Ascension Dot = −4.796266e-07
Crs = 5.3750e + 01 m, Cuc = 1.6691e-04, Crc = 2.7141e + 02 m, Cic = 0.0000e + 00

TABLE 17.5 Almanac Information

SATELLITE ALMANAC STATUS FOR PRN07
Almanac Validity = Valid Almanac at 09:06:07 01/10/1995
Semi-Major Axis = 26560.168 km
Eccentricity = 0.007590
Inclination = 55.258966
Right Ascension = −10.829880
Right Ascension Rate = −4.655885e-07/s
Mean Anomaly = −172.426664
Argument of Perigee = −144.590529
Clock Offset = 7.095337e-04 s, Clock Drift = 0.000000e+00 s/s
Navigation Health Status is: All Data OK
Signal Health Status is: All Signals OK

satellite orbits and to update each satellite's navigation message. Updated information is transmitted to each satellite via the ground antennas.

The GPS system uses time of arrival (TOA) measurements for the determination of user position. A precisely timed clock is not essential for the user because time is obtained in addition to position by the measurement of TOA of four satellites simultaneously in view. If altitude is known (e.g., for a surface user), then three satellites are sufficient. If a stable clock (say, since the last complete coverage) is keeping time, then two satellites in view are sufficient for a fix at known altitude. If the user is, in addition, stationary or has a known speed then, in principle, the position can be obtained by the observation of a complete pass of a single satellite. This could be called the "transit" mode, because the old transit system used this method. In the case of GPS, however, the apparent motion of the satellite is much slower, requiring much more stability of the user clock.

The May 2000 GPS constellation consists of 28 satellites counting 27 operational and one recently launched and soon to become operational. Figure 17.18 and Figure 17.19 illustrate the history of the constellation including both the limiting dates of the operational periods, individual satellites, and the total number available as a function of time.

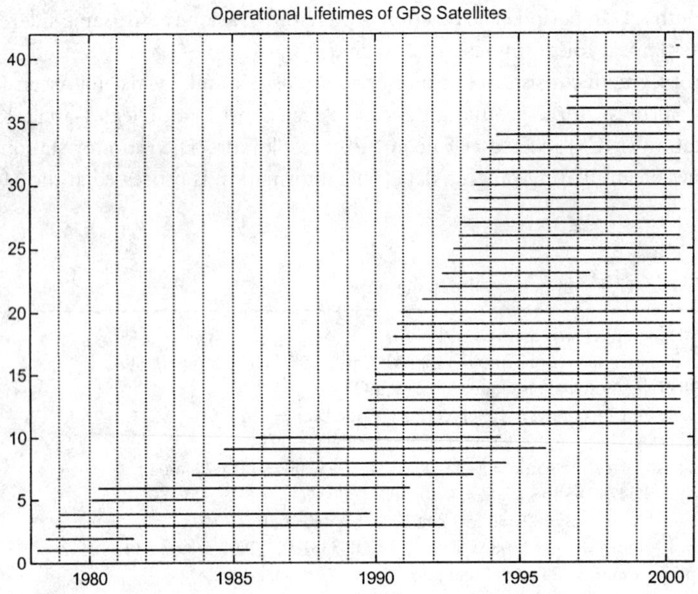

FIGURE 17.18   Operational lifetimes of GPS satellites.

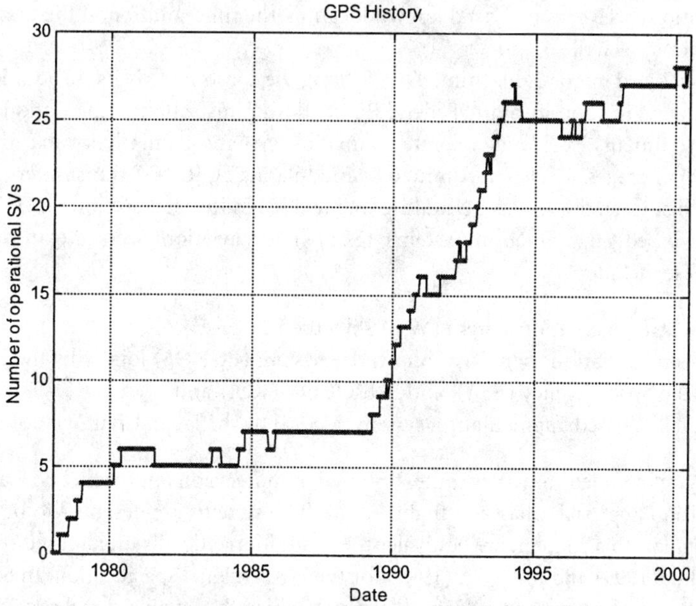

FIGURE 17.19 GPS history.

GPS Augmentations

Many augmentations to GPS exist to improve accuracy and provide integrity. These are both government operated (for safety of life navigation applications) and privately operated. This section will only deal with government operated systems. The recent termination of selective availability has sparked some debate on the need for GPS augmentation, but the most prevalent views are that stand-alone GPS will not satisfy all accuracy and integrity requirements, and that some augmentation will continue to be necessary. Since the dominant post-SA errors vary much more slowly than SA-induced errors did, the information bandwidth requirements to meet specified accuracy levels have certainly been reduced. Since integrity requirements specify maximum times to warn users of system faults, these warning times and time to first fix specifications will now drive information bandwidth requirements.

The simplest and first to become operational were maritime differential GPS stations. Existing radio-beacons in the 285 to 325 kHz band were converted to transmit differential corrections via a minimum shift keyed (MSK) modulation scheme. A base station receiver in a fixed, known location measures pseudo-range errors relative to its own clock and known position. Messages containing the time of observation, these pseudo-range errors, and their rate of change are then transmitted at either 100 or 200 baud. Parity bits are added, but no forward error correction is used. If the user receiver notes a data error, it merely waits for the next message. The user receiver uses the corrections extrapolated ahead in time based on the age of the correction and its rate of change. For a description of the message format the reader is referred to (RTCM, 1992.)

With SA on, satellite clock dithering was the dominant error and was common to all users. It is also assumed that over the few hundred kilometers or less where the signal can be received, that errors such as ephemeris and ionospheric and tropospheric delay are correlated between base station and user receiver as well and can be considerably reduced. The algorithms to predict these delays are disabled in both the base station and user receivers.

Because the corrections are measured relative to the base station clock, they are relative and not absolute corrections. This means they must have corrections for all satellites used for a position and all of these corrections must be from the same base station. For fixed position, precise time users, DGPS provides no improvement in accuracy. For moving users, their solution for time will track the base station clock, and since this base station receiver is in a fixed known location, the time solution in the moving receiver will see improved accuracy as well.

The U. S. Coast Guard operates maritime DGPS along the coasts and rivers. In addition, the system is being expanded inland in support of the Federal Railroad Administration in the National DGPS project. Many foreign governments worldwide operate compatible systems primarily for maritime applications. RTCM type DGPS messages are also transmitted by modulating LORAN C transmitters in Europe. Additional details on this system known as EUROFIX are included in the later section on LORAN C.

Three satellite based augmentation systems (SBASs) for aviation users are in various stages of development. These include

1. the wide area augmentation system (WAAS) by the U.S. FAA
2. the European geostationary navigation overlay system (EGNOS) jointly by the European Union, the European Space Agency (ESA), and EUROCONTROL, and
3. MTSAT satellite based augmentation system (MSAS) by the Japan Civil Aviation Bureau (JCAB)

These systems are intended for the enroute, terminal, nonprecision approach, and Category I (or near Category I) precision approach phases of flight. All of these systems are designed to be compatible, and to the level of detail in this chapter, are equivalent. For additional details on all three systems the reader is referred to Walter (1999) and to RTCA (1999) for the WAAS signal specification. In SBASs corrections need to be applied over the very large geographic area in which the satellite signal can be received. In this case the errors due to satellite ephemeris and ionospheric delay are not the same for all users and cannot be combined into one overall correction. Separate messages are provided for satellite clock corrections, vector corrections of satellite position and velocity, integrity information, and vertical ionospheric delay estimates for selected grid points. These grid points are at 5° increments in latitude and longitude, except larger increments occur in extreme northern or southern latitudes. The user receiver calculates its estimates of ionospheric delay by first determining the ionospheric pierce point of the propagation path from satellite to user receiver, interpolating between grid points, and then correcting for slant angle.

These corrections are or will be transmitted from geostationary satellites at GPS L1 frequencies with signal characteristics similar to GPS. The SBASs will provide additional ranging signals and transmit their own ephemeris information to improve fix availability. Message symbols at 500 symbols/sec will be modulo 2 added to a 1023 bit PRN code. The baseline data rate is 250 bits/sec. These data are rate $^1/_2$ convolutional encoded with a forward error correction code resulting in 500 symbols per second. The data will be sent in 250-bit blocks or 1 block/sec. The data block contains an eight-bit preamble, a six-bit message type, a 212-bit message, and 24 bits of CRC parity.

For Category II and III precision approach, the U.S. FAA will implement the local area augmentation system (LAAS). The transmitted data will include pseudo-range correction data, integrity parameters, approach data, and ground station performance category. The broadcast will be in the 108 to 117.95 MHz band presently used for VOR and ILS systems.

The modulation format is a differentially encoded, eight-phase-shift-keyed (D8PSK) scheme. The broadcast uses an eight-time-slot-per-half-second, fixed-frame, time-division-multiple-access (TDMA) structure. The total data rate of the system is 31,500 bits/sec. After the header and cyclic redundancy check (CRC), the effective rate of the broadcast is 1776 bits (222 bytes) of application data per time slot, or a total of 16×1776 BPS equaling 28,416 application data bits/sec. For additional information on LAAS the reader is referred to Braff (1997) and Skidmore (1999).

17.3.3 Global Navigation Satellite System (GLONASS)

GLONASS is the Russian parallel to GPS and has its origins in the mid-1970s in the former Soviet Union. Like GPS, the system has been primarily developed to support the military, but in recent years has been broadened to include civilian users. In September 1993, Russian President Boris Yeltsin officially proclaimed GLONASS to be an operational system and the basic unit of the Russian Radionavigation Plan. The well-publicized political, economic, and military uncertainties within Russia have undoubtedly hampered the

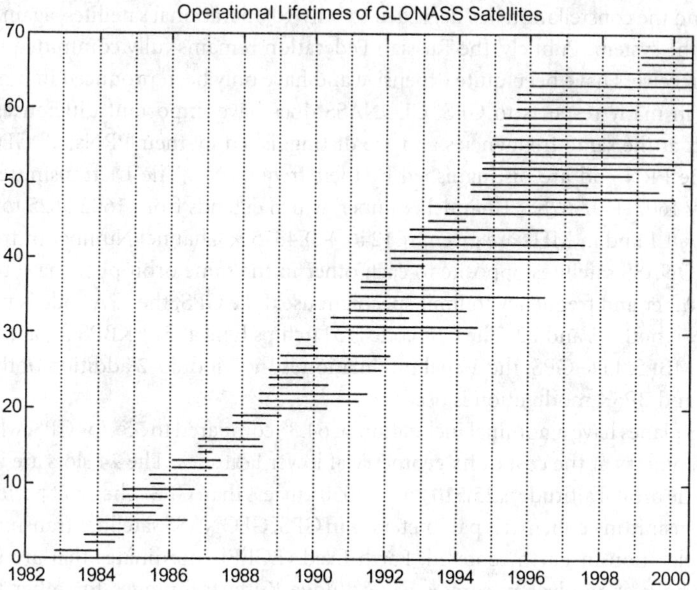

FIGURE 17.20 Operational lifetimes of GLONASS satellites.

implementation and maintenance of the system. These issues combined with the poor reliability record of the early satellites, raised questions relative to its eventual success. Figure 17.20 and Figure 17.21 show data on the operational life of those satellites that became operational from the first one in 1982 through the present. It is also believed that attempts were made to place approximately 10 or more other satellites into operation resulting in failure for a number of reasons (Dale, 1989). However, in the mid-1990s, the reliability of the satellites considerably improved, and many more satellites were put in orbit such that in early 1960 there were 24 operational satellites. The triple launch on December 30, 1998 has been the only launch

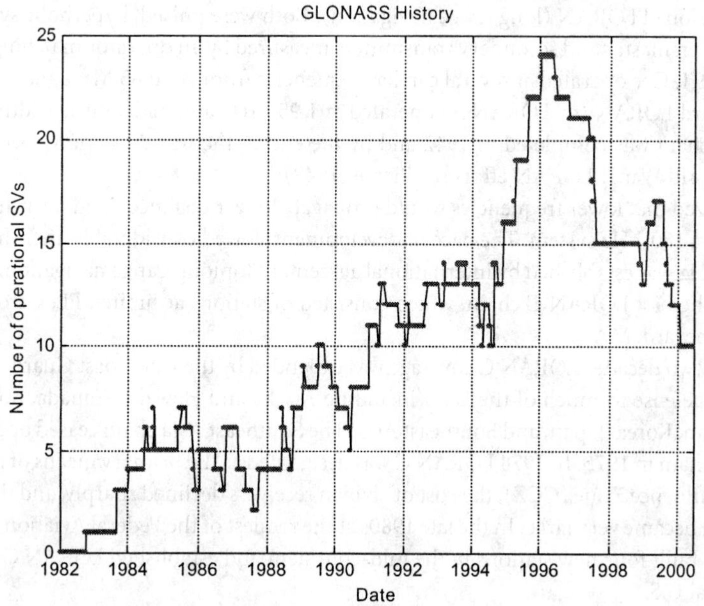

FIGURE 17.21 GLONASS history.

since late 1995, and the constellation has degraded to only 10 operational satellites, again raising concerns on the future of the system. Publicly, the Russian Federation remains fully committed to the system. To date, GLONASS receivers have been quite expensive and have only been produced in limited quantities.

While similar in many respects to GPS, GLONASS does have important differences. While all GPS satellites transmit at the same frequencies and are distinguished by their PRNs, all GLONASS satellites transmit the same PRN and are distinguished by their frequencies. The L1 transmitted frequencies in MHz are given by $1602 + 0.5625 \times$ Channel Number, which extends from 1602.5625 to 1615.5 MHz. L2 frequencies are $^7/_9$ L1 and in MHz are given by $1246 + 0.4375 \times$ Channel Number or from 1246.4375 to 1256.5 MHz. GLONASS satellites opposite to each other in the same orbit plane have been assigned the same channel number and frequency to limit spectrum use. Like GPS, the C/A code is transmitted on L1 only, and P code on both L1 and L2. The C/A code is 511 chips long at 511 KBPS for a length of 1 ms. The P code is at 5.11 MBPS. Like GPS, the actual modulation is the Modulo-2 addition of the PRN sequence and 50 BPS data and BPSK modulation is used.

The GLONASS planes have a nominal inclination of $64.8°$ compared to $55°$ for GPS, which gives slightly better polar fix geometry at the cost of fix geometry at lower latitudes. The 24 slots are in three planes of eight slots each. The orbital altitude is 25,510 km or 1050 km less than GPS. The orbit period is 11 h, 15 min.

Rather than transmitting ephemeris parameters as in GPS, GLONASS satellites transmit actual position, velocity, and acceleration in Earth centered, Earth fixed (ECEF) coordinates that are updated on half-hour intervals. The user receiver integrates using Runge-Kutta techniques for other times. All monitor sites are within the former Soviet Union, which limits accuracy of both ephemeris and time and would delay user notification of satellite failures. The system produces both high accuracy signals for Russian military use only and lesser accuracy for civilian use. The civilian use signals are not degraded to the same extent as GPS was with SA, which resulted in significantly better accuracy than civil GPS during the period GLONASS had a complete constellation. Military accuracy is classified. For more details the reader is referred to the GLONASS Interface Control Document (CSIC, 1998) available at http://www.rssi.ru/sfcsic/sfcsic_main.html

17.3.4   LORAN C History and Future

Early in World War II, both the U.S. and Great Britain recognized the need for an accurate, long range, radionavigation system to support military operations in Europe. As a result the British developed Gee and the U.S. developed LORAN (long range navigation). Both were pulsed, hyperbolic systems with time differences between master and secondary transmitters measured by an operator matching envelopes on a delayed sweep CRT. Gee operated at several carrier frequencies from 20 to 85 MHz and had a pulse width of 6 usec. Standard LORAN (or LORAN A) operated at 1.95 MHz and had a pulse width of 45 usec. The first LORAN A chain was completed in 1942 and by the end of the war 70 stations were operating. For details on the World War II LORAN effort see Pierce (1948).

It was recognized that lower frequencies would propagate longer distances, and near the end of the war testing began on a 180-kHz system. The pace of development slowed considerably after the war. The band from 90 to 110 kHz was established by international agreement for long-range navigation. In 1958, system tests started on the first LORAN C chain, which consisted of stations at Jupiter, FL, Carolina Beach, NC, and Martha's Vineyard, MA.

Over the next two decades LORAN C coverage was expanded by the U.S. Coast Guard in support of the Department of Defense to much of the U.S. (including Alaska and Hawaii), Canada, northwest Europe, the Mediterranean, Korea, Japan, and Southeast Asia. The Southeast Asia chain ceased operations with the fall of South Vietnam in 1975. In 1974 LORAN C was designated as the primary means of navigation in the U.S. Coastal Confluence Zone (CCZ), the cost of civilian receivers declined sharply, and the civil maritime use of the system became very large. In the late 1980s at the request of the Federal Aviation Administration, the Coast Guard built four new stations in the mid-continent and established LORAN C coverage for the entire continental U.S.

Other countries are developing and continuing LORAN C to meet their future navigational needs. Many of the recent initiatives have taken place as a result of the termination of the U.S. DoD requirement for overseas LORAN C. This need came to an end as of December 31, 1994. With the introduction of GPS, many countries have decided that it is in their own best interests not to have their navigational needs met entirely by a U.S. DoD-controlled navigation system. Many of these initiatives have resulted in multilateral agreements between countries which have common navigational interests in those geographic areas where LORAN C previously existed to meet U.S. DoD requirements (e.g., northern Europe and the Far East). The countries of Norway, Denmark, Germany, Ireland, the Netherlands, and France established a common LORAN C system under the designation The Northwest European LORAN C System (NELS). Recently the governments of Italy and the United Kingdom have applied for membership. This system presently comprises eight stations forming four chains. In conjunction with this system, respective foreign governments took over operation of the former USCG stations of B0 and Jan Mayen, Norway; Sylt, Germany; and Ejde, Faroe Islands (Denmark) as of 31 December 1994. The two French stations formerly operated by the French Navy in the rho-rho mode were reconfigured and included and two new stations in Norway were constructed. A planned ninth station at Loop Head, Ireland has yet to be constructed. The former USCG stations at Angissoq, Greenland and Sandur, Iceland were closed.

An important feature of NELS is the transmission of DGPS corrections by pulse position modulation. This concept, called EUROFIX, was developed by Professor Durk Van Willigen and his students at Delft University of Technology in the Netherlands. The last six pulses in each group are modulated in a three-state pattern of prompt and 1 usec advance or retard. Of the possible 729 combinations, 142 of which are balanced, 128 balanced sequences are used to transmit a seven-bit word. Extensive Reed Solomon error correction is used resulting in a very robust data link, although at a somewhat lower data rate than the MSK beacons. With Selective Availability on, the horizontal accuracy was 2 to 3 m, 2 DRMS, or slightly worse than conventional DGPS due to the temporal decorrelation of SA. With SA off, the performance should be virtually comparable. EUROFIX has been operating at the Sylt, Germany transmitter since February, 1997, expansion to other NELS transmitters is planned for the near future. In a joint FAA/USCG effort, preliminary testing is underway of transmitting WAAS data via a much higher data rate LORAN communications channel. If LORAN C survives in the long term, it is likely to be both as a navigation system and as a communications channel to enhance the accuracy and integrity of GPS.

In the Far East, an organization was formed called the Far East Radionavigation Service (FERNS). This organization consists of the countries of Japan, The Peoples Republic of China, The Republic of Korea, and the Russian Federation. Japan took over operation of the former USCG stations in its territory and they are currently being operated by the Japanese Maritime Safety Agency (JMSA). In the Mediterranean Sea area, the four USCG stations were turned over to the host countries. The two stations in Italy, Sellia Marina and Lampedusa, are currently being operated. The stations at Kargaburun, Turkey and Estartit, Spain remain off air. Saudi Arabia and India are currently each operating two LORAN chains.

There is ongoing debate at high levels concerning the future of LORAN C in the U.S. According to the 1999 Federal Radionavigation Plan (DoD/Dot, 1999), "While the administration continues to evaluate the long term need for continuation of the LORAN-C navigation system, the Government will operate the LORAN C system in the short term. The U.S. Government will give users reasonable notice if it concludes that LORAN C is not needed or is not cost effective, so that users will have the opportunity to transition to alternative navigational aids."

LORAN Principles of Operation

Each chain consists of three or more stations, including a master and at least two secondary transmitters. (Each master-secondary pair enables determination of one LOP, and two LOPs are required to determine a position. The algorithms to convert these LOPs into latitude and longitude are discussed in a later section.) Each LORAN C chain provides signals suitable for accurate navigation over a designated geographic area termed a *coverage area*.

The stations in the LORAN chain transmit in a fixed sequence that ensures that time differences (TDs) between a receiving master and secondary can be measured throughout the coverage area. The length of time in usec over which this sequence of transmissions from the master and the secondaries takes place

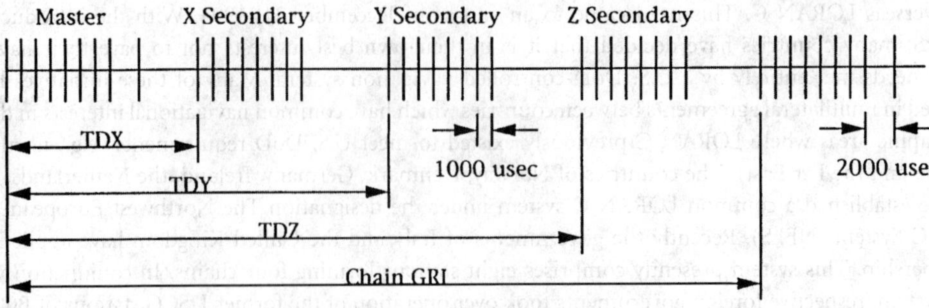

FIGURE 17.22 LORAN C pulse pattern.

is termed the group repetition interval (GRI) of the chain. All LORAN C chains operate on the same frequency (100 kHz), but are distinguished by the GRI of the pulsed transmissions.

The LORAN C system uses pulsed transmission, nine pulses for the master and eight pulses for the secondary transmissions. Figure 17.22 shows this overall pulse pattern for the master and three secondary transmitters (X, Y, and Z). Coding delay is the time between when a secondary receives the master signal and when it transmits. The time differences measured on the baseline extension beyond the secondary should be coding delay. Emission delay is the difference in time of transmission between master and secondary and is the sum of coding delay and baseline length (converted to time). Shown in Fig. 17.23 is an exploded view of the LORAN C pulse shape. It consists of sine waves within an envelope referred to as a t-squared pulse. (The equation for the envelope is also included in Fig. 17.23.) This pulse will rise from zero amplitude to maximum amplitude within the first 65 usec and then slowly trails off or decays over a 200 to 300 usec interval. The pulse shape is designed so that 99% of the radiated power is contained within

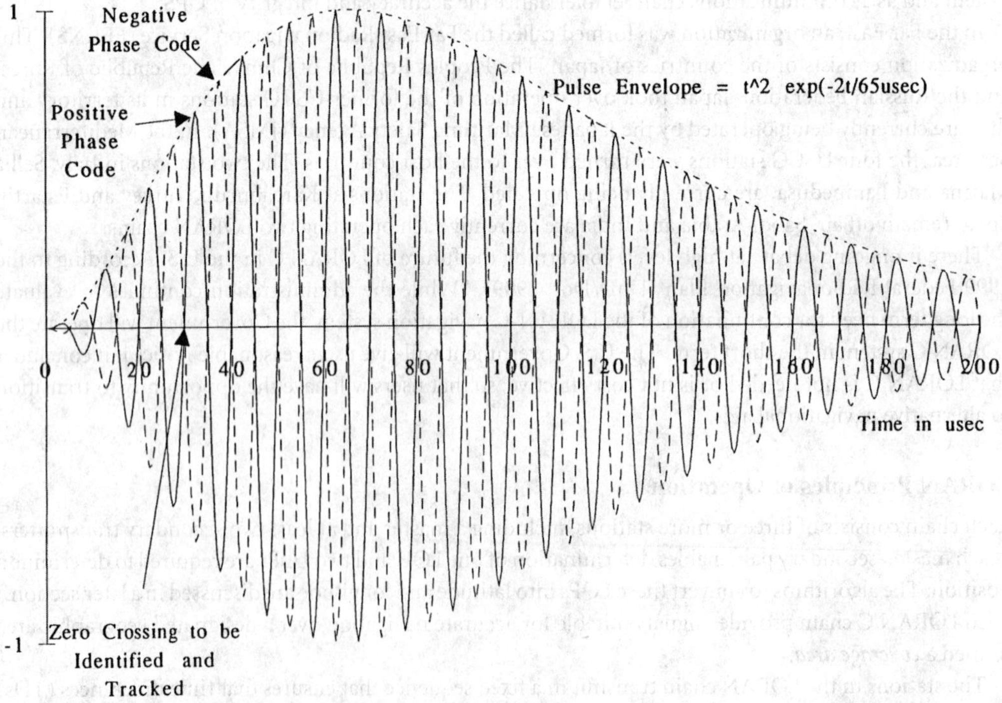

FIGURE 17.23 Ideal LORAN pulse.

TABLE 17.6   LORAN C Phase Codes.

GRI Interval	Master	Secondary
A	+ + − − + − + − +	+ + + + + − − +
B	+ − − + + + + + −	+ − + − + + − −

the allocated frequency band for LORAN C of 90 to 110 kHz. The rapid rise of the pulse allows a receiver to identify one particular cycle of the 100 kHz carrier. Cycles are spaced approximately 10 usec apart. The third cycle of this carrier within the envelope is used when the receiver matches the cycles. The third zero crossing (termed the positive 3rd zero crossing) occurs at 30 usec into the pulse. This time is both late enough in the pulse to ensure an appreciable signal strength and early enough in the pulse to avoid sky wave contamination from those skywaves arriving close after the corresponding ground wave.

Within each pulse group from the master and secondary stations, the phase of the radio frequency (RF) carrier is changed systematically from pulse-to-pulse in the pattern shown in Fig. 17.23 and Table 17.6. This procedure is known as phase coding. The patterns A and B alternate in sequence. The pattern of phase coding differs for the master and secondary transmitters. Thus, the exact sequence of pulses is actually matched every two GRIs, an interval known as a phase code interval (PCI).

Phase coding enables the identification of the pulses in one GRI from those in an earlier or subsequent GRI. Just as selection of the pulse shape and standard zero crossing enable rejection of certain early sky waves interfering with the same pulse, phase coding enables rejection of late sky waves interfering with the next pulse. Since seven of the fourteen pulses that come immediately before another pulse have the same phase code and seven are different, late sky waves interfering with the next pulse will average to zero. Because the master and secondary signals have different phase codes, the LORAN receiver can distinguish between them.

17.3.5   Position Solutions from Radionavigation Data

In general, closed-form solutions that allow direct calculation of position from radionavigation observables do not exist. However, the inverse calculations can be done. Given one's position and clock offset, one can predict GPS or GLONASS pseudo-ranges or LORAN TDs or TOAs. These expressions are nonlinear but can be linearized about a point and the problem solved by iteration. The basic approach can be summarized by

1. Assuming a position
2. Calculating the observables one should have seen at that point
3. Calculating rate of change of observables (partial derivatives) as position is varied
4. Comparing calculated to measured observables
5. Computing a new position based on differences between calculated and measured observables
6. Depending on exit criteria, either exit algorithm or return to "b" above using the calculated position as the new assumed position

We will look at the details of each of these steps and the expected statistics of the results for a number of cases. We will start with the example of processing three LORAN TOAs (or two TDs), but the basic principles apply to GPS as well. We will then consider overdetermined solutions and the techniques for assuring the solutions are optimal in some sense. We will also consider integrated solutions and Kalman filter based solutions.

For LORAN the calculated times of arrival (TOAs) are given by

For Master $\text{TOA}_m = d_i$
and for secondaries $\text{TOA}_i = d_i + \text{ED}_i$, etc.

where d_m and d_i are the propagation times from the master and ith secondary stations to the assumed position and ED_i is emission delay. (Note: Secondary stations are designated by the letters V, W, X, Y, and Z, but because we will also want to use x and y for position variables, in order to avoid confusion, we will use integer subscripts to denote secondaries.) The LORAN propagation times are the sums of three terms:

1. *Primary factor* (PF) or the geodesic distance to the station divided by a nominal phase velocity. See WGA (1982, 57–61) for an algorithm that will work to distances up to 3000 nautical miles.
2. *Secondary factor* (SF) which takes into account the additional phase lag due to the signal following a curved earth surface over an all-seawater path (see COMDT (G-NRN) 1992, II-14 for formula), and
3. *Additional secondary factor* (ASF) which takes into account the difference between the actual path traveled and an all-seawater path. This factor can be calculated based on conductivity maps, or can be obtained by tables (from the U.S. Defense Mapping Agency) or from charts published by the Canadian government. The latter two are based both on calculations and observations.

A typical LORAN receiver measures time differences (TDs) vs. TOAs and the receiver need not necessarily solve for time. In GPS or GLONASS, since the transmitters move rapidly with time, in order to know their location it is essential that an explicit solution for time be made. In GPS each satellite transmits ephemeris parameters which, when substituted in standard equations (NATO, 1991, A-3-26 and 27), give the satellite location as a function of time expressed in Earth centered, Earth fixed (ECEF) coordinates. In this system, the positive x axis goes from the Earth's center to $0°$ latitude and longitude, the positive y axis goes from the Earth's center to $0°$ latitude and $90°$E longitude, and the positive z axis goes from the Earth's center to $90°$N latitude. For GLONASS, the satellites transmit their location, velocity, and acceleration, in ECEF coordinates, and validate at half-hour intervals. The user is expected to numerically integrate the equations on motion to obtain position for a particular time.

The receiver measures the time of arrival of the satellite signal, and since the signal is tagged precisely in time, the difference in time of transmission and time of arrival (converted to distance) is a pseudo-range. "Pseudo" is used as the time of arrival is relative to the receiver clock, which cannot be assumed exact but must be solved for as part of the solution. These pseudo-ranges are then corrected for both ionospheric and tropospheric delay. In civil C/A code receivers, ionospheric delay parameters are transmitted and the delay calculated via an algorithm (NATO, 1991, p A-6-31). For dual frequency receivers, the ionospheric delay is assumed to be inversely proportional to frequency and the delay determined from the difference in pseudo-ranges at the two frequencies. For differential GPS, ionospheric delay is assumed to be the same at the reference station and the user's location and the algorithm is disabled at both locations.

Solution Using Two TDs (LORAN Only)

TDs at the assumed position are calculated ($\mathrm{TD_p}$) and subtracted from observations ($\mathrm{TD_o}$), i.e.,

$$\mathrm{TD}_i = \mathrm{TOA}_i - \mathrm{TOA}_m$$

$$\Delta\mathrm{TD} = \begin{bmatrix} \Delta\mathrm{TD}_1 \\ \Delta\mathrm{TD}_2 \end{bmatrix} = \mathrm{TD}_o - \mathrm{TD}_p$$

This difference in TDs can be linearly related to a difference in position $\Delta P = \begin{bmatrix} \Delta x \\ \Delta y \end{bmatrix}$ by

$$\begin{bmatrix} \sin(\phi_m) - \sin(\phi_1) & \cos(\phi_m) - \cos(\phi_1) \\ \sin(\phi_m) - \sin(\phi_2) & \cos(\phi_m) - \cos(\phi_2) \end{bmatrix} \begin{bmatrix} \Delta x \\ \Delta y \end{bmatrix} = c \begin{bmatrix} \Delta\mathrm{TD}_1 \\ \Delta\mathrm{TD}_2 \end{bmatrix}$$

where ϕ_m, ϕ_1, and ϕ_2 are the azimuths to the transmitters measured in a clockwise direction from the north or positive y direction as shown in Fig. 17.24. While exact calculations taking into account the ellipsoidal nature of the shape of the Earth must be used in the calculation of predicted TDs, only approximate

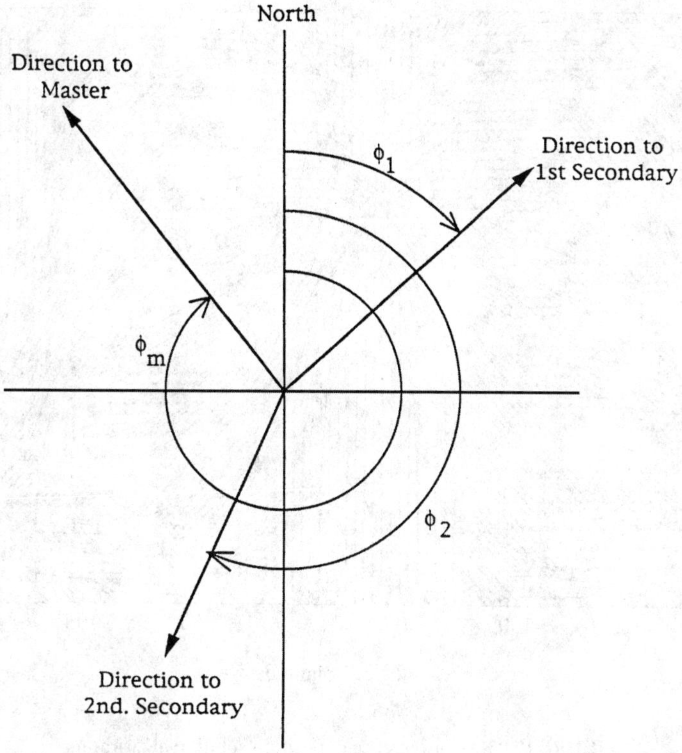

FIGURE 17.24 Definition of LORAN azimuths.

expressions based on a spherical earth are necessary to determine azimuths. The expression above can be expressed in matrix form:

$$A_{\text{TD}}\Delta P = c\Delta\text{TD}$$

and solved by

$$\Delta P = c A_{\text{TD}}^{-1}\Delta\text{TD}$$

At this point a new assumed position would be calculated (using an appropriate conversion of meters to degrees), the predicted TDs at that location determined, and if they were within a selected tolerance of the measured, the algorithm would stop, otherwise a new solution would be determined. An alternative criteria based on the movement of the assumed position, $|\Delta P|$, can be used as well and will be more appropriate in an over-determined solution because exact match of TDs is not possible. Figure 17.25 and Figure 17.26 show typical convergence of this type of algorithm. Since the algorithm is based on the assumption that the TD grid is linear in the vicinity of the assumed position, we need to limit the step size so that the position does not jump to radically different geometry in one step. As can be seen in Fig. 17.25 all lines of constant TD between two transmitters are closed curves on the Earth's surface. Two of these curves intersect, either do not intersect, or intersect exactly twice. Since we assume we are starting with a set of TDs that exist, the latter applies. The initial assumed position is what determines to which of these two ambiguous positions the algorithm will converge. Essentially since the algorithm is based on the gradient of the TDs with respect to change in position and that gradient changes sign on the baseline extension, the position will move to the solution it can reach without crossing a baseline extension. The direction of the gradient, and hence the final solution also changes as the lines of constant TD become parallel as at 30°N and 10°E in Fig. 17.26.

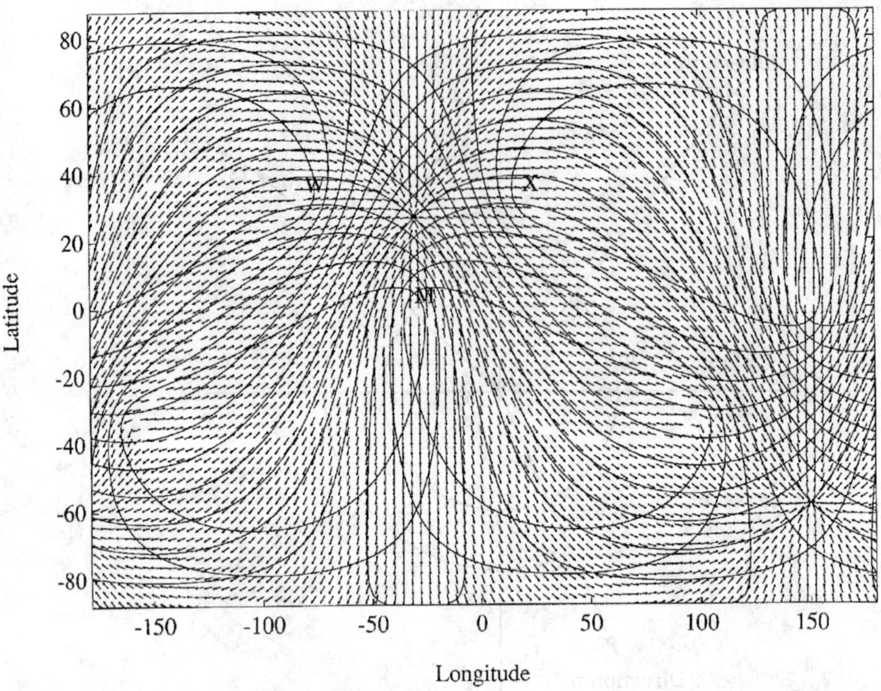

FIGURE 17.25   Convergence of position solution algorithnm.

Frequently, as shown in the Fig. 17.25, one of these ambiguous positions is well beyond the range the signals can be received and can easily be eliminated. This is not true when one of the solutions is near the baseline extension as shown in Fig. 17.26. In this example, it is necessary to have prior knowledge that the actual position is North or South of the master station.

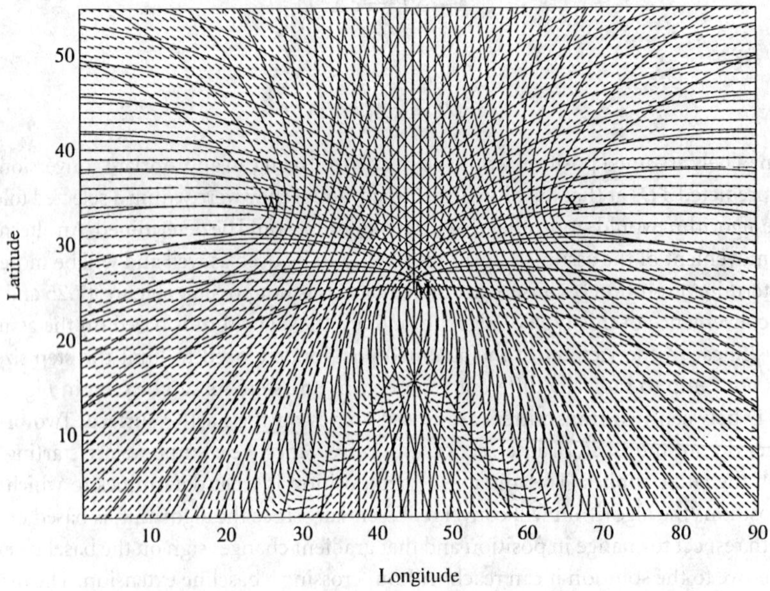

FIGURE 17.26   Example of position solution with close by ambiguous position.

Solution Using TOAs

This method yields a solution for receiver clock bias (modulo one phase code interval) and even though not generally implemented in present receivers, is presented for a number of reasons:

1. It is functionally equivalent to the GPS position solution but is presented first because it is a two vs. three-dimensional and is somewhat easier to visualize.
2. For the exactly determined (two-TD or three-TOA) case it is exactly equivalent to the TD solution and therefore we can extend results from TOA or pseudo-range analysis to the TD case. Early LORAN TD analysis was typically done in scalar form before computer programs such as spreadsheets and MatLab^{TM} made matrix calculations trivial. GPS fix analysis has typically been expressed in matrix form resulting in simpler expressions.
3. For over-determined solutions, since we can assume TOA errors are statistically independent, but cannot make the same assumption for TD errors, the solution is slightly easier to implement and analyze.
4. Solutions using Kalman filters, integrated GPS/LORAN, or a precise clock with only two TOAs, will typically be based on TOAs vs. TDs.

The vector difference between observed (TOA_o) and predicted (TOA_p) TOAs are calculated via

$$\Delta TOA = TOA_o - TOA_p$$

Equation is replaced by

$$- \begin{bmatrix} \sin(\phi_m) & \cos(\phi_m) & 1 \\ \sin(\phi_1) & \cos(\phi_1) & 1 \\ \sin(\phi_2) & \cos(\phi_2) & 1 \end{bmatrix} \begin{bmatrix} \Delta x \\ \Delta y \\ c\,\Delta t \end{bmatrix} = c \begin{bmatrix} \Delta TOA_m \\ \Delta TOA_1 \\ \Delta TOA_2 \end{bmatrix}$$

or

$$a_{TOA}\Delta P = c\,\Delta TOA$$

and solved by

$$\Delta P = c\,A_{TOA}^{-1}\Delta TOA$$

It is useful to think of the sines and cosines in the matrix as the cosines of the angles between the positive direction of the coordinate axes and the directions to the transmitters. This concept can be easily extended to the three-dimensional case in GPS or GLONASS solutions. For a normal GPS-only solution it is usually more convenient to express both satellite and user positions in an Earth centered, Earth fixed (ECEF) coordinate system. In this case the direction cosines are merely the difference of the particular coordinate divided by the range to each satellite. After position solution, an algorithm is used to convert from (xyz) to latitude, longitude, and altitude. An alternative that will prove useful for meaningful dilution of precision (DOP) calculations or for integrated GPS/LORAN receivers is to use a East/North/altitude coordinate system and to solve for the azimuth (AZ) and elevation (EL) of the satellites. In this case the direction cosines for each satellite become

East: $\sin(AZ)\cos(EL)$
North: $\cos(AZ)\cos(EL)$
Altitude: $\sin(EL)$

Error Analysis

With only three TOAs this solution will be exactly the same as the two-TD solution above. In both of these solutions after the algorithm has converged, we can consider ΔTOA or ΔTD as measurement errors and ΔP as the position error due to these errors. The covariance of the errors in position (and in time for TOA processing) can be expressed as functions of the direction cosine matrix and the covariance of the TDs or TOAs.

$$\Delta P \Delta P^T = c^2 A_{TD}^{-1} \Delta TD \Delta TD^T \left(A_{TD}^{-1}\right)^T$$

$$\text{cov}(\Delta P) = E\{\Delta P \Delta P^T\} = c^2 A_{TD}^{-1} E\{\Delta TD \Delta TD^T\}\left(A_{TD}^{-1}\right)^T = c^2 A_{TD}^{-1} \text{cov}(\Delta TD)\left(A_{TD}^{-1}\right)^T$$

$$\text{cov}(\Delta P) = c^2 A_{TOA}^{-1} \text{cov}(\Delta TOA)\left(A_{TOA}^{-1}\right)^T$$

Given that the covariances of the measurements are known it becomes a simple task to evaluate the covariance of the position errors. We will first assume each TOA measurement is uncorrelated, that is,

$$\text{cov}(\Delta TOA) = \begin{bmatrix} \sigma_m^2 & 0 & 0 \\ 0 & \sigma_1^2 & 0 \\ 0 & 0 & \sigma_2^2 \end{bmatrix}$$

and

$$\text{cov}(\Delta TD) = \begin{bmatrix} \sigma_m^2 + \sigma_1^2 & \sigma_m^2 \\ \sigma_m^2 & \sigma_m^2 + \sigma_2^2 \end{bmatrix}$$

For simplicity we will also assume each TOA measurement has the same variance (σ_{TOA}^2). While this assumption is not in general valid, particularly for LORAN C, it is useful in that it allows us to separate out purely fix geometry from signal-in-space issues. Under this assumption

$$\text{cov}(\Delta TOA) = \sigma_{TOA}^2 I$$

and

$$\text{cov}(\Delta TD) = \sigma_{TOA}^2 \begin{bmatrix} 2 & 1 \\ 1 & 2 \end{bmatrix}$$

For the particular case of the same number of measurements and unknowns, these both yield the same results:

$$\text{cov}(\Delta P) = \begin{bmatrix} \sigma_x^2 & \sigma_{xy} & c\sigma_{xt} \\ \sigma_{xy} & \sigma_y^2 & c\sigma_{yt} \\ c\sigma_{xt} & c\sigma_{yt} & c^2\sigma_t^2 \end{bmatrix} = c^2 \sigma_{TOA}^2 A_{TOA}^{-1} \left(A_{TOA}^{-1}\right)^T = c^2 \sigma_{TOA}^2 \left[A_{TOA}^T A_{TOA}\right]^{-1}$$

What is significant is that the matrix $G = [A_{TOA}^T A_{TOA}]^{-1}$ is a dimensionless multiplier that relates our ability to measure the range to a transmitter (LORAN station or GPS satellite). In GPS, (NATO GPS 1991) $\sqrt{\text{trace}(G)}$ is defined as the geometric dilution of precision (GDOP), which includes the time term. (NATO GPS 1991). In LORAN C, since convention has been not to explicitly solve for time, GDOP is normally considered as $\sqrt{G_{11} + G_{22}}$, which only contains horizontal position terms and is equivalent to horizontal dilution of precision (HDOP) in GPS. Since we want to consider both LORAN and GPS in a consistent, integrated way, we will use HDOP and reserve GDOP for expressions including time term. The scalar expression for this quantity is

$$\text{HDOP} = \frac{1}{\sqrt{2}\sin(C)} \sqrt{\frac{1}{\sin^2(A/2)} + \frac{1}{\sin^2(B/2)} + \frac{\cos(C)}{\sin(A/2)\sin(B/2)}}$$

where A and B are the angles from the Master to each secondary and $C = \frac{A+B}{2}$. Figure 17.27 shows contours of constant HDOP as functions of these two angles. While the horizontal accuracy in one dimension is well defined ($\sigma_x^2 = G_{11}c^2\sigma_{TOA}^2$, for example), the two-dimensional accuracy is slightly more complex. Generally the terms drms and 2 drms accuracy are used to refer to the one-sigma and two-sigma accuracies. For example

$$\text{2 drms horizontal accuracy} = 2\,\text{HDOP}\,c\sigma_{TOA}$$

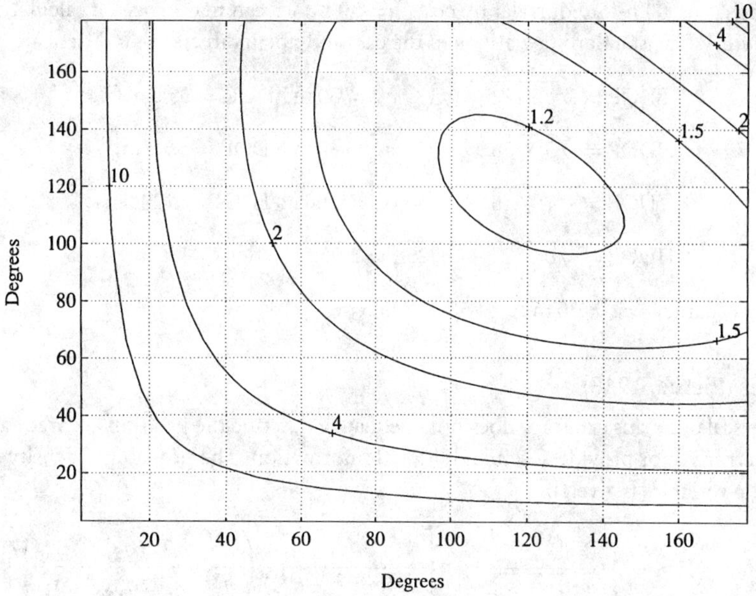

FIGURE 17.27 HDOP as function of angles from master to secondaries.

A circle of this radius should contain between 95.4 and 98.2% of the fixes depending on the eccentricity of the error ellipse describing the distribution of the fixes. For example, if $\sigma_x = \sigma_y$ and $\sigma_{xy} = 0$, the error ellipse becomes a circle and one of radius 2 drms contains $(1 - \exp(-4)) = 98.2\%$. In the other extreme, when the ellipse collapses to a line ($\sigma_{xy} \cong \pm \sigma_x \sigma_y$) the 2 drms circle contains the probability within two standard deviations of the mean of a scalar normal variable or 95.4%.

The relationship between geometry and accuracy can best be explained via numerical example. Shown below are spreadsheet calculations for a master and two secondaries at 0, 80, and 300°, respectively.

Bearing		East (Sine)	North (Cosine)	Time
Master	0	0	1	1
Sec. #1	80	0.9848	0.1736	1
Sec. #2	300	−0.8660	0.5000	1
		INVERSE $(A^T A)$		
TOA		0.7122	0.6748	−0.4047
		0.6748	3.5258	−1.9937
		−0.4047	−1.9937	1.4616
TD		$\sin(M) - \sin(S)$	$\cos(M) - \cos(S)$	
		−0.9848	0.8264	
		0.8660	0.5000	
		R(TD)		
		2	1	
		1	2	
		INV(A) R INV(A^T)		
		0.7122	0.6748	
		0.6748	3.5258	

Assuming $\sigma_{\text{TOA}} = 100$ ns or equivalently $c\sigma_{\text{TOA}} = 30$ m, we can make specific calculations regarding repeatable accuracy. The standard deviations in the east and north directions and time are

$$\text{EDOP} = \sqrt{0.7122} \qquad \sigma_x = 30\,\text{m}\,\text{EDOP} = 25.3\,\text{m},$$

$$\text{NDOP} = \sqrt{3.5258} \qquad \sigma_y = 30\,\text{m}\,\text{NDOP} = 56.3\,\text{m},$$

$$\text{TDOP} = \sqrt{1.4616} \qquad \sigma_t = 100\,\text{ns}\,\text{TDOP} = 120.9\,\text{ns},$$

$$\text{HDOP} = \sqrt{0.7122 + 3.5258} = 2.058$$

and the 2 drms accuracy $= 2 \times 30\,\text{m} \times 2.058 = 123.5\,\text{m}.$

Error Ellipses (Pierce, 1948)

Frequently the scalar 2 drms accuracy does not adequately describe the position accuracy and the distribution of the error vector provides useful additional information. The probability density of two jointly normal random variables is given by

$$p(x, y) = \frac{1}{2\pi \sqrt{\sigma_x^2 \sigma_y^2 - \sigma_{xy}^2}} \exp\left[-\frac{\sigma_x^2 \sigma_y^2}{2(\sigma_x^2 \sigma_y^2 - \sigma_{xy}^2)} \left(\frac{x^2}{\sigma_x^2} - \frac{2xy\sigma_{xy}}{\sigma_x^2 \sigma_y^2} + \frac{y^2}{\sigma_y^2} \right) \right]$$

and the ellipses

$$\frac{x^2}{\sigma_x^2} - \frac{2xy\sigma_{xy}}{\sigma_x^2 \sigma_y^2} + \frac{y^2}{\sigma_y^2} = 2 \left(1 - \frac{\sigma_{xy}^2}{\sigma_x^2 \sigma_y^2} \right) c^2$$

are curves of constant probability density such that the probability a sample lies within the ellipse is $1 - \exp(-c^2)$. For algebraic simplicity this equation will be rewritten:

$$Ax^2 + 2Hxy + By^2 = C$$

which can be rotated by angle (Ω) into a (ξ, η) coordinate system aligned with the ellipse axes

$$\frac{\xi^2}{\alpha^2} + \frac{\eta^2}{\gamma^2} = C$$

where

$$\tan(2\Omega) = -\frac{2H}{B - A}$$

$$\alpha^2 + \gamma^2 = \frac{A + B}{AB - H^2}, \quad \text{and}$$

$$\alpha^2 - \gamma^2 = \frac{(B - A)\sec(2\Omega)}{AB - H^2}$$

Taking the sum and difference of the last two expressions

$$2\alpha^2 = \frac{A[1 - \sec(2\Omega)] + B[1 + \sec(2\Omega)]}{AB - H^2}, \quad \text{and}$$

$$2\gamma^2 = \frac{A[1 + \sec(2\Omega)] + B[1 - \sec(2\Omega)]}{AB - H^2}$$

In the example above, $\Omega = -12.8°$. To draw an ellipse that contains $(1 - \exp(-4)) = 98.17\%$ of the samples, $\alpha = 63.4$ m and $\gamma = 162.8$ m. (Note: α is the length of the ellipse axis rotated by angle Ω from

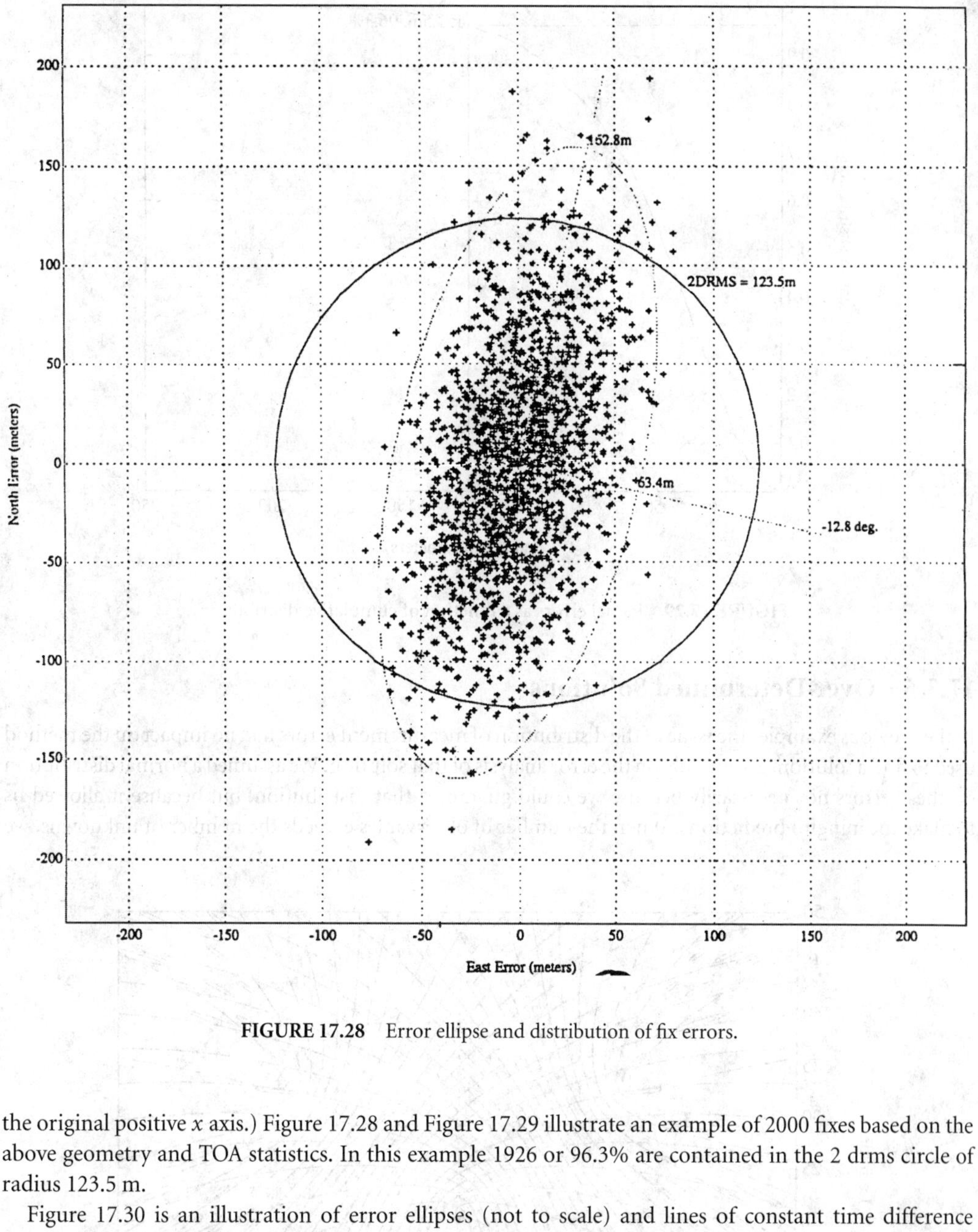

FIGURE 17.28   Error ellipse and distribution of fix errors.

the original positive *x* axis.) Figure 17.28 and Figure 17.29 illustrate an example of 2000 fixes based on the above geometry and TOA statistics. In this example 1926 or 96.3% are contained in the 2 drms circle of radius 123.5 m.

Figure 17.30 is an illustration of error ellipses (not to scale) and lines of constant time difference for a typical triad. The main point is that when geometry becomes poor in the fringes of the coverage area, representing the accuracy by circles of increasing 2 drms radii does not truly reflect one's knowledge of accuracy. In many examples from both air and marine navigation, cross-track accuracy is far more important than along-track accuracy. Knowledge of the orientation of the error ellipse may be important.

In GPS or GLONASS, exactly the same error analysis and ellipses apply, the main difference being that the geometry at any particular point on the earth's surface has constantly changing geometry as satellites move and are added/subtracted from the available constellation. When we look at distributions of GPS or GLONASS fixes they tend to appear circularly symmetric, not because the instantaneous distributions are that way, but because the averaged distributions over long periods such as a day or longer become circularly symmetric.

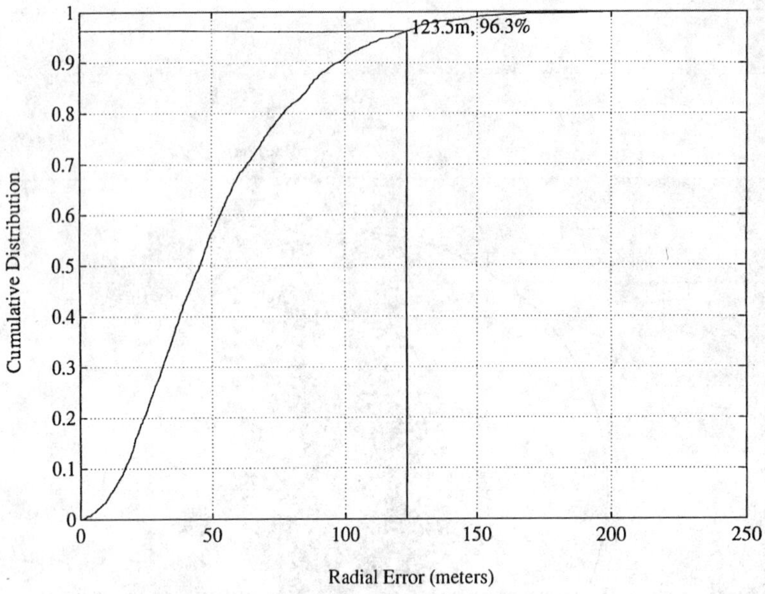

FIGURE 17.29   Radial errors as a function of cumulative distribution.

17.3.6   Over-Determined Solutions

In the previous examples the issue of the distribution of measurement errors had no impact on the method used to find a solution, except only on the error analysis of that solution. We assumed a normal distribution on these errors not necessarily because we could guarantee that distribution, but because it allowed us to make meaningful predictions. When the number of observables exceeds the number of unknowns, we

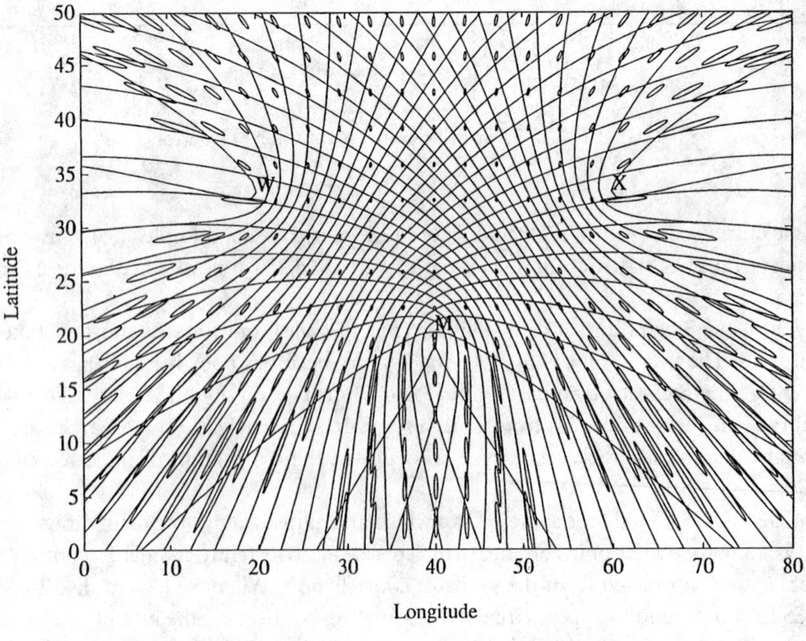

FIGURE 17.30   Lines of constant time difference and error ellipses for typical triad.

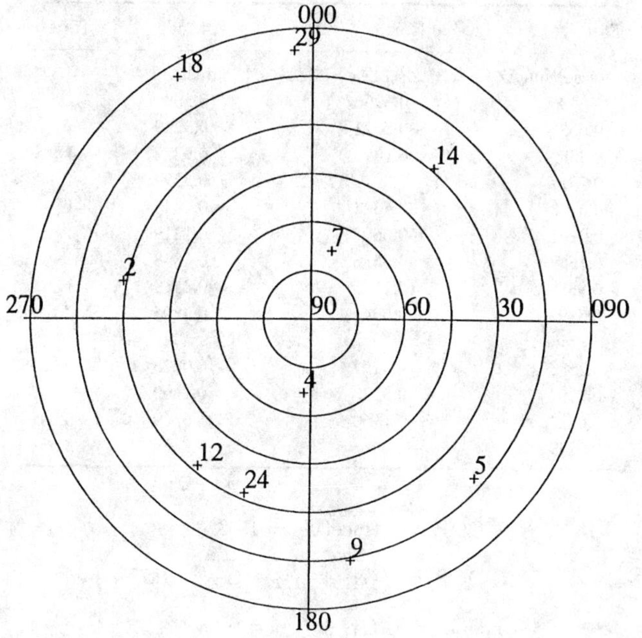

FIGURE 17.31 Azimuth and elevation plot of GPS satellites for 1400Z, Aug. 28, 1995, New London, CT.

have choices in how to process the observations and the optimum choice depends both on the distribution of our measurement errors and what parameter we are trying to optimize. Since the GPS space segment now has 28 satellites, commonly there are more than the minimum of three or four required.

If the errors in the measured pseudo-ranges or TOAs are statistically independent and have the same variance, then the solution that minimizes the sum of the squares of the pseudo-range residuals is the linear least squares solution.

$$\Delta P = (A^T A)^{-1} A^T \Delta PR$$

where A is the matrix of direction cosines and ΔPR is the difference between measured and calculated pseudo-ranges. The covariance of the position error is now

$$\text{cov}(\Delta P) = (A^T A)^{-1} A^T \text{cov}(\Delta PR) A (A^T A)^{-1} = \sigma_{PR}(A^T A)^{-1}$$

where $\text{cov}(\Delta P) = \sigma_{PR} I$.

Figure 17.31 shows the azimuth and elevation plot of GPS satellites for 1400Z, August 28, 1995 in New London CT as a typical example. Eliminating the low elevation satellite (#18) due to potential multipath and using the nine remaining listed in table below, we can calculate expected accuracies.

SVN	Elevation	Azimuth
5	18°	317°
12	32°	232°
7	68°	73°
24	32°	249°
14	29°	50°
29	7°	94°
2	29°	169°
9	14°	280°
4	67°	265°

East	North	Altitude	Time
cos(EL) sin(AZ)	cos(EL) cos(AZ)	sin(EL)	
−0.6486	0.6956	0.309	1
−0.6683	−0.5221	0.5299	1
0.3582	0.1095	0.9272	1
−0.7917	−0.3039	0.5299	1
0.6700	0.5622	0.4848	1
0.9901	−0.0692	0.1219	1
0.1669	−0.8586	0.4848	1
−0.9556	0.1685	0.2419	1
−0.3892	−0.0341	0.9205	1
$G = \text{inv}(A^T A)$			
0.2533	−0.0192	0.0223	0.0239
−0.0192	0.5248	0.1160	−0.0466
0.0223	0.1160	1.6749	−0.8404
0.0239	−0.0466	−0.8404	0.5380

$$\text{GDOP} = \sqrt{\text{trace}(G)} = 1.73$$

$$\text{PDOP} = \sqrt{G_{11} + G_{22} + G_{33}} = 1.57$$

$$\text{HDOP} = \sqrt{G_{11} + G_{22}} = 0.88$$

$$\text{VDOP} = \sqrt{G_{33}} = 1.29$$

Exactly what these DOPs imply in terms of accuracy depends on what version of GPS/DGPS service one is using.

For the standard positioning service (SPS) with selective availability (SA) on, typically the value 32 m was used for the one sigma error. Since this value depended almost entirely on policy, advances in technology did not change its value. In Conley (1999) we can get a glimpse of what accuracies will be expected in the short term, and how they may improve in the future. They report less than 1.4 m rms for the on-orbit clocks and 59 cm for the ephemeris contribution to user range error (URE) in 1999. For SVN 43, the first operational Block IIR satellite, the combined clock and ephemeris errors were 70 cm RMS. This means that the errors in predicting ionospheric delay at about 5 m, the errors in predicting tropospheric delay at about 2 m, and multipath now dominate the statistics. DGPS will remove most of the residual errors due to satellite clock and ephemeris, and ionospheric and tropospheric delay leaving multipath as the most common dominant error source. The multipath errors are highly dependent on receiver and antenna design and antenna placement. Preliminary data after the termination of SA, from fixed sites, with high quality receivers, and where some care has been taken to minimize multipath, is indicating 7 m 2 drms accuracy may be achievable. The performance on typical installations with low cost receivers on moving platforms will be considerably worse. In the GPS modernization program to be implemented over the next decade, the addition of C/A code on L2 and third civil frequency, will mean civil receivers will measure ionospheric delay via differential pseudo-range as in dual frequency, military receivers. This combined with other improvements in technology is expected to improve civil GPS accuracy to better than 5 m 2 drms.

For precise positioning service users, NATO GPS (1991) specifies 13.0 m (95%) URE, which would imply a standard deviation of 6.5 m assuming a normal distribution. The actual performance, while classified, is generally acknowledged to be considerably better than this 1991 value, and will continue to improve with technology.

The negative correlation ($\rho = -0.8404/\sqrt{1.6749 * 0.5380} = -0.885$) between the altitude and time errors is common because the satellites used (by receivers on the Earth's surface) all must lie within a cone starting at some (elevation mask) angle above the horizon. Figure 17.32 shows the error ellipse for the joint time and altitude distribution. By using the first (East), second (North), and fourth (time) columns in the matrix A above we can calculate the expected DOPs for a GPS receiver in a fixed (known) altitude

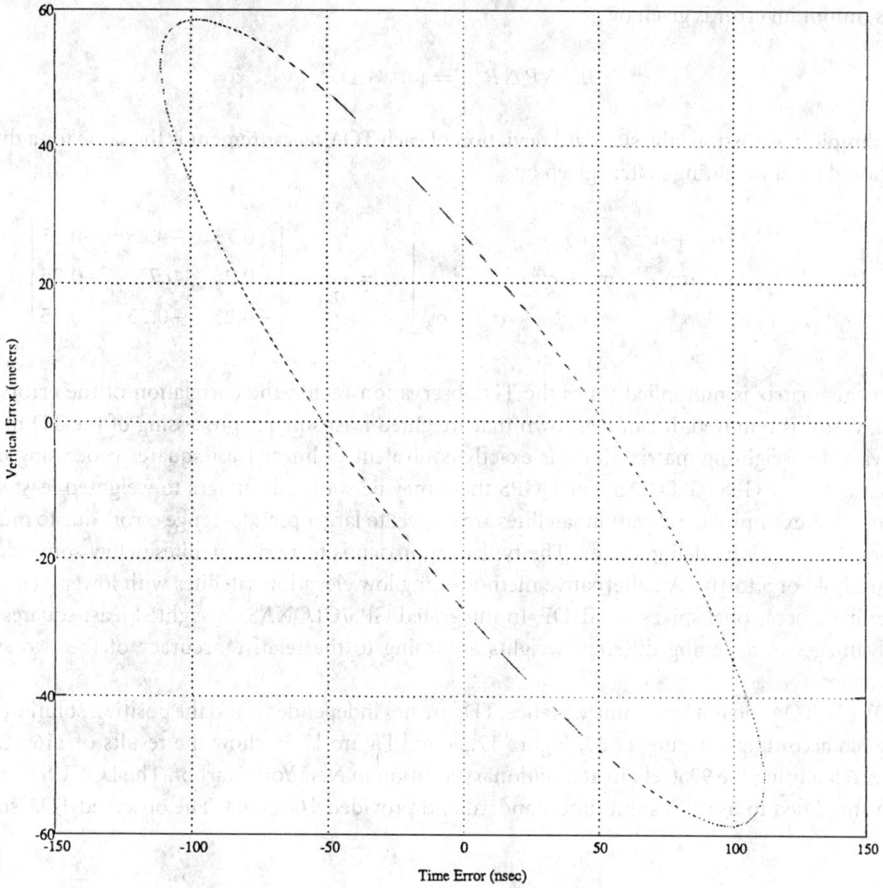

FIGURE 17.32   Typical joint distribution of GPS altitude and time errors.

mode. HDOP is virtually unchanged but TDOP improves from 0.73 to 0.34. Similarly, if we assume we know time accurately and are only solving for three-dimensional position, we use the first three columns and VDOP improves from 1.29 to 0.60. Realistically, we will not know time exactly because we use the GPS to get time. We can improve vertical accuracy somewhat by propagating clock bias ahead via a Kalman filter as described in a later section, but can never get to zero clock error and this VDOP.

17.3.7   Weighted Least Squares

When the covariance of the measurement error is not a constant times an identity matrix, the optimum solution is a weighted vs. linear least squares solution.

$$\Delta P = (A^T W A)^{-1} A^T W \Delta PR$$

Now the covariance of the error is given by

$$E\{\Delta P \Delta P^T\} = E\{(A^T W A)^{-1} A^T W \Delta PR \Delta PR^T W^T A (A^T W A)^{-1}\}$$

This error is minimized if

$$W = R^{-1}$$

And this minimum error is given by

$$E\{\Delta P \Delta P^T\} = (A^T W A)^{-1}$$

For example, if we assume the standard deviation of each TOA measurement is the same for a three TD fix we should use a weighting matrix given by

$$W = R^{-1} = \begin{bmatrix} \sigma_m^2 + \sigma_1^2 & \sigma_m^2 & \sigma_m^2 \\ \sigma_m^2 & \sigma_m^2 + \sigma_2^2 & \sigma_m^2 \\ \sigma_m^2 & \sigma_m^2 & \sigma_m^2 + \sigma_3^2 \end{bmatrix}^{-1} = \frac{1}{\sigma_{\text{TOA}}} \begin{bmatrix} 0.75 & -0.25 & -0.25 \\ -0.25 & 0.75 & -0.25 \\ -0.25 & -0.25 & 0.75 \end{bmatrix}$$

When this matrix is multiplied times the TD observation vector, the correlation of the errors in the resulting vector is removed. It can be shown that weighted least squares processing of the TD measurements with the weighting matrix above is exactly equivalent to linear least squares processing of TOA measurements. For GPS, GLONASS, or DGPS there may be some advantages to weighted least squares processing. For example, low elevation satellites are subject to larger pseudo-range errors due to multipath and ionosphere delay modeling errors. The typical approach is to reject satellites below some elevation mask threshold of 5 to 10°. An alternative method using low elevation satellites with lower weights may help eliminate occasional spikes in GDOP. In integrated GPS/GLONASS, weighted least squares would have advantages in assigning different weights according to the relative accuracy of the two system's pseudoranges.

In LORAN, TOAs do not have same variance, TDs are not independent, and the position solution should be designed accordingly. Figure 17.33, Figure 17.34 and Figure 17.35 show the results of 1166 LORAN fixes over 6.5 h using the 9960 chain at a stationary location in New York Harbor. The LOCUS™ receiver used was modified to use a Cesium time standard and provided TOA data. The observed TOA standard

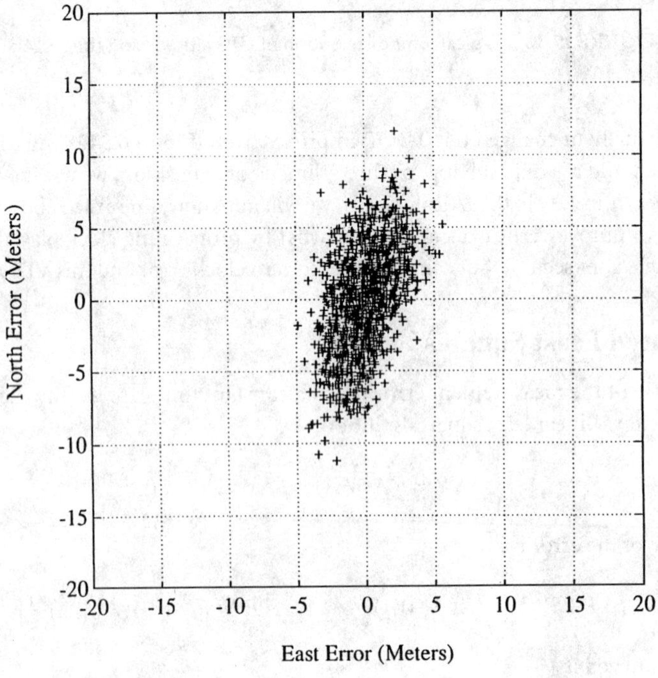

FIGURE 17.33   Example LORAN fixes (error +/− 20 m).

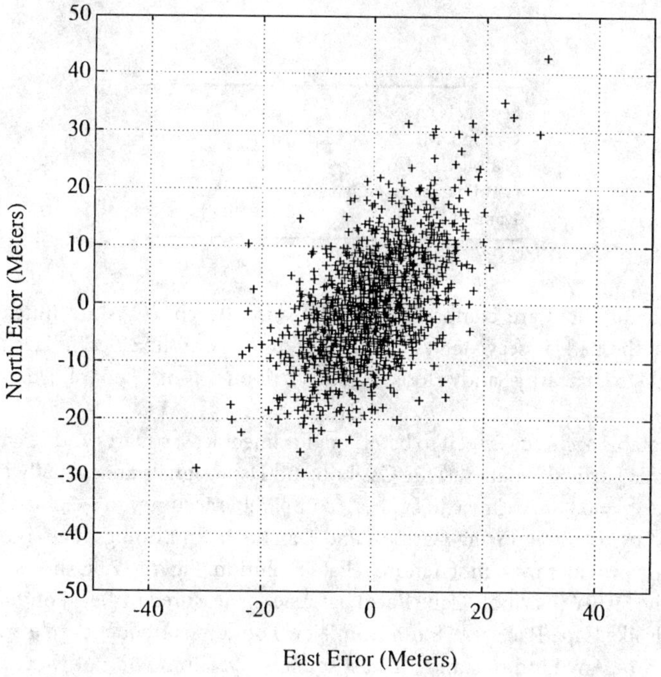

FIGURE 17.34 Example LORAN fixes (error $+/-$ 50 m).

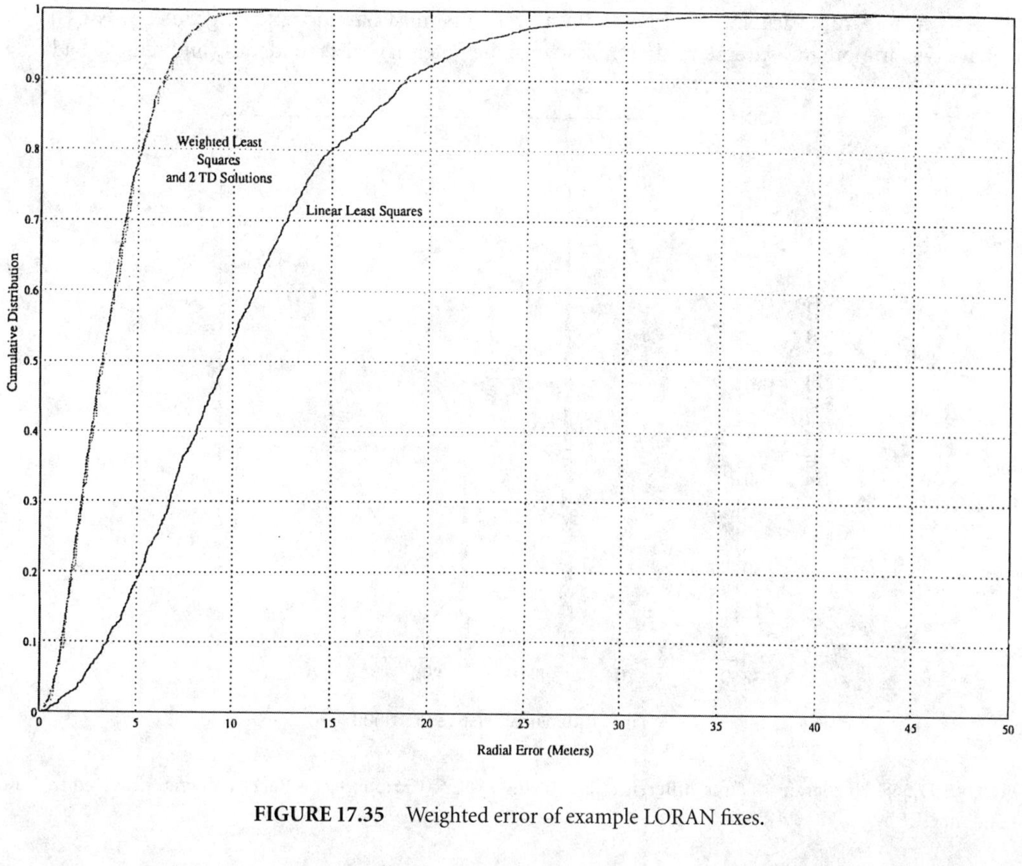

FIGURE 17.35 Weighted error of example LORAN fixes.

deviations were

Seneca, NY	(175 nm)	7.6 ns
Caribou, ME	(452 nm)	100.6 ns
Nantucket, MA	(186 nm)	8.6 ns
Carolina Beach, NC	(438 nm)	21.1 ns
Dana, IN	(618 nm)	60.0 ns

(*Note*: Even though distances are comparable, the Carolina Beach TOAs are much better than those of Caribou because the path is seawater vs. land. In addition, Carolina Beach TDs are monitored and controlled using data from nearby Sandy Hook, NJ and Caribou is controlled using data from a monitor in Maine.)

The 2 drms repeatable accuracy is seen to be 21.8 m for linear least squares and 7.3 m for weighted least squares. Since for this particular data set, the Caribou and Dana signals are virtually ignored (Caribou is weighted $(7.6/100.6)^2 = 0.0057$ relative to Seneca), comparable accuracy to weighted least squares would have been obtained by totally ignoring Caribou and Dana data and doing three-TOA/two-TD fixes. In general using all data provides for a much more reliable solution. Figure 17.36 shows a histogram of time difference data of the 5930Y (Caribou/Cape Race) baseline as measured in New London, CT. What is seen for marginal signals like Cape Race at 875 nm from New London, is frequent cycle errors resulting in 10 and 20 usec TD errors. Any kind of standard least squares algorithm will not work. Assuming an over-determined solution, there are two basic alternatives; first, a maximum likelihood algorithm that takes into account the potential for local maxima in the likelihood function due to cycle errors, and second, receiver autonomous integrity monitoring (RAIM). In RAIM, a fix using all measurements is calculated and the residuals compared to a threshold. For RAIM to be valid, assuming N measurements, each of the N fixes available from each subset of N − 1 measurements must have acceptable geometry. For GPS, if we have two or more measurements than unknowns, the potential exists to identify and remove bad data

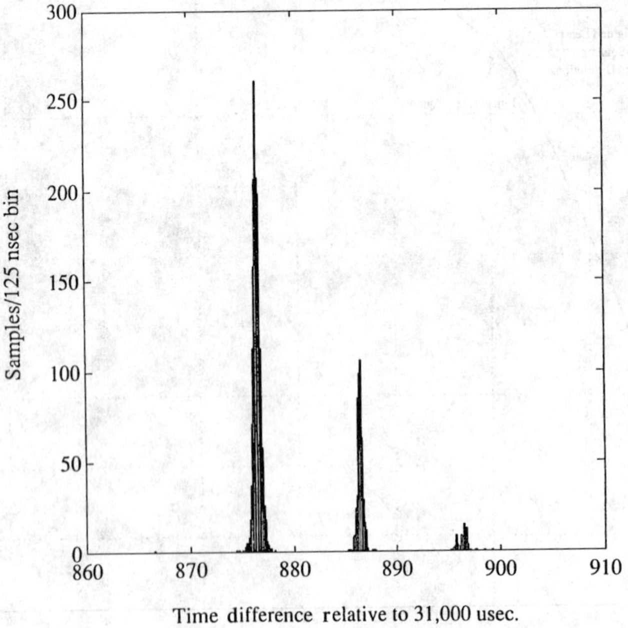

FIGURE 17.36 Histogram of time difference data of the 5930Y (Caribou/Cape Race) baseline measured in New London, CT.

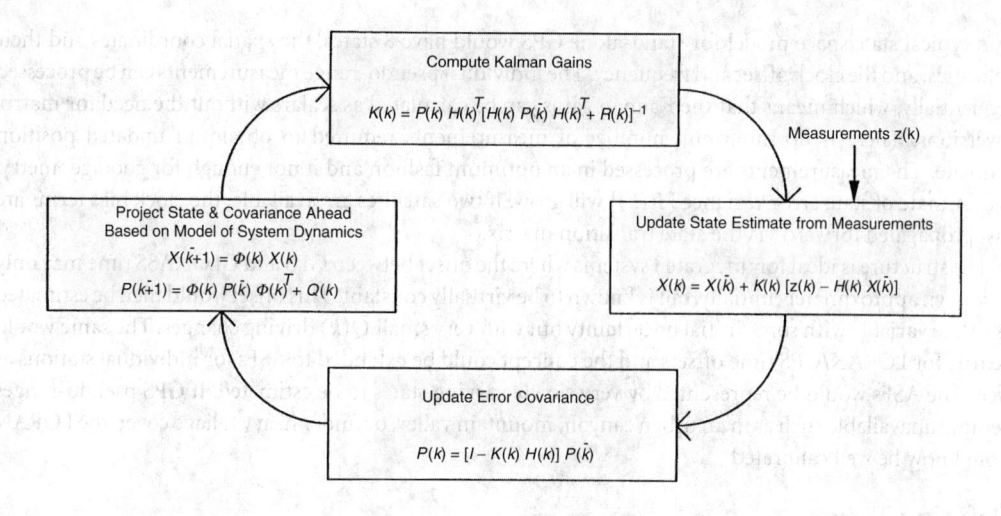

FIGURE 17.37   Kalman filter structure.

via residual analysis on these N over-determined fixes. Since for LORAN, cycle errors have predictable behavior, we may be able to identify and remove them with only one measurement more than the number of unknowns.

17.3.8   Kalman Filters

This section points out the basic principles of Kalman filters and specifically how they can be used in radionavigation systems. All of our analysis above has assumed we process each set of data independently. We know that parameters such as clock frequency and our velocity can only change at finite rates and therefore we ought to be able to do better using all past measurements. When we integrate LORAN TOAs or GLONASS pseudo-ranges with GPS pseudo-ranges we may not be able to assume the various systems are exactly synchronized in time. Therefore we cannot only solve for one time term, but we do know the drift between systems is extremely slow and need to somehow exploit that knowledge. Kalman filters allow us to use past measurements together with *a priori* knowledge of platform and clock dynamics to obtain optimum values of position and velocity. Typically, Kalman filters have found extensive use in the integration of inertial and electronic navigation systems. Here we will concentrate on stand-alone electronic navigation systems. The reader is referred to more extensive references such as Brown (1992) for more detail.

Figure 17.37 shows the basic structure of a discrete Kalman filter. The process starts with initial estimates of the state $[X(0)]$ and the error covariance $[P(0)]$. The other variables are

$R(k)$ is the covariance of the measurement errors. (They are assumed white, this is not valid for SA-dominated pseudo-range errors, which are correlated over minutes. Strictly speaking this correlation should be modeled via additional state variables in system model, but normally it is not.)

$H(k)$ Matrix of direction cosines and ones (as A above) that relate pseudo-range or TOA errors to positions and clock bias and Doppler's to velocity and frequency errors.

$z(k)$ Pseudo-range (or TOA) and Doppler measurements

$K(k)$ Kalman gains

$\Phi(k)$ State transition matrix

$Q(k)$ Covariance of noise driving state changes. This matrix controls allowed accelerations and clock frequency and phase variations. Smaller $Q(k)$ will result in smaller estimates of state error covariance and less weighting of measurements. Larger $Q(k)$ will result in faster system response to measurements. Conventional wisdom suggests using larger Q when uncertain of exact model.

A typical state space model for stand-alone GPS would have 8 states, the spatial coordinates and their velocities, and the clock offset and frequency. The individual pseudo-range measurements can be processed sequentially, which means that the Kalman gains can be calculated as scalars without the need for matrix inversions. There is no minimum number of measurements required to obtain an updated position estimate. The measurements are processed in an optimum fashion and if not enough for good geometry, the estimate of state error variance $[P(k)]$ will grow. If two satellites are available, the clock bias terms are just propagated forward via the state transition matrix.

The structure is ideal for integrated systems where the offset between GPS and GLONASS time may only be known approximately initially, but is known to be virtually constant. This offset would then be estimated as a state variable with some initial uncertainty but with very small $Q(k)$ driving changes. The same would be true for LORAN/GPS time offsets and the concept could be extended to ASFs for individual stations as well. The ASFs would be represented by very slowly varying states to be estimated. If GPS pseudo-ranges became unavailable such as in an urban canyon, mountain valley, or under heavy foliage cover, the LORAN would now be well calibrated.

Defining Terms

Accuracy: The degree of conformance between the estimated or measured position and/or velocity of a platform at a given time and its true position or velocity. Radionavigation system accuracy is usually presented as a statistical measure of system error.

Predictable: The accuracy of a radionavigation system's position solution with respect to the charted solution. Both the position solution and the chart must be based upon the same geodetic datum.

Repeatable: The accuracy with which a user can return to a position whose coordinates have been measured at a previous time with the same navigation system.

Relative: The accuracy with which a user can measure position relative to that of another user of the same navigation system at the same time.

Availability: The availability of a navigation system is the percentage of time that the services of the system are usable. Availability is an indication of the ability of the system to provide usable service within the specified coverage area. Signal availability is the percentage of time that navigational signals transmitted from external sources and are available for use. Availability is a function of both the physical characteristics of the environment and the technical capabilities of the transmitter facilities.

Block I and Block II Satellites: The Block I is a GPS concept validation satellite; it does not have all of the design features and capabilities of the production model GPS satellite, the Block II. The FOC 24 satellite constellation is defined to consist entirely of Block II/IIA satellites.

Coordinated Universal Time (UTC): UTC, an atomic time scale, is the basis for civil time. It is occasionally adjusted by one-second increments to ensure that the difference between the uniform time scale, defined by atomic clocks, does not differ from the earth's rotation by more than 0.9 sec.

Differential: A technique used to improve radionavigation system accuracy by determining positioning error at a known location and subsequently transmitting the determined error, or corrective factors, to users of the same radionavigation system, operating in the same area.

Dilution of precision (DOP): The magnifying effect on radionavigation position error induced by mapping ranging errors into position and time through the position solution. The DOP may be represented in any user local coordinate desired. Examples are HDOP for local horizontal, VDOP for local vertical, PDOP for all three coordinates, TDOP for time, and GDOP for position and time.

Distance root mean square (drms): The root-mean-square value of the distances from the true location point of the position fixes in a collection of measurements. As used in this Chapter 2 drms is the radius of a circle that contains at least 95% of all possible fixes that can be obtained with a system at any one place. Actually, the percentage of fixes contained within 2 drms varies between approximately 95.5 and 98.2%, depending on the degree of ellipticity of the error distribution.

Hyperbolic navigation system: A navigation system that produces hyperbolic lines of position (LOPs) through the measurement of the difference in times of reception (or phase difference) of radio signals from two or more synchronized transmitters.

Integrity: Integrity is the ability of a system to provide timely warnings to users when the system should not be used for navigation.

Multipath transmission: The propagation phenomenon that results in signals reaching the receiving antenna by two or more paths. When two or more signals arrive simultaneously, wave interference results. The received signal fades if the wave interference is time varying or if one of the terminals is in motion.

Pseudo-range: The difference between the ranging signal time of reception (as defined by the receiver's clock) and the time of transmission contained within the satellite's navigation data (as defined by the satellite's clock) multiplied by the speed of light.

Receiver autonomous integrity monitoring (RAIM): A technique whereby a civil GPS receiver/processor determines the integrity of the GPS navigation signals without reference to sensors or non-DoD integrity systems other than the receiver itself. This determination is achieved by a consistency check among redundant pseudo-range measurements.

Selective availability (SA): The denial of full GPS accuracy to civil users by manipulating navigation message orbit data (epsilon) and/or satellite clock frequency (dither).

References

Braff, R. 1997. Description of the FAA's Local Area Augmentation System (LAAS), *Navigation*, 44, 4, 411–424.

Brown, R. G. and Hwang, P. Y. C. 1992. *Introduction to Random Signals and Applied Kalman Filtering*, 2nd Ed., John Wiley & Sons, New York.

Conley, R. and Lavrakas, J. W. 1999. The World after Selective Availability, *Proceedings of ION GPS '99, 14–17*, 1353–1361. Nashville, TN.

CSIC, *GLONASS Interface Control Document, Version 4.0*, The Ministry of Defence of the Russian Federation Coordination Scientific Information Center, 1998. Available in electronic form at http://www.rssi.ru/SFCSIC/SFCSIC_main.html

Dale, S, Daly, P., and Kitching, I. 1989. Understanding signals from GLONASS navigation satellites, *Int. J. Satellite Commun.*, 7, 11–22.

U.S. Departments of Defense and Transportation, (DoD/Dot, 1994) *1994 Federal Radionavigation Plan*, U.S. Departments of Defense and Transportation, NTIS Report DOT-VNTSC-RSPA-95-1/DoD-4650.5, 1995. Available in electronic form (Adobe Acrobat) from USCG NAVCEN (www.navcen,uscg.mil).

Enge, P., Swanson, E., Mullin, R., Ganther, K., Bommarito, A., and Kelly, R. 1995. Terrestrial radionavigation technologies, *Navigation*, 42, 1, 61–108.

Feairheller, S. 1994. The Russian GLONASS system, a U.S. Air Force study, *Proceedings of Institute of Navigation GPS-94*, Salt Lake City.

NATO Navstar GPS Technical Support Group (NATO GPS 1991), *Technical Characteristics of the Navstar GPS*, 1991.

Parkinson, B., Stansell, T., Beard, R., and Gromov, K. 1995. A history of satellite navigation, *Navigation*, 42 ,1, 109–164.

Pierce, J.A., Mckenzie, A.A., and Woodward, R.H. 1948. Eds. *LORAN*, MIT Radiation Laboratory Series. McGraw-Hill, New York.

RTCA. 1998. *Minimum Aviation Performance Standards for Local Area Augmentation System*, RTCA/DO-245.

RTCA. 1999. *Minimum Operational Performance Standards for GPS/Wide Area Augmentation System Airborne Equipment*. RTCA/DO 229B.

RTCM Special Committee 104. 1992. *RTCM Recommended Standard for Differential Navstar GPS Service*, Version 2.0.

Skidmore, T. A., Nyhus, O. K., and Wilson, A. A. 1999. An overview of the LAAS VHF data broadcast, *Proceedings of ION GPS '99, 14–17*, 671–680. Nashville, TN.

U.S. Air Force, GPS Joint Program Office, (USAF JPO, 1995) *Global Positioning System, Standard Positioning Service, Signal Specification*, June, 1995. Available in electronic form (Adobe Acrobat) from USCG NAVCEN (www.navcen,uscg.mil).

U.S. Coast Guard, (USCG, 1992), *LORAN C User Handbook*, COMDTPUB P16562.6, COMDT(G-NRN), 1992. Available in electronic form from USCG NAVCEN (www.navcen,uscg.mil).

Walter, T. and Bakery El-Arini, M. Eds. 1999. *Selected Papers on Satellite based Augmentation Systems (SBASs)*, VI in the GPS Series, Institute of Navigation, Alexandria, VA.

Ward, P. W. 1994–1995. GPS receiver RF interference monitoring, mitigation and analysis techniques, *Navigation*, 41, 4, 367–391.

White House Press Secretary. Statement by the President Regarding the United States Decision to Stop Degrading Global Positioning System Accuracy, White House.

Wild Goose Association (WGA, 1982). On the calculation of geodesic arcs for use with LORAN, *Wild Goose Association Radionavigation Journal*, 57–61.

Further Information

Government Agencies

1. **U.S. Coast Guard Navigation Information Service (NIS)**
   Address: Commanding Officer
                USCG Navigation Center
                7323 Telegraph Road
                Alexandria, VA 22315
   Internet: http://www.navcen.uscg.mil/navcen.htm
   Phone: (703) 313-5900 (NIS Watchstander)

Formerly the GPS Information Center (GPSIC), NIS has been providing information since March, 1990. The NIS is a public information service. At the present time, there is no charge for the information provided. Hours are continuous: 24 h a day, 7 days a week, including federal holidays.

   The mission of the navigation information service (NIS) is to gather, process, and disseminate timely GPS, DGPS, Omega, and LORAN-C status, and other navigation related information to users of these navigation services. Specifically, the functions to be performed by the NIS include the following:

   Provide the Operational Advisory Broadcast Service (OAB).
   Answer questions by telephone or written correspondence.
   Provide information to the public on the NIS services available.
   Provide instruction on the access and use of the information services available.
   Maintain tutorial, instructional, and other relevant handbooks and material for distribution to users.
   Maintain records of GPS, DGPS, and LORAN-C broadcast information, and databases or relevant data
       for reference purposes.
   Maintain bibliography of GPS, DGPS, and LORAN-C publications.

The NIS provides a watchstander to answer radionavigation user inquiries 24 h a day, seven days a week and disseminates general information on GPS, DGPS, Loran-C, and Radiobeacons through various mediums.

2. **NOAA, National Geodetic Survey**
   Address: 1315 East-West Highway, Station 09202
                Silver Spring, MD 20910
   Phone: (301) 713-3242;
   Fax: (301) 713-4172
   Monday through Friday, 7:00 a.m.– 4:30 p.m., Eastern Time
   Internet: http://www.ngs.noaa.gov/

NOAA is modernizing the Nation's spatial reference system, providing horizontal and vertical positions for navigation and engineering purposes. By implementing a new system called CORS (continuously operating reference stations), NOAA will continuously record carrier phase and pseudo-range measurements for all GPS (global positioning system) satellites at each CORS site. A primary objective of CORS is to monitor a particular site, which is determined to millimeter accuracy and to provide local users a tie to the national spatial reference system. NOAA will also use the CORS sites along with other globally distributed tracking stations in computing precise GPS orbits.

NOAA is also the federal agency responsible for providing accurate and timely global positioning system (GPS) satellite ephemerides ("orbits") to the general public. The GPS precise orbits are derived using 24-h data segments from the global GPS network coordinated by the international geodynamics GPS service (IGS). The reference frame used in the computation is the international earth rotation service terrestrial reference frame (ITRF). In addition, an informational summary file is provided to document the computation and to convey relevant information about the observed satellites, such as maneuvers or maintenance. The orbits generally are available two to six days after the date of observation.

Also available

- Software programs to compute, verify, or adjust field surveying observations; convert coordinates from one geodetic datum to another; or assist in other specialized tasks utilizing geodetic data.
- Listings of publications available on geodesy, mapping, charting, photogrammetry, and related topics, such as plane coordinate systems and numerical analysis.
- Information on NGS programs to assist users of geodetic data, including technical workshops and the geodetic advisor program.
- Information on survey data recently incorporated into NSRS.
- Information on the location of NGS field parties.

3. **U.S. Naval Observatory (USNO)**
   Internet: http://www.usno.navy.mil/

USNO is the official source of time used in the U.S. USNO timekeeping is based on an ensemble of cesium beam and hydrogen maser atomic clocks. USNO diseminates information on time offsets of the world's major time services and radionavigation systems.

4. **USSPACECOM GPS Support Center**
   300 O'Malley
   Suite 41
   Schriever AFB, CO 80912-3041
   http://www.peterson.af.mil/usspace/gps_support

The **GPS Support Center** (GSC) is the DoD's focal point for operational issues and questions concerning military use of GPS. The GSC is responsible for

- Receiving reports and coordinating responses to radio frequency interference in the use of GPS in military operations.
- Providing prompt responses to DoD user problems or questions concerning GPS.
- Providing official USSPACECOM monitoring of GPS performance provided to DoD users on a global basis.
- Providing tactical support for planning and assessing military missions involving the use of GPS.

As the DoD's focal point for GPS operational matters, the GSC serves as U.S. Space Command's interface to the civil community, through the U.S. Coast Guard's Navigation Center and Federal Aviation Administration's National Operations Command Center.

5. **The Ministry of Defence of the Russian Federation Coordination Scientific Information Center**
   Internet: http://www.rssi.ru/SFCSIC/SFCSIC_main.html

The mission of the Coordination Scientific Information Center is to plan, manage, and coordinate the activities on

- Use of civil-military space systems (navigation, communications, meteorology etc.)
- Realization of Russian and international scientific and economic space programs
- Realization of programs of international cooperation
- Conversional use of military space facilities, as well as to provide the scientific-informational, contractual and institutional support of these activities.

17.4   Underwater Sonar Systems[1]

Sanjay K. Mehta, Clifford G. Carter, and Bernard E. McTaggart

17.4.1   Introduction

A system that uses sound signals (also known as acoustic signals) propagated through the water to detect, classify, and localize underwater objects is referred to as an underwater sound system (sonar). The system generally consists of four major components. The first component is a transmitter that transmits a signal through water. The second is a receiver that receives the transmitted signal, which has been degraded due to underwater propagation effects, ambient noise, or interference from other signal sources, such as surface warships and fishing vessels. A processing unit that processes the received signal to minimize the degradation effects and to maximize the detection and classification capability of the signal is the third component. The fourth consists of the various displays that aid human operators to detect, classify, and localize signals.

Sound, that is, acoustic energy, propagates better underwater than do other types of energy. For example, both light and radio waves (used for satellite or above ground communications) are attenuated to a far greater degree under water than are sound waves. (Note: there has been some limited success propagating blue–green laser energy in clear water.) For this reason sound waves have generally been used to extract information about underwater objects. A typical underwater acoustical signal processing scenario is shown in Fig. 17.38.

Background

In underwater acoustics, the metric system has not been universally applied, and a number of nonmetric units are still used: distances of nautical miles (1852 m), yards (0.9144 m), and kiloyards; speeds of knots (nautical miles per hour); depths of fathoms (6 feet or 1.8288 m); and bearing in degrees (0.01745 rad). In the past two decades, however, there has been a conscious effort to become totally metric, that is, to use meter–kilogram–second (MKS) or Standard International units.

Underwater sound signals that are processed electronically for detection, classification, and localization can be characterized from a statistical point of view. When time averages of each signal are the same as the ensemble average of signals, the signals are ergodic. When the statistics do not change with time, the signals are said to be stationary. The spatial equivalent to stationary is homogeneous. For many introductory problems, only stationary signals and homogeneous noise are considered; more complex problems involve nonstationary, inhomogeneous environments.

Sound signals have a first-order *probability density function* (PDF); for example, the PDF may be Gaussian, or in the case of clicking, sharp noise spikes or crackling ice noise, the PDF may be non-Gaussian.

---

[1]This work is a revised version from Dorf, R., ed. 1993. *Electrical Engineering Handbook*. CRC Press, Boca Raton, FL.

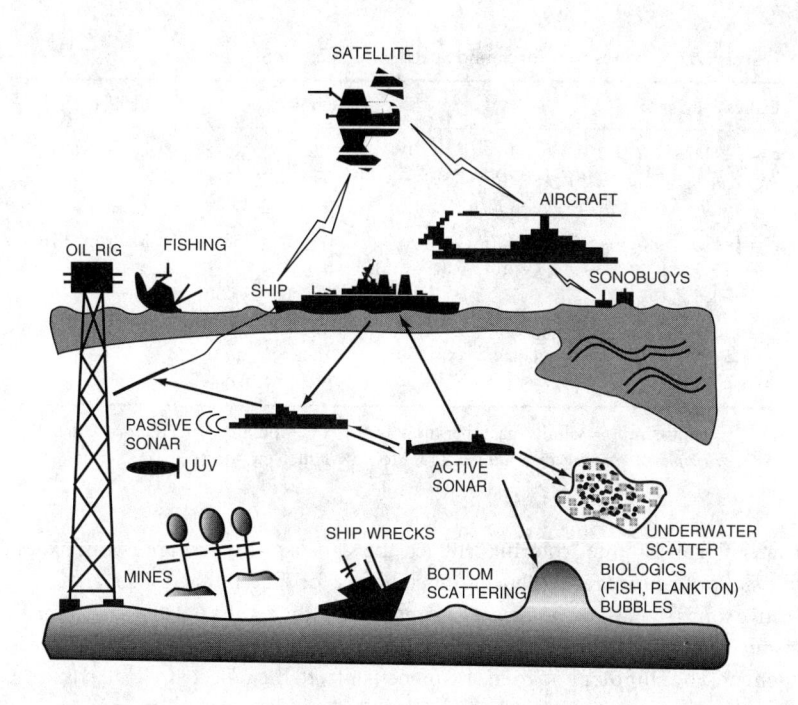

FIGURE 17.38 Active and passive underwater acoustical signal processing.

In addition to being characterized by a PDF, signals can be characterized in the frequency domain by their power spectral density functions, which are Fourier transforms of the autocorrelation functions. White signals, which are uncorrelated from sample to sample, have a delta function autocorrelation or a flat (constant) power spectral density. Ocean signals, in general, are much more colorful and are not limited to being stationary.

Sonar

Sound navigation and ranging, (**SONAR**) (the acronym was adapted in the 1940s similar to the popular acronym for radio detection and ranging (radar)) involves the use of sound to explore the ocean and underwater objects. *Passive sonar* uses sound radiated from the underwater object itself. The duration of the radiated sound may be short or long in time and narrow or broad in frequency. Transmission through the ocean, from the source to a receiving sensor, is one way. *Active sonar* involves echoranging where an acoustical signal is transmitted from a source, and reflected echoes are received back from the object. Here, the transmissions, from a transmitter to an object and back to a receiving sensor are two way. There are three types of active sonar systems

- *Monostatic:* In this most common form, the source and receiver can be identical or distinct, but are located on the same platform (e.g., a surface ship).
- *Bistatic:* In this form, the transmitter and receiver are on different platforms.
- *Multistatic:* Here, one or more transmitters and multiple receivers are located on different receiving platforms or ships.

Passive sonar signals are primarily modeled as random signals. Their first-order PDFs are typically Gaussian; one exception is a stable sinusoidal signal that is non-Gaussian and has a power spectral density function that is a Dirac delta function in the frequency domain. Such sinusoidal signals can be detected by measuring the energy output of narrowband filters. This can be done with fast fourier transform (FFT) electronics and long integration times. In the ocean environment, however, an arbitrarily narrow-frequency width is never observed, and signals have some finite bandwidth. Indeed, the full spectrum of

TABLE 17.7 Expressions for Sound Speed in Meters per Second

Expression	Limits[a]	Reference
$c = 1492.9 + 3(T - 10) - 6 \times 10^{-3}(T - 10)^2$ $- 4 \times 10^{-3}(T - 18)^2 + 1.2(S - 35)$ $- 10^{-2}(T - 18)(S - 35) + D/61$	$-2 \leq T \leq 24.5°$ $30 \leq S \leq 42$ $0 \leq D \leq 1,000$	Leroy 1969
$c = 1449.2 + 4.6T - 5.5 \times 10^{-2}T^2$ $+ 2.9 \times 10^{-4}T^3 + (1.34 - 10^{-2}T)(S - 35)$ $+ 1.6 \times 10^{-2}D$	$0 \leq T \leq 35°$ $0 \leq S \leq 45$ $0 \leq D \leq 1,000$	Medwin 1975
$c = 1448.96 + 4.591T - 5.304 \times 10^{-2}T^2$ $+ 2.374 \times 10^{-4}T^3 + 1.340(S - 35)$ $+ 1.630 \times 10^{-2}D + 1.675 \times 10^{-7}D^2$	$0 \leq T \leq 30°$ $30 \leq S \leq 40$ $0 \leq D \leq 8,000$	Mackenzie 1981

[a] D = depth, m; S = salinity, parts per thousand; T = temperature, °C. (*Source:* Urick, R.J. 1983. *Principles of Underwater Sound*, p. 113. McGraw-Hill, New York.)

most underwater signals is quite "colorful." In fact, the signals of interest are not unlike speech signals, except that the **signal-to-noise ratio (SNR)** is much higher for speech than for sonar.

Received active sonar signals can be viewed as consisting of the results of (1) a deterministic component (known transmit signal) convolved with the medium and reflector transfer functions and (2) a random (noise) component. The **Doppler** imparted (frequency shift) to the reflected signal makes the total system effect nonlinear, thereby complicating analysis and processing of these signals. In addition, in active systems the noise is typically correlated with the signal, making detection of signals more difficult.

17.4.2 Underwater Propagation

Sound speed c in the ocean, in general, lies between 1450–1540 m/s and varies as a function of several physical parameters, such as temperature, salinity, and pressure (depth). Variations in sound speed can significantly affect the propagation (range or quality) of sound in the ocean. Table 17.7 gives approximate expressions for sound speed as a function of these physical parameters.

Sound Velocity Profiles

Sound rays that are normal (perpendicular) to the signal acoustic wavefront can be traced from the source to the receiver by a process called ray tracing. (Ray tracing models are used for high-frequency signals and in deep water. Generally, if the depth-to-wavelength ratio is 100 or more, ray tracing models are accurate. Below that, corrections must be made to these models. In shallow water or low frequencies, that is, when depth-to-wavelength is about 30 or less, mode theory models are used.) In general, the acoustic ray paths are not straight, but bend in a manner analogous to optical rays focused by a lens. In underwater sound, the ray paths are determined by the **sound velocity profile (SVP)** or *sound speed profile* (SSP): that is, the speed of sound in water as a function of water depth. The sound speed not only varies with depth but also varies in different regions of the ocean and with time as well. In deep water, the SVP fluctuates the most in the upper ocean due to variations of temperature and weather. Just below the sea surface is the *surface layer* where the sound speed is greatly affected by temperature and wind action. Below this layer lies the *seasonal thermocline* where the temperature and speed decrease with depth, and the variations are seasonal. In the next layer, the *main thermocline*, the temperature and speed decrease with depth, and surface conditions or seasons have little effect. Finally, there is the *deep isothermal layer* where the temperature is nearly constant at 39°F, and the sound velocity increases almost linearly with depth. A typical deep water sound velocity profile as a function of depth is shown in Fig. 17.39.

If the sound speed is a minimum at a certain depth below the surface, then this depth is called the *axis* of the *underwater sound channel* (also called the *sound fixing and ranging* (SOFAR) channel.) The sound velocity increases both above and below this axis. When the sound wave travels through a medium with a sound speed gradient, the direction of travel of the sound wave is bent toward the area of lower sound speed.

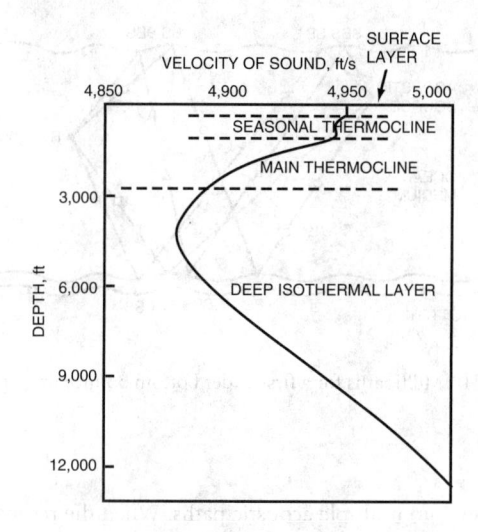

FIGURE 17.39 A Typical sound velocity profile (SVP). (*Source:* Urick, R.J. 1983. *Principles of Underwater Sound*, p. 118. McGraw-Hill, New York.)

Although the definition of shallow water can be signal dependent, in terms of depth-to-wavelength ratio, water depth of less than 100 m is generally referred to as shallow water. In shallow water, the SVP is irregular and difficult to predict because of large surface temperature and salinity variations, wind effects, and multiple reflections of sound from the ocean bottom.

Three Propagation Modes

In general, there are three dominant propagation modes that depend on the distance or range between the sound source and the receiver.

- *Direct path:* Sound energy travels in a (nominal) straight line path between the source and receiver, usually at short ranges.
- *Bottom bounce path:* Sound energy is reflected from the ocean bottom (at intermediate ranges, see Fig. 17.40).
- *Convergence zone (CZ) path:* Sound energy converges at longer ranges where multiple acoustic ray paths add or recombine coherently to reinforce the presence of signal energy from the radiating/reflecting source.

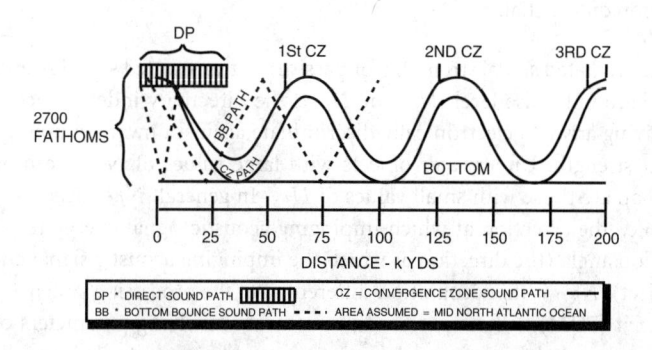

FIGURE 17.40 Typical sound paths between source and receiver, in fathom: unit of length or depth generally used for underwater measurements where 1 fathom = 6 ft. (*Source:* Cox, A.W. 1974. *Sonar and Underwater Sound*, p. 25. Lexington Books, D.C. Heath and Co., Lexington, MA.)

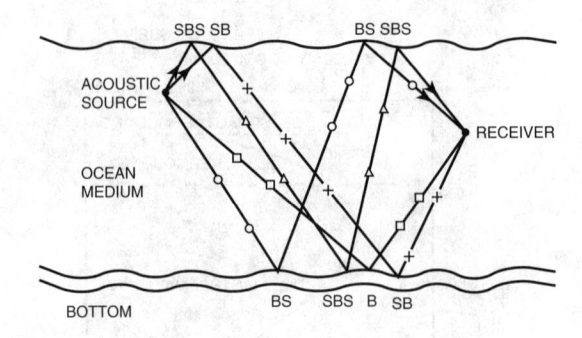

FIGURE 17.41 Multipaths for a first-order bottom bounce propagation model.

Multipaths

The ocean splits signal energy into multiple acoustic paths. When the receiving system can resolve these multiple paths (or multipaths), they should be coherently recombined by optimal signal processing to fully exploit the available signal energy for detection (Chan, 1989). It is also theoretically possible to exploit the geometrical properties of multipaths present in the bottom bounce path by investigation of a virtual aperture that is created by the different path arrivals to localize the energy source. In the case of first-order bottom bounce transmission (i.e., only one bottom interaction), there are four paths (from source to receiver):

- A bottom bounce ray path (B)
- A surface interaction followed by a bottom interaction (SB)
- A bottom bounce followed by a surface interaction (BS)
- A path that first hits the surface, then the bottom, and finally the surface (SBS)

Typical first-order bottom bounce ocean propagation paths are depicted in Fig. 17.41.

Sonar System Performance

The performance of sonar systems can be assessed by the passive and active *sonar equations*. The major parameters in the sonar equation, measured in **decibel**, are as follows:

L_S = source level
L_N = noise level
N_{DI} = directivity index
N_{TS} = echo level or target strength
N_{RD} = recognition differential.

Here, L_S is the target-radiated signal strength (for passive) or transmitted signal strength (for active), and L_N is the total background noise level. N_{DI}, or D_I, is the directivity index, which is a measure of the capability of a receiving array to electronically discriminate against unwanted noise. N_{TS} is the received echo level or target strength. Underwater objects with large values of N_{TS} are more easily detectable with active sonar than are those with small values of N_{TS}. In general, N_{TS} varies as a function of object size, aspect angle (i.e., the direction at which impinging acoustic signal energy reaches the underwater object), and reflection angle (the direction at which the impinging acoustic signal energy is reflected off the underwater object). N_{RD} is the recognition differential of the processing system.

The **figure of merit (FOM)**, a basic performance measure involving parameters of the sonar system, ocean, and target, is computed for active and passive sonar systems (in decibel) as follows:

For passive sonar

$$FOM_P = L_S - (L_N - N_{DI}) - N_{RD}$$

For active sonar

$$\text{FOM}_A = (L_S + N_{TS}) - (L_N - N_{DI}) - N_{RD}$$

Sonar systems are designed so that the FOM exceeds the signal propagation loss for a given set of parameters of the sonar equations. The amount above the FOM is called the *signal excess*. When two sonar systems are compared, the one with the largest signal excess is said to hold the *acoustic advantage*. It should be noted, however, that the set of parameters in the FOM equations given here is simplified. Depending on the design or parameter measurability conditions, parameters can be combined or expanded in terms of such quantities as the frequency dependency of the sonar system in particular ocean conditions, the speed and bearing of the receiving or transmitting platforms, **reverberation** loss, and so forth. Furthermore, due to multipaths, differences in sonar system equipment and operation, and the constantly changing nature of the ocean medium, the FOM parameters fluctuate with time. Thus, the FOM is not an absolute measure of performance but rather an average measure of performance over time.

Sonar System Performance Limitations

In a typical reception of a signal wavefront, noise and interference can degrade the performance of the sonar system and limit the system's capability to detect signals in the underwater environment. The effects of these degradations must be considered when any sonar system is designed. The noise or interference could be from a school of fish, shipping (surface or subsurface), active transmission operations (e.g., jammers), or the use of multiple receivers or sonar systems simultaneously. Also, the ambient noise may have unusual vertical or horizontal directivity, and in some environments, such as the Arctic, the noise due to ice motion may produce unusual interference. Unwanted backscatterers, similar to the headlights of a car driving in fog, can cause a signal-induced and signal-correlated noise that degrades processing gain. Some other performance-limiting factors are

- The loss of signal level and acoustic coherence due to boundary interaction as a function of grazing angle
- The radiated pattern (signal level) of the object and its spatial coherence
- The presence of surface, bottom, and volume reverberation (in active sonar)
- Signal spreading (in time, frequency, or bearing) owing to the modulating effect of surface motion
- Biologic noise as a function of time (both time of day and time of year)
- Statistics of the noise in the medium (e.g., does the noise arrive in the same ray path angles as the signal?).

Imaging and Tomography Systems

Underwater sound and signal processing can be used for bottom imaging and underwater oceanic tomography (Spindel, 1985). Signals are transmitted in succession and the **time delay** measurements between signals and measured multipaths are then used to determine the speed of sound in the ocean. This information along with bathymetry data is used to map depth and temperature variations of the ocean. In addition to mapping ocean bottoms, such information can aid in quantifying global climate and warming trends.

17.4.3 Underwater Sound Systems: Components and Processes

In this section, we describe a generic sonar system and provide a brief summary for some of its components. In Sec. 17.4.4, we describe some of the signal processing functions. A detailed description of the sonar components and various signal processing functions can be found in Knight, Pridham, and Kay (1981), Hueter (1972), and Winder (1975). Figures 17.42 and Figure 17.43 show block diagrams of the major components of a typical active and passive sonar system, respectively. Except for the signal generator, which is present only in active sonar, there are many similarities in the basic components and functions for the active and passive sonar system.

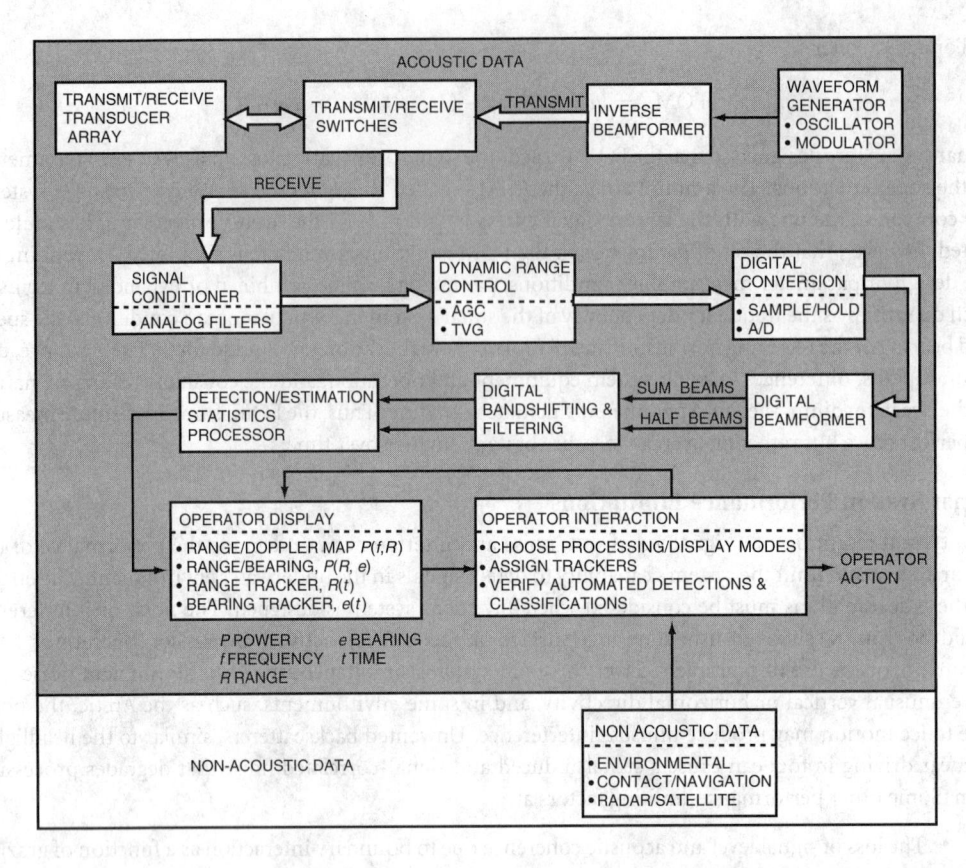

FIGURE 17.42 Generic active sonar system. (*Source:* Adapted from Knight, W., Pridham, R.G., and Kay, S.M. 1981. Digital signal processing for sonar. *Proceedings of the IEEE* 69(11):1454. © IEEE 1981.)

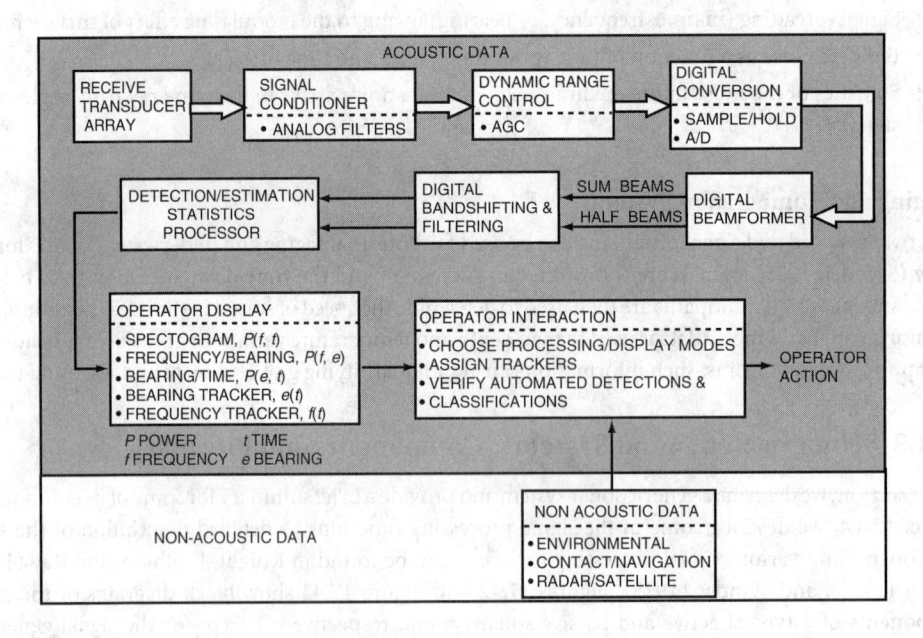

FIGURE 17.43 Generic passive sonar system. (*Source:* Adapted from Knight, W., Pridham, R.G., and Kay, S.M. 1981. Digital signal processing for sonar. *Proceedings of the IEEE* 69(11):1454. © IEEE 1981.)

In an active sonar system, an electronic signal generator produces a signal. The signal is then inverse beamformed by delaying it in time by various amounts. A separate **projector** is used to transmit each of the delayed signals by transforming the electrical signal into an acoustic pressure wave that propagates through water. Thus, an array of projectors is used to transmit the signal and focus it in the desired direction. Depending on the desired range and Doppler resolution, different signal waveforms can be generated and transmitted.

At the receiver (an array of **hydrophones**), the acoustic or pressure waveform is converted back to an electrical signal. The received signal consists of the source signal (usually the transmitted signal in the active sonar case) embedded in ambient noise and interference from other sources present in water. The signal then goes through a number of signal processing functions. In general, each channel of the analog signal is first filtered in a signal conditioner. It is then amplified or attenuated within a specified dynamic range using an automatic gain control (AGC). For active sonar, we can also use a time varied gain (TVG) to amplify or attenuate the signal. The signal, which is analog until this point, is then sampled and digitized by analog-to-digital (A/D) converters. The individual digital sensor outputs are next combined by a digital **beamformer** to form a set of beams. Each beam represents a different search direction of the sonar. The beam output is further processed (bandshifted, filtered, normalized, down sampled, etc.) to obtain *detection*, *classification*, and *localization* (DCL) estimates, which are displayed to the operator on single or multiple displays. Based on the display output (acoustic data) and other nonacoustic data (environmental, contact, navigation, and radar/satellite), the operators make their final decision.

Signal Waveforms

The transmitted signal is an essential part of an active sonar system. The properties of the transmitted signal will strongly affect the quality of the received signal and the information derived from it. The main objective in active sonar is to detect a target and estimate its range and velocity. The range and velocity information is obtained from the reflected signal. To show how the range and velocity information is contained in the received signal, we consider a simple case where $s(t), 0 \leq t \leq T$ is transmitted. Neglecting medium, noise, and other interference effects, we let the transmitted signal be reflected back from a moving target located at range $R = R_0 + vt$, where v is the velocity of the target. The received signal is then given by

$$r(t) = a(R)s[(1-b)t - \tau]$$

$$\tau = 2R_0/c, \qquad b = 2v/c$$

where $a(R)$ is the propagation loss attenuation factor (more generally $a(R)$ is also a function of frequency) and c is the speed of sound. Measuring the delay τ gives the propagation time to and from the target. We can then calculate the range from the time delay. The received signal is also time compressed or expanded by b. The velocity of the target can be estimated by determining the compression or expansion factor of the received signal. Each signal has different range and Doppler (velocity) resolution properties. Some signals are good for range resolution but not for Doppler, some for Doppler but not for range, some for reverberation-limited environments, and some for noise-limited environments (Winder, 1975). Commonly used signals in active sonar are *continuous wave* (CW), *linear frequency modulation* (LFM), *hyperbolic frequency modulation* (HFM), and *pseudo-random noise* (PRN) signals. CW signals have been used in sonar for decades, whereas signals such as *frequency hop codes* (FHC) and Newhall waveforms are recently rediscovered signals (Rihaczek, 1985) that work well in high-reverberation shallow water environments.

Some of the most commonly used signals will be described. So far we have made generalized statements about the effectiveness of signals, which only provide a broad overview. More specifically, the signal has properties that depend on a number of factors not discussed here, such as time duration and frequency bandwidth. Rihaczek (1985) provides a detailed analysis of the properties and effectiveness of signals. The simplest signal is a rectangular CW pulse, which is a single-frequency sinusoid. The CW signal may have high resolution in range (short CW) or Doppler (long CW) but not in both simultaneously. LFM signals are waveforms whose instantaneous frequency varies linearly with time; in HFM signals the instantaneous frequency sweeps monotonically as a hyperbola. Both these signals are good for detecting low Doppler targets in reverberation-limited conditions. PRN signals, which are generated by superimposing binary data on sinusoid carriers, provide simultaneous resolution in range and Doppler. However, such range resolution may not be as good as LFM alone and Doppler resolution is not as good as CW alone. An FHC signal is a waveform that consists of subpulses of equal duration. Each subpulse has a distinct frequency, and these frequencies jump or hop in a defined manner. Similar to the PRN, FHC also provides simultaneous resolution in range and Doppler. Newhall waveforms (also known as coherent pulse trains or saw-tooth frequency modulation), which are trains of repeated modulated subpulses (typically HFM or LFM), allow reverberation suppression and low Doppler target detection.

Sonar Transducers

A transducer is a fundamental element of both receiving hydrophones and projectors. It is a device that converts one form of energy into another. In the case of a sonar transducer, the two forms of energy are electricity and pressure, the pressure being that associated with a signal wavefront in water. The transducer is a reciprocal device, such that when electricity is applied to the transducer, a pressure wave is generated in the water, and when a pressure wave impinges on the transducer, electricity is developed. The heart of a sonar transducer is its active material, which makes the transducer respond to electrical or pressure excitations. These active materials produce electrical charges when subjected to mechanical stress and conversely produce a stress proportional to the applied electrical field strength when subjected to an electrical field. Most sonar transducers employ piezoelectric materials, such as lead zirconate titanate ceramic, as the active material. Magnetostictive materials, such as nickel, can also be used. Figure 17.44 shows a flextensional transducer. In this configuration, the ceramic stack is mounted on the major axis of a metallic elliptical cylinder. Stress is applied by compressing the ellipse along its minor axis, thereby extending the major axis. The ceramic stacks are then inserted into the cylinder and the stress is released, which places the ceramic stacks in compression. This design allows a small change imparted at the ends of the ceramic stack to be converted into a larger change at the major faces of the ellipse.

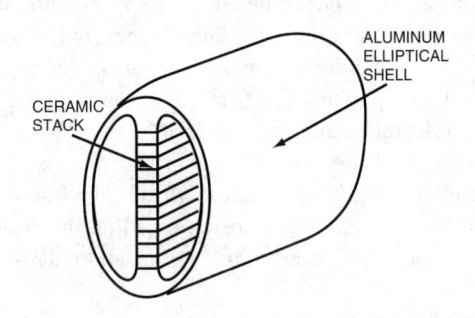

FIGURE 17.44   The flextensional transducer. (*Source:* Burdic, W.S. 1984. *Underwater Acoustic System Analysis*, p. 93. Prentice-Hall, Englewood Cliffs, NJ.)

Hydrophone Receivers

Hydrophone sensors are microphones capable of operating in water under hydrostatic pressure. These sensors receive radiated and reflected sound energy that arrives through the multiple paths of the ocean medium from a variety of sources and reflectors. As with a microphone, hydrophones convert acoustic pressure to electrical voltages or to optical signals. Typical hydrophones are hollow piezoelectric ceramic cylinders with end caps. The cylinders are covered with rubber or polyethylene as water proofing, and an electrical cable exits from one end. These hydrophones are isolation mounted so that they do not pick up stray vibrations from the ship. Most hydrophones are designed to operate below their resonance frequency, thus resulting in a flat or broadband receiving response.

Sonar Projectors (Transmitters)

A sonar projector is a transducer designed principally for transmission, that is, to convert electrical voltage or energy into sound energy. Although sonar projectors are also good receivers (hydrophones), they are too expensive and complicated (it is designed for a specific frequency band of operation) for just receiving signals. Most sonar projectors are of a *tonpilz* (sound mushroom) design with the piezoelectric material sandwiched between a head and tail mass. The head mass is usually very light and stiff (aluminum) and the tail mass very heavy (steel). The combination results in a projector that is basically a half-wavelength

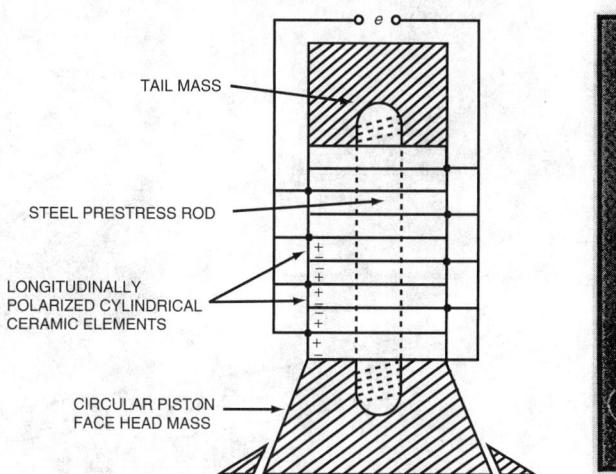

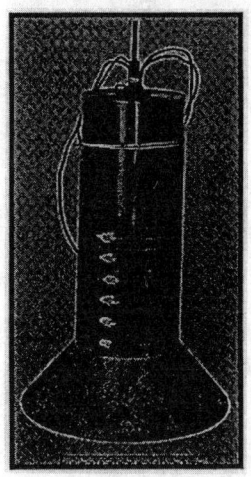

FIGURE 17.45 The tonpilz projector. (*Sources*: Burdic, W.S. 1984. *Underwater Acoustic System Analysis*, p. 92. Prentice-Hall, Englewood Cliffs, NJ; Hueter, T.F., 1972. Twenty years in underwater acoustics: Generation and reception. *J. of the Acoustical Soc. of America* 51(3):1032.)

long at its resonance frequency. The tonpilz resonator is housed in a watertight case that is designed so that the tonpilz is free to vibrate when excited. A pressure release material such as corprene (cork neoprene) is attached to the tail mass so that the housing does not vibrate and all the sound is transmitted from the front face. The piezoelectric material is a dielectric and, as such, acts as a capacitor. To ensure an efficient electrical match to driver amplifiers, a tuning coil or even a transformer is usually contained in the projector housing. A tonpilz projector is shown in Fig. 17.45.

Active Sources

One type of sonar transducer, primarily used in the surveillance community, is a low-frequency active source. The tonpilz design is commonly used for such projectors at frequencies down to about 2 kHz (a tonpilz at this frequency is almost 0.75-m long). For frequencies below 2 kHz, other types of transducer technology are employed, including mechanical transformers such as flexing shells, moving coil (loud speaker) devices, hydraulic sources, and even impulse sources such as spark gap and air. Explosives are a common source for surveillance, and when used with towed arrays make a very sophisticated system.

Receiving Arrays

Most individual hydrophones have an omnidirectional response; that is, sound emanates almost uniformly in all directions. To focus sound in a particular direction requires an array of hydrophones or projectors. The larger the array, the narrower and more focused is the beam of sound energy, and hence the more the signal is isolated from the interfering noise. An array of hydrophones allows discrimination against unwanted background noise by focusing its main response axis (MRA) on the desired signal direction. Arrays of projectors and hydrophones are usually designed with elements spaced one-half wavelength apart. This provides optimum beam structure at the design frequency. As the frequency of operation increases and exceeds three-quarter wavelength spacing, the beam structure, although narrowing, deteriorates. As the frequency decreases, the beam increases in width to the point that focusing diminishes. Some common array configurations are linear (omnidirectional), planar (fan shaped), cylindrical (searchlight) and spherical (steered beams). Arrays can be less than 3.5 in. in diameter and can have on the order of 100 hydrophones or acoustic channels. Some newer arrays have even more channels. Typically, these channels are nested, subgrouped and combined in different configurations to form the low-frequency (LF), mid-frequency

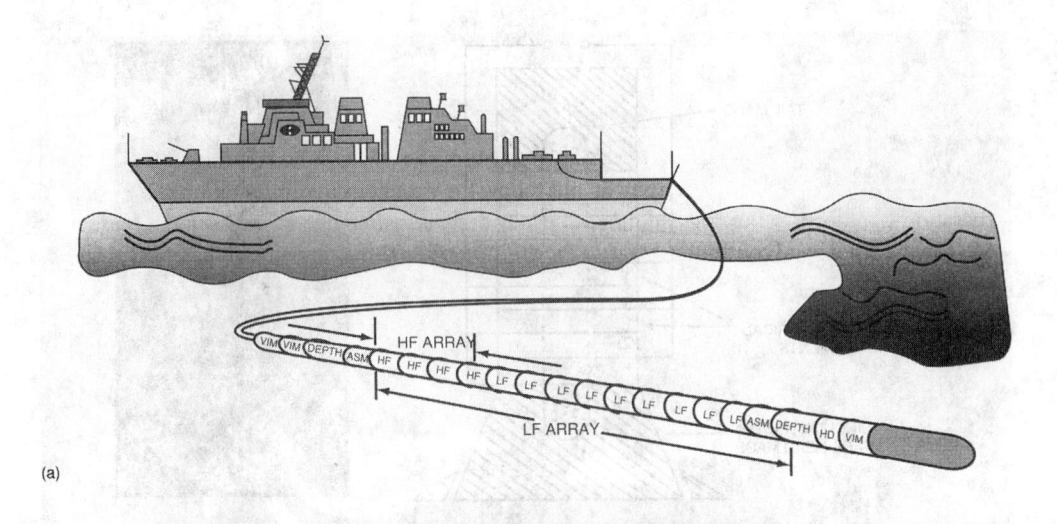

(a)

(b)

FIGURE 17.46   (a) Schematic of a towed line array. (*Source:* Urick, R.J. 1983. *Principles of Underwater Sound*, p. 13. McGraw-Hill, New York.) (b) bow array assembly for the seawolf submarine (SSN-21). (*Source:* NUWC. 1991. Division Newport, BRAC 1991 presentation. Naval Undersea Warfare Center, New London, CT.)

(MF) and high-frequency (HF) apertures of the array. Depending on the frequency of interest, one can use any one of these three apertures to process the data. The data are then prewhitened, amplified, and low-pass filtered before being routed to A/D converters. The A/D converters typically operate or sample the data at about three times the low-pass cutoff frequency.

A common array, shown in Figure 17.46(a), is a single linear line of hydrophones that makes up a device called a *towed array*, which in the oil exploration business are often called *streamers*. The line is towed behind the ship and is effective for searching for low-level and low-frequency signals without interference from the ship's self-noise. Figure 17.46(b) shows a more sophisticated bow array (sphere) assembly.

Sonobuoys

These small expendable sonar devices contain a transmitter to transmit signals and a single hydrophone to receive the signal. Sonobuoys are generally dropped by fixed-wing or rotary wing aircraft for underwater signal detection.

Dynamic Range Control

Analog signals received by the receiving array are converted into a digital format, while ensuring that the dynamic range of the data is within acceptable limits. The receiving array must have sufficient dynamic range so that it can detect the weakest signal but also not saturate upon receiving the largest signal in the presence of noise and interference. To be able to convert the data into digital form and display it, large fluctuations in the data must be eliminated. Not only do these fluctuations overload the digital range capacity, they will affect the background quality and contrast of the displays as well. It has been shown that the optimum display background, a background with uniform fluctuations as a function of range, time, and bearing, for detection is one that has constant temporal variance at a given bearing and a constant spatial variance at a given range. Since the fluctuations (noise, interference, propagation conditions, etc.) are time varying, it is necessary to have a threshold level that is independent of the fluctuations. The concept is to use techniques that can adapt the high dynamic range of the received signal to the limited dynamic range of the computers and displays. Time-varied gain (TVG) for active sonar and automatic gain control (AGC) are two popular techniques to control the dynamic range. TVG controls the receiver gain so that it follows a prescribed variation with time, independent of the background conditions. The disadvantage of TVG is that the variations of gain with time do not follow the variations in reverberation. TVG is sufficient if the reverberation is uniform in bearing and monotonically decreasing in range (which is not the case in shallow water). AGC on the other hand continuously varies the gain according to the current reverberation or interference conditions. Details of how TVG, AGC, and other gain control techniques such as notch filters, reverberation controlled gain, logarithmic receivers, and hard clippers work are presented in Winder (1975).

Beamforming

Beamforming is a process in which outputs from the hydrophone sensors of an array are coherently combined by delaying and summing the outputs to provide enhanced detection and estimation. In underwater applications, we are trying to detect a directional (single direction) signal in the presence of normalized background noise that is ideally isotropic (nondirectional). By arranging the hydrophone (array) sensors in different physical geometries and electronically steering them in a particular direction, we can increase the signal-to-noise ratio in a given direction by rejecting or canceling the noise in other directions. There are many different kinds of arrays that can be beamformed (e.g., equally spaced line, continuous line, circular, cylindrical, spherical, or random sonobuoy arrays). The beam pattern specifies the response of these arrays to the variation in direction. In the simplest case, the increase in SNR due to the beamformer, called the **array gain** (in decibel), is given by

$$AG = 10 \log \frac{\text{SNR}_{\text{array(output)}}}{\text{SNR}_{\text{single sensor(input)}}}$$

Displays

Advancements in processing power and display technologies over the last two decades have made displays an integral and essential part of any sonar system today. Displays have progressed from a single monochrome terminal to very complicated interactive real-time multiterminal color display electronics. The amount of data that can be provided to an operator can be overwhelming; time series, power spectrum, narrowband and broadband lofargrams, time bearing, range bearing, time Doppler, and sector scans are just some of the many available displays. Then add to this a source or contact tracking display for single or multiple sources over multiple (50–100) beams. The most recent displays provide these data in an interactive mode to make it easier for the operator to make a decision.

For passive systems the three main parameters of interest are time, frequency, and bearing. Since three-dimensional data are difficult to visualize and analyze, it is usually displayed in the following formats:

- *Bearing time:* Obtained by integrating over frequency. Useful for targets with significant broadband characteristics; also called bearing time recorder (BTR).
- *Bearing frequency:* Obtained at particular intervals of time or by integrating over time. Effective for targets with strong stable spectral lines; also called *frequency azimuth* (FRAZ).

- *Time frequency:* Usually effective for targets with weak or unstable spectral lines in a particular beam; also called *lofargram or sonogram.*

In active systems, two additional parameters, range and Doppler, can be estimated from the received signals, providing additional displays of range bearing and time Doppler where Doppler provides the speed estimate of the target.

In general, all three formats are required for the operator to make an informed decision. The operator must sort the outputs from all of the displays before classifying them into targets. For example, in passive narrow-band sonar, classification is usually performed on the outputs from spectral/tonal contents of the targets. The operator uses the different tonal content and its harmonic relations of each target for classification. In addition to the acoustic data information and displays, nonacoustic data such as environmental, contact and navigation information, and radar/satellite photographs are also available to the operator.

17.4.4   Signal Processing Functions

Detection of signals in the presence of noise, using classical Bayes or Neyman–Pearson decision criteria, is based on hypothesis testing. In the simplest binary hypothesis case, the detection problem is posed as two hypotheses:

H_0: Signal is not present (referred to as the null hypothesis).
H_1: Signal is present.

For a received wavefront, H_0 relates to the noise-only case and H_1 to the signal-plus-noise case. Complex hypotheses (M hypotheses) can also be formed if detection of a signal among a variety of sources is required.

Probability is a measure, between zero and unity, of how likely an event is to occur. For a received wavefront, the likelihood ratio L, is the ratio of P_{H_1} (probability that hypothesis H_1 is true) to P_{H_0} (probability that hypothesis H_0 is true). A decision (detection) is made by comparing the likelihood to a predetermined threshold h. (Sometimes the logarithm of the likelihood, ratio called the *log-likelihood ratio*, is used.) That is, if $L = P_{H_1}/P_{H_0} > h$, a decision is made that the signal is present.

Probability of detection P_D measures the likelihood of detecting an event or object when the event does occur. *Probability of false alarm* P_{fa} is a measure of the likelihood of saying something happened when the event did *not* occur. **Receiver operating characteristics** (**ROC**) curves plot P_D vs. P_{fa} for a particular (sonar signal) processing system. A single plot of P_D vs. P_{fa} for one system must fix the SNR and processing time. The threshold h is varied to sweep out the ROC curve. The curve is often plotted on either log-log scale or probability scale. In comparing a variety of processing systems, we would like to select the system (or develop one) that maximizes the P_D for every given P_{fa}. Processing systems must operate on their ROC curves, but most processing systems allow the operator to select where on the ROC curve the system is operated by adjusting a threshold; low thresholds ensure a high probability of detection at the expense of high false alarm rate. A sketch of two monotonically increasing ROC curves is given in Fig. 17.47. By proper adjustment of the decision threshold, we can tradeoff detection performance for false alarm performance. Since the points $(0,0)$ and $(1,1)$ are on all ROC curves, we can always guarantee 100% probability of detection with an arbitrarily low threshold (albeit at the expense of 100% probability of false alarm) or 0% probability of false alarm with an arbitrarily high threshold (albeit at the expense of 0% probability of detection). The (log) *likelihood detector* is a detector that

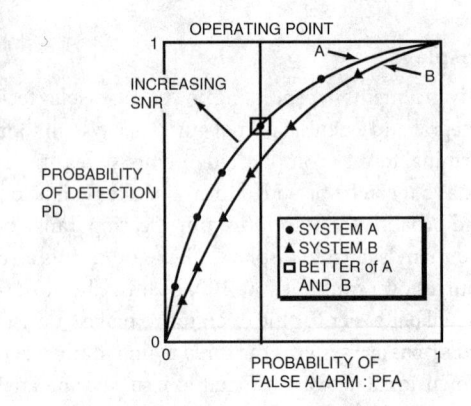

FIGURE 17.47 Typical ROC curves. Note that points $(0,0)$ and $(1,1)$ are on all ROC curves; upper curve represents higher P_D for fixed P_{fa} and, hence, better performance by having higher SNR or processing time.

achieves the maximum probability of detection for fixed probability of false alarm; it is shown in Fig. 17.48 for detecting Gaussian signals reflected or radiated from a stationary object. For spiky non-Gaussian noise, clipping prior to filtering improves detection performance.

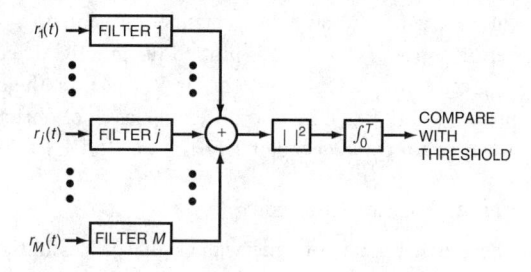

In active sonar, the filters are matched to the known transmitted signals. If the object (acoustic reflector) has motion, it will induce Doppler on the reflected signal, and the receiver will be complicated by the addition of a bank of Doppler compensators. Returns from a moving object are shifted in frequency by $\Delta f = (2v/c)f$, where v is the relative

FIGURE 17.48 Log likelihood detector structure for uncorrelated Gaussian noise in the received signal $r_j(t)$, $j = 1, \ldots, M$.

velocity (range rate) between the source and object, c is the speed of sound in water, and f is the operating frequency of the source transmitter.

In passive sonar, at low SNR, the optimal filters in Fig. 17.48 (so-called Eckart filters) are functions of $G_{ss}^{1/2}(f)/G_{nn}(f)$, where f is frequency in hertz, $G_{ss}(f)$ is the signal power spectrum, and $G_{nn}(f)$ is the noise power spectrum.

Estimation/Localization

The second function of underwater signal processing estimates the parameters that localize the position of the detected object. The source position is estimated in range, bearing, and depth, typically from the underlying parameter of time delay associated with the acoustic signal wave front. The statistical uncertainty of the positional estimates is important. Knowledge of the first-order probability density function or its first- and second-order moments, the mean (expected value), and the variance are vital to understanding the expected performance of the processing system. In the passive case, the ability to estimate range is extremely limited by the geometry of the measurements; indeed, the variance of passive range estimates can be extremely large, especially when the true range to the

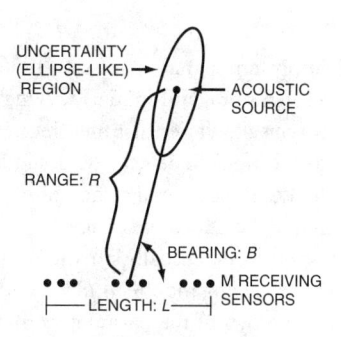

FIGURE 17.49 Array Geometry used to estimate source position. (*Source:* Carter, G.C., 1987. Coherence and time delay estimation. *Proceedings of the IEEE* 75(2):251. © IEEE 1987.)

signal source is long when compared with the aperture length of the receiving array. Figure 17.49 depicts direct path passive ranging uncertainty from a collinear array with sensors clustered so as to minimize the bearing and uncertainty region. Beyond the direct path, multipath signals can be processed to estimate source depth passively. Range estimation accuracy is not difficult with the active sonar, but active sonar is not covert, which for some applications can be important.

Classification

The third function of sonar signal processing is classification. This function determines the type of object that has radiated or reflected signal energy. For example, was the sonar signal return from a school of fish or a reflection from the ocean bottom? The action taken by the operator is highly dependent on this important function. The amount of radiated or reflected signal power relative to the background noise (that is, SNR) necessary to achieve good classification may be many decibels higher than for detection. Also, the type of signal processing required for classification may be different than the type of processing for detection. Processing methods that are developed on the basis of detection might not have the requisite SNR to adequately perform the classification function. Classifiers are, in general, divided into feature (or clue) extractors followed by a classifier decision box. A key to successful classification is feature extraction. Performance of classifiers is plotted (as in ROC detection curves) as the probability of deciding on class A, given A was actually present, or $P(A|A)$, vs. the probability of deciding on class B, given that A was present, or $P(B|A)$, for two different classes of objects, A and B. Of course, for the same class of objects, the operator could also plot $P(B|B)$ vs. $P(A|B)$.

Target Motion Analysis

The fourth function of underwater signal processing is to perform contact or target motion analysis (TMA), that is, to estimate parameters of bearing and speed. Generally, nonlinear filtering methods, including Kalman–Bucy filters, are applied; typically these methods rely on a state-space model for the motion of the contact. For example, the underlying model of motion could assume a straight line course and a constant speed for the contact of interest. When the signal source of interest behaves like the model, then results consistent with the basic theory can be expected. It is also possible to incorporate motion compensation into the signal processing detection function. For example, in the active sonar case, proper signal selection and processing can reduce the degradation of detector performance caused by uncompensated Doppler. Moreover, joint detection and estimation can provide clues to the TMA and classification processes. For example, if the processor simultaneously estimates depth in the process of performing detection, then a submerged object would not be classified as a surface object. Also joint detection and estimation using Doppler for detection can directly improve contact motion estimates.

Normalization

Another important signal processing function for the detection of weak signals in the presence of un-known and (temporal and spatial) varying noise is normalization. The statistics of noise or reverberation for oceans typically varies in time, frequency, and/or bearing from measurement-to-measurement and location-to-location. To detect a weak signal in a broadband, nonstationary, inhomogeneous background, it is usually desirable to make the noise background statistics as uniform as possible for the variations in time, frequency, and/or bearing. The noise background estimates are first obtained from a window of resolution cells (which usually surrounds the test data cell). These estimates are then used to normalize the test cell, thus reducing the effects of the background noise on detection. Window length and distance from the test cell are two of the parameters that can be adjusted to obtain accurate estimates of the different types of stationary or nonstationary noise.

17.4.5   Advanced Signal Processing

Adaptive Beamforming

Beamforming was discussed earlier in Sec. 17.4.3. The cancellation of noise through beamforming can also be done adaptively, which can improve array gain further. Some of the various adaptive beamforming techniques (Knight, Pridham, and Kay, 1981) use Dicanne, sidelobe cancellers, maximum entropy array processing, and maximum-likelihood (ML) array processing.

Coherence Processing

Coherence is a normalized (so that it lies between zero and unity), cross-spectral density function that is a measure of the similarity of received signals and noise between any sensors of the array. The complex coherence functions between two wide-sense-stationary processes x and y are defined by

$$\gamma_{xy}(f) = \frac{G_{xy}(f)}{\sqrt{G_{xx}(f)G_{yy}(f)}}$$

where, as before, f is the frequency in hertz and G is the power spectrum function. Array gain depends on the coherence of the signal and noise between the sensors of the array. To increase the array gain, it is necessary to have high signal coherence but low noise coherence. Coherence of the signal between sensors improves with decreasing separation between the sensors, frequency of the received signal, total bandwidth, and integration time. Loss of coherence of the signal could be due to ocean motion, object motion, multipaths, reverberation, or scattering. The coherence function has many uses, including measurement of SNR or array gain, system identification, and determination of time delays (Carter, 1987, 1993).

Acoustic Data Fusion

Acoustic data fusion is a technique that combines information from multiple receivers or receiving platforms about a common object or channel. Instead of each receiver making a decision, relevant information from the different receivers is sent to a common control unit where the acoustic data is combined and processed (hence the name *data fusion*). After fusion, a decision can be relayed or fed back to each of the receivers. If data transmission is a concern, due to time constraints, cost, or security, other techniques can be used in which each receiver makes a decision and transmits only the decision. The control unit makes a global decision based on the decisions of all of the receivers and relays this global decision back to the receivers. This is called distributed detection. The receivers can then be asked to re-evaluate their individual decisions based on the new global decision. This process could continue until all of the receivers are in agreement or could be terminated whenever an acceptable level of consensus is attained.

An advantage of data fusion is that the receivers can be located at different ranges (e.g., on two different ships), in different mediums (shallow or deep water, or even at the surface), and at different bearings from the object, thus giving comprehensive information about the object or the underwater acoustic channel.

17.4.6   Application

Since World War II, in addition to military applications, there has been an expansion in commercial and industrial underwater acoustic applications. Table 17.8 lists the military and nonmilitary functions of

TABLE 17.8   Underwater Acoustic Applications

Function	Description
Military:	
Detection	Deciding if a target is present or not.
Classification	Deciding if a detected target does or does not belong to a specific class.
Localization	Measuring at least one of the instantaneous positions and velocity components of a target (either relative or absolute), such as range, bearing, range rate, or bearing rate.
Navigation	Determining, controlling, and/or steering a course through a medium (includes avoidance of obstacles and the boundaries of the medium).
Communications	Instead of a wire link, transmitting and receiving acoustic power and information.
Control	Using a sound-activated release mechanism.
Position marking	Transmitting a sound signal continuously (Beacons) or transmitting only when suitably interrogated (Transponders).
Depth sounding	Sending short pulses downward and timing the bottom return.
Acoustic-Speedometers	Using pairs of transducers pointing obliquely downward to obtain speed over the bottom from the Doppler shift of the bottom return.
Commercial Applications:	
Industrial	Oceanographic
Fish finders/fish herding	Subbottom geological mapping
Fish population estimation	Environmental monitoring
Oil and mineral explorations	Ocean topography
River flow meter	Bathyvelocimeter
Acoustic holography	Emergency telephone
Viscosimeter	Seismic simulation and measurement
Acoustic ship docking system	Biological signal and noise measurement
Ultrasonic grinding/drilling	Sonar calibration
Biomedical ultrasound	

Sources: Adapted from Urick, R.J. 1983. *Principles of Underwater Sound*, p. 8. McGraw-Hill, New York; Cox, A.W. 1974. *Sonar and Underwater Sound*, p. 2. Lexington Books, D.C. Heath and Co., Lexington, MA.

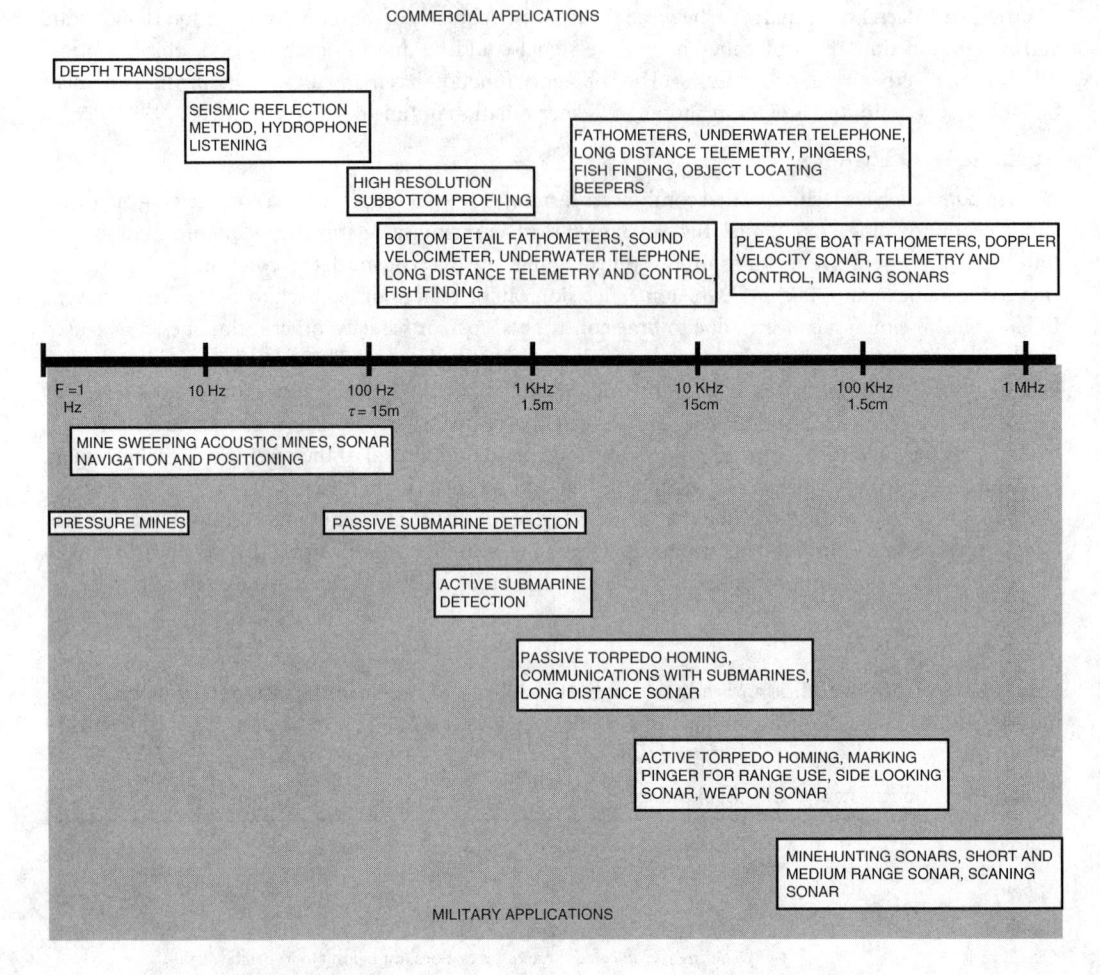

FIGURE 17.50 Sonar frequency allocation. (*Source:* Adapted from Neitzel, E.B. 1973. Civil uses of underwater acoustics. Lectures on Marine Acoustics (AD 156-052).)

sonar along with some of the current applications. Figure 17.50 shows the sonar frequency allocations for the military and commercial applications.

Acknowledgment

The authors thank Louise Miller for her assistance in preparing this chapter.

Defining Terms

Array gain: A measure of how well an array discriminates signals in the presence of noise as a result of beamforming. It is represented as a ratio of the array SNR divided by the SNR of an individual omnidirectional hydrophone. For complex noise environments array gain is also given by the overall signal gain divided by noise gain.

Beamformer: A process in which outputs from individual hydrophone sensors of an array are coherently combined by delaying and summing the outputs to provide enhanced SNR.

Coherence: A normalized cross-spectral density function that is a measure of the similarity of received signals and noise between any sensors of an array.

Decibel (dB): Logarithmic scale representing the ratio of two quantities given as $10 \log_{10}(P_1/P_0)$ for power level ratios and $20 \log_{10}(V_1/V_0)$ for acoustic pressure or voltage ratios. A standard reference pressure or intensity level in SI units is equal to 1 micropascal (1 pascal = 1 newton per square meter = 10 dyne per square centimeter).

Doppler shift: Shift in frequency of transmitted signal due to the relative motion between the source and object.

Figure of merit/sonar equation: Performance evaluation measure for the various target and equipment parameters of a sonar system. It is a subset, which includes reverberation effects, of the broader sonar performance given by *the sonar equations*.

Receiver operating characteristics (ROC) curves: Plots of the probability of detection (likelihood of detecting the object when the object is present) vs. the probability of false alarm (likelihood of detecting the object when the object is not present) for a particular processing system.

Reverberation/clutter: Inhomogeneities, such as dust, sea organisms, schools of fish, and sea mounds on the bottom of the sea, which form mass density discontinuities in the ocean medium. When an acoustic wave strikes these inhomogeneities, some of the acoustic energy is reflected and reradiated. The sum total of all such reradiations is called reverberation. Reverberation is present only in active sonar, and in the case where the object echoes are completely masked by reverberation, the sonar system is said to be reverberation limited.

SONAR: Acronym for sound navigation and ranging, adapted in the 1940s, involves the use of sound to explore the ocean and underwater objects.

Sonar hydrophone: Receiving sensors that convert sound energy into electrical or optical energy (analogous to underwater microphones).

Sonar projector: A transmitting source that converts electrical energy into sound energy.

Sound velocity profile (SVP): Description of the speed of sound in water as a function of water depth.

SNR: The signal-to-noise (power) ratios, usually measured in decibel.

Time Delay: The time (delay) difference in seconds from when an acoustic wavefront impinges on one hydrophone or receiver until it strikes another.

References

Burdic, W.S. 1984. *Underwater Acoustic System Analysis*. Prentice-Hall, Englewood Cliffs, NJ.

Carter, G.C. 1987. Coherence and time delay estimation. *Proceedings of the IEEE* 75(2):236–255.

Carter, G.C., ed. 1993. *Coherence and Time Delay Estimation*. IEEE Press, Piscataway, NJ.

Chan, Y.T., ed. 1989. *Digital Signal Processing for Sonar Underwater Acoustic Signal Processing*, NATO ASI series, series E: Applied Sciences, vol. 161. Kluwer Academic, Norwell, MA.

Cox, A.W. 1974. *Sonar and Underwater Sound*. Lexington Books, D.C. Heath and Co., Lexington, MA.

Hueter, T.F. 1972. Twenty years in underwater acoustics: Generation and reception. *J. Acoust. Soc. Am.* 51(3):1025–1040.

Knight, W.C., Pridham, R.G., and Kay, S.M. 1981. Digital signal processing for sonar. *Proceedings of the IEEE* 69(11):1451–1506.

Leroy, C.C. 1969. Development of simple equations for accurate and more realistic calculation of the speed of sound in sea water. *J. Acoust. Soc. Am.* 46:216.

Mackenzie, K.V. 1981. Nine-term equation for sound speed in the oceans. *J. Acoust. Soc. AM.* 70:807.

Medwin, H. 1975. Speed of sound in water for realistic parameters. *J. Acoust. Soc. Am.* 58:1318.

Neitzel, E.B. 1973. Civil uses of underwater acoustics. Lectures on Marine Acoustics (AD 156-052).

NUWC. 1991. Division Newport BRAC 1991 presentation. Naval Undersea Warfare Center, New London, CT.

Oppenheim, A.V., ed. 1980. *Applications of Digital Signal Processing*. Prentice-Hall, Englewood Cliffs, NJ.

Rihaczek, A.W., ed. 1985. *Principles of High Resolution Radar*. Peninsula, Los Altos, CA.

Spindel, R.C. 1985. Signal processing in ocean tomography. In *Adaptive Methods in Underwater Acoustics*, ed. H.G. Urban, pp. 687–710. D. Reidel.

Urick, R.J. 1983. *Principles of Underwater Sound*. McGraw-Hill Book Company, New York.

Van Trees, H.L. 1968. *Detection, Estimation, and Modulation Theory*. John Wiley & Sons, New York.

Winder, A.A. 1975. II. Sonar system technology. *IEEE Transactions on Sonics and Ultrasonics*, Vol. su-22, No. 5, Sept., pp. 291–332.

Further Information

Journal of Acoustical Society of America (JASA), *IEEE Transactions on Signal Processing* (formerly the *IEEE Transactions on Acoustics, Speech and Signal Processing*), and *IEEE Journal of Oceanic Engineering* are professional journals providing current information on underwater acoustical signal processing.

The annual meetings of the *International Conference on Acoustics, Speech and Signal Processing*, sponsored by the IEEE, and the biannual meetings of the *Acoustical Society of America* are good sources for current trends and technologies.

Digital Signal Processing for Sonar, (Knight, Pridham, and Kay, 1981) and *Sonar System Technology* (Winder, 1975) are informative and detailed tutorials on underwater sound systems. Also the March 1972 issue of *The Journal of Acoustical Society of America*, Vol. 51, No. 3 (Part 2), has historical and review papers on underwater acoustics related topics.

17.5 Electronic Warfare and Countermeasures

Robert D. Hayes

17.5.1 Introduction

Electronic warfare (EW) generally applies to military actions. Specifically, it involves the use of electromagnetic energy to create an advantage for the friendly side of engaged forces while reducing the effectiveness of the opposing hostile armed forces.

Electronic warfare support measures (ESM) or electronic warfare support (ES) are the actions and efforts taken to listen, to search space and time, to intercept and locate the source of radiated energy, and to analyze, identify, and characterize the electromagnetic energy. Tactical employment of forces is established and plans are executed for electronic countermeasures.

Electronic countermeasures (ECM) or electronic attack (EA) are those actions taken to reduce the enemy's effective use of the electromagnetic spectrum. These actions may include jamming, electronic deception, false targets, chaff, flares, transmission disruption, and changing tactics.

Electronic counter-countermeasures (ECCM) or electronic protection (EP) are actions taken to reduce the effectiveness of enemy used ECM to enhance the use of the electromagnetic spectrum for friendly forces (Fig. 17.51). These actions are of a protective nature and are applied to tactical and strategic operations across the equipment used for sonar, radar, command, control, communications, and intelligence.

17.5.2 Radar and Radar Jamming Signal Equations

Electronic jamming is a technique where a false signal of sufficient magnitude and bandwidth is introduced into a receiving device so as to cause confusion with the expected received signal and to create a loss of

GHz	0.1		0.3	0.5	1.0		2	3	4	6	8	10		20		40	60	100
RADAR	VHF		UHF		L		S		C		X	K_u	K	K_a		millimeter		
EW	A		B		C		D	E	F	G	H	I		J		K	L	M
cm	300		100	60	30		15	10		5		3		1.5		.5		.3

FIGURE 17.51 Frequency designation for radar and EW bands.

information. For a communication receiver, it simply overpowers the radio transmission; for radar, the jamming signal overrides the radar echo return from an object under surveillance or track. The jamming equation is the same for both communications and radar systems. In the case of a radar jammer, there are several operational scenarios to consider.

The radar jammer signal level must be determined for the various applications and jammer location. When the jammer is located on the target (self-protection), the jammer signal level must exceed the effective radar cross section of the target to form an adequate screen; when the jammer is located away from the vehicle under protection (stand-off jammer), the jammer should simulate the expected radar cross section observed at a different range and different aspect.

The monostatic radar received signal power equation representing the signal received at an enemy detection radar, and the received jamming power transmitted toward the same enemy detection radar, are presented in the equations given in Table 17.9. Each quantity in the equations is defined in the table. These

TABLE 17.9 Fundamental Radar Equations

Monostatic Radar Equation	Jamming Radar Equation
$$P_r = \frac{P_t G^2 \sigma^2 \lambda^2 F^4 E}{(4\pi)^3 R^4 L_t L_r} e^{-2\alpha R}$$	$$P_r = \frac{P_t G_t G_r \lambda^2 F^2 E}{(4\pi)^2 R^2 L_t L_r} e^{-\alpha R}$$

Quantity	Dependent Function
P_t = transmitter power, W	ΔF = frequency spectrum
	ψ = phase, coherent
L_t = loss, transmitter line	Equipment path
L_r = loss, receiver line	Equipment path
E = receiver gain	Processing techniques
G = antenna gain	p = polarization
	λ = wavelength, RF carrier
	l = length
	w = width
	c = curvature
	ϕ = elevation angle
	θ = azimuth angle
σ = point target RSC, m^2	p
	λ
	l
	w
	c
	ϕ
	θ
	r = rotation
	s = surface roughness
	m = motion
	ϵ = dielectric constant
	μ = permeability
σ_0 = area extended target RCS, m^2/m^2	p
	λ
	ϕ
	θ
	s
	m
	ϵ
	τ = pulse length
	T = temperature
	d = density
σ_v = volume extended target RCS, m^2/m^3	p
	λ
	ϕ
	θ
	s
	m
	ϵ

Quantity	Dependent Function
σ_v = volume extended target RCS, m^2/m^3	p
	μ
	τ
	T
	d
α_0 = clear air loss, Np	λ
	T
	d
	P = pressure
	ρ = gas constant
α = rain, dust, snow, hail, etc., loss, Np	p = polarization
	λ
	c
	m
	ϵ
	T
	d
	P
	ρ
	n = number of particles, or rain rate
	R = range, path length
	β = differential polarization phase shift
F = electric field propagation factor—multipath	p
	λ
	c
	ϕ
	θ
	R
	s
	ϵ
	μ
	ψ
	β
R = range, path length, m	n = index of refraction spherical geometry

Antenna gain = $G = \dfrac{28,000}{\theta\phi}$

where $\theta = 75\dfrac{\lambda}{\omega}$

and $\phi = 75\dfrac{\lambda}{l}$, in degrees

quantities are dependent on electrical, polarization, material, shape, roughness, density, and atmospheric parameters.

Background interference is encountered in every branch of engineering. The received radar/ communication signal power will compete with interfering signals that will impose limitations on performance of the detecting electronic system. There are many types of interference. A few are channel crosstalk, nonlinear harmonic distortion, AM and FM interference, phase interference, polarization cross talk, and noise. The common limiting interference in electronic systems is noise.

There are many types of noise. Every physical body not at absolute zero temperature emits electromagnetic energy at all frequencies, by virtue of the thermal energy the body possesses. This type of radiation is called thermal noise radiation. Johnson noise is a result of random motion of electrons within a resistor; semiconductor noise occurs in transistors; shot noise occurs in both transmitter and receiver vacuum tubes as a result of random fluctuations in electron flow. These fluctuations are controlled by a random mechanism and, thus, in time are random processes described by the Gaussian distribution. The signal is random in phase and polarization, and usually are broad in frequency bandwidth. The average noise power is given by

$$P_N = kTB$$

where
P_N = power, watts
k = Boltzman's constant
T = temperature, absolute degrees.

Noise jamming has been introduced into EW from several modes of operation. *Spot noise jamming* requires the entire output power of the jammer to be concentrated in a narrow bandwidth ideally identical to the bandwidth of the victim radar/communication receiver. It is used to deny range, conversation, and sometimes angle information. *Barrage noise jamming* is a technique applied to wide frequency spectrum to deny the use of multiple channel frequencies to effectively deny range information, or to cover a single wide-band system using pulse compression or spread-spectrum processing. *Blinking* is a technique using two spot noise jammers located in the same cell of angular resolution of the victim system. The jamming transmission is alternated between the two sources causing the radar antenna to oscillate back and forth. Too high a blinking frequency will let the tracker average the data and not oscillate, whereas too low a frequency will allow the tracker to lock on one of the jammers.

In free space the radar echo power returned from the target varies inversely as the fourth power of the range, whereas the power received from a jammer varies as the square of the range from the jammer to the victim radar. There will be some range where the energy for the noise jammer is no longer great enough to hide the friendly vehicle. This range of lost protection is called the burn-through range.

In both EW technology and target identification technology, the received energy, as compared to the received power, better defines the process of signal separation from interference. The received radar energy will be given by

$$S = P\tau$$

where τ is the transmitted pulse length. When the radar receiver bandwidth is matched to the transmitter pulse length, then

$$S = \frac{P_r}{B_r}$$

where B_r is the bandwidth of radar receiver. The transmitted jamming energy will be given by

$$J = \frac{P_j}{B_j}$$

where B_j is the bandwidth of jamming signal.

If for a moment, the system losses are neglected, the system efficiency is 100%, and there are negligible atmospheric losses, then the ratio of jamming energy received to radar energy

$$\frac{J}{S} = \frac{4\pi P_j G_j G_r \lambda^2 F_j^2 R_r^4 B_r}{P_r G_r^2 \sigma \lambda^2 F_r^4 R_j^2 B_j}$$

for the standoff jammer, and

$$\frac{J}{S} = \frac{4\pi P_j G_j B_r R^2}{P_r G_r \sigma F^2 B_j}$$

for the self-protect jammer. For whatever camouflage factor is deemed sufficient (example, $J/S = 10$), the burnthrough range can be determined from this self-protect jamming equation.

Radar Receiver Vulnerable Elements

Range gate pulloff (RGPO) is a deception technique used against pulse tracking radars using range gates for location. The jammer initially repeats the skin echo signal with minimum time delay at a high power to capture the receiver automatic gain control circuitry. The delay is progressively changed, forcing the tracking gates to be pulled away from (walked off) the skin echo. Frequency memory loops (FMLs) or transponders provide the variable delay. The de-
ceptive pulse can be transmitted before the radar pulse is received if the pulse repetition rate is stored and repeated at the proper time. In this technique the range gate can be either pulled off or pulled in.

One implementation is shown in Fig. 17.52. A split gate (early/late gates) range tracking scheme has two gates, which are positioned by an auto-
matic tracking loop so that equal energy appears in each gate. The track point is positioned on the target centroid. First, there is an initial dwell at the target range long enough for the tracker to capture the radar receiver. The received echo has S energy.

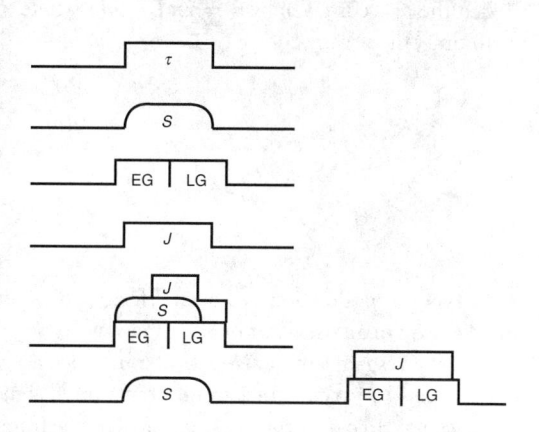

FIGURE 17.52 Range gate pulloff concept.

Then, a jamming signal is introduced which contains J amount of energy. The energy in the early and late gates are compared and the difference is used to reposition the gates so as to keep equal energy in each gate. In this manner, the range gate is continually moved to the desired location. When RGPO is successful, the jamming signal does not compete with the target's skin return, resulting with an almost infinite J/S ratio.

Velocity gate pulloff (VGPO) is used against the radar's received echo modulation frequency tracking gates in a manner analogous to the RGPO used against the received echo signal amplitude tracking gates. Typically in a Doppler or moving target indicator (MTI) system there are a series of fixed filters or a group of frequency tracking filters used to determine the velocity of a detected moving target. By restricting the bandwidth of the tracking filters, the radar can report very accurate speeds, distinguish between approaching and receding targets, and reduce echos from wide-band clutter. Typically, the tracking filters are produced by linearly sweeping across the frequency domain of interest to the detecting radar. The ECM approach to confuse this linear sweep is to transmit a nonlinear frequency modulated sweep which could be parabolic or cubic. The jamming signal will cause the Doppler detector to be walked off of the true frequency as shown in Fig. 17.53.

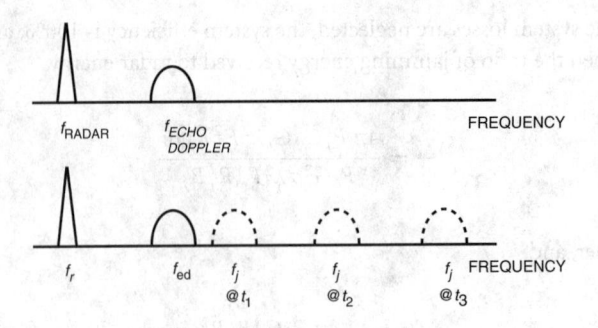

FIGURE 17.53   Frequency gate pulloff concept.

Another technique used to generate a frequency shift is to modulate the phase of the transmitted jamming signal. The instantaneous frequency of any modulated signal is directly related to the change in phase.

$$2\pi f = \frac{d\theta}{dt}$$

When the phase is given by $\theta = 2\pi f_c t + kt$ where f_c is the carrier frequency and k the change of phase with time, then

$$\frac{d\theta}{dt} = 2\pi f_c + k$$

and

$$f = f_c + \frac{k}{2\pi}$$

This type of modulation has been referred to as serrodyne modulation and used with control voltage modulation of an oscillator or traveling wave tube (TWT).

An inverse-gain jammer is a repeater in which a signal is directed back toward the enemy radar to create false targets by varying the jammer transmitted signal level inversely with the magnitude of the received radar signal. To defeat conical scan trackers, the transponder signals have been modulated at the nutation of the scan rate to deceive angle tracking (Fig. 17.54).

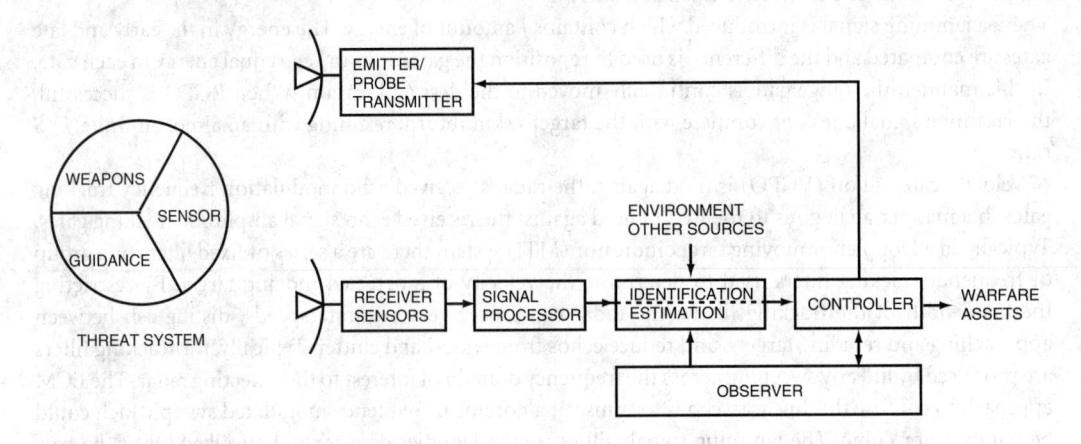

FIGURE 17.54   Surgical countermeasure block diagram.

Most ECM techniques are open-loop control functions, which add signals to the enemy radar return signal so as to confuse or deceive the weapon's ability to track.

17.5.3   Radar Antenna Vulnerable Elements

Radar backscatter from a variety of antennas was measured by Hayes at Georgia Institute of Technology for the U.S. Air Force Avionics Laboratory, Wright-Patterson Air Force Base, Ohio (Hayes and Eaves, 1966).

Typical measured data are presented in Fig. 17.55 and Fig. 17.56 for an AN/APS-3 antenna. This antenna is an 18-in. parabolic dish with a horizontally polarized double-dipole feed built upon a waveguide termination for use at a wavelength of 3.2 cm with an airborne radar. The instrument measured radar operated at a wavelength of 3.2 cm and transmitted either horizontally or vertically polarized signals, while simultaneously receiving both horizontally and vertically polarized signals.

First, the APS-3 antenna was terminated with a matched waveguide load and the backscatter patterns for horizontally polarized and vertically polarized echos obtained for both horizontal polarized transmissions and, second, for vertical polarized transmissions. Then an adjustable short-circuit termination was placed at the end of the waveguide output port. The position of the waveguide termination greatly influenced the backscatter signal from the APS-3 antenna when the polarization of the interrogating measurement radar matched the operational horizontal polarization of the antenna. Tuning of the adjustable short circuit produced returns both larger and smaller than those returns recorded when the matched load was connected to the output port. The waveguide termination had very little effect on the backscatter when the interrogating transmission was vertical polarization (cross pol). In addition to the APS-3, there were several other antennas and pyramidal horns (both in-band and out-of-band) investigated. The W shape in the center of the main beam is typical of backscatter patterns when the waveguide is terminated in

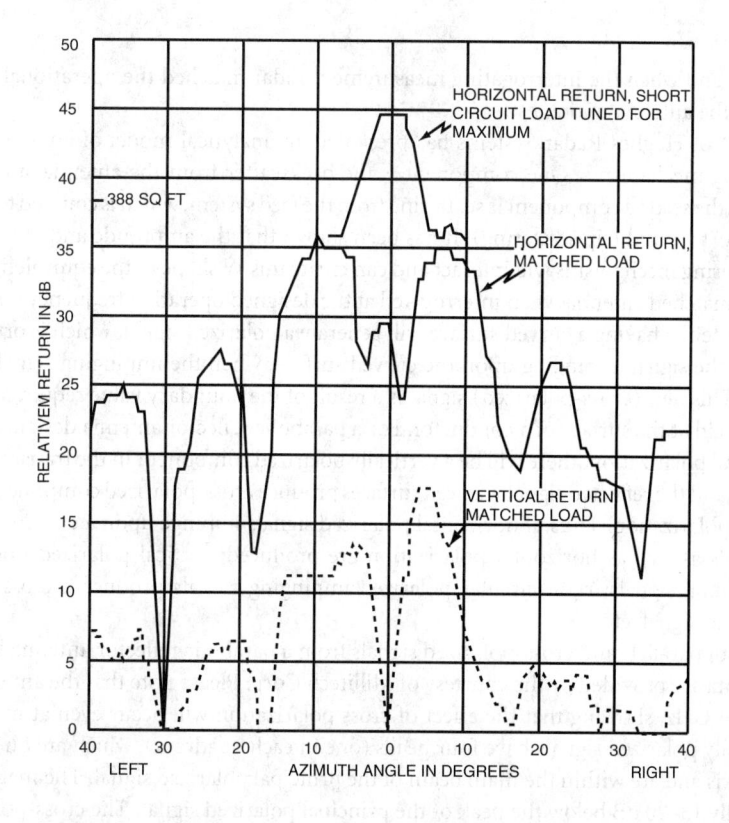

FIGURE 17.55   Backscatter from AN/APS-3 antenna, parallel polarization.

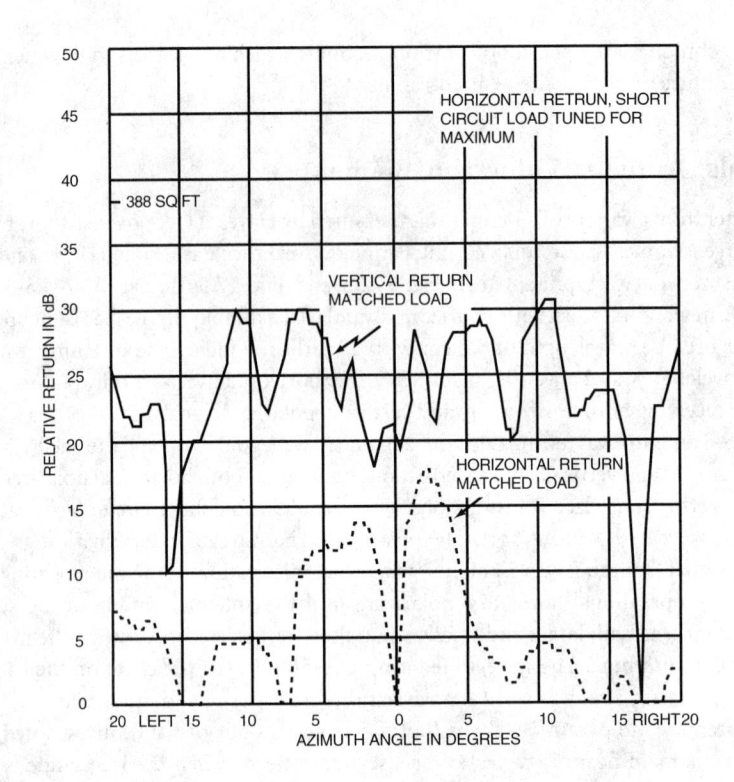

FIGURE 17.56 Backscatter from AN/APS-3 antenna, cross polarization.

a matched load and when the interrogating measurement radar matched the operational frequency and polarization of the antenna under test.

Butler (1981) of Hughes Radar Systems has presented an analytical model of an antenna consisting of two scattering mechanisms. One component is the backscatter from the antenna surfaces (one-way scattering), and the second component is scattering from the feed system, which is coupled to the waveguide input/output port (round trip scattering). It has been shown that the amplitude and phase relationships of the two scattering mechanisms will interact and can create this W shape in the equivalent main beam of an impedance-matched antenna when interrogated at the designed operating frequency and polarization.

A reflector antenna having a curved surface will generate a polarized signal, which is orthogonal to the polarization of the signal impinging upon the curved surface, when the impinging signal is not parallel to the surface. This new (cross-polarized) signal is a result of the boundary value requirement to satisfy a zero electrical field at the surface of a conductor. For a parabolic reflector antenna designed for operation with a horizontal polarization, there will be a vertically polarized component in the transmitted signal, the received signal, and the reflected signal. Curved surfaces produce cross-polarized components regardless of frequency and polarization. Thus, if horizontal polarized signals impinge upon a curved surface, vertical polarized signals as well as horizontal polarization are produced; vertical polarized impinging signals produce horizontal signals; right-circular polarized impinging signals produce a cross-polarized left-circular polarized signal, etc.

An example of parallel- and cross-polarized signals from a parabolic reflector antenna is shown in Fig. 17.57. These data are provided by the courtesy of Millitech Corp. Please note that the antenna is designed to operate at 94 GHz, showing that the effect of cross polarization will occur even at millimeter waves. The orthogonally polarized signals have four peaks (one in each quadrant), which are on the 45° lines to the principal axis and are within the main beam of the principal polarized signal. The magnitude of these peaks is typically 13–20 dB below the peak of the principal polarized signal. The cross-polarized signal is

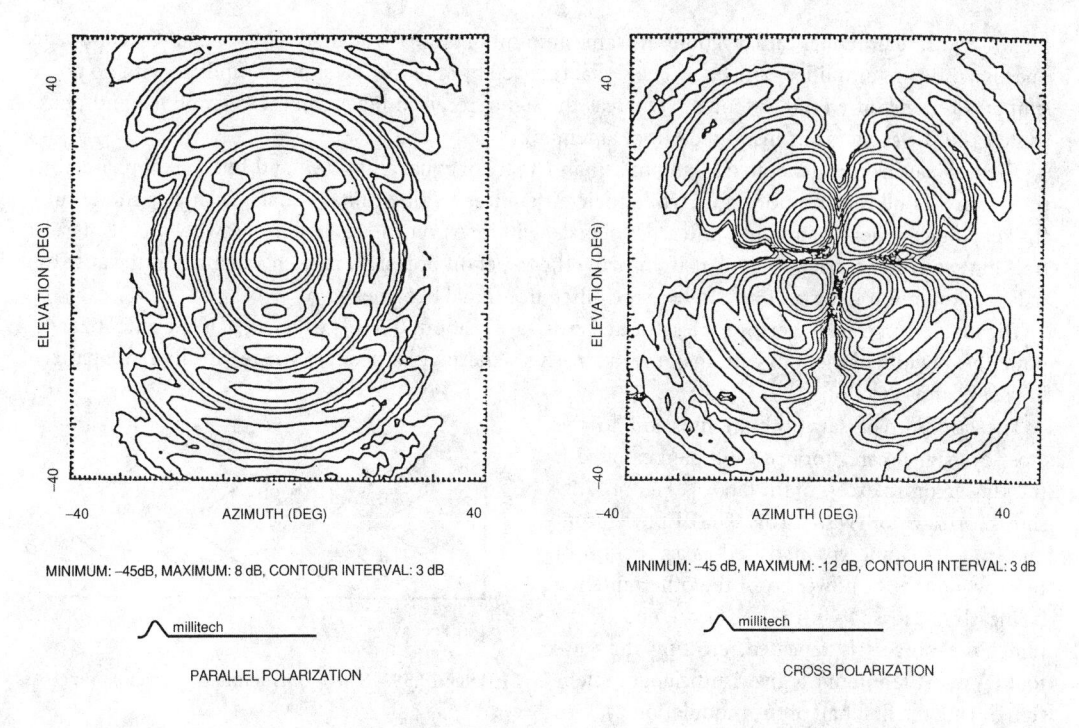

MINIMUM: –45dB, MAXIMUM: 8 dB, CONTOUR INTERVAL: 3 dB   MINIMUM: –45 dB, MAXIMUM: –12 dB, CONTOUR INTERVAL: 3 dB

PARALLEL POLARIZATION   CROSS POLARIZATION

FIGURE 17.57   Antenna radiation patterns for parabola, conical feed, 94 GHz.

theoretically zero on the two principal plane axes, so when perfect track lockup is made, cross polarization will have no effect.

Injecting a cross-polarized signal into a system will have the same effect as entering a monopulse tracker. That is to say, the receiving system cannot detect the difference between parallel and cross-polarized signals and yet the cross-polarized pattern has nulls on the major axes and peaks off the boresight. When the cross-polarized signal dominates, the track should go to an angle at about the 10-dB level of the normal sum main beam.

Cross-eye is a countermeasure technique employed against conical scan and monopulse type angle tracking radars. The basic idea is to intercept the enemy propagated signal, phase shift the signal, and then retransmit the signal to make it appear that the signal represents a reflected return emanating from an object located at an angular position different from the true location. In generating the retransmitted signal, there are two basic approaches to signal level configuration: saturated levels and linear signal levels. A typical implementation of a four antenna cross-eye system is shown in Fig. 17.58. Let us assume that

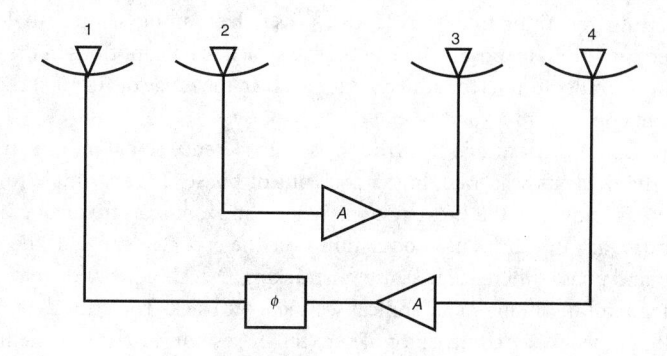

FIGURE 17.58   Cross-eye system block diagram.

the two transmit antennas radiate about the same amount of power (a slight difference could occur due to line loss, different amplifier gain, etc.). The signal received by antenna 2 is simply amplified with no phase shift and retransmitted out antenna 3, whereas the signal received by antenna 4 is amplified and phase shifted 180° before it is retransmitted out of antenna 1.

The two wavefronts (generated by antennas 1 and 3) are of equal amplitude and 180° out of phase, thus their sum will be zero on boresight, while their differences will be finite. Radars, such as a monopulse tracker, which generate sum and difference signals, will interpret the results of the two wave fronts as a tracking error. The track point will drift, moving the target off boresight in such a manner as to cause the radar lockup point to be off target, where the difference signal becomes zero.

The maximum tracking error for a saturated cross-eye is about $0.5\theta_{3\,dB}$ where $\theta_{3\,dB}$ is the antenna half-power beamwidth of the sum pattern of the radar. This is the angular position where the difference pattern has peaks and zero slope.

The addition of sawtooth modulation to the cross-eye signal can produce multiple beamwidth tracking errors in excess of the $0.5\theta_{3dB}$. Figure 17.59 shows two sawtooth waveforms. The RF power from one antenna is linearly increased as a function of time, while the RF power from the other antenna is linearly decreased with time. At the end of the ramp, the process is repeated, creating the saw-tooth. When the antenna is given sufficient momen-tum during the first half of the modulation cycle to enable the antenna to traverse the singularity at

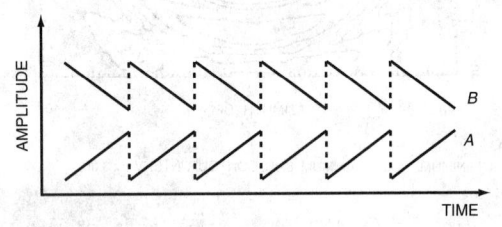

FIGURE 17.59 Saw-tooth modulation diagram for cross-eye.

$0.5\theta_{3dB}$, tracking pulloff will occur. Successive periods of modulation can cause the monopulse tracker to lock onto successive lobes of the sum pattern.

In linear cross-eye, the amplifiers are never allowed to go into saturation. Linear cross-eye differs from saturated cross-eye in that the sum pattern is zero for all pointing angles, and the difference pattern is never zero. No stable track points exist with linear cross-eye and large tracking errors can occur, resulting in large angular breaklocks.

The two most common types of angle tracking used today are sequential lobing and simultaneous lobing.

A common implementation of sequential lobing is to rotate a single antenna lobe about the boresight axis in a cone; the signal amplitude will be modulated at the conical scan frequency. The direction off the boresight is determined by comparing the phase of the scan modulation with an internal reference signal. As the source is brought on to axis, the modulation will go to zero. The source of the RF signal can be either an active source such as a beacon or a communication transmitter, or the target scattering, as in a radar application.

In a typical radar, the conical scan rate is chosen early in the system design. The rate should be slow relative to the radar pulse repetition frequency (PRF), so as to have ample samples to determine the error amplitude. However, the scan rate must be high enough to track source motions in the expected dynamic engagement. The list on tradeoffs puts the scan rates typically in the range of 10–1000 Hz, where the weight and size of the antenna play an important role.

The most effective countermeasure is to introduce an interference signal into the tracker that has the same frequency as the tracker's scan rate, but is 180° out of phase. This is simple to accomplish when the victim's scan rate is known to the ECM system. When the exact scan frequency is not known, then the ECM system must vary the jamming modulation over the expected range of the victim's operation with enough power and phase difference to cause mistracking. A technique known as swept square wave (SSW) has been successful in pulling radar conical scanners off track. The critical item is to keep within the audio bandwidth of the tracker control loop. There can be several variables in attempting to penetrate the tracking loop controls (Fig. 17.60). The sweep frequency may be linear, nonlinear, or triangular, and the sweep time may have a varying time base and dwell times.

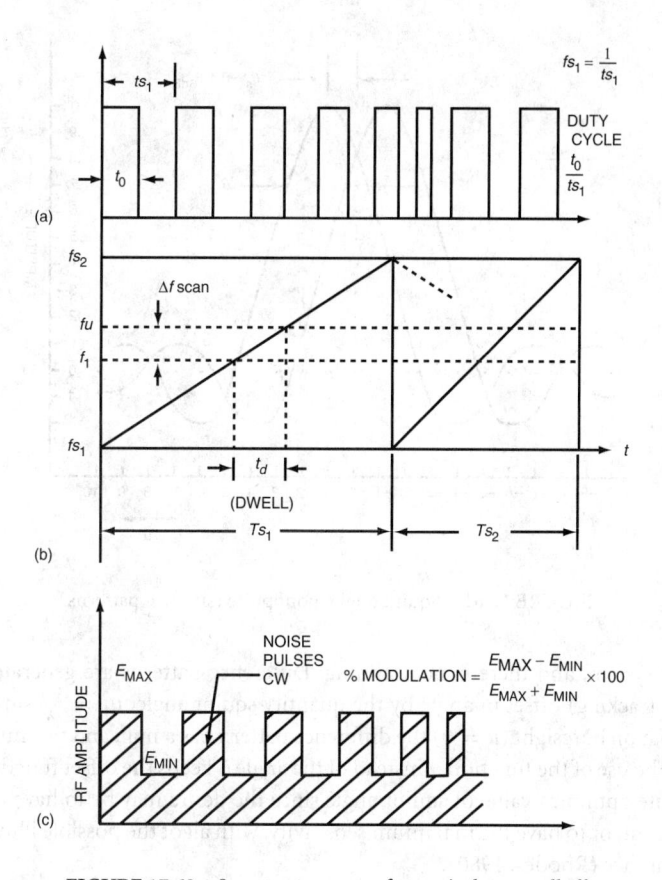

FIGURE 17.60 Swept square wave for conical scan pulloff.

For radar trackers that do not scan the transmitted beam, but instead scan a receiving antenna, the scanning rate is not usually known. It can be assumed with reasonable accurate results that the transmitted beam will be close to boresight of the COSRO tracker, so that a repeated RF signal directed back toward the illuminating beam will intercept the COSRO beam. This energy transmitted toward the victim radar should be at the same RF carrier frequency and amplitude modulated with sufficient noise spectrum so as to cover the audio bandwidth of the tracking control system. The use of pulsed and CW TWTs as jamming noise power sources is common.

In a radar system employing track while scan (TWS), the tracking function is performed in a signal processor/computer by keeping track of the reported position in range, azimuth, and elevation on a scan-to-scan basis. The antenna must move reasonably fast over the field-of-view in order to report target locations in a sufficiently timely basis. The antenna does not produce continuous tracking, and thus does not remain fixed in azimuth or elevation for any long period of time. The countermeasure to a TWS radar is to move fast, turn fast, dispense chaff, noise jam, or some cover technique.

Simultaneous lobing, or monopulse techniques, reduce the number of pulses required to measure the angular error of a target in two dimensions, elevation and azimuth. A minimum of three pulses is required in monopulse tracking, whereas more than four pulses are required for conical scan trackers. In the amplitude comparison monopulse, each pulse is measured in a sum and difference channel to determine the magnitude of the error off boresight, and then the signals are combined in a phase sensitive detector to determine direction. In a four horn antenna system, a sum pattern is generated for transmission and reception to determine range and direction. Two difference patterns, elevation and azimuth, are generated on reception for refining the error off the boresight. The receiving difference patterns produce a passive

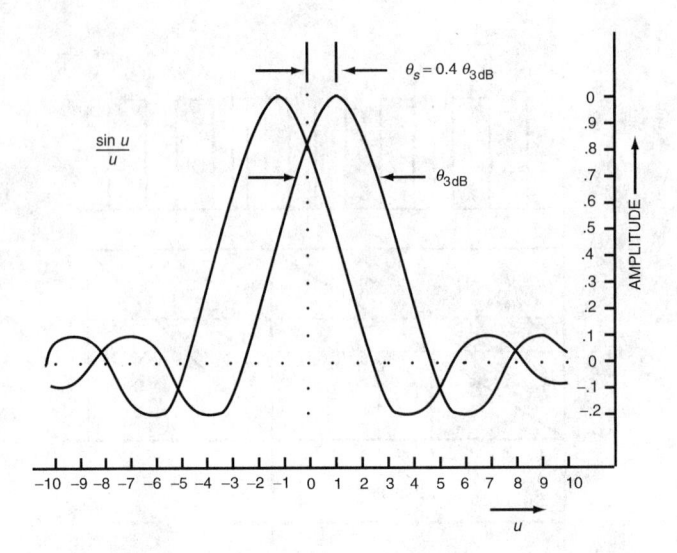

FIGURE 17.61   Squint angle monopulse (sin u)/u patterns.

receiving tracking system, and there is no scanning. Difference patterns are generated by two patterns (for each plane of tracking) offset in angle by the quantity squint angle, $u_s = \frac{\pi d}{\lambda} \sin \theta_s$, and 180° phase difference, such that on boresight ($u = 0$) the difference pattern has a null, and this null is in the center of the sum pattern. The use of the function u instead of the angle θ keeps the offset functions mirror images. There is no absolute optimum value of squint angle since the desire may be to have the best linearity of the track angle output, or to have the maximum sensitivity, with all of the possible illumination functions on the reflecting surface (Rhodes, 1980).

The monopulse tracker is almost completely immune to amplitude modulations.

The two beams, which form the difference pattern (one beam +, and one beam −), cross each other at less than 3 dB below their peaks for maximum error slope-sum pattern product and to give increased angle tracking accuracy. This can be seen in Fig. 17.61 in Skolnik (1962) and in Fig. 17.62 in Rhodes (1980), where a comparison is shown between squint angle and the two difference beam magnitude crossover.

The difference pattern has a minimum on boresight, which coincides with the peak of the sum pattern on boresight. The difference pattern is typically 35–40 dB below the sum pattern. These pattern values will typically improve the identity of the target location by 25:1 over a simple track-the-target in the main beam technique. Some ECM techniques

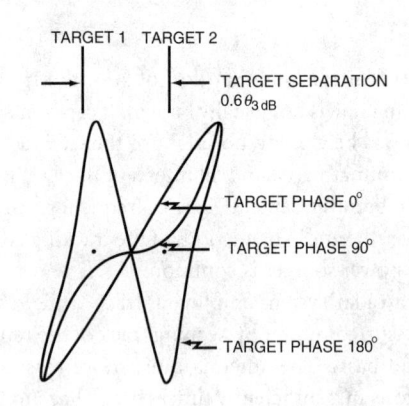

FIGURE 17.62   Two targets under track by monopulse antenna.

are to fill the null of the difference channel so that the tracking accuracy is reduced from the 25:1 expected. If two transmitters are separated by the 3 dB width of the sum pattern (about the same as the peak of the two difference patterns), then a standard ECM technique of blinking of the two transmitters works very well.

Squint angle is a parameter within the control of the victim antenna designer; it may be chosen so as to optimize some desired function such as the best tracking linearity angle output, or maximum sensitivity on boresight. With the many options available to the antenna designer of the victim radar, the ECM designer will have to make some general assumptions. The assumption is that the squint angle off boresight is 0.4

of the half-power antenna beamwidth, given by setting the two difference patterns at a crossover of 2 dB, as shown in Fig. 17.61. (A squint of 0.5 the half-power angle is given by a 3 dB crossover). Two signals of equal amplitude (these may be radar backscatters or electronic sources, it makes no difference) and separated by 0.6 of the sum half-power beamwidth are introduced into the monopulse tracker. A typical monopulse tracker victim radar will seek the positive-going zero crossing of the error signal as a track point. In an example considering these system parameters, when the phase difference of these two signals is 90° or less, the track center position is between the two signals. When the phase difference between the two signals is between 90 and 180° the positive going error signal is both to the left and the right outside the signal sources and the radar track will be outside of the sources. This can be seen in Fig. 17.62. (Dr. E. Rhodes, private communication, Georgia Institute of Technology). Neither target is sought, and wandering of the boresight center is expected. Complex targets, such as multiscattering facet aircraft, can produce 180° differences by slight changes in angle of attack. At X-band, 180° is only 1.5 cm. Thus it is shown that in a buddy system, a monopulse radar tracker can be made to track off of the two sources by slight changes in flight course.

17.5.4   Radar Counter-Countermeasures

The concept of counter-countermeasures is to prevent, or at least decrease, the detection, location, and tracking processes so that the mission can be accomplished before the victim group can defeat the attackers. Reducing the radar scattering cross section has produced real time advantages in combat. This technique, commonly referred to as stealth, has become a part of all advanced systems. The concept is to reduce the cross section of the weapon's delivery system to such a low value that the detection range is so short that the defense system does not have time to react before the mission is completed.

The basic concepts of stealth is to make the object to appear to have a characteristic impedance that is the same as free space, or to absorb all energy that impinges upon our platform, or to reflect the energy in some obscure direction. This is not a new concept. The Germans were using paints and shaping techniques on their U-boats during WW II to reduce 3-GHz radar reflections. A semiflexible material was developed at the Massachusetts Institute of Technology Radiation Laboratories in the U.S. during WW II and it was called HARP. The material consisted of copper, aluminum, and ferromagnetic material in a nonconduction rubber binder and made into cloth, sometimes with conductive backing. HARP-X was used at the X-band. Several techniques are appropriate in this process. First, the impinging signal can be reflected off axis so that no signals are returned in the direction of the transmitter/receiver, or multisignals are generated from multisurfaces, which will be out of phase and appear to cancel in the direction of the transmitter/receiver. These techniques are accomplished by physical configuration and shaping of our platform. The second approach is to bend, absorb, or alter the propagation of the impinging wave in the material covering our platform. Radar absorbing material (RAM) techniques are accomplished by material changes: typically, changing the complex dielectric constant, the dielectric-loss tangent, the magnetic permeability, and/or the magnetic-loss tangent. The Salisbury screen (Salisbury, 1952), single layer of dielectric material backed by a reflecting surface, is often used as a model.

These devices create a resonant impedance to reduce reflections at multiple frequencies corresponding to multiple quarter-wavelengths of the dielectric material. Multilayers, or Jaumann absorbers, are generally more frequency broadband absorbers and may even be graded multidielectric or multimagnetic materials and thus increase the effectiveness at angles off of normal. Knott has shown that the thickness of the absorber can be reduced and the bandwidth considerably expanded if the sheets are allowed to have capacitive reactance in addition to pure resistance (Knott and Lunden, 1995).

The physics of RAM is well described by the Fresnel reflection and transmission equations. The material composite is driven by application, and includes concerns such as weight, weather durability, flexibility, stability, life cycle, heat resistance, and cost.

The application of electronic coding of the transmitter signal and new techniques in processing the received signals have lead to a system concept referred to as low probability of intercept (LPI). There are many subtechnologies under this great umbrella, such as spread spectrum and cover pulses on transmission, pulse compression, frequency-coded and phase-coded chirp, frequency hopping, and variable pulse repetition frequencies. Dave Barton (1975, 1977) has collected a number of papers describing the state of unclassified technology before 1977 in two ready references. A number of publications, primarily U.S. Naval Research Laboratory reports and IEEE papers on coding techniques and codes themselves have been collected by Lewis, Kretschmer, and Shelton (1986).

In developing an LPI system concept, the first item will be to reduce the peak power on transmission so that the standoff detection range is greatly reduced. To have the same amount of energy on target, the transmitter pulse length must be increased. To have the same resolution for target detection, identification, and tracking, the new long pulse can be subdivided on transmission and coherently compressed to construct a much shorter pulse on reception. The compressed pulse produces an effective short pulse for greater range resolution and a greater peak signal, which can exceed noise, clutter, or other interference for improved target detection.

One of the earliest pulse compression techniques was developed at the Bell Telephone Laboratories (Klauder et al., 1960; Rhodes, 1980). The technique was called CHIRP and employs frequency modulation. In a stepped modulation, each separate frequency would be transmitted for a time, which would correspond to a pulse length in a pulsed radar system. Upon reception, the separate frequencies are folded together to effectively form one pulse width containing the total energy transmitted. In a linear modulation, the frequency is continuous swept over a bandwidth in a linear manner during the transmitted pulse length. The received signal can be passed through a matched filter, a delay line inversely related to frequency, an autocorrelater, or any technique to produce an equivalent pulse length equal to the reciprocal of the total bandwidth transmitted. The amplitude of this compressed signal is equal to the square root of the time-frequency-bandwidth product of the transmitted signal. The errors are limited to system component phase and frequency stabilities and linearities over the times to transmit and to receive. When using the Fourier transforms to transform from the frequency domain to the time and thus range domain, sidelobes are generated, and these sidelobes appear as false targets in range. Sidelobe suppression techniques are the same as employed in antenna sidelobe suppression design (Fig. 17.63).

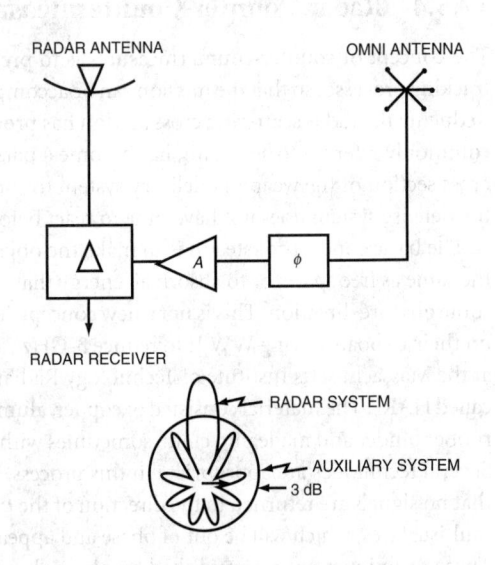

FIGURE 17.63   Sidelobe interference reduction concept.

One of the first biphase coded waveforms proposed for radar pulse compression was the Barker code. See for example, Farnett, Howard, and Stevens (1970). This code changes the phase of the transmitted signal 180° at a time span corresponding to a pulse length of a standard pulse-transmitter radar. The Barker codes are optimum in the sense that the peak magnitude of the autocorrelation function is equal to the length of the coded pulse (or the number of subpulses), and the sidelobes are less than one. When the received signal is reconstructed from the phase domain, the range resolution is the same as the pulse length (or the length of a Barker subpulse) of the standard pulse-transmitter radar.

The maximum length of a Barker code is 13, and this code has sidelobes of −22.3 dB. The phase will look like + + + + +−− + + − + − +, or −−−−− + +−− + − + −.

There are many phase codes discussed by Lewis, Kretschmer, and Shelton (1986), which are unique and thus have unique autocorrelation functions. The compression ratios can be very large and the side lobes very low. The price to be paid is equipment phase stability in polyphase code implementation, length of time to transmit, minimum range due to the long transmitted pulse, and equipment for reception signal processing. The advantages gained are high range resolution, low peak power, forcing the observer to use more power in a noise mode because your radar uses wider bandwidth spread over segments of narrow bandwidths.

Another technique found helpful in reducing interference is to change the transmitted frequency on a pulse-by-pulse basis. This frequency hopping is effective, but does require that the receiver and local oscillator be properly tuned so that the receiver is matched to the return signal for efficient operation. When Doppler or moving target indicators are employed, then the frequency hopping is done in batches. There must be enough pulses at any one frequency to permit enough resolution in the Fourier transform (FT) or the proper number of filter banks to resolve the Doppler components. If the transmitter can handle the total bandwidth and keep each frequency separate so there is no crosstalk in the FTs or in the filters, then this is very effective jamming reduction. Today's digital systems are set up for frequency groups like 32, 64, or 128 and then repeat the frequency hopping in the same frequency direction, or the reverse direction. This reduces velocity gate pulloff.

If the counter measure system has identified your PRF and thus can lock on the your radar's range gate, then there are several techniques to reduce the EW range-gate pulloff. One technique is to place a noise pulse over the standard pulse. This noise pulse is longer than the system range resolution pulse and is not symmetrical around the range pulse. Only your system should know where your own target identification pulse is in the chain of PRF pulses. Another technique is to change the PRF after a batch of pulses have been transmitted. Typically the batch length is controlled in a digital manner. Again, as in the case of frequency hopping, PRF hopping still must permit processing in range for resolution, in frequency for Doppler, and in polarization (or other identifiers) for target identification.

So far as our system is concerned, there is no difference if a jamming or interference signal enters the main beam of our system antenna or a sidelobe of our system antenna. From the jammer's point of view, there needs to be 20–30 dB more power, or 20–30 dB closer in range if the jammer is directed at a sidelobe instead of the main antenna lobe.

Suppression of sidelobe jamming can be accomplished by employing omnidirectional auxiliary antennas having gains greater than the highest sidelobe to be suppressed, and yet lower than the main beam of the antenna. The auxiliary antenna feeds a receiver identical to the receiver used by the radar. The detected output of the auxiliary receiver and the radar receiver are compared, and when the auxiliary receiver output is larger than the radar then the radar is blanked in all range cells where this occurs. The auxiliary antenna should be placed as close as possible to the phase center of the radar antenna to insure that the samples of the interference in the auxiliary antenna can be correlated with interference in the main radar antenna. This requires that the time of arrival of the signals to be compared must be less than the reciprocal of the smaller of the two bandwidths, in the radar receiver or the interference signal.

17.5.5   Chaff

Cylindrical metal rods scatter energy in the same manner as a dipole radiates energy. The same parameters affect the efficiency, beamwidth, loss, and spatial patterns. Thus, in this discussion of chaff as a scattering element, consideration will be given to rod length in terms of excitation frequency, length-to-diameter ratio, impedance or loss scattering pattern as a function of polarization, and interaction (or shielding) between adjacent elements.

The aerodynamic characteristics are determined by material, density, shape, drag coefficient, and atmospheric parameters such as temperature, pressure, and relative humidity. In addition to these parameters, the movement of a rod is strongly influenced by local winds, both horizontal and vertical components, and by wind turbulence and shear.

When all of these variable parameters are defined, the radar scattering characteristic of a collection of rods forming a cloud of chaff can be calculated. The radar parameters of interest in this presentation are the scattering cross section from the cloud, the power spectral density to define the frequency spread, Doppler frequency components, polarization preference, signal attenuation through the cloud, bandwidth, and length of time that these clouds can be expected to exist.

From antenna theory, the tuned half-wave dipole produces the maximum scattering coefficient. There are sources for reference reading; examples are Terman (1943); Kraus (1950); Thourel (1960); Jasik (1961); Eustace (1975–1991); and Van Vleck, Bloch, and Hamermesh (1947).

When the dipole is observed broadside, and with a polarization parallel to the rod, then the maximum radar cross-section is observed and has a value of

$$\sigma = 0.86\lambda^2$$

As the dipole tumbles and is observed over all possible orientations, then the average radar cross section has been calculated and found to be

$$\sigma = 0.18\lambda^2$$

Since the rod has a finite diameter, the maximum scattering cross section occurs for lengths slightly less than $\lambda/2$. A reasonable value of 0.48λ for the length will be assumed to compensate for the end effects of the rod. The first resonance is plotted as a function of the length-to-diameter ratio in the June, 1976 edition of ICH (Eustace, 1975–1991). The equations relating radar cross section to the rod length and diameter in terms of wave length are complicated. The radar cross section has been shown to be directly related to the ratio of physical length-to-radar wave length, and inversely related to the logarithm of the physical length-to-radius ratio in Van Vleck, Bloch, and Hamermesh (1947). Their article is a classic and, along with their references, is the place to begin.

The polarization of the observing radar and the alignment of the chaff material is important since horizontally aligned rods fall more slowly than vertically aligned rods. Under many applications, one end of the chaff filaments is weighted causing these rods to be vertically aligned shortly after release, thus enlarging the chaff cloud rapidly in the vertical elevation plane as well as making the radar cross section of the lower elevation portion of the cloud more susceptible to vertically polarized waves and the upper elevation section of the chaff cloud having a higher cross section to horizontally polarized waves.

Chaff elements have been constructed in all sizes and shapes. Early manufactured elements were cut from sheets of metal, were rectangular in cross section and cut to half-wave dipoles in length; some were bent in V shape, some were long to form ribbons, some were made as ropes; some were cut from copper wire and some from aluminum. In the past decade or two, glass rods coated with aluminum have become very popular, and this configuration will be assumed typical for this discussion. There has been a reasonable amount of experimental work and calculations made for dipoles constructed of both 1-mil- and 1/2-mil-diam glass rods for use as chaff. The metal coating should be at least 3 skin depths at the designed frequency. One skin depth is defined as the thickness where the signal has decreased to $1/e$ of the original value.

There have been limited amounts of work done at 35 and 95 GHz as compared to much work done at lower frequencies. Consider that the wave length at 35 GHz is only 0.857 cm and a half-wave length is only 0.1687 in. Thus, a cloud of dipoles will be alot of very short elements of material. To have a reasonable length-to-diameter ratio and reduce end effects, as well as to provide good electrical bandwidth, it will be assumed that 1/2-mil-diam dipoles would be acceptable. From these dimensions, manufacturing of large bundles of these dipoles is not a trivial matter. Some areas of concern are: (1) the plating of the small diameter glass will not be simple and most likely will produce considerable variations in weight and L/D, with results in varying fall rates and changes in bandwidth of the scattering signal; (2) to consistently cut bundles of pieces of elements this short in length requires considerable ability; and (3) it is required that each dipole be independent and not mashed with several other dipoles or the result is reduction in radar cross section and bird nesting. These effects result in lost efficiency.

There are two types of antennas, which are discussed in the literature on antennas: the thin-wire antenna and the cylindrical antenna. The thin-wire antenna model, where assumed length-to-diameter is greater than 5000, is used to predict general scattering patterns and characteristic impedance as a function of rod length and frequency. The cylindrical antenna will show the effect of wire diameter, which in turn reveals changes in the reactive portion of the terminal impedance and, thus, changes in gain, propagation and scattering diagrams, filling in of pattern nulls, reduction of sidelobes, and wider frequency bandwidth.

It has been shown that a thin-wire half-wave dipole, excited at the center, and at resonance, has an impedance in ohms of

$$Z = 73.2 + j42.5$$

When the length is not half-wave, and the diameter is not negligible, then to a first-order approximation

$$Z = 73.2 + j42.5 \pm j120 \left[\ln \frac{L}{D} - 0.65 \right] \cot \frac{2\pi L}{\lambda}$$

As is the response in all tuned circuits, the band-pass characteristics are primarily determined by the reactive elements as the frequency moves away from the resonant frequency. In low Q circuits, the real part of the impedance also has a notable effect on the total impedance and, thus, also on the bandpass characteristics. To a first-order effect, the real part of the half-wave dipole does not change drastically (less than 10%) from 73.2 Ω; however, the reactive part varies as ($\ln L/D$). As shown by Van Vleck, Bloch, and Hamermash (1947), the radar cross section off of resonance is expressed in reciprocal terms of ln(rod length/rod radius). The effective half-power bandwidth of dipole scatters is, thus, usually referenced to the ratio of dipole length to dipole

PERCENT BANDWIDTH	LENGTH/DIAMETER
12	3000
13	1650
14	1050
15	740
16	520
17	400
18	320
19	260
20	220
21	180
22	150
23	140
24	120
25	110

FIGURE 17.64 Backscatter bandwidth response for chaff.

diameter as shown in Fig. 17.64. As it is desirable to have some bandwidth in television antennas, it is also desirable to have some bandwidth in the radar scattering from chaff, which leads to a measurable diameter of the rods or width of the ribbon strips used for chaff.

Consider an application at X-band in which it is desirable to screen against a wide-band pulse compression radar having a bandwidth of 1.5 GHz. The rods will be cut to 0.48λ. From Fig. 17.64, the value of $L/D = 750$,

$$L/D = 750 \doteq 0.48\lambda/D,$$

where $D = 0.48 \times 3\,\text{cm}/750 = 0.00192\,\text{cm} = 0.000756\,\text{in} = 0.756\,\text{mil}$.

Consider now the generation of a chaff cloud having a radar backscatter cross section per unit volume of -60-dB m^2/m^3. This value cross section is comparable to the radar return from 10 mm/h of rain observed at X-band (Currie, Hayes, and Trebits, 1992). These small aluminum-coated glass cylinders fall very slowly in still air and have been known to stay aloft for hours. The movement is so strongly influenced by local wind conditions that meteorologists in several countries (including the former U.S.S.R., Germany, and the U.S.) have used chaff clouds to determine wind profiles (Nathenson, 1969). Local conditions give extreme wind variations even when the local wind structure is free from perturbations such as squalls, fronts, thunderstorms, etc. Some wind profile models include terrain surface material such as water, grass, dirt, and surface roughness, time of day, time of year, northern hemisphere, and other parameters shown to affect a particular area. At altitudes above 500 ft, the profile of wind has been shown to vary mostly in a horizontal manner, with little vertical constantly defined variations (AFCRC, 1960). The strong jet streams at around 35,000–40,000 ft define the topmost layer typical of zero water vapor levels, zero oxygen levels, definable temperature profiles, wind stability, and pressure profiles. This being the case, it behooves the EW engineer to apply local meteorological records whenever possible.

One-mil-diam aluminum-coated, glass-filament chaff, for example, will adjust to sharp wind gust of up to 50 ft/s within 15 ms (Coleman, 1976–1977). The vertical fall rate of chaff is proportional to the weight and air resistance; thus, the important chaff parameters are diameter, length, and coating. A typical approach is to use a mixture of dipole diameter sizes so that a vertical cloud is generated from different fall rates. For the example above, the diameter mixture could be between 0.5- and 1.0-mil coated glass.

Consider a typical wind condition of a 5-kn horizontal wind, and a wind profile of 1/5 power relationship at a height below 500 ft. Coleman (1976–1977) has shown that a cloud mixture of 0.5–1.0 mil chaff would stay aloft for about 20 min when dispensed at 500 ft. The chaff cloud will smear in both elevation and in horizontal dimensions. The volume of the ellipsoid generated by the horizontal winds and the vertical fall rate is given by

$$\text{volume} = V = \frac{4}{3}\pi abc$$

In this example $a = 100$ ft and $b = c = 300$ ft. The resulting volume is $1{,}068{,}340$ m^3.

The radar cross section of such a cloud is given by

$$\sigma_T = V\sigma_V$$
$$= (1.068 \times 10^6)(10^{-6})$$
$$= 1.068 \text{ m}^2$$

This is not a strong target. But we know that 10 mm/h of rain is not a strong target at X-band. The big problem with rain is a deep volume producing a large amount of attenuation. You notice here that the chaff cloud is only 600 ft across; this would not be a typical rain cloud. Perhaps the comparison should be for the chaff cloud to cover (hide) a 30-dB cross section vehicle.

The number of dipoles in a cloud of chaff is given by

$$N = \frac{\sigma_T}{\sigma_d}$$

where σ_T is the total cross section, and σ_d the cross section of each dipole, by substitution

$$N = \frac{1000}{0.18(3/100)^2}, \frac{\text{m}^2}{\text{m}^2}$$
$$= 6.17 \times 10^6$$

The density of dipoles is given by

$$\frac{N}{V} = \frac{6.17 \times 10^6}{1.068 \times 10^6}$$
$$= 5.78 \text{ dipoles/m}^3$$

This is not a dense chaff cloud, and thus you can see the value of such a cloud to hide a good size target. Of course, there is short fall; the cloud may last only 20 min and will drift several thousand feet in this time. This requires that there be proper area and time planning for chaff use. If the volume to be covered is larger than the $100 \times 300 \times 300$ ft, then, several clouds need to be dispersed.

In the International Countermeasures Handbook (EW Comm., 1975), the author has presented a graph to be used to determine the signal attenuation produced by a chaff cloud. The graph relates normalized two-way attenuation per wave length squared to the number of half-way dipoles per cubic meter, as a function of the dipole orientation K relative to the radar polarization. When the dipoles are uniformly distributed and randomly oriented, K can be as low as 0.115; when the dipoles are aligned with the E-field, $K = 1.0$ and the loss is 10 dB higher; when the dipoles are almost crosspolarized to the radar, $K = 0.0115$, the attenuation will be 10 dB less than the random orientation. For $K = 0.115$, the attenuation relationship is given by

$$L/\lambda^2 = N/\text{m}^3$$

in units of decibel per cubic meter = number/cubic meter. Typical ranges of cloud density range from 10^{-7} to 10^{+2} dipoles per cubic meter. For the example given, a cloud density of 5.78 dipoles/m^3 cube will give a value of $L/\lambda^2 = 5.78$. For a two way loss per meter of chaff cloud, $L = 0.0052$ dB/m at $\lambda = 0.03$ m, giving 0.5 dB of attenuation in a 100-m cloud.

Expendable Jammers

Expendable jammers are a class of electronic devices that are designed to be used only once. These devices are typically released from chaff/flare dispensers much as chaff is deployed. The electronic package can be deployed as a free-falling object, or released with a parachute for slow fall rates. Such packages have been used as screening devices and as simulated false targets. The RF energy can be continuous, pulsed, or modulated to respond as a repeater or a false radar. Expendable jammers have been configured as spot noise jammers and have been deployed as barrage noise jammers, depending upon the mission and victim radars.

Defining Terms

P = power, watts
k = Boltzman constant = 1.38×10^{-23} joules/K
T = temperature = absolute temperature, K
S = energy = power x transmitted pulse length, joules
B = bandwidth, Hz
f = frequency, Hz
Hz = 1 cycle per second
dB = 10 times logarithm of power ratio
Z = impedance, ohms

References

AFCRC. 1960. *Geophysics Handbook*, Chap. 5. Air Force Cambridge Research Center, MA.

Barton, D.K. 1975. Pulse compression. *RADAR*, Vol. 3. ARTECH House, Norwood, MA.

Barton, D.K. 1977. Frequency agility and diversity. *RADAR*, Vol. 6. ARTECH House Norwood, MA.

Butler, W.F. 1981. Antenna backscatter with applications to surgical countermeasures. Contract N00014-81-C-2313, Hughes Aircraft, Sept., El Segundo, CA.

Coleman, E.R. 1976–1977. *International Countermeasures Handbook*. EW Communications, Inc., Palo Alto, CA.

Currie, Hayes, and Trebets. 1992. *Millimeter-Wave Radar Clutter*. ARTECH House, Norwood, MA.

Eustace, H.F., ed. 1975–1991. *The Countermeasures Handbook*.

EW Comm. 1975. *International Countermeasures Handbook*. p. 227. June, EW Communications, Inc., Palo Alto, CA.

Farnett, Howard, and Stevens. 1970. Pulse-compression radar. In *Radar Handbook*, ed. Skolnik, Chap. 20. McGraw-Hill, New York.

Hayes, R.D. and Eaves, J.L. 1966. Study of polarization techniques for target exhancement. Georgia Institute of Technology Project A-871, March, AD 373522.

Jasik, H. 1961. *Antenna Engineering Handbook*. McGraw-Hill, New York.

Klauder, Prince, Darlington, and Albersheim. 1960. *Bell Sys. Tel. J.* 39(4).

Knott, E.F. and Lunden, C.D. 1995. The two-sheet capacitive Jaumann absorber. *IEEE Ant. and Prop. Trans* 43(11).

Kraus, J.D. 1950. *Antennas*. McGraw-Hill, New York.

Lewis, B.L., Kretschmer, F.F., Jr., and Shelton, W.W. 1986. *Aspects of Radar Signal Processing*. ARTECH House, Norwood, MA.

Nathenson. 1969. *Radar Design Principles*. McGraw-Hill, New York.

Rhodes, D.R. 1980. *Introduction to Monopulse*. ARTECH House, Norwood, MA.

Salisbury, W.W. 1952. Absorbent body for electromagnetic waves. U.S. Patent 2,599,944, June 10.

Skolnik, M.I. 1962. *Introduction to Radar Systems*, 2nd ed. McGraw-Hill, New York.

Terman, F.E. 1943. *Radio Engineers' Handbook*. McGraw-Hill, New York.

Thourel, L. 1960. *The Antenna*. Chapman and Hall, New York, NY.

Timberlake, T. 1986. *The International Countermeasure Handbook*, 11th ed. EW Communications, Inc., Palo Alto, CA.

Van Vleck, Bloch, and Hamermesh. 1947. The theory of radar reflection from wires or thin metallic strips. *J. App. Phys.* 18(March).

Further Information

Applied ECM, Vols. 1 and 2, 1978, EW Engineering, Inc., Dunn Loring, VA.

International Countermeasures Handbook, 11th ed. 1986, EW Communications, Inc.

Introduction to Radar Systems, 2nd. ed. 1962, McGraw-Hill.

Modern Radar Systems Analysis, 1988, ARTECH House.

18

Control and
Instrumentation
Technology

Cecil Harrison
University of Southern Mississippi

Edward McConnell
National Instruments

David Jernigan
National Instruments

Thomas F. Edgar
University of Texas

Juergen Hahn
Texas A&M University

John E. McInroy
University of Wyoming

Gibson Morris, Jr.
University of Southern Alabama

18.1 Measurement Techniques: Sensors and Transducers

Cecil Harrison

18.1.1 Introduction

An automatic control system is said to be *error actuated* because the **forward path** components (*comparator, controller, actuator,* and *plant* or *process*) respond to the error signal (Fig. 18.1). The error signal is developed by comparing the measured value of the **controlled output** to some **reference input**, and so the accuracy and precision of the controlled output are largely dependent on the accuracy and precision with which the controlled output is measured. It follows then that measurement of the controlled output, accomplished by a system component called the **transducer**, is arguably the single most important function in an automatic control system.

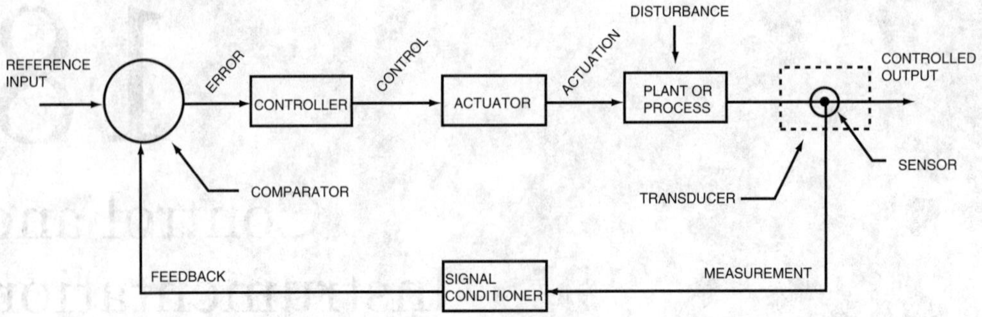

FIGURE 18.1   Functional block diagram of a canonical (standard) automatic control system.

A transducer senses the magnitude or intensity of the controlled output and produces a proportional signal in an energy form suitable for transmission along the feedback path to the comparator. (The term proportional is used loosely here because the output of the transducer may not always be directly proportional to the controlled output; that is, the transducer may not be a linear component. In linear systems, if the output of the transducer (the measurement) is not linear, it is linearized by the signal conditioner.) The element of the transducer which senses the controlled output is called the *sensor;* the remaining elements of a transducer serve to convert the sensor output to the energy form required by the **feedback path**. Possible configurations of the feedback path include

- Mechanical linkage
- Fluid power (pneumatic or hydraulic)
- Electrical, including optical coupling, RF propagation, magnetic coupling, or acoustic propagation

Electrical signals suitable for representing measurement results include

- DC voltage or current amplitude
- AC voltage or current amplitude, frequency, or phase (CW modulated)
- Voltage or current pulses (digital)

In some cases, representation may change (e.g., from a DC amplitude to digital pulses) along the feedback path.

The remainder of this discussion pertains to a large number of automatic control systems in which the feedback signal is electrical and the feedback path consists of wire or cable connections between the feedback path components. The transducers considered hereafter sense the controlled output and produce an electrical signal representative of the magnitude, intensity, or direction of the controlled output.

The **signal conditioner** accepts the electrical output of the transducer and transmits the signal to the comparator in a form compatible with the reference input. The functions of the signal conditioner include

- Amplification/attenuation (scaling)
- Buffering
- Isolation
- Digitizing
- Sampling
- Filtering
- Noise elimination
- Impedance matching
- Linearization
- Wave shaping
- Span and reference shifting
- Phase shifting
- Mathematical manipulation (e.g., differentiation, division, integration, multiplication, root finding, squaring, subtraction, or summation)
- Signal conversion (e.g., DC–AC, AC–DC, frequency–voltage, voltage–frequency, digital–analog, analog–digital, etc.)

In cases in which part or all of the required signal conditioning is accomplished within the transducer, the transducer output may be connected directly to the comparator. (Connection of the transducer output directly to the comparator should not be confused with unity feedback. Unity feedback occurs when the cascaded components of the feedback path (transducer and signal conditioner) have a combined transfer function equal to 1 (unity).) In a digital control system, many of the signal conditioning functions listed here can also be accomplished by software.

Transducers are usually considered in two groups:

- **Motion** and **force transducers**, which are mainly associated with **servomechanisms**
- **Process transducers**, which are mainly associated with **process control** systems

As will be seen, most process transducers incorporate some sort of motion transducer.

18.1.2 Motion and Force Transducers

This section discusses those transducers used in systems that control motion (i.e., *displacement, velocity,* and *acceleration*). Force is closely associated with motion, because motion is the result of unbalanced forces, and so force transducers are discussed concurrently. The discussion is limited to those transducers that measure *rectilinear* motion (straight-line motion within a stationary frame of reference) or *angular* motion (circular motion about a fixed axis). Rectilinear motion is sometimes called *linear* motion, but this leads to confusion in situations where the motion, though along a straight line, really represents a mathematically nonlinear response to input forces. Angular motion is also called *rotation* or *rotary motion* without ambiguity.

The primary theoretical basis for motion transducers is found in rigid-body mechanics. From the equations of motion for rigid-bodies (Table 18.1), it is clear that if any one of displacement, velocity, or

TABLE 18.1 Equations of Motion

Continuous	Discrete $\Delta t = t_i - t_{i-1}$

Rectilinear displacement:

$$x(t) = \int v(t)dt \qquad\qquad x_i = x_{i-1} + \frac{v_i + v_{i-1}}{2} \cdot (\Delta t)$$

$$= \int\int a(t)dt \qquad\qquad = 2x_{i-1} - x_{i-2} + \frac{a_i + 2a_{i-1} + a_{i-2}}{2} \cdot \frac{(\Delta t)}{2}$$

Angular displacement:

$$\theta(t) = \int \omega(t)dt \qquad\qquad \theta_i = \theta_{i-1} + \frac{\omega_i + \omega_{i-1}}{2} \cdot (\Delta t)$$

$$= \int\int \alpha(t)dt \qquad\qquad = 2\theta_{i-1} - \theta_{i-2} + \frac{\alpha_i + 2\alpha_{i-1} + \alpha_{i-2}}{2} \cdot \frac{(\Delta t)}{2}$$

Rectilinear velocity:

$$v(t) = \frac{d}{dt}x(t) \qquad\qquad v_i = \frac{x_i - x_{i-1}}{\Delta t}$$

$$= \int a(t)dt \qquad\qquad = v_{i-1} + \frac{a_i + a_{i-1}}{2} \cdot (\Delta t)$$

Angular velocity:

$$\omega(t) = \frac{d}{dt}\theta(t) \qquad\qquad \omega_i = \frac{\theta_i - \theta_{i-1}}{\Delta t}$$

$$= \int \alpha(t)dt \qquad\qquad = \omega_{i-1} + \frac{\alpha_i + \alpha_{i-1}}{2} \cdot (\Delta t)$$

Rectilinear acceleration:

$$a(t) = \frac{d}{dt}v(t) \qquad\qquad a_i = \frac{v_i - v_{i-1}}{\Delta t}$$

$$= \frac{d^2}{dt^2}x(t) \qquad\qquad = \frac{x_i - 2x_{i-1} + x_{i-2}}{(\Delta t)^2}$$

Angular acceleration:

$$\alpha(t) = \frac{d}{dt}\omega(t) \qquad\qquad \alpha_i = \frac{\omega_i - \omega_{i-1}}{\Delta t}$$

$$= \frac{d^2}{dt^2}\theta(t) \qquad\qquad = \frac{\theta_i - 2\theta_{i-1} + \theta_{i-2}}{(\Delta t)^2}$$

acceleration is measured, the other two can be derived by mathematical manipulation of the signal within an analog signal conditioner or within the controller software of a digital control system.

Position is simply a location within a frame of reference; thus, any measurement of displacement relative to the frame is a measurement of position, and any displacement transducer whose input is referenced to the frame can be used as a position transducer.

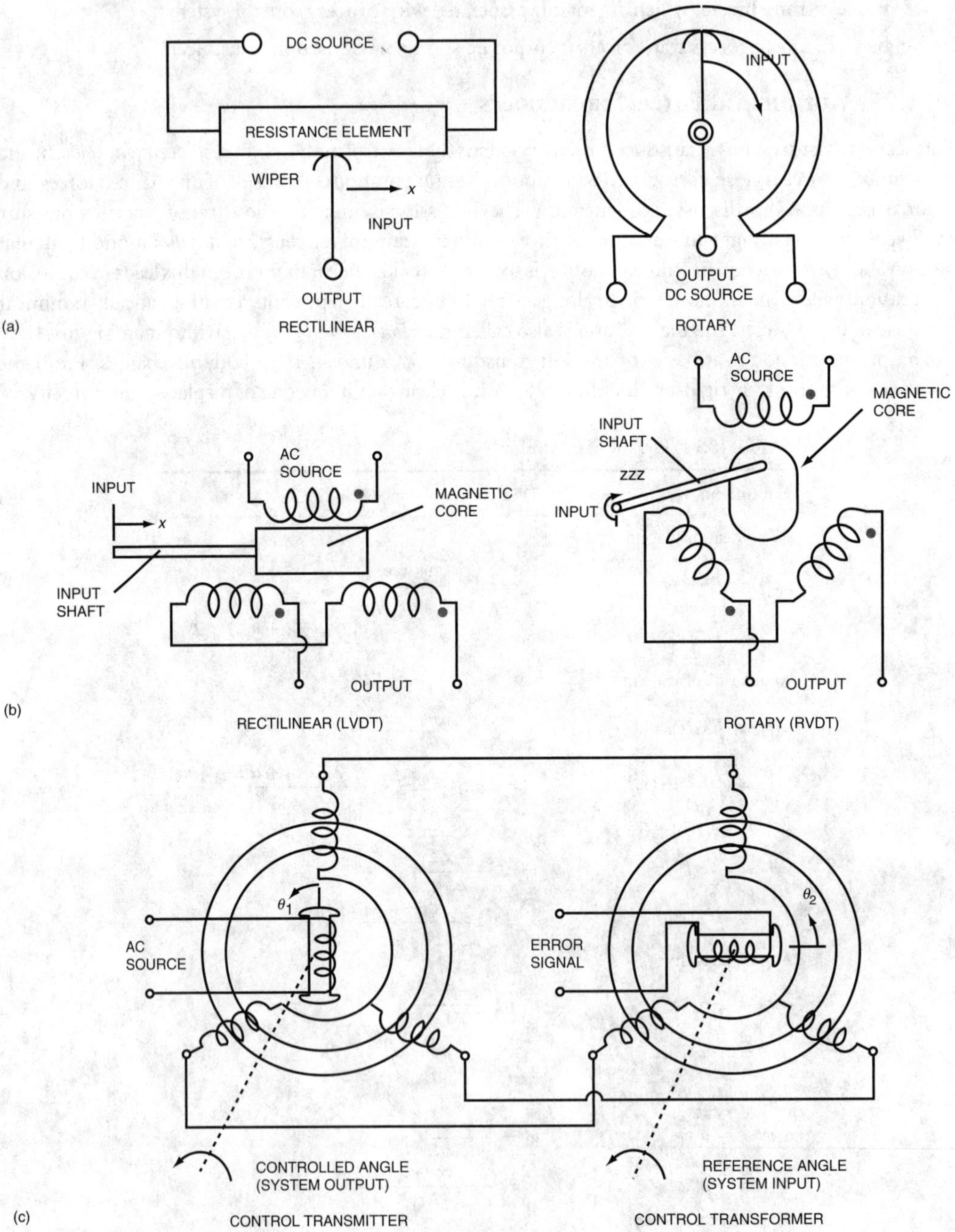

FIGURE 18.2 Gross displacement transducers: (a) potentiometers, (b) variable differential transformers (VDT), (c) synchros (typical connection), (d) resolvers (typical connection), (e) absolute position encoders, (f) code track for incremental position encoder.

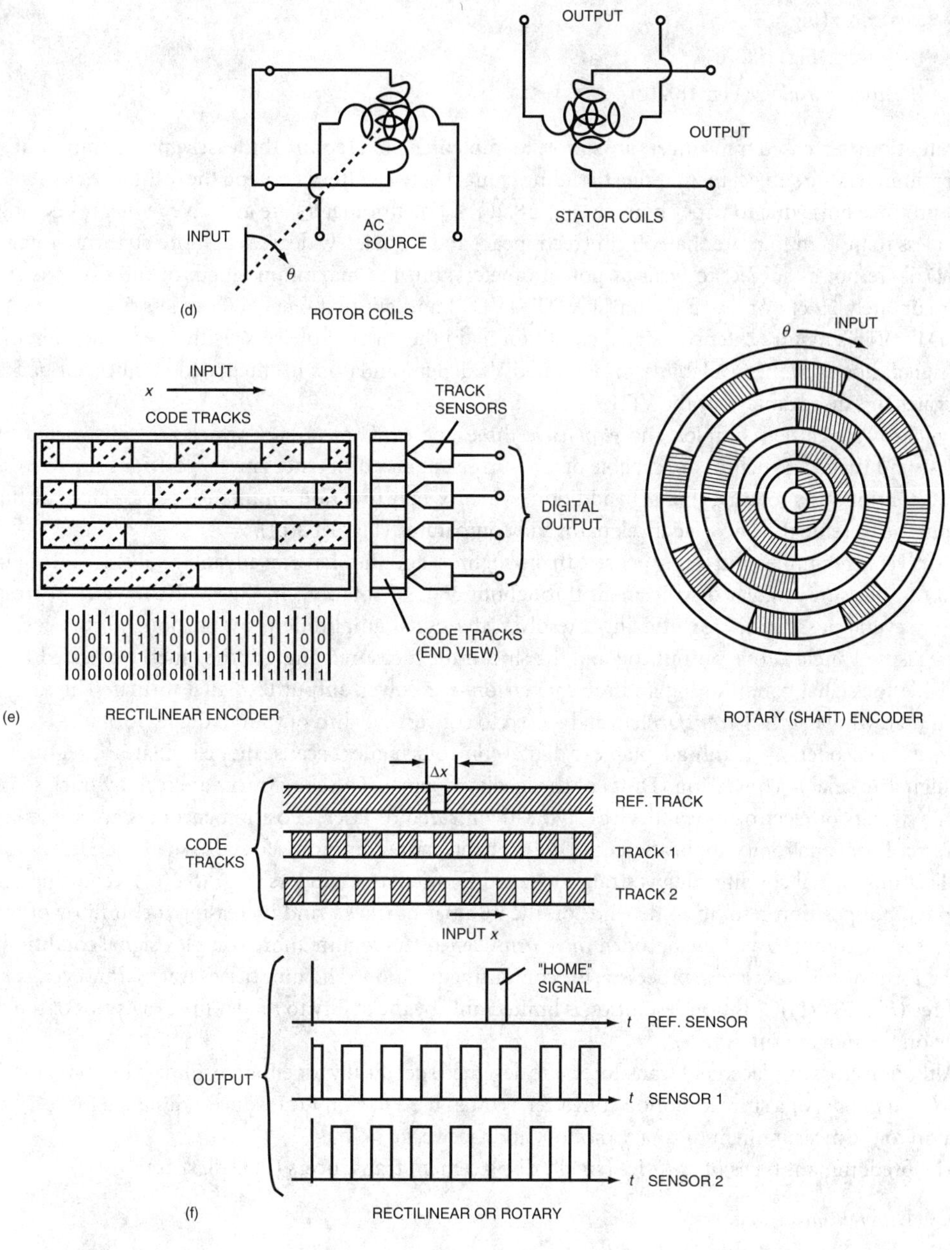

(d) ROTOR COILS

(e) RECTILINEAR ENCODER   ROTARY (SHAFT) ENCODER

(f) RECTILINEAR OR ROTARY

FIGURE 18.2 *(Continued).*

Displacement (Position) Transducers

Displacement transducers may be considered according to application as *gross* (large) displacement transducers or *sensitive* (small) displacement transducers. The demarcation between gross and sensitive displacement is somewhat arbitrary, but may be conveniently taken as approximately 1 mm for rectilinear displacement and approximately 10′ arc ($1/6°$) for angular displacement. The predominant types of gross displacement transducers (Fig. 18.2) are

- *Potentiometers* (Fig. 18.2(a))
- *Variable differential transformers* (VDT) (Fig. 18.2(b))

- *Synchros* (Fig. 18.2(c))
- *Resolvers* (Fig. 18.2(d))
- *Position encoders* (Fig. 18.2(e))

Potentiometer-based transducers are simple to implement and require the least signal conditioning, but potentiometers are subject to wear due to sliding contact between the wiper and the resistance element and may produce noise due to wiper bounce (Fig. 18.2(a)). Potentiometers are available with strokes ranging from less than 1 cm to more than 50 cm (rectilinear) and from a few degrees to more 50 turns (rotary).

VDTs are not as subject to wear as potentiometers, but the maximum length of the stroke is small, approximately 25 cm or less for a linear VDT (LVDT) and approximately 60° or less for a rotary VDT (RVDT). VDTs require extensive signal conditioning in the form of phase-sensitive demodulation of the AC signal; however, the availability of dedicated VDT demodulators in integrated circuit (IC) packages mitigates this disadvantage of the VDT.

Synchros are rather complex and expensive three-phase AC machines, which are constructed to be precise and rugged. Synchros are capable of measuring angular differences in the positions (up to ±180°) of two continuously rotating shafts. In addition, synchros may function simultaneously as reference input, output measurement device, feedback path, and comparator (Fig. 18.2(c)).

Resolvers are simpler and less expensive than synchros, and they have an advantage over RVDTs in their ability to measure angular displacement throughout 360° of rotation. In Fig. 18.2(d), which represents one of several possibilities for utilizing a resolver, the signal amplitude is proportional to the cosine of the measured angle at one output coil and the sine of the measured angle at the other. Dedicated ICs are available for signal conditioning and for conversion of resolver output to digital format. The same IC, when used with a *Scott-T transformer* can be used to convert synchro output to digital format.

Position encoders are highly adaptable to digital control schemes because they eliminate the requirement for digital-to-analog conversion (DAC) of the feedback signal. The code tracks are read by track sensors, usually wipers or electro-optical devices (typically infrared or laser). Position encoders are available for both rectilinear and rotary applications, but are probably more commonly found as shaft encoders in rotary applications. Signal conditioning is straightforward for *absolute* encoders (Fig. 18.2(e)), requiring only a decoder, but position resolution depends on the number of tracks, and increasing the number of tracks increases the complexity of the decoder. *Incremental* encoders require more complex signal conditioning, in the form of counters and a processor for computing position. The number of tracks, however, is fixed at three (Fig. 18.2(f)). Position resolution is limited only by the ability to render finer divisions of the code track on the moving surface.

Although gross displacement transducers are designed specifically for either rectilinear or rotary motion, a *rack* and *pinion*, or a similar motion converter, is often used to adapt transducers designed for rectilinear motion to the measurement of rotary motion, and vice versa.

The predominant types of *sensitive* (small) displacement transducers (Fig. 18.3) are

- *Differential capacitors*
- *Strain gauge resistors*
- *Piezoelectric crystals*

Figure 18.3(a) provides a simplified depiction of a differential capacitor used for sensitive displacement measurements. The motion of the input rod flexes the common plate, which increases the capacitance of one capacitor and decreases the capacitance of the other. In one measurement technique, the two capacitors are made part of an impedance bridge (such as a Schering bridge), and the change in the bridge output is an indication of displacement of the common plate. In another technique, each capacitor is allowed to serve as tuning capacitor for an oscillator, and the difference in frequency between the two oscillators is an indication of displacement.

A strain gauge resistor is used to measure elastic deformation (strain) of materials by bonding the resistor to the material (Fig. 18.3(b)) so that it undergoes the same strain as the material. The resistor is usually incorporated into one of several bridge circuits, and the output of the bridge is taken as an indication of strain.

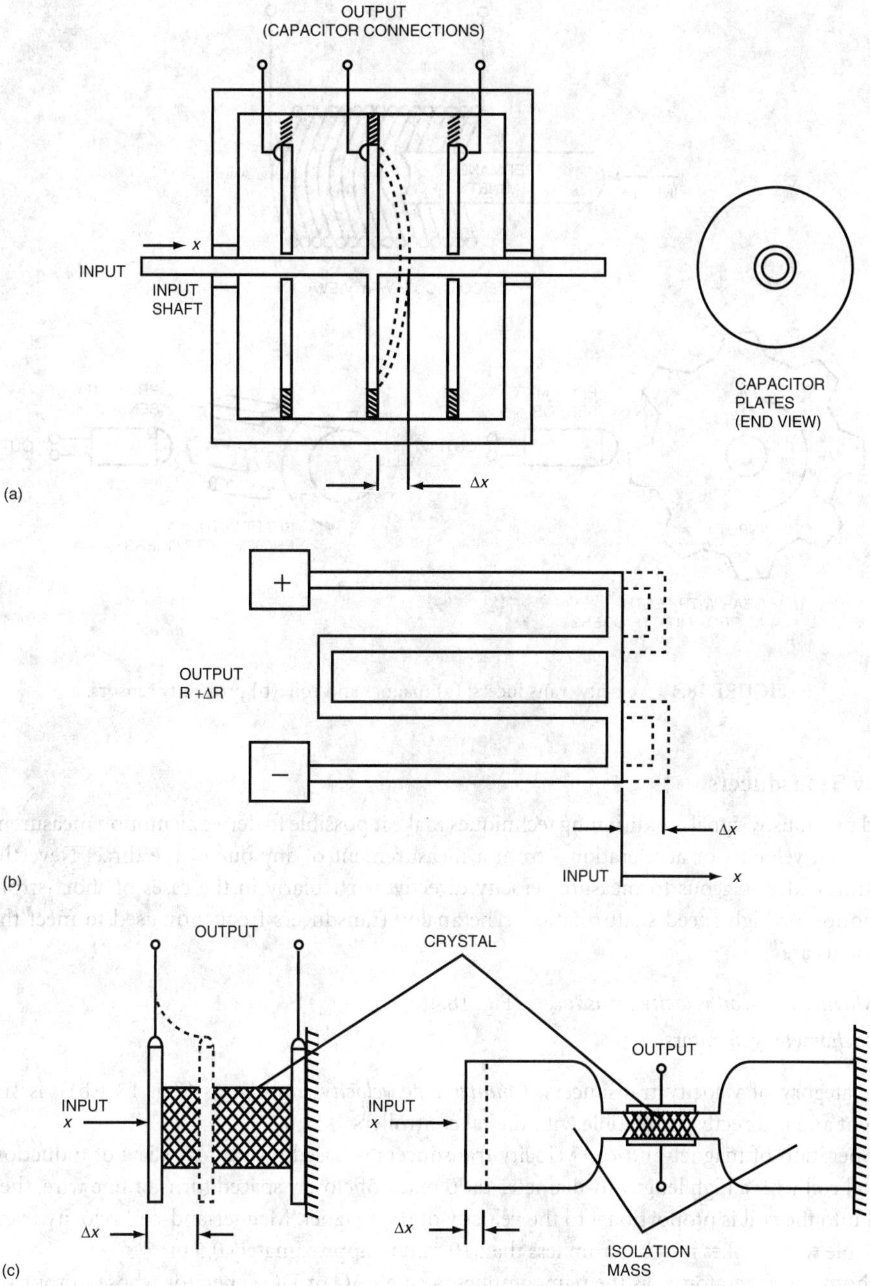

FIGURE 18.3 Sensitive displacement transducers: (a) differential capacitor, (b) strain gauge resistor, (c) piezoelectric crystals.

The piezoelectric effect is used in several techniques for sensitive displacement measurements (Fig. 18.3(c)). In one technique, the input motion deforms the crystal by acting directly on one electrode. In another technique, the crystal is fabricated as part of a larger structure, which is oriented so that input motion bends the structure and deforms the crystal. Deformation of the crystal produces a small output voltage and also alters the resonant frequency of the crystal. In a few situations, the output voltage is taken directly as an indication of motion, but more frequently the crystal is used to control an oscillator, and the oscillator frequency is taken as the indication of strain.

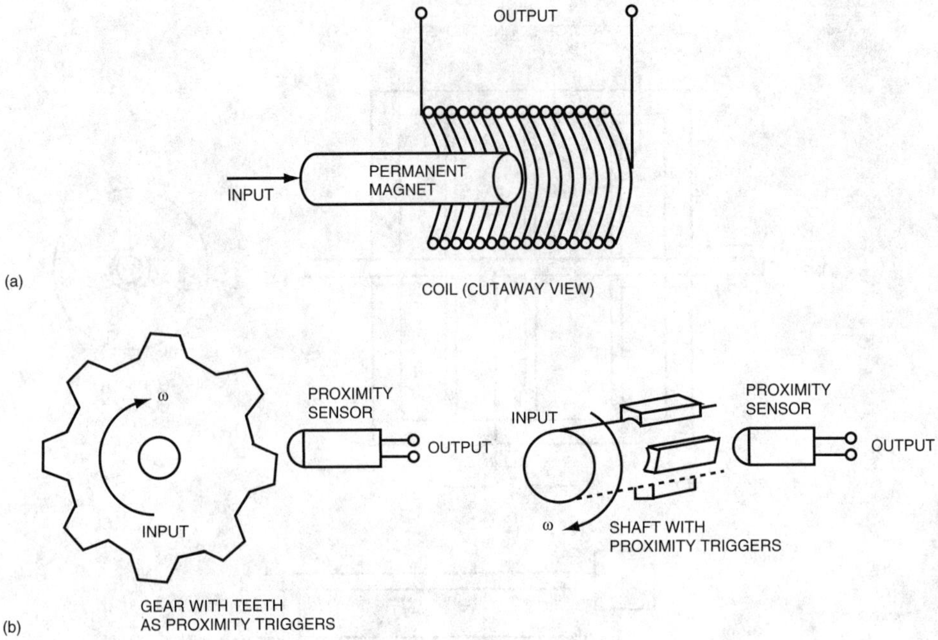

FIGURE 18.4 Velocity transducers: (a) magnet and coil, (b) proximity sensors.

Velocity Transducers

As stated previously, signal conditioning techniques make it possible to derive all motion measurements—displacement, velocity, or acceleration—from a measurement of any one of the three. Nevertheless, it is sometimes advantageous to measure velocity directly, particularly in the cases of short-stroke rectilinear motion or high-speed shaft rotation. The analog transducers frequently used to meet these two requirements are

- *Magnet-and-coil velocity transducers* (Fig. 18.4(a))
- *Tachometer generators*

A third category of velocity transducers, *Counter-type velocity transducers* (Fig. 18.4(b)), is simple to implement and is directly compatible with digital controllers.

The operation of magnet-and-coil velocity transducers is based on Faraday's law of induction. For a solenoidal coil with a high length-to-diameter ratio made of closely spaced turns of fine wire, the voltage induced into the coil is proportional to the velocity of the magnet. Magnet-and-coil velocity transducers are available with strokes ranging from less than 10 mm to approximately 0.5 m.

A tachometer generator is, as the name implies, a small AC or DC generator whose output voltage is directly proportional to the angular velocity of its rotor, which is driven by the controlled output shaft. Tachometer generators are available for shaft speeds of 5000 r/min, or greater, but the output may be nonlinear and there may be an unacceptable output voltage ripple at low speeds.

AC tachometer generators are less expensive and easier to maintain than DC tachometer generators, but DC tachometer generators are directly compatible with analog controllers and the polarity of the output is a direct indication of the direction of rotation. The output of an AC tachometer generator must be demodulated (i.e., rectified and filtered), and the demodulator must be phase sensitive in order to indicate direction of rotation.

Counter-type velocity transducers operate on the principle of counting electrical pulses for a fixed amount of time, then converting the count per unit time to velocity. Counter-type velocity transducers rely on the use of a proximity sensor (*pickup*) or an incremental encoder (Fig. 18.2(f)). Proximity sensors may be one of the following types:

- *Electro-optic*
- *Variable reluctance*
- *Hall effect*
- *Inductance*
- *Capacitance*

Two typical applications of counter-type velocity transducers are shown in Fig. 18.4(b).

Since a digital controller necessarily includes a very accurate electronic clock, both pulse counting and conversion to velocity can be implemented in software (i.e., made a part of the controller program). Hardware implementation of pulse counting may be necessary if time-intensive counting would divert the controller from other necessary control functions. A special-purpose IC, known as a *quadrature decoder/counter interface,* can perform the decoding and counting functions and transmit the count to the controller as a data word.

Acceleration Transducers

As with velocity measurements, it is sometimes preferable to measure acceleration directly, rather than derive acceleration from a displacement or velocity measurement. The majority of acceleration transducers may be categorized as *seismic* accelerometers because the measurement of acceleration is based on measuring the displacement of a mass called the *seismic element* (Fig. 18.5). The configurations shown in Fig. 18.5(a) and Fig. 18.5(b) require a rather precise arrangement of springs for suspension and centering of the seismic mass. One of the disadvantages of a seismic accelerometer is that the seismic mass is displaced during acceleration, and this displacement introduces nonlinearity and bias into the measurement. The force-balance configuration shown in Fig. 18.5(c) uses the core of an electromagnet as the seismic element. A sensitive displacement sensor detects displacement of the core and uses the displacement signal in a negative feedback arrangement to drive the coil, which returns the core to its center position. The output of the force-balance accelerometer is the feedback required to prevent displacement rather than displacement *per se.*

A simpler seismic accelerometer utilizes one electrode of a piezoelectric crystal as the seismic element (Fig. 18.5(d)). Similarly, another simple accelerometer utilizes the common plate of a differential capacitor (Fig. 18.3(a)) as the seismic element.

Force Transducers

Force measurements are usually based on a measurement of the motion which results from the applied force. If the applied force results in gross motion of the controlled output, and the mass of the output element is known, then any appropriate accelerometer attached to the controlled output produces an output proportional to the applied force ($F = Ma$). A simple spring-balance scale (Fig. 18.6(a)) relies on measurement of displacement, which results from the applied force (weight) extending the spring.

Highly precise force measurements in high-value servomechanisms, such as those used in pointing and tracking devices, frequently rely on gyroscope precession as an indication of the applied force. The scheme is shown in Fig. 18.6(b) for a gyroscope with gimbals and a spin element. A motion transducer (either displacement or velocity) on the precession axis provides an output proportional to the applied force. Other types of gyroscopes and precession sensors are also used to implement this force measurement technique.

Static force measurements (in which there is no apparent motion) usually rely on measurement of strain due to the applied force. Figure 18.6(c) illustrates the typical construction of a common force transducer called a *load cell.* The applied force produces a proportional strain in the S-shaped structural member, which is measured with a sensitive displacement transducer, usually a strain gauge resistor or a piezoelectric crystal.

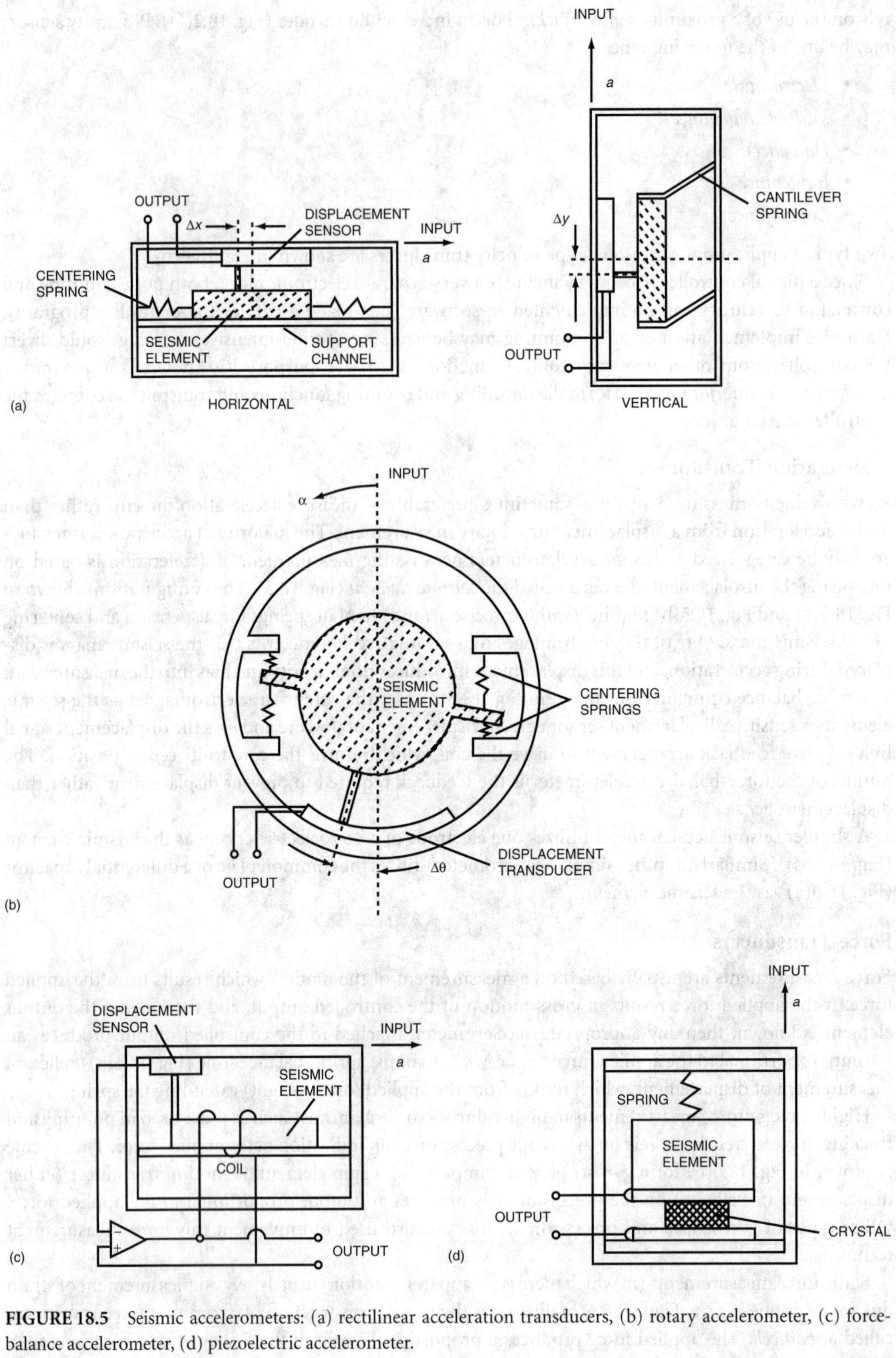

FIGURE 18.5 Seismic accelerometers: (a) rectilinear acceleration transducers, (b) rotary accelerometer, (c) force-balance accelerometer, (d) piezoelectric accelerometer.

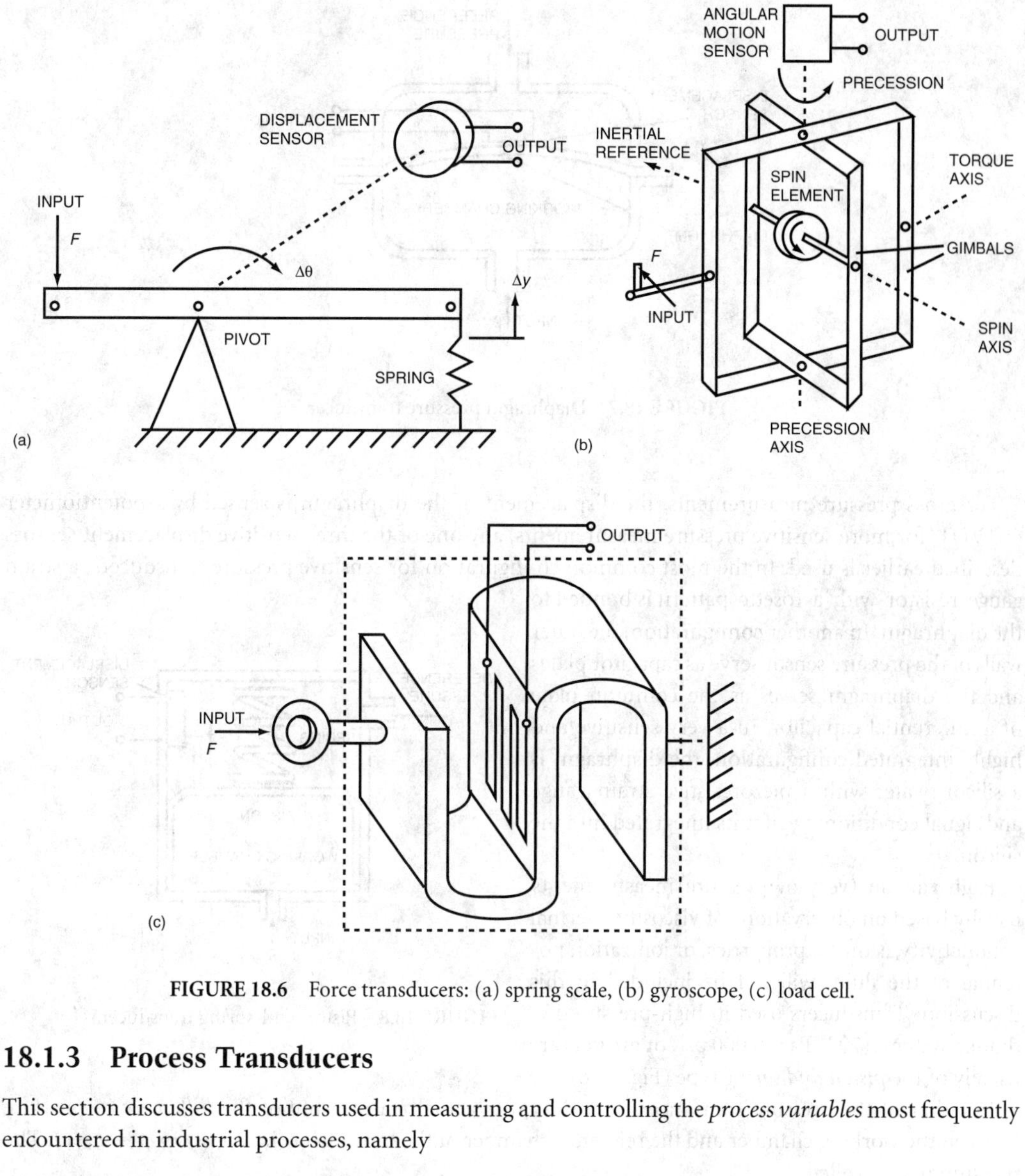

FIGURE 18.6 Force transducers: (a) spring scale, (b) gyroscope, (c) load cell.

18.1.3 Process Transducers

This section discusses transducers used in measuring and controlling the *process variables* most frequently encountered in industrial processes, namely

- *Fluid pressure*
- *Fluid flow*
- *Liquid level*
- *Temperature*

Fluid Pressure Transducers

Most fluid pressure transducers are of the *elastic* type, in which the fluid is confined in a chamber with at least one elastic wall, and the deflection of the elastic wall is taken as an indication of the pressure. The *Bourdon tube* and the *bellows* are examples of elastic pressure transducers, which are used in laboratory-grade transducers and in some industrial process control applications. The fluid pressure transducer depicted in Fig. 18.7, which uses an elastic *diaphragm* to separate two chambers, is the type most frequently encountered in industrial process control. Diaphragms are constructed from one of a variety of elastic materials ranging from thin metal to polymerized fabric.

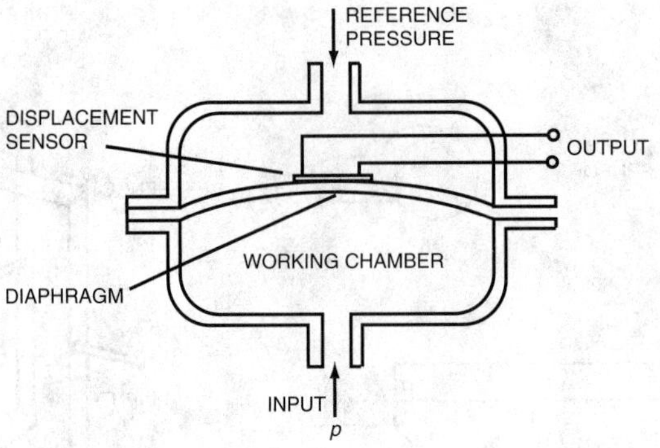

FIGURE 18.7   Diaphragm pressure transducer.

For gross pressure measurements, the displacement of the diaphragm is sensed by a potentiometer or LVDT; for more sensitive pressure measurements, any one of the three sensitive displacement sensors described earlier is used. In the most common configuration for sensitive pressure transducers, a strain gauge resistor with a rosette pattern is bonded to the diaphragm. In another configuration, the outer walls of the pressure sensor serve as capacitor plates and the diaphragm serves as the common plate of a differential capacitor. In a very sensitive and highly integrated configuration, the diaphragm is a silicon wafer with a piezoresistive strain gauge and signal conditioning circuits integrated into the silicon.

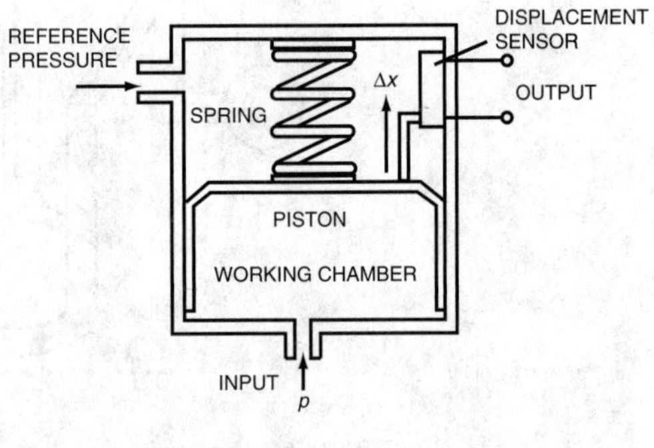

FIGURE 18.8   Piston-and-spring transducer.

High-vacuum (very low pressure) measurements, usually based on observations of viscosity, thermal conductivity, acoustic properties, or ionization potential of the fluid, will not be included in this discussion. Transducers used in high-pressure hydraulic systems (70 MPa (10,000 psi) or greater) are usually of the *piston and spring* type (Fig. 18.8).

In either of the pressure transducers, the output is actually a measure of the *difference* in pressure between the working chamber and the reference chamber of the transducer (i.e., $p_{OUT} = p - p_{REF}$). The measurement is called

- An *absolute pressure* if the reference chamber is sealed and evacuated (i.e., $p_{REF} = 0$ and $p_{OUT} = p$)
- A *gauge pressure* if the reference chamber is vented to the atmosphere (i.e., $p_{OUT} = p - p_{ATM}$)
- A *differential pressure* if any other pressure is applied to the reference chamber

Fluid Flow Transducers (Flowmeters)

Flowmetering, because of the number of variables involved, encompasses a wide range of measurement technology and applications. In industrial processes, the term *fluid* is applied not only to gases and liquids, but also to flowable mixtures (often called *slurries* or *sludges*) such as concrete, sewage, or wood pulp. Control of a fluid flow, and hence the type of measurement required, may involve *volumetric* flow rate, *mass* flow rate, or flow direction. Gas flows may be *compressible*, which also influences the measurement technique. In addition, the condition of the flow—whether or not it is homogenous and clean (free of suspended particles)—has a bearing on flowmeter technology. Another factor to be considered is flow velocity; slow moving laminar flows of viscous material require different measurement techniques than those used for high-velocity turbulent flows. Still another consideration is confinement of the flow. Whereas most fluid flow measurements are concerned with *full flow* through *closed channels* such as ducts and pipes, some applications require measurements of *partial flow* through *open channels* such as troughs and flumes. Only the most widely used flowmeters are considered here.

The major categories of flowmeters are

- *Differential pressure, constriction-type (venturi, orifice, flow nozzle, elbow (or pipe bend),* and *pitot static)* (Fig. 18.9)
- *Fluid-power (gear motors, turbines* and *paddle wheels)* (Fig. 18.10)
- *Ultrasound* (Fig. 18.11)
- *Vortex shedding* (Fig. 18.12)
- *Thermal anemometer* (Fig. 18.13)
- *Electromagnetic* (Fig. 18.14)
- *Rotameter (variable-area in-line flowmeter)* (Fig. 18.15)

Differential pressure flowmeters are suited to high- and moderate-velocity flow of gas and clean, low-viscosity liquids. Venturi flowmeters (Fig. 18.9(a)) are the most accurate, but they are large and expensive. Orifice flowmeters (Fig. 18.9(b)) are smaller, less expensive, and much less accurate than venturi flowmeters. Nozzle flowmeters (Fig. 18.9(c)) are a compromise between venturi and orifice flowmeters. Pipe-bend flowmeters (Fig. 18.9(d)), which can essentially be installed in any bend in an existing piping system, are used primarily for gross flow rate measurements. Pitot-static flowmeters (Fig. 18.9(e)) are used in flows which have a large cross-sectional area, such as in wind tunnels. Pitot-static flowmeters are also used in freestream applications such as airspeed indicators for aircraft.

Fluid-power flowmeters are used in low-velocity, moderately viscous flows. In addition to industrial control applications, turbine flowmeters (Fig. 18.10(a)) are sometime used as speed indicators for ships or boats. Paddle wheel flowmeters (Fig. 18.10(b)) are used both in closed- and open-flow applications such as liquid flow in flumes. Since a fluid-power gear motor (Fig. 18.10(c)) is a constant volume device, motor shaft speed is always a direct indication of fluid flow rate.

Ultrasound flowmeters of the transmission type (Fig. 18.11(a)), which are based on the principle that the sound transmission speed will be increased by the flow rate of the fluid, are used in all types of clean, subsonic flows. Doppler flowmeters (Fig. 18.11(b)) rely on echoes from within the fluid, and are thus only useful in dirty flows that carry suspended particles or turbulent flows that produce bubbles. Ultrasound flowmeters are nonintrusive devices, which can often be retrofitted to existing duct or pipe systems.

Vortex shedding flowmeters (Fig. 18.12) introduce a *shedding body* into the flow to cause production (*shedding*) of *vortices*. The sound accompanying the production and collapse of the vortices is monitored and analyzed. The dominant frequency of the sound is indicative of the rate of vortex production and collapse, and hence an indication of flow rate. Vortex shedding flowmeters are useful in low-velocity, nonturbulent flows.

Thermal anemometers (Fig. 18.13) are used in low-velocity gas flows with large cross-sectional area, such as in heating, ventilation, and air conditioning (HVAC) ducts. Convection cooling of the heating element is related to flow rate. The flow rate measurement is based either on the current required to maintain a constant temperature in the heating element, or alternatively on the change in temperature when the current is held constant.

Electromagnetic flowmeters (Fig. 18.14) are useful for slow moving flows of liquids, sludges, or slurries. The flow material must support electrical conduction between the electrodes, and so in some cases it is necessary to ionize the flow upstream from the measurement point in order to use an electromagnetic flowmeter.

Variable-area in-line flowmeters (Fig. 18.15), or *rotameters,* are sometimes referred to as sight gauges because they provide a visible indication of flow rate. These devices, when fitted with proximity sensors (such as capacitive pickups), which sense the presence of the float, can be used in on-off control applications.

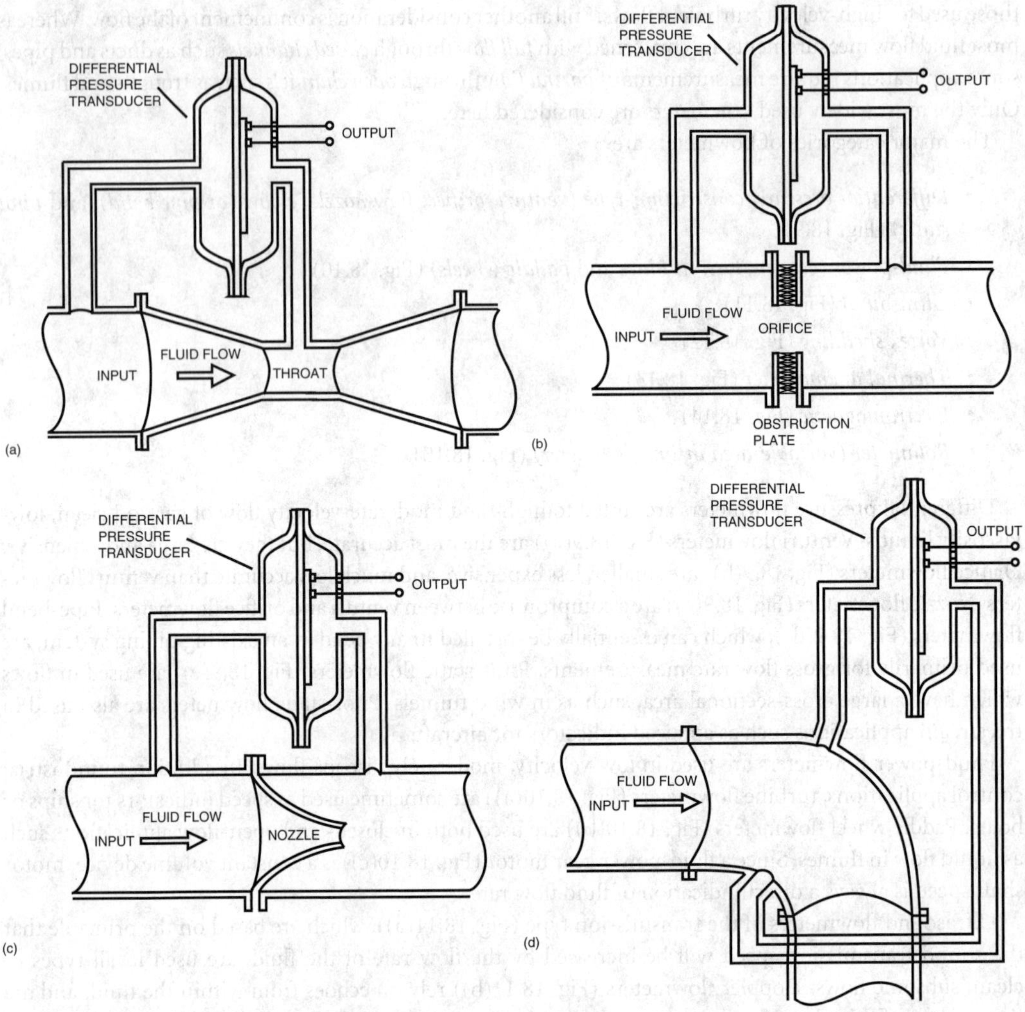

FIGURE 18.9 Differential pressure flowmeters: (a) Venturi flowmeter, (b) orifice flowmeter, (c) nozzle flowmeter, (d) pipebend (elbow) flowmeter, (e) pitot-static-flowmeter.

Liquid Level Transducers

Liquid-level measurements are relatively straightforward, and the transducers fall into the categories of *contact* or *noncontact*. Measurements may be *continuous,* in which the liquid level is monitored continuously throughout its operating range, or *point,* in which the liquid level is determined to be above or below some predetermined level.

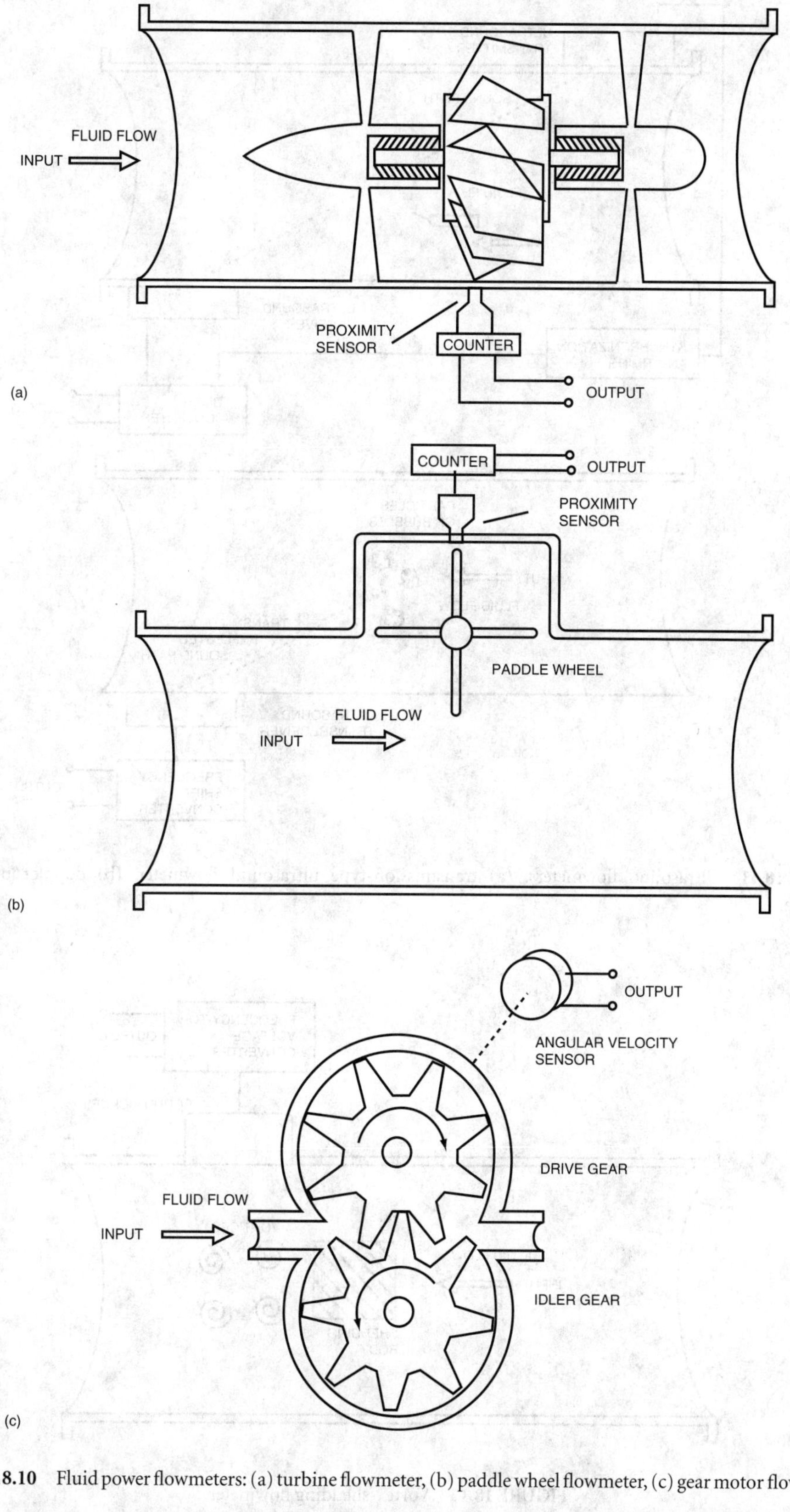

FIGURE 18.10 Fluid power flowmeters: (a) turbine flowmeter, (b) paddle wheel flowmeter, (c) gear motor flowmeter.

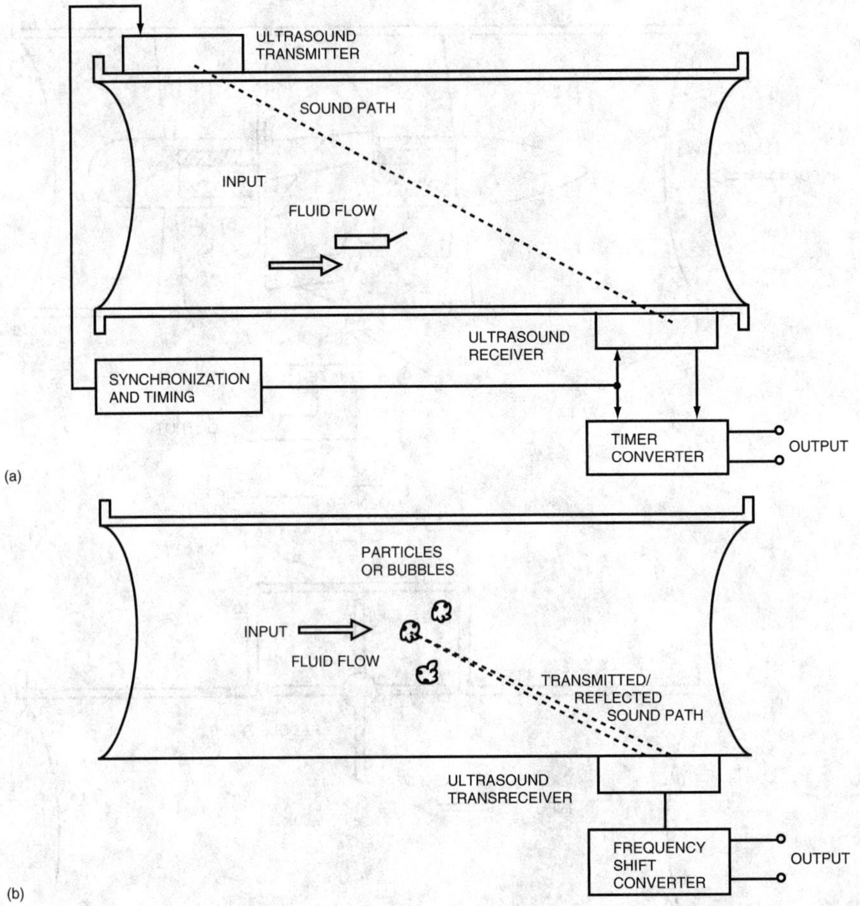

FIGURE 18.11 Ultrasound flowmeters: (a) transmission-type ultrasound flowmeter, (b) doppler ultrasound flowmeter.

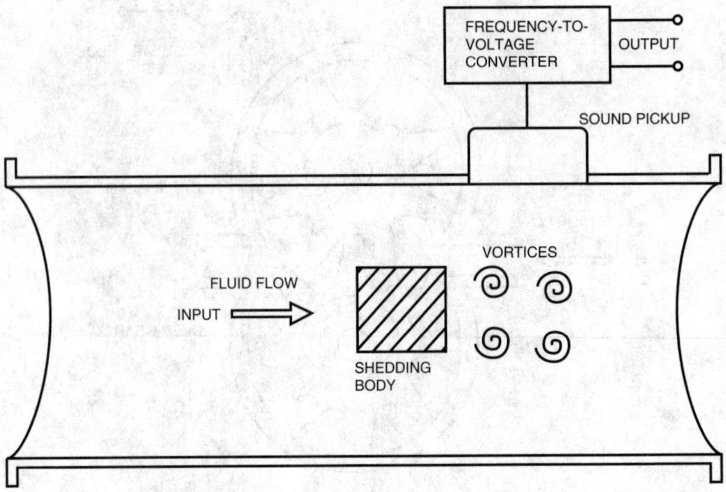

FIGURE 18.12 Vortex-shedding flowmeter.

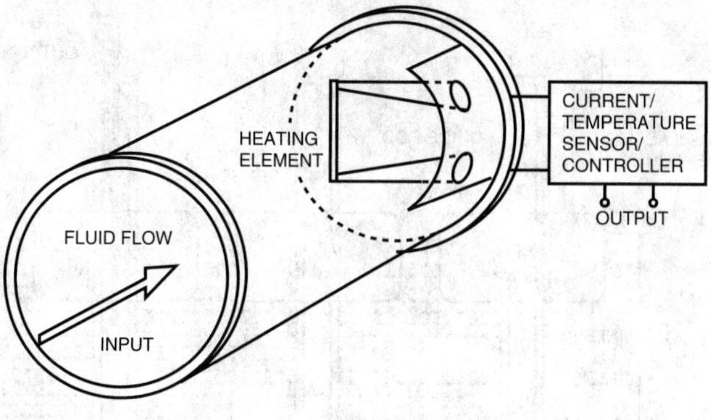

FIGURE 18.13   Thermal anemometer.

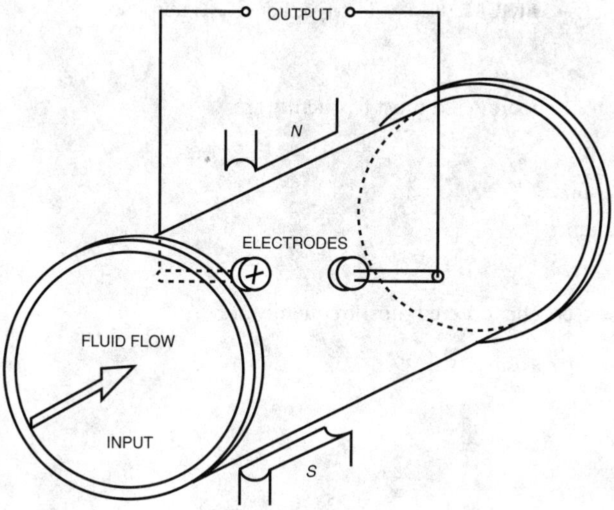

FIGURE 18.14   Electromagnetic flowmeter.

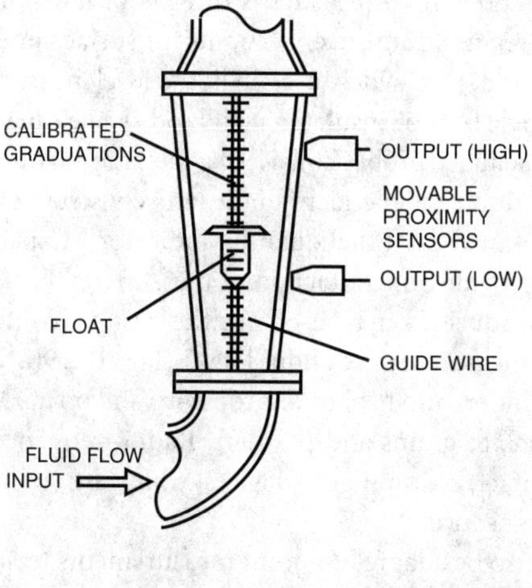

FIGURE 18.15   Variable-area in-line flowmeter (rotameter).

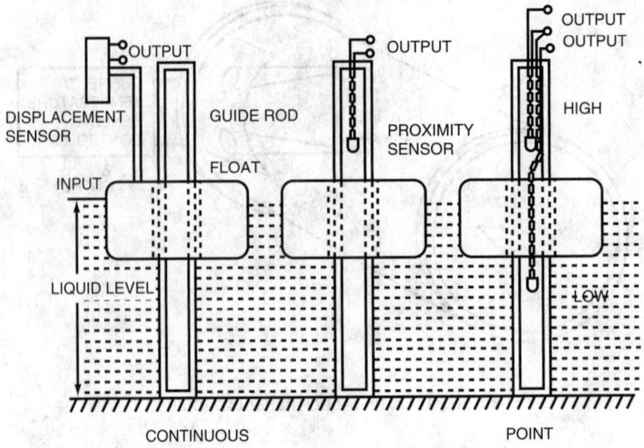

FIGURE 18.16   Float-type liquid level transducers.

The contact transducers encountered most frequently are

- *Float*
- *Hydrostatic pressure*
- *Electrical capacitance*
- *Ultrasound*

The noncontact transducers encountered most frequently are

- *Capacitive proximity sensors*
- *Ultrasound*
- *Radio frequency*
- *Electro-optical*

Float-type liquid level transducers are available in a wide variety of configurations for both continuous and point measurements. One possible configuration is depicted in Fig. 18.16 for continuous measurement and for both single- and dual-point measurements.

Hydrostatic pressure liquid level transducers may be used in either vented or pressurized applications (Fig. 18.17). In either case the differential pressure is directly proportional to the weight of the liquid column, since the differential pressure transducer accounts for surface pressure.

Capacitance probes (Fig. 18.18(a)) are widely used in liquid level measurements. It is possible, when the tank walls are metal, to use a single bare or insulated metal rod as one capacitor plate and the tank walls as the other. More frequently, capacitance probes consist of a metal rod within a concentric cylinder open at the ends, which makes the transducer independent of the tank construction. An interesting application of this type of capacitance probe is in aircraft fuel quantity indicators. Capacitance switches can be utilized as depicted in Fig. 18.18(b) to provide noncontact point measurements of liquid level.

Ultrasound echo ranging transducers can be used in either *wetted* (contact) or *nonwetted* (noncontact) configurations for continuous measurement of liquid level (Fig. 18.19(a)). An interesting application of wetted transducers is as depth finders and fish finders for ships and boats. Nonwetted transducers can also be used with bulk materials such as grains and powders. Radio-frequency and electro-optic liquid level transducers are usually noncontact, echo ranging devices, which are similar in principle and application to the nonwetted ultrasound transducer.

Ultrasonic transducers can also be adapted to point measurements by locating the transmitter and the receiver opposite one another across a gap (Fig. 18.19(b)). When liquid fills the gap, attenuation of the

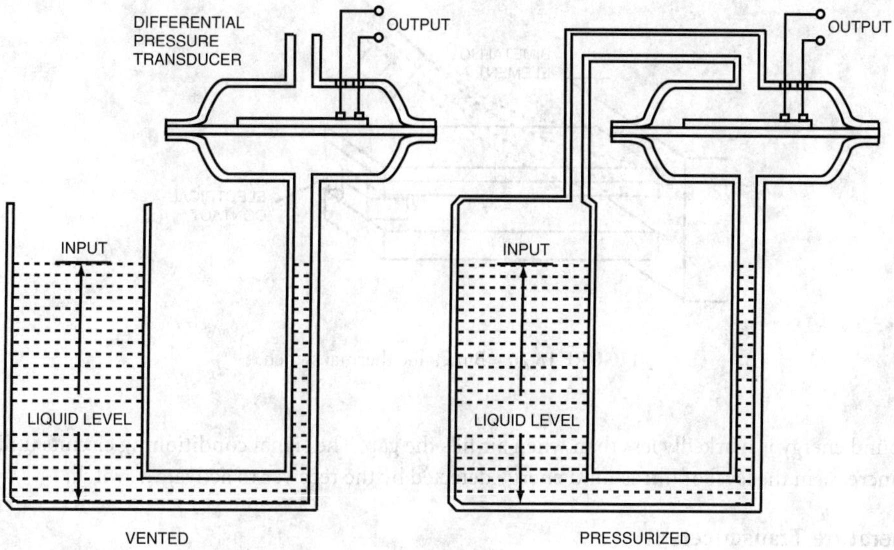

FIGURE 18.17 Hydrostatic pressure liquid level transducers.

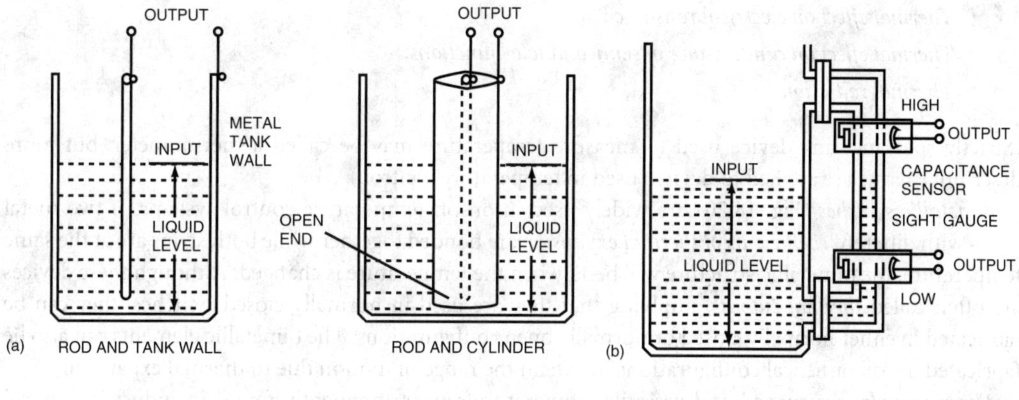

FIGURE 18.18 Capacitive-type liquid level transducers: (a) capacitive probes, (b) capacitive switches.

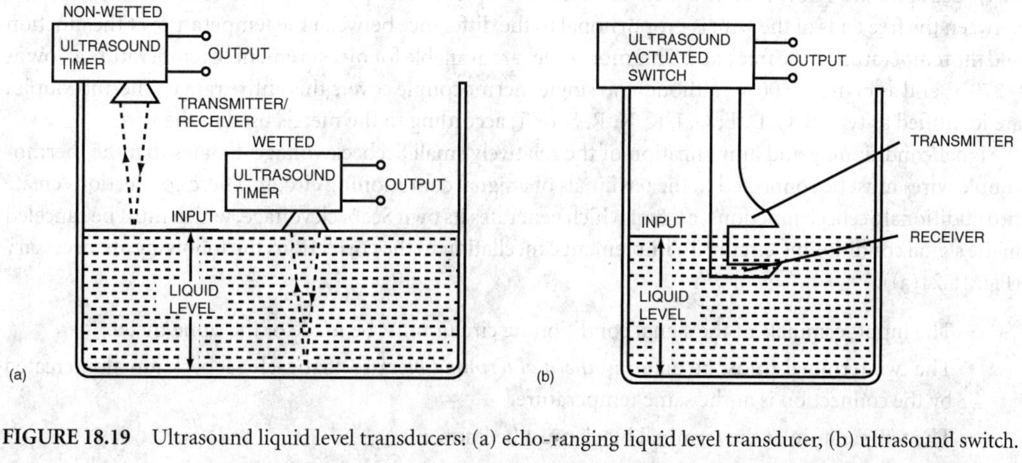

FIGURE 18.19 Ultrasound liquid level transducers: (a) echo-ranging liquid level transducer, (b) ultrasound switch.

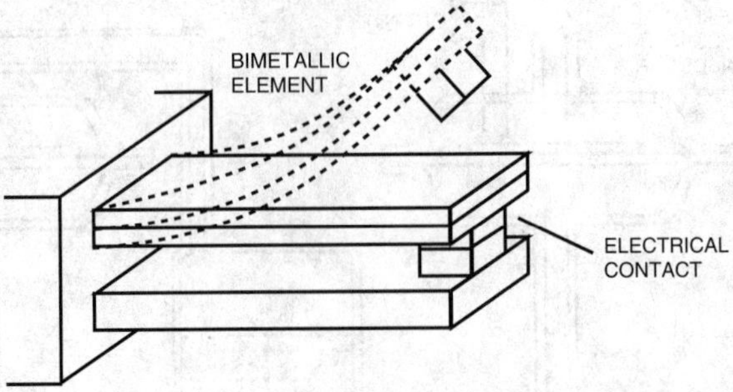

FIGURE 18.20   Bimetallic thermal switch.

ultrasound energy is markedly less than when air fills the gap. The signal conditioning circuits utilize this sharp increase in the level of ultrasound energy detected by the receiver to activate a switch.

Temperature Transducers

Temperature measurement is generally based on one of the following physical principles:

- *Thermal expansion*
- *Thermoelectric phenomena*
- *Thermal effect on electrical resistance*
- *Thermal effect on conductance of semiconductor junctions*
- *Thermal radiation*

(Strictly speaking, any device used to measure temperature may be called a thermometer, but more descriptive terms are applied to devices used in temperature control.)

Bimetallic switches (Fig. 18.20) are widely used in on-off temperature control systems. If two metal strips with different *coefficients of thermal expansion* are bonded together while both strips are at the same temperature, the bimetallic structure will bend when the temperature is changed. Although these devices are often called *thermal cutouts*, implying that they are used in normally closed switches, they can be fabricated in either normally closed or normally open configurations. The bimetallic elements can also be fabricated in coil or helical configurations to extend the range of motion due to thermal expansion.

Thermocouples are rugged and versatile temperature sensors frequently found in industrial control systems. A thermocouple consists of a pair of dissimilar metal wires twisted or otherwise bonded at one end. The *Seebeck effect* is the physical phenomena which accounts for thermocouple operation, so thermocouples are known alternatively as *Seebeck junctions*. The potential difference (*Seebeck voltage*) between the free ends of the wire is proportional to the difference between the temperature at the junction and the temperature at the free ends. Thermocouples are available for measurement of temperature as low as $-270°C$ and as high as $2300°C$, although no single thermocouple covers this entire range. Thermocouples are identified as type B, C, D, E, G, J, K, N, R, S, or T, according to the metals used in the wire.

Signal conditioning and amplification of the relatively small Seebeck voltage dictates that the thermocouple wires must be connected to the terminals of a signal conditioning circuit. These connections create two additional Seebeck junctions, each of which generates its own Seebeck voltage, which must be canceled in the signal conditioning circuit. To implement cancellation to the corrections the following are necessary (Fig. 18.21(a)):

- The input terminals of the signal conditioning circuit must be made of the same metal.
- The two terminals must be on an *isothermal terminal block* so that each Seebeck junction created by the connection is at the same temperature.
- The temperature of the terminal block must be known.

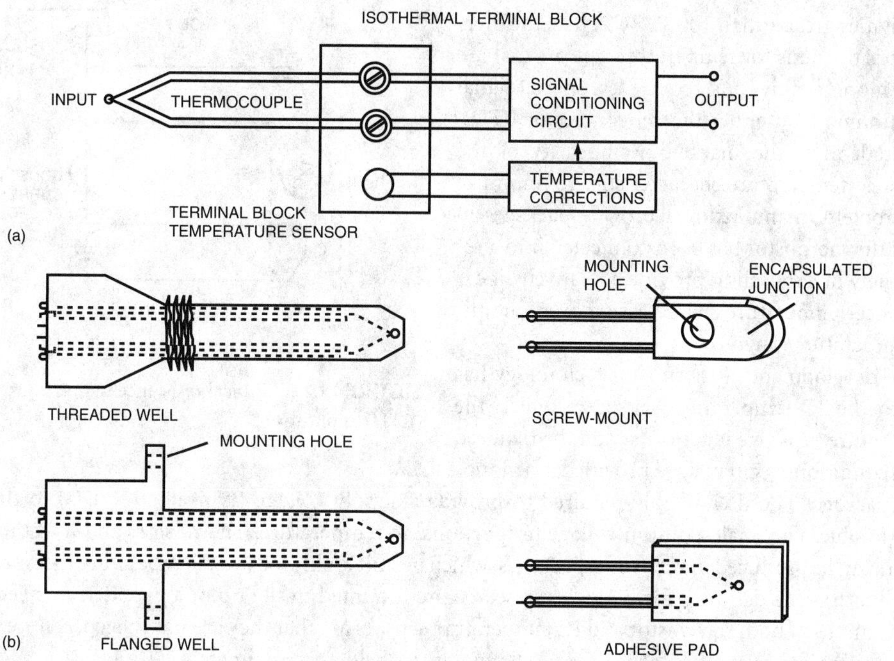

FIGURE 18.21   Thermocouples: (a) typical thermocouple connection, (b) some thermocouple accessories and configurations.

The first two requirements are met by appropriate construction of the signal conditioning circuit. The third requirement is met by using a reference temperature sensor, probably an IC temperature transducer of the type described later.

Thermocouple and thermocouple accessories are fabricated for a variety of applications (Fig. 18.21(b)). Protective shields are used to protect thermocouple junctions in corrosive environments or where conducting liquids can short circuit the thermocouple voltage; however, exposed (bare) junctions are used wherever possible, particularly when fast response is essential.

Resistance temperature detectors (RTDs) are based on the principle that the electrical *resistivity* of most metals increases predictably with temperature. Platinum is the preferred metal for RTDs, although other less expensive metals are used in some applications. The resistivity of platinum is one of the standards by which temperature is measured. The relatively good linearity of the resistivity of platinum over a wide temperature range (-200–$800°$C) makes platinum RTDs suitable for stable, accurate temperature transducers, which are easily adapted to control systems applications.

The disadvantage of the RTD is that the temperature-sensitive element is a rather fragile metal filament wound on a ceramic bobbin or a thin metal film deposited on a ceramic substrate. RTD elements are usually encapsulated and are rarely used as bare elements. The accessories and application packages used with RTDs are similar to those used with thermocouples (Fig. 18.21(b)).

Most platinum RTDs are fabricated to have a nominal resistance of 100 Ω at 0°C. The *resistance temperature coefficient* of platinum is approximately 3–4 m$\Omega/\Omega/°$C, so resolution of temperature to within 1°C for a nominal 100-Ω RTD element requires resolution of the absolute resistance within 0.3–0.4 Ω. These resistance resolution requirements dictate use of special signal conditioning techniques to cancel the lead and contact resistance of the RTD element (Fig. 18.22). The circuit depicted in Fig. 18.22 is a variation of a 4-*wire ohmmeter*. Most RTDs are manufactured with four leads to be compatible with such circuits.

Thermistors are specially prepared metal oxide semiconductors that exhibit a strong *negative* temperature coefficient, in sharp contrast to the weak positive temperature coefficient of RTDs. Nominal thermistor resistance, usually specified for 25°C, ranges from less than 1000 Ω to more than 1 MΩ, with sensitivities greater than 100 $\Omega/°C$. Thus, the thermistor is the basis for temperature sensors that are much more sensitive and require less special signal conditioning than either thermocouples or RTDs. The tradeoff is the marked nonlinearity of the resistance-temperature characteristic. To minimize this problem, manufacturers provide packages in which the thermistor has been connected into a resistor network chosen to provide a relatively linear resistance-temperature characteristic over a nominal temperature range.

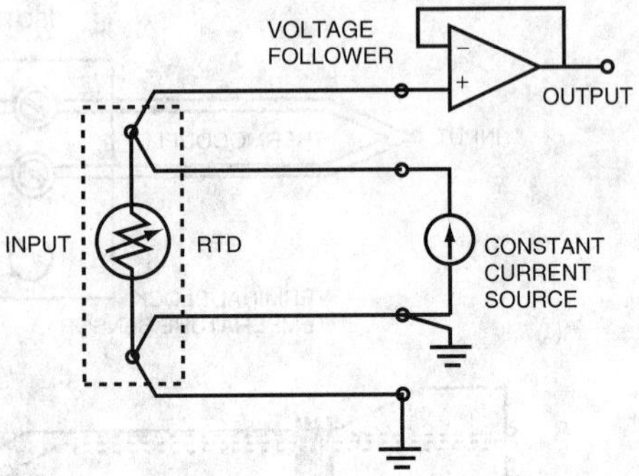

The development of thermistor technology has lead to the IC temperature sensor in which the temperature-sensitive junction(s) and the required signal conditioning circuits are provided in a mono-

FIGURE 18.22 Typical resistance temperature detector (RTD) application.

lithic package. The user is only required to provide a supply voltage (typically 5 V DC) to the IC in order to obtain an analog output voltage proportional to temperature. Thermistors and IC temperature sensors can be produced in very small packages, which permit highly localized temperature measurements. Some thermistors designed for biological research are mounted in the tip of a hypodermic needle. The shortcomings of both thermistor and IC temperature sensors are that they are not rugged, cannot be used in caustic environments, and are limited to temperatures below approximately 200°C.

Radiation thermometers are used for remote (non-contact) sensing of temperature in situations where the contact sensors cannot be used. Operation is based on the principles of heat transfer through thermal radiation. Radiation thermometers focus the *infrared* energy from a heat source onto a *black body* (target) within the radiation thermometer enclosure (Fig. 18.23). One of the contact temperature sensors described previously is incorporated into the target to measure the target temperature. The rise in temperature at the target is related to the source temperature. Typical radiation thermometers have standoff ranges (focal lengths) of 0.5–1.5 m,

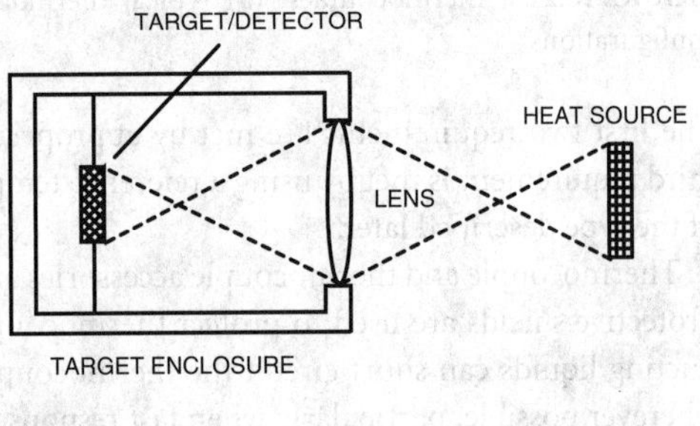

FIGURE 18.23 Schematic of the radiation thermometry scheme.

but instruments with focal length as short as 1 cm or as long as 10 m are available. Radiation thermometers are available for broadband, monochromatic, or two-color thermometry.

18.1.4 Transducer Performance

The operation of a transducer within a control system can be described in terms of its *static performance* and its *dynamic performance*. The static characteristics of greatest interest are

- *Scale factor* (or *sensitivity*)
- *Accuracy, uncertainty, precision,* and *system error* (or *bias*)
- *Threshold, resolution, dead band,* and *hysteresis*
- *Linearity*
- *Analog drift*

The dynamic characteristics of greatest interest are

- *Time constant, response time,* and *rise time*
- *Overshoot, settling time,* and *damped frequency*
- *Frequency response*

Static performance is documented through *calibration,* which consists of applying a known input (quantity or phenomenon to be measured) and observing and recording the transducer output. In a typical calibration procedure, the input is increased in increments from the lower range limit to the upper range limit of the transducer, then decreased to the lower range limit. The range of a component consists of all allowable input values. The difference between the upper and lower range limits is the *input span* of the component; the difference between the output at the upper range limit and the output at the lower range limit is the *output span.*

Dynamic performance is documented by applying a known change, usually a step, in the input and observing and recording the transducer output.

18.1.5   Loading and Transducer Compliance

A prime requirement for an appropriate transducer is that it be *compliant* at its input. Compliance in this sense means that the input energy required for proper operation of the transducer, and hence a correct measurement of the controlled output, does not significantly alter the controlled output. A transducer that does not have this compliance is said to *load* the controlled output. For example, a voltmeter must have a high-impedance input in order that the voltage measurement does not significantly alter circuit current and, hence, alter the voltage being measured.

Defining Terms

Controlled output: The principal product of an automatic control system; the quantity or physical activity to be measured for automatic control.

Feedback path: The cascaded connection of transducer and signal conditioning components in an automatic control system.

Forward path: The cascaded connection of controller, actuator, and plant or process in an automatic control system.

Motion transducer: A transducer used to measure the controlled output of a servomechanism; usually understood to include transducers for static force measurements.

Plant or process: The controlled device that produces the principal output in an automatic control system.

Process control: The term used to refer to the control of industrial processes; most frequently used in reference to control of temperature, fluid pressure, fluid flow, and liquid level.

Process transducer: A transducer used to measure the controlled output of an automatic control system used in process control.

Reference input: The signal provided to an automatic control system to establish the required controlled output; also called *setpoint.*

Servomechanism: A system in which some form of motion is the controlled output.

Signal conditioning: In this context, the term used to refer to the modification of the signal in the feedback path of an automatic control system; signal conditioning converts the sensor output to an electrical signal suitable for comparison to the reference input (setpoint); the term can also be applied to modification of forward path signals.

Transducer: The device used to measure the controlled output in an automatic control system; usually consists of a sensor or pickup and signal conditioning components.

References

Bateson, R.N. 1993. *Introduction to Control System Technology,* 4th ed. Merrill, Columbus, OH.

Berlin, H.M. and Getz, F.C., Jr. 1988. *Principles of Electronic Instrumentation and Measurement.* Merrill, Columbus, OH.

Buchla, D. and McLachlan, W. 1992. *Applied Electronic Instrumentation and Measurement.* Macmillan, New York.

Chaplin, J.W. 1992. *Instrumentation and Automation for Manufacturing.* Delmar, Albany, NY.

Doeblin, E.O. 1990. *Measurement Systems Application and Design,* 4th ed. McGraw-Hill, New York.

Dorf, R.C. and Bishop, R.H. 1995. *Modern Control Systems,* 7th ed. Addison-Wesley, Reading, MA.

O'Dell, T.H. 1991. *Circuits for Electronic Instrumentation.* Cambridge University Press, Cambridge, England, UK.

Seippel, R.G. 1983. *Transducers, Sensors, and Detectors.* Reston Pub., Reston, VA.

Webb, J. and Greshock, K. 1993. *Industrial Control Electronics,* 2nd ed. Macmillan, New York.

Further Information

Manufacturers and vendors catalogs, data documents, handbooks, and applications notes, particularly the handbook series by Omega Engineering, Inc.:

The Flow and Level Handbook
The Pressure, Strain, and Force Handbook
The Temperature Handbook

Trade journals, magazines, and newsletters, particularly:

Instrumentation Newsletter (National Instruments)
Personal Engineering and Instrumentation News
Test and Measurement World

18.2 Data Acquisition

Edward McConnell and David Jernigan

18.2.1 Fundamentals of Data Acquisition

The fundamental task of a data acquisition system is the measurement or generation of real-world physical signals. Before a physical signal can be measured by a computer-based system, you must use a sensor or transducer to convert the physical signal into an electrical signal, such as voltage or current. Often only the plug-in **data acquisition** (**DAQ**) board is considered the data acquisition system; however, it is only one of the components in the system. Unlike stand-alone instruments, signals often cannot be directly connected to the DAQ board. The signals may need to be conditioned by some signal conditioning accessory before they are converted to digital information by the plug-in DAQ board. Finally, software controls the data acquisition system—acquiring the raw data, analyzing the data, and presenting the results. The components are shown in Fig. 18.24.

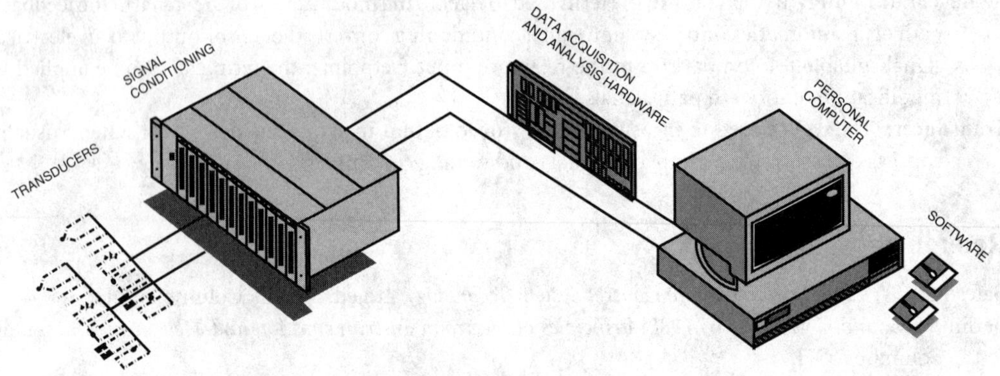

FIGURE 18.24 Components of data acquisition system.

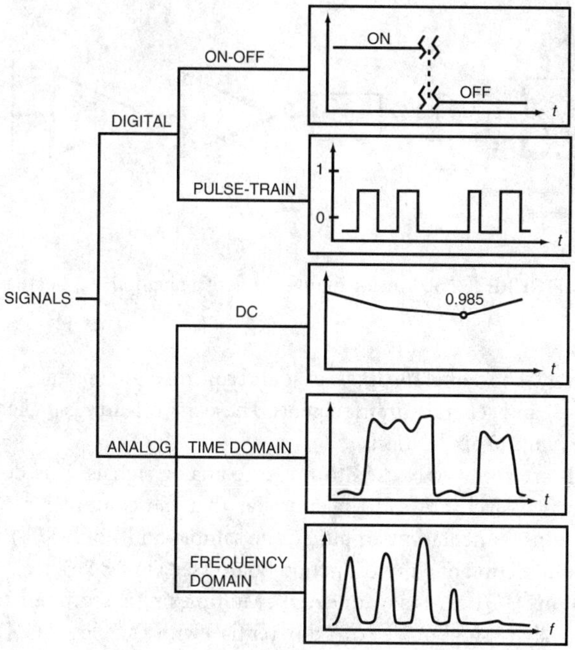

FIGURE 18.25   Classes of signals.

Signals

Signals are defined as any physical variable whose magnitude or variation with time contains information. Signals are measured because they contain some types of useful information. Therefore, the first question you should ask about your signal is: What information does the signal contain, and how is it conveyed? The functionality of the system is determined by the physical characteristics of the signals and the type of information conveyed by the signals. Generally, information is conveyed by a signal through one or more of the following signal parameters: state, rate, level, shape, or frequency content.

All signals are, fundamentally, analog, time-varying signals. For the purpose of discussing the methods of signal measurement using a plug-in DAQ board, a given signal should be classified as one of five signals types. Because the method of signal measurement is determined by the way the signal conveys the needed information, a classification based on these criteria is useful in understanding the fundamental building blocks of a data acquisition system.

As shown in the Fig. 18.25, any signal can generally be classified as analog or digital. A digital, or binary, signal has only two possible discrete levels of interest, a high (on) level and a low (off) level. An analog signal, on the other hand, contains information in the continuous variation of the signal with time. The two digital signal types are the on-off signals and the pulse train signal. The three analog signal types are the DC signal, the time-domain signal, and the frequency-domain signal. The two digital types and three analog types of signals are unique in the information conveyed by each. The category to which a signal belongs depends on the characteristic of the signal to be measured. You can closely parallel the five types of signals with the five basic types of signal information: state, rate, level, shape, and frequency content.

Plug-In DAQ Boards

The fundamental component of a data acquisition system is the plug-in DAQ board. These boards plug directly into a slot in a PC and are available with analog, digital, and timing inputs and outputs. The most versatile of the plug-in DAQ boards is the multifunction input/output (I/O) board. As the name implies, this board typically contains various combinations of analog-to-digital convertors (ADCs), digital-to-analog convertors (DACs), digital I/O lines, and counters/timers. ADCs and DACs measure and generate analog

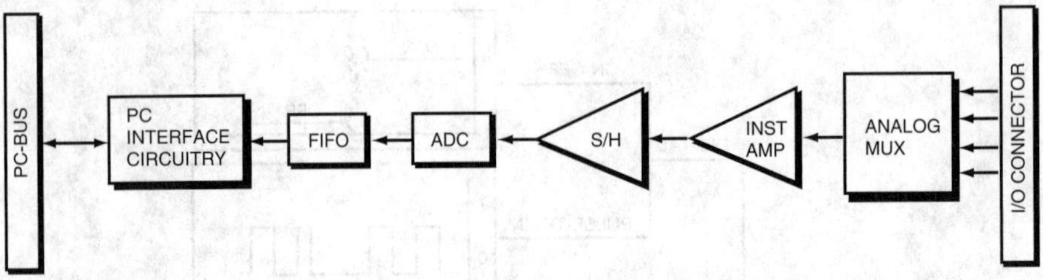

FIGURE 18.26 Analog input section of a plug-in DAQ board.

voltage signals, respectively. The digital I/O lines sense and control digital signals. Counters/timers measure pulse rates, widths, delays, and generate timing signals. These many features make the multifunction DAQ board useful for a wide range of applications.

Multifunction boards are commonly used to measure analog signals. It is done by the ADC, which converts the analog voltage level into a digital number that the computer can interpret. The analog multiplexer (MUX), the instrumentation amplifier, the sample-and-hold (S/H) circuitry, and the ADC comprise the analog input section of a multifunction board (see Fig. 18.26).

Typically, multifunction DAQ boards have one ADC. Multiplexing is a common technique for measuring multiple channels (generally 16 single ended or 8 differential) with a single ADC. The analog mux switches between channels and passes the signal to the instrumentation amplifier and the sample-and-hold circuitry. The multiplexer architecture is the most common approach taken with plug-in DAQ boards. Although plug-in boards typically include up to only 16 single-ended or 8 differential inputs, you can further expanded the number of analog input channels with external multiplexer accessories.

Instrumentation amplifiers typically provide a differential input and selectable gain by jumper or software. The differential input rejects small common-mode voltages. The gain is often software programmable. In addition, many DAQ boards also include the capability to change the amplifier gain while scanning channels at high rates. Therefore, you can easily monitor signals with different ranges of amplitudes. The output of the amplifier is sampled, or held at a constant voltage, by the sample-and-hold device at measurement time so that voltage does not change during digitization.

The ADC digitizes the analog signal into a digital value, which is ultimately sent to computer memory. There are several important parameters of A/D conversion. The fundamental parameter of an ADC is the number of bits. The number of bits of an ADC determines the range of values for the binary output of the ADC conversion. For example, many ADCs are 12 b, so a voltage within the input range of the ADC will produce a binary value that has one of $2^{12} = 4096$ different values. The more bits that an ADC has, the higher the resolution of the measurement. The resolution determines the smallest amount of change that can be detected by the ADC. Depending on your background, you may be more familiar with resolution expressed as number of digits of a voltmeter or dynamic range in decibels, than with bits. Table 18.2 shows the relation between bits, number of digits, and dynamic range in decibels.

The resolution of the A/D conversion is also determined by the input range of the ADC and the gain. DAQ boards usually include an instrumentation amplifier that amplifies the analog signal by a gain factor

TABLE 18.2 Relation Between Bits, Number of Digits, and Dynamic Range

Bits	Digits	dB
20	6.0	120
16	4.5	96
12	3.5	72
8	2.5	48

prior to the conversion. You use this gain to amplify low-level signals so that you can make more accurate measurements.

Together, the input range of the ADC, the gain, and the number of bits of the board determine the maximum accuracy of the measurement. For example, suppose you are measuring a low-level $\pm$-30 mV signal with a 12-b A/D convertor that has a $\pm$-5 V input range. If the system includes an amplifier with a gain of 100, the resulting resolution of the measurement will be range/(gain $\times 2^{bits}$) = resolution, or 10 V/(100 $\times 2^{12}$) = 0.0244 mV.

Finally, an important parameter of digitization is the rate at which A/D conversions are made, referred to as the *sampling rate*. The A/D system must be able to sample the input signal fast enough to accurately measure the important waveform attributes. To meet this criterion, the ADC must be able to convert the analog signal to digital form quickly enough.

When scanning multiple channels with a multiplexing data acquisition system, other factors can affect the throughput of the system. Specifically, the instrumentation amplifier must be able to settle to the needed accuracy before the A/D conversion occurs. With the multiplexed signals, multiple signals are being switched into one instrumentation amplifier. Most amplifiers, especially when amplifying the signals with larger gains, will not be able to settle to the full accuracy of the ADC when scanning channels at high rates. To avoid this situation, consult the specified settling times of the DAQ board for the gains and sampling rates required by your application.

Types of ADCs

Different DAQ boards use different types of ADCs to digitize the signal. The most popular type of ADC on plug-in DAQ boards is the successive approximation ADC, because it offers high speed and high resolution at a modest cost. *Subranging* (also called *half-flash*) ADCs offer very high-speed conversion with sampling speeds up to several million samples per second. *Delta-sigma modulating* ADCs sample at high rates, are able to achieve high resolution, and offer the best linearity of all ADCs. *Integrating and flash* ADCs are mature technologies still used on DAQ boards. Integrating ADCs are able to digitize with high resolution but must sacrifice sampling speed to obtain it. Flash ADCs are able to achieve the highest sampling rate (GHz) but typically with low resolution. The different types of ADCs are summarized in Table 18.3.

Analog Input Architecture

With the typical DAQ board, the multiplexer switches among analog input channels. The analog signal on the channel selected by the multiplexer then passes to the programmable gain instrumentation amplifier (PGIA), which amplifies the signal. After the signal is amplified, the sample and hold keeps the analog signal constant so that the A/D converter can determine the digital representation of the analog signal. A good DAQ board will then place the digital signal in a first-in first-out (FIFO) buffer, so that no data

TABLE 18.3 Types of ADCs

Type of ADC	Advantages
Successive Approximation	High resolution High speed Easily multiplexed
Subranging	Higher speed
Delta–Sigma	High resolution Excellent linearity Built-in antialiasing
Integrated	High resolution Good noise rejection Mature technology
Flash	Highest speed Mature technology

will be lost if the sample cannot transfer immediately over the PC I/O channel to computer memory. Having a FIFO becomes especially important when the board is run under operating systems that have large interrupt latencies.

Basic Analog Specifications

Almost every DAQ board data sheet specifies the number of channels, the maximum sampling rate, the resolution, and the input range and gain.

The number of channels, which is determined by the multiplexer, is usually specified in two forms, differential and single ended. Differential inputs are inputs that have different reference points for each channel, none of which is grounded by the board. Differential inputs are the best way to connect signals to the DAQ board because they provide the best noise immunity.

Single-ended inputs are inputs that are referenced to a common ground point. Because single-ended inputs are referenced to a common ground, they are not as good as differential inputs for rejecting noise. They do have a larger number of channels, however. You can use the single-ended inputs when the input signals are high level (greater than 1 V), the leads from the signal source to the analog input hardware are short (less than 15 ft), and all input signals share a common reference.

Some boards have pseudodifferential inputs, which have all inputs referenced to the same common—like single-ended inputs—but the common is not referenced to ground. Using these boards, you have the benefit of a large number of input channels, like single-ended inputs, and the ability to remove some common mode noise, especially if the common mode noise is consistent across all channels. Differential inputs are still preferable to pseudodifferential, however, because differential is more immune to magnetic noise.

Sampling rate determines how fast the analog signal is converted to a digital signal. If you are measuring AC signals, you will want to sample at least two times faster than the highest frequency of your input signal. Even if you are measuring DC signals, you can sample faster than you need to and then average the samples to increase the accuracy of the signal by reducing the effects of noise.

If you have multiple DC-class signals, you will want to select a board with *interval scanning*. With interval scanning, all channels are scanned at one sample interval (usually the fastest rate of the board), with a second interval (usually slow) determining the time before repeating the scan. Interval scanning gives the effects of simultaneously sampling for slowly varying signals without requiring the addition cost of input circuitry for true simultaneous sampling.

Resolution is the number of bits that are used to represent the analog signal. The higher the resolution, the higher the number of divisions the input range is broken into and, therefore, the smaller the possible detectable voltage. Unfortunately, some data acquisition specifications are misleading when they specify the resolution associated with the DAQ board. Many DAQ boards specifications state the resolution of the ADC without stating the linearity's and noise and, therefore, do not give you the information you need to determine the resolution of the entire board. Resolution of the ADC, combined with the settling time, integral non-linearity, differential non linearity, and noise will give you an understanding of the accuracy of the board.

Input range and gain tell you what level of signal you can connect to the board. Usually, the range and gain are specified separately, so you must combine the two to determine the actual signal input range as

$$signal\ input\ range = range/gain$$

For example, a board using an input range of ±10 V with a gain of 2 will have a signal input range of ±5 V. The closer the signal input range is to the range of your signal, the more accurate your readings from the DAQ board will be. If your signals have different input ranges, you will want to look for a DAQ board that has the capability of different gains per channel.

18.2.2 Data Acquisition Software

The software is often the most critical component of the data acquisition system. Properly chosen software can save you a great deal of time and money. Likewise, poorly chosen software can cost you time and money. A whole spectrum of software options exists, with important tradeoffs and advantages.

Board Register-Level Programming

The first option is not to use vendor-supplied software and program the DAQ board yourself at the hardware level. DAQ boards are typically register based, that is, they include a number of digital registers that control the operation of the board. The developer may use any standard programming language to write series of binary codes to the DAQ board to control its operation.

Driver Software

Driver software typically consists of a library of function calls usable from a standard programming language. These function calls provide a high-level interface to control the standard functions of the plug-in board. For example, a function called SCAN_OP may configure, initiate, and complete a multiple channel scanning data acquisition operation of a predetermined number of points. The function call would include parameters to indicate the channels to be scanned, the amplifier gains to be used, the sampling rate, and the total number of data points to collect. The driver responds to this one function call by programming the plug-in board, the DMA controller, the interrupt controller, and CPU to scan the channels as requested.

18.2.3 Digital Sampling

Every DAQ system has the task of gathering information about analog signals. To do this, the system captures a series of instantaneous snapshots or samples of the signal at definite time intervals. Each sample contains information about the signal at a specific instant. Knowing the exact **conversion time** and the value of the sample, you can reconstruct, analyze, and display the digitized waveform.

Two classifications of sample timing techniques are used to control the ADC conversions, real-time sampling and equivalent-time sampling (ETS). Depending on the type of signal you acquire and the rate of acquisition, one sampling technique may be better than the other.

Real-Time Sampling Techniques

In real-time sampling, you immediately see the changes, as the signal changes (Fig. 18.27). According to the **Nyquist theorem**, you must sample at least twice the rate of the maximum frequency component in that signal to prevent **aliasing**. The frequency at one-half the sampling frequency is referred to as the Nyquist frequency. Theoretically, it is possible to recover information about those signals with frequencies at or below the Nyquist frequency. Frequencies above the Nyquist frequency will alias to appear between DC and the Nyquist frequency.

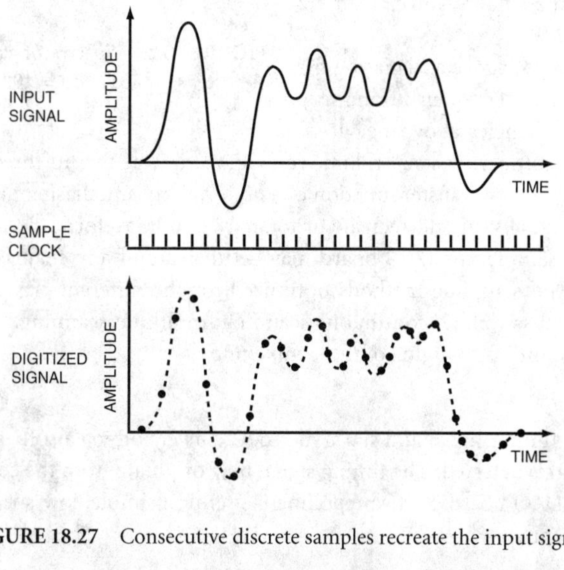

FIGURE 18.27   Consecutive discrete samples recreate the input signal.

For example, assume the sampling frequency fs, is 100 Hz. Also assume the input signal to be sampled contains the following frequencies: 25, 70, 160, and 510 Hz. Figure 18.28 shows a spectral representation of the input signal. The mathematics of sampling theory show us that a sampled signal is shifted in the frequency domain by an amount equal to integer multiples of the sampling frequency, fs. Figure 18.29 shows the spectral content of the input signal after sampling. Frequencies below 50 Hz, the Nyquist frequency ($fs/2$), appear correctly. However, frequencies above the Nyquist appear as aliases below the Nyquist frequency. For example, F1 appears correctly; however, both F2, F3, and F4 have aliases at 30, 40, and 10 Hz, respectively.

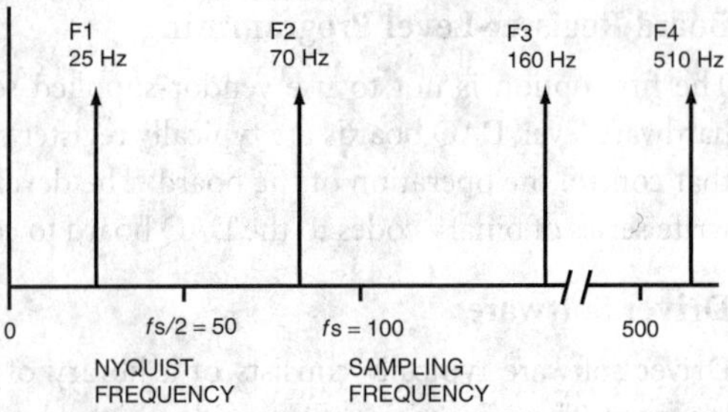

FIGURE 18.28   Spectral of signal with multiple frequencies.

The resulting frequency of aliased signals can be calculated with the following formula:

Apparent (Alias) Freq. = ABS (Closest Integer Multiple of Sampling Freq.-Input Freq.)

For the example of Fig. 18.28 and Fig. 18.29 or 10 Hz

$$\text{alias F2} = |100 - 70| = 30 \text{ Hz}$$
$$\text{alias F3} = |(2)100 - 160| = 40 \text{ Hz}$$
$$\text{alias F4} = |(5)100 - 510| = 10 \text{ Hz}$$

Preventing Aliasing

You can prevent aliasing by using filters on the front end of your DAQ system. These antialiasing filters are set to cut off any frequencies above the Nyquist frequency (half the sampling rate). The perfect filter would reject all frequencies above the Nyquist; however, because perfect filters exist only in textbooks, you must compromise between sampling rate and selecting filters. In many applications, one- or two-pole passive filters are satisfactory. By oversampling (5–10 times) and using these filters, you can sample adequately in most cases.

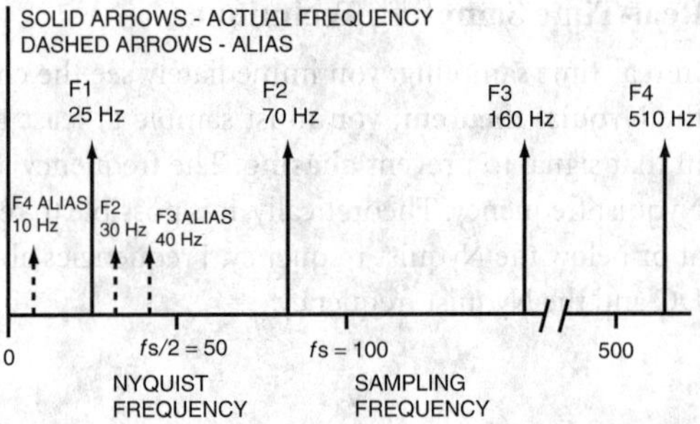

Alternatively, you can use active antialiasing filters with programmable cutoff frequencies and very sharp attenuation of frequencies above the cutoff.

FIGURE 18.29   Spectral of signal with multiple frequencies after sampled at $fs = 100$ Hz.

Because these filters exhibit a very steep rolloff, you can sample at two to three times the filter cutoff frequency. Figure 18.30 shows a transfer function of a high-quality antialiasing filter.

The computer uses digital values to recreate or to analyze the waveform. Because the signal could be anything between each sample, the DAQ board may be unaware of any changes in the signal between samples. There are several sampling methods optimized for the different classes of data; they include software polling, external sampling, continuous scanning, multirate scanning, simultaneous sampling, interval scanning, and seamless changing of the sample rate.

Software Polling

A software loop polls a timing signal and starts the ADC conversion via a software command when the edge of the timing signal is detected. The timing signal may originate from the computer's internal clock or from a clock on the DAQ board. Software polling is useful in simple, low-speed applications, such as temperature measurements.

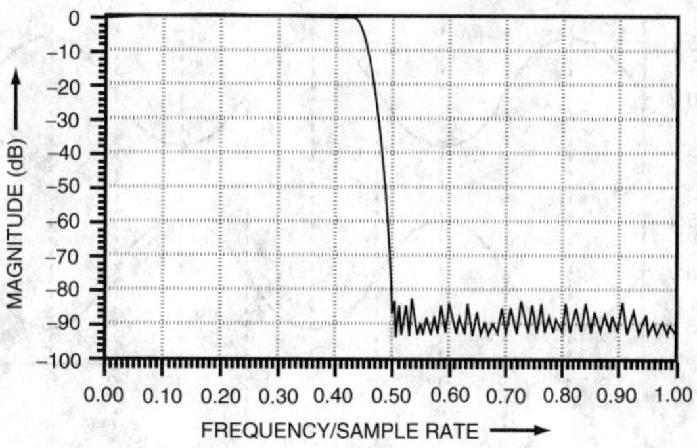

FIGURE 18.30 Transfer function of an antialiasing filter.

The software loop must be fast enough to detect the timing signal and trigger a conversion. Otherwise, a window of uncertainty, also known as *jitter*, will exist between two successive samples. Within the window of uncertainty, the input waveform could change enough to drastically reduce the accuracy of the ADC.

Suppose a 100-Hz, 10-V full-scale sine wave is digitized (Fig. 18.31). If the polling loop takes 5 ms to detect the timing signal and to trigger a conversion, then the voltage of the input sine wave will change as much as 31 mV ($\Delta V - 10 \sin(2\pi \times 100 \times 5 \times 10^{-6})$). For a 12-b ADC operating over an input range of 10 V and a gain of 1, 1 least significant bit (LSB) of error represents 2.44 mV

$$\left(\frac{\text{input range}}{\text{gain} \times 2^n}\right) = \left(\frac{10 \text{ V}}{1 \times 2^{12}}\right) = 2.44 \text{ mV}$$

But because the voltage error due to jitter is 31 mV, the accuracy error is 13 LSB

$$\left(\frac{31 \text{ mV}}{2.44 \text{ mV}}\right)$$

This represents uncertainty in the last 4 b of a 12-b ADC. Thus, the effective accuracy of the system is no longer 12 b, but rather 8 b.

External Sampling

Some DAQ applications must perform a conversion based on another physical event that triggers the data conversion. The event could be a pulse from an optical encoder measuring the rotation of a cylinder. A sample would be taken every time the encoder generates a pulse corresponding to n degrees of rotation. External triggering is advantageous when trying to measure signals whose occurrence is relative to another physical phenomena.

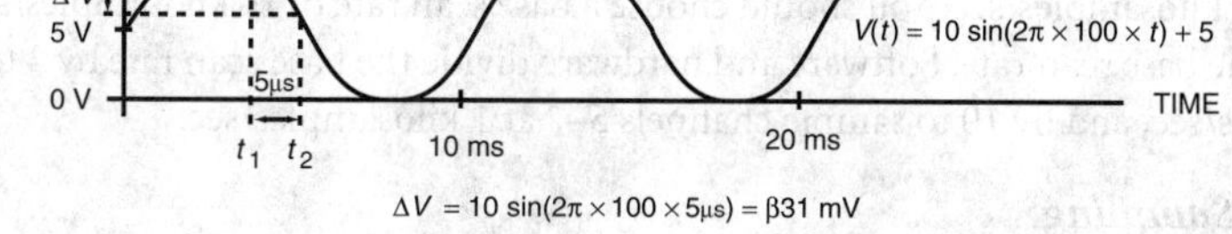

$$\Delta V = 10 \sin(2\pi \times 100 \times 5\mu s) = \beta 31 \text{ mV}$$

FIGURE 18.31 Jitter reduces the effective accuracy of the DAQ board.

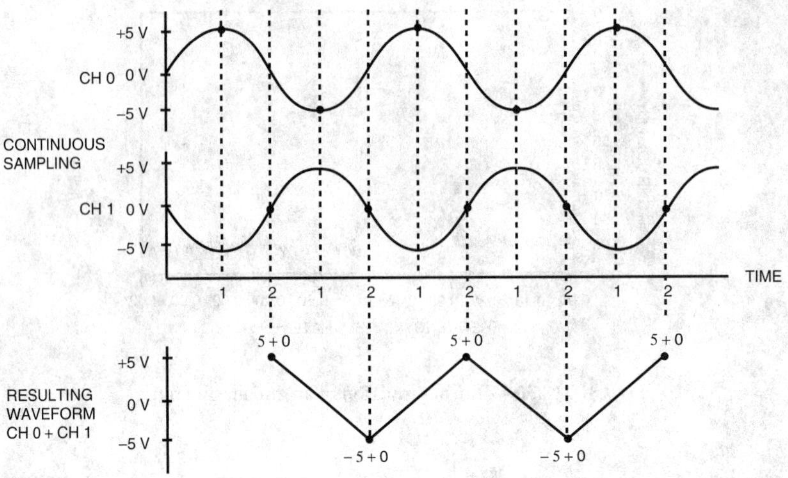

FIGURE 18.32   If the channel skew is large compared to the signal, then erroneous conclusions may result.

Continuous Scanning

When a DAQ board acquires data, several components on the board convert the analog signal to a digital value. These components include the analog multiplexer (mux), the instrumentation amplifier, the sample-and-hold circuitry, and the ADC. When acquiring data from several input channels, the analog mux connects each signal to the ADC at a constant rate. This method, known as *continuous scanning*, is significantly less expensive than having a separate amplifier and ADC for each input channel.

Continuous scanning is advantageous because it eliminates jitter and is easy to implement. However, it is not possible to simultaneously sample multiple channels. Because the mux switches between channels, a time skew occurs between any two successive channel samples. Continuous scanning is appropriate for applications where the time relationship between each sampled point is unimportant or where the skew is relatively negligible compared to the speed of the channel scan.

If you are using samples from two signals to generate a third value, then continuous scanning can lead to significant errors if the time skew is large. In Fig. 18.32, two channels are continuously sampled and added together to produce a third value. Because the two sine waves are 90° out of phase, the sum of the signals should always be zero. But because the skew time between the samples, an erroneous sawtooth signal results.

Multirate Scanning

Multirate scanning, a method that scans multiple channels at different scan rates, is a special case of continuous scanning. Applications that digitize multiple signals with a variety of frequencies use multirate scanning to minimize the amount of buffer space needed to store the sampled signals. You can use channel-independent ADCs to implement hardware multirate scanning; however, this method is extremely expensive. Instead of multiple ADCs, only one ADC is used. A channel/gain configuration register stores the scan rate per channel and software divides down the scan clock based on the per-channel scan rate. Software-controlled multirate scanning works by sampling each input channel at a rate that is a fraction of the specified scan rate.

Suppose you want to scan channels 0–3 at 10 kilosamples/sec, channel 4 at 5 kilosamples/sec, and channels 5–7 at 1 kilosamples/sec. You should choose a base scan rate of 10 kilosamples/sec. Channels 0–3 are acquired at the base scan rate. Software and hardware divide the base scan rate by 2 to sample channel 4 at 5 kilosamples/sec, and by 10 to sample channels 5–7 at 1 kilosamples/sec.

Simultaneous Sampling

For applications where the time relationship between the input signals is important, such as phase analysis of AC signals, you must use simultaneous sampling. DAQ boards capable of simultaneous sampling

typically use independent instrumentation amplifiers and sample-and-hold circuitry for each input
channel, along with an analog mux, which routes the
input signals to the ADC for conversion (as shown
in Fig. 18.33).

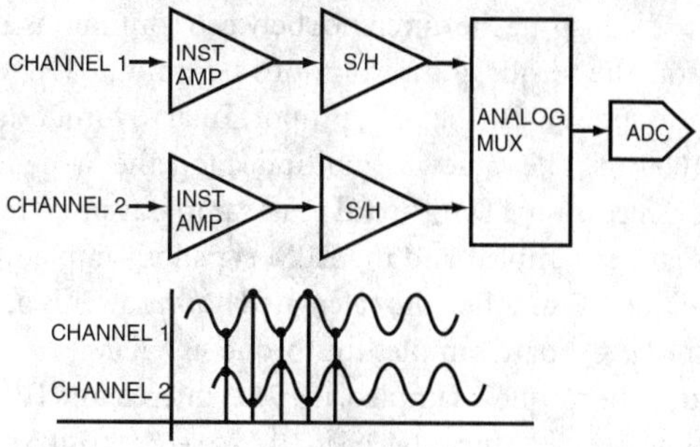

To demonstrate the need for a simultaneous-
sampling DAQ board, consider a system consisting of four 50-kHz input signals sampled at 200
kilosamples/sec. If the DAQ board uses continuous
scanning, the skew between each channel is 5 μs
(1 S/200 kilosamples/sec) which represents a 270°
(15 μs/20 μs × 360°) shift in phase between the
first channel and fourth channel. Alternately, with
a simultaneous-sampling board with a maximum

FIGURE 18.33   Block diagram of DAQ components used
to simultaneously sample multiple channels.

5 ns interchannel time offset, the phase shift is only 0.09° (5 ns/20 μs × 360°). This phenomenon is
illustrated in Fig. 18.34.

Interval Scanning

For low-frequency signals, interval scanning creates the effect of simultaneous sampling, yet maintains the
cost benefits of a continuous scanning system. This method scans the input channels at one rate and uses
a second rate to control when the next scan begins. You can scan the input channels at the fastest rate of
the ADC, creating the effect of simultaneous sampling. Interval scanning is appropriate for slow moving
signals, such as temperature and pressure. Interval scanning results in a jitter-free sample rate and minimal
skew time between channel samples. For example, consider a DAQ system with 10 temperature signals.
Using interval scanning, you can set up the DAQ board to scan all channels with an interchannel delay
of 5 μs, then repeat the scan every second. This method creates the effect of simultaneously sampling 10
channels at 1 S/s, as shown in Fig. 18.35.

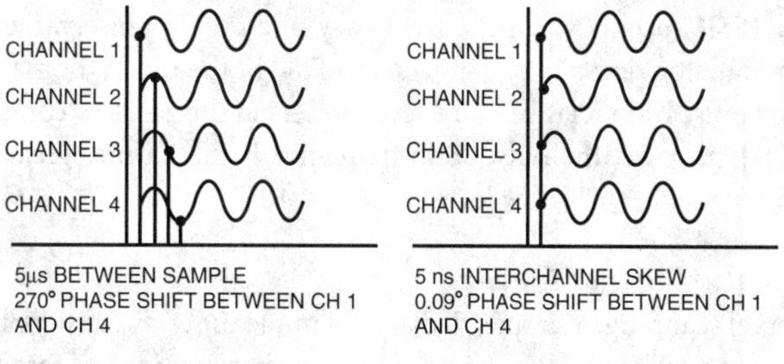

FIGURE 18.34   Comparison of continuous scanning and simultaneous sampling.

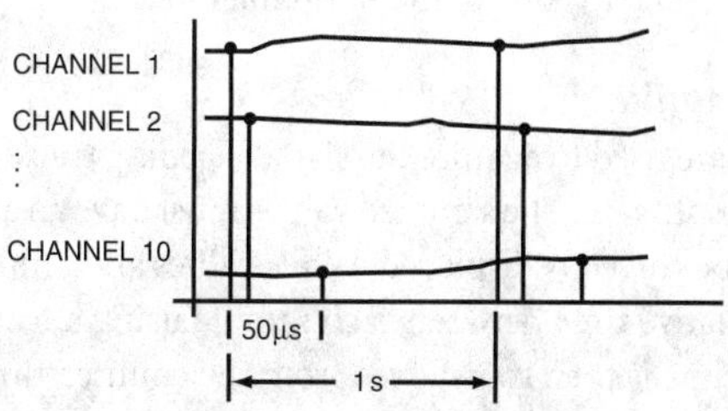

FIGURE 18.35   Interval scanning: all 10 channels are scanned within 45 μs; this is insignificant relative to the overall
acquisition rate of 1 S/s.

To illustrate the difference between continuous and interval scanning, consider an application that monitors the torque and revolutions per minute (RPMs) of an automobile engine and computes the engine horsepower. Two signals, proportional to torque and RPM, are easily sampled by a DAQ board at a rate of 1000 S/s. The values are multiplied together to determine the horsepower as a function of time. A continuously scanning DAQ board must sample at an aggregate rate of 2000 S/s. The time between which the torque signal is sampled and the RPM signal is sampled will always be 0.5 ms (1/2000). If either signal changes within 0.5 ms, then the calculated horsepower is incorrect. But using interval scanning at a rate of 1000 S/s, the DAQ board samples the torque signal every 1 ms, and the RPM signal is sampled as quickly as possible after the torque is sampled. If a 5-μs interchannel delay exists between the torque and RPM samples, then the time skew is reduced by 99% ((0.5 ms − 5 μs))/0.5 ms), and the chance of an incorrect calculation is reduced.

Seamless Changing of the Sampling Rate

This technique, a variation of real-time sampling, is used to vary the sampling rate of the ADC without having to stop and reprogram the counter/timer for different conversion rates. For example, you may want to start sampling slowly, and then, following a trigger, begin sampling quickly; this is particularly useful when performing transient analysis. The ADC samples slowly until the input crosses a voltage level, and then the ADC samples quickly to capture the transient.

Four complex timing and triggering mechanisms are necessary to seamlessly change the sampling rate. They include hardware analog triggering, frequency shift keying (FSK), flexible data acquisition signal routing, and buffered relative-time stamping.

Hardware Analog Triggering

A trigger level serves as the reference point at which the sampling rate changes. Analog trigger circuitry monitors the input voltage of the waveform and generates a transistor-transmitter logic (TTL) high level whenever the voltage input is greater than the trigger voltage Vset; and a TTL low level when the voltage input is less than the trigger voltage. Therefore, as the waveform crosses the trigger level, the analog trigger circuitry signals the counter to count at a different frequency.

Frequency Shift Keying

Frequency shift keying (FSK) occurs when the frequency of a digital pulse varies over time. Frequency modulation (FM) in the analog domain is analogous to FSK in the digital domain. FSK determines the frequency of a generated pulse train relative to the level present at the gate of a counter. For example, when the gate signal is a TTL high level, the pulse train frequency is three times greater than the pulse train frequency when the gate signal is a TTL low level.

Flexible Data Acquisition Signal Routing

For continuous or interval scanning, a simple, dedicated sample timer directly controls ADC conversions. But for more complex sampling techniques, such as seamlessly changing the sampling rate, signals from other parts of the DAQ board control ADC conversions. The DAQ-STC ASIC provides 20 possible signal sources to time each ADC conversion. One of the sources is the output of the general-purpose counter 0, which is more flexible than a dedicated timer. In particular, the general-purpose counter generates the FSK signal that is routed to the ADC to control the sampling rate.

Buffered Relative-Time Stamping

Because different sample rates are used to acquire the signal, keeping track of the various acquisition rates is a challenge for the board and software. The sampled values must have an acquisition time associated with them in order for the signal to be correctly displayed. While values are sampled by the ADC, the DAQ-STC counter/timers measure the relative time between each sample using a technique called *buffered relative-time stamping*. The measured time is then transferred from the counter/timer registers to PC memory via direct memory access.

The counter continuously measures the time interval between successive, same-polarity transitions of the FSK pulse with a measurement resolution of 50 ns. Countings begin at 0. The counter contents are

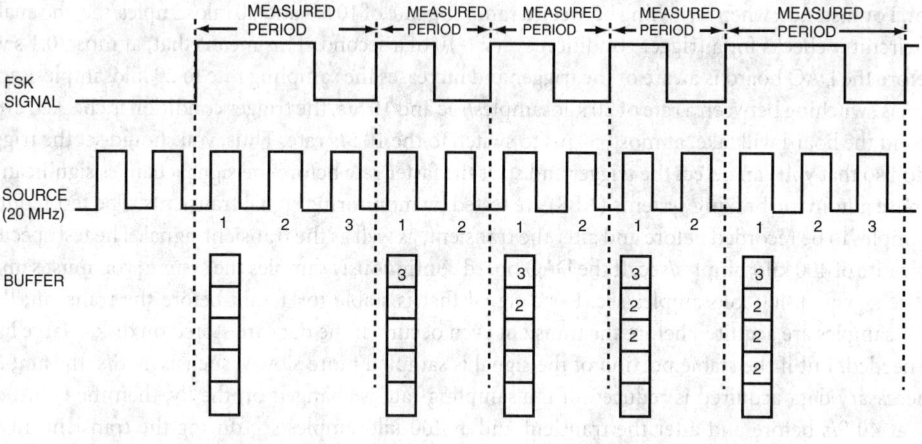

FIGURE 18.36 The number of counts is stored in a buffer. These counts are used to determine the time between conversions.

stored in a buffer after an edge of the appropriate polarity is detected. Then, the counting begins again at 0. Software routines use DMA to transfer data from the counter to a buffer until the buffer is filled.

For example, in Fig. 18.36, the period of an FSK signal is measured. The first period is 150 ns (3 clock cycles × 50-ns resolution); the second, third, and fourth periods are 100 ns (2 clock cycles × 50 ns); the fifth period is 150 ns. For a 10-MHz board that can change its sampling rate seamlessly, the FSK signal determines the ADC conversions; the effective sampling rate is 6.7 MHz for the first part of the signal and 10 MHz for the second part of the signal.

Figure 18.37 details the timing signals necessary to change the sampling rate of the ADC without missing data. As the input waveform crosses the trigger voltage 1 the analog trigger circuitry generates a low on its output. The output signal is routed to a general-purpose counter, which generates pulses at a predetermined frequency 2. Each high-to-low transition of the FSK signal causes the ADC to sample the waveform 3. When the input waveform crosses the trigger level again, the analog trigger circuit generates a high, which causes the general-purpose counter to generate pulses at a second frequency. This timing process continues until a predetermined number of samples has been acquired.

Considerations When Seamlessly Changing the Sampling Rate

The intention of seamlessly changing the sampling rate is to switch between rates, yet not miss significant changes in the signal. Selecting the various rates and the instance at which the rate changes requires some

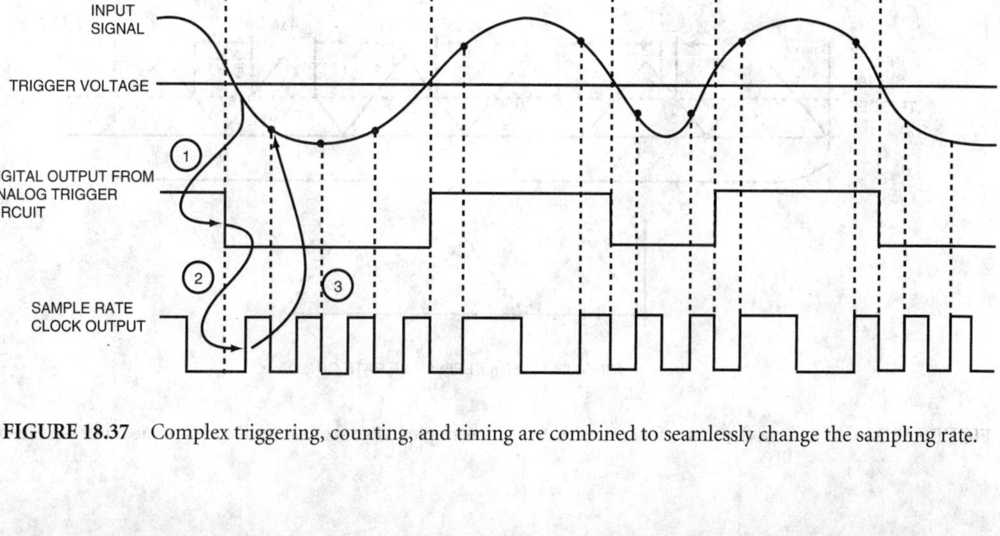

FIGURE 18.37 Complex triggering, counting, and timing are combined to seamlessly change the sampling rate.

thought. For instance, when switching between a sampling rate of 10 S/s and 20 kilosamples/sec, the analog trigger circuitry checks for a trigger condition every 1/10 of a second. This means that, at most, 0.1 s will pass before the DAQ board is aware of the trigger and increases the sampling rate to 20 kilosamples/sec. If the ADC is switching between a rate of 20 kilosamples/sec and 10 S/s, the trigger condition is checked every 50 ms, and the board will take, at most, 50 ms to switch to the 10 S/s rate. Thus, you should set the trigger condition so that you can detect the trigger and start the faster rate before the signal changes significantly.

Suppose automatic braking systems (ABS) are tested by monitoring signal transients. The test requires a few samples to be recorded before and after the transient, as well as the transient signal. The test specifies a sample rate of 400 kilosamples/sec. If the DAQ board continuously samples the system, you must sample the entire signal at 400 kilosamples/sec. For a signal that is stable for 1 min before the transient, then 24×10^6 samples are acquired before the transient even occurs. If the data are stored on disk, a large hard disk is needed. But if the stable portion of the signal is sampled more slowly, such as 40 S/s, the amount of unnecessary data acquired is reduced. If the sampling rate is changed on the fly, then the board can sample at 40 S/s before and after the transient and at 400 kilosamples/sec during the transient. In the 1 min before the transient, only 2400 samples are logged. Once the transient occurs, the board samples at 400 kilosamples/sec. Once the transient passes, the ADC begins to sample at 40 S/s again. Using this method, exactly the number of samples needed relative to the signal are logged to disk.

18.2.4 Equivalent-Time Sampling

If real-time sampling techniques are not fast enough to digitize the signal, then consider ETS as an approach. Unlike continuous, interval, or multirate scanning, complex counter/timers control ETS conversions, and ETS strictly requires that the input waveform be repetitive throughout the entire sampling period.

In ETS mode, an analog trigger arms a counter, which triggers an ADC conversion at progressively increasing time intervals beyond the occurrence of the analog trigger (as shown in Fig. 18.38). Instead of acquiring samples in rapid succession, the ADC digitizes only one sample per cycle. Samples from several cycles of the input waveform are then used to recreate the shape of the signal.

ETS Counter/Timer Operations

Four complex timing and triggering mechanisms are necessary for ETS. They include hardware analog triggering, retriggerable pulse generation, autoincrementing, and flexible data acquisition signal routing.

Hardware Analog Triggering

Analog trigger circuitry monitors the input voltage of the waveform and generates a pulse whenever the trigger conditions are met. Each time the repetitive waveform crosses the trigger level, the analog trigger circuitry arms the board for another acquisition.

MINIMUM $\Delta t = 50$ ns. EFFECTIVE RATE OF 20 MHz.

FIGURE 18.38 In ETS mode, a waveform is recreated from samples taken from several cycles of the input waveform.

Retriggerable Pulse Generation

The ADC samples when it receives a pulse for the conversion counter. In real-time sampling, the conversion counter generates a series of continuous pulses that cause the ADC to successively digitize multiple values. In ETS, however, the conversion counter generates only one pulse, which corresponds to one sample from one cycle of the waveform. This pulse is regenerated in response to a signal from the analog trigger circuitry, which occurs after each new cycle of the waveform.

Autoincrementing

If the ADC sampled every occurrence of the waveform using only retriggerable pulse generation, the same point along the repetitive signal would be digitized, namely, the point corresponding to the analog trigger. A method is needed to make the ADC sample different points along different cycles of the waveform. This method, known as *autoincrementing*, is the most important counter/timer functional for controlling ETS. An autoincrementing counter produces a series of delayed pulses.

Flexible Data Acquisition Signal Routing

As with seamlessly changing the sample rate, ETS uses signals from other parts of the board to control ADC conversions. The DAQ-STC generates autoincrementing retriggerable pulses.

Figure 18.39 details the timing signals used for ETS. As the input waveform crosses the trigger voltage 1, the analog trigger circuitry generates a pulse at the gate input of the autoincrementing counter. This counter generates a conversion pulse 2, which is used to trigger the ADC to take a sample 3. The timing process continues until a predetermined number of samples has been acquired. The delay time Dt is directly related to the relative sampling rate set. For example, for an effective sampling rate of 20 MS/s, Dt is equal to 50 ns. Because the waveform is repetitive over the complete acquisition, the points in the recreated waveform appear as if they were taken every 50 ns.

Considerations When Using ETS

Although ETS is useful in a number of applications, ETS users need to be aware of a few issues. First, the input waveform must be repetitive. ETS will not correctly reproduce a waveform that is nonrepetitive because the analog trigger will never occur at the same place along the waveform. One-shot acquisitions are not possible, because ETS digitizes a sample from several cycles of a repetitive waveform.

The accuracy of the conversion is limited to the accuracy of the counter/timer. Jitter in the counter/timer causes the sample to be taken within a multiple of 50 ns from the trigger. The sample may occur along a

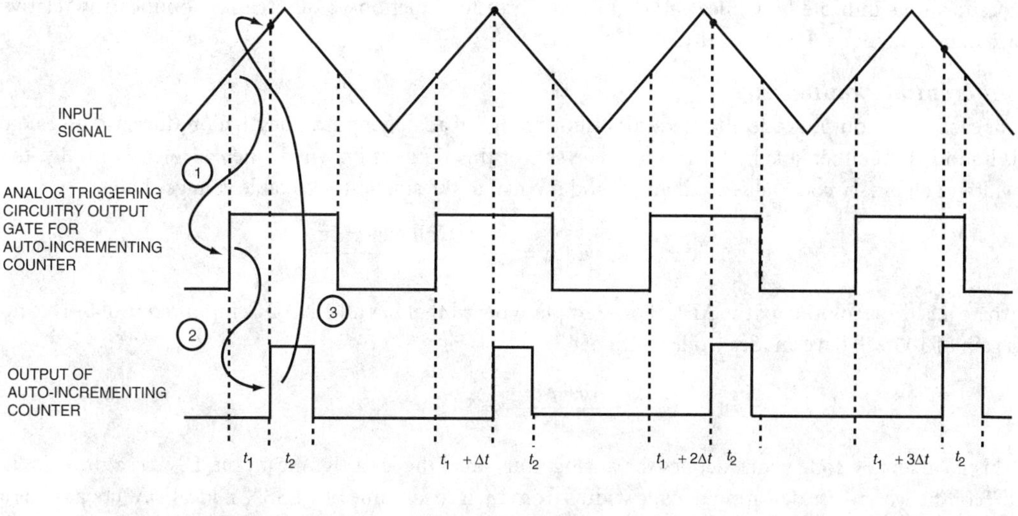

FIGURE 18.39 Complex triggering, counting, and timing are combined to control the ADC in ETS.

slightly different portion of the input waveform than expected. For example, using a 12-b board to sample the 100-Hz, 10-V peak-to-peak sine wave in Fig. 18.31, the accuracy error is 0.13 LSB.

$$\frac{\Delta V}{2.44 \text{ mV}} = \frac{10 \sin(72\pi \times 100 \times 50 \times 10^{-9})}{2.44 \text{ mV}}$$

ETS Application Examples

ETS techniques are useful in measuring the rise and fall times of TTL signals. ETS is also used to measure the impulse response of a 1-MHz, low-pass filter subject to a 2-ms repetitive impulse. The impulse response is a repetitive signal of pulses that also have a duration of 2 ms. Using a 20-MHz ETS conversion rate results in 40 digitized samples per input impulse. This is enough samples to display the impulse response in the time domain. Other applications for ETS include disk drive testing, nondestructive testing, ultrasonic testing, vibration analysis, laser diode characterization, and impact testing.

18.2.5   Factors Influencing the Accuracy of Your Measurements

How do you tell if the plug-in data acquisition board that you already have or the board that you are considering integrating into your system will give you the results you want? With a sophisticated measuring device like a plug-in DAQ board, you can obtain significantly different accuracies depending on which board you are using. For example, you can purchase DAQ products on the market today with 16-b analog to digital converters and get less than 12 b of useful data, or you can purchase a product with a 16-b ADC and actually get 16 b of useful data. This difference in accuracies causes confusion in the PC industry where everyone is used to switching out PCs, video cards, printers, and so on, and experiencing similar results between equipment.

The most important thing to do is to scrutinize more specifications than the resolution of the A/D converter that is used on the DAQ board. For DC-class measurements, you should at least consider the settling time of the instrumentation amplifier, **differential non-linearity** (**DNL**), **relative accuracy, integral nonlinearity** (**INL**), and noise. If the manufacturer of the board you are considering does not supply you with each of these specifications in the data sheets, you can ask the vendor to provide them or you can run tests yourself to determine these specifications of your DAQ board.

Linearity

Ideally, as you increase the level of voltage applied to a DAQ board, the digital codes from the ADC should also increase linearly. If you were to plot the voltage vs. the output code from an ideal ADC, the plot would be a straight line. Deviations from this ideal straight line are specified as the nonlinearity. Three specifications indicate how linear a DAQ boards transfer function is: differential nonlinearity, relative accuracy, and integral nonlinearity.

Differential Nonlinearity

For each digital output code, there is a continuous range of analog input values that produce it. This range is bounded on either side by transitions. The size of this range is known as the *code width*. Ideally, the width of all binary code values is identical and is equal to the smallest detectable voltage change,

$$\text{ideal code width} = \frac{\text{input voltage range}}{2^n}$$

where n is the resolution of the ADC. For example, a board that has a 12-b ADC, input range of 0–10, and a gain of 100 will have an ideal code width of

$$\frac{10 \text{ V}/100}{2^{12}} = 24 \,\mu\text{V}$$

This ideal analog code width defines the analog unit called the least significant bit. DNL is a measure in LSB of the worst-case deviation of code widths from their ideal value of 1 LSB. An ideal DAQ board has a DNL of 0 LSB. Practically, a good DAQ board will have a DNL within ± 0.5 LSB.

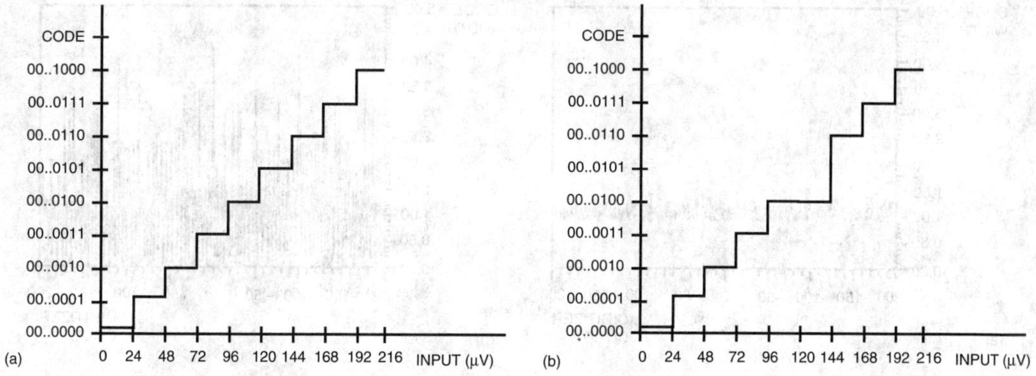

FIGURE 18.40 Demonstration of a missing code: (a) an ideal transfer curve with DNL of 0 and no missing codes, (b) a poor transfer curve with a missing code at 120 μV.

There is no upper limit on how wide a code can be. Codes do not have widths of less than 0 LSB, so the DNL is never worse than −1 LSB. A poorly performing DAQ board may have a code width equal to or very near zero, which indicates a *missing code*. No matter what voltage you input to the DAQ board with a missing code, the board will never quantize the voltage to the value represented by this code. Sometimes DNL is specified by stating that a DAQ board has no missing codes, which means that the DNL is bounded below by −1 LSB but does not make any specifications about the upper bound.

If the DAQ board in the previous example had a missing code at 120 μV, then increasing the voltage from 96 to 120 μV would not be detectable. Only when the voltage is increased another LSB, or in this example, 24 μV, will the voltage change be detectable (Fig. 18.40). As you can see, poor DNL reduces the resolution of the board.

To run your own DNL test:

1. Input a high-resolution, highly linear triangle wave into one channel of the DAQ board. The frequency of the triangle should be low and the amplitude should swing from minus full scale to plus full scale of the input to the DAQ board.
2. Start an acquisition on the plug-in board so that you acquire a large number of points. Recommended values are 1×10^6 samples for a 12-b board and 20×10^6 million samples for a 16-b board.
3. Make a histogram of all the acquired binary codes. This will give you the relative frequency of each code occurrence.
4. Normalize the histogram by dividing by the averaged value of the histogram.
5. The DNL is the greatest deviation from the value of 1 LSB. Because the input was a triangle wave, it had a uniform distribution over the DAQ board codes. The probability of a code occurring is directly proportional to the code width, and therefore shows the DNL.

Figure 18.41 shows a section of the DNL plot of two products with the same 16-b ADC yet significantly different DNL. Ideally, the codewidth plot will be a straight line at 1 LSB. Figure 18.41(a) shows the codewidth plot of a product that has DNL of less than $\pm$0.5 LSB. Because the codewidth is never 0, the product also has no missing codes. Figure 18.41(b) shows the codewidth plot of a product that has poor DNL. This product has many missing codes, a code width that is as much as 4.2 LSB at the code value 0, and is clearly not 16-b linear.

Relative Accuracy

Relative accuracy is a measure in LSBs of the worst-case deviation from the ideal DAQ board transfer function, a straight line. To run your own relative accuracy test:

1. Connect a high-accuracy analog voltage-generation source to one channel of the DAQ board. The source should be very linear and have a higher resolution than the DAQ board that you wish to test.

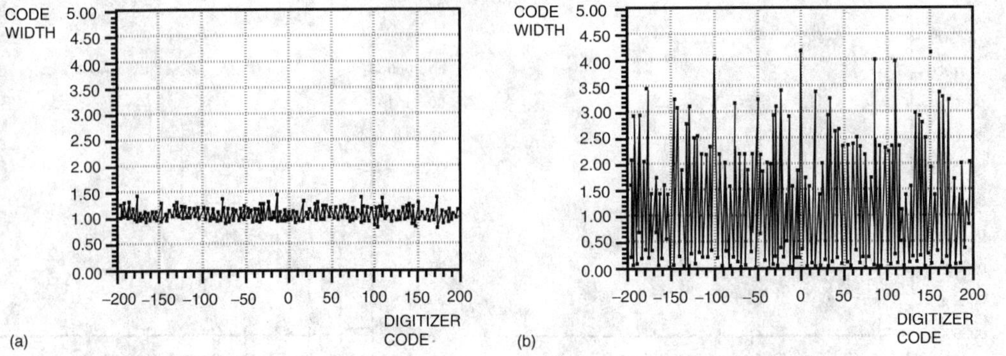

FIGURE 18.41 Section of DNL plots at the critical point (zero crossing) of two products that have the same ADC: (a) the product is the National Instruments AT-MIO-16X, which has a DNL of less than ±0.5 and has no missing codes, (b) the product has very poor DNL and many missing codes.

2. Generate a voltage that is near minus full scale.
3. Acquire at least 100 samples from the DAQ board and average them. The reason for averaging is to reduce the effects of any noise that is present.
4. Increase the voltage slightly and repeat step 3. Continue steps 3 and 4 until you have swept through the input range of the DAQ board.
5. Plot the averaged points on the computer. You will have a straight line, as shown in Fig. 18.42(a), unless the relative accuracy of your DAQ board is astonishingly bad.
6. To see the deviation from a straight line, you must first generate an actual straight line in software that starts at the minus full-scale reading and ends at the plus full-scale reading using a straight-line endpoint fit analysis routine.
7. Subtract the actual straight line from the waveform that you acquired. If you plot the resulting array, you should see a plot similar to Fig. 18.42(b).
8. The maximum deviation from zero is the relative accuracy of the DAQ board.

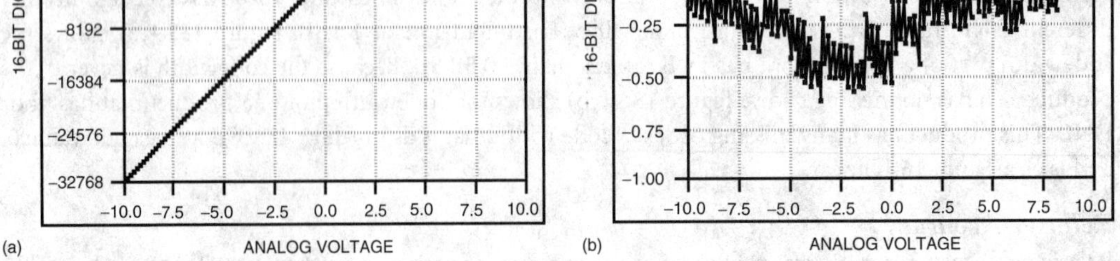

FIGURE 18.42 Determining the relative accuracy of a DAQ board: (a) the plot generated by sweeping the input and how it appears as a straight line, (b) how the straight line of Fig. 18.42(a) is not actually straight by subtracting an actual straight line.

The driver software for a DAQ board will translate the binary code value of the ADC to voltage by multiplying by a constant. Good relative accuracy is important for a DAQ board because it ensures that the translation from the binary code of the ADC to the voltage value is accurate. Obtaining good relative accuracy requires that both the ADC and the surrounding analog support circuitry be designed properly.

Integral Nonlinearity

INL is a measure in LSB of the straightness of a DAQ board's transfer function. Specifically, it indicates how far a plot of the DAQ boards *transitions* deviates from a straight line. Factors that contribute to poor INL on a DAQ board are the ADC, multiplexer, and instrumentation amplifier. The INL test is the most difficult to make:

1. Input a signal from a digital to analog converter that has higher resolution and linearity than the DAQ board that you are testing.
2. Starting at minus full scale, increase the codes to the DAC until the binary reading from the ADC flickers between two consecutive values. When the binary reading flickers evenly between codes, you have found the LSB transition for the DAQ board.
3. Record the DAC codes of all transitions.
4. Using an analysis routine, make an endpoint fit on the recorded DAC codes.
5. Subtract the endpoint fit line from the recorded DAC values.
6. The farthest deviation from zero is the INL.

Even though highly specified, the INL specification does not have as much value as the relative accuracy specification and is much more difficult to make. The relative accuracy specification is showing the deviation from the ideal straight line whereas the INL is showing the deviation of the transitions from an ideal straight line. Therefore, relative accuracy takes the quantization error, INL, and DNL into consideration.

Settling Time

On a typical plug-in DAQ board, an analog signal is first selected by a multiplexer, and then amplified by an instrumentation amplifier before it is converted to a digital signal by the ADC. This instrumentation amplifier must be able to track the output of the multiplexer as the multiplexer switches channels, and the instrumentation amplifier must be able to settle to the accuracy of the ADC. Otherwise, the ADC will convert an analog signal that has not settled to the value that you are trying to measure with your DAQ board. The time required for the instrumentation amplifier to settle to a specified accuracy is called the *settling time*. Poor settling time is a major problem because the amount of inaccuracy usually varies with gain and sampling rate. Because the errors occur in the analog stages of the DAQ board, the board cannot return an error message to the computer when the instrumentation amplifier does not settle.

The instrumentation amplifier is most likely not to settle when you are sampling multiple channels at high gains and high rates. When the application is sampling multiple channels, the multiplexer is switching among different channels that can have significant differences in voltage levels (Fig. 18.43). Instrumentation amplifiers can have difficulty tracking this significant difference in voltage. Typically, the higher the gain and the faster the channel switching time, the less likely it is that the instrumentation amplifier will settle.

It is important to be aware of settling time problems so you know at what multichannel rates and gains you can run your DAQ board and maintain accurate readings. Running your own settling time test is relatively easy:

1. Apply a signal to one channel of your DAQ board that is nearly full scale. Be sure to take the range and gain into account when you select the levels of the signal you will apply. For example, if you are using a gain of 100 on a ± 10 V input range on the DAQ board, apply a signal slightly less than 10 V/100 or $+0.1$-V signal to the channel.
2. Acquire at least 1000 samples from one channel only and average them. The averaged value will be the expected value of the DAQ board when the instrumentation amplifier settles properly.
3. Apply a minus full-scale signal to a second channel. Be sure to take range and gain into account, as described earlier.

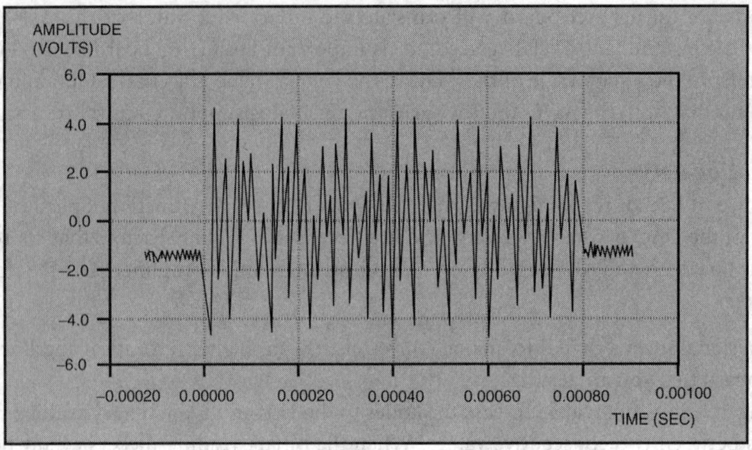

FIGURE 18.43 The input to an instrumentation amplifier that is multiplexing 40 DC signals appears to be a high-frequency AC signal.

4. Acquire at least 1000 samples from the second channel only and average them. This will give you the expected value for the second channel when the instrumentation amplifier settles properly.
5. Have the DAQ board sample both channels at the highest possible sampling rate so that the multiplexer is switching between the first and the second channels.
6. Average at least 100 samples from each channel.
7. The deviation between the values returned from the board when you sampled the channels using a single channel acquisition and the values returned when you sampled the channels using a multichannel acquisition will be the settling time error.

To plot the settling time error, repeat steps 5 and 6, but reduce the sampling rate each time. Then plot the greatest deviation at each sampling rate. A DAQ board settles to the accuracy of the ADC only when the settling time error is less than ±0.5 LSB. By plotting the settling time error, you can determine at what rate you can sample with your DAQ board and settle to the accuracy that your application requires. The settling-time plot will usually vary with gain and so you will want to repeat the test at different gains.

Figure 18.44 shows an example of two DAQ boards that have a 12-b resolution ADC, yet have significantly different settling times. For the tests, both boards were sampling two channels with a gain of 100. The settling-time plot for the first board, shown in Fig. 18.44(a), uses an off-the-shelf instrumentation amplifier, which has 34 LSBs of settling-time error when sampling at 100 kHz. Because the board was using an input range of 20 V at a gain of 100, 1 LSB = 48.8 μV. Therefore, data are inaccurate by as much as 1.7 mV due

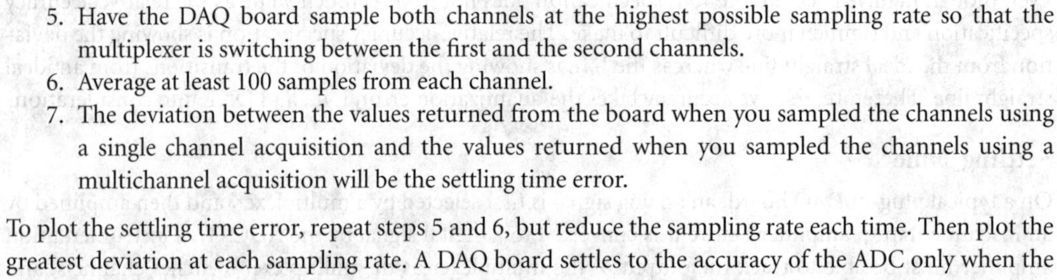

FIGURE 18.44 Settling plots for two DAQ boards that have the same resolution ADC yet significantly different settling times. Both boards were using a gain of 100 (a) settling time of off-the-shelf instrumentation amplifier, (b) settling time of NI-PGIA instrumentation amplifier.

to settling-time when the board is sampling at 100 kHz at a gain of 100. If you use this board you must either increase the sampling period to greater than 20 μs so that the board can settle, or be able to accept the settling-time inaccuracy. The settling-time plot for the second board, shown in Fig. 18.44(b), settles properly because it uses a custom instrumentation amplifier designed specifically to settle in DAQ board applications. Because the board settles to within ±0.5 LSB at the maximum sampling rate, you can run the board at all sampling rates and not have any detectable settling-time error.

Applications that are sampling only one channel, applications that are sampling multiple channels at very slow rates, or applications that are sampling multiple channels at low gains usually do not have a problem with settling time. For the other applications, your best solution is to purchase a DAQ board that will settle at all rates and all gains. Otherwise, you will have to either reduce your sampling rate or realize that your readings are not accurate to the specified value of the ADC.

Noise

Noise is any unwanted signal that appears in the digitized signal of the DAQ board. Because the PC is a noisy digital environment, acquiring data on a plug-in board takes a very careful layout on multilayer DAQ boards by skilled analog designers. Designers can use metal shielding on a DAQ board to help reduce noise. Proper shielding should not only be added around sensitive analog sections on a DAQ board, but must also be built into the DAQ board with ground planes.

Of all of the tests to run, a DC-class signal noise test is the easiest to run:

1. Connect the + and − inputs of the DAQ board directly to ground. If possible, make the connection directly on the I/O connector of the board. By doing this, you can measure only the noise introduced by the DAQ board instead of noise introduced by your cabling.
2. Acquire a large number of points (1×10^6) with the DAQ board at the gain you choose in your application.
3. Make a histogram of the data and normalize the array of samples by dividing by the total number of points acquired.
4. Plot the normalized array.
5. The deviation, in LSB, from the highest probability of occurrence is the noise.

Figure 18.45 shows the DC-noise plot of two DAQ products, both of which use the same ADC. You can determine two qualities of the DAQ board from the noise plots—range of noise and the distribution.

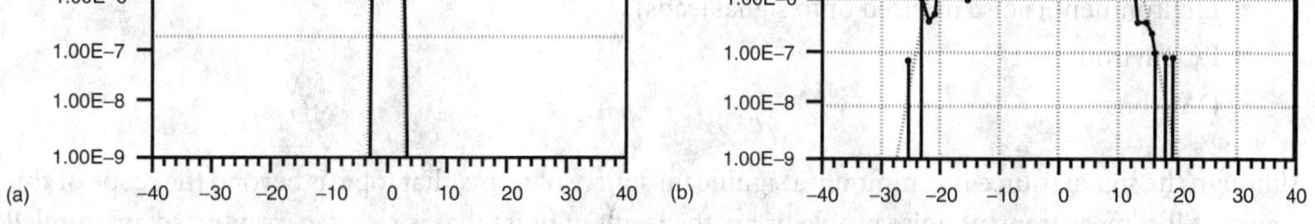

FIGURE 18.45 Noise plots of two DAQ products, which have significantly different noise performance even though they use the same 16-b ADC: (a) The National Instruments AT-MIO-16X, which has codes ranging from −3 LSB to +3 LSB. The codes at ±3 LSB have less than a 10^4 and a 10^{-7} probability of occurrence. (b) The DAQ product has noise as high as ±20 LSB and has a high probability (10^{-4}) of codes occurring as much as 15 LSB from the expected value.

The plot in Fig. 18.45(a) has a high distribution of samples at 0 and a very small number of points occurring at other codes. The distribution is Gaussian, which is what is expected from random noise. From the plot, the noise level is within ±3 LSB. The plot in Fig. 18.45(b) is a very noisy DAQ product, which does not have the expected distribution and has a noise greater than ±20 LSB, with many samples occurring at points other than the expected value. For the DAQ product in Fig. 18.45(b), the tests were run with an input range of ±10 V and a gain of 10. Therefore, 1 LSB = 31 μV, thus a noise level of 20 LSB is equivalent to 620 μV of noise.

18.2.6   Common Issues with PC-based Data Acquisition

Aside from software and programming, the most common problem users run into when putting together DAQ systems is a noisy measurement. Unfortunately, noise in DAQ systems is a complicated issue and difficult to avoid. However, it is useful to understand where noise typically comes from, how much noise is to be expected, and some general techniques to reduce or avoid noise corruption.

The Low-Noise Voltmeter Standard

One of the most common questions asked of technical support personnel of DAQ vendors concerns the voltmeter. Users connect their signal leads to a handy voltmeter or digital multimeter (DMM), without worrying too much about cabling and grounding, and obtain a rock solid reading with little jitter. The user then duplicates the experiment with a DAQ board and is disappointed to find that the readings returned by the DAQ board look very noisy and very unstable.

The user decides there is a problem with the DAQ board and calls the technical support line for help. In fact, the user has just demonstrated the effects of two different measurement techniques, each with advantages and disadvantages. DAQ boards are designed as flexible, general-purpose measurement devices. The DAQ board front end typically consists of a gain amplifier and a *sampling* analog-to-digital converter. A sampling ADC takes an instantaneous measurement of the input signal. If the signal is noisy, the sampling ADC will digitize the signal as well as a noise. The digital voltmeter, on the other hand, will use an *integrating* ADC that integrates the signal over a given time period. This integration effectively filters out any high-frequency noise that is present in the signal.

Although the integrating input of the voltmeter is useful for measuring static, or DC signals, it is not very useful for measuring changing signals or digitizing waveforms, or for capturing transient signals. The plug-in DAQ for board, with its sampling ADC, has the flexibility to perform all of these types of measurements. With a little software, the DAQ board can also emulate the operation of the integrating voltmeter by digitizing the static signal at a higher rate and performing the integration, or averaging, of the signal in software.

Where Does Noise Come from?

There are basically four possible sources of noise in a DAQ system:

- Signal source, or transducer
- Environment (noise induced onto signal leads)
- PC environment
- DAQ board

Although the signal source is commonly a significant source of noise, that topic is beyond the scope of this chapter. Most measurement noise problems are the result of noise that is radiated, conducted, or coupled onto signal wires attaching the sensor or transducer to the DAQ equipment. Signal wires basically act as antennas for noise.

Placing a sensitive analog measurement device, like a plug-in DAQ board, inside a PC chassis might seem like asking for trouble. The high-speed digital traffic and power supplies inside a PC are prime candidates for noise radiation. For example, it is a good idea to not install your DAQ board directly next to your video card.

Probably the most dangerous area for your analog signals is not inside the PC, but on top of it. Keep your signal wires clear of your video monitor, which can radiate large high-frequency noise levels onto your signal.

The DAQ board itself can be a source of measurement noise. Poorly designed boards, for example, may not properly shield the analog sections from the digital logic sections that radiate high-frequency switching noise. Properly designed boards, with well-designed shielding and grounding, can provide very low-noise measurements in the relatively noisy environment of the PC.

Proper Wiring and Signal Connections

In most cases, the major source of noise is the environment through which the signal wires must travel. If your signal leads are relatively long, you will definitely want to pay careful attention to your cabling scheme. A variety of cable types are available for connecting sensors to DAQ systems. Unshielded wires or ribbon cables are inexpensive and work fine for high-level signals and short to moderate cable lengths. For low-level signals or longer signal paths, you will want to consider shielded or twisted-pair wiring. Tie the shield for each signal pair to ground reference at the source. Practically speaking, consider shielded, twisted-pair wiring if the signal is less than 1 V or must travel farther than approximately 1 m. If the signal has a bandwidth greater than 100 kHz, however, you will want to use coaxial cables.

Another useful tip for reducing noise corruption is to use a *differential* measurement. Differential inputs are available on most signal conditioning modules and DAQ boards. Because both the (+) and (−) signal lines travel from signal source to the measurement system, they pick up the same noise. A differential input will reject the voltages that are common to both signal lines. Differential inputs are also best when measuring signals that are referenced to ground. Differential inputs will avoid ground loops and reject any difference in ground potentials. On the other hand, single-ended measurements reference the input to ground, causing ground loops and measurement errors.

Other wiring tips:

- If possible, route your analog signals separately from any digital I/O lines. Separate cables for analog and digital signals are preferred.
- Keep signal cables as far as possible from AC and other power lines.
- Take caution when shielding analog and digital signals together. With a single-ended (not differential) DAQ board, noise coupled from the digital signals to the analog signals via the shield will appear as noise. If using a differential input DAQ board, the coupled noise will be rejected as common-mode noise (assuming the shield is tied to ground at one end only).

Source Impedance Considerations

When time-varying electric fields, such as AC power lines, are in the vicinity of your signal leads, noise is introduced onto the signal leads via capacitive coupling. The capacitive coupling increases in direct proportion to the frequency and amplitude of the noise source and to the impedance of the measurement circuit. Therefore, the source impedance of your sensor or transducer has a direct effect on the susceptibility of your measurement circuit to noise pickup. The higher the source impedance of your sensor or signal source, the larger the amount of capacitive coupling. The best defense against capacitive noise coupling is the shield that is grounded at the source end. Table 18.4 lists some common transducers and their impedance characteristic.

Pinpointing Your Noise Problems

If your system is resulting in noisy measurements, follow these steps to determine the source of the noise and how best to reduce it. The first step can also give you an idea of the noise performance of your DAQ board itself. The steps are

1. Short one of the analog input channels of the DAQ board to ground directly at the I/O connector of the board. Then, take a number of readings and plot the results. The amount of noise present is the amount of noise introduced by the PC and the DAQ board itself with a very low-impedance

TABLE 18.4   Impedance Characteristics of Transducers

Transducer	Impedance Characteristic
Thermocouples	Low ($<20\ \Omega$)
RTDs	Low ($<1\ k\Omega$)
Strain gauges	Low ($<1\ k\Omega$)
Thermistors	Moderate to high ($>1\ k\Omega$)
Solid-state pressure transducer	High ($<1\ k\Omega$)
Potentiometer	High (500 to 100 $k\Omega$)
Glass pH electrode	Very high ($10^9\ \Omega$)

input. Typical results are shown in Fig. 18.46. This plot shows a reading that jumps between 0.00 and 2.44 mV. Because this particular board uses a 12-b ADC and the amplifier was set to a gain of 1, this deviation corresponds to only 1 b, or LSB. In other words, the 12-b ADC toggled between binary values 0 and 1. If this test yields large amount of noise, your DAQ board is not operating properly, or another plug-in board in the PC may be radiating noise onto the DAQ board. Try removing other PC boards to see if the noise level decreases.

2. Attach your signal wires to the DAQ board. At your signal source or signal conditioning unit, ground or short the input leads. Acquire and plot a number of readings as in step 1. If the observed noise levels are roughly the same as those with the actual signal source instead of the short in place, the cabling and/or the environment in which the cabling is run is the culprit. You may need to try relocating your cabling farther from potential noise sources. If the noise source is not known, spectral analysis can help identify the source of the noise.

3. If the noise level in step 2 is less than with the actual signal source, replace the short with a resistor approximately equal to the output impedance of the signal source. This setup will show whether capacitive coupling in the cable due to high impedance is the problem. If the observed noise level is still less than with the actual signal source, then cabling and the environment can be dismissed as the problem. In this case, the culprit is either the signal source itself, or improper grounding configuration.

Noise Removal Strategies

After you have optimized your cabling and hardware setup, you may still need additional techniques to reduce noise that is unavoidable with proper cabling and grounding.

Signal Amplification

If you must pass very low-level signals through long signal leads, you will want to consider amplifying the signals near the source. An amplifying signal conditioner could boost the signal level before it is subject to the noise corruption of the environment. The same amount of noise will be radiated onto the signal, but will have a much smaller effect on the high-level signal.

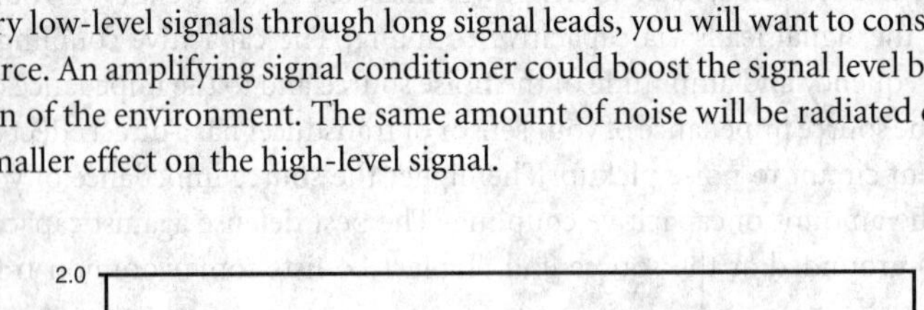

FIGURE 18.46   Measured noise with shorted inputs to DAQ board.

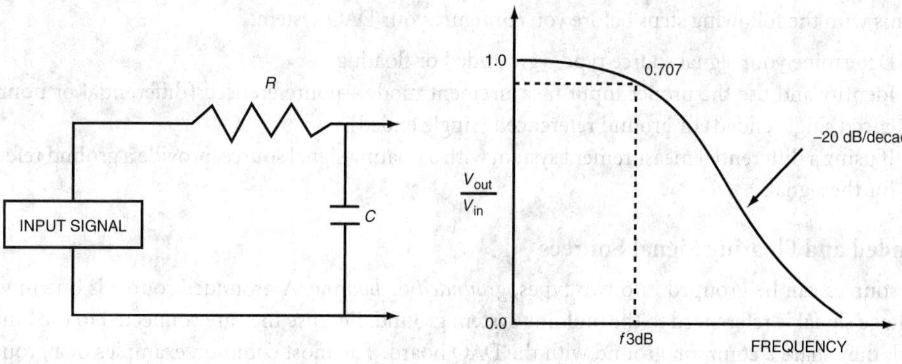

FIGURE 18.47   Simple low-pass RC filter.

Averaging

A very powerful technique for making low-noise measurements of static, or DC, signals is data averaging. For example, suppose you were monitoring the output of a thermocouple in an environment known to contain high amounts of 60-Hz power line noise. For each required temperature reading, therefore, you collect 100 readings over a time period of $i/60$ s, where i is some integer, and average the 100 data readings. Because the data were collected over an integer number of 60-Hz power cycles, the averaging of the data will average out any 60 Hz noise to zero. For 50-Hz power noise, collect the readings over a time period equal to $i/50$ s. This averaging has the same filtering effect as the integrating voltmeter.

Filtering

Of course, one method of removing noise from a electrical signal is with a hardware filter. There are a couple of options. First, you can use commercial signal conditioners that implement low-pass filters. Or, for simple filtering needs (moderate amounts of noise), you might consider building a simple RC filter on the input of your DAQ board. Figure 18.47 shows a single-pole RC filter that you could easily build and would attenuate signals with a frequency higher than the cutoff frequency F_c. F_c will be equal to

$$\frac{1}{2\pi RC}$$

For $R = 5\ k\Omega$ and $C = 4.7\ \mu F$, the cutoff frequency can be calculated as

$$F_c = \frac{1}{2\pi RC} = \frac{1}{2\pi (5000\ \Omega)(4.7 \cdot 10^{-6})} = 6.8\ \text{Hz}$$

Figure 18.47 also shows the transfer function, which shows the attenuation of the signal (defined as V_{out}/V_{in}) as a function of frequency. If a signal of frequency F_C, for example, is applied to the input, the signal output of the RC filter will be equal to $0.707(V_{in})$. The equation that tells us the amount of attenuation as a function of input frequency F_{in}, for the RC filter is

$$\left| \frac{V_{out}}{V_{in}} \right| = \frac{1}{\sqrt{1 + (F_{in}/F_c)^2}}$$

For the single-pole RC filter with $F_c = 6.8$ Hz subject to 60-Hz noise input, the equation above is calculated to be 0.11. Therefore, 60-Hz noise introduced into the signal will be attenuated by 89%.

Grounding

Another very common source of problems in DAQ systems is grounding. In fact, noisy measurement problems are often due to improper grounding of the measurement circuit. You can avoid most grounding problems with the following steps before you configure your DAQ system:

1. Determine your signal source type—grounded or floating.
2. Identify and use the proper input measurement mode—nonreferenced (differential or nonreferenced single ended) or ground referenced (single ended).
3. If using a differential measurement system with a floating signal source, provide a ground reference for the signal.

Grounded and Floating Signal Sources

Signal sources can be grouped into two types, *grounded* or *floating*. A grounded source is one in which the voltage signal is referenced to the building system ground. Because they are connected to the building ground, they share a common ground with the DAQ board. The most common examples of a grounded source are devices that plug into the building ground, such as signal generators and power supplies.

A floating source is a source in which the voltage signal is not referred to an absolute reference, such as Earth or building ground. Some common examples of floating signal sources are batteries, battery-powered signal sources, thermocouples, transformers, isolation amplifiers, and any instrument that explicitly floats its output signal. Notice that neither terminal of the source is referred to the electrical outlet ground. Thus, each terminal is independent of Earth.

Types of Measurement Inputs

In general, with regards to ground-referencing of the inputs, there are three types of measurement systems. Following is a description of each type.

Differential Inputs

A differential, or nonreferenced, measurement system has neither of its inputs tied to a fixed reference such as Earth or building ground. DAQ boards with instrumentation amplifiers can be configured as differential measurement systems.

An ideal differential measurement system responds only to the potential *difference* between its two terminals, the $(+)$ and $(-)$ inputs. Any voltage measured with respect to the instrumentation amplifier ground present at both amplifier inputs is referred to as a common-mode voltage. The term *common-mode* voltage range measures the ability of a DAQ board in differential mode to reject the common-mode voltage signal.

Ground-Referenced Inputs

A grounded or ground-referenced measurement system is similar to a grounded source, in that the measurement is made with respect to ground. This is also referred to as a *ground-referenced single-ended* (GRSE) measurement system.

Nonreferenced Single-Ended Inputs

A variant of the single-ended measurement technique, known as *nonreferenced single-ended* (NRSE) measurement system, is often found in DAQ boards. In an NRSE measurement system, all measurements are still made with respect to a single-node analog sense, but the potential at this node can vary with respect to the measurement system ground.

Now that we have identified the different types of signal sources and measurement systems, we can discuss the proper measurement system for each type of signal source.

Measuring Grounded Signal Sources, Avoiding Loops

A grounded signal source is best measured with a differential or NRSE measurement system. With this configuration, the measured voltage V_m is the sum of the signal voltage V_s and the potential difference ΔV_g that exists between the signal source ground and the measurement system ground. This potential difference is generally not a DC level; thus, the result is a noisy measurement system often showing power-line frequency (50 or 60 Hz) components in the readings. Ground-loop introduced noise may have both AC and DC components, thus introducing offset errors as well as noise in the measurements. The potential difference between the two grounds causes a current to flow in the interconnection. This current is called *ground-loop* current.

The preferable input mode for a grounded signal is differential or NRSE mode. With either of these configurations, any potential difference between references of the source and the measuring device appears as common-mode voltage to the measurement system and is subtracted from the measured signal.

Measuring Floating Signals

You can use differential or single-ended inputs to measure a floating signal source. With a ground-referenced input, the DAQ board provides the one ground reference. When using differential inputs to measure signals that are not ground referenced, however, you must explicitly provide a ground reference to make accurate measurements. The differential input can be referenced by simple grounding of the $(-)$ lead of the signal input. Alternatively, resistors can be connected from each signal lead to ground. This configuration maintains a balanced input and may be desirable for high-impedance signal sources. Many signal conditioning accessories include provisions for installing these resistors or direct connections to ground. Figure 18.48 summarizes the analog input connections.

Basic Signal Conditioning Functions

In general, signal conditioners exist to interface raw signals from transducers to a general-purpose measurement device, such as a plug-in DAQ board, while simultaneously boosting the quality and reliability of the measurement. To accomplish this goal, signal conditioners perform a number of functions, including the following.

Signal Amplification

Many transducers output very small voltages that can be difficult to measure accurately. For example, a J-type thermocouple signal varies only about 50 μV/°C over most of its range. Most signal conditioners, therefore, include amplifiers to boost the signal level to better match the input range of the analog-to-digital converter and improve resolution and sensitivity. Although many DAQ boards and I/O devices include onboard amplifiers for this reason, it may be necessary to locate an additional signal conditioner with amplification near the source of low-level signals, such as thermocouples, to increase their immunity to electrical noise from the environment. Otherwise, any small amount of noise picked up on lead wires can corrupt your data.

Filtering

Additionally, signal conditioners can include filters to reject unwanted noise within a certain frequency range. For example, most conditioners include low-pass filters to reduce high-frequency noise, such as the very common 60- or 50-Hz periodic noise from power systems or machinery. Some signal conditioners that are used for more dynamic measurements, such as vibration monitoring, include special antialiasing that feature programmable bandwidth (variable according to the sampling rate) and very sharp filter rolloff.

Isolation

One of the most common cause of measurement problems, noise, and damaged I/O equipment is improper grounding of the system. These nagging problems tend to disappear when isolated signal conditioners are introduced into the measurement system. Isolated conditioners pass the signal from its source to the measurement device without a galvanic or physical connection. Besides breaking ground loops, isolation blocks high-voltage surges and rejects high common-mode voltage, protecting expensive DAQ instrumentation.

For example, suppose you are to monitor the temperature of an extrusion process. Although you are using thermocouples with output signals of 0 and 50 mV, the thermocouples are soldered to the extruder. The extruder machines are powered by a dedicated power system and your thermocouple leads are actually sitting at 50 V. Connecting the thermocouple leads directly to nonisolated DAQ board would probably damage the board. However, you can connect the thermocouple leads to an isolated signal, which rejects the common-mode voltage (50 V), safely passing the differential 50-mV differential signal on the DAQ board for measurement.

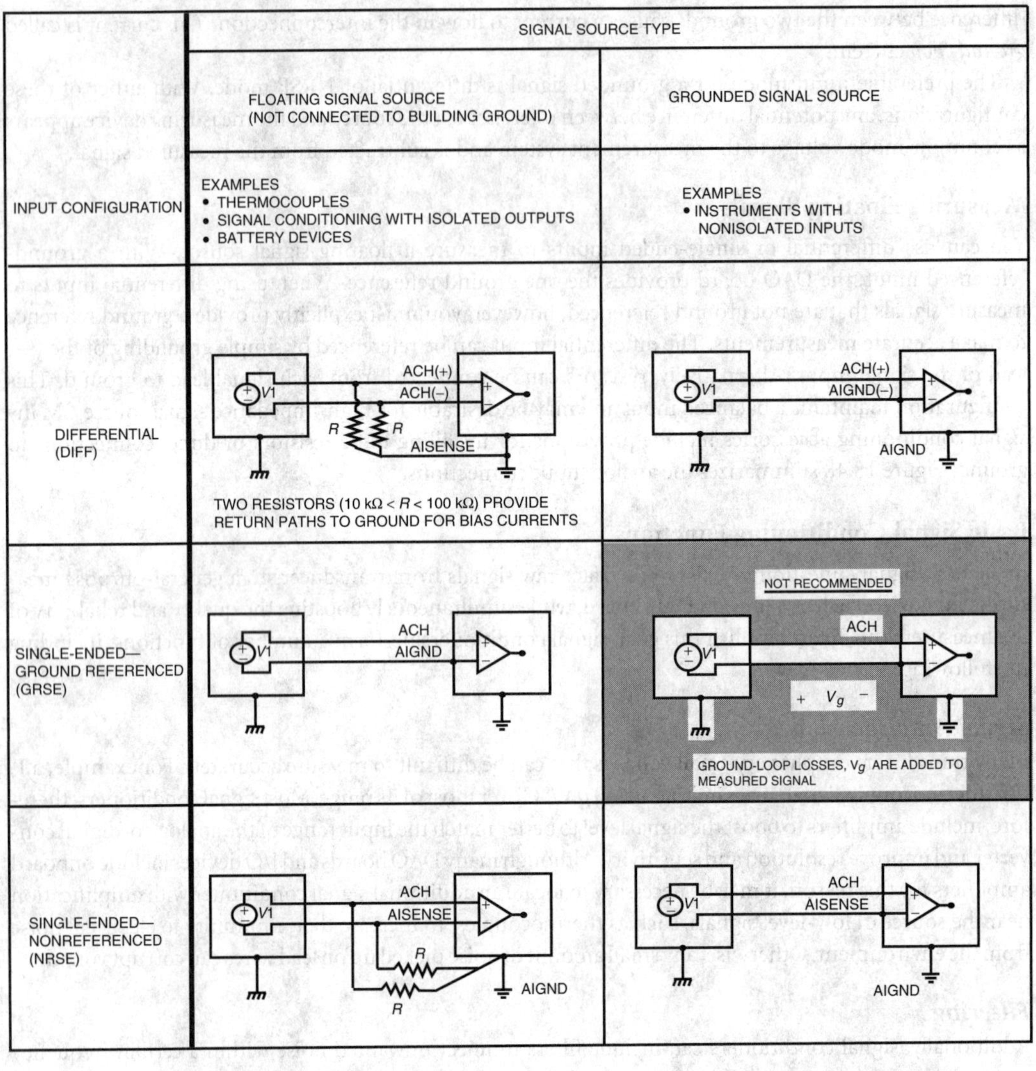

FIGURE 18.48 Summary of connections of grounded and floating signal sources.

A common method for circuit isolation is using optical, capacitive, or transformer isolators. Capacitive and transformer isolators modulate the signal to convert it from a voltage to a frequency value. The frequency signal is then coupled across capacitors or a transformer, where it is then converted back to the proper voltage value. Optical isolators, commonly used for digital signals, use LEDs to convert the voltage on/off information into light signals to couple the signal across the isolation barrier.

Transducer Excitation and Interfacing

Many types of sensors and transducers have particular signal conditioning requirements. For example, thermocouples require cold-junction compensation for the thermoelectric voltages created where the thermocouple wires are connected to the data acquisition equipment. Resistive temperature devices (RTDs) require an accurate current excitation source to convert their small changes in electrical resistance into measurable changes in voltage. To avoid errors caused by the resistance in the lead wires, RTDs are often used in a 4-wire configuration. The 4-wire RTD measurement avoids lead resistance errors because two additional leads carry current to the RTD device, so that current does not flow in the sense, or measurement. Strain gauge transducers, on the other hand, are used in a Wheatstone bridge configuration with a constant voltage or current power source. The signal conditioning requirements for these and other common transducers are listed in Table 18.2.

Linearization

Most sensors exhibit an output that is nonlinear with respect to the measurand. Therefore, many signal conditioners include circuitry or onboard intelligence to linearize the transfer function of the sensor. This onboard linearization is designed to offload some of the processing requirements of the DAQ system. With the increased use of PCs, however, this need is diminished and you can easily perform this linearization function in software. Unlike hardware linearization, software linearization is a very flexible solution, making it possible for a single signal conditioning module to be easily adapted via software for a wide variety of sensors. In fact, you can even implement your own customized transducer linearization routines if necessary.

Variety of Signal Conditioning Architectures

Signal conditioning systems come in all different forms, ranging from single-channel I/O modules to multichannel chassis-based systems with sophisticated signal routing capabilities. In addition, several products commonly classified as signal conditioners include an onboard ADC with digital communications interface. Here, we will concentrate on nondigitizing conditioning systems used as a front-end for data acquisition and control systems, such as plug-in DAQ boards.

The typical single-channel I/O module is a fixed-function conditioner designed for a particular type of transducer and signal range. You cable the conditioned output signal, usually a voltage signal, directly to a DAQ board input channel. Some modules are DIN rail-mountable, whereas others install into a backplane that holds 8–16 modules. Newer versions of the single-channel modules feature programmability and added intelligence for scaling and diagnostics. Because the modules do not incorporate signal multiplexing, they are best suited for applications with fewer I/O channels.

Many DAQ board vendors supply specialized signal conditioning boards for use with their DAQ boards. These signal conditioning boards, usually designed for a particular transducer type, tend to provide a less flexible system. Meanwhile, other DAQ vendors incorporate signal conditioning directly on the PC plug-in DAQ board. Although this approach can provide a low-cost system for simpler applications, the benefits of locating your high-voltage isolation barrier inside the PC are questionable. In addition, you do not have the option of amplifying your low-level sensor signals before they enter the potentially noisy PC.

Signal Conditioners Offer I/O Expansion

A final class of signal conditioners incorporate signal multiplexing to significantly expand the I/O capabilities of the DAQ system. These systems consist of a chassis that houses a variety of signal conditioning modules. Instead of simply passing the conditioned signals to an outgoing connector, each module multiplexes the conditioned signals onto a single analog output channel. You can cable the multiplexed output directly to a DAQ board or pass it to the chassis backplane bus. This backplane bus routes the conditioned analog signals, as well as digital communications and timing control signals, among the modules. Such a system is expandable; adding channels is accomplished by plugging a new multiplexing module into the backplane bus. Because the bus also incorporates a digital communications path, you can also incorporate digital I/O and analog output modules into the same chassis.

The multiplexing architecture of these signal conditioning systems is especially well suited for applications involving larger numbers of channels. For example, some systems can multiplex 3072 channels into a single PC plug-in DAQ board. More importantly, these multiplexing signal conditioners offer significant advantages in cost and physical space requirements. By switching multiple inputs into a single processing block, including amplification, filtering, isolation, and ADC, you can achieve a very low cost per channel not attainable with single-channel modules. Even though single-channel modules are being developed that are smaller and slimmer, systems with a multiplexing architecture will always be able to pack many more I/O channels into a given physical space.

Defining Terms

Alias: A false lower frequency component that appears in sampled data acquired at too low a sampling rate.

Conversion time: The time required, in an analog input or output system, from the moment a channel is interrogated (such as with a read instruction) to the moment that accurate data are available.

Data acquisition (DAQ): (1) Collecting and measuring electrical signals from sensors, transducers, and test probes or fixtures and inputting them to a computer for processing. (2) Collecting and measuring the same kinds of electrical signals with A/D and/or DIO boards plugged into a PC, and possibly generating control signals with D/A and/or DIO boards in the same PC.

Differential nonlinearity (DNL): A measure in LSB of the worst-case deviation of code widths from their ideal value of 1 LSB.

Integral nonlinearity (INL): A measure in LSB of the worst-case deviation from the ideal A/D or D/A transfer characteristic of the analog I/O circuitry.

Nyquist sampling theorem: A law of sampling theory stating that if a continuous bandwidth-limited signal contains no frequency components higher than half the frequency at which it is sampled, then the original signal can be recovered without distortion.

Relative accuracy: A measure in LSB of the accuracy of an ADC. It includes all nonlinearity and quantization errors. It does not include offset and gain errors of the circuitry feeding the ADC.

References

House, R. 1993. Understanding important DA specifications. *Sensors* 10(6):11–16.

House, R. 1994. Understanding inaccuracies due to settling time, linearity, and noise. *National Instruments European User Symposium Proceedings*, Nov. 10–11:17–26.

McConnell, E. 1994. PC-based data acquisition users face numerous challenges. *ECN* 38(8):11–12.

McConnell, E. 1995. Choosing a data-acquisition method. *Electronic Design* 43(13):147–156.

Potter, D. and Razdan, A. 1994. Fundamentals of PC-based data acquisition. *Sensors* 11(2).

Potter, D. 1994. Sensor to PC—Avoiding some common pitfalls. *Sensors Expo Proceedings*, Sept. 20:12–20.

Potter, D. 1995. Signal conditioners expand DAQ system capabilities. *I&CS* 68(8).

Further Information

Johnson, G.W. 1994. *LabVIEW Graphical Programming*. McGraw-Hill, New York.

McConnell, E. 1994. New achievements in counter/timer data acquisition technology. *MessComp 1994 Proceedings*, Sept. 13–15.

McConnell, E. 1995. Equivalent time sampling extends DA performance. *Sensors* 12(6).

18.3   Process Dynamics and Control

Thomas F. Edgar and Juergen Hahn

18.3.1   Introduction

The field of process dynamics and control is concerned with the analysis of dynamic behavior of processes, process simulation, design of automatic controllers, and associated instrumentation. Process control as practiced in the process industries has undergone significant changes since it was first introduced in the 1940s. Perhaps the most significant influence on the changes in process control technology has been the introduction of digital computers and instruments with greater capabilities than their analog predecessors. During the past 25 years automatic control has assumed increased importance in the process industries, which has led to the application of more sophisticated techniques (Seborg, Edgar, and Mellichamp, 2004; Ogunnaike and Ray, 1995).

For instrumenting and controlling a modern plant, it is necessary that an engineer has an understanding of the time-dependent behavior of typical processes. This in turn requires an appreciation of how mathematical tools can be employed in analysis and design of automatic control systems. In this chapter we present the basic ideas involved in developing dynamic models for typical processes and then discuss the use of control valves and proportional-integral-derivative (PID) feedback controllers for process control. Next, various types of advanced control methods that have seen commercial use are reviewed. Finally, some comments on computer control are included. For a thorough discussion of process instrumentation and measurement devices, the reader is referred to other entries in the section, "Control and Instrumentation Technology," in this handbook.

18.3.2   Objectives of Process Control

The main economic objective of process control is to achieve maximum productivity or efficiency while maintaining a satisfactory level of product quality. Manufacturing facilities in the production of chemicals, paper, metals, power, food, and pharmaceuticals require accurate and precise control systems. Although the methods of production vary from industry to industry, the principles of automatic control are generic in nature and can be universally applied, regardless of the size of the plant.

Consider the control of the stirred tank reactor shown in Fig. 18.49. This system can be used to introduce basic concepts and definitions in process control, as given by the following:

1. *Controlled variables:* These are the variables that quantify the performance or quality of the final product, which are also called output variables. In Fig. 18.49, the controlled variables are the tank temperature T and the liquid level in the tank h. Controlling the temperature allows the chemical reaction to proceed at a specified rate. The tank level determines the residence time in the reactor. The desired operating point for each controlled variable is called the *set point*.
2. *Manipulated variables:* These input variables are adjusted dynamically to keep the controlled variables at their setpoints. There are two manipulated variables in Fig. 18.49, namely, the exit flow rate w_o and the steam pressure P_s to the heat transfer coil in the tank.
3. *Disturbance variables:* These are also called load variables and represent input variables that can cause the controlled variables to deviate from their respective set points. For the stirred tank heater, both the feed rate w_i and feed temperature T_i can change, causing the tank level and exit temperature to move away from the set points.

In the design of controllers for this process, two cases can be evaluated:

1. *Set point change:* This involves carrying out a change in the operating conditions, e.g., from a tank exit temperature of 125 to 150°C while holding the level constant. The set-point signal is changed, and the heat transfer rate is adjusted appropriately to achieve the new operating conditions. The changing of set points is also called *servomechanism* (or servo) control, and it can require a time span as short as a few seconds to as long as several hours to carry out, depending on the process design variables. In Fig. 18.49, such design variables could include the flow rate and the tank holdup.

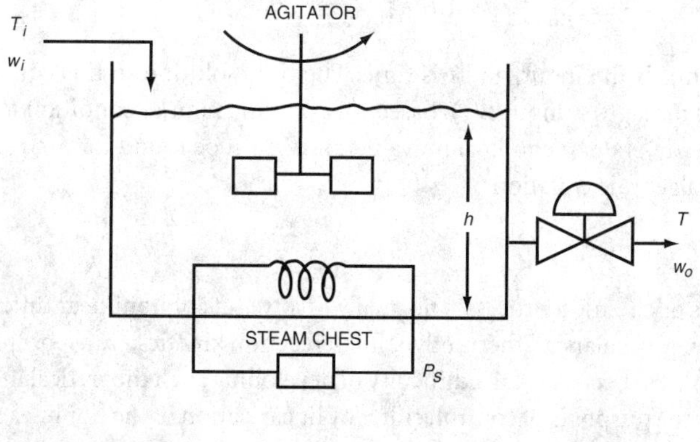

FIGURE 18.49   Flowsheet of a continuous stirred tank reactor.

2. *Disturbance change:* This case relates to the process transient behavior when a disturbance enters, also called *regulatory control*. A control system should be able to return each controlled variable back to its set point.

Several other definitions need to be discussed here. *Steady-state* behavior pertains to the case where there is no variation in the process variables with respect to time. If the system is in equilibrium (at steady state), it can be described by algebraic equations, such as material and energy balances. *Unsteady-state* (dynamic) behavior occurs when the process variables change as a function of time. The required mathematical model for this case includes ordinary differential equations as well as algebraic relationships.

Three further classifications deal with how process control is implemented, namely:

- Open-loop or manual control
- Closed-loop or **feedback control**
- **Feedforward control**

Manual control implies that the operator makes the changes in manipulated variable, which is used occasionally. Feedback control connotes that the manipulated variable is changed automatically in response to the error between the set point and controlled variable. The second approach serves as the basis for most automatic control schemes. Feedforward control is a technique where the manipulated variable is changed as a function of a disturbance variable. Both feedback and feedforward control are discussed in later sections of this chapter.

18.3.3 Development of a Mathematical Model

There are a number of modeling approaches that can be used with process control systems. Whereas mathematical models based on the chemistry and physics of the system represent one alternative, the typical process control model utilizes an empirical input/output relationship, the so-called black-box model. These models are found by experimental tests of the process. Mathematical models of the control system may include not only the process but also the controller, the final control element, and other electronic components such as measurement devices and transducers. Once these component models have been determined, one can proceed to analyze the overall system dynamics, the effect of different controllers in the operating process configuration, and the stability of the system, as well as obtain other useful information.

Mathematical models provide a convenient and compact way of expressing the behavior of a process as a function of process physical parameters and process inputs. The same mathematical model can be expressed in several ways; for example, a continuous-time model based on a differential equation can be converted to a discrete-time system, or it can be transformed to a different type of independent variable altogether (e.g., Laplace transforms or z transforms). Transform models generally feature a simplified notation, which greatly facilitates analysis of complicated systems.

A simple dynamic system with one input and one output can be represented by equations of the form

$$dy/dt = f(y, u) \qquad (18.1)$$

where y is the output, u is the input, and t is time. The time solution to this equation can be found by integrating Eq. (18.1) for a given input $u(t)$, based on a specified initial condition for the output $y(0)$. By letting $dy/dt = 0$, steady-state or equilibrium value(s) of y can be found for each selected value of u, the input, by solving the algebraic equation

$$f(y, u) = 0 \qquad (18.2)$$

Theoretical models of chemical processes normally involve sets of nonlinear differential equations that arise from mass and energy balances, thermodynamics, reaction kinetics, transport phenomena, and physical property relationships. Because of the difficulty of developing such theoretical models, simpler models are usually sought for the purposes of control, either by linearization of the nonlinear models or by making simplifying assumptions. On the other hand, a less time-consuming approach involves developing black box models, which are obtained by fitting experimental input-output data. This latter approach has been the historical basis of process control practice. During the past 10 years, however, there has been increasing use of fundamental models in design and control, largely due to the availability of modeling software packages.

Denoting K as the process gain, the simplest dynamic model used in process control is a first-order linear differential equation:

$$\tau \frac{dy}{dt} = -y + Ku \tag{18.3}$$

where τ is the time constant and K is the process gain.

At steady state, $y_{ss} = Ku$. Linear dynamic models such as Eq. (18.3) provide a theoretical means to determine the time scale of the process. For a step change in $u (= M)$, the solution to Eq. (18.3) can be found analytically,

$$y(t) = KM(1 - e^{-t/\tau}) \tag{18.4}$$

From Eq. (18.4) we can conclude that the response of y reaches 63.2% of its final value when $t = \tau(1 - e^{-1}) = 0.632$. In this way, τ determines the speed of response for the system.

Experience in the process industries indicates that there are a limited number of expected dynamic behaviors that actually influence the controller design step. These behaviors can be categorized using the step response and are based on a transfer function representation of the process model, which is assumed to be linear or a linear approximation of a nonlinear model. A transfer function is found by taking the Laplace transform of the ordinary differential equation that describes the system; the mathematical definition of the Laplace transform is

$$L(f(t)) = \int_0^\infty f(t)e^{-st}dt \tag{18.5}$$

where $f(t)$ is a specified function of time.

The chief mathematical advantage of the Laplace transform is the conversion of the differential equation (such as Eq. (18.1)) to an algebraic equation. The resulting equation can be rearranged as a transfer function,

$$Y(s) = G(s)U(s) \tag{18.6}$$

where

$U(s)$ = transform of the input variable
$Y(s)$ = transform of the output variable
$G(s)$ = transfer function

$G(s)$ describes the dynamic characteristics of the process. For linear systems it is independent of the input variable, so it applies for any time-dependent input signal. For the model in Eq. (18.3), the transfer function is

$$G(s) = \frac{K}{\tau s + 1} = \frac{Y(s)}{U(s)} \tag{18.7}$$

where K is the process gain and τ the time constant already mentioned.

Transfer functions can also be used to describe the dynamic behavior of instruments, controllers, and valves. For a temperature transmitter the steady-state gain is simply the output span divided by the input span; an electronic transmitter which has a 4–20 mA output and a 100–200°F temperature input range gives a gain of $(20 - 4)/(200 - 100) = 0.16$ mA/°F. Sensors usually have simple dynamics, typically described by a first-order transfer function. The characteristics of control valves are more complicated and are discussed in the next section.

The order of the transfer function is the power of the polynomial in s in the denominator of $G(s)$ and is equivalent to the order of the differential equation describing the process. There are a limited number of dynamic transfer functions that are important for process control; Fig. 18.50 shows the dynamic response

DESCRIPTION	TRANSFER FUNCTION	RESPONSE TO UNIT STEP CHANGE AT TIME = 0
(1) FIRST ORDER	$\dfrac{K}{\tau_1 s + 1}$	
(2) SECOND ORDER	$\dfrac{K}{(\tau_1 s + 1)(\tau_2 s + 1)}$	
(3) SECOND ORDER $(\tau_3 > \tau_1 > \tau_2)$	$\dfrac{K(\tau_3 s + 1)}{(\tau_1 s + 1)(\tau_2 s + 1)}$	
(4) SECOND ORDER $(\tau_3 < 0)$	$\dfrac{K(\tau_3 s + 1)}{(\tau_1 s + 1)(\tau_2 s + 1)}$	
(5) DEAD TIME	$e^{-\theta s}$	
(6) FIRST-ORDER LAG WITH DEAD TIME	$\dfrac{Ke^{-\theta s}}{\tau_1 s + 1}$	
(7) INTEGRATOR	$\dfrac{K}{s}$	
(8) SECOND ORDER WITH OSCILLATIONS	$\dfrac{K}{\tau^2 s^2 + 2\tau\zeta s + 1}$	
(9) UNSTABLE $(a > 0)$	$\dfrac{K}{s - a}$	

FIGURE 18.50 Common transfer functions and dynamic step response.

of nine different cases that result from a step change in the input variable. Simple approximate models are usually developed from experimental tests, and the model parameters are fitted either graphically or by nonlinear regression. In Fig. 18.50 only first- and second-order models are used. Some comments on each case are given next:

Case 1

This is the first-order system discussed previously, which is the analog of a resistance-capacitance circuit in electronics. It is mainly applicable to responses of instruments (sensors and valves) but usually is inadequate to describe dynamic behavior of actual chemical processes.

Case 2

The second-order system response provides a better approximation for many real processes dynamics, such as chemical reactors and distillation columns.

Case 3

Overshoot can occur in some open-loop systems as shown here, but this case is relatively rare. Some controllers exhibit this behavior.

Case 4

This behavior is called inverse response, where the output initially goes the opposite direction to the final value. This wrong-way behavior is seen in boilers and distillation columns.

Case 5

Pure delay or dead time usually occurs due to transportation lag, such as flow through a pipe. For a velocity v and distance L, $\theta = L/v$. Another case is a measurement delay, such as for a gas or liquid chromatograph.

Case 6

This is a more realistic response than case 1 because it allows for a time delay as well as first-order behavior, and can be applied to many chemical processes. The delay time may not be clearly identified as transport lag but may be imbedded in higher order dynamics. Hence, this is an approximate but very useful model, especially for staged systems such as distillation or extraction columns. An extension of this model is second order plus time-delay, the response of which is equivalent to case 2 or case 4 with an initial time delay. The advantage over first-order models is that an additional model parameter can give greater accuracy in fitting process data.

Case 7

The integrator describes the behavior of level systems, where the step input (such as a flow change) causes the level to increase with constant slope until the flow is reduced to its original value. There is no time constant here because the level does not reach steady state for a step input but continues to increase until the step input is terminated (or the vessel overflows).

Case 8

Oscillation in the step response is rare for chemical processes under open-loop or manual conditions. When feedback control is implemented, with processes described by cases 2–6, the occurrence of oscillation is quite common. In fact, a well-tuned feedback controller exhibits some degree of oscillation, as discussed later. The time constants for this case are not real values but are imaginary numbers.

Case 9

If any time constant is negative (τ in this case is not physically meaningful), the response is unstable. The output response becomes larger with time and is unbounded. No real system actually behaves this way since some constraint will eventually be reached; for example, a safety valve will open. A linear process is at the limit of instability if it oscillates without decay. Most processes are stable and do not require a controller for stability; however, such self-regulating processes may exhibit very slow responses unless feedback control is applied. On the other hand, when feedback control is used with some processes, instability can occur if the controller is incorrectly designed (see later discussion on controller tuning). There are a few processes, such as chemical reactors, which may be unstable in the open loop, but these can be stabilized by feedback control.

In summary, with all of the various dynamic behaviors shown in Fig. 18.50, it is important for plant engineers, instrument specialists, and even some operators to understand the general dynamic characteristics of a process. It is also crucial that they know how various processes respond under the influence of feedback control.

18.3.4 Final Control Elements

A final control element is a device that receives the manipulated variable from the controller as input and takes action that influences the process in the desired manner. In the process industries valves and pumps are the most common final control elements, because of the necessity to adjust a fluid flow rate such as coolant, steam, or the main process stream.

Control Valves

The control valve is designed to be remotely positioned by the controller output, which is usually an air pressure signal. The valve design contains an opening with variable cross sectional area through which the

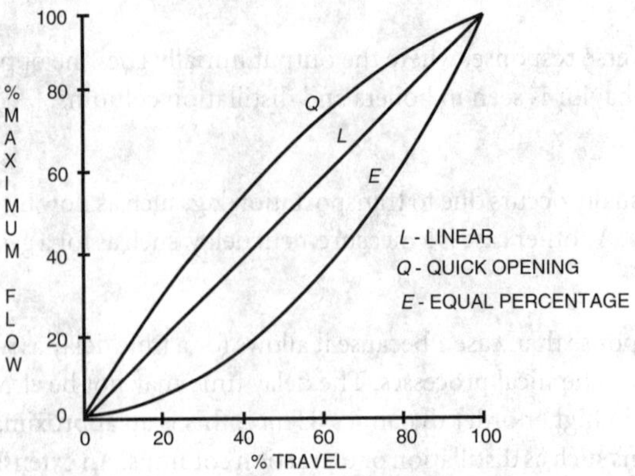

FIGURE 18.51   Inherent control valve characteristics.

fluid can flow. A stem that travels up or down according to a change in the manipulated variable changes the area of the opening and thereby the flowrate. At the end of the stem is a plug of specific shape, which fits into a seat or ring on the perimeter of the valve port. This plug/seat combination is called the valve trim, and it determines the steady-state gain characteristics of the control valve. Based on safety considerations the control valve can be configured to be either air-to-open or air-to-close.

The inherent characteristics of control valves allow classification into three main groups, based on the relationship between valve flow and valve position under constant pressure: linear, equal-percentage, and quick opening (Seborg, Edgar, and Mellichamp, 2004; see Fig. 18.51). Usually in-plant testing is used to determine the actual valve characteristics because the dynamics of the valve can depend on other flow resistances in the process.

The response time of a pneumatic valve is related to the distance from the actuator to the valve, because the pneumatic signal has to travel the distance between the actuator and the valve. It is important for the design of the process that the response time of the valve is at least one order of magnitude faster than the process. Usually valve response times are several seconds, which is small compared to process time constants of minutes and hours. Pneumatic valve performance often suffers from nonideal behavior such as valve hysteresis. In order to achieve better reproducibility and a faster response a valve positioner can be installed to control the actual stem position rather than the diaphragm air pressure (Edgar, Smith, Shinskey, Gassman, Schafbuch, McAvoy, and Seborg, 1997).

Adjustable Speed Pumps

Instead of using fixed speed pumps and then throttling the process with a control valve, an adjustable speed pump can be used as the final control element. In these pumps speed can be varied by using variable-speed drivers such as turbines or electric motors. Adjustable speed pumps offer energy savings as well as performance advantages over throttling valves to offset their higher cost. One of these performance advantages is that, unlike a throttling valve, an adjustable pump does not contain dead time for small amplitude responses. Furthermore, nonlinearities associated with friction in the valve are not present in electronic adjustable speed pumps. However, adjustable speed pumps do not offer shutoff capability like control valves and extra flow check valves or automated on/off valves may be required.

18.3.5   Feedback Control Systems

Feedback control is a fundamental concept that is employed in PID controllers as well as in advanced process control techniques. Figure 18.52 shows a simplified instrumentation diagram for feedback control of the stirred tank discussed earlier, where the inlet flow is equal to the outlet flow (hence, no level control is needed) and outlet temperature is controlled by the steam pressure. Figure 18.53 shows a generic block diagram for a feedback control system. In feedback control the controlled variable is measured

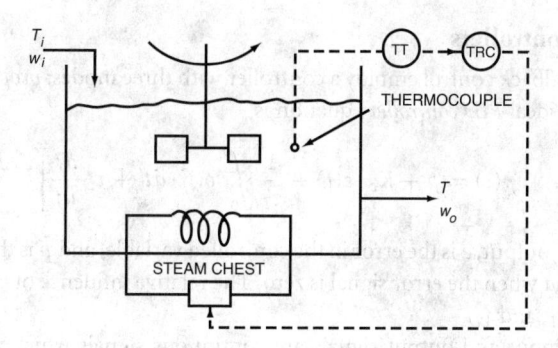

FIGURE 18.52 Instrumentation diagram of a temperature control system for a stirred tank reactor; symbols: ≠ electrical instrument line, TT temperature transmitter, TRC temperature recorder-controller.

and compared to the set point. The resulting error signal is then used by the controller to calculate the appropriate corrective action for the manipulated variable. The manipulated variable influences the controlled variable through the process, which is dynamic in nature.

In many industrial control problems, notably those involving temperature, pressure, and flow control, measurements of the controlled variable are available, and the manipulated variable is adjusted via a control valve. In feedback control, corrective action is taken regardless of the source of the disturbance. Its chief drawback is that no corrective action is taken until after the controlled variable deviates from the set point. Feedback control may also result in undesirable oscillations in the controlled variable if the controller is not tuned properly, that is, if the adjustable controller parameters are not set at appropriate values. The tuning of the controller can be aided by using a mathematical model of the dynamic process, although an experienced control engineer can use trial-and-error tuning to achieve satisfactory performance in many cases. Next, we discuss the two types of controllers used in most commercial applications.

On-Off Controllers

An on-off controller is the simplest type of feedback controller and is widely used in the thermostats of home heating systems and in domestic refrigerators because of its low cost. This type of controller is seldom used in industrial plants, however, because it causes continuous cycling of the controlled variable and excessive wear on the control valve.

In on-off control, the controller output has only two possible values:

$$p(t) = \begin{cases} p_{max} & \text{if } e > 0 \\ p_{min} & \text{if } e < 0 \end{cases} \qquad (18.8)$$

where p_{max} and p_{min} denote the on and off values, respectively, and e is the error or difference between the controlled variable and its set point.

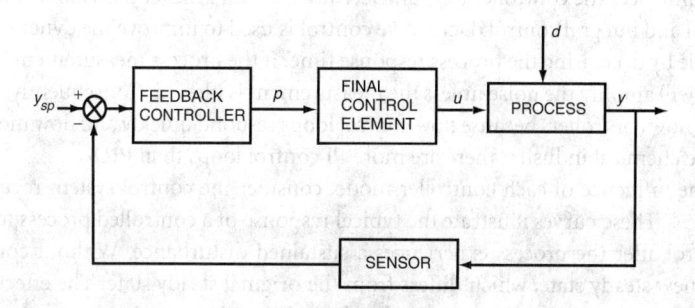

FIGURE 18.53 Block diagram for feedback control structure.

Three Mode (PID) Controllers

Most applications of feedback control employ a controller with three modes: *proportional* (P), *integral* (I), and *derivative* (D). The ideal *PID controller* equation is

$$p(t) = \bar{p} + K_c \left[e(t) + \frac{1}{\tau_I} \int_0^t e(t') \, dt' + \tau_D \frac{de}{dt} \right] \tag{18.9}$$

where p is the controller output, e is the error in the controlled variable, and $\bar{p}$ is the bias, which is set at the desired controller output when the error signal is zero. The relative influence of each mode is determined by the parameters K_c, τ_I, and τ_D.

The PID controller input and output signals are continuous signals which are either pneumatic or electrical. The standard range for pneumatic signals is 3–15 psig, whereas several ranges are available for electronic controllers including 4–20 mA and 1–5 V. Thus, the controllers are designed to be compatible with conventional transmitters and control valves. If the PID controller is implemented as part of a digital control system, the discrete form of the PID equation is used,

$$p_n = \bar{p} + K_c \left[e_n + \frac{\Delta t}{\tau_I} \sum_{k=0}^n e_k + \frac{\tau_D}{\Delta t} (e_n - e_{n-1}) \right] \tag{18.10}$$

where Δt is the sampling period for the control calculations and n denotes the current sampling time. Digital signals are scaled from 0 to 10 V DC.

The principle of proportional action requires that the amount of change in the manipulated variable (e.g., opening or closing of the control valve) vary directly with the size of the error. The controller gain K_c affects the sensitivity of the corrective action and is usually selected after the controller has been installed. The actual input-output behavior of a proportional controller has upper and lower bounds; the controller output saturates when the limits of 3 or 15 psig are reached. The sign of K_c depends on whether the controller is direct acting or reverse acting, that is, whether the control valve is air-to-open or air-to-close.

Integral action in the controller brings the controlled variable back to the set point in the presence of a sustained upset or disturbance. Because the elimination of offset (steady-state error) is an important control objective, integral action is normally employed, even though it may cause more oscillation. A potential difficulty associated with integral action is a phenomenon known as *reset windup*. If a sustained error occurs, then the integral term in Eq. (18.9) becomes quite large and the controller output eventually saturates. Reset windup commonly occurs during the start up of a batch process or after a large set-point change. Reset windup also occurs as a consequence of a large sustained load disturbance that is beyond the range of the manipulated variable. In this situation the physical limitations on the manipulated variable (e.g., control valve fully open or completely shut) prevent the controller from reducing the error signal to zero. Fortunately, many commercial controllers provide antireset windup by disabling the integral mode when the controller output is at a saturation limit.

Derivative control action is also referred to as *rate action, preact,* or *anticipatory control*. Its function is to anticipate the future behavior of the error signal by computing its rate of change; thus, the shape of the error signal influences the controller output. Derivative action is never used alone, but in conjunction with proportional and integral control. Derivative control is used to improve the dynamic response of the controlled variable by decreasing the process response time. If the process measurement is noisy, however, derivative action will amplify the noise unless the measurement is filtered. Consequently, derivative action is seldom used in flow controllers because flow control loops respond quickly and flow measurements tend to be noisy. In the chemical industry, there are more PI control loops than PID.

To illustrate the influence of each controller mode, consider the control system regulatory responses shown in Fig. 18.54. These curves illustrate the typical response of a controlled process for different types of feedback control after the process experiences a sustained disturbance. Without control the process slowly reaches a new steady state, which differs from the original steady state. The effect of proportional control is to speed up the process response and reduce the error from the set point at steady state, or offset. The addition of integral control eliminates offset but tends to make the response more oscillatory.

Adding derivative action reduces the degree of oscillation and the response time, that is, the time it takes the process to reach steady state. Although there are exceptions to Fig. 18.54, the responses shown are typical of what occurs in practice.

Stability Considerations

An important consequence of feedback control is that it can cause oscillations in closed-loop systems. If the oscillations damp out quickly, then the control system performance is generally considered to be acceptable. In some situations, however, the oscillations may persist or their amplitudes may increase with time until a physical bound is reached, such as a control valve being fully open or completely shut. In the latter situation, the closed-loop system is said to be unstable. If the closed-loop response is unstable or too oscillatory, this undesirable behavior can usually be eliminated by proper adjustment of the PID controller constants, K_c, τ_I, and τ_D. Consider the closed-loop response to a unit step change in setpoint for various values of controller gain K_c. For small values of K_c, say K_{c1} and K_{c2}, typical closed-loop responses are shown in Fig. 18.55. If the controller gain is increased to a value of K_{c3}, the sustained oscillation of Fig. 18.56 results; a larger gain, K_{c4} produces the unstable response shown in Fig. 18.56. Note that the amplitude of the unstable oscillation does not continue to grow indefinitely because a physical limit will eventually be reached, as previously noted. In general as the controller gain increases, the closed-loop response typically becomes more oscillatory and, for large values of K_c, can become unstable.

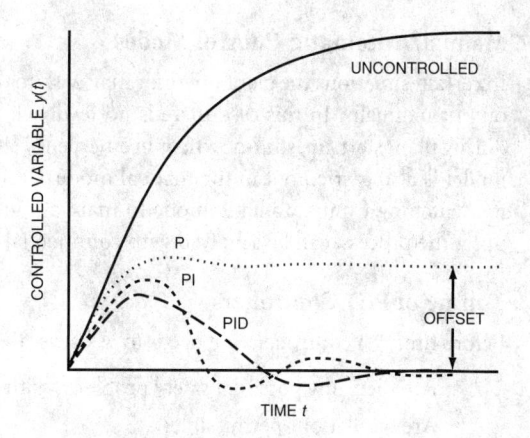

FIGURE 18.54 Response for a step change in disturbance with tuned P, PI, and PID controllers and with no control.

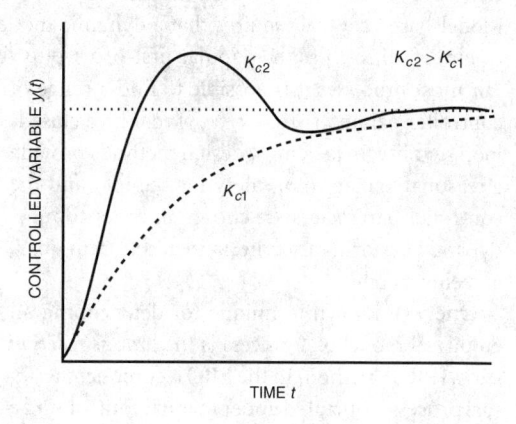

FIGURE 18.55 Closed-loop response for different values of the controller gain.

The conditions under which a feedback control system becomes unstable can be determined theoretically using a number of different techniques (Seborg, Edgar, and Mellichamp, 2004).

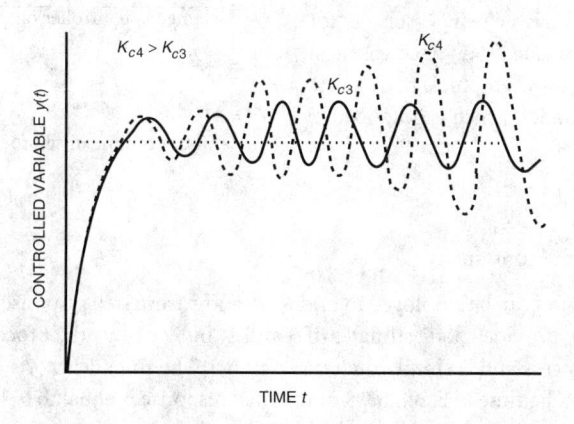

FIGURE 18.56 Transition to instability as controller gain increases.

Manual/Automatic Control Modes

In certain situations the plant operator may wish to override the automatic mode and adjust the controller output manually. In this case there is no feedback loop. This manual mode of operation is very useful during plant start-up, shut-down, or in emergency situations. Testing of a process to obtain a mathematical model is also carried out in the manual mode. Commercial controllers have a manual automatic switch for switching from automatic mode to manual mode and vice versa. Bumpless transfers, which do not upset the process, can be achieved with commercial controllers.

Tuning of PID Controllers

Before tuning a controller, one needs to ask several questions about the process control system:

- Are there any potential safety problems with the nature of the closed-loop response?
- Are oscillations permissible?
- What types of disturbances or set-point changes are expected?

There are several approaches that can be used to tune PID controllers, including model-based correlations, response specification, and frequency response (Seborg, Edgar, and Mellichamp, 2004; Ogunnaike and Ray, 1995). One approach that has recently received much attention is model-based controller design. Model-based control requires that a dynamic model of the process is available. The dynamic model can be empirical, such as the popular first-order plus time delay type of model, or it can be of higher order.

For most processes it is possible to find a range of controller paramters (K_c, τ_I, τ_D) that give closed-loop stability; in fact, most design methods provide a reasonable margin of safety for stability and the controller parameters are chosen in order to meet dynamic performance criteria in addition to guaranteeing stability.

One well-known technique for determining an empirical model of a process is the *process reaction curve* (PRC) method. In the PRC technique, the actual process is operated under manual control, and a step change in the controller output (Δp) is carried out. The size of the step is typically 5% of the span, depending on noise levels for the process variables. One should be careful to take data when other plant fluctuations are minimized. For many processes, the response of the system Δy (change in the measured value of the controlled variable using a sensor) follows the curve shown in Fig. 18.57 (see case 6 in Fig. 18.50). Also shown is the graphical fit by a first-order plus time delay model, which is described by three parameters:

1. the process gain K
2. the time delay θ
3. the dominant time constant τ

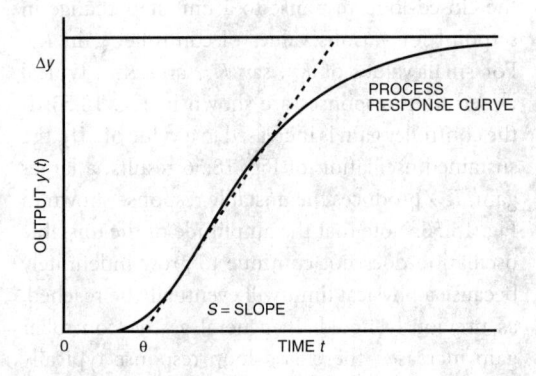

FIGURE 18.57 Process reaction curve for model–based tuning technique; tuning parameters found in Table 18.5 are defined by

$$K = \frac{\text{change in output (at steady state)}}{\text{change in controller output (at steady state)}}$$

$$= \frac{\Delta y}{\Delta p}$$

$$\tau = \Delta y / S$$

θ = amount of time until process responds (delay time).

While nonlinear regression can be employed to find K, θ, and τ from step response data, graphical analysis of the step response can provide good estimates of θ and τ. In Fig. 18.57 the process gain is a steady-state characteristic of the process and is simply the ratio $\Delta y / \Delta p$. The time delay θ is the time elapsed before Δy deviates from zero. The time to reach 63% of the final response is equal to $\theta + \tau$.

A variety of different tuning rules exists which are based on the assumption that the model can be accurately approximated by a first-order plus time delay model. One popular approach, which is not

TABLE 18.5   IMC Tuning Relations for First-Order Plus Time Delay Model

Controller Type	K_C	τ_I	τ_D
PI	$\dfrac{\tau}{K(\tau_C + \theta)}$	τ	—
PID	$\dfrac{\tau + \frac{\theta}{2}}{K\left(\tau_C + \frac{\theta}{2}\right)}$	$\tau + \dfrac{\theta}{2}$	$\dfrac{\tau\theta}{2\tau + \theta}$

restricted to this type of model but nevertheless achieves good performance for it, is internal model control (IMC) (Rivera, Morari, and Skogestad, 1986; Bequette, 2003; Seborg, Edgar, Mellichamp, 2004). Controller tuning using IMC requires only choosing the value of the desired time constant of the closed-loop system, τ_c, in order to compute the values of the three controller tuning parameters, K_C, τ_I, τ_D. The value of τ_c can be further adjusted based on simulation or experimental behavior. The tuning relations for a PI and PID controller for a process which is described by a first-order plus time delay model are shown in Table 18.5 (Chien and Fruehauf, 1990). Tuning relations for other types of process models can be found in Seborg, Edgar, Mellichamp (2004).

Another popular tuning method uses integral error criteria such as the time-weighted integral absolute error (ITAE). The controller parameters are selected in order to minimize the integral of the error. ITAE weights errors at later times more heavily, giving a faster response. Seborg, Edgar, and Mellichamp (2004) have tabulated power law correlations for these PID controller settings for a range of first-order model parameters K, τ, and θ.

Other plant testing and controller design approaches such as frequency response can also be used for more complicated models; refer to Seborg, Edgar, and Mellichamp (2004) for more details.

On-Line Tuning

The model-based tuning of PID controllers presumes that the model is accurate; thus these settings are only a first estimate of the best parameters for an actual operating process. Minor on-line adjustments of K_c, τ_I, and τ_D are usually required. Another technique, called continuous cycling, actually operates the loop at the point of instability, but this approach has obvious safety and operational problems, so it is not used much today. A recent improvement on continuous cycling uses a relay feedback configuration, which puts the loop into a controlled oscillation. The amplitude and period of the oscillation can be used to set proportional and integral settings.

18.3.6   Advanced Control Techniques

Although the single-loop PID controller is satisfactory in many process applications, it does not perform well for processes with slow dynamics, time delays, frequent disturbances, or multivariable interactions. We discuss several advanced control methods next, which can be implemented via computer control.

One of the disadvantages of using conventional feedback control for processes with large time lags or delays is that disturbances are not recognized until after the controlled variable deviates from its setpoint. One way to improve the dynamic response to disturbances is by using a secondary measurement point and a secondary controller; the secondary measurement point is located so that it recognizes the upset condition before the primary controlled variable is affected.

One such approach is called **cascade control**, which is routinely used in most modern computer control systems. Consider a chemical reactor, where reactor temperature is to be controlled by coolant flow to the jacket of the reactor (Fig. 18.58). The reactor temperature can be influenced by changes in disturbance variables such as feed rate or feed temperature; a feedback controller could be employed to compensate for such disturbances by adjusting a valve on the coolant flow to the reactor jacket. However, suppose an increase occurs in the coolant temperature as a result of changes in the plant coolant system. This will cause a change in the reactor temperature measurement, although such a change will not occur quickly, and the corrective action taken by the controller will be delayed.

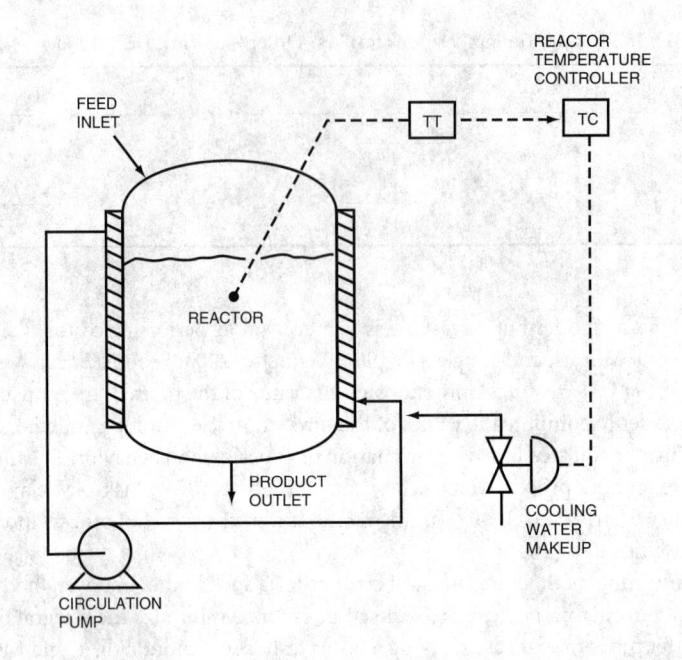

FIGURE 18.58   Conventional control of an exothermic chemical reactor.

Cascade control is one solution to this problem (see Fig. 18.59). Here the jacket temperature is measured and an error signal is sent from this point to the coolant control valve; this reduces coolant flow, maintaining the heat transfer rate to the reactor at a constant level and rejecting the disturbance. The cascade control configuration will also adjust the setting of the coolant control valve when an error occurs in reactor temperature. The cascade control scheme shown in Fig. 18.59 contains two controllers. The primary controller is the reactor temperature coolant temperature controller. It measures the reactor temperature, compares it to the setpoint, and computes an output that is the setpoint for the coolant flow rate controller. This secondary controller (usually a proportional only controller) compares the setpoint to the coolant

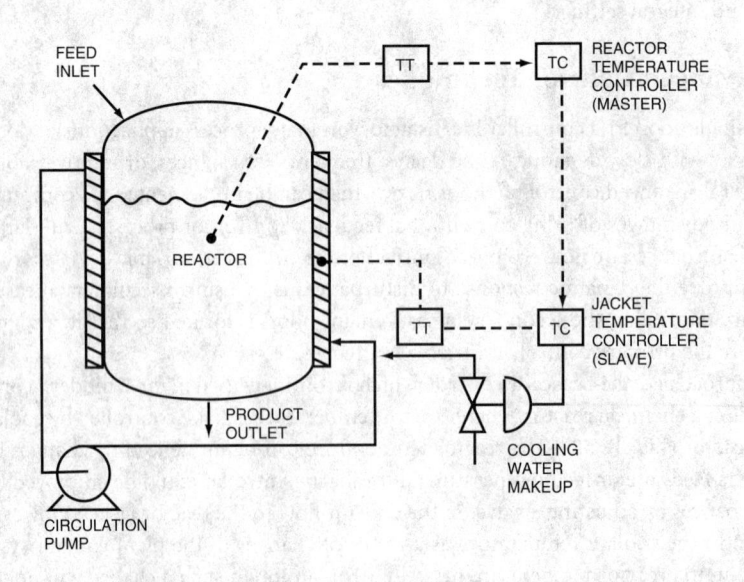

FIGURE 18.59   Cascade control of an exothermic chemical reactor.

temperature measurement and adjusts the valve. The principal advantage of cascade control is that the secondary measurement (jacket temperature) is located closer to a potential disturbance in order to improve the closed-loop response. See Shinskey (1996) for a discussion of other applications of cascade control.

Multivariable Control

Many processes contain several manipulated variables and controlled variables and are called multivariable systems. An example of a multivariable process was shown in Fig. 18.49, where there are two manipulated variables (flow rate and heat transfer rate) and two controlled variables (tank level and temperature). In many applications we can treat a multivariable process with a set of single-loop controllers by pairing inputs and outputs in the most favorable way. Then each loop can be tuned separately using the techniques mentioned previously. One design method employs the relative gain array (McAvoy, 1983). This formula provides guidelines on which variable pairings should be selected for feedback control as well as gives a measure of the potential quality of control for such a multiloop configuration.

If the loop interactions are not severe, then each single-loop controller can be designed using the techniques described earlier in this section. However, the presence of strong interactions requires that the controllers be detuned to reduce oscillations. A better approach is to utilize **multivariable control** techniques, such as model predictive control. Multivariable control is often used in distillation towers as well as in refinery operations such as cracking or reforming.

Model Predictive Control

The model-based control strategy that has been most widely applied in the process industries is **model predictive control** (MPC). It is a general method that is especially well suited for difficult multi-input, multi-output (MIMO) control problems where there are significant interactions between the manipulated inputs and the controlled outputs. Unlike other model-based control strategies, MPC can easily accommodate inequality constraints on input and output variables such as upper and lower limits, or rate of change limits (Ogunnaike and Ray, 1995). These problems are addressed by MPC by solving an optimization problem (Seborg, Edgar, and Mellichamp, 2004). One formulation of an objective function to be minimized is shown below:

$$ J = \sum_{k=0}^{P-1} (y_{sp} - y_k)^T \boldsymbol{Q} (y_{sp} - y_k) + \sum_{k=0}^{M-1} \Delta u_k^T \boldsymbol{R} \Delta u_k \tag{18.11} $$

where y_k refers to the output vector at time k, y_{sp} is the set point for this output, $\Delta u_k = u_k - u_{k-1}$ is the change of the input between time steps $k-1$ and k, and $\boldsymbol{Q}$ and $\boldsymbol{R}$ are weighting matrices. This objective function will penalize deviations of the controlled variable from the desired set point over the prediction horizon of length P as well as changes in the manipulated variable over the control horizon of length M. The optimization determines a future input trajectory that minimizes the objective function subject to constraints on the manipulated as well as the controlled variables. The predictions y_k are made based upon a model of the plant. This plant model will be updated using deviations between the predicted output and the real plant output from past data. An illustration of the trajectories for the manipulated (discrete trajectory) as well as the controlled variables (continuous trajectory) are shown in Figure 18.60.

A variety of different types of models can be used for the prediction. Choosing an appropriate model type is dependent upon the application to be controlled. The model can be based upon first-principles or it can be an empirical model. Also, the supplied model can be either linear or nonlinear, as long as the model predictive control software supports this type of model. Most industrial applications of MPC have relied on linear empirical models, because they can more easily be identified and solved and approximate most processes fairly well. Also, many MPC implementations change set points in order to move the plant to a desired steady state; the actual control changes are implemented by PID controllers in response to the set points. There are over 1000 applications of MPC techniques in oil refineries and petrochemical plants around the world. Thus, MPC has had a substantial impact and is currently the method of choice for difficult constrained multivariable control problems in these industries (Qin and Badgwell, 2003).

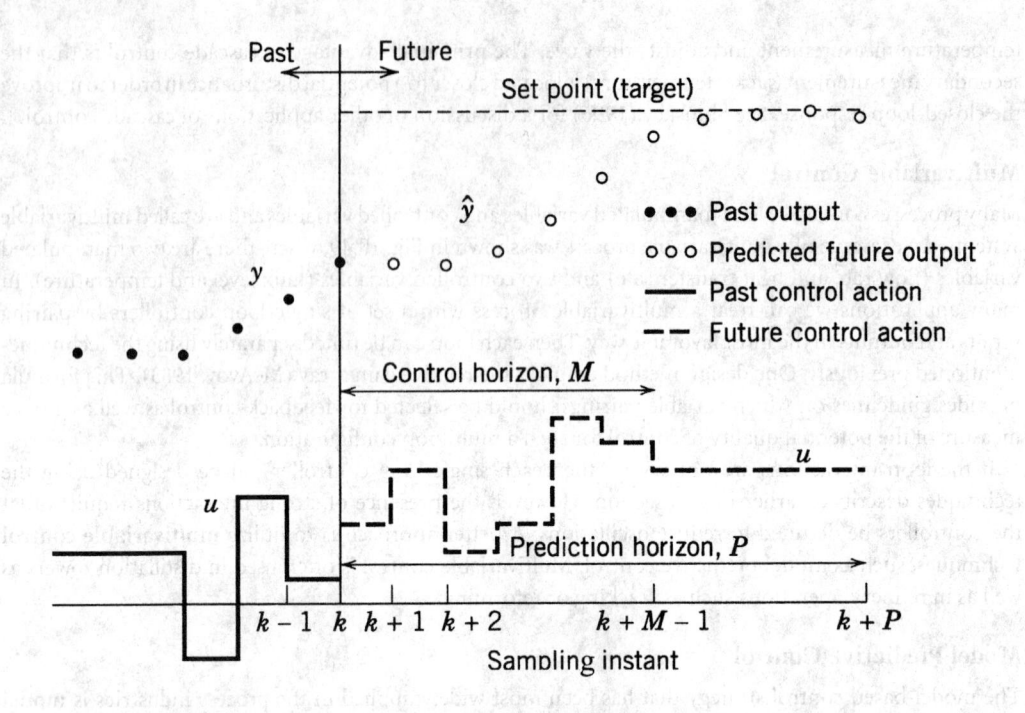

FIGURE 18.60   Past inputs and trajectories are used for forecasting future control action.

A key reason why MPC has become a major commercial and technical success is that there are numerous vendors who are licensed to market MPC products and install them on a turnkey basis. Consequently, even medium-sized companies are able to take advantage of this new technology. Payout times of 3–12 months have been reported. Refer to Maciejowski (2002) for further details on model predictive control.

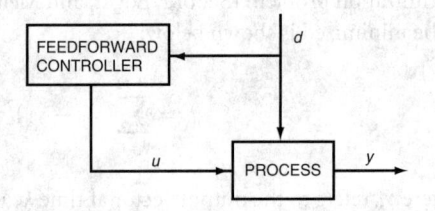

FIGURE 18.61   Block diagram for feedforward control configuration. The controller calculates the manipulated variable u needed to counteract the system upset introduced through the disturbance d. By taking control action based on measured disturbances, rather than controlled variable error, the controller can reject disturbances before they affect the controlled variable y.

Feedforward Control

If the process exhibits slow dynamic response and disturbances are frequent, then the addition of feedforward control with feedback control may be advantageous. Feedforward control differs from feedback control in that the primary disturbance is measured via a sensor and the manipulated variable is adjusted so that the controlled variable does not change (see Fig. 18.61). To determine the appropriate settings for the manipulated variable, one must develop mathematical models which relate:

- The effect of the manipulated variable on the controlled variable.
- The effect of the disturbance on the controlled variable.

These models can be based on steady-state or dynamic analysis. The performance of the feedforward controller depends on the accuracy of both models. If the models are exact, then feedforward control offers the potential of perfect control, that is, holding the controlled variable precisely at the setpoint at all times because of the ability to predict the appropriate control action. However, because most mathematical models are only approximate and not all disturbances are measurable, it is standard practice to utilize

TABLE 18.6 Relative Advantages and Disadvantages of Feedback and Feedforward

Advantages	Disadvantages
Feedforward	
1. Acts before the effect of a disturbance has been felt by the system.	1. Requires identification of all possible disturbances and their direct measurement.
2. Good for systems with large time constant or deadtime.	2. Cannot cope with unmeasured disturbances.
3. It does not introduce instability in the closed-loop response.	3. Sensitive to process/model error.
Feedback	
1. It does not require identification and measurement of any disturbance.	1. Control action is not taken until the effect of the disturbance has been felt by the system.
2. It is relatively insensitive to process/model errors.	2. It is unsatisfactory for processes with large time constants.
	3. It may cause instability in the closed-loop response.

feedforward control in conjunction with feedback control. Table 18.6 lists the relative advantages and disadvantages of feedforward and feedback control. By combining the two control methods, the strengths of both schemes can be utilized.

The tuning of the controller in the feedback loop can be theoretically performed independent of the feedforward loop, that is, the feedforward loop does not introduce instability in the closed-loop response. For more information on feedforward/feedback control applications and design of such controllers, refer to Shinskey (1996).

Adaptive Control and Autotuning

Process control problems inevitably require on-line tuning of the controller constants to achieve a satisfactory degree of control. If the process operating conditions or the environment changes significantly, the controller may have to be retuned. If these changes occur quite frequently, then **adaptive control** techniques should be considered. An adaptive control system is one in which the controller parameters are adjusted automatically to compensate for changing process conditions (Astrom and Wittenmark, 1994). Several adaptive controllers have been field-tested and commercialized in the U.S. and abroad.

Most commercial control systems have an autotuning function that is based on placing the process in a controlled oscillation at very low amplitude, comparable with that of the noise level of the process. This is done via a relay-type step function with hysteresis. The autotuner identifies the dynamic parameters of the process (the ultimate gain and period) and automatically calculates K_c, τ_I, and τ_D using empirical tuning rules. Gain scheduling can also be implemented with this controller, using several sets of PI or PID controller parameters.

Statistical Process Control

Statistical process control (SPC), also called statistical quality control (SQC), has found widespread application in recent years due to the growing focus on increased productivity. Another reason for its increasing use is that feedback control cannot be applied to many processes due to a lack of on-line measurements as is the case in many microelectronics fabrication processes. However, it is important to know if these processes are operating satisfactorily. While SPC is unable to take corrective action while the process is moving away from the desired target, it can serve as an indicator that product quality might not be satisfactory and that corrective action should be taken for future plant operations.

For a process that is operating satisfactorily the variation of product quality will fall within acceptable bounds. These bounds usually correspond to the minimum and maximum values of a specified product property. Normal operating data can be used to compute the mean $\bar{y}$ and the standard deviation σ of a given process variable from a series of n observations $y_1, y_2, \ldots, y_n$ as follows:

$$\bar{y} = \frac{1}{n} \sum_{i=1}^{n} y_i \qquad (18.12)$$

$$\sigma = \left[\frac{1}{n-1} \sum_{i=1}^{n} (y_i - \bar{y})^2 \right]^{1/2} \qquad (18.13)$$

The standard deviation is a measure for how the values of y spread around their mean. A large value of σ indicates that wide variations in y occur. Assuming the process variable follows a normal probability distribution, then 99.7% of all observations should lie within an upper limit of $\bar{y} + 3\sigma$ and a lower limit of $\bar{y} - 3\sigma$. These limits can be used to determine the quality of the control. If all data from a process lie within the $\pm 3\sigma$ limits, then it can be concluded that nothing unusual has happened during the recorded time period, the process environment is relatively unchanged, and the product quality lies within specifications. On the other hand, if repeated violations of the limits occur, then the conclusion can be drawn that the process is out of control and that the process environment has changed. Once this has been determined, the process operator can take action in order to adjust operating conditions to counteract undesired changes that have occurred in the process conditions.

There are several specific rules that determine if a process is out of control. Some of the more widely used ones are the Western-Electric rules that state that a process is out of control if process data include

- One measurement outside the $\pm 3\sigma$ control limit
- Any seven consecutive measurements lying on the same side of the mean
- A decreasing or increasing trend for any seven consecutive measurements
- Any nonrandom pattern in the process measurements

These rules can be applied to data in a control chart, such as a shown in Figure 18.62, where pressure measurements are plotted over a time horizon. It is then possible to read off from the control chart if the process is out of control or if it is operating within normal parameters.

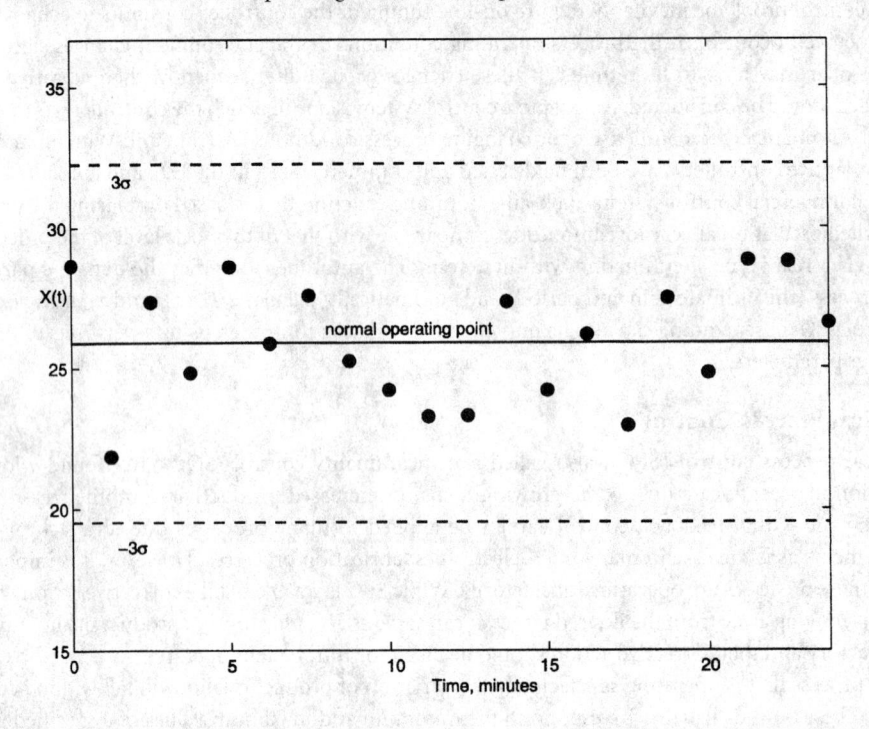

FIGURE 18.62 Control chart for SPC. Dashed lines indicate the $\pm 3\sigma$ levels for the process variables.

Because a process that is out of control can have important economic consequences, such as product waste and customer dissatisfaction, it is important to keep track of the state of the process. Statistical process control provides a convenient method to continuously monitor process performance and product quality. However, it differs from automatic process control (such as feedback control) in that it serves as an indicator that the process is not operating within normal parameters. SPC does not automatically provide a controller setting that will bring the process back to its desired operating point or target.

More details on SPC can be found in References (Seborg, Edgar, and Mellichamp, 2004; Ogunnaike and Ray, 1995; Montgomery, 2001; Grant and Leavenworth, 1996).

18.3.7 Computer Control Systems

Distributed Control Systems (DCS)

Microcomputer-based subsystems are standard in most computer control systems available today. The digital subsystems are interconnected through a digital communications network. Such systems are referred to as distributed digital instrumentation and control systems because of the network approach used to monitor and control the progress.

Figure 18.63 depicts a representative **distributed control system** (Seborg, Edgar, and Mellichamp, 2004). The DCS system consists of many commonly used DCS components, including multiplexers (MUXs), single-loop and multiple-loop controllers, PLCs, and smart devices. A system includes some or all of the following components:

1. Control network. The control network is the communication link between the individual components of a network. Coaxial cable and, more recently, fiber-optic cable have often been used. A redundant pair of cables (dual redundant highway) is normally supplied to reduce the possibility of link failure.
2. Workstations. Workstations are the most powerful computers in the system, and act both as an arbitrator unit to route internodal communications and the database server. Various peripheral devices are coordinated through the workstations. Computationally intensive tasks, such as real-time optimization or model predictive control, are implemented in a workstation.
3. Real-time clocks. Process control systems must respond to events in a timely manner and should have the capability of real-time control.
4. Operator stations. Operator stations typically consist of color graphics monitors with special keyboards to perform dedicated functions. Operators supervise and control processes from these workstations. Operator stations may be connected directly to printers for alarm logging, printing reports, or process graphics.
5. Engineering Workstations. They are similar to operator stations but can also be used as programming terminals to develop system software and applications programs.
6. Remote control units (RCUs). These components are used to implement basic control functions such as PID control. Some RCUs may be configured to acquire or supply set points to single-loop controllers. Radio telemetry (wireless) may be installed to communicate with MUX units located at great distances.
7. Programmable logic controllers (PLCs). These digital devices are used to control batch and sequential processes, but can also implement PID control.
8. Application stations. These separate computers run application software such as databases, spreadsheets, financial software, and simulation software via on OPC interface. OPC is an acronym for object linking and embedding for process control, a software architecture based on standard interfaces. These stations can be used for e-mail and as webservers, for remote diagnosis, configuration, and even for operation of devices that have an IP (Internet protocol) address. Applications stations can communicate with the main database contained in on-line mass storage systems.
9. Mass storage devices. *Typically*, hard disk drives are used to store active data, including on-line and historical databases and nonmemory resident programs. Memory resident programs are also stored to allow loading at system start-up.

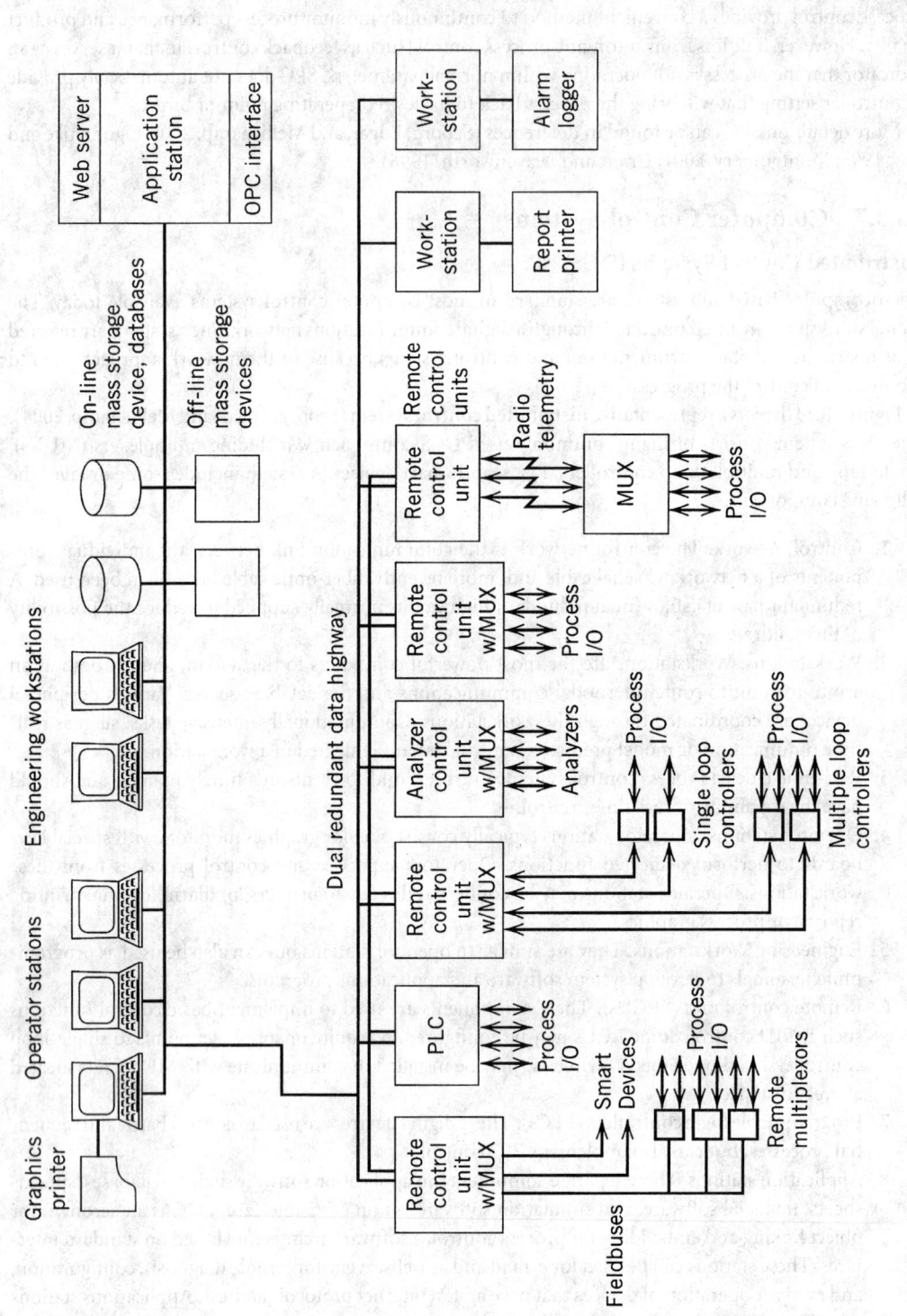

FIGURE 18.63   A typical distributed control system (DCS).

10. Fieldbuses/smart devices. An increasing number of field-mounted devices are available that support digital communication of the process I/O in addition to, or in place of, the traditional 4–20 mA current signal. These devices have greater functionality, resulting in reduced setup time, improved control, combined functionality of separate devices, and control-valve diagnostic capabilities.

Supervisory Control/Real-Time Optimization

The selection of set points in a distributed control network is called **supervisory control**. These set points may be determined in a control computer and then transmitted to specific devices (digital or analog controllers) in each control loop. Most supervisory control strategies are based on real-time optimization (RTO) calculations, wherein the set points (operating conditions) are determined by profitability analysis, that is, maximizing the difference between product value (income) and operating costs. For more details on real-time optimization methods such as linear and nonlinear programming or other search techniques see Seborg, Edgar, and Mellichamp (2004). RTO is often used in conjunction with model predictive control.

Batch Control

The increased availability of digital computer control and the emerging specialty chemical business has made **batch control** a very important component in process plants. Batch operations are of a start-stop nature; however, startup and shutdown of continuous plants must be treated in a similar fashion. Feedback control is of some value for batch processing, but it is more useful for operation near a set point in continuous operations.

The majority of batch steps encompass a wide variety of time-based operating conditions, which are sequential in nature. They can only be managed via a computer control system. Consider the operation of a batch reactor. The operation may include charging the reactor with several reactants (the recipe), applying the heat required to reach the desired reaction temperature, maintaining a specified level of operation until the reaction reaches completion, stopping the reaction, removing the product, and preparing the reactor for another batch.

Discrete functions are implemented via hardware or software to control discrete devices such as on/off valves, pumps, or agitators, based on status (on/off) of equipment or values of process variables. Interlock control can be provided via automatic actuation of a particular device only if certain process conditions sensed by various instruments are met. The two categories of interlocks are safety and permissive interlocks. Safety interlocks are designed to ensure the safety of operating personnel and to protect plant equipment. These types of fail-safe interlocks are associated with equipment malfunction or shutdown. Permissive interlocks establish orderly startup and shutdown of equipment. This prevents accumulation of material in tanks before it is needed. The triggers in the instructions can be time related or process related (temperature, pressure, etc.). Sequencing requires an end condition to be reached before the system can proceed to the next step. More details on batch control system design have been reported by Rosenhof and Ghosh (1987) and Seborg, Edgar, and Mellichamp (2004).

Process Control Software

The introduction of high-level programming languages such as Fortran and Basic in the 1960s was considered a major breakthrough in the area of computer control. For process control applications, some companies have incorporated libraries of software routines for these languages, but others have developed speciality pseudo-languages. These implementations are characterized by their statement-oriented language structure. Although substantial savings in time and efforts can be realized, software development costs can be significant.

The most successful and user-friendly approach, which is now adopted by virtually all commercial systems, is the fill-in-the-forms or table-driven *process control languages* (PCL). The core of these languages is a number of basic functional blocks or software modules. All modules are defined as database points. Using a module is analogous to calling a subroutine in conventional programs.

In general, each module contains some inputs and an output. The programming involves softwiring outputs of blocks to inputs of other blocks. Some modules may require additional parameters to direct module execution. The users are required to fill in the sources of input values, the destinations of output values, and the parameters in the blanks of forms/tables prepared for the modules. The source and destination blanks may be filled with process I/Os when appropriate. To connect modules, some systems require filling the tag names of modules originating or receiving data. The blanks in a pair of interconnecting modules are filled with the tag name of the same data point. A completed control strategy resembles a data flow diagram. All process control languages contain PID controller blocks. The digital PID controller is normally programmed to execute in velocity (difference) form. A pulse duration output may be used to receive the velocity output directly. In addition to the tuning constants, a typical digital PID controller contains some entries not normally found in analog controllers:

- When a process error is below certain tolerable deadband, the controller ceases modifying the output. This is referred to as *gap action*.
- The magnitude of change in a velocity output is limited by a change clamp.
- A pair of output clamps is used to restrict a positional output value from exceeding specified limits.
- The controller action can be disabled by triggering a binary deactivate input signal, during process startup, shutdown, or when some abnormal conditions exist.

Although modules are supplied and their internal configurations are different from system to system, their basic functionalities are the same

Another recent development has been the availability of flexible, yet powerful, software packages that support the controller design, controller testing, and implementation process. Probably, the most widely used program for this purpose is MATLAB® (The MathWorks, Inc.; www.mathworks.com), due to its flexibility. It allows one to implement and test controllers by either solving differential equations, using Laplace transforms, or with block diagrams. MATLAB® also provides a variety of routines that are commonly used for different controller design problems, e.g., optimal control, nonlinear control, optimization, etc. One of the main advantages of MATLAB® is that it is a programming language which provides control-related subroutines. This gives the process engineer flexibility with regard to the use of the software as well as how to extend or reuse already existing routines. It is also possible to exchange data with other software packages from within MATLAB®.

Digital Field Communications

A group of computers can become networked once intercomputer communication is established. Prior to the 1980s, all system suppliers used proprietary protocols to network their systems. The recent introduction of standardized protocols is based on the ISO-OSI* seven-layer model. The manufacturing automation protocol (MAP), which adopted the ISO-OSI standards as its basis, specifies a broadband backbone local area network (LAN). Originally intended for discrete component systems, MAP has evolved to address the integration of DCSs used in process control as well. TCP/IP (transmission control protocol/internet protocol) has been adopted for communication between nodes that have different operating systems.

Microprocessor-based process equipment, such as smart instruments and single-loop controllers, are now available with digital communications capability and are used extensively in process plants. A **fieldbus**, which is a low-cost protocol, provides efficient communication between the DCS and these devices.

Presently, there are several regional and industry-based fieldbus standards, including the French standard (FIP), the German standard (Profibus), and proprietary standards by DCS vendors, generally in the U.S., led by the Fieldbus Foundation. As of 2004, international standards organizations had adopted all of these fieldbus standards rather than a single unifying standard.

Several manufacturers provide fieldbus controllers that reside in the final control element or measurement transmitter. A suitable communications modem is present in the device to interface with a proprietary PC-based, or hybrid analog/digital bus network. Case studies in implementing such digital systems have shown significant reductions in cost of installation (mostly cabling and connections) vs. traditional analog field communication.

An example of a hybrid analog/digital protocol that is open (not proprietary) and in use by several vendors is the highway addressable remote transducer (HART) protocol. Digital communications utilize the same two wires that provide the 4 to 20 mA process control signal without disrupting the actual process signal. This is done by superimposing a frequency-dependent sinusoid ranging from −0.5 to +0.5 mA to represent a digital signal.

Defining Terms

Adaptive control: A control strategy that adjusts its control algorithm as the process changes.

Batch control: Control of a process where there is no inflow/outflow.

Cascade control: Nested multiloop strategy that uses intermediate measured variables to improve control.

Controlled variables: Measurable variables that quantify process performance.

Distributed control system: The standard computer architecture in process control, which involves multiple levels of computers.

Feedback control: A control structure where the controlled variable measurement is compared with the setpoint, generating an error acted upon by the controller.

Feedforward control: A control structure where the disturbance is measured directly and a control action is calculated to counteract it.

Fieldbus: A new communication protocol for digital communication between instruments and a central computer.

Manipulated variables: Input variables that can be adjusted to influence the controlled variables.

Model predictive control: An advanced control strategy that uses a model to optimize the loop performance.

Multivariable control: A control strategy for multiple inputs and multiple outputs.

Statistical process control: A control strategy that keeps a process between upper and lower limits (but not at the setpoint).

Supervisory control: The selection of set points, usually to maximize profitability or minimize costs.

Three-mode (PID) control: A feedback control algorithm that uses proportional, integral, and derivative action on the error signal.

References

Astrom, K.J. and Wittenmark, B. 1994. *Adaptive Control*, 2nd ed. Addison-Wesley, Reading, MA.

Bequette, B.W. 2003. *Process Control: Modeling, Design and Simulation.* Prentice-Hall, Upper Saddle River, NJ.

Chien, I.L. and Fruehauf, P.S. 1990. Consider IMC tuning to improve controller performance. *Chem. Engr. Prog.* 86:33.

Edgar, T.F., Smith, C.L., Shinsky, F.G., Gassman, G.W., Schafbuch, P.J., McAvoy, T.J., and Seborg, D.E. 1997. See. 8, Process control. *Perry's Chemical Engineering Handbook,* 7th ed. McGraw-Hill, New York.

Grant, E.L. and Leavenworth, R.S. 1996. *Statistical Quality Control,* 7th ed. McGraw-Hill, New York.

Maciejowski, J.M. 2002. *Predictive Control with Constraints.* Prentice-Hall, Upper Saddle River, N.J.

McAvoy, T.J. 1983. *Interaction Analysis.* Instrument Society of America, Research Triangle Park, NC.

Montgomery, D.C. 2001. *Introduction to Statistical Quality Control,* 4th ed. Wiley, New York.

Ogunnaike, B. and Ray, W.H. 1995. *Process Dynamics: Modeling and Control.* Oxford Press, New York.

Qin, S.J. and Badgwell, T.A. 2003. A Survey of Industrial Model Predictive Control Technology, *Control Engineering Practice,* 11:733.

Rivera, D.E., Morari, M., and Skogestad, S. 1986. Internal model control. 4. PID controller design. *Ind. Eng. Chem. Process Des. Dev.* 25:252.

Rosenhof, H.P. and Ghosh, A. 1987. *Batch Process Automation.* Van Nostrand Reinhold, New York.

Seborg, D.E., Edgar, T.F., and Mellichamp, D.A. 2004. *Process Dynamics and Control,* 2nd ed. Wiley, New York.

Shinskey, F.G. 1996. *Process Control Systems,* 4th ed. McGraw-Hill, New York.

Smith, C.A. and Corripio, A.B. 1997. *Principles and Practice of Automatic Process Control,* 2nd ed. Wiley, New York.

Further Information

Recent research results on specialized topics in process control are given in the following books and conference proceedings:

Astrom, K.J. and Wittenmark, B. 1984. *Computer Controlled Systems—Theory and Design.* Prentice-Hall, Englewood Cliffs, NJ.

Buckley, P.S., Luyben, W.L., and Shunta, J.P. 1985. *Design of Distillation Column Control Systems.* Instrument Society of America, Research Triangle Park, NC.

Camacho, E.F. and Bordons, C. 1999 *Model Predictive Control.* Springer-Verlag, New York.

Kantor, J.C. Garcia, C.E., and Carnahan, B. 1997. Chemical Process Control, CPC-V, *AIChE Symp. Series,* 93(316).

Liptak, B.G. 1995. *Instrument Engineers Handbook.* Chilton Book, Philadelphia, PA.

Marlin, T. 1995. *Process Control: Designing Processes and Control Systems for Dynamic Performance.* McGraw-Hill, New York.

Rawlings, J.B. and Ogunnaike, B.A., eds. 2002. Chemical Process Control-CPC VI, *AIChE Symp. Series,* 98(208).

18.4   Servo Systems

John E. McInroy

18.4.1   Introduction

During the past 60 years, many methods have been developed for accurately controlling dynamic systems. The vast majority of these systems employ sensors to measure what the system is actually doing. Then, based on these measurements, the input to the system is modified. These concepts were initially used in applications where only a single output was measured, and that measurement was used to modify a single input. For instance, position sensors measured the actual position of a motor, and the current or voltage put into the motor windings was modified based on these position readings. An early definition of these systems is the following:

> A *servosystem* is defined as a combination of elements for the control of a source of power in which the output of the system, or some function of the output, is fed back for comparison with the input and the difference between these quantities is used in controlling the power (James, Nichols, and Phillips, 1947).

This definition is extremely broad and encompasses most of the concepts that are usually referred to today as the technical area of *control systems*. However, the technology and theory of control systems has changed significantly since that 1947 definition. There are now methods for controlling systems with multiple inputs and multiple outputs, many different design and optimization techniques, and intelligent control methodologies sometimes using neural networks and fuzzy logic. As these new theories have evolved, the definition of a servo system has gradually changed. Today, the term servo system often refers to motion control systems, usually employing electric motors. This change may have occurred simply because one of the most important applications of the early servo system theories was in rapid and accurate positioning. For the sake of brevity, this discussion will confine its treatment to motion control systems utilizing electric motors.

18.4.2   Principles of Operation

To obtain rapid and accurate motion control, servo systems employ the same basic principle just outlined, the principle of feedback. By measuring the output (usually the position or velocity of the motor) and feeding it back to the input, the performance can be improved. This output measurement is often subtracted

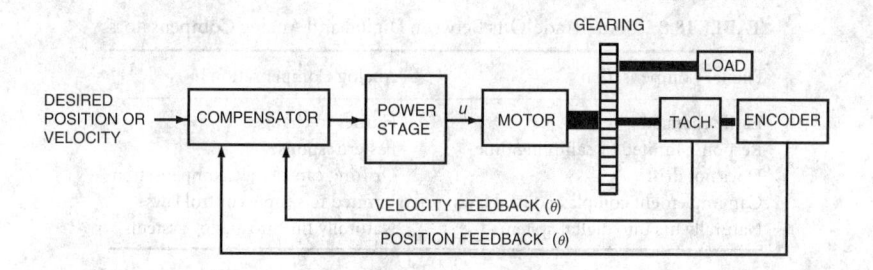

FIGURE 18.64   A typical closed loop motion control or servo system.

from a desired output signal to form an error signal. This error signal is then used to drive the motor. Figure 18.64 illustrates the components present in a typical system.

As the figure illustrates, each servo system consists of several separate components. First, a compensator takes the desired response (usually in terms of a position or velocity of the motor) and compares it to that measured by the sensing system. Based on this measurement, a new input for the motor is calculated. The compensator may be implemented with either analog or digital computing hardware. Since the compensator consists of computing electronics, it rarely has sufficient power to drive the motor directly. Consequently, the compensator's output is used to command a power stage. The power stage amplifies that signal and drives the motor. There are a number of motors that can be used in servo systems. This chapter will briefly discuss DC motors, brushless DC motors, and hybrid stepping motors. To achieve a mechanical advantage, the motor shaft is usually connected to some gears that move the actual load. The performance of the motor is measured by either a single sensor or multiple sensors. Two common sensors, a tachometer (for velocity) and an encoder (for position), are shown. Finally, those measurements are fed back to the compensator.

Servo systems are called **closed-loop** devices because the feedback elements form a loop, as can be seen in Fig. 18.64. In contrast, **open-loop** motion control devices, which do not require sensors, have become popular for some applications. Figure 18.65 depicts an open-loop motion control system. Obviously, it is much easier and less expensive to implement an open-loop control system because far fewer components are involved. The compensator and sensors have been eliminated completely.

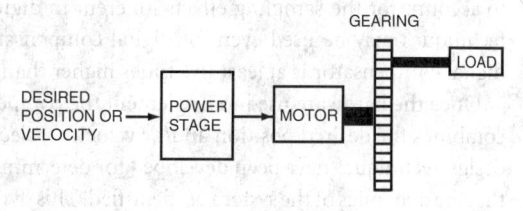

FIGURE 18.65   A typical open loop motion control system.

In addition, the power stage is often simplified because the signals it receives may not be as complex. For instance, switching transistors may replace linear amplifiers. On the other hand, it is difficult to construct an open-loop system with capabilities equal to a servo system's capabilities. Table 18.7 compares open-loop and closed-loop systems.

TABLE 18.7   A Comparison of Open-loop and Closed-loop Control Systems

Open Loop	Closed Loop
Inherently stable	Potentially unstable
Requires no sensing	Requires sensing
Requires no compensator	Requires compensator
Power electronics simplified	Power electronics more complex
Sensitive to modeling errors	Much less sensitive to modeling
Poor disturbance rejection	Rejects unexpected disturbances
May have steady-state error	Steady-state error eliminated

TABLE 18.8   Some Trade-Offs Between Digital and Analog Compensators

Digital Compensation	Analog Compensation
Easily changed	Harder to change
Response limited by sampling time	Faster response
Does not drift	Drifting can change compensation
Can implement complex control laws	Limited to simple control laws
Naturally fits into digital systems	Naturally fits into analog systems

The following sections will briefly describe the components that together are used to create a servo system. The emphasis will be on motor control applications.

Compensator Design

The compensator in Fig. 18.64 is in some ways the brains of a servo system. It takes a desired position or velocity trajectory that the motor system needs to move through, and, by comparing it to the actual movement measured by the sensors, calculates the appropriate input. This calculation can be performed using either digital or analog computing technology. Unlike other areas of computing, in which digital computers long ago replaced analog devices, analog means still are useful today for servo systems. This is mainly because the calculations required are usually simple (a few additions, multiplications, and perhaps integrations) and they must be performed with the utmost speed. On the other hand, digital compensators are increasingly popular because they are easy to change and are capable of more sophisticated control strategies. Table 18.8 compares digital and analog control.

As Table 18.8 indicates, the choice of digital vs. analog compensation depends in part on the control law chosen, since more complex algorithms can be easily implemented on digital computers. If a digital compensator is chosen, then digital control design techniques should be used. These techniques are able to account for the sampling effects inherent in digital compensators. As a rule of thumb, analog design techniques may be used even for digital compensators as long as the sampling frequency used by that digital compensator is at least five times higher than the **bandwidth** of the servo system.

Once the hardware used to implement the compensator has been selected, a control algorithm, which combines the desired position and/or velocity trajectories with the measurements, must be found. Many design techniques have been developed for determining a control law. In general, these techniques require that the dynamics of the system be identified. This dynamic model may come from a physical understanding of the device. For instance, suppose our servo system is used to move a heavy camera. Since the camera is fairly rigid, it might be modeled accurately as a lumped mass. Similarly, the motor's rotor as well as the gearing will yield a mostly inertial load. The motor itself may be modeled as a device that inputs current and outputs torque. By putting these three observations together and writing elementary dynamics equations (e.g., Newton's laws) a set of differential equations that model the actual system can be found. Alternatively, a similar set of differential equations may be found by gathering input and output data (e.g., the motor's current input and the position and velocity output) and employing identification algorithms directly.

Once an adequate dynamic model has been found, it is possible to design a control law that will improve the servo system's behavior in some manner. There are many ways of measuring the performance. Some of the most commonly used and oldest performance measures describe the system's ability to respond to an instantaneous change, or a step. For instance, suppose you are covering a sports event and want to suddenly move the camera to catch the latest action. The camera should reach the desired angle as quickly as possible. The time it does take is termed **rise time**, t_r. In addition, it should stop as quickly as possible so that jitter in the picture will be minimized. The time required to stop is termed **settling time**, t_s. Finally, any excess oscillations will also increase jitter. The magnitude of the worst oscillation is called the **overshoot**. The overshoot divided by the total desired movement and multiplied by 100 gives the *percent overshoot*. These characteristics are depicted in Fig. 18.66.

A vast number of techniques have been developed for achieving these desirable characteristics. As an example, one particular method known as proportional integral derivative (PID) control will be considered.

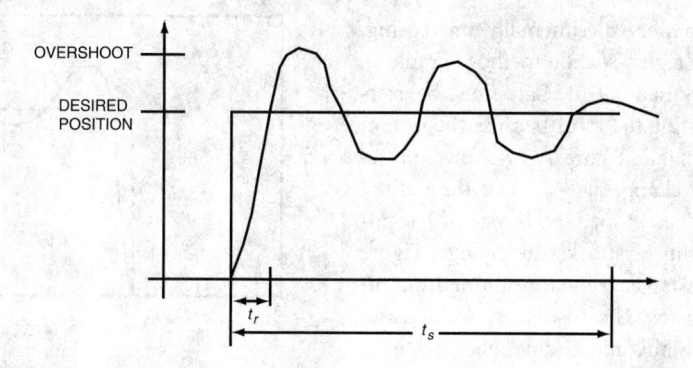

FIGURE 18.66 Performance measures for a step response.

This method is selected because it is intuitive and fairly easy to tune. First, an error signal is formed by subtracting the actual position from the desired position. The error is then

$$e = \theta_d - \theta \tag{18.14}$$

Next, this error signal is operated on mathematically to determine the input to the motor.

The simplest form of this controller uses the following logic:

> If the motor starts falling behind so $\theta_d > \theta$ and $e > 0$, then increase the input to catch up. If the motor is getting ahead, decrease the input to slow down.

The logic can be mathematically written in the formula

$$u = K_P e \tag{18.15}$$

where K_P is a constant number called the *proportional gain* because the input is made proportional to the error.

To make the motor respond faster, it is helpful to know how quickly the motor is falling behind (or getting ahead). This information is contained in the derivative of the error

$$\dot{e} = \dot{\theta}_d - \dot{\theta} \tag{18.16}$$

A derivative term can then be added to the control law to make the input equal

$$u = K_P e + K_D \dot{e} \tag{18.17}$$

Again, K_D is a constant number called the *derivative gain*.

Finally, it is often necessary that the servo system exhibit zero steady-state error. This quality can be achieved by adding a term due to the integral of the error. The reason this works can be understood by considering what would happen should a steady-state error occur. The integrator would keep increasing the input *ad infimum*. Instead, an equilibrium point is reached with zero error. The complete PID control law is then

$$u = K_P e + K_I \int e\, dt + K_D \dot{e} \tag{18.18}$$

with K_I a constant called the *integral gain*.

The choice of the controller gains (K_P, K_I, and K_D) greatly influences the behavior of the servo system. If a good dynamic model has been found and it consists of linear time invariant differential (or difference) equations, then several methods may be employed for finding gains that best meet the performance requirements (the textbooks listed in the Further Information section provide an excellent and detailed treatment).

If not, the gains may be experimentally tuned using, for instance, the Ziegler–Nichols method (Franklin, Powell, and Workman, 1990). Care must be exercised when adjusting the gains because the system may become unstable. Figure 18.67 shows the response to a step change at $t = 5$ in the desired position for a properly tuned servo system. The position response quickly tracks the change. Figure 18.68 plots an unstable response resulting from incorrectly tuned gains. The position error increases rapidly. Obviously, this undesirable behavior can be very destructive.

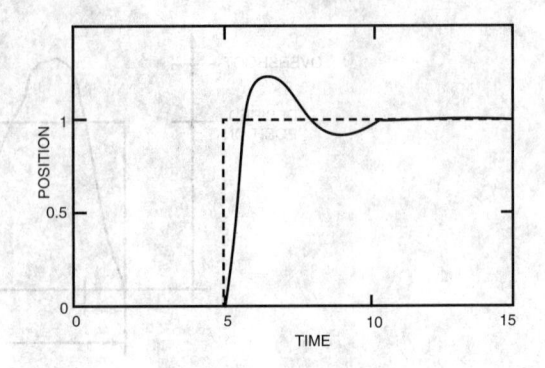

FIGURE 18.67　A stable response to a step command.

Power Stage

A variety of devices are used to implement the power stage, depending on the actuator used to move the servo system. These range from small DC motors to larger AC motors and even more powerful hydraulic systems. This section will briefly discuss the power stage options for the smaller, electric driven servo systems.

There are two distinct methods for electronically implementing the power stage: linear amplifiers and switching amplifiers. Either of these methods typically takes a low-power signal (usually a voltage) as input, and outputs either a desired voltage or current level. This output is regulated so that the voltage or current level remains virtually unchanged despite large changes in the loading caused by the motor. The motor loading does change dramatically as the motor moves because the back electromotive force (EMF) is proportional to the motor velocity ω. For hybrid stepping motors, it is also modulated by either the sine or cosine of the motor position, θ. Figure 18.69 shows the equivalent electric circuit for the two phases of a hybrid stepping motor.

Despite their similarities, linear and switching amplifiers differ markedly in how they achieve voltage or current regulation. Linear amplifiers use a feedback strategy very similar to that described in the previous section. The result is a smooth, continuous output voltage or current. Switching amplifiers, on the other hand, can only turn the voltage to the motor fully on or off (or positive/negative). Consequently, every few microseconds the circuit checks a low-pass filtered version of the motor's voltage or current level. If it is too high, the voltage is turned off. If it is too low, the voltage is turned on. This method works because the sampling is done

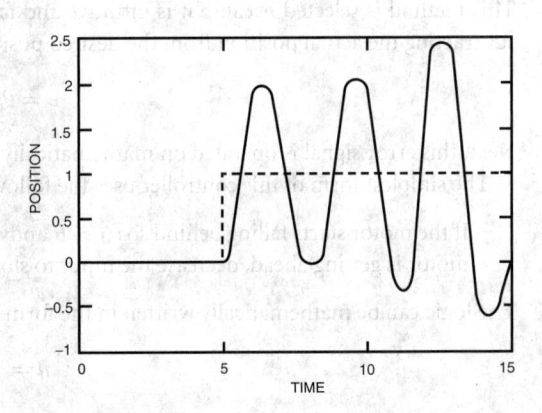

FIGURE 18.68　An unstable response to a step command.

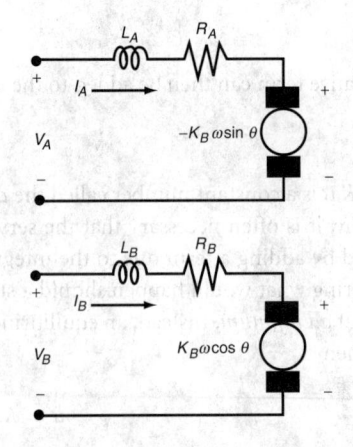

FIGURE 18.69　The electrical equivalent circuit representing the two phases of a hybrid stepping motor.

on a microsecond scale, whereas the motor's current or voltage can only change in a millisecond scale. All motors contain resistance and inductance due to the winding, and this resistor-inductor circuit acts as a low-pass filter. Thus, the motor filters out the high-frequency switching, and responds only to the low-frequency components contained within the switching. Care must be taken when using this technique to perform the switching at a frequency well beyond the bandwidth of the mechanical components so that the motor does not respond to the high-frequency switching.

If the desired output is a voltage, the amplifier is called a **pulse-width-modulated** (PWM) voltage amplifier. If the desired output is a current, the amplifier is usually termed a **current chopper**.

Both linear and switching amplifiers remain in use because each has particular advantages. Linear amplifiers do not introduce high-frequency harmonics to the system; thus, they can provide more accurate control with less vibrations. This is especially true if the servo system contains components that are somewhat flexible. On the other hand, if the system is rigid, the switching amplifier will provide satisfactory results at a lower cost.

Motor Choice

The choice of motor for a servo system can impact cost, complexity, and even the precision. Three motors will be described in this section. Although this is not by any means an exhaustive coverage, it will suffice for a variety of applications.

DC servomotors are the oldest and probably the most commonly used actuator for servo systems. The name arises because a constant (or DC) voltage applied to the motor input will produce motor movement. Over a linear range, these motors exhibit a torque that is proportional to the quantity of current flowing into the device. This can make the design of control laws particularly convenient since torque can often be easily included into the dynamic equations. The back EMF generated by the motor is strictly proportional to the motor velocity, and so a large input voltage is required to overcome this effect at large velocities. The chief disadvantage of the DC motor is caused by the brushes. Inside the motor, mechanical switching is performed with brushes. As you might imagine, these brushes are the first part of the motor to fail, since they are constantly rubbing and switching current. Consequently, they severely limit the reliability of DC motors as compared to other, more modern actuators. The brushes also limit the low-speed performance because the switching effects are not filtered out when the motor is turning slowly. Instead, the motor actually has time to react to each individual switching. For this reason, DC motors are normally run only at high speeds, and gearing is used to convert the speed to that needed for a particular application. Only very specialized designs of DC motors are suitable for highly precise direct drive applications in which no gearing is allowed.

One way to eliminate some of these problems is to use a brushless DC motor rather than an ordinary DC motor. The brushless DC operates in a manner identical to the DC motor. However, the motor construction is quite different. A typical brushless DC motor actually has three phases. Much like the DC motor, these phases are switched either on or off depending on the motor position. Whereas the DC motor accomplishes this switching mechanically with the brushes, the brushless DC contains Hall effect sensors. These sensors determine the position of the motor, then based on these measurements electronically perform the switching. The advantage is that brushes are no longer necessary. As a result, the brushless DC motor will last much longer than an ordinary DC motor. Unfortunately, since Hall effect sensors and more sophisticated switching electronics are required, brushless DC motors are more expensive than DC motors. In addition, since the basic principle still involves switching phases fully on or off, brushless DC motors also exhibit poor performance at low speeds.

Both types of DC motors have become popular because they respond linearly (over a nominal operating range) to an input current command in that the output torque is roughly proportional to the input current. This is convenient because it allows the direct use of linear control design techniques. It was also especially important during early servo system development because it is easy to implement with analog computing devices. Advances in microcontroller computing technology and nonlinear control design techniques now provide many more options. One of these is the use of hybrid stepping motors in servo systems. These motors were originally developed for open-loop operation. They have shown considerable promise for closed-loop operation as well. Hybrid stepping motors typically have two phases. These windings are labeled A and B in Fig. 18.70(a). These two windings are energized in the sequence shown in

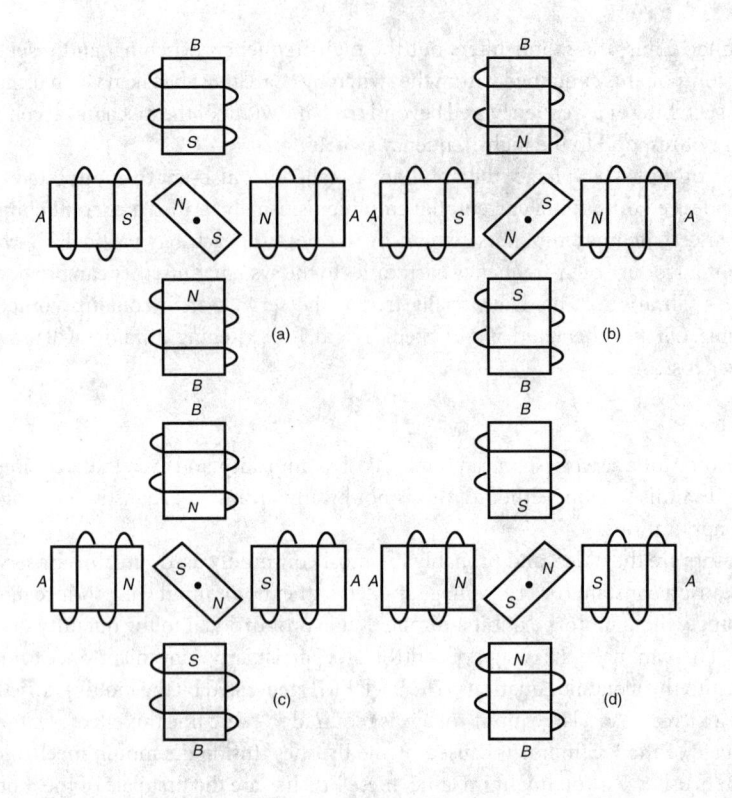

FIGURE 18.70   The stepping sequence for a hybrid stepping motor.

Fig. 18.70(a) to Fig. 70(d), thus causing the permanent magnet rotor to turn. Although Fig. 18.70 shows only four separate energizations (called *steps*) of the windings to complete a revolution, clever mechanical arrangements allow the number of steps per revolution to be greatly increased. Typical hybrid stepping motors have 200 steps/revolution. Consequently, a complete revolution is performed by repeating the four step sequence shown in Fig. 18.70, 50 times.

One advantage of these motors is that they allow fairly accurate positioning in open loop. As long as the motor has sufficient torque capabilities to resist any external torques, the motor position remains within the commanded step. Since open-loop systems are far less complicated (and therefore less expensive), these motors offer an economical alternative to the traditional closed-loop servo system. On the other hand, there are some accompanying disadvantages to open-loop operation. First, since the winding currents are switched on and off to energize the coils in the proper sequence, the motor exhibits a jerky motion at low speeds, much like the DC motors. Second, if an unexpectedly large disturbance torque occurs, the motor can be knocked into the wrong position. Since the operation is open loop and therefore does not employ any position sensor, the system will not correct the position, but will keep running. Finally, the speed of response and damping can be improved using closed-loop techniques.

In addition to the open-loop uses of hybrid stepping motors, methods exist for operating these motors in a closed-in a closed-loop fashion (Lofthus et al., 1994). The principle difficulty arises because hybrid stepping motors are nonlinear devices, so the plethora of design techniques available in linear control theory cannot be directly applied. Fortunately, these motors can be made to behave in a linear fashion by modulating the input currents by sinusoidal functions of the motor position. For instance, the currents through the two windings (I_A and I_B) can be made equal to

$$I_A = -I_C \sin N\theta \qquad (18.19)$$

$$I_B = I_C \cos N\theta \qquad (18.20)$$

where N is the number of motor pole pairs (50 for the typical 200-step/revolution motor) and θ is the motor's angular position. Then the torque output by the motor is proportional to the new input, I_C. I_C can be determined using standard linear control designs such as the PID method outlined earlier.

In contrast to the DC motors, which employ current switching and therefore perform poorly at low speeds, this closed-loop method of controlling stepping motors performs quite accurately even at very low velocities. On the other hand, the generation of the sinusoidally modulated input currents requires both accurate position measurement and separate current choppers for each phase. Thus the implementation cost is somewhat higher than a servo system utilizing DC motors.

Gearing

As alluded to in the previous section, gearing is often used in servo systems so that the motor can operate in its optimum speed range while simultaneously obtaining the required speeds for the application. In addition, it allows the use of a much smaller motor due to the mechanical advantage. Along with these advantages come some disadvantages, the principle disadvantage being **backlash**. Backlash is caused by the small gaps that exist in gearing systems. In gearing systems, there always exist small gaps between a pair of mating gears. As a result of the gaps, the driven gear can wiggle independently from the driving gear. This wiggling is actually a nonlinear phenomena that can cause vibrations and decreases the accuracy that can be obtained. A similar effect occurs when using belt driven gearing because the belt stretches.

These inherent imperfections in gearing can become major problems if extreme accuracy is required in the servo system. Since these effects increase as the gear ratio increases, one way to reduce the effect of the gearing imperfections is to make the gear ratio as small as possible. Unfortunately, DC motors run smoother at higher speeds, so smaller gear ratios tend to accentuate the motor imperfections caused by switching. Consequently, a tradeoff between the inaccuracies caused by the gearing and the inaccuracies caused by the motor must be made. For highly precise systems, the gearing is sometimes completely eliminated. For such applications, a closed-loop hybrid stepping motor or a specialized DC motor are usually necessary to obtain good motion quality.

A Simple Example Application

To clarify some of the engineering choices that must be made when designing a servo system, a simple example of a servo system design for moving a camera will be discussed.

First, suppose the camera will be used in a surveillance application, to constantly pan back and forth across a store. For this application, an actual servo system employing feedback is probably not necessary and would prove too costly. The required precision is fairly low, and low cost is of the utmost importance. Consequently, an open-loop stepping motor powered by transistors switching on or off will probably be adequate. Alternatively, even a DC motor operated in open loop, with microswitches to determine when to reverse direction will suffice.

Now suppose the camera is used to visually inspect components as part of a robotic workcell. In this case, the requirements on precision are significantly higher. Consequently, a servo system using a geared DC motor and powered by a PWM voltage amplifier may be an appropriate choice. If the robotic workcell is used extensively, a more reliable system may be in order. Then a brushless DC motor may be substituted for a higher initial cost.

Finally, suppose the camera system is used to provide feedback to a surgeon performing intricate eye surgeries. In this case, an extremely precise and reliable system is required, and cost is a secondary consideration. To achieve the high accuracy, a direct drive system that does not use any gearing may provide exceptional results. Since the motor must move precisely at low speeds, simultaneously remaining very reliable, a closed-loop hybrid stepping motor may be used. To drive the hybrid stepping motor, either linear current amplifiers or current choppers (one for each of the two phases) will suffice. For this particular application, the hybrid stepping motor offers another advantage because it can be operated fairly accurately in open loop. Thus it offers redundancy if the compensator or sensors fail.

Defining Terms

Backlash: The errors introduced into geared systems due to unavoidable gaps between two mating gears.

Bandwidth: The frequency range over which the Bode plot of a system remains within 3 dB of its nominal value.

Closed-loop system: A system that measures its performance and uses that information to alter the input to the system.

Current chopper: A current amplifier that employs pulse-width-modulation.

Open-loop system: A system that does not use any measurements to alter the input to the system.

Overshoot: The peak magnitude of oscillations when responding to a step command.

Pulse-width-modulation: A means of digital-to-analog conversion that uses the average of high-frequency switchings. This can be performed at high-power levels.

Rise time: The time between an initial step command and the first instance of the system reaching that commanded level.

Settling time: The time between an initial step command and the system reaching equilibrium at the commanded level. Usually this is defined as staying within 2% of the commanded level.

References

Franklin, G.F., Powell, J.D., and Workman, M.L. 1990. *Digital Control of Dynamic Systems*, 2nd ed. Addison-Wesley, Reading, MA.

James, H.M., Nichols, N.B., and Phillips, R.S. 1947. *Theory of Servomechanisms*, Vol. 25, MIT Radiation Laboratory Series. McGraw-Hill, New York.

Lofthus, R.M., Schweid, S.A., McInroy, J.E., and Ota, Y. 1994. Processing back EMF signals of hybrid step motors. *Control Engineering Practice*, a journal of the International Federation of Automatic Control 3(1):1–10. (Updated in Vol. 3, No. 1, (Jan.) 1995, pp. 1–10.)

Further Information

Recent research results on the topic of servo systems and control systems in general is available from the following sources:

Control Systems magazine, a monthly periodical highlighting advanced control systems. It is published by the IEEE Press.

The *IEEE Transactions on Automatic Control* is a scholarly journal featuring new theoretical developments in the field of control systems.

Control Engineering Practice, a journal of the International Federation of Automatic Control, strives to meet the needs of industrial practitioners and industrially related academics and researchers. It is published by Pergamon Press.

In addition, the following books are recommended:

Brogan, W.L. 1985. *Modern Control Theory*. Prentice-Hall, Englewood Cliffs, NJ.

Dorf, R.C. *Modern Control Systems*. Addison-Wesley. Reading, MA.

Jacquot, R.G. 1981. *Modern Digital Control Systems*. Marcel Dekker, New York.

18.5   Power Control and Switching

Gibson Morris, Jr.

18.5.1   Introduction

Control of electric power, or power conditioning, requires the conversion of electric power from one form to another, that is, DC to AC, one voltage level to another, and so on, by the switching characteristics of various electronic power devices. Some of these devices, such as silicon controlled rectifiers (SCRs), can be operated as switches by applying control voltages to their controlling input (gate terminal of gate-turn-off thyristor (GTO)). Others, such as triggered vacuum and gas gap switches rely on over-voltages or triggering. For large power systems, control is exercised by supervisory control and data acquisition (SCADA) computer systems located at a central location. This master station communicates with remote stations in the system to acquire data and system status and to transmit control information. These centers support the two main objectives of power system control, that is, (1) stable (voltage and frequency) power and (2) economical utilization of the generation capabilities in the system. Typically, system control involves both the generation and load ends of a system. At generating plants this is accomplished by direct control to input to the prime mover, and systemwide it is accomplished by the control centers. Distributing control to the load end provides a localized response to changing requirements at subtransmission and distribution circuits. At this level, faults such as lightning strikes and short circuits can be responded to in a timely manner through relays, circuit breakers, and so forth, thereby protecting system assets.

Development of high-power switches was stimulated by the growth of the electric power distribution networks. Control and protection of these networks involved power handling capabilities in excess of 10^9 W. Lasers, X-ray equipment, particle accelerators, fusion research, and so forth, have required the switching of power in the form of fast rising pulses and at levels that can exceed 10^{12} W.

Power control and switching falls into the following general regimes:

- Electric power distribution at low frequencies
- Pulsed power characterized by fast rising pulses

Switches can be divided into two classes, opening switches and closing switches, with the following characteristics:

- Hold-off voltage
- Voltage drop during conduction
- dv/dt during opening or closing
- Trigger voltage requirement, if any
- Maximum current
- di/dt characteristics

Power electronic circuits can be classified as to function as follows:

- Diode rectifiers
- DC–DC converters
- AC–AC converters
- DC–AC converters
- AC–DC converters
- Switches

In general, these power circuits relate two external systems and, therefore, involve current and voltage relationships at the input and output ports of the power circuit. Thus, control of a power circuit must be defined in context to the external systems to which it is connected.

The switching characteristics of semiconductor devices introduce harmonics into the output of converters. These harmonics can affect the power quality of a circuit and reduction of these harmonics by the use of filters is often necessary. With the increasing utilization of semiconductor devices for power control and switching, the issue of power quality is becoming more important.

18.5.2 Electric Power Distribution

Control of a power system is required to keep the speed of machines constant and the voltage within specified limits under changing load conditions. Control schemes fall into several categories: control of excitation systems, turbine control, and faster fault clearing methods. Design strategies that reduce system reactance are: minimum transformer reactance, compensation of transmission lines using series capacitance or shunt inductance, and additional transmission lines.

One of the prinicipal components of the power distribution system is an AC generator driven by a turbine. Changing load conditions require changes on the torque on the shaft from the prime mover to the generator so that the generator can be kept at a constant speed. The field to the alternator must be adjusted to maintain a constant voltage.

This synchronous machine converts mechanical energy to electrical energy and consists of a stator, or armature, with an armature winding. This winding caries the current I_a supplied by the generator to a system. The rotor is mounted on a shaft that rotates inside the stator and has a winding, called the field winding, that is supplied by a DC current from the exciter. The magnetomotive force (mmf) produced by the field winding current combines with the mmf produced by the current in the armature winding to

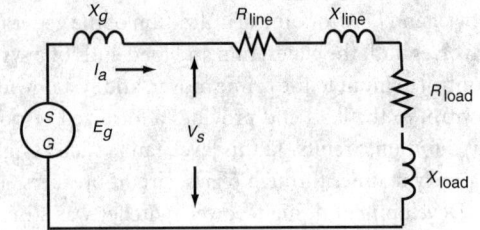

FIGURE 18.71   Basic power circuit.

produce a flux that generates a voltage in the armature winding and provides the electromagnetic torque between the stator and the rotor. This torque opposes the torque of the prime mover, that is, steam or hydraulic turbine.

The basic power circuit is illustrated in Fig. 18.71.

$$V_s = E_g - I_a X_g$$

An important factor in controlling the flow of reactive power is changing the exciter current. Normal excitation of a synchronous generator is given by

$$|E_g| \cos \delta = V_s$$

where δ is the torque or power angle between the generator voltage and the terminal voltage V_s.

Changing the DC exciter current will cause E_g to vary according to

$$I_a = \frac{|E_g| \angle \delta - |V_s| \angle 0}{j X_g} \quad \text{and} \quad I_a^* = \frac{|E_g| \angle -\delta - |V_s| \angle 0}{j X_g}$$

Power from the generator to a system is given by $|V_s||I_a| \cos \theta$, where θ is the power factor angle. Increasing the exciter current increases E_g as shown in Fig. 18.72(a). Note the generator supplies lagging current to the system. In Fig. 18.72(b), the generator is underexcited and supplies leading current to the system, or the generator is drawing lagging current from the system.

If power input to a generator is increased with $|E_g|$ constant, the rotor speed will increase and the angle δ will increase. This increase in δ results in a large I_a and lower θ. The generator delivers more power to

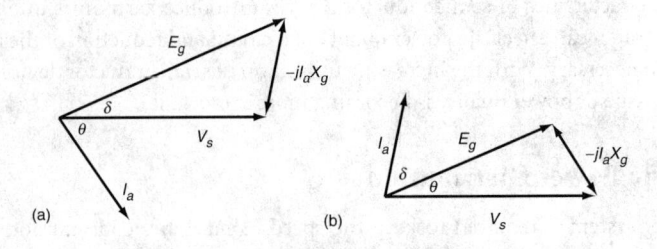

FIGURE 18.72   Generator excitation.

an infinite bus given by

$$S = \frac{|V_s||E_g|\angle 90 - \delta - |V_s|^2 \angle 90}{X_g} = P + jQ = V_s I_a^*$$

where

$$V_s = |V_s|\angle 0 \quad \text{and} \quad E_g = |E_g|\angle \delta$$

Real power

$$P = \mathrm{Re}S = |V_s||E_g|\cos(90 - \delta) - |V_s|\frac{|E_g|}{X_g}\sin\delta$$

Reactive power

$$Q = \mathrm{Im}S = [E_g \cos\delta - |V_s|]\frac{V_s}{X_g}$$

Note that for a small δ, a small change in δ results in a larger P than Q.

Reliable electric power service implies that the loads are fed at a constant voltage and frequency at all times. A stable power system is one in which the synchronous machines, if perturbed, return to their original state if there is no net change in power, or stabilize at a new state without loss of synchronization. The machine rotor angle is used to quantify stability, that is, if the difference in the angle between machines increases or oscillates for an extended period of time, the system is unstable. The swing equation, given by

$$J\delta_m = J\omega_m = T_a$$

governs the motion of the machine rotor. J is moment of inertia, δ_m is mechanical torque angle with respect to a rotating reference, ω_m is shaft angular velocity, and T_a is the accelerating torque. Two factors that act as criteria for the stability of a generating unit are the angular swing of the machine during and following fault conditions, and the time it takes to clear the transient swing.

Mechanical torques of the prime movers, steam or hydraulic, for large generators depend on rotor speed. In unregulated machines the torque speed characteristic is linear over the rated range of speeds. The prime mover speed of a machine will drop in response to an increased load, and the valve position must be opened to increase the speed of the machine. In a regulated machine (governor controlled) the speed control mechanism controls the throttle valves to the steam turbine or the gate position for a water turbine.

Automatic voltage regulation can be used to adjust the field winding current, thus changing E_g as the load on the machine is varied. If the power output of a generator is to be increased while an automatic voltage regulator holds the bus voltage at a constant value, the field winding current must be increased. The maximum voltage output of a machine is limited by the maximum voltage of the exciter supplying the field winding. Figure 18.73 illustrates control of a power-generating unit.

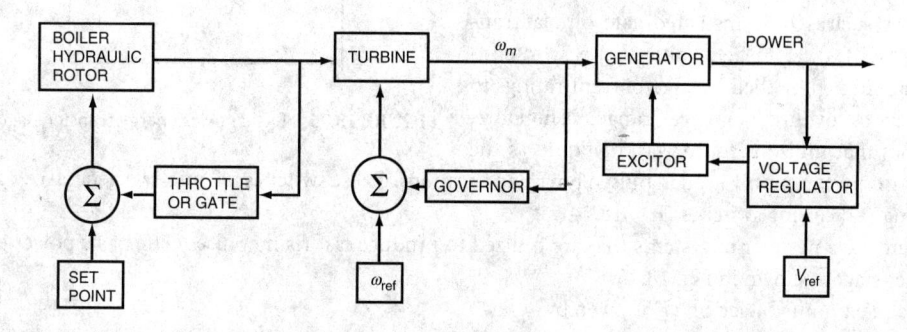

FIGURE 18.73 Generating unit control.

The performance of a transmission system can be improved by reactive compensation of a series or parallel type. Series compensation consists of banks of capacitors placed in series with each phase conductor of the line and is used to reduce the series impedance of the line, which is the principal cause of voltage drop. Shunt compensation consists of inductors placed from each line to neutral and is used to reduce the shunt susceptance of the line.

Circuit interrupting devices in power systems fall into three categories: fuses, switches, and circuit breakers. Design ratings of these devices are

- Power-frequency voltage of the circuit
- Maximum continuous current device can carry
- Maximum short-circuit current device can interrupt

The operation and control of a power system requires continual analysis of the system parameters to ensure optimal service at the cheapest production cost. Computer studies are essential to this goal. The IEEE Brown book (IEEE, 1980) is an excellent reference source for the power engineer. Stability studies, load flow studies, transient studies, and so forth, are detailed in this source, and the reader is encouraged to refer to this document for detailed information about different aspects of power control.

18.5.3 Pulsed Power

The generation of high-power pulses requires transferring energy stored in electric fields (capacitors) and magnetic fields (inductors). Capacitor charging requires high-voltage sources at relatively modest current ratings, whereas inductive charging requires high-current sources at relatively low voltages. Transferring this stored energy to a load requires switches capable of holding off high voltages, conducting high currents, and capable of producing pulses with nanosecond or less rise times. An example of the complexity of controlling high-power pulses is the current effort to generate power with energy derived from the fusion of elements. Ignition of the fusion process requires the simultaneous application of high-energy beams on a single target. Another example of the requirement for fast rising pulses is in the area of a detonator for nuclear weapons. Other applications include lasers, high-energy beams, energy weapons, medical equipment and high-voltage testing of insulators. Solid-state switches, such as SCRs, insulated-gate-bipolar transistors (IGBTs), GTOs, etc., although easy to control, are limited by their voltage/current ratings to the few kilovolt/kiloampere range. Therefore, switches that utilize vacuum, gas, or liquid as the dielectric are commonly used in pulsed power applications. These switches fall into two general categories, opening and closing switches (Fig. 18.74).

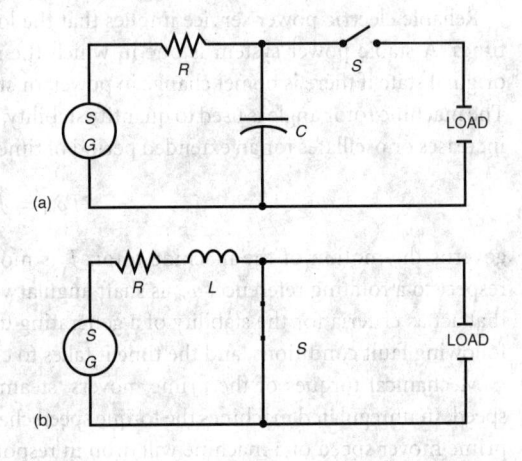

FIGURE 18.74   (a) Closing switch, (b) opening switch.

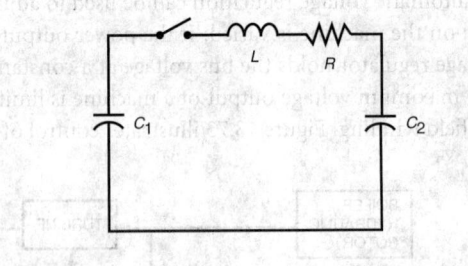

FIGURE 18.75   Basic power energy transfer stage.

There are a number of systems that are utilized to produce fast rising pulses. The basic power energy transfer stage is shown in Fig. 18.75.

The maximum voltage on C_2 is given by

$$V_{c2max} = [V_0 C_1/(C_1 + C_2)][1 - e^{-\pi/2\omega\tau}]$$

where

$$\tau = L/R, \qquad \omega^2 = ABS[1/LC - 1/(2\tau)^2]$$

and V_0 is initial voltage on C_1. The output of this stage is switched to a subsequent stage with a closing switch. Utilizing this technique an energy transfer system can be designed for a particular load.

Another common high-voltage generating system found in almost all high-voltage testing facilities is the marx generator (Fig. 18.76). This system charges capacitors in parallel and then discharges them by switching them to a load in series. The switches are usually spherical electrodes with air as the dielectric. The first stage is the control stage and is triggered into conduction at the desired time.

For capacitive loads

$$V_{\text{max}} = [2NVC/(C + C_L)][1 - e^{-\pi/2\omega\tau}]$$

For resistive loads

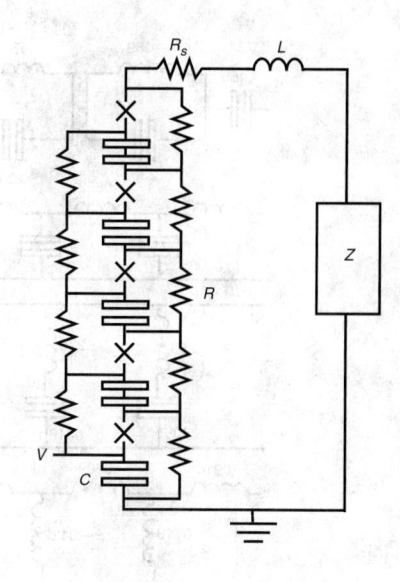

FIGURE 18.76 Marx high-voltage generater.

$$V_{\text{max}} = (NV_0R_Le^{-t/2\tau}/2L\omega)(e^{+\omega t} - e^{-\omega t})$$

where $\tau = L/R$, V is charging voltage, N is number of stages, C_L is capacitance of load, C is capacitance of a single stage

$$\omega^2 = ((NC_L + C)/(NLCC_L)) - 1/(2\tau)^2$$

and the time at which the voltage is at a peak is given by

$$t = T_m = (1/2\omega)\ln[(1 + 2\omega\tau)/(1 - 2\omega\tau)]$$

The breakdown voltage for air gaps is given by $E = 24.5P + 6.7(P/R_{\text{eff}})^{1/2}$ kV/cm where R_{eff} is $0.115R$ for spherical electrodes and $0.23R$ for cylindrical electrodes and P is pressure in atmospheres. To increase the high-voltage characteristics of a switch other dielectrics can be used (Table 18.9).

Another common method of generating high-voltage pulses is to charge a transmission line and

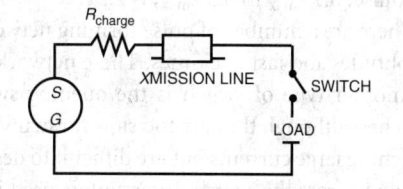

FIGURE 18.77 Transmission line for pulse forming.

then switch the stored energy to a load with a closing switch, see Fig. 18.77. For an ideal pulse line the output voltage is given by

$$V = V_{\text{charge}}(1 - e^{-t/\tau})(Z_c + R)R$$

TABLE 18.9 Relative Breakdown Field Compared to Air

Dielectric	Relative Breakdown Field
Air	1.0
Nitrogen	1.0
SF$_6$	2.7
Freon	2.5
Hydrogen	0.5
30% SF$_6$, 70% air	2.0

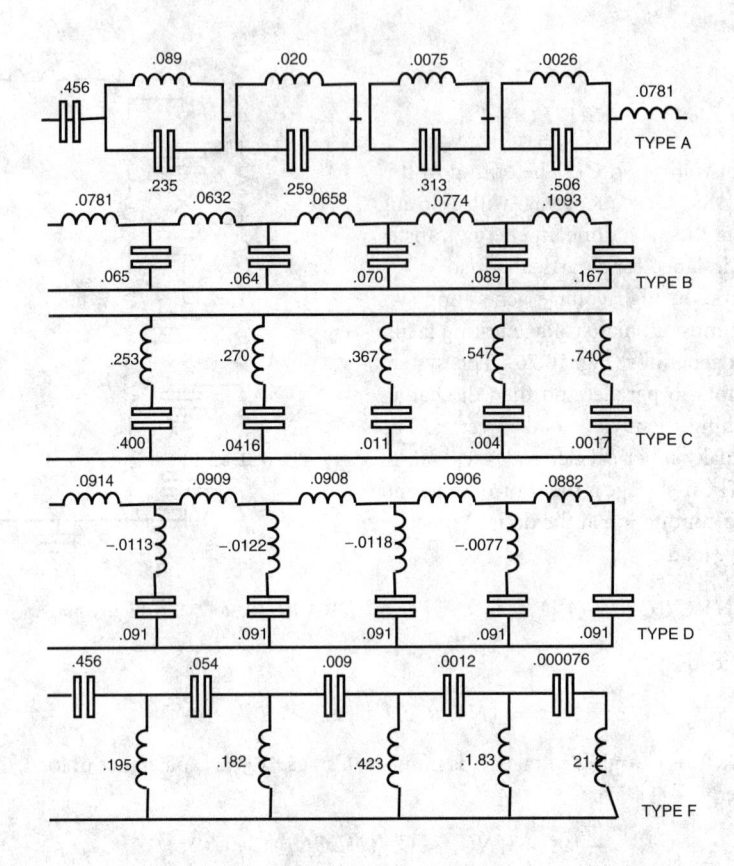

FIGURE 18.78   Pulse forming networks.

where Z_c is the characteristic impedance of the line, R is the load, and $\tau = L/(Z_c + R)$. The rise time of the pulse, $0.1V_{max}$ to $0.9V_{max}$, is 2.2τ.

There are a number of pulse forming networks (Fig. 18.78) that will produce pulses with near constant amplitudes and fast rise times. These networks approximate the ideal transmission line.

Another type of switch is the opening switch. Fuses and circuit breakers are examples of opening switches, although they are too slow for many pulsed power applications. Opening switches are capable of switching large currents but are difficult to design because during switch opening a significant voltage will develop across the switch. Interrupters used in the power industry rely on a current zero occuring while the switch is opening. If the switch configuration is not such that it can withstand the reoccurring voltage it will close. For pulsed power purposes it is necessary to interrupt currents in the kiloampere, range with nanosecond opening. One such switch is the plasma opening switch. In this switch the plasma channel that sustains conduction between electrodes is degraded to the point that it cannot sustain current flow and the switch opens. This switch is utilized in high-energy beam applications. Other methods such as magnetic interruption and explosive devices are used.

Although the power industry is at a mature stage, the pulsed power arena is a new and dynamic challenge for the power engineer. As more exotic requirements develop for the delivery of high energy in very short periods of time to loads, there will be an increasing demand for improvements in switching technology.

References

Anderson, P.M. and Fouad, A.A. 1977. *Power System Control and Stability*. Iowa State University Press, IA.

Eaton, J.R. and Cohen, E. 1983. *Electric Power Transmission Systems*. Prentice-Hall, Englewood Cliffs, NJ.

IEEE. 1980. *IEEE Recommended Practices for Industrial and Commercial Power Systems Analysis*. Wiley, New York.

Kassakian, J.K., Schecht, M.F., and Verghere, G.C. 1991. *Principles of Power Electronics*. Addison-Wesley, New York.

Rashid, M.H. 1988. *Power Electronics*. Prentice-Hall, Englewood Cliffs, NJ.

Russel, B.D. and Council, M.E. 1978. *Power System Control and Protection*. Academic Press, New York.

Stevenson, J.D. 1982. *Elements of Power System Analysis*. McGraw-Hill, New York.

Vitkouitsky, H.M. 1987. *High Power Switching*. Van Nostrand Reinhold, New York.

Weedy, B.M. 1975. *Electric Power Systems*. Wiley, New York.

Further Information

IEEE Power Energy Review

IEEE Recommended Practices for Industrial and Commercial Power Systems Analysis

IEEE Transactions on Power Electronics

IEEE Transactions on Power Systems

19

Computer Systems

Joy S. Shetler
California Polytechnic State University

Margaret H. Hamilton
Hamilton Technologies, Inc.

Bogdan M. Wilamowski
Auburn University

David A. Kosiba
Pennsylvania State University

Rangachar Kasturi
Pennsylvania State University

Yariv Ephraim
George Mason University

Hanoch Lev-Ari
Northeastern University

William J.J. Roberts
Atlantic Coast Technologies, Inc.

Michel D. Yacoub
University of Campinas

Paulo Cardieri
University of Campinas

Élvio João Leonardo
University of Campinas

Álvaro Augusto Machado Medeiros
University of Campinas

James E. Goldman
Purdue University

John D. Meyer
Hewlett-Packard Co.

19.1   Fundamental Architecture

Joy S. Shetler

19.1.1   Introduction

The design space for computer **architectures** is fairly diverse and complicated. Each new architecture strives to fulfill a different set of goals and to carve out a niche in the computer world. A system can be composed of one or several processors. Many concepts apply to both multiple processor and single processor systems. Many researchers have concluded that further advances in computational speed and throughput will come from parallelism rather than relying heavily on technological innovation, as has occurred in the past. But implementation is still an important consideration for any computer system.

19.1.2   Defining a Computer Architecture

Some important attributes of a computer system are as follows:

- Structure of the control path(s)
- Structure of the data path(s)
- The memory organization
- The technology used in the implementation
- The number of clocks, that is, single clocked or dual clocked
- The clock speed(s)
- The degree of pipelining of the units
- The basic topology of the system
- The degree of parallelism within the control, data paths, and interconnection networks

In some cases, the **algorithms** used to execute an instruction are important as in the case of data flow architectures or systolic arrays. Some of these attributes are dependent on the other attributes and can change depending on the implementation of a system. For example, the relative clock speed may be dependent on the technology and the degree of **pipelining**. As more diverse and complicated systems are developed or proposed, the classification of a computer architecture becomes more difficult. The most often cited classification scheme, Flynn's taxonomy (Flynn, 1966) defined four distinct classes of computer architectures: *single instruction, single data* (SISD), *single instruction, multiple data* (SIMD), *multiple instruction, single data* (MISD), and *multiple instruction, multiple data*, (MIMD). A particular classification is distinguished by whether single or multiple paths exist in either the control or data paths. The benefits of this scheme are that it is simple and easy to apply. Its prevalent use to define computer architectures attests to its usefulness. Other classification schemes have been proposed, most of which define the parallelism incorporated in newer computer architectures in greater detail.

19.1.3   Single Processor Systems

A single processor system, depicted in Fig. 19.1 is usually composed of a controller, a data path, a memory, and an input/output unit. The functions of these units can be combined, and there are many names for each unit. The combined controller and data path are sometimes called the central processing unit (CPU). The data path is also called an arithmetic and logical unit (ALU).

Hardware parallelism can be incorporated within the processors through pipelining and the use of multiple functional units, as shown in Fig. 19.2. These methods allow instruction execution to be overlapped in time. Several instructions from the same instruction stream or *thread* could be executing in each pipeline stage for a pipelined system or in separate functional units for a system with multiple functional units.

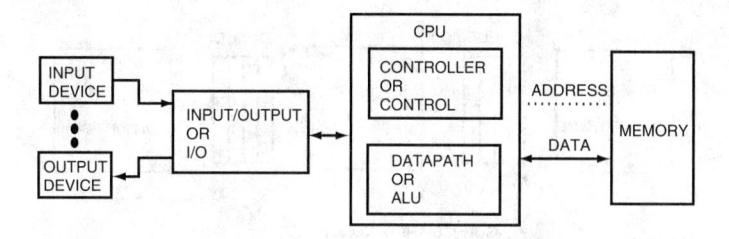

FIGURE 19.1   Block diagram of a single processor system.

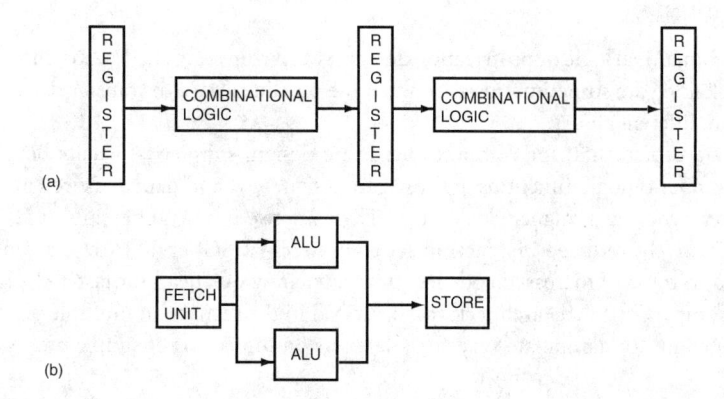

FIGURE 19.2   Methods of incorporating parallelism within a processor system: (a) pipelines, (b) multiple units.

Context is the information needed to execute a process within a processor and can include the contents of a program counter or instruction counter, stack pointers, instruction registers, data registers, or general purpose registers. The instruction, data, and general purpose registers are also sometimes called the register set. Context must be stored in the processor to minimize execution time. Some systems allow multiple instructions from a single stream to be active at the same time. Allowing multiple instructions per stream to be active simultaneously requires considerable hardware overhead to schedule and monitor the separate instructions from each stream.

Pipelined systems came into vogue during the 1970s. In a pipelined system, the operations that an instruction performs are broken up into stages, as shown in Fig. 19.3(a). Several instructions execute simultaneously in the pipeline, each in a different stage. If the total delay through the pipeline is D and there are n stages in the pipeline, then the minimum clock period would be D/n and, optimally, a new instruction would be completed every clock. A deeper pipeline would have a higher value of n and thus a faster clock cycle.

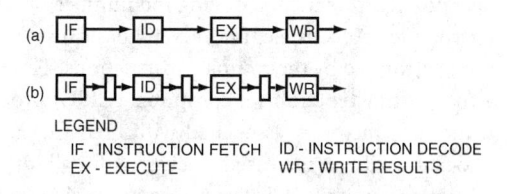

FIGURE 19.3   Comparison of the basic processor pipeline technique: (a) not pipelined, (b) pipelined.

Today most commercial computers use pipelining to increase performance. Significant research has gone into minimizing the clock cycle in a pipeline, determining the problems associated with pipelining an instruction stream, and trying to overcome these problems through techniques such as prefetching instructions and data, compiler techniques, and caching of data and/or instructions. The delay through each stage of a pipeline is also determined by the complexity of the logic in each stage of the pipeline. In many cases, the actual pipeline delay is much larger than the optimal value, D/n, of the logic in each stage of the pipeline. Queues can be added between pipeline stages to absorb any differences in execution time

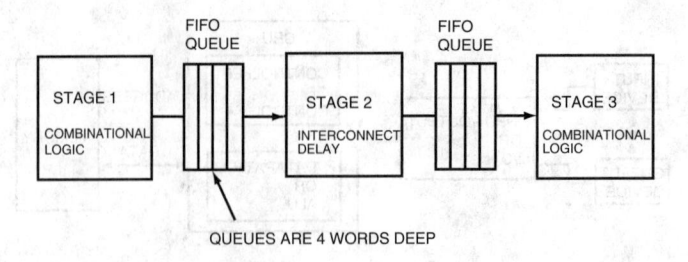

FIGURE 19.4 A pipeline system with queues.

through the combinational logic or propagation delay between chips (Fig. 19.4). Asynchronous techniques including handshaking are sometimes used between the pipeline stages to transfer data between logic or chips running on different clocks.

It is generally accepted that, for computer hardware design, simpler is usually better. Systems that minimize the number of logic functions are easier to design, test, and debug, as well as less power consuming and faster (working at higher clock rates). There are two important architectures that utilize this concept most effectively, **reduced instruction set computers (RISC)** and SIMD machines. RISC architectures are used to tradeoff increased code length and fetching overhead for faster clock cycles and less instruction set complexity. SIMD architectures, described in the section on multiple processors, use single instruction streams to manipulate very large data sets on thousands of simple processors working in parallel.

RISC processors made an entrance during the early 1980s and continue to dominate small processor designs as of this writing. The performance p of a computer can be measured by the relationship

$$p = \left(\frac{\text{computations}}{\text{instruction}} \right) \left(\frac{\text{instructions}}{\text{cycle}} \right) \left(\frac{\text{cycles}}{\text{second}} \right)$$

The first component (computations/instruction) measures the complexity of the instructions being executed and varies according to the structure of the processor and the types of instructions currently being executed. The inverse of the component (instructions/cycle) is commonly quoted for single processor designs as **cycles per instruction** (**CPI**). The inverse of the final component (cycles/second) can also be expressed as the clock period of the processor. In a RISC processor, only hardware for the most common operations is provided, reducing the number of computations per instruction and eliminating complex instructions. At compile time, several basic instructions are used to execute the same operations that had been performed by the complex instructions. Thus, a RISC processor will execute more instructions than a **complex instruction set computer** (**CISC**) processor. By simplifying the hardware, the clock cycle is reduced. If the delay associated with executing more instructions (RISC design) is less than the delay associated with an increased clock cycle for all instructions executed (CISC design), the total system performance is improved. Improved compiler design techniques and large on-chip caching has continued to contribute to higher performance RISC designs (Hennessy and Jouppi, 1991).

One reason that RISC architectures work better than traditional CISC machines is due to the use of large on-chip caches and register sets. Since locality of reference effects (described in the section on memory hierarchy) dominate most instruction and data reference behavior, the use of an on-chip cache and large register sets can reduce the number of instructions and data fetched per instruction execution. Most RISC machines use pipelining to overlap instruction execution, further reducing the clock period. Compiler techniques are used to exploit the natural parallelism inherent in sequentially executed programs.

A register window is a subset of registers used by a particular instruction. These registers are specified as inputs, outputs, or temporary registers to be used by that instruction. One set of instructions outputs become the next inputs in a register window for another instruction. This technique allows more efficient use of the registers and a greater degree of pipelining in some architectures.

Scaling these RISC concepts to large parallel processing systems poses many challenges. As larger, more complex problems are mapped into a RISC-based parallel processing system, communication and allocation of resources significantly affects the ability of the system to utilize its resources efficiently. Unless special routing chips are incorporated into the system, the processors may spend an inordinate amount of time handling requests for other processors or waiting for data and/or instructions. Using a large number of simple RISC processors means that cache accesses must be monitored (*snoopy caches*) or restricted (*directory-based caches*) to maintain data consistency across the system.

The types of problems that are difficult to execute using RISC architectures are those that do an inordinate amount of branching and those that use very large data sets. For these problems, the hit rate of large instruction or data caches may be very low. The overhead needed to fetch large numbers of new instructions or data from memory is significantly higher than the clock cycle, virtually starving the processor. Compiler techniques can only be used to a limited extent when manipulating large data sets.

Further increases in speed for a single stream, pipelined processor will probably come about from either increasing the pipeline depth, **superpipelining** or increasing the width of the data path or control path. The latter can be achieved by either issuing more than one instruction per cycle, **superscalar**, or by using a **very long instruction word** (**VLIW**) architecture in which many operations are performed in parallel by a single instruction. Some researchers have developed the idea of an orthogonal relationship between superscalar and superpipelined designs (Hennessy and Jouppi, 1991). In a superpipelined design, the pipeline depth is increased from the basic pipeline; whereas in a superscalar design, the horizontal width of the pipeline is increased (see Fig. 19.5).

To achieve an overall gain in performance, significant increases in speed due to superpipelining must be accompanied by highly utilized resources. Idle resources contribute little to performance while increasing overall system costs and power consumption. As pipeline depth increases, a single instruction stream cannot keep all of the pipeline stages in a processor fully utilized. Control and data dependencies within

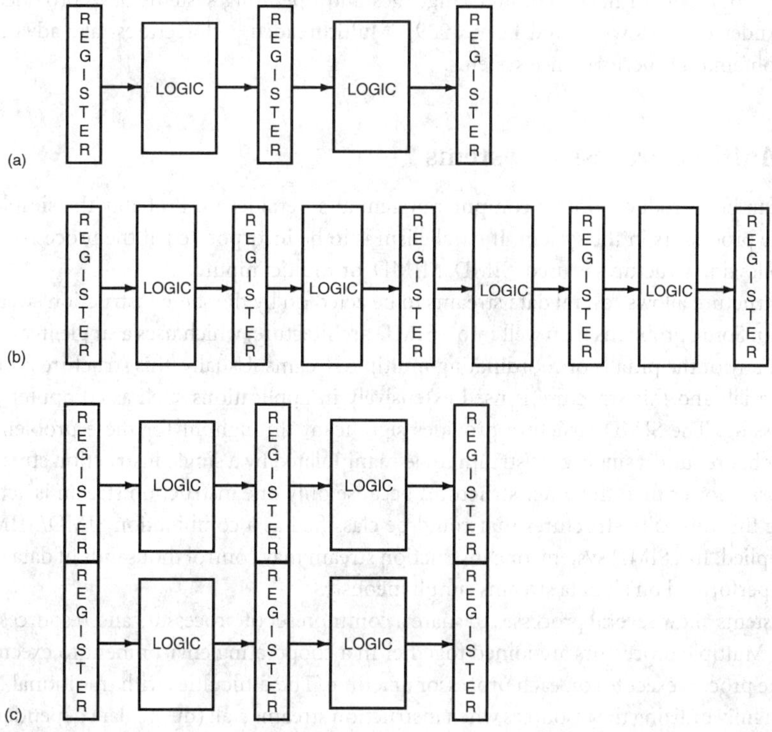

FIGURE 19.5   Superscalar and superpipelined systems: (a) pipelined, (b) superpipelined, (c) superscalar.

the instruction stream limit the number of instructions that can be active for a given instruction stream. *No operation* (NoOps) or null instructions are inserted into the pipeline, creating *bubbles*. Since a NoOp does not perform any useful work, processor cycles are wasted. Some strategies improve pipeline utilization using techniques such as prefetching a number of instructions or data, branch prediction, software pipelining, trace scheduling, alias analysis, and register renaming to keep the memory access overhead to a minimum. An undesirable consequence of this higher level of parallelism is that some prefetched instructions or data might not be used, causing the memory bandwidth to be inefficiently used to fetch useless information. In a single processor system this may be acceptable, but in a multiple processor system, such behavior can decrease the overall performance as the number of memory accesses increases.

Superscalar processors use multiple instruction issue logic to keep the processor busy. Essentially, two or three instructions are issued from a single stream on every clock cycle. This has the effect of widening the control path and part of the datapath in a processor. VLIW processors perform many operations in parallel using different types of functional units. Each instruction is composed of several operation fields and is very complex. The efficient use of these techniques depends on using compilers to partition the instructions in an instruction stream or building extra hardware to perform the partitioning at run time. Again both these techniques are limited by the amount of parallelism inherent in an individual instruction stream.

A solution to fully utilizing a pipeline is to use instructions from independent instruction streams or threads. (The execution of a piece of code specified by parallel constructs is called a thread.) Some machines allow multiple threads per program. A thread can be viewed as a unit of work that is either defined by the programmer or by a parallel compiler. During execution, a thread may spawn or create other threads as required by the parallel execution of a piece of code. Multithreading can mitigate the effects of long memory latencies in uniprocessor systems; the processor executes another thread while the memory system services cache misses for one thread. Multithreading can also be extended to multiprocessor systems, allowing the concurrent use of CPUs, network, and memory.

To get the most performance from **multithreaded** hardware, a compatible software environment is required. Developments in new computer languages and operating systems have provided these environments (Anderson, Lazowska, and Levy, 1989). Multithreaded architectures take advantage of these advances to obtain high-performance systems.

19.1.4   Multiple Processor Systems

Parallelism can be introduced into a computer system at several levels. Probably the simplest level is to have multiple processors in the system. If parallelism is to be incorporated at the processor level, usually one of the following structures is used: SIMD, MIMD, or multicomputers.

A SIMD structure allows several data streams to be acted on by the same instruction stream, as shown in Fig. 19.6(a). Some problems map well into a SIMD architecture, which uses a single instruction stream and avoids many of the pitfalls of coordinating multiple streams. Usually, this structure requires that the data be bit serial, and this structure is used extensively in applications such as computer graphics and image processing. The SIMD structure provides significant throughput for these problems. For many applications that require a single data stream to be manipulated by a single instruction stream, the SIMD structure works slower than the other structures because only one instruction stream is active at a time. To overcome this difficulty structures that could be classified as a combination SIMD/MIMD structure have been applied. In a SIMD system, one instruction stream may control thousands of data streams. Each operation is performed on all data streams simultaneously.

MIMD systems allow several processes to share a common set of processors and resources, as shown in Fig. 19.6(b). Multiple processors are joined together in a cooperating environment to execute programs. Typically, one process executes on each processor at a time. The difficulties with traditional MIMD architectures lie in fully utilizing the resources when instruction streams stall (due to data dependencies, control dependencies, synchronization problems, memory accesses, or I/O accesses) or in assigning new processes quickly once the current process has finished execution. An important problem with this structure is that

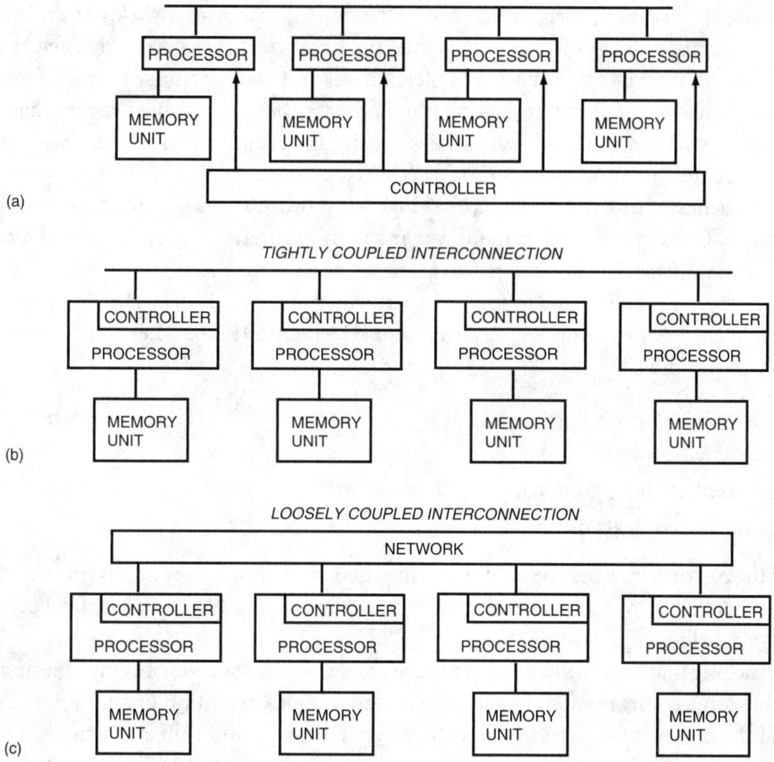

FIGURE 19.6   System level parallelism: (a) SIMD, (b) MIMD, (c) multicomputer.

processors may become idle due to improper load balancing. Implementing an operating system (OS) that can execute on the system without creating imbalances is important to maintain a high utilization of resources.

The next system, which is also popular due to its simple connectivity, is the distributed system or *multicomputer*. A network connects independent processors as shown in Fig. 19.6(c). Each processor is a separate entity, usually running an independent operating system process. A multicomputer will usually use message passing to exchange data and/or instruction streams between the processors. The main difficulties with the multicomputer are the latency involved in passing messages and the difficulty in mapping some algorithms to a distributed memory system.

19.1.5   Memory Hierarchy

High-performance computer systems use a multiple level memory hierarchy ranging from small, fast cache memory to larger, slower main memory to improve performance. Parallelism can be introduced into a system through the memory hierarchy as depicted in Fig. 19.7. A cache is a small, high-speed buffer used to temporarily hold those portions of memory that are currently in use in the processor.

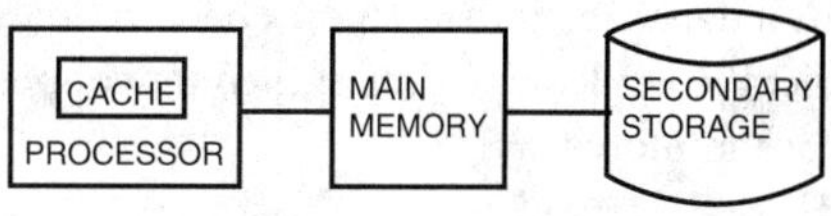

FIGURE 19.7   A common method of incorporating parallelism into a system.

Cache memories take advantage of the property of temporal and spatial locality in program behavior. Temporal locality states that if an item is referenced one point in time, it will tend to be referenced again soon. Spatial locality refers to the likelihood that if an item is referenced, nearby items will tend to be referenced soon.

A cache miss will cause the processor to request the needed data from the slower main memory. The average access time depends on the cache miss rate, the number of cycles needed to fetch the data from the cache, and the number of cycles required to fetch data from main memory in case of a cache miss.

Caches can be divided into separate instruction and data caches or unified caches that contain both instructions and data. The former type of cache is usually associated with a Harvard style architecture in which there are separate memory ports for instructions and data.

In most approaches to instruction fetch, the single chip processor is provided with a simple on-chip instruction cache. On a cache miss, a request is made to external memory for instructions and these off-chip requests compete with data references for access to external memory.

Memory latency is a major problem for high-performance architectures. The disparity in on-chip to off-chip access latencies has prompted the suggestion (Hwang, 1993) that there are four complementary approaches to latency hiding:

- Prefetching techniques
- Using coherent caches
- Using relaxed memory consistency models
- Using multiple contexts or multithreading within a processor

To maximize the performance benefit of multiple instruction issue, memory access time has to be kept low and cache bandwidth has to be increased. Unfortunately, accessing the cache usually becomes a critical timing path for pipelined operation.

Extending caching into a multithreaded processor creates some new problems in cache design. In single-threaded caches, on a context switch (when the processor begins execution of another thread), the cache is flushed and the instructions from the new thread are fetched. A multithreaded cache supports at least two different instruction streams, using different arrays for different threads. Another strategy is to allow the threads to share the cache, making it unnecessary to flush the cache on a context switch. The state of the old thread is stored in the processor, which has different register files for each thread. The cache sends the thread number and the instruction to the processor's instruction buffers. To keep the execution units supplied with instructions, the processor requires several instructions per cycle to be fetched into its instruction buffers. In this variable instruction fetch scheme, an instruction cache access may be misaligned with respect to a cache line. To allow line crossing, an alignment network can be used to select the appropriate instructions to be forwarded to the instruction buffers. Performance is increased by eliminating the need for multiplexing the data bus, thus reducing the I/O delay in a critical timing path and providing simultaneous reads and writes into the cache. Although this operation seems relatively simple, several variables must be considered when designing a multithreaded cache such as: address translation and mapping, variable instruction and data fetching, cache line crossing, request protocols and queuing, line and block replacement strategies, and multiple access ports.

By placing caches, which contain a subset of the main memory, within each processor, several processors could be using the same memory data. When a processor modifies its cache, all other caches within the system with that data are affected. This usually requires that a snooping mechanism be used to update or invalidate data in the caches. These mechanisms usually add considerable hardware to the system and restrict the implementation of the interconnection network to a bus type network so that cache updates can be monitored.

19.1.6 Implementation Considerations

Rather than beginning by building a system and testing the behavior of a breadboard model, the first step is to simulate the behavior using simulation models that approximate the processor structures. The hardware should be modeled at two levels.

High level, hierarchical models provide useful analysis of different types of memory structures within the processor. Flows through the models and utilization parameters can be determined for different structures. Low level, *register transfer level* (RTL) simulation of specific structures usually give a more accurate picture of the actual hardware's performance. These structures may include caches, register files, indexed instruction and data caches, memory controllers, and other elements. *Hardware description languages*, have been developed to represent a design at this level. After simulation and modeling of the design have been completed, the design can be implemented using various technologies.

Although significant technological advances have increased the circuit speed on integrated circuit (IC) chips, this increase in circuit speed is not projected to result in a similar increase in computer performance. The intrachip and interchip propagation delays are not improving at the same rate as the circuit speed. The latency imposed by the physical methods used to transfer the signals between faster logic has limited the computer performance increases that could be practically obtained from incorporating the logic into the designs.

$$\text{Delay}_{\text{circuit}} = \text{Delay}_{\text{interconnect}} + \text{Delay}_{\text{logic}}$$

An example of this phenomena is evident in pipelined systems, where the slowest pipe stage determines the clock rate of the entire system. This effect occurs between the parts of a hardware circuit. The relative speedup (obtained by making one part of the circuit faster) is determined by the slowest part of the circuit. Even if propagation delays through the fastest logic tended toward zero delay, the propagation delay through the interconnect would limit the speedup through the system or block of logic. So that the speedup for a system with zero logic delay compared to a system with some logic delay would be

$$\text{Speedup} = \frac{\text{Delay}_{\text{with logic}}}{\text{Delay}_{\text{without logic}}}$$

$$\text{Speedup} = \frac{\text{Delay}_{\text{interconnect}} + \text{Delay}_{\text{logic}}}{\text{Delay}_{\text{interconnect}}}$$

Rather than suggesting that further speedups due to technological innovation are impossible, these arguments show that new methods must be used that compensate for the latencies imposed by physical limitations. As demonstrated in Fig. 19.8, the time it takes for a signal to be transferred between chips

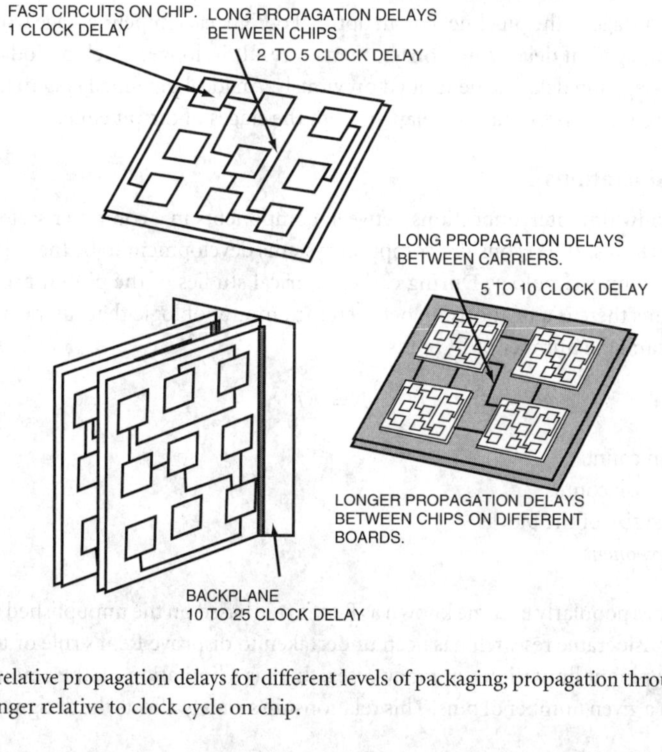

FIGURE 19.8 The relative propagation delays for different levels of packaging; propagation through interconnecting elements becomes longer relative to clock cycle on chip.

might be several times the clock period on a chip implemented in a high-speed circuits. The delay between chips is not only affected by the length of the interconnecting wires between the chips but also by the number and types of interconnections between the wires. Whenever signals must go through a via or pin to another type of material or package, mismatches in impedance and geometry can introduce reflections that affect the propagation delay through the path.

One way of synchronizing logic modules is to use self-timed circuits or **ternary logic** for the signal transfers. Ternary logic is compatible with some optical interconnection methods. Logic must be provided to translate the multiple valued signal back to a digital signal unless the entire system is designed in ternary logic. The overhead of designing the entire system in ternary logic is significant (50% increase in logic).

Pipelining the system so that one instruction is being transferred between chips while another instruction is executing is a promising technique to compensate for this latency. Pipelining techniques (Fig. 19.9) can be used to not only compensate for interconnect delay but also to monitor the inputs and outputs of a chip (when scan paths are designed into the latches), by inserting registers or latches at the inputs and outputs of the chips. This has the effect of increasing the length of the pipeline and the amount of logic, increasing the system latency, and increasing the layout area of the design. The main problems with this use of pipelining are clock synchronization and keeping the propagation delay through each pipeline stage symmetrical. Each register could be running on a different clock signal, each of which is *skewed* from the other, that is, one signal arrives at the pipeline register offset from the time that the other signal arrives at its pipeline register. Clock skew can cause erroneous values to be

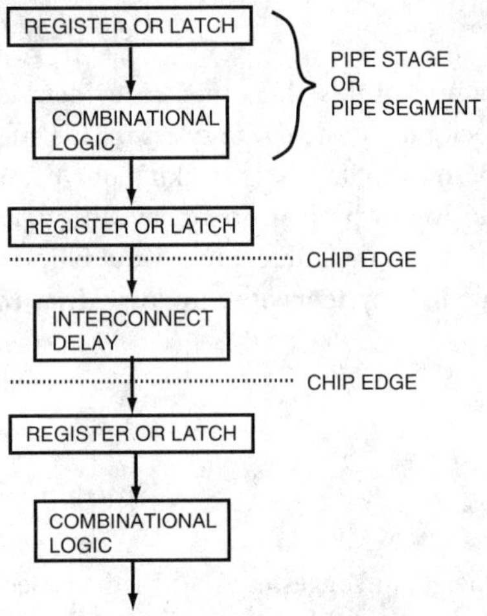

FIGURE 19.9 Using pipelining to compensate for interconnect delays.

loaded into the next stage of the pipeline. For an optimal design, each pipeline stage would be symmetrical with respect to propagation delay. Any nonsymmetries result in longer clock periods to compensate for the worst-case propagation delay as mentioned previously. An ideal pipelined system must allow for clock skew and for differences in propagation delay through the stages of the pipeline.

Packaging Considerations

Ensuring high-bandwidth interconnections between components in a computer system is extremely important and some (Keyes, 1991) consider computer systems development to be the principal driving force behind interconnection technology. During early empirical studies of the pinout problem, it was found that in many designs there is a relationship between the amount of logic that can be implemented within a design partition and the number of I/O pins,

$$I = AB^r$$

where
    I = the I/O pin count
    B = the logic block count
    A = the average size of the blocks
    r = the *Rent exponent*

This relationship has popularly become known as Rent's rule based on the unpublished work by E.F. Rent at IBM. Although considerable research has been undertaken to disprove Rent's rule or to find architectures that overcome the pinout limitations, in many cases, there is a limitation on the amount of logic that can be associated with a given number of pins. This relationship severely affects the designs of some high-speed technologies where the pinout is fairly costly in terms of space and power consumption. Reduced pinout due to packaging constraints means that the logic must be kept simple and sparse for most designs.

RISC designs have benefited from the low number of pins and routing density available in VLSI chips. For a particular architecture certain relations between pinout and logic have been observed. Since the digital logic complexity available per interconnect is determined by empirical relationships, such as Rent's rule, reducing the complexity of a logic block (chip) will usually also reduce the number of interconnects (pins). Because of the simpler implementations, RISC chips have required less routing density and less pins than more complex designs.

Packaging techniques have been developed to provide increased pinout on IC chips. Most of these techniques have been developed for Si chip technology. These packaging techniques are beginning to spread to systems composed of other technologies. Most of the packaging techniques center around providing a higher density pinout by distributing pins across the chip instead of restricting the placement to the edges of the chip. Sometimes this involves using a *flip chip* technology, which turns the chip logic-side down on a board. Heat sinking—removing heat from the chip—becomes an issue as does being able to probe the logic.

Technology Considerations

Silicon semiconductor technology has been the most commonly used technology in computer systems during the past few decades. The types of Si chips available vary according to logic family such as n-channel metal-oxide semiconductor (NMOS), complementary metal-oxide semiconductor (CMOS), transistor transistor logic (TTL), and emitter coupled logic (ECL). Circuit densities and clock rates have continued to improve for these Si logic families, giving the computer designer increased performance without having to cope with entirely new design strategies. Higher speed technologies are preferred for high-end computer systems. Several technologies that hold promise for future computer systems include optical devices, gallium arsenide (GaAs) devices, superconductors, and quantum effect devices.

Many of the high-speed technologies are being developed using III–V materials. GaAs is the current favorite for most devices (sometimes used in combination with AlGaAs layers). Long and Butner (1990) provide a good introduction to the advantages and capabilities of GaAs circuits, including higher electron mobility (faster switching speeds), a semi-insulating substrate, and the major advantage of the suitability of GaAs semiconductors for optical systems. GaAs circuits are steadily increasing in circuit density to the VLSI level and becoming commercially available in large quantities. More designers are incorporating GaAs chips into speed-sensitive areas of their designs.

There are several GaAs circuit families. Some of the circuit families use only depletion mode field effect transistors (FETs), whereas others use both enhancement and depletion (E/D) mode FETs. Very large-scale integrated (VLSI) levels of integration are under development for *direct coupled FET logic* (DCFL) and *source coupled FET logic* (SCFL). By using enhancement and depletion mode FETs with no additional circuitry, DCFL is currently the fastest, simplest commercially available GaAs logic. SCFL (also using E/D FETs) provides high speed with greater drive capabilities but consumes more power and space than DCFL.

Several efforts are underway to harness photons for extremely fast interconnects and logic. Photons have a greater potential bandwidth than electrons, making optical interconnections attractive. Unlike electrons, photons have no charge and do not easily interact in most mediums. There are virtually no effects from crosstalk or electromagnetic fields on even very high densities of optical wires or signals. Although photon beams produce very high-quality signal transmission, the implementation of switching logic is difficult. The same immunity to interference that makes optics ideal for signal transmission makes switching logic difficult to implement. Most optical devices or systems under development exhibit high latencies in exchange for fast switching speeds. Optical interconnects, however, provide wider bandwidth interconnections for interchip or interprocessor communications. Optical interconnections have delays associated with converting signals from electrons to photons and back, increasing the propagation delay.

In a wire connection, the propagation characteristics limit the use of the wire to one pipeline stage. Reflected electrical signals must be dampened before a new signal can enter the wire, precluding the use of the wire by more than one digital signal during a time period. A wire interconnect would only act as a single stage of a pipeline, as shown in Fig. 19.9. An optical interconnect, due to its signal propagation characteristics, allows a finer degree of pipelining. Once a signal enters the optical interconnect, new signals can enter at regular time intervals without causing interference between the signals. Thus, an optical interconnect can represent several pipeline stages in a design, as shown in Fig. 19.10.

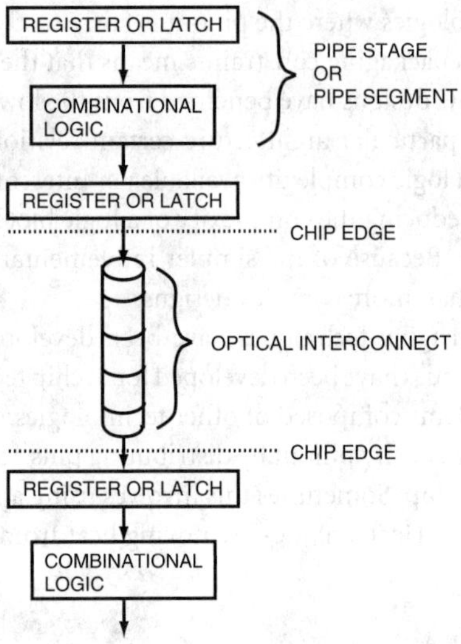

Two types of optical interconnects are under development, fiber optic and free space. Fiber-optic systems provide high bandwidth interconnections at speeds comparable to wire interconnects. Faster switching signals can be transported through fiber-optic systems longer distances than traditional wire and via interconnects. Since the fundamental physical limit for propagation delay between any two locations is the speed of light, free-space optical systems are, the ultimate choice for interconnection technology. The major drawback with optical interconnects is that the technologies that are currently available have a much lower density per area than traditional interconnects.

FIGURE 19.10   Using an optical connection in a pipeline system.

In a fiber-optic system, cables run between source and destination. The cables are usually wide (distances on the order of 250-μm centers between cables are common), difficult to attach, and subject to vibration problems. The lightguide material must maintain its low loss through all subsequent carrier processing (including the fabrication of electrical wiring and the chip-attachment soldering procedures) and the material used must be compatible with the carriers' temperature coefficient of expansion over the IC's operational temperature range. A benefit of using a fiber-optic cable interconnect is that a simple switching function using devices, such as directional optical couplers, can be performed in the cable.

In free-space systems, using just a detector and light source, signals can be transferred from one board to another in free space. The problem of connecting fibers to the boards is removed in a free-space system but the beam destination is limited to next nearest boards. Alignment of detector and sender becomes critical for small feature sizes necessary to provide high-density transfer of signals. For both free-space and fiber-optic systems, lower density or positional limitations makes optical interconnects unattractive for some computer architectures.

The near-term application of optical interconnects are in interprocessor communications with optical storage being the next leading application area. There are significant propagation delays through the translation circuitry when converting signals between the optical and the electrical realms. There are two leading materials for optical interconnect systems, GaAs and InGaAsP/InGaAs/InP. GaAs laser systems are by far the most developed and well understood. InGaAsP/InGaAs/InP *optoelectronic integrated circuits* (OEICs) are the next leading contender with current technology capable of providing laser functions for long-distance applications such as fiber-optic links for metropolitan area networks and CATV distribution.

Lowering the temperature of chips and interconnect results in an improvement in FET circuit switching speed and a reduction in the interconnect resistance. The reduction in resistance causes lower attenuation and greater bandwidth in the interconnect. These effects occur both in normal conductors and super conductors at low temperatures. Room temperature operation is usually preferred because of the problems with maintaining a low-temperature system and the extra power required to lower the temperature. Heat generated by devices on the chip can cause local *hot spots* that affect the operation of other circuits. The development of high-temperature superconductors increased the interest in using high bandwidth superconductor interconnect with high-speed circuits to obtain a significant increase in system clock rate.

Wafer Scale Integration (WSI)

One way to overcome the pinout problem is to reduce the pinout between chips by fabricating the chip set for a design on a single wafer. The logic partitions, which would have been on separate chips, are connected through high-speed metal interconnect on the wafer. Research involving **wafer scale integration (WSI)** dates back to some very early (1960s) work on providing acceptable yield for small Si chips. As processing techniques improved, circuit densities increased to acceptable levels, negating the need for WSI chips in the fledgling Si IC industry.

The driving force behind current WSI efforts is the need to provide high bandwidth communications between logic modules without suffering the degradation in performance due to off-chip interconnections. The problems with WSI are adequate yield, heat sinking, and pinout. The latter two problems can be coped with through using advanced packaging techniques such as flip-chip packages and water-cooled jackets. Obtaining adequate yields for mass production of wafers is probably the largest obstacle for using WSI systems. Processing defects can wipe out large sections of logic on wafers. Most WSI approaches use replicated or redundant logic that can be reconfigured either statically or dynamically to replace defective logic. This technique results in a working wafer if enough spares are available to replace the defective logic.

Architectures that are compatible with WSI are simple with easily replicated structures. Random access memory (RAM) chip designs have had the most successful use of WSI. The rationale for this approach is that GaAs is about at the stage of development that Si was in the early 1960s and, hence, WSI would provide a vehicle for a large gate count implementation of a processor. Large wafers are impractical due to the fragility of GaAs. Packaging techniques using defect-free GaAs chips mounted on another material might be more viable for some designs than WSI, due to the fragility of GaAs, the limited size of the GaAs wafers, and the cost of GaAs material.

In hybrid WSI, defect-free chips are mounted on a substrate and interconnected through normal processing techniques. [Note: the term hybrid is also used to refer to microwave components. A hybrid WSI component is also called a *wafer transmission module* (WTM).] This pseudo-WSI provides the density and performance of on-chip connections without having to use repair mechanisms for yield enhancement. The interconnections are the same feature sizes and defect levels as normal IC processing. Only a few planes are available to route signals, keeping the interconnect density low. Hybrid WSI has been suggested as a way for GaAs chips with optoelectronic components to be integrated with high logic density Si chips.

Multichip Modules (MCMs)

Multichip modules (MCMs) hold great promise for high-performance interconnection of chips, providing high pinout and high manufacturable levels of parts. MCMs are very close in structure to hybrid WSI. Active die are attached directly to a substrate primarily used for interchip wiring, introducing an extra level of packaging between the single die and the printed circuit board (PCB). The main difference between MCMs and hybrid WSI is the size of the substrate and interconnect. The MCM substrate can be as large as a wafer but usually is smaller with several thin-film layers of wiring (up to 35 or more possible).

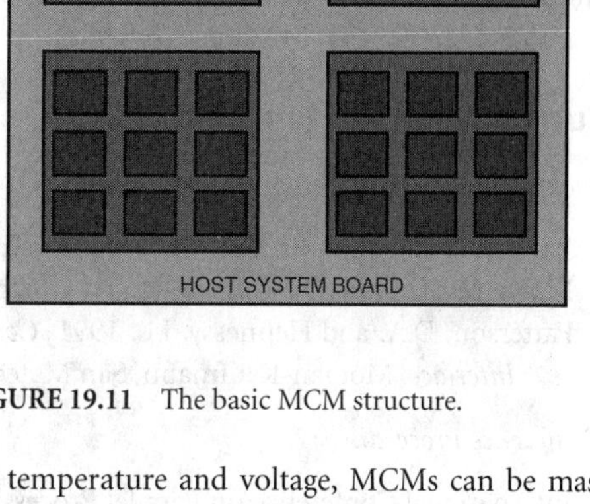

The feature size of the MCM interconnect is about 10 times that of the on-chip interconnect, reducing the interconnect disabled by processing defects to almost nil. By guaranteeing that only good die are

FIGURE 19.11 The basic MCM structure.

attached to the substrates by burning-in the die under temperature and voltage, MCMs can be mass produced at acceptable yield levels. The basis MCM structure is shown in Fig. 19.11.

Defining Terms

Algorithm: A well-defined set of steps or processes used to solve a problem.

Architecture: The physical structure of a computer, its internal components (registers, memory, instruction set, input/output structure, etc.) and the way in which they interact.

Complex instruction set computer (CISC): A design style where there are few constraints on the types of instructions or addressing modes.

Context: The information needed to execute a process including the contents of program counters, address and data registers, and stack pointers

Cycles per instruction (CPI): A performance measurement used to judge the efficiency of a particular design.

Multithreaded: Several instruction streams or threads execute simultaneously.

Pipeline: Dividing up an instruction execution into overlapping steps that can be executed simultaneously with other instructions, each step represents a different pipeline stage.

Reduced instruction set computer (RISC): A design style where instructions are implemented for only the most frequently executed operations; addressing modes are limited to registers and special load/store modes to memory.

Superpipelined: A pipeline whose depth has been increased to allow more overlapped instruction execution.

Superscalar: Hardware capable of dynamically issuing more than one instruction per cycle.

Ternary logic: Digital logic with three valid voltage levels.

Very long instruction word (VLIW): Compiler techniques are used to concatenate many small instructions into a larger instruction word.

Wafer scale integration (WSI): Using an entire semiconductor wafer to implement a design without dicing the wafer into smaller chips.

References

Anderson, T., Lazowska, E., and Levy, H. 1989. The performance implications of thread management alternatives for shared-memory multiprocessors. *IEEE Trans. on Computers* 38(12):1631–1644.

Flynn, M.J. 1966. Very high-speed computing systems. *Proc. of the IEEE* 54(12):1901–1909.

Hennessy, J. and Jouppi, N. 1991. Computer technology and architecture: An evolving interaction. *IEEE Computer* 24(9):18–29.

Hwang, K. 1993. *Advanced Computer Architecture: Parallelism, Scalability, Programmability*. McGraw-Hill, New York.

Keyes, R.W. 1991. The power of connections. *IEEE Circuits and Devices* 7(3):32–35.

Long, S.I. and Butner, S.E. 1990. *Gallium Arsenide Digital Integrated Circuit Design*. McGraw-Hill, New York.

Smith, B. 1978. A pipelined, shared resource MIMD computer. *International Conference on Parallel Processing*, (Bellaire, MI), pp. 6–8. (Aug).

Further Information

Books:

Baron R.J. and Higbee, L. 1992. *Computer Architecture*. Addison-Wesley, Reading, MA.

Mano, M.M. 1993. *Computer System Architecture*. Prentice-Hall, Englewood Cliffs, NJ.

Patterson, D.A. and Hennessy, J.L. 1991. *Computer Organization and Design: The Hardware/Software Interface*. Morgan-Kaufmann, San Mateo, CA.

Conference Proceedings:

International Conference on Parallel Processing
International Symposium on Computer Architecture

Proceedings of the International Conference on VLSI Design
Supercomputing

Magazines and Journals:

ACM Computer Architecture News
Computer Design
IEEE Computer
IEEE Micro
IEEE Transactions on Computers

19.2 Software Design and Development

Margaret H. Hamilton

19.2.1 Introduction

A computer system can be neatly compared with a biological entity called a superorganism. Composed of software, hardware, peopleware and their interconnectivity (such as the internet), and requiring all to survive, the silicon superorganism is itself a part of a larger superorganism, for example, the business. It could be a medical system including patients, drugs, drug companies, doctors, hospitals and health care centers; a space mission including the spacecraft, the laws of the universe, mission control and the astronauts; a system for researching genes including funding organizations, funds, researchers, research subjects, and genes; or a financial system including investors, money, politics, financial institutions, stock markets, and the health of the world economy.

Whether the business be government, academic, or commercial, the computer system, like its biological counterpart, must grow and adapt to meet fast changing requirements. Like other organisms, the business has both physical infrastructure and operational policies which guide, and occasionally constrain, its direction and the rate of evolution, which it can tolerate without becoming dysfunctional.

Unlike a biological superorganism, which may take many generations to effect even a minor hereditary modification, software is immediately modifiable and is, therefore, far superior in this respect to the biological entity in terms of evolutionary adaptability. Continuity of business rules and the physical infrastructure provides a natural tension between "how fast the software can change" and "how rapidly the overall system can accept change."

As the brain of the silicon superorganism, software controls the action of the entire entity, keeping in mind, however, it was a human being that created the software.

In this chapter we will discuss the tenets of software, what it is and how it is developed, as well as the precepts of software engineering, which are the *methodologies* by which ideas are turned into software.

19.2.2 The Notion of Software

Software is the embodiment of logical processes, whether in support of business functions or in control of physical devices. The nature of software as an instantiation of process can apply very broadly, when modeling complex organizations, or very narrowly as when implementing a discrete numerical algorithm. In the former case, there can be significant linkages between re-engineering businesses to accelerate the rate of evolution—even to the point of building operational models which then transition into application suites and thence into narrowly focused implementation of algorithms. Software thus has a potentially wide range of application, and when well designed, it has a potentially long period of utilization.

Whereas some would define software as solely the code generated from programming language statements in the compilation process, a broader and more precise definition includes requirements, specifications, designs, program listings, documentation, procedures, rules, measurements, and data, as well as the tools and reusables used to create, test, optimize, and implement the software.

That there is more than one definition of software is a direct result of the confusion about the very process of software development itself. A 1991 study by the Software Engineering Institute (SEI) (1991) amplifies this rather startling problem. The SEI developed a methodology for classifying an organization's software process maturity into one of five levels that range from Level 1, the *initial level* (where there is no formalization of the software process), to Level 5, the *optimizing level* where methods, procedures, and metrics are in place with a focus toward continuous improvement in software reliability. The result of this study showed that fully 86% of organizations surveyed in the US were at Level 1 where the terms "ad hoc," "dependent on heroes," and "chaotic" are commonly applied. And, given the complexity of today's applications including those that are internet based, it would not be surprising to see the percentage of Level 1 organizations increase. Adding to the mix are the organizations that think the so called order mandated by some techniques serves only to bring about more chaos, confusion, and complexity; or at least spend many more dollars for what they believe delivers little, no, or negative benefit.

Creating order from this chaos requires an insightful understanding into the component parts of software as well as the development process. Borrowing again from the world of natural science, an entelechy is something complex that emerges when you put a large number of simple objects together. For example, one molecule of water is rather boring in its utter lack of activity. But pour a bunch of these molecules into a glass and you have a ring of ripples on the water's surface. If you combine enough of these molecules together, you wind up with an ocean. So, too, software: by itself, a line of code is a rather simple thing. But combine enough of them together and you wind up with the complexity of a program. Add additional programs and you wind up with a system that can put a person on the moon.

Although the whole is indeed bigger than the sum of its parts, one must still understand those parts if the whole is to work in an orderly and controlled fashion. Like a physical entity, software can "wear" as a result of maintenance, changes in the underlying system and updates made to accommodate the requirements of the ongoing user community. Entropy is a significant phenomenon in software, especially for Level 1 organizations.

Software at the lowest programming level is termed *source code*. This differs from *executable code* (i.e., that which can be executed directly by the hardware to perform one or more specified functions) in that it is written in one or more programming languages and cannot, by itself, be executed by the hardware until it is translated into machine executable code. A programming language is a set of words, letters, numerals, and abbreviated mnemonics, regulated by a specific syntax, used to describe a program (made up of lines of source code) to a computer. There are a wide variety of programming languages, many of them tailored for a specific type of application. C, one of today's more popular programming languages, is used in engineering as well as business environments, whereas object-oriented languages, such as C++ (Stroustrup, 1997) and Java (Gosling et al., 1996), have been gaining acceptance in both of these environments. In fact Java has become a language of choice for internet based applications. In the recent past, engineering applications have often used programming languages such as Fortran, Ada (for government applications) and HAL (for space applications) (Lickly, 1974), while commercial business applications have favored common business oriented language (COBOL). For the most part one finds that in any given organization there are no prescribed rules that dictate which languages are to be used. As one might expect, a wide diversity of languages is being deployed.

The programming language, whether that be C (Harbison, 1997), Java, COBOL, C++, C# or something else, provides the capability to code such logical constructs as that having to do with:

- *User interface:* Provides a mechanism whereby the ultimate end user can input, view, manipulate, and query information contained in an organization's computer systems. Studies have shown that productivity increases dramatically when visual user interfaces are provided. Known as *graphical user interfaces (GUIs)*, each operating system provides its own variation. Some common graphical standards are Motif for Unix (including Linux) systems, Aqua for the Macintosh Operating System and Microsoft Windows for PC-based systems.

- *Model calculations:* Perform the calculations or algorithms (step-by-step procedures for solving a problem) intended by a program, for example, process control, payroll calculations, or a Kalman Filter.

- *Program control:* Exerts control in the form of comparisons, branching, calling other programs, and iteration to carry out the logic of the program.

- *Message processing:* There are several varieties of message processing. Help-message processing is the construct by which the program responds to requests for help from the end user. Error-message processing is the automatic capability of the program to notify and then recover from an error during input, output, calculations, reporting, communications, etc. And, in object-oriented development environments, message processing implies the ability of program objects to pass information to other program objects.

- *Moving data:* Programs store data in a data structure. Data can be moved between data structures within a program; moved from an external database or file to an internal data structure or from user input to a program's internal data structures. Alternatively, data can be moved from an internal data structure to a database or even to the user interface of an end user. Sorting, searching, and formatting are data moving and related operations used to prepare the data for further operations.

- *Database:* A collection of data (objects) or information about a subject or related subjects, or a system (for example, an engine in a truck or a personnel department in an organization). A database can include objects such as forms and reports or a set of facts about the system (for example, the information in the personnel department needed about the employees in the company). A database is organized in such a way as to be easily accessible to computer users. Its data is a representation of facts, concepts, or instructions in a manner suitable for processing by computers. It can be displayed, updated, queried, and printed, and reports can be produced from it. A database can store data in several ways including in a relational, hierarchical, network, or object-oriented format.

- *Data declaration:* Describes the organization of a data structure to the program. An example would be associating a particular data structure with its type (for example, data about a particular employee might be of type person).

- *Object:* A person, place, or thing. An object contains data and a set of operations to manipulate data. When brought to life, it knows things (called *attributes*) and can do things (to change itself or interact with other objects). For example, in a robotics system an object may contain the functions to move its own armature to the right, while it is coordinating with another robot to transfer yet another object. Objects can communicate with each other through a communications medium (e.g., message passing, radio waves, internet).

- *Real time:* A software system that satisfies critical timing requirements. The correctness of the software depends on the results of computation, as well as on the time at which the results are produced. Real-time systems can have varied requirements such as performing a task within a specific deadline and processing data in connection with another process outside of the computer. Applications such as transaction processing, avionics, interactive office management, automobile systems, and video games are examples of real-time systems.

- *Distributed system:* Any system in which a number of independent interconnected processes can cooperate (for example processes on more than one computer). The client/server model is one of the most popular forms of distribution in use today. In this model, a client initiates a distributed activity and a server carries out that activity.

- *Simulation:* The representation of selected characteristics of the behavior of one physical or abstract system by another system. For example, a software program can simulate an airplane, an organization, a computer, or another software program.

- *Documentation:* Includes description of requirements, specification and design; it also includs comments that describe the operation of the program that is stored internally in the program, as well as written or generated documentation that describes how each program within the larger system operates.

- *Tools:* The computer programs used to design, develop, test, analyze, or maintain system designs of another computer program and its documentation. They include code generators, compilers, editors, *database management systems (DBMS)*, GUI builders, simulators, debuggers, operating systems and software development, and systems engineering tools (some derived or repackaged from earlier tools that were referred to in the nineties as *computer aided software engineering (CASE)* tools), that combine a set of tools, including some of those listed above.

Although the reader should by now understand the dynamics of a line of source code, where that line of source code fits into the superorganism of software is dependent on many variables. This includes the industry the reader hails from as well as the software development paradigm used by the organization.

As a base unit, a line of code can be joined with other lines of code to form many things. In a traditional software environment many lines of code form a program, sometimes referred to as an application program or just plain application. But lines of source code by themselves cannot be executed. First, the source code must be run through a *compiler* to create *object code*. Next, the object code is run through a *linker*, which is used to construct *executable code*. Compilers are programs themselves. Their function is twofold. The compiler first checks the source code for obvious syntax errors and then, if it finds none, creates the object code for a specific operating system. Unix, Linux (a spinoff of Unix), and Windows are all examples of operating systems. An operating system can be thought of as a supervising program that manages the application programs that run under its control. Since operating systems (as well as computer architectures) can be different from each other, compiled source code for one operating system cannot be executed under a different operating system, without a recompilation.

Solving a complex business or engineering problem often requires more than one program. One or more programs that run in tandem to solve a common problem is known collectively as an application system (or system). The more modern technique of object-oriented development dispenses with the notion of the program altogether and replaces it with the concept of an object (Goldberg and Robson, 1983; Meyer and Bobrow, 1992; Stefik and Bobrow, 1985; Stroustrup, 1994).

Where a program can be considered a critical mass of code that performs many functions in the attempt to solve a problem with little consideration for object boundaries, an object is associated with the code to solve a particular set of functions having to do with just that type of object. By combining objects like molecules, it is possible to create more efficient systems than those created by traditional means. Software development becomes a speedier and less error-prone process as well. Since objects can be reused, once tested and implemented, they can be placed in a library for other developers to reuse. The more the objects in the library, the easier and quicker it is to develop new systems. And since the objects being reused have, in theory, already been warranted (i.e., they have been tested and made error free), there is less possibility that object-oriented systems will have major defects.

The process of building programs and/or objects is known as software development, or software engineering. It is composed of a series of steps or phases, collectively referred to as a development life cycle. The phases include (at a bare minimum): an analysis or requirements phase where the business or engineering problem is dissected and understood, a specification phase where decisions are made as to how the requirements will be fulfilled (e.g., deciding what functions are allocated to software and what functions are allocated to hardware); a design phase where everything from the GUI to the database to the algorithms to the output is designed; a programming (or implementation) phase where one or more tools are used to support the manual process of coding or to automatically generate code, a testing (debugging) phase where the code is tested against a business test case and errors in the program are found and corrected, an installation phase where the systems are placed in production, and a maintenance phase where modifications are made to the system. But different people develop systems in different ways. These different paradigms make up the opposing viewpoints of software engineering.

19.2.3   The Nature of Software Engineering

Engineers often use the term systems engineering to refer to the tasks of specifying, designing, and simulating a non-software system such as a bridge or electronic component. Although software may be used for simulation purposes, it is but one part of the systems engineering process. Software engineering, on the other hand, is concerned with the production of nothing but software. In the 1970s industry pundits began to notice that the cost of producing large-scale systems was growing at a high rate and that many projects were failing or, at the very least, resulting in unreliable products. Dubbed the software crisis, its manifestations were legion, and the most important include

- *Programmer productivity:* In government in the 1980s, an average developer using C was expected to produce only 10 lines of production code per day (an average developer within a commercial organization was expected to produce 30 lines a month); today the benchmark is more like two to five lines a day while at the same time the need is dramatically higher than that, perhaps by several orders of magnitude, ending up with a huge backlog. The National Software Quality Experiment (NSQE) completed a 10-year research project in 2002 that compared the productivity of systems today with that of systems 10 years ago. The assumption at the outset was that the productiviity of today's systems would be much higher given all the more "modern" tools including a proliferation of commercial packages and object-oriented languages that are now available; but they were wrong. Not only did the productivity not increase, it decreased (NSQE, 2002). Further NSQE concluded that there would need to be a significant breakout in the way software is designed and developed in order to achieve a factor of 10 reduction in defect rate.

  Programmer productivity is dependent on a plethora of vagaries: from expertise to complexity of the problem to be coded to size of the program that is generated (Boehm, 1981). The science of measuring the quality and productivity of the software engineering process is called *metrics*. As in the diverse paradigms in software engineering itself, there are many paradigms of software measurement. Today's metric formulas are complex and often take into consideration the following: cost, time to market, productivity on prior projects, data communications, distributed functions, performance, heavily used configurations, transaction rates, online data entry, end user efficiency, online update, processing complexity, reusability, installation ease, operational ease, and multiplicity of operational sites.

- *Defect removal costs:* The same variables that affect programmer productivity affect the cost of debugging the programs and objects generated by those programmers. It has been observed that the testing and correcting of programs consumes a large share of the overall effort.

- *Development environment:* Development tools and development practices greatly affect the quantity and quality of software. Most of today's design and programming environments contain only a fragment of what is really needed to develop a complete system. Life cycle development environments provide a good example of this phenomena. Most of these tools can be described either as addressing the upper part of the life cycle (i.e., they handle the analysis and design) or the lower part of the life cycle (i.e., they handle code generation). There are few integrated tools on the market (i.e., that seamlessly handle both upper and lower functionalities). There are even fewer tools that add simulation, testing and cross-platform generation to the mix. And it would be hard put to find any tools that seamlessly integrate system design to software development.

- *GUI development:* Developing graphical user interfaces is a difficult and expensive process unless the proper tools are used. The movement of systems from a host-based environment to the workstation PC saw the entry of a plethora of GUI development programs onto the marketplace. But, unfortunately, the vast majority of these GUI-based tools do not have the capability of developing the entire system (i.e., the processing component as opposed to merely the front end). This leads to fragmented and error-prone systems. To be efficient, the GUI builder must be well integrated into the software development environment.

The result of these problems is that most of today's systems require more resources allocated to maintenance than to their original development. Lientz and Swanson (1980) demonstrate that the problem is, in fact, larger than the one originally discerned during the 1970s. Software development is indeed complex and the limitations on what can be produced by teams of software engineers given finite amounts of time, budgeted dollars, and talent have been amply documented (Jones, 1977).

Essentially the many paradigms of software engineering attempt to rectify the causes of declining productivity and quality. Unfortunately this fails because current paradigms treat symptoms rather than the root problem. In fact, software engineering is itself extremely dependent on both the underlying systems software (for example, the operating systems) and hardware as well as the business environments upon which they sit.

SEI's process maturity grid very accurately pinpoints the root of most of our software development problems. The fact that a full 86% of the organizations studied, remain at the ad hoc or chaotic level indicates that only a few organizations (the remaining 14%) have adopted any formal process for software engineering. Simply put, 86% of all organizations react to a business problem by just writing code. If they do employ a software engineering discipline, in all likelihood, it is one that no longer fits the requirements of the ever evolving business environment.

In the 1970s, the structured methodology was popularized. Although there were variations on the theme (i.e., different versions of the structured technique included the popular Gane/Sarson method and Yourdon method), for the most part, it provided methodology to develop usable systems in an era of batch computing. In those days, on-line systems with even the dumbest of terminals were a radical concept and graphical user interfaces were as unthinkable as the fall of the Berlin Wall.

Although times have changed and today's hardware is a thousand times more powerful than when structured techniques were introduced, this technique still survives. And survives in spite of the fact that the authors of these techniques have moved on to more adaptable paradigms, and more modern software development and systems engineering environments have entered the market.

In 1981 Finkelstein and Martin (Martin, 1981) popularized information engineering for the more commercially oriented users (i.e., those whose problems to be solved tended to be more database centered), which, to this day, is quite popular amongst mainframe developers with an investment in CASE. Information engineering is essentially a refinement of the structured approach. Instead of focusing on the data so preeminent in the structured approach, however, information engineering focuses on the information needs of the entire organization. Here business experts define high-level information models, as well as detailed data models. Ultimately the system is designed from these models.

Both structured and information engineering methodologies have their roots in mainframe-oriented commercial applications. Today's migration to client/server technologies (where the organization's data can be spread across one or more geographically distributed servers while the end-user uses his or her GUI of choice to perform local processing) disables most of the utility of these methodologies. In fact, many issues now surfacing in more commercial applications are not unlike those addressed earlier in the more engineering oriented environments such as tele-communications and avionics. Client/server environments are characterized by their diversity. One organization may store its data on multiple databases, program in several programming languages, and use more than one operating system and, hence, different GUIs. Since software development complexity is increased a hundredfold in this new environment, a better methodology is required. Today's object-oriented techniques solve some of the problem. Given the complexity of client/server, code trapped in programs is not flexible enough to meet the needs of this type of environment. We have already discussed how coding via objects rather than large programs engenders flexibility as well as productivity and quality through reusability. But object-oriented development is a double-edged sword.

While it is true that to master this technique is to provide dramatic increases in productivity, the sad fact of the matter is that object-oriented development, if done inappropriately, can cause problems far greater than problems generated from structured techniques. The reason for this is simple: the stakes are higher. Object-oriented environments are more complex than any other, the business problems chosen to be solved by object-oriented techniques are far more complex than other types of problems, and there are no conventional object-oriented methodologies and corollary tools to help the development team develop systems that are truly reliable. There are many kinds of object orientation. But with this diversity comes some very real risks. As a result, the following developmental issues must be considered before the computer is even turned on.

Integration is a challenge and needs to be considered at the onset. With traditional systems, developers rely on mismatched modeling methods to capture aspects of even a single definition. Whether it be integration of object to object, module to module, phase to phase, or type of application to type of application, the process can be an arduous one. The mismatch of evolving versions of point products (where a point product is able to be used for developing only an aspect of an application) used in design and development compounds the issue. Integration is usually left to the devices of myriad developers well into development. The resulting system is sometimes hard to understand and objects are difficult to trace. The biggest danger is that there is little correspondence to the real world. Interfaces are often incompatible and errors propagate throughout development. As a result, systems defined in this manner can be ambiguous and just plain incorrect.

Errors need to be minimized. Traditional methods as well as traditional object-oriented methods actually encourage the propagation of errors, such as reuse of unreliable objects with embedded and inherited errors. Internet viruses, worms, packet storms, trojans, and other malicious phenomenon are inexcusable and should be the exception not the rule. To be successful, errors must be eliminated from the very onset of the development process, including in the operating systems and related systems.

Languages need to be more formal. Whereas some languages are formal and others friendly, it is hard to find a language both *formal* and *friendly*. Within environments where more informal languages are used, lack of traceability and an overabundance of interface errors are a common occurrence. Within environments that have more formal languages their application has been found to be impractical for real world systems of any size or complexity. Recently, more modern software requirements languages have been introduced (for example, the Unified Modeling Language, UML (Booch et al., 1999), most of which are informal (or semi-formal); some of these languages were created by "integrating" several languages into one. Unfortunately, the bad comes with the good—often, more of what is not needed and less of what is needed; and since much of the formal part is missing, a common semantics needs to exist to reconcile differences and eliminate redundancies.

Flexibility for change and handling the unpredictable needs to be dealt with up front. Too often it is forgotten that the building of an application must take into account its evolution. In the real world, users change their minds, software development environments change and technologies change. Definitions of requirements in traditional development scenarios concentrate on the application needs of the user, but without consideration of the potential for the user's needs or environment to change. Porting to a new environment becomes a new development for each new architecture, operating system, database, graphics environment, language, or language configuration. Because of this, critical functionality is often avoided for fear of the unknown, and maintenance, the most risky and expensive part of a system's life cycle, is left unaccounted for during development. To address these issues, tools and techniques must be used to allow cross technology and changing technology solutions as well as provide for changing and evolving architectures.

The syndrome of locked-in design needs to be eliminated. Often, developers are forced to develop in terms of an implementation technology that does not have an open architecture, such as a specific database schema or a graphical user interface (GUI). Bad enough is to attempt an evolution of such a system; worse yet is to use parts of it as reusables for a system that does not rely on those technologies. Well thought out and formal business practices and their implementation will help to minimize this problem within an organization.

Developers must prepare for parallelism and distributed environments. Often, when it is not known that a system is targeted for a distributed environment, it is first defined and developed for a single processor environment and then changed/redeveloped for a distributed environment—an unproductive use of resources. Parallelism and distribution need to be dealt with at the very start of the project.

Resource allocation should be transparent to the user. Whether or not a system is allocated to distributed, asynchronous, or synchronous processors and whether or not two processors or ten processors are selected; with traditional methods it is still up to the designer and developer to be concerned with manually incorporating such detail into the application. There is no separation between the specification of what the system is to do vs. how the system does it. This results in far too much implementation detail to be included at the level of design. Once such a resource architecture becomes obsolete, it is necessary to redesign and redevelop those applications which have old designs embedded within them.

The creation of reliable reusable definitions must be promoted, especially those that are inherently provided. Traditional requirements definitions lack the facilities to help find, create, use and ensure commonality in systems. Modelers are forced to use informal and manual methods to find ways to divide a system into components natural for reuse. These components do not lend themselves to integration and, as a result, they tend to be error-prone. Because these systems are not portable or adaptable, there is little incentive for reuse. In conventional methodologies, redundancy becomes a way of doing business. Even when methods are object oriented, developers are often left to their own devices to explicitly make their applications become object oriented; because these methods do not support all that which is inherent in an object.

Automation that minimizes manual work needs to replace "make work" automated solutions. In fact, automation itself is an inherently reusable process. If a system does not exist for reuse, it certainly does not exist for automation. But most of today's development process is needlessly manual. Today's systems are defined with insufficient intelligence for automated tools to use them as input. In fact, these automated tools concentrate on supporting the manual process instead of doing the real work. Typically, developers receive definitions that they manually turn into code. A process that could have been mechanized once for reuse is performed manually again and again. Under this scenario, even when automation attempts to do the real work, it is often incomplete across application domains or even within a domain, resulting in incomplete code, such as shell code (code that is an outline for real code). The generated code is often inefficient or hard wired to an architecture, a language, or even a version of a language. Often partial automations need to be integrated with incompatible partial automations or manual processes. Manual processes are needed to complete unfinished automations.

Run-time performance analysis (decisions between algorithms or architectures) should be based on formal definitions. Conventional system definitions contain insufficient information about a system's run-time performance, including that concerning the decisions between algorithms or architectures. Design decisions where this separation is not taken into account thus depend on analysis of outputs from ad hoc implementations and associated testing scenarios. System definitions must consider how to separate the target system from its target environment.

Design integrity is the first step to usable systems. It is not known if a design is a good one until its implementation has failed or succeeded. Spiral development (an evolutionary approach) as opposed to waterfall development (where each phase of development is completed before the next phase is begun) helps address some of this issue. Usually, a system design is based on short-term considerations because knowledge is not reused from previous lessons learned. Development, ultimately, is driven toward failure.

Once issues like these are addressed, software costs less and takes less time to develop. But, time is of the essence. These issues are becoming more complex and even more critical as developers prepare for the distributed environments that go hand in hand with the increasing predominance of the internet.

Thus far this chapter has explained the derivation of software and attempted to show how it has evolved over time to become the true brains of any automated system. But like a human brain, this software brain must be carefully architected to promote productivity, foster quality, and enforce control and reusability.

Traditional software engineering paradigms fail to see the software development process from the larger perspective of the superorganism described at the beginning of this chapter. It is only when we see the software development process as made up of discrete but well-integrated components, can we begin to develop a methodology that can produce the very benefits that have been promised by the advent of software decades ago.

Software engineering, from this perspective, consists of a methodology as well as a series of tools with which to implement the solution to the business problem at hand. But even before the first tool can be applied, the software engineering methodology must be deployed to assist in specifying the requirements of the problem. How can this be accomplished successfully in the face of the problems outlined? How can it be accomplished in situations where organizations must develop systems that run across diverse and distributed hardware platforms, databases, programming languages, and GUIs when traditional methodologies make no provision for such diversity? And how can software be developed without having to fix or cure those myriad of problems that result "after the fact" of that software's development?

To address these software issues, an organization has several options, ranging from one extreme to the other. The options include (1) keep doing things the same way; (2) add tools and techniques that support business as usual but provide relief in selected areas; (3) bring in more modern but traditional tools and techniques to replace existing ones; (4) use a new paradigm with the most advanced tools and techniques that formalizes the process of software development, while at the same time capitalizing on software already developed; or (5) completely start over with a new paradigm that formalizes the process of software development and uses the most advanced tools and techniques.

19.2.4   A New Approach

The typical reaction to the well known problems and challenges of software and its development has been to lament the fact that this is the way software is and to accept it as a "*fait acccompli.*" But, such a reaction is unacceptable. What is required is a radical revision of the way we build software—an approach that understands how to build systems using the right techniques at the right time. First and foremost, it is a preventative approach. This means it provides a framework for doing things in the right way the first time. Problems associated with traditional methods of design and development would be prevented "before the fact" just by the way a system is defined. Such an approach would concentrate on preventing problems of development from even happening; rather than letting them happen "after the fact" and fixing them after they have surfaced at the most inopportune and expensive point in time.

Consider such an approach in its application to a human system. To fill a tooth before it reaches the stage of a root canal is curative with respect to the cavity, but preventive with respect to the root canal. Preventing the cavity by proper diet prevents not only the root canal, but the cavity as well. To follow a cavity with a root canal is the most expensive alternative; to fill a cavity on time is the next most expensive; and to prevent these cavities in the first place is the least expensive option. Preventiveness is a relative concept. For any given system, be it human or software, one goal is to prevent, to the greatest extent and as early as possible, anything that could go wrong in the life cycle process.

With a preventative philosophy, systems would be carefully constructed to minimize development problems from the very outset. A system that could be developed with properties that controlled its very own design and development. One result would be reusable systems that promote automation. Each system definition would model both its application and its life cycle with built-in constraints—constraints that protect the developer, but yet do not take away his flexibility.

Such an approach could be used throughout a life cycle, starting with requirements and continuing with functional analysis, simulation, specification, analysis, design, system architecture design, algorithm development, implementation, configuration management, testing, maintenance, and reverse engineering. Its users would include end users, managers, system engineers, software engineers, and test engineers.

With this approach, the same language would be used to define any aspect of any system and integrate it with any other aspect. The crucial point is that these aspects are directly related to the real world and, therefore, the same language would be used to define system requirements, specifications, design, and detailed design for functional, resource, and resource allocation architectures throughout all levels and layers of seamless definition, including hardware, software, and peopleware. The same language would be used to define organizations of people, missile or banking systems, cognitive systems, as well as real-time or database environments and is therefore appropriate across industries, academia, or government.

The philosophy behind preventative systems is that reliable systems are defined in terms of reliable systems. Only reliable systems are used as building blocks, and only reliable systems are used as mechanisms to integrate these building blocks to form a new system. The new system becomes a reusable for building other systems.

Effective reuse is a preventative concept. That is, reusing something (e.g., requirements or code) that contains no errors to obtain a desired functionality avoids both the errors and the cost of developing a new system. It allows one to solve a given problem as early as possible, not at the last moment. But to make a system truly reusable, one must start not from the customary end of a life cycle, during the implementation or maintenance phase, but from the very beginning.

Preventative systems are the true realization of the entelechy construct where molecules of software naturally combine to form a whole, which is much greater than the sum of its parts. Or one can think of constructing systems from the tinker toys of our youth. One recalls that the child never errs in building magnificent structures from these tinkertoys. Indeed, tinker toys are built from blocks that are architected to be perpetually reusable, perfectly integratable, and infinitely user-friendly.

There is at least one approach that follows a preventative philosophy. Although not in the mainstream yet, it has been used successfully by research and "trail blazer" organizations and is now being adopted for more commercial use. With this approach there is the potential for users to design systems and build software with seamless integration, including from systems to software, with: no longer a need for the system designer or system engineer to understand the details of programming languages or operating systems; no interface errors; defect rates reduced by at least a factor of 10; correctness by built-in language properties; unambiguous requirements, specifications, design (removing complexity, chaos and confusion); guarantee of function integrity after implementation; complete traceability and evolvability (application to application, architecture to architecture, technology to technology); maximized reuse; generation of complete software solutions from system specifications (full life cycle automation including 100% production ready code for any kind or size of system); a set of tools, each of which is defined and automatically generated by itself; significantly higher reliability, higher productivity, and lower risk.

Most people would say this is not possible, at least in the foreseeable future. This is because it has not been possible with traditional environments. It is possible, however, in major part, because of a non-traditional systems design and software development paradigm and its associated universal systems language that has been derived and evolved over three decades from an empirical study of the Apollo on-board flight software effort. In addition to experience with Apollo and other real world systems, this paradigm also takes its roots from systems theory, formal methods, formal linguistics, and object technologies. We'll describe this technology and its background in order to illustrate by example the potential that preventative approaches have.

19.2.5   Apollo On-Board Flight Software Effort: Lessons Learned

The original purpose of the empirical study, which had its beginnings in 1968, was to learn from Apollo's flight software and its development in order to make use of this effort for future Apollo missions as well as for the then up and coming Skylab and Shuttle missions. A better way was needed to define and develop software systems than the ones being used and available, because the existing ones (just like the traditional ones today) did not solve the pressing problems. There was a driving desire to learn from experience; what could be done better for future systems and what should the designers and developers keep doing because they were doing it right. The search was in particular for a means to build ultra reliable software.

The results of the study took on multiple dimensions, not just for space missions but for systems in general; some of which were not so readily apparent for many years to come. The Apollo software, given what had to be accomplished and the time within which it had to be accomplished, was as complex as it could get; causing other software projects in the future to look (and be) less daunting than they might have been in comparison. The thought was if problems from a worst case scenario could be solved and simplified, the solutions might be able to be applied to all kinds of systems.

On hindsight, this was an ideal setting from which to understand systems and software and the kinds of problems that can occur and issues that need to be addressed. Because of the nature of the Apollo software there was the opportunity to make just about every kind of error possible, especially since the software was being developed concurrently with the planning of the space missions, the hardware, the simulator, and the training of the astronauts (demonstrating how much the flight software was part of a larger system of other software, hardware and peopleware); and since no one had been to the moon before there were many unknowns. In addition the developers were under the gun with what today would have been unrealistic expectations and schedules. This and what was accomplished (or not accomplished) provided a wealth of information from which to learn. Here are some examples:

It was found during the Apollo study that interface errors (errors resulting from ambiguous relationships, mismatches, conflicts in the system, poor communication, lack of coordination, inability to integrate) accounted for approximately 75% of all the errors found in the software during final testing, the Verification and Validation (V&V) phase (using traditional development methods the figure can be as high as 90%); interface errors include data flow, priority and timing errors from the highest levels of a system to the lowest levels of a system, to the lowest level of detail. It was also determined that 44% of the errors were found by manual means (suggesting more areas for automation) and that 60% of the errors had unwittingly existed in earlier missions—missions that had already flown, though no errors occurred (made themselves known) during any flights. The fact that this many errors existed in earlier missions was down-right frightening. It meant lives were at stake during every mission that was flown. It meant more needed to be done in the area of reliability. Although no software problem ever occurred (or was known to occur) on any Apollo mission, it was only because of the dedication of the software team and the methods they used to develop and test the software.

A more detailed analysis of interface errors followed; especially since they not only accounted for the majority of errors but they were often the most subtle errors and therefore the hardest to find. The realization was made that integration problems could be solved if interface problems could be solved and if integration problems could be solved, there would be traceability. Each interface error was placed into a category according to the means that could have been taken to prevent it by the very way a system was defined. It was then established that there were ways to prevent certain errors from happening simply by changing the rules of definition. This work led to a theory and methodology for defining a system that would eliminate all interface errors.

It quickly became apparent that everything including software is a system and the issues of system design were one and the same as those of software. Many things contributed to this understanding of what is meant by the concept of "system." During development of the flight software, once the software group implemented the requirements ("thrown over the wall" by system design experts to the software people), the software people necessarily became the new experts. This phenomenon forced the software experts to become system experts (and vice versa) suggesting that a system was a system whether in the form of higher level algorithms or software that implemented those algorithms.

Everyone learned to always expect the unexpected and this was reflected in every design; that is, they learned to think, plan and design in terms of error detection and recovery, reconfiguration in real time; and to always have backup systems. The recognition of the importance of determining and assigning (and ensuring) unique priorities of processes was established as part of this philosophy. Towards this end it was also established that a "kill and start over again" restart approach to error detection and recovery was far superior as opposed to a "pick up from where you left off" approach; and it simplified both the operation of the software as well as the development and testing of the software. The major impact a system design for one part of the software could have on another part further emphasized that everything involved was part of a system—the "what goes around comes around" syndrome. For example, choosing an asynchronous executive (one where higher priority processes can interrupt lower priority processes) for the flight software not only allowed for more flexibility during the actual flights but it also provided for more flexibility to make a change more safely and in a more modular fashion during development of the flight software.

It became obvious that testing was never over (even with many more levels of testing than with other kinds of applications; and with added testing by independent verification and validation organizations). This prompted a search for better methods of testing. Again, traditional methods were not solving the problem.

With the realization that most of the system design and software development processes could be mechanized (and the same generic processes were being performed throughout each of all phases of development), it became clear they could be automated. This suggested an environment that would do just that, automate what traditional systems still to this day do manually. In addition, for the space missions, software was being developed for many missions all at once and in various stages, leading the way to learn how to successfully perform distributed development.

It was learned during this process of evaluation of space flight software that traditional methods did not support developers in many areas and allowed too much freedom in other areas (such as freedom to make errors, both subtle and serious) and not enough freedom in other areas (such as areas that required an open architecture and the ability to reconfigure in real time to be successful). It further became clear that a new kind of language was needed to define a system having certain "built-in" properties not available in traditional languages; such as inherent reliability, integration, reuse, and open architecture capabilities provided simply by the use of the language itself. In addition it was realized a set of tools could be developed to support such a language as well as take away tedious mechanical processes that could become automated, thus avoiding even further errors, reducing the need for much after the fact testing. It was only later understood the degree to which systems with "built-in reliability" could increase the productivity in their development, resulting in "built-in productivity."

Lessons learned from this effort continue today. Key aspects are that systems are asynchronous in nature and this should be reflected inherently in the language used to define systems. Systems should be assumed to be event driven and every function to be applied should have a unique priority; real time event and priority driven behavior should be part of the way one specifies a system in a systems language and not defined on a case by case basis and in different programming languages with special purpose data types.

Rather than having a language whose purpose is to run a machine the language should naturally describe what objects there are and what actions they are taking. Objects are inherently distributed and their interactions asynchronous with real time, event-driven behavior. This implies that one could define a system and its own definition would have the necessary behaviors to characterize natural behavior in terms of real time execution semantics. Application developers would no longer need to explicitly define schedules of when events were to occur. Events would instead occur when objects interact with other objects. By describing the interactions between objects the schedule of events is inherently defined.

The result of this revelation was that a universal systems language could be used to tackle all aspects of a system: its planning, development, deployment, and evolution. This means that all the systems that work together during the system's lifecycle can be defined and understood using the same semantics.

19.2.6   Development Before the Fact

Once the analysis of the Apollo effort was completed, the next step was to create (and evolve) a new mathematical paradigm from the "heart and soul" of Apollo; one that was preventative instead of curative in its approach. A theory was derived for defining a system such that the entire class of errors, known as interface errors, would be eliminated. The first generation technology derived from this theory concentrated on defining and building reliable systems in terms of functional hierarchies (Hamilton, 1986). Having realized the benefits of addressing one major issue, that is, reliability, just by the way a system is defined, the research effort continued over many years (and still continues) to evolve the philosophy of addressing this issue fur-

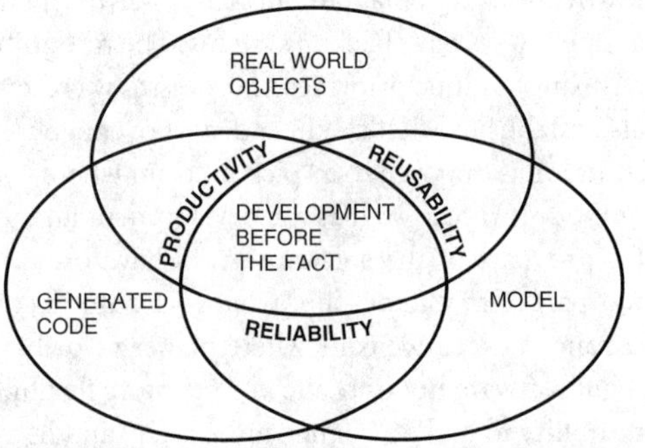

FIGURE 19.12   The development-before-the-fact paradigm.

ther as well as addressing other issues the same way, that is, using language mechanisms that inherently eliminate software problems. The result is a new generation technology called development before the fact (DBTF) (see Fig. 19.12) where systems are designed and built with preventative properties integrating all aspects of a system's definition including the inherent integration of functional and type hierarchical networks, (Hamilton and Hackler, in press; Hamilton, 1994; Hamilton, 1994; Keyes, 2001), <http://www.htius.com/>.

Development before the fact is a system oriented object (SOO) approach based on a concept of control that is lacking in other software engineering paradigms. At the base of the theory that embodies every system are a set of axioms—universally recognized truths—and the design for every DBTF system is based on these axioms and on the assumption of a universal set of objects. Each axiom defines a relation of immediate domination. The union of the relations defined by the axioms is control. Among other things, the axioms establish the control relationships of an object for invocation, input and output, input and output access rights, error detection and recovery, and ordering during its developmental and operational states. Table 19.1 summarizes some of the properties of objects within these systems.

Combined with further research it became clear that the root problem with traditional approaches is that they support users in "fixing wrong things up rather than in "doing things the right way in the first place." Instead of testing software to look for errors *after* the software is developed, with the new paradigm, software could now be defined to not allow errors in, in the first place; correctness could be accomplished by the very way software is defined, by "built-in" language properties; what had been created was a universal semantics for defining not just software systems but systems in general.

Once understood, it became clear that the characteristics of good design can be reused by incorporating them into a language for defining any system (not just a software system). The language is a formalism for representing the mathematics of systems. This language—actually a *meta*-language—is the key to DBTF. Its main attribute is to help the designer reduce the complexity and bring clarity into his thinking process, eventually turning it into the ultimate reusable, wisdom, itself. It is a universal systems language for defining systems, each system of which can be incorporated into the meta-language and then used to define other systems. A system defined with this language has properties that come along "for the ride" that in essence control its own destiny. Based on a theory (DBTF) that extends traditional mathematics of systems with a unique concept of control, this formal but friendly language has embodied within it a natural representation of the physics of time and space. 001AXES evolved as DBTF's formal universal systems language, and the 001 Tool Suite as its automation.

TABLE 19.1 System Oriented Object Properties of Development Before the Fact

Quality	*Automation*
Reliable	Common definitions
Affordable	natural modularity
	natural separation (e.g., functional architecture from
Reliable	its resource architectures);
In control and under control	dumb modules
Based on a set of axioms	an object is integrated with respect to structure,
domain identification (intended, unintended)	behavior and properties of control
ordering (priority and timing)	integration in terms of structure and behavior
access rights: incoming object (or relation),	type of mechanisms
outgoing object (or relation)	function maps (relate an object's function to other
replacement	functions)
Formal	object type maps (relate objects to objects)
consistent, logically complete	structures of functions and types
necessary and sufficient	category
common semantic base	relativity
unique state identification	instantiation
Error free (based on formal definition of "error")	polymorphism
always gets the right answer at the right time and in	parent/child
the right place	being/doing
satisfies users and developers intent	having/not having
Handles the unpredictable	abstraction
Predictable	encapsulation
	replacement
Affordable	relation including function
Reusable	typing including classification
Optimizes resources in operation and development	form including both structure and behavior
in minimum time and space	(for object types and functions)
with best fit of objects to resources	derivation
	deduction
Reusable	inference
Understandable, integratable and maintainable	inheritance
Flexible	
Follows standards	

(continued)

TABLE 19.1 System Oriented Object Properties of Development Before the Fact (*Continued*)

Handles the unpredictable	**Understandable, integratable, and maintainable**
Throughout development and operation	*Reliable*
Without affecting unintended areas	A measurable history
Error detect and recover from the unexpected	Natural correspondence to real world
Interface with, change and reconfigure in asynchronous,	persistence, create and delete
distributed, real time environment	appear and disappear
Flexible	accessibility
Changeable without side effects	reference
Evolvable	assumes existence of objects
Durable	real time and space constraints
Reliable	representation
Extensible	*relativity, abstraction, derivation*
Ability to break up and put together	Provides user friendly definitions
one object to many: modularity, decomposition,	recognizes that one user's friendliness is another user's
instantiation	nightmare
many objects to one: composition, applicative	hides unnecessary detail (*abstraction*)
Operators, integration, abstraction	variable, user selected syntax
Portable	self teaching
secure	derived from a common semantic base
diverse and changing layered developments	common definition mechanisms
open architecture (implementation, resource allocation,	Communicates with common semantics to all entities
and execution independance)	Defined to be simple as possible but not simpler
plug-in (or be plugged into) or reconfiguration of	Defined with integration of all of its objects (and all
different modules	aspects of these objects)
adaptable for different organizations, applications,	Traceability of behavior and structure and their changes
functionality, people, products	(maintenance) throughout its birth, life, and death
Automation	Knows and able to reach the state of completion
The ultimate form of reusable	definition
Formalize, mechanize, then automate	development of itself and that which develops it
it	analysis
its development	design
that which automates its development	implementation
	instantiation
	testing
	maintenance

*All italicized words point to a reusable.

The creation of the concept of reuse definition scenarios during Apollo to save time in development and space in the software was a predecessor to the type of higher level language statements used within the systems language. This included horizontal and vertical reuse that led to a flexible open architecture within the DBTF environment.

Understanding the importance of metrics and the influence it could have on future software has had a major focus throughout this endeavor. What this approach represents, in essence, is the current state of a "reusable" originally based on and learned from Apollo followed by that learned from each of its evolving states. It reuses and builds on that which was (and is) observed to have been beneficial to inherit as well as the avoidance of that observed not to have been beneficial. That learned from history was and will continue to be put into practice for future projects. Someone once said "it is never surprising when something developed empirically turns out to have intimate connections with theory." Such is the case with DBTF .

19.2.7 Development Before the Fact Theory

Mathematical approaches have been known to be difficult to understand and use. They have also been known to be limited in their use for nontrivial systems as well as for much of the life cycle of a given system. What makes DBTF different in this respect is that the mathematics it is based on, has been extended for handling the class of systems that software falls into; the formalism along with its unfriendliness is "hidden" under the covers by language mechanisms derived in terms of that formalism; and the technology based on this formalism has been put to practical use.

With DBTF, all models are defined using its language (and developed using its automated environment) as SOOs. A SOO is understood the same way without ambiguity by all other objects within its environment—including all users, models, and automated tools. Each SOO is made up of other SOOs. Every aspect of a SOO is integrated not the least of which is the integration of its object oriented parts with its function oriented parts and its timing oriented parts. Instead of systems being object-oriented, objects are systems-oriented. All systems are objects. All objects are systems. Because of this, many things heretofore not believed possible with traditional methods are possible, much of what seems counter intuitive with traditional approaches, that tend to be software centric, becomes intuitive with DBTF, which is system centric.

DBTF's automation provides the means to support a system designer or software developer in following its paradigm throughout a system's design or software development life cycle. Take for example, testing. The more a paradigm prevents errors from being made in the first place the less the need for testing and the higher the productivity. Before the fact testing is inherently part of every development step. Most errors are prevented because of that which is inherent or automated.

Unlike a language such as UML or Java that is limited to software, 001AXES is a systems language and as such can be used to state not just software phenomenon but any given system phenomenon. UML is a graphical software specification language. 001AXES is a graphical system specification language. They have different purposes. At some point UML users have to program in some programming language, but 001AXES users do not; a system can be completely specified and automatically generated with all of its functionality, object and timing behavior without a line of code being written. The intent of UML is to run a machine. The intent of 001AXES is to define (and if applicable, execute) a system software or otherwise.

Because 001AXES is a systems language, its semantics is more general than a software language. For example, 001AXES defined systems could run on a computer, person, autonomous robot, or an organization; whereas the content of the types (or classes) in a software language is based on a computer that runs software.

Because 001AXES is systemic in nature it capitalizes upon that which is common to all systems; including software, financial, political, biological, and physical systems. For software the semantics of 001AXES is mapped to appropriate constructs for each of a possible set of underlying programming language implementations. Unlike the DBTF approach, in general these more traditional development approaches emphasize a fragmented approach to the integrated specification of a system and its software, short changing integration and automation in the development process. A more detailed discussion of 001AXES and UML can be found in (Hamilton and Hackler, 2000).

19.2.8 Process

Derived from the combination of steps taken to solve the problems of traditional systems engineering and software development, each DBTF system is defined with built-in quality, built-in productivity and built-in control (like the biological superorganism). The process combines mathematical perfection with engineering precision. Its purpose is to facilitate the "doing things right in the first place" development style, avoiding the "fixing wrong things up" traditional approach. Its automation is developed with the following considerations: error prevention from the early stage of system definition, life cycle control of the system under development, and inherent reuse of highly reliable systems. The development life cycle is divided into a sequence of stages, including: requirements and design modeling by formal specification and analysis; automatic code generation based on consistent and logically complete models; test and execution; and simulation.

The first step is to define a model with the language. This process could be in any phase of the developmental life cycle, including problem analysis, operational scenarios, and design. The model is automatically analyzed to ensure that it was defined properly. This includes static analysis for preventive properties and dynamic analysis for user intent properties. In the next stage, the generic source code generator automatically generates a fully production-ready and fully integrated software implementation for any kind or size of application, consistent with the model, for a selected target environment in the language and architecture of choice. If the selected environment has already been configured, the generator selects that environment directly; otherwise, the generator is first configured for a new language and architecture.

Because of its open architecture, the generator can be configured to reside on any new architecture (or interface to any outside environment), e.g., to a language, communications package, an internet interface, a database package, or an operating system of choice; or it can be configured to interface to the users own legacy code. Once configured for a new environment, an existing system can be automatically regenerated to reside on that new environment. This open architecture approach, which lends itself to true component based development, provides more flexibility to the user when changing requirements or architectures; or when moving from an older technology to a newer one.

It then becomes possible to execute the resulting system. If it is software, the system can undergo testing for further user intent errors. It becomes operational after testing. Application changes are made to the requirements definition—not to the code. Target architecture changes are made to the configuration of the generator environment (which generates one of a possible set of implementations from the model)—not to the code. If the real system is hardware or peopleware, the software system serves as a simulation upon which the real system can be based. Once a system has been developed, the system and the process used to develop it are analyzed to understand how to improve the next round of system development.

Seamless integration is provided throughout from systems to software, requirements to design to code to tests to other requirements and back again; level to level, and layer to layer. The developer is able to trace from requirements to code and back again.

Given an automation that has these capabilities, it should be no surprise that it has been defined with itself and that it continues to automatically generate itself as it evolves with changing architectures and changing technologies. Table 19.2 contains a summary of some of the differences between the more modern preventative paradigm and the traditional approach.

A relatively small set of things is needed to master the concepts behind DBTF. Everything else can be derived, leading to powerful reuse capabilities for building systems. It quickly becomes clear why it is no longer necessary to add features to the language or changes to a developed application in an ad hoc fashion, since each new aspect is ultimately and inherently derived from its mathematical foundations.

Although this approach addresses many of the challenges and solves many of the problems of traditional software environments, it could take time before this paradigm (or one with similar properties) is adopted by the more mainstream users, since it requires a change to the corporate culture. The same has been true in related fields. At the time when the computer was first invented and manual calculators were used for almost every computation; it has been said that it was believed by hardware pioneers like Ken Olson, founder of Digital Equipment Corporation, that there would only be a need for four computers in the world. It took awhile for the idea of computers and what they were capable of to catch on. Such could be the case for a more advanced software paradigm as well. That is, it could take awhile for software to be truly something to manufacture as opposed to being handcrafted as it is in today's traditional development environments.

19.2.9 Select the Right Paradigm and then Automate

Where software engineering fails is in its inability to grasp that not only the right paradigm (out of many paradigms) must be selected, but that the paradigm must be part of an environment that provides an inherent and integrated automated means to solve the problem at hand. What this means is that the paradigm must be coupled with an integrated set of tools with which to implement the results of utilizing the paradigm to design and develop the system. Essentially, the paradigm generates the model and a toolset must be provided to generate the system.

Businesses that expected a big productivity payoff from investing in technology are, in many cases, still waiting to collect. A substantial part of the problem stems from the manner in which organizations are building their automated systems. Although hardware capabilities have increased dramatically, organizations are still mired in the same methodologies that saw the rise of the behemoth mainframes. Obsolete methodologies simply cannot build new systems.

There are other changes as well. Users demand much more functionality and flexibility in their systems. And given the nature of many of the problems to be solved by this new technology, these systems must also be error free.

TABLE 19.2 A Comparison

Traditional (After the Fact)	DBTF (Development Before the Fact)
Integration ad hoc, if at all	*Integration*
Mismatched methods, objects, phases, products, architectures, applications and environment	Seamless life cycle: methods, objects, phases, products, architectures, applications and environment
System not integrated with software	System integrated with software
Function oriented *or* object oriented	System oriented objects: integration of function, timing, and object oriented
GUI not integrated with application	GUI integrated with application
Simulation not integrated with software code	Simulation integrated with software code
Behavior uncertain until after delivery	*Correctness by built-in language properties*
Interface errors abound and infiltrate the system (over 75% of all errors)	*No interface errors*
Most of those found are found after implementation	All found before implementation
Some found manually	All found by automatic and static analysis
Some found by dynamic runs analysis	
Some never found	Always found
Ambiguous requirements, specifications, designs . . . introduce chaos, confusion and complexity	*Unambiguous requirements, specifications, designs . . . remove chaos, confusion and complexity*
Informal or semi-formal language	Formal, but friendly language
Different phases, languages and tools	All phases, same language and tools
Different language for other systems than for software	Same language for software, hardware and any other system
No guarantee of function integrity after implementation	*Guarantee of function integrity after implementation*
Inflexible: Systems not traceable or evolvable	*Flexible: Systems traceable and evolvable*
Locked in bugs, requirements products, architectures, etc.	Open architecture
Painful transition from legacy	Smooth transition from legacy
Maintenance performed at code level	Maintenance performed at spec level
Reuse not inherent	*Inherent Reuse*
Reuse is adhoc	Every object a candidate for reuse
Customization and reuse are mutually exclusive	Customization increases the reuse pool
Automation supports manual process instead of doing real work	*Automation does real work*
Mostly manual: documentation, programming, test generation, traceability, integration	Automatic programming, documentation, test generation, traceability, integration
Limited, incomplete, fragmented, disparate and inefficient	100% code automatically generated for any kind of software
Product x not defined and developed with itself	*001 defined with and generated by itself*
	#1 in all evaluations
Dollars wasted, error prone systems	*Ultra-reliable systems with unprecedented productivity in their development*
High risk	Low risk
Not cost effective	10 to 1, 20 to 1, 50 to 1 . . . dollars saved/dollars made
Difficult to meet schedules	Minimum time to complete
Less of what you need and more of what you do not need	No more, no less of what you need

Where the biological superorganism has built-in control mechanisms fostering quality and productivity, up until now, the silicon superorganism has had none. Hence, the productivity paradox. Often, the only way to solve major issues or to survive tough times is through nontraditional paths or innovation. One must create new methods or new environments for using new methods. Innovation for success often starts with a look at mistakes from traditional systems. The first step is to recognize the true root problems, then categorize them according to how they might be prevented. Derivation of practical solutions is a logical next step. Iterations of the process entail looking for new problem areas in terms of the new solution environment and repeating the scenario. That is how DBTF came into being.

With DBTF, all aspects of system design and development are integrated with one systems language and its associated automation. Reuse naturally takes place throughout the life cycle. Objects, no matter how complex, can be reused and integrated. Environment configurations for different kinds of architectures can be reused. A newly developed system can be safely reused to increase even further the productivity of the systems developed with it.

This preventative approach can support a user in addressing many of the challenges presented in today's software development environments. There will, however, always be more to do to capitalize on this technology. That is part of what makes a technology like this so interesting to work with. Because it is based on a different premise or set of assumptions (axioms), a significant number of things can and will change because of it. There is the continuing opportunity for new research projects and new products. Some problems can be solved, because of the language, that could not be solved before. Software development, as we have known it, will never be the same. Many things will no longer need to exist—they, in fact, will be rendered extinct, just as that phenomenon occurs with the process of natural selection in the biological system. Techniques for bridging the gap from one phase of the life cycle to another become obsolete. Techniques for maintaining the source code as a separate process are no longer needed, since the source is automatically generated from the requirements specification. Verification, too, becomes obsolete. Techniques for managing paper documents give way to entering requirements and their changes directly into the requirements specification database. Testing procedures and tools for finding most errors are no longer needed because those errors no longer exist. Tools to support programming as a manual process are no longer needed.

Compared to development using traditional techniques, the productivity of DBTF developed systems has been shown to be significantly greater (DOD, 1992; Krut, 1993; Ouyang, 1995; Keyes, 2000). Upon further analysis, it was discovered that the productivity was greater the larger and more complex the system—the opposite of what one finds with traditional systems development. This is, in pa2004rt, because of the high degree of DBTF's support of reuse. The larger a system, the more it has the opportunity to capitalize on reuse. As more reuse is employed, productivity *continues* to increase. Measuring productivity becomes a process of relativity—that is, relative to the last system developed.

Capitalizing on reusables within a DBTF environment is an ongoing area of research interest[1] (Hamilton, 2003–2004). An example is understanding the relationship between types of reusables and metrics. This takes into consideration that a reusable can be categorized in many ways. One is according to the manner in which its use saves time (which translates to how it impacts cost and schedules). More intelligent tradeoffs can then be made. The more we know about how some kinds of reusables are used, the more information we have to estimate costs for an overall system. Keep in mind, also, that the traditional methods for estimating time and costs for developing software are no longer valid for estimating systems developed with preventative techniques.

There are other reasons for this higher productivity as well, such as the savings realized and time saved due to tasks and processes that are no longer necessary with the use of this preventative approach. There is less to learn and less to do—less analysis, little or no implementation, less testing, less to manage, less to document, less to maintain, and less to integrate. This is because a major part of these areas has been automated or because of what inherently takes place because of the nature of the formal systems language.

The paradigm shift occurs once a designer realizes that many of the old tools are no longer needed to design and develop a system. For example, with one formal semantic language to define and integrate all aspects of a system, diverse modeling languages (and methodologies for using them), each of which defines only part of a system, are no longer necessary. There is no longer a need to reconcile multiple techniques with semantics that interfere with each other.

---

[1] An initial investigation of the form of a common software interface for use with common guidance systems under contract to ARES Corporation as a part of its contract with TACOM ARDEC, Mr. Nigel Gray, program manager. Mr.Gray is the originator of the Common Guidance Common Sense (CGCS) and Deeply Integrated Guidance and Navigation "smart munitions" and other military and aerospace system applications. The common software interface for these systems will provide a common means for user programming of the internal mission processor in these systems, independent of the embedded operating system (HTI, 2003–2004)

Software is a relatively young technological field that is still in a constant state of change. Changing from a traditional software environment to a preventative one is like going from the typewriter to the word processor. Whenever there is any major change, there is always the initial overhead needed for learning the new way of doing things. But, as with the word processor, progress begets progress.

In the end, it is the combination of the methodology and the technology that executes it that forms the foundation of successful software. Software is so ingrained in our society that its success or failure will dramatically influence both the operation and the success of businesses and government. For that reason, today's decisions about systems engineering and software development will have far reaching effects.

Collective experience strongly confirms that quality and productivity increase with the increased use of properties of preventative systems. In contrast to the "better late than never" after the fact philosophy, the preventative philosophy is to solve—or possible, prevent—a given problem as early as possible. This means finding a problem statically is better than finding it dynamically. Preventing it by the way a system is defined is even better. Better yet, is not having to define (and build) it at all.

Reusing a reliable system is better than reusing one that is not reliable. Automated reuse is better than manual reuse. Inherent reuse is better than automated reuse. Reuse that can evolve is better than one that cannot evolve. Best of all is reuse that approaches wisdom itself. Then, have the wisdom to use it. One such wisdom is that of applying a common systems paradigm to systems and software, unifying their understanding with a commonly held set of system semantics.

DBTF embodies a preventative systems approach, the language supports its representation, and its automation supports its application and use. Each is evolutionary (in fact, recursively so), with experience feeding the theory, the theory feeding the language, which in turn feeds the automation. All are used, in concert, to design systems and build software. The answer continues to be in the results just as in the biological system and the goal is that the systems of tomorrow will inherit the best of the systems of today.

Defining Terms

Data base management system (DBMS): The computer program that is used to control and provide rapid access to a database. A language is used with the DBMS to control the functions that a DBMS provides. For example, SQL is the language that is used to control all of the functions that a relational architecture based DBMS provides for its users, including data definition, data retrieval, data manipulation, access control, data sharing, and data integrity.

Graphical user interface (GUI): The ultimate user interface, by which the deployed system interfaces with the computer most productively, using visual means. Graphical user interfaces provide a series of intuitive, colorful, and graphical mechanisms which enable the end-user to view, update, and manipulate information.

Formal system: A system defined in terms of a known set of axioms (or assumptions); it is therefore mathematically based (for example, a DBTF system is based on a set of axioms of control). Some of its properties are that it is consistent and logically complete. A system is consistent if it can be shown that no assumption of the system contradicts any other assumption of that system. A system is logically complete if the assumptions of the method completely define a given set of properties. This assures that a model of the method has that set of properties. Other properties of the models defined with the method may not be provable from the method's assumptions. A logically complete system has a semantic basis (i.e., a way of expressing the meaning of that system's objects). In terms of the semantics of a DBTF system, this means it has no interface errors and is unambiguous, contains what is necessary and sufficient, and has a unique state identification.

Interface: A seam between objects, or programs, or systems. It is at this juncture that many errors surface. Software can interface with hardware, humans, and other software.

Methodology: A set of procedures, precepts, and constructs for the construction of software.

Metrics: A series of formulas which measure such things as quality and productivity.

Operating system: A program that manages the application programs that run under its control.

Software architecture: The structure and relationships among the components of software.

Acknowledgment

The author is indebted to Jessica Keyes for her helpful suggestions.

References

Boehm, B.W. 1981. *Software Engineering Economics.* Prentice-Hall, Englewood Cliffs, NJ.
Booch, G., Rumbaugh, J., and Jacobson, I. 1999. *The Unified Modeling Language User Guide.* Addison-Wesley, Reading, MA.
Department of Defense. 1992. *Software engineering tools experiment-Final report.* Vol. 1, Experiment Summary, Table 1, p. 9. Strategic Defense Initiative, Washington, DC.
Goldberg, A. and Robson, D. 1983. *Smalltalk-80 the Language and its Implementation.* Addison-Wesley, Reading, MA.
Gosling, J., Joy, B., and Steele, G. 1996. *The Java Language Specification*, Addison-Wesley, Reading, MA.
Hamilton, M. 1994. Development before the fact in action. *Electronic Design*, June 13. ES.
Hamilton, M. 1994. Inside development before the fact. *Electronic Design*, April 4. ES.
Hamilton, M. (In press). *System Oriented Objects: Development Before the Fact.*
Hamilton, M. and Hackler, W.R. 2000. Towards Cost Effective and Timely End-to-End Testing, *HTI*, prepared for Army Research Laboratory, Contract No. DAKF11-99-P-1236.
HTI Technical Report, *Common Software Interface for Common Guidance Systems*, US. Army DAAE30-02-D-1020 and DAAB07-98-D-H502/0180, Precision Guided Munitions, Under the direction of Cliff McLain, ARES, and Nigel Gray, COTR ARDEC, Picatinny Arsenal, NJ, 2003–2004.
Harbison, S.P. and Steele, G.L., Jr. 1997. *C A Reference Manual.* Prentice-Hall, Englewood Cliffs, NJ.
Jones, T.C. 1977. Program quality and programmer productivity. *IBM Tech.* Report TR02.764, Jan.: 80, Santa Teresa Labs., San Jose, CA.
Keyes, J. The Ultimate Internet Developers Sourcebook, Chapter 42, *Developing Web Applications with 001*, AMACOM.
Keyes, J. 2000. *Internet Management*, chapters 30–33, on 001-developed systems for the Internet, Auerbach.
Krut, B. Jr. 1993. *Integrating 001 Tool Support in the Feature-Oriented Domain Analysis Methodology* (CMU/SEI-93-TR-11, ESC-TR-93-188). Pittsburgh, PA: Software Engineering Institute, Carnegie-Mellon University.
Lickly, D. J. 1974. *HAL/S Language Specification,* Intermetrics, prepared for NASA Lyndon Johnson Space Center.
Lientz, B.P. and Swanson, E.B. 1980. *Software Maintenance Management*. Addison-Wesley, Reading, MA.
Martin, J. and Finkelstein, C.B. 1981. *Information Engineering.* Savant Inst., Carnforth, Lancashire, UK.
Meyer, B. 1992. *Eiffel the Language.* Prentice-Hall, New York.
NSQE Experiment: http://hometown.aol.com/ONeillDon/index.html, Director of Defense Research and Engineering (DOD Software Technology Strategy), 1992–2002.
Ouyang, M., Golay, M. W. 1995. *An Integrated Formal Approach for Developing High Quality Software of Safety-Critical Systems*, Massachusetts Institute of Technology, Cambridge, MA, Report No. MIT-ANP-TR-035.
Software Engineering Inst. 1991. *Capability Maturity Model.* Carnegie-Mellon University Pittsburgh, PA.
Stefik, M. and Bobrow, D.J. 1985. Object-Oriented Programming: Themes and Variations. *AI Magazine*, Zerox Corporation, pp. 40–62.
Stroustrup, B. 1994. *The Design and Evolution of C++.* AT&T Bell Laboratories, Murray Hill, NJ.
Stroustrup, B. 1997. *The C++ Programming Language*, Addison-Wesley, Reading, MA.
Yourdon E. and Constantine, L. 1978. *Structured Design: Fundamentals of a Discipline of Computer Program and Systems Design,* Yourdan Press, New York.

Further Information

Hamilton, M. and Hackler, R. 1990. 001: A rapid development approach for rapid prototyping based on a system that supports its own life cycle. *IEEE Proceedings*, First International Workshop on Rapid System Prototyping (Research Triangle Park, NC) pp. 46–62, June 4.

Hamilton, M. 1986. Zero-defect software: The elusive goal. *IEEE Spectrum* 23(3):48–53, March.

Hamilton, M. and Zeldin, S. 1976. Higher Order Software—A Methodology for Defining Software. *IEEE Transactions on Software Engineering*, vol. SE-2, no. 1.

McCauley, B. 1993. Software Development Tools in the 1990s. AIS Security Technology for Space Operations Conference, Houston, Texas.

Schindler, M. 1990. *Computer Aided Software Design*. John Wiley & Sons, New York.

The 001 *Tool Suite Reference Manual*, Version 3, Jan. 1993. Hamilton Technologies Inc., Cambridge, MA.

19.3 Neural Networks and Fuzzy Systems

Bogdan M. Wilamowski

19.3.1 Neural Networks and Fuzzy Systems

New and better electronic devices have inspired researchers to build intelligent machines operating in a fashion similar to the human nervous system. Fascination with this goal started when McCulloch and Pitts (1943) developed their model of an elementary computing neuron and when Hebb (1949) introduced his *learning rules*. A decade latter Rosenblatt (1958) introduced the **perceptron** concept. In the early 1960s Widrow and Holf (1960, 1962) developed intelligent systems such as ADALINE and MADALINE. Nillson (1965) in his book *Learning Machines* summarized many developments of that time. The publication of the Mynsky and Paper (1969) book, with some discouraging results, stopped for sometime the fascination with artificial neural networks, and achievements in the mathematical foundation of the **backpropagation** algorithm by Werbos (1974) went unnoticed. The current rapid growth in the area of neural networks started with the Hopfield (1982, 1984) recurrent network, Kohonen (1982) unsupervised training algorithms, and a description of the backpropagation algorithm by Rumelhart et al. (1986).

19.3.2 Neuron cell

A biological neuron is a complicated structure, which receives trains of pulses on hundreds of *excitatory* and *inhibitory* inputs. Those incoming pulses are summed with different weights (averaged) during the time period of *latent summation*. If the summed value is higher than a threshold, then the neuron itself is generating a pulse, which is sent to neighboring neurons. Because incoming pulses are summed with time, the neuron generates a pulse train with a higher frequency for higher positive excitation. In other words, if the value of the summed weighted inputs is higher, the neuron generates pulses more frequently. At the same time, each neuron is characterized by the nonexcitability for a certain time after the firing pulse. This so-called *refractory period* can be more accurately described as a phenomenon where after excitation the threshold value increases to a very high value and then decreases gradually with a certain time constant. The refractory period sets soft upper limits on the frequency of the output pulse train. In the biological neuron, information is sent in the form of frequency modulated pulse trains.

This description of neuron action leads to a very complex neuron model, which is not practical. McCulloch and Pitts (1943) show that even with a very simple neuron model, it is possible to build logic and memory circuits. Furthermore, these simple neurons with thresholds are usually more powerful than typical logic gates used in computers. The McCulloch-Pitts neuron model assumes that incoming and outgoing signals may have only binary values 0 and 1. If incoming signals summed through positive or negative weights have a value larger than threshold, then the neuron output is set to 1. Otherwise, it is set to 0.

$$T = \begin{cases} 1 & \text{if} \quad net \geq T \\ 0 & \text{if} \quad net < T \end{cases} \tag{19.1}$$

where T is the threshold and *net* value is the weighted sum of all incoming signals

$$net = \sum_{i=1}^{n} w_i x_i \tag{19.2}$$

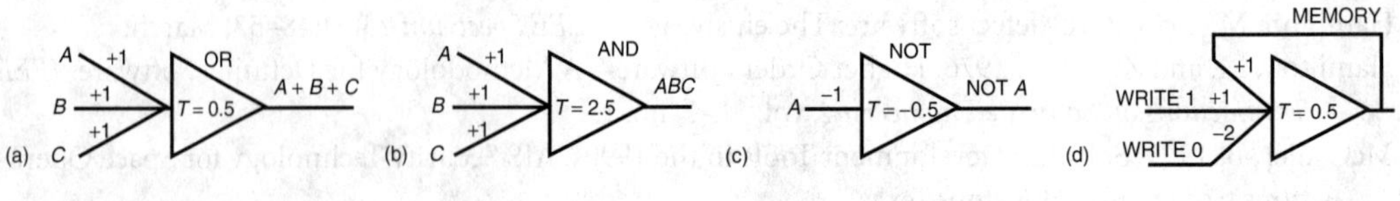

FIGURE 19.13 OR, AND, NOT, and MEMORY operations using networks with McCulloch–Pitts neuron model.

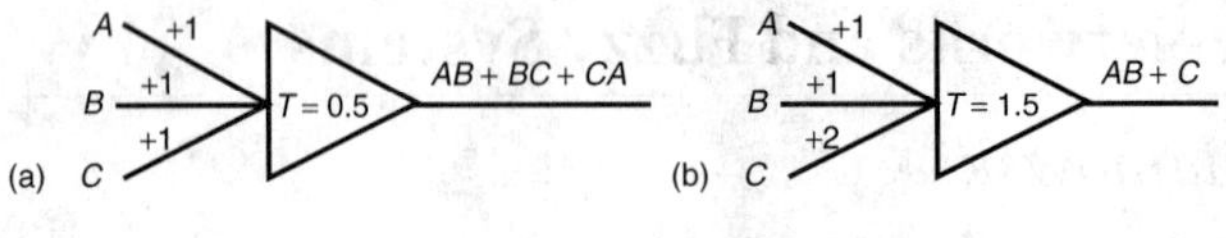

FIGURE 19.14 Other logic functions realized with McCulloch–Pitts neuron model.

Examples of McCulloch-Pitts neurons realizing OR, AND, NOT, and MEMORY operations are shown in Fig. 19.13. Note that the structure of OR and AND gates can be identical. With the same structure, other logic functions can be realized, as Fig. 19.14 shows.

The perceptron model has a similar structure. Its input signals, the weights, and the thresholds could have any positive or negative values. Usually, instead of using variable threshold, one additional constant input with a negative or positive weight can added to each neuron, as Fig. 19.15 shows. In this case, the threshold is always set to be zero and the net value is calculated as

$$net = \sum_{i=1}^{n} w_i x_i + w_{n+1} \tag{19.3}$$

where w_{n+1} has the same value as the required threshold and the opposite sign. Single-layer perceptrons were successfully used to solve many pattern classification problems. The hard threshold activation functions are given by

$$o = f(net) = \frac{\text{sgn}(net) + 1}{2} = \begin{cases} 1 & \text{if} \quad net \geq 0 \\ 0 & \text{if} \quad net < 0 \end{cases} \tag{19.4}$$

for **unipolar** neurons and

$$o = f(net) = \text{sgn}(net) = \begin{cases} 1 & \text{if} \quad net \geq 0 \\ -1 & \text{if} \quad net < 0 \end{cases} \tag{19.5}$$

for **bipolar** neurons. For these types of neurons, most of the known training algorithms are able to adjust weights only in single-layer networks.

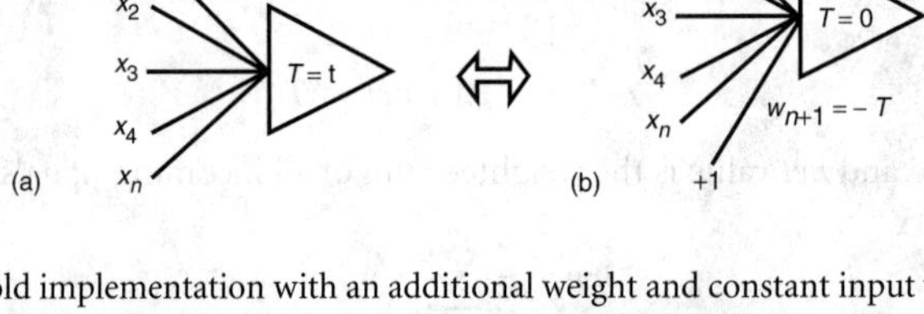

FIGURE 19.15 Threshold implementation with an additional weight and constant input with +1 value: (a) Neuron with threshold T, (b) modified neuron with threshold $T = 0$ and additional weight equal to $-T$.

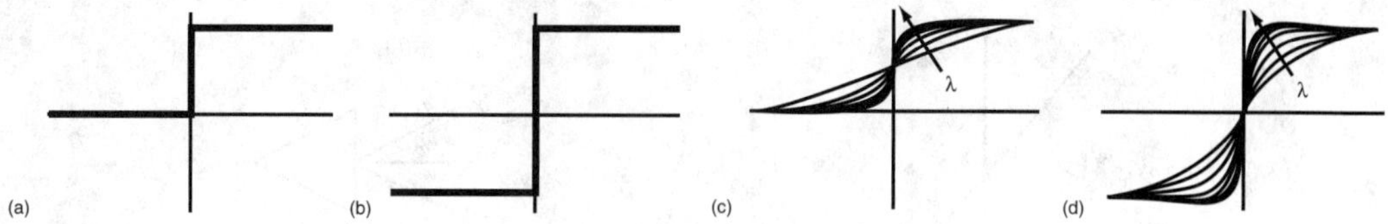

FIGURE 19.16 Typical activation functions: (a) hard threshold unipolar, (b) hard threshold bipolar, (c) continuous unipolar, (d) continuous bipolar.

Multilayer neural networks usually use continuous activation functions, either unipolar

$$o = f(net) = \frac{1}{1 + \exp(-\lambda net)} \tag{19.6}$$

or bipolar

$$o = f(net) = \tanh(0.5\lambda net) = \frac{2}{1 + \exp(-\lambda net)} - 1 \tag{19.7}$$

These continuous activation functions allow for the gradient-based training of multilayer networks. Typical activation functions are shown in Fig. 19.16. In the case when neurons with additional threshold input are used (Fig. 19.15(b)), the λ parameter can be eliminated from Eqs. (19.6) and (19.7) and the steepness of the neuron response can be controlled by the weight scaling only. Therefore, there is no real need to use neurons with variable gains.

Note, that even neuron models with continuous activation functions are far from an actual biological neuron, which operates with frequency modulated pulse trains.

19.3.3 Feedforward Neural Networks

Feedforward neural networks allow only one directional signal flow. Furthermore, most feedforward neural networks are organized in layers. An example of the three-layer feedforward neural network is shown in Fig. 19.17. This network consists of input nodes, two *hidden layers*, and an output layer.

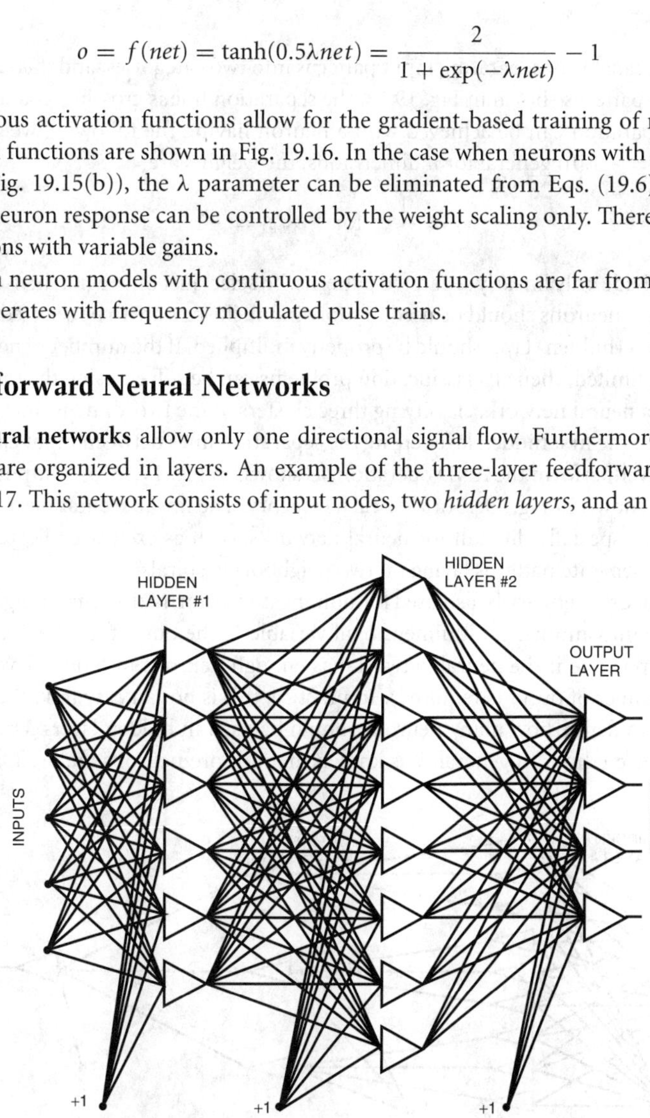

FIGURE 19.17 An example of the three layer feedforward neural network, which is sometimes known also as the backpropagation network.

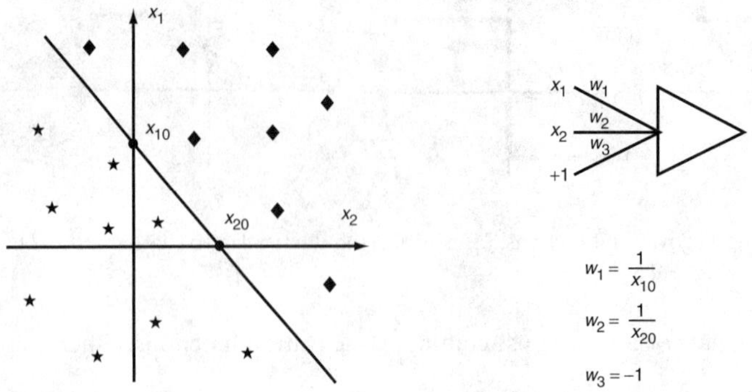

FIGURE 19.18 Illustration of the property of linear separation of patterns in the two-dimensional space by a single neuron.

A single neuron is capable of separating input patterns into two categories, and this separation is linear. For example, for the patterns shown in Fig. 19.18, the separation line is crossing x_1 and x_2 axes at points x_{10} and x_{20}. This separation can be achieved with a neuron having the following weights: $w_1 = 1/x_{10}$, $w_2 = 1/x_{20}$, and $w_3 = -1$. In general for n dimensions, the weights are

$$w_i = \frac{1}{x_{i0}} \quad \text{for} \quad w_{n+1} = -1 \tag{19.8}$$

One neuron can divide only linearly separated patterns. To select just one region in n-dimensional input space, more than $n + 1$ neurons should be used. If more input clusters are to be selected, then the number of neurons in the input (hidden) layer should be properly multiplied. If the number of neurons in the input (hidden) layer is not limited, then all classification problems can be solved using the three-layer network. An example of such a neural network, classifying three clusters in the two-dimensional space, is shown in Fig. 19.19. Neurons in the first hidden layer create the separation lines between input clusters. Neurons in the second hidden layer perform the AND operation, as shown in Fig. 19.13(b). Output neurons perform the OR operation as shown in Fig. 19.13(a), for each category. The linear separation property of neurons makes some problems specially difficult for neural networks, such as exclusive OR, parity computation for several bits, or to separate patterns laying on two neighboring spirals.

The feedforward neural network is also used for nonlinear transformation (mapping) of a multidimensional input variable into another multidimensional variable in the output. In theory, any input-output mapping should be possible if the neural network has enough neurons in hidden layers (size of output layer is set by the number of outputs required). In practice, this is not an easy task. Presently, there is no satisfactory method to define how many neurons should be used in hidden layers. Usually, this is found by the trial-and-error method. In general, it is known that if more neurons are used, more complicated

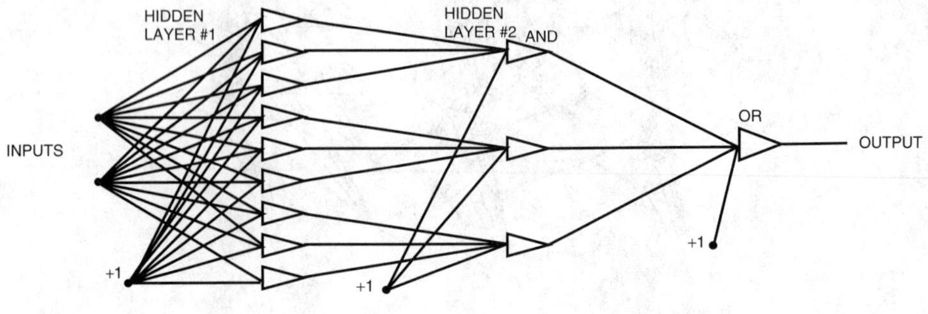

FIGURE 19.19 An example of the three layer neural network with two inputs for classification of three different clusters into one category. This network can be generalized and can be used for solution of all classification problems.

shapes can be mapped. On the other hand, networks with large numbers of neurons lose their ability for generalization, and it is more likely that such networks will also try to map noise supplied to the input.

19.3.4 Learning Algorithms for Neural Networks

Similarly to the biological neurons, the weights in artificial neurons are adjusted during a training procedure. Various learning algorithms were developed, and only a few are suitable for multilayer neuron networks. Some use only local signals in the neurons, others require information from outputs; some require a supervisor who knows what outputs should be for the given patterns, and other unsupervised algorithms need no such information. Common learning rules are described in the following sections.

Hebbian Learning Rule

The Hebb (1949) learning rule is based on the assumption that if two neighbor neurons must be activated and deactivated at the same time, then the weight connecting these neurons should increase. For neurons operating in the opposite phase, the weight between them should decrease. If there is no signal correlation, the weight should remain unchanged. This assumption can be described by the formula

$$\Delta w_{ij} = c x_i o_j \tag{19.9}$$

where

w_{ij} = weight from ith to jth neuron
c = learning constant
x_i = signal on the ith input
o_j = output signal

The training process starts usually with values of all weights set to zero. This learning rule can be used for both soft and hard threshold neurons. Since desired responses of neurons is not used in the learning procedure, this is the **unsupervised learning** rule. The absolute values of the weights are usually proportional to the learning time, which is undesired.

Correlation Learning Rule

The correlation learning rule is based on a similar principle as the Hebbian learning rule. It assumes that weights between simultaneously responding neurons should be largely positive, and weights between neurons with opposite reaction should be largely negative.

Contrary to the Hebbian rule, the correlation rule is the **supervised learning**. Instead of actual response o_j, the desired response d_j is used for the weight change calculation

$$\Delta w_{ij} = c x_i d_j \tag{19.10}$$

This training algorithm usually starts with initialization of weights to zero values.

Instar Learning Rule

If input vectors and weights are normalized, or they have only binary bipolar values (-1 or $+1$), then the *net* value will have the largest positive value when the weights and the input signals are the same. Therefore, weights should be changed only if they are different from the signals

$$\Delta w_i = c(x_i - w_i) \tag{19.11}$$

Note, that the information required for the weight is taken only from the input signals. This is a very local and unsupervised learning algorithm.

Winner Takes All (WTA)

The WTA is a modification of the instar algorithm where weights are modified only for the neuron with the highest *net* value. Weights of remaining neurons are left unchanged. Sometimes this algorithm is modified in such a way that a few neurons with the highest net values are modified at the same time. Although this is an unsupervised algorithm because we do not know what are desired outputs, there is a need for "judge" or "supervisor" to find a winner with a largest net value. The WTA algorithm, developed by Kohonen (1982), is often used for automatic clustering and for extracting statistical properties of input data.

Outstar Learning Rule

In the outstar learning rule, it is required that weights connected to a certain node should be equal to the desired outputs for the neurons connected through those weights

$$\Delta w_{ij} = c(d_j - w_{ij}) \tag{19.12}$$

where d_j is the desired neuron output and c is small learning constant, which further decreases during the learning procedure. This is the supervised training procedure because desired outputs must be known. Both instar and outstar learning rules were developed by Grossberg (1969).

Widrow-Hoff LMS Learning Rule

Widrow and Hoff (1960, 1962) developed a supervised training algorithm which allows training a neuron for the desired response. This rule was derived so that the square of the difference between the net and output value is minimized.

$$Error_j = \sum_{p=1}^{P} (net_{jp} - d_{jp})^2 \tag{19.13}$$

where:
  $Error_j$ = error for jth neuron
  P = number of applied patterns
  d_{jp} = desired output for jth neuron when pth pattern is applied
  net = given by Eq. (19.2).

This rule is also known as the least mean square (LMS) rule. By calculating a derivative of Eq. (19.13) with respect to w_{ij}, a formula for the weight change can be found,

$$\Delta w_{ij} = c x_i \sum_{p=1}^{P} (d_{jp} - net_{jp}) \tag{19.14}$$

Note that weight change Δw_{ij} is a sum of the changes from each of the individual applied patterns. Therefore, it is possible to correct the weight after each individual pattern was applied. This process is known as *incremental updating*; *cumulative updating* is when weights are changed after all patterns have been applied. Incremental updating usually leads to a solution faster, but it is sensitive to the order in which patterns are applied. If the learning constant c is chosen to be small, then both methods give the same result. The LMS rule works well for all types of activation functions. This rule tries to enforce the *net* value to be equal to desired value. Sometimes this is not what the observer is looking for. It is usually not important what the *net* value is, but it is important if the *net* value is positive or negative. For example, a very large *net* value with a proper sign will result in correct output and in large error as defined by Eq. (19.13) and this may be the preferred solution.

Linear Regression

The LMS learning rule requires hundreds or thousands of iterations, using formula (19.14), before it converges to the proper solution. Using the linear regression rule, the same result can be obtained in only one step.

Considering one neuron and using vector notation for a set of the input patterns X applied through weight vector w, the vector of *net* values net is calculated using

$$Xw = net \qquad (19.15)$$

where
   X = rectangular array $(n + 1) \times p$
   n = number of inputs
   p = number of patterns

Note that the size of the input patterns is always augmented by one, and this additional weight is responsible for the threshold (see Fig. 19.3(b)). This method, similar to the LMS rule, assumes a linear activation function, and so the *net* values *net* should be equal to desired output values d

$$Xw = d \qquad (19.16)$$

Usually $p > n + 1$, and the preceding equation can be solved only in the least mean square error sense. Using the vector arithmetic, the solution is given by

$$w = (X^T X)^{-1} X^T d \qquad (19.17)$$

When traditional method is used the set of p equations with $n + 1$ unknowns Eq. (19.16) has to be converted to the set of $n + 1$ equations with $n + 1$ unknowns

$$Yw = z \qquad (19.18)$$

where elements of the Y matrix and the z vector are given by

$$y_{ij} = \sum_{p=1}^{P} x_{ip} x_{jp} \qquad z_i = \sum_{p=1}^{P} x_{ip} d_p \qquad (19.19)$$

Weights are given by Eq. (19.17) or they can be obtained by a solution of Eq. (19.18).

Delta Learning Rule

The LMS method assumes linear activation function $net = o$, and the obtained solution is sometimes far from optimum, as is shown in Fig. 19.20 for a simple two-dimensional case, with four patterns belonging to two categories. In the solution obtained using the LMS algorithm, one pattern is misclassified. If error is defined as

$$Error_j = \sum_{p=1}^{P} (o_{jp} - d_{jp})^2 \qquad (19.20)$$

Then the derivative of the error with respect to the weight w_{ij} is

$$\frac{d\, Error_j}{dw_{ij}} = 2 \sum_{p=1}^{P} (o_{jp} - d_{jp}) \frac{df(net_{jp})}{d\, net_{jp}} x_i \qquad (19.21)$$

since $o = f(net)$ and the *net* is given by Eq. (19.2). Note that this derivative is proportional to the derivative of the activation function $f'(net)$. Thus, this type of approach is possible only for continuous activation

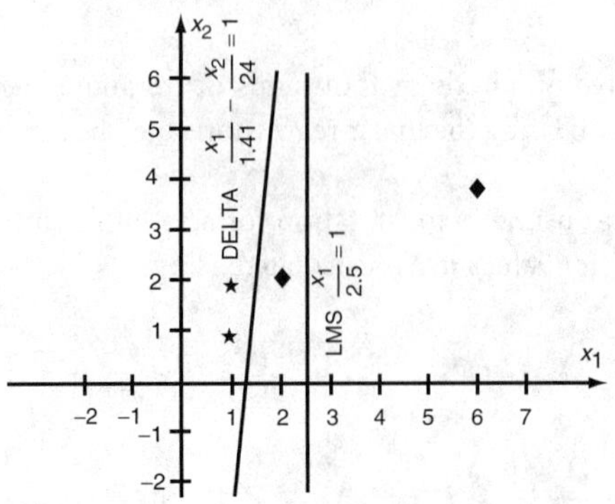

FIGURE 19.20 An example with a comparison of results obtained using LMS and delta training algorithms. Note that LMS is not able to find the proper solution.

functions and this method cannot be used with hard activation functions (19.4) and (19.5). In this respect the LMS method is more general. The derivatives most common continuous activation functions are

$$f' = o(1 - o) \tag{19.22}$$

for the unipolar (Eq. (19.6)) and

$$f' = 0.5(1 - o^2) \tag{19.23}$$

for the bipolar (Eq. (19.7)).

Using the cumulative approach, the neuron weight w_{ij} should be changed with a direction of gradient

$$\Delta w_{ij} = c\, x_i \sum_{p=1}^{P} (d_{jp} - o_{jp}) f'_{jp} \tag{19.24}$$

in case of the incremental training for each applied pattern

$$\Delta w_{ij} = c x_i f'_j (d_j - o_j) \tag{19.25}$$

the weight change should be proportional to input signal x_i, to the difference between desired and actual outputs $d_{jp} - o_{jp}$, and to the derivative of the activation function f'_{jp}. Similar to the LMS rule, weights can be updated in both the incremental and the cumulative methods. In comparison to the LMS rule, the delta rule always leads to a solution close to the optimum. As it is illustrated in Fig. 19.20, when the delta rule is used, all four patterns are classified correctly.

Error Backpropagation Learning

The delta learning rule can be generalized for multilayer networks. Using an approach similiar to the delta rule, the gradient of the global error can be computed with respect to each weight in the network. Interestingly,

$$\Delta w_{ij} = c x_i f'_j E_j \tag{19.26}$$

where
c = learning constant
x_i = signal on the ith neuron input
f'_j = derivative of activation function

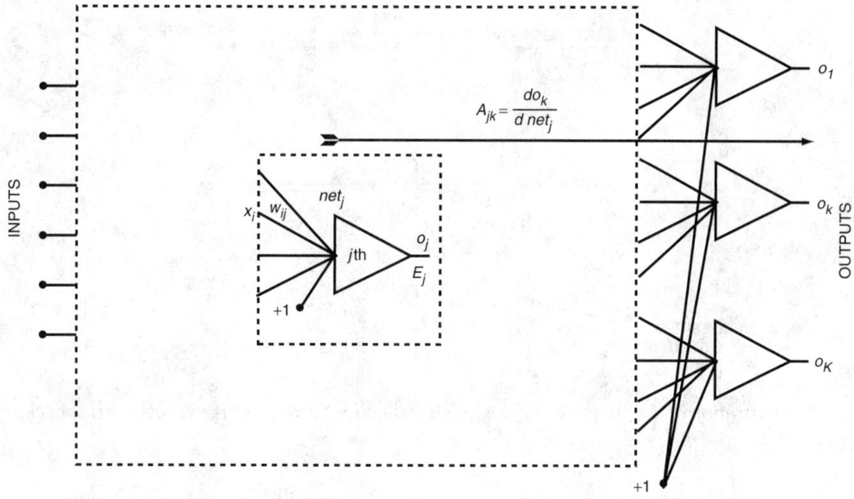

FIGURE 19.21 Illustration of the concept of gain computation in neural networks.

The cumulative error E_j on neuron output is given by

$$E_j = \frac{1}{f'_j} \sum_{k=1}^{K} (o_k - d_k) A_{jk} \tag{19.27}$$

where K is the number of network outputs A_{jk} is the small signal gain from the input of jth neuron to the kth network output, as Fig. 19.21 shows. The calculation of the backpropagating error starts at the output layer and cumulative errors are calculated layer by layer to the input layer. This approach is not practical from the point of view of hardware realization. Instead, it is simpler to find signal gains from the input of the jth neuron to each of the network outputs (Fig. 19.21). In this case, weights are corrected using

$$\Delta w_{ij} = c x_i \sum_{k=1}^{K} (o_k - d_k) A_{jk} \tag{19.28}$$

Note that this formula is general, regardless of if neurons are arranged in layers or not. One way to find gains A_{jk} is to introduce an incremental change on the input of the jth neuron and observe the change in the kth network output. This procedure requires only forward signal propagation, and it is easy to implement in a hardware realization. Another possible way is to calculate gains through each layer and then find the total gains as products of layer gains. This procedure is equally or less computationally intensive than a calculation of cumulative errors in the error backpropagation algorithm.

The backpropagation algorithm has a tendency for oscillation. To smooth the process, the weights increment Δw_{ij} can be modified according to Rumelhart, Hinton, and Wiliams (1986)

$$w_{ij}(n + 1) = w_{ij}(n) + \Delta w_{ij}(n) + \alpha \Delta w_{ij}(n - 1) \tag{19.29}$$

or according to Sejnowski and Rosenberg (1987)

$$w_{ij}(n + 1) = w_{ij}(n) + (1 - \alpha) \Delta w_{ij}(n) + \alpha \Delta w_{ij}(n - 1) \tag{19.30}$$

where α is the momentum term.

The backpropagation algorithm can be significantly sped up, when, after finding components of the gradient, weights are modified along the gradient direction until a minimum is reached. This process can be carried on without the necessity of a computationally intensive gradient calculation at each step. The new gradient components are calculated once a minimum is obtained in the direction of the previous gradient.

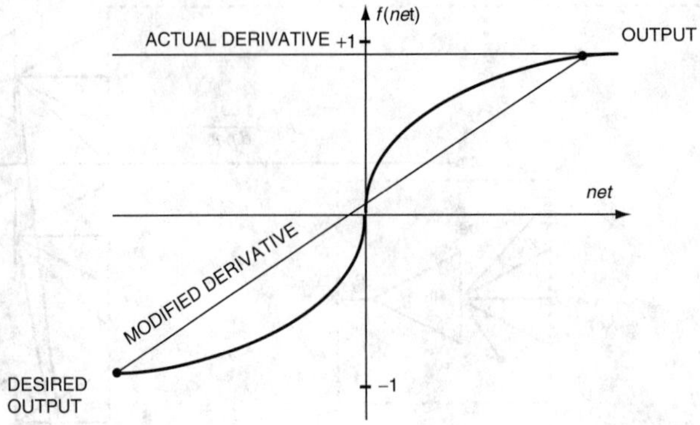

FIGURE 19.22   Illustration of the modified derivative calculation for faster convergency of the error backpropagation algorithm.

This process is only possible for cumulative weight adjustment. One method of finding a minimum along the gradient direction is the *tree step process* of finding error for three points along gradient direction and then, using a parabola approximation, jump directly to the minimum. The fast learning algorithm using the described approach was proposed by Fahlman (1988) and is known as the *quickprop*.

The backpropagation algorithm has many disadvantages, which lead to very slow convergency. One of the most painful is that in the backpropagation algorithm, the learning process almost perishes for neurons responding with the maximally wrong answer. For example, if the value on the neuron output is close to $+1$ and desired output should be close to -1, then the neuron gain $f'(net) \approx 0$ and the error signal cannot backpropagate, and so the learning procedure is not effective. To overcome this difficulty, a modified method for derivative calculation was introduced by Wilamowski and Torvik (1993). The derivative is calculated as the slope of a line connecting the point of the output value with the point of the desired value, as shown in Fig. 19.22.

$$f_{\text{modif}} = \frac{o_{\text{desired}} - o_{\text{actual}}}{net_{\text{desired}} - net_{\text{actual}}} \tag{19.31}$$

Note that for small errors, Eq. (19.31) converges to the derivative of activation function at the point of the output value. With an increase of system dimensionality, the chances for local minima decrease. It is believed that the described phenomenon, rather than a trapping in local minima, is responsible for convergency problems in the error backpropagation algorithm.

19.3.5   Special Feedforward Networks

The multilayer backpropagation network, as shown in Fig. 19.17, is a commonly used feedforward network. This network consists of neurons with the sigmoid type continuous activation function presented in Fig. 19.16(c) and Fig. 19.16(d). In most cases, only one hidden layer is required, and the number of neurons in the hidden layer are chosen to be proportional to the problem complexity. The number of neurons in the hidden layer is usually found by a trial-and-error process. The training process starts with all weights randomized to small values, and the error backpropagation algorithm is used to find a solution. When the learning process does not converge, the training is repeated with a new set of randomly chosen weights. Nguyen and Widrow (1990) proposed an experimental approach for the two-layer network weight initialization. In the second layer, weights are randomly chosen in the range from -0.5 to $+0.5$. In the first layer, initial weights are calculated from

$$w_{ij} = \frac{\beta z_{ij}}{\|z_j\|}; \qquad w_{(n+1)j} = \text{random}(-\beta, +\beta) \tag{19.32}$$

where z_{ij} is the random number from -0.5 to $+0.5$ and the scaling factor β is given by

$$\beta = 0.7 P^{\frac{1}{N}} \tag{19.33}$$

where n is the number of inputs and N is the number of hidden neurons in the first layer. This type of weight initialization usually leads to faster solutions.

For adequate solutions with backpropagation networks, typically many tries are required with different network structures and different initial random weights. It is important that the trained network gains a generalization property. This means that the trained network also should be able to handle correctly patterns that were not used for training. Therefore, in the training procedure, often some data are removed from the training patterns and then these patterns are used for verification. The results with backpropagation networks often depend on luck. This encouraged researchers to develop feedforward networks, which can be more reliable. Some of those networks are described in the following sections.

Functional Link Network

One-layer neural networks are relatively easy to train, but these networks can solve only linearly separated problems. One possible solution for nonlinear problems presented by Nilsson (1965) and elaborated by Pao (1989) using the functional link network is shown in Fig. 19.23. Using nonlinear terms with initially determined functions, the actual number of inputs supplied to the one-layer neural network is increased. In the simplest case, nonlinear elements are higher order terms of input patterns. Note that the functional link network can be treated as a one-layer network, where additional input data are generated off line using nonlinear transformations. The learning procedure for one-layer is easy and fast. Figure 19.24 shows an *XOR* problem solved using functional link networks. Note that when the functional link approach is used, this difficult problem becomes a trivial one. The problem with the functional link network is that proper selection of nonlinear elements is not an easy task. In many practical cases, however, it is not difficult to predict what kind of transformation of input data may linearize the problem, and so the functional link approach can be used.

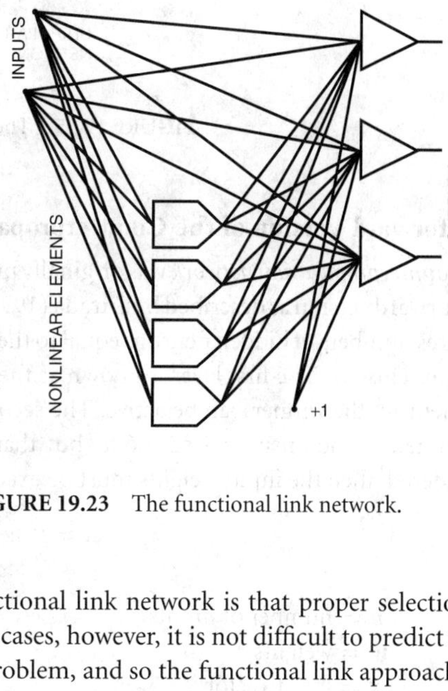

FIGURE 19.23 The functional link network.

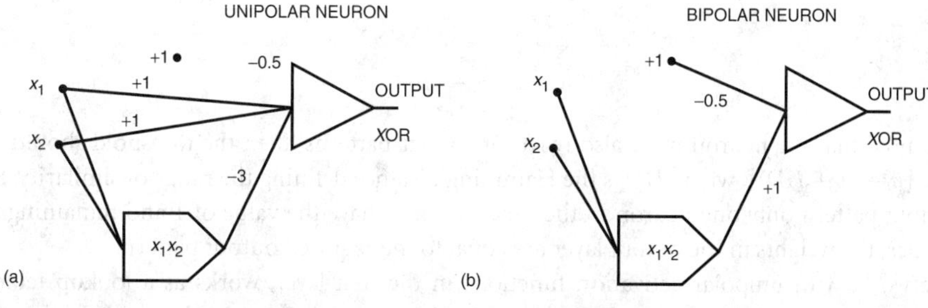

FIGURE 19.24 Functional link networks for solution of the *XOR* problem: (a) using unipolar signals, (b) using bipolar signals.

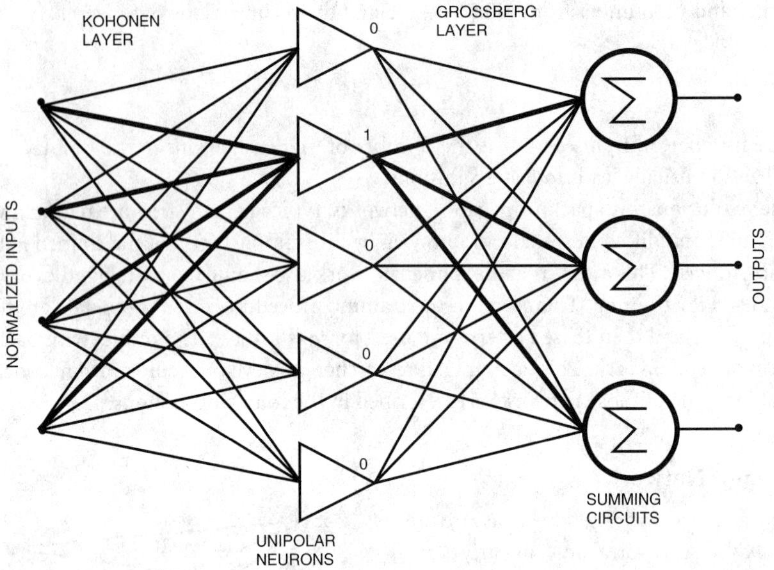

FIGURE 19.25   The counterpropagation network.

Feedforward Version of the Counterpropagation Network

The *counterpropagation network* was originally proposed by Hecht-Nilsen (1987). In this section a modified feedforward version as described by Zurada (1992) is discussed. This network, which is shown in Fig. 19.25, requires numbers of hidden neurons equal to the number of input patterns, or more exactly, to the number of input clusters. The first layer is known as the Kohonen layer with unipolar neurons. In this layer only one neuron, the winner, can be active. The second is the Grossberg outstar layer. The Kohonen layer can be trained in the unsupervised mode, but that need not be the case. When binary input patterns are considered, then the input weights must be exactly equal to the input patterns. In this case,

$$net = x^t w = [n - 2HD(x, w)] \tag{19.34}$$

where

$$n = \text{number of inputs}$$
$$w = \text{weights}$$
$$x = \text{input vector}$$
$$HD(w, x) = Hamming\ distance \text{ between input pattern and weights}$$

For a neuron in the input layer to be reacting just for the stored pattern, the threshold value for this neuron should be

$$w_{(n+1)} = -(n - 1) \tag{19.35}$$

If it is required that the neuron must also react for similar patterns, then the threshold should be set to $w_{n+1} = -[n - (1 + HD)]$, where HD is the Hamming distance defining the range of similarity. Since for a given input pattern only one neuron in the first layer may have the value of 1 and remaining neurons have 0 values, the weights in the output layer are equal to the required output pattern.

The network, with unipolar activation functions in the first layer, works as a lookup table. When the linear activation function (or no activation function at all) is used in the second layer, then the network also can be considered as an analog memory. For the address applied to the input as a binary vector, the stored set of analog values, as weights in the second layer, can be accurately recovered. The feedforward counterpropagation network may also use analog inputs, but in this case all input data should be normalized.

$$w_i = \hat{x}_i = \frac{x_i}{\|x_i\|} \qquad (19.36)$$

The counterpropagation network is very easy to design. The number of neurons in the hidden layer is equal to the number of patterns (clusters). The weights in the input layer are equal to the input patterns, and the weights in the output layer are equal to the output patterns. This simple network can be used for rapid prototyping. The counterpropagation network usually has more hidden neurons than required. However, such an excessive number of hidden neurons are also used in more sophisticated feedforward networks such as the *probabilistic neural network* (PNN) Specht (1990) or the *general regression neural networks* (GRNN) Specht (1992).

WTA Architecture

The winner takes all network was proposed by Kohonen (1988). This is basically a one-layer network used in the unsupervised training algorithm to extract a statistical property of the input data (Fig. 19.26(a)). At the first step, all input data are normalized so that the length of each input vector is the same and, usually, equal to unity (Eq. (19.36)). The activation functions of neurons are unipolar and continuous. The learning process starts with a weight initialization to small random values. During the learning process the weights are changed only for the neuron with the highest value on the output-the winner.

$$\Delta w_w = c(x - w_w) \qquad (19.37)$$

where

w_w = weights of the winning neuron
x = input vector
c = learning constant

Usually, this single-layer network is arranged into a two-dimensional layer shape, as shown in Fig. 19.26(b). The hexagonal shape is usually chosen to secure strong interaction between neurons. Also, the algorithm is modified in such a way that not only the winning neuron but also neighboring neurons are allowed for the weight change. At the same time, the learning constant c in Eq. (19.37) decreases with the distance from the winning neuron. After such a unsupervised training procedure, the Kohonen layer is able to organize data into clusters. Output of the Kohonen layer is then connected to the one- or two-layer feedforward network with the error backpropagation algorithm. This initial data organization in the WTA layer usually leads to rapid training of the following layer or layers.

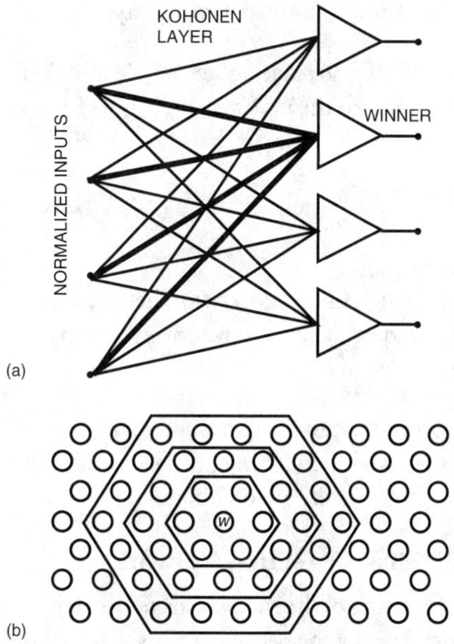

Cascade Correlation Architecture

The cascade correlation architecture was proposed by Fahlman and Lebiere (1990). The process of network building starts with a one-layer neural network and hidden neurons are added as needed. The network architecture is shown in Fig. 19.27. In each training step, a new hidden neuron is added and its weights are adjusted to maximize the magnitude

(a)

(b)

FIGURE 19.26 A winner takes all architecture for cluster extracting in the unsupervised training mode: (a) Network connections, (b) single-layer network arranged into a hexagonal shape.

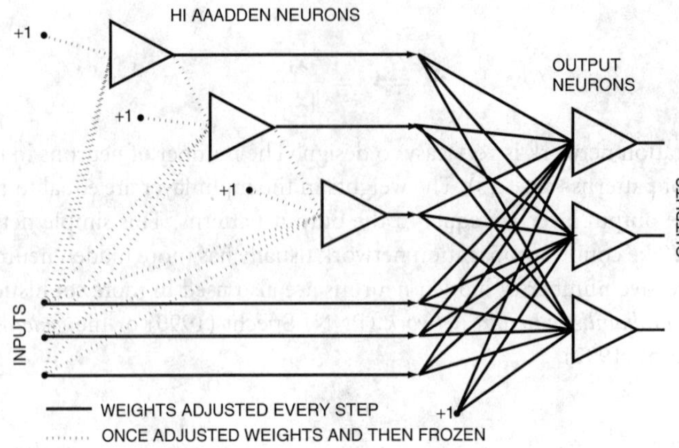

FIGURE 19.27 The cascade correlation architecture.

of the correlation between the new hidden neuron output and the residual error signal on the network output to be eliminated. The correlation parameter S must be maximized.

$$S = \sum_{o=1}^{O} \left| \sum_{p=1}^{P} \left(V_p - \bar{V} \right) \left(E_{po} - \bar{E}_o \right) \right| \tag{19.38}$$

where

O = number of network outputs
P = number of training patterns
V_p = output on the new hidden neuron
E_{po} = error on the network output

$\bar{V}$ and $\bar{E}_o$ are average values of V_p and E_{po}, respectively. By finding the gradient, $\delta S / \delta w_i$, the weight adjustment for the new neuron can be found as

$$\Delta w_i = \sum_{o=1}^{O} \sum_{p=1}^{P} \sigma_o \left(E_{po} - \bar{E}_o \right) f'_p x_{ip} \tag{19.39}$$

where

σ_o = sign of the correlation between the new neuron output value and network output
f'_p = derivative of activation function for pattern p
x_{ip} = input signal

The output neurons are trained using the delta or quickprop algorithms. Each hidden neuron is trained just once and then its weights are frozen. The network learning and building process is completed when satisfactory results are obtained.

Radial Basis Function Networks

The structure of the radial basis network is shown in Fig. 19.28. This type of network usually has only one hidden layer with special neurons. Each of these neurons responds only to the inputs signals close to the stored pattern. The output signal h_i of the ith hidden neuron is computed using formula

$$h_i = \exp\left(-\frac{\|x - s_i\|^2}{2\sigma^2} \right) \tag{19.40}$$

where

x = input vector

s_i = stored pattern representing the center of the i cluster

σ_i = radius of the cluster

Note that the behavior of this "neuron" significantly differs form the biological neuron. In this "neuron," excitation is not a function of the weighted sum of the input signals. Instead, the distance between the input and stored pattern is computed. If this distance is zero then the neuron responds with a maximum output magnitude equal to one. This neuron is capable of recognizing certain patterns and generating output signals that are functions of a similarity. Features of this neuron are much more powerful than a neuron used in the backpropagation networks. As a consequence, a network made of such neurons is also more powerful.

If the input signal is the same as a pattern stored in a neuron, then this neuron responds with 1 and remaining neurons have 0 on the output, as is illustrated in Fig. 19.28. Thus, output signals are exactly equal to the weights coming out from the active neuron. This way, if the number of neurons in the hidden layer is large, then any input/output mapping can be obtained. Unfortunately, it may also happen that for some patterns several neurons in the first layer will respond with a nonzero signal. For a proper approximation, the sum of all signals from the hidden layer should be equal to one. To meet this requirement, output signals are often normalized as shown in Fig. 19.28.

The radial-based networks can be designed or trained. Training is usually carried out in two steps. In the first step, the hidden layer is usually trained in the unsupervised mode by choosing the best patterns for cluster representation. An approach, similar to that used in the WTA architecture can be used. Also in this step, radii σ_i must be found for a proper overlapping of clusters.

The second step of training is the error backpropagation algorithm carried on only for the output layer. Since this is a supervised algorithm for one layer only, the training is very rapid, 100–1000 times faster than in the backpropagation multilayer network. This makes the radial basis-function network very attractive. Also, this network can be easily modeled using computers, however, its hardware implementation would be difficult.

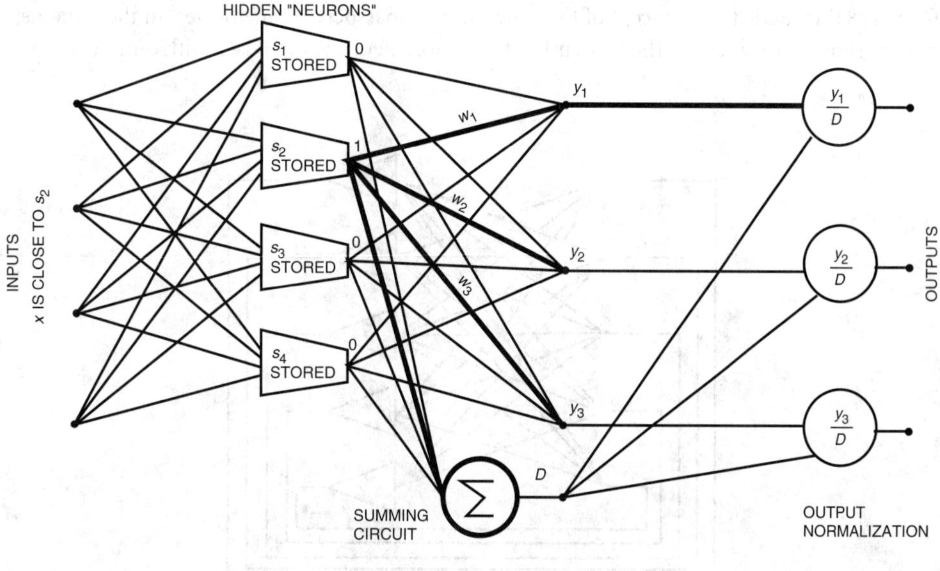

FIGURE 19.28 A typical structure of the radial basis function network.

19.3.6   Recurrent Neural Networks

In contrast to feedforward neural networks, with **recurrent networks** neuron outputs can be connected with their inputs. Thus, signals in the network can continuously circulate. Until recently, only a limited number of recurrent neural networks were described.

Hopfield Network

The single-layer recurrent network was analyzed by Hopfield (1982). This network, shown in Fig. 19.29, has unipolar hard threshold neurons with outputs equal to 0 or 1. Weights are given by a symmetrical square matrix W with zero elements ($w_{ij} = 0$ for $i = j$) on the main diagonal. The stability of the system is usually analyzed by means of the *energy function*

$$E = -\frac{1}{2} \sum_{i=1}^{N} \sum_{j=1}^{N} W_{ij} v_i v_j \tag{19.41}$$

It has been proved that during signal circulation the energy E of the network decreases and the system converges to the stable points. This is especially true when the values of system outputs are updated in the asynchronous mode. This means that at a given cycle, only one random output can be changed to the required values. Hopfield also proved that those stable points which the system converges can be programmed by adjusting the weights using a modified Hebbian rule,

$$\Delta w_{ij} = \Delta w_{ji} = (2v_i - 1)(2v_j - 1)c \tag{19.42}$$

Such memory has limited storage capacity. Based on experiments, Hopfield estimated that the maximum number of stored patterns is $0.15N$, where N is the number of neurons.

Later the concept of energy function was extended by Hopfield (1984) to one-layer recurrent networks having neurons with continuous activation functions. These types of networks were used to solve many optimization and linear programming problems.

Autoassociative Memory

Hopfield (1984) extended the concept of his network to autoassociative memories. In the same network structure as shown in Fig. 19.29, the bipolar hard-threshold neurons were used with outputs equal to -1

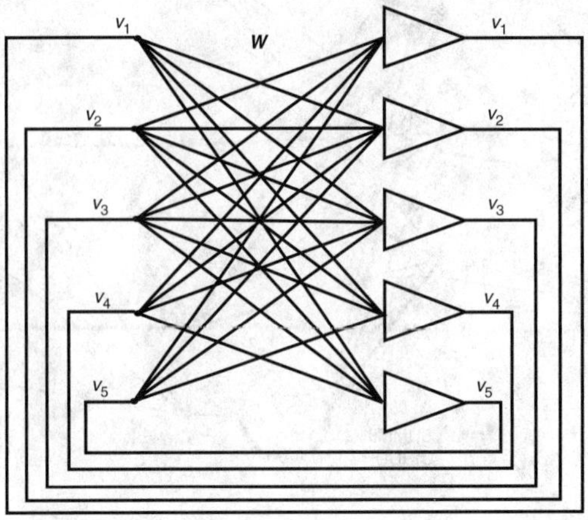

FIGURE 19.29   A Hopfield network or autoassociative memory.

or $+1$. In this network, pattern $\boldsymbol{s}_m$ are stored into the weight matrix $\boldsymbol{W}$ using the autocorrelation algorithm

$$W = \sum_{m=1}^{M} \boldsymbol{s}_m \boldsymbol{s}_m^T - M\boldsymbol{I} \qquad (19.43)$$

where M is the number of stored pattern and $\boldsymbol{I}$ is the unity matrix. Note that $\boldsymbol{W}$ is the square symmetrical matrix with elements on the main diagonal equal to zero (w_{ji} for $i = j$). Using a modified formula (19.42), new patterns can be added or subtracted from memory. When such memory is exposed to a binary bipolar pattern by enforcing the initial network states, after signal circulation the network will converge to the closest (most similar) stored pattern or to its complement. This stable point will be at the closest minimum of the energy

$$E(\boldsymbol{v}) = -\frac{1}{2}\boldsymbol{v}^T \boldsymbol{W} \boldsymbol{v} \qquad (19.44)$$

Like the Hopfield network, the autoassociative memory has limited storage capacity, which is estimated to be about $M_{\max} = 0.15N$. When the number of stored patterns is large and close to the memory capacity, the network has a tendency to converge to spurious states, which were not stored. These spurious states are additional minima of the energy function.

Bidirectional Associative Memories (BAM)

The concept of the autoassociative memory was extended to bidirectional associative memories (BAM) by Kosko (1987, 1988). This memory, shown in Fig. 19.30, is able to associate pairs of the patterns $\boldsymbol{a}$ and $\boldsymbol{b}$. This is the two-layer network with the output of the second layer connected directly to the input of the first layer. The weight matrix of the second layer is $\boldsymbol{W}^T$ and $\boldsymbol{W}$ for the first layer. The rectangular weight matrix $\boldsymbol{W}$ is obtained as a sum of the cross-correlation matrixes

$$W = \sum_{m=1}^{M} \boldsymbol{a}_m \boldsymbol{b}_m \qquad (19.45)$$

where M is the number of stored pairs, and $\boldsymbol{a}_m$ and $\boldsymbol{b}_m$ are the stored vector pairs. If the nodes $\boldsymbol{a}$ or $\boldsymbol{b}$ are initialized with a vector similar to the stored one, then after signal circulations, both stored patterns $\boldsymbol{a}_m$ and $\boldsymbol{b}_m$ should be recovered. The BAM has limited memory capacity and memory corruption problems similar to the autoassociative memory. The BAM concept can be extended for association of three or more vectors.

FIGURE 19.30 An example of the bi-directional autoassociative memory: (a) drawn as a two-layer network with circulating signals, (b) drawn as two-layer network with bi-directional signal flow.

19.3.7   Fuzzy Systems

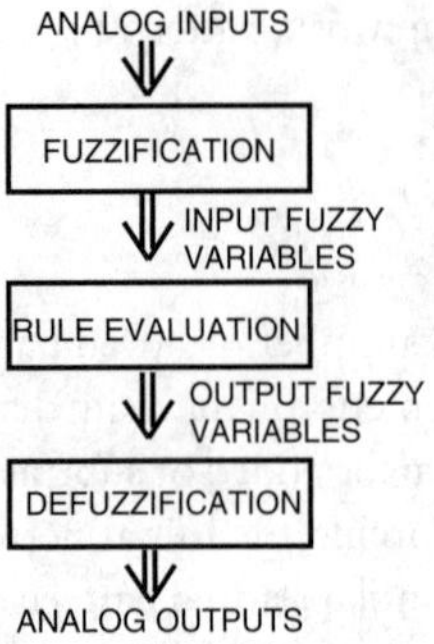

The main applications of neural networks are related to the nonlinear mapping of n-dimensional input variables into m-dimensional output variables. Such a function is often required in control systems, where for specific measured variables certain control variables must be generated. Another approach for nonlinear mapping of one set of variables into another set of variables is the *fuzzy controller*. The principle of operation of the fuzzy controller significantly differs from neural networks. The block diagram of a fuzzy controller is shown in Fig. 19.31. In the first step, analog inputs are converted into a set of fuzzy variables. In this step, for each analog

FIGURE 19.31   The block diagram of the fuzzy controller.

input, 3–9 fuzzy variables typically are generated. Each fuzzy variable has an analog value between zero and one. In the next step, a fuzzy logic is applied to the input fuzzy variables and a resulting set of output variables is generated. In the last step, known as *defuzzification*, from a set of output fuzzy variables, one or more output analog variables are generated, which are used as control variables.

Fuzzification

The purpose of fuzzification is to convert an analog variable input into a set of fuzzy variables. For higher accuracy, more fuzzy variables will be chosen. To illustrate the fuzzification process, consider that the input variable is the temperature and is coded into five fuzzy variables: cold, cool, normal, warm, and hot. Each fuzzy variable should obtain a value between zero and one, which describes a *degree of association* of the analog input (temperature) within the given fuzzy variable. Sometimes, instead of the term degree of association, the term *degree of membership* is used. The process of fuzzification is illustrated in Fig. 19.32. Using Fig. 19.32 we can find the degree of association of each fuzzy variable with the given temperature. For example, for a temperature of 57°F, the following set of fuzzy variables is obtained: (0, 0.5, 0.2, 0, 0), and for $T = 80°F$ it is (0, 0, 0.25, 0.7, 0). Usually only one or two fuzzy variables have a value other than zero. In the example, trapezoidal functions are used for calculation of the degree of association. Various different functions such as triangular or Gaussian, can also be used, as long as the computed value is in the range from zero to one. Each membership function is described by only three or four parameters, which have to be stored in memory.

For proper design of the fuzzification stage, certain practical rules should be used:

- Each point of the input analog variable should belong to at least one and no more than two membership functions.
- For overlapping functions, the sum of two membership functions must not be larger than one. This also means that overlaps must not cross the points of maximum values (ones).
- For higher accuracy, more membership functions should be used. However, very dense functions lead to frequent system reaction and sometimes to system instability.

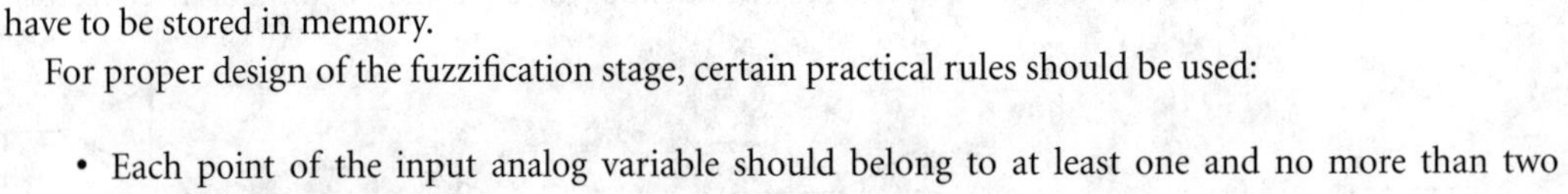

FIGURE 19.32   Fuzzification process: (a) typical membership functions for the fuzzification and the defuzzification processes, (b) example of converting a temperature into fuzzy variables.

Rule Evaluation

Contrary to Boolean logic where variables can have only binary states, in fuzzy logic all variables may have any values between zero and one. The fuzzy logic consists of the same basic $\wedge$ - AND, $\vee$-OR, and NOT operators

$$A \wedge B \wedge C \implies \min\{A, B, C\} \text{—smallest value of } A \text{ or } B \text{ or } C$$
$$A \vee B \vee C \implies \max\{A, B, C\} \text{—largest value of } A \text{ or } B \text{ or } C$$
$$\bar{A} \implies 1 1 - A \quad \text{—one minus value of } A$$

For example $0.1 \wedge 0.7 \wedge 0.3 = 0.1, 0.1 \vee 0.7 \vee 0.3 = 0.7$, and $\bar{0.3} = 0.7$. These rules are also known as *Zadeh* AND, OR, and NOT operators (Zadeh, 1965). Note that these rules are true also for classical binary logic.

Fuzzy rules are specified in the *fuzzy table* as it is shown for a given system. Consider a simple system with two analog input variables x and y, and one output variable z. The goal is to design a fuzzy system generating z as $f(x, y)$. After fuzzification, the analog variable x is represented by five fuzzy variables: x_1, x_2, x_3, x_4, x_5 and an analog variable y is represented by three fuzzy variables: y_1, y_2, y_3. Assume that an analog output variable is represented by four fuzzy variables: z_1, z_2, z_3, z_4. The key issue of the design process is to set proper output fuzzy variables z_k for all combinations of input fuzzy variables, as is shown in the table in Fig. 19.33. The designer has to specify many rules such as if inputs are represented by fuzzy variables x_i and y_j, then the output should be represented by fuzzy variable z_k. Once the fuzzy table is specified, the fuzzy logic computation proceeds in two steps. First each field of the fuzzy table is filled with intermediate fuzzy variables t_{ij}, obtained from AND operator $t_{ij} = \min\{x_i, y_j\}$, as shown in Fig. 19.33(b). This step is independent of the required rules for a given system. In the second step, the OR (max) operator is used to compute each output fuzzy variable z_k. In the given example in Fig. 19.33, $z_1 = \max\{t_{11}, t_{12}, t_{21}, t_{41}, t_{51}\}$, $z_2 = \max\{t_{13}, t_{31}, t_{42}, t_{52}\}$, $z_3 = \max\{t_{22}, t_{23}, t_{43}\}$, $z_4 = \max\{t_{32}, t_{34}, t_{53}\}$. Note that the formulas depend on the specifications given in the fuzzy table shown in Fig. 19.33(a).

	y_1	y_2	y_3
x_1	z_1	z_1	z_2
x_2	z_1	z_3	z_3
x_3	z_1	z_3	z_4
x_4	z_2	z_3	z_4
x_5	z_2	z_3	z_4

(a)

	y_1	y_2	y_3
x_1	t_{11}	t_{12}	t_{13}
x_2	t_{21}	t_{22}	t_{23}
x_3	t_{31}	t_{32}	t_{33}
x_4	t_{41}	t_{42}	t_{43}
x_5	t_{51}	t_{52}	t_{53}

(b)

FIGURE 19.33 Fuzzy tables: (a) table with fuzzy rules, (b) table with the intermediate variables t_{ij}.

Defuzzification

As a result of fuzzy rule evaluation, each analog output variable is represented by several fuzzy variables. The purpose of defuzzification is to obtain analog outputs. This can be done by using a membership function similar to that shown in Fig. 19.32. In the first step, fuzzy variables obtained from rule evaluations are used to modify the membership function employing the formula

$$\mu_k^*(z) = \min\{\mu_k(z), z_k\} \tag{19.46}$$

For example, if the output fuzzy variables are: 0, 0.2, 0.7, 0.0, then the modified membership functions have shapes shown by the thick line in Fig. 19.34. The analog value of the z variable is found as a *center of gravity* of modified membership functions $\mu_k^*(z)$,

$$z_{\text{analog}} = \frac{\left(\sum_{k=1}^{n} \int_{-\infty}^{+\infty} \mu_k^*(z) z \, dz \right)}{\left(\sum_{k=1}^{n} \int_{-\infty}^{+\infty} \mu_k^*(z) \, dz \right)} \tag{19.47}$$

FIGURE 19.34 Illustration of the defuzzification process.

In the case where shapes of the output membership functions $\mu_k(z)$ are the same, the equation can be simplified to

$$z_{\text{analog}} = \frac{\left(\sum_{k=1}^{n} z_k z c_k \right)}{\left(\sum_{k=1}^{n} z_k \right)} \tag{19.48}$$

where

n = number of membership functions of z_{analog} output variable

z_k = fuzzy output variables obtained from rule evaluation

zc_k = analog values corresponding to the center of kth membership function.

Equation (19.47) is usually too complicated to be used in a simple microcontroller based system; therefore, in practical cases, Eq. (19.48) is used more frequently.

19.3.8 Design Example

Consider the design of a simple fuzzy controller for a sprinkler system. The sprinkling time is a function of humidity and temperature. Four membership functions are used for the temperature, three for humidity, and three for the sprinkle time, as shown in Fig. 19.35. Using intuition, the fuzzy table can be developed, as shown in Fig. 19.36(a).

Assume a temperature of 60°F and 70% humidity. Using the membership functions for temperature and humidity the following fuzzy variables can be obtained for the temperature: (0, 0.2, 0.5, 0), and for

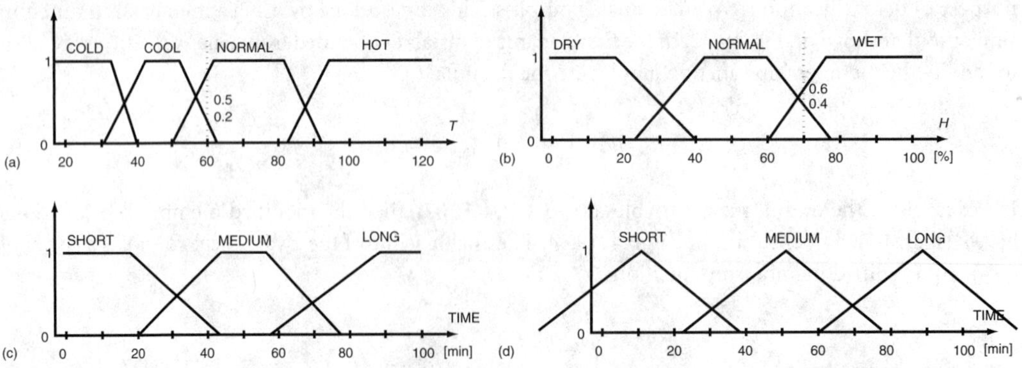

FIGURE 19.35 Membership functions for the presented example: (a) and (b) are membership functions for input variables, (c) and (d) are two possible membership functions for the output variable.

the humidity: (0, 0.4, 0.6). Using the min operator, the fuzzy table can be now filled with temporary fuzzy variables, as shown in Fig. 19.36(b). Note that only four fields have nonzero values. Using fuzzy rules, as shown in Fig. 19.36(a), the max operator can be applied in order to obtain fuzzy output variables: short $\rightarrow o_1 = \max\{0, 0, 0.2, 0.5, 0\} = 0.5$, medium $\rightarrow o_2 = \max\{0, 0, 0.2, 0.4, 0\} = 0.4$, long $\rightarrow o_3 = \max\{0, 0\} = 0$. Using Eq. (19.47) and Fig. 19.35(c) a sprinkle time of 28 min is determined. When the simplified approach is used with Eq. (19.46) and Fig. 19.35(d), then sprinkle time is 27 min.

(a)

	DRY	NORMAL	WET
COLD	M	S	S
COOL	M	M	S
WARM	L	M	S
HOT	L	M	S

(b)

		DRY	NORMAL	WET
		0	0.4	0.6
COLD	0	0	0	0
COOL	0.2	0	0.2	0.2
WARM	0.5	0	0.4	0.5
HOT	0	0	0	0

FIGURE 19.36 Fuzzy tables: (a) fuzzy rules for the design example, (b) fuzzy temporary variables for the design example.

19.3.9 Genetic Algorithms

The success of the artificial neural networks encouraged researchers to search for other patterns in nature to follow. The power of genetics through evolution was able to create such sophisticated machines as the human being. Genetic algorithms follow the evolution process in nature to find better solutions to some complicated problems. The foundations of genetic algorithms are given by Holland (1975) and Goldberg (1989). After initialization, the steps *selection, reproduction with a crossover*, and *mutation* are repeated for each generation. During this procedure, certain strings of symbols, known as chromosomes, evaluate toward a better solution. The genetic algorithm method begins with coding and an initialization. All significant steps of the genetic algorithm will be explained using a simple example of finding a maximum of the function $(\sin^2 x - 0.5x)^2$ (Fig. 19.37) with the range of x from 0 to 1.6. Note that in this range, the function has a global maximum at $x = 1.309$, and a local maximum at $x = 0.262$.

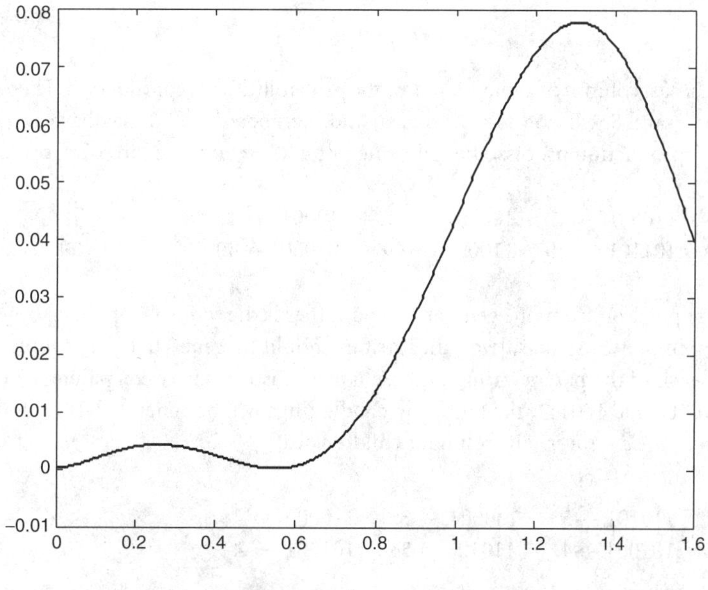

FIGURE 19.37 Function $(\sin^2 x - 0.5x^2)$ for which the minimum must be sought.

TABLE 19.3   Initial Population

String number	String	Decimal value	Variable value	Function value	Fraction of total
1	101101	45	1.125	0.0633	0.2465
2	101000	40	1.000	0.0433	0.1686
3	010100	20	0.500	0.0004	0.0016
4	100101	37	0.925	0.0307	0.1197
5	001010	10	0.250	0.0041	0.0158
6	110001	49	1.225	0.0743	0.2895
7	100111	39	0.975	0.0390	0.1521
8	000100	40	0.100	0.0016	0.0062
Total				0.2568	1.0000

Coding and Initialization

At first, the variable x has to be represented as a string of symbols. With longer strings, the process usually converges faster, so the fewer the symbols for one string field that are used, the better. Although this string may be the sequence of any symbols, the binary symbols 0 and 1 are usually used. In our example, six bit binary numbers are used for coding, having a decimal value of $40x$. The process starts with a random generation of the initial population given in Table 19.3.

Selection and Reproduction

Selection of the best members of the population is an important step in the genetic algorithm. Many different approaches can be used to rank individuals. In this example the ranking function is given. Highest rank has member number 6, and lowest rank has member number 3. Members with higher rank should have higher chances to reproduce. The probability of reproduction for each member can be obtained as a fraction of the sum of all objective function values. This fraction is shown in the last column of Table 19.3. Note that to use this approach, our objective function should always be positive. If it is not, the proper normalization should be introduced at first.

Reproduction

The numbers in the last column of Table 19.3 show the probabilities of reproduction. Therefore, most likely members number 3 and 8 will not be reproduced, and members 1 and 6 may have two or more copies. Using a random reproduction process, the following population, arranged in pairs, could be generated

$101101 \rightarrow 45$	$110001 \rightarrow 49$	$100101 \rightarrow 37$	$110001 \rightarrow 49$
$100111 \rightarrow 39$	$101101 \rightarrow 45$	$110001 \rightarrow 49$	$101000 \rightarrow 40$

If the size of the population from one generation to another is the same, two parents should generate two children. By combining two strings, two other strings should be generated. The simplest way to do this is to split in half each of the parent strings and exchange substrings between parents. For example, from parent strings 010100 and 100111, the following child strings will be generated 010111 and 100100. This process is known as the *crossover*. The resultant children are

$101111 \rightarrow 47$	$110101 \rightarrow 53$	$100001 \rightarrow 33$	$110000 \rightarrow 48$
$100101 \rightarrow 37$	$101001 \rightarrow 41$	$110101 \rightarrow 53$	$101001 \rightarrow 41$

In general, the string need not be split in half. It is usually enough if only selected bits are exchanged between parents. It is only important that bit positions are not changed.

TABLE 19.4 Population of Second Generation

String number	String	Decimal value	Variable value	Function value	Fraction of total
1	010111	47	1.175	0.0696	0.1587
2	100100	37	0.925	0.0307	0.0701
3	110101	53	1.325	0.0774	0.1766
4	010001	41	1.025	0.0475	0.1084
5	100001	33	0.825	0.0161	0.0368
6	110101	53	1.325	0.0774	0.1766
7	110000	48	1.200	0.0722	0.1646
8	101001	41	1.025	0.0475	0.1084
Total				0.4387	1.0000

Mutation

In the evolutionary process, reproduction is enhanced with mutation. In addition to the properties inherited from parents, offspring acquire some new random properties. This process is known as mutation. In most cases mutation generates low-ranked children, which are eliminated in the reproduction process. Sometimes, however, the mutation may introduce a better individual with a new property. This prevents the process of reproduction from degeneration. In genetic algorithms, mutation usually plays a secondary role. For very high-levels of mutation, the process is similar to random pattern generation, and such a searching algorithm is very inefficient. The mutation rate is usually assumed to be at a level well below 1%. In this example, mutation is equivalent to the random bit change of a given pattern. In this simple case, with short strings and a small population, and with a typical mutation rate of 0.1%, the patterns remain practically unchanged by the mutation process. The second generation for this example is shown in Table 19.4.

Note that two identical highest ranking members of the second generation are very close to the solution $x = 1.309$. The randomly chosen parents for the third generation are

$$010111 \rightarrow 47 \quad 110101 \rightarrow 53 \quad 110000 \rightarrow 48 \quad 101001 \rightarrow 41$$
$$110101 \rightarrow 53 \quad 110000 \rightarrow 48 \quad 101001 \rightarrow 41 \quad 110101 \rightarrow 53$$

which produces the following children

$$010101 \rightarrow 21 \quad 110000 \rightarrow 48 \quad 110001 \rightarrow 49 \quad 101101 \rightarrow 45$$
$$110111 \rightarrow 55 \quad 110101 \rightarrow 53 \quad 101000 \rightarrow 40 \quad 110001 \rightarrow 49$$

The best result in the third population is the same as in the second one. By careful inspection of all strings from the second or third generation, it may be concluded that using crossover, where strings are always split in half, the best solution $110100 \rightarrow 52$ will never be reached, regardless of how many generations are created. This is because none of the population in the second generation has a substring ending with 100. For such crossover, a better result can be only obtained due to the mutation process, which may require many generations. Better results in the future generation also can be obtained when strings are split in random places. Another possible solution is that only randomly chosen bits are exchanged between parents.

The genetic algorithm almost always leads to a good solution, but sometimes many generations are required. This solution is usually close to global maximum, but not the best. In the case of a smooth function the gradinet methods are converging much faster and to a better solution. GA are much slower, but more robust.

Defining Terms

Backpropagation: Training technique for multilayer neural networks.
Bipolar neuron: Neuron with output signal between -1 and $+1$.

Feedforward network: Network without feedback.
Perceptron: Network with hard threshold neurons.
Recurrent network: Network with feedback.
Supervised learning: Learning procedure when desired outputs are known.
Unipolar neuron: Neuron with output signal between 0 and +1.
Unsupervised learning: Learning procedure when desired outputs are unknown.

References

Fahlman, S.E. 1988. Faster-learning variations on backpropagation: an empirical study. *Proceedings of the Connectionist Models Summer School*, eds. Touretzky, D. Hinton, G., and Sejnowski, T. Morgan Kaufmann, San Mateo, CA.

Fahlman, S.E. and Lebiere, C. 1990. The cascade correlation learning architecture. *Adv. Ner. Inf. Proc. Syst.*, 2, D.S. Touretzky, ed. pp. 524–532. Morgan, Kaufmann, Los Altos, CA.

Goldberg, D.E. 1989. *Genetic Algorithm in Search, Optimization and Machine Learning*. Addison-Wesley, Reading, MA.

Grossberg, S. 1969. Embedding fields: a theory of learning with physiological implications. *Journal of Mathematical Psychology* 6:209–239.

Hebb, D.O. 1949. *The Organization of Behivior, a Neuropsychological Theory*. John Wiley, New York.

Hecht-Nielsen, R. 1987. Counterpropagation networks. *Appl. Opt.* 26(23):4979–4984.

Hecht-Nielsen, R. 1988. Applications of counterpropagation networks. *Neural Networks* 1:131–139.

Holland, J.H. 1975. *Adaptation in Natural and Artificial Systems*. University. of Michigan Press, Ann Arbor, MI.

Hopfield, J.J. 1982. Neural networks and physical systems with emergent collective computation abilities. *Proceedings of the National Academy of Science* 79:2554–2558.

Hopfield, J.J. 1984. Neurons with graded response have collective computational properties like those of two-state neurons. *Proceedings of the National Academy of Science* 81:3088–3092.

Kohonen, T. 1988. The neural phonetic typerater. *IEEE Computer* 27(3):11–22.

Kohonen, T. 1990. The self-organized map. *Proc. IEEE* 78(9):1464–1480.

Kosko, B. 1987. Adaptive bidirectional associative memories. *App. Opt.* 26:4947–4959.

Kosko, B. 1988. Bidirectional associative memories. *IEEE Trans. Sys. Man, Cyb.* 18:49–60.

McCulloch, W.S. and Pitts, W.H. 1943. A logical calculus of the ideas imminent in nervous activity. *Bull. Math. Biophy.* 5:115–133.

Minsky, M. and Papert, S. 1969. *Perceptrons*. MIT Press, Cambridge, MA.

Nilsson, N.J. 1965. *Learning Machines: Foundations of Trainable Pattern Classifiers*. McGraw-Hill, New York.

Nguyen, D. and Widrow, B. 1990. Improving the learning speed of 2-layer neural networks, by choosing initial values of the adaptive weights. *Proceedings of the International Joint Conference on Neural Networks* (San Diego), CA, June.

Pao, Y.H. 1989. *Adaptive Pattern Recognition and Neural Networks*. Addison-Wesley, Reading, MA.

Rosenblatt, F. 1958. The perceptron: a probabilistic model for information storage and organization in the brain. *Psych. Rev.* 65:386–408.

Rumelhart, D.E., Hinton, G.E., and Williams, R.J. 1986. Learning internal representation by error propagation. *Parallel Distributed Processing*. Vol. 1, pp. 318–362. MIT Press, Cambrige, MA.

Sejnowski, T.J. and Rosenberg, C.R. 1987. Parallel networks that learn to pronounce English text. *Complex Systems* 1:145–168.

Specht, D.F. 1990. Probalistic neural networks. *Neural Networks* 3:109–118.

Specht, D.F. 1992. General regression neural network. *IEEE Trans. Neural Networks* 2:568–576.

Wasserman, P.D. 1989. *Neural Computing Theory and Practice*. Van Nostrand Reinhold, New York.

Werbos, P. 1974. Beyond regression: new tools for prediction and analysis in behavioral sciences. Ph.D. diss., Harvard Universtiy.

Widrow, B. and Hoff, M.E., 1960. Adaptive switching circuits. 1960 IRE Western Electric Show and Convention Record, Part 4 (Aug. 23):96–104.

Widrow, B. 1962. Generalization and information storage in networks of adaline Neurons. In *Self-organizing Systems,* Jovitz, M.C., Jacobi, G.T. and Goldstein, G. eds. pp. 435–461. Sparten Books, Washington, D.C.

Wilamowski, M. and Torvik, L. 1993. Modification of gradient computation in the backpropagation algorithm. *ANNIE'93—Artificial Neural Networks in Engineering.* November 14–17, 1993, St. Louis, Missou.; also in Dagli, C.H. ed. 1993. *Intelligent Engineering Systems Through Artificial Neural Networks* Vol. 3, pp. 175–180. ASME Press, New York.

Zadeh, L.A. 1965. Fuzzy sets. *Information and Control* 8:338–353.

Zurada, J. M. 1992. *Introduction to Artificial Neural Systems.* West Publishing Company, St. Paul. MN.

19.4   Machine Vision

David A. Kosiba and Rangachar Kasturi

19.4.1   Introduction

Machine vision, also known as computer vision, is the scientific discipline whereby explicit, meaningful descriptions of physical objects from the world around us are constructed from their images. Machine vision produces measurements or abstractions from geometrical properties and comprises techniques for estimating features in images, relating feature measurements to the geometry of objects in space, and interpreting this geometric information. This overall task is generally termed *image understanding*.

The goal of a machine vision system is to create a model of the real world from images or sequences of images. Since images are two-dimensional projections of the three-dimensional world, the information is not directly available and must be recovered. This recovery requires the inversion of a many-to-one mapping. To reclaim this information, however, knowledge about the objects in the scene and projection geometry is required.

At every stage in a machine vision system decisions must be made requiring knowledge of the application or goal. Emphasis in machine vision systems is on maximizing the automatic operation at each stage, and these systems should use knowledge to accomplish this. The knowledge used by the system includes models of features, image formation, objects, and relationships among objects. Without explicit use of knowledge, machine vision systems can be designed to work only in a very constrained environment for limited applications. To provide more flexibility and robustness, knowledge is represented explicitly and is directly used by the system. Knowledge is also used by the designers of machine vision systems in many implicit as well as explicit forms. In fact, the efficacy and efficiency of a system is usually governed by the quality of the knowledge used by the system. Difficult problems are often solvable only by identifying the proper source of knowledge and appropriate mechanisms to use it in the system.

Relationship to Other Fields

Machine vision is closely related to many other disciplines and incorporates various techniques adopted from many well-established fields, such as physics, mathematics, and psychology. Techniques developed from many areas are used for recovering information from images. In this section, we briefly describe some very closely related fields.

Image processing techniques generally transform images into other images; the task of information recovery is left to a human user. This field includes topics such as image enhancement, image compression, and correcting blurred or out-of-focus images (Gonzalez and Woods, 1982). On the other hand, machine vision algorithms take images as inputs but produce other types of outputs, such as representations for the object contours in an image or the motion of objects in a series of images. Thus, emphasis in machine vision is on recovering information automatically, with minimal interaction with a human. Image processing algorithms are useful in the early stages of a machine vision system. They are usually used to enhance particular information and suppress noise.

Computer graphics generates images from geometric primitives such as lines, circles, and free-form surfaces. Computer graphics techniques play a significant role in visualization and virtual reality (Foley et al., 1990). Machine vision is the inverse problem: estimating the geometric primitives and other features from the image. Thus, computer graphics is the synthesis of images, and machine vision is the analysis of images. However, these two fields are growing closer. Machine vision is using curve and surface representations and several other techniques from computer graphics, and computer graphics is using many techniques from machine vision to enter models into the computer for creating realistic images. Visualization and virtual reality are bringing these two fields closer.

Pattern recognition classifies numerical and symbolic data. Many statistical and syntactical techniques have been developed for classification of patterns. Techniques from pattern recognition play an important role in machine vision for recognizing objects. In fact, many vision applications rely heavily on pattern recognition. Object recognition in machine vision usually requires many other techniques. For a complete discussion of pattern recognition techniques, see Duda and Hart (1973).

Artificial intelligence is concerned with designing systems that are intelligent and with studying computational aspects of intelligence (Winston, 1992; Tanimoto, 1995). Artificial intelligence is used to analyze scenes by computing a symbolic representation of the scene contents after the images have been processed to obtain features. Artificial intelligence may be viewed as having three stages: perception, cognition, and action. Perception translates signals from the world into symbols, cognition manipulates the symbols, and action translates the symbols into signals that effect changes in the world. Many techniques from artificial intelligence play important roles in all aspects of machine vision. In fact, machine vision is often considered a subfield of artificial intelligence. Directly related to artificial intelligence is the study of *neural networks*. The design and analysis of neural networks has become a very active field in the last decade (Kosko, 1992). Neural networks are being increasingly applied to solve some machine vision problems.

Psychophysics, along with cognitive science, has studied human vision for a long time (Marr, 1982). Some techniques in machine vision are related to what is known about human vision. Many researchers in computer vision are more interested in preparing computational models of human vision than in designing machine vision systems. Therefore, some of the techniques used in machine vision have a strong similarity to those in psychophysics.

Fundamentals of Vision

Like other developing disciplines, machine vision is based on certain fundamental principles and techniques. To design and develop successful machine vision systems, one must have a full understanding of all aspects of the system, from initial image formation to final scene interpretation. The following is a list of the primary topics in machine vision:

- Image formation
- Segmentation
- Feature extraction and matching
- Three-dimensional object recognition
- Dynamic vision

In each of these processes, numerous factors influence the choice of algorithms and techniques to be used in developing a specific system. The system designer should be knowledgeable about the design issues of the system and the tradeoffs between the various algorithms and techniques available. In the following, we very briefly introduce these topics. Throughout the following sections, we make no attempt to cite all original work as the list would be nearly endless. The references, however, do give credit appropriately, and we refer you to the Further Information section for a complete discussion on these and many other topics.

19.4.2   Image Formation

An image is formed when a sensor records received radiation as a two-dimensional function. The brightness or intensity values in an image may represent different physical entities. For example, in a typical image obtained by a video camera, the intensity values represent the reflectance of light from various object surfaces in the scene; in a thermal image, they represent the temperature of corresponding regions in the scene; and in range imaging, they represent the distance from the camera to various points in the scene. Multiple images of the same scene are often captured using different types of sensors to facilitate more robust and reliable interpretation of the scene. Selecting an appropriate image formation system plays a key role in the design of practical machine vision systems. The following sections describe the principles of image formation.

Imaging Geometry

A simple camera-centered imaging model is shown in Fig. 19.38. The coordinate system is chosen such that the xy plane is parallel with the image plane and the z axis passes through the lens center at a distance f, the focal length of the camera, from the image plane. The image of a scene point (x, y, z) forms a point (x', y') on the image plane where

$$x' = x\frac{f}{z} \quad \text{and} \quad y' = y\frac{f}{z} \tag{19.49}$$

These are the *perspective projection* equations of the imaging system.

In a typical imaging situation, the camera may have several degrees of freedom, such as translation, pan, and tilt. Also, more than one camera may be imaging the same scene from different points. In this case, it is convenient to adopt a world coordinate system in reference to which the scene coordinates and camera coordinates are defined. In this situation, however, the imaging equations become more cumbersome and we refer you to the references at the end of this chapter for a more complete discussion of imaging geometry including camera calibration.

Image Intensity

Although imaging geometry uniquely determines the relationship between scene coordinates and image coordinates, the brightness or intensity at each point is determined not only by the imaging geometry but also by several other factors, including scene illumination, reflectance properties and surface orientations of objects in the scene, and radiometric properties of the imaging sensor.

The reflectance properties of a surface are characterized by its *bidirectional reflectance distribution function* (BRDF). The BRDF is the ratio of the scene radiance in the direction of the observer to the irradiance due to a light source from a given direction. It captures how bright a surface will appear when viewed from a given direction and illuminated by another. For example, for a flat Lambertian surface illuminated by a distant point light source, the BRDF is constant; hence, the surface appears equally bright from all directions. For a flat specular (mirrorlike) surface, the BRDF is an impulse function as determined by the laws of reflection.

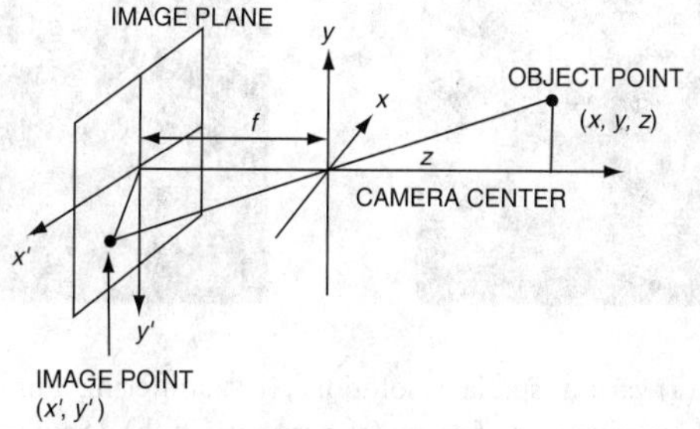

FIGURE 19.38   The camera-centered coordinate system imaging model.

The scene radiance at a point on the surface of an object depends on the reflectance properties of the surface, as well as on the intensity and direction of the illumination sources. For example, for a Lambertian surface illuminated by a point light source, the scene radiance is proportional to the cosine of the angle between the surface normal and the direction of illumination. This relationship between surface orientation and brightness is captured in the *reflectance map*. In the reflectance map for a given surface and illumination, contours of equal brightness are plotted as a function of surface orientation specified by the gradient space coordinates (p, q), where p and q are the surface gradients in the x and y directions, respectively. A typical reflectance map for a Lambertian surface illuminated by a point source is shown in Fig. 19.39. In this figure, the brightest point corresponds to the surface orientation such that its normal points in the direction of the source. Since image brightness is proportional to scene radiance, a direct relationship exists between the image intensity at an image point and the orientation of the surface of the corresponding scene point. *Shape-from-shading* algorithms exploit this relationship to recover the three-dimensional object shape. *Photometric stereo* exploits the same principles to recover the shape of objects from multiple images obtained by illuminating the scene from different directions.

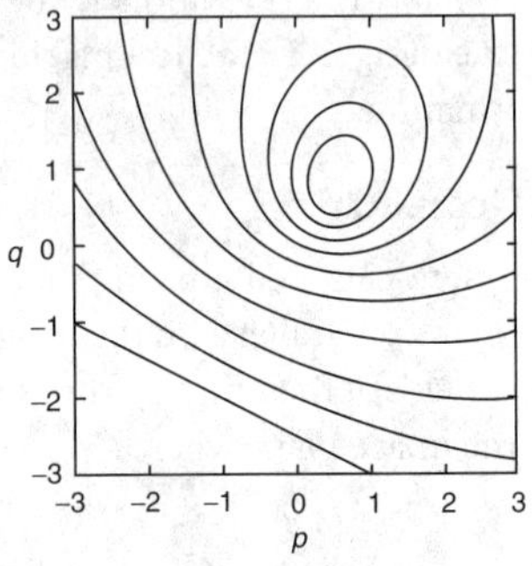

FIGURE 19.39   The reflectance map for a typical Lambertian surface illuminated by a point source from the direction $(p, q) = (0.2, 0.4)$.

Sampling and Quantization

A continuous function cannot be represented exactly in a computer. The interface between the imaging system and the computer must sample the continuous image function at a finite number of points and represent each sample within a finite number of bits. This is called **sampling** and **quantization**. Each image sample is a *pixel*. Generally images are sampled on a regular grid of squares so that the horizontal and vertical distances between pixels are the same throughout the image. Each pixel is represented in the computer as a small integer and represented as a two-dimensional array. Frequently, a pixel is represented as an unsigned 8-b integer in the range of $(0, 255)$, with 0 corresponding to black, 255 corresponding to white, and shades of gray distributed in between.

Many cameras acquire an analog image, which is then sampled and quantized to convert it to a digital image. The sampling rate determines how many pixels the digital image will have (image resolution), and the quantization determines how many intensity levels will be used to represent the intensity value at each sample point. As shown in Fig. 19.40(a) and Fig. 19.40(b), an image looks very different at different sampling

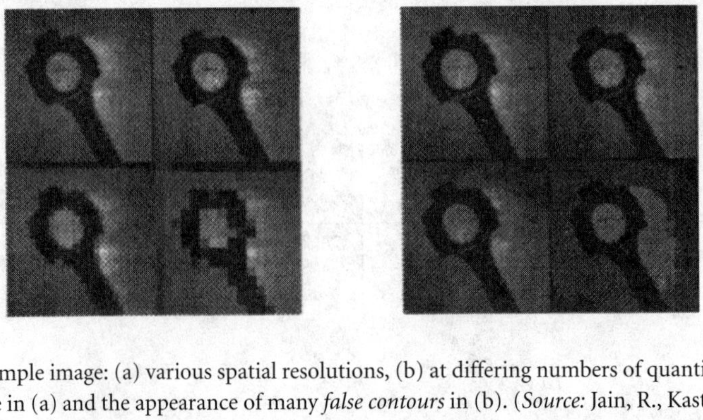

FIGURE 19.40   Sample image: (a) various spatial resolutions, (b) at differing numbers of quantization levels. Notice the blocky structure in (a) and the appearance of many *false contours* in (b). (*Source:* Jain, R., Kasturi, R., and Schunk, B.G. 1995. *Machine Vision.* McGraw-Hill, New York. With permission.)

rates and quantization levels. In most machine vision applications, the sampling rate and quantization are predetermined due to the limited choice available of cameras and image acquisition hardware. In many applications, however, it is important to know the effects of sampling and quantization.

Color Vision

The previous sections were only concerned with the intensity of light; however, light consists of a full spectrum of wavelengths. Images can include samples from a number of different wavelengths, leading to a color image. This perception of color depends on (1) the spectral reflectance of scene surfaces, (2) the spectral content of the illuminating light source, and (3) the spectral response of the sensors in the imaging system. In humans, color perception is attributed to differences in the spectral responses of the millions of neurochemical photoreceptors—rods and cones—in the retina. The rods are responsible for monochromatic vision and the cones are responsible for the sensation of color. The human visual system consists of three types of cones, each with a different spectral response. The total response each of the cones is often modeled by the integral

$$R_i(\lambda) = \int f(\lambda) r(\lambda) h_i(\lambda) d\lambda \qquad i = 1, 2, 3 \tag{19.50}$$

where λ is the wavelength of light incident on the receptor, $f(\lambda)$ the spectral composition of the illumination, $r(\lambda)$ the spectral reflectance function of the reflecting surface, and $h_i(\lambda)$ the spectral response of the ith type of receptor (cone).

In machine vision, color images are typically represented by the output of three different sensors, each having a different spectral response (e.g., the primary colors red, green, and blue). The choice of these primary colors determines the range of possible colors that can be realized by weighted combinations of the primaries. However, it is not necessary for the response of the sensors to be restricted to visible light. The sensors could just as easily respond to wavelengths in the infrared, ultraviolet, or X-ray portions of the electromagnetic spectrum, for example. In either case, the imaging system creates multiple images, called *channels*, one for each sensor. If the outputs from three sensors are used, this translates into a threefold increase in the processing and memory requirements. However, this increase is often offset by exploiting the extra dimensionality that is now available.

Range Imaging

Although three-dimensional information can be extracted from images with two-dimensional intensity—using image cues such as shading, texture, and motion—the problem is greatly simplified by range imaging. Range imaging is aquiring images for which the value at each pixel is a function of the distance of the corresponding point of the object from the sensor. The resulting images are known as *range images* or *depth maps*. A sample range image of a coffee cup is shown in Fig. 19.41. Here we will briefly discuss two common range imaging techniques, namely *imaging radar* and *triangulation*.

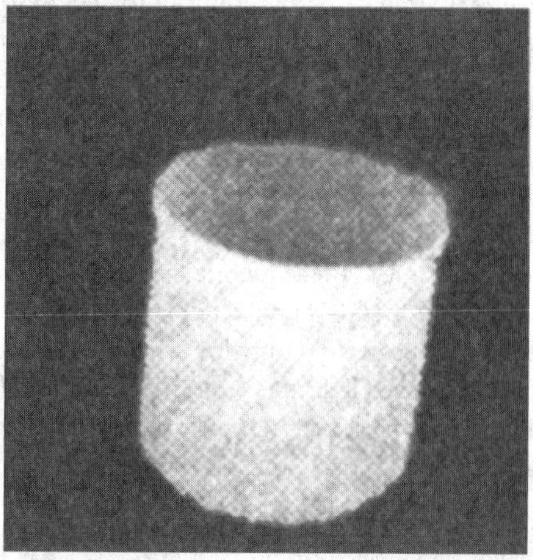

Imaging Radar

In a time-of-flight pulsed radar system, the distance to the object is computed by observing the time difference between transmitted and received electromagnetic pulses. Range information can also be obtained by detecting the phase difference between the transmitted and received waves of an amplitude-modulated beam or by detecting the beat frequency in a coherently mixed transmitted-and-received signal in a frequency-modulated beam.

FIGURE 19.41 A range image of a coffee mug.

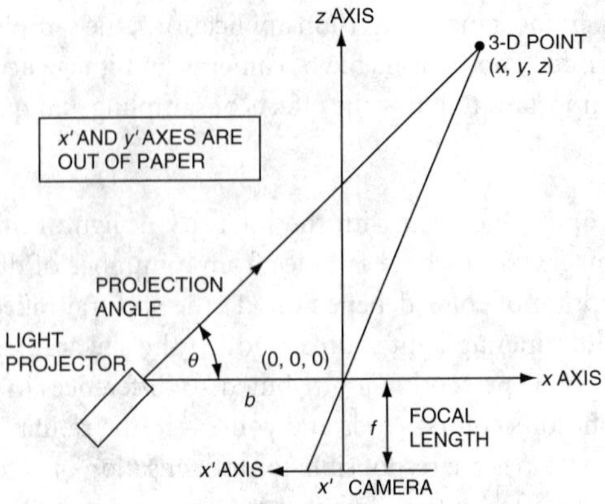

FIGURE 19.42 A triangulation-based range imaging system. (*Source:* Best, P.J. 1988. Active, optical range imaging sensor. *Machine Vision and Applications* 1(2): 127–152.)

Triangulation

In an active triangulation-based range imaging system, a light projector and camera are aligned along the z axis separated by a baseline distance b as shown in Fig. 19.42. The object coordinates (x, y, z) are related to the image coordinates (x', y') and the projection angle θ by the following equation:

$$[x \; y \; z] = \frac{b}{f \cot \theta - x'} [x' y' f] \tag{19.51}$$

The accuracy with which the angle θ and the horizontal position x' can be measured determines the range resolution of such a triangulation system. This system of active triangulation uses a single point light source that is swept across the scene in a raster scan fashion.

Structured Lighting

Imaging using structured lighting refers to systems in which the scene is illuminated by a known geometric pattern of light. In a simple point projection system—like the triangulation system already described—the scene is illuminated, one point at a time, in a two-dimensional grid pattern. The depth at each point is calculated using the previous equation to obtain the range image. In a typical structured lighting system, either planes or two-dimensional patterns of light are projected onto the scene. A camera, which is displaced spatially from the source of illumination, records the patterns of light projected on the object light surfaces. The distortions contained in the observed images of the patterns are determined by the shape and orientation of surfaces of the objects on which the patterns are projected (see Fig. 19.43).

In situations where the scene is changing dynamically, projecting single light stripes in sequence may not be practical due to the sequential nature of the system. In this situation, a set of multiple light stripes—each uniquely encoded—is projected. For example, using a *binary-coded* scheme, a complete set of data can be acquired by projecting only $\log_2 N$ patterns, where $N - 1$ is the total number of stripes in each pattern. Other structured lighting techniques include *color coding*, *grid coding*, and *Fourier domain processing*. The primary drawback of any structured lighting technique is that data cannot be obtained for object points that are not visible to either the light source or the imaging camera.

Active Vision

Most machine vision systems rely on data captured by hardware with fixed characteristics. These systems include passive sensing systems—such as video cameras—and active sensing systems—such as laser range finders. In an active vision system, the parameters and characteristics of data capture are dynamically controlled by the scene interpretation system. Active vision systems may employ either active or passive sensors. In an active vision system, however, the state parameters of the sensors, such as focus, aperature, vergence, and illumination, are controlled to acquire data that will facilitate scene interpretation.

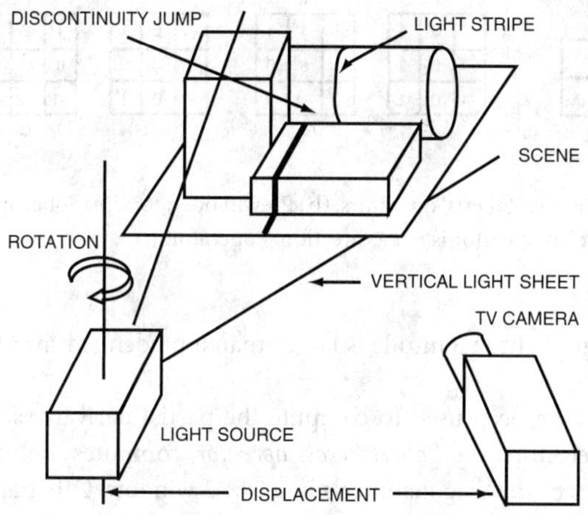

FIGURE 19.43 A typical structured lighting system. (*Source:* Jorvis, R.A. 1983. A perspective on range finding techniques for computer vision. *IEEE Transactions on Pattern Analysis and Machine Intelligence* 5(2):122–139. ©1983 IEEE.)

19.4.3 Segmentation

An image must be analyzed and its relevant features extracted before more abstract representations and descriptions can be generated. Careful selection of these so-called low-level operations is critical for the success of higher level scene interpretation algorithms. One of the first operations that a machine vision system must perform is the separation of objects from the background. This operation, commonly called **segmentation**, is approached in one of two ways: (1) an *edge-based* method for locating discontinuities in certain properties or (2) a *region-based* method for the grouping of pixels according to certain similarities.

Edge-Based Segmentation

In an edge-based approach, the boundaries of objects are used to partition an image. Points that lie on the boundaries of an object must be marked. Such points, called *edge points*, can often be detected by analyzing the local neighborhood of a point. By definition, the regions on either side of an edge point (i.e., the object and background) have dissimilar characteristics. Thus, in edge detection, the emphasis is on detecting dissimilarities in the neighborhoods of points.

Gradient

An *edge* in an image is indicated by a significant local change in image intensity, usually associated with a discontinuity in either the image intensity or the first derivative of the image intensity. Therefore, a direct approach to edge detection is to compute the *gradient* at each pixel in an image. The gradient is the two-dimensional equivalent to the first derivative and is defined as the vector

$$\mathbf{G}[f(x,y)] = \begin{bmatrix} G_x \\ G_y \end{bmatrix} = \begin{bmatrix} \dfrac{df}{dx} \\ \dfrac{df}{dy} \end{bmatrix} \tag{19.52}$$

where $f(x,y)$ represents the two-dimensional image intensity function. The magnitude and direction of the gradient in terms of the first partial derivatives G_x and G_y are given by

$$G(x,y) = \sqrt{G_x^2 + G_y^2} \quad \text{and} \quad \alpha(x,y) = \tan^{-1}\left(\frac{G_y}{G_x}\right) \tag{19.53}$$

$$\text{(a)}\quad G_x = \begin{bmatrix} 1 & 0 \\ 0 & -1 \end{bmatrix} \quad G_y = \begin{bmatrix} 0 & -1 \\ 1 & 0 \end{bmatrix} \qquad \text{(b)}\quad G_x = \begin{bmatrix} -1 & 0 & 1 \\ -1 & 0 & 1 \\ -1 & 0 & 1 \end{bmatrix} \quad G_y = \begin{bmatrix} 1 & 1 & 1 \\ 0 & 0 & 0 \\ -1 & -1 & -1 \end{bmatrix}$$

$$\text{(c)}\quad G_x = \begin{bmatrix} -1 & 0 & 1 \\ -2 & 0 & 2 \\ -1 & 0 & 1 \end{bmatrix} \quad G_y = \begin{bmatrix} 1 & 2 & 1 \\ 0 & 0 & 0 \\ -1 & -2 & -1 \end{bmatrix} \qquad \text{(d)}\quad \begin{bmatrix} 0 & 1 & 0 \\ 0 & -4 & 0 \\ 0 & 1 & 0 \end{bmatrix}$$

FIGURE 19.44 Operators: (a) the Roberts' operators, (b) Prewitt operators, (c) Sobel operators, and (d) the Laplacian operator. Note that the Laplacian operator is *not* a directional operator.

Only those pixels whose gradient magnitude is larger than a predefined threshold are identified as edge pixels.

Many approximations have been used to compute the partial derivatives in the x and y directions. One of the earlier edge detectors, the *Roberts' cross operator*, computes the partial derivatives by taking the difference in intensities of the diagonal pixels in a 2×2 window. Other approximations, such at the *Prewitt* and *Sobel* operators, use a 3×3 window to calculate the partial derivatives. Occasionally, operators that compute the edge strength in a number of specific directions are used in machine vision. These directional operators are defined over 3×3 or larger neighborhoods. For applications where such simple edge detectors are not adequate, the *Canny* edge detector, which attempts to optimize noise suppression with edge localization, is often used. See Fig. 19.44 for a description of the window operators used in each of the mentioned edge operators.

Laplacian

When using gradient-based edge detectors, multiple responses are often obtained at adjacent pixels depending on the rate of transition of the edge. After **thresholding,** thick edges are usually obtained. However, it is desired to obtain only a single response for a single edge; in other words, only keep the maximum of the gradient and not every response that is above a specified threshold. The maximum of the gradient is found by detecting the zero crossings in the second derivative of the image intensity function. The *Laplacian* is defined in terms of the second partial derivatives as follows

$$\nabla^2 f = \frac{\partial^2 f}{\partial x^2} + \frac{\partial^2 f}{\partial y^2} \tag{19.54}$$

The second partial derivatives along the x and y directions are approximated using the difference equations

$$\frac{\partial^2 f}{\partial x^2} = f(x+1, y) - 2f(x, y) + f(x-1, y)$$

and

$$\frac{\partial^2 f}{\partial y^2} = f(x, y+1) - 2f(x, y) + f(x, y-1) \tag{19.55}$$

By combining these two equations, the Laplacian can be implemented in a single window operator (see Fig. 19.44).

The response of the Laplacian operator can be computed in one step by convolving the image with the appropriate window operator. Nevertheless, since the Laplacian is an approximation to the second derivative, it reacts more strongly to lines, line ends, and noise pixels than it does to edges; plus it generates two responses for each edge, one positive and one negative on either side of the edge. Hence, the zero crossing between the two responses is used to localize the edge.

Laplacian of Gaussian

Detecting edge points from the zero crossings of the second derivative of the image intensity is very noisy. Therefore, it is often desirable to filter out the noise before edge detection. To do this, the *Laplacian of Gaussian* (LoG) combines Gaussian filtering with the Laplacian for edge detection. The output of the

LoG operator, $h(x, y)$, is obtained by the convolution operation

$$h(x, y) = \nabla^2(g(x, y) * f(x, y)) \tag{19.56}$$

where

$$g(x, y) = e^{-[\frac{(x^2+y^2)}{2\sigma^2}]} \tag{19.57}$$

is the equation for the Gaussian filter (neglecting the constant factor). Using the derivative rule for convolution, we get the equivalent expression

$$h(x, y) = [\nabla^2 g(x, y)] * f(x, y) \tag{19.58}$$

where the term

$$\nabla^2 g(x, y) = \left(\frac{x^2 + y^2 - 2\sigma^2}{\sigma^4}\right) e^{-[\frac{(x^2+y^2)}{2\sigma^2}]} \tag{19.59}$$

is commonly called the *Mexican hat* operator because of its appearance when plotted (see Fig. 19.45).

This result shows that the Laplacian of Gaussian can be obtained in either of two ways: (1) Convolve the image with a Gaussian smoothing filter and compute the Laplacian of the result. (2) Convolve the image with a filter that is the Laplacian of the Gaussian filter. The zero crossings still must be detected in order to localize the edges; and like the Laplacian, only those edge points whose corresponding first derivative is above a specified threshold are marked as edge points.

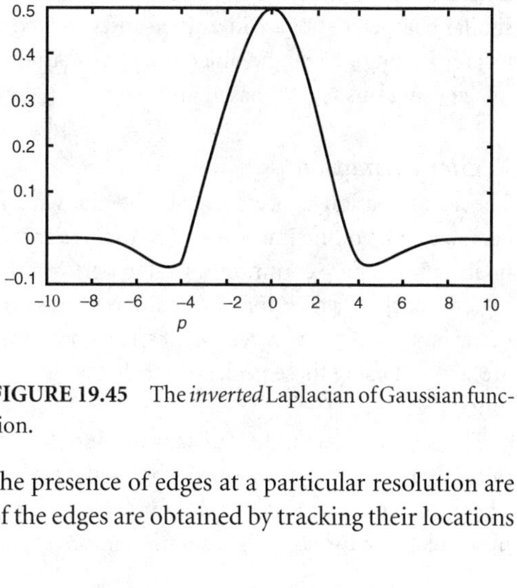

When using the Laplacian of Gaussian approach, edges are detected at a particular resolution (scale) depending on the spread of the Gaussian filter used. The determination of real edges in any image may require the combination of information from operators at several scales using a *scale-space* approach. The presence of edges at a particular resolution are found at the appropriate scale, and the exact locations of the edges are obtained by tracking their locations through lower values of σ.

FIGURE 19.45   The *inverted* Laplacian of Gaussian function.

Surface Fitting

Since a digital image is actually a sampling of a continuous function of two-dimensional spatial variables, another viable approach to edge detection is to approximate and reconstruct the underlying continuous spatial function that the image represents. This process is termed *surface fitting*. The intensity values in the neighborhood of a point can be used to obtain the underlying continuous intensity surface, which then can be used as a best approximation from which properties in the neighborhood of the point can be computed. The continuous intensity function can be described by

$$z = f(x, y) \tag{19.60}$$

If the image represents one surface, the preceeding equation will correctly characterize the image. However, an image generally contains several surfaces, in which case this equation would be satisfied only at local surface points. Therefore, this approach uses the intensity values in the neighborhood about each pixel to approximate a local surface about that point. The *facet model* is a result of this idea. In the facet model, the neighborhood of a point is approximated by the cubic surface patch that best fits the area using

$$f(r, c) = k_1 + k_2 r + k_3 c + k_4 r^2 + k_5 r c + k_6 c^2 + k_7 r^3 + k_8 r^2 c + k_9 r c^2 + k_{10} c^3 \tag{19.61}$$

where r and c are local coordinates about the point in the center of the neighborhood being approximated. From this approximation, the second directional derivative can be calculated and the zero crossings can be detected, and the edge points can be found.

Edge Linking

The detection of edge pixels is only part of the segmentation task. The outputs of typical edge detectors must be linked to form complete boundaries of objects. Edge pixels seldom from closed boundaries; missed edge points will result in breaks in the boundaries. Thus, *edge linking* is an extremely important, but difficult, step in image segmentation. Several methods have been suggested to improve the performance of edge detectors ranging from *relaxation-based edge linking*, which uses the magnitude and direction of an edge point to locate other edges in the neighborhood, to *sequential edge detectors*—also called *edge trackers*—which finds a strong edge point and grows the object boundary by considering neighboring edge strengths and directions.

Region-Based Segmentation

In region-based segmentation, all pixels corresponding to a single object are grouped together and are marked to indicate that they belong to the same object. The grouping of the points is based on some criterion that distinguishes pixels belonging to the same object from all other pixels. Points that have similar characteristics are identified and grouped into a set of connected points, called *regions*. Such points usually belong to a single object or part of an object. Several techniques are available for segmenting an image using this region-based approach.

Region Formation

The segmentation process usually begins with a simple region formation step. In this step, intrinsic characteristics of the image are used to form initial regions. Thresholds obtained from a **histogram** of the image intensity are commonly used to perform this initial grouping. In general, an image will have several regions, each of which may have different intrinsic characteristics. In such cases, the intensity histogram of the image will show several peaks; each peak may correspond to one or more regions. Several thresholds are selected using these peaks. After thresholding, a connected-component algorithm can be used to find initial regions.

Thresholding techniques usually produce too many regions. Since they were formed based only on first-order characteristics, the regions obtained are usually simplistic and do not correspond to complete objects. The regions obtained via thresholding may be considered to be only the first step in segmentation. After the initial histogram-based segmentation, more sophisticated techniques are used to refine the segmentation.

Split and Merge

Automatic refinement of an initial intensity-based segmentation is often done using a combination of *split* and *merge* operations. Split and merge operations eliminate false boundaries and spurious regions by either splitting a region that contains pieces from more than one object or merging adjacent regions that actually belong to the same object. If some property of a region is not uniform, the region should be split. Segmentation based on the split approach starts with large regions. In many cases, the whole image may be used as the starting region. Several decisions must be made before a region is split. One is to decide "when" a property is nonuniform over a region; another is "how" to split a region so that the property for each of the resulting components is uniform. These decisions usually are not easy to make. In some applications, the variance of the intensity values is used as a measure of uniformity.

More difficult than determining the property uniformity is deciding "where" to split a region. Splitting regions based on property values is very difficult. One approach used when trying to determine the best boundaries with which to divide a region is to consider edgeness values within a region. The easiest schemes for splitting region are those that divide a region into a fixed number of equal regions; such methods are called *regular decomposition methods*. For example, in the *quad tree* approach, if a region is considered nonuniform, it is split into four equal quadrants in each step.

Many approaches have been proposed to judge the similarity of regions. Broadly, the approaches are based on either the characteristics of regions or the weakness of edges between them. Two common approaches to judging the similarity of adjacent regions based on region characteristics are to (1) compare their mean intensities and (2) assume their intensity values are drawn from a known probability distribution. In the first approach, if their mean intensities do not differ by more than some predetermined value, the regions are considered to be similar and are candidates for merging. A modified form of this approach uses surface fitting to determine if the regions can be approximated by one surface. In the latter approach, the decision of whether or not to merge adjacent regions is based on considering the probability that they will have the same statistical distribution of intensity values. This approach uses hypothesis testing to judge the similarity of adjacent regions.

Another approach to merging is to combine two regions only if the boundary between them is weak. A weak boundary is one for which the intensities on either side differ by less than a given threshold. This approach attempts to remove weak edges between adjacent regions by considering not only the intensity characteristics, but also the length of the common boundary. The common boundary is dissolved if it is weak and the resulting boundary (of the merged region) does not grow too fast.

Split and merge operations may be used together. After a presegmentation based on thresholding, a succession of splits and merges may be applied as dictated by the properties of the regions. Such schemes have been proposed for segmentation of complex scenes. Domain knowledge with which the split and merge operations can be controlled may also be introduced.

Other Segmentation Methods

In the preceeding discussion, the primary focus was with intensity images. Segmentation techniques that are based on *color*, *texture*, and *motion* have also been developed. Segmentation based on spectral pattern classification techniques is used extensively in remote-sensing applications.

19.4.4 Feature Extraction and Matching

Segmented images are often represented in a compact form to facilitate further abstraction. Representation schemes are chosen to match methods used for object recognition and description. The task of object recognition requires matching an object description in an image to models of known objects. The models, in turn, use certain descriptive features and their relations. Matching also plays an important role in other aspects of information recovery from images. In the following, schemes for representation and description that are popular in machine vision are discussed along with techniques for feature extraction and matching.

Representation and Description

Representation and description of symbolic information can be approached in many ways. One approach is to represent the object in terms of its bounding curve. Popular among several methods developed for boundary representation are chain codes, **polygonalization**, one-dimensional signatures, and representations using dominant points for representing an object. Another approach is to obtain region-based shape descriptors, such as topological or texture descriptors. Representation and shape description schemes are usually chosen such that the descriptors are invariant to rotation, translation, and scale changes.

Chain Codes

One of the earliest methods for representing a boundary uses directional codes, called *chain codes*. The object boundary is resampled at an appropriate scale, and an ordered list of points along the boundary is represented by a string of directions codes (see Fig. 19.46(a)). Often to retain all of the information in the boundary, the resampling step is bypassed. However, resampling eliminates minor fluctuations that typically are due to noise. Chain codes possess some attractive features; namely, the rotation of an object by multiples of 45° is easily implemented; the derivative of the chain code, the *difference code* obtained by computing the first differences of the chain code, is rotation invariant; and other characteristics of a region, such as area and corners, can be computed directly from the chain code. The limitations of this

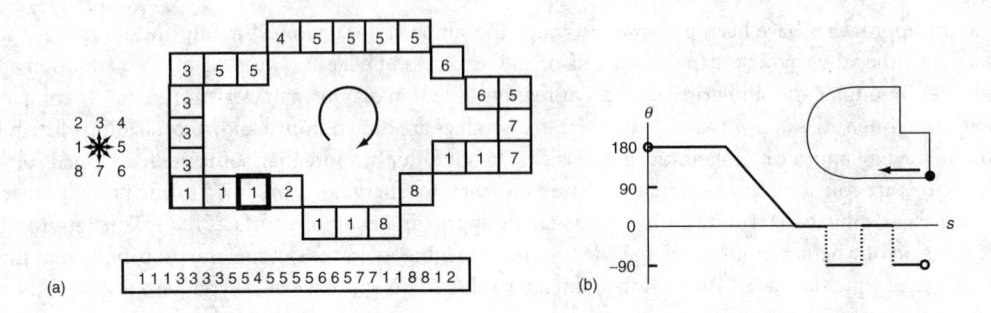

FIGURE 19.46 (a) An object and its chain code. (b) A simple geometric object and its one-dimensional (s, θ) signature.

representation method are attributed to the restricted directions used to represent the tangent at a point. Although codes with larger numbers of directions are occasionally used, the eight-directional chain code is the most commonly used.

Polygonalization

Polygonal approximation of boundaries has been studied extensively and numerous methods have been developed. The fit of a polygon is made to reduce a chosen error criterion between the approximation and the original curve. In an *interative endpoint* scheme, the first step is to connect a straight-line segment between the two farthest points on the boundary. The perpendicular distances from the segment to each point on the curve is measured. If any distance is greater than a chosen threshold, the segment is replaced by two segments; one each from the segment endpoint to the curve point where the distance to the segment is greatest. In another approach, a straight-line fit is constrained to pass within a radius around each data point. The line segment is grown from the first point, and when further extension of the line segment causes it to fall outside the radius point, a new line is started. A *minimax* approach can also be used in which the line segment approximations are chosen to minimize the maximum distance between the data points and the approximating line segment.

Besides polygonal approximation methods, higher order curve- and spline-fitting methods may be used where more precise approximations are required. These are more computationally expensive than most polygonalization methods, and they can be more difficult to apply.

One-Dimensional Signatures

The slope of the tangent to the curve denoted by the angle θ as a function of s, the position along the curve from an arbitrary starting point, is used to represent curves in many applications. Plots of these functions have some interesting and useful characteristics. Horizontal lines in s–θ plots represent straight lines and straight lines at an angle with the horizontal represent circular arcs whose radii are proportional to the slope of the straight line. An s–θ curve can also be treated as a periodic function with a period given by the perimeter of the curve; hence, Fourier techniques can be used. Other functions, such as the distance to the curve from an arbitrary point inside the curve plotted as a function of the angle with reference to the horizontal, are also used as shape signatures. Figure 19.46(b) shows a simple geometric object and its one-dimensional signature.

Boundary Descriptors

Descriptors for objects represented by their boundaries may be generated using the representations already described. Simple descriptors, such as perimeter, length, and orientation of the major axis, shape number, and eccentricity, may be readily computed from the boundary data. The ordered set of points on the boundary having two-dimensional coordinates (x_k, y_k), where $k = 1, \ldots, N$ and N is the total number of points on the boundary, can be treated as a one-dimensional complex function $x_k + iy_k$. The coefficients of the discrete Fourier transform applied to this function can also be used as a shape descriptor.

Other descriptors that use the one-dimensional signature functions may also be obtained. A major drawback of many of these descriptors is that complete data for the object boundary are required. Because of either problems in segmentation or occlusions in the image, however, complete data for the object boundary are often not available. In such instances, recognition strageties based on partial information are needed.

Many region-based representation and description methods analogous to those used to represent object boundaries have been developed. Examples of such methods are the *medial-axis transform, skeletons, convex hulls,* and *deficiencies. Morphological operators* have been developed to extract shape features and generate useful descriptions of object shape. Topological descriptors, such as *Euler number,* are also useful to characterize shape. The facet model has even been used to generate a topographic primal sketch of images using a set of seven descriptive labels, including pit, peak, and ridge to name a few. Such descriptions are useful in object matching. Region-based representation and description are particularly useful to describe objects for which properties within the region, for example, texture, are significant for object recognition. For a complete discussion of these topics, refer to the references at the end of this chapter.

Feature Extraction

If the object to be recognized has unique discriminating features, special-purpose algorithms may be employed to extract such features. Important for object matching and recognition are corners, high-curvature regions, inflection points, or other places along curves at which curvature discontinuities exist. In region-based matching methods, identifying groups of pixels that can be easily distinguished and identified is important. Many methods have been proposed to detect both dominant points in curves and interesting points in regions.

Critical Points

The detection of *critical points,* also called *dominant points,* in curves, such as corners and inflection points, is important for subsequent object matching and recognition. Most algorithms for detecting critical points mark local curvature maxima as dominant. One approach is to analyze the deviations of a curve from a chord to detect dominant points along the curve. Points are marked as either being critical or belonging to a smooth or noisy interval. These markings depend on whether the curve makes a single excursion away from the chord, stays close to the chord, or makes two or more excursions away from the chord, respectively. Other approaches use mathematical expressions to directly compute and plot the curvature values along the contour.

Interesting Points

Points used in matching between two images must be ones that are easily identified and matched. These are known as *interesting points.* Obviously, those points in a uniform region and on edges are not good candidates for matching. Interest operators find image areas with high variance. In applications such as stereo and structure from motion, images should have enough such interesting regions to facilitate matching.

One commonly used interest operator, the *Moravec interest operator,* uses directional variances as a measure of how interesting a point is. A point is considered interesting if it has a local maximum of minimal sums of directional variances. The directional variances in the local neighborhood about a point are calculated by

$$I_1 = \sum_{(x,y)\in s} [f(x,y) - f(x, y+1)]^2$$

$$I_2 = \sum_{(x,y)\in s} [f(x,y) - f(x+1, y)]^2$$

$$I_3 = \sum_{(x,y)\in s} [f(x,y) - f(x+1, y+1)]^2 \tag{19.62}$$

$$I_4 = \sum_{(x,y)\in s} [f(x,y) - f(x+1, y-1)]^2$$

where s represents the neighborhood about the current point. Typical neighborhoods range from 5×5 to 11×11 pixels. The *interestingness* of the point is then given by

$$I(x, y) = \min(I_1, I_2, I_3, I_4) \qquad (19.63)$$

Feature points are chosen where the interestingness is a local maximum. Doing this eliminates the detection of simple edge points since they have no variance in the direction of the edge. Furthermore, a point is considered a good interesting point if its local maximum is above a preset threshold. The Moravec interest operator has found extensive use in stereo matching applications.

Matching

Matching plays a very important role in many phases of machine vision. Object recognition requires matching a description of an object with models of known objects. The goal of matching is to either (1) detect the presence of a known entity, object, or feature or (2) find what an unknown image component is. The difficulty in achieving these matching goals is first encountered with goal-directed matching, wherein the goal is to find a very specific entity in an image. Usually, the location of all instances of such an entity must be found. In stereo and structure-from-motion applications, entities are obtained in one image, and their locations are then determined in a second image. The second problem requires matching an unknown entity with several models to determine which model matches best.

Point Pattern Matching

In matching points in two slightly different images of the same scene (i.e., in a stereo image pair or a motion sequence), interesting points are detected by applying an operator such as the Moravec interest operator. The correspondence process considers the local structure of a selected point in one image in order to assign initial matchable candidate points from the second image. However, this *correspondence problem* is not an easy one to solve, and many constraints are often imposed to ease the process. For example, in stereo-matching applications, the displacements of a point from one image to the other are usually small. Thus, only points within a local neighborhood are considered for matching. To obtain the final correspondence, the set of initial matches is refined by computing the similarity in global structure around each candidate point. In dynamic-scene analysis, one may assume that motions of neighboring points do not differ significantly. To obtain the final matching of points in the two images, relaxation techniques are often employed. An example of the point matching problem is illustrated in Fig. 19.47.

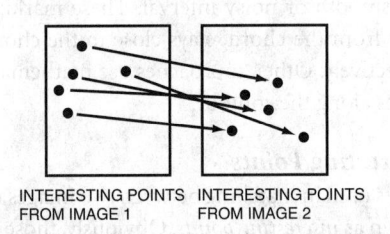

INTERESTING POINTS   INTERESTING POINTS
FROM IMAGE 1   FROM IMAGE 2

FIGURE 19.47   An illustration of the point pattern matching process.

Template Matching

In some applications, a particular pictorial or iconic structure, called a *template*, is to be detected in an image. Templates are usually represented by small two-dimensional intensity functions (typically less than 64×64 pixels). *Template matching* is the process of moving the template over the entire image and detecting the locations at which the template best fits the image. A common measure of similarity to determine a match is the *normalized cross correlation*. The correlation coefficient is given by

$$M(x, y) = \frac{\sum \sum_{(u,v) \in R} g(u, v) f(x + u, y + v)}{\left\{ \sum \sum_{(u,v) \in R} f^2(x + u, y + v) \right\}^{1/2}} \qquad (19.64)$$

where $g(u, v)$ is the template, $f(x, y)$ the image, and R the region spanned by the template. The value M takes maximum value for the point (x, y) at which $g = cf$ (in other words where the pixel values of the template and the image only differ by a constant scaling factor). By thresholding the result of this

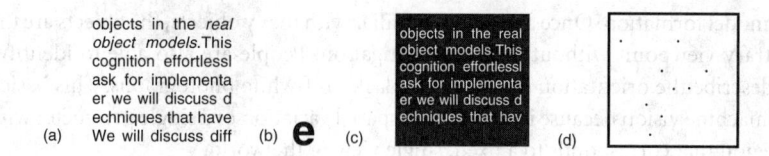

(a)   (b)   (c)   (d)

FIGURE 19.48 (a) A test image. (b) a template (enlarged) of the letter *e*. (c) the results of the normalized cross correlation. (d) the thresholded result indicating the match locations ($T = 240$). (*Source:* Jain, R., Kasturi, R., and Schunck, B.G. 1995. *Machine Vision*. McGraw-Hill, New York. With Permission.)

cross-correlation operation, the locations of a template match can be found. Figure 19.48 shows a test image, a template, and the results of the normalized cross correlation.

A major limitation of the template-matching technique is its sensitivity to the rotation and scaling of objects. To match scaled and rotated objects, separate templates must be constructed. In some approaches, a template is partitioned into several subtemplates, and matching is carried out for these subtemplates. The relationships among the subtemplates are verified in the final matching step.

Hough Transform

Parametric transforms, such as the Hough transform, are useful methods for the recognition of straight lines and curves. For a straight line, the transformation into the parameter space is defined by the parametric representation of a straight line given by

$$\rho = x \cos \theta + y \sin \theta \qquad (19.65)$$

where (ρ, θ) are the variables in the parameter space that represent the length and orientation, respectively, of the normal to the line from the origin. Each point (x, y) in the image plane transforms into a sinusoid in the (ρ, θ) domain. However, all sinusoids corresponding to points that are colinear in the image plane will intersect at a single point in the Hough domain. Thus, pixels belonging to straight lines can be easily detected.

The Hough transform can also be defined to recognize other types of curves. For example, points on a circle can be detected by searching through a three-dimensional parameter space of (x_c, y_c, r) where the first two parameters define the location of the center of the circle and r represents the radius. The Hough transform can also be generalized to detect arbitrary shapes. However, one problem with the Hough transform is the large parameter search space required to detect even simple curves. This problem can be alleviated somewhat by the use of additional information that may be available in the spatial domain. For example, in detecting circles by the brute-force method, searching must be done in a three-dimensional parameter space; however, when the direction of the curve gradient is known, the search is restricted to only one dimension.

Pattern Classification

Features extracted from images may be represented in a multidimensional feature space. A classifier—such as a minimum distance classifier, which is used extensively in statistical pattern recognition—may be used to identify objects. This pattern classification method is especially useful in applications wherein the objective is to assign a label, from among many possible labels, to a region in the image. The classifier may be taught in a supervised-learning mode, by using prototypes of known object classes, or in an unsupervised mode, in which the learning is automatic. Proper selection of features is crucial for the success of this approach. Several methods of pattern classification using structural relationships have also been developed. For a complete discussion of pattern classification, see Duda and Hart (1973).

19.4.5 Three-Dimensional Object Recognition

The real world that we see and touch is primarily composed of three-dimensional solid objects. When an object is viewed for the first time, people typically gather information about that object from many different viewpoints. The process of gathering detailed object information and storing that information is referred to as model formation. Once a person is familiar with many objects, the objects are then identified from an arbitrary viewpoint without further investigation. People are also able to identify, locate, and qualitatively describe the orientation of objects in black-and-white photographs. This basic capability is significant to machine vision because it involves the spatial variation of a single parameter within a framed rectangular region that corresponds to a fixed, single view of the world.

Recognition System Components

Recognition implies awareness of something already known. In modeling real-world objects for recognition purposes, different kinds of schemes have been used. To determine how recognition will take place, a method for matching model data to sensor data must be considered. A straightforward blind-search approach would entail (1) transforming all possible combinations of all possible known object models in all possible distinguishable orientations and locations into a digitized sensor format and (2) matching based on the minimization of a matching-error criterion. Clearly, this approach is impractical. On the other hand, since object models contain more object information than sensor data, we are prohibited from transforming sensor data into complete model data and matching in the model data format. However, this does not prevent us from matching with partial model data. As a result, working with an intermediate domain that is computable from both sensor and model data is advantageous. This domain is referred to as the symbolic scene description domain. A matching procedure is carried out on the quantities in this domain, which are referred to as *features*.

The interactions between the individual components of a recognition system are illustrated in Fig. 19.49. The image formation process creates intensity or range data based purely on physical principles. The description process acts on the sensor data and extracts relevant application-independent features. This process is completely data-driven and includes only the knowledge of the image formation process. The modeling process provides object models for real-world objects. Object reconstruction from sensor data is one method for building models automatically. The understanding, or recognition, process involves an algorithm to perform matching between model and data descriptions. This process might include data- and model-driven subprocesses, where segmented sensor data regions seek explanations in terms of models and hypothesized models seek verification from the data. The rendering process produces synthetic sensor data from object models. Rendering provides an important feedback link by allowing an autonomous system to check on its own understanding of the sensor data by comparing synthetic images to sensed images.

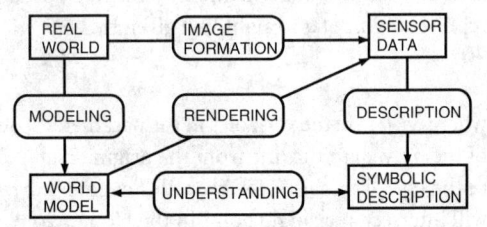

FIGURE 19.49   The components of an object recognition system. (*Source:* Best, P. and Jain, R.C. 1985. Three dimensional object recognition. *ACM Computing Surveys* 17(1).)

Object Representation Schemes

Numerous schemes have been developed for representing three-dimensional objects. The choice of a particular scheme is governed by its intended application. In computer graphics, schemes such as the wire-frame and constructive solid-geometry representations are popular since their data structures are suitable for image rendering. In machine vision systems, other methods such as generalized cones and characteristic views are used extensively. In the following, we briefly describe some commonly used object representation schemes.

Wire Frame

The wire-frame representation of a three-dimensional object consists of a three-dimensional vertex point list and an edge list of vertex pairs. Although this representation is very simple, it is an ambiguous representation for determining such quantities as surface area and volume of an object. Wire-frame

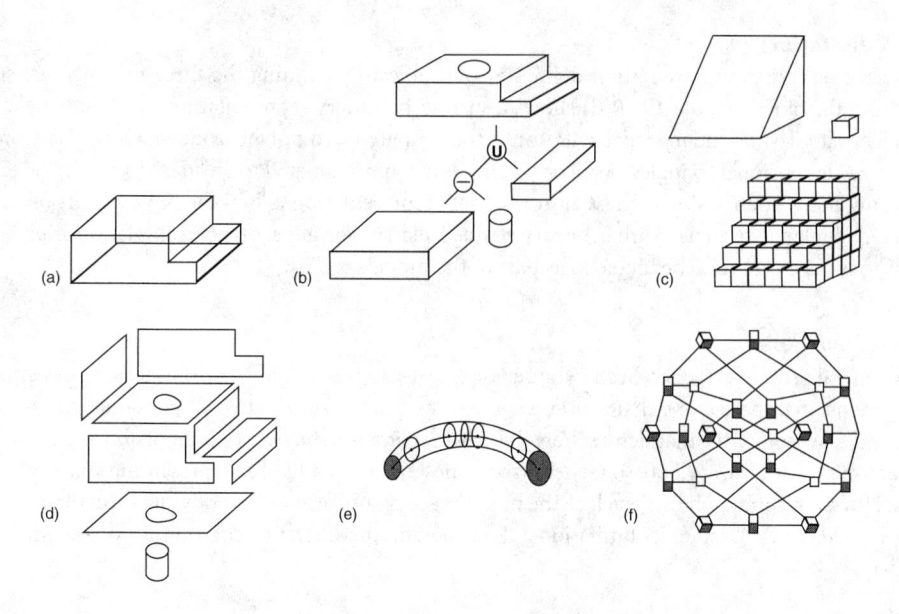

FIGURE 19.50 Various object representation schemes: (a) wire-frame, (b) constructive solid geometry, (c) spatial occupancy, (d) surface-boundary, (e) generalized-cone, (f) aspect graph.

models can sometimes be interpretted as several different solid objects or as different orientations of the same object. Figure 19.50(a) shows the wire-frame representation of a simple three-dimensional object.

Constructive Solid Geometry

The constructive solid-geometry (CSG) representations of an object is specified in terms of a set of three-dimensional volumetric primitives (blocks, cylinders, cones, spheres, etc.) and a set of boolean operators: union, intersection, and difference. Figure 19.50(b) shows the CSG representation for a simple geometric object. The storage data structure is a binary tree, where the terminal nodes are instances of the geometric primitives and the branching nodes represent the Boolean set operations and positioning information. CSG trees define object surface area unambiguously and can, with very little data, represent complex objects. However, the boundary evaluation algorithms required to obtain usable surface information are very computationally expensive. Also, general sculptured surfaces are not easily represented using CSG models.

Spatial Occupancy

Spatial-occupancy representations use nonoverlapping subregions of three-dimensional space occupied by an object to define that object. This method unambiguously defines an object's volume. Some commonly used single primitive representations of this type are the voxel and octree representations. *Voxels* are small volume elements of discretized three-dimensional space (see Fig. 19.50(c)). They are usually fixed-size cubes. Objects are represented by the lists of voxels occupied by the object. Voxel representations tend to be very memory intensive, but algorithms using them tend to be very simple. An *octree* is a hierarchical representation of spatial occupancy. Volumes are decomposed into cubes of different size, where the cube size depends on the distance from the root node. Each branching node of the tree represents a cube and points to eight other nodes, each of which describes object volume occupancy in the corresponding octant subcubes of the branching node cube. The octree representation offers the advantages of the voxel description but is more compact. Beacuse of this compactness, more complicated algorithms are required for many computations than those needed for voxel representation.

Surface Boundary

The surface-boundary representation defines a solid object by defining the three-dimensional surfaces that bound the object. Figure 19.50(d) shows a surface-boundary representation for a simple geometric object. The simplest boundary representation is the triangle-faced polyhedron, which can be stored as a list of three-dimensional triangles. Arbitrary surfaces are approximated to any desired degree of accuracy by utilizing many triangles. A slightly more compact representation allows the replacement of adjacent, connected, coplanar triangles with arbitrary n-sided planar polygons. Structural relationships between bounding surfaces may also be included as part of the model.

Generalized Cone

In the generalized-cone (generalized-cylinder or sweep) representation, an object is represented by a three-dimensional space curve that acts as a spine or axis of the cone, a two-dimensional cross-sectional figure, and a sweeping rule that defines how the cross section is to be swept (and possibly modified) along the space curve (see Fig. 19.50(e)). Generalized cones are well suited for representing many real-world shapes. However, certain objects, such as the human face or an automobile body, are difficult to represent as generalized cones. Despite its limitations, the generalized-cone representation is popular in machine vision.

Skeleton

Skeleton representations use space-curve skeletons. A skeleton can be considered an abstraction of the generalized cone description that consists of only the spines. Skeleton geometry provides useful abstract information. If a radius function is specified at each point on the skeleton, this representation is capable of general-purpose object description.

Multiple Two-Dimensional Projection

For some applications, a library of two-dimensional silhouette projections that represent three-dimensional objects can be conveniently stored. For the recognition of three-dimensional objects with a small number of stable orientations on a flat table, this representation is ideal—if the object silhouettes are sufficiently different. For example, silhouettes have been used to recognize aircraft in any orientation against a well-lit sky background. However, because many different three-dimensional-object shapes can possess the same silhoutte projection, this type of representation is not a general-purpose technique.

Aspect Graphs

In the aspect graph representation, the space of viewpoints is partitioned into maximal regions, where every viewpoint in each region gives the same qualitative view of the object, called the *aspect*. Within each region, **projections** of the object will have the same number and types of features, with identical spatial relationships among them. However, the quantitative properties of these features, such as lengths of edges, vary with the change in viewpoint. Changes in the aspect, called *visual events*, take place at the boundaries between regions. Two aspects are said to be connected by a visual event if their corresponding regions are adjacent in the viewpoint space. An aspect graph is a graph structure whose nodes represent aspects and their associated regions and whose arcs denote the visual events and boundaries between adjacent regions. Figure 19.50(f) shows the aspect graph for a cube.

Characteristic Views

A concept very similar to aspect graphs is that of characteristic views. All of the infinite two-dimensional perspective projection views of an object are grouped into a finite number of topologically equivalent classes. Different views within an equivalence class are related via linear transformations. A representative member of an equivalence class is called the *characteristic view* for that class. In this scheme, objects are assumed to rest on a supporting plane; hence, they are restricted to appear in a number of stable positions. Characteristic views of objects are derived with certain constraints on the camera configuration. It is this use of camera position and orientation information that differentiates the characteristic view representation scheme from the aspect graph. Because characteristic views specify the three-dimensional structure of an object, they also provide a general-purpose object representation.

19.4.6 Dynamic Vision

Early machine vision systems were concerned primarily with static scenes; however, the world is dynamic. Designing machine vision systems capable of analyzing dynamic scenes is increasingly becoming more popular. For a machine vision system engaged in the performance of nontrivial real-world operations and tasks, the ability to cope with moving and changing objects and viewpoints is vital.

The input to a dynamic-scene analysis system is a sequence of image frames. The camera used to acquire the image sequence may itself be in motion. Each frame represents an image of the scene at a particular instant in time. The changes in a scene may be due to camera motion, object motion, illumination changes, or changes in object structure, size, or shape. Changes in a scene are usually assumed to be due to camera and/or object motion; objects are usually assumed to be either rigid or quasirigid. Other changes are not allowed.

A scene usually contains several objects. An image of the scene at a given time represents a projection of part of the scene; the part of the scene depends on the position of the camera. Four cases represent the possibilities for the dynamic-camera/world setup:

1. Stationary camera, stationary objects (SCSO)
2. Stationary camera, moving objects (SCMO)
3. Moving camera, stationary objects (MCSO)
4. Moving camera, moving objects (MCMO)

The first case is simple static-scene analysis. In many applications, processing a single image to obtain the required information may be feasible. However, many more applications exist that require information to be extracted from a dynamic environment.

Clearly, a sequence of image frames offers much more information to aid in scene understanding, but significantly increases the amount of data to be processed by the system. Applying static-scene analysis techniques to each frame of a sequence requires an enormous amount of computation, while still suffering from all of the problems of static-scene analysis. Fortunately, research in dynamic-scene analysis has shown that information recovery may be easier in dynamic scenes than in static scenes. In some cases, the total computational effort may be significantly less and the performance is better.

SCMO scenes have received the most attention in dynamic-scene analysis. The objectives of such scene analysis usually are to detect motion, to extract masks of moving objects with the aim of recognizing them, and to compute their motion characteristics. MCMO is the most general case and possibly presents the most difficult situation in dynamic-scene analysis. Many techniques that have been developed assuming a stationary camera are not applicable to a moving camera. Likewise, techniques developed for a moving camera generally assume a stationary scene and are usually not applicable if the objects are moving. SCMO and MCSO have found uses in many applications and have been studied by researchers in various contexts under various assumptions.

Change Detection

Any perceptible motion in a scene results in some change in the sequence of frames of the scene. If such changes are detected, motion characteristics can be analyzed. A good quantitative estimate of the motion components of an object can be obtained if the motion is restricted to a plane that is parallel to the image plane; for three-dimensional motion, only qualitative estimates are possible. By analyzing frame-to-frame differences, a global analysis of the sequence can be performed. Changes can be detected at different levels: pixel, edge, or region.

Difference Pictures

The most obvious method of detecting change between two frames is to directly compare the corresponding pixels of the frames to determine whether or not they are the same. In the simplest form, a binary *difference picture* $DP_{jk}(x, y)$ between frames j and k is obtained by

$$DP_{jk}(x, y) = \begin{cases} 1 & \text{if } |F(x, y, j) - F(x, y, k)| > \tau \\ 0 & \text{otherwise} \end{cases} \tag{19.66}$$

where τ is a threshold and $F(x, y, j)$ and $F(x, y, k)$ are the image arrays of the jth and kth frames, respectively.

In the difference picture, pixels that have a value of 1 may be considered to be the result of object motion. Because of noise, however, such a simple test on real scenes usually produces unsatisfactory results. A simple size filter may be used to ignore pixels not forming a connected cluster of minimal size. Then only those pixels in the difference picture with the value of 1 that belong to a four-connected (or eight-connected) component larger than a set number of pixels will be attributed to object motion. This filter is very effective in reducing noise, but unfortunately it also filters some desirable signals, such as those from small or slowly moving objects.

Likelihood Ratio

To make change detection more robust, regions or groups of pixels at the same location in two frames may be considered and their intensity characteristics may be compared more rigorously. One method using this approach is based on comparing the frames using the *likelihood ratio*. Thus, we can compute the difference picture by replacing $|F(x, y, j) - F(x, y, k)|$ with λ where

$$\lambda = \frac{\left[\dfrac{\sigma_1^2 + \sigma_2^2}{2} + \left(\dfrac{\mu_1 - \mu_2}{2} \right)^2 \right]^2}{\sigma_1^2 \sigma_2^2} \tag{19.67}$$

and μ and σ denote the mean gray value and the square root of the variance from the frames for the sample area, respectively.

The likelihood ratio can only be applied to regions and not to single pixels. This limitation presents a minor problem, which can be solved by considering corresponding areas of the frames. The likelihood ratio test, combined with a size filter, works well for noise removal in many real-world scenes. The likelihood ratio test can be applied to every point in each frame by considering overlapping areas centered on each pixel of the image, or to a subsampling of points by using nonoverlapping areas, called *superpixels*.

Accumulative Difference Pictures

The problem of missing the detection of small or slowly moving objects can be solved by analyzing change over a series of frames, instead of just between two frames. The *accumulative difference picture* (ADP) is used to detect the motion of small and slowly moving objects. An accumulative difference picture is formed by comparing every frame in an image sequence to a common reference frame. The entry in the accumulative difference picture is increased by one whenever the likelihood ratio for that region exceeds the threshold. Thus, an accumulative difference picture acquired over k frames is given by

$$ADP_0(x, y) = 0$$
$$ADP_k(x, y) = ADP_{k-1}(x, y) + DP_{1k}(x, y) \tag{19.68}$$

where the first frame of a sequence is usually the reference frame.

Time-Varying Edge Detection

As a result of the importance of edge detection in static scenes, it is reasonable to expect that time-varying edge detection may also be important in dynamic-scene analysis. Moving edges can be detected by combining the spatial and temporal gradients using a logical AND operation, which is implemented through multiplication. Thus, the time-varying edgeness of a point in a frame is given by

$$E_t(x, y, t) = \frac{dF(x, y, t)}{dS} \frac{dF(x, y, t)}{dt} \tag{19.69}$$

where dF/dS and dF/dt are the magnitudes of the spatial and temporal gradients of the intensity at point (x, y, t), respectively. Various conventional edge detectors can be used to compute the spatial gradient, and a simple difference can be used to compute the temporal gradient. In most cases, this edge detector works effectively. By applying a threshold to the product—rather than first differencing and then applying an edge detector, or first detecting edges and then computing their temporal gradient—this method overcomes the problem of missing weak or slowly moving edges.

Optical Flow

Optical flow is the distribution of velocity, relative to the observer, over the points of an image. Optical flow carries information that is valuable in dynamic-scene analysis. Optical flow is determined by the velocity vector at each pixel in an image. Several methods have been devised for calculating optical flow based on two or more frames of a sequence. These methods can be classified into two general categories: *feature based* and *gradient based*. If a stationary camera is used, most of the points in an image frame will have zero velocity. This is assuming that a very small subset of the scene is in motion, which is usually true. Thus, most applications for optical flow involve a moving camera.

Feature-Based Methods

Feature-based methods for computing optical flow first select some features in consecutive image frames, match these features between frames, and then calculate the disparities between them. As mentioned earlier, the **correspondence problem** can be solved using relaxation. However, the problem of selecting features and establishing correspondence is not easy. Moreover, this method only produces velocity vectors at sparse points within the image.

Gradient-Based Methods

Gradient-based methods exploit the relationship between the spatial and temporal gradients of image intensity. This relationship can be used to segment images based on the velocity of image points. The relationship between the spatial and temporal gradients and the velocity components is

$$F_x u + F_y v + F_t = 0 \tag{19.70}$$

where $u = dx/dt$ and $v = dy/dt$. In this equation, F_x, F_y, and F_t represent the spatial gradients in the x and y directions and the temporal gradient, respectively, and can be computed directly from the image. Then, at every point in an image, there are two unknowns, u and v, yet there is only one equation. Thus, optical flow cannot be directly determined.

The velocity field, however, can be assumed to vary smoothly over an image. Under this assumption, an iterative approach for computing optical flow using two or more frames can be utilized. The following iterative equations are used for the computation of image flow:

$$u = u_{\text{average}} - F_x \frac{P}{D} \qquad v = v_{\text{average}} - F_y \frac{P}{D} \tag{19.71}$$

where

$$P = F_x u_{\text{average}} + F_y v_{\text{average}} + F_t \tag{19.72}$$

$$D = \lambda^2 + F_x^2 + F_y^2 \tag{19.73}$$

where λ is a constant multiplier. When only two frames are used, the computation is iterated over the same frames many times. For more than two frames, each iteration uses a new frame.

Segmentation Using a Moving Camera

If the camera is moving, then every point in the image has nonzero velocity relative to the camera (except the case where object points are moving with the same velocity as the camera). The velocity relative to the camera depends on both the velocity of the point itself and the distance of the point from the camera. Approaches based on differences may be extended for segmenting moving-camera scenes. If the aim is to extract images of moving objects, however, then additional information is required to decide whether the motion at a point is due solely to its depth or is due to a combination of its depth and its motion. Gradient-based approaches will also require additional information.

If the camera's direction of motion is known, the *focus of expansion* (FOE) with respect to the stationary components in the scene can easily be computed. The FOE will have coordinates (x_f, y_f) where

$$x_f = \frac{dx}{dz} \qquad y_f = \frac{dy}{dz} \tag{19.74}$$

in the image plane. The velocity vectors of all stationary points in a scene project onto the image plane so that they intersect at the FOE. A transformation with respect to the FOE can be used to simplify the task of segmentation. The *ego-motion polar* (EMP) transformation of an image transforms a frame $F(x, y, t)$ into $E(r, \theta, t)$ using

$$E(r, \theta, t) = F(x, y, t) \tag{19.75}$$

where

$$r = \sqrt{(x - x_f)^2 + (y - y_f)^2} \tag{19.76}$$

and

$$\theta = \tan^{-1}\left[\frac{(y - y_f)}{(x - x_f)}\right] \tag{19.77}$$

In EMP space, stationary points are displaced only along the θ axis between the frames of an image sequence, whereas points on moving objects are displaced along the r axis as well as the θ axis. Thus, the displacement in EMP space can be used to segment a scene into its stationary and nonstationary components. Moreover, when *complex logarithmic mapping* (CLM) is performed about the FOE, interesting points in the EMP space can be obtained.

19.4.7 Applications

Machine vision systems have uses in a wide variety of disciplines, from medicine to robotics, from automatic inspection to autonomous navigation, and from document analysis to multimedia systems. New machine vision systems are constantly emerging and becoming a part of everyday life. In this section, we present a brief description some of the various applications of machine vision.

Optical Character Recognition (OCR) and Document Image Analysis

The objective of document image analysis is to recognize the text and graphics components in images and extract the intended information as a human would. Two categories of document processing can be defined, *textual processing* and *graphics processing*. Textual processing deals with the text components of a document image. Some tasks here are recognizing the text by optical character recognition (OCR); skew detection and correction (any tilt at which the documents may have been scanned); and finding the columns, paragraphs, text lines, and words. Graphics processing deals with the nontextual components of a document image, such as lines and symbols that make up line diagrams, delimiting lines between text sections, and company logos.

Document analysis is currently a very active area of research. Researchers have devised systems ranging from automatic engineering drawing interpretation and recognition of tabular drawings to the recognition of zip codes from postal envelopes and the interpretation of musical scores. However, the real success story of document analysis is OCR. This is the one area of machine vision in which scientific study has lead to numerous low-cost marketed products. Many of the current OCR systems report accuracies well in the upper 90th percentile.

Many document analysis publications can be found in the journals and conference proceedings listed in the Further Information section. In addition, the International Conference on Document Analysis and Recognition (ICDAR) and the International Workshop or Graphics Recognition (IWGR) are biannual meetings held in conjunction with each other dedicated entirely to the field of document analysis.

Medical Image Analysis

Medical imaging analysis deals primarily with images such as X-rays, computerized tomograph (CT) scans, and magnetic resonance imaging (MRI) images. Early work in medical image analysis overlapped with that of image processing; the main task was to enhance medical images for viewing by a physician; no automatic interpretation or high-level reasoning by the system was involved. However, more recent work is being conducted on medical imagery, which more closely fits the definition of machine vision; for example, systems to search images for diseased organs or tissues based on known models (images) and features of the diseased samples and systems to generate three-dimensional models of organs based on CT scan and MRI. Some active areas of research include the use of three-dimensional images in surgery and surgical planning (especially neurosurgery), virtual surgery, and generating time-varying three-dimensional models obtained from sequences of images (e.g., pumping heart). These systems encompass many of the aspects of typical machine vision systems plus incorporate many aspects of computer graphics.

In addition to the listings at the end of this section, the following are excellent sources of information on current research in medical imaging: *IEEE Transactions on Medical Imaging*, *IEEE Transactions on Biomedical Engineering*, *IEEE Transactions on Image Processing*, and *IEEE Engineering in Medicine and Biology Magazine*. There are also a number of conferences dedicated to medical imaging research: SPIE Medical Imaging, IEEE Engineering in Medicine and Biology, Medicine Meets Virtual Reality, select sessions/symposium of IEEE Visualization, and SPIE Visualization in Biomedical Computing.

Photogrammetry and Aerial Image Analysis

Photogrammetry deals with the task of making reliable measurements from images. In the early days of photogrammetry, the images were actual printed photographs often taken from balloons. Today, however, the remote sensing process is multispectral using energy in many other parts of the electromagnetic spectrum, such as ultraviolet and infrared. The images are often transmitted directly from satellites orbiting the globe, such as the Landsat satellites first launched in the early 1970s. Some of the applications of photogrammetry and aerial image analysis include atmospheric exploration; thermal image analysis for energy conservation; monitoring natural resources, crop conditions, land cover, and land use; weather prediction; pollution studies; urban planning; military reconnaissance; plus many others in geology, hydrology, and oceanography.

There are several organizations dedicated to the study of these types of remote sensing tasks. The following are very good sources for information on photogrammetry and aerial image analysis: The American Society of Photogrammetry, The International Remote Sensing Institute (ISRI), and The American Congress on Surveying and Mapping. In addition, there are several conferences and symposia dealing with this topic: SPIE Remote Sensing, International Geosciences and Remote Sensing Symposium, *IEEE Transactions on Geoscience and Remote Sensing*, and the Symposium on Remote Sensing sponsored by the American Geological Institute.

Industrial Inspection and Robotics

Unlike many of the already mentioned machine vision tasks, automatic inspection and robotics perform many tasks in real time that significantly increases the complexity of the system. The general goal of such systems is sensor-guided control. Typical industrial applications include automatic inspection of machined parts, solder joint surfaces (welds), silicon wafers, produce, and even candy. Some of the challenges faced by industrial vision system designers include determining the optimal configuration of the camera and lighting, determining the most suitable color space representation of the illumination, modeling various surface reflectance mechanisms, dynamic sensor feedback, real-time manipulator control, real-time operating system interfaces, and neural networks.

Structured lighting techniques have been used extensively in industrial vision applications where the illumination of the scene can be easily controlled. In a typical application, objects on a conveyor belt pass through a plane of light, creating a distortion in the image of the light stripe. The profile of the object at the plane of the light beam is then calculated. This process is repeated at regular intervals as the object moves along the conveyor belt to recover the shape of the object. Then appropriate actions are taken depending on the goal of the system.

The following are sources dedicated primarily to industrial vision applications and robotics: *International Journal of Robotics Research, IEEE Transactions on Robotics and Automation*, IEEE's International Conference on Robotics and Automation, and *IEEE Transactions on Systems, Man, and Cybernetics*.

Autonomous Navigation

Closely related to robotics is the area of autonomous navigation. Much work is being done to develop systems to enable robots or other mobile vehicles to automatically navigate through a specific environment. Techniques involved include active vision (sensor control), neural networks, high-speed stereo vision, three-dimensional vision (range imaging), high-level reasoning for navigational planning, and signal compression and transmission for accurate remote vehicle control.

Visual Information Management Systems

Probably one of the newest areas of machine vision research is in visual information management systems (VIMS). With the recent advances in low-cost computing power and the ever increasing number of multimedia applications, digital imagery is becoming a larger and larger part of every day life. Research in VIMS is providing methods to handle all of this digital information. Such VIMS applications include interactive television, video teleconferencing, digital libraries, video-on-demand, and large-scale video databases. Image processing and machine vision play a very important role in much of this work ranging anywhere from designing video compression schemes, which allow many processing techniques to be performed directly on the compressed datastream and developing more efficient indexing methods for multidimensional data, to automatic scene cut detection for automatically indexing large stockpiles of video data and developing methods to query image databases by image content as opposed to standard structured grey language (SQL) techniques.

Defining Terms

Correspondence problem: The problem of matching points in one image with their corresponding points in a second image.

Histogram: A plot of the frequency of occurrence of the grey levels in an image.

Quantization: The process of representing the continuous range of image intensities by a limited number of discrete grey values.

Sampling: The process of representing the continuous image intensity function as a discrete two-dimensional array.

Segmentation: The process of separating the objects from the background.

Polygonalization: A method of representing a contour by a set of connected straight line segments; for closed curves, these segments form a polygon.

Projection: The transformation and representation of a high-dimensional space in a lesser number of dimensions (i.e., a three-dimensional scene represented as a two-dimensional image).

Thresholding: A method of separating objects from the background by selecting an interval (usually in pixel intensity) and setting any points within the interval to 1 and points outside the interval to 0.

References

Besl, P.J. 1988. Active, Optical range imaging sensors. *Machine Vision and Applications* 1(2):127–152.

Besl, P. and Jain, R.C. 1985. Three dimensional object recognition. *ACM Computing Surveys* 17(1).

Duda, R.O. and Hart, P.E. 1973. *Pattern Classification and Scene Analysis.* Wiley, New York.

Foley, J.D., van Dam, A., Feiner, S.K., and Hughes, J.F. 1990. *Computer Graphics, Principles and Practice.* Addison-Wesley, Reading, MA.

Gonzalez, R.C. and Woods, R.E. 1992. *Digital Image Processing.* Addison-Wesley, Reading, MA.

Javis, R.A. 1983. A perspective on range finding techniques for computer vision. *TEEE Trans. on Pattern Analysis and Machine Intelligence* 5(2):122–139.

Kasturi, R. and Jain, R.C. 1991. *Computer Vision: Principles.* IEEE Computer Society Press.

Kosko, B. 1992. *Neural Networks and Fuzzy Systems.* Prentice-Hall, Englewood Cliffs, NJ.

Jain, R., Kasturi, R., and Schunck, B.G. 1995. *Machine Vision.* McGraw-Hill, New York.

Marr, D. 1982. *Vision: A Computational Investigation into the Human Representation and Processing of Visual Information.* W. H. Freeman, San Francisco, CA.

Tanimoto, S.L. 1995. *The Elements of Artificial Intelligence Using Common Lisp.* Computer Science Press.

Winston, P.H. 1992. *Artificial Intelligence.* Addison-Wesley, Reading, MA.

Further Information

Much of the material presented in this section has been adapted from:

Jain, R., Kasturi, R., and Schunck, B.G. 1995. *Machine Vision.* McGraw-Hill, New York.

Kasturi, R. and Jain, R.C. 1991. *Computer Vision: Principles.* IEEE Computer Society Press, Washington.

In addition the following books are reccommended:

Rosenfeld, A. and Kak, A.C. 1982. *Digital Picture Processing.* Academic Press, Englewood Cliffs, NJ.

Jain, A.K. 1989. *Fundamentals of Digital Image Processing.* Prentice-Hall, New York.

Haralick, R.M. and Shapiro, L.G. 1992–1993. *Computer and Robot Vision.* Addison-Wesley, Reading, MA.

Horn, B. 1986. *Robot Vision.* McGraw-Hill.

Schalkoff, R.J. 1989. *Digital Image Processing and Computer Vision.* Wiley, New York.

Additional information on machine vision can be found in the following technical journals:

Artificial Intelligence (*AI*)
Computer Vision, Graphics, and Image Processing (*CVGIP*)
IEEE Transactions on Image Processing
IEEE Transactions on Pattern Analysis and Machine Intelligence (*PAMI*)
Image and Vision Computing
International Journal of Computer Vision
International Journal on Pattern Recognition and Artificial Intelligence
Machine Vision and Applications (*MVA*)
Pattern Recognition (*PR*)
Pattern Recognition Letters (*PRL*)

The following conference proceedings are also a good source for machine vision information:

A series of conferences by the International Society for Optical Engineering (SPIE)
A series of workshops by the Institute of Electrical and Electronic Engineering (IEEE) Computer Society
A series of workshops by the International Association for Pattern Recognition (IAPR)
European Conference on Computer Vision (ECCV)
IEEE Conference on Computer Vision and Pattern Recognition (CVPR)
International Conference on Computer Vision (ICCV)
International Conference on Pattern Recognition (ICPR)

19.5   A Brief Survey of Speech Enhancement[2]

Yariv Ephraim, Hanoch Lev-Ari, and William J.J. Roberts

19.5.1   Introduction

Speech enhancement aims at improving the performance of speech communication systems in noisy environments. Speech enhancement may be applied, for example, to a mobile radio communication system, a speech recognition system, a set of low quality recordings, or to improve the performance of aids for the hearing impaired. The interference source may be a wide-band noise in the form of a white or colored noise, a periodic signal such as in hum noise, room reverberations, or it can take the form of fading noise. The first two examples represent additive noise sources, while the other two examples represent convolutional and multiplicative noise sources, respectively. The speech signal may be simultaneously attacked by more than one noise source.

There are two principal perceptual criteria for measuring the performance of a speech enhancement system. The *quality* of the enhanced signal measures its clarity, distorted nature, and the level of residual noise in that signal. The quality is a *subjective* measure that is indicative of the extent to which the listener is comfortable with the enhanced signal. The second criterion measures the *intelligibility* of the enhanced signal. This is an *objective* measure which provides the percentage of words that could be correctly identified by listeners. The words in this test need not be meaningful. The two performance measures are not correlated. A signal may be of good quality and poor intelligibility and vice versa. Most speech enhancement systems improve the quality of the signal at the expense of reducing its intelligibility. Listeners can usually extract more information from the noisy signal than from the enhanced signal by careful listening to that signal. This is obvious from the *data processing theorem* of information theory. Listeners, however, experience fatigue over extended listening sessions, a fact that results in reduced intelligibility of the noisy signal. Is such situations, the intelligibility of the enhanced signal may be higher than that of the noisy signal. Less effort would usually be required from the listener to decode portions of the enhanced signal that correspond to high signal to noise ratio segments of the noisy signal.

Both the quality and intelligibility are elaborate and expensive to measure, since they require listening sessions with live subjects. Thus, researchers often resort to less formal listening tests to assess the quality of an enhanced signal, and they use automatic speech recognition tests to assess the intelligibility of that signal. Quality and intelligibility are also hard to quantify and express in a closed form that is amenable to mathematical optimization. Thus, the design of speech enhancement systems is often based on mathematical measures that are somehow believed to be correlated with the quality and intelligibility of the speech signal. A popular example involves estimation of the clean signal by minimizing the mean square error (MSE) between the logarithms of the spectra of the original and estimated signals [5]. This criterion is believed to be more perceptually meaningful than the minimization of the MSE between the original and estimated signal waveforms [13].

Another difficulty in designing efficient speech enhancement systems is the lack of *explicit* statistical models for the speech signal and noise process. In addition, the speech signal, and possibly also the noise process, are not strictly stationary processes. Common parametric models for speech signals, such as an autoregressive process for short-term modeling of the signal, and a hidden Markov process (HMP) for long-term modeling of the signal, have not provided adequate models for speech enhancement applications. A variant of the *expectation-maximization* (EM) algorithm, for maximum likelihood (ML) estimation of the autoregressive parameter from a noisy signal, was developed by Lim and Oppenheim [12] and tested in speech enhancement. Several estimation schemes, which are based on hidden Markov modeling of the clean speech signal and of the noise process, were developed over the years, see, e.g., Ephraim [6]. In each

---

[2] This paper is partially based on "Extension of the Signal Subspace Speech Enhancement Approach to Colored Noise" by Hanoch Lev-Ari and Yariv Ephraim which appeared in *IEEE Sig. Proc. Let.*, vol. 10, pp. 104–106, April 2003 ©2003 IEEE.

case, the HMPs for the speech signal and noise process were designed from training sequences from the two processes, respectively. While autoregressive and hidden Markov models have proved extremely useful in coding and recognition of *clean* speech signals, respectively, they were not found to be sufficiently refined models for speech enhancement applications.

In this chapter we review some common approaches to speech enhancement that were developed primarily for additive wide-band noise sources. Although some of these approaches have been applied to reduction of reverberation noise, we believe that the dereverberation problem requires a completely different approach that is beyond the scope of this capter. Our primary focus is on the spectral subtraction approach [13] and some of its derivatives such as the signal subspace approach [7] [11], and the estimation of the short-term spectral magnitude [16, 4, 5]. This choice is motivated by the fact that some derivatives of the spectral subtraction approach are still the best approaches available to date. These approaches are relatively simple to implement and they usually outperform more elaborate approaches which rely on parametric statistical models and training procedures.

19.5.2   The Signal Subspace Approach

In this section we present the principles of the signal subspace approach and its relations to Wiener filtering and spectral subtraction. Our presentation follows [7] and [11]. This approach assumes that the signal and noise are noncorrelated, and that their second-order statistics are available. It makes no assumptions about the distributions of the two processes.

Let Y and W be k-dimensional random vectors in a Euclidean space $\mathcal{R}^k$ representing the clean signal and noise, respectively. Assume that the expected value of each random variable is zero in an appropriately defined probability space. Let $Z = Y + W$ denote the noisy vector. Let R_y and R_w denote the covariance matrices of the clean signal and noise process, respectively. Assume that R_w is positive definite. Let H denote a $k \times k$ real matrix in the linear space $\mathcal{R}^{k \times k}$, and let $\hat{Y} = HZ$ denote the linear estimator of Y given Z. The residual signal in this estimation is given by

$$Y - \hat{Y} = (I - H)Y - HW \tag{19.78}$$

where I denotes, as usual, the identity matrix. To simplify notation, we shall not explicitly indicate the dimensions of the identity matrix. These dimensions should be clear from the context. In Eq. (19.78), $D = (I - H)Y$ is the signal distortion and $N = HW$ is the residual noise in the linear estimation. Let $(\cdot)'$ denote the transpose of a real matrix or the conjugate transpose of a complex matrix. Let

$$\overline{\epsilon_d^2} = \frac{1}{k}\text{tr}E\{DD'\} = \frac{1}{k}\text{tr}\{(I - H)R_y(I - H)'\} \tag{19.79}$$

denote the average signal distortion power where tr$\{\cdot\}$ denotes the trace of a matrix. Similarly, let

$$\overline{\epsilon_n^2} = \frac{1}{k}\text{tr}E\{NN'\} = \frac{1}{k}\text{tr}\{HR_wH'\} \tag{19.80}$$

denote the average residual noise power.

The matrix H is estimated by minimizing the signal distortion $\overline{\epsilon_d^2}$ subject to a threshold on the residual noise power $\overline{\epsilon_n^2}$. It is obtained from

$$\min_H \overline{\epsilon_d^2}$$
$$\text{subject to}: \overline{\epsilon_n^2} \leq \alpha \tag{19.81}$$

for some given α. Let $\mu \geq 0$ denote the Lagrange multiplier of the inequality constraint. The optimal matrix, say $H = H_1$, is given by

$$H_1 = R_y(R_y + \mu R_w)^{-1} \tag{19.82}$$

The matrix H_1 can be implemented as follows. Let $R_w^{1/2}$ denote the symmetric positive definite square root of R_w and let $R_w^{-1/2} = (R_w^{1/2})^{-1}$. Let U denote an orthogonal matrix of eigenvectors of the symmetric matrix $R_w^{-1/2} R_y R_w^{-1/2}$. Let $\Lambda = \mathrm{diag}(\lambda_1, \ldots, \lambda_k)$ denote the diagonal matrix of nonnegative eigenvalues of $R_w^{-1/2} R_y R_w^{-1/2}$. Then

$$H_1 = R_w^{1/2} U \Lambda (\Lambda + \mu I)^{-1} U' R_w^{-1/2} \tag{19.83}$$

When H_1 in Eq. (19.83) is applied to Z, it first whitens the input noise by applying $R_w^{-1/2}$ to Z. Then, the orthogonal transformation U' corresponding to the covariance matrix of the whitened clean signal is applied, and the transformed signal is modified by a diagonal Wiener-type gain matrix.

In Eq. (19.83), components of the whitened noisy signal that contain noise only are nulled. The indices of these components are given by $\{ j : \lambda_j = 0 \}$. When the noise is white, $R_w = \sigma_w^2 I$, and U and Λ are the matrices of eigenvectors and eigenvalues of R_y / σ_w^2, respectively. The existence of null components $\{ j : \lambda_j = 0 \}$ for the signal means that the signal lies in a subspace of the Euclidean space $\mathcal{R}^k$. At the same time, the eigenvalues of the noise are all equal to σ_w^2 and the noise occupies the entire space $\mathcal{R}^k$. Thus, the signal subspace approach first eliminates the noise components outside the signal subspace and then modifies the signal components inside the signal subspace in accordance with the criterion of Eq. (19.81).

When the signal and noise are *wide-sense stationary*, the matrices R_y and R_w are Toeplitz with associated power spectral densities of $f_y(\theta)$ and $f_w(\theta)$, respectively. The angular frequency θ lies in $[0, 2\pi)$. When the signal and noise are *asymptotically weakly stationary*, the matrices R_y and R_w are asymptotically Toeplitz and have the associated power spectral densities $f_y(\theta)$ and $f_w(\theta)$, respectively [10]. Since the latter represents a somewhat more general situation, we proceed with asymptotically weakly stationary signal and noise. The filter H_1 in Eq. (19.82) is then asymptotically Toeplitz with associated power spectral density

$$h_1(\theta) = \frac{f_y(\theta)}{f_y(\theta) + \mu f_w(\theta)} \tag{19.84}$$

This is the noncausal Wiener filter for the clean signal with an adjustable noise level determined by the constraint α in Eq. (19.81). This filter is commonly implemented using estimates of the two power spectral densities. Let $\hat{f}_y(\theta)$ and $\hat{f}_w(\theta)$ denote the estimates of $f_y(\theta)$ and $f_w(\theta)$, respectively. These estimates could, for example, be obtained from the periodogram or the smoothed periodogram. In that case, the filter is implemented as

$$\hat{h}_1(\theta) = \frac{\hat{f}_y(\theta)}{\hat{f}_y(\theta) + \mu \hat{f}_w(\theta)} \tag{19.85}$$

When $\hat{f}_y(\theta)$ is implemented as

$$\hat{f}_y(\theta) = \begin{cases} \hat{f}_z(\theta) - \hat{f}_w(\theta) & \text{if nonnegative} \\ \varepsilon & \text{otherwise} \end{cases} \tag{19.86}$$

then a *spectral subtraction* estimator for the clean signal results. The constant $\varepsilon \geq 0$ is often referred to as a "spectral floor." Usually $\mu \geq 2$ is chosen for this estimator.

The enhancement filter H could also be designed by imposing constraints on the spectrum of the residual noise. This approach enables shaping of the spectrum of the residual noise to minimize its perceptual effect. Suppose that a set $\{ v_i, i = 1, \ldots, m \}$, $m \leq k$, of k-dimensional real or complex orthonormal vectors, and a set $\{ \alpha_i, i = 1, \ldots, m \}$ of non-negative constants, are chosen. The vectors $\{ v_i \}$ are used to transform the residual noise into the spectral domain, and the constants $\{ \alpha_i \}$ are used as upper bounds on the variances of these spectral components. The matrix H is obtained from

$$\min_H \overline{\epsilon_d^2} \tag{19.87}$$
$$\text{subject to} : E\{|v_i' N|^2\} \leq \alpha_i, \quad i = 1, \ldots, m$$

When the noise is white, the set $\{v_i\}$ could be the set of eigenvectors of R_y and the variances of the residual noise along these coordinate vectors are constrained. Alternatively, the set $\{v_i\}$ could be the set of orthonormal vectors related to the DFT. These vectors are given by $v_i' = k^{-1/2}(1, e^{-j\frac{2\pi}{k}(i-1)\cdot 1}, \ldots, e^{-j\frac{2\pi}{k}(i-1)\cdot(k-1)})$. Here we must choose $\alpha_i = \alpha_{k-i+2}, i = 2, \ldots, k/2$, assuming k is even, for the residual noise power spectrum to be symmetric. This implies that at most $k/2 + 1$ constraints can be imposed. The DFT-related $\{v_i\}$ enable the use of constraints that are consistent with auditory perception of the residual noise.

To present the optimal filter, let e_l denote a unit vector in $\mathcal{R}^k$ for which the lth component is one and all other components are zero. Extend $\{v_1, \ldots, v_m\}$ to a $k \times k$ orthogonal or unitary matrix $V = (v_1, \ldots, v_k)$. Set $\mu_i = 0$ for $m < i \leq k$, and let $M = \text{diag}(k\mu_1, \ldots, k\mu_k)$ denote the matrix of k times the Lagrange multipliers which are assumed nonnegative. Define $Q = R_w^{-1/2}U$ and $T = Q'V$. The optimal estimation matrix, say $H = H_2$, is given by [11]

$$H_2 = R_w^{1/2}U\tilde{H}_2U'R_w^{-1/2} \tag{19.88}$$

where the columns of $\tilde{H}_2$ are given by

$$\tilde{h}_l = T\lambda_l(M + \lambda_l I)^{-1}T^{-1}e_l, \quad l = 1, \ldots, k \tag{19.89}$$

provided that $k\mu_i \neq -\lambda_l$ for all $i.l$. The optimal estimator first whitens the noise, then applies the orthogonal transformation U' obtained from eigendecomposition of the covariance matrix of the whitened signal, and then modifies the resulting components using the matrix $\tilde{H}_2$. This is analogous to the operation of the estimator H_1 in Eq. (19.83). The matrix $\tilde{H}_2$, however, is not diagonal when the input noise is colored.

When the noise is white with variance σ_w^2 and $V = U$ and $m = k$ are chosen, the optimization problem of Eq. (19.87) becomes trivial since knowledge of input and output noise variances uniquely determines the filter H. This filter is given by $H = UGU'$ where $G = \text{diag}(\sqrt{\alpha_1}, \ldots, \sqrt{\alpha_k})$ [7]. For this case, the heuristic choice of

$$\alpha_i = \exp\{-\nu\sigma_w^2/\lambda_i\} \tag{19.90}$$

where $\nu \geq 1$ is an experimental constant and was found useful in practice [7]. This choice is motivated by the observation that for $\nu = 2$, the first order Taylor expansion of $\alpha_i^{-1/2}$ leads to an estimator $H = UGU'$ which coincides with the Wiener estimation matrix in Eq. (19.83) with $\mu = 1$. The estimation matrix using Eq. (19.90) performs significantly better than the Wiener filter in practice.

19.5.3 Short-Term Spectral Estimation

In another earlier approach for speech enhancement, the short-time spectral magnitude of the clean signal is estimated from the noisy signal. The speech signal and noise process are assumed statistically independent, and spectral components of each of these two processes are assumed zero mean statistically independent Gaussian random variables. Let $A_y e^{j\theta_y}$ denote a spectral component of the clean signal Y in a given frame. Let $A_z e^{j\theta_z}$ denote the corresponding spectral component of the noisy signal. Let $\sigma_y^2 = E\{A_y^2\}$ and $\sigma_z^2 = E\{A_z^2\}$ denote, respectively, the variances of the clean and noisy spectral components. If the variance of the corresponding spectral component of the noise process in that frame is denoted by σ_w^2, then we have $\sigma_z^2 = \sigma_y^2 + \sigma_w^2$. Let

$$\xi = \frac{\sigma_y^2}{\sigma_w^2}; \quad \gamma = \frac{A_z^2}{\sigma_w^2}; \quad \vartheta = \frac{\xi}{\xi+1}\gamma \tag{19.91}$$

The MMSE estimation of A_y from $A_z e^{j\theta_z}$ is given by [4]

$$\hat{A}_y = \Gamma(1.5)\frac{\sqrt{\vartheta}}{\gamma}\exp\left(-\frac{\vartheta}{2}\right)\left[(1+\vartheta)I_0\left(\frac{\vartheta}{2}\right) + \vartheta I_1\left(\frac{\vartheta}{2}\right)\right]A_z \tag{19.92}$$

where $\Gamma(1.5) = \frac{\sqrt{\pi}}{2}$, and $I_0(\cdot)$ and $I_1(\cdot)$ denote, respectively, the modified Bessel functions of the zeroth and first order. Similarly to the Wiener filter given in Eq. (19.82), this estimator requires knowledge of second order statistics of each signal and noise spectral components, σ_y^2 and σ_w^2, respectively.

To form an estimator for the spectral component of the clean signal, the spectral magnitude estimator Eq. (19.92) is combined with an estimator of the complex exponential of that component. Let $\widehat{e^{j\theta_y}}$ be an estimator of $e^{j\theta_y}$. This estimator is a function of the noisy spectral component $A_z e^{j\theta_z}$. MMSE estimation of the complex exponential $e^{j\theta_y}$ is obtained from

$$\min_{\widehat{e^{j\theta_y}}} E\{|e^{j\theta_y} - \widehat{e^{j\theta_y}}|^2\}$$
$$\text{subject to} : |\widehat{e^{j\theta_y}}| = 1 \tag{19.93}$$

The constraint in Eq. (19.93) ensures that the estimator $\widehat{e^{j\theta_y}}$ does not affect the optimality of the estimator $\hat{A}_y$ when the two are combined. The constrained minimization problem in Eq. (19.93) results in the estimator $\widehat{e^{j\theta_y}} = e^{j\theta_z}$ which is simply the complex exponential of the noisy signal.

Note that the Wiener filter Eq. (19.85) has zero phase and hence it effectively uses the complex exponential of the noisy signal $e^{j\theta_z}$ in estimating the clean signal spectral component. Thus, both the Wiener estimator of Eq. (19.85) and the MMSE spectral magnitude estimator of Eq. (19.92) use the complex exponential of the noisy phase. The spectral magnitude estimate of the clean signal obtained by the Wiener filter, however, is not optimal in the MMSE sense.

Other criteria for estimating A_y could also be used. For example, A_y could be estimated from

$$\min_{\hat{A}_y} E\{(\log A_y - \log \hat{A}_y)^2\} \tag{19.94}$$

This criterion aims at producing an estimate of A_y whose logarithm is as close as possible to the logarithm of A_y in the MMSE sense. This perceptually motivated criterion results in the estimator given by [5]

$$\hat{A}_y = \frac{\sigma_y^2}{\sigma_y^2 + \sigma_w^2} \exp\left(\frac{1}{2} \int_\vartheta^\infty \frac{e^{-t}}{t} dt\right) A_z \tag{19.95}$$

The integral in Eq. (19.95) is the well known exponential integral of ϑ and it can be numerically evaluated.

In another example, A_y^2 could be estimated from

$$\min_{\hat{A}_y} E\{(A_y^2 - \widehat{A_y^2})^2\} \tag{19.96}$$

and an estimate of A_y could be obtained from $\sqrt{\widehat{A_y^2}}$. The criterion in Eq. (19.96) aims at estimating the magnitude squared of the spectral component of the clean signal in the MMSE sense. This estimator is particularly useful when subsequent processing of the enhanced signal is performed, for example, in autoregressive analysis for low bit rate signal coding applications [13]. In that case, an estimator of the autocorrelation function of the clean signal can be obtained from the estimator $\widehat{A_y^2}$. The optimal estimator in the sense of Eq. (19.96) is well known and is given by (see e.g., [6])

$$\widehat{A_y^2} = \frac{\sigma_y^2 \sigma_w^2}{\sigma_y^2 + \sigma_w^2} + \left|\frac{\sigma_y^2}{\sigma_y^2 + \sigma_w^2} A_z\right|^2 \tag{19.97}$$

19.5.4 Multiple State Speech Model

All estimators presented in Section 19.5.2 and Section 19.5.3 make the implicit assumption that the speech signal is always present in the noisy signal. In the notation of Section 19.5.2, $Z = Y + W$. Since the length of a frame is relatively short, in the order of $30 - 50$ msec, it is more realistic to assume that speech may be present in the noisy signal with some probability, say η, and may be absent from the noisy signal with one minus that probability. Thus we have two hypotheses, one of speech presence, say H_1, and the other of speech absence, say H_0, that occur with probabilities η and $1 - \eta$, respectively. We have

$$Z = \begin{cases} Y + W & \text{under } H_1 \\ W & \text{under } H_0 \end{cases} \tag{19.98}$$

The MMSE estimator of A_y under the uncertainty model can be shown to be

$$\begin{aligned} \tilde{A}_y &= \Pr(H_1|Z) \cdot E\{A_y|Z, H_1\} + \Pr(H_0|Z) \cdot E\{A_y|Z, H_0\} \\ &= \Pr(H_1|Z) \cdot \hat{A}_y \end{aligned} \tag{19.99}$$

since $E\{A_y|Z, H_0\} = 0$ and $E\{A_y|Z, H_1\} = \hat{A}_y$ as given by Eq. (19.92). The more realistic estimator given in Eq. (19.99) was found useful in practice as it improved the performance of the estimator in Eq. (19.92). Other estimators may be derived under this model. Note that the model as in Eq. (19.98) is not applicable to the estimator in Eq. (19.95) since A_y must be positive for the criterion in Eq. (19.94) to be meaningful. Speech enhancement under the speech presence uncertainty model was first proposed and applied by McAulay and Malpass in their pioneering work [16].

An extension of this model leads to the assumption that speech vectors may be in different states at different time instants. The speech presence uncertainty model assumes two states representing speech presence and speech absence. In another model of Drucker [3], five states were proposed representing fricative, stop, vowel, glide, and nasal speech sounds. A speech absence state could be added to that model as a sixth state. This model requires an estimator for each state just as in Eq. (19.99).

A further extension of these ideas, in which multiple states that evolve in time are possible, is obtained when one models the speech signal by a hidden Markov process (HMP) [8]. An HMP is a bivariate random process of states and observations sequences. The state process $\{S_t, t = 1, 2, \ldots\}$ is a finite-state homogeneous Markov chain that is not directly observed. The observation process $\{Y_t, t = 1, 2, \ldots\}$ is conditionally independent given the state process. Thus, each observation depends statistically only on the state of the Markov chain at the same time and not on any other states or observations. Consider, for example, an HMP observed in an additive white noise process $\{W_t, t = 1, 2, \ldots\}$. For each t, let $Z_t = Y_t + W_t$ denote the noisy signal. Let $Z^t = \{Z_1, \ldots, Z_t\}$. Let J denote the number of states of the Markov chain. The causal MMSE estimator of Y_t given $\{Z^t\}$ is given by [6]

$$\begin{aligned} \hat{Y}_t &= E\{Y_t|Z^t\} \\ &= \sum_{j=1}^{J} \Pr(S_t = j|Z^t) E\{Y_t|S_t = j, Z_t\} \end{aligned} \tag{19.100}$$

The estimator as given in Eq. (19.100) reduces to Eq. (19.99) when $J = 2$ and $\Pr(S_t = j|Z^t) = \Pr(S_t = j|Z_t)$, or when the states are statistically independent of each other.

An HMP is a parametric process that depends on the initial distribution and transition matrix of the Markov chain and on the parameter of the conditional distributions of observations given states. The parameter of an HMP can be estimated off-line from training data and then used in constructing the estimator in Eq. (19.100). This approach has a great theoretical appeal since it provides a solid statistical model for speech signals. It also enjoys a great intuitive appeal since speech signals do cluster into sound groups of distinct nature, and dedication of a filter for each group is appropriate. The difficulty in implementing this approach is in achieving low probability of error in mapping vectors of the noisy speech onto states of the HMP. Decoding errors result in wrong association of speech vectors with the set of predesigned estimators $\{E\{Y_t|S_t = j, Z_t\}, j = 1, \ldots, J\}$, and thus in poorly filtered speech vectors. In addition, the complexity of the approach grows with the number of states, since each vector of the signal must be processed by all J filters.

The approach outlined above could be applied based on other models for the speech signal. For example, in [20], a harmonic process based on an estimated pitch period was used to model the speech signal.

19.5.5 Second-Order Statistics Estimation

Each of the estimators presented in Section 19.5.2 and Section 19.5.3 depends on some statistics of the clean signal and noise process which are assumed known *a-priori*. The signal subspace estimators as in Eq. (19.82) and Eq. (19.88) require knowledge of the covariance matrices of the signal and noise. The spectral magnitude estimators in Eq. (19.92), Eq. (19.95), and Eq. (19.97) require knowledge of the variance of each spectral component of the speech signal and of the noise process. In the absence of explicit knowledge of the second-order statistics of the clean signal and noise process, these statistics must be estimated either from training sequences or directly from the noisy signal.

We note that when estimates of the second-order statistics of the signal and noise replace the true second-order statistics in a given estimation scheme, the optimality of that scheme can no longer be guaranteed. The quality of the estimates of these second-order statistics is key for the overall performance of the speech enhancement system. Estimation of the second-order statistics can be performed in various ways usually outlined in the theory of spectral estimation [19]. Some of these approaches are reviewed in this section.

Estimation of speech signals second-order statistics from training data has proven successful in coding and recognition of clean speech signals. This is commonly done in coding applications using vector quantization and in recognition applications using hidden Markov modeling. When only noisy signals are available, however, matching of a given speech frame to a codeword of a vector quantizer or to a state of an HMP is susceptible to decoding errors. The significance of these errors is the creation of a mismatch between the true and estimated second order statistics of the signal. This mismatch results in application of wrong filters to frames of the noisy speech signal, and in unacceptable quality of the processed noisy signal.

On-line estimation of the second-order statistics of a speech signal from a sample function of the noisy signal has proven to be a better choice. Since the analysis frame length is usually relatively small, the covariance matrices of the speech signal may be assumed Toeplitz. Thus, the autocorrelation function of the clean signal in each analysis frame must be estimated. The Fourier transform of a windowed autocorrelation function estimate provides estimates of the variances of the clean signal spectral components.

For a wide-sense stationary noise process, the noise autocorrelation function can be estimated from an initial segment of the noisy signal that contains noise only. If the noise process is not wide-sense stationary, frames of the noisy signal must be classified as in Eq. (19.98), and the autocorrelation function of the noise process must be updated whenever a new noise frame is detected.

In the signal subspace approach [7], the power spectral densities of the noise and noisy processes are first estimated using windowed autocorrelation function estimates. Then, the power spectral density of the clean signal is estimated using Eq. (19.86). That estimate is inverse Fourier transformed to produce an estimate of the desired autocorrelation function of the clean signal. In implementing the MMSE spectral estimator as in Eq. (19.92), a recursive estimator for the variance of each spectral component of the clean signal developed in [4] is often used, see, e.g., [1, 2]. Let $\hat{A}_y(t)$ denote an estimator of the magnitude of a spectral component of the clean signal in a frame at time t. This estimator may be the MMSE estimator as in Eq. (19.92). Let $A_z(t)$ denote the magnitude of the corresponding spectral component of the noisy signal. Let $\widehat{\sigma_w^2}(t)$ denote the estimated variance of the spectral component of the noise process in that frame. Let $\widehat{\sigma_y^2}(t)$ denote the estimated variance of the spectral component of the clean signal in that frame. This estimator is given by

$$\widehat{\sigma_y^2}(t) = \beta \hat{A}_y^2(t-1) + (1-\beta)\max\{A_z^2(t) - \sigma_w^2(t), 0\} \qquad (19.101)$$

where $0 \leq \beta \leq 1$ is an experimental constant. We note that while this estimator was found useful in practice, it is heuristically motivated and its analytical performance is not known.

A parametric approach for estimating the power spectral density of the speech signal from a given sample function of the noisy signal was developed by Musicus and Lim [18]. The clean signal in a given frame is assumed a zero mean Gaussian autoregressive process. The noise is assumed a zero mean white Gaussian process that is independent of the signal. The parameter of the noisy signal comprises the autoregressive coefficients, the gain of the autoregressive process, and the variance of the noise process. This parameter is estimated in the ML sense using the EM algorithm. The EM algorithm and conditions for its convergence were originally developed for this problem by Musicus [17]. The approach starts with an initial estimate of the parameter. This estimate is used to calculate the conditional mean estimates of the clean signal and of the sample covariance matrix of the clean signal given the noisy signal. These estimates are then used to produce a new parameter, and the process is repeated until a fixed point is reached or a stopping criterion is met.

This EM procedure is summarized as follows: Consider a scalar autoregressive process $\{Y_t, t = 1, 2, \ldots\}$ of order r and coefficients $a = (a_1, \ldots, a_r)'$. Let $\{V_t\}$ denote an independent identically distributed (iid) sequence of zero mean unit variance Gaussian random variables. Let σ_v denote a gain factor. A sample function of the autoregressive process is given by

$$y_t = -\sum_{i=1}^{r} a_i y_{t-i} + \sigma_v v_t \tag{19.102}$$

Assume that the initial conditions of Eq. (19.102) are zero, i.e., $y_t = 0$ for $t < 1$. Let $\{W_t, t = 1, 2, \ldots\}$ denote the noise process which comprises an iid sequence of zero mean unit variance Gaussian random variables. Consider a k-dimensional vector of the noisy autoregressive process. Let $Y = (Y_1, \ldots, Y_k)'$, and define similarly the vectors V and W corresponding to the processes $\{V_t\}$ and $\{W_t\}$, respectively. Let A denote a lower triangular Toeplitz matrix with the first column given by $(1, a_1, \ldots, a_r, 0, \ldots, 0)'$. Then

$$Z = \sigma_v A^{-1} V + \sigma_w W \tag{19.103}$$

Suppose $\phi_m = (a(m), \sigma_v(m), \sigma_w(m))$ denotes an estimate of the parameter of Z at the end of the mth iteration. Let $A(m)$ denote the matrix A constructed from the vector $a(m)$. Let $R_y(m) = \sigma_v^2(m) A(m)^{-1} (A(m)^{-1})'$ denote the covariance matrix of the autoregressive process of Eq. (19.102) as obtained from ϕ_m. Let $R_z(m) = R_y(m) + \sigma_w^2(m) I$ denote the covariance matrix of the noisy signal as shown in Eq. (19.103) based on ϕ_m. Let $\hat{Y}$ and $\hat{R}$ denote, respectively, the conditional mean estimates of the clean signal Y and of the sample covariance of Y given Z and the current estimate ϕ_m. Under the Gaussian assumptions of this problem, we have

$$\hat{Y} = E\{Y|Z; \phi_m\}$$
$$= R_y(m) R_z^{-1}(m) Z \tag{19.104}$$

$$\hat{R} = E\{YY'|Z; \phi_m\}$$
$$= R_y(m) - R_y(m) R_z^{-1}(m) R_y(m) + \hat{Y}\hat{Y}' \tag{19.105}$$

The estimation of the statistics of the clean signal in Eq. (19.104) and Eq. (19.105) comprise the E-step of the EM algorithm.

Define the $r + 1 \times r + 1$ covariance matrix S with entries

$$S(i, j) = \sum_{l=\max(i,j)}^{k-1} \hat{R}(l - i, l - j), \quad i, j = 0, \ldots, r \tag{19.106}$$

Define the $r \times 1$ vector $q = (S(1, 0), \ldots, S(r, 0))'$ and the $r \times r$ covariance matrix $Q = \{S(i, j), i, j = 1, \ldots, r\}$. The new estimate of parameter ϕ at the end of the $m + 1$ iteration is given by

$$a(m+1) = -Q^{-1} q \tag{19.107}$$

$$\sigma_v^2(m+1) = (S(0, 0) + 2a'(m)q + a'(m) Q a(m))/k \tag{19.108}$$

$$\sigma_w^2(m+1) = \mathrm{tr}\, (E\{(Z - Y)(Z - Y)'|Z, \phi_m\})/k$$
$$= \mathrm{tr}\, (ZZ' - 2\hat{Y}Z' + \hat{R})/k. \tag{19.109}$$

The calculation of the parameter of the noisy signal in Eq. (19.107) to Eq. (19.109) comprise the M-step of the EM algorithm.

Note that the enhanced signal can be obtained as a by-product of this algorithm using the estimator as in Eq. (19.104) at the last iteration of the algorithm. Formulation of the above parameter estimation problem in a state space assuming colored noise in the form of another autoregressive process, and implementation of the estimators as given in Eq. (19.104) and Eq. (19.105) using Kalman filtering and smoothing was done in [9].

The EM approach presented above differs from an earlier approach pioneered by Lim and Oppenheim [12]. Assume for this comparison that the noise variance is known. In the EM approach, the goal is to derive the ML estimate of the true parameter of the autoregressive process. This is done by local maximization of the likelihood function of the noisy signal over the parameter set of the process. The EM approach results in alternate estimation of the clean signal Y and its sample correlation matrix YY', and of the parameter of the autoregressive process (a, σ_v). The approach taken in [12] aims at estimating the parameter of the autoregressive process that maximizes the joint likelihood function of the clean and noisy signals. Using alternate maximization of the joint likelihood function, the approach resulted in alternate estimation of the clean signal Y and of the parameter of the autoregressive process (a, σ_v). Thus, the main difference between the two approaches is in the estimated statistics from which the autoregressive parameter is estimated in each iteration. This difference impacts the convergence properties of the algorithm [12] which is known to produce an inconsistent parameter estimate. The algorithm [12] is simpler to implement than the EM approach and it is popular among some authors. To overcome the inconsistency, they developed a set of constraints for the parameter estimate in each iteration.

19.5.6 Concluding Remarks

We have surveyed some aspects of the speech enhancement problem and presented state-of-the-art solutions to this problem. In particular, we have identified the difficulties inherent to speech enhancement, and presented statistical models and distortion measures commonly used in designing estimators for the clean signal. We have mainly focused on speech signals degraded by additive noncorrelated wide-band noise. As we have noted earlier, even for this case, a universally accepted solution is not available and more research is required to refine current approaches or alternatively develop new ones. Other noise sources, such as room reverberation noise, present a more significant challenge as the noise is a non-stationary process that is correlated with the signal and it cannot be easily modeled. The speech enhancement problem is expected to attract significant research effort in the future due to challenges that this problem poses, the numerous potential applications, and future advances in computing devices.

References

1. Cappé, O., Elimination of the musical noise phenomenon with the Ephraim and Malah noise suppressor, *IEEE Trans. Speech and Audio Proc.*, vol. 2, pp. 345–349, Apr. 1994.
2. Cohen, I. and Berdugo, B.H., Speech enhancement for non-stationary noise environments, *Signal Processing*, vol. 81, pp. 2403–2418, 2001.
3. Drucker, H., Speech processing in a high ambient noise environment, *IEEE Trans. Audio Electroacoust.*, vol. AU-16, pp. 165–168, Jun. 1968.
4. Ephraim, Y. and Malah, D., Speech enhancement using a minimum mean square error short time spectral amplitude estimator, *IEEE Trans. Acoust., Speech, Signal Processing*, vol. ASSP-32, pp. 1109–1121, Dec. 1984.
5. Ephraim Y. and Malah, D., Speech enhancement using a minimum mean square error Log-spectral amplitude estimator, *IEEE Trans. Acoust., Speech, Signal Processing*, vol. ASSP-33, pp. 443–445, Apr. 1985.
6. Ephraim, Y., Statistical model based speech enhancement systems, *Proc. IEEE*, vol. 80, pp. 1526–1555, Oct. 1992.
7. Ephraim, Y. and Van Trees, H.L., A signal subspace approach for speech enhancement, *IEEE Trans. Speech and Audio Proc.*, vol. 3, pp. 251–266, July 1995.

8. Ephraim, Y. and Merhav, N., Hidden Markov Processes, *IEEE Trans. Inform. Theory*, vol. 48, pp. 1518–1569, June 2002.

9. Gannot, S., Burshtein, D., and Weinstein, E., Iterative and sequential Kalman filter-based speech enhancement algorithms, *IEEE Trans. Speech and Audio Proc.*, vol. 6, pp. 373–385, July 1998.

10. Gray, R.M., *Toeplitz and Circulant Matrices: II.* Stanford Electron. Lab., Tech. Rep. 6504–1, Apr. 1977.

11. Lev-Ari, H. and Ephraim, Y., Extension of the signal subspace speech enhancement approach to colored noise, *IEEE Sig. Proc. Let.*, vol. 10, pp. 104–106, Apr. 2003.

12. Lim, J.S. and Oppenheim, A.V., All-pole modeling of degraded speech, *IEEE Trans. Acoust., Speech, Signal Processing*, vol. ASSP–26, pp. 197–210, June 1978.

13. Lim J.S. and Oppenheim, A.V., Enhancement and bandwidth compression of noisy speech, *Proc. IEEE*, vol. 67, pp. 1586–1604, Dec. 1979.

14. Lim, J.S., ed., *Speech Enhancement*. Prentice-Hall, New Jersey, 1983.

15. Makhoul, J., Crystal, T.H., Green, D.M., Hogan, D., McAulay, R.J., Pisoni, D.B., Sorkin, R.D., and Stockham, T.G., *Removal of Noise From Noise-Degraded Speech Signals.* Panel on removal of noise from a speech/noise National Research Council, National Academy Press, Washington, D.C., 1989.

16. McAulay, R.J. and Malpass, M.L., Speech enhancement using a soft-decision noise suppression filter, *IEEE Trans. Acoust., Speech, Signal Processing*, ASSP-28, pp. 137–145, Apr. 1980.

17. Musicus, B.R., *An Iterative Technique for Maximum Likelihood Parameter Estimation on Noisy Data.* Thesis, S.M., M.I.T., Cambridge, MA, 1979.

18. Musicus, B.R. and Lim, J.S., Maximum likelihood parameter estimation of noisy data, *Proc. IEEE Int. Conf. on Acoust., Speech, Signal Processing*, pp. 224–227, 1979.

19. Priestley, M.B., *Spectral Analysis and Time Series*, Academic Press, London, 1989.

20. Quatieri, T.F. and McAulay, R.J., Noise reduction using a soft-decision sin-wave vector quantizer, *IEEE Int. Conf. Acoust., Speech, Signal Processing*, pp. 821–824, 1990.

Defining Terms

Speech Enhancement: The action of improving perceptual aspects of a given sample of speech signals.

Quality: A subjective measure of the way a speech signal is perceived.

Intelligibility: An objective measure which indicates the percentage of words from a given text that are expected to be correctly understood by listeners.

Signal estimator: A function of the observed noisy signal which approximates the clean signal.

Expectation-maximization: An iterative approach for parameter estimation using alternate estimation and maximization steps.

Autoregressive process: A random process obtained by passing white noise through an all-pole filter.

Wide-sense stationarity: A property of a random process whose second-order statistics do not change with time.

Asymptotically weakly stationarity: A property of a random process indicating eventual wide-sense stationarity.

Hidden Markov Process: A Markov chain observed through a noisy channel.

19.6 Ad Hoc Networks

Michel D. Yacoub, Paulo Cardieri, Élvio João Leonardo, and Álvaro Augusto Machado Medeiros

19.6.1 Introduction

An ad hoc network is a wireless network that is established without the aid of infrastructure or centralized administration. It is formed by a group of wireless terminals (nodes) such that a communication between any two terminals is carried out by means of a store-and-relay mechanism. A terminal wishing to transmit accesses the medium and sends its information to a nearby terminal. Upon receiving such information this terminal determines that this is not addressed to it. It then stores the information in order to relay it to another terminal at an appropriate time, and this process continues until the destination is reached. Note that in ad hoc networks there are no fixed routers. Nodes may be mobile and can be connected dynamically in an arbitrary manner. Nodes function as routers, which discover and maintain routes to other nodes in the network. Ad hoc networks find applications in emergency-and-rescue operations, meeting or conventions, data acquisition operations in inhospitable terrain, sensor networks, and home and office networks. Cheaper hardware, smaller transceivers, and faster processors fuel the increased interest in wireless ad hoc networks. This chapter addresses the ad hoc networks from four main aspects: routing, medium access, TCP/IP issues, and capacity. In routing, the main routing algorithms are illustrated. In medium access, the main medium access protocols are described. In TCP/IP issues, the aspects concerning the performance of TCP/IP in an ad hoc network is discussed. In capacity, some formulation concerning the capacity of the network is tackled.

19.6.2   Routing Algorithms

The design of routing algorithms in ad hoc networks is a challenging task. Algorithms must provide for a high degree of sophistication and intelligence so that the limited resources inherent to the wireless systems can be dealt with efficiently. They must be robust in order to cope with the unkind wireless environment. At the same time they must be flexible in order to adapt to the changing network conditions such as network size, traffic distribution, and mobility. Routing algorithms have long been used in wired systems and they are usually classified into two categories: Distant vector (DV) and Link-state (LS). DV algorithms provide each node with a vector containing the hop distance and the next hop to all the destinations. LS algorithms provide each node with an updated view of the network topology by periodical flooding of link information about its neighbors. A direct application of these algorithms in a wireless and mobile environment may be cumbersome. DV protocols suffer from slow route convergence and may create loops. LS protocols require the frequent use of the resources, thence large bandwidth, in order to maintain the nodes updated.

With the increasing interest in wireless networks, a variety of routing algorithms overcoming the limitations of the DV and LS protocols have been proposed. They are usually classified into three categories: proactive or table-driven, reactive or on-demand, and hybrid. The proactive protocols require the nodes to keep tables with routing information. Updates occur on a periodical basis or as soon as changes in the network topology are perceived. The algorithms differ basically in the type of information is kept and in the way updates occur. The reactive protocols create routes on demand. This is accomplished by means of a route discovery process, which is completed once a route has been found or all possible route permutations have been examined. The discovery process occurs by flooding route request packets through the network. After establishing a route, it is maintained by a route maintenance procedure until either the destination becomes inaccessible along every path from the source or until the route is no longer desired. The hybrid protocols are both proactive and reactive. Nodes with close proximity form a backbone within which proactive protocols are applied. Routes to faraway nodes are found through reactive protocols.

Proactive Algorithms

CGSR—Clusterhead-gateway switch routing (Chiang, 1997). In CGSR, the whole network is partitioned into clusters of nodes. Within each cluster a *clusterhead* is elected among the nodes. A node may belong to one cluster only, in which case it is an internal node, or to more than one cluster, in which case it becomes a *gateway*. Packets are transmitted from the node to the clusterhead and from the clusterhead to the node. Routing within such a network occurs as follows. Assume source and destination belonging to different clusters. The source sends its packets to the clusterhead, which relays these packets to a gateway, which relays them to another clusterhead, and this process continues until the destination is reached.

DREAM—Distance routing effect algorithm for mobility (Basagni, 1998). In DREAM, GPS is used so that each node can maintain a location table with records of locations of all the nodes. The nodes broadcast control packets for location updating purposes. A source having packets to send calculates the direction toward the destination. It then selects a set of one-hop neighbors in the respective direction. (If the set is empty, the data are flooded to the whole network.) The data header encloses the respective set and is sent. Those nodes specified in the header are entitled to receive and process the data. All the nodes of the paths repeat this process until the destination is reached. Upon receiving the packets, the destination issues an ACK, which is transmitted to the source using the same algorithm.

DSDV—Destination-sequential distance-vector routing (Perkins, 1994). In DSDV, each node keeps a routing table containing all of the possible destinations within the network in conjunction with the number of hops to each destination. The entries are marked with a sequence number assigned by the destination node so that mobile nodes can distinguish stable routes from the new ones in order to avoid routing loops. Table consistency is kept by periodical updates transmitted throughout the network.

FSLS—Fuzzy sighted link state (Santivanez, 2001). In FSLS, an optimal link state update algorithm (hazy sighted link state) is used. Updates occur every $2^k T$, in which k is the hop distance, T is the minimum link state update transmission period, and 2^k is the number of nodes to be updated. FSLS operates in a way very similar to FSR, as described below.

FSR—Fisheye state routing (Iwata, 1999; Pei, 2000). In FSR, each node maintains a topology map, and link state information is exchanged with neighbors periodically. The frequency with which it occurs depends on the hop distance to the current node. Nearby destinations are updated more frequently whereas for the faraway ones the updates are less frequent. Therefore, FSR produces accurate distance and path information about immediate neighborhood, and imprecise information about the best path to a distant node. On the other hand, such an imprecision is compensated for as the packet approaches its destination.

GSR—Global state routing (Chen, 1998). In GSR, a link state table based on the update messages from neighboring nodes is kept and periodical exchanges of link state information are carried out. The size of update messages increases with the increase of the size of the network, and a considerable amount of bandwidth is required in this case.

HSR—Hierarchical state routing (Pei, 1999). In HSR, the link state algorithm principle is used in conjunction with a hierarchical addressing and topology map. Clustering algorithms may also be used so as to organize the nodes with close proximity into clusters. Each node has a unique identity, which are typically a MAC address and a hierarchical address. A communication between any two nodes occurs by means of physical and logical links. The physical links support the true communication between nodes whereas the logical links are used for the purposes of the hierarchical structure of the communication. This way, several levels in the hierarchy may be built. The lowest level is always the physical level whereas the higher levels constitute the logical levels. Communications then occur starting from the lowest level up to the higher levels and down again to the lowest level.

MMWN–Multimedia support in mobile wireless networks (Kasera, 1997). In MMWN, a clustering hierarchy is used, each cluster having two types of nodes: switch and endpoint. Endpoints do not communicate with each other but with switches only. Within a cluster, one of the switches is chosen as a location manager, which performs location updating and location finding. This means that routing overhead is drastically reduced as compared to the traditional table-driven algorithms. This way, information in MMWN is stored in a dynamically distributed database.

OLSR—Optimized link state routing (Jacquet, 2001). In OLSR, each node keeps topology information on the network by periodically exchanging link state messages. OLSR uses the multipoint replaying strategy (MPR) in order to minimize the size of the control messages and the number of re-broadcasting nodes. To this end, each node selects a set of neighboring nodes (multipoint relays—MPRs) to retransmit its packets. Those not in the selected set may read and process each packet but not retransmit. The selection of the appropriate set is carried out as follows. Each node periodically broadcasts a list of its one-hop neighbors. From the lists the nodes are able to choose a subset of one-hop neighbors that covers all of its two-hop neighbors. An optimal route to every destination is constructed and stored in a routing table.

STAR—Source-tree adaptive routing (Garci-Luna-Aceves, 1999). In STAR, each node keeps a source tree with the preferred paths to destinations. It uses the least overhead routing approach (LORA) to reduce the amount of routing overhead disseminated into the network. The reduction in the amount of messages is achieved by making update dissemination conditional to the occurrence of certain events.

TBRPF—Topology broadcast based on reverse path forwarding (Bellur, 1999; Ogier, 2002). In TBRPF, two separate modules are implemented: neighbor discovery module and the routing module. The first module performs a differential HELLO messages that reports only the changes (up or lost) of neighbors. The second module operates based on partial topology information. The information is obtained through periodic and differential topology updates. If a node n is to send an update message, then every node in the network selects its next hop (parent) node toward that node. Link state updates are propagated in the reverse direction on the spanning tree formed by the minimum-hop paths from all nodes to the sending node. This means that updates originated at n are only accepted if these updates arrive from the respective parent node. They are then propagated toward the children nodes pertaining to n.

WRP—Wireless routing protocol (Murthy, 1996). In WRP, each node maintains a set of tables as follows: distance table, routing table, link-cost table, message retransmission list (MRL) table. Each entry of the MRL table contains a sequence number of the update message, a retransmission counter, an acknowledgement required flag vector with one entry per neighbor, and a list of updates sent in the update message. It records which updates need to be retransmitted and which neighbors should acknowledge the retransmission. Update messages are sent only between neighboring nodes and they occur after processing updates or detecting a change in the link to the neighbor. Nodes learn of the existence of their neighbors from the receipt of ACK and other messages. In case a node is not sending messages of any kind, then a HELLO message is sent within a specified period of time to ensure connectivity.

Reactive Algorithms

ABR—Associativity-based routing (ABR) (Toh, 1996). In ABR, a query-reply technique is used to determine routes to the required destination. Stable routes are chosen based on an associativity tick that each node maintains with its neighbors, with the links with the higher associativity tick being selected. This may not lead to the shortest paths but rather to paths that last longer. In such a case, fewer route reconstructions are needed thence more bandwidth is available. ABR requires periodical beaconing so as to determine the degree of associativity of the links which requires all nodes to remain active at all time, which result in additional power consumption.

AODV—Ad hoc on demand distance vector (Das, 2002). In AODV, periodic beaconing and sequence numbering procedure are used. The packets convey the destination address rather than the full routing information, this also occurring in the route replies. The advantage of AODV is its adaptability to highly dynamic networks. On the other hand, the nodes may experience large delays in route construction.

ARA—Ant-colony-based routing algorithm (Günes, 2002). In ARA, the food searching behavior of ants is used in order to reduce routing overheads. When searching for food, ants leave a trail behind (pheromone) that is followed by the other ants until it vanishes. In the route discovery procedure, ARA propagates a forwarding ANT (FANT) through the network until it reaches the destination. Then a backward ANT (BANT) is returned, a path is created, and data packet dissemination starts. The route is maintained by means of increasing or decreasing the pheromone value at each node. The pheromone at a given node is increased each time a packet travels through it, and it is decreased overtime until it expires. As can be inferred, the size of FANT and BANT is small; therefore the amount of overhead per control packet is minimized.

CBRP—Cluster-based routing protocol (Jiang, 1999). In CBRP, the nodes are grouped into clusters, a cluster presenting a clusterhead. The advantage of using the hierarchical approach is the decrease in the number of control overhead through the network as compared with the traditional flooding methods. Of course, there are overheads associated with the formation and maintenance of the clusters. The long propagation delay due to the hierarchical structure may render the nodes bearing inconsistent topology information, which may lead to temporary routing loops.

DSR—Dynamic source routing (Johnson, 2002). In DSR, there is no periodic beaconing (HELLO message), an important feature that can be used for battery saving purposes, in which the node may enter the sleep mode. Each packet in DSR conveys the full address of the route, and this is a disadvantage for large or highly dynamic networks. On the other hand, the nodes can store multiple routes in their route cache. The advantage of this is that the node can check its route cache for a valid route before initiating route discovery. A valid route found avoids the need for route discovery. And this is an advantageous feature for low mobility networks.

FORP—Flow oriented routing protocol (Su, 1999). In FORP, routing failure due to mobility is minimized by means of the following algorithm. A Flow-REQ message is disseminated through the network. A node receiving such a message, and based on GPS information, estimates a link expiration time (LET) with the previous hop and append this into its Flow-REQ packet, which is re-transmitted. Upon arrival at the destination, a route expiration time (RET) is estimated using the minimum of all LETs for each node. A Flow-SETUP message is sent back toward the source. Therefore, the destination is able to predict when a link failure may occur. In such a case, a Flow-HANDOFF message is generated and propagated in a similar manner.

LAR—Location aided routing (Ko, 1998). In LAR, using location information routing overhead is minimized, which is commonly present in the traditional flooding algorithms. Assuming each node provided with a GPS, the packets travel in the direction where the relative distance to the destination becomes smaller as they travel from one hop to another.

LMR—Light-weight mobile routing (Corson, 1995). In LMR, the flooding technique is used in order to determine the required routes. Multiple routes are kept at the nodes, the multiplicity being used for reliability purposes as well as to avoid the re-initiation of a route discovery procedure. In addition, the route information concerns the neighborhood only and not the complete route.

RDMAR—Relative distance micro-discovery ad hoc routing (Aggelou, 1999). In RDMAR, a relative-distance micro-discovery procedure is used in order to minimize routing overheads. This is carried out by means of calculating the distance between source and destination, thence limiting each route request packet to a certain number of hops (i.e., the route discovery procedure becomes confined to localized regions). In fact, this is only feasible if previous communications between source and destination has been established. Otherwise, a flooding procedure is applied.

ROAM—Routing on-demand acyclic multi-path (Raju, 1999). In ROAM, inter-nodal coordination and directed acyclic sub-graphs, derived from the distance of the router to the destination, are used. In case the required destination is no longer reachable multiple flood searches stop. In addition, each time the distance of a router to a destination changes by more than a given threshold, the router broadcasts update messages to its neighboring nodes. This increases the network connectivity at the expense of preventing the nodes entering the sleep mode to save battery.

SSA—Signal stability adaptive (Dube, 1997). In SSA, route selection is carried out based on signal strength and location stability, and not on associativity tick. In addition, route requests sent toward a destination cannot be replied by intermediate nodes, which may cause delays before a route is effectively discovered. This is due to the fact that the destination is responsible for selecting the route for data transfer.

TORA—Temporarily ordered routing algorithm (Park, 1997). In TORA, the key design concept is the localization of control messages to a very small set of nodes near the occurrence of a topological change. The nodes maintain routing information about one-hop nodes. Route creation and route maintenance phases use a height metric to establish a directed acyclic graph rooted at the destination. Then, links are assigned as upstream or downstream based on the relative height metric to neighboring nodes. Route's erasure phase involves flooding a broadcast clear packet throughout the network in order to erase invalid routes.

Hybrid Algorithms

DDR—Distributed dynamic routing (Nikaein, 2001). In DDR, a tree-based routing protocol is used but a root node is not required. The trees are set up by means of periodic beaconing messages, exchanged by neighboring nodes only. Different trees, composing a forest, are connected via gateways nodes. Each tree constitutes a zone, which is assigned a zone ID. The routes are determined by hybrid ad hoc protocols.

DST—Distributed spanning trees based routing protocol (Radhakrishnan, 1999). In DST, all nodes are grouped into trees, within which a node becomes a routing node or an internal node. The root, which is also a node, controls the structure of the tree. This may become a disadvantage of DST for the root node creates a single point of failure.

SLURP—Scalable location update routing protocol (Woo, 2001). In SLURP, the nodes are organized into nonoverlapping zones and a home region for each node in the network is assigned. The home region for each node is determined by means of a static mapping function known to all nodes whose inputs are the node ID and the number of nodes. Thus, all nodes are able to determine the home region for each node. The current location of the node within its home region is maintained by unicasting a location update packet toward its region. Once it reaches its home region, it is broadcast to all nodes within its home.

ZHLS—Zone-based hierarchical link state (Joa-Ng, 1999). In ZHLS, hierarchical topology is used such that two levels are established: node level and zone level, for which the use of GPS is required. Each node then has a node ID and a zone ID. In case a route to a node within another zone is required, the source node broadcasts a zone-level location request to all of the other zones. This generates lower overhead as compared to the flooding approach in reactive protocols. Mobility within its own zone maintains the topology of the network such that no further location search is required. In ZHLS, all nodes are supposed to have a pre-programmed static zone map for initial operation.

ZRP—Zone routing protocol (Haas, 1999). In ZRP, a routing zone is established that defines a range in hops within which network connectivity is proactively maintained. This way, nodes within such a zone have the routes available immediately. Outside the zone, routes are determined reactively (on demand) and any reactive algorithm may be used.

19.6.3   Medium Access Protocols

Medium access control (MAC) for wireless ad hoc networks is currently a very active research topic. The characteristics of the network, the diverse physical-layer technologies available, and the range of services envisioned render a difficult task the design of an algorithm to discipline the access to the shared medium that results efficient, fair, power consumption sensitive, and delay bound. A number of issues distinguish wireless MAC protocols from those used in wireline networks (Chandra, 2002), as quoted next.

Half-duplex operation. Due to self-interference (i.e., the energy from the transmitter that leaks into the receiver), it is difficult to construct terminals able to receive while transmitting. Therefore collision detection while sending data is not possible and Ethernet-like protocols cannot be used. Since collisions cannot be detected, wireless MAC protocols use collision avoidance mechanisms to minimize the probability of collision.

Time varying channel. In multipath fading channels, the received signal is the sum of time-shifted and attenuated copies of the transmitted signal. With the change of the channel characteristics as well as in the relative position of terminals, the signal envelope varies as a function of time. The signal experiences fading that may be severe. The nodes establishing a wireless link need to sense the channel so as to assess the communication link conditions.

Burst channel errors. Wireless channels experience higher bit error rate than wireline transmissions. Besides, errors occur in bursts as the signal fades, resulting in high probability of packet loss. Therefore, an acknowledgement mechanism must be implemented so that the packet retransmission may be possible in case of packet loss.

Location-dependent carrier sensing. Because the signal strength decays with distance according to a power law, only nodes within a specific range are able to communicate. This gives rise to the hidden and exposed terminals and the capture effect, as described next.

Hidden terminal. Refer to Fig. 19.51 where the relative positions of terminals A, B, and C are shown. B is within range of both A and C but A and C are out of range of each other. If terminal A is transmitting to B and terminal C wishes to transmit to B, it incorrectly senses that the channel is free because it is out of range of A, the current transmitter. If C starts transmitting it interferes with the reception at B. In this case C is termed the hidden terminal to A. The hidden terminal problem can be minimized with the use of the request-to-sent/clear-to-send (RTS/CTS) handshake protocol (to be explained later) before the data transmission starts.

Exposed terminal. An exposed terminal is one that is within range of the transmitter but out of range of the receiver. In Fig. 19.51, if terminal B is transmitting to A and terminal C senses the channel it perceives it as busy. However, since it is out of range of terminal A it cannot interfere with the current conversation.

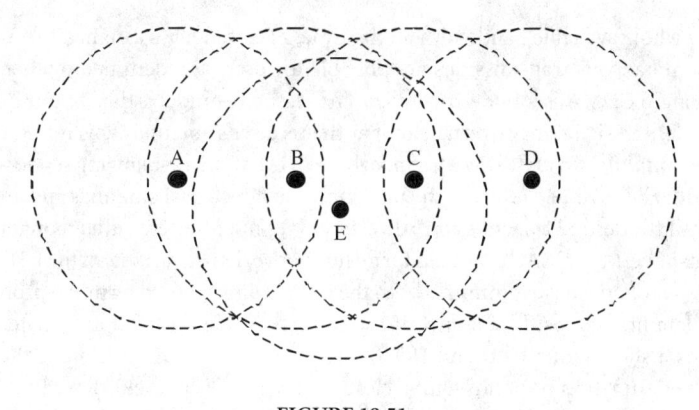

FIGURE 19.51

Therefore it can utilize the channel to establish a parallel link with another terminal that is out of range of B, for instance, terminal D. In this case C is termed the exposed terminal to B. Exposed terminals may result in under-usage of the channel. As in the hidden terminal problem, this also can be minimized with the use of the RTS/CTS handshake protocol.

Capture. Capture at a given terminal occurs in case among several simultaneous signals arriving at it the signal strength of one of them prevails over all of the others combined. In Fig. 19.51, terminals C and E are both within range of terminal B. If C and E are transmitting the interference may result in a collision at B. However, B may be able to receive successfully if one of the signals is much higher than the other, for instance, the signal from E. Capture can improve throughput because it results in less collisions. However, it favors senders that are closer to the intended destination, which may cause unfair allocation of the channel.

From the above considerations, it is promptly inferred that the design of a MAC protocol for ad hoc networks requires a different set of parameters must be considered as compared with those of the wireline ⨍ systems.

Protocols Categories

Jurdak et al. (Jurdak, 2004), after conducting a survey and analysis of a number of current MAC protocol proposals, offer a set of key features that may be used in order to classify MAC protocols for ad hoc networks.

Channel separation and access. The way the medium is organized is an important issue in the protocol design. For instance, all stations may share a single channel, which they use for control and data transmissions. On the other hand, the medium may be divided into multiple channels, in general one for control and the others for data. The single channel approach was favored in earlier MAC designs because of its simplicity. However, it is intrinsically subject to collisions and it does not perform well in medium to heavy traffic conditions. Particularly at heavy loads, simulations show that single channel protocols are prone to increased number of collisions of control packets, for example, RTS and CTS, which cause increased back off delays while the medium is idle (Tseng, 2002). The choice for multiple channels brings the issue of how to separate these channels. The most common ways of separating channels make use of FDMA, TDMA, and CDMA technologies. *Frequency division multiple access* (FDMA) uses multiple carriers to divide the medium into several frequency slots. It allows multiple transmissions to occur simultaneously although each sender can use only the bandwidth of its assigned frequency slot. *Time division multiple access* (TDMA) divides the medium into fixed length time slots. A group of slots forms a time frame and defines the slot repetition rate. Because of its periodic nature, TDMA protocols are suitable to delay sensitive traffic. In TDMA, a sender uses the whole available bandwidth for the duration of a slot assigned to it. In addition, to access the medium terminals need to keep track of frames and slots and, as a result, TDMA protocols require synchronization among terminals. *Code division multiple access* (CDMA) allows senders to use the whole available bandwidth all the time. Each sender is assigned one of several orthogonal codes and simultaneous transmissions are possible for users are identified by their unique code. A general requirement in CDMA is for power control. The reason behind it is that an unwanted signal that is stronger than the desired signal may overwhelm it at the receiver's antenna. This is known as the near-far effect. *space division multiple access* (SDMA), similarly to CDMA, aims at allowing senders to use the whole available bandwidth all the time. However, the terminals use directional antennas and are allowed to start transmission only if the desired transmission's direction does not interfere with an ongoing conversation.

RTS/CTS handshake. Many MAC protocols for ad hoc networks use variants of the RTS/CTS handshake. The original three-way handshake minimizes both the hidden and exposed terminal problems. A terminal wishing to send data first senses the channel. If the channel is idle for the appropriate amount of time, the terminal sends a short request-to-send (RTS) packet. All terminals on hearing the RTS defer their transmissions. The destination responds with a clear-to-send (CTS) packet. All terminals on hearing the CTS also defer their transmissions. The sender, on receiving the CTS assumes the channel is acquired and initiates the data transmission.

Topology. Ad hoc networks have a large degree of flexibility and uncertainty. Terminals may be mobile and have distinct capabilities and resources. The network must take this into account and adapt dynamically while optimizing performance and minimizing power consumption (Jurdak, 2004). A network topology can be centralized, clustered, or flat. *Centralized topologies* have a single terminal or base station that controls and manages the network. The central terminal may be responsible for broadcasting information relevant to the operation of the network. In addition, terminals may only communicate through the central terminal. *Clustered topologies* create a local version of a centralized network where one terminal assumes some or all of the duties expected from the central terminal. *Flat topologies* implement a fully distributed approach where all terminals are at the same level, and central control is not used. Flat topologies are further divided into single-hop and multiple-hop. *Single-hop* assumes that the destination node is within range of the sender. *Multiple-hop* assumes that the destination node may be beyond the sender's reachable neighbors. In this case, intermediate terminals are responsible for relaying the packets until they reach the intended destination. Single-hop protocols are simpler but pose limitations on the size of the network. Multiple-hop adds scalability to the network at the expense of higher complexity.

Power. Power consumption is a relevant issue for all wireless networks. Power conservation is particularly influential for the mobile terminals because of the limited battery power available. An efficient power conservation strategy involves several aspects. The energy used to transmit the signal represents a large share of the power consumption. Ideally the transmit power used should be just enough to reach the intended destination. Another source of wasted energy is the long periods of time terminals need to spend sensing the channel or overhearing irrelevant conversation. If terminals are able to learn in advance about when the medium will be unavailable they may decide to go into a sleep mode for that period of time in order to save energy. The network behavior may be influenced by the terminals' battery power level, for instance, in the selection of a cluster head or in assigning transmission priorities. Terminals aware of their battery level may adjust their behavior accordingly. The exchange of control messages before the data transmission phase also represents power wastage. Reduced control overhead should therefore be pursued for the sake of power efficiency.

Transmission initiation. Intuitively, it is expected that a terminal wishing to start a conversation must initiate the transmission. And, in fact, most of the protocols are organized this way. However, a receiver-initiated protocol may be more suitable to some specialized networks, for instance, a sensor network. In receiver-initiated protocols the receiver polls its neighbors by sending a ready-to-receive (RTR) packet, which indicates its readiness to receive data. If the receiver is able to know or successfully predict when a neighbor wishes to send its data, this class of protocols actually produces better performance. However, for generalized networks and unpredictable traffic, sender-initiated protocols are still a better choice.

Traffic load and scalability. Protocols are usually optimized for the worst expected scenario. Sparse node distribution and light traffic conditions do not pose a challenge for the implementation of ad hoc networks. The protocols are optimized for high traffic load, high node density, and/or real-time traffic, depending on the intended use. Protocols that offer the possibility of channel reservation are those with best performance on both high load and real-traffic situations. Receiver-initiated approaches also tend to work well in high load conditions because there is a high probability that RTR packets reach terminals wishing to send data. If the network ranks terminals and traffic, then it is able to assign priorities based on the traffic nature. Therefore, it can offer favored handling of real-time traffic. Dense networks tend to suffer from higher interference because of the proximity of transmitting nodes. For this reason the use of power control makes a significant difference in the performance of the network.

Range. Transmission range is the distance from the transmitter's antenna that the radio signal strength still remains above the minimum usable level. Protocols can be classified (Jurdak, 2004) as very short-range (range up to 10 m), short-range (from 10 up to 100 m), medium-range (from 100 up to 1000 m), and long-range (from 1000 m). There is a trade-off between increasing the transmission range and achieving high spatial capacity that needs to be negotiated during the protocol design.

Industry Standard Protocols

IEEE 802.11

The family of IEEE 802.11 standards (IEEE, 1999a; IEEE, 1999b; IEEE, 1999c) can be viewed as a wireless version of the local area network (LAN) protocol Ethernet. The 802.11a standard operates in the unlicensed 5 GHz band and offers data rates up to 54 Mb/s. The commercially popular 802.11b operates in the industrial, scientific, and medical (ISM) band at 2.4 GHz and offers data rates up to 11 Mb/s. The current activity of the 802.11-working group is toward quality of service (QoS) (802.11e, described later) and security (802.11i). The 802.11 standards focus on the specification of the MAC and physical (PHY) layers. While their PHY layers differ, existing 802.11 standards rely on the same medium access mechanisms. The basic (and mandatory) access mechanism is referred to as distributed coordination function (DCF). The optional point coordination function (PCF) is an access mechanism in which a central node (the access point) polls terminals according to a list. DCF is available for both flat ad hoc and centralized topologies whereas PCF is only available in centralized configurations. MAC offers two types of traffic services. The mandatory asynchronous data service is based on the best effort and is suited to delay insensitive data. The optional time-bound service is implemented using PCF.

DCF uses the listen-before-talk scheme based on carrier sense multiple access (CSMA). A terminal wishing to transmit a data packet first monitors the medium activity. If the channel is detected idle the terminal waits for a DCF interframe space (DIFS) time interval (34 us in 802.11a). If the channel remains idle during the DIFS period, the terminal starts transmitting its packet immediately after DIFS has expired. The transmission is successfully completed when the sender receives an acknowledgement (ACK) packet from the destination. However, if the channel is sensed busy a collision avoidance procedure is used. In this procedure, after sensing the channel idle again for a DIFS period, the terminal wishing to transmit waits an additional random backoff time. The terminal then initiates its transmission if the channel remains idle during this additional time. The backoff time is a multiple of the slot time (9 us in 802.11a) and it is determined individually and independently by each station. A random number between zero and contention window (CW) is selected for any new transmission attempt. The back off time is decremented while the medium is in contention phase and frozen otherwise. Thus the backoff time may be carried over for several busy cycles of the medium before it expires. Refer to Fig. 19.52 for an example of the backoff

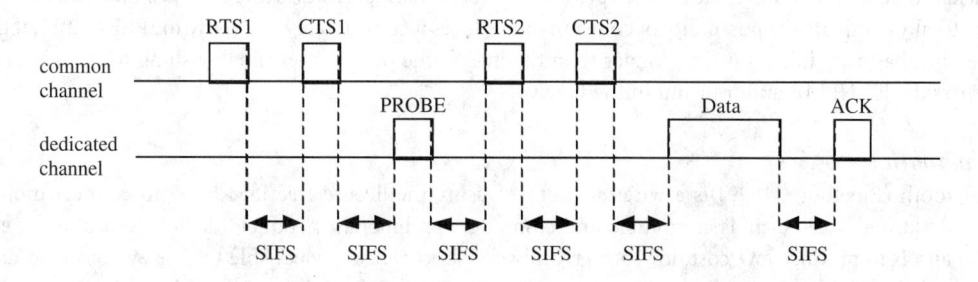

FIGURE 19.52

procedure. The initial value for CW is CWmin (15 for 802.11a) and since all terminals operate with the same CWmin value they all have the same initial medium access priority. After any failed transmission, that is, when the transmitted packet is not acknowledged, the sender doubles its CW up to a maximum defined by CWmax (1023 in 802.11a). A now larger CW decreases the probability of collisions if multiple terminals are trying to access the medium. To reduce the hidden terminal problem, 802.11 optionally uses the RTS/CTS handshake. Both RTS and CTS packets include information on how long the data frame transmission is going to last, including the corresponding ACK. Terminals receiving either the RTS or CTS use this information to start a timer, called network allocation vector (NAV), which informs the period of time the medium is unavailable. Between consecutive frames RTS and CTS, and a data frame and its ACK, the short interframe space (SIFS) (16 us in 802.11) is used. SIFS is shorter than DIFS and therefore gives the terminals sending these frames priority to access the medium.

HIPERLAN 1

High Performance LAN type 1 (HIPERLAN 1) is a wireless LAN standard operating in the 5 GHz band, which offers data rate up to 23.5 Mb/s to mobile users in either clustered ad hoc or centralized topology. HIPERLAN 1 offers asynchronous best effort and time-bound services with hierarchical priorities. There are five priority values defined, from zero (highest) to four (lowest). Each individual MAC protocol data unit (PDU) is assigned a priority that is closely related to its normalized residual lifetime (NRL) value. The NRL is an estimation of the time-to-live the PDU has considering the number of hops it still has to travel. A PDU is discarded if its NRL value reaches zero. In addition, some terminals are designed forwarders and are responsible to relay data to distant nodes in a multi-hop fashion. HIPERLAN 1 allows terminals to go into sleep mode in order to save energy. These terminals, called p-savers, inform support terminals, called p-supporters, of their sleep/wake-up patterns. p-supporters then buffer packets directed to p-savers terminals, as required. Although it has some interesting features, HIPERLAN 1 has not been a commercial success. The channel access mechanism used in HIPERLAN 1 is the elimination-yield non-preemptive priority access (EY-NPMA). It comprises three phases: prioritization (determine the highest priority data packets to be sent); contention (eliminate all contenders except one); and transmission. During the prioritization phase, time is divided in five minislots, numbered sequentially from zero to four. A terminal wishing to transmit has to send a burst during the minislot corresponding to its MAC PDU priority. For example, a terminal with a priority two PDU monitors the medium during minislots zero and one before it can assert its intention by transmitting a burst during minislot two. If the medium becomes busy during either minislot zero or one this terminal defers its transmission. Once a burst is transmitted, the prioritization phase ends and only terminals having PDUs at the same priority level remains in the dispute. The contention phase follows. It starts with the contending terminals transmitting an elimination burst. The individual terminals select the burst length, varying from 0 to 12 minislots, at random and independently. After transmitting the burst the terminals sense the medium. If it is busy they defer their transmissions. Otherwise, the remaining terminals enter the yield listening period. They select at random and independently a value between 0 and 9 and start monitoring the medium. If at the end of this period the medium is still idle the terminal assumes it has won the contention and is allowed to transmit its data. Otherwise, it defers its transmission. It is clear that the mechanism does not allow any lower priority packet to be sent if another with higher priority packet is waiting. At the same time the mechanism does not totally eliminate the possibility of collision but reduces it considerably. Similarly to IEEE 802.11, if the medium has been idle for a time longer than the interframe period a terminal wishing to transmit can bypass the EY-NPMA and transmit immediately.

Bluetooth

Bluetooth (Bluetooth, 1999) is a wireless protocol using the license-free ISM band to connect mobile and desktop devices such as computers and computers peripherals, handheld devices, cell phones, etc. The aim is to produce low-cost, low-power, and very-short range devices able to convey voice and data transmissions at a maximum gross rate of 1 Mb/s. Bluetooth uses frequency hopping spread spectrum (FHSS) with 1600 hops/s. For voice, a 64 kb/s full-duplex link called synchronous connection oriented

(SCO) is used. SCO assigns a periodic single slot to a point-to-point conversation. Data communication uses the best effort asynchronous connectionless (ACL) link in which up to five slots can be assigned. Terminals in Bluetooth are organized in piconets. A piconet contains one terminal identified as the master and up to seven other active slaves. The master determines the hopping pattern and the other terminals need to synchronize to the piconet master. When it joins a piconet, an active terminal is assigned a unique 3-bit long active member address (AMA). It then stays in either transmit state, when it is engaged in a conversation, or connected state. Bluetooth supports three low-power states: park, hold, and sniff. A parked terminal releases its AMA and is assigned one of the 8-bit long parked member address (PMA). Terminals in the hold and sniff states keep their AMA but have limited participation in the piconet. For instance, a terminal in the hold state is unable to communicate using ACL. A terminal not participating in any piconet is in stand-by state. Bluetooth piconets can co-exist in space and time and a terminal may belong to several piconets. A piconet is formed when its future master starts an inquiry process, that is, inquiry messages are broadcast in order to find other terminals in the vicinity. After receiving inquiry responses the master may explicitly page terminals to join the piconet. If a master knows already another terminal's identity it may skip the inquiry phase and page the terminal directly. Bluetooth uses time division duplex (TDD) in which master and slave alternate the opportunity to transmit. A slave can only transmit if the master has just transmitted to it, that is, slaves transmit if polled by the master. Transmissions may last one, three, or five slots although only single-slot transmission is a mandatory feature.

IEEE 802.11e

The IEEE 802.11e is an emerging MAC protocol, which defines a set of QoS features to be added to the 802.11 family of wireless LAN standards. Currently there is a draft version of the specifications (IEEE, 2003). The aim is to better serve delay-sensitive applications, such as voice and multi-media. In 802.11e, the contention-based medium access is referred to as enhanced distributed channel access (EDCA). In order to accommodate different traffic priorities, four access categories (AC) have been introduced. To each AC corresponds a backoff entity. The four distinct parallel backoff entities present in each 802.11e terminal are called (from highest to lowest priority): voice, video, best effort, and background. For the sake of comparison, existing 802.11/a/b standards define only one backoff entity per terminal. Each backoff entity has a distinct set of parameters, such as CWmin, Cwmax, and the arbitration interframe space (AIFS). AIFS is at least equal to DIFS and can be enlarged if desired. Another feature added to 802.11e is referred to as transmission opportunity (TxOP). A TxOP defines a time interval, which a back off entity can use to transmit data. It is specified by its starting time and duration, and the maximum length is AC dependent. The protocol also defines the maximum lifetime of each MAC service data unit (MSDU), which is also AC dependent. Once the maximum lifetime has elapsed, the MSDU is discarded. Finally, the protocol allows for the optional block acknowledgement in which a number of consecutive MSDUs are acknowledged with a single ACK frame.

Other Protocols

PRMA—Packet reservation multiple access (Goodman, 1989). In PRMA, the medium is divided into slots and a group of N slots forms a frame. Slots are either reserved or available. The access to the medium is provided by means of the slotted-ALOHA protocol. Data may be either periodic or sporadic, and this is informed in the header of the packet. Terminals are allowed to reserve a slot when they have periodic data to transmit. Once the central node successfully acknowledges the periodic packet, the terminal assumes the slot is reserved and uses it without contention. When the terminal stops sending periodic information then the reserved slot is released. PRMA assumes the existence of a central node but the mechanism can be adapted to other topologies (Jiang, 2002).

MACA-BI—Multiple access with collision avoidance by invitation (Talucci, 1997). In MACA-BI, the receiver polls a prospective sender by transmitting a ready-to-receive (RTR) packet. (This is an example of a receiver-initiated protocol.) In order to perform the polling in a timely fashion the receiver is required to correctly predict the traffic originated by the sender. Periodic traffic makes this task easier. In case either the data buffer or the delay at the terminal increases above a certain threshold this terminal may trigger a

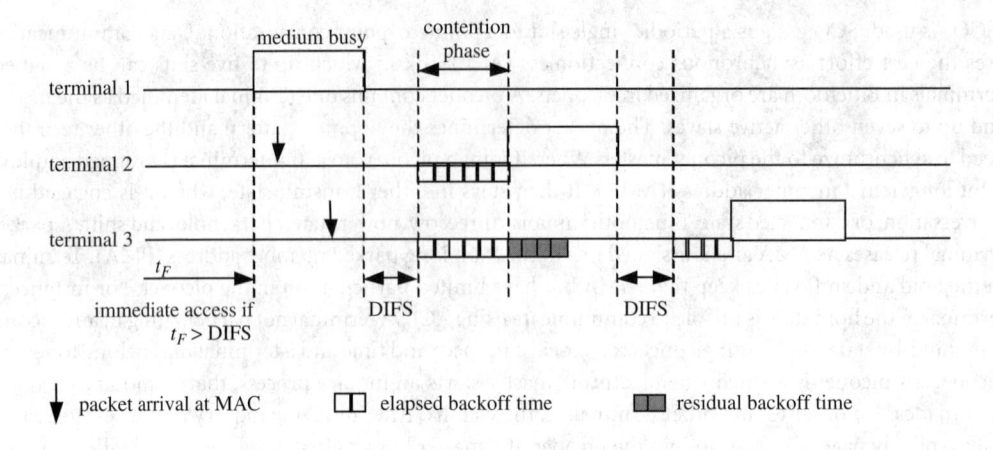

FIGURE 19.53   MAC Handshake Process.

conversation by transmitting an RTS packet. Improvements to MACA-BI are proposed in (Garcia, 1999), in which RIMA-SP–receiver initiated multiple access with simple polling –, and RIMA-DP–Receiver initiated multiple access with dual-purpose polling–are introduced. Both protocols render the RTR-data handshake collision free. RIMA-DP gives an additional purpose to the RTR packet: a request for transmission from the polling terminal. After a reservation phase both terminals can exchange data between them.

DBTMA—Dual busy tone multiple access (Haas, 2002). In DBTMA, the RTS/CTS handshake is replaced by two out-of-band busy tones, namely: BTt (transmit busy tone) and BTr (receive busy tone). When a terminal has data to transmit, it first senses the presence of the BTt and BTr tones. If the medium is free (no busy tone detected), the terminal turns on the BTt, sends an RTS packet, and turns off the BTt. As in other protocols, there is a random backoff time if the medium is busy. The destination terminal, upon receipt of an RTS addressed to it, turns on the BTr and waits for the data. Once the BTr tone is sensed, the sender assumes it has successfully acquired the medium. After waiting a short time (for the BTr to propagate) it transmits the data packet. On successful reception of the data packet the destination terminal turns off the BTr tone, completing the conversation. If no data are received, the BTr tone is turned off after a timer expires at the destination terminal.

Fitzek et al. (Fitzek, 2003) proposes a multi-hop MAC protocol based on the IEEE 802.11. A common channel conveys signaling and dedicated channels carry the data traffic and the ACK packets. Figure 19.53 presents the proposed MAC handshake. The first RTS packet is used to contact the destination and assess its willingness to receive data. The sender includes a list of idle dedicated channels, which is used by the destination terminal to select the dedicated channel. It then transmits this information to the sender in a CTS packet. If no suitable dedicated channel is available the handshake ends. After receiving the CTS packet, the sender transmits a PROBE packet on the dedicated channel. The destination terminal uses this packet to test the channel conditions. It then sends a second CTS packet on the common channel informing about the chosen coding/modulation scheme. The sender to confirm the parameters chosen transmits a second RTS packet. Although at a higher complexity cost, the authors claim that the proposed scheme outperforms the original 802.11.

LA-MAC—Load awareness MAC (Chao, 2003). In LA-MAC, the protocol switches between contention-based and contention-free mechanisms depending on the traffic volume. Contention-based mechanisms are best suited to light traffic conditions in which the probability of collision while attempting to gain the medium is small. For heavy traffic a contention-free mechanism allows higher and more evenly distributed throughput. In (Chao, 2003), the IEEE 802.11 DCF is adopted during contention-based periods while contention-free periods use a token passing protocol. The traffic load is measured by the delay packets are experiencing. Each terminal for the packets it has to transmit computes such delay. During a contention-based period, before a terminal transmits its data packet, it checks the packet's current delay. If the delay is greater than a pre-defined threshold A the terminal creates a token and transmits it attached to the data packet. This indicates to all terminals the start of a contention-free period. Once the delay has fallen below another pre-defined threshold B, the terminal about to transmit removes the token. This indicates the end of the contention-free period and the start of a contention-based period. Threshold A is chosen to be greater than B to give the switching decision some hysteresis.

PCDC—Power controlled dual channel (Muqattash, 2003). In PCDC, the objective is to maintain network connectivity at the lowest possible transmit power. PCDC is a multi-hop protocol that uses the RTS/CTS handshake found in IEEE 802.11 with some modifications. Each terminal is required to keep a list of neighboring terminals and the transmit power needed to reach them. When a packet is received, the list needs to be visited. If the sender is not known an entry is added. Otherwise, the existing entry is updated. In any case, the receiver needs to re-evaluate its connectivity information and confirm that it knows the cheapest way (in a transmit power sense) to reach all terminals that appears in its neighbor list. For instance, for some terminals it might be cheaper to use an intermediate terminal instead of the direct route. At heavy traffic loads there exist enough packets transiting to keep terminals well informed of their neighborhood. For long idle periods terminals are required to broadcast a "hello" packet periodically for this purpose. PCDC achieves space efficiency and simulations carried by the authors indicate an increase in the network's throughput.

MAC ReSerVation—MAC-RSV (Fang, 2003). In MAC-RSV, a reservation-based multihop MAC scheme is proposed. The TDMA frame consists of data and signaling slots. Data slots are marked as follows: reserved for transmission (RT), reserved for reception (RR), free for transmission (FT), free for reception (FR), or free for transmission and reception (FTR). The signaling slot is divided in minislots with each minislot further divided in three parts: request, reply, and confirm. A terminal wishing to transmit sends an RTS packet. In the RTS, the sender informs its own identity, the intended receiver's identity, and the data slots it wishes to reserve. The intended receiver replies with a CTS if any of the requested slots is among its FR or FTR slots. Otherwise, it remains silent. It is possible that the CTS packet accepts reservation of only a subset of the requested slots. Terminals other than the intended receiver replies with a Not CTS (NCTS) if any of the requested slots is among its RR. Any terminal that detects an RTS collision also replies by sending an NCTS. Otherwise, it remains silent. Finally, if the sender successfully receives a CTS it confirms the reservation by sending a confirm packet (CONF). Otherwise, it remains silent. RTS packets are transmitted in the request part of the minislot; CTS and NCTS use the reply part; and CONF packets use the confirm part. Data slots are divided in three parts: receiver beacon (RB), data, and acknowledgement (ACK). A terminal that has a data slot marked RR transmits an RB with the identity of the active data transmitter. In addition, the receiver acknowledges the correct data reception by transmitting an ACK at the end of the data slot. Simulations carried out by the authors indicate that the proposed protocol outperforms the IEEE 802.11 at moderate to heavy traffic loads.

Comments

In (Jurdak, 2004) a set of guidelines is provided that a suitable general-purpose MAC protocol should follow. In particular, it is mentioned that the use of multiple channels to separate control and data is desirable in order to reduce the probability of collisions. The need of flexible channel bandwidth, multiple channels, and the high bandwidth efficiency suggests that CDMA is the optimal choice for channel partition. Multi-hop support is recommended to ensure scalability with flat or clustered topologies depending on the application. In order to favor power efficient terminals, protocols need to be power aware, must control transmission power, and allow for sleep mode. To complete the set of recommendations, the authors include, for the sake of flexibility, short to medium range networks and a sender-initiated approach.

19.6.4   TCP over Ad Hoc Networks

TCP is the prevalent transport protocol in the Internet today and its use over ad hoc networks is a certainty. This has motivated a great deal of research efforts aiming not only at evaluating TCP performance over ad hoc networks, but also at proposing appropriate TCP schemes for this kind of networks. TCP was originally designed for wired network, based on the following assumptions, typical of such an environment: packet losses are mainly caused by congestion, links are reliable (very low bit error rate), round-trip times are stable, and bandwidth is constant (Postel, 1981; Huston, 2001). Based on these assumptions, TCP flow control employs a window-based technique, in which the key idea is to probe the network to determine the available resources. The window is adjusted according to an additive-increase/multiplicative-decrease strategy. When packet loss is detected, the TCP sender retransmits the lost packets and the congestion control mechanisms are invoked, which include exponential backoff of the retransmission timers and reduction of the transmission rate by shrinking the window size. Packet losses are therefore interpreted by TCP as a symptom of congestion (Chandran, 2001). Previous studies on the use of TCP over cellular wireless networks have shown that this protocol suffers from poor performance mainly because the principal cause of packet loss in wireless networks is no longer congestion, but the error-prone wireless medium (Xylomenos, 2001; Balakrishnan, 1997). In addition, multiple users in a wireless network may share the same medium, rendering the transmission delay time-variant. Therefore, packet loss due to transmission error or a delayed packet can be interpreted by TCP as being caused by congestion. When TCP is used over ad hoc networks, additional problems arise. Unlike cellular networks, where only the last hop is wireless, in ad hoc networks the entire path between the TCP sender and the TCP destination may be made up of wireless hops (multihop). Therefore, as discussed earlier in this chapter, appropriate routing protocols and medium access control mechanisms (at the link control layer) are required to establish a path connecting the sender and the destination. The interaction between TCP and the protocols at the physical, link, and network layers can cause serious performance degradation, as discussed in the following.

Physical Layer Impact

Interference and propagation channel effects are the main causes of high bit error rate in wireless networks. Channel induced errors can corrupt TCP data packets or acknowledgement packets (ACK), resulting in packet losses. If an ACK is not received within the retransmit timeout (RTO) interval, the lost packets may be mistakenly interpreted as a symptom of congestion, causing the invocation of TCP congestion control mechanisms. As a consequence, the TCP transmission rate is drastically reduced, degrading the overall performance. Therefore, the reaction of TCP to packet losses due to errors is clearly inappropriate. One approach to avoid this TCP behavior is to make the wireless channel more reliable by employing appropriate forward error correction coding (FEC), at the expense of a reduction of the effective bandwidth (due to the addition of redundancy) and an increase in the transmission delay (Shakkottai, 2003). In addition to FEC, link layer automatic repeat request (ARQ) schemes can be used to provide faster retransmission than that provided at upper layers. ARQ schemes may increase the transmission delay, leading TCP to assume a large round-trip time or to trigger its own retransmission procedure at the same time (Huston, 2001).

MAC Layer Impact

It is well known that the hidden and exposed terminal problems strongly degrade the overall performance of ad hoc networks. Several techniques for avoiding such problems have been proposed, including the RTS/CTS control packets exchange employed in the IEEE 802.11 MAC protocol. However, despite the use of such techniques, hidden and exposed terminal problems can still occur, causing anomalous TCP behavior. The inappropriate interaction between TCP and link control layer mechanisms in multihop scenarios may cause the so-called TCP instability (Xu, 2001). TCP adaptively controls its transmission rate by adjusting its contention window size. The window size determines the number of packets in flight in the network (i.e., the number of packets that can be transmitted before an ACK is received by the TCP sender). Large window sizes increase the contention level at the link layer, as more packets will be trying to make their way to the destination terminal. This increased contention level leads to packet collisions and causes the exposed terminal problem, preventing intermediate nodes from reaching their adjacent terminals (Xu, 2001). When a terminal cannot deliver its packets to its neighbor, it reports a route failure to the source terminal, which reacts by invoking the route reestablishment mechanisms at the routing protocol. If the route reestablishment takes longer than RTO, the TCP congestion control mechanisms are triggered, shrinking the window size and retransmitting the lost packets. The invocation of congestion control mechanisms results in momentary reduction of TCP throughput, causing the mentioned TCP instability. It has been experimentally verified that reducing the TCP contention window size minimizes TCP instability (Xu, 2001). However, reduced window size inhibits spatial channel reuse in multihop scenarios. For the case of IEEE 802.11 MAC, which uses a four-way handshake (RTS-CTS-Data-ACK), it can be shown that, in an H-hop chain configuration, a maximum of $H/4$ terminals can simultaneously transmit (Fu, 2003), assuming ideal scheduling and identical packet sizes. Therefore, a window size smaller than this upper limit degrades the channel utilization efficiency. Another important issue related to the interaction between TCP and the link layer protocols regards the unfairness problem when multiple TCP sessions are active. The unfairness problem (Xu, 2001; Tang, 2001) is also rooted in the hidden (collisions) and exposed terminal problems and can completely shut down one of the TCP sessions. When a terminal is not allowed to send its data packet to its neighbor due to collisions or the exposed terminal problem, its backoff scheme is invoked at the link layer level, increasing (though randomly) its backoff time. If the backoff scheme is repeatedly triggered, the terminal will hardly win a contention, and the winner terminal will eventually capture the medium, shutting down the TCP sessions at the loser terminals.

Mobility Impact

Due to terminal mobility, route failures can frequently occur during the lifetime of a TCP session. As discussed above, when a route failure is detected, the routing protocol invokes its route reestablishment mechanisms, and if the discovery of a new route takes longer than RTO, the TCP sender will interpret the route failure as congestion. Consequently, the TCP congestion control is invoked and the lost packets are retransmitted. However, this reaction of TCP in this situation is clearly inappropriate due to several reasons (Chandran, 2001). Firstly, lost packets should not be retransmitted until the route is reestablished. Secondly, when the route is eventually restored, the TCP slow start strategy will force the throughput to be unnecessarily low immediately after the route reestablishment. In addition, if route failures are frequent, TCP throughput will never reach high rates.

Main TCP Schemes Proposals for Ad Hoc Networks

TCP—Feedback

This TCP scheme is based on explicitly informing the TCP sender of a route failure, such that it does not mistakenly invoke the congestion control (Chandran, 2001). When an intermediate terminal detects a route failure, it sends a route failure notification (RFN) to the TCP sender terminal and records this event. Upon receiving an RFN, the TCP sender transitions to a "snooze" state and (i) stops sending packets, (ii) freezes its flow control window size, as well as all its timers, and (iii) starts a route failure timer, among other actions. When an intermediate terminal that forwarded the RFN finds out a new route, it sends a route reestablishment notification (RRN) to the TCP sender, which in turn leaves the snooze state and resumes its normal operation.

TCP with Explicit Link Failure Notification

Explicit link failure notification (ELFN) technique is based on providing TCP sender with information about link or route failures, preventing TCP from reacting to such failures as if congestions had occurred (Holland, 2002). In this approach, the ELFN message is generated by the routing protocol and a notice to TCP sender about link failure is piggybacked on it. When the TCP sender receives this notice, it disables its retransmission timers and periodically probes the network (by sending packets) to check if the route has been reestablished. When an ACK is received, the TCP sender assumes that a new route has been established and resumes its normal operation.

Ad Hoc TCP

A key feature of this approach is that the standard TCP is not modified, but an intermediate layer, called ad hoc TCP (ATCP), is inserted between IP and TCP (transport) layers. Therefore, ATCP is invisible to TCP and terminals with and without ATCP installed can interoperate. ATCP operates based on the network status information provided by the internet control message protocol (ICMP) and the explicit congestion notification mechanism (ECN) (Floyd, 1994). The ECN mechanism is used to inform the TCP destination of the congestion situation in the network. An ECN bit is included in the TCP header and is set to zero by the TCP sender. Whenever an intermediate router detects congestion, it sets the ECN bit to one. When the TCP destination receives a packet with ECN bit set to one, it informs the TCP sender about the congestion situation, which in turn reduces its transmission rate. ATCP has four possible states: normal, congested, loss, and disconnected. In the normal state ATCP does nothing and is invisible to TCP. In the congested, loss, and disconnected states, ATCP deals with congested network, lossy channel, and partitioned network, respectively. When ATCP sees three duplicate ACKs (likely caused by channel induced errors), ATCP transitions to the loss state and puts TCP into persist mode, ensuring that TCP does not invoke its congestion control mechanisms. In the loss state, ATCP retransmits the unacknowledged segments. When a new ACK arrives, ATCP returns to the normal state and removes TCP from the persist mode, restoring the TCP normal operation. When network congestion occurs, ATCP sees the ECN bit set to one and transitions to congested state. In this state, ATCP does not interfere with TCP congestion control mechanisms. Finally, when a route failure occurs, a destination unreachable message is issued by ICMP. Upon receiving this message, ATCP puts TCP into persist mode and transitions to the disconnected state. While in the persist mode, TCP periodically sends probe packets. When the route is eventually reestablished, TCP is removed from persist mode and ATCP transitions back to the normal state.

19.6.5 Capacity of Ad Hoc Networks

The classical information theory introduced by Shannon (Shannon, 1948) presents the theoretical results on the channel capacity, that is, how much information can be transmitted over a noisy and limited communication channel. In ad-hoc networks, this problem is led to a higher level of difficulty for the capacity now must be investigated in terms of several transmitters and several receivers. The analysis of the capacity of wireless networks has a similar objective as that of the classical information theory: to estimate the limit of how much information can be transmitted and to determine the optimal operation mode, so that this limit can be achieved. A first attempt to calculate these bounds is made by Gupta and Kumar in (Gupta, 2000). In this work, the authors propose a model for studying the capacity of a static ad hoc network (i.e., nodes do not move), based on the following scenario. Suppose that nodes are located in a region of area 1. Each node can transmit at bits per second over a common wireless channel. Packets are sent from node to node in a multi-hop fashion until their final destination is reached and they can be buffered at intermediate nodes while waiting for transmission. Two types of network configurations are considered: arbitrary networks, where the node locations, traffic destinations, rates, and power level are all arbitrary; and random networks, where the node locations and destinations are random, but they have the same transmit power and data rate. Two models of successful reception over one hop are also proposed:

- Protocol Model—in which a transmission from node i to j, with a distance d_{ij} between them, is successful if $d_{kj} \geq (1 + \Delta) d_{ij}$, that is, if the distance between nodes i and j is smaller than that of nodes k and j with both i and k transmitting to j simultaneously over the same channel. The quantity $\Delta > 0$ models the guard zone specified by the protocol to prevent a neighboring node from simultaneous transmission.

- Physical Model—in which, for a subset T of simultaneous transmitting nodes, the transmission from a node $i \in T$ is successfully received by node j if

$$\frac{P_i/d_{ij}^\alpha}{N + \sum_{\substack{k \in T \\ k \neq i}} P_k/d_{kj}^\alpha} \geq \beta \tag{19.110}$$

with a minimum signal-to-interference ratio (SIR) β for successful receptions, noise power level N, transmission power level P_i for node i, and signal power decay α.

TABLE 19.5 Capacity Upper Bounds for Arbitrary and Random Networks

	Protocol Model	Physical Model
Arbitrary Networks (transport capacity in bit-meters/s)	$\Theta\left(W\sqrt{n}\right)$	Feasible $c\,W\sqrt{n}$ Not feasible $c'Wn^{\left(\frac{\alpha-1}{\alpha}\right)}$ for appropriate values of c and c'
Random Networks (node throughput in bits/s)	$\Theta\left(\frac{W}{\sqrt{n\log n}}\right)$	Feasible $c\,W/\sqrt{n\log n}$ Not feasible $c'\,W/\sqrt{n}$ fore appropriate values of c and c'

The transport capacity is so defined as the quantity of bits transported in a certain distance, measured in bit-meter. One bit-meter signifies that one bit has been transported over a distance of one meter toward its destination. With the reception models as described previously, the upper bounds for the transport capacity in arbitrary networks and the throughput per node in random networks were calculated. They are summarized in Table 19.5, where the Knuth's notation has been used, that is, $f(n) = \Theta(g(n))$ denotes that $f(n) = O(g(n))$ as well as $g(n) = O(f(n))$. In Table 19.5, c and c' are constants which are functions of α and β. These results show that, for arbitrary networks, if the transport capacity is divided into equal parts among all nodes, the throughput per node will be $\Theta(W/\sqrt{n})$ bits per second. This shows that, as the number of nodes increases, the throughput capacity for each node diminishes in a square root proportion. The same type of results holds for random networks. These results assume a perfect scheduling algorithm which knows the locations of all nodes and all traffic demands, and which coordinates wireless transmissions temporally and spatially to avoid collisions. Without these assumptions the capacity can be even smaller.

Troumpis and Goldsmith (Toumpis, 2000) extend the analysis of upper limits of (Gupta, 2000) to a three dimensional topology, and incorporated the channel capacity into the link model. In this work, the nodes are assumed as uniformly distributed within a cube of volume $1\ m^3$. The capacity $C(n)$ follows the inequality

$$k_1 \frac{n^{1/3}}{\log(n)} \leq C(n) \leq k_2 \log(n)n^{1/2} \tag{19.111}$$

with probability approaching unity as $n \to \infty$, and k_1, k_2 some positive constants. Equation Eq. (19.111) also suggests that, although the capacity increases with the number of users, the available rate per user decreases.

Case Studies on Capacity of Ad Hoc Networks

IEEE802.11

Li et al (Li, 2001) study the capacity of ad hoc networks through simulations and field tests. Again, the static ad hoc network is the basic scenario, which is justified by the fact that at most mobility scenarios nodes do not move significant distances during packet transmissions. The 802.11 MAC protocol is used to analyze the capacity of different configuration networks.

For the chain of nodes, the ideal capacity is 1/4 of the raw channel bandwidth obtainable from the radio (single-hop throughput). The simulated 802.11-based ad hoc network achieves a capacity of 1/7 of the single-hop throughput, because the 802.11 protocol fails to discover the optimum schedule of transmission and its backoff procedure performs poorly with ad hoc forwarding. The field experiment does not present different results from those obtained in the simulation. The same results are found for the lattice topology. For random networks with random traffic patterns, the 802.11 protocol is less efficient, but the theoretical maximum capacity of $O(1/\sqrt{n})$ per node can be achieved.

It is also shown that the scalability of ad hoc networks is a function of the traffic pattern. In order for the total capacity to scale up with the network size, the average distance between source and destination nodes must remain small as the network grows. Therefore, the key factor deciding whether large networks are feasible is the traffic pattern. For networks with localized traffic the expansion is feasible whereas for networks in which the traffic must traverse it then the expansion is questionable.

Wireless Mesh Networks

A particular case of ad-hoc network, which is drawing significant attention, is the wireless mesh network (WMN). The main characteristic that differentiates a WMN from others ad-hoc networks is the traffic pattern: practically, all traffic is either to or from a node (gateway) that is connected to other networks (e.g. Internet). Consequently, the gateway plays a decisive role in the WMN: the greater the number of gateways the greater the capacity of this network as well as its reliability. Jun and Sichitiu (Jun, (Oct) 2003) analyze the capacity of WMNs with stationary nodes. Their work shows that the capacity of WMNs is extremely dependent on the following aspects:

- Relayed traffic and fairness—Each node in a WMN must transmit relayed traffic as well as its own. Thus, there is an inevitable contention between its own traffic and relayed traffic. In practice, as the offered load at each node increases, the nodes closest to the gateway tends to consume a larger bandwidth, even for a fair MAC layer protocol. The absolute fairness must be forced according to the offered load.
- Nominal capacity of MAC layer (B)—It is defined as the maximum achievable throughput at the MAC layer in one-hop network. It can be calculated as presented in (Jun (April) 2003).
- Link constraints and collision domains—In essence, all MAC protocols are designed to avoid collisions while ensuring that only one node transmits at a time in a given region. The collision domain is the set of links (including the transmitting one) that must be inactive for one link to transmit successfully.

The chain topology is first analyzed. It is observed that the node closer to the gateway has to forward more traffic than nodes farther away. For a n-node network and a per node generated load G, the link between the gateway and the node closer to it has to be able to forward a traffic equal to nG. The link between this node and the next node has to be able to forward traffic equal to $(n-1)G$, and so on. The collision domains are identified and the bottleneck collision domain, which has to transfer the most traffic in the network, is determined. The throughput available for each node is bounded by the nominal capacity B divided by the total traffic of the bottleneck collision domain. The chain topology analysis can be extended to a two-dimensional topology (arbitrary network). The values obtained for the throughput per node are validated with simulation results.

These results lead to an asymptotic throughput per node of $O(1/n)$. This is significantly worse than the results showed in Table 19.5, mainly because of the presence of gateways, which are the network bottlenecks. Clearly, the available throughput improves with the increase of the number of gateways in the network.

Increasing the Capacity of Ad Hoc Networks

The expressions presented in Table 19.5 indicate the best performance achievable considering optimal scheduling, routing, and relaying of packets in the static networks. This is a bad result as far as scalability is concerned and encourages researches to pursue techniques that increase the average throughput. One approach to increase capacity is to add relay-only nodes in the network. The major disadvantage of this scheme is that it requires a large number of pure relay nodes. For random networks under the protocol model with m additional relay nodes, the throughput available per node becomes $\Theta(W(n+m)/n\sqrt{(n+m)\log(n+m)})$ (Gupta, 2000). For example, in a network with 100 senders, at least 4476 relay nodes are needed to quintuplicate the capacity (Gupta, 2000). Another strategy is to introduce mobility into the model. Grossglauser and Tse (Grossglauser, 2001) show that it is possible for each sender-receiver pair to obtain a constant fraction of the total available bandwidth, which is independent of the number of pairs, at the expense of an increasing delay in the transmission of packets and the size of the buffers needed at the intermediate relay nodes. The same results are presented by Bansal and Liu (Bansal, 2003), but with low delay constraints and a particular mobility model similar to the random waypoint model (Bettstetter, 2002). However, mobility introduces new problems such as maintaining connectivity within the ad hoc network, distributing routing information, and establishing access control. (An analysis on the connectivity of ad hoc networks can be found at (Bettstetter, 2002).) The nodes can also be grouped into small clusters, where in each cluster a specific node (clusterhead) is designated to carry all the relaying packets (Lin, 1997). This can increase the capacity and reduce the impact of the transmission overhead due to routing and MAC protocols. On the other hand, the mechanisms to update the information in clusterheads generate additional transmissions, which reduces the effective node throughput.

References

Aggelou, G., Tafazolli, R., RDMAR: A Bandwidth-efficient Routing Protocol for Mobile Ad Hoc Networks, *ACM International Workshop on Wireless Mobile Multimedia (WoWMoM)*, 1999, pp. 26–33.

Balakrishnan, H., Padmanabhan, V.N., Seshan, S., Katz, R.H.A, A Comparison of Mechanisms for Improving TCP Performance over Wireless Links, IEEE/ACM Trans. on Networking, vol. 5, No. 6, pp. 756–769, December 1997.

Bansal, N. and Liu, Z. Capacity, Delay and Mobility in Wireless Ad-Hoc Networks, *Proc. IEE Infocom '03*, April 2003.

Basagni, S. et al., A Distance Routing Effect Algorithm for Mobility (DREAM), *ACM/IEEE Int'l. Conf. Mobile Comp. Net.*, pp. 76–84, 1998.

Bellur, B. and Ogier, R.G., A Reliable, Efficient Topology Broadcast Protocol for Dynamic Networks, *Proc. IEEE INFOCOM '99*, New York, March 1999.

Bettstetter, C., On the Minimum Node Degree and Connectivity of a Wireless Multihop Network, *Proc. ACM Intern. Symp. On Mobile Ad Hoc Networking and Computing (MobiHoc)*, June 2002.

Bluetooth SIG, Specification of the Bluetooth System, vol. 1.0, 1999, available at: http://www.bluetooth.org.

Chandra, A., Gummalla, V., Limb, J.O., Wireless Medium Access Control Protocols, *IEEE Communications Surveys and Tutorials [online]*, vol. 3, no. 2, 2000, available at: http://www.comsoc.org/pubs/surveys/.

Chandran, K., Raghunathan, S., Venkatesan, S., Prakash, R., A Feedback-Based Scheme for Improving TCP Performance in Ad Hoc Wireless Networks, *IEEE Personal Communications*, pp. 34–39, February 2001.

Chao, C.M., Sheu, J.P., Chou, I-C., A load awareness medium access control protocol for wireless ad hoc network, *IEEE International Conference on Communications, ICC'03*, vol. 1, 11–15, pp. 438–442, May 2003.

Chatschik, B., An overview of the Bluetooth wireless technology, *IEEE Communications Magazine*, vol. 39, no. 12, pp. 86–94, Dec. 2001.

Chen, T.-W., Gerla, M., Global State Routing: a New Routing Scheme for Ad-hoc Wireless Networks, *Proceedings of the IEEE ICC*, 1998.

Chiang, C.-C. and Gerla, M., Routing and Multicast in Multihop, Mobile Wireless Networks, *Proc. IEEE ICUPC '97*, San Diego, CA, Oct. 1997.

Corson, M.S. and Ephremides, A., A Distributed Routing Algorithm for Mobile Wireless Networks, *ACM/Baltzer Wireless Networks 1*, vol 1, pp. 61-81, 1995.

Das, S., Perkins, C., and Royer, E., Ad Hoc on Demand Distance Vector (AODV) Routing, Internet Draft, draft-ietf-manet-aodv-11.txt, 2002.

Dube, R., Rais, C., Wang, K., and Tripathi, S., Signal Stability Based Adaptive Routing (SSA) for Ad Hoc Mobile Networks, *IEEE Personal Communication 4*, vol 1, pp. 36–45, 1997.

Fang, J.C. and Kondylis, G.D., A synchronous, reservation based medium access control protocol for multihop wireless networks, *2003 IEEE Wireless Communications and Networking, WCNC 2003*, vol. 2, 16–20. March 2003, pp. 994–998.

Fitzek, F.H.P., Angelini, D., Mazzini, G., and Zorzi, M., Design and performance of an enhanced IEEE 802.11 MAC protocol for multihop coverage extension, *IEEE Wireless Communications*, vol. 10, no. 6, pp. 30–39, Dec. 2003.

Floyd, S., TCP and Explicit Congestion Notification, *ACM Computer Communication Review*, vol. 24, pp. 10–23, October 1994.

Fu, Z., Zerfos, P., Luo, H., Lu, S., Zhang, L., and Gerla, M., The Impact of Multihop Wireless Channel on TCP Throughput and Loss, *IEEE INFOCOM*, pp. 1744–1753, 2003.

Garcia-Luna-Aceves, J.J. and Marcelo Spohn, C., Source-tree Routing in Wireless Networks, *Proceedings of the Seventh Annual International Conference on Network Protocols*, Toronto, Canada, p. 273, October 1999.

Garcia-Luna-Aceves, J.J. and Tzamaloukas, A., Reversing the Collision-Avoidance Handhske in Wireless Networks, *ACM/IEEE MobiCom'99*, pp. 15–20, August 1999.

Goodman, D.J., Valenzuela, R.A., Gayliard, K.T., Ramamurthi, B., Packet reservation multiple access for local wireless communications, *IEEE Transactions on Communications*, vol. 37, no. 8, pp. 885–890, Aug. 1989.

Grossglauser, M. and Tse, D., Mobility Increases the Capacity of Ad Hoc Wireless Networks, *Proc. IEE Infocom '01*, April 2001.

Günes, M., Sorges, U., and Bouazizi, I., ARA–The Ant-colony Based Routing Algorithm for Manets, *ICPP Workshop on Ad Hoc Networks (IWAHN 2002)*, pp. 79–85, August 2002.

Gupta, P. and Kumar, P.R., The Capacity of Wireless Networks, *IEEE Trans. Info. Theory*, vol. 46, March 2000.

Haas, Z.J. and Deng, J., Dual busy tone multiple access (DBTMA)—a multiple access control scheme for ad hoc networks, *IEEE Transactions on Communications*, vol. 50, no. 6, pp. 975–985, June 2002.

Haas, Z.J. and Pearlman, R., Zone Routing Protocol for Ad-hoc Networks, *Internet Draft*, draft-ietf-manet-zrp-02.txt,1999.

Holland, G. and Vaidya, N., Analysis of TCP Performance over Mobile Ad Hoc Networks, *Wireless Networks*, Kluwer Academic Publishers, vol. 8, pp. 275–288, 2002.

Huston, G., TCP in a Wireless World, *IEEE Internet Computing*, pp. 82–84, March–April, 2001.

IEEE 802.11 WG, Wireless LAN Medium Access Control (MAC) and Physical Layer (PHY) Specifications, *IEEE/ANSI Std.* 802-11, 1999 edn.

IEEE 802.11 WG, Wireless LAN Medium Access Control (MAC) and Physical Layer (PHY) Specifications: high-speed physical layer in the 5 GHz band, *IEEE Std.* 802-11a.

IEEE 802.11 WG, Wireless LAN Medium Access Control (MAC) and Physical Layer (PHY) Specifications: higher-speed physical layer extension in the 2.4 GHz band, *IEEE Std.* 802-11b.

IEEE 802.11 WG, Wireless LAN Medium Access Control (MAC) and Physical Layer (PHY) Specifications: Medium Access Control (MAC) Enhancements for Quality of Service (QoS), *IEEE Draft Std.* 802.11e/D5.0, August 2003.

Iwata, A. et al., Scalable Routing Strategies for Ad-hoc Wireless Net-works, *IEEE JSAC*, pp. 1369–179, August. 1999.

Jacquet, P., Muhlethaler, P., Clausen, T., Laouiti, A., Qayyum, A., and Viennot, L., Optimized Link State Routing Protocol for Ad Hoc Networks, *IEEE INMIC*, Pakistan, 2001.

Jiang, M., Ji, J., and Tay, Y.C., Cluster based routing protocol, *Internet Draft*, draft-ietf-manet-cbrp-spec-01.txt, 1999.

Jiang, S., Rao, J., He, D., Ling, X., and Ko, C.C., A simple distributed PRMA for MANETs, *IEEE Transactions on Vehicular Technology*, vol. 51, no. 2, pp. 293–305, March 2002.

Joa-Ng, M. and Lu, I.-T., A Peer-to-peer Zone-based Two-level Link State Routing for Mobile Ad Hoc Networks, *IEEE Journal on Selected Areas in Communications 17*, vol. 8, pp. 1415–1425, 1999.

Johnson, D., Maltz, D., and Jetcheva, J., The Dynamic Source Routing Protocol for Mobile Ad Hoc Networks, *Internet Draft*, draft-ietf-manet-dsr-07.txt, 2002.

Jun, J. and Sichitiu, M.L., The Nominal Capacity of Wireless Mesh Networks, *IEEE Wireless Communications*, October 2003.

Jun, J., Peddabachagari, P., and Sichitiu, M.L., Theoretical Maximum Throughput of IEEE 802.11 and its Applications, *Proc. 2nd IEEE Int'l Symp. Net. Comp. and Applications*, pp. 212–25, April 2003.

Jurdak, R., Lopes, C.V., and Baldi, P., A Survey, Classification and Comparative Analysis of Medium Access Control Protocols for Ad Hoc Networks, *IEEE Communications Surveys and Tutorials* [online], vol. 6, no. 1, 2004. available at: http://www.comsoc.org/pubs/surveys/.

Kasera, K.K. and Ramanathan, R., A Location Management Protocol for Hierarchically Organized Multihop Mobile Wireless Networks, *Proceedings of the IEEE ICUPC'97*, San Diego, CA, pp. 158–162, October 1997.

Ko, Y.-B., Vaidya, N.H., Location-aided Routing (LAR) in Mobile Ad Hoc Networks, *Proceedings of the Fourth Annual ACM/IEEE International Conference on Mobile Computing and Networking (Mobicom'98)*, Dallas, TX, 1998.

Li, J., Blake, C., De Couto, D.S.J., Lee, H.I., and Morris, R., Capacity of Ad Hoc Wireless Networks, *Proc. 7th ACM Int'l Conf. Mobile Comp. and Net.*, pp. 61–69, July 2001.

Lin, C.R. and Gerla, M., Adaptive Clustering for Mobile Wireless Networks, *IEEE Journal on Selected Ares in Communications*, vol. 15, pp. 1265–1275, September 1997.

Mangold, S., Sunghyun Choi, Hiertz, G.R., Klein, O., and Walke, B., Analysis of IEEE 802.11e for QoS support in wireless LANs, *IEEE Wireless Communications*, vol. 10, no. 6, pp. 40–50, December 2003.

Muqattash, A. and Krunz, M., Power controlled dual channel (PCDC) medium access protocol for wireless ad hoc networks, *22nd. Annual Joint Conference of the IEEE Computer and Communications Societies, INFOCOM 2003*, vol. 1, pp. 470–480, 30 March–3 April 2003.

Murthy, S. and Garcia-Lunas-Aceves, J.J., An Efficient Routing Protocol for Wireless Networks, *ACM Mobile Networks and App. J.*, Special Issue on Routing in Mobile Communication Networks, pp. 183–197, Oct. 1996.

Nikaein, N., Laboid, H., and Bonnet, C., Distributed Dynamic Routing Algorithm (DDR) for Mobile Ad Hoc Networks, *Proceedings of the MobiHOC 2000:First Annual Workshop on Mobile Ad Hoc Networking and Computing*, 2000.

Ogier, R.G. et al., Topology Broadcast based on Reverse-Path Forwarding (TBRPF), draft-ietf-manet-tbrpf-05.txt, INTERNET-DRAFT, MANET Working Group, March 2002.

Park, V.D. and Corson, M.S., A Highly Adaptive Distributed Routing Algorithm for Mobile Wireless Networks, *Proceedings of INFOCOM*, April 1997.

Pei, G., Gerla, M., and Chen, T.-W., Fisheye State Routing: A Routing Scheme for Ad Hoc Wireless Networks, *Proc. ICC 2000*, New Orleans, LA, June 2000.

Pei, G., Gerla, M., Hong, X., and Chiang, C., A Wireless Hierarchical Routing Protocol with Group Mobility, *Proceedings of Wireless Communications and Networking*, New Orleans, 1999.

Perkins, C.E. and Bhagwat, P., Highly Dynamic Destination-Sequenced Distance-Vector Routing (DSDV) for Mobile Computers, *Comp. Commun. Rev.*, pp. 234–244, Oct. 1994.

Postel, J., Transmission Control Protocol, *IETF RFC 793*, September 1981.

Radhakrishnan, S., Rao, N.S.V., Racherla, G., Sekharan, C.N., and Batsell, S.G., DST–A Routing Protocol for Ad Hoc Networks Using Distributed Spanning Trees, *IEEE Wireless Communications and Networking Conference*, New Orleans,1999.

Raju, J. and Garcia-Luna-Aceves, J., A New Approach to On-demand Loop-free Multipath Routing, *Proceedings of the 8th Annual IEEE International Conference on Computer Communications and Networks (ICCCN)*, pp. 522–527, Boston, MA, October 1999.

Santivanez, C., Ramanathan, R., and Stavrakakis, I., Making Link-State Routing Scale for Ad Hoc Networks, *Proc. 2001 ACM Int'l. Symp. Mobile Ad Hoc Net. Comp.*, Long Beach, CA, October 2001.

Schiller, J., Mobile Communications, Addison-Wesley, Reading, MA, 2000.

Shakkottai, S., Rappaport, T.S., and Karlsson, P.C, Cross-Layer Design for Wireless Networks, *IEEE Communications Magazine*, pp. 74–80, October 2003.

Shannon, C.E., A mathematical theory of communication, *Bell System Technical Journal*, vol. 79, pp. 379–423, July 1948.

Su, W. and Gerla, M., IPv6 Flow Handoff in Ad-hoc Wireless Networks Using Mobility Prediction, *IEEE Global Communications Conference*, Rio de Janeiro, Brazil, pp. 271–275, December 1999.

Talucci, F., Gerla, M., and Fratta, L., MACA-BI (MACA By Invitation)-a receiver oriented access protocol for wireless multihop networks, *8th IEEE International Symposium on Personal, Indoor and Mobile Radio Communications*, PIMRC'97, Vol. 2, pp. 435–439, 1-4 Sept. 1997.

Tang, K., Gerla, M., and Correa, M., Effects of Ad Hoc MAC Layer Medium Access Mechanisms under TCP, *Mobile Networks and Applications*, Kluwer Academic Publishers, Vol. 6, pp. 317–329, 2001.

Toh, C., A Novel Distributed Routing Protocol to Support Ad-hoc Mobile Computing, *IEEE 15th Annual International Phoenix Conf.*, pp. 480–486, 1996.

Toumpis, S. and Goldsmith, A., Ad Hoc Network Capacity, Conference Record of Thirty-fourth Asilomar Conference on Signals Systems and Computers, Vol 2, pp. 1265–1269, 2000.

Tseng, Y.C. and Hsieh, T.Y., Fully power-aware and location-aware protocols for wireless multi-hop ad hoc networks, *11th. International Conference on Computer Communications and Networks*, pp. 608–613, 14–16 Oct. 2002.

Woo, S.-C. and Singh, S., Scalable Routing Protocol for Ad Hoc Networks, *Wireless Networks 7*, Vol. 5, 513–529, 2001.

Xu, S. and Saadawi, T., Does the IEEE 802.11 MAC Protocol Work Well in Multihop Wireless Ad Hoc Networks? *IEEE Communications Magazine*, pp. 130–137, June 2001.

Xylomenos, G., Polyzos, G. C., Mahonen, P., and Saaranen, M., TCP Performance Issues over Wireless Links, *IEEE Communications Magazine*, Vol. 39, No. 4, pp. 53–58, April 2001.

19.7 Network Communication

James E. Goldman

19.7.1 General Principles of Network Analysis and Design

Use of Top-Down Business Oriented Approach

Network communication is the transport of data, voice, video, image, or facsimile (fax) from one location to another achieved by compatibly combining elements of hardware, software, and media. From a business perspective, network communications is delivering the right information to the right decision maker at the right place and time for the right cost. Because there are so many variables involved in the analysis, design, and implementation of such networks, a structured methodology must be followed in order to assure that the implemented network meets the communications needs of the intended business, organization, or individual.

One such structured methodology is known as the top-down approach. Such an approach can be graphically illustrated in a top-down model as shown in Fig. 19.54. Using a top-down approach as illustrated in the top-down model is relatively straightforward.

One must start with the business level objectives. What is the company (organization, individual) trying to accomplish by installing this network? Without a clear understanding of business level objectives, it is nearly impossible to configure and implement a successful network.

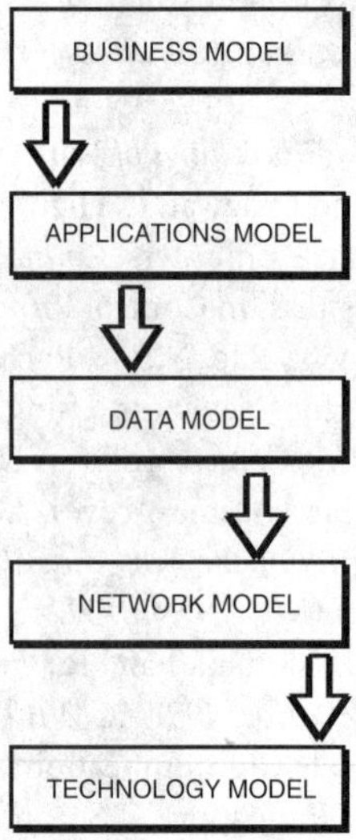

Once business level objectives are understood, one must understand the *applications* which will be running on the computer systems attached to these networks. After all, it is the applications that will be generating the traffic that will travel over the implemented network.

FIGURE 19.54   Top-down design approach: business data comunications analysis.

Data Traffic Analysis Type	Data Traffic Analysis Description
Payload type analysis	Most locations will require at least voice and data service. Videoconferencing and multimedia may need to be supported as well. All payload types should be considered and documented prior to circuit and networking hardware selection.
Transaction analysis	Use process flow analysis or document flow analysis to identify each type of transaction. Analyze detailed data requirements for each transaction type.
Time studies	Once all transaction types have been identified, analyze when and how often each transaction type occurs.
Traffic volume analysis	By combining all types of transactions along with the results of the time study, a time-sensitive traffic volume requirements profile can be produced. This will be a starting point for mapping bandwidth requirements to circuit capacity.
Mission-critical analysis	Results of this analysis may dictate the need for special data security procedures such as encryption or special network design considerations such as redundant links.
Protocol stack analysis	Each corporate location's data traffic is analyzed as to protocols which must be transported across the corporate wide area network. Many alternatives for the transport of numerous protocols exist, but first these protocols must be identified.

FIGURE 19.55   Data traffic analysis.

Once applications are understood and have been documented, the *data* which those applications generate must be examined. In this case, the term data is used in a general sense as today's networks are likely to transport a variety of payloads including voice, video, image, and fax in addition to true data. Data traffic analysis must determine not only the amount of data to be transported, but also must determine important characteristics about the nature of that data. A summarization of data traffic analysis is outlined in Fig. 19.55. It is also during this stage of the top-down analysis in which the geographic proximity of the nodes of the network are examined.

Geographic proximity is one differentiating factor among different categories of networking, which will be examined further subsequently.

Once data traffic analysis has been completed, the following should be known:

1. Physical locations of data (Where?)
2. Data characteristics and compatibility issues (What?)
3. Amount of data generated and transported (How much?)

Given these requirements as determined by the upper layers of the top-down model, the next job is to determine the requirements of the *network* that will possess the capability to deliver this data in a timely, cost-effective manner. Details on the determination of these requirements comprise the remainder of this section on network communications. These network performance criteria could be referred to as *what* the implemented network must do in order to meet the business objectives outlined at the outset of this top-down analysis. These requirements are also sometimes referred to as the *logical network design*.

The *technology* layer analysis, in contrast, will determine *how* various hardware and software components will be combined to build a functional network which will meet predetermined business objectives. The delineation of required technology is often referred to as the *physical network design*.

Overall, the relationship between the layers of the top-down model could be described as follows: analysis at upper layers produces requirements that are passed down to lower layers while solutions meeting these requirements are passed back to upper layers. If this relationship among layers holds true throughout the business oriented network analysis, then the implemented technology (bottom layer) should meet the initially outlined business objectives (top layer).

LAYER	MAJOR FUNCTIONALITY	BLUEPRINT
7 APPLICATION	Layer where Network Operating System and application programs reside. Layer that the user interfaces with.	FURNISHINGS Chairs, couches, tables, paintings
6 PRESENTATION	Assures reliable session transmission between applications. Takes care of differences; data representation.	INTERIOR CARPENTRY Cabinets, shelves, mouldings
5 SESSION	Enables two applications to communicate across the network.	ELECTRICAL Connection between light switch and light wiring.
4 TRANSPORT	Assures reliable transmission from end-to-end, usually across multiple nodes.	HEATING/COOLING/ PLUMBING Furnace, A/C, ductwork
3 NETWORK	This layer sets up the pathways, or end-to-end connections usually across a long distance, or multiple modes.	STRUCTURAL Studs, dry wall
2 DATA LINK	Puts messages together, attaches proper headers to be sent out or received, assures messages are delivered between two points.	FOUNDATION Concrete support upon which entire structure is built.
1 PHYSICAL	Layer that is concerned with transmitting bits of data over a physical medium.	EXCAVATION Site preparation to allow other phases (layers) to proceed according to plans.

FIGURE 19.56   The OSI reference model.

Use of Open Systems Interconnection (OSI) Model

Determining which technology to employ to meet the requirements determined in the logical network design (network layer) requires a structured methodology of its own. Fortunately, a framework for organizing networking technology solutions has been developed by the International Standards Organization (ISO) and is known as the open systems interconnection (OSI) model. The OSI model is illustrated in Fig. 19.56.

The OSI model divides the communication between any two networked computing devices into seven layers or categories. The OSI model allows data communications technology developers as well as standards developers to talk about the interconnection of two networks or computers in common terms without dealing in proprietary vendor jargon.

These common terms are the result of the layered architecture of the seven-layer OSI model. The architecture breaks the task of two computers communicating to each other into separate but interrelated tasks, each represented by its own layer. As can be seen in Fig. 19.56, the top layer (layer 7) represents the application program running on each computer and is therefore aptly named the *application layer*. The bottom layer (layer 1) is concerned with the actual physical connection of the two computers or networks and is therefore named the *physical layer*. The remaining layers (2–6) may not be as obvious but, nonetheless, represent a sufficiently distinct logical group of functions required to connect two computers, as to justify a separate layer.

To use the OSI model, a network analyst lists the known protocols for each computing device or network node in the proper layer of its own seven-layer OSI model. The collection of these known protocols in their proper layers in known as the *protocol stack* of the network node. For example, the physical media employed, such as unshielded twisted pair, coaxial cable, or fiber optic cable, would be entered as a layer 1 protocol, whereas **ethernet** or **token ring** network architectures might be entered as a layer 2 protocol. Other examples of possible protocols in respective layers will be explored in the remainder of this section.

The OSI model allows network analysts to produce an accurate inventory of the protocols present on any given network node. This protocol profile represents a unique personality of each network node and gives the network analyst some insight into what *protocol conversion,* if any, may be necessary in order to get any two network nodes to communicate successfully. Ultimately, the OSI model provides a structured methodology for determining what hardware and software technology will be required in the physical network design in order to meet the requirements of the logical network design.

Perhaps the best analogy for the OSI reference model, which illustrates its architectural or framework purpose, is that of a blueprint for a large office building or skyscraper. The various subcontractors on the job may only be concerned with the layer of the plans that outlines their specific job specifications. However, each specific subcontractor needs to be able to depend on the work of the lower layers subcontractors just as the subcontractors of the upper layers depend on these subcontractors performing their function to specification. Similarly, each layer of the OSI model operates independently of all other layers, while depending on neighboring layers to perform according to specification while cooperating in the attainment of the overall task of communication between two computers or networks.

Differentiation Among Major Categories of Networking

As part of the top-down analysis, geographic proximity of computers or network nodes was mentioned as a key piece of analysis information. Although there are no hard and fast rules for network categorization, following are a few of the more common categories of networking:

- Remote connectivity: A single remote user wishes to access local network resources. This type of networking is particularly important to mobile professionals such as sales representatives, service technicians, field auditors, etc.

- Local area networking: Multiple users' computers are interconnecting for the purpose of sharing applications, data, or networked technology such as printers or CD-ROMs. Local area networks (LANs) may have anywhere from two or three users to several hundred. LANs are often limited to a single department or floor in a building, although technically any single location corporation could be networked via a LAN.

- Internetworking: Also known as LAN-to-LAN networking or connectivity, internetworking involves the connection of multiple LANs and is very common in corporations in which users on departmental LANs now need to share data or otherwise communicate. The challenge of internetworking is in getting departmental LANs of different protocol stacks (as determined by use of the OSI model) to talk to each other, while only allowing authorized users access to the internetwork and other LANs. Variations of internetworking also deal with connecting LANs to mainframes or minicomputers rather than to other LANs.

- Wide area networking: Also known as *enterprise networking*, involves the connection of computers, network nodes, or LANs over a sufficient distance as to require the purchase of wide area network (WAN) service from the phone company or alternative carrier. In some cases, the wide area portion of the network may be owned and operated by the corporation itself. Nonetheless, the geographic distance between nodes is the determining factor in categorizing a wide area network. A subset of WANs known as metropolitan area networks (MANs) are confined to a campus or metropolitan area of usually not more than a few miles in diameter.

The important thing to remember is that categorization of networking is somewhat arbitrary and that what really matters is that the proper networking technology (hardware and software) is specified in any given networking opportunity in order to meet stated business objectives.

19.7.2   Personal Remote Connectivity

Applications

The overall methodology for analysis and design of remote connectivity networking can be summarized as follows:

1. Needs analysis
2. Logical topology choice
3. Physical topology or architecture choice
4. Technology review and specification

Remote Connectivity Needs Analysis

Remote connectivity needs analysis involves documenting the nature and extent of the use of local LAN resources by the remotely connected user. Choices of logical or physical topology for this remote LAN connectivity may be limited depending on the outcome of the remote connectivity needs analysis. Among the possible information sharing needs of remote users are the following: (1) exchange e-mail, (2) upload and download files, (3) run interactive application programs remotely, and (4) utilize LAN attached resources such as printers. One additional question will have a direct impact on topology choice: (5) How many remote users will require simultaneous access to local LAN attached resources?

Remote connectivity architectures comprise the combination of a chosen remote connectivity logical topology and a chosen remote connectivity physical topology.

Logical Topologies

Remote connectivity logical topologies are differentiated by: (1) location of application program execution (local or remote PC) and (2) nature of the data traffic between the local and remote PC. The two most common remote connectivity logical topologies or operations modes are: (1) remote client mode and (2) remote control mode.

Remote Client Mode

Remote client mode, alternatively known as remote access node, executes and often stores applications on the remote PC, using only shared data and other locally attached LAN resources connected to the local LAN **server**. A single local LAN server or specialized communications server can service multiple remote PC clients. The remote node has the same full capabilities as any local node on the network. The fact that the client PC is remote to the local server is transparent. The data traffic between the remote PC and local LAN server are data packets particular to the network operating system, which both the remote PC and local LAN server have installed.

Remote Control Mode

Remote control mode requires a dedicated local PC to be assigned to each remote PC since applications are stored and executed on the local PC. Shared data and other LAN-attached resources are accessed from the LAN server through the local PC. The remote PC is really nothing more than a simple input/output device. All processing is performed on the local PC. Only keystrokes and screen images are transmitted between the remote PC and the local PC. This setup is really just a remote keyboard and monitor for a local PC. The remote PC *controls* the local PC and all of its LAN attached resources, hence the name, remote control mode.

Figure 19.57 outlines some of the details, features and requirements of these two remote PC modes of operation or logical topologies.

Physical Topologies

Remote connectivity physical topologies refer to the physical arrangement of hardware and media which offers access to the local LAN for the remote user. As Fig. 19.58 illustrates, there are three basic ways in which a remote PC user can gain access to the local LAN resources. (1) a LAN attached PC, (2) communications server, and (3) LAN modem.

It is important to understand that the actual implementation of each of these LAN access arrangements may require additional hardware and/or software. They may also be limited in their ability to utilize all LAN attached resources.

19.7.3   Local Area Networks

LAN Applications

Possible business analysis questions for local area networking solutions are listed in Fig. 19.59. This list of business analysis questions is not meant to be exhaustive or all encompassing. Two important things to

Characteristic	Remote Client Operation Mode	Modem Remote Control Operation Mode
Is redirector hardware and/ or software required?	Yes	No
Where do the applications actually execute?	On the remote PC	On the local LAN-attached PC or communications server.
What is the nature of the data traffic over the dial-up line?	All interaction between the LAN server and remote PC travels over the dial-up link.	Only keystrokes from the remote PC keyboard and output for the remote PC monitor travel over the dial-up link.
Network interface	Actually using the remote PC's serial port and modem as a substitute NIC. This option provides a very low speed interface.	Uses installed NIC on local LAN attached PC or communication server to interface to the LAN at a very high speed.
Relative performance	Slower than modem remote control operation mode with substantially more traffic, depending whether application programs are stored remotely or on the LAN.	Faster than remote client mode and generates substantially less traffic. Often not appropriate for Windows-based applications
Also called	Remote LAN node. LAN remote control.	

FIGURE 19.57   Remote LAN connectivity: remote PC operations mode.

remember about any list of business analysis questions are

1. The questions should dig deeper into the required information systems-related business activities.
2. The answers to these questions should provide sufficient insight as to enable the investigation of possible technical solutions.

Next, each of the business analysis questions' categories is explained briefly.

User Issues

User satisfaction is the key to any successful network implementation. To satisfy users, their needs must be first thoroughly understood. Beyond the obvious question of: How many users must the network support? are the more probing questions dealing with specific business activities of individual users. Do users process many short transactions throughout the day? Do users require large file transfers at certain times of day? Are there certain activities which absolutely must be done at certain times of day or within a certain amount of elapsed time? These questions are important in order to establish the amount of network communication required by individual users. Required levels of security should also be addressed. Are payroll files going to be accessed via the network? Who should have access to these files and what security measures will assure authorized access? What is the overall technical ability of the users? Will technical staff need to be hired? Can support be obtained locally from an outside organization?

Local Communication

Remembering that these are business analysis questions and not technical analysis questions, users really cannot be asked how fast their network connections must be. Bits per second or megabits per second have little or no meaning for most users. If users have business activities such as computer aided design/computer aided manufacturing (CAD/CAM) or other three-dimensional modeling or graphics software that will be accessing the network, the network analyst should be aware that these are large consumers of network bandwidth and should document those information system-related business activities which may be large consumers of networking bandwidth.

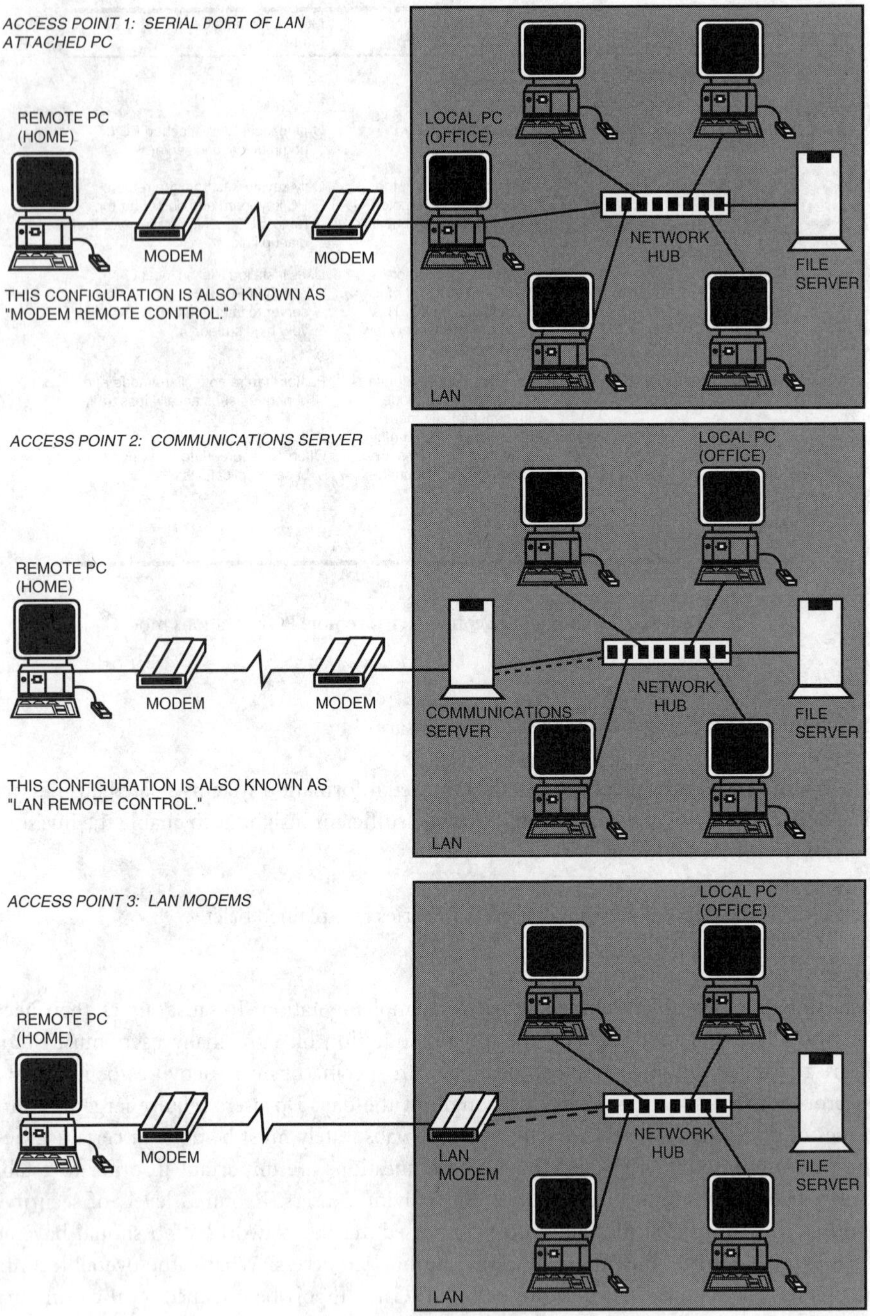

FIGURE 19.58   Remote LAN connectivity: three primary access points.

Resource Sharing

It is important to identify which resources and how many are to be shared: printers, modems, faxes, and the preferred locations of these shared resources. The required distance between shared resources and users can have a bearing on acceptable technical options.

	Current	2–3 years	5 years
User issues:			
How many users?			
What are their business activities?			
What is the budgeted cost/user?			
Comprehensive cost of ownership?			
What are the security needs? (password protection levels, supervisor privileges)			
What are the support issues?			
Local communication:			
Required speed?			
Resource sharing:			
How many printers, modems, and faxes are to be shared?			
What is the greatest distance from the server to each service?			
File sharing:			
Is printer/queue management required?			
How many simultaneous users?			
Application sharing:			
What is the number and type of required applications?			
Are e-mail services required?			
Distributed data access:			
Where will shared data files be stored?			
Lan management/administration:			
Will training be required to manage the network?			
How easy is the network to use?			
Extended communication:			
How many mini/mainframe connections are needed (and what type, IBM, DEC, UNIX based)?			
Will this be an inter-LAN network? [LAN to LAN concerns. Which NOS? Must other protocols be considered? Are the connections local or remote (long-distance?)]			

FIGURE 19.59   LAN and LAN look alikes: business analysis questions.

File Sharing and Application Sharing

Which programs or software packages are users going to need to perform their jobs? Which programs are they currently using? Which new products must be purchased? In many cases, network versions of software packages may cost less than multiple individual licenses of the same software package for individual PCs. The network analyst is really trying at this point to compile a listing of all applications programs which will be shared by users. Not all PC-based software packages are available in network versions and not all PC-based software packages allow simultaneous access by multiple users. Once a complete list of required shared application programs has been completed, it is important to investigate both the availability and capability of the network versions of these programs in order to assure happy, productive users and the successful attainment of business needs.

Distributed Data Access

Although users cannot be expected to be database analysts, sufficient questions must be asked in order to determine which data is to be shared by whom and where these users are located. The major objective of data distribution analysis is to determine the best location on the network for the storage of various data files. That best location is usually the one closest to the greatest number of the most active users of that data.

Some data files that are typically shared, especially in regionalized or multilocation companies, include customer files, employee files, and inventory files. Distributed data access is even more of a concern when the users sharing the data are beyond the reach of a local area network and must share the data via wide area networking solutions. A good starting point for the network analyst might be to ask the question: Has anyone done a comparison of the forms that are used in the various regional and branch offices to determine which data needs to be sent across the network?

Extended Communications

The ability of certain local area networking solutions to communicate beyond the local area network remains a key differentiating factor among local area networking alternatives. Users should be able to articulate connectivity requirements beyond the LAN. The accomplishment of these requirements is the job of the network analyst. Some possible examples of extended communications might include communications to another LAN. If this is the case, the network analyst must investigate all of the technical specifications of this target LAN in order to determine compatibility with the local LAN. The target LAN may be local (within the same building) or remote (across town or around the world). LAN to LAN connection is known as *internetworking* and will be studied in the next section Other examples of extended communications may be the necessity for LAN users to gain access to mainframes, either locally or remotely. Again, users are only asked *what* they need connections to, and *where* those connections must occur, it is the network analyst's job to figure out *how* to make those connections function.

LAN Management and Administration

Another key differentiating factor among LAN alternatives is the level of sophistication required to **manage** and administer the **network**. If the LAN requires a full time, highly trained manager, then that manager's salary should be considered as part of the purchase cost as well as the operational cost of the proposed LAN. Secondly, the users may have requirements for certain management or administration features which must be present. Examples might be user-identification creation or management, or control of access to files or user directories.

Budget Reality

The most comprehensive, well documented, and researched networking proposal is of little value if its costs are beyond the means of the funding organization or business. Initial research into possible networking solutions is often followed by feasibility option reports that outline possible network designs of varying price ranges. Senior management then dictates which options deserve further study based on financial availability. In some cases, senior management may have an approximate project budget in mind which could be shared with network analysts. This acceptable financial range, sometimes expressed as budgeted cost per user, serves as a frame of reference for analysts as technical options are explored. In this sense, budgetary constraints are just another overall, high-level business need or perspective that helps to shape eventual networking proposals.

Anticipated Growth is Key

User needs are not always immediate in nature. These needs can vary dramatically over time. To design networking solutions that will not become obsolete in the near future, it is essential to gain a sense of what the anticipated growth in user demands might be. Imagine the chagrin of the network analyst who must explain to management that the network which was installed last year cannot be expanded and must be replaced due to unanticipated growth of network demand. One method of gaining the necessary insight into future networking requirements, illustrated in Fig. 19.59, is to ask users the same set of business analysis questions with projected time horizons of 2–3 years and 5 years. Incredible as it may seem, 5 years is about the maximum projected lifetime for a given network architecture or design. Of course, there are exceptions. End users may not have the information or knowledge necessary to make these projections. Management can be very helpful in the area of projected growth and informational needs, especially if the company has engaged in any sort of formalized strategic planning methodology.

Network Application	Issues
Network printing	Printers shared; printer requests, buffered and spooled; 2nd generation network printers faster/smarter. Queue/device management, multiple print formats.
Network backup	Match hardware and software to your needs and network. Check out restore features. Scheduling unattended backups. Generation of log reports. Hardware dependencies and capacities.
Network security	Virus control. Additional User ID/password protection. Encryption. User authentication. Automatic logout of inactive terminals.
Network management	Network monitors. Diagnostic and analysis tools, which may be seperated depending on cable and network type. Administration concerns. Software often in NOS for simple LANs.
Network resource optimization	Hardware and Software inventory database. Reporting and query capabilities. Planning and network configuration tools. Monitoring and network alarm capabilities.
Configuration mgmt. Inventory mgmt.	
Remote access-control software	Included in many fully integrated LANs.
Connectivity/gateway software MAC integration software	
Groupware	Workflow automation. Interactive work document review. Information sharing. Group schedules. Enhanced e-mail.

FIGURE 19.60   LAN and LAN look alikes: network applications analysis.

Network Applications: What Can a LAN Do for You ?

Beyond merely being able to share the same application software packages (spreadsheets, word processing, databases), which ran on individuals PCs before they were networked together over a LAN, networking PCs together provides some unique opportunities to run additional networked applications, which can significantly increase worker productivity and/or decrease costs.

Figure 19.60 summarizes the attributes and issues of some of the most popular uses of a LAN. It should be noted that the uses, features, and issues described next apply only to the listed applications functioning on a local area network. Many of these same functions become much more complicated when running across multiple LANs (internetworking) or over long distance WANs. A short description of each of these LAN applications follows

Network Printing

Network printing continues to evolve as user demands change and technology changes to meet those demands.

On a typical LAN, a networked PC would send a request for printing out onto the network through a network interface card. The networked request for printing services would be accepted by a device in charge of organizing the print requests for a networked printer. Depending on the LAN implementation configuration, that device may be a PC with attached printer, a specialized print server with attached printers, or a directly networked printer. Some type of software must manage all of this requesting, spooling, buffering, queuing and printing. The required software may be part of an overall network operating system or may be specifically written for only network printer management.

Network Backup

Backing up data and application files on a network is essential to overall network security and the ability to recover from the inevitable data-destroying disaster. Although the process and overall components are relatively simple, the implementation can be anything but simple. Basically there are only two components to a network backup system: (1) the software, which manages the backup, and (2) the hardware device, which captures the backed-up files.

Some network backup software and hardware work with only certain network operating systems. Other network backup software will work with only the hardware device with which it is sold. The interaction between hardware devices and software such as operating systems or network operating systems is often controlled by specialized software programs (drivers). It is essential to make sure that the necessary drivers are supplied by either the tape backup device vendor or software vendor in order to ensure the operability of the tape backup device. Hardware devices may be various types of tape subsystems, or optical drives. Key differences among hardware devices include

1. How much? What is the storage capacity of the backup device?
2. How fast? How quickly can data be transferred from a PC or server to the backup device? This attribute is important if you are backing up large capacity disk drives.
3. How compressed? Can data be stored on the backup device in compressed form? If so, it may save significant room on the backup media.

Remember, backup is not necessarily a one way process. Restoration of backed up files and the ease with which that restoration can be accomplished is a major purchasing consideration. Being able to schedule unattended backups or restorals as well as the ability to spool or print log reports of backup/restoral activity are also important to **network managers**.

Network Management

The overall task of network management is usually broken down into at least three distinct areas of operation. First, networks must be *monitored*. Therefore, one of the foremost jobs of network management software is to monitor the LAN to detect any irregular activity such as malfunctioning network adapter cards, or an unusually high data transmission rate monopolizing the available network bandwidth to be shared among all attached workstations. Sophisticated LAN monitoring programs can display maps of the network on graphics terminals. Operators can zoom in on a particular node or workstation for more information or performance statistics. Some monitoring programs also have the ability to compare current network activity to preset acceptable parameters and to set off alarms on network monitor terminals, perhaps by turning a node symbol bright red on the screen, when activity goes beyond acceptable limits. Monitoring software is also written specifically for monitoring file servers on a LAN. Besides monitoring server performance and setting off alarms, some monitoring packages have the ability to dial and set off pagers belonging to network managers who may be away from the network console.

Once a problem has been monitored and identified, it must be *analyzed* and *diagnosed*. This is the second major task of network management software. Diagnosis is often done by a class of devices known as protocol analyzers or by the more common name, sniffers. These devices are attached to the LAN and watch, measure, and in some cases, record every bit of data that passes their way. By using multiple sniffers at various points on a LAN, otherwise known as distributed sniffers, bottlenecks can be identified and performance degradation factors can be pinpointed. LAN testing devices must be able to test and isolate the three major segments of any LAN: (1) the wire or cable of the network, (2) the network adapter cards, which interface between the cable and the workstation, and (3) the workstation or PC that generates the network activity.

Having diagnosed the cause of the networking problem, corrective action must be taken against that problem. Perhaps a misbehaving network interface card must be disabled or an application on a workstation that is monopolizing network resources must be logged out. The power to do these things is sometimes called network *administration* or management. In this case, the term administration is preferable in order to provide a contrast to the previous more general use of the term network management. Most often, the required network management software to manage LANs is included in the network operating system itself.

LAN monitoring software and other specialized network management functions are available as an add-on product for most network operating systems. When these add-on products are manufactured by a company other than the original network operating system vendor, they are known as third-party products. These third-party enhancements are often of high quality but should be purchased with caution. Compatibility with associated software or future releases of the network operating system are not necessarily guaranteed.

Network Security

In addition to the typical security features, such as password protection and directory access control supplied with most network operating systems, more sophisticated network security software/hardware is available for LANs. For instance, security software may be added to workstations and or servers which will

1. Require user identification and valid password to be entered before the PC can be booted.
2. Encrypt important data or application files to prevent tampering.
3. Automatically logout inactive terminals to prevent unauthorized access to system resources.
4. Allow certain users to run only certain applications.
5. Require verification of user authenticity by security verifiers.

Another area of network security that is receiving a lot of attention is that of *viruses*. Virus control software is sometimes included in network security packages. Virus control is really a three-step process, implying that effective virus control software should address at least the following three areas:

1. Virus protection: User access to systems is sufficiently controlled as to prevent an unauthorized user from infecting the LAN.
2. Virus detection: Sophisticated software to find viruses regardless of how cleverly they may be disguised.
3. Virus eradication: Sometimes called antibiotic programs, this software eliminates all traces of the virus.

Groupware

Groupware is the name of a category of software that seeks to take advantage of the fact that workers are networked together electronically in order to maximize worker productivity. Groupware is a general term that describes all or some of the following software categories: workflow automation, interactive work, group scheduling, document review, information sharing, electronic whiteboard, and enhanced electronic mail.

Local Area Network Architectures

The choice of a particular network architecture will have a definite bearing on the choice of network adapter cards and less of an impact on the choice of media or network operating system. For instance, an ethernet network architecture requires ethernet adapter cards. As will soon be seen, it is the adapter card which holds the key, or media access control (MAC) layer protocol, which determines whether a network is ethernet, token ring, **fiber distributed data interface (FDDI)** or any other network architecture. Ethernet runs over thick or thin coaxial cable, shielded or unshielded twisted pair, fiber or wireless—clearly a wide choice of media options.

Ethernet

Ethernet, adhering to the IEEE 802.3 standard, is a **carrier sense multiple access with collision detection- (CSMA/CD-)** based network architecture traditionally installed in a bus configuration, but most often installed in a hub-based star physical topology. Every device, most often network adapter cards, attached to an ethernet network has a unique hardware address assigned to it at time of manufacture. As new devices are added to an ethernet network, their addresses become new possible destinations for all other attached ethernet devices.

The media access layer protocol elements of ethernet form data packets for transmission over the shared media according to a standardized format. This ethernet packet format is nearly equivalent to the IEEE 802.3 standard, and the two terms are often used interchangeably.

The potential for collisions and retransmission exists on an ethernet network thanks to its CSMA/CD access methodology. In some cases, ethernet networks with between 100 and 200 users barely use the capacity of the network. However, the nature of the data transmitted is the key to determining potential network capacity problems. Character-based transmissions, such as typical data entry, in which a few characters at a time are typed and sent over the network are much less likely to cause network capacity problems than the transfer of graphical user interface (GUI) screen oriented transmission such as Windows-based applications. CAD/CAM images are even more bandwidth intensive. Simultaneous requests for full screen Windows-based transfers by 30 or more workstations can cause collision and network capacity problems on an ethernet network. As with any data communication problem, there are always solutions or workarounds to these problems. The point in relaying these examples is to provide some assurance that although ethernet is not unlimited in its network capacity, in most cases, it provides more than enough bandwidth.

Token Ring

IBM's token ring network architecture, adhering to the IEEE 802.5 standard, utilizes a star physical topology, sequential message delivery, and a token passing access methodology scheme. Since the sequential logical topology is equivalent to passing messages from neighbor to neighbor around a ring, the token ring network architecture is sometimes referred to as: *logical ring, physical star*. The token ring's use of the token passing access methodology furnishes one of the key positive attributes of this network architecture. The guarantee of no data collisions with assured data delivery afforded by the token passing access methodology is a key selling point in some environments where immediate, guaranteed delivery is essential.

FDDI

As far as trends in network architecture go, as more and more users are attached to LANS, the demand for overall network bandwidth increases. LANs are increasing both in size and overall complexity. Internet-working of LANs of various protocols via bridges and routers means more overall LAN traffic. Network applications are driving the demand for increased bandwidth as well. The concepts of distributed comput-ing, data distribution, and client/server computing all rely on a network architecture foundation of high bandwidth and high reliability. Imaging, multimedia, and data/voice integration all require high amounts of bandwidth in order to transport and display these various data formats in real time. In other words, if full-motion video is to be transported across the LAN as part of a multimedia program, there should be sufficient bandwidth available on that LAN for the video to run at full speed and not in slow motion. Likewise, digitized voice transmission should sound normal when transported across a LAN of sufficient bandwidth.

FDDI supplies not only a great deal of bandwidth, but also a high degree of reliability and security while adhering to standards-based protocols. FDDI's reliability comes not only from the fiber itself which, as we know, is immune to both electromagnetic interference (EMI) and radio frequency interference (RFI). An additional degree of reliability is achieved through the design of the physical topology of FDDI.

FDDI's physical topology comprises not one, but two, separate rings around which data moves simulta-neously in opposite directions. One ring is the primary data ring while the other is a secondary or backup data ring to be used only in the case of the failure of the primary ring. Whereas both rings are attached to a single hub or concentrator, a single point of failure remains in the hub while achieving redundancy in the network media. In addition to speed and reliability, distance is another key feature of an FDDI LAN.

Another positive attribute of FDDI is its ability to interoperate easily with ethernet networks. In this way, a business does not have to scrap its entire existing network in order to upgrade a piece of it to FDDI. An FDDI to ethernet bridge is the specific technology employed in such a setup.

The uses of the FDDI network architecture typically fall into three categories:

1. Campus backbone: Not necessarily implying a college campus, this implementation is used for connecting LANs located throughout a series of closely situated buildings.
2. High bandwidth workgroups: The second application category is when the FDDI LAN is used as a truly local area network, connecting a few PCs or workstations which require high bandwidth communication with each other. Multimedia workstations, engineering workstations, or CAD/CAM workstations are all good examples of high bandwidth workstations.
3. High bandwidth subworkgroup connections: In some cases, only two or three devices, perhaps three servers, have high bandwidth requirements. As distributing computing and data distribution increase, an increasing demand for high-speed server to server data transfer has been seen.

Wireless LANs

Many of the previously mentioned network architectures function over more than one type of media. Another media option is wireless transmission (which is really the absence of any media) for local area networks. There are currently three popular wireless transmission technologies in the local area network technology area. They are microwave transmission, spread spectrum transmission, and infrared transmission.

These are all just radio transmissions at varying frequencies.

Wireless LAN Applications

A primary application of **wireless LANs** optimizes the ease of access of the wireless technology. Portable or notebook PCs equipped with their own wireless LAN adapters can create an instant LAN connection merely by getting within range of a server-based wireless LAN adapter or wireless hub. In this way, a student or employee can sit down anywhere and log into a LAN as long as the user is within range of the wireless hub and has the proper wireless adapter installed in the portable PC.

Meeting rooms can be equipped with wireless hubs to allow spontaneous workgroups to log into network resources without running cables all over the meeting room. Similarly, by quickly installing wireless hubs and portable PCs with wireless adapters, temporary expansion needs or emergency/disaster recovery situations can be handled quickly and with relative ease. No rerunning of wires or finding the proper cross-connects in the wiring closet are necessary.

Finally, wireless LAN technology allows entire LANs to be preconfigured at a central site and shipped ready to run to remote sites. The nontechnical users at the remote site literally just have to plug the power cords into the electrical outlets and they have an instant LAN. For companies with a great number of remote sites and limited technical staff, such a technology is ideal. No preinstallation site visits are necessary. Also avoided are costs and supervision of building wiring jobs and troubleshooting building wiring problems during and after installation.

Local Area Network Hardware

Servers

A server's job is to manage the sharing of networked resources among client PCs. Depending on the number of client PCs and the extent of the shared resources, it may be necessary to have multiple servers and/or to specialize some servers as to the type of resources that they manage. If servers are to be specialized, then shared resources should be grouped in some logical fashion so as to optimize the server performance for managing the sharing of a particular type of resource. A list of potentially shared network resources would probably include: files, application programs, databases, printers, access to other LANs (local), access to other LANs (remote), access to information services, and access to the LAN from remote PCs.

Hubs/Multistation Access Units (MAUs)

The heart of the star physical topology, employed by both ethernet and token ring, is the wiring center, alternatively known as a hub, a concentrator, a repeater or a multistation access unit (MAU).

Repeaters

A repeater, as its name would imply, merely repeats each bit of digital data that it receives. This repeating action actually cleans up the digital signals by retiming and regenerating them before passing this repeated data from one attached device or LAN segment to the next.

Hubs

The terms hub and *concentrator* or *intelligent concentrator* are often used interchangeably. Distinctions can be made, however, between these two broad classes of wiring centers, although there is nothing to stop manufacturers from using the terms as they wish. A hub is often the term reserved for describing a stand-alone device with a fixed number of ports, which offers features beyond that of a simple repeater. The type of media connections and network architecture offered by the hub is determined at the time of manufacture as well. For example, a 10BaseT ethernet hub will offer a fixed number of RJ-45 twisted pair connections for an ethernet network. Additional types of media or network architectures are not usually supported.

MAUs

A MAU is IBM's name for a token ring hub. A MAU would be manufactured with a fixed number of ports and connections for unshielded or shielded twisted pair. IBM uses special connectors for token ring over shielded twisted pair (STP) connections to a MAU. MAUs offer varying degrees of management capability.

Concentrators

The term concentrator or intelligent concentrator (or smart hub) is often reserved for a device characterized by its flexibility and expandability. A concentrator starts with a fairly empty, boxlike device often called a chassis. This chassis contains one or more power supplies and a builtin network backbone. This backbone might be ethernet, token ring, FDDI, or some combination of these. Into this backplane, individual cards or modules are inserted.

For instance, an 8- or 16-port twisted pair ethernet module could be purchased and slid into place in the concentrator chassis. A network management module supporting the SNMP (simple network management protocol) network management protocol could then be purchased and slid into the chassis next to the previously installed 10BaseT Port module. In this mix and match scenario, additional cards could be added for connection of PCs with token ring adapters, PCs, or workstations with FDDI adapters, or dumb asynchronous terminals.

This network in a box is now ready for workstations to be hooked up to it through twisted pair connections to the media interfaces on the network interface cards of the PCs or workstations. Remember that ethernet can run over UTP (unshielded twisted pair), STP, thick and thin coaxial as well as fiber.

Additional modules available for some, but not all, concentrators may allow data traffic from this network in box to travel to other local LANs via bridge or router add-on modules. (Bridges and routers will be discussed in the next section on internetworking.) These combination concentrators are sometimes called *internetworking hubs*. Communication to remote LANs or workstations may be available through the addition of other specialized cards, or modules, designed to provide access to wide area network services purchased from common carriers.

Switching Hubs

The network in a box or backbone in a box offered by concentrators and hubs shrinks the length of the network backbone but does not change the architectural characteristics of a particular network backbone. For instance, in an ethernet concentrator, multiple workstations may access the built in ethernet backbone via a variety of media, but the basic rules of ethernet, such as CSMA/CD access methodology still control performance on this ethernet in a box. Only one workstation at a time can broadcast its message onto the shared backbone.

A *switching hub* seeks to overcome this one at a time broadcast scheme, which can potentially lead to data collisions, retransmissions, and reduced throughput between high-bandwidth demanding devices such as engineering workstations or server-to-server communications.

The ethernet switch is actually able to create connections, or switch, between any two attached ethernet devices on a packet by packet basis. The one-at-time broadcast limitation previously associated with ethernet is overcome with an ethernet switch.

Wiring Centers Technology Analysis

Some of the major technical features to be used for comparative analysis are listed in Fig. 19.61. Before purchasing a hub of any type, consider the implications of these various possible features. To summarize, the following major criteria should be thoroughly considered before a hub or concentrator purchase:

Characteristic/Feature	Options/Implications
Expandability	Options: ports, LANs, slots. Remember, most stand-alone hubs are not expandable, although they may be cascaded. Concentrators may vary in their overall expandability (number of open slots,) or in their ability to add additional ports. Several concentrators allow only one LAN backbone module. Have you anticipated growth in 2- and 5-year horizons?
Network Architecture	Options: ethernet, token ring, FDDI. Some concentrators allow only one type of LAN module, with ethernet being the most widely supported type of LAN module. Before purchasing concentrator, make sure all of the necessary LAN types are supported.
Media	Options: Unshielded twisted pair, shielded twisted pair, thin coax, thick coax, fiber. Remember that each of these different media will connect to the hub with a different type of interface. Make sure that the port modules offer the correct interfaces for attachment of your media; also, remember that a NIC is on the other end of the attached media. Is this concentrator/hub compatible with your network adapter cards?
Extended Communications:	
Internetworking	Options: Internetworking to: ethernet, token ring, FDDI. Internetworking solutions are unique, dependent on the two LANS to be connected. In other words, unless a concentrator offers a module which specifically internetworks ethernet module traffic to token ring LAN, this communication can not take place within the concentrator.
Wide Area Networking	Options: Interfaces to different carrier services.
Management	Options: What level of management is offered? 1. Can individual ports be managed? 2. Is monitoring software included? 3. Can wide area as well as local area link performance be analyzed/ monitored through the concentrator? 4. What level of security in general or regarding management functions specifically is offered? 5. Can ports be remotely enabled/disabled? 6. Can the hub/concentrator be controlled from any attached workstation? Via modem? 7. Can management/monitoring functions be included through a graphical user interface? 8. How are alarm thresholds set? 9. How are faults managed? 10. Can port access be limited by day and/or time? 11. What operating systems can the management software run on? 12. Can a map of the network be displayed?
Reliability	Options: 1. Is integrated UPS included? 2. Are there multiple, redundant power supplies? 3. Can individual modules be swapped out "hot", without powering down the entire hub

FIGURE 19.61 Wiring center technology analysis grid.

(1) expandability (2) supported network architectures (3) supported media types (4) extended communications capabilities, that is, terminal support, internetworking options, and wide area networking options, (5) hub/concentrator management capabilities and (6) reliability features.

Network Interface Cards

Network adapter cards, also known as network interface cards are the physical link between a client or server PC and the shared media of the network. Providing this interface between the network and the PC or workstation requires that the network adapter card have the ability to adhere to the access methodology (CSMA/CD or token passing) of the network architecture to which it is attached. These software rules, implemented by the network adapter card which control the access to the shared network media, are known as *media access control* (MAC) protocols and are represented on the MAC sublayer of the data link layer (layer 2) of the OSI 7-layer reference model.

Since these are MAC layer interface cards and are, therefore, the keepers of the MAC layer interface protocol, it is fair to say that it is the adapter cards themselves that determine network architecture and its constituent protocols more than any other component. Take an ethernet adapter card out of the expansion slot of a PC and replace it with a token ring adapter card and you have a token ring workstation. In this same scenario, the media may not even need to be changed since ethernet, token ring, and FDDI/CDDI often work over the same media.

Role of Adapter Card Drivers

Assuring that the purchased adapter card interfaces successfully to the bus of the CPU as well as the chosen media of the network architecture will assure hardware connectivity. Full interoperability, however, depends on the chosen network adapter card being able to communicate successfully with the network operating system and operating system of the PC into which it is installed.

19.7.4   Internetworking

Applications: What Can Internetworking Do for You?

It is nearly inevitable that sooner or later nearly any organization will need to share information across multiple information platforms or architectures. This sharing may be between LANs that may differ in network architecture or network operating systems. Information systems that combine multiple computing platforms or a variety of network architectures and network operating systems are often referred to as *enterprise computing* environments. Underlying this enterprise computing environment is an *enterprise network* or *internetwork*. The key to successful *internetworking* is that to the end user sitting at a LAN-attached workstation, the connectivity to enterprise computing resources should be completely transparent.

In other words, an end user should not need to know the physical location of a database server or disk drive to which they need access. All the user needs to know is a node name or drive indicator letter and the fact that when these node name or drive letters are entered, data is properly accessed from wherever it may be physically located. That physical location may be across the room or across the country.

As information has become increasingly recognized as a corporate asset to be leveraged to competitive advantage, the delivery of that information in the right form, to the right person at the right place and time has become the goal of information systems and the role of internetworking. Mergers, acquisitions, and enterprise partnerships have accelerated the need to share information seamlessly across geographic or physical LAN boundaries. Intelligent inter-LAN devices perform the task of keeping track of what network attached resources are located and where and how to get a packet of data from one LAN to another. This task is further complicated by the fact that LANS can differ in any of the following categories: (1) network architecture, (2) media, (3) network operating system, and (4) operating system.

This collection of differences is defined by rules or *protocols* that define how certain clearly defined aspects of network communications are to take place. When LANs that need to share information operate according to different protocols, an inter-LAN device that has *protocol conversion* capabilities must be employed. In cases such as this, the inter-LAN device would be transmitting data across logical, as opposed to physical or geographic, boundaries. Among the inter-LAN devices which will be explored in this section

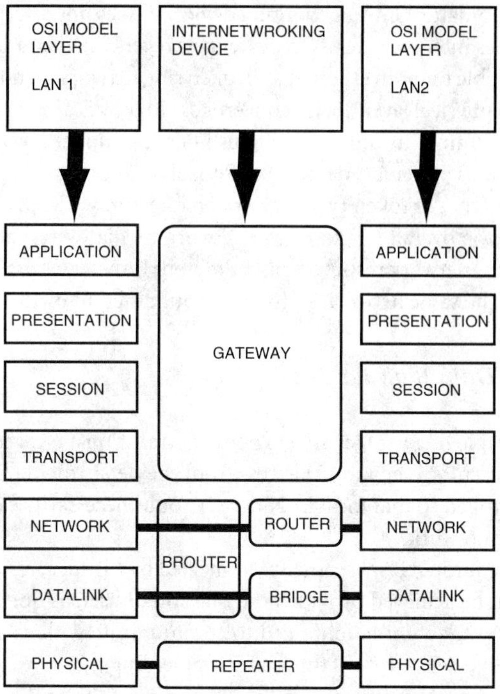

FIGURE 19.62 Relationship between the OSI model and internetworking devices.

are repeaters, bridges, routers, and gateways. Although the differences in functionality of these devices will be distinguished from a technical standpoint, there is no guarantee that manufacturers will follow this differentiation. Functional analysis of data communications devices is the best way to assure that inter-LAN devices will meet internetworking connectivity needs. Internetworking analysis and design is a highly complex and confusing area of study. The purpose of this section is to familiarize the reader with internetworking technology and applications and to provide a list of resources for further study.

The OSI Model and Internetworking Devices

The OSI model was first introduced in Sec. 19.7.1 as a common framework or reference model within which protocols and networking systems could be developed with some assurance of interoperability. In this section, the OSI model provides a most effective framework in which to distinguish between the operational characteristics of the previously mentioned internetworking devices. Figure 19.62 depicts the relationship between each type of internetworking device and its related OSI layer. Each device as well as the related OSI model layer will be explained in detail.

Internetworking Technology

Repeaters: Layer 1—The Physical Layer

Remember that all data traffic on a LAN is in a digital format of discrete voltages of discrete duration traveling over one type of physical media or another. Given this, a repeater's job is fairly simple to understand:

1. Repeat the digital signal by regenerating and retiming the incoming signal.
2. Pass all signals between all attached segments.
3. Do not read destination addresses of data packets.
4. Allow for the connection of different types of media.
5. Effectively extend overall LAN distance by repeating signals between LAN segments.

A repeater is a nondiscriminatory internetworking device. It does not discriminate between data packets. Every signal which comes into one side of a repeater gets regenerated and sent out the other side of the repeater. Repeaters are available for both ethernet and token ring network architectures for a wide variety of media types. A repeater is a physical layer device concerned with physical layer signaling protocols relating to signal voltage levels and timing. It cannot distinguish between upper layer protocols such as between ethernet vs. token ring frames (layer 2 , data link protocols). Therefore, repeaters must be specifically manufactured for either ethernet or token ring network architectures. The primary reasons for employing a repeater are (1) increase the overall length of the network media by repeating signals across multiple LAN segments, (2) isolate key network resources onto different LAN segments, and (3) some repeaters also allow network segments of the same network architecture but different media (layer 1—physical) types to be interconnected.

Bridges: Layer 2—The Data Link Layer
Bridge Functionality
The primary reasons for employing bridges are (1) network traffic on a LAN segment has increased to the point where performance is suffering, and (2) access from the departmental LAN to the corporate LAN backbone needs to be controlled so that local LAN data is not unnecessarily causing congestion problems on the corporate backbone network.

By dividing users across multiple LAN segments connected by a bridge, a substantial reduction in LAN traffic on each segment can be achieved provided the division of users is done in a logical manner. Users should be divided according to job function, need to communicate with each other, and need to access data stored on particular servers. The rule of thumb for segmenting users is that 80% of LAN traffic should remain within the LAN segment and only about 20% should cross the bridge to adjacent LAN segments.

Controlling access to the corporate backbone via a bridge can ensure the viability of enterprise communications by only allowing essential network communication onto the corporate backbone. Servers and other internetworking devices can be connected directly to the corporate backbone while all user's workstations are connected to LAN segments isolated from the corporate backbone by bridges.

When users on one LAN need occasional access to data or resources from another LAN, an internetworking device, which is more sophisticated and discriminating than a repeater, is required. From a comparative standpoint on the functionality of bridges vs repeaters, one could say that bridges are more discriminating. Rather than merely transferring *all* data between LANs or LAN segments like a repeater, a bridge reads the *destination address* of each data frame on a LAN, decides whether the destination is local or remote (on the other side of the bridge), and only allows those data frames with nonlocal destination addresses to cross the bridge to the remote LAN.

How does the bridge know whether a destination is local or not? Data-link protocols such as ethernet contain *source addresses* as well as destination addresses within the predefined ethernet frame layout. A bridge also checks the source address of each frame it receives and adds that source address to a table of known local nodes. After each destination address is read, it is compared with the contents of the known local nodes table in order to determine whether the frame should be allowed to cross the bridge or not (whether the destination is local or not). Bridges are sometimes known as a forward-if-not-local device.

This reading, processing, and discriminating indicates a higher level of sophistication of the bridge, afforded by installed software.

Bridge Categorization
Bridges come in many varieties. Physically, bridges may be cards that can be plugged into an expansion slot of a PC, or they may be standalone devices. Although it is known that the bridge will do the internetwork processing between two LANs, the exact nature of that processing, as well as the bridge's input and output interfaces, will be determined by the characteristics of the two LANs that the bridge is internetworking. In determining the attributes of the input and output bridge one must consider the following issues: MAC sublayer protocol, speed of LANs, local or remote, and wide area network services and media.

1. MAC Sub-Layer Protocol: Depending on the MAC sublayer or network architecture of the LANs to be bridged, any of the following types of bridges may be required: transparent bridges, translating

Bridge Type	LAN Segment Linked
Transparent	• Ethernet to ethernet • Nonsource routing token ring to nonsource routing token ring
Translating	• Ethernet to token ring and vice versa
Encapsulating	• Ethernet to FDDI and vice versa
Source routing	• Source routing token ring to source routing token ring
Source routing transparent	• Source routing token ring to ethernet and vice versa
Adaptive source routing transparent	• Ethernet to ethernet • Source routing token ring to source routing token ring • Source routing token ring to ethernet and vice versa

FIGURE 19.63   Bridges link LAN segments.

bridges, encapsulating bridges, source routing bridges, source routing transparent bridges, and adaptive source routing transparent bridges. First and foremost, are the two LANs which are to be bridged ethernet or token ring? Bridges that connect LANs of similar data link format are known as *transparent bridges*. A special type of bridge that includes a *format converter* can bridge between ethernet and token ring. These special bridges may also be called *multiprotocol bridges* or *translating bridges*. A third type of bridge, somewhat like a translating bridge, is used to bridge between ethernet and FDDI networks. Unlike the translating bridge, which must actually manipulate the data-link layer message before repackaging it, the *encapsulating bridge* merely takes the entire ethernet data link layer message and stuffs it in an envelope (data frame) that conforms to the FDDI data-link layer protocol. *Source routing bridges* are specifically designed for connecting token ring LANs. Bridges that can support links between source routing token ring LANs and nonsource routing LANs, such as ethernet, are known as *source routing transparent bridges*. Finally, bridges that can link transparent bridged ethernet LAN segments to each other, source routing token ring LAN segments to each other, or any combination of the two are known as *adaptive source routing transparent bridges*. Figure 19.63 outlines these various bridge possibilities.

2. Speed of LANs: The speeds of the input and output LANs must be known in order to determine what speed conversion, if any, must be performed by our bridge.
3. Local or Remote: Having determined the MAC layer protocol and speed of the LANs, their geographic proximity to one another must be taken into consideration. If the two LANs are not in close enough proximity to link via traditional LAN media such as UTP, coax, or fiber, the bridge must be equipped with an interface appropriate for linking to wide area carrier services.

Bridge Performance

Bridge performance is generally measured by two criteria:

1. *Filtering rate:* Measured in packets per second or frames per second. When a bridge reads the destination address on an ethernet frame or token ring packet and decides whether or not that packet should be allowed access to the internetwork through the bridge, that process is known as *filtering*.
2. *Forwarding rate:* Also measured in packets per second or frames per second. Having decided whether or not to grant a packet access to the internetwork in the filtering process, the bridge now must perform a separate operation of forwarding the packet onto the internetwork media whether local or remote.

Bridges, Protocols, and the OSI Model

Bridges read the destination addresses within data frames of a predefined structure or protocol. In other words, ethernet and token ring network architectures define a bit-by-bit protocol for formation of data frames. The bridge can rely on this protocol and, therefore, knows just where to look within the ethernet data frames to find the bits, which represent the destination addresses. In terms of the OSI model, ethernet and token ring are considered MAC sublayer protocols. The MAC sublayer is one of two sublayers of OSI model layer 2—the data link layer. The other data link sublayer is known as the *logical link control* sublayer. Because the protocols that a bridge reads and processes are located on the MAC sublayer, bridges are sometimes referred to as MAC layer bridges.

Embedded within the data field of the ethernet frame are all of the higher OSI layer protocols. These higher layer protocols can vary independently of the data-link layer ethernet protocol. In other words, the data-link layer protocols such as ethernet and token ring are *network architectures*, whereas the network layer protocols could be from any one of a number of different *network operating systems*. Bridges only pay attention to network architecture (MAC sublayer) protocols or formats. They completely ignore upper level protocols.

Most network operating systems actually consist of stacks of protocols. In some cases, this *protocol stack* may consist of a separate protocol for each of layers 3–7. Each protocol of a network operating system performs a different networking related function corresponding to the generalized functional definition for the corresponding layer of the OSI model. As an example, the network layer protocol for TCP/IP suite of protocols is known as *internet protocol* (IP).

Routers: The Network Layer Processors

The delivery of data packets to destination addresses across multiple LANs, and perhaps over wide area network links, is the responsibility of a class of internetworking devices known as *routers*. Routers are primarily employed for the following reasons:

1. To build large hierarchical networks. Routers are used to create the backbone network itself.
2. To take part in or gain access to a larger hierarchical network such as the Internet.

Router Functionality

Although they both examine and forward data packets, routers and bridges differ significantly in two key functional areas. First, although a bridge reads the destination address of every data packet on the LAN to which it is attached, a router only examines those data packets that are specifically addressed to it. Second, rather than just merely allowing the data packet access to the internetwork in a manner similar to a bridge, a router is more cautious as well as more helpful. Before indiscriminately forwarding a data packet, a router first confirms the existence of the destination address as well as the latest information on available network paths to reach that destination. Next, based on the latest traffic conditions, the router chooses the best path for the data packet to reach its destination and sends the data packet on its way. The word best is a relative term, controlled by a number of different protocols, which will be examined shortly.

The router itself is a destination address, available to receive, examine, and forward data packets from anywhere on any network to which it is either directly or indirectly internetworked. The destination address on an ethernet or token ring packet must be the address of the router that will handle further internetwork forwarding. Thus, a router is addressed in the data-link layer destination address field. The router then discards this MAC sublayer envelope, which contained its address, and proceeds to read the contents of the data field of the ethernet or token ring frame.

Just as in the case of the data-link layer protocols, network layer protocols dictate a bit-by-bit data frame structure that the router understands. What looked like just data and was ignored by the data-link layer internetworking device (the bridge) is unwrapped by the router and examined thoroughly in order to determine further processing. After reading the network layer destination address and the protocol of the network layer data, the router consults its *routing tables* in order to determine the best path on which to forward this data packet. Having found the best path, the router has the ability to repackage the data packet as required for the delivery route (best path) it has chosen.

As an example, if a packet switched data network was chosen as the wide area link for delivery, then the local router would encapsulate the data packet in compliant envelope. On the other hand, if the best path was over a local ethernet connection, the local router would put the data packet back into a fresh ethernet envelope and send it on its way.

Unlike the bridge, which merely allows access to the internetwork (forward-if-not-local logic), the router specifically addresses the data packet to a distant router. Before a router actually releases a data packet onto the internetwork, however, it confirms the existence of the destination address to which this data packet is bound. Only once the router is satisfied with both the viability of the destination address, as well as with the quality of the intended path, will it release the carefully packaged data packet. This meticulous processing activity on the part of the router is known as *forward-if-proven-remote* logic.

Determination of Best Path

The best path can take into account variables such as

1. Number of intermediate hops. That is, how many other routers will the packet have to be processed by before it reaches its final destination? Every router takes time to process the data packet. Therefore, the fewer the routers, the faster the overall delivery.
2. The speed or condition of the communications circuits. Routers can dynamically maintain their routing tables keeping up to the minute information on network traffic conditions.
3. The protocol of the network operating system, for instance, remembering that multiple protocols can be sealed within ethernet envelopes. We may ask the router to open the ethernet envelopes and forward all NetWare (IPX) traffic to one network and all TCP/IP (IP) to another. In some cases, a certain protocol may require priority handling.

Multiprotocol Routers

Routers are made to read specific network layer protocols in order to maximize filtering and forwarding rates. If a router only has to route one type of network protocol, then it knows exactly where to look for destination addresses every time and can process packets much faster. However, realizing that different network layer protocols will have different packet structures with the destination addresses of various lengths and positions, some more sophisticated routers known as *multiprotocol routers* have the capability to interpret, process, and forward data packets of multiple protocols.

Some common network layer protocols and their associated network operating systems or upper layer protocols as well as other protocols which are actually data-link control protocols processed by some routers are listed in Table 19.6. Remembering that bridges are used to process data-link layer protocols, those routers that can also perform the functionality of a bridge are called bridging routers or *brouters*.

TABLE 19.6 Network and Data-Link Protocols

Network Layer Protocol	Network Operating System or Protocol Stack Name
IPX	NetWare
IP	TCP/IP
VIP	Vines
AFP	Appletalk
XNS	3Com
OSI	Open Systems
Other Protocols	
LAT	Digital DecNet
SNA/SDLC	IBM SNA
Netbios	DOS-based LANs

Router Configuration

Like bridges, routers generally take one of the two physical forms: (1) stand-alone variety, self-contained; and (2) modularized for installation in a slotted chassis. Routers may be installed to link LAN segments either locally or remotely. *Boundary routing* recognizes the need for simple, affordable wide area network devices at remote offices while providing full routing capabilities throughout the wide area network. Boundary routing's physical topology is sometimes referred to as a *hub and spoke topology* due to the fact that each remote branch is connected to a hub office via a single WAN link. If redundant links are a business requirement of a particular node, then it must be a full-function router and not a boundary router in this topology.

Full-function routers are placed at each hub or central node, while less sophisticated boundary routers, or branch-office routers, are placed at each remote or spoke node. Since there are only connected to a single WAN link, these boundary routers make only one decision when examining each piece of data, "If these data are not addressed to a local destination, then it should be forwarded." This forward-if-not-local logic should suggest that, in fact, these boundary routers are acting as bridges.

Gateways

Recalling that in terms of the OSI model, repeaters are considered a physical layer (layer 1) device, bridges are considered a data-link layer (layer 2) device, and routers are considered a network layer (layer 3) device, it could be said that *gateways* provide for interoperability on the session, presentation, and application layers (layers 5–7). Whereas repeaters, bridges and routers provide increasingly more sophisticated connection between two LANs, gateways provide transparent connection between two totally different computing environments. Specialized gateways also translate between different database management systems and are called *database gateways,* or between different e-mail systems and are called e-mail gateways.

The gateway is usually a computer with physical connections to both computing environments to be linked. In addition, the gateway also executes specially written software, which can translate messages between the two computing environments. Unlike other internetworking devices described, gateways are more concerned with translation than with processing destination addresses and delivering messages as efficiently as possible.

19.7.5   Wide Area Networks

Applications

Wide Area Network Architecture

To better understand all of the current and emerging wide area networking technologies and services, a simple model defining the major segments and interrelationships of an overall wide area network architecture is shown in Fig. 19.64. *User demands* are the driving force behind the current and emerging

USER DEMANDS	
INTERFACE SPECIFICATIONS	
NETWORK SERVICES	
NETWORK ARCHITECTURE	
SWITCHING ARCHITECTURE	TRANSMISSION ARCHITECTURE

FIGURE 19.64   Major components of a wide area network architecture.

wide area *network services* which are offered to business and residential customers. Companies offering these services are in business to generate profits by implementing the underlying architectures that will enable them to offer the wide area networking services that users are demanding at the lowest possible cost.

Users are demanding simple, transparent access to variable amounts of bandwidth as required. In addition, this wide area network access must offer support for transmission of data, video, imaging, and fax as well as voice. One of the primary driving forces of increased capacity and sophistication for wide area network services is *LAN interconnection.*

Circuit Switching vs. Packet Switching

Switching of some type or another is necessary in a wide area network because the alternative is unthinkable. To explain, without some type of switching mechanism or architecture, every possible source of data in the world would have to be directly connected to every possible destination of data in the world, not a very likely prospect.

Circuit Switching

Switching allows temporary connections to be established, maintained, and terminated between message sources and message destinations, sometime called sinks in data communications. In the case of the voice-based phone network with which most people are familiar, a call is routed through a central office piece of equipment known as a switch, which creates a temporary circuit between the source phone and the phone of the party to whom one wishes to talk. This connection or circuit only lasts for the duration of the call. This switching technique is known as *circuit switching* and is one of two primary switching techniques employed to deliver messages from here to there. In a circuit switched network, a switched dedicated circuit is created to connect the two or more parties, eliminating the need for source and destination address information such as that provided by packetizing techniques.

The *switched dedicated circuit* established on circuit switched networks makes it appear to the user of the circuit as if a wire has been run directly between the phones of the calling parties. The physical resources required to create this temporary connection are dedicated to that particular circuit for the duration of the connection. If system usage should increase to the point where insufficient resources are available to create additional connections, users would not get a dial tone.

Packet Switching

The other primary switching technique employed to deliver messages from here to there is known as *packet switching.* Packet switching differs from circuit switching in several key areas. First, packets travel one at a time from the message source through a *packet switched network,* otherwise known as a *public data network,* to the message destination. A packet switched network is represented in network diagrams by a symbol which resembles a cloud. Figure 19.65 illustrates such a symbol, as well as the difference between circuit switching and packet switching. The cloud is an appropriate symbol for a packet switched network since all that is known is that the packet of data goes in one side of the public data network (PDN) and comes out the other. The physical path which any packet takes may be different than other packets and in any case, is unknown to the end users. What is beneath the cloud in a packet switched network is a large number of *packet switches,* which pass packets among themselves as the packets are routed from source to destination.

Remember that packets are specially structured groups of data, which include control and address information in addition to the data itself. These packets must be assembled (control and address information added to data) somewhere before entry into the packet switched network and must be subsequently disassembled before delivery of the data to the message destination. This packet assembly and disassembly is done by a device known as a *packet assembler/disassembler (PAD).* PADs may be stand-alone devices or may be integrated into modems or multiplexers.

These PADs may be located at an end-user location, or may be located at the entry point to the packet switched data network. Figure 19.65 illustrates the latter scenario in which the end users employ regular modems to dial up the value added network (VAN) or on-line information service, which provides the PADs to properly assemble the packets prior to transmission over the packet switched network.

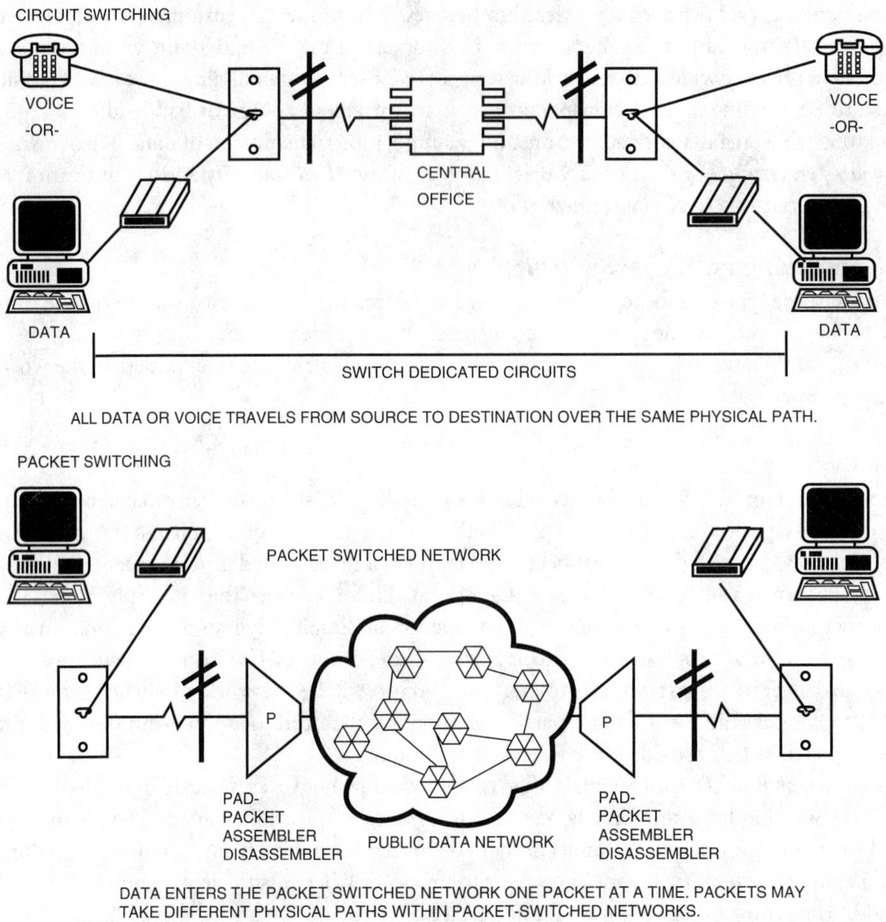

FIGURE 19.65   Circuit-switching vs. packet switching.

Packet Switched Networks

The packet switches illustrated inside the PDN cloud in Fig. 19.65 are generically known as *data switching exchanges (DSEs)*, or *packet switching exchanges (PSEs)*. DSE is the packet switching equivalent of the DCE (data communications equipment) and DTE (data terminal equipment) categorization for modems and dial-up transmission.

Another way in which packet switching differs from circuit switching is that as demand for transmission of data increases on a packet switched network, additional users are not denied access to the packet switched network. Overall performance of the network may suffer, errors and retransmission may occur, or packets of data may be lost, but all users experience the same degradation of service. This is because, in the case of a packet switched network, data travel through the network one packet at a time, traveling over any available path within the network rather than waiting for a switched dedicated path as in the case of the circuit switched network.

For any packet switch to process any packet of data bound for anywhere, it is essential that packet address information be included on each packet. Each packet switch then reads and processes each packet by making routing and forwarding decisions based on the packet's destination address and current network conditions. The full destination address uniquely identifying the ultimate destination of each packet is known as the *global address*.

Because an overall data message is broken up into numerous pieces by the packet assembler, these message pieces may actually arrive out of order at the message destination due to the speed and condition of the alternate paths within the packet switched network over which these message pieces (packets) traveled. The data message must be pieced back together in proper order by the destination PAD before final transmission to the destination address. These self-sufficient packets containing full source and destination address information plus a message segment are known as *datagrams*.

A Business Perspective on Circuit-Switching vs Packet Switching

If the top-down model were applied to an analysis of possible switching methodologies, circuit switching and packet switching could be properly placed on either the network or technology layers. In either case, in order to make the proper switching methodology decision, the top-down model layer directly above the network layer, namely, the data layer must be thoroughly examined. The key data layer question becomes: What is the nature of the data to be transmitted and which switching methodology best supports those data characteristics? The first data-related criteria to examine is the *data source*. What is the nature of the application program (application layer) which will produce this data? Is it a transaction oriented program or more of a batch update or file oriented program?

A transaction oriented program, producing what is sometimes called *interactive data*, is characterized by short *bursts* of data followed by variable length pauses due to users reading screen prompts or pauses between transactions. This *bursty transaction-oriented traffic*, best categorized by banking transactions at an automatic teller machine, must be delivered as quickly and reliably as the network can possibly perform. In addition to data burstiness, time pressures, and reliability constraints are other important data characteristics that will assist in switching methodology decision making.

Applications programs more oriented to large file transfers or batch updates have different data characteristics than transaction oriented programs. Overnight updates from regional offices to corporate headquarters or from local stores to regional offices are typical examples. Rather than occurring in bursts, the data in these types of applications are usually large and flowing steadily. These transfers are important, but often not urgent. If file transfers fail, error detection and correction protocols can retransmit bad data or even restart file transfers at the point of failure.

Defining Terms

Carrier sense multiple access with collision detection (CSMA/CD): A scheme for network communication flow.

Client: The end-users of a network and its resources, typically workstations or personal computers.

Ethernet: A network architecture adhering to IEEE 802.3; a CSMA/CD-based architecture traditionally installed in a bus configuration, but more recently typically in a hub-based star physical topology.

Fiber distributed data interface (FDDI): A networking scheme using separate rings around which data move simultaneously in opposite directions to achieve high speed and operational redundancy.

Gateway: A network device designed to provide a transparent connection between two totally different computing environments.

Hub: The heart of a star physical topology, alternatively known as a concentrator, repeater, or multistation access unit (MAU).

Network interface card (NIC): The physical device or circuit used to interface the network with a local workstation or device.

Network management: The overall task of monitoring and analyzing network traffic and correcting network-related problems.

Open systems interconnection (OSI) model: A framework for organizing networking technology developed by the International Standards Organization.

Router: A device that reads specific network layer protocols in order to maximize filtering and forwarding rates on a network.

Server: The element of a network designed to facilitate and manage the sharing of resources among client devices and workstations.

Token ring: A network architecture adhering to IEEE 802.5, utilizing a star physical topology, sequential message delivery, and a token passing access methodology.

Wireless LAN: An emerging networking system utilizing radio or infrared media as the interconnection method between workstations.

References

Bachus, K. and Longsworth, E. 1993. Road nodes. *Corporate Computing* 2(3):54–61.

Bradner, S. and Greenfield, D. 1993. Routers: Building the highway. *PC Magazine* 12(6):221–270.

Derfler, F. 1993. Ethernet adapters: Fast and efficient. *PC Magazine* 12(3):191.

Derfler, F. 1993. Making the WAN connection: Linking LANs. *PC Magazine* 12(5):183–206.

Derfler, F. 1993. Network printing: Sharing the wealth. *PC Magazine* 12(2):249.

Derfler, F. 1993. To catch a thief. *PC Magazine* 12(16):NE1–NE9.

Derfler, F. 1994. Extend your reach. *PC Magazine* 13(14):315–351.

Derfler, F. 1994. Peer-to-peer LANs: Peer pressure. *PC Magazine* 13(8):237–274.

Donovan, W. 1993. A pain-free approach to SNA internetworking. *Data Communications* 22(16):99.

Gasparro, D. 1994. Putting wireless to work. *Data Communications* 23(5):57–58.

Goldman, J. 1995. *Applied Data Communications: A Business Oriented Approach.* Wiley, New York.

Greenfield, D. 1993. To protect and serve. *PC Magazine* 12(9):179.

Gunnerson, G. 1993. Network operating systems: Playing the odds. *PC Magazine* 12(18):285–333.

Harvey, D. and Santalesa, R. 1994. Wireless gets real. *Byte* 19(5):90.

Held, G. 1993. *Internetworking LANs and WANs.* Wiley, New York.

Held, G. 1994. *The Complete Modem Reference.* Wiley, New York.

Held, G. 1994. *Ethernet Networks: Design Implementation, Operation and Management.* Wiley, New York.

Held, G. 1994. *Local Area Network Performance Issues and Answers.* Wiley, New York.

Heywood, D. et al. 1992. *LAN Connectivity.* New Riders, Carmel, IN.

Jander, M. and Johnson, J. 1993. Managing high speed WANs: Just wait. *Data Communications* 22(7):83.

Johnson, J. 1993. LAN modems: The missing link for remote connectivity. *Data Communications* 22(4):101.

Johnson, J. 1994. Wireless data: Welcome to the enterprise. *Data Communications* 23(5):42–55.

Karney, J. 1993. Network lasers: Built for speed. *PC Magazine* 12(20):199.

Madron, T. 1993. *Peer-to-Peer LANs: Networking Two to Ten PCs.* Wiley, New York.

Mandeville, R. 1994. Ethernet switches evaluated. *Data Communications* 23(4):66–78.

Mathias, C. 1994. New LAN gear naps unseen desktop chains. *Data Communications* 23(5):75–80.

Peterson, M. 1993. Network backup evolves. *PC Magazine* 12(16):277–311.

Pompili, T. 1994. The high speed relay. *PC Magazine* 13(1):NE1–NE12.

Quiat, B. 1994. V.FAST, ISDN, or switched 56 K. *Network Computing* 5(3):70.

Raskin, R. 1993. Antivirus software: Keeping up your guard. *PC Magazine* 12(5):209–264.

Rosen, B. and Fromme, B. 1993. Toppling the SNA internetworking language barrier. *Data Communications* 22(9):79.

Saunders, S. 1993. Choosing high speed LANs: Too many technologies, too little time? *Data Communications* 22(13):58–70.

Saunders, S. 1994. Building a better token ring network. *Data Communications* 23(7):75.

Saunders, S. 1994. Full duplex ethernet: More niche than necessity? *Data Communications* 23(4):87–92.

Schlar, S. 1990. *Inside X.25: A Manager's Guide.* McGraw–Hill, New York.

Shimada, K. 1994. Fast talk about fast ethernet. *Data Communications* 23(5):21–22.

Stallings, W. 1992. *ISDN and Broadband ISDN,* 2nd ed. Macmillan, New York.

Stevenson, T. 1993. Best of a new breed: Groupware—Are we ready? *PC Magazine* 12(11):267–297.

Tabibian, O.R. 1994. Remote access: It all comes down to management. *PC Magazine* 13(14):NE1–NE22.

Thomas, R. 1994. PPP starts to deliver on interoperability promises. *Data Communications* 23(6):83.

Tolly, K. 1993. Can routers be trusted with critical data? *Data Communications* 22(7):58.

Tolly, K. 1993. Checking out channel-attached gateways. *Data Communications* 22(8):75.

Tolly, K. 1993. Token ring adapters: Evaluated for the enterprise. *Data Communications* 22(3):73.

Tolly, K. 1994. FDDI adapters: A sure cure for the bandwidth blues. *Data Communications* 23(10):60.

Tolly, K. 1994. How accurate is your LAN analyzer? *Data Communications* 23(2):42.

Tolly, K. 1994. The new branch-office routers. *Data Communications* 23(11):58.

Tolly, K. 1994. Testing dial up routers: Close, but no cigar. *Data Communications* 23(9):69.

Tolly, K. 1994. Testing remote ethernet bridges. *Data Communications* 23(8):81.

Tolly, K. 1994. Testing remote token ring bridges. *Data Communications* 23(6):93.

Tolly, K. 1994. Testing UNIX-to-SNA gateways. *Data Communications* 23(7):93.

Tolly, K. 1994. Wireless internetworking. *Data Communications* 22(17):60.

Further Information

Books:

The following two books are excellent additions to a professional library for overall coverage of data communications and networking.

Newton, H., *Newton's Telecom Dictionary,* Telecom Library, New York.

Goldman, J. E., *Applied Data Communications: A Business Oriented Approach,* Wiley, New York.

19.8   Printing Technologies and Systems

John D. Meyer

19.8.1   Introduction

The basic parameters of print quality are **resolution**, **addressability**, **gray scale**, and dot microstructure. A real device also has intrinsic variability in the printing process, producing visual artifacts, which come under the general heading of *noise*. Some of the more common manifestations of this are background scatter, dot placement errors, voids (due to nozzle malfunction in ink jet, for example), and banding in images. The significance of any of these aspects of print quality can only be determined by examining them with respect to the properties of the human visual system. The design choices of the basic print quality parameters are, therefore, guided by the properties of the human visual system to determine where improvement needs to be made or where little is to be gained by increasing any one of the specifications.

Resolution and Addressability

Resolution, the most widely used specification to rate print quality, is sometimes confused with the related term addressability. Fundamentally, resolution refers to the ability of the device to render fine detail. This simple definition is complicated by the fact that detail can be regarded as the fineness of the width of a line, the transition between white paper and printed intensity, and/or the smoothness of the edge of a curved line or a line printed at any arbitrary angle. In the simplest case, the resolution of a printer is defined as the spacing of the dots such that full coverage is obtained, that is, no white paper can be seen. For circular dots placed on a square grid, this number would be calculated by dividing the diameter by the square root of two and taking its inverse. For example, an ideal 300 dots per inch (dpi) printer would produce 120-μm-diam dots at an 85 μm spacing. In practice, the dot would be made somewhat larger to allow for dot placement errors. This definition is best understood in terms of the finest line that can be printed by the device. At 300 dpi the line would exhibit a perceptible edge waviness, especially when printed at certain sensitive angles. This would also be true of curved lines. In addition, the range of lines of increasing thickness would have discontinuities since they would consist of an integral number of the basic line, each spaced at 85 μm. These issues have an important bearing on text print quality, which depends on the ability to render both curved and straight lines at variable widths.

The preceding definition is related to the specification of resolution with respect to the human visual system. In this case resolution is determined by the closeness of spacing between alternate black and white lines of equal width and defined contrast. These are known as *line pairs* and for a 300-dpi printer it would result in a value of 150 line pairs per inch. This is not strictly correct since the black lines would be wider than the white spaces due to the roundness of the dot. Since the human visual system has a response that approaches zero near 300 line pairs per inch, gains made in text print quality by increasing resolution alone can be expected to diminish above this value. At this point, issues such as print noise and grayscale enter if further improvement in print quality is desired.

To focus only on the resolution as defined in the previous paragraphs ignores the specific needs of the components of the printed material, that is, text and lines vs. images and area fill. Gains in text print quality may be had if the device can space the dots closer than the fundamental resolution. This can result in substantial dot overlap but allows the line width to be varied more continuously. In addition, at the edge of a curved line, the subpixel adjustments of individual dots increase the perception of smoothness commonly known as getting rid of the *jaggies*. This ultimate dot spacing of the device is called addressability. For example, printers employing this technique are specified as 300×600 dpi indicating a native resolution of 300 dpi in the horizontal direction and a vertical addressability of 600 dpi.

Grayscale

The ability of a printing technology to modulate the printed intensity on the page is referred to as its grayscale capability. There are three ways in which this may be accomplished: variation of the dot size, variation of the intensity of the printed dot, and digital **halftoning** techniques. The first two depend on the intrinsic properties of the technology, whereas digital halftoning can be employed by any printer. A printer that can continuously vary its intensity from white paper through to maximum colorant density is described as having *continuous tone capability*. Other technologies produce a modest number of intensity levels and make use of digital halftoning techniques to create a continuous tone effect. The manner in which gray scale is achieved is of obvious importance in image printing, particularly in the case of color. In recent years considerable effort has gone into the development of sophisticated digital halftoning algorithms to enable binary (single dot size and no intensity modulation) printers to render images. The resulting image quality depends more strongly on resolution than addressability. But the impact of even a few intrinsic gray levels on the print quality achieved by these algorithms can be dramatic.

An important parameter in grayscale considerations is that of the dynamic range, which is simply called range in the graphic arts. This is measured in terms of optical density, the negative logarithm of the reflectance. An optical density of 1.0 represents 10% of reflected light per instant flux, an optical density of 2.0 corresponds to 1% reflectance, and so on. For printed material the smoothness of the printed surface limits the maximum optical density obtainable. If the surface is smooth and mirrorlike, then the print appears glossy and can have optical densities approaching 2.4. The smooth surface reflects light in a specular manner and, therefore, scatters little stray light from the surface into the eye, and the color intensity is not desaturated. It is most noticeable in the case of photographic paper that has a high gloss finish. If the optical density range of the print is high, it is said to have high dynamic range and a very pleasing image will result. Not all papers, however, are designed to have a glossy finish. Papers used in the office are also used in copiers and have a surface which produces diffuse reflection at the interface between the air and the paper. For most uncoated, nonglossy papers this will be between 3–4% and limits the maximum optical density to around 1.4. Image quality on these stocks will depend on the fixing of the colorant to the substrate to produce a smooth surface. The potential image quality for a printer is therefore a complex tradeoff involving the design choices of resolution, addressability, grayscale method, **digital halftoning** algorithm, paper stock, colorant, and fixing technology. Conclusion: for images, resolution alone is not a predictor of print quality.

Dot Microstructure

The microscopic nature of the dot produced by a given technology also has a bearing on final print quality. The most important parameter here relates to the edge gradient of the dot. Known as the *normal-edge profile*, it characterizes the transition between white paper and maximum colorant intensity, that is, the gradient of optical density that occurs at the edge of the dot and measures the steepness of the transition from white paper to full optical density. Some technologies, such as electrophotography, can vary this profile by adjusting various parameters in the imaging and developing process. For ink jet, various paper types will produce different normal-edge profiles. If the profile is very steep, that is, the transition occurs in a very small distance such as $5\,\mu m$, then the dot is described as being a hard dot or having a very sharp edge. This is desirable when printing lines and text that benefit from very sharp transitions between black and white. If this transition is gradual, the dot is described as being soft and produces a blurring of the edge, which can degrade the text quality. In the case of images, where smooth tones and tonal changes are desired, a soft dot can be very beneficial.

Hybrid Methods

From what has been said it should not be inferred that the needs of texts and images are in opposition. In recent years the intrinsic grayscale capability has been used to advantage in improving text print quality. The removal of jaggies can be greatly assisted by the combination of increased addressability and a few gray levels. By the use of gray levels in the region of the jagged stairstep, the transition can be made to take place over several pixels. This is, in essence, blurring the transition to make it less visible to the eye. In the case of certain fonts, there is fine detail requiring resolutions greater than the native resolution of the printer. This fine detail can be rendered through a combination of gray levels and regular pixels. The implementation of these methods requires a complex set of rules to be applied to the data bit stream before it is sent to the marking level of the printer. These rules draw heavily on image processing techniques and a knowledge of the human visual system and are proprietary. Skillfully applied they can have a dramatic effect on the text and line quality. There are a variety of trademarked names for these technologies designed to convey the sense of enhancement of the print quality.

The application of image processing techniques to manipulate the intrinsic properties of electronic printing technologies have made resolution an insufficient measure of print quality. A more comprehensive measure is needed to simplify the identification of the printing technology to serve the design goals for final output quality. Until such a metric is devised, the tradeoff analysis just described, implemented by means of industry standard test charts that separately probe the printer properties, will provide a predictive measure of print quality. Such test charts must also contain test images, which will be subject to the proprietary subjective image enhancement algorithms offered by the manufacturer.

19.8.2 Printing Technologies

The four basic elements of any printing technology are: addressing, marking substance and its storage and delivery, transfer of the marking substance, and fixing. Addressing refers to the communication of electronic data to the marking unit, typically via electronic or optical means. The marking substance contains the colorant, vehicle/carrier material for transport, binders to secure the colorants to the paper, stabilizing agents to resist fading, and technology specific additives such as biocides for liquid inks. The transfer process is the fundamental physical mechanism whereby a specific amount of the marking substance is removed from the bulk and transferred to the paper. Fixing embodies the processes of adhesion, drying, or solidification of the material onto the paper to form a durable image. These fundamental subsystems interact with each other to give each printing technology its own unique characteristics. The common classification of printing technologies today begins with the broad separation into two classes: Impact and nonimpact printing technologies. Impact methods achieve transfer via the direct mechanical application of force or pressure via a marking element, which can be either a fine wire or fully formed character onto a colorant carrying ribbon in contact with the paper; the simplest form of this is a typewriter. Nonimpact methods cover a wide range of technologies that achieve transfer through a variety of means that may be either contact or noncontact in nature.

19.8.3 Nonimpact Printing Technologies

Ink Jet

The transfer process of ink jet printing is one of removing a drop of liquid ink from the bulk and giving it a velocity of sufficient precision and magnitude to place it on a substrate in close proximity to but not touching the printhead. There are three broad techniques: *continuous, electrostatic,* and *drop on demand.* Continuous ink jet printing, because of its intrinsic high drop rate, has tended to find more applications

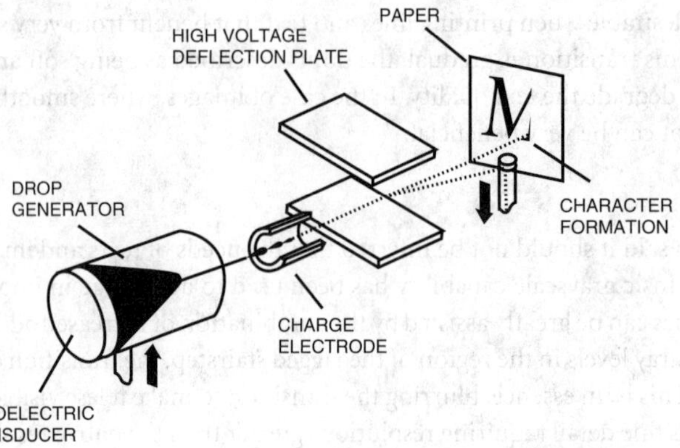

FIGURE 19.66   Character printing with continuous ink jet. The deflection plate applies an analog voltage to steer the drop to the desired location; unwanted droplets are undeflected and captured by the return gutter. (*Source:* Durbeck, R.C. and Sherr, S. 1988. *Hardcopy Output Devices.* Academic Press, San Diego, CA. With permission.)

in commercial systems; electrostatic methods have yet to find widespread application, but have been used for facsimile recording; drop on demand, because of its simplicity and ease of implementation of color, has been widely accepted in the office and home market.

Continuous Ink Jet

The basic principle of continuous ink jet is to take advantage of the natural breakup process due to an instability in the jet that is formed when fluid is forced under pressure through a small orifice. This results from the interplay of surface tension and viscosity and takes place in a quasirandom manner unless external stimulation is applied. This breakup process was first studied by Rayleigh who characterized it via a growth rate for the instability, which depended on the jet diameter D, its velocity V, and the frequency F of any external stimulation. Rayleigh showed that the frequency for maximum growth rate of the instability was $F = V/4.5D$. By stimulating the jet at this frequency it is possible to obtain a uniform stream of droplets. The typical method of providing this stimulation today is via a piezoelectric transducer as an integral part of the printhead.

To make use of these droplets for printing it is necessary to charge them at *breakoff*. This is accomplished by placing electrodes in proximity to the breakup region of the jet. Deflection voltages are then applied farther downstream to direct the droplet to the substrate or into a collector for recirculation and reuse. The earliest techniques involved charging the droplet and applying a variable deflection field to direct it to a specific spot on the paper, enabling full height characters to be printed in one pass (see Fig. 19.66). Later methods focused on producing a stream of charged droplets and using the printing (high-voltage) electrode to deflect unwanted drops for recirculation and reuse (Fig. 19.67). This technique, known as *binary charged continuous ink jet*, lends itself to the construction of multiple nozzle arrays, and there are a number of page-wide implementations in use.

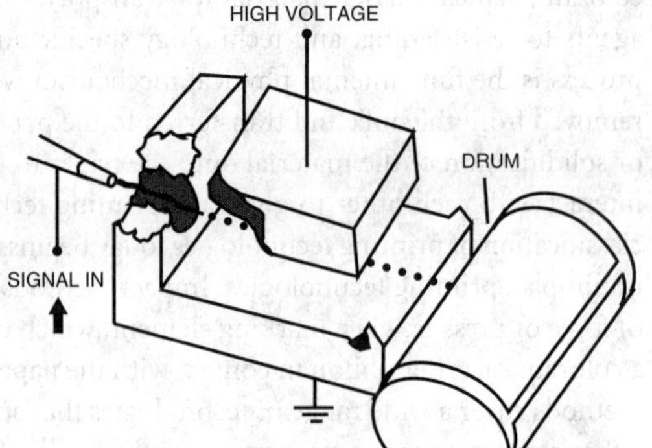

Binary charged continuous ink jet with its high droplet rate makes a simple gray scaling technique

FIGURE 19.67   Binary charged continuous ink jet. Droplets, charged at breakoff, are directed to the paper on a rotating drum. Droplets not selected for printing are diverted by the high-voltage electrode to be collected by a return gutter. (*Source:* Durbeck, R.C. and Sherr, S. 1988. *Hardcopy Output Devices.* Academic Press, San Diego, CA. With permission.

possible. The dot on the paper is modulated in size by printing from one up to N droplets at the same location, where N is the number of different dot sizes desired. By operating at high frequencies and small drop volumes it is possible to produce sufficient gray levels such that full grayscale printing is achieved at reasonable print speeds. The most recent implementation of this method offers 512 gray levels at addressabilities between 200–300 pixels/in. To achieve the small fundamental droplet size, typical implementations employ glass capillaries with diameters of the order of 10 μm and are pressurized from 500–700 lb/in^2. A color printhead will contain four capillaries, one for each color ink plus black.

Drop On Demand (DOD) Ink Jet

For office and home applications the complexities of continuous ink jet technology, such as startup and shutdown procedures, ink recirculation, and the limited nozzle count, have led to the development of drop on demand ink jet technology. These devices employ unpressurized ink delivery systems and, as implied by their name, supply a drop only when requested. The basic technique employed is to produce a volume change in either the ink supply channel or an ink chamber adjacent to the nozzle such that the resulting pressure wave causes drop ejection. Refill is achieved by capillary forces and most DOD systems operate with a slight negative pressure at the ink reservoir. The mechanism for generating the pressure wave dominates the design of these devices, and there are two techniques extant in common DOD printers. One employs the pressure pulse derived from the vaporization of superheated fluid, and the other makes use of piezoelectric materials, which can be deformed by the application of electric potentials.

Devices employing the vaporization of superheated fluid are known concurrently as thermal ink jet or bubble jet printers, the choice of name depending on the manufacturer. Since drop on demand ink jets rely on capillary refill, their operational frequencies are much lower than for continuous ink jet devices. This stresses the importance of the compactness of the actuating system so as to achieve reasonable printing speeds via multiple nozzle printheads. The nozzles must also be precisely registered with respect to each other if systematic print artifacts are to be avoided.

Thermal Ink Jet/Bubble Jet DOD Printers

When fluids are heated at extreme rates (e.g., 500×10^6 W/m^2), they enter a short-lived metastable state where temperatures can far exceed the boiling point at atmospheric pressure. The difference between the elevated temperature and the boiling point is known as the *degree of superheat*. This process does not continue indefinitely, and all fluids have what is known as a *superheat limit*. At this point nucleation and vaporization will occur in the bulk of the fluid. These devices employ an electrically driven planar heater (typically, 50–60 μm^2) in contact with the fluid. Under these conditions vaporization commences at the surface of the heater due to the presence of nucleation sites such as microscopic roughness. With correctly chosen heating rates this can be made very reliable. These heating rates lead to electrical pulse widths of 3–5 μs. In this time frame only a submicron portion of the fluid will be superheated. The net result is a vaporization pulse well in excess of atmospheric pressure and of approximately 3/4-μs duration. By locating a nozzle directly over or alongside the resistor this pressure pulse will eject a droplet (Fig. 19.68).

Within limits of the drop volume desired, it is found that the linear dimensions of the nozzle diameter and planar resistor are comparable. The actuator is therefore optimally compact, and this enables high-nozzle count printheads. The fabrication of the resistors is accomplished by photolithographic techniques common to the IC industry and the resistor substrates are silicon with a thin layer of insulating silicon dioxide. Precise registration from nozzle to nozzle is guaranteed under these circumstances, and electrical drive circuits may be integrated into the head to provide multiplexing capability. This is a valuable attribute for scanning printheads, which employ a flexible printed circuit for interconnect. These features have produced printheads currently numbering 300 or more nozzles for a single color. An additional benefit of the compactness of this technology is that the ink supply can be fully integrated with the printhead. This provides the user with virtually maintenance free operation as the printhead is replaced when the ink supply is consumed. Since the majority of problems arise from paper dust particles finding their way into a nozzle and occasionally becoming lodged there, printhead replacement provides for user service at reasonable cost. Some implementations feature a semipermanent printhead, which is supplied by ink from

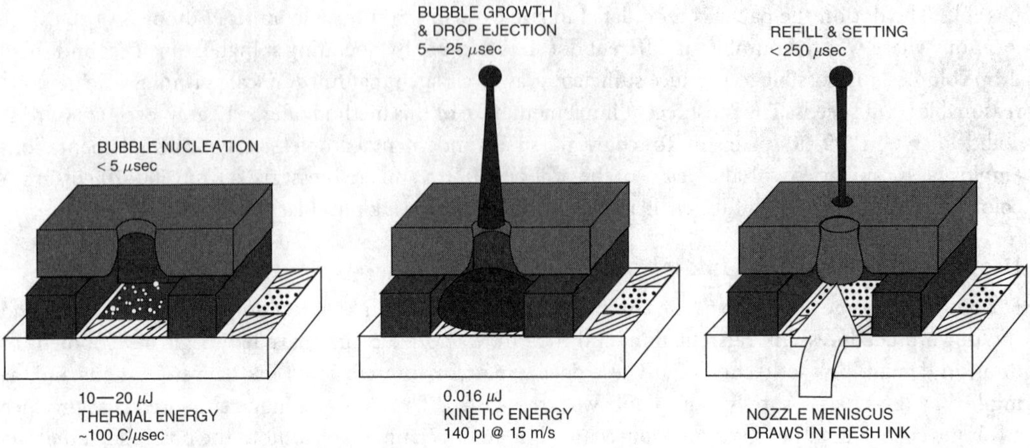

FIGURE 19.68 Drop ejection sequence for thermal ink jet. Nominal time frames and values for various parameters are given to indicate the scale of the three processes of nucleation, bubble growth, and jet formation followed by drop ejection and refill.

replaceable cartridges. The design choice is then a tradeoff involving many factors: frequency of maintenance, cost of operation, how often the printer is to be used, type of material printed, etc. The important subject of ink and paper for these printers will be taken up at the end of the section on DOD technologies.

Piezoelectric DOD Printers

Crystalline structures, which develop a spontaneous dipole moment when mechanically strained, thereby distorting their crystal structures, are called *piezoelectric*. These materials may conversely be caused to be distorted via electrical potentials applied to the appropriate planes of the cyrstal. Piezoceramics have a polarization direction established during the manufacturing process, and the applied fields then interact with this internal polarization to produce mechanical displacement. Depending on the direction of the applied fields, the material can compress or extend longitudinally or transversely. These materials have found widespread use as transducers for DOD printers. An early form was that of a sleeve over a glass capillary, which terminated in a nozzle (Fig. 19.69). Depending on the location of the electrodes either a radial or longitudinal compression could be applied leading to a pressure wave in the enclosed ink sufficient to eject a droplet. Using the diameter of the nozzle as a unit of linear dimension, this approach placed the transducer well upstream from the nozzle (Fig. 19.70). Implementation of this design in a multinozzle printhead required careful matching of transducers and fluid impedance of the individual channels feeding each nozzle. This was a challenging task, and most designs bond a planar transducer to an ink chamber adjacent to a nozzle, as shown in Fig. 19.71.

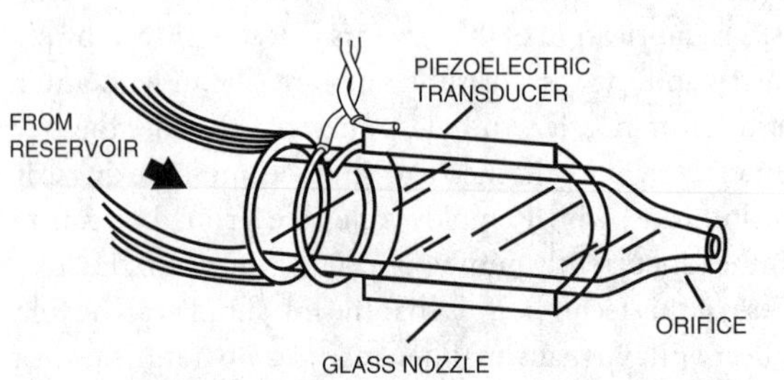

FIGURE 19.69 Squeeze tube piezoelectric ink jet. Early implementation of piezoelectric transducers for drop ejection. (*Source:* Advanced Technology Resources Corporation.)

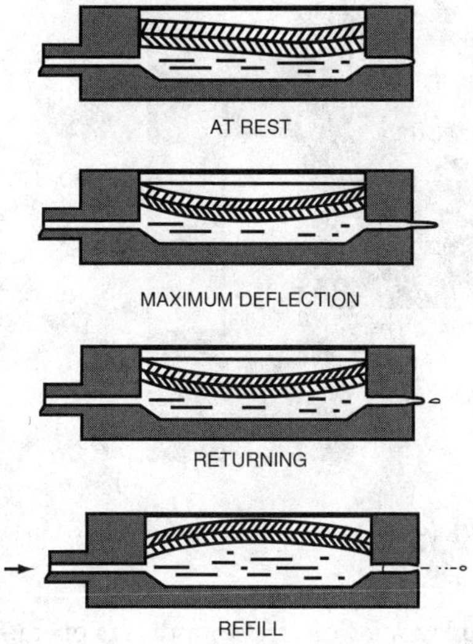

FIGURE 19.70 Drop ejection sequence for piezoelectric printhead. Schematic of drop ejection via deflection of piezocrystal bonded to an ink capillary. In practice the piezodrivers were located well upstream of the nozzle due their size. (*Source:* Advanced Technology Resources Corporation.)

The method of directly coupling the piezoelectric transducer through an ink chamber to an exit nozzle has seen many enhancements and developments since its invention. A feature of some designs is that of air flow channeled at the orifice in such a way as to entrain the droplet as it exits the nozzle and to improve its directional stability, as well as to accelerate the droplet. This enables the device to be operated at lower transducer deflections and, therefore, at higher droplet rate since the settling time of the device has been reduced. Piezodevices can operate at elevated temperatures and are used to eject inks that are solid at room temperature. For solid inks the material is melted to a design temperature for appropriate viscosity and surface tension and then supplied to the piezoelectric-driven ink chamber. The ink then solidifies instantly on contact with the substrate.

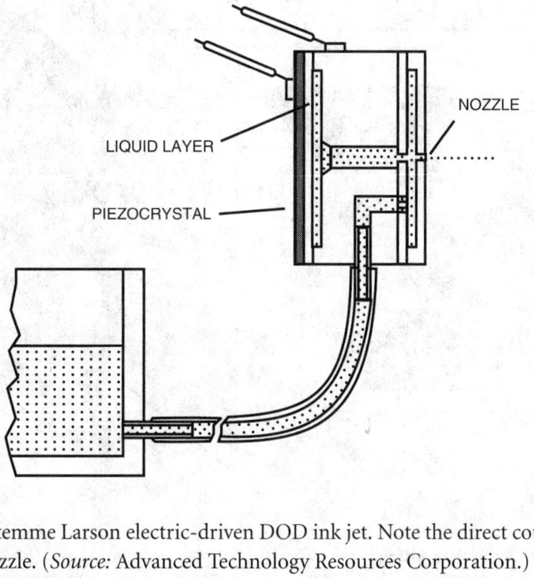

FIGURE 19.71 Design of Stemme Larson electric-driven DOD ink jet. Note the direct coupling of the pressure pulse to the ink chamber at the nozzle. (*Source:* Advanced Technology Resources Corporation.)

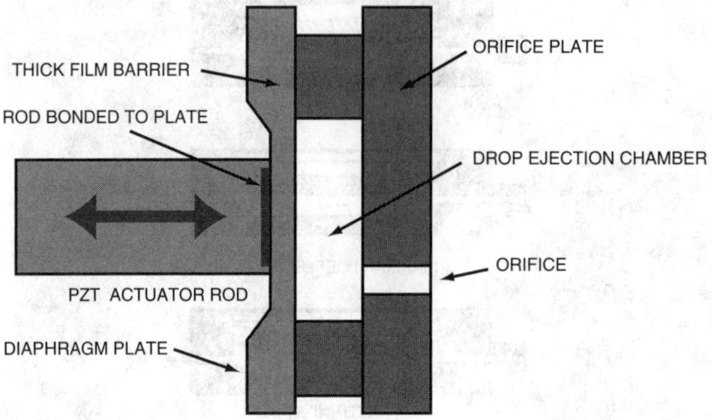

FIGURE 19.72   Schematic of ink chamber and actuator.

A more recent innovation employs piezoelectric transducers operated in the longitudinal mode. The transducers are formed from a single block of piezoceramic material in the form of an array of rods. Suitably placed potentials excite the rods to extend in the longitudinal direction. By bonding one end of the rod in contact with a thin membrane forming the base of an ink chamber, a pressure pulse is generated similar to that of the previous design (Fig. 19.72). To achieve sufficient pressure amplitude a diaphragm is used that is substantially larger than the orifice exit diameter. The consequence of this is that high nozzle density printheads will require multiple rows of nozzles (Fig. 19.73). This design has been implemented to date with liquid inks only.

Grayscale Methods for DOD Ink Jet Printers

The drop rates for DOD devices are typically an order of magnitude less than those of continuous, pressurized systems. This dictates different strategies for the achievement of grayscale. Techniques are based on the generation of a few gray levels that, when incorporated into digital halftoning algorithms, such as error diffusion, clustered, dispersed dot, or blue-noise dither, produce a satisfactory grayscale. The number of levels necessary, their position relative to the maximum modulation achievable (i.e., maximum

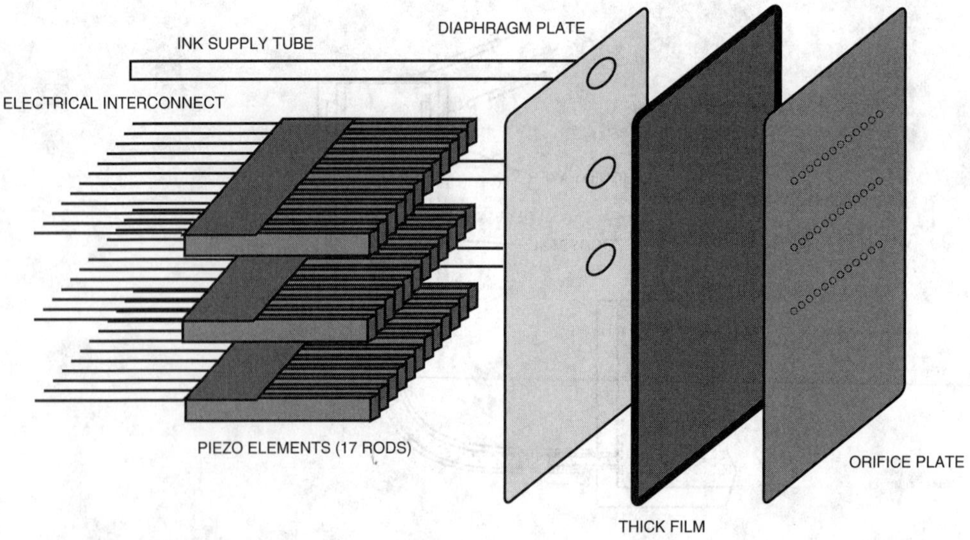

FIGURE 19.73   Exploded view of a multi chamber pen.

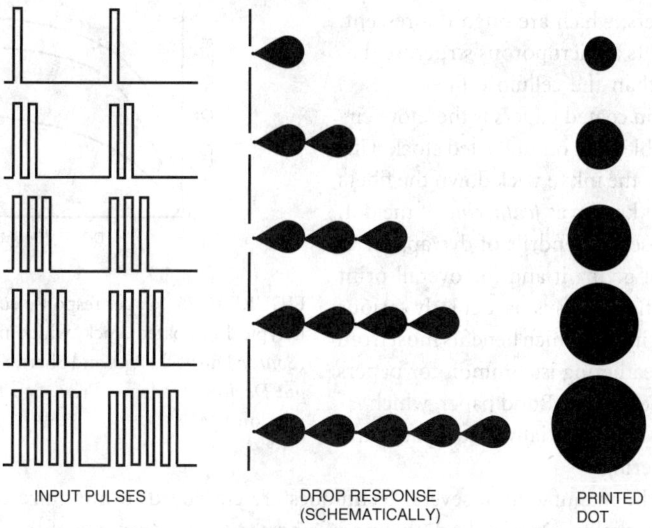

FIGURE 19.74 Stylized representation of multidrop process. Each input pulse consists of a group of drive pulses chosen for the final dot size desired. (*Source:* Durbeck, R.C. and Sherr, S. 1988. *Hardcopy Output Devices.* Academic Press, San Diego, CA. With permission.)

dot size or maximum intensity), and the specialized techniques employed in digital halftoning are an area of active research. There are many patents in the literature governing these techniques, and manufacturers seek to distinguish their devices by the technique offered. When combined with resolution enhancement methods mentioned in the section on print quality, printers with medium resolution, such as 300 dpi and 2 bits of grayscale, can produce remarkable results for both images, text, and graphics.

There are several methods available for DOD devices to modulate either the size or intensity of the dot. For piezodevices, pulse width modulation has been shown to produce droplets of different volumes and, therefore, dot sizes. All DOD ink jet devices have the option of ejecting a droplet repeatedly at the same location by passing over the same swath as many times as desired but this will affect throughput rates. Printheads with sufficient nozzle count can do this and still keep throughput rates within reason. For vapor bubble driven devices, a unique option exits by virtue of the short duration of the actuating bubble. Typical lifetime of bubbles in these devices, from vaporization through to bubble collapse, is of the order of 20 μs. If the resistor is pulsed shortly after bubble collapse, a second droplet can be ejected virtually on the tail of the initial droplet. This technique has been called *multidrop* in the literature. The ink chamber is fired under partial refill conditions, but with proper design several droplets can be ejected by this method at drop rates at around 40 kHz and having substantially the same volume (Fig. 19.74). These merge on the substrate to produce different dot sizes according to the number of droplets ejected for the location. This is not an option for most piezodevices due to the slower settling time of the actuator. Dye dilution methods have also been demonstrated as a way of modulating the intensity of the dot. If no halftone algorithm is employed, this will require many sets of nozzles to accommodate the different dye dilutions.

Ink and Paper for Ink Jet Devices

When liquid inks are employed the paper properties have a major impact on the print quality. The ink droplets will be absorbed by a substrate whose internal structure and surface energy will determine the size, shape, and overall microstructure of the drop. Paper, being a interlocking mesh of cellulose fibers with sizing and binding chemistry, is quite variable in nature. Figure 19.75 is a schematic indication of the response of paper to different volumes of ink. Note that it can be very nonlinear at low drop volumes and either flat or high gain at large volumes. The implication is that by simply changing the paper the print quality is altered. To control this variablity some papers are given a thin coat of claylike material containing whiteners, which are often fluorescent. This coating presents a microporous structure that is more uniform than the cellulose fibers.

Dot formation on coated papers is therefore circular and more stable than on uncoated stock. Uncoated papers allow the ink to wick down the fibers producing an effect known as *feathering* of the dot. In this case, microscopic tendrils of dye appear at the edge of the dot giving it and the overall print quality a blurred effect. This is particularly serious in the case of text printing, which benefits most from sharp dot edges. Feathering is common for papers used in xerographic copiers. Bond paper, which is a popular office stock, is a partially coated paper and exhibits little feathering.

FIGURE 19.75   Paper response curves. The lower curve is typical of coated stock, which minimizes dot spread. (*Source:* Durbeck, R.C. and Sherr, S. 1988. *Hardcopy Output Devices.* Academic Press, San Diego, CA. With permission.)

Depending on the manufacturer, several techniques are employed to minimize the impact of paper variability on print quality. One method, the use of a heater, takes advantage of the fact that absorption into the paper does not commence immediately upon contact. There is a period known as the *wetting time*, which can be as long as 80 μs during which no absorption takes place. The application of heat immediately in the vicinity of the printhead swath can effectively "freeze the dot" by vaporizing carrier. This makes the printer insensitive to change in paper stock and provides uniformly high print quality regardless of substrate. Other methods make use of altering the fluid chemistry by means of surfactants, which increase penetration rate, and high vapor pressure additives to increase removal of fluid into the atmosphere. If the ink penetrates quickly then it is less likely to spread sideways and, thereby, dot size variation is lessened. When choosing a drop on demand ink jet printer, it is advisable to test the performance over the range of paper stocks to be used. In some cases it will be found that high-quality printing can only be obtained when a paper specified by the manufacturer is chosen.

With reference to the section on print quality, it should be kept in mind that the choice of paper will affect the overall dynamic range of print. Text printing, to be pleasing, needs to have an optical density of at least 1.3–1.4. For images, the more dynamic range the better, and special coated stock will always excel over copy paper if image quality is important. Many of the coated papers available still have a matte surface that diffusely reflects the light and limits the dynamic range for reasons previously discussed. Some manufacturers now offer a high-gloss substrate specifically intended for images. These substrates have a plastic base with special coatings to absorb the ink through to the substrate leaving a high gloss finish. This greatly improves the dynamic range to the point of approximating that of photographic paper. These substrates provide ink jet printers with the capability to produce highly saturated brilliant colors with exceptional chromatic and dynamic range and should be used if image printing is the primary objective.

Besides excellent print quality there are other demands placed on the ink. It must provide reliable operation of the device and a durable image. By this it is meant that the image does not fade rapidly, that it is mechanically sound, that it cannot be easily removed from the paper, and that it is impervious to liquids. For liquid ink, this is a challenge since solvent-based color highlighter pens are commonly used to mark up printed documents. These solvents can cause the ink to smear depending on the choice of ink chemistry and the manner in which the colorants are fixed to the substrate. These issues focus on colorant chemistry, and much research is applied to this problem. There are fade-proof dyes, but many are either incompatible with the ink vehicle, typically water, or are toxic or mutagenic.

19.8.4   Thermal Printing Technologies

Printing technologies that employ the controlled application of thermal energy via a contacting printhead to activate either physical or chemical image formation processes come under this general classification.

There are four thermal technologies in current use: direct thermal, direct thermal transfer, dye diffusion thermal transfer, and resistive ribbon thermal transfer.

Direct Thermal

This is the oldest and most prolifically applied thermal technology. The imaging process relies on the application of heat to a thermochromic layer of approximately 10 μm in thickness coated onto a paper substrate. The thermally active layer contains a leuco dye dispersed along with an acid substance in a binder. Upon heating fusion melting occurs resulting in a chemical reaction and conversion of the leuco dye into a visible deeply colored mark. Key to this process is the design of the printhead, which can be either a page-wide array or a vertical array or a scanning printhead. Two technologies are in use, thick film and thin film. Thick-film printheads have resistor material between 10 and 70 μm. The resistive material is covered with a glass layer approximately 10 μm thick for wear resistance. The thin-film printheads bear a strong resemblance to those found in thermal ink jet printheads. They employ resistive material, such as tantalum nitride, at $\frac{1}{10}$ μm thickness and a 7-μm-thick silicon dioxide wear layer. Thin-film heads are manufactured in resolutions up to 400 dpi. In each case the resistors are cycled via electrical heating pulses through temperature ranges from ambient (25°C) up to 400°C. Overall, the thin-film printheads excel in energy efficiency conversion, print quality, response time, and resolution. For these reasons the thin-film printheads are used when high resolution is required, whereas the thick-film printhead excels in commercial applications such as bar coding, airline tickets, fax, etc.

Direct Thermal Transfer

These printers transfer melted wax directly to the paper (Fig. 19.76(a) and Fig. 19.76(b)). The wax that contains the colorant is typically coated at 4 μm thickness onto a polyester film, which, in common implementations, is approximately 6 μm thick. A thermal printhead of the kind described previously presses this ribbon, wax side down, onto the paper. As the individual heating elements are pulsed, the wax melts and transfers by adhesion to the paper. The process is binary in nature; but, by the use of shaped resistors, which produce current crowding via an hourglass shape, for example, the area of wax transferred can be modulated. Common implementations employ page width arrays at 300 dpi with some providing vertical addressability of 600 dpi. The thermal ribbons are also packaged in cassettes for scanning printhead designs in desktop and portable printers. Power consumption is an issue for all thermal printers, and efforts to reduce this for direct thermal transfer have focused on reducing the thickness of the ribbon.

Dye Diffusion Thermal Transfer

This technology involves the transfer of dye from a coated donor ribbon to a receiver sheet via sublimation and diffusion, separately or in combination. The amount of dye transferred is proportional to the

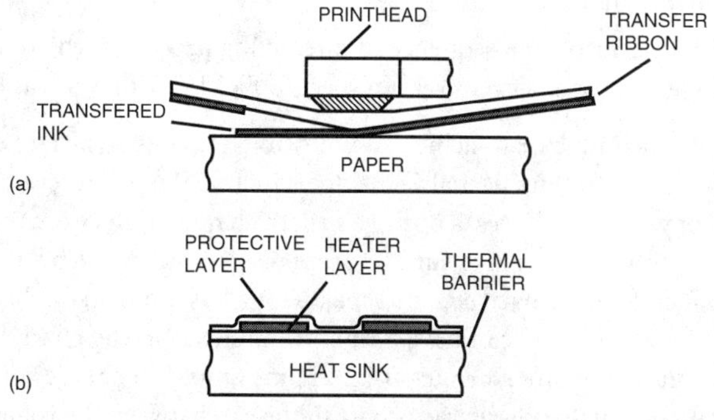

FIGURE 19.76 Schematic of wax transfer process: (a) Intimate contact between printhead, ribbon, and paper is required for successful transfer, (b) design elements of thin-film thermal printhead. Thermal barrier insulates heater for the duration of the heat pulse but allows relation of heater temperature between pulses.

amount of heat energy supplied; therefore, this is a continuous tone technology. It has found application as an alternative to silver halide photography, graphics, and prepress proofing. As with all thermal printers the energy is transferred via a transient heating process. This is governed by a diffusion equation and depending on the length of the heating pulse will produce either large temperature gradients over very short distances or lesser gradients extending well outside the perimeter of the resistor. Much of the design, therefore, focuses on the thicknesses of the various layers through which the heat is to be conducted. In the case of thermal dye sublimation transfer a soft-edged dot results, which is suitable for images but not for text. Shorter heating pulses will lead to sharper dots.

Resistive Ribbon Thermal Transfer

This technology is similar to direct thermal transfer in that a thermoplastic ink is imaged via thermal energy onto the substrate. The ribbon is composed of three layers: An electrically conductive substrate of polycarbonate and carbon black (16 μm thick), an aluminum layer 1000–2000 Å, and an ink layer which is typically 5 μm. The aluminum layer serves as a ground return plane. Heat is generated by passing current from an electrode in the printhead in contact with the ribbon substrate through the polycarbonate/carbon layer to the aluminum layer. The high pressure applied through the printhead ensures intimate contact with the paper, which does not have to be especially smooth for successful transfer. Printed characters can be removed by turning on all electrodes at a reduced energy level and heating the ink to the point that it bonds to the character to be corrected but not to the paper. The unwanted character is removed as the printhead passes over it. This technology does not adapt to color printing in a straightforward way.

19.8.5 Electrophotographic Printing Technology

Electrophotography is a well established and versatile printing technology. Its first application was in 1960 when it was embodied in an office copier. The process itself bears a strong resemblance to offset lithography. The role of the printing plate is played by a cylindrical drum or belt coated with a photoconductor (PC) on which is formed a printing image consisting of charged and uncharged areas. Depending on the implementation of the technology either the charged or uncharged areas will be inked with a charged, pigmented powder known as toner. The image is offset to the paper either by direct contact or indirectly via a silicone-based transfer drum or belt (similar to the blanket cylinder in offset lithography). Early copiers imaged the material to be copied onto the photoconductor by means of geometrical optics. Replacing this optical system with a scanning laser beam, or a linear array of LEDs, which could be electronically modulated, formed the basis of today's laser printers. As a technology it spans the range from desktop office printers (4–10 ppm) to high-speed commercial printers (exceeding 100 ppm). Although capable of E-size printing its broadest application has been in the range of $8\frac{1}{2}$ in to 17 in wide, in color and in black and white.

Printing Process Steps

Electrophotographic printing involves a sequence of interacting processes which must be optimized collectively if quality printing is to be achieved. With respect to Fig. 19.77 they are as follows.

1. Charging of the photoconductor to achieve a uniform electrostatic surface charge can be done by means of a corona in the form of a thin, partially shielded wire maintained at several kilovolts with respect to ground (corotron). For positive voltages, a positive surface charge results from ionization in the vicinity of the wire. For negative voltages, negative surface charge is produced but by a more complex process involving secondary emission, ion impact, etc., that makes for a less uniform discharge. The precise design of the grounded shield for the corona can have a significant effect on the charge uniformity produced. To limit ozone production, many office printers (<20 ppm) employ a charge roller in contact with the photoconductor. A localized, smaller discharge occurs in the gap between the roller and photoconductor, reducing ozone production between two and three orders of magnitude.

2. The charged photoconductor is exposed as described previously to form an image that will be at a significant voltage difference with respect to the background. The particular properties of the

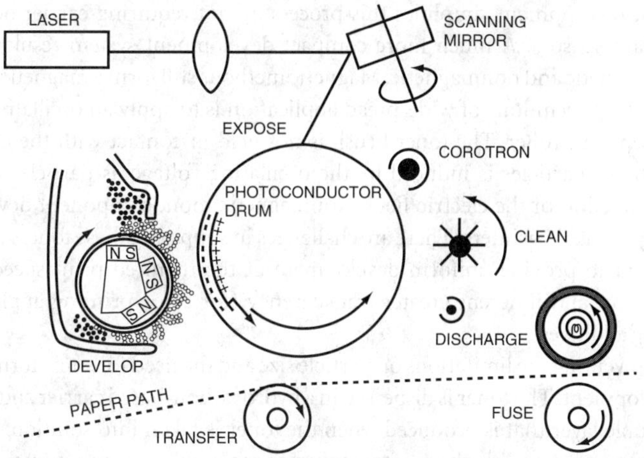

FIGURE 19.77 Schematic of the electrophotographic process. Dual component development is shown with hot roll fusing and coronas for charging and cleaning. (*Source:* Durbeck, R.C. and Sherr, S. 1988. *Hardcopy Output Devices.* Academic Press, San Diego, CA. With permission.)

photoconductor in this step relate to electron hole generation by means of the light and the transport of either electron or hole to the surface to form the image. This process is photographic in nature and has a transfer curve reminiscent of the **H and D** curves for silver halide. The discharge must be swift and as complete as possible to produce a significant difference in voltage between charged and uncharged areas if optimum print quality is to be achieved. Dark decay must be held to a minimum and the PC must be able to sustain repeated voltage cycling without fatigue. In addition to having adequate sensitivity to the dominant wavelength to the exposing light, the PC must also have a wear-resistant surface, be insensitive to fluctuations in temperature and humidity, and release the toner completely to the paper at transfer. It is possible for either the discharged or the charged region to serve as the image to be printed. Widespread practice today, particularly in laser printers, makes use of the discharged area.

Early PCs were sensitive to visible wavelengths and relied on sulfur, selenium, and tellurium alloys. With the use of diode laser scanners, the need for sensitivity in the near infrared has given rise to organic photoconductors (OPC), which in their implementation consist of multiple layers, including a submicron-thick charge generation layer and a charge transport layer in the range of 30 μm thick. This enables the optimization of both processes and is in wide-spread use today. A passivation or *wear layer* is used for OPCs, which are too soft to resist abrasion at the transfer stage. In many desktop devices the photoconductive drum is embodied in a replaceable cartridge containing enough toner for the life of the photoconductor. This provides a level of user servicing similar to that for thermal ink jet printers having replaceable printheads.

3. Image formation is achieved by bringing the exposed photoconductor surface in contact with toner particles, which are themselves charged. Electrostatic attraction attaches these particles to form the image. Once again, uniformity is vital, as well as a ready supply of toner particles to keep pace with the development process. Two methods are in widespread use today: dual component, popular for high-speed printing and monocomponent toners commonly found in desktop printers. Dual component methods employ magnetic toner particles in the 10-μm range and magnetizable carrier beads whose characteristic dimension is around 100 μm. Mechanical agitation of the mixture triboelectrically charges the toner particles, and the combination is made to form threadlike chains by means of imbedded magnets in the development roller. This dense array of threads extending from the development roller is called a magnet brush and is rotated in contact with the charged photoconductor (Fig. 19.75). The toner is then attracted to regions of opposite charge and a sensor-controlled replenishment system is used to maintain the appropriate ratio of toner to carrier beads.

Monocomponent development simplifies this process by not requiring carrier beads, replenishment system, and attendant sensors. A much more compact development system results, and there are two implementations: magnetic and nonmagnetic. Magnetic methods still form a magnetic brush but it consists of toner particles only. A technique of widespread application is to apply an oscillating voltage to a metal sleeve on the development roller. The toner brush is not held in contact with the photoconductor but, rather, a cloud of toner particles is induced by the oscillating voltage as particles detach and reattach depending on the direction of the electric field. Nonmagnetic monocomponent development is equally popular in currently available printers. There are challenges in supplying these toners in charged condition and at rates sufficient to provide uniform development at the required print speed. Their desirability derives from lower cost and inherent greater transparency (for color printing applications) due to the absence of magnetic additives.

One way of circumventing the limitations on particle size and the need for some form of brush technique is to use liquid development. The toner is dispersed in a hydrocarbon-based carrier and is charged by means of the electrical double layer that is produced when the toner is taken into solution. Typically, the liquid toner is brought into contact with the photoconductor via a roller. Upon contact, particle transport mechanisms, such as *electrophoresis*, supply toner to the image regions. Fluid carryout is a major challenge for these printers. To date this has meant commercial use where complex fluid containment systems can be employed. The technique is capable of competing with offset lithography and has also been used for color proofing.

4. The transfer and fuse stage imposes yet more demands on the toner and photoconductor. The toner must be released to the paper cleanly and then fixed to make a durable image (fusing). The majority of fusing techniques employ heat and pressure, although some commercial systems make use of radiant fusing by means of zenon flash tubes. The toner particles must be melted sufficiently to blend together and form a thin film, which will adhere firmly to the substrate. The viscosity of the melted toner, its surface tension, and particle size influence this process. The design challenge for this process step is to avoid excessive use of heat and to limit the pressure so as to avoid smoothing, that is, calendering and/or curling of the paper. Hot-roll fusing passes the toned paper through a nip formed by a heated elastomer-coated roller in contact with an unheated opposing roller that may or may not have an elastomer composition. Some designs also apply a thin film of silicone oil to the heated roller to aid in release of the melted toner from its surface. There is inevitably some fluid carryout under these conditions, as well as a tendency for the silicone oil to permeate the elastomer and degrade its physical properties. Once again materials innovation plays a major role in electrophotography.

5. The final phase involves removal of any remaining toner from the photoconductor prior to charging and imaging for the next impression. Common techniques involve fiber brushes, magnetic brushes, and scraper blades. Coronas to neutralize any residual charge on the PC or background toner are also typical components of the cleaning process. The toner removed in this step is placed in a waste toner hopper to be discarded. The surface hardness of the PC plays a key role in the efficiency of this step. Successful cleaning is especially important for color laser printers since color contrast can make background scatter particularly visible, for example, magenta toner background in a uniform yellow area.

Dot Microstructure

With respect to image microstructure, the design of the toner material, the development technique, and the properties of the photoconductor play key roles. It is desirable to have toner particles as small as possible and in a tightly grouped distribution about their nominal diameter. Composition of toner is the subject of a vast array of publications and patents. Fundamental goals for toner are a high and consistent charge-to-mass ratio, transparency in the case of color, a tightly grouped distribution, and a minimum, preferably no, wrong-sign particles. The latter are primarily responsible for the undesirable background scatter that degrades the print. Recent developments in toner manufacture seek to control this by means of charge control additives which aid in obtaining the appropriate magnitude of charging and its sign. Grayscale in laser printers is achieved by modulating the pulse width of the diode laser. The shape and steepness of the transfer curve, which relates exposure to developed density, is a function of photoconductor properties, development process, and toner properties. It is possible to produce transfer curves of low or high gradient. For text, a steep gradient curve is desirable, but for images a flatter gradient curve provides more control. Since the stability of the development process is subject to ambient temperature and humidity, the production of a stable grayscale color laser printer without print artifacts is most challenging.

19.8.6 Magnetographic and Ionographic Technologies

These two technologies are toner based but utilize different addressing and writing media. The photoconductor is replaced by a thin magnetizable medium or hard dielectric layer, such as anodized aluminum, which is used in ionographic printers. Magnetographic printers employ a printhead that produces magnetic flux transitions in the magnetizable media by changing the field direction in the gap between the poles of the printhead. These magnetic transitions are sources of strong field gradient and field strength. Development is accomplished by means of magnetic toner applied via a magnetic brush. The toner particles are magnetized and attracted to the medium by virtue of the strong field gradient. Transfer and fusing proceed in a similar manner to that of electrophotography. Ionographic printers write onto a dielectric coated drum by means of a printhead containing individual electron sources. The electrons are generated in a miniature cavity by means of air breakdown under the influence of an RF field. The electron beam is focused by a screen electrode, and the cavity functions in a manner similar to that of vacuum tube valves. The role of the plate is played by the dielectric coated metal drum held at ground potential. The charge image is typically developed by monocomponent toner followed by a *transfix*, that is, transfer and fuse operation, often without the influence of heat. Both systems require a cleaning process: mechanical scraping for ionography and magnetic scavenging for magnetography.

19.8.7 System Issues

Processing and communicating data to control today's printers raises significant system issues in view of the material to be printed. Hardcopy output may contain typography, computer-generated graphics, and natural or synthetic images in both color and black and white. The complexity of this information can require a large amount of processing, either in the host computer or in the printer itself. Applications software programs can communicate with the printer in two ways: via a page description language (PDL), or through a printer command set. The choice is driven by the scope of the printed material. If full page layout with text, graphics, and images is the goal, then PDL communication will be needed. For computer generated graphics a graphical language interface will often suffice. However, many graphics programs also provide PDL output capability. Many options exist and a careful analysis of the intended printed material is necessary to determine if a PDL interface is required.

When processing is done in the host computer, it is the function of the printer driver to convert the outline fonts, graphical objects, and images into a stream of bits to be sent to the printer. Functions that the driver may have to perform include digital halftoning, rescaling, color data transformations, and color appearance adjustments among other image processing operations, all designed to enable the printer to deliver its best print quality. Data compression in the host and decompression in the printer may be used to prevent the print speed being limited by the data rate. Printers that do their own internal data processing contain a hardware formatter board whose properties are often quoted as part of the overall specification for the printer. This is typical for printers with a PDL-based interface. Some of the advantages for this approach include speed of communication with the printer and relieving of the host computer of the processing burden, which can be significant for complex documents.

The increase in complexity of printed documents has emphasized several practical system aspects that relate to user needs: visibility and control of the printed process, font management, quick return to the software application, and printer configuration. The degree of visibility and control in the printing process depends on the choice of application and/or operating environment. Fonts, either outline or bit map, may reside on disk, on computer, or printer read-only memory (ROM). To increase speed, outline fonts in use are rasterized and stored in formatter random-access memory (RAM) or computer RAM. Worst cases exist when outline fonts are retrieved at printing and rasterization occurs on a demand basis. This can result in unacceptably slow printing. If quickness of return to the application is important, printers containing their own formatter are an obvious choice. It is necessary, therefore, to take a system view and evaluate the entire configuration (computer hardware; operating system; application program; interconnect; printer formatter, and its CPU, memory, and font storage) to determine if the user needs will be met.

The need to print color images and complex color shaded graphics has brought issues such as color matching, color appearance, and color print quality to the fore. Color printer configuration now includes choices as to halftoning algorithm, color matching method, and, in some cases, smart processing. The latter refers to customized color processing based on whether the object is a text character, image, or graphic. A further complication arises when input devices and software applications also provide some of these services, and it is possible to have color objects suffer redundant processing before being printed. This can severely degrade the print quality and emphasizes the importance of examining the entire image processing chain and turning off the redundant color processing. Color printer configuration choices focus on a tradeoff between print speed and print quality. Halftoning algorithms that minimize visible texture and high print quality modes that require overprinting consume more processing time. For color images and graphics, the relationship between the CRT image and hard copy is a matter of choice and taste. For color graphics, it is common practice to sacrifice accuracy of the hue in the interests of colorfulness or **saturation** of the print. In the case of natural images, hue accuracy, particularly for flesh tones, is more important, and a different tradeoff is made. Some software and hardware vendors provide a default configuration that seeks to make the best processing choice based on a knowledge of the content to be printed. If more precise control is desired, some understanding of the color reproduction issues represented by the combination of color input and output devices linked by a PC having a color monitor is required. This is the domain of color management.

Color Management

The fundamental issue to be addressed by color management is that of enabling the three broad classes of color devices (input, display, output) to communicate with each other in a system configuration. The technical issue in this is one of data representation. Each device has an internal representation of color information that is directly related to the nature in which it either represents or records that information. For printers it is typically amounts of cyan, magenta, yellow, and often black (CMY, K) ink; for displays, digital counts of red, green, and blue (RGB); and for many input devices, digitized values of RGB. These internal spaces are called *device spaces* and map out the volume in three-dimensional color space that can be accessed by the device. To communicate between the devices these internal spaces are converted either by analytical models or three-dimensional lookup tables (LUTs) into a device-independent space. Current practice is to use **Commision Internationale d'Eclairage (CIE)** colorimetric spaces, based on the **CIE 1931 standard observer** for this purpose. This enables the device space to be related to coordinates that are derived from measurements on human color perception. These conversions are known as device profiles, and the common device independent color space is referred to as the *profile connection space* (PCS). When this is done it is found that each device accesses a different volume in human color space. For example, a CRT cannot display yellows at the level of saturation available on most color printers. This problem, in addition to issues relating to viewing conditions and the user state of adaptation, makes it necessary to perform a significant amount of color processing if satisfactory results are to be obtained. Solutions to this problem are known as *color management methods* (CMM) (Fig. 19.78). It is the goal of color management systems to coordinate and perform these operations.

The purpose of a color management system is, therefore, to provide the best possible color preference matching, color editing, and color file transfer capabilities with minimal performance and ease of use penalties. Three levels of color-management solutions are common available, point solutions, application solutions, and operating system solutions. Point solutions perform all processing operations in the device driver and fit transparently into the system. If color matching to the CRT is desired, either information as to the make of CRT or visual calibration tools are provided to calibrate the CRT to the driver. Application solutions contain libraries of device profiles and associated CMMs. This approach is intended

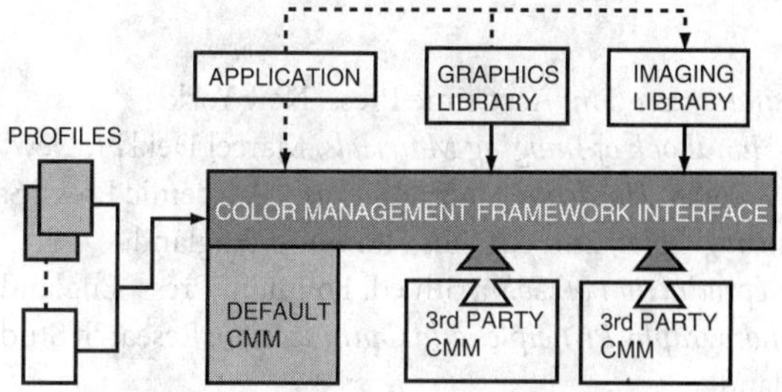

FIGURE 19.78 Proposed ICC color management using ICC profiles. Note fundamental parts: CM framework interface, CMM, third party CMMs, and profiles (which may be resident or embedded in the document).

to be transparent to the peripheral and application vendor. Operating system solutions embed the same functionality within the operating system. These systems provide a default color matching method but also allow vendor-specific CMMs to be used.

Although the creation of a device profile involves straightforward measurement processes, there is much to be done if successful color rendition is to be achieved. It is the property of CIE colorimetry that two colors will match when evaluated under the same viewing conditions. It is rarely the case that viewing conditions are identical and it is necessary to perform a number of adaptations commonly called *color appearance transformations* to allow for this. A simple example is to note that the illuminant in a color scanner will have a different color chromaticity than the white point of the CRT, which will also differ from the white point of the ambient viewing illuminant. In addition, as has been mentioned, different devices access different regions of color space; that is, they have different color gamuts. Colors outside the gamut of a destination device such as a printer must therefore be moved to lie within the printer gamut. This will also apply if the dynamic ranges are mismatched between source and destination. Techniques for performing all of the processes are sophisticated and proprietary and reside in vendor specific CMMs.

Defining Terms

Addressability: The spacing of the dots on the page, specified in dots per unit length. This may be different in horizontal and vertical axes and does not imply a given dot diameter.

CIE 1931 standard observer: Set of curves obtained by averaging the results of color matching experiments performed in 1931 for noncolor defective observers. The relative luminances of the colors of the spectrum were matched by mixtures of three spectral stimuli. The curves are often called *color matching curves*.

Commision Internationale d'Eclairage (CIE): International standards body for lighting and color measurement. Central Bureau of the CIE, a-1033 Vienna, P.O. Box 169, Austria.

Digital halftone: Halftone technique based on patterns of same size dots designed to simulate a shade of gray between white paper and full colorant coverage.

Grayscale: Intrinsic modulation property of the marking technology that enables either dots of different size or intensity to be printed.

Halftone: Technique of simulating continuous tones by varying the amount of area covered by the colorant. Typically accomplished by varying the size of the printed dots in relation to the desired intensity.

H and D curve: Characteristic response curve for a photosensitive material that relates exposure to produced/developed optical density.

Resolution: Spacing of the printer dots such that full ink coverage is just obtained. Calculated from the dot size and represents the fundamental ability of the printer to render fine detail.

Saturation: When applied to color it describes the colorfulness with respect to the achromatic axis. A color is saturated to the degree that it has no achromatic component.

References

Cornsweet, T.N. 1970. *Visual Perception*. Academic Press, New York.

Diamond, A.S., ed. 1991. *Handbook of Imaging Materials*. Marcel Dekker, New York.

Durbeck, R.C. and Sherr, S. 1988. *Hardcopy Output Devices*. Academic Press, San Diego, CA.

Hunt, R.W.G. 1992. *Measuring Color,* 2nd ed. Ellis Horwood, England.

Hunt, R.W.G. 1995. *The Reproduction of Color,* 5th ed. Fountain Press. England.

Scharfe, M. 1984. *Electrophotography Principles and Optimization*. Research Studies Press Ltd., Letchworth, Hertfordshire, England.

Schein, L.B. 1992. *Electrophotography and Development Physics,* 2nd ed. Springer–Verlag, Berlin.

Schreiber, W.F. 1991. *Fundamentals of Electronic Imaging Systems,* 2nd ed. Springer–Verlag, Berlin.

Ulichney, R. 1987. *Digital Halftoning*. MIT Press, Cambridge, MA.

Williams, E.M. 1984. *The Physics and Technology of Xerographic Processes*. Wiley-Interscience, New York.

Further Information

Color Business Report: published by Blackstone Research Associates, P.O. Box 345, Uxbridge, MA 01569-0345. Covers industry issues relating to color, computers, and reprographics.

International Color Consortium: The founding members of this consortium include Adobe Systems Inc., Agfa-Gevaert N.V., Apple Computer, Inc., Eastman Kodak Company, FOGRA (Honorary), Microsoft Corporation, *Hewlett-Packard Journal*, 1985. 36(5); 1988. 39(4) (Entire issues devoted to Thermal Ink Jet).

Journal of Electronic Imaging: co-published by IS&T and SPIE. Publishes papers on the acquisition, display, communication and storage of image data, hardcopy output, image visualization, and related image topics. Source of current research on color processing and digital halftoning for computer printers.

Journal of Imaging Science and Technology: official publication of IS&T, which publishes papers covering a broad range of imaging topics, from silver halide to computer printing technology.

The International Society for Optical Engineering, SPIE, P.O. Box 10, Bellingham, Washington 98227-0010, sponsors conferences in conjunction with IS&T on electronic imaging and publishes topical proceedings from the conference sessions.

The *Hardcopy Observer,* published monthly by Lyra Research Services, P.O. Box 9143, Newtonville, MA 02160. An industry watch magazine providing an overview of the printer industry with a focus on the home and office.

The Society for Imaging Science and Technology, IS&T, 7003 Kilworth Lane, Springfield, VA 22151. Phone (703)642 9090, Fax (703)642 9094. Sponsors wide range of technical conferences on imaging and printing technologies. Publishes conference proceedings, books, *Journal of Electronic Imaging, Journal of Imaging Science and Technology, IS&T Reporter.*

The Society for Information Display, 1526 Brookhollow Drive, Ste 82, Santa Ana, CA 92705-5421, Phone (714)545 1526, Fax (714)545 1547. Cosponsors annual conference on color imaging with IS&T.

20

Signal Measurement, Analysis, and Testing

Jerry C. Whitaker
Editor-in-Chief

Carl Bentz
Intertec Publishing

Samuel O. Agbo
California Polytechnic State University

Jerry C. Hamann
University of Wyoming

John W. Pierre
University of Wyoming

20.1   Audio Frequency Distortion Mechanisms and Analysis

Jerry C. Whitaker

20.1.1   Introduction

Most measurements in the audio field involve characterizing fundamental parameters.[1] These include signal level, phase, and frequency. Most other tests consist of measuring these fundamental parameters and displaying the results in combination by using some convenient format. For example, signal-to-noise ratio (SNR) consists of a pair of level measurements made under different conditions expressed as a logarithmic, or decibel, ratio. When characterizing a device, it is common to view it as a box with input terminals and output terminals. In normal use a signal is applied to the input and the signal, modified in some way, appears at the output. Instruments are necessary to quantify these unintentional changes to the signal. Some measurements are *one-port* tests, such as impedance or noise level, and are not concerned with input/output signals, only with one or the other.

Purpose of Audio Measurements

Measurements are made on audio circuits and equipment to check performance under specified conditions and to assess suitability for use in a particular application. The measurements may be used to verify specified system performance or as a way of comparing several pieces of equipment for use in a system. Measurements may also be used to identify components in need of adjustment or repair. Whatever the application, audio measurements are a key part of audio engineering.

Many parameters are important in audio devices and merit attention in the measurement process. Some common audio measurements are frequency response, gain or loss, harmonic distortion, intermodulation distortion, noise level, phase response, and transient response.

Measurement of level is fundamental to most audio specifications. Level can be measured either in absolute terms or in relative terms. Power output is an example of an absolute level measurement; it does not require any reference. SNR and gain or loss are examples of relative measurements; the result is expressed as a ratio of two measurements. Although it may not appear so at first, frequency response is also a relative measurement. It expresses the gain of the device under test as a function of frequency, with the midband gain, typically, as a reference.

Distortion measurements are a way of quantifying the amount of unwanted components added to a signal by a piece of equipment. The most common technique is total harmonic distortion (THD), but others are often used. Distortion measurements express the amount of unwanted signal components relative to the desired signal, usually as a percentage or decibel value. This is another example of multiple level measurements that are combined to give a new measurement figure.

20.1.2   Level Measurements

The simplest definition of a level measurement is the alternating current (AC) amplitude at a particular place in the audio system. In contrast to direct current measurements, however, there are many ways of specifying AC voltage. The most common methods are average, root-mean-square (RMS), and peak response. Strictly speaking, the term *level* refers to a logarithmic, or decibel, measurement. However, common parlance employs the term for an AC amplitude measurement, and that convention will be followed in this chapter.

---

[1] Portions of this chapter were adapted from: Benson, K.B. and Whitaker, J.C. 1990. *Television and Audio Handbook for Technicians and Engineers.* McGraw-Hill, New York. Used with permission.

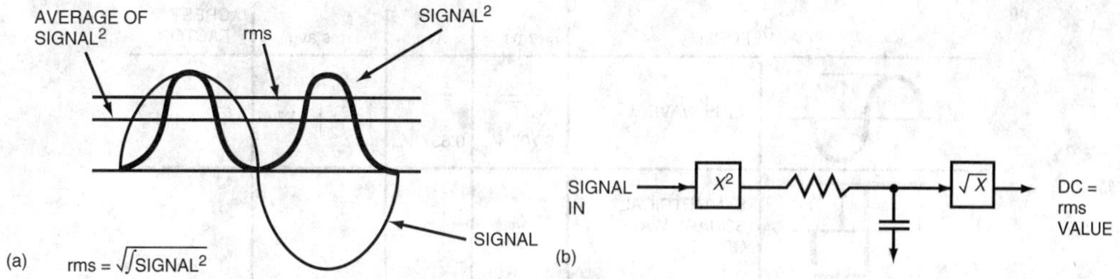

FIGURE 20.1 Root-mean-square (RMS) voltage measurements: (a) the relationship of RMS and average values, (b) RMS measurement circuit.

Root-Mean-Square Measurements

The RMS technique measures the effective power of the AC signal. It specifies the value of the DC equivalent that would dissipate the same power if either were applied to a load resistor. This process is illustrated in Fig. 20.1 for voltage measurements. The input signal is squared, and the average value is found. This is equivalent to finding the average power. The square root of this value is taken to translate the signal from a power value back to a voltage. For the case of a sine wave the RMS value is 0.707 of its maximum value.

Consider the case when the signal is not a sine wave, but rather a sine wave and several of its harmonics. If the RMS amplitude of each harmonic is measured individually and added, the resulting value will be the same as an RMS measurement on the signals together. Because RMS voltages cannot be added directly, it is necessary to perform an RMS addition, as follows:

$$V_{rms} = \sqrt{V_{rms1}^2 + V_{rms2}^2 + V_{rms3}^2 + V_{rmsn}^2}$$

Note that the result is not dependent on the phase relationship of the signal and its harmonics. The RMS value is determined completely by the amplitude of the components. This mathematical predictability is powerful in practical applications of level measurement, enabling measurements made at different places in a system to be correlated. It is also important in correlating measurements with theoretical calculations.

Average-Response Measurements

Average-responding voltmeters were common in audio work some years ago principally because of their low cost. Such devices measure AC voltage by rectifying and filtering the resulting waveform to its average value, which can then be read on a standard DC voltmeter. The average value of a sine wave is 0.637 of its maximum amplitude. Average-responding meters are usually calibrated to read the same as an RMS meter for the case of a single-sine-wave signal. This results in the measurement being scaled by a constant K of 0.707/0.637, or 1.11. Meters of this type are called average-responding, RMS calibrated. For signals other than sine waves, the response will be different and difficult to predict. If multiple sine waves are applied, the reading will depend on the phase shift between the components and will no longer match the RMS measurement. A comparison of RMS and average-response measurements is made in Fig. 20.2 for various waveforms. If the average readings are adjusted as described previously to make the average and RMS values equal for a sine wave, all of the numbers in the *average* column should be increased by 11.1%, whereas the *RMS-average* numbers should be reduced by 11.1%.

Peak-Response Measurements

Peak-responding meters measure the maximum value that the AC signal reaches as a function of time. (See Fig. 20.3.) The signal is full-wave rectified to find its absolute value and then passed through a diode to a storage capacitor. When the absolute value of the voltage rises above the value stored on the capacitor, the diode conducts and increases the stored voltage. When the voltage decreases, the capacitor maintains the old value. Some means for discharging the capacitor is required to allow measuring a new

WAVEFORM		rms	Avg.	rms avg.	CREST FACTOR
	SINE WAVE	$\dfrac{V_m}{\sqrt{2}}$ $0.707\,V_m$	$\dfrac{2}{\pi}\,V_m$ $0.637\,V_m$	$\left\|\dfrac{\pi}{2\sqrt{2}}\right\| = 1.111$	$\sqrt{2} = 1.414$
	SYMMETRICAL SQUARE WAVE OR DC	V_m	V_m	1	1
	TRIANGULAR WAVE OR SAWTOOTH	$\dfrac{V_m}{\sqrt{3}}$	$\dfrac{V_m}{2}$	$\dfrac{2}{\sqrt{3}} = 1.155$	$\sqrt{3} = 1.732$

FIGURE 20.2   Comparison of RMS and average voltage characteristics.

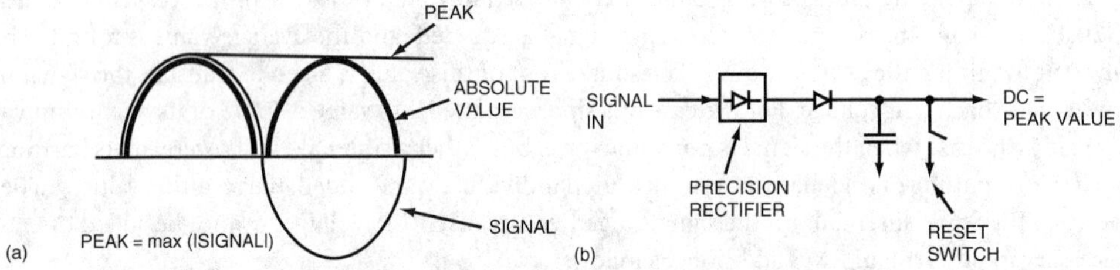

FIGURE 20.3   Peak voltage measurements: (a) illustration of peak detection, (b) peak measurement circuit.

peak value. In a true peak detector, this is accomplished by a switch. Common peak detectors usually include a large resistor to discharge the capacitor gradually after the user has had a chance to read the meter.

The ratio of the true peak to the RMS value is called the **crest factor**. For any signal but an ideal square wave the crest factor will be greater than 1, as illustrated in Fig. 20.4. As the measured signal becomes more peaked, the crest factor increases.

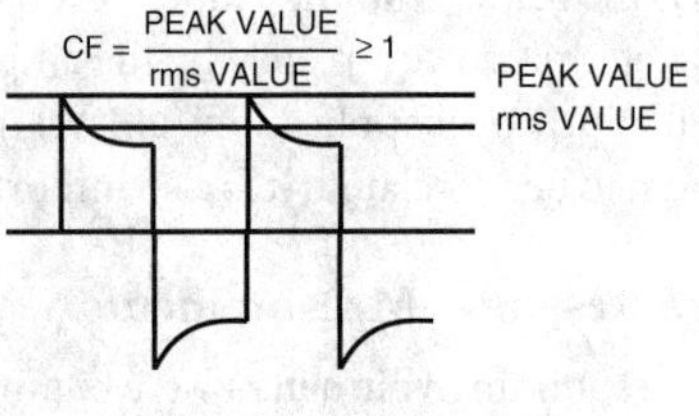

FIGURE 20.4   Illustration of the crest factor in voltage measurements.

By introducing a controlled charge and discharge time, a *quasipeak* detector is achieved. The charge and discharge times may be selected, for example, to simulate the ear's sensitivity to impulsive peaks. International standards define these response times and set requirements for reading accuracy on pulses and sine wave bursts of various durations. The gain of a quasipeak detector is normally calibrated so that it reads the same as an RMS detector for sine waves.

Another method of specifying signal amplitude is the **peak-equivalent sine**, which is the RMS level of a sine wave having the same peak-to-peak amplitude as the signal under consideration. This is the peak value of the waveform scaled by the correction factor 1.414, corresponding to the peak-to-RMS ratio of a sine wave. This is useful when specifying test levels of waveforms in distortion measurements.

Decibel Measurements

Measurements in audio work are usually expressed in decibels. Audio signals span a wide range of level. The sound pressure of a live concert band performance may be one million times that of rustling leaves. This range is too wide to be accommodated on a linear scale. The logarithmic scale of the decibel compresses this wide range down to a more easily handled range. Order-of-magnitude changes result in equal increments on a decibel scale. Furthermore, the human ear perceives changes in amplitude on a logarithmic basis, making measurements with the decibel scale reflect audibility more accurately.

A decibel may be defined as the logarithmic ratio of two power measurements or as the logarithmic ratio of two voltages:

$$dB = 20 \log \frac{E_1}{E_2} \text{ for voltage measurements}$$

$$dB = 10 \log \frac{P_1}{P_2} \text{ for power measurements}$$

There is no difference between decibel values from power measurements and decibel values from voltage measurements if the impedances are equal. In both equations the denominator variable is usually a stated reference. A doubling of voltage will yield a value of 6.02 dB, whereas a doubling of power will yield 3.01 dB. This is true because doubling voltage results in a factor-of-four increase in power.

Audio engineers often express the decibel value of a signal relative to some standard reference instead of another signal. The reference for decibel measurements may be predefined as a power level, as in decibels referenced to 1 mW (dBm), or it may be a voltage reference, as in decibels referenced to 1 V (dBV). When measuring dBm or any power-based decibel value, the reference impedance must be specified or understood. Often it is desirable to specify levels in terms of a reference transmission level somewhere in the system under test. These measurements are designated dBr, where the reference point or level must be separately conveyed.

20.1.3   Noise Measurement

Noise measurements are simply specialized level measurements. It has long been recognized that the ear's sensitivity varies with frequency, especially at low levels. This effect was studied in detail by Fletcher and Munson and later by Robinson and Dadson. The Fletcher–Munson hearing-sensitivity curve for the threshold of hearing and above is given in Fig. 20.5. The ear is most sensitive in the region of 2–4 kHz, with rolloffs above and below these frequencies. To predict how loud something will sound it is necessary to use a filter that duplicates this nonflat behavior electrically. The filter *weights* the signal level on the basis of frequency, thus earning the name *weighting filter*. Various efforts have been made to do this, resulting in several standards for noise measurement. Some of the common weighting filters are shown overlaid on the hearing threshold curve in Fig. 20.6.

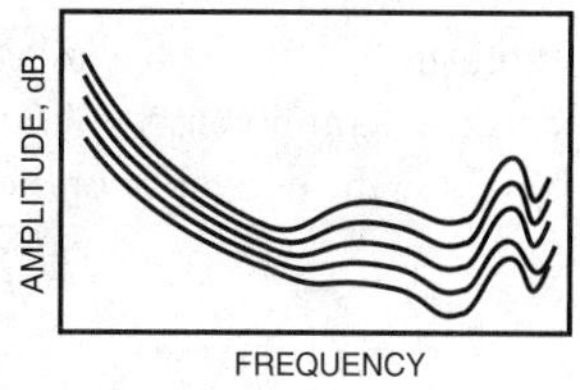

FIGURE 20.5   The Fletcher–Munson curves of hearing sensitivity vs. frequency.

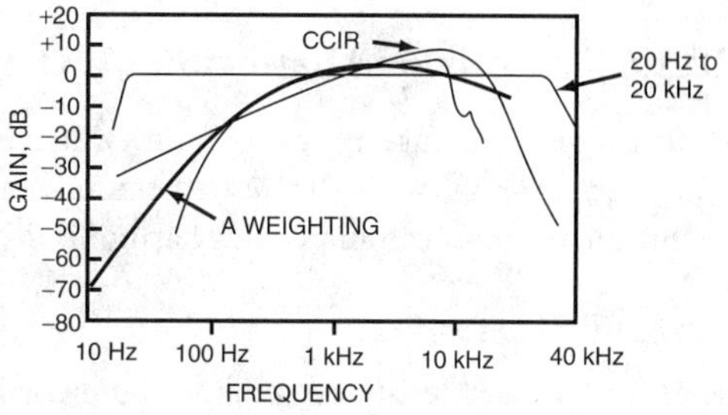

FIGURE 20.6   Response characteristics of several common weighting filters for audio measurements.

The most common filter used in the U.S. for weighted noise measurements is the A-weighting curve. An average-responding meter is often used for A-weighted noise measurements, although RMS meters are also used for this application.

European audio equipment is usually specified with a CCIR filter and a quasipeak detector. The CCIR curve is significantly more peaked than the A curve and has a sharper rolloff at high frequencies. The CCIR quasipeak standard was developed to quantify the noise in telephone systems. The quasipeak detector more accurately represents the ear's sensitivity to impulsive sounds. When used with the CCIR filter curve, it is supposed to correlate better with the subjective level of the noise than A-weighted average-response measurements do.

Some audio equipment manufacturers specify noise with a 20 Hz–20 kHz bandwidth filter and an RMS-responding meter. This is done to specify noise over the audio band without regard to the ear's differing sensitivity with frequency.

20.1.4   Phase Measurement

Phase in an audio system is typically measured and recorded as a function of frequency over the audio range. For most audio devices phase and amplitude responses are closely coupled. Any change in amplitude that varies with frequency produces a corresponding phase shift. A fixed time delay will introduce a phase shift that is a linear function of frequency. This time delay can introduce large values of phase shift at high frequencies that are of no significance in practical applications. The time delay will not distort the waveshape of complex signals and will not be audible. There can be problems, however, with time delay when the delayed signal is used in conjunction with an undelayed signal.

When dealing with complex signals, the meaning of phase can become unclear. Viewing the signal as the sum of its components according to Fourier theory, we find a different value of phase shift at each frequency. With a different phase value on each component, the question is raised as to which one should be used as the reference. If the signal is periodic and the waveshape is unchanged passing through the device under test, then a phase value may still be defined. This may be done by using the shift of the zero crossings as a fraction of the waveform period. If there is differential phase shift with frequency, however, the waveshape will be changed. It is then not possible to define any phase-shift value, and phase must be expressed as a function of frequency.

Group delay is another useful expression of the phase characteristics of an audio device. Group delay is the slope of the phase response. It expresses the relative delay of the spectral components of a complex waveform. If the group delay is flat, all components will arrive together. A peak or rise in the group delay indicates that those components will arrive later by the amount of the peak or rise. Group delay ϕ is computed by taking the derivative of the phase response vs. frequency:

$$\phi = \frac{-(\alpha_{f2} - \alpha_{f1})}{f_2 - f_1}$$

where α_{f1} is the phase at f_1 and α_{f2} phase at f_2. This requires that phase be measured over a range of frequencies to yield a curve that can be differentiated. It also requires that the phase measurements be performed at frequencies close enough together to provide a smooth and accurate derivative.

20.1.5   Nonlinear Distortion Mechanisms

Distortion is a measure of signal impurity. It is usually expressed as a percentage or decibel ratio of the undesired components to the desired components of a signal. There are several methods of measuring distortion, the most common being harmonic distortion and several types of intermodulation distortion.

Harmonic Distortion

The transfer characteristic of a typical amplifier is shown in Fig. 20.7. The transfer characteristic represents the output voltage at any point in the signal waveform for a given input voltage; ideally this is a straight line. The output waveform is the projection of the input sine wave on the device transfer characteristic. A change in the input produces a proportional change in the output. Because the actual transfer characteristic is nonlinear, a distorted version of the input waveshape appears at the output.

Harmonic distortion measurements excite the device under test with a sine wave and measure the spectrum of the output. Because of the nonlinearity of the transfer characteristic, the output is not sinusoidal. By using Fourier series, it can be shown that the output waveform consists of the original input sine wave plus sine waves at integer multiples (harmonics) of the input frequency. The spectrum of the

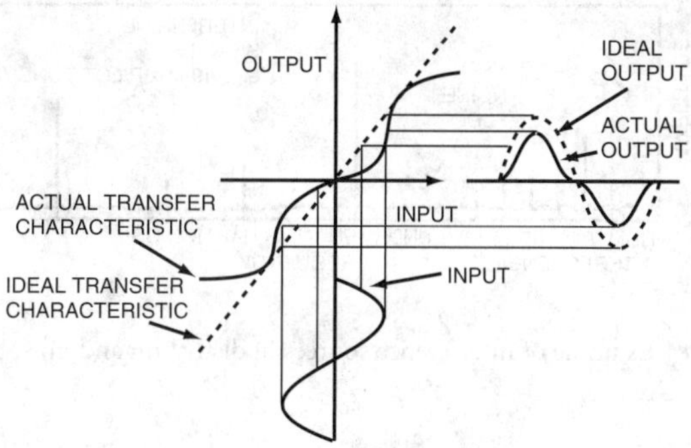

FIGURE 20.7   Illustration of total harmonic distortion (THD) measurement of an amplifier transfer characteristic.

distorted signal is shown in Fig. 20.8 for a 1-kHz input, and output signals consisting of 1, 2, 3 kHz, etc. The harmonic amplitudes are proportional to the amount of distortion in the device under test. The percentage harmonic distortion is the RMS sum of the harmonic amplitudes divided by the RMS amplitude of the fundamental.

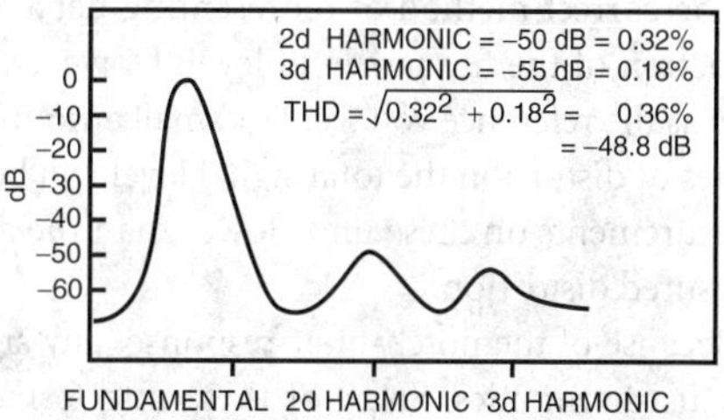

Harmonic distortion may also be measured with a spectrum analyzer. As shown in Fig. 20.8, the fundamental amplitude is adjusted to the 0-dB mark on the display. The amplitudes of the harmonics are

FIGURE 20.8   Example of reading THD from a spectrum analyzer.

then read and converted to linear scale. The RMS sum of these values is taken, which represents the THD.

Notch Filter Analyzer

A simpler approach to the measurement of harmonic distortion can be found in the notch-filter distortion analyzer. This device, commonly referred to as simply a distortion analyzer, removes the fundamental of the signal to be investigated and measures the remainder. A block diagram of such a unit is shown in Fig. 20.9. The fundamental is removed with a notch filter, and the output is measured with an AC voltmeter. Because distortion is normally presented as a percentage of the fundamental signal, this level must be measured or set equal to a predetermined reference value. Additional circuitry (not shown) is required to set the level to the reference value for calibrated measurements. Some analyzers use a series of step attenuators

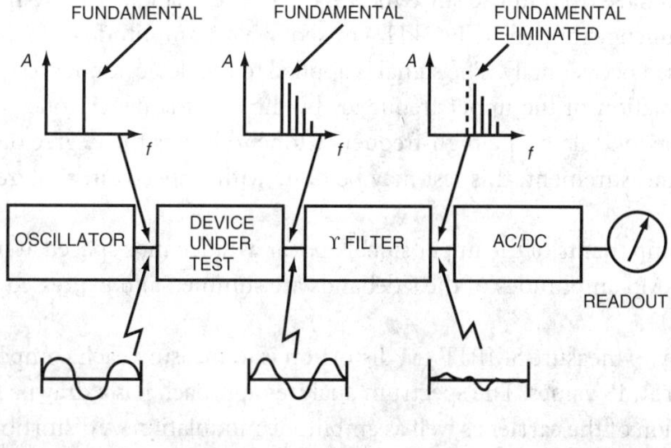

FIGURE 20.9   Simplified block diagram of a harmonic distortion analyzer.

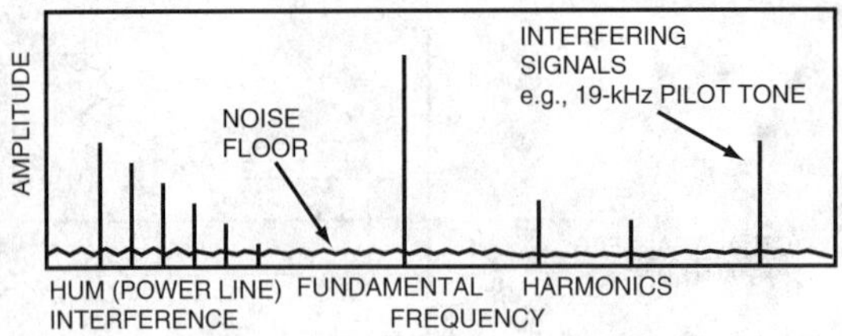

FIGURE 20.10 Example of interference sources in distortion and noise measurements.

and a variable control for setting the input level to the reference value. More sophisticated units eliminate the variable control by using an electronic gain control. Others employ a second AC-to-DC converter to measure the input level and compute the percentage. Automatic units also provide autoranging logic to set the attenuators and ranges, and tabular or graphic display of the results.

The correct method of representing percentage distortion is to express the level of the harmonics as a fraction of the fundamental level. However, most commercial distortion analyzers use the total signal level as the reference voltage. For small amounts of distortion these two quantities are equivalent. At large values of distortion the total signal level will be greater than the fundamental level. This makes distortion measurements on these units lower than the actual value. The errors are not significant until about 20% measured distortion.

Because of the notch-filter response, any signal other than the fundamental will influence the results, not just harmonics. Some of these interfering signals are illustrated in Fig. 20.10. Any practical signal contains some noise, and the distortion analyzer will include these in the reading. Because of these added components, the correct term for this measurement is *total harmonic distortion and noise* (THD + N). Although this fact does limit the reading of very low-THD levels, it is not necessarily bad. Indeed, it can be argued that the ear hears all components present in the signal, not just the harmonics.

Additional filters are included on most distortion analyzers to reduce unwanted hum and noise. These usually consist of one or more high-pass filters (400 Hz) and several low-pass filters. Common low-pass filter frequencies are 22.4, 30, and 80 kHz. A selection of filters eases the tradeoff between limiting bandwidth to reduce noise and the reduction in reading accuracy that results from removing desired components of the signal. When used in conjunction with a good differential input stage on the analyzer, these filters can solve most noise problems.

Intermodulation Distortion

Many methods have been devised to measure the intermodulation (IM) of two or more signals passing through a device simultaneously. The most common of these is **SMPTE IM**, named after the Society of Motion Picture and Television Engineers, which first standardized its use. IM measurements according to the SMPTE method have been in use since the 1930s. The test signal is a low-frequency tone (usually 60 Hz) and a high-frequency tone (usually 7 kHz) mixed in a 4:1 amplitude ratio. Other amplitude ratios and frequencies are used occasionally. The signal is applied to the device under test, and the output signal is examined for modulation of the upper frequency by the low-frequency tone. The amount by which the low-frequency tone modulates the high-frequency tone indicates the degree of nonlinearity. As with harmonic distortion measurement, this test may be done with a spectrum analyzer or with a dedicated distortion analyzer.

The modulation components of the upper signal appear as sidebands spaced at multiples of the lower-frequency tone. The RMS amplitudes of the sidebands are summed and expressed as a percentage of the upper-frequency level.

The most direct way to measure SMPTE IM distortion is to measure each component with a spectrum analyzer and add their RMS values. The spectrum analyzer approach has a drawback in that it is sensitive to frequency modulation of the carrier as well as amplitude modulation. A distortion analyzer for SMPTE testing is quite straightforward. The signal to be analyzed is passed through a high-pass filter to remove

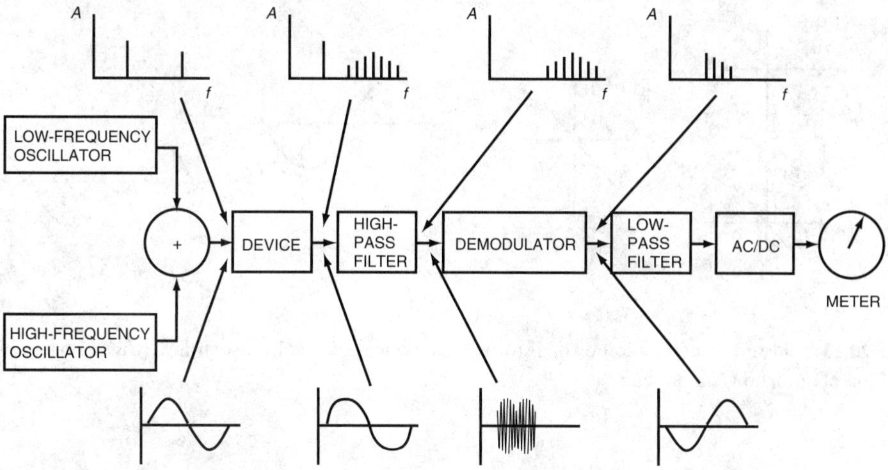

FIGURE 20.11   Simplified block diagram of a SMPTE intermodulation analyzer.

the low-frequency tone, as shown in Fig. 20.11. The high-frequency tone is then demodulated as if it were an amplitude modulated signal to obtain the sidebands. The sidebands pass through a low-pass filter to remove any remaining high-frequency energy. The resulting demodulated low-frequency signal will follow the envelope of the high-frequency tone. This low-frequency fluctuation is the distortion component and is displayed as a percentage of the amplitude of the high-frequency tone. Because low-pass filtering sets the measurement bandwidth, noise has little effect on SMPTE IM measurements.

CCITT IM

Twin-tone intermodulation or CCITT difference frequency distortion is another method of measuring distortion by using two sine waves. The test signal consists of two closely spaced high-frequency tones as shown in Fig. 20.12. When the tones are passed through a nonlinear device, IM products are generated at frequencies related to the difference in frequency between the original tones. For the typical case of signals at 14 and 15 kHz the IM components will be at 1, 2, 3 kHz, etc., and 13, 16, 12, 17 kHz, etc. Even-order, or asymmetrical distortions produce low-frequency difference-frequency components. Odd-order, or symmetrical nonlinearities produce components near the input signal fre-

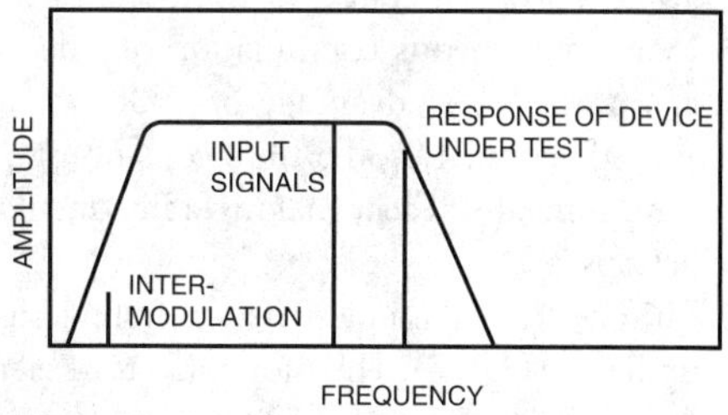

FIGURE 20.12   CCITT intermodulation test of transfer characteristic.

quencies. The most common application of this test measures only the even-order components because they may be measured with a multipole low-pass filter. Measurement of the odd-order components requires a spectrum analyzer or signal analyzer.

The distortion products in the CCITT IM test are usually far removed from the input signal. This positions them outside the range of the auditory system's masking effects. If a test that measures what the ear might hear is desired, the CCITT test is the most likely candidate.

Addition and Cancellation of Distortion Components

A common consideration for system-wide distortion measurements is that of distortion addition and cancellation in the devices under test. Consider the example given in Fig. 20.13. Assume that one

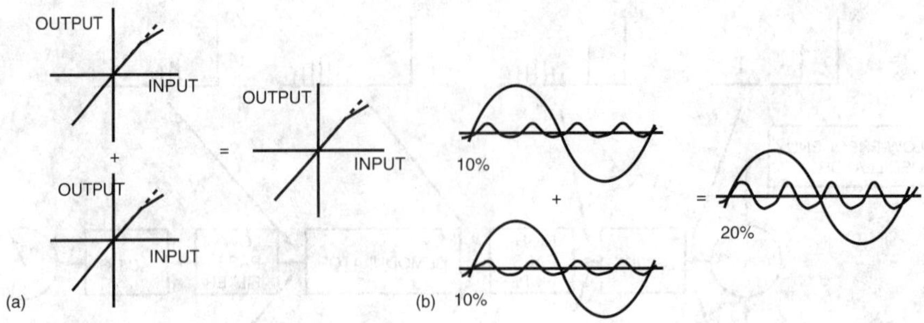

FIGURE 20.13 Illustration of the addition of distortion components: (a) addition of transfer-function nonlinearities, (b) addition of distortion components.

device under test has a transfer characteristic similar to that diagrammed at the top of Fig. 20.13(a) and another has the characteristic diagrammed at the bottom. If the devices are cascaded, the resulting transfer-characteristic nonlinearity will be magnified as shown. The effect on sine waves in the time domain is illustrated in Fig. 20.13(b). The distortion component generated by each nonlinearity is in phase and will sum to a component of twice the original magnitude. If the second device under test has a complementary transfer characteristic, however, quite a different result is obtained. When the devices are cascaded, the effects of the two curves cancel, yielding a straight line for the transfer characteristic. The corresponding distortion products are out of phase with each other, resulting in no measured distortion components in the final output.

20.1.6 Multitone Audio Testing

Multitone testing operates on the premise that audio equipment can be stimulated to the same extent by a simultaneous combination of sine waves as it can be by a series of discrete tones occurring one at a time.[2] The advantage of multitone analysis is clear: it provides complete system evaluation in one step, eliminating the time-consuming process of applying a series of individual tones, allowing a settling time after each, making adjustments after each, and then repeating the process to check for interactions.

Multitone test techniques use carefully designed mixtures of tones applied simultaneously to the device under test (DUT). The individual tone elements have well-defined frequency, phase, and amplitude relationships. The frequencies of multitone components are selected to avoid mathematically predictable harmonic and intermodulation products that would fall on or near any of the fundamentals. The phase of each tone is fixed, but randomly selected relative to each of the other tones. Amplitude relationships may vary, depending on the device under test.

Multitone signals may be created with an inverse FFT for highly accurate and stable signal parameters. Figure 20.14 shows a spectral display of a multitone signal designed for testing wide-band audio recording equipment. Because the frequency, phase, and amplitude relationships of the elements of a multitone signal are well defined, digital signal processor techniques allow easy detection of changes in these relationships. Thus, frequencies not present in the original multitone are the result of distortion and noise. Similarly, changes in phase relationships of the multitone elements can be used to derive phase response information. It follows that deviation from the known amplitude relationships will provide frequency response information.

---

[2]This section was adapted from Thompson, B. 1992. *Multitone Audio Testing. Tektronix Application Note 21* W-7182, Tektronix, Beaverton, OR.

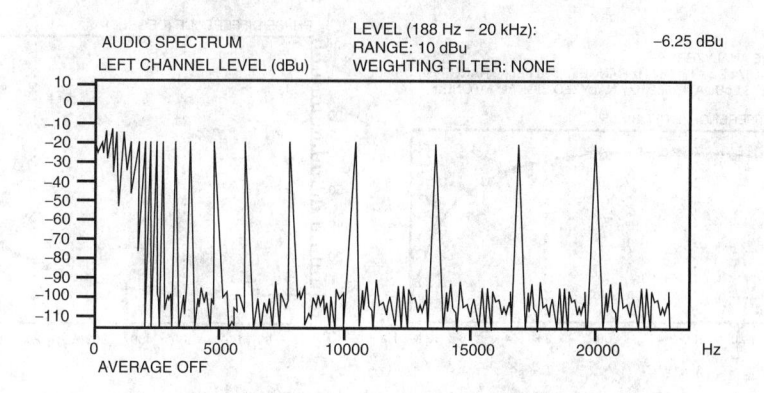

FIGURE 20.14   An audio spectrum display showing the spectral distribution of the numerous fundamental frequencies in a multitone signal. (*Source:* Tektronix, Beaverton, OR.)

Although implementations of multitone analysis vary from one vendor to the next, plots of level vs. frequency, level difference between channels, phase difference between channels vs. frequency, and noise and/or distortion vs. frequency are among those typically provided. Figures 20.15 and 20.16 show two of these multitone displays, levels vs. frequency (frequency response) and level/phase differences between channels plotted against frequency. An audio spectrum display, such as the one shown in Fig. 20.14, is also a valuable tool when combined with the multitone signal.

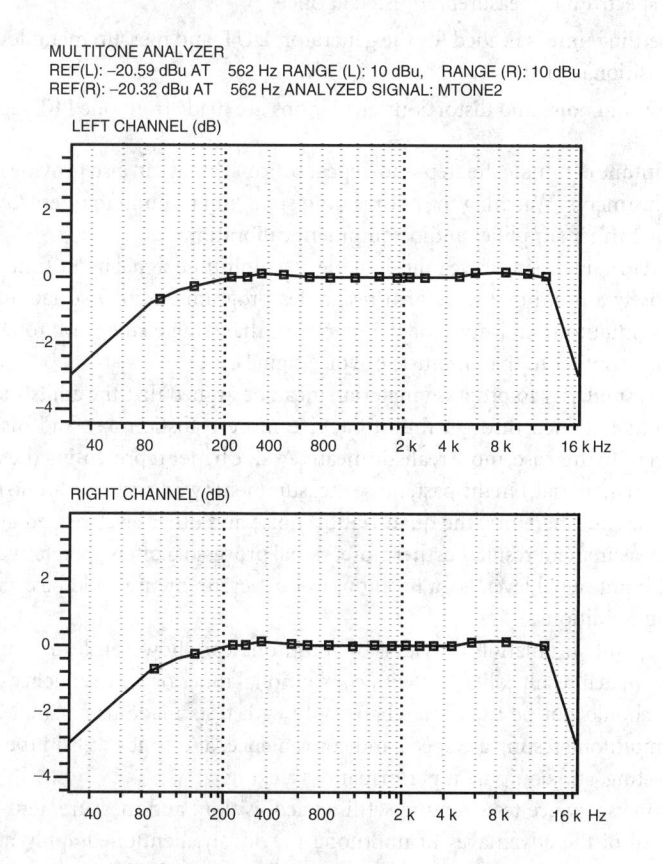

FIGURE 20.15   Level vs. frequency plots resulting from audio system measurement with a multitone signal analyzer. (*Source:* Tektronix, Beaverton, OR.)

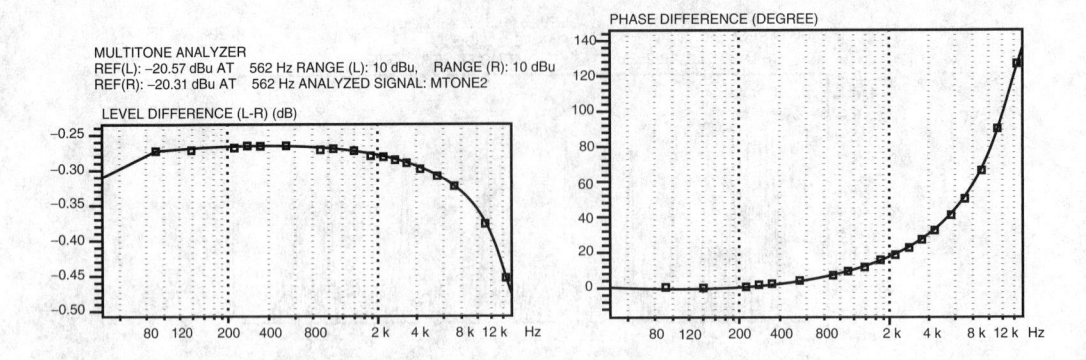

FIGURE 20.16 Level and phase difference measurements taken with a multitone signal analyzer. (*Source:* Tektronix, Beaverton, OR.)

Multitone Versus Discrete Tones

Multitone testing differs from traditional singletone testing in several ways. As previously mentioned, multitone analysis reduces testing and calibration times. This speed advantage stems from several properties:

- All tones are applied at once.
- The entire spectrum is measured/adjusted at once.
- Only one settling time is needed for the generator, DUT, and measurement device because a single signal acquisition is made.
- Level, phase, and noise and distortion calculations are made from one FFT record.

Not only does multitone eliminate the step-and-repeat process, but it can also provide immediate feedback for equipment adjustments. Watching the entire audio spectrum respond to interactive adjustments with little or no time lag further simplifies audio equipment calibration.

When compared to a single sine wave tone, looking at a multitone signal in the time domain (Fig. 20.17) illustrates how closely a multitone waveform resembles program audio. Two factors contribute to this resemblance. First, multitones fill more of the spectrum than single sine wave tones. Second, the crest factor of multitone is similar to that of music or voice signals.

With multitone's similarity to program material, it can be argued that the adjustments performed and measurement results obtained through multitone give more realistic noise and distortion values than discrete tone testing. (In this case, more realistic means a value better representing the noise and distortion present with program material.) In the past, noise measurements were made on audio equipment while no input signal was present. Therefore, the quantization noise and other level-related anomalies were not a part of the noise measurement result. Furthermore, signal processors often operate at different gain ranges depending on the input signal level. So, a no-input noise measurement would be even less representative of actual operating conditions.

The different type of stimulation provided by multitone can, however, lead to measurement results that do not exactly match results obtained with conventional tones or tone sequences. With this in mind, multitone testing should not be used interchangeably with tests made using single-tone input signals. The strengths of multitone testing are speed and convenience, and so it excels in the roles of calibration, production line testing, and equipment performance tracking.

Standardized tone sequence tests require settling, acquisition, and measurement times for each tone in the sequence. All of the advantages of multitone previously mentioned apply here as well. To fully mimic standard tone sequences, a multitone sequence of two or three different levels of multitone can be performed in one-tenth the time of conventional sequences.

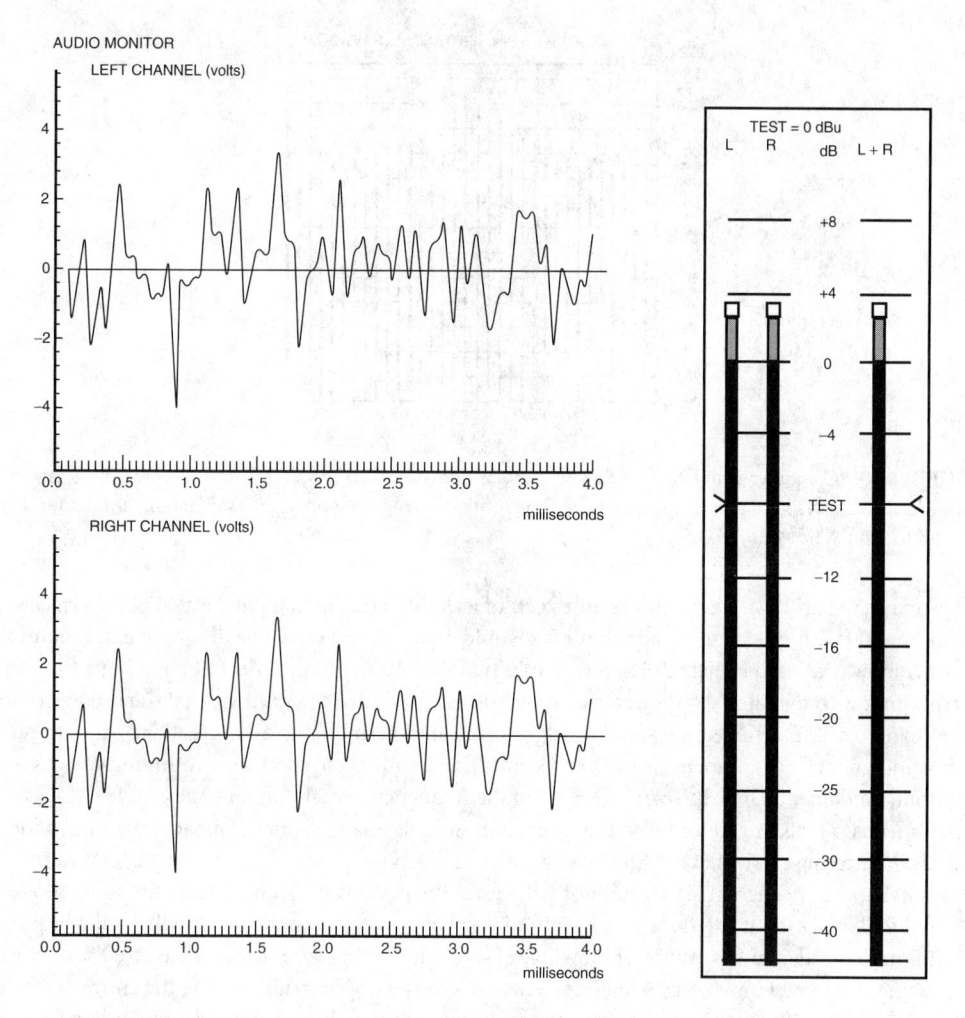

FIGURE 20.17 Time-domain display of a multitone signal. The multitone bears a much stronger resemblance to program material than a single sine wave tone. (*Source:* Tektronix, Beaverton, OR.)

FFT Analysis

A closer look at multitone shows the need for caution when processing the signal. Figures 20.18 and 20.19 illustrate how different processing techniques produce drastically different results, only some of which are meaningful. Both figures are based on a simulated 8192-point (8 K) FFT with a 48-kHz sampling frequency and a 588.87-Hz tone. FFT bin numbers (e.g., bin 100 out of 4096 bins) label the horizontal axes of each graph. The fundamental frequency of 588.87 Hz falls halfway between bin numbers 100 (585.94 Hz) and 101 (591.80 Hz).

Transforming time-domain information into the frequency domain with an FFT requires that the period of the time-domain signal be an integer multiple of the period of the FFT. When such a periodic relationship occurs, a fundamental frequency will fall into a single FFT bin. But this requires that the same sampling clock be used for both signal generation and subsequent resampling of the signal, and further, that no shift in the tone's pitch occurs while passing through the DUT. Because shifts in pitch can occur and because some testing applications require that the generator and analyzer be at separate locations, it is not safe to assume perfect tone-to-bin correspondence.

The result of this imperfect tone-to-bin correspondence is *leakage*, which simply means energy that should fall into a single bin is spread out over a wide frequency range. To reduce the effect of leakage, the time-domain data are multiplied with a window function.

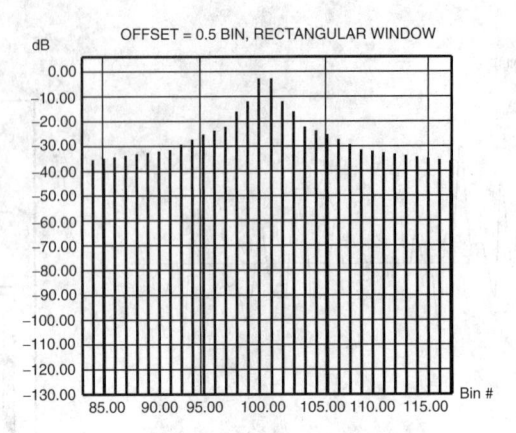

FIGURE 20.18 Computer simulation of FFT without windowing. The 588.87-Hz fundamental frequency in this model does not correspond exactly to an FFT bin. The result is leakage that provides a misleading noise floor display. (*Source:* Tektronix, Beaverton, OR.)

The graph in Fig. 20.18 clearly shows the effects of leakage. If the fundamental was at 585.94 Hz, a single line at bin 100 with no energy in adjacent bins would result. A rectangular window, which is equivalent to no window, was used in Fig. 20.18. The graph in Fig. 20.19 shows the effect of applying a Blackman–Harris window to the same data. After only four bins either side of the fundamental the leakage is down more than 90 dB. In actual application, an algorithm can determine the exact frequency and amplitude of the fundamental. To compensate for significant shifts in pitch, the spectra surrounding each expected multitone fundamental may be searched ±5% of the frequency of each fundamental.

Extraneous signals introduced into the signal path will also have a serious impact on a nonwindowed analysis. Interfering carriers, hum, and sync buzz are a few types of noncoherent signals that fit into this category. These noncoherent signals do not fall within the periodicity requirements for a nonwindowed FFT and will affect the noise floor of a graphical display in the same manner as will a shift in pitch of the multitone fundamentals. With windowing, 60-Hz hum will cause a noise spike at 60 Hz. Without windowing, the same hum energy will be spread across the audio spectrum, raising the entire noise floor.

For noise and distortion measurements, the eight bins surrounding each fundamental in a multitone are set to zero. The remaining bins are plotted to represent the noise (and distortion) floor of the device under test.

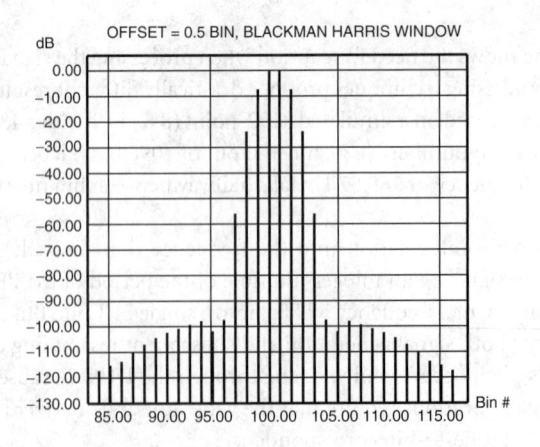

FIGURE 20.19 Computer simulation of FFT with Blackman–Harris windowing. The application of windowing prior to the FFT drastically reduces the effects of leakage. Within 23.44 Hz on either side of the fundamental, leakage is more than 90 dB down. (*Source:* Tektronix, Beaverton, OR.)

Defining Terms

Clipping: A distortion mechanism in an audio system in which the input level to one or more devices or circuits is sufficiently high in amplitude that the maximum input level of the device or circuit is exceeded, resulting in significant distortion of the output waveform.

Crest factor: In an audio system, the ratio of the true peak to the RMS value. For any signal but an ideal square wave the crest factor will be greater than one.

Group delay: The relative delay of the spectral components of a complex waveform. If the group delay is flat, all components arrive together.

Multitone testing: A measurement technique whereby an audio system is characterized by the simultaneous application of a combination of sine waves to a device under test.

Peak-equivalent sine: A method of specifying signal amplitude in an audio system, equal to the RMS level of a sine wave having the same peak-to-peak amplitude as the signal under consideration.

SMPTE IM: A method of measuring intermodulation distortion in an audio system standardized by the Society of Motion Picture and Television Engineers.

Weighting filter: A standardized filter used to impart predetermined characteristics to noise measurements in an audio system.

References

Benson, K.B. and Whitaker, J.C. 1990. *Television and Audio Handbook for Technicians and Engineers.* McGraw-Hill, New York.

Fink, D.G. and Christiansen, D., eds. 1989. *Electronics Engineers' Handbook*, 3rd ed. McGraw-Hill, New York.

Tektronix, 1995. 764 Digital Audio Monitor. Tektronix Application Note 21W7269, Tektronix, Beaverton, OR.

Thompson, B. 1992. Multitone Audio Testing. Tektronix Application Note 21W-7182, Tektronix, Beaverton, OR.

Whitaker, J.C. and Benson, K.B. eds. 2002. *Standard Handbook of Audio and Radio Engineering,* 2nd ed. McGraw-Hill, New York.

Whitaker, J.C. 2001. *Electronic Systems Maintenance Handbook,* 2nd ed. CRC Press, Boca Raton, FL.

Whitaker, J.C. 2000. *The Communications Facility Design Handbook,* CRC Press, Boca Raton, FL.

Further Information

The Audio Engineering Society (AES) conducts trade shows and highly respected technical seminars annually in the U.S. and abroad. The society also publishes a monthly journal (*Journal of the Audio Engineering Society*) and numerous specialized books on the topic of audio technology in general, and digital audio in particular.

20.2 Analog Video Measurements

Carl Bentz and Jerry C. Whitaker

20.2.1 Introduction

Although there are a number of computer-based television signal monitors capable of measuring video, sync, chroma, and burst levels (as well as pulse widths and other timing factors), sometimes the best way to see what a signal is doing is to monitor it visually. The waveform monitor and vectorscope are oscilloscopes specially adapted for the video environment. The waveform monitor, like a traditional oscilloscope, operates in a voltage-vs.-time mode. While an oscilloscope timebase can be set over a wide range of intervals, the waveform monitor timebase triggers automatically on sync pulses in the conventional TV signal, producing line- and field-rate sweeps, as well as multiple lines, multiple fields, and shorter time intervals. Filters, clamps, and other circuits process the video signal for specific monitoring needs.

The vectorscope operates in an $X - Y$ voltage-vs.-voltage mode to display chrominance information. It decodes the signal in much the same way as a television receiver or a video monitor to extract color information and to display phase relationships. These two instruments serve separate, distinct purposes. Some models combine the functions of both types of monitors in one chassis with a single CRT display. Others include a communications link between two separate instruments.

Beyond basic signal monitoring, the waveform monitor and vectorscope provide a means to identify and analyze signal aberrations. If the signal is distorted, these instruments allow a technician to learn the extent of the problem and to locate the offending equipment.

Although designed for analog video measurement and quality control, the waveform monitor and vectorscope still serve valuable roles in the digital video facility, and will continue to do so for some time.

20.2.2   Color Bar Test Patterns

Color bars are the most common pattern for testing encoders, decoders, and other video devices. The test pattern typically contains several bars filled with primary and complementary colors. There are many variants, which differ in color sequence, orientation, saturation, and intensity.

The standard color bar sequence is white, yellow, cyan, green, magenta, red, blue, and black. This sequence can be produced in an RGB format by a simple 3-bit counter. The typical specification of color bar levels (ITU-R Rec. BT.471-1, "Nomenclature and Description of Color Bar Signals") is a set of four numbers separated by slashes or dots and giving RGB levels as a percentage of reference white in the following sequence:

white bar/black bar/max colored bars/min colored bars

For example, 100/0/100/25 means 100% R, G, and B on the white bar; 0% R, G, and B on the black bar; 100% maximum of R, G, and B on colored bars; and 25% minimum of R, G, and B on colored bars. (See Figure 20.20*a*.)

Some color bar patterns merit special names. For example, 100/0/75/0 bars are often called "EBU bars" or "75% bars", and 100/0/100/25 bars are known as "BBC bars." Nevertheless, the ITU four number nomenclature remains the only reliable specification system to designate the exact levels for color bar test patterns.

The SMPTE variation is a matrix test pattern according to SMPTE engineering guideline EG 1, "Alignment Color Bar Test Signal for Television Picture Monitors." The guideline specifies that 67% of the field shall contain seven (without black) 75% color bars, plus 8% of the field with the "new chroma set" bars (blue/black/magenta/black/cyan/black/gray), and the remaining 25% shall contain a combination of $-I$, white, Q, black, and the black set signal (a version of PLUGE). This arrangement is illustrated in Fig. 20.20*b*.

Chroma gain and chroma phase for picture monitors are usually adjusted by observing the standard encoded color bar signal with the red and green CRT guns switched off. The four visible blue bars are set for equal brightness. The use of the chroma set feature greatly increases the accuracy of this adjustment because it provides a signal with the blue bars to be matched vertically adjacent to each other. Because the bars are adjacent, the eye can easily perceive differences in brightness. This also eliminates effects resulting from shading or purity from one part of the screen to another.

The EIA color bar signal is a matrix test pattern according to RS-189-A, which consists of 75% of the field containing seven (without black) 75% color bars (same as SMPTE) and the remaining 25% containing a combination of $-I$, white, Q, and black. (See Figure 20.20*c*.)

HD/SD Video Test Signal

The SMPTE, in recognition of the need for convenient measurement of both high-definition (HD) and standard-definition (SD) video signals within a given plant, developed a recommended practice designed to serve both domains [1]. SMPTE RP-219 specifies a color bar pattern compatible with both HD and SD environments. The multi-format color bar signal is originated as an HD signal with an aspect ratio of 16:9 and may be down converted to an SD color bar signal with an aspect ratio of either 4:3 or 16:9.

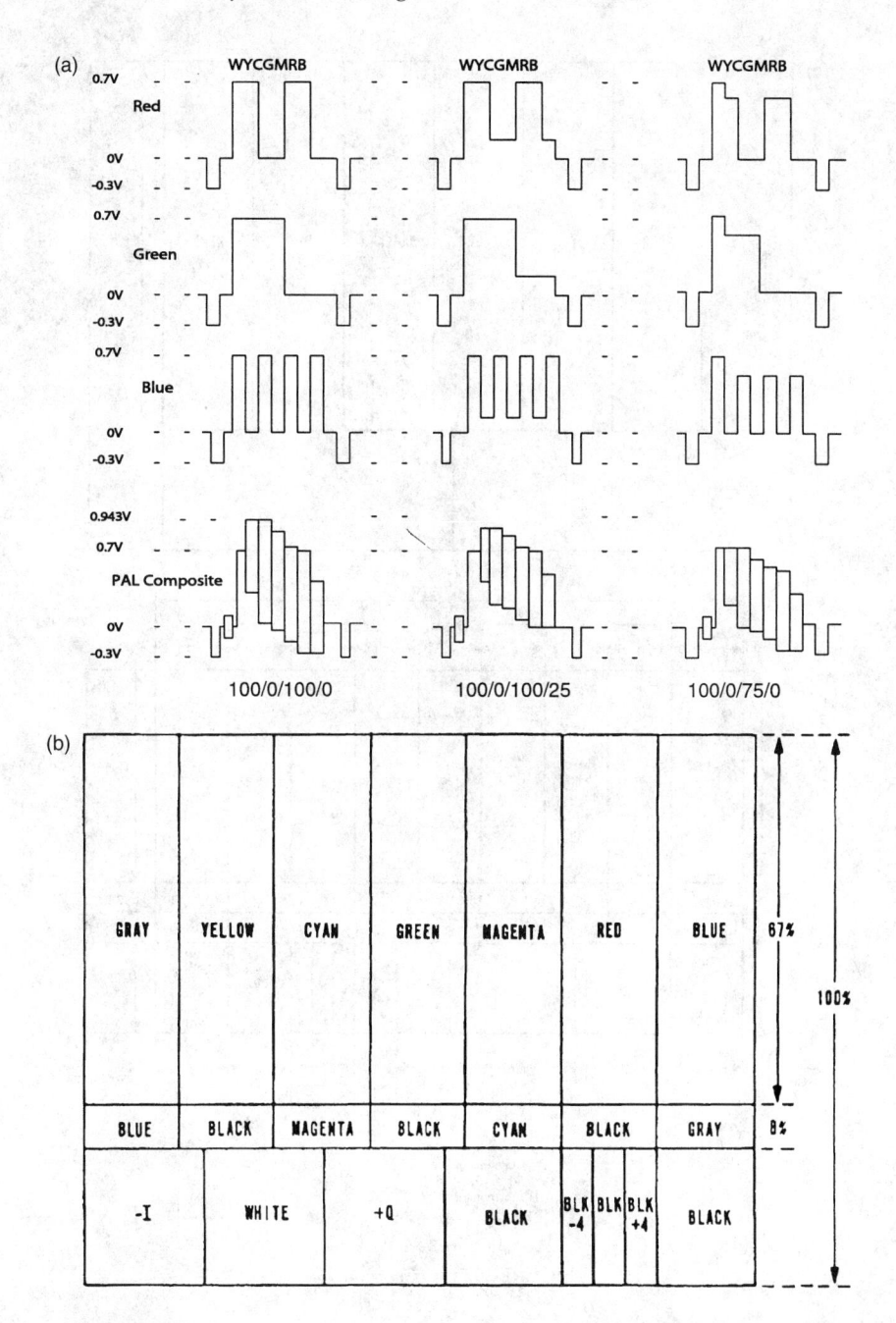

FIGURE 20.20 Examples of color test bars: (a) ITU nomenclature, (b) SMPTE EG 1 color bar test signal, (c) EIA RS-189-A test signal, (d, next page) multi-format color bar test signal.[*Drawing (b)after SMPTE EG 1, (c) after EIA RS-189-A, (d) after SMPTE RP 219.*]

The color bar signal is generated with unconventionally slow rise and fall time value, and is therefore only intended to facilitate video level control and monitor color adjustments of HD and SD equipment. It can be applied to HD video productions, especially in a multi-format environment where HD video sources are frequently converted and used as SD video content either in 525- or 625-line environment with same frame frequencies as in the original HDTV signal.

The multi-format color bar signal is composed of four specific patterns, shown in Fig. 20.20*d*. The first part of the color bar signal represents a signal for the 4:3 aspect ratio; a second part of the total signal

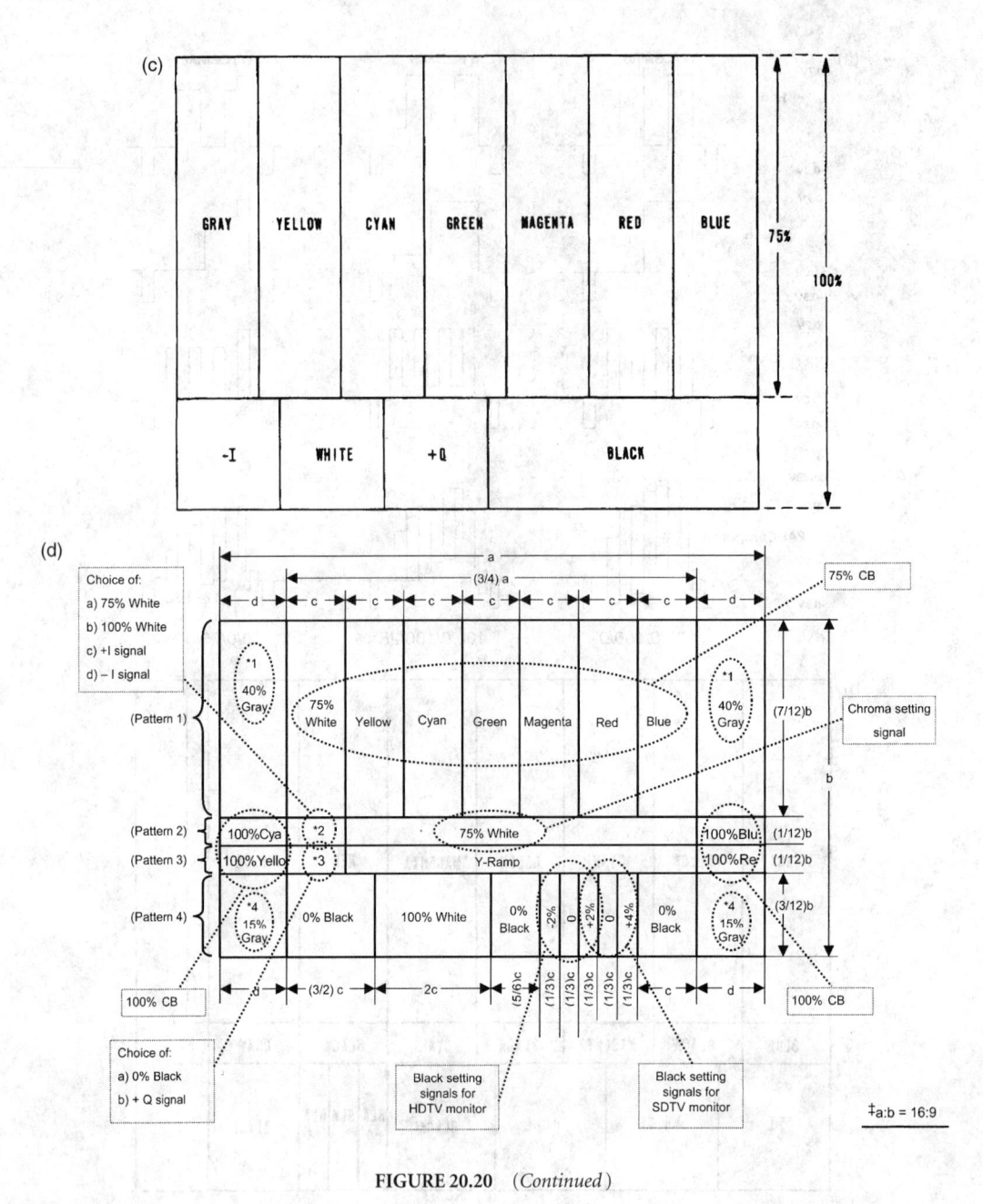

FIGURE 20.20   (*Continued*)

adds side panels for the 16:9 aspect ratio. A third part adds black and white ramps and additional color information, and the last part completes the total signal by adding white and black bars, in addition to a set of near-black-level steps for monitor black level adjustment.

20.2.3   Basic Waveform Measurements

Waveform monitors are used to evaluate the amplitude and timing of video signals and to show timing relationships between two or more signals [2]. The familiar color-bar pattern is the only signal required for these basic tests. Figure 20.21 shows a typical waveform display of color bars. It is important to realize that all color bars are not created equal. Some generators offer a choice of 75% or 100% amplitude bars. Sync, burst, and setup amplitudes remain the same in the two color-bar signals, but the peak-to-peak

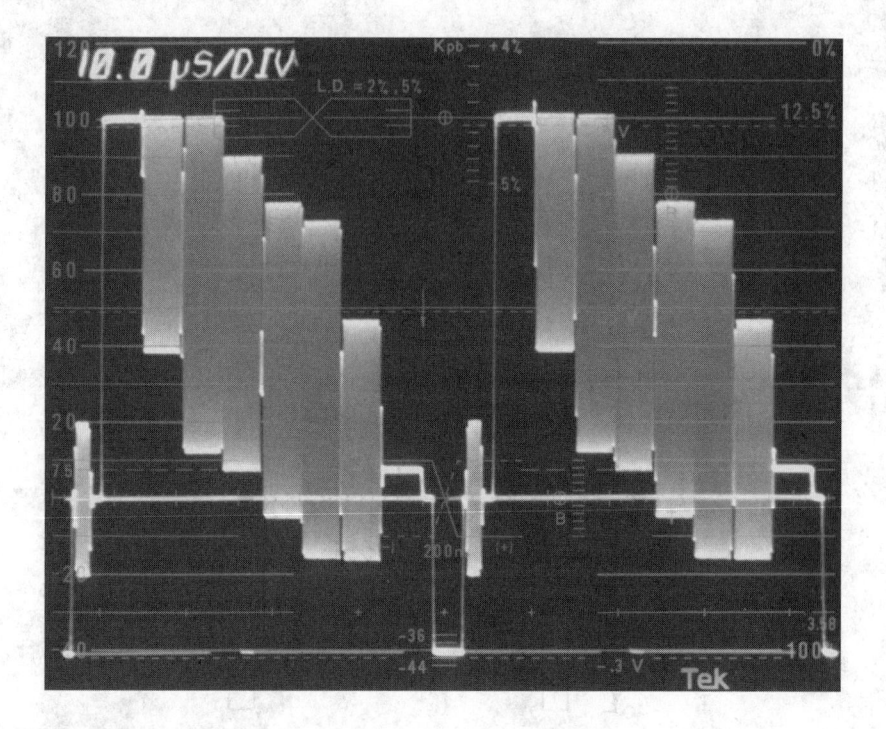

FIGURE 20.21 Waveform monitor display of a color-bar signal at the two-line rate. (*Courtesy of Tektronix.*)

amplitudes of high-frequency chrominance information and low-frequency luminance levels change. The saturation of color, a function of chrominance and luminance amplitudes, remains constant at 100% in both modes. The 75% bar signal has 75% amplitude with 100% saturation. In 100% bars, amplitude and saturation are both 100%.

Chrominance amplitudes in 100% bars exceed the maximum amplitude that should be transmitted via NTSC. Therefore, 75% amplitude color bars, with no chrominance information exceeding 100 IRE, are the standard amplitude bars for conventional television. In the 75% mode, a choice of 100 IRE or 75 IRE white reference level may be offered. Figure 20.22 shows 75% amplitude bars with a 100 IRE white level. Either white level can be used to set levels, but operators must be aware of which signal has been selected. SMPTE bars have a white level of 75 IRE as well as a 100 IRE white flag.

The vertical response of a waveform monitor depends upon filters that process the signal in order to display certain components. The *flat response* mode displays all components of the signal. A *chroma response* filter removes luminance and displays only chrominance. The low-pass filter removes chrominance, leaving only low-frequency luminance levels in the display. Some monitors include an IRE filter, designed to average-out high-level, fine-detail peaks on a monochrome video signal. The IRE filter aids the operator in setting brightness levels. The IRE response removes most, but not all, of the chrominance.

If the waveform monitor has a dual filter mode, the operator can observe luminance levels and overall amplitudes at the same time. The instrument switches between the flat and low-pass filters. The line select mode is another useful feature for monitoring live signals.

Sync Pulses

The duration and frequency of the sync pulses must be monitored closely for conventional video. Most waveform monitors include 0.5 μs or 1 μs per division magnification (MAG) modes, which can be used to verify H-sync width between 4.4 and 5.1 μs. The width is measured at the -4 IRE point. On waveform monitors with good MAG registration, sync appearing in the middle of the screen in the 2-line mode remains centered when the sweep is magnified. Check the rise and fall times of sync, and the widths of the front porch and entire blanking interval. Examine burst and verify there are between 8 and 11 cycles of subcarrier.

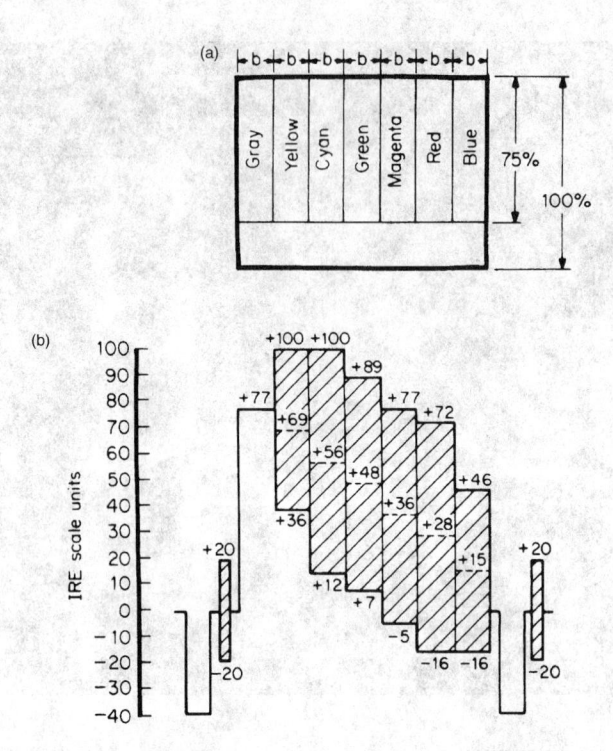

FIGURE 20.22   EIA RS-189-A color-bar displays: (a) color displays of gray and color bars, (b) waveform display of reference gray and primary/complementary colors, plus sync and burst.

Check the vertical intervals for correct format, and measure the timing of the equalizing pulses and vertical sync pulses. The acceptable limits for these parameters are shown in Fig. 22.23.

20.2.4   Basic Vectorscope Measurements

The vectorscope displays chrominance amplitudes, aids hue adjustments, and simplifies the matching of burst phases of multiple signals. These functions require only the color-bar test signal as an input stimulus. To evaluate and adjust chrominance in the TV signal, observe color bars on the vectorscope. The instrument should be in its calibrated gain position. Adjust the vectorscope phase control to place the burst vector at the 9 O'clock position. Note the vector dot positions with respect to the six boxes marked on the graticule. If everything is correctly adjusted, each dot will fall on the crosshairs of its corresponding box, as shown in Fig. 20.24.

The chrominance amplitude of a video signal determines the intensity or brightness of color. If the amplitudes are correct, the color dots fall on the crosshairs in the corresponding graticule boxes. If vectors overshoot the boxes, chrominance amplitude is too high; if they undershoot, it is too low. The boxes at each color location can be used to quantify the error. In the radial direction, the small boxes indicate a ±2.5 IRE error from the standard amplitude. The large boxes indicate a ± 20% error.

Other test signals, including a modulated staircase or multiburst, are used for more advanced tests. It is important to closely observe how these signals appear at the output of the waveform generator. Knowing what the undistorted signal looks like simplifies the subsequent identification of distortions.

20.2.5   Line Select Features

Some waveform monitors and vectorscopes have line select capability. They can display one or two lines out of the entire video frame of 525 lines. (In the normal display, all of the lines are overlaid on top of one another.) The principal use of the single line feature is to monitor VITS (*vertical interval test signals*). VITS

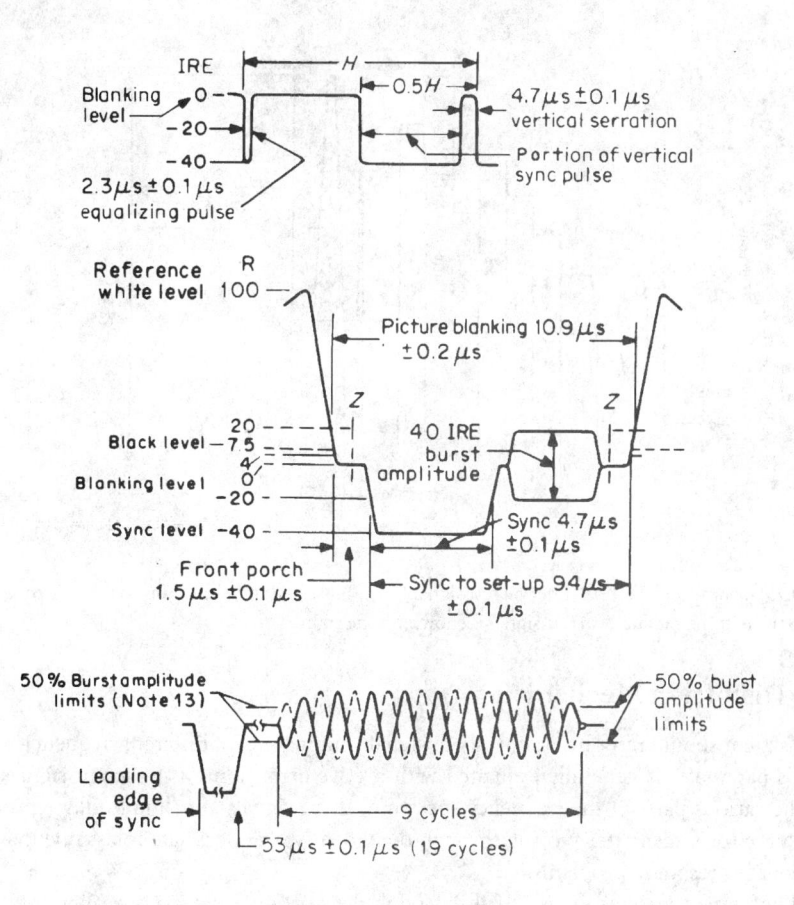

FIGURE 20.23 Sync pulse widths for NTSC.

allows in-service testing of the transmission system. A typical VITS waveform is shown in Fig. 20.25. A full-field line selector drives a picture monitor output with an intensifying pulse. The pulse causes a single horizontal line on a picture monitor to be highlighted. This indicates where the line selector is within the frame to correlate the waveform monitor display with the picture.

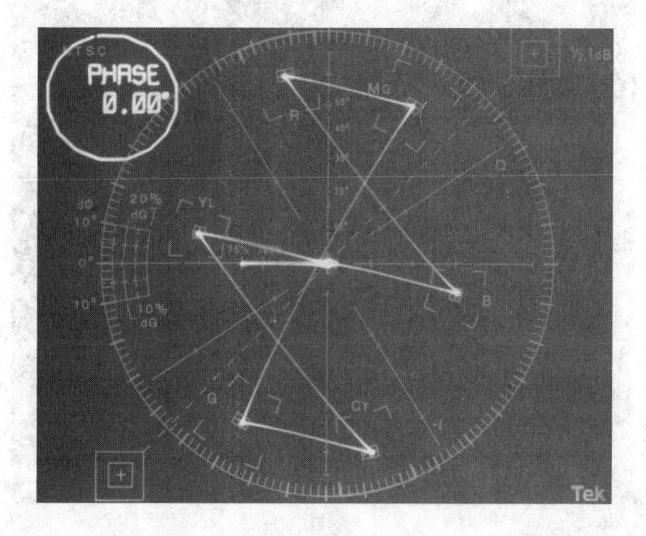

FIGURE 20.24 Vectorscope display of a color-bar signal. (*Courtesy of Tektronix.*)

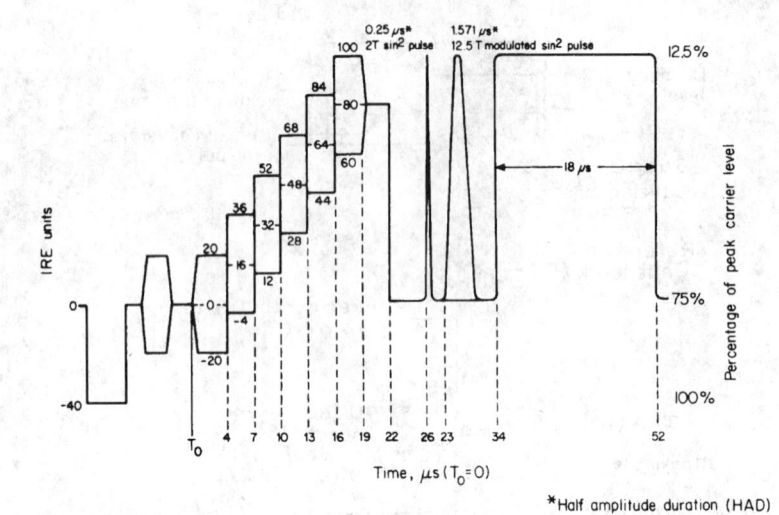

FIGURE 20.25 Composite vertical-interval test signal (VITS) inserted in field 1, line 18. The video level in IRE units is shown on the left; the radiated carrier signal is shown on the right.

20.2.6 Distortion Mechanisms

The video system should respond uniformly to signal components of different frequencies. In an analog system, this parameter is generally evaluated with a waveform monitor. Different signals are required to check the various parts of the frequency spectrum. If the signals are all faithfully reproduced on the waveform monitor screen after passing through the video system, it is safe to assume that there are no serious frequency response problems.

At very low frequencies, look for externally introduced distortions, such as power-line hum or power supply ripple, and distortions resulting from inadequacies in the video processing equipment. Low-frequency distortions usually appear on the video image as flickering or slowly varying brightness. Low-frequency interference can be seen on a waveform monitor when the DC restorer is set to the slow mode and a 2-field sweep is selected. Sine wave distortion from AC power-line hum may be observed in Fig. 20.26. A *bouncing*

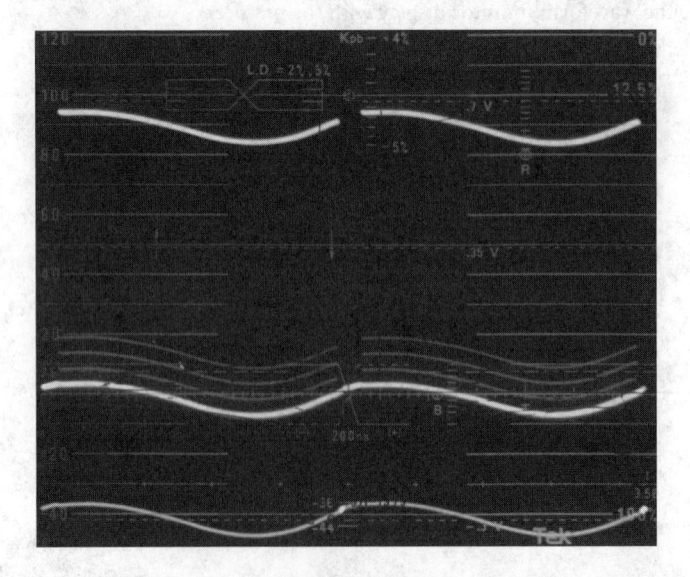

FIGURE 20.26 Waveform monitor display showing additive 60 Hz degradation. (*Courtesy of Tektronix.*)

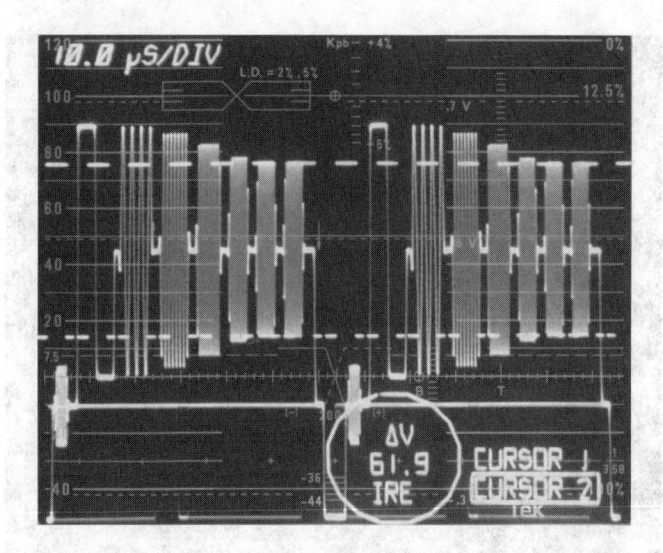

FIGURE 20.27 Waveform monitor display of a multiburst signal showing poor high-frequency response. (*Courtesy of Tektronix.*)

APL signal can be used to detect distortion in the video chain. Vertical shifts in the blanking and sync levels indicate the possibility of low-frequency distortion of the signal.

Field-rate distortions appear as a difference in shading from the top to the bottom of the picture. A field-rate 60 Hz square wave is best suited for measuring field-rate distortion. Distortion of this type is observed as a tilt in the waveform in the 2-field mode with the DC restorer off.

Line-rate distortions appear as streaking, shading, or poor picture stability. To detect such errors, look for tilt in the bar portion of a pulse-and-bar signal. The waveform monitor should be in the 1 or 2H mode with the fast DC restorer selected for the measurement.

The *multiburst* signal is used to test the high-frequency response of a system. The multiburst includes packets of discrete frequencies within the television passband, with the higher frequencies toward the right of each line. The highest frequency packet is at about 4.2 MHz, the upper frequency limit of the NTSC system. The next packet to the left is near the color subcarrier frequency (3.58 MHz, approximately) for checking the chrominance transfer characteristics. Other packets are included at intervals down to 500 kHz. The most common distortion is high-frequency rolloff, seen on the waveform monitor as reduced amplitude packets at higher frequencies. This type of problem is shown in Fig. 20.27. The television picture exhibits loss of fine detail and color intensity when such impairments are present. High frequency peaking, appearing on the waveform as higher amplitude packets at the higher frequencies, causes ghosting on the picture.

Differential Phase

Differential phase ($d\phi$) distortion occurs if a change in luminance level produces a change in the chrominance phase. If the distortion is severe, the hue of an object will change as its brightness changes. A modulated staircase or ramp is used to quantify the problem. Either signal places chrominance of uniform amplitude and phase at different luminance levels. Figure 20.28 shows a 100 IRE modulated ramp. Because $d\phi$ can change with changes in APL, measurements at the center and at the two extremes of the APL range are necessary.

To measure $d\phi$ with a vectorscope, increase the gain control until the vector dot is on the edge of the graticule circle. Use the phase shifter to set the vector to the 9 O'clock position. Phase error appears as circumferential elongation of the dot. The vectorscope graticule has a scale marked with degrees of $d\phi$ error. Figure 20.29 shows a $d\phi$ error of $5°$.

More information can be obtained from a swept R–Y display, which is a common feature of waveform monitor and vectorscope instruments. If one or two lines of demodulated video from the vectorscope are

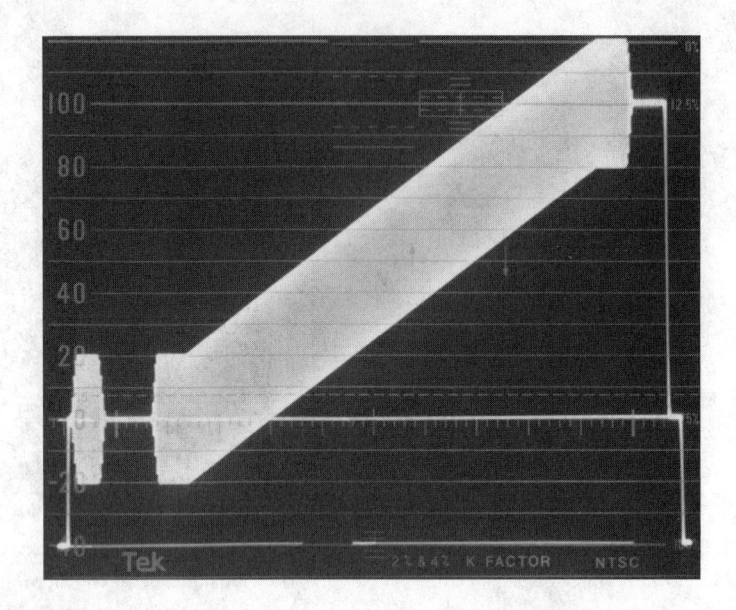

FIGURE 20.28   Waveform monitor display of a modulated ramp signal. (*Courtesy of Tektronix.*)

displayed on a waveform monitor, differential phase appears as tilt across the line. In this mode, the phase control can be adjusted to place the demodulated video on the baseline, which is equivalent in phase to the 9 O'clock position of the vectorscope. Figure 20.30 shows a $d\phi$ error of approximately 6° with the amount of tilt measured against a vertical scale. This mode is useful in troubleshooting applications. By noting where along the line the tilt begins, it is possible to determine at what DC level the problem starts to occur. In addition, field- rate sweeps enable the operator to look at $d\phi$ over the field.

A variation of the swept R–Y display may be available in some instruments for precise measurement of differential phase. Highly accurate measurements can be made with a vectorscope that includes a precision phase shifter and a double-trace mode. This method involves nulling the lowest part of the waveform with the phase shifter, and then using a separate calibrated phase control to null the highest end of the waveform. A readout in tenths of a degree is possible.

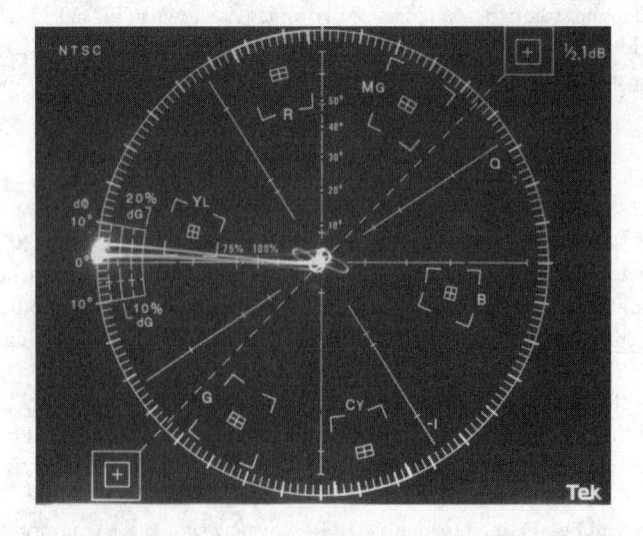

FIGURE 20.29   Vectorscope display showing 5° differential phase error. (*Courtesy of Tektronix.*)

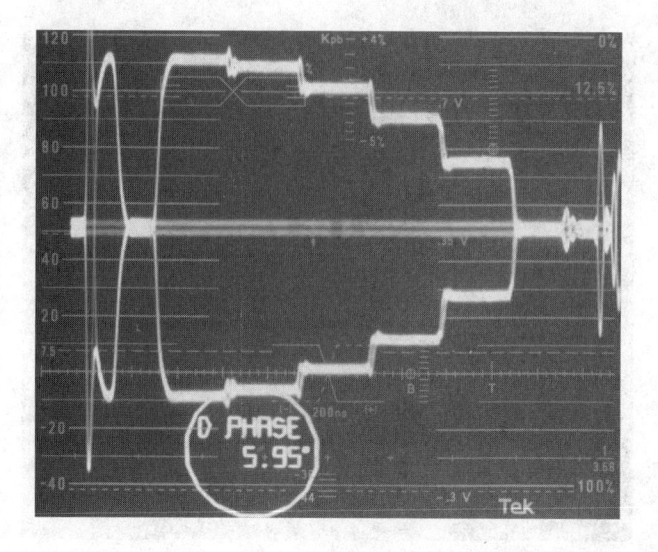

FIGURE 20.30 Vectorscope monitor display showing a differential phase error of 5.95° as a tilt on the vertical scale. (*Courtesy of Tektronix.*)

Differential Gain

Differential gain (dG) distortion refers to a change in chrominance amplitude with changes in luminance level. The vividness of a colored object changes with variations in scene brightness. The modulated ramp or staircase is used to evaluate this impairment with the measurement taken on signals at different APL points.

To measure differential gain with a vectorscope, set the vector to the 9 O'clock position and use the variable gain control to bring it to the edge of the graticule circle. Differential gain error appears as a lengthening of the vector dot in the radial direction. The dG scale at the left side of the graticule can be used to quantify the error. Figure 20.31 shows a dG error of 10%.

Differential gain can be evaluated on a waveform monitor by using the chroma filter and examining the amplitude of the chrominance from a modulated staircase or ramp. With the waveform monitor in 1H sweep, use the variable gain to set the amplitude of the chrominance to 100 IRE. If the chrominance amplitude is not uniform across the line, there is dG error. With the gain normalized to 100 IRE, the error can be expressed as a percentage. Finally, dG can be precisely evaluated with a swept display of

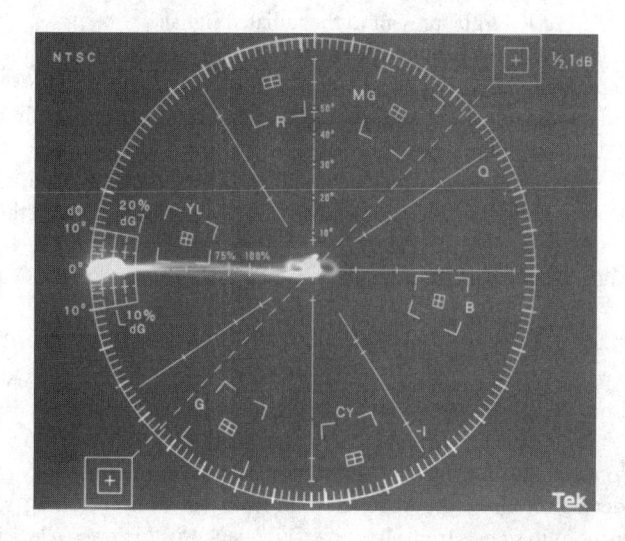

FIGURE 20.31 Vectorscope display of a 10% differential gain error. (*Courtesy of Tektronix.*)

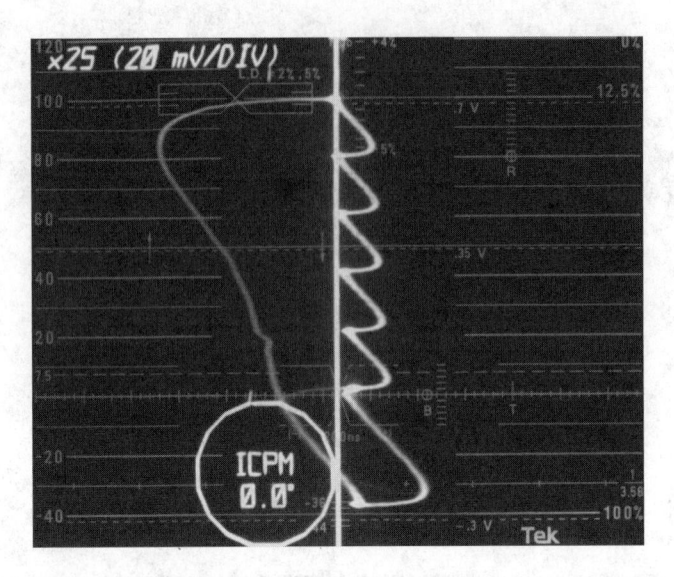

FIGURE 20.32 Waveform monitor display using the ICPM graticule of a five-level stairstep signal with no distortion. (*Courtesy of Tektronix.*)

demodulated video. This is similar to the single trace R–Y methods for differential phase. The B–Y signal is examined for tilt when the phase is set so that the B–Y signal is at its maximum amplitude. The tilt can be quantified against a vertical scale.

ICPM

Television receivers typically use a method known as *intercarrier sound* to reproduce audio information. Sound is recovered by beating the audio carrier against the video carrier, producing a 4.5 MHz IF signal, which is demodulated to produce the sound portion of the transmission. From the interaction between audio and video portions of the signal, certain distortions in the video at the transmitter can produce audio buzz at the receiver. Distortions of this type are referred to as *incidental carrier phase modulation* or ICPM. The widespread use of stereo audio for television increased the importance of measuring this parameter at the transmitter, because the buzz is more objectionable in stereo broadcasts. It is generally suggested that less than 3° of ICPM be present in the radiated signal.

ICPM is measured using a high-quality demodulator with a synchronous detector mode and an oscilloscope operated in a high-gain $X - Y$ mode. Some waveform and vector monitors have such a mode as well. Video from the demodulator is fed to the Y input of the scope and the quadrature output is fed to the X input terminal. Low-pass filters make the display easier to resolve.

An unmodulated 5-step staircase signal produces a polar display, shown in Figure 20.32, on a graticule developed for this purpose. Notice that the bumps all rest in a straight vertical line, if there is no ICPM in the system. Tilt indicates an error, as shown in Figure 20.33. The graticule is calibrated in degrees per radial division for differential gain settings. Adjustment, but not measurement, can be performed without a graticule.

Tilt and Rounding

Good picture quality requires that the characteristics of the visual transmitter be as linear as possible. The transmitted channel bandwidth of conventional television (6 MHz for NTSC and greater for PAL and SECAM) introduces various obstacles for achieving a high degree of linearity. The wideband amplifiers of television transmission systems must have a flat response over the entire frequency range of interest. The design process considers solutions for the problems of spurious harmonics resulting from stray component parameters, and distortions to signals from all sources. Frequency distortion in television systems, ranging from 15 kHz to several hundred kilohertz, is more visible on test equipment if a line-frequency square-wave

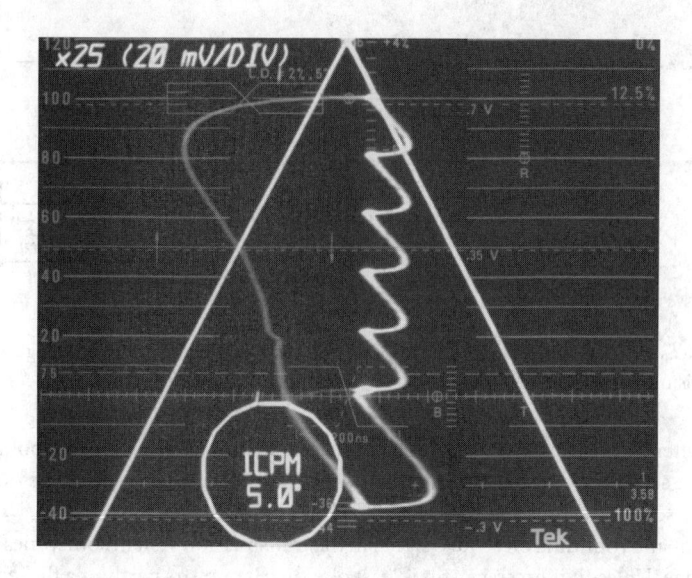

FIGURE 20.33   Waveform monitor display using the ICPM graticule of a five-level stairstep signal with 5° distortion. (*Courtesy of Tektronix.*)

signal (with a rise time of approximately 200 ns) modulates the visual transmitter. Distortion characteristics of this range generally fall into the category of *rounding* and *tilt*.

It is possible to run tests on a transmission system using a number of specific frequencies, monitoring the response to each at various points in the system. However, a more efficient method involves using square-wave test signals with rise times selected to avoid overshoots in transmission. During monitoring of the signals at various test points, tilt and rounding of the square-wave corners may be observed as impaired frequency response. Two square-wave frequencies prove to be useful for this work. The lower frequency of 60 Hz (50 Hz in PAL) aids in detecting response errors in the frequency range below approximately 15 kHz. A 15 kHz square wave serves to determine response to several hundred kilohertz. These two test signals identify low frequency response problems when observed on an oscilloscope as a tilt of the flat part of the square wave (top or bottom) and a rounding of the corners. Ideal response would produce flat tops and bottoms without any rounding at the corners.

Tilt can be determined for both test signal ranges. The measurement is made on a portion of the square wave derived from a demodulator. Particular start and end points of the measurement are suggested to avoid ringing as a result of pulse rise times. To ensure the best measurements, the sweep range and vertical sensitivity of the oscilloscope are set as wide as possible to accommodate the entire length of trace segment of interest. Tops and bottoms of the square waves should be measured separately (see Fig. 20.34.)

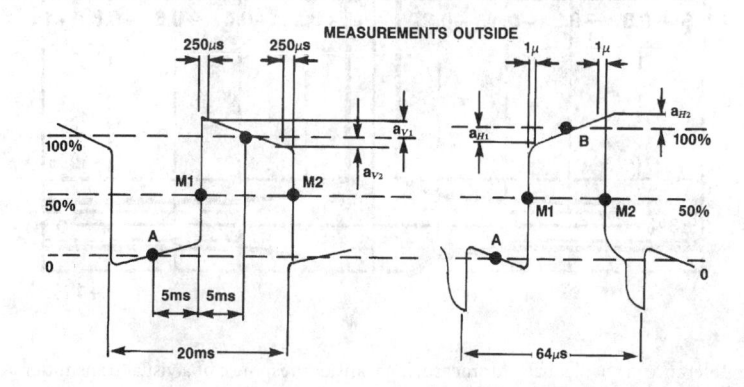

FIGURE 20.34   Measurement of tilt on a visual signal with 50 Hz and 15 kHz signals. (*After* [3].)

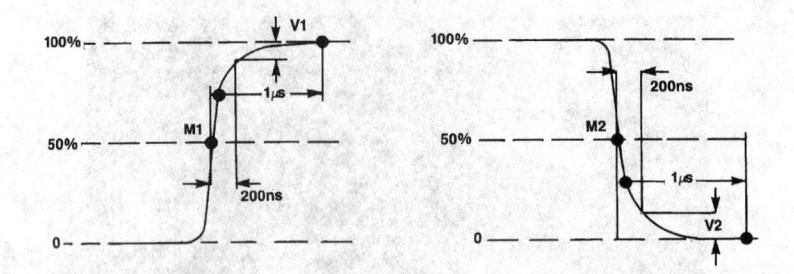

FIGURE 20.35 Measurement of rounding of a 15 kHz square-wave test signal to identify high-frequency-response problems in a television transmitter. (*After* [3].)

Rounding is determined from the 15 kHz signal. It is realistic to delay the start point and to define a duration along the scope trace for the measurement. Separate observations should be made on rising and falling sides of the square wave, as shown in Fig. 20.35.

The effects of tilt and rounding on a video signal in the transmission system sometimes can be seen in the demodulated image. Unless it is excessive, tilt is least obvious. It appears partly as a flickering in the picture and partly as a variation in luminance level from top-to-bottom and/or side-to-side of the image. Rounding, in this case, manifests itself as a reduction of detail in medium-frequency parts of the image. In cases of excessive tilt or rounding error, it may be difficult for some TV receivers to lock on to the degraded signal.

Group Delay

The transients of higher-frequency components (15 to 250 kHz), produce overshoot on the leading edge of transitions and, possibly, ringing at the top and bottom of the waveform area. The shorter duration of

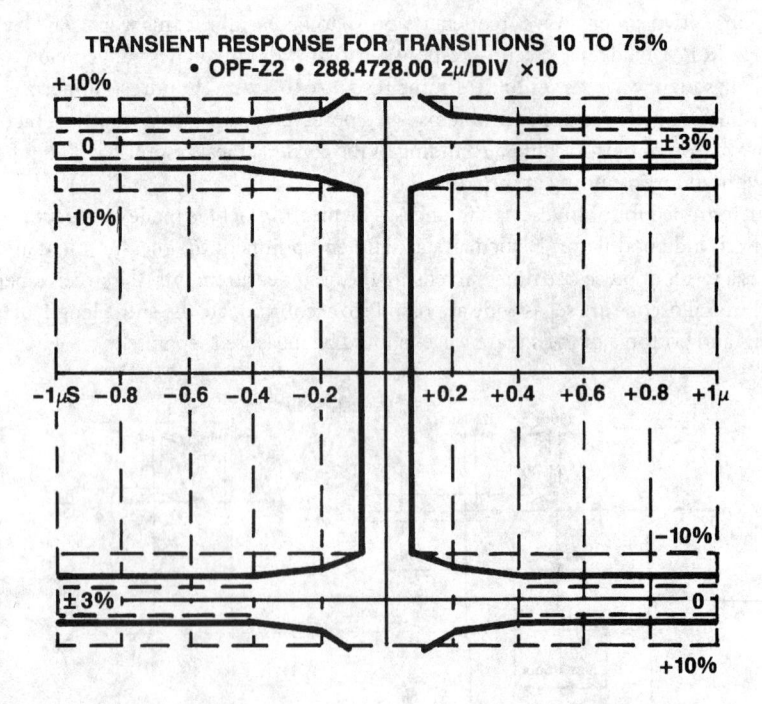

FIGURE 20.36 Tolerance graticule mask for measuring transient response of a visual transmitter. A complete oscillation is first displayed on the screen. The timebase then is expanded, and the signal X-Y position controls are adjusted to shift the trace into the tolerance mask. (*After* [4].)

the "flat" top and bottom lowers concern about tilt and rounding. A square wave contains a fundamental frequency and numerous odd harmonics. The number of the harmonics determines the rise or fall times of the pulses. Square waves with rise times of 100 ns (T) and 200 ns ($2T$) are particularly useful for television measurements. In the $2T$ pulse, significant harmonics approach 5 MHz; a T pulse includes components approaching 10 MHz. Because the TV system should carry as much information as possible, and its response should be as flat as possible, the T pulse is a common test signal for determination of group delay.

Group delay is the effect of time- and frequency-sensitive circuitry on a range of frequencies. Time, frequency, phase shift, and signal delay all are related and can be determined from circuit component values. Excessive group delay in a video signal appears as a loss of image definition. Group delay is a fact of life that cannot be avoided, but its effect can be reduced through *predistortion* of the video signal. Group delay adjustments can be made before the modulator stage of the transmitter, while monitoring of the signal from a feedline test port and adjusting for best performance.

Group delay can be monitored using a special purpose scope or waveform graticule. The goal in making adjustments is to fit all excursions of the signal between the smallest tolerance markings of the graticule, as illustrated in Fig. 20.36. Because quadrature phase errors are caused by the vestigial sideband transmission system, synchronous detection is needed to develop the display.

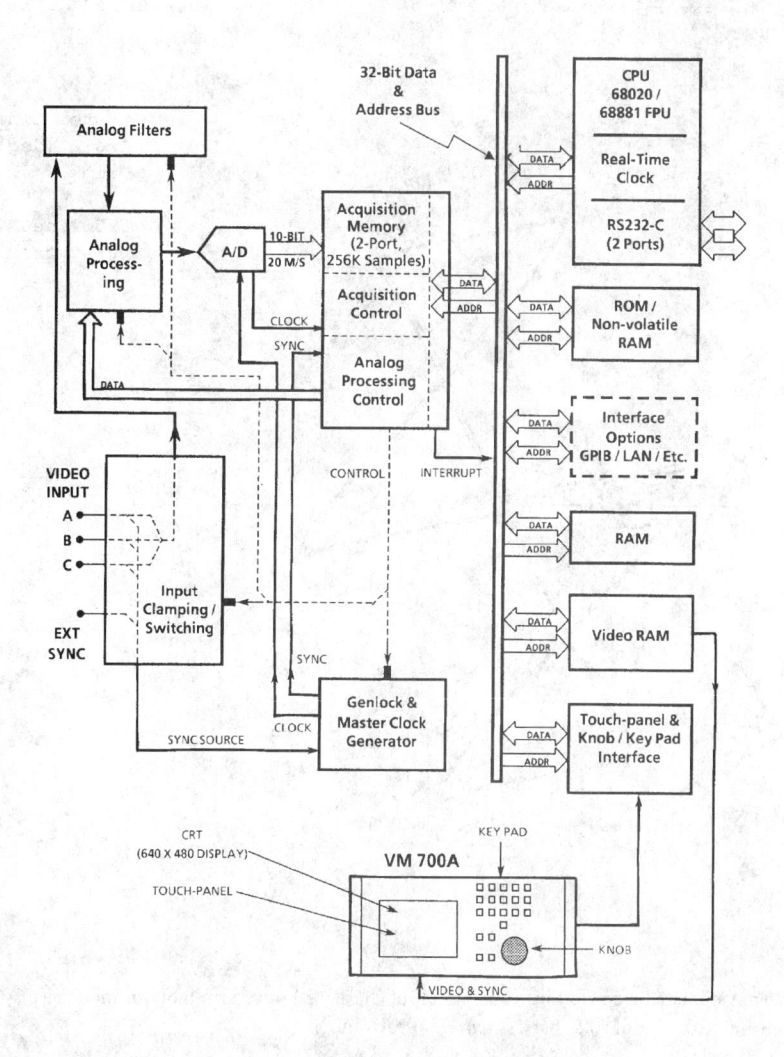

FIGURE 20.37 Block diagram of an automated video test instrument. (*Courtesy of Tektronix.*)

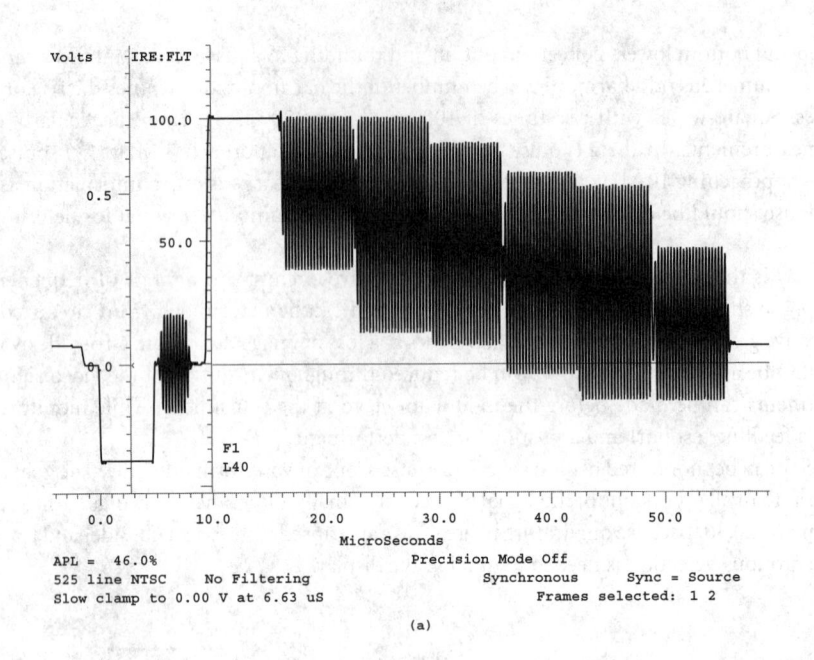

Volts   IRE:FLT

100.0

0.5

50.0

0.0

F1
L40

0.0   10.0   20.0   30.0   40.0   50.0
                MicroSeconds

APL = 46.0%   Precision Mode Off
525 line NTSC   No Filtering   Synchronous   Sync = Source
Slow clamp to 0.00 V at 6.63 uS   Frames selected: 1 2

(a)

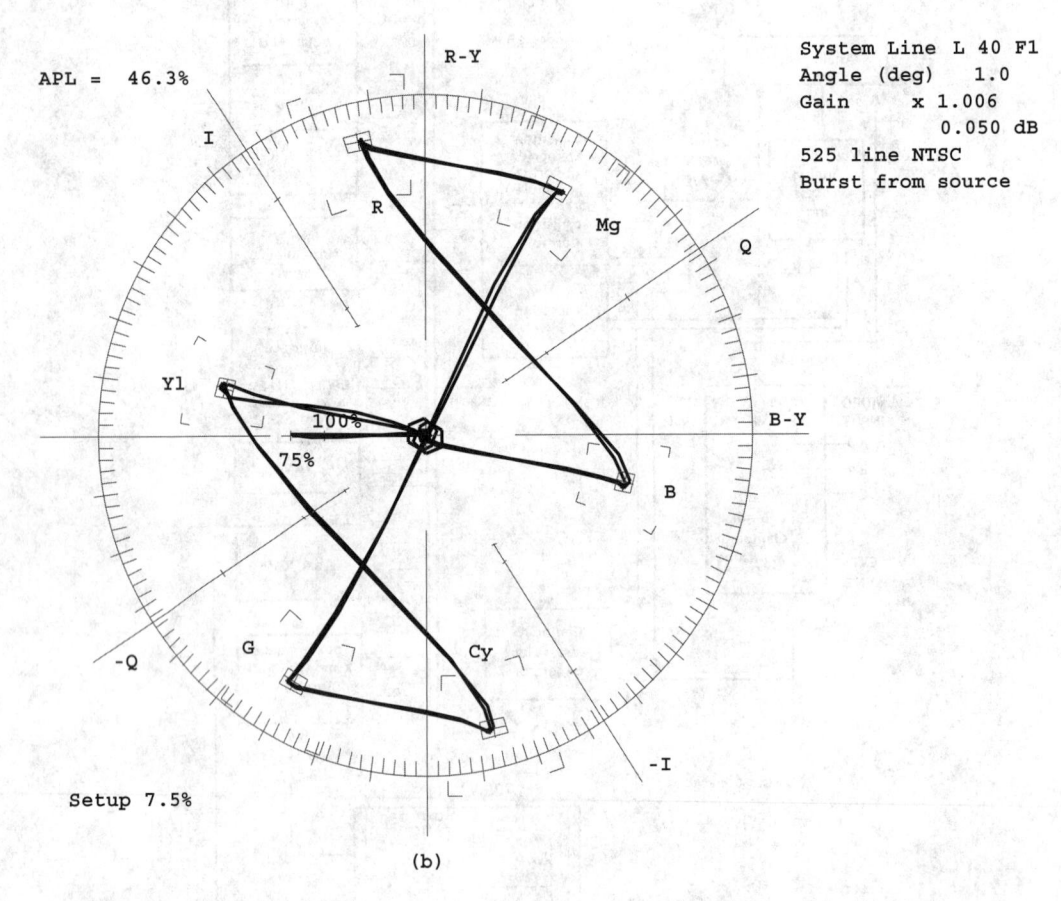

APL = 46.3%

R-Y

System Line L 40 F1
Angle (deg)   1.0
Gain   x 1.006
              0.050 dB
525 line NTSC
Burst from source

I

R

Mg

Q

Yl

100%

B-Y

75%

B

-Q

G

Cy

-I

Setup 7.5%

(b)

FIGURE 20.38   Automated video test instrument output charts: (*a*) waveform monitor mode display of color bars, (*b*) vectorscope mode display of color bars. (*Courtesy of Tektronix.*)

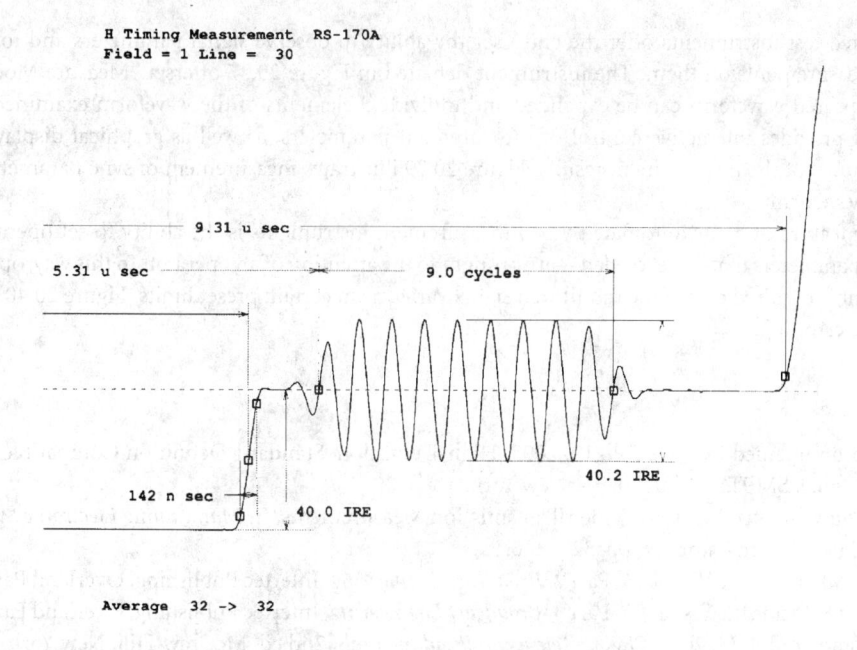

FIGURE 20.39 Expanded view of a sync waveform with measured parameters. (*Courtesy of Tektronix.*)

20.2.7 Automated Video Signal Measurement

Video test instruments based on computer systems provide the maintenance engineer with the ability to rapidly measure a number of parameters with exceptional accuracy. Automated instruments offer a number of benefits, including reduced setup time, test repeatability, waveform storage and transmission capability, and remote control of instrument/measurement functions. Typical features of this class of instrument include the following:

- Waveform monitor functions
- Vectorscope monitor functions
- Picture display capability
- Automatic analysis of an input signals
- RS-232 and/or GPIB I/O ports

Figure 20.37 shows a block diagram of a representative automated video test instrument. Sample output waveforms are shown in Figure 20.38.

			Violated Limits		
			Lower	Upper	
Bar Top	—— % Carr	**	10.0	15.0	Bar Not Found
Blanking Level	—— % Carr	**	72.5	77.5	ZC Pulse Unselected
Bar Amplitude	—— IRE	**	96.0	104.0	Bar Not Found
Sync Variation	—— % Carr	**	0.0	5.0	ZC Pulse Unselected
VIRS Setup	—— % Bar	**	5.0	10.0	Not Found
VIRS Luminance Ref	—— % Bar	**	45.0	55.0	Not Found
VIRS Chroma Ampl	—— % Burst	**	90.0	110.0	Not Found
VIRS Chroma Ampl	—— % Bar	**	36.0	44.0	Not Found
VIRS Chroma Phase	—— Deg	**	-10.0	10.0	Not Found
Line Time Distortion	—— %	**	0.0	2.0	No Composite VITS
Pulse/Bar Ratio	—— %	**	94.0	106.0	No Composite VITS
2T Pulse K-Factor	—— % Kf	**	0.0	2.5	No Composite VITS
IEEE-511 ST Dist	—— % SD	**	0.0	3.0	No Composite VITS

FIGURE 20.40 Example error log for a monitored signal. (*Courtesy of Tektronix.*)

Automated test instruments offer the end-user the ability to observe signal parameters and to make detailed measurements on them. The instrument depicted in Figure 20.37 offers a "Measure Mode," in which a captured waveform can be expanded and individual elements of the waveform examined. The instrument provides interactive control of measurement parameters, as well as graphical displays and digital readouts of the measurement results. Figure 20.39 illustrates measurement of sync parameters on an NTSC waveform.

Another feature of many automated video measurement instruments is the ability to set operational limits and parameters that—if exceeded—are brought to the attention of an operator. In this way, operator involvement is required only if the monitored signal varies from certain preset limits. Figure 20.40 shows an example error log.

References

SMPTE Recommended Practice RP 219. 2002. High-Definition, Standard-Definition Compatible Color Bar Signal. SMPTE, White Plains, New York.

Bentz, C. and Whitaker, J.C. 1991. Video Transmission Measurements, In *Maintaining Electronic Systems*, CRC Press, Boca Raton, FL pp. 328–346.

Bentz, C. 1988. Inside the Visual PA, Part 2. *Broadcast Engineering*, Intertec Publishing, Overland Park, KS.

Bentz, C. 1988. Inside the Visual PA, Part 3. *Broadcast Engineering*, Intertec Publishing, Overland Park, KS.

Robin, M. and Poulin, M. 2001. *Digital Television Fundamentals*, 2nd ed. McGraw-Hill, New York.

Further Information

SMPTE Engineering Guideline EG 1. 1990. Alignment Color Bar Test Signal for Television Picture Monitors. SMPTE, White Plains. New York.

Pank, B. ed. 1988. *The Digital Fact Book*, 9th ed. Quantel Ltd, Newbury, England.

20.3   Radio Frequency Distortion Mechanisms and Analysis

Samuel O. Agbo

20.3.1   Introduction

Radio frequency (RF) communication is usually understood to mean radio and television broadcasting, cellular radio communication, point-to-point radio communication, microwave radio, and other wireless radio communication. For such wireless radio communication, the communication channel is the atmosphere or free space. It is in this sense that RF distortion is treated in this section, although RF signals are also transmitted through other media such as coaxial cables.

Distortion refers to the corruption encountered by a signal during transmission or processing that modifies the signal waveform. A simple model of the atmospheric radio channel, as shown in Fig. 20.41, will help to illustrate the sources of radio frequency distortion. The channel is modeled as a linear time-variant system with additive noise and additive interference. In the figure, $s(t)$ is the RF signal to be transmitted through the channel, $h(t, \tau)$ is the channel impulse response, $n(t)$ is the additive noise, $i(t)$ is the interference, and $r(t)$ is the received signal. Sources of signal corruption, such as signal attenuation, amplitude distortion, phase distortion, multipath effects and time-varying channel characteristics, constitute the linear time-variant channel

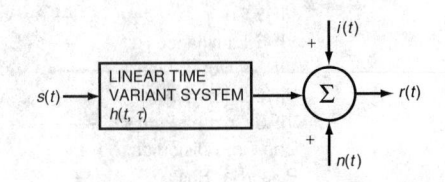

FIGURE 20.41   Model of the atmospheric radio channel as a linear time-variant system with additive noise and additive interference.

impulse response. Noise and interference are additive sources of signal corruption. Corruption by noise may result in a roughness of the output waveform, but the original signal waveform is usually discernible. This is one sense in which noise differs from the distortion mechanisms discussed in this section.

The impulse response $h(t, \tau)$ is the response of the channel at time t due to an impulse $\delta(t - \tau)$ applied at time $t - \tau$. Let $\alpha_k(t)$ represent the time-dependent attenuation factors for the m multipath components of an atmospheric radio channel. Then a simple representation of the impulse response is

$$h(t, \tau) = \sum_{k=1}^{m} \alpha_k(t)\delta(t - \tau_k) \tag{20.1}$$

The received signal, which is the convolution of this impulse response with the signal $s(t)$, plus the additive noise and interference acquired in the channel, is given by

$$r(t) = \sum_{k=1}^{m} \alpha_k s(t - \tau_k) + i(t) + n(t) \tag{20.2}$$

Radio frequency signals are processed in the transmitter, prior to transmission, and in the receiver, prior to demodulation, by electronic circuits. Such processing can be regarded in a generalized way as filtering. Distortion introduced by nonideal filters and the atmospheric radio channel will be discussed. Following this, the effects on demodulation outputs of frequency and phase distortion and interference on radio frequency signals will be discussed.

20.3.2 Types of Distortion

Linear Distortion

The processing of radio frequency signals in transmitter or receiver by electronic circuits can be viewed as forms of filtering. Such processing will introduce distortion unless the filter is ideal or distortionless within a band of frequencies equal to or exceeding the bandwidth of the input signal. A filter is distortionless if its input and output signals have identical waveforms, save for a constant amplitude gain or attenuation and a constant time delay for all of the frequency components. Thus, if $g(t)$ with a bandwidth W is the input signal into an ideal filter of bandwidth B, which is greater than W, the output signal $y(t)$ is given by

$$y(t) = \alpha g(t - \tau) \tag{20.3}$$

In Eq. (20.3) α is the constant gain or attenuation and τ is the time delay in passing through the system. With this definition, $H(f)$, the Fourier transforms of the ideal filter impulse response and $Y(f)$, the filter output are given by, respectively,

$$H(f) = \alpha e^{-j2\pi f\tau} \tag{20.4}$$

$$\begin{aligned} Y(f) &= G(f)H(f) \\ &= \alpha G(f)e^{-j2\pi f\tau} \end{aligned} \tag{20.5}$$

Figure 20.42 shows the frequency-domain representation of an ideal and a nonideal filter and the time-domain representation of the input and output signals. Figures 20.42(a), 20.42(b), and 20.42(c) show, respectively, the frequency response of the ideal filter, the input signal waveform, and the output signal waveform, which is a time delayed version of the input waveform. Both the filter response and the input signal are bandlimited to B Hz.

Figure 20.42(d) shows a nonideal filter frequency response with ideal phase but nonideal amplitude response, which is given by

$$|H(f)| = \begin{cases} \alpha \cos 2\pi f T, & |f| < B \\ 0; & |f| > B \end{cases} \tag{20.6}$$

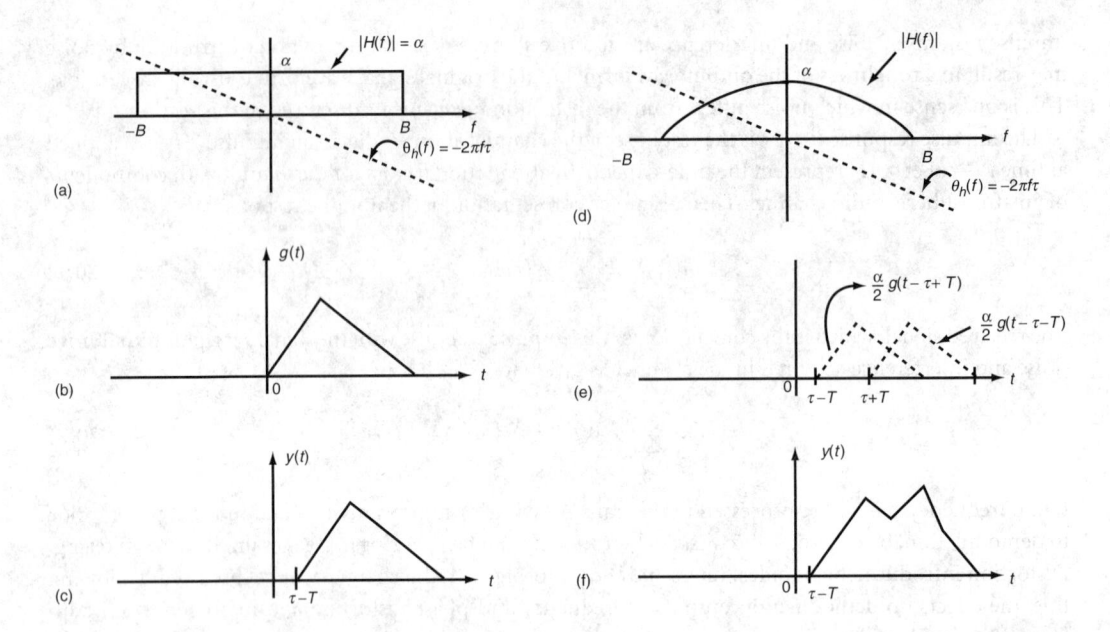

FIGURE 20.42 (a) Ideal filter frequency response, (b) waveform of input signal, (c) waveform of output signal from ideal filter, (d) nonideal filter frequency response with ideal phase but nonideal amplitude response, (e) output waveform of nonideal filter showing its components as time shifted versions of the input signal, (f) output waveform of nonideal filter: a dispersed version of the input signal.

By employing the time-shifting property of the Fourier transform, the time-domain output signal is obtained as

$$y(t) = \frac{\alpha}{2}(g(t - \tau - T) + g(t - \tau + T)) \qquad (20.7)$$

From Eq. 20.7, it is clear that the output signal consists of two scaled and time shifted versions of the input signal as shown in Fig. 20.42(e). The sum of these shifted versions shown in Fig. 20.42(f) indicates that linear distortion results in **dispersion** of the output signal. It is easy to see that nonideal phase response, which implies different delays at different frequencies, also leads to dispersion. Thus, linear distortion causes pulse spreading and interference with neighboring pulses. Consequently, linear distortion should be avoided in time-division multiplexing (TDM) systems where it causes **crosstalk** with neighboring channels. In frequency-division multiplexing (FDM), linear distortion corrupts the signal spectrum, but the pulse spreading does not result in crosstalk because the channels are adjacent in frequency, not in time.

The ideal filter of Fig. 20.42 is a low-pass filter. Bandpass and high-pass filters are more relevant for the processing of RF signals in the transmitter, prior to transmission, and in the receiver, prior to demodulation. For bandpass and high-pass filters, the condition for distortionless filtering is the same as previously indicated: the input signal bandwidth should not exceed the filter bandwidth, and the filter should have a constant gain and the same time delay for all frequencies within its bandwidth. Figure 20.43 illustrates the frequency responses for ideal bandpass and ideal high-pass filters.

Nonlinear Distortion

Nonlinearity in system response usually arises in cases involving large signal amplitudes. For a simple case of memoriless, nonlinear system or channel, with a low-pass input signal $g(t)$, of bandwidth W Hz, the output signal $y(t)$ is given by

$$y(t) = a_0 + a_1 g(t) + a_2 g^2(t) + a_3 g^3(t) + \cdots + a_k g^k(t) + \cdots \qquad (20.8)$$

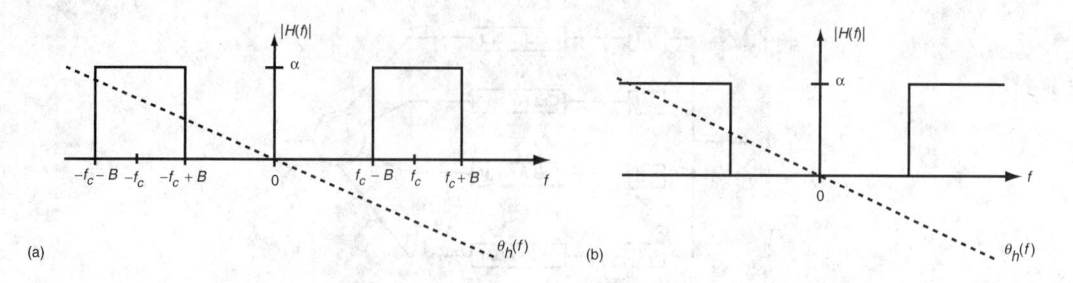

FIGURE 20.43 (a) Ideal bandpass filter, (b) high-pass filters.

Consider the term $a_k g^k(t)$ in the equation. Let $G(f)$ be the Fourier transform of $g(t)$. Then the Fourier transform of $g^k(t)$ is a $k - 1$ fold convolution of $G(f)$. Thus, the bandwidth of $g^k(t)$ is $k - 1$ times the bandwidth of $g(t)$. Suppose m is the last significant term of the power series of the equation, then the bandwidth of the output signal is $(m - 1)W$ Hz. Thus, the bandwidth of the output signal will far exceed the bandwidth of the input signal.

Figure 20.44 illustrates the spectrum of the input and the output signals. If the input signal is a bandpass signal of bandwidth W and carrier frequency f_c, which would be appropriate for a radio frequency signal, the spectrum of the output signal would be much more complex. In this case, the kth term of the power series of the output signal $y(t)$ will, in general, contribute a component of bandwidth $(k - 1)W$ centered at kf_c, in addition to smaller bandwidth components centered at lower harmonics of f_c.

The preceding discussion shows that nonlinear distortion causes spreading in the frequency domain. Such spreading of the spectrum causes interference in FDM systems because it results in frequency-domain overlap of neighboring channels. However, nonlinear distortion does not cause adjacent channel interference in TDM as the channels in that case are not adjacent in frequency, but in time.

Distortion Due to Time-Variant Multipath Channels

Radio waves from a transmitting antenna often reach the receiving antenna through many different paths. Examples include different reflecting layers in the **ionosphere**, numerous scattering points in the **troposphere**, reflections from the Earth's surface, and the direct line-of-sight path. Each of these paths will contribute a different attenuation and time delay. Shown in Fig. 20.45 is a model of the time-variant multipath radio channel. For the sake of simplicity, the effect of noise has been ignored.

This channel is time dispersive because the different time delays entail a spread in time of the composite received signal relative to the transmitted signal. In addition, because of variations with time and locality in weather, temperature and other atmospheric conditions, the time delay and attenuation of these paths vary with time. Because these variations in channel characteristics are generally unpredictable, a statistical characterization of the time-variant multipath channel is appropriate.

To simplify the analysis, consider the transmission of an unmodulated carrier of unit amplitude, $s(t) = \cos 2\pi f_c t$, through an m multipath channel in the absence of noise. Let $\alpha_k(t)$ be the attenuation and $\tau_k(t)$

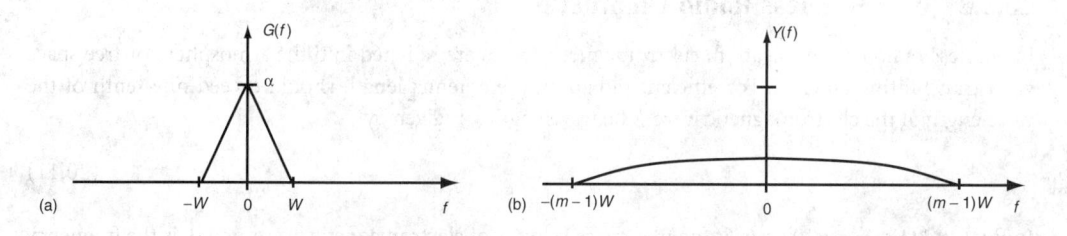

FIGURE 20.44 Input and output spectrum for a channel with nonlinear distortion: (a) amplitude spectrum of output signal, (b) amplitude spectrum of input signal.

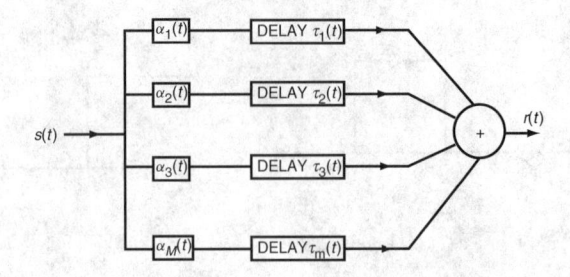

FIGURE 20.45 Model of a noiseless, time-variant, multipath channel.

be time delay contributed by the kth path at the receiver. The received signal is given by

$$r(t) = \sum_{k=1}^{m} \alpha_k(t) \cos[2\pi f_c(t - \tau_k(t))]$$

$$= Re\left[\sum_{k=1}^{m} \alpha_k(t) e^{j2\pi f_c t} e^{-j2\pi f_c \tau_k(t)}\right] \qquad (20.9)$$

$$= [\cos 2\pi f_c t] Re\left[\sum_{k=1}^{m} \alpha_k e^{-j2\pi f_c \tau_k(t)}\right]$$

It is easy to see that the response of the channel to the unmodulated carrier is

$$h(t, \tau) = Re\left[\sum_{k=1}^{m} \alpha_k(t) e^{-j2\pi f_c \tau_k(t)}\right] \qquad (20.10)$$

Note that although the channel input was an unmodulated carrier, the channel output contains a frequency spread around the carrier, resulting from the time-varying delays of the various paths. The bandwidth of this output signal is a measure of how rapidly the channel characteristics are changing. A large change in channel characteristics is required to produce a significant change in the time-varying attenuation $\alpha_k(t)$. Because f_c is a large number, a change of 2π in the phase of the output signal $-2\pi f_c \tau_k(t)$ corresponds to only a change of $1/f_c$ in the time delay $\tau_k(t)$. Thus, the attenuation changes slowly, while the phase (or frequency deviation) changes rapidly.

Because these changes are random in nature, the channel impulse response can be modeled as a random process. Constructive addition of the phases of the various paths result in a strong signal, whereas destructive addition of the phases results in a weak signal. Thus, the multipath propagation results in signal fading, primarily due to the time-variant phase changes. When m, the number of multiple propagation paths, is large, the central limit theorem applies and the channel response will approximate a Gaussian random process.

20.3.3 The Wireless Radio Channel

In wireless radio communication, electromagnetic waves are radiated into the atmosphere or free space via a transmitting antenna. For efficient radiation, the antenna length should exceed one-tenth of the wavelength of the electromagnetic wave. The wavelength λ is given by

$$\lambda = \frac{c}{f} \qquad (20.11)$$

In Eq. (20.11) $c = 3 * 10^8$ m/s, is the free-space velocity of electromagnetic waves and f is the frequency in Hz. Wireless radio communications covers a wide range of frequencies from about 10 kHz in the very low frequency (VLF) band to above 300 GHz in the extra high frequency (EHF) band. Depending on the

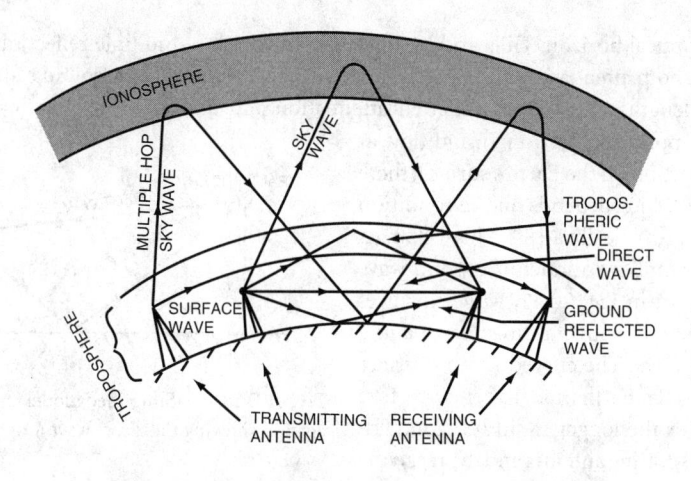

FIGURE 20.46 Wireless radio wave propagation paths.

frequency band and the distance between the transmitter and the receiver, the transmitted wave may reach the receiving antenna by one or a combination of several propagation paths, as shown in Fig. 20.46.

Sky waves are those that reach the receiver after being reflected from the ionosphere, a band of reflecting layers ranging in altitude from 100 to 300 km. Tropospheric waves are those that reach the receiver after being reflected (scattered) by inhomogeneities within the troposphere, the region within 10 km of the Earth's surface. Other waves reach the receiving antenna via ground waves, which consists of space waves and surface waves. Surface waves (in the LF and VLF ranges) have wavelengths comparable to the altitude of the ionosphere. The ionosphere and the ground surface act as guiding planes, so that surface waves follow the curvature of the Earth, permitting transmission over great distances. Space waves consist of direct (line-of-sight) waves and ground reflected waves, which arrive at the receiver after being reflected from the ground. A classification of radio frequency communication by frequency bands, typical uses, and electromagnetic wave propagation modes is given in Table 20.1.

Ground Waves

In the region adjacent to the Earth's surface, free-space conditions are modified by the surface topography, physical structures such as buildings, and the atmosphere. One effect of the atmosphere is to cause a refraction of the wave, which carries it beyond the optical horizon. Consequently, the **radio horizon** is

TABLE 20.1 Classification of Wireless Radio Frequency Channels by Frequency Bands, Typical Uses, and Wave Propagation Modes

Frequency Band	Name	Typical Uses	Propagation Mode
3–30 kHz	Very low frequency (VLF)	Navigation, sonar	Ground waves
30–300 kHz	Low frequency (LF)	Navigation, telephony, telegraphy	Ground waves
0.3–3 MHz	Medium frequency (MF)	AM Broadcasting, amateur, and CB radio	Sky waves and ground waves
3–30 MHz	High frequency (HF)	Mobile, amateur, CB, and military radio	Sky waves
30–300 MHz	Very high frequency (VHF)	VHF TV, FM Broadcasting Police, air traffic control	Sky waves, tropospheric waves
0.3–3 GHz	Ultra high frequency (UHF)	UHF TV, radar, satellite communication	Tropospheric waves, space waves
3–30 GHz	Super high frequency (SHF)	Space and satellite, radar microwave relay	Direct waves, ionospheric penetration waves
30–300 GHz	Extra high frequency (EHF)	Experimental, radar radio astronomy	Direct waves, ionospheric penetration waves

longer than the optical horizon. Hills and building structures cause multiple reflections and shieldings, which contribute both attenuation and phase change to the received signal. Ground reflected waves travel a different path length and undergo different attenuation and time delay relative to direct waves and depending on the point of reflection. In addition, at each point of reflection on the Earth's surface, there is a phase change, which depends on the condition of the surface, but is typically a 180° phase change.

Consider the simplified model of ground wave propagation, shown in Fig. 20.47, which involves only two transmission paths: a direct wave and a ground reflected wave. The curved Earth's surface is represented by a flat Earth model after correction has been made for the longer radio horizon. The height of the transmitting antenna and the receiving antenna are h_T and h_r, respectively.

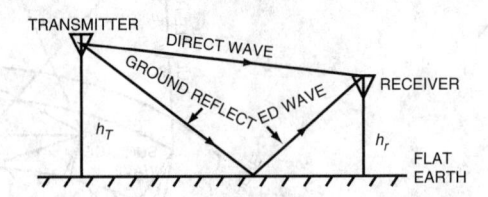

FIGURE 20.47 Simplified model for ground wave propagation showing the direct wave and the ground reflected wave.

The direct wave travels a path length x from the transmitting antenna to the receiving antenna. The attenuation factor for the atmospheric medium is α. Let the corresponding distance traveled by the ground reflected wave be $x + \Delta x$ and the transmitted signal be $s(t)$. Note that a path length equal to the wavelength λ corresponds to a phase change of 2π. The received direct wave $r_d(t)$, the received ground reflected wave $r_g(t)$, and the composite received wave $r(t)$ are given, respectively, by

$$r_d(t) = e^{-\alpha x} s(t) \cos\left(\frac{2\pi x}{\lambda}\right) \tag{20.12}$$

$$r_g(t) = e^{-\alpha(x+\Delta x)} s(t) \cos\left[\frac{2\pi}{\lambda}(x + \Delta x) - \pi\right] \tag{20.13}$$

$$r(t) = r_d(t) + r_g(t) \tag{20.14}$$

From the preceding equations, it is seen that even for small differences in path lengths, the received signals from the direct and the ground reflected waves tend to have large phase differences. Consequently, the composite received signal has significant attenuation and phase error relative to the transmitted signal.

Tropospheric Waves

Waves propagating through the troposphere are only seriously attenuated by adverse atmospheric conditions at frequencies above 10 GHz. Above 10 GHz, heavy rain causes severe attenuation. However, light rain, clouds, and fog cause serious attenuation only at frequencies above 30 GHz. Attenuation due to snow is negligible at all frequencies. On the other hand, the inhomogeneities in the atmosphere due to the weather and other conditions are favorable for tropospheric scatter propagation of radio waves within the 30 MHz–4 GHz frequency band. Within this range of frequencies, the inhomogeneities within the troposphere form effective point scatterers that deflect the transmitted waves downward toward the receiving antenna. By employing large transmitted power and highly directional antennas, long transmission ranges of up to 400 mi can be obtained.

The range of frequencies favored for tropospheric propagation is essentially the domain of microwave transmission, UHF, and VHF radio. Although atmospheric attenuation is higher for frequencies above 10 GHz, it is still quite significant within the 30 MHz–4 GHz range of frequencies. Hence, microwave repeater stations are spaced about 30 mi apart. This spacing is long compared to the spacing of about 3 mi for repeaters for coaxial cable systems. However, such spacing is relatively close, compared with the more recent development of optical fiber transmission in which repeater spacing of above 300 km is practical.

Sky Waves

The ionosphere consists of several layers containing ionized gases at altitudes of 100–300 km. In order of increasing height, these are the C, D, E, F_1, and F_2 layers. The ionization is caused by solar radiation, mainly due to ultraviolet and cosmic rays. Consequently, the electron density in the layers varies with the seasons and with the time of day and night. The gas molecules in the denser, lower layers cause more collisions and hence a higher recombination rate of electrons and ions. At night, the reduced level of radiation and the greater recombination rates cause the lower layers to disappear, so that only the F_2 layer remains.

Electromagnetic wave propagation through the ionosphere is by means of sky waves, mainly in the MF, HF, and VHF bands. The waves are successively refracted at the layers until they are eventually reflected back toward the Earth. At higher frequencies, the waves pass through the ionosphere without being reflected from it. Communication at these higher frequencies depends on a line-of-sight basis. During the daytime, ionospheric propagation is severely attenuated at frequencies below 2 MHz, due to severe absorption in the lower layers, especially the D layer. At night, after the disappearance of these lower layers, powerful AM transmission utilizing the F_2 layer is possible over the range of 100–250 mi. Of course, the maximum transmission range for sky waves depend on the frequency of transmission.

Figure 20.48 shows the geometry relevant to the determination of the range for ionospheric propagation. Because the altitudes involved are much greater than the antenna heights, the latter have been ignored. Rays from the transmitting antenna at angles of elevation greater than the critical angle of elevation for the frequency of transmission result in escape rays, which are not reflected back from the ionosphere. The ray leaving the transmitter at the critical angle, the critical ray, which is reflected from the lowest possible layer, returns to the Earth closest to the transmitter than any other ray, giving the minimum range or the skip zone. The distance between this minimum range and the range of the glancing ray (which is tangential to the Earth at the receiver) is the service range.

The greatest transmission range is obtained when the critical ray is reflected from a layer at so high an altitude that it is also tangential to the Earth's surface at the receiver. The geometry for this case is shown in Fig. 20.48(b) in which the raypaths have been idealized as straight lines, ignoring the curvature of the paths due to refraction. The height of the virtual reflecting point above the Earth's surface is h, and R is the radius of the Earth. The angle ϕ and the maximum range d are given, respectively, by

$$\phi = \cos^{-1}\left(\frac{R}{R+h}\right) \tag{20.15}$$

$$d = 2R\cos^{-1}\left(\frac{R}{R+h}\right) \tag{20.16}$$

This computation is for a single-hop reflection. For multiple hops, the transmission range is considerably greater, although the signal strength would be correspondingly reduced. The condition of the ionosphere within a locality greatly affects the critical angle of elevation and the transmission path. Conditions vary

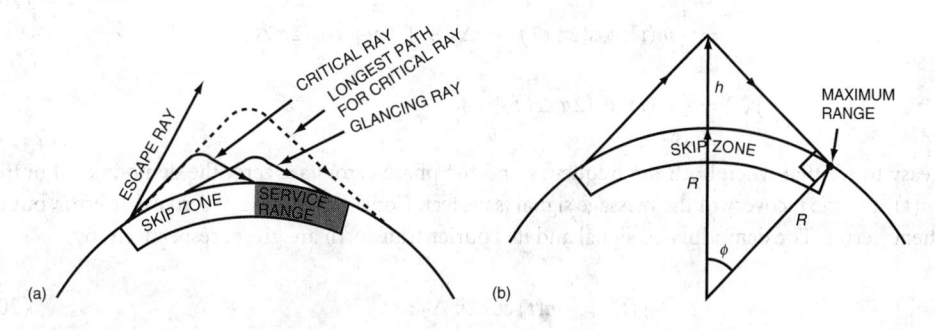

FIGURE 20.48 Geometry of ionospheric single hop: (a) raypaths, skip zone, and service range, (b) geometry for computation of maximum range.

considerably with time and locality within the ionosphere. Such irregularities within the ionosphere can be used for scatter propagation, much like tropospheric scatter propagation, thereby resulting in increased transmission range of up to 1000 mi. However, the undesirable effects of ionospheric inhomogeneities far exceed this favorable effect.

Ionospheric irregularities result from differences in electron densities in the various layers. The irregularities can be quite large in spatial dimensions, ranging from tens of meters to several hundreds of kilometers. Some of these irregularities travel through the ionosphere at speeds in excess of 3000 km/h. These time-varying irregularities are responsible for the time-variant multipath effects and the resultant fading that plagues ionospheric transmission.

20.3.4 Effects of Phase and Frequency Errors in Coherent Demodulation of AM Signals

RF signals acquire phase and amplitude distortion in the wireless radio channel and in processing circuits, through the many mechanisms previously discussed. For coherent (or synchronous) detection, a local carrier, which matches the transmitted signal in phase and frequency, is required at the receiver. In general, such a match is hard to obtain. These mismatch errors are additional to the phase and amplitude errors acquired in the channel.

Double-Sideband Suppressed-Carrier (DSB-SC) Demodulation Errors

The receiver for the synchronous demodulation of double-sideband suppressed-carrier (DSB-SC) waves is shown in Fig. 20.49. In general, it is easy to correct the effect of attenuation or amplitude errors through amplification, filtering, or other processing in the receiver. It is more difficult to correct the effect of phase or frequency errors.

The received signal at the input to the receiver and the local carrier are given, respectively, by

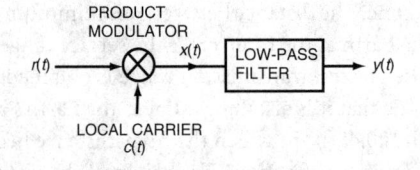

FIGURE 20.49   Receiver for synchronous demodulation of DSB-SC signals.

$$r(t) = m(t) \cos 2\pi f_c t \tag{20.17}$$

$$c(t) = \cos(2\pi(f_c + \Delta f)t + \theta) \tag{20.18}$$

In the last two equations, $m(t)$ is the message signal of bandwidth W, where f_c is the carrier frequency, Δf is the frequency error, and θ is the phase error. The output of the product modulator is $x(t)$. Thus, $x(t)$ and $y(t)$ are given, respectively, by

$$x(t) = m(t) \cos(2\pi f_c t) \cos(2\pi(f_c + \Delta f)t + \theta)$$
$$= \frac{1}{2} m(t) \{ \cos(2\pi(2f_c + \Delta f)t + \theta) + \cos(2\pi \Delta f t + \theta) \} \tag{20.19}$$

$$y(t) = \frac{1}{2} m(t) \cos(2\pi \Delta f t + \theta) \tag{20.20}$$

It is easy to see that when both the frequency and the phase errors are zero, the demodulated output is $1/2m(t)$, and the recovery of the message signal is perfect. Consider the case of zero phase error, but finite frequency error. The demodulated signal and its Fourier transform are given, respectively, by

$$y(t) = \frac{1}{2} m(t) \cos 2\pi \Delta f t \tag{20.21}$$

$$Y(f) = \frac{1}{4}(M(f + \Delta f) + M(f - \Delta f)) \tag{20.22}$$

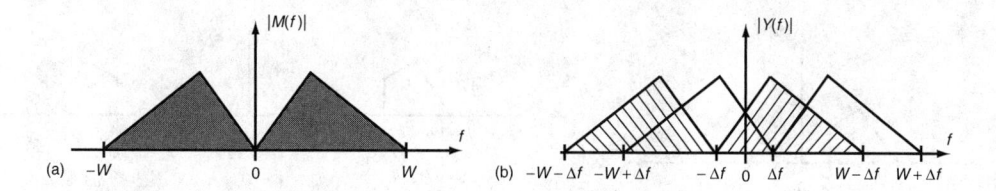

FIGURE 20.50 Frequency spectrum of message signal and demodulated output when there is a Δf frequency error between the local carrier and the carrier in the received DSB-SC signal: (a) message signal spectrum, (b) spectrum of demodulated output signal.

Thus, the demodulated output is a replica of the message signal multiplied by a slowly varying cosine wave of frequency Δf. This is a serious type of distortion, which produces a beating effect. The human ear can tolerate at most about 30 Hz of such frequency drift. Figure 20.50 shows the frequency-domain illustration of this type of distortion. Figure 20.50(a) shows the amplitude spectrum of the original message signal. Figure 20.50(b) shows that the demodulated signal can be viewed as a DSB-SC signal whose carrier is the frequency error Δf. This error is so small that the replicas of the message signal shifted by $+\Delta f$ and $-\Delta f$ overlap in frequency and, therefore, interfere, giving rise to the beating effect.

Next, consider the case in which the frequency error is zero, but the phase error is finite. The demodulated signal is given by

$$y(t) = \frac{1}{2}m(t) \cos \theta \tag{20.23}$$

If θ, the phase error is constant, then it contributes only a constant attenuation, which is easily corrected for through amplification. However, in the extreme case in which $\theta = \pm\pi/2$, the demodulated signal amplitude is reduced to zero. On the other hand, if θ varies randomly with time, as is the case for multipath fading channels, the attenuation induced by the phase error varies randomly and represents a serious problem.

To avoid the undesirable effects of frequency and phase errors between the local and the received carriers, various schemes are employed. One such scheme employs a **pilot carrier** at the transmitter to provide a source of matching carrier at the receiver. Some other schemes employed for providing matching carriers are the *Costas receiver* and signal squaring.

Single-Sideband Suppressed-Carrier (SSB-SC) Demodulation Errors

The receiver for the demodulation of single-sideband suppressed-carrier (SSB-SC) signals is similar to the receiver for DSB-SC signals shown in Fig. 20.49. The local carrier is the same as for the DSB-SC signal, but the received signal is given by

$$r(t) = m(t) \cos 2\pi f_c t \mp m_h(t) \sin 2\pi f_c t \tag{20.24}$$

In Eq. 20.24, the negative $(-)$ sign applies to the upper-sideband (USB) case, whereas the positive sign $(+)$ applies to the lower-sideband (LSB) case. The signal, $m_h(t)$, is the *Hilbert transform* of the message signal $m(t)$. The output of the product modulator is

$$x(t) = (m(t) \cos 2\pi f_c t \mp m_h(t) \sin 2\pi f_c t) \cos(2\pi(f_c + \Delta f)t + \theta)$$

$$= \frac{1}{2}m(t)\{\cos(2\pi \Delta f t + \theta) + \cos(2\pi(2f_c + \Delta f)t + \theta)\} \tag{20.25}$$

$$\pm \frac{1}{2}m_h(t)\{\sin(2\pi \Delta f t + \theta) - \sin(2\pi(2f_c + \Delta f)t + \theta)\}$$

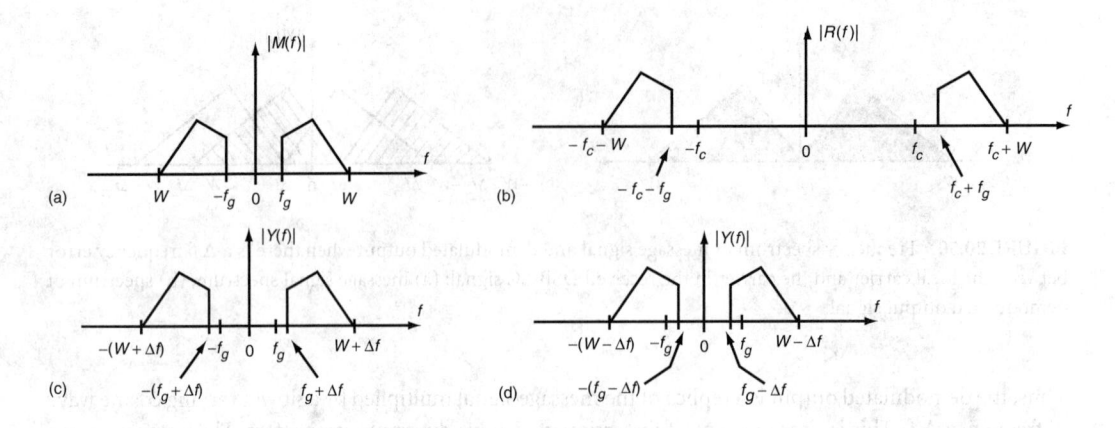

FIGURE 20.51 An illustration of the effect of frequency error in synchronous demodulation of SSB-SC signals: (a) amplitude spectrum of message signal, (b) amplitude spectrum of SSB (USB) signal, (c) amplitude spectrum of demodulated USB signal when $\Delta f < 0$, (d) amplitude spectrum of demodulated USB signal when $\Delta f > 0$.

The low-pass filter suppresses the bandpass components of the product modulator output. Hence, the demodulated output $y(t)$ is given by

$$y(t) = \frac{1}{2}m(t)\cos(2\pi\,\Delta ft + \theta) \pm \frac{1}{2}m_h(t)\sin(2\pi\,\Delta ft + \theta) \tag{20.26}$$

Note that when both the frequency and the phase errors are set to zero in the equation, the demodulator output signal is $1/2m(t)$, an exact replica of the desired message signal. Consider the case of zero phase error, but nonzero frequency error. In this case, the demodulated output signal $y(t)$ can be expressed as

$$y(t) = \frac{1}{2}m(t)\cos 2\pi\,\Delta ft \pm \frac{1}{2}m_h(t)\sin 2\pi\,\Delta ft \tag{20.27}$$

This equation shows that when only frequency error is present between the received and local carriers, the demodulated signal is essentially an SSB-SC signal in which the small frequency error Δf plays the role of the carrier. A comparison of this equation with the expression for the SSB-SC signal shows that if the modulated signal is a USB signal, the demodulated output is and LSB signal, and vice versa.

Figure 20.51 shows the spectra of the message signal, the SSB-SC modulated signal (USB case), and two cases of the demodulated signal when only frequency errors are present. Figure 20.51(c) shows the spectrum of the demodulated output of the USB signal when the frequency error is negative. Figure 20.51(e) shows the spectrum of the demodulated output of the USB signal when the frequency error is positive.

By comparing the demodulated output of an SSB-SC signal with the message signal, it is easy to see that in each case, each frequency component in the demodulated output is shifted by Δf. This does not introduce as serious a distortion as in the DSB-SC carrier case in which each frequency component is shifted by $-\Delta f$ and $+\Delta f$. As shown in Fig. 20.51, $M(f)$, the spectrum of the message signal is zero in the frequency gap from zero to f_g, as should be the case for a well-processed SSB message. By ensuring that f_g exceeds the maximum possible value of Δf, the sidebands for the demodulated output of an SSB-SC wave will never overlap in frequency or interfere with each other as in the DSB-SC case.

For the case of USB modulation in which the frequency error is zero but the phase error is nonzero, the demodulated output is obtained by setting Δf equal to zero in Eq. (20.26). This gives $y(t)$, the demodulated output signal and its frequency domain representation $Y(f)$, respectively, as

$$y(t) = \frac{1}{2}(m(t)\cos\theta + m_h(t)\sin\theta) \tag{20.28}$$

$$Y(f) = \frac{1}{2}(M(f)\cos\theta + M(f)\sin\theta) \tag{20.29}$$

But the frequency domain representation of the Hilbert transform of the message signal, $m_h(f)$, can be expressed as

$$M_h(f) = -j \operatorname{sgn}(f)M(f)$$
$$= \begin{cases} -jM(f); & f > 0 \\ jM(f); & f < 0 \end{cases} \tag{20.30}$$

Thus, $Y(f)$ can be expressed as

$$Y(f) = \begin{cases} \dfrac{1}{4}\left[M(f)e^{-j\theta} - jM(f)\left(\dfrac{-e^{-j\theta}}{j}\right)\right]; & f > 0 \\[3mm] \dfrac{1}{4}\left[M(f)e^{j\theta} + jM(f)\left(\dfrac{e^{j\theta}}{j}\right)\right]; & f < 0 \end{cases}$$

$$= \begin{cases} \dfrac{1}{2}M(f)e^{-j\theta}; & f > 0 \\[3mm] \dfrac{1}{2}M(f)e^{j\theta}; & f < 0 \end{cases} \tag{20.31}$$

From the preceding equation, it is clear that each positive frequency component in the demodulated output signal undergoes a phase shift of $-\theta$, while each negative frequency component undergoes a phase shift of $+\theta$. Thus, phase errors in SSB-SC signals results in phase distortion in the demodulated outputs. For voice signals, the human ear is relatively insensitive to phase distortion. However, this type of phase distortion, which produces a constant phase shift at all message signal frequencies, is quite objectionable in music signals and intolerable in video signals.

20.3.5 Effects of Linear and Nonlinear Distortion in Demodulation of Angle Modulated Waves

The amplitude of an angle modulated wave is supposed to be constant while the information being transmitted is contained in the phase. Linear distortion can be introduced by the channel, non-ideal filters, and other sub-circuits. In general, linear distortion results in time-variant amplitude and phase errors in the angle modulated wave. In effect, the angle modulated wave now contains both amplitude and phase modulation. Nonlinear distortion usually results from channel or power amplifier nonlinearities. Amplitude nonlinearities result in undesirable (or residual) amplitude modulation (AM), whereas phase nonlinearities result in phase distortion of the angle modulated wave. The effect of undesirable amplitude modulation resulting from linear and nonlinear distortion are discussed under separate subheadings. The effects of the time-variant phase errors resulting from linear distortion and from phase nonlinearities are discussed together under phase distortion in angle modulation.

Amplitude Modulation Effects of Linear Distortion

In an angle modulated waveform, the information resides in the phase of the modulated wave. Such an angle modulated wave can be demodulated by first passing it through a circuit whose amplitude response is proportional to the frequency of the input signal. The resulting intermediate signal has both amplitude modulation and angle modulation, and the information also resides in its amplitude. The demodulation process is then completed by passing this intermediate signal through an envelope detector.

A circuit which can implement this process is shown in Fig. 20.52. The differentiator input signal is the angle modulated wave. The amplitude of its output is proportional to the signal frequency, as shown in Fig. 20.52(b). Thus, the differentiator output is an amplitude as well as a frequency modulated waveform. The envelope detector output contains a replica of the message signal. Consider the case in which the

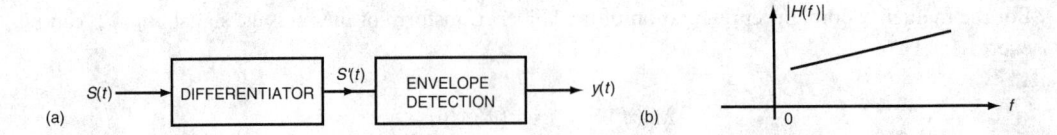

FIGURE 20.52 FM demodulator employing a differentiator and an envelope detector: (a) FM demodulator, (b) amplitude response of an ideal differentiator.

angle modulation involved is frequency modulation (FM). Suppose the FM signal contains no amplitude or phase errors, its amplitude will be constant. For simplicity, it can be assumed to be of unit amplitude. Let f_c be the carrier frequency, k_f be the frequency sensitivity of the modulator, and $m(t)$ be the message signal. Then the FM signal $s(t)$, the differentiator output $s'(t)$, and the envelop detector output $y(t)$, are given, respectively, by

$$s(t) = \cos\left[2\pi f_c t + 2\pi k_f \int_{-\infty}^{t} m(\lambda)d\lambda\right] \tag{20.32}$$

$$s'(t) = -(2\pi f_c + 2\pi k_f m(t)) \cos\left[2\pi f_c t + 2\pi k_f \int_{-\infty}^{t} m(\lambda)d\lambda\right] \tag{20.33}$$

$$y(t) = 2\pi f_c + 2\pi k_f m(t) \tag{20.34}$$

It is easy to see that an exact replica of the message signal can be recovered from the envelope detector output.

Consider the case in which the FM signal contains a time variation in its amplitude due to the channel, nonideal filter, or other system effects. Let $a(t)$ represent the time variation in the amplitude relative to the unit carrier amplitude. Then the FM signal and the output of the differentiator are given, respectively, by

$$s(t) = a(t) \cos\left[2\pi f_c t + 2\pi k_f \int_{-\infty}^{t} m(\lambda)d\lambda\right] \tag{20.35}$$

$$s'(t) = -a(t)(2\pi f_c + 2\pi k_f m(t)) \sin\left[2\pi f_c t + 2\pi k_f \int_{-\infty}^{t} m(\lambda)d\lambda\right]$$

$$+ a'(t) \cos\left[2\pi f_c t + 2\pi k_f \int_{-\infty}^{t} m(\lambda)d\lambda\right] \tag{20.36}$$

The output of the envelope detector, $y(t)$, is the envelope of the differentiator output given in the last equation. Thus, the envelope detector output is

$$y(t) = \left\{a^2(t)(2\pi f_c + 2\pi k_f m(t))^2 + (a'(t))^2\right\}^{\frac{1}{2}} \tag{20.37}$$

From the last equation it is easy to see that because of the time variation in the amplitude of the FM signal, it is not possible to extract an undistorted replica of the message signal from the envelope detector output. Even if the second term in Eq. (20.36) is ignored, the message signal is still multiplied by the time-variant amplitude. Thus, it is necessary to rid the FM wave of variations in its amplitude prior to demodulation. This is usually accomplished by passing the FM signal through a bandpass limiter, such as is illustrated in Fig. 20.53, prior to demodulation.

The FM wave containing time-variant amplitude errors can be expressed as

$$s(t) = a(t) \cos(\phi(t)) \tag{20.38}$$

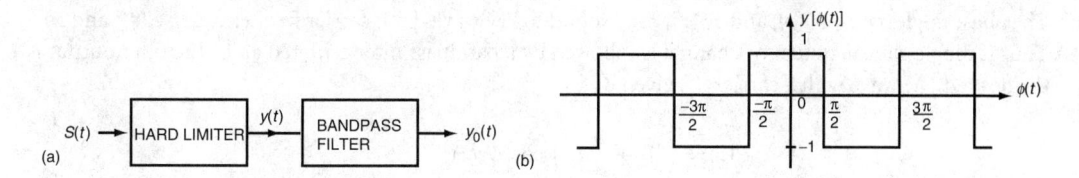

FIGURE 20.53 The bandpass limiter and an illustration of the output waveform of the hard limiter: (a) bandpass limiter, (b) output waveform of the hard limiter.

Its phase $\phi(t)$ is given by

$$\phi(t) = 2\pi f_c t + 2\pi k_f \int_{-\infty}^{t} m(\lambda) d\lambda \qquad (20.39)$$

The hard limiter is a device with a voltage transfer characteristic such that if its input is $s(\phi)$, then its output, $y(\phi)$ is given by

$$y(\phi) = \begin{cases} 1, & \phi > 0 \\ -1, & \phi < 0 \end{cases} \qquad (20.40)$$

When the input to the hard limiter is the FM signal, $a(t)\cos(\phi(t))$, the output of the hard limiter is the square waveform shown in Fig. 20.53(b). This periodic square waveform has a Fourier series expression given by

$$y(\phi(t)) = \frac{4}{\pi} \left[\cos(\phi(t)) - \frac{1}{3}\cos(3\phi(t)) + \frac{1}{5}\cos(5\phi(t)) + \cdots \right] \qquad (20.41)$$

Note that in the last equation, the terms containing $\phi(t), 3\phi(t), 5\phi(t), \ldots$, correspond to FM waves with carrier frequencies $f_c, 3f_c, 5f_c, \ldots$. Hence, the bandpass filter centered at f_c selects only the desired FM signal with carrier frequency f_c. Note that the output of the bandpass limiter, which is the output of the bandpass filter, has a constant amplitude given by

$$y_0(t) = \frac{4}{\pi}\cos\left[2\pi f_c t + 2\pi k_f \int_{-\infty}^{t} m(\lambda) d\lambda \right] \qquad (20.42)$$

Effects of Amplitude Nonlinearities on Angle Modulated Waves

Amplitude nonlinearities can arise from nonlinearities in the channel or from nonlinearities in processing circuits such as power amplifiers. Consider a channel or system with amplitude nonlinearities, whose input signal $x(t)$ and output signal $y(t)$ are related as in the following equation:

$$y(t) = a_0 + a_1 x(t) + a_2 x^2(t) + a_3 x^3(t) + \cdots \qquad (20.43)$$

If the input signal $s(t)$ is the FM wave given by

$$s(t) = \cos(\phi(t)) = \cos\left[2\pi f_c t + 2\pi k_f \int_{-\infty}^{t} m(\lambda) d\lambda \right] \qquad (20.44)$$

then the output signal $y(t)$ is given by

$$\begin{aligned}
y(t) &= a_0 + a_1 \cos(\phi(t)) + a_2 \cos^2(\phi(t)) + a_3 \cos^3(\phi(t)) + \cdots \\
&= a_0 + a_1 \cos(\phi(t)) + \frac{1}{2}a_2(1 + \cos 2\phi(t)) + \frac{1}{2}a_3 \left\{ \cos(\phi(t)) + \frac{1}{2}\cos(\phi(t)) + \frac{1}{2}\cos(3\phi(t)) \right\} + \cdots \\
&= \left(a_0 + \frac{1}{2}a_2 \right) + \left(a_1 + \frac{3}{4}a_3 \right) \cos(\phi(t)) + \frac{1}{2}a_2 \cos(2\phi(t)) + \frac{1}{4}\cos(3\phi(t)) + \cdots \qquad (20.45)
\end{aligned}$$

The phase angles $\phi(t)$, $2\phi(t)$, and $3\phi(t)$, correspond to FM waves with carrier frequencies f_c, $2f_c$, and $3f_c$. Thus if the nonlinear system or channel is followed by a bandpass filter centered at f_c, the filter output is the desired FM wave with a carrier f_c, given by

$$y_0(t) = \left(a_1 + \frac{3}{4}a_3 \right) \cos(\phi(t))$$

$$= \alpha \cos \left[2\pi f_c t + 2\pi k_f \int_{-\infty}^{t} m(\lambda)d\lambda \right]$$

(20.46)

Let B_{FM} be the transmission bandwidth of the FM signal, W the bandwidth of the message signal and Δf the frequency deviation. The FM bandwidth is given for narrowband FM (NBFM) and for wideband FM (WBFM) by

$$B_{\text{FM}} = \begin{cases} 2(\Delta f + W), & \text{for NBFM} \\ 2(\Delta f + 2W), & \text{for WBFM} \end{cases}$$

(20.47)

To ensure that the FM wave with a carrier frequency of $2f_c$ does not interfere with the desired FM signal of carrier frequency f_c, the following condition must be satisfied:

$$f_c + \frac{1}{2}B_{\text{FM}} \leq 2\left(f_c - \frac{1}{2}B_{\text{FM}} \right)$$

(20.48)

This condition is equivalent to ensuring that $f_c \geq 1.5B_{\text{FM}}$. By substituting the expressions for the bandwidths for NBFM and WBFM into the last equation, it is seen that the carrier frequency must satisfy the condition

$$f_c > \begin{cases} 3(\Delta f + W) & \text{for NBFM} \\ 3(\Delta f + 2W) & \text{for WBFM} \end{cases}$$

(20.49)

Note that in the case discussed earlier in which the nonlinearities are not time variant, the effect of passing an FM wave through a nonlinear channel or system, followed by a bandpass filter centered at the carrier frequency, is to produce a gain or attenuation in the constant amplitude. If the nonlinearities are time varying, so that the coefficients in the expression relating the channel input to the output are $a_0(t), a_1(t), a_2(t), \ldots$, a substitution of these time-varying coefficients into the preceding equation shows that the bandpass filter output will be a time-variant amplitude version of the FM signal at the correct carrier frequency. As before, such amplitude variations can be removed with a bandpass limiter, prior to demodulation. Thus, unlike amplitude modulation, FM is fairly tolerant of channel nonlinearities. Hence, FM is widely used in microwave and satellite communication systems where the nonlinearities of the power amplifiers need to be well tolerated.

Phase Distortion in Angle Modulation

Angle modulation systems are very sensitive to phase nonlinearities because in such systems the information resides in the phase of the modulated signal. Phase distortion of the modulated wave eventually results in distortion of the demodulated output. Phase distortion can result from phase nonlinearities of channel, repeater, filter, or amplifier responses. It can also result from nonideal linear filtering of an angle modulated wave with a time-variant amplitude.

Let the input signal to a nonideal linear filter be an FM signal $s_1(t)$ with envelope $E_1(t)$ and phase $\phi_1(t)$. Let the FM wave have a residual amplitude modulation due to distortion previously acquired in the transmission channel. Let the output $s_2(t)$ of the filter have an envelope $E_2(t)$ and a phase $\phi_2(t)$. Then the input and output signal can be expressed, respectively, as

$$s_1(t) = E_1(t) \cos(\phi_1(t))$$

(20.50)

$$s_2(t) = E_2(t) \cos(\phi_2(t))$$

(20.51)

Note that although $s_2(t)$ is a linear function of $s_1(t)$, both the phase and the envelope of the output signal are nonlinear functions of the output signal $s_2(t)$. Hence, they are both nonlinear functions of the input signal $s_1(t)$. Thus, in general, amplitude modulation at the input to a linear filter results in phase modulation at the output. This is known as AM–PM conversion. On the other hand, phase modulation at the input to a linear filter results in residual amplitude modulation at the filter output. This is known as PM–AM conversion. PM–AM conversion does not present much of a problem in angle modulated systems, since the resulting residual amplitude modulation is easily remedied with the aid of bandpass limiters. However, AM–PM conversion presents a great problem in angle modulated systems because it is usually not possible to separate the resulting phase distortion from the demodulated signal.

In addition to the mechanism just described, phase nonlinearities of amplifiers, filters, repeaters, etc., also cause AM–PM conversion. This type of distortion is common in microwave systems where it results from the dependence of the phase characteristic of the repeaters and amplifiers on the envelope of the input signal. Typically, the phase change $\Delta\phi_2(t)$ in the output signal is proportional to the corresponding change, in decibel, of the envelope of the input signal. Let k°/dB be the constant of proportionality relating the change $\Delta E_1(t)$ in the envelope of the input signal $E_1(t)$ to the phase change in the output signal. Then the phase change is given by

$$\Delta\phi_2(t) = k20 \log \frac{\Delta E_1(t)}{E_1(t)} \tag{20.52}$$

20.3.6   Interference as a Radio Frequency Distortion Mechanism

Interference here refers to the corruption of the received signal in the desired channel by adjacent or nearby channels. Although each channel has an assigned bandwidth, adjacent channel interference often occurs because practical transmission filters do not provide infinite signal suppression outside the desired or assigned bandwidth. Adjacent channel interference from a strong channel can be disastrous for a weak channel. The effect of interference will be discussed for AM, DSB-SC AM, SSB-SC AM, and angle modulation. For each case, f_c will represent the carrier frequency of the desired channel, whereas $f_c + \Delta f$ will represent the carrier frequency of the interfering signal. For simplicity, the interference will be regarded as an unmodulated carrier of amplitude A_i and the phase angle between the interference and the desired modulated wave will be ignored. Thus, the interference $r_i(t)$ at the receiver input is

$$r_i(t) = A_i \cos(2\pi(f_c + \Delta f)t) \tag{20.53}$$

Interference in Amplitude Modulation

In the case of AM, demodulation is performed by passing the received signal through an envelope detector. The received signal, which consists of the desired AM signal with carrier amplitude A_c, and the interference is given by

$$
\begin{aligned}
r(t) &= (A_c + m(t)) \cos(2\pi f_c t) + A_i \cos 2\pi(f_c + \Delta f)t \\
&= (A_c + m(t) + A_i \cos 2\pi\Delta f t) \cos(2\pi f_c t) - A_i \sin(2\pi\Delta f t) \sin(2\pi f_c t) \tag{20.54} \\
&= E(t) \cos(2\pi f_c t + \phi(t))
\end{aligned}
$$

In the last equation, $E(t)$ is the envelope and $\phi(t)$ is the phase of the received signal. The phase is not relevant to the present discussion. The envelope is given by

$$E(t) = \left\{ (A_c + m(t) + A_i \cos(2\pi\Delta f t))^2 + A_i^2 \sin^2(2\pi\Delta f t) \right\}^{\frac{1}{2}} \tag{20.55}$$

From the expression for the envelope, it is easy to see that if $A_i \ll A_c$, the envelope is approximated by

$$E(t) \approx A_c + m(t) + A_i \cos 2\pi\Delta f t \tag{20.56}$$

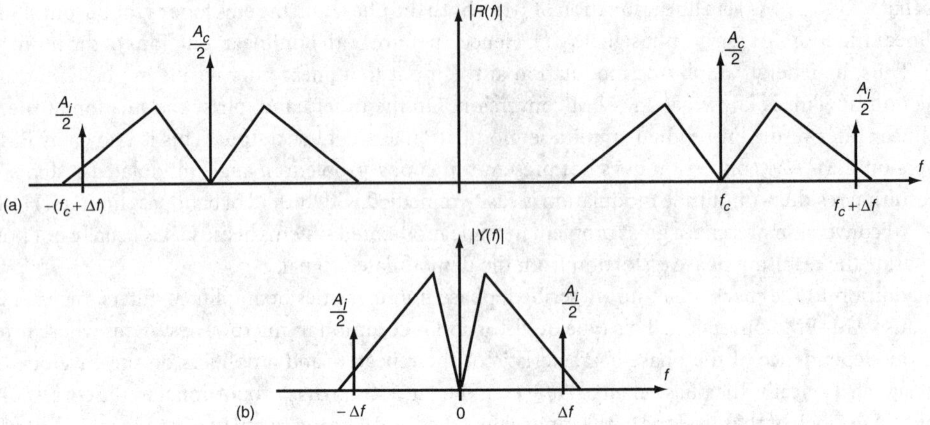

FIGURE 20.54 Illustration of interference in AM: (a) amplitude spectrum of AM wave plus interfrence, (b) amplitude spectrum of demodulated message signal plus interference.

Note that A_c in the last equation can be removed by DC blocking to yield the demodulator output $y(t)$. Let M be the peak amplitude of the message signal. Then it is observed that M/A_i, the ratio of the peak message signal amplitude to the peak interference amplitude, is equal at the input and the output of the receiver. Thus, for the case of small interference, the interference is additive at the demodulator output, and demodulation does not increase or decrease the strength of the interference relative to the message signal. The case for small interference is illustrated in Fig. 20.54. Figure 20.54(a) illustrates the spectrum of the modulated signal and the interference at the receiver input. Figure 20.54(b) illustrates the spectrum of the envelope detector output.

For the case in which the interference amplitude is much larger than the carrier amplitude for the desired channel, the envelope $E(t)$ can be approximated by

$$E(t) \approx (A_c + m(t)) \cos 2\pi \Delta f t + A_i \tag{20.57}$$

In this case, after the constant A_i is blocked with a capacitor, the demodulated output consists of the message signal multiplied by the interference, $\cos 2\pi \Delta f t$ and an additive component of the interference, $A_c \cos 2\pi \Delta f t$. The effect of interference is much worse in this case because the multiplicative interference results in serious distortion.

Interference in DSB-SC AM

Consider the synchronous demodulation of a DSB-SC signal plus the interfering sinusoid previously described. The same receiver of Fig. 20.49 is employed, but in this case, the phase and frequency errors between the received and the local carriers are ignored. The received signal plus interference $r(t)$, the product modulator output $x(t)$, and the demodulated output of the low-pass filter $y(t)$, are given, respectively, by

$$r(t) = m(t) \cos 2\pi f_c t + A_i \cos(2\pi(f_c + \Delta f)t) \tag{20.58}$$

$$x(t) = \frac{1}{2}m(t)(1 + \cos 4\pi f_c t) + \frac{1}{2}A_i(\cos 2\pi \Delta f t + \cos(4\pi f_c t + 2\pi \Delta f t)) \tag{20.59}$$

$$y(t) = \frac{1}{2}m(t) + \frac{1}{2}A_i \cos 2\pi \Delta f t \tag{20.60}$$

The last equation shows that for DSB-SC, the effect of interference is additive to the message signal at the demodulator output. Moreover, the ratio of the peak message output to the interference output M/A_i is equal at both the input and the output of the receiver.

Interference in SSB-SC

The analysis and the results in this case are very similar to those for DSB-SC. The same receiver is employed for both cases. In this case, the received signal plus interference, the product modulator output, and the demodulated output of the low-pass filter are given, respectively, by

$$r(t) = m(t) \cos 2\pi f_c t \mp m_h(t) \sin 2\pi f_c t + A_i \cos 2\pi (f_c + \Delta f)t \qquad (20.61)$$

$$
\begin{aligned}
x(t) &= \frac{1}{2} m(t)(1 + \cos 4\pi f_c t) \mp \frac{1}{2} M_h(t) \sin 4\pi f_c t \\
&\quad + \frac{1}{2} A_i (\cos 2\pi \Delta f t + \cos(4\pi f_c t + 2\pi \Delta f t))
\end{aligned}
\qquad (20.62)
$$

$$y(t) = \frac{1}{2} m(t) + \frac{1}{2} A_i(t) \cos 2\pi \Delta f t \qquad (20.63)$$

The demodulated output in the last equation shows that the effect of interference on SSB-SC is the same as formerly described for DSB-SC. Comparing these results with the AM case, it is easy to see that for small interference, the effect is about the same for AM as for DSB-SC and SSB-SC. However, AM performance is quite inferior to the other two in the case of large interference.

Interference in Angle Modulation

Consider for simplicity, only the unmodulated carrier of an angle modulated wave of carrier frequency f_c given by $A_c \cos 2\pi f_c t$. For the sake of simplicity, the interference is also considered to be the unmodulated sinusoid $A_i \cos 2\pi (f_c + \Delta f)t$. The signal at the input to the receiver is

$$
\begin{aligned}
r(t) &= A_c \cos 2\pi f_c t + A_i \cos 2\pi (f_c + \Delta f)t \\
&= (A_c + A_i \cos 2\pi \Delta f t) \cos 2\pi f_c t - A_i \sin 2\pi \Delta f t \sin 2\pi f_c t \\
&= E(t) \cos(2\pi f_c t + \phi(t))
\end{aligned}
\qquad (20.64)
$$

In the last equation, $E(t)$ is the envelope of the received signal, whereas $\phi(t)$ is its phase. In angle modulated waves, the phase, but not the envelope, is relevant to the demodulator output. The phase is given by

$$
\begin{aligned}
\phi(t) &= \tan^{-1}\left(\frac{-A_i \sin 2\pi \Delta f t}{A_c + A_i \cos 2\pi \Delta f t} \right) \\
&\approx -\frac{A_i \sin 2\pi \Delta f t}{A_c} \quad \text{for } A_i \ll A_c
\end{aligned}
\qquad (20.65)
$$

For PM, the demodulated signal is given by

$$y(t) = \phi(t) = -\frac{A_i}{A_c} \sin 2\pi \Delta f t \qquad (20.66)$$

For FM, the demodulated signal is given by

$$y(t) = \phi'(t) = -\frac{2\pi \Delta f A_i}{A_c} \cos 2\pi \Delta f t \qquad (20.67)$$

The last two equations show that for FM, the amplitude of the interference in the demodulated output is proportional to Δf, the difference between the carrier frequencies of the desired channel and the interference. On the other hand, the interference amplitude is constant for all values of Δf in the case of PM. These differences are illustrated in Fig. 20. 55.

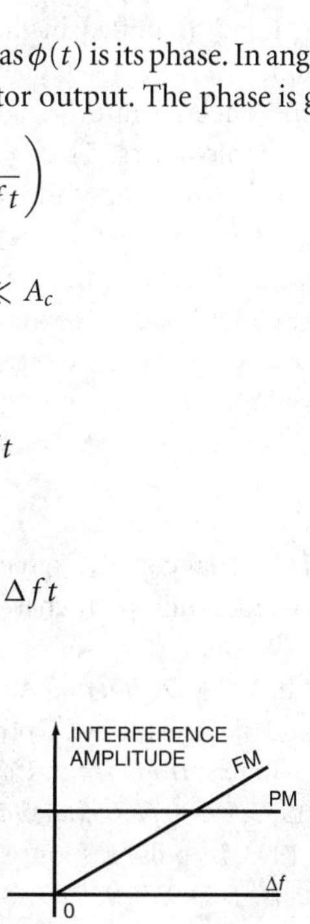

The last two equations also show that for both FM and PM, the interference at the demodulator

FIGURE 20.55 Effect of interference in angle modulation.

output is inversely proportional to A_c, the amplitude of the carrier of the desired channel. Thus, the stronger the desired channel and the weaker the interference, the smaller is the effect of the interference. This is superior in performance to amplitude modulated systems in which, at the very best, the interference is independent of the carrier amplitude. In effect, angle modulated systems suppress weak interference. This statement gives a clue to the behavior of angle modulated systems in the presence of large interference. When the interference is large relative to the desired signal, the so-called **capture effect** sets in. The interference now assumes the role of the desired channel and effectively suppresses the weaker desired channel, in the same way that a weak interference would have been suppressed.

Defining Terms

Capture effect: In angle modulation, the demodulated output of the weaker of two interfering signals is inversely proportional to the amplitude of the stronger signal. Consequently, the stronger signal or channel suppresses the weaker channel. If the stronger signal is noise or an undesired channel, it is said to have captured the desired channel.

Crosstalk: Interference between adjacent communication channels in which an interfering adjacent channel is heard or received in the desired channel.

Dispersion: The variation of wave propagation velocity (or time delay) in a medium with the frequency of the wave, which results in the spreading of the output signal over a time duration greater than the duration of the input signal.

Ionosphere: The ionosphere consists of several layers of the upper atmosphere, ranging in altitude from 100 to 300 km, containing ionized particles. The ionization is due to the effect of solar radiation, particularly ultraviolet and cosmic radiations, on gas particles. The ionosphere serves as reflecting layers for radio (sky) waves.

Pilot carrier: A means of providing a carrier at the receiver, which matches the received carrier in phase and frequency. In this method, which is employed in suppressed carrier modulation systems, a carrier of very small amplitude is inserted into the modulated signal prior to transmission, extracted and amplified in the receiver, and then employed as a matching carrier for coherent detection.

Radio horizon: The maximum range, from transmitter to receiver on the Earth's surface, of direct (line-of-sight) radio waves. This is greater than the optical horizon because the radio waves follow a curved path as a result of the continuous refraction it undergoes in the atmosphere.

Troposphere: The region of the atmosphere within about 10 km of the Earth's surface. Within this region, the wireless radio channel is modified relative to free space conditions by weather conditions, pollution, dust particles, etc. These inhomogeneities act as point scatterers which deflect radio waves downward to reach the receiving antennas, thereby providing tropospheric scatter propagation.

References

Bultitude, R.J.C., Melancon, P., Zaghloul, H., Morrison, G., and Prokiki, M. 1993. The dependence of indoor radio channel multipath characteristics on transmit receive ranges. *IEEE J. Selec. Areas Commun.* 11(7):979–990.

Couch, L.W., II. 1990. *Digital and Analog Communication Systems*, 3rd ed. Macmillan, New York.

Fechtel, S.A. and Meyer, H. 1994. Optimal parametric feedforward estimation of frequency-selective fading radio channels. *IEEE Trans. Commun.* 42(2):1639–1650.

Gibson, J.D. 1993. *Principles of Digital and Analog Communications*, 2nd ed. Macmillan, New York.

Hashemi, H. 1993. Impulse response modeling of indoor radio propagation channels. *IEEE J. Selec. Areas Commun.* 11(7):967–978.

Haykin, S. 1994. *Communication Systems*, 3rd ed. J. Wiley, New York.

Lathi, B.P. 1989. *Modern Digital and Analog Communication Systems*, 2nd ed. Holt, Rinehart and Winston, Philadelphia, PA.

Porakis, J.G. and Salehi, M. 1994. *Communication Systems Engineering.* Prentice-Hall, Englewood Cliffs, NJ.

Roddy, D. and Coolen, J. 1981. *Electronic Communications,* 2nd ed. Reston Pub., Reston, VA.

Serboyakov, I.Y. 1989. An estimate of the effectiveness of using an amplitude-frequency equalizer to compensate selective fading on digital radio relay links. *Telecommun. Radio Eng.* 44(6):112–115.

Stremler, F.G. 1992. *Introduction to Communication Systems,* 3rd ed. Addison-Wesley, Reading, MA.

Taub, H. and Schilling, D. 1986. *Principles of Communication Systems,* 2nd ed. McGraw-Hill, New York.

Voronkov, Y.V. 1992. The generation of an AM oscillation with low nonlinear envelope distortion and low attendant angle modulation. *Telecomm. Radio Eng.* 47(7):10–12.

Ziemer, R.E. and Tranter, W.H. 1990. *Principles of Communications Systems, Modulation and Noise,* 3rd ed. Houghton Mifflin, Boston, MA.

Further Information

Electronics and Communication in Japan

IEEE Transactions on Antenna and Propagation

IEEE Transactions on Communications

IEEE Journal of Selected Areas in Communications

Telecommunications and Radio Engineering, a technical journal published by Scripta Technica, Inc., Wiley Co.

20.4 Oscilloscopes

Jerry C. Whitaker

20.4.1 Introduction

The oscilloscope is one of the most general-purpose of all test instruments. A scope can be used to measure voltages, examine waveshapes, check phase relationships, examine clock pulses, and countless other functions. There has been considerable progress in scope technology within the past few years, most noticeably the move to digital systems. Some instruments offer logic tracing functions, and a hard copy printout.

20.4.2 Specifications

A number of parameters are used to characterize the performance of an oscilloscope. Key parameters include bandwidth and risetime.

Bandwidth

Oscilloscope bandwidth is defined as the frequency at which a sinusoidal signal will be attenuated by a factor of 0.707 (or reduced to 70.7% of its maximum value). This is referred to as the −3 dB point. Bandwidth considerations for sine waves are straightforward. For square waves and other complex waveforms, however, bandwidth considerations become substantially more involved.

Square waves are made up of an infinite number of sine waves: the fundamental frequency plus mostly odd harmonics. Fourier analysis shows that a square wave consists of the fundamental sine wave plus sine waves that are odd multiples of the fundamental. Figure 20.56 illustrates the mechanisms involved. The fundamental frequency contributes about 81.7% of the square wave. The third harmonic contributes about 9.02%, and the fifth harmonic about 3.24%. Higher harmonics contribute less to the shape of the square wave. As outlined here, approximately 94% of the square wave is derived from the fundamental and the third and fifth harmonics. Inaccuracies introduced by the instrument are typically 2% or less. The user's ability to read a conventional CRT display may contribute another 4% error. Thus, a total of 6% error may be introduced by the instrument and the operator. A 96% accurate reproduction of a square wave

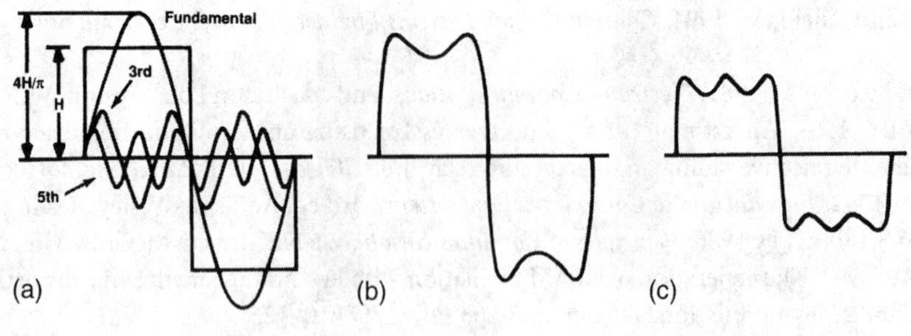

FIGURE 20.56 The mechanisms involved in square waves: (a) individual waveforms involved, (b) waveshape resulting from combining the fundamental and the first odd harmonic (the third harmonic, (c) waveshape resulting from combining the fundamental and the first, third, and fifth harmonics.

should, therefore, be sufficient for all but the most critical applications. It follows that scope bandwidth is important up to and including the fifth harmonic of the desired signal.

Rise Time

The rise time performance required of an oscilloscope depends on the degree of accuracy needed in measuring the input signals. A 2–3% rise time accuracy can be obtained from an instrument that is specified to have approximately five times the rise time of the signal being measured. Rise time can be determined from the bandwidth of the instrument

$$T_r = \frac{0.35}{BW} \tag{20.68}$$

Where

T_r = instrument rise time
BW = instrument bandwidth

Making Measurements

The oscilloscope is a measurement instrument, displaying voltage waveforms within powered circuits. To conduct certain types of waveform analysis, a separate signal generator is required to inject test signals. With a conventional analog oscilloscope, actual measurement of the waveform must be done visually by the operator. The graticule divisions along the trace are counted and multiplied by the time base setting to determine the time interval. The TIME/DIV control determines the value of each increment on the X-axis of the display. Typical sweep ranges vary from 0.05 to 0.5 μs per division. To determine voltage, the operator must estimate the height of a portion of the trace and multiply it by the voltage range selected on the instrument. The VOLT/DIV control determines the value of each increment on the Y-axis of the display. On a dual channel scope, two controls are provided, one for each source. Typical vertical sensitivity ranges vary from 5 mV to 5 V per division.

The need for interpolation introduces inaccuracies in the measurement process. Determination of exact values is especially difficult in the microsecond ranges, where even small increments can make a big difference in the measurement. It is also difficult with a conventional scope to make a close-up examination of specific portions of a waveform. For this purpose, an oscilloscope with two independently adjustable sweep generators is recommended. A small section of the waveform can be selected for closer examination by adjusting the B channel sweep-time and delay-time controls while observing an intensified portion of the waveform. The intensified section can then be expanded across the screen. This type of close-up examination is useful for measuring pulse rise times and viewing details in complex signals, such as video horizontal sync or color burst.

Advanced Features

Modern oscilloscopes remove the potential inaccuracies of previous instruments by incorporating advanced features, including

- Numeric measurement and display. On-screen readouts provide key measurement parameters for the operator. Readout options include peak waveform voltage, frequency, and sweep and sensitivity settings for the instrument. On-screen readout makes waveform photography or screen capture more valuable because the photograph displays not only the waveform, but instrument settings as well.
- Triggering. A wide selection of trigger modes provides operational flexibility. The trigger mode determines what condition starts the horizontal sweep to display the voltage waveform. The trigger signal can be derived from an external source, or from the input channel(s). The trigger is usually continuously adjustable through both positive and negative slopes of the signal. Selectable trigger coupling includes special settings needed in particular applications. A variable hold-off control can be used to obtain a stable display when measuring periodic or complex signals.
- Control and monitoring. Microprocessors have been integrated into scopes to reduce the number of mechanical switch contacts, which are prone to failure, and to permit front-panel setting recall capabilities. Auxiliary signal outputs are available to feed other instruments, such as a distortion analyzer or frequency counter.

20.4.3   Digital Oscilloscope

The digital storage oscilloscope (DSO) offers a number of significant advantages beyond the capabilities of analog instruments. A DSO can store in memory the signal being observed, permitting in-depth analysis impossible with previous technology. Because the waveform resides in memory, the data associated with the waveform can be transferred to a computer for real-time processing, or for processing at a later time.

Operating Principles

Before the introduction of the DSO, the term *storage oscilloscope* referred to a scope that used a *storage CRT*. This type of instrument stored the display waveform as a trace on the scope face. The digital storage oscilloscope operates on a completely different premise. Figure 20.57 shows a block diagram of a DSO. Instead of being amplified and directly applied to the deflection plates of a CRT, the waveform is first converted into a digital form and stored in memory. To reproduce the waveform on the CRT, the data are sequentially read and converted back into an analog signal for display.

The analog-to-digital (A/D) converter transforms the input analog signal into a sequence of digital bits. The amplitude of the analog signal, which varies continuously in time, is sampled at preset intervals. The analog value of the signal at each sample point is converted into a binary number. This *quantization* process is illustrated in Fig. 20.58. The sample rate of a digital oscilloscope must be greater than two times the highest frequency to be measured. The higher the sampling rate relative to the input signal, the greater the measurement accuracy. A sample rate 10 times the input signal is sufficient for most applications. This rule-of-thumb applies for single-shot signals, or signals that are constantly changing. The sample rate may also be expressed as the *sampling interval*, or the period of time between samples. The sample interval is the inverse of the sampling frequency.

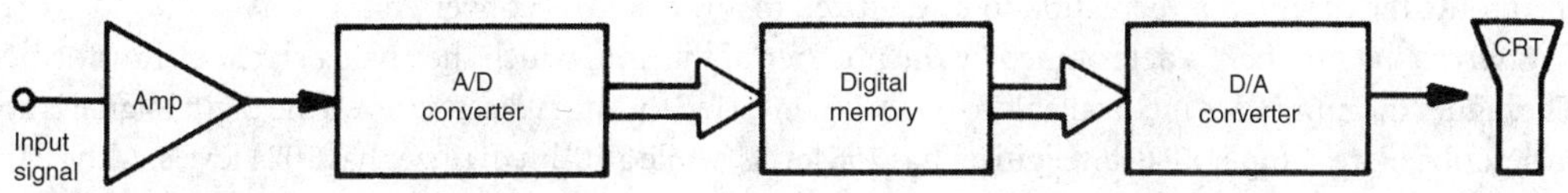

FIGURE 20.57   Simplified block diagram of a digital storage oscilloscope.

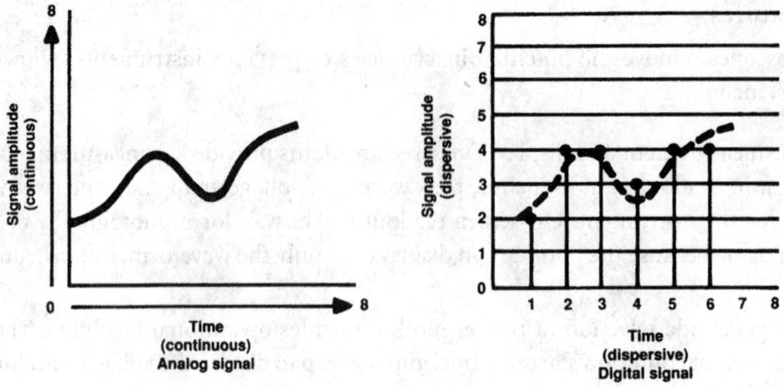

FIGURE 20.58 The quantization process.

Although a DSO is specified by its maximum sampling rate, the actual rate used in acquiring a given waveform is usually dependent on the *time-per-division* setting of the oscilloscope. The *record length* (samples recorded over a given period of time) defines a finite number of sample points available for a given acquisition. The DSO must, therefore, adjust its sampling rate to fill a given record over the period set by the sweep control. To determine the sampling rate for a given sweep speed, the number of displayed points per division is divided into the sweep rate per division. Two additional features can modify the actual sampling rate:

- Use of an external clock for pacing the digitizing rate. With the internal digitizing clock disabled, the digitizer will be paced at a rate defined by the operator.
- Use of a *peak detection* (or *glitch capture*) mode. Peak detection allows the digitizer to sample at the full digitizing rate of the DSO, regardless of the time base setting. The minimum and maximum values found between each normal sample interval are retained in memory. These minimum and maximum values are used to reconstruct the waveform display with the help of an algorithm that recreates a smooth trace along with any captured glitches. Peak detection allows the DSO to capture glitches even at its slowest sweep speed. For higher performance, a technique known as *peak-accumulation* (or *envelope mode*) may be used. With this approach, the instrument accumulates and displays the maximum and minimum excursions of a waveform for a given point in time. This builds an envelope of activity that can reveal infrequent noise spikes, long-term amplitude or time drift, and pulse jitter extremes.

A/D Conversion

The A/D conversion process is critical to the overall accuracy of the oscilloscope. Three types of converters are commonly used:

- Flash converter
- Successive approximation converter
- Charge-coupled device (CCD) converter

The CCD is an analog storage array. The DSO can capture high-speed events in real time using a CCD, then route the captured information to be digitized to low cost A/D converters.

Conversion resolution is determined by the number of bits into which the analog signal is transformed. The higher the number of bits available to the A/D converter, the more discrete levels the digital signal is able to describe. For example, an 8-bit digitizer has 256 levels, while a 10-bit digitizer has 1024 levels. Although a higher number of bits allows greater discrimination between voltage values, the overall accuracy of the DSO may be limited by other factors, including the accuracy of the analog amplifier(s) feeding the digitizer(s). The usefulness of a high conversion digitizer may also be limited by the resolution of the display screen.

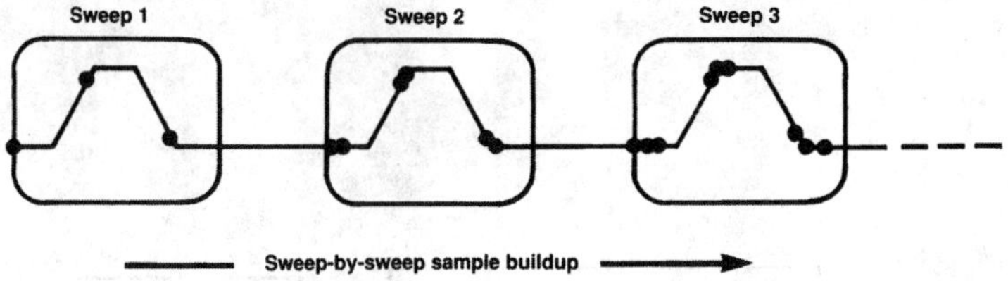

FIGURE 20.59 Increasing sample density through equivalent-time-sampling.

While most DSOs provide 8-bit digitizing, there may be significant differences between individual 8-bit instruments. First among these is the *useful storage bandwidth*. This specification is further divided into *single-shot bandwidth* and *equivalent-time bandwidth*.

Achieving adequate samples for a high frequency waveform places stringent requirements on the sampling rate. High frequency waveforms require high sampling rates. Equivalent-time sampling is often used to provide high-bandwidth capture. This technique relies on sampling a repetitive waveform at a low rate to build up sample density. This concept is illustrated in Fig. 20.59. When the waveform to be acquired triggers the DSO, several samples are taken over the duration of the waveform. The next repetition of the waveform triggers the instrument again, and more samples are taken at different points on the waveform. Over many repetitions, the number of stored samples can be built up to the equivalent of a high sampling rate.

System operating speed also has an effect on the display update rate. Update performance is critical when measuring waveform or voltage changes. If the display does not track the changes, adjustment of the circuit may be difficult.

DSO System Design

A detailed block diagram of a basic DSO is shown in Fig. 20.60. The features included in this design are typical of a digital scope. The major circuit blocks are

- **Data acquisition**—converts the analog input signal into a digital equivalent for storage. Data control signals are routed to a microprocessor through an I/O port.
- **Data processing**—provides interpolation of the stored data for reproduction of the analog signal on the CRT or other display device. The processed data are transferred to the display memory. Except during transfer periods, the contents of the display memory are constantly being sent to the digital-to-analog (D/A) converter and displayed on the CRT.
- **Vertical deflection**—generates the vertical deflection voltages for the CRT. The input signal is impedance-matched and amplified by the Channel 1 and Channel 2 preamplifiers to a level suitable for driving subsequent stages. Channel selection switches operate under control of the microprocessor.
- **Horizontal deflection**—generates the necessary voltages to deflect the beam spot horizontally across the screen. The stage includes trigger and sweep circuits. The horizontal deflection stage also acts as an *X*-axis amplifier for *X*-*Y* operation. Channel selection switches operate under control of the microprocessor.
- **CRT driver**—provides the high voltages necessary to drive the CRT, and controls the *Z*-axis of the CRT. The *Z*-axis circuit sets the intensity and focus of the CRT beam spot. For other types of display devices, such as LCD or plasma units, the appropriate functions are performed in this block.
- **Power supplies**—provides the necessary low voltages for other stages of the instrument.
- **Calibration circuit**—provides a reference signal for calibration of the oscilloscope probe, and for operation of the amplifiers in the noncalibrated mode.

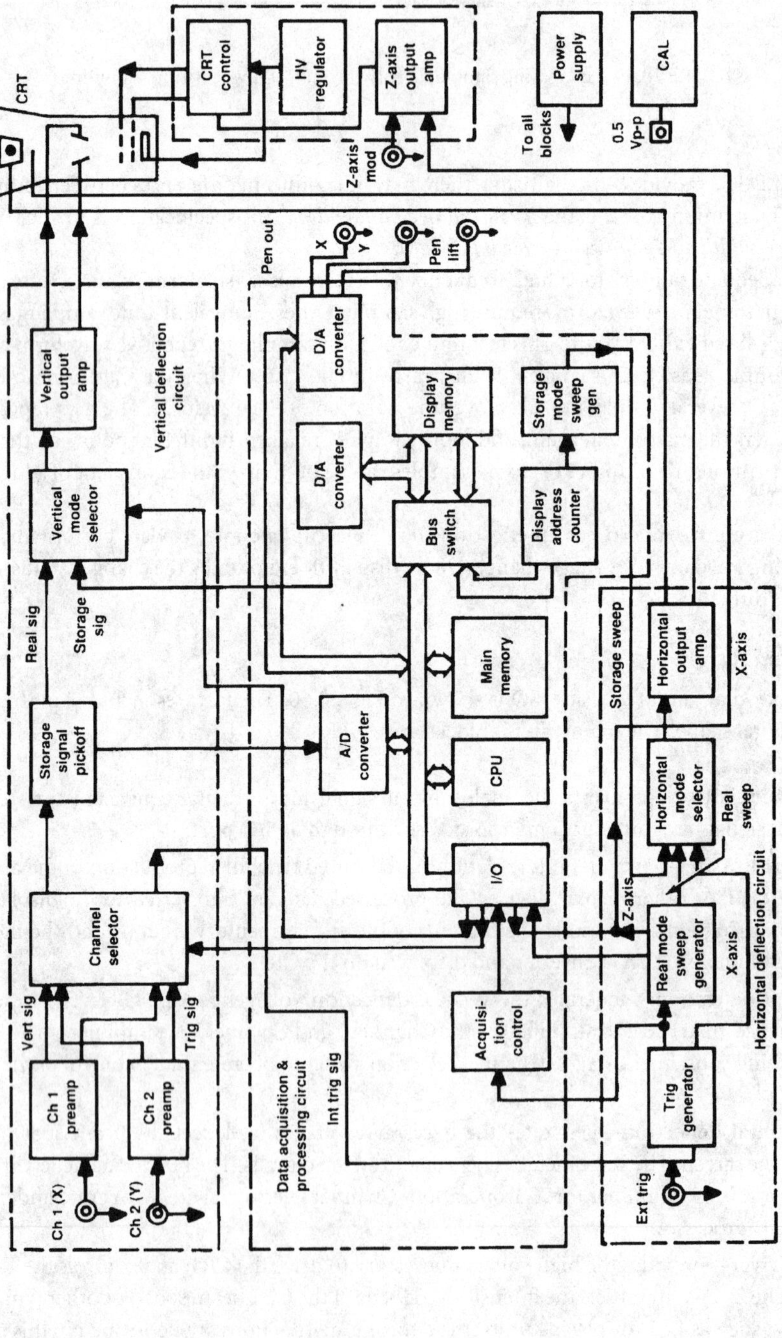

FIGURE 20.60   Block diagram of a DSO showing the conversion, I/O memory, and display circuits (*Courtesy of Kikusui.*)

DSO Features

The digital oscilloscope has become a valuable tool in troubleshooting both analog and digital products. Advanced components and construction techniques have led to lower costs for digital scopes and higher performance. Digital scopes can capture and analyze transient signals, such as race conditions, clock jitter, glitches, dropouts, and intermittent faults. Automated features reduce testing and troubleshooting costs through the use of recallable instrument setups, direct parameter readout, and unattended monitoring. Digital oscilloscopes have inherent benefits not available on most analog oscilloscopes. These benefits include

- Increased resolution (determined by the quality of the analog-to-digital converter).
- Memory storage of digitized waveforms.
- Automatic setup for repetitive signal analysis. For complex multichannel configurations that are used often, front-panel storage/recall features can save dozens of manual selections and adjustments. When multiple memory locations are available, multiple front-panel setups can be stored to save even more time.
- Auto-ranging. Many instruments will automatically adjust for optimum sweep, input sensitivity, and triggering. The instrument's microprocessor automatically configures the front panel for optimum display. Such features permit the operator to concentrate on making measurements, not on adjusting the scope.
- Instant hardcopy output from printers and plotters.
- Remote programmability for automated test applications.
- Trigger flexibility. Single-shot digitizing oscilloscopes capture transient signals and allow the user to view the waveform which *preceded* the trigger point.
- Signal analysis. Intelligent scopes can make key measurements and comparisons. Display capabilities include voltage peak, mean voltage, RMS value, rise time, fall time, and frequency.
- Cursor measurement. Advanced oscilloscopes permit the operator to take measurements or perform comparative operations on data appearing on the display. A measurement cursor consists of a pair of lines or dots that can be moved around the screen as needed.
- Trace quality. Eye fatigue is reduced noticeably with a DSO when viewing low-repetition signals. For example, a 60 Hz waveform can be difficult to view for extended periods of time on a conventional scope because the display tends to flicker. A DSO overcomes this problem by writing waveforms to the screen at the same rate regardless of the input signal.

Capturing Transient Waveforms

Single-shot digitizing makes it possible to capture and clearly display transient and intermittent signals. Waveforms such as signal glitches, dropouts, logic race conditions, intermittent faults, clock jitter, and power-up sequences can be examined with the help of a digital oscilloscope. With single-shot digitizing, the waveform is captured the first time it occurs, on the first trigger. It can then be displayed immediately or held in memory for analysis at a later date.

Triggering

Basic triggering modes available on a digital oscilloscope permit the user to select the desired source, its coupling, level, and slope. More advanced digital scopes contain triggering circuitry similar to that found in a logic analyzer. These powerful features let the user trigger on elusive conditions, such as pulse widths less than or greater than expected, intervals less than or greater than expected, and specified logic conditions. The logic triggering can include digital pattern, state qualified, and time/event qualified conditions. Many trigger modes are further enhanced by allowing the user to hold-off the trigger by a selectable time or number of events. Hold-off is especially useful when the input signal contains bursts of data or follows a repetitive pattern. These features are summarized as follows:

- *Pulse-width triggering* lets the operator quickly check for pulses narrower than expected or wider than expected. The pulse-width trigger circuit checks the time from the trigger source transition of a given slope (typically the rising edge) to the next transition of opposite slope (typically the falling edge). The operator can interactively set the pulse-width threshold for the trigger. For example, a glitch can be considered any signal narrower than 1/2 of a clock period. Conditions preceding the trigger can be displayed to show what events led up to the glitch.

- *Interval triggering* lets the operator quickly check for intervals narrower than expected or wider than expected. Typical applications include monitoring for transmission phase changes in the output of a modem or for signal dropouts, such as missing bits in a transport stream.

- *Pattern triggering* lets the user trigger on the logic state (high, low, or either) of several inputs. The inputs can be external triggers or the input channels themselves. The trigger can be generated either upon entering or exiting the pattern. Applications include triggering on a particular address select or data bus condition. Once the pattern trigger is established, the operator can probe throughout the circuit, taking measurement synchronous with the trigger.

- *State qualified triggering* enables the oscilloscope to trigger on one source, such as the input signal itself, only after the occurrence of a specified logic pattern. The pattern acts as an *enable* or *disable* for the source.

Advanced Features

Some digital oscilloscopes provide enhanced triggering modes that permit the user to select the level and slope for each input. This flexibility makes it easy to look for odd pulse shapes in the pulse-width trigger mode and for subtle dropouts in the interval trigger mode. It also simplifies testing different logic types (TTL, CMOS, and ECL), and testing analog/digital combinational circuits with dual logic triggering. Additional flexibility is available on multiple channel scopes. Once a trigger has been sensed, multiple simultaneously sampled inputs permit the user to monitor conditions at several places in the unit under test, with each channel synchronized to the trigger. Additional useful trigger features for monitoring jitter or drift on a repetitive signal include

- Enveloping
- Extremes
- Waveform delta
- Roof/floor

Each of these functions are related, and in some cases describe the same general operating mode. Various scope manufacturers use different nomenclature to describe proprietary triggering methods. Generally speaking, as the waveshape changes with respect to the trigger, the scope generates upper and lower traces. For every nth sample point with respect to the trigger, the maximum and minimum values are saved. Thus, any jitter or drift is displayed in the envelope.

Advanced triggering features provide the greatest benefit in conjunction with single-shot sampling. Repetitive sampling scopes can only capture and display signals that repeat precisely from cycle to cycle. Several cycles of the waveform are required to create a digitally-reconstructed representation of the input. If the signal varies from cycle to cycle, such scopes can be less effective than a standard analog oscilloscope for accurately viewing a waveform.

References

Albright, J.R. 1989. Waveform Analysis with Professional-Grade Oscilloscopes. *Electronic Servicing and Technology*, Intertec Publishing, Overland Park, KS.

Breya, M. 1988. New Scopes Make Faster Measurements. *Mobile Radio Technology Magazine*, Intertec Publishing, Overland Park, KS.

Harris, B. 1988. The Digital Storage Oscilloscope: Providing the Competitive Edge. *Electronic Servicing and Technology*, Intertec Publishing, Overland Park, KS.

Harris, B. 1989. Understanding DSO Accuracy and Measurement Performance. *Electronic Servicing and Technology*, Intertec Publishing, Overland Park, KS.

Hoyer, M. 1990. Bandwidth and Rise Time: Two Keys to Selecting the Right Oscilloscope. *Electronic Servicing and Technology*, Intertec Publishing, Overland Park, KS.

Montgomery, S. 1989. Advanced in Digital Oscilloscopes. *Broadcast Engineering*, Intertec Publishing, Overland Park, KS.

Persson, C. 1990. Oscilloscope Special Report. *Electronic Servicing and Technology*, Intertec Publishing, Overland Park, KS.

Persson, C. 1987. Oscilloscope: The Eyes of the Technician. *Electronic Servicing and Technology*, Intertec Publishing, Overland Park, KS.

Toorens, H. 1990. Oscilloscopes: From Looking Glass to High-Tech. *Electronic Servicing and Technology*, Intertec Publishing, Overland Park, KS.

Whitaker, J.C. 2001. *Electronic Systems Maintenance Handbook*. CRC Press, Boca Raton, FL.

20.5   Spectrum Analysis

Jerry C. Whitaker

20.5.1   Introduction

An oscilloscope-type instrument displays voltage levels referenced to time and a spectrum analyzer indicates signal levels referenced to frequency. The frequency components of the signal applied to the input of the analyzer are detected and separated for display against a frequency-related time base. Spectrum analyzers are available in a variety of ranges, with some models designed for use with audio or video frequencies and others intended for use with RF frequencies.

The primary application of a spectrum analyzer is the measurement and identification of RF signals. When connected to a small receiving antenna, the analyzer can measure carrier and sideband power levels. By expanding the sweep width of the display, offset or multiple carriers can be observed. By increasing the vertical sensitivity of the analyzer and adjusting the center frequency and sweep width, it is possible to observe the occupied bandwidth of the RF signal. Convention dictates that the vertical axis displays amplitude and the horizontal axis displays frequency. This frequency-domain presentation allows the user to glean more information about the characteristics of an input signal than is possible from an oscilloscope. Figure 20.61 compares the oscilloscope and spectrum analyzer display formats.

Principles of Operation

A spectrum analyzer intended for use at RF frequencies is shown in block diagram form in Fig. 20.62. The instrument includes a superheterodyne receiver with a swept-tuned local oscillator (LO) that feeds a CRT display. The tuning control determines the center frequency (F_c) of the spectrum analyzer, and the *scan-width* selector determines how much of the frequency spectrum around the center frequency will be covered. Full-feature spectrum analyzers also provide front-panel controls for scan-rate selection and bandpass filter selection. Key specifications for a spectrum analyzer include

- *Resolution:* the frequency separation required between two signals so that they may be resolved into two distinct and separate displays on the screen. Resolution is usually specified for equal-level signals. When two signals differ significantly in amplitude and are close together in frequency, greater resolution is required to separate them on the display.

- *Scan width*: the amount of frequency spectrum that can be scanned and shown on the display. Scan width is usually stated in kilohertz or megahertz per division. The minimum scan width available is usually equal to the resolution of the instrument.

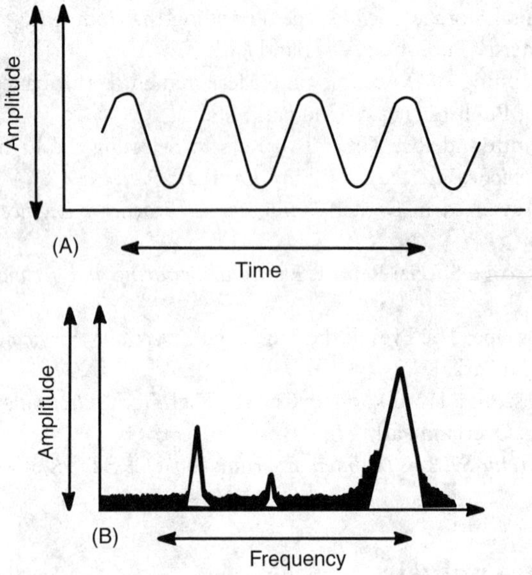

FIGURE 20.61   Comparison of waveform display: (A) oscilloscope, and (B) spectrum analyzer.

- *Dynamic range*: the maximum amplitude difference that two signals can have and still be viewed on the display. Dynamic range is usually stated in decibels.
- *Sensitivity*: the minimum signal level required to produce a usable display on the screen. If low-level signal tracing is planned, as in receiver or off-air monitoring, the sensitivity of the spectrum analyzer is important.

When using the spectrum analyzer, care must be taken not to overload the front-end with a strong input signal. Overloading can cause "false" signals to appear on the display. These false signals are the result of

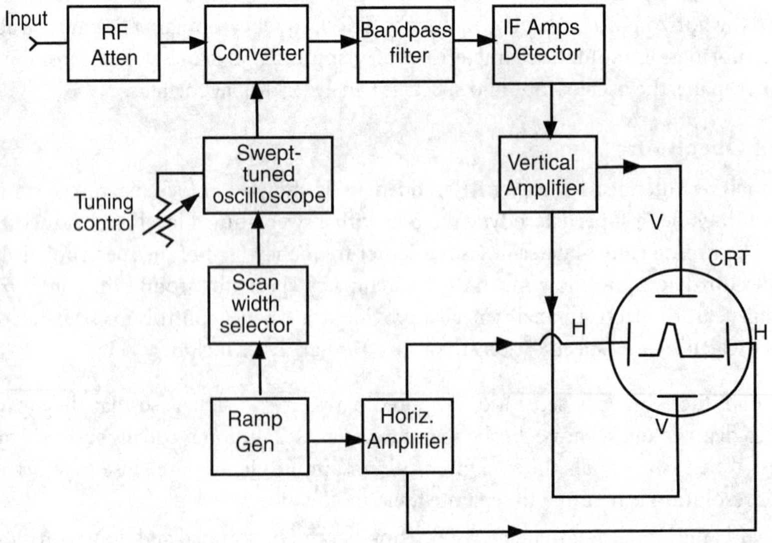

FIGURE 20.62   Block diagram of a spectrum analyzer.

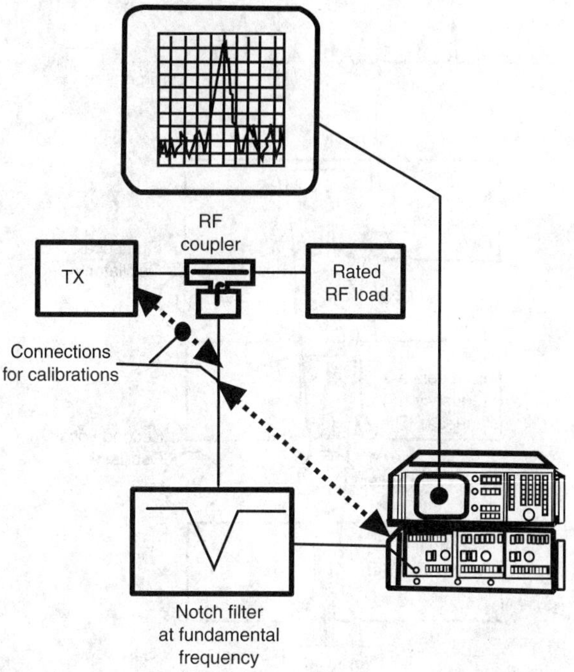

FIGURE 20.63 Typical test setup for measuring the harmonic and spurious output of a transmitter. The notch filer is used to remove the fundamental frequency to prevent overdriving the spectrum analyzer input.

nonlinear mixing in the front end of the instrument. False signals can be identified by changing the RF attenuator setting to a higher level. The amplitude of false signals (caused by overloading) will drop much more than the amount of increased attenuation.

The spectrum analyzer is useful in troubleshooting receivers as well as transmitters. As a tuned signal tracer, it is well adapted to stage-gain measurements and other tests. There is one serious drawback, however. The 50-Ω spectrum analyzer input can load many receiver circuits too heavily, especially high impedance circuits such as FET amplifiers. Isolation probes are available to overcome loading problems. Such probes, however, also attenuate the input signal, and unless the spectrum analyzer has enough reserve gain to overcome the loss caused by the isolation probe, the instrument will fail to provide useful readings. Isolation probes with 20 to 40 dB attenuation are typical. As a rule of thumb, probe impedance should be at least ten times the impedance of the circuit to which it is connected.

Applications

The primary application for a spectrum analyzer centers around measuring the occupied bandwidth of an input signal. Harmonics and spurious signals can be checked and potential causes investigated. Figure 20.63 shows a typical test setup for making transmitter measurements.

The spectrum analyzer is also well-suited for making accurate transmitter FM deviation measurements. This is accomplished using the *Bessel null method*. The Bessel null is a mathematical function that describes the relationship between spectral lines in frequency modulation. The Bessel null technique is highly accurate; it forms the basis for modulation monitor calibration. The concept behind the Bessel null method is to drive the carrier spectral line to zero by changing the modulating frequency. When the carrier amplitude is zero, the modulation index is given by a Bessel function. Deviation can be calculated from

$$\delta f_c = MI \times f$$

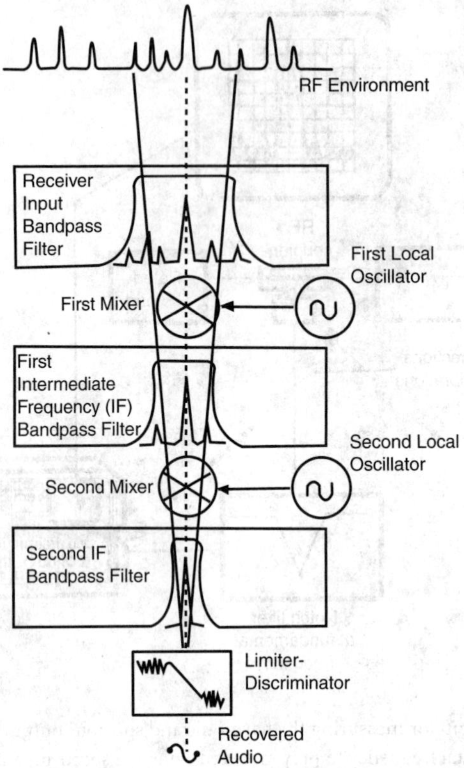

FIGURE 20.64   Using a spectrum analyzer to measure the tunnel effect of bandpass filters in a receiver.

where Δf_c is the deviation frequency, *MI* is the modulation index, and f_m is the modulating frequency. The carrier frequency "disappears" at the Bessel null point, with all power remaining in the FM sidebands.

A *tracking generator* can be used in conjunction with the spectrum analyzer to check the dynamic response of frequency-sensitive devices, such as transmitter isolators, cavities, ring combiners, duplexers, and antenna systems. A tracking generator is a frequency source that is locked in step with the spectrum analyzer horizontal trace rate. The resulting displays shows the relationship of the amplitude-vs.-frequency response of the device under test. The spectrum analyzer can also be used to perform gain-stage measurements. The combination of a spectrum analyzer and a tracking generator makes filter passband measurements possible. As measurements are made along the IF (intermediate frequency) chain of a receiver, the filter passbands become increasingly narrow, as illustrated in Fig. 20.64.

On-Air Measurements

The spectrum analyzer is used for three primary on-air tests:

- *Measuring unknown signal frequencies.* A spectrum analyzer can be coupled to the out-put of a transmitter to determine its exact operating frequency.
- *Intermod and interference signal tracking.* A directional Yagi antenna can be used to identify the source of an interfering signal. If the interference is on-frequency but carries little intelligence, chances are good that it is an intermod being produced by another transmitter. Use the wide dispersion display mode of the analyzer and note signal spikes that appear simultaneously with the interference. A troubleshooter, armed with a spectrum analyzer, a directional antenna, and the

knowledge of how intermod signals are generated, can usually locate a suspected transmitter rapidly. Each suspected unit can then be tested individually by inserting an isolator between the transmitter PA and duplexer (or PA and antenna). When the intermod signal amplitude drops the equivalent of the reverse insert loss of the isolator, the offending transmitter has been located.

• *Field strength measurements.* With an external antenna, a spectrum analyzer can be used for remote field strength measurements. Omnidirectional, broadband antennas are the most versatile because they allow several frequency bands to be checked simultaneously. Obviously, the greater the external antenna height, the greater the testing range.

Spurious Harmonic Distortion

An incorrectly tuned or malfunctioning transmitter can produce spurious harmonics. The spectrum analyzer provides the best way to check for spectral purity. Couple a sample of the transmitter output signal to the analyzer input, either by loop coupling or RF sampling, as illustrated in Fig. 20.65. For low power levels from portable and mobile units, transmit into a dummy load and use a flexible rubber antenna on the analyzer input. For maximum accuracy when measuring larger amounts of power, use an RF sampler to control the input level to the analyzer. Use maximum RF attenuation initially on the analyzer front end to prevent overload damage and internal intermod. False signals on the display can also be observed if covers or shields are not in place on the radio under test. The oscillator, doubler, or trippler levels may radiate sufficient signals to register on the analyzer display.

Energize the transmitter and observe any spurious harmonics on the analyzer display. After centering the main signal of interest and adjusting the input attenuation for maximum display amplitude, calculate the spurious radiation. Spurious signal attenuation is measured in decibels, referenced to the amplitude of the transmitter fundamental. When a radio transmitter has a harmonic distortion problem, the defective stage can usually be isolated by tuning each stage and observing the spurious harmonics on the analyzer display. When the defective stage is tuned, the harmonics either shift frequency or change in amplitude. Signal tracing, with a probe and heavy input attenuation, also helps to determine where the distortion first occurs. Overdriven stages are prime culprits.

Selective-Tuned Filter Alignment

The high input sensitivity and visual frequency-selective display of the spectrum analyzer provide the technician with an efficient analog tuning instrument. Injection circuits can be peaked quickly by loop coupling the analyzer input to the mixer section of a receiver. This is most helpful in tuning older radios when a service manual is not available. Radios without test points, or that require elaborate, specialized test sets, can be tuned in a similar manner. Portable radio transmitters can be tuned directly into their antennas for maximum signal strength. This reveals problems such as improper signal transfer to the antenna or a defective antenna.

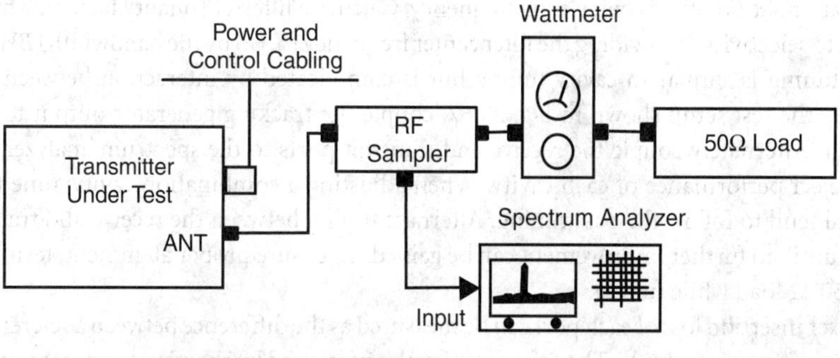

FIGURE 20.65 Common test setup to measure transmitter harmonic distortion with a spectrum analyzer.

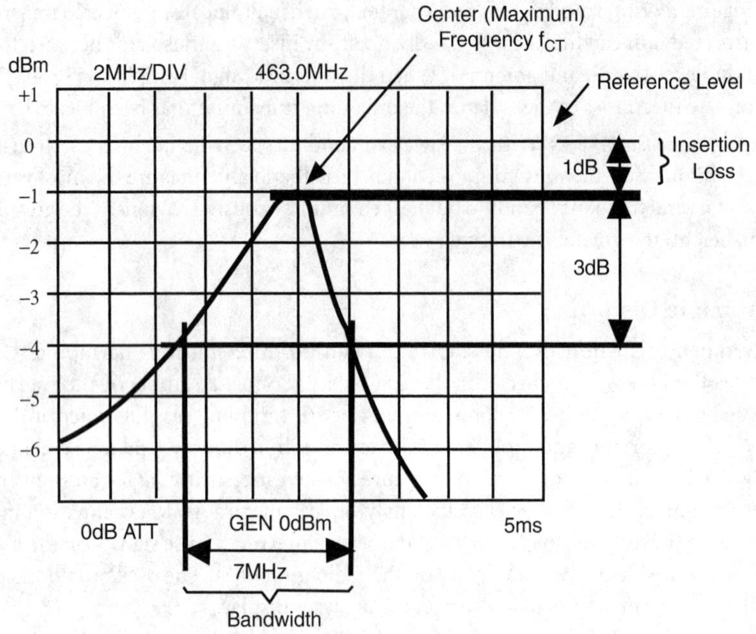

FIGURE 20.66 Typical spectrum analyzer display of a single-cavity filter. Bandwidth (BW) = 458 to 465 MHz = 7 MHz. Filter Q = f_{CT}/BW = 463 MHz/7 MHz = 66. The trace shows 1-dB insertion loss.

Cavities, combiners, and duplexers are all selective-tuned filters. They are composed of three types of passive filters:

- Bandpass
- Band-reject
- Combination pass/reject

The spectrum analyzer is well-suited to tuning these types of filters, particularly duplexers. Couple the cavity between the tracking generator and the spectrum analyzer. Calibrate the tuning frequency in the center of the CRT display and adjust the generator output and analyzer input level for an optimal display trace. Then tune the cavity onto frequency. Measure the various filter characteristics and compare them to the manufacturer's specifications. Figure 20.66 shows a typical spectrum analyzer display of a single-cavity device. Important filter characteristics include frequency, bandwidth, insertion loss, and selectivity. Measure the bandwidth at the points 3 dB down from maximum amplitude. Insertion loss is equal to they cavity's attenuation of the center-tune frequency. Calculate filter Q (quality factor), which is directly proportional to selectivity, by dividing the filter center frequency (f_{CT}) by the bandwidth (BW) of the filter.

Duplexer tuning is similar to cavity tuning but is complicated by interaction between the multiple cavities. Using the test setup shown in Fig. 20.67, couple the tracking generator output to the duplexer antenna input. Alternately couple the receive and transmit ports to the spectrum analyzer while tuning the pass or reject performance of each cavity. When adjusting a combination cavity, tune the reject last because it will tend to follow the pass tuning. Alternate tuning between the receive and transmit sides of the duplexer until no further improvement can be gained. To ensure proper alignment, terminate all open ports into a 50-Ω load while tuning.

The duplexer insertion loss of each port can be measured as the difference between a reference amplitude and the pass frequency amplitude. The reference signal is measured by shorting the generator output cable to the analyzer input cable. This reference level nulls any cable losses to prevent them from being included in duplexer insertion loss measurements.

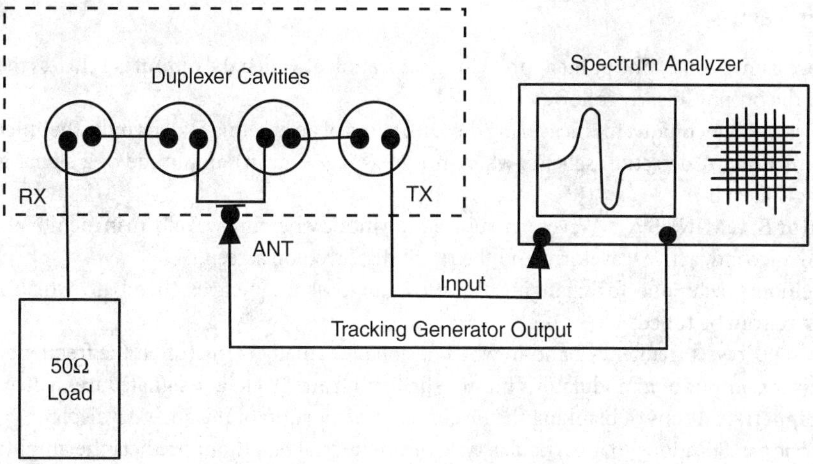

FIGURE 20.67   Test setup for duplexer tuning using a spectrum analyzer.

Small-Signal Troubleshooting

The spectrum analyzer is well-suited for use in small-signal RF troubleshooting. Stage gain, injection level, and signal loss can be easily checked.

Receiver sensitivity loss of less than 6 dB is usually a difficult problem to trace with most test instruments. Such losses are usually associated with the front end of the radio, either as a stage gain or injection amplifier deficiency. Before the spectrum analyzer was commonly available, there was little to aid a technician in tracing such a problem. An RF millivolt meter, with typical usable sensitivity of 1mV, was of little help in troubleshooting such microvolt-level signals. Oscilloscopes are restricted in the same way, as well as having inadequate bandwidth for the high frequencies used in the front-end circuits of most receivers. A spectrum analyzers, with a typical 1 μV sensitivity and broad RF bandwidth, is well-adapted for RF troubleshooting. Microvolt-level signals can be accurately traced, allowing signal losses to be examined at each individual stage.

To conduct measurements, inject a test signal generated by a communications monitor into the antenna input of the radio under test. Use an appropriate probe to trace the signal from stage to stage. Stage gain, mixer output, injection level, and filter loss can be checked from the antenna to the discriminator. Compare measurements taken with those from a correctly functioning unit to isolate the problem. A standard test signal level of 100 μV usually works well. It is large enough to allow testing of the RF amplifier without causing overload. If the radio includes an automatic gain control (AGC) circuit, disable it temporarily during testing to permit measurement of the true stage gain.

When tracing a loss of sensitivity, start with the RF mixer. The mixer provides a junction point for narrowing the direction of troubleshooting. If the output of the mixer is correct, the defective stage lies beyond the mixer, probably in the IF amplifier. If the mixer output is low, check the input signal to the mixer. Check the RF amplifier, input cavities, and receive/transmit (RX/TX) switching circuit (if used). If the mixer IF carrier injection is low, check the multipliers and oscillators along that path.

The bandwidth of IF filters can be measured in circuit by varying the receiver input frequency and measuring the bandwidth of the 3-dB roll-off-points. A crystal bandpass filter can be tested for flaws by injecting an over-deviated signal into the receiver. A cracked filter crystal will produce a sharp, deep notch on the analyzer display. The key to comparative troubleshooting is accurate recording of test signal levels of a radio known to be working correctly prior to troubleshooting.

Remember to protect the front end of the analyzer from DC voltages and overload when tracing signals with a probe. When testing unknown signal levels, use maximum input attenuation and external attenuators as needed to avoid analyzer damage. Isolate the analyzer input from DC voltages using a capacitively coupled probe.

Defining Terms[3]

1 dB compression point: The point approaching saturation at which the output is 1 db less than it should be if the output linearly followed input.

Antenna sweep: A technique for measuring the return loss of an antenna to determine the antenna tuning.

Average detection: A detection scheme wherein the average (mean) amplitude of a signal is measured and displayed.

B-SAVE A (or B, C MINUS A): Waveform subtraction mode wherein a waveform in memory is subtracted from a second, active waveform and the result displayed on screen.

Band switching: Technique for changing the total range of frequencies (band) to which a spectrum analyzer can be tuned.

Baseband: The lowest frequency band in which a signal normally exists (often the frequency range of a receiver's output or a modulator's input); the band from DC to a designated frequency.

Baseline clipper: A means of blanking the bright baseline portion of the analyzer display.

Bessel functions: Solutions to a particular type of differential equation; predicts the amplitudes of FM signal components.

Bassel null method: A technique most often used to calibrate FM deviation meters. A modulating frequency is chosen such that some frequency component of the FM signal nulls at a specified peak deviation.

Calibrator: A signal generator producing a specified output used for calibration purposes.

Carrier-to-noise ratio (C/N): The ratio of carrier signal power to average noise power in a given bandwidth surrounding the carrier; usually expressed in decibels.

Center frequency: The frequency at the center of a given spectrum analyzer display.

Coax bands: The range of frequencies that can be satisfactorily passed via coaxial cable.

Comb generator: A source producing a fundamental frequency component and multiple components at harmonics of the fundamental.

Component: In spectrum analysis, usually denotes one of the constituent sine waves making up electrical signals.

Decibel (dB): Ten times the logarithm of the ratio of one electrical power to another.

Delta F (ΔF): A mode of operation on a spectrum analyzer wherein a difference in frequency may be read out directly.

Distortion: Degradation of a signal, often a result of nonlinear operations, resulting in unwanted signal components. Harmonic and inermodulation distortion are common types.

Dynamic range: The maximum ratio of two simultaneously present signals that can be measured to a specified accuracy.

Emphasis: Deliberate shaping of a signal spectrum or some portion thereof, often used as a means of overcoming system noise. Pre-emphasis is often used before signal transmission and de-emphasis after reception.

Envelope: The limits of an electrical signal or its parameters. For example, the modulation envelope limits the amplitude of an AM carrier.

Equivalent noise bandwidth: The width of a rectangular filter that produces the same noise power at its output as an actual filter when subjected to a spectrally flat input noise signal. Real filters pass different noise power than implied by their nominal bandwidths because their skirts are not infinitely sharp.

External mixers: A mixer, often in a waveguide format, that is used external to a spectrum analyzer.

Filter: A circuit that separates electrical signals or signal components based on their frequencies.

Filter loss: The insertion loss of a filter is the minimum difference, in decibels (dB), between the input signal level and the output level.

---

[3]Definitions courtesy of Tektronix, Beauerton, Oregon.

First mixer input level: Signal amplitude at the input to the first mixer stage of a spectrum analyzer. An optimum value is usually specified by the manufacturer.

Flatness: Unwanted variations in signal amplitude over a specified bandwidth, usually expressed in decibels (dB).

Fourier analysis: A mathematical technique for transforming a signal from the time domain to the frequency domain, and vice versa.

Frequency: The rate at which a signal oscillates, or changes polarity, expressed as hertz or number of cycles per second.

Frequency band: A range of frequencies that can be covered without switching.

Frequency deviation: The maximum difference between the instantaneous frequency and the carrier frequency of an FM signal.

Frequency domain representation: The portrayal of a signal in the frequency domain; representing a signal by displaying its sine wave components; the signal spectrum.

Frequency marker: An intensified or otherwise distinguished spot on a spectrum analyzer display that indicates a specified frequency point.

Frequency range: That range of frequencies over which the performance of the instrument is specified.

Fundamental frequency: The basic rate at which a signal repeats itself.

Grass: Noise or a noise-like signal giving the ragged, hashy appearance of grass seen close-up at eye level.

Graticule: The calibrated grid overlaying the display screen of spectrum analyzers, oscilloscopes, and other test instruments.

Harmonic distortion: The distortion that results when a signal interacts with a itself, often because of nonlinearties in the equipment, to produce sidebands at multiples, or harmonics, of the frequency components of the original signal.

Harmonic mixing: A technique wherein harmonics of the local oscillator signal are deliberately mixed with the input signal to achieve a large total input bandwidth. Enables a spectrum analyzer to function at higher frequencies than would otherwise be possible.

Harmonics: Frequency components of a signal occurring at multiples of the signal's fundamental frequency.

Heterodyne spectrum analyzer: A type of spectrum analyzer that scans the input signal by sweeping the incoming frequency band past one of a set of fixed RBW filters and measuring the signal level at the output of the filter.

Intermediate frequency (IF): In a heterodyne process, the sum of difference frequency at the output of a mixer stage which will be used for further signal processing.

IF gain: The gain of an amplifier stage operating at IF.

Instantaneous frequency: The rate of change of the phase of a sinusoidal signal at a particular instant.

Intermodulation distortion: The distortion that results when two or more signals interact, usually because of nonlinearities in the equipment, to produce new signals.

Linear scale: a scale wherein each increment represents a fixed difference between signal levels.

LO output: A port on a spectrum analyzer where a signal from the local oscillator (LO) is made available; used for tracking generators and external mixing.

Local oscillator (LO): An oscillator that produces the internal signal that is mixed with an incoming signal in a mixer to produce the IF signal.

Logarithmic scale: A scale wherein each scale increment represents a fixed ratio between signal levels.

Magnitude-only measurement: A measurement that responds only to the magnitude of a signal and is insensitive to its phase.

MAX HOLD: A spectrum analyzer feature that captures the maximum signal amplitude at all displayed frequencies over a series of sweeps.

Max span: The maximum frequency span that can be swept and displayed by a spectrum analyzer.

MAX/MIN: A display mode on some spectrum analyzers that shows the maximum and minimum signal levels at alternate frequency points, its advantage is its resemblance to an analog display.

Maximum input level: The maximum input signal amplitude that can be safely handled by a particular instrument.

MIN HOLD: A spectrum analyzer feature that captures the minimum signal amplitude at all displayed frequencies over a series of sweeps.

Mixing: The process whereby two or more signals are combined to produce sum and difference frequencies of the signals and their harmonics.

Noise: Unwanted random disturbances superimposed on a signal that tends to obscure it.

Noise bandwith: The frequency range of a noise-like signal. For white noise, the noise power is directly proportional to the bandwidth of the noise.

Noise floor: The self-noise of an instrument or system that represents the minimum limit at which input signals can be observed. The spectrum analyzer noise floor appears as a "grassy" baseline in the display, even when no signal is present.

Noise sideband: Undesired response caused by noise internal to the spectrum analyzer appearing on the display immediately around a desired response, often having a pedestal-like appearance.

Peak/average cursor: A manually controllable function that enables the user to set the threshold at which the type of signal processing changes prior to display in a digital storage system.

Peak detection: A detection scheme wherein the peak amplitude of a signal is measured and displayed. In spectrum analysis, 20 log(peak) is often displayed.

Peaking: The process of adjusting a circuit for maximum amplitude of a signal by aligning internal filters. In spectrum analysis, peaking is used to align preselectors.

Period: The time interval at which a process recurs; the inverse of the fundamental frequency.

Phase lock: The control of an oscillator or signal generator so as to operate at a constant phase angle relative to a stable reference signal source. Used to ensure frequency stability in spectrum analyzers.

Preselector: A tracking filter located ahead of the first mixer that allows only a narrow band of frequencies to pass into the mixer.

Products: Signal components resulting from mixing or from passing signals through other nonlinear operations such as modulation.

Pulse stretcher: Pulse shaper that produces an output pulse whose duration is greater than that of the input pulse and whose amplitude is proportional to that of the peak amplitude of the input pulse.

Pulse repetition frequency (PRF): The frequency at which a pulsing signal recurs; equal to the fundamental frequency of the pulse train.

Reference level: The signal level required to deflect the CRT display to the top graticule line.

Reference-level control: The control used to vary the reference level on a spectrum analyzer.

Resolution bandwidth (RBW): The width of the narrowest filter in the IF stages of a spectrum analyzer. The RBW determines how well the analyzer can resolve or separate two or more closely spaced signal components.

Return loss: The ration of power sent to a system to that returned by the system. In the case of antennas, the return loss can be used to find the SWR.

Ring: A transient response wherein a signal initially performs a damped oscillation about its steady-state value.

SAVE function: A feature of spectrum analyzers incorporating display storage that enables them to store displayed spectra.

Sensitivity: Measure of a spectrum analyzer's ability to display minimum level signals at a given IF bandwidth, display mode, and any other influencing factors.

Shape factor: In spectrum analysis, the ratio of an RBW filter's 60-dB bandwidth to its 3-dB or 6-dB width (depending on manufacturer).

Sideband: Signal components observable on either or both sides of a carrier as a result of modulation or distortion processes.

Sideband suppression: An amplitude modulation technique wherein one of the AM sidebands is deliberately suppressed, usually to conserve bandwidth.

Single sweep: Operating mode in which the sweep generator must be reset for each sweep. Especially useful for obtaining single examples of a signal spectrum.

Span per division, Span/Div: Frequency difference represented by each major horizontal division of the graticule.

Spectrum: The frequency domain representation of a signal wherein it is represented by displaying its frequency distribution.

Spurious response: An undesired extraneous signal produced by mixing, amplification, or other signal-processing technique.

Stability: The property of retaining defined electrical characteristics for a prescribed time and in specified environments.

Starting frequency: The frequency at the left-hand edge of the spectrum analyzer display.

Sweep speed, sweep rate: The speed or rate, expressed in time per horizontal divisions, at which the electron beam of the CRT sweeps the screen.

Time-domain representation: Representation of signals by displaying the signal amplitude as a function of time. Typical of oscilloscope and waveform monitor displays.

Time-varying signal: A signal whose amplitude changes with time.

Total span: The total width of the displayed spectrum. The Span/Div times the number of divisions.

Tracking generator: A signal generator whose output frequency is synchronized to the frequency being analyzed by the spectrum analyzer.

Ultimate rejection: The ratio, in dB, between a filter's passband response and its response beyond the slopes of its skirts.

Vertical scale factor, vertical display factor: The number of dB, volts, etc. represented by one vertical division of a spectrum analyzer display screen.

Waveform memory: Memory dedicated to storing a digital replica of a spectrum.

Waveform subtraction: A process wherein a saved waveform can be subtracted from a second, active waveform.

Zero hertz peak: A fictitious signal peak occurring at zero hertz that conveniently marks zero frequency. The peak is present regardless of whether or not there is an input signal.

Zero span: A spectrum analyzer mode of operation in which the RBW filter is stationary at the center frequency; the display is essentially a time-domain representation of the signal propagated through the RBW filter.

References

Kinley, H. 1987. Using service monitor/spectrum analyzer combos. *Mobile Radio Technology*, Intertec Publishing, Overland Park, KS.

Pepple, C. 1989. How to use a spectrum analyzer at the cell site. *Cellular Business*, Intertec Publishing, Overland Park, KS.

Whitaker, J. C. 1991. *Maintaining Electronic Systems*. CRC Press, Boca Raton, FL.

Wolf, R. J. 1987. Spectrum analyzer uses for two-way technicians. *Mobile Radio Technology*, Intertec Publishing, Overland Park, KS.

20.6 Fourier Waveform Analysis

Jerry C. Hamann and John W. Pierre

20.6.1 Introduction

Fourier waveform analysis originated in the early 1800s when Baron Jean Baptiste Joseph Fourier (1768–1830), a French mathematical physicist, developed these methods for investigating the conduction of heat in solid bodies. Fourier contended that rather complex **continuous-time waveforms** could be

decomposed into a summation of simple trigonometric functions, sines and cosines of differing frequencies and magnitudes. When Fourier announced his results at a meeting of the French Academy of
Scientists in 1807, his claims drew strong criticism from some of the most prominent mathematicians of
the time, most notably Lagrange. Publication of his work in-full was delayed until 1822 when his seminal
book, *Theorie Analytique de la Chaleur (The Analytical Theory of Heat)*, was finally released. Technical
doubts regarding Fourier's methods were largely dispelled by the late 19th century through the efforts of
Dirichlet, Riemann, and others.

Following this somewhat rocky beginning, the concepts of Fourier analysis have found application in
many fields of science, mathematics, and engineering. Indeed, the study of Fourier techniques is now
a traditional undertaking within undergraduate electrical engineering programs. With the availability of
computers, Fourier techniques have matured dramatically in the analysis of **discrete-time waveforms**. The
now ubiquitous **fast Fourier transform (FFT)** is a somewhat general title applied to efficient algorithms
for the machine computation of *discrete Fourier transforms* (DFTs).

20.6.2   The Mathematical Preliminaries for Fourier Analysis

The analysis techniques described in the following four sections can be conveniently categorized based on
the variety of waveform or function to which they apply. A summary of useful properties of the analysis
methods is provided in Table 20.2. We proceed herein to examine the mathematical preliminaries for the
Fourier methods.

Most if not all **periodic** and **aperiodic** continuous-time **waveforms** $x(t)$, of practical interest, have
Fourier series or Fourier transform counterparts. However, to aid in the existence question, the following
sufficient technical constraints on the function $x(t)$, known generally as the *Dirichlet conditions*, guarantee
convergence of the Fourier technique:

1. The function must be **single-valued**.
2. The function must have a finite number of discontinuities in the periodic interval, or, if aperiodic,
   about the entire real line.
3. The function must remain finite and have a finite number of maxima and minima in the periodic
   interval, or, if aperiodic, about the entire real line.
4. The function must be absolutely integrable, that is, if periodic,

$$\int_{t_0}^{t_0+T} |x(t)|dt$$

must be finite or if aperiodic

$$\int_{-\infty}^{+\infty} |x(t)|dt$$

must be finite.

The techniques of Fourier analysis are perhaps best appreciated initially by an example. Consider the ideal
continuous-time square wave of period T s, having a maximum value of A and a minimum value of 0 as
shown at the top of Fig. 20.68. Because of the periodic nature of the waveform, the technique of Fourier
series (FS) analysis applies, and the resulting decomposition into an infinite sum of simple trigonometric
functions is given by

$$x(t) = \frac{A}{2} + \sum_{k=1,3,5,\dots}^{\infty} \frac{2A}{k\pi}(-1)^{(k-1)/2} \cos(k\Omega_0 t) \qquad (20.69)$$

TABLE 20.2 Summary of Mathematical Properties of Fourier Analysis Methods

Fourier Series (FS)	Fourier Transform (FT)	Discrete-Time Fourier Transform (DTFT)	Discrete-Time Fourier Series (DTFS)
time domain: continuous and periodic	time domain: continuous and aperiodic	time domain: discrete and aperiodic	time domain: discrete and periodic
frequency domain: discrete and aperiodic	frequency domain: continuous and aperiodic	frequency domain: continuous and periodic	frequency domain: discrete and periodic
linearity	linearity	linearity	linearity
$r_1 x_1(t) + r_2 x_2(t) \leftrightarrow r_1 C_{1k} + r_2 C_{2k}$	$r_1 x_1(t) + r_2 x_2(t) \leftrightarrow r_1 X_1(\Omega) + r_2 X_2(\Omega)$	$r_1 x_1(n) + r_2 x_2(n) \leftrightarrow r_1 X_1(\omega) + r_2 X_2(\omega)$	$r_1 x_1(n) + r_2 x_2(n) \leftrightarrow r_1 c_{1k} + r_2 c_{2k}$
shift properties	shift properties	shift properties	shift properties
$x(t - \tau) \leftrightarrow e^{-j\Omega_0 k\tau} C_k$	$x(t - \tau) \leftrightarrow e^{-j\Omega\tau} X(\Omega)$	$x(n - m) \leftrightarrow e^{-j\omega m} X(\omega)$	$x(n - m) \leftrightarrow e^{-j\omega_0 km} c_k$
$e^{j\Omega_0 mt} x(t) \leftrightarrow C_{k-m}$	$e^{j\Omega_0 t} x(t) \leftrightarrow X(\Omega - \Omega_0)$	$e^{j\omega_0 n} x(n) \leftrightarrow X(\omega - \omega_0)$	$e^{j\omega_0 mn} x(n) \leftrightarrow c_{k-m}$
convolution	convolution	convolution	convolution
$x_1(t) * x_2(t) \leftrightarrow C_{1k} C_{2k}$	$x_1(t) * x_2(t) \leftrightarrow X_1(\Omega) X_2(\Omega)$	$x_1(n) * x_2(n) \leftrightarrow X_1(\omega) X_2(\omega)$	$x_1(n) * x_2(n) \leftrightarrow c_{1k} c_{2k}$
$x_1(t) x_2(t) \leftrightarrow C_{1k} * C_{2k}$	$x_1(t) x_2(t) \leftrightarrow 1/2\pi X_1(\Omega) * X_2(\Omega)$	$x_1(n) x_2(n) \leftrightarrow 1/2\pi X_1(\omega) * X_2(\omega)$	$x_1(n) x_2(n) \leftrightarrow c_{1k} * c_{2k}$
modulation	modulation	modulation	modulation
$x(t) \cos(\Omega_0 mt)$	$x(t) \cos(\Omega_0 t)$	$x(n) \cos(\omega_0 n)$	$x(n) \cos(\omega_0 mn)$
$\leftrightarrow \frac{1}{2}[C_{k+m} + C_{k-m}]$	$\leftrightarrow \frac{1}{2}[X(\Omega + \Omega_0) + X(\Omega - \Omega_0)]$	$\leftrightarrow \frac{1}{2}[X(\omega + \omega_0) + X(\omega - \omega_0)]$	$\leftrightarrow \frac{1}{2}[c_{k+m} + c_{k-m}]$
Parseval's Theorem	Parseval's Theorem	Parseval's Theorem	Parseval's Theorem
$\dfrac{1}{T}\displaystyle\int_{t_0}^{t_0+T} x_1(t) x_2^*(t)\,dt$	$\displaystyle\int_{-\infty}^{\infty} x_1(t) x_2^*(t)\,dt$	$\displaystyle\sum_{n=-\infty}^{\infty} x_1(n) x_2^*(n)$	$\dfrac{1}{N}\displaystyle\sum_{n=0}^{N-1} x_1(n) x_2^*(n)$
$= \displaystyle\sum_{k=-\infty}^{\infty} C_{1k} C_{2k}^*$	$= \dfrac{1}{2\pi}\displaystyle\int_{-\infty}^{\infty} X_1(\Omega) X_2^*(\Omega)\,d\Omega$	$= \dfrac{1}{2\pi}\displaystyle\int_{-\pi}^{\pi} X_1(\omega) X_2^*(\omega)\,d\omega$	$= \displaystyle\sum_{k=0}^{N-1} c_{1k} c_{2k}^*$
real signals	real signals	real signals	real signals
$C_k = C_{-k}^*$	$X(\Omega) = X^*(-\Omega)$	$X(\omega) = X^*(-\omega)$	$c_k = c_{-k}^*$
$\text{Re}\, C_k = \text{Re}\, C_{-k}$	$\text{Re}\, X(\Omega) = \text{Re}\, X(-\Omega)$	$\text{Re}\, X(\omega) = \text{Re}\, X(-\omega)$	$\text{Re}\, c_k = \text{Re}\, c_{-k}$
$\text{Im}\, C_k = -\text{Im}\, C_{-k}$	$\text{Im}\, X(\Omega) = -\text{Im}\, X(-\Omega)$	$\text{Im}\, X(\omega) = -\text{Im}\, X(-\omega)$	$\text{Im}\, c_k = -\text{Im}\, c_{-k}$
$|C_k| = |C_{-k}|$	$|X(\Omega)| = |X(-\Omega)|$	$|X(\omega)| = |X(-\omega)|$	$|c_k| = |c_{-k}|$
$\arg C_k = -\arg C_{-k}$	$\arg X(\Omega) = -\arg X(-\Omega)$	$\arg X(\omega) = -\arg X(-\omega)$	$\arg c_k = -\arg c_{-k}$
$x(t)$ real and even C_k real and even	$x(t)$ real and even $X(\Omega)$ real and even	$x(n)$ real and even $X(\omega)$ real and even	$x(n)$ real and even c_k real and even
$x(t)$ real and odd C_k imag. and odd	$x(t)$ real and odd $X(\Omega)$ imag. and odd	$x(n)$ real and odd $X(\omega)$ imag. and odd	$x(n)$ real and odd c_k imag. and odd

In Eq. (20.69), the average value or DC term of the waveform is given by $A/2$, whereas the **fundamental frequency** is described by $\Omega_0 = 2\pi/T$. Because the waveform is an **even function** only cosine terms are present. Because the waveform, less its average value, is characterized by **half-wave symmetry**, only the **odd harmonics** are present. To examine the reconstruction of the function $x(t)$ from the given decomposition, the middle plot of Fig. 20.68 displays the DC term, and the first and third harmonics. The sum of these terms is overlayed with the ideal square wave at the bottom of Fig. 20.68. This progressive reconstruction is continued in Fig. 20.69 with successive odd harmonics added as per Eq. (20.69). The ultimate goal of reconstructing the ideal square wave theoretically demands completion of the infinite sum, an unrealistic task of computation or analog signal summation. By truncating the summation at a large enough value of the harmonic number k, we are assured of having a best approximation of the ideal square wave in a least-square-error sense: that is, the truncated Fourier series is the best fit at any given truncation level $k = k_{max} < \infty$. The ringing or overshoot, which forms near the discontinuity of the ideal square wave, is described by the **Gibbs phenomenon**, in tribute to an early investigator of the effects of truncating Fourier series summations.

In practice, we are seldom presented with ideal functions such as the square wave of Fig. 20.68. Indeed, an analytical description of the waveform of interest, as is implied by the use of $x(t)$ in formulas of the next two sections, is rarely available.

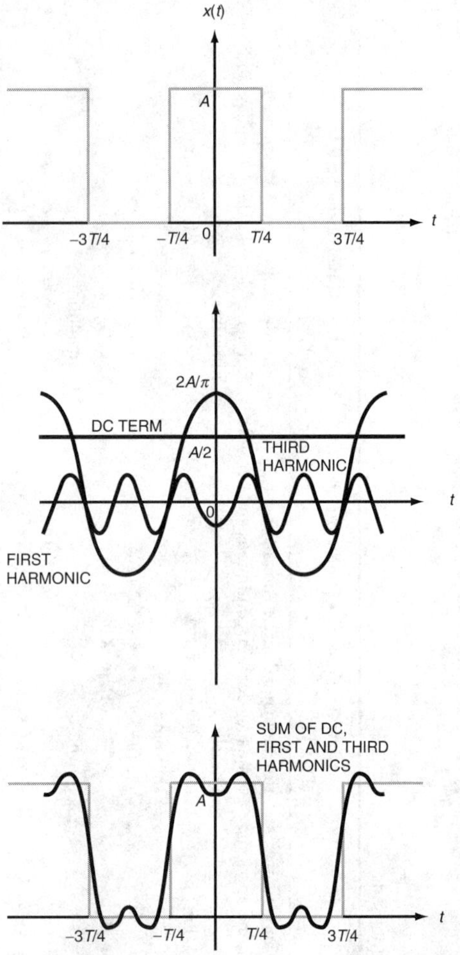

FIGURE 20.68   Example reconstruction of ideal square wave from Fourier series summation.

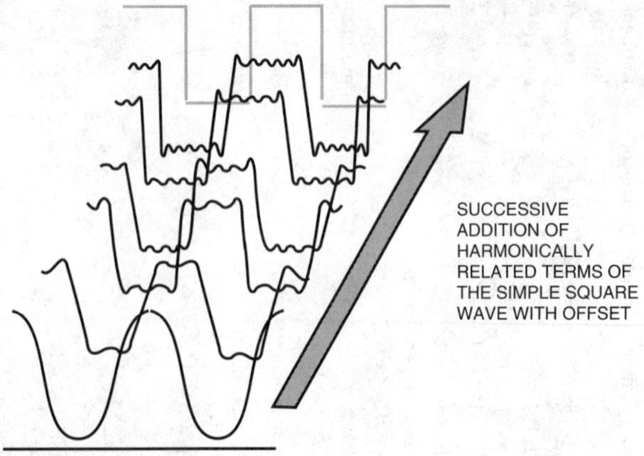

FIGURE 20.69   Continued reconstruction of ideal square wave from Fourier series summation.

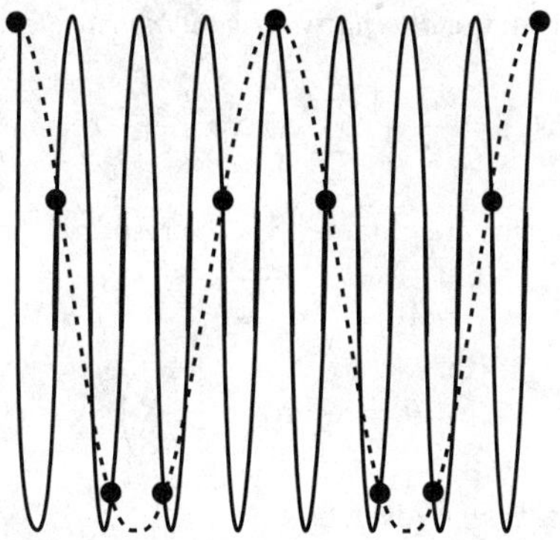

FIGURE 20.70   Example of aliasing wherein uniformly spaced samples from a high-frequency signal are indistinguishable from samples of a low-frequency signal.

The discrete-time waveform counterparts of the Fourier series and Fourier transform provide viable alternatives for estimating the frequency content of signals if discrete measurements of the desired waveforms are available. This avenue of analysis, which is popular in benchtop instrumentation, is considered in further detail in Sec. 20.6.7.

One critical mathematical preliminary for the processing of discrete-time waveforms, which originate from sampling continous-time waveforms, is the issue of **sampling rate** and **aliasing**. The *Nyquist sampling theorem* can be summarized as follows:

> If $x(t)$ is a **bandlimited** continuous-time waveform having no frequency content greater than or equal to F_N Hz (the **Nyquist frequency**), then $x(t)$ is uniquely determined by a discrete-time waveform $x(n)$, which results from uniform sampling of $x(t)$ at a rate F_S, which is greater than or equal to $2F_N$ (the **Nyquist rate**).

Failure to obey the Nyquist sampling theorem results in aliasing, that is, the effect of frequencies greater than the Nyquist frequency are reflected into the band of frequencies between DC and one-half the sampling rate. This situation is demonstrated in Fig. 20.70, where the samples of the high-frequency sinusoid are indistinguishable from a uniform sampling of the low-frequency sinusoid.

20.6.3   The Fourier Series for Continuous-Time Periodic Functions

Three forms of the FS are commonly encountered for continous-time waveforms $x(t)$ of period T. The first is the *trigonometric form*,

$$x(t) = \frac{a_0}{2} + \sum_{k=1}^{\infty} \left[a_k \cos\left(\frac{2\pi k}{T}t\right) + b_k \sin\left(\frac{2\pi k}{T}t\right) \right] \qquad (20.70)$$

where the coefficients a_k and b_k are calculated via the integrals

$$a_k = \frac{2}{T} \int_{t_0}^{t_0+T} x(t) \cos\left(\frac{2\pi k}{T}t\right) dt \qquad (20.71)$$

$$b_k = \frac{2}{T} \int_{t_0}^{t_0+T} x(t) \sin\left(\frac{2\pi k}{T}t\right) dt \qquad (20.72)$$

with the representation for a_0 often appearing without the multiplier and divider of two (and therefore directly representing the average value of $x(t)$). Symmetry properties of the particular function $x(t)$ can lead to simplifications of Eq. (20.71) and Eq. (20.72).

A slight adaptation of this leads to the *alternative trigonometric form,*

$$x(t) = \frac{a_0}{2} + \sum_{k=1}^{\infty} A_k \cos\left(\frac{2\pi k}{T}t + \theta_k\right) \tag{20.73}$$

where the magnitude and phase terms are obtained via the respective identities

$$A_k = \sqrt{a_k^2 + b_k^2} \tag{20.74}$$

$$\theta_k = -\arctan\left(\frac{b_k}{a_k}\right) \tag{20.75}$$

Finally, the *complex exponential form* is given by

$$x(t) = \sum_{k=-\infty}^{\infty} C_k e^{j\frac{2\pi k}{T}t} \tag{20.76}$$

where the coefficients C_k are calculated via

$$C_k = \frac{1}{T}\int_{t_0}^{t_0+T} x(t)e^{-j\frac{2\pi k}{T}t}dt \tag{20.77}$$

It should be noted that the C_k are, in general, complex valued. They may be obtained from the trigonometric form data as follows:

$$|C_k| = |C_{-k}| = \frac{A_k}{2} \quad \text{for } k = 0, 1, 2, \ldots \tag{20.78}$$

$$\arg C_k = -\arg C_{-k} = \theta_k \quad \text{for } k = 0, 1, 2, \ldots \tag{20.79}$$

Common graphical representations of these data include plots of $|C_k|$ (or A_k) and θ_k, which are referred to as the *magnitude* and *phase spectra*, respectively. As is implied by the indexing of the spectral data by the integer k, the Fourier series of a continous-time waveform is itself discrete in character, that is, it takes on nonzero values at only discrete, isolated points of the frequency axis, namely, the frequencies $\Omega_k = 2\pi k/T$.

The average power in a periodic waveform can be found in one of two ways, with the equality noted below commonly attributed to Parseval:

$$P = \frac{1}{T}\int_{t_0}^{t_0+T} |x(t)|^2 dt = \sum_{k=-\infty}^{\infty} |C_k|^2 \tag{20.80}$$

Because of the right-hand side of Eq. (20.80), a plot of $|C_k|^2$ is commonly referred to as the *power spectral density*.

20.6.4   The Fourier Transform for Continuous-Time Aperiodic Functions

The Fourier transform (FT) for an aperiodic, continous-time waveform $x(t)$ is given by

$$X(\Omega) = \mathcal{F}\{x(t)\} = \int_{-\infty}^{\infty} x(t)e^{-j\Omega t}dt \tag{20.81}$$

where the implication is that the frequency-domain counterpart $X(\Omega)$ of $x(t)$ is continuous in frequency and can be used to reconstruct $x(t)$ via

$$x(t) = \mathcal{F}^{-1}\{X(\Omega)\} = \frac{1}{2\pi}\int_{-\infty}^{\infty} X(\Omega)e^{j\Omega t}d\Omega \tag{20.82}$$

If the integration described in Eq. (20.82) is carried out in hertz rather than radians per second, the division by the factor 2π is removed.

The transformation given by Eq. (20.81) can also be applied to some familiar periodic functions as well as functions that are not absolutely integrable. Example Fourier transform pairs are shown in Table 20.3.

TABLE 20.3 Example Fourier Transform Pairs

$x(t)$	$x(t)$	$X(\Omega)$	$|X(\Omega)|$
	$\delta(t)$	1	
	1	$2\pi\delta(\Omega)$	
	$u(t)$	$\pi\delta(\Omega) + \dfrac{1}{j\Omega}$	
	$e^{-at}u(t), a > 0$	$\dfrac{1}{a + j\Omega}$	
	$te^{-at}u(t), a > 0$	$\dfrac{1}{(a + j\Omega)^2}$	
	$e^{-at}\cos(\Omega_0 t)u(t), \; a>0$	$\dfrac{a + j\Omega}{(a + j\Omega)^2 + \Omega_0^2}$	
	$\cos(\Omega_0 t)$	$\pi[\delta(\Omega - \Omega_0) + \delta(\Omega + \Omega_0)]$	
	$1, |t| < \tau/2$ $0, |t| > \tau/2$	$\dfrac{\sin(\Omega\tau/2)}{\Omega/2}$	
	$1 - |t|/\tau, |t| < \tau$ $0, |t| > \tau$	$\dfrac{1}{\tau}\dfrac{\sin^2(\Omega\tau/2)}{(\Omega/2)^2}$	
	$\dfrac{\sin(\Omega_0 t/2)}{\pi t}$	$1, |\Omega| < \Omega_0$ $0, |\Omega| > \Omega_0$	

The total energy in an aperiodic waveform can be found in one of two ways, with the equality noted commonly attributed to Parseval,

$$E = \int_{-\infty}^{\infty} |x(t)|^2 dt = \frac{1}{2\pi}\int_{-\infty}^{\infty} |X(\Omega)|^2 d\Omega \qquad (20.83)$$

Because of the right-hand side of Eq. (20.83), a plot of $|X(\Omega)|^2$ is commonly referred to as the *energy spectral density*. One advantage to the integration vs. frequency is the ability to determine the amount of energy in a signal due to frequency components within a particular range of frequencies.

20.6.5 The Fourier Series for Discrete-Time Periodic Functions

The discrete-time Fourier series (DTFS) for discrete-time waveforms $x(n)$ of period N can also be given in three forms; however, the *complex exponential form* is by far the most common,

$$x(n) = \sum_{k=0}^{N-1} c_k e^{j\frac{2\pi k}{N}n} \qquad (20.84)$$

where the coefficients c_k are calculated via the finite summation

$$c_k = \frac{1}{N} \sum_{n=0}^{N-1} x(n) e^{-j \frac{2\pi k}{N} n} \qquad (20.85)$$

Because of the periodicity of the exponential in Eq. (20.85), the DTFS coefficients c_k form a discrete, periodic sequence of period N if the index k is extended beyond the fundamental range of 0 to $N - 1$.

The average power in a periodic, discrete-time waveform can be found in one of two ways, with the equality noted commonly attributed to Parseval,

$$P = \frac{1}{N} \sum_{n=0}^{N-1} |x(n)|^2 = \sum_{k=0}^{N-1} |c_k|^2 \qquad (20.86)$$

Because of the right-hand side of Eq. (20.86), a plot of $|c_k|^2$ is commonly referred to as the power spectral density.

20.6.6   The Fourier Transform for Discrete-Time Aperiodic Functions

The discrete-time Fourier transform (DTFT) for aperiodic, finite energy discrete-time waveforms $x(n)$ is given by

$$X(\omega) = \sum_{n=-\infty}^{\infty} x(n) e^{-j\omega n} \qquad (20.87)$$

where the implication is that the frequency-domain counterpart $X(\omega)$ of $x(n)$ is continuous in the discrete frequency variable ω and can be used to reconstruct $x(n)$ via

$$x(n) = \frac{1}{2\pi} \int_{-\pi}^{\pi} X(\omega) e^{j\omega n} d\omega \qquad (20.88)$$

If the integration described in Eq. (20.88) is carried out in cycles per sample rather than radian per sample frequency, the division by the factor 2π is removed.

The transformation given by Eq. (20.87) can also be applied to some periodic functions as well as functions that are not absolutely summable. Examples of discrete-time Fourier transform pairs are shown in Table 20.4.

The total energy in an aperiodic, discrete-time waveform can be found in one of two ways, with the equality noted commonly attributed to Parseval,

$$E = \sum_{n=-\infty}^{\infty} |x(n)|^2 = \frac{1}{2\pi} \int_{\pi}^{-\pi} |X(\omega)|^2 d\omega \qquad (20.89)$$

Because of the right-hand side of Eq. (20.89), a plot of $|X(\omega)|^2$ is commonly referred to as the energy spectral density.

20.6.7   Example Applications of Fourier Waveform Techniques

Using the DFT/FFT in Fourier Analysis

The FFT is arguably the most significant advancement in signal analysis in recent decades. The FFT is actually the name given to many algorithms used to efficiently compute the DFT. The DFT of $x(n)$ is defined as

$$X(k) = DFT\{x(n)\} = \sum_{n=0}^{N-1} x(n) e^{-j \frac{2\pi k}{N} n} \quad \text{for } k = 0, 1, 2, \ldots, N-1 \qquad (20.90)$$

TABLE 20.4 Example Discrete-Time Fourier Transform Pairs

$x(n)$	$x(n)$	$X(\omega)$	$|X(\omega)|$
	$\delta(n)$	1	
	1	$\displaystyle\sum_{k=-\infty}^{\infty} 2\pi\delta(\omega + 2\pi k)$	
	$u(n)$	$\displaystyle\frac{1}{1 - e^{-j\omega}} + \sum_{k=-\infty}^{\infty}\pi\delta(\omega + 2\pi k)$	
	$e^{-an}u(n),\, a > 0$	$\displaystyle\frac{1}{1 - e^{-(a+j\omega)}}$	
	$ne^{-an}u(n),\, a > 0$	$\displaystyle\frac{e^{-(a+j\omega)}}{\left(1 - e^{-(a+j\omega)}\right)^2}$	
	$e^{-an}\cos(\omega_0 n)u(n),$ $a > 0$	$\displaystyle\frac{1 - \cos(\omega_0)e^{-(a+j\omega)}}{1 - 2\cos(\omega_0)e^{-(a+jw)} + e^{-2(a+j\omega)}}$	
	$\cos(\omega_0 n)$	$\displaystyle\pi\sum_{k=-\infty}^{\infty}[\delta(\omega - \omega_0 + 2\pi k)$ $+ \delta(\omega + \omega_0 + 2\pi k)]$	
	$1, |n| < L/2$ $0, |n| > (L-1)/2$ L an odd integer	$\displaystyle\frac{\sin(\omega L/2)}{\sin(\omega/2)}$	
	$1 - |n|/L, |n| < L$ $0, |n| \geq L$ L an odd integer	$\displaystyle\frac{1}{L}\frac{\sin^2(\omega L/2)}{\sin^2(\omega/2)}$	
	$\displaystyle\frac{\sin(\omega_0 n)}{\pi n}$	$1, |\omega| < \omega_0$ $0, \omega_0 < |\omega| < \pi$	

and the inverse transform is given by

$$x(n) = IDFT\{X(k)\} = \frac{1}{N}\sum_{k=0}^{N-1} X(k)e^{j\frac{2\pi n}{N}k} \quad \text{for } n = 0, 1, 2, \ldots, N - 1 \tag{20.91}$$

The DFT/FFT can be used to compute or approximate the four Fourier methods described in Sec. 20.6.3 to Sec. 20.6.6, that is the FS, FT, DTFS, and DTFT. Algorithms for the FFT are discussed extensively in the literature (see the **Further Information** entry for this topic).

The DTFS of a periodic, discrete-time waveform can be directly computed using the DFT. If $x(n)$ is one period of the desired signal, then the DTFS coefficients are given by

$$c_k = \frac{1}{N}DFTx(n) \quad \text{for } k = 0, 1, \ldots, N - 1 \tag{20.92}$$

The DTFT of a waveform is a continuous function of the discrete or sample frequency ω. The DFT can be used to compute or approximate the DTFT of an aperiodic waveform at uniformly separated *sample frequencies* $k\omega_0 = k2\pi/N$, where k is any integer between 0 and $N - 1$. If N samples fully describe a finite duration $x(n)$, then

$$X(\omega)|_{\omega = \frac{2\pi}{N}k} = DFTx(n) \quad \text{for } k = 0, 1, \ldots, N - 1 \tag{20.93}$$

If the $x(n)$ in Eq. (20.93) is not of finite duration, then the equality should be changed to an approximately equal. In many practical cases of interest, $x(n)$ is not of finite duration. The literature contains many approaches to truncating and **windowing** such $x(n)$ for Fourier analysis purposes.

The continuous-time FS coefficients can also be approximated using the DFT. If a continuous-time waveform $x(t)$ of period T is sampled at a rate of T_S samples per second to obtain $x(n)$, in accordance with the Nyquist sampling theorem, then the FS coefficients are approximated by

$$C_k \approx \frac{T_s}{T} DFT\{x(n)\} \quad \text{for } k = 0, 1, \ldots, N-1 \tag{20.94}$$

where N samples of $x(n)$ should represent precisely one period of the original $x(t)$. Using the DFT in this manner is equivalent to computing the integral in Eq. (20.77) with a simple rectangular or Euler approximate.

In a similar fashion, the continuous-time FT can be approximated using the DFT. If an aperiodic, continuous-time waveform $x(t)$ is sampled at a rate of T_S samples per second to obtain $x(n)$, in accordance with the Nyquist sampling theorem, then uniformally separated samples of the FT are approximated by

$$X(\Omega)|_{\Omega=\frac{2\pi}{NT_s}k} \approx T_s(DFT\{x(n)\}) \quad \text{for } k = 0, 1, \cdots, N-1 \tag{20.95}$$

For $x(t)$ that are not of finite duration, truncating and windowing can be applied to improve the approximation.

Total Harmonic Distortion Measures

The applications of Fourier analysis are extensive in the electronics and instrumentation industry. One typical application, the computation of total harmonic distortion (THD), is described herein. This application provides a measure of the nonlinear distortion, which is introduced to a pure sinusoidal signal when it passes through a system of interest, perhaps an amplifier. The root-mean-square (rms) total harmonic distortion (THD$_{rms}$) is defined as the ratio of the rms value of the sum of the harmonics, not including the fundamental, to the rms value of the fundamental,

$$\text{THD}_{rms} = \frac{\sqrt{\sum_{k=2}^{\infty} A_k^2}}{A_1} \tag{20.96}$$

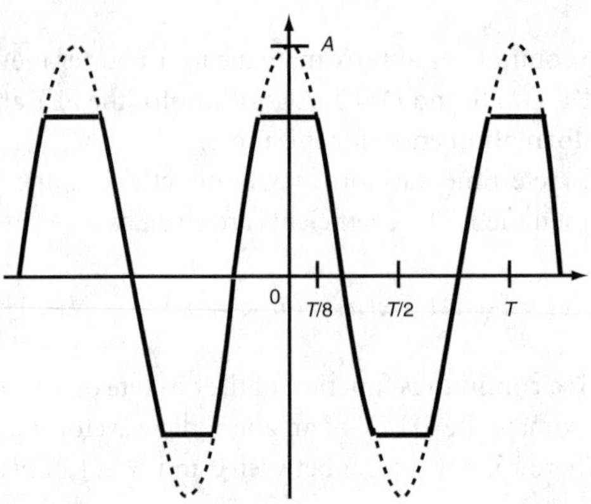

FIGURE 20.71 A clipped sinusoidal waveform for total harmonic distortion example.

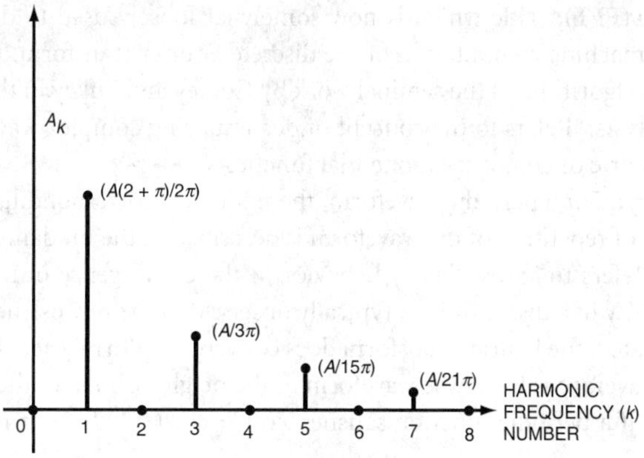

FIGURE 20.72 The magnitude spectrum, through eighth harmonic, for the clipped sinusoid total harmonic distortion example.

As an example, consider the clipped sinusoidal waveform shown in Fig. 20.71. The Fourier series representation of this waveform is given by

$$v(t) = A\left(\frac{1}{2} + \frac{1}{\pi}\right)\cos\left(\frac{2\pi}{T}t\right) + \frac{2\sqrt{2}A}{\pi}\sum_{k=3,5,7,\ldots}^{\infty}\frac{\sin\left(\frac{\pi}{4}k\right) - k\cos\left(\frac{\pi}{4}k\right)}{k(1-k^2)}\cos\left(\frac{2\pi}{T}kt\right) \quad (20.97)$$

The magnitude spectrum for this waveform is shown in Fig. 20.72. Because of symmetry, only the odd harmonics are present. Clearly the fundamental is the dominant component, but due to the clipping, many additional harmonics are present. The THD_{rms} is readily computed from the coefficients in Eq. (20.97), yielding 13.42% for this example. In practice, the coefficients of the Fourier series are typically approximated via the DFT/FFT, as described at the beginning of this section.

Defining Terms

Aliasing: Refers to an often detrimental phenomenon associated with the sampling of continous-time waveforms at a rate below the Nyquist rate. Frequencies greater than one-half the sampling rate become indistinguishable from frequencies in the fundamental bandwidth, that is, between DC and one-half the sampling rate.

Aperiodic waveform: This phrase is used to describe a waveform that does not repeat itself in a uniform, periodic manner. Compare with periodic waveform.

Bandlimited: A waveform is described as bandlimited if the frequency content of the signal is constrained to lie within a finite band of frequencies. This band is often described by an upper limit, the Nyquist frequency, assuming frequencies from DC up to this upper limit may be present. This concept can be extended to frequency bands that do not include DC.

Continuous-time waveform: A waveform, herein represented by $x(t)$, that takes on values over the continuum of time t, the assumed independent variable. The Fourier series and Fourier transform apply to continuous-time waveforms. Compare with discrete-time waveform.

Discrete-time waveform: A waveform, herein represented by $x(n)$, that takes on values at a countable, discrete set of sample times or sample numbers n, the assumed independent variable. The discrete Fourier transform, the discrete-time Fourier series, and discrete-time Fourier transform apply to discrete-time waveforms. Compare with continuous-time waveform.

Even function: If a function $x(t)$ can be characterized as being a mirror image of itself horizontally about the origin, it is described as an even function. Mathematically, this demands that $x(t) = x(-t)$ for all t. The name arises from the standard example of an even function as a polynomial containing only even powers of the independent variable. A pure cosine is also an even function.

Fast Fourier transform (FFT): Title which is now somewhat loosely used to describe any efficient algorithm for the machine computation of the discrete Fourier transform. Perhaps the best known example of these algorithms is the seminal work by Cooley and Tukey in the early 1960s.

Fourier waveform analysis: Refers to the concept of decomposing complex waveforms into the sum of simple trigonometric or complex exponential functions.

Fundamental frequency: For a periodic waveform, the frequency corresponding to the smallest or fundamental period of repetition of the waveform is described as the fundamental frequency.

Gibbs phenomenon: Refers to an oscillatory behavior in the convergence of the Fourier transform or series in the vicinity of a discontinuity, typically observed in the reconstruction of a discontinuous $x(t)$. Formally stated, the Fourier transform does converge uniformly at a discontinuity, but rather, converges to the average value of the waveform in the neighborhood of the discontinuity.

Half-wave symmetry: If a periodic function satisfies $x(t) = -x(t - T/2)$, it is said to have half-wave symmetry.

Harmonic frequency: Refers to any frequency that is a positive integer multiple of the fundamental frequency of a periodic waveform.

Nyquist frequency: For a bandlimited waveform, the width of the band of frequencies contained within the waveform is described by the upper limit known as the Nyquist frequency.

Nyquist rate: To obey the Nyquist sampling theorem, a bandlimited waveform should be sampled at a rate which is at least twice the Nyquist frequency. This minimum sampling rate is known as the Nyquist rate. Failure to follow this restriction results in aliasing.

Odd function: If a function $x(t)$ can be characterized as being a reverse mirror image of itself horizontally about the origin, it is described as an odd function. Mathematically, this demands that $x(t) = -x(-t)$ for all t. The name arises from the standard example of an odd function as a polynomial containing only odd powers of the independent variable. A pure sine is also an odd function.

Periodic waveform: This phrase is used to describe a waveform that repeats itself in a uniform, periodic manner. Mathematically, for the case of a continuous-time waveform, this characteristic is often expressed as $x(t) = x(t \pm kT)$, which implies that the waveform described by the function $x(t)$ takes on the same value for any increment of time kT, where k is any integer and the characteristic value T, a real number greater than zero, describes the fundamental period of $x(t)$. For the case of a discrete-time waveform, we write $x(n) = x(n \pm kN)$, which implies that the waveform $x(n)$ takes on the same value for any increment of sample number kN, where k is any integer and the characteristic value N, an integer greater than zero, describes the fundamental period of $x(n)$.

Quarter-wave symmetry: If a periodic function displays half-wave symmetry and is *even* symmetric about the $\frac{1}{4}$ and $\frac{3}{4}$ period points (the negative and positive lobes of the function), then it is said to have quarter-wave symmetry.

Sampling rate: Refers to the frequency at which a continuous-time waveform is sampled to obtain a corresponding discrete-time waveform. Values are typically given in hertz.

Single-valued: For a function of a single variable, such as $x(t)$, single-valued refers to the quality of having one and only one value $y_0 = x(t_0)$ for any t_0. The square root is an example of a function which is not single-valued.

Windowing: A term used to describe various techniques for preconditioning a discrete-time waveform before processing by algorithms such as the discrete Fourier transform. Typical applications include extracting a finite duration approximation of an infinite duration waveform.

References

Bracewell, R.N. 1989. The Fourier transform. *Sci. Amer.* (June):86–95.

Brigham, E.O. 1974. *The Fast Fourier Transform.* Prentice-Hall, Englewood Cliffs, NJ.

Nilsson, J.W. 1993. *Electric Circuits,* 4th ed. Addison-Wesley, Reading, MA.

Oppenheim, A.V. and Schafer, R. 1989. *Discrete-Time Signal Processing.* Prentice-Hall, Englewood Cliffs, NJ.

Proakis, J.G. and Manolakis, D.G. 1992. *Digital Signal Processing: Principles, Algorithms, and Applications*, 2nd ed. Macmillan, New York.

Ramirez, R.W. 1985. *The FFT, Fundamentals and Concepts*. Prentice-Hall, Englewood Cliffs, NJ.

Further Information

For in-depth descriptions of practical instrumentation incorporating Fourier analysis capability, consult the following sources:

Witte, R.A. 1993. *Spectrum and Network Measurements*. Prentice-Hall, Englewood Cliffs, NJ.

Witte, R.A. 1993. *Electronic Test Instruments: Theory and Applications*. Prentice-Hall, Englewood Cliffs, NJ.

For further investigation of the history of Fourier and his transform, the following source should prove interesting:

Herviel, J. 1975. *Joseph Fourier: The Man and the Physicist*. Clarendon Press.

A thoroughly engaging introduction to Fourier concepts, accessible to the reader with only a fundamental background in trigonometry, can be found in the following publication:

Who is Fourier? A Mathematical Adventure. Transnational College of LEX, English translation by Alan Gleason, Language Research Foundation, Boston, MA, 1995.

Perspectives on the FFT algorithms and their application can be found in the following sources:

Cooley, J.W. and Tukey, J.W. 1965. An algorithm for the machine calculation of complex Fourier series. *Mathematics of Computation* 19(April):297–301.

Cooley, J.W. 1992. How the FFT gained acceptance. *IEEE Signal Processing Magazine* (Jan.):10–13, 1992.

Heideman, M.T., Johnson, D.H., and Burrus, C.S. 1984. Gauss and the history of the fast fourier transform. *IEEE ASSP Magazine* 1:14–21.

Kraniauskas, P. 1994. A plain man's guide to the FFT. *IEEE Signal Processing Magazine* (April):24–35.

Press, W.H., Flannery, B.P., Teukolsky, S.A., and Vetterling, W.T. 1988. *Numerical Recipes in C: The Art of Scientific Computing*. Cambridge University Press.

20.7 Digital Test Instruments

Jerry C. Whitaker

20.7.1 Introduction

As the equipment used by consumers and industry becomes more complex, the requirements for highly-skilled maintenance technicians also increases. Maintenance personnel today require advanced test equipment and must think in a "systems mode" to troubleshoot much of the hardware now in the field. New technologies and changing economic conditions have reshaped the way maintenance professionals view their jobs. As technology drives equipment design forward, maintenance difficulties will continue to increase. Such problems can be met only through improved test equipment and increased technician training.

Servicing computer-based professional equipment typically involves isolating the problem to the board level and then replacing the defective PWB. Taken on a case-by-case basis, this approach seems efficient. The inefficiency in the approach, however (which is readily apparent), is the investment required to keep a stock of spare boards on hand. Furthermore, because of the complex interrelation of circuits today, a PWB that appeared to be faulty may actually turn out to be perfect. The ideal solution is to troubleshoot down to the component level and replace the faulty device instead of swapping boards. In many cases, this approach requires sophisticated and expensive test equipment. In other cases, however, simple test instruments will do the job.

Although the cost of most professional equipment has been going up in recent years, maintenance technicians have seen a buyer's market in test instruments. The semiconductor revolution has done more than given consumers low-cost computers and disposable devices. It has also helped to spawn a broad variety of inexpensive test instruments with impressive measurement capabilities.

20.7.2　Digital Multimeters

Beyond the basic voltage-resistance-current measurements, sophisticated digital multimeters (DMMs) provide a number of unique functions. Large-scale integration has made it possible to pack more features into the limited space inside a DMM, and to do so without significantly increasing the cost of the instrument. In addition, because of competition among manufacturers, the functionality of DMMs is continually improving. The result: increasingly sophisticated products.

Digital multimeters can be divided into two basic classifications:

- **Portable**. Hand-held instruments intended for field applications offer extensive capabilities in a portable, rugged package. Some hand-held meters include true-RMS measurements, 1 msec response times, capacitance and frequency modes, and recording capabilities.
- **Benchtop**. Instruments intended for fixed applications usually offer a wide range of features and interface to other instruments. Options include display devices with multiple readouts, and built-in communications capabilities. A communications interface permits the meter to be controlled over standard computer ports for data storage, analysis, and remote operation.

Evolutionary developments in DMM design have provided a number of helpful features to maintenance technicians, including

- Audible continuity check. The instrument emits a beep when the measured resistance value is less than 5 Ω or so. This permits point-to-point continuity tests to be performed without actually looking at the DMM.
- Audible high voltage warning. The instruments emits a characteristic beep when the measured voltage is dangerously high (100 V or more). This safety feature alerts the technician to use extra care when making measurements.
- Sample-and-hold. The instrument measures and stores the sampled value after it has stabilized. Completion of the sample function is signaled by an audible beep. This feature permits the technician to concentrate on probe placement, not on trying to read the instrument.
- Continuously variable tone. The instrument produces a tone that varies in frequency with the measured parameter. This feature permits a circuit to be adjusted for a peak or null value without actually viewing the DMM. It also may be used to check for intermittent continuity.
- Ruggedized case. Durable cases for portable instruments permit normal physical abuse to be tolerated. Many DMMs in use today may be dropped or exposed to water without failure.
- Input overload hardening. Protection devices are placed at the input circuitry to prevent damage in the event of an accidental overvoltage. Fuses and semiconductor-based protection components are used.
- Logic probe functions. Signal high/low and pulse detection circuitry built into a DMM probe permits the instrument to function as both a conventional multimeter and a logic probe. This eliminates the necessity to change instruments in the middle of a troubleshooting sequence.
- Automatic shutoff. A built-in timer removes power to the DMM after a preset period of non-operation. This feature extends the life of the internal batteries.

DMM functions are further expanded by the wide variety of probes and accessories available for many instruments. Specialized probes include

- High-voltage—used for measuring voltages above 500 V. Many probes permit voltages of 50 kV or more to be measured.
- Clamp-type current—permits current in a conductor to be measured without breaking the circuit. Probes are commonly available in current ranges of 1–2 A and up.
- Demodulator— converts a radio frequency signal into a DC voltage for display on the DMM.

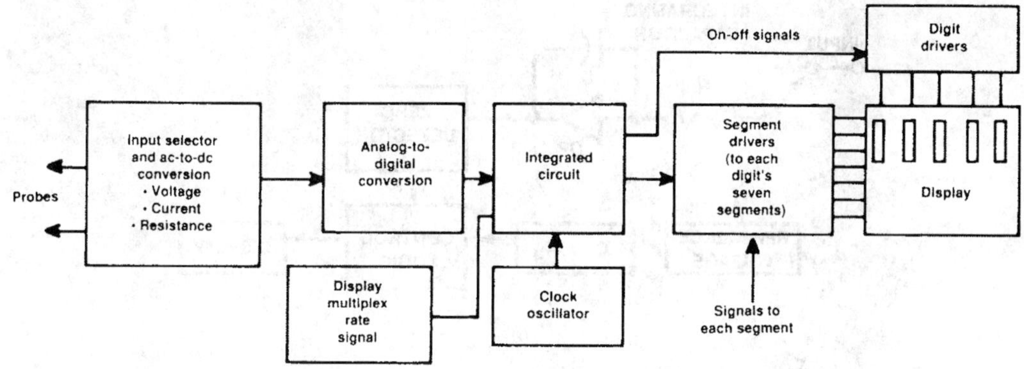

FIGURE 20.73 Functional block diagram of a basic DMM. LSI chip technology has reduced such systems to a single device.

- Temperature—permits the DMM to be used for display of temperature of a probe device. Probes are available for surface measurement of solids and for immersion into liquid or gas.

Advances in DMM technology have resulted in increased reliability. DMMs are engineered to include a minimum of moving parts. Auto-ranging eliminates much of the mechanical switch contacts and movement common with older technology instruments. This approach reduces the number of potential sources of failure, such as corrosion and intermittent switch contacts. Gold plating of the remaining contacts in the DMM further reduces the chance of failure.

Voltage Conversion

Figure 20.73 shows a generalized block diagram of a DMM. Conversion of an input voltage to a digital equivalent can be accomplished in one of several ways. The *dual-slope conversion* method (also know as *double-integration*) is one of the more popular techniques (Figure 20.74). V_{in} represents the input voltage (the voltage to be measured). V_{REF} is a reference voltage with a polarity opposite that of the measured voltage, supplied by the digital conversion circuit. Capacitor C1 and operational amplifier U1 constitute an integrator. S1 is an electronic switch that is initially in the position shown. When the meter probes are first connected to the circuit, the sample voltage is applied to the input of the integrator for a specified period of time, called the *integration period*. The integration period is usually related to the 60 Hz line frequency; integration periods of 1/60th of a second and 1/10th of a second are common. The output signal of the integrator is a voltage determined by the RC time constant of R1 and C1. Because of the nature of the integrator, the maximum voltage (the voltage at the end of the integration period) is proportional to the voltage being measured. At the end of the integration period, switch S1 is moved to the other position (V_{REF}) and a voltage of opposite polarity to the measured voltage is applied. The capacitor is then discharged to zero. As shown in the figure, the discharge interval is directly proportional to the maximum voltage, which in turn is proportional to the applied voltage.

At the same time that the integration interval ends and the discharge interval begins, a counter in the meter begins counting pulses generated by a clock circuit. When the voltage reaches zero, the counter stops. The number of pulses counted is, therefore, proportional to the discharge period. This count is converted to a digital number and displayed as the measured voltage. Although this method works well, it is somewhat slow, so many microcomputer-based meters use a variation called *multislope integration*.

When a DMM is used in the resistance testing mode, it places a low voltage with a constant-current characteristic across the test leads. After the leads are connected, the voltage across the measurement points is determined, which provides the resistance value (voltage divided by the known constant

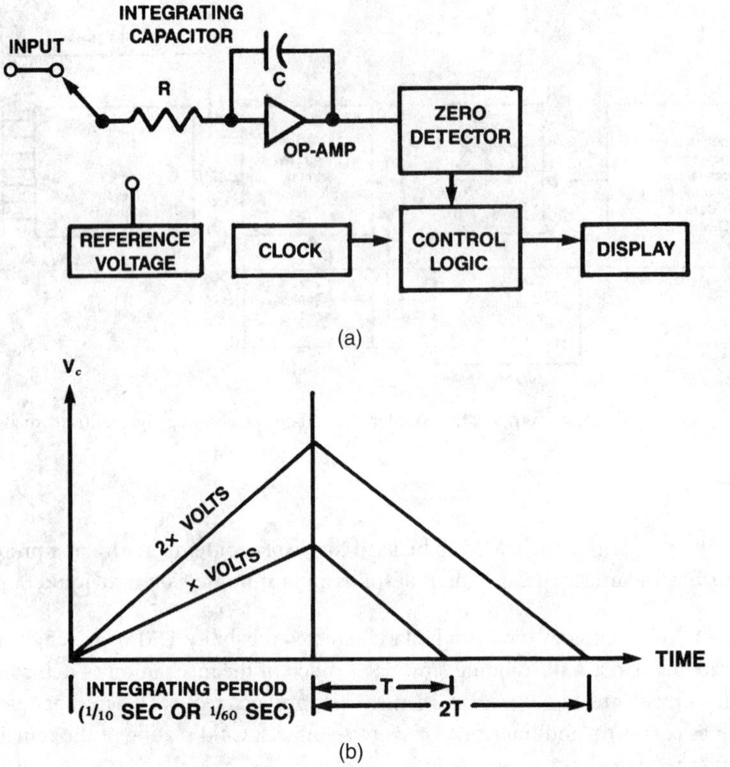

(a)

(b)

FIGURE 20.74　DMM analog-to-digital conversion process: (a) functional block diagram, (b) graph of V_c vs. time.

current = resistance). The meter converts the voltage reading into an equivalent resistance for display. The voltage placed across the circuit under test is kept low to protect semiconductor devices that might be in the loop. The voltage at the ohmmeter probe is about 0.1 V or less, too low to turn on silicon junctions.

The ultimate performance of a DMM in the field is determined by its inherent accuracy. Key specifications include

- **Accuracy**—how closely the meter indicates the actual value of a measured signal, specified in percent of error. Zero percent indicates a perfect meter.
- **Frequency response**—the range of frequencies that can be measured by the meter without exceeding a specified amount of error.
- **Input impedance**—the combined AC and DC resistance at the input terminals of the multimeter. The impedance, if too low, can load critical circuits and result in measurement errors. An input impedance of 10 MΩ or greater will prevent loading.
- **Precision**—the degree to which a measurement is carried out. As a general rule, the more digits a DMM can display, the more precise the measurement will be.

20.7.3　Conventional Test Instruments

Troubleshooting a digital system requires a wide range of test instruments, from simple to complex. Some of the more common conventional test instruments include frequency counters and logic instruments of various types.

Frequency Counter

A frequency counter provides an accurate measure of signal cycles or pulses over a standard period of time. It is used to totalize or to measure frequency, period, frequency ratio, and time intervals. Key specifications for a counter include

- Frequency range—the maximum frequency that the counter can resolve.
- Resolution—the smallest increment of change that can be displayed. The degree of resolution is selectable on many counters. Higher resolution usually requires a longer acquisition time.
- Sensitivity—the lowest amplitude signal that the instrument will count (measured in fractions of a volt).
- Timebase accuracy—a measure of the stability of the timebase. Stability is measured in parts per million (ppm) while the instrument is subjected to temperature and operating voltage variations.

Multi-Function Hand-held Instrument

Although specialized instruments such as a separate frequency counter or capacitance meter may be the best choice for a specific range of tasks, combinations of instruments are available for the convenience of maintenance personnel. Separate analyzers usually offer a broader range of measurement capabilities, but a combined instrument often makes field servicing easier.

Logic Instruments

The simplest of all logic instruments is the logic probe. A basic logic probe tells the technician whether the logic state of the point being checked is high, low, or pulsed. Most probes include a pulse stretcher that will show the presence of a 1-shot pulse. The indicators are usually LEDs. In some cases, a single LED is used to indicate any of the conditions (high, low, or pulsing); generally, individual LEDs are used to indicate each state. Probes usually include a switch to match the level sense circuitry of the unit to the type of logic being checked (TTL, CMOS, ECL, etc.). The probe receives its power from the circuit under test. Connecting the probe in this manner sets the approximate value of signal voltage that constitutes a logic low or high. For example, if the power supply voltage for a CMOS logic circuit is 18 V, a logic low would be about 30 % of 18 V (5.4 V), and a logic high would be about 70% of 18 V (12.6 V).

The logic pulser is the active counterpart to the logic probe. The logic pulser generates a single pulse, or train of pulses into a circuit to check the response.

Logic Current Tracer

The close spacing of pins and conductors on a crowded PWB invites short circuits. On a large board, locating a short can be a difficult and time-consuming task. The logic current tracer can be a valuable tool in troubleshooting such failures.

Even in an extremely thin printed circuit conductor, a change in current induces a magnetic field. The probe of a logic current tracer is designed to sense changes in the minute magnetic field of a PWB trace. An LED indicates the presence of the field. To locate a short circuit, the current tracer is used in conjunction with the logic pulser. As illustrated in Fig. 20.75, pulses are injected at an identifiable failure point. The tracer is then moved slowly along the circuit trace until the LED goes off, indicating that the point of the fault has been passed.

Logic Test Clip

A logic test clip is a device that clips directly over an IC package to display the logic status of all pins on the chip. Each probe of the clip contacts one pin of the IC. The probes are connected to LEDs that indicate whether that point in the circuit is at a logic high or low, or pulsed.

The IC comparator is an extension of the logic test clip. The comparator allows the operation of an in-circuit IC to be compared against a known-good IC of the same type placed in the comparator's "reference" socket. While the circuit containing the IC under test is operating, the comparator matches the responses of the reference IC to the test device.

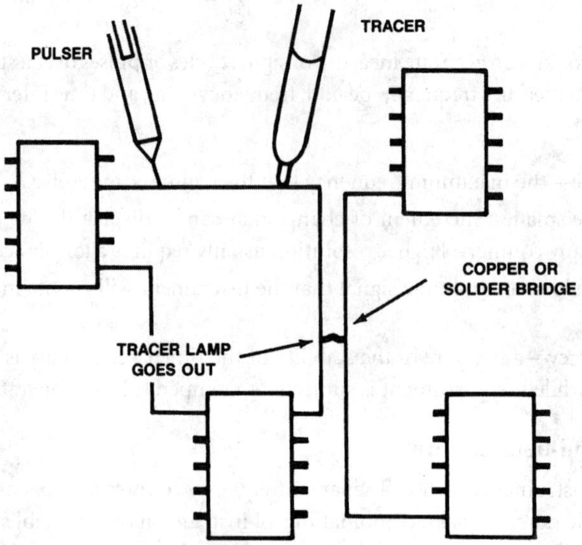

FIGURE 20.75   Use of a logic pulser and current tracer to locate a short-circuit on a PWB.

20.7.4   Logic Analyzer

The logic analyzer is really two instruments in one: a *timing analyzer* and a *state analyzer*. The timing analyzer is analogous to an oscilloscope. It displays information in the same general form as a scope, with the horizontal axis representing time and the vertical axis representing voltage amplitude. The timing analyzer samples the input waveform to determine whether it is high or low. The instrument cares about only one voltage threshold. If the signal is above the threshold when it is sampled, it will be displayed as a 1 (or high); any signal below the threshold is displayed as a 0 (or low). From these sample points, a list of ones and zeros is generated to represent a picture of the input waveform. These data are stored in memory and used to reconstruct the input waveform, as shown in Fig. 20.76. A block diagram of a typical timing analyzer is shown in Fig. 20.77.

Sample points for the timing analyzer are developed by an internal clock. The period of the sample can be selected by the user. Because the analyzer samples asynchronously to the unit-under-test under the direction of the internal clock, a long sample period results in a more accurate picture of the data bus. A sample period should be selected that permits an accurate view of data activity, while not filling up the instrument's memory with unnecessary data.

Accurate sampling of data lines requires a trigger source to begin data acquisition. A fixed delay may be inserted between the trigger point and the *trace point* to allow for bus settling. This concept is illustrated in Fig. 20.78. A variety of trigger modes are commonly available on a timing analyzer, including:

- *Level triggering*—data acquisition and/or display begins when a logic high or low is detected.
- *Edge triggering*—data acquisition/display begins on the rising or falling edge of a selected signal. Although many logic devices are level-dependent, clock and control signals are often edge-sensitive.
- *Bus state triggering*—data acquisition/display is triggered when a specific code is detected (specified in binary or hexadecimal).

The timing analyzer is constantly taking data from the monitored bus. Triggering, and subsequent generation of the trace point, controls the *data window* displayed to the technician. It is possible, therefore, configure the analyzer to display data that precedes the trace point. (See Fig. 20.79) This feature can be a powerful troubleshooting and developmental tool.

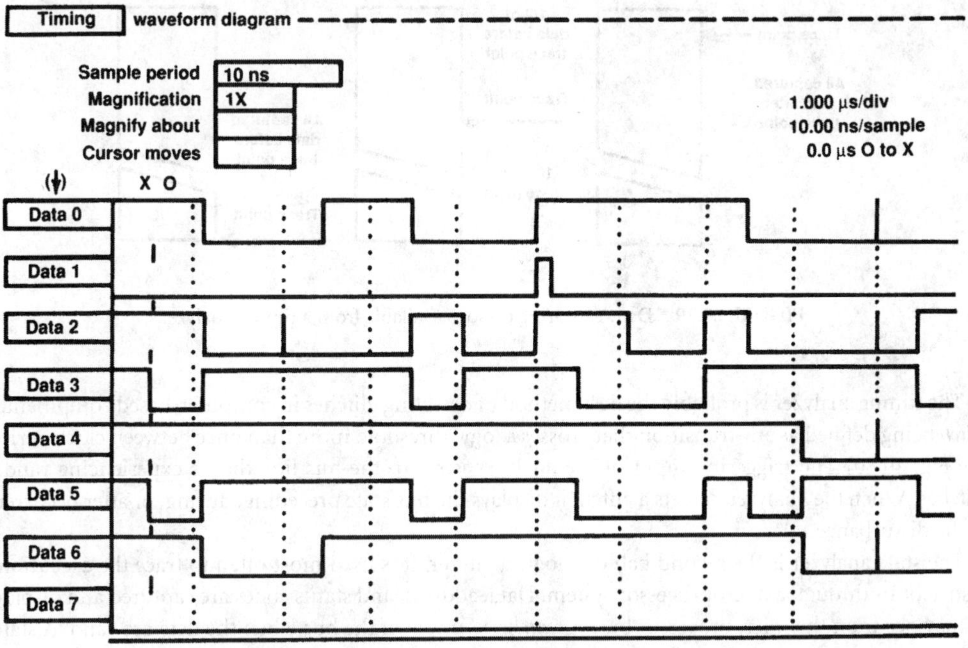

FIGURE 20.76   Typical display of a timing analyzer.

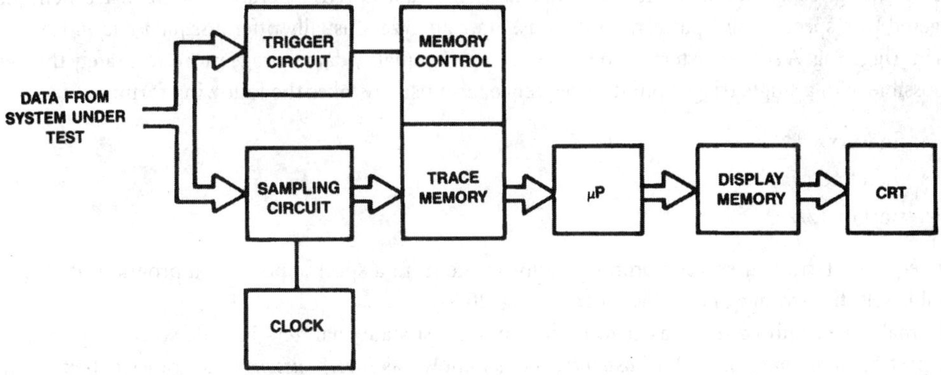

FIGURE 20.77   Block diagram of a timing analyzer.

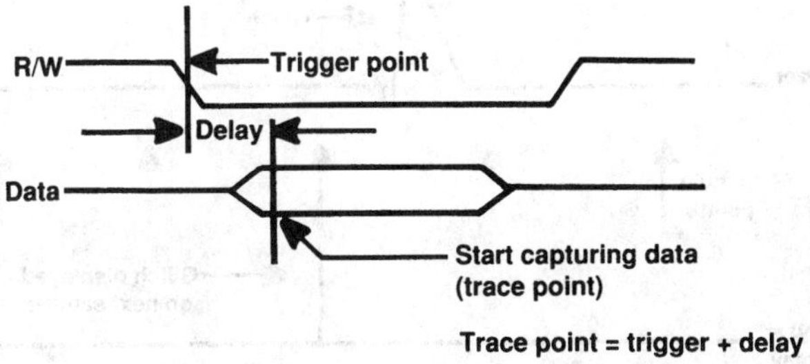

FIGURE 20.78   Use of a delay period between the trigger point and the trace point of a timing analyzer.

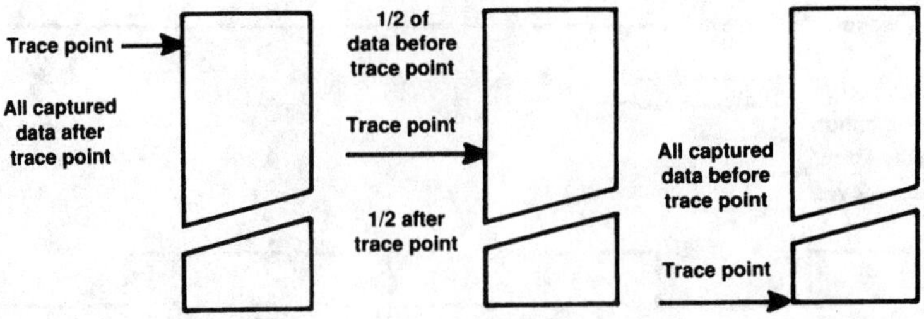

FIGURE 20.79   Data-capturing options available from a timing analyzer.

The timing analyzer is probably the best method of detecting glitches in computer-based equipment. (A *glitch* being defined as any transition that crosses a logic threshold more than once between clock periods. See Fig. 20.80) The triggering input of the analyzer is set to the bus line that is experiencing random glitches. When the analyzer detects a glitch, it displays the bus state preceding, during, or after occurrence of the disturbance.

The state analyzer is the second half of a logic analyzer. It is used most often to trace the execution of instructions through a microprocessor system. Data, address, and status codes are captured and displayed as they occur on the microprocessor bus. A *state* is a sample of the bus when the data are valid. A state is usually displayed in a tabular format of hexadecimal, binary, octal, or assembly language. Because some microprocessors multiplex data and addresses on the same lines, the analyzer must be able to clock-in information at different clock rates. The analyzer, in essence, acts as a demultiplexer to capture an address at the proper time, and then to capture data present on the same bus at a different point in time. A state analyzer also gives the operator the ability to *qualify* the data stored. Operation of the instrument may be triggered by a specific logic pattern on the bus. State analyzers usually offer a *sequence term* feature that aids in triggering. A sequence term allows the operator to qualify data storage more accurately than would be possible with a single trigger point. A sequence term usually takes the following form:

- find xxxx
- then find yyyy
- start on zzzz

A sequence term is useful for probing a subroutine from a specific point in a program. It also makes possible selective storage of data, as shown in Fig. 20.81.

To make the acquired data easier to understand, most state analyzers include software packages that interpret the information. Such *disassemblers* (also known as *inverse assemblers*) translate hex, binary, or octal codes into assembly code (or some other format) to make them easier to read.

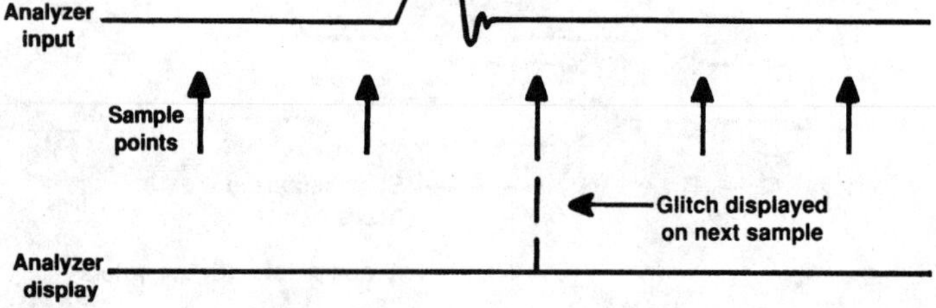

FIGURE 20.80   Use of a timing analyzer for detecting glitches on a monitored line.

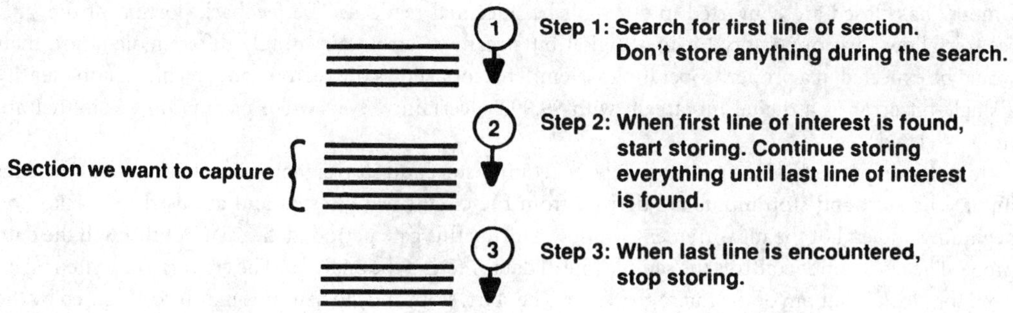
Step 1: Search for first line of section.
Don't store anything during the search.
Step 2: When first line of interest is found, start storing. Continue storing everything until last line of interest is found.
Section we want to capture
Step 3: When last line is encountered, stop storing.
1
2
3

FIGURE 20.81 Illustration of the selective storage capability of a state analyzer.

The logic analyzer is used routinely by design engineers to gain an in-depth look at signals within digital circuits. A logic analyzer can operate at high speeds, making the instrument ideal for detecting glitches resulting from timing problems. Such faults are usually associated with design flaws, not with manufacturing defects or failures in the field.

While the logic analyzer has benefits for the service technician, its use is limited. Designers require the ability to verify hardware and software implementations with test equipment; service technicians simply need to quickly isolate a fault. It is difficult and costly to automate test procedures for a given PWB using a logic analyzer. The technician must examine a long data stream and decide if the data are good or bad. Writing programs to validate state analysis data is possible, but again is time-consuming.

Signature Analyzer

Signature analysis is a common troubleshooting tool based on the old analog method of signal tracing. In analog troubleshooting, the technician followed an annotated schematic that depicted waveforms and voltages that should be observed with an oscilloscope and voltmeter. The signature analyzer captures a complex data stream from a test point on the PWB and converts it into a hexadecimal signature. These hexadecimal signatures are easy to annotate on schematics, permitting their use as references for comparison against signatures obtained from a unit under test. The signature by itself means nothing; the value of the signature analyzer comes from comparing a known-good signature with the UUT.

A block diagram of a simplified signature analyzer is shown in Fig. 20.82. As shown, a digital bit stream, accepted through a data probe during a specified measurement window, is marched through a 16-bit linear feedback shift register. Whatever data are left over in the register after the specified measurement

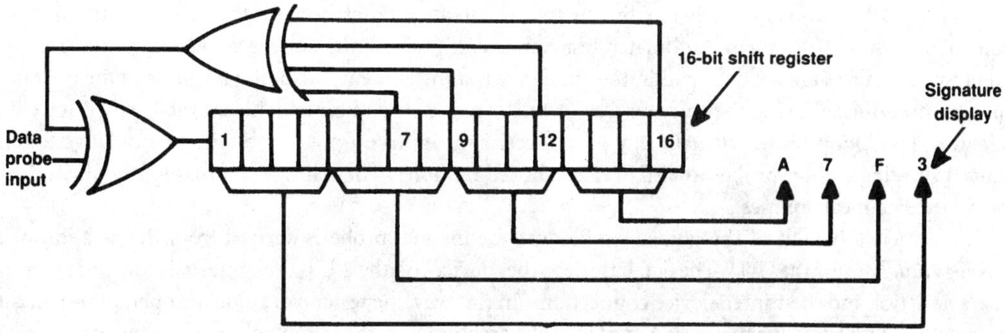
16-bit shift register
Signature display
Data probe input
1
7
9
12
16
A
7
F
3

FIGURE 20.82 Simplified block diagram of a signature analyzer.

window has closed are converted into a 4-bit hexadecimal readout. The feedback portion of the shift register allows just one faulty bit in a digital bit stream to create an entirely different signature than would be expected in a properly operating system. This tool allows the maintenance technician to identify a single bit error in a digital bit stream with 99.998% certainty, even when picking up errors that are timing-related.

To function properly, the signature analyzer requires start and stop signals, a clock input, and the data input. The start and stop inputs are derived from the circuit being tested and are used to bracket the beginning and end of the measurement window. During this gate period, data are input through the data probe. The clock input controls the sample rate of data entering the analyzer. The clock is most often taken from the clock input pin of the microprocessor. The start, stop, and clock inputs may be configured by the technician to trigger on the rising or falling edges of the input signals.

Manual Probe Diagnosis

Manual probe diagnosis integrates the logic probe, logic analyzer, and signature analyzer to troubleshoot computer-based hardware. The manual probe technique employs a database consisting of nodal-level signature measurements. Each node of a known-good unit-under-test is probed and the signature information saved. These known-good data are then compared to the data from a faulty UUT. Efficient use of the manual probe technique requires a skilled operator to determine the proper order for probing circuit nodes.

Emulative Tester

Guided diagnostics can be used on certain types of hardware to facilitate semi-automated circuit testing. The test connector typically taps into one of the system board input/output ports and, armed with the proper software, the instrument goes through a step-by-step test routine that exercises the computer circuits. When the system encounters a fault condition, it outputs an appropriate message.

Emulative testers are a variation on the general guided diagnostics theme. Such an instrument plugs into the microprocessor socket of the computer and runs a series of diagnostic tests to identify the cause of a fault condition. An emulative tester emulates the board's microprocessor while verifying circuit operation. Testing the board "inside-out" allows easy access to all parts of the circuit; synchronization with the various data and address cycles of the PWB is automatic. Test procedures, such as read/write cycles from the microprocessor are generic, so that high quality functional tests can be quickly created for any board. Signature analysis is used to verify that circuits are operating correctly. Even with long streams of data, there is no maximum memory depth.

Several different types of emulative testers are available. Some are designed to check only one brand of computer, or only computers based on one type of microprocessor. Other instruments can check a variety of microcomputer systems using so-called *personality* modules that adapt the instrument to the system being serviced.

Guided-fault isolation (GFI) is practical with an emulative tester because the instrument maintains control over the entire system. Automated tests can isolate faults to the node level. All board information and test procedures are resident within the emulative test instrument, including prompts to the operator on what to do next. Using the microprocessor test connection combined with movable probes or clips allows a closed loop test of virtually any part of a circuit. Input/output capabilities of emulative testers range from single point probes to well over a hundred I/O lines. These lines can provide stimulus as well as measurement capabilities.

The principal benefit of the guided probe over the manual probe is derived from the creation of a *topology database* for the UUT. The database describes devices on the UUT, their internal fault-propagating characteristics, and their inter-device connections. In this way, the tester can guide the operator down the proper logic path to isolate the fault. A guided probe accomplishes the same analysis as a conventional fault tree, but differs in that troubleshooting is based on a generic algorithm that uses the circuit model as its input. First, the system compares measurements taken from the failed UUT against the expected results. Next, it searches for a database representation of the UUT to determine the next logical place to take a measurement. The guided probe system automatically determines which nodes impact other nodes. The algorithm tracks its way through the logic system of the board to the source of the fault. Programming of a guided probe system requires the following steps:

- Development of a stimulus routine to exercise and verify the operating condition of the UUT (go/no-go status).
- Acquisition and storage of measurements for each node affected by the stimulus routine. The programmer divides the system or board into measurement sets (MSETs) of nodes having a common timebase. In this way, the user can take maximum advantage of the time domain feature of the algorithm.
- Programming a representation of the UUT that depicts the interconnection between devices, and the manner in which signals can propagate through the system. Common *connectivity* libraries are developed and reused from one application to another.
- Programming a measurement set cross reference database. This allows the guided probe instrument to cross MSET boundaries automatically without operator intervention.
- Implementation of a test program control routine that executes the stimulus routines. When a failure is detected, the guided probe database is invoked to determine the next step in the troubleshooting process.

Protocol Analyzer

Testing a local area network (LAN)—whether copper, fiber optic, or wireless—presents special challenges to the design engines or maintenance technician. The protocol analyzer is commonly used to service LANs and other communications systems. The protocol analyzer performs data monitoring, terminal simulation, and bit error rate tests (BERTs). Sophisticated analyzers provide *high-level decide* capability. This ability refers to the open system interconnection (OSI) network seven layer model. These functions enable the engineer to observe activity of a communications link and to exercise the facility to verify proper operation. Simulation capability is usually available to emulate almost any data terminal or data communications equipment.

LAN test equipment functions include monitoring of protocols and emulation of a network node, bridge, or terminal. Statistical information is provided to isolate problems to particular devices, or to high periods of activity. Statistical data include use, packet rates, error rates, and collision rates.

20.7.5 Automated Test Instruments

Computers can function as powerful troubleshooting tools. Advanced technology systems, coupled with add-on interface cards and applications software, provide a wide range of testing capabilities. Computers can also be connected to many different instruments to provide for automated testing. There are two basic types of stand-alone automated test instruments (ATEs): *functional* and *in-circuit*.

Functional testing exercises the unit under test to identify faults. Individual PWBs or subassemblies may be checked using this approach. Functional testing provides a fast go/no-go qualification check. Dynamic faults are best discovered through this approach. Functional test instruments are well suited to high volume testing of PWB subassemblies, however, programming is a major undertaking. An intimate knowledge of the subassembly is needed to generate the required test patterns. An in-circuit emulator is one type of functional tester.

In-circuit testing is primarily a diagnostic tool. It verifies the functionality of the individual components of a subassembly. Each device is checked and failing parts are identified. In-circuit testing, while valuable for detailed component checking, does not operate at the clock rate of the subassembly. Propagation delays, race conditions, and other abnormalities may go undetected. Access to key points on the PWB may be accomplished in one of two ways:

- *Bed of nails.* The subassembly is placed on a dedicated test fixture and held in place by a vacuum or mechanical means. Probes access key electrical traces on the subassembly to check individual devices. Through a technique known as *backdriving*, the inputs of each device are isolated from the associated circuitry and the component is tested for functionality. This type of testing is expensive and is only practical for high-volume subassembly qualification testing and rework.
- PWB clips. Intended for lower volume applications than a bed of nails instrument, PWB clips replace the dedicated test fixture. The operator places the clips on the board as directed by the instrument. As access to all components simultaneously is not required, the tester is less hardware-intensive and programming is simplified. Many systems include a library of software routines designed to test various classes of devices. Test instruments using PWB clips tend to be slow. Test times for an average PWB may range from 8–20 min vs. 1 min or less for a bed of nails.

Software Considerations

Most automated test instrument packages include software that permits the user to customize test procedures to meet the required task. The software, in effect, writes software. The user enters the codes needed by each automatic instrument, selects the measurements to be performed, and tells the computer where to store the test data. The program then puts the final software together after the user answers key configuration questions. The software then looks up the operational codes for each instrument and compiles the software to perform the required tests. Automatic generation of programming greatly reduces the need to have experienced programmers on staff. Once installed in the computer, programming becomes as simple as fitting graphic symbols together on the screen.

Applications

The most common applications for computer-controlled testing are data gathering, product go/ no-go qualification, and troubleshooting. All depend on software to control the instruments. Acquiring data can often be accomplished with a computer and a single instrument. The computer collects dozens or hundreds of readings until the occurrence of some event. The event may be a preset elapsed time or the occurrence of some condition at a test point, such as exceeding a preset voltage or dropping below a preset voltage. Readings from a variety of test points are then stored in the computer. Under computer direction, test instruments can also run checks that might be difficult or time-consuming to perform manually. The computer may control several test instruments such as power supplies, signal generators, and frequency counters. The data are stored for later analysis.

References

Carey, G.D. 1989. Automated Test Instruments. *Broadcast Engineering*, Intertec Publishing, Overland Park, KS.

Detwiler, W.L. 1988. Troubleshooting with a Digital Multimeter. *Mobile Radio Technology*, Intertec Publishing, Overland Park, KS.

Gore, G. 1988. Choosing a Hand-Held Test Instrument. *Electronic Servicing and Technology*, Intertec Publishing, Overland Park, KS.

Harju, Rey. 1988. Hands-Free Operation: A New Generation of DMMs. *Microservice Management*, Intertec Publishing, Overland Park, KS.

Persson, C. 1987. Test Equipment for Personal Computers. *Electronic Servicing and Technology*, Intertec Publishing, Overland Park, KS.

Persson, C. 1989. The New Breed of Test Instruments. *Broadcast Engineering*, Intertec Publishing, Overland Park, KS.

Ogden, L. 1989. Choosing the Best DMM for Computer Service. *Microservice Management*, Intertec Publishing, Overland Park, KS.

Siner, T.A. 1987. Guided Probe Diagnosis: Affordable Automation. *Microservice Management*, Intertec Publishing, Overland Park, KS.

Sokol, F. 1989. Specialized Test Equipment. *Microservice Management*, Intertec Publishing, Overland Park, KS.

Whitaker, J.C. 1991. *Electronic Systems Maintenance*, CRC Press, Boca Raton, FL.

Wickstead, M. 1985. Signature Analyzers. *Microservice Management*, Intertec Publishing, Overland Park, KS.

21

Reliability Engineering

Allan White
National Aeronautics and Space Administration

Hagbae Kim
National Aeronautics and Space Administration

Michael Pecht
University of Maryland

Iuliana Bordelon
University of Maryland

Carol Smidts
University of Maryland

21.1 Probability and Statistics

Allan White and Hagbae Kim

21.1.1 Survey of Probability and Statistics

General Introduction to Probability and Statistics

The approach to probability and statistics used in this chapter is the pragmatic one that probability and statistics are methods of operating in the presence of incomplete knowledge.

The most common example is tossing a coin. It will land heads or tails, but the relative frequency is unknown. The two events are often assumed equally likely, but a skilled coin tosser may be able to get heads almost all of the time. The statistical problem in this case is to determine the relative frequency. Given the relative frequency, probability techniques answer such questions as how long can we expect to wait until three heads appear in a row.

Another example is light bulbs. The outcome is known: any bulb will eventually fail. If we had complete knowledge of the local universe, it is conceivable that we might compute the lifetime of any bulb. In reality, the manufacturing variations and future operating conditions for a lightbulb are unknown to us. It may be possible, however, to describe the failure history for the general population. In the absence of any data, we can propose a failure process that is due to the accumulated effect of small events (corrosion, metal evaporation, and cracks) where each of these small events is a random process. It is shown in the statistics section that a failure process that is the sum of many events is closely approximated by a distribution known as the normal (or Gaussian) distribution. This type of curve is displayed in Fig. 21.1(a).

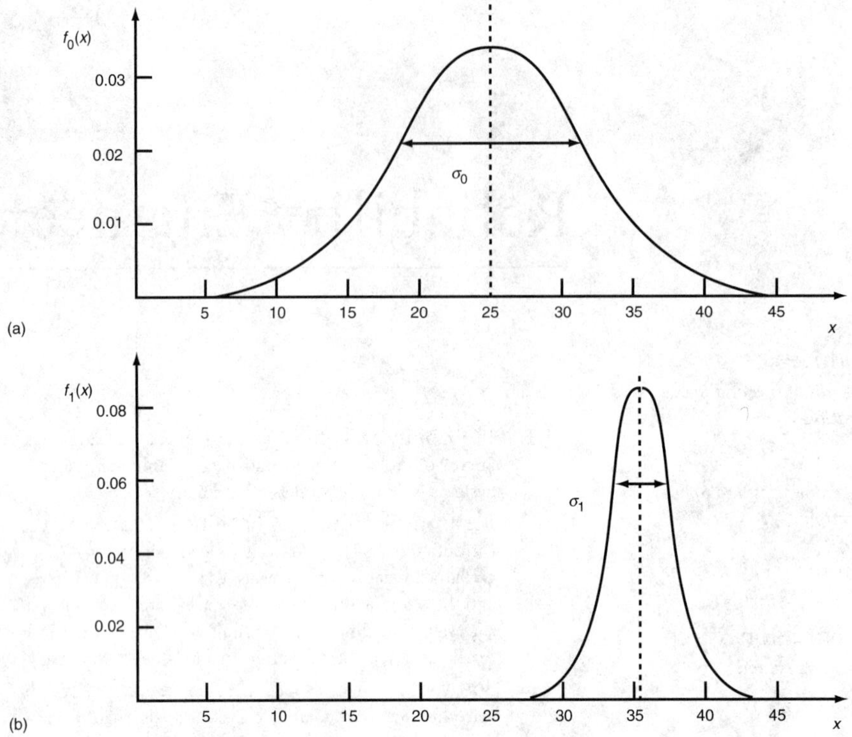

FIGURE 21.1   Two Gaussian distributions having different means and variances.

A manufacturer can try to improve a product by using better materials and requiring closer tolerances in the assembly. If the light bulbs last longer and there are fewer differences between the bulbs, then the failure curve should move to the right and have less dispersion, as shown in Fig. 21.1(b).

There are three types of statistical problems in the light bulb example. One is identifying the shape of the failure distributions. Are they really the normal distribution as conjectured? Another is to estimate such parameters as the average and the deviation from the average for the populations. Still another is to compare the old manufacturing process with the new. Is there a real difference and is the difference significant?

Suppose the failure curve (for light bulbs) and its parameters have been obtained. This information can then be used in probability models. As an example, suppose a city wishes to maintain sufficient illumination for its streets by constructing a certain number of lamp posts and by having a maintenance crew replace failed lights. Since there will be a time lag between light failure and the arrival of the repair crew, the city may construct more than the minimum number of lamp posts needed in order to maintain sufficient illumination. Determining the number of lamp posts depends on the failure curve for the bulbs and the time lag for repair. Furthermore, the city may consider having the repair crew replace all bulbs, whether failed or not, as the crew makes its circuit. This increases the cost for bulbs but decreases the cost for the repair crew. What is the optimum strategy for the number of lamp posts, the repair crew circuit, and the replacement policy? The objective is to maximize the probability of sufficient illumination while minimizing the cost.

Incomplete Knowledge vs. Complete Lack of Knowledge

People unfamiliar with probability sometimes think that random means completely arbitrary, and there is no way to predict what will happen. Physical reasoning, however, can supply a fair amount of knowledge:

- It is reasonable to assume the coin will land heads or tails and the probability of heads remains constant. This gives a unique probability distribution with two possible outcomes.
- It is reasonable to assume that light bulb failure is the accumulation of small effects. This implies that the failure distribution for the light bulb population is approximately the normal distribution.
- It will be shown that the probability distribution for time to failure of a device is uniquely determined by the assumption that the device does not wear out.

Of course, it is possible (for random as for deterministic events) that our assumptions are incorrect. It is conceivable that the coin will land on its edge or disintegrate in the air. Power surges that burn out large numbers of bulbs are not incremental phenomena, and their presence will change the distribution for times to failure. In other words, using a probability distribution implies certain assumptions are being made. It is always good to check for a match between the assumptions underlying the probability model and the phenomenon being modeled.

Probability

This section presents enough technical material to support all our discussions on probability. It covers the (1) definition of a **probability space**, (2) Venn (set) diagrams, and (3) density functions.

Probability Space

The definition of a probability space is based on the intuitive ideas presented in the previous section. We begin with the set of all possible events, call this set X.

- If we are tossing a coin three times, X consists of all sequences of heads (H) and tails (T) of length three.
- If we are testing light bulbs until time to failure, X consists of the positive real numbers.

The next step in the definition of a probability space is to ensure that we can consider simple combinations of events. We want to consider the AND and OR combinations as intersections and unions. We also want to include complements.

In the coin-tossing space, getting at least two heads is the union of the four events, that is

$$\{HHT, HTH, THH, HHH\}$$

In the light-bulb-failure space, a bulb lasting more than 10 h but less than 20 h is the intersection of the two events: (1) bulb lasts more than 10 h and (2) bulb lasts less than 20 h. In the coin-tossing space, getting less than two heads is the complement of the first set, that is

$$\{HTT, THT, TTH, TTT\}$$

When it comes to assigning probabilities to events, the procedure arises from the idea of assigning areas to sets. This idea can be conveyed by pictures called Venn diagrams. For example, consider the disjoint sets of events A and B in the Venn diagram of Fig. 21.2.

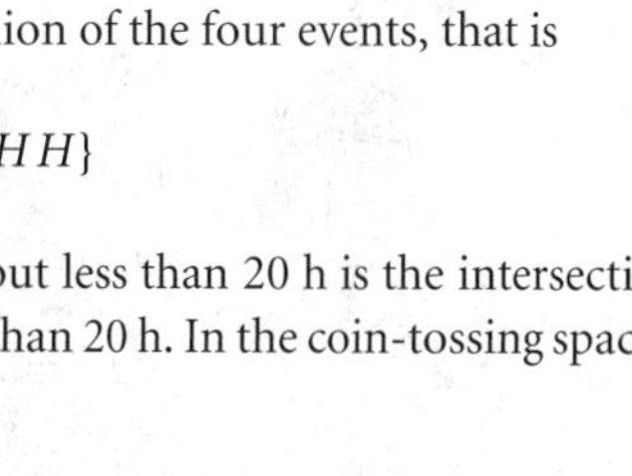

We want the probability of being in either A or B to be the sum of their probabilities. Finally, when considering physical events it is sufficient to consider only finite or countably infinite combinations

FIGURE 21.2 Venn diagram of two disjoint sets of events.

of simpler events. (It is also true that uncountably infinite combinations can produce mathematical pathologies.)

Example

It is necessary to include countably infinite combinations, not just finite combinations. The simplest example is tossing a coin until a head appears. There is no bound for the number of tosses. The sample space consists of all sequences of heads and tails. The combinatorial process in operation for this example is AND; the first toss is T, and the second toss is $T, \ldots$ and the n-th toss is H.

Hence, with this in mind, a probability space is defined as follows:

1. It is a collection of subsets, called events, of a universal space X. This collection of subsets (events) has the following properties:
    - X is in the collection
    - If A is in the collection, its set complement A' is in the collection
    - If $\{A_i\}$ is a finite or countably infinite collection of events, then both $\bigcup A_i$ and $\bigcap A_i$ are in the collection
2. To every event A there is a real number, $Pr[A]$, between 0 and 1 assigned to A such that
    - $Pr[X] = 1$
    - If $\{A_i\}$ is a finite or countably infinite collection of disjoint events, then $Pr[\bigcup A_i] = \sum Pr[A_i]$

As the last axiom suggests, numerous properties about probability are derived by decomposing complex events into unions of disjoint events.

Example

$Pr[A \cup B] = Pr[A] + Pr[B] - Pr[A \cap B]$.

Let $A \backslash B$ be the elements in A that are not in B. Express the sets as the disjoint decompositions

$$A = (A \backslash B) \cup (A \cap B)$$

$$B = (B \backslash A) \cup (A \cap B)$$

$$A \cup B = (A \backslash B) \cup (B \backslash A) \cup (A \cap B)$$

which gives

$$Pr[A] = Pr[A \backslash B] + Pr[A \cap B]$$

$$Pr[B] = Pr[B \backslash A] + Pr[A \cap B]$$

$$Pr[A \cup B] = Pr[A \backslash B] + Pr[B \backslash A] + Pr[A \cap B]$$

and the result follows.

Probability from Density Functions and Integrals

The interpretation of probability as an area, when suitably generalized, gives a universal framework. The formulation given here, in terms of Riemann integrals, is sufficient for many applications. Consider a nonnegative function $f(x)$ defined on the real line with the property that its integral over the entire line is equal to one (Fig. 21.3.)

The probability that the values lie on the interval $[a, b]$ is defined to be

$$Pr[a \leq x \leq b] = \int_a^b f(x)dx$$

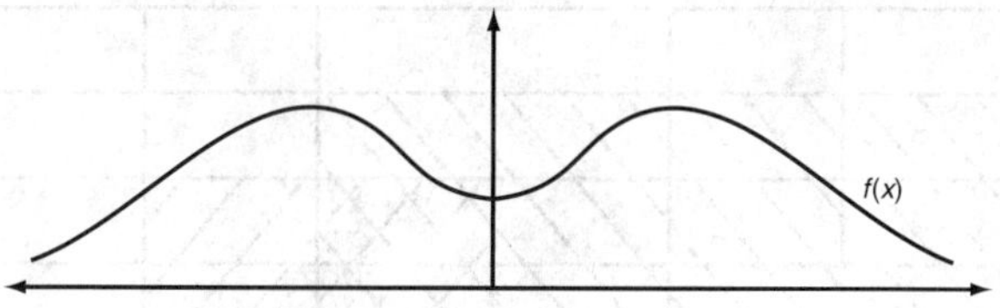

FIGURE 21.3 A nonnegative function $f(x)$ defined on the real line.

The function $f(x)$ is called the *density function*. The probability that the values are less than or equal to t is given by

$$F(t) = Pr[x \le t] = \int_{-\infty}^{t} f(x)dx$$

This function $F(t)$ is called the *distribution function*.

In this formulation by density functions and (Riemann) integrals, the basic events are the intervals $[a, b]$ and countable unions of these disjoint intervals. It is easy to see that this approach satisfies all of the axioms for a probability space listed earlier. This formulation lets us apply all of the techniques of calculus and analysis to probability. This feature is demonstrated repeatedly in the sections to follow.

One last comment should be made. Riemann integration and continuous (or piecewise continuous) density functions are sufficient for a considerable amount of applied probability, but they cannot handle all of the topics in probability because of the highly discontinuous nature of some probability distributions. A generalization of Riemann integration (measure theory and general integration), however, does handle all aspects of probability.

Independence

Conditional probability, and the related topic of independence, give probability a unique character. As already mentioned, mathematical probability can be presented as a topic in measure theory and general integration. The idea of conditional probability distinguishes it from the rest of mathematical analysis. Logically, independence is a consequence of conditional probability, but the exposition to follow attempts an intuitive explanation of independence before introducing conditional events.

Informal Discussion of Independence

In probability, two events are said to be independent if information about one gives no information about the other. Independence is often assumed and used without any explicit discussion, but it is a stringent condition. Furthermore, the probabilistic meaning is different from the usual physical meaning of two events having nothing in common.

Figure 21.4 gives three diagrams illustrating various degrees of correlation and independence. In the first diagram, event A is contained in event B. Hence, if A has occurred then B has occurred, and the events are not independent (since knowledge of A yields some knowledge of B). In the second diagram event A

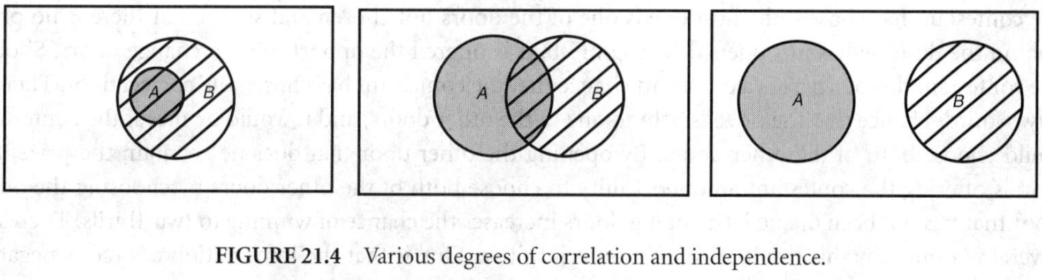

FIGURE 21.4 Various degrees of correlation and independence.

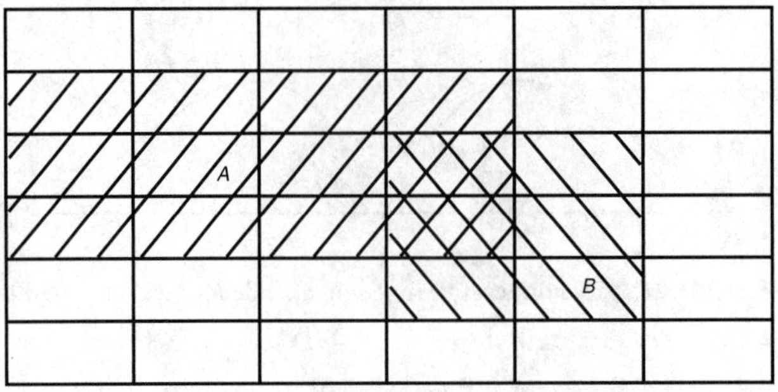

FIGURE 21.5   Pictorial example of conditional probability.

lies half in and half out of event B. Event B has the same relationship to event A. Hence, the two events in the third diagram are independent—knowledge of the occurrence of one gives no information about the occurrence of the other. This mutual relationship is essential. The third diagram illustrates disjoint events. If one event occurs, then the other cannot. Hence, the occurrence of one event yields information about the occurrence of the other event, which means the two events are not independent.

Conditional Probability and the Multiplicative Property

Conditional probability expresses the amount of information that the occurrence of one event gives about the occurrence of another event. The notation is $Pr[A \mid B]$ for the probability of A given B. To motivate the mathematical definition consider the events A and B in Fig. 21.5.

By relative area, $Pr[A] = 12/36$, $Pr B = 6/36$, and $Pr[A \mid B] = 2/6$. The mathematical expression for conditional probability is

$$Pr[A \mid B] = Pr[A \text{ and } B]/Pr[B]$$

From this expression it is easy to derive the multiplication rule for independent events. Suppose A is independent of B. Then information about B gives no information about A. That is, the conditional probability of A given B is the same as the probability of A or (by formula)

$$Pr[A] = Pr[A \mid B] = Pr[A \text{ and } B]/Pr[B]$$

A little algebra gives

$$Pr[A \text{ and } B] = Pr[A]Pr[B]$$

This multiplicative property means that independence is symmetrical. If A is independent of B, then B is independent of A. In its abstract formulation conditional probability is straightforward, but in applications it can be tricky.

Example (The Monte Hall Problem)

A contestant on a quiz show can choose one of three doors, and there is a prize behind one of them. After the contestant has chosen, the host opens one of the doors not chosen and shows that there is no prize behind the door he has just opened. The contestant is offered the opportunity to change doors. Should the contestant do so? There is a one-third chance that the contestant has chosen the correct door. There is a two-thirds chance that the prize is behind one of the other doors, and it would be nice if the contestant could choose both of the other doors. By opening the other door that does not contain the prize, the host is offering the contestant an opportunity to choose both of the other doors by choosing the other door that has not been opened. Changing doors increases the chance of winning to two-thirds. There are several reasons why this result appears counterintuitive. One is that the information offered is negative information (no prize behind this door), and it is hard to interpret negative information. Another is the timing. If the host shows that there is no prize behind a door before the contestant chooses, then the chance of a correct choice is one-half. The timing imposes an additional condition on the host.

The (sometimes) subtle nature of conditional probability is important when conducting an experiment. It is almost trite to say that an event A can be measured only if A or its effect can be observed. It is not quite as trite to say that observability is a special property. Experiments involve the conditional probability:

$$Pr[A \text{ has some value} \mid A \text{ or its effect can be observed}]$$

A primary goal in experimental sampling is to make the observed value of A independent of its observability. One method that accomplishes this is to arrange the experiment so that all events are potentially observable. If X is the set of all events then the equations for this case are

$$Pr[A \mid X] = Pr[A \text{ and } X]/Pr[X] = Pr[A]/1 = Pr[A]$$

For this approach to work it is only necessary that all events are potentially observable, not that all events are actually observed. The last section on statistics will show that a small fraction of the population can give a good estimate, provided it is possible to sample the entire population.

An experiment where part of the population cannot be observed can produce misleading results. Imagine a telephone survey where only half of the population has a telephone. The population with the geographic location and wealth to own a telephone may have opinions that differ from the rest of the population.

These comments apply immediately to fault injection experiments for reconfigurable systems. These experiments attempt to study the diagnostic and recovery properties of a system when a fault arrives. If it is not possible to inject faults into the entire system, then the results can be biased, because the unreachable part of the system may have different diagnostic and recovery characteristics than the reachable part.

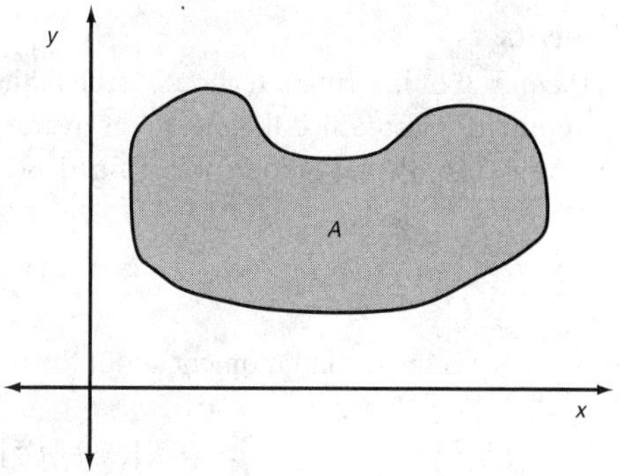

FIGURE 21.6 A set represented by two density functions $f(x)$ and $g(y)$.

Independence and Density Functions

The product formula for the probability of independent events translates into a product formula for the density of independent events. If x and y are random variables with densities $f(x)$ and $g(y)$, then the probability that (x, y) lies in the set A, displayed in Fig. 21.6, is given by

$$\int\int_A f(x)g(y)dy\,dx$$

This property extends to any finite number of random variables. It is used extensively in the statistics section to follow because items in a **sample** are **independent events**.

The Probability Distribution for Memoryless Events

As an example of deriving the density function from the properties of the event, this section will show that memoryless events must have an exponential density function. The derivation uses the result from real analysis that the only continuous function $h(x)$ with the property, $h(a+b) = h(a)h(b)$ is the exponential function $h(t) = e^{\alpha t}$ for some α.

An event is memoryless if the probability of its occurrence in the next time increment does not depend on the amount of time that has already elapsed. That is,

$$Pr[\text{occurrence by time } (s + t) \text{ given has not occurred at time } s] = Pr[\text{occurrence by time } t]$$

Changing to distribution functions and using the formula for conditional probability gives

$$\frac{F(s+t) - F(s)}{1 - F(s)} = F(t)$$

A little algebra yields

$$1 - F(s + t) = [1 - F(s)][1 - F(t)]$$

Let $g(x) = 1 - F(x)$, then $g(s + t) = g(s)g(t)$, which means $g(x) = e^{\alpha t}$. Hence, the density function is

$$f(x) = F'(x) = -\alpha e^{\alpha t}$$

Since the density function must be nonnegative, $\alpha < 0$.

This derivation illustrates a of key activity in probability. If a probability distribution has certain properties, does this imply that it is uniquely determined? If so, what is the density function?

Moments

A fair amount of information about a probability distribution is contained in its first two moments. The first moment is also called the **mean**, the **average**, the **expectation**, or the **expected value**. If the density function is $f(x)$ the first moment is defined as

$$\mu = E(x) = \int_{-\infty}^{+\infty} xf(x)dx$$

The variance is the second moment about the mean

$$\sigma^2 = E[(x - \mu)^2] = \int_{-\infty}^{+\infty} (x - \mu)^2 f(x)dx$$

The variance measures the dispersion of the distribution. If the variance is small then the distribution is concentrated about the mean. A mathematical version of this is given by Chebychev's inequality

$$Pr[\|x - \mu\| \geq \delta] \leq \frac{\sigma^2}{\delta^2}.$$

This inequality says it is unlikely for a member of the population to be a great distance from the mean.

Statistics

The purpose of this section is to present the basic ideas behind statistical techniques by means of concrete examples.

Sampling

Sampling consists of picking individual items from a population in order to measure the parameter in question. For the coin tossing experiment mentioned at the beginning of this chapter, an individual item is a toss of the coin and the parameter measured is heads or tails. If the matter of interest is the failure distribution for a class of devices, then a number of devices are selected and the parameter measured is the time until failure. Selecting and measuring an individual item is known as a trial.

The theoretical formulation arises from the assumptions that (1) selecting an individual from the population is a random event governed by the probability distribution for that population, (2) the results of one trial do not influence the results of another trial, and (3) the population (from which items are chosen) remains the same for each trial. If these assumptions hold, then the samples are said to be independent and identically distributed.

Suppose $f(x)$ is the density function for the item being sampled. The probability that a sample lies in the interval $[a, b]$ is $\int_a^b f(x)dx$. By the assumption of independence, the density function for a sample of size n is the n-fold product

$$f(x_1) f(x_2) \cdots f(x_n)$$

The probability that a sample of size n, say, $x_1, x_2, \ldots, x_n$, is in the set A is

$$\int \int \cdots \int f(x_1) f(x_2) \cdots f(x_n) dx_n \cdots dx_2 dx_1$$

where the integral is taken over the (multidimensional) set A. This formulation for the probability of a sample (in terms of a multiple integral) provides a mathematical basis for statistics.

Estimators: Sample Average and Sample Variance

Estimators are functions of the sample values (and constants). Since samples are random variables, estimators are also random variables. For example, suppose g is a function of the sample $x_1, x_2, \ldots, x_n$, where $f(x)$ is the density function for each of the x. Then the expectation of g is

$$E(g) = \int \int \cdots \int g(x_1, \ldots, x_n) f(x_1) f(x_2) \cdots f(x_n) dx_n \cdots dx_2 dx_1$$

To be useful, of course, estimators must have a good relationship to the unknown parameter we are trying to determine. There are a variety of criteria for estimators. We will only consider the simplest. If g is an estimator for the (unknown) parameter θ, then g is an *unbiased* estimator for θ if the expected value of g equals θ. That is

$$E(g) = \theta$$

In addition, we would like the variance of the estimator to be small since this improves the accuracy. Statistical texts devote considerable time on the efficiency (the **variance**) of estimators.

The most common statistic used is the sample average as an estimator for the mean of the distribution. For a sample of size n it is defined to be

$$\bar{x} = \frac{(x_1 + \cdots + x_n)}{n}$$

Suppose the distribution has mean, variance, and density function of μ, σ^2 and $f(x)$. To illustrate the mathematical formulation of the estimator, we will do a detailed derivation that the sample mean is an unbiased estimator for the mean of the underlying distribution. We have

$$E(\bar{x}) = \int_{-\infty}^{+\infty} \int_{-\infty}^{+\infty} \cdots \int_{-\infty}^{+\infty} \left(\frac{x_1 + \cdots + x_n}{n} \right) f(x_n) \cdots f(x_2) f(x_1) dx_n \cdots dx_2 dx_1$$

$$= \sum_{i=1}^{n} \frac{1}{n} \int_{-\infty}^{+\infty} \int_{-\infty}^{+\infty} \cdots \int_{-\infty}^{+\infty} x_i f(x_n) \cdots f(x_2) f(x_1) dx_n \cdots dx_2 dx_1$$

Writing the ith term in the last sum as an iterated integral and then as a product integral gives

$$\int_{-\infty}^{+\infty} 1 f(x_1) dx_1 \cdots \int_{-\infty}^{+\infty} x_i f(x_i) dx_i \int_{-\infty}^{+\infty} 1 f(x_{i+1}) dx_{i+1} \cdots \int_{-\infty}^{+\infty} 1 f(x_n) dx_n$$

The factors with 1 in the integrand are equal to 1. The factor with x_i in the integrand is equal to μ, the mean of the distribution. Hence

$$E(\bar{x}) = \frac{1}{n} \sum_{i=1}^{n} \mu = \mu$$

The sample average also has a convenient expression for its variance. A derivation similar to, but more complicated than, the preceding one gives

$$\text{Var}(\bar{x}) = \frac{\sigma^2}{n}$$

where σ^2 is the variance of the underlying population. Hence, increasing the sample size n improves the estimate by decreasing the variance of the estimator. This is discussed at greater length in the section on confidence intervals.

We can also estimate the variance of the underlying distribution. Define the sample variance by

$$S^2 = \frac{\sum_{i=1}^{n}(x - \bar{x})^2}{n - 1}$$

A similar derivation shows that

$$E(S^2) = \sigma^2$$

There is also an expression for $\text{Var}(S^2)$ in terms of the higher moments of the underlying distribution.

For emphasis, all of these results about estimators arrive from the formulation of a sample as a collection of independent and identically distributed random variables. Their joint distribution is derived from the n-fold product of the density function of the underlying distribution. Once these ideas are expressed as multiple integrals, deriving results in statistics are exercises in calculus and analysis.

The Density Function of an Estimator

This material motivates the central limit theorem presented in the next section, and it illustrates one of the major ideas in statistics. Since estimators are random variables, it is reasonable to ask about their density functions. If the density function can be computed we can determine the properties of the estimator.

This point of view permeates statistics, but it takes a while to get used to it. There is an underlying distribution. We take a random sample from this distribution. We consider a function of these sample points, which we call an estimator. This estimator is a random variable, and it has a distribution that is related to, but different from, the underlying distribution from which the samples were taken.

We will consider a very simple example for our favorite estimator—the sample average. Suppose the underlying distribution is a constant rate process with the density function $f(z) = e^{-z}$. Suppose the estimator is the average of a sample of size two $\bar{z} = (x + y)/2$. The distribution function for $\bar{z}$ is $G(t) = Pr[x + y \leq 2t]$. The integration is carryed out over the shaded area in Fig. 21.7 bounded by the line $y = -x + 2t$ and the axes.

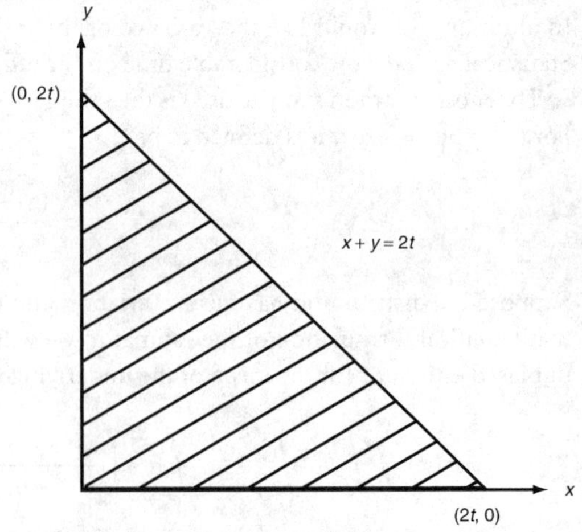

The probability that $x + y \leq 2t$ is given by the double integral

FIGURE 21.7 $x + y \leq 2t$.

$$G(t) = \int_0^{2t} \int_0^{2t-x} e^{-x} e^{-y} dy\, dx = 1 - e^{-2t} - 2te^{-2t}$$

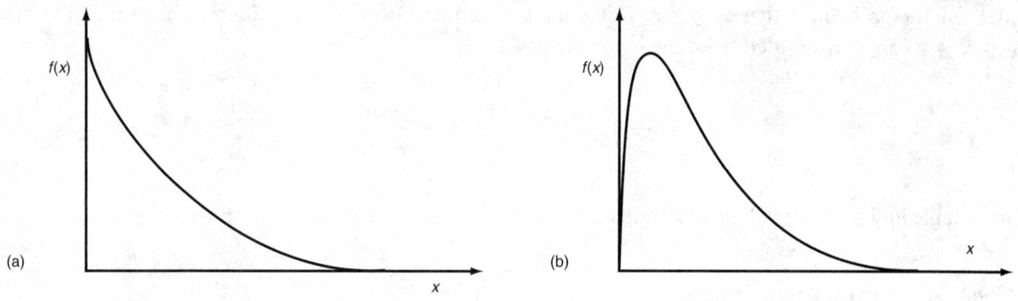

FIGURE 21.8 Density function: (a) underlying exponential distribution, (b) for the estimator of size 2.

which implies the density function is

$$g(t) = G'(t) = 4te^{-2t}$$

Hence, if x and y are sample points from a constant rate process with rate one, then the probability that $a \leq (x+y)/2 \leq b$ is $\int_a^b 4te^{-2t}dt$.

Figure 21.8 displays the density functions for the underlying exponential distribution and for the sample average of size 2. Obviously, we want the density function for a sample of size n, we want the density function of the estimator for other underlying distributions, and we want the density function for estimators other than the sample average. Statisticians have expended considerable effort on this topic, and this material can be found in statistics tests and monographs.

Consider the two density functions in Fig.21.8 from a different point of view. The first is the density function of the sample average when the sample size is 1. The second when the sample size is 2. As the sample size increases, the graph becomes more like the bell-shaped curve for the normal distribution. This is the content of the Central Limit Theorem in the next section. As the sample size increases, the distribution of the sample average becomes approximately normal.

The Central Limit Theorem

A remarkable result in probability is that for a large class of underlying distributions the sample average approaches one type of distribution as the sample size becomes large. The class is those distributions with finite mean and variance. The limiting distribution is the normal (Guassian) distribution. The density function for the normal distribution with mean μ and variance σ^2 (Fig. 21.9) is

$$\frac{1}{\sigma\sqrt{2\pi}}e^{-(x-\mu)^2/2\sigma^2}$$

where $\sigma = \sqrt{\sigma^2}$.

Theorem

Suppose a sample $x_1, \ldots, x_n$ is chosen from a distribution with mean μ and variance σ^2. As n becomes large, the distribution for the sample average $\bar{x}$ approaches the normal distribution with mean μ and variance σ^2/n.

This result is usually used in the following manner. As a consequence of the result, it can be shown that the function, $(\bar{x} - \mu)/\sqrt{\sigma^2/n}$, is (approximately) normal with mean zero and variance one (the standard normal). Extensive tables are available

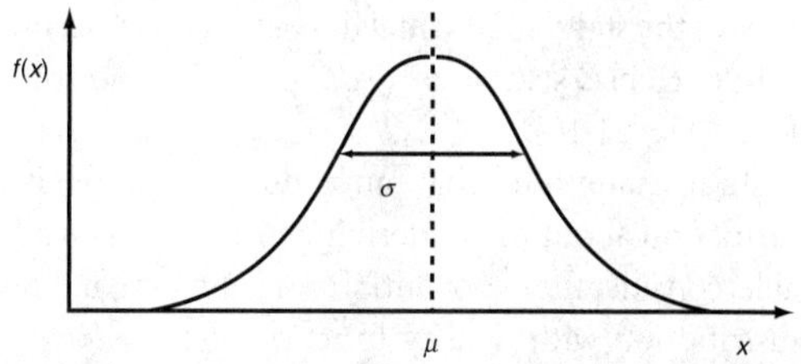

FIGURE 21.9 A Gaussian distribution with mean μ and variance σ^2.

that give the probability that $\alpha \leq Z \leq \beta$ if Z has the standard normal distribution. The probability that $a \leq \bar{x} \leq b$ can be computed from the inequalities

$$\frac{a - \mu}{\sqrt{\sigma^2/n}} \leq \frac{\bar{x} - \mu}{\sqrt{\sigma^2/n}} \leq \frac{b - \mu}{\sqrt{\sigma^2/n}}$$

and a table lookup. This will be illustrated in the next section on confidence intervals.

Confidence Intervals

It is possible for an experiment to mislead us. For example, when estimating the average lifetime for a certain type of device, the sample could contain an unusual number of the better components (or of the poorer components). We need a way to indicate the quality of an experiment—the probability that the experiment has not mislead us. A quantitative method of measuring the quality of an experiment is by using a confidence interval, which usually consists of three parts:

1. The value of the **estimator**: $\hat{x}$
2. An interval (called the **confidence interval**) about the estimator: $[\hat{x} - \alpha, \hat{x} + \beta]$
3. The probability (called the **confidence level**) that this interval contains the true (but unknown) value

The confidence level in part three is usually conservative, the probability that the interval contains the true value is greater than or equal to the stated confidence level.

The confidence interval has the following frequency interpretation. Suppose we obtain a 95% confidence interval from an experiment and statistical techniques. If we performed this experiment many times, then at least 95% of the time the interval will contain the true value of the parameter. This discussion of the confidence interval sounds similar to the previous discussion about estimators; they are random variables, and they have a certain probability of being within certain intervals.

To illustrate the general procedure, suppose we wish a 95% confidence interval for the mean μ of a random variable. Suppose the sample size is n. As described earlier, the sample mean $\bar{x}$ and the sample variance S^2 are unbiased estimators for the mean and variance of the underlying distribution. For a symmetric confidence interval the requirement is

$$Pr[\|\bar{x} - \mu\| \leq \alpha] \geq 0.95$$

It is necessary to solve for α. We will use the central limit theorem. Rewrite the preceding equation

$$Pr\left[\frac{-\alpha}{\sqrt{S^2/n}} \leq \frac{\bar{x} - \mu}{\sqrt{S^2/n}} \leq \frac{\alpha}{\sqrt{S^2/n}}\right]$$

The term in the middle of the inequality has (approximately) the standard normal distribution. Here, 95% of the standard normal lies between -1.96 and $+1.96$. Hence, set $t = \alpha/\sqrt{S^2/n} = 1.96$ and solve for α (Fig. 21.10).

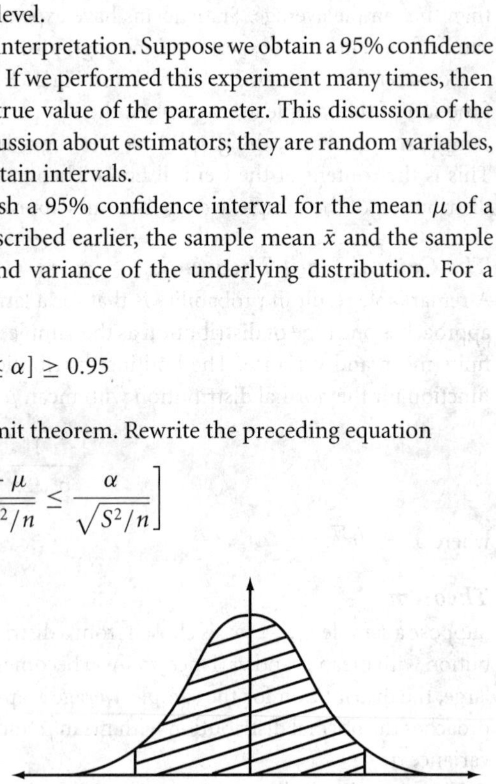

As an example of what can be done with more information about the underlying distribution available, consider the exponential (constant rate) failure distribution with density function $f(t) = \lambda e^{-\lambda t}$. Suppose the requirement is to estimate the mean-time-to-failure within 10% with a 95% confidence level. We will use the central limit theorem to determine how many trials are needed. The underlying

FIGURE 21.10 $-1.96 \leq t \leq 1.96$.

population has mean and variance $\mu = 1/\lambda$ and $\sigma^2 = 1/\lambda^2$. For a sample of size n the sample average $\bar{x}$ has mean and variance:

$$\mu(\bar{x}) = \frac{1}{\lambda} \quad \text{and} \quad \sigma^2(\bar{x}) = \frac{1}{n\lambda^2}$$

The requirement is

$$Pr\left[\|\bar{x} - \mu\| \leq 0.1\mu\right] \geq 0.95$$

Since the estimator is unbiased, $\mu = \mu(\bar{x})$. A little more algebra gives

$$Pr\left[\frac{-0.1\mu(\bar{x})}{\sqrt{\sigma^2(\bar{x})}} \leq \frac{\bar{x} - \mu(\bar{x})}{\sqrt{\sigma^2(\bar{x})}} \leq \frac{0.1\mu(\bar{x})}{\sqrt{\sigma^2(\bar{x})}}\right] \geq 0.95$$

Once again, the middle term of the inequality has (approximately) the standard normal distribution, and 95% percent of the standard normal lies between -1.96 and $+1.96$. Set

$$1.96 = \frac{0.1\mu(\bar{x})}{\sqrt{\sigma^2(\bar{x})}} = 0.1\sqrt{n}$$

write $\mu(\bar{x})$ and $\sigma^2(\bar{x})$ in terms of λ; solve to get $n = 384$.

Opinion Surveys, Monte Carlo Computations, and Binomial Sampling

One of the sources for skepticism about statistical methods is one of the most common applications of statistics, the opinion survey where about 1000 people are polled to determine the preference of the entire population. How can a sample of size 1000 tell us anything about the preference of several hundred million people?

This question is more relevant to engineering applications of statistics than it might appear, because the answer lies in the general nature of binomial sampling. As a similar problem consider a **Monte Carlo** computation of the area enclosed in an irregular graph given in Fig. 21.11. Suppose the area of the square is A. Choose 1000 points at random in the square. If N of the chosen points lie within the enclosed graph, then the area of the graph is estimated to be $NA/1000$. This area estimation poses a greater problem than the opinion survey. Instead of

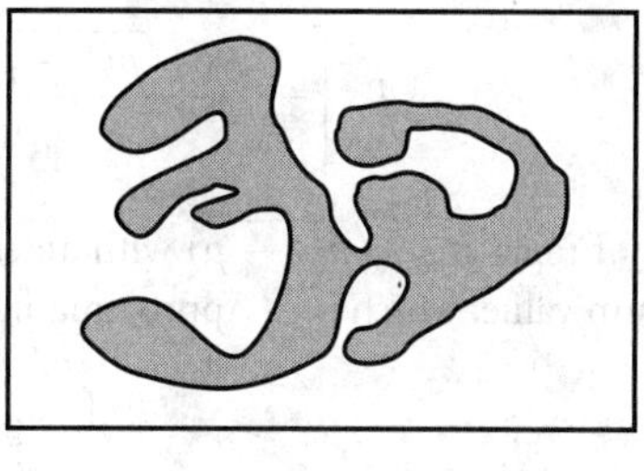

FIGURE 21.11 An area enclosed in an irregular graph.

a total population of several hundred million, the total population of points in the square is uncountably infinite. (How can 1000 points chosen at random tell us anything since 1000 is 0% of the population?) In both the opinion survey and the area problem we are after an unknown probability p, which reflects the global nature of the entire population. For the opinion survey

$$p = \frac{\text{number of yeses}}{\text{total number of population}}$$

For the area problem

$$p = \frac{\text{area enclosed by graph}}{\text{area of square}}$$

Suppose the sampling technique makes it equally likely that any person or point is chosen when a random selection is made. Then p is the probability that the response is yes or inside the curve. Assign a numerical value of 1 to yes or inside the curve. Assign 0 to the complement. The average (the expected value) for the entire population is:

$$1(\text{probability of a 1}) + 0(\text{probability of a 0}) = 1p + 0(1 - p) = p$$

The population variance is

$$\text{(probability of a 1)}(1 - \text{average})^2 + \text{(probability of a 0)}(0 - \text{average})^2$$
$$= p(1 - p)^2 + (1 - p)(0 - p)^2 = p(1 - p)$$

Hence, for a sample of size n, the mean and variance of the sample average are $E(\bar{x}) = p$ and $\text{Var}(\bar{x}) = p(1 - p)/n$.

The last equation answers our question about how a sample that is small compared to the entire population can give an accurate estimate for the entire population. The last equation says the variance of the estimator (which measures the accuracy of the estimator) depends on the sample size, not on the population size. The population size does not appear in the variance, which indicates the accuracy is the same whether the population is ten thousand, or several hundred million, or uncountably infinite.

The next step is to use the **central limit theorem**. Since the binomial distribution has a finite mean and variance, the adjusted sample average

$$\frac{\bar{x} - \mu}{\sqrt{\sigma^2/n}} = \frac{\bar{x} - p}{\sqrt{p(1 - p)/n}}$$

converges to the standard normal distribution (which has mean zero and standard deviation one). The usual confidence level used for opinion surveys is 95%. A table with the normalized Gaussian distribution shows that 95% of the standard normal lies between -1.96 and $+1.96$. Hence

$$Pr\left[-1.96 \le \frac{\bar{x} - p}{\sqrt{p(1 - p)/n}} \le 1.96\right] \ge 0.95$$

or

$$Pr\left[\frac{-1.96\sqrt{p(1 - p)}}{\sqrt{n}} \le \bar{x} - p \le \frac{1.96\sqrt{p(1 - p)}}{\sqrt{n}}\right] \ge 0.95$$

Instead of replacing $\sqrt{p(1 - p)}$ with an estimator, we will be more conservative and replace it with its maximum value, which is $\frac{1}{2}$. Approximating 1.96 by 2 and taking n to be 1000 gives

$$Pr[-0.03 \le \bar{x} - p \le 0.03] \ge 0.95$$

We have arrived at the news announcement that the latest survey gives some percentage, plus or minus 3%. The 3% error comes from a 95% confidence level with a sample size of 1000.

21.1.2 Components and Systems

Basics on Components

Quantitative reliability (the probability of failure by a certain time) begins by considering how fast and in what manner components fail. The source for this information is field data, past experience with classes of devices, and consideration of the manufacturing process. Good data are hard to get. For this reason, reliability analysts tend to be conservative in their assumptions about device quality and to use worst-case analyses when considering failure modes and their effects.

Failure Rates

There is a great body of statistical literature devoted to identifying the failure rate of devices under test. We just present a conceptual survey of rates.

Increasing Rate

There is an increasing failure rate if the device is wearing out. As the device ages, it is more likely to fail in the next increment of time. Most mechanical items have this property. For long-term reliability of a system with such components, there is often a replace before failure strategy as in the streetlight example earlier.

Constant Rate

The device is not wearing out. The most prosaic example is the engineer's coffee mug. It fails by catastrophe (if we drop it), otherwise it is immortal. A constant failure rate model is often used for mathematical convenience when the wear out rate is small. This appears to be accurate for high-quality solid-state electronic devices during most of their useful life.

Decreasing Rate

The device is improving with age. Software is a (sometimes controversial) example, but this requires some discussion. Software does not break, but certain inputs will lead to incorrect output. The frequency of these critical inputs determines the frequency of failure. Debugging enables the software to handle more inputs correctly. Hence, the rate of failures decreases.

A major issue in software is the independence of different versions of the same program. Initially, software reliability was treated similarly to hardware reliability, but software is now recognized as having different characteristics. The main item is that hardware has components that are used over and over (tried and true), whereas most software is custom.

Mathematical Formulation of Rates

The occurrence rate $R(t)$ is defined to be the density function divided by the probability that the event has not yet occurred. Hence, if the density function and distribution functions are $f(t)$ and $F(t)$

$$R(t) = \frac{f(t)}{1 - F(t)}$$

It appears reasonable that a device that is neither wearing out nor improving with age has the memoryless property. This can be shown mathematically. If the failure rate is constant, then

$$\frac{f(t)}{1 - F(t)} = c$$

Since the density function is the derivative of the distribution function, we have the differential equation

$$F'(t) + cF(t) = c$$

with the initial condition $F(0) = 0$. The solution is $F(t) = 1 - e^{-ct}$.

There are infinitely many distributions with increasing or decreasing rates. The Weibull distributions are popular and can be found in any reliability text.

Component Failure Modes

In addition to failing at different rates, devices can exhibit a variety of behaviors once they have failed. One variety is the on–off cycle of faulty behavior. A device, when it fails, can exhibit constant faulty behavior. In some ways, this is best for overall reliability. These devices can be identified and replaced more quickly than devices that exhibit intermittent faulty behavior. There are transient failures, usually due to external shocks, where the device is temporarily faulty, but will perform correctly once it has recovered. It can be difficult to distinguish between a device that is intermittently faulty (which should be replaced) and a device that is transiently faulty (which should not be replaced). In addition, a constantly faulty device in a complicated system may only occasionally cause a measurable performance error.

Devices can exhibit different faulty behavior, some of them malicious. The most famous is the lying clock problem, which can occur in system synchronization when correcting for clock drift.

- Clock A is good, but slow, and sends 11 am to the other two clocks.
- Clock B is good, but fast, and sends 1 pm to the other two clocks.

- Clock *C* is faulty. It sends 6 am to clock A and 6 pm to clock B.
- Clock *A* averages the three times and resets itself to 10 am.
- Clock *B* averages the three times and resets itself to 2 pm.

Even though a majority of the components are good, the system will lose synchronization and fail. The general version of this problem is known as *Byzantine* agreement.

Designing systems to tolerate and diagnose faulty components is an open research area.

Systems

Redundancy

There are two strands in reliability–fault avoidance and fault tolerance. Fault avoidance consists of building high-quality components and designing systems in a conservative manner. Fault tolerance takes the point of view that, despite all our efforts, components will fail and highly reliable systems must function in the presence of these failed components. These systems attempt achieving reliability beyond the reach of any single component by using redundancy.

There is a price in efficiency, however, for using redundancy, especially if the attempt is made to achieve reliability far beyond the reliability of the individual components. This is illustrated in Fig. 21.12, which plots the survival probability of a simple redundant system against the mean-time-to-failure (MTTF) of a single component. The system consists of *n* components, and the system works if a majority of the components are good.

There are several methods of improving the efficiency. One is by reconfiguration, which remove faulty components from the working group. This lets the good components remain in the majority for a longer period of time. Another is to use spares that can replace failed members of the working group. For long-term reliability, the spares can be unpowered, which usually reduces their failure rate. Although reconfiguration improves efficiency, it introduces several problems. One is a more complex system with an increased design cost and an increased possibility of design error. It also introduces a new failure mode. Even if reconfiguration works perfectly, it takes time to detect, identify, and remove a faulty component. Additional component failures during this time span can overwhelm the working group before the faulty components are removed. This failure mode is called a *coverage failure* or a *coincident-fault failure*. Assessing this failure mode requires modeling techniques that reflect system dynamics. One appropriate technique is **Markov models** explained in the next section.

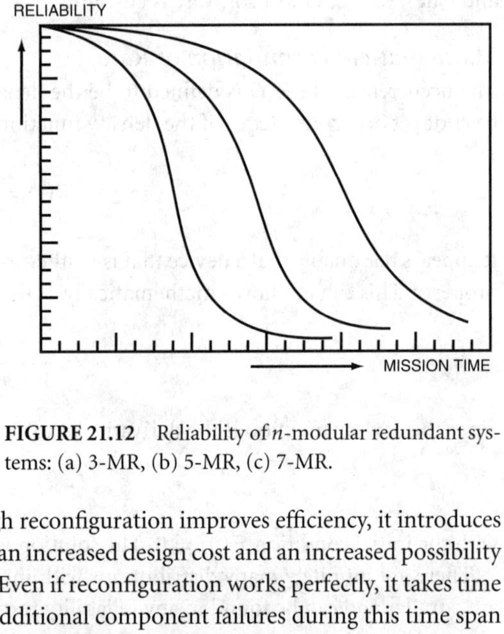

FIGURE 21.12   Reliability of *n*-modular redundant systems: (a) 3-MR, (b) 5-MR, (c) 7-MR.

Periodic Maintenance

In a sense, redundant systems have increasing failure rates. As individual components fail, the system becomes more and more vulnerable. Their reliability can be increased by periodic maintenance that replaces failed components. Figure 21.13 gives the reliability curve of a redundant system. The failure rate is small during the initial period of its life because it is unlikely that a significant number of its components fail in a short period of time. Figure 21.13 also gives the reliability curve of the same system with periodic maintenance. For simplicity, it is assumed that the maintenance restores the system as good as new. The reliability curve for the maintained system repeats the first part of the curve for the original system over and over.

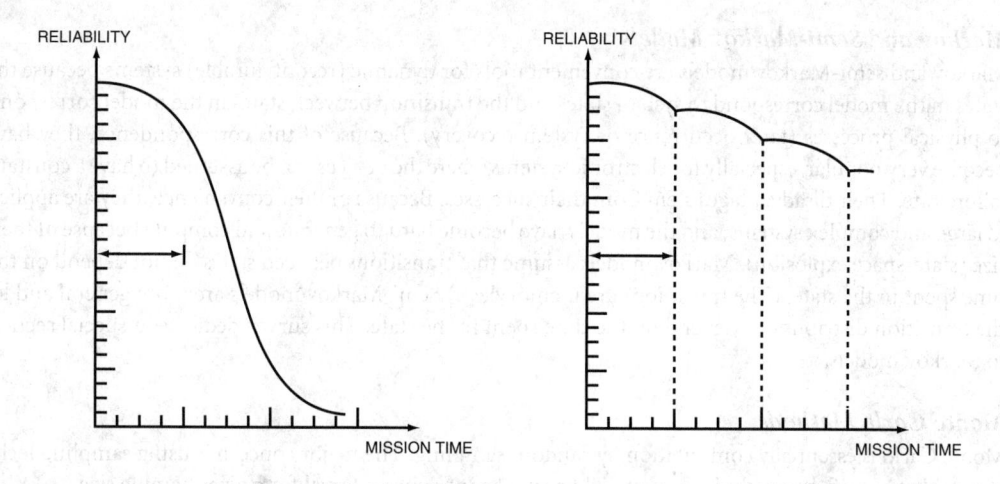

FIGURE 21.13 Reliability curves of an *unmaintained* (left) redundant system and a *maintained* (right) system.

Modeling and Computation Methods

If a system consists of individual components, it is natural to attempt to derive the failure characteristics of the system from the characteristics of the components and the system structure. There are three major modeling approaches.

Combinatorial and Fault Trees

Combinatorial techniques are appropriate if the system is static (nonreconfigurable). The approach is to construct a function that gives the probability of system failure (or survival) in terms of component failure (or survival) and the system structure. Anyone who remembers combinatorial analysis from high school or a course in elementary probability can imagine that the combinatorial expressions for complex systems can become very complicated. Fault trees have been developed as a tool to help express the combinatorial structure of a system. Once the user has described the system, a program computes the combinatorial probability. Fault trees often come with a pictorial input.

Example

A system works if both its subsystems work. Subsystem A works if all three of its components A_1, A_2, and A_3 work. Subsystem B works if either of its subsystems B_1 or B_2 work. Fault trees can be constructed to compute either the probability of success or the probability of failure. Figure 21.14 gives both the success tree and the failure tree for this example.

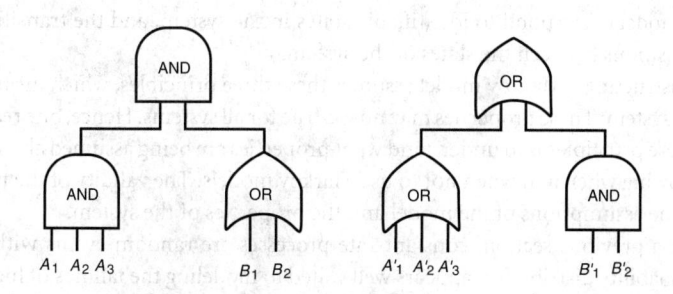

FIGURE 21.14 Success tree (left) and failure tree (right).

Markov and Semi-Markov Models

Markov and semi-Markov models are convenient tools for dynamic (reconfigurable) systems because the states in the model correspond to system states and the transition between states in the model correspond to physical processes (fault occurrence or system recovery). Because of this correspondence, they have become very popular, especially for electronic systems where the devices can be assumed to have a constant failure rate. Their disadvantages stem from their successes. Because of their convenience, they are applied to large and complex systems, and the models have become hard to generate and compute because of their size (state-space explosion). Markov models assume that transitions between states do not depend on the time spent in the state. (The transitions are memoryless.) Semi-Markov models are more general and let the transition distributions depend on the time spent in the state. This survey dedicates a special section to Markov models.

Monte Carlo Methods

Monte Carlo is essentially computation by random sampling. The major concern is using sampling techniques that are efficient techniques that yield a small confidence interval for a given sample size.

Monte Carlo is often the technique of last resort. It is used when the probability distributions are too arbitrary or when the system is too large and complex to be modeled and computed by other approaches. The Monte Carlo approach is usually used when failure rates are time varying. It can be used when maintenance is irregular or imperfect. The last subsection on statistics gave a short presentation of the Monte Carlo approach.

Model Verification

Here is an area for anyone looking for a research topic. Models are based on a inspection of the system and on assumptions about system behavior. There is a possibility that important features of the system have been overlooked and that some of the assumptions are incorrect. There two remedies for these possibilities. One is a detailed examination of the system, a failure modes and effects analysis. The size and complexity of current electronic systems make this an arduous task. Experiments can estimate parameters, but conducting of extensive experiments can be expensive.

21.1.3   Markov Models

Markov models have become popular in both reliability and performance analysis. This section explains this technique from an engineering point of view.

Basic Constructions

It only takes an intuitive understanding of three basic principles to construct Markov models. These are (1) constant rate processes, (2) independent competing events, and (3) independent sequential events. These three principles are described, and it is shown (by several examples) that an intuitive understanding of these principles is sufficient to construct Markov models. A Markov model consists of states and transitions. The states in the model correspond to identifiable states in the system, and the transitions in the model correspond to transitions between the states of the system.

Conversely, constructing a Markov model assumes these three principles, which means assuming these properties about a system. These properties may not be true for all systems. Hence, one reason for a detailed examination of these principles is to understand what properties are being assumed about the system. This is essential for knowing when and when not to use Markov models. The validity of the model depends on a match between the assumptions of the model and the properties of the system.

As discussed in a previous section, constant rate processes are random events with the memoryless property. This probability distribution appears well suited to modeling the failures of high-quality devices operating in a benign environment. The constant rate probability distribution, however, does not appear to be well suited to such events as system reconfiguration. Some of these procedures, for example, are fixed time programs. If the system is halfway through a 10-ms procedure, then it is known that the procedure will be completed in 5 ms. That is, this procedure has memory: how long it takes to complete the procedure depends on how long the procedure has been running. Despite this possible discrepancy, this exposition will describe Markov models because of their simplicity.

Competing events arise naturally in redundant and reconfigurable systems. An important example is the competition between the arrival of a second fault and system recovery from a first fault. Sequential events also arise naturally in redundant and reconfigurable systems. A component failure is followed by system recovery. Another component failure is followed by another system recovery.

The Markov model for an event that has rate λ is given by Fig. 21.15. In state S the event has not occurred. In state F the event has occurred.

Suppose A and B are independent events that occur at rates α and β, respectively. The model for A and B as competing events is given in Fig. 21.16. In state SS neither A nor B has occurred. In state SA, event A has occurred before event B, whereas event B may or may not have occurred. In state SB, event B has occurred before event A, whereas event A may or may not have occurred.

If A and B are independent, then the probability that one or the other or both have occurred can be modeled by adding their rate of occurrence. This is

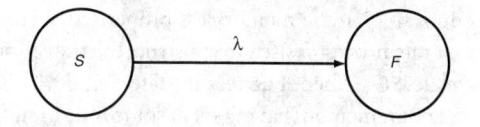

FIGURE 21.15 Model of a device failing at a constant rate λ.

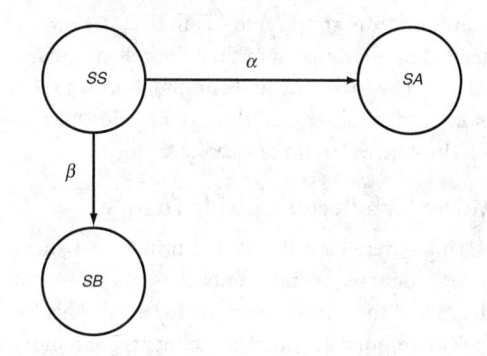

FIGURE 21.16 Model for independent competing events.

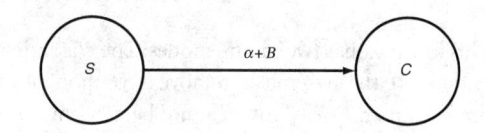

FIGURE 21.17 Model for occurrence of either independent event.

a standard result in probability (Rohatgi 1976; Hoel, Port, and Stone, 1972). For the two events depicted in Fig. 21.16, the sum is $\alpha + \beta$, and the model (for either one or both) is given in Fig. 21.17. The probability of being in state C in Fig. 21.17 is equal to the sum of the probabilities of being in states SA or SB in Fig. 21.16.

The model in Fig. 21.16 tells us whether or not an event has occurred and which event has occurred first. The model does not tell us whether or not both events have occurred. Models with this additional information require considering sequential events.

Figure 21.18 displays a model for two independent sequential events where the first event occurs at rate α and the second rate occurs at rate β. Such a model can arise when a device with failure rate α is operated until it fails, whereupon it is replaced by a device with failure rate β. State S represents the first device operating correctly; state A represents the first device having failed and the second device operating correctly; and state B represents the second device having failed.

Figure 21.19 displays a model with both competing and sequential events. Even though this model is simple, it uses all three of the basic principles. Suppose devices A and B with failure rates α and β are operating. In state S, there are two competing events, the failure of device A and the failure of device B. If device A fails first, the system is in state SA where device A has failed but device B has not failed.

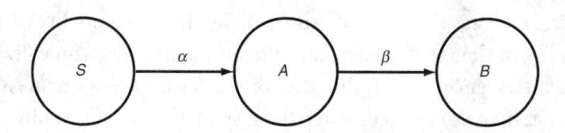

FIGURE 21.18 Model for sequential independent events.

In state SA, the memoryless property of a constant rate process is used to construct the transition
to state SC. When it arrives in state SA, device B
does not remember having spent some time in state
S. Hence, for device B, the transition out of state
SA is the same as the transition out of state S. If
device B were a component that wore out, then the
transition from state SA to SC in Fig. 21.19 would
depend on the amount of time the system spent in
state S. The modeling and computation would be
more difficult. A similar discussion holds for state
SB. In state SC, both devices have failed.

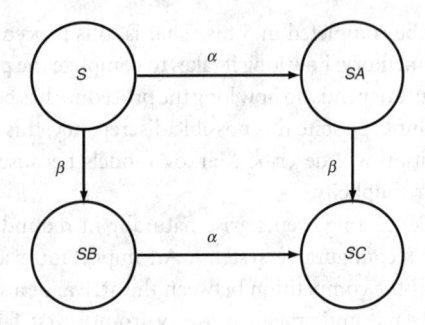

FIGURE 21.19   Model with both completing and sequential events.

Model for a Reconfigurable Fourplex

Having covered the three basic principles (constant rate processes, independent competing events, and
independent sequential events), it is possible to construct Markov reliability models. This section describes
the model for a reconfigurable fourplex. This fourplex uses majority voting to detect the presence of a
faulty component. Faulty components are removed from the system.

The reconfiguration sequence for this fourplex is

$$\text{fourplex} \longrightarrow \text{threeplex} \longrightarrow \text{twoplex} \longrightarrow \text{failure}$$

This fourplex has two failure modes. The coincident-fault failure mode occurs when a second component
fails before the system can remove a component that has previously failed. With two faulty components
present, the majority voter cannot be guaranteed to return the correct answer. A coincident-fault failure
is sometimes called a *coverage failure*. Since a failure due to lack of diagnostics is also called a coverage
failure, this chapter will use the term coincident-fault failure for the system failure mode just described.

The exhaustion of parts failure mode occurs when too many components have failed for the system to
operate correctly. In this reconfigurable fourplex, an exhaustion of parts failure occurs when one component of the final twoplex fails since this fourplex uses only majority voting to detect faulty components. A
system using self-test diagnostics may be able to reconfigure to a simplex before failing. In general, a system
can have a variety of reconfiguration sequences depending on design considerations, the sophistication
of the operating system, and the inclusion of diagnostic routines. This first example will use a simple
reconfiguration sequence.

The Markov reliability model for this fourplex is given in Fig. 21.20. Each component fails at rate λ.
The system recovers from a fault at rate δ. This system recovery includes the detection, identification, and
removal of the faulty component. The mnemonics in Fig. 21.20 are S for a fault free state, R for a recovery
mode state, C for a coincident-fault failure state, and E for the exhaustion of parts failure state.

The initial states and transitions for the model of the fourplex will be examined in detail. In state S_1 there
are four good components each failing at rate λ. Since these components are assumed to fail independently,
the failure rates can be summed to get the rate for the failure of one of the four. This sum is 4λ, which
is the transition rate from state S_1 to R_1. State R_1 has a transition with rate 3λ to state C_1 which represents
the failure of one of the three good components. Hence, in state R_1 precisely one component has failed
because the 4λ transition (from S_1) has been taken, but the 3λ transition (out of R_1) has not been taken.
There is also a δ transition out of state R_1 representing system recovery. System recovery transitions are
usually regarded as independent of component failure transitions because system recovery depends on the
architecture and software, whereas component failure depends on the quality of the devices. The system
recovery procedure would not change if the devices were replaced by electronically equivalent devices with
a higher or lower failure rate. Hence, state R_1 has the competing independent events of component failure
and system recovery. If system recovery occurs first, the system goes to the fault free state S_2 where there are
three good components in the system. The transition 3λ out of state S_2 represents the memoryless failure
rate of these three good components. If, in state R_1, component failure occurs before system recovery, then

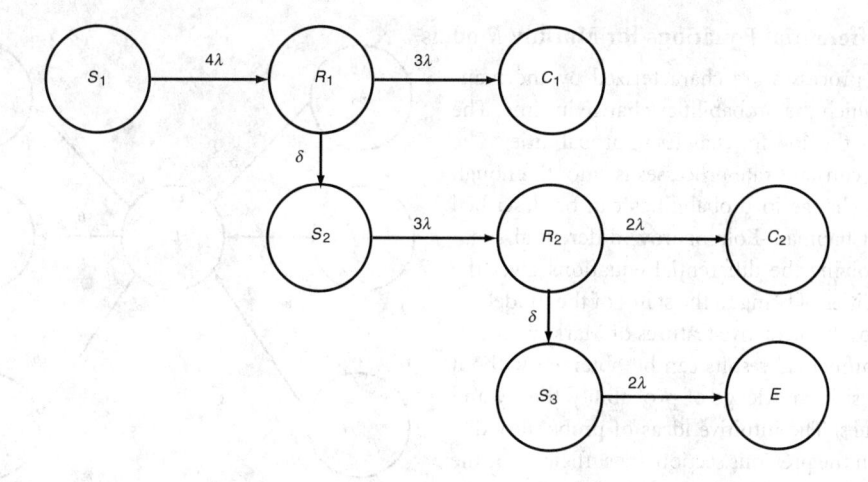

FIGURE 21.20   Reliability model of a reconfigurable fourplex.

the system is in the failed state C_1 where there is the possibility of the faulty components overwhelming the majority voter. The descriptions of the remaining states and transitions in the model are similar to the preceding descriptions.

There are several additional comments about the system failure state C_1. First, it is possible that the fourplex can survive a significant fraction of coincident faults. One possibility is that a fair amount of time elapses between the second fault occurrence and the faulty behavior of the second component. In this case, the system has time to remove the first faulty component before the majority is overwhelmed. Establishing this recovery mode, however, requires experiments with double fault injections, which may be expensive. It may be cheaper and easier to build a system that is more reliable than required instead of requiring an expensive set of experiments to establish the system's reliability, although more extensive experiments are a possible option. Models, such as the one in Fig. 21.20, that overestimate the probability of system failure are said to be conservative. Most models tend to be conservative because of the cost of obtaining detailed information about system performance, especially about system fault recovery.

Correlated Faults

This example demonstrates that Markov models can depict correlated faults. The technique used in this model is to have transitions represent the occurrence of several faults at once. Figure 21.21 displays a model for a fiveplex subjected to shocks, which produce transient faults. These shocks arrive at some constant rate. During a shock, faults appear in 0–5 components with some probability. In Fig. 21.21, the system is in state S_i if it has i faulty components for $i = 0, 1, 2$. The system is in the failure state F if there are three or more faulty components. The system removes all transient faults at rate ρ.

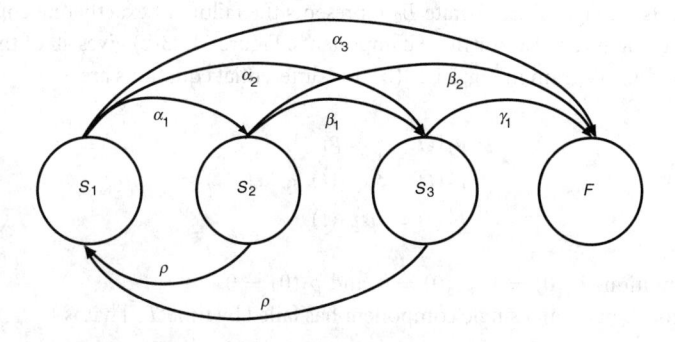

FIGURE 21.21   Fiveplex subjected to shocks.

The Differential Equations for Markov Models

Markov processes are characterized by the manner in which the probabilities change in time. The rates give the flow for changes in probabilities. The flow for constant rate processes is smooth enough that the change in probabilities can be described by the (Chapman–Kolmogorov) differential equations. Solving the differential equations gives the probabilities of being in the states of the model.

One of the attractive features of Markov models is that numerical results can be obtained without an extensive knowledge of probability theory and techniques. The intuitive ideas of probability discussed in the previous section are sufficient for the construction of a Markov model. Once the model is constructed, it can be computed by differential equations. The solution does not require any additional knowledge of probability.

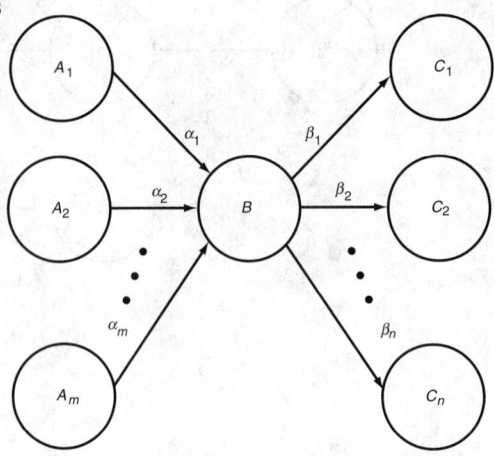

FIGURE 21.22   General diagram for a state in a Markov model.

Writing the differential equations for a model can be done in a local manner. The equation for the derivative of a state depends only on the single step transitions into that state and the single step transitions out of that state. Suppose there are transitions from states $\{A_1, \ldots, A_m\}$ into state B with rates $\{\alpha_1, \ldots, \alpha_m\}$ respectively, and that these are the only transitions into state B. Suppose there are transitions from state B to states $\{C_1, \ldots, C_n\}$ with rates $\{\beta_1, \ldots, \beta_n\}$ respectively, and that these are the only transitions out of state B. This general condition is shown in Fig. 21.22.

The differential equation for state B is

$$p'_B(t) = \alpha_1 p_{A_1}(t) + \alpha_2 p_{A_2}(t) + \cdots + \alpha_m p_{A_m}(t) - (\beta_1 + \cdots + \beta_n) p_B(t)$$

This formula is written for every state in the model, and the equations are solved subject to the initial conditions. While writing the formula for state B, it is possible to ignore all of the other states and transitions.

Computational Examples

This section presents a conceptual example that relates Markov models to combinatorial events and a computational example for a reconfigurable system.

Combinatorial Failure of Components

The next example considers the failure of three identical components with failure rate α. This example uses the basic properties of competing events, sequential events, and memoryless processes. Figure 21.23 presents the model for the failure of at least one component. State A_2 represents the failure of one, two, or three components. In Fig. 21.23(b) state B_2 represents the failure of exactly one component, whereas state B_3 represent the failure of two or three components. Figure 21.23(c) gives all of the details.

Choosing the middle diagram in Fig. 21.23(b), the differential equations are

$$p'_1(t) = -3\alpha p_1(t)$$
$$p'_2(t) = 3\alpha p_1(t) - 2\alpha p_2(t)$$
$$p'_3(t) = 2\alpha p_2(t)$$

with the initial conditions $p_1(0) = 1$, $p_2(0) = 0$, and $p_3(0) = 0$.

Let Q be the probability that a single component has failed by time T. That is

$$Q = 1 - e^{-\alpha T}$$

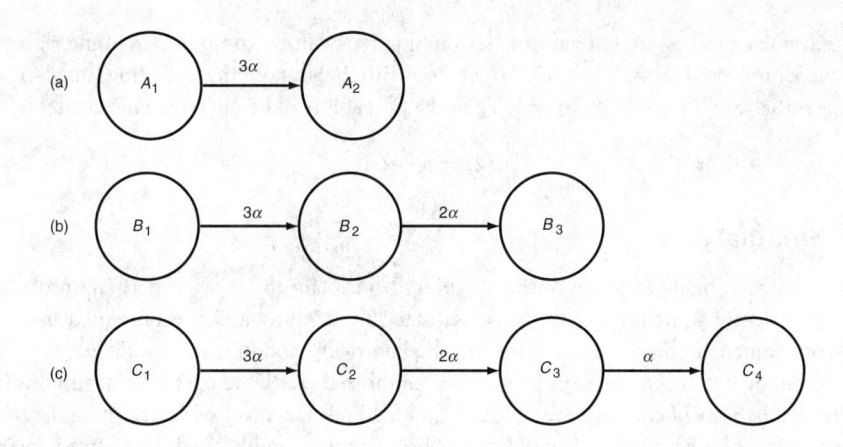

FIGURE 21.23 Markov models for combinatorial failures.

Solving the differential equations gives

$$Pr[B_3] = 3Q^2(1 - Q) + Q^3$$

which is the combinatorial probability that two or three components out of three have failed. A similar result holds for the other models in Fig. 21.23.

Computing the Fourplex

The model of the reconfigurable fourplex using numbers instead of mnemonics to label the states is represented by Fig. 21.24. The differential equations are

$$p_1'(t) = -4\lambda p_1(t)$$
$$p_2'(t) = 4\lambda p_1(t) - (3\lambda + \delta)\lambda p_2(t)$$
$$p_3'(t) = 3\lambda p_2(t)$$
$$p_4'(t) = \delta p_2(t) - 3\lambda p_4(t)$$
$$p_5'(t) = 3\lambda p_4(t) - (2\lambda + \delta)\lambda p_5(t)$$
$$p_6'(t) = 2\lambda p_5(t)$$
$$p_7'(t) = 5\delta p_5(t) - 2\lambda p_7(t)$$
$$p_8'(t) = 2\lambda p_7(t)$$

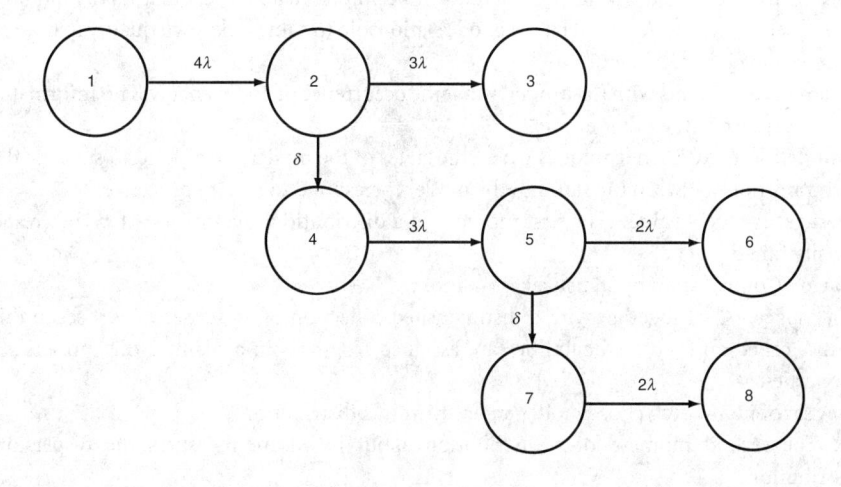

FIGURE 21.24 Original reliability model of a reconfigurable fourplex.

Once the parameter values are known, this set of equations can be computed by numerical methods. Suppose the parameter values are $\lambda = 10^{-4}/h$ and $\delta = 10^4/h$. Suppose the operating time is $T = 2$, and suppose the initial conditional is $p_1(0) = 1$. Then the probability of being in the failed states is

$$p_3(2) = 2.4 \times 10^{11}, \qquad p_6(2) = 4.8 \times 10^{15}, \qquad p_8(2) = 3.2 \times 10^{11}$$

21.1.4 Summary

An attempt has been made to accomplish two things, to explain the ideas underlying probability and statistics and to demonstrate some of the most useful techniques. Probability is presented as a method of describing phenomena in the presence of incomplete knowledge and uncertainty. Elementary events are assigned a likelihood of occurring. Complex events comprise these elementary events, and this is used to compute the probability of complex events. Statistical methods are a way of determining the probability of elementary events by experiment. Sampling and observation are emphasized. These two items are more important than fancy statistical techniques.

The analysis of fault-tolerant systems is presented in light of the view of probability described. Component failure and component behavior when faulty are the elementary events. System failure and behavior is computed according to how the system is built from its components. The text discussed some of the uncertainties and pathologies associated with component failure. It is not easy to design a system that functions as specified when all of the components are working correctly. It is even more difficult to accomplish this when some of the components are faulty.

Markov models of redundant and reconfigurable systems are convenient because states in the model correspond to states in the system and transitions in the model correspond to system processes. The Markov section gives an engineering presentation.

Defining Terms

Confidence interval and confidence level: Indicates the quality of an experiment. A confidence interval is an interval around the estimator. A confidence level is a probability. The interval contains the unknown parameter with this probability.

Central limit theorem: As the sample size becomes large, the distribution of the sample average becomes approximately normal.

Density function of a random variable: Defined implicitly: the probability that the random variable lies between a and b is the area under the density function between a and b. This approach introduces the methods of calculus and analysis into probability.

Estimator: Strictly speaking, any function of the sample points. Hence, an estimator is a random variable. To be useful an estimator must have a good relationship to some unknown quality that we are trying to determine by experiment.

Independent event: Events with the property that the occurrence of one event gives no information about the occurrence of the other events.

Markov model: A modeling techniques where the states of the model correspond to states of the system and transitions between the states in the model correspond to system processes.

Mean, average, expected value: The first moment of a distribution; the integral of x with respect to its density function $f(x)$.

Monte Carlo: Computation by statistical estimation.

Probability space: A set together with a distinguished collection of its subsets. A subset in this collection is called an event. Probabilities are assigned to events in a manner that preserves additive properties.

Sample: A set of items, each chosen independently from a distribution.

Variance: The second moment of a distribution about its mean; measures the dispersion of the distribution.

References

Hoel, P.G., Port, S.C., and Stone, C.J. 1972. *Introduction to Stochastic Processes.* Houghton Mifflin, Boston, MA.

Rohatgi, V.K. 1976. *An Introduction to Probability Theory and Mathematical Statistics.* Wiley, New York.

Ross, S.M. 1990. *A Course in Simulation.* Macmillan, New York.

Siewiorek, D.P. and Swarz, R.S. 1982. *The Theory and Practice of Reliable System Design.* Digital Press, Maynard.

Trivedi, K.S. 1982. *Probability and Statistics with Reliability, Queing, and Computer Science Applications.* Prentice-Hall, Englewood Cliffs.

Further Information

Additional information on the topic of probability and statistics applied to reliability is available from the following journals:

Microelectronics Reliability, Pergamon Journals Ltd.
IEEE Transactions on Computers
IEEE Transactions on Reliability

and proceedings of the following conferences:

IEEE/AIAA Reliability and Maintainability Symposium
IEEE Fault-Tolerant Computing Symposium

21.2 Electronic Hardware Reliability

Michael Pecht and Iuliana Bordelon

21.2.1 Introduction

Reliability is the ability of a product to perform as intended (i.e., without failure and within specified performance limits) for a specified life cycle. Reliability is a characteristic of a product, in the sense that reliability can be designed into a product, controlled in manufacture, measured during test, and sustained in the field.

To achieve **product performance** over time requires an approach that consists of tasks, each having total engineering and management commitment and enforcement. The tasks impact electronic hardware reliability through the selection of materials, structural geometries, design tolerances, manufacturing processes and tolerances, assembly techniques, shipping and handling methods, and maintenance and maintainability guidelines (Pecht, 1994). The tasks are as follows:

1. Define realistic system requirements determined by mission profile, required operating and storage life, performance expectations, size, weight, and cost. The design team must be aware of the environmental and operating conditions for the product. This includes all stress and loading conditions.

2. Characterize the materials and the manufacturing and assembly processes. Variabilities in material properties and manufacturing processes can induce failures. Although knowledge of variability may not be required for some engineering products, due to the inherent strength of the product compared to the stresses to which the product is subjected, concern with product weight, size, and cost often force the design team to consider such extreme margins as wasteful.

3. Identify the potential failure sites and **failure mechanisms**. Critical parts, part details, and the potential failure mechanisms and modes must be identified early in the design, and appropriate

measures must be implemented to assure control. Potential architectural and stress interactions must be defined and assessed.

4. Design to the usage and process capability (i.e., the quality level that can be controlled in manufacturing and assembly) considering the potential failure sites and failure mechanisms. The design stress spectra, the part test spectra, and the full-scale test spectra must be based on the anticipated life cycle usage conditions. Modeling and analysis are steps toward assessment. Tests are conducted to verify the results for complex structures. The goal is to provide a physics-of-failure basis for design decisions with the assessment of possible failure mechanisms for the anticipated product. The proposed product must survive the life cycle profile while being cost effective and available to the market in a timely manner.

5. Qualify the product manufacturing and assembly processes. If all the processes are in control and the design is proper, then product testing is not warranted and is therefore not cost effective. This represents a transition from *product* test, analysis, and screening, to *process* test, analysis, and screening.

6. Control the manufacturing and assembly processes addressed in the design. During manufacturing and assembly, each process must be monitored and controlled so that process shifts do not arise. Each process may involve screens and tests to assess statistical process control.

7. Manage the life cycle usage of the product using closed loop management procedures. This includes realistic inspection and maintenance procedures.

21.2.2 The Life Cycle Usage Environment

The life cycle usage environment or scenario for use of a product goes hand in hand with the product requirements. The life cycle usage information describes the storage, handling, and operating stress profiles and thus contains the necessary load input information for failure assessment and the development of design guidelines, screens, and tests. The stress profile of a product is based on the application profile and the internal stress conditions of the product. Because the performance of a product over time is often highly dependent on the magnitude of the stress cycle, the rate of change of the stress, and even the time and spatial variation of stress, the interaction between the application profile and internal conditions must be specified. Specific information about the product environment includes absolute temperature, temperature ranges, temperature cycles, temperature gradients, vibrational loads and transfer functions, chemically aggressive or inert environments, and electromagnetic conditions. The life cycle usage environment can be divided into three parts: the application and life profile conditions, the external conditions in which the product must operate, and the internal product generated stress conditions.

The application and life profile conditions include the application length, the number of applications in the expected life of the product, the product life expectancy, the product utilization or nonutilization (storage, testing, transportation) profile, the deployment operations, and the maintenance concept or plan. This information is used to group usage platforms (i.e., whether the product will be installed in a car, boat, satellite, underground), develop duty cycles (i.e., on–off cycles, storage cycles, transportation cycles, modes of operation, and repair cycles), determine design criteria, develop screens and test guidelines, and develop support requirements to sustain attainment of reliability and maintainability objectives.

The external operational conditions include the anticipated environment(s) and the associated stresses that the product will be required to survive. The stresses include temperature, vibrations, shock loads, humidity or moisture, chemical contamination, sand, dust and mold, electromagnetic disturbances, radiation, etc.

The internal operational conditions are associated with product generated stresses, such as power consumption and dissipation, internal radiation, and release or outgassing of potential contaminants. If the product is connected to other products or subsystems in a system, stresses associated with the interfaces (i.e., external power consumption, voltage transients, electronic noise, and heat dissipation) must also be included.

The life-cycle application profile is a time-sequential listing of all of the loads that can potentially cause failure. These loads constitute the parameters for quantifying the given application profile. For example, a flight application could be logged at a specified location and could involve engine warm-up, taxi, climb, cruising, high-speed maneuvers, gun firing, ballistic impact, rapid descent, and emergency landing. This information is of little use to a hardware designer unless it is associated with the appropriate application load histories, such as acceleration, vibration, impact force, temperature, humidity, and electrical power cycle.

21.2.3 Characterization of Materials, Parts, and the Manufacturing Processes

Design is intrinsically linked to the materials, parts, interfaces, and manufacturing processes used to establish and maintain functional and structural integrity. It is unrealistic and potentially dangerous to assume defect-free and perfect-tolerance materials, parts, and structures. Materials often have naturally occurring defects and manufacturing processes that can induce additional defects to the materials, parts, and structures. The design team must also recognize that the production lots or vendor sources for parts that comprise the design are subject to change. Even greater variability in parts characteristics is likely to occur during the fielded life of a product as compared to its design or production life cycle phases.

Design decisions involve the selection of components, materials, and controllable process techniques, using tooling and processes appropriate to the scheduled production quantity. Often, the goal is to maximize part and configuration standardization; increase package modularity for ease in fabrication, assembly, and modification; increase flexibility of design adaptation to alternate uses; and to utilize alternate fabrication processes. The design decisions also involve choosing the best materials interfaces and best geometric configurations, given the product requirements and constraints.

21.2.4 Failure Mechanisms

Failure mechanisms are the physical processes by which stresses can damage the materials included in the product. Investigating failure mechanisms aids in the development of failure-free, reliable designs. Numerous studies focusing on material failure mechanisms and physics-based damage models and their role in obtaining reliable electronic products have been extensively illustrated in a series of tutorials comprising all relevant **wearout** and **overstress failures** (Dasgupta and Pecht, 1991; Dasgupta and Hu, 1992a, 1992b, 1992c, 1992d; Dasgupta and Haslach, 1993; Engel, 1993; Li and Dasgupta, 1993, 1994; Dasgupta, 1993; Young and Christou, 1994; Rudra and Jennings, 1994; Al-Sheikhly and Christou, 1994; Diaz, Kang, and Duvvury 1995; Tullmin and Roberge, 1995).

Catastrophic failures that are due to a single occurrence of a stress event, when the intrinsic strength of the material is exceeded, are termed overstress failures. Failure mechanisms due to monotonic accumulation of incremental **damage** beyond the endurance of the material are termed wearout mechanisms. When the damage exceeds the endurance limit of the component, failure occurs. Unanticipated large stress events can either cause an overstress (catastrophic) failure or shorten life by causing the accumulation of wearout damage. Examples of such stresses are accidental abuse and acts of God. On the other hand, in well-designed and high-quality hardware, stresses should cause only uniform accumulation of wearout damage; the threshold of damage required to cause eventual failure should not occur within the usage life of the product.

The design team must be aware of all possible failure mechanisms in order to design hardware capable of withstanding loads without failing. Failure mechanisms and their related models are also important for planning tests and screens to audit the nominal design and manufacturing specifications, as well as the level of defects introduced by excessive variability in manufacturing and material parameters.

Electrical performance failures can be caused when individual components have incorrect resistance, impedance, voltage, current, capacitance, or dielectric properties or by inadequate shielding from electromagnetic interference (EMI) or particle radiation. The failure modes can be manifested as reversible drifts in electrical transient and steady-state responses such as delay time, rise time, attenuation, signal-to-noise ratio, and cross talk. Electrical failures, common in electronic hardware, include overstress mechanisms due to *electrical overstress* (EOS) and *electrostatic discharge* (ESD) such as dielectric breakdown, junction breakdown, hot electron injection, surface and bulk trapping, and surface breakdown and wearout mechanisms such as electromigration.

Thermal performance failures can arise due to incorrect design of thermal paths in an electronic assembly. This includes incorrect conductivity and surface emissivity of individual components, as well as incorrect convective and conductive paths for the heat transfer path. Thermal overstress failures are a result of heating a component beyond such critical temperatures as the glass-transition temperature, melting point, fictive point, or flash point. Some examples of thermal wearout failures are aging due to depolymerization, intermetallic growth, and interdiffusion. Failures due to inadequate thermal design may be manifested as components running too hot or too cold, causing operational parameters to drift beyond specifications, although, the degradation is often reversible upon cooling. Such failures can be caused either by direct thermal loads or by electrical resistive loads, which in turn generate excessive localized thermal stresses. Adequate design checks require proper analysis for thermal stress and should include conductive, convective, and radiative heat paths.

Incorrect product response to mechanical overstress and wearout loads may compromise the product performance, without necessarily causing any irreversible material damage. Such failures include incorrect elastic deformation in response to mechanical static loads, incorrect transient response (such as natural frequency or damping) to dynamic loads, and incorrect time-dependent reversible (anelastic) response. Mechanical failure can also result from buckling, brittle and/or ductile fracture, interfacial separation, fatigue crack initiation and propagation, creep, and creep rupture. To take one example, excessive elastic deformations in slender structures in electronic packages due to overstress loads can sometimes constitute functional failure, such as excessive flexing of interconnection wires, package lids, or flex circuits in electronic devices, causing shorting and/or excessive cross talk. When the load is removed, however, the deformations disappear completely without any permanent damage. Examples of wearout failure mechanism include fatigue damage due to thermomechanical stresses during power cycling of electronic hardware, corrosion rate due to anticipated contaminants, and electromigration in high-power devices.

Radiation failures are principally caused by uranium and thorium contaminants and secondary cosmic rays. Radiation can cause wearout, aging, embrittlement of materials, or overstress soft errors in such electronic hardware as logic chips. Chemical failures occur in adverse chemical environments that result in corrosion, oxidation, or ionic surface dendritic growth. There may also be interactions between different types of stresses. For example, metal migration may be accelerated in the presence of chemical contaminants and composition gradients and a thermal load can accelerate the failure mechanism due to a thermal expansion mismatch.

21.2.5   Design Guidelines and Techniques

Generally, products replace other products. The replaced product can be used for comparisons (i.e., a baseline comparison product). Lessons learned from the baseline comparison product can be used to establish new product parameters, to identify areas of focus in the new product design, and to avoid the mistakes of the past.

Once the parts, materials, and processes are identified along with the stress conditions, the objective is to design a product using parts and materials that have been sufficiently characterized in terms of how they perform over time when subjected to the manufacturing and application profile conditions. Only through a methodical design approach using physics of failure and root cause analysis can a reliable and cost effective product be designed.

In using design guidelines, there may not be a unique path to follow. Instead, there is a general flow in the design process. Multiple branches may exist depending on the input design constraints. The design team should explore enough of the branches to gain confidence that the final design is the best for the prescribed input information. The design team should also assess the use of the guidelines for the complete design and not those limited to specific aspects of an existing design. This statement does not imply that guidelines cannot be used to address only a specific aspect of an existing design, but the design team may have to trace through the implications that a given guideline suggests.

Design guidelines that are based on physics of failure models can also be used to develop tests, screens, and derating factors. Tests can be designed from the physics of failure models to measure specific quantities and to detect the presence of unexpected flaws or manufacturing or maintenance problems. Screens can be designed to precipitate failures in the weak population while not cutting into the design life of the normal population. Derating or safety factors can be determined to lower the stresses for the dominant failure mechanisms.

Preferred Parts

In many cases, a part or a structure much like the required one has already been designed and tested. This "preferred part or structure" is typically mature in the sense that variabilities in manufacturing, assembly, and field operation that can cause problems have been identified and corrected. Many design groups maintain a list of preferred parts and structures of proven performance, cost, availability, and reliability.

Redundancy

Redundancy permits a product to operate even though certain parts and interconnections have failed, thus increasing its reliability and availability. Redundant configurations can be classified as either active or standby. Elements in active redundancy operate simultaneously in performing the same function. Elements in standby redundancy are designed so that an inactive one will or can be switched into service when an active element fails. The reliability of the associated function is increased with the number of standby elements (optimistically assuming that the sensing and switching devices of the redundant configuration are working perfectly, and failed redundant components are replaced before their companion component fails). One preferred design alternative is that a failed redundant component can be repaired without adversely impacting the operation of the product and without placing the maintenance person or the product at risk.

A design team may often find that redundancy is

- The quickest way to improve product reliability if there is insufficient time to explore alternatives or if the part is already designed
- The cheapest solution, if the cost of redundancy is economical in comparison with the cost of redesign
- The only solution, if the reliability requirement is beyond the state-of-the-art

On the other hand, in weighing its disadvantages, the design team may find that redundancy will

- Prove too expensive, if the parts and redundant sensors and switching devices are costly
- Exceed the limitations on size and weight, particularly in avionics, missiles, and satellites
- Exceed the power limitations, particularly in active redundancy
- Attenuate the input signal, requiring additional amplifiers (which increase complexity)
- Require sensing and switching circuitry so complex as to offset the reliability advantage of redundancy.

Protective Architectures

It is generally desirable to include means in a design for preventing a part, structure, or interconnection from failing or from causing further damage if it fails. Protective architectures can be used to sense failure and protect against possible secondary effects. In some cases, self-healing techniques, in that they self-check and self-adjust to effect changes automatically to permit continued operation after a failure, are employed.

Fuses and circuit breakers are examples used in electronic products to sense excessive current drain and disconnect power from a failed part. Fuses within circuits safeguard parts against voltage transients or excessive power dissipation and protect power supplies from shorted parts. Thermostats may be used to sense critical temperature limiting conditions and shutting down the product or a component of the system until the temperature returns to normal. In some products, self-checking circuitry can also be incorporated to sense abnormal conditions and operate adjusting means to restore normal conditions or activate switching means to compensate for the malfunction.

In some instances, means can be provided for preventing a failed part or structure from completely disabling the product. For example, a fuse or circuit breaker can disconnect a failed part from a product in such a way that it is possible to permit partial operation of the product after a part failure, in preference to total product failure. By the same reasoning, degraded performance of a product after failure of a part is often preferable to complete stoppage. An example is the shutting down of a failed circuit whose design function is to provide precise trimming adjustment within a deadband of another control product. Acceptable performance may thus be permitted, perhaps under emergency conditions, with the deadband control product alone.

Sometimes the physical removal of a part from a product can harm or cause failure of another part of the product by removing load, drive, bias, or control. In these cases, the first part should be equipped with some form of interlock to shut down or otherwise protect the second part. The ultimate design, in addition to its ability to act after a failure, would be capable of sensing and adjusting for parametric drifts to avert failures.

In the use of protective techniques, the basic procedure is to take some form of action, after an initial failure or malfunction, to prevent additional or secondary failures. By reducing the number of failures, such techniques can be considered as enhancing product reliability, although they also affect availability and product effectiveness. Another major consideration is the impact of maintenance, repair, and part replacement. If a fuse protecting a circuit is replaced, what is the impact when the product is re-energized? What protective architectures are appropriate for postrepair operations? What maintenance guidance must be documented and followed when fail–safe protective architectures have or have not been included?

Stress Margins

A properly designed product should be capable of operating satisfactorily with parts that drift or change with time, temperature, humidity, and altitude, for as long as the parameters of the parts and the inter-connects are within their rated tolerances. To guard against out-of-tolerance failures, the designer must consider the combined effects of tolerances on parts to be used in manufacture, subsequent changes due to the range of expected environmental conditions, drifts due to aging over the period of time specified in the reliability requirement, and tolerances in parts used in future repair or maintenance functions. Parts and structures should be designed to operate satisfactorily at the extremes of the parameter ranges, and allowable ranges must be included in the procurement or reprocurement specifications.

Methods of dealing with part and structural parameter variations are statistical analysis and worst-case analysis. In statistical design analysis, a functional relationship is established between the output characteristics of the structure and the parameters of one or more of its parts. In worst-case analysis, the effect that a part has on product output is evaluated on the basis of end-of-life performance values or out-of-specification replacement parts.

Derating

Derating is a technique by which either the operational stresses acting on a device or structure are reduced relative to rated strength or the strength is increased relative to allocated operating stress levels. Reducing the stress is achieved by specifying upper limits on the operating loads below the rated capacity of the hardware. For example, manufacturers of electronic hardware often specify limits for supply voltage, output current, power dissipation, junction temperature, and frequency. The equipment designer may decide to select an alternative component or make a design change that ensures that the operational condition for a particular parameter, such as temperature, is always below the rated level. The component is then said to have been derated for thermal stress.

The derating factor, typically defined as the ratio of the rated level of a given stress parameter to its actual operating level, is actually a margin of safety or margin of ignorance, determined by the criticality of any possible failures and by the amount of uncertainty inherent in the reliability model and its inputs. Ideally, this margin should be kept to a minimum to maintain the cost effectiveness of the design. This puts the responsibility on the reliability engineer to identify as unambiguously as possible the rated strength, the relevant operating stresses, and reliability.

To be effective, derating criteria must target the right stress parameter to address modeling of the relevant failure mechanisms. Field measurements may also be necessary, in conjunction with modeling simulations, to identify the actual operating stresses at the failure site. Once the failure models have been quantified, the impact of derating on the effective reliability of the component for a given load can be determined. Quantitative correlations between derating and reliability enable designers and users to effectively tailor the margin of safety to the level of criticality of the component, leading to better and more cost-effective utilization of the functional capacity of the component.

21.2.6 Qualification and Accelerated Testing

Qualification includes all activities that ensure that the nominal design and manufacturing specifications will meet or exceed the reliability targets. The purpose is to define the acceptable range of variabilities for all critical product parameters affected by design and manufacturing, such as geometric dimensions and material properties. Product attributes that are outside the acceptable ranges are termed defects because they have the potential to compromise the product reliability (Pecht et al., 1994).

Qualification validates the capacity of the design and manufacturing specifications of the product to meet customer's expectations. The goal of qualification testing is to verify whether the anticipated reliability is indeed achieved under actual life cycle loads. In other words, qualification tests are intended to assess the performance of survival of a product over the complete life cycle period of the product. Qualification testing thus audits the ability of the design specifications to meet reliability goals. A well-designed qualification procedure provides economic savings and quick turnaround during development of new products or mature products subject to manufacturing and process changes.

Investigating the failure mechanisms and assessing the reliability of products where long life is required may be a challenge because a very long test period under actual operating conditions is necessary to obtain sufficient data to determine actual failure characteristics. One approach to the problem of obtaining meaningful qualification test data for high-reliability devices is **accelerated testing** to achieve test-time compression, sometimes called *accelerated-stress life testing*. When qualifying the reliability for overstress mechanisms, however, a single cycle of the expected overstress load may be adequate, and acceleration of test parameters may not be necessary. This is sometimes called *proof-stress testing*.

Accelerated testing involves measuring the performance of the test product at accelerated conditions of load or stress that are more severe than the normal operating level, in order to induce failures within a reduced time period. The goal of such testing is to accelerate the time-dependent failure mechanisms and the accumulation of damage to reduce the time to failure. A requirement is that the failure mechanisms and modes in the accelerated environment are the same as (or can be quantitatively correlated with) those observed under usage conditions and it is possible to extrapolate quantitatively from the accelerated environment to the usage environment with some reasonable degree of assurance.

A scientific approach to accelerated testing starts with identifying the relevant wearout failure mechanism. The stress parameter that directly causes the time-dependent failure is selected as the acceleration parameter and is commonly called the accelerated stress. Common accelerated stresses include thermal stresses, such as temperature, temperature cycling, or rates of temperature change; chemical stresses, such as humidity, corrosives, acid, or salt; electrical stresses, such as voltage, current, or power; and mechanical stresses, such as vibration loading, mechanical stress cycles, strain cycles, and shock/impulse. The accelerated environment may include one or a combination of these stresses. Interpretation of results for combined stresses requires a very clear and quantitative understanding of their relative interactions and the contribution of each stress to the overall damage.

Once the failure mechanisms are identified, it is necessary to select the appropriate acceleration stress; determine the test procedures and the stress levels; determine the test method, such as constant stress acceleration or step-stress acceleration; perform the tests; and interpret the test data, which includes extrapolating the accelerated test results to normal operating conditions. The test results provide designers with qualitative failure information for improving the hardware through design and/or process changes.

Failure due to a particular mechanism can be induced by several acceleration parameters. For example, corrosion can be accelerated by both temperature and humidity; creep can be accelerated by both mechanical stress and temperature. Furthermore, a single acceleration stress can induce failure by several wearout mechanisms simultaneously. For example, temperature can accelerate wearout damage accumulation not only due to electromigration, but also due to corrosion, creep, and so forth. Failure mechanisms that dominate under usual operating conditions may lose their dominance as stress is elevated. Conversely, failure mechanisms that are dormant under normal use conditions may contribute to device failure under accelerated conditions. Thus, accelerated tests require careful planning in order to represent the actual usage environments and operating conditions without introducing extraneous failure mechanisms or nonrepresentative physical or material behavior. The degree of stress acceleration is usually controlled by an *acceleration factor*, defined as the ratio of the life under normal use conditions to that under the accelerated condition. The acceleration factor should be tailored to the hardware in question and should be estimated from an acceleration transform that gives a functional relationship between the accelerated stress and reduced life, in terms of all of the hardware parameters.

Detailed failure analysis of failed samples is a crucial step in the qualification and validation program. Without such analysis and feedback to designers for corrective action, the purpose of the qualification program is defeated. In other words, it is not adequate simply to collect failure data. The key is to use the test results to provide insights into, and consequent control over, relevant failure mechanisms and ways to prevent them, cost effectively.

21.2.7 Manufacturing Issues

Manufacturing and assembly processes can impact the quality and reliability of hardware. Improper assembly and manufacturing techniques can introduce defects, flaws, and residual stresses that act as potential failure sites or stress raisers later in the life of the product. The fact that the defects and stresses during the assembly and manufacturing process can affect the product reliability during operation necessitates the identification of these defects and stresses to help the design analyst account for them proactively during the design and development phase. The task of auditing the merits of the manufacturing process involves two crucial steps. First, qualification procedures are required, as in design qualification, to ensure that manufacturing specifications do not excessively compromise the long-term reliability of the hardware. Second, lot-to-lot screening is required to ensure that the variabilities of all manufacturing-related parameters are within specified tolerances (Pecht et al., 1994). In other words, screening ensures the quality of the product and improves short-term reliability by precipitating latent defects before they reach the field.

Process Qualification

Like design qualification, process qualification should be conducted at the prototype development phase. The intent is to ensure that the nominal manufacturing specifications and tolerances produce acceptable reliability in the hardware. Once qualified, the process needs requalification only when process parameters, materials, manufacturing specifications, or human factors change.

Process qualification tests can be the same set of accelerated wearout tests used in design qualification. As in design qualification, overstress tests may be used to qualify a product for anticipated field overstress loads. Overstress tests may also be exploited to ensure that manufacturing processes do not degrade the intrinsic material strength of hardware beyond a specified limit. However, such tests should supplement, not replace, the accelerated wearout test program, unless explicit physics-based correlations are available between overstress test results and wearout field-failure data.

Manufacturability

The control and rectification of manufacturing defects has typically been the concern of production and process-control engineers, but not of the designer. In the spirit and context of concurrent product development, however, hardware designers must understand material limits, available processes, and manufacturing process capabilities in order to select materials and construct architectures that promote producibility and reduce the occurrence of defects, and consequently increase yield and quality. Therefore, no specification is complete without a clear discussion of manufacturing defects and acceptability limits. The reliability engineer must have clear definitions of the threshold for acceptable quality, and of what constitutes nonconformance. Nonconformance that compromises hardware performance and reliability is considered a defect. Failure mechanism models provide a convenient vehicle for developing such criteria. It is important for the reliability analyst to understand what deviations from specifications can compromise performance or reliability, and what deviations are benign and can hence be accepted.

A *defect* is any outcome of a process (manufacturing or assembly) that impairs or has the potential to impair the functionality of the product at any time. The defect may arise during a single process or may be the result of a sequence of processes. The *yield* of a process is the fraction of products that are acceptable for use in a subsequent process in the manufacturing sequence or product life cycle. The *cumulative yield* of the process is determined by multiplying the individual yields of each of the individual process steps. The source of defects is not always apparent, because defects resulting from a process can go undetected until the product reaches some downstream point in the process sequence, especially if screening is not employed.

It is often possible to simplify the manufacturing and assembly processes in order to reduce the probability of workmanship defects. As processes become more sophisticated, however, process monitoring and control are necessary to ensure a defect-free product. The bounds that specify whether the process is within tolerance limits, often referred to as the *process window*, are defined in terms of the independent variables to be controlled within the process and the effects of the process on the product, or the dependent product variables. The goal is to understand the effect of each process variable on each product parameter in order to formulate control limits for the process, that is, the points on the variable scale where the defect rate begins to possess a potential for causing failure. In defining the process window, the upper and lower limits of each process variable, beyond which it will produce defects, have to be determined. Manufacturing processes must be contained in the process window by defect testing, analysis of the causes of defects, and elimination of defects by process control, such as closed-loop corrective action systems. The establishment of an effective feedback path to report process-related defect data is critical. Once this is done and the process window is determined, the process window itself becomes a feedback system for the process operator.

Several process parameters may interact to produce a different defect than would have resulted from the individual effects of these parameters acting independently. This complex case may require that the interaction of various process parameters be evaluated in a matrix of experiments. In some cases, a defect cannot be detected until late in the process sequence. Thus, a defect can cause rejection, rework, or failure of the product after considerable value has been added to it. These cost items due to defects can reduce yield and return on investments by adding to hidden factory costs. All critical processes require special attention for defect elimination by process control.

Process Verification Testing

Process verification testing is often called *screening*. Screening involves 100% auditing of all manufactured products to detect or precipitate defects. The aim is to preempt potential quality problems before they reach the field. In principle, this should not be required for a well-controlled process. When uncertainties are likely in process controls, however, screening is often used as a safety net.

Some products exhibit a multimodal probability density function for failures, with a secondary peak during the early period of their service life due to the use of faulty materials, poorly controlled manufacture and assembly technologies, or mishandling. This type of early-life failure is often called *infant mortality*. Properly applied screening techniques can successfully detect or precipitate these failures, eliminating or reducing their occurrence in field use. Screening should only be considered for use during the early stages of production, if at all, and only when products are expected to exhibit infant mortality field failures. Screening will be ineffective and costly if there is only one main peak in the failure probability density function. Further, failures arising due to unanticipated events such as acts of God (lightning, earthquake) may be impossible to screen in a cost effective manner.

Since screening is done on a 100% basis it is important to develop screens that do not harm good components. The best screens, therefore, are nondestructive evaluation techniques, such as microscopic visual exams, X rays, acoustic scans, nuclear magnetic resonance (NMR), electronic paramagnetic resonance (EPR), and so on. Stress screening involves the application of stresses, possibly above the rated operational limits. If stress screens are unavoidable, overstress tests are preferred to accelerated wearout tests, since the latter are more likely to consume some useful life of good components. If damage to good components is unavoidable during stress screening, then quantitative estimates of the screening damage, based on failure mechanism models must be developed to allow the designer to account for this loss of usable life. The appropriate stress levels for screening must be tailored to the specific hardware. As in qualification testing, quantitative models of failure mechanisms can aid in determining screen parameters.

A stress screen need not necessarily simulate the field environment, or even utilize the same failure mechanism as the one likely to be triggered by this defect in field conditions. Instead, a screen should exploit the most convenient and effective failure mechanism to stimulate the defects that would show up in the field as infant mortality. Obviously, this requires an awareness of the possible defects that may occur in the hardware and extensive familiarity with the associated failure mechanisms.

Unlike qualification testing, the effectiveness of screens is maximized when screens are conducted immediately after the operation believed to be responsible for introducing the defect. Qualification testing is preferably conducted on the finished product or as close to the final operation as possible; on the other hand, screening only at the final stage, when all operations have been completed, is less effective, since failure analysis, defect diagnostics, and troubleshooting are difficult and impair corrective actions. Further, if a defect is introduced early in the manufacturing process, subsequent value added through new materials and processes is wasted, which additionally burdens operating costs and reduces productivity. Admittedly, there are also several disadvantages to such an approach. The cost of screening at every manufacturing station may be prohibitive, especially for small batch jobs. Further, components will experience repeated screening loads as they pass through several manufacturing steps, which increases the risk of accumulating wearout damage in good components due to screening stresses. To arrive at a screening matrix that addresses as many defects and failure mechanisms as feasible with each screen test, an optimum situation must be sought through analysis of cost-effectiveness, risk, and the criticality of the defects. All defects must be traced back to the root cause of the variability.

Any commitment to stress screening must include the necessary funding and staff to determine the root cause and appropriate corrective actions for all failed units. The type of stress screening chosen should be derived from the design, manufacturing, and quality teams. Although a stress screen may be necessary during the early stages of production, stress screening carries substantial penalties in capital, operating expense, and cycle time, and its benefits diminish as a product approaches maturity. If almost all of the products fail in a properly designed screen test, the design is probably incorrect. If many products fail, a revision of the manufacturing process is required. If the number of failures in a screen is small, the processes are likely to be within tolerances, and the observed faults may be beyond the resources of the design and production process.

21.2.8 Summary

Hardware reliability is not a matter of chance or good fortune; rather, it is a rational consequence of conscious, systematic, rigorous efforts at every stage of design, development, and manufacture. High product reliability can only be assured through robust product designs, capable processes that are known to be within tolerances, and qualified components and materials from vendors whose processes are also capable and within tolerances. Quantitative understanding and modeling of all relevant failure mechanisms provide a convenient vehicle for formulating effective design and processing specifications and tolerances for high reliability. Scientific reliability assessments may be supplemented by accelerated qualification testing.

The physics-of-failure approach is not only a tool to provide better and more effective designs, but it is also an aid for cost-effective approaches for improving the entire approach to building electronic systems. Proactive improvements can be implemented for defining more realistic performance requirements and environmental conditions, identifying and characterizing key material properties, developing new product architectures and technologies, developing more realistic and effective accelerated stress tests to audit reliability and quality, enhancing manufacturing-for-reliability through mechanistic process modeling and characterization to allow pro-active process optimization, and increasing first-pass yields and reducing hidden factory costs associated with inspection, rework, and scrap.

When utilized early in the concept development stage of a product's development, reliability serves as an aid to determine feasibility and risk. In the design stage of product development, reliability analysis involves methods to enhance performance over time through the selection of materials, design of structures, choice of design tolerance, manufacturing processes and tolerances, assembly techniques, shipping and handling methods, and maintenance and maintainability guidelines. Engineering concepts such as strength, fatigue, fracture, creep, tolerances, corrosion, and aging play a role in these design analyses. The use of physics-of-failure concepts coupled with mechanistic, as well as probabilistic, techniques is often required to understand the potential problems and tradeoffs and to take corrective actions when effective. The use of factors of safety and worst-case studies are as part of the analysis is useful in determining stress screening and burn-in procedures, reliability growth, maintenance modifications, field testing procedures, and various logistics requirements.

Defining Terms

Accelerated testing: Tests conducted at higher stress levels than normal operation but in a shorter period of time for the specific purpose to induce failure faster.

Damage: The failure pattern of an electronic or mechanical product.

Failure mechanism: A physical or chemical defect that results in partial degradation or complete failure of a product.

Overstress failure: Failure mechanisms due to a single occurrence of a stress event when the intrinsic strength of an element of the product is exceeded.

Product performance: The ability of a product to perform as required according to specifications.

Qualification: All activities that ensure that the nominal design and manufacturing specifications will meet or exceed the reliability goals.

Reliability: The ability of a product to perform at required parameters for a specified period of time.

Wearout failure: Failure mechanisms caused by monotonic accumulation of incremental damage beyond the endurance of the product.

References

Al-Sheikhly, M. and Christou, A. 1994. How radiation affects polymeric materials. *IEEE Trans. on Reliability* 43(4):551–556.

Dasgupta, A. 1993. Failure mechanism models for cyclic fatigue. *IEEE Trans. on Reliability* 42(4): 548–555.

Dasgupta, A. and Haslach, H.W., Jr. 1993. Mechanical design failure models for buckling. *IEEE Trans. on Reliability* 42(1):9–16.

Dasgupta, A. and Hu, J.M. 1992a. Failure mechanism models for brittle fracture. *IEEE Trans. on Reliability* 41(3):328–335.

Dasgupta, A. and Hu, J.M. 1992b. Failure mechanism models for ductile fracture. *IEEE Trans. on Reliability* 41(4):489–495.

Dasgupta, A. and Hu, J.M. 1992c. Failure mechanism models for excessive elastic deformation. *IEEE Trans. on Reliability* 41(1):149–154.

Dasgupta, A. and Hu, J.M. 1992d. Failure mechanism models for plastic deformation. *IEEE Trans. on Reliability* 41(2):168–174.

Dasgupta, A. and Pecht, M. 1991. Failure mechanisms and damage models. *IEEE Trans. on Reliability* 40(5):531–536.

Diaz, C., Kang, S.M., and Duvvury, C. 1995. Electrical overstress and electrostatic discharge. *IEEE Trans. on Reliability* 44(1):2–5.

Engel, P.A. 1993. Failure models for mechanical wear modes & mechanisms. *IEEE Trans. on Reliability* 42(2):262–267.

Ghanem and Spanos. 1991. *Stochastic Finite Element Methods*. Springer-Verlag, New York.

Haugen, E.B. 1980. *Probabilistic Mechanical Design*. Wiley, New York.

Hertzberg, R.W. 1989. *Deformation and Fracture Mechanics of Engineering Materials*. Wiley, New York.

Kapur, K.C. and Lamberson, L.R. 1977. *Reliability in Engineering Design*. Wiley, New York.

Lewis, E.E. 1987. *Introduction to Reliability Engineering*. Wiley, New York.

Li, J. and Dasgupta, A. 1994. Failure-mechanism models for creep and creep rupture. *IEEE Trans. on Reliability* 42(3):339–353.

Li, J. and Dasgupta, A. 1994. Failure mechanism models for material aging due to interdiffusion. *IEEE Trans. on Reliability* 43(1):2–10.

Pecht, M. 1994. Physical architecture of VLSI systems. In *Reliability Issues*. Wiley, New York.

Pecht, M. 1994. *Integrated Circuit, Hybrid, and Mutichip Module Package Design Guidelines—A Focus on Reliability*. John Wiley and Sons, New York.

Pecht, M. 1995. *Product Reliability, Maintainability, and Supportability Handbook*. CRC Press, Boca Raton, FL.

Rudra, B. and Jennings, D. 1994. Failure-mechanism models for conductive-filament formation. *IEEE Trans. on Reliability* 43(3):354–360.

Sandor, B. 1972. *Fundamentals of Cyclic Stress and Strain*. City University of Wisconsin Press, WI.

Tullmin, M. and Roberge, P.R. 1995. Corrosion of metallic material. *IEEE Trans. on Reliability* 44(2): 271–278.

Young, D. and Christou, A. 1994. Failure mechanism models for electromigration. *IEEE Trans. on Reliability* 43(2):186–192.

Further Information

IEEE Transactions on Reliability: Corporate Office, New York, NY 10017-2394, phone (212)70-7900.

Pecht, M. 1995. *Product Reliability, Maintainability, and Supportability Handbook*. CRC Press, Boca Raton, FL.

Pecht, M., Dasgupta, A., Evans, J.W., and Evans, J.Y. 1994. *Quality Conformance and Qualification of Microelectronic Packages and Interconnects*. Wiley, New York.

21.3 Software Reliability

Carol Smidts

21.3.1 Introduction

Software is without doubt an important element in our lives. Software systems are present in air traffic control systems, banking systems, manufacturing, nuclear power plants, medical systems, etc. Thus, software failures can have serious consequences: customer annoyance, loss of valuable data in information systems accidents, and so forth, which can result in millions of dollars in lawsuits. Hence, software development needs to be optimized, monitored, and controlled.

Models have been developed to characterize the quality of software development processes in industries. The capability maturity model (Paulk et al., 1993) is one such model. The CMM has five different levels, each of which distinguishes an organization's software process capability:

1. *Initial:* The software development process is characterized as ad hoc; few processes are defined and project outcomes are hard to predict.

2. *Repeatable:* Basic project management processes are established to track cost, schedule, and functionality; processes may vary from project to project, but management controls are standardized; current status can be determined across a project's life; with high probability, the organization can repeat its previous level of performance on similar projects. At level 2, the key process areas are as follows: requirements management, project planning, project tracking and oversight, subcontract management, quality assurance, and configuration management.

3. *Defined:* The software process for both management and engineering activities is documented, standardized, and integrated into an organization-wide software process; all projects use a documented and approved version of the organization's process for developing and maintaining software.

4. *Managed:* Detailed measures of the software process are collected; both the process and the product are quantitatively understood and controlled, using detailed measures.

5. *Optimizing:* Continuous process improvement is made possible by quantitative feedback from the process and from testing innovative ideas and technology.

21.3.2   Software Life Cycle Models

Knowledge of the software development life cycle is needed to understand **software reliability** concepts and techniques. Several life cycle models have been developed over the years: from the *code-and-fix* model to the *spiral model*. We will describe two such models, which have found a wide variety of applications: the *waterfall life cycle model* and the spiral model.

The Waterfall Life Cycle Model

The principal components of this software development process are presented in Fig. 21.25. Together they form the waterfall life cycle model (Royce, 1970, 1987). In actual implementations of the waterfall model variations in the number of phases may exist, naming of phases may differ, and who is responsible for what phase may change.

1. *Requirements definition and analysis* is the phase during which the analyst (usually A software engineer) will determine the user's requirements for the particular piece of software being developed. The notion of requirements will be examined in a later section.

2. *The preliminary design* phase consists of the development of a high-level design of the code based on the requirements.

3. *The detailed design* is characterized by increasing levels of detail and refinement to the preliminary design from the subsystem level to the subroutine level of detail. This implies describing in detail the user's inputs, system outputs, input/output files, and the interfaces at the module level.

4. *The implementation phase* consists of three different activities: coding, testing at the module level (an activity called *unit testing*), and integration of the modules into subsystems. This integration is usually accompanied by integration testing, a testing activity which will ensure compatibility of interfaces between the modules being integrated.

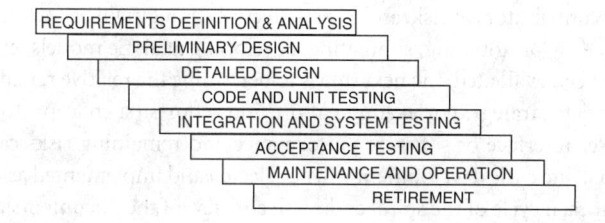

FIGURE 21.25   The waterfall life cycle.

5. *System testing* is defined as when the development team is in charge of testing the entire system from the point of view of the functions the software is supposed to fulfill. The validation process is completed when all tests required in the system test plan are satisfied. The development team will then produce a system description document.

6. *Acceptance testing* is defined as the point at which the development team relinquishes its testing responsibility to an independent acceptance test team. The independent acceptance test team is to determine whether or not the system now satisfies the original system requirements. An acceptance test plan is developed, and the phase ends with successful running of all tests in the acceptance test plan.

7. *Maintenance and operations* is the next phase. If the software requirements are stable, activities will be focused on fixing errors appearing in the exercise of the software or on fine tuning the software to enhance its performance. On the other hand, if the requirements continue to change in response to changing user needs, this stage could resemble to a minilife cycle in itself as software is modified to keep pace with the operational needs.

8. *Retirement* occurs when the user may decide at some point not to use the software anymore. Because of recurrent changes, the software might have become impossible to maintain.

The end of each phase of the life cycle should see the beginning of a new phase. Unfortunately, reality is often very different. Preliminary design will often start before the requirements have been completely analyzed. Actually, the real life cycle is characterized by a series of back-and-forth movements between the eight phases.

The Spiral Model

The waterfall model was extremely influential in the 1970s and still has widespread application today. However, the waterfall model has a number of limitations such as its heavy emphasis on fully developed documents as criteria of completion for early requirements and design phases. This constraint is not practical for many categories of software (for example, interactive end-user applications). Other process models were developed to remedy these weaknesses. The spiral model (Boehm, 1988) (see Fig. 21.26) is one such model. It is a risk-driven software development model that encompasses most existing process models. The software lifecycle model is not known *a priori*, and it is the risk that will dynamically drive selection of a specific process model or of the steps of such process. The spiral ceases when the risk is considered acceptable. In Fig. 21.26, the radial dimension represents the cost of development at that particular stage. The angular dimension describes the progress made in achieving each cycle of the spiral. A new spiral starts with assessing:

1. The objectives of the specific part of the product in development (functionality, required level of performance, etc.)
2. The alternative means of implementing the product (design options, etc.)
3. The constraints on the product development (schedule, cost, etc.)

Given the different objectives and constraints, one will select the optimal alternative. This selection process will usually uncover the principal sources of uncertainty in the development process. These sources of uncertainty are prime contributors to risk, and a cost-effective strategy will be formulated for their prompt resolution (for example, by prototyping, simulating, developing analytic models, etc.).

Once the risks have been evaluated, the next step is to determine the relative remaining risks and resolve them with the appropriate strategy. For example, if previous efforts (such as prototyping) have resolved to issues related to user interface or software performance and remaining risks can be tied to software development, a waterfall life cycle development will be selected and implemented as part of the next spiral. Finally, let us note that each cycle of the spiral ends with a review of the documents that were generated in the cycle. This review ensures that all parties involved in the software development are committed to the phase of the spiral.

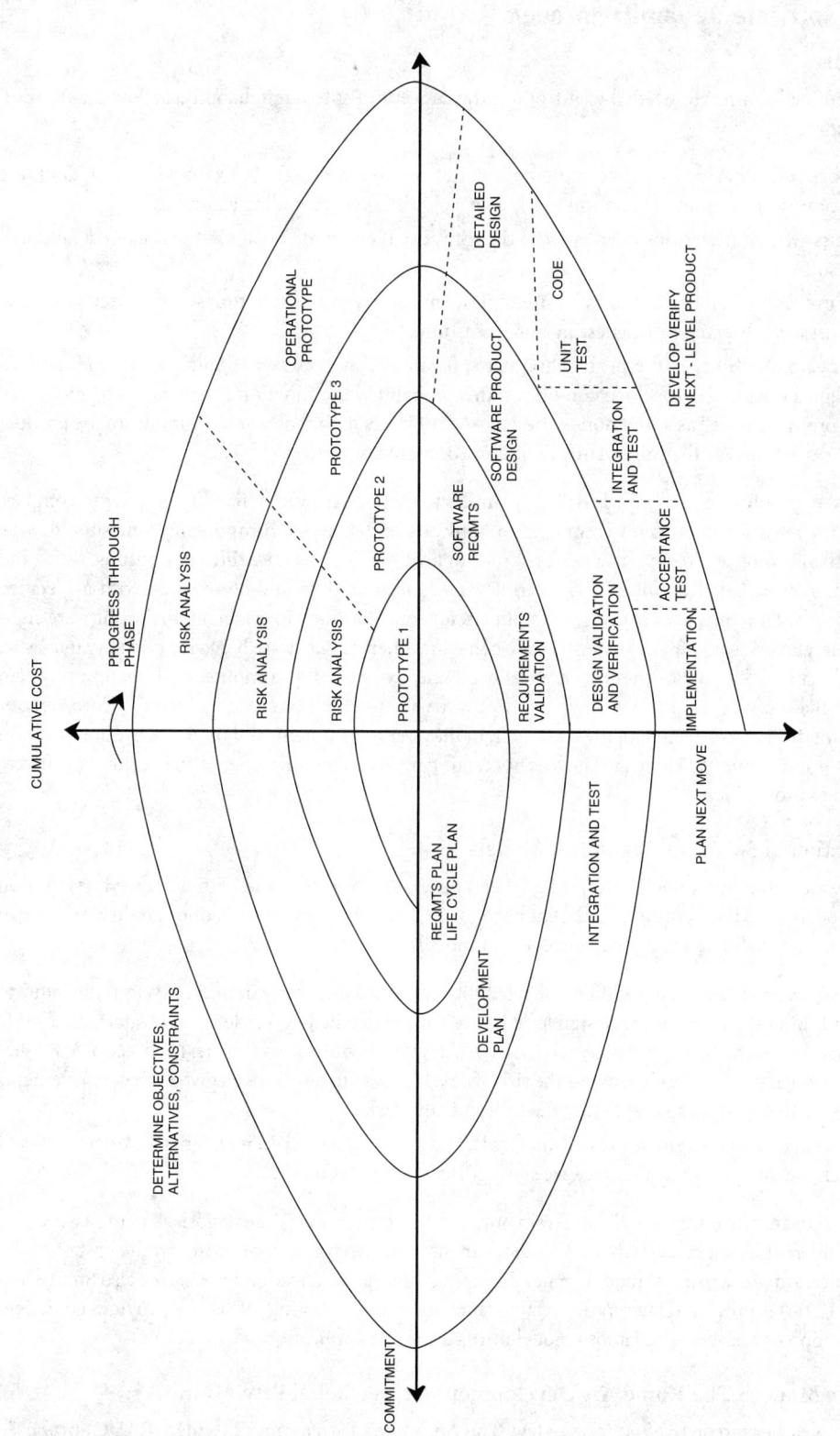

FIGURE 21.26 The spiral model.

21.3.3   Software Reliability Models

Definitions

A number of definitions are needed to introduce the concept of software reliability. Following are IEEE (1983) definitions.

- **Errors** are human actions that result in the software containing a fault. Examples of such faults are the omission or misinterpretation of the user's requirements, a coding error, etc.
- **Faults** are manifestations of an error in the software. If encountered, a fault may cause a failure of the software.
- **Failure** is the inability of the software to perform its mission or function within specified limits. Failures are observed during testing and operation.
- **Software reliability** is the probability that software will not cause the failure of a product for a specified time under specified conditions; this probability is a function of the inputs to and use of the product, as well as a function of the existence of faults in the software; the inputs to the product will determine whether an existing fault is encountered or not.

The definition of software reliability is willingly similar to that of hardware reliability in order to compare and assess the reliability of systems composed of hardware and software components. One should note, however, that in some environments and for some applications such as scientific applications, time (and more precisely time between failures) is an irrelevant concept and should be replaced with a specified number of runs. The concept of software reliability is, for some, difficult to understand. For software engineers and developers, software has deterministic behavior, whereas hardware behavior is partly stochastic. Indeed, once a set of inputs to the software has been selected, once the computer and operating system with which the software will run is known, the software will either fail or execute correctly. However, our knowledge of the inputs selected, of the computer, of the operating system, and of the nature and position of the fault is uncertain. We can translate this uncertainty into probabilities, hence the notion of software reliability as a probability.

Classification of Software Reliability Models

Many software reliability models (SRMs) have been developed over the years. For a detailed description of most models see Musa, Iannino, and Okumoto (1987). Within these models one can distinguish two main categories: predictive models and assessment models.

- *Predictive models* typically address the reliability of the software early in the life cycle at the requirements or at the preliminary design level or even at the detailed design level in a waterfall life cycle process or in the first spiral of a spiral software development process (see next subsection). Predictive models could be used to assess the risk of developing software under a given set of requirements and for specified personnel before the project truly starts.
- *Assessment models* evaluate present and project future software reliability from failure data gathered when the integration of the software starts (discussed subsequently).

Predictive software reliability models are few in number; most models can be categorized in the assessment category. Different classification schemes for assessment models have been proposed in the past.

A classification of assessment models generalized from Goel's classification (1985) (see also Smidts and Kowalski (1995)) is presented later in this section. It includes most existing SRMs and provides guidelines for the selection of a software reliability model fitting a specific application.

Predictive Models: The Rome Air Development Center Reliability Metric

As explained earlier, predictive models are few. The Air Force's Development Center (RADC) proposed one of the first predictive models (ASFC, 1987). A large range of software programs and related failure data were analyzed in order to identify the characteristics that would influence software reliability. The model identifies three characteristics: the application type (A), the development environment (D), and the software characteristics (S). A new software will be examined with reference to these different characteristics. Each characteristic will be quantified, and reliability R in faults per executable lines of code is obtained by multiplying these different metrics

$$R = A \times D \times S \qquad (21.1)$$

A brief description of each characteristic follows:

The *application type (A)* is a basic characteristic of a software: it determines how a software is developed and run. Categories of applications considered by the Air Force were: airborne systems, strategic systems, tactical systems, process control systems, production systems (such as decision aids), and developmental systems (such as software development tools). An initial value for the reliability of the software to be developed is based only on the application type. This initial value is then modified when other factors characterizing the software development process and the product become available.

Different *development environments (D)* are at the origin of variations in software reliability. Boehm (1981) divides development environments into three categories:

- *Organic mode:* Small software teams develop software in a highly familiar, in-house environment. Most software personnel are extremely experienced and knowledgeable about the impact of this software development on the company's objectives.

- *Semidetached mode:* Team members have an intermediate level of expertise with related systems. The team is a mixture of experienced and inexperienced people. Members of the team have experience with some specific aspects of the project but not with others.

- *Embedded mode:* The software needs to operate under tight constraints. In other words, the software will function in a strongly coupled system involving software, hardware, regulations, and procedures. Because of the costs involved in altering other elements of this complex system, it is expected that the software developed will incur all required modifications due to unforeseen problems.

The *software characteristics (S)* metric includes all software characteristics that are likely to impact software reliability. This metric is subdivided into a *requirements and design representation metric (S1)* and *a software implementation metric (S2)* with

$$S = S1 \times S2 \qquad (21.2)$$

The *requirements and design metric (S1)* is the product of the following subset of metrics.

The *anomaly management metric (SA)* degree to which fault tolerance exists in the software studied.

The *traceability metric (ST)* degree to which the software being implemented matches the user requirements.

The effect of traceability on the code is represented by a metric, ST

$$ST = \frac{k_{tc}}{TC} \qquad (21.3)$$

where k_{tc} is a coefficient to be determined by regression and TC is the traceability metric calculated by

$$\frac{NR}{(NR - DR)} \qquad (21.4)$$

where NR is the number of user requirements, and DR is the number of user requirements that cannot be found in design or code.

The last characteristic is the *quality review results metric (SQ)*. During large software developments formal and informal reviews of software requirements and design are conducted. Any problems identified in the documentation or in the software design are recorded in discrepancy reports. Studies show a strong correlation between errors reported at these early phases and existence of unresolved problems that will be found later during test and operation. The quality review results metric is a measure of the effect of the number of discrepancy reports found on reliability

$$SQ = k_q \frac{NR}{(NR - NDR)} \tag{21.5}$$

where k_q is a correlation coefficient to be determined, NR is the number of requirements and NDR is the number of discrepancy reports.

Finally

$$S1 = SA \times ST \times SQ \tag{21.6}$$

The *software implementation metric (S2)* is the product of *SL, SS, SM, SU, SX,* and *SR,* defined as follows.

A *language type metric (SL):* a significant correlation seems to exist between fault density and language type. The software implementation metric identifies two categories of language types: assembly languages ($SL = 1$) and high-order languages ($SL = 1.4$).

Program size metric (SS): Although fault density does account directly for the size of the software (i.e, fault density is the number of faults divided by the number of executable lines of code), nonlinear effects of size on reliability due to factors such as complexity and inability of humans to deal with large systems still need to be assessed. Programs are divided in four broad categories: smaller than 10,000 lines of code, between 10,000 and 50,000 lines of code, between 50,000 and 100,000 lines of code, and larger than 100,000 lines of code.

Modularity metric (SM): It is considered that smaller modules are generally more tractable, can be more easily reviewed, and hence will ultimately be more reliable. Three categories of module size are considered: (1) modules smaller than 200 lines of code, (2) modules between 200 and 3,000 lines of code and (3) modules larger than 3,000 lines of code. For a software composed of α modules in category 1, β modules in category 2, and γ modules in category 3,

$$SM = (\alpha \times SM(1) + \beta \times SM(2) + \gamma \times SM(3))/(\alpha + \beta + \gamma) \tag{21.7}$$

with $SM(i)$ the modularity metric for modules in category i.

Extent of reuse metric (SU): reused code is expected to have a positive impact on reliability because this reused code was part of an application that performed correctly in the field. Two cases need to be considered: (1) The reused code is to be used in an application of the same type as the application from which it originated. (2) The reused code is to be used in an application of a different type requesting different interfaces and so forth. In each case SU is a factor obtained empirically, which characterizes the impact of reuse on reliability.

Complexity metric (SX): it is generally considered that a correlation exists between a module's complexity and its reliability. Although complexity can be measured in different ways, the authors of the model define SX as

$$SX = k_x \cdot \frac{\sum_{i=1,n} SX_i}{n} \tag{21.8}$$

where k_x is a correlation coefficient determined empirically, n is the number of modules, and SX_i is module i McCabe's cyclomatic complexity. (McCabe's cyclomatic complexity (IEEE, 1989b) may be used to determine the structural complexity of a module from a graph of the module's operations. This measure is used to limit the complexity of a module in order to enhance understandability.)

Standards review results metric (SR): Reviews and audits are also performed on the code. The standards review results metric accounts for existence of problem discrepancy reports at the code level. A discrepancy report is generated each time a module meets one of the criteria given in Table 21.1. SR is given by

$$SR = kv \cdot \frac{n}{(n - PR)} \tag{21.9}$$

TABLE 21.1 List of Criteria Determining Generation of a Discrepancy Report for a Code Module

---

- Design is not top down
- Module is not independent
- Module processing is dependent on prior processing
- Module description does not include input, output, processing or limitations
- Module has multiple entrances or multiple exits
- Database is too large
- Database is not compartmentalized
- Functions are duplicated
- Existence of global data

---

where n is the number of modules, PR is the number of modules with severe discrepancies, and kv is a correlation coefficient.

Finally

$$S2 = SL \cdot SS \cdot SM \cdot SU \cdot SX \cdot SR \tag{21.10}$$

The RADC metric was developed from data taken from old projects much smaller than typical software developed today and in a lower level language. Hence its applicability to today's software is questionable.

Models for Software Reliability Assessment

Classification

Most existing SRMs may be grouped into four categories:

- *Times between failures category* includes models that provide estimates of the times between failures (see subsequent subsection, Jelinski and Moranda's model).

- *Failure count category* is interested in the number of faults or failure experienced in specific intervals of time (see subsequent subsection, Musa's basic execution time model and Musa–Okumoto logarithmic Poisson execution time model).

- *Fault seeding category* includes models to assess the number of faults in the program at time 0 via seeding of extraneous faults; (see subsequent subsection, Mills fault seeding model).

- *Input-domain based category* includes models that assess the reliability of a program when the test cases are sampled randomly from a well-known operational distribution of inputs to the program. The **clean room methodology** is an attempt to implement this approach in an industrial environment (the Software Engineering Laboratory, NASA) (Basili and Green, 1993). The reliability estimate is obtained from the number of observed failures during execution (see subsequent subsection, Nelson's model and Ramamoorthi and Bastani).

Table 21.2 lists the key assumptions on which each category of models is based, as well as representative models of the category. Additional assumptions specific to each model are listed in Table 21.3. Table 21.4 examines the validity of some of these assumptions (either the generic assumptions of a category or the additional assumptions of a specific model).

The software development process is environment dependent. Thus, even assumptions that would seem reasonable during the testing of one function or product may not hold true in subsequent testing. The ultimate decision about the appropriateness of the assumptions and the applicability of a model have to be made by the user.

Model Selection

To select an SRM for a specific application, the following practical approach can be used. First, determine to which category the software reliability model you are interested in belongs. Then, assess which specific

TABLE 21.2 Key Assumptions on Which Software Reliability Models Are Based

Key Assumptions	Specific Models
Times between failures models	
• Independent times between failures	• Jelinski–Moranda's de-eutrophication model (1972)
• Equal probability of exposure of each fault	• Schick and Wolverton's model (1973)
• Embedded faults are independent of each other	• Goel and Okumoto's imperfect debugging model (1979)
• No new faults introduced during correction	• Littlewood–Verrall's Bayesian model (1973)
Fault count models	
• Testing intervals are independent of each other	• Shooman's exponential model (1975)
• Testing during intervals is homogeneously distributed	• Goel-Okumoto's nonhomogeneous Poisson process (1979)
• Number of faults detected during nonoverlapping intervals are independent of each other	• Goel's generalized nonhomogeneous Poisson process model (1983)
	• Musa's execution time model (1975)
	• Musa–Okumoto's logarithmic Poisson execution time model (1983)
Fault seeding models	
• Seeded faults are randomly distributed in the program	• Mills seeding model (1972)
• Indigenous and seeded faults have equal probabilities of being detected	
Input-domain based models	
• Input profile distribution is known	• Nelson's model (1978)
• Random testing is used (inputs are selected randomly)	• Ramamoorthy and Bastani's model (1982)
• Input domain can be partitioned into equivalence classes	

model in a given category fits the application. Actually, a choice will only be necessary if the model is in one of the two first categories, that is, times between failures or failure count models. If this is the case, select a software reliability model based on knowledge of the software development process and environment. Collect such failure data as the number of failures, their nature, the time at which the failure occurred, failure severity level, and the time needed to isolate the fault and to make corrections. Plot the cumulative number of failures and the failure intensity as a function of time. Derive the different parameters of the models from the data collected and use the model to predict future behavior. If the future behavior corresponds to the model prediction, keep the model.

Models

Jelinski and Moranda's Model (1972)

Jelinksi and Moranda (1972) developed one of the earliest reliability models. It assumes that

- All faults in a program are equally likely to cause a failure during test.
- The hazard rate is proportional to the number of faults remaining.
- No new defects are incorporated into the software as testing and debugging occur.

Originally, the model assumed only one fault was removed after each failure, but an extension of the model, credited to Lipow (1974), permits more than one fault to be removed.

In this model, the hazard rate for the software is constant between failures and is

$$Z_i(T) = \phi[N - n_{i-1}], \quad i = 1, 2, \ldots, m \tag{21.11}$$

TABLE 21.3 Specific Assumptions Related to Each Specific Software Reliability Model

Specific Representatives of a Model Category	Specific Assumptions
Times between failures models	
• Jelinski and Moranda (JM) de-eutrophication model	• N faults at time 0; detected faults are immediately removed; the hazard rate[a] in the interval of time between two failures is proportional to the number of remaining faults
• Schick and Wolverton model	• Same as above with the hazard rate a function of both the number of faults remaining and the time elapsed since last failure
• Goel and Okumoto imperfect debugging model	• Same as above, but the fault, even if detected, is not removed with certainty
• Littlewood–Verrall Bayesian model	• Initial number of failures is unknown, times between failures are exponentially distributed, and the hazard rate is gamma distributed
Fault count models	
• Shooman's exponential model	• Same assumptions as in the JM model
• Goel–Okumoto's nonhomogeneous Poisson process (GONHPP)	• The cumulative number of failures experienced follows a nonhomogeneous Poisson process (NHPP); the failure rate decreases exponentially with time
• Goel's generalized nonhomogeneous Poisson process model	• Same assumptions as in the GONHPP, but the failure rate attempts better to replicate testing experiments results that show that failure rate first increases then decreases with time
• Musa's execution time model	• Same assumptions as in the JM model
• Musa–Okumoto's logarithmic Poissontion time model	• Same assumptions as in the GONHPP model, but the time element considered in the model is the execution time
Fault seeding models	
• Mills' seeding model	
Input domain based models	
• Nelson's model	• The outcome of a test case provides some stochastic information about the behavior of the program for other inputs close to the inputs used in the test
• Ramamoorthy and Bastani's model	

[a] Software hazard rate is $z(t) = f(t)/R(t)$, where $R(t)$ is the reliability at time t and $f(t) = -dR(t)/d(T)$. The software failure rate density is $\lambda(t) = d\mu(t)/d(t)$, where $\mu(t)$ is the mean value of the cumulative number of failures experienced by time t.

for the interval between the $(i-1)$st ith failures. In this equation, N is the initial number of faults in the software, ϕ is a proportionality constant, and n_{i-1} is the cumulative number of faults removed in the first $(i-1)$ intervals.

The maximum likelihood estimates of N and ϕ are given by the solution of the following equations:

$$\sum_{i=1}^{m} \frac{1}{N - n_{i-1}} - \sum_{i=1}^{m} \phi x_{li} = 0 \tag{21.12}$$

and

$$\frac{n}{\phi} - \sum_{i=1}^{m} [N - n_{i-1}] x_{li} = 0 \tag{21.13}$$

where x_{li} is the length of the interval between the $(i-1)$st and ith failures, and n is the number of errors removed so far.

Once N and ϕ are estimated, the number of remaining errors is given by

$$N(\text{remaining}) = N - n_i \tag{21.14}$$

TABLE 21.4   Validity of Some Software Reliability Models Assumptions

Assumptions	Intrinsic Limitations of Such Assumptions
• Times between failures are independent	• Only true if derivation of test cases is purely random (never the case)
• A detected fault is immediately corrected	• Faults will not be removed immediately; they are usually corrected in batches; however, the assumption is valid as long as further testing avoids the path in which the fault is active
• No new faults are introduced during the fault removal process	• In general, this is not true
• Failure rate decreases with test time	• Reasonable approximation in most cases
• Failure rate is proportional to the number of remaining faults for all faults	• A reasonable assumption if the test cases are chosen to ensure equal probability of testing different parts of code
• Time is used as a basis for failure rate	• Usually time is a good basis for failure rate; if this is not true, the models are valid for other units
• Failure rate increases between failures for a given failure interval	• Generally not the case unless testing intensity increases
• Testing is representative of the operational usage	• Usually not the case since testing usually selects error-prone situations

The mean time to the next software failure is

$$MTTF = \frac{1}{(N - n_i)\phi} \tag{21.15}$$

The reliability is

$$R_{i+1}(t \mid t_i) = \exp[-(N - n_i)\phi(t - t_i)], \quad t \geq t_i \tag{21.16}$$

where $R_{i+1}(t/t_i)$ is the reliability of the software at time t in the interval (t_i, t_{i+1}), given that the ith failure occurred at time t_i.

Musa Basic Execution Time Model (BETM) (Musa, 1975)

This model was first described by Musa in 1975. It assumes that failures occur as a nonhomogeneous Poisson process. The units of failure intensity are failures per central processing unit (CPU) time. This relates failure events to the processor time used by the software. In the BETM, the reduction in the failure intensity function remains constant, irrespective of whether the first or the Nth failure is being fixed.

The failure intensity is a function of failures experienced:

$$\lambda(\mu) = \lambda_0(1 - \mu/v_0) \tag{21.17}$$

where $\lambda(\mu)$ is the failure intensity (failures per CPU hour at μ failures), λ_0 is the initial failure intensity (at $\tau_e = 0$), μ is the mean number of failures experienced at execution time and τ_e, and v_0 is the total number of failures expected to occur in infinite time.

Then, the number of failures that need to occur to move from a present failure intensity, λ_p, to a target intensity, λ_F, is given by

$$\Delta\mu = \frac{v_0}{\lambda_0}(\lambda_P - \lambda_F) \tag{21.18}$$

and the execution time required to reach this objective is

$$\Delta\tau = \frac{v_0}{\lambda_0}\ln(\lambda_P/\lambda_F) \tag{21.19}$$

In practice, ν_0 and λ_0 can be estimated in three ways:

- Use previous experience with similar software. The model can then be applied prior to testing.
- Plot the actual test data to establish or update previous estimates. Plot failure intensity execution time; the y intercept of the straight line fit is an estimate of λ_0. Plot failure intensity failure number; the x intercept of the straight line fits is an estimate of ν_0.
- Use the test data to develop a maximum-likelihood estimate. The details of this approach are described in Musa et al. (1987).

Musa (1987) also developed a method to convert execution time predictions to calendar time. The calendar time component is based on the fact that available resources limit the amount of execution time that is practical in a calendar day.

Musa–Okumoto Logarithmic Poisson Execution Time Model (LPETM)

The logarithmic Poisson execution time model was first described by Musa and Okumoto (1983). In the LPETM, the failure intensity is given by

$$\lambda(\mu) = \lambda_0 \exp(-\theta\mu) \tag{21.20}$$

where θ is the failure intensity decay parameter and λ, μ, and λ_0 are the same as for the BETM. The parameter θ represents the relative change of failure intensity per failure experienced. This model assumes that repair of the first failure has the greatest impact in reducing failure intensity and that the impact of each subsequent repair decreases exponentially.

In the LPETM, no estimate of ν_0 is needed. The expected number of failures that must occur to move from a present failure intensity of λ_P to a target intensity of λ_F is

$$\Delta\mu = (1/\theta)\ln(\lambda_P/\lambda_F) \tag{21.21}$$

and the execution time to reach this objective is given by

$$\Delta\tau = \frac{1}{\theta}\left[\frac{1}{\lambda_F} - \frac{1}{\lambda_P}\right] \tag{21.22}$$

In these equations, λ_0 and θ can be estimated based on previous experience by plotting the test data to make graphical estimates or by making a least-squares fit to the data.

Mills Fault Seeding Model (IEEE, 1989a)

An estimate of the number of defects remaining in a program can be obtained by a seeding process that assumes a homogeneous distribution of a representative class of defects. The variables in this measure are: N_S the number of seeded faults, n_S the number of seeded faults found, and n_F the number of faults found that were not intentionally seeded.

Before seeding, a fault analysis is needed to determine the types of faults expected in the code and their relative frequency of occurrence. An independent monitor inserts into the code N_S faults that are representative of the expected indigenous faults. During reviews (or testing), both seeded and unseeded faults are identified. The number of seeded and indigenous faults discovered permits an estimate of the number of faults remaining for the fault type considered. The measure cannot be computed unless some seeded faults are found. The maximum likelihood estimate of the indigenous (unseeded) faults is given by

$$NF = n_F N_S/n_S \tag{21.23}$$

Example

Here, 20 faults of a given type are seeded and, subsequently, 40 faults of that type are uncovered: 16 seeded and 24 unseeded. Then, $NF = 30$, and the estimate of faults remaining is NF (remaining) $= NF - n_F = 6$.

Input Domain Models

Let us first define the notion of input domain. The input domain of a program is the set of all inputs to this program.

Nelson's Model

Nelson's model (TRW, 1976) is typically used for systems with ultrahigh-reliability requirements, such as software used in nuclear power plants and limited to 1000 lines of code (Goel, 1985; Ramamoorthy and Bastani, 1982). The model is applied to the validation phase of the software (acceptance testing) to estimate reliability. Typically during this phase, if any error is encountered, it will not be corrected (and usually no error will occur).

Nelson defines the reliability of a software run n times (for n test cases) and which failed n_f times as

$$R = 1 - n_f/n$$

where n is the total number of test cases and n_f is the number of failures experienced out of these test cases.

The Nelson model is the only model whose theoretical foundations are really sound. However, it suffers from a number of practical drawbacks.

- We need to run a huge number of test cases in order to reduce our uncertainty on the reliability R.
- Nelson's model assumes random sampling whereas testing strategies tend to use test cases that have a high probability of revealing errors.
- The model does not account for the characteristics of the program. One would expect that to reach an equal level of confidence in the reliability estimate, a logically complex program should be tested more often than a simple program.
- The model does not account for the notion of equivalence classes.

(Equivalence classes are subdomains of the input domain to the program. The expectation is that two sets of input selected from a same equivalence class will lead to similar output conditions.) The model was modified in order to account for these issues. The resulting model (Nelson, 1978) defines the reliability of the program as follows:

$$R = \sum_{j=1,m} P_j(1 - \epsilon_j)$$

where the program is divided in m equivalence classes, P_j is the probability that a set of user's inputs are selected from the equivalence class j and $(1 - \epsilon_j)$ is the probability that the program will execute correctly for any set of inputs in class j knowing that it executed correctly for any input randomly selected in j. Here $(1 - \epsilon_j)$ is the class j correctness probability.

The weakness of the model comes from the fact that values provided for ϵ_j are ad hoc and depend only on the functional structure of the program.

Empirical values for ϵ_j are

- If more than one test case belongs to G_j where G_j is the set of inputs that execute a given logic path L_j, then $\epsilon_j = 0.001$.
- If only one test case belongs to G_j, then $\epsilon_j = 0.01$.
- If no test case belongs to G_j but all segments (i.e., a sequence of executable statements between two branch points) and segment pairs in L_j have been exercised in the testing, then $\epsilon_j = 0.05$.
- If all segments L_j but not all segment pairs have been exercised in the testing, then $\epsilon_j = 0.1$.
- If m segments ($1 < m < 4$) of L_j have not been exercised in testing, then $\epsilon_j = 0.1 + 0.2m$.
- If more than 4 segments of L_j have not been exercised in the testing, then $\epsilon_j = 1$.

Ramamoorthy and Bastani

Ramamoorthy and Bastani (1982) developed an improved version of Nelson's model. The model is an attempt to give an analytical expression of the correctness probability $1 - \epsilon_j$. To achieve this objective the authors introduce the notion of probabilistic equivalence class.

E is a probabilistic equivalence class if $E \sqsubseteq I$, where I is the input domain of the program P and P is correct for all elements in E with probability $P\{X_1, X_2, \ldots, X_d\}$ if P is correct for each $X_i \in E, i = 1, \ldots, d$. Then $P\{E \mid X\}$ is the correctness probability of P based on the set of test cases X. Knowing the correctness probability one derives the reliability of the program easily,

$$R = \sum_{i=1,m} P(E_i \mid X_i) P(E_i)$$

where E_i is a probabilistic equivalence class, $P(E_i)$ is the probability that the user will select inputs from the equivalence class, and $P(E_i \mid X_i)$ is the correctness probability of E_i given the set of test cases X_i.

It is possible to estimate the correctness probability of a program using the continuity assumption (i.e., that closely related points in the input domain are correlated). This assumption holds especially for algebraic equations. Furthermore, we make the assumption that a point in the input domain depends only on its closest neighbors. Figure 21.27 shows the correctness probability for an equivalence class $E_i = (a, a + V)$ where only one test case is selected (Fig. 21.27(a)) and where two test cases are selected (Fig. 21.27(b)).

The authors show that, in general,

$$P\{E_i \text{ is correct} \mid 1 \text{ test case is correct}\} = e^{-\lambda V}$$

$$P(E_i \text{ is correct} \mid n \text{ test cases have successive distances } x_j \text{ are correct}$$

$$j = 1, 2, \ldots, n - 1) = e^{(-\lambda v)} \prod_{j=1}^{n-1} \left(\frac{2}{1 + e^{-\lambda x_j}} \right)$$

where λ is a parameter of the equivalence class (Ramamoorthy and Bastani, 1982).

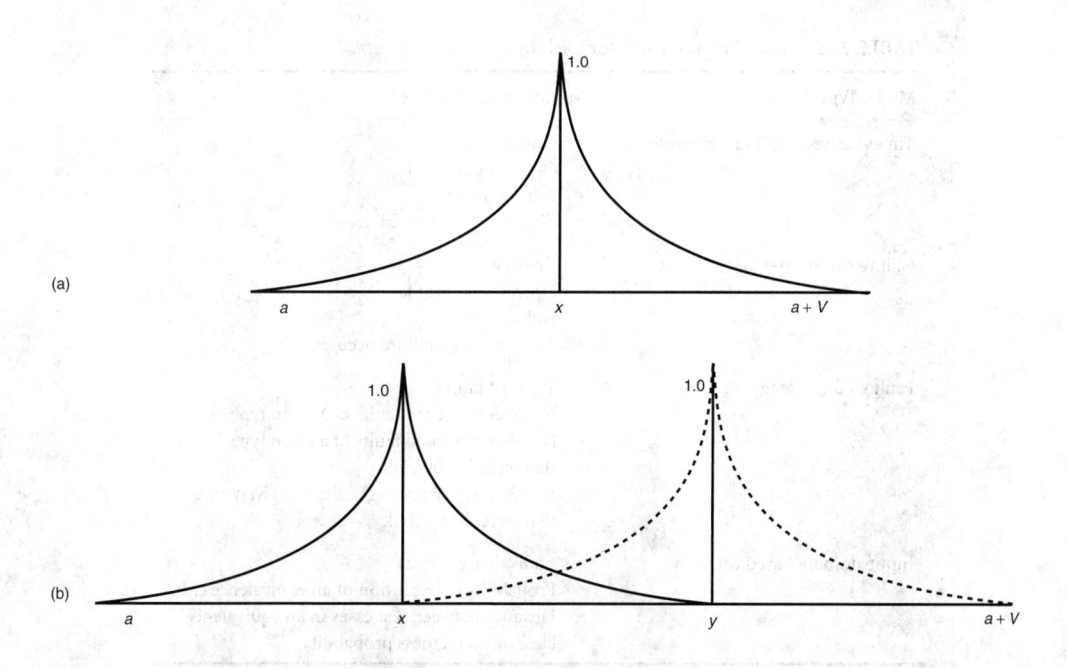

FIGURE 21.27 Interpretation of continuous equivalence classes: (a) single test case, (b) two test cases.

A good approximation of λV is found for

$$\lambda V \simeq (D - 1)//N$$

where N is the number of elements in the class (due to the finite word of the computer) and D is the degree of the equivalence class (i.e., number of distinct test cases that completely validate the class).

Ramamoorthy and Bastani warn that application of the model has several disadvantages. It can be relatively expensive to determine the equivalence classes and their complexity and the probability distribution of the equivalence classes.

Derived Software Reliability Models

From the basic models presented previously, models can be built for applications involving more than one software, such as models to assess the reliability of fault-tolerant software designs (Scott, Gault, and McAllister, 1987) or the reliability of a group of modules assembled during the integration phase.

Littlewood (1979) explicitly takes into account the structure (i.e., the modules) of the software, and models the exchange of control between modules (time of sojourn in a module and target of exchange) using a semi-Markovian process. The failure rates of a given module can be obtained from the basic reliability models applied to the module; interface failures are being modeled explicitly. This model can be used to study the integration process.

Data Collection

This subsection examines the data needed to perform a software reliability assessment with the models just described. In particular, we will define the concepts of criticality level and operational profiles. The complete list of data required by each model can be found in Table 21.5.

Criticality: The faults encountered in a software are usually placed in three or five categories called *severity levels* or *criticality levels*. This categorization helps distinguish those faults that are real contributors to software reliability from others such as mere enhancements. Such distinction is, of course, important

TABLE 21.5 Data Requirements for Each Model Category

Model Type	Minimal Data Required
Times between failures category	• Criticality level • Operational profile • Failure count • Time between two failures
Failure count category	• Criticality • Operational profile • Failure count • Time at which failure occured
Fault seeding category	• Types of faults • Number of seeded faults of a given type • Number of seeded faults of a given type that were discovered • Number of indigenous faults of a given type that were discovered.
Input-domain based category	• Equivalence classes • Probability of execution of an equivalence class • Distance between test cases in an equivalence class or correctness probability

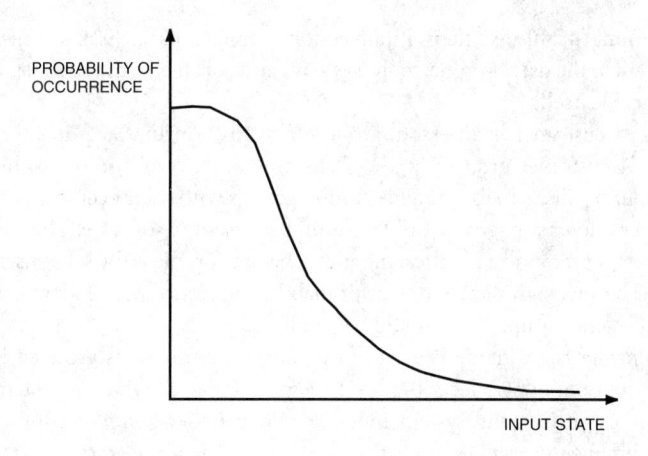

FIGURE 21.28 Input profile for a software program.

if we wish to make a correct assessment of the reliability of our software. The following is a classification with five criticality levels (Neufelder, 1993):

- *Catastrophic:* This fault may cause unrecoverable mission failure, a safety hazard or loss of life.
- *Critical:* This fault may lead to mission failure, loss of data, system downtime with consequences such as lawsuits or loss of business.
- *Moderate:* This fault may lead to a partial loss of functionalities and undesired downtimes. However, work-arounds exist, which can mitigate the effect of the fault.
- *Negligible:* This fault does generally not affect the software functionality. It is merely a nuisance to the user and may never be fixed.
- *All others:* This category includes all other faults.

It will be clear to the reader that the two first categories pertain without question to reliability assessment. But what is it of the three other categories?

Operational profile: A software's reliability depends inherently on its use. If faults exist in a region of the software never executed by its users and no faults exist in regions frequently executed, the software's perceived reliability will be high. On the other hand, if a software contains faults in regions frequently executed, the reliability will be perceived to be low. This remark leads to the concept of *input profile*. An input profile is the distribution obtained by putting the possible inputs to the software on the x-axis and the probability of selection of these inputs on the y-axis (see Fig. 21.28). As the reader will see the notion of input profile is extremely important: input profiles are needed for software reliability assessment but they also drive efficient software testing and even software design. Indeed, testing should start by exercising the software under sets of high probability inputs so that if the software has to be unexpectedly released early in the field, the reliability level achieved will be acceptable. It may even be that a reduced version of the software, the reduced operational software (ROS), constructed from the user's highly demanded functionalities, may be built first to satisfy the customer.

Because of the large number of possible inputs to the software, it would be an horrendous task to build a *pure input* profile (a pure input profile would spell out all combinations of input with their respective occurrence probabilities.) An approach to build a *tractable input profile* is proposed next. This approach assesses the operations that will be performed by the software and their respective occurrence probabilities. This profile is called an *operational profile*.

To build an operational profile five steps should be considered (Musa, 1993): establishing the customer's profile, and refinement to the user's profile, to the system-mode profile, to the functional profile, and finally to the operational profile itself.

Customer's profile: a customer is a person, group, or institution that acquires the software or, more generally, a system. A customer group is a set of customers who will use the software in an identical manner. The customer profile is the set of all customer groups with their occurrence probabilities.

User's profile: a user is a person, group, or institution that uses the software. The user group is then the set of users who use the system in an identical manner. The user profile is the set of users' groups (all users' groups through all customers' groups) and their probability of occurrence. If identical users' groups exist in two different customers' groups, they should be combined.

System mode profile: a system mode is a set of operations or functions grouped in analyzing system execution. The reliability analyst selects the level of granularity at which a system mode is defined. Some reference points for selecting system modes could be: *user group*, administration vs. students in a university; *environmental condition*; overload vs. normal load; *operational architectural structure*, network communications vs. stand-alone computations; *criticality*, normal nuclear power plant operation vs. after trip operations; *hardware components*, functionality is performed on hardware component 1 vs. component 2. The system mode profile is the set of all system modes and their occurrence probabilities.

Functional profile: functions contributing to the execution of a system mode are identified. These functions can be found at the requirements definition and analysis stage (see Fig. 21.25). They may also be identifiable in a prototype of the final software product if a spiral software development model was used. However, in such cases care should be exercised: most likely the prototype will not possess all of the functionalities of the final product. The number of functions to be defined varies from one application to another. It could range from 50 functions to several hundreds. The main criteria governing the definition of a function is the extent to which processing may differ. Let us consider a command C with input parameters X and Y. Three admissible values of X exist, $x1$, $x2$, and $x3$, and five admissible values of parameter Y exist, namely, $y1$, $y2$, $y3$, $y4$, and $y5$. From these different combinations of input parameters we could build 15 different functionalities (i.e., combinations $(x1, y1), (x1, y2), \ldots, (x1, y5); (x2, y1), \ldots, (x2, y5), (x3, y1), \ldots, (x3, y5)$). However, the processing may not differ significantly between these 15 different cases but only for cases that differ in variable X. Each variable defining a function is called a *key input variable*. In our example, the command C and the input X are two key input variables. Please note that even environmental variables could constitute candidate key input variables. The functional profile is the set of all functions identified (for all system modes) and their occurrence probability.

Operational profile: functions are implemented into code. The resulting implementation will lead to one or several operations that need to be executed for a function to be performed. The set of all operations obtained from all functions and their occurrence probability is the operational profile. To derive the operational profile, one will first divide execution into *runs*. A run can be seen as a set of tasks initiated by a given set of inputs. For example, sending e-mail to a specific address can be considered as a run. Note that a run will exercise different functionalities. Identical runs can be grouped into run types. Once the runs have been identified, the input state will be identified. The input state is nothing but the values of the input variables to the program at the initiation of the run. We could then build the input profile as the set of all input states and their occurrence probabilities and test the program for each such input state. However, this would be extremely expensive. Consequently, the run types are grouped into operations. The input space to an operation is called a *domain*. Run types that will be grouped share identical input variables. The set of operations should include all operations of high criticality.

Criticisms and Suggestion for New Approaches

Existing software reliability models will either base reliability predictions on historical data existing for similar ware (prediction models) or on failure data associated with the specific software being developed (software assessment models). A basic criticism held against these models is the fact that they are not

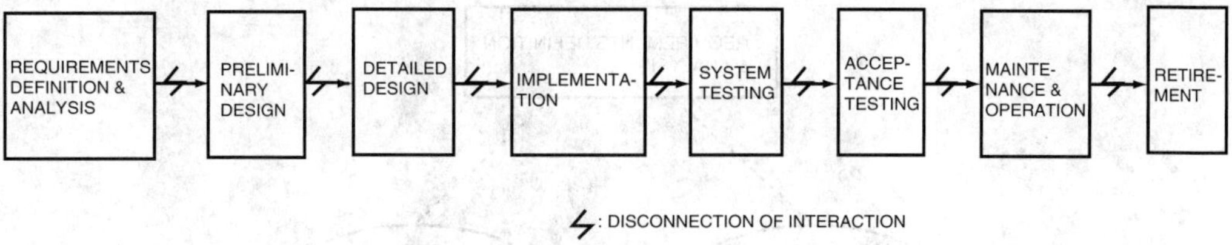

: DISCONNECTION OF INTERACTION

FIGURE 21.29 Sources of failures.

based on a depth of knowledge and understanding of software behavior. The problem resides in the inherent complexity of software. Too many factors seem to influence the final performance of a software. Furthermore, the software engineering process itself is not well defined. For instance, two programmers handed the same set of requirements and following the same detailed design will still generate different software codes.

Given these limitations, how can we modify our approach to the precision, assessment, and measurement of software reliability? How can we integrate our existing knowledge on software behavior in new models? The main issues we need to resolve are

- Why does software fail or how is a specific software failure mode generated?
- How does software fail or what are the different software failure modes?
- What is the likelihood of a given software failure mode?

Software failures are generated through the process of creating software. Consequently, it is in the software development process that lie the roots of software failures. More precisely, failures are generated either within a phase of the software development process or at the interface between two phases (see Fig. 21.29). To determine the ways in which software fail, we will then need to consider each phase of the software development process (for example, the requirements definition and analysis phase, the coding phase, etc.) and for each such phase identify the realm of failures that might occur. Then we will need to consider the interfaces between consequent phases and identify all of the failures that might occur at these interfaces: for example, at the interface between the requirements definition and analysis phase and the preliminary design phase.

To determine how software failure modes are created within a particular software development phase, we should first identify the activities and participants of a specific life cycle phase. For instance, let us consider the requirements definition and analysis phase of software *S*. Some of the activities involved in the software requirements definition phase are elicitation of the following:

1. Functional requirements: the different functionalities of the software.
2. Performance requirements: speed requirements.
3. Reliability and safety requirements: how reliable and how safe the software should be.
4. Portability requirements: the software should be portable from machine A to machine B.
5. Memory requirements.
6. Accuracy requirements.

An exhaustive list of the types of requirements can be found in Sage and Palmer (1990).

The participants of this phase are the user of the software and the software analyst, and generally a systems engineer working for a software development company. Note that the number of people involved in this phase may vary from one for each category (user category and analyst category) to several in each category. We will assume for the sake of simplicity that only one representative of each category is present.

We need now to identify the failure modes. These are the failure modes of the user, the failure modes of the analyst, and the failure modes of the communication between user and analyst (see Fig. 21.30).

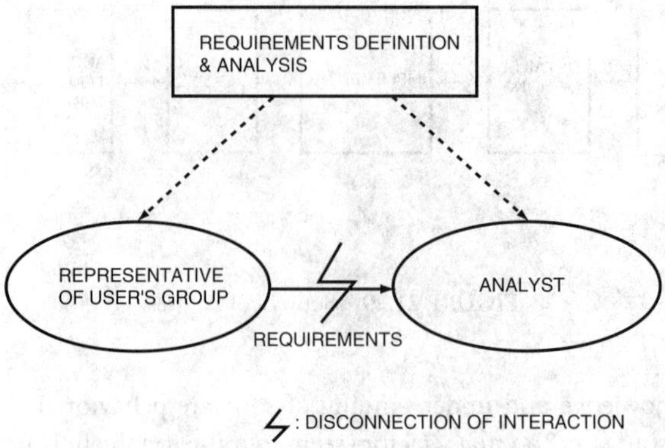

FIGURE 21.30 Sources of failures in the requirements definition and analysis phase.

A preliminary list of user's failure modes are: omission of a known requirement (the user forgets), the user is not aware of a specific requirement, a requirement specified is incorrect (the user misunderstands his/her own needs), and the user specifies a requirement that conflicts with another requirement. Failure modes of the analyst are: the analyst omits (forgets) a specific user requirement, the analyst misunderstands (cognitive error) a specific user requirement, the analyst mistransposes (typographical error) a user requirement, the analyst fails to notice a conflict between two users' requirements (cognitive error). Communication failure is: requirement is misheard. Each such failure is applicable to all types of requirements.

As an example, let us consider the case of a software for which the user defined the following requirements:

- $R(f, 1)$ and $R(f, 2)$, two functional requirements
- A performance requirement $R(p, 1)$ on the whole software system
- A reliability requirement $R(r, 1)$ on the whole software
- One interoperability requirement $R(io, 1)$

Let us further assume for the sake of simplicity that there can only be four categories of requirements.

The failure modes in Table 21.6 are obtained by affecting each requirement category by the failure modes identified previously. Note that we might need to further refine this failure taxonomy in order to correctly assess the impact of each failure mode on the software system. Let us, for instance, consider the case of a reliability requirement. From previous definitions, reliability is defined as the probability that the system will perform its mission under given conditions over a given period of time. Then if the analyst omits a reliability requirement, it might be either too low or too high. If the reliability requirement is too low, the system may still be functional but over a shorter period of time, which may be acceptable or unacceptable. If the reliability requirement is too high, there is an economic impact but the mission itself will be fulfilled.

Once the failure modes are identified, factors that influence their generation should be determined. For example, in the requirements definition and analysis phase, a failure mode such as "user is unaware of a requirement" is influenced by the user's level of knowledge in the task domain.

Once all failure modes and influencing factors have been established, data collection can start. Weakness areas can be identified and corrected. The data collected is meaningful because it ties directly into the failure generation process. Actually, it is likely that a generic set of failure modes can be developed for specific software life cycles and that this set would only need to be tailored for a specific application.

The next step is to find the combination of failure modes that will either fail the software or degrade its operational capabilities in an unacceptable manner. These failure modes are product (software) dependent. Let us focus only on functional requirements. Let us assume that the software will function satisfactorily if any of the two functional requirements $R(f, 1)$ and $R(f, 2)$ is carried out correctly. Hence, the probability

TABLE 21.6 Preliminary List of Failure Modes of the Software Requirements Definition and Analysis Phase

Category: By:	Functional Requirements	Performance Requirements	Reliability Requirements	Interoperability Requirements
User	• Omission of known requirement • Unaware of requirement $R(f, 1)$ is incorrect • $R(f, 2)$ is incorrect • $R(f, 1)$ conflicts with other requirement.	• Omission of known requirement • Unaware of requirement $R(p, 1)$ is incorrect • $R(p, 1)$ conflicts with other requirement.	• Omission of known requirement. • Unaware of requirement $R(r, 1)$ is incorrect • $R(r, 1)$ conflicts with over requirement.	• Omission of known requirement • Unaware of requirement $R(po, 1)$ is incorrect • $R(po, 1)$ conflicts with other requirement.
Analyst	• Analyst omits $R(f, 1)$ • Analyst omits $R(f, 2)$ • Analyst does not notice conflict between $R(f, 1)$ and other requirements • Analyst does not notice conflict between $R(f, 2)$ and other requirements • Analyst misunderstands $R(f, 1)$ • Analyst misunderstands $R(f, 2)$ • Analyst mistransposed $R(f, 1)$ • Analyst mistransposed $R(f, 2)$	• Analyst omits $R(p, 1)$ • Analyst misunderstands $R(p, 1)$ • Analyst mistransposed $R(p, 1)$ • Analyst does not notice conflict between $R(p, 1)$ and other requirements	• Analyst omits $R(r, 1)$ • Analyst misunderstands $R(r, 1)$ • Analyst mistransposed $R(r, 1)$ • Analyst does not notice conflict between $R(r, 1)$ and other requirements	• Analyst omits $R(po, 1)$ • Analyst misunderstands $R(po, 1)$ • Analyst mistransposed $R(po, 1)$ • Analyst does not notice conflict between $R(po, 1)$ and other requirements
Communication between User and Analyst	• User's requirement $R(f, 1)$ is misheard • User's requirement $R(f, 2)$ is misheard	• User's requirement $R(p, 1)$ is misheard	• User's requirement $R(r, 1)$ is misheard	• User's requirement $R(po, 1)$ is misheard

that the software will fail P_f if we take only into account the requirements and definition analysis phase is given by

$$P_f = P(E_{11} \cup (E_{12} \cap E_{13}) \cup E_{14} \cup E_{15} \cup E_{16} \cup E_{21} \cup (E_{22} \cap E_{23}) \cup E_{24} \cup E_{25} \cup E_{26})$$

where $E_{1i}, E_{2i}, i = 1, 6$ are defined in the fault tree given in Fig. 21.31. The actual fault tree will be more complex as failure modes generated by several phases of the software development process contribute to failure of functions 1 and 2. For instance, if we consider a single failure mode for coding such as "software developer fails to code R(f, 1) correctly" and remove, for the sake of the example, contributions for all other phases, a new fault tree is obtained (see Fig. 21.31).

The last question is to determine if we could use such data for reliability prediction. Early in the process, for example, at the requirements definition and analysis phase, our knowledge of the attributes and characteristics of the software development process is limited: coding personnel and coding languages are undetermined, the design approach may not be defined, etc. Hence, our reliability assessment will be uncertain. The more we move into the life cycle process, the more information becomes available, and hence our uncertainty bound reduces. Let us call M the model of the process, that is, the set of characteristics that define the process. The characteristics of such model may include

- The process is a waterfall life cycle
- Top-down design is used
- Language selected is C++
- Number of programmers is 5, etc.

Our uncertainty is in M. Let $p(M)$ be the probability that the model followed is M. This probability is a function of the time into the life cycle, $p(M, t)$. A reasonable approximation of $p(M, t)$ is $p(M, k)$ where k is the kth life cycle phase. Let R_i be the reliability of the software at the end of the software development process if it was developed under model M_i. Initially, at the end of the requirements definition and analysis phase ($k = 1$), the expected reliability $\langle R \rangle$ is

$$\langle R \rangle = \sum_i p(M_i, 1) \cdot R_i$$

and the variance of R due to our uncertainty on the model is

$$\langle R^2 \rangle = \sum_i p(M_i, 1)(R_i - \langle R_i \rangle)^2$$

If information about the process becomes available, such as, for example, at the end of phase 2 (preliminary design), the process can be updated using Bayesian statistics

$$p(M_i, m + 1) = \frac{\mathcal{L}(I \mid M_i) p(M_i, m)}{\sum_j p(M_j, m) \cdot \mathcal{L}(I \mid M_j)}$$

where I is information obtained between phase m and $m + 1$ and where $\mathcal{L}(I \mid M_i)$ is the likelihood of information I if the model is M_i.

Defining Terms

Errors: Human actions that result in the software containing a fault. Examples of such errors are the omission or misinterpretation of the user's requirements, coding error, etc.

Failure: The inability of the software to perform its mission or function within specified limits. Failures are observed during testing and operation.

Faults: Manifestations of an error in the software. If encountered, a fault may cause a failure of the software.

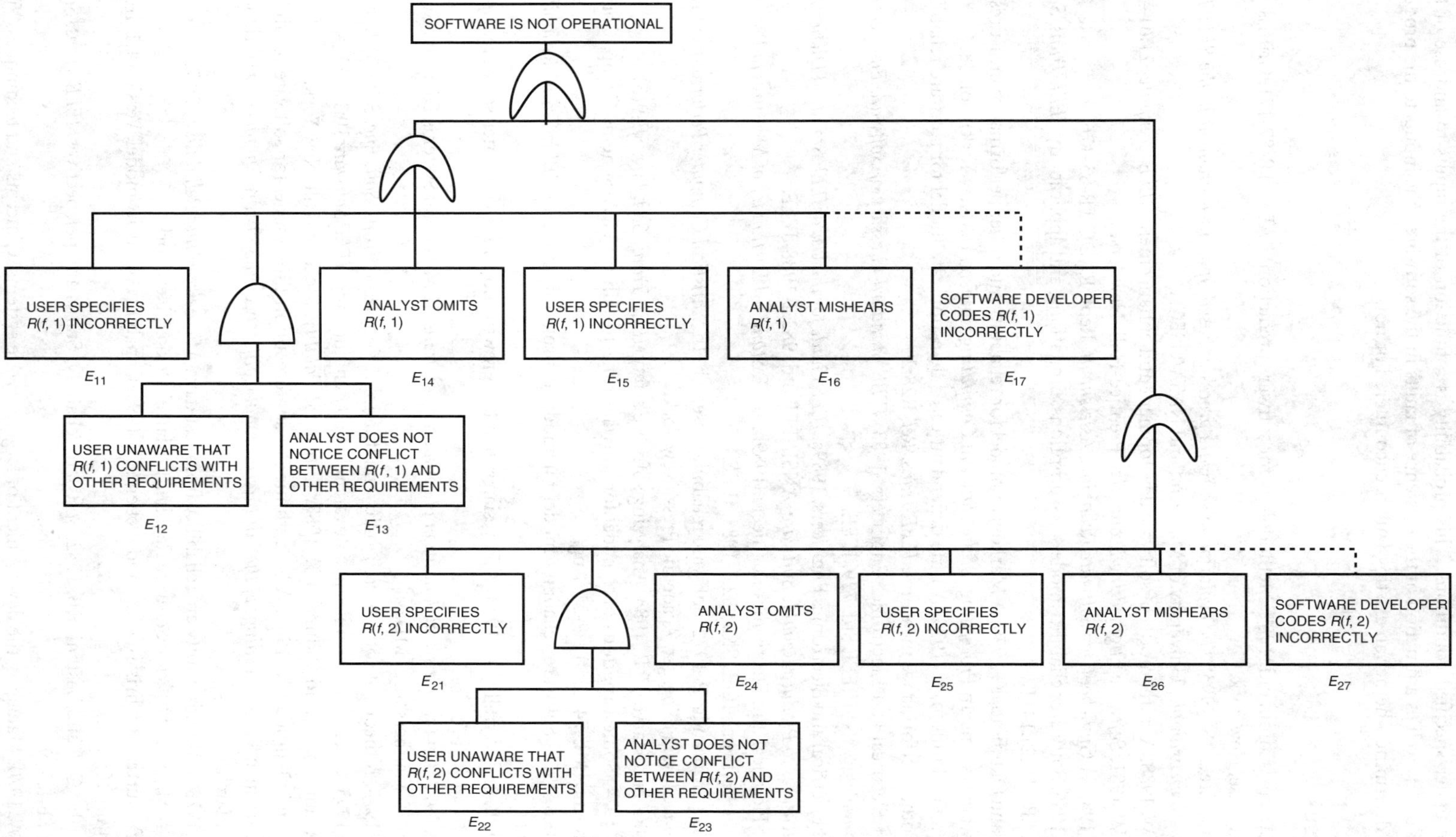

FIGURE 21.31 Software fault tree. The solid lines define the requirements definition and analysis phase fault tree. The requirements definition and analysis phase and coding phase fault tree is obtained by adding solid and dashed contributions.

Software reliability: The probability that software will not cause the failure of a product for a specified time under specified conditions. This probability is a function of the inputs to and use of the product, as well as a function of the existence of faults in the software. The inputs to the product will determine whether an existing fault is encountered or not.

References

Air Force Systems Command. 1987. Methodology for software prediction. RADC-TR-87-171. Griffiss Air Force Base, New York.

Basili, V. and Green, S. 1993. *The evolution of software processes based upon measurement in the SEL: The clean-room example.* University of Maryland and NASA/GSFC.

Boehm, B.W. 1988. A spiral model of software development and enhancement. *IEEE Computer* 21:61–72.

Boehm, B.W. 1981. *Software Engineering Economics.* Prentice-Hall, New Jersey.

Goel, A.L. 1983. *A Guidebook for Software Reliability Assessment,* Rept. RADC TR-83-176.

Goel, A.L. 1985. Software reliability models: Assumptions, limitations, and applicability. *IEEE Trans. Soft. Eng.* SE-11(12):1411.

Goel, A.L. and Okumoto, K. 1979. A Markovian model for reliability and other performance measures of software systems. In *Proceedings of the National Computing Conference* (New York), vol. 48.

Goel, A.L. and Okumoto, K. 1979. A time dependent error detection rate model for software reliability and other performance measures. *IEEE Trans. Rel.* R28:206.

Institute of Electrical and Electronics Engineers. 1983. *IEEE Standard Glossary of Software Engineering Terminology,* ANSI/IEEE Std. 729. IEEE.

Institute of Electrical and Electronics Engineers. 1989a. *IEEE Guide for the Use of IEEE Standard Dictionary of Measures to Produce Reliable Software,* ANSI/IEEE Std. 982.2-1988. IEEE.

Institute of Electrical and Electronics Engineers. 1989b. *IEEE Standard Dictionary of Measures to Produce Reliable Software,* IEEE Std. 982.1-1988. IEEE.

Jelinski, Z. and Moranda, P. 1972. Software reliability research. In *Statistical Computer Performance Evaluation,* ed. W. Freiberger. Academic Press, New York.

Leveson, N.G. and Harvey, P.R. 1983. Analyzing software safety. *IEEE Trans. Soft. Eng.* SE-9(5).

Lipow, M. 1974. Some variations of a model for software time-to-failure. TRW Systems Group, Correspondence ML-74-2260.1.9-21, Aug.

Littlewood, B. 1979. Software reliability model for modular program structure. *IEEE Trans. Reliability* R-28(3).

Littlewood, B. and Verrall, J.K. 1973. A Bayesian reliability growth model for computer software. *Appl. Statist.* 22:332.

Mills, H.D. 1972. *On the statistical validation of computer programs.* IBM Federal Systems Division, Rept. 72-6015, Gaithersburg, MD.

Musa, J.D. 1975. A theory of software reliability and its application. *IEEE Trans. Soft. Eng.* SE-1:312.

Musa, J.D. 1993. Operational profiles in software reliability engineering. *IEEE Software* 10:14–32.

Musa, J.D., Iannino, A., and Okumoto, K. 1987. *Software Reliability.* McGraw-Hill, New York.

Musa, J.D. and Okumoto, K. 1983. A logarithmic Poisson execution time model for software reliability measurement. In *Proceedings of the 7th International Conference on Software Engineering* (Orlando, FL), March.

Nelson, E. 1978. Estimating software reliability from test data. *Microelectronics Reliability* 17:67.

Neufelder, A.M. 1993. *Ensuring Software Reliability.* Marcel Dekker, New York.

Paulk, M.C., Curtis, B., Chrissis, M.B., and Weber, C.V. 1993. Capability maturity model, Version 1.1. *IEEE Software* 10:18–27.

Ramamoorthy, C.V. and Bastani, F.B. 1982. Software reliabiity: Status and perspectives. *IEEE Trans. Soft. Eng.* SE-8:359.

Royce, W.W. 1970. Managing the development of large software systems: Concepts and techniques. Proceedings of Wescon, Aug.; also available in *Proceedings of ICSE9,* 1987. Computer Society Press.

Sage, A.P. and Palmer, J.D. 1990. *Software Systems Engineering*. Wiley Series in Systems Engineering. Wiley, New York.

Saiedian, H. and Kuzara, R. 1995. SEI capability maturity model's impact on contractors. *Computer* 28: 16–26.

Schick, G.J. and Wolverton, R.W. 1973. Assessment of software reliability. 11th Annual Meeting German Oper. Res. Soc., DGOR, Hamburg, Germany; also in *Proc. Oper. Res.*, Physica-Verlag, Wirzberg-Wien.

Scott, R.K., Gault, J. W., and McAllister, D.G. 1987. Fault tolerant software reliability modeling. *IEEE Trans. Soft. Eng.* SE-13(5).

Shooman, M.L. 1975. Software reliability measurement and models. In *Proceedings of the Annual Reliability and Maintainability Symposium* (Washington, DC).

Smidts, C. and Kowalski, R. 1995. Software reliability. In *Product Reliability, Maintainability and Supportability Handbook*, ed. M. Pecht. CRC Press, Boca Raton, FL.

TRW Defense and Space System Group. 1976. Software reliability study. Rept. 76-2260, 1-9.5. TRW, Redondo Beach, CA.

Further Information

Additional information on the topic of software reliability can be found in the following sources:

Computer magazine is a monthly periodical dealing with computer software and hardware. The magazine is published by IEEE Computer Society. IEEE headquarters: 345 E. 47th St. New York, NY 10017-2394.

IBM System Journal is published four times a year by International Business Corporation, Old Orchard Road, Armonk, NY 10504.

IEEE Software is a bimonthly periodical published by the IEEE Computer Society, IEEE headquarters: 345 E. 47th St. New York, NY 10017-2394.

IEEE Transactions on Software is a monthly periodical by the IEEE Computer Society, IEEE headquarters: 345 E. 47th St. New York, NY 10017-2394.

In addition, the following books are recommended.

Friedman, M.A. and Voas, J.M. 1995. *Software Assessment: Reliability, Safety, Testability*. McGraw-Hill, New York.

Littlewood, B. 1987. *Software Reliability*. Blackwell Scientific, Oxford, England, UK.

Rook, P. 1990. *Software Reliability Handbook*. McGraw-Hill, New York.

22

Safety

Clifford D. Ferris
University of Wyoming

William F. Hammett
Hammett and Edison, Inc.

Jerry C. Whitaker
Editor-in-Chief

William. E. DeWitt
Purdue University

22.1   Electric Shock

Clifford D. Ferris

22.1.1   Introduction

Human tissues, such as the skin and the muscles, as well as blood and other body fluids are classified as **electrolytes**. Consequently, they are electrical conductors that may be characterized in terms of ohmic resistances. Electric potential differences applied across human tissues, or at two locations on the external skin surface, generate response currents.

Electric shock can be divided into two classes: microshock and macroshock. Microshock describes an internal shock that may occur as a consequence of certain medical diagnostic procedures (such as cardiac catheterization) or surgical procedures in which electrically operated sensors are introduced into the human body. Electric current levels associated with microshock range from 10 to 100 μA. This phenomenon is specific to medical procedures and will not be addressed further in this chapter.

Macroshock, or simply electric shock, describes simultaneous contact between the body surface and two electrical conductors at different potentials, and the physiological consequences of this contact. The two conductors may be a hot conductor and ground, or two hot conductors, such as two of the phase wires in a three-phase power distribution system.

The severity of the consequences of electric shock depends on a variety of factors. The physiological effects of electrical shock are not produced by electric potential (voltage) *per se*, but rather by the electric current that is driven by the potential difference that is applied externally to the body surface. Consequently,

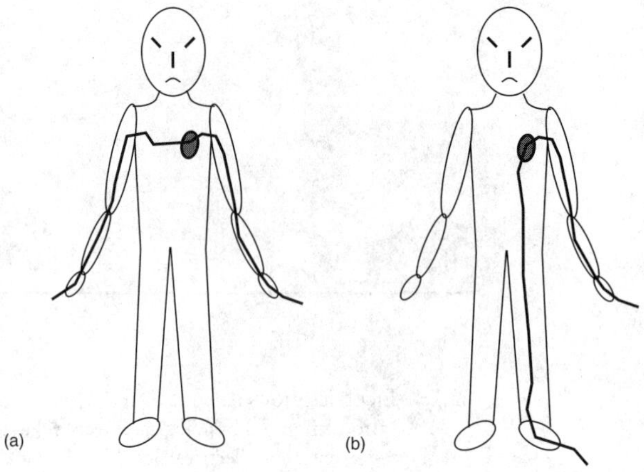

FIGURE 22.1   Most lethal paths for electric shock: (a) hand-to-hand path, (b) hand-to-foot path.

the combined effective electrical resistance of the body volume involved and the intimacy (surface area and pressure) of the skin-conductor contact have a major effect on the severity of electric shock. The paths through the body that the shock currents take determine the potential seriousness of the shock. Generally speaking, current paths that do not include the head or main trunk of the body are not fatal, although permanent damage may be produced in the form of scars, nerve or muscle impairment resulting in partially or fully paralyzed limbs, and loss of limbs. Electric shock currents that pass through the chest in the region of the heart or lungs may cause death, either by direct effect on the heart, or by paralysis of the nerves that control the breathing diaphragm. Severe electric shocks to the head may cause death or permanent brain damage. As illustrated in Fig. 22.1, the two most dangerous current paths are from hand-to-hand, and hand-to-foot (especially left hand to left foot) since the current passes through the chest and potentially through the heart.

Response to electric shock is subjective to some extent, with both physiological and psychological components. The condition of the skin surfaces that come into contact with the electrical conductors is an important factor. For example, if an adult grasps the probes of a digital ohmmeter tightly between the thumb and forefinger of each hand, the resistance displayed will vary widely among individuals. Those individuals who use their hands in activities that build up callouses may produce resistance readings on the order of 1 or 2 MΩ when the hands are dry, and perhaps 100 kΩ when the hands are wet. Other individuals with soft and pliant skin may register 5–10 kΩ with dry hands and significantly lower values when the hands are moist or wet. If the ohmmeter is powered by a 9-V battery, the maximum possible current sustained in the 1-MΩ case is 9 μA, whereas it is 0.9 mA when the resistance is 10 kΩ (a thousandfold increase).

The sensation experienced by individuals exposed to the same low level of electric shock varies widely. Reported sensations range from buzzing, tingling (pins and needles), to burning or a jolt. The emotional state of the individual at the time that the shock was received and the body location of the shock also influence response. Shock response tends to be amplified in persons who are tense. If the shock site is near a nerve or nerve ending that is close to the surface of the skin, the perceived intensity of the shock is also amplified. A parallel situation is bumping one's elbow and triggering the funny-bone response. The physiological consequences of electric shock at different current levels (rms) for alternating current at 60 Hz are presented in Table 22.1. Threshold levels are shown. Refer to the section on Physiological Effects for additional information.

There are various secondary effects of electric shock that can cause serious injury even if the shock itself has produced no direct physiological damage. As with any sudden and unexpected sensory stimulus, the startle reflex may be triggered causing the person to fall down, flail limbs, drop objects, or otherwise sustain injury. High potentials, especially those produced by the discharge of a high-voltage capacitor, can induce

TABLE 22.1　Electric Shock Current Levels and Physiological Effects (AC 60 Hz)

Current Level	Effect
1 mA	Sensation that shock is occurring
5 mA	Upper limit of safe or harmless range
10–20 mA	Let-go threshold: **flexor** muscles are stronger than the **extensor** muscles; subject cannot shake loose from the shock source; perspiration
30–40 mA	**Tetany:** sustained muscle contraction and cramping
50–70 mA	Extreme pain, physical exhaustion, fainting, irreversible nerve damage; possibility of ventricular fibrillation (heart); respiratory arrest with possible asphyxiation
100 mA	Ventricular fibrillation (heart) and death if the current passes through the body trunk
>100 mA	Fibrillation, amnesia, burns, severe electrolysis at contact sites
>5 A	Little likelihood of survival

severe and generalized muscle contractions of sufficient intensity to propel an individual across a room. Because of the violence of the reaction, the person may sustain serious bone fractures (or even death) as a consequence of physical impact with objects in the trajectory path from the site of the shock. Even if the victim does not collide with an object, the force of the muscle contractions alone resulting from electrical shock may fracture spinal vertebrae or cause a shoulder dislocation.

22.1.2　Physiological Effects

As can be inferred from Table 22.1, the medical consequences of electric shock range in severity from no effect through minor burns, nerve and muscle damage, to death. Determining factors are

1. Effective electrical resistance of the body between the contact points
2. Portion of the body involved in the current path
3. Nature of the electrical source producing the shock
4. Body weight

Items 1 and 2 have been addressed to some extent in the Introduction. The lower the contact resistance, the higher is the resulting shock current and the possibility of severe damage. A current path through the body trunk is potentially the most lethal. The electrical resistance of the human body is a function of frequency. By a strange quirk of nature, it is lowest at 60 Hz and rises slightly as frequency decreases to 0 Hz or DC. The resistance also rises as frequency increases above 60 Hz. At frequencies on the order of 100 kHz, skin-surface (capacitance) effects commence such that electric conduction paths include the skin surface as well as bulk effects through the underlying body tissues. A shock that exceeds the let-go threshold is particularly dangerous because the victim begins to perspire, which then reduces the contact resistance between the body and the electrical source. When the contact resistance decreases, the shock current increases. The let-go threshold current level in women is about 30% lower than in men and may relate to differences in skin condition and resistance.

Generally speaking with regard to item 3, the effects of alternating current shock are more severe than direct current shock at the same current levels (rms and DC). The sensation level for DC is about 5 mA, as opposed to 1 mA (rms) for AC at 60 Hz. The let-go threshold shown in Table 22.1 for AC at 60 Hz is 10–20 mA (15 mA average). For DC, the let-go threshold is on the order of 75 mA, a figure that is higher than the corresponding maximum peak-to-peak AC value at the 20-mA rms level. When there is a direct current path through the heart, a momentary current of 60 mA (rms) at 60 Hz can induce **ventricular** fibrillation (defined subsequently), whereas a direct current in the range from 300 to 500 mA is required. A sustained shock at 120 V AC and 60 Hz is especially dangerous. The body resistance is at a minimum and the voltage level is not high enough to cause severe muscle contractions that might otherwise propel the victim away from the electrical source.

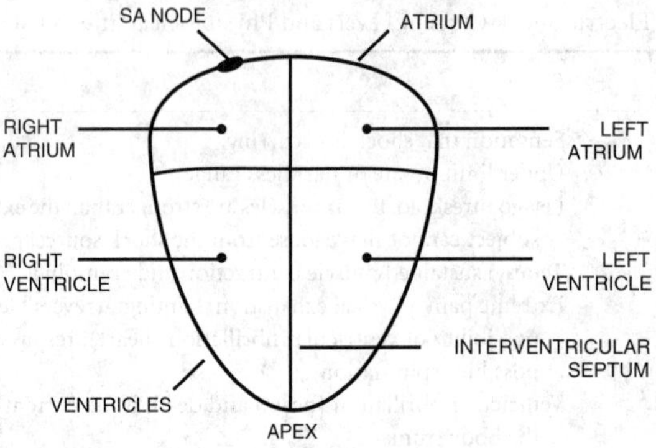

FIGURE 22.2   Stylized anatomical cross section of the heart.

As is discussed in the next section, electrical current that passes through the chest region may cause **cardiac** arrest. The intensity of the current required to produce this condition is a function of the time in the cardiac cycle at which the shock occurs and, to some degree, the victim's body weight. Heavy individuals present larger chest volumes, and the current paths are presumably more diffuse than the current paths through the chest region of a slender person.

Body Trunk (Heart and Lungs)

The human body runs on electricity. Sensory information received from our environment is translated into electrical pulses that are transmitted along nerve pathways to the brain. Muscle activity is initiated by electrical pulses that pass along nerves to sites called myoneural junctions where they trigger a muscle to extend or contract. From a mechanical point of view, the heart is simply a four-chambered pump composed of muscular tissue, as indicated in Fig. 22.2 (as viewed externally from the front).

The two upper chambers (left and right atria) are reservoirs that hold the blood that is returned to the heart from the body organs and the lungs. The main pump chambers are the left and right ventricles. The **atrial** chambers supply blood to the ventricles. When the ventricles (**ventricular** muscles) contract producing a heart beat, blood from the left ventricle is pumped through the aorta to the body organs (the systemic circulation); blood from the right ventricle is pumped through the pulmonary artery to the lungs (the pulmonary circulation). The beating of the heart is controlled by an electric **pacemaker** (analogous to a free-running multivibrator), called the sinoatrial or SA node, which is composed of a small volume of specialized tissue located at a site toward the top of the heart. The pacemaker generates regular pulses that cause the muscles of the ventricles to contract (one pulse per heart beat). In the normal heart, the pacemaker adjusts its output frequency according to the body's demand for oxygen.

The contraction and relaxation of muscles also generates electrical pulses. The beating of the heart is controlled by pulses from the pacemaker, while at the same time the contraction and relaxation of the heart-muscle tissues generate electrical pulses. Because the body is a volume conductor, these pulses can be detected on the surface of the body and recorded as an **electrocardiogram (ECG)**. If a sensing electrode is placed on the inside of each wrist with a third reference electrode on the inside of the right ankle, a signal of the form shown in Fig. 22.3 can be obtained. The signal shown represents one cardiac cycle (heart beat) in a normal heart and is designated as Lead II in the standard method of recording electrocardiograms. The time axis has been normalized to unity, and the major events (called waves) are denoted by P, Q, R, S, T, and U. The maximum peak value (R wave) on the skin surface is approximately 1.5 mV, although values on the order of 1 mV are usually recorded. The P wave represents the muscular contraction of the atrium (two atrial chambers) to force blood into the ventricles. The R wave (in the QRS complex) represents the contraction of the ventricles and the pumping of blood out of the heart into the aorta and pulmonary artery. The T wave is generated by the relaxation (called repolarization) of the ventricular muscles. The electrical signal from the relaxation of the atrium is masked by the high-intensity R wave.

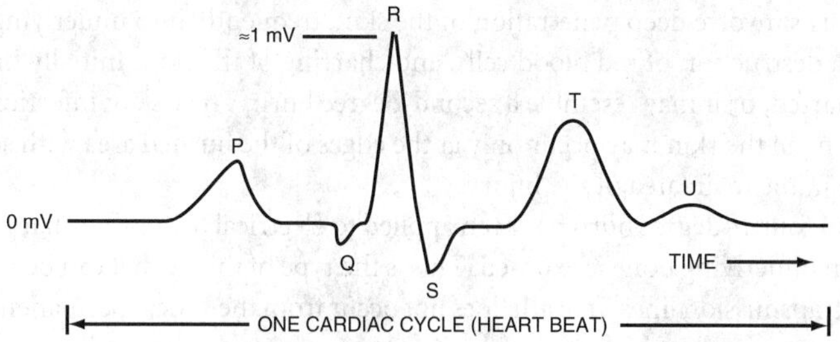

FIGURE 22.3 ECG waveform for one cardiac cycle.

A severe shock current that passes through the chest cavity will cause the entire heart muscle to contract and stop beating while the current persists. If such a current is of short duration, normal heart rhythm begins again with no irreversible cardiac tissue damage. Lower levels of current can disrupt normal heat rhythm. The heart is most susceptible to electric shock during the occurrence of the T wave. Thus a current that just produces a shock response during the T wave typically will not generate any effect during another period of the cardiac cycle. The response of a normally beating heart to an above-threshold electric shock current is cardiac arrest.

The gross symptoms of cardiac arrest produced by electric shock are an unconscious subject who is not breathing, has no detectable pulse (cardiac arrest), and zero blood pressure. The heart, however, may be in either one of two states: standstill or ventricular **fibrillation**. In standstill, as the name implies, the heart is not beating and the ECG signal is a flat line. In fibrillation, the ventricular muscle groups contract randomly and out of unison such that the ventricles flutter. No blood is pumped, but a low-level electrical signal is produced. An ECG recorder is required to differentiate between flat line and ventricular fibrillation (VFIB). The respective ECG signals are illustrated in Fig. 22.4.

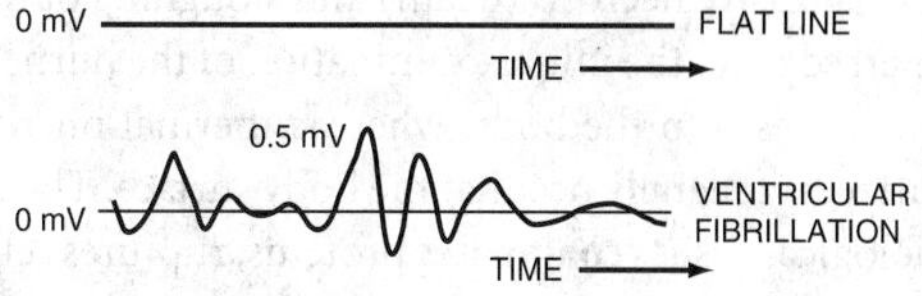

FIGURE 22.4 ECG waveforms in cardiac arrest.

A shock current that passes through the chest region may cause involuntary contraction of the respiratory muscles. The subject can no longer breathe voluntarily, and asphyxiation occurs if the current is not interrupted. Respiratory paralysis (cessation of breathing) may occur as a consequence of electric shock that takes a pathway through the body trunk. Typically, what occurs is the nerve that controls the diaphragm is paralyzed. When the shock occurs, the subject inhales sharply and the chest expands and then becomes rigid, even after the electric current has ceased. The heart may or may not be affected. In some cases, the heart initially continues to beat normally until the beats become erratic because the heart muscle itself is no longer receiving properly oxygenated blood. CO_2 builds up in the blood and produces acidosis, which causes irritability of the heart-muscle tissue and makes restoration of the normal heart beat difficult.

Tissue Damage

Except for very minor electric shocks, burns may be expected at the contact site(s) with the involvement of the skin surface and underlying tissues as a consequence of the penetration by the electric current. Traditional descriptions of the severity of thermal burns includes first-, second-, and third-degree burns. Some medical personnel have designated severe burns produced by electrical shock as fourth degree. First-degree burns are essentially superficial and are described by discoloration or redness of the skin, mild swelling, and associated pain. In subjects without underlying medical problems (such as diabetes), they normally heal quickly and without complications. Second-degree burns penetrate more deeply into the skin tissues than first-degree burns to produce a red or mottled appearance with blister formation and significant pain. Considerable swelling of the affected area occurs over a period of a few days, and the skin surface is usually wet from a loss of plasma from the damaged skin layers.

Third-degree burns involve deep penetration of the skin, frequently into underlying tissue, with coagulation of the skin, destruction of red blood cells, and charring of the skin. Initially the skin surface may appear white or charred, or it may resemble a second-degree burn. The risk of infection at the burn site is high, and full healing of the skin may occur only at the edges of the burned area with scar tissue replacing normal skin tissue in the main area of the injury.

The classification fourth-degree burn has been applied to electrical burns that char the overlying tissues in a manner that an underlying bone is exposed. This is the type of injury that can be produced by contact with high-voltage transmission lines. If death does not occur from the shock, permanent physical disability can be expected.

Whatever the apparent degree, burns resulting from electrical shock should not be taken lightly. With the exception of severe fires and explosions, accidental thermal burns typically affect the skin because they are produced by some hot object or liquid that contacts the skin surface. On the other hand, burns produced by electrical shock result from the passage of current through the skin and underlying tissues. Skin damage may appear to be slight, but there may be extensive tissue damage below the skin surface. Burns associated with electric shock tend to be deep, slow to heal, and subject to infection if not treated immediately. In some cases, both thermal and electrochemical burns can occur simultaneously if the subject accidentally touches an electrically energized hot conductor such as heating coil in an electric oven.

Depending on the nature of the electric shock, the tissues involved can be heated to temperatures as high as $5000°C$, thus converting the water content of the body cells and fluids to superheated steam. There is progressive necrosis (death) and sloughing of the affected tissues. Damage is generally greater than that inferred from the initial examination of the burned area. This situation occurs because the electric current penetrates into the body, whereas thermal burns, such as those from contact with hot liquids or solid surfaces, generally occur at the body surface. The heat produced by the passage of electric current through biological tissues **coagulates** proteins, ruptures cells, and produces tissue death. When nerve and/or muscle tissues are involved, permanent effects may be expected that vary from partial impairment of function of the limb(s) involved to complete paralysis.

Biological tissues may be modeled electrically as parallel resistance-capacitance circuits. For DC and low-frequency AC currents, the resistive component controls the current levels produced by electric shocks. At audio frequencies and above, capacitive reactance decreases to values comparable to the resistive component of the tissue electrical impedance. This situation leads to the potential of very deep and searing burns when electric shock currents pass through tissues. The effects of high-frequency current are put to beneficial use in the now widely used electrosurgical apparatus (electrical scalpel). These instruments typically operate in the 300–400 kHz range, and by modulating the wave form of the signal, surgical cutting, cauterizing, or a combination of the two may be achieved.

22.1.3   First Aid

The best first aid for electric shock is prevention. In the event that an individual has sustained or is sustaining an electric shock in the work place, several guidelines are suggested:

1. *Shock in progress:* For the case when a coworker is receiving an electric shock and cannot let go of the electrical source, the safest action is to trip the circuit breaker that energizes the circuit involved, or to pull the power-line plug on the equipment involved if the latter can be accomplished safely. Under no circumstances should the rescuer touch the individual who is being shocked, since the rescuer's body may then also be in the dangerous current path. If the circuit breaker or equipment plug cannot be located, then an attempt can be made to separate the victim from the electrical source through the use of a nonconducting object such as a wooden stool or a wooden broom handle. Use only an *insulating* object and nothing that contains metal or other electrically conductive material. The rescuer must be very careful not to touch the victim or the electrical source and thus become a second victim. If such equipment is available, hot sticks used in conjunction with lineman's gloves may be applied to push or pull the victim away from the electrical source. Pulling the hot stick normally provides the greatest control over the victim's motion and is the safest action for

the rescuer. Once the the electrical source has been turned off, or the victim can be reached safely, immediate first aid procedures should be implemented as noted in paragraph 2.

2. *Shock no longer in progress:* (a) If the victim is conscious and moving about, have the victim sit down or lie down. Sometimes there is a delayed reaction to an electrical shock that causes the victim to collapse. Call 911 or the appropriate plant-site paramedical team immediately. If there is a delay in the arrival of medical personnel, check for electrical burns. In instances of severe shock, there will normally be burns at a minimum of two sites: the entry point for the current and the exit point(s). Cover the burns with dry (and sterile preferably) dressings. Check for possible bone fractures if the victim was violently thrown away from the electrical source and possibly impacted objects in the vicinity. Apply splints as required if suitable materials are available and you have appropriate training. Cover the victim with a coat or blanket if the environmental temperature is below room temperature, or the victim complains of feeling cold. (b) If the victim is unconscious, call 911 or the appropriate plant-site paramedical team immediately. In the interim, check to see if the victim is breathing and if a pulse can be felt at either the inside of a wrist above the thumb joint (radial pulse) or in the neck above and to either side of the Adam's apple (carotid pulse). It is usually easier to feel the pulse in the neck as opposed to the wrist pulse, which may be weak. The index and middle finger should be used to sense the pulse, and not the thumb. Many individuals have an apparent thumb pulse that can be mistaken for the victim's pulse. If a pulse can be detected but the victim is not breathing, begin mouth-to-mouth respiration if you know how to do so. If no pulse can be detected (presumably the victim will not be breathing), carefully move the victim to a firm surface and begin cardiopulmonary resuscitation if you have been trained in the use of CPR. Respiratory arrest and cardiac arrest are crisis situations. Because of loss of the oxygen supply to the brain, loss of memory or permanent brain damage may occur after several minutes even if the victim is successfully resuscitated. If the victim can be resuscitated and appears to be stable, follow the additional procedures indicated in paragraph 2(a) if the arrival of medical personnel is delayed.

Ironically, the treatment for cardiac arrest induced by an electric shock is a massive countershock, which causes the entire heart muscle to contract. The random and uncoordinated ventricular fibrillation contractions (if present) are thus stilled. Under ideal conditions, normal heart rhythm is restored once the shock current ceases. The countershock is generated by a cardiac defibrillator, various portable models of which are available for use by emergency medical technicians and other *trained* personnel. Although portable defibrillators may be available at industrial sites where there is a high risk of electrical shock to plant personnel, they should be used only by trained personnel. Application of a defibrillator to an unconscious subject whose heart is beating can induce cardiac standstill or ventricular fibrillation, just the conditions that the defibrillator was designed to correct.

22.1.4   Prevention of Electrical Shock

Electric shock is not a matter to be taken lightly, as death may be the consequence. Prevention of electric shock is far superior to remedial medical treatment. The subject of electrical safety is presented in a separate article, but a few items are outlined here to close the present discussion:

1. In the work place, observe all company and local regulations, and Occupational Safety and Health Administration (OHSA) regulations regarding electrical safety matters.
2. Always turn off and *disconnect* electrical equipment from the power mains before opening the equipment cabinet for servicing or calibration procedures. Do not rely on switches to disconnect power, as sometimes they are defective, and sometimes the switch has been miswired or mislabeled such that the off position is actually the on position.
3. Using cables designed for the purpose, always discharge high-voltage capacitors associated with radio transmitters, cathode-ray tubes, or other high-voltage circuits before attempting servicing. The external surface of cathode-ray tubes should also be discharged.

4. If diagnostic tests must be performed on energized equipment, use appropriate electrical safety procedures. Many modern switching/regulating power supplies are potentially lethal and power-line-isolation transformers *must* be used when servicing such equipment. The electrical hazard results from a bridge rectifier being connected directly across the 120-V AC power line when the equipment is energized. Consequently, the ground for the circuit board in the power supply in the equipment is not at power-line neutral or Earth ground potential. If no isolation transformer is used and the service person simultaneously contacts Earth ground and the circuit board ground, a high possibility exists for a fatal electric shock.

5. Do not disable or remove the Earth ground connectors on power cords and within equipment.

6. In the design or modification of electrical equipment and facilities, observe strict adherence to all of the appropriate codes and guidelines such as: the National Electrical Code of the National Fire Protection Association; Medical Device Safety Act of 1971, 1973, and subsequent amendments; industry standards for electronic equipment design and manufacture; other regulations as appropriate.

Defining Terms

Atrial: Pertaining to the atrium or the top reservoir chambers in the upper portion of the heart.

Cardiac: Pertaining to the heart.

Coagulation: The production or formation of a blood clot.

Electrocardiogram (ECG): The visual display of the electrical activity of the heart as recorded from electrodes placed on the skin surface.

Electrolyte: A substance that is electrically conductive and is broken down by the passage of electric current through it.

Extensor muscles: Muscles that extend a joint.

Fibrillation: Spontaneous and individual contraction of muscle fibers.

Flexor muscles: Muscles that flex a joint.

Pacemaker: A small area of specialized tissue near the top of the heart that generates the electrical pulses that initiate heat beats.

Tetany: The state of a muscle when it undergoes sustained contraction.

Ventricular: Pertaining to the ventricles or main pump chambers of the heart.

References

American Academy of Orthopaedic Surgeons. 1981. *Emergency Care and Transportation of the Sick and Injured*, 3rd ed. George Banta, Menasha, WI.

American Red Cross. 1991. *First Aid, Responding to Emergencies*. Mosby Life Line, St. Louis, MO.

Berkow, R., ed. 1992. *The Merck Manual of Diagnosis and Therapy*, 16th ed. Merck, Rahway, NJ.

Ferris, C.D. 1978. *Introduction to Bioinstrumentation*. Humana Press, Clifton, NJ.

Further Information

Additional information on human physiology, the function of the heart, and electric shock is available from the following sources:

Guyton, A.C. 1981. *Textbook of Medical Physiology*, 6th ed. W.B. Saunders, Philadelphia, PA.

Raphael, C.L. et al., eds. 1994. *Electrical Injury: A Multidisciplinary Approach to Therapy, Prevention, and Rehabilitation*. Annals of the New York Academy of Sciences, Vol. 720. The New York Academy of Sciences, New York.

Webster, J.G., ed. 1992. *Medical Instrumentation, Application and Design*, 2nd ed. Houghton Mifflin, Boston, MA.

22.2 Nonionizing Electromagnetic Radiation

William F. Hammett

22.2.1 Definition of Nonionizing Electromagnetic Radiation (NIER)

An electric or magnetic current or potential of alternating polarity will create radiating electromagnetic fields (EMFs). The frequency of alternation is a chief characteristic, and EMF is generated across an enormous range of frequencies. This frequency spectrum is shown in Fig. 22.5, with several common identifiers noted for certain bands.

Electromagnetic radiation is a natural part of our existence, providing light, heat, and, relatively recently, power and communications. Visible light, for instance, occupies a very narrow band of electromagnetic radiation, with characteristic frequencies of $4–8 \times 10^{14}$ Hz. (Sound is not really part of the electromagnetic spectrum, since sound energy is carried by *pressure* waves in air or some other medium. Electromagnetic energy is carried by waves that do not require a medium for transport. Thus, sound can be considered naturally only an Earth-bound phenomenon, whereas EMF literally radiates throughout the universe. Nevertheless, the frequencies of pressure waves audible to humans are shown for interest.)

Electromagnetic waves radiate at, not surprisingly, the speed of light, approximately 300,000,000 m/s, and they behave like streams of particles (called photons). Each EMF has a characteristic quantum energy (also called photon energy), given by the equation

$$QE = hf$$
$$= 4.135 \times 10^{-15} f \text{ eV}$$

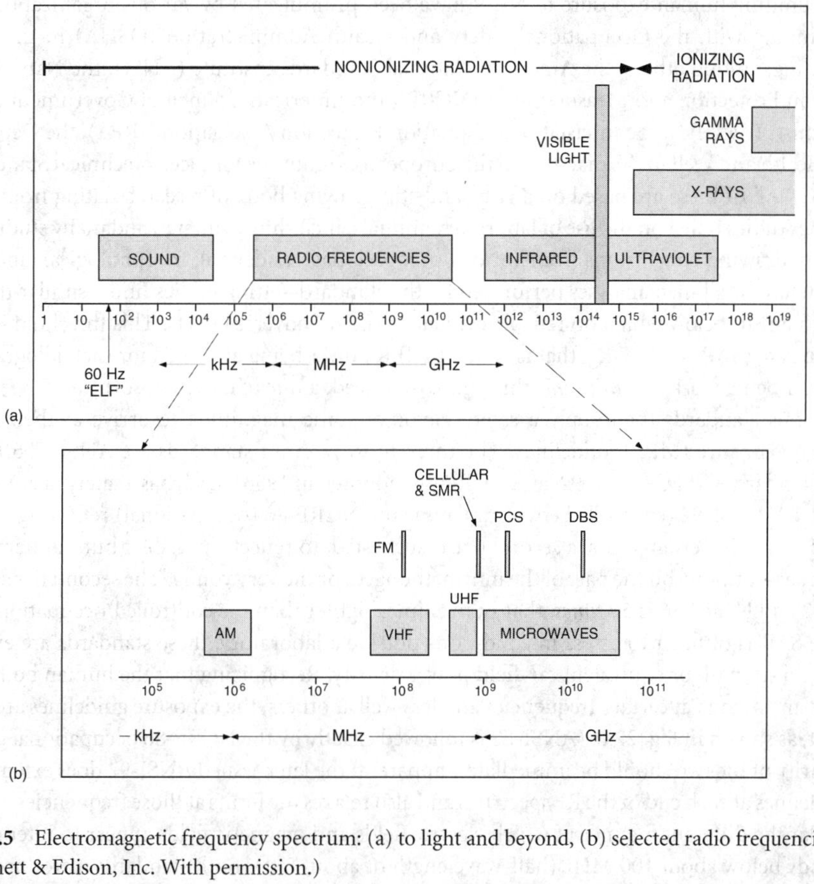

FIGURE 22.5 Electromagnetic frequency spectrum: (a) to light and beyond, (b) selected radio frequencies. (*Source:* 1996 Hammett & Edison, Inc. With permission.)

where h is Planck's constant 6.624×10^{-27} erg-s and f is the frequency of the wave, in hertz (cycles per second).

Quantum energies of at least 4–25 eV are required to break free an electron from an atom of matter, a process known as *ionization*. Therefore, frequencies above about 1–6×10^{15} Hz are capable of this and so are known as *ionizing radiation*. This portion of the spectrum includes ultraviolet light, X-rays, gamma rays, and cosmic rays.

Nonionizing electromagnetic radiation (NIER), by definition, is characterized by quantum energies *below* the ionization threshold and therefore cannot break chemical bonds. Thus, the effects of exposure to NIER are not necessarily destructive or cumulative. This larger portion of the spectrum includes the familiar-sounding TV bands called VHF (very high frequency) and UHF (ultra high frequency). Radio frequencies (RF) are considered those between 300 Hz and 300 GHz.

Extremely low-frequency energy (ELF) is characterized by frequencies below about 100 Hz, which are believed to be similar to human biological electromagnetic activity. The study of ELF and its interaction with the human body is a separate field beyond the scope of this chapter.

Another way to consider EMF, less rigorously but perhaps more intuitively, is by its wavelength, which is proportional to the reciprocal of frequency. Atoms are tiny structures, with diameters of about 1–5 Å. Now, 1 Å equals 10^{-10} m, and so frequencies above about 0.6–3×10^{18} Hz will have wavelengths small enough to get inside an atom and break loose an electron. This is ionizing energy. At frequencies below light, the EMF has wavelengths at least 8000 Å, that is, too big to fit into individual atoms; this is nonionizing energy. In fact, at 2 GHz, a common microwave frequency, the wavelength is 1.5×10^9 Å, or about 6 in.

22.2.2   Exposure Standards for NIER

Standards limiting human exposure to NIER have been promulgated by various organizations over the years, beginning with the Occupational Safety and Health Administration (OSHA) in the 1960s and now including, among others, the American National Standards Institute (ANSI), the National Council on Radiation Protection and Measurements (NCRP), the American Council of Government and Industrial Hygienists (ACGIH), the International Radiation Protection Association (IRPA), the Department of National Health and Welfare (Canada), and the European Committee for Electrotechnical Standardization (CENELEC). All of these are based on a large and still-growing body of research dating from the 1960s. Many of the studies relied on the use of laboratory animals, including primates, and many studies are epistemological, drawing conclusions from various occupational, residential, morphological, and mortality statistics. Tellingly, all metaanalyses performed by the standard-setting bodies find a similar threshold of exposure to NIER, below which no reliable evidence exists of adverse effects. This threshold is a **specific absorption rate (SAR)** of 4 W/kg, that is, 4 W of NIER power being absorbed for each kilogram of body mass. (Power being work *per unit time,* this figure is, indeed, a *rate* of energy absorption.)

The various standards then apply a safety factor of some magnitude to arrive at their **maximum permissible exposure (MPE)** guidelines. The three most relevant standards are ANSI C95.1-1982, the successor standard ANSI C95.1-1992, and NCRP-86 (adopted in 1986). Each has a safety factor of *10 times*, with ANSI-82, ANSI-92 (controlled environments), and NCRP-86 (occupational) reflecting a maximum SAR of 0.4 W/kg. In certain cases, a second tier is suggested, to reflect a possible (but not demonstrated) increase in susceptibility on the part of the infirm, the aged, or the very young. The second tier safety factor in ANSI-92 and NCRP-86 is *50 times,* that is, five times tighter than the controlled/occupational limit.

Because SAR is difficult to assess in conditions outside a laboratory, these standards are expressed in terms of field strength or equivalent **far-field** power density. Recognizing that the human body performs "better" as an antenna at certain frequencies and less well at others, the exposure guidelines are frequency dependent, as shown in Fig. 22.6. (ANSI-82 is followed exactly by the NCRP-86 occupational guidelines.) The similarity of the two should be immediately apparent; the later-issued ANSI-92 does extend the range of the guidelines at each end of the RF spectrum and also relaxes the limits at those frequencies, recognizing in particular the different manner in which electric fields and magnetic fields appear to interact with the human body below about 100 MHz (half-wavelength of about 5 ft). Both standards introduce a limit on

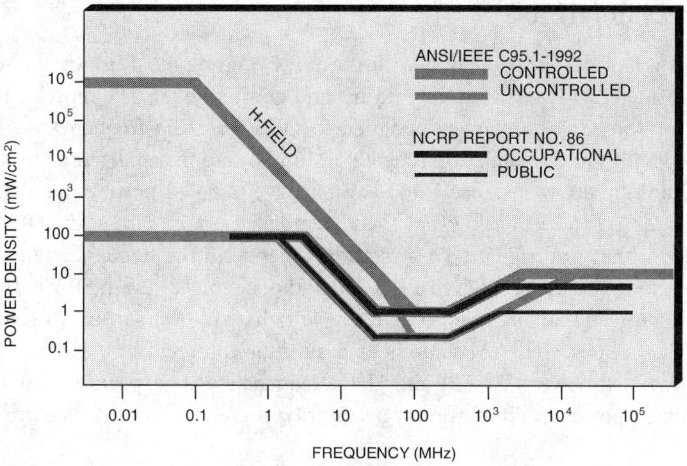

FIGURE 22.6 Comparison of ANSI C95.1-1992 and NCRP-86 (1986) standards. (*Source:* 1996 Hammett & Edison, Inc. With permission.)

current flowing in the human body, ANSI-92 including induced and contact current limits and NCRP-86 just contact current limits. Induced current refers to that flowing in a free-standing body, grounded or ungrounded but not in contact with any conductive object.

Strict compliance with ANSI-92 or NCRP-86 requires not only determination, by calculation or by measurement, of the total far-field equivalent power density from all nearby sources, but also measurement of body current from contact with potentially radiating objects and, for ANSI-92, measurement of body current induced by immersion in fields near such objects. Power density measurements are straightforward to take, using specialized broadband equipment calibrated in power density (or even in percent of some standard); such measurements can also be made using broadband or narrow-band equipment calibrated more traditionally in field strength. The conversion from field strength to equivalent far-field power density S, in **milliwatt per square centimeter (mW/cm^2)**, uses the impedance of free space, 377 Ω, as follows:

$$S = \frac{E^2}{3770}$$

where the electric field strength E is in V/m, and

$$S = 37.7 \, H^2$$

where the magnetic field strength H is in A/m. Figure 22.7 is the tabular form of ANSI-92, providing the derivation for the limits, expressed as E, H, or S, at particular frequencies.

Frequency	Electromagnetic Fields				Body Currents
Applicable Range (MHz)	Electric Field Strength (V/m)	Magnetic Field Strength (A/m)		Equivalent Far-Field Power Density (mW/cm^2)	Induced (Foot) or Contact (mA)
0.003–0.1	614 *614*	163 *163*			1000f *450f*
0.1–1.34	614 *614*	16.3/f *16.3/f*			100 *45*
1.34–3.0	614 *823.8/f*	16.3/f *16.3/f*			100 *45*
3.0–30	1842/f *823.8/f*	16.3/f *16.3/f*			100 *45*
30–100	61.4 *27.5*	16.3/f *158.3/f$^{1.668}$*			100 *45*
100–300	61.4 *27.5*	0.163 *0.0729*		1.0 *0.2*	no limit
300–3,000				f/300 *f/1500*	no limit
3,000–15,000				10 *f/1500*	no limit
15,000–300,000				10 *10*	no limit

Note: f is frequency of emission, in MHz.

FIGURE 22.7 ANSI/IEEE C95.1-1992 radio frequency protection guide. Guidelines for controlled environments in plain text; guidelines for uncontrolled environments in italics.

22.2.3 Sources of NIER

As noted, NIER is a natural phenomenon, with the sun being the most important source. Man-made sources can be considered in two classes: intentional and unintentional. The most important intentional sources of NIER are broadcast antennas, designed to radiate a radio-frequency signal on which data have been encoded with some modulation scheme. FM radio stations are licensed by the FCC at powers up to (and exceeding, in a few instances) 100 kW effective radiated power, and VHF TV stations are licensed at powers up to 316 kW peak visual effective radiated power (or about 160 kW average power, with the inclusion of the aural signal). These stations operate in the frequency range at which the exposure guidelines are tightest. The FCC licenses AM radio stations at up to 50 kW input power, which can become, with a directional antenna array, effective radiated powers of up to several hundred kilowatts, and the FCC licenses UHF TV stations at peak visual powers of 5010 kW (or about 2500 kW average effective radiated power). At AM and UHF frequencies, the exposure guidelines do rise severalfold, so these higher power facilities are not necessarily of greater concern than are FM and VHF TV stations.

Microwave stations are licensed on frequencies from about 1 GHz up to about 40 GHz, although they generally do not send energy over a large area. Instead, this nonbroadcast service uses large dish antennas to focus the NIER tightly and direct it toward a similar receiving dish (known as *point-to-point* service). High radiated powers can be generated at microwave transmitting antennas, but exposures would be of concern generally only to tower riggers.

Other intentional sources include navigational aids for ships and planes, two-way paging, dispatch and emergency services, and signals from satellites. The power density levels experienced by anyone not actually operating that equipment would be very low. There are also household electronic devices, such as nursery monitors, cordless telephones, and garage door openers, that use RF EMF; the powers are generally so low that exposure conditions cannot exceed the standards. Communications devices such as amateur radio stations and cellular and PCS handsets may have the potential for creating localized fields that exceed the standards, but time and space averaging are generally presumed to limit self-exposures to conditions compliant with the standards. Another class of intentional radiators are medical devices, which have been developed for monitoring, for diagnosis, and for treatment.

Unintentional radiators include industrial applications, such as RF sealers and welders, and microwave ovens. Many such devices have emission standards, promulgated by the FCC, the Food and Drug Administration (FDA), or some other agency, limiting radiation from the device generally to avoid electromagnetic interference (EMI). NIER exposure standards apply to unintentional radiators, as well.

22.2.4 Avoidance of NIER

NIER at levels not exceeding the guidelines set forth in the exposure standards need *not* be avoided, as those guidelines are intended to apply for exposure conditions of unlimited duration, that is, 24 hours a day, 365 days a year, for a lifetime. The findings of the standard-setting bodies were that no credible evidence or theory exists for adverse health effects at compliant exposure levels. *Prudent avoidance* is the standard recommendation when findings are inconclusive, and the electric utility companies have taken to including EMF brochures with billing statements, recommending that their customers practice prudent avoidance of the ELF energy those companies supply. With RF radiation, the findings are widely believed to be conclusive already, so avoidance of NIER beyond the safety factors incorporated in the exposure guidelines should not be necessary.

At certain locations, generally near high-power broadcasting antennas or near high-power RF sealing or welding equipment, fields are likely to exceed the exposure guidelines. As NIER power radiates from a source, the spherical surface over which the wave spreads increases as the square of the distance from the source. This leads to the so-called *inverse square law* by which it is known that, in the far field of an antenna, the power density decreases rapidly. Calculation of RF exposure conditions relies on this fact from basic physics, and the incorporation of certain conservative assumptions set forth by the FCC in its

Office of Science and Technology (OST) Bulletin No. 65 allows computers to quickly estimate the NIER exposure conditions at even a multiuser broadcasting site, perhaps avoiding the need for power density measurements. The general form of the OST-65 formula is as follows:

$$\text{power density } S = \frac{2.56 \times 1.64 \times 100 \times \text{RFF}^2 \times [0.4 \times \text{VERP} + \text{AERP}]}{4\pi D^2} \quad \text{in mW/cm}^2$$

where

VERP = total peak visual ERP (all polarizations), kW
AERP = total aural ERP (all polarizations), kW
RFF = relative field factor at direction to actual point of calculation
D = distance from center of radiation to point of calculation, m

The factor of 2.56 accounts for the increase in power density due to ground reflection, assuming a reflection coefficient of 1.6 ($1.6 \times 1.6 = 2.56$). The factor of 1.64 is the gain of a half-wave dipole relative to an isotropic radiator. The factor of 0.4 converts peak visual ERP to an average RMS value; for FM and cellular stations, of course, the value of VERP is zero. The factor of 100 in the numerator converts to the desired units of power density.

Exposure situations exceeding the standards should be mitigated, which can be done in one of three ways: (1) reducing the field, (2) limiting access to the field, or (3) both. Reducing the field can be accomplished, depending on the circumstance, by improved antenna design, improved shielding, relocation of the source, or reduction in power. Limiting access to the field can be accomplished with barricades or, in controlled situations, by signs, markings, and protocols. In the latter situation, care must be exercised to be certain that all persons encountering the fields have been briefed on the appropriate protocols and markings. Figure 22.8 shows a typical warning sign using the ANSI RF symbol. Below this may be added further information such as "Stay Back 10 Feet" or "Check with Supervisor Before Entering Area" that are part of protocols established for ensuring compliance.

FIGURE 22.8 Standard RF warning sign.

Defining Terms

Far field: That region of space in which the electric field and magnetic field components of an electromagnetic wave are related by the impedance of free space. The far field is generally considered to begin no closer to the NIER source than a distance of several wavelengths or several times the antenna aperture.

Maximum permissible exposure (MPE): The limit adopted by a standards-setting body for exposures of unlimited duration. Generally, some allowance is made in the standard for time and/or space averaging of higher exposure conditions.

Milliwatts per square centimeter (mW/cm²): A measure of power density, that is, energy per unit time passing through a unit of surface area. This unit is commonly used to compare electric and magnetic field readings and to assess them against the exposure guidelines.

Nonionizing electromagnetic radiation (NIER): That portion of the electromagnetic spectrum below about 10^{15} Hz, characterized by waves of insufficient quantum energy to break electrons free from atoms.

Specific absorption rate (SAR): The deposition of energy over time into a body. The units are generally watts per kilogram of body mass. This is the attribute on which findings by various researchers can be compared and on which the exposure standards base their guidelines.

References

FCC. 1985. Bulletin No. 65: Evaluating Compliance with FCC-Specified Guidelines for Human Exposure to Radiofrequency Radiation. Office of Science and Technology, Federal Communications Commission, Washington, DC.

Hausmann, E. and Slack, E.P. 1939. *Physics*. D. Van Nostrand, New York.

NCRP. 1986. Biological Effects and Exposure Criteria for Radiofrequency Electromagnetic Fields. Report No. 86, National Council on Radiation Protection and Measurements, Bethesda, MD.

Stein, J., ed. 1967. *The Random House Dictionary*. Random House, New York.

Yost, M.G. 1993. *Nonionizing Radiation Questions and Answers*. San Francisco Press, San Francisco, CA.

Zumdahl, S.S. 1986. *Chemistry*. D.C. Heath, Toronto, Canada.

Further Information

The following texts provide treatments of greater depth on many of the issues introduced concerning NIER:

Hammett, W.F. 1996. *RF EMF—Radiation & Regulation*. McGraw-Hill, New York.

Hitchcock, R.T. and Patterson, R.M. 1995. *Radio-Frequency and ELF Electromagnetic Energies*. Van Nostrand Reinhold, New York.

22.3 PCBs and Other Hazardous Substances

Jerry C. Whitaker

22.3.1 Introduction

Polychlorinated biphenyls (PCBs) belong to a family of organic compounds known as *chlorinated hydrocarbons*. Virtually all PCBs in existence today have been synthetically manufactured. PCBs are of a heavy, oil-like consistency and have a high boiling point, a high degree of chemical stability, low flammability, and low electrical conductivity. These characteristics led to the past widespread use of PCBs in high-voltage capacitors and transformers. Commercial products containing PCBs were distributed widely from 1957 to 1977 under several trade names including

- Aroclor
- Pyroclor
- Sanotherm
- Pyranol
- Askarel

Askarel also is a generic name used for nonflammable dielectric fluids containing PCBs. Table 22.2 lists some common trade names for Askarel. These trade names typically are listed on the nameplate of a PCB transformer or capacitor.

TABLE 22.2 Commonly Used Names for PCB Insulating Material

Apirolio	Abestol	Askarel	Aroclor B	Chlorextol	Chlophen
Chlorinol	Clorphon	Diaclor	DK	Dykanol	EEC-18
Elemex	Eucarel	Fenclor	Hyvol	Inclor	Inerteen
Kanechlor	No-Flamol	Phenodlor	Pydraul	Pyralene	Pyranol
Pyroclor	Sal-T-Kuhl	Santothern FR	Santovac	Solvol	Therminal

22.3.2   Health Risk

PCBs are harmful because, once they are released into the environment, they tend not to break apart into other substances. Instead, PCBs persist, taking several decades to slowly decompose. By remaining in the environment, they can be taken up and stored in the fatty tissues of all organisms, from which they are released slowly into the bloodstream. Therefore, because of the storage in fat, the concentration of PCBs in body tissues can increase with time, even though PCB exposure levels may be quite low. This process is called **bioaccumulation**. Furthermore, as PCBs accumulate in the tissues of simple organisms, which are consumed by progressively higher organisms, the concentration increases. This process is called **biomagnification**. These two factors are especially significant because PCBs are harmful even at low levels. Specifically, PCBs have been shown to cause chronic (long-term) toxic effects in some species of animals and aquatic life. Well-documented tests on laboratory animals show that various levels of PCBs can cause reproductive effects, gastric disorders, skin lesions, and cancerous tumors.

PCBs may enter the body through the lungs, the gastrointestinal tract, and the skin. After absorption, PCBs are circulated in the blood throughout the body and stored in fatty tissues and skin, as well as in a variety of organs, including the liver, kidneys, lungs, adrenal glands, brain, and heart.

The health risk lies not only in the PCB itself, but also in the chemicals developed when PCBs are heated. Laboratory studies have confirmed that PCB by-products, including *polychlorinated dibenzofurans* (PCDFs) and *polychlorinated dibenzo-p-dioxins* (PCDDs), are formed when PCBs or chlorobenzenes are heated to temperatures ranging from approximately 900 to 1300°F. Unfortunately, these products are more toxic than PCBs themselves.

22.3.3   Governmental Action

The Congress took action to control PCBs in October 1975, by passing the **Toxic Substances Control Act** (**TSCA**). A section of this law specifically directed the EPA to regulate PCBs. Three years later, the EPA issued regulations to implement a congressional ban on the manufacture, processing, distribution, and disposal of PCBs. Since then, several revisions and updates have been issued by the EPA. One of these revisions, issued in 1982, specifically addressed the type of equipment used in industrial plants. Failure to properly follow the rules regarding the use and disposal of PCBs has resulted in high fines and some jail sentences.

Although PCBs no longer are being produced for electrical products in the U.S., the EPA estimates that thousands of PCB transformers and millions of PCB small capacitors may still be in use or in storage. The threat of widespread contamination from PCB fire-related incidents is one reason behind the EPA's efforts to reduce the number of PCB products in the environment. The users of high-power equipment are affected by the regulations primarily because of the widespread use of PCB transformers and capacitors in old equipment. Table 22.3 lists the primary classifications of PCB devices.

PCB Components

The two most common PCB components are transformers and capacitors. A PCB transformer is one containing at least 500 ppm (parts per million) PCBs in the dielectric fluid. An Askarel transformer generally has 600,000 ppm or more. A PCB transformer may be converted to a *PCB-contaminated device* (50–500 ppm) or a *non-PCB device* (less than 50 ppm) by being drained, refilled, and tested. The testing must not take place until the transformer has been in service for a minimum of 90 days. Note that this is *not* something that a maintenance technician can do. It is the exclusive domain of specialized remanufacturing companies.

PCB transformers must be inspected quarterly for leaks. However, if an impervious dike (sufficient to contain all the liquid material) is built around the transformer, the inspections can be conducted yearly. Similarly, if the transformer is tested and found to contain less than 60,000 ppm, a yearly inspection is sufficient. Failed PCB transformers cannot be repaired; they must be disposed of properly.

If a leak develops, it must be contained and daily inspections must begin. A cleanup must be initiated as soon as possible, but no later than 48 h after the leak is discovered. Adequate records must be kept of all

TABLE 22.3   Definition of PCB Terms as Identified by the EPA

Term	Definition	Examples
PCB	Any chemical substance that is limited to the biphenyl molecule that has been chlorinated to varying degrees, or any combination of substances that contain such substances.	PCB dielectric fluids, PCB heat-transfer fluids, PCB hydraulic fluids, 2,2′,4-trichlorobiphenyl
PCB article	Any manufactured article, other than a PCB container, that contains PCBs and whose surface has been in direct contact with PCBs.	Capacitors, transformers, electric motors, pumps, pipes
PCB container	A device used to contain PCBs or PCB articles and whose surface has been in direct contact with PCBs.	Packages, cans, bottles, bags, barrels, drums, tanks
PCB article container	A device used to contain PCB articles or equipment and whose surface has not been in direct contact with PCBs.	Packages, cans, bottles bags, barrels, drums, tanks
PCB equipment	Any manufactured item, other than a PCB container or PCB article container, which contains a PCB article or other PCB equipment.	Microwave ovens, fluorescent light ballasts, electronic equipment
PCB item	Any PCB article, PCB article container, PCB container or PCB equipment that deliberately or unintentionally contains, or has as a part of it, any PCBs.	
PCB transformer	Any transformer that contains PCBs in concentrations of 500 ppm or greater.	
PCB contaminated	Any electric equipment that contains more than 50, but less than 500 ppm of PCBs. (Oil-filled electrical equipment other than circuit breakers, reclosers, and cable whose PCB concentration is unknown must be assumed to be PCB-contaminated equipment.)	Transformers, capacitors, circuit breakers, reclosers, voltage regulators, switches, cable, electromagnets

inspections, leaks, and actions taken for 3 years after disposal of the component. Combustible materials must be kept a minimum of 5 m from a PCB transformer and its enclosure.

As of October 1, 1990, the use of PCB transformers (500 ppm or greater) was prohibited in or near commercial buildings when the secondary voltages were 480 V AC or higher. The use of radial PCB transformers was allowed if specified electrical protection was provided. As of October 1, 1993, certain types of PCB transformers must have been removed from service or retrofitted to a lower PCB concentration. EPA regulations required that lower voltage network PCB transformers (those with secondary voltages of less than 480 V) could be used in and around commercial buildings for the rest of their useful life only if they were equipped with electrical protections, such as current-limiting fuses, to prevent ruptures caused by high current faults. If these protections were not installed before October 1, 1990, then the transformer should have been removed from service by October 1, 1993.

The EPA regulations also require that the operator notify others of the possible dangers. All PCB transformers (including those in storage for reuse) must be registered with the local fire department. Supply the following information:

- The location of the PCB transformer(s).
- Address(es) of the building(s). For outdoor PCB transformers, provide the outdoor location.
- Principal constituent of the dielectric fluid in the transformer(s).
- Name and telephone number of the contact person in the event of a fire involving the equipment.

Any PCB transformers used in a commercial building must be registered with the building owner. All owners of buildings within 30 m of such PCB transformers also must be notified. In the event of a fire-related incident involving the release of PCBs, immediately notify the Coast Guard National Spill Response Center at 1-800-424- 8802. Also take appropriate measures to contain and control any possible PCB release into water.

Capacitors are divided into two size classes: *large* and *small*. The following are guidelines for classification:

- A PCB small capacitor contains less than 1.36 kg (3 lb) of dielectric fluid. A capacitor having less than 100 in.^{3} also is considered to contain less than the 3 lb of dielectric fluid.
- A PCB large capacitor has a volume of more than 200 in.^{3} and is considered to contain more than 3 lb of dielectric fluid. Any capacitor having a volume from 100 to 200 in.^{3} is considered to contain 3 lb of dielectric, provided the total weight is less than 9 lb.
- A PCB large *low-voltage capacitor* contains 3 lb or more of dielectric fluid and operates below 2 kV.
- A PCB *large high-voltage capacitor* contains 3 lb or more of dielectric fluid and operates at 2 kV or greater voltages.

The use, servicing, and disposal of PCB small capacitors is not restricted by the EPA unless there is a leak. In that event, the leak must be repaired or the capacitor disposed of properly. Disposal may be performed by an approved incineration facility, or the component may be placed in a specified container and buried in an approved chemical waste landfill. Typically, chemical waste landfills are only for disposal of liquids containing 50–500 ppm PCBs and for solid PCB debris. Items such as capacitors that are leaking oil containing greater than 500-ppm PCBs should be taken to an EPA-approved PCB disposal facility.

Identifying PCB Components

The first task for the facility manager is to identify any PCB items on the premises. Equipment built after 1979 probably does not contain any PCB-filled devices. Even so, inspect all capacitors, transformers, and power switches to be sure. A call to the manufacturer also may help. Older equipment (pre-1979) is more likely to contain PCB transformers and capacitors. A liquid-filled transformer usually has cooling fins, and the nameplate may provide useful information about its contents. If the transformer is not labeled or the fluid is not identified, it must be treated as a PCB transformer. Untested (not analyzed) mineral-oil-filled transformers are assumed to contain at least 50 ppm, but less than 500 ppm PCBs. This places them in the category of PCB-contaminated electric equipment, which has different requirements than PCB transformers. Older high-voltage systems are likely to include both large and small PCB capacitors. Equipment rectifier panels, exciter/modulators, and power-amplifier cabinets may contain a significant number of small capacitors. In older equipment, these capacitors often are Askarel filled. Unless leaking, these devices pose no particular hazard. If a leak does develop, follow proper disposal techniques. Also, liquid-cooled rectifiers may contain Askarel. Even though their use is not regulated, treat them as a PCB article, as if they contain at least 50 ppm PCBs. Never make assumptions about PCB contamination; check with the manufacturer to be sure.

Any PCB article or container being stored for disposal must be date tagged when removed, and inspected for leaks every 30 days. It must be removed from storage and disposed of within 1 year from the date it was placed in storage. Items being stored for disposal must be kept in a storage facility meeting the requirements of 40 Code of Federal Regulations (CFR), Part 761.65(b)(1) unless they fall under alternative regulation provisions. There is a difference between PCB items stored for disposal and those stored for reuse. Once an item has been removed from service and tagged for disposal, it cannot be returned to service.

Labeling PCB Components

After identifying PCB devices, proper labeling is the second step that must be taken by the facility manager. PCB article containers, PCB transformers, and large high-voltage capacitors must be marked with a standard 6- × 6-in. large marking label (ML) as shown in Fig. 22.9. Equipment containing these transformers or capacitors also should be marked. PCB large low-voltage (less than 2 kV) capacitors need not

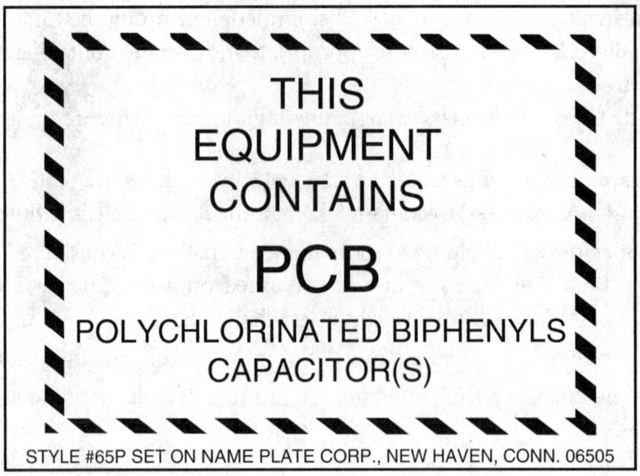

FIGURE 22.9   Marking label (ML) used to identify PCB transformers and PCB large capacitors.

be labeled until removed from service. If the capacitor or transformer is too small to hold the large label, a smaller 1- × 2-in. label is approved for use. Labeling each PCB small capacitor is not required. However, any equipment containing PCB small capacitors should be labeled on the outside of the cabinet or on access panels. Properly label any spare capacitors and transformers that fall under the regulations. Identify with the large label any doors, cabinet panels, or other means of access to PCB transformers. The label must be placed so that it can be read easily by firefighters. All areas used to store PCBs and PCB items for disposal must be marked with the large (6- × 6-in.) PCB label.

Record Keeping

Inspections are a critical component in the management of PCBs. EPA regulations specify a number of steps that must be taken and the information that must recorded. Table 22.4, which summarizes the schedule requirements, can be used as a checklist for each transformer inspection. This record must be retained for 3 years. In addition to the inspection records, some facilities may need to maintain an annual report. This report details the number of PCB capacitors, transformers, and other PCB items on the premises. The report must contain the dates when the items were removed from service, their disposition, and detailed information regarding their characteristics. Such a report must be prepared if the facility uses or stores at least one PCB transformer containing greater than 500 ppm PCBs, 50 or more PCB large capacitors, or at least 45 kg of PCBs in PCB containers. Retain the report for 5 years after the facility ceases using or storing PCBs and PCB items in the prescribed quantities. Table 22.5 lists the information required in the annual PCB report.

TABLE 22.4   The Inspection Schedule Required for PCB Transformers and Other Contaminated Devices

⇒ Standard PCB transformer	Quarterly
⇒ If full-capacity impervious dike added	Annually
⇒ If retrofitted to <60,000 ppm PCB	Annually
⇒ If leak discovered, clean up ASAP	Daily
(retain these records for 3 years)	
⇒ PCB article or container stored for disposal	Monthly
(remove and dispose of within 1 year)	
⇒ Retain all records for 3 years after disposing of transformer	

TABLE 22.5   Required Information for PCB Annual Report

Transformer location: _____

Date of visual inspection: _____

Leak discovered? _____(Yes/No)

If yes, date discovered (if different from inspection date): _____

Location of leak: _____

Person performing inspection: _____

Estimate of the amount of dielectric fluid released from leak: _____

Date of cleanup, containment, repair, or replacement: _____

Description of cleanup, containment, or repair performed:

_____

Results of any containment and daily inspection required for uncorrected active leaks:

_____

Disposal

Disposing of PCBs is not a minor consideration. Before contracting with a company for PCB disposal, verify its license with the area EPA office. That office also can supply background information on the company's compliance and enforcement history.

The fines levied for improper disposal are not mandated by federal regulations. Rather, the local EPA administrator, usually in consultation with local authorities, determines the cleanup procedures and costs. Civil penalties for administrative complaints issued for violations of the PCB regulations are determined according to a matrix provided in the PCB penalty policy. This policy, published in the Federal Register, considers the amount of PCBs involved and the potential for harm posed by the violation.

Proper Management

Properly managing the PCB risk is not difficult. The keys are to understand the regulations and to follow them carefully. A PCB management program should include the following steps:

- Locate and identify all PCB devices. Check all stored or spare devices.
- Properly label PCB transformers and capacitors according to EPA requirements.
- Perform the required inspections, and maintain an accurate log of PCB items, their location, inspection results, and actions taken. These records must be maintained for 3 years after disposal of the PCB component.
- Complete the annual report of PCBs and PCB items by July 1 of each year. This report must be retained for 5 years.
- Arrange for any necessary disposal through a company licensed to handle PCBs. If there are any doubts about the company's license, contact the EPA.
- Report the location of all PCB transformers to the local fire department and owners of any nearby buildings.

The importance of following the EPA regulations cannot be overstated.

22.3.4   Equipment Operating Hazards

Particular hazards exist in the operation and maintenance of high-power equipment. Maintenance personnel must exercise extreme care around such hardware. Consider the following items:

- Use caution around the high voltage stages of the equipment. Many power supplies operate at voltages high enough to kill through electrocution. Always break the primary AC circuit of the power supply and discharge all high-voltage capacitors.
- Minimize exposure to RF radiation. Do not permit personnel in the vicinity of open energized RF generating circuits, RF transmission systems (waveguides, cables, or connectors), or energized

antennas. High levels of radiation can result in severe bodily injury, including blindness. Cardiac pacemakers may also be affected.

- Avoid contact with **beryllium oxide** (BeO) ceramic dust or fumes. BeO ceramic material is often used as a thermal link to carry heat from a tube or transistor to the heat sink. Do not perform any operation on any BeO ceramic that might produce dust or fumes, such as grinding, grit blasting, or acid cleaning. Beryllium oxide dust or fumes are highly toxic and breathing them can result in serious injury or death. BeO ceramics must be disposed of only in a manner prescribed by the device manufacturer.

- Avoid contact with hot surfaces within the equipment. The anode portion of most power tubes are air-cooled. The external surface normally operates at a high temperature (up to 250°C). Other portions of the tube may also reach high temperatures, especially the cathode insulator and the cathode/heater surfaces. All hot surfaces may remain hot for an extended time after the tube is shut off. To prevent serious burns, avoid bodily contact with these surfaces both during and for a reasonable cool-down period after tube operation.

Beryllium Oxide Ceramics

Some power vacuum tubes, both power grid and microwave, contain beryllium oxide (BeO) ceramics, typically at the output waveguide window or around the cathode. Never perform any operations on BeO ceramics that produce dust or fumes; for example, grinding, grit blasting, or acid cleaning. Beryllium oxide dust and fumes are highly toxic and breathing them can result in serious personal injury or death.

If a broken window is suspected on a microwave tube, carefully remove the device from its waveguide and seal the output flange of the tube with tape. Because BeO warning labels may be obliterated or removed, maintenance personnel should contact the tube manufacturer before performing any work on the device. Some tubes have BeO internal to the vacuum envelope.

Take precautions to protect personnel working in the disposal or salvage of tubes containing BeO. All such personnel should be made aware of the deadly hazards involved and the necessity for great care and attention to safety precautions. Some tube manufacturers will dispose of tubes without charge, provided they are returned to the manufacturer prepaid, with a written request for disposal.

Corrosive and Poisonous Compounds

The external output waveguides and cathode high-voltage bushings of microwave tubes are sometimes operated in systems that use a dielectric gas to impede microwave or high-voltage breakdown. If breakdown does occur, the gas may decompose and combine with impurities, such as air or water vapor, to form highly toxic and corrosive compounds. Examples include Freon gas, which may form lethal *phosgene*, and sulfur hexafluoride (SF_6) gas, which may form highly toxic and corrosive sulfur or fluorine compounds such as *beryllium fluoride*. When breakdown does occur in the presence of these gases, proceed as follows:

- Ventilate the area to outside air.
- Avoid breathing any fumes or touching any liquids that develop.
- Take precautions appropriate for beryllium compounds and for other highly toxic and corrosive substances.

FC-75 Toxic Vapor

The decomposition products of FC-75 are highly toxic. Decomposition may occur as a result of any of the following:

- Exposure to temperatures above 200°C
- Exposure to liquid fluorine or alkali metals (lithium, potassium, or sodium)
- Exposure to ionizing radiation

Known thermal decomposition products include *perfluoroisobutylene* (PFIB; $(CF_3)_2 \, C = CF_2$), which is highly toxic in small concentrations.

If FC-75 has been exposed to temperatures above 200°C through fire, electric heating, or prolonged electric arcs, or has been exposed to alkali metals or strong ionizing radiation, take the following steps:

- Strictly avoid breathing any fumes or vapors
- Thoroughly ventilate the area
- Strictly avoid any contact with the FC-75

Under such conditions, promptly replace the FC-75 and handle and dispose the contaminated FC-75 as a toxic waste.

Defining Terms

Askarel: A common trade name for PCB fluid.

Beryllium oxide: A compound commonly used in the production of ceramics for electrical applications and whose dust or fumes are toxic.

Bioaccumulation: The process whereby a harmful substance is increased in concentration in an organism through ingestion of small amounts over a long period of time.

Biomagnification: The process whereby a harmful substance is increased in concentration as contaminated simpler organisms are consumed by progressively higher organisms.

Polychlorinated biphenyls (PCBs): A family of organic compounds known as *chlorinated hydrocarbons* and used as an insulating material for electrical devices until the late 1970s.

Toxic Substances Control Act (TSCA): A federal law passed by the Congress in 1975 to control PCBs and ultimately eliminate them from use.

References

Code of Federal Regulations, 40, Part 761, Washington, DC.

DOL. *Electrical Standards Reference Manual*. U.S. Department of Labor, Washington, DC.

Hammar, W. *Occupational Safety Management and Engineering*. Prentice-Hall, New York.

Pfrimmer, J. 1987. Identifying and Managing PCBs in Broadcast Facilities. In *NAB Engineering Conference Proceedings*. National Association of Broadcasters, Washington, DC.

OSHA. *Handbook for Small Business*. U.S. Department of Labor, Washington, DC.

OSHA. 1985. Occupational Injuries and Illnesses in the U.S. by Industry. *OSHA Bulletin* 2278. U.S. Department of Labor, Washington, DC.

OSHA. 1987. Electrical Hazard Fact Sheets. U.S. Department of Labor, Washington, DC, Jan.

Further Information

The Occupational Safety and Health Administration (OSHA), Washington, DC, offers a number of publications relating to workplace safety. Contact the local or regional OSHA office for a list of available titles.

Specific information relating to PCB reporting and disposal should be directed to the local Environmental Protection Agency (EPA) office.

22.4   Facility Grounding Principles

William. E. DeWitt

22.4.1   Introduction

Grounding is a very broad and complex subject. The material presented here will introduce the reader to facility grounding principles. It is not intended to represent code requirements or recommended practice. Specific code requirements are contained in the current edition of the National Electrical Code and the

National Electrical Code Handbook. Recommended practice is covered in the current edition of the Institute of Electrical and Electronic Engineers (IEEE) and other industry standards.

A facility can be defined as something built, installed, or established to serve a particular purpose [1]. A facility is usually thought of as a single building or group of buildings. The National Electrical Code (NEC) [2] uses the term *premises* to refer to a facility when it defines premises wiring as the interior and exterior wiring, such as power, lighting, control, and signal systems. Premises wiring includes the service and all permanent and temporary wiring between the service and the load equipment. Premises wiring does not include wiring internal to any load equipment.

This article will define grounding, list the reasons for grounding, discuss grounding methods, and present grounding terminology. Applicable codes and industry standards will be referenced.

22.4.2   Grounding Terminology [2]

Grounded conductor—A system or circuit conductor that is intentionally grounded.

Grounding conductor—A conductor used to connect equipment or the grounded circuit of a wiring system to a grounding electrode or electrodes.

Equipment grounding conductor—The conductor used to connect the noncurrent-carrying metal parts of equipment, raceways, and other enclosures to the system grounded conductor, the grounding electrode conductor, or both, at the service equipment or at the source of a separately derived system.

Grounding electrode conductor—The conductor used to connect the grounding electrode(s) to the equipment grounding conductor to the grounded conductor, or to both, at the service, at each building or structure where supplied from a common service, or at the source of a separately derived system.

Ground-fault protection of equipment—A system intended to provide protection of equipment from damaging line-to-ground fault currents by operating to cause a disconnecting means to open all ungrounded conductors of the faulted circuit. This protection is provided at current levels less than those required to protect conductors from damage through the operation of a supply circuit overcurrent device.

Main bonding jumper—The connection between the grounded circuit conductor and the equipment grounding conductor at the service.

Service—The conductors and equipment for delivering energy from the electricity supply system to the wiring system of the premises served.

Service conductors—The conductors from the service point to the service disconnecting means.

Service equipment—The necessary equipment, usually consisting of a circuit breaker(s) or switch(es) and fuse(s) and their accessories, connected to the load end of service conductors to a building or other structure, or an otherwise designated area, and intended to constitute the main control and cutoff of the supply.

Separately derived system—A premises wiring system whose power is derived from a battery, from a solar photovoltaic system, or from a generator, transformer, or converter windings, and that has no direct electrical connection, including a solidly connected grounded circuit conductor, to supply conductors originating in another system.

22.4.3   Reasons for Grounding

The Institute of Electrical and Electronic Engineers (IEEE) defines grounding as a conducting connection, whether intentional or accidental, by which an electric circuit or equipment is connected to the earth, or to some conducting body of relatively large extent that serves in place of the earth. It is used for establishing and maintaining the same or approximately the same potential as that of the earth (or of the conducting body), on conductors connected to it, and for conducting ground current to and from the earth (or the conducting body) [3].

Based on the IEEE definition, the reasons for grounding can be identified as (1) personnel safety by limiting potentials between non-current-carrying metal parts of an electrical distribution system, (2) personnel safety and control of electrostatic discharge by limiting potentials between noncurrent-carrying metal parts of an electrical distribution system and the earth, and (3) fault isolation for equipment and personnel safety by providing a low-impedance fault return path to the power source to facilitate the operation of overcurrent devices during a ground fault.

The IEEE definition implies an important distinction between *ground* and *earth*. *Earth* refers to planet earth, and *ground* refers to the equipment grounding system, which includes equipment grounding conductors, metallic raceways, cable armor, enclosures, cabinets, frames, building steel, and all other noncurrent-carrying metal parts of the electrical distribution system.

Other reasons for grounding are not as implicit in the IEEE definition. Overvoltage control has long been a benefit of power system grounding, and is described in IEEE Std 142, the Green Book [4]. Beeman presents an excellent discussion of overvoltage control in the *Industrial Power Systems Handbook* [5]. With the increasing use of electronic and computer systems, noise control has become associated with the subject of grounding, and is described in IEEE Std 1100, the *Emerald Book*. [6].

22.4.4 Equipment Grounding

Equipment grounding is the process of interconnecting the noncurrent-carrying metal parts of an electrical distribution system, and then connecting the interconnected metal parts to the earth. Equipment grounding insures that there is no difference of potential, and thus no shock hazard between noncurrent-carrying metal parts in the electrical distribution system. Connecting the equipment grounding system to earth insures that there is no difference of potential between the earth and the equipment grounding system. This also prevents static charge build-up. Specific requirements for equipment grounding are contained in Article 250 of the NEC [2].

22.4.5 System Grounding

System grounding is the process of intentionally connecting one of the current-carrying conductors of the electrical distribution system to ground. This conductor is called the *grounded* conductor because it is intentionally grounded. The purpose of system grounding is overvoltage control and equipment and personnel safety through fault isolation. An ungrounded system is subject to serious overvoltages under conditions such as intermittent ground faults, resonant conditions, and contact with higher voltage systems. Fault isolation is achieved by providing a low-impedance ground return path from the load back to the source that will ensure operation of overcurrent devices in case of ground fault. Methods of system grounding include solidly-grounded, ungrounded, and impedance-grounded. Specific requirements for system grounding are also contained in Article 250 of the NEC [2].

Solidly-grounded means that an intentional zero-impedance connection is made between the grounded conductor and the equipment grounding conductor at the service. In a solidly-grounded, three-phase, four-wire, wye system, the neutral is connected directly to ground with no impedance installed in the neutral circuit. The NEC permits this connection to be made at the service only using the main bonding jumper. The advantages of a solidly-grounded wye system include reduced magnitude of transient overvoltages, improved fault protection, and faster location of ground faults. There is one disadvantage of the solidly-grounded wye system. For low-level arcing ground faults, the application of sensitive, properly-coordinated, ground fault protection (GFP) devices is necessary to prevent equipment damage from arcing ground faults. The NEC requires separate arcing ground fault protection at certain points in 480Y/277V systems. Arcing damage is less a problem at the lower voltage 208Y/120V systems, where the arc may be self-extinguishing. The specific GFP requirements are contained in various articles of the NEC [2].

Ungrounded means that there is no intentional connection between a current-carrying conductor and ground. However, charging capacitance will create unintentional capacitive coupling from each phase to ground, making the system essentially a capacitance-grounded system. The most common ungrounded system is the three-phase, three-wire delta system. The most important advantage of an ungrounded system is that an accidental ground fault in one phase does not require immediate removal. This allows for continuity of service that made the ungrounded-delta system very popular in the past. However, ungrounded systems have serious disadvantages. Since there is no fixed system ground point, it is difficult to locate the first ground fault and to sense the magnitude of the fault current. As a result, the fault is often permitted to remain on the system for a long time. If a second fault should occur before the first one is removed, and the second fault is on a different phase, the result will be a double line-to-ground fault causing serious arcing damage. Another problem with the ungrounded-delta system is the occurrence of high transient overvoltages from phase-to-ground. Transient overvoltages are caused by intermittent ground faults, with overvoltages capable of reaching a phase-to-ground voltage of six to eight times the phase-to-neutral voltage. Sustained overvoltages may ultimately result in insulation failure and thus more ground faults.

Impedance-grounded means that an intentional nonzero impedance connection is made between the neutral conductor and the equipment grounding conductor at the service. The high-resistance grounded wye system is an alternative to solidly-grounded and ungrounded systems. High-resistance grounding will limit ground fault current to a few amperes, thus removing the potential for arcing damage inherent in solidly-grounded systems. The ground reference point is fixed, and relaying methods can locate first faults before damages from second faults occur. Internally-generated transient overvoltages are reduced since the neutral to ground resistor dissipates any charge that may build-up on system charging capacitance.

There is no one *best* system grounding method for all applications. In choosing among the various options, the designer must consider the requirements for safety, continuity of service, and cost. Generally, low-voltage systems are operated solidly-grounded. For applications involving continuous processes in industrial plants or where shutdown might create a hazard, a high-resistance grounded wye system, or a solidly-grounded wye system with an alternate power supply, may be used. The high-resistance grounded wye system combines many advantages of the ungrounded-delta system and the solidly-grounded wye system. Recommended practice for system grounding of low and medium-voltage systems is contained in IEEE Std 142 [4].

22.4.6 Earth Connections

The process of connecting the grounding system to earth is called earthing, and consists of immersing a metal electrode or system of electrodes into the earth. The conductor that connects the grounding system to earth is called the *grounding electrode conductor*. Its function is to keep the entire grounding system at earth potential (i.e., voltage equalization during lightning and other transients) rather than for conducting ground fault current. Therefore, the NEC allows reduced sizing requirements for that portion of the grounding electrode conductor that is the sole connection to certain types of electrodes. Refer to the NEC [2] for specific sizing requirements and electrode types.

The basic measure of effectiveness of an earth electrode system is called *earth electrode resistance*. Earth electrode resistance is the resistance, in ohms, between the point of connection and a distant point on the earth called *remote earth*. Remote earth, about 25 ft from a single driven electrode, is a point from the driven electrode where earth electrode resistance does not increase appreciably when this distance is increased. Earth electrode resistance consists of the sum of the resistance of the metal electrode (negligible) plus the contact resistance between the electrode and the soil (negligible) plus the soil resistance itself. Thus, for all practical purposes, earth electrode resistance equals the soil resistance. The soil resistance is nonlinear, with most of the resistance contained within several feet of the electrode. Furthermore, current flows only through the electrolyte portion of the soil, not the soil itself. Thus, soil resistance varies as the electrolyte content (moisture and salts) of the soil varies. Without electrolyte, soil resistance would be essentially infinite.

Soil resistance is a function of soil resistivity. A 1 m^3 sample of soil with a resistivity of one 1 $\Omega\cdot m$ will present a resistance of 1 Ω between opposite faces. A broad variation of soil resistivity occurs as a function of soil types, and soil resistivity can be estimated or measured directly. The most commonly used method of measuring soil resistivity injects a known current into a given volume of soil and measures the resulting voltage drop. When soil resistivity is known, the earth electrode resistance of any given configuration (single rod, multiple rods, or ground ring) may be determined by using standard equations developed by Sunde [7], Schwarz [8], and others.

Earth resistance values should be as low as practicable, but are a function of the application. The NEC requires single rod, pipe, or plate electrodes, that do not have a resistance to ground of 25 ohms or less, to be augmented by one additional approved electrode. Recommended practice for earth electrode resistance is contained in the IEEE and other standards. Methods of reducing earth resistance values include the use of multiple electrodes, the use of ground rings, increased ground rod lengths, installation of ground rods to the permanent water level, increased area of coverage of ground rings, and the use of concrete-encased electrodes, ground wells, and electrolytic electrodes.

22.4.7 Earth Electrodes

Historically, earth electrodes were classified as *made* electrodes, *natural* electrodes, and *special-purpose* electrodes. The term, made electrode, was used in the 1999 NEC but is not used in the 2002 NEC. The term is found once in the 2002 NEC Handbook in reference to a ground rod. Made electrodes include driven rods and pipes, buried conductors, ground mats, buried plates, and ground rings. The electrode selected is a function of the type of soil and the available depth. Driven electrodes are used where the bedrock is well below the earth surface. Mats or buried conductors are used for lesser depths. Buried plates are not widely used because of the higher cost as compared to rods. Ground rings employ equally-spaced driven electrodes interconnected with buried bare conductors. Ground rings are used around buildings, around pad-mounted equipment, and in areas having high soil resistivity.

Natural electrodes could include metal underground water pipe electrodes and concrete-encased electrodes. The NEC lists a metal underground water pipe, available on the premises, in direct contact with the earth for 10-ft or more, and electrically continuous as part of the grounding electrode system. The NEC also lists a concrete-encased electrode (meeting the detailed specifications listed in the code) as part of the grounding electrode system. The application of concrete-encased electrodes is also covered in IEEE Std 142 [4].

A special-purpose electrode called the electrolytic electrode, consists of a ground rod installed in a canister containing an electrolytic salt. The salt absorbs moisture from the atmosphere through *breather* holes at the top of the unit. The salt combines and dissolves with the moisture and forms a constant electrolytic solution. Gravity and changes in atmospheric pressure cause the electrolytic solution to seep out through *weep* holes at the bottom of the unit. The electrolytic solution constantly seeps into the surrounding soil, reducing the soil resistivity, and helps keep the soil resistivity relatively constant. The result is a consistently low and constant earth electrode resistance. Made electrodes can also be backfilled with a material called bentonite to achieve a similar effect.

Specific requirements for the installation of earth electrodes are contained in the NEC [2]. Manufacturer's handling and installation instructions should also be consulted and followed.

22.4.8 Facilities Grounding for Computer and Electronics Equipment

Noise control is an important aspect of computer and electronic systems. The process of noise control through proper grounding techniques is more correctly called *referencing*. IEEE Std 1100 [6] covers the referencing and grounding of sensitive electronics equipment.

22.4.9 Isolated Grounding

Isolated grounding schemes, where the signal reference plane of an electronic system is isolated from equipment ground but connected to an *isolated* electrode in the earth, do not work, are unsafe, and violate the NEC. It was thought by some people that the isolated earth connection is *clean* since there is no connection between it and the *dirty* system ground connection at the service entrance. The clean, isolated earth connection was also viewed (incorrectly) as a point where noise currents can flow into the earth and be dissipated. Kirchoff's current law teaches that any current flowing into the isolated ground must return to the source through another earth connection. Current cannot be dissipated. It must always return to its source. Even lightning current is not dissipated into the earth. It must have a return path (i.e., the electrostatic and electromagnetic fields that created the charge build-up and the lightning strike in the first place). Fine print note (FPN) No. 2 at NEC 250–260 states that the connecting all the grounding electrodes will limit potential differences between them and between their associated wiring systems [2].

Isolated ground (IG) receptacles, which are a version of single-point grounding, are permitted by the NEC where required for the reduction of electrical noise (electromagnetic interference) on grounding circuits. Proper application of IG receptacles is very important, and the specific NEC requirements must be followed.

References

1. Webster's New Collegiate Dictionary.
2. NFPA. 2002. *The National Electrical Code*, Std 70. National Fire Protection Association, Inc., Quincy, MA.
3. IEEE. *Definitions of Electrical and Electronic Terms*, Std 100. The Institute of Electrical and Electronics Engineers, Inc., New York, current edition.
4. IEEE. *Recommended Practice for Grounding Industrial and Commercial Power Systems*, Std 142. Institute of Electrical and Electronics Engineers, Inc., New York, current edition.
5. Beeman, D. R. 1955. *Industrial Power Systems Handbook*, McGraw-Hill, New York.
6. IEEE. *Recommended Practice for Powering and Grounding Sensitive Electronics Equipment*, Std 1100. Institute of Electrical and Electronics Engineers, Inc., New York, current edition.
7. Sunde, E.D. 1949. *Earth Conduction Effects in Transmission Systems*, Van Nonstrand Co., NY. 1949.
8. Schwarz, S.J. 1954. Analytical Expression for Resistance of Grounding Systems, AIEE Trans. Vol. 73 (Part III-B), 1011–1016.
9. Lyncole 1987. XIT ground rod product literature. Lyncole Corp. Torrance, CA.
10. *System Grounding Bulletin*, EESG II-AP-3-1972. General Electric Co.
11. Lazar, I. *System Grounding in Industrial Power Systems*, Specifying Engineer, May 1978–Jan 1979.
12. Kaufmann, R.H. *Some Fundamentals of Equipment Grounding Circuit Design*, GEB No. 957, Industrial Power Systems Applications, General Electric Co.

23

Engineering Management, Standardization, and Regulation

Gene DeSantis
DeSantis Associates

Francis Long
University of Denver

Fred Baumgartner
TCI Technology Ventures

Terrence M. Baun
Criterion Broadcast Services

Richard Rudman
KFWB Radio

Jerry C. Whitaker
Editor-in-Chief

William F. Ames
Georgia Institute of Technology

George Cain
Georgia Institute of Technology

23.1   Systems Engineering Concepts

Gene DeSantis

23.1.1   Introduction

Modern systems engineering emerged during World War II as, due to the degree of complexity in design, development, and deployment, weapons evolved into weapon systems. The complexities of the space program made a systems engineering approach to design and problem solving even more critical. Indeed, the Department of Defense and NASA are two of the staunchest practitioners. With the growth of digital systems, the need for systems engineering has gained increased attention. Today, most large engineering organizations utilize a systems engineering process. Much has been published about system engineering practices in the form of manuals, standards, specifications, and instruction. In 1969, MIL-STD-499 was published to help government and contractor personnel involved in support of defense acquisition programs. In 1974 this standard was updated to MIL-STD-499A, which specifies the application of system engineering principles to military development programs. The tools and techniques of this processes continue to evolve in order to do each job a little better, save time, and cut costs. This chapter will first describe systems theory in a general sense followed by its application in systems engineering and some practical examples and implementations of the process.

23.1.2   Systems Theory

Though there are other areas of application outside of the electronics industry, we will be concerned with systems theory as it applies to electrical systems engineering. Systems theory is applicable to engineering of control, information processing, and computing systems. These systems are made up of component elements that are interconnected and programmed to function together. Many systems theory principles are routinely applied in the aerospace, computer, telecommunications, transportation, and manufacturing industries.

For the purpose of this discussion a *system* is defined as a set of related elements that function together as a single entity.

Systems theory consists of a body of concepts and methods that guide the description, analysis, and design of complex entities.

Decomposition is an essential tool of systems theory. The systems approach attempts to apply an organized methodology to completing large complex projects by breaking them down into simpler and more manageable component elements. These elements are treated separately, analyzed separately, and designed separately. In the end, all of the components are recombined to build the whole.

Holism is an element of systems theory in that the end product is greater than the sum of its component elements. In systems theory, the modeling and analytical methods enable all essential effects and interactions within a system and those between a system and its surroundings to be taken into account. Errors resulting from the idealization and approximation involved in treating parts of a system in isolation, or reducing consideration to a single aspect, are thus avoided.

Another holistic aspect of system theory describes **emergent properties**. Properties that result from the interaction of system components, properties that are not those of the components themselves, are referred to as emergent properties.

Though dealing with concrete systems, **abstraction** is an important feature of systems models. Components are described in terms of their function rather than in terms of their form. Graphical models such as block diagrams, flow diagrams, timing diagrams, and the like are commonly used.

Mathematical models may also be employed. Systems theory shows that, when modeled in abstract formal language, apparently diverse kinds of systems show significant and useful **isomorphisms** of structure and function. Similar interconnection structures occur in different types of systems. Equations that describe the behavior of electrical, thermal, fluid, and mechanical systems are essentially identical in form.

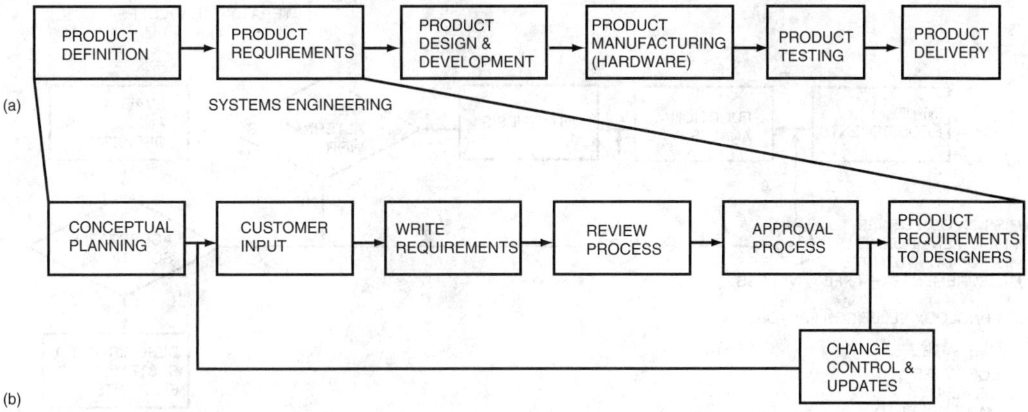

FIGURE 23.1   Systems engineering: (a) product development process, (b) requirements documentation process.

Isomorphism of structure and function implies isomorphism of behavior of a system. Different types of systems exhibit similar dynamic behavior such as response to stimulation.

The concept of **hard** and **soft systems** appears in system theory. In hard systems, the components and their interactions can be described by mathematical models. Soft systems cannot be described as easily. They are mostly human activity systems, which imply unpredictable behavior and nonuniformity. They introduce difficulties and uncertainties of conceptualization, description, and measurement. The kinds of system concepts and methodology described earlier cannot be applied.

23.1.3   Systems Engineering

Systems engineering depends on the use of a process methodology based on systems theory. To deal with the complexity of large projects, systems theory breaks down the process into logical steps.

Even though underlying requirements differ from program to program, there is a consistent, logical process, which can best be used to accomplish system design tasks. The basic product development process is illustrated in Fig. 23.1. The systems engineering starts at the beginning of this process to describe the product to be designed. It includes four activities:

- Functional analysis
- Synthesis
- Evaluation and decision
- Description of system elements

The process is iterative, as shown in Fig. 23.2. That is, with each successive pass, the product element description becomes more detailed. At each stage in the process a decision is made whether to accept,

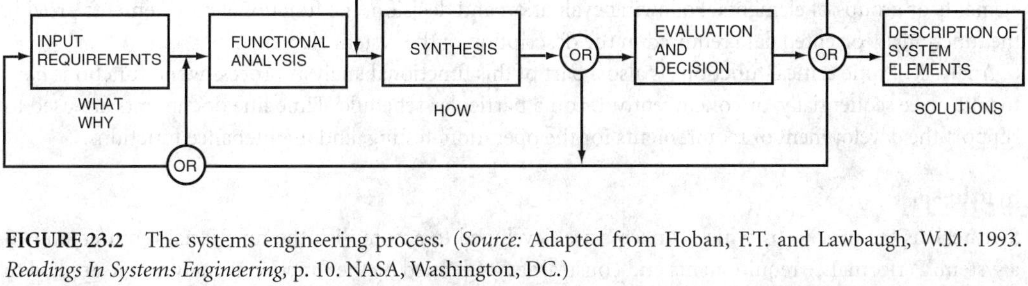

FIGURE 23.2   The systems engineering process. (*Source:* Adapted from Hoban, F.T. and Lawbaugh, W.M. 1993. *Readings In Systems Engineering*, p. 10. NASA, Washington, DC.)

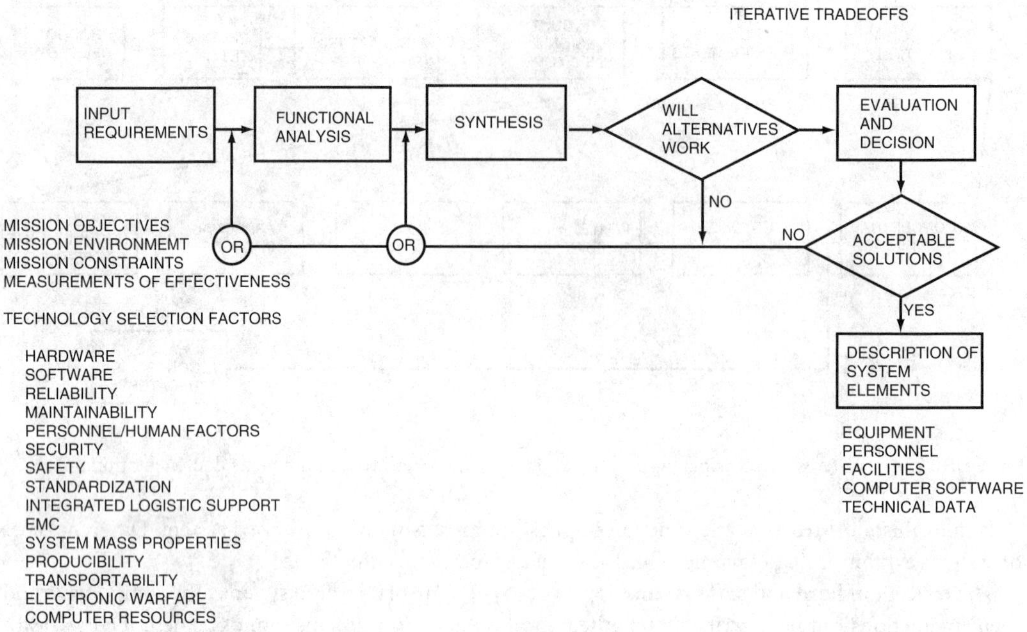

FIGURE 23.3　The systems engineering decision process. (*Source:* Adapted from Hoban, F.T. and Lawbaugh, W.M. 1993. *Readings In Systems Engineering,* p. 12. NASA, Washington, DC.)

make changes, or return to an earlier stage of the process and produce new documentation. The result of this activity is documentation that fully describes all system elements and that can be used to develop and produce the elements of the system. The systems engineering process does not produce the actual system itself.

Functional Analysis

A systematic approach to systems engineering will include elements of systems theory (see Fig. 23.3). To design a product, hardware and software engineers need to develop a vision of the product, the product requirements. These requirements are usually based on customer needs researched by a marketing department. An organized process to identify and validate customer needs will help minimize false starts. System objectives are first defined. This may take the form of a mission statement, which outlines the objectives, the constraints, the mission environment, and the means of measuring mission effectiveness. The purpose of the system is defined, and analysis is carried out to identify the requirements and what essential functions the system must perform and why. The *functional flow block diagram* is a basic tool used to identify functional needs. It shows logical sequences and relationships of operational and support functions at the system level. Other functions, such as maintenance, testing, logistics support, and productivity, may also be required in the functional analysis. The functional requirements will be used during the synthesis phase to show the allocation of the functional performance requirements to individual system elements or groups of elements. Following evaluation and decision, the functional requirements provide the functionally oriented data required in the description of the system elements.

Analysis of time critical functions is also a part of this functional analysis process when functions have to take place sequentially, or concurrently, or on a particular schedule. Time line documents are used to support the development of requirements for the operation, testing, and maintenance functions.

Synthesis

Synthesis is the process by which concepts are developed to accomplish the functional requirements of a system. Performance requirements and constraints, as defined by the functional analysis, are applied to each individual element of the system, and a design approach is proposed for meeting the requirements. Conceptual schematic arrangements of system elements are developed to meet system requirements. These documents can be used to develop a description of the system elements and can be used during the acquisition phase.

Modeling

The concept of **modeling** is the starting point of synthesis. Since we must be able to weigh the effects of different design decisions in order to make choices between alternative concepts, modeling requires the determination of those quantitative features that describe the operation of the system. We would, of course, like a very detailed model with as much detail as possible describing the system. Reality and time constraints, however, dictate that the simplest possible model be selected to improve our chances of design success. The model itself is always a compromise. The model is restricted to those aspects that are important in the evaluation of system operation. A model might start off as a simple block diagram with more detail being added as the need becomes apparent.

Dynamics

Most system problems are **dynamic** in nature. The signals change over time and the components determine the dynamic response of the system. The system behavior depends on the signals at a given instant, as well as on the rates of change of the signals and their past values. The term signals can be replaced by substituting human factors such as the number of users on a computer network, for example.

Optimization

The last concept of synthesis is **optimization**. Every design project involves making a series of compromises and choices based on relative weighting of the merit of important aspects. The best candidate among several alternatives is selected. Decisions are often subjective when it comes to deciding the importance of various features.

Evaluation and Decision

Program costs are determined by the tradeoffs between operational requirements and engineering design. Throughout the design and development phase, decisions must be made based on evaluation of alternatives and their effect on cost. One approach attempts to correlate the characteristics of alternative solutions to the requirements and constraints that make up the selection criteria for a particular element. The rationale for alternative choices in the decision process are documented for review. Mathematical models or computer simulations may be employed to aid in this evaluation decision making process.

Trade Studies

A structured approach is used in the trade study process to guide the selection of alternative configurations and ensure that a logical and unbiased choice is made. Throughout development, **trade studies** are carried out to determine the best configuration that will meet the requirements of the program. In the concept exploration and demonstration phases, trade studies help define the system configuration. Trade studies are used as a detailed design analysis tool for individual system elements in the full-scale development phase. During production, trade studies are used to select alternatives when it is determined that changes need to be made. Figure 23.4 illustrates the relationship of the various types of elements that may be employed in a trade study.

Figure 23.5 is a flow diagram of the trade study process. To provide a basis for the selection criteria, the objectives of the trade study must first be defined. Functional flow diagrams and system block diagrams are used to identify trade study areas that can satisfy certain requirements. Alternative approaches to achieving the defined objectives can then be established.

Complex approaches can be broken down into several simpler areas, and a decision tree constructed to show the relationship and dependences at each level of the selection process. This *trade tree*, as it is called, is illustrated in Fig. 23.6. Several trade study areas may be identified as possible candidates for accomplishing

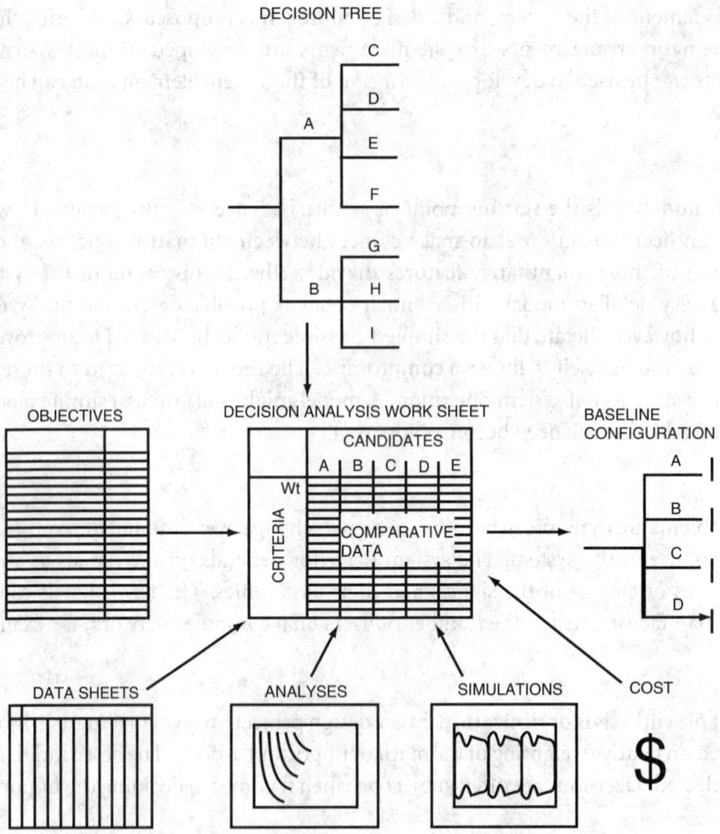

FIGURE 23.4 Trade studies using a systematic approach to decision making. (*Source:* Adapted from Defense Systems Management. 1983. *Systems Engineering Management Guide.* Contract No. MDA 903-82-C-0339. Defense Systems Management College, Fort Belvoir, VA.)

a given function. A trade tree is constructed to show relationships and the path through selected candidate trade areas at each level to arrive at a solution.

Several alternatives may be candidates for solutions in a given area. The selected candidates are then submitted to a systematic evaluation process intended to weed out unacceptable candidates. Criteria are determined, which are intended to reflect the desirable characteristics. Undesirable characteristics may also be included to aid in the evaluation process. Weights are assigned to each criteria to reflect its value or impact on the selection process. This process is subjective. It should also take into account cost, schedule, and hardware availability restraints that may limit the selection.

The criteria data on the candidates are then collected and tabulated on a decision analysis work sheet (Fig. 23.7). The attributes and limitations are listed in the first column and the data for each candidate listed in adjacent columns to the right. The performance data are available from vendor specification sheets or may require laboratory testing and analysis to determine. Each attribute is given a relative score from 1 to 10 based on its comparitive performance relative to the other candidates. Utility function graphs (Fig. 23.8) can be used to assign logical scores for each attribute. The utility curve represents the advantage rating for a particular value of an attribute. A graph is made of ratings on the y axis vs. attribute value on the x axis. Specific scores can then be applied, which correspond to particular performance values. The shape of the curve may take into account requirements, limitations, and any other factor that will influence its value regarding the particular criteria being evaluated. The limits to which the curves should be extended should run from the minimum value, below which no further benefit will accrue, to the maximum value, above which no further benefit will accrue.

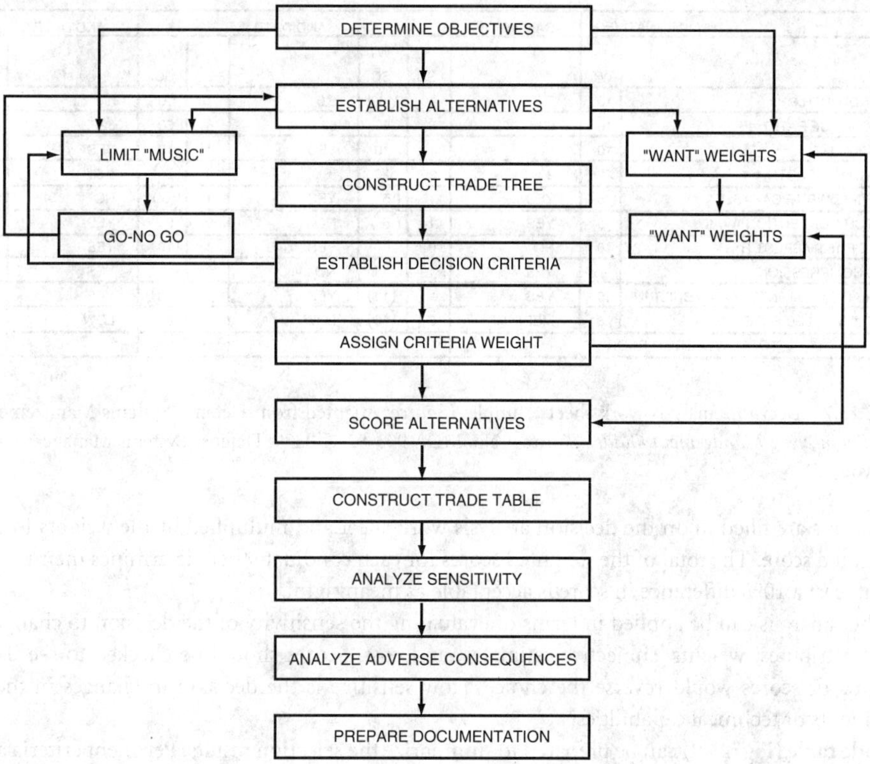

FIGURE 23.5 Trade study process flow chart. (*Source:* Adapted from Defense Systems Management. 1983. *Systems Engineering Management Guide.* Contract No. MDA 903-82-C-0339. Defense Systems Management College, Fort Belvoir, VA.)

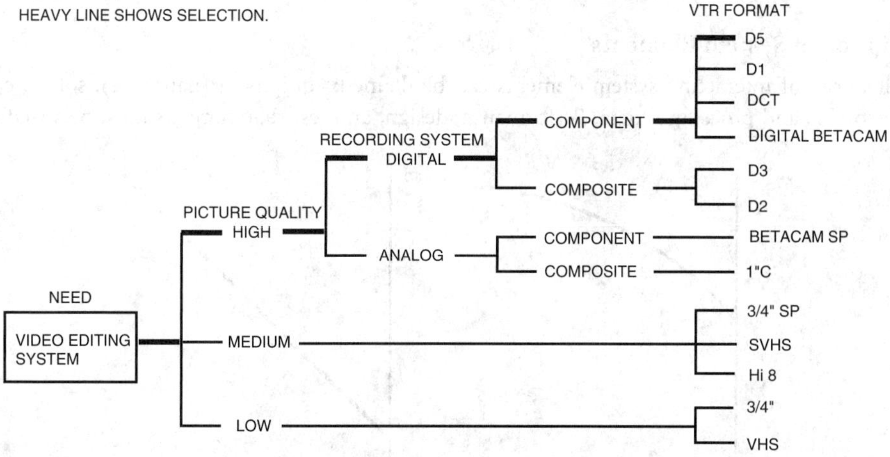

FIGURE 23.6 An example trade tree.

ALTERNATIVES:		CANDIDATE 1			CANDIDATE 2			CANDIDATE 3		
WANTED	WT		SC	WT SC		SC	WT SC		SC	WT SC
VIDEO BANDWIDTH MHz	10	5.6	10	(100)	6.0	10	(100)	5.0	9	90
SIGNAL-TO-NOISE RATIO, dB	10	60	8	80	54	6	60	62	10	(100)
10-bit QUANTIZING	10	YES	1	10	YES	1	10	YES	1	10
MAX PROGRAM LENGTH, h	10	2	2	20	3	3	(30)	1.5	1.5	15
READ BEFORE WRITE CAPABLE	5	YES	1	(5)	YES	1	(5)	NO	0	0
AUDIO PITCH CORRECTION AVAIL	5	YES	1	(5)	NO	0	0	YES	1	(5)
CAPABLE OF 16:9 ASPECT RATIO	10	NO	0	0	YES	1	(10)	YES	1	(10)
EMPLOYS COMPRESSION	−5	YES	1	−5	NO	0	(0)	YES	1	−5
SDIF (SERIAL DIGITAL INTERFACE) BUILT IN	10	YES	1	(10)	YES	1	(10)	YES	1	(10)
CURRENT INSTALLED BASE	8	MEDIUM	2	(16)	LOW	1	8	LOW	1	8
TOTAL WEIGHTED SCORE:				241			234			(243)

FIGURE 23.7 Decision analysis work sheet example. (*Source:* Adapted from Defense Systems Management. 1983. *Systems Engineering Management Guide.* Contract No. MDA 903-82-C-0339. Defense Systems Management College, Fort Belvoir, VA.)

The scores are filled in on the decision analysis work sheet and multiplied by the weights to calculate the weighted score. The total of the weighted scores for each candidate then determines their ranking. As a rule, at least a 10% difference in score is acceptable as meaningful.

Further analysis can be applied in terms of evaluating the sensitivity of the decision to changes in the value of attributes, weights, subjective estimates, and cost. Scores should be checked to see if changes in weights or scores would reverse the choice. How sensitive is the decision to changes in the system requirements or technical capabilities?

A **trade table** (Fig. 23.9) can be prepared to summarize the selection results. Pertinent criteria are listed for each alternative solution. The alternatives may be described in a quantitative manner, such as high, medium, or low.

Finally, the results of the trade study are documented in the form of a report, which discusses the reasons for the selections and may include the trade tree and the trade table.

There has to be a formal system of change control throughout the systems engineering process to prevent changes from being made without proper review and approval by all concerned parties and to keep all parties informed. Change control also ensures that all documentation is kept up to date and can help to eliminate redundant documents. Finally, change control helps to control project costs.

Description of System Elements

Five categories of interacting system elements can be defined: equipment (hardware), software, facilities, personnel, and procedural data. Performance, design, and test requirements must be specified and

FIGURE 23.8 Attribute utility trade curve example.

CRITERIA	COOL ROOM ONLY. ONLY NORMAL CONVECTION COOLING WITHIN ENCLOSURES	FORCED COLD AIR VENTILATION THROUGH RACK THEN DIRECTLY INTO RETURN	FORCED COLD AIR VENTILATION THROUGH RACK, EXHAUSTED INTO THE ROOM, THEN RETURNED THRU THE NORMAL PLENUM
COST	LOWEST. CONVENTIONAL CENTRAL AIR CONDITIONING SYSTEM USED.	HIGH. DEDICATED DUCTING REQUIRED. SEPARATE SYSTEM REQUIRED TO COOL ROOM	MODERATE. DEDICATED DUCTING REQUIRED FOR INPUT AIR.
PERFORMANCE EQUIPMENT TEMPERATURE ROOM TEMPERATURE	POOR $80-120^\circ$ F+ $65-70^\circ$ F TYPICAL AS SET	VERY GOOD $55-70^\circ$ F TYPICAL $65-70^\circ$ F TYPICAL AS SET	VERY GOOD $55-70^\circ$ F TYPICAL $65-70^\circ$ F TYPICAL AS SETF
CONTROL CONTROL OF EQUIPMENT TEMPERATURE	POOR. HOT SPOTS WILL OCCUR WITHIN ENCLOSURES.	VERY GOOD.	VERY GOOD. WHEN THE THERMOSTAT IS SET TO PROVIDE A COMFORTABLE ROOM TEMPERATURE, THE ENCLOSURE WILL BE COOL INSIDE.
CONTROL OF ROOM TEMPERATURE	GOOD. HOT SPOTS MAY STILL EXIST NEAR POWER HUNGRY EQUIPMENT.	GOOD.	GOOD. IF THE ENCLOSURE EXHAUST AIR IS COMFORTABLE FOR OPERATORS, THE INTERNAL EQUIPMENT MUST BE COOL.
OPERATOR COMFORT	GOOD	GOOD. SEPARATE ROOM VENTILATION SYSTEM REQUIRED CAN BE SET FOR COMFORT.	GOOD. WHEN THE THERMOSTAT IS SET TO PROVIDE A COMFORTABLE ROOM TEMPERATURE, THE ENCLOSURE WILL BE COOL INSIDE.

FIGURE 23.9   Trade table example.

documented for equipment, components, and computer software elements of the system. It may be necessary to specify environmental and interface design requirements, which are necessary for proper functioning of system elements within a facility.

The documentation produced by the systems engineering process controls the evolutionary development of the system. Figure 23.10 illustrates the special purpose documentation used by one organization in each step of the systems engineering process.

The requirements are formalized in written specifications. In any organization, there should be clear standards for producing specifications. This can help reduce the variability of technical content and improve product quality as a result. It is also important to make the distinction here that the product should not be overspecified to the point of describing the design or making it too costly. On the other hand, requirements should not be to general or so vague that the product would fail to meet the customer needs. In large departmentalized organizations, commitment to schedules can help assure that other members of the organization can coordinate their time.

The system engineering process does not actually design the system. The system engineering process produces the documentation necessary to define, design, develop, and test the system. The technical integrity provided by this documentation ensures that the design requirements for the system elements reflect the functional performance requirements, that all functional performance requirements are satisfied by the combined system elements, and that such requirements are optimized with respect to system performance requirements and constraints.

23.1.4   Phases of a Typical System Design Project[1]

The television industry has always been a dynamic industry because of the rapid advancement of communications technology. The design of complex modern video facility can be used to illustrate the systems engineering approach.

---

[1]Portions of this section were adapted from: Whitaker, J.C., DeSantis, G., and Paulson, C.R. 1992. *Interconnecting Electronic Systems*, Chap. 1. CRC Press, Boca Raton, FL.

The table in the figure:

	FUNCTIONAL FLOW BLOCK DIAGRAMS (FFBD) IDENTIFY AND SEQUENCE FUNCTIONS THAT MUST BE ACCOMPLISHED TO ACHIEVE SYSTEM OR PROJECT OBEJECTIVES. DEVELOP THE BASICS FOR ESTABLISHING INTER-SYSTEM FUNCTIONAL INTERFACES AND IDENTIFY SYSTEM RELATIONSHIPS.	REQUIREMENTS ALLOCATION SHEETS (RAS) DEFINE THE REQUIREMENTS AND CONSTRAINTS FOR EACH OF THE FUNCTIONS AND RELATE EACH REQUIREMENT TO THE SYSTEM ELEMENTS OF: • EQUIPMENT • FACILITIES • PERSONNEL • PROCEDURAL DATA • COMPUTER SOFTWARE TIME LINE SHEETS (TLS) PRESENT CRITICAL FUNCTIONS AGAINST A TIME BASE IN THE REQUIRED SEQUENCE OF ACCOMPLISHMENT	CONCEPT DESCRIPTION SHEETS (CDS) CONSTRAIN THE DESIGN TO STOP AT A POINT IN THE CYCLE AND CREATE AT THE GROSS LEVEL A DESIGN OR SYNTHESIS MEETING THE FFBD, RAS, TLS, REQUIREMENTS AND CONSTRAINTS. SCHEMATIC BLOCK DIAGRAMS (SBD) DEVELOP AND PORTRAY SCHEMATIC ARRANGE-MENT OF SYSTEM ELEMENTS TO SATISFY SYSTEM REQUIREMENTS.	TRADE STUDY REPORT (TSR) SELECT, EVALUATE, AND OPTIMIZE PROMISING OR ATTRACTIVE CONCEPTS. DOCUMENT THE TRADEOFF AND SUPPORTING RATIONALE. CONSIDER ALL POSSIBLE SOLUTIONS WITHIN THE FRAMEWORK OF REQUIREMENTS.	DESIGN SHEET (DS) DEFINE, DESCRIBE, AND DESIGN, AND TEST CRITERIA FOR THE SYSTEM ELEMENTS • EQUIPMENT • FACILITIES • PERSONNEL • PROCEDURAL DATA • COMPUTER SOFTWARE
BASIC DOCUMENTS					
SPECIAL DOCUMENTS	END ITEM MAINTENANCE SHEET(EIMS), TEST REQT. SHEET (TRS), PRODUCTION SHEETS (PS), LOGISTIC SUPPORT ANALYSIS RECORD (LSAR) IDENTIFY MAINTENANCE, TEST AND PRODUCTION FUNCTIONS ON A SPECIFIC END ITEM, SUBASSEMBLY, AND COMPONENT BASIS.				

FIGURE 23.10 Basic and special purpose documentation for system engineering. (*Source:* Adapted from Hoban, F.T. and Lawbaugh, W.M. 1993. *Readings In Systems Engineering*, p. 10. NASA Science and Technical Information Program, Washington, DC.)

Design Development

System design is carried out in a series of steps that lead to an operational unit. Appropriate research and preliminary design work is completed in the first phase of the project, the design development phase. It is the intent of this phase to fully delineate all requirements of the project and to identify any constraints. Based on initial concepts and information, the design requirements are modified until all concerned parties are satisfied and approval is given for the final design work to proceed. The first objective of this phase is to answer the following questions:

- What are the functional requirements of the product of this work?
- What are the physical requirements of the product of this work?
- What are the performance requirements of the product of this work?
- Are there any constraints limiting design decisions?
- Will existing equipment be used?
- Is the existing equipment acceptable?
- Will this be a new facility or a renovation?
- Will this be a retrofit or upgrade to an existing system?
- Will this be a stand alone system?

Working closely with the customer's representatives, the equipment and functional requirements of each of the major technical areas of the facility are identified. In the case of facility renovation, the systems engineer's first order of business is to analyze existing equipment. A visit is made to the site to gather detailed information about the existing facility. Usually confronted with a mixture of acceptable and unacceptable equipment, the systems engineer must sort out those pieces of equipment that meet current standards and determine which items should be replaced. Then, after soliciting input from the facility's technical personnel, the systems engineer develops a list of needed equipment.

One of the systems engineer's most important contributions is the ability to identify and meet the needs of the customer and do it within the project budget. Based on the customer's initial concepts and any subsequent equipment utilization research conducted by the systems engineer, the desired capabilities are identified as precisely as possible. Design parameters and objectives are defined and reviewed. Functional efficiency is maximized to allow operation by a minimum number of personnel. Future needs are also investigated at this time. Future technical systems expansion is considered.

After the customer approves the equipment list, preliminary system plans are drawn up for review and further development. If architectural drawings of the facility are available, they can be used as a starting point for laying out an equipment floor plan. The systems engineer uses this floor plan to be certain that adequate space is provided for present and future equipment, as well as adequate clearance for maintenance and convenient operation. Equipment identification is then added to the architect's drawings.

Documentation should include, but not be limited to, the generation of a list of major equipment including:

- Equipment prices
- Technical system functional block diagrams
- Custom item descriptions
- Rack and console elevations
- Equipment floor plans

The preliminary drawings and other supporting documents are prepared to record design decisions and to illustrate the design concepts to the customer. Renderings, scale models, or full-size mock ups may also be needed to better illustrate, clarify, or test design ideas.

Ideas and concepts have to be exchanged and understood by all concerned parties. Good communication skills are essential. The bulk of the creative work is carried out in the design development phase. The physical layout—the look and feel—and the functionality of the facility will all have been decided and agreed upon by the completion of this phase of the project. If the design concepts appear feasible, and the cost is within the anticipated budget, management can authorize work to proceed on the final detailed design.

Electronic System Design

Performance standards and specifications have to be established up front in a technical facility project. This will determine the performance level of equipment that will be acceptable for use in the system and effect the size of the budget. Signal quality, stability, reliability, and accuracy are examples of the kinds of parameters that have to be specified. Access and processor speeds are important parameters when dealing with computer-driven products. The systems engineer has to confirm whether selected equipment conforms to the standards.

At this point it must be determined what functions each component in the system will be required to fulfill and how each will function together with other components in the system. The management and operation staff usually know what they would like the system to do and how they can best accomplish it. They have probably selected equipment that they think will do the job. With a familiarity of the capabilities of different equipment, the systems engineer should be able to contribute to this function-definition stage of the process. Questions that need to be answered include

- What functions must be available to the operators?
- What functions are secondary and therefore not necessary?
- What level of automation should be required to perform a function?
- How accessible should the controls be?

Overengineering or overdesign must be avoided. This serious and costly mistake can be made by engineers and company staff when planning technical system requirements. A staff member may, for example, ask for a seemingly simple feature or capability without fully understanding its complexity or the additional cost burden it may impose on a project. Other portions of the system may have to be compromised to implement the additional feature. An experienced systems engineer will be able to spot this and determine if the tradeoffs and added engineering time and cost is really justified.

When existing equipment is going to be used, it will be necessary to make an inventory list. This list will be the starting point for developing a final equipment list. Usually, confronted with a mixture of acceptable and unacceptable equipment, the systems engineer must sort out what meets current standards and what should be replaced. Then, after soliciting input from facility technical personnel, the systems engineer develops a summary of equipment needs, including future acquisitions. One of the systems engineer's most important contributions is the ability to identify and meet these needs within the facility budget.

A list of major equipment is prepared. The systems engineer selects the equipment based on experience with the products, and on customer preferences. Often some existing equipment may be reused. A number of considerations are discussed with the facility customer to arrive at the best product selection. Some of the major points include

- Budget restrictions
- Space limitations
- Performance requirements
- Ease of operation
- Flexibility of use
- Functions and features
- Past performance history
- Manufacturer support

The goal of the systems engineer is the design of equipment to meet the functional requirements of a project efficiently and economically. Simplified block diagrams for the video, audio, control, data, and communication systems are drawn. They are discussed with the customer and presented for approval.

Detailed Design

With the research and preliminary design development completed, the details of the design must now be concluded. The design engineer prepares complete detailed documentation and specifications necessary for the fabrication and installation of the technical systems, including all major and minor components. Drawings must show the final configuration and the relationship of each component to other elements of the system, as well as how they will interface with other building services, such as air conditioning and electrical power. This documentation must communicate the design requirements to the other design professionals, including the construction and installation contractors.

In this phase, the systems engineer develops final, detailed flow diagrams and schematics that show the interconnection of all equipment. Cable interconnection information for each type of signal is taken from the flow diagrams and recorded on the cable schedule. Cable paths are measured and timing calculations performed. Timed cable lengths (used for video and other special services) are entered onto the cable schedule.

The flow diagram is a schematic drawing used to show the interconnections between all equipment that will be installed. It is different from a block diagram in that it contains much more detail. Every wire and cable must be included on the drawings. A typical flow diagram for a video production facility is shown in Fig. 23.11.

The starting point for preparing a flow diagram can vary depending on the information available from the design development phase of the project and on the similarity of the project to previous projects. If a similar system has been designed in the past, the diagrams from that project can be modified to include the equipment and functionality required for the new system. New models of the equipment can be shown

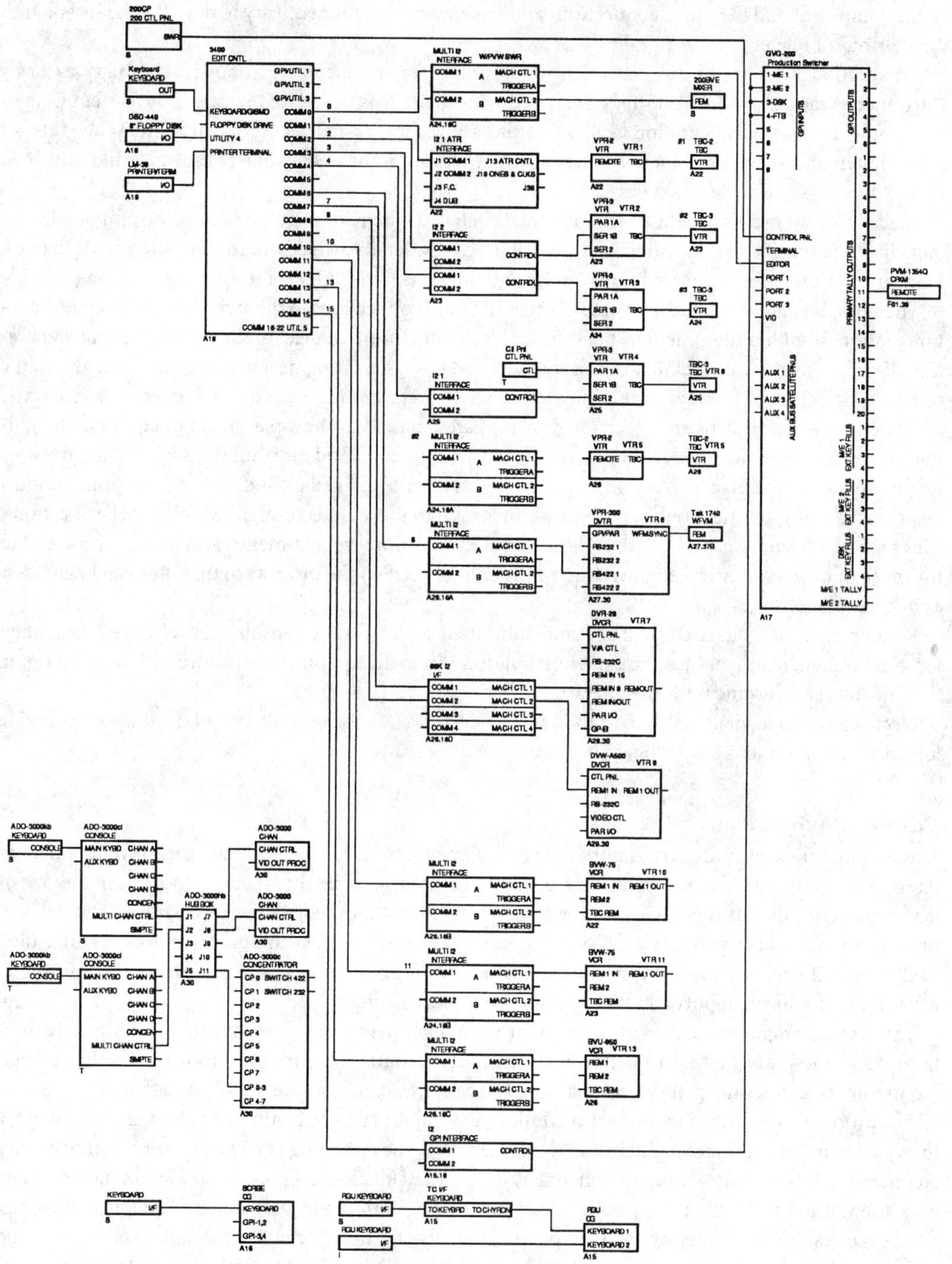

FIGURE 23.11 Video system control flow diagram.

in place of their counterparts on the diagram, and minor wiring changes can be made to reflect the new equipment connections and changes in functional requirements. This method is efficient and easy to complete.

If the facility requirements do not fit any previously completed design, the block diagram and equipment list are used as a starting point. Essentially, the block diagram is expanded and details added to show all of the equipment and their interconnections and to show any details necessary to describe the installation and wiring completely.

An additional design feature that might be desirable for specific applications is the ability to easily disconnect a rack assembly from the system and relocate it. This would be the case if the system were to be prebuilt at a systems integration facility and later moved and installed at the client's site. When this is a requirement, the interconnecting cable harnessing scheme must be well planned in advance and identified on the drawings and cable schedules.

Special custom items are defined and designed. Detailed schematics and assembly diagrams are drawn. Parts lists and specifications are finalized, and all necessary details worked out for these items. Mechanical fabrication drawings are prepared for consoles and other custom-built cabinetry.

The systems engineer provides layouts of cable runs and connections to the architect. Such detailed documentation simplifies equipment installation and facilitates future changes in the system. During preparation of final construction documents, the architect and the systems engineer can firm up the layout of the technical equipment wire ways, including access to flooring, conduits, trenches, and overhead wire ways.

Dimensioned floor plans and elevation drawings are required to show placement of equipment, lighting, electrical cable ways, duct, conduit, and heating, ventilation, and air conditioning (HVAC) ducting. Requirements for special construction, electrical, lighting, HVAC, finishes, and acoustical treatments must be prepared and submitted to the architect for inclusion in the architectural drawings and specifications. This type of information, along with cooling and electrical power requirements, also must be provided to the mechanical and electrical engineering consultants (if used on the project) so that they can begin their design calculations.

Equipment heat loads are calculated and submitted to the HVAC consultant. Steps are taken when locating equipment to avoid any excessive heat buildup within the equipment enclosures, while maintaining a comfortable environment for the operators.

Electrical power loads are calculated and submitted to the electrical consultant and steps taken to provide for sufficient power and proper phase balance.

Customer Support

The systems engineer can assist in purchasing equipment and help to coordinate the move to a new or renovated facility. This can be critical if a great deal of existing equipment is being relocated. In the case of new equipment, the customer will find the systems engineer's knowledge of prices, features, and delivery times to be an invaluable asset. A good systems engineer will see to it that equipment arrives in ample time to allow for sufficient testing and installation. A good working relationship with equipment manufacturers helps guarantee their support and speedy response to the customer's needs.

The systems engineer can also provide engineering management support during planning, construction, installation, and testing to help qualify and select contractors, resolve problems, explain design requirements, and assure quality workmanship by the contractors and the technical staff.

The procedures described in this section outline an ideal scenario. In reality, management may often try to bypass many of the foregoing steps to save money. This, the reasoning goes, will eliminate unnecessary engineering costs and allow construction to start right away. Utilizing in-house personnel, a small company may attempt to handle the job without professional help. With inadequate design detail and planning, which can result when using unqualified people, the job of setting technical standards and making the system work then defaults to the construction contractors, in-house technical staff, or the installation contractor. This can result in costly and uncoordinated work-arounds and, of course, delays and added costs during construction, installation, and testing. It makes the project less manageable and less likely to be completed successfully.

The complexity of a project can be as simple as interconnecting a few pieces of computer equipment together to designing software for the Space Shuttle. The size of a technical facility can vary from a small one room operation to a large multimillion dollar plant or large area network. Where large amounts of money and other resources are going to be involved, management is well advised to recruit the services of qualified system engineers.

Budget Requirements Analysis

The need for a project may originate with customers, management, operations staff, technicians, or engineers. In any case, some sort of logical reasoning or a specific production requirement will justify the cost. On small projects, like the addition of a single piece of equipment, money only has to be available to make the purchase and cover installation costs. When the need may justify a large project, it is not always immediately apparent how much the project will cost to complete. The project has to be analyzed by dividing it up into its constituent elements. These elements include

- Equipment and parts
- Materials
- Resources (including money and time needed to complete the project)

An executive summary or capital project budget request containing a detailed breakdown of these elements can provide the information needed by management to determine the return on investment and to make an informed decision on whether or not to authorize the project.

A capital project budget request containing the minimum information might consist of the following items:

- Project name. Use a name that describes the result of the project, such as control room upgrade.
- Project number (if required). A large organization that does many projects will use a project numbering system of some kind or may use a budget code assigned by the accounting department.
- Project description. A brief description of what the project will accomplish, such as design the technical system upgrade for the renovation of production control room 2.
- Initiation date. The date the request will be submitted.
- Completion date. The date the project will be completed.
- Justification. The reason the project is needed.
- Material cost breakdown. A list of equipment, parts, and materials required for construction, fabrication, and installation of the equipment.
- Total material cost.
- Labor cost breakdown. A list of personnel required to complete the project, their hourly pay rates, the number of hours they will spend on the project, and the total cost for each.
- Total project cost. The sum of material and labor costs.
- Payment schedule. Estimation of individual amounts that will have to be paid out during the course of the project and the approximate dates each will be payable.
- Preparer's name and the date prepared.
- Approval signature(s) and date(s) approved.

More detailed analysis, such as return on investment, can be carried out by an engineer, but financial analysis should be left to the accountants who have access to company financial data.

Feasibility Study and Technology Assessment

Where it is required that an attempt be made to implement new technology and where a determination must be made as to whether certain equipment can perform a desired function, it will be necessary to conduct a feasibility study. The systems engineer may be called upon to assess the state-of-the-art to develop a new application. An executive summary or a more detailed report of evaluation test results may be required, in addition to a budget request, to help management make a decision.

Planning and Control of Scheduling and Resources

Several planning tools have been developed for planning and tracking progress toward the completion of projects and scheduling and controlling resources. The most common graphical project management tools are the Gantt chart and the *critical path method* (CPM) utilizing the *project evaluation and review*

(PERT) technique. Computerized versions of these tools have greatly enhanced the ability of management to control large projects.

Project Tracking and Control

A project team member may be called upon by the project manager to report the status of the work during the course of the project. A standardized project status report form can provide consistent and complete information to the project manager. The purpose is to supply information to the project manager regarding work completed and money spent on resources and materials.

A project status report containing the minimum information might contain the following items:

- Project number (if required)
- Date prepared
- Project name
- Project description
- Start date
- Completion date (the date this part of the project was completed)
- Total material cost
- Labor cost breakdown
- Preparer's name

Change Control

After part or all of a project design has been approved and money allocated to build it, any changes may increase or decrease the cost. Factors that effect the cost include

- Components and material
- Resources, such as labor and special tools or construction equipment
- Costs incurred because of manufacturing or construction delays

Management will want to know about such changes and will want to control them. For this reason, a method of reporting changes to management and soliciting approval to proceed with the change may have to be instituted. The best way to do this is with a *change order request* or *change order*. A change order includes a brief description of the change, the reason for the change and a summary of the effect it will have on costs and on the project schedule.

Management will exercise its authority and approve or disapprove each change based on its understanding of the cost and benefits and the perceived need for the modification of the original plan. Therefore, it is important that the systems engineer provide as much information and explanation as may be necessary to make the change clear and understandable to management.

A change order form containing the minimum information might contain the following items:

- Project number
- Date prepared
- Project name
- Labor cost breakdown
- Preparer's name
- Description of the change
- Reason for the change
- Equipment and materials to be added or deleted
- Material costs or savings
- Labor costs or savings
- Total cost of this change (increase or decrease)
- Impact on the schedule

Program Management

The Defense Systems Management College (Hoban, F.T. and Lawbaugh, W.M. 1993. *Readings in Systems Engineering Management*. NASA Science and Technical Information Program, Washington, D.C., p. 9.) favors the management approach and defines systems engineering as follows:

> Systems engineering is the management function which controls the total system development effort for the purpose of achieving an optimum balance of all system elements. It is a process which transforms an operational need into a description of system parameters and integrates those parameters to optimize the overall system effectiveness.

Systems engineering is both a technical process and a management process. Both processes must be applied throughout a program if it is to be successful. The persons who plan and carry out a project constitute the project team. The makeup of a project team will vary depending on the size of the company and the complexity of the project. It is up to the management to provide the necessary human resources to complete the project.

Executive Management

The executive manager is the person who can authorize that a project be undertaken. This person can allocate funds and delegate authority to others to accomplish the task. Motivation and commitment is toward the goals of the organization. The ultimate responsibility for a project's success is in the hands of the executive manager. This person's job is to get tasks completed through other people by assigning group responsibilities, coordinating activities between groups, and resolving group conflicts. The executive manager establishes policy, provides broad guidelines, approves the project master plan, resolves conflicts, and assures project compliance with commitments.

Executive management delegates the project management functions and assigns authority to qualified professionals, allocates a capital budget for the project, supports the project team, and establishes and maintains a healthy relationship with project team members.

Management has the responsibility to provide clear information and goals, up front, based on their needs and initial research. Before initiating a project, the company executive should be familiar with daily operation of the facility and analyze how the company works, how jobs are done by the staff, and what tools are needed to accomplish the work. Some points that may need to be considered by an executive before initiating a project include

- What is the current capital budget for equipment?
- Why does the staff currently use specific equipment?
- What function of the equipment is the weakest within the organization?
- What functions are needed but cannot be accomplished with current equipment?
- Is the staff satisfied with current hardware?
- Are there any reliability problems or functional weaknesses?
- What is the maintenance budget and is it expected to remain steady?
- How soon must the changes be implemented?
- What is expected from the project team?

Only after answering the appropriate questions will the executive manager be ready to bring in expert project management and engineering assistance. Unless the manager has made a systematic effort to evaluate all of the obvious points about the facility requirements, the not so obvious points may be overlooked. Overall requirements must be broken down into their component parts. Do not try to tackle ideas that have too many branches. Keep the planning as basic as possible. If the company executive does not make a concerted effort to investigate the needs and problems of a facility thoroughly before consulting experts, the expert advice will be shallow and incomplete, no matter how good the engineer.

Engineers work with the information they are given. They put together plans, recommendations, budgets, schedules, purchases, hardware, and installation specifications based on the information they receive from interviewing management and staff. If the management and staff have failed to go through the planning, reflection, and refinement cycle before those interviews, the company will likely waste time and money.

Project Manager

Project management is an outgrowth of the need to accomplish large complex projects in the shortest possible time, within the anticipated cost, and with the required performance and reliability. Project management is based on the realization that modern organizations may be so complex as to preclude effective management using traditional organizational structures and relationships. Project management can be applied to any undertaking that has a specific end objective.

The project manager must be a competent systems engineer, accountant, and manager. As systems engineer, there must be understanding of analysis, simulation, modeling, and reliability and testing techniques. There must be awareness of state-of-the-art technologies and their limitations. As accountant, there must be awareness of the financial implications of planned decisions and knowledge of how to control them. As manager, the planning and control of schedules is an important part of controlling the costs of a project and completing it on time. Also, as manager, there must be the skills necessary to communicate clearly and convincingly with subordinates and superiors to make them aware of problems and their solutions.

The project manager is the person who has the authority to carry out a project. This person has been given the legitimate right to direct the efforts of the project team members. The manager's power comes from the acceptance and respect accorded by superiors and subordinates. The project manager has the power to act and is committed to group goals.

The project manager is responsible for getting the project completed properly, on schedule, and within budget, by utilizing whatever resources are necessary to accomplish the goal in the most efficient manner. The manager provides project schedule, financial, and technical requirement direction and evaluates and reports on project performance. This requires planning, organizing, staffing, directing, and controlling all aspects of the project.

In this leadership role, the project manager is required to perform many tasks including the following:

- Assemble the project organization.
- Develop the project plan.
- Publish the project plan.
- Set measurable and attainable project objectives.
- Set attainable performance standards.
- Determine which scheduling tools (PERT, CPM, and/ or Gantt) are right for the project.
- Using the scheduling tools, develop and coordinate the project plan, which includes the budget, resources, and the project schedule.
- Develop the project schedule.
- Develop the project budget.
- Manage the budget.
- Recruit personnel for the project.
- Select subcontractors.
- Assign work, responsibility, and authority so that team members can make maximum use of their abilities.
- Estimate, allocate, coordinate, and control project resources.
- Deal with specifications and resource needs that are unrealistic.
- Decide on the right level of administrative and computer support.
- Train project members on how to fulfill their duties and responsibilities.

- Supervise project members, giving them day-to-day instructions, guidance, and discipline as required to fulfill their duties and responsibilities.
- Design and implement reporting and briefing information systems or documents that respond to project needs.
- Control the project.

Some basic project management practices can improve the chances for success. Consider the following:

- Secure the necessary commitments from top management to make the project a success.
- Set up an action plan that will be easily adopted by management.
- Use a work breakdown structure that is comprehensive and easy to use.
- Establish accounting practices that help, not hinder, successful completion of the project.
- Prepare project team job descriptions properly up front to eliminate conflict later on.
- Select project team members appropriately the first time.

After the project is under way, follow these steps:

- Manage the project, but make the oversight reasonable and predictable.
- Get team members to accept and participate in the plans.
- Motivate project team members for best performance.
- Coordinate activities so they are carried out in relation to their importance with a minimum of conflict.
- Monitor and minimize interdepartmental conflicts.
- Get the most out of project meetings without wasting the team's productive time. Develop an agenda for each meeting and start on time. Conduct one piece of business at a time. Assign responsibilities where appropriate. Agree on follow up and accountability dates. Indicate the next step for the group. Set the time and place for the next meeting. End on time.
- Spot problems and take corrective action before it is too late.
- Discover the strengths and weaknesses in project team members and manage them to get desired results.
- Help team members solve their own problems.
- Exchange information with subordinates, associates, superiors, and others about plans, progress, and problems.
- Make the best of available resources.
- Measure project performance.
- Determine, through formal and informal reports, the degree to which progress is being made.
- Determine causes of and possible ways to act upon significant deviations from planned performance.
- Take action to correct an unfavorable trend or to take advantage of an unusually favorable trend.
- Look for areas where improvements can be made.
- Develop more effective and economical methods of managing.
- Remain flexible.
- Avoid activity traps.
- Practice effective time management.

When dealing with subordinates, each person must

- Know what is to be done, preferably in terms of an end product.
- Have a clear understanding of the authority and its limits for each individual.
- Know what the relationship with other people is.

- Know what constitutes a job well done in terms of specific results.
- Know when and what is being done exceptionally well.
- Be shown concrete evidence that there are just rewards for work well done and for work exceptionally well done.
- Know where and when expectations are not being met.
- Be made aware of what can and should be done to correct unsatisfactory results.
- Feel that the supervisor has an interest in each person as an individual.
- Feel that the supervisor both believes in each person and is anxious for individual success and progress.

By fostering a good relationship with associates, the manager will have less difficulty communicating with them. The fastest, most effective communication takes place among people with common points of view.

The competent project manager watches what is going on in great detail and can, therefore, perceive problems long before they flow through the paper system. Personal contact is faster than filing out forms. A project manager who spends most of the time in the management office instead of roaming through the places where the work is being done is headed for catastrophe.

Systems Engineer

The term *systems engineer* means different things to different people. The systems engineer is distinguished from the engineering specialist, who is concerned with only one aspect of a well-defined engineering discipline, in that the systems engineer must be able to adapt to the requirements of almost any type of system. The systems engineer provides the employer with a wealth of experience gained from many successful approaches to technical problems developed through hands-on exposure to a variety of situations. This person is a professional with knowledge and experience, possessing skills in a specialized and learned field or fields. The systems engineer is an expert in these fields; highly trained in analyzing problems and developing solutions that satisfy management objectives. The systems engineer takes data from the overall development process and, in return, provides data in the form of requirements and analysis results to the process.

Education in electronics theory is a prerequisite for designing systems that employ electronic components. As a graduate engineer, the systems engineer has the education required to design electronic systems correctly. Mathematics skill acquired in engineering school is one of the tools used by the systems engineer to formulate solutions to design problems and analyze test results. Knowledge of testing techniques and theory enables this individual to specify system components and performance and to measure the results. Drafting and writing skills are required for efficient preparation of the necessary documentation needed to communicate the design to technicians and contractors who will have to build and install the system.

A competent systems engineer has a wealth of technical information that can be used to speed up the design process and help in making cost effective decisions. If necessary information is not at hand, the systems engineer knows where to find it. The experienced systems engineer is familiar with proper fabrication, construction, installation, and wiring techniques and can spot and correct improper work.

Training in personnel relations, a part of the engineering curriculum, helps the systems engineer communicate and negotiate professionally with subordinates and management.

Small in-house projects can be completed on an informal basis and, indeed, this is probably the normal routine where the projects are simple and uncomplicated. In a large project, however, the systems engineer's involvement usually begins with preliminary planning and continues through fabrication, implementation, and testing. The degree to which program objectives are achieved is an important measure of the systems engineer's contribution.

During the design process the systems engineer:

- Concentrates on results and focuses work according to the management objectives.
- Receives input from management and staff.

- Researches the project and develops a workable design.
- Assures balanced influence of all required design specialties.
- Conducts design reviews.
- Performs tradeoff analyses.
- Assists in verifying system performance.
- Resolves technical problems related to the design, interface between system components, and integration of the system into any facility.

Aside from designing a system, the systems engineer has to answer any questions and resolve problems that may arise during fabrication and installation of the hardware. Quality and workmanship of the installation must be monitored. The hardware and software will have to be tested and calibrated upon completion. This, too, is the concern of the systems engineer. During the production or fabrication phase, systems engineering is concerned with

- Verifying system capability
- Verifying system performance
- Maintaining the system baseline
- Forming an analytical framework for producibility analysis

Depending on the complexity of the new installation, the systems engineer may have to provide orientation and operating instruction to the users. During the operational support phase, system engineers

- Receive input from users
- Evaluate proposed changes to the system
- Establish their effectiveness
- Facilitates the effective incorporation of changes, modifications and updates

Depending on the size of the project and the management organization, the systems engineer's duties will vary. In some cases the systems engineer may have to assume the responsibilities of planning and managing smaller projects.

Other Project Team Members

Other key members of the project team where building construction may be involved include the following:

- Architect, responsible design of any structure.
- Electrical engineer, responsible for power system design if not handled by the systems engineer.
- Mechanical engineer, responsible for HVAC, plumbing, and related designs.
- Structural engineer, responsible for concrete and steel structures.
- Construction contractors, responsible for executing the plans developed by the architect, mechanical engineer, and structural engineer.
- Other outside contractors, responsible for certain specialized custom items which can not be developed or fabricated internally or by any of the other contractors.

23.1.5   Conclusion

Systems theory is the theoretical basis of systems engineering that is an organized approach to the design of complex systems. The key components of systems theory applied in systems engineering are a holistic approach, the decomposition of problems, the exploitation of analogies, and the use of models.

A formalized technical project management technique used to define systems includes three major steps.

- Define the requirements in terms of functions to be performed and measurable and testable requirements describing how well each function must be performed.

- Synthesize a way of fulfilling the requirements.
- Study the tradeoffs and select one solution from among the possible alternative solutions.

In the final analysis, beyond any systematic approach to systems engineering, engineers have to engineer. Robert A. Frosch, a former NASA Administrator, in a speech to a group of engineers in New York, urged a common sense approach to systems engineering (Frosch, R.A. 1969. A Classic Look at System Engineering and Management. In *Readings In Systems Engineering.* ed. F.T. Hoban and W.M. Lawbaugh, pp. 1–7. NASA Science and Technical Information Program, Washington, D.C.):

> Systems, even very large systems, are not developed by the tools of systems engineering, but only by the engineers using the tools. . . . I can best describe the spirit of what I have in mind by thinking of a music student who writes a concerto by consulting a check list of the characteristics of the concerto form, being careful to see that all of the canons of the form are observed, but having no flair for the subject, as opposed to someone who just knows roughly what a concerto is like, but has a real feeling for music. The results become obvious upon hearing them. The prescription of technique cannot be a substitute for talent and capability. . . .

Defining Terms

Abstraction: Though dealing with concrete systems, abstraction is an important feature of systems models. Components are described in terms of their function rather than in terms of their form. Graphical models such as block diagrams, flow diagrams, timing diagrams, and the like are commonly used. Mathematical models may also be employed. Systems theory shows that, when modeled in abstract formal language, apparently diverse kinds of systems show significant and useful isomorphisms of structure and function. Similar interconnection structures occur in different types of systems. Equations that describe the behavior of electrical, thermal, fluid, and mechanical systems are essentially identical in form.

Decomposition: The systems approach attempts to apply an organized methodology to completing large complex projects by breaking them down into simpler more manageable component elements. These elements are treated separately, analyzed separately, and designed separately. In the end, all of the components are recombined to build the whole.

Dynamics: Most system problems are dynamic in nature. The signals change over time and the components determine the dynamic response of the system. The system behavior depends on the signals at a given instant, as well as on the rates of change of the signals and their past values. The term signals can be replaced by substituting human factors such as the number of users on a computer network, for example.

Emergent properties: A holistic aspect of system theory describes emergent properties. Properties which result from the interaction of system components, properties which are not those of the components themselves, are referred to as emergent properties.

Hard and soft systems: In hard systems, the components and their interactions can be described by mathematical models. Soft systems can not be described as easily. They are mostly human activity systems, which imply unpredictable behavior and non-uniformity. They introduce difficulties and uncertainties of conceptualization, description, and measurement. Usual system concepts and methodology can not be applied.

Holism: Holism is an element of systems theory in that the end product is greater than the sum of its component elements. In systems theory, the modeling and analytical methods enable all essential effects and interactions within a system and those between a system and its surroundings to be taken into account. Errors resulting from the idealization and approximation involved in treating parts of a system in isolation, or reducing consideration to a single aspect are thus avoided.

Isomorphism: Similarity in elements of different kinds. Similarity of structure and function in elements implies isomorphism of behavior of a system. Different types of systems exhibit similar dynamic behavior such as response to stimulation.

Modeling: The concept of modeling is the starting point of synthesis of a system. Since we must be able to weigh the effects of different design decisions in order to make choices between alternative concepts, modeling requires the determination of those quantitative features that describe the operation of the system. We would, of course, like a very detailed model with as much detail as possible describing the system. Reality and time constraints, however, dictate that the simplest possible model be selected to improve our chances of design success. The model itself is always a compromise. The model is restricted to those aspects that are important in the evaluation of system operation. A model might start off as a simple block diagram with more detail being added as the need becomes apparent.

Optimization: Making an element as perfect, effective, or functional as possible. Every design project involves making a series of compromises and choices based on relative weighting of the merit of important aspects. The best candidate among several alternatives is selected.

Synthesis: This is the process by which concepts are developed to accomplish the functional requirements of a system. Performance requirements and constraints, as defined by the functional analysis, are applied to each individual element of the system, and a design approach is proposed for meeting the requirements. Conceptual schematic arrangements of system elements are developed to meet system requirements. These documents are used to develop a description of the system elements.

Trade Studies: A structured approach to guide the selection of alternative configurations and ensure that a logical and unbiased choice is made. Throughout development, trade studies are carried out to determine the best configuration that will meet the requirements of the program. In the concept exploration and demonstration phases, trade studies help define the system configuration. Trade studies are used as a detailed design analysis tool for individual system elements in the full-scale development phase. During production, trade studies are used to select alternatives when it is determined that changes need to be made. Alternative approaches to achieving the defined objectives can thereby be established.

Trade Table: A trade table is used to summarize selection results of a trade study. Pertinent criteria are listed for each alternative solution. The alternatives may be described in a qualitative manner, such as high, medium, or low.

References

Defense Systems Management. 1983. *Systems Engineering Management Guide.* Defense Systems Management College, Fort Belvoir, VA.

Delatore, J.P., Prell, E.M., and Vora, M.K. 1989. Translating customer needs into product specifications. *Quality Progress* 22(1), Jan.

Finkelstein, L. 1988. Systems theory. *IEE Proceedings*, Pt. A, 135(6), pp. 401–403.

Hoban, F.T. and Lawbaugh, W.M. 1993. *Readings in Systems Engineering.* NASA Science and Technical Information Program, Washington, D.C.

Shinners, S.M. 1976. *A Guide to Systems Engineering and Management.* Lexington Books, Lexington, MA.

Tuxal, J.G. 1972. *Introductory System Engineering.* McGraw-Hill, New York.

Further Information

Additional information on systems engineering concepts is available from the following sources:

MIL-STD-499A is available by writing to The National Council on Systems Engineering (NCOSE) at 333 Cobalt Way, Suite 107, Sunnyvale, CA 94086. NCOSE has set up a working group that specifically deals with commercial systems engineering practices.

Blanchard, B.S. and Fabrycky, W.J. 1990. *Systems Engineering and Analysis.* Prentice-Hall, Englewood Cliffs, NJ. An overview of systems engineering concepts, design methodology, and analytical tools commonly used.

Skytte, K. 1994. Engineering a small system. *IEEE Spectrum* 31(3) describes how systems engineering, once the preserve of large government projects, can benefit commercial products as well.

23.2 Concurrent Engineering

Francis Long

23.2.1 Introduction

Concurrent engineering (CE) is a method used to shorten the time to market for new or improved products. Let it be assumed that a product will, upon reaching the marketplace, be competitive in nearly every respect, such as quality and cost, for example. But the marketplace has shown that products, even though competitive, must not be late to market, because market share, and therefore profitability, will be adversely affected. Concurrent engineering is the technique that, today, is most likely to result in acceptable profits for a given product.

A number of proactive companies began, in the late 1970s and early 1980s, to use what were then innovative techniques to improve their competitive position. But it was not until 1986 that a formal definition of concurrent engineering was published by the Defense Department:

> A systematic approach to the integrated, concurrent design of products and their related processes, including manufacture, and support. This approach is intended to cause the developers, from the outset, to consider all elements of the product life cycle from concept through diposal including quality, cost, schedule, and user requirements.

This definition was printed in the Institute for Defense Analyses Report R-338, 1986. The key words are seen to be *integrated, concurrent design* and *all elements of the product life cycle*. Implicit in this definition is the concept that in addition to input from the originators of the concept, input should come from users of the product, from those who install and maintain the product, from those who manufacture and test the product, as well as from the designers of the product. Such input, as appropriate, should be in every phase of the product life cycle, even the very earliest design work.

This approach is implemented by bringing specialists from manufacturing, test, procurement, field service, etc., into the earliest design considerations. It is very different from the process so long used by industry. The earlier process, now known as the *over the wall* process, was a serial or sequential process. The product concept, formulated at a high level of company management, was then turned over to a design group. The design group completed its design effort, tossed it over the wall to manufacturing, and proceeded to an entirely new and different product design. Manufacturing tossed its product to test, and so on through the chain. The unfortunate result of this sequential process was the necessity for redesign, which happened with great regularity.

Traditional designers too frequently have limited knowledge of a manufacturing process, especially its capabilities and limitations. This may lead to a design that cannot be made economically or made in the time scheduled, or perhaps cannot be made at all. The same can be said of the specialists in the processes of test, marketing, and field service, as well as parts and material procurement. A problem in any of these areas might well require that the design be returned to the design group for redesign. The same result might come from a product that cannot be repaired. The outcome is a redesign effort required to correct the deficiencies found during later processes in the product cycle. Such redesign effort is costly in both economic and time to market terms. Another way to view these redesign efforts is that they are not *value added*. Value added is a particularly useful parameter by which to evaluate a process or practice.

This presence of redesign in the serial process can be illustrated as in Fig. 23.12, showing that even feedback from field service might be needed in a redesign effort. When the process is illustrated in this manner, the presence of redesign can be seen to be less efficient than a process in which little or no redesign is required. A common projection of the added cost of redesign is that changes made in a following process step are about 10 times more costly than correctly designing the product in the first place. If the product should be in the hands of a customer when a failure occurs, the results can be disastrous, both in direct costs to accomplish the repair and in possible lost sales due to a tarnished reputation.

There are two major facets of concurrent engineering that must be kept in mind at all times. The first is that a concurrent engineering process requires team effort. This is more than the customary committee.

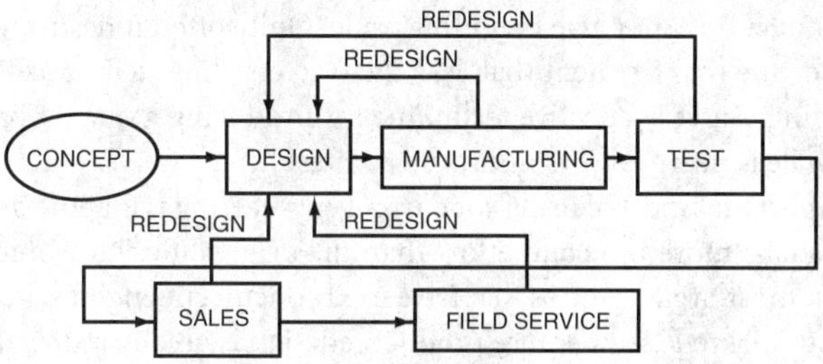

FIGURE 23.12   The serial design process.

Although the team is composed of specialists from the various activities, the team members are not there as representatives of their organizational home. They are there to cooperate in the delivery of product to the market place by contributing their expertise in the task of eliminating the redesign loops shown in Fig. 23.12. Formation of the proper team is critical to the success of most CE endeavors. The second facet is to keep in mind that concurrent engineering is information and communication intensive. There must be no barriers of any kind to complete and rapid communication among all parts of a process, even if they are located at geographically dispersed sites. If top management has access to and uses information relevant to the product or process, this same information must be available to all in the production chain, including the line workers.

An informed and knowledgeable workforce at all levels is essential, so that they may use their efforts to the greatest advantage. The most effective method to accomplish this is to form, as early as possible in the product life cycle, a team composed of knowledgeable people from all aspects of the product life cycle. This team should be able to *anticipate and design out* most if not all possible problems before they actually occur. Figure 23.13 suggests many of the communication pathways that must be freely available to the members of the team. Others will surface as the project progresses. The inputs to the design process are sometimes called the "design for . . . ," inserting the requirement.

The top management that assigns the team members must also be the *coaches* for the team, making certain that the team members are properly trained and then allowing the team to proceed with the project. There is no place here for the traditional *bossism* of the past. The team members must be selected to have the best combination of recognized expertise in their respective fields and the best team skills. It is not always the one most expert in a specialty who will be the best team member. Team members, however, must have the respect for all other team members, not only for their expertise, but also for their interpersonal skills. Only then will the best products result. This concurrency of design, to include these and all other parts of the cycle, can measurably reduce time to market and overall investment in the product.

Preventing downstream problems has another benefit in that employee morale is very likely to be enhanced. People almost universally want to take pride in what they do and produce. Few people want to support or work hard in a process or system that they believe results in a poor product. Producing a

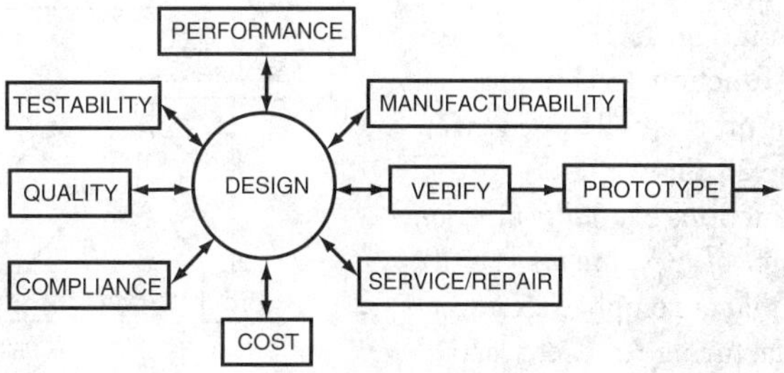

FIGURE 23.13   Concurrency of design is communication intensive.

defective or shoddy product does not give them this pride. Quite often it destroys pride in workmanship and creates a disdain for the management that asks them to employ such processes or systems. The use of concurrent engineering is a very effective technique for producing a quality product in a competitive manner. Employee morale is nearly always improved as a result.

Perhaps the most important aspect of using concurrent engineering is that the design phase of a product cycle will nearly always take more time and effort than the original design would have expended in the serial process. However, most organizations that have used concurrent engineering report that the overall time to market is measurably reduced because product redesign is greatly reduced or eliminated entirely. The time-worn phrase *time is money* takes on added meaning in this context.

Concurrent engineering can be thought of as an evolution rather than a revolution of the product cycle. As such, the principles of **total quality management (TQM)** and *continuous quality improvement* (CQI), involving the ideas of robust design and reduction of variation, are not to be ignored. They continue to be important in process and product improvement. Nor is concurrent engineering, of itself, a type of re-engineering. The concept of re-engineering in today's business generally implies a massive restructuring of the organization of a process, or even a company, probably because the rate of improvement of a process using traditional TQM and CQI is deemed to be too slow or too inefficient or both to remain competitive. Still, the implementation of concurrent engineering does require and demand a certain and often substantial change in the way a company does business.

Concurrent engineering is as much a cultural change as it is a process change. For this reason it is usually achieved with some trauma. The extent of the trauma is dependent on the willingness of people to accept change, which in turn is dependent on the commitment and sales skills of those responsible for installing the concurrent engineering culture. Although it is not usually necessary to re-engineer, that is, to restructure, an entire organization to install concurrent engineering, it is also true that it cannot be installed like an overlay on top of most existing structures. Although some structural changes may be necessary, the most important change is in attitude, in culture. Yet it must also be emphasized that there is no *one size fits all* pattern. Each organization must study itself to determine how best to install concurrent engineering. However, there are some considerations that are helpful in this study. A discussion of many fine ideas can be found in Salomone (1995). The importance of commitment to a concurrent engineering culture from top management to line workers cannot be emphasized too strongly.

23.2.2   The Process View of Production

If production is viewed as a process, the product cycle becomes a seamless movement through the design, manufacture, test, sales, installation, and field maintenance activities. There is no competition within the organization for resources. The team has been charged with the entire product cycle such that allocation of resources is seen from a holistic view rather than from a departmental or specialty view. The needs of each activity are evident to all team members. The product and process are seen as more than the sum of the parts.

The usual divisions of the process cycle can be viewed in a different way. Rather than discuss the obvious activities of manufacturability and testability and the others shown in Fig. 23.13, the process can be viewed in terms of functional techniques that are used to accomplish the process cycle. Such a view might be as shown in Fig. 23.14.

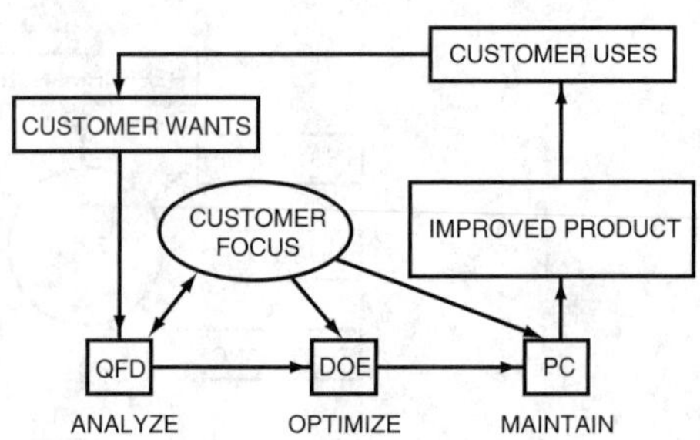

In this view, it is the functions of *quality function deployment* (QFD), *design of experiments* (DOE), and *process control* (PC) that are emphasized rather than the design, manufacturing, test, etc. activities. It is the manner in which the processes of design, manufacturing, and other elements are

FIGURE 23.14   Functional view of the process cycle.

accomplished that is described. In this description, QFD is equated to analysis in the sense that the customers' needs and desires must be the driver in the design of today's products. Through the use of QFD, not only is the customer input, often referred to as the *voice of the customer*, heard, it is translated into a process to produce the product. Thus, both initial product and process design are included in QFD in this view. It is important to note that the product and the process to produce the product are designed together, not just at the same time.

DOE can be used in one of two ways. First is the optimization of an existing process by removing any causes of defects and determining the best target value of the parameters. The purpose of this is to maximize the yield of a process, which frequently involves continuous quality improvement techniques. The second is the determination of a process for a new product by optimization of a proposed process before it is implemented. Today, simulation of processes is becoming increasingly important as the processes become increasingly complex. DOE, combined with simulation, is the problem solving technique, both for running processes and proposed processes.

PC is a monitoring process to insure that the optimized process remains an optimized process. Its primary purpose is to issue an alarm signal when a process is moving away from its optimized state. Often, this makes use of statistical methods and is then called *statistical process control* (SPC). PC is not a problem solving technique, although some have tried to use it for that purpose. When PC signals a problem, then problem solving techniques, possibly involving DOE, must be implemented. The following sections will expand on each of these functional aspects of a product cycle.

Quality Function Deployment

QFD begins with a determination of the customers' needs and desires. There are many ways that raw data can be gathered. Two of these are questionnaires and focus groups. Obtaining the data is a well-developed field. The details of these techniques will not be discussed here as much has been written on the subject. It is important, however, that professionals are involved in the design of such data acquisition because of the many nuances that are inherent in such methods.

The data obtained must be translated into language that is understood by the company and its people. It is this translation that must extract the customers' needs and wants and put them in words that the designers, manufacturers, etc., can use in their tasks. Yet, the intent of the customers' words must not be lost. This is not always an easy task, but is a vitally important one. Another facet of this is the determination of unstated but pleasing qualities of a product that might provide a marketing edge. This idea has been sketched by the Kano model, shown in Fig. 23.15.

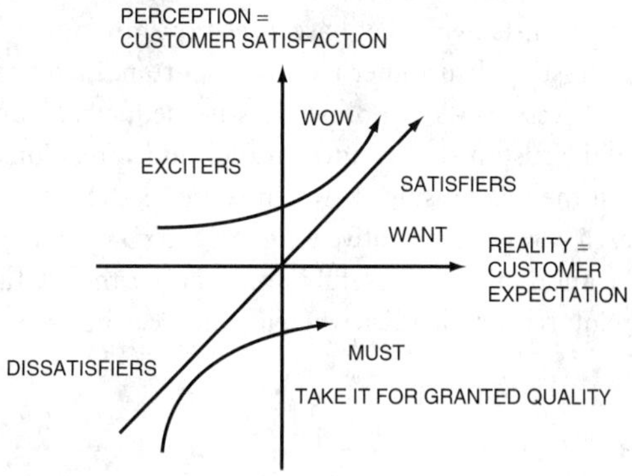

The Kano model, developed by Noriaki Kano, a Japanese professor, describes these pleasers as the *wows*. The model also shows that some characteristics of a product are not even mentioned by customers or potential customers, yet they must be present or the product will be deemed unsatisfactory. The

FIGURE 23.15 The Kano model.

wows are not even thought of by the customers but give the product a competitive advantage. An often cited example is the net in the trunk of the Ford Taurus that can be used to secure loose cargo such as grocery bags.

Translating the customers' responses into useable items is usually accomplished by application of the *house of quality*. The house of quality is a matrix, or perhaps more accurately an augmented matrix. Two important attributes of the house of quality are (1) its capability for ranking the various inputs in terms of perceived importance and (2) the data in the completed house that shows much of the decision making that went into the translation of customers' wants and need into useable task descriptions. The latter attribute is often called the archival characteristic and is especially useful when product up grades are designed or when new products of a similar nature are designed.

Constructing the house of quality matrix begins by identifying each horizontal row of the main or correlation matrix with a customer input, called the *customer attribute* (CA). These CAs are entered on the left side of the matrix row. Each vertical column of the matrix is assigned an activity, called the *engineering characteristic* (EC). The ECs are entered at the top of the columns. Together, the ECs are believed by the team to be able to produce the CAs. Because a certain amount of judgment is required in determining the ECs, such TQM techniques as brainstorming are often used by the team at this time. The team must now decide the relative importance of the ECs in realizing the CAs. Again, the techniques of TQM are used.

The relative importance is indicated by assigning the box that is the intersection of an EC with a CA a numerical value, using an agreed upon rating scale, such as blank for no importance to 5 or 10 as very important. Each box in the matrix is then assigned a number. Note that an EC may affect more than one CA and that one CA may require more than one EC to be realized. An example of a main matrix with ranking information in the boxes and a summary at the bottom is shown in Fig. 23.16. In this example the rankings of the ECs are assigned only three relative values rather than a full range of 0–9. This is frequently done to reduce the uncertainty and lost time as a result of trying to decide, for example, between a 5 or a 6 for the comparisons. Also, weighting the three levels unequally will give emphasis to the more important relationships.

ECs

CAs

	A	B	C	D	E	F	G	H	I	J	
1	9			3	3		1				
2		3	1				9				
3	3	3		1	1		3		9	1	
4				1				1			
5	1	1				3			1		
6			3				1			3	
7					9		3		1		
8	9				3			3			
9		3		1				1		3	
10	3				9						
TOTAL	25	10	20	14	7	11	7	4	11	7	116
%	21	9	17	12	6	10	6	3	10	6	100

FIGURE 23.16 The house of quality main matrix.

Following completion of the main matrix, the augmentation portions are added. The first is usually the planning matrix that is added to the right side of the main matrix. Each new column added by the planning matrix lists items that have a relationship to one or more of the CAs but are not ECs. Such items might be: assumed customer relative importance, current company status, estimated competitor's status, sales positives (wows), improvement needed, etc. Figure 23.17 shows a planning matrix with relative weights of the customer attributes for each added relationship. The range of weights for each column is arbitrary, but the relationships between columns should be such that a row total can be assigned to each CA and each row given a relative weight when compared to other CA rows.

Another item of useful information is the interaction of the ECs because some of these can be positive, reinforcing each other, whereas others can be negative: improving one can lower the effect of another. Again,

CAs

	IMPORTANCE	OUR STATUS	IMPROVEMENT NEEDED	COMPETITIVE ADVANTAGE	OVERALL WEIGHT	%	COMPETITOR
Att 1	5	3	1.7	1.0	5.1	21	4
Att 2	3	3	1.0	1.5	4.5	19	3
Att 3	5	4	1.3	1.0	5.2	22	4
Att 4	2	3	1.0	1.0	3.0	13	2
Att 5	1	4	1.0	1.5	6.0	25	3

23.8 100

FIGURE 23.17 CAs and the planning matrix.

the house of quality can be augmented to indicate these interactions and their relative importance. This is accomplished by adding a roof. Such a roof is shown in Fig. 23.18. This is very important information to the design effort, helping to guide the designers as to where special effort might be needed in the optimization of the product.

It is very likely that the ECs used in the house of quality will need to be translated into other requirements. A useful way to do this is to use the ECs from this first house as the inputs to a second house, whose output might be cost or parts to accomplish the ECs. It is not unusual to have a sequence of several houses of quality, as shown in Fig. 23.19.

The final output of a QFD utilization should be a product description and a first pass at a process description to produce the product. The product description should be traceable to the original cus-

	A	B	C	D	E	F	G	H	I	J	
1	9			3	3		1				
2		3	1			9					
3	3	3		1	1		3		9	1	
4				1				1			
5	1	1				3			1		
6		3				1			3		
7			9				3		1		
8	9				3		3				
9		3		1			1		3		
10	3		9								
TOTAL	25	10	20	14	7	11	7	4	11	7	116
%	21	9	17	12	6	10	6	3	10	6	100

FIGURE 23.18   The house of quality with a roof.

tomer inputs so that this product will be competitive in those terms. The process description should be one that will produce the product in a competitive time and cost framework. It is important to note that the QFD process, to be complete, requires input from all parts of a product cycle.

This brief look at QFD hopefully indicates the power that this tool has. Initial use of this tool will most likely be more expensive than currently used methods because of the familiarization that must take place. It should not be used for small tasks, those with fewer than 10 input statements. It should probably not be used if the number of inputs exceeds 50 because the complexity makes the relative weightings difficult to manage with a satisfactory degree of confidence. Those who have learned to use QFD do find it efficient and valuable as well as cost effective.

Design of Experiments

DOE is an organized procedure for identifying those parts of a process that are causing less than satisfactory product and then optimizing the process. The process might be one already in production or it might be one that is proposed for a new product. A complex process cannot be studied effectively by using the simple technique of varying one parameter while holding all others fixed. Such a procedure, though historically taught in most science and engineering classes, totally ignores the possibility of interactions between parameters, a condition that very often prevails in the real world. However, if all interactions as well as primary parameters are to be tested, the number of experiments rapidly becomes large enough to be out of the question for only a few variables. DOE has been developed to help reduce the number of experiments required to uncover a problem parameter. As DOE developed, it became clear that some method of

FIGURE 23.19   A series of houses of quality.

determining the relative importance of some relationships of parameters needed to be found. Thus, DOE became based in statistics and became the task of statisticians trained in advanced methods of statistics.

Factorial analysis, the study of all parameters and all their interactions, is not time efficient for hand calculations involving more than four parameters. From this, the fractional factorial method of analysis developed, where only the primary parameters and a selected set of their interactions are studied. How to select the interactions, which to study, and which to omit, is the problem. Use of brainstorming and other like techniques from TQM to select the interactions can help but do not remove the nagging questions about those that are left out. Were they left out because of ignorance of them by the team, or by a member of the team dominating the selection because of preconceived, perhaps, erroneous ideas?

An attempt to simplify the selection of parameters for a fractional factorial study was introduced by Taguchi in a technique called *orthogonal arrays* (1993). Much has been written about this technique. A few references are given at the end of this chapter. Use of the technique is not simple and requires more than a passing acquaintance with the method. A major shortcoming of the technique is that there are no guidelines for selecting among the possible interactions, much the same as that for the original fractional factorial method. The response of some to this criticism is that most problems are due to one of the main parameters anyway, leaving the nagging question of the interaction effects that may be important but were excluded.

Because of the large volume written about orthogonal arrays and the difficulties outlined previously, this chapter will discuss a lesser known but simpler system introduced by Dorian Shainen (Bhote, 1991). Only simple hand calculations are required. The techniques are based on valid statistical grounds, but their application require very little knowledge of statistics. To begin, only a knowledge of mean, median, and standard deviation is required. It is important to realize that because of the limited knowledge of statistics required to follow the rules to be described, a professional statistician should be available to answer questions that may arise when application is made to systems slightly different from the simple ones described in the subsequent discussion.

The general procedure to be followed is that of first reducing a large number of possible causes of problems to a few, typically four or fewer if possible, and then use the full factorial method of analysis to identify the most likely culprit or culprits. The principle is that most real life problems are the result of four or fewer primary variables plus their interactions, a situation that can be handled quite nicely by full factorial analysis. Once the culprits have been identified, the next step is to optimize their values so that the process is producing the very best product that it can.

Seven different procedures make up this system. These are shown in Fig. 23.20. In this structure, there are three ways to attack a many variable, more than 20, problem. The basic concept is first to eliminate those

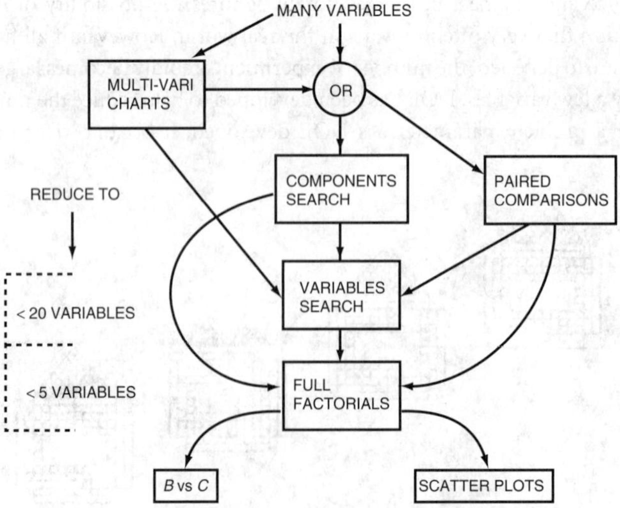

FIGURE 23.20   The procedures of Shainen.

variables that can be shown not to be a cause because they are of the wrong type. For example, *multivari charts* are useful in defining what type, or family, the variation is, eliminating those that are not in this family. *Multivari charts* may be used with one of the other two shown as first level procedures. However, the other two are mutually exclusive. *Components search* requires that the product can be disassembled and reassembled. *Paired comparisons* is the method to use if the product cannot be disassembled but must be studied as a unit.

Colorful language was introduced by Shainen to help users remember the methods. The purpose of the procedures is to find the *red X* or the *pink Xs* or the *pale pink Xs*. The red X is the one primary cause with all other possible causes of much lesser importance. If one cause cannot be found, then the two or more that must be considered are called the pink Xs, or the pale pink Xs if they are of lesser importance.

Multivari Chart

The multivari chart is used to classify the family into which the red X or pink Xs fall. A parameter that is indicative of the problem, and can be measured, is chosen for study. Sets of samples are then taken and the variation noted. The categories used to distinguish the parameter output variation are: (1) variation within sample sets (cyclical variation) is larger than variation within samples or variation over time, (2) time variation (temporal variation) between sample sets is larger than variation within sample sets or variation of the samples, and (3) variations within samples (positional variation) are larger than variation of sample sets over time or variation within the sample sets. These are demonstrated in Fig. 23.21.

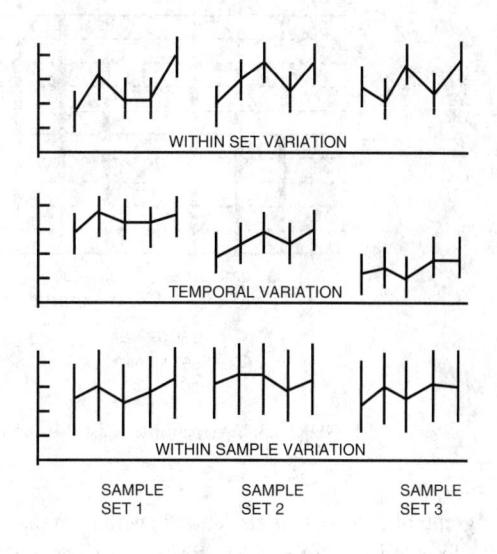

To illustrate, assume a process has been producing defective product at a known historical rate, that is, at an average rate of X ppm, for the past weeks or months. Begin the study by collecting, *consecutively*, three to five products from the process. At a later

FIGURE 23.21 Multivari types of variation.

time, after a number of units have been produced in the interim, collect three to five products again. Repeat this again and again, as often as necessary; three to five times is frequently sufficient to capture at least 80% of the historical defect rate in the samples, that is, these samples should include defects at least at 80% of the historical rate, X, that the process has produced defects in historical samples. This is an important rule to observe, to provide statistical validity to the samples collected.

In the language of statistics, this is a stratified experiment, that is, the samples are grouped according to some criteria. In this case, the criteria is consecutive production of the units. This is not, therefore, a random selection of samples as is required in many statistical methods. It also is not a control chart even though the plot may resemble one. It is a *snapshot* of the process taken at the time of the sampling. Incidentally, the multivari chart is not a new procedure, dating from the 1950s, but it has been incorporated into this system by Shainen.

The purpose of the multivari chart is to discover the *family* of the red X, although on rare occasions the red X itself may become evident. The anticipated result of a successful multivari experiment is a set, the family, of possible causes that includes the red X, the pink Xs, or the pale pink Xs. The family will normally include possible causes numbering from a few up to about 20. Further experiments will be necessary to determine the red X or the pink Xs from this set.

The following example displays in Fig. 23.22 the variations of four 98-Ω nominal valued resistors screen printed on a single ceramic substrate. They will later be separated from each other by scribing. Data recorded over a two shift period, from units from more than 2800 substrate printings, are shown in the chart.

(a)

PRINT	200	201	202		900	901	902		1800	1801	1802		2800	2801	2802
	9:00 A.M.				1:30 P.M.				6:00 P.M.				11:00 P.M.		
R1	92	90	91		98	100	92		97	98	96		102	95	98
R2	94	93	92		96	93	95		102	100	100		98	101	102
R3	95	98	97		92	95	98		95	96	98		96	102	101
R4	93	96	95		94	98	102		102	98	96		99	98	100
GRP MEAN	93.5	94.3	93.8		95.0	96.5	96.8		99.0	98.0	97.5		98.8	99.0	100.3
SET MEAN	93.8				96.1				98.2				99.2		

(b)

● GROUP MEAN
○ SET MEAN

FIGURE 23.22   Example resistor value variation: (a) table of data, (b) graph of data.

A graph of these data is also shown, with the range of data for each substrate indicated and the average of the four resistors for each substrate, as well as the overall average for the three substrates sampled at that time.

Note:

1. Unit-to-unit variation within a group is the largest.
2. Time variation is also serious as the time averages increase sharply.

With this information about the family of the cause(s) of defects, additional experiments to find the cause(s) of the variation can be designed according to the methodology to be described later.

Components Search

Components search is used only when a product can be disassembled and then reassembled. It is a parts swapping procedure that is familiar to many who have done field repair. The first step is to select a performance parameter by which good and bad units can be identified. A good unit is chosen at random, measured, then disassembled and reassembled two times, measuring the performance parameter each time. This establishes a range of variability of the performance parameter that is related to the assembly operation for good units. Repeat this for a randomly selected bad unit, once again establishing the range of variability of the performance parameter for assembly of bad units. The good unit must remain a good unit after disassembly and reassembly, just as the bad unit must remain a bad unit after disassembly and reassembly. If this is not the case, then the parameter chosen as performance indicator needs to be reviewed.

Because there are only three data points for each type of unit, the statistics of small samples is useful. The first requirement here is that the three performance parameter measurements for the good unit must all yield values that are more acceptable than the three for the bad unit. If this is so, then there is only

1 chance in 20 that this ranking of measurements could happen by accident, giving a 95% confidence in this comparison. The second requirement is that the minimum separation between the medians of variability of the good unit and the bad unit exceeds a minimum. This is illustrated in Fig. 23.23, showing the three data points for the good and bad units. The value of 1.25 for the ratio D/d is based on the classical F Table at the 0.05 level. This means that the results of further tests conducted by swapping parts have at least a 95% level of confidence in the results.

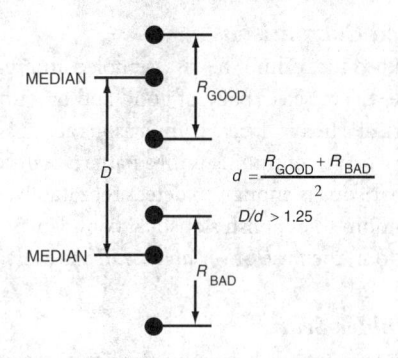

$$d = \frac{R_{GOOD} + R_{BAD}}{2}$$

$$D/d > 1.25$$

FIGURE 23.23 Test of acceptability of data for component search.

The next step is to identify the parts to be swapped and begin doing so, keeping a chart of the results, such as that in Fig. 23.24. In this plot, the left three data points are those of the original good and bad units, plus their disassembly and reassembly two times. The remaining data points represent those measurements for swapping one part for parts labeled *A*, etc.

Three results are possible: (1) no change, indicating the part is not at fault, (2) change in one of the units outside its limits but the other unit remains within its limits; or (3) the units flip-flop, the good unit becoming a bad unit and vice versa. A complete flip-flop indicates a part that is seriously at fault—call it a red X. Parts with measurements that are outside the limits but do not cause a complete reversal of good and bad are deemed pink Xs, worthy of further study. A pink X is a partial cause, so that one or more additional pink Xs should be found. Finally, if pink Xs are found, they should be bundled together, that is, all the parts with a pink X result should be swapped as a block between units. This is called a *capping run* and is illustrated as well. A capping run should result in a complete reversal, as shown, indicating that there are no other causes. Less than a full reversal indicates that other, not identified, causes exist or the performance measure is not the best that could have been chosen.

If a single cause, a red X has been found, then remedial action can be initiated. If two to four pink Xs are found, a full factorial analysis, to be described later, should be done to determine the relative importance of the revealed causes and their interactions. Once the importance has been determined, then allocation of resources can be guided by the relative importance of the causes.

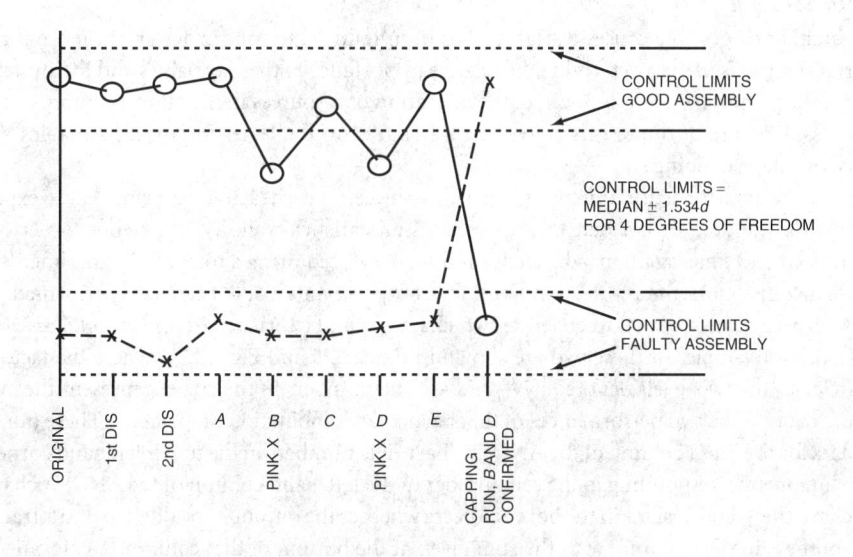

FIGURE 23.24 Components search.

Paired Comparisons

If the product cannot be disassembled and reassembled, the technique to use is paired comparisons. The concept is to select pairs of good and bad units and compare them, using whatever visual, mechanical, electrical, chemical, etc., comparisons are possible, recording whatever differences are noticed. Do this for several pairs, continuing until a pattern of differences becomes evident. In many cases, a half-dozen paired comparisons is enough to detect repeatable differences. The units chosen for this test should be selected at random to establish statistical confidence in the results. If the number of differences detected is more than four, then use of *variables search* is indicated. For four or fewer, a full factorial analysis can be done.

Variables Search

Variables search is best applied when there are 5 or more variables with a practical limit of about 20. It is a binary process. It begins by determining a performance parameter and defining a *best* and a *worst* result. Then a ranking of the variables as possible causes is done, followed by assigning for each variable two levels, call them best and worst or good and bad, or some other distinguishing pair. For all variables simultaneously at the best level, the expected result is the best for the performance parameter chosen, similarly for the worst levels. Run two experiments, one with all variables at their best levels and one with all variables at their worst levels. Do this two more times, randomizing the order of best and worst combinations. Use this set of data in the same manner as that for components search using the same requirements and the same limits formula.

If the results meet the best and worst performance, proceed to the next step. If the results do not meet these requirements, interchange the best and worst levels of one parameter at a time until the requirements are met or until all pair reversals are used. If the requirements are still not met, an important factor has been left out of the original set and additional factors must be added until all important requirements are met.

When the requirements are met, then proceed to run pairs of experiments, choosing first the most likely cause and exchanging it between the two groupings. Let the variables be designated as A, B, etc., and use subscripts B and W to indicate the best and worst levels. Let R designate the remainder of the variables. If A is deemed the most likely cause, then this pair of experiments would use $A_W R_B$ and $A_B R_W$, where R is all remaining variables B, C, etc. Observe whether the results fall within the limits, outside the limits but not reversal, or complete reversal, as before. Use a capping run if necessary. If the red X is found, proceed to remedial efforts. If up to four variables are found, proceed to a full factorial analysis.

Full Factorial Analysis

After the number of possible causes, variables, has been reduced to four or fewer but more than one, a full factorial analysis is used to determine the relative importance of these variables and their interactions. Once again, the purpose of DOE is to direct the allocation of resources in the effort to improve a product and a process. One use of the results is to open tolerances on the lesser important variables if there is economic advantage in doing so.

The simplest four factor factorial analysis is to use two levels for each factor, requiring that 16 experiments be performed in random order. Actually, for reasons of statistical validity, it is better to perform each experiment a second time, again in a different random order, requiring a total of 32 experiments. If there are fewer than four factors, then correspondingly fewer experiments would need to be performed. The data from these experiments are used to generate two charts, a full factorial chart and an *analysis of variance* (ANOVA) chart. Examples of these two are shown in Fig. 23.25 and Fig. 23.26, where the factors are A, B, C, and D with the two levels denoted by $+$ and $-$. The numbers in the circles represent the average or mean of the data for the two performances of that particular combination of variables. These numbers are then the data for the input column of the ANOVA chart. The numbers in the upper left hand corner are the cell or box number corresponding to the cell number in the left-hand column of the ANOVA chart. In the ANOVA chart, the $+$ and $-$ signs in the boxes indicate whether the output is to added to or subtracted from the other outputs in that column, with the sum given at the bottom of that column. A column sum with small net, plus or minus, compared to other columns is deemed to be of little importance. The columns

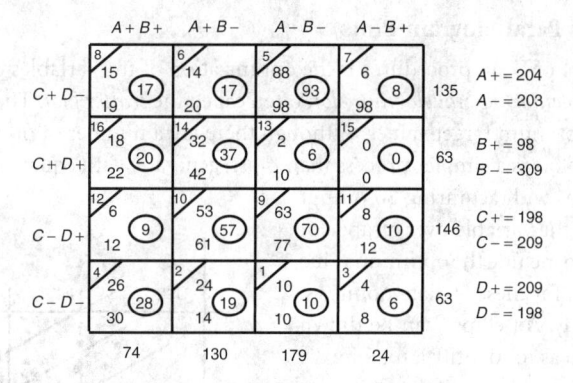

FIGURE 23.25   Full factorial chart.

with large nets, plus or minus, are deemed the ones that require attention. These two charts contain the data necessary to make a determination of resource allocation.

B vs. C Method

At this point it might be desirable to validate these findings by an independent means. The *B* (*better*) vs. *C* (*current*) method is useful for this purpose. There are two parts to this validation: (1) rank a series of samples to see if *B* is better than *C* and (2) determine the degree of risk of assuming that the results are valid. For example, if there are two *B* and two *C*, then there is only one ranking of these four parameters in which the two *B* outrank the two *C*. Therefore, there is only a one in six probability that this ranking occurred by chance. There is a 16.7% risk in assuming that this ranking actually occurred when it should not have. If there are three *B* and three *C*, then requiring that the three *B* outrank the three *C* has only a 1 in 20 probability of happening by chance, a 5% risk. These risk numbers are simply the calculation of the number of combinations of the inputs that can result in the required ranking vs. the total number of combinations that exist. This risk is called the α risk. A definition of an α risk is the risk of assuming improvement when none exists. This is also referred to as a type I error risk. There is also a β risk that is the risk of assuming no improvement when improvement actually does exist, referred to as a type II error risk. It is worthy of note that decreasing one type of risk increases the other for a given sample size. Increasing the sample size may permit decreasing both. It is also true that increasing the sample size may allow some overlap in the *B* vs. *C* ranking, that is, some *C* may be better than some *B* in a larger sample size. Please refer to the references for further discussion.

CELL GROUP	FACTORS				2 FACTOR INTERACTIONS						3 FACTOR INTERACTIONS				4 FACTOR	OUTPUT
	A	B	C	D	AB	AC	AD	BC	BD	CD	ABC	ABD	ACD	BCD	ABCD	
1	−	−	−	−	+	+	+	+	+	+	−	−	−	−	+	10
2	+	−	−	−	−	−	−	+	+	+	+	+	+	−	−	19
3	−	+	−	−	−	+	+	−	−	+	+	+	−	+	−	6
4	+	+	−	−	+	−	−	−	−	+	−	−	+	+	+	28
5	−	−	+	−	+	−	+	−	+	−	+	−	+	+	−	93
6	+	−	+	−	−	+	−	−	+	−	−	+	−	+	−	17
7	−	+	+	−	−	−	+	+	−	−	−	+	+	−	+	8
8	+	+	+	−	+	+	−	+	−	−	+	−	−	−	−	17
9	−	−	−	+	+	+	+	−	−	−	−	+	+	+	−	70
10	+	−	−	+	−	−	+	+	−	+	+	−	−	+	+	57
11	−	+	−	+	−	+	−	−	+	−	+	−	+	−	+	10
12	+	+	−	+	+	−	−	+	+	−	−	+	−	−	−	9
13	−	−	+	+	+	−	−	−	+	+	+	+	−	+	+	6
14	+	−	+	+	−	+	−	+	−	+	−	−	+	−	−	37
15	−	+	+	+	−	−	+	−	+	+	−	−	−	+	−	0
16	+	+	+	+	+	+	+	+	+	+	+	+	+	+	+	20
SUMS	1	−11	11	−211	99	−33	73	−5	−51	−155	49	−97	163	175	−95	

$(+) (+) = +, (−) (−) = +, (+) (−) = (−) (+) = −$

FIGURE 23.26   ANOVA table.

Realistic Tolerances Parallelogram Plots

The final step in this set of DOE procedures is the optimization of the variables of the process. The tool for this is the *realistic tolerances parallelogram plot*, often called the *scatter plot*. The purpose is to establish the variables at their optimum target values. Although there are a number of other techniques for doing this, the use of scatter plots is a simpler process than most, generally with equally acceptable results.

The procedure begins with acquiring 30 output data points by varying the variable over a range of values that is assumed to include the optimum value and plotting the output for these 30 data points vs. the variable under study. An ellipse can be drawn around the data plot so as to identify a major axis. Two lines parallel to the major axis of the ellipse are then drawn on either side of the ellipse to include all but one or one and one-half of the data points. Assuming that specification limits exist for the output, these are drawn on the plot. Then vertical lines are drawn to intersect these specification limit lines at the same point that the parallelogram lines intersect the specification limits, as shown in Fig. 23.27. The intersection of these vertical lines with the variable axis determines the realistic tolerance for the variable.

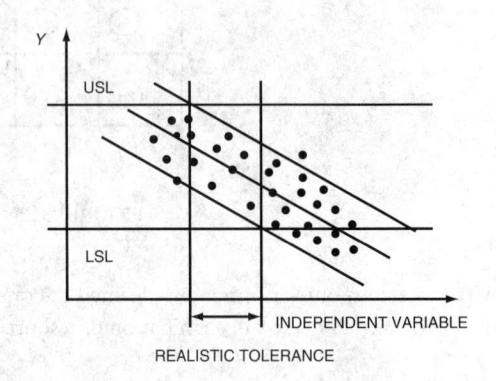

FIGURE 23.27   Scatter plot (realistic tolerance parallelogram plot).

Additional information can be found in this plot. Small vertical scatter of the data indicates that indeed this is an important variable, that most of the variation of the output is caused by this variable. Conversely, large vertical scatter indicates that other variables are largely responsible for variation of the output. Also, little or no slope to the major axis of the ellipse indicates little or no importance of this variable in influencing the output. The actual slope is not important as it depends on the scale of the plot. Scatter plots can be made for each of the variables to determine their optimum, target value.

The techniques presented here are intended to be easy to implement with pencil and paper. As such, they may not be the best, although they are very likely the most economical. The use of recently developed software programs is gaining acceptance, some of these programs able to do very sophisticated data manipulation and plotting. As mentioned previously, the advice or direction of a professional statistician is always to be considered, especially for very complex problems.

Process Control

Once the process has been optimized and is in operation, the task becomes that of maintaining this optimized condition. Here Shainen makes use of two tools, *positrol* and *precontrol*. Again, these are easy to use and understand, and provide in many instances better results than previously used tools, such as control charts from TQM.

Positrol

In simple terms, positrol (short for positive control) is a plan, with appropriate documentation recorded in a log, that identifies *who* is to make *what* measurements, *how* these measurements are to be made, and *when* and *where* they are to be measured. It establishes the responsibility for and the program of measurements that are to be a part of the process control plan. A simple log, though sometimes with detailed entries, is kept so that information about the process is available at any time to operators and managers. Those responsible for keeping the log must be given a short training period on how to keep the log, with emphasis on the importance of making entries promptly and accurately.

An example of a positrol log entry for a surface mount soldering process would be

What	*Who*	*How*	*Where*	*When*
Copper patterns	Etcher	Visual	Etcher	After Etch

Such a positrol log contains complex steps, each of which might well be kept in its own log. For example, the reflow soldering operation could have a number of what steps, such as belt speed; furnace zone temperatures for preheat; soldering zone; cool down zone; atmosphere chemistry control; atmosphere flow speed; visual inspection for missing, misaligned, or tombstoned parts; solder bridges; and solder opens.

A positrol log should contain whatever steps or specifications (whats) in a process are to be monitored. As such, it should be the best log or record of the process that the team, in consultation with the operators, can devise. It should be clear that no important steps are to be omitted. The log provides documentation that could be invaluable should the process develop problems. However, it is not the purpose of the log to identify people to blame, but rather to know the people who might have information to offer in finding a solution to the problem.

Precontrol

The second procedure in process control is precontrol, developed by Frank Satterthwaite and described in the 1950s. One of the problems with control charts of the TQM type is that they are slow to indicate a process that is moving to an out of control state. In part this is because of the way the limits on the variability of a process are determined. Satterthwaite suggested that it would be better if the specification limits were used rather than the traditional control limits. He then divided the range between the *upper specification limit* (USL) and the *lower specification limit* (LSL) into four equal-sized regions. The two in the middle on either side of the center of the region between the limits were called the *green zones*, indicating a satisfactory process. The two regions bordering the green zones on one side and the USL and LSL on the other were called the *yellow zones*. Zones outside the limits were called the *red zones*. For an existing process being studied, simple rules for an operator to follow are

1. If two consecutive samples fall in the green zones, the process may continue.
2. If one sample falls in a green zone and one in a yellow zone, the process is still OK.
3. If both are in the same yellow zone, the process needs adjustment but does not need to be stopped.
4. If the two are in different yellow zones, the process is going out of control and must be stopped.
5. Even one in red zones the process must be stopped.

A stopped process must then be brought back to control.

A new process must be in control before it is brought into production. Here, the procedure is to select five samples. If all five are in green zones, the process may be implemented. If even one is not in the green zones, the process is not yet ready for production and must be further studied.

Another important aspect of precontrol is that, for a process that is in control, the time between samples can be increased the longer it remains in control. The rule is that the time between samples is the time between the prior two consecutive stops divided by six. If it is determined that a more conservative sampling is in order, then divide by a larger number.

Experience has shown that the α risk of precontrol is less than 2%, that is, the risk of stopping a good process. The β risk, not stopping a bad process, is less than 1.5%. These are more than acceptable risks for most real processes.

23.2.3   Robust Design

Because design determines so much of the ultimate expenditure in producing a product, there is increasing interest in process and product designs that can survive in many different environments while maintaining their design performance with quality and production costs that are competitive in the global market. Taguchi (1993) has suggested that the *goal posts* mentality of design, accepting values anywhere within the design specifications, is harmful to society as well as to the producer and the user.

Customers generally want products to be as identical as clones. This implies that the products should have zero variability of a parameter that is important. Its value should be the target value. Any product that does not exhibit this value, even though it remains within specifications, is a loss to society in that the customer/user is not getting from the product the very best that could be delivered.

The Loss Function

Taguchi proposed that a measure of the loss to society, and to the customer, be a quadratic (square law) function of the difference between the target value and the actual value of the parameter of interest. This has become known as *Taguchi's loss function*. The goal posts idea just does not address this loss to society; it does not penalize a product with a parameter that is not the target value. The loss function has been used as a measure of monetary loss associated with less than optimal performance as well as higher maintenance and repair costs.

The formula that is the basis of the loss function is

$$L(y) = (A/\Delta^2)(y - m)^2$$

where L is loss function; A cost of repair or replacement, the failure loss; Δ functional limit of the product, the deviation at which failure occurs; y product value; and m target value. Note that $(y - m)^2$ is the square of the deviation from the target value and could be replaced by the mean squared deviation of the parameter.

Also, Taguchi recognized that any real process has variability, that it is not possible to have ideal clones. To quantify this inherent variability of a process, he chose a measure that he calls signal to noise. (This is not to be confused with the same terminology used in electrical engineering to describe the power in a signal compared to the power in the noise that accompanies the signal.) In the sense used here, noise can be thought of as the uncontrollable factors.

The formulas for signal to noise do use the base 10 logarithm and the units of the signal-to-noise ratio (S/N) are decibel (dB). The purpose is to give a measure of the factor so that comparisons can be made. The basic formula is

$$S/N = 10 \log_{10} (Y^2/\sigma^2)$$

for nominal is best. Taguchi uses S/N in connection with his orthogonal arrays technique. That is a complicated process and will not be discussed here. The reader is referred to the literature for further information on the loss function and the various formula modifications for specific applications.

Process Capability

Process capability C_p is defined as the *specification width* divided by the process width. The main assumption for all of this is that many physical processes have sample distributions that are described by the Gaussian or normal distribution. A second assumption is that the 3σ points of the process normal distribution describe the process width. If the process width coincides with the **specification limits**, then the $C_p = 1.0$. This is somewhat arbitrary but has been the standard usage for many years and continues to be accepted. If $C_p = 1$, the defect rate is approximately 2700 parts per million (ppm), a generally unacceptable defect rate for most products in today's competitive marketplace. If the center of the process is not coincident with the center of the specification width, the process capability is reduced and must be calculated by $C_{pk} = C_p(1 - K)$ where K is the difference between the target value and the process mean divided by one-half the specification width.

Worst Case and Monte Carlo Design

Many designs have been made using the designer's best estimates of the worst possible combination of parameters and attempting to create a design that will still meet the specifications. For those products that demand a high degree of reliability, there is no possibility to be allowed for a defective product. For these, a **worst case design** is mandatory. The major difficulty in doing worst case design is to guarantee that the combination of parameters chosen does in fact result in the worst case. The more the parameters, the more difficult it is to guarantee this. Design of experiments might be used to determine the combinations that will result in the worst case. Then, simulations should be employed to verify the choices. It can be seen that worst case design can be quite expensive, in terms of time as well as dollars. However, most products do not require such high reliability. A method of accepting less than 100% yield might be cost effective for products that experience an occasional failure without catastrophic repercussions, especially if it can be repaired easily.

Still, with the customer demanding higher quality, some form of organized design for quality and reliability must be employed. One method for doing this is to simulate the product variability using **Monte Carlo** statistical analysis of the designs. The specifications are set for the product and tolerances are chosen for the components. Then, the component values of the product are made to vary according to a selected statistical distribution, such as a normal (Gaussian) or a uniform distribution. The components can be varied one at a time or multiple components can be varied at the same time. Generally, varying one component at a time is not the best technique, because interactions between components can be the cause of a defective product. Again, the techniques of design of experiments can aid greatly in selecting the components to be varied. Then, the Monte Carlo analysis becomes a verification of the results obtained from DOE.

The key to successful use of Monte Carlo simulation and analysis is to determine the tolerances to be used and to match the statistical distribution to be used to that of the real component. Then a realistic C_{pk} can be chosen that will guarantee an acceptable level of defect rate. The concurrent engineering team can then determine whether or not such a design can be built and tested.

Monte Carlo simulation and analysis can become expensive if many variables are allowed to vary. The simulation must be run many times. This is another reason to do design of experiments as outlined earlier and incorporate those results with the Monte Carlo analysis. Just as scatter plots can indicate when tolerances might be tightened or loosened or that the original target value is not the best, Monte Carlo analysis can also point out tolerances that might be loosened so as to reduce the costs of the components. Thus, Monte Carlo analysis and design of experiments are seen to be complementary, one yielding verification of the other, as well as presenting a different view of the processes.

Expert Systems and Artificial Intelligence

The advent of **artificial intelligence (AI)** in the world of business in the 1970s and 1980s has begun a new approach to some of the problems of the day. Some companies became acutely aware of the loss of very valuable expertise with the retirement of their most valued employees. Many managers, and often engineers, those who were aware of artificial intelligence research in the university communities and in the fledging consulting companies offering their systems, anticipated the improved productivity that these so far esoteric studies might produce. The most used part of AI became the *expert system*, which could become the repository of the knowledge and expertise of these soon to retire most valued employees.

The history and applications around the world make a fascinating story that is told in the literature. The struggles against the entrenched bureaucratic structures and the doubting managers who must finance the ventures are a classic tale. Some of these pioneers devoted their own time to the venture, while others bootlegged it unbeknownst to their managers. It is the story of many new technologies.

The greatest immediate benefit for many of the companies was the capturing of the knowledge, the heuristics, and the thought processes of the retiring experts. These were then converted to a set of rules that the engine considers in presenting the alternatives to a human decision maker, or even in some instances making a single recommendation. What the expert systems also made possible was the training and assistance of new employees in these critical fields, making the transition from apprentice to expert in a much shorter time than had been possible without the expert system. In other cases, the utility was the uniformity of the decision making process across several workers, such as approving purchases on a credit card or denying the credit, or the increased speed with which such decisions could be made.

Defining Terms

Artificial intelligence: A branch of computer science that attempts to devise computer hardware and software that emulates human intelligence.

Artificial neural networks: Hardware and software that are based on a model of neural synapse data processing.

Control limits: The limits on a process that are determined by running the process, measuring a performance parameter, and determining the limits according to preset formulas based on the output data of the process. A satisfactory process will have control limits that are usually somewhat narrower than the specification limits.

Expert system: A system that uses a computer program containing the knowledge and problem solving techniques of experts in a field.

Knowledge engineers: Those who translate the expertise and knowledge of experts into the software of expert systems.

Monte Carlo design: A method of design that assigns parts a tolerance with variation such as Gaussian or uniform and then studies an output performance parameter to determine if a yield is deemed satisfactory.

Specification limits: The limits on a performance parameter or a part that are set by the designers.

Total quality management (TQM): A philosophy and set of guiding principles required by any organization attempting to compete in today's global-based economy, built upon the work of Deming, Juran, and others.

Worst case design: A method for determining the worst performance values of parameters and assigning all parameters these values while still having the product meet specifications, presumably resulting in a product that will never fail to meet specifications under all operating conditions.

References

Aguayo, R. 1990. *Dr. Deming*. Simon & Schuster, New York.

Bhote, K. 1991. *World Class Quality*. AMACOM, New York.

Brassard, M. and Ritter, D. 1994. *The Memory Jogger II*. GOAL/QPC, Methuen, MA.

Carter, D. and Baker, B. 1992. *Concurrent Engineering*. Addison-Wesley, Reading, MA.

Cohen, L. 1988. Quality function deployment: An application perspective from digital equipment corporation. *National Productivity Review*. Summer.

Department of Defense. 1986. Report R-338. Inst. for Defense Analyses, Washington, DC.

Feigenbaum, E., McCorduck, P., and Nii, H.P. 1988. *The Rise of the Expert Company*. Times Books, New York.

Miller, L. 1993. *Concurrent Engineering Design*. Society of Manufacturing Engineers, SME, Dearborn, MI.

Prybyla, J. 1993. Optimization Doesn't Have To Hurt. *ASIC & EDA*. May.

Ross, P. 1988. *Taguchi Techniques for Quality Engineering*. McGraw-Hill, New York.

Salomone, T. 1995. *What Every Engineer Should Know About Concurrent Engineering*. Marcel Dekker, New York.

Saylor, J. 1995. *TQM Field Manual*. McGraw-Hill, New York.

Shillito, M.L. and De Marle, D. 1992. *Value, Its Measurement, Design and Management*. Wiley, New York.

Shina, S. 1991. *Concurrent Engineering and Design for Manufacture of Electronics Products*. Van Nostrand Reinhold, New York.

Shriver, B. 1994. The concurrent process. *Circuits Assembly* 5(2), Feb.

Taguchi, G. 1993. *Taguchi On Robust Technology Development*. ASME Press, New York.

Walton, M. 1986. *The Deming Management Method*. Putnam. New York.

Further Information

Society of Manufacturing Engineers, headquartered in Dearborn Michigan, publishes a number of books relating to this topic.

ASME Press, headquartered at 345 East 47th St., New York, publishes a number of books relating to this topic.

Concurrrent Engineering Research Center (CERC), located at the University of West Virginia, publishes a newsletter, bibliographies, and other information.

GOAL/QPC, headquartered in Methuen, Massachusetts, publishes a number of monographs and makes other books and publications available.

23.3 Engineering Documentation

Fred Baumgartner and Terrence M. Baun

23.3.1 Introduction

Video facilities are designed to have as little down-time as possible. Yet, inadequate documentation is a major contributor to the high cost of systems maintenance and the resulting widespread replacement of poorly documented facilities. The cost of neglecting *hours* of engineering documentation is paid in *weeks* of reconstruction.

Documentation is a management function every bit as important as project design, budgeting, planning, and quality control; it is often the difference between an efficient and reliable facility and a misadventure. If the engineer does not feel qualified to attempt documentation of a project, the engineer must at the very least oversee and approve the documentation developed by others.

Within the last few years the need for documentation has increased with the complexity of the broadcast systems. Fortunately, the power of documentation tools has kept pace.

23.3.2 Basic Concepts

The first consideration in the documentation process is the complexity of the installation. A basic video editing station may require almost no formal documentation, while a large satellite or network broadcast facility may require computerized databases and a full time staff doing nothing but documentation updates. Most facilities will fall somewhere in the middle of that spectrum.

A second concern is the need for flexibility at the facility. Seldom does a broadcast operation get "completely rewired" because the cabling wears our or fails. More often, it is the supporting documentation that has broken down, frustrating the maintainability of the system. Retroactive documentation is physically difficult and emotionally challenging, and seldom generates the level of commitment required to be entirely successful or accurate; hence, a total rebuild is often the preferred solution to documentation failure. Documentation must be considered a hedge against such unnecessary reconstruction.

Finally, consider efficiency and speed. Documentation is a prepayment of time. Repairs, rerouting, replacements, and reworking all go faster and smoother with proper documentation. If your installation is one in which any downtime or degradation of service is unacceptable, then budgeting sufficient time for the documentation process is critically important.

Because, in essence, documentation is education, knowing "how much is enough" is a difficult decision. We will begin by looking at the basics, and then expand our view of the documentation process to fit specialized situations.

The Manuals

Even if you never take a pen to paper, you do have one source of documentation to care for, since virtually every piece of commercial equipment comes with a manual.

Place those manuals in a centralized location, and arrange them in an order that seems appropriate for your facility. Most engineering shops file manuals alphabetically, but some prefer a filing system based on equipment placement. (For example, production studio equipment manuals would be filed together under a "Production Studio" label and might even be physically located in the referenced studio.) But whatever you do, be consistent. Few things are as frustrating to a technician as being unable to locate a manual when needing something as simple as a part number or the manufacturer's address.

Equipment manuals are the first line of documentation and deserve our attention and respect.

Documentation Conventions

The second essential item of documentation is the statement of "conventions." By this we mean a document containing basic information essential to an understanding of the facility, posted in an obvious location and available to all who maintain the plant. Consider the following examples of conventions:

- Where are the equipment manuals and how are they organized?
- What is the architecture of the ground system? Where is the central station ground and is it a star, grid, or other distribution pattern? Are there separate technical and power grounds? Are audio shield grounded at the source, termination, or both locations?
- What is the standard audio input/output architecture? Is this a +8, +4, or 0 dBm facility? Is equipment sourced at 600 Ω terminated at its destination, or left unterminated? Are unbalanced audio sources wired with the shield as ground or is the low side of a balanced pair used for that purpose? How are XLR-type connectors wired—pin 2 high or low?
- How can a technician disconnect utility power to service line voltage wiring within racks? Where are breakers located, and how are they marked? What equipment is on the UPS power, generator, or utility power? Whom do you call for power and building systems maintenance?
- Where are the keys to the transmitter? Is the site alarmed?

For such an essential information source, you will find it takes very little time to generate the conventions document. Keep it short—it is not meant to be a book. A page or two should be sufficient for most installations and, if located in a obvious place, this document will keep the technical staff on track and will save service personnel from stumbling around searching for basic information. This document is the key to preventing many avoidable embarrassments.

The next step beyond the conventions document is a documentation *system*. There are three primary methods:

- Self documentation
- Database documentation
- Graphic documentation

In most cases, a mixture of all three is necessary and appropriate. In addition to documenting the physical plant and its interconnections, each piece of equipment, whether commercially produced or custom made, must be documented in an organized manner.

Self Documentation

In situations where the facility is small and very routine, self documentation is possible. Self documentation relies on a set of standard practices that are repeated. Telephone installations, for example, appear as a mass of perplexing wires. In reality, the same simple circuits are simply repeated in an organized and universal manner. To the initiated, any telephone installation that follows the rules is easy to understand and repair or modify, no matter where or how large. Such a system is truly self documenting. Familiarity with telephone installations is particularly useful, because the telephone system was the first massive electronic installation. It is the telephone system that gave us "relay" racks, grounding plans, demarcation points, and virtually all of the other concepts that are part of today's electronic control or communications facility.

The organization, color codes, terminology, and layout of telephone systems is recorded in minute detail. Once a technician is familiar with the rules of telephone installations, drawings and written documentation are rarely required for routine expansion and repair. The same is true for many parts of other facilities. Certainly, much of the wiring in any given rack of equipment can be self documenting. For example, a video tape recorder will likely be mounted in a rack with a picture monitor, audio monitor, waveform monitor, and vectorscope. The wiring between each of these pieces of equipment is clearly visible, with all wires short and their purpose obvious to any technician familiar with the rules of video. Furthermore, each video cable will conform to the same standards of level or data configuration. Additional documentation, therefore, is largely unnecessary.

By convention, there are rules of grounding, power, and signal flow in all engineering facilities. In general, it can be assumed that in most communications facilities, the ground will be a star system, the power will be individual 20 amp feeds to each rack, and the signal flow within each rack will be from top to bottom. Rules that might vary from facility to facility include color coding, connector pin outs, and rules for shield and return grounding.

To be self documenting, the rules must be determined and all of the technicians working on the facility must know and follow the conventions. The larger the number of technicians, or the higher the rate of staff turn-over, the more important it is to have a readily available document that clearly covers the conventions in use.

One thing must be very clear: a facility that does not have written documentation is not automatically self documenting. Quite the contrary. A written set of conventions and unfaltering adherence to them are the trade marks of a self documenting facility.

While it is good engineering practice to design all facilities to be as self documenting as possible, there are limits to the power of self documentation. In the practical world, self documentation can greatly reduce the amount of written documentation required, but can seldom replace it entirely.

Database Documentation

As facilities expand in size and complexity, a set of conventions will no longer answer all of the questions. At some point, a wire leaves an equipment rack and its destination is no longer obvious. Likewise, equipment will often require written documentation as to its configuration and purpose, especially if it is utilized in an uncharacteristic way.

Database documentation records the locations of both ends of a given circuit. For this, each cable must be identified individually. There are two common systems for numbering cables: *ascension numbers*, and *from-to coding*. In ascension numbering schemes, each wire or cable is numbered in increasing order, one, two, three, and so on. In from-to coding, the numbers on each cable represent the source location, the destination location, and normally some identification as to purpose and/or a unique identifier. For example, a cable labeled 31–35-B6 might indicated that cable went from a piece of equipment in rack 31 to another unit in rack 35 and carries black, it is also the sixth cable to follow the same route and carry the same class of signal.

Each method has its benefits. Ascension numbering is easier to assign, and commonly available preprinted wire labels can be used. On the other hand, ascension numbers contain no hints as to wire purpose or path, and for that reason "purpose codes" are often added to the markings.

From-to codes can contain a great deal more information without relying on the printed documentation records, but often space does not permit a full delineation on the tag itself. Here again, supplemental information may still be required in a separate document or database.

Whatever numbering system is used, a complete listing must be kept in a database of some type. In smaller installations, this might simply be a spiral notebook that contains a complete list of all cables, their source, destination, any demarcations, and signal parameters.

Because all cabling can be considered as a transmission line, all cabling involves issues of termination. In some data and analog video applications, it is common for a signal to "loop-through" several pieces of equipment. Breaking or tapping into the signal path often has consequences elsewhere, resulting in unterminated or double-terminated lines. While more forgiving, analog audio has similar concerns. Therefore, documentation must include information on such termination.

Analog audio and balanced lines used in instrumentation have special concerns of their own. It is seldom desirable to ground both ends of a shielded cable. Again, the documentation must reflect which end(s) of a given shield is grounded.

In many cases, signal velocity is such that the length of the lines and the resulting propagation delay is critical. In such circumstances, this is significant information that should be retained. In cases where differing signal levels or configurations are used (typical in data and control systems), it is the documentor's obligation to record those circumstances as well.

However or wherever the database documentation is retained, it represents the basic information that defines the facility interconnections and must be available for updating and duplication as required.

Graphic Documentation

Electronics is largely a graphical language. Schematics and flow charts are more understandable than net lists or cable interconnection lists. Drawings, either by hand—done with the aid of drafting machines and tools—or accomplished on CAD (computer aided design) programs are highly useful in conveying overall facility design quickly and clearly. Normally, the wire numbering scheme will follow that used in the database documentation, so that the graphic and text documentation can be used together.

CAD drawings are easy to update and reprint. For this reason, documentation via CAD is becoming more popular, even in smaller installations. Because modern CAD programs not only draw but also store information, they can effectively serve as an electronic file cabinet for documentation. While there have been attempts to provide electronic/telecommunications engineering documentation "templates" and corresponding technical graphics packages for CAD programs, most of the work in this area has been done by engineers working independently to develop their own systems. Obviously, the enormous scope of electronic equipment and telecommunications systems make it impractical for a "standard" CAD package to suit every user.

Update Procedures

Because documentation is a dynamic tool, as the facility changes, so too must the documentation. It is common for a technician to "improve" conditions by reworking a circuit or two. Most often this fixes a problem that should be corrected as a maintenance item. But sometimes, it plants a "time bomb" wherein a future change, based on missing or incomplete documentation of the previous work, will cause problems.

It is essential that there be a means of consistently updating the documentation. The most common way of accomplishing this is the mandatory "change sheet." Here, whenever a technician makes a change it is reported back to those who keep the documentation. If the changes are extensive, the use of the "red-line" drawing and "edit sheet" come into play; the original drawing and database, respectively, are printed and corrected with a "red pen." This document then is used to update the original documentation.

In some cases, the updating process can be tied to the engineering reports or discrepancy process. Most facilities use some form of "trouble ticket" to track equipment and system performance and to report and track maintenance. This same form may be used to report changes required of the documentation or errors in existing documentation.

Equipment Documentation

Plant documentation does not end when all of the circuit paths in the facility are defined. Each circuit begins and ends at a piece of equipment, which can be modified, reconfigured, or removed from service. Keep in mind that unless the lead technician lives forever, never changes jobs, and never takes a day off, someone unfamiliar with the equipment will eventually be asked to return it to service. For this reason, a documentation file for each piece of equipment must be maintained.

Equipment documentation contains these key elements:

- The equipment manual
- Modification record
- Configuration information
- Maintenance record

The equipment manual is the manufacturer's original documentation. As mentioned previously, the manuals must be organized in such a manner that they can be easily located. Typically, the manuals are kept at the site where the equipment is installed (if practical). Remember that equipment with two "ends" such as STLs, RPUs, or remote control systems need manuals at *each* location.

Of course, if a piece of equipment is of custom construction, there must be particular attention paid to creating a manual. For this reason, a copy of the key schematics and documentation is often attached directly to the equipment. This "built-in manual" may be the only documentation to survive over the years.

Many pieces of equipment, over time, will require modification. Typically, the modification is recorded in three ways. First, internally to the equipment. A simple note glued into the chassis may be suitable, or a marker pen is used to write on a printed circuit board or other component. Second, the changes can be recorded in the manual, either inside the cover, or on the schematic or relevant pages. If the manual serves several machines, this may not be appropriate. If this is the case, a third option is to keep the modification information in a separate *equipment file.*

Equipment files are typically kept in standard file folders, and may be filed with the pertinent equipment manuals. Ideally, the equipment file is started when the equipment is purchased, and should contain purchase date, serial number, all modifications, equipment location(s), and a record of service.

The equipment file is the proper place to keep the configuration information. An increasingly large amount of equipment is microprocessor based or otherwise configurable for a specific mode of operation. Having a record of the machine's default configuration is extremely helpful when a power glitch (or an operator) reconfigures a machine unexpectedly.

Equipment files should also contain repair records. With most equipment, documenting failures, major part replacements, operating time, and other service-related events serves a valuable purpose. Nothing is more useful in troubleshooting than a record of previous failures, configuration, modifications, and—of course—a copy of the original manual.

Operator/User Documentation

User documentation provides, at its most basic level, instructions on how to use a system. While most equipment manufacturers provide reasonably good instruction and operations manuals for their products, when those products are integrated into a system another level of documentation may be required. Complex equipment may require interface components that need to be adjusted from time to time, or various machines may be incompatible in some data transmission modes—all of which is essential to the proper operation of the system.

Such information often resides only in the heads of certain key users and is passed on by word of mouth. This level of informality can be dangerous, especially when changes take place in the users or maintenance staff, resulting in differing interpretations between operators and maintenance people about how the system normally operates. Maintenance personnel will then spend considerable time tracking hypothetical errors reported by misinformed users.

A good solution is to have the operators write an operating manual, providing a copy to the maintenance department. Such documentation will go a long way toward improving inter-departmental communications and should result in more efficient maintenance as well.

The True Cost of Documentation

There is no question that documentation is expensive—in some cases it can equal the cost of installation. Still, both installation and documentation expenses pale in comparison to the cost of equipment and potential revenue losses resulting from system down-time.

Documentation must be seen as a management and personnel issue of the highest order. Any lapse in the documentation updating process can result in disaster. Procrastination and the resulting lack of follow through will destroy any documentation system and ultimately result in plant failures, extend down-time, and premature rebuilding of the facility.

Making a business case for documentation is similar to making any business case. Gather together all the costs in time, hardware, and software on one side of the equation, and balance this against the savings in time and lost revenue on the other side. Engineering managers are expected to project costs accurately, and the allocation of sufficient resources for documentation and its requisite updating is an essential part of that responsibility.

References

Baumgartner, F. and Baun, T. 1999. Broadcast Engineering Documentation. In *NAB Engineering Handbook*, 9th ed. J.C. Whitaker, ed., National Association of Broadcasters, Washington, D.C.

Baumgartner, F. and Baun, T. 1996. Engineering Documentation. In *The Electronics Handbook*, J.C. Whitaker, ed., CRC Press, Boca Raton, FL.

23.4 Disaster Planning and Recovery

Richard Rudman

23.4.1 Introduction

Disaster-readiness must be an integral part of the design process. You cannot protect against every possible risk.

Earthquakes, hurricanes, tornados, and floods come to mind when we talk about risks that can disable a facility and its staff. Risk can also be man made. Terrorist attack, arson, utility outage, or even simple accidents that get out of hand are facts of everyday life. Communications facilities, especially those that convey information from government to the public during major catastrophes, must keep their people and systems from becoming disaster victims. No major emergency has yet cut all communications to the public. We should all strive to maintain that record.

Employees all too often become crash dummies when disaster disrupts the workplace. During earthquakes, they sometimes have to dodge falling ceiling tiles, elude objects hurled off shelves, watch computers and work stations dance in different directions, breathe disturbed dust, and suffer the indignities of failed plumbing. This chapter does not have ready-to-implement plans for people, facilities, or systems. It does outline major considerations for disaster planning. It is then up to you to be the champion for emergency preparedness and planning on mission critical projects.

A Short History of Disaster Communications

When man lived in caves, every day was an exercise in disaster planning and recovery. There were no "All Cave News" stations for predator reports. The National Weather Service was not around yet to warn about rain that might put out the community camp fire. We live in an increasingly interdependent information-driven society. Modern information systems managers have an absolute social, technical, and economic responsibility to get information to the public when disaster strikes. Designers of these facilities must do their part to build workplaces that will meet the special needs of the lifeline information mission during major emergencies.

The U.S. recognized the importance of getting information to the public during national emergencies in the 1950s when it devised a national warning system in case of enemy attack. This effort was and still is driven by a belief that a government that loses contact with its citizens during an emergency could fall. The emergency broadcast system (EBS), began life as control of electromagnetic radiation (CONELRAD). It was devised to prevent enemy bombers from homing in on AM broadcast stations. It was conceived amid Cold War fears of nuclear attack, and was born in 1951 to allow government to reach the public rapidly through designated radio stations. EBS replaced CONELRAD and its technical shortcomings. EBS is being replaced by emergency alert system (EAS). EAS introduces multiple alerting paths for redundancy as well as some digital technology. It strengthens the emergency communication partnerships beyond broadcasting, federal government, National Weather Service, local government, and even the Internet.

Major disasters such as Hurricane Andrew and the Northridge (CA) earthquake help us focus on assuring that information systems will work during emergencies. These events expose our weaknesses. They can also be rare windows of opportunity to learn from our mistakes and make improvements for the future.

23.4.2 Emergency Management 101 for Designers

There is a government office of emergency services (or management) in each state. Within each state, *operational areas* for emergency management have been defined. Each operational area has a person or persons assigned to emergency management, usually under the direction of an elected or appointed public safety official. There is usually a mechanism in place to announce when an emergency condition exists. *Declared* has a special meaning when used with the word *emergency*. An operational area normally has a formal method, usually stated in legislation, to announce to the world that it is operating under emergency conditions. That legislation places management of the emergency under a specially trained person or organization. During declared emergencies, operational areas usually open a special management office that is usually called an **emergency operations center (EOC)**. Some large businesses with a lot at stake have emergency management departments with full-time employees and rooms set aside to be their EOC. These departments usually use the same management philosophy as government emergency organizations. They also train their people to understand how government manages emergencies. The unprepared business exists in a vacuum without an understanding of how government is managing the emergency. Misunderstandings are created when people do not know the language. An emergency, when life and property are at stake, is the wrong time to have misunderstandings. Organizations that have to maintain services to the public, have outside plant responsibilities, or have significant responsibilities to the public when on their property should have their plans, people, and purpose in place when disaster strikes.

A growing number of smaller businesses which have experienced serious business disruptions during emergencies, or wish to avoid that unpleasant experience, are writing their own emergency plans and building their own emergency organizations. Broadcasters, information companies, cable systems, and other travellers on the growing information superhighway should do likewise. There are national organizations and possibly local emergency-minded groups you can contact. In Los Angeles, there is the Business and Industry Council for Emergency Planning and Preparedness (BICEPP). A more complete listing of resource organizations follows this chapter.

Figure 23.28 shows the cyclical schematic of emergency management. After an emergency has been declared, *response* deals with immediate issues that have to do with saving lives and property. Finding out what has broken and who has been hurt is called *damage assessment*. Damage assessment establishes the dimensions of the response that are needed. Once you know what is damaged and who is injured, trapped, or killed, you have a much better idea of how much help you need. You also know if you have what you need on hand to meet the challenge. Another aspect of response is to provide the basics of human existence: water, food, and shelter. An entire industry has sprung up to supply such special needs. Water is available packaged to stay

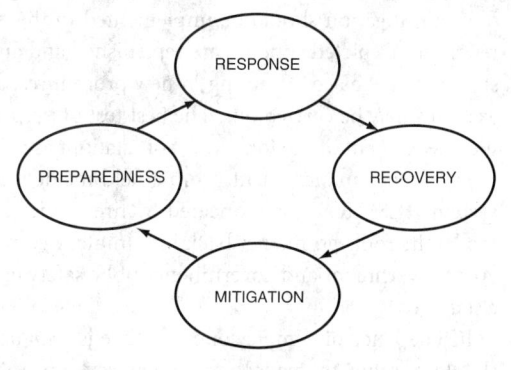

FIGURE 23.28 The response, recovery, mitigation, and preparedness cycle.

drinkable for years. The same is true for food. Some emergency food is based on the old military K rations or the more modern meals ready to eat (MREs). MREs became widely known during the Gulf War.

The second phase of the emergency cycle is called *recovery*. Recovery often begins after immediate threats to life safety have been addressed. This is difficult to determine during an earthquake or flood. A series of life threats may span hours, days, or weeks. It is not uncommon for business and government emergency management organizations to be engaged in recovery and response at the same time.

In business, response covers these broad topics:

- Restoring lost capacity to your production and distribution capabilities
- Getting your people back to work

- Contacting customers and/or alternate vendors
- Contacting insurers and processing claims

A business recovery team's mission is to get the facility back to 100% capacity, or as close to it as possible. Forming a *business recovery team* should be a key action step along with conducting the *business impact analysis,* discussed later in this chapter. Major threats must be eliminated or stabilized so that employees or outside help can work safely. Salvage missions may be necessary to obtain critical records or equipment. Reassurance should also be a part of the mission. Employees may be reluctant to return to work or customer may be loathe to enter even if the building is not *red tagged.* Red tagging is a term for how inspectors identify severely damaged buildings, an instant revocation of a building's certificate of occupancy until satisfactory repairs take place. Entry, even for salvage purposes by owners or tenants, may be prohibited or severely restricted. Recovery will take a whole new turn if the facility is marked unfit for occupancy.

The third stage of the cycle is *mitigation.* Sometimes called lessons learned, mitigation covers a wide range of activities that analyze what went wrong and what went right during response and recovery. Accurate damage assessment records, along with how resources were allocated, are key elements reviewed during mitigation. At the business level, mitigation debriefings might uncover how and why the emergency generator did not start automatically. You might find that a microwave dish was knocked out of alignment by high winds or seismic activity or that a power outage disabled critical components such as security systems. On the *people* side, a review of the emergency might show that no one was available to reset circuit breakers or unlock normally locked doors. There may have been times when people on shift were not fed in a timely manner. Stress, an often overlooked factor that compounds emergencies and emergency response, may have affected performance. In the *what went right* column, you might find that certain people rose above the occasion and averted added disaster, that money spent on preparing paid off, or that you were able to resume 100% capacity much faster than expected.

Be prepared should be the watch-phrase of emergency managers and facilities designers. Lessons learned during mitigation should be implemented in the *preparedness* phase. Procedures that did not work are rewritten. Depleted supplies are replenished and augmented if indicated. New resources are ordered and stored in safe places. Training in new procedures, as well as refresher training, is done. Mock drills and exercises may be carried out. The best test of preparedness is the next actual emergency. Preparedness is the one element of the four you should adopt as a continuous process.

Emergency management, from a field incident to the state level, is based on the **incident command system (ICS)**. ICS was pioneered by fire service agencies in California as an emergency management model. Its roots go further back into military command and control. ICS has been adopted by virtually every government and government public safety emergency management organization in the country as a standard.

ICS depends on simple concepts. One is *span of control.* Span of control theory states that a manager should not directly manage more than seven people. The optimum number is five. Another ICS basic is that an emergency organization builds up in stages, from the bottom up, from the field up, in response to need. For example, a wastebasket fire in your business may call for your fire response team to activate. One trained member on the team may be all that is required to identify the source of the fire, go for a fire extinguisher on the wall, and put it out. A more serious fire may involve the entire team. The team *stages up* according to the situation.

23.4.3   The Planning Process

It is impossible to separate emergency planning from the facility where the plan will be put into action. Emergency planning must be integral to a functional facility. It must support the main mission and the people who must carry it out. It must work when all else fails. Designers first must obtain firm commitment and backing from top management. Commitment is always easier to get if top management has experienced first-hand a major earthquake or powerful storm. Fear is a powerful source of motivation.

Disaster planning and recovery is an art, a science, and a technology. Like entities such as the Institute of Electrical and Electronics Engineers (IEEE) or the Society of Broadcast Engineers (SBE), disaster planners have their own professional groups and certification standards. States such as California provide year-round classroom training for government disaster planners. Many planners work full time for the military or in the public safety arenas of government. Others have found homes in businesses who recognize that staying in business after a major disaster is smart business. Still others offer their skills and services as consultants to entities who need to *jump start* their disaster planning process.

The technical support group should have responsibility or supervision over the environmental infrastructure of a critical communications facility. Without oversight, electronic systems are at the mercy of whomever controls that environment. Local emergencies can be triggered by preventible failures in air supply systems, roof leaks, or uncoordinated telephone, computer, or AC wiring changes. Seemingly harmless acts such as employees plugging electric heaters into the wrong AC outlet have brought down entire facilities. Successful practitioners of systems design and support must take daily emergencies into account in the overall planning process. To do otherwise risks rapid doom, if not swift unemployment.

Starting Your Planning Process: Listing Realistic Risks

Your realistic risk list should contain specific hazards based on local conditions such as

- Regional high water marks for the 100 and 150 year storms
- Regional social, political, and governmental conditions
- Regional commercial electrical power reliability
- Regional weather conditions
- Regional geography
- Regional geology

You should assess specific local hazards that could be triggered by

- Threats from present or former employees who may hold grudges
- External parties who are likely to get mad at your organization
- Other factors that could make you an easy target
- Nearby man-made hazards
- Special on-site hazards
- Neighbors
- Construction of your facility
- Hazardous materials on the premises
- Communications links to the outside world
- Electrical power
- Other utilities
- Buried pipelines

For example, information from the Northridge earthquake could rewrite seismic building codes for many types of structures. The Northridge quake showed that some high rise structures thought to be earthquake safe are not. Designers should be aware that seismic building codes usually allow for safe evacuation. They do not embody design criteria to prevent major structural damage. Earthquake safe is not earthquake proof.

If possible, get help from emergency planning professionals when you finish your list. They can help you devise a well-written and comprehensive emergency plan. They can also help with detailed research on factors such as geology and hazardous materials. For instance, you may be located near a plant that uses hazardous materials. The ultimate expert advice might be to move before something bad happens! Once there is an agreement on the major goals for operations under emergency conditions, you will have a clear direction for *emergency-ready* facilities planning.

Risk Assessment and Business Resumption Planning

Perform a realistic assessment of the risks that your list suggests. Do not overlook the obvious. If computers, transmitters, or telephone equipment depend on cool air, how can they continue to operate during a heat wave when your one air conditioner has malfunctioned?

What level of reliability should a designer build into a lifeline communications facility? Emergencies introduce *chaos* into the reliability equation. Most engineers are quite happy when a system achieves 99.9999% reliability. Although the glass is more than half full, *four nines reliability* still means eight minutes of outage over a one year period! Reliability is an educated prediction based on a number of factors.

Believers in Murphy's law (anything that can go wrong, will go wrong) know that the 0.0001% outage will occur at the worst possible time. Design beyond that stage of reliability so that you can have a greater chance to cheat Murphy and stay on line during major emergencies. Double, triple, and even quadruple redundancy become realistic options when 100% uptime is the goal.

The new Los Angeles County Emergency Operations Center has three diesel generators. Air handling equipment is also built in triplicate. The building rests on huge rubber isolators at the base of each supporting column. These rubber shock mounts are sometimes called *base isolators.* They are built using laminated layers of Neoprene® rubber and metal. The entire structure floats on 26 of these isolators, designed to let the earth move beneath a building during an earthquake, damping transmission of rapid and damaging acceleration. The Los Angeles County EOC is designed to allow 16 in. of lateral movement from normal, or as much as 32 in. of total movement in one axis. LA County emergency planners are hoping there will not be an earthquake that produces more than 16 in. of lateral displacement. The isolators protect the structure and its contents from most violent lateral movement during earthquakes. The design mandate for this structure came directly from the Board of Supervisors and the Sheriff for the County of Los Angeles.

Can that building or a key system still fail? Of course. It is possible, though unlikely, that all three generators will fail to start. Diminishing returns set in beyond a certain point. Designers must balance realistic redundancy with the uptime expectations for the facility.

A facility designer must have a clear understanding of how important continued operation during and after a major catastrophe will be to its future survival. Disaster planning for facilities facing a hurricane with 125-mi/h winds may well entail boarding up the windows and leaving town for the duration. Others facilities will opt for uninterrupted operation, even in the face of nature's fury. Some will do nothing, adding to the list of victims who sap the resources of those who did prepare. The facility's mission may be critical to local government emergency management. For example, the public tunes to radio and television when disaster strikes. Government emergency managers tune in too.

A **business resumption plan (BRP)** is just as important as an organization's disaster plan. Both are just as essential as having an overall strategic business plan. Some experts argue these three elements should be formulated and updated in concert. When disaster strikes, the first concern of an organization must be for the safety of its employees, customers, vendors, and visitors.

Once *life safety* issues have been addressed, the next step is to perform damage assessments, salvage operations, and, if needed, relocation of critical functions. The BRP is activated once these basic requirements have been met and may influence how they are met.

The focus of business resumption planning is maintaining or resuming core business activities following a disaster. The three major goals of the BRP are: resumption of production and delivery, customer service and notification, and cash flow. A comprehensive BRP can accelerate recovery, saving time, money, and jobs. Like any insurance policy or any other investment, a properly designed BRP will cost you both in time and money. A BRP is an insurance policy, if not a major investment. The following actions are the backbone of a BRP:

- Conduct a **business impact analysis (BIA)**
- Promote employee buy-in and participation
- Seek input starting at the lowest staff levels

- Build a recovery strategy and validation process
- Test your BRP
- Assure a continuous update of the BRP[2]

Most companies do not have the staff to do a proper BIA. If a BIA becomes a top management goal, retain an experienced consultant. Qualified consultants can also conduct training and design test exercises. The BIA process does work. Oklahoma City businesses that had BIAs in place recovered faster than their competitors after the terrorist bombing in 1995. First Interstate Bank was able to open for business the very next day after a fire shut down their Los Angeles highrise headquarters in 1988. On January 17, 1994, the day of the Northridge earthquake, Great Western Bank headquarters suffered major structural damage. They made an almost seamless transition to their Florida operations.

23.4.4 Workplace Safety

Employers must always assure safety in the workplace at all times. Some states such as California have passed legislation that mandates that most employers identify hazards and protect their workers from them. Natural emergencies create special hazards that can maim or kill. A *moment magnitude* 7.4 earthquake can hurl heavy objects such as computer monitors and daggerlike shards from plate glass windows lethally through the air. The Richter Scale is no longer used by serious seismic researchers. Moment magnitude calculates energy release based on the surface area of the planes of two adjacent rock structures (an earthquake fault) and the distance these structures will move in relation to one another. Friction across the surface of the fault holds the rocks until enough stress builds to release energy. Some of the released energy travels via low-frequency wave motion through the rock. These low-frequency waves cause the shaking and sometimes violent accelerations that occur during an earthquake. For more on modern seismic research and risk, please refer to the Reference section for this chapter. A strong foundation of day-to-day safety can lessen the impact of major emergencies. For instance, assuring that plate glass in doors has a safety rating could avoid an accidental workplace injury.

Special and dangerous hazards are found in the information and communications workplace. Tall equipment racks are often not secured to floors, much less secured to load-bearing walls. Preventing equipment racks from tipping over during an earthquake may avoid crippling damage to both systems and people. Bookcases and equipment storage shelves should be secured to walls. Certain objects should be *tethered*, rather than firmly bolted. Although securing heavy objects is mostly common sense, consult experts for special cases. Do not forget seismic-rated safety chains for heavy objects like television studio lights and large speakers.

Computers and monitors are usually not secured to work surfaces. A sudden drop from work station height would ruin the day's output for most computers, video monitors, and their managers. An industry has sprung up that provides innovative fasteners for computer and office equipment. Special Velcro® quick-release anchors and fasteners can support the entire weight of a personal computer or printer, even if the work surface falls over.

Bolting work stations to the floor and securing heavy equipment with properly rated fasteners can address major seismic safety issues. *G* forces measured in an upper story of a high rise building during the Northridge quake were greater than 2.7 times the force of gravity; 1 *g* is of course equal to the force of Earth's gravity. An acceleration of 2 *g* doubles the effective force of a person or object in motion, and nullifies the effectiveness of restraints that worked fine just before the earthquake (Force = (mass) × (acceleration)). Seismic accelerations cause objects to make sudden stops: 60 to zero in 1 sec. A room full of unsecured work stations could do a fair imitation of a slam dance contest, even at lower accelerations. Cables can be pulled loose, monitors can implode, and delicate electronics can be smashed into scrap.

---

[2]The author thanks Mary Carrido, President, MLC & Associates of Irvine, CA, for core elements of business resumption. It is far beyond the scope of this chapter to go into more detail on the BIA process

Even regions to the Earth where there has not been recent seismic activity, have been given a *long overdue* rating by respected seismologists. May be you will be the only one on your block to take such warnings seriously. May be you will be the only one left operational on your block!

Maintaining safety standards is difficult in any size organization. A written safety manual that has specific practices and procedures for normal workplace hazards as well as the emergency-related hazards you identify is not only a good idea, it may lower your insurance rates. If outside workers set foot in your facility, prepare a special safety manual for contractors. Include in it installation standards, compliance with *lock-out/tag-out*, and emergency contact names, and phone numbers. (Lock-out/tag-out is a set of standard safety policies that assure that energy is removed from equipment during installation and maintenance. It assures that every member of a work detail is clear before power is reapplied.) Make sure outside contractors carry proper insurance and are qualified, licensed, or certified to do the work for which you contract.

23.4.5   Outside Plant Communications Links

Your facility may be operational, but failure of a wire, microwave, or fiber communications link could be devastating. All outside plant links discussed next presuppose proper installation. For wire and fiber, this means adequate *service loops* (coiled slack) so quake and wind stresses will not snap taut lines. It means that the telephone company has installed terminal equipment so it will not fall over in an earthquake or be easily flooded out. A range of backup options are available.

Outside Plant Wire

Local telephone companies still use a lot of wire. If your facility is served only by wire on telephone poles or underground in flood-prone areas, you may want what the telephone industry calls *alternate routing*. Alternate routing from your location to another **central office (CO)** may be very costly since the next nearest CO is rarely close. Ask to see a map of the proposed alternate route. If it is alternate only to the next block, or duplicates your telephone pole or underground risk, the advantage you gain will be minimal.

Most telephone companies can designate as an *essential service* a limited block of telephone numbers at a given location for lifeline communications. Lines so designated are usually found at hospitals and public safety headquarters. Contact your local phone company representative to see if your facility can qualify. Many broadcasters who have close ties to local government emergency management should easily qualify.

Microwave Links

Wind and seismic activity can cause microwave dishes to go out of alignment. Earthquake-resistant towers and mounts can help prevent alignment failure, even for wind-related problems. Redundant systems should be considered part of your solution. A duplicate microwave system might lead to a false sense of security. Consider a nonmicrowave backup, such as fiber, for a primary microwave link. Smoke, heavy rain, and snow storms can cause enough path loss to disable otherwise sound wireless systems.

Fiber Optics Links

If you are not a fiber customer today, you will be tomorrow. Telephone companies will soon be joined by other providers to seek your fiber business. You may be fortunate enough to be served by separate fiber vendors with separate fiber systems and routing to enhance reliability and uptime. Special installation techniques are essential to make sure fiber links will not be bothered by earth movement, subject to vandalism, or vulnerable to single-point failure. Single-point failure can occur in any system. Single-point failure analysis and prevention is based on simple concepts: A chain is only as strong as its weakest link, but two chains, equally strong, may have the same weak link. The lesson may be make one chain much stronger, or use three chains of a material that has different stress properties.

Fiber should be installed underground in a sturdy plastic sheath, called an interliner. Interliners are usually colored bright orange to make them stand out in trenches, manholes and other places where careless digging and prodding could spell disaster. This sheath offers protection from sharp rocks or other forces that might cause a nick or break in the armor of the cable, or actually sever one or more of the bundled fibers. Cable systems that only have aerial rights-of-way on utility poles for their fiber may not prove as reliable in some areas as underground fiber. Terminal equipment for fiber should be installed in earthquake-secure equipment racks away from flooding hazards. Fiber electronics should have a minimum of two parallel DC power supplies, which are in turn paralleled with rechargeable battery backup.

Sonet® technology is a proven approach you should look for from your fiber vendor. This solution is based on topology that looks like a circle or ring. A fiber optics cable could be cut just like a wire or cable. A ringlike network will automatically provide a path in the other direction, away from the break. Caution! Fiber installations that run through unsealed access points, such as manholes, can be an easy target for terrorism or vandalism.

Satellite

Ku- or C-band satellite is a costly but effective way to link critical communications elements. C band has an added advantage over Ku during heavy rain or snow storms. Liquid or frozen water can disrupt Ku-band satellite transmission. A significant liability of satellite transmission for ultrareliable facilities is the possibility that a storm could cause a deep fade, even for C-band links. Another liability is short but deep semiannual periods of sun outage when a link is lost while the sun is focused directly into a receive dish. Although these periods are predictable and last for only a minute or two, there is nothing that can prevent their effect unless you have alternate service on another satellite with a different sun outage time, or terrestrial backup.

23.4.6 Emergency Power and Batteries

Uninterruptable power supplies (UPS) are common in the information workplace. From small UPS that plug into wall outlets at a personal computer work station, to giant units that can power an entire facility, they all have one thing in common, batteries. UPS batteries have a finite life span. Once exceeded, a UPS is nothing more than an expensive door stop. UPS batteries must be tested regularly. Allow the UPS to go on line to test it. Some UPS test themselves automatically. Routinely pull the UPS AC plug out of the wall for a manual test. Some UPS applications require hours of power, whereas some only need several minutes. Governing factors are

- Availability of emergency power that can be brought on line fast
- A need to keep systems alive long enough for a graceful shutdown
- Systems so critical that they can never go down

Although UPS provide emergency power when the AC mains are dead, many are programmed with another electronic agenda: Protect the devices plugged in from what the UPS thinks is *bad power*. Many diesel generators in emergency service are not sized for the load they have to carry, or may not have proper power factor correction. Computers and other devices with switching power supplies can distort AC power wave forms; the result: bad power.

After a UPS comes on line, it should shut down after the emergency generator picks up the load and charges its batteries. If it senses the AC equivalent of poison, it stays on or cycles on and off. Its battery eventually runs down. Your best defense is to test your entire emergency power system under full load. If a UPS cycles on and off to the point that its batteries run down, you must find out why. Consult your UPS manufacturer, service provider, or service manual to see if your UPS can be adjusted to be more tolerant. Some UPS cycling cannot be avoided with engine-based emergency power, especially if heavy loads such as air conditioner compressors switch on and off line.

Technicians sometimes believe that starting an emergency generator with no equipment load is an adequate weekly test. Even a 30-min test will not get the engine up to proper operating temperature. If your generator is diesel driven, this may lead to *wet stacking*, cylinder glazing, and piston rings that can lose proper seating. Wet stacking occurs when a generator is run repeatedly with no load or a light load.

When the generator is asked to come on line to power a full equipment load, deposits that build up during no-load tests prevent it from developing full power. The engine will also not develop full power if its rings are misseated and there is significant cylinder glazing. The cure is to always test with the load your diesel has to carry during an emergency. If that is not possible, obtain a resistive load bank so you can simulate a full load for an hour or two of hard running several times per year. A really hard run should burn out accumulated carbon, reseat rings, and deglaze cylinder walls.

Fuel stored in tanks gets old and old fuel is unreliable. Gum and varnish can form. Fuel begins to break down. Certain forms of algae can grow in diesel oil, especially at the boundary layer between fuel and the water that can accumulate at the bottom of most tanks. Fuel additives can extend the storage period and prevent algae growth. A good filtering system, and a planned program of cycling fuel through it, can extend storage life dramatically. Individual fuel chemical composition, fuel conditioners, and the age and type of storage tank all affect useful fuel life.

There are companies that will analyze your fuel. They can filter out dirt, water, and debris that can rob your engine of power. The cost of additives and fuel filtering is nominal compared to the cost of new fuel plus hazardous material disposal charges for old fuel. Older fuel tanks can spring leaks that either introduce water into the fuel, or introduce you to a costly hazardous materials clean up project, your tank will be out of service while it is being replaced, and fuel carrying dirt or water can stop the engine.

While you are depending on an emergency generator for your power, you would hate to see it stop. A running generator will consume fuel, crankcase oil, and possibly radiator coolant. You should know your generator's crankcase oil consumption rate so you can add oil well before the engine grinds to a screeching, nonlubricated halt. Water-cooled generators must to be checked periodically to make sure there is enough coolant in the radiator. Make sure you have enough coolant and oil to get the facility through a minimum of one week of constant duty.

Most experts recommend a generator health check every six months. Generators with engine block heaters put special stress on fittings and hoses. Vibration can loosen bolts, crack fittings, and fatigue wires and connectors. If you application is supercritical, a second generator may give you a greater margin of safety. Your generator maintenance technician should take fuel and crankcase oil samples for testing at a qualified laboratory. The fuel report will let you know if your storage conditions are acceptable. The crankcase oil report might find microscopic metal particles, early warning of a major failure. How long will a generator last? Some engine experts say that a properly maintained diesel generator set can run in excess of 9,000 h before it would normally need to be replaced.

Mission dictates need. Need dictates reliability. If the design budget permits, a second or even third emergency generator is a realistic insurance policy. When you are designing a facility you know must never fail, consider redundant UPS wired in parallel. Consult the vendor for details on wiring needs for multiphase parallel UPS installations. During major overhauls and generator work, make sure you have a local source for reliable portable power. High-power diesel generators on wheels are common now to supply field power for events from rock concerts to movie shoots. Check your local telephone directory. If you are installing a new diesel, remember that engines over a certain size may have to be licensed by your local air quality management district and that permits must be obtained to construct and store fuel in an underground tank.

23.4.7 Air Handling Systems

Equipment crashes when it gets too hot. Clean, cool, dry, and pollutant-free air in generous quantities is critical for modern communications facilities. If you lease space in a high-rise, you may not have your own air system. Many building systems often have no backup, are not supervised nights and weekends, and may have uncertain maintenance histories. Your best protection is to get the exact terms for air conditioning nailed down in your lease. You may wish to consider adding your own backup system, a costly but essential strategy if your building air supply is unreliable or has no backup. Several rental companies specialize in emergency portable industrial-strength air conditioning. An emergency contract for **heating ventilating and air conditioning (HVAC)** that can be invoked with a phone call could save you hours or even days of downtime. Consider buying a portable HVAC unit if you are protecting a supercritical facility.

Wherever cooling air comes from, there are times when you need to make sure the system can be forced to recirculate air within the building, temporarily becoming a closed system. Smoke or toxic fumes from a fire in the neighborhood can enter an open system. Toxic air could incapacitate your people in seconds. With some advanced warning, forcing the air system to full recirculation could avoid or forestall calamity. It could buy enough time to arrange an orderly evacuation and transition to an alternate site.

23.4.8   Water Hazards

Water in the wrong place at the wrong time can be part of a larger emergency or be its own emergency. A simple mistake such as locating a water heater where it can flood out electrical equipment can cause short circuits when it finally wears out and begins to leak. Unsecured water heaters can tear away from gas lines, possibly causing an explosion or fire in an earthquake. The water in that heater could be lost, depriving employees of a source of emergency drinking water.

Your facility may be located near a source of water that could flood you out. Many businesses are located in flood plains that see major storms once every 100 or 150 years. If you happen to be on watch at the wrong time of the century, you may wish that you had either located elsewhere, or stocked a very large supply of sand bags.

Remember to include any wet or dry pipe fire sprinkler systems as potential water hazards.

23.4.9   Electromagnetic Pulse Protection (EMP)

The **electromagnetic pulse (EMP)** phenomenon associated with nuclear explosions can disable almost any component in a communications system. EMP energy can enter any component or system coupled to a wire or metal surface directly, capacitively, or inductively. Some chemical weapons can produce EMP, but on a smaller scale. The Federal Emergency Management Agency (FEMA) publishes a three volume set of documents on EMP. They cover the theoretical basis for EMP protection, protection applications, and protection installation. FEMA has been involved in EMP protection since 1970 and is charged at the federal level with the overall direction of the EMP program. FEMA provides detailed guidance and, in some cases, direct assistance on EMP protection to critical communications facilities in the private sector. AM, FM, and television transmitter facilities that need EMP protection should discuss EMP protection tactics with a knowledgeable consultant before installing protection devices on radio frequency (RF) circuitry. EMP devices such as gas discharge tubes can fail in the presence of high-RF voltage conditions and disable facilities through such a failure.

23.4.10   Alternate Sites

No matter how well you plan, something still could happen that will require you to abandon your facility for some period of time. Government emergency planners usually arrange for an alternate site for their EOCs. Communications facilities can sign mutual aid agreements. Sometimes this is the only way to access telephone lines, satellite uplink equipment, microwave, or fiber on short notice. If management shows reluctance to share, respectfully ask what they would do if their own facility is rendered useless.

23.4.11   Security

It is a fact of modern life that man-caused disasters must now enter into the planning and risk assessment process. Events ranging from terrorism to poor training can cause the most mighty organization to tumble. The World Trade Center and Oklahoma City bombings are a warning to us all. Your risk assessment might even prompt you to relocate if you are too close to *ground zero*.

Federal Communications Commission (FCC) rules still state that licensees of broadcast facilities must protect their facilities from hostile takeover. Breaches in basic security have often led to serious incidents at a number of places throughout the county. It has even happened at major market television stations. Here are the basics:

- Approve visits from former employees through their former supervisors.
- Escort nonemployees in critical areas.
- Assure that outside doors are never propped open.
- Secure roof hatches from the inside and have alarm contacts on the hatch.
- Use identification badges when employees will not know each other by sight.
- Check for legislation that may require a written safety and security plan.
- Use video security and card key systems where warranted.
- Repair fences, especially at unmanned sites.
- Install entry alarms at unattended sites; test weekly.
- Redesign to limit the places bombs could be planted.
- Redesign to prevent unauthorized entry.
- Redesign to limit danger from outside windows.
- Plan for fire, bomb threats, hostage situations, terrorist takeovers.
- Plan a safe way to shut the facility down in case of invasion.
- Plan guard patrol schedules to be random, not predictable.
- Plan for off-site relocation and restoration of services on short notice.

California Senate Bill 198 mandates that California businesses with more than 100 employees write an industrial health and safety plan for each facility addressing workplace safety, hazardous materials spills, employee training, and emergency response.

23.4.12   Workplace and Home: Hand-in-Hand Preparedness

A critical facility deprived of its staff will be paralyzed just as surely as if all of the equipment suddenly disappeared. Employees may experience guilt if they are at work when a regional emergency strikes, and they do not know what is happening at home. The first instinct is to go home. This is often the wrong move. Blocked roads, downed bridges, and flooded tunnels are dangerous traps, especially at night. People who leave work during emergencies, especially people experiencing severe stress, often become victims. Encourage employees to prepare their homes, families, and pets for the same types of risks the workplace will face. Emergency food and water and a supply of fresh batteries in the refrigerator are a start. Battery-powered radios and flashlights should be tested regularly. If employees or their families require special foods, prescription drugs, eyewear, oxygen, over-the-counter pharmaceuticals, sun block or bug repellent, remind them to have an adequate supply on hand to tide them over for a lengthy emergency.

Heavy home objects like bookcases should be secured to walls so they will not tip over. Secure or move objects mounted on walls over beds. Make sure someone in the home knows how to shut off natural gas. An extra long hose can help for emergency fire fighting or help drain flooded areas. Suggest family *hazard hunts*. Educate employees on what you are doing to make the workplace safe. The same hazards that can hurt, maim, or kill in the workplace can do the same at home. Personal and company vehicles should all have emergency kits that contain basic home or business emergency supplies. Food, water, comfortable shoes, and old clothes should be added. If their families are prepared at home or on the road, employees will have added peace of mind. It may sustain them until they can get home safely.

An excellent home family preparedness measure is to identify a distant relative or friend who can be the emergency message center. Employees may be able to call a relative from work to find out their family is safe and sound. Disasters that impair telephone communications teach us that it is often possible to make and receive long distance calls when a call across the street will not get through. Business emergency planners should not overlook this hint. A location in another city, or a key customer or supplier may make a good out-of-area emergency contact.

23.4.13 Expectations, 9-1-1, and Emergencies

Television shows depicting 9-1-1 saving lives over the telephone are truly inspirational. But during a major emergency, resources normally available to 9-1-1 services, including their very telephone system, may be unavailable. Emergency experts used to tell us to be prepared to be self-sufficient at the neighborhood and business level for 72 h or more. Some now suggest a week or longer. Government will not respond to every call during a major disaster. That is a fact.

Even experienced communications professionals sometimes forget that an overloaded telephone exchange cannot supply dial tone to all customers at once. Emergency calls will often go through if callers wait patiently for dial tone. If callers do not hear dial tone after 10 or 15 min, it is safe to assume that there is a more significant problem.

23.4.14 Business Rescue Planning for Dire Emergencies

When people are trapped and professionals cannot get through, our first instinct may be to attempt a rescue. Professionals tell us that more people are injured or killed in rescue attempts during major emergencies than are actually saved. Experts in **urban search and rescue (USAR)** not only have the know-how to perform their work safely, but have special tools that make this work possible under impossible conditions. The *jaws of life,* hydraulic cutters used to free victims from wrecked automobiles, is a common USAR tool. Pneumatic jacks that look like large rubber pillows can lift heavy structural members in destroyed buildings to free trapped people.

You, as a facilities designer, may never be faced with a life-or-death decision concerning a rescue when professionals are not available. Those in the facilities you design may be faced with tough decisions. Consider that your design could make their job easier or more difficult. Also consider recommending USAR training for those responsible for on-line management and operations of the facility as a further means to ensure readiness.

23.4.15 Managing Fear

Anyone who says they are not scared while experiencing a hurricane, tornado, flood, or earthquake is either lying or foolish. Normal human reactions when an emergency hits are colored by a number of factors, including fear. As the emergency unfolds, we progress from fear of the unknown, to fear of the known. While preparedness, practice, and experience may help keep fears in check, admitting fear and the normal human response to fear can help us keep calm.

Some people prepare mentally by reviewing their behavior during personal, corporate, or natural emergencies. Then they consider how they could have been better prepared to transition from *normal* human reactions like shock, denial, and panic, to abnormal reactions like grace, acceptance, and steady performance. The latter behaviors reassure those around them and encourage an effective emergency team. Grace under pressure is not a bad goal.

Another normal reaction most people experience is a rapid change of focus toward one's own personal well being. "Am I OK?" is a very normal question at such times. Even the most altruistic people have moments during calamities when they regress. They temporarily become selfish children. Once people know they do not require immediate medical assistance, they can usually start to focus again on others and on the organization.

Defining Terms

Business impact analysis (BIA): A formal study of the impact of a risk or multiple risks on a specific business. A properly conducted BIA becomes critical to the *business recovery plan.*

Business resumption planning (BRP): A blueprint to maintain or resume core business activities following a disaster. Three major goals of BRP are resumption of products and services, customer service, and cash flow.

Central office (CO): Telephone company jargon for the building where local switching is accomplished.

Electromagnetic pulse (EMP): A high burst of energy associated most commonly with nuclear explosions. EMP can instantly destroy many electronic systems and components.

Emergency operations center (EOC): A location where emergency managers receive damage assessments, allocate resources, and begin recovery.

Heating, ventilation, and air conditioning (HVAC): Architectural acronym.

Incident commander (IC): The title of the person in charge at an emergency scene from the street corner to an emergency involving an entire state or region.

Incident command system (ICS): An effective emergency management model invented by fire fighters in California.

Urban Search and Rescue (USAR): Emergency management acronym.

References

Baylus, E. 1992. *Disaster Recovery Handbook.* Chantico, NY.

Fletcher, R. 1990. Federal Response Plan. Federal Emergency Management Agency, Washington, DC.

FEMA. 1991. *Electromagnetic Pulse Protection Guidance*, Vols. 1–3. Federal Emergency Management Agency, Washington, DC.

Handmer, J. and Parker, D. 1993. *Hazard Management and Emergency Planning.* James and James Science, NY.

Rothstein Associates. 1993. *The Rothstein Catalog on Disaster Recovery and Business Resumption Planning.* Rothstein Associates.

Further Information

Associations/Groups:

The Association of Contingency Planners, 14775 Ventura Boulevard, Suite 1-885, Sherman Oaks, CA 91483.

Business and Industry Council for Emergency Planning and Preparedness (BICEPP), P.O. Box 1020, Northridge, CA 91328.

The Disaster Recovery Institute (DRT), 1810 Craig Road, Suite 125, St. Louis, MO 63146.

DRI holds national conferences, and publishes the *Disaster Recovery Journal.*

Earthquake Engineering Research Institute (EERI), 6431 Fairmont Avenue, Suite 7, El Cerritos, CA 94530.

National American Red Cross, 2025 E Street, NW, Washington, DC 20006.

National Center for Earthquake Engineering Research, State University of New York at Buffalo, Science and Engineering Library-304, Capen Hall, Buffalo, NY 14260.

National Coordination Council on Emergency Management (NCCEM), 7297 Lee Highway, Falls Church, VA 22042.

National Hazards Research & Applications Information Center, Campus Box 482, University of Colorado, Boulder, CO 80309.

Business Recovery Planning:

Harris Devlin Associates, 1285 Drummers Lane, Wayne, PA 19087.

Industrial Risk Insurers (IRI), 85 Woodland Street, Hartford, CT 06102.

MLC & Associates, Mary Carrido, President, 15398 Eiffel Circle, Irvine, CA 92714.

Price Waterhouse, Dispute Analysis and Corporate Recovery Dept., 555 California Street, Suite 3130, San Francisco, CA 94104.

Resource Referral Service, P.O. Box 2208, Arlington, VA 22202.

The Workman Group, Janet Gorman, President, P.O. Box 94236, Pasadena, CA 91109.

Life Safety/Disaster Response:

Caroline Pratt & Associates, 24104 Village #14, Camarillo, CA 93013.

Industry Training Associates, 3363 Wrightwood Drive, Suite 100, Studio City, CA 91604.

Emergency Supplies:

BEST Power Technology, P.O. Box 280, Necedah, WI 54646 (UPS).
Exide Electronics Group, Inc., 8521 Six Forks Road, Raleigh, NC 27615.
Extend-A-Life, Inc., 1010 South Arroyo, Parkway #7, Pasadena, CA 91105.
Velcro® USA, P.O. Box 2422, Capistrano Beach, CA 92624.
Worksafe Technologies, 25133 Avenue Tibbets, Building F. Valencia, CA.

Construction/Design/Seismic Bracing:

American Institute of Architects, 1735 New York Avenue, NW, Washington, DC 20006.
DATA Clean Corporation (800-328-2256).

Geotechnical/Environmental Consultants:

H.J. Degenkolb Associates, Engineers, 350 Sansome Street, San Francisco, CA 94104 .
Leighton and Associates, Inc., 17781 Cowan, Irvine, CA 92714.

Miscellaneous:

Commercial Filtering, Inc., 5205 Buffalo Avenue, Sherman Oaks, CA 91423 (Fuel Filtering).
Data Processing Security, Inc., 200 East Loop 820, Forth Worth, TX 76112.
EDP Security, 7 Beaver Brook Road, Littleton, MA 01460.
ENDUR-ALL Glass Coatings, Inc., 23018 Ventura Blvd., Suite 101, Woodland Hills, CA 91464.
Mobile Home Safety Products, 28165 B Front Street, Suite 121, Temecula, CA 92390.

23.5   Conversion Factors

Jerry C. Whitaker

23.5.1   Introduction

Engineers often find it necessary to convert from one unit of measurement to another. The following table of conversion factors provides a convenient method of accomplishing the task. In the following table, a more complete listing of conversion factors is presented, grouped in alphabetical order.

To Convert	Into	Multiply by
	A	
abcoulomb	statcoulombs	2.998×10^{10}
acre	square chain (Gunters)	10
acre	rods	160
acre	square links (Gunters)	1×10^5
acre	hectare or square hectometer	0.4047
acres	square feet	43,560.0
acres	square meters	4,047
acres	square miles	1.562×10^{-3}
acres	square yards	4,840
acre-feet	cubic feet	43,560.0
acre-feet	gallons	3.259×10^5
amperes per square centimeter	ampere per square inch	6.452
amperes per square centimeter	ampere per square meter	10^4
amperes per square inch	ampere per square centimeter	0.1550
amperes per square inch	ampere per square meter	1,550.0
amperes per square meter	ampere per square centimeter	10^{-4}
amperes per square meter	ampere per square inch	6.452×10^{-4}
ampere-hours	coulombs	3,600.0
ampere-hours	faradays	0.03731

(continued)

To Convert	Into	Multiply by
ampere-turns	gilberts	1.257
ampere-turns per centimeter	ampere-turns per inch	2.540
ampere-turns per centimeter	ampere-turns per meter	100.0
ampere-turns per centimeter	gilberts per centimeter	1.257
ampere-turns per inch	ampere-turns per centimeter	0.3937
ampere-turns per inch	ampere-turns per meter	39.37
ampere-turns per inch	gilberts per centimeter	0.4950
ampere-turns per meter	ampere-turns per centimeter	0.01
ampere-turns per meter	ampere-turns per inch	0.0254
ampere-turns per meter	gilberts per centimeter	0.01257
Angstrom unit	inch	3937×10^{-9}
Angstrom unit	meter	1×10^{-10}
Angstrom unit	micron or mu (μ)	1×10^{-4}
are	acre (U.S.)	.02471
ares	square yards	119.60
ares	acres	0.02471
ares	square meters	100.0
astronomical unit	kilometers	1.495×10^{8}
atmospheres	ton per square inch	0.007348
atmospheres	centimeter of mercury	76.0
atmospheres	foot of water (at 4°C)	33.90
atmospheres	inch of mercury (at 0°C)	29.92
atmospheres	kilogram per square centimeter	1.0333
atmospheres	kilogram per square meter	10,332
atmospheres	pounds per square inch	14.70
atmospheres	tons per square foot	1.058
	B	
barrels (U.S., dry)	cubic inch	7056
barrels (U.S., dry)	quarts (dry)	105.0
barrels (U.S., liquid)	gallons	31.5
barrels (oil)	gallons (oil)	42.0
bars	atmospheres	0.9869
bars	dyne per square centimeter	10^{4}
bars	kilogram per square meter	1.020×10^{4}
bars	pound per square foot	2,089
bars	pound per square inch	14.50
Baryl	dyne per square centimeter	1.000
bolt (U.S., cloth)	meter	36.576
British thermal unit	liter-atmosphere	10.409
British thermal unit	erg	1.0550×10^{10}
British thermal unit	foot-pound	778.3
British thermal unit	gram-calorie	252.0
British thermal unit	horsepower-hour	3.931×10^{-4}
British thermal unit	joule	1,054.8
British thermal unit	kilogram-calorie	0.2520
British thermal unit	kilogram-meter	107.5
British thermal unit	kilowatt-hour	2.928×10^{-4}
British thermal unit per hour	foot-pound per second	0.2162
British thermal unit per hour	gram-calorie per second	0.0700
British thermal unit per hour	horsepower-hour	3.929×10^{-4}
British thermal unit per hour	watts	0.2931
British thermal unit per minute	foot-pound per second	12.96
British thermal unit per minute	horsepower	0.02356
British thermal unit per minute	kilowatts	0.01757
British thermal unit per minute	watts	17.57
British thermal unit per square foot per minute	watts per square inch	0.1221
bucket (British, dry)	cubic centimeter	1.818×10^{4}

To Convert	Into	Multiply by
bushels	cubic foot	1.2445
bushels	cubic inch	2,150.4
bushels	cubic meter	0.03524
bushels	liters	35.24
bushels	pecks	4.0
bushels	pints (dry)	64.0
bushels	quarts (dry)	32.0
	C	
calories, gram (mean)	British thermal unit (mean)	3.9685×10^{-3}
candle per square centimeter	lamberts	3.142
candle per square inch	lamberts	.4870
centares (centiares)	square meters	1.0
centigrade	Fahrenheit	$(C° \times 9/5) + 32$
centigrams	grams	0.01
centiliter	ounce fluid (U.S.)	0.3382
centiliter	cubic inch	0.6103
centiliter	drams	2.705
centiliters	liters	0.01
centimeters	feet	3.281×10^{-2}
centimeters	inches	0.3937
centimeters	kilometers	10^{-5}
centimeters	meters	0.01
centimeters	miles	6.214×10^{-6}
centimeters	millimeter	10.0
centimeters	mils	393.7
centimeters	yards	1.094×10^{-2}
centimeter-dynes	centimeter-grams	1.020×10^{-3}
centimeter-dynes	meter-kilogram	1.020×10^{-8}
centimeter-dynes	pound-feet	7.376×10^{-8}
centimeter-grams	centimeter-dynes	980.7
centimeter-grams	meter-kilogram	10^{-5}
centimeter-grams	pound-feet	7.233×10^{-5}
centimeters of mercury	atmospheres	0.01316
centimeters of mercury	feet of water	0.4461
centimeters of mercury	kilogram per square meter	136.0
centimeters of mercury	pounds per square foot	27.85
centimeters of mercury	pounds per square inch	0.1934
centimeters per second	feet per minute	1.9686
centimeters per second	feet per second	0.03281
centimeters per second	kilometer per hour	0.036
centimeters per second	knot	0.1943
centimeters per second	meter per minute	0.6
centimeters per second	mile per hour	0.02237
centimeters per second	mile per minute	3.728×10^{-4}
centimeters per second per second	feet per second per second	0.03281
centimeters per second per second	kilometer per hour per second	0.036
centimeters per second per second	meters per second per second	0.01
centimeters per second per second	miles per hour per second	0.02237
chain	inches	792.00
chain	meters	20.12
chain (surveyors' or Gunter's)	yards	22.00
circular mils	square centimeter	5.067×10^{-6}
circular mils	square mils	0.7854
circular mils	square inches	7.854×10^{-7}
circumference	radians	6.283
cord	cord feet	8
cord feet	cubic feet	16

(*continued*)

To Convert	Into	Multiply by
coulomb	statcoulombs	2.998×10^9
coulombs	faradays	1.036×10^{-5}
coulombs per square centimeter	coulombs per square inch	64.52
coulombs per square centimeter	coulombs per square meter	10^4
coulombs per square inch	coulombs per square centimeter	0.1550
coulombs per square inch	coulombs per square meter	1,550
coulombs per square meter	coulombs per square centimeter	10^{-4}
coulombs per square meter	coulombs per square inch	6.452×10^{-4}
cubic centimeters	cubic feet	3.531×10^{-5}
cubic centimeters	cubic inches	0.06102
cubic centimeters	cubic meters	10^{-6}
cubic centimeters	cubic yards	1.308×10^{-6}
cubic centimeters	gallons (U.S. liquid)	2.642×10^{-4}
cubic centimeters	liters	0.001
cubic centimeters	pints (U.S. liquid)	2.113×10^{-3}
cubic centimeters	quarts (U.S. liquid)	1.057×10^{-3}
cubic feet	bushels (dry)	0.8036
cubic feet	cubic centimeter	28,320.0
cubic feet	cubic inches	1,728.0
cubic feet	cubic meters	0.02832
cubic feet	cubic yards	0.03704
cubic feet	gallons (U.S. liquid)	7.48052
cubic feet	liters	28.32
cubic feet	pints (U.S. liquid)	59.84
cubic feet	quarts (U.S. liquid)	29.92
cubic feet per minute	cubic centimeter per second	472.0
cubic feet per minute	gallons per second	0.1247
cubic feet per minute	liters per second	0.4720
cubic feet per minute	pounds of water per minute	62.43
cubic feet per second	million galllons per day	0.646317
cubic feet per second	gallons per minute	448.831
cubic inches	cubic centimeters	16.39
cubic inches	cubic feet	5.787×10^{-4}
cubic inches	cubic meter	1.639×10^{-5}
cubic inches	cubic yards	2.143×10^{-5}
cubic inches	gallons	4.329×10^{-3}
cubic inches	liters	0.01639
cubic inches	mil-feet	1.061×10^5
cubic inches	pints (U.S. liquid)	0.03463
cubic inches	quarts (U.S. liquid)	0.01732
cubic meters	bushels (dry)	28.38
cubic meters	cubic centimeter	10^6
cubic meters	cubic feet	35.31
cubic meters	cubic inches	61,023.0
cubic meters	cubic yards	1.308
cubic meters	gallons (U.S. liquid)	264.2
cubic meters	liters	1,000.0
cubic meters	pints (U.S. liquid)	2,113.0
cubic meters	quarts (U.S. liquid)	1,057
cubic yards	cubic centimeter	7.646×10^5
cubic yards	cubic feet	27.0
cubic yards	cubic inches	46,656.0
cubic yards	cubic meters	0.7646
cubic yards	gallons (U.S. liquid)	202.0
cubic yards	liters	764.6
cubic yards	pints (U.S. liquid)	1,615.9
cubic yards	quarts (U.S. liquid)	807.9
cubic yards per minute	cubic feet per second	0.45
cubic yards per minute	gallons per second	3.367
cubic yards per minute	liters per second	12.74

To Convert	Into	Multiply by
	D	
Dalton	gram	1.650×10^{-24}
days	seconds	86,400.0
decigrams	grams	0.1
deciliters	liters	0.1
decimeters	meters	0.1
degrees (angle)	quadrants	0.01111
degrees (angle)	radians	0.01745
degrees (angle)	seconds	3,600.0
degrees per second	radians per second	0.01745
degrees per second	revolutions per minute	0.1667
degrees per second	revolutions per second	2.778×10^{-3}
dekagrams	grams	10.0
dekaliters	liters	10.0
dekameters	meters	10.0
drams (apothecaries' or troy)	ounces (avoirdupois)	0.1371429
drams (apothecaries' or troy)	ounces (troy)	0.125
drams (U.S., fluid or apothecaries')	cubic centimeter	3.6967
drams	grams	1.7718
drams	grains	27.3437
drams	ounces	0.0625
dyne per centimeter	erg per square millimeter	0.01
dyne per square centimeter	atmospheres	9.869×10^{-7}
dyne per square centimeter	inch of mercury at 0°C	2.953×10^{-5}
dyne per square centimeter	inch of water at 4°C	4.015×10^{-4}
dynes	grams	1.020×10^{-3}
dynes	joules per centimeter	10^{-7}
dynes	joules per meter (newtons)	10^{-5}
dynes	kilograms	1.020×10^{-6}
dynes	poundals	7.233×10^{-5}
dynes	pounds	2.248×10^{-6}
dynes per square centimeter	bars	10^{-6}
	E	
ell	centimeter	114.30
ell	inches	45
em, pica	inch	0.167
em, pica	centimeter	0.4233
erg per second	dyne–centimeter per second	1.000
ergs	British thermal unit	9.480×10^{-11}
ergs	dyne-centimeters	1.0
ergs	foot-pounds	7.367×10^{-8}
ergs	gram-calories	0.2389×10^{-7}
ergs	gram-centimeter	1.020×10^{-3}
ergs	horsepower-hour	3.7250×10^{-14}
ergs	joules	10^{-7}
ergs	kilogram-calories	2.389×10^{-11}
ergs	kilogram-meters	1.020×10^{-8}
ergs	kilowatt-hour	0.2778×10^{-13}
ergs	watt-hours	0.2778×10^{-10}
ergs per second	British thermal unit per minute	$5,688 \times 10^{-9}$
ergs per second	foot-pound per minute	4.427×10^{-6}
ergs per second	foot-pound per second	7.3756×10^{-8}
ergs per second	horsepower	1.341×10^{-10}
ergs per second	kilogram-calories per minute	1.433×10^{-9}
ergs per second	kilowatts	10^{-10}

(continued)

To Convert	Into	Multiply by
	F	
farads	microfarads	10^6
faraday per second	ampere (absolute)	9.6500×10^4
faradays	ampere-hours	26.80
faradays	coulombs	9.649×10^4
fathom	meter	1.828804
fathoms	feet	6.0
feet	centimeters	30.48
feet	kilometers	3.048×10^{-4}
feet	meters	0.3048
feet	miles (nautical)	1.645×10^{-4}
feet	miles (statute)	1.894×10^{-4}
feet	millimeters	304.8
feet	mils	1.2×10^4
feet of water	atmospheres	0.02950
feet of water	inch of mercury	0.8826
feet of water	kilogram per square centimeter	0.03048
feet of water	kilogram per square meter	304.8
feet of water	pounds per square foot	62.43
feet of water	pounds per square inch	0.4335
feet per minute	centimeter per second	0.5080
feet per minute	feet per second	0.01667
feet per minute	kilometer per hour	0.01829
feet per minute	meters per minute	0.3048
feet per minute	miles per hour	0.01136
feet per second	centimeter per second	30.48
feet per second	kilometer per hour	1.097
feet per second	knots	0.5921
feet per second	meters per minute	18.29
feet per second	miles per hour	0.6818
feet per second	miles per minute	0.01136
feet per second per second	centimeter per second per second	30.48
feet per second per second	kilometer per hour per second	1.097
feet per second per second	meters per second per second	0.3048
feet per second per second	miles per hour per second	0.6818
feet per 100 feet	per centigrade	1.0
foot-candle	Lumen per square meter	10.764
foot-pounds	British thermal unit	1.286×10^{-3}
foot-pounds	ergs	1.356×10^7
foot-pounds	gram-calories	0.3238
foot-pounds	horsepower per hour	5.050×10^{-7}
foot-pounds	joules	1.356
foot-pounds	kilogram-calories	3.24×10^{-4}
foot-pounds	kilogram-meters	0.1383
foot-pounds	kilowatt-hour	3.766×10^{-7}
foot-pounds per minute	British thermal unit per minute	1.286×10^{-3}
foot-pounds per minute	foot-pounds per second	0.01667
foot-pounds per minute	horsepower	3.030×10^{-5}
foot-pounds per minute	kilogram-calories per minute	3.24×10^{-4}
foot-pounds per minute	kilowatts	2.260×10^{-5}
foot-pounds per second	British thermal unit per hour	4.6263
foot-pounds per second	British thermal unit per minute	0.07717
foot-pounds per second	horsepower	1.818×10^{-3}
foot-pounds per second	kilogram-calories per minute	0.01945
foot-pounds per second	kilowatts	1.356×10^{-3}
furlongs	miles (U.S.)	0.125
furlongs	rods	40.0
furlongs	feet	660.0

To Convert	Into	Multiply by
	G	
gallons	cubic centimeter	3,785.0
gallons	cubic feet	0.1337
gallons	cubic inches	231.0
gallons	cubic meters	3.785×10^{-3}
gallons	cubic yards	4.951×10^{-3}
gallons	liters	3.785
gallons (liquid British Imperial)	gallons (U.S. liquid)	1.20095
gallons (U.S.)	gallons (Imperial)	0.83267
gallons of water	pounds of water	8.3453
gallons per minute	cubic foot per second	2.228×10^{-3}
gallons per minute	liters per second	0.06308
gallons per minute	cubic foot per hour	8.0208
gausses	lines per square inch	6.452
gausses	webers per square centimeter	10^{-8}
gausses	webers per square inch	6.452×10^{-8}
gausses	webers per square meter	10^{-4}
gilberts	ampere-turns	0.7958
gilberts per centimeter	ampere-turns per centimeter	0.7958
gilberts per centimeter	ampere-turns per inch	2.021
gilberts per centimeter	ampere-turns per meter	79.58
gills (British)	cubic centimeter	142.07
gills	liters	0.1183
gills	pints (liquid)	0.25
grade	radian	0.01571
grains	drams (avoirdupois)	0.03657143
grains (troy)	grains (avoirdupois)	1.0
grains (troy)	grams	0.06480
grains (troy)	ounces (avoirdupois)	2.0833×10^{-3}
grains (troy)	pennyweight (troy)	0.04167
grains/U.S. gallon	parts per million	17.118
grains/U.S. gallon	pounds per million gallon	142.86
grains/Imp. gallon	parts per million	14.286
grams	dynes	980.7
grams	grains	15.43
grams	joules per centimeter	9.807×10^{-5}
grams	joules per meter (newtons)	9.807×10^{-3}
grams	kilograms	0.001
grams	milligrams	1,000
grams	ounces (avoirdupois)	0.03527
grams	ounces (troy)	0.03215
grams	poundals	0.07093
grams	pounds	2.205×10^{-3}
grams per centimeter	pounds per inch	5.600×10^{-3}
grams per cubic centimeter	pounds per cubic foot	62.43
grams per cubic centimeter	pounds per cubic inch	0.03613
grams per cubic centimeter	pounds per mil-foot	3.405×10^{-7}
grams per liter	grains per gallon	58.417
grams per liter	pounds per 1000 gallon	8.345
grams per liter	pounds per cubic foot	0.062427
grams per liter	parts per million	1,000.0
grams per square centimeter	pounds per square foot	2.0481
gram-calories	British thermal unit	3.9683×10^{-3}
gram-calories	ergs	4.1868×10^{7}
gram-calories	foot-pounds	3.0880
gram-calories	horsepower-hour	1.5596×10^{-6}
gram-calories	kilowatt-hour	1.1630×10^{-6}

(continued)

To Convert	Into	Multiply by
gram-calories	watt-hour	1.1630×10^{-3}
gram-calories per second	British thermal unit per hour	14.286
gram-centimeters	British thermal unit	9.297×10^{-8}
gram-centimeters	ergs	980.7
gram-centimeters	joules	9.807×10^{-5}
gram-centimeters	kilogram-calorie	2.343×10^{-8}
gram-centimeters	kilogram-meters	10^{-5}

H

hand	centimeter	10.16
hectares	acres	2.471
hectares	square feet	1.076×10^{5}
hectograms	grams	100.0
hectoliters	liters	100.0
hectometers	meters	100.0
hectowatts	watts	100.0
henries	millihenries	1,000.0
hogsheads (British)	cubic foot	10.114
hogsheads (U.S.)	cubic foot	8.42184
hogsheads (U.S.)	gallons (U.S.)	63
horsepower	British thermal unit per minute	42.44
horsepower	foot-pound per minute	33,000
horsepower	foot-pound per second	550.0
horsepower (metric) (542.5 foot-pound per second)	horsepower (550 foot-pound per second)	0.9863
horsepower (550 foot-pound per second)	horsepower (metric) (542.5 foot-pound per second)	1.014
horsepower	kilogram-calories per minute	10.68
horsepower	kilowatts	0.7457
horsepower	watts	745.7
horsepower (boiler)	British thermal unit per hour	33.479
horsepower (boiler)	kilowatts	9.803
horsepower-hour	British thermal unit	2,547
horsepower-hour	ergs	2.6845×10^{13}
horsepower-hour	foot-pound	1.98×10^{6}
horsepower-hour	gram-calories	641,190
horsepower-hour	joules	2.684×10^{6}
horsepower-hour	kilogram-calories	641.1
horsepower-hour	kilogram-meters	2.737×10^{5}
horsepower-hour	kilowatt-hour	0.7457
hours	days	4.167×10^{-2}
hours	weeks	5.952×10^{-3}
hundredweights (long)	pounds	112
hundredweights (long)	tons (long)	0.05
hundredweights (short)	ounces (avoirdupois)	1,600
hundredweights (short)	pounds	100
hundredweights (short)	tons (metric)	0.0453592
hundredweights (short)	tons (long)	0.0446429

I

inches	centimeters	2.540
inches	meters	2.540×10^{-2}
inches	miles	1.578×10^{-5}
inches	millimeters	25.40
inches	mils	1,000.0
inches	yards	2.778×10^{-2}
inches of mercury	atmospheres	0.03342
inches of mercury	feet of water	1.133

To Convert	Into	Multiply by
inches of mercury	kilogram per square centimeter	0.03453
inches of mercury	kilogram per square meter	345.3
inches of mercury	pounds per square foot	70.73
inches of mercury	pounds per square inch	0.4912
inches of water (at 4°C)	atmospheres	2.458×10^{-3}
inches of water (at 4°C)	inches of mercury	0.07355
inches of water (at 4°C)	kilogram per square centimeter	2.540×10^{-3}
inches of water (at 4°C)	ounces per square inch	0.5781
inches of water (at 4°C)	pounds per square foot	5.204
inches of water (at 4°C)	pounds per square inch	0.03613
International ampere	ampere (absolute)	.9998
International volt	volts (absolute)	1.0003
International volt	joules (absolute)	1.593×10^{-19}
International volt	joules	9.654×10^{4}

J

To Convert	Into	Multiply by
joules	British thermal unit	9.480×10^{-4}
joules	ergs	10^{7}
joules	foot-pounds	0.7376
joules	kilogram-calories	2.389×10^{-4}
joules	kilogram-meters	0.1020
joules	watt-hour	2.778×10^{-4}
joules per centimeter	grams	1.020×10^{4}
joules per centimeter	dynes	10^{7}
joules per centimeter	joules per meter (newtons)	100.0
joules per centimeter	poundals	723.3
joules per centimeter	pounds	22.48

K

To Convert	Into	Multiply by
kilograms	dynes	980,665
kilograms	grams	1,000.0
kilograms	joules per centimeter	0.09807
kilograms	joules per meter (newtons)	9.807
kilograms	poundals	70.93
kilograms	pounds	2.205
kilograms	tons (long)	9.842×10^{-4}
kilograms	tons (short)	1.102×10^{-3}
kilograms per cubic meter	grams per cubic centimeter	0.001
kilograms per cubic meter	pounds per cubic foot	0.06243
kilograms per cubic meter	pounds per cubic inch	3.613×10^{-5}
kilograms per cubic meter	pounds per mil-foot	3.405×10^{-10}
kilograms per meter	pounds per foot	0.6720
Kilogram per square centimeter	dynes	980,665
kilograms per square centimeter	atmospheres	0.9678
kilograms per square centimeter	feet of water	32.81
kilograms per square centimeter	inches of mercury	28.96
kilograms per square centimeter	pounds per square foot	2,048
kilograms per square centimeter	pounds per square inch	14.22
kilograms per square meter	atmospheres	9.678×10^{-5}
kilograms per square meter	bars	98.07×10^{-6}
kilograms per square meter	feet of water	3.281×10^{-3}
kilograms per square meter	inches of mercury	2.896×10^{-3}
kilograms per square meter	pounds per square foot	0.2048
kilograms per square meter	pounds per square inch	1.422×10^{-3}
kilograms per square millimeter	kilogram per square meter	10^{6}
kilogram-calories	British thermal unit	3.968
kilogram-calories	foot-pounds	3,088

(continued)

To Convert	Into	Multiply by
kilogram-calories	horsepower-hour	1.560×10^{-3}
kilogram-calories	joules	4,186
kilogram-calories	kilogram-meters	426.9
kilogram-calories	kilojoules	4.186
kilogram-calories	kilowatt-hour	1.163×10^{-3}
kilogram meters	British thermal unit	9.294×10^{-3}
kilogram meters	ergs	9.804×10^{7}
kilogram meters	foot-pounds	7.233
kilogram meters	joules	9.804
kilogram meters	kilogram-calories	2.342×10^{-3}
kilogram meters	kilowatt-hour	2.723×10^{-6}
kilolines	maxwells	1,000.0
kiloliters	liters	1,000.0
kilometers	centimeters	10^{5}
kilometers	feet	3,281
kilometers	inches	3.937×10^{4}
kilometers	meters	1,000.0
kilometers	miles	0.6214
kilometers	millimeters	10^{4}
kilometers	yards	1,094
kilometers per hour	centimeter per second	27.78
kilometers per hour	feet per minute	54.68
kilometers per hour	feet per second	0.9113
kilometers per hour	knots	0.5396
kilometers per hour	meters per minute	16.67
kilometers per hour	miles per hour	0.6214
kilometers per second	centimeter per second per second	27.78
kilometers per second	feet per second per second	0.9113
kilometers per second	meters per second per second	0.2778
kilometers per second	miles per hour per second	0.6214
kilowatts	British thermal unit per minute	56.92
kilowatts	foot-pounds per minute	4.426×10^{4}
kilowatts	foot-pounds per second	737.6
kilowatts	horsepower	1.341
kilowatts	kilogram-calories per minute	14.34
kilowatts	watts	1,000.0
kilowatt-hour	British thermal unit	3,413
kilowatt-hour	ergs	3.600×10^{13}
kilowatt-hour	foot-pound	2.655×10^{6}
kilowatt-hour	gram-calories	859,850
kilowatt-hour	horsepower-hour	1.341
kilowatt-hour	joules	3.6×10^{6}
kilowatt-hour	kilogram-calories	860.5
kilowatt-hour	kilogram-meters	3.671×10^{5}
kilowatt-hour	pounds of water evaporated from and at 212°F	8.53
kilowatt-hours	pounds of water raised from 62° to 212°F	22.75
knots	feet per hour	6,080
knots	kilometers per hour	1.8532
knots	nautical miles per hour	1.0
knots	statute miles per hour	1.151
knots	yards per hour	2,027
knots	feet per second	1.689

L

league	miles (approximate)	3.0
Light year	miles	5.9×10^{12}
Light year	kilometers	9.4637×10^{12}
lines per square centimeter	gausses	1.0

To Convert	Into	Multiply by
lines per square inch	gausses	0.1550
lines per square inch	webers per square centimeter	1.550×10^{-9}
lines per square inch	webers per square inch	10^{-8}
lines per square inch	webers per square meter	1.550×10^{-5}
links (engineer's)	inches	12.0
links (surveyor's)	inches	7.92
liters	bushels (U.S. dry)	0.02838
liters	cubic centimeter	1,000.0
liters	cubic feet	0.03531
liters	cubic inches	61.02
liters	cubic meters	0.001
liters	cubic yards	1.308×10^{-3}
liters	gallons (U.S. liquid)	0.2642
liters	pints (U.S. liquid)	2.113
liters	quarts (U.S. liquid)	1.057
liters per minute	cubic foot per second	5.886×10^{-4}
liters per minute	gallons per second	4.403×10^{-3}
Lumen	spherical candle power	.07958
Lumen	watt	.001496
lumens per square foot	foot-candles	1.0
lumen per square foot	lumen per square meter	10.76
lux	foot-candles	0.0929

M

maxwells	kilolines	0.001
maxwells	webers	10^{-8}
megalines	maxwells	10^{6}
megohms	microhms	10^{12}
megohms	ohms	10^{6}
meters	centimeters	100.0
meters	feet	3.281
meters	inches	39.37
meters	kilometers	0.001
meters	miles (nautical)	5.396×10^{-4}
meters	miles (statute)	6.214×10^{-4}
meters	millimeters	1,000.0
meters	yards	1.094
meters	varas	1.179
meters per minute	centimeter per second	1,667
meters per minute	feet per minute	3.281
meters per minute	feet per second	0.05468
meters per minute	kilometer per hour	0.06
meters per minute	knots	0.03238
meters per minute	miles per hour	0.03728
meters per second	feet per minute	196.8
meters per second	feet per second	3.281
meters per second	kilometers per hour	3.6
meters per second	kilometers per minute	0.06
meters per second	miles per hour	2.237
meters per second	miles per minute	0.03728
meters per second per second	centimeter per second per second	100.0
meters per second per second	foot per second per second	3.281
meters per second per second	kilometer per hour per second	3.6
meters per second per second	miles per hour per second	2.237
meter-kilograms	centimeter-dynes	9.807×10^{7}
meter-kilograms	centimeter-grams	10^{5}
meter-kilograms	pound-feet	7.233
microfarad	farads	10^{-6}

(continued)

To Convert	Into	Multiply by
micrograms	grams	10^{-6}
microhms	megohms	10^{-12}
microhms	ohms	10^{-6}
microliters	liters	10^{-6}
Microns	meters	1×10^{-6}
miles (nautical)	feet	6,080.27
miles (nautical)	kilometers	1.853
miles (nautical)	meters	1,853
miles (nautical)	miles (statute)	1.1516
miles (nautical)	yards	2,027
miles (statute)	centimeters	1.609×10^{5}
miles (statute)	feet	5,280
miles (statute)	inches	6.336×10^{4}
miles (statute)	kilometers	1.609
miles (statute)	meters	1,609
miles (statute)	miles (nautical)	0.8684
miles (statute)	yards	1,760
miles per hour	centimeter per second	44.70
miles per hour	feet per minute	88
miles per hour	feet per second	1.467
miles per hour	kilometer per hour	1.609
miles per hour	kilometer per minute	0.02682
miles per hour	knots	0.8684
miles per hour	meters per minute	26.82
miles per hour	miles per minute	0.1667
miles per hour per second	centimeter per second per second	44.70
miles per hour per second	feet per second per second	1.467
miles per hour per second	kilometer per hour per second	1.609
miles per hour per second	meters per second per second	0.4470
miles per minute	centimeter per second	2,682
miles per minute	feet per second	88
miles per minute	kilometer per minute	1.609
miles per minute	knots per minute	0.8684
miles per minute	miles per hour	60
mil-feet	cubic inches	9.425×10^{-6}
milliers	kilograms	1,000
millimicrons	meters	1×10^{-9}
milligrams	grains	0.01543236
milligrams	grams	0.001
milligrams per liter	parts per million	1.0
millihenries	henries	0.001
milliliters	liters	0.001
millimeters	centimeters	0.1
millimeters	feet	3.281×10^{-3}
millimeters	inches	0.03937
millimeters	kilometers	10^{-6}
millimeters	meters	0.001
millimeters	miles	6.214×10^{-7}
millimeters	mils	39.37
millimeters	yards	1.094×10^{-3}
million gallons per day	cubic foot per second	1.54723
mils	centimeters	2.540×10^{-3}
mils	feet	8.333×10^{-5}
mils	inches	0.001
mils	kilometers	2.540×10^{-8}
mils	yards	2.778×10^{-5}
miner's inches	cubic foot per minute	1.5
minims (British)	cubic centimeter	0.059192
minims (U.S., fluid)	cubic centimeter	0.061612

To Convert	Into	Multiply by
minutes (angles)	degrees	0.01667
minutes (angles)	quadrants	1.852×10^{-4}
minutes (angles)	radians	2.909×10^{-4}
minutes (angles)	seconds	60.0
myriagrams	kilograms	10.0
myriameters	kilometers	10.0
myriawatts	kilowatts	10.0

N

nepers	decibels	8.686
newton	dynes	1×105

O

ohm (International)	ohm (absolute)	1.0005
ohms	megohms	10^{-6}
ohms	microhms	10^{6}
ounces	drams	16.0
ounces	grains	437.5
ounces	grams	28.349527
ounces	pounds	0.0625
ounces	ounces (troy)	0.9115
ounces	tons (long)	2.790×10^{-5}
ounces	tons (metric)	2.835×10^{-5}
ounces (fluid)	cu. inches	1.805
ounces (fluid)	liters	0.02957
ounces (troy)	grains	480.0
ounces (troy)	grams	31.103481
ounces (troy)	ounces (avoirdupois)	1.09714
ounces (troy)	pennyweights (troy)	20.0
ounces (troy)	pounds (troy)	0.08333
Ounce per square inch	dynes per square centimeter	4,309
ounces per square inch	pounds per square inch	0.0625

P

parsec	miles	19×10^{12}
parsec	kilometers	3.084×10^{13}
parts per million	grains per U.S. gallon	0.0584
parts per million	grains per Imperial gallon	0.07016
parts per million	pounds per million gallon	8.345
pecks (British)	cubic inches	554.6
pecks (British)	liters	9.091901
pecks (U.S.)	bushels	0.25
pecks (U.S.)	cubic inches	537.605
pecks (U.S.)	liters	8.809582
pecks (U.S.)	quarts (dry)	8
pennyweights (troy)	grains	24.0
pennyweights (troy)	ounces (troy)	0.05
pennyweights (troy)	grams	1.55517
pennyweights (troy)	pounds (troy)	4.1667×10^{-3}
pints (dry)	cubic inches	33.60
pints (liquid)	cubic centimeter	473.2
pints (liquid)	cubic feet	0.01671
pints (liquid)	cubic inches	28.87
pints (liquid)	cubic meters	4.732×10^{-4}
pints (liquid)	cubic yards	6.189×10^{-4}
pints (liquid)	gallons	0.125
pints (liquid)	liters	0.4732
pints (liquid)	quarts (liquid)	0.5

(continued)

To Convert	Into	Multiply by
Planck's quantum	erg-second	6.624×10^{-27}
Poise	Gram per centimeter second	1.00
Pounds (avoirdupois)	ounces (troy)	14.5833
poundals	dynes	13,826
poundals	grams	14.10
poundals	joules per centimeter	1.383×10^{-3}
poundals	joules per meter (newtons)	0.1383
poundals	kilograms	0.01410
poundals	pounds	0.03108
pounds	drams	256
pounds	dynes	44.4823×10^4
pounds	grains	7,000
pounds	grams	453.5924
pounds	joules per centimeter	0.04448
pounds	joules per meter (newtons)	4.448
pounds	kilograms	0.4536
pounds	ounces	16.0
pounds	ounces (troy)	14.5833
pounds	poundals	32.17
pounds	pounds (troy)	1.21528
pounds	tons (short)	0.0005
pounds (troy)	grains	5,760
pounds (troy)	grams	373.24177
pounds (troy)	ounces (avoirdupois)	13.1657
pounds (troy)	ounces (troy)	12.0
pounds (troy)	pennyweights (troy)	240.0
pounds (troy)	pounds (avoirdupois)	0.822857
pounds (troy)	tons (long)	3.6735×10^{-4}
pounds (troy)	tons (metric)	3.7324×10^{-4}
pounds (troy)	tons (short)	4.1143×10^{-4}
pounds of water	cubic foot	0.01602
pounds of water	cubic inches	27.68
pounds of water	gallons	0.1198
pounds of water per minute	cubic foot per second	2.670×10^{-4}
pound-feet	centimeter-dynes	1.356×10^7
pound-feet	centimeter-grams	13,825
pound-feet	meter-kilogram	0.1383
pounds per cubic foot	grams per cubic centimeter	0.01602
pounds per cubic foot	kilogram per cubic meter	16.02
pounds per cubic foot	pounds per cubic inch	5.787×10^{-4}
pounds per cubic foot	pounds per mil-foot	5.456×10^{-9}
pounds per cubic inch	gram per cubic centimeter	27.68
pounds per cubic inch	kilogram per cubic meter	2.768×10^4
pounds per cubic inch	pounds per cubic foot	1,728
pounds per cubic inch	pounds per mil-foot	9.425×10^{-6}
pounds per foot	kilogram per meter	1.488
pounds per inch	gram per centimeter	178.6
pounds per mil-foot	gram per cubic centimeter	2.306×10^6
pounds per square foot	atmospheres	4.725×10^{-4}
pounds per square foot	feet of water	0.01602
pounds per square foot	inches of mercury	0.01414
pounds per square foot	kilogram per square meter	4.882
pounds per square foot	pounds per square inch	6.944×10^{-3}
pounds per square inch	atmospheres	0.06804
pounds per square inch	feet of water	2.307
pounds per square inch	inches of mercury	2.036
pounds per square inch	kilogram per square meter	703.1
pounds per square inch	pounds per square foot	144.0

To Convert	Into	Multiply by
	Q	
quadrants (angle)	degrees	90.0
quadrants (angle)	minutes	5,400.0
quadrants (angle)	radians	1.571
quadrants (angle)	seconds	3.24×10^5
quarts (dry)	cubic inches	67.20
quarts (liquids)	cubic centimeter	946.4
quarts (liquids)	cubic feet	0.03342
quarts (liquids)	cubic inches	57.75
quarts (liquids)	cubic meters	9.464×10^{-4}
quarts (liquids)	cubic yards	1.238×10^{-3}
quarts (liquids)	gallons	0.25
quarts (liquids)	liters	0.9463
	R	
radians	degrees	57.30
radians	minutes	3,438
radians	quadrants	0.6366
radians	seconds	2.063×10^5
radians per second	degrees per second	57.30
radians per second	revolutions per minute	9.549
radians per second	revolutions per second	0.1592
radians per second per second	revolution per minute per minute	573.0
radians per second per second	revolution per minute per second	9.549
radians per second per second	revolution per second per second	0.1592
revolutions	degrees	360.0
revolutions	quadrants	4.0
revolutions	radians	6.283
revolutions per minute	degrees per second	6.0
revolutions per minute	radians per second	0.1047
revolutions per minute	revolution per second	0.01667
revolutions per minute per minute	radians per second per second	1.745×10^{-3}
revolutions per minute per minute	revolution per minute per second	0.01667
revolutions per minute per minute	revolution per second per second	2.778×10^{-4}
revolutions per second	degrees per second	360.0
revolutions per second	radians per second	6.283
revolutions per second	revolution per minute	60.0
revolutions per second per second	radians per second per second	6.283
revolutions per second per second	revolution per minute per minute	3,600.0
revolutions per second per second	revolution per minute per second	60.0
rod	chain (Gunters)	.25
rod	meters	5.029
rods (surveyors' measure)	yards	5.5
rods	feet	16.5
	S	
scruples	grains	20
seconds (angle)	degrees	2.778×10^{-4}
seconds (angle)	minutes	0.01667
seconds (angle)	quadrants	3.087×10^{-6}
seconds (angle)	radians	4.848×10^{-6}
slug	kilogram	14.59
slug	pounds	32.17
sphere	steradians	12.57
square centimeters	circular mils	1.973×10^5
square centimeters	square feet	1.076×10^{-3}
square centimeters	square inches	0.1550
square centimeters	square meters	0.0001

(continued)

To Convert	Into	Multiply by
square centimeters	square miles	3.861×10^{-11}
square centimeters	square millimeters	100.0
square centimeters	square yards	1.196×10^{-4}
square feet	acres	2.296×10^{-5}
square feet	circular mils	1.833×10^{8}
square feet	square centimeter	929.0
square feet	square inches	144.0
square feet	square meters	0.09290
square feet	square miles	3.587×10^{-8}
square feet	square millimeters	9.290×10^{4}
square feet	square yards	0.1111
square inches	circular mils	1.273×10^{6}
square inches	square centimeter	6.452
square inches	square feet	6.944×10^{-3}
square inches	square millimeters	645.2
square inches	square mils	10^{6}
square inches	square yards	7.716×10^{-4}
square kilometers	acres	247.1
square kilometers	square centimeter	10^{10}
square kilometers	square foot	10.76×10^{6}
square kilometers	square inches	1.550×10^{9}
square kilometers	square meters	10^{6}
square kilometers	square miles	0.3861
square kilometers	square yards	1.196×10^{6}
square meters	acres	2.471×10^{-4}
square meters	square centimeter	10^{4}
square meters	square feet	10.76
square meters	square inches	1,550
square meters	square miles	3.861×10^{-7}
square meters	square millimeters	10^{6}
square meters	square yards	1.196
square miles	acres	640.0
square miles	square feet	27.88×10^{6}
square miles	square kilometer	2.590
square miles	square meters	2.590×10^{6}
square miles	square yards	3.098×10^{6}
square millimeters	circular mils	1,973
square millimeters	square centimeter	0.01
square millimeters	square feet	1.076×10^{-5}
square millimeters	square inches	1.550×10^{-3}
square mils	circular mils	1.273
square mils	square centimeter	6.452×10^{-6}
square mils	square inches	10^{-6}
square yards	acres	2.066×10^{-4}
square yards	square centimeter	8,361
square yards	square feet	9.0
square yards	square inches	1,296
square yards	square meters	0.8361
square yards	square miles	3.228×10^{-7}
square yards	square millimeters	8.361×10^{5}

T

To Convert	Into	Multiply by
temperature (°C) + 273	absolute temperature (°C)	1.0
temperature (°C) + 17.78	temperature (°F)	1.8
temperature (°F) + 460	absolute temperature (°F)	1.0
temperature (°F) − 32	temperature (°C)	5/9
tons (long)	kilograms	1,016
tons (long)	pounds	2,240
tons (long)	tons (short)	1.120

To Convert	Into	Multiply by
tons (metric)	kilograms	1,000
tons (metric)	pounds	2,205
tons (short)	kilograms	907.1848
tons (short)	ounces	32,000
tons (short)	ounces (troy)	29,166.66
tons (short)	pounds	2,000
tons (short)	pounds (troy)	2,430.56
tons (short)	tons (long)	0.89287
tons (short)	tons (metric)	0.9078
tons (short) per square foot	kilogram per square meter	9,765
tons (short) per square foot	pounds per square inch	2,000
tons of water per 24 hour	pounds of water per hour	83.333
tons of water per 24 hour	gallons per minute	0.16643
tons of water per 24 hour	cubic foot per hour	1.3349

V

volt/inch	volt per centimeter	.39370
volt (absolute)	statvolts	.003336

W

watts	British thermal unit per hour	3.4129
watts	British thermal unit per minute	0.05688
watts	ergs per second	107
watts	foot-pound per minute	44.27
watts	foot-pound per second	0.7378
watts	horsepower	1.341×10^{-3}
watts	horsepower (metric)	1.360×10^{-3}
watts	kilogram-calories per minute	0.01433
watts	kilowatts	0.001
watts (absolute)	British thermal unit (mean) per minute	0.056884
watts (absolute)	joules per second	1
watt-hours	British thermal unit	3.413
watt-hours	ergs	3.60×10^{10}
watt-hours	foot-pounds	2,656
watt-hours	gram-calories	859.85
watt-hours	horsepower-hour	1.341×10^{-3}
watt-hours	kilogram-calories	0.8605
watt-hours	kilogram-meters	367.2
watt-hours	kilowatt-hour	0.001
watt (International)	watt (absolute)	1.0002
webers	maxwells	10^8
webers	kilolines	10^5
webers per square inch	gausses	1.550×10^7
webers per square inch	lines per square inch	10^8
webers per square inch	webers per square centimeter	0.1550
webers per square inch	webers per square meter	1,550
webers per square meter	gausses	10^4
webers per square meter	lines per square inch	6.452×10^4
webers per square meter	webers per square centimeter	10^{-4}
webers per square meter	webers per square inch	6.452×10^{-4}

Y

yards	centimeters	91.44
yards	kilometers	9.144×10^{-4}
yards	meters	0.9144
yards	miles (nautical)	4.934×10^{-4}
yards	miles (statute)	5.682×10^{-4}
yards	millimeters	914.4

23.5.2 Conversion Constants and Multipliers[3]

Recommended Decimal Multiples and Submultiples

Multiples and Submultiples	Prefixes	Symbols	Multiples and Submultiples	Prefixes	Symbols
10^{18}	exa	E	10^{-1}	deci	d
10^{15}	peta	P	10^{-2}	centi	c
10^{12}	tera	T	10^{-3}	milli	m
10^{9}	giga	G	10^{-6}	micro	μ (Greek mu)
10^{6}	mega	M	10^{-9}	nano	n
10^{3}	kilo	k	10^{-12}	pico	p
10^{2}	hecto	h	10^{-15}	femto	f
10	deca	da	10^{-18}	atto	a

Conversion Factors---Metric to English

To Convert	Into	Multiply By
Inches	Centimeters	0.393700787
Feet	Meters	3.280839895
Yards	Meters	1.093613298
Miles	Kilometers	0.6213711922
Ounces	Grams	$3.527396195 \times 10^{-2}$
Pounds	Kilograms	2.204622622
Gallons (U.S. Liquid)	Liters	0.2641720524
Fluid ounces	Milliliters (cc)	$3.381402270 \times 10^{-2}$
Square inches	Square centimeters	0.1550003100
Square feet	Square meters	10.76391042
Square yards	Square meters	1.195990046
Cubic inches	Milliliters (cc)	$6.102374409 \times 10^{-2}$
Cubic feet	Cubic meters	35.31466672
Cubic yards	Cubic meters	1.307950619

Conversion Factors---English to Metric

To Convert	Into	Multiply By
Microns	Mils	**25.4**
Centimeters	Inches	**2.54**
Meters	Feet	**0.3048**
Meters	Yards	**0.9144**
Kilometers	Miles	**1.609344**
Grams	Ounces	28.34952313
Kilograms	Pounds	**0.45359237**
Liters	Gallons (U.S. Liquid)	**3.785411784**
Millimeters (cc)	Fluid ounces	29.57352956
Square centimeters	Square inches	**6.4516**
Square meters	Square feet	**0.09290304**
Square meters	Square yards	**0.83612736**
Milliliters (cc)	Cubic inches	**16.387064**
Cubic meters	Cubic feet	$2.831684659 \times 10^{-2}$
Cubic meters	Cubic yards	0.764554858

Note: Boldface numbers are exact; others are given to ten significant figures where so indicated by the multiplier factor.

---

[3]Zwillinger, D., ed. 1996. *CRC Standard Mathematical Tables and Formulae,* 30th ed. CRC Press, Boca Raton, FL: Greek alphabet, conversion constants and multipliers (recommended decimal multiples and submultiples, metric to English, English to metric, general, temperature factors).

Conversion Factors---General

To Convert	Into	Multiply By
Atmospheres	Feet of water @ 4°C	2.950×10^{-2}
Atmospheres	Inches of mercury @ 0°C	3.342×10^{-2}
Atmospheres	Pounds per square inch	6.804×10^{-2}
BTU	Foot-pounds	1.285×10^{-3}
BTU	Joules	9.480×10^{-4}
Cubic feet	Cords	**128**
Degree (angle)	Radians	57.2958
Ergs	Foot-pounds	1.356×10^{7}
Feet	Miles	**5280**
Feet of water @ 4°C	Atmospheres	33.90
Foot-pounds	Horsepower-hours	1.98×10^{6}
Foot-pounds	Kilowatt-hours	2.655×10^{6}
Foot-pounds per min	Horsepower	3.3×10^{4}
Horsepower	Foot-pounds per sec	1.818×10^{-3}
Inches of mercury @ 0°C	Pounds per square inch	2.036
Joules	BTU	1054.8
Joules	Foot-pounds	1.35582
Kilowatts	BTU per min	1.758×10^{-2}
Kilowatts	Foot-pounds per min	2.26×10^{-5}
Kilowatts	Horsepower	0.745712
Knots	Miles per hour	0.86897624
Miles	Feet	1.894×10^{-4}
Nautical miles	miles	0.86897624
Radians	Degrees	1.745×10^{-2}
Squares feet	Acres	**43560**
Watts	BTU per min	17.5796

Note: Boldface numbers are exact; others are given to ten significant figures where so indicated by the multiplier factor.

Temperature Factors

$$°F = 9/5(°C) + 32$$

Fahrenheit temperature = 1.8 (temperature in kelvins) − 459.67

$$°C = 5/9[(°F) − 32)]$$

Celsius temperature = temperature in kelvin − 273.15

Fahrenheit temperature = 1.8 (Celsius temperature) + 32

Conversion of Temperatures

From	To	
°Celsius	°Fahrenheit	$t_F = (t_C \times 1.8) + 32$
	Kelvin	$T_K = t_C + 273.15$
	°Rankine	$T_R = (t_C + 273.15) \times 18$
°Fahrenheit	°Celsius	$t_C = \dfrac{t_F - 32}{1.8}$
	Kelvin	$T_K = \dfrac{t_F - 32}{1.8} + 273.15$
	°Rankine	$T_R = t_F + 459.67$
Kelvin	°Celsius	$t_C = T_K - 273.15$
	°Rankine	$T_R = T_K \times 1.8$
°Rankine	°Fahrenheit	$t_F = T_R - 459.67$
	Kelvin	$T_K = \dfrac{T_R}{1.8}$

23.6   General Mathematical Tables[4]

William F. Ames and George Cain

23.6.1   Introduction to Mathematics Chapter

Mathematics has been defined as "the logic of drawing unambiguous conclusions from arbitrary assumptions." The assumptions do not come from the mathematics but arise from the engineering or scientific discipline. Mathematical models of real world phenomena have been remarkably useful. The simplest of these express various physical *laws* in mathematical form. For example, Ohms law ($V = IR$), Newton's second law ($F = ma$) and the ideal gas law ($pv = nRT$) provide starting points for many theories. More complicated models employ difference, differential, or integral equations. Whatever the model, the conclusions drawn from it are unambiguous!

In this mathematics chapter we provide for an extensive review of algebra and geometry in the first section. One feature is the collection of common geometric figures together with their areas, volumes, and other data.

The second section reviews the fundamentals of trigonometry and presents the myriad of identities, which are very useful but very easy to forget. The third section presents tables of various series for a number of useful functions, followed by the theory of how these series are calculated.

The next section gives the elements of the differential calculus, discusses the calculation and theory of maxima and minima, and contains a table of derivatives. The fifth section provides a summary of the integral calculus including formulas for calculating such fundamental physical ideas as arclength, area, volume, work, centroids, etc.

Special functions constitute Sec. 23.6.7. Here are the Bessel functions and various basic polynomials. A special feature found here is the table of three-dimensional quadratic figures.

Because linear algebra and vector calculus are so basic Secs. 23.6.8 and 23.6.9 are devoted to their summaries. The last section presents the various Fourier transforms and their properties in table form.

Throughout the chapter various references are provided.

23.6.2   Elementary Algebra and Geometry

Fundamental Properties (Real Numbers)

$a + b = b + a$	Commutative law for addition
$(a + b) + c = a + (b + c)$	Associative law for addition
$a + 0 = 0 + a$	Identity law for addition
$a + (-a) = (-a) + a = 0$	Inverse law for addition
$a(bc) = (ab)c$	Associative law for multiplication
$a(\frac{1}{a}) = (\frac{1}{a})a = 1,\ a \neq 0$	Inverse law for multiplication
$(a)(1) = (1)(a) = a$	Identity law for multiplication
$ab = ba$	Commutative law for multiplication
$a(b + c) = ab + ac$	Distributive law

Division by zero is not defined.

---

[4]The material in this chapter was previously published by CRC Press in *The Engineering Handbook*, pp. 2037–2079, 2187–2192, and 2196–2202, 1996.

Exponents

For integers m and n

$$a^n a^m = a^{n+m}$$
$$a^n / a^m = a^{n-m}$$
$$(a^n)^m = a^{nm}$$
$$(ab)^m = a^m b^m$$
$$(a/b)^m = a^m / b^m$$

Fractional Exponents

$$a^{p/q} = (a^{1/q})^p$$

where $a^{1/q}$ is the positive qth root of a if $a > 0$ and the negative qth root of a if a is negative and q is odd. Accordingly, the five rules of exponents just given (for integers) are also valid if m and n are fractions, provided a and b are positive.

Irrational Exponents

If an exponent is irrational (e.g., $\sqrt{2}$), the quantity, such as $a^{\sqrt{2}}$, is the limit of the sequence $a^{1.4}, a^{1.41}, a^{1.414}, \ldots$.

Operations with Zero

$$0^m = 0 \qquad a^0 = 1$$

Logarithms

If x, y, and b are positive and $b \neq 1$

$$\log_b(xy) = \log_b x + \log_b y$$
$$\log_b(x/y) = \log_b x - \log_b y$$
$$\log_b x^p = p \log_b x$$
$$\log_b(1/x) = -\log_b x$$
$$\log_b b = 1$$
$$\log_b 1 = 0 \quad \text{Note: } b^{\log_b x} = x$$

Change of Base (a ≠ 1)

$$\log_b x = \log_a x \log_b a$$

Factorials

The factorial of a positive integer n is the product of all of the positive integers less than or equal to the integer n and is denoted $n!$. Thus

$$n! = 1 \cdot 2 \cdot 3 \cdot \cdots \cdot n$$

Factorial 0 is defined: $0! = 1$

Stirling's Approximation

$$\lim_{n \to \infty} (n/e)^n \sqrt{2\pi n} = n!$$

Binomial Theorem

For positive integer n

$$(x + y)^n = x^n + nx^{n-1}y + \frac{n(n-1)}{2!}x^{n-2}y^2 + \frac{n(n-1)(n-2)}{3!}x^{n-3}y^3 + \cdots + nxy^{n-1} + y^n$$

Factors and Expansion

$$(a + b)^2 = a^2 + 2ab + b^2$$
$$(a - b)^2 = a^2 - 2ab + b^2$$
$$(a + b)^3 = a^3 + 3a^2b + 3ab^2 + b^3$$
$$(a - b)^3 = a^3 - 3a^2b + 3ab^2 - b^3$$
$$(a^2 - b^2) = (a - b)(a + b)$$
$$(a^3 - b^3) = (a - b)(a^2 + ab + b^2)$$
$$(a^3 + b^3) = (a + b)(a^2 - ab + b^2)$$

Progression

An *arithmetic progression* is a sequence in which the difference between any term and the preceding term is a constant (d):

$$a, a + d, a + 2d, \ldots, a + (n - 1)d$$

If the last term is denoted $l(= a + (n - 1)d)$, then the sum is

$$s = \frac{n}{2}(a + l)$$

A *geometric progression* is a sequence in which the ratio of any term is a constant r. Thus, for n terms

$$a, ar, ar^2, \ldots, ar^{n-1}$$

The sum is

$$S = \frac{a - ar^n}{1 - r}$$

Complex Numbers

A complex number is an ordered pair of real numbers (a, b).
   Equality:

$$(a, b) = (c, d) \quad \text{if and only if } a = c \text{ and } b = d$$

Addition:

$$(a, b) + (c, d) = (a + c, b + d)$$

Multiplication:

$$(a, b)(c, d) = (ac - bd, ad + bc)$$

The first element (a, b) is called the *real* part, the second the *imaginary* part. An alternative notation for (a, b) is $a + bi$, where $i^2 = (-1, 0)$, and $i = (0, 1)$ or $0 + 1i$ is written for this complex number as a convenience. With this understanding, i behaves as a number, that is, $(2 - 3i)(4 + i) = 8 - 12i + 2i - 3i^2 = 11 - 10i$. The conjugate of $a + bi$ is $a - bi$, and the product of a complex number and its conjugate is $a^2 + b^2$. Thus, *quotients* are computed by multiplying numerator and denominator by the conjugate of the denominator, illustrated as follows:

$$\frac{2+3i}{4+2i} = \frac{(4-2i)(2+3i)}{(4-2i)(4+2i)} = \frac{14+8i}{20} = \frac{7+4i}{10}$$

Polar Form

The complex number $x + iy$ may be represented by a plane vector with components x and y:

$$x + iy = r(\cos\theta + i\sin\theta)$$

(See Fig. 23.29.) Then, given two complex numbers $z_1 = r_1(\cos\theta_1 + i\sin\theta_1)$, and $z_2 = r_2(\cos\theta_2 + i\sin\theta_2)$, the product and quotient are as follows.

Product:

$$z_1z_2 = r_1r_2(\cos(\theta_1 + \theta_2) + i\sin(\theta_1 + \theta_2))$$

Quotient:

$$z_1/z_2 = (r_1/r_2)[\cos(\theta_1 - \theta_2) + i\sin(\theta_1 - \theta_2)]$$

Powers:

$$z^n = [r\cos\theta + i\sin\theta]^n = r^n[\cos n\theta + i\sin n\theta]$$

Roots:

$$z^{1/n} = [r\cos\theta + i\sin\theta]^{1/n}$$
$$= r^{1/n}\left[\cos\frac{\theta + k\cdot 360}{n} + i\sin\frac{\theta + k\cdot 360}{n}\right] \quad k = 0, 1, 2, \ldots, n-1$$

Permutations

A permutation is an ordered arrangement (sequence) of all or part of a set of objects. The number of permutations of n objects taken r at a time is

$$p(n,r) = n(n-1)(n-2)\cdots(n-r+1)$$
$$= \frac{n!}{(n-r)!}$$

A permutation of positive integers is *even* or *odd* if the total number of inversions is an even integer or an odd integer, respectively. Inversions are counted relative to each integer j in the permutation by counting the number of integers that follow j and are less than j. These are summed to give the total number of inversions. For example, the permutation 4132 has four inversions: three relative to 4 and one relative to 3. This permutation is therefore even.

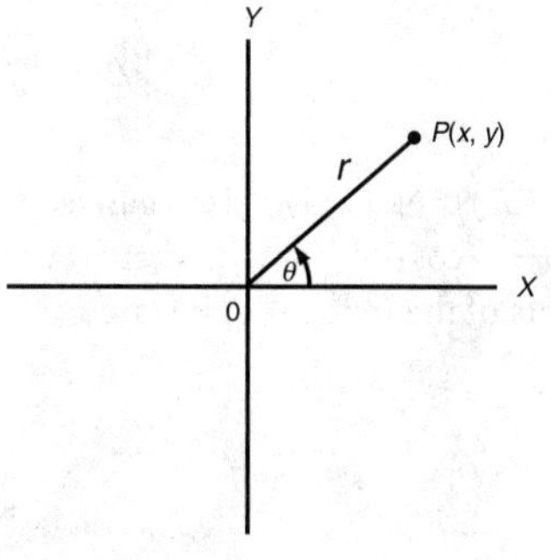

FIGURE 23.29 Polar form of complex number.

Combinations

A combination is a selection of one or more objects from among a set of objects regardless of order. The number of combinations of n different objects taken r at a time is

$$C(n,r) = \frac{P(n,r)}{r!} = \frac{n!}{r!(n-r)!}$$

Algebraic Equations

Quadratic

If $ax^2 + bx + c = 0$, and $a \neq 0$, then roots are

$$x = \frac{-b \pm \sqrt{b^2 - 4ac}}{2a}$$

Cubic

To solve $x^3 + bx^2 + cx + d = 0$, let $x = y - b/3$. Then the *reduced cubic* is obtained

$$y^3 + py + q = 0$$

where $p = c - (1/3)b^2$ and $q = d - (1/3)bc + (2/27)b^3$. Solutions of the original cubic are then in terms of the reduced cubic roots y_1, y_2, y_3

$$x_1 = y_1 - (1/3)b \qquad x_2 = y_2 - (1/3)b \qquad x_3 = y_3 - (1/3)b$$

The three roots of the reduced cubic are

$$y_1 = (A)^{1/3} + (B)^{1/3}$$

$$y_2 = W(A)^{1/3} + W^2(B)^{1/3}$$

$$y_3 = W^2(A)^{1/3} + W(B)^{1/3}$$

where

$$A = -\frac{1}{2}q + \sqrt{(1/27)p^3 + \frac{1}{4}q^2}$$

$$B = -\frac{1}{2}q - \sqrt{(1/27)p^3 + \frac{1}{4}q^2}$$

$$W = \frac{-1 + i\sqrt{3}}{2}, \quad W^2 = \frac{-1 - i\sqrt{3}}{2}$$

When $(1/27)p^3 + (1/4)q^2$ is negative, A is complex; in this case A should be expressed in trigonometric form: $A = r(\cos\theta + i\sin\theta)$ where θ is a first or second quadrant angle, as q is negative or positive. The three roots of the reduced cubic are

$$y_1 = 2(r)^{1/3}\cos(\theta/3)$$

$$y_2 = 2(r)^{1/3}\cos\left(\frac{\theta}{3} + 120°\right)$$

$$y_3 = 2(r)^{1/3}\cos\left(\frac{\theta}{3} + 240°\right)$$

Geometry

Figures 23.30–23.40 are a collection of common geometric figures. Area (A), volume (V), and other measurable features are indicated.

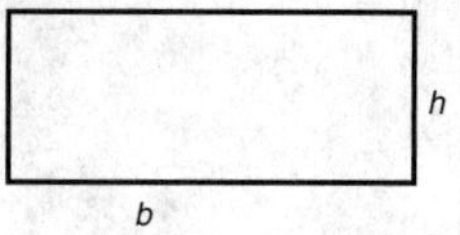

FIGURE 23.30 Rectangle: $A = bh$

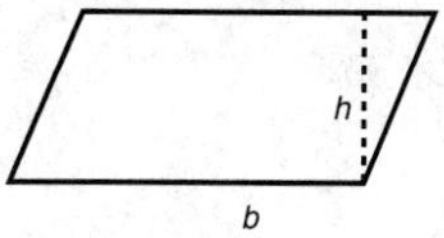

FIGURE 23.31 Parallelogram: $A = bh$

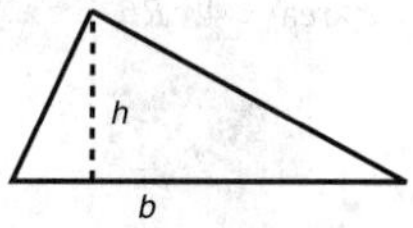

FIGURE 23.32 Triangle: $A = \frac{1}{2}bh$

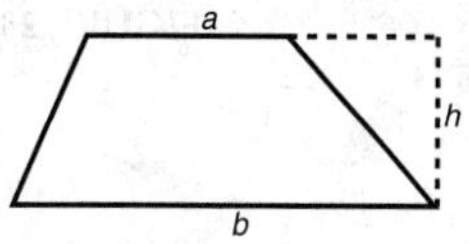

FIGURE 23.33 Trapezoid: $A = \frac{1}{2}(a + b)h$

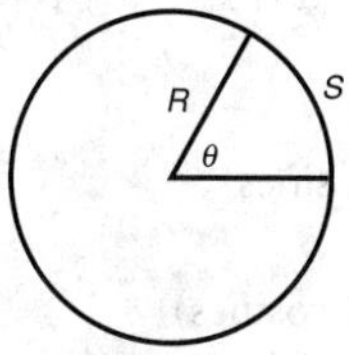

FIGURE 23.34 Circle: $A = \pi R^2$;
circumference $= 2\pi R$;
arc length $S = R\theta$ (θ in radians)

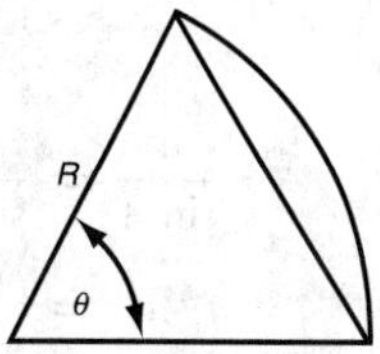

FIGURE 23.35 Sector of a circle:
$A_{\text{sector}} = \frac{1}{2}R^2\theta$;
$A_{\text{segment}} = \frac{1}{2}R^2(\theta - \sin\theta)$

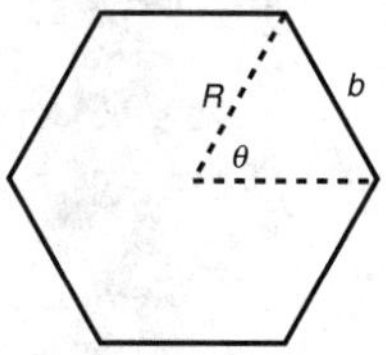

FIGURE 23.36 Regular polygon of n sides:
$A = (n/4)b^2 ctn(\pi/n)$;
$R = (b/2)csc(\pi/n)$

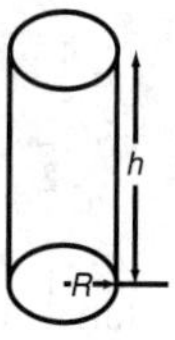

FIGURE 23.37 Right circular cylinder:
$V = \pi R^2 h$;
lateral surface area $= 2\pi Rh$

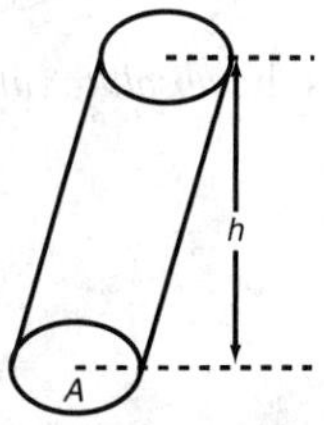

FIGURE 23.38 Cylinder (or prism) with
parallel bases: $V = Ah$

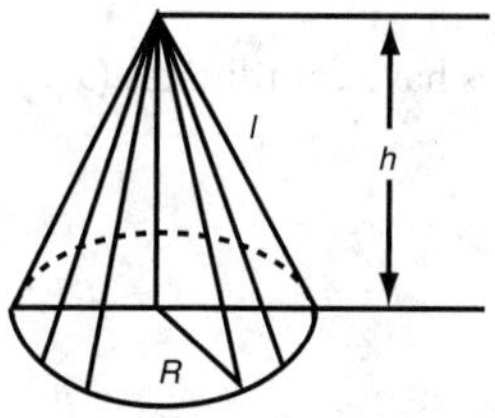

FIGURE 23.39 Right circular cone:
$V = \frac{1}{3}\pi R^2 h$;
lateral surface area $= \pi Rl = \pi R \sqrt{R^2 + h^2}$

FIGURE 23.40   Sphere: $V = \frac{4}{3}\pi R^3$; surface area $= 4\pi R^2$

23.6.3   Trigonometry

Triangles

In any triangle (in a plane) with sides $a, b,$ and c and corresponding opposite angles $A, B,$ and C

$$\frac{a}{\sin A} = \frac{b}{\sin B} = \frac{c}{\sin C} \qquad \text{(Law of sines)}$$

$$a^2 = b^2 + c^2 - 2cb \cos A \qquad \text{(Law of cosines)}$$

$$\frac{a+b}{a-b} = \frac{\tan \frac{1}{2}(A+B)}{\tan \frac{1}{2}(A-B)} \qquad \text{(Law of Tangents)}$$

$$\sin \frac{1}{2}A = \sqrt{\frac{(s-b)(s-c)}{bc}} \qquad \text{where } s = \frac{1}{2}(a+b+c)$$

$$\cos \frac{1}{2}A = \sqrt{\frac{s(s-a)}{bc}}$$

$$\tan \frac{1}{2}A = \sqrt{\frac{(s-b)(s-c)}{s(s-a)}}$$

$$\text{area} = \frac{1}{2}bc \sin A$$

$$= \sqrt{s(s-a)(s-b)(s-c)}$$

If the vertices have coordinates $(x_1, y_1), (x_2, y_2), (x_3, y_3)$, the area is the *absolute value* of the expression

$$\frac{1}{2} \begin{vmatrix} x_1 & y_1 & 1 \\ x_2 & y_2 & 1 \\ x_3 & y_3 & 1 \end{vmatrix}$$

Trigonometric Functions of an Angle

With reference to Fig. 23.41, $P(x, y)$ is a point in any one of the four quadrants and A is an angle whose initial side is coincident with the positive x axis and whose terminal side contains the point $P(x, y)$. The distance from the origin $P(x, y)$ is denoted by r and is positive. The trigonometric functions of the angle

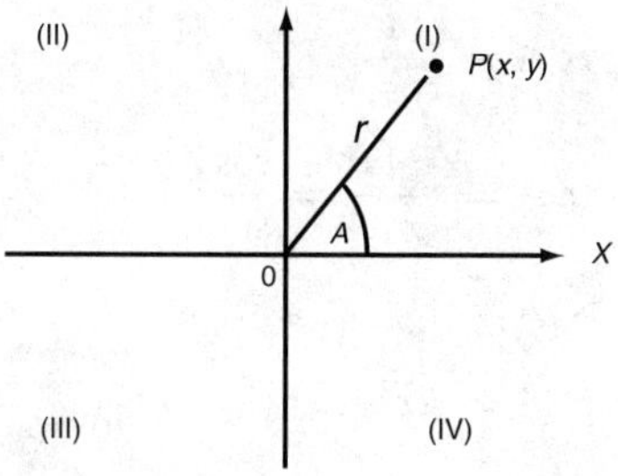

FIGURE 23.41 The trigonometric point: angle A is taken to be positive when the rotation is counterclockwise and negative when the rotation is clockwise. The plane is divided into quadrants as shown.

A are defined as

$$\sin A = \text{sine } A \qquad = y/r$$
$$\cos A = \text{cosine } A \qquad = x/r$$
$$\tan A = \text{tangent } A \qquad = y/x$$
$$\text{ctn } A = \text{cotangent } A = x/y$$
$$\sec A = \text{secant } A \qquad = r/x$$
$$\csc A = \text{cosecant } A \ = r/y$$

Angles are measured in degrees or radians; $180° = \pi$ radians; 1 radian $= 180/\pi°$.
The trigonometric functions of 0, 30, and 45°, and integer multiples of these are directly computed.

	0°	30°	45°	60°	90°	120°	135°	150°	180°
sin	0	$\dfrac{1}{2}$	$\dfrac{\sqrt{2}}{2}$	$\dfrac{\sqrt{3}}{2}$	1	$\dfrac{\sqrt{3}}{2}$	$\dfrac{\sqrt{2}}{2}$	$\dfrac{1}{2}$	0
cos	1	$\dfrac{\sqrt{3}}{2}$	$\dfrac{\sqrt{2}}{2}$	$\dfrac{1}{2}$	0	$-\dfrac{1}{2}$	$-\dfrac{\sqrt{2}}{2}$	$-\dfrac{\sqrt{3}}{2}$	-1
tan	0	$\dfrac{\sqrt{3}}{3}$	1	$\sqrt{3}$	∞	$-\sqrt{3}$	-1	$\dfrac{\sqrt{3}}{3}$	0
ctn	∞	$\sqrt{3}$	1	$\dfrac{\sqrt{3}}{3}$	0	$-\dfrac{\sqrt{3}}{3}$	-1	$-\sqrt{3}$	∞
sec	1	$\dfrac{2\sqrt{3}}{3}$	$\sqrt{2}$	2	∞	-2	$-\sqrt{2}$	$-\dfrac{2\sqrt{3}}{3}$	-1
csc	∞	2	$\sqrt{2}$	$\dfrac{2\sqrt{3}}{3}$	1	$\dfrac{2\sqrt{3}}{3}$	$\sqrt{2}$	2	∞

Trigonometric Identities

$$\sin A = \frac{1}{\csc A}$$

$$\cos A = \frac{1}{\sec A}$$

$$\tan A = \frac{1}{\text{ctn } A} = \frac{\sin A}{\cos A}$$

$$\csc A = \frac{1}{\sin A}$$

$$\sec A = \frac{1}{\cos A}$$

$$\operatorname{ctn} A = \frac{1}{\tan A} = \frac{\cos A}{\sin A}$$

$$\sin^2 A + \cos^2 A = 1$$

$$1 + \tan^2 A = \sec^2 A$$

$$1 + \operatorname{ctn}^2 A = \csc^2 A$$

$$\sin(A \pm B) = \sin A \cos B \pm \cos A \sin B$$

$$\cos(A \pm B) = \cos A \cos B \mp \sin A \sin B$$

$$\tan(A \pm B) = \frac{\tan A \pm \tan B}{1 \mp \tan A \tan B}$$

$$\sin 2A = 2 \sin A \cos A$$

$$\sin 3A = 3 \sin A - 4 \sin^3 A$$

$$\sin nA = 2 \sin(n - 1)A \cos A - \sin(n - 2)A$$

$$\cos 2A = 2 \cos^2 A - 1 = 1 - 2 \sin^2 A$$

$$\cos 3A = 4 \cos^3 A - 3 \cos A$$

$$\cos nA = 2 \cos(n - 1)A \cos A - \cos(n - 2)A$$

$$\sin A + \sin B = 2 \sin \frac{1}{2}(A + B) \cos \frac{1}{2}(A - B)$$

$$\sin A - \sin B = 2 \cos \frac{1}{2}(A + B) \sin \frac{1}{2}(A - B)$$

$$\cos A + \cos B = 2 \cos \frac{1}{2}(A + B) \cos \frac{1}{2}(A - B)$$

$$\cos A - \cos B = -2 \sin \frac{1}{2}(A + B) \sin \frac{1}{2}(A - B)$$

$$\tan A \pm \tan B = \frac{\sin(A \pm B)}{\cos A \cos B}$$

$$\operatorname{ctn} A \pm \operatorname{ctn} B = \pm \frac{\sin(A \pm B)}{\sin A \sin B}$$

$$\sin A \sin B = \frac{1}{2} \cos(A - B) - \frac{1}{2} \cos(A + B)$$

$$\cos A \cos B = \frac{1}{2} \cos(A - B) + \frac{1}{2} \cos(A + B)$$

$$\sin A \cos B = \frac{1}{2} \sin(A + B) + \frac{1}{2} \sin(A - B)$$

$$\sin \frac{A}{2} = \pm \sqrt{\frac{1 - \cos A}{2}}$$

$$\cos \frac{A}{2} = \pm \sqrt{\frac{1 + \cos A}{2}}$$

$$\tan \frac{A}{2} = \frac{1 - \cos A}{\sin A} = \frac{\sin A}{1 + \cos A} = \pm \sqrt{\frac{1 - \cos A}{1 + \cos A}}$$

$$\sin^2 A = \frac{1}{2}(1 - \cos 2A)$$

$$\cos^2 A = \frac{1}{2}(1 + \cos 2A)$$

$$\sin^3 A = \frac{1}{4}(3 \sin A - \sin 3A)$$

$$\cos^3 A = \frac{1}{4}(\cos 3A + 3 \cos A)$$

$$\sin ix = \frac{1}{2}i(e^x - e^{-x}) = i \sinh x$$

$$\cos ix = \frac{1}{2}(e^x + e^{-x}) = \cosh x$$

$$\tan ix = i \frac{(e^x - e^{-x})}{e^x + e^{-x}} = i \tanh x$$

$$e^{x+iy} = e^x(\cos y + i \sin y)$$

$$(\cos x \pm i \sin x)^n = \cos nx \pm i \sin nx$$

Inverse Trigonometric Functions

The inverse trigonometric functions are multiple valued, and this should be taken into account in the use of the following formulas:

$$\sin^{-1} x = \cos^{-1} \sqrt{1 - x^2}$$

$$= \tan^{-1} \frac{x}{\sqrt{1 - x^2}} = \text{ctn}^{-1} \frac{\sqrt{1 - x^2}}{x}$$

$$= \sec^{-1} \frac{1}{\sqrt{1 - x^2}} = \csc^{-1} \frac{1}{x}$$

$$= -\sin^{-1}(-x)$$

$$\cos^{-1} x = \sin^{-1} \sqrt{1 - x^2}$$

$$= \tan^{-1} \frac{\sqrt{1 - x^2}}{x} = \text{ctn}^{-1} \frac{x}{\sqrt{1 - x^2}}$$

$$= \sec^{-1} \frac{1}{x} = \csc^{-1} \frac{1}{\sqrt{1 - x^2}}$$

$$= \pi - \cos^{-1}(-x)$$

$$\tan^{-1} x = \text{ctn}^{-1} \frac{1}{x}$$

$$= \sin^{-1} \frac{x}{\sqrt{1 + x^2}} = \cos^{-1} \frac{1}{\sqrt{1 + x^2}}$$

$$= \sec^{-1} \sqrt{1 + x^2} = \csc^{-1} \frac{\sqrt{1 + x^2}}{x}$$

$$= -\tan^{-1}(-x)$$

23.6.4   Series

Bernoulli and Euler Numbers

A set of numbers, $B_1, B_3, \ldots, B_{2n-1}$ (Bernoulli numbers) and $B_2, B_4, \ldots, B_{2n}$ (Euler numbers) appear in the series expansions of many functions. A partial listing follows; these are computed from the following equations:

$$B_{2n} - \frac{2n(2n-1)}{2!}B_{2n-2} + \frac{2n(2n-1)(2n-2)(2n-3)}{4!}B_{2n-4} - \cdots + (-1)^n = 0$$

and

$$\frac{2^{2n}(2^{2n}-1)}{2n}B_{2n-1} = (2n-1)B_{2n-2} - \frac{(2n-1)(2n-2)(2n-3)}{3!}B_{2n-4} + \cdots + (-1)^{n-1}$$

$$
\begin{array}{ll}
B_1 = 1/6 & B_2 = 1 \\
B_3 = 1/30 & B_4 = 5 \\
B_5 = 1/42 & B_6 = 61 \\
B_7 = 1/30 & B_8 = 1385 \\
B_9 = 5/66 & B_{10} = 50521 \\
B_{11} = 691/2730 & B_{12} = 2\,702\,765 \\
B_{13} = 7/6 & B_{14} = 199\,360\,981 \\
\vdots & \vdots
\end{array}
$$

Series of Functions

In the following, the interval of convergence is indicated; otherwise it is all x. Logarithms are to the base e. Bernoulli and Euler numbers (B_{2n-1} and B_{2n}) appear in certain expressions.

$$(a + x)^n = a^n + na^{n-1}x + \frac{n(n-1)}{2!}a^{n-2}x^2 + \frac{n(n-1)(n-2)}{3!}a^{n-3}x^3 + \cdots$$
$$+ \frac{n!}{(n-j)!j!}a^{n-j}x^j + \cdots \qquad\qquad [x^2 < a^2]$$

$$(a - bx)^{-1} = \frac{1}{a}\left[1 + \frac{bx}{a} + \frac{b^2x^2}{a^2} + \frac{b^3x^3}{a^3} + \cdots\right] \qquad\qquad [b^2x^2 < a^2]$$

$$(1 \pm x)^n = 1 \pm nx + \frac{n(n-1)}{2!}x^2 \pm \frac{n(n-1)(n-2)x^3}{3!} + \cdots \qquad\qquad [x^2 < 1]$$

$$(1 \pm x)^{-n} = 1 \mp nx + \frac{n(n+1)}{2!}x^2 \mp \frac{n(n+1)(n+2)}{3!}x^3 + \cdots \qquad\qquad [x^2 < 1]$$

$$(1 \pm x)^{\frac{1}{2}} = 1 \pm \frac{1}{2}x - \frac{1}{2 \cdot 4}x^2 \pm \frac{1 \cdot 3}{2 \cdot 4 \cdot 6}x^3 - \frac{1 \cdot 3 \cdot 5}{2 \cdot 4 \cdot 6 \cdot 8}x^4 \pm \cdots \qquad\qquad [x^2 < 1]$$

$$(1 \pm x)^{-\frac{1}{2}} = 1 \mp \frac{1}{2}x + \frac{1 \cdot 3}{2 \cdot 4}x^2 \mp \frac{1 \cdot 3 \cdot 5}{2 \cdot 4 \cdot 6}x^3 + \frac{1 \cdot 3 \cdot 5 \cdot 7}{2 \cdot 4 \cdot 6 \cdot 8}x^4 \pm \cdots \qquad\qquad [x^2 < 1]$$

$$(1 \pm x^2)^{\frac{1}{2}} = 1 \pm \frac{1}{2}x^2 - \frac{x^4}{2 \cdot 4} \pm \frac{1 \cdot 3}{2 \cdot 4 \cdot 6}x^6 - \frac{1 \cdot 3 \cdot 5}{2 \cdot 4 \cdot 6 \cdot 8}x^8 \pm \cdots \qquad [x^2 < 1]$$

$$(1 \pm x)^{-1} = 1 \mp x + x^2 \mp x^3 + x^4 \mp x^5 + \cdots \qquad [x^2 < 1]$$

$$(1 \pm x)^{-2} = 1 \mp 2x + 3x^2 \mp 4x^3 + 5x^4 \mp \cdots \qquad [x^2 < 1]$$

$$e^x = 1 + x + \frac{x^2}{2!} + \frac{x^3}{3!} + \frac{x^4}{4!} + \cdots$$

$$e^{-x^2} = 1 - x^2 + \frac{x^4}{2!} - \frac{x^6}{3!} + \frac{x^8}{4!} - \cdots$$

$$a^x = 1 + x \log a + \frac{(x \log a)^2}{2!} + \frac{(x \log a)^3}{3!} + \cdots$$

$$\log x = (x - 1) - \frac{1}{2}(x - 1)^2 + \frac{1}{3}(x - 1)^3 - \cdots \qquad [0 < x < 2]$$

$$\log x = \frac{x-1}{x} + \frac{1}{2}\left(\frac{x-1}{x}\right)^2 + \frac{1}{3}\left(\frac{x-1}{x}\right)^3 + \cdots \qquad \left[x > \frac{1}{2}\right]$$

$$\log x = 2\left[\left(\frac{x-1}{x+1}\right) + \frac{1}{3}\left(\frac{x-1}{x+1}\right)^3 + \frac{1}{5}\left(\frac{x-1}{x+1}\right)^5 + \cdots\right] \qquad [x > 0]$$

$$\log(1 + x) = x - \frac{1}{2}x^2 + \frac{1}{3}x^3 - \frac{1}{4}x^4 + \cdots \qquad [x^2 < 1]$$

$$\log\left(\frac{1+x}{1-x}\right) = 2\left[x + \frac{1}{3}x^3 + \frac{1}{5}x^5 + \frac{1}{7}x^7 + \cdots\right] \qquad [x^2 < 1]$$

$$\log\left(\frac{x+1}{x-1}\right) = 2\left[\frac{1}{x} + \frac{1}{3}\left(\frac{1}{x}\right)^3 + \frac{1}{5}\left(\frac{1}{x}\right)^5 + \cdots\right] \qquad [x^2 < 1]$$

$$\sin x = x - \frac{x^3}{3!} + \frac{x^5}{5!} - \frac{x^7}{7!} + \cdots$$

$$\cos x = 1 - \frac{x^2}{2!} + \frac{x^4}{4!} - \frac{x^6}{6!} + \cdots$$

$$\tan x = x + \frac{x^3}{3} + \frac{2x^5}{15} + \frac{17x^7}{315} + \cdots + \frac{2^{2n}(2^{2n}-1)B_{2n-1}x^{2n-1}}{2n!} \qquad \left[x^2 < \frac{\pi^2}{4}\right]$$

$$\text{ctn } x = \frac{1}{x} - \frac{x}{3} - \frac{x^3}{45} - \frac{2x^5}{945} - \cdots - \frac{B_{2n-1}(2x)^{2n}}{(2n)!x} - \cdots \qquad [x^2 < \pi^2]$$

$$\sec x = 1 + \frac{x^2}{2!} + \frac{5x^4}{4!} + \frac{61x^6}{6!} + \cdots + \frac{B_{2n}x^{2n}}{(2n)!} + \cdots \qquad \left[x^2 < \frac{\pi^2}{4}\right]$$

$$\csc x = \frac{1}{x} + \frac{x}{3!} + \frac{7x^3}{3 \cdot 5!} + \frac{31x^5}{3 \cdot 7!} + \cdots + \frac{2(2^{2n+1}-1)}{(2n+2)!}B_{2n+1}x^{2n+1} + \cdots \qquad [x^2 < \pi^2]$$

$$\sin^{-1} x = x + \frac{x^3}{6} + \frac{(1 \cdot 3)x^5}{(2 \cdot 4)5} + \frac{(1 \cdot 3 \cdot 5)x^7}{(2 \cdot 4 \cdot 6)7} + \cdots \qquad [x^2 < 1]$$

$$\tan^{-1} x = x - \frac{1}{3}x^3 + \frac{1}{5}x^5 - \frac{1}{7}x^7 + \cdots \qquad [x^2 < 1]$$

$$\sec^{-1} x = \frac{\pi}{2} - \frac{1}{x} - \frac{1}{6x^3} - \frac{1 \cdot 3}{(2 \cdot 4)5x^5} - \frac{1 \cdot 3 \cdot 5}{(2 \cdot 4 \cdot 6)7x^7} - \cdots \qquad [x^2 < 1]$$

$$\sinh x = x + \frac{x^3}{3!} + \frac{x^5}{5!} + \frac{x^7}{7!} + \cdots$$

$$\cosh x = 1 + \frac{x^2}{2!} + \frac{x^4}{4!} + \frac{x^6}{6!} + \frac{x^8}{8!} + \cdots$$

$$\tanh x = (2^2 - 1)2^2 B_1 \frac{x}{2!} - (2^4 - 1)2^4 B_3 \frac{x^3}{4!} + (2^6 - 1)2^6 B_5 \frac{x^5}{6!} - \cdots \qquad \left[x^2 < \frac{\pi^2}{4}\right]$$

$$\operatorname{ctnh} x = \frac{1}{x}\left(1 + \frac{2^2 B_1 x^2}{2!} - \frac{2^4 B_3 x^4}{4!} + \frac{2^6 B_5 x^6}{6!} - \cdots\right) \qquad [x^2 < \pi^2]$$

$$\operatorname{sech} x = 1 - \frac{B_2 x^2}{2!} + \frac{B_4 x^4}{4!} - \frac{B_6 x^6}{6!} + \cdots \qquad \left[x^2 < \frac{\pi^2}{4}\right]$$

$$\operatorname{csch} x = \frac{1}{x} - (2 - 1)2B_1 \frac{x}{2!} + (2^3 - 1)2B_3 \frac{x^3}{4!} - \cdots \qquad [x^2 < \pi^2]$$

$$\sinh^{-1} x = x - \frac{1}{2}\frac{x^3}{3} + \frac{1 \cdot 3}{2 \cdot 4}\frac{x^5}{5} - \frac{1 \cdot 3 \cdot 5}{2 \cdot 4 \cdot 6}\frac{x^7}{7} + \cdots \qquad [x^2 < 1]$$

$$\tanh^{-1} x = x + \frac{x^3}{3} + \frac{x^5}{5} + \frac{x^7}{7} + \cdots \qquad [x^2 < 1]$$

$$\operatorname{ctnh}^{-1} x = \frac{1}{x} + \frac{1}{3x^3} + \frac{1}{5x^5} + \cdots \qquad [x^2 > 1]$$

$$\operatorname{csch}^{-1} x = \frac{1}{x} - \frac{1}{2 \cdot 3x^3} + \frac{1 \cdot 3}{2 \cdot 4 \cdot 5x^5} - \frac{1 \cdot 3 \cdot 5}{2 \cdot 4 \cdot 6 \cdot 7x^7} + \cdots \qquad [x^2 > 1]$$

$$\int_0^x e^{-t^2} dt = x - \frac{1}{3}x^3 + \frac{x^5}{5 \cdot 2!} - \frac{x^7}{7 \cdot 3!} + \cdots$$

Error Function

The following function, known as the error function, erf x, arises frequently in applications:

$$\operatorname{erf} x = \frac{2}{\sqrt{\pi}} \int_0^x e^{-t^2} dt$$

The integral cannot be represented in terms of a finite number of elementary functions; therefore, values of erf x have been compiled in tables. The following is the series for erf x:

$$\operatorname{erf} x = \frac{2}{\sqrt{\pi}}\left[x - \frac{x^3}{3} + \frac{x^5}{5 \cdot 2!} - \frac{x^7}{7 \cdot 3!} + \cdots\right]$$

There is a close relation between this function and the area under the standard normal curve. For evaluation it is convenient to use z instead of x; then erf z may be evaluated from the area $F(z)$ by use of the relation

$$\operatorname{erf} z = 2F(\sqrt{2}z)$$

Example

$$\operatorname{erf}(0.5) = 2F[(1.414)(0.5)] = 2F(0.707)$$

By interpolation, $F(0.707) = 0.260$; thus, erf $(0.5) = 0.520$.

Series Expansion

The expression in parentheses following certain series indicates the region of convergence. If not otherwise indicated, it is understood that the series converges for all finite values of x.

Binomial

$$(x + y)^n = x^n + nx^{n-1}y + \frac{n(n-1)}{2!}x^{n-2}y^2 + \frac{n(n-1)(n-2)}{3!}x^{n-3}y^3 + \cdots \qquad (y^2 < x^2)$$

$$(1 \pm x)^n = 1 \pm nx + \frac{n(n-1)x^2}{2!} \pm \frac{n(n-1)(n-2)x^3}{3!} + \cdots \qquad (x^2 < 1)$$

$$(1 \pm x)^{-n} = 1 \mp nx + \frac{n(n+1)x^2}{2!} \mp \frac{n(n+1)(n+2)x^3}{3!} + \cdots \qquad (x^2 < 1)$$

$$(1 \pm x)^{-1} = 1 \mp x + x^2 \mp x^3 + x^4 \mp x^5 + \cdots \qquad (x^2 < 1)$$

$$(1 \pm x)^{-2} = 1 \mp 2x + 3x^2 \mp 4x^3 + 5x^4 \mp 6x^5 + \cdots \qquad (x^2 < 1)$$

Reversion of Series

Let a series be represented by

$$y = a_1 x + a_2 x^2 + a_3 x^3 + a_4 x^4 + a_5 x^5 + a_6 x^6 + \cdots \qquad (a_1 \neq 0)$$

To find the coefficients of the series

$$x = A_1 y + A_2 y^2 + A_3 y^3 + A_4 y^4 + \cdots$$

$$A_1 = \frac{1}{a_1} \qquad A_2 = -\frac{a_2}{a_1^3} \qquad A_3 = \frac{1}{a_1^5}\left(2a_2^2 - a_1 a_3\right)$$

$$A_4 = \frac{1}{a_1^7}\left(5a_1 a_2 a_3 - a_1^2 a_4 - 5a_2^3\right)$$

$$A_5 = \frac{1}{a_1^9}\left(6a_1^2 a_2 a_4 + 3a_1^2 a_3^2 + 14a_2^4 - a_1^3 a_5 - 21a_1 a_2^2 a_3\right)$$

$$A_6 = \frac{1}{a_1^{11}}\left(7a_1^3 a_2 a_5 + 7a_1^3 a_3 a_4 + 84a_1 a_2^3 a_3 - a_1^4 a_6 - 28a_1^2 a_2^2 a_4 - 28a_1^2 a_2 a_3^2 - 42a_2^5\right)$$

$$A_7 = \frac{1}{a_1^{13}}\left(8a_1^4 a_2 a_6 + 8a_1^4 a_3 a_5 + 4a_1^4 a_4^2 + 120a_1^2 a_2^3 a_4 + 180a_1^2 a_2^2 a_3^2 + 132a_2^6 - a_1^5 a_7\right.$$

$$\left. - 36a_1^3 a_2^2 a_5 - 72a_1^3 a_2 a_3 a_4 - 12a_1^3 a_3^3 - 330a_1 a_2^4 a_3\right)$$

Taylor

1.

$$f(x) = f(a) + (x-a)f'(a) + \frac{(x-a)^2}{2!}f''(a) + \frac{(x-a)^3}{3!}f'''(a) + \cdots + \frac{(x-a)^n}{n!}f^{(n)}(a) + \cdots$$

2. (Increment form)

$$f(x+h) = f(x) + hf'(x) + \frac{h^2}{2!}f''(x) + \frac{h^3}{3!}f'''(x) + \cdots$$

$$= f(h) + xf'(h) + \frac{x^2}{2!}f''(h) + \frac{x^3}{3!}f'''(h) + \cdots$$

3. If $f(x)$ is a function possessing derivatives of all orders throughout the interval $a \leq x \leq b$, then there is a value X, with $a < X < b$, such that

$$f(b) = f(a) + (b-a)f'(a) + \frac{(b-a)^2}{2!} f''(a) + \cdots$$

$$+ \frac{(b-a)^{n-1}}{(n-1)!} f^{(n-1)}(a) + \frac{(b-a)^n}{n!} f^{(n)}(X)$$

$$f(a+h) = f(a) + hf'(a) + \frac{h^2}{2!} f''(a) + \cdots + \frac{h^{n-1}}{(n-1)!} f^{(n-1)}(a)$$

$$+ \frac{h^n}{n!} f^{(n)}(a+\theta h), \quad b = a+h, \quad 0 < \theta < 1$$

or

$$f(x) = f(a) + (x-a)f'(a) + \frac{(x-a)^2}{2!} f''(a) + \cdots + (x-a)^{n-1} \frac{f^{(n-1)}(a)}{(n-1)!} + R_n$$

where

$$R_n = \frac{f^{(n)}(a+\theta \cdot (x-a))}{n!}(x-a)^n, \quad 0 < \theta < 1$$

The preceding forms are known as Taylor's series with the remainder term.

4. Taylor's series for a function of two variables: If

$$\left(h\frac{\partial}{\partial x} + k\frac{\partial}{\partial y} \right) f(x,y) = h\frac{\partial f(x,y)}{\partial x} + k\frac{\partial f(x,y)}{\partial y};$$

$$\left(h\frac{\partial}{\partial x} + k\frac{\partial}{\partial y} \right)^2 f(x,y) = h^2 \frac{\partial^2 f(x,y)}{\partial x^2} + 2hk\frac{\partial^2 f(x,y)}{\partial x \, \partial y} + k^2 \frac{\partial^2 f(x,y)}{\partial y^2}$$

and so forth, and if

$$\left(h\frac{\partial}{\partial x} + k\frac{\partial}{\partial y} \right)^n f(x,y) \Big|_{\substack{x=a \\ y=b}}$$

where the bar and subscripts mean that after differentiation we are to replace x by a and y by b,

$$f(a+h, b+k) = f(a,b) + \left(h\frac{\partial}{\partial x} + k\frac{\partial}{\partial y} \right) f(x,y) \Big|_{\substack{x=a \\ y=b}} + \cdots$$

$$+ \frac{1}{n!} \left(h\frac{\partial}{\partial x} + k\frac{\partial}{\partial y} \right)^n f(x,y) \Big|_{\substack{x=a \\ y=b}} + \cdots$$

MacLaurin

$$f(x) = f(0) + xf'(0) + \frac{x^2}{2!} f''(0) + \frac{x^3}{3!} f'''(0) + \cdots + x^{n-1} \frac{f^{n-1}(0)}{(n-1)!} + R_n$$

where

$$R_n = \frac{x^n f^{(n)}(\theta x)}{n!}, \quad 0 < \theta < 1$$

Exponential

$$e = 1 + \frac{1}{1!} + \frac{1}{2!} + \frac{1}{3!} + \frac{1}{4!} + \cdots$$

$$e^x = 1 + x + \frac{x^2}{2!} + \frac{x^3}{3!} + \frac{x^4}{4!} + \cdots \qquad \text{(all real values of } x\text{)}$$

$$a^x = 1 + x \log_e a + \frac{(x \log_e a)^2}{2!} + \frac{(x \log_e a)^3}{3!} + \cdots$$

$$e^x = e^a \left[1 + (x - a) + \frac{(x - a)^2}{2!} + \frac{(x - a)^3}{3!} + \cdots \right]$$

Logarithmic

$$\log_e x = \frac{x-1}{x} + \frac{1}{2}\left(\frac{x-1}{x}\right)^2 + \frac{1}{3}\left(\frac{x-1}{x}\right)^3 + \cdots \qquad \left(x > \frac{1}{2}\right)$$

$$\log_e x = (x-1) - \frac{1}{2}(x-1)^2 + \frac{1}{3}(x-1)^3 - \cdots \qquad (2 \geq x > 0)$$

$$\log_e x = 2\left[\frac{x-1}{x+1} + \frac{1}{3}\left(\frac{x-1}{x+1}\right)^3 \frac{1}{5}\left(\frac{x-1}{x+1}\right)^5 + \cdots\right] \qquad (x > 0)$$

$$\log_e(1+x) = x - \frac{1}{2}x^2 + \frac{1}{3}x^3 - \frac{1}{4}x^4 + \cdots \qquad (-1 < x \leq 1)$$

$$\log_e(n+1) - \log_e(n-1) = 2\left[\frac{1}{n} + \frac{1}{3n^3} + \frac{1}{5n^5} + \cdots\right]$$

$$\log_e(a+x) = \log_e a + 2\left[\frac{x}{2a+x} + \frac{1}{3}\left(\frac{x}{2a+x}\right)^3 + \frac{1}{5}\left(\frac{x}{2a+x}\right)^5 + \cdots\right]$$

$$(a > 0, -a < x < +\infty)$$

$$\log_e\frac{1+x}{1-x} = 2\left[x + \frac{x^3}{3} + \frac{x^5}{5} + \cdots + \frac{x^{2n-1}}{2n-1} + \cdots\right] \qquad (-1 < x < 1)$$

$$\log_e x = \log_e a + \frac{(x-a)}{a} - \frac{(x-a)^2}{2a^2} + \frac{(x-a)^3}{3a^3} - \cdots \qquad (0 < x \leq 2a)$$

Trigonometric

$$\sin x = x - \frac{x^3}{3!} + \frac{x^5}{5!} - \frac{x^7}{7!} + \cdots \qquad \text{(all real values of } x\text{)}$$

$$\cos x = 1 - \frac{x^2}{2!} + \frac{x^4}{4!} - \frac{x^6}{6!} + \cdots \qquad \text{(all real values of } x\text{)}$$

$$\tan x = x + \frac{x^3}{3} + \frac{2x^5}{15} + \frac{17x^7}{315} + \frac{62x^9}{2835} + \cdots$$

$$+ \frac{(-1)^{n-1}2^{2n}(2^{2n}-1)B_{2n}}{(2n)!}x^{2n-1} + \cdots$$

$$(x^2 < \pi^2/4, \text{ and } B_n \text{ represents the } n\text{th Bernoulli number})$$

$$\cot x = \frac{1}{x} - \frac{x}{3} - \frac{x^2}{45} - \frac{2x^5}{945} - \frac{x^7}{4725} - \cdots$$

$$- \frac{(-1)^{n+1}2^{2n}}{(2n)!}B_{2n}x^{2n-1} + \cdots$$

$$(x^2 < \pi^2, \text{ and } B_n \text{ represents the } n\text{th Bernoulli number})$$

23.6.5   Differential Calculus

Notation

For the following equations, the symbols $f(x), g(x)$, and so forth represent functions of x. The value of a function $f(x)$ at $x = a$ is denoted $f(a)$. For the function $y = f(x)$ the derivative of y with respect to x is denoted by one of the following:

$$\frac{dy}{dx}, \quad f'(x), \quad D_x y, \quad y'$$

Higher derivatives are as follows:

$$\frac{d^2 y}{dx^2} = \frac{d}{dx}\left(\frac{dy}{dx}\right) = \frac{d}{dx} f'(x) = f''(x)$$

$$\frac{d^3 y}{dx^3} = \frac{d}{dx}\left(\frac{d^2 y}{dx^2}\right) = \frac{d}{dx} f''(x) = f'''(x)$$

$$\vdots$$

and values of these at $x = a$ are denoted $f''(a), f'''(a)$, and so on (see Table 23.1, Table of Derivatives).

Slope of a Curve

The tangent line at point $P(x, y)$ of the curve $y = f(x)$ has a slope $f'(x)$ provided that $f'(x)$ exists at P. The slope at P is defined to be that of the tangent line at P. The tangent line at $P(x_1, y_1)$ is given by

$$y - y_1 = f'(x_1)(x - x_1)$$

The *normal line* to the curve at $P(x_1, y_1)$ has slope $-1/f'(x_1)$ and thus obeys the equation

$$y - y_1 = [-1/f'(x_1)](x - x_1)$$

(The slope of a vertical line is not defined.)

Angle of Intersection of Two Curves

Two curves, $y = f_1(x)$ and $y = f_2(x)$, that intersect at a point $P(X, Y)$ where derivatives $f_1'(X), f_2'(X)$ exist have an angle (α) of intersection given by

$$\tan \alpha = \frac{f_2'(X) - f_1'(X)}{1 + f_2'(X) \cdot f_1'(X)}$$

If $\tan \alpha > 0$, then α is the acute angle; if $\tan \alpha < 0$, then α is the obtuse angle.

Radius of Curvature

The radius of curvature R of the curve $y = f(x)$ at the point $P(x, y)$ is

$$R = \frac{\{1 + [f'(x)]^2\}^{3/2}}{f''(x)}$$

In polar coordinates (θ, r) the corresponding formula is

$$R = \frac{\left[r^2 + \left(\dfrac{dr}{d\theta}\right)^2\right]^{3/2}}{r^2 + 2\left(\dfrac{dr}{d\theta}\right)^2 - r\dfrac{d^2 r}{d\theta^2}}$$

The *curvature K* is $1/R$.

TABLE 23.1 Table of Derivatives*

1. $\dfrac{d}{dx}(a) = 0$

2. $\dfrac{d}{dx}(x) = 1$

3. $\dfrac{d}{dx}(au) = a\dfrac{du}{dx}$

4. $\dfrac{d}{dx}(u+v) = \dfrac{du}{dx} + \dfrac{dv}{dx}$

5. $\dfrac{d}{dx}(uv) = u\dfrac{dv}{dx} + v\dfrac{du}{dx}$

6. $\dfrac{d}{dx}\dfrac{u}{v} = \dfrac{v\dfrac{du}{dx} - u\dfrac{dv}{dx}}{v^2}$

7. $\dfrac{d}{dx}(u^n) = nu^{n-1}\dfrac{du}{dx}$

8. $\dfrac{d}{dx}e^u = e^u\dfrac{du}{dx}$

9. $\dfrac{d}{dx}a^u = (\log_e a)a^u\dfrac{du}{dx}$

10. $\dfrac{d}{dx}\log_e u = \dfrac{1}{u}\dfrac{du}{dx}$

11. $\dfrac{d}{dx}\log_a u = (\log_a e)\dfrac{1}{u}\dfrac{du}{dx}$

12. $\dfrac{d}{dx}u^v = vu^{v-1}\dfrac{du}{dx} + u^v(\log_e u)\dfrac{dv}{dx}$

13. $\dfrac{d}{dx}\sin u = \cos u\dfrac{du}{dx}$

14. $\dfrac{d}{dx}\cos u = -\sin u\dfrac{du}{dx}$

15. $\dfrac{d}{dx}\tan u = \sec^2 u\dfrac{du}{dx}$

16. $\dfrac{d}{dx}\operatorname{ctn} u = -\csc^2 u\dfrac{du}{dx}$

17. $\dfrac{d}{dx}\sec u = \sec u \tan u\dfrac{du}{dx}$

18. $\dfrac{d}{dx}\csc u = -\csc u \operatorname{ctn} u\dfrac{du}{dx}$

19. $\dfrac{d}{dx}\sin^{-1} u = \dfrac{1}{\sqrt{1-u^2}}\dfrac{du}{dx}$, $\quad\left(-\dfrac{1}{2}\pi \le \sin^{-1} u \le \dfrac{1}{2}\pi\right)$

20. $\dfrac{d}{dx}\cos^{-1} u = \dfrac{-1}{\sqrt{1-u^2}}\dfrac{du}{dx}$, $\quad(0 \le \cos^{-1} u \le \pi)$

21. $\dfrac{d}{dx}\tan^{-1} u = \dfrac{1}{1+u^2}\dfrac{du}{dx}$

22. $\dfrac{d}{dx}\operatorname{ctn}^{-1} u = \dfrac{-1}{1+u^2}\dfrac{du}{dx}$

23. $\dfrac{d}{dx}\sec^{-1} u = \dfrac{1}{u\sqrt{u^2-1}}\dfrac{du}{dx}$,

$\left(-\pi \le \sec^{-1} u < -\dfrac{1}{2}\pi; \quad 0 \le \sec^{-1} u < \dfrac{1}{2}\pi\right)$

24. $\dfrac{d}{dx}\csc^{-1} u = \dfrac{-1}{u\sqrt{u^2-1}}\dfrac{du}{dx}$,

$\left(-\pi \le \csc^{-1} u \le -\dfrac{1}{2}\pi; \quad 0 < \csc^{-1} u \le \dfrac{1}{2}\pi\right)$

25. $\dfrac{d}{dx}\sinh u = \cosh u\dfrac{du}{dx}$

26. $\dfrac{d}{dx}\cosh u = \sinh u\dfrac{du}{dx}$

27. $\dfrac{d}{dx}\tanh u = \operatorname{sech}^2 u\dfrac{du}{dx}$

28. $\dfrac{d}{dx}\operatorname{ctnh} u = -\operatorname{csch}^2 u\dfrac{du}{dx}$

29. $\dfrac{d}{dx}\operatorname{sech} u = -\operatorname{sech} u \tanh u\dfrac{du}{dx}$

30. $\dfrac{d}{dx}\operatorname{csch} u = -\operatorname{csch} u \operatorname{ctnh} u\dfrac{du}{dx}$

31. $\dfrac{d}{dx}\sinh^{-1} u = \dfrac{1}{\sqrt{u^2+1}}\dfrac{du}{dx}$

32. $\dfrac{d}{dx}\cosh^{-1} u = \dfrac{1}{\sqrt{u^2-1}}\dfrac{du}{dx}$

33. $\dfrac{d}{dx}\tanh^{-1} u = \dfrac{1}{1-u^2}\dfrac{du}{dx}$

34. $\dfrac{d}{dx}\operatorname{ctnh}^{-1} u = \dfrac{-1}{u^2-1}\dfrac{du}{dx}$

35. $\dfrac{d}{dx}\operatorname{sech}^{-1} u = \dfrac{-1}{u\sqrt{1-u^2}}\dfrac{du}{dx}$

36. $\dfrac{d}{dx}\operatorname{csch}^{-1} u = \dfrac{1}{u\sqrt{u^2+1}}\dfrac{du}{dx}$

*In this table, a and n are constants, e is the base of the natural logarithms, and u and v denote functions of x.

Relative Maxima and Minima

The function f has a relative maximum at $x = a$ if $f(a) \ge f(a+c)$ for all values of c (positive or negative) that are sufficiently near zero. The function f has a relative minimum at $x = b$ if $f(b) \le f(b+c)$ for all values of c that are sufficiently close to zero. If the function f is defined on the closed interval $x_1 \le x \le x_2$ and has a relative maximum or minimum at $x = a$, where $x_1 < a < x_2$, and if the derivative $f'(x)$ exists at $x = a$, then $f'(a) = 0$. It is noteworthy that a relative maximum or minimum may occur at a point where the derivative does not exist. Further, the derivative may vanish at a point that is neither a maximum nor a minimum for the function. Values of x for which $f'(x) = 0$ are called *critical values*. To determine whether a critical value of x, say

x_c, is a relative maximum or minimum for the function at x_c, one may use the second derivative test:

1. If $f''(x_c)$ is positive, $f(x_c)$ is a minimum.
2. If $f''(x_c)$ is negative, $f(x_c)$ is a maximum.
3. If $f''(x_c)$ is zero, no conclusion may be made.

The sign of the derivative as x advances through x_c may also be used as a test. If $f'(x)$ changes from positive to zero to negative, then a maximum occurs at x_c, whereas a change in $f'(x)$ from negative to zero to positive indicates a minimum. If $f'(x)$ does not change sign as x advances through x_c, then the point is neither a maximum nor a minimum.

Points of Inflection of a Curve

The sign of the second derivative of f indicates whether the graph of $y = f(x)$ is concave upward or concave downward:

$$f''(x) > 0 : \text{concave upward}$$
$$f''(x) < 0 : \text{concave downward}$$

A point of the curve at which the direction of concavity changes is called a point of inflection (Fig. 23.42). Such a point may occur where $f''(x) = 0$ or where $f''(x)$ becomes infinite. More precisely, if the function $y = f(x)$ and its first derivative $y' = f'(x)$ are continuous in the interval $a \leq x \leq b$, and if $y'' = f''(x)$ exists in $a < x < b$, then the graph of $y = f(x)$ for $a < x < b$ is concave upward if $f''(x)$ is positive and concave downward if $f''(x)$ is negative.

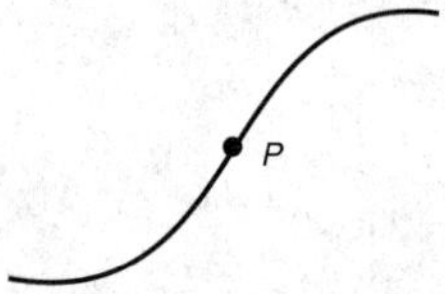

FIGURE 23.42 Point of inflection.

Taylor's Formula

If f is function that is continuous on an interval that contains a and x, and if its first $(n + 1)$ derivatives are continuous on this interval, then

$$f(x) = f(a) + f'(a)(x - a) + \frac{f''(a)}{2!}(x - a)^2$$
$$+ \frac{f'''(a)}{3!}(x - a)^3 + \cdots + \frac{f^{(n)}(a)}{n!}(x - a)^n + R$$

where R is called the *remainder*. There are various common forms of the remainder.

Lagrange's form:

$$R = f^{(n+1)}(\beta) \cdot \frac{(x - a)^{n+1}}{(n + 1)!}, \quad \beta \text{ between } a \text{ and } x$$

Cauchy's form:

$$R = f^{(n+1)}(\beta) \cdot \frac{(x - B)^n (x - a)}{n!}, \quad \beta \text{ between } a \text{ and } x$$

Integral form:

$$R = \int_a^x \frac{(x - t)^n}{n!} f^{(n+1)}(t) dt$$

Indeterminant Forms

If $f(x)$ and $g(x)$ are continuous in an interval that includes $x = a$, and if $f(a) = 0$ and $g(a) = 0$, the limit $\lim_{x \to a}(f(x)/g(x))$ takes the form $0/0$, called an *indeterminant form*. *L'Hôpital's rule* is

$$\lim_{x \to a} \frac{f(x)}{g(x)} = \lim_{x \to a} \frac{f'(x)}{g'(x)}$$

Similarly, it may be shown that if $f(x) \to \infty$ and $g(x) \to \infty$ as $x \to a$, then

$$\lim_{x \to a} \frac{f(x)}{g(x)} = \lim_{x \to a} \frac{f'(x)}{g'(x)}$$

(This holds for $x \to \infty$.)

Examples

$$\lim_{x \to 0} \frac{\sin x}{x} = \lim_{x \to 0} \frac{\cos x}{1} = 1$$

$$\lim_{x \to \infty} \frac{x^2}{e^x} = \lim_{x \to \infty} \frac{2x}{e^x} = \lim_{x \to \infty} \frac{2}{e^x} = 0$$

Numerical Methods

1. *Newton's method* for approximating roots of the equation $f(x) = 0$: A first estimate x_1 of the root is made; then, provided that $f'(x_1) \neq 0$, a better approximation is x_2,

$$x_2 = x_1 - \frac{f(x_1)}{f'(x_1)}$$

The process may be repeated to yield a third approximation, x_3, to the root

$$x_3 = x_2 - \frac{f(x_2)}{f'(x_2)}$$

provided $f'(x_2)$ exists. The process may be repeated. (In certain rare cases the process will not converge.)

2. *Trapezoidal rule for areas* (Fig. 23.43): For the function $y = f(x)$ defined on the interval (a, b) and positive there, take n equal subintervals of width $\Delta x = (b - a)/n$. The area bounded by the curve between $x = a$ and $x = b$ (or definite integral of $f(x)$) is approximately the sum of trapezoidal areas, or

$$A \sim \left(\frac{1}{2}y_0 + y_1 + y_2 + \cdots + y_{n-1} + \frac{1}{2}y_n \right) (\Delta x)^2$$

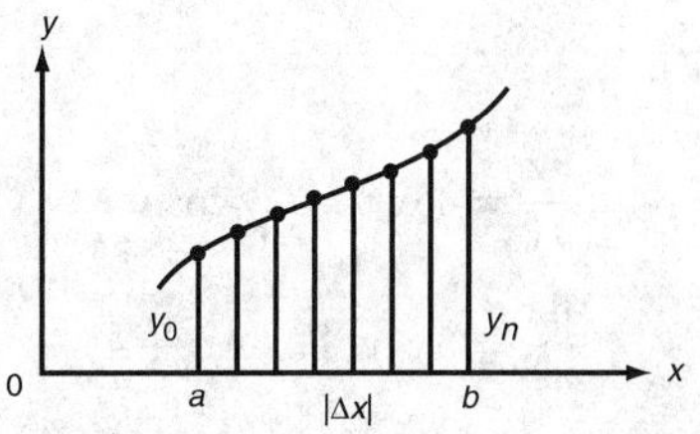

FIGURE 23.43 Trapezoidal rule for area.

Estimation of the error (E) is possible if the second derivative can be obtained:

$$E = \frac{b-a}{12} f''(c)(\Delta x)^2$$

where c is some number between a and b.

Functions of Two Variables

For the function of two variables, denoted $z = f(x, y)$, if y is held constant, say at $y = y_1$, then the resulting function is a function of x only. Similarly, x may be held constant at x_1, to give the resulting function of y.

The Gas Laws

A familiar example is afforded by the ideal gas law relating the pressure p, the volume V, and the absolute temperature T of an ideal gas:

$$pV = nRT$$

where n is the number of moles and R is the gas constant per mole, 8.31 ($\mathrm{J \cdot K^{-1} \cdot mole^{-1}}$). By rearrangement, any one of the three variables may be expressed as a function of the other two. Further, either one of these two may be held constant. If T is held constant, then we get the form known as Boyle's law:

$$p = kV^{-1} \qquad \text{(Boyle's law)}$$

where we have denoted nRT by the constant k and, of course, $V > 0$. If the pressure remains constant, we have Charles' law:

$$V = bT \qquad \text{(Charles' law)}$$

where the constant b denotes nR/p. Similarly, volume may be kept constant:

$$p = aT$$

where now the constant, denoted a, is nR/V.

Partial Derivatives

The physical example afforded by the ideal gas law permits clear interpretations of processes in which one of the variables is held constant. More generally, we may consider a function $z = f(x, y)$ defined over some region of the xy plane in which we hold one of the two coordinates, say y, constant. If the resulting function of x is differentiable at a point (x, y), we denote this derivative by one of the notations.

$$f_x, \quad \frac{\partial f}{\partial x}$$

called the *partial derivative with respect to x*. Similarly, if x is held constant and the resulting function of y is differentiable, we get the *partial derivative with respect to y*, denoted by one of the following:

$$f_y, \quad \frac{\partial f}{\partial y}$$

Example

Given $z = x^4 y^3 - y \sin x + 4y$, then

$$\frac{\partial z}{\partial x} = 4(xy)^3 - y \cos x$$

$$\frac{\partial z}{\partial y} = 3x^4 y^2 - \sin x + 4$$

In Table 23.1, a and n are constants, e is the base of the natural logarithms, and u and v denote functions of x.

Additional Relations with Derivatives

$$\frac{d}{dt} \int_a^t f(x) \, dx = f(t) \qquad \frac{d}{dt} \int_t^a f(x) \, dx = -f(t)$$

If $x = f(y)$, then

$$\frac{dy}{dx} = \frac{1}{\dfrac{dx}{dy}}$$

If $y = f(u)$ and $u = g(x)$, then

$$\frac{dy}{dx} = \frac{dy}{du} \cdot \frac{du}{dx} \qquad \text{(chain rule)}$$

If $x = f(t)$ and $y = g(t)$, then

$$\frac{dy}{dx} = \frac{g'(t)}{f'(t)}, \qquad \text{and} \qquad \frac{d^2}{dx^2} = \frac{f'(t)g''(t) - g'(t)f''(t)}{(f'(t))^3}$$

(*Note*: Exponent in denominator is 3.)

23.6.6 Integral Calculus

Indefinite Integral

If $F(x)$ is differentiable for all values of x in the interval (a, b) and satisfies the equation $dy/dx = f(x)$, then $F(x)$ is an integral of $f(x)$ with respect to x. The notation is $F(x) = \int f(x) \, dx$ or, in differential form, $dF(x) = f(x) \, dx$.

For any function $F(x)$ that is an integral of $f(x)$, it follows that $F(x) + C$ is also an integral. We thus write

$$\int f(x)dx = F(x) + C$$

Definite Integral

Let $f(x)$ be defined on the interval (a, b) which is partitioned by points $x_1, x_2, \ldots, x_j, \ldots, x_{n-1}$ between $a = x_0$ and $b = x_n$. The jth interval has length $\Delta x_j = x_j - x_{j-1}$, which may vary with j. The sum $\sum_{j=1}^n f(v_j) \Delta x_j$, where v_j is arbitrarily chosen in the jth subinterval, depends on the numbers $x_0, \ldots, x_n$ and the choice of the v as well as f; but if such sums approach a common value as all Δx approach zero, then this value is the definite integral of f over the interval (a, b) and is denoted $\int_a^b f(x) \, dx$. The *fundamental theorem of integral calculus* states that

$$\int_a^b f(x) \, dx = F(b) - F(a)$$

where F is any continuous indefinite integral of f in the interval (a, b).

Properties

$$\int_a^b (f_1(x) + f_2(x) + \cdots + f_j(x)) \, dx = \int_a^b f_1(x) \, dx + \int_a^b f_2(x) \, dx + \cdots + \int_a^b f_j(x) \, dx$$

$$\int_a^b cf(x) \, dx = c \int_a^b f(x) \, dx, \qquad \text{if } c \text{ is a constant}$$

$$\int_a^b f(x) \, dx = - \int_b^a f(x) \, dx$$

$$\int_a^b f(x) \, dx = \int_a^c f(x) \, dx + \int_c^b f(x) \, dx$$

Common Applications of the Definite Integral

Area (Rectangular Coordinates)

Given the function $y = f(x)$ such that $y > 0$ for all x between a and b, the area bounded by the curve $y = f(x)$, the x axis, and the vertical lines $x = a$ and $x = b$ is

$$A = \int_a^b f(x)\,dx$$

Length of Arc (Rectangular Coordinates)

Given the smooth curve $f(x, y) = 0$ from point (x_1, y_1) to point (x_2, y_2), the length between these points is

$$L = \int_{x_1}^{x_2} \sqrt{1 + \left(\frac{dy}{dx}\right)^2}\,dx$$

$$L = \int_{y_1}^{y_2} \sqrt{1 + \left(\frac{dx}{dy}\right)^2}\,dy$$

Mean Value of a Function

The mean value of a function $f(x)$ continuous on (a, b) is

$$\frac{1}{(b-a)} \int_a^b f(x)\,dx$$

Area (Polar Coordinates)

Given the curve $r = f(\theta)$, continuous and nonnegative for $\theta_1 \leq \theta \leq \theta_2$, the area enclosed by this curve and the radial lines $\theta = \theta_1$ and $\theta = \theta_2$ is given by

$$A = \int_{\theta_1}^{\theta_2} \frac{1}{2}[f(\theta)]^2 d\theta$$

Length of Arc (Polar Coordinates)

Given the curve $r = f(\theta)$ with continuous derivative $f'(\theta)$ on $\theta_1 \leq \theta \leq \theta_2$, the length of arc from $\theta = \theta_1$ to $\theta = \theta_2$ is

$$L = \int_{\theta_1}^{\theta_2} \sqrt{[f(\theta)]^2 + [f'(\theta)]^2} d\theta$$

Volume of Revolution

Given a function $y = f(x)$ continuous and nonnegative on the interval (a, b), when the region bounded by $f(x)$ between a and b is revolved about the x axis, the volume of revolution is

$$V = \pi \int_a^b [f(x)]^2\,dx$$

Surface Area of Revolution (Revolution about the x axis, between a and b)

If the portion of the curve $y = f(x)$ between $x = a$ and $x = b$ is revolved about the x axis, the area A of the surface generated is given by the following:

$$A = \int_a^b 2\pi f(x)\{1 + [f'(x)]^2\}^{\frac{1}{2}}\,dx$$

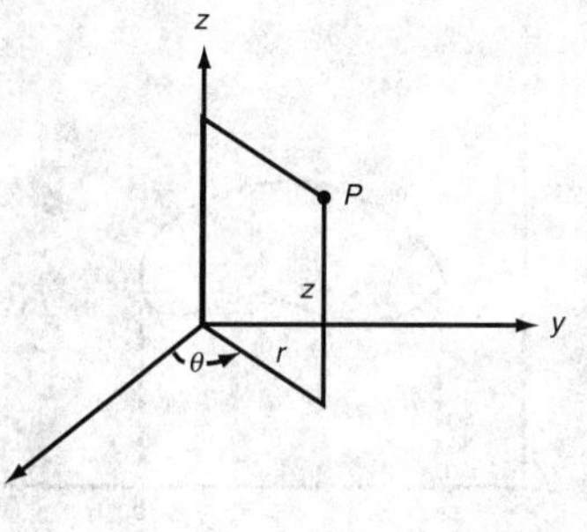

FIGURE 23.44   Cylindrical coordinates.

Work

If a variable force $f(x)$ is applied to an object in the direction of motion along the x axis between $x = a$ and $x = b$, the work done is

$$W = \int_a^b f(x)\,dx$$

Cylindrical and Spherical Coordinates

1. Cylindrical coordinates (Fig. 23.44):

$$x = r\cos\theta$$
$$y = r\sin\theta$$

   Element of volume $dV = r\,dr\,d\theta\,dz$.

2. Spherical coordinates (Fig. 23.45):

$$x = \rho\sin\phi\cos\theta$$
$$y = \rho\sin\phi\sin\theta$$
$$z = \rho\cos\phi$$

   Element of volume $dV = \rho^2\sin\phi\,d\rho\,d\phi\,d\theta$.

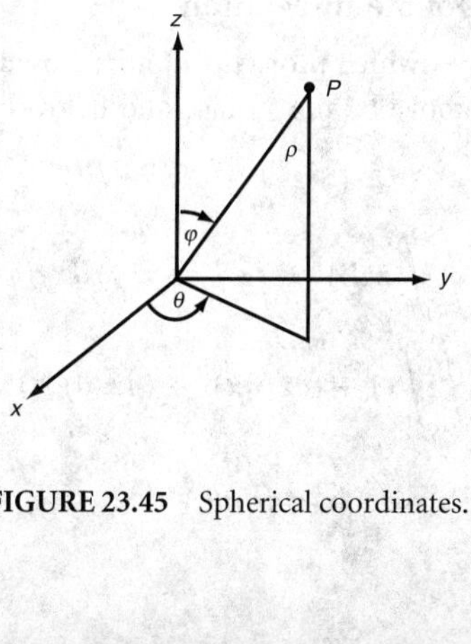

FIGURE 23.45   Spherical coordinates.

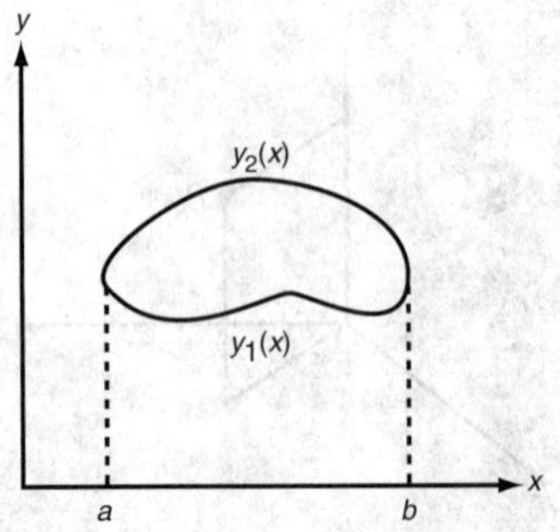

FIGURE 23.46   Region R bounded by $y_2(x)$ and $y_1(x)$.

Double Integration

The evaluation of a double integral of $f(x, y)$ over a plane region R,

$$\int\!\!\int_R f(x, y) \, dA$$

is practically accomplished by iterated (repeated) integration. For example, suppose that a vertical straight line meets the boundary of R in at most two points so that there is an upper boundary, $y = y_2(x)$, and a lower boundary, $y = y_1(x)$. Also, it is assumed that these functions are continuous from a to b (see Fig. 23.46). Then

$$\int\!\!\int_R f(x, y) \, dA = \int_a^b \left(\int_{y_1(x)}^{y_2(y)} f(x, y) \, dy \right) dx$$

If R has left-hand boundary $x = x_1(y)$ and a right-hand boundary $x = x_2(y)$, which are continuous from c to d (the extreme values of y in R), then

$$\int\!\!\int_R f(x, y) \, dA = \int_c^d \left(\int_{x_1(y)}^{x_2(y)} f(x, y) \, dx \right) dy$$

Such integrations are sometimes more convenient in polar coordinates, $x = r \cos \theta$, $y = r \sin \theta$, $dA = r \, dr \, d\theta$.

Surface Area and Volume by Double Integration

For the surface given by $z = f(x, y)$, which projects onto the closed region R of the xy plane, one may calculate the volume V bounded above by the surface and below by R, and the surface area S by the following:

$$V = \int\!\!\int_R z \, dA = \int\!\!\int_R f(x, y) \, dx \, dy$$

$$S = \int\!\!\int_R (1 + (\partial z / \partial x)^2 + (\partial z / \partial y)^2)^{\frac{1}{2}} \, dx \, dy$$

(In polar coordinates, (r, θ), we replace dA by $r \, dr \, d\theta$.)

Centroid

The centroid of a region R of the xy plane is a point (x', y') where

$$x' = \frac{1}{A} \int\!\!\int_R x \, dA, \quad y' = \frac{1}{A} \int\!\!\int_R y \, dA$$

and A is the area of the region.

Example

For the circular sector of angle 2α and radius R, the area A is αR^2; the integral needed for x', expressed in polar coordinates, is

$$\int\!\!\int x \, dA = \int_{-\alpha}^{\alpha} \int_0^R (r\cos\theta) r \, dr \, d\theta = \left[\frac{R^3}{3}\sin\theta\right]_{-\alpha}^{+\alpha} = \frac{2}{3} R^3 \sin\alpha$$

and thus

$$x' = \frac{\dfrac{2}{3}R^3\sin\alpha}{\alpha R^2} = \frac{2}{3}R\frac{\sin\alpha}{\alpha}$$

Centroids of some common regions are shown in Table 23.2.

TABLE 23.2 Centroids

	Area	x'	y'
Rectangle	bh	$b/2$	$h/2$
Isosceles triangle	$bh/2$	$b/2$	$h/3$
($y' = h/3$ for any triangle of altitude h.)			
Semicircle	$\pi R^2/2$	R	$4R/3\pi$
Quarter circle	$\pi R^2/4$	$4R/3\pi$	$4R/3\pi$
Circular sector	$R^2 A$	$2R\sin A/3A$	0

23.6.7   Special Functions

Hyperbolic Functions

$$\sinh x = \frac{e^x - e^{-x}}{2} \qquad\qquad \operatorname{csch} x = \frac{1}{\sinh x}$$

$$\cosh x = \frac{e^x + e^{-x}}{2} \qquad\qquad \operatorname{sech} x = \frac{1}{\cosh x}$$

$$\tanh x = \frac{e^x - e^{-x}}{e^x + e^{-x}} \qquad\qquad \operatorname{ctnh} x = \frac{1}{\tanh x}$$

$$\sinh(-x) = -\sinh x \qquad\qquad \operatorname{ctnh}(-x) = -\operatorname{ctnh} x$$

$$\cosh(-x) = \cosh x \qquad\qquad \operatorname{sech}(-x) = \operatorname{sech} x$$

$$\tanh(-x) = -\tanh x \qquad\qquad \operatorname{csch}(-x) = -\operatorname{csch} x$$

$$\tanh x = \frac{\sinh x}{\cosh x} \qquad\qquad \operatorname{ctnh} x = \frac{\cosh x}{\sinh x}$$

$$\cosh^2 x - \sinh^2 x = 1 \qquad\qquad \cosh^2 x = \frac{1}{2}(\cosh 2x + 1)$$

$$\sinh^2 x = \frac{1}{2}(\cosh 2x - 1) \qquad\qquad \operatorname{ctnh}^2 x - \operatorname{csch}^2 x = 1$$

$$\operatorname{csch}^2 x - \operatorname{sech}^2 x = \operatorname{csch}^2 x \operatorname{sech}^2 x \qquad\qquad \tanh^2 x + \operatorname{sech}^2 x = 1$$

$$\sinh(x + y) = \sinh x \cosh y + \cosh x \sinh y$$

$$\cosh(x + y) = \cosh x \cosh y + \sinh x \sinh y$$

$$\sinh(x - y) = \sinh x \cosh y - \cosh x \sinh y$$

$$\cosh(x - y) = \cosh x \cosh y - \sinh x \sinh y$$

$$\tanh(x + y) = \frac{\tanh x + \tanh y}{1 + \tanh x \tanh y}$$

$$\tanh(x - y) = \frac{\tanh x - \tanh y}{1 - \tanh x \tanh y}$$

Bessel Functions

Bessel functions, also called cylindrical functions, arise in many physical problems as solutions of the differential equation

$$x^2 y'' + xy' + (x^2 - n^2)y = 0$$

which is known as Bessel's equation. Certain solutions, known as *Bessel functions of the first kind of order n*, are given by

$$J_n(x) = \sum_{k=0}^{\infty} \frac{(-1)^k}{k!\Gamma(n + k + 1)} \left(\frac{x}{2}\right)^{n+2k}$$

$$J_{-n}(x) = \sum_{k=0}^{\infty} \frac{(-1)^k}{k!\Gamma(-n + k + 1)} \left(\frac{x}{2}\right)^{-n+2k}$$

see Fig. 23.47a.

In the preceding it is noteworthy that the gamma function must be defined for the negative argument $q : \Gamma(q) = \Gamma(q + 1)/q$, provided that q is not a negative integer. When q is a negative integer, $1/\Gamma(q)$

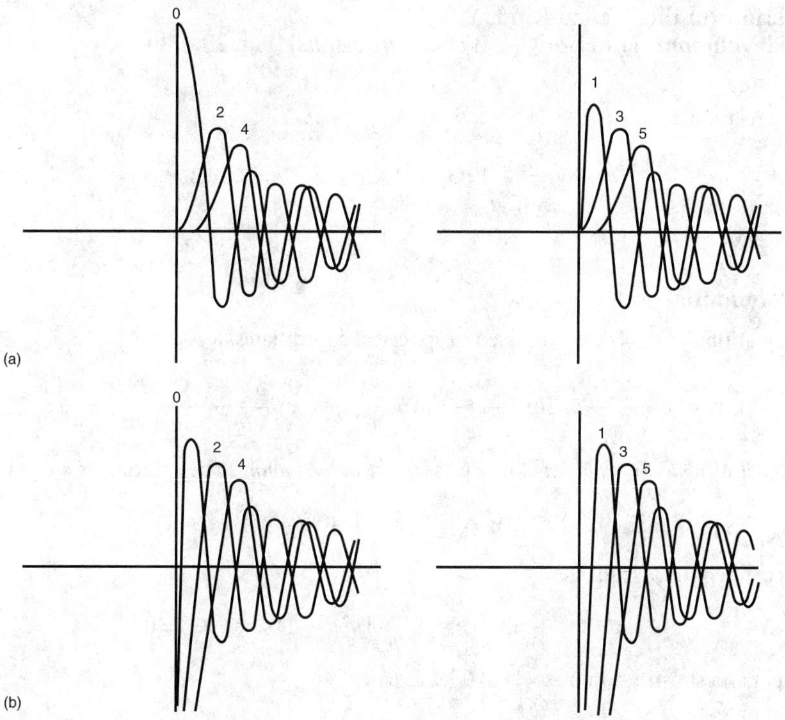

FIGURE 23.47 Bessel Functions: (a) of the first kind, (b) of the second kind.

is defined to be zero. The functions $J_{-n}(x)$ and $J_n(x)$ are solutions of Bessel's equation for all real n. It is seen, for $n = 1, 2, 3, \ldots$, that

$$J_{-n}(x) = (-1)^n J_n(x)$$

and, therefore, these are not independent; hence, a linear combination of these is not a general solution. When, however, n is not a positive integer, a negative integer, or zero, the linear combination with arbitrary constants c_1 and c_2

$$y = c_1 J_n(x) + c_2 J_{-n}(x)$$

is the general solution of the Bessel differential equation.

The zero-order function is especially important as it arises in the solution of the heat equation (for a long cylinder):

$$J_0(x) = 1 - \frac{x^2}{2^2} + \frac{x^4}{2^2 4^2} - \frac{x^6}{2^2 4^2 6^2} + \cdots$$

whereas the following relations show a connection to the trigonometric functions:

$$J_{1/2}(x) = \left[\frac{2}{\pi x} \right]^{1/2} \sin x$$

$$J_{-1/2}(x) = \left[\frac{2}{\pi x} \right]^{1/2} \cos x$$

The following recursion formula gives $J_{n+1}(x)$ for any order in terms of lower order functions:

$$\frac{2n}{x} J_n(x) = J_{n-1}(x) + J_{n+1}(x)$$

Bessel Functions of the Second Kind, $Y_n(x)$
(Also Called *Neumann Functions* or *Weber Functions*) (Fig. 23.47b)

Domain: $[x > 0]$

Recurrence relation:

$$Y_{n+1}(x) = \frac{2n}{x} Y_n(x) - Y_{n-1}(x), \quad n = 0, 1, 2 \ldots$$

Symmetry: $Y_{-n}(x) = (-1)^n Y_n(x)$

Legendre Polynomials

If Laplace's equation, $\nabla^2 V = 0$, is expressed in spherical coordinates, it is

$$r^2 \sin\theta \frac{\delta^2 V}{\delta r^2} + 2r \sin\theta \frac{\delta V}{\delta r} + \sin\theta \frac{\delta^2 V}{\delta\theta^2} + \cos\theta \frac{\delta V}{\delta\theta} + \frac{1}{\sin\theta} \frac{\delta^2 V}{\delta\phi^2} = 0$$

and any of its solutions, $V(r, \theta, \phi)$, are known as *spherical harmonics*. The solution as a product

$$V(r, \theta, \phi) = R(r)\Theta(\theta)$$

which is independent of ϕ, leads to

$$\sin^2\theta\Theta'' + \sin\theta \cos\theta\Theta' + (n(n+1)\sin^2\theta)\Theta = 0$$

Rearrangement and substitution of $x = \cos\theta$ leads to

$$(1 - x^2)\frac{d^2\Theta}{dx^2} - 2x\frac{d\Theta}{dx} + n(n+1)\Theta = 0$$

known as *Legendre's equation*. Important special cases are those in which n is zero or a positive integer, and, for such cases, Legendre's equation is satisfied by polynomials called Legendre polynomials, $P_n(x)$. A short list of Legendre polynomials, expressed in terms of x and $\cos\theta$, is given next. These are given by the following general formula:

$$P_n(x) = \sum_{j=0}^{L} \frac{(-1)^j (2n - 2j)!}{2^n j!(n - j)!(n - 2j)!} x^{n-2j}$$

where $L = n/2$ if n is even and $L = (n - 1)/2$ if n is odd

$$P_0(x) = 1$$

$$P_1(x) = x$$

$$P_2(x) = \frac{1}{2}(3x^2 - 1)$$

$$P_3(x) = \frac{1}{2}(5x^3 - 3x)$$

$$P_4(x) = \frac{1}{8}(35x^4 - 30x^2 + 3)$$

$$P_5(x) = \frac{1}{8}(63x^5 - 70x^3 + 15x)$$

$$P_0(\cos\theta) = 1$$

$$P_1(\cos\theta) = \cos\theta$$

$$P_2(\cos\theta) = \frac{1}{4}(3\cos 2\theta + 1)$$

$$P_3(\cos\theta) = \frac{1}{8}(5\cos 3\theta + 3\cos\theta)$$

$$P_4(\cos\theta) = \frac{1}{64}(35\cos 4\theta + 20\cos 2\theta + 9)$$

Additional Legendre polynomials may be determined from the *recursion formula*

$$(n+1)P_{n+1}(x) - (2n+1)xP_n(x) + nP_{n-1}(x) = 0 \qquad (n = 1, 2, \ldots)$$

or the *Rodrigues formula*

$$P_n(x) = \frac{1}{2^n n!} \frac{d^n}{dx^n}(x^2 - 1)^n$$

Laguerre Polynomials

Laguerre polynomials, denoted $L_n(x)$, are solutions of the differential equation

$$xy'' + (1 - x)y' + ny = 0$$

and are given by

$$L_n(x) = \sum_{j=0}^{n} \frac{(-1)^j}{j!} C_{(n,j)} x^j \qquad (n = 0, 1, 2, \ldots)$$

Thus

$$L_0(x) = 1$$
$$L_1(x) = 1 - x$$
$$L_2(x) = 1 - 2x + \frac{1}{2}x^2$$
$$L_3(x) = 1 - 3x + \frac{3}{2}x^2 - \frac{1}{6}x^3$$

Additional Laguerre polynomials may be obtained from the recursion formula

$$(n+1)L_{n+1}(x) - (2n+1-x)L_n(x) + nL_{n-1}(x) = 0$$

Hermite Polynomials

The Hermite polynomials, denoted $H_n(x)$, are given by

$$H_0 = 1, \quad H_n(x) = (-1)^n e^{x^2} \frac{d^n e^{-x^2}}{dx^n}, \qquad (n = 1, 2, \ldots)$$

and are solutions of the differential equation

$$y'' - 2xy' + 2ny = 0 \quad (n = 0, 1, 2, \ldots)$$

The first few Hermite polynomials are

$$H_0 = 1 \qquad\qquad H_1(x) = 2x$$
$$H_2(x) = 4x^2 - 2 \qquad\qquad H_3(x) = 8x^3 - 12x$$
$$H_4(x) = 16x^4 - 48x^2 + 12$$

Additional Hermite polynomials may be obtained from the relation

$$H_{n+1}(x) = 2xH_n(x) - H_n'(x)$$

where prime denotes differentiation with respect to x.

Orthogonality

A set of functions $\{f_n(x)\}(n = 1, 2, \ldots)$ is orthogonal in an interval (a, b) with respect to a given weight function $w(x)$ if

$$\int_a^b w(x) f_m(x) f_n(x) dx = 0 \quad \text{when } m \neq n$$

The following polynomials are orthogonal on the given interval for the given $w(x)$.

Legendre polynomials:

$$P_n(x)\, w(x) = 1$$
$$a = -1, \quad b = 1$$

Leguerre polynomials:

$$L_n(x)\, w(x) = \exp(-x)$$
$$a = 0, \quad b = \infty$$

Hermite polynomials:

$$H_n(x) w(x) = \exp(-x^2)$$
$$a = -\infty, \quad b = \infty$$

The Bessel functions of *order n*, $J_n(\lambda_1 x), J_n(\lambda_2 x), \ldots$, are orthogonal with respect to $w(x) = x$ over the interval $(0, c)$ provided that the λ_i are the positive roots of $J_n(\lambda c) = 0$

$$\int_0^c x J_n(\lambda_j x) J_n(\lambda_k x)\, dx = 0 \quad (j \neq k)$$

where n is fixed and $n \geq 0$.

Functions with $x^2/a^2 \pm y^2/b^2$

Elliptic Paraboloid (Fig. 23.48)

$$z = c(x^2/a^2 + y^2/b^2)$$
$$x^2/a^2 + y^2/b^2 - z/c = 0$$

Hyperbolic Paraboloid (Commonly Called Saddle) (Fig. 23.49):

$$z = c(x^2/a^2 - y^2/b^2)$$
$$x^2/a^2 - y^2/b^2 - z/c = 0$$

Elliptic Cylinder (Fig. 23.50):

$$1 = x^2/a^2 + y^2/b^2$$
$$x^2/a^2 + y^2/b^2 - 1 = 0$$

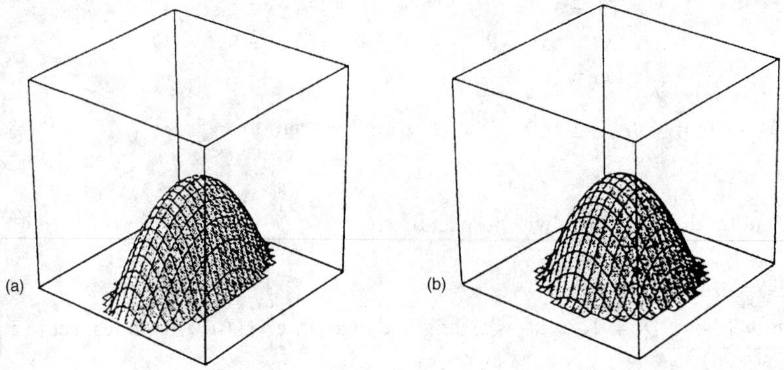

FIGURE 23.48   Elliptic paraboloid: (a) $a = 0.5, b = 1.0, c = -1.0$, viewpoint $= (5, -6, 4)$; (b) $a = 1.0, b = 1.0$, $c = -2.0$, viewpoint $= (5, -6, 4)$.

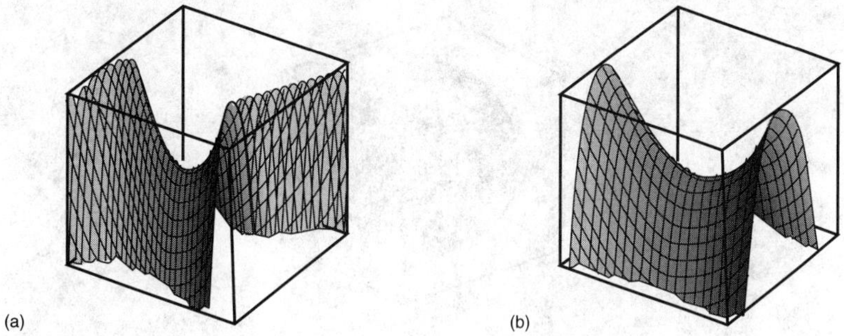

(a)   (b)

FIGURE 23.49   Hyperbolic paraboloid: (a) $a = 0.50, b = 0.5, c = 1.0$, viewpoint $= (4, -6, 4)$; (b) $a = 1.00, b = 0.5, c = 1.0$, viewpoint $= (4, -6.4)$.

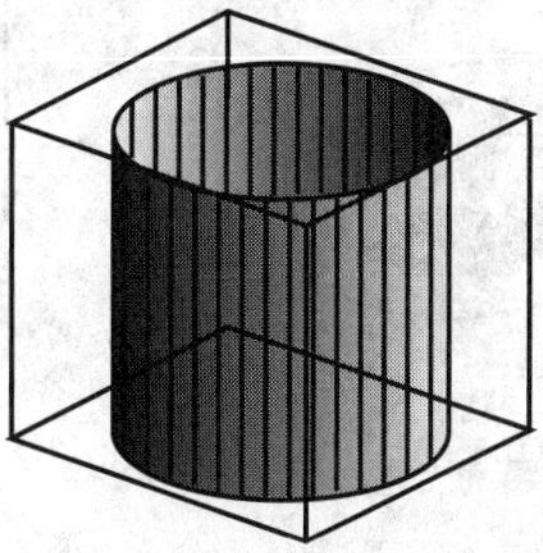

FIGURE 23.50   Elliptic cylinder. $a = 1.0, b = 1.0$, viewpoint $= (4, -5, 2)$.

Hyperbolic Cylinder (Fig. 23.51)

$$1 = x^2/a^2 - y^2/b^2$$
$$x^2/a^2 - y^2/b^2 - 1 = 0$$

Functions with $(x^2/a^2 + y^2/b^2 \pm c^2)^{1/2}$
Sphere (Fig. 23.52)

$$z = (1 - x^2 - y^2)^{1/2}$$
$$x^2 + y^2 + z^2 - 1 = 0$$

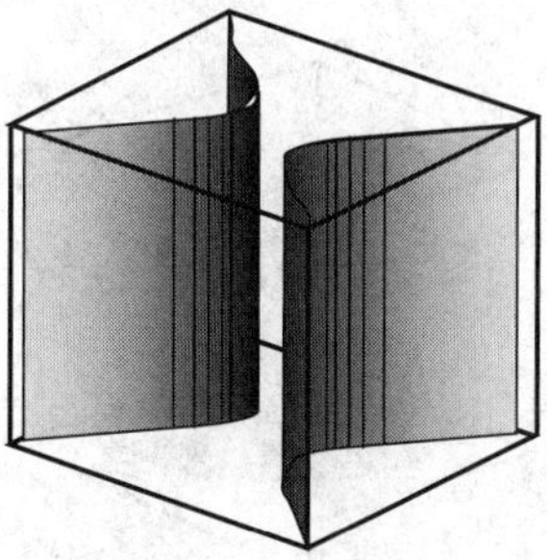

FIGURE 23.51   Hyperbolic cylinder. $a = 1.0, b = 1.0$, viewpoint $= (4, -6, 3)$.

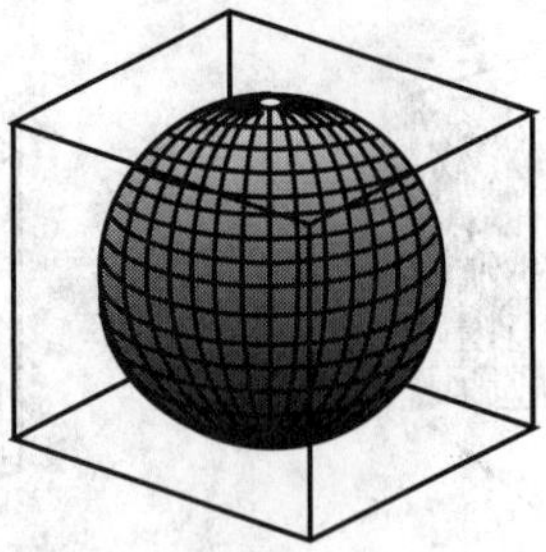

FIGURE 23.52   Sphere: viewpoint = $(4, -5, 2)$.

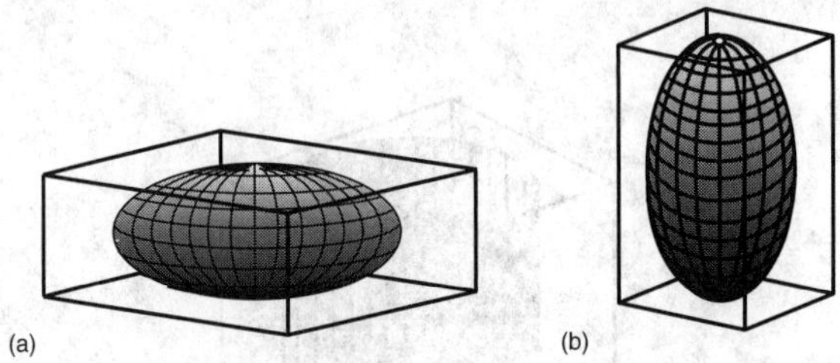

(a)   (b)

FIGURE 23.53   Ellipsoid: (a) $a = 1.00, b = 1.00, c = 0.5$, viewpoint = $(4, -5, 2)$; (b) $a = 0.50, b = 0.50, c = 1.0$, viewpoint = $(4, -5, 2)$.

Ellipsoid (Fig. 23.53)

$$z = c(1 - x^2/a^2 - y^2/b^2)^{1/2}$$
$$x^2/a^2 + y^2/b^2 + z^2/c^2 - 1 = 0$$

Special cases:

$$a = b > c \text{ gives oblate spheroid}$$
$$a = b < c \text{ gives prolate spheroid}$$

Cone (Fig. 23.54)

$$z = (x^2 + y^2)^{1/2}$$
$$x^2 + y^2 - z^2 = 0$$

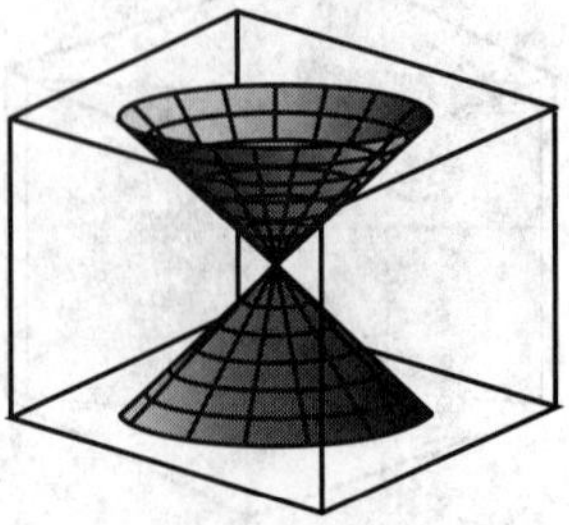

FIGURE 23.54   Cone: view-point = $(4, -5, 2)$.

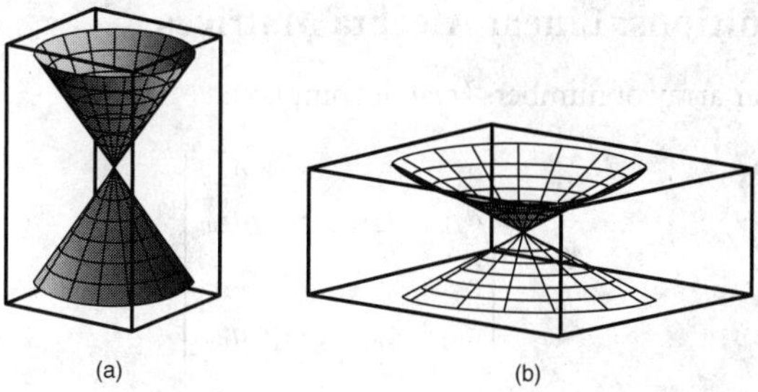

FIGURE 23.55 Elliptic cone: (a) $a = 0.5, b = 0.5, c = 1.00$, viewpoint $= (4, -5, 2)$; (b) $a = 1.0, b = 1.0, c = 0.50$, viewpoint $= (4, -5, 2)$.

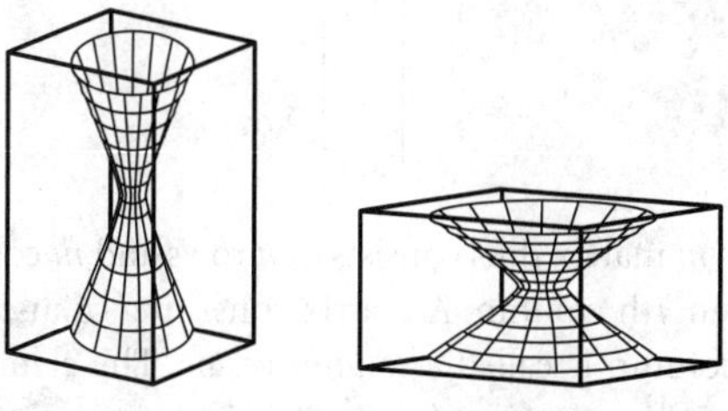

FIGURE 23.56 Hyper-boloid of one sheet: (a) $a = 0.1, b = 0.1, c = 0.2, \pm z = c\sqrt{15}$, viewpoint $= (4, -5, 2)$; (b) $a = 0.2, b = 0.2, c = 0.2, \pm z = c\sqrt{15}$, viewpoint $= (4, -5, 2)$.

Elliptic Cone (Circular Cone if $a = b$) (Fig. 23.55)

$$z = c(x^2/a^2 + y^2/b^2)^{1/2}$$
$$x^2/a^2 + y^2/b^2 - z^2/c^2 = 0$$

Hyperboloid of One Sheet (Fig. 23.56)

$$z = c(x^2/a^2 + y^2/b^2 - 1)^{1/2}$$
$$x^2/a^2 + y^2/b^2 - z^2/c^2 - 1 = 0$$

Hyperboloid of Two Sheets (Fig. 23.57)

$$z = c(x^2/a^2 + y^2/b^2 + 1)^{1/2}$$
$$x^2/a^2 + y^2/b^2 - z^2/c^2 + 1 = 0$$

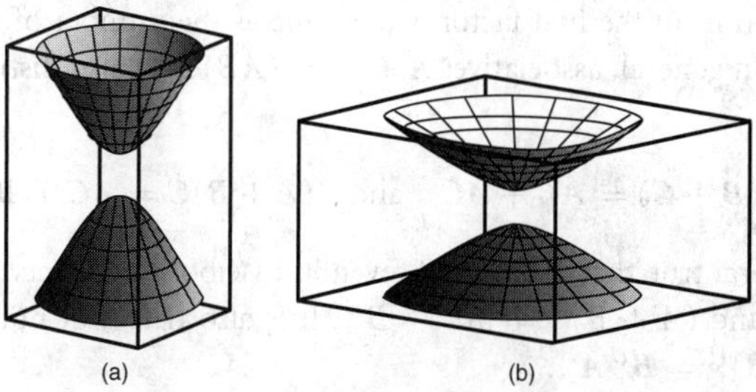

FIGURE 23.57 Hyperboloid of two sheets: (a) $a = 0.125, b = 0.125, c = 0.2, \pm z = c\sqrt{17}$, viewpoint $= (4, -5, 2)$; (b) $a = 0.25, b = 0.25, c = 0.2, \pm z = c\sqrt{17}$, viewpoint $= (4, -5, 2)$.

23.6.8 Basic Definitions: Linear Algebra Matrices

A *matrix* A is a rectangular array of numbers (real or complex):

$$A = \begin{bmatrix} a_{11} & a_{12} & \cdots & a_{1m} \\ a_{21} & a_{22} & \cdots & a_{2m} \\ \vdots & & & \\ a_{n1} & a_{n2} & \cdots & a_{nm} \end{bmatrix}$$

The *size* of the matrix is said to be $n \times m$. The $1 \times m$ matrices $[a_{i1}\, a_{i2} \cdots a_{im}]$ are called *rows of* A, and the $n \times 1$ matrices

$$\begin{bmatrix} a_{1j} \\ a_{2j} \\ \vdots \\ a_{nj} \end{bmatrix}$$

are called *columns of* A. An $n \times m$ matrix thus consists of n rows and m columns; a_{ij} denotes the *element*, or *entry*, of A in the ith row and jth column. A matrix consisting of just one row is called a *row vector*, whereas a matrix of just one column is called a *column vector*. The elements of a vector are frequently called *components* of the vector. When the size of the matrix is clear from the context, we sometimes write $A = (a_{ij})$.

A matrix with the same number of rows as columns is a *square* matrix, and the number of rows and columns is the *order* of the matrix. The diagonal of an $n \times n$ square matrix A from a_{11} to a_{nn} is called the *main*, or *principal, diagonal*. The word *diagonal* with no modifier usually means the main diagonal. The *transpose* of a matrix A is the matrix that results from interchanging the rows and columns of A. It is usually denoted by A^T. A matrix A such that $A = A^T$ is said to be *symmetric*. The *conjugate transpose* of A is the matrix that results from replacing each element of A^T by its complex conjugate, and is usually denoted by A^H. A matrix such that $A = A^H$ is said to be *Hermitian*.

A square matrix $A = (a_{ij})$ is *lower triangular* if $a_{ij} = 0$ for $j > i$ and is *upper triangular* if $a_{ij} = 0$ for $j < i$. A matrix that is both upper and lower triangular is a *diagonal* matrix. The $n \times n$ *identity matrix* is the $n \times n$ diagonal matrix in which each element of the main diagonal is 1. It is traditionally denoted I_n, or simply I when the order is clear from the context.

Algebra of Matrices

The sum and difference of two matrices A and B are defined whenever A and B have the same size. In that case $C = A \pm B$ is defined by $C = (c_{ij}) = (a_{ij} \pm b_{ij})$. The product tA of a scalar t (real or complex number) and a matrix A is defined by $tA = (ta_{ij})$. If A is an $n \times m$ matrix and B is an $m \times p$ matrix, the product $C = AB$ is defined to be the $n \times p$ matrix $C = (c_{ij})$ given by $c_{ij} = \sum_{k=1}^{m} a_{ik} b_{kj}$. Note that the product of an $n \times m$ matrix and an $m \times p$ matrix is an $n \times p$ matrix, and the product is defined only when the number of columns of the first factor is the same as the number of rows of the second factor. Matrix multiplication is, in general, associative: $A(BC) = (AB)C$. It also distributes over addition (and subtraction):

$$A(B + C) = AB + AC \quad \text{and} \quad (A + B)C = AC + BC$$

It is, however, not in general true that $AB = BA$, even in case both products are defined. It is clear that $(A + B)^T = A^T + B^T$ and $(A + B)^H = A^H + B^H$. It is also true, but not so obvious perhaps, that $(AB)^T = B^T A^T$ and $(AB)^H = B^H A^H$.

The $n \times n$ identity matrix I has the property that $IA = AI = A$ for every $n \times n$ matrix A. If A is square, and if there is a matrix B such that $AB = BA = I$, then B is called the *inverse* of A and is denoted A^{-1}. This terminology and notation are justified by the fact that a matrix can have at most one inverse.

A matrix having an inverse is said to be *invertible*, or *nonsingular*, whereas a matrix not having an inverse is said to be *noninvertible*, or *singular*. The product of two invertible matrices is invertible and, in fact, $(AB)^{-1} = B^{-1}A^{-1}$. The sum of two invertible matrices is, obviously, not necessarily invertible.

Systems of Equations

The system of n linear equations in m unknowns

$$a_{11}x_1 + a_{12}x_2 + a_{13}x_3 + \cdots + a_{1m}x_m = b_1$$

$$a_{21}x_1 + a_{22}x_2 + a_{23}x_3 + \cdots + a_{2m}x_m = b_2$$

$$\vdots$$

$$a_{n1}x_1 + a_{n2}x_2 + a_{n3}x_3 + \cdots + a_{nm}x_m = b_n$$

may be written $Ax = b$, where $A = (a_{ij})$, $x = (x_1 \quad x_2 \quad \cdots \quad x_m)^T$ and $b = (b_1 \quad b_2 \quad \cdots \quad b_n)^T$. Thus, A is an $n \times m$ matrix, and x and b are column vectors of the appropriate sizes.

The matrix A is called the *coefficient matrix* of the system. Let us first suppose the coefficient matrix is square; that is, there are an equal number of equations and unknowns. If A is upper triangular, it is quite easy to find all solutions of the system. The ith equation will contain only the unknowns $x_i, x_{i+1}, \ldots, x_n$, and one simply solves the equations in reverse order: the last equation is solved for x_n; the result is substituted into the $(n - 1)$st equation, which is then solved for x_{n-1}; these values of x_n and x_{n-1} are substituted in the $(n - 2)$th equation, which is solved for x_{n-2}, and so on. This procedure is known as *back substitution*.

The strategy for solving an arbitrary system is to find an upper-triangular system equivalent with it and solve this upper-triangular system using back substitution. First, suppose the element $a_{11} \neq 0$. We may rearrange the equations to ensure this, unless, of course the first column of A is all 0s. In this case proceed to the next step, to be described later. For each $i \geq 2$ let $m_{i1} = a_{i1}/a_{11}$. Now replace the ith equation by the result of multiplying the first equation by m_{i1} and subtracting the new equation from the ith equation. Thus

$$a_{i1}x_1 + a_{i2}x_2 + a_{i3}x_3 + \cdots + a_{im}x_m = b_i$$

is replaced by

$$0 \cdot x_1 + (a_{i2} + m_{i1}a_{12})x_2 + (a_{i3} + m_{i1}a_{13})x_3 + \cdots + (a_{im} + m_{i1}a_{1m})x_m = b_i + m_{i1}b_1$$

After this is done for all $i = 2, 3, \ldots, n$, there results the equivalent system

$$a_{11}x_1 + a_{12}x_2 + a_{13}x_3 + \cdots + a_{1n}x_n = b_1$$

$$0 \cdot x_1 + a'_{22}x_2 + a'_{23}x_3 + \cdots + a'_{2n}x_n = b'_2$$

$$0 \cdot x_1 + a'_{32}x_2 + a'_{33}x_3 + \cdots + a'_{3n}x_n = b'_3$$

$$\vdots$$

$$0 \cdot x_1 + a'_{n2}x_2 + a'_{n3}x_3 + \cdots + a'_{nn}x_n = b'_n$$

in which all entries in the first column below a_{11} are 0. (Note that if all entries in the first column were 0 to begin with, then $a_{11} = 0$ also.) This procedure is now repeated for the $(n - 1) \times (n - 1)$ system

$$a'_{22}x_2 + a'_{23}x_3 + \cdots + a'_{2n}x_n = b'_2$$

$$a'_{32}x_2 + a'_{33}x_3 + \cdots + a'_{3n}x_n = b'_3$$

$$\vdots$$

$$a'_{n2}x_2 + a'_{n3}x_3 + \cdots + a'_{nn}x_n = b'_n$$

to obtain an equivalent system in which all entries of the coefficient matrix below a'_{22} are 0. Continuing, we obtain an upper-triangular system $Ux = c$ equivalent with the original system. This procedure is known as *Gaussian elimination*. The numbers m_{ij} are known as the *multipliers*.

Essentially the same procedure may be used in case the coefficient matrix is not square. If the coefficient matrix is not square, we may make it square by appending either rows or columns of 0s as needed. Appending rows of 0s and appending 0s to make b have the appropriate size is equivalent to appending equations $0 = 0$ to the system. Clearly the new system has precisely the same solutions as the original system. Appending columns of 0s and adjusting the size of x appropriately yields a new system with additional unknowns, each appearing only with coefficient 0, thus not affecting the solutions of the original system. In either case we may assume the coefficient matrix is square, and apply the Gauss elimination procedure.

Suppose the matrix A is invertible. Then if there were no row interchanges in carrying out the above Gauss elimination procedure, we have the *LU factorization* of the matrix A:

$$A = LU$$

where U is the upper-triangular matrix produced by elimination and L is the lower-triangular matrix given by

$$L = \begin{bmatrix} 1 & 0 & \cdots & \cdots 0 \\ m_{21} & 1 & 0 & \cdots 0 \\ \vdots & & \ddots & \\ m_{n1} & m_{n2} & \cdots & 1 \end{bmatrix}$$

A *permutation* P_{ij} matrix is an $n \times n$ matrix such that $P_{ij} A$ is the matrix that results from exchanging row i and j of the matrix A. The matrix P_{ij} is the matrix that results from exchanging row i and j of the identity matrix. A product P of such matrices P_{ij} is called a *permutation* matrix. If row interchanges are required in the Gauss elimination procedure, then we have the factorization

$$PA = LU$$

where P is the permutation matrix giving the required row exchanges.

Vector Spaces

The collection of all column vectors with n real components is *Euclidean n-space*, and is denoted R^n. The collection of column vectors with n complex components is denoted C^n. We shall use *vector space* to mean either R^n or C^n. In discussing the space R^n, the word *scalar* will mean a real number, and in discussing the space C^n, it will mean a complex number. A subset S of a vector space is a *subspace* such that if u and v are vectors in S, and if c is any scalar, then $u + v$ and cu are in S. We shall sometimes use the word *space* to mean a subspace. If $B = \{v_1, v_2, \ldots, v_k\}$ is a collection of vectors in a vector space, then the set S consisting of all vectors $c_1 v_1 + c_2 v_2 + \cdots + c_m v_m$ for all scalars $c_1, c_2, \ldots, c_m$ is a subspace, called the *span* of B. A collection $\{v_1, v_2, \ldots, v_m\}$ of vectors $c_1 v_1 + c_2 v_2 + \cdots + c_m v_m\}$ is a *linear combination* of B. If S is a subspace and $B = \{v_1, v_2, \ldots, v_m\}$ is a subset of S such that S is the space of B, then B is said to *span* S.

A collection $\{v_1, v_2, \ldots, v_m\}$ of n vectors is *linearly dependent* if there exist scalars $c_1, c_2, \ldots, c_m$, not all zero, such that $c_1 v_1 + c_2 v_2 + \cdots + c_m v_m = 0$. A collection of vectors that is not linearly dependent is said to be *linearly independent*. The modifier *linearly* is frequently omitted, and we speak simply of dependent and independent collections. A linearly independent collection of vectors in a space S that spans S is a *basis* of S. Every basis of a space S contains the same number of vectors; this number is the *dimension* of S. The dimension of the space consisting of only the zero vector is 0. The collection $B = \{e_1, e_2, \ldots, e_n\}$, where $e_1 = (1, 0, 0, \ldots, 0)^T$, $e_2 = (0, 1, 0, \ldots, 0)^T$, and so forth ($e_i$ has 1 as its ith component and zero for all other components) is a basis for the spaces R^n and C^n. This is the *standard basis* for these spaces. The dimension of these spaces is thus n. In a space S of dimension n, no collection of fewer than n vectors can span S, and no collection of more than n vectors in S can be independent.

Rank and Nullity

The *column space* of an $n \times m$ matrix A is the subspace of R^n or C^n spanned by the columns of A. The *row space* is the subspace of R^m or C^m spanned by the rows of A. Note that for any vector $x = [x_1 x_2 \cdots x_m]^T$

$$Ax = x_1 \begin{bmatrix} a_{11} \\ a_{21} \\ \vdots \\ a_{n1} \end{bmatrix} + x_2 \begin{bmatrix} a_{12} \\ a_{22} \\ \vdots \\ a_{n2} \end{bmatrix} + \cdots + x_m \begin{bmatrix} a_{1m} \\ a_{2m} \\ \vdots \\ a_{nm} \end{bmatrix}$$

so that the column space is the collection of all vectors Ax, and thus the system $Ax = b$ has a solution if and only if b is a member of the column space of A.

The dimension of the column space is the *rank* of A. The row space has the same dimension as the column space. The set of all solutions of the system $Ax = 0$ is a subspace called the *null space* of A, and the dimension of this null space is the *nullity* of A. A fundamental result in matrix theory is the fact that, for an $n \times m$ matrix A

$$\text{rank } A + \text{nullity } A = m$$

The difference of any two solutions of the linear system $Ax = b$ is a member of the null space of A. Thus this system has at most one solution if and only if the nullity of A is zero. If the system is square (that is, if A is $n \times n$), then there will be a solution for every right-hand side b if and only if the collection of columns of A is linearly independent, which is the same as saying the rank of A is n. In this case the nullity must be zero. Thus, for any b, the square system $Ax = b$ has exactly one solution if and only if rank $A = n$. In other words the $n \times n$ matrix A is invertible if and only if rank $A = n$.

Orthogonality and Length

The *inner product* of two vectors x and y is the scalar $x^H y$. The *length*, or *norm*, $\|x\|$, of the vector x is given by $\|x\| = \sqrt{x^H x}$. A *unit vector* is a vector of norm 1. Two vectors x and y are *orthogonal* if $x^H y = 0$. A collection of vectors $\{v_1, v_2, \ldots, v_m\}$ in a space S is said to be an *orthonormal* collection if $v_i^H v_j = 0$ for $i \neq j$ and $v_i^H v_i = 1$. An orthonormal collection is necessarily linearly independent. If S is a subspace (or R^n or C^n) spanned by the orthonormal collection $\{v_1, v_2, \ldots, v_m\}$, then the *projection* of a vector x onto S is the vector

$$\text{proj}(x; S) = (x^H v_1)v_1 + (x^H v_2)v_2 + \cdots + (x^H v_m)v_m$$

The projection of x onto S minimizes the function $f(y) = \|x - y\|^2$ for $y \in S$. In other words the projection of x onto S is the vector in S that is closest to x.

If b is a vector and A is an $n \times m$ matrix, then a vector x minimizes $\|b - Ax\|^2$ if and only if it is a solution of $A^H Ax = A^H b$. This system of equations is called the *system of normal equations* for the least-squares problem of minimizing $\|b - Ax\|^2$.

If A is an $n \times m$ matrix and rank $A = k$, then there is an $n \times k$ matrix Q whose columns form an orthonormal basis for the column space of A and a $k \times m$ upper-triangular matrix R of rank k such that

$$A = QR$$

This is called the *QR factorization* of A. It now follows that x minimizes $\|b - Ax\|^2$ if and only if it is a solution of the upper-triangular system $Rx = Q^H b$.

If $\{w_1, w_2, \ldots, w_m\}$ is a basis for a space S, the following procedure produces an orthonormal basis $\{v_1, v_2, \ldots, v_m\}$ for S:

- Set $v_1 = w_1 / \|w_1\|$
- Let $\tilde{v}_2 = w_2 - \text{proj}(w_2; S_1)$, where S_1 is the span of $\{v_1\}$; set $v_2 = v_2 / \|\tilde{v}_2\|$
- Next, let $\tilde{v}_3 = w_3 - \text{proj}(w_3; s_2)$, where S_2 is the span of $\{v_1, v_2\}$; set $v_3 = \tilde{v}_3 / \|\tilde{v}_3\|$.

And so on: $\tilde{v}_i = w_i - \text{proj}(w_i; S_{i-1})$, where S_{i-1} is the span of $\{v_1, v_2, \ldots, v_{i-1}\}$ set $v_i = \tilde{v}_i / \|\tilde{v}_i\|$. This is the *Gram–Schmidt procedure*.

If the collection of columns of a square matrix is an orthonormal collection, the matrix is called a *unitary matrix*. In case the matrix is a real matrix, it is usually called an *orthogonal matrix*. A unitary matrix U is invertible, and $U^{-1} = U^H$. (In the real case an orthogonal matrix Q is invertible, and $Q^{-1} = Q^T$.)

Determinants

The *determinant* of a square matrix is defined inductively. First, suppose the determinant det A has been defined for all square matrices of order $< n$. Then

$$\det A = a_{11} C_{11} + a_{12} C_{12} + \cdots + a_{1n} C_{1n}$$

where the numbers C_{ij} are *cofactors* of the matrix A

$$C_{ij} = (-1)^{i+j} \det M_{ij}$$

where M_{ij} is the $(n-1) \times (n-1)$ matrix obtained by deleting the ith row and jth column of A. Now det A is defined to be the only entry of a matrix of order 1. Thus, for a matrix of order 2, we have

$$\det \begin{bmatrix} a & b \\ c & d \end{bmatrix} = ad - bc$$

There are many interesting but not obvious properties of determinants. It is true that

$$\det A = a_{i1} C_{i1} + a_{i2} C_{i2} + \cdots + a_{in} C_{in}$$

for any $1 \leq i \leq n$. It is also true that det $A = \det A^T$, so that we have

$$\det A = a_{1j} C_{1j} + a_{2j} C_{2j} + \cdots + a_{nj} C_{nj}$$

for any $1 \leq j \leq n$.

If A and B are matrices of the same order, then det $AB = (\det A)(\det B)$, and the determinant of any identity matrix is 1. Perhaps the most important property of the determinant is the fact that a matrix is invertible if and only if its determinant is not zero.

Eigenvalues and Eigenvectors

If A is a square matrix, and $Av = \lambda v$ for a scalar λ and a nonzero v, then λ is an *eigenvalue* of A and v is an *eigenvector* of A that *corresponds* to λ. Any nonzero linear combination of eigenvectors corresponding to the same eigenvalue λ is also an eigenvector corresponding to λ. The collection of all eigenvectors corresponding to a given eigenvalue λ is thus a subspace, called an *eigenspace* of A. A collection of eigenvectors corresponding to different eigenvalues is necessarily linear independent. It follows that a matrix of order n can have at most n distinct eigenvectors. In fact, the eigenvalues of A are the roots of the nth degree polynomial equation

$$\det(A - \lambda I) = 0$$

called the *characteristic equation* of A. (Eigenvalues and eigenvectors are frequently called *characteristic values* and *characteristic vectors*.)

If the nth order matrix A has an independent collection of n eigenvectors, then A is said to have a *full set* of eigenvectors. In this case there is a set of eigenvectors of A that is a basis for R^n or, in the complex case, C^n. In case there are n distinct eigenvalues of A, then, of course, A has a full set of eigenvectors. If there are fewer than n distinct eigenvalues, then A may or may not have a full set of eigenvectors. If there is a full set of eigenvectors, then

$$D = S^{-1} A S \quad \text{or} \quad A = S D S^{-1}$$

where D is a diagonal matrix the eigenvalues of A on the diagonal, and S is a matrix whose columns are the full set of eigenvectors. If A is symmetric, there are n real distinct eigenvalues of A and the corresponding eigenvectors are orthogonal. There is thus an orthonormal collection of eigenvectors that span R^n, and we have

$$A = Q D Q^T \quad \text{and} \quad D = Q^T A Q$$

where Q is a real orthogonal matrix and D is diagonal. For the complex case, if A is Hermitian, we have

$$A = U D U^H \quad \text{and} \quad D = U^H A U$$

where U is a unitary matrix and D is a *real* diagonal matrix. (A Hermitian matrix also has n distinct real eigenvalues.)

23.6.9 Basic Definitions: Vector Algebra and Calculus

A vector is a directed line segment, with two vectors being equal if they have the same length and the same direction. More precisely, a *vector* is an equivalence class of directed line segments, where two directed segments are equivalent if they have the same length and the same direction. The *length* of a vector is the common length of its directed segments, and the *angle between* vectors is the angle between any of their segments. The length of a vector u is denoted $|u|$. There is defined a distinguished vector having zero length, which is usually denoted 0. It is frequently useful to visualize a directed segment as an arrow; we then speak of the nose and the tail of the segment. The *sum* $u + v$ of two vectors u and v is defined by taking directed segments from u and v and placing the tail of the segment representing v at the nose of the segment representing u and defining $u + v$ to be the vector determined by the segment from the tail of the u representative to the nose of the v representative. It is easy to see that $u + v$ is well defined and that $u + v = v + u$. Subtraction is the inverse operation of addition. Thus the *difference* $u - v$ of two vectors is defined to be the vector that when added to v gives u. In other words, if we take a segment from u and a segment from v and place their tails together, the difference is the segment from the nose of v to the nose of u. The zero vector behaves as one might expect: $u + 0 = u$, and $u - u = 0$. Addition is associative: $u + (v + w) = (u + v) + w$.

To distinguish them from vectors, the real numbers are called *scalars*. The product tu of a scalar t and a vector u is defined to be the vector having length $|t| \, |u|$ and direction the same as u if $t > 0$, the opposite direction if $t < 0$. If $t = 0$, then tu is defined to be the zero vector. Note that $t(u + v) = tu + tv$, and $(t + s)u = tu + su$. From this it follows that $u - v = u + (-1)v$.

The *scalar product* $u \cdot v$ of two vectors is $|u| \, |v| \cos \theta$, where θ is the angle between u and v. The scalar product is frequently called the *dot product*. The scalar product distributes over addition

$$u \cdot (v + w) = u \cdot v + u \cdot w$$

and it is clear that $(tu) \cdot v = t(u \cdot v)$. The *vector product* $u \times v$ of two vectors is defined to be the vector perpendicular to both u and v and having length $|u| \, |v| \sin \theta$, where θ is the angle between u and v. The direction of $u \times v$ is the direction a right-hand threaded bolt advances if the vector u is rotated to v. The vector product is frequently called the *cross product*. The vector product is both associative and distributive, but not commutative: $u \times v = -v \times u$.

Coordinate Systems

Suppose we have a right-handed Cartesian coordinate system in space. For each vector u, we associate a point in space by placing the tail of a representative of u at the origin and associating with u the point at the nose of the segment. Conversely, associated with each point in space is the vector determined by the directed segment from the origin to that point. There is thus a one-to-one correspondence between the points in space and all vectors. The origin corresponds to the zero vector. The coordinates of the point associated with a vector u are called *coordinates* of u. One frequently refers to the vector u and writes

$u = (x, y, z)$, which is, strictly speaking, incorrect, because the left side of this equation is a vector and the right side gives the coordinates of a point in space. What is meant is that (x, y, z) are the coordinates of the point associated with u under the correspondence described. In terms of coordinates, for $u = (u_1, u_2, u_3)$ and $v = (v_1, v_2, v_3)$, we have

$$u + v = (u_1 + v_1, u_2 + v_2, u_3 + v_3)$$

$$tu = (tu_1, tu_2, tu_3)$$

$$u \cdot v = u_1 v_1 + u_2 v_2 + u_3 v_3$$

$$u \times v = (u_2 v_3 - v_2 u_3, u_3 v_1 - v_3 u_1, u_1 v_2 - v_1 u_2)$$

The *coordinate vectors* i, j, and k are the unit vectors $i = (1, 0, 0)$, $j = (0, 1, 0)$, and $k = (0, 0, 1)$. Any vector $u = (u_1, u_2, u_3)$ is thus a linear combination of these coordinate vectors: $u = u_1 i + u_2 j + u_3 k$. A convenient form for the vector product is the formal determinant

$$u \times v = \det \begin{bmatrix} i & j & k \\ u_1 & u_2 & u_3 \\ v_1 & v_2 & v_2 \end{bmatrix}$$

Vector Functions

A *vector function F of one variable* is a rule that associates a vector $F(t)$ with each real number t is some set, called the *domain* of F. The expression $\lim_{t \to t_0} F(t) = a$ means that for any $\varepsilon > 0$, there is a $\delta > 0$ such that $|F(t) - a| < \varepsilon$ whenever $0 < |t - t_0| < \delta$. If $F(t) = (x(t), y(t), z(t))$ and $a = (a_1, a_2, a_3)$, then $\lim_{t \to t_0} F(t) = a$ if and only if

$$\lim_{t \to t_0} x(t) = a_1$$

$$\lim_{t \to t_0} y(t) = a_2$$

$$\lim_{t \to t_0} z(t) = a_3$$

A vector function F is *continuous* at t_0 if $\lim_{t \to t_0} F(t) = F(t_0)$. The vector function F is continuous at t_0 if and only if each of the coordinates $x(t)$, $y(t)$, and $z(t)$ is continuous at t_0.

The function F is *differentiable* at t_0 if the limit

$$\lim_{h \to 0} \frac{1}{h} (F(t + h) - F(t))$$

exists. This limit is called the *derivative* of F at t_0 and is usually written $F'(t_0)$, or $(dF/dt)(t_0)$. The vector function F is differentiable at t_0 if and only if each of its coordinate functions is differentiable at t_0. Moreover, $(dF/dt)(t_0) = ((dx/dt)(t_0), (dy/dt)(t_0), (dz/dt)(t_0))$. The usual rules for derivatives of real valued functions all hold for vector functions. Thus, if F and G are vector functions and s is a scalar function, then

$$\frac{d}{dt}(F + G) = \frac{dF}{dt} + \frac{dG}{dt}$$

$$\frac{d}{dt}(sF) = s\frac{dF}{dt} + \frac{ds}{dt}F$$

$$\frac{d}{dt}(F \cdot G) = F \cdot \frac{dG}{dt} + \frac{dF}{dt} \cdot G$$

$$\frac{d}{dt}(F \times G) = F \times \frac{dG}{dt} + \frac{dF}{dt} \times G$$

If R is a vector function defined for t is some interval, then, as t varies, with the tail of R at the origin, the nose traces out some object C in space. For nice functions R, the object C is a *curve*. If $R(t) = [x(t), y(t), z(t)]$, then the equations

$$x = x(t)$$
$$y = y(t)$$
$$z = z(t)$$

are called *parametric equations* of C. At points where R is differentiable, the derivative dR/dt is a vector *tangent* to the curve. The unit vector $T = (dR/dt)/|dR/dt|$ is called the *unit tangent vector*. If R is differentiable and if the length of the arc of curve described by R between $R(a)$ and $R(t)$ is given by $s(t)$, then

$$\frac{ds}{dt} = \left| \frac{dR}{dt} \right|$$

Thus the length L of the arc from $R(t_0)$ to $R(t_1)$ is

$$L = \int_{t_0}^{t_1} \frac{ds}{dt} dt = \int_{t_0}^{t_1} \left| \frac{dR}{dt} \right| dt$$

The vector $dT/ds = (dT/dt)/(ds/dt)$ is perpendicular to the unit tangent T, and the number $\kappa = |dT/ds|$ is the *curvature* of C. The unit vector $N = (1/\kappa)(dT/ds)$ is the *principal normal*. The vector $B = T \times N$ is the *binormal*, and $dB/ds = -\tau N$. The number τ is the *torsion*. Note that C is a plane curve if and only if τ is zero for all t.

A *vector function F of two variables* is a rule that assigns a vector $F(s,t)$ to each point (s,t) which is some subset of the plane, called the *domain* of F. If $R(s,t)$ is defined for all (s,t) in some region D of the plane, then as the point (s,t) varies over D, with its tail at the origin, the nose of $R(s,t)$ traces out an object in space. For a nice function R, this object is a *surface, S*. The partial derivatives $(\partial R/\partial s)(s,t)$ and $(\partial R/\partial t)(s,t)$ are tangent to the surface at $R(s,t)$ and the vector $(\partial R/\partial s) \times (\partial R/\partial t)$ is thus *normal* to the surface. Of course, $(\partial R/\partial t) \times (\partial R/\partial s) = -(\partial R/\partial s) \times (\partial R/\partial t)$ is also normal to the surface and points in the direction opposite that of $(\partial R/\partial s) \times (\partial R/\partial t)$. By electing one of these normals, we are choosing an *orientation* of the surface. A surface can be oriented only if it has two sides, and the process of orientation consists of choosing which side is positive and which is negative.

Gradient, Curl, and Divergence

If $f(x, y, z)$ is a scalar field defined in some region D, the *gradient* of f is the vector function

$$\text{grad } f = \frac{\partial f}{\partial x} i + \frac{\partial f}{\partial y} j + \frac{\partial f}{\partial z} k$$

If $F(x, y, z) = F_1(x, y, z)i + F_2(x, y, z)j + F_3(x, y, z)k$ is a vector field defined in some region D, then the *divergence* of F is the scalar function

$$\text{div } F = \frac{\partial F_1}{\partial x} + \frac{\partial F_2}{\partial y} + \frac{\partial F_3}{\partial z}$$

The curl is the vector function

$$\text{curl } F = \left(\frac{\partial F_3}{\partial y} - \frac{\partial F_2}{\partial z} \right) i + \left(\frac{\partial F_1}{\partial z} - \frac{\partial F_3}{\partial x} \right) j + \left(\frac{\partial F_2}{\partial x} - \frac{\partial F_1}{\partial y} \right) k$$

In terms of the vector operator *del*, $\nabla = \boldsymbol{i}(\partial/\partial x) + \boldsymbol{j}(\partial/\partial y) + \boldsymbol{k}(\partial/\partial z)$, we can write

$$\text{grad } f = \nabla f$$

$$\text{div } \boldsymbol{F} = \nabla \cdot \boldsymbol{F}$$

$$\text{curl } \boldsymbol{F} = \nabla \times \boldsymbol{F}$$

The *Laplacian operator* is div (grad) $= \nabla \cdot \nabla = \nabla^2 = (\partial^2/\partial x^2) + (\partial^2/\partial y^2) + (\partial^2/\partial z^2)$.

Integration

Suppose C is a curve from the point (x_0, y_0, z_0) to the point (x_1, y_1, z_1) and is described by the vector function $\boldsymbol{R}(t)$ for $t_0 \leq t \leq t_1$. If f is a scalar function (sometimes called a *scalar field*) defined on C, then the integral of f over C is

$$\int_C f(x, y, z) \, ds = \int_{t_0}^{t_1} f[\boldsymbol{R}(t)] \left| \frac{d\boldsymbol{R}}{dt} \right| dt$$

If $\boldsymbol{F}$ is a vector function (sometimes called a *vector field*) defined on C, then the integral of $\boldsymbol{F}$ over C is

$$\int_C \boldsymbol{F}(x, y, z) \cdot d\boldsymbol{R} = \int_{t_0}^{t_1} \boldsymbol{F}[\boldsymbol{R}(t)] \cdot \frac{d\boldsymbol{R}}{dt} \, dt$$

These integrals are called *line integrals*.

In case there is a scalar function f such that $\boldsymbol{F} = \text{grad } f$, then the line integral

$$\int_C \boldsymbol{F}(x, y, z) \cdot d\boldsymbol{R} = f(\boldsymbol{R}(t_1)) - f(\boldsymbol{R}(t_0))$$

The value of the integral thus depends only on the end points of the curve C and not on the curve C itself. The integral is said to be *path independent*. The function f is called a *potential function* for the vector field $\boldsymbol{F}$, and $\boldsymbol{F}$ is said to be a *conservative field*. A vector field $\boldsymbol{F}$ with domain D is conservative if and only if the integral of $\boldsymbol{F}$ around every closed curve in D is zero. If the domain D is simply connected (that is, every closed curve in D can be continuously deformed in D to a point), then $\boldsymbol{F}$ is conservative if and only if curl $\boldsymbol{F} = 0$ in D.

Suppose S is a surface described by $\boldsymbol{R}(s, t)$ for (s, t) in a region D of the plane. If f is a scalar function defined on D, then the integral of f over S is given by

$$\iint_S f(x, y, z) \, dS = \iint_D f(\boldsymbol{R}(s, t)) \left| \frac{\partial \boldsymbol{R}}{\partial s} \times \frac{\partial \boldsymbol{R}}{\partial t} \right| ds \, dt$$

If $\boldsymbol{F}$ is a vector function defined on S, and if an orientation for S is chosen, then the integral of $\boldsymbol{F}$ over S, sometimes called the *flux* of $\boldsymbol{F}$ through S, is

$$\iint_S \boldsymbol{F}(x, y, z) \cdot d\boldsymbol{S} = \iint_D \boldsymbol{F}[\boldsymbol{R}(s, t)] \cdot \left(\frac{\partial \boldsymbol{R}}{\partial s} \times \frac{\partial \boldsymbol{R}}{\partial t} \right) ds \, dt$$

Integral Theorems

Suppose $\boldsymbol{F}$ is a vector field with a closed domain D bounded by the surface S oriented so that the normal points out from D. Then the *divergence theorem* states that

$$\iiint_D \text{div } \boldsymbol{F} dV = \iint_S \boldsymbol{F} \cdot d\boldsymbol{S}$$

If S is an orientable surface bounded by a closed curve C, the orientation of the closed curve C is chosen to be consistent with the orientation of the surface S. Then we have *Stokes's theorem*:

$$\int\int_S (\text{curl } \boldsymbol{F}) \cdot d\boldsymbol{S} = \oint_C \boldsymbol{F} \cdot d\boldsymbol{s}$$

23.6.10 The Fourier Transforms: Overview

For a piecewise continuous function $F(x)$ over a finite interval $0 \le x \le \pi$, the *finite Fourier cosine transform* of $F(x)$ is

$$f_c(n) = \int_0^\pi f(x) \cos nx \, dx \quad (n = 0, 1, 2, \ldots) \tag{23.1}$$

If x ranges over the interval $0 \le x \le L$, the substitution $x' = \pi x / L$ allows the use of this definition also. The inverse transform is written

$$\bar{F}(x) = \frac{1}{\pi} f_c(0) + \frac{2}{\pi} \sum_{n=1}^\infty f_c(n) \cos nx \quad (0 < x < \pi) \tag{23.2}$$

where $\bar{F}(x) = (F(x+0)+F(x-0))/2$. We observe that $\bar{F}(x) = F(x)$ at points of continuity. The formula

$$f_c^{(2)}(n) = \int_0^\pi F''(x) \cos nx \, dx$$
$$= -n^2 f_c(n) - F'(0) + (-1)^n F'(\pi) \tag{23.3}$$

makes the finite Fourier cosine transform useful in certain boundary-value problems.

Analogously, the *finite Fourier sine transform* of $F(x)$ is

$$f_s(n) = \int_0^\pi F(x) \sin nx \, dx \quad (n = 1, 2, 3, \ldots) \tag{23.4}$$

and

$$\bar{F}(x) = \frac{2}{\pi} \sum_{n=1}^\infty f_s(n) \sin nx \quad (0 < x < \pi) \tag{23.5}$$

Corresponding to Eq. (23.3), we have

$$f_s^{(2)}(n) = \int_0^\pi F''(x) \sin nx \, dx \tag{23.6}$$
$$= -n^2 f_s(n) - nF(0) - n(-1)^n F(\pi)$$

Fourier Transforms

If $F(x)$ is defined for $x \ge 0$ and is piecewise continuous over any finite interval, and if

$$\int_0^\infty F(x) \, dx$$

is absolutely convergent, then

$$f_c(\alpha) = \sqrt{\frac{2}{\pi}} \int_0^\infty f(x) \cos(\alpha x) \, dx \tag{23.7}$$

is the *Fourier cosine transform of F(x)*. Furthermore

$$\bar{F}(x) = \sqrt{\frac{2}{\pi}} \int_0^\infty f_c(\alpha) \cos(\alpha x)\, d\alpha \tag{23.8}$$

If $\lim_{x \to \infty} d^n F/dx^n = 0$, an important property of the Fourier cosine transform

$$f_c^{(2r)}(\alpha) = \sqrt{\frac{2}{\pi}} \int_0^\infty \left(\frac{d^{2r}F}{dx^{2r}} \right) \cos(\alpha x)\, dx$$

$$\tag{23.9}$$

$$= -\sqrt{\frac{2}{\pi}} \sum_{n=0}^{r-1} (-1)^n a_{2r-2n-1} \alpha^{2n} + (-1)^r \alpha^{2r} f_c(\alpha)$$

where $\lim_{x \to 0} d^r F/dx^r = a_r$, makes it useful in the solution of many problems.
   Under the same conditions

$$f_s(\alpha) = \sqrt{\frac{2}{\pi}} \int_0^\infty F(x) \sin(\alpha x)\, dx \tag{23.10}$$

defines the *Fourier sine transform of F(x)*, and

$$\bar{F}(x) = \sqrt{\frac{2}{\pi}} \int_0^\infty f_s(\alpha) \sin(\alpha x)\, d\alpha \tag{23.11}$$

Corresponding to Eq. (23.9) we have

$$f_s^{(2r)}(\alpha) = \sqrt{\frac{2}{\pi}} \int_0^\infty \frac{d^{2r}F}{dx^{2r}} \sin(\alpha x)\, dx$$

$$\tag{23.12}$$

$$= -\sqrt{\frac{2}{\pi}} \sum_{n=1}^{r} (-1)^n \alpha^{2n-1} a_{2r-2n} + (-1)^{r-1} \alpha^{2r} f_s(\alpha)$$

Similarly, if $F(x)$ is defined for $-\infty < x < \infty$, and if $\int_{-\infty}^{\infty} F(x)\,dx$ is absolutely convergent, then

$$f(\alpha) = \frac{1}{\sqrt{2\pi}} \int_{-\infty}^{\infty} F(x) e^{i\alpha x}\, dx \tag{23.13}$$

is the *Fourier transform of F(x)*, and

$$\bar{F}(x) = \frac{1}{\sqrt{2\pi}} \int_{-\infty}^{\infty} f(\alpha) e^{-i\alpha x}\, d\alpha \tag{23.14}$$

Also, if

$$\lim_{|x| \to \infty} \left| \frac{d^n F}{dx^n} \right| = 0 \quad (n = 1, 2, \ldots, r-1)$$

then

$$f^{(r)}(\alpha) = \frac{1}{\sqrt{2\pi}} \int_{-\infty}^{\infty} F^{(r)}(x) e^{i\alpha x}\, dx = (-i\alpha)^r f(\alpha) \tag{23.15}$$

The finite and fourier transforms for sine and cosine are given in Table 23.3 to Table 23.6. The fourier transforms for other function are given in Table 23.7.

TABLE 23.3 Finite Sine Transforms

$f_s(n)$	$F(x)$
1. $f_s(n) = \int_0^\pi F(x) \sin nx \, dx \quad (n = 1, 2, 3, \ldots)$	$F(x)$
2. $(-1)^{n+1} f_s(n)$	$F(\pi - x)$
3. $\dfrac{1}{n}$	$\dfrac{\pi - x}{\pi}$
4. $\dfrac{(-1)^{n+1}}{n}$	$\dfrac{x}{\pi}$
5. $\dfrac{1 - (-1)^n}{n}$	1
6. $\dfrac{2}{n^2} \sin \dfrac{n\pi}{2}$	$\begin{cases} x & \text{when} \quad 0 < x < \pi/n \\ \pi - x & \text{when} \quad \pi/2 < x < \pi \end{cases}$
7. $\dfrac{(-1)^{n+1}}{n^3}$	$\dfrac{x(\pi^2 - x^2)}{6\pi}$
8. $\dfrac{1 - (-1)^n}{n^3}$	$\dfrac{x(\pi - x)}{2}$
9. $\dfrac{\pi^2(-1)^{n-1}}{n} - \dfrac{2[1 - (-1)^n]}{n^3}$	x^2
10. $\pi(-1)^n \left(\dfrac{6}{n^3} - \dfrac{\pi^2}{n} \right)$	x^3
11. $\dfrac{n}{n^2 + c^2} [1 - (-1)^n e^{c\pi}]$	e^{cx}
12. $\dfrac{n}{n^2 + c^2}$	$\dfrac{\sin h\, c(\pi - x)}{\sin h\, c\pi}$
13. $\dfrac{n}{n^2 - k^2} \quad (k \neq 0, 1, 2, \ldots)$	$\dfrac{\sin hk(\pi - x)}{\sin k\pi}$
14. $\begin{cases} \dfrac{\pi}{2} & \text{when} \quad n = m \\ 0 & \text{when} \quad n \neq m \end{cases} \quad (m = 1, 2, \ldots)$	$\sin mx$
15. $\dfrac{n}{n^2 - k^2} [1 - (-1)^n \cos k\pi] \quad (k \neq 1, 2, \ldots)$	$\cos kx$
16. $\begin{cases} \dfrac{n}{n^2 - m^2} [1 - (-1)^{n+m}] & \text{when } n \neq m = 1, 2, \ldots \\ 0 & \text{when } n = m \end{cases}$	$\cos mx$
17. $\dfrac{n}{(n^2 - k^2)^2} \quad (k \neq 0, 1, 2, \ldots)$	$\dfrac{\pi \sin kx}{2k \sin^2 k\pi} - \dfrac{x \cos k(\pi - x)}{2k \sin k\pi}$
18. $\dfrac{b^n}{n} \quad (|b| \leq 1)$	$\dfrac{2}{\pi} \arctan \dfrac{b \sin x}{1 - b \cos x}$
19. $\dfrac{1 - (-1)^n}{n} b^n \quad (|b| \leq 1)$	$\dfrac{2}{\pi} \arctan \dfrac{2b \sin x}{1 - b^2}$

TABLE 23.4   Finite Cosine Transforms

$f_c(n)$	$F(x)$
1. $f_c(n) = \int_0^\pi F(x)\cos nx\,dx \quad (n = 0, 1, 2, \ldots)$	$F(x)$
2. $(-1)^n f_c(n)$	$F(\pi - x)$
3. 0 when $n = 1, 2, \ldots;\quad f_c(0) = \pi$	1
4. $\dfrac{2}{\pi}\sin\dfrac{n\pi}{2};\quad f_c(0) = 0$	$\begin{cases} 1 & \text{when } 0 < x < \pi/2 \\ -1 & \text{when } \pi/2 < x < \pi \end{cases}$
5. $-\dfrac{1 - (-1)^n}{n^2};\quad f_c(0) = \dfrac{\pi^2}{2}$	x
6. $\dfrac{(-1)^n}{n^2};\quad f_c(0) = \dfrac{\pi^2}{6}$	$\dfrac{x^2}{2\pi}$
7. $\dfrac{1}{n^2};\quad f_c(0) = 0$	$\dfrac{(\pi - x)^2}{2\pi} - \dfrac{\pi}{6}$
8. $3\pi^2\dfrac{(-1)^n}{n^2} - 6\dfrac{1 - (-1)^n}{n^4};\quad f_c(0) = \dfrac{\pi^4}{4}$	x^3
9. $\dfrac{(-1)^n e^c \pi - 1}{n^2 + c^2}$	$\dfrac{1}{c}e^{cx}$
10. $\dfrac{1}{n^2 + c^2}$	$\dfrac{\cos h\, c(\pi - x)}{c\sin h\, c\pi}$
11. $\dfrac{k}{n^2 - k^2}[(-1)^n \cos \pi k - 1] \quad (k \neq 0, 1, 2, \ldots)$	$\sin kx$
12. $\dfrac{(-1)^{n+m} - 1}{n^2 - m^2}$	$\dfrac{1}{m}\sin mx$
13. $\dfrac{1}{n^2 - k^2} \quad (k \neq 0, 1, 2, \ldots)$	$-\dfrac{\cos k(\pi - x)}{k\sin k\pi}$
14. 0 when $n = 1, 2, \ldots;$	$\cos mx$

TABLE 23.5   Fourier Sine Transforms

$F(x)$	$f_s(\alpha)$
1. $\begin{cases} 1 & (0 < x < a) \\ 0 & (x > a) \end{cases}$	$\sqrt{\dfrac{2}{\pi}}\left[\dfrac{1 - \cos\alpha}{\alpha}\right]$
2. $x^{p-1} \quad (0 < p < 1)$	$\sqrt{\dfrac{2}{\pi}}\dfrac{\Gamma(p)}{\alpha^p}\sin\dfrac{p\pi}{2}$
3. $\begin{cases} \sin x & (0 < x < a) \\ 0 & (x > a) \end{cases}$	$\dfrac{1}{\sqrt{2\pi}}\left[\dfrac{\sin[a(1 - \alpha)]}{1 - \alpha} - \dfrac{\sin[a(1 + \alpha)]}{1 + \alpha}\right]$
4. e^{-x}	$\sqrt{\dfrac{2}{\pi}}\left[\dfrac{\alpha}{1 + \alpha^2}\right]$
5. $xe^{-x^2/2}$	$\alpha e^{-\alpha^2/2}$
6. $\cos\dfrac{x^2}{2}$	$\sqrt{2}\left[\sin\dfrac{\alpha^2}{2}C\left(\dfrac{\alpha^2}{2}\right) - \cos\dfrac{\alpha^2}{2}S\left(\dfrac{\alpha^2}{2}\right)\right]^{a}$
7. $\sin\dfrac{x^2}{2}$	$\sqrt{2}\left[\cos\dfrac{\alpha^2}{2}C\left(\dfrac{\alpha^2}{2}\right) + \sin\dfrac{\alpha^2}{2}S\left(\dfrac{\alpha^2}{2}\right)\right]^{a}$

[a] $C(y)$ and $S(y)$ are the Fresnel integrals

$$C(y) = \frac{1}{\sqrt{2\pi}}\int_0^y \frac{1}{\sqrt{t}}\cos t\,dt$$

$$S(y) = \frac{1}{\sqrt{2\pi}}\int_0^y \frac{1}{\sqrt{t}}\sin t\,dt$$

TABLE 23.6 Fourier Cosine Transforms

$F(x)$	$f_c(\alpha)$
1. $\begin{cases} 1 & (0 < x < a) \\ 0 & (x < a) \end{cases}$	$\sqrt{\dfrac{2}{\pi}}\dfrac{\sin a\alpha}{\alpha}$
2. $x^{p-1} \quad (0 < p < 1)$	$\sqrt{\dfrac{2}{\pi}}\dfrac{\Gamma(p)}{\alpha^p}\cos\dfrac{p\pi}{2}$
3. $\begin{cases} \cos x & (0 < x < a) \\ 0 & (x > a) \end{cases}$	$\dfrac{1}{\sqrt{2\pi}}\left[\dfrac{\sin[a(1-\alpha)]}{1-\alpha}+\dfrac{\sin[a(1+\alpha)]}{1+\alpha}\right]$
4. e^{-x}	$\sqrt{\dfrac{2}{\pi}}\left(\dfrac{1}{1+\alpha^2}\right)$
5. $e^{-x^2/2}$	$e^{-\alpha^2/2}$
6. $\cos\dfrac{x^2}{2}$	$\cos\left(\dfrac{\alpha^2}{2}-\dfrac{\pi}{4}\right)$
7. $\sin\dfrac{x^2}{2}$	$\cos\left(\dfrac{\alpha^2}{2}-\dfrac{\pi}{4}\right)$

TABLE 23.7 Fourier Transforms

$F(x)$	$f(\alpha)$
1. $\dfrac{\sin ax}{x}$	$\begin{cases} \sqrt{\dfrac{\pi}{2}} & \lvert\alpha\rvert < a \\ 0 & \lvert\alpha\rvert > a \end{cases}$
2. $\begin{cases} e^{iwx} & (p < x < q) \\ 0 & (x < p, x > q) \end{cases}$	$\dfrac{i}{\sqrt{2\pi}}\dfrac{e^{ip(w+\alpha)}-e^{iq(w+\alpha)}}{(w+\alpha)}$
3. $\begin{cases} e^{-cx+iwx} & (x > 0) \\ & (c > 0) \\ 0 & (x < 0) \end{cases}$	$\dfrac{i}{\sqrt{2\pi}(w+\alpha+ic)}$
4. $e^{-px^2} \quad R(p) > 0$	$\dfrac{1}{\sqrt{2p}}e^{-\alpha^2/4p}$
5. $\cos px^2$	$\dfrac{1}{\sqrt{2p}}\cos\left[\dfrac{\alpha^2}{4p}-\dfrac{\pi}{4}\right]$
6. $\sin px^2$	$\dfrac{1}{\sqrt{2p}}\cos\left[\dfrac{\alpha^2}{4p}+\dfrac{\pi}{4}\right]$
7. $\lvert x\rvert^{-p} \quad (0 < p < 1)$	$\sqrt{\dfrac{2}{\pi}}\dfrac{\Gamma(1-p)\sin\dfrac{p\pi}{2}}{\lvert\alpha\rvert^{(1-p)}}$
8. $\dfrac{e^{-a\lvert x\rvert}}{\sqrt{\lvert x\rvert}}$	$\dfrac{\sqrt{\sqrt{(a^2+\alpha^2)}+a}}{\sqrt{a^2+\alpha^2}}$
9. $\dfrac{\cosh ax}{\cosh \pi x} \quad (-\pi < a < \pi)$	$\sqrt{\dfrac{2}{\pi}}\dfrac{\cos\dfrac{a}{2}\cosh\dfrac{\alpha}{2}}{\cosh a + \cos a}$
10. $\dfrac{\sinh ax}{\sinh \pi x} \quad (-\pi < a < \pi)$	$\dfrac{1}{\sqrt{2\pi}}\dfrac{\sin a}{\cosh a + \cos a}$
11. $\begin{cases} \dfrac{1}{\sqrt{a^2-x^2}} & (\lvert x\rvert < a) \\ 0 & (\lvert x\rvert > a) \end{cases}$	$\sqrt{\dfrac{\pi}{2}}J_0(a\alpha)$

TABLE 23.7 Fourier Transforms (*Continued*)

	$F(x)$	$f(\alpha)$
12.	$\dfrac{\sin[b\sqrt{a^2+x^2}]}{\sqrt{a^2+x^2}}$	$\begin{cases} 0 & (\|\alpha\| > b) \\ \sqrt{\dfrac{\pi}{2}}J_0(a\sqrt{b^2-\alpha^2}) & (\|\alpha\| < b) \end{cases}$
13.	$\begin{cases} P_n(x) & (\|x\| < 1) \\ 0 & (\|x\| > 1) \end{cases}$	$\dfrac{i^n}{\sqrt{\alpha}}J_{n+1/2}(\alpha)$
14.	$\begin{cases} \dfrac{\cos[b\sqrt{a^2-x^2}]}{\sqrt{a^2-x^2}} & (\|x\| < a) \\ 0 & (\|x\| > a) \end{cases}$	$\sqrt{\dfrac{\pi}{2}}J_0(a\sqrt{a^2+b^2})$
15.	$\begin{cases} \dfrac{\cosh[b\sqrt{a^2-x^2}]}{\sqrt{a^2-x^2}} & (\|x\| < a) \\ 0 & (\|x\| > a) \end{cases}$	$\sqrt{\dfrac{\pi}{2}}J_0(a\sqrt{a^2-b^2})$

TABLE 23.8 Functions Among Transforms Tables Entries

Function	Definition	Name
$\mathrm{Ei}(x)$	$\displaystyle\int_{-\infty}^{x}\frac{e^v}{v}\,dv;$ or sometimes defined as $$-Ei(-x) = \int_{x}^{\infty}\frac{e^{-v}}{v}\,dv$$	Exponential integral function
$\mathrm{Si}(x)$	$\displaystyle\int_{0}^{x}\frac{\sin v}{v}\,dv$	Sine integral function
$\mathrm{Ci}(x)$	$\displaystyle\int_{\infty}^{x}\frac{\cos v}{v}\,dv;$ or sometimes defined as negative of this integral	Cosine integral function
$\mathrm{erf}(x)$	$\dfrac{2}{\sqrt{\pi}}\displaystyle\int_{0}^{x}e^{-v^2}\,dv$	Error function
$\mathrm{erfc}(x)$	$1-\mathrm{erf}(x)=\dfrac{2}{\pi}\displaystyle\int_{x}^{\infty}e^{-v^2}\,dv$	Complementary function to error function
$L_n(x)$	$\dfrac{e^x}{n!}\dfrac{d^n}{dx^n}(x^n e^{-x}),\quad n=0,1,\dots$	Laguerre polynomial of degree n

Table 23.8 provided function among transforms table entries.

References

Daniel, J.W. and Noble, B. 1988. *Applied Linear Algebra.* Prentice-Hall, Englewood Cliffs, NJ.
Davis, H.F. and Snider, A.D. 1991. *Introduction to Vector Analysis,* 6th ed. Wm. C. Brown, Dubuque, IA.
Strang, G. 1993. *Introduction to Linear Algebra.* Wellesley-Cambridge Press, Wellesley, MA.
Wylie, C.R. 1975. *Advanced Engineering Mathematics,* 4th ed. McGraw-Hill, New York.

Further Information

More advanced topics leading into the theory and applications of tensors may be found in Simmonds, J.G. *A Brief on Tensor Analysis* (1982, Springer-Verlag, New York).

23.7 Glossary of Terms

Jerry C. Whitaker

A

Absolute delay: The amount of time a signal is delayed. The delay may be expressed in time or number of pulse events.

Absolute zero: The lowest temperature theoretically possible, $-273.16°C$. *Absolute zero* is equal to zero degrees Kelvin.

Absorption: The transference of some or all of the energy contained in an electromagnetic wave to the substance or medium in which it is propagating or upon which it is incident.

Absorption auroral: The loss of energy in a radio wave passing through an area affected by solar auroral activity.

Ac coupling: A method of coupling one circuit to another through a capacitor or transformer so as to transmit the varying (ac) characteristics of the signal while blocking the static (dc) characteristics.

Ac/dc coupling: Coupling between circuits that accommodates the passing of both ac and dc signals (may also be referred to as simply dc coupling).

Accelerated life test: A special form of reliability testing performed by an equipment manufacturer. The unit under test is subjected to stresses that exceed those typically experienced in normal operation. The goal of an *accelerated life test* is to improve the reliability of products shipped by forcing latent failures in components to become evident before the unit leaves the factory.

Accelerating electrode: The electrode that causes electrons emitted from an electron gun to accelerate in their journey to the screen of a cathode ray tube.

Accelerating voltage: The voltage applied to an electrode that accelerates a beam of electrons or other charged particles.

Acceptable reliability level: The maximum number of failures allowed per thousand operating hours of a given component or system.

Acceptance test: The process of testing newly purchased equipment to ensure that it is fully compliant with contractual specifications.

Access: The point at which entry is gained to a circuit or facility.

Acquisition time: In a communication system, the amount of time required to attain synchronism.

Active: Any device or circuit that introduces gain or uses a source of energy other than that inherent in the signal to perform its function.

Adapter: A fitting or electrical connector that links equipment that cannot be connected directly.

Adaptive: A device able to adjust or react to a condition or application, as an *adaptive circuit*. This term usually refers to filter circuits.

Adaptive system: A general name for a system that is capable of reconfiguring itself to meet new requirements.

Adder: A device the output of which represents the sum of its inputs.

Adjacent channel interference: Interference to communications caused by a transmitter operating on an adjacent radio channel. The sidebands of the transmitter mix with the carrier being received on the desired channel, resulting in noise.

Admittance: A measure of how well alternating current flows in a conductor. It is the reciprocal of *impedance* and is expressed in *siemens*. The real part of admittance is *conductance*; the imaginary part is *susceptance*.

AFC (automatic frequency control): A circuit that automatically keeps an oscillator on frequency by comparing the output of the oscillator with a standard frequency source or signal.

Air core: An inductor with no magnetic material in its core.

Algorithm: A prescribed finite set of well-defined rules or processes for the solution of a problem in a finite number of steps.

Alignment: The adjustment of circuit components so that an entire system meets minimum performance values. For example, the stages in a radio are aligned to ensure proper reception.

Allocation: The planned use of certain facilities and equipment to meet current, pending, and/or fore-casted circuit- and carrier-system requirements.

Alternating current (ac): A continuously variable current, rising to a maximum in one direction, falling to zero, then reversing direction and rising to a maximum in the other direction, then falling to zero and repeating the cycle. Alternating current usually follows a sinusoidal growth and decay curve. Note that the correct usage of the term *ac* is lower case.

Alternator: A generator that produces alternating current electric power.

Ambient electromagnetic environment: The radiated or conducted electromagnetic signals and noise at a specific location and time.

Ambient level: The magnitude of radiated or conducted electromagnetic signals and noise at a specific test location when equipment-under-test is not powered.

Ambient temperature: The temperature of the surrounding medium, typically air, that comes into contact with an apparatus. Ambient temperature may also refer simply to room temperature.

American National Standards Institute (ANSI): A nonprofit organization that coordinates voluntary standards activities in the U.S.

American Wire Gauge (AWG): The standard American method of classifying wire diameter.

Ammeter: An instrument that measures and records the amount of current in amperes flowing in a circuit.

Amp (A): An abbreviation of the term *ampere*.

Ampacity: A measure of the current carrying capacity of a power cable. *Ampacity* is determined by the maximum continuous-performance temperature of the insulation, by the heat generated in the cable (as a result of conductor and insulation losses), and by the heat-dissipating properties of the cable and its environment.

Ampere (amp): The standard unit of electric current.

Ampere per meter: The standard unit of magnetic field strength.

Ampere-hour: The energy that is consumed when a current of one ampere flows for a period of one hour.

Ampere-turns: The product of the number of turns of a coil and the current in amperes flowing through the coil.

Amplification: The process that results when the output of a circuit is an enlarged reproduction of the input signal. Amplifiers may be designed to provide amplification of voltage, current, or power, or a combination of these quantities.

Amplification factor: In a vacuum tube, the ratio of the change in plate voltage to the change in grid voltage that causes a corresponding change in plate current. Amplification factor is expressed by the Greek letter μ (*mu*).

Amplifier (1—general): A device that receives an input signal and provides as an output a magnified replica of the input waveform. (**2—audio**) An amplifier designed to cover the normal audio frequency range (20 Hz to 20 kHz). (**3—balanced**) A circuit with two identical connected signal branches that operate in phase opposition, with input and output connections each balanced to ground. (**4—bridging**) An amplifying circuit featuring high input impedance to prevent loading of the source. (**5—broadband**) An amplifier capable of operating over a specified broad band of frequencies with acceptably small amplitude variations as a function of frequency. (**6—buffer**) An amplifier stage used to isolate a frequency-sensitive circuit from variations in the load presented by following stages. (**7—linear**) An amplifier in which the instantaneous output signal is a linear function of the corresponding input signal. (**8—magnetic**) An amplifier incorporating a control device dependent on magnetic saturation. A small dc signal applied to a control circuit triggers a large change in operating impedance and, hence, in the output of the circuit. (**9—microphone**) A circuit that amplifies the low level output from a microphone to make it sufficient to be used as an input signal to a power amplifier or another stage in a modulation circuit. Such a circuit is commonly known as a *preamplifier*. (**10—push-pull**) A balanced amplifier with two similar amplifying units connected in phase opposition in order to cancel undesired harmonics and minimize distortion.

(**11—tuned radio frequency**) An amplifier tuned to a particular radio frequency or band so that only selected frequencies are amplified.

Amplifier operating class (1—general): The operating point of an amplifying stage. The operating point, termed the operating *class*, determines the period during which current flows in the output. (**2—class A**) An amplifier in which output current flows during the whole of the input current cycle. (**3—class AB**) An amplifier in which the output current flows for more than half but less than the whole of the input cycle. (**4—class B**) An amplifier in which output current is cut off at zero input signal; a half-wave rectified output is produced. (**5—class C**) An amplifier in which output current flows for less than half the input cycle. (**6—class D**) An amplifier operating in a pulse-only mode.

Amplitude: The magnitude of a signal in voltage or current, frequently expressed in terms of *peak*, *peak-to-peak*, or *root-mean-square* (RMS). The actual amplitude of a quantity at a particular instant often varies in a sinusoidal manner.

Amplitude distortion: A distortion mechanism occurring in an amplifier or other device when the output amplitude is not a linear function of the input amplitude under specified conditions.

Amplitude equalizer: A corrective network that is designed to modify the amplitude characteristics of a circuit or system over a desired frequency range.

Amplitude-versus-frequency distortion: The distortion in a transmission system caused by the nonuniform attenuation or gain of the system with respect to frequency under specified conditions.

Analog carrier system: A carrier system whose signal amplitude, frequency, or phase is varied continuously as a function of a modulating input.

Anode (1—general): A positive pole or element. (**2—vacuum tube**) The outermost positive element in a vacuum tube, also called the *plate*. (**3—battery**) The positive element of a battery or cell.

Anodize: The formation of a thin film of oxide on a metallic surface, usually to produce an insulating layer.

Antenna (1—general): A device used to transmit or receive a radio signal. An antenna is usually designed for a specified frequency range and serves to couple electromagnetic energy from a transmission line to and/or from the free space through which it travels. Directional antennas concentrate the energy in a particular horizontal or vertical direction. (**2—aperiodic**) An antenna that is not periodic or resonant at particular frequencies, and so can be used over a wide band of frequencies. (**3—artificial**) A device that behaves, so far as the transmitter is concerned, like a proper antenna, but does not radiate any power at radio frequencies. (**4—broadband**) An antenna that operates within specified performance limits over a wide band of frequencies, without requiring retuning for each individual frequency. (**5—Cassegrain**) A double reflecting antenna, often used for ground stations in satellite systems. (**6—coaxial**) A dipole antenna made by folding back on itself a quarter wavelength of the outer conductor of a coaxial line, leaving a quarter wavelength of the inner conductor exposed. (**7—corner**) An antenna within the angle formed by two plane-reflecting surfaces. (**8—dipole**) A center-fed antenna, one half-wavelength long. (**9—directional**) An antenna designed to receive or emit radiation more efficiently in a particular direction. (**10—dummy**) An artificial antenna, designed to accept power from the transmitter but not to radiate it. (**11—ferrite**) A common AM broadcast receive antenna that uses a small coil mounted on a short rod of ferrite material. (**12—flat top**) An antenna in which all the horizontal components are in the same horizontal plane. (**13—folded dipole**) A radiating device consisting of two ordinary half-wave dipoles joined at their outer ends and fed at the center of one of the dipoles. (**14—horn reflector**) A radiator in which the feed horn extends into a parabolic reflector, and the power is radiated through a window in the horn. (**15—isotropic**) A theoretical antenna in free space that transmits or receives with the same efficiency in all directions. (**16—log-periodic**) A broadband directional antenna incorporating an array of dipoles of different lengths, the length and spacing between dipoles increasing logarithmically away from the feeder element. (**17—long wire**) An antenna made up of one or more conductors in a straight line pointing in the required direction with a total length of several wavelengths at the operating frequency. (**18—loop**) An antenna consisting of one or more turns of wire in the same or parallel planes. (**19—nested rhombic**) An assembly of two rhombic antennas, one smaller than the other, so that the complete diamond-shaped antenna fits inside the area occupied by the larger unit. (**20—omnidirectional**) An antenna whose radiating or receiving properties are the same in all horizontal plane directions. (**21—periodic**) A resonant antenna designed for use at a particular frequency. (**22—quarter-wave**) A dipole antenna whose length is equal to one quarter of a wavelength at the operating frequency. (**23—rhombic**) A large diamond-shaped antenna, with sides of the diamond several wavelengths long. The rhombic antenna is fed at one of the corners, with directional efficiency in the direction of the diagonal. (**24—series fed**) A vertical antenna that is fed at its lower end. (**25—shunt fed**) A vertical antenna whose base is grounded, and is fed at a specified point above ground. The point at which the antenna is fed above ground determines the operating impedance. (**26—steerable**) An antenna so constructed that its major lobe may readily be changed in direction. (**27—top-loaded**)A vertical antenna capacitively loaded at its upper end, often by simple enlargement or the attachment of a disc or plate. (**28—turnstile**) An antenna with one or more tiers of horizontal dipoles, crossed at right angles to each other and with excitation of the dipoles in phase quadrature. (**29—whip**) An antenna constructed of a thin semiflexible metal rod or tube, fed at its base. (**30—Yagi**) A directional antenna constructed of a series of dipoles cut to specific lengths. *Director* elements are placed in front of the active dipole and *reflector* elements are placed behind the active element.

Antenna array: A group of several antennas coupled together to yield a required degree of directivity.

Antenna beamwidth: The angle between the *half-power* points (3 dB points) of the main lobe of the antenna pattern when referenced to the peak power point of the antenna pattern. *Antenna beamwidth* is measured in degrees and normally refers to the horizontal radiation pattern.

Antenna directivity factor: The ratio of the power flux density in the desired direction to the average value of power flux density at crests in the antenna directivity pattern in the interference section.

Antenna factor: A factor that, when applied to the voltage appearing at the terminals of measurement equipment, yields the electrical field strength at an antenna. The unit of antenna factor is volts per meter per measured volt.

Antenna gain: The ratio of the power required at the input of a theoretically perfect omnidirectional reference antenna to the power supplied to the input of the given antenna to produce the same field at the same distance. When not specified otherwise, the figure expressing the gain of an antenna refers to the gain in the direction of the radiation main lobe. In services using *scattering* modes of propagation, the full gain of an antenna may not be realizable in practice and the apparent gain may vary with time.

Antenna gain-to-noise temperature: For a satellite earth terminal receiving system, a figure of merit that equals G/T, where G is the gain in dB of the earth terminal antenna at the receive frequency, and T is the equivalent noise temperature of the receiving system in Kelvins.

Antenna matching: The process of adjusting an antenna matching circuit (or the antenna itself) so that the input impedance of the antenna is equal to the characteristic impedance of the transmission line.

Antenna monitor: A device used to measure the ratio and phase between the currents flowing in the towers of a directional AM broadcast station.

Antenna noise temperature: The temperature of a resistor having an available noise power per unit bandwidth equal to that at the antenna output at a specified frequency.

Antenna pattern: A diagram showing the efficiency of radiation in all directions from the antenna.

Antenna power rating: The maximum continuous-wave power that can be applied to an antenna without degrading its performance.

Antenna preamplifier: A small amplifier, usually mast-mounted, for amplifying weak signals to a level sufficient to compensate for down-lead losses.

Apparent power: The product of the root-mean-square values of the voltage and current in an alternating-current circuit without a correction for the phase difference between the voltage and current.

Arc: A sustained luminous discharge between two or more electrodes.

Arithmetic mean: The sum of the values of several quantities divided by the number of quantities, also referred to as the *average*.

Armature winding: The winding of an electrical machine, either a motor or generator, in which current is induced.

Array (1—antenna): An assembly of several directional antennas so placed and interconnected that directivity may be enhanced. (**2—broadside**) An antenna array the elements of which are all in the same plane, producing a major lobe perpendicular to the plane. (**3—colinear**) An antenna array the elements of which are in the same line, either horizontal or vertical. (**4—end-fire**) An antenna array the elements of which are in parallel rows, one behind the other, producing a major lobe perpendicular to the plane in which individual elements are placed. (**5—linear**) An antenna array the elements of which are arranged end-to-end. (**6—stacked**) An antenna array the elements of which are stacked, one above the other.

Artificial line: An assembly of resistors, inductors, and capacitors that simulates the electrical characteristics of a transmission line.

Assembly: A manufactured part made by combining several other parts or subassemblies.

Assumed values: A range of values, parameters, levels, and other elements assumed for a mathematical model, hypothetical circuit, or network, from which analysis, additional estimates, or calculations will be made. The range of values, while not measured, represents the best engineering judgment and is generally derived from values found or measured in real circuits or networks of the same generic type, and includes projected improvements.

Atmosphere: The gaseous envelope surrounding the earth, composed largely of oxygen, carbon dioxide, and water vapor. The atmosphere is divided into four primary layers: *troposphere*, *stratosphere*, *ionosphere*, and *exosphere*.

Atmospheric noise: Radio noise caused by natural atmospheric processes, such as lightning.

Attack time: The time interval in seconds required for a device to respond to a control stimulus.

Attenuation: The decrease in amplitude of an electrical signal traveling through a transmission medium caused by dielectric and conductor losses.

Attenuation coefficient: The rate of decrease in the amplitude of an electrical signal caused by attenuation. The *attenuation coefficient* can be expressed in decibels or nepers per unit length. It may also be referred to as the *attenuation constant*.

Attenuation distortion: The distortion caused by attenuation that varies over the frequency range of a signal.

Attenuation-limited operation: The condition prevailing when the received signal amplitude (rather than distortion) limits overall system performance.

Attenuator: A fixed or adjustable component that reduces the amplitude of an electrical signal without causing distortion.

Atto: A prefix meaning one *quintillionth*.

Attraction: The attractive force between two unlike magnetic poles (N/S) or electrically charged bodies (+/−).

Attributes: The characteristics of equipment that aid planning and circuit design.

Automatic frequency control (AFC): A system designed to maintain the correct operating frequency of a receiver. Any drift in tuning results in the production of a control voltage, which is used to adjust the frequency of a local oscillator so as to minimize the tuning error.

Automatic gain control (AGC): An electronic circuit that compares the level of an incoming signal with a previously defined standard and automatically amplifies or attenuates the signal so it arrives at its destination at the correct level.

Autotransformer: A transformer in which both the primary and secondary currents flow through one common part of the coil.

Auxiliary power: An alternate source of electric power, serving as a back-up for the primary utility company ac power.

Availability: A measure of the degree to which a system, subsystem, or equipment is operable and not in a stage of congestion or failure at any given point in time.

Avalanche effect: The effect obtained when the electric field across a barrier region is sufficiently strong for electrons to collide with *valence electrons*, thereby releasing more electrons and giving a cumulative multiplication effect in a semiconductor.

Average life: The mean value for a normal distribution of product or component lives, generally applied to mechanical failures resulting from "wear-out."

B

Back emf: A voltage induced in the reverse direction when current flows through an inductance. *Back emf* is also known as *counter-emf*.

Back scattering: A form of wave scattering in which at least one component of the scattered wave is deflected opposite to the direction of propagation of the incident wave.

Background noise: The total system noise in the absence of information transmission, independent of the presence or absence of a signal.

Backscatter: The deflection or reflection of radiant energy through angles greater than 90° with respect to the original angle of travel.

Backscatter range: The maximum distance from which backscattered radiant energy can be measured.

Backup: A circuit element or facility used to replace an element that has failed.

Backup supply: A redundant power supply that takes over if the primary power supply fails.

Balance: The process of equalizing the voltage, current, or other parameter between two or more circuits or systems.

Balanced: A circuit having two sides (conductors) carrying voltages that are symmetrical about a common reference point, typically ground.

Balanced circuit: A circuit whose two sides are electrically equal in all transmission respects.

Balanced line: A transmission line consisting of two conductors in the presence of ground capable of being operated in such a way that when the voltages of the two conductors at all transverse planes are equal in magnitude and opposite in polarity with respect to ground, the currents in the two conductors are equal in magnitude and opposite in direction.

Balanced modulator: A modulator that combines the information signal and the carrier so that the output contains the two sidebands without the carrier.

Balanced three-wire system: A power distribution system using three conductors, one of which is balanced to have a potential midway between the potentials of the other two.

Balanced-to-ground: The condition when the impedance to ground on one wire of a two-wire circuit is equal to the impedance to ground on the other wire.

Balun (balanced/unbalanced): A device used to connect balanced circuits with unbalanced circuits.

Band: A range of frequencies between a specified upper and lower limit.

Band elimination filter: A filter having a single continuous attenuation band, with neither the upper nor lower cut-off frequencies being zero or infinite. A *band elimination filter* may also be referred to as a *band-stop*, *notch*, or *band reject* filter.

Bandpass filter: A filter having a single continuous transmission band with neither the upper nor the lower cut-off frequencies being zero or infinite. A bandpass filter permits only a specific band of frequencies to pass; frequencies above or below are attenuated.

Bandwidth: The range of signal frequencies that can be transmitted by a communications channel with a defined maximum loss or distortion. Bandwidth indicates the information-carrying capacity of a channel.

Bandwidth expansion ratio: The ratio of the necessary bandwidth to the baseband bandwidth.

Bandwidth-limited operation: The condition prevailing when the frequency spectrum or bandwidth, rather than the amplitude (or power) of the signal, is the limiting factor in communication capability. This condition is reached when the system distorts the shape of the waveform beyond tolerable limits.

Bank: A group of similar items connected together in a specified manner and used in conjunction with one another.

Bare: A wire conductor that is not enameled or enclosed in an insulating sheath.

Baseband: The band of frequencies occupied by a signal before it modulates a carrier wave to form a transmitted radio or line signal.

Baseband channel: A channel that carries a signal without modulation, in contrast to a *passband* channel.

Baseband signal: The original form of a signal, unchanged by modulation.

Bath tub: The shape of a typical graph of component failure rates: high during an initial period of operation, falling to an acceptable low level during the normal usage period, and then rising again as the components become time-expired.

Battery: A group of several cells connected together to furnish current by conversion of chemical, thermal, solar, or nuclear energy into electrical energy. A single cell is itself sometimes also called a battery.

Bay: A row or suite of racks on which transmission, switching, and/or processing equipment is mounted.

Bel: A unit of power measurement, named in honor of Alexander Graham Bell. The commonly used unit is one tenth of a Bel, or a decibel (dB). One Bel is defined as a tenfold increase in power. If an amplifier increases the power of a signal by 10 times, the power gain of the amplifier is equal to 1 Bel or 10 *decibels* (dB). If power is increased by 100 times, the power gain is 2 Bels or 20 decibels.

Bend: A transition component between two elements of a transmission waveguide.

Bending radius: The smallest bend that may be put into a cable under a stated pulling force. The bending radius is typically expressed in inches.

Bias: A dc voltage difference applied between two elements of an active electronic device, such as a vacuum tube, transistor, or integrated circuit. Bias currents may or may not be drawn, depending on the device and circuit type.

Bidirectional: An operational qualification which implies that the transmission of information occurs in both directions.

Bifilar winding: A type of winding in which two insulated wires are placed side by side. In some components, bifilar winding is used to produce balanced circuits.

Bipolar: A signal that contains both positive-going and negative-going amplitude components. A bipolar signal may also contain a zero amplitude state.

Bleeder: A high resistance connected in parallel with one or more filter capacitors in a high voltage dc system. If the power supply load is disconnected, the capacitors discharge through the bleeder.

Block diagram: An overview diagram that uses geometric figures to represent the principal divisions or sections of a circuit, and lines and arrows to show the path of a signal, or to show program functionalities. It is not a *schematic*, which provides greater detail.

Blocking capacitor: A capacitor included in a circuit to stop the passage of direct current.

BNC: An abbreviation for *bayonet Neill-Concelman*, a type of cable connector used extensively in RF applications (named for its inventor).

Boltzmann's constant: 1.38×10^{-23} joules.

Bridge: A type of network circuit used to match different circuits to each other, ensuring minimum transmission impairment.

Bridging: The shunting or paralleling of one circuit with another.

Broadband: The quality of a communications link having essentially uniform response over a given range of frequencies. A communications link is said to be *broadband* if it offers no perceptible degradation to the signal being transported.

Buffer: A circuit or component that isolates one electrical circuit from another.

Burn-in: The operation of a device, sometimes under extreme conditions, to stabilize its characteristics and identify latent component failures before bringing the device into normal service.

Bus: A central conductor for the primary signal path. The term bus may also refer to a signal path to which a number of inputs may be connected for feed to one or more outputs.

Busbar: A main dc power bus.

Bypass capacitor: A capacitor that provides a signal path that effectively shunts or bypasses other components.

Bypass relay: A switch used to bypass the normal electrical route of a signal or current in the event of power, signal, or equipment failure.

C

Cable: An electrically and/or optically conductive interconnecting device.

Cable loss: Signal loss caused by passing a signal through a coaxial cable. Losses are the result of resistance, capacitance, and inductance in the cable.

Cable splice: The connection of two pieces of cable by joining them mechanically and closing the joint with a weather-tight case or sleeve.

Cabling: The wiring used to interconnect electronic equipment.

Calibrate: The process of checking, and adjusting if necessary, a test instrument against one known to be set correctly.

Calibration: The process of identifying and measuring errors in instruments and/or procedures.

Capacitance: The property of a device or component that enables it to store energy in an electrostatic field and to release it later. A capacitor consists of two conductors separated by an insulating material. When the conductors have a voltage difference between them, a charge will be stored in the electrostatic field between the conductors.

Capacitor: A device that stores electrical energy. A capacitor allows the apparent flow of alternating current, while blocking the flow of direct current. The degree to which the device permits ac current flow depends on the frequency of the signal and the size of the capacitor. Capacitors are used in filters, delay-line components, couplers, frequency selectors, timing elements, voltage transient suppression, and other applications.

Carrier: A single frequency wave that, prior to transmission, is modulated by another wave containing information. A carrier may be modulated by manipulating its amplitude and/or frequency in direct relation to one or more applied signals.

Carrier frequency: The frequency of an unmodulated oscillator or transmitter. Also, the average frequency of a transmitter when a signal is frequency modulated by a symmetrical signal.

Cascade connection: A tandem arrangement of two or more similar component devices or circuits, with the output of one connected to the input of the next.

Cascaded: An arrangement of two or more circuits in which the output of one circuit is connected to the input of the next circuit.

Cathode ray tube (CRT): A vacuum tube device, usually glass, that is narrow at one end and widens at the other to create a surface onto which images can be projected. The narrow end contains the necessary circuits to generate and focus an electron beam on the luminescent screen at the other end. CRTs are used to display pictures in TV receivers, video monitors, oscilloscopes, computers, and other systems.

Cell: An elementary unit of communication, of power supply, or of equipment.

Celsius: A temperature measurement scale, expressed in degrees C, in which water freezes at $0°C$ and boils at $100°C$. To convert to degrees Fahrenheit, multiply by 0.555 and add 32. To convert to Kelvins add 273 (approximately).

Center frequency: In frequency modulation, the resting frequency or initial frequency of the carrier before modulation.

Center tap: A connection made at the electrical center of a coil.

Channel: The smallest subdivision of a circuit that provides a single type of communication service.

Channel decoder: A device that converts an incoming modulated signal on a given channel back into the source-encoded signal.

Channel encoder: A device that takes a given signal and converts it into a form suitable for transmission over the communications channel.

Channel noise level: The ratio of the channel noise at any point in a transmission system to some arbitrary amount of circuit noise chosen as a reference. This ratio is usually expressed in *decibels above reference noise*, abbreviated *dBrn*.

Channel reliability: The percent of time a channel is available for use in a specific direction during a specified period.

Channelization: The allocation of communication circuits to channels and the forming of these channels into groups for higher order multiplexing.

Characteristic: The property of a circuit or component.

Characteristic impedance: The impedance of a transmission line, as measured at the driving point, if the line were of infinite length. In such a line, there would be no standing waves. The *characteristic impedance* may also be referred to as the *surge impedance*.

Charge: The process of replenishing or replacing the electrical charge in a secondary cell or storage battery.

Charger: A device used to recharge a battery. Types of charging include: (1) *constant voltage charge*, (2) *equalizing charge*, and (3) *trickle charge*.

Chassis ground: A connection to the metal frame of an electronic system that holds the components in a place. The chassis ground connection serves as the ground return or electrical common for the system.

Circuit: Any closed path through which an electrical current can flow. In a *parallel circuit*, components are connected between common inputs and outputs such that all paths are parallel to each other. The same voltage appears across all paths. In a *series circuit*, the same current flows through all components.

Circuit noise level: The ratio of the circuit noise at some given point in a transmission system to an established reference, usually expressed in decibels above the reference.

Circuit reliability: The percentage of time a circuit is available to the user during a specified period of scheduled availability.

Circular mil: The measurement unit of the cross-sectional area of a circular conductor. A *circular mil* is the area of a circle whose diameter is one mil, or 0.001 inch.

Clear channel: A transmission path wherein the full bandwidth is available to the user, with no portions of the channel used for control, framing, or signaling. Can also refer to a classification of AM broadcast station.

Clipper: A limiting circuit which ensures that a specified output level is not exceeded by restricting the output waveform to a maximum peak amplitude.

Clipping: The distortion of a signal caused by removing a portion of the waveform through restriction of the amplitude of the signal by a circuit or device.

Coax: A short-hand expression for *coaxial cable*, which is used to transport high-frequency signals.

Coaxial cable: A transmission line consisting of an inner conductor surrounded first by an insulating material and then by an outer conductor, either solid or braided. The mechanical dimensions of the cable determine its *characteristic impedance*.

Coherence: The correlation between the phases of two or more waves.

Coherent: The condition characterized by a fixed phase relationship among points on an electromagnetic wave.

Coherent pulse: The condition in which a fixed phase relationship is maintained between consecutive pulses during pulse transmission.

Cold joint: A soldered connection that was inadequately heated, with the result that the wire is held in place by rosin flux, not solder. A cold joint is sometimes referred to as a *dry joint*.

Comb filter: An electrical filter circuit that passes a series of frequencies and rejects the frequencies in between, producing a frequency response similar to the teeth of a comb.

Common: A point that acts as a reference for circuits, often equal in potential to the local ground.

Common mode: Signals identical with respect to amplitude, frequency, and phase that are applied to both terminals of a cable and/or both the input and reference of an amplifier.

Common return: A return path that is common to two or more circuits, and returns currents to their source or to ground.

Common return offset: The dc common return potential difference of a line.

Communications system: A collection of individual communications networks, transmission systems, relay stations, tributary stations, and terminal equipment capable of interconnection and inter-operation to form an integral whole. The individual components must serve a common purpose, be technically compatible, employ common procedures, respond to some form of control, and, in general, operate in unison.

Commutation: A successive switching process carried out by a commutator.

Commutator: A circular assembly of contacts, insulated one from another, each leading to a different portion of the circuit or machine.

Compatibility: The ability of diverse systems to exchange necessary information at appropriate levels of command directly and in usable form. Communications equipment items are compatible if signals can be exchanged between them without the addition of buffering or translation for the specific purpose of achieving workable interface connections, and if the equipment or systems being inter-connected possess comparable performance characteristics, including the suppression of undesired radiation.

Complex wave: A waveform consisting of two or more sinewave components. At any instant of time, a complex wave is the algebraic sum of all its sinewave components.

Compliance: For mechanical systems, a property which is the reciprocal of stiffness.

Component: An assembly, or part thereof, that is essential to the operation of some larger circuit or system. A *component* is an immediate subdivision of the assembly to which it belongs.

COMSAT: The *Communications Satellite Corporation*, an organization established by an act of Congress in 1962. COMSAT launches and operates the international satellites for the INTELSAT consortium of countries.

Concentricity: A measure of the deviation of the center conductor position relative to its ideal location in the exact center of the dielectric cross-section of a coaxial cable.

Conditioning: The adjustment of a channel in order to provide the appropriate transmission character-istics needed for data or other special services.

Conditioning equipment: The equipment used to match transmission levels and impedances, and to provide equalization between facilities.

Conductance: A measure of the capability of a material to conduct electricity. It is the reciprocal of *resistance* (ohm) and is expressed in *siemens*. (Formerly expressed as *mho*.)

Conducted emission: An electromagnetic energy propagated along a conductor.

Conduction: The transfer of energy through a medium, such as the conduction of electricity by a wire, or of heat by a metallic frame.

Conduction band: A partially filled or empty atomic energy band in which electrons are free to move easily, allowing the material to carry an electric current.

Conductivity: The conductance per unit length.

Conductor: Any material that is capable of carrying an electric current.

Configuration: A relative arrangement of parts.

Connection: A point at which a junction of two or more conductors is made.

Connector: A device mounted on the end of a wire or fiber optic cable that mates to a similar device on a specific piece of equipment or another cable.

Constant-current source: A source with infinitely high output impedance so that output current is in-dependent of voltage, for a specified range of output voltages.

Constant-voltage charge: A method of charging a secondary cell or storage battery during which the terminal voltage is kept at a constant value.

Constant-voltage source: A source with low, ideally zero, internal impedance, so that voltage will remain constant, independent of current supplied.

Contact: The points that are brought together or separated to complete or break an electrical circuit.

Contact bounce: The rebound of a contact, which temporarily opens the circuit after its initial *make*.

Contact form: The configuration of a contact assembly on a relay. Many different configurations are possible from simple *single-make* contacts to complex arrangements involving *breaks* and *makes*.

Contact noise: A noise resulting from current flow through an electrical contact that has a rapidly varying resistance, as when the contacts are corroded or dirty.

Contact resistance: The resistance at the surface when two conductors make contact.

Continuity: A continuous path for the flow of current in an electrical circuit.

Continuous wave: An electromagnetic signal in which successive oscillations of the waves are identical.

Control: The supervision that an operator or device exercises over a circuit or system.

Control grid: The grid in an electron tube that controls the flow of current from the cathode to the anode.

Convention: A generally acceptable symbol, sign, or practice in a given industry.

Coordinated Universal Time (UTC): The time scale, maintained by the BIH (Bureau International de l'Heure), that forms the basis of a coordinated dissemination of standard frequencies and time signals.

Copper loss: The loss resulting from the heating effect of current.

Corona: A bluish luminous discharge resulting from ionization of the air near a conductor carrying a voltage gradient above a certain *critical level*.

Corrective maintenance: The necessary tests, measurements, and adjustments required to remove or correct a fault.

Cosmic noise: The random noise originating outside the earth's atmosphere.

Coulomb: The standard unit of electric quantity or charge. One *coulomb* is equal to the quantity of electricity transported in 1 second by a current of 1 ampere.

Coulomb's Law: The attraction and repulsion of electric charges act on a line between them. The charges are inversely proportional to the square of the distance between them, and proportional to the product of their magnitudes. (Named for the French physicist Charles-Augustine de Coulomb, 1736–1806.)

Counter-electromotive force: The effective electromotive force within a system that opposes the passage of current in a specified direction.

Couple: The process of linking two circuits by inductance, so that energy is transferred from one circuit to another.

Coupled mode: The selection of either ac or dc coupling.

Coupling: The relationship between two components that enables the transfer of energy between them. Included are *direct coupling* through a direct electrical connection, such as a wire; *capacitive coupling* through the capacitance formed by two adjacent conductors; and *inductive coupling* in which energy is transferred through a magnetic field. Capacitive coupling is also called *electrostatic coupling*. Inductive coupling is often referred to as *electromagnetic coupling*.

Coupling coefficient: A measure of the electrical coupling that exists between two circuits. The *coupling coefficient* is equal to the ratio of the mutual impedance to the square root of the product of the self impedances of the coupled circuits.

Cross coupling: The coupling of a signal from one channel, circuit, or conductor to another, where it becomes an undesired signal.

Crossover distortion: A distortion that results in an amplifier when an irregularity is introduced into the signal as it crosses through a zero reference point. If an amplifier is properly designed and biased, the upper half cycle and lower half cycle of the signal coincide at the zero crossover reference.

Crossover frequency: The frequency at which output signals pass from one channel to the other in a *crossover network*. At the *crossover frequency* itself, the outputs to each side are equal.

Crossover network: A type of filter that divides an incoming signal into two or more outputs, with higher frequencies directed to one output, and lower frequencies to another.

Crosstalk: Undesired transmission of signals from one circuit into another circuit in the same system. Crosstalk is usually caused by unintentional capacitive (ac) coupling.

Crosstalk coupling: The ratio of the power in a disturbing circuit to the induced power in the disturbed circuit, observed at a particular point under specified conditions. Crosstalk coupling is typically expressed in dB.

Crowbar: A short-circuit or low resistance path placed across the input to a circuit, usually for protective purposes.

CRT(cathode ray tube): A vacuum tube device that produces light when energized by the electron beam generated inside the tube. A CRT includes an electron gun, deflection mechanism, and phosphor-covered faceplate.

Crystal: A solidified form of a substance that has atoms and molecules arranged in a symmetrical pattern.

Crystal filter: A filter that uses piezoelectric crystals to create resonant or antiresonant circuits.

Crystal oscillator: An oscillator using a piezoelectric crystal as the tuned circuit that controls the resonant frequency.

Crystal-controlled oscillator: An oscillator in which a piezoelectric-effect crystal is coupled to a tuned oscillator circuit in such a way that the crystal pulls the oscillator frequency to its own natural frequency and does not allow frequency drift.

Current (1—general): A general term for the transfer of electricity, or the movement of electrons or *holes*. **(2—alternating):]** An electric current that is constantly varying in amplitude and periodically reversing direction. **(3—average):]** The arithmetic mean of the instantaneous values of current, averaged over one complete half cycle. **(4—charging)** The current that flows in to charge a capacitor when it is first connected to a source of electric potential. **(5—direct)** Electric current that flows in one direction only. **(6—eddy)** A wasteful current that flows in the core of a transformer and produces heat. *Eddy currents* are largely eliminated through the use of laminated cores. **(7—effective)** The ac current that will produce the same effective heat in a resistor as is produced by dc. If the ac is sinusoidal, the *effective current* value is 0.707 times the peak ac value. **(8—fault)** The current that flows between conductors or to ground during a fault condition. **(9—ground fault)** A fault current that flows to ground. **(10—ground return)** A current that returns through the earth. **(11—lagging)** A phenomenon observed in an inductive circuit where alternating current lags behind the voltage that produces it. **(12—leading)** A phenomenon observed in a capacitive circuit where alternating current leads the voltage that produces it. **(13—magnetizing)** The current in a transformer primary winding that is just sufficient to magnetize the core and offset iron losses. **(14—neutral)** The current that flows in the neutral conductor of an unbalanced polyphase power circuit. If correctly balanced, the neutral would carry no net current. **(15—peak)** The maximum value reached by a varying current during one cycle. **(16—pick-up)** The minimum current at which a relay just begins to operate. **(17—plate)** The anode current of an electron tube. **(18—residual)** The vector sum of the currents in the phase wires of an unbalanced polyphase power circuit. **(19—space)** The total current flowing through an electron tube.

Current amplifier: A low output impedance amplifier capable of providing high current output.

Current probe: A sensor, clamped around an electrical conductor, in which an induced current is developed from the magnetic field surrounding the conductor. For measurements, the current probe is connected to a suitable test instrument.

Current transformer: A transformer-type of instrument in which the primary carries the current to be measured and the secondary is in series with a low current ammeter. A current transformer is used to measure high values of alternating current.

Current-carrying capacity: A measure of the maximum current that can be carried continuously without damage to components or devices in a circuit.

Cut-off frequency: The frequency above or below which the output current in a circuit is reduced to a specified level.

Cycle: The interval of time or space required for a periodic signal to complete one period.

Cycles per second: The standard unit of frequency, expressed in Hertz (one cycle per second).

D

Damped oscillation: An oscillation exhibiting a progressive diminution of amplitude with time.

Damping: The dissipation and resultant reduction of any type of energy, such as electromagnetic waves.

DB (decibel): A measure of voltage, current, or power gain equal to 0.1 Bel. Decibels are given by the equations

$$20 \log \frac{V_{out}}{V_{in}}, \ 20 \log \frac{I_{out}}{I_{in}}, \quad \text{or} \quad 10 \log \frac{P_{out}}{P_{in}}.$$

DBk: A measure of power relative to 1 kilowatt. 0 dBk equals 1 kW.

DBm (decibels above 1 milliwatt): A logarithmic measure of power with respect to a reference power of one milliwatt.

DBmv: A measure of voltage gain relative to 1 millivolt at 75 ohms.

DBr: The power difference expressed in dB between any point and a reference point selected as the *zero relative transmission level* point. A power expressed in *dBr* does not specify the absolute power; it is a relative measurement only.

DBu: A term that reflects comparison between a measured value of voltage and a reference value of 0.775 V, expressed under conditions in which the impedance at the point of measurement (and of the reference source) are not considered.

DbV: A measure of voltage gain relative to 1 V.

DBW: A measure of power relative to 1 watt. 0 dBW equals 1 W.

Dc: An abbreviation for *direct current*. Note that the preferred usage of the term *dc* is lower case.

Dc amplifier: A circuit capable of amplifying dc and slowly varying alternating current signals.

Dc component: The portion of a signal that consists of direct current. This term may also refer to the average value of a signal.

Dc coupled: A connection configured so that both the signal (ac component) and the constant voltage on which it is riding (dc component) are passed from one stage to the next.

Dc coupling: A method of coupling one circuit to another so as to transmit the static (dc) characteristics of the signal as well as the varying (ac) characteristics. Any dc offset present on the input signal is maintained and will be present in the output.

Dc offset: The amount that the dc component of a given signal has shifted from its correct level.

Dc signal bounce: Overshoot of the proper dc voltage level resulting from multiple ac couplings in a signal path.

De-energized: A system from which sources of power have been disconnected.

Deca: A prefix meaning *ten*.

Decay: The reduction in amplitude of a signal on an exponential basis.

Decay time: The time required for a signal to fall to a certain fraction of its original value.

Decibel (dB): One tenth of a Bel. The decibel is a logarithmic measure of the ratio between two powers.

Decode: The process of recovering information from a signal into which the information has been encoded.

Decoder: A device capable of deciphering encoded signals. A decoder interprets input instructions and initiates the appropriate control operations as a result.

Decoupling: The reduction or removal of undesired coupling between two circuits or stages.

Deem phasis: The reduction of the high-frequency components of a received signal to reverse the pre-emphasis that was placed on them to overcome attenuation and noise in the transmission process.

Defect: An error made during initial planning that is normally detected and corrected during the development phase. Note that a *fault* is an error that occurs in an in-service system.

Deflection: The control placed on electron direction and motion in CRTs and other vacuum tube devices by varying the strengths of electrostatic (electrical) or electromagnetic fields.

Degradation: In susceptibility testing, any undesirable change in the operational performance of a test specimen. This term does not necessarily mean malfunction or catastrophic failure.

Degradation failure: A failure that results from a gradual change in performance characteristics of a system or part with time.

Delay: The amount of time by which a signal is delayed or an event is retarded.

Delay circuit: A circuit designed to delay a signal passing through it by a specified amount.

Delay distortion: The distortion resulting from the difference in phase delays at two frequencies of interest.

Delay equalizer: A network that adjusts the velocity of propagation of the frequency components of a complex signal to counteract the delay distortion characteristics of a transmission channel.

Delay line: A transmission network that increases the propagation time of a signal traveling through it.

Delta connection: A common method of joining together a three-phase power supply, with each phase across a different pair of the three wires used.

Delta-connected system: A 3-phase power distribution system where a single-phase output can be derived from each of the adjacent pairs of an equilateral triangle formed by the service drop transformer secondary windings.

Demodulator: Any device that recovers the original signal after it has modulated a high-frequency carrier. The output from the unit may be in baseband composite form.

Demultiplexer (demux): A device used to separate two or more signals that were previously combined by a compatible multiplexer and are transmitted over a single channel.

Derating factor: An operating safety margin provided for a component or system to ensure reliable performance. A *derating allowance* also is typically provided for operation under extreme environmental conditions, or under stringent reliability requirements.

Desiccant: A drying agent used for drying out cable splices or sensitive equipment.

Design: A layout of all the necessary equipment and facilities required to make a special circuit, piece of equipment, or system work.

Design objective: The desired electrical or mechanical performance characteristic for electronic circuits and equipment.

Detection: The rectification process that results in the modulating signal being separated from a modulated wave.

Detectivity: The reciprocal of *noise equivalent power*.

Detector: A device that converts one type of energy into another.

Device: A functional circuit, component, or network unit, such as a vacuum tube or transistor.

Dewpoint: The temperature at which moisture will condense out.

Diagnosis: The process of locating errors in software, or equipment faults in hardware.

Diagnostic routine: A software program designed to trace errors in software, locate hardware faults, or identify the cause of a breakdown.

Dielectric: An insulating material that separates the elements of various components, including capacitors and transmission lines. Dielectric materials include air, plastic, mica, ceramic, and Teflon. A dielectric material must be an insulator. (*Teflon* is a registered trademark of Du Pont.)

Dielectric constant: The ratio of the capacitance of a capacitor with a certain dielectric material to the capacitance with a vacuum as the dielectric. The *dielectric constant* is considered a measure of the capability of a dielectric material to store an electrostatic charge.

Dielectric strength: The potential gradient at which electrical breakdown occurs.

Differential amplifier: An input circuit that rejects voltages that are the same at both input terminals but amplifies any voltage difference between the inputs. Use of a differential amplifier causes any signal present on both terminals, such as common mode hum, to cancel itself.

Differential dc: The maximum dc voltage that can be applied between the differential inputs of an amplifier while maintaining linear operation.

Differential gain: The difference in output amplitude (expressed in percent or dB) of a small high frequency sinewave signal at two stated levels of a low frequency signal on which it is superimposed.

Differential phase: The difference in output phase of a small high frequency sinewave signal at two stated levels of a low frequency signal on which it is superimposed.

Differential-mode interference: An interference source that causes a change in potential of one side of a signal transmission path relative to the other side.

Diffuse reflection: The scattering effect that occurs when light, radio, or sound waves strike a rough surface.

Diffusion: The spreading or scattering of a wave, such as a radio wave.

Diode: A semiconductor or vacuum tube with two electrodes that passes electric current in one direction only. Diodes are used in rectifiers, gates, modulators, and detectors.

Direct coupling: A coupling method between stages that permits dc current to flow between the stages.

Direct current: An electrical signal in which the direction of current flow remains constant.

Discharge: The conversion of stored energy, as in a battery or capacitor, into an electric current.

Discontinuity: An abrupt nonuniform point of change in a transmission circuit that causes a disruption of normal operation.

Discrete: An individual circuit component.

Discrete component: A separately contained circuit element with its own external connections.

Discriminator: A device or circuit whose output amplitude and polarity vary according to how much the input signal varies from a standard or from another signal. A discriminator can be used to recover the modulating waveform in a frequency modulated signal.

Dish: An antenna system consisting of a parabolic shaped reflector with a signal feed element at the focal point. Dish antennas commonly are used for transmission and reception from microwave stations and communications satellites.

Dispersion: The wavelength dependence of a parameter.

Display: The representation of text and images on a cathode-ray tube, an array of light-emitting diodes, a liquid-crystal readout, or another similar device.

Display device: An output unit that provides a visual representation of data.

Distortion: The difference between the wave shape of an original signal and the signal after it has traversed a transmission circuit.

Distortion-limited operation: The condition prevailing when the shape of the signal, rather than the amplitude (or power), is the limiting factor in communication capability. This condition is reached when the system distorts the shape of the waveform beyond tolerable limits. For linear systems, *distortion-limited* operation is equivalent to *bandwidth-limited* operation.

Disturbance: The interference with normal conditions and communications by some external energy source.

Disturbance current: The unwanted current of any irregular phenomenon associated with transmission that tends to limit or interfere with the interchange of information.

Disturbance power: The unwanted power of any irregular phenomenon associated with transmission that tends to limit or interfere with the interchange of information.

Disturbance voltage: The unwanted voltage of any irregular phenomenon associated with transmission that tends to limit or interfere with the interchange of information.

Diversity receiver: A receiver using two antennas connected through circuitry that senses which antenna is receiving the stronger signal. Electronic gating permits the stronger source to be routed to the receiving system.

Documentation: A written description of a program. *Documentation* can be considered as any record that has permanence and can be read by humans or machines.

Down-lead: A lead-in wire from an antenna to a receiver.

Downlink: The portion of a communication link used for transmission of signals from a satellite or airborne platform to a surface terminal.

Downstream: A specified signal modification occurring after other given devices in a signal path.

Downtime: The time during which equipment is not capable of doing useful work because of malfunction. This does not include preventive maintenance time. In other words, *downtime* is measured from the occurrence of a malfunction to the correction of that malfunction.

Drift: A slow change in a nominally constant signal characteristic, such as frequency.

Drift-space: The area in a klystron tube in which electrons drift at their entering velocities and form electron *bunches*.

Drive: The input signal to a circuit, particularly to an amplifier.

Driver: An electronic circuit that supplies an isolated output to drive the input of another circuit.

Drop-out value: The value of current or voltage at which a relay will cease to be operated.

Dropout: The momentary loss of a signal.

Dropping resistor: A resistor designed to carry current that will make a required voltage available.

Duplex separation: The frequency spacing required in a communications system between the *forward* and *return* channels to maintain interference at an acceptably low level.

Duplex signaling: A configuration permitting signaling in both transmission directions simultaneously.

Duty cycle: The ratio of operating time to total elapsed time of a device that operates intermittently, expressed in percent.

Dynamic: A situation in which the operating parameters and/or requirements of a given system are continually changing.

Dynamic range: The maximum range or extremes in amplitude, from the lowest to the highest (noise floor to system clipping), that a system is capable of reproducing. The dynamic range is expressed in dB against a reference level.

Dynamo: A rotating machine, normally a dc generator.

Dynamotor: A rotating machine used to convert dc into ac.

E

Earth: A large conducting body with no electrical potential, also called *ground*.

Earth capacitance: The capacitance between a given circuit or component and a point at ground potential.

Earth current: A current that flows to earth/ground, especially one that follows from a fault in the system. *Earth current* may also refer to a current that flows in the earth, resulting from ionospheric disturbances, lightning, or faults on power lines.

Earth fault: A fault that occurs when a conductor is accidentally grounded/earthed, or when the resistance to earth of an insulator falls below a specified value.

Earth ground: A large conducting body that represents *zero level* in the scale of electrical potential. An *earth ground* is a connection made either accidentally or by design between a conductor and earth.

Earth potential: The potential taken to be the arbitrary zero in a scale of electric potential.

Effective ground: A connection to ground through a medium of sufficiently low impedance and adequate current-carrying capacity to prevent the buildup of voltages that might be hazardous to equipment or personnel.

Effective resistance: The increased resistance of a conductor to an alternating current resulting from the *skin effect*, relative to the direct-current resistance of the conductor. Higher frequencies tend to travel only on the outer skin of the conductor, whereas dc flows uniformly through the entire area.

Efficiency: The useful power output of an electrical device or circuit divided by the total power input, expressed in percent.

Electric: Any device or circuit that produces, operates on, transmits, or uses electricity.

Electric charge: An excess of either electrons or protons within a given space or material.

Electric field strength: The magnitude, measured in volts per meter, of the electric field in an electromagnetic wave.

Electric flux: The amount of electric charge, measured in coulombs, across a dielectric of specified area. *Electric flux* may also refer simply to electric lines of force.

Electricity: An energy force derived from the movement of negative and positive electric charges.

Electrode: An electrical terminal that emits, collects, or controls an electric current.

Electrolysis: A chemical change induced in a substance resulting from the passage of electric current through an electrolyte.

Electrolyte: A nonmetallic conductor of electricity in which current is carried by the physical movement of ions.

Electromagnet: An iron or steel core surrounded by a wire coil. The core becomes magnetized when current flows through the coil but loses its magnetism when the current flow is stopped.

Electromagnetic compatibility: The capability of electronic equipment or systems to operate in a specific electromagnetic environment, at designated levels of efficiency and within a defined margin of safety, without interfering with itself or other systems.

Electromagnetic field: The electric and magnetic fields associated with radio and light waves.

Electromagnetic induction: An electromotive force created with a conductor by the relative motion between the conductor and a nearby magnetic field.

Electromagnetism: The study of phenomena associated with varying magnetic fields, electromagnetic radiation, and moving electric charges.

Electromotive force (EMF): An electrical potential, measured in volts, that can produce the movement of electrical charges.

Electron: A stable elementary particle with a negative charge that is mainly responsible for electrical conduction. Electrons move when under the influence of an electric field. This movement constitutes an *electric current*.

Electron beam: A stream of emitted electrons, usually in a vacuum.

Electron gun: A hot cathode that produces a finely focused stream of fast electrons, which are necessary for the operation of a vacuum tube, such as a cathode ray tube. The gun is made up of a hot cathode electron source, a control grid, accelerating anodes, and (usually) focusing electrodes.

Electron lens: A device used for focusing an electron beam in a cathode ray tube. Such focusing can be accomplished by either magnetic forces, in which external coils are used to create the proper magnetic field within the tube, or electrostatic forces, where metallic plates within the tube are charged electrically in such a way as to control the movement of electrons in the beam.

Electron volt: The energy acquired by an electron in passing through a potential difference of one volt in a vacuum.

Electronic: A description of devices (or systems) that are dependent on the flow of electrons in electron tubes, semiconductors, and other devices, and not solely on electron flow in ordinary wires, inductors, capacitors, and similar passive components.

Electronic Industries Association (EIA): A trade organization, based in Washington, DC, representing the manufacturers of electronic systems and parts, including communications systems. The association develops standards for electronic components and systems.

Electronic switch: A transistor, semiconductor diode, or a vacuum tube used as an on/off switch in an electrical circuit. Electronic switches can be controlled manually, by other circuits, or by computers.

Electronics: The field of science and engineering that deals with electron devices and their utilization.

Electroplate: The process of coating a given material with a deposit of metal by electrolytic action.

Electrostatic: The condition pertaining to electric charges that are at rest.

Electrostatic field: The space in which there is electric stress produced by static electric charges.

Electrostatic induction: The process of inducing static electric charges on a body by bringing it near other bodies that carry high electrostatic charges.

Element: A substance that consists of atoms of the same atomic number. Elements are the basic units in all chemical changes other than those in which *atomic changes*, such as fusion and fission, are involved.

EMI (electromagnetic interference): Undesirable electromagnetic waves that are radiated unintentionally from an electronic circuit or device into other circuits or devices, disrupting their operation.

Emission (1—radiation): The radiation produced, or the production of radiation by a radio transmitting system. The emission is considered to be a *single emission* if the modulating signal and other characteristics are the same for every transmitter of the radio transmitting system and the spacing between antennas is not more than a few wavelengths. (**2—cathode**) The release of electrons from the cathode of a vacuum tube. (**3—parasitic**) A spurious radio frequency emission unintentionally generated at frequencies that are independent of the carrier frequency being amplified or modulated. (**4—secondary**) In an electron tube, emission of electrons by a plate or grid because of bombardment by *primary emission* electrons from the cathode of the tube. (**5—spurious**) An emission outside the radio frequency band authorized for a transmitter. (**6—thermonic**) An emission from a cathode resulting from high temperature.

Emphasis: The intentional alteration of the frequency-amplitude characteristics of a signal to reduce the adverse effects of noise in a communication system.

Empirical: A conclusion not based on pure theory, but on practical and experimental work.

Emulation: The use of one system to imitate the capabilities of another system.

Enable: To prepare a circuit for operation or to allow an item to function.

Enabling signal: A signal that permits the occurrence of a specified event.

Encode: The conversion of information from one form into another to obtain characteristics required by a transmission or storage system.

Encoder: A device that processes one or more input signals into a specified form for transmission and/or storage.

Energized: The condition when a circuit is switched on, or powered up.

Energy spectral density: A frequency-domain description of the energy in each of the frequency components of a pulse.

Envelope: The boundary of the family of curves obtained by varying a parameter of a wave.

Envelope delay: The difference in absolute delay between the fastest and slowest propagating frequencies within a specified bandwidth.

Envelope delay distortion: The maximum difference or deviation of the envelope-delay characteristic between any two specified frequencies.

Envelope detection: A demodulation process that senses the shape of the modulated RF envelope. A diode detector is one type of envelope detection device.

Environmental: An equipment specification category relating to temperature and humidity.

EQ (equalization) network: A network connected to a circuit to correct or control its transmission frequency characteristics.

Equalization (EQ): The reduction of frequency distortion and/or phase distortion of a circuit through the introduction of one or more networks to compensate for the difference in attenuation, time delay, or both, at the various frequencies in the transmission band.

Equalize: The process of inserting in a line a network with complementary transmission characteristics to those of the line, so that when the loss or delay in the line and that in the equalizer are combined, the overall loss or delay is approximately equal at all frequencies.

Equalizer: A network that corrects the transmission-frequency characteristics of a circuit to allow it to transmit selected frequencies in a uniform manner.

Equatorial orbit: The plane of a satellite orbit which coincides with that of the equator of the primary body.

Equipment: A general term for electrical apparatus and hardware, switching systems, and transmission components.

Equipment failure: The condition when a hardware fault stops the successful completion of a task.

Equipment ground: A protective ground consisting of a conducting path to ground of noncurrent carrying metal parts.

Equivalent circuit: A simplified network that emulates the characteristics of the real circuit it replaces. An equivalent circuit is typically used for mathematical analysis.

Equivalent noise resistance: A quantitative representation in resistance units of the spectral density of a noise voltage generator at a specified frequency.

Error: A collective term that includes all types of inconsistencies, transmission deviations, and control failures.

Excitation: The current that energizes field coils in a generator.

Expandor: A device with a nonlinear gain characteristic that acts to increase the gain more on larger input signals than it does on smaller input signals.

Extremely high frequency (EHF): The band of microwave frequencies between the limits of 30 GHz and 300 GHz (wavelengths between 1 cm and 1 mm).

Extremely low frequency: The radio signals with operating frequencies below 300 Hz (wavelengths longer than 1000 km).

F

Fail-safe operation: A type of control architecture for a system that prevents improper functioning in the event of circuit or operator failure.

Failure: A detected cessation of ability to perform a specified function or functions within previously established limits. A *failure* is beyond adjustment by the operator by means of controls normally accessible during routine operation of the system. (This requires that measurable limits be established to define "satisfactory performance".)

Failure effect: The result of the malfunction or failure of a device or component.

Failure in time (FIT): A unit value that indicates the reliability of a component or device. One failure in time corresponds to a failure rate of 10^{-9} per hour.

Failure mode and effects analysis (FMEA): An iterative documented process performed to identify basic faults at the component level and determine their effects at higher levels of assembly.

Failure rate: The ratio of the number of actual failures to the number of times each item has been subjected to a set of specified stress conditions.

Fall time: The length of time during which a pulse decreases from 90 percent to 10 percent of its maximum amplitude.

Farad: The standard unit of capacitance equal to the value of a capacitor with a potential of one volt between its plates when the charge on one plate is one coulomb and there is an equal and opposite charge on the other plate. The farad is a large value and is more commonly expressed in *microfarads* or *picofarads*. The *farad* is named for the English chemist and physicist Michael Faraday (1791–1867).

Fast frequency shift keying (FFSK): A system of digital modulation where the digits are represented by different frequencies that are related to the baud rate, and where transitions occur at the zero crossings.

Fatigue: The reduction in strength of a metal caused by the formation of crystals resulting from repeated flexing of the part in question.

Fault: A condition that causes a device, a component, or an element to fail to perform in a required manner. Examples include a short-circuit, broken wire, or intermittent connection.

Fault to ground: A fault caused by the failure of insulation and the consequent establishment of a direct path to ground from a part of the circuit that should not normally be grounded.

Fault tree analysis (FTA): An iterative documented process of a systematic nature performed to identify basic faults, determine their causes and effects, and establish their probabilities of occurrence.

Feature: A distinctive characteristic or part of a system or piece of equipment, usually visible to end users and designed for their convenience.

Federal Communications Commission (FCC): The federal agency empowered by law to regulate all interstate radio and wireline communications services originating in the United States, including radio, television, facsimile, telegraph, data transmission, and telephone systems. The agency was established by the Communications Act of 1934.

Feedback: The return of a portion of the output of a device to the input. *Positive feedback* adds to the input; *negative feedback* subtracts from the input.

Feedback amplifier: An amplifier with the components required to feed a portion of the output back into the input to alter the characteristics of the output signal.

Feedline: A transmission line, typically coaxial cable, that connects a high frequency energy source to its load.

Femto: A prefix meaning *one quadrillionth* (10^{-15}).

Ferrite: A ceramic material made of powdered and compressed ferric oxide, plus other oxides (mainly cobalt, nickel, zinc, yttrium-iron, and manganese). These materials have low eddy current losses at high frequencies.

Ferromagnetic material: A material with low relative permeability and high coercive force so that it is difficult to magnetize and demagnetize. Hard ferromagnetic materials retain magnetism well, and are commonly used in permanent magnets.

Fidelity: The degree to which a system, or a portion of a system, accurately reproduces at its output the essential characteristics of the signal impressed upon its input.

Field strength: The strength of an electric, magnetic, or electromagnetic field.

Filament: A wire that becomes hot when current is passed through it, used either to emit light (for a light bulb) or to heat a cathode to enable it to emit electrons (for an electron tube).

Film resistor: A type of resistor made by depositing a thin layer of resistive material on an insulating core.

Filter: A network that passes desired frequencies but greatly attenuates other frequencies.

Filtered noise: White noise that has been passed through a filter. The power spectral density of filtered white noise has the same shape as the transfer function of the filter.

Fitting: A coupling or other mechanical device that joins one component with another.

Fixed: A system or device that is not changeable or movable.

Flashover: An arc or spark between two conductors.

Flashover voltage: The voltage between conductors at which flashover just occurs.

Flat face tube: The design of CRT tube with almost a flat face, giving improved legibility of text and reduced reflection of ambient light.

Flat level: A signal that has an equal amplitude response for all frequencies within a stated range.

Flat loss: A circuit, device, or channel that attenuates all frequencies of interest by the same amount, also called *flat slope*.

Flat noise: A noise whose power per unit of frequency is essentially independent of frequency over a specified frequency range.

Flat response: The performance parameter of a system in which the output signal amplitude of the system is a faithful reproduction of the input amplitude over some range of specified input frequencies.

Floating: A circuit or device that is not connected to any source of potential or to ground.

Fluorescence: The characteristic of a material to produce light when excited by an external energy source. Minimal or no heat results from the process.

Flux: The electric or magnetic lines of force resulting from an applied energy source.

Flywheel effect: The characteristic of an oscillator that enables it to sustain oscillations after removal of the control stimulus. This characteristic may be desirable, as in the case of a phase-locked loop employed in a synchronous system, or undesirable, as in the case of a voltage-controlled oscillator.

Focusing: A method of making beams of radiation converge on a target, such as the face of a CRT.

Fourier analysis: A mathematical process for transforming values between the frequency domain and the time domain. This term also refers to the decomposition of a time-domain signal into its frequency components.

Fourier transform: An integral that performs an actual transformation between the frequency domain and the time domain in Fourier analysis.

Frame: A segment of an analog or digital signal that has a repetitive characteristic, in that corresponding elements of successive *frames* represent the same things.

Free electron: An electron that is not attached to an atom and is, thus, mobile when an electromotive force is applied.

Free running: An oscillator that is not controlled by an external synchronizing signal.

Free-running oscillator: An oscillator that is not synchronized with an external timing source.

Frequency: The number of complete cycles of a periodic waveform that occur within a given length of time. Frequency is usually specified in cycles per second (*Hertz*). Frequency is the reciprocal of wavelength. The higher the frequency, the shorter the wavelength. In general, the higher the frequency of a signal, the more capacity it has to carry information, the smaller an antenna is required, and the more susceptible the signal is to absorption by the atmosphere and by physical structures. At microwave frequencies, radio signals take on a *line-of-sight* characteristic and require highly directional and focused antennas to be used successfully.

Frequency accuracy: The degree of conformity of a given signal to the specified value of a frequency.

Frequency allocation: The designation of radio-frequency bands for use by specific radio services.

Frequency content: The band of frequencies or specific frequency components contained in a signal.

Frequency converter: A circuit or device used to change a signal of one frequency into another of a different frequency.

Frequency coordination: The process of analyzing frequencies in use in various bands of the spectrum to achieve reliable performance for current and new services.

Frequency counter: An instrument or test set used to measure the frequency of a radio signal or any other alternating waveform.

Frequency departure: An unintentional deviation from the nominal frequency value.

Frequency difference: The algebraic difference between two frequencies. The two frequencies can be of identical or different nominal values.

Frequency displacement: The end-to-end shift in frequency that may result from independent frequency translation errors in a circuit.

Frequency distortion: The distortion of a multifrequency signal caused by unequal attenuation or amplification at the different frequencies of the signal. This term may also be referred to as *amplitude distortion*.

Frequency domain: A representation of signals as a function of frequency, rather than of time.

Frequency modulation (FM): The modulation of a carrier signal so that its instantaneous frequency is proportional to the instantaneous value of the modulating wave.

Frequency multiplier: A circuit that provides as an output an exact multiple of the input frequency.

Frequency offset: A frequency shift that occurs when a signal is sent over an analog transmission facility in which the modulating and demodulating frequencies are not identical. A channel with frequency offset does not preserve the waveform of a transmitted signal.

Frequency response: The measure of system linearity in reproducing signals across a specified bandwidth. Frequency response is expressed as a frequency range with a specified amplitude tolerance in dB.

Frequency response characteristic: The variation in the transmission performance (gain or loss) of a system with respect to variations in frequency.

Frequency reuse: A technique used to expand the capacity of a given set of frequencies or channels by separating the signals either geographically or through the use of different polarization techniques. Frequency reuse is a common element of the *frequency coordination* process.

Frequency selectivity: The ability of equipment to separate or differentiate between signals at different frequencies.

Frequency shift: The difference between the frequency of a signal applied at the input of a circuit and the frequency of that signal at the output.

Frequency shift keying (FSK): A commonly used method of digital modulation in which a one and a zero (the two possible states) are each transmitted as separate frequencies.

Frequency stability: A measure of the variations of the frequency of an oscillator from its mean frequency over a specified period of time.

Frequency standard: An oscillator with an output frequency sufficiently stable and accurate that it is used as a reference.

Frequency-division multiple access (FDMA): The provision of multiple access to a transmission facility, such as an earth satellite, by assigning each transmitter its own frequency band.

Frequency-division multiplexing (FDM): The process of transmitting multiple analog signals by an orderly assignment of frequency slots, that is, by dividing transmission bandwidth into several narrow bands, each of which carries a single communication and is sent simultaneously with others over a common transmission path.

Full duplex: A communications system capable of transmission simultaneously in two directions.

Full-wave rectifier: A circuit configuration in which both positive and negative half-cycles of the incoming ac signal are rectified to produce a unidirectional (dc) current through the load.

Functional block diagram: A diagram illustrating the definition of a device, system, or problem on a logical and functional basis.

Functional unit: An entity of hardware and/or software capable of accomplishing a given purpose.

Fundamental frequency: The lowest frequency component of a complex signal.

Fuse: A protective device used to limit current flow in a circuit to a specified level. The fuse consists of a metallic link that melts and opens the circuit at a specified current level.

Fuse wire: A fine-gauge wire made of an alloy that overheats and melts at the relatively low temperatures produced when the wire carries overload currents. When used in a fuse, the wire is called a fuse (or fusible) link.

G

Gain: An increase or decrease in the level of an electrical signal. Gain is measured in terms of decibels or number-of-times of magnification. Strictly speaking, *gain* refers to an increase in level. Negative numbers, however, are commonly used to denote a decrease in level.

Gain-bandwidth: The gain times the frequency of measurement when a device is biased for maximum obtainable gain.

Gain/frequency characteristic: The gain-versus-frequency characteristic of a channel over the bandwidth provided, also referred to as *frequency response*.

Gain/frequency distortion: A circuit defect in which a change in frequency causes a change in signal amplitude.

Galvanic: A device that produces direct current by chemical action.

Gang: The mechanical connection of two or more circuit devices so that they can all be adjusted simultaneously.

Gang capacitor: A variable capacitor with more than one set of moving plates linked together.

Gang tuning: The simultaneous tuning of several different circuits by turning a single shaft on which ganged capacitors are mounted.

Ganged: One or more devices that are mechanically coupled, normally through the use of a shared shaft.

Gas breakdown: The ionization of a gas between two electrodes caused by the application of a voltage that exceeds a threshold value. The ionized path has a low impedance. Certain types of circuit and line protectors rely on gas breakdown to divert hazardous currents away from protected equipment.

Gas tube: A protection device in which a sufficient voltage across two electrodes causes a gas to ionize, creating a low impedance path for the discharge of dangerous voltages.

Gas-discharge tube: A gas-filled tube designed to carry current during gas breakdown. The gas-discharge tube is commonly used as a protective device, preventing high voltages from damaging sensitive equipment.

Gauge: A measure of wire diameter. In measuring wire gauge, the lower the number, the thicker the wire.

Gaussian distribution: A statistical distribution, also called the *normal* distribution. The graph of a Gaussian distribution is a bell-shaped curve.

Gaussian noise: Noise in which the distribution of amplitude follows a Gaussian model; that is, the noise is random but distributed about a reference voltage of zero.

Gaussian pulse: A pulse that has the same form as its own Fourier transform.

Generator: A machine that converts mechanical energy into electrical energy, or one form of electrical energy into another form.

Geosynchronous: The attribute of a satellite in which the relative position of the satellite as viewed from the surface of a given planet is stationary. For earth, the geosynchronous position is 22,300 miles above the planet.

Getter: A metal used in vaporized form to remove residual gas from inside an electron tube during manufacture.

Giga: A prefix meaning one billion.

Gigahertz (GHz): A measure of frequency equal to one billion cycles per second. Signals operating above 1 gigahertz are commonly known as *microwaves*, and begin to take on the characteristics of visible light.

Glitch: A general term used to describe a wide variety of momentary signal discontinuities.

Graceful degradation: An equipment failure mode in which the system suffers reduced capability, but does not fail altogether.

Graticule: A fixed pattern of reference markings used with oscilloscope CRTs to simplify measurements. The graticule may be etched on a transparent plate covering the front of the CRT or, for greater accuracy in readings, may be electrically generated within the CRT itself.

Grid (1—general): A mesh electrode within an electron tube that controls the flow of electrons between the cathode and plate of the tube. **(2—bias)** The potential applied to a grid in an electron tube to control its center operating point. **(3—control)** The grid in an electron tube to which the input signal is usually applied. **(4—screen)** The grid in an electron tube, typically held at a steady potential, that screens the control grid from changes in anode potential. **(5—suppressor)** The grid in an electron tube near the anode (plate) that suppresses the emission of secondary electrons from the plate.

Ground: An electrical connection to earth or to a common conductor usually connected to earth.

Ground clamp: A clamp used to connect a ground wire to a ground rod or system.

Ground loop: An undesirable circulating ground current in a circuit grounded via multiple connections or at multiple points.

Ground plane: A conducting material at ground potential, physically close to other equipment, so that connections may be made readily to ground the equipment at the required points.

Ground potential: The point at zero electric potential.

Ground return: A conductor used as a path for one or more circuits back to the ground plane or central facility ground point.

Ground rod: A metal rod driven into the earth and connected into a mesh of interconnected rods so as to provide a low resistance link to ground.

Ground window: A single-point interface between the integrated ground plane of a building and an isolated ground plane.

Ground wire: A copper conductor used to extend a good low-resistance earth ground to protective devices in a facility.

Grounded: The connection of a piece of equipment to earth via a low resistance path.

Grounding: The act of connecting a device or circuit to ground or to a conductor that is grounded.

Group delay: A condition where the different frequency elements of a given signal suffer differing propagation delays through a circuit or a system. The delay at a lower frequency is different from the delay at a higher frequency, resulting in a time-related distortion of the signal at the receiving point.

Group delay time: The rate of change of the total phase shift of a waveform with angular frequency through a device or transmission facility.

Group velocity: The speed of a pulse on a transmission line.

Guard band: A narrow bandwidth between adjacent channels intended to reduce interference or crosstalk.

H

Half-wave rectifier: A circuit or device that changes only positive or negative half-cycle inputs of alternating current into direct current.

Hall effect: The phenomenon by which a voltage develops between the edges of a current-carrying metal strip whose faces are perpendicular to an external magnetic field.

Hard-wired: Electrical devices connected through physical wiring.

Harden: The process of constructing military telecommunications facilities so as to protect them from damage by enemy action, especially *electromagnetic pulse* (EMP) radiation.

Hardware: Physical equipment, such as mechanical, magnetic, electrical, or electronic devices or components.

Harmonic: A periodic wave having a frequency that is an integral multiple of the fundamental frequency. For example, a wave with twice the frequency of the fundamental is called the *second harmonic*.

Harmonic analyzer: A test set capable of identifying the frequencies of the individual signals that make up a complex wave.

Harmonic distortion: The production of harmonics at the output of a circuit when a periodic wave is applied to its input. The level of the distortion is usually expressed as a percentage of the level of the input.

Hazard: A condition that could lead to danger for operating personnel.

Headroom: The difference, in decibels, between the typical operating signal level and a peak overload level.

Heat loss: The loss of useful electrical energy resulting from conversion into unwanted heat.

Heat sink: A device that conducts heat away from a heat-producing component so that it stays within a safe working temperature range.

Heater: In an electron tube, the filament that heats the cathode to enable it to emit electrons.

Hecto: A prefix meaning 100.

Henry: The standard unit of electrical inductance, equal to the self-inductance of a circuit or the mutual inductance of two circuits when there is an induced electromotive force of one volt and a current change of one ampere per second. The symbol for inductance is H, named for the American physicist Joseph Henry (1797–1878).

Hertz (Hz): The unit of frequency that is equal to one cycle per second. Hertz is the reciprocal of the *period*, the interval after which the same portion of a periodic waveform recurs. Hertz was named for the German physicist Heinrich R. Hertz (1857–1894).

Heterodyne: The mixing of two signals in a nonlinear device in order to produce two additional signals at frequencies that are the sum and difference of the original frequencies.

Heterodyne frequency: The sum of, or the difference between, two frequencies, produced by combining the two signals together in a modulator or similar device.

Heterodyne wavemeter: A test set that uses the heterodyne principle to measure the frequencies of incoming signals.

High-frequency loss: Loss of signal amplitude at higher frequencies through a given circuit or medium. For example, high frequency loss could be caused by passing a signal through a coaxial cable.

High: Q: An inductance or capacitance whose ratio of reactance to resistance is high.

High tension: A high voltage circuit.

High-pass filter: A network that passes signals of higher than a specified frequency but attenuates signals of all lower frequencies.

Homochronous: Signals whose corresponding significant instants have a constant but uncontrolled phase relationship with each other.

Horn gap: A lightning arrester utilizing a gap between two horns. When lightning causes a discharge between the horns, the heat produced lengthens the arc and breaks it.

Horsepower: The basic unit of mechanical power. One horsepower (hp) equals 550 foot-pounds per second or 746 watts.

Hot: A charged electrical circuit or device.

Hot dip galvanized: The process of galvanizing steel by dipping it into a bath of molten zinc.

Hot standby: System equipment that is fully powered but not in service. A *hot standby* can rapidly replace a primary system in the event of a failure.

Hum: Undesirable coupling of the 60 Hz power sine wave into other electrical signals and/or circuits.

HVAC: An abbreviation for *heating, ventilation, and air conditioning* system.

Hybrid system: A communication system that accommodates both digital and analog signals.

Hydrometer: A testing device used to measure specific gravity, particularly the specific gravity of the dilute sulphuric acid in a lead-acid storage battery, to learn the state of charge of the battery.

Hygrometer: An instrument that measures the relative humidity of the atmosphere.

Hygroscopic: The ability of a substance to absorb moisture from the air.

Hysteresis: The property of an element evidenced by the dependence of the value of the output, for a given excursion of the input, upon the history of prior excursions and direction of the input. Originally, *hysteresis* was the name for magnetic phenomena only—the lagging of flux density behind the change in value of the magnetizing flux—but now, the term is also used to describe other inelastic behavior.

Hysteresis loop: The plot of magnetizing current against magnetic flux density (or of other similarly related pairs of parameters), which appears as a loop. The area within the loop is proportional to the power loss resulting from hysteresis.

Hysteresis loss: The loss in a magnetic core resulting from hysteresis.

I

$I^2 R$ **loss:** The power lost as a result of the heating effect of current passing through resistance.

Idling current: The current drawn by a circuit, such as an amplifier, when no signal is present at its input.

Image frequency: A frequency on which a carrier signal, when heterodyned with the local oscillator in a superheterodyne receiver, will cause a sum or difference frequency that is the same as the intermediate frequency of the receiver. Thus, a signal on an *image frequency* will be demodulated along with the desired signal and will interfere with it.

Impact ionization: The ionization of an atom or molecule as a result of a high energy collision.

Impedance: The total passive opposition offered to the flow of an alternating current.
Impedance consists of a combination of resistance, inductive reactance, and capacitive reactance. It is the vector sum of resistance and reactance $(R+jX)$ or the vector of magnitude Z at an angle θ.

Impedance characteristic: A graph of the impedance of a circuit showing how it varies with frequency.

Impedance irregularity: A discontinuity in an impedance characteristic caused, for example, by the use of different coaxial cable types.

Impedance matching: The adjustment of the impedances of adjoining circuit components to a common value so as to minimize reflected energy from the junction and to maximize energy transfer across it. Incorrect adjustment results in an *impedance mismatch*.

Impedance matching transformer: A transformer used between two circuits of different impedances with a turns ratio that provides for maximum power transfer and minimum loss by reflection.

Impulse: A short high energy surge of electrical current in a circuit or on a line.

Impulse current: A current that rises rapidly to a peak then decays to zero without oscillating.

Impulse excitation: The production of an oscillatory current in a circuit by impressing a voltage for a relatively short period compared with the duration of the current produced.

Impulse noise: A noise signal consisting of random occurrences of energy spikes, having random amplitude and bandwidth.

Impulse response: The amplitude-versus-time output of a transmission facility or device in response to an impulse.

Impulse voltage: A unidirectional voltage that rises rapidly to a peak and then falls to zero, without any appreciable oscillation.

In-phase: The property of alternating current signals of the same frequency that achieve their peak positive, peak negative, and zero amplitude values simultaneously.

Incidence angle: The angle between the perpendicular to a surface and the direction of arrival of a signal.

Increment: A small change in the value of a quantity.

Induce: To produce an electrical or magnetic effect in one conductor by changing the condition or position of another conductor.

Induced current: The current that flows in a conductor because a voltage has been induced across two points in, or connected to, the conductor.

Induced voltage: A voltage developed in a conductor when the conductor passes through magnetic lines of force.

Inductance: The property of an inductor that opposes any change in a current that flows through it. The standard unit of inductance is the *Henry*.

Induction: The electrical and magnetic interaction process by which a changing current in one circuit produces a voltage change not only in its own circuit (*self inductance*) but also in other circuits to which it is linked magnetically.

Inductive: A circuit element exhibiting inductive reactance.

Inductive kick: A voltage surge produced when a current flowing through an inductance is interrupted.

Inductive load: A load that possesses a net inductive reactance.

Inductive reactance: The reactance of a circuit resulting from the presence of inductance and the phenomenon of induction.

Inductor: A coil of wire, usually wound on a core of high permeability, that provides high inductance without necessarily exhibiting high resistance.

Inert: An inactive unit, or a unit that has no power requirements.

Infinite line: A transmission line that appears to be of infinite length. There are no reflections back from the far end because it is terminated in its characteristic impedance.

Infra low frequency (ILF): The frequency band from 300 Hz to 3000 Hz.

Inhibit: A control signal that prevents a device or circuit from operating.

Injection: The application of a signal to an electronic device.

Input: The waveform fed into a circuit, or the terminals that receive the input waveform.

Insertion gain: The gain resulting from the insertion of a transducer in a transmission system, expressed as the ratio of the power delivered to that part of the system following the transducer to the power delivered to that same part before insertion. If more than one component is involved in the input or output, the particular component used must be specified. This ratio is usually expressed in decibels. If the resulting number is negative, an *insertion loss* is indicated.

Insertion loss: The signal loss within a circuit, usually expressed in decibels as the ratio of input power to output power.

Insertion loss-vs.-frequency characteristic: The amplitude transfer characteristic of a system or component as a function of frequency. The amplitude response may be stated as actual gain, loss, amplification, or attenuation, or as a ratio of any one of these quantities at a particular frequency, with respect to that at a specified reference frequency.

Inspection lot: A collection of units of product from which a sample is drawn and inspected to determine conformance with acceptability criteria.

Instantaneous value: The value of a varying waveform at a given instant of time. The value can be in volts, amperes, or phase angle.

Institute of Electrical and Electronics Engineers (IEEE): The organization of electrical and electronics scientists and engineers formed in 1963 by the merger of the Institute of Radio Engineers (IRE) and the American Institute of Electrical Engineers (AIEE).

Instrument multiplier: A measuring device that enables a high voltage to be measured using a meter with only a low voltage range.

Instrument rating: The range within which an instrument has been designed to operate without damage.

Insulate: The process of separating one conducting body from another conductor.

Insulation: The material that surrounds and insulates an electrical wire from other wires or circuits. *Insulation* may also refer to any material that does not ionize easily and thus presents a large impedance to the flow of electrical current.

Insulator: A material or device used to separate one conducting body from another.

Intelligence signal: A signal containing information.

Intensity: The strength of a given signal under specified conditions.

Interconnect cable: A short distance cable intended for use between equipment (generally less than 3 m in length).

Interface: A device or circuit used to interconnect two pieces of electronic equipment.

Interface device: A unit that joins two interconnecting systems.

Interference emission: An emission that results in an electrical signal being propagated into and interfering with the proper operation of electrical or electronic equipment.

Interlock: A protection device or system designed to remove all dangerous voltages from a machine or piece of equipment when access doors or panels are opened or removed.

Intermediate frequency: A frequency that results from combining a signal of interest with a signal generated within a radio receiver. In superheterodyne receivers, all incoming signals are converted to a single intermediate frequency for which the amplifiers and filters of the receiver have been optimized.

Intermittent: A noncontinuous recurring event, often used to denote a problem that is difficult to find because of its unpredictable nature.

Intermodulation: The production, in a nonlinear transducer element, of frequencies corresponding to the sums and differences of the fundamentals and harmonics of two or more frequencies that are transmitted through the transducer.

Intermodulation distortion (IMD): The distortion that results from the mixing of two input signals in a nonlinear system. The resulting output contains new frequencies that represent the sum and difference of the input signals and the sums and differences of their harmonics. IMD is also called *intermodulation noise.*

Intermodulation noise: In a transmission path or device, the noise signal that is contingent upon modulation and demodulation, resulting from nonlinear characteristics in the path or device.

Internal resistance: The actual resistance of a source of electric power. The total electromotive force produced by a power source is not available for external use; some of the energy is used in driving current through the source itself.

International Standards Organization (ISO): An international body concerned with worldwide standardization for a broad range of industrial products, including telecommunications equipment. Members are represented by national standards organizations, such as ANSI (American National Standards Institute) in the United States. ISO was established in 1947 as a specialized agency of the United Nations.

International Telecommunications Satellite Consortium (Intelsat): A nonprofit cooperative of member nations that owns and operates a satellite system for international and, in many instances, domestic communications.

International Telecommunications Union (ITU): A specialized agency of the United Nations established to maintain and extend international cooperation for the maintenance, development, and efficient use of telecommunications. The union does this through standards and recommended regulations, and through technical and telecommunications studies.

Interoperability: The condition achieved among communications and electronics systems or equipment when information or services can be exchanged directly between them or their users, or both.

Interpolate: The process of estimating unknown values based on a knowledge of comparable data that falls on both sides of the point in question.

Interrupting capacity: The rating of a circuit breaker or fuse that specifies the maximum current the device is designed to interrupt at its rated voltage.

Interval: The points or numbers lying between two specified endpoints.

Inverse voltage: The effective value of voltage across a rectifying device, which conducts a current in one direction during one half cycle of the alternating input, during the half cycle when current is not flowing.

Inversion: The change in the polarity of a pulse, such as from positive to negative.

Inverter: A circuit or device that converts a direct current into an alternating current.

Ionizing radiation: The form of electromagnetic radiation that can turn an atom into an ion by knocking one or more of its electrons loose. Examples of ionizing radiation include X rays, gamma rays, and cosmic rays.

IR drop: A drop in voltage because of the flow of current (I) through a resistance (R), also called *resistance drop.*

IR loss: The conversion of electrical power to heat caused by the flow of electrical current through a resistance.

Isochronous: A signal in which the time interval separating any two significant instants is theoretically equal to a specified unit interval or to an integral multiple of the unit interval.

Isolated ground: A ground circuit that is isolated from all equipment framework and any other grounds, except for a single-point external connection.

Isolated ground plane: A set of connected frames that are grounded through a single connection to a ground reference point. That point and all parts of the frames are insulated from any other ground system in a building.

Isolated pulse: A pulse uninfluenced by other pulses in the same signal.

Isophasing amplifier: A timing device that corrects for small timing errors.

Isotropic: A quantity exhibiting the same properties in all planes and directions.

J

Jack: Areceptacle or connector that makes electrical contact with the mating contacts of a plug. In combination, the plug and jack provide a ready means for making connections in electrical circuits.

Jacket: An insulating layer of material surrounding a wire in a cable.

Jitter: Small, rapid variations in a waveform resulting from fluctuations in a supply voltage or other causes.

Joule: The standard unit of work that is equal to the work done by one newton of force when the point at which the force is applied is displaced a distance of one meter in the direction of the force. The *joule* is named for the English physicist James Prescott Joule (1818–1889).

Julian date: A chronological date in which days of the year are numbered in sequence. For example, the first day is 001, the second is 002, and the last is 365 (or 366 in a leap year).

K

Kelvin (K): The standard unit of thermodynamic temperature. Zero degrees Kelvin represents *absolute zero*. Water freezes at 273 K and water boils at 373 K under standard pressure conditions.

Kilo: A prefix meaning one thousand.

Kilohertz (kHz): A unit of measure of frequency equal to 1,000 Hz.

Kilovar: A unit equal to one thousand volt-amperes.

Kilovolt (kV): A unit of measure of electrical voltage equal to 1,000 V.

Kilowatt: A unit equal to one thousand watts.

Kirchoff's Law: At any point in a circuit, there is as much current flowing into the point as there is flowing away from it.

Klystron (1—general): A family of electron tubes that function as microwave amplifiers and oscillators. Simplest in form are two-cavity klystrons in which an electron beam passes through a cavity that is excited by a microwave input, producing a velocity-modulated beam which passes through a second cavity a precise distance away that is coupled to a tuned circuit, thereby producing an amplified output of the original input signal frequency. If part of the output is fed back to the input, an oscillator can be the result. (2—**multi-cavity**) An amplifier device for UHF and microwave signals based on velocity modulation of an electron beam. The beam is directed through an input cavity, where the input RF signal polarity initializes a *bunching effect* on electrons in the beam. The bunching effect excites subsequent cavities, which increase the bunching through an energy flywheel concept. Finally, the beam passes to an output cavity that couples the amplified signal to the load (antenna system). The beam falls onto a collector element that forms the return path for the current and dissipates the heat resulting from electron beam bombardment. (3—**reflex**) A klystron with only one cavity. The action is the same as in a two-cavity klystron but the beam is reflected back into the cavity in which it was first excited, after being sent out to a reflector. The one cavity, therefore, acts both as the original exciter (or buncher) and as the collector from which the output is taken.

Knee: In a response curve, the region of maximum curvature.

Ku band: Radio frequencies in the range of 15.35 GHz to 17.25 GHz, typically used for satellite telecommunications.

L

Ladder network: A type of filter with components alternately across the line and in the line.

Lag: The difference in phase between a current and the voltage that produced it, expressed in electrical degrees.

Lagging current: A current that lags behind the alternating electromotive force that produced it. A circuit that produces a *lagging current* is one containing inductance alone, or whose effective impedance is inductive.

Lagging load: A load whose combined inductive reactance exceeds its capacitive reactance. When an alternating voltage is applied, the current lags behind the voltage.

Laminate: A material consisting of layers of the same or different materials bonded together and built up to the required thickness.

Latitude: An angular measurement of a point on the earth above or below the equator. The equator represents $0°$, the north pole $+90°$, and the south pole $-90°$.

Layout: A proposed or actual arrangement or allocation of equipment.

LC circuit: An electrical circuit with both inductance (L) and capacitance (C) that is resonant at a particular frequency.

LC ratio: The ratio of inductance to capacitance in a given circuit.

Lead: An electrical wire, usually insulated.

Leading edge: The initial portion of a pulse or wave in which voltage or current rise rapidly from zero to a final value.

Leading load: A reactive load in which the reactance of capacitance is greater than that of inductance. Current through such a load *leads* the applied voltage causing the current.

Leakage: The loss of energy resulting from the flow of electricity past an insulating material, the escape of electromagnetic radiation beyond its shielding, or the extension of magnetic lines of force beyond their intended working area.

Leakage resistance: The resistance of a path through which leakage current flows.

Level: The strength or intensity of a given signal.

Level alignment: The adjustment of transmission levels of single links and links in tandem to prevent overloading of transmission subsystems.

Life cycle: The predicted useful life of a class of equipment, operating under normal (specified) working conditions.

Life safety system: A system designed to protect life and property, such as emergency lighting, fire alarms, smoke exhaust and ventilating fans, and site security.

Life test: A test in which random samples of a product are checked to see how long they can continue to perform their functions satisfactorily. A form of *stress testing* is used, including temperature, current, voltage, and/or vibration effects, cycled at many times the rate that would apply in normal usage.

Limiter: An electronic device in which some characteristic of the output is automatically prevented from exceeding a predetermined value.

Limiter circuit: A circuit of nonlinear elements that restricts the electrical excursion of a variable in accordance with some specified criteria.

Limiting: A process by which some characteristic at the output of a device is prevented from exceeding a predetermined value.

Line loss: The total end-to-end loss in decibels in a transmission line.

Line-up: The process of adjusting transmission parameters to bring a circuit to its specified values.

Linear: A circuit, device, or channel whose output is directly proportional to its input.

Linear distortion: A distortion mechanism that is independent of signal amplitude.

Linearity: A constant relationship, over a designated range, between the input and output characteristics of a circuit or device.

Lines of force: A group of imaginary lines indicating the direction of the electric or magnetic field at all points along it.

Lissajous pattern: The looping patterns generated by a CRT spot when the horizontal (X) and vertical (Y) deflection signals are sinusoids. The lissajous pattern is useful for evaluating the delay or phase of two sinusoids of the same frequency.

Live: A device or system connected to a source of electric potential.

Load: The work required of an electrical or mechanical system.

Load factor: The ratio of the average load over a designated period of time to the peak load occurring during the same period.

Load line: A straight line drawn across a grouping of plate current/plate voltage characteristic curves showing the relationship between grid voltage and plate current for a particular plate load resistance of an electron tube.

Logarithm: The power to which a base must be raised to produce a given number. Common logarithms are to base 10.

Logarithmic scale: A meter scale with displacement proportional to the logarithm of the quantity represented.

Long persistence: The quality of a cathode ray tube that has phosphorescent compounds on its screen (in addition to fluorescent compounds) so that the image continues to glow after the original electron beam has ceased to create it by producing the usual fluorescence effect. Long persistence is often used in radar screens or where photographic evidence is needed of a display. Most such applications, however, have been superseded through the use of digital storage techniques.

Longitude: The angular measurement of a point on the surface of the earth in relation to the meridian of Greenwich (London). The earth is divided into 360° of longitude, beginning at the Greenwich mean. As one travels west around the globe, the longitude increases.

Longitudinal current: A current that travels in the same direction on both wires of a pair. The return current either flows in another pair or via a ground return path.

Loss: The power dissipated in a circuit, usually expressed in decibels, that performs no useful work.

Loss deviation: The change of actual loss in a circuit or system from a designed value.

Loss variation: The change in actual measured loss over time.

Lossy: The condition when the line loss per unit length is significantly greater than some defined normal parameter.

Lossy cable: A coaxial cable constructed to have high transmission loss so it can be used as an artificial load or as an attenuator.

Lot size: A specific quantity of similar material or a collection of similar units from a common source; in inspection work, the quantity offered for inspection and acceptance at any one time. The **lot size** may be a collection of raw material, parts, subassemblies inspected during production, or a consignment of finished products to be sent out for service.

Low tension: A low voltage circuit.

Low-pass filter: A filter network that passes all frequencies below a specified frequency with little or no loss, but that significantly attenuates higher frequencies.

Lug: A tag or projecting terminal onto which a wire may be connected by wrapping, soldering, or crimping.

Lumped constant: A resistance, inductance, or capacitance connected at a point, and not distributed uniformly throughout the length of a route or circuit.

M

MA: An abbreviation for *milliamperes* (0.001 A).

Magnet: A device that produces a magnetic field and can attract iron, and attract or repel other magnets.

Magnetic field: An energy field that exists around magnetic materials and current-carrying conductors. Magnetic fields combine with electric fields in light and radio waves.

Magnetic flux: The field produced in the area surrounding a magnet or electric current. The standard unit of flux is the *Weber*.

Magnetic flux density: A vector quantity measured by a standard unit called the *Tesla*. The *magnetic flux density* is the number of magnetic lines of force per unit area, at right angles to the lines.

Magnetic leakage: The magnetic flux that does not follow a useful path.

Magnetic pole: A point that appears from the outside to be the center of magnetic attraction or repulsion at or near one end of a magnet.

Magnetic storm: A violent local variation in the earth's magnetic field, usually the result of sunspot activity.

Magnetism: A property of iron and some other materials by which external magnetic fields are maintained, other magnets being thereby attracted or repelled.

Magnetization: The exposure of a magnetic material to a magnetizing current, field, or force.

Magnetizing force: The force producing magnetization.

Magnetomotive force: The force that tends to produce lines of force in a magnetic circuit. The *magnetomotive force* bears the same relationship to a magnetic circuit that voltage does to an electrical circuit.

Magnetron: A high-power, ultra high frequency electron tube oscillator that employs the interaction of a strong electric field between an anode and cathode with the field of a strong permanent magnet to cause oscillatory electron flow through multiple internal cavity resonators. The magnetron may operate in a continuous or pulsed mode.

Maintainability: The probability that a failure will be repaired within a specified time after the failure occurs.

Maintenance: Any activity intended to keep a functional unit in satisfactory working condition. The term includes the tests, measurements, replacements, adjustments, and repairs necessary to keep a device or system operating properly.

Malfunction: An equipment failure or a fault.

Manometer: A test device for measuring gas pressure.

Margin: The difference between the value of an operating parameter and the value that would result in unsatisfactory operation. Typical *margin* parameters include signal level, signal-to-noise ratio, distortion, crosstalk coupling, and/or undesired emission level.

Markov model: A statistical model of the behavior of a complex system over time in which the probabilities of the occurrence of various future states depend only on the present state of the system, and not on the path by which the present state was achieved. This term was named for the Russian mathematician Andrei Andreevich Markov (1856–1922).

Master clock: An accurate timing device that generates a synchronous signal to control other clocks or equipment.

Master oscillator: A stable oscillator that provides a standard frequency signal for other hardware and/or systems.

Matched termination: A termination that absorbs all the incident power and so produces no reflected waves or mismatch loss.

Matching: The connection of channels, circuits, or devices in a manner that results in minimal reflected energy.

Matrix: A logical network configured in a rectangular array of intersections of input/output signals.

Maxwell's equations: Four differential equations that relate electric and magnetic fields to electromagnetic waves. The equations are a basis of electrical and electronic engineering.

Mean: An arithmetic average in which values are added and divided by the number of such values.

Mean time between failures (MTBF): For a particular interval, the total functioning life of a population of an item divided by the total number of failures within the population during the measurement interval.

Mean time to failure (MTTF): The measured operating time of a single piece of equipment divided by the total number of failures during the measured period of time. This measurement is normally made during that period between early life and wear-out failures.

Mean time to repair (MTTR): The total corrective maintenance time on a component or system divided by the total number of corrective maintenance actions during a given period of time.

Measurement: A procedure for determining the amount of a quantity.

Median: A value in a series that has as many readings or values above it as below.

Medium: An electronic pathway or mechanism for passing information from one point to another.

Mega: A prefix meaning one million.

Megahertz (MHz): A quantity equal to one million Hertz (cycles per second).

Megohm: A quantity equal to one million ohms.

Metric system: A decimal system of measurement based on the meter, the kilogram, and the second.

Micro: A prefix meaning one millionth.

Micron: A unit of length equal to one millionth of a meter (1/25,000 of an inch).

Microphonic(s): Unintended noise introduced into an electronic system by mechanical vibration of electrical components.

Microsecond: One millionth of a second (0.000001 s).

Microvolt: A quantity equal to one-millionth of a volt.

Milli: A prefix meaning one thousandth.

Milliammeter: A test instrument for measuring electrical current, often part of a _multimeter_.

Millihenry: A quantity equal to one-thousandth of a henry.

Milliwatt: A quantity equal to one thousandth of a watt.

Minimum discernible signal: The smallest input that will produce a discernible change in the output of a circuit or device.

Mixer: A circuit used to combine two or more signals to produce a third signal that is a function of the input waveforms.

Mixing ratio: The ratio of the mass of water vapor to the mass of dry air in a given volume of air. The _mixing ratio_ affects radio propagation.

Mode: An electromagnetic field distribution that satisfies theoretical requirements for propagation in a waveguide or oscillation in a cavity.

Modified refractive index: The sum of the refractive index of the air at a given height above sea level, and the ratio of this height to the radius of the earth.

Modular: An equipment design in which major elements are readily separable, and which the user may replace, reducing the mean-time-to-repair.

Modulation: The process whereby the amplitude, frequency, or phase of a single-frequency wave (the _carrier_) is varied in step with the instantaneous value of, or samples of, a complex wave (the _modulating wave_).

Modulator: A device that enables the intelligence in an information-carrying modulating wave to be conveyed by a signal at a higher frequency. A _modulator_ modifies a carrier wave by amplitude, phase, and/or frequency as a function of a control signal that carries intelligence. Signals are _modulated_ in this way to permit more efficient and/or reliable transmission over any of several media.

Module: An assembly replaceable as an entity, often as an interchangeable plug-in item. A _module_ is not normally capable of being disassembled.

Monostable: A device that is stable in one state only. An input pulse causes the device to change state, but it reverts immediately to its stable state.

Motor: A machine that converts electrical energy into mechanical energy.

Motor effect: The repulsion force exerted between adjacent conductors carrying currents in opposite directions.

Moving coil: Any device that utilizes a coil of wire in a magnetic field in such a way that the coil is made to move by varying the applied current, or itself produces a varying voltage because of its movement.

Ms: An abbreviation for _millisecond_ (0.001 s).

Multimeter: A test instrument fitted with several ranges for measuring voltage, resistance, and current, and equipped with an analog meter or digital display readout. The _multimeter_ is also known as a _volt-ohm-milliammeter_, or _VOM_.

Multiplex (MUX): The use of a common channel to convey two or more channels. This is done either by splitting of the common channel frequency band into narrower bands, each of which is used to constitute a distinct channel (*frequency division multiplex*), or by allotting this common channel to multiple users in turn to constitute different intermittent channels (*time division multiplex*).

Multiplexer: A device or circuit that combines several signals onto a single signal.

Multiplexing: A technique that uses a single transmission path to carry multiple channels. In *time division multiplexing* (TDM), path time is shared. For *frequency division multiplexing* (FDM) or *wavelength division multiplexing* (WDM), signals are divided into individual channels sent along the same path but at different frequencies.

Multiplication: Signal mixing that occurs within a multiplier circuit.

Multiplier: A circuit in which one or more input signals are mixed under the direction of one or more control signals. The resulting output is a composite of the input signals, the characteristics of which are determined by the scaling specified for the circuit.

Mutual induction: The property of the magnetic flux around a conductor that induces a voltage in a nearby conductor. The voltage generated in the secondary conductor in turn induces a voltage in the primary conductor. The inductance of two conductors so coupled is referred to as *mutual inductance*.

MV: An abbreviation for *millivolt* (0.001 V).

MW: An abbreviation for *milliwatt* (0.001 W).

N

Nano: A prefix meaning one billionth.

Nanometer: 1×10^{-9} meter.

Nanosecond (ns): One billionth of a second (1×10^{-9} s).

Narrowband: A communications channel of restricted bandwidth, often resulting in degradation of the transmitted signal.

Narrowband emission: An emission having a spectrum exhibiting one or more sharp peaks that are narrow in width compared to the nominal bandwidth of the measuring instrument, and are far enough apart in frequency to be resolvable by the instrument.

National Electrical Code (NEC): A document providing rules for the installation of electric wiring and equipment in public and private buildings, published by the National Fire Protection Association. The NEC has been adopted as law by many states and municipalities in the U.S.

National Institute of Standards and Technology (NIST): A nonregulatory agency of the Department of Commerce that serves as a national reference and measurement laboratory for the physical and engineering sciences. Formerly called the *National Bureau of Standards*, the agency was renamed in 1988 and given the additional responsibility of aiding U.S. companies in adopting new technologies to increase their international competitiveness.

Negative: In a conductor or semiconductor material, an excess of electrons or a deficiency of positive charge.

Negative feedback: The return of a portion of the output signal from a circuit to the input but 180° out of phase. This type of feedback decreases signal amplitude but stabilizes the amplifier and reduces distortion and noise.

Negative impedance: An impedance characterized by a decrease in voltage drop across a device as the current through the device is increased, or a decrease in current through the device as the voltage across it is increased.

Neutral: A device or object having no electrical charge.

Neutral conductor: A conductor in a power distribution system connected to a point in the system that is designed to be at neutral potential. In a balanced system, the neutral conductor carries no current.

Neutral ground: An intentional ground applied to the neutral conductor or neutral point of a circuit, transformer, machine, apparatus, or system.

Newton: The standard unit of force. One *newton* is the force that, when applied to a body having a mass of 1 kg, gives it an acceleration of 1 m/s^2.

Nitrogen: A gas widely used to pressurize radio frequency transmission lines. If a small puncture occurs in the cable sheath, the nitrogen keeps moisture out so that service is not adversely affected.

Node: The points at which the current is at minimum in a transmission system in which standing waves are present.

Noise: Any random disturbance or unwanted signal in a communication system that tends to obscure the clarity or usefulness of a signal in relation to its intended use.

Noise factor (NF): The ratio of the noise power measured at the output of a receiver to the noise power that would be present at the output if the thermal noise resulting from the resistive component of the source impedance were the only source of noise in the system.

Noise figure: A measure of the noise in dB generated at the input of an amplifier, compared with the noise generated by an impedance-method resistor at a specified temperature.

Noise filter: A network that attenuates noise frequencies.

Noise generator: A generator of wideband random noise.

Noise power ratio (NPR): The ratio, expressed in decibels, of signal power to intermodulation product power plus residual noise power, measured at the baseband level.

Noise suppressor: A filter or digital signal processing circuit in a receiver or transmitter that automatically reduces or eliminates noise.

Noise temperature: The temperature, expressed in Kelvin, at which a resistor will develop a particular noise voltage. The noise temperature of a radio receiver is the value by which the temperature of the resistive component of the source impedance should be increased—if it were the only source of noise in the system—to cause the noise power at the output of the receiver to be the same as in the real system.

Nominal: The most common value for a component or parameter that falls between the maximum and minimum limits of a tolerance range.

Nominal value: A specified or intended value independent of any uncertainty in its realization.

Nomogram: A chart showing three or more scales across which a straight edge may be held in order to read off a graphical solution to a three-variable equation.

Nonconductor: A material that does not conduct energy, such as electricity, heat, or sound.

Noncritical technical load: That part of the technical power load for a facility not required for minimum acceptable operation.

Noninductive: A device or circuit without significant inductance.

Nonionizing radiation: Electromagnetic radiation that does not turn an atom into an ion. Examples of nonionizing radiation include visible light and radio waves.

Nonlinearity: A distortion in which the output of a circuit or system does not rise or fall in direct proportion to the input.

Nontechnical load: The part of the total operational load of a facility used for such purposes as general lighting, air conditioning, and ventilating equipment during normal operation.

Normal: A line perpendicular to another line or to a surface.

Normal-mode noise: Unwanted signals in the form of voltages appearing in line-to-line and line-to-neutral signals.

Normalized frequency: The ratio between the actual frequency and its nominal value.

Normalized frequency departure: The frequency departure divided by the nominal frequency value.

Normalized frequency difference: The algebraic difference between two normalized frequencies.

Normalized frequency drift: The frequency drift divided by the nominal frequency value.

Normally closed: Switch contacts that are closed in their nonoperated state, or relay contacts that are closed when the relay is de-energized.

Normally open: Switch contacts that are open in their nonoperated state, or relay contacts that are open when the relay is de-energized.

North pole: The pole of a magnet that seeks the north magnetic pole of the earth.

Notch filter: A circuit designed to attenuate a specific frequency band; also known as a *band stop filter*.

Notched noise: A noise signal in which a narrow band of frequencies has been removed.

Ns: An abbreviation for *nanosecond*.

Null: A zero or minimum amount or position.

O

Octave: Any frequency band in which the highest frequency is twice the lowest frequency.

Off-line: A condition wherein devices or subsystems are not connected into, do not form a part of, and are not subject to the same controls as an operational system.

Offset: An intentional difference between the realized value and the nominal value.

Ohm: The unit of electric resistance through which one ampere of current will flow when there is a difference of one volt. The quantity is named for the German physicist Georg Simon Ohm (1787–1854).

Ohm's law: A law that sets forth the relationship between voltage (E), current (I), and resistance (R). The law states that $E = I \times R$, $I = \frac{E}{R}$, and $R = \frac{E}{I}$. *Ohm's Law* is named for the German physicist Georg Simon Ohm (1787–1854).

Ohmic loss: The power dissipation in a line or circuit caused by electrical resistance.

Ohmmeter: A test instrument used for measuring resistance, often part of a *multimeter*.

Ohms-per-volt: A measure of the sensitivity of a voltmeter.

On-line: A device or system that is energized and operational, and ready to perform useful work.

Open: An interruption in the flow of electrical current, as caused by a broken wire or connection.

Open-circuit: A defined loop or path that closes on itself and contains an infinite impedance.

Open-circuit impedance: The input impedance of a circuit when its output terminals are open, that is, not terminated.

Open-circuit voltage: The voltage measured at the terminals of a circuit when there is no load and, hence, no current flowing.

Operating lifetime: The period of time during which the principal parameters of a component or system remain within a prescribed range.

Optimize: The process of adjusting for the best output or maximum response from a circuit or system.

Orbit: The path, relative to a specified frame of reference, described by the center of mass of a satellite or other object in space, subjected solely to natural forces (mainly gravitational attraction).

Order of diversity: The number of independently fading propagation paths or frequencies, or both, used in a diversity reception system.

Original equipment manufacturer (OEM): A manufacturer of equipment that is used in systems assembled and sold by others.

Oscillation: A variation with time of the magnitude of a quantity with respect to a specified reference when the magnitude is alternately greater than and smaller than the reference.

Oscillator: A nonrotating device for producing alternating current, the output frequency of which is determined by the characteristics of the circuit.

Oscilloscope: A test instrument that uses a display, usually a cathode-ray tube, to show the instantaneous values and waveforms of a signal that varies with time or some other parameter.

Out-of-band energy: Energy emitted by a transmission system that falls outside the frequency spectrum of the intended transmission.

Outage duration: The average elapsed time between the start and the end of an outage period.

Outage probability: The probability that an outage state will occur within a specified time period. In the absence of specific known causes of outages, the *outage probability* is the sum of all outage durations divided by the time period of measurement.

Outage threshold: A defined value for a supported performance parameter that establishes the minimum operational service performance level for that parameter.

Output impedance: The impedance presented at the output terminals of a circuit, device, or channel.

Output stage: The final driving circuit in a piece of electronic equipment.

Ovenized crystal oscillator (OXO): A crystal oscillator enclosed within a temperature regulated heater (oven) to maintain a stable frequency despite external temperature variations.

Overcoupling: A degree of coupling greater than the *critical coupling* between two resonant circuits. *Overcoupling* results in a wide bandwidth circuit with two peaks in the response curve.

Overload: In a transmission system, a power greater than the amount the system was designed to carry. In a power system, an overload could cause excessive heating. In a communications system, distortion of a signal could result.

Overshoot: The first maximum excursion of a pulse beyond the 100% level. Overshoot is the portion of the pulse that exceeds its defined level temporarily before settling to the correct level. Overshoot amplitude is expressed as a percentage of the defined level.

P

Pentode: An electron tube with five electrodes, the cathode, control grid, screen grid, suppressor grid, and plate.

Photocathode: An electrode in an electron tube that will emit electrons when bombarded by photons of light.

Picture tube: A cathode-ray tube used to produce an image by variation of the intensity of a scanning beam on a phosphor screen.

Pin: A terminal on the base of a component, such as an electron tube.

Plasma (1—arc): An ionized gas in an arc-discharge tube that provides a conducting path for the discharge. **(2—solar)** The ionized gas at extremely high temperature found in the sun.

Plate (1—electron tube): The anode of an electron tube. **(2—battery)** An electrode in a storage battery. **(3—capacitor)** One of the surfaces in a capacitor. **(4—chassis)** A mounting surface to which equipment may be fastened.

Propagation time delay: The time required for a signal to travel from one point to another.

Protector: A device or circuit that prevents damage to lines or equipment by conducting dangerously high voltages or currents to ground. Protector types include spark gaps, semiconductors, varistors, and gas tubes.

Proximity effect: A nonuniform current distribution in a conductor, caused by current flow in a nearby conductor.

Pseudonoise: In a spread-spectrum system, a seemingly random series of pulses whose frequency spectrum resembles that of continuous noise.

Pseudorandom: A sequence of signals that appears to be completely random but have, in fact, been carefully drawn up and repeat after a significant time interval.

Pseudorandom noise: A noise signal that satisfies one or more of the standard tests for statistical randomness. Although it seems to lack any definite pattern, there is a sequence of pulses that repeats after a long time interval.

Pseudorandom number sequence: A sequence of numbers that satisfies one or more of the standard tests for statistical randomness. Although it seems to lack any definite pattern, there is a sequence that repeats after a long time interval.

Pulsating direct current: A current changing in value at regular or irregular intervals but which has the same direction at all times.

Pulse: One of the elements of a repetitive signal characterized by the rise and decay in time of its magnitude. A *pulse* is usually short in relation to the time span of interest.

Pulse decay time: The time required for the trailing edge of a pulse to decrease from 90 percent to 10 percent of its peak amplitude.

Pulse duration: The time interval between the points on the leading and trailing edges of a pulse at which the instantaneous value bears a specified relation to the peak pulse amplitude.

Pulse duration modulation (PDM): The modulation of a pulse carrier by varying the width of the pulses according to the instantaneous values of the voltage samples of the modulating signal (also called *pulse width modulation*).

Pulse edge: The leading or trailing edge of a pulse, defined as the 50 percent point of the pulse rise or fall time.

Pulse fall time: The interval of time required for the edge of a pulse to fall from 90 percent to 10 percent of its peak amplitude.

Pulse interval: The time between the start of one pulse and the start of the next.

Pulse length: The duration of a pulse (also called *pulse width*).

Pulse level: The voltage amplitude of a pulse.

Pulse period: The time between the start of one pulse and the start of the next.

Pulse ratio: The ratio of the length of any pulse to the total pulse period.

Pulse repetition period: The time interval from the beginning of one pulse to the beginning of the next pulse.

Pulse repetition rate: The number of times each second that pulses are transmitted.

Pulse rise time: The time required for the leading edge of a pulse to rise from 10 percent to 90 percent of its peak amplitude.

Pulse train: A series of pulses having similar characteristics.

Pulse width: The measured interval between the 50 percent amplitude points of the leading and trailing edges of a pulse.

Puncture: A breakdown of insulation or of a dielectric, such as in a cable sheath or in the insulant around a conductor.

PW: An abbreviation for picowatt, a unit of power equal to 10^{-12} W (-90 dBm).

Q

Q(quality factor): A figure of merit that defines how close a coil comes to functioning as a pure inductor. *High Q* describes an inductor with little energy loss resulting from resistance. *Q* is found by dividing the inductive reactance of a device by its resistance.

Quadrature: A state of alternating current signals separated by one quarter of a cycle (90°).

Quadrature amplitude modulation (QAM): A process that allows two different signals to modulate a single carrier frequency. The two signals of interest amplitude modulate two samples of the carrier that are of the same frequency, but differ in phase by 90°. The two resultant signals can be added and transmitted. Both signals may be recovered at a decoder when they are demodulated 90° apart.

Quadrature component: The component of a voltage or current at an angle of 90° to a reference signal, resulting from inductive or capacitive reactance.

Quadrature phase shift keying (QPSK): A type of phase shift keying using four phase states.

Quality: The absence of objectionable distortion.

Quality assurance (QA): All those activities, including surveillance, inspection, control, and documentation, aimed at ensuring that a given product will meet its performance specifications.

Quality control (QC): A function whereby management exercises control over the quality of raw material or intermediate products in order to prevent the production of defective devices or systems.

Quantum noise: Any noise attributable to the discrete nature of electromagnetic radiation. Examples include shot noise, photon noise, and recombination noise.

Quantum-limited operation: An operation wherein the minimum detectable signal is limited by quantum noise.

Quartz: A crystalline mineral that when electrically excited vibrates with a stable period. Quartz is typically used as the frequency-determining element in oscillators and filters.

Quasi-peak detector: A detector that delivers an output voltage that is some fraction of the peak value of the regularly repeated pulses applied to it. The fraction increases toward unity as the pulse repetition rate increases.

Quick-break fuse: A fuse in which the fusible link is under tension, providing for rapid operation.

Quiescent: An inactive device, signal, or system.

Quiescent current: The current that flows in a device in the absence of an applied signal.

R

Rack: An equipment rack, usually measuring 19 in (48.26 cm) wide at the front mounting rails.

Rack unit (RU): A unit of measure of vertical space in an equipment enclosure. One rack unit is equal to 1.75 in (4.45 cm).

Radiate: The process of emitting electromagnetic energy.

Radiation: The emission and propagation of electromagnetic energy in the form of waves. *Radiation* is also called *radiant energy*.

Radiation scattering: The diversion of thermal, electromagnetic, or nuclear radiation from its original path as a result of interactions or collisions with atoms, molecules, or large particles in the atmosphere or other media between the source of radiation and a point some distance away. As a result of scattering, radiation (especially gamma rays and neutrons) will be received at such a point from many directions, rather than only from the direction of the source.

Radio: The transmission of signals over a distance by means of electromagnetic waves in the approximate frequency range of 150 kHz to 300 GHz. The term may also be used to describe the equipment used to transmit or receive electromagnetic waves.

Radio detection: The detection of the presence of an object by radio location without precise determination of its position.

Radio frequency interference (RFI): The intrusion of unwanted signals or electromagnetic noise into various types of equipment resulting from radio frequency transmission equipment or other devices using radio frequencies.

Radio frequency spectrum: Those frequency bands in the electromagnetic spectrum that range from several hundred thousand cycles per second (*very low frequency*) to several billion cycles per second (*microwave frequencies*).

Radio recognition: In military communications, the determination by radio means of the "friendly" or "unfriendly" character of an aircraft or ship.

Random noise: Electromagnetic signals that originate in transient electrical disturbances and have random time and amplitude patterns. Random noise is generally undesirable; however, it may also be generated for testing purposes.

Rated output power: The power available from an amplifier or other device under specified conditions of operation.

RC constant: The time constant of a resistor-capacitor circuit. The *RC constant* is the time in seconds required for current in an RC circuit to rise to 63 percent of its final steady value or fall to 37 percent of its original steady value, obtained by multiplying the resistance value in ohms by the capacitance value in farads.

RC network: A circuit that contains resistors and capacitors, normally connected in series.

Reactance: The part of the impedance of a network resulting from inductance or capacitance. The *reactance* of a component varies with the frequency of the applied signal.

Reactive power: The power circulating in an ac circuit. It is delivered to the circuit during part of the cycle and is returned during the other half of the cycle. The *reactive power* is obtained by multiplying the voltage, current, and the sine of the phase angle between them.

Reactor: A component with inductive reactance.

Received signal level (RSL): The value of a specified bandwidth of signals at the receiver input terminals relative to an established reference.

Receiver: Any device for receiving electrical signals and converting them to audible sound, visible light, data, or some combination of these elements.

Receptacle: An electrical socket designed to receive a mating plug.

Reception: The act of receiving, listening to, or watching information-carrying signals.

Rectification: The conversion of alternating current into direct current.

Rectifier: A device for converting alternating current into direct current. A *rectifier* normally includes filters so that the output is, within specified limits, smooth and free of ac components.

Rectify: The process of converting alternating current into direct current.

Redundancy: A system design that provides a back-up for key circuits or components in the event of a failure. Redundancy improves the overall reliability of a system.

Redundant: A configuration when two complete systems are available at one time. If the online system fails, the backup will take over with no loss of service.

Reference voltage: A voltage used for control or comparison purposes.

Reflectance: The ratio of reflected power to incident power.

Reflection: An abrupt change, resulting from an impedance mismatch, in the direction of propagation of an electromagnetic wave. For light, at the interface of two dissimilar materials, the incident wave is returned to its medium of origin.

Reflection coefficient: The ratio between the amplitude of a reflected wave and the amplitude of the incident wave. For large smooth surfaces, the reflection coefficient may be near unity.

Reflection gain: The increase in signal strength that results when a reflected wave combines, in phase, with an incident wave.

Reflection loss: The apparent loss of signal strength caused by an impedance mismatch in a transmission line or circuit. The loss results from the reflection of part of the signal back toward the source from the point of the impedance discontinuity. The greater the mismatch, the greater the loss.

Reflectometer: A device that measures energy traveling in each direction in a waveguide, used in determining the standing wave ratio.

Refraction: The bending of a sound, radio, or light wave as it passes obliquely from a medium of one density to a medium of another density that varies its speed.

Regulation: The process of adjusting the level of some quantity, such as circuit gain, by means of an electronic system that monitors an output and feeds back a controlling signal to constantly maintain a desired level.

Regulator: A device that maintains its output voltage at a constant level.

Relative envelope delay: The difference in envelope delay at various frequencies when compared with a reference frequency that is chosen as having zero delay.

Relative humidity: The ratio of the quantity of water vapor in the atmosphere to the quantity that would cause saturation at the ambient temperature.

Relative transmission level: The ratio of the signal power in a transmission system to the signal power at some point chosen as a reference. The ratio is usually determined by applying a standard test signal at the input to the system and measuring the gain or loss at the location of interest.

Relay: A device by which current flowing in one circuit causes contacts to operate that control the flow of current in another circuit.

Relay armature: The movable part of an electromechanical relay, usually coupled to spring sets on which contacts are mounted.

Relay bypass: A device that, in the event of a loss of power or other failure, routes a critical signal around the equipment that has failed.

Release time: The time required for a pulse to drop from steady-state level to zero, also referred to as the *decay time*.

Reliability: The ability of a system or subsystem to perform within the prescribed parameters of quality of service. *Reliability* is often expressed as the probability that a system or subsystem will perform its intended function for a specified interval under stated conditions.

Reliability growth: The action taken to move a hardware item toward its reliability potential, during development or subsequent manufacturing or operation.

Reliability predictions: The compiled failure rates for parts, components, subassemblies, assemblies, and systems. These generic failure rates are used as basic data to predict the reliability of a given device or system.

Remote control: A system used to control a device from a distance.

Remote station: A station or terminal that is physically remote from a main station or center but can gain access through a communication channel.

Repeater: The equipment between two circuits that receives a signal degraded by normal factors during transmission and amplifies the signal to its original level for retransmission.

Repetition rate: The rate at which regularly recurring pulses are repeated.

Reply: A transmitted message that is a direct response to an original message.

Repulsion: The mechanical force that tends to separate like magnetic poles, like electric charges, or conductors carrying currents in opposite directions.

Reset: The act of restoring a device to its default or original state.

Residual flux: The magnetic flux that remains after a magnetomotive force has been removed.

Residual magnetism: The magnetism or flux that remains in a core after current ceases to flow in the coil producing the magnetomotive force.

Residual voltage: The vector sum of the voltages in all the phase wires of an unbalanced polyphase power system.

Resistance: The opposition of a material to the flow of electrical current. Resistance is equal to the voltage drop through a given material divided by the current flow through it. The standard unit of resistance is the *ohm*, named for the German physicist Georg Simon Ohm (1787–1854).

Resistance drop: The fall in potential (volts) between two points, the product of the current and resistance.

Resistance-grounded: A circuit or system grounded for safety through a resistance, which limits the value of the current flowing through the circuit in the event of a fault.

Resistive load: A load in which the voltage is in phase with the current.

Resistivity: The resistance per unit volume or per unit area.

Resistor: A device the primary function of which is to introduce a specified resistance into an electrical circuit.

Resonance: A tuned condition conducive to oscillation, when the reactance resulting from capacitance in a circuit is equal in value to the reactance resulting from inductance.

Resonant frequency: The frequency at which the inductive reactance and capacitive reactance of a circuit are equal.

Resonator: A resonant cavity.

Return: A return path for current, sometimes through ground.

Reversal: A change in magnetic polarity, in the direction of current flow.

Reverse current: A small current that flows through a diode when the voltage across it is such that normal forward current does not flow.

Reverse voltage: A voltage in the reverse direction from that normally applied.

Rheostat: A two-terminal variable resistor, usually constructed with a sliding or rotating shaft that can be used to vary the resistance value of the device.

Ripple: An ac voltage superimposed on the output of a dc power supply, usually resulting from imperfect filtering.

Rise time: The time required for a pulse to rise from 10 percent to 90 percent of its peak value.

Roll-off: A gradual attenuation of gain-frequency response at either or both ends of a transmission pass band.

Root-mean-square (RMS): The square root of the average value of the squares of all the instantaneous values of current or voltage during one half-cycle of an alternating current. For an alternating current, the RMS voltage or current is equal to the amount of direct current or voltage that would produce the same heating effect in a purely resistive circuit. For a sinewave, the root-mean-square value is equal to 0.707 times the peak value. RMS is also called the *effective value*.

Rotor: The rotating part of an electric generator or motor.

RU: An abbreviation for *rack unit*.

S

Scan: One sweep of the target area in a camera tube, or of the screen in a picture tube.

Screen grid: A grid in an electron tube that improves performance of the device by shielding the control grid from the plate.

Self-bias: The provision of bias in an electron tube through a voltage drop in the cathode circuit.

Shot noise: The noise developed in a vacuum tube or photoconductor resulting from the random number and velocity of emitted charge carriers.

Slope: The rate of change, with respect to frequency, of transmission line attenuation over a given frequency spectrum.

Slope equalizer: A device or circuit used to achieve a specified slope in a transmission line.

Smoothing circuit: A filter designed to reduce the amount of ripple in a circuit, usually a dc power supply.

Snubber: An electronic circuit used to suppress high frequency noise.

Solar wind: Charged particles from the sun that continuously bombard the surface of the earth.

Solid: A single wire conductor, as contrasted with a stranded, braided, or rope-type wire.

Solid-state: The use of semiconductors rather than electron tubes in a circuit or system.

Source: The part of a system from which signals or messages are considered to originate.

Source terminated: A circuit whose output is terminated for correct impedance matching with standard cable.

Spare: A system that is available but not presently in use.

Spark gap: A gap between two electrodes designed to produce a spark under given conditions.

Specific gravity: The ratio of the weight of a volume, liquid, or solid to the weight of the same volume of water at a specified temperature.

Spectrum: A continuous band of frequencies within which waves have some common characteristics.

Spectrum analyzer: A test instrument that presents a graphic display of signals over a selected frequency bandwidth. A cathode-ray tube is often used for the display.

Spectrum designation of frequency: A method of referring to a range of communication frequencies. In American practice, the designation is a two or three letter acronym for the name. The ranges are: below 300 Hz, ELF (extremely low frequency); 300 Hz–3000 Hz, ILF (infra low frequency); 3 kHz–30 kHz, VLF (very low frequency); 30 kHz–300 kHz, LF (low frequency); 300 kHz–3000 kHz, MF (medium frequency); 3 MHz–30 MHz, HF (high frequency); 30 MHz–300 MHz, VHF (very high frequency); 300 MHz–3000 MHz, UHF (ultra high frequency); 3 GHz–30 GHz, SHF (super high frequency); 30 GHz–300 GHz, EHF (extremely high frequency); 300 GHz–3000 GHz, THF (tremendously high frequency).

Spherical antenna: A type of satellite receiving antenna that permits more than one satellite to be accessed at any given time. A spherical antenna has a broader angle of acceptance than a parabolic antenna.

Spike: A high amplitude, short duration pulse superimposed on an otherwise regular waveform.

Split-phase: A device that derives a second phase from a single phase power supply by passing it through a capacitive or inductive reactor.

Splitter: A circuit or device that accepts one input signal and distributes it to several outputs.

Splitting ratio: The ratio of the power emerging from the output ports of a coupler.

Sporadic: An event occurring at random and infrequent intervals.

Spread spectrum: A communications technique in which the frequency components of a narrowband signal are spread over a wide band. The resulting signal resembles white noise. The technique is used to achieve signal security and privacy, and to enable the use of a common band by many users.

Spurious signal: Any portion of a given signal that is not part of the fundamental waveform. Spurious signals include transients, noise, and hum.

Square wave: A square or rectangular-shaped periodic wave that alternately assumes two fixed values for equal lengths of time, the transition being negligible in comparison with the duration of each fixed value.

Square wave testing: The use of a square wave containing many odd harmonics of the fundamental frequency as an input signal to a device. Visual examination of the output signal on an oscilloscope indicates the amount of distortion introduced.

Stability: The ability of a device or circuit to remain stable in frequency, power level, and/or other specified parameters.

Standard: The specific signal configuration, reference pulses, voltage levels, and other parameters that describe the input/output requirements for a particular type of equipment.

Standard time and frequency signal: A time-controlled radio signal broadcast at scheduled intervals on a number of different frequencies by government-operated radio stations to provide a method for calibrating instruments.

Standing wave ratio (SWR): The ratio of the maximum to the minimum value of a component of a wave in a transmission line or waveguide, such as the maximum voltage to the minimum voltage.

Static charge: An electric charge on the surface of an object, particularly a dielectric.

Station: One of the input or output points in a communications system.

Stator: The stationary part of a rotating electric machine.

Status: The present condition of a device.

Statute mile: A unit of distance equal to 1,609 km or 5,280 ft.

Steady-state: A condition in which circuit values remain essentially constant, occurring after all initial transients or fluctuating conditions have passed.

Steady-state condition: A condition occurring after all initial transient or fluctuating conditions have damped out in which currents, voltages, or fields remain essentially constant or oscillate uniformly without changes in characteristics such as amplitude, frequency, or wave shape.

Steep wavefront: A rapid rise in voltage of a given signal, indicating the presence of high frequency odd harmonics of a fundamental wave frequency.

Step up (or down): The process of increasing (or decreasing) the voltage of an electrical signal, as in a step-up (or step-down) transformer.

Straight-line capacitance: A capacitance employing a variable capacitor with plates so shaped that capacitance varies directly with the angle of rotation.

Stray capacitance: An unintended—and usually undesired—capacitance between wires and components in a circuit or system.

Stray current: A current through a path other than the intended one.

Stress: The force per unit of cross-sectional area on a given object or structure.

Subassembly: A functional unit of a system.

Subcarrier (SC): A carrier applied as modulation on another carrier, or on an intermediate subcarrier.

Subharmonic: A frequency equal to the fundamental frequency of a given signal divided by a whole number.

Submodule: A small circuit board or device that mounts on a larger module or device.

Subrefraction: A refraction for which the refractivity gradient is greater than standard.

Subsystem: A functional unit of a system.

Superheterodyne receiver: A radio receiver in which all signals are first converted to a common frequency for which the intermediate stages of the receiver have been optimized, both for tuning and filtering. Signals are converted by mixing them with the output of a local oscillator whose output is varied in accordance with the frequency of the received signals so as to maintain the desired *intermediate frequency*.

Suppressor grid: The fifth grid of a pentode electron tube, which provides screening between plate and screen grid.

Surface leakage: A leakage current from line to ground over the face of an insulator supporting an open wire route.

Surface refractivity: The refractive index, calculated from observations of pressure, temperature, and humidity at the surface of the earth.

Surge: A rapid rise in current or voltage, usually followed by a fall back to the normal value.

Susceptance: The reciprocal of reactance, and the imaginary component of admittance, expressed in siemens.

Sweep: The process of varying the frequency of a signal over a specified bandwidth.

Sweep generator: A test oscillator, the frequency of which is constantly varied over a specified bandwidth.

Switching: The process of making and breaking (connecting and disconnecting) two or more electrical circuits.

Synchronization: The process of adjusting the corresponding significant instants of signals—for example, the zero-crossings—to make them synchronous. The term *synchronization* is often abbreviated as *sync*.

Synchronize: The process of causing two systems to operate at the same speed.

Synchronous: In step or in phase, as applied to two or more devices; a system in which all events occur in a predetermined timed sequence.

Synchronous detection: A demodulation process in which the original signal is recovered by multiplying the modulated signal by the output of a synchronous oscillator locked to the carrier.

Synchronous system: A system in which the transmitter and receiver are operating in a fixed time relationship.

System standards: The minimum required electrical performance characteristics of a specific collection of hardware and/or software.

Systems analysis: An analysis of a given activity to determine precisely what must be accomplished and how it is to be done.

T

Tetrode: A four element electron tube consisting of a cathode, control grid, screen grid, and plate.

Thyratron: A gas-filled electron tube in which plate current flows when the grid voltage reaches a predetermined level. At that point, the grid has no further control over the current, which continues to flow until it is interrupted or reversed.

Tolerance: The permissible variation from a standard.

Torque: A moment of force acting on a body and tending to produce rotation about an axis.

Total harmonic distortion (THD): The ratio of the sum of the amplitudes of all signals harmonically related to the fundamental versus the amplitude of the fundamental signal. THD is expressed in percent.

Trace: The pattern on an oscilloscope screen when displaying a signal.

Tracking: The locking of tuned stages in a radio receiver so that all stages are changed appropriately as the receiver tuning is changed.

Trade-off: The process of weighing conflicting requirements and reaching a compromise decision in the design of a component or a subsystem.

Transceiver: Any circuit or device that receives and transmits signals.

Transconductance: The mutual conductance of an electron tube expressed as the change in plate current divided by the change in control grid voltage that produced it.

Transducer: A device that converts energy from one form to another.

Transfer characteristics: The intrinsic parameters of a system, subsystem, or unit of equipment which, when applied to the input of the system, subsystem, or unit of equipment, will fully describe its output.

Transformer: A device consisting of two or more windings wrapped around a single core or linked by a common magnetic circuit.

Transformer ratio: The ratio of the number of turns in the secondary winding of a transformer to the number of turns in the primary winding, also known as the *turns ratio*.

Transient: A sudden variance of current or voltage from a steady-state value. A transient normally results from changes in load or effects related to switching action.

Transient disturbance: A voltage pulse of high energy and short duration impressed upon the ac waveform. The overvoltage pulse can be one to 100 times the normal ac potential (or more) and can last up to 15 ms. Rise times measure in the nanosecond range.

Transient response: The time response of a system under test to a stated input stimulus.

Transition: A sequence of actions that occurs when a process changes from one state to another in response to an input.

Transmission: The transfer of electrical power, signals, or an intelligence from one location to another by wire, fiber optic, or radio means.

Transmission facility: A transmission medium and all the associated equipment required to transmit information.

Transmission loss: The ratio, in decibels, of the power of a signal at a point along a transmission path to the power of the same signal at a more distant point along the same path. This value is often used as a measure of the quality of the transmission medium for conveying signals. Changes in power level are normally expressed in decibels by calculating ten times the logarithm (base 10) of the ratio of the two powers.

Transmission mode: One of the field patterns in a waveguide in a plane transverse to the direction of propagation.

Transmission system: The set of equipment that provides single or multichannel communications facilities capable of carrying audio, video, or data signals.

Transmitter: The device or circuit that launches a signal into a passive medium, such as the atmosphere.

Transparency: The property of a communications system that enables it to carry a signal without altering or otherwise affecting the electrical characteristics of the signal.

Tray: The metal cabinet that holds circuit boards.

Tremendously high frequency (THF): The frequency band from 300 GHz to 3000 GHz.

Triangular wave: An oscillation, the values of which rise and fall linearly, and immediately change upon reaching their peak maximum and minimum. A graphical representation of a triangular wave resembles a triangle.

Trim: The process of making fine adjustments to a circuit or a circuit element.

Trimmer: A small mechanically-adjustable component connected in parallel or series with a major component so that the net value of the two can be finely adjusted for tuning purposes.

Triode: A three-element electron tube, consisting of a cathode, control grid, and plate.

Triple beat: A third-order beat whose three beating carriers all have different frequencies, but are spaced at equal frequency separations.

Troposphere: The layer of the earth's atmosphere, between the surface and the stratosphere, in which about 80 percent of the total mass of atmospheric air is concentrated and in which temperature normally decreases with altitude.

Trouble: A failure or fault affecting the service provided by a system.

Troubleshoot: The process of investigating, localizing, and (if possible) correcting a fault.

Tube (1—electron): An evacuated or gas-filled tube enclosed in a glass or metal case in which the electrodes are maintained at different voltages, giving rise to a controlled flow of electrons from the cathode to the anode. **(2—cathode ray, CRT)** An electron beam tube used for the display of changing electrical phenomena, generally similar to a television picture tube. **(3—cold-cathode)** An electron tube whose cathode emits electrons without the need of a heating filament. **(4—gas)** A gas-filled electron tube in which the gas plays an essential role in operation of the device. **(5—mercury-vapor)** A tube filled with mercury vapor at low pressure, used as a rectifying device. **(6—metal)** An electron tube enclosed in a metal case. **(7—traveling wave, TWT)** A wide band microwave amplifier in which a stream of electrons interacts with a guided electromagnetic wave moving substantially in synchronism with the electron stream, resulting in a net transfer of energy from the electron stream to the wave. **(8—velocity-modulated)** An electron tube in which the velocity of the electron stream is continually changing, as in a klystron.

Tune: The process of adjusting the frequency of a device or circuit, such as for resonance or for maximum response to an input signal.

Tuned trap: A series resonant network bridged across a circuit that eliminates ("traps") the frequency of the resonant network.

Tuner: The radio frequency and intermediate frequency parts of a radio receiver that produce a low level output signal.

Tuning: The process of adjusting a given frequency; in particular, to adjust for resonance or for maximum response to a particular incoming signal.

Turns ratio: In a transformer, the ratio of the number of turns on the secondary to the number of turns on the primary.

Tweaking: The process of adjusting an electronic circuit to optimize its performance.

Twin-line: A feeder cable with two parallel, insulated conductors.

Two-phase: A source of alternating current circuit with two sinusoidal voltages that are 90° apart.

U

Ultra high frequency (UHF): The frequency range from 300 MHz to 3000 MHz.

Ultraviolet radiation: Electromagnetic radiation in a frequency range between visible light and high-frequency X-rays.

Unattended: A device or system designed to operate without a human attendant.

Unattended operation: A system that permits a station to receive and transmit messages without the presence of an attendant or operator.

Unavailability: A measure of the degree to which a system, subsystem, or piece of equipment is not operable and not in a committable state at the start of a mission, when the mission is called for at a random point in time.

Unbalanced circuit: A two-wire circuit with legs that differ from one another in resistance, capacity to earth or to other conductors, leakage, or inductance.

Unbalanced line: A transmission line in which the magnitudes of the voltages on the two conductors are not equal with respect to ground. A coaxial cable is an example of an unbalanced line.

Unbalanced modulator: A modulator whose output includes the carrier signal.

Unbalanced output: An output with one leg at ground potential.

Unbalanced wire circuit: A circuit whose two sides are inherently electrically unlike.

Uncertainty: An expression of the magnitude of a possible deviation of a measured value from the true value. Frequently, it is possible to distinguish two components: the *systematic uncertainty* and the *random uncertainty*. The random uncertainty is expressed by the standard deviation or by a multiple of the standard deviation. The systematic uncertainty is generally estimated on the basis of the parameter characteristics.

Undamped wave: A signal with constant amplitude.

Underbunching: A condition in a traveling wave tube wherein the tube is not operating at its optimum bunching rate.

Underwriters Laboratories, Inc.: A laboratory established by the National Board of Fire Underwriters which tests equipment, materials, and systems that may affect insurance risks, with special attention to fire dangers and other hazards to life.

Ungrounded: A circuit or line not connected to ground.

Unicoupler: A device used to couple a balanced circuit to an unbalanced circuit.

Unidirectional: A signal or current flowing in one direction only.

Uniform transmission line: A transmission line with electrical characteristics that are identical, per unit length, over its entire length.

Unit: An assembly of equipment and associated wiring that together forms a complete system or independent subsystem.

Unity coupling: In a theoretically perfect transformer, complete electromagnetic coupling between the primary and secondary windings with no loss of power.

Unity gain: An amplifier or active circuit in which the output amplitude is the same as the input amplitude.

Unity power factor: A power factor of 1.0, which means that the load is—in effect—a pure resistance, with ac voltage and current completely in phase.

Unterminated: A device or system that is not terminated.

Up-converter: A frequency translation device in which the frequency of the output signal is greater than that of the input signal. Such devices are commonly found in microwave radio and satellite systems.

Uplink: A transmission system for sending radio signals from the ground to a satellite or aircraft.

Upstream: A device or system placed ahead of other devices or systems in a signal path.

Useful life: The period during which a low, constant failure rate can be expected for a given device or system. The *useful life* is the portion of a product life cycle between break-in and wear out.

User: A person, organization, or group that employs the services of a system for the transfer of information or other purposes.

V

VA: An abbreviation for *volt-amperes*, volts times amperes.

Vacuum relay: A relay the contacts of which are enclosed in an evacuated space, usually to provide reliable long-term operation.

Vacuum switch: A switch the contacts of which are enclosed in an evacuated container so that spark formation is discouraged.

Vacuum tube: An electron tube. The most common vacuum tubes include the diode, triode, tetrode, and pentode.

Validity check: A test designed to ensure that the quality of transmission is maintained over a given system.

Varactor: A semiconductor that behaves like a capacitor under the influence of an external control voltage.

Varactor diode: A semiconductor device with capacitance as a function of the applied voltage. A varactor diode, also called a *variable reactance diode* or simply a *varactor*, is often used to tune the operating frequency of a radio circuit.

Variable frequency oscillator (VFO): An oscillator the frequency of which can be set to any required value in a given range of frequencies.

Variable impedance: A capacitor, inductor, or resistor that is adjustable in value.

Variable-gain amplifier: An amplifier the gain of which can be controlled by an external signal source.

Variable-reluctance: A transducer in which the input (usually a mechanical movement) varies the magnetic reluctance of a device.

Variation monitor: A device used for sensing a deviation in voltage, current, or frequency, which is capable of providing an alarm and/or initiating transfer to another power source when programmed limits of voltage, frequency, current, or time are exceeded.

Varicap: A diode used as a variable capacitor.

VCXO (voltage controlled crystal oscillator): A device the output frequency of which is determined by an input control voltage.

Vector: A quantity having both magnitude and direction.

Vector diagram: A diagram using vectors to indicate the relationship between voltage and current in a circuit.

Vector sum: The sum of two vectors which, when they are at right angles to each other, equal the length of the hypotenuse of the right triangle so formed. In the general case, the vector sum of the two vectors equals the diagonal of the parallelogram formed on the two vectors.

Velocity of light: The speed of propagation of electromagnetic waves in a vacuum, equal to 299,792,458 m/s, or approximately 186,000 mi/s. For rough calculations, the figure of 300,000 km/s is used.

Velocity of propagation: The velocity of signal transmission. In free space, electromagnetic waves travel at the speed of light. In a cable, the velocity is substantially lower.

Vernier: A device that enables precision reading of a measuring set or gauge, or the setting of a dial with precision.

Very low frequency (VLF): A radio frequency in the band 3 kHz to 30 kHz.

Vestigial sideband: A form of transmission in which one sideband is significantly attenuated. The carrier and the other sideband are transmitted without attenuation.

Vibration testing: A testing procedure whereby subsystems are mounted on a test base that vibrates, thereby revealing any faults resulting from badly soldered joints or other poor mechanical design features.

Volt: The standard unit of electromotive force, equal to the potential difference between two points on a conductor that is carrying a constant current of one ampere when the power dissipated between the two points is equal to one watt. One *volt* is equivalent to the potential difference across a resistance of one ohm when one ampere is flowing through it. The volt is named for the Italian physicist Alessandro Volta (1745–1827).

Volt-ampere (VA): The apparent power in an ac circuit (volts times amperes).

Volt-ohm-milliammeter (VOM): A general purpose multirange test meter used to measure voltage, resistance, and current.

Voltage: The potential difference between two points.

Voltage drop: A decrease in electrical potential resulting from current flow through a resistance.

Voltage gradient: The continuous drop in electrical potential, per unit length, along a uniform conductor or thickness of a uniform dielectric.

Voltage level: The ratio of the voltage at a given point to the voltage at an arbitrary reference point.

Voltage reference circuit: A stable voltage reference source.

Voltage regulation: The deviation from a nominal voltage, expressed as a percentage of the nominal voltage.

Voltage regulator: A circuit used for controlling and maintaining a voltage at a constant level.

Voltage stabilizer: A device that produces a constant or substantially constant output voltage despite variations in input voltage or output load current.

Voltage to ground: The voltage between any given portion of a piece of equipment and the ground potential.

Voltmeter: An instrument used to measure differences in electrical potential.

Vox: A voice-operated relay circuit that permits the equivalent of push-to-talk operation of a transmitter by the operator.

VSAT (very small aperture terminal): A satellite Ku-band earth station intended for fixed or portable use. The antenna diameter of a VSAT is on the order of 1.5 m or less.

W

Watt: The unit of power equal to the work done at one joule per second, or the rate of work measured as a current of one ampere under an electric potential of one volt. Designated by the symbol W, the watt is named after the Scottish inventor James Watt (1736–1819).

Watt meter: A meter indicating in watts the rate of consumption of electrical energy.

Watt-hour: The work performed by one watt over a one hour period.

Wave: A disturbance that is a function of time or space, or both, and is propagated in a medium or through space.

Wave number: The reciprocal of wavelength; the number of wave lengths per unit distance in the direction of propagation of a wave.

Waveband: A band of wavelengths defined for some given purpose.

Waveform: The characteristic shape of a periodic wave, determined by the frequencies present and their amplitudes and relative phases.

Wavefront: A continuous surface that is a locus of points having the same phase at a given instant. A *wavefront* is a surface at right angles to rays that proceed from the wave source. The surface passes through those parts of the wave that are in the same phase and travel in the same direction. For parallel rays the wavefront is a plane; for rays that radiate from a point, the wavefront is spherical.

Waveguide: Generally, a rectangular or circular pipe that constrains the propagation of an acoustic or electromagnetic wave along a path between two locations. The dimensions of a waveguide determine the frequencies for optimum transmission.

Wavelength: For a sinusoidal wave, the distance between points of corresponding phase of two consecutive cycles.

Weber: The unit of magnetic flux equal to the flux that, when linked to a circuit of one turn, produces an electromotive force of one volt as the flux is reduced at a uniform rate to zero in one second. The *weber* is named for the German physicist Wilhelm Eduard Weber (1804–1891).

Weighted: The condition when a correction factor is applied to a measurement.

Weighting: The adjustment of a measured value to account for conditions that would otherwise be different or appropriate during a measurement.

Weighting network: A circuit, used with a test instrument, that has a specified amplitude-versus-frequency characteristic.

Wideband: The passing or processing of a wide range of frequencies. The meaning varies with the context.

Wien bridge: An ac bridge used to measure capacitance or inductance.

Winding: A coil of wire used to form an inductor.

Wire: A single metallic conductor, usually solid-drawn and circular in cross section.

Working range: The permitted range of values of an analog signal over which transmitting or other processing equipment can operate.

Working voltage: The rated voltage that may safely be applied continuously to a given circuit or device.

X

X-band: A microwave frequency band from 5.2 GHz to 10.9 GHz.

X-cut: A method of cutting a quartz plate for an oscillator, with the x-axis of the crystal perpendicular to the faces of the plate.

X ray: An electromagnetic radiation of approximately 100 nm to 0.1 nm, capable of penetrating non-metallic materials.

Y

Y-cut: A method of cutting a quartz plate for an oscillator, with the y-axis of the crystal perpendicular to the faces of the plate.

Yield strength: The magnitude of mechanical stress at which a material will begin to deform. Beyond the *yield strength* point, extension is no longer proportional to stress and rupture is possible.

Yoke: A material that interconnects magnetic cores. *Yoke* can also refer to the deflection windings of a CRT.

Yttrium-iron garnet (YIG): A crystalline material used in microwave devices.

23.8   Abbreviations and Acronyms

Jerry C. Whitaker

A

a	atto (10^{-18})
Å	angstrom
A	ampere
AAR	automatic alternate routing
AARTS	automatic audio remote test set
ac	alternating current

This chapter is adapted from: General Services Administration, Information Technology Service, *National Communications System: Glossary of Telecommunication Terms*, Technology and Standards Division, Federal Standard 1037C, General Services Administration, Washington, D.C., August, 7, 1996.

ACA	automatic circuit assurance
ACC	automatic callback calling
ACD	automatic call distributor
ac-dc	alternating current-direct current
ACK	acknowledge character
ACTS	Advanced Communications Technology Satellite
ACU	automatic calling unit
A-D	analog-to-digital
ADC	analog-to-digital converter; analog-to-digital conversion
ADCCP	Advanced Data Communication Control Procedures
ADH	automatic data handling
ADP	automatic data processing
ADPCM	adaptive differential pulse-code modulation
ADPE	automatic data processing equipment
ADU	automatic dialing unit
ADX	automatic data exchange
AECS	Aeronautical Emergency Communications System [Plan]
AF	audio frequency
AFC	area frequency coordinator; automatic frequency control
AFRS	Armed Forces Radio Service
AGC	automatic gain control
AGE	aerospace ground equipment
AI	artificial intelligence
AIG	address indicator group; address indicating group
AIM	amplitude intensity modulation
AIN	advanced intelligent network
AIOD	automatic identified outward dialing
AIS	automated information system
AJ	anti-jamming
ALC	automatic level control; automatic load control
ALE	automatic link establishment
ALU	arithmetic and logic unit
AM	amplitude modulation
AMA	automatic message accounting
AMC	administrative management complex
AME	amplitude modulation equivalent; automatic message exchange
AMI	alternate mark inversion [signal]
AM/PM/VSB	amplitude modulation/phase modulation/vestigial sideband
AMPS	automatic message processing system
AMPSSO	automated message processing system security officer
AMSC	American Mobile Satellite Corporation
AMTS	automated maritime telecommunications system
ANI	automatic number identification
ANL	automatic noise limiter
ANMCC	Alternate National Military Command Center
ANS	American National Standard
ANSI	American National Standards Institute
AP	anomalous propagation
APC	adaptive predictive coding
API	application program interface
APK	amplitude phase-shift keying

APL	average picture level
ARP	address resolution protocol
ARPA	Advanced Research Projects Agency [now DARPA]
ARPANET	Advanced Research Projects Agency Network
ARQ	automatic repeat-request
ARS	automatic route selection
ARSR	air route surveillance radar
ARU	audio response unit
ASCII	American Standard Code for Information Interchange
ASIC	application-specific integrated circuits
ASP	Aggregated Switch Procurement; adjunct service point
ASR	automatic send and receive; airport surveillance radar
AT	access tandem
ATACS	Army Tactical Communications System
ATB	all trunks busy
ATCRBS	air traffic control radar beacon system
ATDM	asynchronous time-division multiplexing
ATE	automatic test equipment
ATM	asynchronous transfer mode
ATV	advanced television
au	astronomical unit
AUI	attachment unit interface
AUTODIN	Automatic Digital Network
AUTOVON	Automatic Voice Network
AVD	alternate voice/data
AWG	American wire gauge
AWGN	additive white Gaussian noise
AZ	azimuth

B

b	bit
B	bel; byte
balun	balanced to unbalanced
basecom	base communications
BASIC	beginners' all-purpose symbolic instruction code
BCC	block check character
BCD	binary coded decimal; binary-coded decimal notation
BCI	bit-count integrity
Bd	baud
B8ZS	bipolar with eight-zero substitution
bell	BEL character
BER	bit error ratio
BERT	bit error ratio tester
BETRS	basic exchange telecommunications radio service
BEX	broadband exchange
BIH	International Time Bureau
B-ISDN	broadband ISDN
bi-sync	binary synchronous [communication]
bit	binary digit
BIT	built-in test

BITE	built-in test equipment
BIU	bus interface unit
BNF	Backus Naur form
BOC	Bell Operating Company
bpi	bits per inch; bytes per inch
BPSK	binary phase-shift keying
b/s	bits per second
b/in	bits per inch
BPSK	binary phase-shift keying
BR	bit rate
BRI	basic rate interface
BSA	basic serving arrangement
BSE	basic service element
BSI	British Standards Institution
B6ZS	bipolar with six-zero substitution
B3ZS	bipolar with three-zero substitution
BTN	billing telephone number
BW	bandwidth

C

c	centi (10^{-2})
CACS	centralized alarm control system
CAM	computer-aided manufacturing
CAMA	centralized automatic message accounting
CAN	cancel character
CAP	competitive access provider; customer administration panel
CARS	cable television relay service [station]
CAS	centralized attendant services
CASE	computer-aided software engineering; computer aided system engineering; computer-assisted software engineering
CATV	cable TV; cable television; community antenna television
CBX	computer branch exchange
C^2	command and control
C^3	command, control, and communications
C^3CM	C^3 countermeasures
C^3I	command, control, communications and intelligence
CCA	carrier-controlled approach
CCD	charge-coupled device
CCH	connections per circuit hour
CCIF	International Telephone Consultative Committee
CCIR	International Radio Consultative Committee
CCIS	common-channel interoffice signaling
CCIT	International Telegraph Consultative Committee
CCITT	International Telegraph and Telephone Consultative Committee
CCL	continuous communications link
CCS	hundred call-seconds
CCSA	common control switching arrangement
CCTV	closed-circuit television
CCW	cable cutoff wavelength
cd	candela

CD	collision detection; compact disk
CDF	combined distribution frame; cumulative distribution function
CDMA	code-division multiple access
CDPSK	coherent differential phase-shift keying
CDR	call detail recording
CD ROM	compact disk read-only memory
CDT	control data terminal
CDU	central display unit
C-E	communications-electronics
CEI	comparably efficient interconnection
CELP	code-excited linear prediction
CEP	circular error probable
CFE	contractor-furnished equipment
cgs	centimeter-gram-second
ChR	channel reliability
CIAS	circuit inventory and analysis system
CIC	content indicator code
CIF	common intermediate format
CIFAX	ciphered facsimile
CiR	circuit reliability
C/kT	carrier-to-receiver noise density
CLASS	custom local area signaling service
cm	centimeter
CMI	coded mark inversion
CMIP	Common Management Information Protocol
CMIS	common management information service
CMOS	complementary metal oxide substrate
CMRR	common-mode rejection ratio
CNR	carrier-to-noise ratio; combat-net radio
CNS	complementary network service
C.O.	central office
COAM	customer owned and maintained equipment
COBOL	common business oriented language
codec	coder-decoder
COG	centralized ordering group
COMINT	communications intelligence
COMJAM	communications jamming
compandor	compressor-expander
COMPUSEC	computer security
COMSAT	Communications Satellite Corporation
COMSEC	communications security
CONEX	connectivity exchange
CONUS	Continental United States
COP	Committee of Principals
COR	Council of Representatives
COT	customer office terminal
CPAS	cellular priority access services
CPE	customer premises equipment
cpi	characters per inch
cpm	counts per minute
cps	characters per second

CPU	central processing unit; communications processor unit
CR	channel reliability; circuit reliability
CRC	cyclic redundancy check
CRITICOM	Critical Intelligence Communications
CROM	control read-only memory
CRT	cathode ray tube
c/s	cycles per second
CSA	Canadian Standards Association
CSC	circuit-switching center; common signaling channel
CSMA	carrier sense multiple access
CSMA/CA	carrier sense multiple access with collision avoidance
CSMA/CD	carrier sense multiple access with collision detection
CSU	channel service unit; circuit switching unit; customer service unit
CTS	clear to send
CTX	Centrex® [service]; clear to transmit
CVD	chemical vapor deposition
CVSD	continuously variable slope delta [modulation]
cw	carrier wave; composite wave; continuous wave
CX	composite signaling
cxr	carrier

D

d	deci (10^{-1})
da	deka (10)
D-A	digital-to-analog; digital-to-analog converter
D/L	downlink
DACS	digital access and cross-connect system
DAMA	demand assignment multiple access
DARPA	Defense Advanced Research Projects Agency
dB	decibel
dBa	decibels adjusted
dBa0	noise power measured at zero transmission level point
dBc	dB relative to carrier power
dBm	dB referred to 1 milliwatt
dBm(psoph)	noise power in dBm measured by a set with psophometric weighting
DBMS	database management system
dBmV	dB referred to 1 millivolt across 75 ohms
dBm0	noise power in dBm referred to or measured at 0TLP
dBm0p	noise power in dBm0 measured by a psophometric or noise measuring set having psophometric weighting
dBr	power difference in dB between any point and a reference point
dBrn	dB above reference noise
dBrnC	noise power in dBrn measured by a set with C-message weighting
dBrnC0	noise power in dBrnC referred to or measured at 0TLP
dBrn($f_1 - f_2$)	flat noise power in dBrn
dBrn(144)	noise power in dBrn measured by a set with 144-line weighting
dBv	dB relative to 1 V (volt) peak-to-peak
dBW	dB referred to 1 W (watt)
dBx	dB above reference coupling
dc	direct current

DCA	Defense Communications Agency
DCE	data circuit-terminating equipment
DCL	direct communications link
DCPSK	differentially coherent phase-shift keying
DCS	Defense Communications System
DCTN	Defense Commercial Telecommunications Network
DCWV	direct-current working volts
DDD	direct distance dialing
DDN	Defense Data Network
DDS	digital data service
DEL	delete character
demarc	demarcation point
demux	demultiplex; demultiplexer; demultiplexing
dequeue	double-ended queue
DES	Data Encryption Standard
detem	detector/emitter
DFSK	double-frequency shift keying
DIA	Defense Intelligence Agency
DID	direct inward dialing
DIN	Deutsches Institut für Normung
DIP	dual in-line package
DISA	Defense Information Systems Agency
DISC	disconnect command
DISN	Defense Information System Network
DISNET	Defense Integrated Secure Network
DLA	Defense Logistic Agency
DLC	digital loop carrier
DLE	data link escape character
DM	delta modulation
DMA	Defense Mapping Agency; direct memory access
DME	distance measuring equipment
DMS	Defense Message System
DNA	Defense Nuclear Agency
DNIC	data network identification code
DNPA	data numbering plan area
DNS	Domain Name System
DO	design objective
DoC	Department of Commerce
DOD	Department of Defense; direct outward dialing
DODD	Department of Defense Directive
DODISS	Department of Defense Index of Specifications and Standards
DOD-STD	Department of Defense Standard
DOS	Department of State
DPCM	differential pulse-code modulation
DPSK	differential phase-shift keying
DQDB	distributed-queue dual-bus [network]
DRAM	dynamic random access memory
DRSI	destination station routing indicator
DS	digital signal; direct support
DS0	digital signal 0
DS1	digital signal 1

DS1C	digital signal 1C
DS2	digital signal 2
DS3	digital signal 3
DS4	digital signal 4
DSA	dial service assistance
DSB	double sideband (transmission); Defense Science Board
DSB-RC	double-sideband reduced carrier transmission
DSB-SC	double-sideband suppressed-carrier transmission
DSC	digital selective calling
DSCS	Defense Satellite Communications System
DSE	data switching exchange
DSI	digital speech interpolation
DSL	digital subscriber line
DSN	Defense Switched Network
DSR	data signaling rate
DSS	direct station selection
DSSCS	Defense Special Service Communications System
DSTE	data subscriber terminal equipment
DSU	data service unit
DTE	data terminal equipment
DTG	date-time group
DTMF	dual-tone multifrequency (signaling)
DTN	data transmission network
DTS	Diplomatic Telecommunications Service
DTU	data transfer unit; data tape unit; digital transmission unit; direct to user
DVL	direct voice link
DX signaling	direct current signaling; duplex signaling

E

E	exa (10^{18})
E-MAIL	electronic mail
EAS	extended area service
EBCDIC	extended binary coded decimal interchange code
E_b/N_0	signal energy per bit per hertz of thermal noise
EBO	embedded base organization
EBS	Emergency Broadcast System
EBX	electronic branch exchange
EC	Earth coverage; Earth curvature
ECC	electronically controlled coupling; enhance call completion
ECCM	electronic counter-countermeasures
ECM	electronic countermeasures
EDC	error detection and correction
EDI	electronic data interchange
EDTV	extended-definition television
EHF	extremely high frequency
EIA	Electronic Industries Association
eirp	effective isotropically radiated power; equivalent isotopically radiated power
EIS	Emergency Information System
el	elevation
ELF	extremely low frequency

ELINT	electronics intelligence; electromagnetic intelligence
ELSEC	electronics security
ELT	emergency locator transmitter
EMC	electromagnetic compatibility
EMCON	emission control
EMD	equilibrium mode distribution
EME	electromagnetic environment
emf	electromotive force
EMI	electromagnetic interference; electromagnetic interference control
EMP	electromagnetic pulse
EMR	electromagnetic radiation
EMR hazards	electromagnetic radiation hazards
e.m.r.p.	effective monopole radiated power; equivalent monopole radiated power
EMS	electronic message system
EMSEC	emanations security
emu	electromagnetic unit
EMV	electromagnetic vulnerability
EMW	electromagnetic warfare; electromagnetic wave
ENQ	enquiry character
EO	end office
E.O.	Executive Order
EOD	end of data
EOF	end of file
EOL	end of line
EOM	end of message
EOP	end of program; end output
EOS	end-of-selection character
EOT	end-of-transmission character; end of tape
EOW	engineering orderwire
EPROM	erasable programmable read-only memory
EPSCS	enhanced private switched communications system
ERL	echo return loss
ERLINK	emergency response link
ERP, e.r.p.	effective radiated power
ES	end system; expert system
ESC	escape character; enhanced satellite capability
ESF	extended superframe
ESM	electronic warfare support measures
ESP	enhanced service provider
ESS	electronic switching system
ETB	end-of-transmission-block character
ETX	end-of-text character
EW	electronic warfare
EXCSA	Exchange Carriers Standards Association

F

f	femto (10^{-15})
f	frequency
FAA	Federal Aviation Administration
FAQ file	Frequently Asked Questions file

FAX	facsimile
FC	functional component
FCC	Federal Communications Commission
FCS	frame check sequence
FDDI	fiber distributed data interface
FDDI-2	fiber distributed data interface-2
FDHM	full duration at half maximum
FDM	frequency-division multiplexing
FDMA	frequency-division multiple access
FDX	full duplex
FEC	forward error correction
FECC	Federal Emergency Communications Coordinators
FED-STD	Federal Standard
FEMA	Federal Emergency Management Agency
FEP	front-end processor
FET	field effect transistor
FIFO	first-in first-out
FIP	Federal Information Processing
FIPS	Federal Information Processing Standards
FIR	finite impulse response
FIRMR	Federal Information Resources Management Regulations
FISINT	foreign instrumentation signals intelligence
flops	floating-point operations per second
FM	frequency modulation
FO	fiber optics
FOC	final operational capability; full operational capability
FOT	frequency of optimum traffic; frequency of optimum transmission
FPIS	forward propagation ionospheric scatter
fps	foot-pound-second
FPS	frames per second; focus projection and scanning
FRP	Federal Response Plan
FSDPSK	filtered symmetric differential phase-shift keying
FSK	frequency-shift keying
FSS	fully separate subsidiary
FT	fiber optic T-carrier
FTAM	file transfer, access, and management
FTF	Federal Telecommunications Fund
ft/min	feet per minute
FTP	file transfer protocol
ft/s	feet per second
FTS	Federal Telecommunications System
FTS2000	Federal Telecommunications System 2000
FTSC	Federal Telecommunications Standards Committee
FWHM	full width at half maximum
FX	fixed service; foreign exchange service
FYDP	Five Year Defense Plan

G

g	profile parameter
G	giga (10^9)

GBH	group busy hour
GCT	Greenwich Civil Time
GDF	group distribution frame
GETS	Government Emergency Telecommunications Service
GFE	Government-furnished equipment
GGCL	government-to-government communications link
GHz	gigahertz
GII	Global Information Infrastructure
GMT	Greenwich Mean Time
GOS	grade of service
GOSIP	Government Open Systems Interconnection Profile
GSA	General Services Administration
GSTN	general switched telephone network
G/T	antenna gain-to-noise-temperature
GTP	Government Telecommunications Program
GTS	Government Telecommunications System
GUI	graphical user interface

H

h	hecto (10^2); hour; Planck's constant
HCS	hard clad silica (fiber)
HDLC	high-level data link control
HDTV	high-definition television
HDX	half-duplex (operation)
HE^{11} mode	the fundamental hybrid mode (of an optical fiber)
HEMP	high-altitude electromagnetic pulse
HERF	hazards of electromagnetic radiation to fuel
HERO	hazards of electromagnetic radiation to ordnance
HERP	hazards of electromagnetic radiation to personnel
HF	high frequency
HFDF	high-frequency distribution frame
HLL	high-level language
HPC	high probability of completion
HV	high voltage
Hz	hertz

I

IA	International Alphabet
I&C	installation and checkout
IC	integrated circuit
ICI	incoming call identification
ICNI	Integrated Communications, Navigation, and Identification
ICW	interrupted continuous wave
IDDD	International Direct Distance Dialing
IDF	intermediate distribution frame
IDN	integrated digital network
IDTV	improved-definition television
IEC	International Electrotechnical Commission
IEEE	Institute of Electrical and Electronics Engineers

IES	Industry Executive Subcommittee
IF	intermediate frequency
I/F	interface
IFF	identification, friend or foe
IFRB	International Frequency Registration Board
IFS	ionospheric forward scatter
IIR	infinite impulse response
IITF	Information Infrastructure Task Force
ILD	injection laser diode
ILS	instrument landing system
IM	intensity modulation; intermodulation
I&M	installation and maintenance
IMD	intermodulation distortion
IMP	interface message processor
IN	intelligent network
INFOSEC	information systems security
INS	inertial navigation system
INTELSAT	International Telecommunications Satellite Consortium
INWATS	Inward Wide-Area Telephone Service
I/O	input/output (device)
IOC	integrated optical circuit; initial operational capability; input-output controller
IP	Internet protocol; intelligent peripheral
IPA	intermediate power amplifier
IPC	information processing center
IPM	impulses per minute; interference prediction model; internal polarization modulation; interruptions per minute
in/s	inches per second
ips	interruptions per second
IPX	Internet Packet Exchange
IQF	intrinsic quality factor
IR	infrared
IRAC	Interdepartment Radio Advisory Committee
IRC	international record carrier; Interagency Radio Committee
ISB	independent-sideband (transmission)
ISDN	Integrated Services Digital Network
ISM	industrial, scientific, and medical (applications)
ISO	International Organization for Standardization
ITA	International Telegraph Alphabet
ITA-5	International Telegraph Alphabet Number 5
ITC	International Teletraffic Congress
ITS	Institute for Telecommunication Sciences
ITSO	International Telecommunications Satellite Organization
ITU	International Telecommunication Union
IVDT	integrated voice data terminal
IXC	interexchange carrier

J

JANAP	Joint Army-Navy-Air Force Publication(s)
JCS	Joint Chiefs of Staff
JPL	Jet Propulsion Laboratory

JSC   Joint Steering Committee; Joint Spectrum Center
JTC³A   Joint Tactical Command, Control and Communications Agency
JTIDS   Joint Tactical Information Distribution System
JTRB   Joint Telecommunications Resources Board
JTSSG   Joint Telecommunications Standards Steering Group
JWID   Joint Warrior Interoperability Demonstration

K

k   kilo (10^3); Boltzmann's constant
K   coefficient of absorption; kelvin
KDC   key distribution center
KDR   keyboard data recorder
KDT   keyboard display terminal
kg   kilogram
kg·m·s   kilogram-meter-second
kHz   kilohertz
km   kilometer
kΩ, k   kilohm
KSR   keyboard send/receive device
kT   noise power density
KTS   key telephone system
KTU   key telephone unit

L

LAN   local area network
LAP-B   Data Link Layer protocol (CCITT Recommendation X.25 [1989])
LAP-D   link access procedure D
laser   light amplification by stimulated emission of radiation
LASINT   laser intelligence
LATA   local access and transport area
LBO   line buildout
LC   limited capability
LCD   liquid crystal display
LD   long distance
LDM   limited distance modem
LEC   local exchange carrier
LED   light-emitting diode
LF   low frequency
LFB   look-ahead-for-busy (information)
LIFO   last-in first-out
LLC   logical link control (sublayer)
l/m   lines per minute
LMF   language media format
LMR   land mobile radio
LNA   launch numerical aperture
LOF   lowest operating frequency
loran   long-range aid to navigation system; long-range radio navigation; long-range
                  radio aid to navigation system
LOS   line of sight, loss of signal

LP	linearly polarized (mode); linear programming; linking protection; log-periodic (antenna); log-periodic (array)
LPA	linear power amplifier
LPC	linear predictive coding
LPD	low probability of detection
LPI	low probability of interception
lpi	lines per inch
lpm	lines per minute
LP_{01}	the fundamental mode (of an optical fiber)
LQA	link quality analysis
LRC	longitudinal redundancy check
LSB	lower sideband, least significant bit
LSI	large scale integrated (circuit); large scale integration; line status indication
LTC	line traffic coordinator
LUF	lowest usable high frequency
LULT	line-unit-line termination
LUNT	line-unit-network termination
LV	low voltage

M

m	meter
M	mega (10^6)
MAC	medium access control [sublayer]
MACOM	major command
MAN	metropolitan area network
MAP	manufacturers' automation protocol
maser	microwave amplification by the stimulated emission of radiation
MAU	medium access unit
MCC	maintenance control circuit
MCEB	Military Communications-Electronics Board
MCM	multicarrier modulation
MCS	Master Control System
MCW	modulated continuous wave
MCXO	microcomputer compensated crystal oscillator
MDF	main distribution frame
MDT	mean downtime
MEECN	Minimum Essential Emergency Communications Network
MERCAST	merchant-ship broadcast system
MF	medium frequency; multifrequency (signaling)
MFD	mode field diameter
MFJ	Modification of Final Judgment
MFSK	multiple frequency-shift keying
MHF	medium high frequency
MHS	message handling service; message handling system
MHz	megahertz
mi	mile
MIC	medium interface connector; microphone; microwave integrated circuit; minimum ignition current; monolithic integrated circuit; mutual interface chart
MILNET	military network
MIL-STD	Military Standard
min	minute

MIP	medium interface point
MIPS, mips	million instructions per second
MIS	management information system
MKS	meter-kilogram-second
MLPP	multilevel precedence and preemption
MMW	millimeter wave
modem	modulator-demodulator
mol	mole
ms	millisecond (10^{-3} second)
MSB	most significant bit
MSK	minimum-shift keying
MTBF	mean time between failures
MTBM	mean time between maintenance
MTBO	mean time between outages
MTBPM	mean time between preventive maintenance
MTF	modulation transfer function
MTSO	mobile telephone switching office
MTSR	mean time to service restoration
MTTR	mean time to repair
μ	micro (10^{-6})
μs	microsecond
MUF	maximum usable frequency
MUX	multiplex; multiplexer
MUXing	multiplexing
mw	microwave
MWI	message waiting indicator
MWV	maximum working voltage

N

n	nano (10^{-9}); refractive index
N_0	sea level refractivity; spectral noise density
NA	numerical aperture
NACSEM	National Communications Security Emanation Memorandum
NACSIM	National Communications Security Information Memorandum
NAK	negative-acknowledge character
NASA	National Aeronautics and Space Administration
NATA	North American Telecommunications Association
NATO	North Atlantic Treaty Organization
NAVSTAR	Navigational Satellite Timing and Ranging
NBFM	narrowband frequency modulation
NBH	network busy hour
NBRVF	narrowband radio voice frequency
NBS	National Bureau of Standards
NBSV	narrowband secure voice
NCA	National Command Authorities
NCC	National Coordinating Center for Telecommunications
NCS	National Communications System; net control station
NCSC	National Communications Security Committee
NDCS	network data control system
NDER	National Defense Executive Reserve

NEACP	National Emergency Airborne Command Post
NEC	National Electric Code®
NEP	noise equivalent power
NES	noise equivalent signal
NF	noise figure
NFS	Network File System
NIC	network interface card
NICS	NATO Integrated Communications System
NID	network interface device; network inward dialing; network information database
NII	National Information Infrastructure
NIOD	network inward/outward dialing
NIST	National Institute of Standards and Technology
NIU	network interface unit
NLP	National Level Program
nm	nanometer
NMCS	National Military Command System
nmi	nautical mile
NOD	network outward dialing
Np	neper
NPA	numbering plan area
NPR	noise power ratio
NRI	net radio interface
NRM	network resource manager
NRRC	Nuclear Risk Reduction Center
NRZ	non-return-to-zero
NRZI	non-return-to-zero inverted
NRZ-M	non-return-to-zero mark
NRZ-S	non-return-to-zero space
NRZ1	non-return-to-zero, change on ones
NRZ-1	non-return-to-zero mark
ns	nanosecond
NSA	National Security Agency
NSC	National Security Council
NS/EP	National Security or Emergency Preparedness telecommunications
NSTAC	National Security Telecommunications Advisory Committee
NTCN	National Telecommunications Coordinating Network
NTDS	Naval Tactical Data System
NT1	Network termination 1
NT2	Network termination 2
NTI	network terminating interface
NTIA	National Telecommunications and Information Administration
NTMS	National Telecommunications Management Structure
NTN	network terminal number
NTSC	National Television Standards Committee; National Television Standards Committee (standard)
NUL	null character
NVIS	near vertical incidence skywave

O

O&M	operations and maintenance
OC	operations center

OCC	other common carrier
OCR	optical character reader; optical character recognition
OCU	orderwire control unit
OCVCXO	oven controlled-voltage controlled crystal oscillator
OCXO	oven controlled crystal oscillator
OD	optical density; outside diameter
OFC	optical fiber, conductive
OFCP	optical fiber, conductive, plenum
OFCR	optical fiber, conductive, riser
OFN	optical fiber, nonconductive
OFNP	optical fiber, nonconductive, plenum
OFNR	optical fiber, nonconductive, riser
OMB	Office of Management and Budget
ONA	open network architecture
opm	operations per minute
OPMODEL	operations model
OPSEC	operations security
OPX	off-premises extension
OR	off-route service; off-route aeronautical mobile service
OSHA	Occupational Safety and Health Administration
OSI	open switching interval; Open Systems Interconnection
OSI-RM	Open Systems Interconnection—Reference Model
OSRI	originating stations routing indicator
OSSN	originating stations serial number
OTAM	over-the-air management of automated HF network nodes
OTAR	over-the-air rekeying
OTDR	optical time domain reflectometer; optical time domain reflectometry
OW	orderwire [circuit]

P

p	pico (10^{-12})
P	peta (10^{15})
PABX	private automatic branch exchange
PAD	packet assembler/disassembler
PAL	phase alternation by line
PAL-M	phase alternation by line—modified
PAM	pulse-amplitude modulation
PAMA	pulse-address multiple access
p/a r	peak-to-average ratio
PAR	performance analysis and review
PARAMP	parametric amplifier
par meter	peak-to-average ratio meter
PAX	private automatic exchange
PBER	pseudo-bit-error-ratio
PBX	private branch exchange
PC	carrier power (of a radio transmitter); personal computer
PCB	power circuit breaker; printed circuit board
PCM	pulse-code modulation; plug compatible module; process control module
PCS	Personal Communications Services; personal communications system; plastic-clad silica (fiber)

PCSR	parallel channels signaling rate
PD	photodetector
PDM	pulse delta modulation; pulse-duration modulation
PDN	public data network
PDS	protected distribution system; power distribution system; program data source
PDT	programmable data terminal
PDU	protocol data unit
PE	phase-encoded (recording)
PEP	peak envelope power (of a radio transmitter)
pF	picofarad
PF	power factor
PFM	pulse-frequency modulation
PI	protection interval
PIC	plastic insulated cable
ping	packet Internet groper
PIV	peak inverse voltage
PLA	programmable logic array
PL/I	programming language 1
PLL	phase-locked loop
PLN	private line network
PLR	pulse link repeater
PLS	physical signaling sublayer
pm	phase modulation
PM	mean power; polarization-maintaining (optical fiber); preventive maintenance; pulse modulation
PMB	pilot-make-busy (circuit)
PMO	program management office
POI	point of interface
POP	point of presence
POSIX	portable operating system interface for computer environments
POTS	plain old telephone service
PP	polarization-preserving (optical fiber)
P-P	peak-to-peak (value)
P/P	point-to-point
PPM	pulse-position modulation
pps	pulses per second
PR	pulse rate
PRF	pulse-repetition frequency
PRI	primary rate interface
PRM	pulse-rate modulation
PROM	programmable read-only memory
PRR	pulse repetition rate
PRSL	primary area switch locator
PS	permanent signal
psi	pounds (force) per square inch
PSK	phase-shift keying
PSN	public switched network
p-static	precipitation static
PSTN	public switched telephone network
PTF	patch and test facility
PTM	pulse-time modulation

PTT	postal, telephone, and telegraph; push-to-talk (operation)
PTTC	paper tape transmission code
PTTI	precise time and time interval
PU	power unit
PUC	public utility commission; public utilities commission
PVC	permanent virtual circuit; polyvinyl chloride (insulation)
pW	picowatt
PWM	pulse-width modulation
PX	private exchange

Q

QA	quality assurance
QAM	quadrature amplitude modulation
QC	quality control
QCIF	quarter common intermediate format
QMR	qualitative material requirement
QOS	quality of service
QPSK	quadrature phase-shift keying
QRC	quick reaction capability

R

racon	radar beacon
rad	radian; radiation absorbed dose
radar	radio detection and ranging
RADHAZ	electromagnetic radiation hazards
RADINT	radar intelligence
RAM	random access memory; reliability, availability, and maintainability
R&D	research and development
RATT	radio teletypewriter system
RBOC	Regional Bell Operating Company
RbXO	rubidium-crystal oscillator
RC	reflection coefficient; resource controller
RCC	radio common carrier
RCVR	receiver
RDF	radio-direction finding
REA	Rural Electrification Administration
REN	ringer equivalency number
RF	radio frequency; range finder
RFI	radio frequency interference
RFP	request for proposal
RFQ	request for quotation
RGB	red-green-blue
RH	relative humidity
RHR	radio horizon range
RI	routing indicator
RISC	reduced instruction set chip
RJ	registered jack
RJE	remote job entry
rms	root-mean-square (deviation)
RO	read only; receive only

ROA	recognized operating agency
ROC	required operational capability
ROM	read-only memory
ROSE	remote operations service element protocol
rpm	revolutions per minute
RPM	rate per minute
RPOA	recognized private operating agency
rps	revolutions per second
RQ	repeat-request
RR	repetition rate
RSL	received signal level
rss	root-sum-square
R/T	real time
RTA	remote trunk arrangement
RTS	request to send
RTTY	radio teletypewriter
RTU	remote terminal unit
RTX	request to transmit
RVA	reactive volt-ampere
RVWG	Reliability and Vulnerability Working Group
RWI	radio and wire integration
RX	receive; receiver
RZ	return-to-zero

S

s	second
SCC	specialized common carrier
SCE	service creation environment
SCF	service control facility
SCP	service control point
SCPC	single channel per carrier
SCR	semiconductor-controlled rectifier; silicon-controlled rectifier
SCSR	single channel signaling rate
SDLC	synchronous data link control
SDM	space-division multiplexing
SDN	software-defined network
SECDEF	Secretary of Defense
SECORD	secure voice cord board
SECTEL	secure telephone
SETAMS	systems engineering, technical assistance, and management services
SEVAS	Secure Voice Access System
S-F	store-and-forward
SF	single-frequency (signaling)
SGDF	supergroup distribution frame
S/H	sample and hold
SHA	sidereal hour angle
SHARES	Shared Resources (SHARES) HF Radio Program
SHF	super high frequency
SI	International System of Units
SID	sudden ionospheric disturbance

SIGINT	signals intelligence
SINAD	signal-plus-noise-plus-distortion to noise-plus-distortion ratio
SLD	superluminescent diode
SLI	service logic interpreter
SLP	service logic program
SMDR	station message-detail recording
SMSA	standard metropolitan statistical area
SNR	signal-to-noise ratio
SOH	start-of-heading character
SOM	start of message
sonar	sound navigation and ranging
SONET	synchronous optical network
SOP	standard operating procedure
SOR	start of record
SOW	statement of work
$(S+N)/N$	signal-plus-noise-to-noise ratio
sr	steradian
S/R	send and receive
SSB	single-sideband (transmission)
SSB-SC	single-sideband suppressed carrier (transmission)
SSN	station serial number
SSP	service switching point
SSUPS	solid-state uninterruptible power system
STALO	stabilized local oscillator
STD	subscriber trunk dialing
STFS	standard time and frequency signal (service); standard time and frequency service
STL	standard telegraph level; studio-to-transmitter link
STP	standard temperature and pressure; signal transfer point
STU	secure telephone unit
STX	start-of-text character
SUB	substitute character
SWR	standing wave ratio
SX	simplex signaling
SXS	step-by-step switching system
SYN	synchronous idle character
SYSGEN	system generation

T

T	tera (10^{12})
TADIL	tactical data information link
TADIL-A	tactical data information link-A
TADIL-B	tactical data information link-B
TADS	teletypewriter automatic dispatch system
TADSS	Tactical Automatic Digital Switching System
TAI	International Atomic Time
TASI	time-assignment speech interpolation
TAT	trans-Atlantic telecommunication (cable)
TC	toll center
TCB	trusted computing base
TCC	telecommunications center

TCCF	Tactical Communications Control Facility
TCF	technical control facility
TCP	transmission control protocol
TCS	trusted computer system
TCU	teletypewriter control unit
TCVXO	temperature compensated-voltage controlled crystal oscillator
TCXO	temperature controlled crystal oscillator
TD	time delay; transmitter distributor
TDD	Telecommunications Device for the Deaf
TDM	time-division multiplexing
TDMA	time-division multiple access
TE	transverse electric [mode]
TED	trunk encryption device
TEK	traffic encryption key
TEM	transverse electric and magnetic [mode]
TEMPEST	compromising emanations
TEMS	telecommunications management system
TGM	trunk group multiplexer
THD	total harmonic distortion
THF	tremendously high frequency
THz	terahertz
TIA	Telecommunications Industry Association
TIE	time interval error
TIFF	tag image file format
TIP	terminal interface processor
T_K	response timer
TLP	transmission level point
TM	transverse magnetic [mode]
TP	toll point
TRANSEC	transmission security
TRC	transverse redundancy check
TRF	tuned radio frequency
TRI-TAC	tri-services tactical [equipment]
TSK	transmission security key
TSP	Telecommunications Service Priority [system]
TSPS	traffic service position system
TSR	telecommunications service request; terminate and stay resident
TTL	transistor-transistor logic
TTTN	tandem tie trunk network
TTY	teletypewriter
TTY/TDD	Telecommunications Device for the Deaf
TV	television
TW	traveling wave
TWT	traveling wave tube
TWTA	traveling wave tube amplifier
TWX®	teletypewriter exchange service
TX	transmit; transmitter

U

UDP	User Datagram Protocol
UHF	ultra high frequency

U/L	uplink
ULF	ultra low frequency
UPS	uninterruptible power supply
UPT	Universal Personal Telecommunications service
USB	upper sideband
USDA	U.S. Department of Agriculture
USFJ	U.S. Forces, Japan
USFK	U.S. Forces, Korea
USNO	U.S. Naval observatory
USTA	U.S. Telephone Association
UT	Universal Time
UTC	Coordinated Universal Time
uv	ultraviolet

V

V	volt
VA	volt-ampere
VAN	value-added network
VAR	value added reseller
VARISTAR	variable resistor
vars	volt-amperes reactive
VC	virtual circuit
VCO	voltage-controlled oscillator
VCXO	voltage-controlled crystal oscillator
V/D	voice/data
Vdc	volts direct current
VDU	video display unit; visual display unit
VF	voice frequency
VFCT	voice frequency carrier telegraph
VFDF	voice frequency distribution frame
VFO	variable-frequency oscillator
VFTG	voice-frequency telegraph
VHF	very high frequency
VLF	very low frequency
V/m	volts per meter
VNL	via net loss
VNLF	via net loss factor
vocoder	voice-coder
vodas	voice-operated device anti-sing
vogad	voice-operated gain-adjusting device
volcas	voice-operated loss control and echo/signaling suppression
vox	voice-operated relay circuit; voice operated transmit
VRC	vertical redundancy check
VSAT	very small aperture terminal
VSB	vestigial sideband [transmission]
VSM	vestigial sideband modulation
VSWR	voltage standing wave ratio
VT	virtual terminal
VTU	video teleconferencing unit
vu	volume unit

W

WADS	wide area data service
WAIS	Wide Area Information Servers
WAN	wide area network
WARC	World Administrative Radio Conference
WATS	Wide Area Telecommunications Service; Wide Area Telephone Service
WAWS	Washington Area Wideband System
WDM	wavelength-division multiplexing
WHSR	White House Situation Room
WIN	WWMCCS Intercomputer Network
WITS	Washington Integrated Telecommunications System
WORM	write once, read many times
wpm	words per minute
wps	words per second
wv	working voltage
WVDC	working voltage direct current
WWDSA	worldwide digital system architecture
WWMCCS	Worldwide Military Command and Control System
WWW	World Wide Web

X

XMIT	transmit
XMSN	transmission
XMTD	transmitted
XMTR	transmitter
XO	crystal oscillator
XOFF	transmitter off
XON	transmitter on
XT	crosstalk
XTAL	crystal

Z

Z	Zulu time
ZD	zero defects
Z_o	characteristic impedance
0TLP	zero transmission level point

Indexes

Author Index

A

Agbo, Samuel O.
    Integrated Circuit Design, 716
    Radio Frequency Distortion Mechanisms and Analysis, 2194
Alkin, Oktay
    Digital Coding Schemes, 1449
Allen, Brent
    SONET, 1647
Ames, William F.
    General Mathematical Tables, 2420
Anagnostopoulos, Constantine N.
    Application-Specific Integrated Circuits, 791
Aronhime, Peter
    Applications of Operational Amplifiers, 641
    Operational Amplifiers, 611
Asthana, Praveen
    Optical Storage Systems, 1592

B

Baumgartner, Fred
    Engineering Documentation, 2383
Baun, Terrence M.
    Engineering Documentation, 2383
Belcher, Jr., Melvin L.
    Radar System Implementation, 1820
Bentz, Carl
    Analog Video Measurements, 2177
Besch, David F.
    Thermal Properties, 144
Bhat, Ashoka K.S.
    DC-to-DC Conversion, 1081
    Inverters, 1067

Blackwell, Glenn R.
    Surface Mount Technology, 1297
Bomar, Bruce W.
    Digital Filters, 808
Bordelon, Iuliana
    Electronic Hardware Reliability, 2281
Breitenbach, Jerome R.
    Data Compression, 1619
Brews, John R.
    Metal-Oxide-Semiconductor Field-Effect Transistor, 545
Buchmann, Isidor
    Batteries, 1246
Burgess, Stuart K.
    Circuit Fundamentals, 112

C

Cain, George
    General Mathematical Tables, 2420
Cardieri, Paulo
    Ad Hoc Networks, 2097
Carter, Clifford G.
    Underwater Sonar Systems, 1878
Chambers, Jonathon
    Digital Filters, 808
Chan, Curtis J.
    Television and Video Production Systems, 1708
Chauvin, Ken
    Fiber Optic Cable, 905
    Optical Fiber, 875
Chen, Tom
    Integrated Circuits, 708

Subject Index

B

F

H

J

K

M

O